SECOND EDITION

Compilation of EPA's Sampling and Analysis Methods

Edited by **Lawrence H. Keith**

LEWIS PUBLISHERS

A CRC Press Company
Boca Raton   London   New York   Washington, D.C.

Library of Congress Cataloging-in-Publication Data

Compilation of EPA's sampling and analysis methods / edited by Lawrence H. Keith — 2nd ed.
    p. cm.
    Includes index.
    ISBN 1-56670-170-8 (alk. paper)
    1. Pollutants—Analysis—Handbooks, manuals, etc. 2. Environmental sampling—Handbooks, manuals, etc.
  I. Keith, Lawrence H., 1938–.
TD193.C65 1996
628.1'61—dc21
                                                                       96-25999
                                                                                     CIP

This book contains information obtained from authentic and highly regarded sources. Reprinted material is quoted with permission, and sources are indicated. A wide variety of references are listed. Reasonable efforts have been made to publish reliable data and information, but the authors and the publisher cannot assume responsibility for the validity of all materials or for the consequences of their use.

Neither this book nor any part may be reproduced or transmitted in any form or by any means, electronic or mechanical, including photocopying, microfilming, and recording, or by any information storage or retrieval system, without prior permission in writing from the publisher.

The consent of CRC Press LLC does not extend to copying for general distribution, for promotion, for creating new works, or for resale. Specific permission must be obtained in writing from CRC Press LLC for such copying.

Direct all inquiries to CRC Press LLC, 2000 N.W. Corporate Blvd., Boca Raton, Florida 33431.

Trademark Notice: Product or corporate names may be trademarks or registered trademarks, and are used only for identification and explanation, without intent to infringe.

Disclaimer: Although the information in this document has been funded in part by the United States Environmental Protection Agency (some data from the first edition), it has not been subject to Agency review and, therefore, does not necessarily reflect the views of the Agency and no official endorsement should be inferred.

Visit the CRC Press Web site at www.crcpress.com

© 1996 by CRC Press LLC
Lewis Publishers is an imprint of CRC Press LLC

No claim to original U.S. Government works
International Standard Book Number 1-56670-170-8
Library of Congress Card Number 96-25999
Printed in the United States of America   3 4 5 6 7 8 9 0
Printed on acid-free paper

The Editor

Dr. Lawrence A. Keith is a Corporate Fellow at Radian International LLC, in Austin, Texas. Before joining Radian in 1977, he was a research scientist with the U.S. Environmental Protection Agency laboratory in Athens, Georgia. A pioneer in environmental sampling and analysis, method development, and handling hazardous compounds, Dr. Keith has published many books and technical articles on these subjects. Recent publications have employed electronic books and expert systems as a medium rather than traditional paper publications like this one. He also teaches American Chemical Society (ACS) short courses on environmental sampling and analysis and has lectured extensively in the U.S. and abroad. He is the current editor of the ACS Division of Environmental Chemistry newsletter, *EnvirofACS,* and chairman of the subcommittee on Environmental Sampling and Analysis within the ACS Committee on Environmental Improvement. Other offices within the ACS have included Chairman, Secretary, Program Chairman, and Alternate Councilor for the Division of Environmental Chemistry. He also serves on several academic, industrial and government advisory boards.

Preface

The popular first edition of this book contained about 600 analyte/method summaries and was published in 1991. It was a printed version of EPA's *Sampling and Analysis Methods Database* and was intended to be a convenient reference for use with that database so an index was omitted; this edition, however, contains a comprehensive double table of contents which serves that purpose. The purpose of the database, and this book, is to help people to rapidly and easily select the most appropriate methods of sampling and analysis for a particular situation without having to become an expert in using these methods or searching through many volumes of published EPA methods.

This edition contains twice as many new analyte method summaries as in the first edition plus all of those in the first edition (for a total of about 1,700 analyte/method summaries). In addition, many of those in the first edition were updated and expanded. A comprehensive double table of contents has been included, listing:

- chemical names with CAS Registry Numbers in alphabetical order, and
- CAS Registry Numbers in numerical order with chemical names.

Each analyte/method summary focuses on the information that people require in order to decide which, among several methods, will be the best for their particular needs. And, an important feature of the format used is that each analyte/method summary contains all of the information required to stand alone as a reference for that analyte analyzed by that method. This format was a popular feature of the first edition and it has been maintained in this edition. Thus, in addition to a brief summary of each method, descriptions include the instrumentation required, interferences, sampling containers, preservation techniques, maximum holding times, detection levels, accuracy, precision, quality control requirements, EPA reference, and, when available, EPA contacts with telephone numbers.

The first edition was also limited in its coverage of semivolatile analytes. This edition greatly increases coverage of this important category of pollutants. It also provides coverage of plasma emission methods (ICP) for elemental analysis and more extensive coverage of pesticides and herbicides. And, like the first edition, this book is primarily a printed version of the second edition of EPA's *Sampling and Analysis Methods Database*.

A software program named *DQO-PRO* may be used with this book or with EPA's *Sampling and Analysis Methods Database*. *DQO-PRO* helps to estimate how many samples are needed to meet your objectives of a given confidence level with your desired quality objectives (such as maximum rates of false positive or false negative results or maximum tolerable error). It also provides estimates of numbers of samples needed for remediation projects using systematic sampling with grids. The program runs in Microsoft® Windows™ with a user interface of a simple calculator. *DQO-PRO* was developed by Radian International LLC for the U.S. Government at a cost of about $10,000 and it is free. You may download *DQL-PRO* from the Internet at http://www.instantref.com/inst-ref.htm or by contacting me by mail, e-mail, phone or fax and I'll be glad to send you a copy.

You may reach me at:

Dr. Lawrence H. Keith
Radian International LLC
P. O. Box 201088
Austin, TX 78720-1088

e-mail: larry_keith@radian.com
Fax: + (512) 345-2386
Telephone: + (512) 419-5412

I hope you'll find this book to be useful in your work. Please feel free to contact me if you have any comments or questions. I'll always be glad to try to help you solve a problem. And, if I don't know the answer, we've got 3,000 other experts at Radian who can help me find what you need.

Lawrence H. Keith, Ph.D.
February 1996

Chemical Table of Contents

Chemical Name	CAS Number	Page	Chemical Name	CAS Number	Page
A			Aramite	140-57-8	64
			Aramite	140-57-8	65
Acenaphthene	83-32-9	1	Aroclor 1016	12674-11-2	66
Acenaphthene	83-32-9	2	Aroclor 1016	12674-11-2	67
Acenaphthene	83-32-9	3	Aroclor 1016	12674-11-2	68
Acenaphthene	83-32-9	4	Aroclor 1016	12674-11-2	69
Acenaphthene	83-32-9	5	Aroclor 1221	11104-28-2	71
Acenaphthylene	208-96-8	5	Aroclor 1221	11104-28-2	72
Acenaphthylene	208-96-8	6	Aroclor 1221	11104-28-2	73
Acenaphthylene	208-96-8	7	Aroclor 1221	11104-28-2	74
Acenaphthylene	208-96-8	9	Aroclor 1232	11141-16-5	76
Acenaphthylene	208-96-8	9	Aroclor 1232	11141-16-5	77
Acetone	67-64-1	10	Aroclor 1232	11141-16-5	78
Acetonitrile	75-05-8	11	Aroclor 1232	11141-16-5	79
Acetonitrile	75-05-8	13	Aroclor 1242	53469-21-9	80
Acetophenone	98-86-2	13	Aroclor 1242	53469-21-9	82
Acetophenone	98-86-2	15	Aroclor 1242	53469-21-9	82
1-Acetyl-2-thiourea	591-08-2	16	Aroclor 1242	53469-21-9	84
2-Acetylaminofluorene	53-96-3	18	Aroclor 1248	12672-29-6	85
Acidity (Titrimetric)		19	Aroclor 1248	12672-29-6	86
Acifluorfen	50594-66-6	19	Aroclor 1248	12672-29-6	87
Acrolein	107-02-8	21	Aroclor 1248	12672-29-6	88
Acrolein	107-02-8	22	Aroclor 1254	11097-69-1	90
Acrolein	107-02-8	23	Aroclor 1254	11097-69-1	91
Acrylamide	79-06-1	24	Aroclor 1254	11097-69-1	92
Acrylonitrile	107-13-1	24	Aroclor 1254	11097-69-1	93
Acrylonitrile	107-13-1	25	Aroclor 1260	11096-82-5	94
Acrylonitrile	107-13-1	27	Aroclor 1260	11096-82-5	96
Alachlor	15972-60-8	27	Aroclor 1260	11096-82-5	97
Alachlor	15972-60-8	29	Aroclor 1260	11096-82-5	98
Aldrin	309-00-2	30	Arsenic	7440-38-2	99
Aldrin	309-00-2	31	Arsenic	7440-38-2	100
Aldrin	309-00-2	32	Arsenic	7440-38-2	101
Aldrin	309-00-2	33	Arsenic	7440-38-2	101
Alkalinity		34	Arsenic	7440-38-2	102
Alkalinity (Titrimetric, pH 4.5)		35	Atraton	1610-17-9	102
Allyl alcohol	107-18-6	35	Atrazine	1912-24-9	103
Allyl alcohol	107-18-6	36	Atrazine	1912-24-9	104
Allyl chloride	107-05-1	38	Azinphos-methyl	86-50-0	105
Aluminum	7429-90-5	39	Azinphos-methyl	86-50-0	107
Aluminum	7429-90-5	40	Azinphos-methyl	86-50-0	108
Aluminum	7429-90-5	41			
Ametryn	834-12-8	41	**B**		
3-Amino-9-ethylcarbazole	132-32-1	42			
2-Aminoanthraquinone	117-79-3	44	Barban	101-27-9	111
Aminoazobenzene	60-09-3	45	Barium	7440-39-3	112
4-Aminobiphenyl	92-67-1	47	Barium	7440-39-3	113
4-Aminobiphenyl	92-67-1	48	Barium	7440-39-3	114
Anilazine	101-05-3	50	Barium	7440-39-3	114
Aniline	62-53-3	51	Barium	7440-39-3	114
Aniline	62-53-3	52	Bentazon	25057-89-0	115
o-Anisidine	90-04-0	54	Benzal chloride	98-87-3	116
o-Anisidine	90-04-0	55	Benzanthrone	82-05-3	118
Anthracene	120-12-7	57	Benzene	71-43-2	119
Anthracene	120-12-7	58	Benzene	71-43-2	120
Anthracene	120-12-7	59	Benzene	71-43-2	121
Anthracene	120-12-7	60	Benzene	71-43-2	122
Anthracene	120-12-7	61	Benzene	71-43-2	124
Antimony	7440-36-0	61	Benzene	71-43-2	125
Antimony	7440-36-0	62	Benzene	71-43-2	127
Antimony	7440-36-0	63	Benzene	71-43-2	127
Antimony	7440-36-0	63	Benzene	71-43-2	128

Chemical Name	CAS Number	Page	Chemical Name	CAS Number	Page
1,3-Benzenediol (Resorcinol)	108-46-3	128	Bis(2-chloroethoxy)methane	111-91-1	203
Benzenethiol	108-98-5	129	Bis(2-chloroethoxy)methane	111-91-1	204
Benzidine	92-87-5	131	Bis(2-chloroethoxy)methane	111-91-1	205
Benzidine	92-87-5	132	Bis(2-chloroethyl) ether	111-44-4	206
Benzidine	92-87-5	133	Bis(2-chloroethyl) ether	111-44-4	208
Benz(a)anthracene	56-55-3	134	Bis(2-chloroethyl) ether	111-44-4	209
Benz(a)anthracene	56-55-3	135	Bis(2-chloroethyl) ether	111-44-4	210
Benz(a)anthracene	56-55-3	136	Bis(2-chloroisopropyl) ether	108-60-1	211
Benz(a)anthracene	56-55-3	138	Bis(2-chloroisopropyl) ether	108-60-1	212
Benz(a)anthracene	56-55-3	138	Bis(2-chloroisopropyl) ether	108-60-1	214
Benzo(a)pyrene	50-32-8	139	Bis(2-chloroisopropyl) ether	108-60-1	214
Benzo(a)pyrene	50-32-8	140	Bis(2-ethoxyethyl) phthalate	605-54-9	216
Benzo(a)pyrene	50-32-8	141	Bis(2-ethylhexl) phthalate	117-81-7	218
Benzo(a)pyrene	50-32-8	142	Bis(2-ethylhexyl) phthalate	117-81-7	219
Benzo(a)pyrene	50-32-8	143	Bis(2-ethylhexyl) phthalate	117-81-7	220
Benzo(b)fluoranthene	205-99-2	143	Bis(2-ethylhexyl) phthalate	117-81-7	221
Benzo(b)fluoranthene	205-99-2	145	Bis(2-ethylhexyl) phthalate	117-81-7	223
Benzo(b)fluoranthene	205-99-2	146	Bis(2-ethylhexyl) phthalate	117-81-7	225
Benzo(b)fluoranthene	205-99-2	147	Bis(2-methoxyethyl) phthalate	117-82-8	225
Benzo(b)fluoranthene	205-99-2	148	Bis(2-n-butoxyethyl) phthalate	117-83-9	227
Benzo(g,h,i)perylene	191-24-2	148	Bis(4-methyl-2-pentyl) phthalate	146-50-9	229
Benzo(g,h,i)perylene	191-24-2	149	Bolstar (Sulprofos)	35400-43-2	231
Benzo(g,h,i)perylene	191-24-2	150	Bolstar (Sulprofos)	35400-43-2	232
Benzo(g,h,i)perylene	191-24-2	152	Boron	7440-42-8	233
Benzo(g,h,i)perylene	191-24-2	152	Boron	7440-42-8	234
Benzo(k)fluoranthene	207-08-9	153	Bromide (Titrimetric)		235
Benzo(k)fluoranthene	207-08-9	154	Bromoacetone	598-31-2	235
Benzo(k)fluoranthene	207-08-9	155	Bromobenzene	108-86-1	237
Benzo(k)fluoranthene	207-08-9	156	Bromobenzene	108-86-1	238
Benzo(k)fluoranthene	207-08-9	157	Bromobenzene	108-86-1	239
Benzo(k)fluoranthene	207-08-9	158	Bromobenzene	108-86-1	240
2,3-Benzofluorene	243-17-4	158	Bromobenzene	108-86-1	242
2,3-Benzofluorene	243-17-4	160	2-Bromochlorobenzene	694-80-4	243
Benzoic acid	65-85-0	161	3-Bromochlorobenzene	108-37-2	244
Benzoic acid	65-85-0	162	Bromochloromethane	74-97-5	245
Benzoic acid	65-85-0	163	Bromochloromethane	74-97-5	246
p-Benzoquinone	106-51-4	165	Bromochloromethane	74-97-5	247
Benzotrichloride	98-07-7	166	Bromochloromethane	74-97-5	248
Benzyl alcohol	100-51-6	168	Bromocil	314-40-9	250
Benzyl alcohol	100-51-6	169	Bromodichloromethane	75-27-4	251
Benzyl alcohol	100-51-6	170	Bromodichloromethane	75-27-4	252
Benzyl butyl phthalate	85-68-7	172	Bromodichloromethane	75-27-4	253
Benzyl chloride	100-44-7	173	Bromodichloromethane	75-27-4	254
Benzyl chloride	100-44-7	174	Bromodichloromethane	75-27-4	256
Beryllium	7440-41-7	176	Bromodichloromethane	75-27-4	257
Beryllium	7440-41-7	177	Bromodichloromethane	75-27-4	259
Beryllium	7440-41-7	178	Bromodichloromethane	75-27-4	259
Beryllium	7440-41-7	178	Bromodichloromethane	75-27-4	260
β-BHC	319-85-7	178	Bromoform	75-25-2	260
δ-BHC	319-86-8	179	Bromoform	75-25-2	261
α-BHC	319-84-6	180	Bromoform	75-25-2	262
β-BHC	319-85-7	181	Bromoform	75-25-2	263
δ-BHC	319-86-8	183	Bromoform	75-25-2	265
α-BHC	319-84-6	184	Bromoform	75-25-2	266
β-BHC	319-85-7	185	Bromoform	75-25-2	268
δ-BHC	319-86-8	187	Bromoform	75-25-2	268
α-BHC	319-84-6	189	Bromoform	75-25-2	269
β-BHC	319-85-7	190	Bromomethane	74-83-9	269
δ-BHC	319-86-8	192	Bromomethane	74-83-9	270
γ-BHC (Lindane)	58-89-9	193	Bromomethane	74-83-9	271
γ-BHC (Lindane)	58-89-9	195	Bromomethane	74-83-9	272
γ-BHC (Lindane)	58-89-9	196	Bromomethane	74-83-9	274
γ-BHC (Lindane)	58-89-9	197	Bromomethane	74-83-9	275
Biochemical Oxygen Demand (BOD)		199	Bromomethane	74-83-9	277
Biphenyl	92-52-4	199	Bromomethane	74-83-9	277
Biphenyl	92-52-4	200	Bromomethane	74-83-9	278
Bis(2-chloroethoxy)methane	111-91-1	202	4-Bromophenyl phenyl ether	101-55-3	278

Chemical Name	CAS Number	Page	Chemical Name	CAS Number	Page
4-Bromophenyl phenyl ether	101-55-3	279	Chloride		350
4-Bromophenyl phenyl ether	101-55-3	280	Chloride		350
Bromoxynil	1689-84-5	282	Chloride		351
Butachlor	23184-66-9	283	Chloride		351
2-Butanone	78-93-3	284	Chlorine		351
Butyl benzyl phthalate	85-68-7	286	Chlorine		352
Butyl benzyl phthalate	85-68-7	287	Chlorine		352
Butyl benzyl phthalate	85-68-7	289	Chlorine		352
Butyl benzyl phthalate	85-68-7	290	Chlorine		353
2-sec-Butyl-4,6-dinitrophenol	88-85-7	291	4-Chloro-1,2-phenylenediamine	95-83-0	353
Butylate	2008-41-5	293	2-Chloro-1,3-butadiene	126-99-8	355
n-Butylbenzene	104-51-8	294	4-Chloro-1,3-phenylenediamine	5131-60-2	355
sec-Butylbenzene	135-98-8	295	5-Chloro-2-methylaniline	95-79-4	357
tert-Butylbenzene	98-06-6	296	4-Chloro-2-nitroaniline	89-63-4	358
n-Butylbenzene	104-51-8	297	4-Chloro-3-methylphenol	59-50-7	360
sec-Butylbenzene	135-98-8	298	4-Chloro-3-methylphenol	59-50-7	361
tert-Butylbenzene	98-06-6	300	4-Chloro-3-methylphenol	59-50-7	362
n-Butylbenzene	104-51-8	301	4-Chloro-3-methylphenol	59-50-7	363
sec-Butylbenzene	135-98-8	302	5-Chloro-o-toluidine	95-79-4	364
tert-Butylbenzene	98-06-6	303	Chloroacetonitrile	107-14-2	365
n-Butylbenzene	104-51-8	304	4-Chloroaniline	106-47-8	366
sec-Butylbenzene	135-98-8	306	4-Chloroaniline	106-47-8	367
tert-Butylbenzene	98-06-6	308	Chlorobenzene	108-90-7	369
n-Butylbenzene	104-51-8	309	Chlorobenzene	108-90-7	370
sec-Butylbenzene	135-98-8	310	Chlorobenzene	108-90-7	371
tert-Butylbenzene	98-06-6	310	Chlorobenzene	108-90-7	372
			Chlorobenzene	108-90-7	373
C			Chlorobenzene	108-90-7	375
			Chlorobenzene	108-90-7	376
Cadmium	7440-43-9	313	Chlorobenzene	108-90-7	377
Cadmium	7440-43-9	314	Chlorobenzene	108-90-7	377
Cadmium	7440-43-9	314	Chlorobenzene	108-90-7	377
Cadmium	7440-43-9	315	Chlorobenzene	108-90-7	378
Calcium	7440-70-2	315	Chlorobenzene	108-90-7	378
Calcium	7440-70-2	316	Chlorobenzilate	510-15-6	379
Calcium	7440-70-2	317	Chlorodibromomethane	124-48-1	380
Calcium	7440-70-2	317	Chlorodibromomethane	124-48-1	381
Captafol	2425-06-1	317	Chlorodibromomethane	124-48-1	382
Captan	133-06-2	319	Chloroethane	75-00-3	384
Carbaryl	63-25-2	320	Chloroethane	75-00-3	385
Carbazole	86-74-8	322	Chloroethane	75-00-3	386
Carbofuran	1563-66-2	323	Chloroethane	75-00-3	387
Carbon disulfide	75-15-0	325	Chloroethane	75-00-3	388
Carbon disulfide	75-15-0	326	Chloroethane	75-00-3	390
Carbon tetrachloride	56-23-5	327	Chloroethane	75-00-3	392
Carbon tetrachloride	56-23-5	328	Chloroethane	75-00-3	392
Carbon tetrachloride	56-23-5	329	Chloroethane	75-00-3	392
Carbon tetrachloride	56-23-5	330	2-Chloroethanol	107-07-3	393
Carbon tetrachloride	56-23-5	332	2-Chloroethyl vinyl ether	110-75-8	395
Carbon tetrachloride	56-23-5	333	2-Chloroethyl vinyl ether	110-75-8	395
Carbon tetrachloride	56-23-5	335	2-Chloroethyl vinyl ether	110-75-8	397
Carbon tetrachloride	56-23-5	335	2-Chloroethyl vinyl ether	110-75-8	397
Carbon tetrachloride	56-23-5	336	2-Chloroethyl vinyl ether	110-75-8	398
Carbophenothion	786-19-6	336	Chloroform	67-66-3	398
Carboxin	5234-68-5	338	Chloroform	67-66-3	399
Chemical Oxygen Demand (COD)		339	Chloroform	67-66-3	400
Chemical Oxygen Demand (COD)		339	Chloroform	67-66-3	401
Chemical Oxygen Demand (COD)		339	Chloroform	67-66-3	403
Chemical Oxygen Demand (COD)		340	Chloroform	67-66-3	404
Chloramben	133-90-4	340	Chloroform	67-66-3	406
Chlordane	57-74-9	342	Chloroform	67-66-3	406
α-Chlordane	5103-71-9	343	Chloroform	67-66-3	407
γ-Chlordane	5103-74-2	344	Chloromethane	74-87-3	407
Chlordane	57-74-9	345	Chloromethane	74-87-3	408
Chlordane	57-74-9	346	Chloromethane	74-87-3	409
Chlordane (technical)	12789-03-6	348	Chloromethane	74-87-3	410
Chlorfenvinphos	470-90-6	349	Chloromethane	74-87-3	412

Chemical Name	CAS Number	Page	Chemical Name	CAS Number	Page
Chloromethane	74-87-3	413	Cyanide	None	478
Chloromethane	74-87-3	415	Cycloate	1134-23-2	479
Chloromethane	74-87-3	415	2-Cyclohexyl-4,6-dinitrophenol	131-89-5	480
Chloromethane	74-87-3	416	p-Cymene	99-87-6	482
3-(Chloromethyl)pyridine hydrochloride	6959-48-4	416			
			D		
1-Chloronaphthalene	90-13-1	418			
2-Chloronaphthalene	91-58-7	419	2,4-D	94-75-7	485
2-Chloronaphthalene	91-58-7	420	2,4-D	94-75-7	486
2-Chloronaphthalene	91-58-7	421	Dalapon	75-99-0	487
2-Chloronaphthalene	91-58-7	422	Dalapon	75-99-0	488
2-Chloronaphthalene	91-58-7	424	2,4-DB	94-82-6	489
3-Chloronitrobenzene	121-73-3	425	2,4-DB	94-82-6	490
2-Chlorophenol	95-57-8	427	DCPA diacid	2136-79-0	491
2-Chlorophenol	95-57-8	428	4,4'-DDD	72-54-8	492
2-Chlorophenol	95-57-8	429	4,4'-DDD	72-54-8	493
2-Chlorophenol	95-57-8	430	4,4'-DDD	72-54-8	494
4-Chlorophenyl phenyl ether	7005-72-3	431	4,4'-DDE	72-55-9	496
4-Chlorophenyl phenyl ether	7005-72-3	432	4,4'-DDE	72-55-9	497
4-Chlorophenyl phenyl ether	7005-72-3	433	4,4'-DDE	72-55-9	498
Chloroprene	126-99-8	434	4,4'-DDT	50-29-3	499
3-Chloropropene	107-05-1	436	4,4'-DDT	50-29-3	500
3-Chloropropionitrile	542-76-7	437	4,4'-DDT	50-29-3	501
2-Chlorotoluene	95-49-8	438	n-Decane	124-18-5	503
4-Chlorotoluene	106-43-4	440	Demeton, -O and -S	8065-48-3	504
2-Chlorotoluene	95-49-8	441	Demeton-O	298-03-3	506
4-Chlorotoluene	106-43-4	442	Demeton-O	298-03-3	507
2-Chlorotoluene	95-49-8	443	Demeton-S	126-75-0	508
4-Chlorotoluene	106-43-4	444	Demeton-S	126-75-0	509
2-Chlorotoluene	95-49-8	446	Di-n-butyl phthalate	84-74-2	510
4-Chlorotoluene	106-43-4	447	Di-n-butyl phthalate	84-74-2	511
2-Chlorotoluene	95-49-8	449	Di-n-butyl phthalate	84-74-2	512
4-Chlorotoluene	106-43-4	449	Di-n-butyl phthalate	84-74-2	513
Chlorpropham	101-21-3	450	Di-n-butyl phthalate	84-74-2	515
Chlorpyrifos	2921-88-2	451	Di-n-octyl phthalate	117-84-0	516
Chlorpyrifos	2921-88-2	452	Di-n-octyl phthalate	117-84-0	517
Chromium	7440-47-3	453	Di-n-octyl phthalate	117-84-0	518
Chromium	7440-47-3	454	Di-n-octyl phthalate	117-84-0	519
Chromium	7440-47-3	454	Di-n-octyl phthalate	117-84-0	521
Chromium (total)	7440-47-3	454	2,6-Di-tert-butyl-p-benzoquinone	719-22-2	522
Chromium VI, Hexavalent		455	Diallate (cis or trans)	2303-16-4	523
Chromium VI, Hexavalent	7440-47-3	456	2,4-Diaminotoluene	95-80-7	525
Chromium VI, Hexavalent	7440-47-3	456	2,4-Diaminotoluene	95-80-7	526
Chromium VI, Hexavalent	7440-47-3	457	Diamyl phthalate	131-18-0	527
Chrysene	218-01-9	457	Diazinon	333-41-5	529
Chrysene	218-01-9	458	Diazinon	333-41-5	530
Chrysene	218-01-9	459	Diazinon	333-41-5	532
Chrysene	218-01-9	461	Dibenz(a,h)anthracene	53-70-3	533
Chrysene	218-01-9	461	Dibenz(a,h)anthracene	53-70-3	534
Cobalt	7440-48-4	462	Dibenz(a,j)acridine	224-42-0	535
Cobalt	7440-48-4	463	Dibenzo(a,e)pyrene	192-65-4	537
Cobalt	7440-48-4	463	Dibenz(a,h)anthracene	53-70-3	538
Cobalt	7440-48-4	464	Dibenz(a,h)anthracene	53-70-3	539
Conductance		464	Dibenz(a,h)anthracene	53-70-3	540
Copper	7440-50-8	464	Dibenzofuran	132-64-9	540
Copper	7440-50-8	466	Dibenzofuran	132-64-9	541
Copper	7440-50-8	466	Dibenzothiophene	132-65-0	543
Copper	7440-50-8	467	1,2-Dibromo-3-chloropropane	96-12-8	544
Coumaphos	56-72-4	467	1,2-Dibromo-3-chloropropane	96-12-8	545
Coumaphos	56-72-4	469	1,2-Dibromo-3-chloropropane	96-12-8	546
Coumaphos	56-72-4	470	1,2-Dibromo-3-chloropropane	96-12-8	548
p-Cresidine	120-71-8	471	1,2-Dibromo-3-chloropropane	96-12-8	549
o-Cresol	95-48-7	472	1,2-Dibromo-3-chloropropane	96-12-8	550
p-Cresol	106-44-5	473	1,2-Dibromo-3-chloropropane	96-12-8	552
Crotonaldehyde	123-73-9	474	1,2-Dibromo-3-chloropropane	96-12-8	554
Crotoxyphos	7700-17-6	475	3,5-Dibromo-4-hydroxybenzonitrile	1689-84-5	554
Crotoxyphos	7700-17-6	477	Dibromochloromethane	124–48-1	555

Chemical Name	CAS Number	Page	Chemical Name	CAS Number	Page
Dibromochloromethane	124–48-1	556	1,3-Dichlorobenzene	541-73-1	633
Dibromochloromethane	124–48-1	557	1,4-Dichlorobenzene	106-46-7	634
Dibromochloromethane	124–48-1	559	1,2-Dichlorobenzene	95-50-1	634
Dibromochloromethane	124–48-1	560	1,3-Dichlorobenzene	541-73-1	634
Dibromochloromethane	124–48-1	561	1,4-Dichlorobenzene	106-46-7	635
Dibromochloromethane	124–48-1	563	1,2-Dichlorobenzene	95-50-1	635
Dibromochloromethane	124–48-1	563	1,3-Dichlorobenzene	541-73-1	635
Dibromochloromethane	124–48-1	564	1,4-Dichlorobenzene	106-46-7	636
1,2-Dibromoethane	106-93-4	564	1,2-Dichlorobenzene	95-50-1	636
1,2-Dibromoethane	106-93-4	565	1,3-Dichlorobenzene	541-73-1	637
1,2-Dibromoethane	106-93-4	567	1,4-Dichlorobenzene	106-46-7	638
1,2-Dibromoethane	106-93-4	568	3,3′-Dichlorobenzidine	91-94-1	638
1,2-Dibromoethane	106-93-4	569	3,3′-Dichlorobenzidine	91-94-1	639
1,2-Dibromoethane	106-93-4	571	3,3′-Dichlorobenzidine	91-94-1	640
1,2-Dibromoethane	106-93-4	572	3,5-Dichlorobenzoic acid	51-36-5	642
Dibromomethane	74-95-3	572	Dichlorodifluoromethane	75-71-8	643
Dibromomethane	74-95-3	573	Dichlorodifluoromethane	75-71-8	644
Dibromomethane	74-95-3	575	Dichlorodifluoromethane	75-71-8	645
Dibromomethane	74-95-3	576	Dichlorodifluoromethane	75-71-8	647
Dibromomethane	74-95-3	577	Dichlorodifluoromethane	75-71-8	648
Dibromomethane	74-95-3	578	Dichlorodifluoromethane	75-71-8	650
Dicamba	1918-00-9	580	1,1-Dichloroethane	75-34-3	650
Dicamba	1918-00-9	582	1,2-Dichloroethane	107-06-2	651
Dichlone	117-80-6	582	1,1-Dichloroethane	75-34-3	652
trans-1,4-Dichloro-2-butene	110-57-6	584	1,2-Dichloroethane	107-06-2	653
1,4-Dichloro-2-butene	764-41-0	585	1,1-Dichloroethane	75-34-3	655
1,3-Dichloro-2-propanol	96-23-1	586	1,2-Dichloroethane	107-06-2	656
1,3-Dichloro-2-propanol	96-23-1	587	1,1-Dichloroethane	75-34-3	657
2,6-Dichloro-4-nitroaniline	99-30-9	589	1,2-Dichloroethane	107-06-2	658
2,3-Dichloroaniline	608-27-5	590	1,1-Dichloroethane	75-34-3	659
1,2-Dichlorobenzene	95-50-1	591	1,2-Dichloroethane	107-06-2	661
1,3-Dichlorobenzene	541-73-1	593	1,1-Dichloroethane	75-34-3	662
1,4-Dichlorobenzene	106-46-7	594	1,2-Dichloroethane	107-06-2	664
1,2-Dichlorobenzene	95-50-1	595	1,1-Dichloroethane	75-34-3	666
1,3-Dichlorobenzene	541-73-1	596	1,2-Dichloroethane	107-06-2	666
1,4-Dichlorobenzene	106-46-7	598	1,1-Dichloroethane	75-34-3	666
1,2-Dichlorobenzene	95-50-1	599	1,2-Dichloroethane	107-06-2	667
1,3-Dichlorobenzene	541-73-1	600	1,1-Dichloroethane	75-34-3	667
1,4-Dichlorobenzene	106-46-7	601	1,2-Dichloroethane	107-06-2	668
1,2-Dichlorobenzene	95-50-1	602	1,1-Dichloroethene	75-35-4	668
1,3-Dichlorobenzene	541-73-1	603	trans-1,2-Dichloroethene	156-60-5	669
1,4-Dichlorobenzene	106-46-7	604	1,1-Dichloroethene	75-35-4	670
1,2-Dichlorobenzene	95-50-1	605	cis-1,2-Dichloroethene	156-59-4	671
1,3-Dichlorobenzene	541-73-1	606	trans-1,2-Dichloroethene	156-60-5	672
1,4-Dichlorobenzene	106-46-7	607	1,1-Dichloroethene	75-35-4	674
1,2-Dichlorobenzene	95-50-1	608	cis-1,2-Dichloroethene	156-59-4	675
1,3-Dichlorobenzene	541-73-1	609	trans-1,2-Dichloroethene	156-60-5	676
1,4-Dichlorobenzene	106-46-7	610	1,1-Dichloroethene	75-35-4	677
1,2-Dichlorobenzene	95-50-1	611	cis-1,2-Dichloroethene	156-59-4	678
1,3-Dichlorobenzene	541-73-1	613	trans-1,2-Dichloroethene	156-60-5	679
1,4-Dichlorobenzene	106-46-7	615	1,1-Dichloroethene	75-35-4	681
1,2-Dichlorobenzene	95-50-1	616	trans-1,2-Dichloroethene	156-60-5	682
1,3-Dichlorobenzene	541-73-1	618	1,1-Dichloroethene	75-35-4	684
1,4-Dichlorobenzene	106-46-7	620	cis-1,2-Dichloroethene	156-59-4	686
1,2-Dichlorobenzene	95-50-1	621	trans-1,2-Dichloroethene	156-60-5	687
1,3-Dichlorobenzene	541-73-1	623	1,1-Dichloroethene	75-35-4	689
1,4-Dichlorobenzene	106-46-7	625	1,1-Dichloroethene	75-35-4	689
1,2-Dichlorobenzene	95-50-1	626	1,1-Dichloroethene	75-35-4	690
1,3-Dichlorobenzene	541-73-1	628	trans-1,2-Dichloroethene	156-60-5	690
1,4-Dichlorobenzene	106-46-7	629	trans-1,2-Dichloroethene	156-60-5	690
1,2-Dichlorobenzene	95-50-1	631	trans-1,2-Dichloroethene	156-60-5	691
1,3-Dichlorobenzene	541-73-1	631	2,3-Dichloronitrobenzene	3209-22-1	691
1,4-Dichlorobenzene	106-46-7	632	2,4-Dichlorophenol	120-83-2	693
1,2-Dichlorobenzene	95-50-1	632	2,6-Dichlorophenol	87-65-0	694
1,3-Dichlorobenzene	541-73-1	632	2,4-Dichlorophenol	120-83-2	695
1,4-Dichlorobenzene	106-46-7	633	2,4-Dichlorophenol	120-83-2	696
1,2-Dichlorobenzene	95-50-1	633	2,6-Dichlorophenol	87-65-0	697

Chemical Name	CAS Number	Page	Chemical Name	CAS Number	Page
2,4-Dichlorophenol	120-83-2	699	3,3'-Dimethoxybenzidine	119-90-4	777
Dichloroprop	120-36-5	699	Dimethyl phthalate	131-11-3	779
1,2-Dichloropropane	78-87-5	700	Dimethyl phthalate	131-11-3	780
1,3-Dichloropropane	142-28-9	701	Dimethyl phthalate	131-11-3	781
1,2-Dichloropropane	78-87-5	702	Dimethyl phthalate	131-11-3	783
1,3-Dichloropropane	142-28-9	703	Dimethyl phthalate	131-11-13	784
2,2-Dichloropropane	590-20-7	704	Dimethyl sulfone	67-71-0	785
1,2-Dichloropropane	78-87-5	705	p-Dimethylaminoazobenzene	60-11-7	786
1,3-Dichloropropane	142-28-9	707	Dimethylaminoazobenzene	60-11-7	787
2,2-Dichloropropane	590-20-7	708	7,12-Dimethylbenz(a)anthracene	57-97-6	789
1,2-Dichloropropane	78-87-5	709	7,12-Dimethylbenz(a)anthracene	57-97-6	790
1,3-Dichloropropane	142-28-9	710	3,3'-Dimethylbenzidine	119-93-7	792
2,2-Dichloropropane	590-20-7	711	N,N-Dimethylformamide	68-12-2	793
1,2-Dichloropropane	78-87-5	712	3,6-Dimethylphenanthrene	1576-67-6	794
1,2-Dichloropropane	78-87-5	714	α,α-Dimethylphenethylamine	122-09-8	795
1,3-Dichloropropane	142-28-9	716	2,4-Dimethylphenol	105-67-9	797
2,2-Dichloropropane	590-20-7	717	2,4-Dimethylphenol	105-67-9	798
1,2-Dichloropropane	78-87-5	719	2,4-Dimethylphenol	105-67-9	799
1,2-Dichloropropane	78-87-5	719	2,4-Dimethylphenol	105-67-9	801
1,2-Dichloropropane	78-87-5	720	4,6-Dinitro-2-methylphenol	534-52-1	801
cis-1,3-Dichloropropene	10061-01-5	720	1,4-Dinitrobenzene	100-25-4	803
trans-1,3-Dichloropropene	10061-02-6	721	1,2-Dinitrobenzene	528-29-0	804
1,1-Dichloropropene	563-58-6	722	1,3-Dinitrobenzene	99-65-0	805
cis-1,3-Dichloropropene	10061-01-5	723	1,4-Dinitrobenzene	100-25-4	807
trans-1,3-Dichloropropene	10061-02-6	725	2,4-Dinitrophenol	51-28-5	808
1,1-Dichloropropene	563-58-6	726	2,4-Dinitrophenol	51-28-5	809
cis-1,3-Dichloropropene	10061-01-5	727	2,4-Dinitrophenol	51-28-5	810
trans-1,3-Dichloropropene	10061-02-6	728	2,4-Dinitrophenol	51-28-5	812
1,1-Dichloropropene	563-58-6	729	2,4-Dinitrotoluene	121-14-2	812
cis-1,3-Dichloropropene	10061-01-5	730	2,6-Dinitrotoluene	606-20-2	814
trans-1,3-Dichloropropene	10061-02-6	732	2,4-Dinitrotoluene	121-14-2	815
cis-1,3-Dichloropropene	10061-01-5	733	2,6-Dinitrotoluene	606-20-2	816
trans-1,3-Dichloropropene	10061-02-6	735	2,4-Dinitrotoluene	121-14-2	817
1,1-Dichloropropene	563-58-6	736	2,6-Dinitrotoluene	606-20-2	818
cis-1,3-Dichloropropene	10061-01-5	738	2,4-Dinitrotoluene	121-14-2	820
cis-1,3-Dichloropropene	10061-01-5	738	2,6-Dinitrotoluene	606-20-2	820
cis-1,3-Dichloropropene	10061-01-5	739	Dinocap	39300-45-3	821
trans-1,3-Dichloropropene	10061-02-6	739	Dinonyl phthalate	84-76-4	822
trans-1,3-Dichloropropene	10061-02-6	739	Dinoseb	88-85-7	824
trans-1,3-Dichloropropene	10061-02-6	740	Dinoseb	88-85-7	826
Dichlorovos	62-73-7	740	Dinoseb	88-85-7	827
Dichlorovos	62-73-7	741	1,4-Dioxane (p-Dioxane)	123-91-1	828
Dichlorovos	62-73-7	743	1,4-Dioxane (p-Dioxane)	123-91-1	829
Dichlorovos	62-73-7	744	Dioxathion	78-34-2	830
Dichlorprop	120-36-5	745	Diphenamid	957-51-7	832
Dicrotophos	141-66-2	746	Diphenyl ether	101-84-8	833
Dicyclohexyl phthalate	84-61-7	748	Diphenylamine	122-39-4	834
Dieldrin	60-57-1	750	Diphenylamine	122-39-4	835
Dieldrin	60-57-1	751	Diphenyldisulfide	882-33-7	837
Dieldrin	60-57-1	752	5,5-Diphenylhydantoin	57-41-0	838
Dieldrin	60-57-1	753	1,2-Diphenylhydrazine	122-66-7	839
1,2:3,4-Diepoxybutane	1464-53-5	755	1,2-Diphenylhydrazine	122-66-7	841
1,2:3,4-Diepoxybutane	1464-53-5	756	Dissolved Oxygen Uptake Rate		842
Diethyl ether	60-29-7	757	Disulfoton	298-04-4	842
Diethyl ether	60-29-7	758	Disulfoton	298-04-4	843
Diethyl phthalate	84-66-2	759	Disulfoton	298-04-4	845
Diethyl phthalate	84-66-2	760	Disulfoton	298-04-4	847
Diethyl phthalate	84-66-2	761	Disulfoton sulfone	2497-06-5	847
Diethyl phthalate	84-66-2	763	Disulfoton sulfoxide	2497-07-6	848
Diethyl phthalate	84-66-2	764	n-Docosane	629-97-0	849
Diethyl sulfate	64-67-5	765	n-Dodecane	112-40-3	850
Diethylstilbestrol	56-53-1	766			
Dihexyl phthalate	84-75-3	768	**E**		
Dihydrosaffrole	56312-13-1	770			
Diisobutyl phthalate	84-69-5	771	n-Eicosane	112-95-8	853
Dimethoate	60-51-5	773	Endosulfan I	959-98-8	854
Dimethoate	60-51-5	775	Endosulfan I	959-98-8	855
3,3'-Dimethoxybenzidine	119-90-4	776	Endosulfan I	959-98-8	856

Chemical Name	CAS Number	Page	Chemical Name	CAS Number	Page
Endosulfan II	33213-65-9	857	Fluorene	86-73-7	934
Endosulfan II	33213-65-9	858	Fluoride		934
Endosulfan II	33213-65-9	859	Fluoride		935
Endosulfan sulfate	1031-07-8	861	Fluoride		935
Endosulfan sulfate	1031-07-8	862	Fluoride		935
Endosulfan sulfate	1031-07-8	863	Fluoride, Total		936
Endrin	72-20-8	864	Fluridone	59756-60-4	936
Endrin	72-20-8	866			
Endrin	72-20-8	866	**H**		
Endrin	72-20-8	868			
Endrin aldehyde	7421-93-4	869	Hardness, Total (mg/L as CaCO3)		939
Endrin aldehyde	7421-93-4	870	Hardness, Total (mg/L as CaCO3)		939
Endrin aldehyde	7421-93-4	871	Heptachlor	76-44-8	939
Endrin ketone	53494-70-5	873	Heptachlor	76-44-8	940
Epichlorohydrin	106-89-8	874	Heptachlor	76-44-8	941
EPN	2104-64-5	876	Heptachlor	76-44-8	942
EPN	2104-64-5	877	Heptachlor epoxide	1024-57-3	944
EPTC	759-94-4	879	Heptachlor epoxide	1024-57-3	945
Ethanol	64-17-5	880	Heptachlor epoxide	1024-57-3	946
Ethanol	64-17-5	881	Heptachlor epoxide	1024-57-3	947
Ethion	563-12-2	882	Hexachlorobenzene	118-74-1	949
Ethoprop	13194-48-4	883	Hexachlorobenzene	118-74-1	950
Ethoprop	13194-48-4	884	Hexachlorobenzene	118-74-1	951
Ethoprop	13194-48-4	886	Hexachlorobenzene	118-74-1	952
Ethyl carbamate	51-79-6	887	Hexachlorobenzene	118-74-1	953
Ethyl cyanide	107-12-0	888	Hexachlorobenzene	118-74-1	954
Ethyl methacrylate	97-63-2	889	Hexachlorobutadiene	87-68-3	956
Ethyl methacrylate	97-63-2	890	Hexachlorobutadiene	87-68-3	957
Ethyl methanesulfonate	62-50-0	891	Hexachlorobutadiene	87-68-3	959
Ethyl methanesulfonate	62-50-0	893	Hexachlorobutadiene	87-88-3	959
Ethyl parathion	56-38-2	894	Hexachlorobutadiene	87-68-3	960
Ethyl parathion	56-38-2	896	Hexachlorobutadiene	87-68-3	961
Ethylbenzene	100-41-4	897	Hexachlorobutadiene	87-68-3	962
Ethylbenzene	100-41-4	898	Hexachlorobutadiene	87-68-3	964
Ethylbenzene	100-41-4	899	Hexachlorobutadiene	87-68-3	966
Ethylbenzene	100-41-4	900	Hexachlorobutadiene	87-68-3	967
Ethylbenzene	100-41-4	902	Hexachlorocyclohexane	608-73-1	968
Ethylbenzene	100-41-4	903	Hexachlorocyclopentadiene	77-47-4	969
Ethylbenzene	100-41-4	905	Hexachlorocyclopentadiene	77-47-4	970
Ethylbenzene	100-41-4	905	Hexachlorocyclopentadiene	77-47-4	971
Ethylbenzene	100-41-4	906	Hexachlorocyclopentadiene	77-47-4	972
Ethylbenzene	100-41-4	906	Hexachlorocyclopentadiene	77-47-4	973
Ethylene oxide	75-21-8	907	Hexachloroethane	67-72-1	975
Ethylenethiourea	96-45-7	908	Hexachloroethane	67-72-1	976
Ethynylestradiol-3 methyl ether	72-33-3	909	Hexachloroethane	67-72-1	977
			Hexachloroethane	67-72-1	978
			Hexachloroethane	67-72-1	980
F			Hexachlorophene	70-30-4	981
Famphur	52-85-7	913	Hexachloropropene	1888-71-7	983
Fenamiphos	22224-92-6	914	Hexachloropropene	1888-71-7	984
Fenarimol	60168-88-9	915	n-Hexacosane	630-01-3	985
Fensulfothion	115-90-2	916	n-Hexadecane	544-76-3	987
Fensulfothion	115-90-2	918	Hexamethyl phosphoramide	680-31-9	988
Fensulfothion	115-90-2	919	Hexanoic acid	142-62-1	989
Fenthion	55-38-9	920	2-Hexanone	591-78-6	991
Fenthion	55-38-9	921	2-Hexanone	591-78-6	991
Fenthion	55-38-9	923	Hexazinone	51235-04-2	993
Fluchloralin	33245-39-5	923	Hexyl 2-ethylhexyl phthalate	75673-16-4	994
Fluoranthene	206-44-0	925	1,2,3,4,6,7,8-HpCDD	35822-46-9	996
Fluoranthene	206-44-0	926	1,2,3,4,6,7,8-HpCDD	35822-46-9	998
Fluoranthene	206-44-0	927	1,2,3,4,6,7,8-HpCDF	67562-39-4	999
Fluoranthene	206-44-0	928	1,2,3,4,6,7,8-HpCDF	67562-39-4	1001
Fluoranthene	206-44-0	929	1,2,3,4,7,8,9-HpCDF	55673-89-7	1002
Fluorene	86-73-7	929	1,2,3,4,7,8-HxCDD	57653-85-7	1003
Fluorene	86-73-7	931	1,2,3,6,7,8-HxCDD	34465-46-8	1005
Fluorene	86-73-7	932	1,2,3,4,7,8-HxCDD	57653-85-7	1007
Fluorene	86-73-7	933	1,2,3,6,7,8-HxCDD	34465-46-8	1008

Chemical Name	CAS Number	Page	Chemical Name	CAS Number	Page
1,2,3,7,8,9-HxCDD	19408-74-3	1009	Magnesium		1070
1,2,3,4,7,8-HxCDD	39227-28-6	1011	Magnesium	7439-95-4	1071
1,2,3,6,7,8-HxCDD	57653-85-7	1012	Malachite green	569-64-2	1071
1,2,3,4,7,8-HxCDF	70648-26-9	1012	Malathion	121-75-5	1072
1,2,3,4,7,8-HxCDF	70648-26-9	1014	Malathion	121-75-5	1074
1,2,3,6,7,8-HxCDF	57117-44-9	1015	Maleic anhydride	108-31-6	1075
1,2,3,7,8,9-HxCDF	72918-21-9	1017	Malononitrile	109-77-3	1077
2,3,4,6,7,8-HxCDF	60851-34-5	1018	Manganese	7439-96-5	1078
1,2,3,4,7,8-HxCDF	70648-26-9	1019	Manganese	7439-96-5	1080
Hydroquinone	123-31-9	1020	Manganese	7439-96-5	1080
5-Hydroxydicamba	7600-50-2	1022	Manganese	7439-96-5	1080
2-Hydroxypropionitrile	78-97-7	1023	Manganese	7439-96-5	1081
			MCPA	94-74-6	1081
I			MCPA	94-74-6	1083
			MCPP	93-65-2	1083
Indeno(1,2,3-cd)pyrene	193-39-5	1027	MCPP	93-65-2	1085
Indeno(1,2,3-cd)pyrene	193-39-5	1027	Mercury	7439-97-6	1085
Indeno(1,2,3-cd)pyrene	193-39-5	1029	Mercury	7439-97-6	1086
Indeno(1,2,3-cd)pyrene	193-39-5	1030	Merphos	150-50-5	1087
Indeno(1,2,3-cd)pyrene	193-39-5	1031	Merphos	150-50-5	1088
Iodide (Titrimetric)		1031	Merphos	150-50-5	1090
Iodomethane	74-88-4	1032	Mestranol	72-33-3	1090
Iodomethane	74-88-4	1032	Methacrylonitrile	126-98-7	1092
Iron	7439-89-6	1034	Methacrylonitrile	126-98-7	1093
Iron	7439-89-6	1035	Methapyrilene	91-80-5	1094
Iron	7439-89-6	1036	Methapyrilene	91-80-5	1096
Iron	7439-89-6	1036	Metholachlor	51218-45-2	1097
Iron	7439-89-6	1036	Methoxychlor	72-43-5	1098
Iron	7439-89-6	1037	4,4'-Methoxychlor	72-43-5	1099
Isobutyl alcohol	78-83-1	1037	Methoxychlor	72-43-5	1100
Isobutyl alcohol	78-83-1	1038	Methyl ethyl ketone (MEK)	78-93-3	1102
Isodrin	465-73-6	1040	Methyl ethyl ketone (MEK)	78-93-3	1103
Isophorone	78-59-1	1041	Methyl iodide	74-88-4	1103
Isophorone	78-59-1	1043	Methyl isobutyl ketone (MIBK)	108-10-1	1105
Isophorone	78-59-1	1043	Methyl methacrylate	80-62-6	1105
Isophorone	78-59-1	1045	Methyl methacrylate	80-62-6	1106
Isopropylbenzene	98-82-8	1045	Methyl methanesulfonate	66-27-3	1108
Isopropylbenzene	98-82-8	1047	Methyl methanesulfonate	66-27-3	1109
Isopropylbenzene	98-82-8	1048	Methyl paraoxon	950-35-6	1111
Isopropylbenzene	98-82-8	1050	Methyl parathion	298-00-0	1112
Isopropylbenzene	98-82-8	1051	Methyl parathion	298-00-0	1113
2-Isopropylnaphthalene	2027-17-0	1051	Methyl parathion	298-00-0	1115
4-Isopropyltoluene	99-87-6	1052	4-Methyl-2-pentanone	108-10-1	1115
4-Isopropyltoluene	99-87-6	1053	4-Methyl-2-pentanone	108-10-1	1116
4-Isopropyltoluene	99-87-6	1054	2-Methyl-4,6-dinitrophenol	534-52-1	1118
4-Isopropyltoluene	99-87-6	1056	2-Methyl-4,6-dinitrophenol	534-52-1	1119
p-Isopropyltoluene	99-87-6	1057	2-Methyl-4,6-dinitrophenol	534-52-1	1120
4-Isopropyltoluene	99-87-6	1057	2-Methyl-5-nitroaniline	99-55-8	1120
Isosafrole	120-58-1	1058	2-Methylbenzothioazole	120-75-2	1122
Isosafrole	120-58-1	1059	3-Methylcholanthrene	56-49-5	1123
			3-Methylcholanthrene	56-49-5	1124
K			Methylene chloride	75-09-2	1126
Kepone	143-50-0	1061	Methylene chloride	75-09-2	1127
			Methylene chloride	75-09-2	1128
L			Methylene chloride	75-09-2	1129
			Methylene chloride	75-09-2	1130
Lead	7439-92-1	1063	Methylene chloride	75-09-2	1132
Lead	7439-92-1	1064	Methylene chloride	75-09-2	1134
Lead	7439-92-1	1064	Methylene chloride	75-09-2	1134
Lead	7439-92-1	1065	Methylene chloride	75-09-2	1134
Leptophos	21609-90-5	1065	4,4'-Methylenebis(2-chloroaniline)	101-14-4	1135
Longifolene	475-20-7	1067	4,4'-Methylenebis(2-chloroaniline)	101-14-4	1136
			4,4'-Methylenebis(N,N-dimethylaniline)	101-61-1	1137
M			4,5-Methylenephenanthrene	203-64-5	1139
			1-Methylfluorene	1730-37-6	1140
Magnesium	7439-95-4	1069	2-Methylnaphthalene	91-57-6	1141
Magnesium		1070	2-Methylnaphthalene	91-57-6	1142

Chemical Name	CAS Number	Page
1-Methylphenanthrene	832-69-9	1144
2-Methylphenol	95-48-7	1145
3-Methylphenol	108-39-4	1147
4-Methylphenol	106-44-5	1148
2-Methylpyridine	109-06-8	1150
2-(Methylthio)-benzothiazole	615-22-5	1151
Metribuzin	21087-64-9	1152
Mevinphos	7786-34-7	1153
Mevinphos	7786-34-7	1154
Mevinphos	7786-34-7	1156
Mevinphos	7786-34-7	1157
Mexacarbate	315-18-4	1158
MGK 264	113-48-4	1159
Mirex	2385-85-5	1161
Molinate	2212-67-1	1162
Molybdenum	7439-98-7	1163
Molybdenum	7439-98-7	1164
Molybdenum	7439-98-7	1165
Molybdenum	7439-98-7	1165
Monocrotophos	6923-22-4	1166
Monocrotophos	6923-22-4	1167

N

Chemical Name	CAS Number	Page
Naled	300-76-5	1171
Naled	300-76-5	1172
Naled	300-76-5	1174
Naphthalene	91-20-3	1174
Naphthalene	91-20-3	1175
Naphthalene	91-20-3	1177
Naphthalene	91-20-3	1178
Naphthalene	91-20-3	1179
Naphthalene	91-20-3	1180
Naphthalene	91-20-3	1182
Naphthalene	91-20-3	1183
Naphthalene	91-20-3	1183
Naphthalene	91-20-3	1184
1,5-Naphthalenediamine	2243-62-1	1184
1,4-Naphthoquinone	130-15-4	1186
1,4-Naphthoquinone	130-15-4	1187
1-Naphthylamine (α-Naphthylamine)	134-32-7	1188
2-Naphthylamine (β-Naphthylamine)	91-59-8	1190
1-Naphthylaminel (α-Naphthylamine)	134-32-7	1191
2-Naphthylamine (β-Naphthylamine)	91-59-8	1192
Napropamide	15299-99-7	1194
Nickel	7440-02-0	1195
Nickel	7440-02-0	1196
Nickel	7440-02-0	1197
Nicotine	54-11-5	1197
Nitrate-N (Total)		1198
5-Nitro-o-anisidine	99-59-2	1199
5-Nitro-o-toluidine	99-55-8	1200
5-Nitro-o-toluidine	99-55-8	1201
5-Nitroacenaphthene	602-87-9	1203
2-Nitroaniline	88-74-4	1204
3-Nitroaniline	99-09-2	1206
4-Nitroaniline	100-01-6	1207
2-Nitroaniline	88-74-4	1208
3-Nitroaniline	99-09-2	1209
4-Nitroaniline	100-01-6	1211
Nitrobenzene	98-95-3	1212
Nitrobenzene	98-95-3	1214
Nitrobenzene	98-95-3	1215
Nitrobenzene	98-95-3	1216
4-Nitrobiphenyl	92-93-3	1217
Nitrofen	1836-75-5	1218
Nitrogen, Ammonia		1219

Chemical Name	CAS Number	Page
Nitrogen, Ammonia		1220
Nitrogen, Ammonia		1220
Nitrogen, Kjeldahl, Total (TKN)		1221
Nitrogen, Kjeldahl, Total (TKN)		1221
Nitrogen, Kjeldahl, Total (TKN)		1221
Nitrogen, Kjeldahl, Total (TKN)		1222
Nitrogen, Nitrate		1222
Nitrogen, Nitrate-Nitrite		1222
Nitrogen, Nitrate-Nitrite		1223
Nitrogen, Nitrate-Nitrite		1223
Nitrogen, Nitrite		1223
2-Nitrophenol	88-75-5	1224
4-Nitrophenol	100-02-7	1225
2-Nitrophenol	88-75-5	1226
4-Nitrophenol	100-02-7	1227
4-Nitrophenol	100-02-7	1228
2-Nitrophenol	88-75-5	1229
4-Nitrophenol	100-02-7	1231
2-Nitrophenol	88-75-5	1232
4-Nitrophenol	100-02-7	1233
Nitroquinoline-1-oxide	56-57-5	1233
N-Nitrosodi-n-butylamine	924-16-3	1235
N-Nitrosodi-n-propylamine	621-64-7	1236
N-Nitrosodi-n-propylamine	621-64-7	1237
N-Nitrosodi-n-propylamine	621-64-7	1238
N-Nitrosodibutylamine	924-16-3	1240
N-Nitrosodiethylamine	55-18-5	1241
N-Nitrosodiethylamine	55-18-5	1242
N-Nitrosodimethylamine	62-75-9	1244
N-Nitrosodimethylamine	62-75-9	1245
N-Nitrosodimethylamine	62-75-9	1246
N-Nitrosodiphenylamine	86-30-6	1247
N-Nitrosodiphenylamine	86-30-6	1249
N-Nitrosodiphenylamine	86-30-6	1249
N-Nitrosomethylethylamine	10595-95-6	1251
N-Nitrosomethylethylamine	10595-95-6	1252
N-Nitrosomethylphenylamine	614-00-6	1254
N-Nitrosomorpholine	59-89-2	1255
N-Nitrosomorpholine	59-89-2	1256
N-Nitrosopiperidine	100-75-4	1257
N-Nitrosopiperidine	100-75-4	1259
N-Nitrosopyrrolidine	930-55-2	1260
cis-Nonachlor	39765-80-5	1262
trans-Nonachlor	39765-80-5	1263
Norflurazon	27314-13-2	1264

O

Chemical Name	CAS Number	Page
OCDD	3268-87-9	1267
OCDD	3268-87-9	1268
OCDF	39001-02-0	1270
OCDF	39001-02-0	1271
n-Octacosane	630-02-4	1273
n-Octadecane	593-45-3	1274
Organic Carbon, Total		1275
Osmium	7440-04-2	1275
Oxygen		1276
Oxygen		1276
Oxygen, Dissolved (DO)		1276
Oxygen, Dissolved (DO)		1277

P

Chemical Name	CAS Number	Page
Paraldehyde	123-3-7	1279
Pebulate	1114-71-2	1279
1,2,3,4,7-PeCDD	39227-61-7	1280
1,2,3,7,8-PeCDD	40321-76-4	1282

Chemical Name	CAS Number	Page	Chemical Name	CAS Number	Page
1,2,3,7,8-PeCDD	40321-76-4	1283	Propionitrile	107-12-0	1359
1,2,3,7,8-PeCDF	57117-41-6	1285	n-Propylamine	107-10-8	1361
1,2,3,7,8-PeCDF	57117-41-6	1286	n-Propylbenzene	103-65-1	1363
2,3,4,7,8-PeCDF	57117-31-4	1288	n-Propylbenzene	103-65-1	1364
1,2,3,7,8-PeCDF	57117-41-6	1289	n-Propylbenzene	103-65-1	1365
Pentachlorobenzene	608-93-5	1290	n-Propylbenzene	103-65-1	1366
Pentachlorobenzene	608-93-5	1291	n-Propylbenzene	103-65-1	1368
Pentachlorobenzene	608-93-5	1293	Propylthiouracil	51-52-5	1368
Pentachloroethane	76-01-7	1294	Pyrene	129-00-0	1370
Pentachloroethane	76-01-7	1295	Pyrene	129-00-0	1371
Pentachlorohexane	96989-91-2	1297	Pyrene	129-00-0	1372
Pentachloronitrobenzene	82-68-8	1298	Pyrene	129-00-0	1373
Pentachlorophenol	87-86-5	1299	Pyrene	129-00-0	1375
Pentachlorophenol	87-86-5	1300	Pyrene	129-00-0	1375
Pentachlorophenol	87-86-5	1301	Pyridine	110-86-1	1376
Pentachlorophenol	87-86-5	1303	Pyridine	110-86-1	1377
Pentachlorophenol	87-86-5	1304	Pyridine	110-86-1	1378
Pentamethylbenzene	700-12-9	1305	Pyridine	110-86-1	1380
Perylene	198-55-0	1306			
pH		1307	**R**		
pH (electrometric)		1308			
Phenacetin	62-44-2	1308	Residue, Filterable (TDS)		1383
Phenacetin	62-44-2	1309	Residue, Volatile (VSS) and (VS)		1383
Phenanthrene	85-01-8	1311	Resorcinol	108-46-3	1383
Phenanthrene	85-01-8	1312	Ronnel	299-84-3	1385
Phenanthrene	85-01-8	1313	Ronnel	299-84-3	1386
Phenanthrene	85-01-8	1314			
Phenanthrene	85-01-8	1315	**S**		
Phenobarbital	50-06-6	1315			
Phenol	108-95-2	1317	Safrole	94-59-7	1389
Phenol	108-95-2	1318	Safrole	94-59-7	1390
Phenol	108-95-2	1319	Safrole	94-59-7	1391
Phenol	108-95-2	1320	Selenium	7782-49-2	1392
Phenothiazine	92-84-2	1321	Selenium	7782-49-2	1394
1,4-Phenylenediamine	106-50-3	1322	Selenium	7782-49-2	1394
1-Phenylnaphthalene	605-02-7	1324	Selenium	7782-49-2	1395
2-Phenylnaphthalene	612-94-2	1325	Silicon	7440-21-3	1395
Phorate	298-02-2	1326	Silicon	7440-21-3	1396
Phorate	298-02-2	1328	Silver	7440-22-4	1397
Phorate	298-02-2	1329	Silver	7440-22-4	1398
Phosalone	2310-17-0	1330	Silver	7440-22-4	1398
Phosmet	732-11-6	1331	Silver	7440-22-4	1399
Phosphamidon	13171-21-6	1333	Simazine	122-34-9	1399
Phosphorus	7723-14-0	1334	Simazine	122-34-9	1400
Phosphorus	7723-14-0	1335	Simetryn	1014-70-6	1401
Phosphorus	7723-14-0	1336	Sodium	7440-23-5	1402
Phosphorus	7723-14-0	1336	Sodium	7440-23-5	1404
Phosphorus	7723-14-0	1336	Sodium	7440-23-5	1404
Phthalic anhydride	85-44-9	1337	Sodium	7440-23-5	1404
Picloram	1918-02-1	1338	Specific Conductance	None	1405
2-Picoline	109-06-8	1340	Squalene	7683-64-9	1405
2-Picoline	109-06-8	1341	Stirophos (Tetrachlorovinphos)	22248-79-9	1407
2-Picoline	109-06-8	1342	Stirophos (Tetrachlorovinphos)	22248-79-9	1408
2-Picoline	109-06-8	1344	Stirophos (Tetrachlorovinphos)	22248-79-9	1409
Piperonyl sulfoxide	120-62-7	1345	Strychnine	60-41-3	1410
Potassium	7440-09-7	1347	Styrene	100-42-5	1411
Potassium	7440-09-7	1347	Styrene	100-42-5	1413
Potassium	7440-09-7	1348	Styrene	100-42-5	1414
Prometon	1610-18-0	1348	Styrene	100-42-5	1415
Prometryn	7287-19-6	1349	Styrene	100-42-5	1416
Pronamide	23950-58-5	1350	Styrene	100-42-5	1417
Pronamide	23950-58-5	1351	Styrene	100-42-5	1419
Pronamide	23950-58-5	1353	Styrene	100-42-5	1421
Pronamide	23950-58-5	1354	Sulfallate	95-06-7	1421
Propargyl alcohol	107-19-7	1355	Sulfate		1423
Propazine	139-40-2	1357	Sulfate		1423
β-Propiolactone	57-57-8	1358	Sulfate (Gravimetric)		1423

Chemical Name	CAS Number	Page	Chemical Name	CAS Number	Page
Sulfate (Total)		1424	Thallium	7440-28-0	1500
Sulfate (Turbidometric)		1424	Thianaphthene(2,3-benzothiophene)	95-15-8	1500
Sulfotep	3689-24-5	1424	Thioacetamide	62-55-5	1502
			Thionazine	297-97-2	1503
T			Thiophenol (Benzenethiol)	108-98-5	1504
			Thioxanthone	492-22-8	1506
2,4,5-T	93-76-5	1427	Tin	7440-31-5	1507
1,2,3,4-TCDD	30746-58-8	1428	Tin	7440-31-5	1508
1,2,7,8-TCDD	34816-53-0	1430	Tokuthion (Protothiofos)	34643-46-4	1508
1,2,8,9-TCDD	62470-54-6	1431	Tokuthion (Protothiofos)	34643-46-4	1509
1,3,6,8-TCDD	33423-92-6	1433	Toluene	108-88-3	1511
1,3,7,8-TCDD	50585-46-1	1434	Toluene	108-88-3	1511
1,3,7,9-TCDD	62470-53-5	1436	Toluene	108-88-3	1513
2,3,7,8-TCDD	1746-01-6	1437	Toluene	108-88-3	1514
2,3,7,8-TCDD	1746-01-6	1439	Toluene	108-88-3	1515
1,2,7,8-TCDF	58802-20-3	1440	Toluene	108-88-3	1517
2,3,7,8-TCDF	51207-31-9	1442	Toluene	108-88-3	1518
2,3,7,8-TCDF	51207-31-9	1443	Toluene	108-88-3	1519
Tebuthiuron	34014-18-1	1445	Toluene	108-88-3	1519
Temperature (Thermometric)		1446	Toluene	108-88-3	1519
TEPP	21646-99-1	1446	Toluene diisocyanate	584-84-9	1520
Terbacil	5902-51-2	1448	o-Toluidine	95-53-4	1521
Terbufos	13071-79-9	1449	o-Toluidine	95-53-4	1523
Terbufos	13071-79-9	1450	Total Organic Halides (TOX)		1524
Terbutryn	886-50-0	1451	Total Solids (TS)		1524
α-Terpineol	98-55-5	1452	Total Suspended Solids (TSS)		1525
1,2,4,5-Tetrachlorobenzene	95-94-3	1453	Toxaphene	8001-35-2	1525
1,2,3,4-Tetrachlorobenzene	634-66-2	1455	Toxaphene	8001-35-2	1526
1,2,3,5-Tetrachlorobenzene	634-90-2	1456	Toxaphene	8001-35-2	1527
1,2,4,5-Tetrachlorobenzene	95-94-2	1458	Toxaphene	8001-35-2	1528
1,2,4,5-Tetrachlorobenzene	95-94-3	1461	2,4,5-TP (Silvex)	93-72-1	1530
Tetrachlorobenzenes (3 isomers)	Various	1462	2,4,5-TP (Silvex)	93-72-1	1530
1,1,1,2-Tetrachloroethane	630-20-6	1462	Tri-p-tolyl phosphate	78-32-0	1532
1,1,2,2-Tetrachloroethane	79-34-5	1463	n-Triacontane	638-68-6	1533
1,1,1,2-Tetrachloroethane	630-20-6	1464	Triademefon	43121-43-3	1534
1,1,2,2-Tetrachloroethane	79-34-5	1465	1,2,3-Trichlorobenzene	87-61-6	1536
1,1,1,2-Tetrachloroethane	630-20-6	1467	1,2,4-Trichlorobenzene	120-82-1	1537
1,1,2,2-Tetrachloroethane	79-34-5	1468	1,2,3-Trichlorobenzene	87-61-6	1538
1,1,1,2-Tetrachloroethane	630-20-6	1469	1,2,4-Trichlorobenzene	120-82-1	1539
1,1,2,2-Tetrachloroethane	79-34-5	1470	1,2,3-Trichlorobenzene	87-61-6	1540
1,1,1,2-Tetrachloroethane	630-20-6	1471	1,2,4-Trichlorobenzene	120-82-1	1541
1,1,2,2-Tetrachloroethane	79-34-5	1473	1,2,4-Trichlorobenzene	120-82-1	1543
1,1,1,2-Tetrachloroethane	630-20-6	1474	1,2,3-Trichlorobenzene	87-61-6	1543
1,1,2,2-Tetrachloroethane	79-34-5	1476	1,2,4-Trichlorobenzene	120-82-1	1545
1,1,2,2-Tetrachloroethane	79-34-5	1478	1,2,4-Trichlorobenzene	120-82-1	1546
1,1,2,2-Tetrachloroethane	79-34-5	1478	1,2,3-Trichlorobenzene	87-61-6	1547
1,1,2,2-Tetrachloroethane	79-34-5	1478	1,2,4-Trichlorobenzene	120-82-1	1548
Tetrachloroethylene	127-18-4	1479	1,3,5-Trichlorobenzene	108-70-3	1550
Tetrachloroethylene	127-18-4	1480	1,2,3-Trichlorobenzene	87-61-6	1552
Tetrachloroethylene	127-18-4	1481	1,2,4-Trichlorobenzene	120-82-1	1554
Tetrachloroethylene	127-18-4	1482	1,2,4-Trichlorobenzene	120-82-1	1555
Tetrachloroethylene	127-18-4	1483	1,2,3-Trichlorobenzene	87-61-6	1557
Tetrachloroethylene	127-18-4	1485	1,2,4-Trichlorobenzene	120-82-1	1557
Tetrachloroethylene	127-18-4	1487	1,1,1-Trichloroethane	71-55-6	1558
Tetrachloroethylene	127-18-4	1487	1,1,2-Trichloroethane	79-00-5	1558
Tetrachloroethylene	127-18-4	1487	1,1,1-Trichloroethane	71-55-6	1559
Tetrachloroethylene	127-18-4	1488	1,1,2-Trichloroethane	79-00-5	1561
2,3,4,6-Tetrachlorophenol	58-90-2	1488	1,1,1-Trichloroethane	71-55-6	1562
2,3,4,6-Tetrachlorophenol	58-90-2	1489	1,1,2-Trichloroethane	79-00-5	1563
Tetrachlorvinphos	961-11-5	1491	1,1,1-Trichloroethane	71-55-6	1564
n-Tetracosane	646-31-1	1492	1,1,2-Trichloroethane	79-00-5	1565
n-Tetradecane	629-59-4	1494	1,1,1-Trichloroethane	71-55-6	1566
Tetraethyl dithiopyrophosphate	3689-24-5	1495	1,1,2-Trichloroethane	79-00-5	1568
Tetraethyl pyrophosphate	107-49-3	1496	1,1,1-Trichloroethane	71-55-6	1570
Thallium	7440-28-0	1498	1,1,2-Trichloroethane	79-00-5	1571
Thallium	7440-28-0	1499	1,1,1-Trichloroethane	71-55-6	1573
Thallium	7440-28-0	1500	1,1,2-Trichloroethane	79-00-5	1573

Chemical Name	CAS Number	Page	Chemical Name	CAS Number	Page
1,1,1-Trichloroethane	71-55-6	1574	Tris(2,3-dibromopropyl) phosphate	126-72-7	1636
1,1,2-Trichloroethane	79-00-5	1574	1,3,5-Trithiane	291-21-4	1638
1,1,1-Trichloroethane	71-55-6	1574	Turbidity (Nephelometric)		1639
1,1,2-Trichloroethane	79-00-5	1575			
Trichloroethylene	79-01-6	1575	**V**		
Trichloroethylene	79-01-6	1576			
Trichloroethylene	79-01-6	1577	Vanadium	7440-62-2	1641
Trichloroethylene	79-01-6	1579	Vanadium	7440-62-2	1642
Trichloroethylene	79-01-6	1580	Vanadium	7440-62-2	1642
Trichloroethylene	79-01-6	1581	Vanadium	7440-62-2	1643
Trichloroethylene	79-01-6	1583	Vernolate	1929-77-7	1643
Trichloroethylene	79-01-6	1583	Vinyl acetate	108-05-4	1644
Trichloroethylene	79-01-6	1584	Vinyl acetate	108-05-4	1645
Trichloroethylene	79-01-6	1584	Vinyl chloride	75-01-4	1647
Trichlorofluoromethane	75-69-4	1585	Vinyl chloride	75-01-4	1648
Trichlorofluoromethane	75-69-4	1586	Vinyl chloride	75-01-4	1649
Trichlorofluoromethane	75-69-4	1587	Vinyl chloride	75-01-4	1650
Trichlorofluoromethane	75-69-4	1588	Vinyl chloride	75-01-4	1651
Trichlorofluoromethane	75-69-4	1589	Vinyl chloride	75-01-4	1653
Trichlorofluoromethane	75-69-4	1591	Vinyl chloride	75-01-4	1654
Trichlorofluoromethane	75-69-4	1592	Vinyl chloride	75-01-4	1656
Trichlorofluoromethane	75-69-4	1593	Vinyl chloride	75-01-4	1656
Trichlorofluoromethane	75-69-4	1593	Vinyl chloride	75-01-4	1657
Trichloronate	327-98-0	1594			
Trichloronate	327-98-0	1595	**X**		
2,3,6-Trichlorophenol	933-75-5	1596			
2,4,5-Trichlorophenol	95-95-4	1597	m-Xylene	108-38-3	1659
2,4,6-Trichlorophenol	88-06-2	1598	o-Xylene	95-47-6	1660
2,4,6-Trichlorophenol	88-06-2	1599	p-Xylene	106-42-3	1660
2,4,5-Trichlorophenol	95-95-4	1600	m-Xylene	108-38-3	1661
2,4,6-Trichlorophenol	88-06-2	1602	o-Xylene	95-47-6	1663
2,4,6-Trichlorophenol	88-06-2	1603	p-Xylene	106-42-3	1664
1,2,3-Trichloropropane	96-18-4	1604	m-Xylene	108-38-3	1665
1,2,3-Trichloropropane	96-18-4	1605	o-Xylene	95-47-6	1666
1,2,3-Trichloropropane	96-18-4	1606	o-Xylene	95-47-6	1667
1,2,3-Trichloropropane	96-18-4	1607	p-Xylene	106-42-3	1668
1,2,3-Trichloropropane	96-18-4	1608	p-Xylene	106-42-3	1669
1,2,3-Trichloropropane	96-18-4	1611	m-Xylene	108-38-3	1671
Tricyclozole	41814-78-2	1612	o-Xylene	95-47-6	1671
O,O,O-Triethyl phosphorothioate	126-68-1	1613	p-Xylene	106-42-3	1673
Trifluralin	1582-09-8	1614	m-Xylene	108-38-3	1674
1,2,3-Trimethoxybenzene	634-36-6	1616	m-Xylene	108-38-3	1676
Trimethyl phosphate	512-56-1	1617	o-Xylene	95-47-6	1678
2,4,5-Trimethylaniline	137-17-7	1618	o-Xylene	95-47-6	1679
2,4,5-Trimethylaniline	137-17-7	1619	p-Xylene	106-42-3	1681
1,2,4-Trimethylbenzene	95-63-6	1621	p-Xylene	106-42-3	1683
1,3,5-Trimethylbenzene	108-67-8	1622	m-Xylene	108-38-3	1684
1,2,4-Trimethylbenzene	95-63-6	1623	o-Xylene	95-47-6	1685
1,3,5-Trimethylbenzene	108-67-8	1624	p-Xylene	106-42-3	1685
1,2,4-Trimethylbenzene	95-63-6	1626	Xylenes	1330-20-7	1686
1,3,5-Trimethylbenzene	108-67-8	1627	Xylenes	1330-20-7	1687
1,2,4-Trimethylbenzene	95-63-6	1628			
1,3,5-Trimethylbenzene	108-67-8	1631	**Z**		
1,2,4-Trimethylbenzene	95-63-6	1631			
1,3,5-Trimethylbenzene	108-67-8	1632	Zinc	7440-66-6	1689
1,3,5-Trinitrobenzene	99-35-4	1632	Zinc	7440-66-6	1690
Triphenylene	217-59-4	1634	Zinc	7440-66-6	1690
Tripropyleneglycolmethyl ether	20324-33-8	1635	Zinc	7440-66-6	1691

CAS Number Table of Contents

CAS Number	Chemical Name	Page	CAS Number	Chemical Name	Page
50-06-6	Phenobarbital	1317	58-89-9	γ-BHC (Lindane)	195
50-06-6	Phenobarbital	1315	58-89-9	γ-BHC (Lindane)	196
50-29-3	4,4'-DDT	499	58-89-9	γ-BHC (Lindane)	197
50-29-3	4,4'-DDT	500	58-90-2	2,3,4,6-Tetrachlorophenol	1488
50-29-3	4,4'-DDT	501	58-90-2	2,3,4,6-Tetrachlorophenol	1489
50-32-8	Benzo(a)pyrene	139	59-50-7	4-Chloro-3-methylphenol	360
50-32-8	Benzo(a)pyrene	140	59-50-7	4-Chloro-3-methylphenol	361
50-32-8	Benzo(a)pyrene	141	59-50-7	4-Chloro-3-methylphenol	362
50-32-8	Benzo(a)pyrene	142	59-50-7	4-Chloro-3-methylphenol	363
50-32-8	Benzo(a)pyrene	143	59-89-2	N-Nitrosomorpholine	1255
51-28-5	2,4-Dinitrophenol	808	59-89-2	N-Nitrosomorpholine	1256
51-28-5	2,4-Dinitrophenol	809	60-09-3	Aminoazobenzene	45
51-28-5	2,4-Dinitrophenol	810	60-11-7	Dimethylaminoazobenzene	787
51-28-5	2,4-Dinitrophenol	812	60-11-7	p-Dimethylaminoazobenzene	786
51-36-5	3,5-Dichlorobenzoic acid	642	60-29-7	Diethyl ether	757
51-52-5	Propylthiouracil	1368	60-29-7	Diethyl ether	758
51-79-6	Ethyl carbamate	887	60-41-3	Strychnine	1410
52-85-7	Famphur	913	60-51-5	Dimethoate	773
53-70-3	Dibenz(a,h)anthracene	533	60-51-5	Dimethoate	775
53-70-3	Dibenz(a,h)anthracene	534	60-57-1	Dieldrin	750
53-70-3	Dibenz(a,h)anthracene	538	60-57-1	Dieldrin	751
53-70-3	Dibenz(a,h)anthracene	539	60-57-1	Dieldrin	752
53-70-3	Dibenz(a,h)anthracene	540	60-57-1	Dieldrin	753
53-96-3	2-Acetylaminofluorene	18	62-44-2	Phenacetin	1308
54-11-5	Nicotine	1197	62-44-2	Phenacetin	1309
55-18-5	N-Nitrosodiethylamine	1241	62-50-0	Ethyl methanesulfonate	891
55-18-5	N-Nitrosodiethylamine	1242	62-50-0	Ethyl methanesulfonate	893
55-38-9	Fenthion	920	62-53-3	Aniline	51
55-38-9	Fenthion	921	62-53-3	Aniline	52
55-38-9	Fenthion	923	62-55-5	Thioacetamide	1502
56-23-5	Carbon tetrachloride	327	62-73-7	Dichlorovos	740
56-23-5	Carbon tetrachloride	328	62-73-7	Dichlorovos	741
56-23-5	Carbon tetrachloride	329	62-73-7	Dichlorovos	743
56-23-5	Carbon tetrachloride	330	62-73-7	Dichlorovos	744
56-23-5	Carbon tetrachloride	332	62-75-9	N-Nitrosodimethylamine	1244
56-23-5	Carbon tetrachloride	333	62-75-9	N-Nitrosodimethylamine	1245
56-23-5	Carbon tetrachloride	335	62-75-9	N-Nitrosodimethylamine	1246
56-23-5	Carbon tetrachloride	335	63-25-2	Carbaryl	320
56-23-5	Carbon tetrachloride	336	64-17-5	Ethanol	880
56-38-2	Ethyl parathion	894	64-17-5	Ethanol	881
56-38-2	Ethyl parathion	896	64-67-5	Diethyl sulfate	765
56-49-5	3-Methylcholanthrene	1123	65-85-0	Benzoic acid	161
56-49-5	3-Methylcholanthrene	1124	65-85-0	Benzoic acid	162
56-53-1	Diethylstilbestrol	766	65-85-0	Benzoic acid	163
56-55-3	Benz(a)anthracene	134	66-27-3	Methyl methanesulfonate	1108
56-55-3	Benz(a)anthracene	135	66-27-3	Methyl methanesulfonate	1109
56-55-3	Benz(a)anthracene	136	67-64-1	Acetone	10
56-55-3	Benz(a)anthracene	138	67-66-3	Chloroform	398
56-55-3	Benz(a)anthracene	138	67-66-3	Chloroform	399
56-57-5	Nitroquinoline-1-oxide	1233	67-66-3	Chloroform	400
56-72-4	Coumaphos	467	67-66-3	Chloroform	401
56-72-4	Coumaphos	469	67-66-3	Chloroform	403
56-72-4	Coumaphos	470	67-66-3	Chloroform	404
57-41-0	5,5-Diphenylhydantoin	838	67-66-3	Chloroform	406
57-57-8	β-Propiolactone	1358	67-66-3	Chloroform	406
57-74-9	Chlordane	342	67-66-3	Chloroform	407
57-74-9	Chlordane	345	67-71-0	Dimethyl sulfone	785
57-74-9	Chlordane	346	67-72-1	Hexachloroethane	975
57-97-6	7,12-Dimethylbenz(a)anthracene	789	67-72-1	Hexachloroethane	976
57-97-6	7,12-Dimethylbenz(a)anthracene	790	67-72-1	Hexachloroethane	977
58-89-9	γ-BHC (Lindane)	193	67-72-1	Hexachloroethane	978

CAS Number	Chemical Name	Page	CAS Number	Chemical Name	Page
67-72-1	Hexachloroethane	980	75-00-3	Chloroethane	384
68-12-2	N,N-Dimethylformamide	793	75-00-3	Chloroethane	385
70-30-4	Hexachlorophene	981	75-00-3	Chloroethane	386
71-43-2	Benzene	119	75-00-3	Chloroethane	387
71-43-2	Benzene	120	75-00-3	Chloroethane	388
71-43-2	Benzene	121	75-00-3	Chloroethane	390
71-43-2	Benzene	122	75-00-3	Chloroethane	392
71-43-2	Benzene	124	75-00-3	Chloroethane	392
71-43-2	Benzene	125	75-00-3	Chloroethane	392
71-43-2	Benzene	127	75-01-4	Vinyl chloride	1647
71-43-2	Benzene	127	75-01-4	Vinyl chloride	1648
71-43-2	Benzene	128	75-01-4	Vinyl chloride	1649
71-55-6	1,1,1-Trichloroethane	1558	75-01-4	Vinyl chloride	1650
71-55-6	1,1,1-Trichloroethane	1559	75-01-4	Vinyl chloride	1651
71-55-6	1,1,1-Trichloroethane	1562	75-01-4	Vinyl chloride	1653
71-55-6	1,1,1-Trichloroethane	1564	75-01-4	Vinyl chloride	1654
71-55-6	1,1,1-Trichloroethane	1566	75-01-4	Vinyl chloride	1656
71-55-6	1,1,1-Trichloroethane	1570	75-01-4	Vinyl chloride	1656
71-55-6	1,1,1-Trichloroethane	1573	75-01-4	Vinyl chloride	1657
71-55-6	1,1,1-Trichloroethane	1574	75-05-8	Acetonitrile	11
71-55-6	1,1,1-Trichloroethane	1574	75-05-8	Acetonitrile	13
72-20-8	Endrin	864	75-09-2	Methylene chloride	1126
72-20-8	Endrin	866	75-09-2	Methylene chloride	1127
72-20-8	Endrin	866	75-09-2	Methylene chloride	1128
72-20-8	Endrin	868	75-09-2	Methylene chloride	1129
72-33-3	Ethynylestradiol-3 methyl ether	909	75-09-2	Methylene chloride	1130
72-33-3	Mestranol	1090	75-09-2	Methylene chloride	1132
72-43-5	4,4'-Methoxychlor	1099	75-09-2	Methylene chloride	1134
72-43-5	Methoxychlor	1098	75-09-2	Methylene chloride	1134
72-43-5	Methoxychlor	1100	75-09-2	Methylene chloride	1134
72-54-8	4,4'-DDD	492	75-15-0	Carbon disulfide	325
72-54-8	4,4'-DDD	493	75-15-0	Carbon disulfide	326
72-54-8	4,4'-DDD	494	75-21-8	Ethylene oxide	907
72-55-9	4,4'-DDE	496	75-25-2	Bromoform	260
72-55-9	4,4'-DDE	497	75-25-2	Bromoform	261
72-55-9	4,4'-DDE	498	75-25-2	Bromoform	262
74-83-9	Bromomethane	269	75-25-2	Bromoform	263
74-83-9	Bromomethane	270	75-25-2	Bromoform	265
74-83-9	Bromomethane	271	75-25-2	Bromoform	266
74-83-9	Bromomethane	272	75-25-2	Bromoform	268
74-83-9	Bromomethane	274	75-25-2	Bromoform	268
74-83-9	Bromomethane	275	75-25-2	Bromoform	269
74-83-9	Bromomethane	277	75-27-4	Bromodichloromethane	251
74-83-9	Bromomethane	277	75-27-4	Bromodichloromethane	252
74-83-9	Bromomethane	278	75-27-4	Bromodichloromethane	253
74-87-3	Chloromethane	407	75-27-4	Bromodichloromethane	254
74-87-3	Chloromethane	408	75-27-4	Bromodichloromethane	256
74-87-3	Chloromethane	409	75-27-4	Bromodichloromethane	257
74-87-3	Chloromethane	410	75-27-4	Bromodichloromethane	259
74-87-3	Chloromethane	412	75-27-4	Bromodichloromethane	259
74-87-3	Chloromethane	413	75-27-4	Bromodichloromethane	260
74-87-3	Chloromethane	415	75-34-3	1,1-Dichloroethane	650
74-87-3	Chloromethane	415	75-34-3	1,1-Dichloroethane	652
74-87-3	Chloromethane	416	75-34-3	1,1-Dichloroethane	655
74-88-4	Iodomethane	1032	75-34-3	1,1-Dichloroethane	657
74-88-4	Iodomethane	1032	75-34-3	1,1-Dichloroethane	659
74-88-4	Methyl iodide	1103	75-34-3	1,1-Dichloroethane	662
74-95-3	Dibromomethane	572	75-34-3	1,1-Dichloroethane	666
74-95-3	Dibromomethane	573	75-34-3	1,1-Dichloroethane	666
74-95-3	Dibromomethane	575	75-34-3	1,1-Dichloroethane	667
74-95-3	Dibromomethane	576	75-35-4	1,1-Dichloroethene	668
74-95-3	Dibromomethane	577	75-35-4	1,1-Dichloroethene	670
74-95-3	Dibromomethane	578	75-35-4	1,1-Dichloroethene	674
74-97-5	Bromochloromethane	245	75-35-4	1,1-Dichloroethene	677
74-97-5	Bromochloromethane	246	75-35-4	1,1-Dichloroethene	681
74-97-5	Bromochloromethane	247	75-35-4	1,1-Dichloroethene	684
74-97-5	Bromochloromethane	248	75-35-4	1,1-Dichloroethene	689

CAS Number	Chemical Name	Page	CAS Number	Chemical Name	Page
75-35-4	1,1-Dichloroethene	689	79-01-6	Trichloroethylene	1583
75-35-4	1,1-Dichloroethene	690	79-01-6	Trichloroethylene	1584
75-69-4	Trichlorofluoromethane	1585	79-01-6	Trichloroethylene	1584
75-69-4	Trichlorofluoromethane	1586	79-06-1	Acrylamide	24
75-69-4	Trichlorofluoromethane	1587	79-34-5	1,1,2,2-Tetrachloroethane	1463
75-69-4	Trichlorofluoromethane	1588	79-34-5	1,1,2,2-Tetrachloroethane	1465
75-69-4	Trichlorofluoromethane	1589	79-34-5	1,1,2,2-Tetrachloroethane	1468
75-69-4	Trichlorofluoromethane	1591	79-34-5	1,1,2,2-Tetrachloroethane	1470
75-69-4	Trichlorofluoromethane	1592	79-34-5	1,1,2,2-Tetrachloroethane	1473
75-69-4	Trichlorofluoromethane	1593	79-34-5	1,1,2,2-Tetrachloroethane	1476
75-69-4	Trichlorofluoromethane	1593	79-34-5	1,1,2,2-Tetrachloroethane	1478
75-71-8	Dichlorodifluoromethane	643	79-34-5	1,1,2,2-Tetrachloroethane	1478
75-71-8	Dichlorodifluoromethane	644	79-34-5	1,1,2,2-Tetrachloroethane	1478
75-71-8	Dichlorodifluoromethane	645	80-62-6	Methyl methacrylate	1105
75-71-8	Dichlorodifluoromethane	647	80-62-6	Methyl methacrylate	1106
75-71-8	Dichlorodifluoromethane	648	82-05-3	Benzanthrone	118
75-71-8	Dichlorodifluoromethane	650	82-68-8	Pentachloronitrobenzene	1298
75-99-0	Dalapon	487	83-32-9	Acenaphthene	1
75-99-0	Dalapon	488	83-32-9	Acenaphthene	2
76-01-7	Pentachloroethane	1294	83-32-9	Acenaphthene	3
76-01-7	Pentachloroethane	1295	83-32-9	Acenaphthene	4
76-44-8	Heptachlor	939	83-32-9	Acenaphthene	5
76-44-8	Heptachlor	940	84-61-7	Dicyclohexyl phthalate	748
76-44-8	Heptachlor	941	84-66-2	Diethyl phthalate	759
76-44-8	Heptachlor	942	84-66-2	Diethyl phthalate	760
77-47-4	Hexachlorocyclopentadiene	969	84-66-2	Diethyl phthalate	761
77-47-4	Hexachlorocyclopentadiene	970	84-66-2	Diethyl phthalate	763
77-47-4	Hexachlorocyclopentadiene	971	84-66-2	Diethyl phthalate	764
77-47-4	Hexachlorocyclopentadiene	972	84-69-5	Diisobutyl phthalate	771
77-47-4	Hexachlorocyclopentadiene	973	84-74-2	Di-n-butyl phthalate	510
78-32-0	Tri-p-tolyl phosphate	1532	84-74-2	Di-n-butyl phthalate	511
78-34-2	Dioxathion	830	84-74-2	Di-n-butyl phthalate	512
78-59-1	Isophorone	1041	84-74-2	Di-n-butyl phthalate	513
78-59-1	Isophorone	1043	84-74-2	Di-n-butyl phthalate	515
78-59-1	Isophorone	1043	84-75-3	Dihexyl phthalate	768
78-59-1	Isophorone	1045	84-76-4	Dinonyl phthalate	822
78-83-1	Isobutyl alcohol	1037	85-01-8	Phenanthrene	1311
78-83-1	Isobutyl alcohol	1038	85-01-8	Phenanthrene	1312
78-87-5	1,2-Dichloropropane	700	85-01-8	Phenanthrene	1313
78-87-5	1,2-Dichloropropane	702	85-01-8	Phenanthrene	1314
78-87-5	1,2-Dichloropropane	705	85-01-8	Phenanthrene	1315
78-87-5	1,2-Dichloropropane	709	85-44-9	Phthalic anhydride	1337
78-87-5	1,2-Dichloropropane	712	85-68-7	Benzyl butyl phthalate	172
78-87-5	1,2-Dichloropropane	714	85-68-7	Butyl benzyl phthalate	286
78-87-5	1,2-Dichloropropane	719	85-68-7	Butyl benzyl phthalate	287
78-87-5	1,2-Dichloropropane	719	85-68-7	Butyl benzyl phthalate	289
78-87-5	1,2-Dichloropropane	720	85-68-7	Butyl benzyl phthalate	290
78-93-3	2-Butanone	284	86-30-6	N-Nitrosodiphenylamine	1247
78-93-3	Methyl ethyl ketone (MEK)	1102	86-30-6	N-Nitrosodiphenylamine	1249
78-93-3	Methyl ethyl ketone (MEK)	1103	86-30-6	N-Nitrosodiphenylamine	1249
78-97-7	2-Hydroxypropionitrile	1023	86-50-0	Azinphos-methyl	105
79-00-5	1,1,2-Trichloroethane	1558	86-50-0	Azinphos-methyl	107
79-00-5	1,1,2-Trichloroethane	1561	86-50-0	Azinphos-methyl	108
79-00-5	1,1,2-Trichloroethane	1563	86-73-7	Fluorene	929
79-00-5	1,1,2-Trichloroethane	1565	86-73-7	Fluorene	931
79-00-5	1,1,2-Trichloroethane	1568	86-73-7	Fluorene	932
79-00-5	1,1,2-Trichloroethane	1571	86-73-7	Fluorene	933
79-00-5	1,1,2-Trichloroethane	1573	86-73-7	Fluorene	934
79-00-5	1,1,2-Trichloroethane	1574	86-74-8	Carbazole	322
79-00-5	1,1,2-Trichloroethane	1575	87-61-6	1,2,3-Trichlorobenzene	1536
79-01-6	Trichloroethylene	1575	87-61-6	1,2,3-Trichlorobenzene	1538
79-01-6	Trichloroethylene	1576	87-61-6	1,2,3-Trichlorobenzene	1540
79-01-6	Trichloroethylene	1577	87-61-6	1,2,3-Trichlorobenzene	1543
79-01-6	Trichloroethylene	1579	87-61-6	1,2,3-Trichlorobenzene	1547
79-01-6	Trichloroethylene	1580	87-61-6	1,2,3-Trichlorobenzene	1552
79-01-6	Trichloroethylene	1581	87-61-6	1,2,3-Trichlorobenzene	1557
79-01-6	Trichloroethylene	1583	87-65-0	2,6-Dichlorophenol	694

CAS Number	Chemical Name	Page	CAS Number	Chemical Name	Page
87-65-0	2,6-Dichlorophenol	697	93-65-2	MCPP	1083
87-68-3	Hexachlorobutadiene	956	93-65-2	MCPP	1085
87-68-3	Hexachlorobutadiene	957	93-72-1	2,4,5-TP (Silvex)	1530
87-68-3	Hexachlorobutadiene	959	93-72-1	2,4,5-TP (Silvex)	1530
87-68-3	Hexachlorobutadiene	960	93-76-5	2,4,5-T	1427
87-68-3	Hexachlorobutadiene	961	94-59-7	Safrole	1389
87-68-3	Hexachlorobutadiene	962	94-59-7	Safrole	1390
87-68-3	Hexachlorobutadiene	964	94-59-7	Safrole	1391
87-68-3	Hexachlorobutadiene	966	94-74-6	MCPA	1081
87-68-3	Hexachlorobutadiene	967	94-74-6	MCPA	1083
87-86-5	Pentachlorophenol	1299	94-75-7	2,4-D	485
87-86-5	Pentachlorophenol	1300	94-75-7	2,4-D	486
87-86-5	Pentachlorophenol	1301	94-82-6	2,4-DB	489
87-86-5	Pentachlorophenol	1303	94-82-6	2,4-DB	490
87-86-5	Pentachlorophenol	1304	95-06-7	Sulfallate	1421
87-88-3	Hexachlorobutadiene	959	95-15-8	Thianaphthene(2,3-benzothiophene)	1500
88-06-2	2,4,6-Trichlorophenol	1598	95-47-6	o-Xylene	1660
88-06-2	2,4,6-Trichlorophenol	1599	95-47-6	o-Xylene	1663
88-06-2	2,4,6-Trichlorophenol	1602	95-47-6	o-Xylene	1666
88-06-2	2,4,6-Trichlorophenol	1603	95-47-6	o-Xylene	1667
88-74-4	2-Nitroaniline	1204	95-47-6	o-Xylene	1671
88-74-4	2-Nitroaniline	1208	95-47-6	o-Xylene	1678
88-75-5	2-Nitrophenol	1224	95-47-6	o-Xylene	1679
88-75-5	2-Nitrophenol	1226	95-47-6	o-Xylene	1685
88-75-5	2-Nitrophenol	1229	95-48-7	2-Methylphenol	1145
88-75-5	2-Nitrophenol	1232	95-48-7	o-Cresol	472
88-85-7	2-sec-Butyl-4,6-dinitrophenol	291	95-49-8	2-Chlorotoluene	438
88-85-7	Dinoseb	824	95-49-8	2-Chlorotoluene	441
88-85-7	Dinoseb	826	95-49-8	2-Chlorotoluene	443
88-85-7	Dinoseb	827	95-49-8	2-Chlorotoluene	446
89-63-4	4-Chloro-2-nitroaniline	358	95-49-8	2-Chlorotoluene	449
90-04-0	o-Anisidine	54	95-50-1	1,2-Dichlorobenzene	591
90-04-0	o-Anisidine	55	95-50-1	1,2-Dichlorobenzene	595
90-13-1	1-Chloronaphthalene	418	95-50-1	1,2-Dichlorobenzene	599
91-20-3	Naphthalene	1174	95-50-1	1,2-Dichlorobenzene	602
91-20-3	Naphthalene	1175	95-50-1	1,2-Dichlorobenzene	605
91-20-3	Naphthalene	1177	95-50-1	1,2-Dichlorobenzene	608
91-20-3	Naphthalene	1178	95-50-1	1,2-Dichlorobenzene	611
91-20-3	Naphthalene	1179	95-50-1	1,2-Dichlorobenzene	616
91-20-3	Naphthalene	1180	95-50-1	1,2-Dichlorobenzene	621
91-20-3	Naphthalene	1182	95-50-1	1,2-Dichlorobenzene	626
91-20-3	Naphthalene	1183	95-50-1	1,2-Dichlorobenzene	631
91-20-3	Naphthalene	1183	95-50-1	1,2-Dichlorobenzene	632
91-20-3	Naphthalene	1184	95-50-1	1,2-Dichlorobenzene	633
91-57-6	2-Methylnaphthalene	1141	95-50-1	1,2-Dichlorobenzene	634
91-57-6	2-Methylnaphthalene	1142	95-50-1	1,2-Dichlorobenzene	635
91-58-7	2-Chloronaphthalene	419	95-50-1	1,2-Dichlorobenzene	636
91-58-7	2-Chloronaphthalene	420	95-53-4	o-Toluidine	1521
91-58-7	2-Chloronaphthalene	421	95-53-4	o-Toluidine	1523
91-58-7	2-Chloronaphthalene	422	95-57-8	2-Chlorophenol	427
91-58-7	2-Chloronaphthalene	424	95-57-8	2-Chlorophenol	428
91-59-8	2-Naphthylamine (β-Naphthylamine)	1190	95-57-8	2-Chlorophenol	429
91-59-8	2-Naphthylamine (β-Naphthylamine)	1192	95-57-8	2-Chlorophenol	430
91-80-5	Methapyrilene	1094	95-63-6	1,2,4-Trimethylbenzene	1621
91-80-5	Methapyrilene	1096	95-63-6	1,2,4-Trimethylbenzene	1623
91-94-1	3,3'-Dichlorobenzidine	638	95-63-6	1,2,4-Trimethylbenzene	1626
91-94-1	3,3'-Dichlorobenzidine	639	95-63-6	1,2,4-Trimethylbenzene	1628
91-94-1	3,3'-Dichlorobenzidine	640	95-63-6	1,2,4-Trimethylbenzene	1631
92-52-4	Biphenyl	199	95-79-4	5-Chloro-2-methylaniline	357
92-52-4	Biphenyl	200	95-79-4	5-Chloro-o-toluidine	364
92-67-1	4-Aminobiphenyl	47	95-80-7	2,4-Diaminotoluene	525
92-67-1	4-Aminobiphenyl	48	95-80-7	2,4-Diaminotoluene	526
92-84-2	Phenothiazine	1321	95-83-0	4-Chloro-1,2-phenylenediamine	353
92-87-5	Benzidine	131	95-94-2	1,2,4,5-Tetrachlorobenzene	1458
92-87-5	Benzidine	132	95-94-3	1,2,4,5-Tetrachlorobenzene	1453
92-87-5	Benzidine	133	95-94-3	1,2,4,5-Tetrachlorobenzene	1461
92-93-3	4-Nitrobiphenyl	1217	95-95-4	2,4,5-Trichlorophenol	1597

CAS Number	Chemical Name	Page	CAS Number	Chemical Name	Page
95-95-4	2,4,5-Trichlorophenol	1600	100-41-4	Ethylbenzene	902
96-12-8	1,2-Dibromo-3-chloropropane	544	100-41-4	Ethylbenzene	903
96-12-8	1,2-Dibromo-3-chloropropane	545	100-41-4	Ethylbenzene	905
96-12-8	1,2-Dibromo-3-chloropropane	546	100-41-4	Ethylbenzene	905
96-12-8	1,2-Dibromo-3-chloropropane	548	100-41-4	Ethylbenzene	906
96-12-8	1,2-Dibromo-3-chloropropane	549	100-41-4	Ethylbenzene	906
96-12-8	1,2-Dibromo-3-chloropropane	550	100-42-5	Styrene	1411
96-12-8	1,2-Dibromo-3-chloropropane	552	100-42-5	Styrene	1413
96-12-8	1,2-Dibromo-3-chloropropane	554	100-42-5	Styrene	1414
96-18-4	1,2,3-Trichloropropane	1604	100-42-5	Styrene	1415
96-18-4	1,2,3-Trichloropropane	1605	100-42-5	Styrene	1416
96-18-4	1,2,3-Trichloropropane	1606	100-42-5	Styrene	1417
96-18-4	1,2,3-Trichloropropane	1607	100-42-5	Styrene	1419
96-18-4	1,2,3-Trichloropropane	1608	100-42-5	Styrene	1421
96-18-4	1,2,3-Trichloropropane	1611	100-44-7	Benzyl chloride	173
96-23-1	1,3-Dichloro-2-propanol	586	100-44-7	Benzyl chloride	174
96-23-1	1,3-Dichloro-2-propanol	587	100-51-6	Benzyl alcohol	168
96-45-7	Ethylenethiourea	908	100-51-6	Benzyl alcohol	169
97-63-2	Ethyl methacrylate	889	100-51-6	Benzyl alcohol	170
97-63-2	Ethyl methacrylate	890	100-75-4	N-Nitrosopiperidine	1257
98-06-6	tert-Butylbenzene	296	100-75-4	N-Nitrosopiperidine	1259
98-06-6	tert-Butylbenzene	300	101-05-3	Anilazine	50
98-06-6	tert-Butylbenzene	303	101-14-4	4,4′-Methylenebis(2-chloraniline)	1135
98-06-6	tert-Butylbenzene	308	101-14-4	4,4′-Methylenebis(2-chloroaniline)	1136
98-06-6	tert-Butylbenzene	310	101-21-3	Chlorpropham	450
98-07-7	Benzotrichloride	166	101-27-9	Barban	111
98-55-5	α-Terpineol	1452	101-55-3	4-Bromophenyl phenyl ether	278
98-82-8	Isopropylbenzene	1045	101-55-3	4-Bromophenyl phenyl ether	279
98-82-8	Isopropylbenzene	1047	101-55-3	4-Bromophenyl phenyl ether	280
98-82-8	Isopropylbenzene	1048	101-61-1	4,4′-Methylenebis(N,N-dimethylaniline)	1137
98-82-8	Isopropylbenzene	1050	101-84-8	Diphenyl ether	833
98-82-8	Isopropylbenzene	1051	103-65-1	n-Propylbenzene	1363
98-86-2	Acetophenone	13	103-65-1	n-Propylbenzene	1364
98-86-2	Acetophenone	15	103-65-1	n-Propylbenzene	1365
98-87-3	Benzal chloride	116	103-65-1	n-Propylbenzene	1366
98-95-3	Nitrobenzene	1212	103-65-1	n-Propylbenzene	1368
98-95-3	Nitrobenzene	1214	104-51-8	n-Butylbenzene	294
98-95-3	Nitrobenzene	1215	104-51-8	n-Butylbenzene	297
98-95-3	Nitrobenzene	1216	104-51-8	n-Butylbenzene	301
99-09-2	3-Nitroaniline	1206	104-51-8	n-Butylbenzene	304
99-09-2	3-Nitroaniline	1209	104-51-8	n-Butylbenzene	309
99-30-9	2,6-Dichloro-4-nitroaniline	589	105-67-9	2,4-Dimethylphenol	797
99-35-4	1,3,5-Trinitrobenzene	1632	105-67-9	2,4-Dimethylphenol	798
99-55-8	2-Methyl-5-nitroaniline	1120	105-67-9	2,4-Dimethylphenol	799
99-55-8	5-Nitro-o-toluidine	1200	105-67-9	2,4-Dimethylphenol	801
99-55-8	5-Nitro-o-toluidine	1201	106-42-3	p-Xylene	1660
99-59-2	5-Nitro-o-anisidine	1199	106-42-3	p-Xylene	1664
99-65-0	1,3-Dinitrobenzene	805	106-42-3	p-Xylene	1668
99-87-6	4-Isopropyltoluene	1052	106-42-3	p-Xylene	1669
99-87-6	4-Isopropyltoluene	1053	106-42-3	p-Xylene	1673
99-87-6	4-Isopropyltoluene	1054	106-42-3	p-Xylene	1681
99-87-6	4-Isopropyltoluene	1057	106-42-3	p-Xylene	1683
99-87-6	p-Cymene	482	106-42-3	p-Xylene	1685
99-87-6	p-Isopropyltoluene	1056	106-43-4	4-Chlorotoluene	440
100-01-6	4-Nitroaniline	1207	106-43-4	4-Chlorotoluene	442
100-01-6	4-Nitroaniline	1211	106-43-4	4-Chlorotoluene	444
100-02-7	4-Nitrophenol	1225	106-43-4	4-Chlorotoluene	447
100-02-7	4-Nitrophenol	1227	106-43-4	4-Chlorotoluene	449
100-02-7	4-Nitrophenol	1228	106-44-5	4-Methylphenol	1148
100-02-7	4-Nitrophenol	1231	106-44-5	p-Cresol	473
100-02-7	4-Nitrophenol	1233	106-46-7	1,4-Dichlorobenzene	594
100-25-4	1,4-Dinitrobenzene	803	106-46-7	1,4-Dichlorobenzene	598
100-25-4	1,4-Dinitrobenzene	807	106-46-7	1,4-Dichlorobenzene	601
100-41-4	Ethylbenzene	897	106-46-7	1,4-Dichlorobenzene	604
100-41-4	Ethylbenzene	898	106-46-7	1,4-Dichlorobenzene	607
100-41-4	Ethylbenzene	899	106-46-7	1,4-Dichlorobenzene	610
100-41-4	Ethylbenzene	900	106-46-7	1,4-Dichlorobenzene	615

CAS Number	Chemical Name	Page	CAS Number	Chemical Name	Page
106-46-7	1,4-Dichlorobenzene	620	108-60-1	Bis(2-chloroisopropyl) ether	214
106-46-7	1,4-Dichlorobenzene	625	108-67-8	1,3,5-Trimethylbenzene	1622
106-46-7	1,4-Dichlorobenzene	629	108-67-8	1,3,5-Trimethylbenzene	1624
106-46-7	1,4-Dichlorobenzene	632	108-67-8	1,3,5-Trimethylbenzene	1627
106-46-7	1,4-Dichlorobenzene	633	108-67-8	1,3,5-Trimethylbenzene	1631
106-46-7	1,4-Dichlorobenzene	634	108-67-8	1,3,5-Trimethylbenzene	1632
106-46-7	1,4-Dichlorobenzene	635	108-70-3	1,3,5-Trichlorobenzene	1550
106-46-7	1,4-Dichlorobenzene	636	108-86-1	Bromobenzene	237
106-46-7	1,4-Dichlorobenzene	638	108-86-1	Bromobenzene	238
106-47-8	4-Chloroaniline	366	108-86-1	Bromobenzene	239
106-47-8	4-Chloroaniline	367	108-86-1	Bromobenzene	240
106-50-3	1,4-Phenylenediamine	1322	108-86-1	Bromobenzene	242
106-51-4	p-Benzoquinone	165	108-88-3	Toluene	1511
106-89-8	Epichlorohydrin	874	108-88-3	Toluene	1511
106-93-4	1,2-Dibromoethane	564	108-88-3	Toluene	1513
106-93-4	1,2-Dibromoethane	565	108-88-3	Toluene	1514
106-93-4	1,2-Dibromoethane	567	108-88-3	Toluene	1515
106-93-4	1,2-Dibromoethane	568	108-88-3	Toluene	1517
106-93-4	1,2-Dibromoethane	569	108-88-3	Toluene	1518
106-93-4	1,2-Dibromoethane	571	108-88-3	Toluene	1519
106-93-4	1,2-Dibromoethane	572	108-88-3	Toluene	1519
107-02-8	Acrolein	21	108-88-3	Toluene	1519
107-02-8	Acrolein	22	108-90-7	Chlorobenzene	369
107-02-8	Acrolein	23	108-90-7	Chlorobenzene	370
107-05-1	3-Chloropropene	436	108-90-7	Chlorobenzene	371
107-05-1	Allyl chloride	38	108-90-7	Chlorobenzene	372
107-06-2	1,2-Dichloroethane	651	108-90-7	Chlorobenzene	373
107-06-2	1,2-Dichloroethane	653	108-90-7	Chlorobenzene	375
107-06-2	1,2-Dichloroethane	656	108-90-7	Chlorobenzene	376
107-06-2	1,2-Dichloroethane	658	108-90-7	Chlorobenzene	377
107-06-2	1,2-Dichloroethane	661	108-90-7	Chlorobenzene	377
107-06-2	1,2-Dichloroethane	664	108-90-7	Chlorobenzene	377
107-06-2	1,2-Dichloroethane	666	108-90-7	Chlorobenzene	378
107-06-2	1,2-Dichloroethane	667	108-90-7	Chlorobenzene	378
107-06-2	1,2-Dichloroethane	668	108-95-2	Phenol	1317
107-07-3	2-Chloroethanol	393	108-95-2	Phenol	1318
107-10-8	n-Propylamine	1361	108-95-2	Phenol	1319
107-12-0	Ethyl cyanide	888	108-95-2	Phenol	1320
107-12-0	Propionitrile	1359	108-98-5	Benzenethiol	129
107-13-1	Acrylonitrile	24	108-98-5	Thiophenol (Benzenethiol)	1504
107-13-1	Acrylonitrile	25	109-06-8	2-Methylpyridine	1150
107-13-1	Acrylonitrile	27	109-06-8	2-Picoline	1340
107-14-2	Chloroacetonitrile	365	109-06-8	2-Picoline	1341
107-18-6	Allyl alcohol	35	109-06-8	2-Picoline	1342
107-18-6	Allyl alcohol	36	109-06-8	2-Picoline	1344
107-19-7	Propargyl alcohol	1355	109-77-3	Malononitrile	1077
107-49-3	Tetraethyl pyrophosphate	1496	110-57-6	trans-1,4-Dichloro-2-butene	584
108-05-4	Vinyl acetate	1644	110-75-8	2-Chloroethyl vinyl ether	395
108-05-4	Vinyl acetate	1645	110-75-8	2-Chloroethyl vinyl ether	395
108-10-1	4-Methyl-2-pentanone	1115	110-75-8	2-Chloroethyl vinyl ether	397
108-10-1	4-Methyl-2-pentanone	1116	110-75-8	2-Chloroethyl vinyl ether	397
108-10-1	Methyl isobutyl ketone (MIBK)	1105	110-75-8	2-Chloroethyl vinyl ether	398
108-31-6	Maleic anhydride	1075	110-86-1	Pyridine	1376
108-37-2	3-Bromochlorobenzene	244	110-86-1	Pyridine	1377
108-38-3	m-Xylene	1659	110-86-1	Pyridine	1378
108-38-3	m-Xylene	1661	110-86-1	Pyridine	1380
108-38-3	m-Xylene	1665	111-44-4	Bis(2-chloroethyl) ether	206
108-38-3	m-Xylene	1671	111-44-4	Bis(2-chloroethyl) ether	208
108-38-3	m-Xylene	1674	111-44-4	Bis(2-chloroethyl) ether	209
108-38-3	m-Xylene	1676	111-44-4	Bis(2-chloroethyl) ether	210
108-38-3	m-Xylene	1684	111-91-1	Bis(2-chloroethoxy)methane	202
108-39-4	3-Methylphenol	1147	111-91-1	Bis(2-chloroethoxy)methane	203
108-46-3	1,3-Benzenediol (Resorcinol)	128	111-91-1	Bis(2-chloroethoxy)methane	204
108-46-3	Resorcinol	1383	111-91-1	Bis(2-chloroethoxy)methane	205
108-60-1	Bis(2-chloroisopropyl) ether	211	112-40-3	n-Dodecane	850
108-60-1	Bis(2-chloroisopropyl) ether	212	112-95-8	n-Eicosane	853
108-60-1	Bis(2-chloroisopropyl) ether	214	113-48-4	MGK 264	1159

CAS Number	Chemical Name	Page	CAS Number	Chemical Name	Page
115-90-2	Fensulfothion	916	123-3-7	Paraldehyde	1279
115-90-2	Fensulfothion	918	123-31-9	Hydroquinone	1020
115-90-2	Fensulfothion	919	123-73-9	Crotonaldehyde	474
117-79-3	2-Aminoanthraquinone	44	123-91-1	1,4-Dioxane (p-Dioxane)	828
117-80-6	Dichlone	582	123-91-1	1,4-Dioxane (p-Dioxane)	829
117-81-7	Bis(2-ethylhexl) phthalate	218	124-18-5	n-Decane	503
117-81-7	Bis(2-ethylhexyl) phthalate	219	124–48-1	Chlorodibromomethane	380
117-81-7	Bis(2-ethylhexyl) phthalate	220	124–48-1	Chlorodibromomethane	381
117-81-7	Bis(2-ethylhexyl) phthalate	221	124–48-1	Chlorodibromomethane	382
117-81-7	Bis(2-ethylhexyl) phthalate	223	124–48-1	Dibromochloromethane	555
117-81-7	Bis(2-ethylhexyl) phthalate	225	124–48-1	Dibromochloromethane	556
117-82-8	Bis(2-methoxyethyl) phthalate	225	124–48-1	Dibromochloromethane	557
117-83-9	Bis(2-n-butoxyethyl) phthalate	227	124–48-1	Dibromochloromethane	559
117-84-0	Di-n-octyl phthalate	516	124–48-1	Dibromochloromethane	560
117-84-0	Di-n-octyl phthalate	517	124–48-1	Dibromochloromethane	561
117-84-0	Di-n-octyl phthalate	518	124–48-1	Dibromochloromethane	563
117-84-0	Di-n-octyl phthalate	519	124–48-1	Dibromochloromethane	563
117-84-0	Di-n-octyl phthalate	521	124–48-1	Dibromochloromethane	564
118-74-1	Hexachlorobenzene	949	126-68-1	O,O,O-Triethyl phosphorothioate	1613
118-74-1	Hexachlorobenzene	950	126-72-7	Tris(2,3-dibromopropyl) phosphate	1636
118-74-1	Hexachlorobenzene	951	126-75-0	Demeton-S	508
118-74-1	Hexachlorobenzene	952	126-75-0	Demeton-S	509
118-74-1	Hexachlorobenzene	953	126-98-7	Methacrylonitrile	1092
118-74-1	Hexachlorobenzene	954	126-98-7	Methacrylonitrile	1093
119-90-4	3,3'-Dimethoxybenzidine	776	126-99-8	2-Chloro-1,3-butadiene	355
119-90-4	3,3'-Dimethoxybenzidine	777	126-99-8	Chloroprene	434
119-93-7	3,3'-Dimethylbenzidine	792	127-18-4	Tetrachloroethylene	1479
120-12-7	Anthracene	57	127-18-4	Tetrachloroethylene	1480
120-12-7	Anthracene	58	127-18-4	Tetrachloroethylene	1481
120-12-7	Anthracene	59	127-18-4	Tetrachloroethylene	1482
120-12-7	Anthracene	60	127-18-4	Tetrachloroethylene	1483
120-12-7	Anthracene	61	127-18-4	Tetrachloroethylene	1485
120-36-5	Dichloroprop	699	127-18-4	Tetrachloroethylene	1487
120-36-5	Dichlorprop	745	127-18-4	Tetrachloroethylene	1487
120-58-1	Isosafrole	1058	127-18-4	Tetrachloroethylene	1487
120-58-1	Isosafrole	1059	127-18-4	Tetrachloroethylene	1488
120-62-7	Piperonyl sulfoxide	1345	129-00-0	Pyrene	1370
120-71-8	p-Cresidine	471	129-00-0	Pyrene	1371
120-75-2	2-Methylbenzothioazole	1122	129-00-0	Pyrene	1372
120-82-1	1,2,4-Trichlorobenzene	1537	129-00-0	Pyrene	1373
120-82-1	1,2,4-Trichlorobenzene	1539	129-00-0	Pyrene	1375
120-82-1	1,2,4-Trichlorobenzene	1541	129-00-0	Pyrene	1375
120-82-1	1,2,4-Trichlorobenzene	1543	130-15-4	1,4-Naphthoquinone	1186
120-82-1	1,2,4-Trichlorobenzene	1545	130-15-4	1,4-Naphthoquinone	1187
120-82-1	1,2,4-Trichlorobenzene	1546	131-11-13	Dimethyl phthalate	784
120-82-1	1,2,4-Trichlorobenzene	1548	131-11-3	Dimethyl phthalate	779
120-82-1	1,2,4-Trichlorobenzene	1554	131-11-3	Dimethyl phthalate	780
120-82-1	1,2,4-Trichlorobenzene	1555	131-11-3	Dimethyl phthalate	781
120-82-1	1,2,4-Trichlorobenzene	1557	131-11-3	Dimethyl phthalate	783
120-83-2	2,4-Dichlorophenol	693	131-18-0	Diamyl phthalate	527
120-83-2	2,4-Dichlorophenol	695	131-89-5	2-Cyclohexyl-4,6-dinitrophenol	480
120-83-2	2,4-Dichlorophenol	696	132-32-1	3-Amino-9-ethylcarbazole	42
120-83-2	2,4-Dichlorophenol	699	132-64-9	Dibenzofuran	540
121-14-2	2,4-Dinitrotoluene	812	132-64-9	Dibenzofuran	541
121-14-2	2,4-Dinitrotoluene	815	132-65-0	Dibenzothiophene	543
121-14-2	2,4-Dinitrotoluene	817	133-06-2	Captan	319
121-14-2	2,4-Dinitrotoluene	820	133-90-4	Chloramben	340
121-73-3	3-Chloronitrobenzene	425	134-32-7	1-Naphthylamine (α-Naphthylamine)	1188
121-75-5	Malathion	1072	134-32-7	1-Naphthylaminel (α-Naphthylamine)	1191
121-75-5	Malathion	1074	135-98-8	sec-Butylbenzene	295
122-09-8	α,α-Dimethylphenethylamine	795	135-98-8	sec-Butylbenzene	298
122-34-9	Simazine	1399	135-98-8	sec-Butylbenzene	302
122-34-9	Simazine	1400	135-98-8	sec-Butylbenzene	306
122-39-4	Diphenylamine	834	135-98-8	sec-Butylbenzene	310
122-39-4	Diphenylamine	835	137-17-7	2,4,5-Trimethylaniline	1618
122-66-7	1,2-Diphenylhydrazine	839	137-17-7	2,4,5-Trimethylaniline	1619
122-66-7	1,2-Diphenylhydrazine	841	139-40-2	Propazine	1357

CAS Number	Chemical Name	Page	CAS Number	Chemical Name	Page
140-57-8	Aramite	64	224-42-0	Dibenz(a,j)acridine	535
140-57-8	Aramite	65	243-17-4	2,3-Benzofluorene	158
141-66-2	Dicrotophos	746	243-17-4	2,3-Benzofluorene	160
142-28-9	1,3-Dichloropropane	701	291-21-4	1,3,5-Trithiane	1638
142-28-9	1,3-Dichloropropane	703	297-97-2	Thionazine	1503
142-28-9	1,3-Dichloropropane	707	298-00-0	Methyl parathion	1112
142-28-9	1,3-Dichloropropane	710	298-00-0	Methyl parathion	1113
142-28-9	1,3-Dichloropropane	716	298-00-0	Methyl parathion	1115
142-62-1	Hexanoic acid	989	298-02-2	Phorate	1326
143-50-0	Kepone	1061	298-02-2	Phorate	1328
146-50-9	Bis(4-methyl-2-pentyl) phthalate	229	298-02-2	Phorate	1329
150-50-5	Merphos	1087	298-03-3	Demeton-O	506
150-50-5	Merphos	1088	298-03-3	Demeton-O	507
150-50-5	Merphos	1090	298-04-4	Disulfoton	842
156-59-4	cis-1,2-Dichloroethene	671	298-04-4	Disulfoton	843
156-59-4	cis-1,2-Dichloroethene	675	298-04-4	Disulfoton	845
156-59-4	cis-1,2-Dichloroethene	678	298-04-4	Disulfoton	847
156-59-4	cis-1,2-Dichloroethene	686	299-84-3	Ronnel	1385
156-60-5	trans-1,2-Dichloroethene	669	299-84-3	Ronnel	1386
156-60-5	trans-1,2-Dichloroethene	672	300-76-5	Naled	1171
156-60-5	trans-1,2-Dichloroethene	676	300-76-5	Naled	1172
156-60-5	trans-1,2-Dichloroethene	679	300-76-5	Naled	1174
156-60-5	trans-1,2-Dichloroethene	682	309-00-2	Aldrin	30
156-60-5	trans-1,2-Dichloroethene	687	309-00-2	Aldrin	31
156-60-5	trans-1,2-Dichloroethene	690	309-00-2	Aldrin	32
156-60-5	trans-1,2-Dichloroethene	690	309-00-2	Aldrin	33
156-60-5	trans-1,2-Dichloroethene	691	314-40-9	Bromocil	250
191-24-2	Benzo(g,h,i)perylene	148	315-18-4	Mexacarbate	1158
191-24-2	Benzo(g,h,i)perylene	149	319-84-6	α-BHC	180
191-24-2	Benzo(g,h,i)perylene	150	319-84-6	α-BHC	184
191-24-2	Benzo(g,h,i)perylene	152	319-84-6	α-BHC	189
191-24-2	Benzo(g,h,i)perylene	152	319-85-7	β-BHC	178
192-65-4	Dibenzo(a,e)pyrene	537	319-85-7	β-BHC	181
193-39-5	Indeno(1,2,3-cd)pyrene	1027	319-85-7	β-BHC	185
193-39-5	Indeno(1,2,3-cd)pyrene	1027	319-85-7	β-BHC	190
193-39-5	Indeno(1,2,3-cd)pyrene	1029	319-86-8	δ-BHC	179
193-39-5	Indeno(1,2,3-cd)pyrene	1030	319-86-8	δ-BHC	183
193-39-5	Indeno(1,2,3-cd)pyrene	1031	319-86-8	δ-BHC	187
198-55-0	Perylene	1306	319-86-8	δ-BHC	192
203-64-5	4,5-Methylenephenanthrene	1139	327-98-0	Trichloronate	1594
205-99-2	Benzo(b)fluoranthene	143	327-98-0	Trichloronate	1595
205-99-2	Benzo(b)fluoranthene	145	333-41-5	Diazinon	529
205-99-2	Benzo(b)fluoranthene	146	333-41-5	Diazinon	530
205-99-2	Benzo(b)fluoranthene	147	333-41-5	Diazinon	532
205-99-2	Benzo(b)fluoranthene	148	465-73-6	Isodrin	1040
206-44-0	Fluoranthene	925	470-90-6	Chlorfenvinphos	349
206-44-0	Fluoranthene	926	475-20-7	Longifolene	1067
206-44-0	Fluoranthene	927	492-22-8	Thioxanthone	1506
206-44-0	Fluoranthene	928	510-15-6	Chlorobenzilate	379
206-44-0	Fluoranthene	929	512-56-1	Trimethyl phosphate	1617
207-08-9	Benzo(k)fluoranthene	153	528-29-0	1,2-Dinitrobenzene	804
207-08-9	Benzo(k)fluoranthene	154	534-52-1	2-Methyl-4,6-dinitrophenol	1118
207-08-9	Benzo(k)fluoranthene	155	534-52-1	2-Methyl-4,6-dinitrophenol	1119
207-08-9	Benzo(k)fluoranthene	156	534-52-1	2-Methyl-4,6-dinitrophenol	1120
207-08-9	Benzo(k)fluoranthene	157	534-52-1	4,6-Dinitro-2-methylphenol	801
207-08-9	Benzo(k)fluoranthene	158	541-73-1	1,3-Dichlorobenzene	593
208-96-8	Acenaphthylene	5	541-73-1	1,3-Dichlorobenzene	596
208-96-8	Acenaphthylene	6	541-73-1	1,3-Dichlorobenzene	600
208-96-8	Acenaphthylene	7	541-73-1	1,3-Dichlorobenzene	603
208-96-8	Acenaphthylene	9	541-73-1	1,3-Dichlorobenzene	606
208-96-8	Acenaphthylene	9	541-73-1	1,3-Dichlorobenzene	609
217-59-4	Triphenylene	1634	541-73-1	1,3-Dichlorobenzene	613
218-01-9	Chrysene	457	541-73-1	1,3-Dichlorobenzene	618
218-01-9	Chrysene	458	541-73-1	1,3-Dichlorobenzene	623
218-01-9	Chrysene	459	541-73-1	1,3-Dichlorobenzene	628
218-01-9	Chrysene	461	541-73-1	1,3-Dichlorobenzene	631
218-01-9	Chrysene	461	541-73-1	1,3-Dichlorobenzene	632

CAS Number	Chemical Name	Page	CAS Number	Chemical Name	Page
541-73-1	1,3-Dichlorobenzene	633	924-16-3	N-Nitrosodi-n-butylamine	1235
541-73-1	1,3-Dichlorobenzene	634	924-16-3	N-Nitrosodibutylamine	1240
541-73-1	1,3-Dichlorobenzene	635	930-55-2	N-Nitrosopyrrolidine	1260
541-73-1	1,3-Dichlorobenzene	637	933-75-5	2,3,6-Trichlorophenol	1596
542-76-7	3-Chloropropionitrile	437	950-35-6	Methyl paraoxon	1111
544-76-3	n-Hexadecane	987	957-51-7	Diphenamid	832
563-12-2	Ethion	882	959-98-8	Endosulfan I	854
563-58-6	1,1-Dichloropropene	722	959-98-8	Endosulfan I	855
563-58-6	1,1-Dichloropropene	726	959-98-8	Endosulfan I	856
563-58-6	1,1-Dichloropropene	729	961-11-5	Tetrachlorvinphos	1491
563-58-6	1,1-Dichloropropene	736	1014-70-6	Simetryn	1401
569-64-2	Malachite green	1071	1024-57-3	Heptachlor epoxide	944
584-84-9	Toluene diisocyanate	1520	1024-57-3	Heptachlor epoxide	945
590-20-7	2,2-Dichloropropane	704	1024-57-3	Heptachlor epoxide	946
590-20-7	2,2-Dichloropropane	708	1024-57-3	Heptachlor epoxide	947
590-20-7	2,2-Dichloropropane	711	1031-07-8	Endosulfan sulfate	861
590-20-7	2,2-Dichloropropane	717	1031-07-8	Endosulfan sulfate	862
591-08-2	1-Acetyl-2-thiourea	16	1031-07-8	Endosulfan sulfate	863
591-78-6	2-Hexanone	991	1114-71-2	Pebulate	1279
591-78-6	2-Hexanone	991	1134-23-2	Cycloate	479
593-45-3	n-Octadecane	1274	1330-20-7	Xylenes	1686
598-31-2	Bromoacetone	235	1330-20-7	Xylenes	1687
602-87-9	5-Nitroacenaphthene	1203	1464-53-5	1,2:3,4-Diepoxybutane	755
605-02-7	1-Phenylnaphthalene	1324	1464-53-5	1,2:3,4-Diepoxybutane	756
605-54-9	Bis(2-ethoxyethyl) phthalate	216	1563-66-2	Carbofuran	323
606-20-2	2,6-Dinitrotoluene	814	1576-67-6	3,6-Dimethylphenanthrene	794
606-20-2	2,6-Dinitrotoluene	816	1582-09-8	Trifluralin	1614
606-20-2	2,6-Dinitrotoluene	818	1610-17-9	Atraton	102
606-20-2	2,6-Dinitrotoluene	820	1610-18-0	Prometon	1348
608-27-5	2,3-Dichloroaniline	590	1689-84-5	3,5-Dibromo-4-hydroxybenzonitrile	554
608-73-1	Hexachlorocyclohexane	968	1689-84-5	Bromoxynil	282
608-93-5	Pentachlorobenzene	1290	1730-37-6	1-Methylfluorene	1140
608-93-5	Pentachlorobenzene	1291	1746-01-6	2,3,7,8-TCDD	1437
608-93-5	Pentachlorobenzene	1293	1746-01-6	2,3,7,8-TCDD	1439
612-94-2	2-Phenylnaphthalene	1325	1836-75-5	Nitrofen	1218
614-00-6	N-Nitrosomethylphenylamine	1254	1888-71-7	Hexachloropropene	983
615-22-5	2-(Methylthio)-benzothiazole	1151	1888-71-7	Hexachloropropene	984
621-64-7	N-Nitrosodi-n-propylamine	1236	1912-24-9	Atrazine	103
621-64-7	N-Nitrosodi-n-propylamine	1237	1912-24-9	Atrazine	104
621-64-7	N-Nitrosodi-n-propylamine	1238	1918-00-9	Dicamba	580
629-59-4	n-Tetradecane	1494	1918-00-9	Dicamba	582
629-97-0	n-Docosane	849	1918-02-1	Picloram	1338
630-01-3	n-Hexacosane	985	1929-77-7	Vernolate	1643
630-02-4	n-Octacosane	1273	2008-41-5	Butylate	293
630-20-6	1,1,1,2-Tetrachloroethane	1462	2027-17-0	2-Isopropylnaphthalene	1051
630-20-6	1,1,1,2-Tetrachloroethane	1464	2104-64-5	EPN	876
630-20-6	1,1,1,2-Tetrachloroethane	1467	2104-64-5	EPN	877
630-20-6	1,1,1,2-Tetrachloroethane	1469	2136-79-0	DCPA diacid	491
630-20-6	1,1,1,2-Tetrachloroethane	1471	2212-67-1	Molinate	1162
630-20-6	1,1,1,2-Tetrachloroethane	1474	2243-62-1	1,5-Naphthalenediamine	1184
634-36-6	1,2,3-Trimethoxybenzene	1616	2303-16-4	Diallate (cis or trans)	523
634-66-2	1,2,3,4-Tetrachlorobenzene	1455	2310-17-0	Phosalone	1330
634-90-2	1,2,3,5-Tetrachlorobenzene	1456	2385-85-5	Mirex	1161
638-68-6	n-Triacontane	1533	2425-06-1	Captafol	317
646-31-1	n-Tetracosane	1492	2497-06-5	Disulfoton sulfone	847
680-31-9	Hexamethyl phosphoramide	988	2497-07-6	Disulfoton sulfoxide	848
694-80-4	2-Bromochlorobenzene	243	2921-88-2	Chlorpyrifos	451
700-12-9	Pentamethylbenzene	1305	2921-88-2	Chlorpyrifos	452
719-22-2	2,6-Di-tert-butyl-p-benzoquinone	522	3209-22-1	2,3-Dichloronitrobenzene	691
732-11-6	Phosmet	1331	3268-87-9	OCDD	1267
759-94-4	EPTC	879	3268-87-9	OCDD	1268
764-41-0	1,4-Dichloro-2-butene	585	3689-24-5	Sulfotep	1424
786-19-6	Carbophenothion	336	3689-24-5	Tetraethyl dithiopyrophosphate	1495
832-69-9	1-Methylphenanthrene	1144	5103-71-9	α-Chlordane	343
834-12-8	Ametryn	41	5103-74-2	γ-Chlordane	344
882-33-7	Diphenyldisulfide	837	5131-60-2	4-Chloro-1,3-phenylenediamine	355
886-50-0	Terbutryn	1451	5234-68-5	Carboxin	338

CAS Number	Chemical Name	Page	CAS Number	Chemical Name	Page
5902-51-2	Terbacil	1448	7440-38-2	Arsenic	101
6923-22-4	Monocrotophos	1166	7440-38-2	Arsenic	102
6923-22-4	Monocrotophos	1167	7440-39-3	Barium	112
6959-48-4	3-(Chloromethyl)pyridine hydrochloride	416	7440-39-3	Barium	113
7005-72-3	4-Chlorophenyl phenyl ether	431	7440-39-3	Barium	114
7005-72-3	4-Chlorophenyl phenyl ether	432	7440-39-3	Barium	114
7005-72-3	4-Chlorophenyl phenyl ether	433	7440-39-3	Barium	114
7287-19-6	Prometryn	1349	7440-41-7	Beryllium	176
7421-93-4	Endrin aldehyde	869	7440-41-7	Beryllium	177
7421-93-4	Endrin aldehyde	870	7440-41-7	Beryllium	178
7421-93-4	Endrin aldehyde	871	7440-41-7	Beryllium	178
7429-90-5	Aluminum	39	7440-42-8	Boron	233
7429-90-5	Aluminum	40	7440-42-8	Boron	234
7429-90-5	Aluminum	41	7440-43-9	Cadmium	313
7439-89-6	Iron	1034	7440-43-9	Cadmium	314
7439-89-6	Iron	1035	7440-43-9	Cadmium	314
7439-89-6	Iron	1036	7440-43-9	Cadmium	315
7439-89-6	Iron	1036	7440-47-3	Chromium (total)	454
7439-89-6	Iron	1036	7440-47-3	Chromium	453
7439-89-6	Iron	1037	7440-47-3	Chromium	454
7439-92-1	Lead	1063	7440-47-3	Chromium	454
7439-92-1	Lead	1064	7440-47-3	Chromium VI, Hexavalent	456
7439-92-1	Lead	1064	7440-47-3	Chromium VI, Hexavalent	456
7439-92-1	Lead	1065	7440-47-3	Chromium VI, Hexavalent	457
7439-95-4	Magnesium	1069	7440-48-4	Cobalt	462
7439-95-4	Magnesium	1071	7440-48-4	Cobalt	463
7439-96-5	Manganese	1078	7440-48-4	Cobalt	463
7439-96-5	Manganese	1080	7440-48-4	Cobalt	464
7439-96-5	Manganese	1080	7440-50-8	Copper	464
7439-96-5	Manganese	1080	7440-50-8	Copper	466
7439-96-5	Manganese	1081	7440-50-8	Copper	466
7439-97-6	Mercury	1085	7440-50-8	Copper	467
7439-97-6	Mercury	1086	7440-62-2	Vanadium	1641
7439-98-7	Molybdenum	1163	7440-62-2	Vanadium	1642
7439-98-7	Molybdenum	1164	7440-62-2	Vanadium	1642
7439-98-7	Molybdenum	1165	7440-62-2	Vanadium	1643
7439-98-7	Molybdenum	1165	7440-66-6	Zinc	1689
7440-02-0	Nickel	1195	7440-66-6	Zinc	1690
7440-02-0	Nickel	1196	7440-66-6	Zinc	1690
7440-02-0	Nickel	1197	7440-66-6	Zinc	1691
7440-04-2	Osmium	1275	7440-70-2	Calcium	315
7440-09-7	Potassium	1347	7440-70-2	Calcium	316
7440-09-7	Potassium	1347	7440-70-2	Calcium	317
7440-09-7	Potassium	1348	7440-70-2	Calcium	317
7440-21-3	Silicon	1395	7600-50-2	5-Hydroxydicamba	1022
7440-21-3	Silicon	1396	7683-64-9	Squalene	1405
7440-22-4	Silver	1397	7700-17-6	Crotoxyphos	475
7440-22-4	Silver	1398	7700-17-6	Crotoxyphos	477
7440-22-4	Silver	1398	7723-14-0	Phosphorus	1334
7440-22-4	Silver	1399	7723-14-0	Phosphorus	1335
7440-23-5	Sodium	1402	7723-14-0	Phosphorus	1336
7440-23-5	Sodium	1404	7723-14-0	Phosphorus	1336
7440-23-5	Sodium	1404	7723-14-0	Phosphorus	1336
7440-23-5	Sodium	1404	7782-49-2	Selenium	1392
7440-28-0	Thallium	1498	7782-49-2	Selenium	1394
7440-28-0	Thallium	1499	7782-49-2	Selenium	1394
7440-28-0	Thallium	1500	7782-49-2	Selenium	1395
7440-28-0	Thallium	1500	7786-34-7	Mevinphos	1153
7440-31-5	Tin	1507	7786-34-7	Mevinphos	1154
7440-31-5	Tin	1508	7786-34-7	Mevinphos	1156
7440-36-0	Antimony	61	7786-34-7	Mevinphos	1157
7440-36-0	Antimony	62	8001-35-2	Toxaphene	1525
7440-36-0	Antimony	63	8001-35-2	Toxaphene	1526
7440-36-0	Antimony	63	8001-35-2	Toxaphene	1527
7440-38-2	Arsenic	99	8001-35-2	Toxaphene	1528
7440-38-2	Arsenic	100	8065-48-3	Demeton, -O and -S	504
7440-38-2	Arsenic	101	10061-01-5	*cis*-1,3-Dichloropropene	720

CAS Number	Chemical Name	Page	CAS Number	Chemical Name	Page
10061-01-5	*cis*-1,3-Dichloropropene	723	25057-89-0	Bentazon	115
10061-01-5	*cis*-1,3-Dichloropropene	727	27314-13-2	Norflurazon	1264
10061-01-5	*cis*-1,3-Dichloropropene	730	30746-58-8	1,2,3,4-TCDD	1428
10061-01-5	*cis*-1,3-Dichloropropene	733	33213-65-9	Endosulfan II	857
10061-01-5	*cis*-1,3-Dichloropropene	738	33213-65-9	Endosulfan II	858
10061-01-5	*cis*-1,3-Dichloropropene	738	33213-65-9	Endosulfan II	859
10061-01-5	*cis*-1,3-Dichloropropene	739	33245-39-5	Fluchloralin	923
10061-02-6	*trans*-1,3-Dichloropropene	721	33423-92-6	1,3,6,8-TCDD	1433
10061-02-6	*trans*-1,3-Dichloropropene	725	34014-18-1	Tebuthiuron	1445
10061-02-6	*trans*-1,3-Dichloropropene	728	34465-46-8	1,2,3,6,7,8-HxCDD	1005
10061-02-6	*trans*-1,3-Dichloropropene	732	34465-46-8	1,2,3,6,7,8-HxCDD	1008
10061-02-6	*trans*-1,3-Dichloropropene	735	34643-46-4	Tokuthion (Protothiofos)	1508
10061-02-6	*trans*-1,3-Dichloropropene	739	34643-46-4	Tokuthion (Protothiofos)	1509
10061-02-6	*trans*-1,3-Dichloropropene	739	34816-53-0	1,2,7,8-TCDD	1430
10061-02-6	*trans*-1,3-Dichloropropene	740	35400-43-2	Bolstar (Sulprofos)	231
10595-95-6	N-Nitrosomethylethylamine	1251	35400-43-2	Bolstar (Sulprofos)	232
10595-95-6	N-Nitrosomethylethylamine	1252	35822-46-9	1,2,3,4,6,7,8-HpCDD	996
11096-82-5	Aroclor 1260	94	35822-46-9	1,2,3,4,6,7,8-HpCDD	998
11096-82-5	Aroclor 1260	96	39001-02-0	OCDF	1270
11096-82-5	Aroclor 1260	97	39001-02-0	OCDF	1271
11096-82-5	Aroclor 1260	98	39227-28-6	1,2,3,4,7,8-HxCDD	1011
11097-69-1	Aroclor 1254	90	39227-61-7	1,2,3,4,7-PeCDD	1280
11097-69-1	Aroclor 1254	91	39300-45-3	Dinocap	821
11097-69-1	Aroclor 1254	92	39765-80-5	*cis*-Nonachlor	1262
11097-69-1	Aroclor 1254	93	39765-80-5	*trans*-Nonachlor	1263
11104-28-2	Aroclor 1221	71	40321-76-4	1,2,3,7,8-PeCDD	1282
11104-28-2	Aroclor 1221	72	40321-76-4	1,2,3,7,8-PeCDD	1283
11104-28-2	Aroclor 1221	73	41814-78-2	Tricyclozole	1612
11104-28-2	Aroclor 1221	74	43121-43-3	Triademefon	1534
11141-16-5	Aroclor 1232	76	50585-46-1	1,3,7,8-TCDD	1434
11141-16-5	Aroclor 1232	77	50594-66-6	Acifluorfen	19
11141-16-5	Aroclor 1232	78	51207-31-9	2,3,7,8-TCDF	1442
11141-16-5	Aroclor 1232	79	51207-31-9	2,3,7,8-TCDF	1443
12672-29-6	Aroclor 1248	85	51218-45-2	Metholachlor	1097
12672-29-6	Aroclor 1248	86	51235-04-2	Hexazinone	993
12672-29-6	Aroclor 1248	87	53469-21-9	Aroclor 1242	80
12672-29-6	Aroclor 1248	88	53469-21-9	Aroclor 1242	82
12674-11-2	Aroclor 1016	66	53469-21-9	Aroclor 1242	82
12674-11-2	Aroclor 1016	67	53469-21-9	Aroclor 1242	84
12674-11-2	Aroclor 1016	68	53494-70-5	Endrin ketone	873
12674-11-2	Aroclor 1016	69	55673-89-7	1,2,3,4,7,8,9-HpCDF	1002
12789-03-6	Chlordane (technical)	348	56312–13-1	Dihydrosaffrole	770
13071-79-9	Terbufos	1449	57117-31-4	2,3,4,7,8-PeCDF	1288
13071-79-9	Terbufos	1450	57117-41-6	1,2,3,7,8-PeCDF	1285
13171-21-6	Phosphamidon	1333	57117-41-6	1,2,3,7,8-PeCDF	1286
13194-48-4	Ethoprop	883	57117-41-6	1,2,3,7,8-PeCDF	1289
13194-48-4	Ethoprop	884	57117-44-9	1,2,3,6,7,8-HxCDF	1015
13194-48-4	Ethoprop	886	57653-85-7	1,2,3,4,7,8-HxCDD	1003
15299-99-7	Napropamide	1194	57653-85-7	1,2,3,4,7,8-HxCDD	1007
15972-60-8	Alachlor	27	57653-85-7	1,2,3,6,7,8-HxCDD	1012
15972-60-8	Alachlor	29	58802-20-3	1,2,7,8-TCDF	1440
19408-74-3	1,2,3,7,8,9-HxCDD	1009	59756-60-4	Fluridone	936
20324-33-8	Tripropyleneglycolmethyl ether	1635	60168-88-9	Fenarimol	915
21087-64-9	Metribuzin	1152	60851-34-5	2,3,4,6,7,8-HxCDF	1018
21609-90-5	Leptophos	1065	62470-53-5	1,3,7,9-TCDD	1436
21646-99-1	TEPP	1446	62470-54-6	1,2,8,9-TCDD	1431
22224-92-6	Fenamiphos	914	67562-39-4	1,2,3,4,6,7,8-HpCDF	999
22248-79-9	Stirophos (Tetrachlorovinphos)	1407	67562-39-4	1,2,3,4,6,7,8-HpCDF	1001
22248-79-9	Stirophos (Tetrachlorovinphos)	1408	70648-26-9	1,2,3,4,7,8-HxCDF	1012
22248-79-9	Stirophos (Tetrachlorovinphos)	1409	70648-26-9	1,2,3,4,7,8-HxCDF	1014
23184-66-9	Butachlor	283	70648-26-9	1,2,3,4,7,8-HxCDF	1019
23950-58-5	Pronamide	1350	72918-21-9	1,2,3,7,8,9-HxCDF	1017
23950-58-5	Pronamide	1351	75673-16-4	Hexyl 2-ethylhexyl phthalate	994
23950-58-5	Pronamide	1353	96989-91-2	Pentachlorohexane	1297
23950-58-5	Pronamide	1354			

A

Acenaphthene
CAS #83-32-9

EPA Method 1625

TITLE Semivolatile Organic Compounds by Isotope Dilution GC/MS

MATRIX The compounds may be determined in waters, soils, and municipal sludges by this method.

METHOD SUMMARY This method is used to determine 176 semivolatile toxic organic pollutants associated with the CWA (as amended 1987); the RCRA (as amended 1986); the CERCLA (as amended 1986); and other compounds amenable to extraction and analysis by capillary column gas chromatography-mass spectrometry (GC/MS).

Stable isotopically-labeled analogs of the compounds of interest are added to the sample. If the solids content is less than 1%, a 1-L sample is extracted at pH 12–13, then at pH <2 with methylene chloride using continuous extraction techniques.

If the solids content is 30% or less, the sample is diluted to 1% solids with reagent water, homogenized ultrasonically, and extracted at pH 12–13, then at pH <2 with methylene chloride using continuous extraction techniques. If the solids content is greater than 30%, the sample is extracted using ultrasonic techniques.

Each extract is dried over sodium sulfate, concentrated to a volume of 5 mL, cleaned up using GPC, if necessary, and concentrated. Extracts are concentrated to 1 mL if GPC is not performed, and to 0.5 mL if GPC is performed.

An internal standard is added to the extract, and a 1-mL aliquot of the extract is injected into the GC. The compounds are separated by GC and detected by a MS. The labeled compounds serve to correct the variability of the analytical technique.

INTERFERENCES Solvents, reagents, glassware, and other sample processing hardware may yield artifacts and/or elevated baselines causing misinterpretation of chromatograms and spectra. Materials used in the analysis must be demonstrated to be free from interferences under the conditions of analysis by running method blanks initially and with each sample lot (sample started through the extraction process on a given 8-h shift, to a maximum of 20). Specific selection of reagents and purification of solvents by distillation in all glass systems may be required. Glassware and, where possible, reagents are cleaned by solvent rinse and baking at 450°C for 1-h minimum. Interferences coextracted from samples will vary considerably from source to source, depending on the diversity of the site being sampled.

INSTRUMENTATION Major instrumentation includes a GC with a splitless or on-column injection port for capillary column, a MS with 70 eV electron impact ionization, and a data system to collect and record MS data, and process it. A K-D apparatus is used to concentrate extracts.

GC Column: 30 m × 0.25 mm I.D. 5% phenyl, 94% methyl, 1% vinyl silicone bonded phased fused silica capillary column.

PRECISION & ACCURACY The detection limits of the method are usually dependent on the level of interferences rather than instrumental limitations. The limits typify the minimum quantities that can be detected with no interferences present.

The minimum level (in µg/mL) was 10. This is defined as a minimum level at which the analytical system shall give recognizable mass spectra (background corrected) and acceptable calibration points. The MDL (in µg/kg) in low solids was 64 and in high solids was 55; these were determined in digested sludge (low solids) and in filter cake or compost (high solids).

The labeled and native compound initial precision as standard deviation (in µg/L) was 21.
The labeled and native compound initial accuracy as average recovery (in µg/L) was 79–134.

SAMPLE COLLECTION, PRESERVATION, & HANDLING Collect samples in glass containers. Aqueous samples which flow freely are collected in refrigerated bottles using automatic sampling equipment. Solid samples are collected as grab samples using widemouth jars. Maintain samples at 0 to 4°C from the time of collection until extraction. If residual chlorine is present in aqueous samples, add 80 mg sodium thiosulfate/L of water. Begin sample extraction within 7 days of collection, and analyze all extracts within 40 days of extraction.

SAMPLE PREPARATION Samples containing 1% solids or less are extracted directly using continuous liquid-liquid extraction techniques. Samples containing 1 to 30% solids are diluted to the 1% level with reagent water and extracted using continuous liquid-liquid extraction techniques. Samples containing greater than 30% solids are extracted using ultrasonic techniques.

Base/neutral extraction — Adjust the pH of the waters in the extractors to 12–13 with 6 N NaOH. Extract with methylene chloride for 24–48 h.

Acid extraction — Adjust the pH of the waters in the extractors to 2 or less using 6 N sulfuric acid. Extract with methylene chloride for 24–48 h.

Ultrasonic extraction of high solids samples — Add anhydrous sodium sulfate to the sample and QC aliquot(s). Add acetone:methylene chloride (1:1) to the sample and mix thoroughly

Concentrate extracts using a K-D apparatus.

QUALITY CONTROL The analyst is permitted to modify this method to improve separations or lower the costs of measurements, provided all performance specifications are met. Analyses of blanks are required to demonstrate freedom from contamination. When results of spikes indicate atypical method performance for samples, the samples are diluted to bring method performance within acceptable limits.

For low solids (aqueous samples), extract, concentrate, and analyze two sets of four 1-L aliquots (8 aliquots total) of the precision and recovery standard. For high solids samples, two sets of four 30-g aliquots of the high solids reference matrix are used.

Spike all samples with labeled compounds to assess method performance. Compute percent recovery of the labeled compounds using the internal standard method. Compare the labeled compound recovery for each compound with the corresponding labeled compound recovery.

Reagent water and high solids reference matrix blanks are analyzed to demonstrate freedom from contamination. Extract and concentrate a 1-L reagent water blank or a high solids reference matrix blank with each sample's lot (samples started through the extraction process on the same 8-h shift, to a maximum of 20 samples).

Field replicates may be collected to determine the precision of the sampling technique, and spiked samples may be required to determine the accuracy of the analysis when the internal standard method is used.

REFERENCE Semivolatile Organic Compounds by Isotope Dilution GC/MS. Office of Water Regulation and Standards, U.S. EPA Industrial Technology Division, Washington, DC, EPA Method 1625, Rev. C, June 1989 (contact W.A. Telliard, U.S. EPA, Office of Water Regulations and Standards, 401 M St., SW, Washington, DC, 20460. Phone: 202-382-7131).

Acenaphthene **EPA Method 625**
CAS #83-32-9

TITLE Base/Neutrals and Acids, U.S. EPA Method 625

MATRIX This methods covers municipal and industrial wastewater.

METHOD SUMMARY Approximately 1 L of sample is serially extracted with methylene chloride at a pH greater than 11 and again at a pH less than 2 using a separatory funnel or a continuous extractor. The methylene chloride extract is dried, concentrated to a volume of 1 mL, and analyzed by GC/MS. Qualitative identification of the parameters in the extract is performed using the retention time and the relative abundance of three characteristic masses (m/z). Qualitative analysis is performed using external or internal standard techniques with a single characteristic m/z.

INTERFERENCES Method interferences may be caused by contaminants in solvents, reagents, glassware, and other sample processing hardware. Glassware must be scrupulously cleaned. Glassware should be heated in a muffle furnace at 400°C for 5 to 30 min. Some thermally stable materials, such as PCBs, may not be eliminated by this treatment. Solvent rinses with acetone and pesticide quality hexane may be substituted for the muffle furnace heating. Matrix interferences may be caused by contaminants that are coextracted from the sample. The base/neutral extraction may cause significantly reduced recovery of phenols. The packed gas chromatographic columns recommended for the basic fraction may not exhibit sufficient resolution for some analytes.

INSTRUMENTATION A GC/MS system with an injection port designed for on-column injection when using packed columns and for splitless injection when using capillary columns.

Column for base/neutrals: 1.8 m long × 2 mm I.D. glass, packed with 3% SP-2550 on Supelcoport (100/120 mesh) or equivalent.

Column for acids: 1.8 m long × 2 mm I.D. glass, packed with 1% SP-1240DA on Supelcoport (100/120 mesh) or equivalent.

PRECISION & ACCURACY The MDL concentrations were obtained using reagent water. The MDL actually achieved in a given analysis will vary depending on instrument sensitivity and matrix effects. This method was tested by 15 laboratories using reagent water, drinking water, surface water, and industrial wastewater spiked at six concentrations over the range 5 to 100 µg/L. Single operator precision, overall precision, and method accuracy were found to be directly related to the concentration of the parameter matrix.

The MDL (in µg/L) in reagent water was 1.9.

The standard deviation (in µg/L based on 4 recovery measurements) was 27.6.

The range (in µg/L) for average recovery for 4 measurements was 60.1–132.3.

The range (in %) for percent recovery was 47–145.

Accuracy (in µg/L) as expected recovery for one or more measurements of a sample containing a true concentration of C was 0.96C+0.19.

Precision (in µg/L) as expected single analyst standard deviation of measurements at an average concentration found at X was 0.15X–0.12.

Overall precision (in µg/L) as expected interlaboratory standard deviation of measurements in an average concentration found at X was 0.21X–0.67.

$C =$ *True value of the concentration in µg/L.*
$X =$ *Average recovery found for measurements of samples containing a concentration at C in µg/L.*

SAMPLE PREPARATION Adjust the pH to >11 with sodium hydroxide and serially extract in a separatory funnel with methylene chloride or else in a continuous extractor. Next, adjust the pH to <2 with sulfuric acid and serially extract in a separatory funnel with methylene chloride or else in a continuous extractor. Dry the extracts separately through a column of anhydrous sodium sulfate and then concentrate each of the extracts to 1.0 mL using a K-D apparatus.

SAMPLE COLLECTION, PRESERVATION, & HANDLING Grab samples must be collected in glass containers. All samples must be refrigerated at 4°C from the time of collection until extraction. If residual chlorine is present, add 80 mg of sodium thiosulfate/L of sample and mix well. All samples must be extracted within 7 days of collection and completely analyzed within 40 days of extraction.

QUALITY CONTROL Make an initial, one-time, demonstration of the ability to generate acceptable accuracy and precision with this method. Before processing any samples, the analyst must analyze a reagent water blank to demonstrate that interferences from the analytical system and glassware are under control. Each time a set of samples is extracted or reagents are changed, a reagent water blank must be processed. Spike and analyze a minimum of 5% of all samples to monitor and evaluate lab data quality. A QC check sample concentrate that contains each parameter of interest at a concentration of 100 µg/mL in acetone is required. PCBs and multicomponent pesticides may be omitted from this test.

After analysis of five spiked wastewater samples, calculate the average percent recovery and the standard deviation of the percent recovery. Spike all samples with the surrogate standard spiking solution and calculate the percent recovery of each surrogate compound.

REFERENCE *Federal Register*, Vol. 49, No. 209. Friday, Oct. 26, 1984.

Acenaphthene   **EPA Method 8270**
CAS #83-32-9

TITLE Semivolatile Organic Compounds by GC/MS

MATRIX This method is used to determine the concentration of semivolatile organic compounds in extracts prepared from all types of solid waste matrices, soils, and groundwater. Although surface waters are not specifically mentioned, this method should be applicable to water samples from rivers, lakes, etc.

METHOD SUMMARY This method covers 259 semivolatile organic compounds. In very limited applications direct injection of the sample into the GC/MS system may be appropriate, but this results in very high detection limits (approximately 10,000 µg/L). Typically, a 1-L liquid sample, containing surrogate, and matrix spiking standards, is extracted in a continuous extractor first under acid conditions and then under basic conditions. Typically 30 g of a solid sample, containing surrogate, and matrix spiking standards, is extracted ultrasonically. After concentrating the extract to 1 mL it is spiked with 10 µL of an internal standard solution just prior to analysis by GC/MS. The volume injected should contain about 100 ng of base/neutral and 200 ng of acid surrogates (for a 1-µL injection). Analysis is performed by GC/MS using a capillary GC column.

INTERFERENCES Raw GC/MS data from all blanks, samples, and spikes must be evaluated for interferences. Contamination by carryover can occur whenever high-concentration and low-concentration samples are sequentially analyzed. To reduce carryover, the sample syringe must be rinsed out between samples with solvent. Whenever an unusually concentrated sample is encountered, it should be followed by the analysis of blank solvent to check for cross-contamination.

INSTRUMENTATION A GC/MS and a data system are required. The GC column used is a 30 m × 0.25 mm I.D. (or 0.32 mm I.D.) 1-µm film thickness silicone-coated fused silica capillary column. A continuous liquid-liquid extractor equipped with Teflon® or glass connection joints and stopcocks requiring no lubrication, a K-D concentrating apparatus, water bath, and an ultrasonic disrupter with a minimum power of 300 W and with pulsing capability are also required.

PRECISION & ACCURACY The estimated quantitation limit (EQL) of Method 8270B for determining an individual compound is approximately 1 mg/kg (wet weight) for soil or sediment samples, 1–200 mg/kg for wastes (dependent on matrix and method of preparation), and 10 µg/L for groundwater samples. EQLs will be proportionately higher for sample extracts that require dilution to avoid saturation of the detector.

The EQL(b) for groundwater in µg/L is 10.
The EQL (a, b) for low concentrations in soil and sediment in µg/kg is 660.
Accuracy as µg/L is 0.96C +0.19.
Overall precision in µg/L is 0.21X–0.67.

(a) *EQLs listed for soil/sediment are based on wet weight. Normally data is reported in a dry-weight basis; therefore, EQLs will be higher based on the % dry weight of each sample. This calculation is based on a 30-g sample and gel permeation chromatography cleanup.*
(b) *Sample EQLs are highly matrix-dependent. The EQLs are provided for guidance and may not always be achievable.*
C = *True value for concentration, in µg/L.*
X = *Average recovery found for measurements of samples containing a concentration of C, in µg/L.*

ESTIMATED QUANTITATION LIMIT

Other Matrices	Factor (a)
High-concentration soil and sludges by sonicator	7.5
Non-water miscible waste	75

(a) *EQL for other matrices = [EQL for low soil/sediment] × [Factor]. This estimated EQL is similar to an EPA "Practical Quantitation Limit."*

SAMPLING METHOD
Liquid samples — Use a 1 or 2½ gallon amber glass bottle with a screw-top Teflon®-lined cover that has been prewashed with detergent and rinsed with distilled water and methanol (or isopropanol).

Soils, sediments, or sludges — Use an 8-oz. widemouth glass with a screw-top Teflon®-lined cover that has been prewashed with detergent and rinsed with distilled water and methanol (or isopropanol).

SAMPLE PRESERVATION
Liquid samples — If residual chlorine is present, add 3 mL of 10% sodium thiosulfate per gallon, cool to 4°C and store in a solvent-free refrigerator until analysis; if chlorine is not present, then eliminate the sodium thiosulfate addition.

Soils, sediments, or sludges — Cool samples to 4°C and store in a solvent-free refrigerator.

MHT Liquid samples must be extracted within 7 days and the extracts analyzed within 40 days. Soils, sediments, or sludges may be stored for a maximum of 14 days and the extracts analyzed within 40 days.

SAMPLE PREPARATION
Liquid samples — Transfer 1 L quantitatively to a continuous extractor. If high concentrations are anticipated, a smaller volume may be used and then diluted with organic-free reagent

water to 1 L. Adjust pH, if necessary, to pH <2 using 1:1 (V/V) sulfuric acid. Pipette 1.0 mL of a surrogate standard spiking solution into each sample. For the sample in each analytical batch selected for spiking, add 1.0 mL of a matrix spiking standard. For base/neutral acid analysis, the amount of the surrogates and matrix spiking compounds added to the sample should result in a final concentration of 100 ng/μL of each analyte in the extract to be analyzed (assuming a 1-μL injection). Extract with methylene chloride for 18–24 h. Next, adjust the pH of the aqueous phase to pH >11 using 10 N sodium hydroxide and extract it with methylene chloride again for 18–24 h. Dry the extract through a column containing anhydrous sodium sulfate and concentrate it to 1 mL using a K-D concentrator.

Soils, sediments, or sludges — Use 30 g of sample. Nonporous or wet samples (gummy or clay type) that do not have a free-flowing sandy texture must be mixed with anhydrous sodium sulfate until the sample is free flowing. Add 1 mL of surrogate standards to all samples, spikes, standards, and blanks. For the sample in each analytical batch selected for spiking, add 1.0 mL of a matrix spiking standard. For base/neutral acid analysis, the amount added of the surrogates and matrix spiking compounds should result in a final concentration of 100 ng/μL of each base/neutral analyte and 200 ng/μL of each acid analyte in the extract to be analyzed (assuming a 1-μL injection). Immediately add a 100-mL mixture of 1:1 methylene chloride:acetone and extract the sample ultrasonically for 3 min and then decant or filter the extracts. Repeat the extraction two or more times. Dry the extract using a column with anhydrous sodium sulfate and concentrate it to 1 mL in a K-D concentrator.

QUALITY CONTROL A methylene chloride solution containing 50 ng/μL of decafluorotriphenylphosphine (DFTPP) is used for tuning the GC/MS system each 12-h shift. A system performance check also must be made during every 12-h shift. A standard containing 50 ng/μL each of 4,4′-DDT, pentachlorophenol, and benzidine is required to verify injection port inertness and GC column performance. A calibration standard at mid-concentration, containing each compound of interest, including all required surrogates, must be performed every 12 h during analysis. After the system performance check is met, calibration check compounds (CCCs) are used to check the validity of the initial calibration.

The internal standard responses and retention times in the calibration check standard must be evaluated immediately after or during data acquisition. If the retention time for any internal standard changes by more than 30 seconds from the last check calibration (12 h), the chromatographic system must be inspected for malfunctions and corrections must be made, as required. If the electron ionization current plot (EICP) area for any of the internal standards changes by a factor of two from the last daily calibration standard check, the mass spectrometer must be inspected for malfunctions and corrections must be made, as appropriate.

Demonstrate, through the analysis of a reagent water blank, that interferences from the analytical system, glassware, and reagents are under control. The blank samples should be carried through all stages of the sample preparation and measurement steps. For each analytical batch (up to 20 samples), a reagent blank, matrix spike, and matrix spike duplicate/duplicate must be analyzed (the frequency of the spikes may be different for different monitoring programs). The blank and spiked samples must be carried through all stages of the sample preparation and measurement steps. A QC reference sample concentrate containing each analyte at a concentration of 100 mg/L in methanol is required.

REFERENCE Test Methods for Evaluating Solid Waste (SW-846). U.S. EPA 1983, Method 8270B, Rev. 2, Nov. 1990. Office of Solid Waste, Washington, DC.

Acenaphthene   **EPA Method 8100**
CAS #83-32-9

TITLE Polynuclear Aromatic Hydrocarbons

MATRIX Groundwater, soils, sludges, water miscible liquid wastes, and non-water miscible wastes.

APPLICATION This method is used for the analysis of various PAHs. Samples are extracted, concentrated, and analyzed using direct injection of both neat and diluted organic liquids. The method provides two optional GC columns that are better than Column 1 and that may help resolve analytes from interferences.

INTERFERENCES Solvents, reagents, and glassware may introduce artifacts. Other interferences may come from coextracted compounds from samples.

INSTRUMENTATION GC capable of on-column injections and a flame ionization detector (FID). Column 1: a 1.8 m by 2 mm % OV-17 on Chromosorb W-AW-DCMS column. Column 2: a 30 m by 0.25 mm SE-54 fused silica capillary column. Column 3: a 30 m by 0.32 mm SE-54 fused silica capillary column.

RANGE 0.1–425 μg/L

MDL Not reported.

PQL FACTORS FOR MULTIPLYING × FID MDL VALUE Not available.

PRECISION $0.53X + 1.32$ μg/L (overall precision).

ACCURACY $0.52C + 0.54$ μg/L (as recovery).

SAMPLING METHOD Use 8-oz. widemouth glass bottles with Teflon®-lined caps for concentrated waste samples, soils, sediments, and sludges. Use 1 or 2½ gallon amber glass bottles with Teflon®-lined caps for liquid (water) samples.

STABILITY Cool soil, sediment, sludge, and liquid samples to 4°C. If residual chlorine is present in liquid samples add 3 mL of 0% sodium thiosulfate per gallon of sample and cool to 4°C.

MHT 14 days for concentrated waste, soil, sediment, or sludge; 7 days for liquid samples; all extracts must be analyzed within 40 days.

QUALITY CONTROL A quality control check sample concentrate containing each analyte of interest is required. The QC check sample concentrate may be prepared from pure standard materials or purchased as certified solutions. Use appropriate trip, matrix, control site, method, reagent, and solvent blanks. Internal, surrogate, and five concentration level calibration standards are used. The quality control check sample concentrate should contain acenaphthene at 100 µg/mL in acetonitrile.

REFERENCE Method 8100, SW-846, 3rd ed., Nov. 1986.

Acenaphthene   **EPA Method 8310**
CAS #83-32-9

TITLE Polynuclear Aromatic Hydrocarbons

MATRIX Groundwater, soils, sludges, water miscible liquid wastes, and non-water miscible wastes.

APPLICATION This method is used for the analysis of 16 polynuclear aromatic hydrocarbons (PAHs). Samples are extracted, concentrated, and analyzed using HPLC with detection by UV and fluorescence detectors.

INTERFERENCES Solvents, reagents, and glassware may introduce artifacts. Other interferences may come from coextracted compounds from samples.

INSTRUMENTATION HPLC with a gradient pumping system and a 250 mm by 2.6 mm reverse phase HC-ODS Sil-X 5-micron particle size column. The UV detector uses an excitation wavelength of 254 nm coupled to the fluorescence detector. The fluorescence detector uses an excitation wavelength of 280 nm and emission greater than 389 nm cutoff with dispersive optics.

RANGE 0.1–425 µg/L

MDL 1.8 µg/L (UV detector; reagent water).

PQL FACTORS FOR MULTIPLYING × FID MDL VALUE

Matrix	Multiplication Factor
Groundwater	10
Low-level soil by sonication with GPC cleanup	670
High-level soil and sludge by sonication	10,000
Non-water miscible waste	100,000

PRECISION 0.53X + 1.32 µg/L (overall precision).

ACCURACY 0.52C + 0.54 µg/L (as recovery).

SAMPLING METHOD Use 8-oz. widemouth glass bottles with Teflon®-lined caps for concentrated waste samples, soils, sediments, and sludges. Use 1 or 2½ gallon amber glass bottles with Teflon®-lined caps for liquid (water) samples.

STABILITY Cool soil, sediment, sludge, and liquid samples to 4°C. If residual chlorine is present in liquid samples add 3 mL of 10% sodium thiosulfate per gallon of sample and cool to 4°C.

MHT 14 days for concentrated waste, soil, sediment, or sludge; 7 days for liquid samples; all extracts must be analyzed within 40 days.

QUALITY CONTROL Internal, surrogate, and five concentration level calibration standards are used. The calibration standards must be used with the analytical method blank. A quality control check sample concentrate containing acenaphthene at 100 µg/mL is required. The QC check sample concentrate may be prepared from pure standard materials or purchased as certified solutions. Use appropriate trip, matrix, control site, method, reagent, and solvent blanks.

REFERENCE Method 8310, SW-846, 3rd ed., Nov. 1986.

Acenaphthylene   **EPA Method 1625**
CAS #208-96-8

TITLE Semivolatile Organic Compounds by Isotope Dilution GC/MS

MATRIX The compounds may be determined in waters, soils, and municipal sludges by this method.

METHOD SUMMARY This method is used to determine 176 semivolatile toxic organic pollutants associated with the CWA (as amended 1987); the RCRA (as amended 1986); the CERCLA (as amended 1986); and other compounds amenable to extraction and analysis by capillary column gas chromatography-mass spectrometry (GC/MS).

Stable isotopically-labeled analogs of the compounds of interest are added to the sample. If the solids content is less than 1%, a 1-L sample is extracted at pH 12–13, then at pH <2 with methylene chloride using continuous extraction techniques.

If the solids content is 30% or less, the sample is diluted to 1% solids with reagent water, homogenized ultrasonically, and extracted at pH 12–13, then at pH <2 with methylene chloride using continuous extraction techniques. If the solids content is greater than 30%, the sample is extracted using ultrasonic techniques.

Each extract is dried over sodium sulfate, concentrated to a volume of 5 mL, cleaned up using GPC, if necessary, and concentrated. Extracts are concentrated to 1 mL if GPC is not performed, and to 0.5 mL if GPC is performed.

An internal standard is added to the extract, and a 1-mL aliquot of the extract is injected into the GC. The compounds are separated by GC and detected by a MS. The labeled compounds serve to correct the variability of the analytical technique.

INTERFERENCES Solvents, reagents, glassware, and other sample processing hardware may yield artifacts and/or elevated baselines causing misinterpretation of chromatograms and spectra. Materials used in the analysis must be demonstrated to be free from interferences under the conditions of analysis by running method blanks initially and with each sample lot (sample started through the extraction process on a given 8-h shift, to a maximum of 20). Specific selection of reagents and purification of solvents by distillation in all glass systems may be required. Glassware and, where possible, reagents are cleaned by solvent rinse and baking at 450°C for 1-h minimum. Interferences coextracted from samples will vary considerably from source to source, depending on the diversity of the site being sampled.

INSTRUMENTATION Major instrumentation includes a GC with a splitless or on-column injection port for capillary column, a MS with 70 eV electron impact ionization, and a data system to collect and record MS data, and process it. A K-D apparatus is used to concentrate extracts.

GC Column: 30 m × 0.25 mm I.D. 5% phenyl, 94% methyl, 1% vinyl silicone bonded phased fused silica capillary column.

PRECISION & ACCURACY The detection limits of the method are usually dependent on the level of interferences rather than instrumental limitations. The limits typify the minimum quantities that can be detected with no interferences present.

The minimum level (in µg/mL) was 10. This is defined as a minimum level at which the analytical system shall give recognizable mass spectra (background corrected) and acceptable calibration points.

The MDL (in µg/kg) in low solids was 57 and in high solids was 18; these were determined in digested sludge (low solids) and in filter cake or compost (high solids).

The labeled and native compound initial precision as standard deviation (in µg/L) was 38.
The labeled and native compound initial accuracy as average recovery (in µg/L) was 69–186.

SAMPLE COLLECTION, PRESERVATION & HANDLING Collect samples in glass containers. Aqueous samples which flow freely are collected in refrigerated bottles using automatic sampling equipment. Solid samples are collected as grab samples using widemouth jars. Maintain samples at 0 to 4°C from the time of collection until extraction. If residual chlorine is present in aqueous samples, add 80 mg sodium thiosulfate/L of water. Begin sample extraction within 7 days of collection, and analyze all extracts within 40 days of extraction.

SAMPLE PREPARATION Samples containing 1% solids or less are extracted directly using continuous liquid-liquid extraction techniques. Samples containing 1 to 30% solids are diluted to the 1% level with reagent water and extracted using continuous liquid-liquid extraction techniques. Samples containing greater than 30% solids are extracted using ultrasonic techniques.

Base/neutral extraction — Adjust the pH of the waters in the extractors to 12–13 with 6 N NaOH. Extract with methylene chloride for 24–48 h.
Acid extraction — Adjust the pH of the waters in the extractors to 2 or less using 6 N sulfuric acid. Extract with methylene chloride for 24–48 h.
Ultrasonic extraction of high solids samples — Add anhydrous sodium sulfate to the sample and QC aliquot(s). Add acetone:methylene chloride (1:1) to the sample and mix thoroughly

Concentrate extracts using a K-D apparatus.

QUALITY CONTROL The analyst is permitted to modify this method to improve separations or lower the costs of measurements, provided all performance specifications are met. Analyses of blanks are required to demonstrate freedom from contamination. When results of spikes indicate atypical method performance for samples, the samples are diluted to bring method performance within acceptable limits.

For low solids (aqueous samples), extract, concentrate, and analyze two sets of four 1-L aliquots (8 aliquots total) of the precision and recovery standard. For high solids samples, two sets of four 30-g aliquots of the high solids reference matrix are used.

Spike all samples with labeled compounds to assess method performance. Compute percent recovery of the labeled compounds using the internal standard method. Compare the labeled compound recovery for each compound with the corresponding labeled compound recovery.

Reagent water and high solids reference matrix blanks are analyzed to demonstrate freedom from contamination. Extract and concentrate a 1-L reagent water blank or a high solids reference matrix blank with each sample's lot (samples started through the extraction process on the same 8-h shift, to a maximum of 20 samples).

Field replicates may be collected to determine the precision of the sampling technique, and spiked samples may be required to determine the accuracy of the analysis when the internal standard method is used.

REFERENCE Semivolatile Organic Compounds by Isotope Dilution GC/MS. Office of Water Regulation and Standards, U.S. EPA Industrial Technology Division, Washington, DC, EPA Method 1625, Rev. C, June 1989 (contact W.A. Telliard, U.S. EPA, Office of Water Regulations and Standards, 401 M St., SW, Washington, DC, 20460. Phone: 202-382-7131).

Acenaphthylene   **EPA Method 625**
CAS #208-96-8

TITLE Base/Neutrals and Acids, U.S. EPA Method 625

MATRIX This methods covers municipal and industrial wastewater.

METHOD SUMMARY Approximately 1 L of sample is serially extracted with methylene chloride at a pH greater than 11 and again at a pH less than 2 using a separatory funnel or a continuous extractor. The methylene chloride extract is dried, concentrated to a volume of 1 mL, and analyzed by GC/MS. Qualitative identification of the parameters in the extract is performed using the retention time and the relative abundance of three characteristic masses (m/z). Qualitative analysis is performed using either external or internal standard techniques with a single characteristic m/z.

INTERFERENCES Method interferences may be caused by contaminants in solvents, reagents, glassware, and other sample processing hardware. Glassware must be scrupulously cleaned. Glassware should be heated in a muffle furnace at 400°C for 5 to 30 min. Some thermally stable materials, such as PCBs, may not be eliminated by this treatment. Solvent rinses with acetone and pesticide quality hexane may be substituted for the muffle furnace heating. Matrix interferences may be caused by contaminants that are coextracted from the sample. The base/neutral extraction may cause significantly reduced recovery of phenols. The packed gas chromatographic columns recommended for the basic fraction may not exhibit sufficient resolution for some analytes.

INSTRUMENTATION A GC/MS system with an injection port designed for on-column injection when using packed columns and for splitless injection when using capillary columns.

Column for base/neutrals: 1.8 m long × 2 mm I.D. glass, packed with 3% SP-2550 on Supelcoport (100/120 mesh) or equivalent.

Column for acids: 1.8 m long × 2 mm I.D. glass, packed with 1% SP-1240DA on Supelcoport (100/120 mesh) or equivalent.

PRECISION & ACCURACY The MDL concentrations were obtained using reagent water. The MDL actually achieved in a given analysis will vary depending on instrument sensitivity and matrix effects. This method was tested by 15 laboratories using reagent water, drinking water, surface water, and industrial wastewater spiked at six concentrations over the range 5 to 100 μg/L. Single operator precision, overall precision, and method accuracy were found to be directly related to the concentration of the parameter matrix.

The MDL (in μg/L) in reagent water was not reported.
The standard deviation (in μg/L based on 4 recovery measurements) was 40.2.
The range (in μg/L) for average recovery for 4 measurements was 53.5–126.0.
The range (in %) for percent recovery was 33–145.
Accuracy (in μg/L) as expected recovery for one or more measurements of a sample containing a true concentration of C was 0.89C+0.74.
Precision (in μg/L) as expected single analyst standard deviation of measurements at an average concentration found at X was 0.24X–1.06.
Overall precision (in μg/L) as expected interlaboratory standard deviation of measurements in an average concentration found at X was 0.26X–0.54.

C = *True value of the concentration in μg/L.*
X = *Average recovery found for measurements of samples containing a concentration at C in μg/L.*

SAMPLE PREPARATION Adjust the pH to >11 with sodium hydroxide and serially extract in a separatory funnel with methylene chloride or else in a continuous extractor. Next, adjust the pH to <2 with sulfuric acid and serially extract in a separatory funnel with methylene chloride or else in a continuous extractor. Dry the extracts separately through a column of anhydrous sodium sulfate and then concentrate each of the extracts to 1.0 mL using a K-D apparatus.

SAMPLE COLLECTION, PRESERVATION & HANDLING
Grab samples must be collected in glass containers. All samples must be refrigerated at 4°C from the time of collection until extraction. If residual chlorine is present, add 80 mg of sodium thiosulfate/L of sample and mix well. All samples must be extracted within 7 days of collection and completely analyzed within 40 days of extraction.

QUALITY CONTROL Make an initial, one-time, demonstration of the ability to generate acceptable accuracy and precision with this method. Before processing any samples, the analyst must analyze a reagent water blank to demonstrate that interferences from the analytical system and glassware are under control. Each time a set of samples is extracted or reagents are changed, a reagent water blank must be processed. Spike and analyze a minimum of 5% of all samples to monitor and evaluate lab data quality. A QC check sample concentrate that contains each parameter of interest at a concentration of 100 μg/mL in acetone is required. PCBs and multicomponent pesticides may be omitted from this test.

After analysis of five spiked wastewater samples, calculate the average percent recovery and the standard deviation of the percent recovery. Spike all samples with the surrogate standard spiking solution and calculate the percent recovery of each surrogate compound.

REFERENCE *Federal Register*, Vol. 49, No. 209. Friday, Oct. 26, 1984.

Acenaphthylene   **EPA Method 8270**
CAS #208-96-8

TITLE Semivolatile Organic Compounds by GC/MS

MATRIX This method is used to determine the concentration of semivolatile organic compounds in extracts prepared from all types of solid waste matrices, soils, and groundwater. Although surface waters are not specifically mentioned, this method should be applicable to water samples from rivers, lakes, etc.

METHOD SUMMARY This method covers 259 semivolatile organic compounds. In very limited applications direct injection of the sample into the GC/MS system may be appropriate, but this results in very high detection limits (approximately 10,000 μg/L). Typically, a 1-L liquid sample, containing surrogate, and matrix spiking standards, is extracted in a continuous extractor first under acid conditions and then under basic conditions. Typically 30 g of a solid sample, containing surrogate, and matrix spiking standards, is extracted ultrasonically. After concentrating the extract to 1 mL it is spiked with 10 μL of an internal standard solution just prior to analysis by GC/MS. The volume injected should contain about 100 ng of base/neutral and 200 ng of acid surrogates (for a 1-μL injection). Analysis is performed by GC/MS using a capillary GC column.

INTERFERENCES Raw GC/MS data from all blanks, samples, and spikes must be evaluated for interferences. Contamination by carryover can occur whenever high-concentration and low-concentration samples are sequentially analyzed. To reduce carryover, the sample syringe must be rinsed out between samples with solvent. Whenever an unusually concentrated

sample is encountered, it should be followed by the analysis of blank solvent to check for cross-contamination.

INSTRUMENTATION A GC/MS and a data system are required. The GC column used is a 30 m × 0.25 mm I.D. (or 0.32 mm I.D.) 1-µm film thickness silicone-coated fused silica capillary column. A continuous liquid-liquid extractor equipped with Teflon® or glass connection joints and stopcocks requiring no lubrication, a K-D concentrating apparatus, water bath, and an ultrasonic disrupter with a minimum power of 300 W and with pulsing capability are also required.

PRECISION & ACCURACY The estimated quantitation limit (EQL) of Method 8270B for determining an individual compound is approximately 1 mg/kg (wet weight) for soil or sediment samples, 1–200 mg/kg for wastes (dependent on matrix and method of preparation), and 10 µg/L for groundwater samples. EQLs will be proportionately higher for sample extracts that require dilution to avoid saturation of the detector.

The EQL(b) for groundwater in µg/L is 10.
The EQL (a, b) for low concentrations in soil and sediment in µg/kg is 660.
Accuracy as µg/L is $0.89C + 0.74$.
Overall precision in µg/L is $0.26X - 0.54$.

(a) EQLs listed for soil/sediment are based on wet weight. Normally data is reported in a dry-weight basis; therefore, EQLs will be higher based on the % dry weight of each sample. This calculation is based on a 30-g sample and gel permeation chromatography cleanup.

(b) Sample EQLs are highly matrix-dependent. The EQLs are provided for guidance and may not always be achievable.

C = True value for concentration, in µg/L.

X = Average recovery found for measurements of samples containing a concentration of C, in µg/L.

ESTIMATED QUANTITATION LIMIT

Other Matrices	Factor (a)
High-concentration soil and sludges by sonicator	7.5
Non-water miscible waste	75

(a) EQL for other matrices = [EQL for low soil/sediment] × [Factor]. This estimated EQL is similar to an EPA "Practical Quantitation Limit."

SAMPLING METHOD

Liquid samples — Use a 1 or 2½ gallon amber glass bottle with a screw-top Teflon®-lined cover that has been prewashed with detergent and rinsed with distilled water and methanol (or isopropanol).

Soils, sediments, or sludges — Use an 8-oz. widemouth glass with a screw-top Teflon®-lined cover that has been prewashed with detergent and rinsed with distilled water and methanol (or isopropanol).

SAMPLE PRESERVATION

Liquid samples — If residual chlorine is present, add 3 mL of 10% sodium thiosulfate per gallon, cool to 4°C and store in a solvent-free refrigerator until analysis; if chlorine is not present, then eliminate the sodium thiosulfate addition.

Soils, sediments, or sludges — Cool samples to 4°C and store in a solvent-free refrigerator.

MHT Liquid samples must be extracted within 7 days and the extracts analyzed within 40 days. Soils, sediments, or sludges may be stored for a maximum of 14 days and the extracts analyzed within 40 days.

SAMPLE PREPARATION

Liquid samples — Transfer 1 L quantitatively to a continuous extractor. If high concentrations are anticipated, a smaller volume may be used and then diluted with organic-free reagent water to 1 L. Adjust pH, if necessary, to pH <2 using 1:1 (V/V) sulfuric acid. Pipette 1.0 mL of a surrogate standard spiking solution into each sample. For the sample in each analytical batch selected for spiking, add 1.0 mL of a matrix spiking standard. For base/neutral acid analysis, the amount of the surrogates and matrix spiking compounds added to the sample should result in a final concentration of 100 ng/µL of each analyte in the extract to be analyzed (assuming a 1-µL injection). Extract with methylene chloride for 18–24 h. Next, adjust the pH of the aqueous phase to pH >11 using 10 N sodium hydroxide and extract it with methylene chloride again for 18–24 h. Dry the extract through a column containing anhydrous sodium sulfate and concentrate it to 1 mL using a K-D concentrator.

Soils, sediments, or sludges — Use 30 g of sample. Nonporous or wet samples (gummy or clay type) that do not have a free-flowing sandy texture must be mixed with anhydrous sodium sulfate until the sample is free flowing. Add 1 mL of surrogate standards to all samples, spikes, standards, and blanks. For the sample in each analytical batch selected for spiking, add 1.0 mL of a matrix spiking standard. For base/neutral acid analysis, the amount added of the surrogates and matrix spiking compounds should result in a final concentration of 100 ng/µL of each base/neutral analyte and 200 ng/µL of each acid analyte in the extract to be analyzed (assuming a 1-µL injection). Immediately add a 100-mL mixture of 1:1 methylene chloride:acetone and extract the sample ultrasonically for 3 min and then decant or filter the extracts. Repeat the extraction two or more times. Dry the extract using a column with anhydrous sodium sulfate and concentrate it to 1 mL in a K-D concentrator.

QUALITY CONTROL A methylene chloride solution containing 50 ng/µL of decafluorotriphenylphosphine (DFTPP) is used for tuning the GC/MS system each 12-h shift. A system performance check also must be made during every 12-h shift. A standard containing 50 ng/µL each of 4,4'-DDT, pentachlorophenol, and benzidine is required to verify injection port inertness and GC column performance. A calibration standard at mid-concentration, containing each compound of interest, including all required surrogates, must be performed every 12 h during analysis. After the system performance check is met, calibration check compounds (CCCs) are used to check the validity of the initial calibration.

The internal standard responses and retention times in the calibration check standard must be evaluated immediately after or during data acquisition. If the retention time for any internal standard changes by more than 30 seconds from the last check calibration (12 h), the chromatographic system must be

inspected for malfunctions and corrections must be made, as required. If the electron ionization current plot (EICP) area for any of the internal standards changes by a factor of two from the last daily calibration standard check, the mass spectrometer must be inspected for malfunctions and corrections must be made, as appropriate.

Demonstrate, through the analysis of a reagent water blank, that interferences from the analytical system, glassware, and reagents are under control. The blank samples should be carried through all stages of the sample preparation and measurement steps. For each analytical batch (up to 20 samples), a reagent blank, matrix spike, and matrix spike duplicate/duplicate must be analyzed (the frequency of the spikes may be different for different monitoring programs). The blank and spiked samples must be carried through all stages of the sample preparation and measurement steps. A QC reference sample concentrate containing each analyte at a concentration of 100 mg/L in methanol is required.

REFERENCE Test Methods for Evaluating Solid Waste (SW-846). U.S. EPA 1983, Method 8270B, Rev. 2, Nov. 1990. Office of Solid Waste, Washington, DC.

Acenaphthylene   **EPA Method 8100**
CAS #208-96-8

TITLE Polynuclear Aromatic Hydrocarbons

MATRIX Groundwater, soils, sludges, water miscible liquid wastes, and non-water miscible wastes.

APPLICATION This method is used for the analysis of various PAHs. Samples are extracted, concentrated, and analyzed using direct injection of both neat and diluted organic liquids. The method provides two optional GC columns that are better than Column 1 and that may help resolve analytes from interferences.

INTERFERENCES Solvents, reagents, and glassware may introduce artifacts. Other interferences may come from coextracted compounds from samples.

INSTRUMENTATION GC capable of on-column injections and a flame ionization detector (FID). Column 1: a 1.8 m by 2 mm 3% OV-17 on Chromosorb W-AW-DCMS column. Column 2: a 30 m by 0.25 mm SE-54 fused silica capillary column. Column 3: a 30 m by 0.32 mm SE-54 fused silica capillary column.

RANGE 0.1–425 µg/L

MDL Not reported.

PQL FACTORS FOR MULTIPLYING × FID MDL VALUE Not available.

PRECISION 0.42X + 0.52 µg/L (overall precision).

ACCURACY 0.69C–1.89 µg/L (as recovery).

SAMPLING METHOD Use 8-oz. widemouth glass bottles with Teflon®-lined caps for concentrated waste samples, soils, sediments, and sludges. Use 1 or 2½ gallon amber glass bottles with Teflon®-lined caps for liquid (water) samples.

STABILITY Cool soil, sediment, sludge, and liquid samples to 4°C. If residual chlorine is present in liquid samples add 3 mL of 10% sodium thiosulfate per gallon of sample and cool to 4°C.

MHT 14 days for concentrated waste, soil, sediment, or sludge; 7 days for liquid samples; all extracts must be analyzed within 40 days.

QUALITY CONTROL A quality control check sample concentrate containing each analyte of interest is required. The QC check sample concentrate may be prepared from pure standard materials or purchased as certified solutions. Use appropriate trip, matrix, control site, method, reagent, and solvent blanks. Internal, surrogate, and five concentration level calibration standards are used. The quality control check sample concentrate should contain acenaphthylene at 100 µg/mL in acetonitrile.

REFERENCE Method 8100, SW-846, 3rd ed., Nov. 1986.

Acenaphthylene   **EPA Method 8310**
CAS #208-96-8

TITLE Polynuclear Aromatic Hydrocarbons

MATRIX Groundwater, soils, sludges, water miscible liquid wastes, and non-water miscible wastes.

APPLICATION This method is used for the analysis of 16 polynuclear aromatic hydrocarbons (PAHs). Samples are extracted, concentrated, and analyzed using HPLC with detection by UV and fluorescence detectors.

INTERFERENCES Solvents, reagents, and glassware may introduce artifacts. Other interferences may come from coextracted compounds from samples.

INSTRUMENTATION HPLC with a gradient pumping system and a 250 mm by 2.6 mm reverse phase HC-ODS Sil-X 5-micron particle size column. The UV detector uses an excitation wavelength of 254 nm coupled to the fluorescence detector. The fluorescence detector uses an excitation wavelength of 280 nm and emission greater than 389 nm cutoff with dispersive optics.

RANGE 0.1–425 µg/L

MDL 2.3 µg/L (UV detector; reagent water).

PQL FACTORS FOR MULTIPLYING × FID MDL VALUE

Matrix	Multiplication Factor
Groundwater	10
Low-level soil by sonication with GPC cleanup	670
High-level soil and sludge by sonication	10,000
Non-water miscible waste	100,000

PRECISION 0.42X + 0.52 µg/L (overall precision).

ACCURACY 0.69C–1.89 µg/L (as recovery).

SAMPLING METHOD Use 8-oz. widemouth glass bottles with Teflon®-lined caps for concentrated waste samples, soils, sediments, and sludges. Use 1 or 2½ gallon amber glass bottles with Teflon®-lined caps for liquid (water) samples.

STABILITY Cool soil, sediment, sludge, and liquid samples to 4°C. If residual chlorine is present in liquid samples add 3 mL of 10% sodium thiosulfate per gallon of sample and cool to 4°C.

MHT 14 days for concentrated waste, soil, sediment, or sludge; 7 days for liquid samples; all extracts must be analyzed within 40 days.

QUALITY CONTROL Internal, surrogate, and five concentration level calibration standards are used. The calibration standards must be used with the analytical method blank. A quality control check sample concentrate containing acenaphthylene at 100 µg/mL is required. The QC check sample concentrate may be prepared from pure standard materials or purchased as certified solutions. Use appropriate trip, matrix, control site, method, reagent, and solvent blanks.

REFERENCE Method 8310, SW-846, 3rd ed., Nov. 1986.

Acetone
CAS #67-64-1

EPA Method 8240

TITLE Volatile Organics By GC/MS: Packed Column Technique

MATRIX Nearly all types of sample matrices, regardless of water content, can be analyzed using this method. This includes groundwater, aqueous sludges, caustic liquors, acid liquors, waste solvents, oily wastes, mousses, tars, fibrous wastes, polymetric emulsions, filter cakes, spent carbons, spent catalysts, soils, and sediments.

METHOD SUMMARY Method 8240B covers 80 volatile organic compounds that are introduced into a gas chromatograph by the purge-and-trap method or by direct injection (in limited applications). For the purge-and-trap method an inert gas (zero grade nitrogen or helium) is bubbled through a 5-mL solution at ambient temperature. Purged sample components are trapped in a tube of sorbent materials. When purging is complete, the sorbent tube is heated and backflushed with inert gas to desorb the trapped components onto a GC column.

INTERFERENCES Impurities in the purge gas and from organic compounds outgassing from the plumbing ahead of the trap account for many contamination problems. Interferences purged or coextracted from the samples will vary considerably from source to source. Cross-contamination can occur whenever high-level and low-level samples are analyzed sequentially. Whenever an unusually concentrated sample is analyzed, it should be followed by the analysis of organic-free reagent water to check for cross-contamination. Samples also can be contam-inated by diffusion of volatile organics (particularly methylene chloride and fluorocarbons) through the septum seal into the sample during shipment and storage. A trip blank can serve as a check on such contamination. The lab where volatile analysis is performed and also the refrigerated storage area should be completely free of solvents.

INSTRUMENTATION A gas chromatograph/mass spectrometry/data system (GC/MS) equipped with a 6 ft × 0.1 in I.D. glass column packed with 1% SP-1000 on Carbopack-B (60/80 mesh) is required. Also needed is a 5-mL purging device, a sorbent trap, and a thermal desorption apparatus.

PRECISION & ACCURACY This method is reported to have been tested by 15 laboratories using organic-free reagent water, drinking water, surface water, and industrial wastewater (not specified) fortified at six concentrations over the range 5–600 µg/L.

Sample estimated quantitation limits (EQLs) are highly matrix dependent. The EQLs listed may not always be achievable. EQLs listed for soils or sediments are based on wet weight. Normally, data is reported on a dry-weight basis; therefore, EQLs will be higher, based on the percent dry weight of each sample. Note that EQLs are even more variable than MDLs and that they are highly variable depending on the matrix being analyzed.

EQL in groundwater in µg/L was 100.
EQL in low soil or sediment in µg/kg was 100.
Accuracy (a) in µg/L was not listed.
Precision (b) in µg/L was not listed.

(a) Average recovery found for measurements of samples containing a concentration of C, in µg/L.
(b) Overall precision found for measurements of samples with average recovery X for samples containing a concentration of C in µg/L.
X = Average recovery found for measurement of samples containing a concentration of C in µg/L.

ESTIMATED QUANTITATION LIMITS FOR MATRICES OTHER THAN WATER, SOIL, AND SEDIMENTS

Other Matrices	Multiplication Factor (a)
Waste miscible liquid waste	50
High-concentration soil and sludge	125
Non-water miscible waste	500

(a) EQL = [EQL for low soil/sediment] × [Factor]. For non-aqueous samples, the factor is on a wet-weight basis.

SAMPLING METHOD

Liquid samples — Use a 40-mL glass screw-cap VOA vial with a Teflon®-faced silicone septum that has been prewashed, rinsed with distilled deionized water, and oven dried. However, if residual chlorine is present, collect sample in a 40-oz. soil VOA container which has been pre-preserved with 4 drops of 10% sodium thiosulfate, mix gently, and then transfer the sample to a 40-mL VOA vial. Collect bubble-free samples in duplicate and seal them in separate plastic bags.

Soils or sediments, and sludges — Use an 8 oz. widemouth glass bottle with a Teflon®-faced silicone septum that has been prewashed with detergent, rinsed with distilled deionized water, and oven dried. Tap slightly to eliminate free air space. Collect samples in duplicate and seal them in separate plastic bags.

SAMPLE PRESERVATION

Liquid samples — Add 4 drops of concentrated HCL and immediately cool samples to 4°C and store in a solvent-free refrigerator.

Soils or sediments, and sludges — Cool samples to 4°C and store in a solvent-free refrigerator.

MHT Maximum holding time is 14 days from the date of sample collection.

SAMPLE PREPARATION

Liquid samples — Remove the plunger from a 5-ml syringe and carefully pour the sample into the syringe barrel to just short of overflowing. Replace the syringe plunger and compress the sample. Open the syringe valve and vent any residual air while adjusting the sample volume to 5.0 mL. If there is only one volatile organic analysis (VOA) vial, a second syringe should be filled at this time to protect against possible loss of sample integrity. Add 10 µL of surrogate spiking solution and 10 µL of internal standard spiking solution through the valve bore of the 5-ml syringe, then close the valve. The surrogate and internal standards may be mixed and added as a single spiking solution.

Sediments, soils, and waste samples — All samples of this type should be screened by GC analysis using a headspace method (EPA Method 3810) or the hexadecane extraction and screening method (EPA Method 3820). Use the screening data to determine whether to use the low-concentration method (0.005–1 mg/kg) or the high-concentration method (>1 mg/kg).

Low-concentration method — The low-concentration method is based on purging a heated sediment or soil sample mixed with organic-free reagent water containing the surrogate and internal standards. Analyze all reagent blanks and standards under the same conditions as the samples.

Use a 5-g sample if the expected concentration is <0.1 mg/kg or a 1-g sample for expected concentrations between 0.1 and 1 mg/kg. Mix the contents of the sample container with a narrow metal spatula. Weigh the amount of the sample into a tared purge device. Add the spiked water to the purge device, which contains the weighed amount of sample, and connect the device to the purge-and-trap system.

High-concentration method — This method is based on extracting the sediment or soil with methanol. A waste sample is either extracted or diluted, depending on its solubility in methanol. Wastes that are insoluble in methanol are diluted with reagent tetraglyme or possibly polyethylene glycol (PEG). An aliquot of the extract is added to organic-free reagent water containing surrogate and internal standards. This is purged at ambient temperature. All samples with an expected concentration of >1.0 mg/kg should be analyzed by this method.

Mix the contents of the sample container with a narrow metal spatula. For sediments or soils and solid wastes that are insoluble in methanol, weigh 4 g (wet weight) of sample into a tared 20-mL vial. For waste that is soluble in methanol, tetraglyme, or PEG, weigh 1 g (wet weight) into a tared scintillation vial or culture tube or a 10-mL volumetric flask. Quickly add 9.0 mL of appropriate solvent then add 1.0 mL of a surrogate spiking solution to the vial, cap it, and shake it for 2 min.

QUALITY CONTROL Demonstrate, through the analysis of a reagent water blank, that interferences from the analytical system, glassware, and reagents are under control. Blank samples should be carried through all stages of the sample preparation and measurement steps. For each analytical batch (up to 20 samples), a reagent blank, matrix spike, and matrix spike duplicate must be analyzed (the frequency of the spikes may be different for different monitoring programs). The blank and spiked samples must be carried through all stages of the sample preparation and measurement steps. QC samples mentioned in the section on Interferences will also be needed as appropriate to those situations.

METHANOL EXTRACT REQUIRED FOR ANALYSIS OF HIGH-CONCENTRATION SOILS OR SEDIMENTS

Approximate Concentration Range	Volume of Methanol Extract (a)
500–10,000 µg/kg	100 µL
1,000–20,000 µg/kg	50 µL
5,000–100,000 µg/kg	10 µL
25,000–500,000 µg/kg	100 µL of 1/50 dilution (b)

Calculate appropriate dilution factor for concentrations exceeding this table.

(a) The volume of methanol added to 5 mL of water being purged should be kept constant. Therefore, add to the 5-mL syringe whatever volume of methanol is necessary to maintain a volume of 100 µL added to the syringe.
(b) Dilute an aliquot of the methanol extract and then take 100 µL for analysis.

REFERENCE Test Methods for Evaluating Solid Waste (SW-846). U.S. EPA. 1983. Method 8240B, Rev. 2, Nov. 1990. Office of Solid Wastes, Washington, DC.

Acetonitrile   EPA Method 8240
CAS #75-05-8

TITLE Volatile Organics By GC/MS: Packed Column Technique

MATRIX Nearly all types of sample matrices, regardless of water content, can be analyzed using this method. This includes groundwater, aqueous sludges, caustic liquors, acid liquors, waste solvents, oily wastes, mousses, tars, fibrous wastes, polymetric emulsions, filter cakes, spent carbons, spent catalysts, soils, and sediments.

METHOD SUMMARY Method 8240B covers 80 volatile organic compounds that are introduced into a gas chromatograph by the purge-and-trap method or by direct injection (in limited applications). For the purge-and-trap method an inert gas (zero grade nitrogen or helium) is bubbled through a 5-mL solution at ambient temperature. Purged sample components are trapped in a tube of sorbent materials. When purging is complete, the sorbent tube is heated and backflushed with inert gas to desorb the trapped components onto a GC column.

INTERFERENCES Impurities in the purge gas and from organic compounds outgassing from the plumbing ahead of the trap account for many contamination problems. Interferences purged or coextracted from the samples will vary considerably from source to source. Cross-contamination can occur whenever high-level and low-level samples are analyzed sequentially. Whenever an unusually concentrated sample is analyzed, it should be followed by the analysis of organic-free reagent water to check for cross-contamination. Samples also can be contaminated by diffusion of volatile organics (particularly methylene chloride and fluorocarbons) through the septum seal into the sample during shipment and storage. A trip blank can serve as a check on such contamination. The lab where volatile analysis is performed and also the refrigerated storage area should be completely free of solvents.

INSTRUMENTATION A gas chromatograph/mass spectrometry/data system (GC/MS) equipped with a 6 ft × 0.1 in I.D. glass column packed with 1% SP-1000 on Carbopack-B (60/80 mesh) is required. Also needed is a 5-mL purging device, a sorbent trap, and a thermal desorption apparatus.

PRECISION & ACCURACY This method is reported to have been tested by 15 laboratories using organic-free reagent water, drinking water, surface water, and industrial wastewater (not specified) fortified at six concentrations over the range 5–600 µg/L.

Sample estimated quantitation limits (EQLs) are highly matrix dependent. The EQLs listed may not always be achievable. EQLs listed for soils or sediments are based on wet weight. Normally, data is reported on a dry-weight basis; therefore, EQLs will be higher, based on the percent dry weight of each sample. Note that EQLs are even more variable than MDLs and that they are highly variable depending on the matrix being analyzed.

EQL in groundwater in µg/L was not listed.
EQL in low soil or sediment in µg/kg was not listed.
Accuracy (a) in µg/L was not listed.
Precision (b) in µg/L was not listed.

(a) *Average recovery found for measurements of samples containing a concentration of C, in µg/L.*
(b) *Overall precision found for measurements of samples with average recovery X for samples containing a concentration of C in µg/L.*
X = *Average recovery found for measurement of samples containing a concentration of C in µg/L.*

ESTIMATED QUANTITATION LIMITS FOR MATRICES OTHER THAN WATER, SOIL, AND SEDIMENTS

Other Matrices	Multiplication Factor (a)
Waste miscible liquid waste	50
High-concentration soil and sludge	125
Non-water miscible waste	500

(a) *EQL = [EQL for low soil/sediment] × [Factor]. For non-aqueous samples, the factor is on a wet-weight basis.*

SAMPLING METHOD
Liquid samples — Use a 40-mL glass screw-cap VOA vial with a Teflon®-faced silicone septum that has been prewashed, rinsed with distilled deionized water, and oven dried. However, if residual chlorine is present, collect sample in a 40-oz. soil VOA container which has been pre-preserved with 4 drops of 10% sodium thiosulfate, mix gently, and then transfer the sample to a 40-mL VOA vial. Collect bubble-free samples in duplicate and seal them in separate plastic bags.

Soils or sediments, and sludges — Use an 8 oz. widemouth glass bottle with a Teflon®-faced silicone septum that has been prewashed with detergent, rinsed with distilled deionized water, and oven dried. Tap slightly to eliminate free air space. Collect samples in duplicate and seal them in separate plastic bags.

SAMPLE PRESERVATION
Liquid samples — Add 4 drops of concentrated HCL and immediately cool samples to 4°C and store in a solvent-free refrigerator.

Soils or sediments, and sludges — Cool samples to 4°C and store in a solvent-free refrigerator.

MHT Maximum holding time is 14 days from the sample collection date.

SAMPLE PREPARATION
Liquid samples — Remove the plunger from a 5-mL syringe and carefully pour the sample into the syringe barrel to just short of overflowing. Replace the syringe plunger and compress the sample. Open the syringe valve and vent any residual air while adjusting the sample volume to 5.0 mL. If there is only one volatile organic analysis (VOA) vial, a second syringe should be filled at this time to protect against possible loss of sample integrity. Add 10 µL of surrogate spiking solution and 10 µL of internal standard spiking solution through the valve bore of the 5-mL syringe, then close the valve. The surrogate and internal standards may be mixed and added as a single spiking solution.

Sediments, soils, and waste samples — All samples of this type should be screened by GC analysis using a headspace method (EPA Method 3810) or the hexadecane extraction and screening method (EPA Method 3820). Use the screening data to determine whether to use the low-concentration method (0.005–1 mg/kg) or the high-concentration method (>1 mg/kg).

Low-concentration method — The low-concentration method is based on purging a heated sediment or soil sample mixed with organic-free reagent water containing the surrogate and internal standards. Analyze all reagent blanks and standards under the same conditions as the samples.

Use a 5-g sample if the expected concentration is <0.1 mg/kg or a 1-g sample for expected concentrations between 0.1 and 1 mg/kg. Mix the contents of the sample container with a narrow metal spatula. Weigh the amount of the sample into a tared purge device. Add the spiked water to the purge device, which contains the weighed amount of sample, and connect the device to the purge-and-trap system.

High-concentration method — This method is based on extracting the sediment or soil with methanol. A waste sample is either extracted or diluted, depending on its solubility in methanol. Wastes that are insoluble in methanol are diluted with reagent tetraglyme or possibly polyethylene glycol (PEG).

An aliquot of the extract is added to organic-free reagent water containing surrogate and internal standards. This is purged at ambient temperature. All samples with an expected concentration of >1.0 mg/kg should be analyzed by this method.

Mix the contents of the sample container with a narrow metal spatula. For sediments or soils and solid wastes that are insoluble in methanol, weigh 4 g (wet weight) of sample into a tared 20-mL vial. For waste that is soluble in methanol, tetraglyme, or PEG, weigh 1 g (wet weight) into a tared scintillation vial or culture tube or a 10-mL volumetric flask. Quickly add 9.0 mL of appropriate solvent then add 1.0 mL of a surrogate spiking solution to the vial, cap it, and shake it for 2 min.

METHANOL EXTRACT REQUIRED FOR ANALYSIS OF HIGH-CONCENTRATION SOILS OR SEDIMENTS

Approximate Concentration Range	Volume of Methanol Extract (a)
500–10,000 µg/kg	100 µL
1,000–20,000 µg/kg	50 µL
5,000–100,000 µg/kg	10 µL
25,000–500,000 µg/kg	100 µL of 1/50 dilution (b)

Calculate appropriate dilution factor for concentrations exceeding this table.

(a) The volume of methanol added to 5 mL of water being purged should be kept constant. Therefore, add to the 5-mL syringe whatever volume of methanol is necessary to maintain a volume of 100 µL added to the syringe.

(b) Dilute an aliquot of the methanol extract and then take 100 µL for analysis.

QUALITY CONTROL Demonstrate, through the analysis of a reagent water blank, that interferences from the analytical system, glassware, and reagents are under control. Blank samples should be carried through all stages of the sample preparation and measurement steps. For each analytical batch (up to 20 samples), a reagent blank, matrix spike, and matrix spike duplicate must be analyzed (the frequency of the spikes may be different for different monitoring programs). The blank and spiked samples must be carried through all stages of the sample preparation and measurement steps. QC samples mentioned in the section on Interferences will also be needed as appropriate to those situations.

REFERENCE Test Methods for Evaluating Solid Waste (SW-846). U.S. EPA. 1983. Method 8240B, Rev. 2, Nov. 1990. Office of Solid Wastes, Washington, DC.

Acetonitrile   **EPA Method 8030**
CAS #75-05-8

TITLE Other Nonhalogenated VOCs

MATRIX Groundwater, soils, sludges, water miscible liquid wastes, and non-water miscible wastes.

APPLICATION This method is used for the analysis of 3 nonhalogenated VOCs. Samples are analyzed using direct injection or purge-and-trap methods. Groundwater must be analyzed by the purge and trap method. The method provides an optional GC column which is used for analyte confirmation and that may help resolve analytes from interferences.

INTERFERENCES There can be carryover contamination with high- and low-level samples. Impurities may come from the purge and trap apparatus, organic compounds outgassing from the plumbing ahead of trap, diffusion of VOCs through the sample bottle septum during shipping or storage, or from solvent vapors in the lab.

INSTRUMENTATION GC capable of on-column injections or purge-and-trap sample introduction and a flame ionization detector (FID). Column 1: 10 ft by 2 mm with Porapak-QS. Column 2: 6 ft by 0.1 in with Chromosorb 101.

RANGE 5 to 100 µg/L

MDL 0.7 µg/L (reagent water).

PQL FACTORS FOR MULTIPLYING × FID MDL VALUE

Matrix	Multiplication Factor
Groundwater	10
Low-level soil	10
Water miscible liquid waste	500
High-level soil and sludge	1250
Non-water miscible waste	1250

PRECISION & ACCURACY Not determined.

SAMPLING METHOD Use two glass 40-mL vials with Teflon®-lined septum caps per sample location and collect samples with no headspace.

STABILITY Adjust to pH 4–5 and cool to 4°C.

MHT 14 days.

QUALITY CONTROL Analyze a reagent blank, matrix spike, and matrix spike duplicate/duplicate for each analytical batch (up to 20 Samples). Demonstrate the purity of glassware and reagents by analyzing a reagent water method blank. Internal, surrogate, and five concentration level calibration standards are used. The QC check sample concentrate should contain this compound at 25 µg/mL in reagent water.

REFERENCE Method 8030, SW-846, 3rd ed., Nov. 1986.

Acetophenone   **EPA Method 1625**
CAS #98-86-2

TITLE Semivolatile Organic Compounds by Isotope Dilution GC/MS

MATRIX The compounds may be determined in waters, soils, and municipal sludges by this method.

METHOD SUMMARY This method is used to determine 176 semivolatile toxic organic pollutants associated with the CWA (as amended 1987); the RCRA (as amended 1986); the CERCLA (as amended 1986); and other compounds amenable to extraction and analysis by capillary column gas chromatography-mass spectrometry (GC/MS).

Stable isotopically-labeled analogs of the compounds of interest are added to the sample. If the solids content is less than 1%, a 1-L sample is extracted at pH 12–13, then at pH <2 with methylene chloride using continuous extraction techniques.

If the solids content is 30% or less, the sample is diluted to 1% solids with reagent water, homogenized ultrasonically, and extracted at pH 12–13, then at pH <2 with methylene chloride using continuous extraction techniques. If the solids content is greater than 30%, the sample is extracted using ultrasonic techniques.

Each extract is dried over sodium sulfate, concentrated to a volume of 5 mL, cleaned up using GPC, if necessary, and concentrated. Extracts are concentrated to 1 mL if GPC is not performed, and to 0.5 mL if GPC is performed.

An internal standard is added to the extract, and a 1-mL aliquot of the extract is injected into the GC. The compounds are separated by GC and detected by a MS. The labeled compounds serve to correct the variability of the analytical technique.

INTERFERENCES Solvents, reagents, glassware, and other sample processing hardware may yield artifacts and/or elevated baselines causing misinterpretation of chromatograms and spectra. Materials used in the analysis must be demonstrated to be free from interferences under the conditions of analysis by running method blanks initially and with each sample lot (sample started through the extraction process on a given 8-h shift, to a maximum of 20). Specific selection of reagents and purification of solvents by distillation in all glass systems may be required. Glassware and, where possible, reagents are cleaned by solvent rinse and baking at 450°C for 1-h minimum. Interferences coextracted from samples will vary considerably from source to source, depending on the diversity of the site being sampled.

INSTRUMENTATION Major instrumentation includes a GC with a splitless or on-column injection port for capillary column, a MS with 70 eV electron impact ionization, and a data system to collect and record MS data, and process it. A K-D apparatus is used to concentrate extracts.

GC Column: 30 m × 0.25 mm I.D. 5% phenyl, 94% methyl, 1% vinyl silicone bonded phased fused silica capillary column.

PRECISION & ACCURACY The detection limits of the method are usually dependent on the level of interferences rather than instrumental limitations. The limits typify the minimum quantities that can be detected with no interferences present.

The minimum level (in µg/mL) was not listed. This is defined as a minimum level at which the analytical system shall give recognizable mass spectra (background corrected) and acceptable calibration points.

The MDL (in µg/kg) in low solids was not listed and in high solids was not listed; these were determined in digested sludge (low solids) and in filter cake or compost (high solids).

The labeled and native compound initial precision as standard deviation (in µg/L) was not listed.

The labeled and native compound initial accuracy as average recovery (in µg/L) was not listed.

SAMPLE COLLECTION, PRESERVATION & HANDLING
Collect samples in glass containers. Aqueous samples which flow freely are collected in refrigerated bottles using automatic sampling equipment. Solid samples are collected as grab samples using widemouth jars. Maintain samples at 0 to 4°C from the time of collection until extraction. If residual chlorine is present in aqueous samples, add 80 mg sodium thiosulfate/L of water. Begin sample extraction within 7 days of collection, and analyze all extracts within 40 days of extraction.

SAMPLE PREPARATION Samples containing 1% solids or less are extracted directly using continuous liquid-liquid extraction techniques. Samples containing 1 to 30% solids are diluted to the 1% level with reagent water and extracted using continuous liquid-liquid extraction techniques. Samples containing greater than 30% solids are extracted using ultrasonic techniques.

Base/neutral extraction — Adjust the pH of the waters in the extractors to 12–13 with 6 *N* NaOH. Extract with methylene chloride for 24–48 h.

Acid extraction — Adjust the pH of the waters in the extractors to 2 or less using 6 *N* sulfuric acid. Extract with methylene chloride for 24–48 h.

Ultrasonic extraction of high solids samples — Add anhydrous sodium sulfate to the sample and QC aliquot(s). Add acetone:methylene chloride (1:1) to the sample and mix thoroughly

Concentrate extracts using a K-D apparatus.

QUALITY CONTROL The analyst is permitted to modify this method to improve separations or lower the costs of measurements, provided all performance specifications are met. Analyses of blanks are required to demonstrate freedom from contamination. When results of spikes indicate atypical method performance for samples, the samples are diluted to bring method performance within acceptable limits.

For low solids (aqueous samples), extract, concentrate, and analyze two sets of four 1-L aliquots (8 aliquots total) of the precision and recovery standard. For high solids samples, two sets of four 30-g aliquots of the high solids reference matrix are used.

Spike all samples with labeled compounds to assess method performance. Compute percent recovery of the labeled compounds using the internal standard method. Compare the labeled compound recovery for each compound with the corresponding labeled compound recovery.

Reagent water and high solids reference matrix blanks are analyzed to demonstrate freedom from contamination. Extract and concentrate a 1-L reagent water blank or a high solids reference matrix blank with each sample's lot (samples started through the extraction process on the same 8-h shift, to a maximum of 20 samples).

Field replicates may be collected to determine the precision of the sampling technique, and spiked samples may be required

to determine the accuracy of the analysis when the internal standard method is used.

REFERENCE Semivolatile Organic Compounds by Isotope Dilution GC/MS. Office of Water Regulation and Standards, U.S. EPA Industrial Technology Div., Washington, DC, EPA Method 1625, Rev. C, June 1989 (contact W.A. Telliard, U.S. EPA, Office of Water Regulations and Stan-dards, 401 M St., SW, Washington, DC, 20460. Phone: 202-382-7131).

Acetophenone **EPA Method 8270**
CAS #98-86-2

TITLE Semivolatile Organic Compounds by GC/MS

MATRIX This method is used to determine the concentration of semivolatile organic compounds in extracts prepared from all types of solid waste matrices, soils, and groundwater. Although surface waters are not specifically mentioned, this method should be applicable to water samples from rivers, lakes, etc.

METHOD SUMMARY This method covers 259 semivolatile organic compounds. In very limited applications direct injection of the sample into the GC/MS system may be appropriate, but this results in very high detection limits (approximately 10,000 µg/L). Typically, a 1-L liquid sample, containing surrogate, and matrix spiking standards, is extracted in a continuous extractor first under acid conditions and then under basic conditions. Typically 30 g of a solid sample, containing surrogate, and matrix spiking standards, is extracted ultrasonically. After concentrating the extract to 1 mL it is spiked with 10 µL of an internal standard sol-ution just prior to analysis by GC/MS. The volume injected should contain about 100 ng of base/neutral and 200 ng of acid surrogates (for a 1-µL injection). Analysis is performed by GC/MS using a capillary GC column.

INTERFERENCES Raw GC/MS data from all blanks, samples, and spikes must be evaluated for interferences. Contamination by carryover can occur whenever high-concentration and low-concentration samples are sequentially analyzed. To reduce carryover, the sample syringe must be rinsed out between samples with solvent. Whenever an unusually concentrated sample is encountered, it should be followed by the analysis of blank solvent to check for cross-contamination.

INSTRUMENTATION A GC/MS and a data system are required. The GC column used is a 30 m × 0.25 mm I.D. (or 0.32 mm I.D.) 1-µm film thickness silicone-coated fused silica capillary column. A continuous liquid-liquid extractor equipped with Teflon® or glass connection joints and stopcocks requiring no lubrication, a K-D concentrating apparatus, water bath, and an ultrasonic disrupter with a minimum power of 300 W and with pulsing capability are also required.

PRECISION & ACCURACY The estimated quantitation limit (EQL) of Method 8270B for determining an individual compound is approximately 1 mg/kg (wet weight) for soil or sediment samples, 1–200 mg/kg for wastes (dependent on matrix and method of preparation), and 10 µg/L for groundwater samples. EQLs will be proportionately higher for sample extracts that require dilution to avoid saturation of the detector.

The EQL(b) for groundwater in µg/L is 10.
The EQL (a, b) for low concentrations in soil and sediment in µg/kg is not determined.
Accuracy as µg/L is not listed.
Overall precision in µg/L is not listed.

(a) *EQLs listed for soil/sediment are based on wet weight. Normally data is reported in a dry-weight basis; therefore, EQLs will be higher based on the % dry weight of each sample. This calculation is based on a 30-g sample and gel permeation chromatography cleanup.*
(b) *Sample EQLs are highly matrix-dependent. The EQLs are provided for guidance and may not always be achievable.*
C = *True value for concentration, in µg/L.*
X = *Average recovery found for measurements of samples containing a concentration of C, in µg/L.*

ESTIMATED QUANTITATION LIMIT

Other Matrices	Factor (a)
High-concentration soil and sludges by sonicator	7.5
Non-water miscible waste	75

(a) *EQL for other matrices = [EQL for low soil/sediment] × [Factor]. This estimated EQL is similar to an EPA "Practical Quantitation Limit."*

SAMPLING METHOD
Liquid samples — Use a 1 or 2½ gallon amber glass bottle with a screw-top Teflon®-lined cover that has been prewashed with detergent and rinsed with distilled water and methanol (or isopropanol).

Soils, sediments, or sludges — Use an 8-oz. widemouth glass with a screw-top Teflon®-lined cover that has been prewashed with detergent and rinsed with distilled water and methanol (or isopropanol).

SAMPLE PRESERVATION
Liquid samples — If residual chlorine is present, add 3 mL of 10% sodium thiosulfate per gallon, cool to 4°C and store in a solvent-free refrigerator until analysis; if chlorine is not present, then eliminate the sodium thiosulfate addition.

Soils, sediments, or sludges — Cool samples to 4°C and store in a solvent-free refrigerator.

MHT Liquid samples must be extracted within 7 days and the extracts analyzed within 40 days. Soils, sediments, or sludges may be stored for a maximum of 14 days and the extracts analyzed within 40 days.

SAMPLE PREPARATION
Liquid samples — Transfer 1 L quantitatively to a continuous extractor. If high concentrations are anticipated, a smaller volume may be used and then diluted with organic-free reagent water to 1 L. Adjust pH, if necessary, to pH <2 using 1:1 (V/V) sulfuric acid. Pipette 1.0 mL of a surrogate standard spiking solution into each sample. For the sample in each analytical batch selected for spiking, add 1.0 mL of a matrix spiking standard. For base/neutral acid analysis, the amount of the surrogates

and matrix spiking compounds added to the sample should result in a final concentration of 100 ng/μL of each analyte in the extract to be analyzed (assuming a 1-μL injection). Extract with methylene chloride for 18–24 h. Next, adjust the pH of the aqueous phase to pH >11 using 10 N sodium hydroxide and extract it with methylene chloride again for 18–24 h. Dry the extract through a column containing anhydrous sodium sulfate and concentrate it to 1 mL using a K-D concentrator.

Soils, sediments, or sludges — Use 30 g of sample. Nonporous or wet samples (gummy or clay type) that do not have a free-flowing sandy texture must be mixed with anhydrous sodium sulfate until the sample is free flowing. Add 1 mL of surrogate standards to all samples, spikes, standards, and blanks. For the sample in each analytical batch selected for spiking, add 1.0 mL of a matrix spiking standard. For base/neutral acid analysis, the amount added of the surrogates and matrix spiking compounds should result in a final concentration of 100 ng/μL of each base/neutral analyte and 200 ng/μL of each acid analyte in the extract to be analyzed (assuming a 1-μL injection). Immediately add a 100-mL mixture of 1:1 methylene chloride:acetone and extract the sample ultrasonically for 3 min and then decant or filter the extracts. Repeat the extraction two or more times. Dry the extract using a column with anhydrous sodium sulfate and concentrate it to 1 mL in a K-D concentrator.

QUALITY CONTROL   A methylene chloride solution containing 50 ng/μL of decafluorotriphenylphosphine (DFTPP) is used for tuning the GC/MS system each 12-h shift. A system performance check also must be made during every 12-h shift. A standard containing 50 ng/μL each of 4,4'-DDT, pentachlorophenol, and benzidine is required to verify injection port inertness and GC column performance. A calibration standard at mid-concentration, containing each compound of interest, including all required surrogates, must be performed every 12 h during analysis. After the system performance check is met, calibration check compounds are used to check the validity of the initial calibration.

The internal standard responses and retention times in the calibration check standard must be evaluated immediately after or during data acquisition. If the retention time for any internal standard changes by more than 30 seconds from the last check calibration (12 h), the chromatographic system must be inspected for malfunctions and corrections must be made, as required. If the electron ionization current plot (EICP) area for any of the internal standards changes by a factor of two from the last daily calibration standard check, the mass spectrometer must be inspected for malfunctions and corrections must be made, as appropriate.

Demonstrate, through the analysis of a reagent water blank, that interferences from the analytical system, glassware, and reagents are under control. The blank samples should be carried through all stages of the sample preparation and measurement steps. For each analytical batch (up to 20 samples), a reagent blank, matrix spike, and matrix spike duplicate/duplicate must be analyzed (the frequency of the spikes may be different for different monitoring programs). The blank and spiked samples must be carried through all stages of the sample preparation and measurement steps. A QC reference sample concentrate containing each analyte at a concentration of 100 mg/L in methanol is required.

REFERENCE   Test Methods for Evaluating Solid Waste (SW-846). U.S. EPA 1983, Method 8270B, Rev. 2, Nov. 1990. Office of Solid Waste, Washington, DC.

1-Acetyl-2-thiourea **EPA Method 8270**
CAS #591-08-2

TITLE   Semivolatile Organic Compounds by GC/MS

MATRIX   This method is used to determine the concentration of semivolatile organic compounds in extracts prepared from all types of solid waste matrices, soils, and groundwater. Although surface waters are not specifically mentioned, this method should be applicable to water samples from rivers, lakes, etc.

METHOD SUMMARY   This method covers 259 semivolatile organic compounds. In very limited applications direct injection of the sample into the GC/MS system may be appropriate, but this results in very high detection limits (approximately 10,000 μg/L). Typically, a 1-L liquid sample, containing surrogate, and matrix spiking standards, is extracted in a continuous extractor first under acid conditions and then under basic conditions. Typically 30 g of a solid sample, containing surrogate, and matrix spiking standards, is extracted ultrasonically. After concentrating the extract to 1 mL it is spiked with 10 μL of an internal standard solution just prior to analysis by GC/MS. The volume injected should contain about 100 ng of base/neutral and 200 ng of acid surrogates (for a 1-μL injection). Analysis is performed by GC/MS using a capillary GC column.

INTERFERENCES   Raw GC/MS data from all blanks, samples, and spikes must be evaluated for interferences. Contamination by carryover can occur whenever high-concentration and low-concentration samples are sequentially analyzed. To reduce carryover, the sample syringe must be rinsed out between samples with solvent. Whenever an unusually concentrated sample is encountered, it should be followed by the analysis of blank solvent to check for cross-contamination.

INSTRUMENTATION   A GC/MS and a data system are required. The GC column used is a 30 m × 0.25 mm I.D. (or 0.32 mm I.D.) 1-μm film thickness silicone-coated fused silica capillary column. A continuous liquid-liquid extractor equipped with Teflon® or glass connection joints and stopcocks requiring no lubrication, a K-D concentrating apparatus, water bath, and an ultrasonic disrupter with a minimum power of 300 W and with pulsing capability are also required.

PRECISION & ACCURACY   The estimated quantitation limit (EQL) of Method 8270B for determining an individual compound is approximately 1 mg/kg (wet weight) for soil or sediment samples, 1–200 mg/kg for wastes (dependent on matrix and method of preparation), and 10 μg/L for groundwater samples. EQLs will be proportionately higher for sample extracts that require dilution to avoid saturation of the detector.

The EQL(b) for groundwater in μg/L is 1000.

The EQL (a, b) for low concentrations in soil and sediment in µg/kg is not determined.
Accuracy as µg/L is not listed.
Overall precision in µg/L is not listed.

(a) EQLs listed for soil/sediment are based on wet weight. Normally data is reported in a dry-weight basis; therefore, EQLs will be higher based on the % dry weight of each sample. This calculation is based on a 30-g sample and gel permeation chromatography cleanup.
(b) Sample EQLs are highly matrix-dependent. The EQLs are provided for guidance and may not always be achievable.
C = *True value for concentration, in µg/L.*
X = *Average recovery found for measurements of samples containing a concentration of C, in µg/L.*

ESTIMATED QUANTITATION LIMIT

Other Matrices	Factor (a)
High-concentration soil and sludges by sonicator	7.5
Non-water miscible waste	75

(a) EQL for other matrices = [EQL for low soil/sediment] × [Factor]. This estimated EQL is similar to an EPA "Practical Quantitation Limit."

SAMPLING METHOD
Liquid samples — Use a 1 or 2½ gallon amber glass bottle with a screw-top Teflon®-lined cover that has been prewashed with detergent and rinsed with distilled water and methanol (or isopropanol).X

Soils, sediments, or sludges — Use an 8-oz. widemouth glass with a screw-top Teflon®-lined cover that has been prewashed with detergent and rinsed with distilled water and methanol (or isopropanol).

SAMPLE PRESERVATION
Liquid samples — If residual chlorine is present, add 3 mL of 10% sodium thiosulfate per gallon, cool to 4°C and store in a solvent-free refrigerator until analysis; if chlorine is not present, then eliminate the sodium thiosulfate addition.

Soils, sediments, or sludges — Cool samples to 4°C and store in a solvent-free refrigerator.

MHT Liquid samples must be extracted within 7 days and the extracts analyzed within 40 days. Soils, sediments, or sludges may be stored for a maximum of 14 days and the extracts analyzed within 40 days.

SAMPLE PREPARATION
Liquid samples — Transfer 1 L quantitatively to a continuous extractor. If high concentrations are anticipated, a smaller volume may be used and then diluted with organic-free reagent water to 1 L. Adjust pH, if necessary, to pH <2 using 1:1 (V/V) sulfuric acid. Pipette 1.0 mL of a surrogate standard spiking solution into each sample. For the sample in each analytical batch selected for spiking, add 1.0 mL of a matrix spiking standard. For base/neutral acid analysis, the amount of the surrogates and matrix spiking compounds added to the sample should result in a final concentration of 100 ng/µL of each analyte in the extract to be analyzed (assuming a 1-µL injection). Extract with methylene chloride for 18–24 h. Next, adjust the pH of the aqueous phase to pH >11 using 10 N sodium hydroxide and extract it with methylene chloride again for 18–24 h. Dry the extract through a column containing anhydrous sodium sulfate and concentrate it to 1 mL using a K-D concentrator.

Soils, sediments, or sludges — Use 30 g of sample. Nonporous or wet samples (gummy or clay type) that do not have a free-flowing sandy texture must be mixed with anhydrous sodium sulfate until the sample is free flowing. Add 1 mL of surrogate standards to all samples, spikes, standards, and blanks. For the sample in each analytical batch selected for spiking, add 1.0 mL of a matrix spiking standard. For base/neutral acid analysis, the amount added of the surrogates and matrix spiking compounds should result in a final concentration of 100 ng/µL of each base/neutral analyte and 200 ng/µL of each acid analyte in the extract to be analyzed (assuming a 1-µL injection). Immediately add a 100-mL mixture of 1:1 methylene chloride:acetone and extract the sample ultrasonically for 3 min and then decant or filter the extracts. Repeat the extraction two or more times. Dry the extract using a column with anhydrous sodium sulfate and concentrate it to 1 mL in a K-D concentrator.

QUALITY CONTROL A methylene chloride solution containing 50 ng/µL of decafluorotriphenylphosphine (DFTPP) is used for tuning the GC/MS system each 12-h shift. A system performance check also must be made during every 12-h shift. A standard containing 50 ng/µL each of 4,4′-DDT, pentachlorophenol, and benzidine is required to verify injection port inertness and GC column performance. A calibration standard at mid-concentration, containing each compound of interest, including all required surrogates, must be performed every 12 h during analysis. After the system performance check is met, calibration check compounds (CCCs) are used to check the validity of the initial calibration.

The internal standard responses and retention times in the calibration check standard must be evaluated immediately after or during data acquisition. If the retention time for any internal standard changes by more than 30 seconds from the last check calibration (12 h), the chromatographic system must be inspected for malfunctions and corrections must be made, as required. If the electron ionization current plot (EICP) area for any of the internal standards changes by a factor of two from the last daily calibration standard check, the mass spectrometer must be inspected for malfunctions and corrections must be made, as appropriate.

Demonstrate, through the analysis of a reagent water blank, that interferences from the analytical system, glassware, and reagents are under control. The blank samples should be carried through all stages of the sample preparation and measurement steps. For each analytical batch (up to 20 samples), a reagent blank, matrix spike, and matrix spike duplicate/duplicate must be analyzed (the frequency of the spikes may be different for different monitoring programs). The blank and spiked samples must be carried through all stages of the sample preparation and measurement steps. A QC reference sample concentrate containing each analyte at a concentration of 100 mg/L in methanol is required.

REFERENCE Test Methods for Evaluating Solid Waste (SW-846). U.S. EPA 1983, Method 8270B, Rev. 2, Nov. 1990. Office of Solid Waste, Washington, DC.

2-Acetylaminofluorene **EPA Method 8270**
CAS #53-96-3

TITLE Semivolatile Organic Compounds by GC/MS

MATRIX This method is used to determine the concentration of semivolatile organic compounds in extracts prepared from all types of solid waste matrices, soils, and groundwater. Although surface waters are not specifically mentioned, this method should be applicable to water samples from rivers, lakes, etc.

METHOD SUMMARY This method covers 259 semivolatile organic compounds. In very limited applications direct injection of the sample into the GC/MS system may be appropriate, but this results in very high detection limits (approximately 10,000 µg/L). Typically, a 1-L liquid sample, containing surrogate, and matrix spiking standards, is extracted in a continuous extractor first under acid conditions and then under basic conditions. Typically 30 g of a solid sample, containing surrogate, and matrix spiking standards, is extracted ultrasonically. After concentrating the extract to 1 mL it is spiked with 10 µL of an internal standard solution just prior to analysis by GC/MS. The volume injected should contain about 100 ng of base/neutral and 200 ng of acid surrogates (for a 1-µL injection). Analysis is performed by GC/MS using a capillary GC column.

INTERFERENCES Raw GC/MS data from all blanks, samples, and spikes must be evaluated for interferences. Contamination by carryover can occur whenever high-concentration and low-concentration samples are sequentially analyzed. To reduce carryover, the sample syringe must be rinsed out between samples with solvent. Whenever an unusually concentrated sample is encountered, it should be followed by the analysis of blank solvent to check for cross-contamination.

INSTRUMENTATION A GC/MS and a data system are required. The GC column used is a 30 m × 0.25 mm I.D. (or 0.32 mm I.D.) 1-µm film thickness silicone-coated fused silica capillary column. A continuous liquid-liquid extractor equipped with Teflon® or glass connection joints and stopcocks requiring no lubrication, a K-D concentrating apparatus, water bath, and an ultrasonic disrupter with a minimum power of 300 W and with pulsing capability are also required.

PRECISION & ACCURACY The estimated quantitation limit (EQL) of Method 8270B for determining an individual compound is approximately 1 mg/kg (wet weight) for soil or sediment samples, 1–200 mg/kg for wastes (dependent on matrix and method of preparation), and 10 µg/L for groundwater samples. EQLs will be proportionately higher for sample extracts that require dilution to avoid saturation of the detector.

The EQL(b) for groundwater in µg/L is 20.
The EQL (a, b) for low concentrations in soil and sediment in µg/kg is not determined.
Accuracy as µg/L is not listed.

Overall precision in µg/L is not listed.

(a) *EQLs listed for soil/sediment are based on wet weight. Normally data is reported in a dry-weight basis; therefore, EQLs will be higher based on the % dry weight of each sample. This calculation is based on a 30-g sample and gel permeation chromatography cleanup.*
(b) *Sample EQLs are highly matrix-dependent. The EQLs are provided for guidance and may not always be achievable.*
C = *True value for concentration, in µg/L.*
X = *Average recovery found for measurements of samples containing a concentration of C, in µg/L.*

ESTIMATED QUANTITATION LIMIT

Other Matrices	Factor (a)
High-concentration soil and sludges by sonicator	7.5
Non-water miscible waste	75

(a) EQL for other matrices = [EQL for low soil/sediment] × [Factor]. This estimated EQL is similar to an EPA "Practical Quantitation Limit."

SAMPLING METHOD
Liquid samples — Use a 1 or 2½ gallon amber glass bottle with a screw-top Teflon®-lined cover that has been prewashed with detergent and rinsed with distilled water and methanol (or isopropanol).

Soils, sediments, or sludges — Use an 8-oz. widemouth glass with a screw-top Teflon®-lined cover that has been prewashed with detergent and rinsed with distilled water and methanol (or isopropanol).

SAMPLE PRESERVATION
Liquid samples — If residual chlorine is present, add 3 mL of 10% sodium thiosulfate per gallon, cool to 4°C and store in a solvent-free refrigerator until analysis; if chlorine is not present, then eliminate the sodium thiosulfate addition.

Soils, sediments, or sludges — Cool samples to 4°C and store in a solvent-free refrigerator.

MHT Liquid samples must be extracted within 7 days and the extracts analyzed within 40 days. Soils, sediments, or sludges may be stored for a maximum of 14 days and the extracts analyzed within 40 days.

SAMPLE PREPARATION
Liquid samples — Transfer 1 L quantitatively to a continuous extractor. If high concentrations are anticipated, a smaller volume may be used and then diluted with organic-free reagent water to 1 L. Adjust pH, if necessary, to pH <2 using 1:1 (V/V) sulfuric acid. Pipette 1.0 mL of a surrogate standard spiking solution into each sample. For the sample in each analytical batch selected for spiking, add 1.0 mL of a matrix spiking standard. For base/neutral acid analysis, the amount of the surrogates and matrix spiking compounds added to the sample should result in a final concentration of 100 ng/µL of each analyte in the extract to be analyzed (assuming a 1-µL injection). Extract with methylene chloride for 18–24 h. Next, adjust the pH of the aqueous phase to pH >11 using 10 N sodium hydroxide and extract it with methylene chloride again for 18–24 h. Dry

the extract through a column containing anhydrous sodium sulfate and concentrate it to 1 mL using a K-D concentrator.

Soils, sediments, or sludges — Use 30 g of sample. Nonporous or wet samples (gummy or clay type) that do not have a free-flowing sandy texture must be mixed with anhydrous sodium sulfate until the sample is free flowing. Add 1 mL of surrogate standards to all samples, spikes, standards, and blanks. For the sample in each analytical batch selected for spiking, add 1.0 mL of a matrix spiking standard. For base/neutral acid analysis, the amount added of the surrogates and matrix spiking compounds should result in a final concentration of 100 ng/μL of each base/neutral analyte and 200 ng/μL of each acid analyte in the extract to be analyzed (assuming a 1-μL injection). Immediately add a 100-mL mixture of 1:1 methylene chloride:acetone and extract the sample ultrasonically for 3 min and then decant or filter the extracts. Repeat the extraction two or more times. Dry the extract using a column with anhydrous sodium sulfate and concentrate it to 1 mL in a K-D concentrator.

QUALITY CONTROL A methylene chloride solution containing 50 ng/μL of decafluorotriphenylphosphine (DFTPP) is used for tuning the GC/MS system each 12-h shift. A system performance check also must be made during every 12-h shift. A standard containing 50 ng/μL each of 4,4′-DDT, pentachlorophenol, and benzidine is required to verify injection port inertness and GC column performance. A calibration standard at mid-concentration, containing each compound of interest, including all required surrogates, must be performed every 12 h during analysis. After the system performance check is met, calibration check compounds (CCCs) are used to check the validity of the initial calibration.

The internal standard responses and retention times in the calibration check standard must be evaluated immediately after or during data acquisition. If the retention time for any internal standard changes by more than 30 seconds from the last check calibration (12 h), the chromatographic system must be inspected for malfunctions and corrections must be made, as required. If the electron ionization current plot (EICP) area for any of the internal standards changes by a factor of two from the last daily calibration standard check, the mass spectrometer must be inspected for malfunctions and corrections must be made, as appropriate.

Demonstrate, through the analysis of a reagent water blank, that interferences from the analytical system, glassware, and reagents are under control. The blank samples should be carried through all stages of the sample preparation and measurement steps. For each analytical batch (up to 20 samples), a reagent blank, matrix spike, and matrix spike duplicate/duplicate must be analyzed (the frequency of the spikes may be different for different monitoring programs). The blank and spiked samples must be carried through all stages of the sample preparation and measurement steps. A QC reference sample concentrate containing each analyte at a concentration of 100 mg/L in methanol is required.

REFERENCE Test Methods for Evaluating Solid Waste (SW-846). U.S. EPA 1983, Method 8270B, Rev. 2, Nov. 1990. Office of Solid Waste, Washington, DC.

Acidity (Titrimetric) — EPA Method 305.1

TITLE Inorganics, Nonmetallics

MATRIX Surface and waste waters.

APPLICATION Date issued 1971. Technical Rev. 1974. The pH of the sample is determined and a measured amount of standard acid is added, as needed, to lower pH to 4 or less. Hydrogen peroxide is added, the solution boiled for several min, cooled and tirated to pH 8.2. Method measures mineral acidity of a sample plus acidity from oxidation and hydrolysis of polyvalent cations, including salts of iron and aluminum.

INTERFERENCES Suspended matter present in the sample, or precipitates formed during the titration may cause sluggish electrode response. (This is overcome by allowing 15–20 second pauses between titrant additions and drop by drop titrant additions near end point).

INSTRUMENTATION pH m suitable for electrometric titrations.

RANGE 10–1000 mg/L as $CaCO_3$, on (50 mL)

MDL standard acid = (0.02 N sulfuric)

PRECISION ±10 mg/L on 4 sample concentrations (up to 2000 mg/L).

ACCURACY Calculate acidity as mg/L $CaCO_3$ or as meq/L.

SAMPLING METHOD Plastic or glass (100 mL).

STABILITY Cool, 4°C.

MHT 14 days.

QUALITY CONTROL Cool (boiled sample) to room temperature before titrating electrometrically with standard sodium hydroxide (0.02 N) to pH 8.2

REFERENCE EPA Methods for the Chemical Analysis of Water and Wastes, EPA-600/4-79-020, U.S. EPA, EMSL, 1979.

Acifluorfen — EPA Method 8151
CAS #50594-66-6

TITLE Chlorinated Herbicides by GC Using Methylation or Pentafluorobenzylation Derivatization: Capillary Column Technique.

MATRIX This method covers aqueous and solid matrices. This includes a wide variety such as drinking water, groundwater, industrial wastewater, surface waters, soils, solids and sediments.

METHOD SUMMARY This is a GC method for determining 19 chlorinated acid herbicides in aqueous, soil, and waste matrices. Because these compounds are produced and used in various forms (i.e., acid, salt, ester, etc.) a hydrolysis step is included to convert the herbicide to the acid form prior to analysis. This method provides hydrolysis, extraction, derivatization, and GC conditions for the analysis of chlorinated acid herbicides in water, soil, and waste samples. Water samples are hydrolyzed *in situ*, extracted with diethyl ether, and then esterified with either diazomethane or pentafluorobenzyl bromide. The derivatives are determined by GC with an electron capture detector (GC/ECD). The results are reported as acid equivalents. The sensitivity of this method depends on the level of interferences in addition to instrumental limitations.

INTERFERENCES Method interferences may be caused by contaminants in solvents, reagents, glassware, and other sample processing hardware. Immediately prior to use, glassware should be rinsed with the next solvent to be used. Matrix interferences may be caused by contaminants that are coextracted from the sample. Organic acids, especially chlorinated acids, cause the most direct interference with the determination by methylation. Phenols, including chlorophenols, may also interfere with this procedure. The determination using pentafluorobenzylation is more sensitive, and more prone to interferences from the presence of organic acids of phenols than by methylation. Alkaline hydrolysis and subsequent extraction of the basic solution removes many chlorinated hydrocarbons and phthalate esters that might otherwise interfere with the ECD analysis. The herbicides, being strong organic acids, react readily with alkaline substances and may be lost during analysis. Therefore, glassware must be acid-rinsed and then rinsed to constant pH with organic-free reagent water.

INSTRUMENTATION A GC suitable for Grob-type injection using capillary columns. A data system for measuring peak heights and/or peak areas is recommended. An electron capture detector (ECD) is used. Also a K-D apparatus, a diazomethane generator, a centrifuge, and an ultrasonic disrupter will be required.

Narrow Bore Columns:
Primary Column 1: 30 m × 0.25 mm, 5% phenyl/95% methyl silicone (DB-5), 0.25 μm film thickness.
Primary Column 1a (GC/MS): 30 m × 0.32 mm, 5% phenyl/95% methyl silicone (DB-5), 1-μm film thickness.
Column 2: 30 m × 0.25 mm DB-608 with a 25 μm film thickness.
Confirmation Column: 30 m × 0.25 mm, 14% cyanopropyl phenyl silicone (DB-1701), 0.25 μm film thickness.

Megabore Columns:
Primary Column: 30 m × 0.53 mm DB-608 with 0.83 μm film thickness.
Confirmation Column: 30 m × 0.53 mm, 14% cyanopropyl phenyl silicone (DB-1701), 1.0 μm film thickness.

PRECISION & ACCURACY Method detection limits (MDLs) are compound-dependent and vary with derivitization efficiency, derivative recovery, the matrix sampled, and herbicide concentration.

The estimated MDL (in μg/L) was 0.096 for aqueous samples using GC/ECD.

The estimated MDL (in μg/kg) was not reported for soil samples using GC/ECD when corrected back to 50-g samples extracted and concentrated to 10 mL with 5-μL injections.

The estimated GC/MS identification limit (in ng) was not reported for soil samples using GC/MS.

Mean percent recovery, calculated from 7–8 determinations of spiked reagent water, after diazomethane derivatization, from a spike concentration (in μg/L) of 0.2 was 121 with a standard deviation of the percent recovery of 15.7.

Mean percent recovery, calculated from 10 determinations of spiked clay and clay/still bottom samples over the linear concentration range (in ng/g) of no data, was none reported with a percent relative standard deviation of none. The RSD % was calculated on 10 samples high in the linear concentration range and 10 low in the range. The linear concentration range was determined using standard solutions and corrected to 50-g soil samples.

SAMPLE COLLECTION, PRESERVATION & HANDLING
Containers used to collect samples for the determination of semivolatile organic compounds should be soap and water washed followed by methanol (or isopropanol) rinsing. The sample containers should be of glass or Teflon® and have screw-top covers with Teflon® liners.

No preservation is used with concentrated waste samples. With liquid samples containing no residual chlorine and with soil, sediment, and sludge samples, immediately cooling to 4°C is the only preservation used. When residual chlorine is present then 3 mL of 10% aqueous sodium sulfate is added for each gallon of sample collected, followed by cooling to 4°C.

The holding time for all volatile organics samples is 14 days. Liquid samples must be extracted within 7 days and their extracts analyzed within 40 days. Concentrated waste, soil, sediment, and sludge samples must be extracted within 14 days and their extracts analyzed within 40 days.

SAMPLE PREPARATION
Preparation of soil, sediment, and other solid samples — Acidify 30 g (dry weight) solids with 0.1 M phosphate buffer (pH = 2.5) and thoroughly mix the contents. Spike the sample with surrogate compound(s). The ultrasonic extraction of solids must be optimized for each type of sample. In order for the ultrasonic extractor to efficiently extract solid samples, the sample must be free flowing when the solvent is added. Acidified anhydrous sodium sulfate should be added to clay-type soils, or any other solid that is not a free-flowing sandy texture, until a free flowing mixture is obtained. Add methylene chloride and perform ultrasonic extraction. Combine organic extracts from the repetitive extractings of the sample and centrifuge. Add aqueous potassium hydroxide, water, and methanol to the extract and reflux the mixture on a water bath. Extract the solution three times with methylene chloride and discard the methylene chloride phase. The basic solution contains the herbicide salts. Adjust the pH of the solution to <2 with cold sulfuric acid and extract three times with methylene chloride. Combine the extracts and pour them through a pre-rinsed drying column containing acidified anhydrous sodium sulfate. Collect the dried extracts in a K-D flask and concentrate them.

Preparation of aqueous samples — Measure 1 L of sample into a 2-L separatory funnel and spike it with surrogate compound(s). Add NaCl to the sample, then add 6 N NaOH to the sample to a pH of 12 or more and let the sample sit at room temperature for 1 h to hydrolyze esters. Extract the sample three times with methylene chloride and discard the extracts. Then add cold 12 N sulfuric acid to a pH less than or equal to 2, and extract the sample three times with ethyl ether. Collect the ether phase in a flask containing acidified anhydrous sodium sulfate and allow it to remain in contact with the sodium sulfate for a minimum of 2 h. The drying step is very critical to ensuring complete esterification; any moisture remaining in the ether will result in low herbicide recoveries.

Extract concentration and derivatization — The combined ether extract is concentrated to about 1 mL using a K-D apparatus followed by using a micro Snyder column or nitrogen gas blowdown. If methyl esters are to be produced, then dilute the concentrated ether extract with 1 mL of isooctane and 0.5 mL of methanol, dilute to a final volume of 4 mL, and esterify with diazomethane. If pentafluorobenzene esters are to be produced, then dilute concentrated ether extract with acetone to a final volume of 4 mL and esterify with pentafluorobenzyl bromide.

QUALITY CONTROL Select a representative spike concentration for each compound (acid or ester) to be measured. Using stock standard, prepare a quality control check sample concentrate, in acetone, that is 1000 times more concentrated than the selected concentrations. Use this quality control check sample concentrate to prepare quality control check samples. Calculate surrogate standard recovery on all standards, samples, blanks, and spikes. GC/MS techniques should be judiciously employed to support qualitative identifications made with this method. When available, chemical ionization mass spectra may be employed to aid the qualitative identification process.

REFERENCE Test Methods for Evaluating Solid Waste, Physical/Chemical Methods, SW-846, 3rd Edition, U.S. EPA, Office of Solid Waste, Washington, DC, EPA Method 8151, Nov. 1990.

Acrolein   **EPA Method 1624**
CAS #107-02-8

TITLE Volatile Organic Compounds by Isotope Dilution GC/MS

MATRIX Compounds may be determined in waters, soils, and municipal sludges by this method.

METHOD SUMMARY This method is used to determine 58 volatile toxic organic pollutants associated with the CWA (as amended 1987); the RCRA (as amended 1986); the CERCLA (as amended 1986); and other compounds amenable to purge and trap gas chromatography-mass spectrometry (GC/MS).

If the solids content is less than 1%, stable isotopically-labeled analogs of the compounds of interest are added to a 5-mL sample and the sample is purged with an inert gas at 20–25°C in a chamber designed for soil or water samples. If the solids content is greater than 1%, 5 mL of reagent water and the labeled compounds are added to a 5 g aliquot of sample and the mixture is purged at 40°C. Compounds that will not purge at 20–25°C or at 40°C are purged at 78–85°C. In the purging process, the volatile compounds are transferred from the aqueous phase into the gaseous phase where they are passed into a sorbent column and trapped. After purging is completed, the trap is backflushed and heated rapidly to desorb the compounds into a GC. The compounds are separated by the GC and detected by a MS. The labeled compounds serve to correct the variability of the analytical technique.

INTERFERENCES Impurities in the purge gas, organic compounds outgassing from the plumbing upstream of the trap, and solvent vapors in the lab account for most problems. Samples can be contaminated by diffusion of volatile organic compounds (particularly methylene chloride) through the bottle seal during shipment and storage. Contamination by carryover can occur when high-level and low-level samples are analyzed sequentially. When an unusually concentrated sample is encountered, follow it by analysis of a reagent water blank to check for carryover.

INSTRUMENTATION Major equipment includes a GC with linear temperature programming and a glass jet separator as the MS interface, a MS with 70 eV electron impact ionization, and a data system to collect and record response factors.

Column: 2.8 m × 2 mm I.D. glass, packed with 1% SP-1000 on Carbopak B, 60/80 mesh, or equivalent.

PRECISION & ACCURACY The detection limits of the method are usually dependent on the level of interferences rather than instrumental limitations. The method detection limits were determined in digested sludge (low solids) and in filter cake or compost (high solids).

The MDL (in µg/kg) for low solids is 377 and for high solids is 18.

Background levels of this compound were present in the sludge with low solids, resulting in a higher than expected MDL.

Labeled and native compound precision (in µg/L) as standard deviation was 72.0.

Labeled and native compound accuracy (in µg/L) as average recovery was 32–168.

Acceptance criteria are at 100 µg/L for this compound.

SAMPLE COLLECTION, PRESERVATION & HANDLING Grab samples are collected in glass containers having a total volume greater than 20 mL. Fill and seal each bottle so that no air bubbles are entrapped. Samples are maintained at 0 to 4°C from the time of collection until analysis. If an aqueous sample contains residual chlorine, add sodium thiosulfate preservative (10 mg/40 mL) to the empty sample bottles just prior to shipment to the sample site. All samples must be analyzed within 14 days of collection.

SAMPLE PREPARATION Samples containing less than 1% solids are analyzed directly as aqueous samples. Samples containing 1% solids or greater are analyzed as solid samples utilizing one of two methods, depending on the levels of pollutants, in the sample. Samples containing 1% solids or greater, and low to moderate levels of pollutants are analyzed by purging a known weight of sample added to 5 mL of reagent water. Samples containing 1% solids or greater, and high levels of pollutants, are extracted with methanol, and an aliquot of the methanol extract is added to reagent water and purged.

QUALITY CONTROL A field blank prepared from reagent water and carried through the sampling and handling protocol may serve as a check on contamination from shipment and storage.

The analyst is permitted to modify this method to improve separations or lower the costs of measurements, provided all performance specifications are met. Analyses of blanks are required. When results of spikes indicate atypical method performance for samples, the samples are diluted to bring method performance within acceptable limits. Analyze two sets of four 5-mL aliquots (8 aliquots total) of the aqueous performance standard. Spike all samples with labeled compounds to assess method performance on the sample matrix. Compute the percent recovery of the labeled compounds using the internal standard method. Compare the percent recovery for each compound with the corresponding labeled compound recovery. Reagent water blanks are analyzed to demonstrate freedom from carryover contamination. Field replicates may be collected to determine the precision of the sampling technique, and spiked samples may be required to determine the accuracy of the analysis when the internal method is used.

REFERENCE Volatile Organic Compounds by Isotope Dilution GC/MS. Office of Water Regulation and Standards, U.S. EPA Industrial Technology Division, Washington, DC, EPA Method 1624, Rev. C, June 1989 (contact W.A. Telliard, U.S. EPA, Office of Water Regulations and Standards, 401 M St., SW, Washington, DC, 20460. Phone: 202-382-7131).

Acrolein EPA Method 8240
CAS #107-02-8

TITLE Volatile Organics By GC/MS: Packed Column Technique

MATRIX Nearly all types of sample matrices, regardless of water content, can be analyzed using this method. This includes groundwater, aqueous sludges, caustic liquors, acid liquors, waste solvents, oily wastes, mousses, tars, fibrous wastes, polymetric emulsions, filter cakes, spent carbons, spent catalysts, soils, and sediments.

METHOD SUMMARY Method 8240B covers 80 volatile organic compounds that are introduced into a gas chromatograph by the purge-and-trap method or by direct injection (in limited applications). For the purge-and-trap method an inert gas (zero grade nitrogen or helium) is bubbled through a 5-mL solution at ambient temperature. Purged sample components are trapped in a tube of sorbent materials. When purging is complete, the sorbent tube is heated and backflushed with inert gas to desorb the trapped components onto a GC column.

INTERFERENCES Impurities in the purge gas and from organic compounds outgassing from the plumbing ahead of the trap account for many contamination problems. Interferences purged or coextracted from the samples will vary considerably from source to source. Cross-contamination can occur whenever high-level and low-level samples are analyzed sequentially. Whenever an unusually concentrated sample is analyzed, it should be followed by the analysis of organic-free reagent water to check for cross-contamination. Samples also can be contaminated by diffusion of volatile organics (particularly methylene chloride and fluorocarbons) through the septum seal into the sample during shipment and storage. A trip blank can serve as a check on such contamination. The lab where volatile analysis is performed and also the refrigerated storage area should be completely free of solvents.

INSTRUMENTATION A gas chromatograph/mass spectrometry/data system (GC/MS) equipped with a 6 ft × 0.1 in I.D. glass column packed with 1% SP-1000 on Carbopack-B (60/80 mesh) is required. Also needed is a 5-mL purging device, a sorbent trap, and a thermal desorption apparatus.

PRECISION & ACCURACY This method is reported to have been tested by 15 laboratories using organic-free reagent water, drinking water, surface water, and industrial wastewater (not specified) fortified at six concentrations over the range 5–600 μg/L.

Sample estimated quantitation limits (EQLs) are highly matrix dependent. The EQLs listed may not always be achievable. EQLs listed for soils or sediments are based on wet weight. Normally, data is reported on a dry-weight basis; therefore, EQLs will be higher, based on the percent dry weight of each sample. Note that EQLs are even more variable than MDLs and that they are highly variable depending on the matrix being analyzed.

EQL in groundwater in μg/L was not listed.
EQL in low soil or sediment in μg/kg was not listed.
Accuracy (a) in μg/L was not listed.
Precision (b) in μg/L was not listed.

(a) Average recovery found for measurements of samples containing a concentration of C, in μg/L.
(b) Overall precision found for measurements of samples with average recovery X for samples containing a concentration of C in μg/L.
X = Average recovery found for measurement of samples containing a concentration of C in μg/L.

ESTIMATED QUANTITATION LIMITS FOR MATRICES OTHER THAN WATER, SOIL, AND SEDIMENTS

Other Matrices	Factor (a)
Waste miscible liquid waste	50
High-concentration soil and sludge	125
Non-water miscible waste	500

(a) EQL = [EQL for low soil/sediment] × [Factor]. For non-aqueous samples, the factor is on a wet-weight basis.

SAMPLING METHOD

Liquid samples — Use a 40-mL glass screw-cap VOA vial with a Teflon®-faced silicone septum that has been prewashed, rinsed with distilled deionized water, and oven dried. However, if residual chlorine is present, collect sample in a 40-oz. soil VOA container which has been pre-preserved with 4 drops of 10% sodium thiosulfate, mix gently, and then transfer the sample to a 40-mL VOA vial. Collect bubble-free samples in duplicate and seal them in separate plastic bags.

Soils or sediments, and sludges — Use an 8 oz. widemouth glass bottle with a Teflon®-faced silicone septum that has been prewashed with detergent, rinsed with distilled deionized water, and oven dried. Tap slightly to eliminate free air space. Collect samples in duplicate and seal them in separate plastic bags.

SAMPLE PRESERVATION

Liquid samples — Add 4 drops of concentrated HCL and immediately cool samples to 4°C and store in a solvent-free refrigerator.

Soils or sediments, and sludges — Cool samples to 4°C and store in a solvent-free refrigerator.

MHT Maximum holding time is 14 days from the date of sample collection.

SAMPLE PREPARATION

Liquid samples — Remove the plunger from a 5-mL syringe and carefully pour the sample into the syringe barrel to just short of overflowing. Replace the syringe plunger and compress the sample. Open the syringe valve and vent any residual air while adjusting the sample volume to 5.0 mL. If there is only one volatile organic analysis (VOA) vial, a second syringe should be filled at this time to protect against possible loss of sample integrity. Add 10 µL of surrogate spiking solution and 10 µL of internal standard spiking solution through the valve bore of the 5-mL syringe, then close the valve. The surrogate and internal standards may be mixed and added as a single spiking solution.

Sediments, soils, and waste samples — All samples of this type should be screened by GC analysis using a headspace method (EPA Method 3810) or the hexadecane extraction and screening method (EPA Method 3820). Use the screening data to determine whether to use the low-concentration method (0.005–1 mg/kg) or the high-concentration method (>1 mg/kg).

Low-concentration method — The low-concentration method is based on purging a heated sediment or soil sample mixed with organic-free reagent water containing the surrogate and internal standards. Analyze all reagent blanks and standards under the same conditions as the samples.

Use a 5-g sample if the expected concentration is <0.1 mg/kg or a 1-g sample for expected concentrations between 0.1 and 1 mg/kg. Mix the contents of the sample container with a narrow metal spatula. Weigh the amount of sample into a tared purge device. Add the spiked water to the purge device, which contains the weighed amount of sample, and connect the device to the purge-and-trap system.

High-concentration method — This method is based on extracting the sediment or soil with methanol. A waste sample is either extracted or diluted, depending on its solubility in methanol. Wastes that are insoluble in methanol are diluted with reagent tetraglyme or possibly polyethylene glycol (PEG). An aliquot of the extract is added to organic-free reagent water containing surrogate and internal standards. This is purged at ambient temperature. All samples with an expected concentration of >1.0 mg/kg should be analyzed by this method.

Mix the contents of the sample container with a narrow metal spatula. For sediments or soils and solid wastes that are insoluble in methanol, weigh 4 g (wet weight) of sample into a tared 20-mL vial. For waste that is soluble in methanol, tetraglyme, or PEG, weigh 1 g (wet weight) into a tared scintillation vial or culture tube or a 10-mL volumetric flask. Quickly add 9.0 mL of appropriate solvent then add 1.0 mL of a surrogate spiking solution to the vial, cap it, and shake it for 2 min.

METHANOL EXTRACT REQUIRED FOR ANALYSIS OF HIGH-CONCENTRATION SOILS OR SEDIMENTS

Approximate Concentration Range	Volume of Methanol Extract (a)
500–10,000 µg/kg	100 µL
1,000–20,000 µg/kg	50 µL
5,000–100,000 µg/kg	10 µL
25,000–500,000 µg/kg	100 µL of 1/50 dilution (b)

Calculate appropriate dilution factor for concentrations exceeding this table.

(a) The volume of methanol added to 5 mL of water being purged should be kept constant. Therefore, add to the 5-mL syringe whatever volume of methanol is necessary to maintain a volume of 100 µL added to the syringe.
(b) Dilute an aliquot of the methanol extract and then take 100 µL for analysis.

QUALITY CONTROL Demonstrate, through the analysis of a reagent water blank, that interferences from the analytical system, glassware, and reagents are under control. Blank samples should be carried through all stages of the sample preparation and measurement steps. For each analytical batch (up to 20 samples), a reagent blank, matrix spike, and matrix spike duplicate must be analyzed (the frequency of the spikes may be different for different monitoring programs). The blank and spiked samples must be carried through all stages of the sample preparation and measurement steps. QC samples mentioned in the section on Interferences will also be needed as appropriate to those situations.

REFERENCE Test Methods for Evaluating Solid Waste (SW-846). U.S. EPA. 1983. Method 8240B, Rev. 2, Nov. 1990. Office of Solid Wastes, Washington, DC.

Acrolein	EPA Method 8030
CAS #107-02-8	

TITLE Other Nonhalogenated VOCs

MATRIX Groundwater, soils, sludges, water miscible liquid wastes, and non-water miscible wastes.

APPLICATION This method is used for the analysis of 3 nonhalogenated VOCs. Samples are analyzed using direct injection or purge-and-trap methods. Groundwater must be analyzed by the purge and trap method. The method provides an optional GC column which is used for analyte confirmation and that may help resolve analytes from interferences.

INTERFERENCES There can be carryover contamination with high- and low-level samples. Impurities may come from the purge and trap apparatus, organic compounds outgassing from the plumbing ahead of trap, diffusion of VOCs through the sample bottle septum during shipping or storage, or from solvent vapors in the lab.

INSTRUMENTATION GC capable of on-column injections or purge-and-trap sample introduction and a flame ionization detector (FID). Column 1: 10 ft by 2 mm with Porapak-QS. Column 2: 6 ft by 0.1 in with Chromosorb 101.

RANGE 5 to 100 µg/L

MDL 0.7 µg/L (reagent water).

PQL FACTORS FOR MULTIPLYING × FID MDL VALUE

Matrix	Multiplication Factor
Groundwater	10
Low-level soil	10
Water miscible liquid waste	500
High-level soil and sludge	1250
Non-water miscible waste	1250

PRECISION (as standard deviation): 0.7 µg/L, reagent water and 0.8 µg/L, POTW at 50 µg/L spike, 1.1 µg/L, industrial water at 100 µg/L spike.

ACCURACY (as % recovery): 103%, reagent water and 89%, POTW at 50 µg/L spike, 9%, industrial wastewater at 100 µg/L spike.

SAMPLING METHOD Use two glass 40-mL vials with Teflon®-lined septum caps per sample location and collect samples with no headspace.

STABILITY Adjust to pH 4–5 and cool to 4°C.

MHT 14 days.

QUALITY CONTROL Analyze a reagent blank, matrix spike, and matrix spike duplicate/duplicate for each analytical batch (up to 20 samples). Demonstrate the purity of glassware and reagents by analyzing a reagent water method blank. Internal, surrogate, and five concentration level calibration standards are used. The QC check sample concentrate should contain this compound at 25 µg/mL in reagent water.

REFERENCE Method 8030, SW-846, 3rd ed., Nov. 1986.

Acrylamide   **EPA Method 8015**
CAS #79-06-1

TITLE Nonhalogenated Volatile Organics

MATRIX Groundwater, soils, sludges, water miscible liquid wastes, and non-water miscible wastes.

APPLICATION This method is used for the analysis of 6 nonhalogenated VOCs. Samples are analyzed using direct injection or purge-and-trap methods. Groundwater must be analyzed by the purge and trap method. The method provides an optional GC column that may help resolve analytes from interferences and which is also used for analyte confirmation.

INTERFERENCES There can be carryover contamination with high- and low-level samples. Impurities may come from the purge and trap apparatus, organic compounds outgassing from the plumbing ahead of trap, diffusion of VOCs through the sample bottle septum during shipping or storage, or from solvent vapors in the lab.

INSTRUMENTATION GC capable of on-column injections or purge-and-trap sample introduction and a flame ionization detector (FID). Column 1: an 8 ft by 0.1 in 1% SP-1000 on Carbopack-B. Column 2: a 6 ft by 0.1 in bonded n-octane on Porasil-C.

RANGE Not available.

MDL Not available.

PRECISION Not available.

ACCURACY Not available.

SAMPLING METHOD For water and liquid samples, use glass 40-mL vials with Teflon®-lined septum caps and collect two vials per sample location with no headspace. For solids and concentrated waste samples, use widemouth glass bottles with Teflon® liners. Cool all samples to 4°C.

STABILITY For concentrated wastes, soils, sediments, or sludges, cool to 4°C. For liquids, add 4 drops of concentrated hydrochloric acid, cool to 4°C.

MHT 14 days.

QUALITY CONTROL Analyze a reagent blank, matrix spike, and matrix spike duplicate/duplicate for each analytical batch (up to 20 Samples). Demonstrate the purity of glassware and reagents by analyzing a reagent water method blank. Internal, surrogate, and five concentration level calibration standards are used.

REFERENCE Method 8015, SW-846, 3rd ed., Nov. 1986.

Acrylonitrile   **EPA Method 1624**
CAS #107-13-1

TITLE Volatile Organic Compounds by Isotope Dilution GC/MS

MATRIX Compounds may be determined in waters, soils, and municipal sludges by this method.

METHOD SUMMARY This method is used to determine 58 volatile toxic organic pollutants associated with the CWA (as amended 1987); the RCRA (as amended 1986); the CERCLA (as amended 1986); and other compounds amenable to purge and trap gas chromatography-mass spectrometry (GC/MS).

If the solids content is less than 1%, stable isotopically-labeled analogs of the compounds of interest are added to a 5-mL sample and the sample is purged with an inert gas at 20–25°C in a chamber designed for soil or water samples. If the solids content is greater than 1%, 5 mL of reagent water and the labeled compounds are added to a 5 g aliquot of sample and the mixture is purged at 40°C. Compounds that will not purge at 20–25°C or at 40°C are purged at 78–85°C. In the purging process, the volatile compounds are transferred from the aqueous phase into the gaseous phase where they are passed into a sorbent column and trapped. After purging is completed, the trap is backflushed and heated rapidly to desorb the compounds into a GC. The compounds are separated by the GC and detected by a MS. The labeled compounds serve to correct the variability of the analytical technique.

INTERFERENCES Impurities in the purge gas, organic compounds outgassing from the plumbing upstream of the trap, and solvent vapors in the lab account for most problems. Samples can be contaminated by diffusion of volatile organic compounds (particularly methylene chloride) through the bottle seal during shipment and storage. Contamination by carryover can occur when high-level and low-level samples are analyzed sequentially. When an unusually concentrated sample is encountered, follow it by analysis of a reagent water blank to check for carryover.

INSTRUMENTATION Major equipment includes a GC with linear temperature programming and a glass jet separator as the MS interface, a MS with 70 eV electron impact ionization, and a data system to collect and record response factors.

Column: 2.8 m × 2 mm I.D. glass, packed with 1% SP-1000 on Carbopak B, 60/80 mesh, or equivalent.

PRECISION & ACCURACY The detection limits of the method are usually dependent on the level of interferences rather than instrumental limitations. The method detection limits were determined in digested sludge (low solids) and in filter cake or compost (high solids).

The MDL (in µg/kg) for low solids is 360 and for high solids is 9.

Background levels of this compound were present in the sludge with low solids, resulting in a higher than expected MDL.

Labeled and native compound precision (in µg/L) as standard deviation was 16.0.
Labeled and native compound accuracy (in µg/L) as average recovery was 70–132.
Acceptance criteria are at 100 µg/L for this compound.

SAMPLE COLLECTION, PRESERVATION & HANDLING Grab samples are collected in glass containers having a total volume greater than 20 mL. Fill and seal each bottle so that no air bubbles are entrapped. Samples are maintained at 0 to 4°C from the time of collection until analysis. If an aqueous sample contains residual chlorine, add sodium thiosulfate preservative (10 mg/40 mL) to the empty sample bottles just prior to shipment to the sample site. All samples must be analyzed within 14 days of collection.

SAMPLE PREPARATION Samples containing less than 1% solids are analyzed directly as aqueous samples. Samples containing 1% solids or greater are analyzed as solid samples utilizing one of two methods, depending on the levels of pollutants in the sample. Samples containing 1% solids or greater, and low to moderate levels of pollutants are analyzed by purging a known weight of sample added to 5 mL of reagent water. Samples containing 1% solids or greater, and high levels of pollutants, are extracted with methanol, and an aliquot of the methanol extract is added to reagent water and purged.

QUALITY CONTROL A field blank prepared from reagent water and carried through the sampling and handling protocol may serve as a check on contamination from shipment and storage.

The analyst is permitted to modify this method to improve separations or lower the costs of measurements, provided all performance specifications are met. Analyses of blanks are required. When results of spikes indicate atypical method performance for samples, the samples are diluted to bring method performance within acceptable limits. Analyze two sets of four 5-mL aliquots (8 aliquots total) of the aqueous performance standard. Spike all samples with labeled compounds to assess method performance on the sample matrix. Compute the percent recovery of the labeled compounds using the internal standard method. Compare the percent recovery for each compound with the corresponding labeled compound recovery. Reagent water blanks are analyzed to demonstrate freedom from carryover contamination. Field replicates may be collected to determine the precision of the sampling technique, and spiked samples may be required to determine the accuracy of the analysis when the internal method is used.

REFERENCE Volatile Organic Compounds by Isotope Dilution GC/MS. Office of Water Regulation and Standards, U.S. EPA Industrial Technology Division, Washington, DC, EPA Method 1624, Rev. C, June 1989 (contact W.A. Telliard, U.S. EPA, Office of Water Regulations and Standards, 401 M St., SW, Washington, DC, 20460. Phone: 202-382-7131).

Acrylonitrile **EPA Method 8240**
CAS #107-13-1

TITLE Volatile Organics By GC/MS: Packed Column Technique

MATRIX Nearly all types of sample matrices, regardless of water content, can be analyzed using this method. This includes groundwater, aqueous sludges, caustic liquors, acid liquors, waste solvents, oily wastes, mousses, tars, fibrous wastes, polymetric emulsions, filter cakes, spent carbons, spent catalysts, soils, and sediments.

METHOD SUMMARY Method 8240B covers 80 volatile organic compounds that are introduced into a gas chromatograph by the purge-and-trap method or by direct injection (in limited applications). For the purge-and-trap method an inert gas (zero grade nitrogen or helium) is bubbled through a 5-mL solution at ambient temperature. Purged sample components are trapped in a tube of sorbent materials. When purging is complete, the sorbent tube is heated and backflushed with inert gas to desorb the trapped components onto a GC column.

INTERFERENCES Impurities in the purge gas and from organic compounds outgassing from the plumbing ahead of the trap account for many contamination problems. Interferences purged or coextracted from the samples will vary considerably from source to source. Cross-contamination can occur whenever high-level and low-level samples are analyzed sequentially. Whenever an unusually concentrated sample is analyzed, it should be followed by the analysis of organic-free reagent water to check for cross-contamination. Samples also can be contaminated by diffusion of volatile organics (particularly methylene chloride and fluorocarbons) through the septum seal into the sample during shipment and storage. A trip blank can serve as a check on such contamination. The lab where volatile analysis is performed and also the refrigerated storage area should be completely free of solvents.

INSTRUMENTATION A gas chromatograph/mass spectrometry/data system (GC/MS) equipped with a 6 ft × 0.1 in I.D. glass column packed with 1% SP-1000 on Carbopack-B (60/80 mesh) is required. Also needed is a 5-mL purging device, a sorbent trap, and a thermal desorption apparatus.

PRECISION & ACCURACY This method is reported to have been tested by 15 laboratories using organic-free reagent water, drinking water, surface water, and industrial wastewater (not specified) fortified at six concentrations over the range 5–600 µg/L.

Sample estimated quantitation limits (EQLs) are highly matrix dependent. The EQLs listed may not always be achievable. EQLs listed for soils or sediments are based on wet weight. Normally, data is reported on a dry-weight basis; therefore, EQLs will be higher, based on the percent dry weight of each sample. Note that EQLs are even more variable than MDLs and that they are highly variable depending on the matrix being analyzed.

EQL in groundwater in µg/L was not listed.
EQL in low soil or sediment in µg/kg was not listed.
Accuracy (a) in µg/L was not listed.
Precision (b) in µg/L was not listed.

(a) *Average recovery found for measurements of samples containing a concentration of C, in µg/L.*
(b) *Overall precision found for measurements of samples with average recovery X for samples containing a concentration of C in µg/L.*
X = *Average recovery found for measurement of samples containing a concentration of C in µg/L.*

ESTIMATED QUANTITATION LIMITS FOR MATRICES OTHER THAN WATER, SOIL, AND SEDIMENTS

Other Matrices	Factor (a)
Waste miscible liquid waste	50
High-concentration soil and sludge	125
Non-water miscible waste	500

(a) *EQL = [EQL for low soil/sediment] × [Factor]. For non-aqueous samples, the factor is on a wet-weight basis.*

SAMPLING METHOD
Liquid samples — Use a 40-mL glass screw-cap VOA vial with a Teflon®-faced silicone septum that has been prewashed, rinsed with distilled deionized water, and oven dried. However, if residual chlorine is present, collect sample in a 40-oz. soil VOA container which has been pre-preserved with 4 drops of 10% sodium thiosulfate, mix gently, and then transfer the sample to a 40-mL VOA vial. Collect bubble-free samples in duplicate and seal them in separate plastic bags.

Soils or sediments, and sludges — Use an 8 oz. widemouth glass bottle with a Teflon®-faced silicone septum that has been prewashed with detergent, rinsed with distilled deionized water, and oven dried. Tap slightly to eliminate free air space. Collect samples in duplicate and seal them in separate plastic bags.

SAMPLE PRESERVATION
Liquid samples — Add 4 drops of concentrated HCL and immediately cool samples to 4°C and store in a solvent-free refrigerator.

Soils or sediments, and sludges — Cool samples to 4°C and store in a solvent-free refrigerator.

MHT Maximum holding time is 14 days from the sample collection date.

SAMPLE PREPARATION
Liquid samples — Remove the plunger from a 5-mL syringe and carefully pour the sample into the syringe barrel to just short of overflowing. Replace the syringe plunger and compress the sample. Open the syringe valve and vent any residual air while adjusting the sample volume to 5.0 mL. If there is only one volatile organic analysis (VOA) vial, a second syringe should be filled at this time to protect against possible loss of sample integrity. Add 10 µL of surrogate spiking solution and 10 µL of internal standard spiking solution through the valve bore of the 5-mL syringe, then close the valve. The surrogate and internal standards may be mixed and added as a single spiking solution.

Sediments, soils, and waste samples — All samples of this type should be screened by GC analysis using a headspace method (EPA Method 3810) or the hexadecane extraction and screening method (EPA Method 3820). Use the screening data to determine whether to use the low-concentration method (0.005–1 mg/kg) or the high-concentration method (>1 mg/kg).

Low-concentration method — The low-concentration method is based on purging a heated sediment or soil sample mixed with organic-free reagent water containing the surrogate and internal standards. Analyze all reagent blanks and standards under the same conditions as the samples.

Use a 5-g sample if the expected concentration is <0.1 mg/kg or a 1-g sample for expected concentrations between 0.1 and 1 mg/kg. Mix the contents of the sample container with a narrow metal spatula. Weigh the amount of the sample into a tared purge device. Add the spiked water to the purge device, which contains the weighed amount of sample, and connect the device to the purge-and-trap system.

High-concentration method — This method is based on extracting the sediment or soil with methanol. A waste sample is either extracted or diluted, depending on its solubility in methanol. Wastes that are insoluble in methanol are diluted with reagent tetraglyme or possibly polyethylene glycol (PEG). An aliquot of the extract is added to organic-free reagent water containing surrogate and internal standards. This is purged at ambient temperature. All samples with an expected concentration of >1.0 mg/kg should be analyzed by this method.

Mix the contents of the sample container with a narrow metal spatula. For sediments or soils and solid wastes that are insoluble in methanol, weigh 4 g (wet weight) of sample into a tared 20-mL vial. For waste that is soluble in methanol, tetraglyme, or PEG, weigh 1 g (wet weight) into a tared scintillation vial or culture tube or a 10-mL volumetric flask. Quickly add 9.0 mL of appropriate solvent then add 1.0 mL of a surrogate spiking solution to the vial, cap it, and shake it for 2 min.

QUALITY CONTROL Demonstrate, through the analysis of a reagent water blank, that interferences from the analytical system, glassware, and reagents are under control. Blank samples should be carried through all stages of the sample preparation and measurement steps. For each analytical batch (up to 20 samples), a reagent blank, matrix spike, and matrix spike duplicate must be analyzed (the frequency of the spikes may be different for different monitoring programs). The blank and spiked samples must be carried through all stages of the sample preparation and measurement steps. QC samples mentioned in the section on Interferences will also be needed as appropriate to those situations.

METHANOL EXTRACT REQUIRED FOR ANALYSIS OF HIGH-CONCENTRATION SOILS OR SEDIMENTS

Approximate Concentration Range	Volume of Methanol Extract (a)
500–10,000 µg/kg	100 µL
1,000–20,000 µg/kg	50 µL
5,000–100,000 µg/kg	10 µL
25,000–500,000 µg/kg	100 µL of 1/50 dilution (b)

Calculate appropriate dilution factor for concentrations exceeding this table.

(a) The volume of methanol added to 5 mL of water being purged should be kept constant. Therefore, add to the 5-mL syringe whatever volume of methanol is necessary to maintain a volume of 100 µL added to the syringe.
(b) Dilute an aliquot of the methanol extract and then take 100 µL for analysis.

REFERENCE Test Methods for Evaluating Solid Waste (SW-846). U.S. EPA 1983. Method 8240B, Rev. 2, Nov. 1990. Office of Solid Wastes, Washington, DC.

Acrylonitrile   **EPA Method 8030**
CAS #107-13-1

TITLE Other Nonhalogenated VOCs

MATRIX Groundwater, soils, sludges, water miscible liquid wastes, and non-water miscible wastes.

APPLICATION This method is used for the analysis of 3 nonhalogenated VOCs. Samples are analyzed using direct injection or purge-and-trap methods. Groundwater must be analyzed by the purge and trap method. The method provides an optional GC column which is used for analyte confirmation and that may help resolve analytes from interferences.

INTERFERENCES There can be carryover contamination with high- and low-level samples. Impurities may come from the purge and trap apparatus, organic compounds outgassing from the plumbing ahead of trap, diffusion of VOCs through the sample bottle septum during shipping or storage, or from solvent vapors in the lab.

INSTRUMENTATION GC capable of on-column injections or purge-and-trap sample introduction and a flame ionization detector (FID). Column 1: 10 ft by 2 mm with Porapak-QS. Column 2: 6 ft by 0.1 in with Chromosorb 101.

RANGE 5 to 100 µg/L

MDL 0.5 µg/L (reagent water).

PQL FACTORS FOR MULTIPLYING × FID MDL VALUE

Matrix	Multiplication Factor
Groundwater	10
Low-level soil	10
Water miscible liquid waste	500
High-level soil and sludge	1250
Non-water miscible waste	1250

PRECISION (as standard deviation): 1.5 µg/L, reagent water at 50 µg/L spike; 1.5 µg/L, POTW and 3.2 µg/L industrial water at 100 µg/L spike.

ACCURACY (as % recovery): 103%, reagent water at 50 µg/L spike; 101% POTW and 104% industrial water at 100 µg/L spike.

SAMPLING METHOD Use two glass 40-mL vials with Teflon®-lined septum caps per sample location and collect samples with no headspace.

STABILITY Adjust to pH 4–5 and cool to 4°C.

MHT 14 days.

QUALITY CONTROL Analyze a reagent blank, matrix spike, and matrix spike duplicate/duplicate for each analytical batch (up to 20 Samples). Demonstrate the purity of glassware and reagents by analyzing a reagent water method blank. Internal, surrogate, and five concentration level calibration standards are used. The QC check sample concentrate should contain this compound at 25 µg/mL in reagent water.

REFERENCE Method 8030, SW-846, 3rd ed., Nov. 1986.

Alachlor   **EPA Method 505**
CAS #15972-60-8

TITLE Analysis of Organohalide Pesticides and Commercial Polychlorinated Biphenyl (PCB) Products in Water by

Microextraction and Gas Chromatography. U.S. EPA Method 505, Rev. 2.0, 1989.

MATRIX This method is applicable to drinking water and raw source water. The latter should include most surface water and groundwater sources.

METHOD SUMMARY Method 505 covers 25 pesticides and commercial PCB products. This is a very sensitive method that is more useful for monitoring than for exploratory analyses. 5-mL samples of water are saturated with sodium chloride and then extracted by shaking with 2 mL of hexane. The sample extracts are transferred to an auto sampler setup to inject 1- to 2-µL portions into a gas chromatograph (GC) for analysis. Alternatively, 1- to 2-µL portions of samples, blanks, and standards may be manually injected. Each extract is analyzed by capillary GC/ECD with confirmation using either a second capillary column or GC/MS. The electron capture detector is easy to use, but it is a nonselective detector. The microextraction technique also eliminates the expensive costs of other methods, but it has the disadvantage of being less sensitive than most because the extracts are not concentrated.

INTERFERENCES Method interferences may be caused by contaminants in solvents, reagents, glassware, and other sample processing apparatus that lead to discrete artifacts or elevated baselines. Interfering contamination may occur when a sample containing low concentrations of analytes is analyzed immediately following a sample containing relatively high concentrations of the analytes. Matrix interferences also may be caused by contaminants that are coextracted from the sample; cleanup of sample extracts may be necessary in these cases. Some pesticides and commercial PCB products from aqueous solutions adhere to glass surfaces, so sample transfers and contact with glass surfaces should be minimized. Some pesticides are rapidly oxidized by chlorine so dechlorination with sodium thiosulfate at the time of sample collection is important. Also, splitless injectors may cause degradation of some pesticides.

INSTRUMENTATION A gas chromatograph/electron capture detector/data system, with temperature programming and split/splitless injector suitable for use with capillary columns is needed.

Column 1: 0.32 mm I.D. × 30 m fused silica capillary with chemically bond methyl polysiloxane phase (DB-1, 1.0 µm film, or equivalent).
Column 2: 0.32 mm I.D. × 30 m fused silica capillary with 1:1 mixed phase of dimethyl silicone and polyethylene glycol (Durawax-DX3, 0.25 µm film, or equivalent).
Column 3: 0.32 mm I.D. × 25 m fused silica capillary with chemically bonded 50:50 methyl-phenyl silicone (OV-17, 1.5 µm film, or equivalent).

Column 1 should be used as the primary analytical column. Columns 2 and 3 are recommended for use as confirmatory columns when GC/MS confirmation is not available.

PRECISION & ACCURACY Method detection limits are dependent upon the characteristics of the gas chromatographic system used. Analytes that are not separated chromatographically cannot be individually identified and used in the same calibration mixture or water samples unless an alternative technique for identification and quantification, such as mass spectrometry, is used.

The concentration(s) (in µg/L) used for these QC measurements was 0.50.
The MDL (in µg/L) was 0.225.
The accuracy (% recovery) for reagent water at the above concentration(s) was 102 and the precision (%) was 13.4.
The accuracy (% recovery) for groundwater at the above concentration(s) was not listed and the precision (%) was not listed.
The accuracy (% recovery) for tap water at the above concentration(s) was not listed and the precision (5) was not listed.

Note: No range of concentrations is provided with this method.

SAMPLING METHOD Collect samples using a 40-mL screw-cap vial (prewashed with detergent, rinsed with distilled water and oven dried at 400°C for one h) with a Teflon®-faced silicone septum. Collect bubble-free samples and place the septum with the Teflon® side down on the water.

SAMPLE PRESERVATION If residual chlorine is present in the water add about 3 mg of sodium thiosulfate to each vial before samples are collected to remove the chlorine. Alternatively, add 75 µL of 0.04 g/mL solution of sodium thiosulfate to each vial just prior to sampling. Immediately cool samples to 4°C and store them in a solvent-free refrigerator at 4°C until analysis.

MHT The maximum holding time is 14 days from the time the sample was collected until it must be analyzed.

SAMPLE PREPARATION Remove the sample from storage and allow it to come to room temperature. Remove a 5-mL volume from each container and weigh the container to the nearest 0.1 g. Add 6 g of sodium chloride and 2.0 mL of hexane to each sample bottle. Recap the sample and shake it vigorously for one min. Allow the water and hexane phases to separate, remove the cap, and transfer 0.5 mL of hexane into an auto sampler vial using a disposable glass pipette. Transfer the remaining hexane phase into a second auto sampler vial and store at 4°C for reanalysis, if necessary. Discard the remaining sample/hexane mixture and reweigh the empty container to determine net weight of sample.

QUALITY CONTROL Minimum quality control requirements are initial demonstration of lab capability, analysis of lab reagent blanks, fortified blanks, fortified sample matrix, and quality control samples. The lab must analyze at least one fortified blank per sample set, or at least one for every 20 samples. The fortifying concentration of each analyte should be 10 times the method detection limit or the maximum calibration limit (MCL), whichever is less. Calculate accuracy as percent recovery and develop control limits from the mean percent recovery and standard deviation.

The lab must add a known concentration of the analytes to a minimum of 10% of the routine samples, or one lab fortified sample matrix per sample set. Calculate the percent recovery for each analyte and compare to the control limits established from the analyses of the fortified blanks.

EPA CONTACT & HOTLINE For technical questions contact Dr. Baldev Bathija, U.S. EPA, Office of Ground Water and Drinking Water (WH-550D), 401 M St. SW, Washington, DC 20460. Tel. (202) 260-3040. For further information the EPA Safe Drinking Water Hotline may be called at: (800) 426-4791.

REFERENCE Methods for the Determination of Organic Compounds in Drinking Water, EPA/600/4-88/039 (revised July 1991). U.S. EPA Environmental Monitoring Systems Laboratory, Cincinnati, OH, 45268, U.S.A. Available from the National Technical Information Service (NTIS), 5285 Port Royal Road, Springfield, VA 22161; Tel. 800-553-6847. NTIS Order Number is PB91-231480.

Alachlor **EPA Method 507**
CAS #15972-60-8

TITLE Determination of Nitrogen and Phosphorus-Containing Pesticides in Water by GC/NPD

MATRIX This method is applicable to the determination of certain nitrogen and phosphorus-containing pesticides in finished drinking water and groundwater.

METHOD SUMMARY Method 507 covers 46 nitrogen- and phosphorus-containing pesticides. A 1-L sample is fortified with a surrogate standard, salted, buffered, extracted with methylene chloride, and concentrated; then the solvent is exchanged with methyl tert-butyl ether (MTBE) and concentrated again, and a 2-µL aliquot of a sample extract is injected into a GC system equipped with a selective nitrogen-phosphorus detector and a capillary column for analysis.

INTERFERENCES Method interferences may be caused by contaminants in solvents, reagents, glassware, and other sample processing apparatus. Interfering contamination may occur when a sample containing low concentrations of analytes is analyzed immediately following a sample containing relatively high concentrations. One or more injections of MTBE should be made following the analysis of a sample with high concentrations of analytes to check for analyte carryover. Matrix interferences may be caused by contaminants that are coextracted from the sample. The extent of matrix interferences will vary considerably from source to source, depending upon the water sampled.

INSTRUMENTATION A gas chromatograph system (GC) equipped with a nitrogen-phosphorus detector (NPD) is needed.

Column 1: 30 m × 0.25 mm I.D. DB-5 bonded fused silica column, 0.25 µm film thickness, or equivalent.

Column 2: 30 M × 0.25 mm I.D. DB-1701 bonded fused silica column, 0.25 µm film thickness, or equivalent.

PRECISION & ACCURACY This method has been validated in a single lab and estimated detection limits (EDLs) have been determined for each analyte. Observed detection limits may vary among waters, depending upon the nature of the interferences in the sample matrix and the specific instrumentation used. Analytes that are not separated chromatographically cannot be individually identified and measured unless an alternative technique for identification and quantification exist.

The estimated detection limit (in µg/L) was 0.38. The EDL is defined as either method detection limit or a level of compound in a sample yielding a peak in the final extract with signal-to-noise ratio of approximately 5, whichever value is higher.

The concentration used for these measurements (in µg/L) was 3.8.

The accuracy (as % recovery) was 95.

The precision (% RSD) was 11.

SAMPLING METHOD Grab samples are collected in 1-L glass sample bottles (prewashed with detergent and hot tap water, rinsed with reagent water, and dried in an oven at 400°C for 1 h) with screw caps lined with PTFE-fluorocarbon.

SAMPLE PRESERVATION Add mercuric chloride to the sample bottle in amounts to produce a concentration of 10 mg/L. If residual chlorine is present, add 80 mg of sodium thiosulfate/L of sample to the sample bottle prior to collection. After collection, seal bottle and shake vigorously for 1 min, then cool the sample to 4°C immediately and store it at 4°C in the dark until extraction.

MHT Maximum holding time of the samples, and in some cases the extracts, is 14 days.

SAMPLE PREPARATION Fortify the sample with 50 µL of the surrogate standard solution, adjust to pH 7 with phosphate buffer, add 100 g NaCl to the sample, and seal and shake to dissolve the salt; then extract with methylene chloride in a separatory funnel or in a mechanical tumbler bottle. Dry the extract by pouring it through a solvent-rinsed drying column containing about 10 cm of anhydrous sodium sulfate. Collect the extract in a Kuderna-Danish (K-D) concentrator and rinse the column with 20–30 mL methylene chloride. Concentrate the extract to about 2 mL and rinse the flask and its lower joint into the concentrator tube with 1 to 2 mL of methyl t-butyl ether (MTBE). Add 5–10 mL of MTBE and concentrate the extract twice (adding more MTBE) to a final volume of 5.0 mL and store it at 4°C until analysis.

Note: If methylene chloride is not completely removed from the final extract, it may cause detector problems.

QUALITY CONTROL Minimum quality control requirements are initial demonstration of lab capability, determination of surrogate compound recoveries in each sample and blank, monitoring internal standard peak area or height in each sample and blank, analysis of lab reagent blanks, lab fortified samples, lab fortified blanks, and other QC samples. A lab reagent blank is analyzed to demonstrate that all glassware and reagent interferences are under control.

Initial demonstration of capability is fulfilled by analyzing four fortified reagent water samples with the recovery value for each analyte falling within the acceptable range (±30% average recovery). Surrogate recoveries from samples or method blanks must be 70–130%. The internal standard response for any sample chromatogram should not deviate from the daily calibration check standard's internal standard response by more than

30% or lab fortified blanks and sample matrices are used to assess lab performance and analyte recovery, respectively.

If the response for the target analyte peak exceeds the working range of the system, dilute the extract and reanalyze. Alternative techniques such as an alternate detector or second chromatography column should be used to confirm peak identification when sample components are not resolved adequately.

EPA CONTACT & HOTLINE For technical questions contact Dr. Baldev Bathija, U.S. EPA, Office of Ground Water and Drinking Water (WH-550D), 401 M St. SW, Washington, DC 20460. Tel. (202) 260-3040. For further information the EPA Safe Drinking Water Hotline may be called at: (800) 426-4791.

REFERENCE Methods for the Determination of Organic Compounds in Drinking Water, EPA/600/4-88/039 (revised July 1991). U.S. EPA Environmental Monitoring Systems Laboratory, Cincinnati, OH, 45268, U.S.A. Available from the National Technical Information Service (NTIS), 5285 Port Royal Road, Springfield, VA 22161; Tel. 800-553-6847. NTIS Order Number is PB91-231480.

Aldrin **EPA Method 505**
CAS #309-00-2

TITLE Analysis of Organohalide Pesticides and Commercial Polychlorinated Biphenyl (PCB) Products in Water by Microextraction and Gas Chromatography. U.S. EPA Method 505, Rev. 2.0, 1989.

MATRIX This method is applicable to drinking water and raw source water. The latter should include most surface water and groundwater sources.

METHOD SUMMARY Method 505 covers 25 pesticides and commercial PCB products. This is a very sensitive method that is more useful for monitoring than for exploratory analyses. 5-mL samples of water are saturated with sodium chloride and then extracted by shaking with 2 mL of hexane. The sample extracts are transferred to an auto sampler setup to inject 1- to 2-µL portions into a gas chromatograph (GC) for analysis. Alternatively, 1- to 2-µL portions of samples, blanks, and standards may be manually injected. Each extract is analyzed by capillary GC/ECD with confirmation using either a second capillary column or GC/MS. The electron capture detector is easy to use, but it is a nonselective detector. The microextraction technique also eliminates the expensive sample preparation costs of other methods, but it has the disadvantage of being less sensitive than most because the extracts are not concentrated.

INTERFERENCES Method interferences may be caused by contaminants in solvents, reagents, glassware, and other sample processing apparatus that lead to discrete artifacts or elevated baselines. Interfering contamination may occur when a sample containing low concentrations of analytes is analyzed immediately following a sample containing relatively high concentrations of the analytes. Matrix interferences also may be caused by contaminants that are coextracted from the sample; cleanup of sample extracts may be necessary in these cases. Some pesticides and commercial PCB products from aqueous solutions adhere to glass surfaces, so sample transfers and contact with glass surfaces should be minimized. Some pesticides are rapidly oxidized by chlorine so dechlorination with sodium thiosulfate at the time of sample collection is important. Also, splitless injectors may cause degradation of some pesticides.

INSTRUMENTATION A gas chromatograph/electron capture detector/data system, with temperature programming and split/splitless injector suitable for use with capillary columns is needed.

Column 1: 0.32 mm I.D. × 30 m fused silica capillary with chemically bond methyl polysiloxane phase (DB-1, 1.0 µm film, or equivalent).
Column 2: 0.32 mm I.D. × 30 m fused silica capillary with 1:1 mixed phase of dimethyl silicone and polyethylene glycol (Durawax-DX3, 0.25 µm film, or equivalent).
Column 3: 0.32 mm I.D. × 25 m fused silica capillary with chemically bonded 50:50 methyl-phenyl silicone (OV-17, 1.5 µm film, or equivalent).

Column 1 should be used as the primary analytical column. Columns 2 and 3 are recommended for use as confirmatory columns when GC/MS confirmation is not available.

PRECISION & ACCURACY Method detection limits are dependent upon the characteristics of the gas chromatographic system used. Analytes that are not separated chromatographically cannot be individually identified and used in the same calibration mixture or water samples unless an alternative technique for identification and quantification, such as mass spectrometry, is used.

The concentration(s) (in µg/L) used for these QC measurements was 0.15 and 0.05.
The MDL (in µg/L) was 0.075 and 0.007.
The accuracy (% recovery) for reagent water at the above concentration(s) was 86 and 106 and the precision (%) was 95 and 20.
The accuracy (% recovery) for groundwater at the above concentration(s) was 100 and 86 and the precision (%) was 11.0 and 16.3.
The accuracy (% recovery) for tap water at the above concentration(s) was 69 and not listed and the precision (5) was 9 and not listed.

Note: No range of concentrations is provided with this method.

SAMPLING METHOD Collect samples using a 40-mL screw-cap vial (prewashed with detergent, rinsed with distilled water and oven dried at 400°C for one h) with a Teflon®-faced silicone septum. Collect bubble-free samples and place the septum with the Teflon® side down on the water.

SAMPLE PRESERVATION If residual chlorine is present in the water add about 3 mg of sodium thiosulfate to each vial before samples are collected to remove the chlorine. Alternatively, add 75 µL of 0.04 g/mL solution of sodium thiosulfate to each vial just prior to sampling. Immediately cool sample to 4°C and store them in a solvent-free refrigerator at 4°C until analysis.

MHT The maximum holding time is 14 days from the time the sample was collected until it must be analyzed.

SAMPLE PREPARATION Remove the sample from storage and allow it to come to room temperature. Remove a 5-mL volume from each container and weigh the container to the nearest 0.1 g. Add 6 g of sodium chloride and 2.0 mL of hexane to each sample bottle. Recap the sample and shake it vigorously for one min. Allow the water and hexane phases to separate, remove the cap, and transfer 0.5 mL of hexane into an auto sampler vial using a disposable glass pipette. Transfer the remaining hexane phase into a second auto sampler vial and store at 4°C for reanalysis, if necessary. Discard the remaining sample/hexane mixture and reweigh the empty container to determine net weight of sample.

QUALITY CONTROL Minimum quality control requirements are initial demonstration of lab capability, analysis of lab reagent blanks, fortified blanks, fortified sample matrix, and quality control samples. The lab must analyze at least one fortified blank per sample set, or at least one for every 20 samples. The fortifying concentration of each analyte should be 10 times the method detection limit or the maximum calibration limit (MCL), whichever is less. Calculate accuracy as percent recovery and develop control limits from the mean percent recovery and standard deviation.

The lab must add a known concentration of the analytes to a minimum of 10% of the routine samples, or one lab fortified sample matrix per sample set. Calculate the percent recovery for each analyte and compare to the control limits established from the analyses of the fortified blanks.

EPA CONTACT & HOTLINE For technical questions contact Dr. Baldev Bathija, U.S. EPA, Office of Ground Water and Drinking Water (WH-550D), 401 M St. SW, Washington, DC 20460. Tel. (202) 260-3040. For further information the EPA Safe Drinking Water Hotline may be called at: (800) 426-4791.

REFERENCE Methods for the Determination of Organic Compounds in Drinking Water, EPA/600/4-88/039 (revised July 1991). U.S. EPA Environmental Monitoring Systems Laboratory, Cincinnati, OH, 45268, U.S.A. Available from the National Technical Information Service (NTIS), 5285 Port Royal Road, Springfield, VA 22161; Tel. 800-553-6847. NTIS Order Number is PB91-231480.

Aldrin EPA Method 625
CAS #309-00-2

TITLE Base/Neutrals and Acids, U.S. EPA Method 625

MATRIX This methods covers municipal and industrial wastewater.

METHOD SUMMARY Approximately 1 L of sample is serially extracted with methylene chloride at a pH greater than 11 and again at a pH less than 2 using a separatory funnel or a continuous extractor. The methylene chloride extract is dried, concentrated to a volume of 1 mL, and analyzed by GC/MS. Qualitative identification of the parameters in the extract is performed using the retention time and the relative abundance of three characteristic masses (m/z). Qualitative analysis is performed using either external or internal standard techniques with a single characteristic m/z.

INTERFERENCES Method interferences may be caused by contaminants in solvents, reagents, glassware, and other sample processing hardware. Glassware must be scrupulously cleaned. Glassware should be heated in a muffle furnace at 400°C for 5 to 30 min. Some thermally stable materials, such as PCBs, may not be eliminated by this treatment. Solvent rinses with acetone and pesticide quality hexane may be substituted for the muffle furnace heating. Matrix interferences may be caused by contaminants that are coextracted from the sample. The base/neutral extraction may cause significantly reduced recovery of phenols. The packed gas chromatographic columns recommended for the basic fraction may not exhibit sufficient resolution for some analytes.

INSTRUMENTATION A GC/MS system with an injection port designed for on-column injection when using packed columns and for splitless injection when using capillary columns.

Column for base/neutrals: 1.8 m long × 2 mm I.D. glass, packed with 3% SP-2550 on Supelcoport (100/120 mesh) or equivalent.

Column for acids: 1.8 m long × 2 mm I.D. glass, packed with 1% SP-1240DA on Supelcoport (100/120 mesh) or equivalent.

PRECISION & ACCURACY The MDL concentrations were obtained using reagent water. The MDL actually achieved in a given analysis will vary depending on instrument sensitivity and matrix effects. This method was tested by 15 laboratories using reagent water, drinking water, surface water, and industrial wastewater spiked at six concentrations over the range 5 to 100 µg/L. Single operator precision, overall precision, and method accuracy were found to be directly related to the concentration of the parameter matrix.

The MDL (in µg/L) in reagent water was 1.9.
The standard deviation (in µg/L based on 4 recovery measurements) was 39.0.
The range (in µg/L) for average recovery for 4 measurements was 7.2–152.2.
The range (in %) for percent recovery was D-166.
Accuracy (in µg/L) as expected recovery for one or more measurements of a sample containing a true concentration of C was 0.78C+1.66.
Precision (in µg/L) as expected single analyst standard deviation of measurements at an average concentration found at X was 0.27X−1.28.
Overall precision (in µg/L) as expected interlaboratory standard deviation of measurements in an average concentration found at X was 0.43X−1.13.

C = *True value of the concentration in µg/L.*
X = *Average recovery found for measurements of samples containing a concentration at C in µg/L.*

SAMPLE PREPARATION Adjust the pH to >11 with sodium hydroxide and serially extract in a separatory funnel with methylene chloride or else in a continuous extractor. Next, adjust the pH to <2 with sulfuric acid and serially extract in a separatory

funnel with methylene chloride or else in a continuous extractor. Dry the extracts separately through a column of anhydrous sodium sulfate and then concentrate each of the extracts to 1.0 mL using a K-D apparatus.

SAMPLE COLLECTION, PRESERVATION & HANDLING
Grab samples must be collected in glass containers. All samples must be refrigerated at 4°C from the time of collection until extraction. If residual chlorine is present, add 80 mg of sodium thiosulfate/L of sample and mix well. All samples must be extracted within 7 days of collection and completely analyzed within 40 days of extraction.

QUALITY CONTROL Make an initial, one-time, demonstration of the ability to generate acceptable accuracy and precision with this method. Before processing any samples, the analyst must analyze a reagent water blank to demonstrate that interferences from the analytical system and glassware are under control. Each time a set of samples is extracted or reagents are changed, a reagent water blank must be processed. Spike and analyze a minimum of 5% of all samples to monitor and evaluate lab data quality. A QC check sample concentrate that contains each parameter of interest at a concentration of 100 µg/mL in acetone is required. PCBs and multicomponent pesticides may be omitted from this test.

After analysis of five spiked wastewater samples, calculate the average percent recovery and the standard deviation of the percent recovery. Spike all samples with the surrogate standard spiking solution and calculate the percent recovery of each surrogate compound.

REFERENCE *Federal Register*, Vol. 49, No. 209. Friday, Oct. 26, 1984.

Aldrin   **EPA Method 8080**
CAS #309-00-2

TITLE Organochlorine Pesticides and Polychlorinated Biphenyls By Gas Chromatography

MATRIX This method is used to determine the concentration of various organochlorine pesticides and polychlorinated biphenyls in extracts prepared from water, groundwater, soils, and sediments.

METHOD SUMMARY This method covers 26 pesticides and Aroclor (PCB) mixtures and it is suitable for monitoring-type analyses. After extraction, concentration and solvent exchange to hexane, a 2- to 5-µL sample aliquot is injected into a GC using the solvent flush technique, and the analytes are detected by an electron capture detector (ECD) or an electrolytic conductivity detector in the halogen mode (HECD). Both neat and diluted organic liquids may be analyzed by direct injection.

INTERFERENCES Interferences coextracted from the samples will vary considerably from source to source. Interferences by phthalate esters can pose a major problem in pesticide determinations when using the ECD. Cross-contamination of clean glassware routinely occurs when plastics are handled during extraction steps, especially when solvent-wetted surfaces are handled. The contamination from phthalate esters can be completely eliminated with a microcoulometric or electrolytic conductivity detector. Solvents, reagent, glassware, and other sample processing hardware may yield artifacts and/or interferences to sample analysis.

INSTRUMENTATION A gas chromatograph capable of on-column injections is needed. It must be equipped with an ECD or a HECD and one of the following GC columns:

Column 1: Supelcoport (100/120 mesh) coated with 1.5% SP-2250/1.95% SP-2401 packed in a 1.8 m × 4 mm I.D. glass column.

Column 2: Supelcoport (100/120 mesh) coated with 3% OV-1 in a 1.8 m × 4 mm I.D. glass column.

PRECISION & ACCURACY The method was tested by 20 laboratories using organic-free reagent water, drinking water, surface water, and three industrial wastewater spiked at six concentrations. Concentrations used in the study ranged from 0.5 to 30 µg/L for single-component pesticides and from 8.5 to 400 µg/L for multicomponent parameters. Overall precision and method accuracy were found to be directly related to the concentration of the analyte and essentially independent of the sample matrix. The sensitivity of this method usually depends on the concentration of interferences rather than on instrumental limitations.

MDL in µg/L was 0.004.
Concentration range in µg/L was 0.5–30.
Accuracy as recovery (x*) in µg/L was 0.81C+0.04.
Overall precision (S*) in µg/L was 0.20x–0.01.

x^* *Expected recovery for one or more measurements of a sample containing concentration C, in µg/L.*
$S^* =$ *Expected interlaboratory standard deviation of measurements at an average concentration found of the analyte in µg/L.*
$C =$ *True value for the concentration, in µg/L.*
$X =$ *Average recovery found for measurements of samples containing a concentration of C, in µg/L.*

SAMPLING METHOD
Liquid samples — Use a 1 or 2½ gallon amber glass bottle with a screw-top Teflon®-lined cover. Pre-wash the bottle with detergent, rinse with distilled water and methanol (or isopropanol).

Soil, sediments, and sludges — Use an 8-oz. widemouth glass with a screw-top Teflon®-lined cover. Pre-wash the bottle with detergent, rinse with distilled water and methanol (or isopropanol).

SAMPLE PRESERVATION Cool water, soil, sediment, or sludge samples immediately to 4°C.

Water samples — If residual chlorine is present, add 3 mL of 10% sodium thiosulfate per gallon and cool to 4°C. All extracts and samples should be stored under refrigeration.

MHT Liquid samples must be extracted within 7 days and the extracts must be analyzed within 40 days. Soils, sediments, and sludges may be stored for a maximum of 14 days prior to extraction.

SAMPLE PREPARATION

Liquid samples — Extract 1-L samples in a continuous extractor at pH 5–9 with methylene chloride after adding 1.0 mL of surrogate spiking solution to each sample. Pass the extract through a column of anhydrous sodium sulfate to dry and concentrate it in a K-D apparatus to 1 mL volume.

Soils, sediments and sludges — Rapidly weigh approximately 30 g of sample into a 400-mL beaker to avoid loss of the more volatile extractables. Nonporous or wet samples (gummy or clay type) that do not have a free-flowing sandy texture must be mixed with anhydrous sodium sulfate until the sample is free flowing. Add 1 mL of surrogate standards to all samples, spikes, standards, and blanks. Add 100 mL of 1:1 methylene chloride:acetone and extract ultrasonically. Decant and filter extracts, dry the extract by passing it through a drying column containing anhydrous sodium sulfate, and concentrate to 1 mL in a K-D apparatus.

Hexane eolvent exchange — Add 50 mL of hexane, a new boiling chip, and concentrate until the apparent volume of liquid reaches 1 mL. Adjust the extract volume to 10.0 mL. Stopper the concentration tube and store refrigerated at 4°C if further processing will not be performed immediately. If the extract will be stored longer than two days, transfer it to a vial with Teflon®-lined screw-cap or crimp top.

QUALITY CONTROL
Demonstrate through the analysis of a reagent water blank, that all glassware and reagents are interference free. Each time a set of samples is processed, a method blank should be processed as a safeguard against chronic lab contamination. A reagent blank, a matrix spike, and a duplicate or matrix spike duplicate must be performed for each analytical batch (up to a maximum of 20 samples) analyzed.

Analytical system performance must be verified by analyzing QC check samples. The QC check sample concentration should contain each single-component analyte at the following concentrations in acetone: 4,4'-DDD, 10 µg/mL; 4,4'-DDT, 10 µg/mL; endosulfan II, 10 µg/mL; endosulfan sulfate, 10 µg/mL; and any other single-component pesticide at 2 µg/mL. If the method is only to be used to analyze PCBs, Chlordane, or Toxaphene, the QC check sample concentrate should contain the most representative multicomponent parameter at a concentration of 50 µg/mL in acetone.

REFERENCE
Test Methods for Evaluating Solid Waste (SW-846). U.S. EPA. 1983. Method 8080B, Rev. 2, Nov. 1990. Office of Solid Wastes, Washington, DC.

Aldrin
CAS #309-00-2

EPA Method 8270

TITLE
Semivolatile Organic Compounds by GC/MS

MATRIX
This method is used to determine the concentration of semivolatile organic compounds in extracts prepared from all types of solid waste matrices, soils, and groundwater. Although surface waters are not specifically mentioned, this method should be applicable to water samples from rivers, lakes, etc.

METHOD SUMMARY
This method covers 259 semivolatile organic compounds. In very limited applications direct injection of the sample into the GC/MS system may be appropriate, but this results in very high detection limits (approximately 10,000 µg/L). Typically, a 1-L liquid sample, containing surrogate, and matrix spiking standards, is extracted in a continuous extractor first under acid conditions and then under basic conditions. Typically 30 g of a solid sample, containing surrogate, and matrix spiking standards, is extracted ultrasonically. After concentrating the extract to 1 mL it is spiked with 10 µL of an internal standard solution just prior to analysis by GC/MS. The volume injected should contain about 100 ng of base/neutral and 200 ng of acid surrogates (for a 1-µL injection). Analysis is performed by GC/MS using a capillary GC column.

INTERFERENCES
Raw GC/MS data from all blanks, samples, and spikes must be evaluated for interferences. Contamination by carryover can occur whenever high-concentration and low-concentration samples are sequentially analyzed. To reduce carryover, the sample syringe must be rinsed out between samples with solvent. Whenever an unusually concentrated sample is encountered, it should be followed by the analysis of blank solvent to check for cross-contamination.

INSTRUMENTATION
A GC/MS and a data system are required. The GC column used is a 30 m × 0.25 mm I.D. (or 0.32 mm I.D.) 1-µm film thickness silicone-coated fused silica capillary column. A continuous liquid-liquid extractor equipped with Teflon® or glass connection joints and stopcocks requiring no lubrication, a K-D concentrating apparatus, water bath, and an ultrasonic disrupter with a minimum power of 300 W and with pulsing capability are also required.

PRECISION & ACCURACY
The estimated quantitation limit (EQL) of Method 8270B for determining an individual compound is approximately 1 mg/kg (wet weight) for soil or sediment samples, 1–200 mg/kg for wastes (dependent on matrix and method of preparation), and 10 µg/L for groundwater samples. EQLs will be proportionately higher for sample extracts that require dilution to avoid saturation of the detector.

The EQL(b) for groundwater in µg/L is not listed.
The EQL (a, b) for low concentrations in soil and sediment in µg/kg is not listed.
Accuracy as µg/L is $0.78C + 1.66$.
Overall precision in µg/L is $0.43X + 1.13$.

(a) *EQLs listed for soil/sediment are based on wet weight. Normally data is reported in a dry-weight basis; therefore, EQLs will be higher based on the % dry weight of each sample. This calculation is based on a 30-g sample and gel permeation chromatography cleanup.*
(b) *Sample EQLs are highly matrix-dependent. The EQLs are provided for guidance and may not always be achievable.*
$C = $ *True value for concentration, in µg/L.*
$X = $ *Average recovery found for measurements of samples containing a concentration of C, in µg/L.*

ESTIMATED QUANTITATION LIMIT

Other Matrices	Factor (a)
High-concentration soil and sludges by sonicator	7.5
Non-water miscible waste	75

(a) EQL for other matrices = [EQL for low soil/sediment] × [Factor]. This estimated EQL is similar to an EPA "Practical Quantitation Limit."

SAMPLING METHOD

Liquid samples — Use a 1 or 2½ gallon amber glass bottle with a screw-top Teflon®-lined cover that has been prewashed with detergent and rinsed with distilled water and methanol (or isopropanol).

Soils, sediments, or sludges — Use an 8-oz. widemouth glass with a screw-top Teflon®-lined cover that has been prewashed with detergent and rinsed with distilled water and methanol (or isopropanol).

SAMPLE PRESERVATION

Liquid samples — If residual chlorine is present, add 3 mL of 10% sodium thiosulfate per gallon, cool to 4°C and store in a solvent-free refrigerator until analysis; if chlorine is not present, then eliminate the sodium thiosulfate addition.

Soils, sediments, or sludges — Cool samples to 4°C and store in a solvent-free refrigerator.

MHT Liquid samples must be extracted within 7 days and the extracts analyzed within 40 days. Soils, sediments, or sludges may be stored for a maximum of 14 days and the extracts analyzed within 40 days.

SAMPLE PREPARATION

Liquid samples — Transfer 1 L quantitatively to a continuous extractor. If high concentrations are anticipated, a smaller volume may be used and then diluted with organic-free reagent water to 1 L. Adjust pH, if necessary, to pH <2 using 1:1 (V/V) sulfuric acid. Pipette 1.0 mL of a surrogate standard spiking solution into each sample. For the sample in each analytical batch selected for spiking, add 1.0 mL of a matrix spiking standard. For base/neutral acid analysis, the amount of the surrogates and matrix spiking compounds added to the sample should result in a final concentration of 100 ng/μL of each analyte in the extract to be analyzed (assuming a 1-μL injection). Extract with methylene chloride for 18–24 h. Next, adjust the pH of the aqueous phase to pH >11 using 10 N sodium hydroxide and extract it with methylene chloride again for 18–24 h. Dry the extract through a column containing anhydrous sodium sulfate and concentrate it to 1 mL using a K-D concentrator.

Soils, sediments, or sludges — Use 30 g of sample. Nonporous or wet samples (gummy or clay type) that do not have a free-flowing sandy texture must be mixed with anhydrous sodium sulfate until the sample is free flowing. Add 1 mL of surrogate standards to all samples, spikes, standards, and blanks. For the sample in each analytical batch selected for spiking, add 1.0 mL of a matrix spiking standard. For base/neutral acid analysis, the amount added of the surrogates and matrix spiking compounds should result in a final concentration of 100 ng/μL of each base/neutral analyte and 200 ng/μL of each acid analyte in the extract to be analyzed (assuming a 1-μL injection). Immediately add a 100-mL mixture of 1:1 methylene chloride:acetone and extract the sample ultrasonically for 3 min and then decant or filter the extracts. Repeat the extraction two or more times. Dry the extract using a column with anhydrous sodium sulfate and concentrate it to 1 mL in a K-D concentrator.

QUALITY CONTROL A methylene chloride solution containing 50 ng/μL of decafluorotriphenylphosphine (DFTPP) is used for tuning the GC/MS system each 12-h shift. A system performance check also must be made during every 12-h shift. A standard containing 50 ng/μL each of 4,4'-DDT, pentachlorophenol, and benzidine is required to verify injection port inertness and GC column performance. A calibration standard at mid-concentration, containing each compound of interest, including all required surrogates, must be performed every 12 h during analysis. After the system performance check is met, calibration check compounds (CCCs) are used to check the validity of the initial calibration.

The internal standard responses and retention times in the calibration check standard must be evaluated immediately after or during data acquisition. If the retention time for any internal standard changes by more than 30 seconds from the last check calibration (12 h), the chromatographic system must be inspected for malfunctions and corrections must be made, as required. If the electron ionization current plot (EICP) area for any of the internal standards changes by a factor of two from the last daily calibration standard check, the mass spectrometer must be inspected for malfunctions and corrections must be made, as appropriate.

Demonstrate, through the analysis of a reagent water blank, that interferences from the analytical system, glassware, and reagents are under control. The blank samples should be carried through all stages of the sample preparation and measurement steps. For each analytical batch (up to 20 samples), a reagent blank, matrix spike, and matrix spike duplicate/duplicate must be analyzed (the frequency of the spikes may be different for different monitoring programs). The blank and spiked samples must be carried through all stages of the sample preparation and measurement steps. A QC reference sample concentrate containing each analyte at a concentration of 100 mg/L in methanol is required.

REFERENCE Test Methods for Evaluating Solid Waste (SW-846). U.S. EPA 1983, Method 8270B, Rev. 2, Nov. 1990. Office of Solid Waste, Washington, DC.

Alkalinity   EPA Method 310.2

TITLE Inorganics, Non-Metallics

MATRIX Drinking, Surface and Saline waters. Wastewater.

APPLICATION Date issued 1971. Editorial Rev. 1974. (Calorimetric, automated, methyl orange). Methyl orange indicator is dissolved in a weak buffer at pH 3.1. Just below equivalent point. Any alkali addition causes a loss of color directly proportional to amount of alkalinity.

INTERFERENCES Sample turbidity and color may interfere with this method. Turbidity must be removed by filtration prior to analysis. If sample is filtered, method is not approved for NPDES monitoring. Sample color absorbed in photometric range interferes.

INSTRUMENTATION Technicon auto analyzer. 550 nm Filters. 15 mm Tubular flow cells.

RANGE 10–200 mg/L as $CaCO_3$.

MDL Not listed.

PRECISION SD = ±0.5 concentrations; 15, 57, 154, and 193 mg/L as $CaCO_3$.

ACCURACY Recoveries = 100 and 99% at 31 and 149 mg/L as $CaCO_3$.

SAMPLING METHOD Plastic or glass (100 mL).

STABILITY Cool, 4°C.

MHT 14 days.

QUALITY CONTROL Place working standards in sampler in order of decreasing concentration. (Methyl orange is used as indicator because its pH range is in the same range as equivalent point for total alkalinity and has distinct color change that can be easily measured).

REFERENCE Methods for the Chemical Analysis of Water and Wastes, EPA-600/4-79-020. USEPA, EMSL, 1979.

Alkalinity (Titrimetric, pH 4.5)   EPA Method 310.1

TITLE Inorganics, Non-Metallics

MATRIX Drinking, Surface and Saline waters. Wastewater.

APPLICATION Date issued 1971. Editorial Rev. 1978. (Titrimetric, pH 4.5). An unaltered sample is titrated, using standard acid, to an electrometrically determined end point of pH 4.5. The sample must not be filtered, diluted, concentrated, or altered in any way.

INTERFERENCES Substances such as salts of weak organic and inorganic acids present in large amounts, may cause interference in the electrometric pH measurements. Oil and grease, by coating the pH electrode, may also interfere, causing sluggish response.

INSTRUMENTATION pH m or electrically operated titrator.

RANGE For all alkalinity ranges.

MDL Keep titration volume <50 mL.

PRECISION A standard deviation of 1 mg $CaCO_3$ per L can be achieved.

ACCURACY Not listed.

SAMPLING METHOD Plastic or glass. (100 mL).

STABILITY Cool, 4°C.

MHT 14 days.

QUALITY CONTROL Standardize and calibrate pH m according to manufacturer's instructions. If automatic temperature compensation is not provided, make titration at 25 ± 2 C. (For <1000 mg $CaCO_3$ per L use 0.02 N titrant. For >1000 mg $CaCO_3$ per L use 0.1 N titrant).

REFERENCE EPA Methods for the Chemical Analysis of Water and Wastes, EPA-600/4-79-020, USEPA, EMSL, 1979.

Allyl alcohol   EPA Method 1624
CAS #107-18-6

TITLE Volatile Organic Compounds by Isotope Dilution GC/MS

MATRIX Compounds may be determined in waters, soils, and municipal sludges by this method.

METHOD SUMMARY This method is used to determine 58 volatile toxic organic pollutants associated with the CWA (as amended 1987); the RCRA (as amended 1986); the CERCLA (as amended 1986); and other compounds amenable to purge and trap gas chromatography-mass spectrometry (GC/MS).

If the solids content is less than 1%, stable isotopically-labeled analogs of the compounds of interest are added to a 5-mL sample and the sample is purged with an inert gas at 20–25°C in a chamber designed for soil or water samples. If the solids content is greater than 1%, 5 mL of reagent water and the labeled compounds are added to a 5 g aliquot of sample and the mixture is purged at 40°C. Compounds that will not purge at 20–25°C or at 40°C are purged at 78–85°C. In the purging process, the volatile compounds are transferred from the aqueous phase into the gaseous phase where they are passed into a sorbent column and trapped. After purging is completed, the trap is backflushed and heated rapidly to desorb the compounds into a GC. The compounds are separated by the GC and detected by a MS. The labeled compounds serve to correct the variability of the analytical technique.

INTERFERENCES Impurities in the purge gas, organic compounds outgassing from the plumbing upstream of the trap, and solvent vapors in the lab account for most problems. Samples can be contaminated by diffusion of volatile organic compounds (particularly methylene chloride) through the bottle seal during shipment and storage. Contamination by carryover can occur when high-level and low-level samples are analyzed sequentially. When an unusually concentrated sample is encountered, follow it by analysis of a reagent water blank to check for carryover.

INSTRUMENTATION Major equipment includes a GC with linear temperature programming and a glass jet separator as the MS interface, a MS with 70 eV electron impact ionization, and a data system to collect and record response factors.

Column: 2.8 m × 2 mm I.D. glass, packed with 1% SP-1000 on Carbopak B, 60/80 mesh, or equivalent.

PRECISION & ACCURACY The detection limits of the method are usually dependent on the level of interferences rather than instrumental limitations. The method detection

limits were determined in digested sludge (low solids) and in filter cake or compost (high solids).

The MDL (in μg/kg) for low solids is not listed and for high solids is not listed.

Labeled and native compound precision (in μg/L) as standard deviation was not listed.
Labeled and native compound accuracy (in μg/L) as average recovery was not listed.

Acceptance criteria are at 20 μg/L for this compound.

SAMPLE COLLECTION, PRESERVATION & HANDLING Grab samples are collected in glass containers having a total volume greater than 20 mL. Fill and seal each bottle so that no air bubbles are entrapped. Samples are maintained at 0 to 4°C from the time of collection until analysis. If an aqueous sample contains residual chlorine, add sodium thiosulfate preservative (10 mg/40 mL) to the empty sample bottles just prior to shipment to the sample site. All samples must be analyzed within 14 days of collection.

SAMPLE PREPARATION Samples containing less than 1% solids are analyzed directly as aqueous samples. Samples containing 1% solids or greater are analyzed as solid samples utilizing one of two methods, depending on the levels of pollutants in the sample. Samples containing 1% solids or greater, and low to moderate levels of pollutants are analyzed by purging a known weight of sample added to 5 mL of reagent water. Samples containing 1% solids or greater, and high levels of pollutants, are extracted with methanol, and an aliquot of the methanol extract is added to reagent water and purged.

QUALITY CONTROL A field blank prepared from reagent water and carried through the sampling and handling protocol may serve as a check on contamination from shipment and storage.

The analyst is permitted to modify this method to improve separations or lower the costs of measurements, provided all performance specifications are met. Analyses of blanks are required. When results of spikes indicate atypical method performance for samples, the samples are diluted to bring method performance within acceptable limits. Analyze two sets of four 5-mL aliquots (8 aliquots total) of the aqueous performance standard. Spike all samples with labeled compounds to assess method performance on the sample matrix. Compute the percent recovery of the labeled compounds using the internal standard method. Compare the percent recovery for each compound with the corresponding labeled compound recovery. Reagent water blanks are analyzed to demonstrate freedom from carryover contamination. Field replicates may be collected to determine the precision of the sampling technique, and spiked samples may be required to determine the accuracy of the analysis when the internal method is used.

REFERENCE Volatile Organic Compounds by Isotope Dilution GC/MS. Office of Water Regulation and Standards, U.S. EPA Industrial Technology Division, Washington, DC, EPA Method 1624, Rev. C, June 1989 (contact W.A. Telliard, U.S. EPA, Office of Water Regulations and Standards, 401 M St., SW, Washington, DC, 20460. Phone: 202-382-7131).

Allyl alcohol   **EPA Method 8240**
CAS #107-18-6

TITLE Volatile Organics By GC/MS: Packed Column Technique

MATRIX Nearly all types of sample matrices, regardless of water content, can be analyzed using this method. This includes groundwater, aqueous sludges, caustic liquors, acid liquors, waste solvents, oily wastes, mousses, tars, fibrous wastes, polymetric emulsions, filter cakes, spent carbons, spent catalysts, soils, and sediments.

METHOD SUMMARY Method 8240B covers 80 volatile organic compounds that are introduced into a gas chromatograph by the purge-and-trap method or by direct injection (in limited applications). For the purge-and-trap method an inert gas (zero grade nitrogen or helium) is bubbled through a 5-mL solution at ambient temperature. Purged sample components are trapped in a tube of sorbent materials. When purging is complete, the sorbent tube is heated and backflushed with inert gas to desorb the trapped components onto a GC column.

INTERFERENCES Impurities in the purge gas and from organic compounds outgassing from the plumbing ahead of the trap account for many contamination problems. Interferences purged or coextracted from the samples will vary considerably from source to source. Cross-contamination can occur whenever high-level and low-level samples are analyzed sequentially. Whenever an unusually concentrated sample is analyzed, it should be followed by the analysis of organic-free reagent water to check for cross-contamination. Samples also can be contaminated by diffusion of volatile organics (particularly methylene chloride and fluorocarbons) through the septum seal into the sample during shipment and storage. A trip blank can serve as a check on such contamination. The lab where volatile analysis is performed and also the refrigerated storage area should be completely free of solvents.

INSTRUMENTATION A gas chromatograph/mass spectrometry/data system (GC/MS) equipped with a 6 ft × 0.1 in I.D. glass column packed with 1% SP-1000 on Carbopack-B (60/80 mesh) is required. Also needed is a 5-mL purging device, a sorbent trap, and a thermal desorption apparatus.

PRECISION & ACCURACY This method is reported to have been tested by 15 laboratories using organic-free reagent water, drinking water, surface water, and industrial wastewater (not specified) fortified at six concentrations over the range 5–600 μg/L.

Sample estimated quantitation limits (EQLs) are highly matrix dependent. The EQLs listed may not always be achievable. EQLs listed for soils or sediments are based on wet weight. Normally, data is reported on a dry-weight basis; therefore, EQLs will be higher, based on the percent dry weight of each sample. Note that EQLs are even more variable than MDLs and that they are highly variable depending on the matrix being analyzed.

EQL in groundwater in μg/L was not listed.
EQL in low soil or sediment in μg/kg was not listed.

Accuracy (a) in µg/L was not listed.
Precision (b) in µg/L was not listed.

(a) *Average recovery found for measurements of samples containing a concentration of C, in µg/L.*
(b) *Overall precision found for measurements of samples with average recovery X for samples containing a concentration of C in µg/L.*
X = *Average recovery found for measurement of samples containing a concentration of C in µg/L.*

ESTIMATED QUANTITATION LIMITS FOR MATRICES OTHER THAN WATER, SOIL, AND SEDIMENTS

Other Matrices	Factor (a)
Waste miscible liquid waste	50
High-concentration soil and sludge	125
Non-water miscible waste	500

(a) *EQL = [EQL for low soil/sediment] × [Factor]. For non-aqueous samples, the factor is on a wet-weight basis.*

SAMPLING METHOD
Liquid samples — Use a 40-mL glass screw-cap VOA vial with a Teflon®-faced silicone septum that has been prewashed, rinsed with distilled deionized water, and oven dried. However, if residual chlorine is present, collect sample in a 40-oz. soil VOA container which has been pre-preserved with 4 drops of 10% sodium thiosulfate, mix gently, and then transfer the sample to a 40-mL VOA vial. Collect bubble-free samples in duplicate and seal them in separate plastic bags.

Soils or sediments, and sludges — Use an 8 oz. widemouth glass bottle with a Teflon®-faced silicone septum that has been prewashed with detergent, rinsed with distilled deionized water, and oven dried. Tap slightly to eliminate free air space. Collect samples in duplicate and seal them in separate plastic bags.

SAMPLE PRESERVATION
Liquid samples — Add 4 drops of concentrated HCL and immediately cool samples to 4°C and store in a solvent-free refrigerator.

Soils or sediments, and sludges — Cool samples to 4°C and store in a solvent-free refrigerator.

MHT Maximum holding time is 14 days from the date of sample collection.

SAMPLE PREPARATION
Liquid samples - - Remove the plunger from a 5-mL syringe and carefully pour the sample into the syringe barrel to just short of overflowing. Replace the syringe plunger and compress the sample. Open the syringe valve and vent any residual air while adjusting the sample volume to 5.0 mL. If there is only one volatile organic analysis (VOA) vial, a second syringe should be filled at this time to protect against possible loss of sample integrity. Add 10 µL of surrogate spiking solution and 10 µL of internal standard spiking solution through the valve bore of the 5-mL syringe, then close the valve. The surrogate and internal standards may be mixed and added as a single spiking solution.

Sediments, soils, and waste samples — All samples of this type should be screened by GC analysis using a headspace method (EPA Method 3810) or the hexadecane extraction and screening method (EPA Method 3820). Use the screening data to determine whether to use the low-concentration method (0.005–1 mg/kg) or the high-concentration method (>1 mg/kg).

Low-concentration method — The low-concentration method is based on purging a heated sediment or soil sample mixed with organic-free reagent water containing the surrogate and internal standards. Analyze all reagent blanks and standards under the same conditions as the samples.

Use a 5-g sample if the expected concentration is <0.1 mg/kg or a 1-g sample for expected concentrations between 0.1 and 1 mg/kg. Mix the contents of the sample container with a narrow metal spatula. Weigh the amount of the sample into a tared purge device. Add the spiked water to the purge device, which contains the weighed amount of sample, and connect the device to the purge-and-trap system.

High-concentration method — This method is based on extracting the sediment or soil with methanol. A waste sample is either extracted or diluted, depending on its solubility in methanol. Wastes that are insoluble in methanol are diluted with reagent tetraglyme or possibly polyethylene glycol (PEG). An aliquot of the extract is added to organic-free reagent water containing surrogate and internal standards. This is purged at ambient temperature. All samples with an expected concentration of >1.0 mg/kg should be analyzed by this method.

Mix the contents of the sample container with a narrow metal spatula. For sediments or soils and solid wastes that are insoluble in methanol, weigh 4 g (wet weight) of sample into a tared 20-mL vial. For waste that is soluble in methanol, tetraglyme, or PEG, weigh 1 g (wet weight) into a tared scintillation vial or culture tube or a 10-mL volumetric flask. Quickly add 9.0 mL of appropriate solvent then add 1.0 mL of a surrogate spiking solution to the vial, cap it, and shake it for 2 min.

METHANOL EXTRACT REQUIRED FOR ANALYSIS OF HIGH-CONCENTRATION SOILS OR SEDIMENTS

Approximate Concentration Range	Volume of Methanol Extract (a)
500–10,000 µg/kg	100 µL
1,000–20,000 µg/kg	50 µL
5,000–100,000 µg/kg	10 µL
25,000–500,000 µg/kg	100 µL of 1/50 dilution (b)

Calculate appropriate dilution factor for concentrations exceeding this table.

(a) *The volume of methanol added to 5 mL of water being purged should be kept constant. Add to the 5-mL syringe whatever volume of methanol is necessary to maintain a volume of 100 µL added to the syringe.*
(b) *Dilute an aliquot of the methanol extract and then take 100 µL for analysis.*

QUALITY CONTROL Demonstrate, through the analysis of a reagent water blank, that interferences from the analytical system, glassware, and reagents are under control. Blank samples should be carried through all stages of the sample preparation and measurement steps. For each analytical batch (up to 20 samples), a reagent blank, matrix spike, and matrix spike duplicate must be analyzed (the frequency of the spikes may be different for different monitoring programs). The blank and spiked samples must be carried through all stages of the sample preparation and measurement steps. QC samples mentioned in the section on Interferences will also be needed as appropriate to those situations.

REFERENCE Test Methods for Evaluating Solid Waste (SW-846). U.S. EPA. 1983. Method 8240B, Rev. 2, Nov. 1990. Office of Solid Wastes, Washington, DC.

Allyl chloride   EPA Method 8240
CAS #107-05-1

TITLE Volatile Organics By GC/MS: Packed Column Technique

MATRIX Nearly all types of sample matrices, regardless of water content, can be analyzed using this method. This includes groundwater, aqueous sludges, caustic liquors, acid liquors, waste solvents, oily wastes, mousses, tars, fibrous wastes, polymetric emulsions, filter cakes, spent carbons, spent catalysts, soils, and sediments.

METHOD SUMMARY Method 8240B covers 80 volatile organic compounds that are introduced into a gas chromatograph by the purge-and-trap method or by direct injection (in limited applications). For the purge-and-trap method an inert gas (zero grade nitrogen or helium) is bubbled through a 5-mL solution at ambient temperature. Purged sample components are trapped in a tube of sorbent materials. When purging is complete, the sorbent tube is heated and backflushed with inert gas to desorb the trapped components onto a GC column.

INTERFERENCES Impurities in the purge gas and from organic compounds outgassing from the plumbing ahead of the trap account for many contamination problems. Interferences purged or coextracted from the samples will vary considerably from source to source. Cross-contamination can occur whenever high-level and low-level samples are analyzed sequentially. Whenever an unusually concentrated sample is analyzed, it should be followed by the analysis of organic-free reagent water to check for cross-contamination. Samples also can be contaminated by diffusion of volatile organics (particularly methylene chloride and fluorocarbons) through the septum seal into the sample during shipment and storage. A trip blank can serve as a check on such contamination. The lab where volatile analysis is performed and also the refrigerated storage area should be completely free of solvents.

INSTRUMENTATION A gas chromatograph/mass spectrometry/data system (GC/MS) equipped with a 6 ft × 0.1 in I.D. glass column packed with 1% SP-1000 on Carbopack-B (60/80 mesh) is required. Also needed is a 5-mL purging device, a sorbent trap, and a thermal desorption apparatus.

PRECISION & ACCURACY This method is reported to have been tested by 15 laboratories using organic-free reagent water, drinking water, surface water, and industrial wastewater (not specified) fortified at six concentrations over the range 5–600 µg/L.

Sample estimated quantitation limits (EQLs) are highly matrix dependent. The EQLs listed may not always be achievable. EQLs listed for soils or sediments are based on wet weight. Normally, data is reported on a dry-weight basis; therefore, EQLs will be higher, based on the percent dry weight of each sample. Note that EQLs are even more variable than MDLs and that they are highly variable depending on the matrix being analyzed.

EQL in groundwater in µg/L was 5.
EQL in low soil or sediment in µg/kg was 5.
Accuracy (a) in µg/L was 0.93C+2.00.
Precision (b) in µg/L was 0.25x–1.13.

(a) *Average recovery found for measurements of samples containing a concentration of C, in µg/L.*
(b) *Overall precision found for measurements of samples with average recovery X for samples containing a concentration of C in µg/L.*
X = *Average recovery found for measurement of samples containing a concentration of C in µg/L.*

ESTIMATED QUANTITATION LIMITS FOR MATRICES OTHER THAN WATER, SOIL, AND SEDIMENTS

Other Matrices	Factor (a)
Waste miscible liquid waste	50
High-concentration soil and sludge	125
Non-water miscible waste	500

(a) *EQL = [EQL for low soil/sediment] × [Factor]. For non-aqueous samples, the factor is on a wet-weight basis.*

SAMPLING METHOD
Liquid samples — Use a 40-mL glass screw-cap VOA vial with a Teflon®-faced silicone septum that has been prewashed, rinsed with distilled deionized water, and oven dried. However, if residual chlorine is present, collect sample in a 40-oz. soil VOA container which has been pre-preserved with 4 drops of 10% sodium thiosulfate, mix gently, and then transfer the sample to a 40-mL VOA vial. Collect bubble-free samples in duplicate and seal them in separate plastic bags.

Soils or sediments, and sludges — Use an 8 oz. widemouth glass bottle with a Teflon®-faced silicone septum that has been prewashed with detergent, rinsed with distilled deionized water, and oven dried. Tap slightly to eliminate free air space. Collect samples in duplicate and seal them in separate plastic bags.

SAMPLE PRESERVATION
Liquid samples — Add 4 drops of concentrated HCL and immediately cool samples to 4°C and store in a solvent-free refrigerator.

Soils or sediments, and sludges — Cool samples to 4°C and store in a solvent-free refrigerator.

MHT Maximum holding time is 14 days from the date of sample collection.

SAMPLE PREPARATION

Liquid samples — Remove the plunger from a 5-mL syringe and carefully pour the sample into the syringe barrel to just short of overflowing. Replace the syringe plunger and compress the sample. Open the syringe valve and vent any residual air while adjusting the sample volume to 5.0 mL. If there is only one volatile organic analysis (VOA) vial, a second syringe should be filled at this time to protect against possible loss of sample integrity. Add 10 µL of surrogate spiking solution and 10 µL of internal standard spiking solution through the valve bore of the 5-mL syringe, then close the valve. The surrogate and internal standards may be mixed and added as a single spiking solution.

Sediments, soils, and waste samples — All samples of this type should be screened by GC analysis using a headspace method (EPA Method 3810) or the hexadecane extraction and screening method (EPA Method 3820). Use the screening data to determine whether to use the low-concentration method (0.005–1 mg/kg) or the high-concentration method (>1 mg/kg).

Low-concentration method — The low-concentration method is based on purging a heated sediment or soil sample mixed with organic-free reagent water containing the surrogate and internal standards. Analyze all reagent blanks and standards under the same conditions as the samples.

Use a 5-g sample if the expected concentration is <0.1 mg/kg or a 1-g sample for expected concentrations between 0.1 and 1 mg/kg. Mix the contents of the sample container with a narrow metal spatula. Weigh the amount of the sample into a tared purge device. Add the spiked water to the purge device, which contains the weighed amount of sample, and connect the device to the purge-and-trap system.

High-concentration method — This method is based on extracting the sediment or soil with methanol. A waste sample is either extracted or diluted, depending on its solubility in methanol. Wastes that are insoluble in methanol are diluted with reagent tetraglyme or possibly polyethylene glycol (PEG). An aliquot of the extract is added to organic-free reagent water containing surrogate and internal standards. This is purged at ambient temperature. All samples with an expected concentration of >1.0 mg/kg should be analyzed by this method.

Mix the contents of the sample container with a narrow metal spatula. For sediments or soils and solid wastes that are insoluble in methanol, weigh 4 g (wet weight) of sample into a tared 20-mL vial. For waste that is soluble in methanol, tetraglyme, or PEG, weigh 1 g (wet weight) into a tared scintillation vial or culture tube or a 10-mL volumetric flask. Quickly add 9.0 mL of appropriate solvent then add 1.0 mL of a surrogate spiking solution to the vial, cap it, and shake it for 2 min.

METHANOL EXTRACT REQUIRED FOR ANALYSIS OF HIGH-CONCENTRATION SOILS OR SEDIMENTS

Approximate Concentration Range	Volume of Methanol Extract (a)
500–10,000 µg/kg	100 µL
1,000–20,000 µg/kg	50 µL
5,000–100,000 µg/kg	10 µL
25,000–500,000 µg/kg	100 µL of 1/50 dilution (b)

Calculate appropriate dilution factor for concentrations exceeding this table.

(a) The volume of methanol added to 5 mL of water being purged should be kept constant. Therefore, add to the 5-mL syringe whatever volume of methanol is necessary to maintain a volume of 100 µL added to the syringe.

(b) Dilute an aliquot of the methanol extract and then take 100 µL for analysis.

QUALITY CONTROL Demonstrate, through the analysis of a reagent water blank, that interferences from the analytical system, glassware, and reagents are under control. Blank samples should be carried through all stages of the sample preparation and measurement steps. For each analytical batch (up to 20 samples), a reagent blank, matrix spike, and matrix spike duplicate must be analyzed (the frequency of the spikes may be different for different monitoring programs). The blank and spiked samples must be carried through all stages of the sample preparation and measurement steps. QC samples mentioned in the section on Interferences will also be needed as appropriate to those situations.

REFERENCE Test Methods for Evaluating Solid Waste (SW-846). U.S. EPA. 1983. Method 8240B, Rev. 2, Nov. 1990. Office of Solid Wastes, Washington, DC.

Aluminum
CAS #7429-90-5

EPA Method 6010

TITLE Inductively Coupled Plasma-Atomic Emission Spectroscopy

MATRIX This method is applicable to the determination of trace elements, including metals, in groundwater, soils, sludges, sediments, and other solid wastes. All matrices require digestion prior to analysis. The method of standard addition must be used for the analysis of all sample digests unless either serial dilution or matrix spike addition demonstrates it is not required.

METHOD SUMMARY Method 6010 covers 25 elements using ICP analysis. It measures element-emitted light by optical spectrometry. Samples, following an appropriate acid digestion, are nebulized and the resulting aerosol is transported to the plasma torch. Element-specific atomic line emission spectra are produced by a radio-frequency inductively coupled plasma.

INTERFERENCES Interferences may be categorized as spectral or non-spectral. Spectral interferences are caused by overlap of a spectral line from another element, unresolved overlap of molecular band spectra, background contribution from continuous or recombination phenomena, and stray light from the line emission of high concentration elements. Non-spectral interferences include physical and chemical interferences. Physical interferences are effects associated with the sample nebulization and transport processes. Changes in viscosity and surface tension can cause significant inaccuracies. Chemical interferences include molecular compound formation, ionization effects, and solute vaporization effects. Normally these effects are not significant and can be minimized by careful selection of operating conditions. Chemical interferences are highly dependent on matrix type and the specific analyte element.

INSTRUMENTATION An inductively coupled argon plasma emission spectrometer (ICP) capable of background correction is required.

PRECISION & ACCURACY Detection limits, sensitivity, and optimum ranges of the metals will vary with the matrices and model of the spectrometer. In a single lab evaluation, seven wastes were analyzed for 22 elements. The mean percent relative standard deviation from triplicate analyses for all elements and wastes was 9 ± 2%. The mean percent recovery of spiked elements for all wastes was 93 ± 6%. Spike levels ranged from 100 µg/L to 100 mg/L. The wastes included sludges and industrial wastewater.

Estimated instrument detection limit in µg/L is 45.
Spiked concentration in µg/L is 60.
Mean reported value in µg/L is 62.
Precision as RSD % is 33.

SAMPLING METHOD Samples should be collected in borosilicate glass, linear polyethylene, polypropylene, or Teflon® bottles that have been prewashed with detergent and tap water, and rinsed with 1:1 nitric acid and tap water or 1:1 hydrochloric acid and tap water. Collect at least 2 g of solids and 200 mL of aqueous samples.

SAMPLE PRESERVATION Add nitric acid to make the samples pH <2.

MHT The maximum holding time for properly preserved samples is 6 months.

SAMPLE PREPARATION Preliminary treatment of most matrices is necessary because of the complexity and variability of sample matrices. Water samples that have been prefiltered and acidified will not need acid digestion. Methods for acid digestion of waters for total recoverable or dissolved metals, acid digestions of aqueous samples and extracts for total metals, and acid digestion of sediments, sludges, and soils are summarized below.

Total recoverable or dissolved metals in water — To prepare surface and groundwater samples for determination of total recoverable and dissolved metals, a 100 mL aliquot of well-mixed sample is acidified with concentrated nitric acid and concentrated hydrochloric acid, then heated until the volume is reduced to 15–20 mL. Adjust the final volume to 100 mL with reagent water.

Total metals in aqueous samples, soil and sediment extracts — To prepare aqueous samples, soil and sediment extracts, and wastes that contain suspended solids, a 100 mL aliquot is made acidic with concentrated nitric acid and the solution is evaporated to about 5 mL on a hot plate. Continue heating and adding additional acid until sample digestion is complete, which is usually indicated when the digestate is light in color or does not change in appearance. Evaporate the solution to about 3 mL and cool it and add a small quantity of 1:1 hydrochloric acid (10 mL100 mL of final solution). Cover the beaker and reflux for 15 min. Wash down the beaker walls and filters or centrifuge the sample to remove silicates and other insoluble material. Filter the sample and adjust the final volume to 100 mL with reagent water and the final acid concentration to 10%.

Sediments, sludges, and soil — To prepare sediments, sludges and soil samples, transfer 1–2 g to a conical beaker and add 10 mL of 1:1 nitric acid, mix the slurry, and cover it with a watch glass. Heat the sample and reflux for 10 to 15 min without boiling. Allow it to cool, then add 5 mL of concentrated nitric acid and reflux for 30 min. Repeat last step and then allow the solution to evaporate to 5 mL without boiling. Cool and add 2 mL of water and 3 mL of 30% hydrogen peroxide. Cover and place the beaker on the hot plate. Heat and add 30% hydrogen peroxide in 1-mL aliquots with warming until the effervescence is minimal but do not add more than a total of 10 mL of 30% hydrogen peroxide. If the sample is being prepared for the analysis of Ag, Al, As, Ba, Be, Ca, Cd, Co, Cr, Cu, Fe, K, Mg, Mn, Mo, Na, Ni, Os, Pb, Se, Tl, V, and Zn, then add 5 mL of concentrated hydrochloric acid and 10 mL of water and return the covered beaker to a hot plate for 15 min of additional refluxing without boiling. Dilute the sample to a 100 mL volume with water after cooling and filter or centrifuge to remove particulates.

QUALITY CONTROL Laboratory control samples must be analyzed for each analytical method. A method blank should be analyzed with each batch of samples. The effect of the matrix on method performance must be demonstrated: when appropriate, there should be at least one matrix spike and either one matrix duplicate or one matrix spike duplicate per analytical batch. The bias and precision of the method, as well as the method detection limit for each specific matrix type, must be measured.

Dilute and reanalyze samples that are more concentrated than the linear calibration limit. Employ a minimum of one reagent blank per sample batch to determine if contamination or any memory effects are occurring. Whenever a new or unusual sample matrix is encountered, perform either a serial dilution test or a matrix spike addition test to ensure that neither positive or negative interferences are operating on any of the analyte elements. Check the instrument standardization by verifying calibration every 10 samples using a calibration blank and a check standard.

REFERENCE Test Methods for Evaluating Solid Waste (SW-846). U.S. EPA. 1983. Method 6010, Rev. 0, Sept. 1986. Office of Solid Wastes, Washington, DC.

Aluminum — EPA Method 200.7
CAS #7429-90-5

TITLE Inductively Coupled Plasma

MATRIX Dissolved, suspended or (ICP) total element in drinking and surface waters and in domestic and industrial wastewater.

APPLICATION The method covers the determination of 25 metals. Dissolved elements are determined in filtered and acidified samples after appropriate digestion (which increases dissolved solids). Its primary advantage is that ICP instruments allow simultaneous or rapid sequential determination of many elements in a short time. Samples are first nebulized and the aerosol is transported to a plasma torch in which element specific atomic line emission spectra are produced by a radio frequency inductively coupled plasma. Background correction is required for trace element detection except in the case of line broadening.

INTERFERENCES There are spectral, physical, and chemical interferences. The primary disadvantage of ICP instruments is background radiation from other elements and the plasma gases (spectral interferences). Changes in sample viscosity and surface tension with samples containing high dissolved solids (especially those exceeding 1500 mg/L) or high acid concentrations can cause physical interferences. Ionization effects, solute vaporization, and molecular compound formation can cause chemical interferences. Manganese and vanadium can cause interference at the 100 mg/L level.

INSTRUMENTATION Inductively coupled argon plasma emission spectroscopy. 308.215 nm Wavelength

RANGE Not listed.

MDL 45 µg/L.

PRECISION SD = 5.6% Mean at true value 700 µg/L.

ACCURACY Mean Recovery = 93% ± 6% of spiked elements for all wastes.

SAMPLING METHOD Wash sample container with detergent and tap water, rinse with 1+1 nitric acid and tap water, then rinse with 1+1 hydrochloric acid and tap water, then rinse with deionized, distilled water in that order. Perform any filtration or acid preservation steps when the sample is collected or as soon as possible thereafter.

STABILITY Cool samples to 4°C.

MHT 24 h.

QUALITY CONTROL Mixed calibration standards, an instrument check standard, and an interference check solution are used in addition to a quality control sample. The quality control sample should be prepared in the same acid matrix as the calibration standards at 10 times the instrumental detection limits and in accordance with the instructions provided by the supplier. Furthermore, two types of blanks are required: a calibration blank and a reagent blank.

REFERENCE Method 200.7, U.S. EPA, EMSL-Cincinnati, OH, Nov. 1980

Aluminum **EPA Method 7020**
CAS #7429-90-5

TITLE Atomic Absorption, (AA)

MATRIX Drinking, Surface and Direct Aspiration Saline Waters, Wastewater.

APPLICATION Sample is aspirated and atomized in a flame. A light beam from an aluminum hollow cathode lamp is directed through the flame into a monochromator and onto a detector. Since wavelength of light beam is specific for aluminum, light energy absorbed by detector is measure of aluminum concentration.

INTERFERENCES The most troublesome type of interference is chemical, and caused by lack of absorption of atoms bound in molecular combination in the flame. High dissolved solids in a sample may result in nonatomic absorbance interference. Ionization and spectral interferences can occur.

INSTRUMENTATION Atomic absorption spectrometer. Aluminum hollow cathode lamp or electrodeless discharge lamp. (309.3 nm wavelength).

RANGE 5–50 mg/L.

MDL 0.1 mg/L.

PRECISION deviation = 299 µg/L at 1205 µg/L (true value) 38 labs.

ACCURACY As bias = +6.3% at 1205 µg/L (true value) 38 labs.

SAMPLING METHOD Use glass or plastic containers. Collect 200 g of solids and 600 mL of liquid samples.

STABILITY Cool solid samples to 4°C and analyze as soon as possible. Add nitric acid to liquid samples to pH <2.

MHT 6 months.

QUALITY CONTROL At least one duplicate and one spike sample should be run every 20 samples or with each matrix type to verify precision of the method. For 20 or more samples per day, verify working standard curve. Run an additional standard at or near mid-range every 10 samples.

REFERENCE Method 7020, SW-846, 3rd ed., Nov. 1986.

Ametryn **EPA Method 507**
CAS #834-12-8

TITLE Determination of Nitrogen and Phosphorus-Containing Pesticides in Water by GC/NPD

MATRIX This method is applicable to the determination of certain nitrogen and phosphorus-containing pesticides in finished drinking water and groundwater.

METHOD SUMMARY Method 507 covers 46 nitrogen- and phosphorus-containing pesticides. A 1-L sample is fortified with a surrogate standard, salted, buffered, extracted with methylene chloride, and concentrated; then the solvent is exchanged with methyl tert-butyl ether (MTBE) and concentrated again, and a 2-µL aliquot of a sample extract is injected into a GC system equipped with a selective nitrogen-phosphorus detector and a capillary column for analysis.

INTERFERENCES Method interferences may be caused by contaminants in solvents, reagents, glassware, and other sample processing apparatus. Interfering contamination may occur when a sample containing low concentrations of analytes is analyzed immediately following a sample containing relatively high concentrations. One or more injections of MTBE should be made following the analysis of a sample with high concentrations of analytes to check for analyte carryover. Matrix interferences may be caused by contaminants that are coextracted from the sample. The extent of matrix interferences will vary considerably from source to source, depending upon the water sampled.

INSTRUMENTATION A gas chromatograph system (GC) equipped with a nitrogen-phosphorus detector (NPD) is needed.

Column 1: 30 m × 0.25 mm I.D. DB-5 bonded fused silica column, 0.25 μm film thickness, or equivalent.

Column 2: 30 M × 0.25 mm I.D. DB-1701 bonded fused silica column, 0.25 μm film thickness, or equivalent.

PRECISION & ACCURACY This method has been validated in a single lab and estimated detection limits (EDLs) have been determined for each analyte. Observed detection limits may vary among waters, depending upon the nature of the interferences in the sample matrix and the specific instrumentation used. Analytes that are not separated chromatographically cannot be individually identified and measured unless an alternative technique for identification and quantification exist.

The estimated detection limit (in μg/L) was 2. The EDL is defined as either method detection limit or a level of compound in a sample yielding a peak in the final extract with signal-to-noise ratio of approximately 5, whichever value is higher.

The concentration used for these measurements (in μg/L) was 20.
The accuracy (as % recovery) was 91.
The precision (% RSD) was 10.

SAMPLING METHOD Grab samples are collected in 1-L glass sample bottles (prewashed with detergent and hot tap water, rinsed with reagent water, and dried in an oven at 400°C for 1 h) with screw caps lined with PTFE-fluorocarbon.

SAMPLE PRESERVATION Add mercuric chloride to the sample bottle in amounts to produce a concentration of 10 mg/L. If residual chlorine is present, add 80 mg of sodium thiosulfate/L of sample to the sample bottle prior to collection. After collection, seal bottle and shake vigorously for 1 min, then cool the sample to 4°C immediately and store it at 4°C in the dark until extraction.

MHT Maximum holding time of the samples and, in some cases the extracts, is 14 days.

SAMPLE PREPARATION Fortify the sample with 50 μL of the surrogate standard solution, adjust to pH 7 with phosphate buffer, add 100 g NaCl to the sample, and seal and shake to dissolve the salt; then extract with methylene chloride in a separatory funnel or in a mechanical tumbler bottle. Dry the extract by pouring it through a solvent-rinsed drying column containing about 10 cm of anhydrous sodium sulfate. Collect the extract in a Kuderna-Danish (K-D) concentrator and rinse the column with 20–30 mL methylene chloride. Concentrate the extract to about 2 mL and rinse the flask and its lower joint into the concentrator tube with 1 to 2 mL of methyl t-butyl ether (MTBE). Add 5–10 mL of MTBE and concentrate the extract twice (adding more MTBE) to a final volume of 5.0 mL and store it at 4°C until analysis.

Note: If methylene chloride is not completely removed from the final extract, it may cause detector problems.

QUALITY CONTROL Minimum quality control requirements are initial demonstration of lab capability, determination of surrogate compound recoveries in each sample and blank, monitoring internal standard peak area or height in each sample and blank, analysis of lab reagent blanks, lab fortified samples, lab fortified blanks, and other QC samples. A lab reagent blank is analyzed to demonstrate that all glassware and reagent interferences are under control.

Initial demonstration of capability is fulfilled by analyzing four fortified reagent water samples with the recovery value for each analyte falling within the acceptable range (±30% average recovery). Surrogate recoveries from samples or method blanks must be 70–130%. The internal standard response for any sample chromatogram should not deviate from the daily calibration check standard's internal standard response by more than 30% or lab fortified blanks and sample matrices are used to assess lab performance and analyte recovery, respectively.

If the response for the target analyte peak exceeds the working range of the system, dilute the extract and reanalyze. Alternative techniques such as an alternate detector or second chromatography column should be used to confirm peak identification when sample components are not resolved adequately.

EPA CONTACT & HOTLINE For technical questions contact Dr. Baldev Bathija, U.S. EPA, Office of Ground Water and Drinking Water (WH-550D), 401 M St. SW, Washington, DC 20460. Tel. (202) 260-3040. For further information the EPA Safe Drinking Water Hotline may be called at: (800) 426-4791.

REFERENCE Methods for the Determination of Organic Compounds in Drinking Water, EPA/600/4-88/039 (revised July 1991). U.S. EPA Environmental Monitoring Systems Laboratory, Cincinnati, OH, 45268, U.S.A. Available from the National Technical Information Service (NTIS), 5285 Port Royal Road, Springfield, VA 22161; Tel. 800-553-6847. NTIS Order Number is PB91-231480.

3-Amino-9-ethylcarbazole   **EPA Method 8270**
CAS #132-32-1

TITLE Semivolatile Organic Compounds by GC/MS

MATRIX This method is used to determine the concentration of semivolatile organic compounds in extracts prepared from all types of solid waste matrices, soils, and groundwater. Although surface waters are not specifically mentioned, this method should be applicable to water samples from rivers, lakes, etc.

METHOD SUMMARY This method covers 259 semivolatile organic compounds. In very limited applications direct injection of the sample into the GC/MS system may be appropriate, but this results in very high detection limits (approximately 10,000 µg/L). Typically, a 1-L liquid sample, containing surrogate, and matrix spiking standards, is extracted in a continuous extractor first under acid conditions and then under basic conditions. Typically 30 g of a solid sample, containing surrogate, and matrix spiking standards, is extracted ultrasonically. After concentrating the extract to 1 mL it is spiked with 10 µL of an internal standard solution just prior to analysis by GC/MS. The volume injected should contain about 100 ng of base/neutral and 200 ng of acid surrogates (for a 1-µL injection). Analysis is performed by GC/MS using a capillary GC column.

INTERFERENCES Raw GC/MS data from all blanks, samples, and spikes must be evaluated for interferences. Contamination by carryover can occur whenever high-concentration and low-concentration samples are sequentially analyzed. To reduce carryover, the sample syringe must be rinsed out between samples with solvent. Whenever an unusually concentrated sample is encountered, it should be followed by the analysis of blank solvent to check for cross-contamination.

INSTRUMENTATION A GC/MS and a data system are required. The GC column used is a 30 m × 0.25 mm I.D. (or 0.32 mm I.D.) 1-µm film thickness silicone-coated fused silica capillary column. A continuous liquid-liquid extractor equipped with Teflon® or glass connection joints and stopcocks requiring no lubrication, a K-D concentrating apparatus, water bath, and an ultrasonic disrupter with a minimum power of 300 W and with pulsing capability are also required.

PRECISION & ACCURACY The estimated quantitation limit (EQL) of Method 8270B for determining an individual compound is approximately 1 mg/kg (wet weight) for soil or sediment samples, 1–200 mg/kg for wastes (dependent on matrix and method of preparation), and 10 µg/L for groundwater samples. EQLs will be proportionately higher for sample extracts that require dilution to avoid saturation of the detector.

The EQL(b) for groundwater in µg/L is not listed.
The EQL (a, b) for low concentrations in soil and sediment in µg/kg is not listed.
Accuracy as µg/L is not listed.
Overall precision in µg/L is not listed.

(a) *EQLs listed for soil/sediment are based on wet weight. Normally data is reported in a dry-weight basis; therefore, EQLs will be higher based on the % dry weight of each sample. This calculation is based on a 30-g sample and gel permeation chromatography cleanup.*
(b) *Sample EQLs are highly matrix-dependent. The EQLs are provided for guidance and may not always be achievable.*
C = *True value for concentration, in µg/L.*
X = *Average recovery found for measurements of samples containing a concentration of C, in µg/L.*

ESTIMATED QUANTITATION LIMIT

Other Matrices	Factor (a)
High-concentration soil and sludges by sonicator	7.5
Non-water miscible waste	75

(a) *EQL for other matrices = [EQL for low soil/sediment] × [Factor]. This estimated EQL is similar to an EPA "Practical Quantitation Limit."*

SAMPLING METHOD
Liquid samples — Use a 1 or 2½ gallon amber glass bottle with a screw-top Teflon®-lined cover that has been prewashed with detergent and rinsed with distilled water and methanol (or isopropanol).

Soils, sediments, or sludges — Use an 8-oz. widemouth glass with a screw-top Teflon®-lined cover that has been prewashed with detergent and rinsed with distilled water and methanol (or isopropanol).

SAMPLE PRESERVATION
Liquid samples — If residual chlorine is present, add 3 mL of 10% sodium thiosulfate per gallon, cool to 4°C and store in a solvent-free refrigerator until analysis; if chlorine is not present, then eliminate the sodium thiosulfate addition.

Soils, sediments, or sludges — Cool samples to 4°C and store in a solvent-free refrigerator.

MHT Liquid samples must be extracted within 7 days and the extracts analyzed within 40 days. Soils, sediments, or sludges may be stored for a maximum of 14 days and the extracts analyzed within 40 days.

SAMPLE PREPARATION
Liquid samples — Transfer 1 L quantitatively to a continuous extractor. If high concentrations are anticipated, a smaller volume may be used and then diluted with organic-free reagent water to 1 L. Adjust pH, if necessary, to pH <2 using 1:1 (V/V) sulfuric acid. Pipette 1.0 mL of a surrogate standard spiking solution into each sample. For the sample in each analytical batch selected for spiking, add 1.0 mL of a matrix spiking standard. For base/neutral acid analysis, the amount of the surrogates and matrix spiking compounds added to the sample should result in a final concentration of 100 ng/µL of each analyte in the extract to be analyzed (assuming a 1-µL injection). Extract with methylene chloride for 18–24 h. Next, adjust the pH of the aqueous phase to pH >11 using 10 N sodium hydroxide and extract it with methylene chloride again for 18–24 h. Dry the extract through a column containing anhydrous sodium sulfate and concentrate it to 1 mL using a K-D concentrator.

Soils, sediments, or sludges — Use 30 g of sample. Nonporous or wet samples (gummy or clay type) that do not have a free-flowing sandy texture must be mixed with anhydrous sodium sulfate until the sample is free flowing. Add 1 mL of surrogate standards to all samples, spikes, standards, and blanks. For the sample in each analytical batch selected for spiking, add 1.0 mL of a matrix spiking standard. For base/neutral acid analysis, the amount added of the surrogates and matrix spiking compounds should result in a final concentration of 100 ng/µL of

each base/neutral analyte and 200 ng/µL of each acid analyte in the extract to be analyzed (assuming a 1-µL injection). Immediately add a 100-mL mixture of 1:1 methylene chloride:acetone and extract the sample ultrasonically for 3 min and then decant or filter the extracts. Repeat the extraction two or more times. Dry the extract using a column with anhydrous sodium sulfate and concentrate it to 1 mL in a K-D concentrator.

QUALITY CONTROL A methylene chloride solution containing 50 ng/µL of decafluorotriphenylphosphine (DFTPP) is used for tuning the GC/MS system each 12-h shift. A system performance check also must be made during every 12-h shift. A standard containing 50 ng/µL each of 4,4′-DDT, pentachlorophenol, and benzidine is required to verify injection port inertness and GC column performance. A calibration standard at mid-concentration, containing each compound of interest, including all required surrogates, must be performed every 12 h during analysis. After the system performance check is met, calibration check compounds are used to check the validity of the initial calibration.

The internal standard responses and retention times in the calibration check standard must be evaluated immediately after or during data acquisition. If the retention time for any internal standard changes by more than 30 seconds from the last check calibration (12 h), the chromatographic system must be inspected for malfunctions and corrections must be made, as required. If the electron ionization current plot (EICP) area for any of the internal standards changes by a factor of two from the last daily calibration standard check, the mass spectrometer must be inspected for malfunctions and corrections must be made, as appropriate.

Demonstrate, through the analysis of a reagent water blank, that interferences from the analytical system, glassware, and reagents are under control. The blank samples should be carried through all stages of the sample preparation and measurement steps. For each analytical batch (up to 20 samples), a reagent blank, matrix spike, and matrix spike duplicate/duplicate must be analyzed (the frequency of the spikes may be different for different monitoring programs). The blank and spiked samples must be carried through all stages of the sample preparation and measurement steps. A QC reference sample concentrate containing each analyte at a concentration of 100 mg/L in methanol is required.

REFERENCE Test Methods for Evaluating Solid Waste (SW-846). U.S. EPA 1983, Method 8270B, Rev. 2, Nov. 1990. Office of Solid Waste, Washington, DC.

2-Aminoanthraquinone **EPA Method 8270**
CAS #117-79-3

TITLE Semivolatile Organic Compounds by GC/MS

MATRIX This method is used to determine the concentration of semivolatile organic compounds in extracts prepared from all types of solid waste matrices, soils, and groundwater. Although surface waters are not specifically mentioned, this method should be applicable to water samples from rivers, lakes, etc.

METHOD SUMMARY This method covers 259 semivolatile organic compounds. In very limited applications direct injection of the sample into the GC/MS system may be appropriate, but this results in very high detection limits (approximately 10,000 µg/L). Typically, a 1-L liquid sample, containing surrogate, and matrix spiking standards, is extracted in a continuous extractor first under acid conditions and then under basic conditions. Typically 30 g of a solid sample, containing surrogate, and matrix spiking standards, is extracted ultrasonically. After concentrating the extract to 1 mL it is spiked with 10 µL of an internal standard solution just prior to analysis by GC/MS. The volume injected should contain about 100 ng of base/neutral and 200 ng of acid surrogates (for a 1-µL injection). Analysis is performed by GC/MS using a capillary GC column.

INTERFERENCES Raw GC/MS data from all blanks, samples, and spikes must be evaluated for interferences. Contamination by carryover can occur whenever high-concentration and low-concentration samples are sequentially analyzed. To reduce carryover, the sample syringe must be rinsed out between samples with solvent. Whenever an unusually concentrated sample is encountered, it should be followed by the analysis of blank solvent to check for cross-contamination.

INSTRUMENTATION A GC/MS and a data system are required. The GC column used is a 30 m × 0.25 mm I.D. (or 0.32 mm I.D.) 1-µm film thickness silicone-coated fused silica capillary column. A continuous liquid-liquid extractor equipped with Teflon® or glass connection joints and stopcocks requiring no lubrication, a K-D concentrating apparatus, water bath, and an ultrasonic disrupter with a minimum power of 300 W and with pulsing capability are also required.

PRECISION & ACCURACY The estimated quantitation limit (EQL) of Method 8270B for determining an individual compound is approximately 1 mg/kg (wet weight) for soil or sediment samples, 1–200 mg/kg for wastes (dependent on matrix and method of preparation), and 10 µg/L for groundwater samples. EQLs will be proportionately higher for sample extracts that require dilution to avoid saturation of the detector.

The EQL(b) for groundwater in µg/L is 20.
The EQL (a, b) for low concentrations in soil and sediment in µg/kg is not determined.
Accuracy as µg/L is not listed.
Overall precision in µg/L is not listed.

(a) *EQLs listed for soil/sediment are based on wet weight. Normally data is reported in a dry-weight basis; therefore, EQLs will be higher based on the % dry weight of each sample. This calculation is based on a 30-g sample and gel permeation chromatography cleanup.*
(b) *Sample EQLs are highly matrix-dependent. The EQLs are provided for guidance and may not always be achievable.*
C = *True value for concentration, in µg/L.*
X = *Average recovery found for measurements of samples containing a concentration of C, in µg/L.*

ESTIMATED QUANTITATION LIMIT

Other Matrices	Factor (a)
High-concentration soil and sludges by sonicator	7.5
Non-water miscible waste	75

(a) EQL for other matrices = [EQL for low soil/sediment] × [Factor]. This estimated EQL is similar to an EPA "Practical Quantitation Limit."

SAMPLING METHOD

Liquid samples — Use a 1 or 2½ gallon amber glass bottle with a screw-top Teflon®-lined cover that has been prewashed with detergent and rinsed with distilled water and methanol (or isopropanol).

Soils, sediments, or sludges — Use an 8-oz. widemouth glass with a screw-top Teflon®-lined cover that has been prewashed with detergent and rinsed with distilled water and methanol (or isopropanol).

SAMPLE PRESERVATION

Liquid samples — If residual chlorine is present, add 3 mL of 10% sodium thiosulfate per gallon, cool to 4°C and store in a solvent-free refrigerator until analysis; if chlorine is not present, then eliminate the sodium thiosulfate addition.

Soils, sediments, or sludges — Cool samples to 4°C and store in a solvent-free refrigerator.

MHT Liquid samples must be extracted within 7 days and the extracts analyzed within 40 days. Soils, sediments, or sludges may be stored for a maximum of 14 days and the extracts analyzed within 40 days.

SAMPLE PREPARATION

Liquid samples — Transfer 1 L quantitatively to a continuous extractor. If high concentrations are anticipated, a smaller volume may be used and then diluted with organic-free reagent water to 1 L. Adjust pH, if necessary, to pH <2 using 1:1 (V/V) sulfuric acid. Pipette 1.0 mL of a surrogate standard spiking solution into each sample. For the sample in each analytical batch selected for spiking, add 1.0 mL of a matrix spiking standard. For base/neutral acid analysis, the amount of the surrogates and matrix spiking compounds added to the sample should result in a final concentration of 100 ng/μL of each analyte in the extract to be analyzed (assuming a 1-μL injection). Extract with methylene chloride for 18–24 h. Next, adjust the pH of the aqueous phase to pH >11 using 10 N sodium hydroxide and extract it with methylene chloride again for 18–24 h. Dry the extract through a column containing anhydrous sodium sulfate and concentrate it to 1 mL using a K-D concentrator.

Soils, sediments, or sludges — Use 30 g of sample. Nonporous or wet samples (gummy or clay type) that do not have a free-flowing sandy texture must be mixed with anhydrous sodium sulfate until the sample is free flowing. Add 1 mL of surrogate standards to all samples, spikes, standards, and blanks. For the sample in each analytical batch selected for spiking, add 1.0 mL of a matrix spiking standard. For base/neutral acid analysis, the amount added of the surrogates and matrix spiking compounds should result in a final concentration of 100 ng/μL of each base/neutral analyte and 200 ng/μL of each acid analyte in the extract to be analyzed (assuming a 1-μL injection). Immediately add a 100-mL mixture of 1:1 methylene chloride:acetone and extract the sample ultrasonically for 3 min and then decant or filter the extracts. Repeat the extraction two or more times. Dry the extract using a column with anhydrous sodium sulfate and concentrate it to 1 mL in a K-D concentrator.

QUALITY CONTROL A methylene chloride solution containing 50 ng/μL of decafluorotriphenylphosphine (DFTPP) is used for tuning the GC/MS system each 12-h shift. A system performance check also must be made during every 12-h shift. A standard containing 50 ng/μL each of 4,4′-DDT, pentachlorophenol, and benzidine is required to verify injection port inertness and GC column performance. A calibration standard at mid-concentration, containing each compound of interest, including all required surrogates, must be performed every 12 h during analysis. After the system performance check is met, calibration check compounds (CCCs) are used to check the validity of the initial calibration.

The internal standard responses and retention times in the calibration check standard must be evaluated immediately after or during data acquisition. If the retention time for any internal standard changes by more than 30 seconds from the last check calibration (12 h), the chromatographic system must be inspected for malfunctions and corrections must be made, as required. If the electron ionization current plot (EICP) area for any of the internal standards changes by a factor of two from the last daily calibration standard check, the mass spectrometer must be inspected for malfunctions and corrections must be made, as appropriate.

Demonstrate, through the analysis of a reagent water blank, that interferences from the analytical system, glassware, and reagents are under control. The blank samples should be carried through all stages of the sample preparation and measurement steps. For each analytical batch (up to 20 samples), a reagent blank, matrix spike, and matrix spike duplicate/duplicate must be analyzed (the frequency of the spikes may be different for different monitoring programs). The blank and spiked samples must be carried through all stages of the sample preparation and measurement steps. A QC reference sample concentrate containing each analyte at a concentration of 100 mg/L in methanol is required.

REFERENCE Test Methods for Evaluating Solid Waste (SW-846). U.S. EPA 1983, Method 8270B, Rev. 2, Nov. 1990. Office of Solid Waste, Washington, DC.

Aminoazobenzene **EPA Method 8270**
CAS #60-09-3

TITLE Semivolatile Organic Compounds by GC/MS

MATRIX This method is used to determine the concentration of semivolatile organic compounds in extracts prepared from all types of solid waste matrices, soils, and groundwater. Although surface waters are not specifically mentioned, this method should be applicable to water samples from rivers, lakes, etc.

METHOD SUMMARY This method covers 259 semivolatile organic compounds. In very limited applications direct injection of the sample into the GC/MS system may be appropriate, but this results in very high detection limits (approximately 10,000 µg/L). Typically, a 1-L liquid sample, containing surrogate, and matrix spiking standards, is extracted in a continuous extractor first under acid conditions and then under basic conditions. Typically 30 g of a solid sample, containing surrogate, and matrix spiking standards, is extracted ultrasonically. After concentrating the extract to 1 mL it is spiked with 10 µL of an internal standard solution just prior to analysis by GC/MS. The volume injected should contain about 100 ng of base/neutral and 200 ng of acid surrogates (for a 1-µL injection). Analysis is performed by GC/MS using a capillary GC column.

INTERFERENCES Raw GC/MS data from all blanks, samples, and spikes must be evaluated for interferences. Contamination by carryover can occur whenever high-concentration and low-concentration samples are sequentially analyzed. To reduce carryover, the sample syringe must be rinsed out between samples with solvent. Whenever an unusually concentrated sample is encountered, it should be followed by the analysis of blank solvent to check for cross-contamination.

INSTRUMENTATION A GC/MS and a data system are required. The GC column used is a 30 m × 0.25 mm I.D. (or 0.32 mm I.D.) 1-µm film thickness silicone-coated fused silica capillary column. A continuous liquid-liquid extractor equipped with Teflon® or glass connection joints and stopcocks requiring no lubrication, a K-D concentrating apparatus, water bath, and an ultrasonic disrupter with a minimum power of 300 W and with pulsing capability are also required.

PRECISION & ACCURACY The estimated quantitation limit (EQL) of Method 8270B for determining an individual compound is approximately 1 mg/kg (wet weight) for soil or sediment samples, 1–200 mg/kg for wastes (dependent on matrix and method of preparation), and 10 µg/L for groundwater samples. EQLs will be proportionately higher for sample extracts that require dilution to avoid saturation of the detector.

The EQL(b) for groundwater in µg/L is 10.
The EQL (a, b) for low concentrations in soil and sediment in µg/kg is not determined.
Accuracy as µg/L is not listed.
Overall precision in µg/L is not listed.

(a) EQLs listed for soil/sediment are based on wet weight. Normally data is reported in a dry-weight basis; therefore, EQLs will be higher based on the % dry weight of each sample. This calculation is based on a 30-g sample and gel permeation chromatography cleanup.
(b) Sample EQLs are highly matrix-dependent. The EQLs are provided for guidance and may not always be achievable.
C = *True value for concentration, in µg/L.*
X = *Average recovery found for measurements of samples containing a concentration of C, in µg/L.*

ESTIMATED QUANTITATION LIMIT

Other Matrices	Factor (a)
High-concentration soil and sludges by sonicator	7.5
Non-water miscible waste	75

(a) EQL for other matrices = [EQL for low soil/sediment] × [Factor]. This estimated EQL is similar to an EPA "Practical Quantitation Limit."

SAMPLING METHOD
Liquid samples — Use a 1 or 2½ gallon amber glass bottle with a screw-top Teflon®-lined cover that has been prewashed with detergent and rinsed with distilled water and methanol (or isopropanol).

Soils, sediments, or sludges — Use an 8-oz. widemouth glass with a screw-top Teflon®-lined cover that has been prewashed with detergent and rinsed with distilled water and methanol (or isopropanol).

SAMPLE PRESERVATION
Liquid samples — If residual chlorine is present, add 3 mL of 10% sodium thiosulfate per gallon, cool to 4°C and store in a solvent-free refrigerator until analysis; if chlorine is not present, then eliminate the sodium thiosulfate addition.

Soils, sediments, or sludges — Cool samples to 4°C and store in a solvent-free refrigerator.

MHT Liquid samples must be extracted within 7 days and the extracts analyzed within 40 days. Soils, sediments, or sludges may be stored for a maximum of 14 days and the extracts analyzed within 40 days.

SAMPLE PREPARATION
Liquid samples — Transfer 1 L quantitatively to a continuous extractor. If high concentrations are anticipated, a smaller volume may be used and then diluted with organic-free reagent water to 1 L. Adjust pH, if necessary, to pH <2 using 1:1 (V/V) sulfuric acid. Pipette 1.0 mL of a surrogate standard spiking solution into each sample. For the sample in each analytical batch selected for spiking, add 1.0 mL of a matrix spiking standard. For base/neutral acid analysis, the amount of the surrogates and matrix spiking compounds added to the sample should result in a final concentration of 100 ng/µL of each analyte in the extract to be analyzed (assuming a 1-µL injection). Extract with methylene chloride for 18–24 h. Next, adjust the pH of the aqueous phase to pH >11 using 10 N sodium hydroxide and extract it with methylene chloride again for 18–24 h. Dry the extract through a column containing anhydrous sodium sulfate and concentrate it to 1 mL using a K-D concentrator.

Soils, sediments, or sludges — Use 30 g of sample. Nonporous or wet samples (gummy or clay type) that do not have a free-flowing sandy texture must be mixed with anhydrous sodium sulfate until the sample is free flowing. Add 1 mL of surrogate standards to all samples, spikes, standards, and blanks. For the sample in each analytical batch selected for spiking, add 1.0 mL of a matrix spiking standard. For base/neutral acid analysis, the amount added of the surrogates and matrix spiking compounds

should result in a final concentration of 100 ng/µL of each base/neutral analyte and 200 ng/µL of each acid analyte in the extract to be analyzed (assuming a 1-µL injection). Immediately add a 100-mL mixture of 1:1 methylene chloride:acetone and extract the sample ultrasonically for 3 min and then decant or filter the extracts. Repeat the extraction two or more times. Dry the extract using a column with anhydrous sodium sulfate and concentrate it to 1 mL in a K-D concentrator.

QUALITY CONTROL A methylene chloride solution containing 50 ng/µL of decafluorotriphenylphosphine (DFTPP) is used for tuning the GC/MS system each 12-h shift. A system performance check also must be made during every 12-h shift. A standard containing 50 ng/µL each of 4,4'-DDT, pentachlorophenol, and benzidine is required to verify injection port inertness and GC column performance. A calibration standard at mid-concentration, containing each compound of interest, including all required surrogates, must be performed every 12 h during analysis. After the system performance check is met, calibration check compounds (CCCs) are used to check the validity of the initial calibration.

The internal standard responses and retention times in the calibration check standard must be evaluated immediately after or during data acquisition. If the retention time for any internal standard changes by more than 30 seconds from the last check calibration (12 h), the chromatographic system must be inspected for malfunctions and corrections must be made, as required. If the electron ionization current plot (EICP) area for any of the internal standards changes by a factor of two from the last daily calibration standard check, the mass spectrometer must be inspected for malfunctions and corrections must be made, as appropriate.

Demonstrate, through the analysis of a reagent water blank, that interferences from the analytical system, glassware, and reagents are under control. The blank samples should be carried through all stages of the sample preparation and measurement steps. For each analytical batch (up to 20 samples), a reagent blank, matrix spike, and matrix spike duplicate/duplicate must be analyzed (the frequency of the spikes may be different for different monitoring programs). The blank and spiked samples must be carried through all stages of the sample preparation and measurement steps. A QC reference sample concentrate containing each analyte at a concentration of 100 mg/L in methanol is required.

REFERENCE Test Methods for Evaluating Solid Waste (SW-846). U.S. EPA 1983, Method 8270B, Rev. 2, Nov. 1990. Office of Solid Waste, Washington, DC.

4-Aminobiphenyl **EPA Method 1625**
CAS #92-67-1

TITLE Semivolatile Organic Compounds by Isotope Dilution GC/MS

MATRIX The compounds may be determined in waters, soils, and municipal sludges by this method.

METHOD SUMMARY This method is used to determine 176 semivolatile toxic organic pollutants associated with the CWA (as amended 1987); the RCRA (as amended 1986); the CERCLA (as amended 1986); and other compounds amenable to extraction and analysis by capillary column gas chromatography-mass spectrometry (GC/MS).

Stable isotopically-labeled analogs of the compounds of interest are added to the sample. If the solids content is less than 1%, a 1-L sample is extracted at pH 12–13, then at pH <2 with methylene chloride using continuous extraction techniques.

If the solids content is 30% or less, the sample is diluted to 1% solids with reagent water, homogenized ultrasonically, and extracted at pH 12–13, then at pH <2 with methylene chloride using continuous extraction techniques. If the solids content is greater than 30%, the sample is extracted using ultrasonic techniques.

Each extract is dried over sodium sulfate, concentrated to a volume of 5 mL, cleaned up using GPC, if necessary, and concentrated. Extracts are concentrated to 1 mL if GPC is not performed, and to 0.5 mL if GPC is performed.

An internal standard is added to the extract, and a 1-mL aliquot of the extract is injected into the GC. The compounds are separated by GC and detected by a MS. The labeled compounds serve to correct the variability of the analytical technique.

INTERFERENCES Solvents, reagents, glassware, and other sample processing hardware may yield artifacts and/or elevated baselines causing misinterpretation of chromatograms and spectra. Materials used in the analysis must be demonstrated to be free from interferences under the conditions of analysis by running method blanks initially and with each sample lot (sample started through the extraction process on a given 8-h shift, to a maximum of 20). Specific selection of reagents and purification of solvents by distillation in all glass systems may be required. Glassware and, where possible, reagents are cleaned by solvent rinse and baking at 450°C for 1-h minimum. Interferences coextracted from samples will vary considerably from source to source, depending on the diversity of the site being sampled.

INSTRUMENTATION Major instrumentation includes a GC with a splitless or on-column injection port for capillary column, a MS with 70 eV electron impact ionization, and a data system to collect and record MS data, and process it. A K-D apparatus is used to concentrate extracts.

GC Column: 30 m × 0.25 mm I.D. 5% phenyl, 94% methyl, 1% vinyl silicone bonded phased fused silica capillary column.

PRECISION & ACCURACY The detection limits of the method are usually dependent on the level of interferences rather than instrumental limitations. The limits typify the minimum quantities that can be detected with no interferences present.

The minimum level (in µg/mL) was not listed. This is defined as a minimum level at which the analytical system shall give recognizable mass spectra (background corrected) and acceptable calibration points.

The MDL (in µg/kg) in low solids was not listed and in high solids was not listed; these were determined in digested sludge (low solids) and in filter cake or compost (high solids).

The labeled and native compound initial precision as standard deviation (in µg/L) was not listed.

The labeled and native compound initial accuracy as average recovery (in µg/L) was not listed.

SAMPLE COLLECTION, PRESERVATION & HANDLING
Collect samples in glass containers. Aqueous samples which flow freely are collected in refrigerated bottles using automatic sampling equipment. Solid samples are collected as grab samples using widemouth jars. Maintain samples at 0 to 4°C from the time of collection until extraction. If residual chlorine is present in aqueous samples, add 80 mg sodium thiosulfate/L of water. Begin sample extraction within 7 days of collection, and analyze all extracts within 40 days of extraction.

SAMPLE PREPARATION Samples containing 1% solids or less are extracted directly using continuous liquid-liquid extraction techniques. Samples containing 1 to 30% solids are diluted to the 1% level with reagent water and extracted using continuous liquid-liquid extraction techniques. Samples containing greater than 30% solids are extracted using ultrasonic techniques.

Base/neutral extraction — Adjust the pH of the waters in the extractors to 12–13 with 6 N NaOH. Extract with methylene chloride for 24–48 h.

Acid extraction — Adjust the pH of the waters in the extractors to 2 or less using 6 N sulfuric acid. Extract with methylene chloride for 24–48 h.

Ultrasonic extraction of high solids samples — Add anhydrous sodium sulfate to the sample and QC aliquot(s). Add acetone:methylene chloride (1:1) to the sample and mix thoroughly

Concentrate extracts using a K-D apparatus.

QUALITY CONTROL The analyst is permitted to modify this method to improve separations or lower the costs of measurements, provided all performance specifications are met. Analyses of blanks are required to demonstrate freedom from contamination. When results of spikes indicate atypical method performance for samples, the samples are diluted to bring method performance within acceptable limits.

For low solids (aqueous samples), extract, concentrate, and analyze two sets of four 1-L aliquots (8 aliquots total) of the precision and recovery standard. For high solids samples, two sets of four 30-g aliquots of the high solids reference matrix are used.

Spike all samples with labeled compounds to assess method performance. Compute percent recovery of the labeled compounds using the internal standard method. Compare the labeled compound recovery for each compound with the corresponding labeled compound recovery.

Reagent water and high solids reference matrix blanks are analyzed to demonstrate freedom from contamination. Extract and concentrate a 1-L reagent water blank or a high solids reference matrix blank with each sample's lot (samples started through the extraction process on the same 8-h shift, to a maximum of 20 samples).

Field replicates may be collected to determine the precision of the sampling technique, and spiked samples may be required to determine the accuracy of the analysis when the internal standard method is used.

REFERENCE Semivolatile Organic Compounds by Isotope Dilution GC/MS. Office of Water Regulation and Standards, U.S. EPA Industrial Technology Division, Washington, DC, EPA Method 1625, Rev. C, June 1989 (contact W.A. Telliard, U.S. EPA, Office of Water Regulations and Standards, 401 M St., SW, Washington, DC, 20460. Phone: 202-382-7131).

4-Aminobiphenyl   **EPA Method 8270**
CAS #92-67-1

TITLE Semivolatile Organic Compounds by GC/MS

MATRIX This method is used to determine the concentration of semivolatile organic compounds in extracts prepared from all types of solid waste matrices, soils, and groundwater. Although surface waters are not specifically mentioned, this method should be applicable to water samples from rivers, lakes, etc.

METHOD SUMMARY This method covers 259 semivolatile organic compounds. In very limited applications direct injection of the sample into the GC/MS system may be appropriate, but this results in very high detection limits (approximately 10,000 µg/L). Typically, a 1-L liquid sample, containing surrogate, and matrix spiking standards, is extracted in a continuous extractor first under acid conditions and then under basic conditions. Typically 30 g of a solid sample, containing surrogate, and matrix spiking standards, is extracted ultrasonically. After concentrating the extract to 1 mL it is spiked with 10 µL of an internal standard solution just prior to analysis by GC/MS. The volume injected should contain about 100 ng of base/neutral and 200 ng of acid surrogates (for a 1-µL injection). Analysis is performed by GC/MS using a capillary GC column.

INTERFERENCES Raw GC/MS data from all blanks, samples, and spikes must be evaluated for interferences. Contamination by carryover can occur whenever high-concentration and low-concentration samples are sequentially analyzed. To reduce carryover, the sample syringe must be rinsed out between samples with solvent. Whenever an unusually concentrated sample is encountered, it should be followed by the analysis of blank solvent to check for cross-contamination.

INSTRUMENTATION A GC/MS and a data system are required. The GC column used is a 30 m × 0.25 mm I.D. (or 0.32 mm I.D.) 1-µm film thickness silicone-coated fused silica capillary column. A continuous liquid-liquid extractor equipped with Teflon® or glass connection joints and stopcocks requiring no lubrication, a K-D concentrating apparatus, water bath, and an ultrasonic disrupter with a minimum power of 300 W and with pulsing capability are also required.

PRECISION & ACCURACY The estimated quantitation limit (EQL) of Method 8270B for determining an individual compound is approximately 1 mg/kg (wet weight) for soil or sediment samples, 1–200 mg/kg for wastes (dependent on matrix and method of preparation), and 10 µg/L for groundwater samples. EQLs will be proportionately higher for sample extracts that require dilution to avoid saturation of the detector.

The EQL(b) for groundwater in µg/L is 20.
The EQL (a, b) for low concentrations in soil and sediment in µg/kg is not determined.
Accuracy as µg/L is not listed.
Overall precision in µg/L is not listed.

(a) *EQLs listed for soil/sediment are based on wet weight. Normally data is reported in a dry-weight basis; therefore, EQLs will be higher based on the % dry weight of each sample. This calculation is based on a 30-g sample and gel permeation chromatography cleanup.*
(b) *Sample EQLs are highly matrix-dependent. The EQLs are provided for guidance and may not always be achievable.*
C = *True value for concentration, in µg/L.*
X = *Average recovery found for measurements of samples containing a concentration of C, in µg/L.*

ESTIMATED QUANTITATION LIMIT

Other Matrices	Factor (a)
High-concentration soil and sludges by sonicator	7.5
Non-water miscible waste	75

(a) *EQL for other matrices = [EQL for low soil/sediment] × [Factor]. This estimated EQL is similar to an EPA "Practical Quantitation Limit."*

SAMPLING METHOD
Liquid samples — Use a 1 or 2½ gallon amber glass bottle with a screw-top Teflon®-lined cover that has been prewashed with detergent and rinsed with distilled water and methanol (or isopropanol).

Soils, sediments, or sludges — Use an 8-oz. widemouth glass with a screw-top Teflon®-lined cover that has been prewashed with detergent and rinsed with distilled water and methanol (or isopropanol).

SAMPLE PRESERVATION
Liquid samples — If residual chlorine is present, add 3 mL of 10% sodium thiosulfate per gallon, cool to 4°C and store in a solvent-free refrigerator until analysis; if chlorine is not present, then eliminate the sodium thiosulfate addition.

Soils, sediments, or sludges — Cool samples to 4°C and store in a solvent-free refrigerator.

MHT Liquid samples must be extracted within 7 days and the extracts analyzed within 40 days. Soils, sediments, or sludges may be stored for a maximum of 14 days and the extracts analyzed within 40 days.

SAMPLE PREPARATION
Liquid samples — Transfer 1 L quantitatively to a continuous extractor. If high concentrations are anticipated, a smaller volume may be used and then diluted with organic-free reagent water to 1 L. Adjust pH, if necessary, to pH <2 using 1:1 (V/V) sulfuric acid. Pipette 1.0 mL of a surrogate standard spiking solution into each sample. For the sample in each analytical batch selected for spiking, add 1.0 mL of a matrix spiking standard. For base/neutral acid analysis, the amount of the surrogates and matrix spiking compounds added to the sample should result in a final concentration of 100 ng/µL of each analyte in the extract to be analyzed (assuming a 1-µL injection). Extract with methylene chloride for 18–24 h. Next, adjust the pH of the aqueous phase to pH >11 using 10 N sodium hydroxide and extract it with methylene chloride again for 18–24 h. Dry the extract through a column containing anhydrous sodium sulfate and concentrate it to 1 mL using a K-D concentrator.

Soils, sediments, or sludges — Use 30 g of sample. Nonporous or wet samples (gummy or clay type) that do not have a free-flowing sandy texture must be mixed with anhydrous sodium sulfate until the sample is free flowing. Add 1 mL of surrogate standards to all samples, spikes, standards, and blanks. For the sample in each analytical batch selected for spiking, add 1.0 mL of a matrix spiking standard. For base/neutral acid analysis, the amount added of the surrogates and matrix spiking compounds should result in a final concentration of 100 ng/µL of each base/neutral analyte and 200 ng/µL of each acid analyte in the extract to be analyzed (assuming a 1-µL injection). Immediately add a 100-mL mixture of 1:1 methylene chloride:acetone and extract the sample ultrasonically for 3 min and then decant or filter the extracts. Repeat the extraction two or more times. Dry the extract using a column with anhydrous sodium sulfate and concentrate it to 1 mL in a K-D concentrator.

QUALITY CONTROL A methylene chloride solution containing 50 ng/µL of decafluorotriphenylphosphine (DFTPP) is used for tuning the GC/MS system each 12-h shift. A system performance check also must be made during every 12-h shift. A standard containing 50 ng/µL each of 4,4'-DDT, pentachlorophenol, and benzidine is required to verify injection port inertness and GC column performance. A calibration standard at mid-concentration, containing each compound of interest, including all required surrogates, must be performed every 12 h during analysis. After the system performance check is met, calibration check compounds (CCCs) are used to check the validity of the initial calibration.

The internal standard responses and retention times in the calibration check standard must be evaluated immediately after or during data acquisition. If the retention time for any internal standard changes by more than 30 seconds from the last check calibration (12 h), the chromatographic system must be inspected for malfunctions and corrections must be made, as required. If the electron ionization current plot (EICP) area for any of the internal standards changes by a factor of two from the last daily calibration standard check, the mass spectrometer must be inspected for malfunctions and corrections must be made, as appropriate.

Demonstrate, through the analysis of a reagent water blank, that interferences from the analytical system, glassware, and reagents are under control. The blank samples should be carried through all stages of the sample preparation and measurement

steps. For each analytical batch (up to 20 samples), a reagent blank, matrix spike, and matrix spike duplicate/duplicate must be analyzed (the frequency of the spikes may be different for different monitoring programs). The blank and spiked samples must be carried through all stages of the sample preparation and measurement steps. A QC reference sample concentrate containing each analyte at a concentration of 100 mg/L in methanol is required.

REFERENCE Test Methods for Evaluating Solid Waste (SW-846). U.S. EPA 1983, Method 8270B, Rev. 2, Nov. 1990. Office of Solid Waste, Washington, DC.

Anilazine   EPA Method 8270
CAS #101-05-3

TITLE Semivolatile Organic Compounds by GC/MS

MATRIX This method is used to determine the concentration of semivolatile organic compounds in extracts prepared from all types of solid waste matrices, soils, and groundwater. Although surface waters are not specifically mentioned, this method should be applicable to water samples from rivers, lakes, etc.

METHOD SUMMARY This method covers 259 semivolatile organic compounds. In very limited applications direct injection of the sample into the GC/MS system may be appropriate, but this results in very high detection limits (approximately 10,000 µg/L). Typically, a 1-L liquid sample, containing surrogate, and matrix spiking standards, is extracted in a continuous extractor first under acid conditions and then under basic conditions. Typically 30 g of a solid sample, containing surrogate, and matrix spiking standards, is extracted ultrasonically. After concentrating the extract to 1 mL it is spiked with 10 µL of an internal standard solution just prior to analysis by GC/MS. The volume injected should contain about 100 ng of base/neutral and 200 ng of acid surrogates (for a 1-µL injection). Analysis is performed by GC/MS using a capillary GC column.

INTERFERENCES Raw GC/MS data from all blanks, samples, and spikes must be evaluated for interferences. Contamination by carryover can occur whenever high-concentration and low-concentration samples are sequentially analyzed. To reduce carryover, the sample syringe must be rinsed out between samples with solvent. Whenever an unusually concentrated sample is encountered, it should be followed by the analysis of blank solvent to check for cross-contamination.

INSTRUMENTATION A GC/MS and a data system are required. The GC column used is a 30 m × 0.25 mm I.D. (or 0.32 mm I.D.) 1-µm film thickness silicone-coated fused silica capillary column. A continuous liquid-liquid extractor equipped with Teflon® or glass connection joints and stopcocks requiring no lubrication, a K-D concentrating apparatus, water bath, and an ultrasonic disrupter with a minimum power of 300 W and with pulsing capability are also required.

PRECISION & ACCURACY The estimated quantitation limit (EQL) of Method 8270B for determining an individual compound is approximately 1 mg/kg (wet weight) for soil or sediment samples, 1–200 mg/kg for wastes (dependent on matrix and method of preparation), and 10 µg/L for groundwater samples. EQLs will be proportionately higher for sample extracts that require dilution to avoid saturation of the detector.

The EQL(b) for groundwater in µg/L is 100.
The EQL (a, b) for low concentrations in soil and sediment in µg/kg is not determined.
Accuracy as µg/L is not listed.
Overall precision in µg/L is not listed.

(a) *EQLs listed for soil/sediment are based on wet weight. Normally data is reported in a dry-weight basis; therefore, EQLs will be higher based on the % dry weight of each sample. This calculation is based on a 30-g sample and gel permeation chromatography cleanup.*
(b) *Sample EQLs are highly matrix-dependent. The EQLs are provided for guidance and may not always be achievable.*
C = *True value for concentration, in µg/L.*
X = *Average recovery found for measurements of samples containing a concentration of C, in µg/L.*

ESTIMATED QUANTITATION LIMIT

Other Matrices	Factor (a)
High-concentration soil and sludges by sonicator	7.5
Non-water miscible waste	75

(a) *EQL for other matrices = [EQL for low soil/sediment] × [Factor]. This estimated EQL is similar to an EPA "Practical Quantitation Limit."*

SAMPLING METHOD

Liquid samples — Use a 1 or 2½ gallon amber glass bottle with a screw-top Teflon®-lined cover that has been prewashed with detergent and rinsed with distilled water and methanol (or isopropanol).

Soils, sediments, or sludges — Use an 8-oz. widemouth glass with a screw-top Teflon®-lined cover that has been prewashed with detergent and rinsed with distilled water and methanol (or isopropanol).

SAMPLE PRESERVATION

Liquid samples — If residual chlorine is present, add 3 mL of 10% sodium thiosulfate per gallon, cool to 4°C and store in a solvent-free refrigerator until analysis; if chlorine is not present, then eliminate the sodium thiosulfate addition.

Soils, sediments, or sludges — Cool samples to 4°C and store in a solvent-free refrigerator.

MHT Liquid samples must be extracted within 7 days and the extracts analyzed within 40 days. Soils, sediments, or sludges may be stored for a maximum of 14 days and the extracts analyzed within 40 days.

SAMPLE PREPARATION

Liquid samples — Transfer 1 L quantitatively to a continuous extractor. If high concentrations are anticipated, a smaller volume may be used and then diluted with organic-free reagent water to 1 L. Adjust pH, if necessary, to pH <2 using 1:1 (V/V) sulfuric acid. Pipette 1.0 mL of a surrogate standard spiking solution into each sample. For the sample in each analytical

batch selected for spiking, add 1.0 mL of a matrix spiking standard. For base/neutral acid analysis, the amount of the surrogates and matrix spiking compounds added to the sample should result in a final concentration of 100 ng/µL of each analyte in the extract to be analyzed (assuming a 1-µL injection). Extract with methylene chloride for 18–24 h. Next, adjust the pH of the aqueous phase to pH >11 using 10 N sodium hydroxide and extract it with methylene chloride again for 18–24 h. Dry the extract through a column containing anhydrous sodium sulfate and concentrate it to 1 mL using a K-D concentrator.

Soils, sediments, or sludges — Use 30 g of sample. Nonporous or wet samples (gummy or clay type) that do not have a free-flowing sandy texture must be mixed with anhydrous sodium sulfate until the sample is free flowing. Add 1 mL of surrogate standards to all samples, spikes, standards, and blanks. For the sample in each analytical batch selected for spiking, add 1.0 mL of a matrix spiking standard. For base/neutral acid analysis, the amount added of the surrogates and matrix spiking compounds should result in a final concentration of 100 ng/µL of each base/neutral analyte and 200 ng/µL of each acid analyte in the extract to be analyzed (assuming a 1-µL injection). Immediately add a 100-mL mixture of 1:1 methylene chloride:acetone and extract the sample ultrasonically for 3 min and then decant or filter the extracts. Repeat the extraction two or more times. Dry the extract using a column with anhydrous sodium sulfate and concentrate it to 1 mL in a K-D concentrator.

QUALITY CONTROL A methylene chloride solution containing 50 ng/µL of decafluorotriphenylphosphine (DFTPP) is used for tuning the GC/MS system each 12-h shift. A system performance check also must be made during every 12-h shift. A standard containing 50 ng/µL each of 4,4′-DDT, pentachlorophenol, and benzidine is required to verify injection port inertness and GC column performance. A calibration standard at mid-concentration, containing each compound of interest, including all required surrogates, must be performed every 12 h during analysis. After the system performance check is met, calibration check compounds (CCCs) are used to check the validity of the initial calibration.

The internal standard responses and retention times in the calibration check standard must be evaluated immediately after or during data acquisition. If the retention time for any internal standard changes by more than 30 seconds from the last check calibration (12 h), the chromatographic system must be inspected for malfunctions and corrections must be made, as required. If the electron ionization current plot (EICP) area for any of the internal standards changes by a factor of two from the last daily calibration standard check, the mass spectrometer must be inspected for malfunctions and corrections must be made, as appropriate.

Demonstrate, through the analysis of a reagent water blank, that interferences from the analytical system, glassware, and reagents are under control. The blank samples should be carried through all stages of the sample preparation and measurement steps. For each analytical batch (up to 20 samples), a reagent blank, matrix spike, and matrix spike duplicate/duplicate must be analyzed (the frequency of the spikes may be different for different monitoring programs). The blank and spiked samples must be carried through all stages of the sample preparation and measurement steps. A QC reference sample concentrate containing each analyte at a concentration of 100 mg/L in methanol is required.

REFERENCE Test Methods for Evaluating Solid Waste (SW-846). U.S. EPA 1983, Method 8270B, Rev. 2, Nov. 1990. Office of Solid Waste, Washington, DC.

Aniline — EPA Method 1625
CAS #62-53-3

TITLE Semivolatile Organic Compounds by Isotope Dilution GC/MS

MATRIX The compounds may be determined in waters, soils, and municipal sludges by this method.

METHOD SUMMARY This method is used to determine 176 semivolatile toxic organic pollutants associated with the CWA (as amended 1987); the RCRA (as amended 1986); the CERCLA (as amended 1986); and other compounds amenable to extraction and analysis by capillary column gas chromatography-mass spectrometry (GC/MS).

Stable isotopically-labeled analogs of the compounds of interest are added to the sample. If the solids content is less than 1%, a 1-L sample is extracted at pH 12–13, then at pH <2 with methylene chloride using continuous extraction techniques.

If the solids content is 30% or less, the sample is diluted to 1% solids with reagent water, homogenized ultrasonically, and extracted at pH 12–13, then at pH <2 with methylene chloride using continuous extraction techniques. If the solids content is greater than 30%, the sample is extracted using ultrasonic techniques.

Each extract is dried over sodium sulfate, concentrated to a volume of 5 mL, cleaned up using GPC, if necessary, and concentrated. Extracts are concentrated to 1 mL if GPC is not performed, and to 0.5 mL if GPC is performed.

An internal standard is added to the extract, and a 1-mL aliquot of the extract is injected into the GC. The compounds are separated by GC and detected by a MS. The labeled compounds serve to correct the variability of the analytical technique.

INTERFERENCES Solvents, reagents, glassware, and other sample processing hardware may yield artifacts and/or elevated baselines causing misinterpretation of chromatograms and spectra. Materials used in the analysis must be demonstrated to be free from interferences under the conditions of analysis by running method blanks initially and with each sample lot (sample started through the extraction process on a given 8-h shift, to a maximum of 20). Specific selection of reagents and purification of solvents by distillation in all glass systems may be required. Glassware and, where possible, reagents are cleaned by solvent rinse and baking at 450°C for 1-h minimum. Interferences coextracted from samples will vary considerably from source to source, depending on the diversity of the site being sampled.

INSTRUMENTATION Major instrumentation includes a GC with a splitless or on-column injection port for capillary column, a MS with 70 eV electron impact ionization, and a data system to collect and record MS data, and process it. A K-D apparatus is used to concentrate extracts.

GC Column: 30 m × 0.25 mm I.D. 5% phenyl, 94% methyl, 1% vinyl silicone bonded phased fused silica capillary column.

PRECISION & ACCURACY The detection limits of the method are usually dependent on the level of interferences rather than instrumental limitations. The limits typify the minimum quantities that can be detected with no interferences present.

The minimum level (in µg/mL) was not listed. This is defined as a minimum level at which the analytical system shall give recognizable mass spectra (background corrected) and acceptable calibration points.

The MDL (in µg/kg) in low solids was not listed and in high solids was not listed; these were determined in digested sludge (low solids) and in filter cake or compost (high solids).

The labeled and native compound initial precision as standard deviation (in µg/L) was not listed.

The labeled and native compound initial accuracy as average recovery (in µg/L) was not listed.

SAMPLE COLLECTION, PRESERVATION & HANDLING Collect samples in glass containers. Aqueous samples which flow freely are collected in refrigerated bottles using automatic sampling equipment. Solid samples are collected as grab samples using widemouth jars. Maintain samples at 0 to 4°C from the time of collection until extraction. If residual chlorine is present in aqueous samples, add 80 mg sodium thiosulfate/L of water. Begin sample extraction within 7 days of collection, and analyze all extracts within 40 days of extraction.

SAMPLE PREPARATION Samples containing 1% solids or less are extracted directly using continuous liquid-liquid extraction techniques. Samples containing 1 to 30% solids are diluted to the 1% level with reagent water and extracted using continuous liquid-liquid extraction techniques. Samples containing greater than 30% solids are extracted using ultrasonic techniques.

Base/neutral extraction — Adjust the pH of the waters in the extractors to 12–13 with 6 N NaOH. Extract with methylene chloride for 24–48 h.

Acid extraction — Adjust the pH of the waters in the extractors to 2 or less using 6 N sulfuric acid. Extract with methylene chloride for 24–48 h.

Ultrasonic extraction of high solids samples — Add anhydrous sodium sulfate to the sample and QC aliquot(s). Add acetone:methylene chloride (1:1) to the sample and mix thoroughly

Concentrate extracts using a K-D apparatus.

QUALITY CONTROL The analyst is permitted to modify this method to improve separations or lower the costs of measurements, provided all performance specifications are met. Analyses of blanks are required to demonstrate freedom from contamination. When results of spikes indicate atypical method performance for samples, the samples are diluted to bring method performance within acceptable limits.

For low solids (aqueous samples), extract, concentrate, and analyze two sets of four 1-L aliquots (8 aliquots total) of the precision and recovery standard. For high solids samples, two sets of four 30-g aliquots of the high solids reference matrix are used.

Spike all samples with labeled compounds to assess method performance. Compute percent recovery of the labeled compounds using the internal standard method. Compare the labeled compound recovery for each compound with the corresponding labeled compound recovery.

Reagent water and high solids reference matrix blanks are analyzed to demonstrate freedom from contamination. Extract and concentrate a 1-L reagent water blank or a high solids reference matrix blank with each sample's lot (samples started through the extraction process on the same 8-h shift, to a maximum of 20 samples).

Field replicates may be collected to determine the precision of the sampling technique, and spiked samples may be required to determine the accuracy of the analysis when the internal standard method is used.

REFERENCE Semivolatile Organic Compounds by Isotope Dilution GC/MS. Office of Water Regulation and Standards, U.S. EPA Industrial Technology Division, Washington, DC, EPA Method 1625, Rev. C, June 1989 (contact W.A. Telliard, U.S. EPA, Office of Water Regulations and Standards, 401 M St., SW, Washington, DC, 20460. Phone: 202-382-7131).

Aniline   **EPA Method 8270**
CAS #62-53-3

TITLE Semivolatile Organic Compounds by GC/MS

MATRIX This method is used to determine the concentration of semivolatile organic compounds in extracts prepared from all types of solid waste matrices, soils, and groundwater. Although surface waters are not specifically mentioned, this method should be applicable to water samples from rivers, lakes, etc.

METHOD SUMMARY This method covers 259 semivolatile organic compounds. In very limited applications direct injection of the sample into the GC/MS system may be appropriate, but this results in very high detection limits (approximately 10,000 µg/L). Typically, a 1-L liquid sample, containing surrogate, and matrix spiking standards, is extracted in a continuous extractor first under acid conditions and then under basic conditions. Typically 30 g of a solid sample, containing surrogate, and matrix spiking standards, is extracted ultrasonically. After concentrating the extract to 1 mL it is spiked with 10 µL of an internal standard solution just prior to analysis by GC/MS. The volume injected should contain about 100 ng of base/neutral

and 200 ng of acid surrogates (for a 1-μL injection). Analysis is performed by GC/MS using a capillary GC column.

INTERFERENCES Raw GC/MS data from all blanks, samples, and spikes must be evaluated for interferences. Contamination by carryover can occur whenever high-concentration and low-concentration samples are sequentially analyzed. To reduce carryover, the sample syringe must be rinsed out between samples with solvent. Whenever an unusually concentrated sample is encountered, it should be followed by the analysis of blank solvent to check for cross-contamination.

INSTRUMENTATION A GC/MS and a data system are required. The GC column used is a 30 m × 0.25 mm I.D. (or 0.32 mm I.D.) 1-μm film thickness silicone-coated fused silica capillary column. A continuous liquid-liquid extractor equipped with Teflon® or glass connection joints and stopcocks requiring no lubrication, a K-D concentrating apparatus, water bath, and an ultrasonic disrupter with a minimum power of 300 W and with pulsing capability are also required.

PRECISION & ACCURACY The estimated quantitation limit (EQL) of Method 8270B for determining an individual compound is approximately 1 mg/kg (wet weight) for soil or sediment samples, 1–200 mg/kg for wastes (dependent on matrix and method of preparation), and 10 μg/L for groundwater samples. EQLs will be proportionately higher for sample extracts that require dilution to avoid saturation of the detector.

The EQL(b) for groundwater in μg/L is not listed.
The EQL (a, b) for low concentrations in soil and sediment in μg/kg is not listed.
Accuracy as μg/L is not listed.
Overall precision in μg/L is not listed.

(a) *EQLs listed for soil/sediment are based on wet weight. Normally data is reported in a dry-weight basis; therefore, EQLs will be higher based on the % dry weight of each sample. This calculation is based on a 30-g sample and gel permeation chromatography cleanup.*
(b) *Sample EQLs are highly matrix-dependent. The EQLs are provided for guidance and may not always be achievable.*
C = *True value for concentration, in μg/L.*
X = *Average recovery found for measurements of samples containing a concentration of C, in μg/L.*

ESTIMATED QUANTITATION LIMIT

Other Matrices	Factor (a)
High-concentration soil and sludges by sonicator	7.5
Non-water miscible waste	75

(a) *EQL for other matrices = [EQL for low soil/sediment] × [Factor]. This estimated EQL is similar to an EPA "Practical Quantitation Limit."*

SAMPLING METHOD
Liquid samples — Use a 1 or 2½ gallon amber glass bottle with a screw-top Teflon®-lined cover that has been prewashed with detergent and rinsed with distilled water and methanol (or isopropanol).

Soils, sediments, or sludges — Use an 8-oz. widemouth glass with a screw-top Teflon®-lined cover that has been prewashed with detergent and rinsed with distilled water and methanol (or isopropanol).

SAMPLE PRESERVATION
Liquid samples — If residual chlorine is present, add 3 mL of 10% sodium thiosulfate per gallon, cool to 4°C and store in a solvent-free refrigerator until analysis; if chlorine is not present, then eliminate the sodium thiosulfate addition.

Soils, sediments, or sludges — Cool samples to 4°C and store in a solvent-free refrigerator.

MHT Liquid samples must be extracted within 7 days and the extracts analyzed within 40 days. Soils, sediments, or sludges may be stored for a maximum of 14 days and the extracts analyzed within 40 days.

SAMPLE PREPARATION
Liquid samples — Transfer 1 L quantitatively to a continuous extractor. If high concentrations are anticipated, a smaller volume may be used and then diluted with organic-free reagent water to 1 L. Adjust pH, if necessary, to pH <2 using 1:1 (V/V) sulfuric acid. Pipette 1.0 mL of a surrogate standard spiking solution into each sample. For the sample in each analytical batch selected for spiking, add 1.0 mL of a matrix spiking standard. For base/neutral acid analysis, the amount of the surrogates and matrix spiking compounds added to the sample should result in a final concentration of 100 ng/μL of each analyte in the extract to be analyzed (assuming a 1-μL injection). Extract with methylene chloride for 18–24 h. Next, adjust the pH of the aqueous phase to pH >11 using 10 N sodium hydroxide and extract it with methylene chloride again for 18–24 h. Dry the extract through a column containing anhydrous sodium sulfate and concentrate it to 1 mL using a K-D concentrator.

Soils, sediments, or sludges — Use 30 g of sample. Nonporous or wet samples (gummy or clay type) that do not have a free-flowing sandy texture must be mixed with anhydrous sodium sulfate until the sample is free flowing. Add 1 mL of surrogate standards to all samples, spikes, standards, and blanks. For the sample in each analytical batch selected for spiking, add 1.0 mL of a matrix spiking standard. For base/neutral acid analysis, the amount added of the surrogates and matrix spiking compounds should result in a final concentration of 100 ng/μL of each base/neutral analyte and 200 ng/μL of each acid analyte in the extract to be analyzed (assuming a 1-μL injection). Immediately add a 100-mL mixture of 1:1 methylene chloride:acetone and extract the sample ultrasonically for 3 min and then decant or filter the extracts. Repeat the extraction two or more times. Dry the extract using a column with anhydrous sodium sulfate and concentrate it to 1 mL in a K-D concentrator.

QUALITY CONTROL A methylene chloride solution containing 50 ng/μL of decafluorotriphenylphosphine (DFTPP) is used for tuning the GC/MS system each 12-h shift. A system performance check also must be made during every 12-h shift. A standard containing 50 ng/μL each of 4,4′-DDT, pentachlorophenol, and benzidine is required to verify injection port inertness and GC column performance. A calibration standard at mid-concentration, containing each compound of interest, including all required surrogates, must be performed every 12 h

during analysis. After the system performance check is met, calibration check compounds (CCCs) are used to check the validity of the initial calibration.

The internal standard responses and retention times in the calibration check standard must be evaluated immediately after or during data acquisition. If the retention time for any internal standard changes by more than 30 seconds from the last check calibration (12 h), the chromatographic system must be inspected for malfunctions and corrections must be made, as required. If the electron ionization current plot (EICP) area for any of the internal standards changes by a factor of two from the last daily calibration standard check, the mass spectrometer must be inspected for malfunctions and corrections must be made, as appropriate.

Demonstrate, through the analysis of a reagent water blank, that interferences from the analytical system, glassware, and reagents are under control. The blank samples should be carried through all stages of the sample preparation and measurement steps. For each analytical batch (up to 20 samples), a reagent blank, matrix spike, and matrix spike duplicate/duplicate must be analyzed (the frequency of the spikes may be different for different monitoring programs). The blank and spiked samples must be carried through all stages of the sample preparation and measurement steps. A QC reference sample concentrate containing each analyte at a concentration of 100 mg/L in methanol is required.

REFERENCE Test Methods for Evaluating Solid Waste (SW-846). U.S. EPA 1983, Method 8270B, Rev. 2, Nov. 1990. Office of Solid Waste, Washington, DC.

---

o-Anisidine **EPA Method 1625**
CAS #90-04-0

TITLE Semivolatile Organic Compounds by Isotope Dilution GC/MS

MATRIX The compounds may be determined in waters, soils, and municipal sludges by this method.

METHOD SUMMARY This method is used to determine 176 semivolatile toxic organic pollutants associated with the CWA (as amended 1987); the RCRA (as amended 1986); the CERCLA (as amended 1986); and other compounds amenable to extraction and analysis by capillary column gas chromatography-mass spectrometry (GC/MS).

Stable isotopically-labeled analogs of the compounds of interest are added to the sample. If the solids content is less than 1%, a 1-L sample is extracted at pH 12–13, then at pH <2 with methylene chloride using continuous extraction techniques.

If the solids content is 30% or less, the sample is diluted to 1% solids with reagent water, homogenized ultrasonically, and extracted at pH 12–13, then at pH <2 with methylene chloride using continuous extraction techniques. If the solids content is greater than 30%, the sample is extracted using ultrasonic techniques.

Each extract is dried over sodium sulfate, concentrated to a volume of 5 mL, cleaned up using GPC, if necessary, and concentrated. Extracts are concentrated to 1 mL if GPC is not performed, and to 0.5 mL if GPC is performed.

An internal standard is added to the extract, and a 1-mL aliquot of the extract is injected into the GC. The compounds are separated by GC and detected by a MS. The labeled compounds serve to correct the variability of the analytical technique.

INTERFERENCES Solvents, reagents, glassware, and other sample processing hardware may yield artifacts and/or elevated baselines causing misinterpretation of chromatograms and spectra. Materials used in the analysis must be demonstrated to be free from interferences under the conditions of analysis by running method blanks initially and with each sample lot (sample started through the extraction process on a given 8-h shift, to a maximum of 20). Specific selection of reagents and purification of solvents by distillation in all glass systems may be required. Glassware and, where possible, reagents are cleaned by solvent rinse and baking at 450°C for 1-h minimum. Interferences coextracted from samples will vary considerably from source to source, depending on the diversity of the site being sampled.

INSTRUMENTATION Major instrumentation includes a GC with a splitless or on-column injection port for capillary column, a MS with 70 eV electron impact ionization, and a data system to collect and record MS data, and process it. A K-D apparatus is used to concentrate extracts.

GC Column: 30 m × 0.25 mm I.D. 5% phenyl, 94% methyl, 1% vinyl silicone bonded phased fused silica capillary column.

PRECISION & ACCURACY The detection limits of the method are usually dependent on the level of interferences rather than instrumental limitations. The limits typify the minimum quantities that can be detected with no interferences present.

The minimum level (in µg/mL) was not listed. This is defined as a minimum level at which the analytical system shall give recognizable mass spectra (background corrected) and acceptable calibration points.

The MDL (in µg/kg) in low solids was not listed and in high solids was not listed; these were determined in digested sludge (low solids) and in filter cake or compost (high solids).

The labeled and native compound initial precision as standard deviation (in µg/L) was not listed.
The labeled and native compound initial accuracy as average recovery (in µg/L) was not listed.

SAMPLE COLLECTION, PRESERVATION & HANDLING Collect samples in glass containers. Aqueous samples which flow freely are collected in refrigerated bottles using automatic sampling equipment. Solid samples are collected as grab samples using widemouth jars. Maintain samples at 0 to 4°C from the time of collection until extraction. If residual chlorine is present in aqueous samples, add 80 mg sodium thiosulfate/L of water. Begin sample extraction within 7 days of collection, and analyze all extracts within 40 days of extraction.

SAMPLE PREPARATION Samples containing 1% solids or less are extracted directly using continuous liquid-liquid extraction techniques. Samples containing 1 to 30% solids are diluted to the 1% level with reagent water and extracted using continuous liquid-liquid extraction techniques. Samples containing greater than 30% solids are extracted using ultrasonic techniques.

- Base/neutral extraction — Adjust the pH of the waters in the extractors to 12–13 with 6 N NaOH. Extract with methylene chloride for 24–48 h.
- Acid extraction — Adjust the pH of the waters in the extractors to 2 or less using 6 N sulfuric acid. Extract with methylene chloride for 24–48 h.
- Ultrasonic extraction of high solids samples — Add anhydrous sodium sulfate to the sample and QC aliquot(s). Add acetone:methylene chloride (1:1) to the sample and mix thoroughly

Concentrate extracts using a K-D apparatus.

QUALITY CONTROL The analyst is permitted to modify this method to improve separations or lower the costs of measurements, provided all performance specifications are met. Analyses of blanks are required to demonstrate freedom from contamination. When results of spikes indicate atypical method performance for samples, the samples are diluted to bring method performance within acceptable limits.

For low solids (aqueous samples), extract, concentrate, and analyze two sets of four 1-L aliquots (8 aliquots total) of the precision and recovery standard. For high solids samples, two sets of four 30-g aliquots of the high solids reference matrix are used.

Spike all samples with labeled compounds to assess method performance. Compute percent recovery of the labeled compounds using the internal standard method. Compare the labeled compound recovery for each compound with the corresponding labeled compound recovery.

Reagent water and high solids reference matrix blanks are analyzed to demonstrate freedom from contamination. Extract and concentrate a 1-L reagent water blank or a high solids reference matrix blank with each sample's lot (samples started through the extraction process on the same 8-h shift, to a maximum of 20 samples).

Field replicates may be collected to determine the precision of the sampling technique, and spiked samples may be required to determine the accuracy of the analysis when the internal standard method is used.

REFERENCE Semivolatile Organic Compounds by Isotope Dilution GC/MS. Office of Water Regulation and Standards, U.S. EPA Industrial Technology Division, Washington, DC, EPA Method 1625, Rev. C, June 1989 (contact W.A. Telliard, U.S. EPA, Office of Water Regulations and Standards, 401 M St., SW, Washington, DC, 20460. Phone: 202-382-7131).

o-Anisidine　　　　　　　　　　　　　　EPA Method 8270
CAS #90-04-0

TITLE Semivolatile Organic Compounds by GC/MS

MATRIX This method is used to determine the concentration of semivolatile organic compounds in extracts prepared from all types of solid waste matrices, soils, and groundwater. Although surface waters are not specifically mentioned, this method should be applicable to water samples from rivers, lakes, etc.

METHOD SUMMARY This method covers 259 semivolatile organic compounds. In very limited applications direct injection of the sample into the GC/MS system may be appropriate, but this results in very high detection limits (approximately 10,000 µg/L). Typically, a 1-L liquid sample, containing surrogate, and matrix spiking standards, is extracted in a continuous extractor first under acid conditions and then under basic conditions. Typically 30 g of a solid sample, containing surrogate, and matrix spiking standards, is extracted ultrasonically. After concentrating the extract to 1 mL it is spiked with 10 µL of an internal standard solution just prior to analysis by GC/MS. The volume injected should contain about 100 ng of base/neutral and 200 ng of acid surrogates (for a 1-µL injection). Analysis is performed by GC/MS using a capillary GC column.

INTERFERENCES Raw GC/MS data from all blanks, samples, and spikes must be evaluated for interferences. Contamination by carryover can occur whenever high-concentration and low-concentration samples are sequentially analyzed. To reduce carryover, the sample syringe must be rinsed out between samples with solvent. Whenever an unusually concentrated sample is encountered, it should be followed by the analysis of blank solvent to check for cross-contamination.

INSTRUMENTATION A GC/MS and a data system are required. The GC column used is a 30 m × 0.25 mm I.D. (or 0.32 mm I.D.) 1-µm film thickness silicone-coated fused silica capillary column. A continuous liquid-liquid extractor equipped with Teflon® or glass connection joints and stopcocks requiring no lubrication, a K-D concentrating apparatus, water bath, and an ultrasonic disrupter with a minimum power of 300 W and with pulsing capability are also required.

PRECISION & ACCURACY The estimated quantitation limit (EQL) of Method 8270B for determining an individual compound is approximately 1 mg/kg (wet weight) for soil or sediment samples, 1–200 mg/kg for wastes (dependent on matrix and method of preparation), and 10 µg/L for groundwater samples. EQLs will be proportionately higher for sample extracts that require dilution to avoid saturation of the detector.

The EQL(b) for groundwater in µg/L is 10.
The EQL (a, b) for low concentrations in soil and sediment in µg/kg is not determined.
Accuracy as µg/L is not listed.
Overall precision in µg/L is not listed.

(a)　EQLs listed for soil/sediment are based on wet weight. Normally data is reported in a dry-weight basis; therefore, EQLs

will be higher based on the % dry weight of each sample. This calculation is based on a 30-g sample and gel permeation chromatography cleanup.

(b) Sample EQLs are highly matrix-dependent. The EQLs are provided for guidance and may not always be achievable.

C = *True value for concentration, in µg/L.*

X = *Average recovery found for measurements of samples containing a concentration of C, in µg/L.*

ESTIMATED QUANTITATION LIMIT

Other Matrices	Factor (a)
High-concentration soil and sludges by sonicator	7.5
Non-water miscible waste	75

(a) EQL for other matrices = [EQL for low soil/sediment] × [Factor]. This estimated EQL is similar to an EPA "Practical Quantitation Limit."

SAMPLING METHOD

Liquid samples — Use a 1 or 2½ gallon amber glass bottle with a screw-top Teflon®-lined cover that has been prewashed with detergent and rinsed with distilled water and methanol (or isopropanol).

Soils, sediments, or sludges — Use an 8-oz. widemouth glass with a screw-top Teflon®-lined cover that has been prewashed with detergent and rinsed with distilled water and methanol (or isopropanol).

SAMPLE PRESERVATION

Liquid samples — If residual chlorine is present, add 3 mL of 10% sodium thiosulfate per gallon, cool to 4°C and store in a solvent-free refrigerator until analysis; if chlorine is not present, then eliminate the sodium thiosulfate addition.

Soils, sediments, or sludges — Cool samples to 4°C and store in a solvent-free refrigerator.

MHT Liquid samples must be extracted within 7 days and the extracts analyzed within 40 days. Soils, sediments, or sludges may be stored for a maximum of 14 days and the extracts analyzed within 40 days.

SAMPLE PREPARATION

Liquid samples — Transfer 1 L quantitatively to a continuous extractor. If high concentrations are anticipated, a smaller volume may be used and then diluted with organic-free reagent water to 1 L. Adjust pH, if necessary, to pH <2 using 1:1 (V/V) sulfuric acid. Pipette 1.0 mL of a surrogate standard spiking solution into each sample. For the sample in each analytical batch selected for spiking, add 1.0 mL of a matrix spiking standard. For base/neutral acid analysis, the amount of the surrogates and matrix spiking compounds added to the sample should result in a final concentration of 100 ng/µL of each analyte in the extract to be analyzed (assuming a 1-µL injection). Extract with methylene chloride for 18–24 h. Next, adjust the pH of the aqueous phase to pH >11 using 10 N sodium hydroxide and extract it with methylene chloride again for 18–24 h. Dry the extract through a column containing anhydrous sodium sulfate and concentrate it to 1 mL using a K-D concentrator.

Soils, sediments, or sludges — Use 30 g of sample. Nonporous or wet samples (gummy or clay type) that do not have a free-flowing sandy texture must be mixed with anhydrous sodium sulfate until the sample is free flowing. Add 1 mL of surrogate standards to all samples, spikes, standards, and blanks. For the sample in each analytical batch selected for spiking, add 1.0 mL of a matrix spiking standard. For base/neutral acid analysis, the amount added of the surrogates and matrix spiking compounds should result in a final concentration of 100 ng/µL of each base/neutral analyte and 200 ng/µL of each acid analyte in the extract to be analyzed (assuming a 1-µL injection). Immediately add a 100-mL mixture of 1:1 methylene chloride:acetone and extract the sample ultrasonically for 3 min and then decant or filter the extracts. Repeat the extraction two or more times. Dry the extract using a column with anhydrous sodium sulfate and concentrate it to 1 mL in a K-D concentrator.

QUALITY CONTROL A methylene chloride solution containing 50 ng/µL of decafluorotriphenylphosphine (DFTPP) is used for tuning the GC/MS system each 12-h shift. A system performance check also must be made during every 12-h shift. A standard containing 50 ng/µL each of 4,4'-DDT, pentachlorophenol, and benzidine is required to verify injection port inertness and GC column performance.

A calibration standard at mid-concentration, containing each compound of interest, including all required surrogates, must be performed every 12 h during analysis. After the system performance check is met, calibration check compounds (CCCs) are used to check the validity of the initial calibration.

The internal standard responses and retention times in the calibration check standard must be evaluated immediately after or during data acquisition. If the retention time for any internal standard changes by more than 30 seconds from the last check calibration (12 h), the chromatographic system must be inspected for malfunctions and corrections must be made, as required. If the electron ionization current plot (EICP) area for any of the internal standards changes by a factor of two from the last daily calibration standard check, the mass spectrometer must be inspected for malfunctions and corrections must be made, as appropriate.

Demonstrate, through the analysis of a reagent water blank, that interferences from the analytical system, glassware, and reagents are under control. The blank samples should be carried through all stages of the sample preparation and measurement steps. For each analytical batch (up to 20 samples), a reagent blank, matrix spike, and matrix spike duplicate/duplicate must be analyzed (the frequency of the spikes may be different for different monitoring programs). The blank and spiked samples must be carried through all stages of the sample preparation and measurement steps. A QC reference sample concentrate containing each analyte at a concentration of 100 mg/L in methanol is required.

REFERENCE Test Methods for Evaluating Solid Waste (SW-846). U.S. EPA 1983, Method 8270B, Rev. 2, Nov. 1990. Office of Solid Waste, Washington, DC.

Anthracene
CAS #120-12-7

EPA Method 1625

TITLE Semivolatile Organic Compounds by Isotope Dilution GC/MS

MATRIX The compounds may be determined in waters, soils, and municipal sludges by this method.

METHOD SUMMARY This method is used to determine 176 semivolatile toxic organic pollutants associated with the CWA (as amended 1987); the RCRA (as amended 1986); the CERCLA (as amended 1986); and other compounds amenable to extraction and analysis by capillary column gas chromatography-mass spectrometry (GC/MS).

Stable isotopically-labeled analogs of the compounds of interest are added to the sample. If the solids content is less than 1%, a 1-L sample is extracted at pH 12–13, then at pH <2 with methylene chloride using continuous extraction techniques.

If the solids content is 30% or less, the sample is diluted to 1% solids with reagent water, homogenized ultrasonically, and extracted at pH 12–13, then at pH <2 with methylene chloride using continuous extraction techniques. If the solids content is greater than 30%, the sample is extracted using ultrasonic techniques.

Each extract is dried over sodium sulfate, concentrated to a volume of 5 mL, cleaned up using GPC, if necessary, and concentrated. Extracts are concentrated to 1 mL if GPC is not performed, and to 0.5 mL if GPC is performed.

An internal standard is added to the extract, and a 1-mL aliquot of the extract is injected into the GC. The compounds are separated by GC and detected by a MS. The labeled compounds serve to correct the variability of the analytical technique.

INTERFERENCES Solvents, reagents, glassware, and other sample processing hardware may yield artifacts and/or elevated baselines causing misinterpretation of chromatograms and spectra. Materials used in the analysis must be demonstrated to be free from interferences under the conditions of analysis by running method blanks initially and with each sample lot (sample started through the extraction process on a given 8-h shift, to a maximum of 20). Specific selection of reagents and purification of solvents by distillation in all glass systems may be required. Glassware and, where possible, reagents are cleaned by solvent rinse and baking at 450°C for 1-h minimum. Interferences coextracted from samples will vary considerably from source to source, depending on the diversity of the site being sampled.

INSTRUMENTATION Major instrumentation includes a GC with a splitless or on-column injection port for capillary column, a MS with 70 eV electron impact ionization, and a data system to collect and record MS data, and process it. A K-D apparatus is used to concentrate extracts.

GC Column: 30 m × 0.25 mm I.D. 5% phenyl, 94% methyl, 1% vinyl silicone bonded phased fused silica capillary column.

PRECISION & ACCURACY The detection limits of the method are usually dependent on the level of interferences rather than instrumental limitations. The limits typify the minimum quantities that can be detected with no interferences present.

The minimum level (in µg/mL) was 10. This is defined as a minimum level at which the analytical system shall give recognizable mass spectra (background corrected) and acceptable calibration points.

The MDL (in µg/kg) in low solids was 52 and in high solids was 21; these were determined in digested sludge (low solids) and in filter cake or compost (high solids).

The labeled and native compound initial precision as standard deviation (in µg/L) was 41.

The labeled and native compound initial accuracy as average recovery (in µg/L) was 58–174.

SAMPLE COLLECTION, PRESERVATION & HANDLING Collect samples in glass containers. Aqueous samples which flow freely are collected in refrigerated bottles using automatic sampling equipment. Solid samples are collected as grab samples using widemouth jars. Maintain samples at 0 to 4°C from the time of collection until extraction. If residual chlorine is present in aqueous samples, add 80 mg sodium thiosulfate/L of water. Begin sample extraction within 7 days of collection, and analyze all extracts within 40 days of extraction.

SAMPLE PREPARATION Samples containing 1% solids or less are extracted directly using continuous liquid-liquid extraction techniques. Samples containing 1 to 30% solids are diluted to the 1% level with reagent water and extracted using continuous liquid-liquid extraction techniques. Samples containing greater than 30% solids are extracted using ultrasonic techniques.

Base/neutral extraction — Adjust the pH of the waters in the extractors to 12–13 with 6 N NaOH. Extract with methylene chloride for 24–48 h.

Acid extraction — Adjust the pH of the waters in the extractors to 2 or less using 6 N sulfuric acid. Extract with methylene chloride for 24–48 h.

Ultrasonic extraction of high solids samples — Add anhydrous sodium sulfate to the sample and QC aliquot(s). Add acetone:methylene chloride (1:1) to the sample and mix thoroughly

Concentrate extracts using a K-D apparatus.

QUALITY CONTROL The analyst is permitted to modify this method to improve separations or lower the costs of measurements, provided all performance specifications are met. Analyses of blanks are required to demonstrate freedom from contamination. When results of spikes indicate atypical method performance for samples, the samples are diluted to bring method performance within acceptable limits.

For low solids (aqueous samples), extract, concentrate, and analyze two sets of four 1-L aliquots (8 aliquots total) of the precision and recovery standard. For high solids samples, two sets of four 30-g aliquots of the high solids reference matrix are used.

Spike all samples with labeled compounds to assess method performance. Compute percent recovery of the labeled compounds using the internal standard method. Compare the labeled compound recovery for each compound with the corresponding labeled compound recovery.

Reagent water and high solids reference matrix blanks are analyzed to demonstrate freedom from contamination. Extract and concentrate a 1-L reagent water blank or a high solids reference matrix blank with each sample's lot (samples started through the extraction process on the same 8-h shift, to a maximum of 20 samples).

Field replicates may be collected to determine the precision of the sampling technique, and spiked samples may be required to determine the accuracy of the analysis when the internal standard method is used.

REFERENCE Semivolatile Organic Compounds by Isotope Dilution GC/MS. Office of Water Regulation and Standards, U.S. EPA Industrial Technology Div., Washington, DC, EPA Method 1625, Rev. C, June 1989 (contact W.A. Telliard, U.S. EPA, Office of Water Regulations and Standards, 401 M St., SW, Washington, DC, 20460. Phone: 202-382-7131).

Anthracene   **EPA Method 625**
CAS #120-12-7

TITLE Base/Neutrals and Acids, U.S. EPA Method 625

MATRIX This methods covers municipal and industrial wastewater.

METHOD SUMMARY Approximately 1 L of sample is serially extracted with methylene chloride at a pH greater than 11 and again at a pH less than 2 using a separatory funnel or a continuous extractor. The methylene chloride extract is dried, concentrated to a volume of 1 mL, and analyzed by GC/MS. Qualitative identification of the parameters in the extract is performed using the retention time and the relative abundance of three characteristic masses (m/z). Qualitative analysis is performed using external or internal standard techniques with a single characteristic m/z.

INTERFERENCES Method interferences may be caused by contaminants in solvents, reagents, glassware, and other sample processing hardware. Glassware must be scrupulously cleaned. Glassware should be heated in a muffle furnace at 400°C for 5 to 30 min. Some thermally stable materials, such as PCBs, may not be eliminated by this treatment. Solvent rinses with acetone and pesticide quality hexane may be substituted for the muffle furnace heating. Matrix interferences may be caused by contaminants that are coextracted from the sample. The base/neutral extraction may cause significantly reduced recovery of phenols. The packed gas chromatographic columns recommended for the basic fraction may not exhibit sufficient resolution for some analytes.

INSTRUMENTATION A GC/MS system with an injection port designed for on-column injection when using packed columns and for splitless injection when using capillary columns. Column for base/neutrals: 1.8 m long × 2 mm I.D. glass, packed with 3% SP-2550 on Supelcoport (100/120 mesh) or equivalent.
Column for acids: 1.8 m long × 2 mm I.D. glass, packed with 1% SP-1240DA on Supelcoport (100/120 mesh) or equivalent.

PRECISION & ACCURACY The MDL concentrations were obtained using reagent water. The MDL actually achieved in a given analysis will vary depending on instrument sensitivity and matrix effects. This method was tested by 15 laboratories using reagent water, drinking water, surface water, and industrial wastewater spiked at six concentrations over the range 5 to 100 µg/L. Single operator precision, overall precision, and method accuracy were found to be directly related to the concentration of the parameter matrix.

The MDL (in µg/L) in reagent water was 1.9.
The standard deviation (in µg/L based on 4 recovery measurements) was 32.0.
The range (in µg/L) for average recovery for 4 measurements was 43.4–118.0.
The range (in %) for percent recovery was 27–133.
Accuracy (in µg/L) as expected recovery for one or more measurements of a sample containing a true concentration of C was $0.80C+0.68$.
Precision (in µg/L) as expected single analyst standard deviation of measurements at an average concentration found at X was $0.21X-0.32$.
Overall precision (in µg/L) as expected interlaboratory standard deviation of measurements in an average concentration found at X was $0.27X-0.64$.

C = *True value of the concentration in µg/L.*
X = *Average recovery found for measurements of samples containing a concentration at C in µg/L.*

SAMPLE PREPARATION Adjust the pH to >11 with sodium hydroxide and serially extract in a separatory funnel with methylene chloride or else in a continuous extractor. Next, adjust the pH to <2 with sulfuric acid and serially extract in a separatory funnel with methylene chloride or else in a continuous extractor. Dry the extracts separately through a column of anhydrous sodium sulfate and then concentrate each of the extracts to 1.0 mL using a K-D apparatus.

SAMPLE COLLECTION, PRESERVATION & HANDLING Grab samples must be collected in glass containers. All samples must be refrigerated at 4°C from the time of collection until extraction. If residual chlorine is present, add 80 mg of sodium thiosulfate/L of sample and mix well. All samples must be extracted within 7 days of collection and completely analyzed within 40 days of extraction.

QUALITY CONTROL Make an initial, one-time, demonstration of the ability to generate acceptable accuracy and precision with this method. Before processing any samples, the analyst must analyze a reagent water blank to demonstrate that interferences from the analytical system and glassware are under control. Each time a set of samples is extracted or reagents are changed, a reagent water blank must be processed. Spike and analyze a minimum of 5% of all samples to monitor and evaluate lab data quality. A QC check sample concentrate that

contains each parameter of interest at a concentration of 100 µg/mL in acetone is required. PCBs and multicomponent pesticides may be omitted from this test.

After analysis of five spiked wastewater samples, calculate the average percent recovery and the standard deviation of the percent recovery. Spike all samples with the surrogate standard spiking solution and calculate the percent recovery of each surrogate compound.

REFERENCE *Federal Register*, Vol. 49, No. 209. Friday, Oct. 26, 1984.

Anthracene   **EPA Method 8270**
CAS #120-12-7

TITLE Semivolatile Organic Compounds by GC/MS

MATRIX This method is used to determine the concentration of semivolatile organic compounds in extracts prepared from all types of solid waste matrices, soils, and groundwater. Although surface waters are not specifically mentioned, this method should be applicable to water samples from rivers, lakes, etc.

METHOD SUMMARY This method covers 259 semivolatile organic compounds. In very limited applications direct injection of the sample into the GC/MS system may be appropriate, but this results in very high detection limits (approximately 10,000 µg/L). Typically, a 1-L liquid sample, containing surrogate, and matrix spiking standards, is extracted in a continuous extractor first under acid conditions and then under basic conditions. Typically 30 g of a solid sample, containing surrogate, and matrix spiking standards, is extracted ultrasonically. After concentrating the extract to 1 mL it is spiked with 10 µL of an internal standard solution just prior to analysis by GC/MS. The volume injected should contain about 100 ng of base/neutral and 200 ng of acid surrogates (for a 1-µL injection). Analysis is performed by GC/MS using a capillary GC column.

INTERFERENCES Raw GC/MS data from all blanks, samples, and spikes must be evaluated for interferences. Contamination by carryover can occur whenever high-concentration and low-concentration samples are sequentially analyzed. To reduce carryover, the sample syringe must be rinsed out between samples with solvent. Whenever an unusually concentrated sample is encountered, it should be followed by the analysis of blank solvent to check for cross-contamination.

INSTRUMENTATION A GC/MS and a data system are required. The GC column used is a 30 m × 0.25 mm I.D. (or 0.32 mm I.D.) 1-µm film thickness silicone-coated fused silica capillary column. A continuous liquid-liquid extractor equipped with Teflon® or glass connection joints and stopcocks requiring no lubrication, a K-D concentrating apparatus, water bath, and an ultrasonic disrupter with a minimum power of 300 W and with pulsing capability are also required.

PRECISION & ACCURACY The estimated quantitation limit (EQL) of Method 8270B for determining an individual compound is approximately 1 mg/kg (wet weight) for soil or sediment samples, 1–200 mg/kg for wastes (dependent on matrix and method of preparation), and 10 µg/L for groundwater samples. EQLs will be proportionately higher for sample extracts that require dilution to avoid saturation of the detector.

The EQL(b) for groundwater in µg/L is 10.
The EQL (a, b) for low concentrations in soil and sediment in µg/kg is 660.
Accuracy as µg/L is 0.80C +0.68.
Overall precision in µg/L is 0.27X–0.64.

(a) *EQLs listed for soil/sediment are based on wet weight. Normally data is reported in a dry-weight basis; therefore, EQLs will be higher based on the % dry weight of each sample. This calculation is based on a 30-g sample and gel permeation chromatography cleanup.*
(b) *Sample EQLs are highly matrix-dependent. The EQLs are provided for guidance and may not always be achievable.*
$C =$ *True value for concentration, in µg/L.*
$X =$ *Average recovery found for measurements of samples containing a concentration of C, in µg/L.*

ESTIMATED QUANTITATION LIMIT

Other Matrices	Factor (a)
High-concentration soil and sludges by sonicator	7.5
Non-water miscible waste	75

(a) *EQL for other matrices = [EQL for low soil/sediment] × [Factor]. This estimated EQL is similar to an EPA "Practical Quantitation Limit."*

SAMPLING METHOD
Liquid samples — Use a 1 or 2½ gallon amber glass bottle with a screw-top Teflon®-lined cover that has been prewashed with detergent and rinsed with distilled water and methanol (or isopropanol).

Soils, sediments, or sludges — Use an 8-oz. widemouth glass with a screw-top Teflon®-lined cover that has been prewashed with detergent and rinsed with distilled water and methanol (or isopropanol).

SAMPLE PRESERVATION
Liquid samples — If residual chlorine is present, add 3 mL of 10% sodium thiosulfate per gallon, cool to 4°C and store in a solvent-free refrigerator until analysis; if chlorine is not present, then eliminate the sodium thiosulfate addition.

Soils, sediments, or sludges — Cool samples to 4°C and store in a solvent-free refrigerator.

MHT Liquid samples must be extracted within 7 days and the extracts analyzed within 40 days. Soils, sediments, or sludges may be stored for a maximum of 14 days and the extracts analyzed within 40 days.

SAMPLE PREPARATION
Liquid samples — Transfer 1 L quantitatively to a continuous extractor. If high concentrations are anticipated, a smaller volume may be used and then diluted with organic-free reagent water to 1 L. Adjust pH, if necessary, to pH <2 using 1:1 (V/V) sulfuric acid. Pipette 1.0 mL of a surrogate standard spiking solution into each sample. For the sample in each analytical

batch selected for spiking, add 1.0 mL of a matrix spiking standard. For base/neutral acid analysis, the amount of the surrogates and matrix spiking compounds added to the sample should result in a final concentration of 100 ng/µL of each analyte in the extract to be analyzed (assuming a 1-µL injection). Extract with methylene chloride for 18–24 h. Next, adjust the pH of the aqueous phase to pH >11 using 10 N sodium hydroxide and extract it with methylene chloride again for 18–24 h. Dry the extract through a column containing anhydrous sodium sulfate and concentrate it to 1 mL using a K-D concentrator.

Soils, sediments, or sludges — Use 30 g of sample. Nonporous or wet samples (gummy or clay type) that do not have a free-flowing sandy texture must be mixed with anhydrous sodium sulfate until the sample is free flowing. Add 1 mL of surrogate standards to all samples, spikes, standards, and blanks. For the sample in each analytical batch selected for spiking, add 1.0 mL of a matrix spiking standard. For base/neutral acid analysis, the amount added of the surrogates and matrix spiking compounds should result in a final concentration of 100 ng/µL of each base/neutral analyte and 200 ng/µL of each acid analyte in the extract to be analyzed (assuming a 1-µL injection). Immediately add a 100-mL mixture of 1:1 methylene chloride:acetone and extract the sample ultrasonically for 3 min and then decant or filter the extracts. Repeat the extraction two or more times. Dry the extract using a column with anhydrous sodium sulfate and concentrate it to 1 mL in a K-D concentrator.

QUALITY CONTROL A methylene chloride solution containing 50 ng/µL of decafluorotriphenylphosphine (DFTPP) is used for tuning the GC/MS system each 12-h shift. A system performance check also must be made during every 12-h shift. A standard containing 50 ng/µL each of 4,4′-DDT, pentachlorophenol, and benzidine is required to verify injection port inertness and GC column performance. A calibration standard at mid-concentration, containing each compound of interest, including all required surrogates, must be performed every 12 h during analysis. After the system performance check is met, calibration check compounds (CCCs) are used to check the validity of the initial calibration.

The internal standard responses and retention times in the calibration check standard must be evaluated immediately after or during data acquisition. If the retention time for any internal standard changes by more than 30 seconds from the last check calibration (12 h), the chromatographic system must be inspected for malfunctions and corrections must be made, as required. If the electron ionization current plot (EICP) area for any of the internal standards changes by a factor of two from the last daily calibration standard check, the mass spectrometer must be inspected for malfunctions and corrections must be made, as appropriate.

Demonstrate, through the analysis of a reagent water blank, that interferences from the analytical system, glassware, and reagents are under control. The blank samples should be carried through all stages of the sample preparation and measurement steps. For each analytical batch (up to 20 samples), a reagent blank, matrix spike, and matrix spike duplicate/duplicate must be analyzed (the frequency of the spikes may be different for different monitoring programs). The blank and spiked samples must be carried through all stages of the sample preparation and measurement steps. A QC reference sample concentrate containing each analyte at a concentration of 100 mg/L in methanol is required.

REFERENCE Test Methods for Evaluating Solid Waste (SW-846). U.S. EPA 1983, Method 8270B, Rev. 2, Nov. 1990. Office of Solid Waste, Washington, DC.

Anthracene   **EPA Method 8100**
CAS #120-12-7

TITLE Polynuclear Aromatic Hydrocarbons

MATRIX Groundwater, soils, sludges, water miscible liquid wastes, and non-water miscible wastes.

APPLICATION This method is used for the analysis of various PAHs. Samples are extracted, concentrated, and analyzed using direct injection of both neat and diluted organic liquids. The method provides two optional GC columns that are better than Column 1 and that may help resolve analytes from interferences.

INTERFERENCES Solvents, reagents, and glassware may introduce artifacts. Other interferences may come from coextracted compounds from samples.

INSTRUMENTATION GC capable of on-column injections and a flame ionization detector (FID). Column 1: a 1.8 m by 2 mm 3% OV-17 on Chromosorb W-AW-DCMS column. Column 2: a 30 m by 0.25 mm SE-54 fused silica capillary column. Column 3: a 30 m by 0.32 mm SE-54 fused silica capillary column.

RANGE   0.1–425 µg/L

MDL   Not reported.

PQL FACTORS FOR MULTIPLYING × FID MDL VALUE
Not available.

PRECISION   0.41X + 0.45 µg/L (overall precision).

ACCURACY   0.63C — 1.26 µg/L (as recovery).

SAMPLING METHOD Use 8-oz. widemouth glass bottles with Teflon®-lined caps for concentrated waste samples, soils, sediments, and sludges. Use 1 or 2½ gallon amber glass bottles with Teflon®-lined caps for liquid (water) samples.

STABILITY Cool soil, sediment, sludge, and liquid samples to 4°C. If residual chlorine is present in liquid samples add 3 mL of 10% sodium thiosulfate per gallon of sample and cool to 4°C.

MHT 14 days for concentrated waste, soil, sediment, or sludge; 7 days for liquid samples; all extracts must be analyzed within 40 days.

QUALITY CONTROL A quality control check sample concentrate containing each analyte of interest is required. The QC check sample concentrate may be prepared from pure standard materials or purchased as certified solutions. Use appropriate trip, matrix, control site, method, reagent, and solvent blanks. Internal, surrogate, and five concentration level calibration standards are used. The quality control check sample concentrate should contain anthracene at 100 µg/mL in acetonitrile.

REFERENCE   Method 8100, SW-846, 3rd ed., Nov. 1986.

Anthracene   EPA Method 8310
CAS #120-12-7

TITLE   Polynuclear Aromatic Hydrocarbons

MATRIX   Groundwater, soils, sludges, water miscible liquid wastes, and non-water miscible wastes.

APPLICATION   This method is used for the analysis of 16 polynuclear aromatic hydrocarbons (PAHs). Samples are extracted, concentrated, and analyzed using HPLC with detection by UV and fluorescence detectors.

INTERFERENCES   Solvents, reagents, and glassware may introduce artifacts. Other interferences may come from coextracted compounds from samples.

INSTRUMENTATION   HPLC with a gradient pumping system and a 250 mm by 2.6 mm reverse phase HC-ODS Sil-X 5-micron particle size column. The fluorescence detector uses an excitation wavelength of 280 nm and emission greater than 389 nm cutoff with dispersive optics.

RANGE   0.1–425 µg/L

MDL   0.66 µg/L (Fluorescence; reagent water).

PQL FACTORS FOR MULTIPLYING × FID MDL VALUE

Matrix	Multiplication Factor
Groundwater	10
Low-level soil by sonication with GPC cleanup	670
High-level soil and sludge by sonication	10,000
Non-water miscible waste	100,000

PRECISION $0.41X + 0.45$ µg/L (overall precision).

ACCURACY $0.63C - 1.26$ µg/L (as recovery).

SAMPLING METHOD   Use 8-oz. widemouth glass bottles with Teflon®-lined caps for concentrated waste samples, soils, sediments, and sludges. Use 1 or 2½ gallon amber glass bottles with Teflon®-lined caps for liquid (water) samples.

STABILITY   Cool soil, sediment, sludge, and liquid samples to 4°C. If residual chlorine is present in liquid samples add 3 mL of 10% sodium thiosulfate per gallon of sample and cool to 4°C.

MHT   14 days for concentrated waste, soil, sediment, or sludge; 7 days for liquid samples; all extracts must be analyzed within 40 days.

QUALITY CONTROL   Internal, surrogate, and five concentration level calibration standards are used. The calibration standards must be used with the analytical method blank. A quality control check sample concentrate containing anthracene at 100 µg/mL is required. The QC check sample concentrate may be prepared from pure standard materials or purchased as certified solutions. Use appropriate trip, matrix, control site, method, reagent, and solvent blanks.

REFERENCE   Method 8310, SW-846, 3rd ed., Nov. 1986.

Antimony   EPA Method 6010
CAS #7440-36-0

TITLE   Inductively Coupled Plasma-Atomic Emission Spectroscopy

MATRIX   This method is applicable to the determination of trace elements, including metals, in groundwater, soils, sludges, sediments, and other solid wastes. All matrices require digestion prior to analysis. The method of standard addition must be used for the analysis of all sample digests unless either serial dilution or matrix spike addition demonstrates it is not required.

METHOD SUMMARY   Method 6010 covers 25 elements using ICP analysis. It measures element-emitted light by optical spectrometry. Samples, following an appropriate acid digestion, are nebulized and the resulting aerosol is transported to the plasma torch. Element-specific atomic line emission spectra are produced by a radio-frequency inductively coupled plasma.

INTERFERENCES   Interferences may be categorized as spectral or non-spectral. Spectral interferences are caused by overlap of a spectral line from another element, unresolved overlap of molecular band spectra, background contribution from continuous or recombination phenomenon, and stray light from the line emission of high concentration elements. Non-spectral interferences include physical and chemical interferences. Physical interferences are effects associated with the sample nebulization and transport processes. Changes in viscosity and surface tension can cause significant inaccuracies. Chemical interferences include molecular compound formation, ionization effects, and solute vaporization effects. Normally these effects are not significant and can be minimized by careful selection of operating conditions. Chemical interferences are highly dependent on matrix type and the specific analyte element.

INSTRUMENTATION   An inductively coupled argon plasma emission spectrometer (ICP) capable of background correction is required.

PRECISION & ACCURACY   Detection limits, sensitivity, and optimum ranges of the metals will vary with the matrices and model of the spectrometer. In a single lab evaluation, seven wastes were analyzed for 22 elements. The mean percent relative standard deviation from triplicate analyses for all elements and wastes was 9 ± 2%. The mean percent recovery of spiked elements for all wastes was 93 ± 6%. Spike levels ranged from 100 µg/L to 100 mg/L. The wastes included sludges and industrial wastewater.

Estimated instrument detection limit in µg/L is 32.
Spiked concentration in µg/L is not listed.

Mean reported value in μg/L is not listed.
Precision as RSD % is not listed.

SAMPLING METHOD Samples should be collected in borosilicate glass, linear polyethylene, polypropylene, or Teflon® bottles that have been prewashed with detergent and tap water, and rinsed with 1:1 nitric acid and tap water or 1:1 hydrochloric acid and tap water. Collect at least 2 g of solids and 200 mL of aqueous samples.

SAMPLE PRESERVATION Add nitric acid to make the samples pH <2.

MHT The maximum holding time for properly preserved samples is 6 months.

SAMPLE PREPARATION Preliminary treatment of most matrices is necessary because of the complexity and variability of sample matrices. Water samples which have been prefiltered and acidified will not need acid digestion. Methods for acid digestion of waters for total recoverable or dissolved metals, acid digestions of aqueous samples and extracts for total metals, and acid digestion of sediments, sludges, and soils are summarized below.

Total recoverable or dissolved metals in water — To prepare surface and groundwater samples for determination of total recoverable and dissolved metals, a 100 mL aliquot of well-mixed sample is acidified with concentrated nitric acid and concentrated hydrochloric acid, then heated until the volume is reduced to 15–20 mL. Adjust the final volume to 100 mL with reagent water.

Total metals in aqueous samples, soil and sediment extracts — To prepare aqueous samples, soil and sediment extracts, and wastes that contain suspended solids, a 100 mL aliquot is made acidic with concentrated nitric acid and the solution is evaporated to about 5 mL on a hot plate. Continue heating and adding additional acid until sample digestion is complete, which is usually indicated when the digestate is light in color or does not change in appearance. Evaporate the solution to about 3 mL and cool it and add a small quantity of 1:1 hydrochloric acid (10 mL100 mL of final solution). Cover the beaker and reflux for 15 min. Wash down the beaker walls and filters or centrifuge the sample to remove silicates and other insoluble material. Filter the sample and adjust the final volume to 100 mL with reagent water and the final acid concentration to 10%.

Sediments, sludges, and soils — To prepare sediments, sludges and soil samples, transfer 1–2 g to a conical beaker and add 10 mL of 1:1 nitric acid, mix the slurry, and cover it with a watch glass. Heat the sample and reflux for 10 to 15 min without boiling. Allow it to cool, then add 5 mL of concentrated nitric acid and reflux for 30 min. Repeat last step and then allow the solution to evaporate to 5 mL without boiling. Cool and add 2 mL of water and 3 mL of 30% hydrogen peroxide. Cover and place the beaker on the hot plate. Heat and add 30% hydrogen peroxide in 1-mL aliquots with warming until the effervescence is minimal but do not add more than a total of 10 mL of 30% hydrogen peroxide. If the sample is being prepared for the analysis of Ag, Al, As, Ba, Be, Ca, Cd, Co, Cr, Cu, Fe, K, Mg, Mn, Mo, Na, Ni, Os, Pb, Se, Tl, V, and Zn, then add 5 mL of concentrated hydrochloric acid and 10 mL of water and return the covered beaker to a hot plate for 15 min of additional refluxing without boiling. Dilute the sample to a 100 mL volume with water after cooling and filter or centrifuge to remove particulates.

QUALITY CONTROL Laboratory control samples must be analyzed for each analytical method. A method blank should be analyzed with each batch of samples. The effect of the matrix on method performance must be demonstrated: when appropriate, there should be at least one matrix spike and either one matrix duplicate or one matrix spike duplicate per analytical batch. The bias and precision of the method, as well as the method detection limit for each specific matrix type, must be measured.

Dilute and reanalyze samples that are more concentrated than the linear calibration limit. Employ a minimum of one reagent blank per sample batch to determine if contamination or any memory effects are occurring. Whenever a new or unusual sample matrix is encountered, perform either a serial dilution test or a matrix spike addition test to ensure that neither positive or negative interferences are operating on any of the analyte elements. Check the instrument standardization by verifying calibration every 10 samples using a calibration blank and a check standard.

REFERENCE Test Methods for Evaluating Solid Waste (SW-846). U.S. EPA. 1983. Method 6010, Rev. 0, Sept. 1986. Office of Solid Wastes, Washington, DC.

Antimony   **EPA Method 200.7**
CAS #7440-36-0

TITLE Inductively Coupled Plasma (ICP)

MATRIX Dissolved, suspended or total element in drinking and surface waters and in domestic and industrial wastewater.

APPLICATION The method covers the determination of 25 metals. Dissolved elements are determined in filtered and acidified samples after appropriate digestion (which increases dissolved solids). Its primary advantage is that ICP instruments allow simultaneous or rapid sequential determination of many elements in a short time. Samples are first nebulized and the aerosol is transported to a plasma torch in which element specific atomic line emission spectra are produced by a radio frequency inductively coupled plasma. Background correction is required for trace element detection except in the case of line roadening.

INTERFERENCES There are spectral, physical, and chemical interferences. The primary disadvantage of ICP instruments is background radiation from other elements and the plasma gases (spectral interferences). Changes in sample viscosity and surface tension with samples containing high dissolved solids (especially those exceeding 1500 mg/L) or high acid concentrations can cause physical interferences. Ionization effects, solute vaporization and molecular compound formation can cause chemical interferences. Aluminum, chromium, iron, thallium, and vanadium can cause interferences at the 100 mg/L level.

INSTRUMENTATION Inductively coupled argon plasma emission spectroscopy. 206.833 nm wavelength.

RANGE Not listed.

MDL 32 µg/L

PRECISION Not listed.

ACCURACY Mean Recovery = 93% ± 6% of spiked elements for all wastes.

SAMPLING METHOD Wash sample container with detergent and tap water, rinse with 1+1 nitric acid and tap water, then rinse with 1+1 hydrochloric acid and tap water, then rinse with deionized, distilled water in that order. Perform any filtration or acid preservation steps when the sample is collected or as soon as possible thereafter.

STABILITY Cool samples to 4°C.

MHT 24 h.

QUALITY CONTROL Mixed calibration standards, an instrument check standard, and an interference check solution are used in addition to a quality control sample. The quality control sample should be prepared in the same acid matrix as the calibration standards at 10 times the instrumental detection limits and in accordance with the instructions provided by the supplier. Furthermore, two types of blanks are required: a calibration blank and a reagent blank.

REFERENCE Method 200.7, U.S. EPA, EMSL-Cincinnati, OH, Nov. 1980

Antimony   EPA Method 7040
CAS #7440-36-0

TITLE Atomic Absorption, (AA)

MATRIX Drinking, surface and direct aspiration saline waters, wastewater.

APPLICATION Sample is aspirated and atomized in a flame. A light beam from an Sb hollow cathode lamp is directed through the flame into a monochromator and onto a detector. Since wavelength of light beam is specific for Sb, light energy absorbed by detector is measure of Sb.

INTERFERENCES The most troublesome type is chemical, caused by lack of absorption of atoms bound in molecular combination in the flame. High dissolved solids in sample may result in nonatomic absorbance interference. Lead, copper, nickel, and excess acid can interfere.

INSTRUMENTATION Atomic absorption spectrometer. Antimony hollow cathode lamp or electrodeless discharge lamp. (217.6 nm Wavelength).

RANGE 1–40 mg/L

MDL 0.2 mg/L

PRECISION standard deviation = ±0.08 and 0.10% at 5.0 and 15 mg Sb/L

ACCURACY recoveries = 96% and 97% at 5.0 mg and 15 mg Sb/L

SAMPLING METHOD Use glass or plastic containers. Collect 200 g of solids and 600 mL of liquid samples.

STABILITY Cool solid samples to 4°C and analyze as soon as possible. Add nitric acid to liquid samples to pH <2.

MHT 6 months.

QUALITY CONTROL At least one duplicate and one spike sample should be run every 20 samples or with each matrix type to verify precision of the method. For 20 or more samples per day, verify working standard curve. Run an additional standard at or near mid-range every 10 samples.

REFERENCE Method 7040, SW-846, 3rd ed., Nov. 1986.

Antimony EPA Method 7041
CAS #7440-36-0

TITLE Atomic Absorption, (AA) Furnace Technique

MATRIX Wastes, mobility procedure extracts, soils, and groundwater

APPLICATION Aqueous samples, EP extracts, industrial wastes, soils, sludges, sediments and solid wastes require digestion before analysis. An aliquot of sample is placed in the graphite tube in the furnace and slowly evaporated, charred and atomized. Absorption of lamp radiation during atomization is proportional to (Sb) concentration.

INTERFERENCES The furnace technique is subject to chemical interferences. Composition of sample matrix can have major effect on analysis. Modify matrix to remove interferences. High lead concentration may cause spectral interference at 217.6 nm Line. (With interference, use 231.1 nm).

INSTRUMENTATION Atomic absorption spectrometer. (Sb) hollow cathode lamp or electrodeless discharge lamp. Graphite furnace. Strip-chart recorder.

RANGE 20–300 µg/L

MDL 3 µg/L [217.6 nm line (primary)]

PRECISION Not listed.

ACCURACY Not listed.

SAMPLING METHOD Use glass or plastic containers. Collect 200 g of solids and 600 mL of liquid samples.

STABILITY Cool solid samples to 4°C and analyze as soon as possible.

Add nitric acid to liquid samples to pH <2.

MHT 6 months.

QUALITY CONTROL At least one duplicate and one spike sample should be run every 20 samples, or with each matrix type to verify method precision. If 20 or more samples are run a day, run a standard (at or near mid-range) every 10 samples.

REFERENCE Method 7041, SW-846, 3rd ed., Nov. 1986.

Aramite **EPA Method 1625**
CAS #140-57-8

TITLE Semivolatile Organic Compounds by Isotope Dilution GC/MS

MATRIX The compounds may be determined in waters, soils, and municipal sludges by this method.

METHOD SUMMARY This method is used to determine 176 semivolatile toxic organic pollutants associated with the CWA (as amended 1987); the RCRA (as amended 1986); the CERCLA (as amended 1986); and other compounds amenable to extraction and analysis by capillary column gas chromatography-mass spectrometry (GC/MS).

Stable isotopically-labeled analogs of the compounds of interest are added to the sample. If the solids content is less than 1%, a 1-L sample is extracted at pH 12–13, then at pH <2 with methylene chloride using continuous extraction techniques.

If the solids content is 30% or less, the sample is diluted to 1% solids with reagent water, homogenized ultrasonically, and extracted at pH 12–13, then at pH <2 with methylene chloride using continuous extraction techniques. If the solids content is greater than 30%, the sample is extracted using ultrasonic techniques.

Each extract is dried over sodium sulfate, concentrated to a volume of 5 mL, cleaned up using GPC, if necessary, and concentrated. Extracts are concentrated to 1 mL if GPC is not performed, and to 0.5 mL if GPC is performed.

An internal standard is added to the extract, and a 1-mL aliquot of the extract is injected into the GC. The compounds are separated by GC and detected by a MS. The labeled compounds serve to correct the variability of the analytical technique.

INTERFERENCES Solvents, reagents, glassware, and other sample processing hardware may yield artifacts and/or elevated baselines causing misinterpretation of chromatograms and spectra. Materials used in the analysis must be demonstrated to be free from interferences under the conditions of analysis by running method blanks initially and with each sample lot (sample started through the extraction process on a given 8-h shift, to a maximum of 20). Specific selection of reagents and purification of solvents by distillation in all glass systems may be required. Glassware and, where possible, reagents are cleaned by solvent rinse and baking at 450°C for 1-h minimum. Interferences coextracted from samples will vary considerably from source to source, depending on the diversity of the site being sampled.

INSTRUMENTATION Major instrumentation includes a GC with a splitless or on-column injection port for capillary column, a MS with 70 eV electron impact ionization, and a data system to collect and record MS data, and process it. A K-D apparatus is used to concentrate extracts.

GC Column: 30 m × 0.25 mm I.D. 5% phenyl, 94% methyl, 1% vinyl silicone bonded phased fused silica capillary column.

PRECISION & ACCURACY The detection limits of the method are usually dependent on the level of interferences rather than instrumental limitations. The limits typify the minimum quantities that can be detected with no interferences present.

The minimum level (in µg/mL) was not listed. This is defined as a minimum level at which the analytical system shall give recognizable mass spectra (background corrected) and acceptable calibration points.

The MDL (in µg/kg) in low solids was not listed and in high solids was not listed; these were determined in digested sludge (low solids) and in filter cake or compost (high solids).

The labeled and native compound initial precision as standard deviation (in µg/L) was not listed.
The labeled and native compound initial accuracy as average recovery (in µg/L) was not listed.

SAMPLE COLLECTION, PRESERVATION & HANDLING Collect samples in glass containers. Aqueous samples which flow freely are collected in refrigerated bottles using automatic sampling equipment. Solid samples are collected as grab samples using widemouth jars. Maintain samples at 0 to 4°C from the time of collection until extraction. If residual chlorine is present in aqueous samples, add 80 mg sodium thiosulfate/L of water. Begin sample extraction within 7 days of collection, and analyze all extracts within 40 days of extraction.

SAMPLE PREPARATION Samples containing 1% solids or less are extracted directly using continuous liquid-liquid extraction techniques. Samples containing 1 to 30% solids are diluted to the 1% level with reagent water and extracted using continuous liquid-liquid extraction techniques. Samples containing greater than 30% solids are extracted using ultrasonic techniques.

Base/neutral extraction — Adjust the pH of the waters in the extractors to 12–13 with 6 *N* NaOH. Extract with methylene chloride for 24–48 h.
Acid extraction — Adjust the pH of the waters in the extractors to 2 or less using 6 *N* sulfuric acid. Extract with methylene chloride for 24–48 h.
Ultrasonic extraction of high solids samples — Add anhydrous sodium sulfate to the sample and QC aliquot(s). Add acetone:methylene chloride (1:1) to the sample and mix thoroughly

Concentrate extracts using a K-D apparatus.

QUALITY CONTROL The analyst is permitted to modify this method to improve separations or lower the costs of measurements, provided all performance specifications are met. Analyses of blanks are required to demonstrate freedom from contamination. When results of spikes indicate atypical method performance for samples, the samples are diluted to bring method performance within acceptable limits.

For low solids (aqueous samples), extract, concentrate, and analyze two sets of four 1-L aliquots (8 aliquots total) of the

precision and recovery standard. For high solids samples, two sets of four 30-g aliquots of the high solids reference matrix are used.

Spike all samples with labeled compounds to assess method performance. Compute percent recovery of the labeled compounds using the internal standard method. Compare the labeled compound recovery for each compound with the corresponding labeled compound recovery.

Reagent water and high solids reference matrix blanks are analyzed to demonstrate freedom from contamination. Extract and concentrate a 1-L reagent water blank or a high solids reference matrix blank with each sample's lot (samples started through the extraction process on the same 8-h shift, to a maximum of 20 samples).

Field replicates may be collected to determine the precision of the sampling technique, and spiked samples may be required to determine the accuracy of the analysis when the internal standard method is used.

REFERENCE Semivolatile Organic Compounds by Isotope Dilution GC/MS. Office of Water Regulation and Standards, U.S. EPA Industrial Technology Div., Washington, DC, EPA Method 1625, Rev. C, June 1989 (contact W.A. Telliard, U.S. EPA, Office of Water Regulations and Standards, 401 M St., SW, Washington, DC, 20460. Phone: 202-382-7131).

Aramite
CAS #140-57-8   **EPA Method 8270**

TITLE Semivolatile Organic Compounds by GC/MS

MATRIX This method is used to determine the concentration of semivolatile organic compounds in extracts prepared from all types of solid waste matrices, soils, and groundwater. Although surface waters are not specifically mentioned, this method should be applicable to water samples from rivers, lakes, etc.

METHOD SUMMARY This method covers 259 semivolatile organic compounds. In very limited applications direct injection of the sample into the GC/MS system may be appropriate, but this results in very high detection limits (approximately 10,000 µg/L). Typically, a 1-L liquid sample containing surrogate and matrix spiking standards is extracted in a continuous extractor first under acid conditions and then under basic conditions. Typically 30 g of a solid sample containing surrogate and matrix spiking standards is extracted ultrasonically. After concentrating the extract to 1 mL it is spiked with 10 µL of an internal standard solution just prior to analysis by GC/MS. The volume injected should contain about 100 ng of base/neutral and 200 ng of acid surrogates (for a 1-µL injection). Analysis is performed by GC/MS using a capillary GC column.

INTERFERENCES Raw GC/MS data from all blanks, samples, and spikes must be evaluated for interferences. Contamination by carryover can occur whenever high-concentration and low-concentration samples are sequentially analyzed. To reduce carryover, the sample syringe must be rinsed out between samples with solvent. Whenever an unusually concentrated sample is encountered, it should be followed by the analysis of blank solvent to check for cross-contamination.

INSTRUMENTATION A GC/MS and a data system are required. The GC column used is a 30 m × 0.25 mm I.D. (or 0.32 mm I.D.) 1-µm film thickness silicone-coated fused silica capillary column. A continuous liquid-liquid extractor equipped with Teflon® or glass connection joints and stopcocks requiring no lubrication, a K-D concentrating apparatus, water bath, and an ultrasonic disrupter with a minimum power of 300 W and with pulsing capability are also required.

PRECISION & ACCURACY The estimated quantitation limit (EQL) of Method 8270B for determining an individual compound is approximately 1 mg/kg (wet weight) for soil or sediment samples, 1–200 mg/kg for wastes (dependent on matrix and method of preparation), and 10 µg/L for groundwater samples. EQLs will be proportionately higher for sample extracts that require dilution to avoid saturation of the detector.

The EQL(b) for groundwater in µg/L is 20.
The EQL (a, b) for low concentrations in soil and sediment in µg/kg is not determined.
Accuracy as µg/L is not listed.
Overall precision in µg/L is not listed.

(a) *EQLs listed for soil/sediment are based on wet weight. Normally data is reported in a dry-weight basis; therefore, EQLs will be higher based on the % dry weight of each sample. This calculation is based on a 30-g sample and gel permeation chromatography cleanup.*
(b) *Sample EQLs are highly matrix-dependent. The EQLs are provided for guidance and may not always be achievable.*
$C =$ *True value for concentration, in µg/L.*
$X =$ *Average recovery found for measurements of samples containing a concentration of C, in µg/L.*

ESTIMATED QUANTITATION LIMIT

Other Matrices	Factor (a)
High-concentration soil and sludges by sonicator	7.5
Non-water miscible waste	75

(a) *EQL for other matrices = [EQL for low soil/sediment] × [Factor]. This estimated EQL is similar to an EPA "Practical Quantitation Limit."*

SAMPLING METHOD

Liquid samples — Use a 1 or 2½ gallon amber glass bottle with a screw-top Teflon®-lined cover that has been prewashed with detergent and rinsed with distilled water and methanol (or isopropanol).

Soils, sediments, or sludges — Use an 8-oz. widemouth glass with a screw-top Teflon®-lined cover that has been prewashed with detergent and rinsed with distilled water and methanol (or isopropanol).

SAMPLE PRESERVATION

Liquid samples — If residual chlorine is present, add 3 mL of 10% sodium thiosulfate per gallon, cool to 4°C and store in a solvent-free refrigerator until analysis; if chlorine is not present, then eliminate the sodium thiosulfate addition.

Soils, sediments, or sludges — Cool samples to 4°C and store in a solvent-free refrigerator.

MHT Liquid samples must be extracted within 7 days and the extracts analyzed within 40 days. Soils, sediments, or sludges may be stored for a maximum of 14 days and the extracts analyzed within 40 days.

SAMPLE PREPARATION

Liquid samples — Transfer 1 L quantitatively to a continuous extractor. If high concentrations are anticipated, a smaller volume may be used and then diluted with organic-free reagent water to 1 L. Adjust pH, if necessary, to pH <2 using 1:1 (V/V) sulfuric acid. Pipette 1.0 mL of a surrogate standard spiking solution into each sample. For the sample in each analytical batch selected for spiking, add 1.0 mL of a matrix spiking standard. For base/neutral acid analysis, the amount of the surrogates and matrix spiking compounds added to the sample should result in a final concentration of 100 ng/µL of each analyte in the extract to be analyzed (assuming a 1-µL injection). Extract with methylene chloride for 18–24 h. Next, adjust the pH of the aqueous phase to pH >11 using 10 *N* sodium hydroxide and extract it with methylene chloride again for 18–24 h. Dry the extract through a column containing anhydrous sodium sulfate and concentrate it to 1 mL using a K-D concentrator.

Soils, sediments, or sludges — Use 30 g of sample. Nonporous or wet samples (gummy or clay type) that do not have a free-flowing sandy texture must be mixed with anhydrous sodium sulfate until the sample is free flowing. Add 1 mL of surrogate standards to all samples, spikes, standards, and blanks. For the sample in each analytical batch selected for spiking, add 1.0 mL of a matrix spiking standard. For base/neutral acid analysis, the amount added of the surrogates and matrix spiking compounds should result in a final concentration of 100 ng/µL of each base/neutral analyte and 200 ng/µL of each acid analyte in the extract to be analyzed (assuming a 1-µL injection). Immediately add a 100-mL mixture of 1:1 methylene chloride:acetone and extract the sample ultrasonically for 3 min and then decant or filter the extracts. Repeat the extraction two or more times. Dry the extract using a column with anhydrous sodium sulfate and concentrate it to 1 mL in a K-D concentrator.

QUALITY CONTROL A methylene chloride solution containing 50 ng/µL of decafluorotriphenylphosphine (DFTPP) is used for tuning the GC/MS system each 12-h shift. A system performance check also must be made during every 12-h shift. A standard containing 50 ng/µL each of 4,4′-DDT, pentachlorophenol, and benzidine is required to verify injection port inertness and GC column performance. A calibration standard at mid-concentration, containing each compound of interest, including all required surrogates, must be performed every 12 h during analysis. After the system performance check is met, calibration check compounds (CCCs) are used to check the validity of the initial calibration.

The internal standard responses and retention times in the calibration check standard must be evaluated immediately after or during data acquisition. If the retention time for any internal standard changes by more than 30 seconds from the last check calibration (12 h), the chromatographic system must be inspected for malfunctions and corrections must be made, as required. If the electron ionization current plot (EICP) area for any of the internal standards changes by a factor of two from the last daily calibration standard check, the mass spectrometer must be inspected for malfunctions and corrections must be made, as appropriate.

Demonstrate, through the analysis of a reagent water blank, that interferences from the analytical system, glassware, and reagents are under control. The blank samples should be carried through all stages of the sample preparation and measurement steps. For each analytical batch (up to 20 samples), a reagent blank, matrix spike, and matrix spike duplicate/duplicate must be analyzed (the frequency of the spikes may be different for different monitoring programs). The blank and spiked samples must be carried through all stages of the sample preparation and measurement steps. A QC reference sample concentrate containing each analyte at a concentration of 100 mg/L in methanol is required.

REFERENCE Test Methods for Evaluating Solid Waste (SW-846). U.S. EPA 1983, Method 8270B, Rev. 2, Nov. 1990. Office of Solid Waste, Washington, DC.

Aroclor 1016   EPA Method 505
CAS #12674-11-2

TITLE Analysis of Organohalide Pesticides and Commercial Polychlorinated Biphenyl (PCB) Products in Water by Microextraction and Gas Chromatography. U.S. EPA Method 505, Rev. 2.0, 1989.

MATRIX This method is applicable to drinking water and raw source water. The latter should include most surface water and groundwater sources.

METHOD SUMMARY Method 505 covers 25 pesticides and commercial PCB products. This is a very sensitive method that is more useful for monitoring than for exploratory analyses. 5-mL samples of water are saturated with sodium chloride and then extracted by shaking with 2 mL of hexane. The sample extracts are transferred to an auto sampler setup to inject 1- to 2-µL portions into a gas chromatograph (GC) for analysis. Alternatively, 1- to 2-µL portions of samples, blanks, and standards may be manually injected. Each extract is analyzed by capillary GC/ECD with confirmation using either a second capillary column or GC/MS. The electron capture detector is easy to use, but it is a nonselective detector. The microextraction technique also eliminates the expensive sample preparation costs of other methods, but it has the disadvantage of being less sensitive than most because the extracts are not concentrated.

INTERFERENCES Method interferences may be caused by contaminants in solvents, reagents, glassware, and other sample processing apparatus that lead to discrete artifacts or elevated baselines. Interfering contamination may occur when a sample containing low concentrations of analytes is analyzed immediately following a sample containing relatively high concentrations of the analytes. Matrix interferences also may be caused by contaminants that are coextracted from the sample; cleanup of sample extracts may be necessary in these cases. Some pesticides and commercial PCB products from aqueous solutions adhere to glass surfaces, so sample transfers and contact with glass surfaces should be minimized. Some pesticides are rapidly oxidized by chlorine so dechlorination with sodium thiosulfate at the time of sample collection is important. Also, splitless injectors may cause degradation of some pesticides.

INSTRUMENTATION A gas chromatograph/electron capture detector/data system, with temperature programming and split/splitless injector suitable for use with capillary columns is needed.

Column 1: 0.32 mm I.D. × 30 m fused silica capillary with chemically bond methyl polysiloxane phase (DB-1, 1.0 μm film, or equivalent).

Column 2: 0.32 mm I.D. × 30 m fused silica capillary with 1:1 mixed phase of dimethyl silicone and polyethylene glycol (Durawax-DX3, 0.25 μm film, or equivalent).

Column 3: 0.32 mm I.D. × 25 m fused silica capillary with chemically bonded 50:50 methyl-phenyl silicone (OV-17, 1.5 μm film, or equivalent).

Column 1 should be used as the primary analytical column. Columns 2 and 3 are recommended for use as confirmatory columns when GC/MS confirmation is not available.

PRECISION & ACCURACY Method detection limits are dependent upon the characteristics of the gas chromatographic system used. Analytes that are not separated chromatographically cannot be individually identified and used in the same calibration mixture or water samples unless an alternative technique for identification and quantification, such as mass spectrometry, is used.

The concentration(s) (in μg/L) used for these QC measurements was 1.0.

The MDL (in μg/L) was 0.08.

The accuracy (% recovery) for reagent water at the above concentration(s) was not available and the precision (%) was 6.6.

The accuracy (% recovery) for groundwater at the above concentration(s) was not listed and the precision (%) was not listed.

The accuracy (% recovery) for tap water at the above concentration(s) was 97 and the precision (5) was 7.5.

Note: No range of concentrations is provided with this method.

SAMPLING METHOD Collect samples using a 40-mL screw-cap vial (prewashed with detergent, rinsed with distilled water and oven dried at 400°C for one h) with a Teflon®-faced silicone septum. Collect bubble-free samples and place the septum with the Teflon® side down on the water.

SAMPLE PRESERVATION If residual chlorine is present in the water add about 3 mg of sodium thiosulfate to each vial before samples are collected to remove the chlorine. Alternatively, add 75 μL of 0.04 g/mL solution of sodium thiosulfate to each vial just prior to sampling. Immediately cool samples to 4°C and store them in a solvent-free refrigerator at 4°C until analysis.

MHT The maximum holding time is 14 days from the time the sample was collected until it must be analyzed.

SAMPLE PREPARATION Remove the sample from storage and allow it to come to room temperature. Remove a 5-mL volume from each container and weigh the container to the nearest 0.1 g. Add 6 g of sodium chloride and 2.0 mL of hexane to each sample bottle. Recap the sample and shake it vigorously for one min. Allow the water and hexane phases to separate, remove the cap, and transfer 0.5 mL of hexane into an auto sampler vial using a disposable glass pipette. Transfer the remaining hexane phase into a second auto sampler vial and store at 4°C for reanalysis, if necessary. Discard the remaining sample/hexane mixture and reweigh the empty container to determine net weight of sample.

QUALITY CONTROL Minimum quality control requirements are initial demonstration of lab capability, analysis of lab reagent blanks, fortified blanks, fortified sample matrix, and quality control samples. The lab must analyze at least one fortified blank per sample set, or at least one for every 20 samples. The fortifying concentration of each analyte should be 10 times the method detection limit or the maximum calibration limit (MCL), whichever is less. Calculate accuracy as percent recovery and develop control limits from the mean percent recovery and standard deviation.

The lab must add a known concentration of the analytes to a minimum of 10% of the routine samples, or one lab fortified sample matrix per sample set. Calculate the percent recovery for each analyte and compare to the control limits established from the analyses of the fortified blanks.

EPA CONTACT & HOTLINE For technical questions contact Dr. Baldev Bathija, U.S. EPA, Office of Ground Water and Drinking Water (WH-550D), 401 M St. SW, Washington, DC 20460. Tel. (202) 260-3040. For further information the EPA Safe Drinking Water Hotline may be called at: (800) 426-4791.

REFERENCE Methods for the Determination of Organic Compounds in Drinking Water, EPA/600/4-88/039 (revised July 1991). U.S. EPA Environmental Monitoring Systems Laboratory, Cincinnati, OH, 45268, U.S.A. Available from the National Technical Information Service (NTIS), 5285 Port Royal Road, Springfield, VA 22161; Tel. 800-553-6847. NTIS Order Number is PB91-231480.

Aroclor 1016 **EPA Method 625**
CAS #12674-11-2

TITLE Base/Neutrals and Acids, U.S. EPA Method 625

MATRIX This methods covers municipal and industrial wastewater.

METHOD SUMMARY Approximately 1 L of sample is serially extracted with methylene chloride at a pH greater than 11 and again at a pH less than 2 using a separatory funnel or a continuous extractor. The methylene chloride extract is dried, concentrated to a volume of 1 mL, and analyzed by GC/MS. Qualitative identification of the parameters in the extract is performed using the retention time and the relative abundance of three characteristic masses (m/z). Qualitative analysis is performed using either external or internal standard techniques with a single characteristic m/z

INTERFERENCES Method interferences may be caused by contaminants in solvents, reagents, glassware, and other sample processing hardware. Glassware must be scrupulously cleaned. Glassware should be heated in a muffle furnace at 400°C for 5 to 30 min. Some thermally stable materials, such as PCBs, may not be eliminated by this treatment. Solvent rinses with acetone and pesticide quality hexane may be substituted for the muffle furnace heating. Matrix interferences may be caused by contaminants that are coextracted from the sample. The base/neutral extraction may cause significantly reduced recovery of phenols. The packed gas chromatographic columns recommended for the basic fraction may not exhibit sufficient resolution for some analytes.

INSTRUMENTATION A GC/MS system with an injection port designed for on-column injection when using packed columns and for splitless injection when using capillary columns.

Column for base/neutrals: 1.8 m long × 2 mm I.D. glass, packed with 3% SP-2550 on Supelcoport (100/120 mesh) or equivalent.
Column for acids: 1.8 m long × 2 mm I.D. glass, packed with 1% SP-1240DA on Supelcoport (100/120 mesh) or equivalent.

PRECISION & ACCURACY The MDL concentrations were obtained using reagent water. The MDL actually achieved in a given analysis will vary depending on instrument sensitivity and matrix effects. This method was tested by 15 laboratories using reagent water, drinking water, surface water, and industrial wastewater spiked at six concentrations over the range 5 to 100 µg/L. Single operator precision, overall precision, and method accuracy were found to be directly related to the concentration of the parameter matrix.

The MDL (in µg/L) in reagent water was not detected.
The standard deviation (in µg/L based on 4 recovery measurements) was not reported.
The range (in µg/L) for average recovery for 4 measurements was not reported.
The range (in %) for percent recovery was not reported.
Accuracy (in µg/L) as expected recovery for one or more measurements of a sample containing a true concentration of C was not reported.
Precision (in µg/L) as expected single analyst standard deviation of measurements at an average concentration found at X was not reported.
Overall precision (in µg/L) as expected interlaboratory standard deviation of measurements in an average concentration found at X was not reported.

C = *True value of the concentration in µg/L.*
X = *Average recovery found for measurements of samples containing a concentration at C in µg/L.*

SAMPLE PREPARATION Adjust the pH to >11 with sodium hydroxide and serially extract in a separatory funnel with methylene chloride or else in a continuous extractor. Next, adjust the pH to <2 with sulfuric acid and serially extract in a separatory funnel with methylene chloride or else in a continuous extractor. Dry the extracts separately through a column of anhydrous sodium sulfate and then concentrate each of the extracts to 1.0 mL using a K-D apparatus.

SAMPLE COLLECTION, PRESERVATION & HANDLING Grab samples must be collected in glass containers. All samples must be refrigerated at 4°C from the time of collection until extraction. If residual chlorine is present, add 80 mg of sodium thiosulfate/L of sample and mix well. All samples must be extracted within 7 days of collection and completely analyzed within 40 days of extraction.

QUALITY CONTROL Make an initial, one-time, demonstration of the ability to generate acceptable accuracy and precision with this method. Before processing any samples, the analyst must analyze a reagent water blank to demonstrate that interferences from the analytical system and glassware are under control. Each time a set of samples is extracted or reagents are changed, a reagent water blank must be processed. Spike and analyze a minimum of 5% of all samples to monitor and evaluate lab data quality. A QC check sample concentrate that contains each parameter of interest at a concentration of 100 µg/mL in acetone is required. PCBs and multicomponent pesticides may be omitted from this test.

After analysis of five spiked wastewater samples, calculate the average percent recovery and the standard deviation of the percent recovery. Spike all samples with the surrogate standard spiking solution and calculate the percent recovery of each surrogate compound.

REFERENCE Federal Register, Vol. 49, No. 209. Friday, Oct. 26, 1984.

Aroclor 1016   **EPA Method 8080**
CAS #12674-11-2

TITLE Organochlorine Pesticides and Polychlorinated Biphenyls By Gas Chromatography

MATRIX This method is used to determine the concentration of various organochlorine pesticides and polychlorinated biphenyls in extracts prepared from water, groundwater, soils, and sediments.

METHOD SUMMARY This method covers 26 pesticides and Aroclor (PCB) mixtures and it is suitable for monitoring-type analyses. After extraction, concentration and solvent exchange to hexane, a 2- to 5-µL sample aliquot is injected into a GC using the solvent flush technique, and the analytes are detected by an electron capture detector (ECD) or an electrolytic conductivity detector in the halogen mode (HECD). Both neat and diluted organic liquids may be analyzed by direct injection.

INTERFERENCES Interferences coextracted from the samples will vary considerably from source to source. Interferences by phthalate esters can pose a major problem in pesticide determinations when using the ECD. Cross-contamination of clean glassware routinely occurs when plastics are handled during extraction steps, especially when solvent-wetted surfaces are handled. The contamination from phthalate esters can be completely eliminated with a microcoulometric or electrolytic conductivity detector. Solvents, reagent, glassware, and other sample processing hardware may yield artifacts and/or interferences to sample analysis.

INSTRUMENTATION A gas chromatograph capable of on-column injections is needed. It must be equipped with an ECD or a HECD and one of the following GC columns:

Column 1: Supelcoport (100/120 mesh) coated with 1.5% SP-2250/1.95% SP-2401 packed in a 1.8 m × 4 mm I.D. glass column.

Column 2: Supelcoport (100/120 mesh) coated with 3% OV-1 in a 1.8 m × 4 mm I.D. glass column.

PRECISION & ACCURACY The method was tested by 20 laboratories using organic-free reagent water, drinking water, surface water, and three industrial wastewater spiked at six concentrations. Concentrations used in the study ranged from 0.5 to 30 µg/L for single-component pesticides and from 8.5 to 400 µg/L for multicomponent parameters. Overall precision and method accuracy were found to be directly related to the concentration of the analyte and essentially independent of the sample matrix. The sensitivity of this method usually depends on the concentration of interferences rather than on instrumental limitations.

MDL in µg/L was ND.
Concentration range in µg/L was 8.5–40.
Accuracy as recovery (x*) in µg/L was 0.81C+0.50.
Overall precision (S*) in µg/L was 0.15x +0.45.

x^* *Expected recovery for one or more measurements of a sample containing concentration C, in µg/L.*
$S^* =$ *Expected interlaboratory standard deviation of measurements at an average concentration found of the analyte in µg/L.*
$C =$ *True value for the concentration, in µg/L.*
$X =$ *Average recovery found for measurements of samples containing a concentration of C, in µg/L.*

SAMPLING METHOD
Liquid samples — Use a 1 or 2½ gallon amber glass bottle with a screw-top Teflon®-lined cover. Pre-wash the bottle with detergent, rinse with distilled water and methanol (or isopropanol).

Soil, sediments, and sludges — Use an 8-oz. widemouth glass with a screw-top Teflon®-lined cover. Pre-wash the bottle with detergent, rinse with distilled water and methanol (or isopropanol).

SAMPLE PRESERVATION Cool water, soil, sediment, or sludge samples immediately to 4°C.

Water samples — If residual chlorine is present, add 3 mL of 10% sodium thiosulfate per gallon and cool to 4°C. All extracts and samples should be stored under refrigeration.

MHT Liquid samples must be extracted within 7 days and the extracts must be analyzed within 40 days. Soils, sediments, and sludges may be stored for a maximum of 14 days prior to extraction.

SAMPLE PREPARATION
Liquid samples — Extract 1-L samples in a continuous extractor at pH 5–9 with methylene chloride after adding 1.0 mL of surrogate spiking solution to each sample. Pass the extract through a column of anhydrous sodium sulfate to dry and concentrate it in a K-D apparatus to 1 mL volume.

Soils, sediments and sludges — Rapidly weigh approximately 30 g of sample into a 400-mL beaker to avoid loss of the more volatile extractables. Nonporous or wet samples (gummy or clay type) that do not have a free-flowing sandy texture must be mixed with anhydrous sodium sulfate until the sample is free flowing. Add 1 mL of surrogate standards to all samples, spikes, standards, and blanks. Add 100 mL of 1:1 methylene chloride:acetone and extract ultrasonically. Decant and filter extracts, dry the extract by passing it through a drying column containing anhydrous sodium sulfate and concentrate to 1 mL in a K-D apparatus.

Hexane eolvent exchange — Add 50 mL of hexane, a new boiling chip, and concentrate until the apparent volume of liquid reaches 1 mL. Adjust the extract volume to 10.0 mL. Stopper the concentration tube and store refrigerated at 4°C if further processing will not be performed immediately. If the extract will be stored longer than two days, transfer it to a vial with Teflon®-lined screw-cap or crimp top.

QUALITY CONTROL Demonstrate through the analysis of a reagent water blank, that all glassware and reagents are interference free. Each time a set of samples is processed, a method blank should be processed as a safeguard against chronic lab contamination. A reagent blank, a matrix spike, and a duplicate or matrix spike duplicate must be performed for each analytical batch (up to a maximum of 20 samples) analyzed.

Analytical system performance must be verified by analyzing QC check samples. The QC check sample concentration should contain each single-component analyte at the following concentrations in acetone: 4,4'-DDD, 10 µg/mL; 4,4'-DDT, 10 µg/mL; endosulfan II, 10 µg/mL; endosulfan sulfate, 10 µg/mL; and any other single-component pesticide at 2 µg/mL. If the method is only to be used to analyze PCBs, Chlordane, or Toxaphene, the QC check sample concentrate should contain the most representative multicomponent parameter at a concentration of 50 µg/mL in acetone.

REFERENCE Test Methods for Evaluating Solid Waste (SW-846). U.S. EPA. 1983. Method 8080B, Rev. 2, Nov. 1990. Office of Solid Wastes, Washington, DC.

Aroclor 1016
CAS #12674-11-2

EPA Method 8270

TITLE Semivolatile Organic Compounds by GC/MS

MATRIX This method is used to determine the concentration of semivolatile organic compounds in extracts prepared from all types of solid waste matrices, soils, and groundwater. Although surface waters are not specifically mentioned, this method should be applicable to water samples from rivers, lakes, etc.

METHOD SUMMARY This method covers 259 semivolatile organic compounds. In very limited applications direct injection of the sample into the GC/MS system may be appropriate, but this results in very high detection limits (approximately 10,000 µg/L). Typically, a 1-L liquid sample, containing surrogate, and matrix spiking standards, is extracted in a continuous extractor first under acid conditions and then under basic conditions. Typically 30 g of a solid sample containing surrogate and matrix spiking standards is extracted ultrasonically. After concentrating the extract to 1 mL it is spiked with 10 µL of an internal standard solution just prior to analysis by GC/MS. The volume injected should contain about 100 ng of base/neutral and 200 ng of acid surrogates (for a 1-µL injection). Analysis is performed by GC/MS using a capillary GC column.

INTERFERENCES Raw GC/MS data from all blanks, samples, and spikes must be evaluated for interferences. Contamination by carryover can occur whenever high-concentration and low-concentration samples are sequentially analyzed. To reduce carryover, the sample syringe must be rinsed out between samples with solvent. Whenever an unusually concentrated sample is encountered, it should be followed by the analysis of blank solvent to check for cross-contamination.

INSTRUMENTATION A GC/MS and a data system are required. The GC column used is a 30 m × 0.25 mm I.D. (or 0.32 mm I.D.) 1-µm film thickness silicone-coated fused silica capillary column. A continuous liquid-liquid extractor equipped with Teflon® or glass connection joints and stopcocks requiring no lubrication, a K-D concentrating apparatus, water bath, and an ultrasonic disrupter with a minimum power of 300 W and with pulsing capability are also required.

PRECISION & ACCURACY The estimated quantitation limit (EQL) of Method 8270B for determining an individual compound is approximately 1 mg/kg (wet weight) for soil or sediment samples, 1–200 mg/kg for wastes (dependent on matrix and method of preparation), and 10 µg/L for groundwater samples. EQLs will be proportionately higher for sample extracts that require dilution to avoid saturation of the detector.

The EQL(b) for groundwater in µg/L is not listed.
The EQL (a, b) for low concentrations in soil and sediment in µg/kg is not listed.
Accuracy as µg/L is not listed.
Overall precision in µg/L is not listed.

(a) *EQLs listed for soil/sediment are based on wet weight. Normally data is reported in a dry-weight basis; therefore, EQLs will be higher based on the % dry weight of each sample. This calculation is based on a 30-g sample and gel permeation chromatography cleanup.*
(b) *Sample EQLs are highly matrix-dependent. The EQLs are provided for guidance and may not always be achievable.*
C = *True value for concentration, in µg/L.*
X = *Average recovery found for measurements of samples containing a concentration of C, in µg/L.*

ESTIMATED QUANTITATION LIMIT

Other Matrices	Factor (a)
High-concentration soil and sludges by sonicator	7.5
Non-water miscible waste	75

(a) *EQL for other matrices = [EQL for low soil/sediment] × [Factor]. This estimated EQL is similar to an EPA "Practical Quantitation Limit."*

SAMPLING METHOD
Liquid samples — Use a 1 or 2½ gallon amber glass bottle with a screw-top Teflon®-lined cover that has been prewashed with detergent and rinsed with distilled water and methanol (or isopropanol).

Soils, sediments, or sludges — Use an 8-oz. widemouth glass with a screw-top Teflon®-lined cover that has been prewashed with detergent and rinsed with distilled water and methanol (or isopropanol).

SAMPLE PRESERVATION
Liquid samples — If residual chlorine is present, add 3 mL of 10% sodium thiosulfate per gallon, cool to 4°C and store in a solvent-free refrigerator until analysis; if chlorine is not present, then eliminate the sodium thiosulfate addition.

Soils, sediments, or sludges — Cool samples to 4°C and store in a solvent-free refrigerator.

MHT Liquid samples must be extracted within 7 days and the extracts analyzed within 40 days. Soils, sediments, or sludges may be stored for a maximum of 14 days and the extracts analyzed within 40 days.

SAMPLE PREPARATION
Liquid samples — Transfer 1 L quantitatively to a continuous extractor. If high concentrations are anticipated, a smaller volume may be used and then diluted with organic-free reagent water to 1 L. Adjust pH, if necessary, to pH <2 using 1:1 (V/V) sulfuric acid. Pipette 1.0 mL of a surrogate standard spiking solution into each sample. For the sample in each analytical batch selected for spiking, add 1.0 mL of a matrix spiking standard. For base/neutral acid analysis, the amount of the surrogates and matrix spiking compounds added to the sample should result in a final concentration of 100 ng/µL of each analyte in the extract to be analyzed (assuming a 1-µL injection). Extract with methylene chloride for 18–24 h. Next, adjust the pH of the aqueous phase to pH >11 using 10 N sodium hydroxide and extract it with methylene chloride again for 18–24 h. Dry the extract through a column containing anhydrous sodium sulfate and concentrate it to 1 mL using a K-D concentrator.

Soils, sediments, or sludges — Use 30 g of sample. Nonporous or wet samples (gummy or clay type) that do not have a free-flowing sandy texture must be mixed with anhydrous sodium sulfate until the sample is free flowing. Add 1 mL of surrogate standards to all samples, spikes, standards, and blanks. For the sample in each analytical batch selected for spiking, add 1.0 mL of a matrix spiking standard. For base/neutral acid analysis, the amount added of the surrogates and matrix spiking compounds

should result in a final concentration of 100 ng/μL of each base/neutral analyte and 200 ng/μL of each acid analyte in the extract to be analyzed (assuming a 1-μL injection). Immediately add a 100 mL mixture of 1:1 methylene chloride:acetone and extract the sample ultrasonically for 3 min and then decant or filter the extracts. Repeat the extraction two or more times. Dry the extract using a column with anhydrous sodium sulfate and concentrate it to 1 mL in a K-D concentrator.

QUALITY CONTROL A methylene chloride solution containing 50 ng/μL of decafluorotriphenylphosphine (DFTPP) is used for tuning the GC/MS system each 12-h shift. A system performance check also must be made during every 12-h shift. A standard containing 50 ng/μL each of 4,4'-DDT, pentachlorophenol, and benzidine is required to verify injection port inertness and GC column performance. A calibration standard at mid-concentration, containing each compound of interest, including all required surrogates, must be performed every 12 h during analysis. After the system performance check is met, calibration check compounds (CCCs) are used to check the validity of the initial calibration.

The internal standard responses and retention times in the calibration check standard must be evaluated immediately after or during data acquisition. If the retention time for any internal standard changes by more than 30 seconds from the last check calibration (12 h), the chromatographic system must be inspected for malfunctions and corrections must be made, as required. If the electron ionization current plot (EICP) area for any of the internal standards changes by a factor of two from the last daily calibration standard check, the mass spectrometer must be inspected for malfunctions and corrections must be made, as appropriate.

Demonstrate, through the analysis of a reagent water blank, that interferences from the analytical system, glassware, and reagents are under control. The blank samples should be carried through all stages of the sample preparation and measurement steps. For each analytical batch (up to 20 samples), a reagent blank, matrix spike, and matrix spike duplicate/duplicate must be analyzed (the frequency of the spikes may be different for different monitoring programs). The blank and spiked samples must be carried through all stages of the sample preparation and measurement steps. A QC reference sample concentrate containing each analyte at a concentration of 100 mg/L in methanol is required.

REFERENCE Test Methods for Evaluating Solid Waste (SW-846). U.S. EPA 1983, Method 8270B, Rev. 2, Nov. 1990. Office of Solid Waste, Washington, DC.

Aroclor 1221   EPA Method 505
CAS #11104-28-2

TITLE Analysis of Organohalide Pesticides and Commercial Polychlorinated Biphenyl (PCB) Products in Water by Microextraction and Gas Chromatography. U.S. EPA Method 505, Rev. 2.0, 1989.

MATRIX This method is applicable to drinking water and raw source water. The latter should include most surface water and groundwater sources.

METHOD SUMMARY Method 505 covers 25 pesticides and commercial PCB products. This is a very sensitive method that is more useful for monitoring than for exploratory analyses. 5-mL samples of water are saturated with sodium chloride and then extracted by shaking with 2 mL of hexane. The sample extracts are transferred to an autosampler setup to inject 1- to 2-μL portions into a gas chromatograph (GC) for analysis. Alternatively, 1- to 2-μL portions of samples, blanks, and standards may be manually injected. Each extract is analyzed by capillary GC/ECD with confirmation using either a second capillary column or GC/MS. The electron capture detector is easy to use, but it is a nonselective detector. The microextraction technique also eliminates the expensive sample preparation costs of other methods, but it has the disadvantage of being less sensitive than most because the extracts are not concentrated.

INTERFERENCES Method interferences may be caused by contaminants in solvents, reagents, glassware, and other sample processing apparatus that lead to discrete artifacts or elevated baselines. Interfering contamination may occur when a sample containing low concentrations of analytes is analyzed immediately following a sample containing relatively high concentrations of the analytes. Matrix interferences also may be caused by contaminants that are coextracted from the sample; cleanup of sample extracts may be necessary in these cases. Some pesticides and commercial PCB products from aqueous solutions adhere to glass surfaces, so sample transfers and contact with glass surfaces should be minimized. Some pesticides are rapidly oxidized by chlorine so dechlorination with sodium thiosulfate at the time of sample collection is important. Also, splitless injectors may cause degradation of some pesticides.

INSTRUMENTATION A gas chromatograph/electron capture detector/data system, with temperature programming and split/splitless injector suitable for use with capillary columns is needed.

Column 1: 0.32 mm I.D. × 30 m fused silica capillary with chemically bond methyl polysiloxane phase (DB-1, 1.0 μm film, or equivalent).

Column 2: 0.32 mm I.D. × 30 m fused silica capillary with 1:1 mixed phase of dimethyl silicone and polyethylene glycol (Durawax-DX3, 0.25 μm film, or equivalent).

Column 3: 0.32 mm I.D. × 25 m fused silica capillary with chemically bonded 50:50 methyl-phenyl silicone (OV-17, 1.5 μm film, or equivalent).

Column 1 should be used as the primary analytical column. Columns 2 and 3 are recommended for use as confirmatory columns when GC/MS confirmation is not available.

PRECISION & ACCURACY Method detection limits are dependent upon the characteristics of the gas chromatographic system used. Analytes that are not separated chromatographically cannot be individually identified and used in the same calibration mixture or water samples unless an alternative technique

for identification and quantification, such as mass spectrometry, is used.

The concentration(s) (in µg/L) used for these QC measurements was 180.

The MDL (in µg/L) was 15.0.

The accuracy (% recovery) for reagent water at the above concentration(s) was not available and the precision (%) was 8.3.

The accuracy (% recovery) for groundwater at the above concentration(s) was not listed and the precision (%) was not listed.

The accuracy (% recovery) for tap water at the above concentration(s) was 92 and the precision (5) was 9.6.

Note: No range of concentrations is provided with this method.

SAMPLING METHOD Collect samples in a 40-mL screw-cap vial (prewashed with detergent, rinsed in distilled water and oven dried at 400°C for one h) with a Teflon®-faced silicone septum. Collect bubble-free samples and place the septum with the Teflon® side down on the water.

SAMPLE PRESERVATION If residual chlorine is present in the water add about 3 mg of sodium thiosulfate to each vial before samples are collected to remove the chlorine. Alternatively, add 75 µL of 0.04 g/mL solution of sodium thiosulfate to each vial just prior to sampling. Immediately cool samples to 4°C and store them in a solvent-free refrigerator at 4°C until analysis.

MHT The maximum holding time is 14 days from the time the sample was collected until it must be analyzed.

SAMPLE PREPARATION Remove the sample from storage and allow it to come to room temperature. Remove a 5-mL volume from each container and weigh the container to the nearest 0.1 g. Add 6 g of sodium chloride and 2.0 mL of hexane to each sample bottle. Recap the sample and shake it vigorously for one min. Allow the water and hexane phases to separate, remove the cap, and transfer 0.5 mL of hexane into an autosampler vial using a disposable glass pipette. Transfer the remaining hexane phase into a second autosampler vial and store at 4°C for reanalysis, if necessary. Discard the remaining sample/hexane mixture and reweigh the empty container to determine net weight of sample.

QUALITY CONTROL Minimum quality control requirements are initial demonstration of lab capability, analysis of lab reagent blanks, fortified blanks, fortified sample matrix, and quality control samples. The lab must analyze at least one fortified blank per sample set, or at least one for every 20 samples. The fortifying concentration of each analyte should be 10 times the method detection limit or the maximum calibration limit (MCL), whichever is less. Calculate accuracy as percent recovery and develop control limits from the mean percent recovery and standard deviation.

The lab must add a known concentration of the analytes to a minimum of 10% of the routine samples, or one lab fortified sample matrix per sample set. Calculate the percent recovery for each analyte and compare to the control limits established from the analyses of the fortified blanks.

EPA CONTACT & HOTLINE For technical questions contact Dr. Baldev Bathija, U.S. EPA, Office of Ground Water and Drinking Water (WH-550D), 401 M St. SW, Washington, DC 20460. Tel. (202) 260-3040. For further information the EPA Safe Drinking Water Hotline may be called at: (800) 426-4791.

REFERENCE Methods for the Determination of Organic Compounds in Drinking Water, EPA/600/4-88/039 (revised July 1991). U.S. EPA Environmental Monitoring Systems Laboratory, Cincinnati, OH, 45268, U.S.A. Available from the National Technical Information Service (NTIS), 5285 Port Royal Road, Springfield, VA 22161; Tel. 800-553-6847. NTIS Order Number is PB91-231480.

Aroclor 1221 **EPA Method 625**
CAS #11104-28-2

TITLE Base/Neutrals and Acids, U.S. EPA Method 625

MATRIX This methods covers municipal and industrial wastewaters.

METHOD SUMMARY Approximately 1 L of sample is serially extracted with methylene chloride at a pH greater than 11 and again at a pH less than 2 using a separatory funnel or a continuous extractor. The methylene chloride extract is dried, concentrated to a volume of 1 mL, and analyzed by GC/MS. Qualitative identification of the parameters in the extract is performed using the retention time and the relative abundance of three characteristic masses (m/z). Qualitative analysis is performed using either external or internal standard techniques with a single characteristic m/z.

INTERFERENCES Method interferences may be caused by contaminants in solvents, reagents, glassware, and other sample processing hardware. Glassware must be scrupulously cleaned. Glassware should be heated in a muffle furnace at 400°C for 5 to 30 min. Some thermally stable materials, such as PCBs, may not be eliminated by this treatment. Solvent rinses with acetone and pesticide quality hexane may be substituted for the muffle furnace heating. Matrix interferences may be caused by contaminants that are coextracted from the sample. The base/neutral extraction may cause significantly reduced recovery of phenols. The packed gas chromatographic columns recommended for the basic fraction may not exhibit sufficient resolution for some analytes.

INSTRUMENTATION A GC/MS system with an injection port designed for on-column injection when using packed columns and for splitless injection when using capillary columns.

Column for base/neutrals: 1.8 m long × 2 mm I.D. glass, packed with 3% SP-2550 on Supelcoport (100/120 mesh) or equivalent.

Column for acids: 1.8 m long × 2 mm I.D. glass, packed with 1% SP-1240DA on Supelcoport (100/120 mesh) or equivalent.

PRECISION & ACCURACY The MDL concentrations were obtained using reagent water. The MDL actually achieved in a given analysis will vary depending on instrument sensitivity and matrix effects. This method was tested by 15 laboratories

using reagent water, drinking water, surface water, and industrial wastewaters spiked at six concentrations over the range 5 to 100 µg/L. Single operator precision, overall precision, and method accuracy were found to be directly related to the concentration of the parameter matrix.

The MDL (in µg/L) in reagent water was 30.
The standard deviation (in µg/L based on 4 recovery measurements) was not reported.
The range (in µg/L) for average recovery for 4 measurements was not reported.
The range (in %) for percent recovery was not reported.
Accuracy (in µg/L) as expected recovery for one or more measurements of a sample containing a true concentration of C was not reported.
Precision (in µg/L) as expected single analyst standard deviation of measurements at an average concentration found at X was not reported.
Overall precision (in µg/L) as expected interlaboratory standard deviation of measurements in an average concentration found at X was not reported.

C = *True value of the concentration in µg/L.*
X = *Average recovery found for measurements of samples containing a concentration at C in µg/L.*

SAMPLE PREPARATION Adjust the pH to >11 with sodium hydroxide and serially extract in a separatory funnel with methylene chloride or else in a continuous extractor. Next, adjust the pH to <2 with sulfuric acid and serially extract in a separatory funnel with methylene chloride or else in a continuous extractor. Dry the extracts separately through a column of anhydrous sodium sulfate and then concentrate each of the extracts to 1.0 mL using a K-D apparatus.

SAMPLE COLLECTION, PRESERVATION & HANDLING
Grab samples must be collected in glass containers. All samples must be refrigerated at 4°C from the time of collection until extraction. If residual chlorine is present, add 80 mg of sodium thiosulfate/L of sample and mix well. All samples must be extracted within 7 days of collection and completely analyzed within 40 days of extraction.

QUALITY CONTROL Make an initial, one-time, demonstration of the ability to generate acceptable accuracy and precision with this method. Before processing any samples, the analyst must analyze a reagent water blank to demonstrate that interferences from the analytical system and glassware are under control. Each time a set of samples is extracted or reagents are changed, a reagent water blank must be processed. Spike and analyze a minimum of 5% of all samples to monitor and evaluate lab data quality. A QC check sample concentrate that contains each parameter of interest at a concentration of 100 µg/mL in acetone is required. PCBs and multicomponent pesticides may be omitted from this test.

After analysis of five spiked wastewater samples, calculate the average percent recovery and the standard deviation of the percent recovery. Spike all samples with the surrogate standard spiking solution and calculate the percent recovery of each surrogate compound.

REFERENCE *Federal Register*, Vol. 49, No. 209. Friday, Oct. 26, 1984.

Aroclor 1221 **EPA Method 8080**
CAS #11104-28-2

TITLE Organochlorine Pesticides and Polychlorinated Biphenyls By Gas Chromatography

MATRIX This method is used to determine the concentration of various organochlorine pesticides and polychlorinated biphenyls in extracts prepared from water, groundwater, soils, and sediments.

METHOD SUMMARY This method covers 26 pesticides and Aroclor (PCB) mixtures and it is suitable for monitoring-type analyses. After extraction, concentration and solvent exchange to hexane, a 2- to 5-µL sample aliquot is injected into a GC using the solvent flush technique, and the analytes are detected by an electron capture detector (ECD) or an electrolytic conductivity detector in the halogen mode (HECD). Both neat and diluted organic liquids may be analyzed by direct injection.

INTERFERENCES Interferences coextracted from the samples will vary considerably from source to source. Interferences by phthalate esters can pose a major problem in pesticide determinations when using the ECD. Cross-contamination of clean glassware routinely occurs when plastics are handled during extraction steps, especially when solvent-wetted surfaces are handled. The contamination from phthalate esters can be completely eliminated with a microcoulometric or electrolytic conductivity detector. Solvents, reagent, glassware, and other sample processing hardware may yield artifacts and/or interferences to sample analysis.

INSTRUMENTATION A gas chromatograph capable of on-column injections is needed. It must be equipped with an ECD or a HECD and one of the following GC columns:

Column 1: Supelcoport (100/120 mesh) coated with 1.5% SP-2250/1.95% SP-2401 packed in a 1.8 m × 4 mm I.D. glass column.

Column 2: Supelcoport (100/120 mesh) coated with 3% OV-1 in a 1.8 m × 4 mm I.D. glass column.

PRECISION & ACCURACY The method was tested by 20 laboratories using organic-free reagent water, drinking water, surface water, and three industrial wastewaters spiked at six concentrations. Concentrations used in the study ranged from 0.5 to 30 µg/L for single-component pesticides and from 8.5 to 400 µg/L for multicomponent parameters. Overall precision and method accuracy were found to be directly related to the concentration of the analyte and essentially independent of the sample matrix. The sensitivity of this method usually depends on the concentration of interferences rather than on instrumental limitations.

MDL in µg/L was ND.
Concentration range in µg/L was 8.5–40.
Accuracy as recovery (x^*) in µg/L was $0.96C+0.65$.
Overall precision (S^*) in µg/L was $0.35x-0.62$.

x*   *Expected recovery for one or more measurements of a sample containing concentration C, in μg/L.*

S* = *Expected interlaboratory standard deviation of measurements at an average concentration found of the analyte in μg/L.*

C = *True value for the concentration, in μg/L.*

X = *Average recovery found for measurements of samples containing a concentration of C, in μg/L.*

SAMPLING METHOD

Liquid samples — Use a 1 or 2½ gallon amber glass bottle with a screw-top Teflon®-lined cover. Pre-wash the bottle with detergent, rinse with distilled water and methanol (or isopropanol).

Soil, sediments, and sludges — Use an 8-oz. widemouth glass with a screw-top Teflon®-lined cover. Pre-wash the bottle with detergent, rinse with distilled water and methanol (or isopropanol).

SAMPLE PRESERVATION   Cool water, soil, sediment, or sludge samples immediately to 4°C.

Water samples — If residual chlorine is present, add 3 mL of 10% sodium thiosulfate per gallon and cool to 4°C. All extracts and samples should be stored under refrigeration.

MHT   Liquid samples must be extracted within 7 days and the extracts must be analyzed within 40 days. Soils, sediments, and sludges may be stored for a maximum of 14 days prior to extraction.

SAMPLE PREPARATION

Liquid samples — Extract 1-L samples in a continuous extractor at pH 5–9 with methylene chloride after adding 1.0 mL of surrogate spiking solution to each sample. Pass the extract through a column of anhydrous sodium sulfate to dry and concentrate it in a K-D apparatus to 1 mL volume.

Soils, sediments and sludges — Rapidly weigh approximately 30 g of sample into a 400-mL beaker to avoid loss of the more volatile extractables. Nonporous or wet samples (gummy or clay type) that do not have a free-flowing sandy texture must be mixed with anhydrous sodium sulfate until the sample is free flowing. Add 1 mL of surrogate standards to all samples, spikes, standards, and blanks. Add 100 mL of 1:1 methylene chloride:acetone and extract ultrasonically. Decant and filter extracts, dry the extract by passing it through a drying column containing anhydrous sodium sulfate and concentrate to 1 mL in a K-D apparatus.

Hexane eolvent exchange — Add 50 mL of hexane, a new boiling chip, and concentrate until the apparent volume of liquid reaches 1 Ml. Adjust the extract volume to 10.0 Ml. Stopper the concentration tube and store refrigerated at 4°C if further processing will not be performed immediately. If the extract will be stored longer than two days, transfer it to a vial with Teflon®-lined screw-cap or crimp top.

QUALITY CONTROL   Demonstrate through the analysis of a reagent water blank, that all glassware and reagents are interference free. Each time a set of samples is processed, a method blank should be processed as a safeguard against chronic lab contamination. A reagent blank, a matrix spike, and a duplicate or matrix spike duplicate must be performed for each analytical batch (up to a maximum of 20 samples) analyzed.

Analytical system performance must be verified by analyzing QC check samples. The QC check sample concentration should contain each single-component analyte at the following concentrations in acetone: 4,4'-DDD, 10 μg/mL; 4,4'-DDT, 10 μg/mL; endosulfan II, 10 μg/mL; endosulfan sulfate, 10 μg/mL; and any other single-component pesticide at 2 μg/mL. If the method is only to be used to analyze PCBs, Chlordane, or Toxaphene, the QC check sample concentrate should contain the most representative multicomponent parameter at a concentration of 50 μg/mL in acetone.

REFERENCE   Test Methods for Evaluating Solid Waste (SW-846). U.S. EPA. 1983. Method 8080B, Rev. 2, Nov. 1990. Office of Solid Wastes, Washington, DC.

Aroclor 1221   **EPA Method 8270**
CAS #11104-28-2

TITLE   Semivolatile Organic Compounds by GC/MS

MATRIX   This method is used to determine the concentration of semivolatile organic compounds in extracts prepared from all types of solid waste matrices, soils, and groundwater. Although surface waters are not specifically mentioned, this method should be applicable to water samples from rivers, lakes, etc.

METHOD SUMMARY   This method covers 259 semivolatile organic compounds. In very limited applications direct injection of the sample into the GC/MS system may be appropriate, but this results in very high detection limits (approximately 10,000 μg/L). Typically, a 1-L liquid sample, containing surrogate, and matrix spiking standards, is extracted in a continuous extractor first under acid conditions and then under basic conditions. Typically 30 g of a solid sample, containing surrogate, and matrix spiking standards, is extracted ultrasonically. After concentrating the extract to 1 mL it is spiked with 10 μL of an internal standard solution just prior to analysis by GC/MS. The volume injected should contain about 100 ng of base/neutral and 200 ng of acid surrogates (for a 1-μL injection). Analysis is performed by GC/MS using a capillary GC column.

INTERFERENCES   Raw GC/MS data from all blanks, samples, and spikes must be evaluated for interferences. Contamination by carryover can occur whenever high-concentration and low-concentration samples are sequentially analyzed. To reduce carryover, the sample syringe must be rinsed out between samples with solvent. Whenever an unusually concentrated sample is encountered, it should be followed by the analysis of blank solvent to check for cross-contamination.

INSTRUMENTATION   A GC/MS and a data system are required. The GC column used is a 30 m × 0.25 mm I.D. (or 0.32 mm I.D.) 1-μm film thickness silicone-coated fused silica capillary column. A continuous liquid-liquid extractor equipped with Teflon® or glass connection joints and stopcocks requiring no lubrication, a K-D concentrating apparatus, water

bath, and an ultrasonic disrupter with a minimum power of 300 W and with pulsing capability are also required.

PRECISION & ACCURACY The estimated quantitation limit (EQL) of Method 8270B for determining an individual compound is approximately 1 mg/kg (wet weight) for soil or sediment samples, 1–200 mg/kg for wastes (dependent on matrix and method of preparation), and 10 µg/L for groundwater samples. EQLs will be proportionately higher for sample extracts that require dilution to avoid saturation of the detector.

The EQL(b) for groundwater in µg/L is not listed.
The EQL (a, b) for low concentrations in soil and sediment in µg/kg is not listed.
Accuracy as µg/L is not listed.
Overall precision in µg/L is not listed.

(a) *EQLs listed for soil/sediment are based on wet weight. Normally data is reported in a dry-weight basis; therefore, EQLs will be higher based on the % dry weight of each sample. This calculation is based on a 30-g sample and gel permeation chromatography cleanup.*
(b) *Sample EQLs are highly matrix-dependent. The EQLs are provided for guidance and may not always be achievable.*
$C =$ *True value for concentration, in µg/L.*
$X =$ *Average recovery found for measurements of samples containing a concentration of C, in µg/L.*

ESTIMATED QUANTITATION LIMIT

Other Matrices	Factor (a)
High-concentration soil and sludges by sonicator	7.5
Non-water miscible waste	75

(a) EQL for other matrices = [EQL for low soil/sediment] × [Factor]. This estimated EQL is similar to an EPA "Practical Quantitation Limit."

SAMPLING METHOD
Liquid samples — Use a 1 or 2½ gallon amber glass bottle with a screw-top Teflon®-lined cover that has been prewashed with detergent and rinsed with distilled water and methanol (or isopropanol).

Soils, sediments, or sludges — Use an 8-oz. widemouth glass with a screw-top Teflon®-lined cover that has been prewashed with detergent and rinsed with distilled water and methanol (or isopropanol).

SAMPLE PRESERVATION
Liquid samples — If residual chlorine is present, add 3 mL of 10% sodium thiosulfate per gallon, cool to 4°C and store in a solvent-free refrigerator until analysis; if chlorine is not present, then eliminate the sodium thiosulfate addition.

Soils, sediments, or sludges — Cool samples to 4°C and store in a solvent-free refrigerator.

MHT Liquid samples must be extracted within 7 days and the extracts analyzed within 40 days. Soils, sediments, or sludges may be stored for a maximum of 14 days and the extracts analyzed within 40 days.

SAMPLE PREPARATION
Liquid samples — Transfer 1 L quantitatively to a continuous extractor. If high concentrations are anticipated, a smaller volume may be used and then diluted with organic-free reagent water to 1 L. Adjust pH, if necessary, to pH <2 using 1:1 (V/V) sulfuric acid. Pipette 1.0 mL of a surrogate standard spiking solution into each sample. For the sample in each analytical batch selected for spiking, add 1.0 mL of a matrix spiking standard. For base/neutral acid analysis, the amount of the surrogates and matrix spiking compounds added to the sample should result in a final concentration of 100 ng/µL of each analyte in the extract to be analyzed (assuming a 1-µL injection). Extract with methylene chloride for 18–24 h. Next, adjust the pH of the aqueous phase to pH >11 using 10 N sodium hydroxide and extract it with methylene chloride again for 18–24 h. Dry the extract through a column containing anhydrous sodium sulfate and concentrate it to 1 mL using a K-D concentrator.

Soils, sediments, or sludges — Use 30 g of sample. Nonporous or wet samples (gummy or clay type) that do not have a free-flowing sandy texture must be mixed with anhydrous sodium sulfate until the sample is free flowing. Add 1 mL of surrogate standards to all samples, spikes, standards, and blanks. For the sample in each analytical batch selected for spiking, add 1.0 mL of a matrix spiking standard. For base/neutral acid analysis, the amount added of the surrogates and matrix spiking compounds should result in a final concentration of 100 ng/µL of each base/neutral analyte and 200 ng/µL of each acid analyte in the extract to be analyzed (assuming a 1-µL injection). Immediately add a 100 mL mixture of 1:1 methylene chloride:acetone and extract the sample ultrasonically for 3 min and then decant or filter the extracts. Repeat the extraction two or more times. Dry the extract using a column with anhydrous sodium sulfate and concentrate it to 1 mL in a K-D concentrator.

QUALITY CONTROL A methylene chloride solution containing 50 ng/µL of decafluorotriphenylphosphine (DFTPP) is used for tuning the GC/MS system each 12-h shift. A system performance check also must be made during every 12-h shift. A standard containing 50 ng/µL each of 4,4'-DDT, pentachlorophenol, and benzidine is required to verify injection port inertness and GC column performance. A calibration standard at mid-concentration, containing each compound of interest, including all required surrogates, must be performed every 12 h during analysis. After the system performance check is met, calibration check compounds (CCCs) are used to check the validity of the initial calibration.

The internal standard responses and retention times in the calibration check standard must be evaluated immediately after or during data acquisition. If the retention time for any internal standard changes by more than 30 seconds from the last check calibration (12 h), the chromatographic system must be inspected for malfunctions and corrections must be made, as required. If the electron ionization current plot (EICP) area for any of the internal standards changes by a factor of two from the last daily calibration standard check, the mass spectrometer must be inspected for malfunctions and corrections must be made, as appropriate.

Demonstrate, through the analysis of a reagent water blank, that interferences from the analytical system, glassware, and reagents are under control. The blank samples should be carried through all stages of the sample preparation and measurement steps. For each analytical batch (up to 20 samples), a reagent blank, matrix spike, and matrix spike duplicate/duplicate must be analyzed (the frequency of the spikes may be different for different monitoring programs). The blank and spiked samples must be carried through all stages of the sample preparation and measurement steps. A QC reference sample concentrate containing each analyte at a concentration of 100 mg/L in methanol is required.

REFERENCE Test Methods for Evaluating Solid Waste (SW-846). U.S. EPA 1983, Method 8270B, Rev. 2, Nov. 1990. Office of Solid Waste, Washington, DC.

Aroclor 1232   EPA Method 505
CAS #11141-16-5

TITLE Analysis of Organohalide Pesticides and Commercial Polychlorinated Biphenyl (PCB) Products in Water by Microextraction and Gas Chromatography. U.S. EPA Method 505, Rev. 2.0, 1989.

MATRIX This method is applicable to drinking water and raw source water. The latter should include most surface water and groundwater sources.

METHOD SUMMARY Method 505 covers 25 pesticides and commercial PCB products. This is a very sensitive method that is more useful for monitoring than for exploratory analyses. 5-mL samples of water are saturated with sodium chloride and then extracted by shaking with 2 mL of hexane. The sample extracts are transferred to an autosampler setup to inject 1- to 2-µL portions into a gas chromatograph (GC) for analysis. Alternatively, 1- to 2-µL portions of samples, blanks, and standards may be manually injected. Each extract is analyzed by capillary GC/ECD with confirmation using either a second capillary column or GC/MS. The electron capture detector is easy to use, but it is a nonselective detector. The microextraction technique also eliminates the expensive sample preparation costs of other methods, but it has the disadvantage of being less sensitive than most because the extracts are not concentrated.

INTERFERENCES Method interferences may be caused by contaminants in solvents, reagents, glassware, and other sample processing apparatus that lead to discrete artifacts or elevated baselines. Interfering contamination may occur when a sample containing low concentrations of analytes is analyzed immediately following a sample containing relatively high concentrations of the analytes. Matrix interferences also may be caused by contaminants that are coextracted from the sample; cleanup of sample extracts may be necessary in these cases. Some pesticides and commercial PCB products from aqueous solutions adhere to glass surfaces, so sample transfers and contact with glass surfaces should be minimized. Some pesticides are rapidly oxidized by chlorine so dechlorination with sodium thiosulfate at the time of sample collection is important. Also, splitless injectors may cause degradation of some pesticides.

INSTRUMENTATION A gas chromatograph/electron capture detector/data system, with temperature programming and split/splitless injector suitable for use with capillary columns is needed.

Column 1: 0.32 mm I.D. × 30 m fused silica capillary with chemically bond methyl polysiloxane phase (DB-1, 1.0 µm film, or equivalent).

Column 2: 0.32 mm I.D. × 30 m fused silica capillary with 1:1 mixed phase of dimethyl silicone and polyethylene glycol (Durawax-DX3, 0.25 µm film, or equivalent).

Column 3: 0.32 mm I.D. × 25 m fused silica capillary with chemically bonded 50:50 methyl-phenyl silicone (OV-17, 1.5 µm film, or equivalent).

Column 1 should be used as the primary analytical column. Columns 2 and 3 are recommended for use as confirmatory columns when GC/MS confirmation is not available.

PRECISION & ACCURACY Method detection limits are dependent upon the characteristics of the gas chromatographic system used. Analytes that are not separated chromatographically cannot be individually identified and used in the same calibration mixture or water samples unless an alternative technique for identification and quantification, such as mass spectrometry, is used.

The concentration(s) (in µg/L) used for these QC measurements was 3.9.

The MDL (in µg/L) was 0.48.

The accuracy (% recovery) for reagent water at the above concentration(s) was not available and the precision (%) was 13.5.

The accuracy (% recovery) for groundwater at the above concentration(s) was not listed and the precision (%) was not listed.

The accuracy (% recovery) for tap water at the above concentration(s) was 86 and the precision (5) was 7.3.

Note: No range of concentrations is provided with this method.

SAMPLING METHOD Collect samples using a 40-mL screw-cap vial (prewashed with detergent, rinsed with distilled water and oven dried at 400°C for one h) with a Teflon®-faced silicone septum. Collect bubble-free samples and place the septum with the Teflon® side down on the water.

SAMPLE PRESERVATION If residual chlorine is present in the water add about 3 mg of sodium thiosulfate to each vial before samples are collected to remove the chlorine. Alternatively, add 75 µL of 0.04 g/mL solution of sodium thiosulfate to each vial just prior to sampling. Immediately cool samples to 4°C and store them in a solvent-free refrigerator at 4°C until analysis.

MHT The maximum holding time is 14 days from the time the sample was collected until it must be analyzed.

SAMPLE PREPARATION Remove the sample from storage and allow it to come to room temperature. Remove a 5-mL volume from each container and weigh the container to the nearest 0.1 g. Add 6 g of sodium chloride and 2.0 mL of hexane to each sample bottle. Recap the sample and shake it vigorously for one min. Allow the water and hexane phases to separate, remove the cap, and transfer 0.5 mL of hexane into an autosampler vial using a disposable glass pipette. Transfer the remaining hexane phase into a second autosampler vial and store at 4°C for reanalysis, if necessary. Discard the remaining sample/hexane mixture and reweigh the empty container to determine net weight of sample.

QUALITY CONTROL Minimum quality control requirements are initial demonstration of lab capability, analysis of lab reagent blanks, fortified blanks, fortified sample matrix, and quality control samples. The lab must analyze at least one fortified blank per sample set, or at least one for every 20 samples. The fortifying concentration of each analyte should be 10 times the method detection limit or the maximum calibration limit (MCL), whichever is less. Calculate accuracy as percent recovery and develop control limits from the mean percent recovery and standard deviation.

The lab must add a known concentration of the analytes to a minimum of 10% of the routine samples, or one lab fortified sample matrix per sample set. Calculate the percent recovery for each analyte and compare to the control limits established from the analyses of the fortified blanks.

EPA CONTACT & HOTLINE For technical questions contact Dr. Baldev Bathija, U.S. EPA, Office of Ground Water and Drinking Water (WH-550D), 401 M St. SW, Washington, DC 20460. Tel. (202) 260-3040. For further information the EPA Safe Drinking Water Hotline may be called at: (800) 426-4791.

REFERENCE Methods for the Determination of Organic Compounds in Drinking Water, EPA/600/4-88/039 (revised July 1991). U.S. EPA Environmental Monitoring Systems Laboratory, Cincinnati, OH, 45268, U.S.A. Available from the National Technical Information Service (NTIS), 5285 Port Royal Road, Springfield, VA 22161; Tel. 800-553-6847. NTIS Order Number is PB91-231480.

Aroclor 1232   **EPA Method 625**
CAS #11141-16-5

TITLE Base/Neutrals and Acids, U.S. EPA Method 625

MATRIX This methods covers municipal and industrial wastewaters.

METHOD SUMMARY Approximately 1 L of sample is serially extracted with methylene chloride at a pH greater than 11 and again at a pH less than 2 using a separatory funnel or a continuous extractor. The methylene chloride extract is dried, concentrated to a volume of 1 mL, and analyzed by GC/MS. Qualitative identification of the parameters in the extract is performed using the retention time and the relative abundance of three characteristic masses (m/z). Qualitative analysis is performed using either external or internal standard techniques with a single characteristic m/z.

INTERFERENCES Method interferences may be caused by contaminants in solvents, reagents, glassware, and other sample processing hardware. Glassware must be scrupulously cleaned. Glassware should be heated in a muffle furnace at 400°C for 5 to 30 min. Some thermally stable materials, such as PCBs, may not be eliminated by this treatment. Solvent rinses with acetone and pesticide quality hexane may be substituted for the muffle furnace heating. Matrix interferences may be caused by contaminants that are coextracted from the sample. The base/neutral extraction may cause significantly reduced recovery of phenols. The packed gas chromatographic columns recommended for the basic fraction may not exhibit sufficient resolution for some analytes.

INSTRUMENTATION A GC/MS system with an injection port designed for on-column injection when using packed columns and for splitless injection when using capillary columns.

Column for base/neutrals: 1.8 m long × 2 mm I.D. glass, packed with 3% SP-2550 on Supelcoport (100/120 mesh) or equivalent.

Column for acids: 1.8 m long × 2 mm I.D. glass, packed with 1% SP-1240DA on Supelcoport (100/120 mesh) or equivalent.

PRECISION & ACCURACY The MDL concentrations were obtained using reagent water. The MDL actually achieved in a given analysis will vary depending on instrument sensitivity and matrix effects. This method was tested by 15 laboratories using reagent water, drinking water, surface water, and industrial wastewaters spiked at six concentrations over the range 5 to 100 µg/L. Single operator precision, overall precision, and method accuracy were found to be directly related to the concentration of the parameter matrix.

The MDL (in µg/L) in reagent water was not detected.
The standard deviation (in µg/L based on 4 recovery measurements) was not reported.
The range (in µg/L) for average recovery for 4 measurements was not reported.
The range (in %) for percent recovery was not reported.
Accuracy (in µg/L) as expected recovery for one or more measurements of a sample containing a true concentration of C was not reported.
Precision (in µg/L) as expected single analyst standard deviation of measurements at an average concentration found at X was not reported.
Overall precision (in µg/L) as expected interlaboratory standard deviation of measurements in an average concentration found at X was not reported.

$C =$ *True value of the concentration in µg/L.*
$X =$ *Average recovery found for measurements of samples containing a concentration at C in µg/L.*

SAMPLE PREPARATION Adjust the pH to >11 with sodium hydroxide and serially extract in a separatory funnel with methylene chloride or else in a continuous extractor. Next, adjust the pH to <2 with sulfuric acid and serially extract in a separatory funnel with methylene chloride or else in a continuous extractor. Dry the extracts separately through a column of anhydrous sodium sulfate and then concentrate each of the extracts to 1.0 mL using a K-D apparatus.

SAMPLE COLLECTION, PRESERVATION & HANDLING
Grab samples must be collected in glass containers. All samples must be refrigerated at 4°C from the time of collection until extraction. If residual chlorine is present, add 80 mg of sodium thiosulfate/L of sample and mix well. All samples must be extracted within 7 days of collection and completely analyzed within 40 days of extraction.

QUALITY CONTROL Make an initial, one-time, demonstration of the ability to generate acceptable accuracy and precision with this method. Before processing any samples, the analyst must analyze a reagent water blank to demonstrate that interferences from the analytical system and glassware are under control. Each time a set of samples is extracted or reagents are changed, a reagent water blank must be processed. Spike and analyze a minimum of 5% of all samples to monitor and evaluate lab data quality. A QC check sample concentrate that contains each parameter of interest at a concentration of 100 µg/mL in acetone is required. PCBs and multicomponent pesticides may be omitted from this test.

After analysis of five spiked wastewater samples, calculate the average percent recovery and the standard deviation of the percent recovery. Spike all samples with the surrogate standard spiking solution and calculate the percent recovery of each surrogate compound.

REFERENCE Federal Register, Vol. 49, No. 209. Friday, Oct. 26, 1984.

Aroclor 1232 **EPA Method 8080**
CAS #11141-16-5

TITLE Organochlorine Pesticides and Polychlorinated Biphenyls By Gas Chromatography

MATRIX This method is used to determine the concentration of various organochlorine pesticides and polychlorinated biphenyls in extracts prepared from water, groundwater, soils, and sediments.

METHOD SUMMARY This method covers 26 pesticides and Aroclor (PCB) mixtures and it is suitable for monitoring-type analyses. After extraction, concentration and solvent exchange to hexane, a 2- to 5-µL sample aliquot is injected into a GC using the solvent flush technique, and the analytes are detected by an electron capture detector (ECD) or an electrolytic conductivity detector in the halogen mode (HECD). Both neat and diluted organic liquids may be analyzed by direct injection.

INTERFERENCES Interferences coextracted from the samples will vary considerably from source to source. Interferences by phthalate esters can pose a major problem in pesticide determinations when using the ECD. Cross-contamination of clean glassware routinely occurs when plastics are handled during extraction steps, especially when solvent-wetted surfaces are handled. The contamination from phthalate esters can be completely eliminated with a microcoulometric or electrolytic conductivity detector. Solvents, reagent, glassware, and other sample processing hardware may yield artifacts and/or interferences to sample analysis.

INSTRUMENTATION A gas chromatograph capable of on-column injections is needed. It must be equipped with an ECD or a HECD and one of the following GC columns:

Column 1: Supelcoport (100/120 mesh) coated with 1.5% SP-2250/1.95% SP-2401 packed in a 1.8 m × 4 mm I.D. glass column.

Column 2: Supelcoport (100/120 mesh) coated with 3% OV-1 in a 1.8 m × 4 mm I.D. glass column.

PRECISION & ACCURACY The method was tested by 20 laboratories using organic-free reagent water, drinking water, surface water, and three industrial wastewaters spiked at six concentrations. Concentrations used in the study ranged from 0.5 to 30 µg/L for single-component pesticides and from 8.5 to 400 µg/L for multicomponent parameters. Overall precision and method accuracy were found to be directly related to the concentration of the analyte and essentially independent of the sample matrix. The sensitivity of this method usually depends on the concentration of interferences rather than on instrumental limitations.

MDL in µg/L was ND.
Concentration range in µg/L was 8.5–40.
Accuracy as recovery (x^*) in µg/L was $0.91C+10.79$.
Overall precision (S^*) in µg/L was $0.31x +3.50$.

x^* = *Expected recovery for one or more measurements of a sample containing concentration C, in µg/L.*
S^* = *Expected interlaboratory standard deviation of measurements at an average concentration found of the analyte in µg/L.*
C = *True value for the concentration, in µg/L.*
X = *Average recovery found for measurements of samples containing a concentration of C, in µg/L.*

SAMPLING METHOD
Liquid samples — Use a 1 or 2½ gallon amber glass bottle with a screw-top Teflon®-lined cover. Pre-wash the bottle with detergent, rinse with distilled water and methanol (or isopropanol).

Soil, sediments and sludges — Use an 8-oz. widemouth glass with a screw-top Teflon®-lined cover. Pre-wash the bottle with detergent, rinse with distilled water and methanol (or isopropanol).

SAMPLE PRESERVATION Cool water, soil, sediment, or sludge samples immediately to 4°C.

Water samples — If residual chlorine is present, add 3 mL of 10% sodium thiosulfate per gallon and cool to 4°C. All extracts and samples should be stored under refrigeration.

MHT Liquid samples must be extracted within 7 days and the extracts must be analyzed within 40 days. Soils, sediments, and sludges may be stored for a maximum of 14 days prior to extraction.

SAMPLE PREPARATION
Liquid samples — Extract 1-L samples in a continuous extractor at pH 5–9 with methylene chloride after adding 1.0 mL of surrogate spiking solution to each sample. Pass the extract through a column of anhydrous sodium sulfate to dry and concentrate it in a K-D apparatus to 1 mL volume.

Soils, sediments and sludges — Rapidly weigh approximately 30 g of sample into a 400-mL beaker to avoid loss of the more volatile extractables. Nonporous or wet samples (gummy or clay type) that do not have a free-flowing sandy texture must be mixed with anhydrous sodium sulfate until the sample is free flowing. Add 1 mL of surrogate standards to all samples, spikes, standards, and blanks. Add 100 mL of 1:1 methylene chloride:acetone and extract ultrasonically. Decant and filter extracts, dry the extract by passing it through a drying column containing anhydrous sodium sulfate and concentrate to 1 mL in a K-D apparatus.

Hexane eolvent exchange — Add 50 mL of hexane, a new boiling chip, and concentrate until the apparent volume of liquid reaches 1 Ml. Adjust the extract volume to 10.0 Ml. Stopper the concentration tube and store refrigerated at 4°C if further processing will not be performed immediately. If the extract will be stored longer than two days, transfer it to a vial with Teflon®-lined screw-cap or crimp top.

QUALITY CONTROL Demonstrate through the analysis of a reagent water blank, that all glassware and reagents are interference free. Each time a set of samples is processed, a method blank should be processed as a safeguard against chronic lab contamination. A reagent blank, a matrix spike, and a duplicate or matrix spike duplicate must be performed for each analytical batch (up to a maximum of 20 samples) analyzed.

Analytical system performance must be verified by analyzing QC check samples. The QC check sample concentration should contain each single-component analyte at the following concentrations in acetone: 4,4'-DDD, 10 µg/mL; 4,4'-DDT, 10 µg/mL; endosulfan II, 10 µg/mL; endosulfan sulfate, 10 µg/mL; and any other single-component pesticide at 2 µg/mL. If the method is only to be used to analyze PCBs, Chlordane, or Toxaphene, the QC check sample concentrate should contain the most representative multicomponent parameter at a concentration of 50 µg/mL in acetone.

REFERENCE Test Methods for Evaluating Solid Waste (SW-846). U.S. EPA. 1983. Method 8080B, Rev. 2, Nov. 1990. Office of Solid Wastes, Washington, DC.

Aroclor 1232   **EPA Method 8270**
CAS #11141-16-5

TITLE Semivolatile Organic Compounds by GC/MS

MATRIX This method is used to determine the concentration of semivolatile organic compounds in extracts prepared from all types of solid waste matrices, soils, and groundwater. Although surface waters are not specifically mentioned, this method should be applicable to water samples from rivers, lakes, etc.

METHOD SUMMARY This method covers 259 semivolatile organic compounds. In very limited applications direct injection of the sample into the GC/MS system may be appropriate, but this results in very high detection limits (approximately 10,000 µg/L). Typically, a 1-L liquid sample, containing surrogate, and matrix spiking standards, is extracted in a continuous extractor first under acid conditions and then under basic conditions. Typically 30 g of a solid sample, containing surrogate, and matrix spiking standards, is extracted ultrasonically. After concentrating the extract to 1 mL it is spiked with 10 µL of an internal standard solution just prior to analysis by GC/MS. The volume injected should contain about 100 ng of base/neutral and 200 ng of acid surrogates (for a 1-µL injection). Analysis is performed by GC/MS using a capillary GC column.

INTERFERENCES Raw GC/MS data from all blanks, samples, and spikes must be evaluated for interferences. Contamination by carryover can occur whenever high-concentration and low-concentration samples are sequentially analyzed. To reduce carryover, the sample syringe must be rinsed out between samples with solvent. Whenever an unusually concentrated sample is encountered, it should be followed by the analysis of blank solvent to check for cross-contamination.

INSTRUMENTATION A GC/MS and a data system are required. The GC column used is a 30 m × 0.25 mm I.D. (or 0.32 mm I.D.) 1-µm film thickness silicone-coated fused silica capillary column. A continuous liquid-liquid extractor equipped with Teflon® or glass connection joints and stopcocks requiring no lubrication, a K-D concentrating apparatus, water bath, and an ultrasonic disrupter with a minimum power of 300 W and with pulsing capability are also required.

PRECISION & ACCURACY The estimated quantitation limit (EQL) of Method 8270B for determining an individual compound is approximately 1 mg/kg (wet weight) for soil or sediment samples, 1–200 mg/kg for wastes (dependent on matrix and method of preparation), and 10 µg/L for groundwater samples. EQLs will be proportionately higher for sample extracts that require dilution to avoid saturation of the detector.

The EQL(b) for groundwater in µg/L is not listed.
The EQL (a, b) for low concentrations in soil and sediment in µg/kg is not listed.
Accuracy as µg/L is not listed.
Overall precision in µg/L is not listed.

(a) *EQLs listed for soil/sediment are based on wet weight. Normally data is reported in a dry-weight basis; therefore, EQLs will be higher based on the % dry weight of each sample. This calculation is based on a 30-g sample and gel permeation chromatography cleanup.*
(b) *Sample EQLs are highly matrix-dependent. The EQLs are provided for guidance and may not always be achievable.*
C = *True value for concentration, in µg/L.*
X = *Average recovery found for measurements of samples containing a concentration of C, in µg/L.*

ESTIMATED QUANTITATION LIMIT

Other Matrices	Factor (a)
High-concentration soil and sludges by sonicator	7.5
Non-water miscible waste	75

(a) *EQL for other matrices = [EQL for low soil/sediment] × [Factor]. This estimated EQL is similar to an EPA "Practical Quantitation Limit."*

SAMPLING METHOD

Liquid samples — Use a 1 or 2½ gallon amber glass bottle with a screw-top Teflon®-lined cover that has been prewashed with detergent and rinsed with distilled water and methanol (or isopropanol).

Soils, sediments, or sludges — Use an 8-oz. widemouth glass with a screw-top Teflon®-lined cover that has been prewashed with detergent and rinsed with distilled water and methanol (or isopropanol).

SAMPLE PRESERVATION

Liquid samples — If residual chlorine is present, add 3 mL of 10% sodium thiosulfate per gallon, cool to 4°C and store in a solvent-free refrigerator until analysis; if chlorine is not present, then eliminate the sodium thiosulfate addition.

Soils, sediments, or sludges — Cool samples to 4°C and store in a solvent-free refrigerator.

MHT Liquid samples must be extracted within 7 days and the extracts analyzed within 40 days. Soils, sediments, or sludges may be stored for a maximum of 14 days and the extracts analyzed within 40 days.

SAMPLE PREPARATION

Liquid samples — Transfer 1 L quantitatively to a continuous extractor. If high concentrations are anticipated, a smaller volume may be used and then diluted with organic-free reagent water to 1 L. Adjust pH, if necessary, to pH <2 using 1:1 (V/V) sulfuric acid. Pipette 1.0 mL of a surrogate standard spiking solution into each sample. For the sample in each analytical batch selected for spiking, add 1.0 mL of a matrix spiking standard. For base/neutral acid analysis, the amount of the surrogates and matrix spiking compounds added to the sample should result in a final concentration of 100 ng/μL of each analyte in the extract to be analyzed (assuming a 1-μL injection). Extract with methylene chloride for 18–24 h. Next, adjust the pH of the aqueous phase to pH >11 using 10 N sodium hydroxide and extract it with methylene chloride again for 18–24 h. Dry the extract through a column containing anhydrous sodium sulfate and concentrate it to 1 mL using a K-D concentrator.

Soils, sediments, or sludges — Use 30 g of sample. Nonporous or wet samples (gummy or clay type) that do not have a free-flowing sandy texture must be mixed with anhydrous sodium sulfate until the sample is free flowing. Add 1 mL of surrogate standards to all samples, spikes, standards, and blanks. For the sample in each analytical batch selected for spiking, add 1.0 mL of a matrix spiking standard. For base/neutral acid analysis, the amount added of the surrogates and matrix spiking compounds should result in a final concentration of 100 ng/μL of each base/neutral analyte and 200 ng/μL of each acid analyte in the extract to be analyzed (assuming a 1-μL injection). Immediately add a 100 mL mixture of 1:1 methylene chloride:acetone and extract the sample ultrasonically for 3 min and then decant or filter the extracts. Repeat the extraction two or more times. Dry the extract using a column with anhydrous sodium sulfate and concentrate it to 1 mL in a K-D concentrator.

QUALITY CONTROL

A methylene chloride solution containing 50 ng/μL of decafluorotriphenylphosphine (DFTPP) is used for tuning the GC/MS system each 12-h shift. A system performance check also must be made during every 12-h shift. A standard containing 50 ng/μL each of 4,4′-DDT, pentachlorophenol, and benzidine is required to verify injection port inertness and GC column performance. A calibration standard at mid-concentration, containing each compound of interest, including all required surrogates, must be performed every 12 h during analysis. After the system performance check is met, calibration check compounds are used to check the validity of the initial calibration.

The internal standard responses and retention times in the calibration check standard must be evaluated immediately after or during data acquisition. If the retention time for any internal standard changes by more than 30 seconds from the last check calibration (12 h), the chromatographic system must be inspected for malfunctions and corrections must be made, as required. If the electron ionization current plot (EICP) area for any of the internal standards changes by a factor of two from the last daily calibration standard check, the mass spectrometer must be inspected for malfunctions and corrections must be made, as appropriate.

Demonstrate, through the analysis of a reagent water blank, that interferences from the analytical system, glassware, and reagents are under control. The blank samples should be carried through all stages of the sample preparation and measurement steps. For each analytical batch (up to 20 samples), a reagent blank, matrix spike, and matrix spike duplicate/duplicate must be analyzed (the frequency of the spikes may be different for different monitoring programs). The blank and spiked samples must be carried through all stages of the sample preparation and measurement steps. A QC reference sample concentrate containing each analyte at a concentration of 100 mg/L in methanol is required.

REFERENCE Test Methods for Evaluating Solid Waste (SW-846). U.S. EPA 1983, Method 8270B, Rev. 2, Nov. 1990. Office of Solid Waste, Washington, DC.

Aroclor 1242 EPA Method 505
CAS #53469-21-9

TITLE Analysis of Organohalide Pesticides and Commercial Polychlorinated Biphenyl (PCB) Products in Water by Microextraction and Gas Chromatography. U.S. EPA Method 505, Rev. 2.0, 1989.

MATRIX This method is applicable to drinking water and raw source water. The latter should include most surface water and groundwater sources.

METHOD SUMMARY Method 505 covers 25 pesticides and commercial PCB products. This is a very sensitive method that is more useful for monitoring than for exploratory analyses. 5-mL samples of water are saturated with sodium chloride and then extracted by shaking with 2 mL of hexane. The sample extracts are transferred to an autosampler setup to inject 1- to 2-μL portions into a gas chromatograph (GC) for analysis. Alternatively, 1- to 2-μL portions of samples, blanks, and standards may be manually injected. Each extract is analyzed by capillary GC/ECD with confirmation using either a second capillary column or GC/MS. The electron capture detector is easy to use, but it is a nonselective detector. The microextraction technique also eliminates the expensive sample preparation costs of other methods, but it has the disadvantage of being less sensitive than most because the extracts are not concentrated.

INTERFERENCES Method interferences may be caused by contaminants in solvents, reagents, glassware, and other sample processing apparatus that lead to discrete artifacts or elevated baselines. Interfering contamination may occur when a sample containing low concentrations of analytes is analyzed immediately following a sample containing relatively high concentrations of the analytes. Matrix interferences also may be caused by contaminants that are coextracted from the sample; cleanup of sample extracts may be necessary in these cases. Some pesticides and commercial PCB products from aqueous solutions adhere to glass surfaces, so sample transfers and contact with glass surfaces should be minimized. Some pesticides are rapidly oxidized by chlorine so dechlorination with sodium thiosulfate at the time of sample collection is important. Also, splitless injectors may cause degradation of some pesticides.

INSTRUMENTATION A gas chromatograph/electron capture detector/data system, with temperature programming and split/splitless injector suitable for use with capillary columns is needed.

Column 1: 0.32 mm I.D. × 30 m fused silica capillary with chemically bond methyl polysiloxane phase (DB-1, 1.0 µm film, or equivalent).

Column 2: 0.32 mm I.D. × 30 m fused silica capillary with 1:1 mixed phase of dimethyl silicone and polyethylene glycol (Durawax-DX3, 0.25 µm film, or equivalent).

Column 3: 0.32 mm I.D. × 25 m fused silica capillary with chemically bonded 50:50 methyl-phenyl silicone (OV-17, 1.5 µm film, or equivalent).

Column 1 should be used as the primary analytical column. Columns 2 and 3 are recommended for use as confirmatory columns when GC/MS confirmation is not available.

PRECISION & ACCURACY Method detection limits are dependent upon the characteristics of the gas chromatographic system used. Analytes that are not separated chromatographically cannot be individually identified and used in the same calibration mixture or water samples unless an alternative technique for identification and quantification, such as mass spectrometry, is used.

The concentration(s) (in µg/L) used for these QC measurements was 4.7.
The MDL (in µg/L) was 0.31.
The accuracy (% recovery) for reagent water at the above concentration(s) was not available and the precision (%) was 6.0.
The accuracy (% recovery) for groundwater at the above concentration(s) was not listed and the precision (%) was not listed.
The accuracy (% recovery) for tap water at the above concentration(s) was 96 and the precision (5) was 7.4.

Note: No range of concentrations is provided with this method.

SAMPLING METHOD Collect samples using a 40-mL screw-cap vial (prewashed with detergent, rinsed with distilled water and oven dried at 400°C for one h) with a Teflon®-faced silicone septum. Collect bubble-free samples and place the septum with the Teflon® side down on the water.

SAMPLE PRESERVATION If residual chlorine is present in the water add about 3 mg of sodium thiosulfate to each vial before samples are collected to remove the chlorine. Alternatively, add 75 µL of 0.04 g/mL solution of sodium thiosulfate to each vial just prior to sampling. Immediately cool samples to 4°C and store them in a solvent-free refrigerator at 4°C until analysis.

MHT The maximum holding time is 14 days from the time the sample was collected until it must be analyzed.

SAMPLE PREPARATION Remove the sample from storage and allow it to come to room temperature. Remove a 5-mL volume from each container and weigh the container to the nearest 0.1 g. Add 6 g of sodium chloride and 2.0 mL of hexane to each sample bottle. Recap the sample and shake it vigorously for one min. Allow the water and hexane phases to separate, remove the cap, and transfer 0.5 mL of hexane into an autosampler vial using a disposable glass pipette. Transfer the remaining hexane phase into a second autosampler vial and store at 4°C for reanalysis, if necessary. Discard the remaining sample/hexane mixture and reweigh the empty container to determine net weight of sample.

QUALITY CONTROL Minimum quality control requirements are initial demonstration of lab capability, analysis of lab reagent blanks, fortified blanks, fortified sample matrix, and quality control samples. The lab must analyze at least one fortified blank per sample set, or at least one for every 20 samples. The fortifying concentration of each analyte should be 10 times the method detection limit or the maximum calibration limit (MCL), whichever is less. Calculate accuracy as percent recovery and develop control limits from the mean percent recovery and standard deviation.

The lab must add a known concentration of the analytes to a minimum of 10% of the routine samples, or one lab fortified sample matrix per sample set. Calculate the percent recovery for each analyte and compare to the control limits established from the analyses of the fortified blanks.

EPA CONTACT & HOTLINE For technical questions contact Dr. Baldev Bathija, U.S. EPA, Office of Ground Water and Drinking Water (WH-550D), 401 M St. SW, Washington, DC 20460. Tel. (202) 260-3040. For further information the EPA Safe Drinking Water Hotline may be called at: (800) 426-4791.

REFERENCE Methods for the Determination of Organic Compounds in Drinking Water, EPA/600/4-88/039 (revised July 1991). U.S. EPA Environmental Monitoring Systems Laboratory, Cincinnati, OH, 45268, U.S.A. Available from the National Technical Information Service (NTIS), 5285 Port Royal Road, Springfield, VA 22161; Tel. 800-553-6847. NTIS Order Number is PB91-231480.

Aroclor 1242
CAS #53469-21-9
EPA Method 625

TITLE Base/Neutrals and Acids, U.S. EPA Method 625

MATRIX This methods covers municipal and industrial wastewaters.

METHOD SUMMARY Approximately 1 L of sample is serially extracted with methylene chloride at a pH greater than 11 and again at a pH less than 2 using a separatory funnel or a continuous extractor. The methylene chloride extract is dried, concentrated to a volume of 1 mL, and analyzed by GC/MS. Qualitative identification of the parameters in the extract is performed using the retention time and the relative abundance of three characteristic masses (m/z). Qualitative analysis is performed using either external or internal standard techniques with a single characteristic m/z.

INTERFERENCES Method interferences may be caused by contaminants in solvents, reagents, glassware, and other sample processing hardware. Glassware must be scrupulously cleaned. Glassware should be heated in a muffle furnace at 400°C for 5 to 30 min. Some thermally stable materials, such as PCBs, may not be eliminated by this treatment. Solvent rinses with acetone and pesticide quality hexane may be substituted for the muffle furnace heating. Matrix interferences may be caused by contaminants that are coextracted from the sample. The base/neutral extraction may cause significantly reduced recovery of phenols. The packed gas chromatographic columns recommended for the basic fraction may not exhibit sufficient resolution for some analytes.

INSTRUMENTATION A GC/MS system with an injection port designed for on-column injection when using packed columns and for splitless injection when using capillary columns.

Column for base/neutrals: 1.8 m long × 2 mm I.D. glass, packed with 3% SP-2550 on Supelcoport (100/120 mesh) or equivalent.

Column for acids: 1.8 m long × 2 mm I.D. glass, packed with 1% SP-1240DA on Supelcoport (100/120 mesh) or equivalent.

PRECISION & ACCURACY The MDL concentrations were obtained using reagent water. The MDL actually achieved in a given analysis will vary depending on instrument sensitivity and matrix effects. This method was tested by 15 laboratories using reagent water, drinking water, surface water, and industrial wastewaters spiked at six concentrations over the range 5 to 100 µg/L. Single operator precision, overall precision, and method accuracy were found to be directly related to the concentration of the parameter matrix.

The MDL (in µg/L) in reagent water was not detected.
The standard deviation (in µg/L based on 4 recovery measurements) was not reported.
The range (in µg/L) for average recovery for 4 measurements was not reported.
The range (in %) for percent recovery was not reported.
Accuracy (in µg/L) as expected recovery for one or more measurements of a sample containing a true concentration of C was not reported.
Precision (in µg/L) as expected single analyst standard deviation of measurements at an average concentration found at X was not reported.
Overall precision (in µg/L) as expected interlaboratory standard deviation of measurements in an average concentration found at X was not reported.

C = *True value of the concentration in µg/L.*
X = *Average recovery found for measurements of samples containing a concentration at C in µg/L.*

SAMPLE PREPARATION Adjust the pH to >11 with sodium hydroxide and serially extract in a separatory funnel with methylene chloride or else in a continuous extractor. Next, adjust the pH to <2 with sulfuric acid and serially extract in a separatory funnel with methylene chloride or else in a continuous extractor. Dry the extracts separately through a column of anhydrous sodium sulfate and then concentrate each of the extracts to 1.0 mL using a K-D apparatus.

SAMPLE COLLECTION, PRESERVATION & HANDLING Grab samples must be collected in glass containers. All samples must be refrigerated at 4°C from the time of collection until extraction. If residual chlorine is present, add 80 mg of sodium thiosulfate/L of sample and mix well. All samples must be extracted within 7 days of collection and completely analyzed within 40 days of extraction.

QUALITY CONTROL Make an initial, one-time, demonstration of the ability to generate acceptable accuracy and precision with this method. Before processing any samples, the analyst must analyze a reagent water blank to demonstrate that interferences from the analytical system and glassware are under control. Each time a set of samples is extracted or reagents are changed, a reagent water blank must be processed. Spike and analyze a minimum of 5% of all samples to monitor and evaluate lab data quality. A QC check sample concentrate that contains each parameter of interest at a concentration of 100 µg/mL in acetone is required. PCBs and multicomponent pesticides may be omitted from this test.

After analysis of five spiked wastewater samples, calculate the average percent recovery and the standard deviation of the percent recovery. Spike all samples with the surrogate standard spiking solution and calculate the percent recovery of each surrogate compound.

REFERENCE Federal Register, Vol. 49, No. 209. Friday, Oct. 26, 1984.

Aroclor 1242
CAS #53469-21-9
EPA Method 8080

TITLE Organochlorine Pesticides and Polychlorinated Biphenyls By Gas Chromatography

MATRIX This method is used to determine the concentration of various organochlorine pesticides and polychlorinated biphenyls in extracts prepared from water, groundwater, soils, and sediments.

METHOD SUMMARY This method covers 26 pesticides and Aroclor (PCB) mixtures and it is suitable for monitoring-type analyses. After extraction, concentration and solvent exchange to hexane, a 2- to 5-μL sample aliquot is injected into a GC using the solvent flush technique, and the analytes are detected by an electron capture detector (ECD) or an electrolytic conductivity detector in the halogen mode (HECD). Both neat and diluted organic liquids may be analyzed by direct injection.

INTERFERENCES Interferences coextracted from the samples will vary considerably from source to source. Interferences by phthalate esters can pose a major problem in pesticide determinations when using the ECD. Cross-contamination of clean glassware routinely occurs when plastics are handled during extraction steps, especially when solvent-wetted surfaces are handled. The contamination from phthalate esters can be completely eliminated with a microcoulometric or electrolytic conductivity detector. Solvents, reagent, glassware, and other sample processing hardware may yield artifacts and/or interferences to sample analysis.

INSTRUMENTATION A gas chromatograph capable of on-column injections is needed. It must be equipped with an ECD or a HECD and one of the following GC columns:

Column 1: Supelcoport (100/120 mesh) coated with 1.5% SP-2250/1.95% SP-2401 packed in a 1.8 m × 4 mm I.D. glass column.

Column 2: Supelcoport (100/120 mesh) coated with 3% OV-1 in a 1.8 m × 4 mm I.D. glass column.

PRECISION & ACCURACY The method was tested by 20 laboratories using organic-free reagent water, drinking water, surface water, and three industrial wastewaters spiked at six concentrations. Concentrations used in the study ranged from 0.5 to 30 μg/L for single-component pesticides and from 8.5 to 400 μg/L for multicomponent parameters. Overall precision and method accuracy were found to be directly related to the concentration of the analyte and essentially independent of the sample matrix. The sensitivity of this method usually depends on the concentration of interferences rather than on instrumental limitations.

MDL in μg/L was 0.065.
Concentration range in μg/L was 8.5–40.
Accuracy as recovery (x^*) in μg/L was $0.91C+10.79$.
Overall precision (S^*) in μg/L was $0.31x +3.50$.

x^* *Expected recovery for one or more measurements of a sample containing concentration C, in μg/L.*
$S^* =$ *Expected interlaboratory standard deviation of measurements at an average concentration found of the analyte in μg/L.*
$C =$ *True value for the concentration, in μg/L.*
$X =$ *Average recovery found for measurements of samples containing a concentration of C, in μg/L.*

SAMPLING METHOD
Liquid samples — Use a 1 or 2½ gallon amber glass bottle with a screw-top Teflon®-lined cover. Pre-wash the bottle with detergent, rinse with distilled water and methanol (or isopropanol).

Soil, sediments and sludges — Use an 8-oz. widemouth glass with a screw-top Teflon®-lined cover. Pre-wash the bottle with detergent, rinse with distilled water and methanol (or isopropanol).

SAMPLE PRESERVATION Cool water, soil, sediment, or sludge samples immediately to 4°C.

Water samples — If residual chlorine is present, add 3 mL of 10% sodium thiosulfate per gallon and cool to 4°C. All extracts and samples should be stored under refrigeration.

MHT Liquid samples must be extracted within 7 days and the extracts must be analyzed within 40 days. Soils, sediments, and sludges may be stored for a maximum of 14 days prior to extraction.

SAMPLE PREPARATION
Liquid samples — Extract 1-L samples in a continuous extractor at pH 5–9 with methylene chloride after adding 1.0 mL of surrogate spiking solution to each sample. Pass the extract through a column of anhydrous sodium sulfate to dry and concentrate it in a K-D apparatus to 1 mL volume.

Soils, sediments and sludges — Rapidly weigh approximately 30 g of sample into a 400-mL beaker to avoid loss of the more volatile extractables. Nonporous or wet samples (gummy or clay type) that do not have a free-flowing sandy texture must be mixed with anhydrous sodium sulfate until the sample is free flowing. Add 1 mL of surrogate standards to all samples, spikes, standards, and blanks. Add 100 mL of 1:1 methylene chloride:acetone and extract ultrasonically. Decant and filter extracts, dry the extract by passing it through a drying column containing anhydrous sodium sulfate and concentrate to 1 mL in a K-D apparatus.

Hexane eolvent exchange — Add 50 mL of hexane, a new boiling chip, and concentrate until the apparent volume of liquid reaches 1 Ml. Adjust the extract volume to 10.0 Ml. Stopper the concentration tube and store refrigerated at 4°C if further processing will not be performed immediately. If the extract will be stored longer than two days, transfer it to a vial with Teflon®-lined screw-cap or crimp top.

QUALITY CONTROL Demonstrate through the analysis of a reagent water blank, that all glassware and reagents are interference free. Each time a set of samples is processed, a method blank should be processed as a safeguard against chronic lab contamination. A reagent blank, a matrix spike, and a duplicate or matrix spike duplicate must be performed for each analytical batch (up to a maximum of 20 samples) analyzed.

Analytical system performance must be verified by analyzing QC check samples. The QC check sample concentration should contain each single-component analyte at the following concentrations in acetone: 4,4'-DDD, 10 μg/mL; 4,4'-DDT, 10 μg/mL; endosulfan II, 10 μg/mL; endosulfan sulfate, 10 μg/mL; and any other single-component pesticide at 2 μg/mL. If the method is only to be used to analyze PCBs, Chlordane, or Toxaphene, the QC check sample concentrate should contain the most representative multicomponent parameter at a concentration of 50 μg/mL in acetone.

REFERENCE Test Methods for Evaluating Solid Waste (SW-846). U.S. EPA. 1983. Method 8080B, Rev. 2, Nov. 1990. Office of Solid Wastes, Washington, DC.

Aroclor 1242   **EPA Method 8270**
CAS #53469-21-9

TITLE Semivolatile Organic Compounds by GC/MS

MATRIX This method is used to determine the concentration of semivolatile organic compounds in extracts prepared from all types of solid waste matrices, soils, and groundwater. Although surface waters are not specifically mentioned, this method should be applicable to water samples from rivers, lakes, etc.

METHOD SUMMARY This method covers 259 semivolatile organic compounds. In very limited applications direct injection of the sample into the GC/MS system may be appropriate, but this results in very high detection limits (approximately 10,000 µg/L). Typically, a 1-L liquid sample, containing surrogate, and matrix spiking standards, is extracted in a continuous extractor first under acid conditions and then under basic conditions. Typically 30 g of a solid sample, containing surrogate, and matrix spiking standards, is extracted ultrasonically. After concentrating the extract to 1 mL it is spiked with 10 µL of an internal standard solution just prior to analysis by GC/MS. The volume injected should contain about 100 ng of base/neutral and 200 ng of acid surrogates (for a 1-µL injection). Analysis is performed by GC/MS using a capillary GC column.

INTERFERENCES Raw GC/MS data from all blanks, samples, and spikes must be evaluated for interferences. Contamination by carryover can occur whenever high-concentration and low-concentration samples are sequentially analyzed. To reduce carryover, the sample syringe must be rinsed out between samples with solvent. Whenever an unusually concentrated sample is encountered, it should be followed by the analysis of blank solvent to check for cross-contamination.

INSTRUMENTATION A GC/MS and a data system are required. The GC column used is a 30 m × 0.25 mm I.D. (or 0.32 mm I.D.) 1-µm film thickness silicone-coated fused silica capillary column. A continuous liquid-liquid extractor equipped with Teflon® or glass connection joints and stopcocks requiring no lubrication, a K-D concentrating apparatus, water bath, and an ultrasonic disrupter with a minimum power of 300 W and with pulsing capability are also required.

PRECISION & ACCURACY The estimated quantitation limit (EQL) of Method 8270B for determining an individual compound is approximately 1 mg/kg (wet weight) for soil or sediment samples, 1–200 mg/kg for wastes (dependent on matrix and method of preparation), and 10 µg/L for groundwater samples. EQLs will be proportionately higher for sample extracts that require dilution to avoid saturation of the detector.

The EQL(b) for groundwater in µg/L is not listed.
The EQL (a, b) for low concentrations in soil and sediment in µg/kg is not listed.
Accuracy as µg/L is not listed.

Overall precision in µg/L is not listed.

(a) *EQLs listed for soil/sediment are based on wet weight. Normally data is reported in a dry-weight basis; therefore, EQLs will be higher based on the % dry weight of each sample. This calculation is based on a 30-g sample and gel permeation chromatography cleanup.*
(b) *Sample EQLs are highly matrix-dependent. The EQLs are provided for guidance and may not always be achievable.*
C = True value for concentration, in µg/L.
X = Average recovery found for measurements of samples containing a concentration of C, in µg/L.

ESTIMATED QUANTITATION LIMIT

Other Matrices	Factor (a)
High-concentration soil and sludges by sonicator	7.5
Non-water miscible waste	75

(a) *EQL for other matrices = [EQL for low soil/sediment] × [Factor]. This estimated EQL is similar to an EPA "Practical Quantitation Limit."*

SAMPLING METHOD
Liquid samples — Use a 1 or 2½ gallon amber glass bottle with a screw-top Teflon®-lined cover that has been prewashed with detergent and rinsed with distilled water and methanol (or isopropanol).

Soils, sediments, or sludges — Use an 8-oz. widemouth glass with a screw-top Teflon®-lined cover that has been prewashed with detergent and rinsed with distilled water and methanol (or isopropanol).

SAMPLE PRESERVATION
Liquid samples — If residual chlorine is present, add 3 mL of 10% sodium thiosulfate per gallon, cool to 4°C and store in a solvent-free refrigerator until analysis; if chlorine is not present, then eliminate the sodium thiosulfate addition.

Soils, sediments, or sludges — Cool samples to 4°C and store in a solvent-free refrigerator.

MHT Liquid samples must be extracted within 7 days and the extracts analyzed within 40 days. Soils, sediments, or sludges may be stored for a maximum of 14 days and the extracts analyzed within 40 days.

SAMPLE PREPARATION
Liquid samples — Transfer 1 L quantitatively to a continuous extractor. If high concentrations are anticipated, a smaller volume may be used and then diluted with organic-free reagent water to 1 L. Adjust pH, if necessary, to pH <2 using 1:1 (V/V) sulfuric acid. Pipette 1.0 mL of a surrogate standard spiking solution into each sample. For the sample in each analytical batch selected for spiking, add 1.0 mL of a matrix spiking standard. For base/neutral acid analysis, the amount of the surrogates and matrix spiking compounds added to the sample should result in a final concentration of 100 ng/µL of each analyte in the extract to be analyzed (assuming a 1-µL injection). Extract with methylene chloride for 18–24 h. Next, adjust the pH of the aqueous phase to pH >11 using 10 N sodium hydroxide and extract it with methylene chloride again for 18–24 h. Dry the extract through a column containing anhydrous

sodium sulfate and concentrate it to 1 mL using a K-D concentrator.

Soils, sediments, or sludges — Use 30 g of sample. Nonporous or wet samples (gummy or clay type) that do not have a free-flowing sandy texture must be mixed with anhydrous sodium sulfate until the sample is free flowing. Add 1 mL of surrogate standards to all samples, spikes, standards, and blanks. For the sample in each analytical batch selected for spiking, add 1.0 mL of a matrix spiking standard. For base/neutral acid analysis, the amount added of the surrogates and matrix spiking compounds should result in a final concentration of 100 ng/µL of each base/neutral analyte and 200 ng/µL of each acid analyte in the extract to be analyzed (assuming a 1-µL injection). Immediately add a 100 mL mixture of 1:1 methylene chloride:acetone and extract the sample ultrasonically for 3 min and then decant or filter the extracts. Repeat the extraction two or more times. Dry the extract using a column with anhydrous sodium sulfate and concentrate it to 1 mL in a K-D concentrator.

QUALITY CONTROL A methylene chloride solution containing 50 ng/µL of decafluorotriphenylphosphine (DFTPP) is used for tuning the GC/MS system each 12-h shift. A system performance check also must be made during every 12-h shift. A standard containing 50 ng/µL each of 4,4'-DDT, pentachlorophenol, and benzidine is required to verify injection port inertness and GC column performance. A calibration standard at mid-concentration, containing each compound of interest, including all required surrogates, must be performed every 12 h during analysis. After the system performance check is met, calibration check compounds (CCCs) are used to check the validity of the initial calibration.

The internal standard responses and retention times in the calibration check standard must be evaluated immediately after or during data acquisition. If the retention time for any internal standard changes by more than 30 seconds from the last check calibration (12 h), the chromatographic system must be inspected for malfunctions and corrections must be made, as required. If the electron ionization current plot (EICP) area for any of the internal standards changes by a factor of two from the last daily calibration standard check, the mass spectrometer must be inspected for malfunctions and corrections must be made, as appropriate.

Demonstrate, through the analysis of a reagent water blank, that interferences from the analytical system, glassware, and reagents are under control. The blank samples should be carried through all stages of the sample preparation and measurement steps. For each analytical batch (up to 20 samples), a reagent blank, matrix spike, and matrix spike duplicate/duplicate must be analyzed (the frequency of the spikes may be different for different monitoring programs). The blank and spiked samples must be carried through all stages of the sample preparation and measurement steps. A QC reference sample concentrate containing each analyte at a concentration of 100 mg/L in methanol is required.

REFERENCE Test Methods for Evaluating Solid Waste (SW-846). U.S. EPA 1983, Method 8270B, Rev. 2, Nov. 1990. Office of Solid Waste, Washington, DC.

Aroclor 1248 — EPA Method 505
CAS #12672-29-6

TITLE Analysis of Organohalide Pesticides and Commercial Polychlorinated Biphenyl (PCB) Products in Water by Microextraction and Gas Chromatography. U.S. EPA Method 505, Rev. 2.0, 1989.

MATRIX This method is applicable to drinking water and raw source water. The latter should include most surface water and groundwater sources.

METHOD SUMMARY Method 505 covers 25 pesticides and commercial PCB products. This is a very sensitive method that is more useful for monitoring than for exploratory analyses. 5-mL samples of water are saturated with sodium chloride and then extracted by shaking with 2 mL of hexane. The sample extracts are transferred to an autosampler setup to inject 1- to 2-µL portions into a gas chromatograph (GC) for analysis. Alternatively, 1- to 2-µL portions of samples, blanks, and standards may be manually injected. Each extract is analyzed by capillary GC/ECD with confirmation using either a second capillary column or GC/MS. The electron capture detector is easy to use, but it is a nonselective detector. The microextraction technique also eliminates the expensive sample preparation costs of other methods, but it has the disadvantage of being less sensitive than most because the extracts are not concentrated.

INTERFERENCES Method interferences may be caused by contaminants in solvents, reagents, glassware, and other sample processing apparatus that lead to discrete artifacts or elevated baselines. Interfering contamination may occur when a sample containing low concentrations of analytes is analyzed immediately following a sample containing relatively high concentrations of the analytes. Matrix interferences also may be caused by contaminants that are coextracted from the sample; cleanup of sample extracts may be necessary in these cases. Some pesticides and commercial PCB products from aqueous solutions adhere to glass surfaces, so sample transfers and contact with glass surfaces should be minimized. Some pesticides are rapidly oxidized by chlorine so dechlorination with sodium thiosulfate at the time of sample collection is important. Also, splitless injectors may cause degradation of some pesticides.

INSTRUMENTATION A gas chromatograph/electron capture detector/data system, with temperature programming and split/splitless injector suitable for use with capillary columns is needed.

Column 1: 0.32 mm I.D. × 30 m fused silica capillary with chemically bond methyl polysiloxane phase (DB-1, 1.0 µm film, or equivalent).
Column 2: 0.32 mm I.D. × 30 m fused silica capillary with 1:1 mixed phase of dimethyl silicone and polyethylene glycol (Durawax-DX3, 0.25 µm film, or equivalent).
Column 3: 0.32 mm I.D. × 25 m fused silica capillary with chemically bonded 50:50 methyl-phenyl silicone (OV-17, 1.5 µm film, or equivalent).

Column 1 should be used as the primary analytical column. Columns 2 and 3 are recommended for use as confirmatory columns when GC/MS confirmation is not available.

PRECISION & ACCURACY Method detection limits are dependent upon the characteristics of the gas chromatographic system used. Analytes that are not separated chromatographically cannot be individually identified and used in the same calibration mixture or water samples unless an alternative technique for identification and quantification, such as mass spectrometry, is used.

The concentration(s) (in µg/L) used for these QC measurements was 3.6 and 3.4.
The MDL (in µg/L) was 0.102.
The accuracy (% recovery) for reagent water at the above concentration(s) was not available and not listed and the precision (%) was 115 and not listed.
The accuracy (% recovery) for groundwater at the above concentration(s) was not listed and not listed and the precision (%) was not listed and not listed.
The accuracy (% recovery) for tap water at the above concentration(s) was not listed and 84 and the precision (5) was not listed and 9.9.

Note: No range of concentrations is provided with this method.

SAMPLING METHOD Collect samples using a 40-mL screw-cap vial (prewashed with detergent, rinsed with distilled water and oven dried at 400°C for one h) with a Teflon®-faced silicone septum. Collect bubble-free samples and place the septum with the Teflon® side down on the water.

SAMPLE PRESERVATION If residual chlorine is present in the water add about 3 mg of sodium thiosulfate to each vial before samples are collected to remove the chlorine. Alternatively, add 75 µL of 0.04 g/mL solution of sodium thiosulfate to each vial just prior to sampling. Immediately cool samples to 4°C and store them in a solvent-free refrigerator at 4°C until analysis.

MHT The maximum holding time is 14 days from the time the sample was collected until it must be analyzed.

SAMPLE PREPARATION Remove the sample from storage and allow it to come to room temperature. Remove a 5-mL volume from each container and weigh the container to the nearest 0.1 g. Add 6 g of sodium chloride and 2.0 mL of hexane to each sample bottle. Recap the sample and shake it vigorously for one min. Allow the water and hexane phases to separate, remove the cap, and transfer 0.5 mL of hexane into an autosampler vial using a disposable glass pipette. Transfer the remaining hexane phase into a second autosampler vial and store at 4°C for reanalysis, if necessary. Discard the remaining sample/hexane mixture and reweigh the empty container to determine net weight of sample.

QUALITY CONTROL Minimum quality control requirements are initial demonstration of lab capability, analysis of lab reagent blanks, fortified blanks, fortified sample matrix, and quality control samples. The lab must analyze at least one fortified blank per sample set, or at least one for every 20 samples. The fortifying concentration of each analyte should be 10 times the method detection limit or the maximum calibration limit (MCL), whichever is less. Calculate accuracy as percent recovery and develop control limits from the mean percent recovery and standard deviation.

The lab must add a known concentration of the analytes to a minimum of 10% of the routine samples, or one lab fortified sample matrix per sample set. Calculate the percent recovery for each analyte and compare to the control limits established from the analyses of the fortified blanks.

EPA CONTACT & HOTLINE For technical questions contact Dr. Baldev Bathija, U.S. EPA, Office of Ground Water and Drinking Water (WH-550D), 401 M St. SW, Washington, DC 20460. Tel. (202) 260-3040. For further information the EPA Safe Drinking Water Hotline may be called at: (800) 426-4791.

REFERENCE Methods for the Determination of Organic Compounds in Drinking Water, EPA/600/4-88/039 (revised July 1991). U.S. EPA Environmental Monitoring Systems Laboratory, Cincinnati, OH, 45268, U.S.A. Available from the National Technical Information Service (NTIS), 5285 Port Royal Road, Springfield, VA 22161; Tel. 800-553-6847. NTIS Order Number is PB91-231480.

Aroclor 1248   EPA Method 625
CAS #12672-29-6

TITLE Base/Neutrals and Acids, U.S. EPA Method 625

MATRIX This methods covers municipal and industrial wastewaters.

METHOD SUMMARY Approximately 1 L of sample is serially extracted with methylene chloride at a pH greater than 11 and again at a pH less than 2 using a separatory funnel or a continuous extractor. The methylene chloride extract is dried, concentrated to a volume of 1 mL, and analyzed by GC/MS. Qualitative identification of the parameters in the extract is performed using the retention time and the relative abundance of three characteristic masses (m/z). Qualitative analysis is performed using either external or internal standard techniques with a single characteristic m/z.

INTERFERENCES Method interferences may be caused by contaminants in solvents, reagents, glassware, and other sample processing hardware. Glassware must be scrupulously cleaned. Glassware should be heated in a muffle furnace at 400°C for 5 to 30 min. Some thermally stable materials, such as PCBs, may not be eliminated by this treatment. Solvent rinses with acetone and pesticide quality hexane may be substituted for the muffle furnace heating. Matrix interferences may be caused by contaminants that are coextracted from the sample. The base/neutral extraction may cause significantly reduced recovery of phenols. The packed gas chromatographic columns recommended for the basic fraction may not exhibit sufficient resolution for some analytes.

INSTRUMENTATION A GC/MS system with an injection port designed for on-column injection when using packed columns and for splitless injection when using capillary columns.

Column for base/neutrals: 1.8 m long × 2 mm I.D. glass, packed with 3% SP-2550 on Supelcoport (100/120 mesh) or equivalent.

Column for acids: 1.8 m long × 2 mm I.D. glass, packed with 1% SP-1240DA on Supelcoport (100/120 mesh) or equivalent.

PRECISION & ACCURACY The MDL concentrations were obtained using reagent water. The MDL actually achieved in a given analysis will vary depending on instrument sensitivity and matrix effects. This method was tested by 15 laboratories using reagent water, drinking water, surface water, and industrial wastewaters spiked at six concentrations over the range 5 to 100 µg/L. Single operator precision, overall precision, and method accuracy were found to be directly related to the concentration of the parameter matrix.

The MDL (in µg/L) in reagent water was not detected.

The standard deviation (in µg/L based on 4 recovery measurements) was not reported.

The range (in µg/L) for average recovery for 4 measurements was not reported.

The range (in %) for percent recovery was not reported.

Accuracy (in µg/L) as expected recovery for one or more measurements of a sample containing a true concentration of C was not reported.

Precision (in µg/L) as expected single analyst standard deviation of measurements at an average concentration found at X was not reported.

Overall precision (in µg/L) as expected interlaboratory standard deviation of measurements in an average concentration found at X was not reported.

C = *True value of the concentration in µg/L.*
X = *Average recovery found for measurements of samples containing a concentration at C in µg/L.*

SAMPLE PREPARATION Adjust the pH to >11 with sodium hydroxide and serially extract in a separatory funnel with methylene chloride or else in a continuous extractor. Next, adjust the pH to <2 with sulfuric acid and serially extract in a separatory funnel with methylene chloride or else in a continuous extractor. Dry the extracts separately through a column of anhydrous sodium sulfate and then concentrate each of the extracts to 1.0 mL using a K-D apparatus.

SAMPLE COLLECTION, PRESERVATION & HANDLING Grab samples must be collected in glass containers. All samples must be refrigerated at 4°C from the time of collection until extraction. If residual chlorine is present, add 80 mg of sodium thiosulfate/L of sample and mix well. All samples must be extracted within 7 days of collection and completely analyzed within 40 days of extraction.

QUALITY CONTROL Make an initial, one-time, demonstration of the ability to generate acceptable accuracy and precision with this method. Before processing any samples, the analyst must analyze a reagent water blank to demonstrate that interferences from the analytical system and glassware are under control. Each time a set of samples is extracted or reagents are changed, a reagent water blank must be processed. Spike and analyze a minimum of 5% of all samples to monitor and evaluate lab data quality. A QC check sample concentrate that contains each parameter of interest at a concentration of 100 µg/mL in acetone is required. PCBs and multicomponent pesticides may be omitted from this test.

After analysis of five spiked wastewater samples, calculate the average percent recovery and the standard deviation of the percent recovery. Spike all samples with the surrogate standard spiking solution and calculate the percent recovery of each surrogate compound.

REFERENCE *Federal Register*, Vol. 49, No. 209. Friday, Oct. 26, 1984.

Aroclor 1248   **EPA Method 8080**
CAS #12672-29-6

TITLE Organochlorine Pesticides and Polychlorinated Biphenyls By Gas Chromatography

MATRIX This method is used to determine the concentration of various organochlorine pesticides and polychlorinated biphenyls in extracts prepared from water, groundwater, soils, and sediments.

METHOD SUMMARY This method covers 26 pesticides and Aroclor (PCB) mixtures and it is suitable for monitoring-type analyses. After extraction, concentration and solvent exchange to hexane, a 2- to 5-µL sample aliquot is injected into a GC using the solvent flush technique, and the analytes are detected by an electron capture detector (ECD) or an electrolytic conductivity detector in the halogen mode (HECD). Both neat and diluted organic liquids may be analyzed by direct injection.

INTERFERENCES Interferences coextracted from the samples will vary considerably from source to source. Interferences by phthalate esters can pose a major problem in pesticide determinations when using the ECD. Cross-contamination of clean glassware routinely occurs when plastics are handled during extraction steps, especially when solvent-wetted surfaces are handled. The contamination from phthalate esters can be completely eliminated with a microcoulometric or electrolytic conductivity detector. Solvents, reagent, glassware, and other sample processing hardware may yield artifacts and/or interferences to sample analysis.

INSTRUMENTATION A gas chromatograph capable of on-column injections is needed. It must be equipped with an ECD or a HECD and one of the following GC columns:

Column 1: Supelcoport (100/120 mesh) coated with 1.5% SP-2250/1.95% SP-2401 packed in a 1.8 m × 4 mm I.D. glass column.

Column 2: Supelcoport (100/120 mesh) coated with 3% OV-1 in a 1.8 m × 4 mm I.D. glass column.

PRECISION & ACCURACY The method was tested by 20 laboratories using organic-free reagent water, drinking water, surface water, and three industrial wastewaters spiked at six concentrations. Concentrations used in the study ranged from 0.5 to 30 µg/L for single-component pesticides and from 8.5 to 400 µg/L for multicomponent parameters. Overall precision and method accuracy were found to be directly related to the concentration of the analyte and essentially independent of the sample matrix. The sensitivity of this method usually depends on the concentration of interferences rather than on instrumental limitations.

MDL in µg/L was ND.
Concentration range in µg/L was 8.5–40.
Accuracy as recovery (x*) in µg/L was 0.91C+10.79 .
Overall precision (S*) in µg/L was 0.31x +3.50.

x*   *Expected recovery for one or more measurements of a sample containing concentration C, in µg/L.*
S* = *Expected interlaboratory standard deviation of measurements at an average concentration found of the analyte in µg/L.*
C = *True value for the concentration, in µg/L.*
X = *Average recovery found for measurements of samples containing a concentration of C, in µg/L.*

SAMPLING METHOD

Liquid samples — Use a 1 or 2½ gallon amber glass bottle with a screw-top Teflon®-lined cover. Pre-wash the bottle with detergent, rinse with distilled water and methanol (or isopropanol).

Soil, sediments, and sludges — Use an 8-oz. widemouth glass with a screw-top Teflon®-lined cover. Pre-wash the bottle with detergent, rinse with distilled water and methanol (or isopropanol).

SAMPLE PRESERVATION   Cool water, soil, sediment, or sludge samples immediately to 4°C.

Water samples — If residual chlorine is present, add 3 mL of 10% sodium thiosulfate per gallon and cool to 4°C. All extracts and samples should be stored under refrigeration.

MHT   Liquid samples must be extracted within 7 days and the extracts must be analyzed within 40 days. Soils, sediments, and sludges may be stored for a maximum of 14 days prior to extraction.

SAMPLE PREPARATION

Liquid samples — Extract 1-L samples in a continuous extractor at pH 5–9 with methylene chloride after adding 1.0 mL of surrogate spiking solution to each sample. Pass the extract through a column of anhydrous sodium sulfate to dry and concentrate it in a K-D apparatus to 1 mL volume.

Soils, sediments and sludges — Rapidly weigh approximately 30 g of sample into a 400-mL beaker to avoid loss of the more volatile extractables. Nonporous or wet samples (gummy or clay type) that do not have a free-flowing sandy texture must be mixed with anhydrous sodium sulfate until the sample is free flowing. Add 1 mL of surrogate standards to all samples, spikes, standards, and blanks. Add 100 mL of 1:1 methylene chloride:acetone and extract ultrasonically. Decant and filter extracts, dry the extract by passing it through a drying column containing anhydrous sodium sulfate and concentrate to 1 mL in a K-D apparatus.

Hexane eolvent exchange — Add 50 mL of hexane, a new boiling chip, and concentrate until the apparent volume of liquid reaches 1 Ml. Adjust the extract volume to 10.0 Ml. Stopper the concentration tube and store refrigerated at 4°C if further processing will not be performed immediately. If the extract will be stored longer than two days, transfer it to a vial with Teflon®-lined screw-cap or crimp top.

QUALITY CONTROL   Demonstrate through the analysis of a reagent water blank, that all glassware and reagents are interference free. Each time a set of samples is processed, a method blank should be processed as a safeguard against chronic lab contamination. A reagent blank, a matrix spike, and a duplicate or matrix spike duplicate must be performed for each analytical batch (up to a maximum of 20 samples) analyzed.

Analytical system performance must be verified by analyzing QC check samples. The QC check sample concentration should contain each single-component analyte at the following concentrations in acetone: 4,4'-DDD, 10 µg/mL; 4,4'-DDT, 10 µg/mL; endosulfan II, 10 µg/mL; endosulfan sulfate, 10 µg/mL; and any other single-component pesticide at 2 µg/mL. If the method is only to be used to analyze PCBs, Chlordane, or Toxaphene, the QC check sample concentrate should contain the most representative multicomponent parameter at a concentration of 50 µg/mL in acetone.

REFERENCE   Test Methods for Evaluating Solid Waste (SW-846). U.S. EPA. 1983. Method 8080B, Rev. 2, Nov. 1990. Office of Solid Wastes, Washington, DC.

Aroclor 1248   **EPA Method 8270**
CAS #12672-29-6

TITLE   Semivolatile Organic Compounds by GC/MS

MATRIX   This method is used to determine the concentration of semivolatile organic compounds in extracts prepared from all types of solid waste matrices, soils, and groundwater. Although surface waters are not specifically mentioned, this method should be applicable to water samples from rivers, lakes, etc.

METHOD SUMMARY   This method covers 259 semivolatile organic compounds. In very limited applications direct injection of the sample into the GC/MS system may be appropriate, but this results in very high detection limits (approximately 10,000 µg/L). Typically, a 1-L liquid sample, containing surrogate, and matrix spiking standards, is extracted in a continuous extractor first under acid conditions and then under basic conditions. Typically 30 g of a solid sample, containing surrogate, and matrix spiking standards, is extracted ultrasonically. After concentrating the extract to 1 mL it is spiked with 10 µL of an internal standard solution just prior to analysis by GC/MS. The volume injected should contain about 100 ng of base/neutral and 200 ng of acid surrogates (for a 1-µL injection). Analysis is performed by GC/MS using a capillary GC column.

INTERFERENCES   Raw GC/MS data from all blanks, samples, and spikes must be evaluated for interferences. Contamination by carryover can occur whenever high-concentration and low-concentration samples are sequentially analyzed. To reduce carryover, the sample syringe must be rinsed out between samples with solvent. Whenever an unusually concentrated

sample is encountered, it should be followed by the analysis of blank solvent to check for cross-contamination.

INSTRUMENTATION A GC/MS and a data system are required. The GC column used is a 30 m × 0.25 mm I.D. (or 0.32 mm I.D.) 1-μm film thickness silicone-coated fused silica capillary column. A continuous liquid-liquid extractor equipped with Teflon® or glass connection joints and stopcocks requiring no lubrication, a K-D concentrating apparatus, water bath, and an ultrasonic disrupter with a minimum power of 300 W and with pulsing capability are also required.

PRECISION & ACCURACY The estimated quantitation limit (EQL) of Method 8270B for determining an individual compound is approximately 1 mg/kg (wet weight) for soil or sediment samples, 1–200 mg/kg for wastes (dependent on matrix and method of preparation), and 10 μg/L for groundwater samples. EQLs will be proportionately higher for sample extracts that require dilution to avoid saturation of the detector.

The EQL(b) for groundwater in μg/L is not listed.
The EQL (a, b) for low concentrations in soil and sediment in μg/kg is not listed.
Accuracy as μg/L is not listed.
Overall precision in μg/L is not listed.

(a) EQLs listed for soil/sediment are based on wet weight. Normally data is reported in a dry-weight basis; therefore, EQLs will be higher based on the % dry weight of each sample. This calculation is based on a 30-g sample and gel permeation chromatography cleanup.
(b) Sample EQLs are highly matrix-dependent. The EQLs are provided for guidance and may not always be achievable.
C = True value for concentration, in μg/L.
X = Average recovery found for measurements of samples containing a concentration of C, in μg/L.

ESTIMATED QUANTITATION LIMIT

Other Matrices	Factor (a)
High-concentration soil and sludges by sonicator	7.5
Non-water miscible waste	75

(a) EQL for other matrices = [EQL for low soil/sediment] × [Factor]. This estimated EQL is similar to an EPA "Practical Quantitation Limit."

SAMPLING METHOD
Liquid samples — Use a 1 or 2½ gallon amber glass bottle with a screw-top Teflon®-lined cover that has been prewashed with detergent and rinsed with distilled water and methanol (or isopropanol).

Soils, sediments, or sludges — Use an 8-oz. widemouth glass with a screw-top Teflon®-lined cover that has been prewashed with detergent and rinsed with distilled water and methanol (or isopropanol).

SAMPLE PRESERVATION
Liquid samples — If residual chlorine is present, add 3 mL of 10% sodium thiosulfate per gallon, cool to 4°C and store in a solvent-free refrigerator until analysis; if chlorine is not present, then eliminate the sodium thiosulfate addition.

Soils, sediments, or sludges — Cool samples to 4°C and store in a solvent-free refrigerator.

MHT Liquid samples must be extracted within 7 days and the extracts analyzed within 40 days. Soils, sediments, or sludges may be stored for a maximum of 14 days and the extracts analyzed within 40 days.

SAMPLE PREPARATION
Liquid samples — Transfer 1 L quantitatively to a continuous extractor. If high concentrations are anticipated, a smaller volume may be used and then diluted with organic-free reagent water to 1 L. Adjust pH, if necessary, to pH <2 using 1:1 (V/V) sulfuric acid. Pipette 1.0 mL of a surrogate standard spiking solution into each sample. For the sample in each analytical batch selected for spiking, add 1.0 mL of a matrix spiking standard. For base/neutral acid analysis, the amount of the surrogates and matrix spiking compounds added to the sample should result in a final concentration of 100 ng/μL of each analyte in the extract to be analyzed (assuming a 1-μL injection). Extract with methylene chloride for 18–24 h. Next, adjust the pH of the aqueous phase to pH >11 using 10 N sodium hydroxide and extract it with methylene chloride again for 18–24 h. Dry the extract through a column containing anhydrous sodium sulfate and concentrate it to 1 mL using a K-D concentrator.

Soils, sediments, or sludges — Use 30 g of sample. Nonporous or wet samples (gummy or clay type) that do not have a free-flowing sandy texture must be mixed with anhydrous sodium sulfate until the sample is free flowing. Add 1 mL of surrogate standards to all samples, spikes, standards, and blanks. For the sample in each analytical batch selected for spiking, add 1.0 mL of a matrix spiking standard.

For base/neutral acid analysis, the amount added of the surrogates and matrix spiking compounds should result in a final concentration of 100 ng/μL of each base/neutral analyte and 200 ng/μL of each acid analyte in the extract to be analyzed (assuming a 1-μL injection). Immediately add a 100 mL mixture of 1:1 methylene chloride:acetone and extract the sample ultrasonically for 3 min and then decant or filter the extracts. Repeat the extraction two or more times. Dry the extract using a column with anhydrous sodium sulfate and concentrate it to 1 mL in a K-D concentrator.

QUALITY CONTROL A methylene chloride solution containing 50 ng/μL of decafluorotriphenylphosphine (DFTPP) is used for tuning the GC/MS system each 12-h shift. A system performance check also must be made during every 12-h shift. A standard containing 50 ng/μL each of 4,4′-DDT, pentachlorophenol, and benzidine is required to verify injection port inertness and GC column performance. A calibration standard at mid-concentration, containing each compound of interest, including all required surrogates, must be performed every 12 h during analysis. After the system performance check is met, calibration check compounds (CCCs) are used to check the validity of the initial calibration.

The internal standard responses and retention times in the calibration check standard must be evaluated immediately after or during data acquisition. If the retention time for any internal standard changes by more than 30 seconds from the last check calibration (12 h), the chromatographic system must be inspected for malfunctions and corrections must be made, as required. If the electron ionization current plot (EICP) area for any of the internal standards changes by a factor of two from the last daily calibration standard check, the mass spectrometer must be inspected for malfunctions and corrections must be made, as appropriate.

Demonstrate, through the analysis of a reagent water blank, that interferences from the analytical system, glassware, and reagents are under control. The blank samples should be carried through all stages of the sample preparation and measurement steps. For each analytical batch (up to 20 samples), a reagent blank, matrix spike, and matrix spike duplicate/duplicate must be analyzed (the frequency of the spikes may be different for different monitoring programs). The blank and spiked samples must be carried through all stages of the sample preparation and measurement steps. A QC reference sample concentrate containing each analyte at a concentration of 100 mg/L in methanol is required.

REFERENCE Test Methods for Evaluating Solid Waste (SW-846). U.S. EPA 1983, Method 8270B, Rev. 2, Nov. 1990. Office of Solid Waste, Washington, DC.

Aroclor 1254   EPA Method 505
CAS #11097-69-1

TITLE Analysis of Organohalide Pesticides and Commercial Polychlorinated Biphenyl (PCB) Products in Water by Microextraction and Gas Chromatography. U.S. EPA Method 505, Rev. 2.0, 1989.

MATRIX This method is applicable to drinking water and raw source water. The latter should include most surface water and groundwater sources.

METHOD SUMMARY Method 505 covers 25 pesticides and commercial PCB products. This is a very sensitive method that is more useful for monitoring than for exploratory analyses. 5-mL samples of water are saturated with sodium chloride and then extracted by shaking with 2 mL of hexane. The sample extracts are transferred to an autosampler setup to inject 1- to 2-μL portions into a gas chromatograph (GC) for analysis. Alternatively, 1- to 2-μL portions of samples, blanks, and standards may be manually injected. Each extract is analyzed by capillary GC/ECD with confirmation using either a second capillary column or GC/MS. The electron capture detector is easy to use, but it is a nonselective detector. The microextraction technique also eliminates the expensive sample preparation costs of other methods, but it has the disadvantage of being less sensitive than most because the extracts are not concentrated.

INTERFERENCES Method interferences may be caused by contaminants in solvents, reagents, glassware, and other sample processing apparatus that lead to discrete artifacts or elevated baselines. Interfering contamination may occur when a sample containing low concentrations of analytes is analyzed immediately following a sample containing relatively high concentrations of the analytes. Matrix interferences also may be caused by contaminants that are coextracted from the sample; cleanup of sample extracts may be necessary in these cases. Some pesticides and commercial PCB products from aqueous solutions adhere to glass surfaces, so sample transfers and contact with glass surfaces should be minimized. Some pesticides are rapidly oxidized by chlorine so dechlorination with sodium thiosulfate at the time of sample collection is important. Also, splitless injectors may cause degradation of some pesticides.

INSTRUMENTATION A gas chromatograph/electron capture detector/data system, with temperature programming and split/splitless injector suitable for use with capillary columns is needed.

Column 1: 0.32 mm I.D. × 30 m fused silica capillary with chemically bond methyl polysiloxane phase (DB-1, 1.0 μm film, or equivalent).
Column 2: 0.32 mm I.D. × 30 m fused silica capillary with 1:1 mixed phase of dimethyl silicone and polyethylene glycol (Durawax-DX3, 0.25 μm film, or equivalent).
Column 3: 0.32 mm I.D. × 25 m fused silica capillary with chemically bonded 50:50 methyl-phenyl silicone (OV-17, 1.5 μm film, or equivalent).

Column 1 should be used as the primary analytical column. Columns 2 and 3 are recommended for use as confirmatory columns when GC/MS confirmation is not available.

PRECISION & ACCURACY Method detection limits are dependent upon the characteristics of the gas chromatographic system used. Analytes that are not separated chromatographically cannot be individually identified and used in the same calibration mixture or water samples unless an alternative technique for identification and quantification, such as mass spectrometry, is used.

The concentration(s) (in μg/L) used for these QC measurements was 1.8 and 1.7.
The MDL (in μg/L) was 0.102.
The accuracy (% recovery) for reagent water at the above concentration(s) was not available and not listed and the precision (%) was 10.4 and not listed.
The accuracy (% recovery) for groundwater at the above concentration(s) was not listed and not listed and the precision (%) was not listed and not listed.
The accuracy (% recovery) for tap water at the above concentration(s) was not listed and 85 and the precision (5) was not listed and 11.8.

Note: No range of concentrations is provided with this method.

SAMPLING METHOD Collect samples using a 40-mL screw-cap vial (prewashed with detergent, rinsed with distilled water and oven dried at 400°C for one h) with a Teflon®-faced silicone septum. Collect bubble-free samples and place the septum with the Teflon® side down on the water.

SAMPLE PRESERVATION If residual chlorine is present in the water add about 3 mg of sodium thiosulfate to each vial before samples are collected to remove the chlorine. Alternatively, add 75 μL of 0.04 g/mL solution of sodium thiosulfate to each vial just prior to sampling. Immediately cool samples to 4°C and store them in a solvent-free refrigerator at 4°C until analysis.

MHT The maximum holding time is 14 days from the time the sample was collected until it must be analyzed.

SAMPLE PREPARATION Remove the sample from storage and allow it to come to room temperature. Remove a 5-mL volume from each container and weigh the container to the nearest 0.1 g. Add 6 g of sodium chloride and 2.0 mL of hexane to each sample bottle. Recap the sample and shake it vigorously for one min. Allow the water and hexane phases to separate, remove the cap, and transfer 0.5 mL of hexane into an autosampler vial using a disposable glass pipette. Transfer the remaining hexane phase into a second autosampler vial and store at 4°C for reanalysis, if necessary. Discard the remaining sample/hexane mixture and reweigh the empty container to determine net weight of sample.

QUALITY CONTROL Minimum quality control requirements are initial demonstration of lab capability, analysis of lab reagent blanks, fortified blanks, fortified sample matrix, and quality control samples. The lab must analyze at least one fortified blank per sample set, or at least one for every 20 samples. The fortifying concentration of each analyte should be 10 times the method detection limit or the maximum calibration limit (MCL), whichever is less. Calculate accuracy as percent recovery and develop control limits from the mean percent recovery and standard deviation.

The lab must add a known concentration of the analytes to a minimum of 10% of the routine samples, or one lab fortified sample matrix per

sample set. Calculate the percent recovery for each analyte and compare to the control limits established from the analyses of the fortified blanks.

EPA CONTACT & HOTLINE For technical questions contact Dr. Baldev Bathija, U.S. EPA, Office of Ground Water and Drinking Water (WH-550D), 401 M St. SW, Washington, DC 20460. Tel. (202) 260-3040. For further information the EPA Safe Drinking Water Hotline may be called at: (800) 426-4791.

REFERENCE Methods for the Determination of Organic Compounds in Drinking Water, EPA/600/4-88/039 (revised July 1991). U.S. EPA Environmental Monitoring Systems Laboratory, Cincinnati, OH, 45268, U.S.A. Available from the National Technical Information Service (NTIS), 5285 Port Royal Road, Springfield, VA 22161; Tel. 800-553-6847. NTIS Order Number is PB91-231480.

Aroclor 1254   **EPA Method 625**
CAS #11097-69-1

TITLE Base/Neutrals and Acids, U.S. EPA Method 625

MATRIX This methods covers municipal and industrial wastewaters.

METHOD SUMMARY Approximately 1 L of sample is serially extracted with methylene chloride at a pH greater than 11 and again at a pH less than 2 using a separatory funnel or a continuous extractor. The methylene chloride extract is dried, concentrated to a volume of 1 mL, and analyzed by GC/MS. Qualitative identification of the parameters in the extract is performed using the retention time and the relative abundance of three characteristic masses (m/z). Qualitative analysis is performed using either external or internal standard techniques with a single characteristic m/z.

INTERFERENCES Method interferences may be caused by contaminants in solvents, reagents, glassware, and other sample processing hardware. Glassware must be scrupulously cleaned. Glassware should be heated in a muffle furnace at 400°C for 5 to 30 min. Some thermally stable materials, such as PCBs, may not be eliminated by this treatment. Solvent rinses with acetone and pesticide quality hexane may be substituted for the muffle furnace heating. Matrix interferences may be caused by contaminants that are coextracted from the sample. The base/neutral extraction may cause significantly reduced recovery of phenols. The packed gas chromatographic columns recommended for the basic fraction may not exhibit sufficient resolution for some analytes.

INSTRUMENTATION A GC/MS system with an injection port designed for on-column injection when using packed columns and for splitless injection when using capillary columns.

Column for base/neutrals: 1.8 m long × 2 mm I.D. glass, packed with 3% SP-2550 on Supelcoport (100/120 mesh) or equivalent.
Column for acids: 1.8 m long × 2 mm I.D. glass, packed with 1% SP-1240DA on Supelcoport (100/120 mesh) or equivalent.

PRECISION & ACCURACY The MDL concentrations were obtained using reagent water. The MDL actually achieved in a given analysis will vary depending on instrument sensitivity and matrix effects. This method was tested by 15 laboratories using reagent water, drinking water, surface water, and industrial wastewaters spiked at six concentrations over the range 5 to 100 µg/L. Single operator precision, overall precision, and method accuracy were found to be directly related to the concentration of the parameter matrix.

The MDL (in µg/L) in reagent water was 36.
The standard deviation (in µg/L based on 4 recovery measurements) was not reported.
The range (in µg/L) for average recovery for 4 measurements was not reported.
The range (in %) for percent recovery was not reported.
Accuracy (in µg/L) as expected recovery for one or more measurements of a sample containing a true concentration of C was not reported.
Precision (in µg/L) as expected single analyst standard deviation of measurements at an average concentration found at X was not reported.
Overall precision (in µg/L) as expected interlaboratory standard deviation of measurements in an average concentration found at X was not reported.

$C =$ *True value of the concentration in µg/L.*

$X =$ *Average recovery found for measurements of samples containing a concentration at C in µg/L.*

SAMPLE PREPARATION Adjust the pH to >11 with sodium hydroxide and serially extract in a separatory funnel with methylene chloride or else in a continuous extractor. Next, adjust the pH to <2 with sulfuric acid and serially extract in a separatory funnel with methylene chloride or else in a continuous extractor. Dry the extracts separately through a column of anhydrous sodium sulfate and then concentrate each of the extracts to 1.0 mL using a K-D apparatus.

SAMPLE COLLECTION, PRESERVATION & HANDLING Grab samples must be collected in glass containers. All samples must be refrigerated at 4°C from the time of collection until extraction. If residual chlorine is present, add 80 mg of sodium thiosulfate/L of sample and mix well. All samples must be extracted within 7 days of collection and completely analyzed within 40 days of extraction.

QUALITY CONTROL Make an initial, one-time, demonstration of the ability to generate acceptable accuracy and precision with this method. Before processing any samples, the analyst must analyze a reagent water blank to demonstrate that interferences from the analytical system and glassware are under control. Each time a set of samples is extracted or reagents are changed, a reagent water blank must be processed. Spike and analyze a minimum of 5% of all samples to monitor and evaluate lab data quality. A QC check sample concentrate that contains each parameter of interest at a concentration of 100 µg/mL in acetone is required. PCBs and multicomponent pesticides may be omitted from this test.

After analysis of five spiked wastewater samples, calculate the average percent recovery and the standard deviation of the percent recovery. Spike all samples with the surrogate standard spiking solution and calculate the percent recovery of each surrogate compound.

REFERENCE *Federal Register*, Vol. 49, No. 209. Friday, Oct. 26, 1984.

Aroclor 1254   **EPA Method 8080**
CAS #11097-69-1

TITLE Organochlorine Pesticides and Polychlorinated Biphenyls By Gas Chromatography

MATRIX This method is used to determine the concentration of various organochlorine pesticides and polychlorinated biphenyls in extracts prepared from water, groundwater, soils, and sediments.

METHOD SUMMARY This method covers 26 pesticides and Aroclor (PCB) mixtures and it is suitable for monitoring-type analyses. After extraction, concentration and solvent exchange to hexane, a 2- to 5-µL sample aliquot is injected into a GC using the solvent flush technique, and the analytes are detected by an electron capture detector (ECD) or an electrolytic conductivity detector in the halogen mode (HECD). Both neat and diluted organic liquids may be analyzed by direct injection.

INTERFERENCES Interferences coextracted from the samples will vary considerably from source to source. Interferences by phthalate esters can pose a major problem in pesticide determinations when using the ECD. Cross-contamination of clean glassware routinely occurs when plastics are handled during extraction steps, especially when solvent-wetted surfaces are handled. The contamination from phthalate esters can be completely eliminated with a microcoulometric or electrolytic conductivity detector. Solvents, reagent, glassware, and other sample processing hardware may yield artifacts and/or interferences to sample analysis.

INSTRUMENTATION A gas chromatograph capable of on-column injections is needed. It must be equipped with an ECD or a HECD and one of the following GC columns:

Column 1: Supelcoport (100/120 mesh) coated with 1.5% SP-2250/1.95% SP-2401 packed in a 1.8 m × 4 mm I.D. glass column.
Column 2: Supelcoport (100/120 mesh) coated with 3% OV-1 in a 1.8 m × 4 mm I.D. glass column.

PRECISION & ACCURACY The method was tested by 20 laboratories using organic-free reagent water, drinking water, surface water, and three industrial wastewaters spiked at six concentrations. Concentrations used in the study ranged from 0.5 to 30 µg/L for single-component pesticides and from 8.5 to 400 µg/L for multicomponent parameters. Overall precision and method accuracy were found to be directly related to the concentration of the analyte and essentially independent of the sample matrix. The sensitivity of this method usually depends on the concentration of interferences rather than on instrumental limitations.

MDL in µg/L was ND.
Concentration range in µg/L was 8.5–40.
Accuracy as recovery (x^*) in µg/L was $0.91C+10.79$.
Overall precision (S^*) in µg/L was $0.31x +3.50$.

x^* *Expected recovery for one or more measurements of a sample containing concentration C, in µg/L.*
$S^* =$ *Expected interlaboratory standard deviation of measurements at an average concentration found of the analyte in µg/L.*
$C =$ *True value for the concentration, in µg/L.*
$X =$ *Average recovery found for measurements of samples containing a concentration of C, in µg/L.*

SAMPLING METHOD
Liquid samples — Use a 1 or 2½ gallon amber glass bottle with a screw-top Teflon®-lined cover. Pre-wash the bottle with detergent, rinse with distilled water and methanol (or isopropanol).

Soil, sediments and sludges — Use an 8-oz. widemouth glass with a screw-top Teflon®-lined cover. Pre-wash the bottle with detergent, rinse with distilled water and methanol (or isopropanol).

SAMPLE PRESERVATION Cool water, soil, sediment, or sludge samples immediately to 4°C.

Water samples — If residual chlorine is present, add 3 mL of 10% sodium thiosulfate per gallon and cool to 4°C. All extracts and samples should be stored under refrigeration.

MHT Liquid samples must be extracted within 7 days and the extracts must be analyzed within 40 days. Soils, sediments, and sludges may be stored for a maximum of 14 days prior to extraction.

SAMPLE PREPARATION

Liquid samples — Extract 1-L samples in a continuous extractor at pH 5–9 with methylene chloride after adding 1.0 mL of surrogate spiking solution to each sample. Pass the extract through a column of anhydrous sodium sulfate to dry and concentrate it in a K-D apparatus to 1 mL volume.

Soils, sediments and sludges — Rapidly weigh approximately 30 g of sample into a 400-mL beaker to avoid loss of the more volatile extractables. Nonporous or wet samples (gummy or clay type) that do not have a free-flowing sandy texture must be mixed with anhydrous sodium sulfate until the sample is free flowing. Add 1 mL of surrogate standards to all samples, spikes, standards, and blanks. Add 100 mL of 1:1 methylene chloride:acetone and extract ultrasonically. Decant and filter extracts, dry the extract by passing it through a drying column containing anhydrous sodium sulfate and concentrate to 1 mL in a K-D apparatus.

Hexane eolvent exchange — Add 50 mL of hexane, a new boiling chip, and concentrate until the apparent volume of liquid reaches 1 Ml. Adjust the extract volume to 10.0 Ml. Stopper the concentration tube and store refrigerated at 4°C if further processing will not be performed immediately. If the extract will be stored longer than two days, transfer it to a vial with Teflon®-lined screw-cap or crimp top.

QUALITY CONTROL Demonstrate through the analysis of a reagent water blank, that all glassware and reagents are interference free. Each time a set of samples is processed, a method blank should be processed as a safeguard against chronic lab contamination. A reagent blank, a matrix spike, and a duplicate or matrix spike duplicate must be performed for each analytical batch (up to a maximum of 20 samples) analyzed.

Analytical system performance must be verified by analyzing QC check samples. The QC check sample concentration should contain each single-component analyte at the following concentrations in acetone: 4,4'-DDD, 10 µg/mL; 4,4'-DDT, 10 µg/mL; endosulfan II, 10 µg/mL; endosulfan sulfate, 10 µg/mL; and any other single-component pesticide at 2 µg/mL. If the method is only to be used to analyze PCBs, Chlordane, or Toxaphene, the QC check sample concentrate should contain the most representative multicomponent parameter at a concentration of 50 µg/mL in acetone.

REFERENCE Test Methods for Evaluating Solid Waste (SW-846). U.S. EPA. 1983. Method 8080B, Rev. 2, Nov. 1990. Office of Solid Wastes, Washington, DC.

Aroclor 1254 **EPA Method 8270**
CAS #11097-69-1

TITLE Semivolatile Organic Compounds by GC/MS

MATRIX This method is used to determine the concentration of semivolatile organic compounds in extracts prepared from all types of solid waste matrices, soils, and groundwater. Although surface waters are not specifically mentioned, this method should be applicable to water samples from rivers, lakes, etc.

METHOD SUMMARY This method covers 259 semivolatile organic compounds. In very limited applications direct injection of the sample into the GC/MS system may be appropriate, but this results in very high detection limits (approximately 10,000 µg/L). Typically, a 1-L liquid sample, containing surrogate, and matrix spiking standards, is extracted in a continuous extractor first under acid conditions and then under basic conditions.

Typically 30 g of a solid sample, containing surrogate, and matrix spiking standards, is extracted ultrasonically. After concentrating the extract to 1 mL it is spiked with 10 µL of an internal standard solution just prior to analysis by GC/MS. The volume injected should contain about 100 ng of base/neutral and 200 ng of acid surrogates (for a 1-µL injection). Analysis is performed by GC/MS using a capillary GC column.

INTERFERENCES Raw GC/MS data from all blanks, samples, and spikes must be evaluated for interferences. Contamination by carryover can occur whenever high-concentration and low-concentration samples are sequentially analyzed. To reduce carryover, the sample syringe must be rinsed out between samples with solvent. Whenever an unusually concentrated sample is encountered, it should be followed by the analysis of blank solvent to check for cross-contamination.

INSTRUMENTATION A GC/MS and a data system are required. The GC column used is a 30 m × 0.25 mm I.D. (or 0.32 mm I.D.) 1-µm film thickness silicone-coated fused silica capillary column. A continuous liquid-liquid extractor equipped with Teflon® or glass connection joints and stopcocks requiring no lubrication, a K-D concentrating apparatus, water bath, and an ultrasonic disrupter with a minimum power of 300 W and with pulsing capability are also required.

PRECISION & ACCURACY The estimated quantitation limit (EQL) of Method 8270B for determining an individual compound is approximately 1 mg/kg (wet weight) for soil or sediment samples, 1–200 mg/kg for wastes (dependent on matrix and method of preparation), and 10 µg/L for groundwater samples. EQLs will be proportionately higher for sample extracts that require dilution to avoid saturation of the detector.

The EQL(b) for groundwater in µg/L is not listed.
The EQL (a, b) for low concentrations in soil and sediment in µg/kg is not listed.
Accuracy as µg/L is not listed.
Overall precision in µg/L is not listed.

(a) *EQLs listed for soil/sediment are based on wet weight. Normally data is reported in a dry-weight basis; therefore, EQLs will be higher based on the % dry weight of each sample. This calculation is based on a 30-g sample and gel permeation chromatography cleanup.*

(b) *Sample EQLs are highly matrix-dependent. The EQLs are provided for guidance and may not always be achievable.*

C = True value for concentration, in µg/L.
X = Average recovery found for measurements of samples containing a concentration of C, in µg/L.

ESTIMATED QUANTITATION LIMIT

Other Matrices	Factor (a)
High-concentration soil and sludges by sonicator	7.5
Non-water miscible waste	75

(a) EQL for other matrices = [EQL for low soil/sediment] × [Factor]. This estimated EQL is similar to an EPA "Practical Quantitation Limit."

SAMPLING METHOD

Liquid samples — Use a 1 or 2½ gallon amber glass bottle with a screw-top Teflon®-lined cover that has been prewashed with detergent and rinsed with distilled water and methanol (or isopropanol).

Soils, sediments, or sludges — Use an 8-oz. widemouth glass with a screw-top Teflon®-lined cover that has been prewashed with detergent and rinsed with distilled water and methanol (or isopropanol).

SAMPLE PRESERVATION

Liquid samples — If residual chlorine is present, add 3 mL of 10% sodium thiosulfate per gallon, cool to 4°C and store in a solvent-free refrigerator until analysis; if chlorine is not present, then eliminate the sodium thiosulfate addition.

Soils, sediments, or sludges — Cool samples to 4°C and store in a solvent-free refrigerator.

MHT Liquid samples must be extracted within 7 days and the extracts analyzed within 40 days. Soils, sediments, or sludges may be stored for a maximum of 14 days and the extracts analyzed within 40 days.

SAMPLE PREPARATION

Liquid samples — Transfer 1 L quantitatively to a continuous extractor. If high concentrations are anticipated, a smaller volume may be used and then diluted with organic-free reagent water to 1 L. Adjust pH, if necessary, to pH <2 using 1:1 (V/V) sulfuric acid. Pipette 1.0 mL of a surrogate standard spiking solution into each sample. For the sample in each analytical batch selected for spiking, add 1.0 mL of a matrix spiking standard. For base/neutral acid analysis, the amount of the surrogates and matrix spiking compounds added to the sample should result in a final concentration of 100 ng/µL of each analyte in the extract to be analyzed (assuming a 1-µL injection). Extract with methylene chloride for 18–24 h. Next, adjust the pH of the aqueous phase to pH >11 using 10 N sodium hydroxide and extract it with methylene chloride again for 18–24 h. Dry the extract through a column containing anhydrous sodium sulfate and concentrate it to 1 mL using a K-D concentrator.

Soils, sediments, or sludges — Use 30 g of sample. Nonporous or wet samples (gummy or clay type) that do not have a free-flowing sandy texture must be mixed with anhydrous sodium sulfate until the sample is free flowing. Add 1 mL of surrogate standards to all samples, spikes, standards, and blanks. For the sample in each analytical batch selected for spiking, add 1.0 mL of a matrix spiking standard. For base/neutral acid analysis, the amount added of the surrogates and matrix spiking compounds should result in a final concentration of 100 ng/µL of each base/neutral analyte and 200 ng/µL of each acid analyte in the extract to be analyzed (assuming a 1-µL injection). Immediately add a 100 mL mixture of 1:1 methylene chloride:acetone and extract the sample ultrasonically for 3 min and then decant or filter the extracts. Repeat the extraction two or more times. Dry the extract using a column with anhydrous sodium sulfate and concentrate it to 1 mL in a K-D concentrator.

QUALITY CONTROL A methylene chloride solution containing 50 ng/µL of decafluorotriphenylphosphine (DFTPP) is used for tuning the GC/MS system each 12-h shift. A system performance check also must be made during every 12-h shift. A standard containing 50 ng/µL each of 4,4'-DDT, pentachlorophenol, and benzidine is required to verify injection port inertness and GC column performance. A calibration standard at mid-concentration, containing each compound of interest, including all required surrogates, must be performed every 12 h during analysis. After the system performance check is met, calibration check compounds (CCCs) are used to check the validity of the initial calibration.

The internal standard responses and retention times in the calibration check standard must be evaluated immediately after or during data acquisition. If the retention time for any internal standard changes by more than 30 seconds from the last check calibration (12 h), the chromatographic system must be inspected for malfunctions and corrections must be made, as required. If the electron ionization current plot (EICP) area for any of the internal standards changes by a factor of two from the last daily calibration standard check, the mass spectrometer must be inspected for malfunctions and corrections must be made, as appropriate.

Demonstrate, through the analysis of a reagent water blank, that interferences from the analytical system, glassware, and reagents are under control. The blank samples should be carried through all stages of the sample preparation and measurement steps. For each analytical batch (up to 20 samples), a reagent blank, matrix spike, and matrix spike duplicate/duplicate must be analyzed (the frequency of the spikes may be different for different monitoring programs). The blank and spiked samples must be carried through all stages of the sample preparation and measurement steps. A QC reference sample concentrate containing each analyte at a concentration of 100 mg/L in methanol is required.

REFERENCE Test Methods for Evaluating Solid Waste (SW-846). U.S. EPA 1983, Method 8270B, Rev. 2, Nov. 1990. Office of Solid Waste, Washington, DC.

Aroclor 1260   **EPA Method 505**
CAS #11096-82-5

TITLE Analysis of Organohalide Pesticides and Commercial Polychlorinated Biphenyl (PCB) Products in Water by Microextraction and Gas Chromatography. U.S. EPA Method 505, Rev. 2.0, 1989.

MATRIX This method is applicable to drinking water and raw source water. The latter should include most surface water and groundwater sources.

METHOD SUMMARY Method 505 covers 25 pesticides and commercial PCB products. This is a very sensitive method that is more useful for monitoring than for exploratory analyses. 5-mL samples of water are saturated with sodium chloride and then extracted by shaking with 2 mL of hexane. The sample extracts are transferred to an autosampler setup to inject 1- to 2-µL portions into a gas chromatograph (GC) for analysis. Alternatively, 1- to 2-µL portions of samples, blanks, and standards may be manually injected. Each extract is analyzed by capillary GC/ECD with confirmation using either a second capillary column or GC/MS. The electron capture detector is easy to use, but it is a nonselective detector. The microextraction technique also eliminates the expensive sample preparation costs of other methods, but it has the disadvantage of being less sensitive than most because the extracts are not concentrated.

INTERFERENCES Method interferences may be caused by contaminants in solvents, reagents, glassware, and other sample processing apparatus that lead to discrete artifacts or elevated baselines. Interfering contamination may occur when a sample containing low concentrations of analytes is analyzed immediately following a sample containing relatively high concentrations of the analytes. Matrix interferences also may be caused by contaminants that are coextracted from the sample; cleanup of sample extracts may be necessary in these cases. Some pesticides and commercial PCB products from aqueous solutions adhere to glass surfaces, so sample transfers and contact with glass surfaces should be minimized. Some pesticides are rapidly oxidized by chlorine so dechlorination with sodium thiosulfate at the time of sample collection is important. Also, splitless injectors may cause degradation of some pesticides.

INSTRUMENTATION A gas chromatograph/electron capture detector/data system, with temperature programming and split/splitless injector suitable for use with capillary columns is needed.

Column 1: 0.32 mm I.D. × 30 m fused silica capillary with chemically bond methyl polysiloxane phase (DB-1, 1.0 µm film, or equivalent).

Column 2: 0.32 mm I.D. × 30 m fused silica capillary with 1:1 mixed phase of dimethyl silicone and polyethylene glycol (Durawax-DX3, 0.25 µm film, or equivalent).

Column 3: 0.32 mm I.D. × 25 m fused silica capillary with chemically bonded 50:50 methyl-phenyl silicone (OV-17, 1.5 µm film, or equivalent).

Column 1 should be used as the primary analytical column. Columns 2 and 3 are recommended for use as confirmatory columns when GC/MS confirmation is not available.

PRECISION & ACCURACY Method detection limits are dependent upon the characteristics of the gas chromatographic system used. Analytes that are not separated chromatographically cannot be individually identified and used in the same calibration mixture or water samples unless an alternative technique for identification and quantification, such as mass spectrometry, is used.

The concentration(s) (in µg/L) used for these QC measurements was 2.0 and 1.8.

The MDL (in µg/L) was 0.189.

The accuracy (% recovery) for reagent water at the above concentration(s) was not available and not listed and the precision (%) was 20.7 and not available.

The accuracy (% recovery) for groundwater at the above concentration(s) was not listed and not listed and the precision (%) was not listed and not listed.

The accuracy (% recovery) for tap water at the above concentration(s) was not listed and 88 and the precision (5) was not listed and 19.8.

Note: No range of concentrations is provided with this method.

SAMPLING METHOD Collect samples using a 40-mL screw-cap vial (prewashed with detergent, rinsed with distilled water and oven dried at 400°C for one h) with a Teflon®-faced silicone septum. Collect bubble-free samples and place the septum with the Teflon® side down on the water.

SAMPLE PRESERVATION If residual chlorine is present in the water add about 3 mg of sodium thiosulfate to each vial before samples are collected to remove the chlorine. Alternatively, add 75 µL of 0.04 g/mL solution of sodium thiosulfate to each vial just prior to sampling. Immediately cool samples to 4°C and store them in a solvent-free refrigerator at 4°C until analysis.

MHT The maximum holding time is 14 days from the time the sample was collected until it must be analyzed.

SAMPLE PREPARATION Remove the sample from storage and allow it to come to room temperature. Remove a 5-mL volume from each container and weigh the container to the nearest 0.1 g. Add 6 g of sodium chloride and 2.0 mL of hexane to each sample bottle. Recap the sample and shake it vigorously for one min. Allow the water and hexane phases to separate, remove the cap, and transfer 0.5 mL of hexane into an autosampler vial using a disposable glass pipette. Transfer the remaining hexane phase into a second autosampler vial and store at 4°C for reanalysis, if necessary. Discard the remaining sample/hexane mixture and reweigh the empty container to determine net weight of sample.

QUALITY CONTROL Minimum quality control requirements are initial demonstration of lab capability, analysis of lab reagent blanks, fortified blanks, fortified sample matrix, and quality control samples. The lab must analyze at least one fortified blank per sample set, or at least one for every 20 samples. The fortifying concentration of each analyte should be 10 times the method detection limit or the maximum calibration limit (MCL), whichever is less. Calculate accuracy as percent recovery and develop control limits from the mean percent recovery and standard deviation.

The lab must add a known concentration of the analytes to a minimum of 10% of the routine samples, or one lab fortified sample matrix per sample set. Calculate the percent recovery for each analyte and compare to the control limits established from the analyses of the fortified blanks.

EPA CONTACT & HOTLINE For technical questions contact Dr. Baldev Bathija, U.S. EPA, Office of Ground Water and Drinking Water (WH-550D), 401 M St. SW, Washington, DC 20460. Tel. (202) 260-3040. For further information the EPA Safe Drinking Water Hotline may be called at: (800) 426-4791.

REFERENCE Methods for the Determination of Organic Compounds in Drinking Water, EPA/600/4-88/039 (revised July 1991). U.S. EPA Environmental Monitoring Systems Laboratory, Cincinnati, OH, 45268, U.S.A. Available from the National Technical Information Service (NTIS), 5285 Port Royal Road, Springfield, VA 22161; Tel. 800-553-6847. NTIS Order Number is PB91-231480.

Aroclor 1260   **EPA Method 625**
CAS #11096-82-5

TITLE Base/Neutrals and Acids, U.S. EPA Method 625

MATRIX This methods covers municipal and industrial wastewaters.

METHOD SUMMARY Approximately 1 L of sample is serially extracted with methylene chloride at a pH greater than 11 and again at a pH less than 2 using a separatory funnel or a continuous extractor. The methylene chloride extract is dried, concentrated to a volume of 1 mL, and analyzed by GC/MS. Qualitative identification of the parameters in the extract is performed using the retention time and the relative abundance of three characteristic masses (m/z). Qualitative analysis is performed using either external or internal standard techniques with a single characteristic m/z.

INTERFERENCES Method interferences may be caused by contaminants in solvents, reagents, glassware, and other sample processing hardware. Glassware must be scrupulously cleaned. Glassware should be heated in a muffle furnace at 400°C for 5 to 30 min. Some thermally stable materials, such as PCBs, may not be eliminated by this treatment. Solvent rinses with acetone and pesticide quality hexane may be substituted for the muffle furnace heating. Matrix interferences may be caused by contaminants that are coextracted from the sample. The base/neutral extraction may cause significantly reduced recovery of phenols. The packed gas chromatographic columns recommended for the basic fraction may not exhibit sufficient resolution for some analytes.

INSTRUMENTATION A GC/MS system with an injection port designed for on-column injection when using packed columns and for splitless injection when using capillary columns.

Column for base/neutrals: 1.8 m long × 2 mm I.D. glass, packed with 3% SP-2550 on Supelcoport (100/120 mesh) or equivalent.
Column for acids: 1.8 m long × 2 mm I.D. glass, packed with 1% SP-1240DA on Supelcoport (100/120 mesh) or equivalent.

PRECISION & ACCURACY The MDL concentrations were obtained using reagent water. The MDL actually achieved in a given analysis will vary depending on instrument sensitivity and matrix effects. This method was tested by 15 laboratories using reagent water, drinking water, surface water, and industrial wastewaters spiked at six concentrations over the range 5 to 100 µg/L. Single operator precision, overall precision, and method accuracy were found to be directly related to the concentration of the parameter matrix.

The MDL (in µg/L) in reagent water was not detected.
The standard deviation (in µg/L based on 4 recovery measurements) was 54.2.
The range (in µg/L) for average recovery for 4 measurements was 19.3–121.0.
The range (in %) for percent recovery was D-164.
Accuracy (in µg/L) as expected recovery for one or more measurements of a sample containing a true concentration of C was 0.81C-10.86.
Precision (in µg/L) as expected single analyst standard deviation of measurements at an average concentration found at X was 0.35X+3.61.
Overall precision (in µg/L) as expected interlaboratory standard deviation of measurements in an average concentration found at X was 0.43X+1.82.

$C =$ *True value of the concentration in µg/L.*
$X =$ *Average recovery found for measurements of samples containing a concentration at C in µg/L.*

SAMPLE PREPARATION Adjust the pH to >11 with sodium hydroxide and serially extract in a separatory funnel with methylene chloride or else in a continuous extractor. Next, adjust the pH to <2 with sulfuric acid and serially extract in a separatory funnel with methylene chloride or else in a continuous extractor. Dry the extracts separately through a column of anhydrous sodium sulfate and then concentrate each of the extracts to 1.0 mL using a K-D apparatus.

SAMPLE COLLECTION, PRESERVATION & HANDLING Grab samples must be collected in glass containers. All samples must be refrigerated at 4°C from the time of collection until extraction. If residual chlorine is present, add 80 mg of sodium thiosulfate/L of sample and mix well. All samples must be extracted within 7 days of collection and completely analyzed within 40 days of extraction.

QUALITY CONTROL Make an initial, one-time, demonstration of the ability to generate acceptable accuracy and precision with this method. Before processing any samples, the analyst must analyze a reagent water blank to demonstrate that interferences from the analytical system and glassware are under control. Each time a set of samples is extracted or reagents are changed, a reagent water blank must be processed. Spike and analyze a minimum of 5% of all samples to monitor and evaluate lab data quality. A QC check sample concentrate that contains each parameter of interest at a concentration of 100 µg/mL in acetone is required. PCBs and multicomponent pesticides may be omitted from this test.

After analysis of five spiked wastewater samples, calculate the average percent recovery and the standard deviation of the percent recovery. Spike all samples with the surrogate standard spiking solution and calculate the percent recovery of each surrogate compound.

REFERENCE Federal Register, Vol. 49, No. 209. Friday, Oct. 26, 1984.

Aroclor 1260 — EPA Method 8080
CAS #11096-82-5

TITLE Organochlorine Pesticides and Polychlorinated Biphenyls By Gas Chromatography

MATRIX This method is used to determine the concentration of various organochlorine pesticides and polychlorinated biphenyls in extracts prepared from water, groundwater, soils, and sediments.

METHOD SUMMARY This method covers 26 pesticides and Aroclor (PCB) mixtures and it is suitable for monitoring-type analyses. After extraction, concentration and solvent exchange to hexane, a 2- to 5-µL sample aliquot is injected into a GC using the solvent flush technique, and the analytes are detected by an electron capture detector (ECD) or an electrolytic conductivity detector in the halogen mode (HECD). Both neat and diluted organic liquids may be analyzed by direct injection.

INTERFERENCES Interferences coextracted from the samples will vary considerably from source to source. Interferences by phthalate esters can pose a major problem in pesticide determinations when using the ECD.

Cross-contamination of clean glassware routinely occurs when plastics are handled during extraction steps, especially when solvent-wetted surfaces are handled. The contamination from phthalate esters can be completely eliminated with a microcoulometric or electrolytic conductivity detector. Solvents, reagent, glassware, and other sample processing hardware may yield artifacts and/or interferences to sample analysis.

INSTRUMENTATION A gas chromatograph capable of on-column injections is needed. It must be equipped with an ECD or a HECD and one of the following GC columns:

Column 1: Supelcoport (100/120 mesh) coated with 1.5% SP-2250/1.95% SP-2401 packed in a 1.8 m × 4 mm I.D. glass column.

Column 2: Supelcoport (100/120 mesh) coated with 3% OV-1 in a 1.8 m × 4 mm I.D. glass column.

PRECISION & ACCURACY The method was tested by 20 laboratories using organic-free reagent water, drinking water, surface water, and three industrial wastewaters spiked at six concentrations. Concentrations used in the study ranged from 0.5 to 30 µg/L for single-component pesticides and from 8.5 to 400 µg/L for multicomponent parameters. Overall precision and method accuracy were found to be directly related to the concentration of the analyte and essentially independent of the sample matrix. The sensitivity of this method usually depends on the concentration of interferences rather than on instrumental limitations.

MDL in µg/L was ND.

Concentration range in µg/L was 8.5–40.
Accuracy as recovery (x^*) in µg/L was $0.91C+10.79$.
Overall precision (S^*) in µg/L was $0.31x +3.50$.

x^* *Expected recovery for one or more measurements of a sample containing concentration C, in µg/L.*
$S^* =$ *Expected interlaboratory standard deviation of measurements at an average concentration found of the analyte in µg/L.*
$C =$ *True value for the concentration, in µg/L.*
$X =$ *Average recovery found for measurements of samples containing a concentration of C, in µg/L.*

SAMPLING METHOD
Liquid samples — Use a 1 or 2½ gallon amber glass bottle with a screw-top Teflon®-lined cover. Pre-wash the bottle with detergent, rinse with distilled water and methanol (or isopropanol).

Soil, sediments and sludges — Use an 8-oz. widemouth glass with a screw-top Teflon®-lined cover. Pre-wash the bottle with detergent, rinse with distilled water and methanol (or isopropanol).

SAMPLE PRESERVATION Cool water, soil, sediment, or sludge samples immediately to 4°C.

Water samples — If residual chlorine is present, add 3 mL of 10% sodium thiosulfate per gallon and cool to 4°C. All extracts and samples should be stored under refrigeration.

MHT Liquid samples must be extracted within 7 days and the extracts must be analyzed within 40 days. Soils, sediments, and sludges may be stored for a maximum of 14 days prior to extraction.

SAMPLE PREPARATION
Liquid samples — Extract 1-L samples in a continuous extractor at pH 5–9 with methylene chloride after adding 1.0 mL of surrogate spiking solution to each sample. Pass the extract through a column of anhydrous sodium sulfate to dry and concentrate it in a K-D apparatus to 1 mL volume.

Soils, sediments and sludges — Rapidly weigh approximately 30 g of sample into a 400-mL beaker to avoid loss of the more volatile extractables. Nonporous or wet samples (gummy or clay type) that do not have a free-flowing sandy texture must be mixed with anhydrous sodium sulfate until the sample is free flowing. Add 1 mL of surrogate standards to all samples, spikes, standards, and blanks. Add 100 mL of 1:1 methylene chloride:acetone and extract ultrasonically. Decant and filter extracts, dry the extract by passing it through a drying column containing anhydrous sodium sulfate and concentrate to 1 mL in a K-D apparatus.

Hexane eolvent exchange — Add 50 mL of hexane, a new boiling chip, and concentrate until the apparent volume of liquid reaches 1 Ml. Adjust the extract volume to 10.0 Ml. Stopper the concentration tube and store refrigerated at 4°C if further processing will not be performed immediately. If the extract will be stored longer than two days, transfer it to a vial with Teflon®-lined screw-cap or crimp top.

QUALITY CONTROL Demonstrate through the analysis of a reagent water blank, that all glassware and reagents are interference free. Each time a set of samples is processed, a method blank should be processed as a safeguard against chronic lab contamination. A reagent blank, a matrix spike, and a duplicate or matrix spike duplicate must be performed for each analytical batch (up to a maximum of 20 samples) analyzed.

Analytical system performance must be verified by analyzing QC check samples. The QC check sample concentration should contain each single-component analyte at the following concentrations in acetone: 4,4'-DDD, 10 µg/mL; 4,4'-DDT, 10 µg/mL; endosulfan II, 10 µg/mL; endosulfan sulfate, 10 µg/mL; and any other single-component pesticide at 2 µg/mL. If the method is only to be used to analyze PCBs, Chlordane, or Toxaphene, the QC check sample concentrate should contain the most representative multicomponent parameter at a concentration of 50 µg/mL in acetone.

REFERENCE Test Methods for Evaluating Solid Waste (SW-846). U.S. EPA. 1983. Method 8080B, Rev. 2, Nov. 1990. Office of Solid Wastes, Washington, DC.

Aroclor 1260   **EPA Method 8270**
CAS #11096-82-5

TITLE Semivolatile Organic Compounds by GC/MS

MATRIX This method is used to determine the concentration of semivolatile organic compounds in extracts prepared from all types of solid waste matrices, soils, and groundwater. Although surface waters are not specifically mentioned, this method should be applicable to water samples from rivers, lakes, etc.

METHOD SUMMARY This method covers 259 semivolatile organic compounds. In very limited applications direct injection of the sample into the GC/MS system may be appropriate, but this results in very high detection limits (approximately 10,000 µg/L). Typically, a 1-L liquid sample, containing surrogate, and matrix spiking standards, is extracted in a continuous extractor first under acid conditions and then under basic conditions. Typically 30 g of a solid sample, containing surrogate, and matrix spiking standards, is extracted ultrasonically. After concentrating the extract to 1 mL it is spiked with 10 µL of an internal standard solution just prior to analysis by GC/MS. The volume injected should contain about 100 ng of base/neutral and 200 ng of acid surrogates (for a 1-µL injection). Analysis is performed by GC/MS using a capillary GC column.

INTERFERENCES Raw GC/MS data from all blanks, samples, and spikes must be evaluated for interferences. Contamination by carryover can occur whenever high-concentration and low-concentration samples are sequentially analyzed. To reduce carryover, the sample syringe must be rinsed out between samples with solvent. Whenever an unusually concentrated sample is encountered, it should be followed by the analysis of blank solvent to check for cross-contamination.

INSTRUMENTATION A GC/MS and a data system are required. The GC column used is a 30 m × 0.25 mm I.D. (or 0.32 mm I.D.) 1-µm film thickness silicone-coated fused silica capillary column. A continuous liquid-liquid extractor equipped with Teflon® or glass connection joints and stopcocks requiring no lubrication, a K-D concentrating apparatus, water bath, and an ultrasonic disrupter with a minimum power of 300 W and with pulsing capability are also required.

PRECISION & ACCURACY The estimated quantitation limit (EQL) of Method 8270B for determining an individual compound is approximately 1 mg/kg (wet weight) for soil or sediment samples, 1–200 mg/kg for wastes (dependent on matrix and method of preparation), and 10 µg/L for groundwater samples. EQLs will be proportionately higher for sample extracts that require dilution to avoid saturation of the detector.

The EQL(b) for groundwater in µg/L is not listed.
The EQL (a, b) for low concentrations in soil and sediment in µg/kg is not listed.
Accuracy as µg/L is $0.81C - 10.86$.
Overall precision in µg/L is $0.43X + 1.82$.

(a) EQLs listed for soil/sediment are based on wet weight. Normally data is reported in a dry-weight basis; therefore, EQLs will be higher based on the % dry weight of each sample. This calculation is based on a 30-g sample and gel permeation chromatography cleanup.

(b) Sample EQLs are highly matrix-dependent. The EQLs are provided for guidance and may not always be achievable.

$C =$ True value for concentration, in µg/L.
$X =$ Average recovery found for measurements of samples containing a concentration of C, in µg/L.

ESTIMATED QUANTITATION LIMIT

Other Matrices	Factor (a)
High-concentration soil and sludges by sonicator	7.5
Non-water miscible waste	75

(a) EQL for other matrices = [EQL for low soil/sediment] × [Factor]. This estimated EQL is similar to an EPA "Practical Quantitation Limit."

SAMPLING METHOD

Liquid samples — Use a 1 or 2½ gallon amber glass bottle with a screw-top Teflon®-lined cover that has been prewashed with detergent and rinsed with distilled water and methanol (or isopropanol).

Soils, sediments, or sludges — Use an 8-oz. widemouth glass with a screw-top Teflon®-lined cover that has been prewashed with detergent and rinsed with distilled water and methanol (or isopropanol).

SAMPLE PRESERVATION

Liquid samples — If residual chlorine is present, add 3 mL of 10% sodium thiosulfate per gallon, cool to 4°C and store in a solvent-free refrigerator until analysis; if chlorine is not present, then eliminate the sodium thiosulfate addition.

Soils, sediments, or sludges — Cool samples to 4°C and store in a solvent-free refrigerator.

MHT Liquid samples must be extracted within 7 days and the extracts analyzed within 40 days. Soils, sediments, or sludges may

be stored for a maximum of 14 days and the extracts analyzed within 40 days.

SAMPLE PREPARATION

Liquid samples — Transfer 1 L quantitatively to a continuous extractor. If high concentrations are anticipated, a smaller volume may be used and then diluted with organic-free reagent water to 1 L. Adjust pH, if necessary, to pH <2 using 1:1 (V/V) sulfuric acid. Pipette 1.0 mL of a surrogate standard spiking solution into each sample. For the sample in each analytical batch selected for spiking, add 1.0 mL of a matrix spiking standard. For base/neutral acid analysis, the amount of the surrogates and matrix spiking compounds added to the sample should result in a final concentration of 100 ng/µL of each analyte in the extract to be analyzed (assuming a 1-µL injection). Extract with methylene chloride for 18–24 h. Next, adjust the pH of the aqueous phase to pH >11 using 10 N sodium hydroxide and extract it with methylene chloride again for 18–24 h. Dry the extract through a column containing anhydrous sodium sulfate and concentrate it to 1 mL using a K-D concentrator.

Soils, sediments, or sludges — Use 30 g of sample. Nonporous or wet samples (gummy or clay type) that do not have a free-flowing sandy texture must be mixed with anhydrous sodium sulfate until the sample is free flowing. Add 1 mL of surrogate standards to all samples, spikes, standards, and blanks. For the sample in each analytical batch selected for spiking, add 1.0 mL of a matrix spiking standard. For base/neutral acid analysis, the amount added of the surrogates and matrix spiking compounds should result in a final concentration of 100 ng/µL of each base/neutral analyte and 200 ng/µL of each acid analyte in the extract to be analyzed (assuming a 1-µL injection). Immediately add a 100 mL mixture of 1:1 methylene chloride:acetone and extract the sample ultrasonically for 3 min and then decant or filter the extracts. Repeat the extraction two or more times. Dry the extract using a column with anhydrous sodium sulfate and concentrate it to 1 mL in a K-D concentrator.

QUALITY CONTROL

A methylene chloride solution containing 50 ng/µL of decafluorotriphenylphosphine (DFTPP) is used for tuning the GC/MS system each 12-h shift. A system performance check also must be made during every 12-h shift. A standard containing 50 ng/µL each of 4,4′-DDT, pentachlorophenol, and benzidine is required to verify injection port inertness and GC column performance. A calibration standard at mid-concentration, containing each compound of interest, including all required surrogates, must be performed every 12 h during analysis. After the system performance check is met, calibration check compounds (CCCs) are used to check the validity of the initial calibration.

The internal standard responses and retention times in the calibration check standard must be evaluated immediately after or during data acquisition. If the retention time for any internal standard changes by more than 30 seconds from the last check calibration (12 h), the chromatographic system must be inspected for malfunctions and corrections must be made, as required. If the electron ionization current plot (EICP) area for any of the internal standards changes by a factor of two from the last daily calibration standard check, the mass spectrometer must be inspected for malfunctions and corrections must be made, as appropriate.

Demonstrate, through the analysis of a reagent water blank, that interferences from the analytical system, glassware, and reagents are under control. The blank samples should be carried through all stages of the sample preparation and measurement steps. For each analytical batch (up to 20 samples), a reagent blank, matrix spike, and matrix spike duplicate/duplicate must be analyzed (the frequency of the spikes may be different for different monitoring programs). The blank and spiked samples must be carried through all stages of the sample preparation and measurement steps. A QC reference sample concentrate containing each analyte at a concentration of 100 mg/L in methanol is required.

REFERENCE Test Methods for Evaluating Solid Waste (SW-846). U.S. EPA 1983, Method 8270B, Rev. 2, Nov. 1990. Office of Solid Waste, Washington, DC.

Arsenic
CAS #7440-38-2
EPA Method 6010

TITLE Inductively Coupled Plasma-Atomic Emission Spectroscopy

MATRIX This method is applicable to the determination of trace elements, including metals, in groundwater, soils, sludges, sediments, and other solid wastes. All matrices require digestion prior to analysis. The method of standard addition must be used for the analysis of all sample digests unless either serial dilution or matrix spike addition demonstrates it is not required.

METHOD SUMMARY Method 6010 covers 25 elements using ICP analysis. It measures element-emitted light by optical spectrometry. Samples, following an appropriate acid digestion, are nebulized and the resulting aerosol is transported to the plasma torch. Element-specific atomic line emission spectra are produced by a radio-frequency inductively coupled plasma.

INTERFERENCES Interferences may be categorized as spectral or non-spectral. Spectral interferences are caused by overlap of a spectral line from another element, unresolved overlap of molecular band spectra, background contribution from continuous or recombination phenomenon, and stray light from the line emission of high concentration elements. Non-spectral interferences include physical and chemical interferences. Physical interferences are effects associated with the sample nebulization and transport processes. Changes in viscosity and surface tension can cause significant inaccuracies. Chemical interferences include molecular compound formation, ionization effects, and solute vaporization effects. Normally these effects are not significant and can be minimized by careful selection of operating conditions. Chemical interferences are highly dependent on matrix type and the specific analyte element.

INSTRUMENTATION An inductively coupled argon plasma emission spectrometer (ICP) capable of background correction is required.

PRECISION & ACCURACY Detection limits, sensitivity, and optimum ranges of the metals will vary with the matrices and model of the spectrometer. In a single lab evaluation, seven wastes were analyzed for 22 elements. The mean percent relative standard deviation from triplicate analyses for all elements and wastes was 9 ± 2%. The mean percent recovery of spiked elements for all wastes was 93 ± 6%. Spike levels ranged from 100 µg/L to 100 mg/L. The wastes included sludges and industrial wastewaters.

Estimated instrument detection limit in µg/L is 53.
Spiked concentration in µg/L is 22.
Mean reported value in µg/L is 19.
Precision as RSD % is 23.

SAMPLING METHOD Samples should be collected in borosilicate glass, linear polyethylene, polypropylene, or Teflon® bottles that have been prewashed with detergent and tap water, and rinsed with 1:1 nitric acid and tap water or 1:1 hydrochloric acid and tap water. Collect at least 2 g of solids and 200 mL of aqueous samples.

SAMPLE PRESERVATION Add nitric acid to make the samples pH <2.

MHT The maximum holding time for properly preserved samples is 6 months.

SAMPLE PREPARATION Preliminary treatment of most matrices is necessary because of the complexity and variability of sample matrices. Water samples which have been prefiltered and acidified will not need acid digestion. Methods for acid digestion of waters for total recoverable or dissolved metals, acid digestions of aqueous samples and extracts for total metals, and acid digestion of sediments, sludges, and soils are summarized below.

Total recoverable or dissolved metals in water — To prepare surface and groundwater samples for determination of total recoverable and dissolved metals, a 100 mL aliquot of well-mixed sample is acidified with concentrated nitric acid and concentrated hydrochloric acid, then heated until the volume is reduced to 15–20 Ml. Adjust the final volume to 100 mL with reagent water.

Total metals in aqueous samples, soil and sediment extracts — To prepare aqueous samples, soil and sediment extracts, and wastes that contain suspended solids, a 100 mL aliquot is made acidic with concentrated nitric acid and the solution is evaporated to about 5 mL on a hot plate. Continue heating and adding additional acid until sample digestion is complete, which is usually indicated when the digestate is light in color or does not change in appearance. Evaporate the solution to about 3 mL and cool it and add a small quantity of 1:1 hydrochloric acid (10 Ml/100 mL of final solution). Cover the beaker and reflux for 15 min. Wash down the beaker walls and filters or centrifuge the sample to remove silicates and other insoluble material. Filter the sample and adjust the final volume to 100 mL with reagent water and the final acid concentration to 10%.

Sediments, sludges, and soils — To prepare sediments, sludges, and soil samples, transfer 1–2 g to a conical beaker and add 10 mL of 1:1 nitric acid, mix the slurry, and cover it with a watch glass. Heat the sample and reflux for 10 to 15 min without boiling. Allow it to cool, then add 5 mL of concentrated nitric acid and reflux for 30 min. Repeat last step and then allow the solution to evaporate to 5 mL without boiling. Cool and add 2 mL of water and 3 mL of 30% hydrogen peroxide. Cover and place the beaker on the hot plate. Heat and add 30% hydrogen peroxide in 1 mL aliquots with warming until the effervescence is minimal but do not add more than a total of 10 mL of 30% hydrogen peroxide. If the sample is being prepared for the analysis of Ag, Al, As, Ba, Be, Ca, Cd, Co, Cr, Cu, Fe, K, Mg, Mn, Mo, Na, Ni, Os, Pb, Se, Tl, V, and Zn, then add 5 mL of concentrated hydrochloric acid and 10 mL of water and return the covered beaker to a hot plate for 15 min of additional refluxing without boiling. Dilute the sample to a 100 mL volume with water after cooling and filter or centrifuge to remove particulates.

QUALITY CONTROL Laboratory control samples must be analyzed for each analytical method. A method blank should be analyzed with each batch of samples. The effect of the matrix on method performance must be demonstrated: when appropriate, there should be at least one matrix spike and either one matrix duplicate or one matrix spike duplicate per analytical batch. The bias and precision of the method, as well as the method detection limit for each specific matrix type, must be measured.

Dilute and reanalyze samples that are more concentrated than the linear calibration limit. Employ a minimum of one reagent blank per sample batch to determine if contamination or any memory effects are occurring. Whenever a new or unusual sample matrix is encountered, perform either a serial dilution test or a matrix spike addition test to ensure that neither positive or negative interferences are operating on any of the analyte elements. Check the instrument standardization by verifying calibration every 10 samples using a calibration blank and a check standard.

REFERENCE Test Methods for Evaluating Solid Waste (SW-846). U.S. EPA. 1983. Method 6010, Rev. 0, Sept. 1986. Office of Solid Wastes, Washington, DC.

Arsenic EPA Method 200.7
CAS #7440-38-2

TITLE Inductively Coupled Plasma (ICP)

MATRIX Dissolved, suspended or total element in drinking and surface waters and in domestic and industrial wastewaters.

APPLICATION The method covers the determination of 25 metals. Dissolved elements are determined in filtered and acidified samples after appropriate digestion (which increases dissolved solids). Its primary advantage is that ICP instruments allow simultaneous or rapid sequential determination of many elements in a short time. Samples are first nebulized and the aerosol is transported to a plasma torch in which element specific atomic line emission spectra are produced by a radio frequency inductively coupled plasma. Background correction is required for trace element detection except in the case of line broadening.

INTERFERENCES There are spectral, physical, and chemical interferences. The primary disadvantage of ICP instruments is background radiation from other elements and the plasma gases (spectral interferences). Changes in sample viscosity and surface tension with samples containing high dissolved solids (especially those exceeding 1500 mg/L) or high acid concentrations can cause physical interferences. Ionization effects, solute vaporization and molecular compound formation can cause chemical interferences. Aluminum, chromium, and vanadium can cause interference at the 100 mg/L level.

INSTRUMENTATION Inductively coupled argon plasma emission spectroscopy. 193.696 nm wavelength.

RANGE Not listed.

MDL 53 µg/L.

PRECISION SD = 7.5% Mean at true value 200 µg/L.

ACCURACY Mean Recovery = 93% ± 6% of spiked elements for all wastes.

SAMPLING METHOD Wash sample container with detergent and tap water, rinse with 1+1 nitric acid and tap water, then rinse with 1+1 hydrochloric acid and tap water, then rinse with deionized, distilled water in that order. Perform any filtration or acid preservation steps when the sample is collected or as soon as possible thereafter.

STABILITY Cool samples to 4°C.

MHT = 24 h.

QUALITY CONTROL Mixed calibration standards, an instrument check standard, and an interference check solution are used in addition to a quality control sample. The quality control sample should be prepared in the same acid matrix as the calibration standards at 10 times the instrumental detection limits and in accordance with the instructions provided by the supplier. Furthermore, two types of blanks are required: a calibration blank and a reagent blank.

REFERENCE Method 200.7, U.S. EPA, EMSL-Cincinnati, OH, Nov. 1980

Arsenic   **EPA Method 206.2**
CAS #7440-38-2

TITLE Metals (Total, Dissolved, Suspended) AAS, Furnace Technique

MATRIX drinking, surface, and saline waters. Wastewater.

APPLICATION Date issued 1978. A representative sample aliquot (containing nickel nitrate) is placed in graphite tube in furnace, evaporated to dryness, charred and atomized. Radiation from excited element is passed through vapor and decreases proportional to amount in vapor.

INTERFERENCES Furnace technique subject to chemical and matrix interferences. Furnace gases may have molecular absorption bands enclosing analytical wavelength. Smoke-producing sample matrix can interfere. If As isn't volatilized and removed from furnace, memory effects occur.

INSTRUMENTATION AAS. Arsenic(As) hollow cathode lamp or EDL. Graphite furnace. Pipets.

RANGE 5–100 µg/L

MDL 1 µg/L

PRECISION SD = ±1.6 at 100 µg As/L

ACCURACY Recovery = 101% at 100 µg As/L

SAMPLING METHOD Plastic or glass (prewashed).

STABILITY HNO3 to pH <2.

MHT 6 months.

QUALITY CONTROL A check standard should be run approximately after every 10 sample injections. Standards are run in part to monitor the life and performance of the graphite tube. Lack of reproducibility or significant change in the signal for the standard indicates tube should be replaced.

REFERENCE EPA Methods for the Chemical Analysis of Water and Wastes, EPA-600/4-79-020, USEPA, EMSL, 1979.

Arsenic   **EPA Method 7061**
CAS #7440-38-2

TITLE Atomic Absorption (AA)

MATRIX Wastes, mobility procedure. Gaseous hydride extracts, soils, and groundwater.

APPLICATION Method approved only for sample matrices without high concentrations of Cr, Cu, Hg, Ni, Ag, Co and Mo after sample preparation with HNO3/H2SO4 digestion, Arsenic in digestate is reduced to trivalent form with SnCl2. As(III) is then converted to a volatile hydride using hydrogen and is swept into an argon-hydrogen flame located in the optical path of an atomic absorption spectrometer. Absorption of the lamp radiation is proportional to the arsenic concentration.

INTERFERENCES Traces of nitric acid left following sample work-up can result in analytical interferences. Elemental As and many of its compounds are volatilized, thus samples may be subject to losses during sample preparation. High concentrations of Cr, Cu, Hg, Ni, Ag, Co and Mo cause analytical interferences.

INSTRUMENTATION Atomic absorption spectrometer. Arsenic (As) hollow cathode lamp or electrodeless discharge lamp. Burner (for argon-hydrogen flame). 193.7 nm wavelength.

RANGE 2–20 µg/L

MDL 0.002 mg/L

PRECISION standard deviation = ±0.9 at 10 µg/L on o-arsenilic acid solution.

ACCURACY Recovery = 93% at 10 µg/L on o-arsenilic acid solution.

SAMPLING METHOD Use plastic or glass containers (prewashed). Collect 100 mL of sample.

STABILITY Add nitric acid to pH <2.

MHT 6 months.

QUALITY CONTROL Run one spike duplicate sample for every 20 samples. Verify calibration with an independently prepared check standard every 15 samples.

REFERENCE Method 7061, SW-846, 3rd ed., Nov. 1986.

Arsenic   **EPA Method 7060**
CAS #7440-38-2

TITLE Atomic Absorption (AA)

MATRIX Wastes, mobility procedure. Furnace technique extracts, soils, and groundwater.

APPLICATION Sample preparation converts organic forms of As to inorganic forms. Sample preparation varies with matrix. Following appropriate dissolution of sample, a representative aliquot of digestate is spiked with nickel nitrate solution and placed in a graphite tube furnace.

INTERFERENCES Elemental As and many of its compounds are volatilized. There may be losses in As during sample preparation. There can be severe nonspecific absorption and light scattering caused by matrix components during atomization. Memory effects occur if As isn't volatilized and removed from furnace. Aluminum is a severe positive interferent.

INSTRUMENTATION Atomic absorption spectrometer, arsenic (As) hollow cathode lamp or electrodeless discharge lamp. Graphite furnace.

RANGE 5–100 µg/L

MDL 1 µg/L

PRECISION Standard deviation = ±1.6 at 100 µg As/L.

ACCURACY Recovery = 101% at 100 µg As/L.

SAMPLING METHOD Use plastic or glass containers (prewashed). Collect 100 mL of sample.

STABILITY Add nitric acid to pH <2.

MHT 6 months.

QUALITY CONTROL Run one spike duplicate sample for every 20 samples. Verify calibration with an independently prepared check standard every 15 samples. (Low wavelength, 193.7 nm, makes As analysis susceptible to problems).

REFERENCE Method 7060, SW-846, 3rd ed., Nov. 1986.

Atraton   **EPA Method 507**
CAS #1610-17-9

TITLE Determination of Nitrogen and Phosphorus-Containing Pesticides in Water by GC/NPD

MATRIX This method is applicable to the determination of certain nitrogen and phosphorus-containing pesticides in finished drinking water and groundwater.

METHOD SUMMARY Method 507 covers 46 nitrogen- and phosphorus-containing pesticides. A 1-L sample is fortified with a surrogate standard, salted, buffered, extracted with methylene chloride, and concentrated; then the solvent is exchanged with methyl tert-butyl ether (MTBE) and concentrated again, and a 2-µL aliquot of a sample extract is injected into a GC system equipped with a selective nitrogen-phosphorus detector and a capillary column for analysis.

INTERFERENCES Method interferences may be caused by contaminants in solvents, reagents, glassware, and other sample processing apparatus. Interfering contamination may occur when a sample containing low concentrations of analytes is analyzed immediately following a sample containing relatively high concentrations. One or more injections of MTBE should be made following the analysis of a sample with high concentrations of analytes to check for analyte carryover. Matrix interferences may be caused by contaminants that are coextracted from the sample. The extent of matrix interferences will vary considerably from source to source, depending upon the water sampled.

INSTRUMENTATION A gas chromatograph system (GC) equipped with a nitrogen-phosphorus detector (NPD) is needed.

Column 1: 30 m × 0.25 mm I.D. DB-5 bonded fused silica column, 0.25 µm film thickness, or equivalent.
Column 2: 30 M × 0.25 mm I.D. DB-1701 bonded fused silica column, 0.25 µm film thickness, or equivalent.

PRECISION & ACCURACY This method has been validated in a single lab and estimated detection limits (EDLs) have been determined for each analyte. Observed detection limits may vary among waters, depending upon the nature of the interferences in the sample matrix and the specific instrumentation used. Analytes that are not separated chromatographically cannot be individually identified and measured unless an alternative technique for identification and quantification exist.

The estimated detection limit (in µg/L) was 0.6. The EDL is defined as either method detection limit or a level of compound in a sample yielding a peak in the final extract with signal-to-noise ratio of approximately 5, whichever value is higher.

The concentration used for these measurements (in µg/L) was 6.
The accuracy (as % recovery) was 91.
The precision (% RSD) was 11.

SAMPLING METHOD Grab samples are collected in 1-L glass sample bottles (prewashed with detergent and hot tap water, rinsed with reagent water, and dried in an oven at 400°C for 1 h) with screw caps lined with PTFE-fluorocarbon.

SAMPLE PRESERVATION Add mercuric chloride to the sample bottle in amounts to produce a concentration of 10 mg/L. If residual chlorine is present, add 80 mg of sodium thiosulfate/L of sample to the sample bottle prior to collection. After collection, seal bottle and shake vigorously for 1 min, then cool the sample to 4°C immediately and store it at 4°C in the dark until extraction.

MHT Maximum holding time of the samples, and in some cases the extracts, is 14 days.

SAMPLE PREPARATION Fortify the sample with 50 µL of the surrogate standard solution, adjust to pH 7 with phosphate buffer, add 100 g NaCl to the sample, and seal and shake to dissolve the salt; then extract with methylene chloride in a separatory funnel or in a mechanical tumbler bottle. Dry the extract by pouring it through a solvent-rinsed drying column containing about 10 cm of anhydrous sodium sulfate. Collect the extract in a Kuderna-Danish (K-D) concentrator and rinse the column with 20–30 mL methylene chloride. Concentrate the extract to about 2 mL and rinse the flask and its lower joint into the concentrator tube with 1 to 2 mL of methyl t-butyl ether (MTBE). Add 5–10 mL of MTBE and concentrate the extract twice (adding more MTBE) to a final volume of 5.0 mL and store it at 4°C until analysis.

Note: If methylene chloride is not completely removed from the final extract, it may cause detector problems.

QUALITY CONTROL Minimum quality control requirements are initial demonstration of lab capability, determination of surrogate compound recoveries in each sample and blank, monitoring internal standard peak area or height in each sample and blank, analysis of lab reagent blanks, lab fortified samples, lab fortified blanks, and other QC samples. A lab reagent blank is analyzed to demonstrate that all glassware and reagent interferences are under control.

Initial demonstration of capability is fulfilled by analyzing four fortified reagent water samples with the recovery value for each analyte falling within the acceptable range (±30% average recovery). Surrogate recoveries from samples or method blanks must be 70–130%. The internal standard response for any sample chromatogram should not deviate from the daily calibration check standard's internal standard response by more than 30% or lab fortified blanks and sample matrices are used to assess lab performance and analyte recovery, respectively.

If the response for the target analyte peak exceeds the working range of the system, dilute the extract and reanalyze. Alternative techniques such as an alternate detector or second chromatography column should be used to confirm peak identification when sample components are not resolved adequately.

EPA CONTACT & HOTLINE For technical questions contact Dr. Baldev Bathija, U.S. EPA, Office of Ground Water and Drinking Water (WH-550D), 401 M St. SW, Washington, DC 20460. Tel. (202) 260-3040. For further information the EPA Safe Drinking Water Hotline may be called at: (800) 426-4791.

REFERENCE Methods for the Determination of Organic Compounds in Drinking Water, EPA/600/4-88/039 (revised July 1991). U.S. EPA Environmental Monitoring Systems Laboratory, Cincinnati, OH, 45268, U.S.A. Available from the National Technical Information Service (NTIS), 5285 Port Royal Road, Springfield, VA 22161; Tel. 800-553-6847. NTIS Order Number is PB91-231480.

Atrazine EPA Method 505
CAS #1912-24-9

TITLE Analysis of Organohalide Pesticides and Commercial Polychlorinated Biphenyl (PCB) Products in Water by Microextraction and Gas Chromatography. U.S. EPA Method 505, Rev. 2.0, 1989.

MATRIX This method is applicable to drinking water and raw source water. The latter should include most surface water and groundwater sources.

METHOD SUMMARY Method 505 covers 25 pesticides and commercial PCB products. This is a very sensitive method that is more useful for monitoring than for exploratory analyses. 5-mL samples of water are saturated with sodium chloride and then extracted by shaking with 2 mL of hexane. The sample extracts are transferred to an autosampler setup to inject 1- to 2-µL portions into a gas chromatograph (GC) for analysis. Alternatively, 1- to 2-µL portions of samples, blanks, and standards may be manually injected. Each extract is analyzed by capillary GC/ECD with confirmation using either a second capillary column or GC/MS. The electron capture detector is easy to use, but it is a nonselective detector. The microextraction technique also eliminates the expensive sample preparation costs of other methods, but it has the disadvantage of being less sensitive than most because the extracts are not concentrated.

INTERFERENCES Method interferences may be caused by contaminants in solvents, reagents, glassware, and other sample processing apparatus that lead to discrete artifacts or elevated baselines. Interfering contamination may occur when a sample containing low concentrations of analytes is analyzed immediately following a sample containing relatively high concentrations of the analytes. Matrix interferences also may be caused by contaminants that are coextracted from the sample; cleanup of sample extracts may be necessary in these cases. Some pesticides and commercial PCB products from aqueous solutions adhere to glass surfaces, so sample transfers and contact with glass surfaces should be minimized. Some pesticides are rapidly oxidized by chlorine so dechlorination with sodium thiosulfate at the time of sample collection is important. Also, splitless injectors may cause degradation of some pesticides.

INSTRUMENTATION A gas chromatograph/electron capture detector/data system, with temperature programming and split/splitless injector suitable for use with capillary columns is needed.

Column 1: 0.32 mm I.D. × 30 m fused silica capillary with chemically bond methyl polysiloxane phase (DB-1, 1.0 µm film, or equivalent).

Column 2: 0.32 mm I.D. × 30 m fused silica capillary with 1:1 mixed phase of dimethyl silicone and polyethylene glycol (Durawax-DX3, 0.25 μm film, or equivalent).

Column 3: 0.32 mm I.D. × 25 m fused silica capillary with chemically bonded 50:50 methyl-phenyl silicone (OV-17, 1.5 μm film, or equivalent).

Column 1 should be used as the primary analytical column. Columns 2 and 3 are recommended for use as confirmatory columns when GC/MS confirmation is not available.

PRECISION & ACCURACY Method detection limits are dependent upon the characteristics of the gas chromatographic system used. Analytes that are not separated chromatographically cannot be individually identified and used in the same calibration mixture or water samples unless an alternative technique for identification and quantification, such as mass spectrometry, is used.

The concentration(s) (in μg/L) used for these QC measurements was 5.0 and 20.0.

The MDL (in μg/L) was 2.4.

The accuracy (% recovery) for reagent water at the above concentration(s) was 85 and 95 and the precision (%) was 16.2 and 5.2.

The accuracy (% recovery) for groundwater at the above concentration(s) was 95 and 86 and the precision (%) was 7.3 and 9.1.

The accuracy (% recovery) for tap water at the above concentration(s) was 108 and 91 and the precision (5) was 7.9 and 3.1.

Note: No range of concentrations is provided with this method.

SAMPLING METHOD Collect samples using a 40-mL screw-cap vial (prewashed with detergent, rinsed with distilled water and oven dried at 400°C for one h) with a Teflon®-faced silicone septum. Collect bubble-free samples and place the septum with the Teflon® side down on the water.

SAMPLE PRESERVATION If residual chlorine is present in the water add about 3 mg of sodium thiosulfate to each vial before samples are collected to remove the chlorine. Alternatively, add 75 μL of 0.04 g/mL solution of sodium thiosulfate to each vial just prior to sampling. Immediately cool samples to 4°C and store them in a solvent-free refrigerator at 4°C until analysis.

MHT The maximum holding time is 14 days from the time the sample was collected until it must be analyzed.

SAMPLE PREPARATION Remove the sample from storage and allow it to come to room temperature. Remove a 5-mL volume from each container and weigh the container to the nearest 0.1 g. Add 6 g of sodium chloride and 2.0 mL of hexane to each sample bottle. Recap the sample and shake it vigorously for one min. Allow the water and hexane phases to separate, remove the cap, and transfer 0.5 mL of hexane into an autosampler vial using a disposable glass pipette. Transfer the remaining hexane phase into a second autosampler vial and store at 4°C for reanalysis, if necessary. Discard the remaining sample/hexane mixture and reweigh the empty container to determine net weight of sample.

QUALITY CONTROL Minimum quality control requirements are initial demonstration of lab capability, analysis of lab reagent blanks, fortified blanks, fortified sample matrix, and quality control samples. The lab must analyze at least one fortified blank per sample set, or at least one for every 20 samples. The fortifying concentration of each analyte should be 10 times the method detection limit or the maximum calibration limit (MCL), whichever is less. Calculate accuracy as percent recovery and develop control limits from the mean percent recovery and standard deviation.

The lab must add a known concentration of the analytes to a minimum of 10% of the routine samples, or one lab fortified sample matrix per sample set. Calculate the percent recovery for each analyte and compare to the control limits established from the analyses of the fortified blanks.

EPA CONTACT & HOTLINE For technical questions contact Dr. Baldev Bathija, U.S. EPA, Office of Ground Water and Drinking Water (WH-550D), 401 M St. SW, Washington, DC 20460. Tel. (202) 260-3040. For further information the EPA Safe Drinking Water Hotline may be called at: (800) 426-4791.

REFERENCE Methods for the Determination of Organic Compounds in Drinking Water, EPA/600/4-88/039 (revised July 1991). U.S. EPA Environmental Monitoring Systems Laboratory, Cincinnati, OH, 45268, U.S.A. Available from the National Technical Information Service (NTIS), 5285 Port Royal Road, Springfield, VA 22161; Tel. 800-553-6847. NTIS Order Number is PB91-231480.

Atrazine   EPA Method 507
CAS #1912-24-9

TITLE Determination of Nitrogen and Phosphorus-Containing Pesticides in Water by GC/NPD

MATRIX This method is applicable to the determination of certain nitrogen and phosphorus-containing pesticides in finished drinking water and groundwater.

METHOD SUMMARY Method 507 covers 46 nitrogen- and phosphorus-containing pesticides. A 1-L sample is fortified with a surrogate standard, salted, buffered, extracted with methylene chloride, and concentrated; then the solvent is exchanged with methyl tert-butyl ether (MTBE) and concentrated again, and a 2-μL aliquot of a sample extract is injected into a GC system equipped with a selective nitrogen-phosphorus detector and a capillary column for analysis.

INTERFERENCES Method interferences may be caused by contaminants in solvents, reagents, glassware, and other sample processing apparatus. Interfering contamination may occur when a sample containing low concentrations of analytes is analyzed immediately following a sample containing relatively high concentrations. One or more injections of MTBE should be made following the analysis of a sample with high concentrations of analytes to check for analyte carryover. Matrix interferences may be caused by contaminants that are coextracted from the sample. The extent of matrix interferences will vary considerably from source to source, depending upon the water sampled.

INSTRUMENTATION A gas chromatograph system (GC) equipped with a nitrogen-phosphorus detector (NPD) is needed.

Column 1: 30 m × 0.25 mm I.D. DB-5 bonded fused silica column, 0.25 μm film thickness, or equivalent.
Column 2: 30 M × 0.25 mm I.D. DB-1701 bonded fused silica column, 0.25 μm film thickness, or equivalent.

PRECISION & ACCURACY This method has been validated in a single lab and estimated detection limits (EDLs) have been determined for each analyte. Observed detection limits may vary among waters, depending upon the nature of the interferences in the sample matrix and the specific instrumentation used. Analytes that are not separated chromatographically cannot be individually identified and measured unless an alternative technique for identification and quantification exist.

The estimated detection limit (in μg/L) was 0.13. The EDL is defined as either method detection limit or a level of compound in a sample yielding a peak in the final extract with signal-to-noise ratio of approximately 5, whichever value is higher.

The concentration used for these measurements (in μg/L) was 1.3.
The accuracy (as % recovery) was 92.
The precision (% RSD) was 8.

SAMPLING METHOD Grab samples are collected in 1-L glass sample bottles (prewashed with detergent and hot tap water, rinsed with reagent water, and dried in an oven at 400°C for 1 h) with screw caps lined with PTFE-fluorocarbon.

SAMPLE PRESERVATION Add mercuric chloride to the sample bottle in amounts to produce a concentration of 10 mg/L. If residual chlorine is present, add 80 mg of sodium thiosulfate/L of sample to the sample bottle prior to collection. After collection, seal bottle and shake vigorously for 1 min, then cool the sample to 4°C immediately and store it at 4°C in the dark until extraction.

MHT Maximum holding time of the samples, and in some cases the extracts, is 14 days.

SAMPLE PREPARATION Fortify the sample with 50 μL of the surrogate standard solution, adjust to pH 7 with phosphate buffer, add 100 g NaCl to the sample, and seal and shake to dissolve the salt; then extract with methylene chloride in a separatory funnel or in a mechanical tumbler bottle. Dry the extract by pouring it through a solvent-rinsed drying column containing about 10 cm of anhydrous sodium sulfate. Collect the extract in a Kuderna-Danish (K-D) concentrator and rinse the column with 20–30 mL methylene chloride. Concentrate the extract to about 2 mL and rinse the flask and its lower joint into the concentrator tube with 1 to 2 mL of methyl t-butyl ether (MTBE). Add 5–10 mL of MTBE and concentrate the extract twice (adding more MTBE) to a final volume of 5.0 mL and store it at 4°C until analysis.

Note: If methylene chloride is not completely removed from the final extract, it may cause detector problems.

QUALITY CONTROL Minimum quality control requirements are initial demonstration of lab capability, determination of surrogate compound recoveries in each sample and blank, monitoring internal standard peak area or height in each sample and blank, analysis of lab reagent blanks, lab fortified samples, lab fortified blanks, and other QC samples. A lab reagent blank is analyzed to demonstrate that all glassware and reagent interferences are under control.

Initial demonstration of capability is fulfilled by analyzing four fortified reagent water samples with the recovery value for each analyte falling within the acceptable range (±30% average recovery). Surrogate recoveries from samples or method blanks must be 70–130%. The internal standard response for any sample chromatogram should not deviate from the daily calibration check standard's internal standard response by more than 30% or lab fortified blanks and sample matrices are used to assess lab performance and analyte recovery, respectively.

If the response for the target analyte peak exceeds the working range of the system, dilute the extract and reanalyze. Alternative techniques such as an alternate detector or second chromatography column should be used to confirm peak identification when sample components are not resolved adequately.

EPA CONTACT & HOTLINE For technical questions contact Dr. Baldev Bathija, U.S. EPA, Office of Ground Water and Drinking Water (WH-550D), 401 M St. SW, Washington, DC 20460. Tel. (202) 260-3040. For further information the EPA Safe Drinking Water Hotline may be called at: (800) 426-4791.

REFERENCE Methods for the Determination of Organic Compounds in Drinking Water, EPA/600/4-88/039 (revised July 1991). U.S. EPA Environmental Monitoring Systems Laboratory, Cincinnati, OH, 45268, U.S.A. Available from the National Technical Information Service (NTIS), 5285 Port Royal Road, Springfield, VA 22161; Tel. 800-553-6847. NTIS Order Number is PB91-231480.

Azinphos-methyl **EPA Method 8141**
CAS #86-50-0

TITLE Organophosphorus Compounds by Gas Chromatography: Capillary Column Technique

MATRIX This method covers aqueous and solid matrices. This includes a wide variety such as drinking water, groundwater, industrial wastewaters, surface waters, soils, solids and sediments.

METHOD SUMMARY This is a GC method used to determine the concentration of 28 organophosphorus pesticides.

The use of gel permeation cleanup (EPA Method 3640) for sample cleanup has been demonstrated to yield recoveries of less than 85% for many method analytes and is therefore not recommended for use with this method.

This method provides GC conditions for the detection of ppb concentrations of organophosphorus compounds. Prior to the use of this method, appropriate sample preparation techniques must be used. Water samples are extracted at a neutral pH with methylene chloride as a solvent by using a separatory funnel (EPA Method 3510) or a continuous liquid-liquid extractor

(EPA Method 3520). Soxhlet extraction (EPA Method 3540) or ultrasonic extraction (EPA Method 3550) using methylene chloride/acetone (1:1) are used for solid samples. Both neat and diluted organic liquids (EPA Method 3580) may be analyzed by direct injection. Spiked samples are used to verify the applicability of the chosen extraction technique to each new sample type. A GC with a flame photometric (FPD) or nitrogen-phosphorus detector (NPD) is used for this multiresidue procedure.

INTERFERENCES The use of Florisil cleanup materials (EPA Method 3620) for some of the compounds in this method has been demonstrated to yield recoveries less than 85% and is therefore not recommended for all compounds. Use of phosphorus or halogen specific detectors, however, often obviates the necessity for cleanup for relatively clean sample matrices. If particular circumstances demand the use of an alternative cleanup procedure, the analyst must determine the elution profile and demonstrate that the recovery of each analyte is no less than 85%.

Use of a flame photometric detector (FPD) in the phosphorus mode will minimize interferences from materials that do not contain phosphorus. Elemental sulfur, however, may interfere with the determination of certain organophosphorus compounds by flame photometric gas chromatography. Sulfur cleanup using EPA Method 3660 may alleviate this interference. A nitrogen phosphorus detector (NPD) is also recommended.

A few analytes coelute on certain columns. Therefore, select a second column for confirmation where coelution of the analytes of interest does not occur.

Method interferences may be caused by contaminants in solvents, reagents, glassware, and other sample processing hardware that lead to discrete artifacts or elevated baselines in gas chromatograms. All these materials must be routinely demonstrated to be free from interferences under the conditions of the analysis by analyzing reagent blanks.

INSTRUMENTATION A GC with a NPD or a FPD will be needed. A data system or integrator is recommended for measuring peak areas and/or peak heights. A Kuderna-Danish (K-D) apparatus will be needed for extract concentration.

Column 1: 15 m × 0.53 mm megabore capillary column, 1.0 µm film thickness, DB-210.
Column 2: 15 m × 0.53 mm megabore capillary column, 1.5 µm film thickness, SPB-608.
Column 3: 15 m × 0.53 mm megabore capillary column, 1.0 µm film thickness, DB-5.

Three megabore capillary columns are included for analysis of organophosphates by this method. Column 1 (DB-210 or equivalent) and Column 2 (SPB-608 or equivalent) are recommended if a large number of organophosphorus analytes are to be determined. If the superior resolution offered by Column 1 and Column 2 is not required, Column 3 (DB-5 or equivalent) may be used. For megabore capillary columns, automatic injections of 1 µL are recommended.

PRECISION & ACCURACY The MDL actually achieved in a given analysis will vary, as it is dependent on instrument sensitivity and matrix effects. Single operator accuracy and precision studies have been conducted with spiked water and soil samples.

MULTIPLICATION FACTORS FOR OTHER MATRICES (a)

Matrix	Factor (b)
Groundwater	
(EPA Method 3510 or EPA Method 3520)	10(c)
Low-concentration soil by Soxhlet and no cleanup	10 (c)
Low-concentration soil by ultrasonic extraction with GPC cleanup	6.7 (c)
High-concentration soil and sludges by ultrasonic extraction	500 (c)
Non-Water miscible waste (EPA Method 3580)	1000 (c)

(a) Sample EQLs are highly matrix dependent. The EQLs listed here are provided for guidance and may not always be achievable.
(b) EQL = [Method Detection Limit] × [Factor]. For non-aqueous samples the factor is on a wet-weight basis.
(c) Multiply this factory times the soil MDL.

The MDL (in µg/L) when reagent water was extracted using a separatory funnel was 0.10.
The MDL (in µg/kg) when soil was extracted using Soxhlet extraction (EPA Method 3540) was 5.0.
Accuracy (as % recovery) with separatory funnel extraction ranged from 126 (with low spikes) to 101 (with high spikes).
Accuracy (as % recovery) with continuous liquid-liquid extraction ranged from not recovered (with low spikes) to 122 (with high spikes).
Accuracy (as % recovery) with Soxhlet extraction of soils ranged from 156 (with low spikes to 87 (with high spikes).
Accuracy (as % recovery) with ultrasonic extraction of soils ranged from not recovered (with low spikes) to 21 (with high spikes).

SAMPLE COLLECTION, PRESERVATION & HANDLING
Containers used to collect samples for the determination of semivolatile organic compounds should be soap and water washed followed by methanol (or isopropanol) rinsing. The sample containers should be of glass or Teflon® and have screw-top covers with Teflon® liners.

No preservation is used with concentrated waste samples. With liquid samples containing no residual chlorine and with soil, sediment, and sludge samples, immediately cooling to 4°C is the only preservation used. When residual chlorine is present then 3 mL of 10% aqueous sodium sulfate is added for each gallon of sample collected, followed by cooling to 4°C.

Liquid samples must be extracted within 7 days and their extracts analyzed within 40 days. Concentrated waste, soil, sediment, and sludge samples must be extracted within 14 days and their extracts analyzed within 40 days.

SAMPLE PREPARATION In general, water samples are extracted at a neutral pH with methylene chloride, using either EPA Method 3510 or EPA Method 3520. Solid samples are extracted using either EPA Method 3540 or EPA Method 3550 with methylene chloride/acetone (1:1) as the extraction solvent.

Prior to GC analysis, the extraction solvent may be exchanged to hexane. Single lab data indicates that samples should not be transferred with 100% hexane during sample workup as the more water soluble organophosphorus compounds may be lost.

If cleanup is performed on the samples, the analyst should analyze the samples by GC. This will confirm elution patterns and the absence of interferences from the reagents. If peak detection and identification is prevented by the presence of interferences, further cleanup is required.

QUALITY CONTROL The analyst should monitor the performance of the extraction, cleanup (when used), and analytical system and the effectiveness of the method in dealing with each sample matrix by spiking each sample, standard, and blank with one or two surrogates (e.g., organophosphorus compounds not expected to be present in the sample). Deuterated analogs of analytes should not be used as surrogates for gas chromatographic analysis due to coelution problems.

A minimum of five concentrations for each analyte of interest should be prepared through dilution of the stock standards with isooctane. One of the concentrations should be at a concentration near, but above, the MDL.

Include a mid-level check standard after each group of 10 samples in the analysis sequence. GC/MS techniques should be judiciously employed to support qualitative identifications made with this method. Follow the GC/MS operating requirements specified in EPA Method 8270.

When available, chemical ionization mass spectra may be employed to aid in the qualitative identification process. To confirm an identification of a compound, the background-corrected mass spectrum of the compound must be obtained from the sample extract and must be compared with a mass spectrum from a stock or calibration standard analyzed under the same chromatographic conditions. The molecular ion and all other ions present above 20% relative abundance in the mass spectrum of the standard must be present in the mass spectrum of the sample with agreement to ± 20%. The retention time of the compound in the sample must be within six seconds of the retention time for the same compound in the standard solution.

Should the MS procedure fail to provide satisfactory results, additional steps may be taken before reanalysis. These steps may include the use of alternate packed or capillary GC columns or additional sample cleanup.

REFERENCE Test Methods for Evaluating Solid Waste, Physical/Chemical Methods, SW-846, 3rd Edition, U.S. EPA, Office of Solid Waste, Washington, DC, EPA Method 8141 July 1992.

Azinphos-methyl   **EPA Method 8270**
CAS #86-50-0

TITLE Semivolatile Organic Compounds by GC/MS

MATRIX This method is used to determine the concentration of semivolatile organic compounds in extracts prepared from all types of solid waste matrices, soils, and groundwater. Although surface waters are not specifically mentioned, this method should be applicable to water samples from rivers, lakes, etc.

METHOD SUMMARY This method covers 259 semivolatile organic compounds. In very limited applications direct injection of the sample into the GC/MS system may be appropriate, but this results in very high detection limits (approximately 10,000 µg/L). Typically, a 1-L liquid sample, containing surrogate, and matrix spiking standards, is extracted in a continuous extractor first under acid conditions and then under basic conditions. Typically 30 g of a solid sample, containing surrogate, and matrix spiking standards, is extracted ultrasonically. After concentrating the extract to 1 mL it is spiked with 10 µL of an internal standard solution just prior to analysis by GC/MS. The volume injected should contain about 100 ng of base/neutral and 200 ng of acid surrogates (for a 1-µL injection). Analysis is performed by GC/MS using a capillary GC column.

INTERFERENCES Raw GC/MS data from all blanks, samples, and spikes must be evaluated for interferences. Contamination by carryover can occur whenever high-concentration and low-concentration samples are sequentially analyzed. To reduce carryover, the sample syringe must be rinsed out between samples with solvent. Whenever an unusually concentrated sample is encountered, it should be followed by the analysis of blank solvent to check for cross-contamination.

INSTRUMENTATION A GC/MS and a data system are required. The GC column used is a 30 m × 0.25 mm I.D. (or 0.32 mm I.D.) 1-µm film thickness silicone-coated fused silica capillary column. A continuous liquid-liquid extractor equipped with Teflon® or glass connection joints and stopcocks requiring no lubrication, a K-D concentrating apparatus, water bath, and an ultrasonic disrupter with a minimum power of 300 W and with pulsing capability are also required.

PRECISION & ACCURACY The estimated quantitation limit (EQL) of Method 8270B for determining an individual compound is approximately 1 mg/kg (wet weight) for soil or sediment samples, 1–200 mg/kg for wastes (dependent on matrix and method of preparation), and 10 µg/L for groundwater samples. EQLs will be proportionately higher for sample extracts that require dilution to avoid saturation of the detector.

The EQL(b) for groundwater in µg/L is 100.
The EQL (a, b) for low concentrations in soil and sediment in µg/kg is not determined.
Accuracy as µg/L is not listed.
Overall precision in µg/L is not listed.

(a) *EQLs listed for soil/sediment are based on wet weight. Normally data is reported in a dry-weight basis; therefore, EQLs will be higher based on the % dry weight of each sample. This calculation is based on a 30-g sample and gel permeation chromatography cleanup.*
(b) *Sample EQLs are highly matrix-dependent. The EQLs are provided for guidance and may not always be achievable.*
$C =$ *True value for concentration, in µg/L.*
$X =$ *Average recovery found for measurements of samples containing a concentration of C, in µg/L.*

ESTIMATED QUANTITATION LIMIT

Other Matrices	Factor (a)
High-concentration soil and sludges by sonicator	7.5
Non-water miscible waste	75

(a) EQL for other matrices = [EQL for low soil/sediment] × [Factor]. This estimated EQL is similar to an EPA "Practical Quantitation Limit."

SAMPLING METHOD

Liquid samples — Use a 1 or 2½ gallon amber glass bottle with a screw-top Teflon®-lined cover that has been prewashed with detergent and rinsed with distilled water and methanol (or isopropanol).

Soils, sediments, or sludges — Use an 8-oz. widemouth glass with a screw-top Teflon®-lined cover that has been prewashed with detergent and rinsed with distilled water and methanol (or isopropanol).

SAMPLE PRESERVATION

Liquid samples — If residual chlorine is present, add 3 mL of 10% sodium thiosulfate per gallon, cool to 4°C and store in a solvent-free refrigerator until analysis; if chlorine is not present, then eliminate the sodium thiosulfate addition.

Soils, sediments, or sludges — Cool samples to 4°C and store in a solvent-free refrigerator.

MHT Liquid samples must be extracted within 7 days and the extracts analyzed within 40 days. Soils, sediments, or sludges may be stored for a maximum of 14 days and the extracts analyzed within 40 days.

SAMPLE PREPARATION

Liquid samples — Transfer 1 L quantitatively to a continuous extractor. If high concentrations are anticipated, a smaller volume may be used and then diluted with organic-free reagent water to 1 L. Adjust pH, if necessary, to pH <2 using 1:1 (V/V) sulfuric acid. Pipette 1.0 mL of a surrogate standard spiking solution into each sample. For the sample in each analytical batch selected for spiking, add 1.0 mL of a matrix spiking standard. For base/neutral acid analysis, the amount of the surrogates and matrix spiking compounds added to the sample should result in a final concentration of 100 ng/µL of each analyte in the extract to be analyzed (assuming a 1-µL injection). Extract with methylene chloride for 18–24 h. Next, adjust the pH of the aqueous phase to pH >11 using 10 N sodium hydroxide and extract it with methylene chloride again for 18–24 h. Dry the extract through a column containing anhydrous sodium sulfate and concentrate it to 1 mL using a K-D concentrator.

Soils, sediments, or sludges — Use 30 g of sample. Nonporous or wet samples (gummy or clay type) that do not have a free-flowing sandy texture must be mixed with anhydrous sodium sulfate until the sample is free flowing. Add 1 mL of surrogate standards to all samples, spikes, standards, and blanks. For the sample in each analytical batch selected for spiking, add 1.0 mL of a matrix spiking standard. For base/neutral acid analysis, the amount added of the surrogates and matrix spiking compounds should result in a final concentration of 100 ng/µL of each base/neutral analyte and 200 ng/µL of each acid analyte in the extract to be analyzed (assuming a 1-µL injection). Immediately add a 100-mL mixture of 1:1 methylene chloride:acetone and extract the sample ultrasonically for 3 min and then decant or filter the extracts. Repeat the extraction two or more times. Dry the extract using a column with anhydrous sodium sulfate and concentrate it to 1 mL in a K-D concentrator.

QUALITY CONTROL A methylene chloride solution containing 50 ng/µL of decafluorotriphenylphosphine (DFTPP) is used for tuning the GC/MS system each 12-h shift. A system performance check also must be made during every 12-h shift. A standard containing 50 ng/µL each of 4,4'-DDT, pentachlorophenol, and benzidine is required to verify injection port inertness and GC column performance. A calibration standard at mid-concentration, containing each compound of interest, including all required surrogates, must be performed every 12 h during analysis. After the system performance check is met, calibration check compounds (CCCs) are used to check the validity of the initial calibration.

The internal standard responses and retention times in the calibration check standard must be evaluated immediately after or during data acquisition. If the retention time for any internal standard changes by more than 30 seconds from the last check calibration (12 h), the chromatographic system must be inspected for malfunctions and corrections must be made, as required. If the electron ionization current plot (EICP) area for any of the internal standards changes by a factor of two from the last daily calibration standard check, the mass spectrometer must be inspected for malfunctions and corrections must be made, as appropriate.

Demonstrate, through the analysis of a reagent water blank, that interferences from the analytical system, glassware, and reagents are under control. The blank samples should be carried through all stages of the sample preparation and measurement steps. For each analytical batch (up to 20 samples), a reagent blank, matrix spike, and matrix spike duplicate/duplicate must be analyzed (the frequency of the spikes may be different for different monitoring programs). The blank and spiked samples must be carried through all stages of the sample preparation and measurement steps. A QC reference sample concentrate containing each analyte at a concentration of 100 mg/L in methanol is required.

REFERENCE Test Methods for Evaluating Solid Waste (SW-846). U.S. EPA 1983, Method 8270B, Rev. 2, Nov. 1990. Office of Solid Waste, Washington, DC.

Azinphos-methyl   **EPA Method 8140**
CAS #86-50-0

TITLE Organophosphorus Pesticides

MATRIX Groundwater, soils, sludges, water miscible liquid wastes, and non-water miscible wastes.

APPLICATION This method is used for the analysis of 21 organophosphorus pesticides. Samples are extracted, concentrated, and analyzed using direct injection of both neat and diluted organic liquid into a gas chromatograph (GC).

INTERFERENCES Solvents, reagents, and glassware may introduce artifacts. Other interferences may come from coextracted compounds from samples. The use of Florisil cleanup materials may produce low recoveries. Elemental sulfur may interfere with some compounds when using a flame photometric detector. Sulfur cleanup (Method 3660) may alleviate sulfur interference.

INSTRUMENTATION GC capable of on-column injections and a flame photometric detector (FPD) or a thermionic detector. Column 1: 1.8 m by 2 mm with 5% SP-2401 on Supelcoport. Column 2: 1.8 m by 2 mm with 3% SP-2401 on Supelcoport. Column 3: 50 cm by ⅛ in Teflon® with 15% SE-54 on Gas Chrom Q. The preferred column is Column Number 1.

RANGE 21 to 250 µg/L

MDL 1.5 µg/L (in reagent water).

PQL FACTORS FOR MULTIPLYING × FID MDL VALUE

Matrix	Multiplication Factor
Groundwater	10
Low-level soil by sonication with GPC cleanup	670
High-level soil and sludge by sonication	10,000
Non-water miscible waste	100,000

PRECISION 18.8% (single operator standard deviation)

ACCURACY 72.7% (single operator average recovery)

SAMPLING METHOD Use 8-oz. widemouth glass bottles with Teflon®-lined caps for concentrated waste samples, soils, sediments, and sludges. Use 1 or 2½ gallon amber glass bottles with Teflon®-lined caps for liquid (water) samples.

STABILITY Cool soil, sediment, sludge, and liquid samples to 4°C. If residual chlorine is present in liquid samples add 3 mL of 10% sodium thiosulfate per gallon of sample and cool to 4°C.

MHT 14 days for concentrated waste, soil, sediment, or sludge; 7 days for liquid samples; all extracts must be analyzed within 40 days.

QUALITY CONTROL A quality control check sample concentrate containing this compound in acetone at a concentration 1,000 times more concentrated than the selected spike concentration is required. The QC check sample concentrate may be prepared from pure standard materials or purchased as certified solutions. Use appropriate trip, matrix, control site, method, reagent, and solvent blanks. Internal, surrogate, and five concentration level calibration standards are used.

REFERENCE Method 8140, SW-846, 3rd ed., Sept. 1986.

B

Barban
CAS #101-27-9

EPA Method 8270

TITLE Semivolatile Organic Compounds by GC/MS

MATRIX This method is used to determine the concentration of semivolatile organic compounds in extracts prepared from all types of solid waste matrices, soils, and groundwater. Although surface waters are not specifically mentioned, this method should be applicable to water samples from rivers, lakes, etc.

METHOD SUMMARY This method covers 259 semivolatile organic compounds. In very limited applications direct injection of the sample into the GC/MS system may be appropriate, but this results in very high detection limits (approximately 10,000 µg/L). Typically, a 1-L liquid sample, containing surrogate, and matrix spiking standards, is extracted in a continuous extractor first under acid conditions and then under basic conditions. Typically 30 g of a solid sample, containing surrogate, and matrix spiking standards, is extracted ultrasonically. After concentrating the extract to 1 mL it is spiked with 10 µL of an internal standard solution just prior to analysis by GC/MS. The volume injected should contain about 100 ng of base/neutral and 200 ng of acid surrogates (for a 1-µL injection). Analysis is performed by GC/MS using a capillary GC column.

INTERFERENCES Raw GC/MS data from all blanks, samples, and spikes must be evaluated for interferences. Contamination by carryover can occur whenever high-concentration and low-concentration samples are sequentially analyzed. To reduce carryover, the sample syringe must be rinsed out between samples with solvent. Whenever an unusually concentrated sample is encountered, it should be followed by the analysis of blank solvent to check for cross-contamination.

INSTRUMENTATION A GC/MS and a data system are required. The GC column used is a 30 m × 0.25 mm I.D. (or 0.32 mm I.D.) 1-µm film thickness silicone-coated fused silica capillary column. A continuous liquid-liquid extractor equipped with Teflon® or glass connection joints and stopcocks requiring no lubrication, a K-D concentrating apparatus, water bath, and an ultrasonic disrupter with a minimum power of 300 W and with pulsing capability are also required.

PRECISION & ACCURACY The estimated quantitation limit (EQL) of Method 8270B for determining an individual compound is approximately 1 mg/kg (wet weight) for soil or sediment samples, 1–200 mg/kg for wastes (dependent on matrix and method of preparation), and 10 µg/L for groundwater samples. EQLs will be proportionately higher for sample extracts that require dilution to avoid saturation of the detector.

The EQL (b) for groundwater in µg/L is 200.
The EQL (a, b) for low concentrations in soil and sediment in µg/kg is not determined.
Accuracy as µg/L is not listed.
Overall precision in µg/L is not listed.

(a) EQLs listed for soil/sediment are based on wet weight. Normally data is reported in a dry-weight basis; therefore, EQLs will be higher based on the % dry weight of each sample. This calculation is based on a 30-g sample and gel permeation chromatography cleanup.
(b) Sample EQLs are highly matrix-dependent. The EQLs are provided for guidance and may not always be achievable.
C = True value for concentration, in µg/L.
X = Average recovery found for measurements of samples containing a concentration of C, in µg/L.

ESTIMATED QUANTITATION LIMIT

Other Matrices	Factor (a)
High-concentration soil and sludges by sonicator	7.5
Non-water miscible waste	75

(a) EQL for other matrices = [EQL for low soil/sediment] × [Factor]. This estimated EQL is similar to an EPA "Practical Quantitation Limit."

SAMPLING METHOD
Liquid samples — Use a 1 or 2½ gallon amber glass bottle with a screw-top Teflon®-lined cover that has been prewashed with detergent and rinsed with distilled water and methanol (or isopropanol).

Soils, sediments, or sludges — Use an 8-oz. widemouth glass with a screw-top Teflon®-lined cover that has been prewashed with detergent and rinsed with distilled water and methanol (or isopropanol).

SAMPLE PRESERVATION
Liquid samples — If residual chlorine is present, add 3 mL of 10% sodium thiosulfate per gallon, cool to 4°C and store in a solvent-free refrigerator until analysis; if chlorine is not present, then eliminate the sodium thiosulfate addition.

Soils, sediments, or sludges — Cool samples to 4°C and store in a solvent-free refrigerator.

MHT Liquid samples must be extracted within 7 days and the extracts analyzed within 40 days. Soils, sediments, or sludges may be stored for a maximum of 14 days and the extracts analyzed within 40 days.

SAMPLE PREPARATION
Liquid samples — Transfer 1 L quantitatively to a continuous extractor. If high concentrations are anticipated, a smaller volume may be used and then diluted with organic-free reagent water to 1 L. Adjust pH, if necessary, to pH <2 using 1:1 (V/V) sulfuric acid. Pipette 1.0 mL of a surrogate standard spiking solution into each sample. For the sample in each analytical batch selected for spiking, add 1.0 mL of a matrix spiking standard. For base/neutral acid analysis, the amount of the surrogates and matrix spiking compounds added to the sample should result in a final concentration of 100 ng/µL of each analyte in the extract to be analyzed (assuming a 1-µL injection). Extract with methylene chloride for 18–24 h. Next, adjust the pH of the aqueous phase to pH >11 using 10 N sodium hydroxide and extract it with methylene chloride again for

18–24 h. Dry the extract through a column containing anhydrous sodium sulfate and concentrate it to 1 mL using a K-D concentrator.

Soils, sediments, or sludges — Use 30 g of sample. Nonporous or wet samples (gummy or clay type) that do not have a free-flowing sandy texture must be mixed with anhydrous sodium sulfate until the sample is free flowing. Add 1 mL of surrogate standards to all samples, spikes, standards, and blanks. For the sample in each analytical batch selected for spiking, add 1.0 mL of a matrix spiking standard. For base/neutral acid analysis, the amount added of the surrogates and matrix spiking compounds should result in a final concentration of 100 ng/µL of each base/neutral analyte and 200 ng/µL of each acid analyte in the extract to be analyzed (assuming a 1-µL injection). Immediately add a 100 mL mixture of 1:1 methylene chloride:acetone and extract the sample ultrasonically for 3 min and then decant or filter the extracts. Repeat the extraction two or more times. Dry the extract using a column with anhydrous sodium sulfate and concentrate it to 1 mL in a K-D concentrator.

QUALITY CONTROL A methylene chloride solution containing 50 ng/µL of decafluorotriphenylphosphine (DFTPP) is used for tuning the GC/MS system each 12-h shift. A system performance check also must be made during every 12-h shift. A standard containing 50 ng/µL each of 4,4′-DDT, pentachlorophenol, and benzidine is required to verify injection port inertness and GC column performance. A calibration standard at mid-concentration, containing each compound of interest, including all required surrogates, must be performed every 12 h during analysis. After the system performance check is met, calibration check compounds (CCCs) are used to check the validity of the initial calibration.

The internal standard responses and retention times in the calibration check standard must be evaluated immediately after or during data acquisition. If the retention time for any internal standard changes by more than 30 seconds from the last check calibration (12 h), the chromatographic system must be inspected for malfunctions and corrections must be made, as required. If the electron ionization current plot (EICP) area for any of the internal standards changes by a factor of two from the last daily calibration standard check, the mass spectrometer must be inspected for malfunctions and corrections must be made, as appropriate.

Demonstrate, through the analysis of a reagent water blank, that interferences from the analytical system, glassware, and reagents are under control. The blank samples should be carried through all stages of the sample preparation and measurement steps. For each analytical batch (up to 20 samples), a reagent blank, matrix spike, and matrix spike duplicate/duplicate must be analyzed (the frequency of the spikes may be different for different monitoring programs). The blank and spiked samples must be carried through all stages of the sample preparation and measurement steps. A QC reference sample concentrate containing each analyte at a concentration of 100 mg/L in methanol is required.

REFERENCE Test Methods for Evaluating Solid Waste (SW-846). U.S. EPA 1983, Method 8270B, Rev. 2, Nov. 1990. Office of Solid Waste, Washington, DC.

Barium   **EPA Method 6010**
CAS #7440-39-3

TITLE Inductively Coupled Plasma-Atomic Emission Spectroscopy

MATRIX This method is applicable to the determination of trace elements, including metals, in groundwater, soils, sludges, sediments, and other solid wastes. All matrices require digestion prior to analysis. The method of standard addition must be used for the analysis of all sample digests unless either serial dilution or matrix spike addition demonstrates it is not required.

METHOD SUMMARY Method 6010 covers 25 elements using ICP analysis. It measures element-emitted light by optical spectrometry. Samples, following an appropriate acid digestion, are nebulized and the resulting aerosol is transported to the plasma torch. Element-specific atomic line emission spectra are produced by a radio-frequency inductively coupled plasma.

INTERFERENCES Interferences may be categorized as spectral or non-spectral. Spectral interferences are caused by overlap of a spectral line from another element, unresolved overlap of molecular band spectra, background contribution from continuous or recombination phenomenon, and stray light from the line emission of high concentration elements. Non-spectral interferences include physical and chemical interferences. Physical interferences are effects associated with the sample nebulization and transport processes. Changes in viscosity and surface tension can cause significant inaccuracies. Chemical interferences include molecular compound formation, ionization effects, and solute vaporization effects. Normally these effects are not significant and can be minimized by careful selection of operating conditions. Chemical interferences are highly dependent on matrix type and the specific analyte element.

INSTRUMENTATION An inductively coupled argon plasma emission spectrometer (ICP) capable of background correction is required.

PRECISION & ACCURACY Detection limits, sensitivity, and optimum ranges of the metals will vary with the matrices and model of the spectrometer. In a single lab evaluation, seven wastes were analyzed for 22 elements. The mean percent relative standard deviation from triplicate analyses for all elements and wastes was 9 ± 2%. The mean percent recovery of spiked elements for all wastes was 93 ± 6%. Spike levels ranged from 100 µg/L to 100 mg/L. The wastes included sludges and industrial wastewaters.

Estimated instrument detection limit in µg/L is 2.
Spiked concentration in µg/L is not listed.
Mean reported value in µg/L is not listed.
Precision as RSD % is not listed.

SAMPLING METHOD Samples should be collected in borosilicate glass, linear polyethylene, polypropylene, or Teflon® bottles that have been prewashed with detergent and tap water, and rinsed with 1:1 nitric acid and tap water or 1:1 hydrochloric acid and tap water. Collect at least 2 g of solids and 200 mL of aqueous samples.

SAMPLE PRESERVATION Add nitric acid to make the samples pH <2.

MHT The maximum holding time for properly preserved samples is 6 months.

SAMPLE PREPARATION Preliminary treatment of most matrices is necessary because of the complexity and variability of sample matrices. Water samples that have been prefiltered and acidified will not need acid digestion. Methods for acid digestion of waters for total recoverable or dissolved metals, acid digestions of aqueous samples and extracts for total metals, and acid digestion of sediments, sludges, and soils are summarized below.

Total recoverable or dissolved metals in water — To prepare surface and groundwater samples for determination of total recoverable and dissolved metals, a 100-mL aliquot of well-mixed sample is acidified with concentrated nitric acid and concentrated hydrochloric acid, then heated until the volume is reduced to 15–20 mL. Adjust the final volume to 100 mL with reagent water.

Total metals in aqueous samples, soil and sediment extracts — To prepare aqueous samples, soil and sediment extracts, and wastes that contain suspended solids, a 100-mL aliquot is made acidic with concentrated nitric acid and the solution is evaporated to about 5 mL on a hot plate. Continue heating and adding additional acid until sample digestion is complete, which is usually indicated when the digestate is light in color or does not change in appearance. Evaporate the solution to about 3 mL and cool it and add a small quantity of 1:1 hydrochloric acid (10 mL/100 mL of final solution). Cover the beaker and reflux for 15 min. Wash down the beaker walls and filters or centrifuge the sample to remove silicates and other insoluble material. Filter the sample and adjust the final volume to 100 mL with reagent water and the final acid concentration to 10%.

Sediments, sludges, and soils — To prepare sediments, sludges and soil samples, transfer 1–2 g to a conical beaker and add 10 mL of 1:1 nitric acid, mix the slurry, and cover it with a watch glass. Heat the sample and reflux for 10 to 15 min without boiling. Allow it to cool, then add 5 mL of concentrated nitric acid and reflux for 30 min. Repeat last step and then allow the solution to evaporate to 5 mL without boiling. Cool and add 2 mL of water and 3 mL of 30% hydrogen peroxide. Cover and place the beaker on the hot plate. Heat and add 30% hydrogen peroxide in 1-mL aliquots with warming until the effervescence is minimal but do not add more than a total of 10 mL of 30% hydrogen peroxide. If the sample is being prepared for the analysis of Ag, Al, As, Ba, Be, Ca, Cd, Co, Cr, Cu, Fe, K, Mg, Mn, Mo, Na, Ni, Os, Pb, Se, Tl, V, and Zn, then add 5 mL of concentrated hydrochloric acid and 10 mL of water and return the covered beaker to a hot plate for 15 min of additional refluxing without boiling. Dilute the sample to a 100 mL volume with water after cooling and filter or centrifuge to remove particulates.

QUALITY CONTROL Laboratory control samples must be analyzed for each analytical method. A method blank should be analyzed with each batch of samples. The effect of the matrix on method performance must be demonstrated: when appropriate, there should be at least one matrix spike and either one matrix duplicate or one matrix spike duplicate per analytical batch. The bias and precision of the method, as well as the method detection limit for each specific matrix type, must be measured.

Dilute and reanalyze samples that are more concentrated than the linear calibration limit. Employ a minimum of one reagent blank per sample batch to determine if contamination or any memory effects are occurring. Whenever a new or unusual sample matrix is encountered, perform either a serial dilution test or a matrix spike addition test to ensure that neither positive or negative interferences are operating on any of the analyte elements. Check the instrument standardization by verifying calibration every 10 samples using a calibration blank and a check standard.

REFERENCE Test Methods for Evaluating Solid Waste (SW-846). U.S. EPA. 1983. Method 6010, Rev. 0, Sept. 1986. Office of Solid Wastes, Washington, DC.

Barium   EPA Method 200.7
CAS #7440-39-3

TITLE Inductively Coupled Plasma (ICP)

MATRIX Dissolved, suspended or total element in drinking and surface waters and in domestic and industrial wastewaters.

APPLICATION The method covers the determination of 25 metals. Dissolved elements are determined in filtered and acidified samples after appropriate digestion (which increases dissolved solids). Its primary advantage is that ICP instruments allow simultaneous or rapid sequential determination of many elements in a short time. Samples are first nebulized and the aerosol is transported to a plasma torch in which element specific atomic line emission spectra are produced by a radio frequency inductively coupled plasma. Background correction is required for trace element detection except in the case of line broadening.

INTERFERENCES There are spectral, physical, and chemical interferences. The primary disadvantage of ICP instruments is background radiation from other elements and the plasma gases (spectral interferences). Changes in sample viscosity and surface tension with samples containing high dissolved solids (especially those exceeding 1500 mg/L) or high acid concentrations can cause physical interferences. Ionization effects, solute vaporization and molecular compound formation can cause chemical interferences. Other metals do not cause interference at the 100 mg/L level.

INSTRUMENTATION Inductively coupled argon plasma emission spectroscopy. 455.403 nm wavelength.

RANGE Not listed.

MDL 2 µg/L

PRECISION Not listed.

ACCURACY Mean recovery = 93% ± 6% of spiked elements for all wastes.

SAMPLING METHOD Wash sample container with detergent and tap water, rinse with 1+1 nitric acid and tap water, then rinse with 1+1 hydrochloric acid and tap water, then rinse with deionized, distilled water in that order. Perform any filtration or acid preservation steps when the sample is collected or as soon as possible thereafter.

STABILITY Cool samples to 4°C.

MHT 4 h.

QUALITY CONTROL Mixed calibration standards, an instrument check standard, and an interference check solution are used in addition to a quality control sample. The quality control sample should be prepared in the same acid matrix as the calibration standards at 10 times the instrumental detection limits and in accordance with the instructions provided by the supplier. Furthermore, two types of blanks are required: a calibration blank and a reagent blank.

REFERENCE Method 200.7, U.S. EPA, EMSL-Cincinnati, OH, Nov. 1980

Barium   EPA Method 208.2
CAS #7440-39-3

TITLE Metals (Total, Dissolved, Suspended) AAS, Furnace Technique

MATRIX Drinking, surface and saline waters. Wastewater.

APPLICATION Date issued 1978. A representative sample aliquot is placed in graphite tube in the furnace, evaporated to dryness, charred and atomized. Radiation from excited element is passed through vapor and radiation intensity decreases proportional to amount of Ba in vapor.

INTERFERENCES Furnace technique subject to chemical and matrix interferences. Furnace gases may have molecular absorption bands enclosing analytical wavelength. Smoke-producing sample matrix can interfere. If Ba isn't volatilized and removed from furnace, memory effects occur.

INSTRUMENTATION AAS. Barium (Ba) hollow cathode lamp or EDL. Graphite furnace. Pipets.

RANGE 10–200 µg/L

MDL 2 µg/L.

PRECISION SD = ±2.2 µg at 1000 µg Ba/L

ACCURACY Recovery = 102% at 1000 µg Ba/L

SAMPLING METHOD Plastic or glass (prewashed).

STABILITY HNO3 to pH <2.

MHT 6 months.

QUALITY CONTROL A check standard should be run approximately after every 10 sample injections. Standards are run in part to monitor the life and performance of the graphite tube. Lack of reproducibility or significant change in the signal for the standard indicates tube should be replaced.

REFERENCE Methods for the chemical analysis of water and wastes, EPA-600/4-79-020, U.S. EPA, EMSL, 1979.

Barium   EPA Method 7080
CAS #7440-39-3

TITLE Atomic Absorption, (AA) Direct Aspiration

MATRIX Drinking, surface and saline waters, Wastewater

APPLICATION Sample is aspirated and atomized in a flame. A light beam from a Ba hollow cathode lamp is directed through the flame into a monochromator and onto a detector. Since wavelength of light beam is specific for barium, light energy absorbed by detector is a measure of barium.

INTERFERENCES The most troublesome type is chemical, caused by lack of absorption of atoms bound in molecular combination in the flame. High dissolved solids in sample may result in nonatomic absorbance interference. Narrow spectral band pass must be used.

INSTRUMENTATION Atomic absorption spectrometer. Barium hollow cathode lamp. (553.6 nm wavelength; calcium also emits here).

RANGE 1–20 mg/L.

MDL 0.1 mg/L

PRECISION Standard deviation = ±(0.043 and 0.13) at 0.4 and 2 mg Ba/L

ACCURACY Recoveries = 94 and 113% at 0.4 and 2 mg Ba/L

SAMPLING METHOD Use glass or plastic containers. Collect 200 g of solids and 600 mL of liquid samples.

STABILITY Cool solid samples to 4°C and analyze as soon as possible. Add nitric acid to liquid samples to pH <2.

MHT 6 months.

QUALITY CONTROL At least one duplicate and one spike sample should be run every 20 samples or with each matrix type to verify precision of the method. For 20 or more samples per day, verify working standard curve. Run an additional standard at or near mid-range every 10 samples.

REFERENCE Method 7080, SW-846, 3rd ed., Nov. 1986.

Barium   EPA Method 7081
CAS #7440-39-3

TITLE Atomic Absorption, (AA) Furnace Technique

MATRIX Wastes, mobility procedure, extracts, soils and groundwater.

APPLICATION Aqueous samples, EP extracts, industrial wastes, soils, sludges, sediments, and solid wastes require digestion before analysis. An aliquot of sample is placed in the graphite tube in the furnace and slowly evaporated, charred and atomized. Absorption of lamp radiation during atomization is proportional to barium concentration.

INTERFERENCES The furnace technique is subject to chemical interferences. Composition of sample matrix can have major effect on analysis. Modify matrix to remove interferences. Derived barium carbide causes loss of sensitivity and memory effects. Avoid halide acids.

INSTRUMENTATION Atomic absorption spectrometer. Barium hollow cathode lamp or electrodeless discharge lamp. Graphite furnace. Strip-chart recorder.

RANGE 10–200 µg/L

MDL 2 µg/L (553.6 nm wavelength).

PRECISION & ACCURACY Not listed.

SAMPLING METHOD Use glass or plastic containers. Collect 200 g of solids and 600 mL of liquid samples.

STABILITY Cool solid samples to 4°C and analyze as soon as possible. Add nitric acid to liquid samples to pH <2.

MHT 6 months.

QUALITY CONTROL At least one duplicate and one spike sample should be run every 20 samples, or with each matrix type to verify method precision. If 20 or more samples are run a day, run a standard (at or near mid-range) every 10 samples.

REFERENCE Method 7081, SW-846, 3rd ed., (Included as Rev. 0, Dec. 1987)

Bentazon EPA Method 8151
CAS #25057-89-0

TITLE Chlorinated Herbicides by GC Using Methylation or Pentafluorobenzylation Derivatization: Capillary Column Technique.

MATRIX This method covers aqueous and solid matrices. This includes a wide variety such as drinking water, groundwater, industrial wastewaters, surface waters, soils, solids, and sediments.

METHOD SUMMARY This is a GC method for determining 19 chlorinated acid herbicides in aqueous, soil, and waste matrices. Because these compounds are produced and used in various forms (i.e., acid, salt, ester, etc.) a hydrolysis step is included to convert the herbicide to the acid form prior to analysis. This method provides hydrolysis, extraction, derivatization and GC conditions for the analysis of chlorinated acid herbicides in water, soil, and waste samples. Water samples are hydrolyzed *in situ*, extracted with diethyl ether, and then esterified with either diazomethane or pentafluorobenzyl bromide. The derivatives are determined by gas chromatography with an electron capture detector (GC/ECD). The results are reported as acid equivalents. The sensitivity of this method depends on the level of interferences in addition to instrumental limitations.

INTERFERENCES Method interferences may be caused by contaminants in solvents, reagents, glassware, and other sample processing hardware. Immediately prior to use, glassware should be rinsed with the next solvent to be used. Matrix interferences may be caused by contaminants that are coextracted from the sample. Organic acids, especially chlorinated acids, cause the most direct interference with the determination by methylation. Phenols, including chlorophenols, may also interfere with this procedure. The determination using pentafluorobenzylation is more sensitive, and more prone to interferences from the presence of organic acids of phenols than by methylation. Alkaline hydrolysis and subsequent extraction of the basic solution removes many chlorinated hydrocarbons and phthalate esters that might otherwise interfere with the ECD analysis. The herbicides, being strong organic acids, react readily with alkaline substances and may be lost during analysis. Therefore, glassware must be acid-rinsed and then rinsed to constant pH with organic-free reagent water.

INSTRUMENTATION A GC suitable for Grob-type injection using capillary columns. A data system for measuring peak heights and/or peak areas is recommended. An electron capture detector (ECD) is used. Also a K-D apparatus, a diazomethane generator, a centrifuge and an ultrasonic disrupter will be required.

Narrow Bore Columns:
Primary Column 1: 30 m × 0.25 mm, 5% phenyl/95% methyl silicone (DB-5), 0.25 µm film thickness.
Primary Column 1a (GC/MS): 30 m × 0.32 mm, 5% phenyl/95% methyl silicone (DB-5), 1-µm film thickness.
Column 2: 30 m × 0.25 mm DB-608 with a 25 µm film thickness.
Confirmation Column: 30 m × 0.25 mm, 14% cyanopropyl phenyl silicone (DB-1701), 0.25 µm film thickness.

Megabore Columns:
Primary Column: 30 m × 0.53 mm DB-608 with 0.83 µm film thickness.
Confirmation Column: 30 m × 0.53 mm, 14% cyanopropyl phenyl silicone (DB-1701), 1.0 µm film thickness.

PRECISION & ACCURACY Method detection limits (MDLs) are compound-dependent and vary with derivitization efficiency, derivative recovery, the matrix sampled, and herbicide concentration.

The estimated MDL (in µg/L) was 0.2 for aqueous samples using GC/ECD.
The estimated MDL (in µg/kg) was not reported for soil samples using GC/ECD when corrected back to 50-g samples extracted and concentrated to 10 mL with 5-µL injections.
The estimated GC/MS identification limit (in ng) was not reported for soil samples using GC/MS.

Mean percent recovery, calculated from 7–8 determinations of spiked reagent water, after diazomethane derivatization, from a spike concentration (in µg/L) of 1 was 120 with a standard deviation of the percent recovery of 16.8.

Mean percent recovery, calculated from 10 determinations of spiked clay and clay/still bottom samples over the linear concentration range (in ng/g) of no data was none reported with a percent relative standard deviation of none. The RSD % was calculated on 10 samples high in the linear concentration range and 10 low in the range. The linear concentration range was determined using standard solutions and corrected to 50 g soil samples.

SAMPLE COLLECTION, PRESERVATION & HANDLING
Containers used to collect samples for the determination of semivolatile organic compounds should be soap and water washed followed by methanol (or isopropanol) rinsing. The sample containers should be of glass or Teflon® and have screw-top covers with Teflon® liners.

No preservation is used with concentrated waste samples. With liquid samples containing no residual chlorine and with soil, sediment, and sludge samples, immediately cooling to 4°C is the only preservation used. When residual chlorine is present then 3 mL of 10% aqueous sodium sulfate is added for each gallon of sample collected, followed by cooling to 4°C.

The holding time for all volatile organics samples is 14 days. Liquid samples must be extracted within 7 days and their extracts analyzed within 40 days. Concentrated waste, soil, sediment, and sludge samples must be extracted within 14 days and their extracts analyzed within 40 days.

SAMPLE PREPARATION
Preparation of soil, sediment, and other solid samples — Acidify 30 g (dry weight) solids with 0.1 M phosphate buffer (pH = 2.5) and thoroughly mix the contents. Spike the sample with surrogate compound(s). The ultrasonic extraction of solids must be optimized for each type of sample. In order for the ultrasonic extractor to efficiently extract solid samples, the sample must be free flowing when the solvent is added. Acidified anhydrous sodium sulfate should be added to clay-type soils, or any other solid that is not a free-flowing sandy texture, until a free flowing mixture is obtained. Add methylene chloride and perform ultrasonic extraction. Combine organic extracts from the repetitive extractings of the sample and centrifuge. Add aqueous potassium hydroxide, water, and methanol to the extract and reflux the mixture on a water bath. Extract the solution three times with methylene chloride and discard the methylene chloride phase. The basic solution contains the herbicide salts. Adjust the pH of the solution to <2 with cold sulfuric acid and extract three times with methylene chloride. Combine the extracts and pour them through a pre-rinsed drying column containing acidified anhydrous sodium sulfate. Collect the dried extracts in a K-D flask and concentrate them.

Preparation of aqueous samples — Measure 1 L of sample into a 2-L separatory funnel and spike it with surrogate compound(s). Add NaCl to the sample, then add 6 N NaOH to the sample to a pH of 12 or more and let the sample sit at room temperature for 1 h to hydrolyze esters. Extract the sample three times with methylene chloride and discard the extracts. Then add cold 12 N sulfuric acid to a pH less than or equal to 2, and extract the sample three times with ethyl ether. Collect the ether phase in a flask containing acidified anhydrous sodium sulfate and allow it to remain in contact with the sodium sulfate for a minimum of 2 h. The drying step is very critical to ensuring complete esterification; any moisture remaining in the ether will result in low herbicide recoveries.

Extract Concentration and Derivatization — The combined ether extract is concentrated to about 1 mL using a K-D apparatus followed by using a micro Snyder column or nitrogen gas blowdown. If methyl esters are to be produced, then dilute the concentrated ether extract with 1 mL of isooctane and 0.5 mL of methanol, dilute to a final volume of 4 mL, and esterify with diazomethane. If pentafluorobenzene esters are to be produced, then dilute concentrated ether extract with acetone to a final volume of 4 mL and esterify with pentafluorobenzyl bromide.

QUALITY CONTROL Select a representative spike concentration for each compound (acid or ester) to be measured. Using stock standard, prepare a quality control check sample concentrate, in acetone, that is 1000 times more concentrated than the selected concentrations. Use this quality control check sample concentrate to prepare quality control check samples. Calculate surrogate standard recovery on all standards, samples, blanks, and spikes. GC/MS techniques should be judiciously employed to support qualitative identifications made with this method. When available, chemical ionization mass spectra may be employed to aid the qualitative identification process.

REFERENCE Test Methods for Evaluating Solid Waste, Physical/Chemical Methods, SW-846, 3rd Edition, U.S. EPA, Office of Solid Waste, Washington, DC, EPA Method 8151, Nov. 1990.

Benzal chloride   **EPA Method 8121**
CAS #98-87-3

TITLE Chlorinated Hydrocarbons by GC: Capillary Column Technique

MATRIX This method covers aqueous and solid matrices. This includes a wide variety such as drinking water, groundwater, industrial wastewaters, surface waters, soils, solids, and sediments.

METHOD SUMMARY This method provides procedures for the determination of 22 chlorinated hydrocarbons in water, soil/sediment, and waste matrices. A measured volume or weight of sample is extracted by using one of the appropriate sample extraction techniques specified in EPA Method 3510, EPA Method 3520, EPA Method 3540, or EPA Method 3550, or diluted using EPA Method 3580. Aqueous samples are extracted at neutral pH with methylene chloride by using either a separatory funnel (EPA Method 3510) or a continuous liquid-liquid extractor (EPA Method 3520). Solid samples are extracted with hexane/acetone (1:1) by using a Soxhlet extractor (EPA Method 3540) or with methylene chloride/acetone (1:1) by using an ultrasonic extractor (EPA Method 3550). After cleanup, the extract or diluted sample is analyzed by gas chromatography with electron capture detection (GC/ECD).

The sensitivity level of this method usually depends on the level of interferences rather than on instrumental limitations. This method may be used in conjunction with EPA Method 3620, Florisil Column Cleanup, EPA Method 3660, Sulfur Cleanup, and EPA Method 3640, Gel Permeation Chromatography, to aid in the elimination of interferences.

INTERFERENCES Solvents, reagents, glassware, and other hardware used in sample processing may introduce artifacts which may result in elevated baselines, causing misinterpretation of gas chromatograms. Interferants coextracted from the samples will vary considerably from waste to waste. Glassware must be scrupulously clean. Phthalate esters, if present in a sample, will interfere only with the BHC isomers. The presence of elemental sulfur will result in large peaks, and can often mask the region of compounds eluting after 1,2,4,5-tetrachlorobenzene. The tetrabutylammonium (TBA)-sulfite procedure (EPA Method 3660) works well for the removal of elemental sulfur. Waxes and lipids can be removed by gel permeation chromatography (EPA Method 3640).

INSTRUMENTATION A GC suitable for on-column injections and all required accessories, including and electron capture detector (ECD), analytical columns, recorder, gases, and syringes are needed. A data system for measuring peak heights and/or peak areas is recommended. A Kuderna-Danish (K-D) apparatus will also be needed to concentrate extracts.

Column 1: 30 m × 0.53 mm I.D. fused-silica capillary column chemically bonded with trifluoropropyl methyl silicone (DB-210 or equivalent).

Column 2: 30 m × 0.53 mm I.D. fused-silica capillary column chemically bonded with polyethylene glycol (DB-WAX or equivalent).

PRECISION & ACCURACY This method has been tested in a single lab by using organic-free reagent water, sandy loam samples, and extracts which were spiked with the test compounds at one concentration. Single-operator precision and method accuracy were found to be related to the concentration of compound and the type of matrix. The accuracy and precision technique will be determined by the sample matrix, sample preparation technique, optional cleanup techniques, and calibration procedures used.

ESTIMATED QUANTITATION LIMIT (EQL) FACTOR FOR VARIOUS MATRICES (a)

Matrix	Factor (b)
Groundwater	10
Low-concentration soil by ultrasonic extraction with GPC cleanup	670
High-concentration soil and sludges by ultrasonic extraction	10,000
Waste not miscible with water	100,000

(a) Sample EQLs are highly matrix-dependent. The EQLs listed herein are provided for guidance and may not always be achievable. (b) EQL = [Method detection limit] × [Factor]. For nonaqueous samples, the factor is on a wet-weight basis.

PRECISION & ACCURACY MDL is the method detection limit for organic-free reagent water. MDL was determined from the analysis of eight replicate aliquots processed through the entire analytical method (extraction, Florisil cartridge cleanup, and GC/ECD analysis).

The MDL (in ng/L) was 2–5 (estimated from the instrument detection limit).

The accuracy (as average % recovery using 5 determinations and no Florisil cleanup) from a spike concentration of 10 µg/L and separatory funnel extraction was 95% with a final volume of 10 mL.

The precision (as RSD% using 5 determinations and no Florisil cleanup) from a spike concentration of 10 µg/L and separatory funnel extraction was 3.0% with a final volume of 10 mL.

The accuracy (as average % recovery using 5 determinations and no Florisil cleanup), from a spike concentration of 3300 µg/L and ultrasonic extraction of solid samples using 1:1 methylene chloride and acetone, was 89% with a final volume of 10 mL.

The precision (as RSD% using 5 determinations and no Florisil cleanup), from a spike concentration of 3300 µg/L and ultrasonic extraction of solid samples using 1:1 methylene chloride and acetone, was 2.7% with a final volume of 10 mL.

SAMPLE COLLECTION, PRESERVATION & HANDLING
Volatile Organics — Standard 40-mL glass screw-cap VOA vials with Teflon®-faced silicone septum may be used for both liquid and solid matrices. When collecting samples, liquids and solids should be introduced into the vials gently to reduce agitation which might drive off volatile compounds. If there are any air bubbles present the sample must be retaken. The vials with solids should be tapped slightly as they are filled to try and eliminate as much free air space as possible. Two vials from each sampling location should be sealed in separate plastic bags to prevent cross-contamination between samples.

Semivolatile organics — Containers used to collect samples for the determination of semivolatile organic compounds should be soap and water washed followed by methanol (or isopropanol) rinsing. The sample containers should be of glass or Teflon® and have screw-top covers with Teflon® liners.

Preservation for volatile organics — No preservation is used with concentrated waste samples. With liquid samples containing no residual chlorine, 4 drops of concentrated hydrochloric acid are added and the samples are immediately cooled to 4°C. When liquid samples contain residual chlorine, they are treated as above and, in addition, 4 drops of 4% aqueous sodium thiosulfate are added to remove the residual chlorine. Soil, sediment, and sludge samples are only cooled to 4°C.

Preservation for semivolatile organics — No preservation is used with concentrated waste samples. With liquid samples containing no residual chlorine and with soil, sediment, and sludge samples, immediately cooling to 4°C is the only preservation used. When residual chlorine is present then 3 mL of 10% aqueous sodium sulfate is added for each gallon of sample collected, followed by cooling to 4°C.

Holding times — The holding time for all volatile organics samples is 14 days. Liquid samples must be extracted within 7 days and their extracts analyzed within 40 days. Concentrated waste, soil, sediment, and sludge samples must be extracted within 14 days and their extracts analyzed within 40 days.

SAMPLE PREPARATION Prepare stock standard solutions in hexane. Calibration standards at a minimum of five concentrations should be prepared through dilution of the stock standards with hexane. The suggested internal standards are: 2,5-dibromotoluene, 1,3,5-tribromobenzene, and α,α-dibromo-m-xylene. The analyst can use any of the three compounds provided that they are resolved from matrix interferences. Recommended surrogate compounds are α-2,6-trichlorotoluene, 1,4-dichloronaphthalene, and 2,3,4,5,6-pentachlorotoluene.

In general, water samples are extracted at a neutral pH with methylene chloride using a separatory funnel (EPA Method 3510) or a continuous liquid-liquid extractor (EPA Method 3520). Solid samples are extracted with hexane/acetone (1:1 v:v) using a Soxhlet extractor (EPA Method 3540) or with methylene chloride/acetone (1:1 v:v) using an ultrasonic extractor (EPA Method 3550). Non-aqueous waste samples may be diluted using EPA Method 3580. Prior to Florisil cleanup or gas chromatographic analysis, the extraction solvent must be exchanged to hexane. Sample extracts that will be subjected to gel permeation chromatography do not need solvent exchange.

Cleanup procedures may not be necessary for a relatively clean matrix. If removal of interferences such as chlorinated phenols, phthalate esters, etc., is required, proceed with the procedure outlined in EPA Method 3620.

QUALITY CONTROL Analyze a quality control check standard to demonstrate that the operation of the GC is in control. The frequency of the check standard analysis is equivalent to 10% of the samples analyzed. If the recovery of any compound found in the check standard is less than 80% of the certified value, the problem must be corrected and a new set of calibration standards must be prepared and analyzed. Calculate surrogate standard recoveries for all samples, blanks, and spikes. An internal standard peak area check must be performed on all samples. The internal standard must be evaluated for acceptance by determining whether the measured area for the internal standard deviates by more than 30% from the average area for the internal standard in the calibration standards. When the internal standard peak area is outside that limit, all samples that fall outside the QC criteria must be reanalyzed. Any compound confirmed by two columns may also be confirmed by GC/MS (EPA Method 8270). The GC/MS would normally require a minimum concentration of 1 ng/μL in the final extract for each compound. Include a mid-concentration calibration standard after each group of 20 samples in the analysis sequence. The response factors for the mid-concentration calibration must be within 15% of the average values for the multiconcentration calibration.

REFERENCE Test Methods for Evaluating Solid Waste, Physical/Chemical Methods, SW-846, 3rd Edition, U.S. EPA, Office of Solid Waste, Washington, DC, 1990. EPA Method 8121, Rev. 0, Nov. 1990.

Benzanthrone   **EPA Method 1625**
CAS #82-05-3

TITLE Semivolatile Organic Compounds by Isotope Dilution GC/MS

MATRIX The compounds may be determined in waters, soils, and municipal sludges by this method.

METHOD SUMMARY This method is used to determine 176 semivolatile toxic organic pollutants associated with the CWA (as amended 1987); the RCRA (as amended 1986); the CERCLA (as amended 1986); and other compounds amenable to extraction and analysis by capillary column gas chromatography-mass spectrometry (GC/MS).

Stable isotopically-labeled analogs of the compounds of interest are added to the sample. If the solids content is less than 1%, a 1-L sample is extracted at pH 12–13, then at pH <2 with methylene chloride using continuous extraction techniques.

If the solids content is 30% or less, the sample is diluted to 1% solids with reagent water, homogenized ultrasonically, and extracted at pH 12–13, then at pH <2 with methylene chloride using continuous extraction techniques. If the solids content is greater than 30%, the sample is extracted using ultrasonic techniques.

Each extract is dried over sodium sulfate, concentrated to a volume of 5 mL, cleaned up using GPC, if necessary, and concentrated. Extracts are concentrated to 1 mL if GPC is not performed, and to 0.5 mL if GPC is performed.

An internal standard is added to the extract, and a 1-mL aliquot of the extract is injected into the GC. The compounds are separated by GC and detected by a MS. The labeled compounds serve to correct the variability of the analytical technique.

INTERFERENCES Solvents, reagents, glassware, and other sample processing hardware may yield artifacts and/or elevated baselines causing misinterpretation of chromatograms and spectra. Materials used in the analysis must be demonstrated to be free from interferences under the conditions of analysis by running method blanks initially and with each sample lot (sample started through the extraction process on a given 8-h shift, to a maximum of 20). Specific selection of reagents and purification of solvents by distillation in all glass systems may be required. Glassware and, where possible, reagents are cleaned by solvent rinse and baking at 450°C for 1-h minimum. Interferences coextracted from samples will vary considerably from source to source, depending on the diversity of the site being sampled.

INSTRUMENTATION Major instrumentation includes a GC with a splitless or on-column injection port for capillary column, a MS with 70 eV electron impact ionization, and a data system to collect and record MS data, and process it. A K-D apparatus is used to concentrate extracts.

GC Column: 30 m × 0.25 mm I.D. 5% phenyl, 94% methyl, 1% vinyl silicone bonded phased fused silica capillary column.

PRECISION & ACCURACY The detection limits of the method are usually dependent on the level of interferences rather than instrumental limitations. The limits typify the minimum quantities that can be detected with no interferences present.

The minimum level (in μg/mL) was not listed. This is defined as a minimum level at which the analytical system shall give

recognizable mass spectra (background corrected) and acceptable calibration points.

The MDL (in µg/kg) in low solids was not listed and in high solids was not listed; these were determined in digested sludge (low solids) and in filter cake or compost (high solids).

The labeled and native compound initial precision as standard deviation (in µg/L) was not listed.

The labeled and native compound initial accuracy as average recovery (in µg/L) was not listed.

SAMPLE COLLECTION, PRESERVATION & HANDLING

Collect samples in glass containers. Aqueous samples which flow freely are collected in refrigerated bottles using automatic sampling equipment. Solid samples are collected as grab samples using widemouth jars. Maintain samples at 0 to 4°C from the time of collection until extraction. If residual chlorine is present in aqueous samples, add 80 mg sodium thiosulfate/L of water. Begin sample extraction within 7 days of collection, and analyze all extracts within 40 days of extraction.

SAMPLE PREPARATION
Samples containing 1% solids or less are extracted directly using continuous liquid-liquid extraction techniques. Samples containing 1 to 30% solids are diluted to the 1% level with reagent water and extracted using continuous liquid-liquid extraction techniques. Samples containing greater than 30% solids are extracted using ultrasonic techniques.

- Base/neutral extraction — Adjust the pH of the waters in the extractors to 12–13 with 6 N NaOH. Extract with methylene chloride for 24–48 h.
- Acid extraction — Adjust the pH of the waters in the extractors to 2 or less using 6 N sulfuric acid. Extract with methylene chloride for 24–48 h.
- Ultrasonic extraction of high solids samples — Add anhydrous sodium sulfate to the sample and QC aliquot(s). Add acetone:methylene chloride (1:1) to the sample and mix thoroughly

Concentrate extracts using a K-D apparatus.

QUALITY CONTROL
The analyst is permitted to modify this method to improve separations or lower the costs of measurements, provided all performance specifications are met. Analyses of blanks are required to demonstrate freedom from contamination. When results of spikes indicate atypical method performance for samples, the samples are diluted to bring method performance within acceptable limits.

For low solids (aqueous samples), extract, concentrate, and analyze two sets of four 1-L aliquots (8 aliquots total) of the precision and recovery standard. For high solids samples, two sets of four 30-g aliquots of the high solids reference matrix are used.

Spike all samples with labeled compounds to assess method performance. Compute percent recovery of the labeled compounds using the internal standard method. Compare the labeled compound recovery for each compound with the corresponding labeled compound recovery.

Reagent water and high solids reference matrix blanks are analyzed to demonstrate freedom from contamination. Extract and concentrate a 1-L reagent water blank or a high solids reference matrix blank with each sample's lot (samples started through the extraction process on the same 8-h shift, to a maximum of 20 samples).

Field replicates may be collected to determine the precision of the sampling technique, and spiked samples may be required to determine the accuracy of the analysis when the internal standard method is used.

REFERENCE
Semivolatile Organic Compounds by Isotope Dilution GC/MS. Office of Water Regulation and Standards, U.S. EPA Industrial Technology Division, Washington, DC, EPA Method 1625, Rev. C, June 1989 (contact W.A. Telliard, U.S. EPA, Office of Water Regulations and Standards, 401 M St., SW, Washington, DC, 20460. Phone: 202-382-7131).

Benzene   **EPA Method 1624**
CAS #71-43-2

TITLE
Volatile Organic Compounds by Isotope Dilution GC/MS

MATRIX
Compounds may be determined in waters, soils, and municipal sludges by this method.

METHOD SUMMARY
This method is used to determine 58 volatile toxic organic pollutants associated with the CWA (as amended 1987); the RCRA (as amended 1986); the CERCLA (as amended 1986); and other compounds amenable to purge-and-trap gas chromatography-mass spectrometry (GC/MS).

If the solids content is less than 1%, stable isotopically-labeled analogs of the compounds of interest are added to a 5-mL sample and the sample is purged with an inert gas at 20–25°C in a chamber designed for soil or water samples. If the solids content is greater than 1%, 5 mL of reagent water and the labeled compounds are added to a 5-g aliquot of sample and the mixture is purged at 40°C. Compounds that will not purge at 20–25°C or at 40°C are purged at 78–85°C. In the purging process, the volatile compounds are transferred from the aqueous phase into the gaseous phase where they are passed into a sorbent column, and trapped. After purging is completed, the trap is backflushed and heated rapidly to desorb the compounds into a GC. The compounds are separated by the GC and detected by a MS. The labeled compounds serve to correct the variability of the analytical technique.

INTERFERENCES
Impurities in the purge gas, organic compounds outgassing from the plumbing upstream of the trap, and solvent vapors in the lab account for most problems. Samples can be contaminated by diffusion of volatile organic compounds (particularly methylene chloride) through the bottle seal during shipment and storage. Contamination by carryover can occur when high-level and low-level samples are analyzed sequentially. When an unusually concentrated sample is encountered, follow it by analysis of a reagent water blank to check for carryover.

INSTRUMENTATION
Major equipment includes a GC with linear temperature programming and a glass jet separator as

the MS interface, a MS with 70 eV electron impact ionization, and a data system to collect and record response factors.

Column: 2.8 m × 2 mm I.D. glass, packed with 1% SP-1000 on Carbopak B, 60/80 mesh, or equivalent.

PRECISION & ACCURACY The detection limits of the method are usually dependent on the level of interferences rather than instrumental limitations. The method detection limits were determined in digested sludge (low solids) and in filter cake or compost (high solids).

The MDL (in µg/kg) for low solids is 23 and for high solids is 8.

Labeled and native compound precision (in µg/L) as standard deviation was 9.0.
Labeled and native compound accuracy (in µg/L) as average recovery was 13–28.

Acceptance criteria are at 20 µg/L for this compound.

SAMPLE COLLECTION, PRESERVATION & HANDLING Grab samples are collected in glass containers having a total volume greater than 20 mL. Fill and seal each bottle so that no air bubbles are entrapped. Samples are maintained at 0 to 4°C from the time of collection until analysis. If an aqueous sample contains residual chlorine, add sodium thiosulfate preservative (10 mg/40 mL) to the empty sample bottles just prior to shipment to the sample site. All samples must be analyzed within 14 days of collection.

SAMPLE PREPARATION Samples containing less than 1% solids are analyzed directly as aqueous samples. Samples containing 1% solids or greater are analyzed as solid samples utilizing one of two methods, depending on the levels of pollutants, in the sample. Samples containing 1% solids or greater, and low to moderate levels of pollutants are analyzed by purging a known weight of sample added to 5 mL of reagent water. Samples containing 1% solids or greater, and high levels of pollutants, are extracted with methanol, and an aliquot of the methanol extract is added to reagent water and purged.

QUALITY CONTROL A field blank prepared from reagent water and carried through the sampling and handling protocol may serve as a check on contamination from shipment and storage.

The analyst is permitted to modify this method to improve separations or lower the costs of measurements, provided all performance specifications are met. Analyses of blanks are required. When results of spikes indicate atypical method performance for samples, the samples are diluted to bring method performance within acceptable limits. Analyze two sets of four 5-mL aliquots (8 aliquots total) of the aqueous performance standard. Spike all samples with labeled compounds to assess method performance on the sample matrix. Compute the percent recovery of the labeled compounds using the internal standard method. Compare the percent recovery for each compound with the corresponding labeled compound recovery. Reagent water blanks are analyzed to demonstrate freedom from carryover contamination. Field replicates may be collected to determine the precision of the sampling technique, and spiked samples may be required to determine the accuracy of the analysis when the internal method is used.

REFERENCE Volatile Organic Compounds by Isotope Dilution GC/MS. Office of Water Regulation and Standards, U.S. EPA Industrial Technology Division, Washington, DC, EPA Method 1624, Rev. C, June 1989 (contact W.A. Telliard, U.S. EPA, Office of Water Regulations and Standards, 401 M St., SW, Washington, DC, 20460. Phone: 202-382-7131).

Benzene — EPA Method 502
CAS #71-43-2

TITLE Volatile Organic Compounds in Water By Purge and Trap Capillary Column Gas Chromatography with Photoionization and Electrolytic Conductivity Detectors in Series. U.S. EPA Method 502.2, Rev. 2.0, 1989.

MATRIX Drinking water and raw source water. The latter should include most surface water and groundwater sources.

METHOD SUMMARY This method covers 60 volatile organic compounds that contain halogen atoms and/or that are aromatic. An inert gas (zero grade nitrogen or helium) is bubbled through a 25-mL or a 5-mL water sample (depending on the expected concentration of the analytes). Purged sample components are trapped in a tube of sorbent materials. When purging is complete, the sorbent tube is heated and backflushed with helium to desorb the trapped sample onto a capillary GC column. The column is temperature programmed to separate the method analytes which are then detected with a photoionization detector (PID) and a Hall electrolytic conductivity (HECD) placed in series. The PID is selective for aromatic compounds and the HECD is selective for halogenated compounds.

INTERFERENCES Impurities in the purge gas and from organic compounds outgassing from the plumbing ahead of the trap account for many contamination problems. Interferences purged or coextracted from the samples will vary considerably from source to source, depending upon the particular sample or extract being tested. Cross-contamination can occur whenever high-level and low-level samples are analyzed sequentially. Samples also can be contaminated by diffusion of volatile organics (particularly methylene chloride and fluorocarbons) through the septum seal into the sample during shipment and storage. The lab where volatile analysis is performed and also the refrigerated storage area should be completely free of solvents.

INSTRUMENTATION A GC containing a series configuration of a high temperature photoionization detector (PID) equipped with 10.0 eV (nominal) lamp and Hall electrolytic conductivity detector (HECD) is required. Also required is an all-glass 5-mL purging device, a sorbent trap, and a thermal desorption apparatus which is connected to the GC system.

Column 1: VOCOL glass wide-bore capillary column.
Column 2: RTX–502.2 mega-bore capillary column.
Column 3: DB-62 mega-bore capillary column.

PRECISION & ACCURACY Method detection limits are dependent upon the characteristics of the gas chromatographic system used. Analytes that are not separated chromatographically cannot be individually identified and used in the same calibration mixture or water samples unless an alternative technique for identification and quantification, such as mass spectrometry, is used.

Electrolytic Conductivity Detetor (c) range in µg/L (a) was 0.02–200.
Electrolytic Conductivity Detetor (c) MDL in µg/L (b) was not listed.
Electrolytic Conductivity Detetor (c) accuracy as % Recovery was not listed.
Electrolytic Conductivity Detetor (c) precision as % RSD was not listed.
Photoionization Detector (d) range in µg/L (a) was 0.02–200.
Photoionization Detector (d) MDL in µg/L (b) was 0.01.
Photoionization Detector (d) accuracy as % Recovery was 99.
Photoionization Detector (d) precision as % RSD was 1.2.

(a) The applicable concentration range of this method is compound, instrument, and matrix-dependent. It is listed as being approximately 0.02 to 200 µg/L but no specific information is provided so caution should be observed.

(b) The method detection limits reports with this method are compound, instrument, and matrix-dependent. The values reported were calculated using reagent water fortified with the corresponding compounds at 10 µg/L and a GC-equipped with a 60 m × 0.75 mm VOLCOL wide bore capillary column with 1.5 µm film thickness and using helium carrier gas.

(c) Recoveries and relative standard deviations were determined from seven samples of reagent water fortified with 10 µg/L of each compound. 2-Bromo-1-chloropropane was used as the internal standard for calculating average recoveries.

(d) Recoveries and relative standard deviations were determined from seven samples of reagent water fortified with 10 µg/L of each compound. Fluorobenzene was used as the internal standard for calculating average recoveries.

SAMPLING METHOD Collect samples using a 40- to 120-mL screw-cap vial (prewashed with detergent, rinsed with distilled water and oven dried at 105°C) with a Teflon®-faced silicone septum. Collect bubble-free samples and place the septum with the Teflon® side down on the water.

SAMPLE PRESERVATION If residual chlorine is present in the water add about 25 mg of ascorbic acid to each vial before samples are collected to remove the chlorine. Add hydrochloric acid to reduce pH to <2, immediately cool samples to 4°C, and store them in a solvent-free refrigerator at 4°C until analysis.

MHT The maximum holding time for samples is 14 days from the time they were collected.

SAMPLE PREPARATION Remove the plungers from two 5-mL syringes and attach a closed syringe valve to each. Warm the sample to room temperature, open the sample bottle, and carefully pour the sample into one of the syringe barrels to just short of overflowing. Replace the syringe plunger, invert the syringe, and compress the sample. Open the syringe valve and vent any residual air while adjusting the sample volume to 5.0 mL. Add 10 µL of the internal calibration standard to the sample through the syringe valve. Close the valve. Fill the second syringe in an identical manner from the same sample bottle. Reserve this second syringe for a reanalysis if necessary.

QUALITY CONTROL As an initial demonstration of lab accuracy and precision, analyze 4 to 7 replicates of a lab fortified blank containing analyte at 0.1–5 µg/L. Collect all samples in duplicate. Surrogate analytes (similar to those of the analytes of interest), whose concentration is known in every sample, are measured using the same internal standard calibration procedure. Duplicate field reagent water blanks (trip blanks) must be analyzed with each set of samples, lab reagent blanks (method blanks) must be analyzed with each batch of samples processed as a group within a work shift. Also, a single lab-fortified blank that contains each of the analytes of interest should be analyzed with each batch of samples processed as a group within a work shift. A 3- to 5-point calibration curve is needed depending on the calibration range factor required.

EPA CONTACT & HOTLINE For technical questions contact Dr. Baldev Bathija, U.S. EPA, Office of Ground Water and Drinking Water (WH-550D), 401 M St. SW, Washington, DC 20460. Tel. (202) 260-3040. For further information the EPA Safe Drinking Water Hotline may be called at: (800) 426-4791.

REFERENCE Methods for the Determination of Organic Compounds in Drinking Water, EPA/600/4-88/039 (revised July 1991; Final Rule for determination of compliance with the MCL for Total Trihalomethanes under 141.30, in 40 CFR Part 141, Vol. 58, No. 147, Fed. Reg., Tuesday Aug. 3, 1993). U.S. EPA Environmental Monitoring Systems Laboratory, Cincinnati, OH, 45268, U.S.A. Available from the National Technical Information Service (NTIS), 5285 Port Royal Road, Springfield, VA 22161; Tel. 800-553-6847. NTIS Order Number is PB91-231480.

Benzene   **EPA Method 524**
CAS #71-43-2

TITLE Measurement of Purgeable Organic Compounds in Water by Capillary Column GC/MS.

MATRIX Drinking water and raw source water; the latter should include most surface water and groundwater sources.

METHOD SUMMARY Method 524.2 covers 60 volatile organic compounds. An inert gas (zero grade nitrogen or helium) is bubbled through a 25-mL or a 5-mL water sample (depending on the expected concentration of the analytes). Purged sample components are trapped in a tube of sorbent materials. When purging is complete, the sorbent tube is heated and backflushed with helium to desorb the trapped sample onto a capillary GC column.

INTERFERENCES Impurities in the purge gas and from organic compounds outgassing from the plumbing ahead of the trap account for many contamination problems. Interferences purged or coextracted from the samples will vary considerably from source to source, depending upon the particular sample or extract being tested. Cross-contamination can occur whenever high-level and low-level samples are analyzed sequentially. Samples also can be contaminated by diffusion of volatile organics (particularly methylene chloride and fluorocarbons) through the septum seal into the sample during shipment and storage.

INSTRUMENTATION A GC/MS with a data system equipped with one of the following capillary GC columns:

Column 1: VOCOL glass wide bore capillary column.
Column 2: DB-624 fused silica capillary column.
Column 3: DB-5 fused silica capillary column.

Also required is an all-glass 25-mL or 5-mL purging device, a sorbent trap, and a thermal desorption apparatus which is connected to the GC/MS system.

PRECISION & ACCURACY Method detection limits are compound- and instrument-dependent, and may vary from approximately 0.02–0.35 µg/L. Note in the table below that the "true" concentration range used for accuracy and precision measurements was quite narrow. However, the applicable concentration range of this method is primarily column dependent and is approximately 0.02 to 200 µg/L for the wide-bore thick-film columns. Narrow-bore thin-film columns may have a capacity which limits the range to about 0.02 to 20 µg/L. Analytes that are inefficiently purged from water will not be detected when present at low concentrations, but they can be measured with acceptable accuracy and precision when present in sufficient amounts.

Analytes that are not separated chromatographically, but which have different mass spectra and non-interfering quantification ions, can be identified and measured in the same calibration mixture or water sample. Analytes which have very similar mass spectra cannot be individually identified and measured in the same calibration mixture or water samples unless they have different retention times. Co-eluting compounds with very similar mass spectra, typically many structural isomers, must be reported as an isomeric group or pair.

The range (in µg/L) was 0.1–10.
The method detection limit (in µg/L) was 0.04.
The accuracy (as % recovery) was 97.
The precision (in %) was 5.7.

Note: Data were obtained from 16–31 determinations using a wide-bore capillary column and a jet separator interfaced to a quadrupole mass spectrometer. All analytes were in a reagent water matrix.

SAMPLING METHOD Collect samples using a 40- to 120-mL screw-cap vial (prewashed with detergent, rinsed with distilled water and oven dried at 105°C) with a Teflon®-faced silicone septum. Collect bubble-free samples and place the septum with the Teflon® side down on the water.

SAMPLE PRESERVATION If residual chlorine is present in the water add about 25 mg of ascorbic acid to each vial before samples are collected to remove the chlorine. Add hydrochloric acid to reduce pH to <2, and immediately cool samples to 4°C, and store them in a solvent-free refrigerator at 4°C until analysis.

MHT The maximum holding time for samples is 14 days from the time they were collected.

SAMPLE PREPARATION Remove the plungers from two 25-mL (or 5-mL depending on sample size) syringes and attach a closed syringe valve to each. Warm the sample to room temperature, open the sample bottle, and carefully pour the sample into one of the syringe barrels to just short of overflowing. Replace the syringe plunger, invert the syringe, and compress the sample. Open the syringe valve and vent any residual air while adjusting the sample volume to 25.0 mL (or 5 mL). For samples and blanks, add 5 µL of the fortification solution containing the internal standard and the surrogates to the sample through the syringe valve. For calibration standards and lab fortified blanks, add 5 µL of the fortification solution containing the internal standard only. Close the valve. Fill the second syringe in an identical manner from the same sample bottle. Reserve this second syringe for a reanalysis if necessary.

QUALITY CONTROL As an initial demonstration of lab accuracy and precision, analyze 4 to 7 replicates of a lab fortified blank containing analyte at 0.2–5 µg/L. Collect all samples in duplicate. Surrogate analytes (similar to those of the analytes of interest), whose concentration is known in every sample, are measured using the same internal standard calibration procedure. Duplicate field reagent water blanks (trip blanks) must be analyzed with each set of samples, lab reagent blanks (method blanks) must be analyzed with each batch of samples processed as a group within a work shift. Also, a single lab-fortified blank that contains each of the analytes of interest should be analyzed with each batch of samples processed as a group within a work shift. A 3- to 5-point calibration curve is needed depending on the calibration range factor required.

EPA CONTACT & HOTLINE For technical questions contact Dr. Baldev Bathija, U.S. EPA, Office of Ground Water and Drinking Water (WH-550D), 401 M St. SW, Washington, DC 20460. Tel. (202) 260-3040. For further information the EPA Safe Drinking Water Hotline may be called at: (800) 426-4791.

REFERENCE Methods for the Determination of Organic Compounds in Drinking Water, EPA/600/4-88/039 (revised July 1991; Final Rule for determination of compliance with the MCL for Total Trihalomethanes under 141.30, in 40 CFR Part 141, Vol. 58, No. 147, Fed. Reg., Tuesday Aug. 3, 1993). U.S. EPA Environmental Monitoring Systems Laboratory, Cincinnati, OH, 45268, U.S.A. Available from the National Technical Information Service (NTIS), 5285 Port Royal Road, Springfield, VA 22161; Tel. 800-553-6847. NTIS Order Number is PB91-231480.

Benzene   **EPA Method 8021**
CAS #71-43-2

VOAs by Capillary Column GC/PID-HECD

TITLE Halogenated Volatile by Gas Chromatography Using Photoionization and Electrolytic Conductivity Detectors in Series: Capillary Column Technique

MATRIX This method is applicable to nearly all types of samples, regardless of water content, including groundwater, aqueous sludges, caustic liquors, acid liquors, waste solvents, oily wastes, mousses, tars, fibrous wastes, polymeric emulsions, filter cakes, spent carbons, spent catalysts, soils, and sediments.

METHOD SUMMARY This method is used to determine 60 volatile organic compounds in a variety of solid waste matrices. It provides GC conditions for the detection of halogenated and aromatic volatile organic compounds. Samples can be analyzed using direct injection or purge-and-trap (EPA Method 5030). Groundwater samples must be analyzed using EPA Method 5030 (where applicable). A temperature program is used with the GC. Detection is achieved by a photoionization detector (PID) and a Hall electrolytic conductivity detector (HECD) in series.

INTERFERENCES Samples can be contaminated by diffusion of volatile organics (particularly chlorofluorocarbons and methylene chloride) through the sample container septum during shipment and storage.

INSTRUMENTATION A GC-equipped with variable-constant differential flow controllers, subambient oven controller, PID and HECD detectors connected with a short piece of uncoated capillary tubing and a data system.

Column: 60 m × 0.75 mm I.D. VOCOL wide-bore capillary column with 1.5 µm film thickness.

PRECISION & ACCURACY MDLs are compound-dependent and vary with purging efficiency and concentration. The applicable concentration range of this method is compound- and instrument-dependent but is approximately 0.1 to 200 µg/L. Analytes that are inefficiently purged from water will not be detected when present at low concentrations, but they can be measured with acceptable accuracy and precision when present in sufficient amounts. The estimated quantitation limit (EQL) for an individual compound is approximately 1 µg/kg (wet weight) for soil/sediment samples, 100 µg/kg (wet weight) for wastes, and 1 µg/L for groundwater. EQLs will be proportionately higher for sample extracts and samples that require dilution or reduced sample size to avoid saturation of the detector.

MULTIPLICATION FACTORS FOR OTHER MATRICES (a)

Matrix	Factor (b)
Groundwater	10
Low-concentration soil	10
Water miscible liquid waste	500
High-concentration soil and sludge	1250
Non-water miscible waste	1250

(a) Sample EQLs are highly matrix-dependent. The EQLs listed herein are provided for guidance and may not always be achievable. (b) EQL = [Method detection limit] × [Factor]. For non-aqueous samples, the factor is on a wet-weight basis.

SINGLE LABORATORY ACCURACY & PRECISION DATA FOR VOCs IN WATER

This method was tested in a single lab using water spiked at 10 µg/L and the following data was reported:

Recoveries and standard deviations were determined from seven samples and spiked at 10 µg/L of each analyte. Recoveries were determined by the internal standard method. Internal standards were: Fluorobenzene for PID and 2-Bromo-1-chloropropane for HECD.

The average recovery (in percent) for the PID was 99.
The standard deviation of the recovery for the PID was 1.2.
The MDL (in µg/mL) for the PID was 0.009.
The average recovery (in percent) for the HECD was none (no response for this detector).
The standard deviation of the recovery for the HECD was none (no response for this detector).
The MDL (in µg/mL) for the HECD was none (no response for this detector).

SAMPLE COLLECTION, PRESERVATION & HANDLING
Volatile Organics — Standard 40-mL glass screw-cap VOA vials with Teflon®-faced silicone septum may be used for both liquid and solid matrices. When collecting samples, liquids and solids should be introduced into the vials gently to reduce agitation which might drive off volatile compounds. If there are any air bubbles present the sample must be retaken. Tap slightly as they are filled to try and eliminate as much free air space as possible. The two vials from each sampling locations should be sealed in separate plastic bags to prevent cross-contamination between samples particularly if the sampled waste is suspected of containing high levels of volatile organics.

Semivolatile organics — Containers used to collect samples for the determination of semivolatile organic compounds should be soap and water washed followed by methanol (or isopropanol) rinsing. The sample containers should be of glass or Teflon® and have screw-top covers with Teflon® liners.

Preservation for volatile organics — No preservation is used with concentrated waste samples. With liquid samples containing no residual chlorine, 4 drops of concentrated hydrochloric acid are added and the samples are immediately cooled to 4°C. When liquid samples contain residual chlorine, they are treated as above and, in addition, 4 drops of 4% aqueous sodium thiosulfate are added. Soil, sediment, and sludge samples are only cooled to 4°C.

Preservation for semivolatile organics — No preservation is used with concentrated waste samples. With liquid samples containing no residual chlorine and with soil, sediment, and sludge samples, immediately cooling to 4°C is the only preservation used. When residual chlorine is present then 3 mL of 10% aqueous sodium sulfate is added for each gallon of sample collected, followed by cooling to 4°C.

MHT The holding time for all volatile organics samples is 14 days. Liquid samples must be extracted within 7 days and their extracts analyzed within 40 days. Concentrated waste, soil, sediment, and sludge samples must be extracted within 14 days and their extracts analyzed within 40 days.

SAMPLE PREPARATION Volatile compounds are introduced into the gas chromatograph either by direct injector or purge-and-trap (EPA Method 5030). EPA Method 5030 may be used directly on groundwater samples or low-concentration contaminated soils and sediments. For medium-concentration soils or sediments, methanolic extraction, as described in EPA Method 5030, may be necessary prior to purge-and-trap analysis.

QUALITY CONTROL Calculate surrogate standard recovery on all samples, blanks, and spikes. A trip blank is recommended to check on sampling, storage, and handling contamination. Calibration standards, at a minimum of five concentration levels, are prepared in organic-free reagent water. One of the concentration levels should be at a concentration near, but above, the method detection limit.

A combination of bromochloromethane, 2-bromo-1-chloropropane, 1,4-dichlorobutane, and bromochlorobenzene are recommended as surrogate standards to encompass the range of the temperature program used in this method.

REFERENCE Test Methods for Evaluating Solid Waste, Physical/Chemical Methods, SW-846, 3rd Edition, U.S. EPA, Office of Solid Waste, Washington, DC, EPA Method 8021A, Rev. 1, Nov. 1992.

Benzene **EPA Method 8240**
CAS #71-43-2

TITLE Volatile Organics By GC/MS: Packed Column Technique

MATRIX Nearly all types of sample matrices, regardless of water content, can be analyzed using this method. This includes groundwater, aqueous sludges, caustic liquors, acid liquors, waste solvents, oily wastes, mousses, tars, fibrous wastes, polymetric emulsions, filter cakes, spent carbons, spent catalysts, soils, and sediments.

METHOD SUMMARY Method 8240B covers 80 volatile organic compounds that are introduced into a gas chromatograph by the purge-and-trap method or by direct injection (in limited applications). For the purge-and-trap method an inert gas (zero grade nitrogen or helium) is bubbled through a 5-mL solution at ambient temperature. Purged sample components are trapped in a tube of sorbent materials. When purging is complete, the sorbent tube is heated and backflushed with inert gas to desorb the trapped components onto a GC column.

INTERFERENCES Impurities in the purge gas and from organic compounds outgassing from the plumbing ahead of the trap account for many contamination problems. Interferences purged or coextracted from the samples will vary considerably from source to source. Cross-contamination can occur whenever high-level and low-level samples are analyzed sequentially. Whenever an unusually concentrated sample is analyzed, it should be followed by the analysis of organic-free reagent water to check for cross-contamination. Samples also can be contaminated by diffusion of volatile organics (particularly methylene chloride and fluorocarbons) through the septum seal into the sample during shipment and storage. A trip blank can serve as a check on such contamination. The lab where volatile analysis is performed and also the refrigerated storage area should be completely free of solvents.

INSTRUMENTATION A gas chromatograph/mass spectrometry/data system (GC/MS) equipped with a 6 ft × 0.1 in I.D. glass column packed with 1% SP-1000 on Carbopack-B (60/80 mesh) is required. Also needed is a 5-mL purging device, a sorbent trap, and a thermal desorption apparatus.

PRECISION & ACCURACY This method is reported to have been tested by 15 laboratories using organic-free reagent water, drinking water, surface water, and industrial wastewaters (not specified) fortified at six concentrations over the range 5–600 µg/L.

Sample estimated quantitation limits (EQLs) are highly matrix-dependent. The EQLs listed may not always be achievable. EQLs listed for soils or sediments are based on wet weight. Normally, data is reported on a dry-weight basis; therefore, EQLs will be higher, based on the percent dry weight of each sample. Note that EQLs are even more variable than MDLs and that they are highly variable depending on the matrix being analyzed.

EQL in groundwater in µg/L was 5.
EQL in low soil or sediment in µg/kg was 5.
Accuracy (a) in µg/L was not listed.
Precision (b) in µg/L was not listed.

(a) *Average recovery found for measurements of samples containing a concentration of C, in µg/L.*
(b) *Overall precision found for measurements of samples with average recovery X for samples containing a concentration of C in µg/L.*
X = *Average recovery found for measurement of samples containing a concentration of C in µg/L.*

ESTIMATED QUANTITATION LIMITS FOR MATRICES OTHER THAN WATER, SOIL, AND SEDIMENTS

Other Matrices	Factor (a)
Waste miscible liquid waste	50
High-concentration soil and sludge	125
Non-water miscible waste	500

(a) *EQL = [EQL for low soil/sediment] × [Factor]. For non-aqueous samples, the factor is on a wet-weight basis.*

SAMPLING METHOD
Liquid samples — Use a 40-mL glass screw-cap VOA vial with a Teflon®-faced silicone septum that has been prewashed, rinsed with distilled deionized water, and oven dried. However, if residual chlorine is present, collect sample in a 40-oz. soil VOA container which has been pre-preserved with 4 drops of 10% sodium thiosulfate, mix gently, and then transfer the sample to a 40-mL VOA vial. Collect bubble-free samples in duplicate and seal them in separate plastic bags.

Soils or sediments, and sludges — Use an 8-oz. widemouth glass bottle with a Teflon®-faced silicone septum that has been prewashed with detergent, rinsed with distilled deionized water, and oven dried. Tap slightly to eliminate free air space. Collect samples in duplicate and seal them in separate plastic bags.

SAMPLE PRESERVATION
Liquid samples — Add 4 drops of concentrated HCL and immediately cool samples to 4°C and store in a solvent-free refrigerator.

Soils or sediments, and sludges — Cool samples to 4°C and store in a solvent-free refrigerator.

MHT Maximum holding time is 14 days from the date of sample collection.

SAMPLE PREPARATION

Liquid samples — Remove the plunger from a 5-mL syringe and carefully pour the sample into the syringe barrel to just short of overflowing. Replace the syringe plunger and compress the sample. Open the syringe valve and vent any residual air while adjusting the sample volume to 5.0 mL. If there is only one volatile organic analysis (VOA) vial, a second syringe should be filled at this time to protect against possible loss of sample integrity. Add 10 μL of surrogate spiking solution and 10 μL of internal standard spiking solution through the valve bore of the 5-mL syringe, then close the valve. The surrogate and internal standards may be mixed and added as a single spiking solution.

Sediments, soils, and waste samples — All samples of this type should be screened by GC analysis using a headspace method (EPA Method 3810) or the hexadecane extraction and screening method (EPA Method 3820). Use the screening data to determine whether to use the low-concentration method (0.005–1 mg/kg) or the high-concentration method (>1 mg/kg).

Low-concentration method — The low-concentration method is based on purging a heated sediment or soil sample mixed with organic-free reagent water containing the surrogate and internal standards. Analyze all reagent blanks and standards under the same conditions as the samples.

Use a 5-g sample if the expected concentration is <0.1 mg/kg or a 1-g sample for expected concentrations between 0.1 and 1 mg/kg. Mix the contents of the sample container with a narrow metal spatula. Weigh the amount of the sample into a tared purge device. Add the spiked water to the purge device, which contains the weighed amount of sample, and connect the device to the purge-and-trap system.

High-concentration method — This method is based on extracting the sediment or soil with methanol. A waste sample is either extracted or diluted, depending on its solubility in methanol. Wastes that are insoluble in methanol are diluted with reagent tetraglyme or possibly polyethylene glycol (PEG). An aliquot of the extract is added to organic-free reagent water containing surrogate and internal standards. This is purged at ambient temperature. All samples with an expected concentration of >1.0 mg/kg should be analyzed by this method.

Mix the contents of the sample container with a narrow metal spatula. For sediments or soils and solid wastes that are insoluble in methanol, weigh 4 g (wet weight) of sample into a tared 20-mL vial. For waste that is soluble in methanol, tetraglyme, or PEG, weigh 1 g (wet weight) into a tared scintillation vial or culture tube or a 10-mL volumetric flask. Quickly add 9.0 mL of appropriate solvent then add 1.0 mL of a surrogate spiking solution to the vial, cap it, and shake it for 2 min.

METHANOL EXTRACT REQUIRED FOR ANALYSIS OF HIGH-CONCENTRATION SOILS OR SEDIMENTS

Approximate Concentration Range	Volume of Methanol Extract (a)
500–10,000 μg/kg	100 μL
1,000–20,000 μg/kg	50 μL
5,000–100,000 μg/kg	10 μL
25,000–500,000 μg/kg	100 μL of 1/50 dilution (b)

Calculate appropriate dilution factor for concentrations exceeding this table.

(a) The volume of methanol added to 5 mL of water being purged should be kept constant. Therefore, add to the 5-mL syringe whatever volume of methanol is necessary to maintain a volume of 100 μL added to the syringe.

(b) Dilute an aliquot of the methanol extract and then take 100 μL for analysis.

QUALITY CONTROL Demonstrate, through the analysis of a reagent water blank, that interferences from the analytical system, glassware, and reagents are under control. Blank samples should be carried through all stages of the sample preparation and measurement steps. For each analytical batch (up to 20 samples), a reagent blank, matrix spike, and matrix spike duplicate must be analyzed (the frequency of the spikes may be different for different monitoring programs). The blank and spiked samples must be carried through all stages of the sample preparation and measurement steps. QC samples mentioned in the section on Interferences will also be needed as appropriate to those situations.

REFERENCE Test Methods for Evaluating Solid Waste (SW-846). U.S. EPA. 1983. Method 8240B, Rev. 2, Nov. 1990. Office of Solid Wastes, Washington, DC.

Benzene   **EPA Method 8260**
CAS #71-43-2

TITLE Volatile Organic Compounds by GC/MS: Capillary Column Technique

MATRIX This method is applicable to nearly all types of samples, regardless of water content, including groundwater, soils, and sediments.

METHOD SUMMARY Method 8260A covers 58 volatile organic compounds that are introduced into a gas chromatograph by the purge-and-trap method or by direct injection (in limited applications). Zero-grade helium is bubbled through a 5-mL solution at ambient temperature. Purged sample components are trapped in a tube containing suitable sorbent materials. When purging is complete, the sorbent tube is heated and backflushed with helium to desorb trapped sample components. The analytes are desorbed directly to a large bore capillary or cryofocussed on a capillary precolumn before being flash evaporated to a narrow bore capillary for analysis.

INTERFERENCES Major contaminant sources are volatile materials in the lab and impurities in the inert purging gas and in the sorbent trap. Interfering contamination may occur when a sample containing low concentrations of volatile organic compounds is analyzed immediately after a sample containing high concentrations of volatile organic compounds. After analysis of a sample containing high concentrations of volatile organic compounds, one or more calibration blanks should be analyzed to check for cross-contamination. Screening of the samples prior to purge-and-trap GC/MS analysis is highly recommended to prevent contamination of the system. This is especially true for soil and waste samples.

Special precautions must be taken to analyze for methylene chloride. The analytical and sample storage area should be isolated from all atmospheric sources of methylene chloride. All gas chromatography carrier gas lines and purge gas plumbing should be constructed from stainless steel or copper tubing. Laboratory clothing previously exposed to methylene chloride fumes during liquid-liquid extraction procedures can contribute to sample contamination.

Samples can also be contaminated by diffusion of volatile organics (particularly methylene chloride and fluorocarbons) through the septum seal during shipment and storage. A trip blank can serve as a check on such contamination.

INSTRUMENTATION GC/MS with a temperature-programmable chromatograph suitable for splitless injection equipped with variable constant differential flow controllers, a subambient oven controller, a purging device, sorbent trap, a thermal desorption apparatus and a capillary precolumn interface when using cryogenic cooling will be needed. The following GC columns may be used:

Column 1: 60 m × 0.75 mm I.D. capillary column coated with VOCOL, 1.5 μm film thickness.
Column 2: 30 m × 0.53 mm I.D. capillary column coated with DB-624 or VOCOL, 3 μm film thickness.
Column 3: 30 m × 0.32 mm I.D. capillary column coated with DB-5 or SE-54, 1-μm film thickness.

PRECISION & ACCURACY This method has been tested in a single lab using spiked water. Using a wide-bore capillary column, water was spiked at concentrations between 0.5 and 10 μg/L. Single lab accuracy and precision data are presented. The MDL actually achieved in a given analysis will vary depending on instrument sensitivity and matrix effects.

The MDL (a) in μg/L was 0.04.
The concentration range in μg/L was 0.1–10.
The mean accuracy (% of true value) was 97.
The precision as relative standard deviation was 5.7.

Note: The MDL is based on a 25-mL sample volume instead of a 5-mL sample volume.

SAMPLING METHOD
Liquid samples — Use a 40-mL glass screw-cap VOA vial with a Teflon®-faced silicone septum that has been prewashed, rinsed with distilled deionized water, and oven dried. If residual chlorine is present, collect the sample in a 4-oz soil VOA container which has been pre-preserved with 4 drops of 10% sodium thiosulfate. Mix gently and transfer the sample to a 40-mL VOA vial. Collect bubble-free samples in duplicate and seal each sample in a separate plastic bag.

Soils, sediments, and sludges — Use an 8-oz widemouth glass bottle with Teflon®-faced silicone septum that has been prewashed, rinsed with distilled deionized water, and oven dried. **Do not** heat the septum for more than 1 h. Tap slightly to eliminate any free air space. Collect samples in duplicate and seal each one in a separate plastic bag.

SAMPLE PRESERVATION
Liquid samples — Add 4 drops of concentrated HCL, cool to 4°C and store in a solvent-free refrigerator.

Soils, sediments, and sludges — Cool samples to 4°C and store in a solvent-free refrigerator.

MHT The maximum holding time of any sample (liquids, soils, sediments, and sludges) is 14 days.

SAMPLE PREPARATION
Liquid samples — Remove the plunger from a 5-mL syringe and carefully pour the sample into the syringe barrel to just short of overflowing. Replace the syringe plunger and compress the sample. Open the syringe valve and vent any residual air while adjusting the sample volume to 5.0 mL. If there is only one volatile organic analysis (VOA) vial, a second syringe should be filled at this time to protect against possible loss of sample integrity. Add 10 μL of surrogate spiking solution and 10 μL of internal standard spiking solution through the valve bore of the 5-mL syringe, then close the valve. The surrogate and internal standards may be mixed and added as a single spiking solution.

Sediments, soils, and waste samples — All samples of this type should be screened by GC analysis using a headspace method (EPA Method 3810) or the hexadecane extraction and screening method (EPA Method 3820). Use the screening data to determine whether to use the low-concentration method (0.005–1 mg/kg) or the high-concentration method (>1 mg/kg).

Low-concentration method — The low-concentration method is based on purging a heated sediment or soil sample mixed with organic-free reagent water containing the surrogate and internal standards. Analyze all reagent blanks and standards under the same conditions as the samples.

Use a 5-g sample if the expected concentration is <0.1 mg/kg or a 1-g sample for expected concentrations between 0.1 and 1 mg/kg. Mix the contents of the sample container with a narrow metal spatula. Weigh the amount of the sample into a tared purge device. Add the spiked water to the purge device, which contains the weighed amount of sample, and connect the device to the purge-and-trap system.

High-concentration method — This method is based on extracting the sediment or soil with methanol. A waste sample is either extracted or diluted, depending on its solubility in methanol. Wastes that are insoluble in methanol are diluted with reagent tetraglyme or possibly polyethylene glycol (PEG). An aliquot of the extract is added to organic-free reagent water containing surrogate and internal standards. This is purged at ambient temperature. All samples with an expected concentration of >1.0 mg/kg should be analyzed by this method.

Mix the contents of the sample container with a narrow metal spatula. For sediments or soils and solid wastes that are insoluble in methanol, weigh 4 g (wet weight) of sample into a tared 20-mL vial. For waste that is soluble in methanol, tetraglyme, or PEG, weigh 1 g (wet weight) into a tared scintillation vial or culture tube or a 10-mL volumetric flask. Quickly add 9.0 mL of appropriate solvent then add 1.0 mL of a surrogate spiking solution to the vial, cap it, and shake it for 2 min.

METHANOL EXTRACT REQUIRED FOR ANALYSIS OF HIGH-CONCENTRATION SOILS OR SEDIMENTS

Approximate Concentration Range	Volume of Methanol Extract (a)
500–10,000 µg/kg	100 µL
1,000–20,000 µg/kg	50 µL
5,000–100,000 µg/kg	10 µL
25,000–500,000 µg/kg	100 µL of 1/50 dilution (b)

Calculate appropriate dilution factor for concentrations exceeding this table.

(a) The volume of methanol added to 5 mL of water being purged should be kept constant. Therefore, add to the 5-mL syringe whatever volume of methanol is necessary to maintain a volume of 100 µL added to the syringe.
(b) Dilute an aliquot of the methanol extract and then take 100 µL for analysis.

QUALITY CONTROL Demonstrate, through the analysis of a reagent water blank, that interferences from the analytical system, glassware, and reagents are under control. Blank samples should be carried through all stages of the sample preparation and measurement steps. For each analytical batch (up to 20 samples), a reagent blank, matrix spike, and matrix spike duplicate must be analyzed (the frequency of the spikes may be different for different monitoring programs). The blank and spiked samples must be carried through all stages of the sample preparation and measurement steps. QC samples mentioned in the section on Interferences will also be needed as appropriate to those situations.

Matrix spiking standards should be prepared from volatile organic compounds which will be representative of the compounds being investigated. The recommended internal standards are chlorobenzene-d5, 1,4-difluorobenzene, 1,4-dichlorobenzene-d4, and pentafluorobenzene. Using stock standard solutions, prepare secondary dilution standards containing the compounds of interest, either singly or mixed together in methanol. Store them in a vial with no headspace for no more than one week. Surrogates recommended are toluene-d8, 4-bromofluorobenzene, and dibromofluoromethane. Each sample undergoing GC/MS analysis must be spiked with 10 µL of the surrogate spiking solution prior to analysis.

REFERENCE Test Methods for Evaluating Solid Waste (SW-846). U.S. EPA 1983, Method 8260A, Rev. 1, Nov. 1990. Office of Solid Waste, Washington, DC.

Benzene   EPA Method 602
CAS #71-43-2

TITLE Purgeable Aromatics

MATRIX Wastewater.

APPLICATION Method covers 7 purgeable aromatics. (Method 624 provides GC/MS conditions appropriate for the qualitative and quantitative confirmation of results). Method describes conditions for a 2nd GC column to confirm measurements made with primary column.

INTERFERENCES Impurities in the purge gas and organic compounds outgassing from the plumbing ahead of the trap. With high- and low-level samples, there can be carryover contamination. Diffusion of volatile organics through the septum seal into the sample.

INSTRUMENTATION GC-equipped with photoionization detector. (With purge-and-trap unit.)

RANGE 2.1–550 µg/L

MDL 0.2 µg/L

PRECISION 0.21X+0.56 µg/L (overall precision).

ACCURACY 0.92C+0.57 µg/L (as recovery).

SAMPLING METHOD 25-mL glass vial. Teflon®-lined septum.

STABILITY cool, 4°C, 0.008% Na2S2O3, HCl to pH 2.

MHT 14 days.

QUALITY CONTROL The lab must on an ongoing basis, spike at least 10% of the samples from each sample site being monitored to assess accuracy.

REFERENCE Method 602, *Federal Register* Part VIII 40 CFR Part 136, Oct. 26, 1984.

Benzene   EPA Method 624
CAS #71-43-2

TITLE Purgeables

MATRIX Wastewater.

APPLICATION Method covers 31 purgeable organics. An inert gas is bubbled through a 5-mL water sample in a specially designed purging chamber. Here, purgeables are transferred from aqueous to gaseous phase, passed onto a sorbent column, and trapped. Trap is heated and backflushed with inert gas to desorb purgeables onto a GC column, where purgeables are separated.

INTERFERENCES Impurities in the purge gas, organic compounds outgassing from the plumbing ahead of the trap, and solvent vapors in the lab. With high- and low-level samples, there can be carryover contamination.

INSTRUMENTATION GC/MS with purge-and-trap unit.

RANGE 5–600 µg/L

MDL 4.4 µg/L

PRECISION 0.25X–1.33 µg/L (overall precision).

ACCURACY 0.93C+2.00 µg/L (as recovery).

SAMPLING METHOD 25-mL glass vial. Teflon®-lined septum.

STABILITY cool, 4°C, 0.008% Na2S2O3. HCl to pH 2.

MHT 14 days.

QUALITY CONTROL The lab must on an ongoing basis, spike at least 5% of the samples from each sample site being monitored to assess accuracy.

REFERENCE Method 624, *Federal Register* Part VIII 40 CFR Part 136, Oct. 26, 1984.

Benzene   **EPA Method 8020**
CAS #71-43-2

TITLE Aromatic Volatile Organics

MATRIX Groundwater, soils, sludges, water miscible liquid wastes, and non-water miscible wastes.

APPLICATION This method is used to analyze for 8 aromatic VOCs. Samples are analyzed using direct injection or purge-and-trap methods. Groundwater must be analyzed by the purge-and-trap method. The method provides an optional GC column that is used for analyte confirmation and may also help resolve analytes from interferences.

INTERFERENCES There can be carryover contamination with high- and low-level samples. Impurities may come from the purge-and-trap apparatus, organic compounds outgassing from the plumbing ahead of trap, diffusion of VOCs through the sample bottle septum during shipping or storage, or from solvent vapors in the lab.

INSTRUMENTATION GC capable of on-column injections or purge-and-trap sample introduction and a photoionization detector (PID). Column 1: 6 ft by 0.082 in with 5% SP-1200 and 1.75% Bentone-34 on Supelcoport. Column 2: 8 ft by 0.1 in with 5% 1,2,3-tris(2-cyanoethoxy)propane on Chromosorb W-AW.

RANGE 2.1 to 500 µg/L

MDL 0.2 µg/L (reagent water).

PQL FACTORS FOR MULTIPLYING × FID MDL VALUE

Matrix	Multiplication Factor
Groundwater	10
Low-level soil	10
Water miscible liquid waste	500
High-level soil and sludge	1250
Non-water miscible waste	1250

PRECISION 0.21X + 0.56 µg/L (overall precision).

ACCURACY 0.92C + 0.57 µg/L (as recovery).

SAMPLING METHOD For water and liquid samples use glass 40-mL vials with Teflon®-lined septum caps and collect two vials per sample location with no headspace. For solids and concentrated waste samples use widemouth glass bottles with Teflon® liners. Cool all samples to 4°C

STABILITY For concentrated wastes, soils, sediments, or sludges cool to 4°C. For liquids, add 4 drops of concentrated hydrochloric acid and cool to 4°C.

MHT 14 days.

QUALITY CONTROL Analyze a reagent blank, matrix spike, and matrix spike duplicate/duplicate for each analytical batch (up to 20 samples). Demonstrate the purity of glassware and reagents by analyzing a reagent water method blank. Internal, surrogate, and five concentration level calibration standards are used. The QC check sample concentrate should contain this compound at 10 µg/mL in methanol.

REFERENCE Method 8020, SW-846, 3rd ed., Nov. 1986.

1,3-Benzenediol (Resorcinol)   **EPA Method 1625**
CAS #108-46-3

TITLE Semivolatile Organic Compounds by Isotope Dilution GC/MS

MATRIX The compounds may be determined in waters, soils, and municipal sludges by this method.

METHOD SUMMARY This method is used to determine 176 semivolatile toxic organic pollutants associated with the CWA (as amended 1987); the RCRA (as amended 1986); the CERCLA (as amended 1986); and other compounds amenable to extraction and analysis by capillary column gas chromatography-mass spectrometry (GC/MS).

Stable isotopically-labeled analogs of the compounds of interest are added to the sample. If the solids content is less than 1%, a 1-L sample is extracted at pH 12–13, then at pH <2 with methylene chloride using continuous extraction techniques.

If the solids content is 30% or less, the sample is diluted to 1% solids with reagent water, homogenized ultrasonically, and extracted at pH 12–13, then at pH <2 with methylene chloride using continuous extraction techniques. If the solids content is greater than 30%, the sample is extracted using ultrasonic techniques.

Each extract is dried over sodium sulfate, concentrated to a volume of 5 mL, cleaned up using GPC, if necessary, and concentrated. Extracts are concentrated to 1 mL if GPC is not performed, and to 0.5 mL if GPC is performed.

An internal standard is added to the extract, and a 1-mL aliquot of the extract is injected into the GC. The compounds are separated by GC and detected by a MS. The labeled compounds serve to correct the variability of the analytical technique.

INTERFERENCES Solvents, reagents, glassware, and other sample processing hardware may yield artifacts and/or elevated baselines causing misinterpretation of chromatograms and spectra. Materials used in the analysis must be demonstrated to be free from interferences under the conditions of analysis by running method blanks initially and with each sample lot (sample started through the extraction process on a given 8-h shift, to a maximum of 20). Specific selection of reagents and purification of solvents by distillation in all glass systems may be required. Glassware and, where possible, reagents are cleaned by solvent rinse and baking at 450°C for 1-h minimum. Interferences coextracted from samples will vary considerably from source to source, depending on the diversity of the site being sampled.

INSTRUMENTATION Major instrumentation includes a GC with a splitless or on-column injection port for capillary column, a MS with 70 eV electron impact ionization, and a data system to collect and record MS data, and process it. A K-D apparatus is used to concentrate extracts.

GC Column: 30 m × 0.25 mm I.D. 5% phenyl, 94% methyl, 1% vinyl silicone bonded phased fused silica capillary column.

PRECISION & ACCURACY The detection limits of the method are usually dependent on the level of interferences rather than instrumental limitations. The limits typify the minimum quantities that can be detected with no interferences present.

- The minimum level (in µg/mL) was not listed. This is defined as a minimum level at which the analytical system shall give recognizable mass spectra (background corrected) and acceptable calibration points.
- The MDL (in µg/kg) in low solids was not listed and in high solids was not listed; these were determined in digested sludge (low solids) and in filter cake or compost (high solids).
- The labeled and native compound initial precision as standard deviation (in µg/L) was not listed.
- The labeled and native compound initial accuracy as average recovery (in µg/L) was not listed.

SAMPLE COLLECTION, PRESERVATION & HANDLING Collect samples in glass containers. Aqueous samples which flow freely are collected in refrigerated bottles using automatic sampling equipment. Solid samples are collected as grab samples using widemouth jars. Maintain samples at 0 to 4°C from the time of collection until extraction. If residual chlorine is present in aqueous samples, add 80 mg sodium thiosulfate/L of water. Begin sample extraction within 7 days of collection, and analyze all extracts within 40 days of extraction.

SAMPLE PREPARATION Samples containing 1% solids or less are extracted directly using continuous liquid-liquid extraction techniques. Samples containing 1 to 30% solids are diluted to the 1% level with reagent water and extracted using continuous liquid-liquid extraction techniques. Samples containing greater than 30% solids are extracted using ultrasonic techniques.

- Base/neutral extraction — Adjust the pH of the waters in the extractors to 12–13 with 6 N NaOH. Extract with methylene chloride for 24–48 h.
- Acid extraction — Adjust the pH of the waters in the extractors to 2 or less using 6 N sulfuric acid. Extract with methylene chloride for 24–48 h.
- Ultrasonic extraction of high solids samples — Add anhydrous sodium sulfate to the sample and QC aliquot(s). Add acetone:methylene chloride (1:1) to the sample and mix thoroughly

Concentrate extracts using a K-D apparatus.

QUALITY CONTROL The analyst is permitted to modify this method to improve separations or lower the costs of measurements, provided all performance specifications are met. Analyses of blanks are required to demonstrate freedom from contamination. When results of spikes indicate atypical method performance for samples, the samples are diluted to bring method performance within acceptable limits.

For low solids (aqueous samples), extract, concentrate, and analyze two sets of four 1-L aliquots (8 aliquots total) of the precision and recovery standard. For high solids samples, two sets of four 30-g aliquots of the high solids reference matrix are used.

Spike all samples with labeled compounds to assess method performance. Compute percent recovery of the labeled compounds using the internal standard method. Compare the labeled compound recovery for each compound with the corresponding labeled compound recovery.

Reagent water and high solids reference matrix blanks are analyzed to demonstrate freedom from contamination. Extract and concentrate a 1-L reagent water blank or a high solids reference matrix blank with each sample's lot (samples started through the extraction process on the same 8-h shift, to a maximum of 20 samples).

Field replicates may be collected to determine the precision of the sampling technique, and spiked samples may be required to determine the accuracy of the analysis when the internal standard method is used.

REFERENCE Semivolatile Organic Compounds by Isotope Dilution GC/MS. Office of Water Regulation and Standards, U.S. EPA Industrial Technology Division, Washington, DC, EPA Method 1625, Rev. C, June 1989 (contact W.A. Telliard, U.S. EPA, Office of Water Regulations and Standards, 401 M St., SW, Washington, DC, 20460. Phone: 202-382-7131).

Benzenethiol **EPA Method 1625**
CAS #108-98-5

TITLE Semivolatile Organic Compounds by Isotope Dilution GC/MS

MATRIX The compounds may be determined in waters, soils, and municipal sludges by this method.

METHOD SUMMARY This method is used to determine 176 semivolatile toxic organic pollutants associated with the CWA (as amended 1987); the RCRA (as amended 1986); the CERCLA (as amended 1986); and other compounds amenable

to extraction and analysis by capillary column gas chromatography-mass spectrometry (GC/MS).

Stable isotopically-labeled analogs of the compounds of interest are added to the sample. If the solids content is less than 1%, a 1-L sample is extracted at pH 12–13, then at pH <2 with methylene chloride using continuous extraction techniques.

If the solids content is 30% or less, the sample is diluted to 1% solids with reagent water, homogenized ultrasonically, and extracted at pH 12–13, then at pH <2 with methylene chloride using continuous extraction techniques. If the solids content is greater than 30%, the sample is extracted using ultrasonic techniques.

Each extract is dried over sodium sulfate, concentrated to a volume of 5 mL, cleaned up using GPC, if necessary, and concentrated. Extracts are concentrated to 1 mL if GPC is not performed, and to 0.5 mL if GPC is performed.

An internal standard is added to the extract, and a 1-mL aliquot of the extract is injected into the GC. The compounds are separated by GC and detected by a MS. The labeled compounds serve to correct the variability of the analytical technique.

INTERFERENCES Solvents, reagents, glassware, and other sample processing hardware may yield artifacts and/or elevated baselines causing misinterpretation of chromatograms and spectra. Materials used in the analysis must be demonstrated to be free from interferences under the conditions of analysis by running method blanks initially and with each sample lot (sample started through the extraction process on a given 8-h shift, to a maximum of 20). Specific selection of reagents and purification of solvents by distillation in all glass systems may be required. Glassware and, where possible, reagents are cleaned by solvent rinse and baking at 450°C for 1-h minimum. Interferences coextracted from samples will vary considerably from source to source, depending on the diversity of the site being sampled.

INSTRUMENTATION Major instrumentation includes a GC with a splitless or on-column injection port for capillary column, a MS with 70 eV electron impact ionization, and a data system to collect and record MS data, and process it. A K-D apparatus is used to concentrate extracts.

GC Column: 30 m × 0.25 mm I.D. 5% phenyl, 94% methyl, 1% vinyl silicone bonded phased fused silica capillary column.

PRECISION & ACCURACY The detection limits of the method are usually dependent on the level of interferences rather than instrumental limitations. The limits typify the minimum quantities that can be detected with no interferences present.

The minimum level (in µg/mL) was not listed. This is defined as a minimum level at which the analytical system shall give recognizable mass spectra (background corrected) and acceptable calibration points.

The MDL (in µg/kg) in low solids was not listed and in high solids was not listed; these were determined in digested sludge (low solids) and in filter cake or compost (high solids).

The labeled and native compound initial precision as standard deviation (in µg/L) was not listed.

The labeled and native compound initial accuracy as average recovery (in µg/L) was not listed.

SAMPLE COLLECTION, PRESERVATION & HANDLING
Collect samples in glass containers. Aqueous samples which flow freely are collected in refrigerated bottles using automatic sampling equipment. Solid samples are collected as grab samples using widemouth jars. Maintain samples at 0 to 4°C from the time of collection until extraction. If residual chlorine is present in aqueous samples, add 80 mg sodium thiosulfate/L of water. Begin sample extraction within 7 days of collection, and analyze all extracts within 40 days of extraction.

SAMPLE PREPARATION Samples containing 1% solids or less are extracted directly using continuous liquid-liquid extraction techniques. Samples containing 1 to 30% solids are diluted to the 1% level with reagent water and extracted using continuous liquid-liquid extraction techniques. Samples containing greater than 30% solids are extracted using ultrasonic techniques.

Base/neutral extraction — Adjust the pH of the waters in the extractors to 12–13 with 6 N NaOH. Extract with methylene chloride for 24–48 h.

Acid extraction — Adjust the pH of the waters in the extractors to 2 or less using 6 N sulfuric acid. Extract with methylene chloride for 24–48 h.

Ultrasonic extraction of high solids samples — Add anhydrous sodium sulfate to the sample and QC aliquot(s). Add acetone:methylene chloride (1:1) to the sample and mix thoroughly

Concentrate extracts using a K-D apparatus.

QUALITY CONTROL The analyst is permitted to modify this method to improve separations or lower the costs of measurements, provided all performance specifications are met. Analyses of blanks are required to demonstrate freedom from contamination. When results of spikes indicate atypical method performance for samples, the samples are diluted to bring method performance within acceptable limits.

For low solids (aqueous samples), extract, concentrate, and analyze two sets of four 1-L aliquots (8 aliquots total) of the precision and recovery standard. For high solids samples, two sets of four 30-g aliquots of the high solids reference matrix are used.

Spike all samples with labeled compounds to assess method performance. Compute percent recovery of the labeled compounds using the internal standard method. Compare the labeled compound recovery for each compound with the corresponding labeled compound recovery.

Reagent water and high solids reference matrix blanks are analyzed to demonstrate freedom from contamination. Extract and concentrate a 1-L reagent water blank or a high solids reference matrix blank with each sample's lot (samples started through the extraction process on the same 8-h shift, to a maximum of 20 samples).

Field replicates may be collected to determine the precision of the sampling technique, and spiked samples may be required

to determine the accuracy of the analysis when the internal standard method is used.

REFERENCE Semivolatile Organic Compounds by Isotope Dilution GC/MS. Office of Water Regulation and Standards, U.S. EPA Industrial Technology Division, Washington, DC, EPA Method 1625, Rev. C, June 1989 (contact W.A. Telliard, U.S. EPA, Office of Water Regulations and Standards, 401 M St., SW, Washington, DC, 20460. Phone: 202-382-7131).

Benzidine   EPA Method 1625
CAS #92-87-5

TITLE Semivolatile Organic Compounds by Isotope Dilution GC/MS

MATRIX The compounds may be determined in waters, soils, and municipal sludges by this method.

METHOD SUMMARY This method is used to determine 176 semivolatile toxic organic pollutants associated with the CWA (as amended 1987); the RCRA (as amended 1986); the CERCLA (as amended 1986); and other compounds amenable to extraction and analysis by capillary column gas chromatography-mass spectrometry (GC/MS).

Stable isotopically-labeled analogs of the compounds of interest are added to the sample. If the solids content is less than 1%, a 1-L sample is extracted at pH 12–13, then at pH <2 with methylene chloride using continuous extraction techniques.

If the solids content is 30% or less, the sample is diluted to 1% solids with reagent water, homogenized ultrasonically, and extracted at pH 12–13, then at pH <2 with methylene chloride using continuous extraction techniques. If the solids content is greater than 30%, the sample is extracted using ultrasonic techniques.

Each extract is dried over sodium sulfate, concentrated to a volume of 5 mL, cleaned up using GPC, if necessary, and concentrated. Extracts are concentrated to 1 mL if GPC is not performed, and to 0.5 mL if GPC is performed.

An internal standard is added to the extract, and a 1-mL aliquot of the extract is injected into the GC. The compounds are separated by GC and detected by a MS. The labeled compounds serve to correct the variability of the analytical technique.

INTERFERENCES Solvents, reagents, glassware, and other sample processing hardware may yield artifacts and/or elevated baselines causing misinterpretation of chromatograms and spectra. Materials used in the analysis must be demonstrated to be free from interferences under the conditions of analysis by running method blanks initially and with each sample lot (sample started through the extraction process on a given 8-h shift, to a maximum of 20). Specific selection of reagents and purification of solvents by distillation in all glass systems may be required. Glassware and, where possible, reagents are cleaned by solvent rinse and baking at 450°C for 1-h minimum. Interferences coextracted from samples will vary considerably from source to source, depending on the diversity of the site being sampled.

INSTRUMENTATION Major instrumentation includes a GC with a splitless or on-column injection port for capillary column, a MS with 70 eV electron impact ionization, and a data system to collect and record MS data, and process it. A K-D apparatus is used to concentrate extracts.

GC Column: 30 m × 0.25 mm I.D. 5% phenyl, 94% methyl, 1% vinyl silicone bonded phased fused silica capillary column.

PRECISION & ACCURACY The detection limits of the method are usually dependent on the level of interferences rather than instrumental limitations. The limits typify the minimum quantities that can be detected with no interferences present.

The minimum level (in µg/mL) was 50. This is defined as a minimum level at which the analytical system shall give recognizable mass spectra (background corrected) and acceptable calibration points.
The MDL (in µg/kg) in low solids was not listed and in high solids was not listed; these were determined in digested sludge (low solids) and in filter cake or compost (high solids).
The labeled and native compound initial precision as standard deviation (in µg/L) was 119.
The labeled and native compound initial accuracy as average recovery (in µg/L) was 16–518.

SAMPLE COLLECTION, PRESERVATION & HANDLING Collect samples in glass containers. Aqueous samples which flow freely are collected in refrigerated bottles using automatic sampling equipment. Solid samples are collected as grab samples using widemouth jars. Maintain samples at 0 to 4°C from the time of collection until extraction. If residual chlorine is present in aqueous samples, add 80 mg sodium thiosulfate/L of water. Begin sample extraction within 7 days of collection, and analyze all extracts within 40 days of extraction.

SAMPLE PREPARATION Samples containing 1% solids or less are extracted directly using continuous liquid-liquid extraction techniques. Samples containing 1 to 30% solids are diluted to the 1% level with reagent water and extracted using continuous liquid-liquid extraction techniques. Samples containing greater than 30% solids are extracted using ultrasonic techniques.

Base/neutral extraction — Adjust the pH of the waters in the extractors to 12–13 with 6 *N* NaOH. Extract with methylene chloride for 24–48 h.
Acid extraction — Adjust the pH of the waters in the extractors to 2 or less using 6 *N* sulfuric acid. Extract with methylene chloride for 24–48 h.
Ultrasonic extraction of high solids samples — Add anhydrous sodium sulfate to the sample and QC aliquot(s). Add acetone:methylene chloride (1:1) to the sample and mix thoroughly

Concentrate extracts using a K-D apparatus.

QUALITY CONTROL The analyst is permitted to modify this method to improve separations or lower the costs of measurements, provided all performance specifications are met. Analyses of blanks are required to demonstrate freedom from contamination. When results of spikes indicate atypical

method performance for samples, the samples are diluted to bring method performance within acceptable limits.

For low solids (aqueous samples), extract, concentrate, and analyze two sets of four 1-L aliquots (8 aliquots total) of the precision and recovery standard. For high solids samples, two sets of four 30-g aliquots of the high solids reference matrix are used.

Spike all samples with labeled compounds to assess method performance. Compute percent recovery of the labeled compounds using the internal standard method. Compare the labeled compound recovery for each compound with the corresponding labeled compound recovery.

Reagent water and high solids reference matrix blanks are analyzed to demonstrate freedom from contamination. Extract and concentrate a 1-L reagent water blank or a high solids reference matrix blank with each sample's lot (samples started through the extraction process on the same 8-h shift, to a maximum of 20 samples).

Field replicates may be collected to determine the precision of the sampling technique, and spiked samples may be required to determine the accuracy of the analysis when the internal standard method is used.

REFERENCE Semivolatile Organic Compounds by Isotope Dilution GC/MS. Office of Water Regulation and Standards, U.S. EPA Industrial Technology Division, Washington, DC, EPA Method 1625, Rev. C, June 1989 (contact W.A. Telliard, U.S. EPA, Office of Water Regulations and Standards, 401 M St., SW, Washington, DC, 20460. Phone: 202-382-7131).

Benzidine   **EPA Method 625**
CAS #92-87-5

TITLE Base/Neutrals and Acids, U.S. EPA Method 625

MATRIX This methods covers municipal and industrial wastewaters.

METHOD SUMMARY Approximately 1 L of sample is serially extracted with methylene chloride at a pH greater than 11 and again at a pH less than 2 using a separatory funnel or a continuous extractor. The methylene chloride extract is dried, concentrated to a volume of 1 mL, and analyzed by GC/MS. Qualitative identification of the parameters in the extract is performed using the retention time and the relative abundance of three characteristic masses (m/z). Qualitative analysis is performed using either external or internal standard techniques with a single characteristic m/z.

INTERFERENCES Method interferences may be caused by contaminants in solvents, reagents, glassware, and other sample processing hardware. Glassware must be scrupulously cleaned. Glassware should be heated in a muffle furnace at 400°C for 5 to 30 min. Some thermally stable materials, such as PCBs, may not be eliminated by this treatment. Solvent rinses with acetone and pesticide quality hexane may be substituted for the muffle furnace heating. Matrix interferences may be caused by contaminants that are coextracted from the sample. The base-neutral extraction may cause significantly reduced recovery of phenols. The packed gas chromatographic columns recommended for the basic fraction may not exhibit sufficient resolution for some analytes.

INSTRUMENTATION A GC/MS system with an injection port designed for on-column injection when using packed columns and for splitless injection when using capillary columns.

Column for base/neutrals: 1.8 m long × 2 mm I.D. glass, packed with 3% SP-2550 on Supelcoport (100/120 mesh) or equivalent.

Column for acids: 1.8 m long × 2 mm I.D. glass, packed with 1% SP-1240DA on Supelcoport (100/120 mesh) or equivalent.

PRECISION & ACCURACY The MDL concentrations were obtained using reagent water. The MDL actually achieved in a given analysis will vary depending on instrument sensitivity and matrix effects. This method was tested by 15 laboratories using reagent water, drinking water, surface water, and industrial wastewaters spiked at six concentrations over the range 5 to 100 µg/L. Single operator precision, overall precision, and method accuracy were found to be directly related to the concentration of the parameter matrix.

The MDL (in µg/L) in reagent water was 44.
The standard deviation (in µg/L based on 4 recovery measurements) was not reported.
The range (in µg/L) for average recovery for 4 measurements was not reported.
The range (in %) for percent recovery was not reported.
Accuracy (in µg/L) as expected recovery for one or more measurements of a sample containing a true concentration of C was not reported.
Precision (in µg/L) as expected single analyst standard deviation of measurements at an average concentration found at X was not reported.
Overall precision (in µg/L) as expected interlaboratory standard deviation of measurements in an average concentration found at X was not reported.

C = *True value of the concentration in µg/L.*
X = *Average recovery found for measurements of samples containing a concentration at C in µg/L.*

SAMPLE PREPARATION Adjust the pH to >11 with sodium hydroxide and serially extract in a separatory funnel with methylene chloride or else in a continuous extractor. Next, adjust the pH to <2 with sulfuric acid and serially extract in a separatory funnel with methylene chloride or else in a continuous extractor. Dry the extracts separately through a column of anhydrous sodium sulfate and then concentrate each of the extracts to 1.0 mL using a K-D apparatus.

SAMPLE COLLECTION, PRESERVATION & HANDLING Grab samples must be collected in glass containers. All samples must be refrigerated at 4°C from the time of collection until extraction. If residual chlorine is present, add 80 mg of sodium thiosulfate/L of sample and mix well. All samples must be extracted within 7 days of collection and completely analyzed within 40 days of extraction.

QUALITY CONTROL Make an initial, one-time, demonstration of the ability to generate acceptable accuracy and precision with this method. Before processing any samples, the analyst must analyze a reagent water blank to demonstrate that interferences from the analytical system and glassware are under control. Each time a set of samples is extracted or reagents are changed, a reagent water blank must be processed. Spike and analyze a minimum of 5% of all samples to monitor and evaluate lab data quality. A QC check sample concentrate that contains each parameter of interest at a concentration of 100 µg/mL in acetone is required. PCBs and multicomponent pesticides may be omitted from this test.

After analysis of five spiked wastewater samples, calculate the average percent recovery and the standard deviation of the percent recovery. Spike all samples with the surrogate standard spiking solution and calculate the percent recovery of each surrogate compound.

REFERENCE *Federal Register*, Vol. 49, No. 209. Friday, Oct. 26, 1984.

Benzidine   **EPA Method 8270**
CAS #92-87-5

TITLE Semivolatile Organic Compounds by GC/MS

MATRIX This method is used to determine the concentration of semivolatile organic compounds in extracts prepared from all types of solid waste matrices, soils, and groundwater. Although surface waters are not specifically mentioned, this method should be applicable to water samples from rivers, lakes, etc.

METHOD SUMMARY This method covers 259 semivolatile organic compounds. In very limited applications direct injection of the sample into the GC/MS system may be appropriate, but this results in very high detection limits (approximately 10,000 µg/L). Typically, a 1-L liquid sample, containing surrogate, and matrix spiking standards, is extracted in a continuous extractor first under acid conditions and then under basic conditions. Typically 30 g of a solid sample, containing surrogate, and matrix spiking standards, is extracted ultrasonically. After concentrating the extract to 1 mL it is spiked with 10 µL of an internal standard solution just prior to analysis by GC/MS. The volume injected should contain about 100 ng of base/neutral and 200 ng of acid surrogates (for a 1-µL injection). Analysis is performed by GC/MS using a capillary GC column.

INTERFERENCES Raw GC/MS data from all blanks, samples, and spikes must be evaluated for interferences. Contamination by carryover can occur whenever high-concentration and low-concentration samples are sequentially analyzed. To reduce carryover, the sample syringe must be rinsed out between samples with solvent. Whenever an unusually concentrated sample is encountered, it should be followed by the analysis of blank solvent to check for cross-contamination.

INSTRUMENTATION A GC/MS and a data system are required. The GC column used is a 30 m × 0.25 mm I.D. (or 0.32 mm I.D.) 1-µm film thickness silicone-coated fused silica capillary column. A continuous liquid-liquid extractor equipped with Teflon® or glass connection joints and stopcocks requiring no lubrication, a K-D concentrating apparatus, water bath, and an ultrasonic disrupter with a minimum power of 300 W and with pulsing capability are also required.

PRECISION & ACCURACY The estimated quantitation limit (EQL) of Method 8270B for determining an individual compound is approximately 1 mg/kg (wet weight) for soil or sediment samples, 1–200 mg/kg for wastes (dependent on matrix and method of preparation), and 10 µg/L for groundwater samples. EQLs will be proportionately higher for sample extracts that require dilution to avoid saturation of the detector.

The EQL(b) for groundwater in µg/L is not listed.
The EQL (a, b) for low concentrations in soil and sediment in µg/kg is not listed.
Accuracy as µg/L is not listed.
Overall precision in µg/L is not listed.

(a) EQLs listed for soil/sediment are based on wet weight. Normally data is reported in a dry-weight basis; therefore, EQLs will be higher based on the % dry weight of each sample. This calculation is based on a 30-g sample and gel permeation chromatography cleanup.
(b) Sample EQLs are highly matrix-dependent. The EQLs are provided for guidance and may not always be achievable.
C = True value for concentration, in µg/L.
X = Average recovery found for measurements of samples containing a concentration of C, in µg/L.

ESTIMATED QUANTITATION LIMIT

Other Matrices	Factor (a)
High-concentration soil and sludges by sonicator	7.5
Non-water miscible waste	75

(a) EQL for other matrices = [EQL for low soil/sediment] × [Factor]. This estimated EQL is similar to an EPA "Practical Quantitation Limit."

SAMPLING METHOD
Liquid samples — Use a 1 or 2½ gallon amber glass bottle with a screw-top Teflon®-lined cover that has been prewashed with detergent and rinsed with distilled water and methanol (or isopropanol).

Soils, sediments, or sludges — Use an 8-oz. widemouth glass with a screw-top Teflon®-lined cover that has been prewashed with detergent and rinsed with distilled water and methanol (or isopropanol).

SAMPLE PRESERVATION
Liquid samples — If residual chlorine is present, add 3 mL of 10% sodium thiosulfate per gallon, cool to 4°C and store in a solvent-free refrigerator until analysis; if chlorine is not present, then eliminate the sodium thiosulfate addition.

Soils, sediments, or sludges — Cool samples to 4°C and store in a solvent-free refrigerator.

MHT Liquid samples must be extracted within 7 days and the extracts analyzed within 40 days. Soils, sediments, or sludges may

be stored for a maximum of 14 days and the extracts analyzed within 40 days.

SAMPLE PREPARATION

Liquid samples — Transfer 1 L quantitatively to a continuous extractor. If high concentrations are anticipated, a smaller volume may be used and then diluted with organic-free reagent water to 1 L. Adjust pH, if necessary, to pH <2 using 1:1 (V/V) sulfuric acid. Pipette 1.0 mL of a surrogate standard spiking solution into each sample. For the sample in each analytical batch selected for spiking, add 1.0 mL of a matrix spiking standard. For base/neutral acid analysis, the amount of the surrogates and matrix spiking compounds added to the sample should result in a final concentration of 100 ng/µL of each analyte in the extract to be analyzed (assuming a 1-µL injection). Extract with methylene chloride for 18–24 h. Next, adjust the pH of the aqueous phase to pH >11 using 10 N sodium hydroxide and extract it with methylene chloride again for 18–24 h. Dry the extract through a column containing anhydrous sodium sulfate and concentrate it to 1 mL using a K-D concentrator.

Soils, sediments, or sludges — Use 30 g of sample. Nonporous or wet samples (gummy or clay type) that do not have a free-flowing sandy texture must be mixed with anhydrous sodium sulfate until the sample is free flowing. Add 1 mL of surrogate standards to all samples, spikes, standards, and blanks. For the sample in each analytical batch selected for spiking, add 1.0 mL of a matrix spiking standard. For base/neutral acid analysis, the amount added of the surrogates and matrix spiking compounds should result in a final concentration of 100 ng/µL of each base/neutral analyte and 200 ng/µL of each acid analyte in the extract to be analyzed (assuming a 1-µL injection). Immediately add a 100 mL mixture of 1:1 methylene chloride:acetone and extract the sample ultrasonically for 3 min and then decant or filter the extracts. Repeat the extraction two or more times. Dry the extract using a column with anhydrous sodium sulfate and concentrate it to 1 mL in a K-D concentrator.

Note: Benzidine can be subject to oxidative losses during solvent concentration and chromatographic separation of it may be poor.

QUALITY CONTROL A methylene chloride solution containing 50 ng/µL of decafluorotriphenylphosphine (DFTPP) is used for tuning the GC/MS system each 12-h shift. A system performance check also must be made during every 12-h shift. A standard containing 50 ng/µL each of 4,4′-DDT, pentachlorophenol, and benzidine is required to verify injection port inertness and GC column performance. A calibration standard at mid-concentration, containing each compound of interest, including all required surrogates, must be performed every 12 h during analysis. After the system performance check is met, calibration check compounds (CCCs) are used to check the validity of the initial calibration.

The internal standard responses and retention times in the calibration check standard must be evaluated immediately after or during data acquisition. If the retention time for any internal standard changes by more than 30 seconds from the last check calibration (12 h), the chromatographic system must be inspected for malfunctions and corrections must be made, as required. If the electron ionization current plot (EICP) area for any of the internal standards changes by a factor of two from the last daily calibration standard check, the mass spectrometer must be inspected for malfunctions and corrections must be made, as appropriate.

Demonstrate, through the analysis of a reagent water blank, that interferences from the analytical system, glassware, and reagents are under control. The blank samples should be carried through all stages of the sample preparation and measurement steps. For each analytical batch (up to 20 samples), a reagent blank, matrix spike, and matrix spike duplicate/duplicate must be analyzed (the frequency of the spikes may be different for different monitoring programs). The blank and spiked samples must be carried through all stages of the sample preparation and measurement steps. A QC reference sample concentrate containing each analyte at a concentration of 100 mg/L in methanol is required.

REFERENCE Test Methods for Evaluating Solid Waste (SW-846). U.S. EPA 1983, Method 8270B, Rev. 2, Nov. 1990. Office of Solid Waste, Washington, DC.

Benz(a)anthracene   EPA Method 1625
CAS #56-55-3

TITLE Semivolatile Organic Compounds by Isotope Dilution GC/MS

MATRIX The compounds may be determined in waters, soils, and municipal sludges by this method.

METHOD SUMMARY This method is used to determine 176 semivolatile toxic organic pollutants associated with the CWA (as amended 1987); the RCRA (as amended 1986); the CERCLA (as amended 1986); and other compounds amenable to extraction and analysis by capillary column gas chromatography-mass spectrometry (GC/MS).

Stable isotopically-labeled analogs of the compounds of interest are added to the sample. If the solids content is less than 1%, a 1-L sample is extracted at pH 12–13, then at pH <2 with methylene chloride using continuous extraction techniques.

If the solids content is 30% or less, the sample is diluted to 1% solids with reagent water, homogenized ultrasonically, and extracted at pH 12–13, then at pH <2 with methylene chloride using continuous extraction techniques. If the solids content is greater than 30%, the sample is extracted using ultrasonic techniques.

Each extract is dried over sodium sulfate, concentrated to a volume of 5 mL, cleaned up using GPC, if necessary, and concentrated. Extracts are concentrated to 1 mL if GPC is not performed, and to 0.5 mL if GPC is performed.

An internal standard is added to the extract, and a 1-mL aliquot of the extract is injected into the GC. The compounds are separated by GC and detected by a MS. The labeled compounds serve to correct the variability of the analytical technique.

INTERFERENCES Solvents, reagents, glassware, and other sample processing hardware may yield artifacts and/or elevated baselines causing misinterpretation of chromatograms and spectra. Materials used in the analysis must be demonstrated to be free from interferences under the conditions of analysis by running method blanks initially and with each sample lot (sample started through the extraction process on a given 8-h shift, to a maximum of 20). Specific selection of reagents and purification of solvents by distillation in all glass systems may be required. Glassware and, where possible, reagents are cleaned by solvent rinse and baking at 450°C for 1-h minimum. Interferences coextracted from samples will vary considerably from source to source, depending on the diversity of the site being sampled.

INSTRUMENTATION Major instrumentation includes a GC with a splitless or on-column injection port for capillary column, a MS with 70 eV electron impact ionization, and a data system to collect and record MS data, and process it. A K-D apparatus is used to concentrate extracts.

GC Column: 30 m × 0.25 mm I.D. 5% phenyl, 94% methyl, 1% vinyl silicone bonded phased fused silica capillary column.

PRECISION & ACCURACY The detection limits of the method are usually dependent on the level of interferences rather than instrumental limitations. The limits typify the minimum quantities that can be detected with no interferences present.

The minimum level (in µg/mL) was 10. This is defined as a minimum level at which the analytical system shall give recognizable mass spectra (background corrected) and acceptable calibration points.

The MDL (in µg/kg) in low solids was 61 and in high solids was 47; these were determined in digested sludge (low solids) and in filter cake or compost (high solids).

The labeled and native compound initial precision as standard deviation (in µg/L) was 20.

The labeled and native compound initial accuracy as average recovery (in µg/L) was 65–168.

SAMPLE COLLECTION, PRESERVATION & HANDLING Collect samples in glass containers. Aqueous samples which flow freely are collected in refrigerated bottles using automatic sampling equipment. Solid samples are collected as grab samples using widemouth jars. Maintain samples at 0 to 4°C from the time of collection until extraction. If residual chlorine is present in aqueous samples, add 80 mg sodium thiosulfate/L of water. Begin sample extraction within 7 days of collection, and analyze all extracts within 40 days of extraction.

SAMPLE PREPARATION Samples containing 1% solids or less are extracted directly using continuous liquid-liquid extraction techniques. Samples containing 1 to 30% solids are diluted to the 1% level with reagent water and extracted using continuous liquid-liquid extraction techniques. Samples containing greater than 30% solids are extracted using ultrasonic techniques.

Base/neutral extraction — Adjust the pH of the waters in the extractors to 12–13 with 6 *N* NaOH. Extract with methylene chloride for 24–48 h.

Acid extraction — Adjust the pH of the waters in the extractors to 2 or less using 6 *N* sulfuric acid. Extract with methylene chloride for 24–48 h.

Ultrasonic extraction of high solids samples — Add anhydrous sodium sulfate to the sample and QC aliquot(s). Add acetone:methylene chloride (1:1) to the sample and mix thoroughly

Concentrate extracts using a K-D apparatus.

QUALITY CONTROL The analyst is permitted to modify this method to improve separations or lower the costs of measurements, provided all performance specifications are met. Analyses of blanks are required to demonstrate freedom from contamination. When results of spikes indicate atypical method performance for samples, the samples are diluted to bring method performance within acceptable limits.

For low solids (aqueous samples), extract, concentrate, and analyze two sets of four 1-L aliquots (8 aliquots total) of the precision and recovery standard. For high solids samples, two sets of four 30-g aliquots of the high solids reference matrix are used.

Spike all samples with labeled compounds to assess method performance. Compute percent recovery of the labeled compounds using the internal standard method. Compare the labeled compound recovery for each compound with the corresponding labeled compound recovery.

Reagent water and high solids reference matrix blanks are analyzed to demonstrate freedom from contamination. Extract and concentrate a 1-L reagent water blank or a high solids reference matrix blank with each sample's lot (samples started through the extraction process on the same 8-h shift, to a maximum of 20 samples).

Field replicates may be collected to determine the precision of the sampling technique, and spiked samples may be required to determine the accuracy of the analysis when the internal standard method is used.

REFERENCE Semivolatile Organic Compounds by Isotope Dilution GC/MS. Office of Water Regulation and Standards, U.S. EPA Industrial Technology Division, Washington, DC, EPA Method 1625, Rev. C, June 1989 (contact W.A. Telliard, U.S. EPA, Office of Water Regulations and Standards, 401 M St., SW, Washington, DC, 20460. Phone: 202-382-7131).

Benz(a)anthracene   **EPA Method 625**
CAS #56-55-3

TITLE Base/Neutrals and Acids, U.S. EPA Method 625

MATRIX This methods covers municipal and industrial wastewaters.

METHOD SUMMARY Approximately 1 L of sample is serially extracted with methylene chloride at a pH greater than 11 and again at a pH less than 2 using a separatory funnel or a continuous extractor. The methylene chloride extract is dried, concentrated to a volume of 1 mL, and analyzed by GC/MS.

Qualitative identification of the parameters in the extract is performed using the retention time and the relative abundance of three characteristic masses (m/z). Qualitative analysis is performed using either external or internal standard techniques with a single characteristic m/z.

INTERFERENCES Method interferences may be caused by contaminants in solvents, reagents, glassware, and other sample processing hardware. Glassware must be scrupulously cleaned. Glassware should be heated in a muffle furnace at 400°C for 5 to 30 min. Some thermally stable materials, such as PCBs, may not be eliminated by this treatment. Solvent rinses with acetone and pesticide quality hexane may be substituted for the muffle furnace heating. Matrix interferences may be caused by contaminants that are coextracted from the sample. The base-neutral extraction may cause significantly reduced recovery of phenols. The packed gas chromatographic columns recommended for the basic fraction may not exhibit sufficient resolution for some analytes.

INSTRUMENTATION A GC/MS system with an injection port designed for on-column injection when using packed columns and for splitless injection when using capillary columns.

Column for base/neutrals: 1.8 m long × 2 mm I.D. glass, packed with 3% SP-2550 on Supelcoport (100/120 mesh) or equivalent.
Column for acids: 1.8 m long × 2 mm I.D. glass, packed with 1% SP-1240DA on Supelcoport (100/120 mesh) or equivalent.

PRECISION & ACCURACY The MDL concentrations were obtained using reagent water. The MDL actually achieved in a given analysis will vary depending on instrument sensitivity and matrix effects. This method was tested by 15 laboratories using reagent water, drinking water, surface water, and industrial wastewaters spiked at six concentrations over the range 5 to 100 µg/L. Single operator precision, overall precision, and method accuracy were found to be directly related to the concentration of the parameter matrix.

The MDL (in µg/L) in reagent water was 7.8.
The standard deviation (in µg/L based on 4 recovery measurements) was 27.6.
The range (in µg/L) for average recovery for 4 measurements was 41.8–133.0.
The range (in %) for percent recovery was 33–143.
Accuracy (in µg/L) as expected recovery for one or more measurements of a sample containing a true concentration of C was 0.88C–0.60.
Precision (in µg/L) as expected single analyst standard deviation of measurements at an average concentration found at X was 0.15+0.93.
Overall precision (in µg/L) as expected interlaboratory standard deviation of measurements in an average concentration found at X was 0.26X–0.28.

C = True value of the concentration in µg/L.
X = Average recovery found for measurements of samples containing a concentration at C in µg/L.

SAMPLE PREPARATION Adjust the pH to >11 with sodium hydroxide and serially extract in a separatory funnel with methylene chloride or else in a continuous extractor. Next, adjust the pH to <2 with sulfuric acid and serially extract in a separatory funnel with methylene chloride or else in a continuous extractor. Dry the extracts separately through a column of anhydrous sodium sulfate and then concentrate each of the extracts to 1.0 mL using a K-D apparatus.

SAMPLE COLLECTION, PRESERVATION & HANDLING Grab samples must be collected in glass containers. All samples must be refrigerated at 4°C from the time of collection until extraction. If residual chlorine is present, add 80 mg of sodium thiosulfate/L of sample and mix well. All samples must be extracted within 7 days of collection and completely analyzed within 40 days of extraction.

QUALITY CONTROL Make an initial, one-time, demonstration of the ability to generate acceptable accuracy and precision with this method. Before processing any samples, the analyst must analyze a reagent water blank to demonstrate that interferences from the analytical system and glassware are under control. Each time a set of samples is extracted or reagents are changed, a reagent water blank must be processed. Spike and analyze a minimum of 5% of all samples to monitor and evaluate lab data quality. A QC check sample concentrate that contains each parameter of interest at a concentration of 100 µg/mL in acetone is required. PCBs and multicomponent pesticides may be omitted from this test.

After analysis of five spiked wastewater samples, calculate the average percent recovery and the standard deviation of the percent recovery. Spike all samples with the surrogate standard spiking solution and calculate the percent recovery of each surrogate compound.

REFERENCE Federal Register, Vol. 49, No. 209. Friday, Oct. 26, 1984.

Benz(a)anthracene   **EPA Method 8270**
CAS #56-55-3

TITLE Semivolatile Organic Compounds by GC/MS

MATRIX This method is used to determine the concentration of semivolatile organic compounds in extracts prepared from all types of solid waste matrices, soils, and groundwater. Although surface waters are not specifically mentioned, this method should be applicable to water samples from rivers, lakes, etc.

METHOD SUMMARY This method covers 259 semivolatile organic compounds. In very limited applications direct injection of the sample into the GC/MS system may be appropriate, but this results in very high detection limits (approximately 10,000 µg/L). Typically, a 1-L liquid sample, containing surrogate, and matrix spiking standards, is extracted in a continuous extractor first under acid conditions and then under basic conditions. Typically 30 g of a solid sample, containing surrogate, and matrix spiking standards, is extracted ultrasonically. After concentrating the extract to 1 mL it is spiked with 10 µL of an internal standard solution just prior to analysis by GC/MS. The volume injected should contain about 100 ng of base/neutral

and 200 ng of acid surrogates (for a 1-μL injection). Analysis is performed by GC/MS using a capillary GC column.

INTERFERENCES Raw GC/MS data from all blanks, samples, and spikes must be evaluated for interferences. Contamination by carryover can occur whenever high-concentration and low-concentration samples are sequentially analyzed. To reduce carryover, the sample syringe must be rinsed out between samples with solvent. Whenever an unusually concentrated sample is encountered, it should be followed by the analysis of blank solvent to check for cross-contamination.

INSTRUMENTATION A GC/MS and a data system are required. The GC column used is a 30 m × 0.25 mm I.D. (or 0.32 mm I.D.) 1-μm film thickness silicone-coated fused silica capillary column. A continuous liquid-liquid extractor equipped with Teflon® or glass connection joints and stopcocks requiring no lubrication, a K-D concentrating apparatus, water bath, and an ultrasonic disrupter with a minimum power of 300 W and with pulsing capability are also required.

PRECISION & ACCURACY The estimated quantitation limit (EQL) of Method 8270B for determining an individual compound is approximately 1 mg/kg (wet weight) for soil or sediment samples, 1–200 mg/kg for wastes (dependent on matrix and method of preparation), and 10 μg/L for groundwater samples. EQLs will be proportionately higher for sample extracts that require dilution to avoid saturation of the detector.

The EQL(b) for groundwater in μg/L is 10.
The EQL (a, b) for low concentrations in soil and sediment in μg/kg is 660.
Accuracy as μg/L is $0.88C-0.60$.
Overall precision in μg/L is $0.26X-0.21$.

(a) EQLs listed for soil/sediment are based on wet weight. Normally data is reported in a dry-weight basis; therefore, EQLs will be higher based on the % dry weight of each sample. This calculation is based on a 30-g sample and gel permeation chromatography cleanup.
(b) Sample EQLs are highly matrix-dependent. The EQLs are provided for guidance and may not always be achievable.
C = *True value for concentration, in μg/L.*
X = *Average recovery found for measurements of samples containing a concentration of C, in μg/L.*

ESTIMATED QUANTITATION LIMIT

Other Matrices	Factor (a)
High-concentration soil and sludges by sonicator	7.5
Non-water miscible waste	75

(a) EQL for other matrices = [EQL for low soil/sediment] × [Factor]. This estimated EQL is similar to an EPA "Practical Quantitation Limit."

SAMPLING METHOD
Liquid samples — Use a 1 or 2½ gallon amber glass bottle with a screw-top Teflon®-lined cover that has been prewashed with detergent and rinsed with distilled water and methanol (or isopropanol).

Soils, sediments, or sludges — Use an 8-oz. widemouth glass with a screw-top Teflon®-lined cover that has been prewashed with detergent and rinsed with distilled water and methanol (or isopropanol).

SAMPLE PRESERVATION
Liquid samples — If residual chlorine is present, add 3 mL of 10% sodium thiosulfate per gallon, cool to 4°C and store in a solvent-free refrigerator until analysis; if chlorine is not present, then eliminate the sodium thiosulfate addition.

Soils, sediments, or sludges — Cool samples to 4°C and store in a solvent-free refrigerator.

MHT Liquid samples must be extracted within 7 days and the extracts analyzed within 40 days. Soils, sediments, or sludges may be stored for a maximum of 14 days and the extracts analyzed within 40 days.

SAMPLE PREPARATION
Liquid samples — Transfer 1 L quantitatively to a continuous extractor. If high concentrations are anticipated, a smaller volume may be used and then diluted with organic-free reagent water to 1 L. Adjust pH, if necessary, to pH <2 using 1:1 (V/V) sulfuric acid. Pipette 1.0 mL of a surrogate standard spiking solution into each sample. For the sample in each analytical batch selected for spiking, add 1.0 mL of a matrix spiking standard. For base/neutral acid analysis, the amount of the surrogates and matrix spiking compounds added to the sample should result in a final concentration of 100 ng/μL of each analyte in the extract to be analyzed (assuming a 1-μL injection). Extract with methylene chloride for 18–24 h. Next, adjust the pH of the aqueous phase to pH >11 using 10 N sodium hydroxide and extract it with methylene chloride again for 18–24 h. Dry the extract through a column containing anhydrous sodium sulfate and concentrate it to 1 mL using a K-D concentrator.

Soils, sediments, or sludges — Use 30 g of sample. Nonporous or wet samples (gummy or clay type) that do not have a free-flowing sandy texture must be mixed with anhydrous sodium sulfate until the sample is free flowing. Add 1 mL of surrogate standards to all samples, spikes, standards, and blanks. For the sample in each analytical batch selected for spiking, add 1.0 mL of a matrix spiking standard. For base/neutral acid analysis, the amount added of the surrogates and matrix spiking compounds should result in a final concentration of 100 ng/μL of each base/neutral analyte and 200 ng/μL of each acid analyte in the extract to be analyzed (assuming a 1-μL injection). Immediately add a 100 mL mixture of 1:1 methylene chloride:acetone and extract the sample ultrasonically for 3 min and then decant or filter the extracts. Repeat the extraction two or more times. Dry the extract using a column with anhydrous sodium sulfate and concentrate it to 1 mL in a K-D concentrator.

QUALITY CONTROL A methylene chloride solution containing 50 ng/μL of decafluorotriphenylphosphine (DFTPP) is used for tuning the GC/MS system each 12-h shift. A system performance check also must be made during every 12-h shift. A standard containing 50 ng/μL each of 4,4'-DDT, pentachlorophenol, and benzidine is required to verify injection port inertness and GC column performance. A calibration standard at mid-concentration, containing each compound of interest, including all required surrogates, must be performed every 12 h during analysis. After the system performance check is met, calibration check compounds (CCCs) are used to check the validity of the initial calibration.

The internal standard responses and retention times in the calibration check standard must be evaluated immediately after or during data acquisition. If the retention time for any internal standard changes by more than 30 seconds from the last check calibration (12 h), the chromatographic system must be inspected for malfunctions and corrections must be made, as required. If the electron ionization current plot (EICP) area for any of the internal standards changes by a factor of two from the last daily calibration standard check, the mass spectrometer must be inspected for malfunctions and corrections must be made, as appropriate.

Demonstrate, through the analysis of a reagent water blank, that interferences from the analytical system, glassware, and reagents are under control. The blank samples should be carried through all stages of the sample preparation and measurement steps. For each analytical batch (up to 20 samples), a reagent blank, matrix spike, and matrix spike duplicate/duplicate must be analyzed (the frequency of the spikes may be different for different monitoring programs). The blank and spiked samples must be carried through all stages of the sample preparation and measurement steps. A QC reference sample concentrate containing each analyte at a concentration of 100 mg/L in methanol is required.

REFERENCE Test Methods for Evaluating Solid Waste (SW-846). U.S. EPA 1983, Method 8270B, Rev. 2, Nov. 1990. Office of Solid Waste, Washington, DC.

Benz(a)anthracene EPA Method 8100
CAS #56-55-3

TITLE Polynuclear Aromatic Hydrocarbons

MATRIX Groundwater, soils, sludges, water miscible liquid wastes, and non-water miscible wastes.

APPLICATION This method is used for the analysis of various PAHs. Samples are extracted, concentrated, and analyzed using direct injection of both neat and diluted organic liquids. The method provides two optional GC columns that are better than Column 1 and that may help resolve analytes from interferences.

INTERFERENCES Solvents, reagents, and glassware may introduce artifacts. Other interferences may come from coextracted compounds from samples.

INSTRUMENTATION GC capable of on-column injections and a flame with detector (FID). Column 1: a 1.8 m by 2 mm 3% OV-17 on Chromosorb W-AW-DCMS column. Column 2: a 30 m by 0.25 mm SE-54 fused silica capillary column. Column 3: a 30 m by 0.32 mm SE-54 fused silica capillary column.

RANGE 0.1–425 µg/L

MDL Not reported.

PQL FACTORS FOR MULTIPLYING × FID MDL VALUE Not available.

PRECISION 0.34X + 0.02 µg/L (overall precision).

ACCURACY 0.73C + 0.05 µg/L (as recovery).

SAMPLING METHOD Use 8-oz. widemouth glass bottles with Teflon®-lined caps for concentrated waste samples, soils, sediments, and sludges. Use 1 or 2½ gallon amber glass bottles with Teflon®-lined caps for liquid (water) samples.

STABILITY Cool soil, sediment, sludge, and liquid samples to 4°C. If residual chlorine is present in liquid samples add 3 mL of 10% sodium thiosulfate per gallon of sample and cool to 4°C.

MHT 14 days for concentrated waste, soil, sediment, or sludge; 7 days for liquid samples; all extracts must be analyzed within 40 days.

QUALITY CONTROL A quality control check sample concentrate containing each analyte of interest is required. The QC check sample concentrate may be prepared from pure standard materials or purchased as certified solutions. Use appropriate trip, matrix, control site, method, reagent, and solvent blanks. Internal, surrogate, and five concentration level calibration standards are used. The quality control check sample concentrate should contain Benz(a)anthracene at 10 µg/mL in acetonitrile.

REFERENCE Method 8100, SW-846, 3rd ed., Nov. 1986.

Benz(a)anthracene EPA Method 8310
CAS #56-55-3

TITLE Polynuclear Aromatic Hydrocarbons

MATRIX Groundwater, soils, sludges, water miscible liquid wastesand non-water miscible wastes.

APPLICATION This method is used for the analysis of 16 polynuclear aromatic hydrocarbons (PAHs). Samples are extracted, concentrated, and analyzed using HPLC with detection by UV and fluorescence detectors.

INTERFERENCES Solvents, reagents, and glassware may introduce artifacts. Other interferences may come from coextracted compounds from samples.

INSTRUMENTATION HPLC with a gradient pumping system and a 250 mm by 2.6 mm reverse phase HC-ODS Sil-X 5-micron particle-size column. The fluorescence detector uses an excitation wavelength of 280 nm and emission greater than 389 nm cutoff with dispersive optics.

RANGE 0.1–425 µg/L

MDL 0.013 µg/L (fluorescence; reagent water).

PQL FACTORS FOR MULTIPLYING × FID MDL VALUE

Matrix	Multiplication Factor
Groundwater	10
Low-level soil by sonication with GPC cleanup	670
High-level soil and sludge by sonication	10,000
Non-water miscible waste	100,000

PRECISION $0.34X + 0.02$ µg/L (overall precision).

ACCURACY $0.73C + 0.05$ µg/L (as recovery).

SAMPLING METHOD Use 8-oz. widemouth glass bottles with Teflon®-lined caps for concentrated waste samples, soils, sediments, and sludges. Use 1 or 2½ gallon amber glass bottles with Teflon®-lined caps for liquid (water) samples.

STABILITY Cool soil, sediment, sludge, and liquid samples to 4°C. If residual chlorine is present in liquid samples add 3 mL of 10% sodium thiosulfate per gallon of sample and cool to 4°C.

MHT 14 days for concentrated waste, soil, sediment, or sludge; 7 days for liquid samples; all extracts must be analyzed within 40 days.

QUALITY CONTROL Internal, surrogate, and five concentration level calibration standards are used. The calibration standards must be used with the analytical method blank. A quality control check sample concentrate containing benz(a)anthracene at 10 µg/mL is required. The QC check sample concentrate may be prepared from pure standard materials or purchased as certified solutions. Use appropriate trip, matrix, control site, method, reagent, and solvent blanks.

REFERENCE Method 8310, SW-846, 3rd ed., Nov. 1986.

Benzo(a)pyrene **EPA Method 1625**
CAS #50-32-8

TITLE Semivolatile Organic Compounds by Isotope Dilution GC/MS

MATRIX The compounds may be determined in waters, soils, and municipal sludges by this method.

METHOD SUMMARY This method is used to determine 176 semivolatile toxic organic pollutants associated with the CWA (as amended 1987); the RCRA (as amended 1986); the CERCLA (as amended 1986); and other compounds amenable to extraction and analysis by capillary column gas chromatography-mass spectrometry (GC/MS).

Stable isotopically-labeled analogs of the compounds of interest are added to the sample. If the solids content is less than 1%, a 1-L sample is extracted at pH 12–13, then at pH <2 with methylene chloride using continuous extraction techniques.

If the solids content is 30% or less, the sample is diluted to 1% solids with reagent water, homogenized ultrasonically, and extracted at pH 12–13, then at pH <2 with methylene chloride using continuous extraction techniques. If the solids content is greater than 30%, the sample is extracted using ultrasonic techniques.

Each extract is dried over sodium sulfate, concentrated to a volume of 5 mL, cleaned up using GPC, if necessary, and concentrated. Extracts are concentrated to 1 mL if GPC is not performed, and to 0.5 mL if GPC is performed.

An internal standard is added to the extract, and a 1-mL aliquot of the extract is injected into the GC. The compounds are separated by GC and detected by a MS. The labeled compounds serve to correct the variability of the analytical technique.

INTERFERENCES Solvents, reagents, glassware, and other sample processing hardware may yield artifacts and/or elevated baselines causing misinterpretation of chromatograms and spectra. Materials used in the analysis must be demonstrated to be free from interferences under the conditions of analysis by running method blanks initially and with each sample lot (sample started through the extraction process on a given 8-h shift, to a maximum of 20). Specific selection of reagents and purification of solvents by distillation in all glass systems may be required. Glassware and, where possible, reagents are cleaned by solvent rinse and baking at 450°C for 1-h minimum. Interferences coextracted from samples will vary considerably from source to source, depending on the diversity of the site being sampled.

INSTRUMENTATION Major instrumentation includes a GC with a splitless or on-column injection port for capillary column, a MS with 70 eV electron impact ionization, and a data system to collect and record MS data, and process it. A K-D apparatus is used to concentrate extracts.

GC Column: 30 m × 0.25 mm I.D. 5% phenyl, 94% methyl, 1% vinyl silicone bonded phased fused silica capillary column.

PRECISION & ACCURACY The detection limits of the method are usually dependent on the level of interferences rather than instrumental limitations. The limits typify the minimum quantities that can be detected with no interferences present.

- The minimum level (in µg/mL) was 10. This is defined as a minimum level at which the analytical system shall give recognizable mass spectra (background corrected) and acceptable calibration points.
- The MDL (in µg/kg) in low solids was 52 and in high solids was 15; these were determined in digested sludge (low solids) and in filter cake or compost (high solids).
- The labeled and native compound initial precision as standard deviation (in µg/L) was 26.
- The labeled and native compound initial accuracy as average recovery (in µg/L) was 62–195.

SAMPLE COLLECTION, PRESERVATION & HANDLING Collect samples in glass containers. Aqueous samples which flow freely are collected in refrigerated bottles using automatic sampling equipment. Solid samples are collected as grab samples using widemouth jars. Maintain samples at 0 to 4°C from the time of collection until extraction. If residual chlorine is present in aqueous samples, add 80 mg sodium thiosulfate/L

of water. Begin sample extraction within 7 days of collection, and analyze all extracts within 40 days of extraction.

SAMPLE PREPARATION Samples containing 1% solids or less are extracted directly using continuous liquid-liquid extraction techniques. Samples containing 1 to 30% solids are diluted to the 1% level with reagent water and extracted using continuous liquid-liquid extraction techniques. Samples containing greater than 30% solids are extracted using ultrasonic techniques.

Base/neutral extraction — Adjust the pH of the waters in the extractors to 12–13 with 6 N NaOH. Extract with methylene chloride for 24–48 h.

Acid extraction — Adjust the pH of the waters in the extractors to 2 or less using 6 N sulfuric acid. Extract with methylene chloride for 24–48 h.

Ultrasonic extraction of high solids samples — Add anhydrous sodium sulfate to the sample and QC aliquot(s). Add acetone:methylene chloride (1:1) to the sample and mix thoroughly

Concentrate extracts using a K-D apparatus.

QUALITY CONTROL The analyst is permitted to modify this method to improve separations or lower the costs of measurements, provided all performance specifications are met. Analyses of blanks are required to demonstrate freedom from contamination. When results of spikes indicate atypical method performance for samples, the samples are diluted to bring method performance within acceptable limits.

For low solids (aqueous samples), extract, concentrate, and analyze two sets of four 1-L aliquots (8 aliquots total) of the precision and recovery standard. For high solids samples, two sets of four 30-g aliquots of the high solids reference matrix are used.

Spike all samples with labeled compounds to assess method performance. Compute percent recovery of the labeled compounds using the internal standard method. Compare the labeled compound recovery for each compound with the corresponding labeled compound recovery.

Reagent water and high solids reference matrix blanks are analyzed to demonstrate freedom from contamination. Extract and concentrate a 1-L reagent water blank or a high solids reference matrix blank with each sample's lot (samples started through the extraction process on the same 8-h shift, to a maximum of 20 samples).

Field replicates may be collected to determine the precision of the sampling technique, and spiked samples may be required to determine the accuracy of the analysis when the internal standard method is used.

REFERENCE Semivolatile Organic Compounds by Isotope Dilution GC/MS. Office of Water Regulation and Standards, U.S. EPA Industrial Technology Division, Washington, DC, EPA Method 1625, Rev. C, June 1989 (contact W.A. Telliard, U.S. EPA, Office of Water Regulations and Standards, 401 M St., SW, Washington, DC, 20460. Phone: 202-382-7131).

Benzo(a)pyrene   EPA Method 625
CAS #50-32-8

TITLE Base/Neutrals and Acids, U.S. EPA Method 625

MATRIX This methods covers municipal and industrial wastewaters.

METHOD SUMMARY Approximately 1 L of sample is serially extracted with methylene chloride at a pH greater than 11 and again at a pH less than 2 using a separatory funnel or a continuous extractor. The methylene chloride extract is dried, concentrated to a volume of 1 mL, and analyzed by GC/MS. Qualitative identification of the parameters in the extract is performed using the retention time and the relative abundance of three characteristic masses (m/z). Qualitative analysis is performed using either external or internal standard techniques with a single characteristic m/z.

INTERFERENCES Method interferences may be caused by contaminants in solvents, reagents, glassware, and other sample processing hardware. Glassware must be scrupulously cleaned. Glassware should be heated in a muffle furnace at 400°C for 5 to 30 min. Some thermally stable materials, such as PCBs, may not be eliminated by this treatment. Solvent rinses with acetone and pesticide quality hexane may be substituted for the muffle furnace heating. Matrix interferences may be caused by contaminants that are coextracted from the sample. The base-neutral extraction may cause significantly reduced recovery of phenols. The packed gas chromatographic columns recommended for the basic fraction may not exhibit sufficient resolution for some analytes.

INSTRUMENTATION A GC/MS system with an injection port designed for on-column injection when using packed columns and for splitless injection when using capillary columns.

Column for base/neutrals: 1.8 m long × 2 mm I.D. glass, packed with 3% SP-2550 on Supelcoport (100/120 mesh) or equivalent.

Column for acids: 1.8 m long × 2 mm I.D. glass, packed with 1% SP-1240DA on Supelcoport (100/120 mesh) or equivalent.

PRECISION & ACCURACY The MDL concentrations were obtained using reagent water. The MDL actually achieved in a given analysis will vary depending on instrument sensitivity and matrix effects. This method was tested by 15 laboratories using reagent water, drinking water, surface water, and industrial wastewaters spiked at six concentrations over the range 5 to 100 µg/L. Single operator precision, overall precision, and method accuracy were found to be directly related to the concentration of the parameter matrix.

The MDL (in µg/L) in reagent water was 2.5.

The standard deviation (in µg/L based on 4 recovery measurements) was 39.0.

The range (in µg/L) for average recovery for 4 measurements was 31.7–148.0.

The range (in %) for percent recovery was 17–163.

Accuracy (in µg/L) as expected recovery for one or more measurements of a sample containing a true concentration of C was 0.90C–0.13.

Precision (in µg/L) as expected single analyst standard deviation of measurements at an average concentration found at X was 0.22X+0.48.

Overall precision (in µg/L) as expected interlaboratory standard deviation of measurements in an average concentration found at X was 0.32X+1.35.

C = *True value of the concentration in µg/L.*
X = *Average recovery found for measurements of samples containing a concentration at C in µg/L.*

SAMPLE PREPARATION Adjust the pH to >11 with sodium hydroxide and serially extract in a separatory funnel with methylene chloride or else in a continuous extractor. Next, adjust the pH to <2 with sulfuric acid and serially extract in a separatory funnel with methylene chloride or else in a continuous extractor. Dry the extracts separately through a column of anhydrous sodium sulfate and then concentrate each of the extracts to 1.0 mL using a K-D apparatus.

SAMPLE COLLECTION, PRESERVATION & HANDLING Grab samples must be collected in glass containers. All samples must be refrigerated at 4°C from the time of collection until extraction. If residual chlorine is present, add 80 mg of sodium thiosulfate/L of sample and mix well. All samples must be extracted within 7 days of collection and completely analyzed within 40 days of extraction.

QUALITY CONTROL Make an initial, one-time, demonstration of the ability to generate acceptable accuracy and precision with this method. Before processing any samples, the analyst must analyze a reagent water blank to demonstrate that interferences from the analytical system and glassware are under control. Each time a set of samples is extracted or reagents are changed, a reagent water blank must be processed. Spike and analyze a minimum of 5% of all samples to monitor and evaluate lab data quality. A QC check sample concentrate that contains each parameter of interest at a concentration of 100 µg/mL in acetone is required. PCBs and multicomponent pesticides may be omitted from this test.

After analysis of five spiked wastewater samples, calculate the average percent recovery and the standard deviation of the percent recovery. Spike all samples with the surrogate standard spiking solution and calculate the percent recovery of each surrogate compound.

REFERENCE *Federal Register*, Vol. 49, No. 209. Friday, Oct. 26, 1984.

Benzo(a)pyrene EPA Method 8270
CAS #50-32-8

TITLE Semivolatile Organic Compounds by GC/MS

MATRIX This method is used to determine the concentration of semivolatile organic compounds in extracts prepared from all types of solid waste matrices, soils, and groundwater. Although surface waters are not specifically mentioned, this method should be applicable to water samples from rivers, lakes, etc.

METHOD SUMMARY This method covers 259 semivolatile organic compounds. In very limited applications direct injection of the sample into the GC/MS system may be appropriate, but this results in very high detection limits (approximately 10,000 µg/L). Typically, a 1-L liquid sample, containing surrogate, and matrix spiking standards, is extracted in a continuous extractor first under acid conditions and then under basic conditions. Typically 30 g of a solid sample, containing surrogate, and matrix spiking standards, is extracted ultrasonically. After concentrating the extract to 1 mL it is spiked with 10 µL of an internal standard solution just prior to analysis by GC/MS. The volume injected should contain about 100 ng of base/neutral and 200 ng of acid surrogates (for a 1-µL injection). Analysis is performed by GC/MS using a capillary GC column.

INTERFERENCES Raw GC/MS data from all blanks, samples, and spikes must be evaluated for interferences. Contamination by carryover can occur whenever high-concentration and low-concentration samples are sequentially analyzed. To reduce carryover, the sample syringe must be rinsed out between samples with solvent. Whenever an unusually concentrated sample is encountered, it should be followed by the analysis of blank solvent to check for cross-contamination.

INSTRUMENTATION A GC/MS and a data system are required. The GC column used is a 30 m × 0.25 mm I.D. (or 0.32 mm I.D.) 1-µm film thickness silicone-coated fused silica capillary column. A continuous liquid-liquid extractor equipped with Teflon® or glass connection joints and stopcocks requiring no lubrication, a K-D concentrating apparatus, water bath, and an ultrasonic disrupter with a minimum power of 300 W and with pulsing capability are also required.

PRECISION & ACCURACY The estimated quantitation limit (EQL) of Method 8270B for determining an individual compound is approximately 1 mg/kg (wet weight) for soil or sediment samples, 1–200 mg/kg for wastes (dependent on matrix and method of preparation), and 10 µg/L for groundwater samples. EQLs will be proportionately higher for sample extracts that require dilution to avoid saturation of the detector.

The EQL(b) for groundwater in µg/L is 10.
The EQL (a, b) for low concentrations in soil and sediment in µg/kg is 660.
Accuracy as µg/L is 0.90C -0.13.
Overall precision in µg/L is 0.32X +1.35.

(a) *EQLs listed for soil/sediment are based on wet weight. Normally data is reported in a dry-weight basis; therefore, EQLs will be higher based on the % dry weight of each sample. This calculation is based on a 30-g sample and gel permeation chromatography cleanup.*
(b) *Sample EQLs are highly matrix-dependent. The EQLs are provided for guidance and may not always be achievable.*
C = *True value for concentration, in µg/L.*
X = *Average recovery found for measurements of samples containing a concentration of C, in µg/L.*

ESTIMATED QUANTITATION LIMIT

Other Matrices	Factor (a)
High-concentration soil and sludges by sonicator	7.5
Non-water miscible waste	75

(a) EQL for other matrices = [EQL for low soil/sediment] × [Factor]. This estimated EQL is similar to an EPA "Practical Quantitation Limit."

SAMPLING METHOD
Liquid samples — Use a 1 or 2½ gallon amber glass bottle with a screw-top Teflon®-lined cover that has been prewashed with detergent and rinsed with distilled water and methanol (or isopropanol).

Soils, sediments, or sludges — Use an 8-oz. widemouth glass with a screw-top Teflon®-lined cover that has been prewashed with detergent and rinsed with distilled water and methanol (or isopropanol).

SAMPLE PRESERVATION
Liquid samples — If residual chlorine is present, add 3 mL of 10% sodium thiosulfate per gallon, cool to 4°C and store in a solvent-free refrigerator until analysis; if chlorine is not present, then eliminate the sodium thiosulfate addition.

Soils, sediments, or sludges — Cool samples to 4°C and store in a solvent-free refrigerator.

MHT Liquid samples must be extracted within 7 days and the extracts analyzed within 40 days. Soils, sediments, or sludges may be stored for a maximum of 14 days and the extracts analyzed within 40 days.

SAMPLE PREPARATION
Liquid samples — Transfer 1 L quantitatively to a continuous extractor. If high concentrations are anticipated, a smaller volume may be used and then diluted with organic-free reagent water to 1 L. Adjust pH, if necessary, to pH <2 using 1:1 (V/V) sulfuric acid. Pipette 1.0 mL of a surrogate standard spiking solution into each sample. For the sample in each analytical batch selected for spiking, add 1.0 mL of a matrix spiking standard. For base/neutral acid analysis, the amount of the surrogates and matrix spiking compounds added to the sample should result in a final concentration of 100 ng/µL of each analyte in the extract to be analyzed (assuming a 1-µL injection). Extract with methylene chloride for 18–24 h. Next, adjust the pH of the aqueous phase to pH >11 using 10 N sodium hydroxide and extract it with methylene chloride again for 18–24 h. Dry the extract through a column containing anhydrous sodium sulfate and concentrate it to 1 mL using a K-D concentrator.

Soils, sediments, or sludges — Use 30 g of sample. Nonporous or wet samples (gummy or clay type) that do not have a free-flowing sandy texture must be mixed with anhydrous sodium sulfate until the sample is free flowing. Add 1 mL of surrogate standards to all samples, spikes, standards, and blanks. For the sample in each analytical batch selected for spiking, add 1.0 mL of a matrix spiking standard. For base/neutral acid analysis, the amount added of the surrogates and matrix spiking compounds should result in a final concentration of 100 ng/µL of each base/neutral analyte and 200 ng/µL of each acid analyte in the extract to be analyzed (assuming a 1-µL injection). Immediately add a 100 mL mixture of 1:1 methylene chloride:acetone and extract the sample ultrasonically for 3 min and then decant or filter the extracts. Repeat the extraction two or more times. Dry the extract using a column with anhydrous sodium sulfate and concentrate it to 1 mL in a K-D concentrator.

QUALITY CONTROL A methylene chloride solution containing 50 ng/µL of decafluorotriphenylphosphine (DFTPP) is used for tuning the GC/MS system each 12-h shift. A system performance check also must be made during every 12-h shift. A standard containing 50 ng/µL each of 4,4′-DDT, pentachlorophenol, and benzidine is required to verify injection port inertness and GC column performance. A calibration standard at mid-concentration, containing each compound of interest, including all required surrogates, must be performed every 12 h during analysis. After the system performance check is met, calibration check compounds (CCCs) are used to check the validity of the initial calibration.

The internal standard responses and retention times in the calibration check standard must be evaluated immediately after or during data acquisition. If the retention time for any internal standard changes by more than 30 seconds from the last check calibration (12 h), the chromatographic system must be inspected for malfunctions and corrections must be made, as required. If the electron ionization current plot (EICP) area for any of the internal standards changes by a factor of two from the last daily calibration standard check, the mass spectrometer must be inspected for malfunctions and corrections must be made, as appropriate.

Demonstrate, through the analysis of a reagent water blank, that interferences from the analytical system, glassware, and reagents are under control. The blank samples should be carried through all stages of the sample preparation and measurement steps. For each analytical batch (up to 20 samples), a reagent blank, matrix spike, and matrix spike duplicate/duplicate must be analyzed (the frequency of the spikes may be different for different monitoring programs). The blank and spiked samples must be carried through all stages of the sample preparation and measurement steps. A QC reference sample concentrate containing each analyte at a concentration of 100 mg/L in methanol is required.

REFERENCE Test Methods for Evaluating Solid Waste (SW-846). U.S. EPA 1983, Method 8270B, Rev. 2, Nov. 1990. Office of Solid Waste, Washington, DC.

Benzo(a)pyrene   **EPA Method 8100**
CAS #50-32-8

TITLE Polynuclear Aromatic Hydrocarbons

MATRIX Groundwater, soils, sludges, water miscible liquid wastes, and non-water miscible wastes.

APPLICATION This method is used for the analysis of various PAHs. Samples are extracted, concentrated, and analyzed using direct injection of bothneat and diluted organic liquids. The method provides two optional GC columns that are better than Column 1 and that may help resolve analytes from interferences.

INTERFERENCES Solvents, reagents, and glassware may introduce artifacts. Other interferences may come from coextracted compounds from samples.

INSTRUMENTATION GC capable of on-column injections and a flame with detector (FID). Column 1: a 1.8 m by 2 mm 3% OV-17 on Chromosorb W-AW-DCMS column. Column 2: a 30 m by 0.25 mm SE-54 fused silica capillary column. Column 3: a 30 m by 0.32 mm SE-54 fused silica capillary column.

RANGE 0.1–425 µg/L

MDL Not reported.

PQL FACTORS FOR MULTIPLYING × FID MDL VALUE
Not available.

PRECISION 0.53X + 0.01 µg/L (overall precision).

ACCURACY 0.56C + 0.01 µg/L (as recovery).

SAMPLING METHOD Use 8-oz. widemouth glass bottles with Teflon®-lined caps for concentrated waste samples, soils, sediments, and sludges. Use 1 or 2½ gallon amber glass bottles with Teflon®-lined caps for liquid (water) samples.

STABILITY Cool soil, sediment, sludge, and liquid samples to 4°C. If residual chlorine is present in liquid samples add 3 mL of 10% sodium thiosulfate per gallon of sample and cool to 4°C.

MHT 14 days for concentrated waste, soil, sediment, or sludge.

MHT 7 days for liquid samples; all extracts must be analyzed within 40 days.

QUALITY CONTROL A quality control check sample concentrate containing each analyte of interest is required. The QC check sample concentrate may be prepared from pure standard materials or purchased ascertified solutions. Use appropriate trip, matrix, controlsite, method, reagent, and solvent blanks. Internal, surrogate, and five concentration level calibration standards are used. The quality control check sample concentrate should contain benzo(a)pyrene at 10 µg/mL in acetonitrile.

REFERENCE Method 8100, SW-846, 3rd ed., Nov. 1986.

Benzo(a)pyrene EPA Method 8310
CAS #50-32-8

TITLE Polynuclear Aromatic Hydrocarbons

MATRIX Groundwater, soils, sludges, water miscible liquid wastes, and non-water miscible wastes.

APPLICATION This method is used for the analysis of 16 polynuclear aromatic hydrocarbons(PAHs). Samples are extracted, concentrated, and analyzed using HPLC with detection by UV and fluorescence detectors.

INTERFERENCES Solvents, reagents, and glassware may introduce artifacts. Other interferences may come from coextracted compounds from samples.

INSTRUMENTATION HPLC with a gradient pumping system and a 250 mm by 2.6mm reverse phase HC-ODSSil-X 5-micron particle-size column. The fluorescence detector uses an excitation wavelength of 280 nm and emission greater than 389 nm cutoff with dispersive optics.

RANGE 0.1–425 µg/L

MDL 0.023 µg/L (fluorescence; reagent water).

PQL FACTORS FOR MULTIPLYING × FID MDL VALUE

Matrix	Multiplication Factor
Groundwater	10
Low-level soil by sonication with GPC cleanup	670
High-level soil and sludge by sonication	10,000
Non-water miscible waste	100,000

PRECISION 0.53X–0.01 µg/L (overall precision).

ACCURACY 0.56C + 0.01 µg/L (as recovery).

SAMPLING METHOD Use 8-oz. widemouth glass bottles with Teflon®-lined caps for concentrated waste samples, soils, sediments, and sludges. Use 1 or 2½ gallon amber glass bottles with Teflon®-lined caps for liquid (water) samples.

STABILITY Cool soil, sediment, sludge, and liquid samples to 4°C. If residual chlorine is present in liquid samples add 3 mL of 10% sodium thiosulfate per gallon of sample and cool to 4°C.

MHT 14 days for concentrated waste, soil, sediment, or sludge; 7 days for liquid samples; all extracts must be analyzed within 40 days.

QUALITY CONTROL Internal, surrogate, and five concentration level calibration standards are used. The calibration standards must be used with the analytical method blank. Aquality control check sample concentrate containing benzo(a)pyrene at 10 µg/mL is required. The QC check sample concentrate may be prepared from pure standard materials or purchased as certified solutions. Use appropriate trip, matrix, control site, method, reagent, and solvent blanks.

REFERENCE Method 8310, SW-846, 3rd ed., Nov. 1986.

Benzo(b)fluoranthene EPA Method 1625
CAS #205-99-2

TITLE Semivolatile Organic Compounds by Isotope Dilution GC/MS

MATRIX The compounds may be determined in waters, soils, and municipal sludges by this method.

METHOD SUMMARY This method is used to determine 176 semivolatile toxic organic pollutants associated with the CWA (as amended 1987); the RCRA (as amended 1986); the CERCLA (as amended 1986); and other compounds amenable

to extraction and analysis by capillary column gas chromatography-mass spectrometry (GC/MS).

Stable isotopically-labeled analogs of the compounds of interest are added to the sample. If the solids content is less than 1%, a 1-L sample is extracted at pH 12–13, then at pH <2 with methylene chloride using continuous extraction techniques.

If the solids content is 30% or less, the sample is diluted to 1% solids with reagent water, homogenized ultrasonically, and extracted at pH 12–13, then at pH <2 with methylene chloride using continuous extraction techniques. If the solids content is greater than 30%, the sample is extracted using ultrasonic techniques.

Each extract is dried over sodium sulfate, concentrated to a volume of 5 mL, cleaned up using GPC, if necessary, and concentrated. Extracts are concentrated to 1 mL if GPC is not performed, and to 0.5 mL if GPC is performed.

An internal standard is added to the extract, and a 1-mL aliquot of the extract is injected into the GC. The compounds are separated by GC and detected by a MS. The labeled compounds serve to correct the variability of the analytical technique.

INTERFERENCES Solvents, reagents, glassware, and other sample processing hardware may yield artifacts and/or elevated baselines causing misinterpretation of chromatograms and spectra. Materials used in the analysis must be demonstrated to be free from interferences under the conditions of analysis by running method blanks initially and with each sample lot (sample started through the extraction process on a given 8-h shift, to a maximum of 20). Specific selection of reagents and purification of solvents by distillation in all glass systems may be required. Glassware and, where possible, reagents are cleaned by solvent rinse and baking at 450°C for 1-h minimum. Interferences coextracted from samples will vary considerably from source to source, depending on the diversity of the site being sampled.

INSTRUMENTATION Major instrumentation includes a GC with a splitless or on-column injection port for capillary column, a MS with 70 eV electron impact ionization, and a data system to collect and record MS data, and process it. A K-D apparatus is used to concentrate extracts.

GC Column: 30 m × 0.25 mm I.D. 5% phenyl, 94% methyl, 1% vinyl silicone bonded phased fused silica capillary column.

PRECISION & ACCURACY The detection limits of the method are usually dependent on the level of interferences rather than instrumental limitations. The limits typify the minimum quantities that can be detected with no interferences present.

- The minimum level (in µg/mL) was 10. This is defined as a minimum level at which the analytical system shall give recognizable mass spectra (background corrected) and acceptable calibration points.
- The MDL (in µg/kg) in low solids was 54 and in high solids was 30; these were determined in digested sludge (low solids) and in filter cake or compost (high solids).
- The labeled and native compound initial precision as standard deviation (in µg/L) was 183.
- The labeled and native compound initial accuracy as average recovery (in µg/L) was 32–545.

SAMPLE COLLECTION, PRESERVATION & HANDLING
Collect samples in glass containers. Aqueous samples which flow freely are collected in refrigerated bottles using automatic sampling equipment. Solid samples are collected as grab samples using widemouth jars. Maintain samples at 0 to 4°C from the time of collection until extraction. If residual chlorine is present in aqueous samples, add 80 mg sodium thiosulfate/L of water. Begin sample extraction within 7 days of collection, and analyze all extracts within 40 days of extraction.

SAMPLE PREPARATION Samples containing 1% solids or less are extracted directly using continuous liquid-liquid extraction techniques. Samples containing 1 to 30% solids are diluted to the 1% level with reagent water and extracted using continuous liquid-liquid extraction techniques. Samples containing greater than 30% solids are extracted using ultrasonic techniques.

Base/neutral extraction — Adjust the pH of the waters in the extractors to 12–13 with 6 *N* NaOH. Extract with methylene chloride for 24–48 h.

Acid extraction — Adjust the pH of the waters in the extractors to 2 or less using 6 *N* sulfuric acid. Extract with methylene chloride for 24–48 h.

Ultrasonic extraction of high solids samples — Add anhydrous sodium sulfate to the sample and QC aliquot(s). Add acetone:methylene chloride (1:1) to the sample and mix thoroughly

Concentrate extracts using a K-D apparatus.

QUALITY CONTROL The analyst is permitted to modify this method to improve separations or lower the costs of measurements, provided all performance specifications are met. Analyses of blanks are required to demonstrate freedom from contamination. When results of spikes indicate atypical method performance for samples, the samples are diluted to bring method performance within acceptable limits.

For low solids (aqueous samples), extract, concentrate, and analyze two sets of four 1-L aliquots (8 aliquots total) of the precision and recovery standard. For high solids samples, two sets of four 30-g aliquots of the high solids reference matrix are used.

Spike all samples with labeled compounds to assess method performance. Compute percent recovery of the labeled compounds using the internal standard method. Compare the labeled compound recovery for each compound with the corresponding labeled compound recovery.

Reagent water and high solids reference matrix blanks are analyzed to demonstrate freedom from contamination. Extract and concentrate a 1-L reagent water blank or a high solids reference matrix blank with each sample's lot (samples started through the extraction process on the same 8-h shift, to a maximum of 20 samples).

Field replicates may be collected to determine the precision of the sampling technique, and spiked samples may be required

to determine the accuracy of the analysis when the internal standard method is used.

REFERENCE Semivolatile Organic Compounds by Isotope Dilution GC/MS. Office of Water Regulation and Standards, U.S. EPA Industrial Technology Division, Washington, DC, EPA Method 1625, Rev. C, June 1989 (contact W.A. Telliard, U.S. EPA, Office of Water Regulations and Standards, 401 M St., SW, Washington, DC, 20460. Phone: 202-382-7131).

Benzo(b)fluoranthene   EPA Method 625
CAS #205-99-2

TITLE Base/Neutrals and Acids, U.S. EPA Method 625

MATRIX This methods covers municipal and industrial wastewaters.

METHOD SUMMARY Approximately 1 L of sample is serially extracted with methylene chloride at a pH greater than 11 and again at a pH less than 2 using a separatory funnel or a continuous extractor. The methylene chloride extract is dried, concentrated to a volume of 1 mL, and analyzed by GC/MS. Qualitative identification of the parameters in the extract is performed using the retention time and the relative abundance of three characteristic masses (m/z). Qualitative analysis is performed using either external or internal standard techniques with a single characteristic m/z.

INTERFERENCES Method interferences may be caused by contaminants in solvents, reagents, glassware, and other sample processing hardware. Glassware must be scrupulously cleaned. Glassware should be heated in a muffle furnace at 400°C for 5 to 30 min. Some thermally stable materials, such as PCBs, may not be eliminated by this treatment. Solvent rinses with acetone and pesticide quality hexane may be substituted for the muffle furnace heating. Matrix interferences may be caused by contaminants that are coextracted from the sample. The base-neutral extraction may cause significantly reduced recovery of phenols. The packed gas chromatographic columns recommended for the basic fraction may not exhibit sufficient resolution for some analytes.

INSTRUMENTATION A GC/MS system with an injection port designed for on-column injection when using packed columns and for splitless injection when using capillary columns.

Column for base/neutrals: 1.8 m long × 2 mm I.D. glass, packed with 3% SP-2550 on Supelcoport (100/120 mesh) or equivalent.
Column for acids: 1.8 m long × 2 mm I.D. glass, packed with 1% SP-1240DA on Supelcoport (100/120 mesh) or equivalent.

PRECISION & ACCURACY The MDL concentrations were obtained using reagent water. The MDL actually achieved in a given analysis will vary depending on instrument sensitivity and matrix effects. This method was tested by 15 laboratories using reagent water, drinking water, surface water, and industrial wastewaters spiked at six concentrations over the range 5 to 100 µg/L. Single operator precision, overall precision, and method accuracy were found to be directly related to the concentration of the parameter matrix.

The MDL (in µg/L) in reagent water was 4.8.
The standard deviation (in µg/L based on 4 recovery measurements) was 38.8.
The range (in µg/L) for average recovery for 4 measurements was 42.0–140.4.
The range (in %) for percent recovery was 24–159.
Accuracy (in µg/L) as expected recovery for one or more measurements of a sample containing a true concentration of C was 0.93C–1.60.
Precision (in µg/L) as expected single analyst standard deviation of measurements at an average concentration found at X was 0.22X+0.43.
Overall precision (in µg/L) as expected interlaboratory standard deviation of measurements in an average concentration found at X was 0.29X–0.96.

$C =$ *True value of the concentration in µg/L.*
$X =$ *Average recovery found for measurements of samples containing a concentration at C in µg/L.*

SAMPLE PREPARATION Adjust the pH to >11 with sodium hydroxide and serially extract in a separatory funnel with methylene chloride or else in a continuous extractor. Next, adjust the pH to <2 with sulfuric acid and serially extract in a separatory funnel with methylene chloride or else in a continuous extractor. Dry the extracts separately through a column of anhydrous sodium sulfate and then concentrate each of the extracts to 1.0 mL using a K-D apparatus.

SAMPLE COLLECTION, PRESERVATION & HANDLING- Grab samples must be collected in glass containers. All samples must be refrigerated at 4°C from the time of collection until extraction. If residual chlorine is present, add 80 mg of sodium thiosulfate/L of sample and mix well. All samples must be extracted within 7 days of collection and completely analyzed within 40 days of extraction.

QUALITY CONTROL Make an initial, one-time, demonstration of the ability to generate acceptable accuracy and precision with this method. Before processing any samples, the analyst must analyze a reagent water blank to demonstrate that interferences from the analytical system and glassware are under control. Each time a set of samples is extracted or reagents are changed, a reagent water blank must be processed. Spike and analyze a minimum of 5% of all samples to monitor and evaluate lab data quality. A QC check sample concentrate that contains each parameter of interest at a concentration of 100 µg/mL in acetone is required. PCBs and multicomponent pesticides may be omitted from this test.

After analysis of five spiked wastewater samples, calculate the average percent recovery and the standard deviation of the percent recovery. Spike all samples with the surrogate standard spiking solution and calculate the percent recovery of each surrogate compound.

REFERENCE *Federal Register*, Vol. 49, No. 209. Friday, Oct. 26, 1984.

Benzo(b)fluoranthene
CAS #205-99-2
EPA Method 8270

TITLE Semivolatile Organic Compounds by GC/MS

MATRIX This method is used to determine the concentration of semivolatile organic compounds in extracts prepared from all types of solid waste matrices, soils, and groundwater. Although surface waters are not specifically mentioned, this method should be applicable to water samples from rivers, lakes, etc.

METHOD SUMMARY This method covers 259 semivolatile organic compounds. In very limited applications direct injection of the sample into the GC/MS system may be appropriate, but this results in very high detection limits (approximately 10,000 μg/L). Typically, a 1-L liquid sample, containing surrogate, and matrix spiking standards, is extracted in a continuous extractor first under acid conditions and then under basic conditions. Typically 30 g of a solid sample, containing surrogate, and matrix spiking standards, is extracted ultrasonically. After concentrating the extract to 1 mL it is spiked with 10 μL of an internal standard solution just prior to analysis by GC/MS. The volume injected should contain about 100 ng of base/neutral and 200 ng of acid surrogates (for a 1-μL injection). Analysis is performed by GC/MS using a capillary GC column.

INTERFERENCES Raw GC/MS data from all blanks, samples, and spikes must be evaluated for interferences. Contamination by carryover can occur whenever high-concentration and low-concentration samples are sequentially analyzed. To reduce carryover, the sample syringe must be rinsed out between samples with solvent. Whenever an unusually concentrated sample is encountered, it should be followed by the analysis of blank solvent to check for cross-contamination.

INSTRUMENTATION A GC/MS and a data system are required. The GC column used is a 30 m × 0.25 mm I.D. (or 0.32 mm I.D.) 1-μm film thickness silicone-coated fused silica capillary column. A continuous liquid-liquid extractor equipped with Teflon® or glass connection joints and stopcocks requiring no lubrication, a K-D concentrating apparatus, water bath, and an ultrasonic disrupter with a minimum power of 300 W and with pulsing capability are also required.

PRECISION & ACCURACY The estimated quantitation limit (EQL) of Method 8270B for determining an individual compound is approximately 1 mg/kg (wet weight) for soil or sediment samples, 1–200 mg/kg for wastes (dependent on matrix and method of preparation), and 10 μg/L for groundwater samples. EQLs will be proportionately higher for sample extracts that require dilution to avoid saturation of the detector.

The EQL(b) for groundwater in μg/L is 10.
The EQL (a, b) for low concentrations in soil and sediment in μg/kg is 660.
Accuracy as μg/L is $0.93C - 1.80$.
Overall precision in μg/L is $0.29X + 0.96$.

(a) *EQLs listed for soil/sediment are based on wet weight. Normally data is reported in a dry-weight basis; therefore, EQLs will be higher based on the % dry weight of each sample. This calculation is based on a 30-g sample and gel permeation chromatography cleanup.*

(b) *Sample EQLs are highly matrix-dependent. The EQLs are provided for guidance and may not always be achievable.*

$C =$ *True value for concentration, in μg/L.*
$X =$ *Average recovery found for measurements of samples containing a concentration of C, in μg/L.*

ESTIMATED QUANTITATION LIMIT FOR OTHER MATRICES

Other Matrices	Factor (a)
High-concentration soil and sludges by sonicator	7.5
Non-water miscible waste	75

(a) *EQL for other matrices = [EQL for low soil/sediment] × [Factor]. This estimated EQL is similar to an EPA "Practical Quantitation Limit."*

SAMPLING METHOD
Liquid samples — Use a 1 or 2½ gallon amber glass bottle with a screw-top Teflon®-lined cover that has been prewashed with detergent and rinsed with distilled water and methanol (or isopropanol).

Soils, sediments, or sludges — Use an 8-oz. widemouth glass with a screw-top Teflon®-lined cover that has been prewashed with detergent and rinsed with distilled water and methanol (or isopropanol).

SAMPLE PRESERVATION
Liquid samples — If residual chlorine is present, add 3 mL of 10% sodium thiosulfate per gallon, cool to 4°C and store in a solvent-free refrigerator until analysis; if chlorine is not present, then eliminate the sodium thiosulfate addition.

Soils, sediments, or sludges — Cool samples to 4°C and store in a solvent-free refrigerator.

MHT Liquid samples must be extracted within 7 days and the extracts analyzed within 40 days. Soils, sediments, or sludges may be stored for a maximum of 14 days and the extracts analyzed within 40 days.

SAMPLE PREPARATION
Liquid samples — Transfer 1 L quantitatively to a continuous extractor. If high concentrations are anticipated, a smaller volume may be used and then diluted with organic-free reagent water to 1 L. Adjust pH, if necessary, to pH <2 using 1:1 (V/V) sulfuric acid. Pipette 1.0 mL of a surrogate standard spiking solution into each sample. For the sample in each analytical batch selected for spiking, add 1.0 mL of a matrix spiking standard. For base/neutral acid analysis, the amount of the surrogates and matrix spiking compounds added to the sample should result in a final concentration of 100 ng/μL of each analyte in the extract to be analyzed (assuming a 1-μL injection). Extract with methylene chloride for 18–24 h. Next, adjust the pH of the aqueous phase to pH >11 using 10 N sodium hydroxide and extract it with methylene chloride again for 18–24 h. Dry the extract through a column containing anhydrous sodium sulfate and concentrate it to 1 mL using a K-D concentrator.

Soils, sediments, or sludges — Use 30 g of sample. Nonporous or wet samples (gummy or clay type) that do not have a free-flowing sandy texture must be mixed with anhydrous sodium sulfate until the sample is free flowing. Add 1 mL of surrogate standards to all samples, spikes, standards, and blanks. For the sample in each analytical batch selected for spiking, add 1.0 mL of a matrix spiking standard. For base/neutral acid analysis, the amount added of the surrogates and matrix spiking compounds should result in a final concentration of 100 ng/µL of each base/neutral analyte and 200 ng/µL of each acid analyte in the extract to be analyzed (assuming a 1-µL injection). Immediately add a 100 mL mixture of 1:1 methylene chloride:acetone and extract the sample ultrasonically for 3 min and then decant or filter the extracts. Repeat the extraction two or more times. Dry the extract using a column with anhydrous sodium sulfate and concentrate it to 1 mL in a K-D concentrator.

QUALITY CONTROL A methylene chloride solution containing 50 ng/µL of decafluorotriphenylphosphine (DFTPP) is used for tuning the GC/MS system each 12-h shift. A system performance check also must be made during every 12-h shift. A standard containing 50 ng/µL each of 4,4'-DDT, pentachlorophenol, and benzidine is required to verify injection port inertness and GC column performance. A calibration standard at mid-concentration, containing each compound of interest, including all required surrogates, must be performed every 12 h during analysis. After the system performance check is met, calibration check compounds (CCCs) are used to check the validity of the initial calibration.

The internal standard responses and retention times in the calibration check standard must be evaluated immediately after or during data acquisition. If the retention time for any internal standard changes by more than 30 seconds from the last check calibration (12 h), the chromatographic system must be inspected for malfunctions and corrections must be made, as required. If the electron ionization current plot (EICP) area for any of the internal standards changes by a factor of two from the last daily calibration standard check, the mass spectrometer must be inspected for malfunctions and corrections must be made, as appropriate.

Demonstrate, through the analysis of a reagent water blank, that interferences from the analytical system, glassware, and reagents are under control. The blank samples should be carried through all stages of the sample preparation and measurement steps. For each analytical batch (up to 20 samples), a reagent blank, matrix spike, and matrix spike duplicate/duplicate must be analyzed (the frequency of the spikes may be different for different monitoring programs). The blank and spiked samples must be carried through all stages of the sample preparation and measurement steps. A QC reference sample concentrate containing each analyte at a concentration of 100 mg/L in methanol is required.

REFERENCE Test Methods for Evaluating Solid Waste (SW-846). U.S. EPA 1983, Method 8270B, Rev. 2, Nov. 1990. Office of Solid Waste, Washington, DC.

Benzo(b)fluoranthene — EPA Method 8100
CAS #205-99-2

TITLE Polynuclear Aromatic Hydrocarbons

MATRIX Groundwater, soils, sludges, water miscible liquid wastes, and non-water miscible wastes.

APPLICATION This methodis used for the analysis of various PAHs. Samples are extracted, concentrated, and analyzed using direct injection of bothneat and diluted organic liquids. The method provides two optional GC columns that are better than Column 1 and that may help resolve analytes from interferences.

INTERFERENCES Solvents, reagents, and glassware may introduce artifacts. Other interferences may come from coextracted compounds from samples.

INSTRUMENTATION GC capable of on-column injections and a flame with detector (FID). Column 1: a 1.8 m by 2 mm 3% OV-17 on Chromosorb W-AW-DCMS column. Column 2: a 30 m by 0.25 mm SE-54 fused silica capillary column. Column 3: a 30 m by 0.32 mm SE-54 fused silica capillary column.

RANGE 0.1–425 µg/L

MDL Not reported.

PQL FACTORS FOR MULTIPLYING × FID MDL VALUE
Not available.

PRECISION 0.38X–0.00 µg/L (overall precision).

ACCURACY 0.78C + 0.01 µg/L (as recovery).

SAMPLING METHOD Use 8-oz. widemouth glass bottles with Teflon®-lined caps for concentrated waste samples, soils. sediments, and sludges. Use 1 or 2½ gallon amber glass bottles with Teflon®-lined caps for liquid (water) samples.

STABILITY Cool soil, sediment, sludge, and liquid samples to 4°C. If residual chlorine is present in liquid samples add 3 mL of 10% sodium thiosulfate per gallon of sample and cool to 4°C.

MHT 14 days for concentrated waste, soil, sediment, or sludge; 7 days for liquid samples; all extracts must be analyzed within 40 days.

QUALITY CONTROL A quality control check sample concentrate containing each analyte of interest is required. The QC check sample concentrate may be prepared from pure standard materials or purchased as certified solutions. Use appropriate trip, matrix, controlsite, method, reagent, and solvent blanks. Internal, surrogate, and five concentration level calibration standards are used. The quality control check sample concentrate should contain benzo(b)fluoranthene at 10 µg/mL in acetonitrile.

REFERENCE Method 8100, SW-846, 3rd ed., Nov. 1986.

Benzo(b)fluoranthene
CAS #205-99-2

EPA Method 8310

TITLE Polynuclear Aromatic Hydrocarbons

MATRIX Groundwater, soils, sludges, water miscible liquid wastes, and non-water miscible wastes.

APPLICATION This method is used for the analysis of 16 polynucleararomatic hydrocarbons(PAHs). Samples are extracted, concentrated, and analyzed using HPLC with detection by UV and fluorescence detectors.

INTERFERENCES Solvents, reagents, and glassware may introduce artifacts. Other interferences may come from coextracted compounds from samples.

INSTRUMENTATION HPLCwith a gradient pumping system and a 250 mm by 2.6 mm reverse phase HC-ODS Sil-X 5-micron particle sizecolumn. Thefluorescence detectorusesanexcitation wavelength of 280 nm and emission greater than 389 nm cutoff with dispersive optics.

RANGE 0.1–425 µg/L

MDL 0.018 µg/L (fluorescence; reagent water).

PQL FACTORS FOR MULTIPLYING × FID MDL VALUE

Matrix	Multiplication Factor
Groundwater	10
Low-level soil by sonication with GPC cleanup	670
High-level soil and sludge by sonication	10,000
Non-water miscible waste	100,000

PRECISION 0.38X–0.00 µg/L (overall precision).

ACCURACY 0.78C + 0.01 µg/L (as recovery).

SAMPLING METHOD Use 8-oz. widemouth glass bottles with Teflon®-lined caps for concentrated waste samples, soils, sediments, and sludges. Use 1 or 2½ gallon amber glass bottles with Teflon®-lined caps for liquid (water) samples.

STABILITY Cool soil, sediment, sludge, and liquid samples to 4°C. If residual chlorine is present in liquid samples add 3 mL of 10% sodium thiosulfate per gallon of sample and cool to 4°C.

MHT 14 days for concentrated waste, soil, sediment, or sludge; 7 days for liquid samples; all extracts must be analyzed within 40 days.

QUALITY CONTROL Internal, surrogate, and five concentration level calibration standardsareused. The calibration standards must be used with the analytical method blank. Aquality control check sample concentrate containing benzo(b)fluoranthene at 10 µg/mL is required. The QC check sample concentrate may be prepared from pure standard materials or purchased as certified solutions. Use appropriate trip, matrix, control site, method, reagent, and solvent blanks.

REFERENCE Method 8310, SW-846, 3rd ed., Nov. 1986.

Benzo(g,h,i)perylene
CAS #191-24-2

EPA Method 1625

TITLE Semivolatile Organic Compounds by Isotope Dilution GC/MS

MATRIX The compounds may be determined in waters, soils, and municipal sludges by this method.

METHOD SUMMARY This method is used to determine 176 semivolatile toxic organic pollutants associated with the CWA (as amended 1987); the RCRA (as amended 1986); the CERCLA (as amended 1986); and other compounds amenable to extraction and analysis by capillary column gas chromatography-mass spectrometry (GC/MS).

Stable isotopically-labeled analogs of the compounds of interest are added to the sample. If the solids content is less than 1%, a 1-L sample is extracted at pH 12–13, then at pH <2 with methylene chloride using continuous extraction techniques.

If the solids content is 30% or less, the sample is diluted to 1% solids with reagent water, homogenized ultrasonically, and extracted at pH 12–13, then at pH <2 with methylene chloride using continuous extraction techniques. If the solids content is greater than 30%, the sample is extracted using ultrasonic techniques.

Each extract is dried over sodium sulfate, concentrated to a volume of 5 mL, cleaned up using GPC, if necessary, and concentrated. Extracts are concentrated to 1 mL if GPC is not performed, and to 0.5 mL if GPC is performed.

An internal standard is added to the extract, and a 1-mL aliquot of the extract is injected into the GC. The compounds are separated by GC and detected by a MS. The labeled compounds serve to correct the variability of the analytical technique.

INTERFERENCES Solvents, reagents, glassware, and other sample processing hardware may yield artifacts and/or elevated baselines causing misinterpretation of chromatograms and spectra. Materials used in the analysis must be demonstrated to be free from interferences under the conditions of analysis by running method blanks initially and with each sample lot (sample started through the extraction process on a given 8-h shift, to a maximum of 20). Specific selection of reagents and purification of solvents by distillation in all glass systems may be required. Glassware and, where possible, reagents are cleaned by solvent rinse and baking at 450°C for 1-h minimum. Interferences coextracted from samples will vary considerably from source to source, depending on the diversity of the site being sampled.

INSTRUMENTATION Major instrumentation includes a GC with a splitless or on-column injection port for capillary column, a MS with 70 eV electron impact ionization, and a data system to collect and record MS data, and process it. A K-D apparatus is used to concentrate extracts.

GC Column: 30 m × 0.25 mm I.D. 5% phenyl, 94% methyl, 1% vinyl silicone bonded phased fused silica capillary column.

PRECISION & ACCURACY The detection limits of the method are usually dependent on the level of interferences rather than instrumental limitations. The limits typify the minimum quantities that can be detected with no interferences present.

The minimum level (in µg/mL) was 20. This is defined as a minimum level at which the analytical system shall give recognizable mass spectra (background corrected) and acceptable calibration points.

The MDL (in µg/kg) in low solids was 44 and in high solids was not detected; these were determined in digested sludge (low solids) and in filter cake or compost (high solids).

The labeled and native compound initial precision as standard deviation (in µg/L) was 21.

The labeled and native compound initial accuracy as average recovery (in µg/L) was 72–160.

SAMPLE COLLECTION, PRESERVATION & HANDLING Collect samples in glass containers. Aqueous samples which flow freely are collected in refrigerated bottles using automatic sampling equipment. Solid samples are collected as grab samples using widemouth jars. Maintain samples at 0 to 4°C from the time of collection until extraction. If residual chlorine is present in aqueous samples, add 80 mg sodium thiosulfate/L of water. Begin sample extraction within 7 days of collection, and analyze all extracts within 40 days of extraction.

SAMPLE PREPARATION Samples containing 1% solids or less are extracted directly using continuous liquid-liquid extraction techniques. Samples containing 1 to 30% solids are diluted to the 1% level with reagent water and extracted using continuous liquid-liquid extraction techniques. Samples containing greater than 30% solids are extracted using ultrasonic techniques.

Base/neutral extraction — Adjust the pH of the waters in the extractors to 12–13 with 6 N NaOH. Extract with methylene chloride for 24–48 h.

Acid extraction — Adjust the pH of the waters in the extractors to 2 or less using 6 N sulfuric acid. Extract with methylene chloride for 24–48 h.

Ultrasonic extraction of high solids samples — Add anhydrous sodium sulfate to the sample and QC aliquot(s). Add acetone:methylene chloride (1:1) to the sample and mix thoroughly

Concentrate extracts using a K-D apparatus.

QUALITY CONTROL The analyst is permitted to modify this method to improve separations or lower the costs of measurements, provided all performance specifications are met. Analyses of blanks are required to demonstrate freedom from contamination. When results of spikes indicate atypical method performance for samples, the samples are diluted to bring method performance within acceptable limits.

For low solids (aqueous samples), extract, concentrate, and analyze two sets of four 1-L aliquots (8 aliquots total) of the precision and recovery standard. For high solids samples, two sets of four 30-g aliquots of the high solids reference matrix are used.

Spike all samples with labeled compounds to assess method performance. Compute percent recovery of the labeled compounds using the internal standard method. Compare the labeled compound recovery for each compound with the corresponding labeled compound recovery.

Reagent water and high solids reference matrix blanks are analyzed to demonstrate freedom from contamination. Extract and concentrate a 1-L reagent water blank or a high solids reference matrix blank with each sample's lot (samples started through the extraction process on the same 8-h shift, to a maximum of 20 samples).

Field replicates may be collected to determine the precision of the sampling technique, and spiked samples may be required to determine the accuracy of the analysis when the internal standard method is used.

REFERENCE Semivolatile Organic Compounds by Isotope Dilution GC/MS. Office of Water Regulation and Standards, U.S. EPA Industrial Technology Division, Washington, DC, EPA Method 1625, Rev. C, June 1989 (contact W.A. Telliard, U.S. EPA, Office of Water Regulations and Standards, 401 M St., SW, Washington, DC, 20460. Phone: 202-382-7131).

Benzo(g,h,i)perylene   EPA Method 625
CAS #191-24-2

TITLE Base/Neutrals and Acids, U.S. EPA Method 625

MATRIX This methods covers municipal and industrial wastewaters.

METHOD SUMMARY Approximately 1 L of sample is serially extracted with methylene chloride at a pH greater than 11 and again at a pH less than 2 using a separatory funnel or a continuous extractor. The methylene chloride extract is dried, concentrated to a volume of 1 mL, and analyzed by GC/MS. Qualitative identification of the parameters in the extract is performed using the retention time and the relative abundance of three characteristic masses (m/z). Qualitative analysis is performed using either external or internal standard techniques with a single characteristic m/z.

INTERFERENCES Method interferences may be caused by contaminants in solvents, reagents, glassware, and other sample processing hardware. Glassware must be scrupulously cleaned. Glassware should be heated in a muffle furnace at 400°C for 5 to 30 min. Some thermally stable materials, such as PCBs, may not be eliminated by this treatment. Solvent rinses with acetone and pesticide quality hexane may be substituted for the muffle furnace heating. Matrix interferences may be caused by contaminants that are coextracted from the sample. The base-neutral extraction may cause significantly reduced recovery of phenols. The packed gas chromatographic columns recommended for the basic fraction may not exhibit sufficient resolution for some analytes.

INSTRUMENTATION A GC/MS system with an injection port designed for on-column injection when using packed columns and for splitless injection when using capillary columns.

Column for base/neutrals: 1.8 m long × 2 mm I.D. glass, packed with 3% SP-2550 on Supelcoport (100/120 mesh) or equivalent.

Column for acids: 1.8 m long × 2 mm I.D. glass, packed with 1% SP-1240DA on Supelcoport (100/120 mesh) or equivalent.

PRECISION & ACCURACY The MDL concentrations were obtained using reagent water. The MDL actually achieved in a given analysis will vary depending on instrument sensitivity and matrix effects. This method was tested by 15 laboratories using reagent water, drinking water, surface water, and industrial wastewaters spiked at six concentrations over the range 5 to 100 µg/L. Single operator precision, overall precision, and method accuracy were found to be directly related to the concentration of the parameter matrix.

The MDL (in µg/L) in reagent water was 4.1.

The standard deviation (in µg/L based on 4 recovery measurements) was 58.9.

The range (in µg/L) for average recovery for 4 measurements was D-195.0.

The range (in %) for percent recovery was D-219.

Accuracy (in µg/L) as expected recovery for one or more measurements of a sample containing a true concentration of C was $0.96C-0.66$.

Precision (in µg/L) as expected single analyst standard deviation of measurements at an average concentration found at X was $0.29X+2.40$.

Overall precision (in µg/L) as expected interlaboratory standard deviation of measurements in an average concentration found at X was $0.51X-0.44$.

C = *True value of the concentration in µg/L.*

X = *Average recovery found for measurements of samples containing a concentration at C in µg/L.*

SAMPLE PREPARATION Adjust the pH to >11 with sodium hydroxide and serially extract in a separatory funnel with methylene chloride or else in a continuous extractor. Next, adjust the pH to <2 with sulfuric acid and serially extract in a separatory funnel with methylene chloride or else in a continuous extractor. Dry the extracts separately through a column of anhydrous sodium sulfate and then concentrate each of the extracts to 1.0 mL using a K-D apparatus.

SAMPLE COLLECTION, PRESERVATION & HANDLING Grab samples must be collected in glass containers. All samples must be refrigerated at 4°C from the time of collection until extraction. If residual chlorine is present, add 80 mg of sodium thiosulfate/L of sample and mix well. All samples must be extracted within 7 days of collection and completely analyzed within 40 days of extraction.

QUALITY CONTROL Make an initial, one-time, demonstration of the ability to generate acceptable accuracy and precision with this method. Before processing any samples, the analyst must analyze a reagent water blank to demonstrate that interferences from the analytical system and glassware are under control. Each time a set of samples is extracted or reagents are changed, a reagent water blank must be processed. Spike and analyze a minimum of 5% of all samples to monitor and evaluate lab data quality. A QC check sample concentrate that contains each parameter of interest at a concentration of 100 µg/mL in acetone is required. PCBs and multicomponent pesticides may be omitted from this test.

After analysis of five spiked wastewater samples, calculate the average percent recovery and the standard deviation of the percent recovery. Spike all samples with the surrogate standard spiking solution and calculate the percent recovery of each surrogate compound.

REFERENCE Federal Register, Vol. 49, No. 209. Friday, Oct. 26, 1984.

Benzo(g,h,i)perylene   **EPA Method 8270**
CAS #191-24-2

TITLE Semivolatile Organic Compounds by GC/MS

MATRIX This method is used to determine the concentration of semivolatile organic compounds in extracts prepared from all types of solid waste matrices, soils, and groundwater. Although surface waters are not specifically mentioned, this method should be applicable to water samples from rivers, lakes, etc.

METHOD SUMMARY This method covers 259 semivolatile organic compounds. In very limited applications direct injection of the sample into the GC/MS system may be appropriate, but this results in very high detection limits (approximately 10,000 µg/L). Typically, a 1-L liquid sample, containing surrogate, and matrix spiking standards, is extracted in a continuous extractor first under acid conditions and then under basic conditions. Typically 30 g of a solid sample, containing surrogate, and matrix spiking standards, is extracted ultrasonically. After concentrating the extract to 1 mL it is spiked with 10 µL of an internal standard solution just prior to analysis by GC/MS. The volume injected should contain about 100 ng of base/neutral and 200 ng of acid surrogates (for a 1-µL injection). Analysis is performed by GC/MS using a capillary GC column.

INTERFERENCES Raw GC/MS data from all blanks, samples, and spikes must be evaluated for interferences. Contamination by carryover can occur whenever high-concentration and low-concentration samples are sequentially analyzed. To reduce carryover, the sample syringe must be rinsed out between samples with solvent. Whenever an unusually concentrated sample is encountered, it should be followed by the analysis of blank solvent to check for cross-contamination.

INSTRUMENTATION A GC/MS and a data system are required. The GC column used is a 30 m × 0.25 mm I.D. (or 0.32 mm I.D.) 1-µm film thickness silicone-coated fused silica capillary column. A continuous liquid-liquid extractor equipped with Teflon® or glass connection joints and stopcocks requiring no lubrication, a K-D concentrating apparatus, water bath, and an ultrasonic disrupter with a minimum power of 300 W and with pulsing capability are also required.

PRECISION & ACCURACY The estimated quantitation limit (EQL) of Method 8270B for determining an individual compound is approximately 1 mg/kg (wet weight) for soil or sediment samples, 1–200 mg/kg for wastes (dependent on matrix and method of preparation), and 10 μg/L for groundwater samples. EQLs will be proportionately higher for sample extracts that require dilution to avoid saturation of the detector.

The EQL(b) for groundwater in μg/L is 10.
The EQL (a, b) for low concentrations in soil and sediment in μg/kg is 660.
Accuracy as μg/L is 0.98C −0.86.
Overall precision in μg/L is 0.51X−0.44.

(a) *EQLs listed for soil/sediment are based on wet weight. Normally data is reported in a dry-weight basis; therefore, EQLs will be higher based on the % dry weight of each sample. This calculation is based on a 30-g sample and gel permeation chromatography cleanup.*
(b) *Sample EQLs are highly matrix-dependent. The EQLs are provided for guidance and may not always be achievable.*
C = *True value for concentration, in μg/L.*
X = *Average recovery found for measurements of samples containing a concentration of C, in μg/L.*

ESTIMATED QUANTITATION LIMIT FOR OTHER MATRICES

Other Matrices	Factor (a)
High-concentration soil and sludges by sonicator	7.5
Non-water miscible waste	75

(a) *EQL for other matrices = [EQL for low soil/sediment] × [Factor]. This estimated EQL is similar to an EPA "Practical Quantitation Limit."*

SAMPLING METHOD

Liquid samples — Use a 1 or 2½ gallon amber glass bottle with a screw-top Teflon®-lined cover that has been prewashed with detergent and rinsed with distilled water and methanol (or isopropanol).

Soils, sediments, or sludges — Use an 8-oz. widemouth glass with a screw-top Teflon®-lined cover that has been prewashed with detergent and rinsed with distilled water and methanol (or isopropanol).

SAMPLE PRESERVATION

Liquid samples — If residual chlorine is present, add 3 mL of 10% sodium thiosulfate per gallon, cool to 4°C and store in a solvent-free refrigerator until analysis; if chlorine is not present, then eliminate the sodium thiosulfate addition.

Soils, sediments, or sludges — Cool samples to 4°C and store in a solvent-free refrigerator.

MHT Liquid samples must be extracted within 7 days and the extracts analyzed within 40 days. Soils, sediments, or sludges may be stored for a maximum of 14 days and the extracts analyzed within 40 days.

SAMPLE PREPARATION

Liquid samples — Transfer 1 L quantitatively to a continuous extractor. If high concentrations are anticipated, a smaller volume may be used and then diluted with organic-free reagent water to 1 L. Adjust pH, if necessary, to pH <2 using 1:1 (V/V) sulfuric acid. Pipette 1.0 mL of a surrogate standard spiking solution into each sample. For the sample in each analytical batch selected for spiking, add 1.0 mL of a matrix spiking standard. For base/neutral acid analysis, the amount of the surrogates and matrix spiking compounds added to the sample should result in a final concentration of 100 ng/μL of each analyte in the extract to be analyzed (assuming a 1-μL injection). Extract with methylene chloride for 18–24 h. Next, adjust the pH of the aqueous phase to pH >11 using 10 N sodium hydroxide and extract it with methylene chloride again for 18–24 h. Dry the extract through a column containing anhydrous sodium sulfate and concentrate it to 1 mL using a K-D concentrator.

Soils, sediments, or sludges — Use 30 g of sample. Nonporous or wet samples (gummy or clay type) that do not have a free-flowing sandy texture must be mixed with anhydrous sodium sulfate until the sample is free flowing. Add 1 mL of surrogate standards to all samples, spikes, standards, and blanks. For the sample in each analytical batch selected for spiking, add 1.0 mL of a matrix spiking standard. For base/neutral acid analysis, the amount added of the surrogates and matrix spiking compounds should result in a final concentration of 100 ng/μL of each base/neutral analyte and 200 ng/μL of each acid analyte in the extract to be analyzed (assuming a 1-μL injection). Immediately add a 100 mL mixture of 1:1 methylene chloride:acetone and extract the sample ultrasonically for 3 min and then decant or filter the extracts. Repeat the extraction two or more times. Dry the extract using a column with anhydrous sodium sulfate and concentrate it to 1 mL in a K-D concentrator.

QUALITY CONTROL A methylene chloride solution containing 50 ng/μL of decafluorotriphenylphosphine (DFTPP) is used for tuning the GC/MS system each 12-h shift. A system performance check also must be made during every 12-h shift. A standard containing 50 ng/μL each of 4,4′-DDT, pentachlorophenol, and benzidine is required to verify injection port inertness and GC column performance. A calibration standard at mid-concentration, containing each compound of interest, including all required surrogates, must be performed every 12 h during analysis. After the system performance check is met, calibration check compounds (CCCs) are used to check the validity of the initial calibration.

The internal standard responses and retention times in the calibration check standard must be evaluated immediately after or during data acquisition. If the retention time for any internal standard changes by more than 30 seconds from the last check calibration (12 h), the chromatographic system must be inspected for malfunctions and corrections must be made, as required. If the electron ionization current plot (EICP) area for any of the internal standards changes by a factor of two from the last daily calibration standard check, the mass spectrometer must be inspected for malfunctions and corrections must be made, as appropriate.

Demonstrate, through the analysis of a reagent water blank, that interferences from the analytical system, glassware, and reagents are under control. The blank samples should be carried through all stages of the sample preparation and measurement steps. For each analytical batch (up to 20 samples), a reagent blank, matrix spike, and matrix spike duplicate/duplicate must be analyzed (the frequency of the spikes may be different for different monitoring programs). The blank and spiked samples must be carried through all stages of the sample preparation and measurement steps. A QC reference sample concentrate containing each analyte at a concentration of 100 mg/L in methanol is required.

REFERENCE Test Methods for Evaluating Solid Waste (SW-846). U.S. EPA 1983, Method 8270B, Rev. 2, Nov. 1990. Office of Solid Waste, Washington, DC.

Benzo(g,h,i)perylene   **EPA Method 8100**
CAS #191-24-2

TITLE Polynuclear Aromatic Hydrocarbons

MATRIX Groundwater, soils, sludges, water miscible liquid wastes, and non-water miscible wastes.

APPLICATION This method is used for the analysis of various PAHs. Samples are extracted, concentrated, and analyzed using direct injection of both neat and diluted organic liquids. The method provides two optional GC columns that are better than Column 1 and that may help resolve analytes from interferences. The recovery of this compound at 80 and 800 times the MDL is reported as low (35% and 45% respectively) although the MDL was not reported.

INTERFERENCES Solvents, reagents, and glassware may introduce artifacts. Other interferences may come from coextracted compounds from samples.

INSTRUMENTATION GC capable of on-column injections and a flame with detector (FID). Column 1: a 1.8 m by 2 mm 3% OV-17 on Chromosorb W-AW-DCMS column. Column 2: a 30 m by 0.25 mm SE-54 fused silica capillary column. Column 3: a 30 m by 0.32 mm SE-54 fused silica capillary column.

RANGE 0.1–425 µg/L

MDL Not reported.

PQL FACTORS FOR MULTIPLYING × FID MDL VALUE
Not available.

PRECISION 0.58X + 0.10 µg/L (overall precision).

ACCURACY 0.44C + 0.30 µg/L (as recovery).

SAMPLING METHOD Use 8-oz. widemouth glass bottles with Teflon®-lined caps for concentrated waste samples, soils, sediments, and sludges. Use 1 or 2½ gallon amber glass bottles with Teflon®-lined caps for liquid (water) samples.

STABILITY Cool soil, sediment, sludge, and liquid samples to 4°C. If residual chlorine is present in liquid samples add 3 mL of 10% sodium thiosulfate per gallon of sample and cool to 4°C.

MHT 14 days for concentrated waste, soil, sediment, or sludge; 7 days for liquid samples; all extracts must be analyzed within 40 days.

QUALITY CONTROL A quality control check sample concentrate containing each analyte of interest is required. The QC check sample concentrate may be prepared from pure standard materials or purchased as certified solutions. Use appropriate trip, matrix, control site, method, reagent, and solvent blanks. Internal, surrogate, and five concentration level calibration standards are used. The quality control check sample concentrate should contain benzo(ghi)perylene at 10 µg/mL in acetonitrile.

REFERENCE Method 8100, SW-846, 3rd ed., Nov. 1986.

Benzo(g,h,i)perylene   **EPA Method 8310**
CAS #191-24-2

TITLE Polynuclear Aromatic Hydrocarbons

MATRIX Groundwater, soils, sludges, water miscible liquid wastes, and non-water miscible wastes.

APPLICATION This method is used for the analysis of 16 polynuclear aromatic hydrocarbons (PAHs). Samples are extracted, concentrated, and analyzed using HPLC with detection by UV and fluorescence detectors.

INTERFERENCES Solvents, reagents, and glassware may introduce artifacts. Other interferences may come from coextracted compounds from samples.

INSTRUMENTATION HPLC with a gradient pumping system and a 250 mm by 2.6 mm reversephase HC-ODSSil-X 5-micron particle-size column. The fluorescence detector uses an excitation wavelength of 280 nm and emission greater than 389 nm cutoff with dispersive optics.

RANGE 0.1–425 µg/L

MDL 0.076 µg/L (fluorescence; reagent water).

PQL FACTORS FOR MULTIPLYING × FID MDL VALUE

Matrix	Multiplication Factor
Groundwater	10
Low-level soil by sonication with GPC cleanup	670
High-level soil and sludge by sonication	10,000
Non-water miscible waste	100,000

PRECISION 0.58X + 0.10 µg/L (overall precision).

ACCURACY 0.44C + 0.30 µg/L (as recovery).

SAMPLING METHOD Use 8-oz. widemouth glass bottles with Teflon®-lined caps for concentrated waste samples, soils, sediments, and sludges. Use 1 or 2½ gallon amber glass bottles with Teflon®-lined caps for liquid (water) samples.

STABILITY Cool soil, sediment, sludge, and liquid samples to 4°C. If residual chlorine is present in liquid samples add 3 mL of 10% sodium thiosulfate per gallon of sample and cool to 4°C.

MHT 14 days for concentrated waste, soil, sediment, or sludge; 7 days for liquid samples; all extracts must be analyzed within 40 days.

QUALITY CONTROL Internal, surrogate, and five concentration level calibration standards are used. The calibration standards must be used with the analytical method blank. A quality control check sample concentrate containing benzo(ghi)perylene at 10 µg/mL is required. The QC check sample concentrate may be prepared from pure standard materials or purchased as certified solutions. Use appropriate trip, matrix, control site, method, reagent, and solvent blanks.

REFERENCE Method 8310, SW-846, 3rd ed., Nov. 1986.

Benzo(k)fluoranthene   **EPA Method 1625**
CAS #207-08-9

TITLE Semivolatile Organic Compounds by Isotope Dilution GC/MS

MATRIX The compounds may be determined in waters, soils, and municipal sludges by this method.

METHOD SUMMARY This method is used to determine 176 semivolatile toxic organic pollutants associated with the CWA (as amended 1987); the RCRA (as amended 1986); the CERCLA (as amended 1986); and other compounds amenable to extraction and analysis by capillary column gas chromatography-mass spectrometry (GC/MS).

Stable isotopically-labeled analogs of the compounds of interest are added to the sample. If the solids content is less than 1%, a 1-L sample is extracted at pH 12–13, then at pH <2 with methylene chloride using continuous extraction techniques.

If the solids content is 30% or less, the sample is diluted to 1% solids with reagent water, homogenized ultrasonically, and extracted at pH 12–13, then at pH <2 with methylene chloride using continuous extraction techniques. If the solids content is greater than 30%, the sample is extracted using ultrasonic techniques.

Each extract is dried over sodium sulfate, concentrated to a volume of 5 mL, cleaned up using GPC, if necessary, and concentrated. Extracts are concentrated to 1 mL if GPC is not performed, and to 0.5 mL if GPC is performed.

An internal standard is added to the extract, and a 1-mL aliquot of the extract is injected into the GC. The compounds are separated by GC and detected by a MS. The labeled compounds serve to correct the variability of the analytical technique.

INTERFERENCES Solvents, reagents, glassware, and other sample processing hardware may yield artifacts and/or elevated baselines causing misinterpretation of chromatograms and spectra. Materials used in the analysis must be demonstrated to be free from interferences under the conditions of analysis by running method blanks initially and with each sample lot (sample started through the extraction process on a given 8-h shift, to a maximum of 20). Specific selection of reagents and purification of solvents by distillation in all glass systems may be required. Glassware and, where possible, reagents are cleaned by solvent rinse and baking at 450°C for 1-h minimum. Interferences coextracted from samples will vary considerably from source to source, depending on the diversity of the site being sampled.

INSTRUMENTATION Major instrumentation includes a GC with a splitless or on-column injection port for capillary column, a MS with 70 eV electron impact ionization, and a data system to collect and record MS data, and process it. A K-D apparatus is used to concentrate extracts.

GC Column: 30 m × 0.25 mm I.D. 5% phenyl, 94% methyl, 1% vinyl silicone bonded phased fused silica capillary column.

PRECISION & ACCURACY The detection limits of the method are usually dependent on the level of interferences rather than instrumental limitations. The limits typify the minimum quantities that can be detected with no interferences present.

The minimum level (in µg/mL) was 10. This is defined as a minimum level at which the analytical system shall give recognizable mass spectra (background corrected) and acceptable calibration points.

The MDL (in µg/kg) in low solids was 95 and in high solids was 20; these were determined in digested sludge (low solids) and in filter cake or compost (high solids).

The labeled and native compound initial precision as standard deviation (in µg/L) was 26.

The labeled and native compound initial accuracy as average recovery (in µg/L) was 59–143.

SAMPLE COLLECTION, PRESERVATION & HANDLING Collect samples in glass containers. Aqueous samples which flow freely are collected in refrigerated bottles using automatic sampling equipment. Solid samples are collected as grab samples using widemouth jars. Maintain samples at 0 to 4°C from the time of collection until extraction. If residual chlorine is present in aqueous samples, add 80 mg sodium thiosulfate/L of water. Begin sample extraction within 7 days of collection, and analyze all extracts within 40 days of extraction.

SAMPLE PREPARATION Samples containing 1% solids or less are extracted directly using continuous liquid-liquid extraction techniques. Samples containing 1 to 30% solids are diluted to the 1% level with reagent water and extracted using continuous liquid-liquid extraction techniques. Samples containing greater than 30% solids are extracted using ultrasonic techniques.

Base/neutral extraction — Adjust the pH of the waters in the extractors to 12–13 with 6 N NaOH. Extract with methylene chloride for 24–48 h.

Acid extraction — Adjust the pH of the waters in the extractors to 2 or less using 6 N sulfuric acid. Extract with methylene chloride for 24–48 h.

Ultrasonic extraction of high solids samples — Add anhydrous sodium sulfate to the sample and QC aliquot(s). Add acetone:methylene chloride (1:1) to the sample and mix thoroughly

Concentrate extracts using a K-D apparatus.

QUALITY CONTROL The analyst is permitted to modify this method to improve separations or lower the costs of measurements, provided all performance specifications are met.

Analyses of blanks are required to demonstrate freedom from contamination. When results of spikes indicate atypical method performance for samples, the samples are diluted to bring method performance within acceptable limits.

For low solids (aqueous samples), extract, concentrate, and analyze two sets of four 1-L aliquots (8 aliquots total) of the precision and recovery standard. For high solids samples, two sets of four 30-g aliquots of the high solids reference matrix are used.

Spike all samples with labeled compounds to assess method performance. Compute percent recoveryof the labeled compounds using the internal standard method. Compare the labeled compound recovery for each compound with the corresponding labeled compound recovery.

Reagent water and high solids reference matrix blanks are analyzed to demonstrate freedom from contamination. Extract and concentrate a 1-L reagent water blank or a high solids reference matrix blank with each sample's lot (samples started through the extraction process on the same 8-h shift, to a maximum of 20 samples).

Field replicates may be collected to determine the precision of the sampling technique, and spiked samples may be required to determine the accuracy of the analysis when the internal standard method is used.

REFERENCE Semivolatile Organic Compounds by Isotope Dilution GC/MS. Office of Water Regulation and Standards, U.S. EPA Industrial Technology Division, Washington, DC, EPA Method 1625, Rev. C, June 1989 (contact W.A. Telliard, U.S. EPA, Office of Water Regulations and Standards, 401 M St., SW, Washington, DC, 20460. Phone: 202-382-7131).

Benzo(k)fluoranthene   **EPA Method 1625**
CAS #207-08-9

TITLE Semivolatile Organic Compounds by Isotope Dilution GC/MS

MATRIX The compounds may be determined in waters, soils, and municipal sludges by this method.

METHOD SUMMARY This method is used to determine 176 semivolatile toxic organic pollutants associated with the CWA (as amended 1987); the RCRA (as amended 1986); the CERCLA (as amended 1986); and other compounds amenable to extraction and analysis by capillary column gas chromatography-mass spectrometry (GC/MS).

Stable isotopically-labeled analogs of the compounds of interest are added to the sample. If the solids content is less than 1%, a 1-L sample is extracted at pH 12–13, then at pH <2 with methylene chloride using continuous extraction techniques.

If the solids content is 30% or less, the sample is diluted to 1% solids with reagent water, homogenized ultrasonically, and extracted at pH 12–13, then at pH <2 with methylene chloride using continuous extraction techniques. If the solids content is greater than 30%, the sample is extracted using ultrasonic techniques.

Each extract is dried over sodium sulfate, concentrated to a volume of 5 mL, cleaned up using GPC, if necessary, and concentrated. Extracts are concentrated to 1 mL if GPC is not performed, and to 0.5 mL if GPC is performed.

An internal standard is added to the extract, and a 1-mL aliquot of the extract is injected into the GC. The compounds are separated by GC and detected by a MS. The labeled compounds serve to correct the variability of the analytical technique.

INTERFERENCES Solvents, reagents, glassware, and other sample processing hardware may yield artifacts and/or elevated baselines causing misinterpretation of chromatograms and spectra. Materials used in the analysis must be demonstrated to be free from interferences under the conditions of analysis by running method blanks initially and with each sample lot (sample started through the extraction process on a given 8-h shift, to a maximum of 20). Specific selection of reagents and purification of solvents by distillation in all glass systems may be required. Glassware and, where possible, reagents are cleaned by solvent rinse and baking at 450°C for 1-h minimum. Interferences coextracted from samples will vary considerably from source to source, depending on the diversity of the site being sampled.

INSTRUMENTATION Major instrumentation includes a GC with a splitless or on-column injection port for capillary column, a MS with 70 eV electron impact ionization, and a data system to collect and record MS data, and process it. A K-D apparatus is used to concentrate extracts.

GC Column: 30 m × 0.25 mm I.D. 5% phenyl, 94% methyl, 1% vinyl silicone bonded phased fused silica capillary column.

PRECISION & ACCURACY The detection limits of the method are usually dependent on the level of interferences rather than instrumental limitations. The limits typify the minimum quantities that can be detected with no interferences present.

The minimum level (in µg/mL) was 10. This is defined as a minimum level at which the analytical system shall give recognizable mass spectra (background corrected) and acceptable calibration points.

The MDL (in µg/kg) in low solids was 95 and in high solids was 20; these were determined in digested sludge (low solids) and in filter cake or compost (high solids).

The labeled and native compound initial precision as standard deviation (in µg/L) was 26.

The labeled and native compound initial accuracyas average recovery (in µg/L) was 59–143.

SAMPLE COLLECTION, PRESERVATION & HANDLING Collect samples in glass containers. Aqueous samples which flow freely are collected in refrigerated bottles using automatic sampling equipment. Solid samples are collected as grab samples using widemouth jars. Maintain samples at 0 to 4°C from the time of collection until extraction. If residual chlorine is present in aqueous samples, add 80 mg sodium thiosulfate/L

of water. Begin sample extraction within 7 days of collection, and analyze all extracts within 40 days of extraction.

SAMPLE PREPARATION Samples containing 1% solids or less are extracted directly using continuous liquid-liquid extraction techniques. Samples containing 1 to 30% solids are diluted to the 1% level with reagent water and extracted using continuous liquid-liquid extraction techniques. Samples containing greater than 30% solids are extracted using ultrasonic techniques.

Base/neutral extraction — Adjust the pH of the waters in the extractors to 12–13 with 6 N NaOH. Extract with methylene chloride for 24–48 h.

Acid extraction — Adjust the pH of the waters in the extractors to 2 or less using 6 N sulfuric acid. Extract with methylene chloride for 24–48 h.

Ultrasonic extraction of high solids samples — Add anhydrous sodium sulfate to the sample and QC aliquot(s). Add acetone:methylene chloride (1:1) to the sample and mix thoroughly

Concentrate extracts using a K-D apparatus.

QUALITY CONTROL The analyst is permitted to modify this method to improve separations or lower the costs of measurements, provided all performance specifications are met. Analyses of blanks are required to demonstrate freedom from contamination. When results of spikes indicate atypical method performance for samples, the samples are diluted to bring method performance within acceptable limits.

For low solids (aqueous samples), extract, concentrate, and analyze two sets of four 1-L aliquots (8 aliquots total) of the precision and recovery standard. For high solids samples, two sets of four 30-g aliquots of the high solids reference matrix are used.

Spike all samples with labeled compounds to assess method performance. Compute percent recovery of the labeled compounds using the internal standard method. Compare the labeled compound recovery for each compound with the corresponding labeled compound recovery.

Reagent water and high solids reference matrix blanks are analyzed to demonstrate freedom from contamination. Extract and concentrate a 1-L reagent water blank or a high solids reference matrix blank with each sample's lot (samples started through the extraction process on the same 8-h shift, to a maximum of 20 samples).

Field replicates may be collected to determine the precision of the sampling technique, and spiked samples may be required to determine the accuracy of the analysis when the internal standard method is used.

REFERENCE Semivolatile Organic Compounds by Isotope Dilution GC/MS. Office of Water Regulation and Standards, U.S. EPA Industrial Technology Division, Washington, DC, EPA Method 1625, Rev. C, June 1989 (contact W.A. Telliard, U.S. EPA, Office of Water Regulations and Standards, 401 M St., SW, Washington, DC, 20460. Phone: 202-382-7131).

Benzo(k)fluoranthene   EPA Method 625
CAS #207-08-9

TITLE Base/Neutrals and Acids, U.S. EPA Method 625

MATRIX This methods covers municipal and industrial wastewaters.

METHOD SUMMARY Approximately 1 L of sample is serially extracted with methylene chloride at a pH greater than 11 and again at a pH less than 2 using a separatory funnel or a continuous extractor. The methylene chloride extract is dried, concentrated to a volume of 1 mL, and analyzed by GC/MS. Qualitative identification of the parameters in the extract is performed using the retention time and the relative abundance of three characteristic masses (m/z). Qualitative analysis is performed using either external or internal standard techniques with a single characteristic m/z.

INTERFERENCES Method interferences may be caused by contaminants in solvents, reagents, glassware, and other sample processing hardware. Glassware must be scrupulously cleaned. Glassware should be heated in a muffle furnace at 400°C for 5 to 30 min. Some thermally stable materials, such as PCBs, may not be eliminated by this treatment. Solvent rinses with acetone and pesticide quality hexane may be substituted for the muffle furnace heating. Matrix interferences may be caused by contaminants that are coextracted from the sample. The base-neutral extraction may cause significantly reduced recovery of phenols. The packed gas chromatographic columns recommended for the basic fraction may not exhibit sufficient resolution for some analytes.

INSTRUMENTATION A GC/MS system with an injection port designed for on-column injection when using packed columns and for splitless injection when using capillary columns.

Column for base/neutrals: 1.8 m long × 2 mm I.D. glass, packed with 3% SP-2550 on Supelcoport (100/120 mesh) or equivalent.

Column for acids: 1.8 m long × 2 mm I.D. glass, packed with 1% SP-1240DA on Supelcoport (100/120 mesh) or equivalent.

PRECISION & ACCURACY The MDL concentrations were obtained using reagent water. The MDL actually achieved in a given analysis will vary depending on instrument sensitivity and matrix effects. This method was tested by 15 laboratories using reagent water, drinking water, surface water, and industrial wastewaters spiked at six concentrations over the range 5 to 100 µg/L. Single operator precision, overall precision, and method accuracy were found to be directly related to the concentration of the parameter matrix.

The MDL (in µg/L) in reagent water was 2.5.

The standard deviation (in µg/L based on 4 recovery measurements) was 32.3.

The range (in µg/L) for average recovery for 4 measurements was 25.2–145.7.

The range (in %) for percent recovery was 11–162.

Accuracy (in µg/L) as expected recovery for one or more measurements of a sample containing a true concentration of C was 0.87C–1.58.

Precision (in µg/L) as expected single analyst standard deviation of measurements at an average concentration found at X was 0.19X+1.03.

Overall precision (in µg/L) as expected interlaboratory standard deviation of measurements in an average concentration found at X was 0.35X+0.40.

$C =$ *True value of the concentration in µg/L.*
$X =$ *Average recovery found for measurements of samples containing a concentration at C in µg/L.*

SAMPLE PREPARATION   Adjust the pH to >11 with sodium hydroxide and serially extract in a separatory funnel with methylene chloride or else in a continuous extractor. Next, adjust the pH to <2 with sulfuric acid and serially extract in a separatory funnel with methylene chloride or else in a continuous extractor. Dry the extracts separately through a column of anhydrous sodium sulfate and then concentrate each of the extracts to 1.0 mL using a K-D apparatus.

SAMPLE COLLECTION, PRESERVATION & HANDLING-Grab samples must be collected in glass containers. All samples must be refrigerated at 4°C from the time of collection until extraction. If residual chlorine is present, add 80 mg of sodium thiosulfate/L of sample and mix well. All samples must be extracted within 7 days of collection and completely analyzed within 40 days of extraction.

QUALITY CONTROL   Make an initial, one-time, demonstration of the ability to generate acceptable accuracy and precision with this method. Before processing any samples, the analyst must analyze a reagent water blank to demonstrate that interferences from the analytical system and glassware are under control. Each time a set of samples is extracted or reagents are changed, a reagent water blank must be processed. Spike and analyze a minimum of 5% of all samples to monitor and evaluate lab data quality. A QC check sample concentrate that contains each parameter of interest at a concentration of 100 µg/mL in acetone is required. PCBs and multicomponent pesticides may be omitted from this test.

After analysis of five spiked wastewater samples, calculate the average percent recovery and the standard deviation of the percent recovery. Spike all samples with the surrogate standard spiking solution and calculate the percent recovery of each surrogate compound.

REFERENCE   *Federal Register*, Vol. 49, No. 209. Friday, Oct. 26, 1984.

Benzo(k)fluoranthene　　　　　　　　　　**EPA Method 8270**
CAS #207-08-9

TITLE   Semivolatile Organic Compounds by GC/MS

MATRIX   This method is used to determine the concentration of semivolatile organic compounds in extracts prepared from all types of solid waste matrices, soils, and groundwater. Although surface waters are not specifically mentioned, this method should be applicable to water samples from rivers, lakes, etc.

METHOD SUMMARY   This method covers 259 semivolatile organic compounds. In very limited applications direct injection of the sample into the GC/MS system may be appropriate, but this results in very high detection limits (approximately 10,000 µg/L). Typically, a 1-L liquid sample, containing surrogate, and matrix spiking standards, is extracted in a continuous extractor first under acid conditions and then under basic conditions. Typically 30 g of a solid sample, containing surrogate, and matrix spiking standards, is extracted ultrasonically. After concentrating the extract to 1 mL it is spiked with 10 µL of an internal standard solution just prior to analysis by GC/MS. The volume injected should contain about 100 ng of base/neutral and 200 ng of acid surrogates (for a 1-µL injection). Analysis is performed by GC/MS using a capillary GC column.

INTERFERENCES   Raw GC/MS data from all blanks, samples, and spikes must be evaluated for interferences. Contamination by carryover can occur whenever high-concentration and low-concentration samples are sequentially analyzed. To reduce carryover, the sample syringe must be rinsed out between samples with solvent. Whenever an unusually concentrated sample is encountered, it should be followed by the analysis of blank solvent to check for cross-contamination.

INSTRUMENTATION   A GC/MS and a data system are required. The GC column used is a 30 m × 0.25 mm I.D. (or 0.32 mm I.D.) 1-µm film thickness silicone-coated fused silica capillary column. A continuous liquid-liquid extractor equipped with Teflon® or glass connection joints and stopcocks requiring no lubrication, a K-D concentrating apparatus, water bath, and an ultrasonic disrupter with a minimum power of 300 W and with pulsing capability are also required.

PRECISION & ACCURACY   The estimated quantitation limit (EQL) of Method 8270B for determining an individual compound is approximately 1 mg/kg (wet weight) for soil or sediment samples, 1–200 mg/kg for wastes (dependent on matrix and method of preparation), and 10 µg/L for groundwater samples. EQLs will be proportionately higher for sample extracts that require dilution to avoid saturation of the detector.

The EQL(b) for groundwater in µg/L is 10.
The EQL (a, b) for low concentrations in soil and sediment in µg/kg is 660.
Accuracy as µg/L is 0.87C–1.56.
Overall precision in µg/L is 0.035X +0.40.

(a) *EQLs listed for soil/sediment are based on wet weight. Normally data is reported in a dry-weight basis; therefore, EQLs will be higher based on the % dry weight of each sample. This calculation is based on a 30-g sample and gel permeation chromatography cleanup.*
(b) *Sample EQLs are highly matrix-dependent. The EQLs are provided for guidance and may not always be achievable.*

$C =$ *True value for concentration, in µg/L.*
$X =$ *Average recovery found for measurements of samples containing a concentration of C, in µg/L.*

ESTIMATED QUANTITATION LIMIT FOR OTHER MATRICES

Other Matrices	Factor (a)
High-concentration soil and sludges by sonicator	7.5
Non-water miscible waste	75

(a) EQL for other matrices = [EQL for low soil/sediment] × [Factor]. This estimated EQL is similar to an EPA "Practical Quantitation Limit."

SAMPLING METHOD

Liquid samples — Use a 1 or 2½ gallon amber glass bottle with a screw-top Teflon®-lined cover that has been prewashed with detergent and rinsed with distilled water and methanol (or isopropanol).

Soils, sediments, or sludges — Use an 8-oz. widemouth glass with a screw-top Teflon®-lined cover that has been prewashed with detergent and rinsed with distilled water and methanol (or isopropanol).

SAMPLE PRESERVATION

Liquid samples — If residual chlorine is present, add 3 mL of 10% sodium thiosulfate per gallon, cool to 4°C and store in a solvent-free refrigerator until analysis; if chlorine is not present, then eliminate the sodium thiosulfate addition.

Soils, sediments, or sludges — Cool samples to 4°C and store in a solvent-free refrigerator.

MHT Liquid samples must be extracted within 7 days and the extracts analyzed within 40 days. Soils, sediments, or sludges may be stored for a maximum of 14 days and the extracts analyzed within 40 days.

SAMPLE PREPARATION

Liquid samples — Transfer 1 L quantitatively to a continuous extractor. If high concentrations are anticipated, a smaller volume may be used and then diluted with organic-free reagent water to 1 L. Adjust pH, if necessary, to pH <2 using 1:1 (V/V) sulfuric acid. Pipette 1.0 mL of a surrogate standard spiking solution into each sample. For the sample in each analytical batch selected for spiking, add 1.0 mL of a matrix spiking standard. For base/neutral acid analysis, the amount of the surrogates and matrix spiking compounds added to the sample should result in a final concentration of 100 ng/µL of each analyte in the extract to be analyzed (assuming a 1-µL injection). Extract with methylene chloride for 18–24 h. Next, adjust the pH of the aqueous phase to pH >11 using 10 N sodium hydroxide and extract it with methylene chloride again for 18–24 h. Dry the extract through a column containing anhydrous sodium sulfate and concentrate it to 1 mL using a K-D concentrator.

Soils, sediments, or sludges — Use 30 g of sample. Nonporous or wet samples (gummy or clay type) that do not have a free-flowing sandy texture must be mixed with anhydrous sodium sulfate until the sample is free flowing. Add 1 mL of surrogate standards to all samples, spikes, standards, and blanks. For the sample in each analytical batch selected for spiking, add 1.0 mL of a matrix spiking standard. For base/neutral acid analysis, the amount added of the surrogates and matrix spiking compounds should result in a final concentration of 100 ng/µL of each base/neutral analyte and 200 ng/µL of each acid analyte in the extract to be analyzed (assuming a 1-µL injection). Immediately add a 100 mL mixture of 1:1 methylene chloride:acetone and extract the sample ultrasonically for 3 min and then decant or filter the extracts. Repeat the extraction two or more times. Dry the extract using a column with anhydrous sodium sulfate and concentrate it to 1 mL in a K-D concentrator.

QUALITY CONTROL A methylene chloride solution containing 50 ng/µL of decafluorotriphenylphosphine (DFTPP) is used for tuning the GC/MS system each 12-h shift. A system performance check also must be made during every 12-h shift. A standard containing 50 ng/µL each of 4,4'-DDT, pentachlorophenol, and benzidine is required to verify injection port inertness and GC column performance. A calibration standard at mid-concentration, containing each compound of interest, including all required surrogates, must be performed every 12 h during analysis. After the system performance check is met, calibration check compounds (CCCs) are used to check the validity of the initial calibration.

The internal standard responses and retention times in the calibration check standard must be evaluated immediately after or during data acquisition. If the retention time for any internal standard changes by more than 30 seconds from the last check calibration (12 h), the chromatographic system must be inspected for malfunctions and corrections must be made, as required. If the electron ionization current plot (EICP) area for any of the internal standards changes by a factor of two from the last daily calibration standard check, the mass spectrometer must be inspected for malfunctions and corrections must be made, as appropriate.

Demonstrate, through the analysis of a reagent water blank, that interferences from the analytical system, glassware, and reagents are under control. The blank samples should be carried through all stages of the sample preparation and measurement steps. For each analytical batch (up to 20 samples), a reagent blank, matrix spike, and matrix spike duplicate/duplicate must be analyzed (the frequency of the spikes may be different for different monitoring programs). The blank and spiked samples must be carried through all stages of the sample preparation and measurement steps. A QC reference sample concentrate containing each analyte at a concentration of 100 mg/L in methanol is required.

REFERENCE Test Methods for Evaluating Solid Waste (SW-846). U.S. EPA 1983, Method 8270B, Rev. 2, Nov. 1990. Office of Solid Waste, Washington, DC.

Benzo(k)fluoranthene   **EPA Method 8100**
CAS #207-08-9

TITLE Polynuclear Aromatic Hydrocarbons

MATRIX Groundwater, soils, sludges, water miscible liquid wastes, and non-water miscible wastes.

APPLICATION This method is used for the analysis of various PAHs. Samples are extracted, concentrated, and analyzed using direct injection of both neat and diluted organic liquids.

The method provides two optional GC columns that are better than Column 1 and that may help resolve analytes from interferences.

INTERFERENCES Solvents, reagents, and glassware may introduce artifacts. Other interferences may come from coextracted compounds from samples.

INSTRUMENTATION GC capable of on-column injections and a flame with detector (FID). Column 1: a 1.8 m by 2 mm 3% OV-17 on Chromosorb W-AW-DCMS column. Column 2: a 30 m by 0.25 mm SE-54 fused silica capillary column. Column 3: a 30 m by 0.32 mm SE-54 fused silica capillary column.

RANGE 0.1–425 µg/L

MDL Not reported.

PQL FACTORS FOR MULTIPLYING × FID MDL VALUE
Not available.

PRECISION 0.69X + 0.10 µg/L (overall precision).

ACCURACY 0.59C + 0.00 µg/L (as recovery).

SAMPLING METHOD Use 8-oz. widemouth glass bottles with Teflon®-lined caps for concentrated waste samples, soils, sediments, and sludges. Use 1 or 2½ gallon amber glass bottles with Teflon®-lined caps for liquid (water) samples.

STABILITY Cool soil, sediment, sludge, and liquid samples to 4°C. If residual chlorine is present in liquid samples add 3 mL of 10% sodium thiosulfate per gallon of sample and cool to 4°C.

MHT 14 days for concentrated waste, soil, sediment, or sludge; 7 days for liquid samples; all extracts must be analyzed within 40 days.

QUALITY CONTROL A quality control check sample concentrate containing each analyte of interest is required. The QC check sample concentrate may be prepared from pure standard materials or purchased as certified solutions. Use appropriate trip, matrix, control site, method, reagent, and solvent blanks. Internal, surrogate, and five concentration level calibration standards are used. The quality control check sample concentrate should contain benzo(k)fluoranthene at 5 µg/mL in acetonitrile.

REFERENCE Method 8100, SW-846, 3rd ed., Nov. 1986.

Benzo(k)fluoranthene   **EPA Method 8310**
CAS #207-08-9

TITLE Polynuclear Aromatic Hydrocarbons

MATRIX Groundwater, soils, sludges, water miscible liquid wastes, and non-water miscible wastes.

APPLICATION This method is used for the analysis of 16 polynucleararomatic hydrocarbons(PAHs). Samples are extracted, concentrated, and analyzed using HPLC with detection by UV and fluorescence detectors.

INTERFERENCES Solvents, reagents, and glassware may introduce artifacts. Other interferences may come from coextracted compounds from samples.

INSTRUMENTATION HPLC with a gradient pumping system and a 250 mm by 2.6 mm reverse phase HC-ODS Sil-X 5-micron particle size column. The fluorescence detector uses an excitation wavelength of 280 nm and emission greater than 389 nm cutoff with dispersive optics.

RANGE 0.1–425 µg/L

MDL 0.017 µg/L (fluorescence; reagent water).

PQL FACTORS FOR MULTIPLYING × FID MDL VALUE
Matrix Multiplication Factor

Groundwater	10
Low-level soil by sonication with GPC cleanup	670
High-level soil and sludge by sonication	10,000
Non-water miscible waste	100,000

PRECISION 0.69X + 0.10 µg/L (overall precision).

ACCURACY 0.59C + 0.00 µg/L (as recovery).

SAMPLING METHOD Use 8-oz. widemouth glass bottles with Teflon®-lined caps for concentrated waste samples, soils, sediments, and sludges. Use 1 or 2½ gallon amber glass bottles with Teflon®-lined caps for liquid (water) samples.

STABILITY Cool soil, sediment, sludge, and liquid samples to 4°C. If residual chlorine is present in liquid samples add 3 mL of 10% sodium thiosulfate per gallon of sample and cool to 4°C.

MHT 14 days for concentrated waste, soil, sediment, or sludge; 7 days for liquid samples; all extracts must be analyzed within 40 days.

QUALITY CONTROL Internal, surrogate, and five concentration level calibration standardsareused. The calibration standards must be used with the analytical method blank. Aquality control check sample concentrate containingbenzo(k)fluoranthene at 5 µg/mL is required. The QC check sample concentrate may be prepared from pure standard materials or purchased as certified solutions. Use appropriate trip, matrix, control site, method, reagent, and solvent blanks.

REFERENCE Method 8310, SW-846, 3rd ed., Nov. 1986.

2,3-Benzofluorene   **EPA Method 1625**
CAS #243-17-4

TITLE Semivolatile Organic Compounds by Isotope Dilution GC/MS

MATRIX The compounds may be determined in waters, soils, and municipal sludges by this method.

METHOD SUMMARY This method is used to determine 176 semivolatile toxic organic pollutants associated with the CWA (as amended 1987); the RCRA (as amended 1986); the CERCLA (as amended 1986); and other compounds amenable

to extraction and analysis by capillary column gas chromatography-mass spectrometry (GC/MS).

Stable isotopically-labeled analogs of the compounds of interest are added to the sample. If the solids content is less than 1%, a 1-L sample is extracted at pH 12–13, then at pH <2 with methylene chloride using continuous extraction techniques.

If the solids content is 30% or less, the sample is diluted to 1% solids with reagent water, homogenized ultrasonically, and extracted at pH 12–13, then at pH <2 with methylene chloride using continuous extraction techniques. If the solids content is greater than 30%, the sample is extracted using ultrasonic techniques.

Each extract is dried over sodium sulfate, concentrated to a volume of 5 mL, cleaned up using GPC, if necessary, and concentrated. Extracts are concentrated to 1 mL if GPC is not performed, and to 0.5 mL if GPC is performed.

An internal standard is added to the extract, and a 1-mL aliquot of the extract is injected into the GC. The compounds are separated by GC and detected by a MS. The labeled compounds serve to correct the variability of the analytical technique.

INTERFERENCES Solvents, reagents, glassware, and other sample processing hardware may yield artifacts and/or elevated baselines causing misinterpretation of chromatograms and spectra. Materials used in the analysis must be demonstrated to be free from interferences under the conditions of analysis by running method blanks initially and with each sample lot (sample started through the extraction process on a given 8-h shift, to a maximum of 20). Specific selection of reagents and purification of solvents by distillation in all glass systems may be required. Glassware and, where possible, reagents are cleaned by solvent rinse and baking at 450°C for 1-h minimum. Interferences coextracted from samples will vary considerably from source to source, depending on the diversity of the site being sampled.

INSTRUMENTATION Major instrumentation includes a GC with a splitless or on-column injection port for capillary column, a MS with 70 eV electron impact ionization, and a data system to collect and record MS data, and process it. A K-D apparatus is used to concentrate extracts.

GC Column: 30 m × 0.25 mm I.D. 5% phenyl, 94% methyl, 1% vinyl silicone bonded phased fused silica capillary column.

PRECISION & ACCURACY The detection limits of the method are usually dependent on the level of interferences rather than instrumental limitations. The limits typify the minimum quantities that can be detected with no interferences present.

The minimum level (in µg/mL) was not listed. This is defined as a minimum level at which the analytical system shall give recognizable mass spectra (background corrected) and acceptable calibration points.
The MDL (in µg/kg) in low solids was not listed and in high solids was not listed; these were determined in digested sludge (low solids) and in filter cake or compost (high solids).
The labeled and native compound initial precision as standard deviation (in µg/L) was not listed.

The labeled and native compound initial accuracy as average recovery (in µg/L) was not listed.

SAMPLE COLLECTION, PRESERVATION & HANDLING
Collect samples in glass containers. Aqueous samples which flow freely are collected in refrigerated bottles using automatic sampling equipment. Solid samples are collected as grab samples using widemouth jars. Maintain samples at 0 to 4°C from the time of collection until extraction. If residual chlorine is present in aqueous samples, add 80 mg sodium thiosulfate/L of water. Begin sample extraction within 7 days of collection, and analyze all extracts within 40 days of extraction.

SAMPLE PREPARATION Samples containing 1% solids or less are extracted directly using continuous liquid-liquid extraction techniques. Samples containing 1 to 30% solids are diluted to the 1% level with reagent water and extracted using continuous liquid-liquid extraction techniques. Samples containing greater than 30% solids are extracted using ultrasonic techniques.

Base/neutral extraction — Adjust the pH of the waters in the extractors to 12–13 with 6 N NaOH. Extract with methylene chloride for 24–48 h.
Acid extraction — Adjust the pH of the waters in the extractors to 2 or less using 6 N sulfuric acid. Extract with methylene chloride for 24–48 h.
Ultrasonic extraction of high solids samples — Add anhydrous sodium sulfate to the sample and QC aliquot(s). Add acetone:methylene chloride (1:1) to the sample and mix thoroughly

Concentrate extracts using a K-D apparatus.

QUALITY CONTROL The analyst is permitted to modify this method to improve separations or lower the costs of measurements, provided all performance specifications are met. Analyses of blanks are required to demonstrate freedom from contamination. When results of spikes indicate atypical method performance for samples, the samples are diluted to bring method performance within acceptable limits.

For low solids (aqueous samples), extract, concentrate, and analyze two sets of four 1-L aliquots (8 aliquots total) of the precision and recovery standard. For high solids samples, two sets of four 30-g aliquots of the high solids reference matrix are used.

Spike all samples with labeled compounds to assess method performance. Compute percent recovery of the labeled compounds using the internal standard method. Compare the labeled compound recovery for each compound with the corresponding labeled compound recovery.

Reagent water and high solids reference matrix blanks are analyzed to demonstrate freedom from contamination. Extract and concentrate a 1-L reagent water blank or a high solids reference matrix blank with each sample's lot (samples started through the extraction process on the same 8-h shift, to a maximum of 20 samples).

Field replicates may be collected to determine the precision of the sampling technique, and spiked samples may be required

to determine the accuracy of the analysis when the internal standard method is used.

REFERENCE Semivolatile Organic Compounds by Isotope Dilution GC/MS. Office of Water Regulation and Standards, U.S. EPA Industrial Technology Division, Washington, DC, EPA Method 1625, Rev. C, June 1989 (contact W.A. Telliard, U.S. EPA, Office of Water Regulations and Standards, 401 M St., SW, Washington, DC, 20460. Phone: 202-382-7131).

2,3-Benzofluorene   **EPA Method 1625**
CAS #243-17-4

TITLE Semivolatile Organic Compounds by Isotope Dilution GC/MS

MATRIX The compounds may be determined in waters, soils, and municipal sludges by this method.

METHOD SUMMARY This method is used to determine 176 semivolatile toxic organic pollutants associated with the CWA (as amended 1987); the RCRA (as amended 1986); the CERCLA (as amended 1986); and other compounds amenable to extraction and analysis by capillary column gas chromatography-mass spectrometry (GC/MS).

Stable isotopically-labeled analogs of the compounds of interest are added to the sample. If the solids content is less than 1%, a 1-L sample is extracted at pH 12–13, then at pH <2 with methylene chloride using continuous extraction techniques.

If the solids content is 30% or less, the sample is diluted to 1% solids with reagent water, homogenized ultrasonically, and extracted at pH 12–13, then at pH <2 with methylene chloride using continuous extraction techniques. If the solids content is greater than 30%, the sample is extracted using ultrasonic techniques.

Each extract is dried over sodium sulfate, concentrated to a volume of 5 mL, cleaned up using GPC, if necessary, and concentrated. Extracts are concentrated to 1 mL if GPC is not performed, and to 0.5 mL if GPC is performed.

An internal standard is added to the extract, and a 1-mL aliquot of the extract is injected into the GC. The compounds are separated by GC and detected by a MS. The labeled compounds serve to correct the variability of the analytical technique.

INTERFERENCES Solvents, reagents, glassware, and other sample processing hardware may yield artifacts and/or elevated baselines causing misinterpretation of chromatograms and spectra. Materials used in the analysis must be demonstrated to be free from interferences under the conditions of analysis by running method blanks initially and with each sample lot (sample started through the extraction process on a given 8-h shift, to a maximum of 20). Specific selection of reagents and purification of solvents by distillation in all glass systems may be required. Glassware and, where possible, reagents are cleaned by solvent rinse and baking at 450°C for 1-h minimum. Interferences coextracted from samples will vary considerably from source to source, depending on the diversity of the site being sampled.

INSTRUMENTATION Major instrumentation includes a GC with a splitless or on-column injection port for capillary column, a MS with 70 eV electron impact ionization, and a data system to collect and record MS data, and process it. A K-D apparatus is used to concentrate extracts.

GC Column: 30 m × 0.25 mm I.D. 5% phenyl, 94% methyl, 1% vinyl silicone bonded phased fused silica capillary column.

PRECISION & ACCURACY The detection limits of the method are usually dependent on the level of interferences rather than instrumental limitations. The limits typify the minimum quantities that can be detected with no interferences present.

The minimum level (in µg/mL) was not listed. This is defined as a minimum level at which the analytical system shall give recognizable mass spectra (background corrected) and acceptable calibration points.
The MDL (in µg/kg) in low solids was not listed and in high solids was not listed; these were determined in digested sludge (low solids) and in filter cake or compost (high solids).
The labeled and native compound initial precision as standard deviation (in µg/L) was not listed.
The labeled and native compound initial accuracy as average recovery (in µg/L) was not listed.

SAMPLE COLLECTION, PRESERVATION & HANDLING Collect samples in glass containers. Aqueous samples which flow freely are collected in refrigerated bottles using automatic sampling equipment. Solid samples are collected as grab samples using widemouth jars. Maintain samples at 0 to 4°C from the time of collection until extraction. If residual chlorine is present in aqueous samples, add 80 mg sodium thiosulfate/L of water. Begin sample extraction within 7 days of collection, and analyze all extracts within 40 days of extraction.

SAMPLE PREPARATION Samples containing 1% solids or less are extracted directly using continuous liquid-liquid extraction techniques. Samples containing 1 to 30% solids are diluted to the 1% level with reagent water and extracted using continuous liquid-liquid extraction techniques. Samples containing greater than 30% solids are extracted using ultrasonic techniques.

Base/neutral extraction — Adjust the pH of the waters in the extractors to 12–13 with 6 *N* NaOH. Extract with methylene chloride for 24–48 h.
Acid extraction — Adjust the pH of the waters in the extractors to 2 or less using 6 *N* sulfuric acid. Extract with methylene chloride for 24–48 h.
Ultrasonic extraction of high solids samples — Add anhydrous sodium sulfate to the sample and QC aliquot(s). Add acetone:methylene chloride (1:1) to the sample and mix thoroughly

Concentrate extracts using a K-D apparatus.

QUALITY CONTROL The analyst is permitted to modify this method to improve separations or lower the costs of measurements, provided all performance specifications are met. Analyses of blanks are required to demonstrate freedom from contamination. When results of spikes indicate atypical

method performance for samples, the samples are diluted to bring method performance within acceptable limits.

For low solids (aqueous samples), extract, concentrate, and analyze two sets of four 1-L aliquots (8 aliquots total) of the precision and recovery standard. For high solids samples, two sets of four 30-g aliquots of the high solids reference matrix are used.

Spike all samples with labeled compounds to assess method performance. Compute percent recovery of the labeled compounds using the internal standard method. Compare the labeled compound recovery for each compound with the corresponding labeled compound recovery.

Reagent water and high solids reference matrix blanks are analyzed to demonstrate freedom from contamination. Extract and concentrate a 1-L reagent water blank or a high solids reference matrix blank with each sample's lot (samples started through the extraction process on the same 8-h shift, to a maximum of 20 samples).

Field replicates may be collected to determine the precision of the sampling technique, and spiked samples may be required to determine the accuracy of the analysis when the internal standard method is used.

REFERENCE Semivolatile Organic Compounds by Isotope Dilution GC/MS. Office of Water Regulation and Standards, U.S. EPA Industrial Technology Division, Washington, DC, EPA Method 1625, Rev. C, June 1989 (contact W.A. Telliard, U.S. EPA, Office of Water Regulations and Standards, 401 M St., SW, Washington, DC, 20460. Phone: 202-382-7131).

Benzoic acid   **EPA Method 1625**
CAS #65-85-0

TITLE Semivolatile Organic Compounds by Isotope Dilution GC/MS

MATRIX The compounds may be determined in waters, soils, and municipal sludges by this method.

METHOD SUMMARY This method is used to determine 176 semivolatile toxic organic pollutants associated with the CWA (as amended 1987); the RCRA (as amended 1986); the CERCLA (as amended 1986); and other compounds amenable to extraction and analysis by capillary column gas chromatography-mass spectrometry (GC/MS).

Stable isotopically-labeled analogs of the compounds of interest are added to the sample. If the solids content is less than 1%, a 1-L sample is extracted at pH 12–13, then at pH <2 with methylene chloride using continuous extraction techniques.

If the solids content is 30% or less, the sample is diluted to 1% solids with reagent water, homogenized ultrasonically, and extracted at pH 12–13, then at pH <2 with methylene chloride using continuous extraction techniques. If the solids content is greater than 30%, the sample is extracted using ultrasonic techniques.

Each extract is dried over sodium sulfate, concentrated to a volume of 5 mL, cleaned up using GPC, if necessary, and concentrated. Extracts are concentrated to 1 mL if GPC is not performed, and to 0.5 mL if GPC is performed.

An internal standard is added to the extract, and a 1-mL aliquot of the extract is injected into the GC. The compounds are separated by GC and detected by a MS. The labeled compounds serve to correct the variability of the analytical technique.

INTERFERENCES Solvents, reagents, glassware, and other sample processing hardware may yield artifacts and/or elevated baselines causing misinterpretation of chromatograms and spectra. Materials used in the analysis must be demonstrated to be free from interferences under the conditions of analysis by running method blanks initially and with each sample lot (sample started through the extraction process on a given 8-h shift, to a maximum of 20). Specific selection of reagents and purification of solvents by distillation in all glass systems may be required. Glassware and, where possible, reagents are cleaned by solvent rinse and baking at 450°C for 1-h minimum. Interferences coextracted from samples will vary considerably from source to source, depending on the diversity of the site being sampled.

INSTRUMENTATION Major instrumentation includes a GC with a splitless or on-column injection port for capillary column, a MS with 70 eV electron impact ionization, and a data system to collect and record MS data, and process it. A K-D apparatus is used to concentrate extracts.

GC Column: 30 m × 0.25 mm I.D. 5% phenyl, 94% methyl, 1% vinyl silicone bonded phased fused silica capillary column.

PRECISION & ACCURACY The detection limits of the method are usually dependent on the level of interferences rather than instrumental limitations. The limits typify the minimum quantities that can be detected with no interferences present.

The minimum level (in µg/mL) was not listed. This is defined as a minimum level at which the analytical system shall give recognizable mass spectra (background corrected) and acceptable calibration points.
The MDL (in µg/kg) in low solids was not listed and in high solids was not listed; these were determined in digested sludge (low solids) and in filter cake or compost (high solids).
The labeled and native compound initial precision as standard deviation (in µg/L) was not listed.
The labeled and native compound initial accuracy as average recovery (in µg/L) was not listed.

SAMPLE COLLECTION, PRESERVATION & HANDLING
Collect samples in glass containers. Aqueous samples which flow freely are collected in refrigerated bottles using automatic sampling equipment. Solid samples are collected as grab samples using widemouth jars. Maintain samples at 0 to 4°C from the time of collection until extraction. If residual chlorine is present in aqueous samples, add 80 mg sodium thiosulfate/L of water. Begin sample extraction within 7 days of collection, and analyze all extracts within 40 days of extraction.

SAMPLE PREPARATION Samples containing 1% solids or less are extracted directly using continuous liquid-liquid extraction techniques. Samples containing 1 to 30% solids are diluted to the 1% level with reagent water and extracted using continuous liquid-liquid extraction techniques. Samples containing greater than 30% solids are extracted using ultrasonic techniques.

Base/neutral extraction — Adjust the pH of the waters in the extractors to 12–13 with 6 *N* NaOH. Extract with methylene chloride for 24–48 h.

Acid extraction — Adjust the pH of the waters in the extractors to 2 or less using 6 *N* sulfuric acid. Extract with methylene chloride for 24–48 h.

Ultrasonic extraction of high solids samples — Add anhydrous sodium sulfate to the sample and QC aliquot(s). Add acetone:methylene chloride (1:1) to the sample and mix thoroughly

Concentrate extracts using a K-D apparatus.

QUALITY CONTROL The analyst is permitted to modify this method to improve separations or lower the costs of measurements, provided all performance specifications are met. Analyses of blanks are required to demonstrate freedom from contamination. When results of spikes indicate atypical method performance for samples, the samples are diluted to bring method performance within acceptable limits.

For low solids (aqueous samples), extract, concentrate, and analyze two sets of four 1-L aliquots (8 aliquots total) of the precision and recovery standard. For high solids samples, two sets of four 30-g aliquots of the high solids reference matrix are used.

Spike all samples with labeled compounds to assess method performance. Compute percent recoveryof the labeled compounds using the internal standard method. Compare the labeled compound recovery for each compound with the corresponding labeled compound recovery.

Reagent water and high solids reference matrix blanks are analyzed to demonstrate freedom from contamination. Extract and concentrate a 1-L reagent water blank or a high solids reference matrix blank with each sample's lot (samples started through the extraction process on the same 8-h shift, to a maximum of 20 samples).

Field replicates may be collected to determine the precision of the sampling technique, and spiked samples may be required to determine the accuracy of the analysis when the internal standard method is used.

REFERENCE Semivolatile Organic Compounds by Isotope Dilution GC/MS. Office of Water Regulation and Standards, U.S. EPA Industrial Technology Division, Washington, DC, EPA Method 1625, Rev. C, June 1989 (contact W.A. Telliard, U.S. EPA, Office of Water Regulations and Standards, 401 M St., SW, Washington, DC, 20460. Phone: 202-382-7131).

Benzoic acid   **EPA Method 1625**
CAS #65-85-0

TITLE Semivolatile Organic Compounds by Isotope Dilution GC/MS

MATRIX The compounds may be determined in waters, soils, and municipal sludges by this method.

METHOD SUMMARY This method is used to determine 176 semivolatile toxic organic pollutants associated with the CWA (as amended 1987); the RCRA (as amended 1986); the CERCLA (as amended 1986); and other compounds amenable to extraction and analysis by capillary column gas chromatography-mass spectrometry (GC/MS).

Stable isotopically-labeled analogs of the compounds of interest are added to the sample. If the solids content is less than 1%, a 1-L sample is extracted at pH 12–13, then at pH <2 with methylene chloride using continuous extraction techniques.

If the solids content is 30% or less, the sample is diluted to 1% solids with reagent water, homogenized ultrasonically, and extracted at pH 12–13, then at pH <2 with methylene chloride using continuous extraction techniques. If the solids content is greater than 30%, the sample is extracted using ultrasonic techniques.

Each extract is dried over sodium sulfate, concentrated to a volume of 5 mL, cleaned up using GPC, if necessary, and concentrated. Extracts are concentrated to 1 mL if GPC is not performed, and to 0.5 mL if GPC is performed.

An internal standard is added to the extract, and a 1-mL aliquot of the extract is injected into the GC. The compounds are separated by GC and detected by a MS. The labeled compounds serve to correct the variability of the analytical technique.

INTERFERENCES Solvents, reagents, glassware, and other sample processing hardware may yield artifacts and/or elevated baselines causing misinterpretation of chromatograms and spectra. Materials used in the analysis must be demonstrated to be free from interferences under the conditions of analysis by running method blanks initially and with each sample lot (sample started through the extraction process on a given 8-h shift, to a maximum of 20). Specific selection of reagents and purification of solvents by distillation in all glass systems may be required. Glassware and, where possible, reagents are cleaned by solvent rinse and baking at 450°C for 1-h minimum. Interferences coextracted from samples will vary considerably from source to source, depending on the diversity of the site being sampled.

INSTRUMENTATION Major instrumentation includes a GC with a splitless or on-column injection port for capillary column, a MS with 70 eV electron impact ionization, and a data system to collect and record MS data, and process it. A K-D apparatus is used to concentrate extracts.

GC Column: 30 m × 0.25 mm I.D. 5% phenyl, 94% methyl, 1% vinyl silicone bonded phased fused silica capillary column.

PRECISION & ACCURACY The detection limits of the method are usually dependent on the level of interferences rather than instrumental limitations. The limits typify the minimum quantities that can be detected with no interferences present.

The minimum level (in µg/mL) was not listed. This is defined as a minimum level at which the analytical system shall give recognizable mass spectra (background corrected) and acceptable calibration points.

The MDL (in µg/kg) in low solids was not listed and in high solids was not listed; these were determined in digested sludge (low solids) and in filter cake or compost (high solids).

The labeled and native compound initial precision as standard deviation (in µg/L) was not listed.

The labeled and native compound initial accuracy as average recovery (in µg/L) was not listed.

SAMPLE COLLECTION, PRESERVATION & HANDLING Collect samples in glass containers. Aqueous samples which flow freely are collected in refrigerated bottles using automatic sampling equipment. Solid samples are collected as grab samples using widemouth jars. Maintain samples at 0 to 4°C from the time of collection until extraction. If residual chlorine is present in aqueous samples, add 80 mg sodium thiosulfate/L of water. Begin sample extraction within 7 days of collection, and analyze all extracts within 40 days of extraction.

SAMPLE PREPARATION Samples containing 1% solids or less are extracted directly using continuous liquid-liquid extraction techniques. Samples containing 1 to 30% solids are diluted to the 1% level with reagent water and extracted using continuous liquid-liquid extraction techniques. Samples containing greater than 30% solids are extracted using ultrasonic techniques.

Base/neutral extraction — Adjust the pH of the waters in the extractors to 12–13 with 6 N NaOH. Extract with methylene chloride for 24–48 h.

Acid extraction — Adjust the pH of the waters in the extractors to 2 or less using 6 N sulfuric acid. Extract with methylene chloride for 24–48 h.

Ultrasonic extraction of high solids samples — Add anhydrous sodium sulfate to the sample and QC aliquot(s). Add acetone:methylene chloride (1:1) to the sample and mix thoroughly

Concentrate extracts using a K-D apparatus.

QUALITY CONTROL The analyst is permitted to modify this method to improve separations or lower the costs of measurements, provided all performance specifications are met. Analyses of blanks are required to demonstrate freedom from contamination. When results of spikes indicate atypical method performance for samples, the samples are diluted to bring method performance within acceptable limits.

For low solids (aqueous samples), extract, concentrate, and analyze two sets of four 1-L aliquots (8 aliquots total) of the precision and recovery standard. For high solids samples, two sets of four 30-g aliquots of the high solids reference matrix are used.

Spike all samples with labeled compounds to assess method performance. Compute percent recovery of the labeled compounds using the internal standard method. Compare the labeled compound recovery for each compound with the corresponding labeled compound recovery.

Reagent water and high solids reference matrix blanks are analyzed to demonstrate freedom from contamination. Extract and concentrate a 1-L reagent water blank or a high solids reference matrix blank with each sample's lot (samples started through the extraction process on the same 8-h shift, to a maximum of 20 samples).

Field replicates may be collected to determine the precision of the sampling technique, and spiked samples may be required to determine the accuracy of the analysis when the internal standard method is used.

REFERENCE Semivolatile Organic Compounds by Isotope Dilution GC/MS. Office of Water Regulation and Standards, U.S. EPA Industrial Technology Division, Washington, DC, EPA Method 1625, Rev. C, June 1989 (contact W.A. Telliard, U.S. EPA, Office of Water Regulations and Standards, 401 M St., SW, Washington, DC, 20460. Phone: 202-382-7131).

Benzoic acid   **EPA Method 8270**
CAS #65-85-0

TITLE Semivolatile Organic Compounds by GC/MS

MATRIX This method is used to determine the concentration of semivolatile organic compounds in extracts prepared from all types of solid waste matrices, soils, and groundwater. Although surface waters are not specifically mentioned, this method should be applicable to water samples from rivers, lakes, etc.

METHOD SUMMARY This method covers 259 semivolatile organic compounds. In very limited applications direct injection of the sample into the GC/MS system may be appropriate, but this results in very high detection limits (approximately 10,000 µg/L). Typically, a 1-L liquid sample, containing surrogate, and matrix spiking standards, is extracted in a continuous extractor first under acid conditions and then under basic conditions. Typically 30 g of a solid sample, containing surrogate, and matrix spiking standards, is extracted ultrasonically. After concentrating the extract to 1 mL it is spiked with 10 µL of an internal standard solution just prior to analysis by GC/MS. The volume injected should contain about 100 ng of base/neutral and 200 ng of acid surrogates (for a 1-µL injection). Analysis is performed by GC/MS using a capillary GC column.

INTERFERENCES Raw GC/MS data from all blanks, samples, and spikes must be evaluated for interferences. Contamination by carryover can occur whenever high-concentration and low-concentration samples are sequentially analyzed. To reduce carryover, the sample syringe must be rinsed out between samples with solvent. Whenever an unusually concentrated sample is encountered, it should be followed by the analysis of blank solvent to check for cross-contamination.

INSTRUMENTATION A GC/MS and a data system are required. The GC column used is a 30 m × 0.25 mm I.D. (or 0.32 mm I.D.) 1-μm film thickness silicone-coated fused silica capillary column. A continuous liquid-liquid extractor equipped with Teflon® or glass connection joints and stopcocks requiring no lubrication, a K-D concentrating apparatus, water bath, and an ultrasonic disrupter with a minimum power of 300 W and with pulsing capability are also required.

PRECISION & ACCURACY The estimated quantitation limit (EQL) of Method 8270B for determining an individual compound is approximately 1 mg/kg (wet weight) for soil or sediment samples, 1–200 mg/kg for wastes (dependent on matrix and method of preparation), and 10 μg/L for groundwater samples. EQLs will be proportionately higher for sample extracts that require dilution to avoid saturation of the detector.

The EQL(b) for groundwater in μg/L is 50.
The EQL (a, b) for low concentrations in soil and sediment in μg/kg is 3300.
Accuracy as μg/L is not listed.
Overall precision in μg/L is not listed.

(a) *EQLs listed for soil/sediment are based on wet weight. Normally data is reported in a dry-weight basis; therefore, EQLs will be higher based on the % dry weight of each sample. This calculation is based on a 30-g sample and gel permeation chromatography cleanup.*
(b) *Sample EQLs are highly matrix-dependent. The EQLs are provided for guidance and may not always be achievable.*
C = *True value for concentration, in μg/L.*
X = *Average recovery found for measurements of samples containing a concentration of C, in μg/L.*

ESTIMATED QUANTITATION LIMIT FOR OTHER MATRICES

Other Matrices	Factor (a)
High-concentration soil and sludges by sonicator	7.5
Non-water miscible waste	75

(a) *EQL for other matrices = [EQL for low soil/sediment] × [Factor]. This estimated EQL is similar to an EPA "Practical Quantitation Limit."*

SAMPLING METHOD
Liquid samples — Use a 1 or 2½ gallon amber glass bottle with a screw-top Teflon®-lined cover that has been prewashed with detergent and rinsed with distilled water and methanol (or isopropanol).

Soils, sediments, or sludges — Use an 8-oz. widemouth glass with a screw-top Teflon®-lined cover that has been prewashed with detergent and rinsed with distilled water and methanol (or isopropanol).

SAMPLE PRESERVATION
Liquid samples — If residual chlorine is present, add 3 mL of 10% sodium thiosulfate per gallon, cool to 4°C and store in a solvent-free refrigerator until analysis; if chlorine is not present, then eliminate the sodium thiosulfate addition.

Soils, sediments, or sludges — Cool samples to 4°C and store in a solvent-free refrigerator.

MHT Liquid samples must be extracted within 7 days and the extracts analyzed within 40 days. Soils, sediments, or sludges may be stored for a maximum of 14 days and the extracts analyzed within 40 days.

SAMPLE PREPARATION
Liquid samples — Transfer 1 L quantitatively to a continuous extractor. If high concentrations are anticipated, a smaller volume may be used and then diluted with organic-free reagent water to 1 L. Adjust pH, if necessary, to pH <2 using 1:1 (V/V) sulfuric acid. Pipette 1.0 mL of a surrogate standard spiking solution into each sample. For the sample in each analytical batch selected for spiking, add 1.0 mL of a matrix spiking standard. For base/neutral acid analysis, the amount of the surrogates and matrix spiking compounds added to the sample should result in a final concentration of 100 ng/μL of each analyte in the extract to be analyzed (assuming a 1-μL injection). Extract with methylene chloride for 18–24 h. Next, adjust the pH of the aqueous phase to pH >11 using 10 N sodium hydroxide and extract it with methylene chloride again for 18–24 h. Dry the extract through a column containing anhydrous sodium sulfate and concentrate it to 1 mL using a K-D concentrator.

Soils, sediments, or sludges — Use 30 g of sample. Nonporous or wet samples (gummy or clay type) that do not have a free-flowing sandy texture must be mixed with anhydrous sodium sulfate until the sample is free flowing. Add 1 mL of surrogate standards to all samples, spikes, standards, and blanks. For the sample in each analytical batch selected for spiking, add 1.0 mL of a matrix spiking standard. For base/neutral acid analysis, the amount added of the surrogates and matrix spiking compounds should result in a final concentration of 100 ng/μL of each base/neutral analyte and 200 ng/μL of each acid analyte in the extract to be analyzed (assuming a 1-μL injection). Immediately add a 100 mL mixture of 1:1 methylene chloride:acetone and extract the sample ultrasonically for 3 min and then decant or filter the extracts. Repeat the extraction two or more times. Dry the extract using a column with anhydrous sodium sulfate and concentrate it to 1 mL in a K-D concentrator.

Note: This compound may be exhibit erratic chromatographic behavior, especially if the GC system is contaminated with high boiling material.

QUALITY CONTROL A methylene chloride solution containing 50 ng/μL of decafluorotriphenylphosphine (DFTPP) is used for tuning the GC/MS system each 12-h shift. A system performance check also must be made during every 12-h shift. A standard containing 50 ng/μL each of 4,4'-DDT, pentachlorophenol, and benzidine is required to verify injection port inertness and GC column performance. A calibration standard at mid-concentration, containing each compound of interest, including all required surrogates, must be performed every 12 h during analysis. After the system performance check is met, calibration check compounds (CCCs) are used to check the validity of the initial calibration.

The internal standard responses and retention times in the calibration check standard must be evaluated immediately after or during data acquisition. If the retention time for any internal standard changes by more than 30 seconds from the last check calibration (12 h), the chromatographic system must be inspected for malfunctions and corrections must be made, as required. If the electron ionization current plot (EICP) area for any of the internal standards changes by a factor of two from the last daily calibration standard check, the mass spectrometer must be inspected for malfunctions and corrections must be made, as appropriate.

Demonstrate, through the analysis of a reagent water blank, that interferences from the analytical system, glassware, and reagents are under control. The blank samples should be carried through all stages of the sample preparation and measurement steps. For each analytical batch (up to 20 samples), a reagent blank, matrix spike, and matrix spike duplicate/duplicate must be analyzed (the frequency of the spikes may be different for different monitoring programs). The blank and spiked samples must be carried through all stages of the sample preparation and measurement steps. A QC reference sample concentrate containing each analyte at a concentration of 100 mg/L in methanol is required.

REFERENCE Test Methods for Evaluating Solid Waste (SW-846). U.S. EPA 1983, Method 8270B, Rev. 2, Nov. 1990. Office of Solid Waste, Washington, DC.

---

p-Benzoquinone   EPA Method 8270
CAS #106-51-4

TITLE Semivolatile Organic Compounds by GC/MS

MATRIX This method is used to determine the concentration of semivolatile organic compounds in extracts prepared from all types of solid waste matrices, soils, and groundwater. Although surface waters are not specifically mentioned, this method should be applicable to water samples from rivers, lakes, etc.

METHOD SUMMARY This method covers 259 semivolatile organic compounds. In very limited applications direct injection of the sample into the GC/MS system may be appropriate, but this results in very high detection limits (approximately 10,000 µg/L). Typically, a 1-L liquid sample, containing surrogate, and matrix spiking standards, is extracted in a continuous extractor first under acid conditions and then under basic conditions. Typically 30 g of a solid sample, containing surrogate, and matrix spiking standards, is extracted ultrasonically. After concentrating the extract to 1 mL it is spiked with 10 µL of an internal standard solution just prior to analysis by GC/MS. The volume injected should contain about 100 ng of base/neutral and 200 ng of acid surrogates (for a 1-µL injection). Analysis is performed by GC/MS using a capillary GC column.

INTERFERENCES Raw GC/MS data from all blanks, samples, and spikes must be evaluated for interferences. Contamination by carryover can occur whenever high-concentration and low-concentration samples are sequentially analyzed. To reduce carryover, the sample syringe must be rinsed out between samples with solvent. Whenever an unusually concentrated sample is encountered, it should be followed by the analysis of blank solvent to check for cross-contamination.

INSTRUMENTATION A GC/MS and a data system are required. The GC column used is a 30 m × 0.25 mm I.D. (or 0.32 mm I.D.) 1-µm film thickness silicone-coated fused silica capillary column. A continuous liquid-liquid extractor equipped with Teflon® or glass connection joints and stopcocks requiring no lubrication, a K-D concentrating apparatus, water bath, and an ultrasonic disrupter with a minimum power of 300 W and with pulsing capability are also required.

PRECISION & ACCURACY The estimated quantitation limit (EQL) of Method 8270B for determining an individual compound is approximately 1 mg/kg (wet weight) for soil or sediment samples, 1–200 mg/kg for wastes (dependent on matrix and method of preparation), and 10 µg/L for groundwater samples. EQLs will be proportionately higher for sample extracts that require dilution to avoid saturation of the detector.

The EQL(b) for groundwater in µg/L is 10.
The EQL (a, b) for low concentrations in soil and sediment in µg/kg is not determined.
Accuracy as µg/L is not listed.
Overall precision in µg/L is not listed.

(a) EQLs listed for soil/sediment are based on wet weight. Normally data is reported in a dry-weight basis; therefore, EQLs will be higher based on the % dry weight of each sample. This calculation is based on a 30-g sample and gel permeation chromatography cleanup.
(b) Sample EQLs are highly matrix-dependent. The EQLs are provided for guidance and may not always be achievable.
C = True value for concentration, in µg/L.
X = Average recovery found for measurements of samples containing a concentration of C, in µg/L.

ESTIMATED QUANTITATION LIMIT FOR OTHER MATRICES

Other Matrices	Factor (a)
High-concentration soil and sludges by sonicator	7.5
Non-water miscible waste	75

(a) EQL for other matrices = [EQL for low soil/sediment] × [Factor]. This estimated EQL is similar to an EPA "Practical Quantitation Limit."

SAMPLING METHOD

Liquid samples — Use a 1 or 2½ gallon amber glass bottle with a screw-top Teflon®-lined cover that has been prewashed with detergent and rinsed with distilled water and methanol (or isopropanol).

Soils, sediments, or sludges — Use an 8-oz. widemouth glass with a screw-top Teflon®-lined cover that has been prewashed with detergent and rinsed with distilled water and methanol (or isopropanol).

SAMPLE PRESERVATION

Liquid samples — If residual chlorine is present, add 3 mL of 10% sodium thiosulfate per gallon, cool to 4°C and store in a solvent-free refrigerator until analysis; if chlorine is not present, then eliminate the sodium thiosulfate addition.

Soils, sediments, or sludges — Cool samples to 4°C and store in a solvent-free refrigerator.

MHT Liquid samples must be extracted within 7 days and the extracts analyzed within 40 days. Soils, sediments, or sludges may be stored for a maximum of 14 days and the extracts analyzed within 40 days.

SAMPLE PREPARATION
Liquid samples — Transfer 1 L quantitatively to a continuous extractor. If high concentrations are anticipated, a smaller volume may be used and then diluted with organic-free reagent water to 1 L. Adjust pH, if necessary, to pH <2 using 1:1 (V/V) sulfuric acid. Pipette 1.0 mL of a surrogate standard spiking solution into each sample. For the sample in each analytical batch selected for spiking, add 1.0 mL of a matrix spiking standard. For base/neutral acid analysis, the amount of the surrogates and matrix spiking compounds added to the sample should result in a final concentration of 100 ng/µL of each analyte in the extract to be analyzed (assuming a 1-µL injection). Extract with methylene chloride for 18–24 h. Next, adjust the pH of the aqueous phase to pH >11 using 10 N sodium hydroxide and extract it with methylene chloride again for 18–24 h. Dry the extract through a column containing anhydrous sodium sulfate and concentrate it to 1 mL using a K-D concentrator.

Soils, sediments, or sludges — Use 30 g of sample. Nonporous or wet samples (gummy or clay type) that do not have a free-flowing sandy texture must be mixed with anhydrous sodium sulfate until the sample is free flowing. Add 1 mL of surrogate standards to all samples, spikes, standards, and blanks. For the sample in each analytical batch selected for spiking, add 1.0 mL of a matrix spiking standard. For base/neutral acid analysis, the amount added of the surrogates and matrix spiking compounds should result in a final concentration of 100 ng/µL of each base/neutral analyte and 200 ng/µL of each acid analyte in the extract to be analyzed (assuming a 1-µL injection). Immediately add a 100 mL mixture of 1:1 methylene chloride:acetone and extract the sample ultrasonically for 3 min and then decant or filter the extracts. Repeat the extraction two or more times. Dry the extract using a column with anhydrous sodium sulfate and concentrate it to 1 mL in a K-D concentrator.

QUALITY CONTROL A methylene chloride solution containing 50 ng/µL of decafluorotriphenylphosphine (DFTPP) is used for tuning the GC/MS system each 12-h shift. A system performance check also must be made during every 12-h shift. A standard containing 50 ng/µL each of 4,4'-DDT, pentachlorophenol, and benzidine is required to verify injection port inertness and GC column performance. A calibration standard at mid-concentration, containing each compound of interest, including all required surrogates, must be performed every 12 h during analysis. After the system performance check is met, calibration check compounds (CCCs) are used to check the validity of the initial calibration.

The internal standard responses and retention times in the calibration check standard must be evaluated immediately after or during data acquisition. If the retention time for any internal standard changes by more than 30 seconds from the last check calibration (12 h), the chromatographic system must be inspected for malfunctions and corrections must be made, as required. If the electron ionization current plot (EICP) area for any of the internal standards changes by a factor of two from the last daily calibration standard check, the mass spectrometer must be inspected for malfunctions and corrections must be made, as appropriate.

Demonstrate, through the analysis of a reagent water blank, that interferences from the analytical system, glassware, and reagents are under control. The blank samples should be carried through all stages of the sample preparation and measurement steps. For each analytical batch (up to 20 samples), a reagent blank, matrix spike, and matrix spike duplicate/duplicate must be analyzed (the frequency of the spikes may be different for different monitoring programs). The blank and spiked samples must be carried through all stages of the sample preparation and measurement steps. A QC reference sample concentrate containing each analyte at a concentration of 100 mg/L in methanol is required.

REFERENCE Test Methods for Evaluating Solid Waste (SW-846). U.S. EPA 1983, Method 8270B, Rev. 2, Nov. 1990. Office of Solid Waste, Washington, DC.

Benzotrichloride   **EPA Method 8121**
CAS #98-07-7

TITLE Chlorinated Hydrocarbons by GC: Capillary Column Technique

MATRIX This method covers aqueous and solid matrices. This includes a wide variety such as drinking water, groundwater, industrial wastewaters, surface waters, soils, solids, and sediments.

METHOD SUMMARY This method provides procedures for the determination of 22 chlorinated hydrocarbons in water, soil/sediment, and waste matrices. A measured volume or weight of sample is extracted by using one of the appropriate sample extraction techniques specified in EPA Method 3510, EPA Method 3520, EPA Method 3540, or EPA Method 3550, or diluted using EPA Method 3580. Aqueous samples are extracted at neutral pH with methylene chloride by using either a separatory funnel (EPA Method 3510) or a continuous liquid-liquid extractor (EPA Method 3520). Solid samples are extracted with hexane/acetone (1:1) by using a Soxhlet extractor (EPA Method 3540) or with methylene chloride/acetone (1:1) by using an ultrasonic extractor (EPA Method 3550). After cleanup, the extract or diluted sample is analyzed by gas chromatography with electron capture detection (GC/ECD).

The sensitivity level of this method usually depends on the level of interferences rather than on instrumental limitations. This method may be used in conjunction with EPA Method 3620, Florisil Column Cleanup, EPA Method 3660, Sulfur Cleanup, and EPA Method 3640, Gel Permeation Chromatography, to aid in the elimination of interferences.

INTERFERENCES Solvents, reagents, glassware, and other hardware used in sample processing may introduce artifacts which may result in elevated baselines, causing misinterpretation of gas chromatograms. Interferants coextracted from the samples will vary considerably from waste to waste. Glassware must be scrupulously clean. Phthalate esters, if present in a sample, will interfere only with the BHC isomers. The presence of elemental sulfur will result in large peaks, and can often mask the region of compounds eluting after 1,2,4,5-tetrachlorobenzene. The tetrabutylammonium (TBA)-sulfite procedure (EPA Method 3660) works well for the removal of elemental sulfur. Waxes and lipids can be removed by gel permeation chromatography (EPA Method 3640).

INSTRUMENTATION A GC suitable for on-column injections and all required accessories, including and electron capture detector (ECD), analytical columns, recorder, gases, and syringes are needed. A data system for measuring peak heights and/or peak areas is recommended. A Kuderna-Danish (K-D) apparatus will also be needed to concentrate extracts.

Column 1: 30 m × 0.53 mm I.D. fused-silica capillary column chemically bonded with trifluoropropyl methyl silicone (DB-210 or equivalent).
Column 2: 30 m × 0.53 mm I.D. fused-silica capillary column chemically bonded with polyethylene glycol (DB-WAX or equivalent).

PRECISION & ACCURACY This method has been tested in a single lab by using organic-free reagent water, sandy loam samples, and extracts which were spiked with the test compounds at one concentration. Single-operator precision and method accuracy were found to be related to the concentration of compound and the type of matrix. The accuracy and precision technique will be determined by the sample matrix, sample preparation technique, optional cleanup techniques, and calibration procedures used.

ESTIMATED QUANTITATION LIMIT (EQL) FACTORS FOR VARIOUS MATRICES (a)

Matrix	Factor (b)
Groundwater	10
Low-concentration soil by ultrasonic extraction with GPC cleanup	670
High-concentration soil and sludges by ultrasonic extraction	10,000
Waste not miscible with water	100,000

(a) Sample EQLs are highly matrix-dependent. The EQLs listed herein are provided for guidance and may not always be achievable. (b) EQL = [Method detection limit] × [Factor]. For nonaqueous samples, the factor is on a wet-weight basis.

PRECISION & ACCURACY MDL is the method detection limit for organic-free reagent water. MDL was determined from the analysis of eight replicate aliquots processed through the entire analytical method (extraction, Florisil cartridge cleanup, and GC/ECD analysis).

The MDL (in ng/L) was 6.0.
The accuracy (as average % recovery using 5 determinations and no Florisil cleanup) from a spike concentration of 1.0 µg/L and separatory funnel extraction was 97% with a final volume of 10 mL.
The precision (as RSD% using 5 determinations and no Florisil cleanup) from a spike concentration of 1.0 µg/L and separatory funnel extraction was 2.1% with a final volume of 10 mL.

The accuracy (as average % recovery using 5 determinations and no Florisil cleanup), from a spike concentration of 3300 µg/L and ultrasonic extraction of solid samples using 1:1 methylene chloride and acetone, was 90% with a final volume of 10 mL.
The precision (as RSD% using 5 determinations and no Florisil cleanup), from a spike concentration of 3300 µg/L and ultrasonic extraction of solid samples using 1:1 methylene chloride and acetone, was 2.9% with a final volume of 10 mL.

SAMPLE COLLECTION, PRESERVATION & HANDLING
Volatile Organics — Standard 40-mL glass screw-cap VOA vials with Teflon®-faced silicone septum may be used for both liquid and solid matrices. When collecting samples, liquids and solids should be introduced into the vials gently to reduce agitation which might drive off volatile compounds. If there are any air bubbles present the sample must be retaken. The vials with solids should be tapped slightly as they are filled to try and eliminate as much free air space as possible. Two vials from each sampling location should be sealed in separate plastic bags to prevent cross-contamination between samples.

Semivolatile organics — Containers used to collect samples for the determination of semivolatile organic compounds should be soap and water washed followed by methanol (or isopropanol) rinsing. The sample containers should be of glass or Teflon® and have screw-top covers with Teflon® liners.

Preservation for volatile organics — No preservation is used with concentrated waste samples. With liquid samples containing no residual chlorine, 4 drops of concentrated hydrochloric acid are added and the samples are immediately cooled to 4°C. When liquid samples contain residual chlorine, they are treated as above and, in addition, 4 drops of 4% aqueous sodium thiosulfate are added to remove the residual chlorine. Soil, sediment, and sludge samples are only cooled to 4°C.

Preservation for semivolatile organics — No preservation is used with concentrated waste samples. With liquid samples containing no residual chlorine and with soil, sediment, and sludge samples, immediately cooling to 4°C is the only preservation used. When residual chlorine is present then 3 mL of 10% aqueous sodium sulfate is added for each gallon of sample collected, followed by cooling to 4°C.

Holding times — The holding time for all volatile organics samples is 14 days. Liquid samples must be extracted within 7 days and their extracts analyzed within 40 days. Concentrated waste, soil, sediment, and sludge samples must be extracted within 14 days and their extracts analyzed within 40 days.

SAMPLE PREPARATION Prepare stock standard solutions in hexane. Calibration standards at a minimum of five concentrations should be prepared through dilution of the stock standards with hexane. The suggested internal standards are: 2,5-dibromotoluene, 1,3,5-tribromobenzene, and α, α-dibromom-xylene. The analyst can use any of the three compounds provided that they are resolved from matrix interferences. Recommended surrogate compounds are α-2,6-trichlorotoluene, 1,4-dichloronaphthalene, and 2,3,4,5,6-pentachlorotoluene.

In general, water samples are extracted at a neutral pH with methylene chloride using a separatory funnel (EPA Method

3510) or a continuous liquid-liquid extractor (EPA Method 3520). Solid samples are extracted with hexane/acetone (1:1 v:v) using a Soxhlet extractor (EPA Method 3540) or with methylene chloride/acetone (1:1 v:v) using an ultrasonic extractor (EPA Method 3550). Non-aqueous waste samples may be diluted using EPA Method 3580. Prior to Florisil cleanup or gas chromatographic analysis, the extraction solvent must be exchanged to hexane. Sample extracts that will be subjected to gel permeation chromatography do not need solvent exchange.

Cleanup procedures may not be necessary for a relatively clean matrix. If removal of interferences such as chlorinated phenols, phthalate esters, etc., is required, proceed with the procedure outlined in EPA Method 3620.

QUALITY CONTROL Analyze a quality control check standard to demonstrate that the operation of the GC is in control. The frequency of the check standard analysis is equivalent to 10% of the samples analyzed. If the recovery of any compound found in the check standard is less than 80% of the certified value, the problem must be corrected and a new set of calibration standards must be prepared and analyzed. Calculate surrogate standard recoveries for all samples, blanks, and spikes. An internal standard peak area check must be performed on all samples. The internal standard must be evaluated for acceptance by determining whether the measured area for the internal standard deviates by more than 30% from the average area for the internal standard in the calibration standards. When the internal standard peak area is outside that limit, all samples that fall outside the QC criteria must be reanalyzed. Any compound confirmed by two columns may also be confirmed by GC/MS (EPA Method 8270). The GC/MS would normally require a minimum concentration of 1 ng/μL in the final extract for each compound. Include a mid-concentration calibration standard after each group of 20 samples in the analysis sequence. The response factors for the mid-concentration calibration must be within 15% of the average values for the multiconcentration calibration.

REFERENCE Test Methods for Evaluating Solid Waste, Physical/Chemical Methods, SW-846, 3rd Edition, U.S. EPA, Office of Solid Waste, Washington, DC, 1990. EPA Method 8121, Rev. 0, Nov. 1990.

Benzyl alcohol   **EPA Method 1625**
CAS #100-51-6

TITLE Semivolatile Organic Compounds by Isotope Dilution GC/MS

MATRIX The compounds may be determined in waters, soils, and municipal sludges by this method.

METHOD SUMMARY This method is used to determine 176 semivolatile toxic organic pollutants associated with the CWA (as amended 1987); the RCRA (as amended 1986); the CERCLA (as amended 1986); and other compounds amenable to extraction and analysis by capillary column gas chromatography-mass spectrometry (GC/MS).

Stable isotopically-labeled analogs of the compounds of interest are added to the sample. If the solids content is less than 1%, a 1-L sample is extracted at pH 12–13, then at pH <2 with methylene chloride using continuous extraction techniques.

If the solids content is 30% or less, the sample is diluted to 1% solids with reagent water, homogenized ultrasonically, and extracted at pH 12–13, then at pH <2 with methylene chloride using continuous extraction techniques. If the solids content is greater than 30%, the sample is extracted using ultrasonic techniques.

Each extract is dried over sodium sulfate, concentrated to a volume of 5 mL, cleaned up using GPC, if necessary, and concentrated. Extracts are concentrated to 1 mL if GPC is not performed, and to 0.5 mL if GPC is performed.

An internal standard is added to the extract, and a 1-mL aliquot of the extract is injected into the GC. The compounds are separated by GC and detected by a MS. The labeled compounds serve to correct the variability of the analytical technique.

INTERFERENCES Solvents, reagents, glassware, and other sample processing hardware may yield artifacts and/or elevated baselines causing misinterpretation of chromatograms and spectra. Materials used in the analysis must be demonstrated to be free from interferences under the conditions of analysis by running method blanks initially and with each sample lot (sample started through the extraction process on a given 8-h shift, to a maximum of 20). Specific selection of reagents and purification of solvents by distillation in all glass systems may be required. Glassware and, where possible, reagents are cleaned by solvent rinse and baking at 450°C for 1-h minimum. Interferences coextracted from samples will vary considerably from source to source, depending on the diversity of the site being sampled.

INSTRUMENTATION Major instrumentation includes a GC with a splitless or on-column injection port for capillary column, a MS with 70 eV electron impact ionization, and a data system to collect and record MS data, and process it. A K-D apparatus is used to concentrate extracts.

GC Column: 30 m × 0.25 mm I.D. 5% phenyl, 94% methyl, 1% vinyl silicone bonded phased fused silica capillary column.

PRECISION & ACCURACY The detection limits of the method are usually dependent on the level of interferences rather than instrumental limitations. The limits typify the minimum quantities that can be detected with no interferences present.

The minimum level (in μg/mL) wasnot listed. This is defined as a minimum level at which the analytical system shall give recognizable mass spectra (background corrected) and acceptable calibration points.
The MDL (in μg/kg) in low solids was not listed and in high solids was not listed; these were determined in digested sludge (low solids) and in filter cake or compost (high solids).
The labeled and native compound initial precision as standard deviation (in μg/L) was not listed.
The labeled and native compound initial accuracyas average recovery (in μg/L) was not listed.

SAMPLE COLLECTION, PRESERVATION & HANDLING

Collect samples in glass containers. Aqueous samples which flow freely are collected in refrigerated bottles using automatic sampling equipment. Solid samples are collected as grab samples using widemouth jars. Maintain samples at 0 to 4°C from the time of collection until extraction. If residual chlorine is present in aqueous samples, add 80 mg sodium thiosulfate/L of water. Begin sample extraction within 7 days of collection, and analyze all extracts within 40 days of extraction.

SAMPLE PREPARATION

Samples containing 1% solids or less are extracted directly using continuous liquid-liquid extraction techniques. Samples containing 1 to 30% solids are diluted to the 1% level with reagent water and extracted using continuous liquid-liquid extraction techniques. Samples containing greater than 30% solids are extracted using ultrasonic techniques.

Base/neutral extraction — Adjust the pH of the waters in the extractors to 12–13 with 6 N NaOH. Extract with methylene chloride for 24–48 h.

Acid extraction — Adjust the pH of the waters in the extractors to 2 or less using 6 N sulfuric acid. Extract with methylene chloride for 24–48 h.

Ultrasonic extraction of high solids samples — Add anhydrous sodium sulfate to the sample and QC aliquot(s). Add acetone:methylene chloride (1:1) to the sample and mix thoroughly

Concentrate extracts using a K-D apparatus.

QUALITY CONTROL

The analyst is permitted to modify this method to improve separations or lower the costs of measurements, provided all performance specifications are met. Analyses of blanks are required to demonstrate freedom from contamination. When results of spikes indicate atypical method performance for samples, the samples are diluted to bring method performance within acceptable limits.

For low solids (aqueous samples), extract, concentrate, and analyze two sets of four 1-L aliquots (8 aliquots total) of the precision and recovery standard. For high solids samples, two sets of four 30-g aliquots of the high solids reference matrix are used.

Spike all samples with labeled compounds to assess method performance. Compute percent recoveryof the labeled compounds using the internal standard method. Compare the labeled compound recovery for each compound with the corresponding labeled compound recovery.

Reagent water and high solids reference matrix blanks are analyzed to demonstrate freedom from contamination. Extract and concentrate a 1-L reagent water blank or a high solids reference matrix blank with each sample's lot (samples started through the extraction process on the same 8-h shift, to a maximum of 20 samples).

Field replicates may be collected to determine the precision of the sampling technique, and spiked samples may be required to determine the accuracy of the analysis when the internal standard method is used.

REFERENCE

Semivolatile Organic Compounds by Isotope Dilution GC/MS. Office of Water Regulation and Standards, U.S. EPA Industrial Technology Division, Washington, DC, EPA Method 1625, Rev. C, June 1989 (contact W.A. Telliard, U.S. EPA, Office of Water Regulations and Standards, 401 M St., SW, Washington, DC, 20460. Phone: 202-382-7131).

Benzyl alcohol   **EPA Method 1625**
CAS #100-51-6

TITLE

Semivolatile Organic Compounds by Isotope Dilution GC/MS

MATRIX

The compounds may be determined in waters, soils, and municipal sludges by this method.

METHOD SUMMARY

This method is used to determine 176 semivolatile toxic organic pollutants associated with the CWA (as amended 1987); the RCRA (as amended 1986); the CERCLA (as amended 1986); and other compounds amenable to extraction and analysis by capillary column gas chromatography-mass spectrometry (GC/MS).

Stable isotopically-labeled analogs of the compounds of interest are added to the sample. If the solids content is less than 1%, a 1-L sample is extracted at pH 12–13, then at pH <2 with methylene chloride using continuous extraction techniques.

If the solids content is 30% or less, the sample is diluted to 1% solids with reagent water, homogenized ultrasonically, and extracted at pH 12–13, then at pH <2 with methylene chloride using continuous extraction techniques. If the solids content is greater than 30%, the sample is extracted using ultrasonic techniques.

Each extract is dried over sodium sulfate, concentrated to a volume of 5 mL, cleaned up using GPC, if necessary, and concentrated. Extracts are concentrated to 1 mL if GPC is not performed, and to 0.5 mL if GPC is performed.

An internal standard is added to the extract, and a 1-mL aliquot of the extract is injected into the GC. The compounds are separated by GC and detected by a MS. The labeled compounds serve to correct the variability of the analytical technique.

INTERFERENCES

Solvents, reagents, glassware, and other sample processing hardware may yield artifacts and/or elevated baselines causing misinterpretation of chromatograms and spectra. Materials used in the analysis must be demonstrated to be free from interferences under the conditions of analysis by running method blanks initially and with each sample lot (sample started through the extraction process on a given 8-h shift, to a maximum of 20). Specific selection of reagents and purification of solvents by distillation in all glass systems may be required. Glassware and, where possible, reagents are cleaned by solvent rinse and baking at 450°C for 1-h minimum. Interferences coextracted from samples will vary considerably from source to source, depending on the diversity of the site being sampled.

INSTRUMENTATION

Major instrumentation includes a GC with a splitless or on-column injection port for capillary column,

a MS with 70 eV electron impact ionization, and a data system to collect and record MS data, and process it. A K-D apparatus is used to concentrate extracts.

GC Column: 30 m × 0.25 mm I.D. 5% phenyl, 94% methyl, 1% vinyl silicone bonded phased fused silica capillary column.

PRECISION & ACCURACY The detection limits of the method are usually dependent on the level of interferences rather than instrumental limitations. The limits typify the minimum quantities that can be detected with no interferences present.

The minimum level (in µg/mL) wasnot listed. This is defined as a minimum level at which the analytical system shall give recognizable mass spectra (background corrected) and acceptable calibration points.

The MDL (in µg/kg) in low solids was not listed and in high solids was not listed; these were determined in digested sludge (low solids) and in filter cake or compost (high solids).

The labeled and native compound initial precision as standard deviation (in µg/L) was not listed.

The labeled and native compound initial accuracyas average recovery (in µg/L) was not listed.

SAMPLE COLLECTION, PRESERVATION & HANDLING Collect samples in glass containers. Aqueous samples which flow freely are collected in refrigerated bottles using automatic sampling equipment. Solid samples are collected as grab samples using widemouth jars. Maintain samples at 0 to 4°C from the time of collection until extraction. If residual chlorine is present in aqueous samples, add 80 mg sodium thiosulfate/L of water. Begin sample extraction within 7 days of collection, and analyze all extracts within 40 days of extraction.

SAMPLE PREPARATION Samples containing 1% solids or less are extracted directly using continuous liquid-liquid extraction techniques. Samples containing 1 to 30% solids are diluted to the 1% level with reagent water and extracted using continuous liquid-liquid extraction techniques. Samples containing greater than 30% solids are extracted using ultrasonic techniques.

- Base/neutral extraction — Adjust the pH of the waters in the extractors to 12–13 with 6 N NaOH. Extract with methylene chloride for 24–48 h.
- Acid extraction — Adjust the pH of the waters in the extractors to 2 or less using 6 N sulfuric acid. Extract with methylene chloride for 24–48 h.
- Ultrasonic extraction of high solids samples — Add anhydrous sodium sulfate to the sample and QC aliquot(s). Add acetone:methylene chloride (1:1) to the sample and mix thoroughly

Concentrate extracts using a K-D apparatus.

QUALITY CONTROL The analyst is permitted to modify this method to improve separations or lower the costs of measurements, provided all performance specifications are met. Analyses of blanks are required to demonstrate freedom from contamination. When results of spikes indicate atypical method performance for samples, the samples are diluted to bring method performance within acceptable limits.

For low solids (aqueous samples), extract, concentrate, and analyze two sets of four 1-L aliquots (8 aliquots total) of the precision and recovery standard. For high solids samples, two sets of four 30-g aliquots of the high solids reference matrix are used.

Spike all samples with labeled compounds to assess method performance. Compute percent recoveryof the labeled compounds using the internal standard method. Compare the labeled compound recovery for each compound with the corresponding labeled compound recovery.

Reagent water and high solids reference matrix blanks are analyzed to demonstrate freedom from contamination. Extract and concentrate a 1-L reagent water blank or a high solids reference matrix blank with each sample's lot (samples started through the extraction process on the same 8-h shift, to a maximum of 20 samples).

Field replicates may be collected to determine the precision of the sampling technique, and spiked samples may be required to determine the accuracy of the analysis when the internal standard method is used.

REFERENCE Semivolatile Organic Compounds by Isotope Dilution GC/MS. Office of Water Regulation and Standards, U.S. EPA Industrial Technology Division, Washington, DC, EPA Method 1625, Rev. C, June 1989 (contact W.A. Telliard, U.S. EPA, Office of Water Regulations and Standards, 401 M St., SW, Washington, DC, 20460. Phone: 202-382-7131).

Benzyl alcohol   **EPA Method 8270**
CAS #100-51-6

TITLE Semivolatile Organic Compounds by GC/MS

MATRIX This method is used to determine the concentration of semivolatile organic compounds in extracts prepared from all types of solid waste matrices, soils, and groundwater. Although surface waters are not specifically mentioned, this method should be applicable to water samples from rivers, lakes, etc.

METHOD SUMMARY This method covers 259 semivolatile organic compounds. In very limited applications direct injection of the sample into the GC/MS system may be appropriate, but this results in very high detection limits (approximately 10,000 µg/L). Typically, a 1-L liquid sample, containing surrogate, and matrix spiking standards, is extracted in a continuous extractor first under acid conditions and then under basic conditions. Typically 30 g of a solid sample, containing surrogate, and matrix spiking standards, is extracted ultrasonically. After concentrating the extract to 1 mL it is spiked with 10 µL of an internal standard solution just prior to analysis by GC/MS. The volume injected should contain about 100 ng of base/neutral and 200 ng of acid surrogates (for a 1-µL injection). Analysis is performed by GC/MS using a capillary GC column.

INTERFERENCES Raw GC/MS data from all blanks, samples, and spikes must be evaluated for interferences. Contamination by carryover can occur whenever high-concentration

and low-concentration samples are sequentially analyzed. To reduce carryover, the sample syringe must be rinsed out between samples with solvent. Whenever an unusually concentrated sample is encountered, it should be followed by the analysis of blank solvent to check for cross-contamination.

INSTRUMENTATION A GC/MS and a data system are required. The GC column used is a 30 m × 0.25 mm I.D. (or 0.32 mm I.D.) 1-µm film thickness silicone-coated fused silica capillary column. A continuous liquid-liquid extractor equipped with Teflon® or glass connection joints and stopcocks requiring no lubrication, a K-D concentrating apparatus, water bath, and an ultrasonic disrupter with a minimum power of 300 W and with pulsing capability are also required.

PRECISION & ACCURACY The estimated quantitation limit (EQL) of Method 8270B for determining an individual compound is approximately 1 mg/kg (wet weight) for soil or sediment samples, 1–200 mg/kg for wastes (dependent on matrix and method of preparation), and 10 µg/L for groundwater samples. EQLs will be proportionately higher for sample extracts that require dilution to avoid saturation of the detector.

The EQL(b) for groundwater in µg/L is 20.
The EQL (a, b) for low concentrations in soil and sediment in µg/kg is 1300.
Accuracy as µg/L is not listed.
Overall precision in µg/L is not listed.

(a) EQLs listed for soil/sediment are based on wet weight. Normally data is reported in a dry-weight basis; therefore, EQLs will be higher based on the % dry weight of each sample. This calculation is based on a 30-g sample and gel permeation chromatography cleanup.
(b) Sample EQLs are highly matrix-dependent. The EQLs are provided for guidance and may not always be achievable.
C = *True value for concentration, in µg/L.*
X = *Average recovery found for measurements of samples containing a concentration of C, in µg/L.*

ESTIMATED QUANTITATION LIMIT FOR OTHER MATRICES

Other Matrices	Factor (a)
High-concentration soil and sludges by sonicator	7.5
Non-water miscible waste	75

(a) EQL for other matrices = [EQL for low soil/sediment] × [Factor]. This estimated EQL is similar to an EPA "Practical Quantitation Limit."

SAMPLING METHOD
Liquid samples — Use a 1 or 2½ gallon amber glass bottle with a screw-top Teflon®-lined cover that has been prewashed with detergent and rinsed with distilled water and methanol (or isopropanol).

Soils, sediments, or sludges — Use an 8-oz. widemouth glass with a screw-top Teflon®-lined cover that has been prewashed with detergent and rinsed with distilled water and methanol (or isopropanol).

SAMPLE PRESERVATION
Liquid samples — If residual chlorine is present, add 3 mL of 10% sodium thiosulfate per gallon, cool to 4°C and store in a solvent-free refrigerator until analysis; if chlorine is not present, then eliminate the sodium thiosulfate addition.

Soils, sediments, or sludges — Cool samples to 4°C and store in a solvent-free refrigerator.

MHT Liquid samples must be extracted within 7 days and the extracts analyzed within 40 days. Soils, sediments, or sludges may be stored for a maximum of 14 days and the extracts analyzed within 40 days.

SAMPLE PREPARATION
Liquid samples — Transfer 1 L quantitatively to a continuous extractor. If high concentrations are anticipated, a smaller volume may be used and then diluted with organic-free reagent water to 1 L. Adjust pH, if necessary, to pH <2 using 1:1 (V/V) sulfuric acid. Pipette 1.0 mL of a surrogate standard spiking solution into each sample. For the sample in each analytical batch selected for spiking, add 1.0 mL of a matrix spiking standard. For base/neutral acid analysis, the amount of the surrogates and matrix spiking compounds added to the sample should result in a final concentration of 100 ng/µL of each analyte in the extract to be analyzed (assuming a 1-µL injection). Extract with methylene chloride for 18–24 h. Next, adjust the pH of the aqueous phase to pH >11 using 10 N sodium hydroxide and extract it with methylene chloride again for 18–24 h. Dry the extract through a column containing anhydrous sodium sulfate and concentrate it to 1 mL using a K-D concentrator.

Soils, sediments, or sludges — Use 30 g of sample. Nonporous or wet samples (gummy or clay type) that do not have a free-flowing sandy texture must be mixed with anhydrous sodium sulfate until the sample is free flowing. Add 1 mL of surrogate standards to all samples, spikes, standards, and blanks. For the sample in each analytical batch selected for spiking, add 1.0 mL of a matrix spiking standard. For base/neutral acid analysis, the amount added of the surrogates and matrix spiking compounds should result in a final concentration of 100 ng/µL of each base/neutral analyte and 200 ng/µL of each acid analyte in the extract to be analyzed (assuming a 1-µL injection). Immediately add a 100 mL mixture of 1:1 methylene chloride:acetone and extract the sample ultrasonically for 3 min and then decant or filter the extracts. Repeat the extraction two or more times. Dry the extract using a column with anhydrous sodium sulfate and concentrate it to 1 mL in a K-D concentrator.

Note: This compound may be exhibit erratic chromatographic behavior, especially if the GC system is contaminated with high boiling material.

QUALITY CONTROL A methylene chloride solution containing 50 ng/µL of decafluorotriphenylphosphine (DFTPP) is used for tuning the GC/MS system each 12-h shift. A system performance check also must be made during every 12-h shift. A standard containing 50 ng/µL each of 4,4'-DDT, pentachlorophenol, and benzidine is required to verify injection port inertness and GC column performance. A calibration standard at mid-concentration, containing each compound of interest,

including all required surrogates, must be performed every 12 h during analysis. After the system performance check is met, calibration check compounds (CCCs) are used to check the validity of the initial calibration.

The internal standard responses and retention times in the calibration check standard must be evaluated immediately after or during data acquisition. If the retention time for any internal standard changes by more than 30 seconds from the last check calibration (12 h), the chromatographic system must be inspected for malfunctions and corrections must be made, as required. If the electron ionization current plot (EICP) area for any of the internal standards changes by a factor of two from the last daily calibration standard check, the mass spectrometer must be inspected for malfunctions and corrections must be made, as appropriate.

Demonstrate, through the analysis of a reagent water blank, that interferences from the analytical system, glassware, and reagents are under control. The blank samples should be carried through all stages of the sample preparation and measurement steps. For each analytical batch (up to 20 samples), a reagent blank, matrix spike, and matrix spike duplicate/duplicate must be analyzed (the frequency of the spikes may be different for different monitoring programs). The blank and spiked samples must be carried through all stages of the sample preparation and measurement steps. A QC reference sample concentrate containing each analyte at a concentration of 100 mg/L in methanol is required.

REFERENCE Test Methods for Evaluating Solid Waste (SW-846). U.S. EPA 1983, Method 8270B, Rev. 2, Nov. 1990. Office of Solid Waste, Washington, DC.

Benzyl butyl phthalate   **EPA Method 625**
CAS #85-68-7

TITLE Base/Neutrals and Acids, U.S. EPA Method 625

MATRIX This methods covers municipal and industrial wastewaters.

METHOD SUMMARY Approximately 1 L of sample is serially extracted with methylene chloride at a pH greater than 11 and again at a pH less than 2 using a separatory funnel or a continuous extractor. The methylene chloride extract is dried, concentrated to a volume of 1 mL, and analyzed by GC/MS. Qualitative identification of the parameters in the extract is performed using the retention time and the relative abundance of three characteristic masses (m/z). Qualitative analysis is performed using either external or internal standard techniques with a single characteristic m/z.

INTERFERENCES Method interferences may be caused by contaminants in solvents, reagents, glassware, and other sample processing hardware. Glassware must be scrupulously cleaned. Glassware should be heated in a muffle furnace at 400°C for 5 to 30 min. Some thermally stable materials, such as PCBs, may not be eliminated by this treatment. Solvent rinses with acetone and pesticide quality hexane may be substituted for the muffle furnace heating. Matrix interferences may be caused by contaminants that are coextracted from the sample. The base-neutral extraction may cause significantly reduced recovery of phenols. The packed gas chromatographic columns recommended for the basic fraction may not exhibit sufficient resolution for some analytes.

INSTRUMENTATION A GC/MS system with an injection port designed for on-column injection when using packed columns and for splitless injection when using capillary columns.

Column for base/neutrals: 1.8 m long × 2 mm I.D. glass, packed with 3% SP-2550 on Supelcoport (100/120 mesh) or equivalent.

Column for acids: 1.8 m long × 2 mm I.D. glass, packed with 1% SP-1240DA on Supelcoport (100/120 mesh) or equivalent.

PRECISION & ACCURACY The MDL concentrations were obtained using reagent water. The MDL actually achieved in a given analysis will vary depending on instrument sensitivity and matrix effects. This method was tested by 15 laboratories using reagent water, drinking water, surface water, and industrial wastewaters spiked at six concentrations over the range 5 to 100 µg/L. Single operator precision, overall precision, and method accuracy were found to be directly related to the concentration of the parameter matrix.

The MDL (in µg/L) in reagent water was 2.5.
The standard deviation (in µg/L based on 4 recovery measurements) was 23.4.
The range (in µg/L) for average recovery for 4 measurements was D–139.9.
The range (in %) for percent recovery was D–152.
Accuracy (in µg/L) as expected recovery for one or more measurements of a sample containing a true concentration of C was 0.66C–1.68.
Precision (in µg/L) as expected single analyst standard deviation of measurements at an average concentration found at X was 0.18X+0.94.
Overall precision (in µg/L) as expected interlaboratory standard deviation of measurements in an average concentration found at X was 0.53X+0.92.

$C =$ *True value of the concentration in µg/L.*
$X =$ *Average recovery found for measurements of samples containing a concentration at C in µg/L.*

SAMPLE PREPARATION Adjust the pH to >11 with sodium hydroxide and serially extract in a separatory funnel with methylene chloride or else in a continuous extractor. Next, adjust the pH to <2 with sulfuric acid and serially extract in a separatory funnel with methylene chloride or else in a continuous extractor. Dry the extracts separately through a column of anhydrous sodium sulfate and then concentrate each of the extracts to 1.0 mL using a K-D apparatus.

SAMPLE COLLECTION, PRESERVATION & HANDLING- Grab samples must be collected in glass containers. All samples must be refrigerated at 4°C from the time of collection until extraction. If residual chlorine is present, add 80 mg of sodium thiosulfate/L of sample and mix well. All samples must be extracted within 7 days of collection and completely analyzed within 40 days of extraction.

QUALITY CONTROL Make an initial, one-time, demonstration of the ability to generate acceptable accuracy and precision with this method. Before processing any samples, the analyst must analyze a reagent water blank to demonstrate that interferences from the analytical system and glassware are under control. Each time a set of samples is extracted or reagents are changed, a reagent water blank must be processed. Spike and analyze a minimum of 5% of all samples to monitor and evaluate lab data quality. A QC check sample concentrate that contains each parameter of interest at a concentration of 100 µg/mL in acetone is required. PCBs and multicomponent pesticides may be omitted from this test.

After analysis of five spiked wastewater samples, calculate the average percent recovery and the standard deviation of the percent recovery. Spike all samples with the surrogate standard spiking solution and calculate the percent recovery of each surrogate compound.

REFERENCE *Federal Register*, Vol. 49, No. 209. Friday, Oct. 26, 1984.

Benzyl chloride   **EPA Method 8121**
CAS #100-44-7

TITLE Chlorinated Hydrocarbons by GC: Capillary Column Technique

MATRIX This method covers aqueous and solid matrices. This includes a wide variety such as drinking water, groundwater, industrial wastewaters, surface waters, soils, solids, and sediments.

METHOD SUMMARY This method provides procedures for the determination of 22 chlorinated hydrocarbons in water, soil/sediment, and waste matrices. A measured volume or weight of sample is extracted by using one of the appropriate sample extraction techniques specified in EPA Method 3510, EPA Method 3520, EPA Method 3540, or EPA Method 3550, or diluted using EPA Method 3580. Aqueous samples are extracted at neutral pH with methylene chloride by using either a separatory funnel (EPA Method 3510) or a continuous liquid-liquid extractor (EPA Method 3520). Solid samples are extracted with hexane/acetone (1:1) by using a Soxhlet extractor (EPA Method 3540) or with methylene chloride/acetone (1:1) by using an ultrasonic extractor (EPA Method 3550). After cleanup, the extract or diluted sample is analyzed by gas chromatography with electron capture detection (GC/ECD).

The sensitivity level of this method usually depends on the level of interferences rather than on instrumental limitations. This method may be used in conjunction with EPA Method 3620, Florisil Column Cleanup, EPA Method 3660, Sulfur Cleanup, and EPA Method 3640, Gel Permeation Chromatography, to aid in the elimination of interferences.

INTERFERENCES Solvents, reagents, glassware, and other hardware used in sample processing may introduce artifacts which may result in elevated baselines, causing misinterpretation of gas chromatograms. Interferants coextracted from the samples will vary considerably from waste to waste. Glassware must be scrupulously clean. Phthalate esters, if present in a sample, will interfere only with the BHC isomers. The presence of elemental sulfur will result in large peaks, and can often mask the region of compounds eluting after 1,2,4,5-tetrachlorobenzene. The tetrabutylammonium (TBA)-sulfite procedure (EPA Method 3660) works well for the removal of elemental sulfur. Waxes and lipids can be removed by gel permeation chromatography (EPA Method 3640).

INSTRUMENTATION A GC suitable for on-column injections and all required accessories, including and electron capture detector (ECD), analytical columns, recorder, gases, and syringes are needed. A data system for measuring peak heights and/or peak areas is recommended. A Kuderna-Danish (K-D) apparatus will also be needed to concentrate extracts.

Column 1: 30 m × 0.53 mm I.D. fused-silica capillary column chemically bonded with trifluoropropyl methyl silicone (DB-210 or equivalent).

Column 2: 30 m × 0.53 mm I.D. fused-silica capillary column chemically bonded with polyethylene glycol (DB-WAX or equivalent).

PRECISION & ACCURACY This method has been tested in a single lab by using organic-free reagent water, sandy loam samples, and extracts which were spiked with the test compounds at one concentration. Single-operator precision and method accuracy were found to be related to the concentration of compound and the type of matrix. The accuracy and precision technique will be determined by the sample matrix, sample preparation technique, optional cleanup techniques, and calibration procedures used.

ESTIMATED QUANTITATION LIMIT (EQL) FACTORS FOR VARIOUS MATRICES (a)

Matrix	Factor (b)
Groundwater	10
Low-concentration soil by ultrasonic extraction with GPC cleanup	670
High-concentration soil and sludges by ultrasonic extraction	10,000
Waste not miscible with water	100,000

(a) Sample EQLs are highly matrix-dependent. The EQLs listed herein are provided for guidance and may not always be achievable. (b) EQL = [Method detection limit] × [Factor]. For nonaqueous samples, the factor is on a wet-weight basis.

PRECISION & ACCURACY MDL is the method detection limit for organic-free reagent water. MDL was determined from the analysis of eight replicate aliquots processed through the entire analytical method (extraction, Florisil cartridge cleanup, and GC/ECD analysis).

The MDL (in ng/L) was 180.

The accuracy (as average % recovery using 5 determinations and no Florisil cleanup) from a spike concentration of 100 µg/L and separatory funnel extraction was 90% with a final volume of 10 mL.

The precision (as RSD% using 5 determinations and no Florisil cleanup) from a spike concentration of 100 µg/L and separatory funnel extraction was 6.2% with a final volume of 10 mL.

The accuracy (as average % recovery using 5 determinations and no Florisil cleanup), from a spike concentration of 33,00 µg/L and ultrasonic extraction of solid samples using 1:1 methylene chloride and acetone, was 121% with a final volume of 10 mL.

The precision (as RSD% using 5 determinations and no Florisil cleanup), from a spike concentration of 33,00 µg/L and ultrasonic extraction of solid samples using 1:1 methylene chloride and acetone, was 5.9% with a final volume of 10 mL.

SAMPLE COLLECTION, PRESERVATION & HANDLING
Volatile Organics — Standard 40-mL glass screw-cap VOA vials with Teflon®-faced silicone septum may be used for both liquid and solid matrices. When collecting samples, liquids and solids should be introduced into the vials gently to reduce agitation which might drive off volatile compounds. If there are any air bubbles present the sample must be retaken. The vials with solids should be tapped slightly as they are filled to try and eliminate as much free air space as possible. Two vials from each sampling location should be sealed in separate plastic bags to prevent cross-contamination between samples.

Semivolatile organics — Containers used to collect samples for the determination of semivolatile organic compounds should be soap and water washed followed by methanol (or isopropanol) rinsing. The sample containers should be of glass or Teflon® and have screw-top covers with Teflon® liners.

Preservation for volatile organics — No preservation is used with concentrated waste samples. With liquid samples containing no residual chlorine, 4 drops of concentrated hydrochloric acid are added and the samples are immediately cooled to 4°C. When liquid samples contain residual chlorine, they are treated as above and, in addition, 4 drops of 4% aqueous sodium thiosulfate are added to remove the residual chlorine. Soil, sediment, and sludge samples are only cooled to 4°C.

Preservation for semivolatile organics — No preservation is used with concentrated waste samples. With liquid samples containing no residual chlorine and with soil, sediment, and sludge samples, immediately cooling to 4°C is the only preservation used. When residual chlorine is present then 3 mL of 10% aqueous sodium sulfate is added for each gallon of sample collected, followed by cooling to 4°C.

Holding times — The holding time for all volatile organics samples is 14 days. Liquid samples must be extracted within 7 days and their extracts analyzed within 40 days. Concentrated waste, soil, sediment, and sludge samples must be extracted within 14 days and their extracts analyzed within 40 days.

SAMPLE PREPARATION Prepare stock standard solutions in hexane. Calibration standards at a minimum of five concentrations should be prepared through dilution of the stock standards with hexane. The suggested internal standards are: 2,5-dibromotoluene, 1,3,5-tribromobenzene, and α, α-dibromo-m-xylene. The analyst can use any of the three compounds provided that they are resolved from matrix interferences. Recommended surrogate compounds are α-2,6-trichlorotoluene, 1,4-dichloronaphthalene, and 2,3,4,5,6-pentachlorotoluene.

In general, water samples are extracted at a neutral pH with methylene chloride using a separatory funnel (EPA Method 3510) or a continuous liquid-liquid extractor (EPA Method 3520). Solid samples are extracted with hexane/acetone (1:1 v:v) using a Soxhlet extractor (EPA Method 3540) or with methylene chloride/acetone (1:1 v:v) using an ultrasonic extractor (EPA Method 3550). Non-aqueous waste samples may be diluted using EPA Method 3580. Prior to Florisil cleanup or gas chromatographic analysis, the extraction solvent must be exchanged to hexane. Sample extracts that will be subjected to gel permeation chromatography do not need solvent exchange.

Cleanup procedures may not be necessary for a relatively clean matrix. If removal of interferences such as chlorinated phenols, phthalate esters, etc., is required, proceed with the procedure outlined in EPA Method 3620.

QUALITY CONTROL Analyze a quality control check standard to demonstrate that the operation of the GC is in control. The frequency of the check standard analysis is equivalent to 10% of the samples analyzed. If the recovery of any compound found in the check standard is less than 80% of the certified value, the problem must be corrected and a new set of calibration standards must be prepared and analyzed. Calculate surrogate standard recoveries for all samples, blanks, and spikes. An internal standard peak area check must be performed on all samples. The internal standard must be evaluated for acceptance by determining whether the measured area for the internal standard deviates by more than 30% from the average area for the internal standard in the calibration standards. When the internal standard peak area is outside that limit, all samples that fall outside the QC criteria must be reanalyzed. Any compound confirmed by two columns may also be confirmed by GC/MS (EPA Method 8270). The GC/MS would normally require a minimum concentration of 1 ng/µL in the final extract for each compound. Include a mid-concentration calibration standard after each group of 20 samples in the analysis sequence. The response factors for the mid-concentration calibration must be within 15% of the average values for the multiconcentration calibration.

REFERENCE Test Methods for Evaluating Solid Waste, Physical/Chemical Methods, SW-846, 3rd Edition, U.S. EPA, Office of Solid Waste, Washington, DC, 1990. EPA Method 8121, Rev. 0, Nov. 1990.

Benzyl chloride   EPA Method 8240
CAS #100-44-7

TITLE Volatile Organics By GC/MS: Packed Column Technique

MATRIX Nearly all types of sample matrices, regardless of water content, can be analyzed using this method. This includes groundwater, aqueous sludges, caustic liquors, acid liquors, waste solvents, oily wastes, mousses, tars, fibrous wastes, polymetric

emulsions, filter cakes, spent carbons, spent catalysts, soils, and sediments.

METHOD SUMMARY Method 8240B covers 80 volatile organic compounds that are introduced into a gas chromatograph by the purge-and-trap method or by direct injection (in limited applications). For the purge-and-trap method an inert gas (zero grade nitrogen or helium) is bubbled through a 5-mL solution at ambient temperature. Purged sample components are trapped in a tube of sorbent materials. When purging is complete, the sorbent tube is heated and backflushed with inert gas to desorb the trapped components onto a GC column.

INTERFERENCES Impurities in the purge gas and from organic compounds outgassing from the plumbing ahead of the trap account for many contamination problems. Interferences purged or coextracted from the samples will vary considerably from source to source. Cross-contamination can occur whenever high-level and low-level samples are analyzed sequentially. Whenever an unusually concentrated sample is analyzed, it should be followed by the analysis of organic-free reagent water to check for cross-contamination. Samples also can be contaminated by diffusion of volatile organics (particularly methylene chloride and fluorocarbons) through the septum seal into the sample during shipment and storage. A trip blank can serve as a check on such contamination. The lab where volatile analysis is performed and also the refrigerated storage area should be completely free of solvents.

INSTRUMENTATION A gas chromatograph/mass spectrometry/data system (GC/MS) equipped with a 6 ft × 0.1 in I.D. glass column packed with 1% SP-1000 on Carbopack-B (60/80 mesh) is required. Also needed is a 5-mL purging device, a sorbent trap, and a thermal desorption apparatus.

PRECISION & ACCURACY This method is reported to have been tested by 15 laboratories using organic-free reagent water, drinking water, surface water, and industrial wastewaters (not specified) fortified at six concentrations over the range 5–600 µg/L.

Sample estimated quantitation limits (EQLs) are highly matrix-dependent. The EQLs listed may not always be achievable. EQLs listed for soils or sediments are based on wet weight. Normally, data is reported on a dry-weight basis; therefore, EQLs will be higher, based on the percent dry weight of each sample. Note that EQLs are even more variable than MDLs and that they are highly variable depending on the matrix being analyzed.

EQL in groundwater in µg/L was 100.
EQL in low soil or sediment in µg/kg was 100.
Accuracy (a) in µg/L was not listed.
Precision (b) in µg/L was not listed.

(a) *Average recovery found for measurements of samples containing a concentration of C, in µg/L.*
(b) *Overall precision found for measurements of samples with average recovery X for samples containing a concentration of C in µg/L.*
X = *Average recovery found for measurement of samples containing a concentration of C in µg/L.*

MULTIPLICATION FACTORS FOR OTHER MATRICES

Other Matrices	Factor (a)
Waste miscible liquid waste	50
High-concentration soil and sludge	125
Non-water miscible waste	500

(a) *EQL = [EQL for low soil/sediment] × [Factor]. For non-aqueous samples, the factor is on a wet-weight basis.*

SAMPLING METHOD

Liquid samples — Use a 40-mL glass screw-cap VOA vial with a Teflon®-faced silicone septum that has been prewashed, rinsed with distilled deionized water, and oven dried. However, if residual chlorine is present, collect sample in a 40-oz. soil VOA container which has been pre-preserved with 4 drops of 10% sodium thiosulfate, mix gently, and then transfer the sample to a 40-mL VOA vial. Collect bubble-free samples in duplicate and seal them in separate plastic bags.

Soils or sediments and sludges — Use an 8-oz. widemouth glassbottle with a Teflon®-faced silicone septum that has been prewashed with detergent, rinsed with distilled deionized water, and oven dried. Tap slightly to eliminate free air space. Collect samples in duplicate and seal them in separate plastic bags.

SAMPLE PRESERVATION

Liquid samples — Add 4 drops of concentrated HCL and immediately coolsamples to 4°C and store in a solvent-free refrigerator.

Soils or sediments, and sludges — Cool samples to 4°C and store in a solvent-free refrigerator.

MHT Maximum holding time is 14 days from the date of sample collection.

SAMPLE PREPARATION

Liquid samples — Remove the plunger from a 5-mL syringe and carefully pour the sample into the syringe barrel to just short of overflowing. Replace the syringe plunger and compress the sample. Open the syringe valve and vent any residual air while adjusting the sample volume to 5.0 mL. If there is only one volatile organic analysis (VOA) vial, a second syringe should be filled at this time to protect against possible loss of sample integrity. Add 10 µL of surrogate spiking solution and 10 µL of internal standard spiking solution through the valve bore of the 5-mL syringe, then close the valve. The surrogate and internal standards may be mixed and added as a single spiking solution.

Sediments, soils, and waste samples — All samples of this type should be screened by GC analysis using a headspace method (EPA Method 3810) or the hexadecane extraction and screening method (EPA Method 3820). Use the screening data to determine whether to use the low-concentration method (0.005–1 mg/kg) or the high-concentration method (>1 mg/kg).

Low-concentration method — The low-concentration method is based on purging a heated sediment or soil sample mixed with organic-free reagent water containing the surrogate and internal standards. Analyze all reagent blanks and standards under the same conditions as the samples.

Use a 5-g sample if the expected concentration is <0.1 mg/kg or a 1-g sample for expected concentrations between 0.1 and 1 mg/kg. Mix the contents of the sample container with a narrow metal spatula. Weigh the amount of the sample into a tared purge device. Add the spiked water to the purge device, which contains the weighed amount of sample, and connect the device to the purge-and-trap system.

High-concentration method — This method is based on extracting the sediment or soil with methanol. A waste sample is either extracted or diluted, depending on its solubility in methanol. Wastes that are insoluble in methanol are diluted with reagent tetraglyme or possibly polyethylene glycol (PEG). An aliquot of the extract is added to organic-free reagent water containing surrogate and internal standards. This is purged at ambient temperature. All samples with an expected concentration of >1.0 mg/kg should be analyzed by this method.

Mix the contents of the sample container with a narrow metal spatula. For sediments or soils and solid wastes that are insoluble in methanol, weigh 4 g (wet weight) of sample into a tared 20-mL vial. For waste that is soluble in methanol, tetraglyme, or PEG, weigh 1 g (wet weight) into a tared scintillation vial or culture tube or a 10-mL volumetric flask. Quickly add 9.0 mL of appropriate solvent then add 1.0 mL of a surrogate spiking solution to the vial, cap it, and shake it for 2 min.

METHANOL EXTRACT REQUIRED FOR ANALYSIS OF HIGH-CONCENTRATION SOILS OR SEDIMENTS

Approximate Concentration Range	Volume of Methanol Extract (a)
500–10,000 µg/kg	100 µL
1,000–20,000 µg/kg	50 µL
5,000–100,000 µg/kg	10 µL
25,000–500,000 µg/kg	100 µL of 1/50 dilution (b)

Calculate appropriate dilution factor for concentrations exceeding this table.

(a) The volume of methanol added to 5 mL of water being purged should be kept constant. Therefore, add to the 5-mL syringe whatever volume of methanol is necessary to maintain a volume of 100 µL added to the syringe.
(b) Dilute an aliquot of the methanol extract and then take 100 µL for analysis.

QUALITY CONTROL Demonstrate, through the analysis of a reagent water blank, that interferences from the analytical system, glassware, and reagents are under control. Blank samples should be carried through all stages of the sample preparation and measurement steps. For each analytical batch (up to 20 samples), a reagent blank, matrix spike, and matrix spike duplicate must be analyzed (the frequency of the spikes may be different for different monitoring programs). The blank and spiked samples must be carried through all stages of the sample preparation and measurement steps. QC samples mentioned in the section on Interferences will also be needed as appropriate to those situations.

REFERENCE Test Methods for Evaluating Solid Waste (SW-846). U.S. EPA. 1983. Method 8240B, Rev. 2, Nov. 1990. Office of Solid Wastes, Washington, DC.

Beryllium — EPA Method 6010
CAS #7440-41-7

TITLE Inductively Coupled Plasma-Atomic Emission Spectroscopy

MATRIX This method is applicable to the determination of trace elements, including metals, in groundwater, soils, sludges, sediments, and other solid wastes. All matrices require digestion prior to analysis. The method of standard addition must be used for the analysis of all sample digests unless either serial dilution or matrix spike addition demonstrates it is not required.

METHOD SUMMARY Method 6010 covers 25 elements using ICP analysis. It measures element-emitted light by optical spectrometry. Samples, following an appropriate acid digestion, are nebulized and the resulting aerosol is transported to the plasma torch. Element-specific atomic line emission spectra are produced by a radio-frequency inductively coupled plasma.

INTERFERENCES Interferences may be categorized as spectral or non-spectral. Spectral interferences are caused by overlap of a spectral line from another element, unresolved overlap of molecular band spectra, background contribution from continuous or recombination phenomenon, and stray light from the line emission of high concentration elements. Non-spectral interferences include physical and chemical interferences. Physical interferences are effects associated with the sample nebulization and transport processes. Changes in viscosity and surface tension can cause significant inaccuracies. Chemical interferences include molecular compound formation, ionization effects, and solute vaporization effects. Normally these effects are not significant and can be minimized by careful selection of operating conditions. Chemical interferences are highly dependent on matrix type and the specific analyte element.

INSTRUMENTATION An inductively coupled argon plasma emission spectrometer (ICP) capable of background correction is required.

PRECISION & ACCURACY Detection limits, sensitivity, and optimum ranges of the metals will vary with the matrices and model of the spectrometer. In a single lab evaluation, seven wastes were analyzed for 22 elements. The mean percent relative standard deviation from triplicate analyses for all elements and wastes was 9 ± 2%. The mean percent recovery of spiked elements for all wastes was 93 ± 6%. Spike levels ranged from 100 µg/L to 100 mg/L. The wastes included sludges and industrial wastewaters.

Estimated instrument detection limit in µg/L is 0.3.
Spiked concentration in µg/L is 20.
Mean reported value in µg/L is 20.
Precision as RSD % is 9.8.

SAMPLING METHOD Samples should be collected in borosilicate glass, linear polyethylene, polypropylene, or Teflon® bottles that have been prewashed with detergent and tap water, and rinsed with 1:1 nitric acid and tap water or 1:1 hydrochloric acid and tap water. Collect at least 2 g of solids and 200 mL of aqueous samples.

SAMPLE PRESERVATION Add nitric acid to make the samples pH <2.

MHT The maximum holding time for properly preserved samples is 6 months.

SAMPLE PREPARATION Preliminary treatment of most matrices is necessary because of the complexity and variability of sample matrices. Water samples that have been prefiltered and acidified will not need acid digestion. Methods for acid digestion of waters for total recoverable or dissolved metals, acid digestions of aqueous samples and extracts for total metals, and acid digestion of sediments, sludges, and soils are summarized below.

Total recoverable or dissolved metals in water — To prepare surface and groundwater samples for determination of total recoverable and dissolved metals, a 100-mL aliquot of well-mixed sample is acidified with concentrated nitric acid and concentrated hydrochloric acid, then heated until the volume is reduced to 15–20 mL. Adjust the final volume to 100 mL with reagent water.

Total metals in aqueous samples, soil and sediment extracts — To prepare aqueous samples, soil and sediment extracts, and wastes that contain suspended solids, a 100-mL aliquot is made acidic with concentrated nitric acid and the solution is evaporated to about 5 mL on a hot plate. Continue heating and adding additional acid until sample digestion is complete, which is usually indicated when the digestate is light in color or does not change in appearance. Evaporate the solution to about 3 mL andcool it and add a small quantity of 1:1 hydrochloric acid (10 mL/100 mL of final solution). Cover the beaker and reflux for 15 min. Wash down the beaker walls and filters or centrifuge the sample to remove silicates and other insoluble material. Filter the sample and adjust the final volume to 100 mL with reagent water and the final acid concentration to 10%.

Sediments, sludges, and soils — To prepare sediments, sludges and soil samples, transfer 1–2 g to a conical beaker and add 10 mL of 1:1 nitric acid, mix the slurry, and cover it with a watch glass. Heat the sample and reflux for 10 to 15 min without boiling. Allow it to cool, then add 5 mL of concentrated nitric acid and reflux for 30 min. Repeat last step and then allow the solution to evaporate to 5 mL without boiling. Cool and add 2 mL of water and 3 mL of 30% hydrogen peroxide. Cover and place the beaker on the hot plate. Heat and add 30% hydrogen peroxide in 1-mL aliquots with warming until the effervescence is minimal but do not add more than a total of 10 mL of 30% hydrogen peroxide. If the sample is being prepared for the analysis of Ag, Al, As, Ba, Be, Ca, Cd, Co, Cr, Cu, Fe, K, Mg, Mn, Mo, Na, Ni, Os, Pb, Se, Tl, V, and Zn, then add 5 mL of concentrated hydrochloric acid and 10 mL of water and return the covered beaker to a hot plate for 15 min of additional refluxing without boiling. Dilute the sample to a 100 mL volume with water after cooling and filter or centrifuge to remove particulates.

QUALITY CONTROL Laboratory control samples must be analyzed for each analytical method. A method blank should be analyzed with each batch of samples. The effect of the matrix on method performance must be demonstrated: when appropriate, there should be at least one matrix spike and either one matrix duplicate or one matrix spike duplicate per analytical batch. The bias and precision of the method, as well as the method detection limit for each specific matrix type, must be measured.

Dilute and reanalyze samples that are more concentrated than the linear calibration limit. Employ a minimum of one reagent blank per sample batch to determine if contamination or any memory effects are occurring. Whenever a new or unusual sample matrix is encountered, perform either a serial dilution test or a matrix spike addition test to ensure that neither positive or negative interferences are operating on any of the analyte elements. Check the instrument standardization by verifying calibration every 10 samples using a calibration blank and a check standard.

REFERENCE Test Methods for Evaluating Solid Waste (SW-846). U.S. EPA. 1983. Method 6010, Rev. 0, Sept. 1986. Office of Solid Wastes, Washington, DC.

Beryllium **EPA Method 200.7**
CAS #7440-41-7

TITLE Inductively Coupled Plasma

MATRIX Dissolved, suspended or(ICP) total element in drinking and surfacewaters and in domestic and industrial waste waters.

APPLICATION The method covers the determination of 25 metals. Dissolved elements are determined in filtered and acidified samples after appropriate digestion (which increases dissolved solids). Its primary advantage is that ICP instruments allow simultaneous or rapid sequential determination of many elements in a short time. Samples are first nebulized and theaerosolis transported to a plasma torch in which element specific atomic line emission spectra are produced by a radio frequency inductively coupled plasma. Background correctionis required for trace element detection except in the case of line broadening.

INTERFERENCES There are spectral, physical, and chemical interferences. The primary disadvantage of ICP instruments is background radiation from other elements and the plasma gases (spectral interferences). Changes in sample viscosity and surface tension with samples containing high dissolved solids (especially those exceeding 1500 mg/L) or high acid concentrations can cause physical interferences. Ionization effects, solute vaporization and molecular compound formation can cause chemical interferences. Thallium and vanadium can cause interference at the 100 mg/L level.

INSTRUMENTATION Inductively Coupled Argon Plasma Emission Spectroscopy. 313.042 nm Wavelength

RANGE Not listed.

MDL 0.3 µg/L.

PRECISION SD = 6.2% Mean at true value 750 µg/L.

ACCURACY Mean recovery = 93% ± 6% of spiked elements for all wastes.

SAMPLING METHOD Wash sample container with detergent and tap water, rinse with 1+1 nitric acid and tap water, then rinse with 1+1 hydrochloric acid and tap water, then rinse with deionized, distilled water in that order. Perform any filtration or acid preservation steps when the sample is collected or as soon as possible thereafter.

STABILITY Cool samples to 4°C.

MHT 24 h.

QUALITY CONTROL Mixed calibration standards, an instrument check standard and an interference check solution are used in addition to a quality control sample. The quality control sample should be prepared in the same acid matrix as the calibration standards at 10 times the instrumental detection limits and in accordance with the instructions provided by the supplier. Furthermore, two types of blanks are required: a calibration blank and a reagent blank.

REFERENCE Method 200.7, U.S. EPA, EMSL-Cincinnati, OH, Nov. 1980

Beryllium EPA Method 7090
CAS #7440-41-7

TITLE Atomic Absorption, (AA)

MATRIX Drinking, surface and direct aspiration saline waters, wastewater.

APPLICATION Sample is aspirated and atomized in a flame. A light beam from a be hollow cathode lamp is directed through the flame into a monochromator and onto a detector. Since wavelength of lightbeam is specific for beryllium, light energy absorbed by detector is measure of be.

INTERFERENCES The most troublesome type is chemical, caused by lack of absorption of atoms bound in molecular combination in the flame. High dissolved solids in sample may result in nonatomic absorbance interference. Aluminum concentrations greater than 500 ppm can interfere.

INSTRUMENTATION Atomic absorption spectrometer. Berylliumhollowcathode lamp. (234.9 nm wavelength).

RANGE 0.05–2 mg/L

MDL 0.005 mg/L

PRECISION Standard deviation = ±0.001 and 0.002 at 0.01 and 0.25 mg Be/L

ACCURACY Recoveries = 100 and 97% at 0.01 and 0.25 mg Be/L

SAMPLING METHOD Use glass or plastic containers. Collect 200 g of solids and 600 mL of liquid samples.

STABILITY Cool solid samples to 4°C and analyze as soon as possible.

Add nitric acid to liquid samples to pH <2.

MHT 6 months.

QUALITY CONTROL At least one duplicate and one spike sampleshouldberun every20 samples or with each matrix type to verify precision of the method. For 20 or more samples perday,verifyworkingstandardcurve. Runan additional standard at or near mid-range every 10 samples.

REFERENCE Method 7090, SW-846, 3rd ed., Nov. 1986.

Beryllium EPA Method 7091
CAS #7440-41-7

TITLE Atomic Absorption, (AA)

MATRIX Wastes, mobility procedure Furnace Technique extracts, soils, and groundwater.

APPLICATION Aqueous samples, EP extracts, industrial wastes, soils, sludges, sediments, and solid wastes require digestion before analysis. An aliquot of sample is placed in the graphite tube in the furnace and slowly evaporated, charred and atomized. Absorption of lamp radiation during atomization is proportional to beryllium concentration.

INTERFERENCES The furnace technique is subject to chemical interferences.

Composition of sample matrix can have major effect on analysis. The long residence time and high concentrations of atomized sample in optical path can result in severe physical and chemical interferences.

INSTRUMENTATION Atomic absorption spectrometer. Beryllium hollow cathode lamp or electrodeless discharge lamp. Graphite furnace. Strip-chart recorder.

RANGE 1–30 µg/L (234.9 nm wavelength)

MDL 0.2 µg/L

PRECISION Not listed.

ACCURACY Not listed.

SAMPLING METHOD Use glass or plastic containers. Collect 200 g of solids and 600 mL of liquid samples.

STABILITY Cool solid samples to 4°C and analyze as soon as possible. Add nitric acid to liquid samples to pH <2.

MHT 6 months.

QUALITY CONTROL At least one duplicate and one spike sampleshouldberun every 20 samples, or with each matrix type to verify method precision. If 20 or more samples are run a day, run a standard (at or near mid-range) every 10 samples.

REFERENCE Method 7091, SW-846, 3rd ed., Nov. 1986.

β-BHC EPA Method 625
CAS #319-85-7

TITLE Base/Neutrals and Acids, U.S. EPA Method 625

MATRIX This methods covers municipal and industrial wastewaters.

METHOD SUMMARY Approximately 1 L of sample is serially extracted with methylene chloride at a pH greater than 11 and again at a pH less than 2 using a separatory funnel or a continuous extractor. The methylene chloride extract is dried, concentrated to a volume of 1 mL, and analyzed by GC/MS. Qualitative identification of the parameters in the extract is performed using the retention time and the relative abundance of three characteristic masses (m/z). Qualitative analysis is performed using either external or internal standard techniques with a single characteristic m/z.

INTERFERENCES Method interferences may be caused by contaminants in solvents, reagents, glassware, and other sample processing hardware. Glassware must be scrupulously cleaned. Glassware should be heated in a muffle furnace at 400°C for 5 to 30 min. Some thermally stable materials, such as PCBs, may not be eliminated by this treatment. Solvent rinses with acetone and pesticide quality hexane may be substituted for the muffle furnace heating. Matrix interferences may be caused by contaminants that are coextracted from the sample. The base-neutral extraction may cause significantly reduced recovery of phenols. The packed gas chromatographic columns recommended for the basic fraction may not exhibit sufficient resolution for some analytes.

INSTRUMENTATION A GC/MS system with an injection port designed for on-column injection when using packed columns and for splitless injection when using capillary columns.

Column for base/neutrals: 1.8 m long × 2 mm I.D. glass, packed with 3% SP-2550 on Supelcoport (100/120 mesh) or equivalent.

Column for acids: 1.8 m long × 2 mm I.D. glass, packed with 1% SP-1240DA on Supelcoport (100/120 mesh) or equivalent.

PRECISION & ACCURACY The MDL concentrations were obtained using reagent water. The MDL actually achieved in a given analysis will vary depending on instrument sensitivity and matrix effects. This method was tested by 15 laboratories using reagent water, drinking water, surface water, and industrial wastewaters spiked at six concentrations over the range 5 to 100 µg/L. Single operator precision, overall precision, and method accuracy were found to be directly related to the concentration of the parameter matrix.

The MDL (in µg/L) in reagent water was 4.2.
The standard deviation (in µg/L based on 4 recovery measurements) was 31.5.
The range (in µg/L) for average recovery for 4 measurements was 41.5–130.6.
The range (in %) for percent recovery was 24–149.
Accuracy (in µg/L) as expected recovery for one or more measurements of a sample containing a true concentration of C was $0.87C - 0.94$.
Precision (in µg/L) as expected single analyst standard deviation of measurements at an average concentration found at X was $0.20X - 0.58$.

Overall precision (in µg/L) as expected interlaboratory standard deviation of measurements in an average concentration found at X was $0.30C - 1.94$.

$C =$ *True value of the concentration in µg/L.*
$X =$ *Average recovery found for measurements of samples containing a concentration at C in µg/L.*

SAMPLE PREPARATION Adjust the pH to >11 with sodium hydroxide and serially extract in a separatory funnel with methylene chloride or else in a continuous extractor. Next, adjust the pH to <2 with sulfuric acid and serially extract in a separatory funnel with methylene chloride or else in a continuous extractor. Dry the extracts separately through a column of anhydrous sodium sulfate and then concentrate each of the extracts to 1.0 mL using a K-D apparatus.

SAMPLE COLLECTION, PRESERVATION & HANDLING Grab samples must be collected in glass containers. All samples must be refrigerated at 4°C from the time of collection until extraction. If residual chlorine is present, add 80 mg of sodium thiosulfate/L of sample and mix well. All samples must be extracted within 7 days of collection and completely analyzed within 40 days of extraction.

QUALITY CONTROL Make an initial, one-time, demonstration of the ability to generate acceptable accuracy and precision with this method. Before processing any samples, the analyst must analyze a reagent water blank to demonstrate that interferences from the analytical system and glassware are under control. Each time a set of samples is extracted or reagents are changed, a reagent water blank must be processed. Spike and analyze a minimum of 5% of all samples to monitor and evaluate lab data quality. A QC check sample concentrate that contains each parameter of interest at a concentration of 100 µg/mL in acetone is required. PCBs and multicomponent pesticides may be omitted from this test.

After analysis of five spiked wastewater samples, calculate the average percent recovery and the standard deviation of the percent recovery. Spike all samples with the surrogate standard spiking solution and calculate the percent recovery of each surrogate compound.

REFERENCE *Federal Register*, Vol. 49, No. 209. Friday, Oct. 26, 1984.

δ-BHC   EPA Method 625
CAS #319-86-8

TITLE Base/Neutrals and Acids, U.S. EPA Method 625

MATRIX This methods covers municipal and industrial wastewaters.

METHOD SUMMARY Approximately 1 L of sample is serially extracted with methylene chloride at a pH greater than 11 and again at a pH less than 2 using a separatory funnel or a continuous extractor. The methylene chloride extract is dried, concentrated to a volume of 1 mL, and analyzed by GC/MS. Qualitative identification of the parameters in the extract is performed using the retention time and the relative abundance of three characteristic masses (m/z). Qualitative analysis is performed using either external or internal standard techniques with a single characteristic m/z.

INTERFERENCES Method interferences may be caused by contaminants in solvents, reagents, glassware, and other sample processing hardware. Glassware must be scrupulously cleaned. Glassware should be heated in a muffle furnace at 400°C for 5 to 30 min. Some thermally stable materials, such as PCBs, may not be eliminated by this treatment. Solvent rinses with acetone and pesticide quality hexane may be substituted for the muffle furnace heating. Matrix interferences may be caused by contaminants that are coextracted from the sample. The base-neutral extraction may cause significantly reduced recovery of phenols. The packed gas chromatographic columns recommended for the basic fraction may not exhibit sufficient resolution for some analytes.

INSTRUMENTATION A GC/MS system with an injection port designed for on-column injection when using packed columns and for splitless injection when using capillary columns.

Column for base/neutrals: 1.8 m long × 2 mm I.D. glass, packed with 3% SP-2550 on Supelcoport (100/120 mesh) or equivalent.

Column for acids: 1.8 m long × 2 mm I.D. glass, packed with 1% SP-1240DA on Supelcoport (100/120 mesh) or equivalent.

PRECISION & ACCURACY The MDL concentrations were obtained using reagent water. The MDL actually achieved in a given analysis will vary depending on instrument sensitivity and matrix effects. This method was tested by 15 laboratories using reagent water, drinking water, surface water, and industrial wastewaters spiked at six concentrations over the range 5 to 100 µg/L. Single operator precision, overall precision, and method accuracy were found to be directly related to the concentration of the parameter matrix.

The MDL (in µg/L) in reagent water was 3.1.
The standard deviation (in µg/L based on 4 recovery measurements) was 21.6.
The range (in µg/L) for average recovery for 4 measurements was D-100.0.
The range (in %) for percent recovery was D-110.
Accuracy (in µg/L) as expected recovery for one or more measurements of a sample containing a true concentration of C was $0.29C-1.09$.
Precision (in µg/L) as expected single analyst standard deviation of measurements at an average concentration found at X was $0.34X+0.86$.
Overall precision (in µg/L) as expected interlaboratory standard deviation of measurements in an average concentration found at X was $0.93X-0.17$.

$C = $ *True value of the concentration in µg/L.*
$X = $ *Average recovery found for measurements of samples containing a concentration at C in µg/L.*

SAMPLE PREPARATION Adjust the pH to >11 with sodium hydroxide and serially extract in a separatory funnel with methylene chloride or else in a continuous extractor. Next, adjust the pH to <2 with sulfuric acid and serially extract in a separatory funnel with methylene chloride or else in a continuous extractor. Dry the extracts separately through a column of anhydrous sodium sulfate and then concentrate each of the extracts to 1.0 mL using a K-D apparatus.

SAMPLE COLLECTION, PRESERVATION & HANDLING- Grab samples must be collected in glass containers. All samples must be refrigerated at 4°C from the time of collection until extraction. If residual chlorine is present, add 80 mg of sodium thiosulfate/L of sample and mix well. All samples must be extracted within 7 days of collection and completely analyzed within 40 days of extraction.

QUALITY CONTROL Make an initial, one-time, demonstration of the ability to generate acceptable accuracy and precision with this method. Before processing any samples, the analyst must analyze a reagent water blank to demonstrate that interferences from the analytical system and glassware are under control. Each time a set of samples is extracted or reagents are changed, a reagent water blank must be processed. Spike and analyze a minimum of 5% of all samples to monitor and evaluate lab data quality. A QC check sample concentrate that contains each parameter of interest at a concentration of 100 µg/mL in acetone is required. PCBs and multicomponent pesticides may be omitted from this test.

After analysis of five spiked wastewater samples, calculate the average percent recovery and the standard deviation of the percent recovery. Spike all samples with the surrogate standard spiking solution and calculate the percent recovery of each surrogate compound.

REFERENCE *Federal Register*, Vol. 49, No. 209. Friday, Oct. 26, 1984.

α-BHC   EPA Method 8080
CAS #319-84-6

TITLE Organochlorine Pesticides and Polychlorinated Biphenyls By Gas Chromatography

MATRIX This method is used to determine the concentration of various organochlorine pesticides and polychlorinated biphenyls in extracts prepared from water, groundwater, soils, and sediments.

METHOD SUMMARY This method covers 26 pesticides and Aroclor (PCB) mixtures and it is suitable for monitoring-type analyses. After extraction, concentration and solvent exchange to hexane, a 2- to 5-µL sample aliquot is injected into a GC using the solvent flush technique, and the analytes are detected by an electron capture detector (ECD) or an electrolytic conductivity detector in the halogen mode (HECD). Both neat and diluted organic liquids may be analyzed by direct injection.

INTERFERENCES Interferences coextracted from the samples will vary considerably from source to source. Interferences by phthalate esters can pose a major problem in pesticide determinations when using the ECD. Cross-contamination of clean glassware routinely occurs when plastics are handled during extraction steps, especially when solvent-wetted surfaces are handled. The contamination from phthalate esters can be completely eliminated with a microcoulometric or electrolytic conductivity detector. Solvents, reagent, glassware, and other sample processing hardware may yield artifacts and/or interferences to sample analysis.

INSTRUMENTATION A gas chromatograph capable of on-column injections is needed. It must be equipped with an ECD or a HECD and one of the following GC columns:

Column 1: Supelcoport (100/120 mesh) coated with 1.5% SP-2250/1.95% SP-2401 packed in a 1.8 m × 4 mm I.D. glass column.

Column 2: Supelcoport (100/120 mesh) coated with 3% OV-1 in a 1.8 m × 4 mm I.D. glass column.

PRECISION & ACCURACY The method was tested by 20 laboratories using organic-free reagent water, drinking water, surface water, and three industrial wastewaters spiked at six concentrations. Concentrations used in the study ranged from 0.5 to 30 µg/L for single-component pesticides and from 8.5 to 400 µg/L for multicomponent parameters. Overall precision and method accuracy were found to be directly related to the concentration of the analyte and essentially independent of the sample matrix. The sensitivity of this method usually depends on the concentration of interferences rather than on instrumental limitations.

MDL in µg/L was 0.003.
Concentration range in µg/L was 0.5–30.
Accuracy as recovery (x^*) in µg/L was 0.84C+0.03 .
Overall precision (S^*) in µg/L was 0.23x–0.00.

x^* *Expected recovery for one or more measurements of a sample containing concentration C, in µg/L.*
S^* = *Expected interlaboratory standard deviation of measurements at an average concentration found of the analyte in µg/L.*
C = *True value for the concentration, in µg/L.*
X = *Average recovery found for measurements of samples containing a concentration of C, in µg/L.*

SAMPLING METHOD
Liquid samples — Use a 1 or 2½ gallon amber glass bottle with a screw-top Teflon®-lined cover. Pre-wash the bottle with detergent, rinse with distilled water and methanol (or isopropanol).

Soil, sediments, and sludges — Use an 8-oz. widemouth glass with a screw-top Teflon®-lined cover. Pre-wash the bottle with detergent, rinse with distilled water and methanol (or isopropanol).

SAMPLE PRESERVATION Cool water, soil, sediment, or sludge samples immediately to 4°C.

Water samples — If residual chlorine is present, add 3 mL of 10% sodium thiosulfate per gallon and cool to 4°C. All extracts and samples should be stored under refrigeration.

MHT Liquid samples must be extracted within 7 days and the extracts must be analyzed within 40 days. Soils, sediments, and sludges may be stored for a maximum of 14 days prior to extraction.

SAMPLE PREPARATION
Liquid samples — Extract 1-L sample in a continuous extractor at pH 5–9 with methylene chloride after adding 1.0 mL of surrogate spiking solution to each sample. Pass the extract through a column of anhydrous sodium sulfate to dry and concentrate it in a K-D apparatus to 1 mL volume.

Soils, sediments, and sludges — Rapidly weigh approximately 30 g of sample into a 400-mL beaker to avoid loss of the more volatile extractables. Nonporous or wet samples (gummy or clay type) that do not have a free-flowing sandy texture must be mixed with anhydrous sodium sulfate until the sample is free flowing. Add 1 mL of surrogate standards to all samples, spikes, standards, and blanks. Add 100 mL of 1:1 methylene chloride:acetone and extract ultrasonically. Decant and filter extracts, dry the extract by passing it through a drying column containing anhydrous sodium sulfate and concentrate to 1 mL in a K-D apparatus.

Hexane solvent exchange — Add 50 mL of hexane, a new boiling chip, and concentrate until the apparent volume of liquid reaches 1 mL. Adjust the extract volume to 10.0 mL. Stopper the concentration tube and store refrigerated at 4°C if further processing will not be performed immediately. If the extract will be stored longer than two days, transfer it to a vial with Teflon®-lined screw-cap or crimp top.

QUALITY CONTROL Demonstrate through the analysis of a reagent water blank, that all glassware and reagents are interference free. Each time a set of samples is processed, a method blank should be processed as a safeguard against chronic lab contamination. A reagent blank, a matrix spike, and a duplicate or matrix spike duplicate must be performed for each analytical batch (up to a maximum of 20 samples) analyzed.

Analytical system performance must be verified by analyzing QC check samples. The QC check sample concentration should contain each single-component analyte at the following concentrations in acetone: 4,4'-DDD, 10 µg/mL; 4,4'-DDT, 10 µg/mL; endosulfan II, 10 µg/mL; endosulfan sulfate, 10 µg/mL; and any other single-component pesticide at 2 µg/mL. If the method is only to be used to analyze PCBs, Chlordane, or Toxaphene, the QC check sample concentrate should contain the most representative multicomponent parameter at a concentration of 50 µg/mL in acetone.

REFERENCE Test Methods for Evaluating Solid Waste (SW-846). U.S. EPA. 1983. Method 8080B, Rev. 2, Nov. 1990. Office of Solid Wastes, Washington, DC.

β-BHC   EPA Method 8080
CAS #319-85-7

TITLE Organochlorine Pesticides and Polychlorinated Biphenyls By Gas Chromatography

MATRIX This method is used to determine the concentration of various organochlorine pesticides and polychlorinated biphenyls in extracts prepared from water, groundwater, soils, and sediments.

METHOD SUMMARY This method covers 26 pesticides and Aroclor (PCB) mixtures and it is suitable for monitoring-type analyses. After extraction, concentration and solvent exchange to hexane, a 2- to 5-μL sample aliquot is injected into a GC using the solvent flush technique, and the analytes are detected by an electron capture detector (ECD) or an electrolytic conductivity detector in the halogen mode (HECD). Both neat and diluted organic liquids may be analyzed by direct injection.

INTERFERENCES Interferences coextracted from the samples will vary considerably from source to source. Interferences by phthalate esters can pose a major problem in pesticide determinations when using the ECD. Cross-contamination of clean glassware routinely occurs when plastics are handled during extraction steps, especially when solvent-wetted surfaces are handled. The contamination from phthalate esters can be completely eliminated with a microcoulometric or electrolytic conductivity detector. Solvents, reagent, glassware, and other sample processing hardware may yield artifacts and/or interferences to sample analysis.

INSTRUMENTATION A gas chromatograph capable of on-column injections is needed. It must be equipped with an ECD or a HECD and one of the following GC columns:

Column 1: Supelcoport (100/120 mesh) coated with 1.5% SP-2250/1.95% SP-2401 packed in a 1.8 m × 4 mm I.D. glass column.

Column 2: Supelcoport (100/120 mesh) coated with 3% OV-1 in a 1.8 m × 4 mm I.D. glass column.

PRECISION & ACCURACY The method was tested by 20 laboratories using organic-free reagent water, drinking water, surface water, and three industrial wastewaters spiked at six concentrations. Concentrations used in the study ranged from 0.5 to 30 μg/L for single-component pesticides and from 8.5 to 400 μg/L for multicomponent parameters. Overall precision and method accuracy were found to be directly related to the concentration of the analyte and essentially independent of the sample matrix. The sensitivity of this method usually depends on the concentration of interferences rather than on instrumental limitations.

MDL in μg/L was 0.006.
Concentration range in μg/L was 0.5–30.
Accuracy as recovery (x^*) in μg/L was $0.81C+0.07$.
Overall precision (S^*) in μg/L was $0.33x-0.95$.

x^* *Expected recovery for one or more measurements of a sample containing concentration C, in μg/L.*

$S^* =$ *Expected interlaboratory standard deviation of measurements at an average concentration found of the analyte in μg/L.*

$C =$ *True value for the concentration, in μg/L.*

$X =$ *Average recovery found for measurements of samples containing a concentration of C, in μg/L.*

SAMPLING METHOD
Liquid samples — Use a 1 or 2½ gallon amber glass bottle with a screw-top Teflon®-lined cover. Pre-wash the bottle with detergent, rinse with distilled water and methanol (or isopropanol).

Soil, sediments, and sludges — Use an 8-oz. widemouth glass with a screw-top Teflon®-lined cover. Pre-wash the bottle with detergent, rinse with distilled water and methanol (or isopropanol).

SAMPLE PRESERVATION Cool water, soil, sediment, or sludge samples immediately to 4°C.

Water samples — If residual chlorine is present, add 3 mL of 10% sodium thiosulfate per gallon and cool to 4°C. All extracts and samples should be stored under refrigeration.

MHT Liquid samples must be extracted within 7 days and the extracts must be analyzed within 40 days. Soils, sediments, and sludges may be stored for a maximum of 14 days prior to extraction.

SAMPLE PREPARATION
Liquid samples — Extract 1-L sample in a continuous extractor at pH 5–9 with methylene chloride after adding 1.0 mL of surrogate spiking solution to each sample. Pass the extract through a column of anhydrous sodium sulfate to dry and concentrate it in a K-D apparatus to 1 mL volume.

Soils, sediments, and sludges — Rapidly weigh approximately 30 g of sample into a 400-mL beaker to avoid loss of the more volatile extractables. Nonporous or wet samples (gummy or clay type) that do not have a free-flowing sandy texture must be mixed with anhydrous sodium sulfate until the sample is free flowing. Add 1 mL of surrogate standards to all samples, spikes, standards, and blanks. Add 100 mL of 1:1 methylene chloride:acetone and extract ultrasonically. Decant and filter extracts, dry the extract by passing it through a drying column containing anhydrous sodium sulfate and concentrate to 1 mL in a K-D apparatus.

Hexane solvent exchange — Add 50 mL of hexane, a new boiling chip, and concentrate until the apparent volume of liquid reaches 1 mL. Adjust the extract volume to 10.0 mL. Stopper the concentration tube and store refrigerated at 4°C if further processing will not be performed immediately. If the extract will be stored longer than two days, transfer it to a vial with Teflon®-lined screw-cap or crimp top.

QUALITY CONTROL Demonstrate through the analysis of a reagent water blank, that all glassware and reagents are interference free. Each time a set of samples is processed, a method blank should be processed as a safeguard against chronic lab contamination. A reagent blank, a matrix spike, and a duplicate or matrix spike duplicate must be performed for each analytical batch (up to a maximum of 20 samples) analyzed.

Analytical system performance must be verified by analyzing QC check samples. The QC check sample concentration should contain each single-component analyte at the following concentrations in acetone: 4,4'-DDD, 10 μg/mL; 4,4'-DDT, 10 μg/mL; endosulfan II, 10 μg/mL; endosulfan sulfate, 10 μg/mL; and any other single-component pesticide at 2 μg/mL. If the method is only to be used to analyze PCBs, Chlordane, or Toxaphene, the QC check sample concentrate should contain the most representative multicomponent parameter at a concentration of 50 μg/mL in acetone.

REFERENCE Test Methods for Evaluating Solid Waste (SW-846). U.S. EPA. 1983. Method 8080B, Rev. 2, Nov. 1990. Office of Solid Wastes, Washington, DC.

δ-BHC   EPA Method 8080
CAS #319-86-8

TITLE Organochlorine Pesticides and Polychlorinated Biphenyls By Gas Chromatography

MATRIX This method is used to determine the concentration of various organochlorine pesticides and polychlorinated biphenyls in extracts prepared from water, groundwater, soils, and sediments.

METHOD SUMMARY This method covers 26 pesticides and Aroclor (PCB) mixtures and it is suitable for monitoring-type analyses. After extraction, concentration and solvent exchange to hexane, a 2- to 5-μL sample aliquot is injected into a GC using the solvent flush technique, and the analytes are detected by an electron capture detector (ECD) or an electrolytic conductivity detector in the halogen mode (HECD). Both neat and diluted organic liquids may be analyzed by direct injection.

INTERFERENCES Interferences coextracted from the samples will vary considerably from source to source. Interferences by phthalate esters can pose a major problem in pesticide determinations when using the ECD. Cross-contamination of clean glassware routinely occurs when plastics are handled during extraction steps, especially when solvent-wetted surfaces are handled. The contamination from phthalate esters can be completely eliminated with a microcoulometric or electrolytic conductivity detector. Solvents, reagent, glassware, and other sample processing hardware may yield artifacts and/or interferences to sample analysis.

INSTRUMENTATION A gas chromatograph capable of on-column injections is needed. It must be equipped with an ECD or a HECD and one of the following GC columns:

Column 1: Supelcoport (100/120 mesh) coated with 1.5% SP-2250/1.95% SP-2401 packed in a 1.8 m × 4 mm I.D. glass column.

Column 2: Supelcoport (100/120 mesh) coated with 3% OV-1 in a 1.8 m × 4 mm I.D. glass column.

PRECISION & ACCURACY The method was tested by 20 laboratories using organic-free reagent water, drinking water, surface water, and three industrial wastewaters spiked at six concentrations. Concentrations used in the study ranged from 0.5 to 30 μg/L for single-component pesticides and from 8.5 to 400 μg/L for multicomponent parameters. Overall precision and method accuracy were found to be directly related to the concentration of the analyte and essentially independent of the sample matrix. The sensitivity of this method usually depends on the concentration of interferences rather than on instrumental limitations.

MDL in μg/L was 0.009.
Concentration range in μg/L was 0.5–30.
Accuracy as recovery (x^*) in μg/L was $0.81C+0.07$.
Overall precision (S^*) in μg/L was $0.25x +0.03$.

x^* *Expected recovery for one or more measurements of a sample containing concentration C, in μg/L.*
S^* = *Expected interlaboratory standard deviation of measurements at an average concentration found of the analyte in μg/L.*
C = *True value for the concentration, in μg/L.*
X = *Average recovery found for measurements of samples containing a concentration of C, in μg/L.*

SAMPLING METHOD
Liquid samples — Use a 1 or 2½ gallon amber glass bottle with a screw-top Teflon®-lined cover. Pre-wash the bottle with detergent, rinse with distilled water and methanol (or isopropanol).

Soil, sediments, and sludges — Use an 8-oz. widemouth glass with a screw-top Teflon®-lined cover. Pre-wash the bottle with detergent, rinse with distilled water and methanol (or isopropanol).

SAMPLE PRESERVATION Cool water, soil, sediment, or sludge samples immediately to 4°C.

Water samples — If residual chlorine is present, add 3 mL of 10% sodium thiosulfate per gallon and cool to 4°C. All extracts and samples should be stored under refrigeration.

MHT Liquid samples must be extracted within 7 days and the extracts must be analyzed within 40 days. Soils, sediments, and sludges may be stored for a maximum of 14 days prior to extraction.

SAMPLE PREPARATION
Liquid samples — Extract 1-L sample in a continuous extractor at pH 5–9 with methylene chloride after adding 1.0 mL of surrogate spiking solution to each sample. Pass the extract through a column of anhydrous sodium sulfate to dry and concentrate it in a K-D apparatus to 1 mL volume.

Soils, sediments, and sludges — Rapidly weigh approximately 30 g of sample into a 400-mL beaker to avoid loss of the more volatile extractables. Nonporous or wet samples (gummy or clay type) that do not have a free-flowing sandy texture must be mixed with anhydrous sodium sulfate until the sample is free flowing. Add 1 mL of surrogate standards to all samples, spikes, standards, and blanks. Add 100 mL of 1:1 methylene chloride:acetone and extract ultrasonically. Decant and filter extracts, dry the extract by passing it through a drying column containing anhydrous sodium sulfate and concentrate to 1 mL in a K-D apparatus.

Hexane solvent exchange — Add 50 mL of hexane, a new boiling chip, and concentrate until the apparent volume of liquid reaches 1 mL. Adjust the extract volume to 10.0 mL. Stopper the concentration tube and store refrigerated at 4°C if further processing will not be performed immediately. If the extract will be stored longer than two days, transfer it to a vial with Teflon®-lined screw-cap or crimp top.

QUALITY CONTROL Demonstrate through the analysis of a reagent water blank, that all glassware and reagents are interference free. Each time a set of samples is processed, a method blank should be processed as a safeguard against chronic lab contamination. A reagent blank, a matrix spike, and a duplicate or matrix spike duplicate must be performed for each analytical batch (up to a maximum of 20 samples) analyzed.

Analytical system performance must be verified by analyzing QC check samples. The QC check sample concentration should contain each single-component analyte at the following concentrations in acetone: 4,4′-DDD, 10 µg/mL; 4,4′-DDT, 10 µg/mL; endosulfan II, 10 µg/mL; endosulfan sulfate, 10 µg/mL; and any other single-component pesticide at 2 µg/mL. If the method is only to be used to analyze PCBs, Chlordane, or Toxaphene, the QC check sample concentrate should contain the most representative multicomponent parameter at a concentration of 50 µg/mL in acetone.

REFERENCE Test Methods for Evaluating Solid Waste (SW-846). U.S. EPA. 1983. Method 8080B, Rev. 2, Nov. 1990. Office of Solid Wastes, Washington, DC.

α-BHC   EPA Method 8121
CAS #319-84-6

TITLE Chlorinated Hydrocarbons by GC: Capillary Column Technique

MATRIX This method covers aqueous and solid matrices. This includes a wide variety such as drinking water, groundwater, industrial wastewaters, surface waters, soils, solids, and sediments.

METHOD SUMMARY This method provides procedures for the determination of 22 chlorinated hydrocarbons in water, soil/sediment, and waste matrices. A measured volume or weight of sample is extracted by using one of the appropriate sample extraction techniques specified in EPA Method 3510, EPA Method 3520, EPA Method 3540, or EPA Method 3550, or diluted using EPA Method 3580. Aqueous samples are extracted at neutral pH with methylene chloride by using either a separatory funnel (EPA Method 3510) or a continuous liquid-liquid extractor (EPA Method 3520). Solid samples are extracted with hexane/acetone (1:1) by using a Soxhlet extractor (EPA Method 3540) or with methylene chloride/acetone (1:1) by using an ultrasonic extractor (EPA Method 3550). After cleanup, the extract or diluted sample is analyzed by gas chromatography with electron capture detection (GC/ECD).

The sensitivity level of this method usually depends on the level of interferences rather than on instrumental limitations. This method may be used in conjunction with EPA Method 3620, Florisil Column Cleanup, EPA Method 3660, Sulfur Cleanup, and EPA Method 3640, Gel Permeation Chromatography, to aid in the elimination of interferences.

INTERFERENCES Solvents, reagents, glassware, and other hardware used in sample processing may introduce artifacts which may result in elevated baselines, causing misinterpretation of gas chromatograms. Interferants coextracted from the samples will vary considerably from waste to waste. Glassware must be scrupulously clean. Phthalate esters, if present in a sample, will interfere only with the BHC isomers. The presence of elemental sulfur will result in large peaks, and can often mask the region of compounds eluting after 1,2,4,5-tetrachlorobenzene. The tetrabutylammonium (TBA)-sulfite procedure (EPA Method 3660) works well for the removal of elemental sulfur. Waxes and lipids can be removed by gel permeation chromatography (EPA Method 3640).

INSTRUMENTATION A GC suitable for on-column injections and all required accessories, including and electron capture detector (ECD), analytical columns, recorder, gases, and syringes are needed. A data system for measuring peak heights and/or peak areas is recommended. A Kuderna-Danish (K-D) apparatus will also be needed to concentrate extracts.

Column 1: 30 m × 0.53 mm I.D. fused-silica capillary column chemically bonded with trifluoropropyl methyl silicone (DB-210 or equivalent).

Column 2: 30 m × 0.53 mm I.D. fused-silica capillary column chemically bonded with polyethylene glycol (DB-WAX or equivalent).

PRECISION & ACCURACY This method has been tested in a single lab by using organic-free reagent water, sandy loam samples, and extracts which were spiked with the test compounds at one concentration. Single-operator precision and method accuracy were found to be related to the concentration of compound and the type of matrix. The accuracy and precision technique will be determined by the sample matrix, sample preparation technique, optional cleanup techniques, and calibration procedures used.

ESTIMATED QUANTITATION LIMIT (EQL) FACTORS FOR VARIOUS MATRICES (a)

Matrix	Factor(b)
Groundwater	10
Low-concentration soil by ultrasonic extraction with GPC cleanup	670
High-concentration soil and sludges by ultrasonic extraction	10,000
Waste not miscible with water	100,000

(a) Sample EQLs are highly matrix-dependent. The EQLs listed herein are provided for guidance and may not always be achievable. (b) EQL = [Method detection limit] × [Factor]. For nonaqueous samples, the factor is on a wet-weight basis.

PRECISION & ACCURACY MDL is the method detection limit for organic-free reagent water. MDL was determined from the analysis of eight replicate aliquots processed through the entire analytical method (extraction, Florisil cartridge cleanup, and GC/ECD analysis).

The MDL (in ng/L) was 11.

The accuracy (as average % recovery using 5 determinations and no Florisil cleanup) from a spike concentration of 10 µg/L and separatory funnel extraction was 96% with a final volume of 10 mL.

The precision (as RSD% using 5 determinations and no Florisil cleanup) from a spike concentration of 10 µg/L and separatory funnel extraction was 2.6% with a final volume of 10 mL.

The accuracy (as average % recovery using 5 determinations and no Florisil cleanup), from a spike concentration of 3300 µg/L and ultrasonic extraction of solid samples using 1:1 methylene chloride and acetone, was 100% with a final volume of 10 mL.

The precision (as RSD% using 5 determinations and no Florisil cleanup), from a spike concentration of 3300 µg/L and ultrasonic extraction of solid samples using 1:1 methylene chloride and acetone, was 2.9% with a final volume of 10 mL.

SAMPLE COLLECTION, PRESERVATION & HANDLING
Volatile Organics — Standard 40-mL glass screw-cap VOA vials with Teflon®-faced silicone septum may be used for both liquid and solid matrices. When collecting samples, liquids and solids should be introduced into the vials gently to reduce agitation which might drive off volatile compounds. If there are any air bubbles present the sample must be retaken. The vials with solids should be tapped slightly as they are filled to try and eliminate as much free air space as possible. Two vials from each sampling location should be sealed in separate plastic bags to prevent cross-contamination between samples.

Semivolatile organics — Containers used to collect samples for the determination of semivolatile organic compounds should be soap and water washed followed by methanol (or isopropanol) rinsing. The sample containers should be of glass or Teflon® and have screw-top covers with Teflon® liners.

Preservation for volatile organics — No preservation is used with concentrated waste samples. With liquid samples containing no residual chlorine, 4 drops of concentrated hydrochloric acid are added and the samples are immediately cooled to 4°C. When liquid samples contain residual chlorine, they are treated as above and, in addition, 4 drops of 4% aqueous sodium thiosulfate are added to remove the residual chlorine. Soil, sediment, and sludge samples are only cooled to 4°C.

Preservation for semivolatile organics — No preservation is used with concentrated waste samples. With liquid samples containing no residual chlorine and with soil, sediment, and sludge samples, immediately cooling to 4°C is the only preservation used. When residual chlorine is present then 3 mL of 10% aqueous sodium sulfate is added for each gallon of sample collected, followed by cooling to 4°C.

Holding times — The holding time for all volatile organics samples is 14 days. Liquid samples must be extracted within 7 days and their extracts analyzed within 40 days. Concentrated waste, soil, sediment, and sludge samples must be extracted within 14 days and their extracts analyzed within 40 days.

SAMPLE PREPARATION Prepare stock standard solutions in hexane. Calibration standards at a minimum of five concentrations should be prepared through dilution of the stock standards with hexane. The suggested internal standards are: 2,5-dibromotoluene, 1,3,5-tribromobenzene, and α,α-dibromo-m-xylene. The analyst can use any of the three compounds provided that they are resolved from matrix interferences. Recommended surrogate compounds are α-2,6-trichlorotoluene, 1,4-dichloronaphthalene, and 2,3,4,5,6-pentachlorotoluene.

In general, water samples are extracted at a neutral pH with methylene chloride using a separatory funnel (EPA Method 3510) or a continuous liquid-liquid extractor (EPA Method 3520). Solid samples are extracted with hexane/acetone (1:1 v:v) using a Soxhlet extractor (EPA Method 3540) or with methylene chloride/acetone (1:1 v:v) using an ultrasonic extractor (EPA Method 3550). Non-aqueous waste samples may be diluted using EPA Method 3580. Prior to Florisil cleanup or gas chromatographic analysis, the extraction solvent must be exchanged to hexane. Sample extracts that will be subjected to gel permeation chromatography do not need solvent exchange.

Cleanup procedures may not be necessary for a relatively clean matrix. If removal of interferences such as chlorinated phenols, phthalate esters, etc., is required, proceed with the procedure outlined in EPA Method 3620.

QUALITY CONTROL Analyze a quality control check standard to demonstrate that the operation of the GC is in control. The frequency of the check standard analysis is equivalent to 10% of the samples analyzed. If the recovery of any compound found in the check standard is less than 80% of the certified value, the problem must be corrected and a new set of calibration standards must be prepared and analyzed. Calculate surrogate standard recoveries for all samples, blanks, and spikes. An internal standard peak area check must be performed on all samples. The internal standard must be evaluated for acceptance by determining whether the measured area for the internal standard deviates by more than 30% from the average area for the internal standard in the calibration standards. When the internal standard peak area is outside that limit, all samples that fall outside the QC criteria must be reanalyzed. Any compound confirmed by two columns may also be confirmed by GC/MS (EPA Method 8270). The GC/MS would normally require a minimum concentration of 1 ng/µL in the final extract for each compound. Include a mid-concentration calibration standard after each group of 20 samples in the analysis sequence. The response factors for the mid-concentration calibration must be within 15% of the average values for the multiconcentration calibration.

REFERENCE Test Methods for Evaluating Solid Waste, Physical/Chemical Methods, SW-846, 3rd Edition, U.S. EPA, Office of Solid Waste, Washington, DC, 1990. EPA Method 8121, Rev. 0, Nov. 1990.

β-BHC   **EPA Method 8121**
CAS #319-85-7

TITLE Chlorinated Hydrocarbons by GC: Capillary Column Technique

MATRIX This method covers aqueous and solid matrices. This includes a wide variety such as drinking water, groundwater, industrial wastewaters, surface waters, soils, solids, and sediments.

METHOD SUMMARY This method provides procedures for the determination of 22 chlorinated hydrocarbons in water, soil/sediment, and waste matrices. A measured volume or weight of sample is extracted by using one of the appropriate sample extraction techniques specified in EPA Method 3510, EPA Method 3520, EPA Method 3540, or EPA Method 3550, or diluted using EPA Method 3580. Aqueous samples are extracted at neutral pH with methylene chloride by using either a separatory funnel (EPA Method 3510) or a continuous liquid-liquid extractor (EPA Method 3520). Solid samples are extracted with hexane/acetone (1:1) by using a Soxhlet extractor (EPA Method 3540) or with methylene chloride/acetone (1:1) by using an ultrasonic extractor (EPA Method 3550). After cleanup, the extract or diluted sample is analyzed by gas chromatography with electron capture detection (GC/ECD).

The sensitivity level of this method usually depends on the level of interferences rather than on instrumental limitations. This method may be used in conjunction with EPA Method 3620, Florisil Column Cleanup, EPA Method 3660, Sulfur Cleanup, and EPA Method 3640, Gel Permeation Chromatography, to aid in the elimination of interferences.

INTERFERENCES Solvents, reagents, glassware, and other hardware used in sample processing may introduce artifacts which may result in elevated baselines, causing misinterpretation of gas chromatograms. Interferants coextracted from the samples will vary considerably from waste to waste. Glassware must be scrupulously clean. Phthalate esters, if present in a sample, will interfere only with the BHC isomers. The presence of elemental sulfur will result in large peaks, and can often mask the region of compounds eluting after 1,2,4,5-tetrachlorobenzene. The tetrabutylammonium (TBA)-sulfite procedure (EPA Method 3660) works well for the removal of elemental sulfur. Waxes and lipids can be removed by gel permeation chromatography (EPA Method 3640).

INSTRUMENTATION A GC suitable for on-column injections and all required accessories, including and electron capture detector (ECD), analytical columns, recorder, gases, and syringes are needed. A data system for measuring peak heights and/or peak areas is recommended. A Kuderna-Danish (K-D) apparatus will also be needed to concentrate extracts.

Column 1: 30 m × 0.53 mm I.D. fused-silica capillary column chemically bonded with trifluoropropyl methyl silicone (DB-210 or equivalent).
Column 2: 30 m × 0.53 mm I.D. fused-silica capillary column chemically bonded with polyethylene glycol (DB-WAX or equivalent).

PRECISION & ACCURACY This method has been tested in a single lab by using organic-free reagent water, sandy loam samples, and extracts which were spiked with the test compounds at one concentration. Single-operator precision and method accuracy were found to be related to the concentration of compound and the type of matrix. The accuracy and precision technique will be determined by the sample matrix, sample preparation technique, optional cleanup techniques, and calibration procedures used.

ESTIMATED QUANTITATION LIMIT (EQL) FACTORS FOR VARIOUS MATRICES (a)

Matrix	Factor (b)
Groundwater	10
Low-concentration soil by ultrasonic extraction with GPC cleanup	670
High-concentration soil and sludges by ultrasonic extraction	10,000
Waste not miscible with water	100,000

(a) Sample EQLs are highly matrix-dependent. The EQLs listed herein are provided for guidance and may not always be achievable. (b) EQL = [Method detection limit] × [Factor]. For nonaqueous samples, the factor is on a wet-weight basis.

PRECISION & ACCURACY MDL is the method detection limit for organic-free reagent water. MDL was determined from the analysis of eight replicate aliquots processed through the entire analytical method (extraction, Florisil cartridge cleanup, and GC/ECD analysis).

The MDL (in ng/L) was 31.

The accuracy (as average % recovery using 5 determinations and no Florisil cleanup) from a spike concentration of 10 µg/L and separatory funnel extraction was 103% with a final volume of 10 mL.

The precision (as RSD% using 5 determinations and no Florisil cleanup) from a spike concentration of 10 µg/L and separatory funnel extraction was 3.6% with a final volume of 10 mL.

The accuracy (as average % recovery using 5 determinations and no Florisil cleanup), from a spike concentration of 3300 µg/L and ultrasonic extraction of solid samples using 1:1 methylene chloride and acetone, was 923% with a final volume of 10 mL.

The precision (as RSD% using 5 determinations and no Florisil cleanup), from a spike concentration of 3300 µg/L and ultrasonic extraction of solid samples using 1:1 methylene chloride and acetone, was 2.4% with a final volume of 10 mL.

SAMPLE COLLECTION, PRESERVATION & HANDLING
Volatile Organics — Standard 40-mL glass screw-cap VOA vials with Teflon®-faced silicone septum may be used for both liquid and solid matrices. When collecting samples, liquids and solids should be introduced into the vials gently to reduce agitation which might drive off volatile compounds. If there are any air bubbles present the sample must be retaken. The vials with solids should be tapped slightly as they are filled to try and eliminate as much free air space as possible. Two vials from each sampling location should be sealed in separate plastic bags to prevent cross-contamination between samples.

Semivolatile organics — Containers used to collect samples for the determination of semivolatile organic compounds should be soap and water washed followed by methanol (or isopropanol) rinsing. The sample containers should be of glass or Teflon® and have screw-top covers with Teflon® liners.

Preservation for volatile organics — No preservation is used with concentrated waste samples. With liquid samples containing no

residual chlorine, 4 drops of concentrated hydrochloric acid are added and the samples are immediately cooled to 4°C. When liquid samples contain residual chlorine, they are treated as above and, in addition, 4 drops of 4% aqueous sodium thiosulfate are added to remove the residual chlorine. Soil, sediment, and sludge samples are only cooled to 4°C.

Preservation for semivolatile organics — No preservation is used with concentrated waste samples. With liquid samples containing no residual chlorine and with soil, sediment, and sludge samples, immediately cooling to 4°C is the only preservation used. When residual chlorine is present then 3 mL of 10% aqueous sodium sulfate is added for each gallon of sample collected, followed by cooling to 4°C.

Holding times — The holding time for all volatile organics samples is 14 days. Liquid samples must be extracted within 7 days and their extracts analyzed within 40 days. Concentrated waste, soil, sediment, and sludge samples must be extracted within 14 days and their extracts analyzed within 40 days.

SAMPLE PREPARATION Prepare stock standard solutions in hexane. Calibration standards at a minimum of five concentrations should be prepared through dilution of the stock standards with hexane. The suggested internal standards are: 2,5-dibromotoluene, 1,3,5-tribromobenzene, and α, α-dibromo-m-xylene. The analyst can use any of the three compounds provided that they are resolved from matrix interferences. Recommended surrogate compounds are α-2,6-trichlorotoluene, 1,4-dichloronaphthalene, and 2,3,4,5,6-pentachlorotoluene.

In general, water samples are extracted at a neutral pH with methylene chloride using a separatory funnel (EPA Method 3510) or a continuous liquid-liquid extractor (EPA Method 3520). Solid samples are extracted with hexane/acetone (1:1 v:v) using a Soxhlet extractor (EPA Method 3540) or with methylene chloride/acetone (1:1 v:v) using an ultrasonic extractor (EPA Method 3550). Non-aqueous waste samples may be diluted using EPA Method 3580. Prior to Florisil cleanup or gas chromatographic analysis, the extraction solvent must be exchanged to hexane. Sample extracts that will be subjected to gel permeation chromatography do not need solvent exchange.

Cleanup procedures may not be necessary for a relatively clean matrix. If removal of interferences such as chlorinated phenols, phthalate esters, etc., is required, proceed with the procedure outlined in EPA Method 3620.

QUALITY CONTROL Analyze a quality control check standard to demonstrate that the operation of the GC is in control. The frequency of the check standard analysis is equivalent to 10% of the samples analyzed. If the recovery of any compound found in the check standard is less than 80% of the certified value, the problem must be corrected and a new set of calibration standards must be prepared and analyzed. Calculate surrogate standard recoveries for all samples, blanks, and spikes. An internal standard peak area check must be performed on all samples. The internal standard must be evaluated for acceptance by determining whether the measured area for the internal standard deviates by more than 30% from the average area for the internal standard in the calibration standards. When the internal standard peak area is outside that limit, all samples that fall outside the QC criteria must be reanalyzed. Any compound confirmed by two columns may also be confirmed by GC/MS (EPA Method 8270). The GC/MS would normally require a minimum concentration of 1 ng/µL in the final extract for each compound. Include a mid-concentration calibration standard after each group of 20 samples in the analysis sequence. The response factors for the mid-concentration calibration must be within 15% of the average values for the multiconcentration calibration.

REFERENCE Test Methods for Evaluating Solid Waste, Physical/Chemical Methods, SW-846, 3rd Edition, U.S. EPA, Office of Solid Waste, Washington, DC, 1990. EPA Method 8121, Rev. 0, Nov. 1990.

δ-BHC EPA Method 8121
CAS #319-86-8

TITLE Chlorinated Hydrocarbons by GC: Capillary Column Technique

MATRIX This method covers aqueous and solid matrices. This includes a wide variety such as drinking water, groundwater, industrial wastewaters, surface waters, soils, solids, and sediments.

METHOD SUMMARY This method provides procedures for the determination of 22 chlorinated hydrocarbons in water, soil/sediment, and waste matrices. A measured volume or weight of sample is extracted by using one of the appropriate sample extraction techniques specified in EPA Method 3510, EPA Method 3520, EPA Method 3540, or EPA Method 3550, or diluted using EPA Method 3580. Aqueous samples are extracted at neutral pH with methylene chloride by using either a separatory funnel (EPA Method 3510) or a continuous liquid-liquid extractor (EPA Method 3520). Solid samples are extracted with hexane/acetone (1:1) by using a Soxhlet extractor (EPA Method 3540) or with methylene chloride/acetone (1:1) by using an ultrasonic extractor (EPA Method 3550). After cleanup, the extract or diluted sample is analyzed by gas chromatography with electron capture detection (GC/ECD).

The sensitivity level of this method usually depends on the level of interferences rather than on instrumental limitations. This method may be used in conjunction with EPA Method 3620, Florisil Column Cleanup, EPA Method 3660, Sulfur Cleanup, and EPA Method 3640, Gel Permeation Chromatography, to aid in the elimination of interferences.

INTERFERENCES Solvents, reagents, glassware, and other hardware used in sample processing may introduce artifacts which may result in elevated baselines, causing misinterpretation of gas chromatograms. Interferants coextracted from the samples will vary considerably from waste to waste. Glassware must be scrupulously clean. Phthalate esters, if present in a sample, will interfere only with the BHC isomers. The presence of elemental sulfur will result in large peaks, and can often mask the region of compounds eluting after 1,2,4,5-tetrachlorobenzene. The tetrabutylammonium (TBA)-sulfite procedure

(EPA Method 3660) works well for the removal of elemental sulfur. Waxes and lipids can be removed by gel permeation chromatography (EPA Method 3640).

INSTRUMENTATION A GC suitable for on-column injections and all required accessories, including and electron capture detector (ECD), analytical columns, recorder, gases, and syringes are needed. A data system for measuring peak heights and/or peak areas is recommended. A Kuderna-Danish (K-D) apparatus will also be needed to concentrate extracts.

Column 1: 30 m × 0.53 mm I.D. fused-silica capillary column chemically bonded with trifluoropropyl methyl silicone (DB-210 or equivalent).

Column 2: 30 m × 0.53 mm I.D. fused-silica capillary column chemically bonded with polyethylene glycol (DB-WAX or equivalent).

PRECISION & ACCURACY This method has been tested in a single lab by using organic-free reagent water, sandy loam samples, and extracts which were spiked with the test compounds at one concentration. Single-operator precision and method accuracy were found to be related to the concentration of compound and the type of matrix. The accuracy and precision technique will be determined by the sample matrix, sample preparation technique, optional cleanup techniques, and calibration procedures used.

ESTIMATED QUANTITATION LIMIT (EQL) FACTORS FOR VARIOUS MATRICES (a)

Matrix	Factor (b)
Groundwater	10
Low-concentration soil by ultrasonic extraction with GPC cleanup	670
High-concentration soil and sludges by ultrasonic extraction	10,000
Waste not miscible with water	100,000

(a) Sample EQLs are highly matrix-dependent. The EQLs listed herein are provided for guidance and may not always be achievable. (b) EQL = [Method detection limit] × [Factor]. For nonaqueous samples, the factor is on a wet-weight basis.

PRECISION & ACCURACY MDL is the method detection limit for organic-free reagent water. MDL was determined from the analysis of eight replicate aliquots processed through the entire analytical method (extraction, Florisil cartridge cleanup, and GC/ECD analysis).

The MDL (in ng/L) was 20.

The accuracy (as average % recovery using 5 determinations and no Florisil cleanup) from a spike concentration of 10 µg/L and separatory funnel extraction was 103% with a final volume of 10 mL.

The precision (as RSD% using 5 determinations and no Florisil cleanup) from a spike concentration of 10 µg/L and separatory funnel extraction was 2.7% with a final volume of 10 mL.

The accuracy (as average % recovery using 5 determinations and no Florisil cleanup), from a spike concentration of 3300 µg/L and ultrasonic extraction of solid samples using 1:1 methylene chloride and acetone, was 973% with a final volume of 10 mL.

The precision (as RSD% using 5 determinations and no Florisil cleanup), from a spike concentration of 3300 µg/L and ultrasonic extraction of solid samples using 1:1 methylene chloride and acetone, was 1.5% with a final volume of 10 mL.

SAMPLE COLLECTION, PRESERVATION & HANDLING
Volatile Organics — Standard 40-mL glass screw-cap VOA vials with Teflon®-faced silicone septum may be used for both liquid and solid matrices. When collecting samples, liquids and solids should be introduced into the vials gently to reduce agitation which might drive off volatile compounds. If there are any air bubbles present the sample must be retaken. The vials with solids should be tapped slightly as they are filled to try and eliminate as much free air space as possible. Two vials from each sampling location should be sealed in separate plastic bags to prevent cross-contamination between samples.

Semivolatile organics — Containers used to collect samples for the determination of semivolatile organic compounds should be soap and water washed followed by methanol (or isopropanol) rinsing. The sample containers should be of glass or Teflon® and have screw-top covers with Teflon® liners.

Preservation for volatile organics — No preservation is used with concentrated waste samples. With liquid samples containing no residual chlorine, 4 drops of concentrated hydrochloric acid are added and the samples are immediately cooled to 4°C. When liquid samples contain residual chlorine, they are treated as above and, in addition, 4 drops of 4% aqueous sodium thiosulfate are added to remove the residual chlorine. Soil, sediment, and sludge samples are only cooled to 4°C.

Preservation for semivolatile organics — No preservation is used with concentrated waste samples. With liquid samples containing no residual chlorine and with soil, sediment, and sludge samples, immediately cooling to 4°C is the only preservation used. When residual chlorine is present then 3 mL of 10% aqueous sodium sulfate is added for each gallon of sample collected, followed by cooling to 4°C.

Holding times — The holding time for all volatile organics samples is 14 days. Liquid samples must be extracted within 7 days and their extracts analyzed within 40 days. Concentrated waste, soil, sediment, and sludge samples must be extracted within 14 days and their extracts analyzed within 40 days.

SAMPLE PREPARATION Prepare stock standard solutions in hexane. Calibration standards at a minimum of five concentrations should be prepared through dilution of the stock standards with hexane. The suggested internal standards are: 2,5-dibromotoluene, 1,3,5-tribromobenzene, and α, α-dibromo-m-xylene. The analyst can use any of the three compounds provided that they are resolved from matrix interferences. Recommended surrogate compounds are α-2,6-trichlorotoluene, 1,4-dichloronaphthalene, and 2,3,4,5,6-pentachlorotoluene.

In general, water samples are extracted at a neutral pH with methylene chloride using a separatory funnel (EPA Method 3510) or a continuous liquid-liquid extractor (EPA Method 3520). Solid samples are extracted with hexane/acetone (1:1

v:v) using a Soxhlet extractor (EPA Method 3540) or with methylene chloride/acetone (1:1 v:v) using an ultrasonic extractor (EPA Method 3550). Non-aqueous waste samples may be diluted using EPA Method 3580. Prior to Florisil cleanup or gas chromatographic analysis, the extraction solvent must be exchanged to hexane. Sample extracts that will be subjected to gel permeation chromatography do not need solvent exchange.

Cleanup procedures may not be necessary for a relatively clean matrix. If removal of interferences such as chlorinated phenols, phthalate esters, etc., is required, proceed with the procedure outlined in EPA Method 3620.

QUALITY CONTROL Analyze a quality control check standard to demonstrate that the operation of the GC is in control. The frequency of the check standard analysis is equivalent to 10% of the samples analyzed. If the recovery of any compound found in the check standard is less than 80% of the certified value, the problem must be corrected and a new set of calibration standards must be prepared and analyzed. Calculate surrogate standard recoveries for all samples, blanks, and spikes. An internal standard peak area check must be performed on all samples. The internal standard must be evaluated for acceptance by determining whether the measured area for the internal standard deviates by more than 30% from the average area for the internal standard in the calibration standards. When the internal standard peak area is outside that limit, all samples that fall outside the QC criteria must be reanalyzed. Any compound confirmed by two columns may also be confirmed by GC/MS (EPA Method 8270). The GC/MS would normally require a minimum concentration of 1 ng/μL in the final extract for each compound. Include a mid-concentration calibration standard after each group of 20 samples in the analysis sequence. The response factors for the mid-concentration calibration must be within 15% of the average values for the multiconcentration calibration.

REFERENCE Test Methods for Evaluating Solid Waste, Physical/Chemical Methods, SW-846, 3rd Edition, U.S. EPA, Office of Solid Waste, Washington, DC, 1990. EPA Method 8121, Rev. 0, Nov. 1990.

α-BHC   EPA Method 8270
CAS #319-84-6

TITLE Semivolatile Organic Compounds by GC/MS

MATRIX This method is used to determine the concentration of semivolatile organic compounds in extracts prepared from all types of solid waste matrices, soils, and groundwater. Although surface waters are not specifically mentioned, this method should be applicable to water samples from rivers, lakes, etc.

METHOD SUMMARY This method covers 259 semivolatile organic compounds. In very limited applications direct injection of the sample into the GC/MS system may be appropriate, but this results in very high detection limits (approximately 10,000 μg/L). Typically, a 1-L liquid sample, containing surrogate, and matrix spiking standards, is extracted in a continuous extractor first under acid conditions and then under basic conditions. Typically 30 g of a solid sample, containing surrogate, and matrix spiking standards, is extracted ultrasonically. After concentrating the extract to 1 mL it is spiked with 10 μL of an internal standard solution just prior to analysis by GC/MS. The volume injected should contain about 100 ng of base/neutral and 200 ng of acid surrogates (for a 1-μL injection). Analysis is performed by GC/MS using a capillary GC column.

INTERFERENCES Raw GC/MS data from all blanks, samples, and spikes must be evaluated for interferences. Contamination by carryover can occur whenever high-concentration and low-concentration samples are sequentially analyzed. To reduce carryover, the sample syringe must be rinsed out between samples with solvent. Whenever an unusually concentrated sample is encountered, it should be followed by the analysis of blank solvent to check for cross-contamination.

INSTRUMENTATION A GC/MS and a data system are required. The GC column used is a 30 m × 0.25 mm I.D. (or 0.32 mm I.D.) 1-μm film thickness silicone-coated fused silica capillary column. A continuous liquid-liquid extractor equipped with Teflon® or glass connection joints and stopcocks requiring no lubrication, a K-D concentrating apparatus, water bath, and an ultrasonic disrupter with a minimum power of 300 W and with pulsing capability are also required.

PRECISION & ACCURACY The estimated quantitation limit (EQL) of Method 8270B for determining an individual compound is approximately 1 mg/kg (wet weight) for soil or sediment samples, 1–200 mg/kg for wastes (dependent on matrix and method of preparation), and 10 μg/L for groundwater samples. EQLs will be proportionately higher for sample extracts that require dilution to avoid saturation of the detector.

The EQL(b) for groundwater in μg/L is not listed.
The EQL (a, b) for low concentrations in soil and sediment in μg/kg is not listed.
Accuracy as μg/L is not listed.
Overall precision in μg/L is not listed.

(a) EQLs listed for soil/sediment are based on wet weight. Normally data is reported in a dry-weight basis; therefore, EQLs will be higher based on the % dry weight of each sample. This calculation is based on a 30-g sample and gel permeation chromatography cleanup.
(b) Sample EQLs are highly matrix-dependent. The EQLs are provided for guidance and may not always be achievable.
C = *True value for concentration, in μg/L.*
X = *Average recovery found for measurements of samples containing a concentration of C, in μg/L.*

ESTIMATED QUANTITATION LIMIT FOR OTHER MATRICES

Other Matrices	Factor (a)
High-concentration soil and sludges by sonicator	7.5
Non-water miscible waste	75

(a) EQL for other matrices = [EQL for low soil/sediment] × [Factor]. This estimated EQL is similar to an EPA "Practical Quantitation Limit."

SAMPLING METHOD

Liquid samples — Use a 1 or 2½ gallon amber glass bottle with a screw-top Teflon®-lined cover that has been prewashed with detergent and rinsed with distilled water and methanol (or isopropanol).

Soils, sediments, or sludges — Use an 8-oz. widemouth glass with a screw-top Teflon®-lined cover that has been prewashed with detergent and rinsed with distilled water and methanol (or isopropanol).

SAMPLE PRESERVATION

Liquid samples — If residual chlorine is present, add 3 mL of 10% sodium thiosulfate per gallon, cool to 4°C and store in a solvent-free refrigerator until analysis; if chlorine is not present, then eliminate the sodium thiosulfate addition.

Soils, sediments, or sludges — Cool samples to 4°C and store in a solvent-free refrigerator.

MHT Liquid samples must be extracted within 7 days and the extracts analyzed within 40 days. Soils, sediments, or sludges may be stored for a maximum of 14 days and the extracts analyzed within 40 days.

SAMPLE PREPARATION

Liquid samples — Transfer 1 L quantitatively to a continuous extractor. If high concentrations are anticipated, a smaller volume may be used and then diluted with organic-free reagent water to 1 L. Adjust pH, if necessary, to pH <2 using 1:1 (V/V) sulfuric acid. Pipette 1.0 mL of a surrogate standard spiking solution into each sample. For the sample in each analytical batch selected for spiking, add 1.0 mL of a matrix spiking standard. For base/neutral acid analysis, the amount of the surrogates and matrix spiking compounds added to the sample should result in a final concentration of 100 ng/μL of each analyte in the extract to be analyzed (assuming a 1-μL injection). Extract with methylene chloride for 18–24 h. Next, adjust the pH of the aqueous phase to pH >11 using 10 N sodium hydroxide and extract it with methylene chloride again for 18–24 h. Dry the extract through a column containing anhydrous sodium sulfate and concentrate it to 1 mL using a K-D concentrator.

Soils, sediments, or sludges — Use 30 g of sample. Nonporous or wet samples (gummy or clay type) that do not have a free-flowing sandy texture must be mixed with anhydrous sodium sulfate until the sample is free flowing. Add 1 mL of surrogate standards to all samples, spikes, standards, and blanks. For the sample in each analytical batch selected for spiking, add 1.0 mL of a matrix spiking standard. For base/neutral acid analysis, the amount added of the surrogates and matrix spiking compounds should result in a final concentration of 100 ng/μL of each base/neutral analyte and 200 ng/μL of each acid analyte in the extract to be analyzed (assuming a 1-μL injection). Immediately add a 100 mL mixture of 1:1 methylene chloride:acetone and extract the sample ultrasonically for 3 min and then decant or filter the extracts. Repeat the extraction two or more times. Dry the extract using a column with anhydrous sodium sulfate and concentrate it to 1 mL in a K-D concentrator.

Note: Under the alkaline conditions of the extraction step α-BHC is subject to decomposition so neutral extraction should be performed if this compound is expected.

QUALITY CONTROL

A methylene chloride solution containing 50 ng/μL of decafluorotriphenylphosphine (DFTPP) is used for tuning the GC/MS system each 12-h shift. A system performance check also must be made during every 12-h shift. A standard containing 50 ng/μL each of 4,4'-DDT, pentachlorophenol, and benzidine is required to verify injection port inertness and GC column performance. A calibration standard at mid-concentration, containing each compound of interest, including all required surrogates, must be performed every 12 h during analysis. After the system performance check is met, calibration check compounds (CCCs) are used to check the validity of the initial calibration.

The internal standard responses and retention times in the calibration check standard must be evaluated immediately after or during data acquisition. If the retention time for any internal standard changes by more than 30 seconds from the last check calibration (12 h), the chromatographic system must be inspected for malfunctions and corrections must be made, as required. If the electron ionization current plot (EICP) area for any of the internal standards changes by a factor of two from the last daily calibration standard check, the mass spectrometer must be inspected for malfunctions and corrections must be made, as appropriate.

Demonstrate, through the analysis of a reagent water blank, that interferences from the analytical system, glassware, and reagents are under control. The blank samples should be carried through all stages of the sample preparation and measurement steps. For each analytical batch (up to 20 samples), a reagent blank, matrix spike, and matrix spike duplicate/duplicate must be analyzed (the frequency of the spikes may be different for different monitoring programs). The blank and spiked samples must be carried through all stages of the sample preparation and measurement steps. A QC reference sample concentrate containing each analyte at a concentration of 100 mg/L in methanol is required.

REFERENCE Test Methods for Evaluating Solid Waste (SW-846). U.S. EPA 1983, Method 8270B, Rev. 2, Nov. 1990. Office of Solid Waste, Washington, DC.

β-BHC
CAS #319-85-7

EPA Method 8270

TITLE Semivolatile Organic Compounds by GC/MS

MATRIX This method is used to determine the concentration of semivolatile organic compounds in extracts prepared from all types of solid waste matrices, soils, and groundwater. Although surface waters are not specifically mentioned, this method should be applicable to water samples from rivers, lakes, etc.

METHOD SUMMARY This method covers 259 semivolatile organic compounds. In very limited applications direct injection of the sample into the GC/MS system may be appropriate, but this results in very high detection limits (approximately 10,000 μg/L). Typically, a 1-L liquid sample, containing surrogate, and matrix spiking standards, is extracted in a continuous extractor first under acid conditions and then under basic conditions.

Typically 30 g of a solid sample, containing surrogate, and matrix spiking standards, is extracted ultrasonically. After concentrating the extract to 1 mL it is spiked with 10 µL of an internal standard solution just prior to analysis by GC/MS. The volume injected should contain about 100 ng of base/neutral and 200 ng of acid surrogates (for a 1-µL injection). Analysis is performed by GC/MS using a capillary GC column.

INTERFERENCES Raw GC/MS data from all blanks, samples, and spikes must be evaluated for interferences. Contamination by carryover can occur whenever high-concentration and low-concentration samples are sequentially analyzed. To reduce carryover, the sample syringe must be rinsed out between samples with solvent. Whenever an unusually concentrated sample is encountered, it should be followed by the analysis of blank solvent to check for cross-contamination.

INSTRUMENTATION A GC/MS and a data system are required. The GC column used is a 30 m × 0.25 mm I.D. (or 0.32 mm I.D.) 1-µm film thickness silicone-coated fused silica capillary column. A continuous liquid-liquid extractor equipped with Teflon® or glass connection joints and stopcocks requiring no lubrication, a K-D concentrating apparatus, water bath, and an ultrasonic disrupter with a minimum power of 300 W and with pulsing capability are also required.

PRECISION & ACCURACY The estimated quantitation limit (EQL) of Method 8270B for determining an individual compound is approximately 1 mg/kg (wet weight) for soil or sediment samples, 1–200 mg/kg for wastes (dependent on matrix and method of preparation), and 10 µg/L for groundwater samples. EQLs will be proportionately higher for sample extracts that require dilution to avoid saturation of the detector.

The EQL(b) for groundwater in µg/L is not listed.
The EQL (a, b) for low concentrations in soil and sediment in µg/kg is not listed.
Accuracy as µg/L is $0.87C - 0.94$.
Overall precision in µg/L is $0.30X + 1.94$.

(a) EQLs listed for soil/sediment are based on wet weight. Normally data is reported in a dry-weight basis; therefore, EQLs will be higher based on the % dry weight of each sample. This calculation is based on a 30-g sample and gel permeation chromatography cleanup.
(b) Sample EQLs are highly matrix-dependent. The EQLs are provided for guidance and may not always be achievable.
C = True value for concentration, in µg/L.
X = Average recovery found for measurements of samples containing a concentration of C, in µg/L.

ESTIMATED QUANTITATION LIMIT FOR OTHER MATRICES

Other Matrices	Factor (a)
High-concentration soil and sludges by sonicator	7.5
Non-water miscible waste	75

(a) EQL for other matrices = [EQL for low soil/sediment] × [Factor]. This estimated EQL is similar to an EPA "Practical Quantitation Limit."

SAMPLING METHOD
Liquid samples — Use a 1 or 2½ gallon amber glass bottle with a screw-top Teflon®-lined cover that has been prewashed with detergent and rinsed with distilled water and methanol (or isopropanol).

Soils, sediments, or sludges — Use an 8-oz. widemouth glass with a screw-top Teflon®-lined cover that has been prewashed with detergent and rinsed with distilled water and methanol (or isopropanol).

SAMPLE PRESERVATION
Liquid samples — If residual chlorine is present, add 3 mL of 10% sodium thiosulfate per gallon, cool to 4°C and store in a solvent-free refrigerator until analysis; if chlorine is not present, then eliminate the sodium thiosulfate addition.

Soils, sediments, or sludges — Cool samples to 4°C and store in a solvent-free refrigerator.

MHT Liquid samples must be extracted within 7 days and the extracts analyzed within 40 days. Soils, sediments, or sludges may be stored for a maximum of 14 days and the extracts analyzed within 40 days.

SAMPLE PREPARATION
Liquid samples — Transfer 1 L quantitatively to a continuous extractor. If high concentrations are anticipated, a smaller volume may be used and then diluted with organic-free reagent water to 1 L. Adjust pH, if necessary, to pH <2 using 1:1 (V/V) sulfuric acid. Pipette 1.0 mL of a surrogate standard spiking solution into each sample. For the sample in each analytical batch selected for spiking, add 1.0 mL of a matrix spiking standard. For base/neutral acid analysis, the amount of the surrogates and matrix spiking compounds added to the sample should result in a final concentration of 100 ng/µL of each analyte in the extract to be analyzed (assuming a 1-µL injection). Extract with methylene chloride for 18–24 h. Next, adjust the pH of the aqueous phase to pH >11 using 10 N sodium hydroxide and extract it with methylene chloride again for 18–24 h. Dry the extract through a column containing anhydrous sodium sulfate and concentrate it to 1 mL using a K-D concentrator.

Soils, sediments, or sludges — Use 30 g of sample. Nonporous or wet samples (gummy or clay type) that do not have a free-flowing sandy texture must be mixed with anhydrous sodium sulfate until the sample is free flowing. Add 1 mL of surrogate standards to all samples, spikes, standards, and blanks. For the sample in each analytical batch selected for spiking, add 1.0 mL of a matrix spiking standard. For base/neutral acid analysis, the amount added of the surrogates and matrix spiking compounds should result in a final concentration of 100 ng/µL of each base/neutral analyte and 200 ng/µL of each acid analyte in the extract to be analyzed (assuming a 1-µL injection). Immediately add a 100 mL mixture of 1:1 methylene chloride:acetone and extract the sample ultrasonically for 3 min and then decant or filter the extracts. Repeat the extraction two or more times. Dry the extract using a column with anhydrous sodium sulfate and concentrate it to 1 mL in a K-D concentrator.

QUALITY CONTROL A methylene chloride solution containing 50 ng/µL of decafluorotriphenylphosphine (DFTPP) is used for tuning the GC/MS system each 12-h shift. A system performance check also must be made during every 12-h shift. A standard containing 50 ng/µL each of 4,4'-DDT, pentachlorophenol, and benzidine is required to verify injection port inertness and GC column performance. A calibration standard at mid-concentration, containing each compound of interest, including all required surrogates, must be performed every 12 h during analysis. After the system performance check is met, calibration check compounds (CCCs) are used to check the validity of the initial calibration.

The internal standard responses and retention times in the calibration check standard must be evaluated immediately after or during data acquisition. If the retention time for any internal standard changes by more than 30 seconds from the last check calibration (12 h), the chromatographic system must be inspected for malfunctions and corrections must be made, as required. If the electron ionization current plot (EICP) area for any of the internal standards changes by a factor of two from the last daily calibration standard check, the mass spectrometer must be inspected for malfunctions and corrections must be made, as appropriate.

Demonstrate, through the analysis of a reagent water blank, that interferences from the analytical system, glassware, and reagents are under control. The blank samples should be carried through all stages of the sample preparation and measurement steps. For each analytical batch (up to 20 samples), a reagent blank, matrix spike, and matrix spike duplicate/duplicate must be analyzed (the frequency of the spikes may be different for different monitoring programs). The blank and spiked samples must be carried through all stages of the sample preparation and measurement steps. A QC reference sample concentrate containing each analyte at a concentration of 100 mg/L in methanol is required.

REFERENCE Test Methods for Evaluating Solid Waste (SW-846). U.S. EPA 1983, Method 8270B, Rev. 2, Nov. 1990. Office of Solid Waste, Washington, DC.

δ-BHC   EPA Method 8270
CAS #319-86-8

TITLE Semivolatile Organic Compounds by GC/MS

MATRIX This method is used to determine the concentration of semivolatile organic compounds in extracts prepared from all types of solid waste matrices, soils, and groundwater. Although surface waters are not specifically mentioned, this method should be applicable to water samples from rivers, lakes, etc.

METHOD SUMMARY This method covers 259 semivolatile organic compounds. In very limited applications direct injection of the sample into the GC/MS system may be appropriate, but this results in very high detection limits (approximately 10,000 μg/L). Typically, a 1-L liquid sample, containing surrogate, and matrix spiking standards, is extracted in a continuous extractor first under acid conditions and then under basic conditions. Typically 30 g of a solid sample, containing surrogate, and matrix spiking standards, is extracted ultrasonically. After concentrating the extract to 1 mL it is spiked with 10 μL of an internal standard solution just prior to analysis by GC/MS. The volume injected should contain about 100 ng of base/neutral and 200 ng of acid surrogates (for a 1-μL injection). Analysis is performed by GC/MS using a capillary GC column.

INTERFERENCES Raw GC/MS data from all blanks, samples, and spikes must be evaluated for interferences. Contamination by carryover can occur whenever high-concentration and low-concentration samples are sequentially analyzed. To reduce carryover, the sample syringe must be rinsed out between samples with solvent. Whenever an unusually concentrated sample is encountered, it should be followed by the analysis of blank solvent to check for cross-contamination.

INSTRUMENTATION A GC/MS and a data system are required. The GC column used is a 30 m × 0.25 mm I.D. (or 0.32 mm I.D.) 1-μm film thickness silicone-coated fused silica capillary column. A continuous liquid-liquid extractor equipped with Teflon® or glass connection joints and stopcocks requiring no lubrication, a K-D concentrating apparatus, water bath, and an ultrasonic disrupter with a minimum power of 300 W and with pulsing capability are also required.

PRECISION & ACCURACY The estimated quantitation limit (EQL) of Method 8270B for determining an individual compound is approximately 1 mg/kg (wet weight) for soil or sediment samples, 1–200 mg/kg for wastes (dependent on matrix and method of preparation), and 10 μg/L for groundwater samples. EQLs will be proportionately higher for sample extracts that require dilution to avoid saturation of the detector.

The EQL(b) for groundwater in μg/L is not listed.
The EQL (a, b) for low concentrations in soil and sediment in μg/kg is not listed.
Accuracy as μg/L is 0.29C–1.09.
Overall precision in μg/L is 0.93X–0.17.

(a) *EQLs listed for soil/sediment are based on wet weight. Normally data is reported in a dry-weight basis; therefore, EQLs will be higher based on the % dry weight of each sample. This calculation is based on a 30-g sample and gel permeation chromatography cleanup.*
(b) *Sample EQLs are highly matrix-dependent. The EQLs are provided for guidance and may not always be achievable.*
C = *True value for concentration, in μg/L.*
X = *Average recovery found for measurements of samples containing a concentration of C, in μg/L.*

ESTIMATED QUANTITATION LIMIT FOR OTHER MATRICES

Other Matrices	Factor (a)
High-concentration soil and sludges by sonicator	7.5
Non-water miscible waste	75

(a) *EQL for other matrices = [EQL for low soil/sediment] × [Factor]. This estimated EQL is similar to an EPA "Practical Quantitation Limit."*

SAMPLING METHOD
Liquid samples — Use a 1 or 2½ gallon amber glass bottle with a screw-top Teflon®-lined cover that has been prewashed with detergent and rinsed with distilled water and methanol (or isopropanol).

Soils, sediments, or sludges — Use an 8-oz. widemouth glass with a screw-top Teflon®-lined cover that has been prewashed with detergent and rinsed with distilled water and methanol (or isopropanol).

SAMPLE PRESERVATION

Liquid samples — If residual chlorine is present, add 3 mL of 10% sodium thiosulfate per gallon, cool to 4°C and store in a solvent-free refrigerator until analysis; if chlorine is not present, then eliminate the sodium thiosulfate addition.

Soils, sediments, or sludges — Cool samples to 4°C and store in a solvent-free refrigerator.

MHT Liquid samples must be extracted within 7 days and the extracts analyzed within 40 days. Soils, sediments, or sludges may be stored for a maximum of 14 days and the extracts analyzed within 40 days.

SAMPLE PREPARATION

Liquid samples — Transfer 1 L quantitatively to a continuous extractor. If high concentrations are anticipated, a smaller volume may be used and then diluted with organic-free reagent water to 1 L. Adjust pH, if necessary, to pH <2 using 1:1 (V/V) sulfuric acid. Pipette 1.0 mL of a surrogate standard spiking solution into each sample. For the sample in each analytical batch selected for spiking, add 1.0 mL of a matrix spiking standard. For base/neutral acid analysis, the amount of the surrogates and matrix spiking compounds added to the sample should result in a final concentration of 100 ng/µL of each analyte in the extract to be analyzed (assuming a 1-µL injection). Extract with methylene chloride for 18–24 h. Next, adjust the pH of the aqueous phase to pH >11 using 10 N sodium hydroxide and extract it with methylene chloride again for 18–24 h. Dry the extract through a column containing anhydrous sodium sulfate and concentrate it to 1 mL using a K-D concentrator.

Soils, sediments, or sludges — Use 30 g of sample. Nonporous or wet samples (gummy or clay type) that do not have a free-flowing sandy texture must be mixed with anhydrous sodium sulfate until the sample is free flowing. Add 1 mL of surrogate standards to all samples, spikes, standards, and blanks. For the sample in each analytical batch selected for spiking, add 1.0 mL of a matrix spiking standard. For base/neutral acid analysis, the amount added of the surrogates and matrix spiking compounds should result in a final concentration of 100 ng/µL of each base/neutral analyte and 200 ng/µL of each acid analyte in the extract to be analyzed (assuming a 1-µL injection). Immediately add a 100 mL mixture of 1:1 methylene chloride:acetone and extract the sample ultrasonically for 3 min and then decant or filter the extracts. Repeat the extraction two or more times. Dry the extract using a column with anhydrous sodium sulfate and concentrate it to 1 mL in a K-D concentrator.

QUALITY CONTROL

A methylene chloride solution containing 50 ng/µL of decafluorotriphenylphosphine (DFTPP) is used for tuning the GC/MS system each 12-h shift. A system performance check also must be made during every 12-h shift. A standard containing 50 ng/µL each of 4,4'-DDT, pentachlorophenol, and benzidine is required to verify injection port inertness and GC column performance. A calibration standard at mid-concentration, containing each compound of interest, including all required surrogates, must be performed every 12 h during analysis. After the system performance check is met, calibration check compounds (CCCs) are used to check the validity of the initial calibration.

The internal standard responses and retention times in the calibration check standard must be evaluated immediately after or during data acquisition. If the retention time for any internal standard changes by more than 30 seconds from the last check calibration (12 h), the chromatographic system must be inspected for malfunctions and corrections must be made, as required. If the electron ionization current plot (EICP) area for any of the internal standards changes by a factor of two from the last daily calibration standard check, the mass spectrometer must be inspected for malfunctions and corrections must be made, as appropriate.

Demonstrate, through the analysis of a reagent water blank, that interferences from the analytical system, glassware, and reagents are under control. The blank samples should be carried through all stages of the sample preparation and measurement steps. For each analytical batch (up to 20 samples), a reagent blank, matrix spike, and matrix spike duplicate/duplicate must be analyzed (the frequency of the spikes may be different for different monitoring programs). The blank and spiked samples must be carried through all stages of the sample preparation and measurement steps. A QC reference sample concentrate containing each analyte at a concentration of 100 mg/L in methanol is required.

REFERENCE Test Methods for Evaluating Solid Waste (SW-846). U.S. EPA 1983, Method 8270B, Rev. 2, Nov. 1990. Office of Solid Waste, Washington, DC.

γ-BHC (Lindane) EPA Method 505
CAS #58-89-9

TITLE Analysis of Organohalide Pesticides and Commercial Polychlorinated Biphenyl (PCB) Products in Water by Microextraction and Gas Chromatography. U.S. EPA Method 505, Rev. 2.0, 1989.

MATRIX This method is applicable to drinking water and raw source water. The latter should include most surface water and groundwater sources.

METHOD SUMMARY Method 505 covers 25 pesticides and commercial PCB products. This is a very sensitive method that is more useful for monitoring than for exploratory analyses. 5-mL of water are saturated with sodium chloride and then extracted by shaking with 2 mL of hexane. The sample extracts are transferred to an autosampler setup to inject 1- to 2-µL portions into a gas chromatograph (GC) for analysis. Alternatively, 1- to 2-µL portions of samples, blanks, and standards may be manually injected. Each extract is analyzed by capillary GC/ECD with confirmation using either a second capillary column or GC/MS. The electron capture detector is easy to use, but it is a nonselective detector. The microextraction technique also eliminates the expensive sample preparation costs of other methods, but it has the disadvantage of being less sensitive than most because the extracts are not concentrated.

INTERFERENCES Method interferences may be caused by contaminants in solvents, reagents, glassware, and other sample processing apparatus that lead to discrete artifacts or elevated baselines. Interfering contamination may occur when a sample containing low concentrations of analytes is analyzed immediately following a sample containing relatively high concentrations of the analytes. Matrix interferences also may be caused by contaminants that are coextracted from the sample; cleanup of sample extracts may be necessary in these cases. Some pesticides and commercial PCB products from aqueous solutions adhere to glass surfaces, so sample transfers and contact with glass surfaces should be minimized. Some pesticides are rapidly oxidized by chlorine so dechlorination with sodium thiosulfate at the time of sample collection is important. Also, splitless injectors may cause degradation of some pesticides.

INSTRUMENTATION A gas chromatograph/electron capture detector/data system, with temperature programming and split/splitless injector suitable for use with capillary columns is needed.

Column 1: 0.32 mm I.D. × 30 m fused silica capillary with chemically bond methyl polysiloxane phase (DB-1, 1.0 μm film, or equivalent).

Column 2: 0.32 mm I.D. × 30 m fused silica capillary with 1:1 mixed phase of dimethyl silicone and polyethylene glycol (Durawax-DX3, 0.25 μm film, or equivalent).

Column 3: 0.32 mm I.D. × 25 m fused silica capillary with chemically bonded 50:50 methyl-phenyl silicone (OV-17, 1.5 μm film, or equivalent).

Column 1 should be used as the primary analytical column. Columns 2 and 3 are recommended for use as confirmatory columns when GC/MS confirmation is not available.

PRECISION & ACCURACY Method detection limits are dependent upon the characteristics of the gas chromatographic system used. Analytes that are not separated chromatographically cannot be individually identified and used in the same calibration mixture or water samples unless an alternative technique for identification and quantification, such as mass spectrometry, is used.

The concentration(s) (in μg/L) used for these QC measurements was 0.03 and 1.2.
The MDL (in μg/L) was 0.003.
The accuracy (% recovery) for reagent water at the above concentration(s) was 91 and 111 and the precision (%) was 6.5 and 5.0.
The accuracy (% recovery) for groundwater at the above concentration(s) was 88 and 109 and the precision (%) was 7.7 and 3.4.
The accuracy (% recovery) for tap water at the above concentration(s) was 103 and 93 and the precision (5) was 8.1 and 18.4.

Note: No range of concentrations is provided with this method.

SAMPLING METHOD Collect samples using a 40-mL screw-cap vial (prewashed with detergent, rinsed with distilled water and oven dried at 400°C for one h) with a Teflon®-faced silicone septum. Collect bubble-free samples and place the septum with the Teflonside down on the water.

SAMPLE PRESERVATION If residual chlorine is present in the water add about 3 mg of sodium thiosulfate to each vial before samples are collected to remove thechlorine. Alternatively, add 75 μL of 0.04 g/mL solution of sodium thiosulfate to each vial just prior to sampling. Immediately coolsamples to 4°C, and store them in a solvent-free refrigerator at 4°C until analysis.

MHT The maximum holding time is 14 days from the time the sample was collected until it must be analyzed.

SAMPLE PREPARATION Remove the sample from storage and allow it to come to room temperature. Remove a 5 mL volume from each container and weigh the container to the nearest 0.1 g. Add 6 g of sodium chloride and 2.0 mL of hexane to each sample bottle. Recap the sample and shake it vigorously for one min. Allow the water and hexane phases to separate, remove the cap, and transfer 0.5 mL of hexane into an autosampler vial using a disposable glass pipette. Transfer the remaining hexane phase into a second autosampler vial and store at 4°C for reanalysis, if necessary. Discard the remaining sample/hexane mixture and reweigh the empty container to determine net weight of sample.

QUALITY CONTROL Minimum quality control requirements are initial demonstration of lab capability, analysis of lab reagent blanks, fortified blanks, fortified sample matrix, and quality control samples. The lab must analyze at least one fortified blank per sample set, or at least one for every 20 samples. The fortifying concentration of each analyte should be 10 times the method detection limit or the maximum calibration limit (MCL), whichever is less. Calculate accuracy as percent recovery and develop control limits from the mean percent recovery and standard deviation.

The lab must add a known concentration of the analytes to a minimum of 10% of the routine samples, or one lab fortified sample matrix per sample set. Calculate the percent recovery for each analyte and compare to the control limits established from the analyses of the fortified blanks.

EPA CONTACT & HOTLINE For technical questions contact Dr. Baldev Bathija, U.S. EPA,Office of Ground Water and Drinking Water (WH-550D), 401 M St. SW, Washington, DC 20460. Tel. (202) 260-3040. For further information the EPA Safe Drinking Water Hotline may be called at: (800) 426-4791.

REFERENCE Methods for the Determination of Organic Compounds in Drinking Water, EPA/600/4-88/039 (revised July 1991). U.S. EPA Environmental Monitoring Systems Laboratory, Cincinnati, OH, 45268, U.S.A. Available from the National Technical Information Service (NTIS), 5285 Port Royal Road, Springfield, VA 22161; Tel. 800-553-6847. NTIS Order Number is PB91-231480.

γ-BHC (Lindane) EPA Method 8080
CAS #58-89-9

TITLE Organochlorine Pesticides and Polychlorinated Biphenyls By Gas Chromatography

MATRIX This method is used to determine the concentration of various organochlorine pesticides and polychlorinated biphenyls in extracts prepared from water, groundwater, soils, and sediments.

METHOD SUMMARY This method covers 26 pesticides and Aroclor (PCB) mixtures and it is suitable for monitoring-type analyses. After extraction, concentration and solvent exchange to hexane, a 2- to 5-μL sample aliquot is injected into a GC using the solvent flush technique, and the analytes are detected by an electron capture detector (ECD) or an electrolytic conductivity detector in the halogen mode (HECD). Both neat and diluted organic liquids may be analyzed by direct injection.

INTERFERENCES Interferences coextracted from the samples will vary considerably from source to source. Interferences by phthalate esters can pose a major problem in pesticide determinations when using the ECD. Cross-contamination of clean glassware routinely occurs when plastics are handled during extraction steps, especially when solvent-wetted surfaces are handled. The contamination from phthalate esters can be completely eliminated with a microcoulometric or electrolytic conductivity detector. Solvents, reagent, glassware, and other sample processing hardware may yield artifacts and/or interferences to sample analysis.

INSTRUMENTATION A gas chromatograph capable of on-column injections is needed. It must be equipped with an ECD or a HECD and one of the following GC columns:

Column 1: Supelcoport (100/120 mesh) coated with 1.5% SP-2250/1.95% SP-2401 packed in a 1.8 m × 4 mm I.D. glass column.
Column 2: Supelcoport (100/120 mesh) coated with 3% OV-1 in a 1.8 m × 4 mm I.D. glass column.

PRECISION & ACCURACY The method was tested by 20 laboratories using organic-free reagent water, drinking water, surface water, and three industrial wastewaters spiked at six concentrations. Concentrations used in the study ranged from 0.5 to 30 μg/L for single-component pesticides and from 8.5 to 400 μg/L for multicomponent parameters. Overall precision and method accuracy were found to be directly related to the concentration of the analyte and essentially independent of the sample matrix. The sensitivity of this method usually depends on the concentration of interferences rather than on instrumental limitations.

MDL in μg/L was 0.004.
Concentration range in μg/L was 0.5–30.
Accuracy as recovery (x^*) in μg/L was $0.82C - 0.05$.
Overall precision (S^*) in μg/L was $0.22x + 0.04$.

x^* = Expected recovery for one or more measurements of a sample containing concentration C, in μg/L.
S^* = Expected interlaboratory standard deviation of measurements at an average concentration found of the analyte in μg/L.
C = True value for the concentration, in μg/L.
X = Average recovery found for measurements of samples containing a concentration of C, in μg/L.

SAMPLING METHOD
Liquid samples — Use a 1 or 2½ gallon amber glass bottle with a screw-top Teflon®-lined cover. Pre-wash the bottle with detergent, rinse with distilled water and methanol (or isopropanol).

Soil, sediments, and sludges — Use an 8-oz. widemouth glass with a screw-top Teflon®-lined cover. Pre-wash the bottle with detergent, rinse with distilled water and methanol (or isopropanol).

SAMPLE PRESERVATION Cool water, soil, sediment, or sludge samples immediately to 4°C.

Water samples — If residual chlorine is present, add 3 mL of 10% sodium thiosulfate per gallon and cool to 4°C. All extracts and samples should be stored under refrigeration.

MHT Liquid samples must be extracted within 7 days and the extracts must be analyzed within 40 days. Soils, sediments, and sludges may be stored for a maximum of 14 days prior to extraction.

SAMPLE PREPARATION
Liquid samples — Extract 1-L sample in a continuous extractor at pH 5–9 with methylene chloride after adding 1.0 mL of surrogate spiking solution to each sample. Pass the extract through a column of anhydrous sodium sulfate to dry and concentrate it in a K-D apparatus to 1 mL volume.

Soils, sediments, and sludges — Rapidly weigh approximately 30 g of sample into a 400-mL beaker to avoid loss of the more volatile extractables. Nonporous or wet samples (gummy or clay type) that do not have a free-flowing sandy texture must be mixed with anhydrous sodium sulfate until the sample is free flowing. Add 1 mL of surrogate standards to all samples, spikes, standards, and blanks. Add 100 mL of 1:1 methylene chloride:acetone and extract ultrasonically. Decant and filter extracts, dry the extract by passing it through a drying column containing anhydrous sodium sulfate and concentrate to 1 mL in a K-D apparatus.

Hexane solvent exchange — Add 50 mL of hexane, a new boiling chip, and concentrate until the apparent volume of liquid reaches 1 mL. Adjust the extract volume to 10.0 mL. Stopper the concentration tube and store refrigerated at 4°C if further processing will not be performed immediately. If the extract will be stored longer than two days, transfer it to a vial with Teflon®-lined screw-cap or crimp top.

QUALITY CONTROL Demonstrate through the analysis of a reagent water blank, that all glassware and reagents are interference free. Each time a set of samples is processed, a method blank should be processed as a safeguard against chronic lab contamination. A reagent blank, a matrix spike, and a duplicate or matrix spike duplicate must be performed for each analytical batch (up to a maximum of 20 samples) analyzed.

Analytical system performance must be verified by analyzing QC check samples. The QC check sample concentration should contain each single-component analyte at the following concentrations in acetone: 4,4'-DDD, 10 µg/mL; 4,4'-DDT, 10 µg/mL; endosulfan II, 10 µg/mL; endosulfan sulfate, 10 µg/mL; and any other single-component pesticide at 2 µg/mL. If the method is only to be used to analyze PCBs, Chlordane, or Toxaphene, the QC check sample concentrate should contain the most representative multicomponent parameter at a concentration of 50 µg/mL in acetone.

REFERENCE Test Methods for Evaluating Solid Waste (SW-846). U.S. EPA. 1983. Method 8080B, Rev. 2, Nov. 1990. Office of Solid Wastes, Washington, DC.

γ-BHC (Lindane)   **EPA Method 8121**
CAS #58-89-9

TITLE Chlorinated Hydrocarbons by GC: Capillary Column Technique

MATRIX This method covers aqueous and solid matrices. This includes a wide variety such as drinking water, groundwater, industrial wastewaters, surface waters, soils, solids, and sediments.

METHOD SUMMARY This method provides procedures for the determination of 22 chlorinated hydrocarbons in water, soil/sediment, and waste matrices. A measured volume or weight of sample is extracted by using one of the appropriate sample extraction techniques specified in EPA Method 3510, EPA Method 3520, EPA Method 3540, or EPA Method 3550, or diluted using EPA Method 3580. Aqueous samples are extracted at neutral pH with methylene chloride by using either a separatory funnel (EPA Method 3510) or a continuous liquid-liquid extractor (EPA Method 3520). Solid samples are extracted with hexane/acetone (1:1) by using a Soxhlet extractor (EPA Method 3540) or with methylene chloride/acetone (1:1) by using an ultrasonic extractor (EPA Method 3550). After cleanup, the extract or diluted sample is analyzed by gas chromatography with electron capture detection (GC/ECD).

The sensitivity level of this method usually depends on the level of interferences rather than on instrumental limitations. This method may be used in conjunction with EPA Method 3620, Florisil Column Cleanup, EPA Method 3660, Sulfur Cleanup, and EPA Method 3640, Gel Permeation Chromatography, to aid in the elimination of interferences.

INTERFERENCES Solvents, reagents, glassware, and other hardware used in sample processing may introduce artifacts which may result in elevated baselines, causing misinterpretation of gas chromatograms. Interferants coextracted from the samples will vary considerably from waste to waste. Glassware must be scrupulously clean. Phthalate esters, if present in a sample, will interfere only with the BHC isomers. The presence of elemental sulfur will result in large peaks, and can often mask the region of compounds eluting after 1,2,4,5-tetrachlorobenzene. The tetrabutylammonium (TBA)-sulfite procedure (EPA Method 3660) works well for the removal of elemental sulfur. Waxes and lipids can be removed by gel permeation chromatography (EPA Method 3640).

INSTRUMENTATION A GC suitable for on-column injections and all required accessories, including and electron capture detector (ECD), analytical columns, recorder, gases, and syringes are needed. A data system for measuring peak heights and/or peak areas is recommended. A Kuderna-Danish (K-D) apparatus will also be needed to concentrate extracts.

Column 1: 30 m × 0.53 mm I.D. fused-silica capillary column chemically bonded with trifluoropropyl methyl silicone (DB-210 or equivalent).

Column 2: 30 m × 0.53 mm I.D. fused-silica capillary column chemically bonded with polyethylene glycol (DB-WAX or equivalent).

PRECISION & ACCURACY This method has been tested in a single lab by using organic-free reagent water, sandy loam samples, and extracts which were spiked with the test compounds at one concentration. Single-operator precision and method accuracy were found to be related to the concentration of compound and the type of matrix. The accuracy and precision technique will be determined by the sample matrix, sample preparation technique, optional cleanup techniques, and calibration procedures used.

ESTIMATED QUANTITATION LIMIT (EQL) FACTORS FOR VARIOUS MATRICES (a)

Matrix	Factor (b)
Groundwater	10
Low-concentration soil by ultrasonic extraction with GPC cleanup	670
High-concentration soil and sludges by ultrasonic extraction	10,000
Waste not miscible with water	100,000

(a) Sample EQLs are highly matrix-dependent. The EQLs listed herein are provided for guidance and may not always be achievable. (b) EQL = [Method detection limit] × [Factor]. For nonaqueous samples, the factor is on a wet-weight basis.

PRECISION & ACCURACY MDL is the method detection limit for organic-free reagent water. MDL was determined from the analysis of eight replicate aliquots processed through the entire analytical method (extraction, Florisil cartridge cleanup, and GC/ECD analysis).

The MDL (in ng/L) was 23.

The accuracy (as average % recovery using 5 determinations and no Florisil cleanup) from a spike concentration of 10 µg/L and separatory funnel extraction was 96% with a final volume of 10 mL.

The precision (as RSD% using 5 determinations and no Florisil cleanup) from a spike concentration of 10 µg/L and separatory funnel extraction was 2.8% with a final volume of 10 mL.

The accuracy (as average % recovery using 5 determinations and no Florisil cleanup), from a spike concentration of 3300 µg/L and ultrasonic extraction of solid samples using 1:1

methylene chloride and acetone, was 99% with a final volume of 10 mL.

The precision (as RSD% using 5 determinations and no Florisil cleanup), from a spike concentration of 3300 µg/L and ultrasonic extraction of solid samples using 1:1 methylene chloride and acetone, was 4.1% with a final volume of 10 mL.

SAMPLE COLLECTION, PRESERVATION & HANDLING
Volatile Organics — Standard 40-mL glass screw-cap VOA vials with Teflon®-faced silicone septum may be used for both liquid and solid matrices. When collecting samples, liquids and solids should be introduced into the vials gently to reduce agitation which might drive off volatile compounds. If there are any air bubbles present the sample must be retaken. The vials with solids should be tapped slightly as they are filled to try and eliminate as much free air space as possible. Two vials from each sampling location should be sealed in separate plastic bags to prevent cross-contamination between samples.

Semivolatile organics — Containers used to collect samples for the determination of semivolatile organic compounds should be soap and water washed followed by methanol (or isopropanol) rinsing. The sample containers should be of glass or Teflon® and have screw-top covers with Teflon® liners.

Preservation for volatile organics — No preservation is used with concentrated waste samples. With liquid samples containing no residual chlorine, 4 drops of concentrated hydrochloric acid are added and the samples are immediately cooled to 4°C. When liquid samples contain residual chlorine, they are treated as above and, in addition, 4 drops of 4% aqueous sodium thiosulfate are added to remove the residual chlorine. Soil, sediment, and sludge samples are only cooled to 4°C.

Preservation for semivolatile organics — No preservation is used with concentrated waste samples. With liquid samples containing no residual chlorine and with soil, sediment, and sludge samples, immediately cooling to 4°C is the only preservation used. When residual chlorine is present then 3 mL of 10% aqueous sodium sulfate is added for each gallon of sample collected, followed by cooling to 4°C.

Holding times — The holding time for all volatile organics samples is 14 days. Liquid samples must be extracted within 7 days and their extracts analyzed within 40 days. Concentrated waste, soil, sediment, and sludge samples must be extracted within 14 days and their extracts analyzed within 40 days.

SAMPLE PREPARATION Prepare stock standard solutions in hexane. Calibration standards at a minimum of five concentrations should be prepared through dilution of the stock standards with hexane. The suggested internal standards are: 2,5-dibromotoluene, 1,3,5-tribromobenzene, and α, α-dibromom-xylene. The analyst can use any of the three compounds provided that they are resolved from matrix interferences. Recommended surrogate compounds are α-2,6-trichlorotoluene, 1,4-dichloronaphthalene, and 2,3,4,5,6-pentachlorotoluene.

In general, water samples are extracted at a neutral pH with methylene chloride using a separatory funnel (EPA Method 3510) or a continuous liquid-liquid extractor (EPA Method 3520). Solid samples are extracted with hexane/acetone (1:1 v:v) using a Soxhlet extractor (EPA Method 3540) or with methylene chloride/acetone (1:1 v:v) using an ultrasonic extractor (EPA Method 3550). Non-aqueous waste samples may be diluted using EPA Method 3580. Prior to Florisil cleanup or gas chromatographic analysis, the extraction solvent must be exchanged to hexane. Sample extracts that will be subjected to gel permeation chromatography do not need solvent exchange.

Cleanup procedures may not be necessary for a relatively clean matrix. If removal of interferences such as chlorinated phenols, phthalate esters, etc., is required, proceed with the procedure outlined in EPA Method 3620.

QUALITY CONTROL Analyze a quality control check standard to demonstrate that the operation of the GC is in control. The frequency of the check standard analysis is equivalent to 10% of the samples analyzed. If the recovery of any compound found in the check standard is less than 80% of the certified value, the problem must be corrected and a new set of calibration standards must be prepared and analyzed. Calculate surrogate standard recoveries for all samples, blanks, and spikes. An internal standard peak area check must be performed on all samples. The internal standard must be evaluated for acceptance by determining whether the measured area for the internal standard deviates by more than 30% from the average area for the internal standard in the calibration standards. When the internal standard peak area is outside that limit, all samples that fall outside the QC criteria must be reanalyzed. Any compound confirmed by two columns may also be confirmed by GC/MS (EPA Method 8270). The GC/MS would normally require a minimum concentration of 1 ng/µL in the final extract for each compound. Include a mid-concentration calibration standard after each group of 20 samples in the analysis sequence. The response factors for the mid-concentration calibration must be within 15% of the average values for the multiconcentration calibration.

REFERENCE Test Methods for Evaluating Solid Waste, Physical/Chemical Methods, SW-846, 3rd Edition, U.S. EPA, Office of Solid Waste, Washington, DC, 1990. EPA Method 8121, Rev. 0, Nov. 1990.

γ-BHC (Lindane)   **EPA Method 8270**
CAS #58-89-9

TITLE Semivolatile Organic Compounds by GC/MS

MATRIX This method is used to determine the concentration of semivolatile organic compounds in extracts prepared from all types of solid waste matrices, soils, and groundwater. Although surface waters are not specifically mentioned, this method should be applicable to water samples from rivers, lakes, etc.

METHOD SUMMARY This method covers 259 semivolatile organic compounds. In very limited applications direct injection of the sample into the GC/MS system may be appropriate, but this results in very high detection limits (approximately 10,000 µg/L). Typically, a 1-L liquid sample, containing surrogate, and matrix spiking standards, is extracted in a continuous extractor first under acid conditions and then under basic conditions. Typically 30 g of a solid sample, containing surrogate, and matrix spiking standards, is extracted ultrasonically. After concentrating the extract to 1 mL it is spiked with 10 µL of an internal standard solution just prior to analysis by GC/MS. The volume injected should contain about 100 ng of base/neutral and 200 ng of acid surrogates (for a 1-µL injection). Analysis is performed by GC/MS using a capillary GC column.

INTERFERENCES Raw GC/MS data from all blanks, samples, and spikes must be evaluated for interferences. Contamination by carryover can occur whenever high-concentration and low-concentration samples are sequentially analyzed. To reduce carryover, the sample syringe must be rinsed out between samples with solvent. Whenever an unusually concentrated sample is encountered, it should be followed by the analysis of blank solvent to check for cross-contamination.

INSTRUMENTATION A GC/MS and a data system are required. The GC column used is a 30 m × 0.25 mm I.D. (or 0.32 mm I.D.) 1-µm film thickness silicone-coated fused silica capillary column. A continuous liquid-liquid extractor equipped with Teflon® or glass connection joints and stopcocks requiring no lubrication, a K-D concentrating apparatus, water bath, and an ultrasonic disrupter with a minimum power of 300 W and with pulsing capability are also required.

PRECISION & ACCURACY The estimated quantitation limit (EQL) of Method 8270B for determining an individual compound is approximately 1 mg/kg (wet weight) for soil or sediment samples, 1–200 mg/kg for wastes (dependent on matrix and method of preparation), and 10 µg/L for groundwater samples. EQLs will be proportionately higher for sample extracts that require dilution to avoid saturation of the detector.

The EQL(b) for groundwater in µg/L is not listed.
The EQL (a, b) for low concentrations in soil and sediment in µg/kg is not listed.
Accuracy as µg/L is not listed.
Overall precision in µg/L is not listed.

(a) *EQLs listed for soil/sediment are based on wet weight. Normally data is reported in a dry-weight basis; therefore, EQLs will be higher based on the % dry weight of each sample. This calculation is based on a 30-g sample and gel permeation chromatography cleanup.*
(b) *Sample EQLs are highly matrix-dependent. The EQLs are provided for guidance and may not always be achievable.*
C = *True value for concentration, in µg/L.*
X = *Average recovery found for measurements of samples containing a concentration of C, in µg/L.*

ESTIMATED QUANTITATION LIMIT FOR OTHER MATRICES

Other Matrices	Factor (a)
High-concentration soil and sludges by sonicator	7.5
Non-water miscible waste	75

(a) *EQL for other matrices = [EQL for low soil/sediment] × [Factor]. This estimated EQL is similar to an EPA "Practical Quantitation Limit."*

SAMPLING METHOD
Liquid samples — Use a 1 or 2½ gallon amber glass bottle with a screw-top Teflon®-lined cover that has been prewashed with detergent and rinsed with distilled water and methanol (or isopropanol).

Soils, sediments, or sludges — Use an 8-oz. widemouth glass with a screw-top Teflon®-lined cover that has been prewashed with detergent and rinsed with distilled water and methanol (or isopropanol).

SAMPLE PRESERVATION
Liquid samples — If residual chlorine is present, add 3 mL of 10% sodium thiosulfate per gallon, cool to 4°C and store in a solvent-free refrigerator until analysis; if chlorine is not present, then eliminate the sodium thiosulfate addition.

Soils, sediments, or sludges — Cool samples to 4°C and store in a solvent-free refrigerator.

MHT Liquid samples must be extracted within 7 days and the extracts analyzed within 40 days. Soils, sediments, or sludges may be stored for a maximum of 14 days and the extracts analyzed within 40 days.

SAMPLE PREPARATION
Liquid samples — Transfer 1 L quantitatively to a continuous extractor. If high concentrations are anticipated, a smaller volume may be used and then diluted with organic-free reagent water to 1 L. Adjust pH, if necessary, to pH <2 using 1:1 (V/V) sulfuric acid. Pipette 1.0 mL of a surrogate standard spiking solution into each sample. For the sample in each analytical batch selected for spiking, add 1.0 mL of a matrix spiking standard. For base/neutral acid analysis, the amount of the surrogates and matrix spiking compounds added to the sample should result in a final concentration of 100 ng/µL of each analyte in the extract to be analyzed (assuming a 1-µL injection). Extract with methylene chloride for 18–24 h. Next, adjust the pH of the aqueous phase to pH >11 using 10 N sodium hydroxide and extract it with methylene chloride again for 18–24 h. Dry the extract through a column containing anhydrous sodium sulfate and concentrate it to 1 mL using a K-D concentrator.

Soils, sediments, or sludges — Use 30 g of sample. Nonporous or wet samples (gummy or clay type) that do not have a free-flowing sandy texture must be mixed with anhydrous sodium sulfate until the sample is free flowing. Add 1 mL of surrogate standards to all samples, spikes, standards, and blanks. For the sample in each analytical batch selected for spiking, add 1.0 mL of a matrix spiking standard. For base/neutral acid analysis, the amount added of the surrogates and matrix spiking compounds should result in a final concentration of 100 ng/µL of each base/neutral analyte and 200 ng/µL of each acid analyte in the extract to be analyzed (assuming a 1-µL injection). Immediately add a 100 mL mixture of 1:1 methylene chloride:acetone and extract the sample ultrasonically for 3 min and then decant or filter the extracts. Repeat the extraction two or more times. Dry the extract using a column with anhydrous sodium sulfate and concentrate it to 1 mL in a K-D concentrator.

QUALITY CONTROL A methylene chloride solution containing 50 ng/µL of decafluorotriphenylphosphine (DFTPP) is

used for tuning the GC/MS system each 12-h shift. A system performance check also must be made during every 12-h shift. A standard containing 50 ng/μL each of 4,4'-DDT, pentachlorophenol, and benzidine is required to verify injection port inertness and GC column performance. A calibration standard at mid-concentration, containing each compound of interest, including all required surrogates, must be performed every 12 h during analysis. After the system performance check is met, calibration check compounds (CCCs) are used to check the validity of the initial calibration.

The internal standard responses and retention times in the calibration check standard must be evaluated immediately after or during data acquisition. If the retention time for any internal standard changes by more than 30 seconds from the last check calibration (12 h), the chromatographic system must be inspected for malfunctions and corrections must be made, as required. If the electron ionization current plot (EICP) area for any of the internal standards changes by a factor of two from the last daily calibration standard check, the mass spectrometer must be inspected for malfunctions and corrections must be made, as appropriate.

Demonstrate, through the analysis of a reagent water blank, that interferences from the analytical system, glassware, and reagents are under control. The blank samples should be carried through all stages of the sample preparation and measurement steps. For each analytical batch (up to 20 samples), a reagent blank, matrix spike, and matrix spike duplicate/duplicate must be analyzed (the frequency of the spikes may be different for different monitoring programs). The blank and spiked samples must be carried through all stages of the sample preparation and measurement steps. A QC reference sample concentrate containing each analyte at a concentration of 100 mg/L in methanol is required.

REFERENCE Test Methods for Evaluating Solid Waste (SW-846). U.S. EPA 1983, Method 8270B, Rev. 2, Nov. 1990. Office of Solid Waste, Washington, DC.

Biochemical Oxygen Demand (BOD)   EPA Method 405.1

TITLE Organics

MATRIX Municipal and industrial wastewaters.

APPLICATION Date issued 1971. Editorial Rev. 1974. (5 Days, 20c). The BOD test is an empirical bioassay-type procedure which measures the dissolved oxygen consumed by microbial life while assimilating and oxidizing the organic matter present.

INTERFERENCES The actual environment conditions of temperature, biological population, water movement, sunlight, and oxygen concentration can not be accurately reproduced in the lab. Results obtained must take above factors into account when relating BOD results.

INSTRUMENTATION Modified Winkler with full-bottle technique or probe method.

RANGE Not listed.

MDL Not listed.

PRECISION At mean of 2.1 and 175 mg/L BOD, S.D. = 0.7 and 26 mg/L.

ACCURACY No acceptable procedure to determine accuracy.

SAMPLING METHOD Plastic or glass. (1000 mL).

STABILITY cool, 4°C.

MHT 48 h.

QUALITY CONTROL Sample of waste, or an appropriate dilution, is incubated for 5 days @ 20°C in the dark. The reduction in dissolved oxygen concentration during the incubation period yields a measure of the biochemical oxygen demand.

REFERENCE Methods for the Chemical Analysis of Water and Wastes, EPA-600/4-79-020, U.S. EPA, EMSL, 1979.

Biphenyl   EPA Method 1625
CAS #92-52-4

TITLE Semivolatile Organic Compounds by Isotope Dilution GC/MS

MATRIX The compounds may be determined in waters, soils, and municipal sludges by this method.

METHOD SUMMARY This method is used to determine 176 semivolatile toxic organic pollutants associated with the CWA (as amended 1987); the RCRA (as amended 1986); the CERCLA (as amended 1986); and other compounds amenable to extraction and analysis by capillary column gas chromatography-mass spectrometry (GC/MS).

Stable isotopically-labeled analogs of the compounds of interest are added to the sample. If the solids content is less than 1%, a 1-L sample is extracted at pH 12–13, then at pH <2 with methylene chloride using continuous extraction techniques.

If the solids content is 30% or less, the sample is diluted to 1% solids with reagent water, homogenized ultrasonically, and extracted at pH 12–13, then at pH <2 with methylene chloride using continuous extraction techniques. If the solids content is greater than 30%, the sample is extracted using ultrasonic techniques.

Each extract is dried over sodium sulfate, concentrated to a volume of 5 mL, cleaned up using GPC, if necessary, and concentrated. Extracts are concentrated to 1 mL if GPC is not performed, and to 0.5 mL if GPC is performed.

An internal standard is added to the extract, and a 1-mL aliquot of the extract is injected into the GC. The compounds are separated by GC and detected by a MS. The labeled compounds serve to correct the variability of the analytical technique.

INTERFERENCES Solvents, reagents, glassware, and other sample processing hardware may yield artifacts and/or elevated baselines causing misinterpretation of chromatograms and spectra. Materials used in the analysis must be demonstrated to be free from interferences under the conditions of analysis

by running method blanks initially and with each sample lot (sample started through the extraction process on a given 8-h shift, to a maximum of 20). Specific selection of reagents and purification of solvents by distillation in all glass systems may be required. Glassware and, where possible, reagents are cleaned by solvent rinse and baking at 450°C for 1-h minimum. Interferences coextracted from samples will vary considerably from source to source, depending on the diversity of the site being sampled.

INSTRUMENTATION Major instrumentation includes a GC with a splitless or on-column injection port for capillary column, a MS with 70 eV electron impact ionization, and a data system to collect and record MS data, and process it. A K-D apparatus is used to concentrate extracts.

GC Column: 30 m × 0.25 mm I.D. 5% phenyl, 94% methyl, 1% vinyl silicone bonded phased fused silica capillary column.

PRECISION & ACCURACY The detection limits of the method are usually dependent on the level of interferences rather than instrumental limitations. The limits typify the minimum quantities that can be detected with no interferences present.

The minimum level (in µg/mL) was 10. This is defined as a minimum level at which the analytical system shall give recognizable mass spectra (background corrected) and acceptable calibration points.

The MDL (in µg/kg) in low solids was 67 and in high solids was 55; these were determined in digested sludge (low solids) and in filter cake or compost (high solids).

The labeled and native compound initial precision as standard deviation (in µg/L) was 41.

The labeled and native compound initial accuracyas average recovery (in µg/L) was 75–148.

SAMPLE COLLECTION, PRESERVATION & HANDLING Collect samples in glass containers. Aqueous samples which flow freely are collected in refrigerated bottles using automatic sampling equipment. Solid samples are collected as grab samples using widemouth jars. Maintain samples at 0 to 4°C from the time of collection until extraction. If residual chlorine is present in aqueous samples, add 80 mg sodium thiosulfate/L of water. Begin sample extraction within 7 days of collection, and analyze all extracts within 40 days of extraction.

SAMPLE PREPARATION Samples containing 1% solids or less are extracted directly using continuous liquid-liquid extraction techniques. Samples containing 1 to 30% solids are diluted to the 1% level with reagent water and extracted using continuous liquid-liquid extraction techniques. Samples containing greater than 30% solids are extracted using ultrasonic techniques.

Base/neutral extraction — Adjust the pH of the waters in the extractors to 12–13 with 6 N NaOH. Extract with methylene chloride for 24–48 h.

Acid extraction — Adjust the pH of the waters in the extractors to 2 or less using 6 N sulfuric acid. Extract with methylene chloride for 24–48 h.

Ultrasonic extraction of high solids samples — Add anhydrous sodium sulfate to the sample and QC aliquot(s). Add acetone:methylene chloride (1:1) to the sample and mix thoroughly

Concentrate extracts using a K-D apparatus.

QUALITY CONTROL The analyst is permitted to modify this method to improve separations or lower the costs of measurements, provided all performance specifications are met. Analyses of blanks are required to demonstrate freedom from contamination. When results of spikes indicate atypical method performance for samples, the samples are diluted to bring method performance within acceptable limits.

For low solids (aqueous samples), extract, concentrate, and analyze two sets of four 1-L aliquots (8 aliquots total) of the precision and recovery standard. For high solids samples, two sets of four 30-g aliquots of the high solids reference matrix are used.

Spike all samples with labeled compounds to assess method performance. Compute percent recoveryof the labeled compounds using the internal standard method. Compare the labeled compound recovery for each compound with the corresponding labeled compound recovery.

Reagent water and high solids reference matrix blanks are analyzed to demonstrate freedom from contamination. Extract and concentrate a 1-L reagent water blank or a high solids reference matrix blank with each sample's lot (samples started through the extraction process on the same 8-h shift, to a maximum of 20 samples).

Field replicates may be collected to determine the precision of the sampling technique, and spiked samples may be required to determine the accuracy of the analysis when the internal standard method is used.

REFERENCE Semivolatile Organic Compounds by Isotope Dilution GC/MS. Office of Water Regulation and Standards, U.S. EPA Industrial Technology Division, Washington, DC, EPA Method 1625, Rev. C, June 1989 (contact W.A. Telliard, U.S. EPA, Office of Water Regulations and Standards, 401 M St., SW, Washington, DC, 20460. Phone: 202-382-7131).

Biphenyl
CAS #92-52-4

EPA Method 1625

TITLE Semivolatile Organic Compounds by Isotope Dilution GC/MS

MATRIX The compounds may be determined in waters, soils, and municipal sludges by this method.

METHOD SUMMARY This method is used to determine 176 semivolatile toxic organic pollutants associated with the CWA (as amended 1987); the RCRA (as amended 1986); the CERCLA (as amended 1986); and other compounds amenable to extraction and analysis by capillary column gas chromatography-mass spectrometry (GC/MS).

Stable isotopically-labeled analogs of the compounds of interest are added to the sample. If the solids content is less than 1%, a 1-L sample is extracted at pH 12–13, then at pH <2 with methylene chloride using continuous extraction techniques.

If the solids content is 30% or less, the sample is diluted to 1% solids with reagent water, homogenized ultrasonically, and extracted at pH 12–13, then at pH <2 with methylene chloride using continuous extraction techniques. If the solids content is greater than 30%, the sample is extracted using ultrasonic techniques.

Each extract is dried over sodium sulfate, concentrated to a volume of 5 mL, cleaned up using GPC, if necessary, and concentrated. Extracts are concentrated to 1 mL if GPC is not performed, and to 0.5 mL if GPC is performed.

An internal standard is added to the extract, and a 1-mL aliquot of the extract is injected into the GC. The compounds are separated by GC and detected by a MS. The labeled compounds serve to correct the variability of the analytical technique.

INTERFERENCES Solvents, reagents, glassware, and other sample processing hardware may yield artifacts and/or elevated baselines causing misinterpretation of chromatograms and spectra. Materials used in the analysis must be demonstrated to be free from interferences under the conditions of analysis by running method blanks initially and with each sample lot (sample started through the extraction process on a given 8-h shift, to a maximum of 20). Specific selection of reagents and purification of solvents by distillation in all glass systems may be required. Glassware and, where possible, reagents are cleaned by solvent rinse and baking at 450°C for 1-h minimum. Interferences coextracted from samples will vary considerably from source to source, depending on the diversity of the site being sampled.

INSTRUMENTATION Major instrumentation includes a GC with a splitless or on-column injection port for capillary column, a MS with 70 eV electron impact ionization, and a data system to collect and record MS data, and process it. A K-D apparatus is used to concentrate extracts.

GC Column: 30 m × 0.25 mm I.D. 5% phenyl, 94% methyl, 1% vinyl silicone bonded phased fused silica capillary column.

PRECISION & ACCURACY The detection limits of the method are usually dependent on the level of interferences rather than instrumental limitations. The limits typify the minimum quantities that can be detected with no interferences present.

The minimum level (in µg/mL) was 10. This is defined as a minimum level at which the analytical system shall give recognizable mass spectra (background corrected) and acceptable calibration points.
The MDL (in µg/kg) in low solids was 67 and in high solids was 55; these were determined in digested sludge (low solids) and in filter cake or compost (high solids).
The labeled and native compound initial precision as standard deviation (in µg/L) was 41.
The labeled and native compound initial accuracy as average recovery (in µg/L) was 75–148.

SAMPLE COLLECTION, PRESERVATION & HANDLING Collect samples in glass containers. Aqueous samples which flow freely are collected in refrigerated bottles using automatic sampling equipment. Solid samples are collected as grab samples using widemouth jars. Maintain samples at 0 to 4°C from the time of collection until extraction. If residual chlorine is present in aqueous samples, add 80 mg sodium thiosulfate/L of water. Begin sample extraction within 7 days of collection, and analyze all extracts within 40 days of extraction.

SAMPLE PREPARATION Samples containing 1% solids or less are extracted directly using continuous liquid-liquid extraction techniques. Samples containing 1 to 30% solids are diluted to the 1% level with reagent water and extracted using continuous liquid-liquid extraction techniques. Samples containing greater than 30% solids are extracted using ultrasonic techniques.

Base/neutral extraction — Adjust the pH of the waters in the extractors to 12–13 with 6 N NaOH. Extract with methylene chloride for 24–48 h.
Acid extraction — Adjust the pH of the waters in the extractors to 2 or less using 6 N sulfuric acid. Extract with methylene chloride for 24–48 h.
Ultrasonic extraction of high solids samples — Add anhydrous sodium sulfate to the sample and QC aliquot(s). Add acetone:methylene chloride (1:1) to the sample and mix thoroughly

Concentrate extracts using a K-D apparatus.

QUALITY CONTROL The analyst is permitted to modify this method to improve separations or lower the costs of measurements, provided all performance specifications are met. Analyses of blanks are required to demonstrate freedom from contamination. When results of spikes indicate atypical method performance for samples, the samples are diluted to bring method performance within acceptable limits.

For low solids (aqueous samples), extract, concentrate, and analyze two sets of four 1-L aliquots (8 aliquots total) of the precision and recovery standard. For high solids samples, two sets of four 30-g aliquots of the high solids reference matrix are used.

Spike all samples with labeled compounds to assess method performance. Compute percent recovery of the labeled compounds using the internal standard method. Compare the labeled compound recovery for each compound with the corresponding labeled compound recovery.

Reagent water and high solids reference matrix blanks are analyzed to demonstrate freedom from contamination. Extract and concentrate a 1-L reagent water blank or a high solids reference matrix blank with each sample's lot (samples started through the extraction process on the same 8-h shift, to a maximum of 20 samples).

Field replicates may be collected to determine the precision of the sampling technique, and spiked samples may be required to determine the accuracy of the analysis when the internal standard method is used.

REFERENCE Semivolatile Organic Compounds by Isotope Dilution GC/MS. Office of Water Regulation and Standards, U.S. EPA Industrial Technology Division, Washington, DC,

EPA Method 1625, Rev. C, June 1989 (contact W.A. Telliard, U.S. EPA, Office of Water Regulations and Standards, 401 M St., SW, Washington, DC, 20460. Phone: 202-382-7131).

Bis(2-chloroethoxy)methane   **EPA Method 8270**
CAS #111-91-1

TITLE Semivolatile Organic Compounds by GC/MS

MATRIX This method is used to determine the concentration of semivolatile organic compounds in extracts prepared from all types of solid waste matrices, soils, and groundwater. Although surface waters are not specifically mentioned, this method should be applicable to water samples from rivers, lakes, etc.

METHOD SUMMARY This method covers 259 semivolatile organic compounds. In very limited applications direct injection of the sample into the GC/MS system may be appropriate, but this results in very high detection limits (approximately 10,000 µg/L). Typically, a 1-L liquid sample, containing surrogate, and matrix spiking standards, is extracted in a continuous extractor first under acid conditions and then under basic conditions. Typically 30 g of a solid sample, containing surrogate, and matrix spiking standards, is extracted ultrasonically. After concentrating the extract to 1 mL it is spiked with 10 µL of an internal standard solution just prior to analysis by GC/MS. The volume injected should contain about 100 ng of base/neutral and 200 ng of acid surrogates (for a 1-µL injection). Analysis is performed by GC/MS using a capillary GC column.

INTERFERENCES Raw GC/MS data from all blanks, samples, and spikes must be evaluated for interferences. Contamination by carryover can occur whenever high-concentration and low-concentration samples are sequentially analyzed. To reduce carryover, the sample syringe must be rinsed out between samples with solvent. Whenever an unusually concentrated sample is encountered, it should be followed by the analysis of blank solvent to check for cross-contamination.

INSTRUMENTATION A GC/MS and a data system are required. The GC column used is a 30 m × 0.25 mm I.D. (or 0.32 mm I.D.) 1-µm film thickness silicone-coated fused silica capillary column. A continuous liquid-liquid extractor equipped with Teflon® or glass connection joints and stopcocks requiring no lubrication, a K-D concentrating apparatus, water bath, and an ultrasonic disrupter with a minimum power of 300 W and with pulsing capability are also required.

PRECISION & ACCURACY The estimated quantitation limit (EQL) of Method 8270B for determining an individual compound is approximately 1 mg/kg (wet weight) for soil or sediment samples, 1–200 mg/kg for wastes (dependent on matrix and method of preparation), and 10 µg/L for groundwater samples. EQLs will be proportionately higher for sample extracts that require dilution to avoid saturation of the detector.

The EQL(b) for groundwater in µg/L is 10.
The EQL (a, b) for low concentrations in soil and sediment in µg/kg is 660.
Accuracy as µg/L is 1.12C -5.04.

Overall precision in µg/L is 0.26X +2.01.

(a) *EQLs listed for soil/sediment are based on wet weight. Normally data is reported in a dry-weight basis; therefore, EQLs will be higher based on the % dry weight of each sample. This calculation is based on a 30-g sample and gel permeation chromatography cleanup.*
(b) *Sample EQLs are highly matrix-dependent. The EQLs are provided for guidance and may not always be achievable.*
C = *True value for concentration, in µg/L.*
X = *Average recovery found for measurements of samples containing a concentration of C, in µg/L.*

ESTIMATED QUANTITATION LIMIT FOR OTHER MATRICES

Other Matrices	Factor (a)
High-concentration soil and sludges by sonicator	7.5
Non-water miscible waste	75

(a) *EQL for other matrices = [EQL for low soil/sediment] × [Factor]. This estimated EQL is similar to an EPA "Practical Quantitation Limit."*

SAMPLING METHOD
Liquid samples — Use a 1 or 2½ gallon amber glass bottle with a screw-top Teflon®-lined cover that has been prewashed with detergent and rinsed with distilled water and methanol (or isopropanol).

Soils, sediments, or sludges — Use an 8-oz. widemouth glass with a screw-top Teflon®-lined cover that has been prewashed with detergent and rinsed with distilled water and methanol (or isopropanol).

SAMPLE PRESERVATION
Liquid samples — If residual chlorine is present, add 3 mL of 10% sodium thiosulfate per gallon, cool to 4°C and store in a solvent-free refrigerator until analysis; if chlorine is not present, then eliminate the sodium thiosulfate addition.

Soils, sediments, or sludges — Cool samples to 4°C and store in a solvent-free refrigerator.

MHT Liquid samples must be extracted within 7 days and the extracts analyzed within 40 days. Soils, sediments, or sludges may be stored for a maximum of 14 days and the extracts analyzed within 40 days.

SAMPLE PREPARATION
Liquid samples — Transfer 1 L quantitatively to a continuous extractor. If high concentrations are anticipated, a smaller volume may be used and then diluted with organic-free reagent water to 1 L. Adjust pH, if necessary, to pH <2 using 1:1 (V/V) sulfuric acid. Pipette 1.0 mL of a surrogate standard spiking solution into each sample. For the sample in each analytical batch selected for spiking, add 1.0 mL of a matrix spiking standard. For base/neutral acid analysis, the amount of the surrogates and matrix spiking compounds added to the sample should result in a final concentration of 100 ng/µL of each analyte in the extract to be analyzed (assuming a 1-µL injection). Extract with methylene chloride for 18–24 h. Next, adjust the pH of the aqueous phase to pH >11 using 10 N sodium hydroxide and extract it with methylene chloride again for

18–24 h. Dry the extract through a column containing anhydrous sodium sulfate and concentrate it to 1 mL using a K-D concentrator.

Soils, sediments, or sludges — Use 30 g of sample. Nonporous or wet samples (gummy or clay type) that do not have a free-flowing sandy texture must be mixed with anhydrous sodium sulfate until the sample is free flowing. Add 1 mL of surrogate standards to all samples, spikes, standards, and blanks. For the sample in each analytical batch selected for spiking, add 1.0 mL of a matrix spiking standard. For base/neutral acid analysis, the amount added of the surrogates and matrix spiking compounds should result in a final concentration of 100 ng/µL of each base/neutral analyte and 200 ng/µL of each acid analyte in the extract to be analyzed (assuming a 1-µL injection). Immediately add a 100 mL mixture of 1:1 methylene chloride:acetone and extract the sample ultrasonically for 3 min and then decant or filter the extracts. Repeat the extraction two or more times. Dry the extract using a column with anhydrous sodium sulfate and concentrate it to 1 mL in a K-D concentrator.

QUALITY CONTROL A methylene chloride solution containing 50 ng/µL of decafluorotriphenylphosphine (DFTPP) is used for tuning the GC/MS system each 12-h shift. A system performance check also must be made during every 12-h shift. A standard containing 50 ng/µL each of 4,4′-DDT, pentachlorophenol, and benzidine is required to verify injection port inertness and GC column performance. A calibration standard at mid-concentration, containing each compound of interest, including all required surrogates, must be performed every 12 h during analysis. After the system performance check is met, calibration check compounds (CCCs) are used to check the validity of the initial calibration.

The internal standard responses and retention times in the calibration check standard must be evaluated immediately after or during data acquisition. If the retention time for any internal standard changes by more than 30 seconds from the last check calibration (12 h), the chromatographic system must be inspected for malfunctions and corrections must be made, as required. If the electron ionization current plot (EICP) area for any of the internal standards changes by a factor of two from the last daily calibration standard check, the mass spectrometer must be inspected for malfunctions and corrections must be made, as appropriate.

Demonstrate, through the analysis of a reagent water blank, that interferences from the analytical system, glassware, and reagents are under control. The blank samples should be carried through all stages of the sample preparation and measurement steps. For each analytical batch (up to 20 samples), a reagent blank, matrix spike, and matrix spike duplicate/duplicate must be analyzed (the frequency of the spikes may be different for different monitoring programs). The blank and spiked samples must be carried through all stages of the sample preparation and measurement steps. A QC reference sample concentrate containing each analyte at a concentration of 100 mg/L in methanol is required.

REFERENCE Test Methods for Evaluating Solid Waste (SW-846). U.S. EPA 1983, Method 8270B, Rev. 2, Nov. 1990. Office of Solid Waste, Washington, DC.

Bis(2-chloroethoxy)methane   EPA Method 1625
CAS #111-91-1

TITLE Semivolatile Organic Compounds by Isotope Dilution GC/MS

MATRIX The compounds may be determined in waters, soils, and municipal sludges by this method.

METHOD SUMMARY This method is used to determine 176 semivolatile toxic organic pollutants associated with the CWA (as amended 1987); the RCRA (as amended 1986); the CERCLA (as amended 1986); and other compounds amenable to extraction and analysis by capillary column gas chromatography-mass spectrometry (GC/MS).

Stable isotopically-labeled analogs of the compounds of interest are added to the sample. If the solids content is less than 1%, a 1-L sample is extracted at pH 12–13, then at pH <2 with methylene chloride using continuous extraction techniques.

If the solids content is 30% or less, the sample is diluted to 1% solids with reagent water, homogenized ultrasonically, and extracted at pH 12–13, then at pH <2 with methylene chloride using continuous extraction techniques. If the solids content is greater than 30%, the sample is extracted using ultrasonic techniques.

Each extract is dried over sodium sulfate, concentrated to a volume of 5 mL, cleaned up using GPC, if necessary, and concentrated. Extracts are concentrated to 1 mL if GPC is not performed, and to 0.5 mL if GPC is performed.

An internal standard is added to the extract, and a 1-mL aliquot of the extract is injected into the GC. The compounds are separated by GC and detected by a MS. The labeled compounds serve to correct the variability of the analytical technique.

INTERFERENCES Solvents, reagents, glassware, and other sample processing hardware may yield artifacts and/or elevated baselines causing misinterpretation of chromatograms and spectra. Materials used in the analysis must be demonstrated to be free from interferences under the conditions of analysis by running method blanks initially and with each sample lot (sample started through the extraction process on a given 8-h shift, to a maximum of 20). Specific selection of reagents and purification of solvents by distillation in all glass systems may be required. Glassware and, where possible, reagents are cleaned by solvent rinse and baking at 450°C for 1-h minimum. Interferences coextracted from samples will vary considerably from source to source, depending on the diversity of the site being sampled.

INSTRUMENTATION Major instrumentation includes a GC with a splitless or on-column injection port for capillary column, a MS with 70 eV electron impact ionization, and a data system to collect and record MS data, and process it. A K-D apparatus is used to concentrate extracts.

GC Column: 30 m × 0.25 mm I.D. 5% phenyl, 94% methyl, 1% vinyl silicone bonded phased fused silica capillary column.

PRECISION & ACCURACY The detection limits of the method are usually dependent on the level of interferences

rather than instrumental limitations. The limits typify the minimum quantities that can be detected with no interferences present.

The minimum level (in μg/mL) was 20. This is defined as a minimum level at which the analytical system shall give recognizable mass spectra (background corrected) and acceptable calibration points.

The MDL (in μg/kg) in low solids was 26 and in high solids was 23; these were determined in digested sludge (low solids) and in filter cake or compost (high solids).

The labeled and native compound initial precision as standard deviation (in μg/L) was 27.

The labeled and native compound initial accuracy as average recovery (in μg/L) was 43–153.

SAMPLE COLLECTION, PRESERVATION & HANDLING
Collect samples in glass containers. Aqueous samples which flow freely are collected in refrigerated bottles using automatic sampling equipment. Solid samples are collected as grab samples using widemouth jars. Maintain samples at 0 to 4°C from the time of collection until extraction. If residual chlorine is present in aqueous samples, add 80 mg sodium thiosulfate/L of water. Begin sample extraction within 7 days of collection, and analyze all extracts within 40 days of extraction.

SAMPLE PREPARATION Samples containing 1% solids or less are extracted directly using continuous liquid-liquid extraction techniques. Samples containing 1 to 30% solids are diluted to the 1% level with reagent water and extracted using continuous liquid-liquid extraction techniques. Samples containing greater than 30% solids are extracted using ultrasonic techniques.

- Base/neutral extraction — Adjust the pH of the waters in the extractors to 12–13 with 6 N NaOH. Extract with methylene chloride for 24–48 h.
- Acid extraction — Adjust the pH of the waters in the extractors to 2 or less using 6 N sulfuric acid. Extract with methylene chloride for 24–48 h.
- Ultrasonic extraction of high solids samples — Add anhydrous sodium sulfate to the sample and QC aliquot(s). Add acetone:methylene chloride (1:1) to the sample and mix thoroughly

Concentrate extracts using a K-D apparatus.

QUALITY CONTROL The analyst is permitted to modify this method to improve separations or lower the costs of measurements, provided all performance specifications are met. Analyses of blanks are required to demonstrate freedom from contamination. When results of spikes indicate atypical method performance for samples, the samples are diluted to bring method performance within acceptable limits.

For low solids (aqueous samples), extract, concentrate, and analyze two sets of four 1-L aliquots (8 aliquots total) of the precision and recovery standard. For high solids samples, two sets of four 30-g aliquots of the high solids reference matrix are used.

Spike all samples with labeled compounds to assess method performance. Compute percent recovery of the labeled compounds using the internal standard method. Compare the labeled compound recovery for each compound with the corresponding labeled compound recovery.

Reagent water and high solids reference matrix blanks are analyzed to demonstrate freedom from contamination. Extract and concentrate a 1-L reagent water blank or a high solids reference matrix blank with each sample's lot (samples started through the extraction process on the same 8-h shift, to a maximum of 20 samples).

Field replicates may be collected to determine the precision of the sampling technique, and spiked samples may be required to determine the accuracy of the analysis when the internal standard method is used.

REFERENCE Semivolatile Organic Compounds by Isotope Dilution GC/MS. Office of Water Regulation and Standards, U.S. EPA Industrial Technology Division, Washington, DC, EPA Method 1625, Rev. C, June 1989 (contact W.A. Telliard, U.S. EPA, Office of Water Regulations and Standards, 401 M St., SW, Washington, DC, 20460. Phone: 202-382-7131).

Bis(2-chloroethoxy)methane   **EPA Method 1625**
CAS #111-91-1

TITLE Semivolatile Organic Compounds by Isotope Dilution GC/MS

MATRIX The compounds may be determined in waters, soils, and municipal sludges by this method.

METHOD SUMMARY This method is used to determine 176 semivolatile toxic organic pollutants associated with the CWA (as amended 1987); the RCRA (as amended 1986); the CERCLA (as amended 1986); and other compounds amenable to extraction and analysis by capillary column gas chromatography-mass spectrometry (GC/MS).

Stable isotopically-labeled analogs of the compounds of interest are added to the sample. If the solids content is less than 1%, a 1-L sample is extracted at pH 12–13, then at pH <2 with methylene chloride using continuous extraction techniques.

If the solids content is 30% or less, the sample is diluted to 1% solids with reagent water, homogenized ultrasonically, and extracted at pH 12–13, then at pH <2 with methylene chloride using continuous extraction techniques. If the solids content is greater than 30%, the sample is extracted using ultrasonic techniques.

Each extract is dried over sodium sulfate, concentrated to a volume of 5 mL, cleaned up using GPC, if necessary, and concentrated. Extracts are concentrated to 1 mL if GPC is not performed, and to 0.5 mL if GPC is performed.

An internal standard is added to the extract, and a 1-mL aliquot of the extract is injected into the GC. The compounds are separated by GC and detected by a MS. The labeled compounds serve to correct the variability of the analytical technique.

INTERFERENCES Solvents, reagents, glassware, and other sample processing hardware may yield artifacts and/or elevated

baselines causing misinterpretation of chromatograms and spectra. Materials used in the analysis must be demonstrated to be free from interferences under the conditions of analysis by running method blanks initially and with each sample lot (sample started through the extraction process on a given 8-h shift, to a maximum of 20). Specific selection of reagents and purification of solvents by distillation in all glass systems may be required. Glassware and, where possible, reagents are cleaned by solvent rinse and baking at 450°C for 1-h minimum. Interferences coextracted from samples will vary considerably from source to source, depending on the diversity of the site being sampled.

INSTRUMENTATION Major instrumentation includes a GC with a splitless or on-column injection port for capillary column, a MS with 70 eV electron impact ionization, and a data system to collect and record MS data, and process it. A K-D apparatus is used to concentrate extracts.

GC Column: 30 m × 0.25 mm I.D. 5% phenyl, 94% methyl, 1% vinyl silicone bonded phased fused silica capillary column.

PRECISION & ACCURACY The detection limits of the method are usually dependent on the level of interferences rather than instrumental limitations. The limits typify the minimum quantities that can be detected with no interferences present.

The minimum level (in µg/mL) was 20. This is defined as a minimum level at which the analytical system shall give recognizable mass spectra (background corrected) and acceptable calibration points.

The MDL (in µg/kg) in low solids was 26 and in high solids was 23; these were determined in digested sludge (low solids) and in filter cake or compost (high solids).

The labeled and native compound initial precision as standard deviation (in µg/L) was 27.

The labeled and native compound initial accuracy as average recovery (in µg/L) was 43–153.

SAMPLE COLLECTION, PRESERVATION & HANDLING Collect samples in glass containers. Aqueous samples which flow freely are collected in refrigerated bottles using automatic sampling equipment. Solid samples are collected as grab samples using widemouth jars. Maintain samples at 0 to 4°C from the time of collection until extraction. If residual chlorine is present in aqueous samples, add 80 mg sodium thiosulfate/L of water. Begin sample extraction within 7 days of collection, and analyze all extracts within 40 days of extraction.

SAMPLE PREPARATION Samples containing 1% solids or less are extracted directly using continuous liquid-liquid extraction techniques. Samples containing 1 to 30% solids are diluted to the 1% level with reagent water and extracted using continuous liquid-liquid extraction techniques. Samples containing greater than 30% solids are extracted using ultrasonic techniques.

Base/neutral extraction — Adjust the pH of the waters in the extractors to 12–13 with 6 N NaOH. Extract with methylene chloride for 24–48 h.

Acid extraction — Adjust the pH of the waters in the extractors to 2 or less using 6 N sulfuric acid. Extract with methylene chloride for 24–48 h.

Ultrasonic extraction of high solids samples — Add anhydrous sodium sulfate to the sample and QC aliquot(s). Add acetone:methylene chloride (1:1) to the sample and mix thoroughly

Concentrate extracts using a K-D apparatus.

QUALITY CONTROL The analyst is permitted to modify this method to improve separations or lower the costs of measurements, provided all performance specifications are met. Analyses of blanks are required to demonstrate freedom from contamination. When results of spikes indicate atypical method performance for samples, the samples are diluted to bring method performance within acceptable limits.

For low solids (aqueous samples), extract, concentrate, and analyze two sets of four 1-L aliquots (8 aliquots total) of the precision and recovery standard. For high solids samples, two sets of four 30-g aliquots of the high solids reference matrix are used.

Spike all samples with labeled compounds to assess method performance. Compute percent recovery of the labeled compounds using the internal standard method. Compare the labeled compound recovery for each compound with the corresponding labeled compound recovery.

Reagent water and high solids reference matrix blanks are analyzed to demonstrate freedom from contamination. Extract and concentrate a 1-L reagent water blank or a high solids reference matrix blank with each sample's lot (samples started through the extraction process on the same 8-h shift, to a maximum of 20 samples).

Field replicates may be collected to determine the precision of the sampling technique, and spiked samples may be required to determine the accuracy of the analysis when the internal standard method is used.

REFERENCE Semivolatile Organic Compounds by Isotope Dilution GC/MS. Office of Water Regulation and Standards, U.S. EPA Industrial Technology Division, Washington, DC, EPA Method 1625, Rev. C, June 1989 (contact W.A. Telliard, U.S. EPA, Office of Water Regulations and Standards, 401 M St., SW, Washington, DC, 20460. Phone: 202-382-7131).

Bis(2-chloroethoxy)methane   **EPA Method 625**
CAS #111-91-1

TITLE Base/Neutrals and Acids, U.S. EPA Method 625

MATRIX This methods covers municipal and industrial wastewaters.

METHOD SUMMARY Approximately 1 L of sample is serially extracted with methylene chloride at a pH greater than 11 and again at a pH less than 2 using a separatory funnel or a continuous extractor. The methylene chloride extract is dried, concentrated to a volume of 1 mL, and analyzed by GC/MS.

Qualitative identification of the parameters in the extract is performed using the retention time and the relative abundance of three characteristic masses (m/z). Qualitative analysis is performed using either external or internal standard techniques with a single characteristic m/z.

INTERFERENCES Method interferences may be caused by contaminants in solvents, reagents, glassware, and other sample processing hardware. Glassware must be scrupulously cleaned. Glassware should be heated in a muffle furnace at 400°C for 5 to 30 min. Some thermally stable materials, such as PCBs, may not be eliminated by this treatment. Solvent rinses with acetone and pesticide quality hexane may be substituted for the muffle furnace heating. Matrix interferences may be caused by contaminants that are coextracted from the sample. The base-neutral extraction may cause significantly reduced recovery of phenols. The packed gas chromatographic columns recommended for the basic fraction may not exhibit sufficient resolution for some analytes.

INSTRUMENTATION A GC/MS system with an injection port designed for on-column injection when using packed columns and for splitless injection when using capillary columns.

Column for base/neutrals: 1.8 m long × 2 mm I.D. glass, packed with 3% SP-2550 on Supelcoport (100/120 mesh) or equivalent.
Column for acids: 1.8 m long × 2 mm I.D. glass, packed with 1% SP-1240DA on Supelcoport (100/120 mesh) or equivalent.

PRECISION & ACCURACY The MDL concentrations were obtained using reagent water. The MDL actually achieved in a given analysis will vary depending on instrument sensitivity and matrix effects. This method was tested by 15 laboratories using reagent water, drinking water, surface water, and industrial wastewaters spiked at six concentrations over the range 5 to 100 µg/L. Single operator precision, overall precision, and method accuracy were found to be directly related to the concentration of the parameter matrix.

The MDL (in µg/L) in reagent water was not reported.
The standard deviation (in µg/L based on 4 recovery measurements) was 34.5.
The range (in µg/L) for average recovery for 4 measurements was 49.2–164.7.
The range (in %) for percent recovery was 33–184.
Accuracy (in µg/L) as expected recovery for one or more measurements of a sample containing a true concentration of C was 1.12C-5.04.
Precision (in µg/L) as expected single analyst standard deviation of measurements at an average concentration found at X was 0.16X+1.34.
Overall precision (in µg/L) as expected interlaboratory standard deviation of measurements in an average concentration found at X was 0.26X+2.01.

C = *True value of the concentration in µg/L.*
X = *Average recovery found for measurements of samples containing a concentration at C in µg/L.*

SAMPLE PREPARATION Adjust the pH to >11 with sodium hydroxide and serially extract in a separatory funnel with methylene chloride or else in a continuous extractor. Next, adjust the pH to <2 with sulfuric acid and serially extract in a separatory funnel with methylene chloride or else in a continuous extractor. Dry the extracts separately through a column of anhydrous sodium sulfate and then concentrate each of the extracts to 1.0 mL using a K-D apparatus.

SAMPLE COLLECTION, PRESERVATION & HANDLING Grab samples must be collected in glass containers. All samples must be refrigerated at 4°C from the time of collection until extraction. If residual chlorine is present, add 80 mg of sodium thiosulfate/L of sample and mix well. All samples must be extracted within 7 days of collection and completely analyzed within 40 days of extraction.

QUALITY CONTROL Make an initial, one-time, demonstration of the ability to generate acceptable accuracy and precision with this method. Before processing any samples, the analyst must analyze a reagent water blank to demonstrate that interferences from the analytical system and glassware are under control. Each time a set of samples is extracted or reagents are changed, a reagent water blank must be processed. Spike and analyze a minimum of 5% of all samples to monitor and evaluate lab data quality. A QC check sample concentrate that contains each parameter of interest at a concentration of 100 µg/mL in acetone is required. PCBs and multicomponent pesticides may be omitted from this test.

After analysis of five spiked wastewater samples, calculate the average percent recovery and the standard deviation of the percent recovery. Spike all samples with the surrogate standard spiking solution and calculate the percent recovery of each surrogate compound.

REFERENCE *Federal Register*, Vol. 49, No. 209. Friday, Oct. 26, 1984.

Bis(2-chloroethyl) ether   **EPA Method 1625**
CAS #111-44-4

TITLE Semivolatile Organic Compounds by Isotope Dilution GC/MS

MATRIX The compounds may be determined in waters, soils, and municipal sludges by this method.

METHOD SUMMARY This method is used to determine 176 semivolatile toxic organic pollutants associated with the CWA (as amended 1987); the RCRA (as amended 1986); the CERCLA (as amended 1986); and other compounds amenable to extraction and analysis by capillary column gas chromatography-mass spectrometry (GC/MS).

Stable isotopically-labeled analogs of the compounds of interest are added to the sample. If the solids content is less than 1%, a 1-L sample is extracted at pH 12–13, then at pH <2 with methylene chloride using continuous extraction techniques.

If the solids content is 30% or less, the sample is diluted to 1% solids with reagent water, homogenized ultrasonically, and extracted at pH 12–13, then at pH <2 with methylene chloride using continuous extraction techniques. If the solids content is greater than 30%, the sample is extracted using ultrasonic techniques.

Each extract is dried over sodium sulfate, concentrated to a volume of 5 mL, cleaned up using GPC, if necessary, and concentrated. Extracts are concentrated to 1 mL if GPC is not performed, and to 0.5 mL if GPC is performed.

An internal standard is added to the extract, and a 1-mL aliquot of the extract is injected into the GC. The compounds are separated by GC and detected by a MS. The labeled compounds serve to correct the variability of the analytical technique.

INTERFERENCES Solvents, reagents, glassware, and other sample processing hardware may yield artifacts and/or elevated baselines causing misinterpretation of chromatograms and spectra. Materials used in the analysis must be demonstrated to be free from interferences under the conditions of analysis by running method blanks initially and with each sample lot (sample started through the extraction process on a given 8-h shift, to a maximum of 20). Specific selection of reagents and purification of solvents by distillation in all glass systems may be required. Glassware and, where possible, reagents are cleaned by solvent rinse and baking at 450°C for 1-h minimum. Interferences coextracted from samples will vary considerably from source to source, depending on the diversity of the site being sampled.

INSTRUMENTATION Major instrumentation includes a GC with a splitless or on-column injection port for capillary column, a MS with 70 eV electron impact ionization, and a data system to collect and record MS data, and process it. A K-D apparatus is used to concentrate extracts.

GC Column: 30 m × 0.25 mm I.D. 5% phenyl, 94% methyl, 1% vinyl silicone bonded phased fused silica capillary column.

PRECISION & ACCURACY The detection limits of the method are usually dependent on the level of interferences rather than instrumental limitations. The limits typify the minimum quantities that can be detected with no interferences present.

- The minimum level (in µg/mL) was 10. This is defined as a minimum level at which the analytical system shall give recognizable mass spectra (background corrected) and acceptable calibration points.
- The MDL (in µg/kg) in low solids was 32 and in high solids was 22; these were determined in digested sludge (low solids) and in filter cake or compost (high solids).
- The labeled and native compound initial precision as standard deviation (in µg/L) was 34.
- The labeled and native compound initial accuracyas average recovery (in µg/L) was 55–196.

SAMPLE COLLECTION, PRESERVATION & HANDLING Collect samples in glass containers. Aqueous samples which flow freely are collected in refrigerated bottles using automatic sampling equipment. Solid samples are collected as grab samples using widemouth jars. Maintain samples at 0 to 4°C from the time of collection until extraction. If residual chlorine is present in aqueous samples, add 80 mg sodium thiosulfate/L of water. Begin sample extraction within 7 days of collection, and analyze all extracts within 40 days of extraction.

SAMPLE PREPARATION Samples containing 1% solids or less are extracted directly using continuous liquid-liquid extraction techniques. Samples containing 1 to 30% solids are diluted to the 1% level with reagent water and extracted using continuous liquid-liquid extraction techniques. Samples containing greater than 30% solids are extracted using ultrasonic techniques.

Base/neutral extraction — Adjust the pH of the waters in the extractors to 12–13 with 6 N NaOH. Extract with methylene chloride for 24–48 h.

Acid extraction — Adjust the pH of the waters in the extractors to 2 or less using 6 N sulfuric acid. Extract with methylene chloride for 24–48 h.

Ultrasonic extraction of high solids samples — Add anhydrous sodium sulfate to the sample and QC aliquot(s). Add acetone:methylene chloride (1:1) to the sample and mix thoroughly

Concentrate extracts using a K-D apparatus.

QUALITY CONTROL The analyst is permitted to modify this method to improve separations or lower the costs of measurements, provided all performance specifications are met. Analyses of blanks are required to demonstrate freedom from contamination. When results of spikes indicate atypical method performance for samples, the samples are diluted to bring method performance within acceptable limits.

For low solids (aqueous samples), extract, concentrate, and analyze two sets of four 1-L aliquots (8 aliquots total) of the precision and recovery standard. For high solids samples, two sets of four 30-g aliquots of the high solids reference matrix are used.

Spike all samples with labeled compounds to assess method performance. Compute percent recoveryof the labeled compounds using the internal standard method. Compare the labeled compound recovery for each compound with the corresponding labeled compound recovery.

Reagent water and high solids reference matrix blanks are analyzed to demonstrate freedom from contamination. Extract and concentrate a 1-L reagent water blank or a high solids reference matrix blank with each sample's lot (samples started through the extraction process on the same 8-h shift, to a maximum of 20 samples).

Field replicates may be collected to determine the precision of the sampling technique, and spiked samples may be required to determine the accuracy of the analysis when the internal standard method is used.

REFERENCE Semivolatile Organic Compounds by Isotope Dilution GC/MS. Office of Water Regulation and Standards, U.S. EPA Industrial Technology Division, Washington, DC, EPA Method 1625, Rev. C, June 1989 (contact W.A. Telliard, U.S. EPA, Office of Water Regulations and Standards, 401 M St., SW, Washington, DC, 20460. Phone: 202-382-7131).

Bis(2-chloroethyl) ether
CAS #111-44-4

EPA Method 1625

TITLE Semivolatile Organic Compounds by Isotope Dilution GC/MS

MATRIX The compounds may be determined in waters, soils, and municipal sludges by this method.

METHOD SUMMARY This method is used to determine 176 semivolatile toxic organic pollutants associated with the CWA (as amended 1987); the RCRA (as amended 1986); the CERCLA (as amended 1986); and other compounds amenable to extraction and analysis by capillary column gas chromatography-mass spectrometry (GC/MS).

Stable isotopically-labeled analogs of the compounds of interest are added to the sample. If the solids content is less than 1%, a 1-L sample is extracted at pH 12–13, then at pH <2 with methylene chloride using continuous extraction techniques.

If the solids content is 30% or less, the sample is diluted to 1% solids with reagent water, homogenized ultrasonically, and extracted at pH 12–13, then at pH <2 with methylene chloride using continuous extraction techniques. If the solids content is greater than 30%, the sample is extracted using ultrasonic techniques.

Each extract is dried over sodium sulfate, concentrated to a volume of 5 mL, cleaned up using GPC, if necessary, and concentrated. Extracts are concentrated to 1 mL if GPC is not performed, and to 0.5 mL if GPC is performed.

An internal standard is added to the extract, and a 1-mL aliquot of the extract is injected into the GC. The compounds are separated by GC and detected by a MS. The labeled compounds serve to correct the variability of the analytical technique.

INTERFERENCES Solvents, reagents, glassware, and other sample processing hardware may yield artifacts and/or elevated baselines causing misinterpretation of chromatograms and spectra. Materials used in the analysis must be demonstrated to be free from interferences under the conditions of analysis by running method blanks initially and with each sample lot (sample started through the extraction process on a given 8-h shift, to a maximum of 20). Specific selection of reagents and purification of solvents by distillation in all glass systems may be required. Glassware and, where possible, reagents are cleaned by solvent rinse and baking at 450°C for 1-h minimum. Interferences coextracted from samples will vary considerably from source to source, depending on the diversity of the site being sampled.

INSTRUMENTATION Major instrumentation includes a GC with a splitless or on-column injection port for capillary column, a MS with 70 eV electron impact ionization, and a data system to collect and record MS data, and process it. A K-D apparatus is used to concentrate extracts.

GC Column: 30 m × 0.25 mm I.D. 5% phenyl, 94% methyl, 1% vinyl silicone bonded phased fused silica capillary column.

PRECISION & ACCURACY The detection limits of the method are usually dependent on the level of interferences rather than instrumental limitations. The limits typify the minimum quantities that can be detected with no interferences present.

The minimum level (in µg/mL) was 10. This is defined as a minimum level at which the analytical system shall give recognizable mass spectra (background corrected) and acceptable calibration points.

The MDL (in µg/kg) in low solids was 32 and in high solids was 22; these were determined in digested sludge (low solids) and in filter cake or compost (high solids).

The labeled and native compound initial precision as standard deviation (in µg/L) was 34.

The labeled and native compound initial accuracyas average recovery (in µg/L) was 55–196.

SAMPLE COLLECTION, PRESERVATION & HANDLING
Collect samples in glass containers. Aqueous samples which flow freely are collected in refrigerated bottles using automatic sampling equipment. Solid samples are collected as grab samples using widemouth jars. Maintain samples at 0 to 4°C from the time of collection until extraction. If residual chlorine is present in aqueous samples, add 80 mg sodium thiosulfate/L of water. Begin sample extraction within 7 days of collection, and analyze all extracts within 40 days of extraction.

SAMPLE PREPARATION Samples containing 1% solids or less are extracted directly using continuous liquid-liquid extraction techniques. Samples containing 1 to 30% solids are diluted to the 1% level with reagent water and extracted using continuous liquid-liquid extraction techniques. Samples containing greater than 30% solids are extracted using ultrasonic techniques.

Base/neutral extraction — Adjust the pH of the waters in the extractors to 12–13 with 6 N NaOH. Extract with methylene chloride for 24–48 h.

Acid extraction — Adjust the pH of the waters in the extractors to 2 or less using 6 N sulfuric acid. Extract with methylene chloride for 24–48 h.

Ultrasonic extraction of high solids samples — Add anhydrous sodium sulfate to the sample and QC aliquot(s). Add acetone:methylene chloride (1:1) to the sample and mix thoroughly

Concentrate extracts using a K-D apparatus.

QUALITY CONTROL The analyst is permitted to modify this method to improve separations or lower the costs of measurements, provided all performance specifications are met. Analyses of blanks are required to demonstrate freedom from contamination. When results of spikes indicate atypical method performance for samples, the samples are diluted to bring method performance within acceptable limits.

For low solids (aqueous samples), extract, concentrate, and analyze two sets of four 1-L aliquots (8 aliquots total) of the precision and recovery standard. For high solids samples, two sets of four 30-g aliquots of the high solids reference matrix are used.

Spike all samples with labeled compounds to assess method performance. Compute percent recoveryof the labeled compounds

using the internal standard method. Compare the labeled compound recovery for each compound with the corresponding labeled compound recovery.

Reagent water and high solids reference matrix blanks are analyzed to demonstrate freedom from contamination. Extract and concentrate a 1-L reagent water blank or a high solids reference matrix blank with each sample's lot (samples started through the extraction process on the same 8-h shift, to a maximum of 20 samples).

Field replicates may be collected to determine the precision of the sampling technique, and spiked samples may be required to determine the accuracy of the analysis when the internal standard method is used.

REFERENCE Semivolatile Organic Compounds by Isotope Dilution GC/MS. Office of Water Regulation and Standards, U.S. EPA Industrial Technology Division, Washington, DC, EPA Method 1625, Rev. C, June 1989 (contact W.A. Telliard, U.S. EPA, Office of Water Regulations and Standards, 401 M St., SW, Washington, DC, 20460. Phone: 202-382-7131).

Bis(2-chloroethyl) ether EPA Method 625
CAS #111-44-4

TITLE Base/Neutrals and Acids, U.S. EPA Method 625

MATRIX This methods covers municipal and industrial wastewaters.

METHOD SUMMARY Approximately 1 L of sample is serially extracted with methylene chloride at a pH greater than 11 and again at a pH less than 2 using a separatory funnel or a continuous extractor. The methylene chloride extract is dried, concentrated to a volume of 1 mL, and analyzed by GC/MS. Qualitative identification of the parameters in the extract is performed using the retention time and the relative abundance of three characteristic masses (m/z). Qualitative analysis is performed using either external or internal standard techniques with a single characteristic m/z.

INTERFERENCES Method interferences may be caused by contaminants in solvents, reagents, glassware, and other sample processing hardware. Glassware must be scrupulously cleaned. Glassware should be heated in a muffle furnace at 400°C for 5 to 30 min. Some thermally stable materials, such as PCBs, may not be eliminated by this treatment. Solvent rinses with acetone and pesticide quality hexane may be substituted for the muffle furnace heating. Matrix interferences may be caused by contaminants that are coextracted from the sample. The base-neutral extraction may cause significantly reduced recovery of phenols. The packed gas chromatographic columns recommended for the basic fraction may not exhibit sufficient resolution for some analytes.

INSTRUMENTATION A GC/MS system with an injection port designed for on-column injection when using packed columns and for splitless injection when using capillary columns.

Column for base/neutrals: 1.8 m long × 2 mm I.D. glass, packed with 3% SP-2550 on Supelcoport (100/120 mesh) or equivalent.

Column for acids: 1.8 m long × 2 mm I.D. glass, packed with 1% SP-1240DA on Supelcoport (100/120 mesh) or equivalent.

PRECISION & ACCURACY The MDL concentrations were obtained using reagent water. The MDL actually achieved in a given analysis will vary depending on instrument sensitivity and matrix effects. This method was tested by 15 laboratories using reagent water, drinking water, surface water, and industrial wastewaters spiked at six concentrations over the range 5 to 100 µg/L. Single operator precision, overall precision, and method accuracy were found to be directly related to the concentration of the parameter matrix.

The MDL (in µg/L) in reagent water was not reported.
The standard deviation (in µg/L based on 4 recovery measurements) was 55.0.
The range (in µg/L) for average recovery for 4 measurements was 42.9–126.0.
The range (in %) for percent recovery was 12–158.
Accuracy (in µg/L) as expected recovery for one or more measurements of a sample containing a true concentration of C was $0.86C - 1.54$.
Precision (in µg/L) as expected single analyst standard deviation of measurements at an average concentration found at X was $0.35X - 0.99$.
Overall precision (in µg/L) as expected interlaboratory standard deviation of measurements in an average concentration found at X was $0.35X + 0.10$.

$C =$ *True value of the concentration in µg/L.*
$X =$ *Average recovery found for measurements of samples containing a concentration at C in µg/L.*

SAMPLE PREPARATION Adjust the pH to >11 with sodium hydroxide and serially extract in a separatory funnel with methylene chloride or else in a continuous extractor. Next, adjust the pH to <2 with sulfuric acid and serially extract in a separatory funnel with methylene chloride or else in a continuous extractor. Dry the extracts separately through a column of anhydrous sodium sulfate and then concentrate each of the extracts to 1.0 mL using a K-D apparatus.

SAMPLE COLLECTION, PRESERVATION & HANDLING Grab samples must be collected in glass containers. All samples must be refrigerated at 4°C from the time of collection until extraction. If residual chlorine is present, add 80 mg of sodium thiosulfate/L of sample and mix well. All samples must be extracted within 7 days of collection and completely analyzed within 40 days of extraction.

QUALITY CONTROL Make an initial, one-time, demonstration of the ability to generate acceptable accuracy and precision with this method. Before processing any samples, the analyst must analyze a reagent water blank to demonstrate that interferences from the analytical system and glassware are under control. Each time a set of samples is extracted or reagents are changed, a reagent water blank must be processed. Spike and analyze a minimum of 5% of all samples to monitor and evaluate lab data quality. A QC check sample concentrate that

contains each parameter of interest at a concentration of 100 µg/mL in acetone is required. PCBs and multicomponent pesticides may be omitted from this test.

After analysis of five spiked wastewater samples, calculate the average percent recovery and the standard deviation of the percent recovery. Spike all samples with the surrogate standard spiking solution and calculate the percent recovery of each surrogate compound.

REFERENCE *Federal Register*, Vol. 49, No. 209. Friday, Oct. 26, 1984.

Bis(2-chloroethyl) ether   **EPA Method 8270**
CAS #111-44-4

TITLE Semivolatile Organic Compounds by GC/MS

MATRIX This method is used to determine the concentration of semivolatile organic compounds in extracts prepared from all types of solid waste matrices, soils, and groundwater. Although surface waters are not specifically mentioned, this method should be applicable to water samples from rivers, lakes, etc.

METHOD SUMMARY This method covers 259 semivolatile organic compounds. In very limited applications direct injection of the sample into the GC/MS system may be appropriate, but this results in very high detection limits (approximately 10,000 µg/L). Typically, a 1-L liquid sample, containing surrogate, and matrix spiking standards, is extracted in a continuous extractor first under acid conditions and then under basic conditions. Typically 30 g of a solid sample, containing surrogate, and matrix spiking standards, is extracted ultrasonically. After concentrating the extract to 1 mL it is spiked with 10 µL of an internal standard solution just prior to analysis by GC/MS. The volume injected should contain about 100 ng of base/neutral and 200 ng of acid surrogates (for a 1-µL injection). Analysis is performed by GC/MS using a capillary GC column.

INTERFERENCES Raw GC/MS data from all blanks, samples, and spikes must be evaluated for interferences. Contamination by carryover can occur whenever high-concentration and low-concentration samples are sequentially analyzed. To reduce carryover, the sample syringe must be rinsed out between samples with solvent. Whenever an unusually concentrated sample is encountered, it should be followed by the analysis of blank solvent to check for cross-contamination.

INSTRUMENTATION A GC/MS and a data system are required. The GC column used is a 30 m × 0.25 mm I.D. (or 0.32 mm I.D.) 1-µm film thickness silicone-coated fused silica capillary column. A continuous liquid-liquid extractor equipped with Teflon® or glass connection joints and stopcocks requiring no lubrication, a K-D concentrating apparatus, water bath, and an ultrasonic disrupter with a minimum power of 300 W and with pulsing capability are also required.

PRECISION & ACCURACY The estimated quantitation limit (EQL) of Method 8270B for determining an individual compound is approximately 1 mg/kg (wet weight) for soil or sediment samples, 1–200 mg/kg for wastes (dependent on matrix and method of preparation), and 10 µg/L for groundwater samples. EQLs will be proportionately higher for sample extracts that require dilution to avoid saturation of the detector.

The EQL(b) for groundwater in µg/L is 10.
The EQL (a, b) for low concentrations in soil and sediment in µg/kg is 660.
Accuracy as µg/L is 0.86C–1.54.
Overall precision in µg/L is 0.35X +0.10.

(a) *EQLs listed for soil/sediment are based on wet weight. Normally data is reported in a dry-weight basis; therefore, EQLs will be higher based on the % dry weight of each sample. This calculation is based on a 30-g sample and gel permeation chromatography cleanup.*
(b) *Sample EQLs are highly matrix-dependent. The EQLs are provided for guidance and may not always be achievable.*
$C =$ *True value for concentration, in µg/L.*
$X =$ *Average recovery found for measurements of samples containing a concentration of C, in µg/L.*

ESTIMATED QUANTITATION LIMIT
FOR OTHER MATRICES

Other Matrices	Factor (a)
High-concentration soil and sludges by sonicator	7.5
Non-water miscible waste	75

(a) *EQL for other matrices = [EQL for low soil/sediment] × [Factor]. This estimated EQL is similar to an EPA "Practical Quantitation Limit."*

SAMPLING METHOD
Liquid samples — Use a 1 or 2½ gallon amber glass bottle with a screw-top Teflon®-lined cover that has been prewashed with detergent and rinsed with distilled water and methanol (or isopropanol).

Soils, sediments, or sludges — Use an 8-oz. widemouth glass with a screw-top Teflon®-lined cover that has been prewashed with detergent and rinsed with distilled water and methanol (or isopropanol).

SAMPLE PRESERVATION
Liquid samples — If residual chlorine is present, add 3 mL of 10% sodium thiosulfate per gallon, cool to 4°C and store in a solvent-free refrigerator until analysis; if chlorine is not present, then eliminate the sodium thiosulfate addition.

Soils, sediments, or sludges — Cool samples to 4°C and store in a solvent-free refrigerator.

MHT Liquid samples must be extracted within 7 days and the extracts analyzed within 40 days. Soils, sediments, or sludges may be stored for a maximum of 14 days and the extracts analyzed within 40 days.

SAMPLE PREPARATION
Liquid samples — Transfer 1 L quantitatively to a continuous extractor. If high concentrations are anticipated, a smaller volume may be used and then diluted with organic-free reagent water to 1 L. Adjust pH, if necessary, to pH <2 using 1:1 (V/V) sulfuric acid. Pipette 1.0 mL of a surrogate standard spiking

solution into each sample. For the sample in each analytical batch selected for spiking, add 1.0 mL of a matrix spiking standard. For base/neutral acid analysis, the amount of the surrogates and matrix spiking compounds added to the sample should result in a final concentration of 100 ng/µL of each analyte in the extract to be analyzed (assuming a 1-µL injection). Extract with methylene chloride for 18–24 h. Next, adjust the pH of the aqueous phase to pH >11 using 10 N sodium hydroxide and extract it with methylene chloride again for 18–24 h. Dry the extract through a column containing anhydrous sodium sulfate and concentrate it to 1 mL using a K-D concentrator.

Soils, sediments, or sludges — Use 30 g of sample. Nonporous or wet samples (gummy or clay type) that do not have a free-flowing sandy texture must be mixed with anhydrous sodium sulfate until the sample is free flowing. Add 1 mL of surrogate standards to all samples, spikes, standards, and blanks. For the sample in each analytical batch selected for spiking, add 1.0 mL of a matrix spiking standard. For base/neutral acid analysis, the amount added of the surrogates and matrix spiking compounds should result in a final concentration of 100 ng/µL of each base/neutral analyte and 200 ng/µL of each acid analyte in the extract to be analyzed (assuming a 1-µL injection). Immediately add a 100 mL mixture of 1:1 methylene chloride:acetone and extract the sample ultrasonically for 3 min and then decant or filter the extracts. Repeat the extraction two or more times. Dry the extract using a column with anhydrous sodium sulfate and concentrate it to 1 mL in a K-D concentrator.

QUALITY CONTROL A methylene chloride solution containing 50 ng/µL of decafluorotriphenylphosphine (DFTPP) is used for tuning the GC/MS system each 12-h shift. A system performance check also must be made during every 12-h shift. A standard containing 50 ng/µL each of 4,4′-DDT, pentachlorophenol, and benzidine is required to verify injection port inertness and GC column performance. A calibration standard at mid-concentration, containing each compound of interest, including all required surrogates, must be performed every 12 h during analysis. After the system performance check is met, calibration check compounds (CCCs) are used to check the validity of the initial calibration.

The internal standard responses and retention times in the calibration check standard must be evaluated immediately after or during data acquisition. If the retention time for any internal standard changes by more than 30 seconds from the last check calibration (12 h), the chromatographic system must be inspected for malfunctions and corrections must be made, as required. If the electron ionization current plot (EICP) area for any of the internal standards changes by a factor of two from the last daily calibration standard check, the mass spectrometer must be inspected for malfunctions and corrections must be made, as appropriate.

Demonstrate, through the analysis of a reagent water blank, that interferences from the analytical system, glassware, and reagents are under control. The blank samples should be carried through all stages of the sample preparation and measurement steps. For each analytical batch (up to 20 samples), a reagent blank, matrix spike, and matrix spike duplicate/duplicate must be analyzed (the frequency of the spikes may be different for different monitoring programs). The blank and spiked samples must be carried through all stages of the sample preparation and measurement steps. A QC reference sample concentrate containing each analyte at a concentration of 100 mg/L in methanol is required.

REFERENCE Test Methods for Evaluating Solid Waste (SW-846). U.S. EPA 1983, Method 8270B, Rev. 2, Nov. 1990. Office of Solid Waste, Washington, DC.

Bis(2-chloroisopropyl) ether   **EPA Method 1625**
CAS #108-60-1

TITLE Semivolatile Organic Compounds by Isotope Dilution GC/MS.

MATRIX The compounds may be determined in waters, soils, and municipal sludges by this method.

METHOD SUMMARY This method is used to determine 176 semivolatile toxic organic pollutants associated with the CWA (as amended 1987); the RCRA (as amended 1986); the CERCLA (as amended 1986); and other compounds amenable to extraction and analysis by capillary column gas chromatography-mass spectrometry (GC/MS).

Stable isotopically-labeled analogs of the compounds of interest are added to the sample. If the solids content is less than 1%, a 1-L sample is extracted at pH 12–13, then at pH <2 with methylene chloride using continuous extraction techniques.

If the solids content is 30% or less, the sample is diluted to 1% solids with reagent water, homogenized ultrasonically, and extracted at pH 12–13, then at pH <2 with methylene chloride using continuous extraction techniques. If the solids content is greater than 30%, the sample is extracted using ultrasonic techniques.

Each extract is dried over sodium sulfate, concentrated to a volume of 5 mL, cleaned up using GPC, if necessary, and concentrated. Extracts are concentrated to 1 mL if GPC is not performed, and to 0.5 mL if GPC is performed.

An internal standard is added to the extract, and a 1-mL aliquot of the extract is injected into the GC. The compounds are separated by GC and detected by a MS. The labeled compounds serve to correct the variability of the analytical technique.

INTERFERENCES Solvents, reagents, glassware, and other sample processing hardware may yield artifacts and/or elevated baselines causing misinterpretation of chromatograms and spectra. Materials used in the analysis must be demonstrated to be free from interferences under the conditions of analysis by running method blanks initially and with each sample lot (sample started through the extraction process on a given 8-h shift, to a maximum of 20). Specific selection of reagents and purification of solvents by distillation in all glass systems may be required. Glassware and, where possible, reagents are cleaned by solvent rinse and baking at 450°C for 1-h minimum. Interferences coextracted from samples will vary considerably from source to source, depending on the diversity of the site being sampled.

INSTRUMENTATION Major instrumentation includes a GC with a splitless or on-column injection port for capillary column, a MS with 70 eV electron impact ionization, and a data system to collect and record MS data, and process it. A K-D apparatus is used to concentrate extracts.

GC Column: 30 m × 0.25 mm I.D. 5% phenyl, 94% methyl, 1% vinyl silicone bonded phased fused silica capillary column.

PRECISION & ACCURACY The detection limits of the method are usually dependent on the level of interferences rather than instrumental limitations. The limits typify the minimum quantities that can be detected with no interferences present.

The minimum level (in µg/mL) was 10. This is defined as a minimum level at which the analytical system shall give recognizable mass spectra (background corrected) and acceptable calibration points.

The MDL (in µg/kg) in low solids was 24 and in high solids was 39; these were determined in digested sludge (low solids) and in filter cake or compost (high solids).

The labeled and native compound initial precision as standard deviation (in µg/L) was 17.

The labeled and native compound initial accuracyas average recovery (in µg/L) was 81–138.

SAMPLE COLLECTION, PRESERVATION & HANDLING Collect samples in glass containers. Aqueous samples which flow freely are collected in refrigerated bottles using automatic sampling equipment. Solid samples are collected as grab samples using widemouth jars. Maintain samples at 0 to 4°C from the time of collection until extraction. If residual chlorine is present in aqueous samples, add 80 mg sodium thiosulfate/L of water. Begin sample extraction within 7 days of collection, and analyze all extracts within 40 days of extraction.

SAMPLE PREPARATION Samples containing 1% solids or less are extracted directly using continuous liquid-liquid extraction techniques. Samples containing 1 to 30% solids are diluted to the 1% level with reagent water and extracted using continuous liquid-liquid extraction techniques. Samples containing greater than 30% solids are extracted using ultrasonic techniques.

- Base/neutral extraction — Adjust the pH of the waters in the extractors to 12–13 with 6 N NaOH. Extract with methylene chloride for 24–48 h.
- Acid extraction — Adjust the pH of the waters in the extractors to 2 or less using 6 N sulfuric acid. Extract with methylene chloride for 24–48 h.
- Ultrasonic extraction of high solids samples — Add anhydrous sodium sulfate to the sample and QC aliquot(s). Add acetone:methylene chloride (1:1) to the sample and mix thoroughly

Concentrate extracts using a K-D apparatus.

QUALITY CONTROL The analyst is permitted to modify this method to improve separations or lower the costs of measurements, provided all performance specifications are met. Analyses of blanks are required to demonstrate freedom from contamination. When results of spikes indicate atypical method performance for samples, the samples are diluted to bring method performance within acceptable limits.

For low solids (aqueous samples), extract, concentrate, and analyze two sets of four 1-L aliquots (8 aliquots total) of the precision and recovery standard. For high solids samples, two sets of four 30-g aliquots of the high solids reference matrix are used.

Spike all samples with labeled compounds to assess method performance. Compute percent recoveryof the labeled compounds using the internal standard method. Compare the labeled compound recovery for each compound with the corresponding labeled compound recovery.

Reagent water and high solids reference matrix blanks are analyzed to demonstrate freedom from contamination. Extract and concentrate a 1-L reagent water blank or a high solids reference matrix blank with each sample's lot (samples started through the extraction process on the same 8-h shift, to a maximum of 20 samples).

Field replicates may be collected to determine the precision of the sampling technique, and spiked samples may be required to determine the accuracy of the analysis when the internal standard method is used.

REFERENCE Semivolatile Organic Compounds by Isotope Dilution GC/MS. Office of Water Regulation and Standards, U.S. EPA Industrial Technology Division, Washington, DC, EPA Method 1625, Rev. C, June 1989 (contact W.A. Telliard, U.S. EPA, Office of Water Regulations and Standards, 401 M St., SW, Washington, DC, 20460. Phone: 202-382-7131).

Bis(2-chloroisopropyl) ether   **EPA Method 1625**
CAS #108-60-1

TITLE Semivolatile Organic Compounds by Isotope Dilution GC/MS

MATRIX The compounds may be determined in waters, soils, and municipal sludges by this method.

METHOD SUMMARY This method is used to determine 176 semivolatile toxic organic pollutants associated with the CWA (as amended 1987); the RCRA (as amended 1986); the CERCLA (as amended 1986); and other compounds amenable to extraction and analysis by capillary column gas chromatography-mass spectrometry (GC/MS).

Stable isotopically-labeled analogs of the compounds of interest are added to the sample. If the solids content is less than 1%, a 1-L sample is extracted at pH 12–13, then at pH <2 with methylene chloride using continuous extraction techniques.

If the solids content is 30% or less, the sample is diluted to 1% solids with reagent water, homogenized ultrasonically, and extracted at pH 12–13, then at pH <2 with methylene chloride using continuous extraction techniques. If the solids content is greater than 30%, the sample is extracted using ultrasonic techniques.

Each extract is dried over sodium sulfate, concentrated to a volume of 5 mL, cleaned up using GPC, if necessary, and concentrated. Extracts are concentrated to 1 mL if GPC is not performed, and to 0.5 mL if GPC is performed.

An internal standard is added to the extract, and a 1-mL aliquot of the extract is injected into the GC. The compounds are separated by GC and detected by a MS. The labeled compounds serve to correct the variability of the analytical technique.

INTERFERENCES Solvents, reagents, glassware, and other sample processing hardware may yield artifacts and/or elevated baselines causing misinterpretation of chromatograms and spectra. Materials used in the analysis must be demonstrated to be free from interferences under the conditions of analysis by running method blanks initially and with each sample lot (sample started through the extraction process on a given 8-h shift, to a maximum of 20). Specific selection of reagents and purification of solvents by distillation in all glass systems may be required. Glassware and, where possible, reagents are cleaned by solvent rinse and baking at 450°C for 1-h minimum. Interferences coextracted from samples will vary considerably from source to source, depending on the diversity of the site being sampled.

INSTRUMENTATION Major instrumentation includes a GC with a splitless or on-column injection port for capillary column, a MS with 70 eV electron impact ionization, and a data system to collect and record MS data, and process it. A K-D apparatus is used to concentrate extracts.

GC Column: 30 m × 0.25 mm I.D. 5% phenyl, 94% methyl, 1% vinyl silicone bonded phased fused silica capillary column.

PRECISION & ACCURACY The detection limits of the method are usually dependent on the level of interferences rather than instrumental limitations. The limits typify the minimum quantities that can be detected with no interferences present.

The minimum level (in µg/mL) was 10. This is defined as a minimum level at which the analytical system shall give recognizable mass spectra (background corrected) and acceptable calibration points.
The MDL (in µg/kg) in low solids was 24 and in high solids was 39; these were determined in digested sludge (low solids) and in filter cake or compost (high solids).
The labeled and native compound initial precision as standard deviation (in µg/L) was 17.
The labeled and native compound initial accuracy as average recovery (in µg/L) was 81–138.

SAMPLE COLLECTION, PRESERVATION & HANDLING Collect samples in glass containers. Aqueous samples which flow freely are collected in refrigerated bottles using automatic sampling equipment. Solid samples are collected as grab samples using widemouth jars. Maintain samples at 0 to 4°C from the time of collection until extraction. If residual chlorine is present in aqueous samples, add 80 mg sodium thiosulfate/L of water. Begin sample extraction within 7 days of collection, and analyze all extracts within 40 days of extraction.

SAMPLE PREPARATION Samples containing 1% solids or less are extracted directly using continuous liquid-liquid extraction techniques. Samples containing 1 to 30% solids are diluted to the 1% level with reagent water and extracted using continuous liquid-liquid extraction techniques. Samples containing greater than 30% solids are extracted using ultrasonic techniques.

Base/neutral extraction — Adjust the pH of the waters in the extractors to 12–13 with 6 N NaOH. Extract with methylene chloride for 24–48 h.
Acid extraction — Adjust the pH of the waters in the extractors to 2 or less using 6 N sulfuric acid. Extract with methylene chloride for 24–48 h.
Ultrasonic extraction of high solids samples — Add anhydrous sodium sulfate to the sample and QC aliquot(s). Add acetone:methylene chloride (1:1) to the sample and mix thoroughly

Concentrate extracts using a K-D apparatus.

QUALITY CONTROL The analyst is permitted to modify this method to improve separations or lower the costs of measurements, provided all performance specifications are met. Analyses of blanks are required to demonstrate freedom from contamination. When results of spikes indicate atypical method performance for samples, the samples are diluted to bring method performance within acceptable limits.

For low solids (aqueous samples), extract, concentrate, and analyze two sets of four 1-L aliquots (8 aliquots total) of the precision and recovery standard. For high solids samples, two sets of four 30-g aliquots of the high solids reference matrix are used.

Spike all samples with labeled compounds to assess method performance. Compute percent recovery of the labeled compounds using the internal standard method. Compare the labeled compound recovery for each compound with the corresponding labeled compound recovery.

Reagent water and high solids reference matrix blanks are analyzed to demonstrate freedom from contamination. Extract and concentrate a 1-L reagent water blank or a high solids reference matrix blank with each sample's lot (samples started through the extraction process on the same 8-h shift, to a maximum of 20 samples).

Field replicates may be collected to determine the precision of the sampling technique, and spiked samples may be required to determine the accuracy of the analysis when the internal standard method is used.

REFERENCE Semivolatile Organic Compounds by Isotope Dilution GC/MS. Office of Water Regulation and Standards, U.S. EPA Industrial Technology Division, Washington, DC, EPA Method 1625, Rev. C, June 1989 (contact W.A. Telliard,

U.S. EPA, Office of Water Regulations and Standards, 401 M St., SW, Washington, DC, 20460. Phone: 202-382-7131).

Bis(2-chloroisopropyl) ether   **EPA Method 625**
CAS #108-60-1

TITLE Base/Neutrals and Acids, U.S. EPA Method 625

MATRIX This methods covers municipal and industrial wastewaters.

METHOD SUMMARY Approximately 1 L of sample is serially extracted with methylene chloride at a pH greater than 11 and again at a pH less than 2 using a separatory funnel or a continuous extractor. The methylene chloride extract is dried, concentrated to a volume of 1 mL, and analyzed by GC/MS. Qualitative identification of the parameters in the extract is performed using the retention time and the relative abundance of three characteristic masses (m/z). Qualitative analysis is performed using either external or internal standard techniques with a single characteristic m/z.

INTERFERENCES Method interferences may be caused by contaminants in solvents, reagents, glassware, and other sample processing hardware. Glassware must be scrupulously cleaned. Glassware should be heated in a muffle furnace at 400°C for 5 to 30 min. Some thermally stable materials, such as PCBs, may not be eliminated by this treatment. Solvent rinses with acetone and pesticide quality hexane may be substituted for the muffle furnace heating. Matrix interferences may be caused by contaminants that are coextracted from the sample. The base-neutral extraction may cause significantly reduced recovery of phenols. The packed gas chromatographic columns recommended for the basic fraction may not exhibit sufficient resolution for some analytes.

INSTRUMENTATION A GC/MS system with an injection port designed for on-column injection when using packed columns and for splitless injection when using capillary columns.

Column for base/neutrals: 1.8 m long × 2 mm I.D. glass, packed with 3% SP-2550 on Supelcoport (100/120 mesh) or equivalent.
Column for acids: 1.8 m long × 2 mm I.D. glass, packed with 1% SP-1240DA on Supelcoport (100/120 mesh) or equivalent.

PRECISION & ACCURACY The MDL concentrations were obtained using reagent water. The MDL actually achieved in a given analysis will vary depending on instrument sensitivity and matrix effects. This method was tested by 15 laboratories using reagent water, drinking water, surface water, and industrial wastewaters spiked at six concentrations over the range 5 to 100 µg/L. Single operator precision, overall precision, and method accuracy were found to be directly related to the concentration of the parameter matrix.

The MDL (in µg/L) in reagent water was not reported.
The standard deviation (in µg/L based on 4 recovery measurements) was 46.3.
The range (in µg/L) for average recovery for 4 measurements was 62.8–138.6.
The range (in %) for percent recovery was 36–166.
Accuracy (in µg/L) as expected recovery for one or more measurements of a sample containing a true concentration of C was 1.03C-2.31.
Precision (in µg/L) as expected single analyst standard deviation of measurements at an average concentration found at X was 0.24X+0.28.
Overall precision (in µg/L) as expected interlaboratory standard deviation of measurements in an average concentration found at X was 0.28X+1.04.

C = *True value of the concentration in µg/L.*
X = *Average recovery found for measurements of samples containing a concentration at C in µg/L.*

SAMPLE PREPARATION Adjust the pH to >11 with sodium hydroxide and serially extract in a separatory funnel with methylene chloride or else in a continuous extractor. Next, adjust the pH to <2 with sulfuric acid and serially extract in a separatory funnel with methylene chloride or else in a continuous extractor. Dry the extracts separately through a column of anhydrous sodium sulfate and then concentrate each of the extracts to 1.0 mL using a K-D apparatus.

SAMPLE COLLECTION, PRESERVATION & HANDLING- Grab samples must be collected in glass containers. All samples must be refrigerated at 4°C from the time of collection until extraction. If residual chlorine is present, add 80 mg of sodium thiosulfate/L of sample and mix well. All samples must be extracted within 7 days of collection and completely analyzed within 40 days of extraction.

QUALITY CONTROL Make an initial, one-time, demonstration of the ability to generate acceptable accuracy and precision with this method. Before processing any samples, the analyst must analyze a reagent water blank to demonstrate that interferences from the analytical system and glassware are under control. Each time a set of samples is extracted or reagents are changed, a reagent water blank must be processed. Spike and analyze a minimum of 5% of all samples to monitor and evaluate lab data quality. A QC check sample concentrate that contains each parameter of interest at a concentration of 100 µg/mL in acetone is required. PCBs and multicomponent pesticides may be omitted from this test.

After analysis of five spiked wastewater samples, calculate the average percent recovery and the standard deviation of the percent recovery. Spike all samples with the surrogate standard spiking solution and calculate the percent recovery of each surrogate compound.

REFERENCE Federal Register, Vol. 49, No. 209. Friday, Oct. 26, 1984.

Bis(2-chloroisopropyl) ether   **EPA Method 8270**
CAS #108-60-1

TITLE Semivolatile Organic Compounds by GC/MS

MATRIX This method is used to determine the concentration of semivolatile organic compounds in extracts prepared from all types of solid waste matrices, soils, and groundwater. Although surface waters are not specifically mentioned, this method should be applicable to water samples from rivers, lakes, etc.

METHOD SUMMARY This method covers 259 semivolatile organic compounds. In very limited applications direct injection of the sample into the GC/MS system may be appropriate, but this results in very high detection limits (approximately 10,000 µg/L). Typically, a 1-L liquid sample, containing surrogate, and matrix spiking standards, is extracted in a continuous extractor first under acid conditions and then under basic conditions. Typically 30 g of a solid sample, containing surrogate, and matrix spiking standards, is extracted ultrasonically. After concentrating the extract to 1 mL it is spiked with 10 µL of an internal standard solution just prior to analysis by GC/MS. The volume injected should contain about 100 ng of base/neutral and 200 ng of acid surrogates (for a 1-µL injection). Analysis is performed by GC/MS using a capillary GC column.

INTERFERENCES Raw GC/MS data from all blanks, samples, and spikes must be evaluated for interferences. Contamination by carryover can occur whenever high-concentration and low-concentration samples are sequentially analyzed. To reduce carryover, the sample syringe must be rinsed out between samples with solvent. Whenever an unusually concentrated sample is encountered, it should be followed by the analysis of blank solvent to check for cross-contamination.

INSTRUMENTATION A GC/MS and a data system are required. The GC column used is a 30 m × 0.25 mm I.D. (or 0.32 mm I.D.) 1-µm film thickness silicone-coated fused silica capillary column. A continuous liquid-liquid extractor equipped with Teflon® or glass connection joints and stopcocks requiring no lubrication, a K-D concentrating apparatus, water bath, and an ultrasonic disrupter with a minimum power of 300 W and with pulsing capability are also required.

PRECISION & ACCURACY The estimated quantitation limit (EQL) of Method 8270B for determining an individual compound is approximately 1 mg/kg (wet weight) for soil or sediment samples, 1–200 mg/kg for wastes (dependent on matrix and method of preparation), and 10 µg/L for groundwater samples. EQLs will be proportionately higher for sample extracts that require dilution to avoid saturation of the detector.

The EQL(b) for groundwater in µg/L is 10.
The EQL (a, b) for low concentrations in soil and sediment in µg/kg is 660.
Accuracy as µg/L is $1.03C - 2.31$.
Overall precision in µg/L is $0.25X + 1.04$.

(a) EQLs listed for soil/sediment are based on wet weight. Normally data is reported in a dry-weight basis; therefore, EQLs will be higher based on the % dry weight of each sample. This calculation is based on a 30-g sample and gel permeation chromatography cleanup.

(b) Sample EQLs are highly matrix-dependent. The EQLs are provided for guidance and may not always be achievable.

$C =$ *True value for concentration, in µg/L.*

$X =$ *Average recovery found for measurements of samples containing a concentration of C, in µg/L.*

ESTIMATED QUANTITATION LIMIT FOR OTHER MATRICES

Other Matrices	Factor (a)
High-concentration soil and sludges by sonicator	7.5
Non-water miscible waste	75

(a) EQL for other matrices = [EQL for low soil/sediment] × [Factor]. This estimated EQL is similar to an EPA "Practical Quantitation Limit."

SAMPLING METHOD
Liquid samples — Use a 1 or 2½ gallon amber glass bottle with a screw-top Teflon®-lined cover that has been prewashed with detergent and rinsed with distilled water and methanol (or isopropanol).

Soils, sediments, or sludges — Use an 8-oz. widemouth glass with a screw-top Teflon®-lined cover that has been prewashed with detergent and rinsed with distilled water and methanol (or isopropanol).

SAMPLE PRESERVATION
Liquid samples — If residual chlorine is present, add 3 mL of 10% sodium thiosulfate per gallon, cool to 4°C and store in a solvent-free refrigerator until analysis; if chlorine is not present, then eliminate the sodium thiosulfate addition.

Soils, sediments, or sludges — Cool samples to 4°C and store in a solvent-free refrigerator.

MHT Liquid samples must be extracted within 7 days and the extracts analyzed within 40 days. Soils, sediments, or sludges may be stored for a maximum of 14 days and the extracts analyzed within 40 days.

SAMPLE PREPARATION
Liquid samples — Transfer 1 L quantitatively to a continuous extractor. If high concentrations are anticipated, a smaller volume may be used and then diluted with organic-free reagent water to 1 L. Adjust pH, if necessary, to pH <2 using 1:1 (V/V) sulfuric acid. Pipette 1.0 mL of a surrogate standard spiking solution into each sample. For the sample in each analytical batch selected for spiking, add 1.0 mL of a matrix spiking standard. For base/neutral acid analysis, the amount of the surrogates and matrix spiking compounds added to the sample should result in a final concentration of 100 ng/µL of each analyte in the extract to be analyzed (assuming a 1-µL injection). Extract with methylene chloride for 18–24 h. Next, adjust the pH of the aqueous phase to pH >11 using 10 N sodium hydroxide and extract it with methylene chloride again for 18–24 h. Dry the extract through a column containing anhydrous sodium sulfate and concentrate it to 1 mL using a K-D concentrator.

Soils, sediments, or sludges — Use 30 g of sample. Nonporous or wet samples (gummy or clay type) that do not have a free-flowing sandy texture must be mixed with anhydrous sodium sulfate until the sample is free flowing. Add 1 mL of surrogate standards to all samples, spikes, standards, and blanks. For the sample in each analytical batch selected for spiking, add 1.0 mL of a matrix spiking standard. For base/neutral acid analysis, the amount added of the surrogates and matrix spiking compounds should result in a final concentration of 100 ng/µL of each base/neutral analyte and 200 ng/µL of each acid analyte in the extract to be analyzed (assuming a 1-µL injection). Immediately add a 100 mL mixture of 1:1 methylene chloride:acetone and extract the sample ultrasonically for 3 min and then decant or filter the extracts. Repeat the extraction two or more times. Dry the extract using a column with anhydrous sodium sulfate and concentrate it to 1 mL in a K-D concentrator.

QUALITY CONTROL A methylene chloride solution containing 50 ng/µL of decafluorotriphenylphosphine (DFTPP) is used for tuning the GC/MS system each 12-h shift. A system performance check also must be made during every 12-h shift. A standard containing 50 ng/µL each of 4,4'-DDT, pentachlorophenol, and benzidine is required to verify injection port inertness and GC column performance. A calibration standard at mid-concentration, containing each compound of interest, including all required surrogates, must be performed every 12 h during analysis. After the system performance check is met, calibration check compounds (CCCs) are used to check the validity of the initial calibration.

The internal standard responses and retention times in the calibration check standard must be evaluated immediately after or during data acquisition. If the retention time for any internal standard changes by more than 30 seconds from the last check calibration (12 h), the chromatographic system must be inspected for malfunctions and corrections must be made, as required. If the electron ionization current plot (EICP) area for any of the internal standards changes by a factor of two from the last daily calibration standard check, the mass spectrometer must be inspected for malfunctions and corrections must be made, as appropriate.

Demonstrate, through the analysis of a reagent water blank, that interferences from the analytical system, glassware, and reagents are under control. The blank samples should be carried through all stages of the sample preparation and measurement steps. For each analytical batch (up to 20 samples), a reagent blank, matrix spike, and matrix spike duplicate/duplicate must be analyzed (the frequency of the spikes may be different for different monitoring programs). The blank and spiked samples must be carried through all stages of the sample preparation and measurement steps. A QC reference sample concentrate containing each analyte at a concentration of 100 mg/L in methanol is required.

REFERENCE Test Methods for Evaluating Solid Waste (SW-846). U.S. EPA 1983, Method 8270B, Rev. 2, Nov. 1990. Office of Solid Waste, Washington, DC.

Bis(2-ethoxyethyl) phthalate   **EPA Method 8061**
CAS #605-54-9

TITLE Phthalate Esters by Capillary Gas Chromatography With Electron Capture Detection (GC/ECD)

MATRIX This method covers aqueous and solid matrices. This includes a wide variety such as drinking water, groundwater, industrial wastewaters, surface waters, soils, solids, and sediments.

METHOD SUMMARY This method is used to determine the identities and concentrations of phthalate esters in liquid, solid and sludge matrices. When used to analyze for any or all of the target analytes, compound identification should be supported by at least one additional qualitative technique. This method describes conditions for parallel column, dual electron capture detector analysis, which fulfills the above requirement. Alternatively, GC/MS could be used for compound confirmation.

A measured volume or weight of sample (approximately 1 L for liquids, 10 to 30 g for solids and sludges) is extracted by using the appropriate sample extraction technique specified in EPA Method 3510, EPA Method 3540, and EPA Method 3550. After cleanup, the extract is analyzed by GC/ECD.

INTERFERENCES The sensitivity of this method usually depends on the level of interferences rather than on instrumental limitations. If interferences prevent detection of the analytes, cleanup of the sample extracts is necessary. Either EPA Method 3610 or EPA Method 3620 alone or followed by EPA Method 3660, Sulfur Cleanup, may be used to eliminate interferences in the analysis. EPA Method 3640, Gel Permeation Cleanup, is applicable for samples that contain high amounts of lipids and waxes.

Interferences coextracted from the samples will vary considerably from waste to waste. Glassware must be scrupulously clean. All glassware require treatment in a muffle furnace at 400°C for 2 to 4 h, or thorough rinsing with pesticide-grade solvent, prior to use. Volumetric glassware should not be heated in a muffle furnace. Storage of glassware in the lab introduces contamination, even if the glassware is wrapped in aluminum foil. Sodium sulfate, Florisil, and alumina may be contaminated with phthalate esters and, therefore, use of these materials in sample cleanup should be employed cautiously. If these materials are used, they must be obtained packaged in glass. Heating at 400°C for sodium sulfate, 320°C for Florisil, and 210°C for alumina is recommended. Glass wool used in any step of sample preparation should be a specially treated Pyrex wool, pesticide grade, and must be baked at 400°C for 4 h immediately prior to use.

Paper thimbles and filter paper must be exhaustively washed with the solvent that will be used in the sample extraction. Soxhlet extraction of paper thimbles and filter paper for 12 h with fresh solvent should be repeated for a minimum of three times. Method blanks should be obtained before any of the precleaned thimbles or filter papers are used.

INSTRUMENTATION Gas chromatograph suitable for on-column and split/splitless injections.

Column 1: 30 m × 0.53 mm ID, 5% phenyl/95% methyl silicone fused-silica open tubular column, DB-5, 1.5 µg film thickness.
Column 2: 30 m × 0.53 mm ID, 14% cyanopropyl phenyl silicone fused-silica open tubular column, DB-1701, 1.0 µg film thickness.

A dual electron capture detector (ECD) is used. A Kuderna-Danish (K-D) apparatus is required along with a vacuum manifold consisting of individually adjustable, easily accessible flow-control valves for up to 24 cartridges, sample rack, chemically resistant cover and seals, heavy-duty glass basin, removable stainless steel solvent guides, built-in vacuum gauge and valve. Also, 6-mL, 1-g solid-phase extraction cartridges, LC-Florisil or equivalent, prepackaged, ready to use will be needed.

PRECISION & ACCURACY The MDL actually achieved in a given analysis will vary, as it is dependent on instrument sensitivity and matrix effects. This method has been tested in a single lab. Single-operator precision, overall precision, and method accuracy were found to be related to the concentration of the compounds and the type of matrix.

ESTIMATED QUANTITATION LIMIT (EQL) FACTORS FOR VARIOUS MATRICES (a)

Matrix	Factor (b)
Groundwater	10
Low-concentration soil by ultrasonic extraction with GPC cleanup	670
High-concentration soil and sludges by ultrasonic extraction	10,000
Waste not miscible with water	100,000

(a) Sample EQLs are highly matrix-dependent.
(b) EQL = [Method detection limit] × [Factor]. For non-aqueous samples, the factor is on a wet-weight basis.

The MDL using 7 replicate determinations and a spike concentration of 100 µg/L was 270 ng/L.

The average recovery from HPLC-grade water using 4 determinatons and a spike concentration of 100 µg/L was 108%.

The precision (as RSD) from HPLC-grade water using 4 determinatons and a spike concentration of 100 µg/L was 8.9%.

The average recovery from groundwater using 4 determinatons and a spike concentration of 100 µg/L was 102%.

The precision (as RSD) from groundwater using 4 determinatons and a spike concentration of 100 µg/L was 4.0%.

The average recovery (in %) with %RSD (in parenthesis) from 3 determinations and a spike concentration of 20 µg/L in water was 75.6 (3.3) using 3M Empore Disks and EPA Method 8061.

The average recovery (in %) with %RSD (in parenthesis) from 3 determinations and a spike concentration of 20 µg/L in leachate was 90.8 (22.4) using 3M Empore Disks and EPA Method 8061.

The average recovery (in %) with %RSD (in parenthesis) from 3 determinations and a spike concentration of 20 µg/L in estuarine groundwater was 86.4 (5.8) using 3M Empore Disks and EPA Method 8061.

The average recovery (in %) with %RSD (in parenthesis) from 3 determinations and a spike concentration of 1 mg/kg in estuarine sediment was not determined (matrix interferant) after sulfur cleanup with EPA Method 3660.

The average recovery (in %) with %RSD (in parenthesis) from 3 determinations and a spike concentration of 1 mg/kg in municipal sludge was 66.6 (4.9).

The average recovery (in %) with %RSD (in parenthesis) from 3 determinations and a spike concentration of 1 mg/kg in sandy loam soil was not determined (matrix interferant).

SAMPLE COLLECTION, PRESERVATION & HANDLING
Containers used to collect samples for the determination of semivolatile organic compounds should be soap and water washed followed by methanol (or isopropanol) rinsing. The sample containers should be of glass or Teflon® and have screw-top covers with Teflon® liners. Sample containers should be filled with care to prevent any portion of the collected sample coming in contact with the sampler's gloves.

No preservation is used with concentrated waste samples. With liquid samples containing no residual chlorine and with soil, sediment, and sludge samples, immediately cooling to 4°C is the only preservation used. When residual chlorine is present then 3 mL of 10% aqueous sodium sulfate is added for each gallon of sample collected, followed by cooling to 4°C.

MHT Liquid samples must be extracted within 7 days and their extracts analyzed within 40 days. Concentrated waste, soil, sediment, and sludge samples must be extracted within 14 days and their extracts analyzed within 40 days.

SAMPLE PREPARATION In general, water samples are extracted at a pH of 5 to 7 with methylene chloride in a separatory funnel (EPA Method 3510). EPA Method 3520 is not recommended for the extraction of aqueous samples because the longer chain esters tend to adsorb to the glassware and consequently, their extraction recoveries may be poor. Solid samples are extracted with hexane/acetone (1:) or methylene chloride/acetone (1:1) in a Soxhlet extractor (EPA Method 3540) or with an ultrasonic extractor (EPA Method 3550). Immediately prior to extraction, spike 500 µL of the surrogate standard spiking solution into 1-L aqueous sample or 30-g solid sample. Extraction of particulate-free aqueous samples using C–18 extraction disks is an optional method that can be used.

Prior to Florisil cleanup or GC analysis, the methylene chloride and methylene chloride/acetone extracts must be exchanged to hexane. Exchange is not required for the acetonitrile extracts. Cleanup may not be necessary for extracts from a relatively clean sample matrix. Florisil Cartridge Cleanup may be used for extract cleanup.

If PCBs and organochlorine pesticides are known to be present in the sample, and if Florisil Cartridge Cleanup is considered, then two fractions are collected: Fraction 1 is eluted with 5 mL of 20% methylene chloride in hexane and Fraction 2 is eluted with 5 mL of 10% acetone in hexane. Fraction 1 contains the organochlorine pesticides and PCBs, and can be discarded. Fraction 2 contains the phthalate esters and is analyzed by GC/ECD.

QUALITY CONTROL Identify compounds in the sample by comparing the retention times of the peaks in the sample chromatogram with those of the peaks in standard chromatograms. The retention time window used to make identification is based upon measurements of actual retention time variations over the course of 10 consecutive injections.

Calibration standards are prepared at a minimum of five concentrations for each parameter of interest through dilution of the stock standard solutions with hexane. One of the concentrations should be at a concentration near, but above, the method detection limit. Prepare stock standard solutions in hexane. Stock standards should be checked frequently for signs of degradation or evaporation, especially just prior to preparing calibration standards from them. Stock standard solutions must be replaced after one year, or sooner if comparison with check standards indicates a problem. The suggested internal standards are: 2,5-dibromotoluene, 1,3,5-tribromobenzene, and α, α'-dibromo-m-xylene. The analyst can use any of the three compounds provided that they are resolved from matrix interferences. Recommended surrogate compounds are α-2,6-trichlorotoluene, 1,4-dichloronaphthalene, and 2,3,4,5,6-pentachlorotoluene.

Spike each sample, standard, and blank with surrogate compounds. Three surrogates are suggested for this method: diphenyl phthalate, diphenyl isophthalate, and dibenzyl phthalate.

The quality control check sample concentrate should contain the test compounds at 5 to 10 ng/µL. An internal standard peak area check must be performed on all samples. The internal standard must be evaluated for acceptance by determining whether the measured area for the internal standard deviates by more than 30% from the average are for the internal standard in the calibration standards. When the internal standard peak area is outside that limit, all samples that fall outside the QC criteria must be reanalyzed. Benzyl benzoate has been tested and found appropriate as an internal standard for this method.

Any compounds confirmed by two columns may also be confirmed by GC/MS. The sample extract and associated blank should be analyzed by GC/MS. A reference standard of the compound must also be analyzed by GC/MS. Include a mid-concentration calibration standard after each group of 20 samples. The response factors for the mid-concentration calibration must be within ± 15% of the average values for the multiconcentration calibration. Demonstrate through the analyses of standards that the Florisil fractionation scheme is reproducible.

REFERENCE Test Methods for Evaluating Solid Waste, Physical/Chemical Methods, SW-846, 3rd Edition, U.S. EPA, Office of Solid Waste, Washington, DC, EPA Method 8061, Nov. 1990.

Bis(2-ethylhexl) phthalate   **EPA Method 625**
CAS #117-81-7

TITLE Base/Neutrals and Acids, U.S. EPA Method 625

MATRIX This methods covers municipal and industrial wastewaters.

METHOD SUMMARY Approximately 1 L of sample is serially extracted with methylene chloride at a pH greater than 11 and again at a pH less than 2 using a separatory funnel or a continuous extractor. The methylene chloride extract is dried, concentrated to a volume of 1 mL, and analyzed by GC/MS. Qualitative identification of the parameters in the extract is performed using the retention time and the relative abundance of three characteristic masses (m/z). Qualitative analysis is performed using either external or internal standard techniques with a single characteristic m/z.

INTERFERENCES Method interferences may be caused by contaminants in solvents, reagents, glassware, and other sample processing hardware. Glassware must be scrupulously cleaned. Glassware should be heated in a muffle furnace at 400°C for 5 to 30 min. Some thermally stable materials, such as PCBs, may not be eliminated by this treatment. Solvent rinses with acetone and pesticide quality hexane may be substituted for the muffle furnace heating. Matrix interferences may be caused by contaminants that are coextracted from the sample. The base-neutral extraction may cause significantly reduced recovery of phenols. The packed gas chromatographic columns recommended for the basic fraction may not exhibit sufficient resolution for some analytes.

INSTRUMENTATION A GC/MS system with an injection port designed for on-column injection when using packed columns and for splitless injection when using capillary columns.

Column for base/neutrals: 1.8 m long × 2 mm I.D. glass, packed with 3% SP-2550 on Supelcoport (100/120 mesh) or equivalent.
Column for acids: 1.8 m long × 2 mm I.D. glass, packed with 1% SP-1240DA on Supelcoport (100/120 mesh) or equivalent.

PRECISION & ACCURACY The MDL concentrations were obtained using reagent water. The MDL actually achieved in a given analysis will vary depending on instrument sensitivity and matrix effects. This method was tested by 15 laboratories using reagent water, drinking water, surface water, and industrial wastewaters spiked at six concentrations over the range 5 to 100 µg/L. Single operator precision, overall precision, and method accuracy were found to be directly related to the concentration of the parameter matrix.

The MDL (in µg/L) in reagent water was 2.5.
The standard deviation (in µg/L based on 4 recovery measurements) was 41.1.
The range (in µg/L) for average recovery for 4 measurements was 28.9–136.8.
The range (in %) for percent recovery was 8–158.
Accuracy (in µg/L) as expected recovery for one or more measurements of a sample containing a true concentration of C was 0.84C–1.18.
Precision (in µg/L) as expected single analyst standard deviation of measurements at an average concentration found at X was 0.26X+0.73.
Overall precision (in µg/L) as expected interlaboratory standard deviation of measurements in an average concentration found at X was 0.36X+0.67.

$C =$ *True value of the concentration in µg/L.*
$X =$ *Average recovery found for measurements of samples containing a concentration at C in µg/L.*

SAMPLE PREPARATION Adjust the pH to >11 with sodium hydroxide and serially extract in a separatory funnel with methylene chloride or else in a continuous extractor. Next, adjust the pH to <2 with sulfuric acid and serially extract in a separatory

funnel with methylene chloride or else in a continuous extractor. Dry the extracts separately through a column of anhydrous sodium sulfate and then concentrate each of the extracts to 1.0 mL using a K-D apparatus.

SAMPLE COLLECTION, PRESERVATION & HANDLING- Grab samples must be collected in glass containers. All samples must be refrigerated at 4°C from the time of collection until extraction. If residual chlorine is present, add 80 mg of sodium thiosulfate/L of sample and mix well. All samples must be extracted within 7 days of collection and completely analyzed within 40 days of extraction.

QUALITY CONTROL Make an initial, one-time, demonstration of the ability to generate acceptable accuracy and precision with this method. Before processing any samples, the analyst must analyze a reagent water blank to demonstrate that interferences from the analytical system and glassware are under control. Each time a set of samples is extracted or reagents are changed, a reagent water blank must be processed. Spike and analyze a minimum of 5% of all samples to monitor and evaluate lab data quality. A QC check sample concentrate that contains each parameter of interest at a concentration of 100 µg/mL in acetone is required. PCBs and multicomponent pesticides may be omitted from this test.

After analysis of five spiked wastewater samples, calculate the average percent recovery and the standard deviation of the percent recovery. Spike all samples with the surrogate standard spiking solution and calculate the percent recovery of each surrogate compound.

REFERENCE *Federal Register,* Vol. 49, No. 209. Friday, Oct. 26, 1984.

Bis(2-ethylhexyl) phthalate   **EPA Method 1625**
CAS #117-81-7

TITLE Semivolatile Organic Compounds by Isotope Dilution GC/MS

MATRIX The compounds may be determined in waters, soils, and municipal sludges by this method.

METHOD SUMMARY This method is used to determine 176 semivolatile toxic organic pollutants associated with the CWA (as amended 1987); the RCRA (as amended 1986); the CERCLA (as amended 1986); and other compounds amenable to extraction and analysis by capillary column gas chromatography-mass spectrometry (GC/MS).

Stable isotopically-labeled analogs of the compounds of interest are added to the sample. If the solids content is less than 1%, a 1-L sample is extracted at pH 12–13, then at pH <2 with methylene chloride using continuous extraction techniques.

If the solids content is 30% or less, the sample is diluted to 1% solids with reagent water, homogenized ultrasonically, and extracted at pH 12–13, then at pH <2 with methylene chloride using continuous extraction techniques. If the solids content is greater than 30%, the sample is extracted using ultrasonic techniques.

Each extract is dried over sodium sulfate, concentrated to a volume of 5 mL, cleaned up using GPC, if necessary, and concentrated. Extracts are concentrated to 1 mL if GPC is not performed, and to 0.5 mL if GPC is performed.

An internal standard is added to the extract, and a 1-mL aliquot of the extract is injected into the GC. The compounds are separated by GC and detected by a MS. The labeled compounds serve to correct the variability of the analytical technique.

INTERFERENCES Solvents, reagents, glassware, and other sample processing hardware may yield artifacts and/or elevated baselines causing misinterpretation of chromatograms and spectra. Materials used in the analysis must be demonstrated to be free from interferences under the conditions of analysis by running method blanks initially and with each sample lot (sample started through the extraction process on a given 8-h shift, to a maximum of 20). Specific selection of reagents and purification of solvents by distillation in all glass systems may be required. Glassware and, where possible, reagents are cleaned by solvent rinse and baking at 450°C for 1-h minimum. Interferences coextracted from samples will vary considerably from source to source, depending on the diversity of the site being sampled.

INSTRUMENTATION Major instrumentation includes a GC with a splitless or on-column injection port for capillary column, a MS with 70 eV electron impact ionization, and a data system to collect and record MS data, and process it. A K-D apparatus is used to concentrate extracts.

GC Column: 30 m × 0.25 mm I.D. 5% phenyl, 94% methyl, 1% vinyl silicone bonded phased fused silica capillary column.

PRECISION & ACCURACY The detection limits of the method are usually dependent on the level of interferences rather than instrumental limitations. The limits typify the minimum quantities that can be detected with no interferences present.

The minimum level (in µg/mL) was 10. This is defined as a minimum level at which the analytical system shall give recognizable mass spectra (background corrected) and acceptable calibration points.

The MDL (in µg/kg) in low solids was 553 and in high solids was 1310; these were determined in digested sludge (low solids) and in filter cake or compost (high solids).

Note: Background levels of this compound were present in the sludge tested, resulting in higher than expected MDLs. The MDL for this compound is expected to be approximately 50 µg/kg with no interferences present.

The labeled and native compound initial precision as standard deviation (in µg/L) was 31.

The labeled and native compound initial accuracyas average recovery (in µg/L) was 69–220.

SAMPLE COLLECTION, PRESERVATION & HANDLING
Collect samples in glass containers. Aqueous samples which

flow freely are collected in refrigerated bottles using automatic sampling equipment. Solid samples are collected as grab samples using widemouth jars. Maintain samples at 0 to 4°C from the time of collection until extraction. If residual chlorine is present in aqueous samples, add 80 mg sodium thiosulfate/L of water. Begin sample extraction within 7 days of collection, and analyze all extracts within 40 days of extraction.

SAMPLE PREPARATION Samples containing 1% solids or less are extracted directly using continuous liquid-liquid extraction techniques. Samples containing 1 to 30% solids are diluted to the 1% level with reagent water and extracted using continuous liquid-liquid extraction techniques. Samples containing greater than 30% solids are extracted using ultrasonic techniques.

Base/neutral extraction — Adjust the pH of the waters in the extractors to 12–13 with 6 N NaOH. Extract with methylene chloride for 24–48 h.

Acid extraction — Adjust the pH of the waters in the extractors to 2 or less using 6 N sulfuric acid. Extract with methylene chloride for 24–48 h.

Ultrasonic extraction of high solids samples — Add anhydrous sodium sulfate to the sample and QC aliquot(s). Add acetone:methylene chloride (1:1) to the sample and mix thoroughly

Concentrate extracts using a K-D apparatus.

QUALITY CONTROL The analyst is permitted to modify this method to improve separations or lower the costs of measurements, provided all performance specifications are met. Analyses of blanks are required to demonstrate freedom from contamination. When results of spikes indicate atypical method performance for samples, the samples are diluted to bring method performance within acceptable limits.

For low solids (aqueous samples), extract, concentrate, and analyze two sets of four 1-L aliquots (8 aliquots total) of the precision and recovery standard. For high solids samples, two sets of four 30-g aliquots of the high solids reference matrix are used.

Spike all samples with labeled compounds to assess method performance. Compute percent recoveryof the labeled compounds using the internal standard method. Compare the labeled compound recovery for each compound with the corresponding labeled compound recovery.

Reagent water and high solids reference matrix blanks are analyzed to demonstrate freedom from contamination. Extract and concentrate a 1-L reagent water blank or a high solids reference matrix blank with each sample's lot (samples started through the extraction process on the same 8-h shift, to a maximum of 20 samples).

Field replicates may be collected to determine the precision of the sampling technique, and spiked samples may be required to determine the accuracy of the analysis when the internal standard method is used.

REFERENCE Semivolatile Organic Compounds by Isotope Dilution GC/MS. Office of Water Regulation and Standards, U.S. EPA Industrial Technology Division, Washington, DC, EPA Method 1625, Rev. C, June 1989 (contact W.A. Telliard, U.S. EPA, Office of Water Regulations and Standards, 401 M St., SW, Washington, DC, 20460. Phone: 202-382-7131).

Bis(2-ethylhexyl) phthalate   **EPA Method 1625**
CAS #117-81-7

TITLE Semivolatile Organic Compounds by Isotope Dilution GC/MS

MATRIX The compounds may be determined in waters, soils, and municipal sludges by this method.

METHOD SUMMARY This method is used to determine 176 semivolatile toxic organic pollutants associated with the CWA (as amended 1987); the RCRA (as amended 1986); the CERCLA (as amended 1986); and other compounds amenable to extraction and analysis by capillary column gas chromatography-mass spectrometry (GC/MS).

Stable isotopically-labeled analogs of the compounds of interest are added to the sample. If the solids content is less than 1%, a 1-L sample is extracted at pH 12–13, then at pH <2 with methylene chloride using continuous extraction techniques.

If the solids content is 30% or less, the sample is diluted to 1% solids with reagent water, homogenized ultrasonically, and extracted at pH 12–13, then at pH <2 with methylene chloride using continuous extraction techniques. If the solids content is greater than 30%, the sample is extracted using ultrasonic techniques.

Each extract is dried over sodium sulfate, concentrated to a volume of 5 mL, cleaned up using GPC, if necessary, and concentrated. Extracts are concentrated to 1 mL if GPC is not performed, and to 0.5 mL if GPC is performed.

An internal standard is added to the extract, and a 1-mL aliquot of the extract is injected into the GC. The compounds are separated by GC and detected by a MS. The labeled compounds serve to correct the variability of the analytical technique.

INTERFERENCES Solvents, reagents, glassware, and other sample processing hardware may yield artifacts and/or elevated baselines causing misinterpretation of chromatograms and spectra. Materials used in the analysis must be demonstrated to be free from interferences under the conditions of analysis by running method blanks initially and with each sample lot (sample started through the extraction process on a given 8-h shift, to a maximum of 20). Specific selection of reagents and purification of solvents by distillation in all glass systems may be required. Glassware and, where possible, reagents are cleaned by solvent rinse and baking at 450°C for 1-h minimum. Interferences coextracted from samples will vary considerably from source to source, depending on the diversity of the site being sampled.

INSTRUMENTATION Major instrumentation includes a GC with a splitless or on-column injection port for capillary column, a MS with 70 eV electron impact ionization, and a data system to collect and record MS data, and process it. A K-D apparatus is used to concentrate extracts.

GC Column: 30 m × 0.25 mm I.D. 5% phenyl, 94% methyl, 1% vinyl silicone bonded phased fused silica capillary column.

PRECISION & ACCURACY The detection limits of the method are usually dependent on the level of interferences rather than instrumental limitations. The limits typify the minimum quantities that can be detected with no interferences present.

The minimum level (in µg/mL) was 10. This is defined as a minimum level at which the analytical system shall give recognizable mass spectra (background corrected) and acceptable calibration points.

The MDL (in µg/kg) in low solids was 553 and in high solids was 1310; these were determined in digested sludge (low solids) and in filter cake or compost (high solids).

Note: Background levels of this compound were present in the sludge tested, resulting in higher than expected MDLs. The MDL for this compound is expected to be approximately 50 µg/kg with no interferences present.

The labeled and native compound initial precision as standard deviation (in µg/L) was 31.

The labeled and native compound initial accuracy as average recovery (in µg/L) was 69–220.

SAMPLE COLLECTION, PRESERVATION & HANDLING
Collect samples in glass containers. Aqueous samples which flow freely are collected in refrigerated bottles using automatic sampling equipment. Solid samples are collected as grab samples using widemouth jars. Maintain samples at 0 to 4°C from the time of collection until extraction. If residual chlorine is present in aqueous samples, add 80 mg sodium thiosulfate/L of water. Begin sample extraction within 7 days of collection, and analyze all extracts within 40 days of extraction.

SAMPLE PREPARATION Samples containing 1% solids or less are extracted directly using continuous liquid-liquid extraction techniques. Samples containing 1 to 30% solids are diluted to the 1% level with reagent water and extracted using continuous liquid-liquid extraction techniques. Samples containing greater than 30% solids are extracted using ultrasonic techniques.

Base/neutral extraction — Adjust the pH of the waters in the extractors to 12–13 with 6 *N* NaOH. Extract with methylene chloride for 24–48 h.

Acid extraction — Adjust the pH of the waters in the extractors to 2 or less using 6 *N* sulfuric acid. Extract with methylene chloride for 24–48 h.

Ultrasonic extraction of high solids samples — Add anhydrous sodium sulfate to the sample and QC aliquot(s). Add acetone:methylene chloride (1:1) to the sample and mix thoroughly

Concentrate extracts using a K-D apparatus.

QUALITY CONTROL The analyst is permitted to modify this method to improve separations or lower the costs of measurements, provided all performance specifications are met. Analyses of blanks are required to demonstrate freedom from contamination. When results of spikes indicate atypical method performance for samples, the samples are diluted to bring method performance within acceptable limits.

For low solids (aqueous samples), extract, concentrate, and analyze two sets of four 1-L aliquots (8 aliquots total) of the precision and recovery standard. For high solids samples, two sets of four 30-g aliquots of the high solids reference matrix are used.

Spike all samples with labeled compounds to assess method performance. Compute percent recovery of the labeled compounds using the internal standard method. Compare the labeled compound recovery for each compound with the corresponding labeled compound recovery.

Reagent water and high solids reference matrix blanks are analyzed to demonstrate freedom from contamination. Extract and concentrate a 1-L reagent water blank or a high solids reference matrix blank with each sample's lot (samples started through the extraction process on the same 8-h shift, to a maximum of 20 samples).

Field replicates may be collected to determine the precision of the sampling technique, and spiked samples may be required to determine the accuracy of the analysis when the internal standard method is used.

REFERENCE Semivolatile Organic Compounds by Isotope Dilution GC/MS. Office of Water Regulation and Standards, U.S. EPA Industrial Technology Division, Washington, DC, EPA Method 1625, Rev. C, June 1989 (contact W.A. Telliard, U.S. EPA, Office of Water Regulations and Standards, 401 M St., SW, Washington, DC, 20460. Phone: 202-382-7131).

Bis(2-ethylhexyl) phthalate   EPA Method 8270
CAS #117-81-7

TITLE Semivolatile Organic Compounds by GC/MS

MATRIX This method is used to determine the concentration of semivolatile organic compounds in extracts prepared from all types of solid waste matrices, soils, and groundwater. Although surface waters are not specifically mentioned, this method should be applicable to water samples from rivers, lakes, etc.

METHOD SUMMARY This method covers 259 semivolatile organic compounds. In very limited applications direct injection of the sample into the GC/MS system may be appropriate, but this results in very high detection limits (approximately 10,000 µg/L). Typically, a 1-L liquid sample, containing surrogate, and matrix spiking standards, is extracted in a continuous extractor first under acid conditions and then under basic conditions. Typically 30 g of a solid sample, containing surrogate, and matrix spiking standards, is extracted ultrasonically. After concentrating the extract to 1 mL it is spiked with 10 µL of an internal standard solution just prior to analysis by GC/MS. The volume injected should contain about 100 ng of base/neutral and 200 ng of acid surrogates (for a 1-µL injection). Analysis is performed by GC/MS using a capillary GC column.

INTERFERENCES Raw GC/MS data from all blanks, samples, and spikes must be evaluated for interferences. Contamination by carryover can occur whenever high-concentration and low-concentration samples are sequentially analyzed. To reduce carryover, the sample syringe must be rinsed out between samples with solvent. Whenever an unusually concentrated sample is encountered, it should be followed by the analysis of blank solvent to check for cross-contamination.

INSTRUMENTATION A GC/MS and a data system are required. The GC column used is a 30 m × 0.25 mm I.D. (or 0.32 mm I.D.) 1um film thickness silicone-coated fused silica capillary column. A continuous liquid-liquid extractor equipped with Teflon® or glass connection joints and stopcocks requiring no lubrication, a K-D concentrating apparatus, water bath, and an ultrasonic disrupter with a minimum power of 300 W and with pulsing capability are also required.

PRECISION & ACCURACY The estimated quantitation limit (EQL) of Method 8270B for determining an individual compound is approximately 1 mg/kg (wet weight) for soil or sediment samples, 1–200 mg/kg for wastes (dependent on matrix and method of preparation), and 10 µg/L for groundwater samples. EQLs will be proportionately higher for sample extracts that require dilution to avoid saturation of the detector.

The EQL(b) for groundwater in µg/L is not listed.
The EQL (a, b) for low concentrations in soil and sediment in µg/kg is not listed.
Accuracy as µg/L is $0.84C - 1.18$.
Overall precision in µg/L is $0.36X + 0.67$.

(a) *EQLs listed for soil/sediment are based on wet weight. Normally data is reported in a dry-weight basis; therefore, EQLs will be higher based on the % dry weight of each sample. This calculation is based on a 30-g sample and gel permeation chromatography cleanup.*
(b) *Sample EQLs are highly matrix-dependent. The EQLs are provided for guidance and may not always be achievable.*
$C =$ *True value for concentration, in µg/L.*
$X =$ *Average recovery found for measurements of samples containing a concentration of C, in µg/L.*

ESTIMATED QUANTITATION LIMIT FOR OTHER MATRICES

Other Matrices	Factor (a)
High-concentration soil and sludges by sonicator	7.5
Non-water miscible waste	75

(a) *EQL for other matrices = [EQL for low soil/sediment] × [Factor]. This estimated EQL is similar to an EPA "Practical Quantitation Limit."*

SAMPLING METHOD
Liquid samples — Use a 1 or 2½ gallon amber glass bottle with a screw-top Teflon®-lined cover that has been prewashed with detergent and rinsed with distilled water and methanol (or isopropanol).

Soils, sediments, or sludges — Use an 8-oz. widemouth glass with a screw-top Teflon®-lined cover that has been prewashed with detergent and rinsed with distilled water and methanol (or isopropanol).

SAMPLE PRESERVATION
Liquid samples — If residual chlorine is present, add 3 mL of 10% sodium thiosulfate per gallon, cool to 4°C and store in a solvent-free refrigerator until analysis; if chlorine is not present, then eliminate the sodium thiosulfate addition.

Soils, sediments, or sludges — Cool samples to 4°C and store in a solvent-free refrigerator.

MHT Liquid samples must be extracted within 7 days and the extracts analyzed within 40 days. Soils, sediments, or sludges may be stored for a maximum of 14 days and the extracts analyzed within 40 days.

SAMPLE PREPARATION
Liquid samples — Transfer 1 L quantitatively to a continuous extractor. If high concentrations are anticipated, a smaller volume may be used and then diluted with organic-free reagent water to 1 L. Adjust pH, if necessary, to pH <2 using 1:1 (V/V) sulfuric acid. Pipette 1.0 mL of a surrogate standard spiking solution into each sample. For the sample in each analytical batch selected for spiking, add 1.0 mL of a matrix spiking standard. For base/neutral acid analysis, the amount of the surrogates and matrix spiking compounds added to the sample should result in a final concentration of 100 ng/µL of each analyte in the extract to be analyzed (assuming a 1-µL injection). Extract with methylene chloride for 18–24 h. Next, adjust the pH of the aqueous phase to pH >11 using 10 N sodium hydroxide and extract it with methylene chloride again for 18–24 h. Dry the extract through a column containing anhydrous sodium sulfate and concentrate it to 1 mL using a K-D concentrator.

Soils, sediments, or sludges — Use 30 g of sample. Nonporous or wet samples (gummy or clay type) that do not have a free-flowing sandy texture must be mixed with anhydrous sodium sulfate until the sample is free flowing. Add 1 mL of surrogate standards to all samples, spikes, standards, and blanks. For the sample in each analytical batch selected for spiking, add 1.0 mL of a matrix spiking standard. For base/neutral acid analysis, the amount added of the surrogates and matrix spiking compounds should result in a final concentration of 100 ng/µL of each base/neutral analyte and 200 ng/µL of each acid analyte in the extract to be analyzed (assuming a 1-µL injection). Immediately add a 100 mL mixture of 1:1 methylene chloride:acetone and extract the sample ultrasonically for 3 min and then decant or filter the extracts. Repeat the extraction two or more times. Dry the extract using a column with anhydrous sodium sulfate and concentrate it to 1 mL in a K-D concentrator.

QUALITY CONTROL A methylene chloride solution containing 50 ng/µL of decafluorotriphenylphosphine (DFTPP) is used for tuning the GC/MS system each 12-h shift. A system performance check also must be made during every 12-h shift. A standard containing 50 ng/µL each of 4,4′-DDT, pentachlorophenol, and benzidine is required to verify injection port inertness and GC column performance. A calibration standard at mid-concentration, containing each compound of interest,

including all required surrogates, must be performed every 12 h during analysis. After the system performance check is met, calibration check compounds (CCCs) are used to check the validity of the initial calibration.

The internal standard responses and retention times in the calibration check standard must be evaluated immediately after or during data acquisition. If the retention time for any internal standard changes by more than 30 seconds from the last check calibration (12 h), the chromatographic system must be inspected for malfunctions and corrections must be made, as required. If the electron ionization current plot (EICP) area for any of the internal standards changes by a factor of two from the last daily calibration standard check, the mass spectrometer must be inspected for malfunctions and corrections must be made, as appropriate.

Demonstrate, through the analysis of a reagent water blank, that interferences from the analytical system, glassware, and reagents are under control. The blank samples should be carried through all stages of the sample preparation and measurement steps. For each analytical batch (up to 20 samples), a reagent blank, matrix spike, and matrix spike duplicate/duplicate must be analyzed (the frequency of the spikes may be different for different monitoring programs). The blank and spiked samples must be carried through all stages of the sample preparation and measurement steps. A QC reference sample concentrate containing each analyte at a concentration of 100 mg/L in methanol is required.

REFERENCE Test Methods for Evaluating Solid Waste (SW-846). U.S. EPA 1983, Method 8270B, Rev. 2, Nov. 1990. Office of Solid Waste, Washington, DC.

Bis(2-ethylhexyl) phthalate   EPA Method 8061
CAS #117-81-7

TITLE Phthalate Esters by Capillary Gas Chromatography With Electron Capture Detection (GC/ECD)

MATRIX This method covers aqueous and solid matrices. This includes a wide variety such as drinking water, groundwater, industrial wastewaters, surface waters, soils, solids, and sediments.

METHOD SUMMARY This method is used to determine the identities and concentrations of phthalate esters in liquid, solid and sludge matrices. When used to analyze for any or all of the target analytes, compound identification should be supported by at least one additional qualitative technique. This method describes conditions for parallel column, dual electron capture detector analysis, which fulfills the above requirement. Alternatively, GC/MS could be used for compound confirmation.

A measured volume or weight of sample (approximately 1 L for liquids, 10 to 30 g for solids and sludges) is extracted by using the appropriate sample extraction technique specified in EPA Method 3510, EPA Method 3540, and EPA Method 3550. After cleanup, the extract is analyzed by GC/ECD.

INTERFERENCES The sensitivity of this method usually depends on the level of interferences rather than on instrumental limitations. If interferences prevent detection of the analytes, cleanup of the sample extracts is necessary. Either EPA Method 3610 or EPA Method 3620 alone or followed by EPA Method 3660, Sulfur Cleanup, may be used to eliminate interferences in the analysis. EPA Method 3640, Gel Permeation Cleanup, is applicable for samples that contain high amounts of lipids and waxes.

Interferences coextracted from the samples will vary considerably from waste to waste. Glassware must be scrupulously clean. All glassware require treatment in a muffle furnace at 400°C for 2 to 4 h, or thorough rinsing with pesticide-grade solvent, prior to use. Volumetric glassware should not be heated in a muffle furnace. Storage of glassware in the lab introduces contamination, even if the glassware is wrapped in aluminum foil. Sodium sulfate, Florisil, and alumina may be contaminated with phthalate esters and, therefore, use of these materials in sample cleanup should be employed cautiously. If these materials are used, they must be obtained packaged in glass. Heating at 400°C for sodium sulfate, 320°C for Florisil, and 210°C for alumina is recommended. Glass wool used in any step of sample preparation should be a specially treated pyrex wool, pesticide grade, and must be baked at 400°C for 4 h immediately prior to use.

Paper thimbles and filter paper must be exhaustively washed with the solvent that will be used in the sample extraction. Soxhlet extraction of paper thimbles and filter paper for 12 h with fresh solvent should be repeated for a minimum of three times. Method blanks should be obtained before any of the precleaned thimbles or filter papers are used.

INSTRUMENTATION Gas chromatograph suitable for on-column and split/splitless injections.

- Column 1: 30 m × 0.53 mm ID, 5% phenyl/95% methyl silicone fused-silica open tubular column, DB-5, 1.5 µg film thickness.
- Column 2: 30 m × 0.53 mm ID, 14% cyanopropyl phenyl silicone fused-silica open tubular column, DB-1701, 1.0 µg film thickness.

A dual electron capture detector (ECD) is used. A Kuderna-Danish (K-D) apparatus is required along with a vacuum manifold consisting of individually adjustable, easily accessible flow-control valves for up to 24 cartridges, sample rack, chemically resistant cover and seals, heavy-duty glass basin, removable stainless steel solvent guides, built-in vacuum gauge and valve. Also, 6-mL, 1-g solid-phase extraction cartridges, LC-Florisil or equivalent, prepackaged, ready to use will be needed.

PRECISION & ACCURACY The MDL actually achieved in a given analysis will vary, as it is dependent on instrument sensitivity and matrix effects. This method has been tested in a single lab. Single-operator precision, overall precision, and method accuracy were found to be related to the concentration of the compounds and the type of matrix.

ESTIMATED QUANTITATION LIMIT (EQL) FACTORS FOR VARIOUS MATRICES (a)

Matrix	Factor (b)
Groundwater	10
Low-concentration soil by ultrasonic extraction with GPC cleanup	670
High-concentration soil and sludges by ultrasonic extraction	10,000
Waste not miscible with water	100,000

(a) Sample EQLs are highly matrix-dependent.
(b) EQL = [Method detection limit] × [Factor]. For non-aqueous samples, the factor is on a wet-weight basis.

The MDL using 7 replicate determinations and a spike concentration of 100 µg/L was 270 ng/L.

The average recovery from HPLC-grade water using 4 determinatons and a spike concentration of 100 µg/L was 91.3%.

The precision (as RSD) from HPLC-grade water using 4 determinatons and a spike concentration of 100 µg/L was 7.4%.

The average recovery from groundwater using 4 determinatons and a spike concentration of 100 µg/L was 96.3%.

The precision (as RSD) from groundwater using 4 determinatons and a spike concentration of 100 µg/L was 7.9%.

The average recovery (in %) with %RSD (in parenthesis) from 3 determinations and a spike concentration of 20 µg/L in water was 81.4 (4.1) using 3M Empore Disks and EPA Method 8061.

The average recovery (in %) with %RSD (in parenthesis) from 3 determinations and a spike concentration of 20 µg/L in leachate was 93.0 15.0) using 3M Empore Disks and EPA Method 8061.

The average recovery (in %) with %RSD (in parenthesis) from 3 determinations and a spike concentration of 20 µg/L in estuarine groundwater was 90.5 (4.9) using 3M Empore Disks and EPA Method 8061.

The average recovery (in %) with %RSD (in parenthesis) from 3 determinations and a spike concentration of 1 mg/kg in estuarine sediment was not determined (matrix interferant) after sulfur cleanup with EPA Method 3660.

The average recovery (in %) with %RSD (in parenthesis) from 3 determinations and a spike concentration of 1 mg/kg in municipal sludge was76.6 (10.6).

The average recovery (in %) with %RSD (in parenthesis) from 3 determinations and a spike concentration of 1 mg/kg in sandy loam soil was 99.2 (25.3).

SAMPLE COLLECTION, PRESERVATION & HANDLING

Containers used to collect samples for the determination of semivolatile organic compounds should be soap and water washed followed by methanol (or isopropanol) rinsing. The sample containers should be of glass or Teflon® and have screw-top covers with Teflon® liners. Sample containers should be filled with careto prevent any portion of the collected sample coming in contact with the sampler's gloves.

No preservation is used with concentrated waste samples. With liquid samples containing no residual chlorine and with soil, sediment, and sludge samples, immediately cooling to 4°C is the only preservation used. When residual chlorine is present then 3 mL of 10% aqueous sodium sulfate is added for each gallon of sample collected, followed by cooling to 4°C.

MHT Liquid samples must be extracted within 7 days and their extracts analyzed within 40 days. Concentrated waste, soil, sediment, and sludge samples must be extracted within 14 days and their extracts analyzed within 40 days.

SAMPLE PREPARATION In general, water samples are extracted at a pH of 5 to 7 with methylene chloride in a separatory funnel (EPA Method 3510). EPA Method 3520 is not recommended for the extraction of aqueous samples because the longer chain esters tend to adsorb to the glassware and consequently, their extraction recoveries may be poor. Solid samples are extracted with hexane/acetone (1:) or methylene chloride/acetone (1:1) in a Soxhlet extractor (EPA Method 3540) or with an ultrasonic extractor (EPA Method 3550). Immediately prior to extraction, spike 500 µL of the surrogate standard spiking solution into 1-L aqueous sample or 30-g solid sample. Extraction of particulate-free aqueous samples using C–18 extraction disks is an optional method that can be used.

Prior to Florisil cleanup or GC analysis, the methylene chloride and methylene chloride/acetone extracts must be exchanged to hexane. Exchange is not required for the acetonitrile extracts. Cleanup may not be necessary for extracts from a relatively clean sample matrix. Florisil Cartridge Cleanup may be used for extract cleanup.

If PCBs and organochlorine pesticides are known to be present in the sample, and if Florisil Cartridge Cleanup is considered, then two fractions are collected: Fraction 1 is eluted with 5 mL of 20% methylene chloride in hexane and Fraction 2 is eluted with 5 mL of 10% acetone in hexane. Fraction 1 contains the organochlorine pesticides and PCBs, and can be discarded. Fraction 2 contains the phthalate esters and is analyzed by GC/ECD.

QUALITY CONTROL Identify compounds in the sample by comparing the retention times of the peaks in the sample chromatogram with those of the peaks in standard chromatograms. The retention time window used to make identification is based upon measurements of actual retention time variations over the course of 10 consecutive injections.

Calibration standards are prepared at a minimum of five concentrations for each parameter of interest through dilution of the stock standard solutions with hexane. One of the concentrations should be at a concentration near, but above, the method detection limit. Prepare stock standard solutions in hexane. Stock standards should be checked frequently for signs of degradation or evaporation, especially just prior to preparing calibration standards from them. Stock standard solutions must be replaced after one year, or sooner if comparison with check standards indicates a problem. The suggested internal standards are: 2,5-dibromotoluene, 1,3,5-tribromobenzene, and α, α'-dibromo-m-xylene. The analyst can use any of the three compounds provided that they are resolved from matrix interferences. Recommended surrogate compounds are α-2,6-trichlorotoluene, 1,4-dichloronaphthalene, and 2,3,4,5,6-pentachlorotoluene.

Spike each sample, standard, and blank with surrogate compounds. Three surrogates are suggested for this method: diphenyl phthalate, diphenyl isophthalate, and dibenzyl phthalate.

The quality control check sample concentrate should contain the test compounds at 5 to 10 ng/µL. An internal standard peak area check must be performed on all samples. The internal standard must be evaluated for acceptance by determining whether the measured area for the internal standard deviates by more than 30% from the average are for the internal standard in the calibration standards. When the internal standard peak area is outside that limit, all samples that fall outside the QC criteria must be reanalyzed. Benzyl benzoate has been tested and found appropriate as an internal standard for this method.

Any compounds confirmed by two columns may also be confirmed by GC/MS. The sample extract and associated blank should be analyzed by GC/MS. A reference standard of the compound must also be analyzed by GC/MS. Include a mid-concentration calibration standard after each group of 20 samples. The response factors for the mid-concentration calibration must be within ± 15% of the average values for the multiconcentration calibration. Demonstrate through the analyses of standards that the Florisil fractionation scheme is reproducible.

REFERENCE Test Methods for Evaluating Solid Waste, Physical/Chemical Methods, SW-846, 3rd Edition, U.S. EPA, Office of Solid Waste, Washington, DC, EPA Method 8061, Nov. 1990.

Bis(2-ethylhexyl) phthalate   **EPA Method 8060**
CAS #117-81-7

TITLE Phthalate Esters

MATRIX Groundwater, soils, sludges, water miscible liquid wastes, and non-water miscible wastes.

APPLICATION This method is used for theanalys is of 6 phthalate esters.

Samples are extracted, concentrated and analyzed using direct injection of both neat and diluted organic liquids into a gas chromatograph. Analytes are detected by a flame ionization detector (FID) or an electron capture detector (ECD). Groundwater samples should be determined by ECD. The method provides an optional GC column which is used for analyte confirmation and that may help resolve analytes from interferences.

INTERFERENCES Solvents, reagents and glassware may introduce artifacts. Plastics, in particular, must be avoided. Other interferences may come from coextracted compounds from samples. There can be carryover contamination with high- and low-level samples.

INSTRUMENTATION GC capable of on-column injections and a flame with detector (FID) or electron capture detector (ECD). Column 1: 1.8 m by 4 mm with 1.5% SP-2250/1.95% SP-2401 on Supelcoport. Column 2: 1.8 m by 4 mm with 3% OV-1 on supelcoport.

RANGE 0.7 to 106 µg/L

MDL 20 µg/L (FID) and 2.0 µg/L (ECD)

PQL FACTORS FOR MULTIPLYING × FID MDL VALUE

Matrix	Multiplication Factor
Groundwater	10
Low-level soil by sonication with GPC cleanup	670
High-level soil and sludge by sonication	10,000
Non-water miscible waste	100,000

PRECISION 0.73–0.17 µg/L (overall precision using FID)

ACCURACY 0.53C + 2.02 µg/L (as recovery using FID)

SAMPLING METHOD Use 8-oz. widemouth glass bottles with Teflon®-lined caps for concentrated waste samples, soils, sediments, and sludges. Use 1 or 2½ gallon amber glass bottles with Teflon®-lined caps for liquid (water) samples.

STABILITY Cool soil, sediment, sludge, and liquid samples to 4°C. If residual chlorine is present in liquid samples add 3 mL of 10% sodium thiosulfate per gallon of sample and cool to 4°C.

MHT 14 days for concentrated waste, soil, sediment, or sludge; 7 days for liquid samples; all extracts must be analyzed within 40 days.

QUALITY CONTROL A quality control check sample concentrate containing each analyte of interest is required. The QC check sample concentrate may be prepared from pure standard materials or purchased as certified solutions. Use appropriate trip, matrix, control site, method, reagent, and solvent blanks. Internal, surrogate, and five concentration level calibration standards are used. The quality control check sample concentrate should contain benzo(ghi)perylene at 10 µg/mL in acetonitrile.d solvent blanks. Internal, surrogate and five concentration level calibration standards are used. The QC check sample concentrate should contain this compound at 50 µg/mL in acetone.

REFERENCE Method 8060, SW-846, 3rd ed., Nov.1986.

Bis(2-methoxyethyl) phthalate   **EPA Method 8061**
CAS #117-82-8

TITLE Phthalate Esters by Capillary Gas Chromatography With Electron Capture Detection (GC/ECD)

MATRIX This method covers aqueous and solid matrices. This includes a wide variety such as drinking water, groundwater, industrial wastewaters, surface waters, soils, solids, and sediments.

METHOD SUMMARY This method is used to determine the identities and concentrations of phthalate esters in liquid, solid and sludge matrices. When used to analyze for any or all of the target analytes, compound identification should be supported by at least one additional qualitative technique. This method describes conditions for parallel column, dual electron capture detector analysis, which fulfills the above requirement. Alternatively, GC/MS could be used for compound confirmation.

A measured volume or weight of sample (approximately 1 L for liquids, 10 to 30 g for solids and sludges) is extracted by using the appropriate sample extraction technique specified in EPA Method 3510, EPA Method 3540, and EPA Method 3550. After cleanup, the extract is analyzed by GC/ECD.

INTERFERENCES The sensitivity of this method usually depends on the level of interferences rather than on instrumental limitations. If interferences prevent detection of the analytes, cleanup of the sample extracts is necessary. Either EPA Method 3610 or EPA Method 3620 alone or followed by EPA Method 3660, Sulfur Cleanup, may be used to eliminate interferences in the analysis. EPA Method 3640, Gel Permeation Cleanup, is applicable for samples that contain high amounts of lipids and waxes.

Interferences coextracted from the samples will vary considerably from waste to waste. Glassware must be scrupulously clean. All glassware require treatment in a muffle furnace at 400°C for 2 to 4 h, or thorough rinsing with pesticide-grade solvent, prior to use. Volumetric glassware should not be heated in a muffle furnace. Storage of glassware in the lab introduces contamination, even if the glassware is wrapped in aluminum foil. Sodium sulfate, Florisil, and alumina may be contaminated with phthalate esters and, therefore, use of these materials in sample cleanup should be employed cautiously. If these materials are used, they must be obtained packaged in glass. Heating at 400°C for sodium sulfate, 320°C for Florisil, and 210°C for alumina is recommended. Glass wool used in any step of sample preparation should be a specially treated pyrex wool, pesticide grade, and must be baked at 400°C for 4 h immediately prior to use.

Paper thimbles and filter paper must be exhaustively washed with the solvent that will be used in the sample extraction. Soxhlet extraction of paper thimbles and filter paper for 12 h with fresh solvent should be repeated for a minimum of three times. Method blanks should be obtained before any of the precleaned thimbles or filter papers are used.

INSTRUMENTATION Gas chromatograph suitable for on-column and split/splitless injections.

Column 1: 30 m × 0.53 mm ID, 5% phenyl/95% methyl silicone fused-silica open tubular column, DB-5, 1.5 µg film thickness.
Column 2: 30 m × 0.53 mm ID, 14% cyanopropyl phenyl silicone fused-silica open tubular column, DB-1701, 1.0 µg film thickness.

A dual electron capture detector (ECD) is used. A Kuderna-Danish (K-D) apparatus is required along with a vacuum manifold consisting of individually adjustable, easily accessible flow-control valves for up to 24 cartridges, sample rack, chemically resistant cover and seals, heavy-duty glass basin, removable stainless steel solvent guides, built-in vacuum gauge and valve. Also, 6-mL, 1-g solid-phase extraction cartridges, LC-Florisil or equivalent, prepackaged, ready to use will be needed.

PRECISION & ACCURACY The MDL actually achieved in a given analysis will vary, as it is dependent on instrument sensitivity and matrix effects. This method has been tested in a single lab. Single-operator precision, overall precision, and method accuracy were found to be related to the concentration of the compounds and the type of matrix.

ESTIMATED QUANTITATION LIMIT (EQL) FACTORS FOR VARIOUS MATRICES (a)

Matrix	Factor (b)
Groundwater	10
Low-concentration soil by ultrasonic extraction with GPC cleanup	670
High-concentration soil and sludges by ultrasonic extraction	10,000
Waste not miscible with water	100,000

(a) Sample EQLs are highly matrix-dependent.
(b) EQL = [Method detection limit] × [Factor]. For non-aqueous samples, the factor is on a wet-weight basis.

The MDL using 7 replicate determinations and a spike concentration of 100 µg/L was 510 ng/L.
The average recovery from HPLC-grade water using 4 determinatons and a spike concentration of 100 µg/L was 107%.
The precision (as RSD) from HPLC-grade water using 4 determinatons and a spike concentration of 100 µg/L was 13.6%.
The average recovery from groundwater using 4 determinatons and a spike concentration of 100 µg/L was 113%.
The precision (as RSD) from groundwater using 4 determinatons and a spike concentration of 100 µg/L was 2.8%.
The average recovery (in %) with %RSD (in parenthesis) from 3 determinations and a spike concentration of 20 µg/L in water was 73.8 (1.0) using 3M Empore Disks and EPA Method 8061.
The average recovery (in %) with %RSD (in parenthesis) from 3 determinations and a spike concentration of 20 µg/L in leachate was 87.2 (21.7) using 3M Empore Disks and EPA Method 8061.
The average recovery (in %) with %RSD (in parenthesis) from 3 determinations and a spike concentration of 20 µg/L in estuarine groundwater was 82.4 (4.4) using 3M Empore Disks and EPA Method 8061.
The average recovery (in %) with %RSD (in parenthesis) from 3 determinations and a spike concentration of 1 mg/kg in estuarine sediment was 26.6 (26.8) after sulfur cleanup with EPA Method 3660.
The average recovery (in %) with %RSD (in parenthesis) from 3 determinations and a spike concentration of 1 mg/kg in municipal sludge was 72.7 (8.3).
The average recovery (in %) with %RSD (in parenthesis) from 3 determinations and a spike concentration of 1 mg/kg in sandy loam soil was not determined (matrix interferant).

SAMPLE COLLECTION, PRESERVATION & HANDLING
Containers used to collect samples for the determination of semivolatile organic compounds should be soap and water washed followed by methanol (or isopropanol) rinsing. The sample containers should be of glass or Teflon® and have screw-top covers with Teflon® liners. Sample containers should be

filled with careto prevent any portion of the collected sample coming in contact with the sampler's gloves.

No preservation is used with concentrated waste samples. With liquid samples containing no residual chlorine and with soil, sediment, and sludge samples, immediately cooling to 4°C is the only preservation used. When residual chlorine is present then 3 mL of 10% aqueous sodium sulfate is added for each gallon of sample collected, followed by cooling to 4°C.

MHT   Liquid samples must be extracted within 7 days and their extracts analyzed within 40 days. Concentrated waste, soil, sediment, and sludge samples must be extracted within 14 days and their extracts analyzed within 40 days.

SAMPLE PREPARATION   In general, water samples are extracted at a pH of 5 to 7 with methylene chloride in a separatory funnel (EPA Method 3510). EPA Method 3520 is not recommended for the extraction of aqueous samples because the longer chain esters tend to adsorb to the glassware and consequently, their extraction recoveries may be poor. Solid samples are extracted with hexane/acetone (1:) or methylene chloride/acetone (1:1) in a Soxhlet extractor (EPA Method 3540) or with an ultrasonic extractor (EPA Method 3550). Immediately prior to extraction, spike 500 µL of the surrogate standard spiking solution into 1-L aqueous sample or 30-g solid sample. Extraction of particulate-free aqueous samples using C–18 extraction disks is an optional method that can be used.

Prior to Florisil cleanup or GC analysis, the methylene chloride and methylene chloride/acetone extracts must be exchanged to hexane. Exchange is not required for the acetonitrile extracts. Cleanup may not be necessary for extracts from a relatively clean sample matrix. Florisil Cartridge Cleanup may be used for extract cleanup.

If PCBs and organochlorine pesticides are known to be present in the sample, and if Florisil Cartridge Cleanup is considered, then two fractions are collected: Fraction 1 is eluted with 5 mL of 20% methylene chloride in hexane and Fraction 2 is eluted with 5 mL of 10% acetone in hexane. Fraction 1 contains the organochlorine pesticides and PCBs, and can be discarded. Fraction 2 contains the phthalate esters and is analyzed by GC/ECD.

QUALITY CONTROL   Identify compounds in the sample by comparing the retention times of the peaks in the sample chromatogram with those of the peaks in standard chromatograms. The retention time window used to make identification is based upon measurements of actual retention time variations over the course of 10 consecutive injections.

Calibration standards are prepared at a minimum of five concentrations for each parameter of interest through dilution of the stock standard solutions with hexane. One of the concentrations should be at a concentration near, but above, the method detection limit. Prepare stock standard solutions in hexane. Stock standards should be checked frequently for signs of degradation or evaporation, especially just prior to preparing calibration standards from them. Stock standard solutions must be replaced after one year, or sooner if comparison with check standards indicates a problem. The suggested internal standards are: 2,5-dibromotoluene, 1,3,5-tribromobenzene, and α, α'-dibromo-m-xylene. The analyst can use any of the three compounds provided that they are resolved from matrix interferences. Recommended surrogate compounds are α-2,6-trichlorotoluene, 1,4-dichloronaphthalene, and 2,3,4,5,6-pentachlorotoluene.

Spike each sample, standard, and blank with surrogate compounds. Three surrogates are suggested for this method: diphenyl phthalate, diphenyl isophthalate, and dibenzyl phthalate.

The quality control check sample concentrate should contain the test compounds at 5 to 10 ng/µL. An internal standard peak area check must be performed on all samples. The internal standard must be evaluated for acceptance by determining whether the measured area for the internal standard deviates by more than 30% from the average are for the internal standard in the calibration standards. When the internal standard peak area is outside that limit, all samples that fall outside the QC criteria must be reanalyzed. Benzyl benzoate has been tested and found appropriate as an internal standard for this method.

Any compounds confirmed by two columns may also be confirmed by GC/MS. The sample extract and associated blank should be analyzed by GC/MS. A reference standard of the compound must also be analyzed by GC/MS. Include a mid-concentration calibration standard after each group of 20 samples. The response factors for the mid-concentration calibration must be within ± 15% of the average values for the multiconcentration calibration. Demonstrate through the analyses of standards that the Florisil fractionation scheme is reproducible.

REFERENCE   Test Methods for Evaluating Solid Waste, Physical/Chemical Methods, SW-846, 3rd Edition, U.S. EPA, Office of Solid Waste, Washington, DC, EPA Method 8061, Nov. 1990.

Bis(2-n-butoxyethyl) phthalate   EPA Method 8061
CAS #117-83-9

TITLE   Phthalate Esters by Capillary Gas Chromatography With Electron Capture Detection (GC/ECD)

MATRIX   This method covers aqueous and solid matrices. This includes a wide variety such as drinking water, groundwater, industrial wastewaters, surface waters, soils, solids, and sediments.

METHOD SUMMARY   This method is used to determine the identities and concentrations of phthalate esters in liquid, solid and sludge matrices. When used to analyze for any or all of the target analytes, compound identification should be supported by at least one additional qualitative technique. This method describes conditions for parallel column, dual electron capture detector analysis, which fulfills the above requirement. Alternatively, GC/MS could be used for compound confirmation.

A measured volume or weight of sample (approximately 1 L for liquids, 10 to 30 g for solids and sludges) is extracted by using the appropriate sample extraction technique specified in

EPA Method 3510, EPA Method 3540, and EPA Method 3550. After cleanup, the extract is analyzed by GC/ECD.

INTERFERENCES The sensitivity of this method usually depends on the level of interferences rather than on instrumental limitations. If interferences prevent detection of the analytes, cleanup of the sample extracts is necessary. Either EPA Method 3610 or EPA Method 3620 alone or followed by EPA Method 3660, Sulfur Cleanup, may be used to eliminate interferences in the analysis. EPA Method 3640, Gel Permeation Cleanup, is applicable for samples that contain high amounts of lipids and waxes.

Interferences coextracted from the samples will vary considerably from waste to waste. Glassware must be scrupulously clean. All glassware require treatment in a muffle furnace at 400°C for 2 to 4 h, or thorough rinsing with pesticide-grade solvent, prior to use. Volumetric glassware should not be heated in a muffle furnace. Storage of glassware in the lab introduces contamination, even if the glassware is wrapped in aluminum foil. Sodium sulfate, Florisil, and alumina may be contaminated with phthalate esters and, therefore, use of these materials in sample cleanup should be employed cautiously. If these materials are used, they must be obtained packaged in glass. Heating at 400°C for sodium sulfate, 320°C for Florisil, and 210°C for alumina is recommended. Glass wool used in any step of sample preparation should be a specially treated pyrex wool, pesticide grade, and must be baked at 400°C for 4 h immediately prior to use.

Paper thimbles and filter paper must be exhaustively washed with the solvent that will be used in the sample extraction. Soxhlet extraction of paper thimbles and filter paper for 12 h with fresh solvent should be repeated for a minimum of three times. Method blanks should be obtained before any of the precleaned thimbles or filter papers are used.

INSTRUMENTATION Gas chromatograph suitable for on-column and split/splitless injections.

Column 1: 30 m × 0.53 mm ID, 5% phenyl/95% methyl silicone fused-silica open tubular column, DB-5, 1.5 µg film thickness.
Column 2: 30 m × 0.53 mm ID, 14% cyanopropyl phenyl silicone fused-silica open tubular column, DB-1701, 1.0 µg film thickness.

A dual electron capture detector (ECD) is used. A Kuderna-Danish (K-D) apparatus is required along with a vacuum manifold consisting of individually adjustable, easily accessible flow-control valves for up to 24 cartridges, sample rack, chemically resistant cover and seals, heavy-duty glass basin, removable stainless steel solvent guides, built-in vacuum gauge and valve. Also, 6-mL, 1-g solid-phase extraction cartridges, LC-Florisil or equivalent, prepackaged, ready to use will be needed.

PRECISION & ACCURACY The MDL actually achieved in a given analysis will vary, as it is dependent on instrument sensitivity and matrix effects. This method has been tested in a single lab. Single-operator precision, overall precision, and method accuracy were found to be related to the concentration of the compounds and the type of matrix.

ESTIMATED QUANTITATION LIMIT (EQL) FACTORS FOR VARIOUS MATRICES (a)

Matrix	Factor (b)
Groundwater	10
Low-concentration soil by ultrasonic extraction with GPC cleanup	670
High-concentration soil and sludges by ultrasonic extraction	10,000
Waste not miscible with water	100,000

(a) Sample EQLs are highly matrix-dependent.
(b) EQL = [Method detection limit] × [Factor]. For non-aqueous samples, the factor is on a wet-weight basis.

The MDL using 7 replicate determinations and a spike concentration of 100 µg/L was 84 ng/L.
The average recovery from HPLC-grade water using 4 determinatons and a spike concentration of 100 µg/L was 94.8%.
The precision (as RSD) from HPLC-grade water using 4 determinatons and a spike concentration of 100 µg/L was 6.3%.
The average recovery from groundwater using 4 determinatons and a spike concentration of 100 µg/L was 98.7%.
The precision (as RSD) from groundwater using 4 determinatons and a spike concentration of 100 µg/L was 6.0%.
The average recovery (in %) with %RSD (in parenthesis) from 3 determinations and a spike concentration of 20 µg/L in water was 78.5 (3.5) using 3M Empore Disks and EPA Method 8061.
The average recovery (in %) with %RSD (in parenthesis) from 3 determinations and a spike concentration of 20 µg/L in leachate was 92.3 (16.1) using 3M Empore Disks and EPA Method 8061.
The average recovery (in %) with %RSD (in parenthesis) from 3 determinations and a spike concentration of 20 µg/L in estuarine groundwater was 89.3 (3.6) using 3M Empore Disks and EPA Method 8061.
The average recovery (in %) with %RSD (in parenthesis) from 3 determinations and a spike concentration of 1 mg/kg in estuarine sediment was 114 (21.1) after sulfur cleanup with EPA Method 3660.
The average recovery (in %) with %RSD (in parenthesis) from 3 determinations and a spike concentration of 1 mg/kg in municipal sludge was74.0 (15.6).
The average recovery (in %) with %RSD (in parenthesis) from 3 determinations and a spike concentration of 1 mg/kg in sandy loam soil was not determined (matrix interferant).

SAMPLE COLLECTION, PRESERVATION & HANDLING
Containers used to collect samples for the determination of semivolatile organic compounds should be soap and water washed followed by methanol (or isopropanol) rinsing. The sample containers should be of glass or Teflon® and have screw-top covers with Teflon® liners. Sample containers should be filled with careto prevent any portion of the collected sample coming in contact with the sampler's gloves.

No preservation is used with concentrated waste samples. With liquid samples containing no residual chlorine and with soil, sediment, and sludge samples, immediately cooling to 4°C is the only preservation used. When residual chlorine is present then 3 mL of 10% aqueous sodium sulfate is added for each gallon of sample collected, followed by cooling to 4°C.

MHT Liquid samples must be extracted within 7 days and their extracts analyzed within 40 days. Concentrated waste, soil, sediment, and sludge samples must be extracted within 14 days and their extracts analyzed within 40 days.

SAMPLE PREPARATION In general, water samples are extracted at a pH of 5 to 7 with methylene chloride in a separatory funnel (EPA Method 3510). EPA Method 3520 is not recommended for the extraction of aqueous samples because the longer chain esters tend to adsorb to the glassware and consequently, their extraction recoveries may be poor. Solid samples are extracted with hexane/acetone (1:) or methylene chloride/acetone (1:1) in a Soxhlet extractor (EPA Method 3540) or with an ultrasonic extractor (EPA Method 3550). Immediately prior to extraction, spike 500 µL of the surrogate standard spiking solution into 1-L aqueous sample or 30-g solid sample. Extraction of particulate-free aqueous samples using C–18 extraction disks is an optional method that can be used.

Prior to Florisil cleanup or GC analysis, the methylene chloride and methylene chloride/acetone extracts must be exchanged to hexane. Exchange is not required for the acetonitrile extracts. Cleanup may not be necessary for extracts from a relatively clean sample matrix. Florisil Cartridge Cleanup may be used for extract cleanup.

If PCBs and organochlorine pesticides are known to be present in the sample, and if Florisil Cartridge Cleanup is considered, then two fractions are collected: Fraction 1 is eluted with 5 mL of 20% methylene chloride in hexane and Fraction 2 is eluted with 5 mL of 10% acetone in hexane. Fraction 1 contains the organochlorine pesticides and PCBs, and can be discarded. Fraction 2 contains the phthalate esters and is analyzed by GC/ECD.

QUALITY CONTROL Identify compounds in the sample by comparing the retention times of the peaks in the sample chromatogram with those of the peaks in standard chromatograms. The retention time window used to make identification is based upon measurements of actual retention time variations over the course of 10 consecutive injections.

Calibration standards are prepared at a minimum of five concentrations for each parameter of interest through dilution of the stock standard solutions with hexane. One of the concentrations should be at a concentration near, but above, the method detection limit. Prepare stock standard solutions in hexane. Stock standards should be checked frequently for signs of degradation or evaporation, especially just prior to preparing calibration standards from them. Stock standard solutions must be replaced after one year, or sooner if comparison with check standards indicates a problem. The suggested internal standards are: 2,5-dibromotoluene, 1,3,5-tribromobenzene, and α, α'-dibromo-m-xylene. The analyst can use any of the three compounds provided that they are resolved from matrix interferences. Recommended surrogate compounds are α-2,6-trichlorotoluene, 1,4-dichloronaphthalene, and 2,3,4,5,6-pentachlorotoluene.

Spike each sample, standard, and blank with surrogate compounds. Three surrogates are suggested for this method: diphenyl phthalate, diphenyl isophthalate, and dibenzyl phthalate.

The quality control check sample concentrate should contain the test compounds at 5 to 10 ng/µL. An internal standard peak area check must be performed on all samples. The internal standard must be evaluated for acceptance by determining whether the measured area for the internal standard deviates by more than 30% from the average are for the internal standard in the calibration standards. When the internal standard peak area is outside that limit, all samples that fall outside the QC criteria must be reanalyzed. Benzyl benzoate has been tested and found appropriate as an internal standard for this method.

Any compounds confirmed by two columns may also be confirmed by GC/MS. The sample extract and associated blank should be analyzed by GC/MS. A reference standard of the compound must also be analyzed by GC/MS. Include a mid-concentration calibration standard after each group of 20 samples. The response factors for the mid-concentration calibration must be within ± 15% of the average values for the multiconcentration calibration. Demonstrate through the analyses of standards that the Florisil fractionation scheme is reproducible.

REFERENCE Test Methods for Evaluating Solid Waste, Physical/Chemical Methods, SW-846, 3rd Edition, U.S. EPA, Office of Solid Waste, Washington, DC, EPA Method 8061, Nov. 1990.

Bis(4-methyl-2-pentyl) phthalate   **EPA Method 8061**
CAS #146-50-9

TITLE Phthalate Esters by Capillary Gas Chromatography With Electron Capture Detection (GC/ECD)

MATRIX This method covers aqueous and solid matrices. This includes a wide variety such as drinking water, groundwater, industrial wastewaters, surface waters, soils, solids, and sediments.

METHOD SUMMARY This method is used to determine the identities and concentrations of phthalate esters in liquid, solid and sludge matrices. When used to analyze for any or all of the target analytes, compound identification should be supported by at least one additional qualitative technique. This method describes conditions for parallel column, dual electron capture detector analysis, which fulfills the above requirement. Alternatively, GC/MS could be used for compound confirmation.

A measured volume or weight of sample (approximately 1 L for liquids, 10 to 30 g for solids and sludges) is extracted by using the appropriate sample extraction technique specified in EPA Method 3510, EPA Method 3540, and EPA Method 3550. After cleanup, the extract is analyzed by GC/ECD.

INTERFERENCES The sensitivity of this method usually depends on the level of interferences rather than on instrumental limitations. If interferences prevent detection of the analytes, cleanup of the sample extracts is necessary. Either EPA Method 3610 or EPA Method 3620 alone or followed by EPA Method 3660, Sulfur Cleanup, may be used to eliminate interferences in the analysis. EPA Method 3640, Gel Permeation Cleanup, is applicable for samples that contain high amounts of lipids and waxes.

Interferences coextracted from the samples will vary considerably from waste to waste. Glassware must be scrupulously clean. All glassware require treatment in a muffle furnace at 400°C for 2 to 4 h, or thorough rinsing with pesticide-grade solvent, prior to use. Volumetric glassware should not be heated in a muffle furnace. Storage of glassware in the lab introduces contamination, even if the glassware is wrapped in aluminum foil. Sodium sulfate, Florisil, and alumina may be contaminated with phthalate esters and, therefore, use of these materials in sample cleanup should be employed cautiously. If these materials are used, they must be obtained packaged in glass. Heating at 400°C for sodium sulfate, 320°C for Florisil, and 210°C for alumina is recommended. Glass wool used in any step of sample preparation should be a specially treated pyrex wool, pesticide grade, and must be baked at 400°C for 4 h immediately prior to use.

Paper thimbles and filter paper must be exhaustively washed with the solvent that will be used in the sample extraction. Soxhlet extraction of paper thimbles and filter paper for 12 h with fresh solvent should be repeated for a minimum of three times. Method blanks should be obtained before any of the precleaned thimbles or filter papers are used.

INSTRUMENTATION Gas chromatograph suitable for on-column and split/splitless injections.

Column 1: 30 m × 0.53 mm ID, 5% phenyl/95% methyl silicone fused-silica open tubular column, DB-5, 1.5 µg film thickness.

Column 2: 30 m × 0.53 mm ID, 14% cyanopropyl phenyl silicone fused-silica open tubular column, DB-1701, 1.0 µg film thickness.

A dual electron capture detector (ECD) is used. A Kuderna-Danish (K-D) apparatus is required along with a vacuum manifold consisting of individually adjustable, easily accessible flow-control valves for up to 24 cartridges, sample rack, chemically resistant cover and seals, heavy-duty glass basin, removable stainless steel solvent guides, built-in vacuum gauge and valve. Also, 6-mL, 1-g solid-phase extraction cartridges, LC-Florisil or equivalent, prepackaged, ready to use will be needed.

PRECISION & ACCURACY The MDL actually achieved in a given analysis will vary, as it is dependent on instrument sensitivity and matrix effects. This method has been tested in a single lab. Single-operator precision, overall precision, and method accuracy were found to be related to the concentration of the compounds and the type of matrix.

ESTIMATED QUANTITATION LIMIT (EQL) FACTORS FOR VARIOUS MATRICES (a)

Matrix	Factor (b)
Groundwater	10
Low-concentration soil by ultrasonic extraction with GPC cleanup	670
High-concentration soil and sludges by ultrasonic extraction	10,000
Waste not miscible with water	100,000

(a) Sample EQLs are highly matrix-dependent.
(b) EQL = [Method detection limit] × [Factor]. For non-aqueous samples, the factor is on a wet-weight basis.

The MDL using 7 replicate determinations and a spike concentration of 100 µg/L was 370 ng/L.

The average recovery from HPLC-grade water using 4 determinatons and a spike concentration of 100 µg/L was 87.2%.

The precision (as RSD) from HPLC-grade water using 4 determinatons and a spike concentration of 100 µg/L was 9.5%.

The average recovery from groundwater using 4 determinatons and a spike concentration of 100 µg/L was 86.7%.

The precision (as RSD) from groundwater using 4 determinatons and a spike concentration of 100 µg/L was 4.9%.

The average recovery (in %) with %RSD (in parenthesis) from 3 determinations and a spike concentration of 20 µg/L in water was 78.6 (2.6) using 3M Empore Disks and EPA Method 8061.

The average recovery (in %) with %RSD (in parenthesis) from 3 determinations and a spike concentration of 20 µg/L in leachate was 87.3 (18.2) using 3M Empore Disks and EPA Method 8061.

The average recovery (in %) with %RSD (in parenthesis) from 3 determinations and a spike concentration of 20 µg/L in estuarine groundwater was 92.6 (13.7) using 3M Empore Disks and EPA Method 8061.

The average recovery (in %) with %RSD (in parenthesis) from 3 determinations and a spike concentration of 1 mg/kg in estuarine sediment was 108 (57.4) after sulfur cleanup with EPA Method 3660.

The average recovery (in %) with %RSD (in parenthesis) from 3 determinations and a spike concentration of 1 mg/kg in municipal sludge was97.3 (7.4).

The average recovery (in %) with %RSD (in parenthesis) from 3 determinations and a spike concentration of 1 mg/kg in sandy loam soil was not determined (matrix interferant).

SAMPLE COLLECTION, PRESERVATION & HANDLING
Containers used to collect samples for the determination of semivolatile organic compounds should be soap and water washed followed by methanol (or isopropanol) rinsing. The sample containers should be of glass or Teflon® and have screw-top covers with Teflon® liners. Sample containers should be filled with careto prevent any portion of the collected sample coming in contact with the sampler's gloves.

No preservation is used with concentrated waste samples. With liquid samples containing no residual chlorine and with soil, sediment, and sludge samples, immediately cooling to 4°C is the only preservation used. When residual chlorine is present then 3 mL of 10% aqueous sodium sulfate is added for each gallon of sample collected, followed by cooling to 4°C.

MHT Liquid samples must be extracted within 7 days and their extracts analyzed within 40 days. Concentrated waste, soil, sediment, and sludge samples must be extracted within 14 days and their extracts analyzed within 40 days.

SAMPLE PREPARATION In general, water samples are extracted at a pH of 5 to 7 with methylene chloride in a separatory funnel (EPA Method 3510). EPA Method 3520 is not recommended for the extraction of aqueous samples because the longer chain esters tend to adsorb to the glassware and consequently, their extraction recoveries may be poor. Solid samples are extracted with hexane/acetone (1:) or methylene chloride/acetone (1:1) in a Soxhlet extractor (EPA Method 3540) or with an ultrasonic extractor (EPA Method 3550).

Immediately prior to extraction, spike 500 µL of the surrogate standard spiking solution into 1-L aqueous sample or 30-g solid sample. Extraction of particulate-free aqueous samples using C–18 extraction disks is an optional method that can be used.

Prior to Florisil cleanup or GC analysis, the methylene chloride and methylene chloride/acetone extracts must be exchanged to hexane. Exchange is not required for the acetonitrile extracts. Cleanup may not be necessary for extracts from a relatively clean sample matrix. Florisil Cartridge Cleanup may be used for extract cleanup.

If PCBs and organochlorine pesticides are known to be present in the sample, and if Florisil Cartridge Cleanup is considered, then two fractions are collected: Fraction 1 is eluted with 5 mL of 20% methylene chloride in hexane and Fraction 2 is eluted with 5 mL of 10% acetone in hexane. Fraction 1 contains the organochlorine pesticides and PCBs, and can be discarded. Fraction 2 contains the phthalate esters and is analyzed by GC/ECD.

QUALITY CONTROL Identify compounds in the sample by comparing the retention times of the peaks in the sample chromatogram with those of the peaks in standard chromatograms. The retention time window used to make identification is based upon measurements of actual retention time variations over the course of 10 consecutive injections.

Calibration standards are prepared at a minimum of five concentrations for each parameter of interest through dilution of the stock standard solutions with hexane. One of the concentrations should be at a concentration near, but above, the method detection limit. Prepare stock standard solutions in hexane. Stock standards should be checked frequently for signs of degradation or evaporation, especially just prior to preparing calibration standards from them. Stock standard solutions must be replaced after one year, or sooner if comparison with check standards indicates a problem. The suggested internal standards are: 2,5-dibromotoluene, 1,3,5-tribromobenzene, and α, α′-dibromo-m-xylene. The analyst can use any of the three compounds provided that they are resolved from matrix interferences. Recommended surrogate compounds are α-2,6-trichlorotoluene, 1,4-dichloronaphthalene, and 2,3,4,5,6-pentachlorotoluene.

Spike each sample, standard, and blank with surrogate compounds. Three surrogates are suggested for this method: diphenyl phthalate, diphenyl isophthalate, and dibenzyl phthalate.

The quality control check sample concentrate should contain the test compounds at 5 to 10 ng/µL. An internal standard peak area check must be performed on all samples. The internal standard must be evaluated for acceptance by determining whether the measured area for the internal standard deviates by more than 30% from the average are for the internal standard in the calibration standards. When the internal standard peak area is outside that limit, all samples that fall outside the QC criteria must be reanalyzed. Benzyl benzoate has been tested and found appropriate as an internal standard for this method.

Any compounds confirmed by two columns may also be confirmed by GC/MS. The sample extract and associated blank should be analyzed by GC/MS. A reference standard of the compound must also be analyzed by GC/MS. Include a mid-concentration calibration standard after each group of 20 samples. The response factors for the mid-concentration calibration must be within ± 15% of the average values for the multiconcentration calibration. Demonstrate through the analyses of standards that the Florisil fractionation scheme is reproducible.

REFERENCE Test Methods for Evaluating Solid Waste, Physical/Chemical Methods, SW-846, 3rd Edition, U.S. EPA, Office of Solid Waste, Washington, DC, EPA Method 8061, Nov. 1990.

Bolstar (Sulprofos)   EPA Method 8140
CAS #35400-43-2

TITLE Organophosphorus Pesticides

MATRIX Groundwater, soils, sludges, water miscible liquid wastes, and non-water miscible wastes.

APPLICATION This method is used for the analysis of 21 organ ophosphorus pesticides. Samples are extracted, concentrated, and analyzed using direct injection of both neat and diluted organic liquid into a gas chromatograph (GC).

INTERFERENCES Solvents, reagents and glassware may introduce artifacts. Other interferences may come from coextracted compounds from samples. The use of Florisil cleanup materials may produce low recoveries. Elemental sulfur may interfere with some compounds when using a flame photometric detector. Sulfur cleanup (Method 3660) may alleviate sulfur interference.

INSTRUMENTATION GC capable of on-column injections and a flame photometric detector (FPD) or a thermionic detector. Column 1: 1.8 m by 2 mm with 5% SP-2401on Supelcoport. Column 2: 1.8 m by 2 mm with 3% SP-2401 on Supelcoport. Column 3: 50 cm by ⅛ in Teflon® with 15% SE-54 on Gas Chrom Q. The preferred column is Column Number 1.

RANGE 4.9 to 46 µg/L

MDL 0.15 µg/L (in reagent water).

PQL FACTORS FOR MULTIPLYING × FID MDL VALUE

Matrix	Multiplication Factor
Groundwater	10
Low-level soil by sonication with GPC cleanup	670
High-level soil and sludge by sonication	10,000
Non-water miscible waste	100,000

PRECISION 6.3% (single operator standard deviation).

ACCURACY 64.6% (single operator average recovery).

SAMPLING METHOD Use 8-oz. widemouth glass bottles with Teflon®-lined caps for concentrated waste samples, soils, sediments, and sludges. Use 1 or 2½ gallon amber glass bottles with Teflon®-lined caps for liquid (water) samples.

STABILITY Cool soil, sediment, sludge, and liquid samples to 4°C. If residual chlorine is present in liquid samples add 3 mL of 10% sodium thiosulfate per gallon of sample and cool to 4°C.

MHT 14 days for concentrated waste, soil, sediment, or sludge; 7 days for liquid samples; all extracts must be analyzed within 40 days.

QUALITY CONTROL A quality control check sample concentrate containing this compound in acetone at a concentration 1,000 times more concentrated than the selected spike concentration is required. The QC check sample concentrate may be prepared from pure standard materials or purchased as certified solutions. Use appropriate trip, matrix, control site, method, reagent, and solvent blanks. Internal, surrogate, and five concentration level calibration standards are used.

REFERENCE Method 8140, SW-846, 3rd ed., Sept. 1986.

Bolstar (Sulprofos)   EPA Method 8141
CAS #35400-43-2

TITLE Organophosphorus Compounds by Gas Chromatography: Capillary Column Technique

MATRIX This method covers aqueous and solid matrices. This includes a wide variety such as drinking water, groundwater, industrial wastewaters, surface waters, soils, solids, and sediments.

METHOD SUMMARY This is a GC method used to determine the concentration of 28 organophosphorus pesticides.

The use of Gel Permeation Cleanup (EPA Method 3640) for sample cleanup has been demonstrated to yield recoveries of less than 85% for many method analytes and is therefore not recommended for use with this method.

This method provides GC conditions for the detection of ppb concentrations of organophosphorus compounds. Prior to the use of this method, appropriate sample preparation techniques must be used. Water samples are extracted at a neutral pH with methylene chloride as a solvent by using a separatory funnel (EPA Method 3510) or a continuous liquid-liquid extractor (EPA Method 3520). Soxhlet extraction (EPA Method 3540) or ultrasonic extraction (EPA Method 3550) using methylene chloride/acetone (1:1) are used for solid samples. Both neat and diluted organic liquids (EPA Method 3580) may be analyzed by direct injection. Spiked samples are used to verify the applicability of the chosen extraction technique to each new sample type. A GC with a flame photometric (FPD) or nitrogen-phosphorus detector (NPD) is used for this multiresidue procedure.

INTERFERENCES The use of Florisil cleanup materials (EPA Method 3620) for some of the compounds in this method has been demonstrated to yield recoveries less than 85% and is therefore not recommended for all compounds. Use of phosphorus or halogen specific detectors, however, often obviates the necessity for cleanup for relatively clean sample matrices. If particular circumstances demand the use of an alternative cleanup procedure, the analyst must determine the elution profile and demonstrate that the recovery of each analyte is no less than 85%.

Use of a flame photometric detector (FPD) in the phosphorus mode will minimize interferences from materials that do not contain phosphorus. Elemental sulfur, however, may interfere with the determination of certain organophosphorus compounds by flame photometric gas chromatography. Sulfur cleanup using EPA Method 3660 may alleviate this interference. A nitrogen phosphorus detector (NPD) is also recommended.

A few analytes coelute on certain columns. Therefore, select a second column for confirmation where coelution of the analytes of interest does not occur.

Method interferences may be caused by contaminants in solvents, reagents, glassware, and other sample processing hardware that lead to discrete artifacts or elevated baselines in gas chromatograms. All these materials must be routinely demonstrated to be free from interferences under the conditions of the analysis by analyzing reagent blanks.

INSTRUMENTATION A GC with a NPD or a FPD will be needed. A data system or integrator is recommended for measuring peak areas and/or peak heights. A Kuderna-Danish (K-D) apparatus will be needed for extract concentration.

Column 1: 15 m × 0.53 mm megabore capillary column, 1.0 μm film thickness, DB-210.
Column 2: 15 m × 0.53 mm megabore capillary column, 1.5 μm film thickness, SPB-608.
Column 3: 15 m × 0.53 mm megabore capillary column, 1.0 μm film thickness, DB-5.

Three megabore capillary columns are included for analysis of organophosphates by this method. Column 1 (DB-210 or equivalent) and Column 2 (SPB-608 or equivalent) are recommended if a large number of organophosphorus analytes are to be determined. If the superior resolution offered by Column 1 and Column 2 is not required, Column 3 (DB-5 or equivalent) may be used. For megabore capillary columns, automatic injections of 1 μL are recommended.

PRECISION & ACCURACY The MDL actually achieved in a given analysis will vary, as it is dependent on instrument sensitivity and matrix effects. Single operator accuracy and precision studies have been conducted with spiked water and soil samples.

MULTIPLICATION FACTORS FOR OTHER MATRICES (a)

Matrix	Factor (b)
Groundwater (EPA Method 3510 or EPA Method 3520)	10
Low-concentration soil by Soxhlet and no cleanup	10 (c)
Low-concentration soil by ultrasonic extraction with GPC cleanup	6.7 (c)
High-concentration soil and sludges by ultrasonic extraction	500 (c)
Non-Water miscible waste (EPA Method 3580)	1000 (c)

(a) Sample EQLs are highly matrix-dependent. The EQLs listed here are provided for guidance and may not always be achievable.
(b) EQL = [Method detection limit] × [Factor]. For non-aqueous samples the factor is on a wet-weight basis.
(c) Multiply this factory times the soil MDL.

The MDL (in μg/L) when reagent water was extracted using a separatory funnel was 0.07.

The MDL (in μg/kg) when soil was extracted using Soxhlet extraction (EPA Method 3540) was 3.5.

Accuracy (as % recovery) with separatory funnel extraction ranged from 134 (with low spikes) to 101 (with high spikes).

Accuracy (as % recovery) with continuous liquid-liquid extraction ranged from not recovered (with low spikes) to 128 (with high spikes).

Accuracy (as % recovery) with Soxhlet extraction of soils ranged from 102 (with low spikes to 79 (with high spikes).

Accuracy (as % recovery) with ultrasonic extraction of soils ranged from not recovered (with low spikes) to 114 (with high spikes).

SAMPLE COLLECTION, PRESERVATION & HANDLING

Containers used to collect samples for the determination of semivolatile organic compounds should be soap and water washed followed by methanol (or isopropanol) rinsing. The sample containers should be of glass or Teflon® and have screw-top covers with Teflon® liners.

No preservation is used with concentrated waste samples. With liquid samples containing no residual chlorine and with soil, sediment, and sludge samples, immediately cooling to 4°C is the only preservation used. When residual chlorine is present then 3 mL of 10% aqueous sodium sulfate is added for each gallon of sample collected, followed by cooling to 4°C.

Liquid samples must be extracted within 7 days and their extracts analyzed within 40 days. Concentrated waste, soil, sediment, and sludge samples must be extracted within 14 days and their extracts analyzed within 40 days.

SAMPLE PREPARATION

In general, water samples are extracted at a neutral pH with methylene chloride, using either EPA Method 3510 or EPA Method 3520. Solid samples are extracted using either EPA Method 3540 or EPA Method 3550 with methylene chloride/acetone (1:1) as the extraction solvent.

Prior to GC analysis, the extraction solvent may be exchanged to hexane. Single lab data indicates that samples should not be transferred with 100% hexane during sample workup as the more water soluble organophosphorus compounds may be lost.

If cleanup is performed on the samples, the analyst should analyze the samples by GC. This will confirm elution patterns and the absence of interferences from the reagents. If peak detection and identification is prevented by the presence of interferences, further cleanup is required.

QUALITY CONTROL

The analyst should monitor the performance of the extraction, cleanup (when used), and analytical system and the effectiveness of the method in dealing with each sample matrix by spiking each sample, standard, and blank with one or two surrogates (e.g., organophosphorus compounds not expected to be present in the sample). Deuterated analogs of analytes should not be used as surrogates for gas chromatographic analysis due to coelution problems.

A minimum of five concentrations for each analyte of interest should be prepared through dilution of the stock standards with isooctane. One of the concentrations should be at a concentration near, but above, the MDL.

Include a mid-level check standard after each group of 10 samples in the analysis sequence. GC/MS techniques should be judiciously employed to support qualitative identifications made with this method. Follow the GC/MS operating requirements specified in EPA Method 8270.

When available, chemical ionization mass spectra may be employed to aid in the qualitative identification process. To confirm an identification of a compound, the background-corrected mass spectrum of the compound must be obtained from the sample extract and must be compared with a mass spectrum from a stock or calibration standard analyzed under the same chromatographic conditions. The molecular ion and all other ions present above 20% relative abundance in the mass spectrum of the standard must be present in the mass spectrum of the sample with agreement to ± 20%. The retention time of the compound in the sample must be within six seconds of the retention time for the same compound in the standard solution.

Should the MS procedure fail to provide satisfactory results, additional steps may be taken before reanalysis. These steps may include the use of alternate packed or capillary GC columns or additional sample cleanup.

REFERENCE Test Methods for Evaluating Solid Waste, Physical/Chemical Methods, SW-846, 3rd Edition, U.S. EPA, Office of Solid Waste, Washington, DC, EPA Method 8141 July 1992.

Boron EPA Method 6010
CAS #7440-42-8

TITLE Inductively Coupled Plasma-Atomic Emission Spectroscopy

MATRIX This method is applicable to the determination of trace elements, including metals, in groundwater, soils, sludges, sediments, and other solid wastes. All matrices require digestion prior to analysis. The method of standard addition must be used for the analysis of all sample digests unless either serial dilution or matrix spike addition demonstrates it is not required.

METHOD SUMMARY Method 6010 covers 25 elements using ICP analysis. It measures element-emitted light by optical spectrometry. Samples, following an appropriate acid digestion, are nebulized and the resulting aerosol is transported to the plasma torch. Element-specific atomic line emission spectra are produced by a radio-frequency inductively coupled plasma.

INTERFERENCES Interferences may be categorized as spectral or non-spectral. Spectral interferences are caused by overlap of a spectral line from another element, unresolved overlap of molecular band spectra, background contribution from continuous or recombination phenomenon, and stray light from the line emission of high concentration elements. Non-spectral interferences include physical and chemical interferences. Physical interferences are effects associated with the sample nebulization and transport processes. Changes in viscosity and surface tension can cause significant inaccuracies. Chemical interferences include molecular compound formation, ionization effects, and solute vaporization effects. Normally these effects are not significant and can be minimized by careful selection of operating conditions. Chemical interferences are highly dependent on matrix type and the specific analyte element.

INSTRUMENTATION An inductively coupled argon plasma emission spectrometer (ICP) capable of background correction is required.

PRECISION & ACCURACY Detection limits, sensitivity, and optimum ranges of the metals will vary with the matrices and model of the spectrometer. In a single lab evaluation, seven wastes were analyzed for 22 elements. The mean percent relative standard deviation from triplicate analyses for all elements and wastes was 9 ± 2%. The mean percent recovery of spiked elements for all wastes was 93 ± 6%. Spike levels ranged from 100 µg/L to 100 mg/L. The wastes included sludges and industrial wastewaters.

Estimated instrument detection limit in µg/L is 5.
Spiked concentration in µg/L is not listed.
Mean reported value in µg/L is not listed.
Precision as RSD % is not listed.

SAMPLING METHOD Samples should be collected in borosilicate glass, linear polyethylene, polypropylene, or Teflon® bottles that have been prewashed with detergent and tap water, and rinsed with 1:1 nitric acid and tap water or 1:1 hydrochloric acid and tap water. Collect at least 2 g of solids and 200 mL of aqueous samples.

SAMPLE PRESERVATION Add nitric acid to make the samples pH <2.

MHT The maximum holding time for properly preserved samples is 6 months.

SAMPLE PREPARATION Preliminary treatment of most matrices is necessary because of the complexity and variability of sample matrices. Water samples that have been prefiltered and acidified will not need acid digestion. Methods for acid digestion of waters for total recoverable or dissolved metals, acid digestions of aqueous samples and extracts for total metals, and acid digestion of sediments, sludges, and soils are summarized below.

Total recoverable or dissolved metals in water — To prepare surface and groundwater samples for determination of total recoverable and dissolved metals, a 100-mL aliquot of well-mixed sample is acidified with concentrated nitric acid and concentrated hydrochloric acid, then heated until the volume is reduced to 15–20 mL. Adjust the final volume to 100 mL with reagent water.

Total metals in aqueous samples, soil and sediment extracts — To prepare aqueous samples, soil and sediment extracts, and wastes that contain suspended solids, a 100-mL aliquot is made acidic with concentrated nitric acid and the solution is evaporated to about 5 mL on a hot plate. Continue heating and adding additional acid until sample digestion is complete, which is usually indicated when the digestate is light in color or does not change in appearance. Evaporate the solution to about 3 mL andcool it and add a small quantity of 1:1 hydrochloric acid (10 mL/100 mL of final solution). Cover the beaker and reflux for 15 min. Wash down the beaker walls and filters or centrifuge the sample to remove silicates and other insoluble material. Filter the sample and adjust the final volume to 100 mL with reagent water and the final acid concentration to 10%.

Sediments, sludges, and soils — To prepare sediments, sludges and soil samples, transfer 1–2 g to a conical beaker and add 10 mL of 1:1 nitric acid, mix the slurry, and cover it with a watch glass. Heat the sample and reflux for 10 to 15 min without boiling. Allow it to cool, then add 5 mL of concentrated nitric acid and reflux for 30 min. Repeat last step and then allow the solution to evaporate to 5 mL without boiling. Cool and add 2 mL of water and 3 mL of 30% hydrogen peroxide. Cover and place the beaker on the hot plate. Heat and add 30% hydrogen peroxide in 1-mL aliquots with warming until the effervescence is minimal but do not add more than a total of 10 mL of 30% hydrogen peroxide. If the sample is being prepared for the analysis of Ag, Al, As, Ba, Be, Ca, Cd, Co, Cr, Cu, Fe, K, Mg, Mn, Mo, Na, Ni, Os, Pb, Se, Tl, V, and Zn, then add 5 mL of concentrated hydrochloric acid and 10 mL of water and return the covered beaker to a hot plate for 15 min of additional refluxing without boiling. Dilute the sample to a 100 mL volume with water after cooling and filter or centrifuge to remove particulates.

QUALITY CONTROL Laboratory control samples must be analyzed for each analytical method. A method blank should be analyzed with each batch of samples. The effect of the matrix on method performance must be demonstrated: when appropriate, there should be at least one matrix spike and either one matrix duplicate or one matrix spike duplicate per analytical batch. The bias and precision of the method, as well as the method detection limit for each specific matrix type, must be measured.

Dilute and reanalyze samples that are more concentrated than the linear calibration limit. Employ a minimum of one reagent blank per sample batch to determine if contamination or any memory effects are occurring. Whenever a new or unusual sample matrix is encountered, perform either a serial dilution test or a matrix spike addition test to ensure that neither positive or negative interferences are operating on any of the analyte elements. Check the instrument standardization by verifying calibration every 10 samples using a calibration blank and a check standard.

REFERENCE Test Methods for Evaluating Solid Waste (SW-846). U.S. EPA. 1983. Method 6010, Rev. 0, Sept. 1986. Office of Solid Wastes, Washington, DC.

BoronEPA Method 200.7
CAS #7440-42-8

TITLE Inductively Coupled Plasma (ICP)

MATRIX Dissolved, suspended or total element in drinking and surfacewaters and in domestic and industrial waste waters.

APPLICATION The method covers the determination of 25 metals. Dissolved elements are determined in filtered and acidified samples after appropriate digestion (which increases dissolved solids). Its primary advantage is that ICP instruments allow simultaneous or rapid sequential determination of many elements in a short time. Samples are first nebulized and the aerosol is transported to a plasma torch in which element specific atomic line emission spectra are produced by a radio frequency inductively coupled plasma. Background correction is required for trace element detection except in the case of line broadening.

INTERFERENCES There are spectral, physical, and chemical interferences. The primary disadvantage of ICP instruments is background radiation from other elements and the plasma gases (spectral interferences). Changes in sample viscosity and surface tension with samples containing high dissolved solids (especially those exceeding 1500 mg/L) or high acid concentrations can cause physical interferences. Ionization effects, solute vaporization and molecular compound formation can cause chemical interferences. Thallium and vanadium can cause interference at the 100 mg/L level.

INSTRUMENTATION Inductively Coupled Argon Plasma Emission Spectroscopy. 249.773 nm wavelength.

RANGE Not listed.

MDL 5 µg/L.

PRECISION Not listed.

ACCURACY Mean recovery = 93% ± 6% of spiked elements for all wastes.

SAMPLING METHOD Wash sample container with detergent and tap water, rinse with 1+1 nitric acid and tap water, then rinse with 1+1 hydrochloric acid and tap water, then rinse with deionized, distilled water in that order. Perform any filtration or acid preservation steps when the sample is collected or as soon as possible thereafter.

STABILITY Cool samples to 4°C.

MHT 24 h.

QUALITY CONTROL Mixed calibration standards, an instrument check standard and an interference check solution are used in addition to a quality control sample. The quality control sample should be prepared in the same acid matrix as the calibration standards at 10 times the instrumental detection limits and in accordance with the instructions provided by the supplier. Furthermore, two types of blanks are required: a calibration blank and a reagent blank.

REFERENCE Method 200.7, U.S. EPA, EMSL-Cincinnati, OH, Nov. 1980

Bromide (Titrimetric)   **EPA Method 320.1**

TITLE Inorganics, Non-metallics

MATRIX Drinking, surface, and saline waters. Wastewaters.

APPLICATION Date issued 1974. After pretreatment, sample is divided. One aliquot is run by converting iodide to iodate and titrating with Phenylarsine oxide(pao) or sodium thiosulfate. Other aliquot is run for iodide+bromide by conversion to iodate and bromate and titrating.

INTERFERENCES Iron, manganese and organic matter can interfere. (Calcium oxide pretreatment nullifies this interference). Color interferes with observation of indicator and bromine-water color changes. (Eliminate, using a pH m and standard amounts of oxidant and oxidant-quencher).

INSTRUMENTATION Laboratory iodometric titration equipment and glassware.

RANGE 2–20 mg bromide/L.

MDL Not listed.

PRECISION SD = ±0.42 mg/L of bromide at 20.3 mg/L of bromide.

ACCURACY Recovery = 99% at 20.3 mg/L of bromide.

SAMPLING METHOD Plastic or glass (100 mL).

STABILITY No preservation required.

MHT 28 Days.

QUALITY CONTROL When titrating either aliquot, run a distilled water blank with each sample set because of iodide, iodate, bromide and/or bromate in reagents. Calculate bromide by difference. [(iodide + bromide) — (iodide) = bromide].

REFERENCE Methods for the Chemical Analysis of Water and Wastes, EPA-600/4-79-020, U.S. EPA, EMSL, 1979.

Bromoacetone   **EPA Method 8240**
CAS #598-31-2

TITLE Volatile Organics By GC/MS: Packed Column Technique

MATRIX Nearly all types of sample matarices, regardless of water content, can be analyzed using this method. This includes groundwater, aqueous sludges, caustic liquors, acid liquors, waste solvents, oily wastes, mousses, tars, fibrous wastes, polymetric emulsions, filter cakes, spent carbons, spent catalysts, soils, and sediments.

METHOD SUMMARY Method 8240B covers 80 volatile organic compounds that are introduced into a gas chromatograph by the purge-and-trap method or by direct injection (in limited applications). For the purge-and-trap method an inert gas (zero grade nitrogen or helium) is bubbled through a 5-mL solution at ambient temperature. Purged sample components are trapped in a tube of sorbent materials. When purging is complete, the sorbent tube is heated and backflushed with inert gas to desorb the trapped components onto a GC column.

INTERFERENCES Impurities in the purge gas and from organic compounds outgassing from the plumbing ahead of the trap account for many contamination problems. Interferences purged or coextracted from the samples will vary considerably from source to source. Cross-contamination can occur whenever high-level and low-level samples are analyzed sequentially. Whenever an unusually concentrated sample is analyzed, it should be followed by the analysis of organic-free reagent water to check for cross-contamination. Samples also can be contaminated by diffusion of volatile organics (particularly methylene chloride and fluorocarbons) through the septum seal into the sample during shipment and storage. A trip blank can serve as a check on such contamination. The lab where volatile analysis is performed and also the refrigerated storage area should be completely free of solvents.

INSTRUMENTATION A gas chromatograph/mass spectrometry/data system (GC/MS) equipped with a 6 ft × 0.1 in I.D. glass column packed with 1% SP-1000 on Carbopack-B (60/80 mesh) is required. Also needed is a 5-mL purging device, a sorbent trap, and a thermal desorption apparatus.

PRECISION & ACCURACY This method is reported to have been tested by 15 laboratories using organic-free reagent water, drinking water, surface water, and industrial wastewaters (not specified) fortified at six concentrations over the range 5–600 µg/L.

Sample estimated quantitation limits (EQLs) are highly matrix-dependent. The EQLs listed may not always be achievable. EQLs listed for soils or sediments are based on wet weight. Normally, data is reported on a dry-weight basis; therefore, EQLs will be higher, based on the percent dry weight of each sample. Note that EQLs are even more variable than MDLs and that they are highly variable depending on the matrix being analyzed.

EQL in groundwater in µg/L was not listed.
EQL in low soil or sediment in µg/kg was not listed.
Accuracy (a) in µg/L was not listed.
Precision (b) in µg/L was not listed.

(a) *Average recovery found for measurements of samples containing a concentration of C, in µg/L.*
(b) *Overall precision found for measurements of samples with average recovery X for samples containing a concentration of C in µg/L.*
X = *Average recovery found for measurement of samples containing a concentration of C in µg/L.*

MULTIPLICATION FACTORS FOR OTHER MATRICES

Other Matrices	Factor (a)
Waste miscible liquid waste	50
High-concentration soil and sludge	125
Non-water miscible waste	500

(a) *EQL = [EQL for low soil/sediment] × [Factor]. For non-aqueous samples, the factor is on a wet-weight basis.*

SAMPLING METHOD

Liquid samples — Use a 40-mL glass screw-cap VOA vial with a Teflon®-faced silicone septum that has been prewashed, rinsed with distilled deionized water, and oven dried. However, if residual chlorine is present, collect sample in a 40-oz. soil VOA container which has been pre-preserved with 4 drops of 10% sodium thiosulfate, mix gently, and then transfer the sample to a 40-mL VOA vial. Collect bubble-free samples in duplicate and seal them in separate plastic bags.

Soils or sediments and sludges — Use an 8-oz. widemouth glassbottle with a Teflon®-faced silicone septum that has been prewashed with detergent, rinsed with distilled deionized water, and oven dried. Tap slightly to eliminate free air space. Collect samples in duplicate and seal them in separate plastic bags.

SAMPLE PRESERVATION

Liquid samples — Add 4 drops of concentrated HCL and immediately coolsamples to 4°C and store in a solvent-free refrigerator.

Soils or sediments, and sludges — Cool samples to 4°C and store in a solvent-free refrigerator.

MHT Maximum holding time is 14 days from the date of sample collection.

SAMPLE PREPARATION

Liquid samples — Remove the plunger from a 5-mL syringe and carefully pour the sample into the syringe barrel to just short of overflowing. Replace the syringe plunger and compress the sample. Open the syringe valve and vent any residual air while adjusting the sample volume to 5.0 mL. If there is only one volatile organic analysis (VOA) vial, a second syringe should be filled at this time to protect against possible loss of sample integrity. Add 10 µL of surrogate spiking solution and 10 µL of internal standard spiking solution through the valve bore of the 5-mL syringe, then close the valve. The surrogate and internal standards may be mixed and added as a single spiking solution.

Sediments, soils, and waste samples — All samples of this type should be screened by GC analysis using a headspace method (EPA Method 3810) or the hexadecane extraction and screening method (EPA Method 3820). Use the screening data to determine whether to use the low-concentration method (0.005–1 mg/kg) or the high-concentration method (>1 mg/kg).

Low-concentration method — The low-concentration method is based on purging a heated sediment or soil sample mixed with organic-free reagent water containing the surrogate and internal standards. Analyze all reagent blanks and standards under the same conditions as the samples.

Use a 5-g sample if the expected concentration is <0.1 mg/kg or a 1-g sample for expected concentrations between 0.1 and 1 mg/kg. Mix the contents of the sample container with a narrow metal spatula. Weigh the amount of the sample into a tared purge device. Add the spiked water to the purge device, which contains the weighed amount of sample, and connect the device to the purge-and-trap system.

High-concentration method — This method is based on extracting the sediment or soil with methanol. A waste sample is either extracted or diluted, depending on its solubility in methanol. Wastes that are insoluble in methanol are diluted with reagent tetraglyme or possibly polyethylene glycol (PEG). An aliquot of the extract is added to organic-free reagent water containing surrogate and internal standards. This is purged at

ambient temperature. All samples with an expected concentration of >1.0 mg/kg should be analyzed by this method.

Mix the contents of the sample container with a narrow metal spatula. For sediments or soils and solid wastes that are insoluble in methanol, weigh 4 g (wet weight) of sample into a tared 20-mL vial. For waste that is soluble in methanol, tetraglyme, or PEG, weigh 1 g (wet weight) into a tared scintillation vial or culture tube or a 10-mL volumetric flask. Quickly add 9.0 mL of appropriate solvent then add 1.0 mL of a surrogate spiking solution to the vial, cap it, and shake it for 2 min.

METHANOL EXTRACT REQUIRED FOR ANALYSIS OF HIGH-CONCENTRATION SOILS OR SEDIMENTS

Approximate Concentration Range	Volume of Methanol Extract (a)
500–10,000 µg/kg	100 µL
1,000–20,000 µg/kg	50 µL
5,000–100,000 µg/kg	10 µL
25,000–500,000 µg/kg	100 µL of 1/50 dilution (b)

Calculate appropriate dilution factor for concentrations exceeding this table.

(a) The volume of methanol added to 5 mL of water being purged should be kept constant. Therefore, add to the 5-mL syringe whatever volume of methanol is necessary to maintain a volume of 100 µL added to the syringe.
(b) Dilute an aliquot of the methanol extract and then take 100 µL for analysis.

QUALITY CONTROL Demonstrate, through the analysis of a reagent water blank, that interferences from the analytical system, glassware, and reagents are under control. Blank samples should be carried through all stages of the sample preparation and measurement steps. For each analytical batch (up to 20 samples), a reagent blank, matrix spike, and matrix spike duplicate must be analyzed (the frequency of the spikes may be different for different monitoring programs). The blank and spiked samples must be carried through all stages of the sample preparation and measurement steps. QC samples mentioned in the section on Interferences will also be needed as appropriate to those situations.

REFERENCE Test Methods for Evaluating Solid Waste (SW-846). U.S. EPA. 1983. Method 8240B, Rev. 2, Nov. 1990. Office of Solid Wastes, Washington, DC.

Bromobenzene   **EPA Method 502**
CAS #108-86-1

TITLE Volatile Organic Compounds in Water By Purge and Trap Capillary Column Gas Chromatography with Photoionization and Electrolytic Conductivity Detectors in Series. U.S. EPA Method 502.2, Rev. 2.0, 1989.

MATRIX Drinking water and raw source water. The latter should include most surface water and groundwater sources.

METHOD SUMMARY This method covers 60 volatile organic compounds that contain halogen atoms and/or that are aromatic. An inert gas (zero grade nitrogen or helium) is bubbled through a 25-mL or a 5-mL water sample (depending on the expected concentration of the analytes). Purged sample components are trapped in a tube of sorbent materials. When purging is complete, the sorbent tube is heated and backflushed with helium to desorb the trapped sample onto a capillary GC column. The column is temperature programmed to separate the method analytes which are then detected with a photoionization detector (PID) and a Hall electrolytic conductivity (HECD) placed in series. The PID is selective for aromatic compounds and the HECD is selective for halogenated compounds.

INTERFERENCES Impurities in the purge gas and from organic compounds outgassing from the plumbing ahead of the trap account for many contamination problems. Interferences purged or coextracted from the samples will vary considerably from source to source, depending upon the particular sample or extract being tested. Cross-contamination can occur whenever high-level and low-level samples are analyzed sequentially. Samples also can be contaminated by diffusion of volatile organics (particularly methylene chloride and fluorocarbons) through the septum seal into the sample during shipment and storage. The lab where volatile analysis is performed and also the refrigerated storage area should be completely free of solvents.

INSTRUMENTATION A GC containing a series configuration of a high temperature photoionization detector (PID) equipped with 10.0 eV (nominal) lamp and Hall electrolytic conductivity detector (HECD) is required. Also required is an all-glass 5-mL purging device, a sorbent trap, and a thermal desorption apparatus which is connected to the GC system.

Column 1: VOCOL glass wide-bore capillary column.
Column 2: RTX–502.2 mega-bore capillary column.
Column 3: DB-62 mega-bore capillary column.

PRECISION & ACCURACY Method detection limits are dependent upon the characteristics of the gas chromatographic system used. Analytes that are not separated chromatographically cannot be individually identified and used in the same calibration mixture or water samples unless an alternative technique for identification and quantification, such as mass spectrometry, is used.

Electrolytic conductivity detetor (c) range in µg/L (a) was 0.02–200.
Electrolytic conductivity detetor (c) MDL in µg/L (b) was 0.03.
Electrolytic conductivity detetor (c) accuracy as % recovery was 97.
Electrolytic conductivity detetor (c) precision as % RSD was 2.7.
Photoionization detector (d) range in µg/L (a) was 0.02–200.
Photoionization detector (d) MDL in µg/L (b) was 0.01.
Photoionization detector (d) accuracy as % recovery was 99.
Photoionization detector (d) precision as % RSD was 1.7.

(a) The applicable concentration range of this method is compound, instrument, and matrix-dependent. It is listed as being approximately 0.02 to 200 µg/L but no specific information is provided so caution should be observed.

(b) The method detection limits reports with this method are compound, instrument, and matrix-dependent. The values reported were calculated using reagent water fortified with the corresponding compounds at 10 µg/L and a GC-equipped with a 60 m × 0.75 mm VOLCOL wide bore capillary column with 1.5 µm film thickness and using helium carrier gas.

(c) Recoveries and relative standard deviations were determined from seven samples of reagent water fortified with 10 µg/L of each compound. 2-Bromo-1-chloropropane was used as the internal standard for calculating average recoveries.

(d) Recoveries and relative standard deviations were determined from seven samples of reagent water fortified with 10 µg/L of each compound. Fluorobenzene was used as the internal standard for calculating average recoveries.

SAMPLING METHOD Collect samples using a 40- to 120-mL screw-cap vial (prewashed with detergent, rinsed with distilled water and oven dried at 105°C) with a Teflon®-faced silicone septum. Collect bubble-free samples and place the septum with the Teflon® side down on the water.

SAMPLE PRESERVATION If residual chlorine is present in the water add about 25 mg of ascorbic acid to each vial before samples are collected to remove thechlorine. Add hydrochloric acid to reduce pH to <2, immediately coolsamples to 4°C, and store them in a solvent-free refrigerator at 4°C until analysis.

MHT The maximum holding time for samples is 14 days from the time they were collected.

SAMPLE PREPARATION Remove the plungers from two 5-mL syringes and attach a closed syringe valve to each. Warm the sample to room temperature, open the sample bottle, and carefully pour the sample into one of the syringe barrels to just short of overflowing. Replace the syringe plunger, invert the syringe, and compress the sample. Open the syringe valve and vent any residual air while adjusting the sample volume to 5.0 mL. Add 10 µL of the internal calibration standard to the sample through the syringe valve. Close the valve. Fill the second syringe in an identical manner from the same sample bottle. Reserve this second syringe for a reanalysis if necessary.

QUALITY CONTROL As an initial demonstration of lab accuracy and precision, analyze 4 to 7 replicates of a lab fortified blank containing analyte at 0.1–5 µg/L. Collect all samples in duplicate. Surrogate analytes (similar to those of the analytes of interest), whose concentration is known in every sample, are measured using the same internal standard calibration procedure. Duplicate field reagent water blanks (trip blanks) must be analyzed with each set of samples, lab reagent blanks (method blanks) must be analyzed with each batch of samples processed as a group within a work shift. Also, a single lab-fortified blank that contains each of the analytes of interest should be analyzed with each batch of samples processed as a group within a work shift. A 3- to 5-point calibration curve is needed depending on the calibration range factor required.

EPA CONTACT & HOTLINE For technical questions contact Dr. Baldev Bathija, U.S. EPA,Office of Ground Water and Drinking Water (WH-550D), 401 M St. SW, Washington, DC 20460. Tel. (202) 260-3040. For further information the EPA Safe Drinking Water Hotline may be called at: (800) 426-4791.

REFERENCE Methods for the Determination of Organic Compounds in Drinking Water, EPA/600/4-88/039 (revised July 1991; Final Rule for determination of compliance with the MCL for Total Trihalomethanes under 141.30, in 40 CFR Part 141, Vol. 58, No. 147, Fed. Reg., Tuesday Aug. 3, 1993). U.S. EPA Environmental Monitoring Systems Laboratory, Cincinnati, OH, 45268, U.S.A. Available from the National Technical Information Service (NTIS), 5285 Port Royal Road, Springfield, VA 22161; Tel. 800-553-6847. NTIS Order Number is PB91-231480.

Bromobenzene   **EPA Method 524**
CAS #108-86-1

TITLE Measurement of Purgeable Organic Compounds in Water by Capillary Column GC/MS.

MATRIX Drinking water and raw source water; the latter should include most surface water and groundwater sources.

METHOD SUMMARY Method 524.2 covers 60 volatile organic compounds. An inert gas (zero grade nitrogen or helium) is bubbled through a 25-mL or a 5-mL water sample (depending on the expected concentration of the analytes). Purged sample components are trapped in a tube of sorbent materials. When purging is complete, the sorbent tube is heated and backflushed with helium to desorb the trapped sample onto a capillary GC column.

INTERFERENCES Impurities in the purge gas and from organic compounds outgassing from the plumbing ahead of the trap account for many contamination problems. Interferences purged or coextracted from the samples will vary considerably from source to source, depending upon the particular sample or extract being tested. Cross-contamination can occur whenever high-level and low-level samples are analyzed sequentially. Samples also can be contaminated by diffusion of volatile organics (particularly methylene chloride and fluorocarbons) through the septum seal into the sample during shipment and storage.

INSTRUMENTATION A GC/MS with a data system equipped with one of the following capillary GC columns:

Column 1: VOCOL glass wide bore capillary column.
Column 2: DB-624 fused silica capillary column.
Column 3: DB-5 fused silica capillary column.

Also required is an all-glass 25-mL or 5-mL purging device, a sorbent trap, and a thermal desorption apparatus which is connected to the GC/MS system.

PRECISION & ACCURACY Method detection limits are compound- and instrument-dependent, and may vary from approximately 0.02–0.35 µg/L. Note in the table below that the "true" concentration range used for accuracy and precision measurements was quite narrow. However, the applicable concentration range of this method is primarily column dependent and is approximately 0.02 to 200 µg/L for the wide-bore thick-film

columns. Narrow-bore thin-film columns may have a capacity which limits the range to about 0.02 to 20 µg/L. Analytes that are inefficiently purged from water will not be detected when present at low concentrations, but they can be measured with acceptable accuracy and precision when present in sufficient amounts.

Analytes that are not separated chromatographically, but which have different mass spectra and non-interfering quantification ions, can be identified and measured in the same calibration mixture or water sample. Analytes which have very similar mass spectra cannot be individually identified and measured in the same calibration mixture or water samples unless they have different retention times. Co-eluting compounds with very similar mass spectra, typically many structural isomers, must be reported as an isomeric group or pair.

The range (in µg/L) was 0.1–10.
The method detection limit (in µg/L) was 0.03.
The accuracy (as % recovery) was 100.
The precision (in %) was 5.5.

Note: Data were obtained from 16–31 determinations using a wide-bore capillary column and a jet separator interfaced to a quadrupole mass spectrometer. All analytes were in a reagent water matrix.

SAMPLING METHOD Collect samples using a 40- to 120-mL screw-cap vial (prewashed with detergent, rinsed with distilled water and oven dried at 105°C) with a Teflon®-faced silicone septum. Collect bubble-free samples and place the septum with the Teflonside down on the water.

SAMPLE PRESERVATION If residual chlorine is present in the water add about 25 mg of ascorbic acid to each vial before samples are collected to remove thechlorine. Add hydrochloric acid to reduce pH to <2, and immediately coolsamples to 4°C, and store them in a solvent-free refrigerator at 4°C until analysis.

MHT The maximum holding time for samples is 14 days from the time they were collected.

SAMPLE PREPARATION Remove the plungers from two 25-mL (or 5-mL depending on sample size) syringes and attach a closed syringe valve to each. Warm the sample to room temperature, open the sample bottle, and carefully pour the sample into one of the syringe barrels to just short of overflowing. Replace the syringe plunger, invert the syringe, and compress the sample. Open the syringe valve and vent any residual air while adjusting the sample volume to 25.0 mL (or 5 mL). For samples and blanks, add 5 µL of the fortification solution containing the internal standard and the surrogates to the sample through the syringe valve. For calibration standards and lab fortified blanks, add 5 µL of the fortification solution containing the internal standard only. Close the valve. Fill the second syringe in an identical manner from the same sample bottle. Reserve this second syringe for a reanalysis if necessary.

QUALITY CONTROL As an initial demonstration of lab accuracy and precision, analyze 4 to 7 replicates of a lab fortified blank containing analyte at 0.2–5 µg/L. Collect all samples in duplicate. Surrogate analytes (similar to those of the analytes of interest), whose concentration is known in every sample, are measured using the same internal standard calibration procedure. Duplicate field reagent water blanks (trip blanks) must be analyzed with each set ofsamples, lab reagent blanks (method blanks) must be analyzed with each batch ofsamples processed as a group within a work shift. Also, a singlelab-fortified blank that contains each of the analytes of interestshould be analyzed with each batch of samples processed as a group within a work shift. A 3- to 5-point calibration curve is needed depending on the calibration range factor required.

EPA CONTACT & HOTLINE For technical questions contact Dr. Baldev Bathija, U.S. EPA,Office of Ground Water and Drinking Water (WH-550D), 401 M St. SW, Washington, DC 20460. Tel. (202) 260-3040. For further information the EPA Safe Drinking Water Hotline may be called at: (800) 426-4791.

REFERENCE Methods for the Determination of Organic Compounds in Drinking Water, EPA/600/4-88/039 (revised July 1991; Final Rule for determination of compliance with the MCL for Total Trihalomethanes under 141.30, in 40 CFR Part 141, Vol. 58, No. 147, Fed. Reg., Tuesday Aug. 3, 1993). U.S. EPA Environmental Monitoring Systems Laboratory, Cincinnati, OH, 45268, U.S.A. Available from the National Technical Information Service (NTIS), 5285 Port Royal Road, Springfield, VA 22161; Tel. 800-553-6847. NTIS Order Number is PB91-231480.

Bromobenzene **EPA Method 8021**
CAS #108-86-1

TITLE Halogenated Volatile by Gas Chromatography Using Photoionization and Electrolytic Conductivity Detectors in Series: Capillary Column Technique

MATRIX This method is applicable to nearly all types of samples, regardless of water content, including groundwater, aqueous sludges, caustic liquors, acid liquors, waste solvents, oily wastes, mousses, tars, fibrous wastes, polymeric emulsions, filter cakes, spent carbons, spent catalysts, soils, and sediments.

METHOD SUMMARY This method is used to determine 60 volatile organic compounds in a variety of solid waste matrices. It provides GC conditions for the detection of halogenated and aromatic volatile organic compounds. Samples can be analyzed using direct injection or purge-and-trap (EPA Method 5030). Groundwater samples must be analyzed using EPA Method 5030 (where applicable). A temperature program is used with the GC. Detection is achieved by a photoionization detector (PID) and a Hall electrolytic conductivity detector (HECD) in series.

INTERFERENCES Samples can be contaminated by diffusion of volatile organics (particularly chlorofluorocarbons and methylene chloride) through the sample container septum during shipment and storage.

INSTRUMENTATION A GC-equipped with variable-constant differential flow controllers, subambient oven controller,

PID and HECD detectors connected with a short piece of uncoated capillary tubing and a data system.

Column: 60 m × 0.75 mm I.D. VOCOL wide-bore capillary column with 1.5 μm film thickness.

PRECISION & ACCURACY MDLs are compound-dependent and vary with purging efficiency and concentration. The applicable concentration range of this method is compound- and instrument-dependent but is approximately 0.1–200 μg/L. Analytes that are inefficiently purged from water will not be detected when present at low concentrations, but they can be measured with acceptable accuracy and precision when present in sufficient amounts. The estimated quantitation limit (EQL) for an individual compound is approximately 1 μg/kg (wet weight) for soil/sediment samples, 100 μg/kg (wet weight) for wastes, and 1 μg/L for groundwater. EQLs will be proportionately higher for sample extracts and samples that require dilution or reduced sample size to avoid saturation of the detector.

MULTIPLICATION FACTORS FOR OTHER MATRICES (a)

Matrix	Factor (b)
Groundwater	10
Low-concentration soil	10
Water miscible liquid waste	500
High-concentration soil and sludge	1250
Non-water miscible waste	1250

(a) Sample EQLs are highly matrix-dependent. The EQLs listed herein are provided for guidance and may not always be achievable. (b) EQL = [Method detection limit] × [Factor]. For non-aqueous samples, the factor is on a wet-weight basis.

SINGLE LABORATORY ACCURACY & PRECISION DATA FOR VOCs IN WATER

This method was tested in a single lab using water spiked at 10 μg/L and the following data was reported:

Recoveries and standard deviations were determined from seven samples and spiked at 10 μg/L of each analyte. Recoveries were determined by the internal standard method. Internal standards were: Fluorobenzene for PID and 2-Bromo-1-chloropropane for HECD.

The average recovery (in percent) for the PID was 99.
The standard deviation of the recovery for the PID was 1.7.
The MDL (in μg/mL) for the PID was 0.006.
The average recovery (in percent) for the HECD was 97.
The standard deviation of the recovery for the HECD was 2.7.
The MDL (in μg/mL) for the HECD was 0.03.

SAMPLE COLLECTION, PRESERVATION & HANDLING
Volatile Organics — Standard 40-mL glass screw-cap VOA vials with Teflon®-faced silicone septum may be used for both liquid and solid matrices. When collecting samples, liquids and solids should be introduced into the vials gently to reduce agitation which might drive off volatile compounds. If there are any air bubbles present the sample must be retaken. Tap slightly as they are filled to try and eliminate as much free air space as possible. The two vials from each sampling locations should be sealed in separate plastic bags to prevent cross-contamination between samples particularly if the sampled waste is suspected of containing high levels of volatile organics.

Semivolatile organics — Containers used to collect samples for the determination of semivolatile organic compounds should be soap and water washed followed by methanol (or isopropanol) rinsing. The sample containers should be of glass or Teflon® and have screw-top covers with Teflon® liners.

Preservation for volatile organics — No preservation is used with concentrated waste samples. With liquid samples containing no residual chlorine, 4 drops of concentrated hydrochloric acid are added and the samples are immediately cooled to 4°C. When liquid samples contain residual chlorine, they are treated as above and, in addition, 4 drops of 4% aqueous sodium thiosulfate are added. Soil, sediment, and sludge samples are only cooled to 4°C.

Preservation for semivolatile organics — No preservation is used with concentrated waste samples. With liquid samples containing no residual chlorine and with soil, sediment, and sludge samples, immediately cooling to 4°C is the only preservation used. When residual chlorine is present then 3 mL of 10% aqueous sodium sulfate is added for each gallon of sample collected, followed by cooling to 4°C.

MHT The holding time for all volatile organics samples is 14 days. Liquid samples must be extracted within 7 days and their extracts analyzed within 40 days. Concentrated waste, soil, sediment, and sludge samples must be extracted within 14 days and their extracts analyzed within 40 days.

SAMPLE PREPARATION Volatile compounds are introduced into the gas chromatograph either by direct injector or purge-and-trap (EPA Method 5030). EPA Method 5030 may be used directly on groundwater samples or low-concentration contaminated soils and sediments. For medium-concentration soils or sediments, methanolic extraction, as described in EPA Method 5030, may be necessary prior to purge-and-trap analysis.

QUALITY CONTROL Calculate surrogate standard recovery on all samples, blanks, and spikes. A trip blank is recommended to check on sampling, storage, and handling contamination. Calibration standards, at a minimum of five concentration levels, are prepared in organic-free reagent water. One of the concentration levels should be at a concentration near, but above, the method detection limit.

A combination of bromochloromethane, 2-bromo-1-chloropropane, 1,4-dichlorobutane, and bromochlorobenzene are recommended as surrogate standards to encompass the range of the temperature program used in this method.

REFERENCE Test Methods for Evaluating Solid Waste, Physical/Chemical Methods, SW-846, 3rd Edition, U.S. EPA, Office of Solid Waste, Washington, DC, EPA Method 8021A, Rev. 1, Nov. 1992.

Bromobenzene   **EPA Method 8260**
CAS #108-86-1

TITLE Volatile Organic Compounds by GC/MS: Capillary Column Technique

MATRIX This method is applicable to nearly all types of samples, regardless of water content, including groundwater, soils, and sediments.

METHOD SUMMARY Method 8260A covers 58 volatile organic compounds that are introduced into a gas chromatograph by the purge-and-trap method or by direct injection (in limited applications). Zero-grade helium is bubbled through a 5-mL solution at ambient temperature. Purged sample components are trapped in a tube containing suitable sorbent materials. When purging is complete, the sorbent tube is heated and backflushed with helium to desorb trapped sample components. The analytes are desorbed directly to a large bore capillary or cryofocussed on a capillary precolumn before being flash evaporated to a narrow bore capillary for analysis.

INTERFERENCES Major contaminant sources are volatile materials in the lab and impurities in the inert purging gas and in the sorbent trap. Interfering contamination may occur when a sample containing low concentrations of volatile organic compounds is analyzed immediately after a sample containing high concentrations of volatile organic compounds. After analysis of a sample containing high concentrations of volatile organic compounds, one or more calibration blanks should be analyzed to check for cross-contamination. Screening of the samples prior to purge-and-trap GC/MS analysis is highly recommended to prevent contamination of the system. This is especially true for soil and waste samples.

Special precautions must be taken to analyze for methylene chloride. The analytical and sample storage area should be isolated from all atmospheric sources of methylene chloride. All gas chromatography carrier gas lines and purge gas plumbing should be constructed from stainless steel or copper tubing. Laboratory clothing previously exposed to methylene chloride fumes during liquid-liquid extraction procedures can contribute to sample contamination.

Samples can also be contaminated by diffusion of volatile organics (particularly methylene chloride and fluorocarbons) through the septum seal during shipment and storage. A trip blank can serve as a check on such contamination.

INSTRUMENTATION GC/MS with a temperature-programmable chromatograph suitable for splitless injection equipped with variable constant differential flow controllers, a subambient oven controller, a purging device, sorbent trap, a thermal desorption apparatus and a capillary precolumn interface when using cryogenic cooling will be needed. The following GC columns may be used:

Column 1: 60 m × 0.75 mm I.D. capillary column coated with VOCOL, 1.5 μm film thickness.
Column 2: 30 m × 0.53 mm I.D. capillary column coated with DB-624 or VOCOL, 3 μm film thickness.
Column 3: 30 m × 0.32 mm I.D. capillary column coated with DB-5 or SE-54, 1-μm film thickness.

PRECISION & ACCURACY This method has been tested in a single lab using spiked water. Using a wide-bore capillary column, water was spiked at concentrations between 0.5 and 10 μg/L. Single lab accuracy and precision data are presented. The MDL actually achieved in a given analysis will vary depending on instrument sensitivity and matrix effects.

The MDL (a) in μg/L was 0.03.
The concentration range in μg/L was 0.1–10.
The mean accuracy (% of true value) was 100.
The precision as relative standard deviation was 5.5.

Note: The MDL is based on a 25-mL sample volume instead of a 5-mL sample volume.

SAMPLING METHOD
Liquid samples — Use a 40-mL glass screw-cap VOA vial with a Teflon®-faced silicone septum that has been prewashed, rinsed with distilled deionized water, and oven dried. If residual chlorine is present, collect the sample in a 4-oz soil VOA container which has been pre-preserved with 4 drops of 10% sodium thiosulfate. Mix gently and transfer the sample to a 40-mL VOA vial. Collect bubble-free samples in duplicate and seal each sample in a separate plastic bag.

Soils, sediments, and sludges — Use an 8-oz widemouth glass bottle with Teflon®-faced silicone septum that has been prewashed, rinsed with distilled deionized water, and oven dried. **Do not** heat the septum for more than 1 h. Tap slightly to eliminate any free air space. Collect samples in duplicate and seal each one in a separate plastic bag.

SAMPLE PRESERVATION
Liquid samples — Add 4 drops of concentrated HCL, cool to 4°C and store in a solvent-free refrigerator.

Soils, sediments, and sludges — Cool samples to 4°C and store in a solvent-free refrigerator.

MHT The maximum holding time of any sample (liquids, soils, sediments, and sludges) is 14 days.

SAMPLE PREPARATION
Liquid samples — Remove the plunger from a 5-mL syringe and carefully pour the sample into the syringe barrel to just short of overflowing. Replace the syringe plunger and compress the sample. Open the syringe valve and vent any residual air while adjusting the sample volume to 5.0 mL. If there is only one volatile organic analysis (VOA) vial, a second syringe should be filled at this time to protect against possible loss of sample integrity. Add 10 μL of surrogate spiking solution and 10 μL of internal standard spiking solution through the valve bore of the 5-mL syringe, then close the valve. The surrogate and internal standards may be mixed and added as a single spiking solution.

Sediments, soils, and waste samples — All samples of this type should be screened by GC analysis using a headspace method (EPA Method 3810) or the hexadecane extraction and screening method (EPA Method 3820). Use the screening data to determine whether to use the low-concentration method (0.005–1 mg/kg) or the high-concentration method (>1 mg/kg).

Low-concentration method — The low-concentration method is based on purging a heated sediment or soil sample mixed with organic-free reagent water containing the surrogate and internal standards. Analyze all reagent blanks and standards under the same conditions as the samples.

Use a 5-g sample if the expected concentration is <0.1 mg/kg or a 1-g sample for expected concentrations between 0.1 and 1 mg/kg. Mix the contents of the sample container with a narrow metal spatula. Weigh the amount of the sample into a tared purge device. Add the spiked water to the purge device, which contains the weighed amount of sample, and connect the device to the purge-and-trap system.

High-concentration method — This method is based on extracting the sediment or soil with methanol. A waste sample is either extracted or diluted, depending on its solubility in methanol. Wastes that are insoluble in methanol are diluted with reagent tetraglyme or possibly polyethylene glycol (PEG). An aliquot of the extract is added to organic-free reagent water containing surrogate and internal standards. This is purged at ambient temperature. All samples with an expected concentration of >1.0 mg/kg should be analyzed by this method.

Mix the contents of the sample container with a narrow metal spatula. For sediments or soils and solid wastes that are insoluble in methanol, weigh 4 g (wet weight) of sample into a tared 20-mL vial. For waste that is soluble in methanol, tetraglyme, or PEG, weigh 1 g (wet weight) into a tared scintillation vial or culture tube or a 10-mL volumetric flask. Quickly add 9.0 mL of appropriate solvent then add 1.0 mL of a surrogate spiking solution to the vial, cap it, and shake it for 2 min.

METHANOL EXTRACT REQUIRED FOR ANALYSIS OF HIGH-CONCENTRATION SOILS OR SEDIMENTS

Approximate Concentration Range	Volume of Methanol Extract (a)
500–10,000 µg/kg	100 µL
1,000–20,000 µg/kg	50 µL
5,000–100,000 µg/kg	10 µL
25,000–500,000 µg/kg	100 µL of 1/50 dilution (b)

Calculate appropriate dilution factor for concentrations exceeding this table.

(a) The volume of methanol added to 5 mL of water being purged should be kept constant. Therefore, add to the 5-mL syringe whatever volume of methanol is necessary to maintain a volume of 100 µL added to the syringe.
(b) Dilute an aliquot of the methanol extract and then take 100 µL for analysis.

QUALITY CONTROL Demonstrate, through the analysis of a reagent water blank, that interferences from the analytical system, glassware, and reagents are under control. Blank samples should be carried through all stages of the sample preparation and measurement steps. For each analytical batch (up to 20 samples), a reagent blank, matrix spike, and matrix spike duplicate must be analyzed (the frequency of the spikes may be different for different monitoring programs). The blank and spiked samples must be carried through all stages of the sample preparation and measurement steps. QC samples mentioned in the section on Interferences will also be needed as appropriate to those situations.

Matrix spiking standards should be prepared from volatile organic compounds which will be representative of the compounds being investigated. The recommended internal standards are chlorobenzene-d5, 1,4-difluorobenzene, 1,4-dichlorobenzene-d4, and pentafluorobenzene. Using stock standard solutions, prepare secondary dilution standards containing the compounds of interest, either singly or mixed together in methanol. Store them in a vial with no headspace for no more than one week. Surrogates recommended are toluene-d8, 4-bromofluorobenzene, and dibromofluoromethane. Each sample undergoing GC/MS analysis must be spiked with 10 µL of the surrogate spiking solution prior to analysis.

REFERENCE Test Methods for Evaluating Solid Waste (SW-846). U.S. EPA 1983, Method 8260A, Rev. 1, Nov. 1990. Office of Solid Waste, Washington, DC.

Bromobenzene   **EPA Method 503.1**
CAS #108-86-1

TITLE Aromatic and Unsaturated VOCs in Water

MATRIX Drinking water (finished or any treatment stage) and raw source water.

APPLICATION Method covers 28 aromatic and unsaturated VOCs. An inert gas is bubbled through a 5-mL water sample. Purged sample components are trapped in tube of sorbent materials. When purging is complete, sorbent tube is heated and backflushed with inert gas to desorb trapped sample onto a packed GC column.

INTERFERENCES During analysis, major contaminant sources are volatile materials in the lab and impurities in purging gas and sorbent trap. With high and low level samples, there can be carry over contamination. Excess water causes a negative baseline deflection.

INSTRUMENTATION Purge and Trap GC w/photoionization detector. (Two GC columns are recommended); Column 1: 5% SP-1200 and 1.75% Bentone 34 on Supelcoport; 5% 1,2,3-tris(2-cyanoethoxy)propane on Chromosorb W.

RANGE 2.2–600 µg/L (drinking water).

MDL 0.002 µg/L in water.

PRECISION RSD = 6.2% at 0.50 µg/L; 19 samples.

ACCURACY Average recovery = 93% at 0.50 µg/L; 19 samples.

SAMPLING METHOD Use a 40- to 120-mL screw-cap vial (prewashed with detergent, rinsed with distilled water and oven dried at 105°C) with a PTFE-faced silicone septum. If residual chlorine is in the water add about 25 mg of ascorbic acid to each vial before sample collection. Collect bubble-free samples.

STABILITY Cool to 4°C; HCl to pH <2.

MHT 14 days.

QUALITY CONTROL As an initial demonstration of lab accuracy and precision, analyze 4 to 7 replicates of a lab fortified blank containing analyte at 0.1–5 µg/L. Collect all samples in duplicate.

REFERENCE Method 503.1, Volatile aromatic and unsaturated organic compounds in H2O by Purge and Trap GC, EPA 600/4-88/039.

2-Bromochlorobenzene   **EPA Method 1625**
CAS #694-80-4

TITLE Semivolatile Organic Compounds by Isotope Dilution GC/MS

MATRIX The compounds may be determined in waters, soils, and municipal sludges by this method.

METHOD SUMMARY This method is used to determine 176 semivolatile toxic organic pollutants associated with the CWA (as amended 1987); the RCRA (as amended 1986); the CERCLA (as amended 1986); and other compounds amenable to extraction and analysis by capillary column gas chromatography-mass spectrometry (GC/MS).

Stable isotopically-labeled analogs of the compounds of interest are added to the sample. If the solids content is less than 1%, a 1-L sample is extracted at pH 12–13, then at pH <2 with methylene chloride using continuous extraction techniques.

If the solids content is 30% or less, the sample is diluted to 1% solids with reagent water, homogenized ultrasonically, and extracted at pH 12–13, then at pH <2 with methylene chloride using continuous extraction techniques. If the solids content is greater than 30%, the sample is extracted using ultrasonic techniques.

Each extract is dried over sodium sulfate, concentrated to a volume of 5 mL, cleaned up using GPC, if necessary, and concentrated. Extracts are concentrated to 1 mL if GPC is not performed, and to 0.5 mL if GPC is performed.

An internal standard is added to the extract, and a 1-mL aliquot of the extract is injected into the GC. The compounds are separated by GC and detected by a MS. The labeled compounds serve to correct the variability of the analytical technique.

INTERFERENCES Solvents, reagents, glassware, and other sample processing hardware may yield artifacts and/or elevated baselines causing misinterpretation of chromatograms and spectra. Materials used in the analysis must be demonstrated to be free from interferences under the conditions of analysis by running method blanks initially and with each sample lot (sample started through the extraction process on a given 8-h shift, to a maximum of 20). Specific selection of reagents and purification of solvents by distillation in all glass systems may be required. Glassware and, where possible, reagents are cleaned by solvent rinse and baking at 450°C for 1-h minimum. Interferences coextracted from samples will vary considerably from source to source, depending on the diversity of the site being sampled.

INSTRUMENTATION Major instrumentation includes a GC with a splitless or on-column injection port for capillary column, a MS with 70 eV electron impact ionization, and a data system to collect and record MS data, and process it. A K-D apparatus is used to concentrate extracts.

GC Column: 30 m × 0.25 mm I.D. 5% phenyl, 94% methyl, 1% vinyl silicone bonded phased fused silica capillary column.

PRECISION & ACCURACY The detection limits of the method are usually dependent on the level of interferences rather than instrumental limitations. The limits typify the minimum quantities that can be detected with no interferences present.

The minimum level (in µg/mL) was not listed. This is defined as a minimum level at which the analytical system shall give recognizable mass spectra (background corrected) and acceptable calibration points.

The MDL (in µg/kg) in low solids was not listed and in high solids was not listed; these were determined in digested sludge (low solids) and in filter cake or compost (high solids).

The labeled and native compound initial precision as standard deviation (in µg/L) was not listed.

The labeled and native compound initial accuracy as average recovery (in µg/L) was not listed.

SAMPLE COLLECTION, PRESERVATION & HANDLING Collect samples in glass containers. Aqueous samples which flow freely are collected in refrigerated bottles using automatic sampling equipment. Solid samples are collected as grab samples using widemouth jars. Maintain samples at 0 to 4°C from the time of collection until extraction. If residual chlorine is present in aqueous samples, add 80 mg sodium thiosulfate/L of water. Begin sample extraction within 7 days of collection, and analyze all extracts within 40 days of extraction.

SAMPLE PREPARATION Samples containing 1% solids or less are extracted directly using continuous liquid-liquid extraction techniques. Samples containing 1 to 30% solids are diluted to the 1% level with reagent water and extracted using continuous liquid-liquid extraction techniques. Samples containing greater than 30% solids are extracted using ultrasonic techniques.

Base/neutral extraction — Adjust the pH of the waters in the extractors to 12–13 with 6 N NaOH. Extract with methylene chloride for 24–48 h.

Acid extraction — Adjust the pH of the waters in the extractors to 2 or less using 6 N sulfuric acid. Extract with methylene chloride for 24–48 h.

Ultrasonic extraction of high solids samples — Add anhydrous sodium sulfate to the sample and QC aliquot(s). Add acetone:methylene chloride (1:1) to the sample and mix thoroughly

Concentrate extracts using a K-D apparatus.

QUALITY CONTROL The analyst is permitted to modify this method to improve separations or lower the costs of measurements, provided all performance specifications are met. Analyses of blanks are required to demonstrate freedom from contamination. When results of spikes indicate atypical method performance for samples, the samples are diluted to bring method performance within acceptable limits.

For low solids (aqueous samples), extract, concentrate, and analyze two sets of four 1-L aliquots (8 aliquots total) of the precision and recovery standard. For high solids samples, two sets of four 30-g aliquots of the high solids reference matrix are used.

Spike all samples with labeled compounds to assess method performance. Compute percent recovery of the labeled compounds using the internal standard method. Compare the labeled compound recovery for each compound with the corresponding labeled compound recovery.

Reagent water and high solids reference matrix blanks are analyzed to demonstrate freedom from contamination. Extract and concentrate a 1-L reagent water blank or a high solids reference matrix blank with each sample's lot (samples started through the extraction process on the same 8-h shift, to a maximum of 20 samples).

Field replicates may be collected to determine the precision of the sampling technique, and spiked samples may be required to determine the accuracy of the analysis when the internal standard method is used.

REFERENCE Semivolatile Organic Compounds by Isotope Dilution GC/MS. Office of Water Regulation and Standards, U.S. EPA Industrial Technology Division, Washington, DC, EPA Method 1625, Rev. C, June 1989 (contact W.A. Telliard, U.S. EPA, Office of Water Regulations and Standards, 401 M St., SW, Washington, DC, 20460. Phone: 202-382-7131).

3-Bromochlorobenzene   **EPA Method 1625**
CAS #108-37-2

TITLE Semivolatile Organic Compounds by Isotope Dilution GC/MS

MATRIX The compounds may be determined in waters, soils, and municipal sludges by this method.

METHOD SUMMARY This method is used to determine 176 semivolatile toxic organic pollutants associated with the CWA (as amended 1987); the RCRA (as amended 1986); the CERCLA (as amended 1986); and other compounds amenable to extraction and analysis by capillary column gas chromatography-mass spectrometry (GC/MS).

Stable isotopically-labeled analogs of the compounds of interest are added to the sample. If the solids content is less than 1%, a 1-L sample is extracted at pH 12–13, then at pH <2 with methylene chloride using continuous extraction techniques.

If the solids content is 30% or less, the sample is diluted to 1% solids with reagent water, homogenized ultrasonically, and extracted at pH 12–13, then at pH <2 with methylene chloride using continuous extraction techniques. If the solids content is greater than 30%, the sample is extracted using ultrasonic techniques.

Each extract is dried over sodium sulfate, concentrated to a volume of 5 mL, cleaned up using GPC, if necessary, and concentrated. Extracts are concentrated to 1 mL if GPC is not performed, and to 0.5 mL if GPC is performed.

An internal standard is added to the extract, and a 1-mL aliquot of the extract is injected into the GC. The compounds are separated by GC and detected by a MS. The labeled compounds serve to correct the variability of the analytical technique.

INTERFERENCES Solvents, reagents, glassware, and other sample processing hardware may yield artifacts and/or elevated baselines causing misinterpretation of chromatograms and spectra. Materials used in the analysis must be demonstrated to be free from interferences under the conditions of analysis by running method blanks initially and with each sample lot (sample started through the extraction process on a given 8-h shift, to a maximum of 20). Specific selection of reagents and purification of solvents by distillation in all glass systems may be required. Glassware and, where possible, reagents are cleaned by solvent rinse and baking at 450°C for 1-h minimum. Interferences coextracted from samples will vary considerably from source to source, depending on the diversity of the site being sampled.

INSTRUMENTATION Major instrumentation includes a GC with a splitless or on-column injection port for capillary column, a MS with 70 eV electron impact ionization, and a data system to collect and record MS data, and process it. A K-D apparatus is used to concentrate extracts.

GC Column: 30 m × 0.25 mm I.D. 5% phenyl, 94% methyl, 1% vinyl silicone bonded phased fused silica capillary column.

PRECISION & ACCURACY The detection limits of the method are usually dependent on the level of interferences rather than instrumental limitations. The limits typify the minimum quantities that can be detected with no interferences present.

The minimum level (in µg/mL) was not listed. This is defined as a minimum level at which the analytical system shall give recognizable mass spectra (background corrected) and acceptable calibration points.
The MDL (in µg/kg) in low solids was not listed and in high solids was not listed; these were determined in digested sludge (low solids) and in filter cake or compost (high solids).
The labeled and native compound initial precision as standard deviation (in µg/L) was not listed.
The labeled and native compound initial accuracy as average recovery (in µg/L) was not listed.

SAMPLE COLLECTION, PRESERVATION & HANDLING Collect samples in glass containers. Aqueous samples which flow freely are collected in refrigerated bottles using automatic sampling equipment. Solid samples are collected as grab samples using widemouth jars. Maintain samples at 0 to 4°C from the time of collection until extraction. If residual chlorine is present in aqueous samples, add 80 mg sodium thiosulfate/L of water. Begin sample extraction within 7 days of collection, and analyze all extracts within 40 days of extraction.

SAMPLE PREPARATION Samples containing 1% solids or less are extracted directly using continuous liquid-liquid extraction techniques. Samples containing 1 to 30% solids are diluted to the 1% level with reagent water and extracted using continuous liquid-liquid extraction techniques. Samples containing greater than 30% solids are extracted using ultrasonic techniques.

Base/neutral extraction — Adjust the pH of the waters in the extractors to 12–13 with 6 N NaOH. Extract with methylene chloride for 24–48 h.

Acid extraction — Adjust the pH of the waters in the extractors to 2 or less using 6 N sulfuric acid. Extract with methylene chloride for 24–48 h.

Ultrasonic extraction of high solids samples — Add anhydrous sodium sulfate to the sample and QC aliquot(s). Add acetone:methylene chloride (1:1) to the sample and mix thoroughly

Concentrate extracts using a K-D apparatus.

QUALITY CONTROL The analyst is permitted to modify this method to improve separations or lower the costs of measurements, provided all performance specifications are met. Analyses of blanks are required to demonstrate freedom from contamination. When results of spikes indicate atypical method performance for samples, the samples are diluted to bring method performance within acceptable limits.

For low solids (aqueous samples), extract, concentrate, and analyze two sets of four 1-L aliquots (8 aliquots total) of the precision and recovery standard. For high solids samples, two sets of four 30-g aliquots of the high solids reference matrix are used.

Spike all samples with labeled compounds to assess method performance. Compute percent recoveryof the labeled compounds using the internal standard method. Compare the labeled compound recovery for each compound with the corresponding labeled compound recovery.

Reagent water and high solids reference matrix blanks are analyzed to demonstrate freedom from contamination. Extract and concentrate a 1-L reagent water blank or a high solids reference matrix blank with each sample's lot (samples started through the extraction process on the same 8-h shift, to a maximum of 20 samples).

Field replicates may be collected to determine the precision of the sampling technique, and spiked samples may be required to determine the accuracy of the analysis when the internal standard method is used.

REFERENCE Semivolatile Organic Compounds by Isotope Dilution GC/MS. Office of Water Regulation and Standards, U.S. EPA Industrial Technology Division, Washington, DC, EPA Method 1625, Rev. C, June 1989 (contact W.A. Telliard, U.S. EPA, Office of Water Regulations and Standards, 401 M St., SW, Washington, DC, 20460. Phone: 202-382-7131).

Bromochloromethane   **EPA Method 502**
CAS #74-97-5

TITLE Volatile Organic Compounds in Water By Purge and Trap Capillary Column Gas Chromatography with Photoionization and Electrolytic Conductivity Detectors in Series. U.S. EPA Method 502.2, Rev. 2.0, 1989.

MATRIX Drinking water and raw source water. The latter should include most surface water and groundwater sources.

METHOD SUMMARY This method covers 60 volatile organic compounds that contain halogen atoms and/or that are aromatic. An inert gas (zero grade nitrogen or helium) is bubbled through a 25-mL or a 5-mL water sample (depending on the expected concentration of the analytes). Purged sample components are trapped in a tube of sorbent materials. When purging is complete, the sorbent tube is heated and backflushed with helium to desorb the trapped sample onto a capillary GC column. The column is temperature programmed to separate the method analytes which are then detected with a photoionization detector (PID) and a Hall electrolytic conductivity (HECD) placed in series. The PID is selective for aromatic compounds and the HECD is selective for halogenated compounds.

INTERFERENCES Impurities in the purge gas and from organic compounds outgassing from the plumbing ahead of the trap account for many contamination problems. Interferences purged or coextracted from the samples will vary considerably from source to source, depending upon the particular sample or extract being tested. Cross-contamination can occur whenever high-level and low-level samples are analyzed sequentially. Samples also can be contaminated by diffusion of volatile organics (particularly methylene chloride and fluorocarbons) through the septum seal into the sample during shipment and storage. The lab where volatile analysis is performed and also the refrigerated storage area should be completely free of solvents.

INSTRUMENTATION A GC containing a series configuration of a high temperature photoionization detector (PID) equipped with 10.0 eV (nominal) lamp and Hall electrolytic conductivity detector (HECD) is required. Also required is an all-glass 5-mL purging device, a sorbent trap, and a thermal desorption apparatus which is connected to the GC system.

Column 1: VOCOL glass wide-bore capillary column.
Column 2: RTX–502.2 mega-bore capillary column.
Column 3: DB-62 mega-bore capillary column.

PRECISION & ACCURACY Method detection limits are dependent upon the characteristics of the gas chromatographic system used. Analytes that are not separated chromatographically cannot be individually identified and used in the same calibration mixture or water samples unless an alternative technique for identification and quantification, such as mass spectrometry, is used.

Electrolytic conductivity detetor (c) range in µg/L (a) was 0.02–200.
Electrolytic conductivity detetor (c) MDL in µg/L (b) was 0.01.
Electrolytic conductivity detetor (c) accuracy as % recovery was 96.
Electrolytic conductivity detetor (c) precision as % RSD was 3.0.
Photoionization detector (D) range in µg/L (a) was 0.02–200.
Photoionization detector (D) MDL in µg/L (b) was not listed.
Photoionization detector (D) accuracy as % recovery was not listed.
Photoionization detector (D) precision as % RSD was not listed.

(a) The applicable concentration range of this method is compound, instrument, and matrix-dependent. It is listed as being approximately 0.02 to 200 µg/L but no specific information is provided so caution should be observed.

(b) The method detection limits reports with this method are compound, instrument, and matrix-dependent. The values reported were calculated using reagent water fortified with the corresponding compounds at 10 µg/L and a GC-equipped with a 60 m × 0.75 mm VOLCOL wide bore capillary column with 1.5 µm film thickness and using helium carrier gas.

(c) Recoveries and relative standard deviations were determined from seven samples of reagent water fortified with 10 µg/L of each compound. 2-Bromo-1-chloropropane was used as the internal standard for calculating average recoveries.

(d) Recoveries and relative standard deviations were determined from seven samples of reagent water fortified with 10 µg/L of each compound. Fluorobenzene was used as the internal standard for calculating average recoveries.

SAMPLING METHOD Collect samples using a 40- to 120-mL screw-cap vial (prewashed with detergent, rinsed with distilled water and oven dried at 105°C) with a Teflon®-faced silicone septum. Collect bubble-free samples and place the septum with the Teflon® side down on the water.

SAMPLE PRESERVATION If residual chlorine is present in the water add about 25 mg of ascorbic acid to each vial before samples are collected to remove the chlorine. Add hydrochloric acid to reduce pH to <2, immediately cool samples to 4°C, and store them in a solvent-free refrigerator at 4°C until analysis.

MHT The maximum holding time for samples is 14 days from the time they were collected.

SAMPLE PREPARATION Remove the plungers from two 5-mL syringes and attach a closed syringe valve to each. Warm the sample to room temperature, open the sample bottle, and carefully pour the sample into one of the syringe barrels to just short of overflowing. Replace the syringe plunger, invert the syringe, and compress the sample. Open the syringe valve and vent any residual air while adjusting the sample volume to 5.0 mL. Add 10 µL of the internal calibration standard to the sample through the syringe valve. Close the valve. Fill the second syringe in an identical manner from the same sample bottle. Reserve this second syringe for a reanalysis if necessary.

QUALITY CONTROL As an initial demonstration of lab accuracy and precision, analyze 4 to 7 replicates of a lab fortified blank containing analyte at 0.1–5 µg/L. Collect all samples in duplicate. Surrogate analytes (similar to those of the analytes of interest), whose concentration is known in every sample, are measured using the same internal standard calibration procedure. Duplicate field reagent water blanks (trip blanks) must be analyzed with each set of samples, lab reagent blanks (method blanks) must be analyzed with each batch of samples processed as a group within a work shift. Also, a single lab-fortified blank that contains each of the analytes of interest should be analyzed with each batch of samples processed as a group within a work shift. A 3- to 5-point calibration curve is needed depending on the calibration range factor required.

EPA CONTACT & HOTLINE For technical questions contact Dr. Baldev Bathija, U.S. EPA, Office of Ground Water and Drinking Water (WH-550D), 401 M St. SW, Washington, DC 20460. Tel. (202) 260-3040. For further information the EPA Safe Drinking Water Hotline may be called at: (800) 426-4791.

REFERENCE Methods for the Determination of Organic Compounds in Drinking Water, EPA/600/4-88/039 (revised July 1991; Final Rule for determination of compliance with the MCL for Total Trihalomethanes under 141.30, in 40 CFR Part 141, Vol. 58, No. 147, Fed. Reg., Tuesday Aug. 3, 1993). U.S. EPA Environmental Monitoring Systems Laboratory, Cincinnati, OH, 45268, U.S.A. Available from the National Technical Information Service (NTIS), 5285 Port Royal Road, Springfield, VA 22161; Tel. 800-553-6847. NTIS Order Number is PB91-231480.

Bromochloromethane   EPA Method 524
CAS #74-97-5

TITLE Measurement of Purgeable Organic Compounds in Water by Capillary Column GC/MS.

MATRIX Drinking water and raw source water; the latter should include most surface water and groundwater sources.

METHOD SUMMARY Method 524.2 covers 60 volatile organic compounds. An inert gas (zero grade nitrogen or helium) is bubbled through a 25-mL or a 5-mL water sample (depending on the expected concentration of the analytes). Purged sample components are trapped in a tube of sorbent materials. When purging is complete, the sorbent tube is heated and backflushed with helium to desorb the trapped sample onto a capillary GC column.

INTERFERENCES Impurities in the purge gas and from organic compounds outgassing from the plumbing ahead of the trap account for many contamination problems. Interferences purged or coextracted from the samples will vary considerably from source to source, depending upon the particular sample or extract being tested. Cross-contamination can occur whenever high-level and low-level samples are analyzed sequentially. Samples also can be contaminated by diffusion of volatile organics (particularly methylene chloride and fluorocarbons) through the septum seal into the sample during shipment and storage.

INSTRUMENTATION A GC/MS with a data system equipped with one of the following capillary GC columns:

Column 1: VOCOL glass wide bore capillary column.
Column 2: DB-624 fused silica capillary column.
Column 3: DB-5 fused silica capillary column.

Also required is an all-glass 25-mL or 5-mL purging device, a sorbent trap, and a thermal desorption apparatus which is connected to the GC/MS system.

PRECISION & ACCURACY Method detection limits are compound- and instrument-dependent, and may vary from approximately 0.02–0.35 µg/L. Note in the table below that the "true" concentration range used for accuracy and precision measurements was quite narrow. However, the applicable concentration range of this method is primarily column dependent and is approximately 0.02 to 200 µg/L for the wide-bore thick-film columns. Narrow-bore thin-film columns may have a capacity which limits the range to about 0.02 to 20 µg/L. Analytes that are inefficiently purged from water will not be detected when present at low concentrations, but they can be measured with acceptable accuracy and precision when present in sufficient amounts.

Analytes that are not separated chromatographically, but which have different mass spectra and non-interfering quantification ions, can be identified and measured in the same calibration mixture or water sample. Analytes which have very similar mass spectra cannot be individually identified and measured in the same calibration mixture or water samples unless they have different retention times. Co-eluting compounds with very similar mass spectra, typically many structural isomers, must be reported as an isomeric group or pair.

The range (in µg/L) was 0.5–10.
The method detection limit (in µg/L) was 0.04.
The accuracy (as % recovery) was 90.
The precision (in %) was 6.4.

Note: Data were obtained from 16–31 determinations using a wide-bore capillary column and a jet separator interfaced to a quadrupole mass spectrometer. All analytes were in a reagent water matrix.

SAMPLING METHOD Collect samples using a 40- to 120-mL screw-cap vial (prewashed with detergent, rinsed with distilled water and oven dried at 105°C) with a Teflon®-faced silicone septum. Collect bubble-free samples and place the septum with the Teflon side down on the water.

SAMPLE PRESERVATION If residual chlorine is present in the water add about 25 mg of ascorbic acid to each vial before samples are collected to remove the chlorine. Add hydrochloric acid to reduce pH to <2, and immediately cool samples to 4°C, and store them in a solvent-free refrigerator at 4°C until analysis.

MHT The maximum holding time for samples is 14 days from the time they were collected.

SAMPLE PREPARATION Remove the plungers from two 25-mL (or 5-mL depending on sample size) syringes and attach a closed syringe valve to each. Warm the sample to room temperature, open the sample bottle, and carefully pour the sample into one of the syringe barrels to just short of overflowing. Replace the syringe plunger, invert the syringe, and compress the sample. Open the syringe valve and vent any residual air while adjusting the sample volume to 25.0 mL (or 5 mL). For samples and blanks, add 5 µL of the fortification solution containing the internal standard and the surrogates to the sample through the syringe valve. For calibration standards and lab fortified blanks, add 5 µL of the fortification solution containing the internal standard only. Close the valve. Fill the second syringe in an identical manner from the same sample bottle. Reserve this second syringe for a reanalysis if necessary.

QUALITY CONTROL As an initial demonstration of lab accuracy and precision, analyze 4 to 7 replicates of a lab fortified blank containing analyte at 0.2–5 µg/L. Collect all samples in duplicate. Surrogate analytes (similar to those of the analytes of interest), whose concentration is known in every sample, are measured using the same internal standard calibration procedure. Duplicate field reagent water blanks (trip blanks) must be analyzed with each set of samples, lab reagent blanks (method blanks) must be analyzed with each batch of samples processed as a group within a work shift. Also, a single lab-fortified blank that contains each of the analytes of interest should be analyzed with each batch of samples processed as a group within a work shift. A 3- to 5-point calibration curve is needed depending on the calibration range factor required.

EPA CONTACT & HOTLINE For technical questions contact Dr. Baldev Bathija, U.S. EPA, Office of Ground Water and Drinking Water (WH-550D), 401 M St. SW, Washington, DC 20460. Tel. (202) 260-3040. For further information the EPA Safe Drinking Water Hotline may be called at: (800) 426-4791.

REFERENCE Methods for the Determination of Organic Compounds in Drinking Water, EPA/600/4-88/039 (revised July 1991; Final Rule for determination of compliance with the MCL for Total Trihalomethanes under 141.30, in 40 CFR Part 141, Vol. 58, No. 147, Fed. Reg., Tuesday Aug. 3, 1993). U.S. EPA Environmental Monitoring Systems Laboratory, Cincinnati, OH, 45268, U.S.A. Available from the National Technical Information Service (NTIS), 5285 Port Royal Road, Springfield, VA 22161; Tel. 800-553-6847. NTIS Order Number is PB91-231480.

Bromochloromethane   **EPA Method 8021**
CAS #74-97-5

TITLE Halogenated Volatile by Gas Chromatography Using Photoionization and Electrolytic Conductivity Detectors in Series: Capillary Column Technique

MATRIX This method is applicable to nearly all types of samples, regardless of water content, including groundwater, aqueous sludges, caustic liquors, acid liquors, waste solvents, oily wastes, mousses, tars, fibrous wastes, polymeric emulsions, filter cakes, spent carbons, spent catalysts, soils, and sediments.

METHOD SUMMARY This method is used to determine 60 volatile organic compounds in a variety of solid waste matrices. It provides GC conditions for the detection of halogenated and aromatic volatile organic compounds. Samples can be analyzed using direct injection or purge-and-trap (EPA Method 5030). Groundwater samples must be analyzed using EPA Method 5030 (where applicable). A temperature program is used with the GC. Detection is achieved by a photoionization detector (PID) and a Hall electrolytic conductivity detector (HECD) in series.

INTERFERENCES Samples can be contaminated by diffusion of volatile organics (particularly chlorofluorocarbons and methylene chloride) through the sample container septum during shipment and storage.

INSTRUMENTATION A GC-equipped with variable-constant differential flow controllers, subambient oven controller, PID, and HECD detectors connected with a short piece of uncoated capillary tubing and a data system.

Column: 60 m × 0.75 mm I.D. VOCOL wide-bore capillary column with 1.5 μm film thickness.

PRECISION & ACCURACY MDLs are compound-dependent and vary with purging efficiency and concentration. The applicable concentration range of this method is compound- and instrument-dependent but is approximately 0.1–200 μg/L. Analytes that are inefficiently purged from water will not be detected when present at low concentrations, but they can be measured with acceptable accuracy and precision when present in sufficient amounts. The estimated quantitation limit (EQL) for an individual compound is approximately 1 μg/kg (wet weight) for soil/sediment samples, 100 μg/kg (wet weight) for wastes, and 1 μg/L for groundwater. EQLs will be proportionately higher for sample extracts and samples that require dilution or reduced sample size to avoid saturation of the detector.

MULTIPLICATION FACTORS FOR OTHER MATRICES (a)

Matrix	Factor (b)
Groundwater	10
Low-concentration soil	10
Water miscible liquid waste	500
High-concentration soil and sludge	1250
Non-water miscible waste	1250

(a) Sample EQLs are highly matrix-dependent. The EQLs listed herein are provided for guidance and may not always be achievable.
(b) EQL = [Method detection limit] × [Factor]. For non-aqueous samples, the factor is on a wet-weight basis.

SINGLE LABORATORY ACCURACY & PRECISION DATA FOR VOCs IN WATER

This method was tested in a single lab using water spiked at 10 μg/L and the following data was reported:

Recoveries and standard deviations were determined from seven samples and spiked at 10 μg/L of each analyte. Recoveries were determined by the internal standard method. Internal standards were: Fluorobenzene for PID and 2-Bromo-1-chloropropane for HECD.

The average recovery (in percent) for the PID was none (no response for this detector).
The standard deviation of the recovery for the PID was none (no response for this detector).
The MDL (in μg/mL) for the PID was none (no response for this detector).
The average recovery (in percent) for the HECD was 96.
The standard deviation of the recovery for the HECD was 3.0.
The MDL (in μg/mL) for the HECD was 0.01.

SAMPLE COLLECTION, PRESERVATION & HANDLING
Volatile Organics — Standard 40-mL glass screw-cap VOA vials with Teflon®-faced silicone septum may be used for both liquid and solid matrices. When collecting samples, liquids and solids should be introduced into the vials gently to reduce agitation which might drive off volatile compounds. If there are any air bubbles present the sample must be retaken. Tap slightly as they are filled to try and eliminate as much free air space as possible. The two vials from each sampling locations should be sealed in separate plastic bags to prevent cross-contamination between samples particularly if the sampled waste is suspected of containing high levels of volatile organics.

Semivolatile organics — Containers used to collect samples for the determination of semivolatile organic compounds should be soap and water washed followed by methanol (or isopropanol) rinsing. The sample containers should be of glass or Teflon® and have screw-top covers with Teflon® liners.

Preservation for volatile organics — No preservation is used with concentrated waste samples. With liquid samples containing no residual chlorine, 4 drops of concentrated hydrochloric acid are added and the samples are immediately cooled to 4°C. When liquid samples contain residual chlorine, they are treated as above and, in addition, 4 drops of 4% aqueous sodium thiosulfate are added. Soil, sediment, and sludge samples are only cooled to 4°C.

Preservation for semivolatile organics — No preservation is used with concentrated waste samples. With liquid samples containing no residual chlorine and with soil, sediment, and sludge samples, immediately cooling to 4°C is the only preservation used. When residual chlorine is present then 3 mL of 10% aqueous sodium sulfate is added for each gallon of sample collected, followed by cooling to 4°C.

MHT The holding time for all volatile organics samples is 14 days. Liquid samples must be extracted within 7 days and their extracts analyzed within 40 days. Concentrated waste, soil, sediment, and sludge samples must be extracted within 14 days and their extracts analyzed within 40 days.

SAMPLE PREPARATION Volatile compounds are introduced into the gas chromatograph either by direct injector or purge-and-trap (EPA Method 5030). EPA Method 5030 may be used directly on groundwater samples or low-concentration contaminated soils and sediments. For medium-concentration soils or sediments, methanolic extraction, as described in EPA Method 5030, may be necessary prior to purge-and-trap analysis.

QUALITY CONTROL Calculate surrogate standard recovery on all samples, blanks, and spikes. A trip blank is recommended to check on sampling, storage, and handling contamination. Calibration standards, at a minimum of five concentration levels, are prepared in organic-free reagent water. One of the concentration levels should be at a concentration near, but above, the method detection limit.

A combination of bromochloromethane, 2-bromo-1-chloropropane, 1,4-dichlorobutane, and bromochlorobenzene are recommended as surrogate standards to encompass the range of the temperature program used in this method.

REFERENCE Test Methods for Evaluating Solid Waste, Physical/Chemical Methods, SW-846, 3rd Edition, U.S. EPA, Office of Solid Waste, Washington, DC, EPA Method 8021A, Rev. 1, Nov. 1992.

Bromochloromethane   **EPA Method 8260**
CAS #74-97-5

TITLE Volatile Organic Compounds by GC/MS: Capillary Column Technique

MATRIX This method is applicable to nearly all types of samples, regardless of water content, including groundwater, soils, and sediments.

METHOD SUMMARY Method 8260A covers 58 volatile organic compounds that are introduced into a gas chromatograph by the purge-and-trap method or by direct injection (in limited applications). Zero-grade helium is bubbled through a 5-mL solution at ambient temperature. Purged sample components are trapped in a tube containing suitable sorbent materials. When purging is complete, the sorbent tube is heated and backflushed with helium to desorb trapped sample components. The analytes are desorbed directly to a large bore capillary or cryofocussed on a capillary precolumn before being flash evaporated to a narrow bore capillary for analysis.

INTERFERENCES Major contaminant sources are volatile materials in the lab and impurities in the inert purging gas and in the sorbent trap. Interfering contamination may occur when a sample containing low concentrations of volatile organic compounds is analyzed immediately after a sample containing high concentrations of volatile organic compounds. After analysis of a sample containing high concentrations of volatile organic compounds, one or more calibration blanks should be analyzed to check for cross-contamination. Screening of the samples prior to purge-and-trap GC/MS analysis is highly recommended to prevent contamination of the system. This is especially true for soil and waste samples.

Special precautions must be taken to analyze for methylene chloride. The analytical and sample storage area should be isolated from all atmospheric sources of methylene chloride. All gas chromatography carrier gas lines and purge gas plumbing should be constructed from stainless steel or copper tubing. Laboratory clothing previously exposed to methylene chloride fumes during liquid-liquid extraction procedures can contribute to sample contamination.

Samples can also be contaminated by diffusion of volatile organics (particularly methylene chloride and fluorocarbons) through the septum seal during shipment and storage. A trip blank can serve as a check on such contamination.

INSTRUMENTATION GC/MS with a temperature-programmable chromatograph suitable for splitless injection equipped with variable constant differential flow controllers, a subambient oven controller, a purging device, sorbent trap, a thermal desorption apparatus and a capillary precolumn interface when using cryogenic cooling will be needed. The following GC columns may be used:

Column 1: 60 m × 0.75 mm I.D. capillary column coated with VOCOL, 1.5 µm film thickness.
Column 2: 30 m × 0.53 mm I.D. capillary column coated with DB-624 or VOCOL, 3 µm film thickness.
Column 3: 30 m × 0.32 mm I.D. capillary column coated with DB-5 or SE-54, 1-µm film thickness.

PRECISION & ACCURACY This method has been tested in a single lab using spiked water. Using a wide-bore capillary column, water was spiked at concentrations between 0.5 and 10 µg/L. Single lab accuracy and precision data are presented. The MDL actually achieved in a given analysis will vary depending on instrument sensitivity and matrix effects.

The MDL (a) in µg/L was 0.04.
The concentration range in µg/L was 0.5–10.
The mean accuracy (% of true value) was 90.
The precision as relative standard deviation was 6.4.

Note: The MDL is based on a 25-mL sample volume instead of a 5-mL sample volume.

SAMPLING METHOD
Liquid samples — Use a 40-mL glass screw-cap VOA vial with a Teflon®-faced silicone septum that has been prewashed, rinsed with distilled deionized water, and oven dried. If residual chlorine is present, collect the sample in a 4-oz soil VOA container which has been pre-preserved with 4 drops of 10% sodium thiosulfate. Mix gently and transfer the sample to a 40-mL VOA vial. Collect bubble-free samples in duplicate and seal each sample in a separate plastic bag.

Soils, sediments, and sludges — Use an 8-oz widemouth glass bottle with Teflon®-faced silicone septumthat has been prewashed, rinsed with distilled deionized water, and oven dried. **Do not** heat the septum for more than 1 h. Tap slightly to eliminate any free air space. Collect samples in duplicate and seal each one in a separate plastic bag.

SAMPLE PRESERVATION
Liquid samples — Add 4 drops of concentrated HCL, cool to 4°C and store in a solvent-free refrigerator.

Soils, sediments, and sludges — Cool samples to 4°C and store in a solvent-free refrigerator.

MHT The maximum holding time of any sample (liquids, soils, sediments, and sludges) is 14 days.

SAMPLE PREPARATION
Liquid samples — Remove the plunger from a 5-mL syringe and carefully pour the sample into the syringe barrel to just short of overflowing. Replace the syringe plunger and compress the sample. Open the syringe valve and vent any residual air while adjusting the sample volume to 5.0 mL. If there is only one volatile organic analysis (VOA) vial, a second syringe should be filled at this time to protect against possible loss of sample integrity. Add 10 µL of surrogate spiking solution and 10 µL of internal standard spiking solution through the valve bore of the 5-mL syringe, then close the valve. The surrogate and internal standards may be mixed and added as a single spiking solution.

Sediments, soils, and waste samples — All samples of this type should be screened by GC analysis using a headspace method (EPA Method 3810) or the hexadecane extraction and screening method (EPA Method 3820). Use the screening data to determine whether to use the low-concentration method (0.005–1 mg/kg) or the high-concentration method (>1 mg/kg).

Low-concentration method — The low-concentration method is based on purging a heated sediment or soil sample mixed with organic-free reagent water containing the surrogate and internal standards. Analyze all reagent blanks and standards under the same conditions as the samples.

Use a 5-g sample if the expected concentration is <0.1 mg/kg or a 1-g sample for expected concentrations between 0.1 and 1 mg/kg. Mix the contents of the sample container with a narrow metal spatula. Weigh the amount of the sample into a tared purge device. Add the spiked water to the purge device, which contains the weighed amount of sample, and connect the device to the purge-and-trap system.

High-concentration method — This method is based on extracting the sediment or soil with methanol. A waste sample is either extracted or diluted, depending on its solubility in methanol. Wastes that are insoluble in methanol are diluted with reagent tetraglyme or possibly polyethylene glycol (PEG). An aliquot of the extract is added to organic-free reagent water containing surrogate and internal standards. This is purged at ambient temperature. All samples with an expected concentration of >1.0 mg/kg should be analyzed by this method.

Mix the contents of the sample container with a narrow metal spatula. For sediments or soils and solid wastes that are insoluble in methanol, weigh 4 g (wet weight) of sample into a tared 20-mL vial. For waste that is soluble in methanol, tetraglyme, or PEG, weigh 1 g (wet weight) into a tared scintillation vial or culture tube or a 10-mL volumetric flask. Quickly add 9.0 mL of appropriate solvent then add 1.0 mL of a surrogate spiking solution to the vial, cap it, and shake it for 2 min.

METHANOL EXTRACT REQUIRED FOR ANALYSIS OF HIGH-CONCENTRATION SOILS OR SEDIMENTS

Approximate Concentration Range	Volume of Methanol Extract (a)
500–10,000 µg/kg	100 µL
1,000–20,000 µg/kg	50 µL
5,000–100,000 µg/kg	10 µL
25,000–500,000 µg/kg	100 µL of 1/50 dilution (b)

Calculate appropriate dilution factor for concentrations exceeding this table.

(a) The volume of methanol added to 5 mL of water being purged should be kept constant. Therefore, add to the 5-mL syringe whatever volume of methanol is necessary to maintain a volume of 100 µL added to the syringe.
(b) Dilute an aliquot of the methanol extract and then take 100 µL for analysis.

QUALITY CONTROL Demonstrate, through the analysis of a reagent water blank, that interferences from the analytical system, glassware, and reagents are under control. Blank samples should be carried through all stages of the sample preparation and measurement steps. For each analytical batch (up to 20 samples), a reagent blank, matrix spike, and matrix spike duplicate must be analyzed (the frequency of the spikes may be different for different monitoring programs). The blank and spiked samples must be carried through all stages of the sample preparation and measurement steps. QC samples mentioned in the section on Interferences will also be needed as appropriate to those situations.

Matrix spiking standards should be prepared from volatile organic compounds which will be representative of the compounds being investigated. The recommended internal standards are chlorobenzene-d5, 1,4-difluorobenzene, 1,4-dichlorobenzene-d4, and pentafluorobenzene. Using stock standard solutions, prepare secondary dilution standards containing the compounds of interest, either singly or mixed together in methanol. Store them in a vial with no headspace for no more than one week. Surrogates recommended are toluene-d8, 4-bromofluorobenzene, and dibromofluoromethane. Each sample undergoing GC/MS analysis must be spiked with 10 µL of the surrogate spiking solution prior to analysis.

REFERENCE Test Methods for Evaluating Solid Waste (SW-846). U.S. EPA 1983, Method 8260A, Rev. 1, Nov. 1990. Office of Solid Waste, Washington, DC.

Bromocil   **EPA Method 507**
CAS #314-40-9

TITLE Determination of Nitrogen and Phosphorus-Containing Pesticides in Water by GC/NPD

MATRIX This method is applicable to the determination of certain nitrogen and phosphorus-containing pesticides in finished drinking water and groundwater.

METHOD SUMMARY Method 507 covers 46 nitrogen- and phosphorus-containing pesticides. A 1 L sample is fortified with a surrogate standard, salted, buffered, extracted with methylene chloride, and concentrated; then the solvent is exchanged with methyl tert-butyl ether (MTBE) and concentrated again, and a 2-µL aliquot of a sample extract is injected into a GC system equipped with a selective nitrogen-phosphorus detector and a capillary column for analysis.

INTERFERENCES Method interferences may be caused by contaminants in solvents, reagents, glassware, and other sample processing apparatus. Interfering contamination may occur when a sample containing low concentrations of analytes is analyzed immediately following a sample containing relatively high concentrations. One or more injections of MTBE should be made following the analysis of a sample with high concentrations of analytes to check for analyte carryover. Matrix interferences may be caused by contaminants that are coextracted from the sample. The extent of matrix interferences will vary considerably from source to source, depending upon the water sampled.

INSTRUMENTATION A gas chromatograph system (GC) equipped with a nitrogen-phosphorus detector (NPD) is needed.

Column 1: 30 m × 0.25 mm I.D. DB-5 bonded fused silica column, 0.25 µm film thickness, or equivalent.
Column 2: 30 m × 0.25 mm I.D. DB-1701 bonded fused silica column, 0.25 µm film thickness, or equivalent.

PRECISION & ACCURACY This method has been validated in a single lab and estimated detection limits (EDLs) have been determined for each analyte. Observed detection limits may vary among waters, depending upon the nature of the interferences in the sample matrix and the specific instrumentation used. Analytes that are not separated chromatographically cannot be individually identified and measured unless an alternative technique for identification and quantification exist.

The estimated detection limit (in µg/L) was 2.5. The EDL is defined as either method detection limit or a level of compound in a sample yielding a peak in the final extract with signal-to-noise ratio of approximately 5, whichever value is higher.

The concentration used for these measurements (in µg/L) was 25.
The accuracy (as % recovery) was 91.
The precision (% RSD) was 9.

SAMPLING METHOD Grab samples are collected in 1-L glass sample bottles (prewashed with detergent and hot tap water, rinsed with reagent water, and dried in an oven at 400°C for 1 h) with screw caps lined with PTFE-fluorocarbon.

SAMPLE PRESERVATION Add mercuric chloride to the sample bottle in amounts to produce a concentration of 10 mg/L. If residual chlorine is present, add 80 mg of sodium thiosulfate/L of sample to the sample bottle prior to collection. After collection, seal bottle and shake vigorously for 1 min, then cool the sample to 4°C immediately and store it at 4°C in the dark until extraction.

MHT Maximum holding time of the samples, and in some cases the extracts, is 14 days.

SAMPLE PREPARATION Fortify the sample with 50 µL of the surrogate standard solution, adjust to pH 7 with phosphate buffer, add 100 g NaCl to the sample, and seal and shake to dissolve the salt; then extract with methylene chloride in a separatory funnel or in a mechanical tumbler bottle. Dry the extract by pouring it through a solvent-rinsed drying column containing about 10 cm of anhydrous sodium sulfate. Collect the extract in a Kuderna-Danish (K-D) concentrator and rinse the column with 20-30 mL methylene chloride. Concentrate the extract to about 2 mL and rinse the flask and its lower joint into the concentrator tube with 1 to 2 mL of methyl t-butyl ether (MTBE). Add 5–10 mL of MTBE and concentrate the extract twice (adding more MTBE) to a final volume of 5.0 mL and store it at 4°C until analysis.

Note: If methylene chloride is not completely removed from the final extract, it may cause detector problems.

QUALITY CONTROL Minimum quality control requirements are initial demonstration of lab capability, determination of surrogate compound recoveries in each sample and blank, monitoring internal standard peak area or height in each sample and blank, analysis of lab reagent blanks, lab fortified samples, lab fortified blanks, and other QC samples. A lab reagent blank is analyzed to demonstrate that all glassware and reagent interferences are under control.

Initial demonstration of capability is fulfilled by analyzing four fortified reagent water samples with the recovery value for each analyte falling within the acceptable range (±30% average recovery). Surrogate recoveries from samples or method blanks must be 70–130%. The internal standard response for any sample chromatogram should not deviate from the daily calibration check standard's internal standard response by more than 30% or lab fortified blanks and sample matrices are used to assess lab performance and analyte recovery, respectively.

If the response for the target analyte peak exceeds the working range of the system, dilute the extract and reanalyze. Alternative techniques such as an alternate detector or second chromatography column should be used to confirm peak identification when sample components are not resolved adequately.

EPA CONTACT & HOTLINE For technical questions contact Dr. Baldev Bathija, U.S. EPA, Office of Ground Water and Drinking Water (WH-550D), 401 M St. SW, Washington, DC 20460. Tel. (202) 260-3040. For further information the EPA Safe Drinking Water Hotline may be called at: (800) 426-4791.

REFERENCE Methods for the Determination of Organic Compounds in Drinking Water, EPA/600/4-88/039 (revised July 1991). U.S. EPA Environmental Monitoring Systems Laboratory, Cincinnati, OH, 45268, U.S.A. Available from the National Technical Information Service (NTIS), 5285 Port Royal Road, Springfield, VA 22161; Tel. 800-553-6847. NTIS Order Number is PB91-231480.

Bromodichloromethane — EPA Method 1624
CAS #75-27-4

TITLE Volatile Organic Compounds by Isotope Dilution GC/MS

MATRIX Compounds may be determined in waters, soils, and municipal sludges by this method.

METHOD SUMMARY This method is used to determine 58 volatile toxic organic pollutants associated with the CWA (as amended 1987); the RCRA (as amended 1986); the CERCLA (as amended 1986); and other compounds amenable to purge-and-trap gas chromatography-mass spectrometry (GC/MS).

If the solids content is less than 1%, stable isotopically-labeled analogs of the compounds of interest are added to a 5-mL sample and the sample is purged with an inert gas at 20–25°C in a chamber designed for soil or water samples. If the solids content is greater than 1%, 5 mL of reagent water and the labeled compounds are added to a 5-g aliquot of sample and the mixture is purged at 40°C. Compounds that will not purge at 20–25°C or at 40°C are purged at 78–85°C. In the purging process, the volatile compounds are transferred from the aqueous phase into the gaseous phase where they are passed into a sorbent column, and trapped. After purging is completed, the trap is backflushed and heated rapidly to desorb the compounds into a GC. The compounds are separated by the GC and detected by a MS. The labeled compounds serve to correct the variability of the analytical technique.

INTERFERENCES Impurities in the purge gas, organic compounds outgassing from the plumbing upstream of the trap, and solvent vapors in the lab account for most problems. Samples can be contaminated by diffusion of volatile organic compounds (particularly methylene chloride) through the bottle seal during shipment and storage. Contamination by carryover can occur when high-level and low-level samples are analyzed sequentially. When an unusually concentrated sample is encountered, follow it by analysis of a reagent water blank to check for carryover.

INSTRUMENTATION Major equipment includes a GC with linear temperature programming and a glass jet separator as the MS interface, a MS with 70 eV electron impact ionization, and a data system to collect and record response factors.

Column: 2.8 m × 2 mm I.D. glass, packed with 1% SP-1000 on Carbopak B, 60/80 mesh, or equivalent.

PRECISION & ACCURACY The detection limits of the method are usually dependent on the level of interferences rather than instrumental limitations. The Method detection limitswere determined in digested sludge (low solids) and in filter cake or compost (high solids).

The MDL (in µg/kg) for low solids is 28and for high solids is 33.
Labeled and native compound precision (in µg/L) as standard deviation was 8.2.
Labeled and native compound accuracy (in µg/L) as average recovery was 7–32.
Acceptance criteria are at 20 µg/L for this compound.

SAMPLE COLLECTION, PRESERVATION & HANDLING Grab samples are collected in glass containers having a total volume greater than 20 mL. Fill and seal each bottle so that no air bubbles are entrapped. Samples are maintained at 0 to 4°C from the time of collection until analysis. If an aqueous sample contains residual chlorine, add sodium thiosulfate preservative (10 mg/40 mL) to the empty sample bottles just prior to shipment to the sample site. All samples must be analyzed within 14 days of collection.

SAMPLE PREPARATION Samples containing less than 1% solids are analyzed directly as aqueous samples. Samples containing 1% solids or greater are analyzed as solid samples utilizing one of two methods, depending on the levels of pollutants, in the sample. Samples containing 1% solids or greater, and low to moderate levels of pollutants are analyzed by purging a known weight of sample added to 5 mL of reagent water. Samples containing 1% solids or greater, and high levels of pollutants, are extracted with methanol, and an aliquot of the methanol extract is added to reagent water and purged.

QUALITY CONTROL A field blank prepared from reagent water and carried through the sampling and handling protocol may serve as a check on contamination from shipment and storage.

The analyst is permitted to modify this method to improve separations or lower the costs of measurements, provided all performance specifications are met. Analyses of blanks are required. When results of spikes indicate atypical method performance for samples, the samples are diluted to bring method performance within acceptable limits. Analyze two sets of four 5-mL aliquots (8 aliquots total) of the aqueous performance standard. Spike all samples with labeled compounds to assess method performance on the sample matrix. Compute the percent recovery of the labeled compounds using the internal standard method. Compare the percent recovery for each compound with the corresponding labeled compound recovery. Reagent water blanks are analyzed to demonstrate freedom from carryover contamination. Field replicates may be collected to determine the precision of the sampling technique, and spiked samples may be required to determine the accuracy of the analysis when the internal method is used.

REFERENCE Volatile Organic Compounds by Isotope Dilution GC/MS. Office of Water Regulation and Standards, U.S. EPA Industrial Technology Division, Washington, DC, EPA Method 1624, Rev. C, June 1989 (contact W.A. Telliard, U.S. EPA, Office of Water Regulations and Standards, 401 M St., SW, Washington, DC, 20460. Phone: 202-382-7131).

Bromodichloromethane   **EPA Method 502**
CAS #75-27-4

TITLE Volatile Organic Compounds in Water By Purge and Trap Capillary Column Gas Chromatography with Photoionization and Electrolytic Conductivity Detectors in Series. U.S. EPA Method 502.2, Rev. 2.0, 1989.

MATRIX Drinking water and raw source water. The latter should include most surface water and groundwater sources.

METHOD SUMMARY This method covers 60 volatile organic compounds that contain halogen atoms and/or that are aromatic. An inert gas (zero grade nitrogen or helium) is bubbled through a 25-mL or a 5-mL water sample (depending on the expected concentration of the analytes). Purged sample components are trapped in a tube of sorbent materials. When purging is complete, the sorbent tube is heated and backflushed with helium to desorb the trapped sample onto a capillary GC column. The column is temperature programmed to separate the method analytes which are then detected with a photoionization detector (PID) and a Hall electrolytic conductivity (HECD) placed in series. The PID is selective for aromatic compounds and the HECD is selective for halogenated compounds.

INTERFERENCES Impurities in the purge gas and from organic compounds outgassing from the plumbing ahead of the trap account for many contamination problems. Interferences purged or coextracted from the samples will vary considerably from source to source, depending upon the particular sample or extract being tested. Cross-contamination can occur whenever high-level and low-level samples are analyzed sequentially. Samples also can be contaminated by diffusion of volatile organics (particularly methylene chloride and fluorocarbons) through the septum seal into the sample during shipment and storage. The lab where volatile analysis is performed and also the refrigerated storage area should be completely free of solvents.

INSTRUMENTATION A GC containing a series configuration of a high temperature photoionization detector (PID) equipped with 10.0 eV (nominal) lamp and Hall electrolytic conductivity detector (HECD) is required. Also required is an all-glass 5-mL purging device, a sorbent trap, and a thermal desorption apparatus which is connected to the GC system.

Column 1: VOCOL glass wide-bore capillary column.

Column 2: RTX–502.2 mega-bore capillary column.
Column 3: DB-62 mega-bore capillary column.

PRECISION & ACCURACY Method detection limits are dependent upon the characteristics of the gas chromatographic system used. Analytes that are not separated chromatographically cannot be individually identified and used in the same calibration mixture or water samples unless an alternative technique for identification and quantification, such as mass spectrometry, is used.

Electrolytic conductivity detetor (c) range in µg/L (a) was 0.02–200.
Electrolytic conductivity detetor (c) MDL in µg/L (b) was 0.02.
Electrolytic conductivity detetor (c) accuracy as % recovery was 97.
Electrolytic conductivity detetor (c) precision as % RSD was 2.9.
Photoionization detector (d) range in µg/L (a) was 0.02–200.
Photoionization detector (d) MDL in µg/L (b) was not listed.
Photoionization detector (d) accuracy as % recovery was not listed.
Photoionization detector (d) precision as % RSD was not listed.

(a) The applicable concentration range of this method is compound, instrument, and matrix-dependent. It is listed as being approximately 0.02 to 200 µg/L but no specific information is provided so caution should be observed.
(b) The method detection limits reports with this method are compound, instrument, and matrix-dependent. The values reported were calculated using reagent water fortified with the corresponding compounds at 10 µg/L and a GC-equipped with a 60 m × 0.75 mm VOLCOL wide bore capillary column with 1.5 µm film thickness and using helium carrier gas.
(c) Recoveries and relative standard deviations were determined from seven samples of reagent water fortified with 10 µg/L of each compound. 2-Bromo-1-chloropropane was used as the internal standard for calculating average recoveries.
(d) Recoveries and relative standard deviations were determined from seven samples of reagent water fortified with 10 µg/L of each compound. Fluorobenzene was used as the internal standard for calculating average recoveries.

SAMPLING METHOD Collect samples using a 40- to 120-mL screw-cap vial (prewashed with detergent, rinsed with distilled water and oven dried at 105°C) with a Teflon®-faced silicone septum. Collect bubble-free samples and place the septum with the Teflon® side down on the water.

SAMPLE PRESERVATION If residual chlorine is present in the water add about 25 mg of ascorbic acid to each vial before samples are collected to remove the chlorine. Add hydrochloric acid to reduce pH to <2, immediately coolsamples to 4°C, and store them in a solvent-free refrigerator at 4°C until analysis.

MHT The maximum holding time for samples is 14 days from the time they were collected.

SAMPLE PREPARATION Remove the plungers from two 5-mL syringes and attach a closed syringe valve to each. Warm the sample to room temperature, open the sample bottle, and carefully pour the sample into one of the syringe barrels to just short of overflowing. Replace the syringe plunger, invert the syringe, and compress the sample. Open the syringe valve and vent any residual air while adjusting the sample volume to 5.0 mL. Add 10 µL of the internal calibration standard to the sample through the syringe valve. Close the valve. Fill the second syringe in an identical manner from the same sample bottle. Reserve this second syringe for a reanalysis if necessary.

QUALITY CONTROL As an initial demonstration of lab accuracy and precision, analyze 4 to 7 replicates of a lab fortified blank containing analyte at 0.1–5 µg/L. Collect all samples in duplicate. Surrogate analytes (similar to those of the analytes of interest), whose concentration is known in every sample, are measured using the same internal standard calibration procedure. Duplicate field reagent water blanks (trip blanks) must be analyzed with each set of samples, lab reagent blanks (method blanks) must be analyzed with each batch of samples processed as a group within a work shift. Also, a single lab-fortified blank that contains each of the analytes of interest should be analyzed with each batch of samples processed as a group within a work shift. A 3- to 5-point calibration curve is needed depending on the calibration range factor required.

EPA CONTACT & HOTLINE For technical questions contact Dr. Baldev Bathija, U.S. EPA, Office of Ground Water and Drinking Water (WH-550D), 401 M St. SW, Washington, DC 20460. Tel. (202) 260-3040. For further information the EPA Safe Drinking Water Hotline may be called at: (800) 426-4791.

REFERENCE Methods for the Determination of Organic Compounds in Drinking Water, EPA/600/4-88/039 (revised July 1991; Final Rule for determination of compliance with the MCL for Total Trihalomethanes under 141.30, in 40 CFR Part 141, Vol. 58, No. 147, Fed. Reg., Tuesday Aug. 3, 1993). U.S. EPA Environmental Monitoring Systems Laboratory, Cincinnati, OH, 45268, U.S.A. Available from the National Technical Information Service (NTIS), 5285 Port Royal Road, Springfield, VA 22161; Tel. 800-553-6847. NTIS Order Number is PB91-231480.

Bromodichloromethane   **EPA Method 524**
CAS #75-27-4

TITLE Measurement of Purgeable Organic Compounds in Water by Capillary Column GC/MS.

MATRIX Drinking water and raw source water; the latter should include most surface water and groundwater sources.

METHOD SUMMARY Method 524.2 covers 60 volatile organic compounds. An inert gas (zero grade nitrogen or helium) is bubbled through a 25-mL or a 5-mL water sample (depending on the expected concentration of the analytes). Purged sample components are trapped in a tube of sorbent materials. When purging is complete, the sorbent tube is heated and backflushed with helium to desorb the trapped sample onto a capillary GC column.

INTERFERENCES Impurities in the purge gas and from organic compounds outgassing from the plumbing ahead of

the trap account for many contamination problems. Interferences purged or coextracted from the samples will vary considerably from source to source, depending upon the particular sample or extract being tested. Cross-contamination can occur whenever high-level and low-level samples are analyzed sequentially. Samples also can be contaminated by diffusion of volatile organics (particularly methylene chloride and fluorocarbons) through the septum seal into the sample during shipment and storage.

INSTRUMENTATION A GC/MS with a data system equipped with one of the following capillary GC columns:

Column 1: VOCOL glass wide bore capillary column.
Column 2: DB-624 fused silica capillary column.
Column 3: DB-5 fused silica capillary column.

Also required is an all-glass 25-mL or 5-mL purging device, a sorbent trap, and a thermal desorption apparatus which is connected to the GC/MS system.

PRECISION & ACCURACY Method detection limits are compound- and instrument-dependent, and may vary from approximately 0.02–0.35 µg/L. Note in the table below that the "true" concentration range used for accuracy and precision measurements was quite narrow. However, the applicable concentration range of this method is primarily column dependent and is approximately 0.02 to 200 µg/L for the wide-bore thick-film columns. Narrow-bore thin-film columns may have a capacity which limits the range to about 0.02 to 20 µg/L. Analytes that are inefficiently purged from water will not be detected when present at low concentrations, but they can be measured with acceptable accuracy and precision when present in sufficient amounts.

Analytes that are not separated chromatographically, but which have different mass spectra and non-interfering quantification ions, can be identified and measured in the same calibration mixture or water sample. Analytes which have very similar mass spectra cannot be individually identified and measured in the same calibration mixture or water samples unless they have different retention times. Co-eluting compounds with very similar mass spectra, typically many structural isomers, must be reported as an isomeric group or pair.

The range (in µg/L) was 0.1–10.
The method detection limit (in µg/L) was 0.08.
The accuracy (as % recovery) was 95.
The precision (in %) was 6.1.

Note: Data were obtained from 16–31 determinations using a wide-bore capillary column and a jet separator interfaced to a quadrupole mass spectrometer. All analytes were in a reagent water matrix.

SAMPLING METHOD Collect samples using a 40- to 120-mL screw-cap vial (prewashed with detergent, rinsed with distilled water and oven dried at 105°C) with a Teflon®-faced silicone septum. Collect bubble-free samples and place the septum with the Teflon side down on the water.

SAMPLE PRESERVATION If residual chlorine is present in the water add about 25 mg of ascorbic acid to each vial before samples are collected to remove the chlorine. Add hydrochloric acid to reduce pH to <2, and immediately cool samples to 4°C, and store them in a solvent-free refrigerator at 4°C until analysis.

MHT The maximum holding time for samples is 14 days from the time they were collected.

SAMPLE PREPARATION Remove the plungers from two 25-mL (or 5-mL depending on sample size) syringes and attach a closed syringe valve to each. Warm the sample to room temperature, open the sample bottle, and carefully pour the sample into one of the syringe barrels to just short of overflowing. Replace the syringe plunger, invert the syringe, and compress the sample. Open the syringe valve and vent any residual air while adjusting the sample volume to 25.0 mL (or 5 mL). For samples and blanks, add 5 µL of the fortification solution containing the internal standard and the surrogates to the sample through the syringe valve. For calibration standards and lab fortified blanks, add 5 µL of the fortification solution containing the internal standard only. Close the valve. Fill the second syringe in an identical manner from the same sample bottle. Reserve this second syringe for a reanalysis if necessary.

QUALITY CONTROL As an initial demonstration of lab accuracy and precision, analyze 4 to 7 replicates of a lab fortified blank containing analyte at 0.2–5 µg/L. Collect all samples in duplicate. Surrogate analytes (similar to those of the analytes of interest), whose concentration is known in every sample, are measured using the same internal standard calibration procedure. Duplicate field reagent water blanks (trip blanks) must be analyzed with each set of samples, lab reagent blanks (method blanks) must be analyzed with each batch of samples processed as a group within a work shift. Also, a single lab-fortified blank that contains each of the analytes of interest should be analyzed with each batch of samples processed as a group within a work shift. A 3- to 5-point calibration curve is needed depending on the calibration range factor required.

EPA CONTACT & HOTLINE For technical questions contact Dr. Baldev Bathija, U.S. EPA, Office of Ground Water and Drinking Water (WH-550D), 401 M St. SW, Washington, DC 20460. Tel. (202) 260-3040. For further information the EPA Safe Drinking Water Hotline may be called at: (800) 426-4791.

REFERENCE Methods for the Determination of Organic Compounds in Drinking Water, EPA/600/4-88/039 (revised July 1991; Final Rule for determination of compliance with the MCL for Total Trihalomethanes under 141.30, in 40 CFR Part 141, Vol. 58, No. 147, Fed. Reg., Tuesday Aug. 3, 1993). U.S. EPA Environmental Monitoring Systems Laboratory, Cincinnati, OH, 45268, U.S.A. Available from the National Technical Information Service (NTIS), 5285 Port Royal Road, Springfield, VA 22161; Tel. 800-553-6847. NTIS Order Number is PB91-231480.

Bromodichloromethane   **EPA Method 8021**
CAS #75-27-4

TITLE Halogenated Volatile by Gas Chromatography Using Photoionization and Electrolytic Conductivity Detectors in Series: Capillary Column Technique

MATRIX This method is applicable to nearly all types of samples, regardless of water content, including groundwater, aqueous sludges, caustic liquors, acid liquors, waste solvents, oily wastes, mousses, tars, fibrous wastes, polymeric emulsions, filter cakes, spent carbons, spent catalysts, soils, and sediments.

METHOD SUMMARY This method is used to determine 60 volatile organic compounds in a variety of solid waste matrices. It provides GC conditions for the detection of halogenated and aromatic volatile organic compounds. Samples can be analyzed using direct injection or purge-and-trap (EPA Method 5030). Groundwater samples must be analyzed using EPA Method 5030 (where applicable). A temperature program is used with the GC. Detection is achieved by a photoionization detector (PID) and a Hall electrolytic conductivity detector (HECD) in series.

INTERFERENCES Samples can be contaminated by diffusion of volatile organics (particularly chlorofluorocarbons and methylene chloride) through the sample container septum during shipment and storage.

INSTRUMENTATION A GC-equipped with variable-constant differential flow controllers, subambient oven controller, PID and HECD detectors connected with a short piece of uncoated capillary tubing and a data system.

Column: 60 m × 0.75 mm I.D. VOCOL wide-bore capillary column with 1.5 µm film thickness.

PRECISION & ACCURACY MDLs are compound-dependent and vary with purging efficiency and concentration. The applicable concentration range of this method is compound- and instrument-dependent but is approximately 0.1–200 µg/L. Analytes that are inefficiently purged from water will not be detected when present at low concentrations, but they can be measured with acceptable accuracy and precision when present in sufficient amounts. The estimated quantitation limit (EQL) for an individual compound is approximately 1 µg/kg (wet weight) for soil/sediment samples, 100 µg/kg (wet weight) for wastes, and 1 µg/L for groundwater. EQLs will be proportionately higher for sample extracts and samples that require dilution or reduced sample size to avoid saturation of the detector.

MULTIPLICATION FACTORS FOR OTHER MATRICES (a)

Matrix	Factor (b)
Groundwater	10
Low-concentration soil	10
Water miscible liquid waste	500
High-concentration soil and sludge	1250
Non-water miscible waste	1250

(a) Sample EQLs are highly matrix-dependent. The EQLs listed herein are provided for guidance and may not always be achievable. (b) EQL = [Method detection limit] × [Factor]. For non-aqueous samples, the factor is on a wet-weight basis.

SINGLE LABORATORY ACCURACY & PRECISION DATA FOR VOCs IN WATER

This method was tested in a single lab using water spiked at 10 µg/L and the following data was reported:

Recoveries and standard deviations were determined from seven samples and spiked at 10 µg/L of each analyte. Recoveries were determined by the internal standard method. Internal standards were: Fluorobenzene for PID and 2-Bromo-1-chloropropane for HECD.

The average recovery (in percent) for the PID was none (no response for this detector).
The standard deviation of the recovery for the PID was none (no response for this detector).
The MDL (in µg/mL) for the PID was none (no response for this detector).
The average recovery (in percent) for the HECD was 97.
The standard deviation of the recovery for the HECD was 2.9.
The MDL (in µg/mL) for the HECD was 0.02.

SAMPLE COLLECTION, PRESERVATION & HANDLING
Volatile Organics — Standard 40-mL glass screw-cap VOA vials with Teflon®-faced silicone septum may be used for both liquid and solid matrices. When collecting samples, liquids and solids should be introduced into the vials gently to reduce agitation which might drive off volatile compounds. If there are any air bubbles present the sample must be retaken. Tap slightly as they are filled to try and eliminate as much free air space as possible. The two vials from each sampling locations should be sealed in separate plastic bags to prevent cross-contamination between samples particularly if the sampled waste is suspected of containing high levels of volatile organics.

Semivolatile organics — Containers used to collect samples for the determination of semivolatile organic compounds should be soap and water washed followed by methanol (or isopropanol) rinsing. The sample containers should be of glass or Teflon® and have screw-top covers with Teflon® liners.

Preservation for volatile organics — No preservation is used with concentrated waste samples. With liquid samples containing no residual chlorine, 4 drops of concentrated hydrochloric acid are added and the samples are immediately cooled to 4°C. When liquid samples contain residual chlorine, they are treated as above and, in addition, 4 drops of 4% aqueous sodium thiosulfate are added. Soil, sediment, and sludge samples are only cooled to 4°C.

Preservation for semivolatile organics — No preservation is used with concentrated waste samples. With liquid samples containing no residual chlorine and with soil, sediment, and sludge samples, immediately cooling to 4°C is the only preservation used. When residual chlorine is present then 3 mL of 10% aqueous sodium sulfate is added for each gallon of sample collected, followed by cooling to 4°C.

MHT The holding time for all volatile organics samples is 14 days. Liquid samples must be extracted within 7 days and their extracts analyzed within 40 days. Concentrated waste, soil, sediment, and sludge samples must be extracted within 14 days and their extracts analyzed within 40 days.

SAMPLE PREPARATION Volatile compounds are introduced into the gas chromatograph either by direct injector or purge-and-trap (EPA Method 5030). EPA Method 5030 may be used directly on groundwater samples or low-concentration contaminated soils and sediments. For medium-concentration

soils or sediments, methanolic extraction, as described in EPA Method 5030, may be necessary prior to purge-and-trap analysis.

QUALITY CONTROL Calculate surrogate standard recovery on all samples, blanks, and spikes. A trip blank is recommended to check on sampling, storage, and handling contamination. Calibration standards, at a minimum of five concentration levels, are prepared in organic-free reagent water. One of the concentration levels should be at a concentration near, but above, the method detection limit.

A combination of bromochloromethane, 2-bromo-1-chloropropane, 1,4-dichlorobutane, and bromochlorobenzene are recommended as surrogate standards to encompass the range of the temperature program used in this method.

REFERENCE Test Methods for Evaluating Solid Waste, Physical/Chemical Methods, SW-846, 3rd Edition, U.S. EPA, Office of Solid Waste, Washington, DC, EPA Method 8021A, Rev. 1, Nov. 1992.

Bromodichloromethane   EPA Method 8240
CAS #75-27-4

TITLE Volatile Organics By GC/MS: Packed Column Technique

MATRIX Nearly all types of sample matarices, regardless of water content, can be analyzed using this method. This includes groundwater, aqueous sludges, caustic liquors, acid liquors, waste solvents, oily wastes, mousses, tars, fibrous wastes, polymetric emulsions, filter cakes, spent carbons, spent catalysts, soils, and sediments.

METHOD SUMMARY Method 8240B covers 80 volatile organic compounds that are introduced into a gas chromatograph by the purge-and-trap method or by direct injection (in limited applications). For the purge-and-trap method an inert gas (zero grade nitrogen or helium) is bubbled through a 5-mL solution at ambient temperature. Purged sample components are trapped in a tube of sorbent materials. When purging is complete, the sorbent tube is heated and backflushed with inert gas to desorb the trapped components onto a GC column.

INTERFERENCES Impurities in the purge gas and from organic compounds outgassing from the plumbing ahead of the trap account for many contamination problems. Interferences purged or coextracted from the samples will vary considerably from source to source. Cross-contamination can occur whenever high-level and low-level samples are analyzed sequentially. Whenever an unusually concentrated sample is analyzed, it should be followed by the analysis of organic-free reagent water to check for cross-contamination. Samples also can be contaminated by diffusion of volatile organics (particularly methylene chloride and fluorocarbons) through the septum seal into the sample during shipment and storage. A trip blank can serve as a check on such contamination. The lab where volatile analysis is performed and also the refrigerated storage area should be completely free of solvents.

INSTRUMENTATION A gas chromatograph/mass spectrometry/data system (GC/MS) equipped with a 6 ft × 0.1 in I.D. glass column packed with 1% SP-1000 on Carbopack-B (60/80 mesh) is required. Also needed is a 5-mL purging device, a sorbent trap, and a thermal desorption apparatus.

PRECISION & ACCURACY This method is reported to have been tested by 15 laboratories using organic-free reagent water, drinking water, surface water, and industrial wastewaters (not specified) fortified at six concentrations over the range 5–600 µg/L.

Sample estimated quantitation limits (EQLs) are highly matrix-dependent. The EQLs listed may not always be achievable. EQLs listed for soils or sediments are based on wet weight. Normally, data is reported on a dry-weight basis; therefore, EQLs will be higher, based on the percent dry weight of each sample. Note that EQLs are even more variable than MDLs and that they are highly variable depending on the matrix being analyzed.

EQL in groundwater in µg/L was 5.
EQL in low soil or sediment in µg/kg was 5.
Accuracy (a) in µg/L was 1.03C–1.58.
Precision (b) in µg/L was 0.20x+1.13.

(a) Average recovery found for measurements of samples containing a concentration of C, in µg/L.
(b) Overall precision found for measurements of samples with average recovery X for samples containing a concentration of C in µg/L.
X = Average recovery found for measurement of samples containing a concentration of C in µg/L.

MULTIPLICATION FACTORS FOR OTHER MATRICES

Other Matrices	Factor (a)
Waste miscible liquid waste	50
High-concentration soil and sludge	125
Non-water miscible waste	500

(a) EQL = [EQL for low soil/sediment] × [Factor]. For non-aqueous samples, the factor is on a wet-weight basis.

SAMPLING METHOD

Liquid samples — Use a 40-mL glass screw-cap VOA vial with a Teflon®-faced silicone septum that has been prewashed, rinsed with distilled deionized water, and oven dried. However, if residual chlorine is present, collect sample in a 40-oz. soil VOA container which has been pre-preserved with 4 drops of 10% sodium thiosulfate, mix gently, and then transfer the sample to a 40-mL VOA vial. Collect bubble-free samples in duplicate and seal them in separate plastic bags.

Soils or sediments and sludges — Use an 8-oz. widemouth glassbottle with a Teflon®-faced silicone septum that has been prewashed with detergent, rinsed with distilled deionized water, and oven dried. Tap slightly to eliminate free air space. Collect samples in duplicate and seal them in separate plastic bags.

SAMPLE PRESERVATION

Liquid samples — Add 4 drops of concentrated HCL and immediately coolsamples to 4°C and store in a solvent-free refrigerator.

Soils or sediments, and sludges — Cool samples to 4°C and store in a solvent-free refrigerator.

MHT Maximum holding time is 14 days from the date of sample collection.

SAMPLE PREPARATION

Liquid samples — Remove the plunger from a 5-mL syringe and carefully pour the sample into the syringe barrel to just short of overflowing. Replace the syringe plunger and compress the sample. Open the syringe valve and vent any residual air while adjusting the sample volume to 5.0 mL. If there is only one volatile organic analysis (VOA) vial, a second syringe should be filled at this time to protect against possible loss of sample integrity. Add 10 µL of surrogate spiking solution and 10 µL of internal standard spiking solution through the valve bore of the 5-mL syringe, then close the valve. The surrogate and internal standards may be mixed and added as a single spiking solution.

Sediments, soils, and waste samples — All samples of this type should be screened by GC analysis using a headspace method (EPA Method 3810) or the hexadecane extraction and screening method (EPA Method 3820). Use the screening data to determine whether to use the low-concentration method (0.005–1 mg/kg) or the high-concentration method (>1 mg/kg).

Low-concentration method — The low-concentration method is based on purging a heated sediment or soil sample mixed with organic-free reagent water containing the surrogate and internal standards. Analyze all reagent blanks and standards under the same conditions as the samples.

Use a 5-g sample if the expected concentration is <0.1 mg/kg or a 1-g sample for expected concentrations between 0.1 and 1 mg/kg. Mix the contents of the sample container with a narrow metal spatula. Weigh the amount of the sample into a tared purge device. Add the spiked water to the purge device, which contains the weighed amount of sample, and connect the device to the purge-and-trap system.

High-concentration method — This method is based on extracting the sediment or soil with methanol. A waste sample is either extracted or diluted, depending on its solubility in methanol. Wastes that are insoluble in methanol are diluted with reagent tetraglyme or possibly polyethylene glycol (PEG). An aliquot of the extract is added to organic-free reagent water containing surrogate and internal standards. This is purged at ambient temperature. All samples with an expected concentration of >1.0 mg/kg should be analyzed by this method.

Mix the contents of the sample container with a narrow metal spatula. For sediments or soils and solid wastes that are insoluble in methanol, weigh 4 g (wet weight) of sample into a tared 20-mL vial. For waste that is soluble in methanol, tetraglyme, or PEG, weigh 1 g (wet weight) into a tared scintillation vial or culture tube or a 10-mL volumetric flask. Quickly add 9.0 mL of appropriate solvent then add 1.0 mL of a surrogate spiking solution to the vial, cap it, and shake it for 2 min.

METHANOL EXTRACT REQUIRED FOR ANALYSIS OF HIGH-CONCENTRATION SOILS OR SEDIMENTS

Approximate Concentration Range	Volume of Methanol Extract (a)
500–10,000 µg/kg	100 µL
1,000–20,000 µg/kg	50 µL
5,000–100,000 µg/kg	10 µL
25,000–500,000 µg/kg	100 µL of 1/50 dilution (b)

Calculate appropriate dilution factor for concentrations exceeding this table.

(a) The volume of methanol added to 5 mL of water being purged should be kept constant. Therefore, add to the 5-mL syringe whatever volume of methanol is necessary to maintain a volume of 100 µL added to the syringe.

(b) Dilute an aliquot of the methanol extract and then take 100 µL for analysis.

QUALITY CONTROL Demonstrate, through the analysis of a reagent water blank, that interferences from the analytical system, glassware, and reagents are under control. Blank samples should be carried through all stages of the sample preparation and measurement steps. For each analytical batch (up to 20 samples), a reagent blank, matrix spike, and matrix spike duplicate must be analyzed (the frequency of the spikes may be different for different monitoring programs). The blank and spiked samples must be carried through all stages of the sample preparation and measurement steps. QC samples mentioned in the section on Interferences will also be needed as appropriate to those situations.

REFERENCE Test Methods for Evaluating Solid Waste (SW-846). U.S. EPA. 1983. Method 8240B, Rev. 2, Nov. 1990. Office of Solid Wastes, Washington, DC.

Bromodichloromethane   **EPA Method 8260**
CAS #75-27-4

TITLE Volatile Organic Compounds by GC/MS: Capillary Column Technique

MATRIX This method is applicable to nearly all types of samples, regardless of water content, including groundwater, soils, and sediments.

METHOD SUMMARY Method 8260A covers 58 volatile organic compounds that are introduced into a gas chromatograph by the purge-and-trap method or by direct injection (in limited applications). Zero-grade helium is bubbled through a 5-mL solution at ambient temperature. Purged sample components are trapped in a tube containing suitable sorbent materials. When purging is complete, the sorbent tube is heated and backflushed with helium to desorb trapped sample components. The analytes are desorbed directly to a large bore capillary or cryofocussed on a capillary precolumn before being flash evaporated to a narrow bore capillary for analysis.

INTERFERENCES Major contaminant sources are volatile materials in the lab and impurities in the inert purging gas and in the sorbent trap. Interfering contamination may occur when a sample containing low concentrations of volatile organic compounds is analyzed immediately after a sample containing high concentrations of volatile organic compounds. After analysis of a sample containing high concentrations of volatile organic compounds, one or more calibration blanks should be analyzed to check for cross-contamination. Screening of the samples prior to purge-and-trap GC/MS analysis is highly recommended to prevent contamination of the system. This is especially true for soil and waste samples.

Special precautions must be taken to analyze for methylene chloride. The analytical and sample storage area should be isolated from all atmospheric sources of methylene chloride. All gas chromatography carrier gas lines and purge gas plumbing should be constructed from stainless steel or copper tubing. Laboratory clothing previously exposed to methylene chloride fumes during liquid-liquid extraction procedures can contribute to sample contamination.

Samples can also be contaminated by diffusion of volatile organics (particularly methylene chloride and fluorocarbons) through the septum seal during shipment and storage. A trip blank can serve as a check on such contamination.

INSTRUMENTATION GC/MS with a temperature-programmable chromatograph suitable for splitless injection equipped with variable constant differential flow controllers, a subambient oven controller, a purging device, sorbent trap, a thermal desorption apparatus and a capillary precolumn interface when using cryogenic cooling will be needed. The following GC columns may be used:

Column 1: 60 m × 0.75 mm I.D. capillary column coated with VOCOL, 1.5 µm film thickness.
Column 2: 30 m × 0.53 mm I.D. capillary column coated with DB-624 or VOCOL, 3 µm film thickness.
Column 3: 30 m × 0.32 mm I.D. capillary column coated with DB-5 or SE-54, 1-µm film thickness.

PRECISION & ACCURACY This method has been tested in a single lab using spiked water. Using a wide-bore capillary column, water was spiked at concentrations between 0.5 and 10 µg/L. Single lab accuracy and precision data are presented. The MDL actually achieved in a given analysis will vary depending on instrument sensitivity and matrix effects.

The MDL (a) in µg/L was 0.08.
The concentration range in µg/L was 0.1–10.
The mean accuracy (% of true value) was 95.
The precision as relative standard deviation was 6.1.

Note: The MDL is based on a 25-mL sample volume instead of a 5-mL sample volume.

SAMPLING METHOD
Liquid samples — Use a 40-mL glass screw-cap VOA vial with a Teflon®-faced silicone septum that has been prewashed, rinsed with distilled deionized water, and oven dried. If residual chlorine is present, collect the sample in a 4-oz soil VOA container which has been pre-preserved with 4 drops of 10% sodium thiosulfate. Mix gently and transfer the sample to a 40-mL VOA vial. Collect bubble-free samples in duplicate and seal each sample in a separate plastic bag.

Soils, sediments, and sludges — Use an 8-oz widemouth glass bottle with Teflon®-faced silicone septum that has been prewashed, rinsed with distilled deionized water, and oven dried. **Do not** heat the septum for more than 1 h. Tap slightly to eliminate any free air space. Collect samples in duplicate and seal each one in a separate plastic bag.

SAMPLE PRESERVATION
Liquid samples — Add 4 drops of concentrated HCL, cool to 4°C and store in a solvent-free refrigerator.

Soils, sediments, and sludges — Cool samples to 4°C and store in a solvent-free refrigerator.

MHT The maximum holding time of any sample (liquids, soils, sediments, and sludges) is 14 days.

SAMPLE PREPARATION
Liquid samples — Remove the plunger from a 5-mL syringe and carefully pour the sample into the syringe barrel to just short of overflowing. Replace the syringe plunger and compress the sample. Open the syringe valve and vent any residual air while adjusting the sample volume to 5.0 mL. If there is only one volatile organic analysis (VOA) vial, a second syringe should be filled at this time to protect against possible loss of sample integrity. Add 10 µL of surrogate spiking solution and 10 µL of internal standard spiking solution through the valve bore of the 5-mL syringe, then close the valve. The surrogate and internal standards may be mixed and added as a single spiking solution.

Sediments, soils, and waste samples — All samples of this type should be screened by GC analysis using a headspace method (EPA Method 3810) or the hexadecane extraction and screening method (EPA Method 3820). Use the screening data to determine whether to use the low-concentration method (0.005–1 mg/kg) or the high-concentration method (>1 mg/kg).

Low-concentration method — The low-concentration method is based on purging a heated sediment or soil sample mixed with organic-free reagent water containing the surrogate and internal standards. Analyze all reagent blanks and standards under the same conditions as the samples.

Use a 5-g sample if the expected concentration is <0.1 mg/kg or a 1-g sample for expected concentrations between 0.1 and 1 mg/kg. Mix the contents of the sample container with a narrow metal spatula. Weigh the amount of the sample into a tared purge device. Add the spiked water to the purge device, which contains the weighed amount of sample, and connect the device to the purge-and-trap system.

High-concentration method — This method is based on extracting the sediment or soil with methanol. A waste sample is either extracted or diluted, depending on its solubility in methanol. Wastes that are insoluble in methanol are diluted with reagent tetraglyme or possibly polyethylene glycol (PEG). An aliquot of the extract is added to organic-free reagent water containing surrogate and internal standards. This is purged at

ambient temperature. All samples with an expected concentration of >1.0 mg/kg should be analyzed by this method.

Mix the contents of the sample container with a narrow metal spatula. For sediments or soils and solid wastes that are insoluble in methanol, weigh 4 g (wet weight) of sample into a tared 20-mL vial. For waste that is soluble in methanol, tetraglyme, or PEG, weigh 1 g (wet weight) into a tared scintillation vial or culture tube or a 10-mL volumetric flask. Quickly add 9.0 mL of appropriate solvent then add 1.0 mL of a surrogate spiking solution to the vial, cap it, and shake it for 2 min.

METHANOL EXTRACT REQUIRED FOR ANALYSIS OF HIGH-CONCENTRATION SOILS OR SEDIMENTS

Approximate Concentration Range	Volume of Methanol Extract (a)
500–10,000 µg/kg	100 µL
1,000–20,000 µg/kg	50 µL
5,000–100,000 µg/kg	10 µL
25,000–500,000 µg/kg	100 µL of 1/50 dilution (b)

Calculate appropriate dilution factor for concentrations exceeding this table.

(a) The volume of methanol added to 5 mL of water being purged should be kept constant. Therefore, add to the 5-mL syringe whatever volume of methanol is necessary to maintain a volume of 100 µL added to the syringe.
(b) Dilute an aliquot of the methanol extract and then take 100 µL for analysis.

QUALITY CONTROL Demonstrate, through the analysis of a reagent water blank, that interferences from the analytical system, glassware, and reagents are under control. Blank samples should be carried through all stages of the sample preparation and measurement steps. For each analytical batch (up to 20 samples), a reagent blank, matrix spike, and matrix spike duplicate must be analyzed (the frequency of the spikes may be different for different monitoring programs). The blank and spiked samples must be carried through all stages of the sample preparation and measurement steps. QC samples mentioned in the section on Interferences will also be needed as appropriate to those situations.

Matrix spiking standards should be prepared from volatile organic compounds which will be representative of the compounds being investigated. The recommended internal standards are chlorobenzene-d5, 1,4-difluorobenzene, 1,4-dichlorobenzene-d4, and pentafluorobenzene. Using stock standard solutions, prepare secondary dilution standards containing the compounds of interest, either singly or mixed together in methanol. Store them in a vial with no headspace for no more than one week. Surrogates recommended are toluene-d8, 4-bromofluorobenzene, and dibromofluoromethane. Each sample undergoing GC/MS analysis must be spiked with 10 µL of the surrogate spiking solution prior to analysis.

REFERENCE Test Methods for Evaluating Solid Waste (SW-846). U.S. EPA 1983, Method 8260A, Rev. 1, Nov. 1990. Office of Solid Waste, Washington, DC.

Bromodichloromethane EPA Method 601
CAS #75-27-4

TITLE Purgeable Halocarbons

MATRIX Wastewater

APPLICATION Method covers 29 purgeable halocarbons. (Method 624 provides GC/MS conditions appropriate for the qualitative and quantitative confirmation of results). Method describes conditions for a 2nd GC column to confirm measurements made with primary column.

INTERFERENCES Impurities in the purge gas and organic compounds outgassing from the plumbing ahead of the trap. With high- and low-level samples, there can be carryover contamination. Diffusion of volatile organics through the septum seal into the sample.

INSTRUMENTATION GC-equipped with halide-specific detector. (With purge-and-trap unit).

RANGE 8.0–500 µg/L.

MDL 0.10 µg/L.

PRECISION 0.20X+1.00 µg/L (overall precision).

ACCURACY 1.12C–1.02 µg/L (as recovery).

SAMPLING METHOD 25-mL glass vial. Teflon®-lined septum.

STABILITY cool, 4°C, 0.008% Na2S2O3.

MHT 14 days.

QUALITY CONTROL The lab must on an ongoing basis, spike at least 10% of the samples from each sample site being monitored to assess accuracy.

REFERENCE Method 601, *Federal Register* Part VIII 40 CFR Part 136, Oct 26, 1984.

Bromodichloromethane EPA Method 624
CAS #75-27-4

TITLE Purgeables

MATRIX Wastewater

APPLICATION Method covers 31 purgeable organics. An inert gas is bubbled through a 5-mL water sample in a specially designed purging chamber. Here, purgeables are transferred from aqueous to gaseous phase, passed onto a sorbent column, and trapped. Trap is heated and backflushed with inert gas to desorb purgeables onto a GC column, where purgeables are separated.

INTERFERENCES Impurities in the purge gas, organic compounds outgassing from the plumbing ahead of the trap, and solvent vapors in the lab. With high- and low-level samples, there can be carryover contamination.

INSTRUMENTATION GC/MS with purge-and-trap unit.

RANGE 5–600 µg/L

MDL 2.2 µg/L

PRECISION 0.20X+1.13 µg/L (overall precision).

ACCURACY 1.03C–1.58 µg/L (as recovery).

SAMPLING METHOD 25-mL glass vial. Teflon®-lined septum.

STABILITY Cool, 4°C, 0.008% Na2S2O3.

MHT 14 days.

QUALITY CONTROL The lab must on an ongoing basis, spike at least 5% of the samples from each sample site being monitored to assess accuracy.

REFERENCE Method 624, *Federal Register* Part VIII 40 CFR Part 136, Oct 26, 1984.

Bromodichloromethane **EPA Method 8010**
CAS #75-27-4

TITLE Halogenated Volatile Organics

MATRIX Groundwater, soils, sludges, water miscible liquid wastes, and non-water miscible wastes.

APPLICATION This method is used for the analysis of 39 halogenated VOCs.

Samples are analyzed using direct injection or purge-and-trap methods.

Groundwater must be analyzed by the purge-and-trap method. The method provides an optional GC column which is used for analyte confirmation and that may help resolve analytes from interferences.

INTERFERENCES There can be carryover contamination with high- and low-level samples. Impurities may come from the purge-and-trap apparatus, organic compounds outgassing from the plumbing ahead of trap, diffusion of VOCs through the sample bottle septum during shipping or storage, or from solvent vapors in the lab.

INSTRUMENTATION GC capable of on-column injections or purge-and-trap sample introduction and a halogen specific detector. Column 1: 8 ft by 0.1 in 1%. SP-1000 on Carbopack-B. Column 2: 6 ft by 0.1 in bonded n-octane on Porasil-C.

RANGE 8 to 500 µg/L (reagent water).

MDL 0.10 µg/L (reagent water).

PQL FACTORS FOR MULTIPLYING × FID MDL VALUE

Matrix	Multiplication Factor
Groundwater	10
Low-level soil	10
Water miscible liquid waste	500
High-level soil and sludge	1250
Non-water miscible waste	1250

PRECISION 0.20X + 1.00 µg/L (overall precision).

ACCURACY 1.12C — 1.02 µg/L (as recovery).

SAMPLING METHOD For water and liquid samples; use glass 40-mL vials with Teflon®-lined septum caps and collect two vials per sample location with no headspace. For solids and concentrated waste samples; use widemouth glass bottles with Teflon® liners.

STABILITY For concentrated wastes, soils, sediments, or sludges: cool to 4°C. For liquids: add 4 drops of concentrated hydrochloric acid and cool to 4°C.

MHT 14 days.

QUALITY CONTROL Analyze a reagent blank, matrix spike, and matrix spike duplicate/duplicate for each analytical batch (up to 20 samples). Demonstrate the purity of glassware and reagents by analyzing a reagent water method blank. Internal, surrogate, and five concentration level calibration standards are used.

REFERENCE Test Methods for Evaluating Solid Waste (SW-846), U.S. EPA Office of Solid Waste, Washington, DC, Method 8010B, Rev. 2, Nov. 1992.

Bromoform **EPA Method 1624**
CAS #75-25-2

TITLE Volatile Organic Compounds by Isotope Dilution GC/MS

MATRIX Compounds may be determined in waters, soils, and municipal sludges by this method.

METHOD SUMMARY This method is used to determine 58 volatile toxic organic pollutants associated with the CWA (as amended 1987); the RCRA (as amended 1986); the CERCLA (as amended 1986); and other compounds amenable to purge-and-trap gas chromatography-mass spectrometry (GC/MS).

If the solids content is less than 1%, stable isotopically-labeled analogs of the compounds of interest are added to a 5-mL sample and the sample is purged with an inert gas at 20–25°C in a chamber designed for soil or water samples. If the solids content is greater than 1%, 5 mL of reagent water and the labeled compounds are added to a 5-g aliquot of sample and the mixture is purged at 40°C. Compounds that will not purge at 20–25°C or at 40°C are purged at 78–85°C. In the purging process, the volatile compounds are transferred from the aqueous phase into the gaseous phase where they are passed into a sorbent column, and trapped. After purging is completed, the trap is backflushed and heated rapidly to desorb the compounds into a GC. The compounds are separated by the GC and detected by a MS. The labeled compounds serve to correct the variability of the analytical technique.

INTERFERENCES Impurities in the purge gas, organic compounds outgassing from the plumbing upstream of the trap, and solvent vapors in the lab account for most problems. Samples can be contaminated by diffusion of volatile organic compounds (particularly methylene chloride) through the bottle seal during shipment and storage. Contamination by carryover can occur when high-level and low-level samples are analyzed sequentially. When an unusually concentrated sample is encountered, follow it by analysis of a reagent water blank to check for carryover.

INSTRUMENTATION Major equipment includes a GC with linear temperature programming and a glass jet separator as the MS interface, a MS with 70 eV electron impact ionization, and a data system to collect and record response factors.

Column: 2.8 m × 2 mm I.D. glass, packed with 1% SP-1000 on Carbopak B, 60/80 mesh, or equivalent.

PRECISION & ACCURACY The detection limits of the method are usually dependent on the level of interferences rather than instrumental limitations. The method detection limits were determined in digested sludge (low solids) and in filter cake or compost (high solids).

The MDL (in µg/kg) for low solids is 917 and for high solids is not listed.

Labeled and native compound precision (in µg/L) as standard deviation was 7.0.

Labeled and native compound accuracy (in µg/L) as average recovery was 7–35.

Acceptance criteria are at 20 µg/L for this compound.

SAMPLE COLLECTION, PRESERVATION & HANDLING Grab samples are collected in glass containers having a total volume greater than 20 mL. Fill and seal each bottle so that no air bubbles are entrapped. Samples are maintained at 0 to 4°C from the time of collection until analysis. If an aqueous sample contains residual chlorine, add sodium thiosulfate preservative (10 mg/40 mL) to the empty sample bottles just prior to shipment to the sample site. All samples must be analyzed within 14 days of collection.

SAMPLE PREPARATION Samples containing less than 1% solids are analyzed directly as aqueous samples. Samples containing 1% solids or greater are analyzed as solid samples utilizing one of two methods, depending on the levels of pollutants, in the sample. Samples containing 1% solids or greater, and low to moderate levels of pollutants are analyzed by purging a known weight of sample added to 5 mL of reagent water. Samples containing 1% solids or greater, and high levels of pollutants, are extracted with methanol, and an aliquot of the methanol extract is added to reagent water and purged.

QUALITY CONTROL A field blank prepared from reagent water and carried through the sampling and handling protocol may serve as a check on contamination from shipment and storage.

The analyst is permitted to modify this method to improve separations or lower the costs of measurements, provided all performance specifications are met. Analyses of blanks are required. When results of spikes indicate atypical method performance for samples, the samples are diluted to bring method performance within acceptable limits. Analyze two sets of four 5-mL aliquots (8 aliquots total) of the aqueous performance standard. Spike all samples with labeled compounds to assess method performance on the sample matrix. Compute the percent recovery of the labeled compounds using the internal standard method. Compare the percent recovery for each compound with the corresponding labeled compound recovery. Reagent water blanks are analyzed to demonstrate freedom from carryover contamination. Field replicates may be collected to determine the precision of the sampling technique, and spiked samples may be required to determine the accuracy of the analysis when the internal method is used.

REFERENCE Volatile Organic Compounds by Isotope Dilution GC/MS. Office of Water Regulation and Standards, U.S. EPA Industrial Technology Division, Washington, DC, EPA Method 1624, Rev. C, June 1989 (contact W.A. Telliard, U.S. EPA, Office of Water Regulations and Standards, 401 M St., SW, Washington, DC, 20460. Phone: 202-382-7131).

Bromoform   **EPA Method 502**
CAS #75-25-2

TITLE Volatile Organic Compounds in Water By Purge and Trap Capillary Column Gas Chromatography with Photoionization and Electrolytic Conductivity Detectors in Series. U.S. EPA Method 502.2, Rev. 2.0, 1989.

MATRIX Drinking water and raw source water. The latter should include most surface water and groundwater sources.

METHOD SUMMARY This method covers 60 volatile organic compounds that contain halogen atoms and/or that are aromatic. An inert gas (zero grade nitrogen or helium) is bubbled through a 25-mL or a 5-mL water sample (depending on the expected concentration of the analytes). Purged sample components are trapped in a tube of sorbent materials. When purging is complete, the sorbent tube is heated and backflushed with helium to desorb the trapped sample onto a capillary GC column. The column is temperature programmed to separate the method analytes which are then detected with a photoionization detector (PID) and a Hall electrolytic conductivity (HECD) placed in series. The PID is selective for aromatic compounds and the HECD is selective for halogenated compounds.

INTERFERENCES Impurities in the purge gas and from organic compounds outgassing from the plumbing ahead of the trap account for many contamination problems. Interferences purged or coextracted from the samples will vary considerably from source to source, depending upon the particular sample or extract being tested. Cross-contamination can occur whenever high-level and low-level samples are analyzed sequentially. Samples also can be contaminated by diffusion of volatile organics (particularly methylene chloride and fluorocarbons) through the septum seal into the sample during shipment and storage. The lab where volatile analysis is performed and also the refrigerated storage area should be completely free of solvents.

INSTRUMENTATION A GC containing a series configuration of a high temperature photoionization detector (PID) equipped with 10.0 eV (nominal) lamp and Hall electrolytic conductivity detector (HECD) is required. Also required is an

all-glass 5-mL purging device, a sorbent trap, and a thermal desorption apparatus which is connected to the GC system.

Column 1: VOCOL glass wide-bore capillary column.
Column 2: RTX–502.2 mega-bore capillary column.
Column 3: DB-62 mega-bore capillary column.

PRECISION & ACCURACY Method detection limits are dependent upon the characteristics of the gas chromatographic system used. Analytes that are not separated chromatographically cannot be individually identified and used in the same calibration mixture or water samples unless an alternative technique for identification and quantification, such as mass spectrometry, is used.

Electrolytic conductivity detetor (c) range in µg/L (a) was 0.02–200.
Electrolytic conductivity detetor (c) MDL in µg/L (b) was 1.6.
Electrolytic conductivity detetor (c) accuracy as % recovery was 106.
Electrolytic conductivity detetor (c) precision as % RSD was 5.2.
Photoionization detector (d) range in µg/L (a) was 0.02–200.
Photoionization detector (d) MDL in µg/L (b) was not listed.
Photoionization detector (d) accuracy as % recovery was not listed.
Photoionization detector (d) precision as % RSD was not listed.

(a) *The applicable concentration range of this method is compound, instrument, and matrix-dependent. It is listed as being approximately 0.02 to 200 µg/L but no specific information is provided so caution should be observed.*

(b) *The method detection limits reports with this method are compound, instrument, and matrix-dependent. The values reported were calculated using reagent water fortified with the corresponding compounds at 10 µg/L and a GC-equipped with a 60 m × 0.75 mm VOLCOL wide bore capillary column with 1.5 µm film thickness and using helium carrier gas.*

(c) *Recoveries and relative standard deviations were determined from seven samples of reagent water fortified with 10 µg/L of each compound. 2-Bromo-1-chloropropane was used as the internal standard for calculating average recoveries.*

(d) *Recoveries and relative standard deviations were determined from seven samples of reagent water fortified with 10 µg/L of each compound. Fluorobenzene was used as the internal standard for calculating average recoveries.*

SAMPLING METHOD Collect samples using a 40- to 120-mL screw-cap vial (prewashed with detergent, rinsed with distilled water and oven dried at 105°C) with a Teflon®-faced silicone septum. Collect bubble-free samples and place the septum with the Teflon® side down on the water.

SAMPLE PRESERVATION If residual chlorine is present in the water add about 25 mg of ascorbic acid to each vial before samples are collected to remove the chlorine. Add hydrochloric acid to reduce pH to <2, immediately cool samples to 4°C, and store them in a solvent-free refrigerator at 4°C until analysis.

MHT The maximum holding time for samples is 14 days from the time they were collected.

SAMPLE PREPARATION Remove the plungers from two 5-mL syringes and attach a closed syringe valve to each. Warm the sample to room temperature, open the sample bottle, and carefully pour the sample into one of the syringe barrels to just short of overflowing. Replace the syringe plunger, invert the syringe, and compress the sample. Open the syringe valve and vent any residual air while adjusting the sample volume to 5.0 mL. Add 10 µL of the internal calibration standard to the sample through the syringe valve. Close the valve. Fill the second syringe in an identical manner from the same sample bottle. Reserve this second syringe for a reanalysis if necessary.

QUALITY CONTROL As an initial demonstration of lab accuracy and precision, analyze 4 to 7 replicates of a lab fortified blank containing analyte at 0.1–5 µg/L. Collect all samples in duplicate. Surrogate analytes (similar to those of the analytes of interest), whose concentration is known in every sample, are measured using the same internal standard calibration procedure. Duplicate field reagent water blanks (trip blanks) must be analyzed with each set of samples, lab reagent blanks (method blanks) must be analyzed with each batch of samples processed as a group within a work shift. Also, a single lab-fortified blank that contains each of the analytes of interest should be analyzed with each batch of samples processed as a group within a work shift. A 3- to 5-point calibration curve is needed depending on the calibration range factor required.

EPA CONTACT & HOTLINE For technical questions contact Dr. Baldev Bathija, U.S. EPA, Office of Ground Water and Drinking Water (WH-550D), 401 M St. SW, Washington, DC 20460. Tel. (202) 260-3040. For further information the EPA Safe Drinking Water Hotline may be called at: (800) 426-4791.

REFERENCE Methods for the Determination of Organic Compounds in Drinking Water, EPA/600/4-88/039 (revised July 1991; Final Rule for determination of compliance with the MCL for Total Trihalomethanes under 141.30, in 40 CFR Part 141, Vol. 58, No. 147, Fed. Reg., Tuesday Aug. 3, 1993). U.S. EPA Environmental Monitoring Systems Laboratory, Cincinnati, OH, 45268, U.S.A. Available from the National Technical Information Service (NTIS), 5285 Port Royal Road, Springfield, VA 22161; Tel. 800-553-6847. NTIS Order Number is PB91-231480.

Bromoform   **EPA Method 524**
CAS #75-25-2

TITLE Measurement of Purgeable Organic Compounds in Water by Capillary Column GC/MS.

MATRIX Drinking water and raw source water; the latter should include most surface water and groundwater sources.

METHOD SUMMARY Method 524.2 covers 60 volatile organic compounds. An inert gas (zero grade nitrogen or helium) is bubbled through a 25-mL or a 5-mL water sample (depending on the expected concentration of the analytes). Purged sample components are trapped in a tube of sorbent materials. When purging is complete, the sorbent tube is heated and backflushed with helium to desorb the trapped sample onto a capillary GC column.

INTERFERENCES Impurities in the purge gas and from organic compounds outgassing from the plumbing ahead of the trap account for many contamination problems. Interferences purged or coextracted from the samples will vary considerably from source to source, depending upon the particular sample or extract being tested. Cross-contamination can occur whenever high-level and low-level samples are analyzed sequentially. Samples also can be contaminated by diffusion of volatile organics (particularly methylene chloride and fluorocarbons) through the septum seal into the sample during shipment and storage.

INSTRUMENTATION A GC/MS with a data system equipped with one of the following capillary GC columns:

Column 1: VOCOL glass wide bore capillary column.
Column 2: DB-624 fused silica capillary column.
Column 3: DB-5 fused silica capillary column.

Also required is an all-glass 25-mL or 5-mL purging device, a sorbent trap, and a thermal desorption apparatus which is connected to the GC/MS system.

PRECISION & ACCURACY Method detection limits are compound- and instrument-dependent, and may vary from approximately 0.02–0.35 µg/L. Note in the table below that the "true" concentration range used for accuracy and precision measurements was quite narrow. However, the applicable concentration range of this method is primarily column dependent and is approximately 0.02 to 200 µg/L for the wide-bore thick-film columns. Narrow-bore thin-film columns may have a capacity which limits the range to about 0.02 to 20 µg/L. Analytes that are inefficiently purged from water will not be detected when present at low concentrations, but they can be measured with acceptable accuracy and precision when present in sufficient amounts.

Analytes that are not separated chromatographically, but which have different mass spectra and non-interfering quantification ions, can be identified and measured in the same calibration mixture or water sample. Analytes which have very similar mass spectra cannot be individually identified and measured in the same calibration mixture or water samples unless they have different retention times. Co-eluting compounds with very similar mass spectra, typically many structural isomers, must be reported as an isomeric group or pair.

The range (in µg/L) was 0.5–10.
The method detection limit (in µg/L) was 0.12.
The accuracy (as % recovery) was 101.
The precision (in %) was 6.3.

Note: Data were obtained from 16–31 determinations using a wide-bore capillary column and a jet separator interfaced to a quadrupole mass spectrometer. All analytes were in a reagent water matrix.

SAMPLING METHOD Collect samples using a 40- to 120-mL screw-cap vial (prewashed with detergent, rinsed with distilled water and oven dried at 105°C) with a Teflon®-faced silicone septum. Collect bubble-free samples and place the septum with the Teflon® side down on the water.

SAMPLE PRESERVATION If residual chlorine is present in the water add about 25 mg of ascorbic acid to each vial before samples are collected to remove the chlorine. Add hydrochloric acid to reduce pH to <2, and immediately cool samples to 4°C, and store them in a solvent-free refrigerator at 4°C until analysis.

MHT The maximum holding time for samples is 14 days from the time they were collected.

SAMPLE PREPARATION Remove the plungers from two 25-mL (or 5-mL depending on sample size) syringes and attach a closed syringe valve to each. Warm the sample to room temperature, open the sample bottle, and carefully pour the sample into one of the syringe barrels to just short of overflowing. Replace the syringe plunger, invert the syringe, and compress the sample. Open the syringe valve and vent any residual air while adjusting the sample volume to 25.0 mL (or 5 mL). For samples and blanks, add 5 µL of the fortification solution containing the internal standard and the surrogates to the sample through the syringe valve. For calibration standards and lab fortified blanks, add 5 µL of the fortification solution containing the internal standard only. Close the valve. Fill the second syringe in an identical manner from the same sample bottle. Reserve this second syringe for a reanalysis if necessary.

QUALITY CONTROL As an initial demonstration of lab accuracy and precision, analyze 4 to 7 replicates of a lab fortified blank containing analyte at 0.2–5 µg/L. Collect all samples in duplicate. Surrogate analytes (similar to those of the analytes of interest), whose concentration is known in every sample, are measured using the same internal standard calibration procedure. Duplicate field reagent water blanks (trip blanks) must be analyzed with each set of samples, lab reagent blanks (method blanks) must be analyzed with each batch of samples processed as a group within a work shift. Also, a single lab-fortified blank that contains each of the analytes of interest should be analyzed with each batch of samples processed as a group within a work shift. A 3- to 5-point calibration curve is needed depending on the calibration range factor required.

EPA CONTACT & HOTLINE For technical questions contact Dr. Baldev Bathija, U.S. EPA, Office of Ground Water and Drinking Water (WH-550D), 401 M St. SW, Washington, DC 20460. Tel. (202) 260-3040. For further information the EPA Safe Drinking Water Hotline may be called at: (800) 426-4791.

REFERENCE Methods for the Determination of Organic Compounds in Drinking Water, EPA/600/4-88/039 (revised July 1991; Final Rule for determination of compliance with the MCL for Total Trihalomethanes under 141.30, in 40 CFR Part 141, Vol. 58, No. 147, Fed. Reg., Tuesday Aug. 3, 1993). U.S. EPA Environmental Monitoring Systems Laboratory, Cincinnati, OH, 45268, U.S.A. Available from the National Technical Information Service (NTIS), 5285 Port Royal Road, Springfield, VA 22161; Tel. 800-553-6847. NTIS Order Number is PB91-231480.

Bromoform **EPA Method 8021**
CAS #75-25-2

VOAs by Capillary Column GC/PID-HECD

TITLE Halogenated Volatile by Gas Chromatography Using Photoionization and Electrolytic Conductivity Detectors in Series: Capillary Column Technique

MATRIX This method is applicable to nearly all types of samples, regardless of water content, including groundwater, aqueous sludges, caustic liquors, acid liquors, waste solvents, oily wastes, mousses, tars, fibrous wastes, polymeric emulsions, filter cakes, spent carbons, spent catalysts, soils, and sediments.

METHOD SUMMARY This method is used to determine 60 volatile organic compounds in a variety of solid waste matrices. It provides GC conditions for the detection of halogenated and aromatic volatile organic compounds. Samples can be analyzed using direct injection or purge-and-trap (EPA Method 5030). Groundwater samples must be analyzed using EPA Method 5030 (where applicable). A temperature program is used with the GC. Detection is achieved by a photoionization detector (PID) and a Hall electrolytic conductivity detector (HECD) in series.

INTERFERENCES Samples can be contaminated by diffusion of volatile organics (particularly chlorofluorocarbons and methylene chloride) through the sample container septum during shipment and storage.

INSTRUMENTATION A GC-equipped with variable-constant differential flow controllers, subambient oven controller, PID and HECD detectors connected with a short piece of uncoated capillary tubing and a data system.

Column: 60 m × 0.75 mm I.D. VOCOL wide-bore capillary column with 1.5 µm film thickness.

PRECISION & ACCURACY MDLs are compound-dependent and vary with purging efficiency and concentration. The applicable concentration range of this method is compound- and instrument-dependent but is approximately 0.1 to 200 µg/L. Analytes that are inefficiently purged from water will not be detected when present at low concentrations, but they can be measured with acceptable accuracy and precision when present in sufficient amounts. The estimated quantitation limit (EQL) for an individual compound is approximately 1 µg/kg (wet weight) for soil/sediment samples, 100 µg/kg (wet weight) for wastes, and 1 µg/L for groundwater. EQLs will be proportionately higher for sample extracts and samples that require dilution or reduced sample size to avoid saturation of the detector.

MULTIPLICATION FACTORS FOR OTHER MATRICES (a)

Matrix	Factor (b)
Groundwater	10
Low-concentration soil	10
Water miscible liquid waste	500
High-concentration soil and sludge	1250
Non-water miscible waste	1250

(a) Sample EQLs are highly matrix-dependent. The EQLs listed herein are provided for guidance and may not always be achievable. (b) EQL = [Method detection limit] × [Factor]. For non-aqueous samples, the factor is on a wet-weight basis.

SINGLE LABORATORY ACCURACY & PRECISION DATA FOR VOCs IN WATER

This method was tested in a single lab using water spiked at 10 µg/L and the following data was reported:

Recoveries and standard deviations were determined from seven samples and spiked at 10 µg/L of each analyte. Recoveries were determined by the internal standard method. Internal standards were: Fluorobenzene for PID and 2-Bromo-1-chloropropane for HECD.

The average recovery (in percent) for the PID was none (no response for this detector).
The standard deviation of the recovery for the PID was none (no response for this detector).
The MDL (in µg/mL) for the PID was none (no response for this detector).
The average recovery (in percent) for the HECD was 106.
The standard deviation of the recovery for the HECD was 5.5.
The MDL (in µg/mL) for the HECD was 1.6.

SAMPLE COLLECTION, PRESERVATION & HANDLING
Volatile Organics — Standard 40-mL glass screw-cap VOA vials with Teflon®-faced silicone septum may be used for both liquid and solid matrices. When collecting samples, liquids and solids should be introduced into the vials gently to reduce agitation which might drive off volatile compounds. If there are any air bubbles present the sample must be retaken. Tap slightly as they are filled to try and eliminate as much free air space as possible. The two vials from each sampling locations should be sealed in separate plastic bags to prevent cross-contamination between samples particularly if the sampled waste is suspected of containing high levels of volatile organics.

Semivolatile organics — Containers used to collect samples for the determination of semivolatile organic compounds should be soap and water washed followed by methanol (or isopropanol) rinsing. The sample containers should be of glass or Teflon® and have screw-top covers with Teflon® liners.

Preservation for volatile organics — No preservation is used with concentrated waste samples. With liquid samples containing no residual chlorine, 4 drops of concentrated hydrochloric acid are added and the samples are immediately cooled to 4°C. When liquid samples contain residual chlorine, they are treated as above and, in addition, 4 drops of 4% aqueous sodium thiosulfate are added. Soil, sediment, and sludge samples are only cooled to 4°C.

Preservation for semivolatile organics — No preservation is used with concentrated waste samples. With liquid samples containing no residual chlorine and with soil, sediment, and sludge samples, immediately cooling to 4°C is the only preservation used. When residual chlorine is present then 3 mL of 10% aqueous sodium sulfate is added for each gallon of sample collected, followed by cooling to 4°C.

MHT The holding time for all volatile organics samples is 14 days. Liquid samples must be extracted within 7 days and their extracts analyzed within 40 days. Concentrated waste, soil, sediment, and sludge samples must be extracted within 14 days and their extracts analyzed within 40 days.

SAMPLE PREPARATION Volatile compounds are introduced into the gas chromatograph either by direct injector or purge-and-trap (EPA Method 5030). EPA Method 5030 may be used directly on groundwater samples or low-concentration contaminated soils and sediments. For medium-concentration soils or sediments, methanolic extraction, as described in EPA Method 5030, may be necessary prior to purge-and-trap analysis.

QUALITY CONTROL Calculate surrogate standard recovery on all samples, blanks, and spikes. A trip blank is recommended to check on sampling, storage, and handling contamination. Calibration standards, at a minimum of five concentration levels, are prepared in organic-free reagent water. One of the concentration levels should be at a concentration near, but above, the method detection limit.

A combination of bromochloromethane, 2-bromo-1-chloropropane, 1,4-dichlorobutane, and bromochlorobenzene are recommended as surrogate standards to encompass the range of the temperature program used in this method.

REFERENCE Test Methods for Evaluating Solid Waste, Physical/Chemical Methods, SW-846, 3rd Edition, U.S. EPA, Office of Solid Waste, Washington, DC, EPA Method 8021A, Rev. 1, Nov. 1992.

Bromoform   **EPA Method 8240**
CAS #75-25-2

TITLE Volatile Organics By GC/MS: Packed Column Technique

MATRIX Nearly all types of sample matarices, regardless of water content, can be analyzed using this method. This includes groundwater, aqueous sludges, caustic liquors, acid liquors, waste solvents, oily wastes, mousses, tars, fibrous wastes, polymetric emulsions, filter cakes, spent carbons, spent catalysts, soils, and sediments.

METHOD SUMMARY Method 8240B covers 80 volatile organic compounds that are introduced into a gas chromatograph by the purge-and-trap method or by direct injection (in limited applications). For the purge-and-trap method an inert gas (zero grade nitrogen or helium) is bubbled through a 5-mL solution at ambient temperature. Purged sample components are trapped in a tube of sorbent materials. When purging is complete, the sorbent tube is heated and backflushed with inert gas to desorb the trapped components onto a GC column.

INTERFERENCES Impurities in the purge gas and from organic compounds outgassing from the plumbing ahead of the trap account for many contamination problems. Interferences purged or coextracted from the samples will vary considerably from source to source. Cross-contamination can occur whenever high-level and low-level samples are analyzed sequentially. Whenever an unusually concentrated sample is analyzed, it should be followed by the analysis of organic-free reagent water to check for cross-contamination. Samples also can be contaminated by diffusion of volatile organics (particularly methylene chloride and fluorocarbons) through the septum seal into the sample during shipment and storage. A trip blank can serve as a check on such contamination. The lab where volatile analysis is performed and also the refrigerated storage area should be completely free of solvents.

INSTRUMENTATION A gas chromatograph/mass spectrometry/data system (GC/MS) equipped with a 6 ft × 0.1 in I.D. glass column packed with 1% SP-1000 on Carbopack-B (60/80 mesh) is required. Also needed is a 5-mL purging device, a sorbent trap, and a thermal desorption apparatus.

PRECISION & ACCURACY This method is reported to have been tested by 15 laboratories using organic-free reagent water, drinking water, surface water, and industrial wastewaters (not specified) fortified at six concentrations over the range 5–600 µg/L.

Sample estimated quantitation limits (EQLs) are highly matrix-dependent. The EQLs listed may not always be achievable. EQLs listed for soils or sediments are based on wet weight. Normally, data is reported on a dry-weight basis; therefore, EQLs will be higher, based on the percent dry weight of each sample. Note that EQLs are even more variable than MDLs and that they are highly variable depending on the matrix being analyzed.

EQL in groundwater in µg/L was 5.
EQL in low soil or sediment in µg/kg was 5.
Accuracy (a) in µg/L was $1.18C - 2.35$.
Precision (b) in µg/L was $0.17x + 1.38$.

(a) *Average recovery found for measurements of samples containing a concentration of C, in µg/L.*
(b) *Overall precision found for measurements of samples with average recovery X for samples containing a concentration of C in µg/L.*
X = *Average recovery found for measurement of samples containing a concentration of C in µg/L.*

MULTIPLICATION FACTORS FOR OTHER MATRICES

Other Matrices	Factor (a)
Waste miscible liquid waste	50
High-concentration soil and sludge	125
Non-water miscible waste	500

(a) *EQL = [EQL for low soil/sediment] × [Factor]. For non-aqueous samples, the factor is on a wet-weight basis.*

SAMPLING METHOD

Liquid samples — Use a 40-mL glass screw-cap VOA vial with a Teflon®-faced silicone septum that has been prewashed, rinsed with distilled deionized water, and oven dried. However, if residual chlorine is present, collect sample in a 40-oz. soil VOA container which has been pre-preserved with 4 drops of 10% sodium thiosulfate, mix gently, and then transfer the sample to a 40-mL VOA vial. Collect bubble-free samples in duplicate and seal them in separate plastic bags.

Soils or sediments, and sludges — Use an 8-oz. widemouth glass bottle with a Teflon®-faced silicone septum that has been prewashed with detergent, rinsed with distilled deionized water, and oven dried. Tap slightly to eliminate free air space. Collect samples in duplicate and seal them in separate plastic bags.

SAMPLE PRESERVATION

Liquid samples — Add 4 drops of concentrated HCL and immediately cool samples to 4°C and store in a solvent-free refrigerator.

Soils or sediments, and sludges — Cool samples to 4°C and store in a solvent-free refrigerator.

MHT Maximum holding time is 14 days from the date of sample collection.

SAMPLE PREPARATION

Liquid samples — Remove the plunger from a 5-mL syringe and carefully pour the sample into the syringe barrel to just short of overflowing. Replace the syringe plunger and compress the sample. Open the syringe valve and vent any residual air while adjusting the sample volume to 5.0 mL. If there is only one volatile organic analysis (VOA) vial, a second syringe should be filled at this time to protect against possible loss of sample integrity. Add 10 µL of surrogate spiking solution and 10 µL of internal standard spiking solution through the valve bore of the 5-mL syringe, then close the valve. The surrogate and internal standards may be mixed and added as a single spiking solution.

Sediments, soils, and waste samples — All samples of this type should be screened by GC analysis using a headspace method (EPA Method 3810) or the hexadecane extraction and screening method (EPA Method 3820). Use the screening data to determine whether to use the low-concentration method (0.005–1 mg/kg) or the high-concentration method (>1 mg/kg).

Low-concentration method — The low-concentration method is based on purging a heated sediment or soil sample mixed with organic-free reagent water containing the surrogate and internal standards. Analyze all reagent blanks and standards under the same conditions as the samples.

Use a 5-g sample if the expected concentration is <0.1 mg/kg or a 1-g sample for expected concentrations between 0.1 and 1 mg/kg. Mix the contents of the sample container with a narrow metal spatula. Weigh the amount of the sample into a tared purge device. Add the spiked water to the purge device, which contains the weighed amount of sample, and connect the device to the purge-and-trap system.

High-concentration method — This method is based on extracting the sediment or soil with methanol. A waste sample is either extracted or diluted, depending on its solubility in methanol. Wastes that are insoluble in methanol are diluted with reagent tetraglyme or possibly polyethylene glycol (PEG). An aliquot of the extract is added to organic-free reagent water containing surrogate and internal standards. This is purged at ambient temperature. All samples with an expected concentration of >1.0 mg/kg should be analyzed by this method.

Mix the contents of the sample container with a narrow metal spatula. For sediments or soils and solid wastes that are insoluble in methanol, weigh 4 g (wet weight) of sample into a tared 20-mL vial. For waste that is soluble in methanol, tetraglyme, or PEG, weigh 1 g (wet weight) into a tared scintillation vial or culture tube or a 10-mL volumetric flask. Quickly add 9.0 mL of appropriate solvent then add 1.0 mL of a surrogate spiking solution to the vial, cap it, and shake it for 2 min.

METHANOL EXTRACT REQUIRED FOR ANALYSIS OF HIGH-CONCENTRATION SOILS OR SEDIMENTS

Approximate Concentration Range	Volume of Methanol Extract (a)
500–10,000 µg/kg	100 µL
1,000–20,000 µg/kg	50 µL
5,000–100,000 µg/kg	10 µL
25,000–500,000 µg/kg	100 µL of 1/50 dilution (b)

Calculate appropriate dilution factor for concentrations exceeding this table.

(a) The volume of methanol added to 5 mL of water being purged should be kept constant. Therefore, add to the 5-mL syringe whatever volume of methanol is necessary to maintain a volume of 100 µL added to the syringe.

(b) Dilute an aliquot of the methanol extract and then take 100 µL for analysis.

QUALITY CONTROL Demonstrate, through the analysis of a reagent water blank, that interferences from the analytical system, glassware, and reagents are under control. Blank samples should be carried through all stages of the sample preparation and measurement steps. For each analytical batch (up to 20 samples), a reagent blank, matrix spike, and matrix spike duplicate must be analyzed (the frequency of the spikes may be different for different monitoring programs). The blank and spiked samples must be carried through all stages of the sample preparation and measurement steps. QC samples mentioned in the section on Interferences will also be needed as appropriate to those situations.

REFERENCE Test Methods for Evaluating Solid Waste (SW-846). U.S. EPA. 1983. Method 8240B, Rev. 2, Nov. 1990. Office of Solid Wastes, Washington, DC.

Bromoform   **EPA Method 8260**
CAS #75-25-2

TITLE Volatile Organic Compounds by GC/MS: Capillary Column Technique

MATRIX This method is applicable to nearly all types of samples, regardless of water content, including groundwater, soils, and sediments.

METHOD SUMMARY Method 8260A covers 58 volatile organic compounds that are introduced into a gas chromatograph by the purge-and-trap method or by direct injection (in limited applications). Zero-grade helium is bubbled through a 5-mL solution at ambient temperature. Purged sample components are trapped in a tube containing suitable sorbent materials. When purging is complete, the sorbent tube is heated and backflushed with helium to desorb trapped sample components. The analytes are desorbed directly to a large bore capillary or cryofocussed on a capillary precolumn before being flash evaporated to a narrow bore capillary for analysis.

INTERFERENCES Major contaminant sources are volatile materials in the lab and impurities in the inert purging gas and in the sorbent trap. Interfering contamination may occur when a sample containing low concentrations of volatile organic compounds is analyzed immediately after a sample containing high concentrations of volatile organic compounds. After analysis of a sample containing high concentrations of volatile organic compounds, one or more calibration blanks should be analyzed to check for cross-contamination. Screening of the samples prior to purge-and-trap GC/MS analysis is highly recommended to prevent contamination of the system. This is especially true for soil and waste samples.

Special precautions must be taken to analyze for methylene chloride. The analytical and sample storage area should be isolated from all atmospheric sources of methylene chloride. All gas chromatography carrier gas lines and purge gas plumbing should be constructed from stainless steel or copper tubing. Laboratory clothing previously exposed to methylene chloride fumes during liquid-liquid extraction procedures can contribute to sample contamination.

Samples can also be contaminated by diffusion of volatile organics (particularly methylene chloride and fluorocarbons) through the septum seal during shipment and storage. A trip blank can serve as a check on such contamination.

INSTRUMENTATION GC/MS with a temperature-programmable chromatograph suitable for splitless injection equipped with variable constant differential flow controllers, a subambient oven controller, a purging device, sorbent trap, a thermal desorption apparatus and a capillary precolumn interface when using cryogenic cooling will be needed. The following GC columns may be used:

Column 1: 60 m × 0.75 mm I.D. capillary column coated with VOCOL, 1.5 µm film thickness.
Column 2: 30 m × 0.53 mm I.D. capillary column coated with DB-624 or VOCOL, 3 µm film thickness.
Column 3: 30 m × 0.32 mm I.D. capillary column coated with DB-5 or SE-54, 1-µm film thickness.

PRECISION & ACCURACY This method has been tested in a single lab using spiked water. Using a wide-bore capillary column, water was spiked at concentrations between 0.5 and 10 µg/L. Single lab accuracy and precision data are presented. The MDL actually achieved in a given analysis will vary depending on instrument sensitivity and matrix effects.

The MDL (a) in µg/L was 0.12.
The concentration range in µg/L was 0.5–10.
The mean accuracy (% of true value) was 101.
The precision as relative standard deviation was 6.3.

Note: The MDL is based on a 25-mL sample volume instead of a 5-mL sample volume.

SAMPLING METHOD
Liquid samples — Use a 40-mL glass screw-cap VOA vial with a Teflon®-faced silicone septum that has been prewashed, rinsed with distilled deionized water, and oven dried. If residual chlorine is present, collect the sample in a 4-oz soil VOA container which has been pre-preserved with 4 drops of 10% sodium thiosulfate. Mix gently and transfer the sample to a 40-mL VOA vial. Collect bubble-free samples in duplicate and seal each sample in a separate plastic bag.

Soils, sediments, and sludges — Use an 8-oz widemouth glass bottle with Teflon®-faced silicone septum that has been prewashed, rinsed with distilled deionized water, and oven dried. **Do not** heat the septum for more than 1 h. Tap slightly to eliminate any free air space. Collect samples in duplicate and seal each one in a separate plastic bag.

SAMPLE PRESERVATION
Liquid samples — Add 4 drops of concentrated HCL, cool to 4°C and store in a solvent-free refrigerator.

Soils, sediments, and sludges — Cool samples to 4°C and store in a solvent-free refrigerator.

MHT The maximum holding time of any sample (liquids, soils, sediments, and sludges) is 14 days.

SAMPLE PREPARATION
Liquid samples — Remove the plunger from a 5-mL syringe and carefully pour the sample into the syringe barrel to just short of overflowing. Replace the syringe plunger and compress the sample. Open the syringe valve and vent any residual air while adjusting the sample volume to 5.0 mL. If there is only one volatile organic analysis (VOA) vial, a second syringe should be filled at this time to protect against possible loss of sample integrity. Add 10 µL of surrogate spiking solution and 10 µL of internal standard spiking solution through the valve bore of the 5-mL syringe, then close the valve. The surrogate and internal standards may be mixed and added as a single spiking solution.

Sediments, soils, and waste samples — All samples of this type should be screened by GC analysis using a headspace method (EPA Method 3810) or the hexadecane extraction and screening method (EPA Method 3820). Use the screening data to determine whether to use the low-concentration method (0.005–1 mg/kg) or the high-concentration method (>1 mg/kg).

Low-concentration method — The low-concentration method is based on purging a heated sediment or soil sample mixed with organic-free reagent water containing the surrogate and internal standards. Analyze all reagent blanks and standards under the same conditions as the samples.

Use a 5-g sample if the expected concentration is <0.1 mg/kg or a 1-g sample for expected concentrations between 0.1 and 1 mg/kg. Mix the contents of the sample container with a narrow metal spatula. Weigh the amount of the sample into a tared purge device. Add the spiked water to the purge device, which contains the weighed amount of sample, and connect the device to the purge-and-trap system.

High-concentration method — This method is based on extracting the sediment or soil with methanol. A waste sample is either extracted or diluted, depending on its solubility in methanol. Wastes that are insoluble in methanol are diluted with reagent tetraglyme or possibly polyethylene glycol (PEG). An aliquot of the extract is added to organic-free reagent water containing surrogate and internal standards. This is purged at

ambient temperature. All samples with an expected concentration of >1.0 mg/kg should be analyzed by this method.

Mix the contents of the sample container with a narrow metal spatula. For sediments or soils and solid wastes that are insoluble in methanol, weigh 4 g (wet weight) of sample into a tared 20-mL vial. For waste that is soluble in methanol, tetraglyme, or PEG, weigh 1 g (wet weight) into a tared scintillation vial or culture tube or a 10-mL volumetric flask. Quickly add 9.0 mL of appropriate solvent then add 1.0 mL of a surrogate spiking solution to the vial, cap it, and shake it for 2 min.

METHANOL EXTRACT REQUIRED FOR ANALYSIS OF HIGH-CONCENTRATION SOILS OR SEDIMENTS

Approximate Concentration Range	Volume of Methanol Extract (a)
500–10,000 µg/kg	100 µL
1,000–20,000 µg/kg	50 µL
5,000–100,000 µg/kg	10 µL
25,000–500,000 µg/kg	100 µL of 1/50 dilution (b)

Calculate appropriate dilution factor for concentrations exceeding this table.

(a) The volume of methanol added to 5 mL of water being purged should be kept constant. Therefore, add to the 5-mL syringe whatever volume of methanol is necessary to maintain a volume of 100 µL added to the syringe.
(b) Dilute an aliquot of the methanol extract and then take 100 µL for analysis.

QUALITY CONTROL Demonstrate, through the analysis of a reagent water blank, that interferences from the analytical system, glassware, and reagents are under control. Blank samples should be carried through all stages of the sample preparation and measurement steps. For each analytical batch (up to 20 samples), a reagent blank, matrix spike, and matrix spike duplicate must be analyzed (the frequency of the spikes may be different for different monitoring programs). The blank and spiked samples must be carried through all stages of the sample preparation and measurement steps. QC samples mentioned in the section on Interferences will also be needed as appropriate to those situations.

Matrix spiking standards should be prepared from volatile organic compounds which will be representative of the compounds being investigated. The recommended internal standards are chlorobenzene-d5, 1,4-difluorobenzene, 1,4-dichlorobenzene-d4, and pentafluorobenzene. Using stock standard solutions, prepare secondary dilution standards containing the compounds of interest, either singly or mixed together in methanol. Store them in a vial with no headspace for no more than one week. Surrogates recommended are toluene-d8, 4-bromofluorobenzene, and dibromofluoromethane. Each sample undergoing GC/MS analysis must be spiked with 10 µL of the surrogate spiking solution prior to analysis.

REFERENCE Test Methods for Evaluating Solid Waste (SW-846). U.S. EPA 1983, Method 8260A, Rev. 1, Nov. 1990. Office of Solid Waste, Washington, DC.

Bromoform   **EPA Method 601**
CAS #75-25-2

TITLE Purgeable Halocarbons

MATRIX Wastewater

APPLICATION Method covers 29 purgeable halocarbons. (Method 624 provides GC/MS conditions appropriate for the qualitative and quantitative confirmation of results). Method describes conditions for a 2nd GC column to confirm measurements made with primary column.

INTERFERENCES Impurities in the purge gas and organic compounds outgassing from the plumbing ahead of the trap. With high- and low-level samples, there can be carryover contamination. Diffusion of volatile organics through the septum seal into the sample.

INSTRUMENTATION GC-equipped with halide-specific detector. (With purge-and-trap unit).

RANGE 8.0–500 µg/L.

MDL 0.20 µg/L.

PRECISION 0.21X+2.41 µg/L (overall precision).

ACCURACY 0.96C-2.05 µg/L (as recovery).

SAMPLING METHOD 25-mL glass vial. Teflon®-lined septum.

STABILITY cool, 4°C, 0.008% Na2S2O3.

MHT 14 days.

QUALITY CONTROL The lab must on an ongoing basis, spike at least 10% of the samples from each sample site being monitored to assess accuracy.

REFERENCE Method 601, *Federal Register* Part VIII 40 CFR Part 136, Oct 26, 1984.

Bromoform   **EPA Method 624**
CAS #75-25-2

TITLE Purgeables

MATRIX Wastewater

APPLICATION Method covers 31 purgeable organics. An inert gas is bubbled through a 5-mL water sample in a specially designed purging chamber. Here, purgeables are transferred from aqueous to gaseous phase, passed onto a sorbent column, and trapped. Trap is heated and backflushed with inert gas to desorb purgeables onto a GC column, where purgeables are separated.

INTERFERENCES Impurities in the purge gas, organic compounds outgassing from the plumbing ahead of the trap, and solvent vapors in the lab. With high- and low-level samples, there can be carryover contamination.

INSTRUMENTATION GC/MS with purge-and-trap unit.

RANGE 5–600 µg/L

MDL 4.7 µg/L

PRECISION 0.17X+1.38 µg/L (overall precision).

ACCURACY 1.18C-2.35 µg/L (as recovery).

SAMPLING METHOD 25-mL glass vial. Teflon®-lined septum.

STABILITY cool, 4°C, 0.008% Na2S2O3.

MHT 14 days.

QUALITY CONTROL The lab must on an ongoing basis, spike at least 5% of the samples from each sample site being monitored to assess accuracy.

REFERENCE Method 624, *Federal Register* Part VIII 40 CFR Part 136, Oct 26, 1984.

Bromoform — EPA Method 8010
CAS #75-25-2

TITLE Halogenated Volatile Organics

MATRIX Groundwater, soils, sludges, water miscible liquid wastes, and non-water miscible wastes.

APPLICATION This method is used for the analysis of 39 halogenated VOCs. Samples are analyzed using direct injection or purge-and-trap methods. Groundwater must be analyzed by the purge-and-trap method. The method provides an optional GC column which is used for analyte confirmation and that may help resolve analytes from interferences.

INTERFERENCES There can be carryover contamination with high- and low-level samples. Impurities may come from the purge-and-trap apparatus, organic compounds outgassing from the plumbing ahead of trap, diffusion of VOCs through the sample bottle septum during shipping or storage, or from solvent vapors in the lab.

INSTRUMENTATION GC capable of on-column injections or purge-and-trap sample introduction and a halogen specific detector. Column 1: 8 ft by 0.1 in 1%. SP-1000 on Carbopack-B. Column 2: 6 ft by 0.1 in bonded n-octane on Porasil-C.

RANGE 8 to 500 µg/L (reagent water)

MDL 0.20 µg/L (reagent water).

PQL FACTORS FOR MULTIPLYING × FID MDL VALUE
Matrix Multiplication Factor

Groundwater	10
Low-level soil	10
Water miscible liquid waste	500
High-level soil and sludge	1250
Non-water miscible waste	1250

PRECISION 0.21X + 2.41 µg/L (overall precision).

ACCURACY 0.96C–2.05 µg/L (as recovery).

SAMPLING METHOD For water and liquid samples; use glass 40-mL vials with Teflon®-lined septum caps and collect two vials per sample location with no headspace. For solids and concentrated waste samples; use widemouth glass bottles with Teflon® liners.

STABILITY For concentrated wastes, soils, sediments, or sludges: cool to 4°C. For liquids: add 4 drops of concentrated hydrochloric acid and cool to 4°C.

MHT 14 days.

QUALITY CONTROL Analyze a reagent blank, matrix spike, and matrix spike duplicate/duplicate for each analytical batch (up to 20 samples). Demonstrate the purity of glassware and reagents by analyzing a reagent water method blank. Internal, surrogate, and five concentration level calibration standards are used.

REFERENCE Test Methods for Evaluating Solid Waste (SW-846), U.S. EPA Office of Solid Waste, Washington, DC, Method 8010B, Rev. 2, Nov. 1992.

Bromomethane — EPA Method 1624
CAS #74-83-9

TITLE Volatile Organic Compounds by Isotope Dilution GC/MS

MATRIX Compounds may be determined in waters, soils, and municipal sludges by this method.

METHOD SUMMARY This method is used to determine 58 volatile toxic organic pollutants associated with the CWA (as amended 1987); the RCRA (as amended 1986); the CERCLA (as amended 1986); and other compounds amenable to purge-and-trap gas chromatography-mass spectrometry (GC/MS).

If the solids content is less than 1%, stable isotopically-labeled analogs of the compounds of interest are added to a 5-mL sample and the sample is purged with an inert gas at 20–25°C in a chamber designed for soil or water samples. If the solids content is greater than 1%, 5 mL of reagent water and the labeled compounds are added to a 5-g aliquot of sample and the mixture is purged at 40°C. Compounds that will not purge at 20–25°C or at 40°C are purged at 78–85°C. In the purging process, the volatile compounds are transferred from the aqueous phase into the gaseous phase where they are passed into a sorbent column, and trapped. After purging is completed, the trap is backflushed and heated rapidly to desorb the compounds into a GC. The compounds are separated by the GC and detected by a MS. The labeled compounds serve to correct the variability of the analytical technique.

INTERFERENCES Impurities in the purge gas, organic compounds outgassing from the plumbing upstream of the trap, and solvent vapors in the lab account for most problems. Samples can be contaminated by diffusion of volatile organic compounds (particularly methylene chloride) through the bottle seal during shipment and storage. Contamination by carryover can occur when high-level and low-level samples are analyzed sequentially. When an unusually concentrated sample is encountered, follow it by analysis of a reagent water blank to check for carryover.

INSTRUMENTATION Major equipment includes a GC with linear temperature programming and a glass jet separator as the MS interface, a MS with 70 eV electron impact ionization, and a data system to collect and record response factors.

Column: 2.8 m × 2 mm I.D. glass, packed with 1% SP-1000 on Carbopak B, 60/80 mesh, or equivalent.

PRECISION & ACCURACY The detection limits of the method are usually dependent on the level of interferences rather than instrumental limitations. The method detection limits were determined in digested sludge (low solids) and in filter cake or compost (high solids).

The MDL (in µg/kg) for low solids is 148 and for high solids is 11.
Background levels of this compound were present in the sludge with low solids, resulting in a higher than expected MDL.
Labeled and native compound precision (in µg/L) as standard deviation was 25.0.
Labeled and native compound accuracy (in µg/L) as average recovery was detected to 54.
Acceptance criteria are at 20 µg/L for this compound.

SAMPLE COLLECTION, PRESERVATION & HANDLING Grab samples are collected in glass containers having a total volume greater than 20 mL. Fill and seal each bottle so that no air bubbles are entrapped. Samples are maintained at 0 to 4°C from the time of collection until analysis. If an aqueous sample contains residual chlorine, add sodium thiosulfate preservative (10 mg/40 mL) to the empty sample bottles just prior to shipment to the sample site. All samples must be analyzed within 14 days of collection.

SAMPLE PREPARATION Samples containing less than 1% solids are analyzed directly as aqueous samples. Samples containing 1% solids or greater are analyzed as solid samples utilizing one of two methods, depending on the levels of pollutants, in the sample. Samples containing 1% solids or greater, and low to moderate levels of pollutants are analyzed by purging a known weight of sample added to 5 mL of reagent water. Samples containing 1% solids or greater, and high levels of pollutants, are extracted with methanol, and an aliquot of the methanol extract is added to reagent water and purged.

QUALITY CONTROL A field blank prepared from reagent water and carried through the sampling and handling protocol may serve as a check on contamination from shipment and storage.

The analyst is permitted to modify this method to improve separations or lower the costs of measurements, provided all performance specifications are met. Analyses of blanks are required. When results of spikes indicate atypical method performance for samples, the samples are diluted to bring method performance within acceptable limits. Analyze two sets of four 5-mL aliquots (8 aliquots total) of the aqueous performance standard. Spike all samples with labeled compounds to assess method performance on the sample matrix. Compute the percent recovery of the labeled compounds using the internal standard method. Compare the percent recovery for each compound with the corresponding labeled compound recovery. Reagent water blanks are analyzed to demonstrate freedom from carryover contamination. Field replicates may be collected to determine the precision of the sampling technique, and spiked samples may be required to determine the accuracy of the analysis when the internal method is used.

REFERENCE Volatile Organic Compounds by Isotope Dilution GC/MS. Office of Water Regulation and Standards, U.S. EPA Industrial Technology Division, Washington, DC, EPA Method 1624, Rev. C, June 1989 (contact W.A. Telliard, U.S. EPA, Office of Water Regulations and Standards, 401 M St., SW, Washington, DC, 20460. Phone: 202-382-7131).

Bromomethane   EPA Method 502
CAS #74-83-9

TITLE Volatile Organic Compounds in Water By Purge and Trap Capillary Column Gas Chromatography with Photoionization and Electrolytic Conductivity Detectors in Series. U.S. EPA Method 502.2, Rev. 2.0, 1989.

MATRIX Drinking water and raw source water. The latter should include most surface water and groundwater sources.

METHOD SUMMARY This method covers 60 volatile organic compounds that contain halogen atoms and/or that are aromatic. An inert gas (zero grade nitrogen or helium) is bubbled through a 25-mL or a 5-mL water sample (depending on the expected concentration of the analytes). Purged sample components are trapped in a tube of sorbent materials. When purging is complete, the sorbent tube is heated and backflushed with helium to desorb the trapped sample onto a capillary GC column. The column is temperature programmed to separate the method analytes which are then detected with a photoionization detector (PID) and a Hall electrolytic conductivity (HECD) placed in series. The PID is selective for aromatic compounds and the HECD is selective for halogenated compounds.

INTERFERENCES Impurities in the purge gas and from organic compounds outgassing from the plumbing ahead of the trap account for many contamination problems. Interferences purged or coextracted from the samples will vary considerably from source to source, depending upon the particular sample or extract being tested. Cross-contamination can occur whenever high-level and low-level samples are analyzed sequentially. Samples also can be contaminated by diffusion of volatile organics (particularly methylene chloride and fluorocarbons) through the septum seal into the sample during shipment and storage. The lab where volatile analysis is performed and also the refrigerated storage area should be completely free of solvents.

INSTRUMENTATION A GC containing a series configuration of a high temperature photoionization detector (PID) equipped with 10.0 eV (nominal) lamp and Hall electrolytic conductivity detector (HECD) is required. Also required is an all-glass 5-mL purging device, a sorbent trap, and a thermal desorption apparatus which is connected to the GC system.

Column 1: VOCOL glass wide-bore capillary column.
Column 2: RTX–502.2 mega-bore capillary column.
Column 3: DB-62 mega-bore capillary column.

PRECISION & ACCURACY Method detection limits are dependent upon the characteristics of the gas chromatographic system used. Analytes that are not separated chromatographically cannot be individually identified and used in the same calibration mixture or water samples unless an alternative technique for identification and quantification, such as mass spectrometry, is used.

Electrolytic conductivity detetor (c) range in µg/L (a) was 0.02–200.
Electrolytic conductivity detetor (c) MDL in µg/L (b) was 1.1.
Electrolytic conductivity detetor (c) accuracy as % recovery was 97.
Electrolytic conductivity detetor (c) precision as % RSD was 3.8.
Photoionization detector (d) range in µg/L (a) was 0.02–200.
Photoionization detector (d) MDL in µg/L (b) was not listed.
Photoionization detector (d) accuracy as % recovery was not listed.
Photoionization detector (d) precision as % RSD was not listed.

(a) *The applicable concentration range of this method is compound, instrument, and matrix-dependent. It is listed as being approximately 0.02 to 200 µg/L but no specific information is provided so caution should be observed.*

(b) *The method detection limits reports with this method are compound, instrument, and matrix-dependent. The values reported were calculated using reagent water fortified with the corresponding compounds at 10 µg/L and a GC-equipped with a 60 m × 0.75 mm VOLCOL wide bore capillary column with 1.5 µm film thickness and using helium carrier gas.*

(c) *Recoveries and relative standard deviations were determined from seven samples of reagent water fortified with 10 µg/L of each compound. 2-Bromo-1-chloropropane was used as the internal standard for calculating average recoveries.*

(d) *Recoveries and relative standard deviations were determined from seven samples of reagent water fortified with 10 µg/L of each compound. Fluorobenzene was used as the internal standard for calculating average recoveries.*

SAMPLING METHOD Collect samples using a 40- to 120-mL screw-cap vial (prewashed with detergent, rinsed with distilled water and oven dried at 105°C) with a Teflon®-faced silicone septum. Collect bubble-free samples and place the septum with the Teflon® side down on the water.

SAMPLE PRESERVATION If residual chlorine is present in the water add about 25 mg of ascorbic acid to each vial before samples are collected to remove the chlorine. Add hydrochloric acid to reduce pH to <2, immediately cool samples to 4°C, and store them in a solvent-free refrigerator at 4°C until analysis.

MHT The maximum holding time for samples is 14 days from the time they were collected.

SAMPLE PREPARATION Remove the plungers from two 5-mL syringes and attach a closed syringe valve to each. Warm the sample to room temperature, open the sample bottle, and carefully pour the sample into one of the syringe barrels to just short of overflowing. Replace the syringe plunger, invert the syringe, and compress the sample. Open the syringe valve and vent any residual air while adjusting the sample volume to 5.0 mL. Add 10 µL of the internal calibration standard to the sample through the syringe valve. Close the valve. Fill the second syringe in an identical manner from the same sample bottle. Reserve this second syringe for a reanalysis if necessary.

QUALITY CONTROL As an initial demonstration of lab accuracy and precision, analyze 4 to 7 replicates of a lab fortified blank containing analyte at 0.1–5 µg/L. Collect all samples in duplicate. Surrogate analytes (similar to those of the analytes of interest), whose concentration is known in every sample, are measured using the same internal standard calibration procedure. Duplicate field reagent water blanks (trip blanks) must be analyzed with each set of samples, lab reagent blanks (method blanks) must be analyzed with each batch of samples processed as a group within a work shift. Also, a single lab-fortified blank that contains each of the analytes of interest should be analyzed with each batch of samples processed as a group within a work shift. A 3- to 5-point calibration curve is needed depending on the calibration range factor required.

EPA CONTACT & HOTLINE For technical questions contact Dr. Baldev Bathija, U.S. EPA, Office of Ground Water and Drinking Water (WH-550D), 401 M St. SW, Washington, DC 20460. Tel. (202) 260-3040. For further information the EPA Safe Drinking Water Hotline may be called at: (800) 426-4791.

REFERENCE Methods for the Determination of Organic Compounds in Drinking Water, EPA/600/4-88/039 (revised July 1991; Final Rule for determination of compliance with the MCL for Total Trihalomethanes under 141.30, in 40 CFR Part 141, Vol. 58, No. 147, Fed. Reg., Tuesday Aug. 3, 1993). U.S. EPA Environmental Monitoring Systems Laboratory, Cincinnati, OH, 45268, U.S.A. Available from the National Technical Information Service (NTIS), 5285 Port Royal Road, Springfield, VA 22161; Tel. 800-553-6847. NTIS Order Number is PB91-231480.

Bromomethane   **EPA Method 524**
CAS #74-83-9

TITLE Measurement of Purgeable Organic Compounds in Water by Capillary Column GC/MS.

MATRIX Drinking water and raw source water; the latter should include most surface water and groundwater sources.

METHOD SUMMARY Method 524.2 covers 60 volatile organic compounds. An inert gas (zero grade nitrogen or helium) is bubbled through a 25-mL or a 5-mL water sample (depending on the expected concentration of the analytes). Purged sample components are trapped in a tube of sorbent materials. When purging is complete, the sorbent tube is heated and backflushed with helium to desorb the trapped sample onto a capillary GC column.

INTERFERENCES Impurities in the purge gas and from organic compounds outgassing from the plumbing ahead of the trap account for many contamination problems. Interferences purged or coextracted from the samples will vary considerably from source to source, depending upon the particular sample or extract being tested. Cross-contamination can occur whenever high-level and low-level samples are analyzed sequentially. Samples also can be contaminated by diffusion of volatile organics (particularly methylene chloride and fluorocarbons) through the septum seal into the sample during shipment and storage.

INSTRUMENTATION A GC/MS with a data system equipped with one of the following capillary GC columns:

Column 1: VOCOL glass wide bore capillary column.
Column 2: DB-624 fused silica capillary column.
Column 3: DB-5 fused silica capillary column.

Also required is an all-glass 25-mL or 5-mL purging device, a sorbent trap, and a thermal desorption apparatus which is connected to the GC/MS system.

PRECISION & ACCURACY Method detection limits are compound- and instrument-dependent, and may vary from approximately 0.02–0.35 µg/L. Note in the table below that the "true" concentration range used for accuracy and precision measurements was quite narrow. However, the applicable concentration range of this method is primarily column dependent and is approximately 0.02 to 200 µg/L for the wide-bore thick-film columns. Narrow-bore thin-film columns may have a capacity which limits the range to about 0.02 to 20 µg/L. Analytes that are inefficiently purged from water will not be detected when present at low concentrations, but they can be measured with acceptable accuracy and precision when present in sufficient amounts.

Analytes that are not separated chromatographically, but which have different mass spectra and non-interfering quantification ions, can be identified and measured in the same calibration mixture or water sample. Analytes which have very similar mass spectra cannot be individually identified and measured in the same calibration mixture or water samples unless they have different retention times. Co-eluting compounds with very similar mass spectra, typically many structural isomers, must be reported as an isomeric group or pair.

The range (in µg/L) was 0.5–10.
The method detection limit (in µg/L) was 0.11.
The accuracy (as % recovery) was 95.
The precision (in %) was 8.2.

Note: Data were obtained from 16–31 determinations using a wide-bore capillary column and a jet separator interfaced to a quadrupole mass spectrometer. All analytes were in a reagent water matrix.

SAMPLING METHOD Collect samples using a 40- to 120-mL screw-cap vial (prewashed with detergent, rinsed with distilled water and oven dried at 105°C) with a Teflon®-faced silicone septum. Collect bubble-free samples and place the septum with the Teflon® side down on the water.

SAMPLE PRESERVATION If residual chlorine is present in the water add about 25 mg of ascorbic acid to each vial before samples are collected to remove the chlorine. Add hydrochloric acid to reduce pH to <2, and immediately cool samples to 4°C, and store them in a solvent-free refrigerator at 4°C until analysis.

MHT The maximum holding time for samples is 14 days from the time they were collected.

SAMPLE PREPARATION Remove the plungers from two 25-mL (or 5-mL depending on sample size) syringes and attach a closed syringe valve to each. Warm the sample to room temperature, open the sample bottle, and carefully pour the sample into one of the syringe barrels to just short of overflowing. Replace the syringe plunger, invert the syringe, and compress the sample. Open the syringe valve and vent any residual air while adjusting the sample volume to 25.0 mL (or 5 mL). For samples and blanks, add 5 µL of the fortification solution containing the internal standard and the surrogates to the sample through the syringe valve. For calibration standards and lab fortified blanks, add 5 µL of the fortification solution containing the internal standard only. Close the valve. Fill the second syringe in an identical manner from the same sample bottle. Reserve this second syringe for a reanalysis if necessary.

QUALITY CONTROL As an initial demonstration of lab accuracy and precision, analyze 4 to 7 replicates of a lab fortified blank containing analyte at 0.2–5 µg/L. Collect all samples in duplicate. Surrogate analytes (similar to those of the analytes of interest), whose concentration is known in every sample, are measured using the same internal standard calibration procedure. Duplicate field reagent water blanks (trip blanks) must be analyzed with each set of samples, lab reagent blanks (method blanks) must be analyzed with each batch of samples processed as a group within a work shift. Also, a single lab-fortified blank that contains each of the analytes of interest should be analyzed with each batch of samples processed as a group within a work shift. A 3- to 5-point calibration curve is needed depending on the calibration range factor required.

EPA CONTACT & HOTLINE For technical questions contact Dr. Baldev Bathija, U.S. EPA, Office of Ground Water and Drinking Water (WH-550D), 401 M St. SW, Washington, DC 20460. Tel. (202) 260-3040. For further information the EPA Safe Drinking Water Hotline may be called at: (800) 426-4791.

REFERENCE Methods for the Determination of Organic Compounds in Drinking Water, EPA/600/4-88/039 (revised July 1991; Final Rule for determination of compliance with the MCL for Total Trihalomethanes under 141.30, in 40 CFR Part 141, Vol. 58, No. 147, Fed. Reg., Tuesday Aug. 3, 1993). U.S. EPA Environmental Monitoring Systems Laboratory, Cincinnati, OH, 45268, U.S.A. Available from the National Technical Information Service (NTIS), 5285 Port Royal Road, Springfield, VA 22161; Tel. 800-553-6847. NTIS Order Number is PB91-231480.

Bromomethane EPA Method 8021
CAS #74-83-9

VOAs by Capillary Column GC/PID-HECD

TITLE Halogenated Volatile by Gas Chromatography Using Photoionization and Electrolytic Conductivity Detectors in Series: Capillary Column Technique

MATRIX This method is applicable to nearly all types of samples, regardless of water content, including groundwater, aqueous sludges, caustic liquors, acid liquors, waste solvents, oily wastes, mousses, tars, fibrous wastes, polymeric emulsions, filter cakes, spent carbons, spent catalysts, soils, and sediments.

METHOD SUMMARY This method is used to determine 60 volatile organic compounds in a variety of solid waste matrices. It provides GC conditions for the detection of halogenated and aromatic volatile organic compounds. Samples can be analyzed using direct injection or purge-and-trap (EPA Method 5030). Groundwater samples must be analyzed using EPA Method 5030 (where applicable). A temperature program is used with the GC. Detection is achieved by a photoionization detector (PID) and a Hall electrolytic conductivity detector (HECD) in series.

INTERFERENCES Samples can be contaminated by diffusion of volatile organics (particularly chlorofluorocarbons and methylene chloride) through the sample container septum during shipment and storage.

INSTRUMENTATION A GC-equipped with variable-constant differential flow controllers, subambient oven controller, PID and HECD detectors connected with a short piece of uncoated capillary tubing and a data system.

Column: 60 m × 0.75 mm I.D. VOCOL wide-bore capillary column with 1.5 µm film thickness.

PRECISION & ACCURACY MDLs are compound-dependent and vary with purging efficiency and concentration. The applicable concentration range of this method is compound- and instrument-dependent but is approximately 0.1 to 200 µg/L. Analytes that are inefficiently purged from water will not be detected when present at low concentrations, but they can be measured with acceptable accuracy and precision when present in sufficient amounts. The estimated quantitation limit (EQL) for an individual compound is approximately 1 µg/kg (wet weight) for soil/sediment samples, 100 µg/kg (wet weight) for wastes, and 1 µg/L for groundwater. EQLs will be proportionately higher for sample extracts and samples that require dilution or reduced sample size to avoid saturation of the detector.

MULTIPLICATION FACTORS FOR OTHER MATRICES (a)

Matrix	Factor (b)
Groundwater	10
Low-concentration soil	10
Water miscible liquid waste	500
High-concentration soil and sludge	1250
Non-water miscible waste	1250

(a) Sample EQLs are highly matrix-dependent. The EQLs listed herein are provided for guidance and may not always be achievable. (b) EQL = [Method detection limit] × [Factor]. For non-aqueous samples, the factor is on a wet-weight basis.

SINGLE LABORATORY ACCURACY & PRECISION DATA FOR VOCs IN WATER

This method was tested in a single lab using water spiked at 10 µg/L and the following data was reported:

Recoveries and standard deviations were determined from seven samples and spiked at 10 µg/L of each analyte. Recoveries were determined by the internal standard method. Internal standards were: Fluorobenzene for PID and 2-Bromo-1-chloropropane for HECD.

The average recovery (in percent) for the PID was none (no response for this detector).
The standard deviation of the recovery for the PID was none (no response for this detector).
The MDL (in µg/mL) for the PID was none (no response for this detector).
The average recovery (in percent) for the HECD was 97.
The standard deviation of the recovery for the HECD was 3.7.
The MDL (in µg/mL) for the HECD was 1.1.

SAMPLE COLLECTION, PRESERVATION & HANDLING
Volatile Organics — Standard 40-mL glass screw-cap VOA vials with Teflon®-faced silicone septum may be used for both liquid and solid matrices. When collecting samples, liquids and solids should be introduced into the vials gently to reduce agitation which might drive off volatile compounds. If there are any air bubbles present the sample must be retaken. Tap slightly as they are filled to try and eliminate as much free air space as possible. The two vials from each sampling locations should be sealed in separate plastic bags to prevent cross-contamination between samples particularly if the sampled waste is suspected of containing high levels of volatile organics.

Semivolatile organics — Containers used to collect samples for the determination of semivolatile organic compounds should be soap and water washed followed by methanol (or isopropanol) rinsing. The sample containers should be of glass or Teflon® and have screw-top covers with Teflon® liners.

Preservation for volatile organics — No preservation is used with concentrated waste samples. With liquid samples containing no residual chlorine, 4 drops of concentrated hydrochloric acid are added and the samples are immediately cooled to 4°C. When liquid samples contain residual chlorine, they are treated as above and, in addition, 4 drops of 4% aqueous sodium thiosulfate are added. Soil, sediment, and sludge samples are only cooled to 4°C.

Preservation for semivolatile organics — No preservation is used with concentrated waste samples. With liquid samples containing no residual chlorine and with soil, sediment, and sludge samples, immediately cooling to 4°C is the only preservation used. When residual chlorine is present then 3 mL of 10% aqueous sodium sulfate is added for each gallon of sample collected, followed by cooling to 4°C.

MHT The holding time for all volatile organics samples is 14 days. Liquid samples must be extracted within 7 days and their extracts analyzed within 40 days. Concentrated waste, soil, sediment, and sludge samples must be extracted within 14 days and their extracts analyzed within 40 days.

SAMPLE PREPARATION Volatile compounds are introduced into the gas chromatograph either by direct injector or purge-and-trap (EPA Method 5030). EPA Method 5030 may be used directly on groundwater samples or low-concentration contaminated soils and sediments. For medium-concentration soils or sediments, methanolic extraction, as described in EPA Method 5030, may be necessary prior to purge-and-trap analysis.

QUALITY CONTROL Calculate surrogate standard recovery on all samples, blanks, and spikes. A trip blank is recommended to check on sampling, storage, and handling contamination. Calibration standards, at a minimum of five concentration levels, are prepared in organic-free reagent water. One of the concentration levels should be at a concentration near, but above, the method detection limit.

A combination of bromochloromethane, 2-bromo-1-chloropropane, 1,4-dichlorobutane, and bromochlorobenzene are recommended as surrogate standards to encompass the range of the temperature program used in this method.

REFERENCE Test Methods for Evaluating Solid Waste, Physical/Chemical Methods, SW-846, 3rd Edition, U.S. EPA, Office of Solid Waste, Washington, DC, EPA Method 8021A, Rev. 1, Nov. 1992.

Bromomethane   **EPA Method 8240**
CAS #74-83-9

TITLE Volatile Organics By GC/MS: Packed Column Technique

MATRIX Nearly all types of sample matarices, regardless of water content, can be analyzed using this method. This includes groundwater, aqueous sludges, caustic liquors, acid liquors, waste solvents, oily wastes, mousses, tars, fibrous wastes, polymetric emulsions, filter cakes, spent carbons, spent catalysts, soils, and sediments.

METHOD SUMMARY Method 8240B covers 80 volatile organic compounds that are introduced into a gas chromatograph by the purge-and-trap method or by direct injection (in limited applications). For the purge-and-trap method an inert gas (zero grade nitrogen or helium) is bubbled through a 5-mL solution at ambient temperature. Purged sample components are trapped in a tube of sorbent materials. When purging is complete, the sorbent tube is heated and backflushed with inert gas to desorb the trapped components onto a GC column.

INTERFERENCES Impurities in the purge gas and from organic compounds outgassing from the plumbing ahead of the trap account for many contamination problems. Interferences purged or coextracted from the samples will vary considerably from source to source. Cross-contamination can occur whenever high-level and low-level samples are analyzed sequentially. Whenever an unusually concentrated sample is analyzed, it should be followed by the analysis of organic-free reagent water to check for cross-contamination. Samples also can be contaminated by diffusion of volatile organics (particularly methylene chloride and fluorocarbons) through the septum seal into the sample during shipment and storage. A trip blank can serve as a check on such contamination. The lab where volatile analysis is performed and also the refrigerated storage area should be completely free of solvents.

INSTRUMENTATION A gas chromatograph/mass spectrometry/data system (GC/MS) equipped with a 6 ft × 0.1 in I.D. glass column packed with 1% SP-1000 on Carbopack-B (60/80 mesh) is required. Also needed is a 5-mL purging device, a sorbent trap, and a thermal desorption apparatus.

PRECISION & ACCURACY This method is reported to have been tested by 15 laboratories using organic-free reagent water, drinking water, surface water, and industrial wastewaters (not specified) fortified at six concentrations over the range 5–600 µg/L.

Sample estimated quantitation limits (EQLs) are highly matrix-dependent. The EQLs listed may not always be achievable. EQLs listed for soils or sediments are based on wet weight. Normally, data is reported on a dry-weight basis; therefore, EQLs will be higher, based on the percent dry weight of each sample. Note that EQLs are even more variable than MDLs and that they are highly variable depending on the matrix being analyzed.

EQL in groundwater in µg/L was 10.
EQL in low soil or sediment in µg/kg was 10.
Accuracy (a) in µg/L was 1.00C.
Precision (b) in µg/L was 0.58x.

(a) *Average recovery found for measurements of samples containing a concentration of C, in µg/L.*
(b) *Overall precision found for measurements of samples with average recovery X for samples containing a concentration of C in µg/L.*
X = *Average recovery found for measurement of samples containing a concentration of C in µg/L.*

MULTIPLICATION FACTORS FOR OTHER MATRICES

Other Matrices	Factor (a)
Waste miscible liquid waste	50
High-concentration soil and sludge	125
Non-water miscible waste	500

(a) *EQL = [EQL for low soil/sediment] × [Factor]. For non-aqueous samples, the factor is on a wet-weight basis.*

SAMPLING METHOD

Liquid samples — Use a 40-mL glass screw-cap VOA vial with a Teflon®-faced silicone septum that has been prewashed, rinsed with distilled deionized water, and oven dried. However, if residual chlorine is present, collect sample in a 40-oz. soil VOA container which has been pre-preserved with 4 drops of 10% sodium thiosulfate, mix gently, and then transfer the sample to a 40-mL VOA vial. Collect bubble-free samples in duplicate and seal them in separate plastic bags.

Soils or sediments, and sludges — Use an 8-oz. widemouth glass bottle with a Teflon®-faced silicone septum that has been prewashed with detergent, rinsed with distilled deionized water, and oven dried. Tap slightly to eliminate free air space. Collect samples in duplicate and seal them in separate plastic bags.

SAMPLE PRESERVATION

Liquid samples — Add 4 drops of concentrated HCL and immediately cool samples to 4°C and store in a solvent-free refrigerator.

Soils or sediments, and sludges — Cool samples to 4°C and store in a solvent-free refrigerator.

MHT Maximum holding time is 14 days from the date of sample collection.

SAMPLE PREPARATION

Liquid samples — Remove the plunger from a 5-mL syringe and carefully pour the sample into the syringe barrel to just short of overflowing. Replace the syringe plunger and compress the sample. Open the syringe valve and vent any residual air while adjusting the sample volume to 5.0 mL. If there is only one volatile organic analysis (VOA) vial, a second syringe should be filled at this time to protect against possible loss of sample integrity. Add 10 μL of surrogate spiking solution and 10 μL of internal standard spiking solution through the valve bore of the 5-mL syringe, then close the valve. The surrogate and internal standards may be mixed and added as a single spiking solution.

Sediments, soils, and waste samples — All samples of this type should be screened by GC analysis using a headspace method (EPA Method 3810) or the hexadecane extraction and screening method (EPA Method 3820). Use the screening data to determine whether to use the low-concentration method (0.005–1 mg/kg) or the high-concentration method (>1 mg/kg).

Low-concentration method — The low-concentration method is based on purging a heated sediment or soil sample mixed with organic-free reagent water containing the surrogate and internal standards. Analyze all reagent blanks and standards under the same conditions as the samples.

Use a 5-g sample if the expected concentration is <0.1 mg/kg or a 1-g sample for expected concentrations between 0.1 and 1 mg/kg. Mix the contents of the sample container with a narrow metal spatula. Weigh the amount of the sample into a tared purge device. Add the spiked water to the purge device, which contains the weighed amount of sample, and connect the device to the purge-and-trap system.

High-concentration method — This method is based on extracting the sediment or soil with methanol. A waste sample is either extracted or diluted, depending on its solubility in methanol. Wastes that are insoluble in methanol are diluted with reagent tetraglyme or possibly polyethylene glycol (PEG). An aliquot of the extract is added to organic-free reagent water containing surrogate and internal standards. This is purged at ambient temperature. All samples with an expected concentration of >1.0 mg/kg should be analyzed by this method.

Mix the contents of the sample container with a narrow metal spatula. For sediments or soils and solid wastes that are insoluble in methanol, weigh 4 g (wet weight) of sample into a tared 20-mL vial. For waste that is soluble in methanol, tetraglyme, or PEG, weigh 1 g (wet weight) into a tared scintillation vial or culture tube or a 10-mL volumetric flask. Quickly add 9.0 mL of appropriate solvent then add 1.0 mL of a surrogate spiking solution to the vial, cap it, and shake it for 2 min.

METHANOL EXTRACT REQUIRED FOR ANALYSIS OF HIGH-CONCENTRATION SOILS OR SEDIMENTS

Approximate Concentration Range	Volume of Methanol Extract (a)
500–10,000 μg/kg	100 μL
1,000–20,000 μg/kg	50 μL
5,000–100,000 μg/kg	10 μL
25,000–500,000 μg/kg	100 μL of 1/50 dilution (b)

Calculate appropriate dilution factor for concentrations exceeding this table.

(a) The volume of methanol added to 5 mL of water being purged should be kept constant. Therefore, add to the 5-mL syringe whatever volume of methanol is necessary to maintain a volume of 100 μL added to the syringe.

(b) Dilute an aliquot of the methanol extract and then take 100 μL for analysis.

QUALITY CONTROL Demonstrate, through the analysis of a reagent water blank, that interferences from the analytical system, glassware, and reagents are under control. Blank samples should be carried through all stages of the sample preparation and measurement steps. For each analytical batch (up to 20 samples), a reagent blank, matrix spike, and matrix spike duplicate must be analyzed (the frequency of the spikes may be different for different monitoring programs). The blank and spiked samples must be carried through all stages of the sample preparation and measurement steps. QC samples mentioned in the section on Interferences will also be needed as appropriate to those situations.

REFERENCE Test Methods for Evaluating Solid Waste (SW-846). U.S. EPA. 1983. Method 8240B, Rev. 2, Nov. 1990. Office of Solid Wastes, Washington, DC.

Bromomethane **EPA Method 8260**
CAS #74-83-9

TITLE Volatile Organic Compounds by GC/MS: Capillary Column Technique

MATRIX This method is applicable to nearly all types of samples, regardless of water content, including groundwater, soils, and sediments.

METHOD SUMMARY Method 8260A covers 58 volatile organic compounds that are introduced into a gas chromatograph by the purge-and-trap method or by direct injection (in limited applications). Zero-grade helium is bubbled through a 5-mL solution at ambient temperature. Purged sample components are trapped in a tube containing suitable sorbent materials. When purging is complete, the sorbent tube is heated and backflushed with helium to desorb trapped sample components. The analytes are desorbed directly to a large bore capillary or cryofocussed on a capillary precolumn before being flash evaporated to a narrow bore capillary for analysis.

INTERFERENCES Major contaminant sources are volatile materials in the lab and impurities in the inert purging gas and in the sorbent trap. Interfering contamination may occur when a sample containing low concentrations of volatile organic compounds is analyzed immediately after a sample containing high concentrations of volatile organic compounds. After analysis of a sample containing high concentrations of volatile organic compounds, one or more calibration blanks should be analyzed to check for cross-contamination. Screening of the samples prior to purge-and-trap GC/MS analysis is highly recommended to prevent contamination of the system. This is especially true for soil and waste samples.

Special precautions must be taken to analyze for methylene chloride. The analytical and sample storage area should be isolated from all atmospheric sources of methylene chloride. All gas chromatography carrier gas lines and purge gas plumbing should be constructed from stainless steel or copper tubing. Laboratory clothing previously exposed to methylene chloride fumes during liquid-liquid extraction procedures can contribute to sample contamination.

Samples can also be contaminated by diffusion of volatile organics (particularly methylene chloride and fluorocarbons) through the septum seal during shipment and storage. A trip blank can serve as a check on such contamination.

INSTRUMENTATION GC/MS with a temperature-programmable chromatograph suitable for splitless injection equipped with variable constant differential flow controllers, a subambient oven controller, a purging device, sorbent trap, a thermal desorption apparatus and a capillary precolumn interface when using cryogenic cooling will be needed. The following GC columns may be used:

Column 1: 60 m × 0.75 mm I.D. capillary column coated with VOCOL, 1.5 µm film thickness.
Column 2: 30 m × 0.53 mm I.D. capillary column coated with DB-624 or VOCOL, 3 µm film thickness.
Column 3: 30 m × 0.32 mm I.D. capillary column coated with DB-5 or SE-54, 1-µm film thickness.

PRECISION & ACCURACY This method has been tested in a single lab using spiked water. Using a wide-bore capillary column, water was spiked at concentrations between 0.5 and 10 µg/L. Single lab accuracy and precision data are presented. The MDL actually achieved in a given analysis will vary depending on instrument sensitivity and matrix effects.

The MDL (a) in µg/L was 0.11.
The concentration range in µg/L was 0.5–10.
The mean accuracy (% of true value) was 95.
The precision as relative standard deviation was 8.2.

Note: The MDL is based on a 25-mL sample volume instead of a 5-mL sample volume.

SAMPLING METHOD
Liquid samples — Use a 40-mL glass screw-cap VOA vial with a Teflon®-faced silicone septum that has been prewashed, rinsed with distilled deionized water, and oven dried. If residual chlorine is present, collect the sample in a 4-oz soil VOA container which has been pre-preserved with 4 drops of 10% sodium thiosulfate. Mix gently and transfer the sample to a 40-mL VOA vial. Collect bubble-free samples in duplicate and seal each sample in a separate plastic bag.

Soils, sediments, and sludges — Use an 8-oz widemouth glass bottle with Teflon®-faced silicone septum that has been prewashed, rinsed with distilled deionized water, and oven dried. **Do not** heat the septum for more than 1 h. Tap slightly to eliminate any free air space. Collect samples in duplicate and seal each one in a separate plastic bag.

SAMPLE PRESERVATION
Liquid samples — Add 4 drops of concentrated HCL, cool to 4°C and store in a solvent-free refrigerator.

Soils, sediments, and sludges — Cool samples to 4°C and store in a solvent-free refrigerator.

MHT The maximum holding time of any sample (liquids, soils, sediments, and sludges) is 14 days.

SAMPLE PREPARATION
Liquid samples — Remove the plunger from a 5-mL syringe and carefully pour the sample into the syringe barrel to just short of overflowing. Replace the syringe plunger and compress the sample. Open the syringe valve and vent any residual air while adjusting the sample volume to 5.0 mL. If there is only one volatile organic analysis (VOA) vial, a second syringe should be filled at this time to protect against possible loss of sample integrity. Add 10 µL of surrogate spiking solution and 10 µL of internal standard spiking solution through the valve bore of the 5-mL syringe, then close the valve. The surrogate and internal standards may be mixed and added as a single spiking solution.

Sediments, soils, and waste samples — All samples of this type should be screened by GC analysis using a headspace method (EPA Method 3810) or the hexadecane extraction and screening method (EPA Method 3820). Use the screening data to determine whether to use the low-concentration method (0.005–1 mg/kg) or the high-concentration method (>1 mg/kg).

Low-concentration method — The low-concentration method is based on purging a heated sediment or soil sample mixed with organic-free reagent water containing the surrogate and internal standards. Analyze all reagent blanks and standards under the same conditions as the samples.

Use a 5-g sample if the expected concentration is <0.1 mg/kg or a 1-g sample for expected concentrations between 0.1 and 1 mg/kg. Mix the contents of the sample container with a narrow metal spatula. Weigh the amount of the sample into a tared purge device. Add the spiked water to the purge device, which contains the weighed amount of sample, and connect the device to the purge-and-trap system.

High-concentration method — This method is based on extracting the sediment or soil with methanol. A waste sample is either extracted or diluted, depending on its solubility in methanol. Wastes that are insoluble in methanol are diluted with reagent tetraglyme or possibly polyethylene glycol (PEG). An aliquot of the extract is added to organic-free reagent water containing surrogate and internal standards. This is purged at

ambient temperature. All samples with an expected concentration of >1.0 mg/kg should be analyzed by this method.

Mix the contents of the sample container with a narrow metal spatula. For sediments or soils and solid wastes that are insoluble in methanol, weigh 4 g (wet weight) of sample into a tared 20-mL vial. For waste that is soluble in methanol, tetraglyme, or PEG, weigh 1 g (wet weight) into a tared scintillation vial or culture tube or a 10-mL volumetric flask. Quickly add 9.0 mL of appropriate solvent then add 1.0 mL of a surrogate spiking solution to the vial, cap it, and shake it for 2 min.

METHANOL EXTRACT REQUIRED FOR ANALYSIS OF HIGH-CONCENTRATION SOILS OR SEDIMENTS

Approximate Concentration Range	Volume of Methanol Extract (a)
500–10,000 µg/kg	100 µL
1,000–20,000 µg/kg	50 µL
5,000–100,000 µg/kg	10 µL
25,000–500,000 µg/kg	100 µL of 1/50 dilution (b)

Calculate appropriate dilution factor for concentrations exceeding this table.

(a) The volume of methanol added to 5 mL of water being purged should be kept constant. Therefore, add to the 5-mL syringe whatever volume of methanol is necessary to maintain a volume of 100 µL added to the syringe.
(b) Dilute an aliquot of the methanol extract and then take 100 µL for analysis.

QUALITY CONTROL Demonstrate, through the analysis of a reagent water blank, that interferences from the analytical system, glassware, and reagents are under control. Blank samples should be carried through all stages of the sample preparation and measurement steps. For each analytical batch (up to 20 samples), a reagent blank, matrix spike, and matrix spike duplicate must be analyzed (the frequency of the spikes may be different for different monitoring programs). The blank and spiked samples must be carried through all stages of the sample preparation and measurement steps. QC samples mentioned in the section on Interferences will also be needed as appropriate to those situations.

Matrix spiking standards should be prepared from volatile organic compounds which will be representative of the compounds being investigated. The recommended internal standards are chlorobenzene-d5, 1,4-difluorobenzene, 1,4-dichlorobenzene-d4, and pentafluorobenzene. Using stock standard solutions, prepare secondary dilution standards containing the compounds of interest, either singly or mixed together in methanol. Store them in a vial with no headspace for no more than one week. Surrogates recommended are toluene-d8, 4-bromofluorobenzene, and dibromofluoromethane. Each sample undergoing GC/MS analysis must be spiked with 10 µL of the surrogate spiking solution prior to analysis.

REFERENCE Test Methods for Evaluating Solid Waste (SW-846). U.S. EPA 1983, Method 8260A, Rev. 1, Nov. 1990. Office of Solid Waste, Washington, DC.

Bromomethane EPA Method 601
CAS #74-83-9

TITLE Purgeable Halocarbons

MATRIX Wastewater

APPLICATION Method covers 29 purgeable halocarbons. (Method 624 provides GC/MS conditions appropriate for the qualitative and quantitative confirmation of results). Method describes conditions for a 2nd GC column to confirm measurements made with primary column.

INTERFERENCES Impurities in the purge gas and organic compounds outgassing from the plumbing ahead of the trap. With high- and low-level samples, there can be carryover contamination. Diffusion of volatile organics through the septum seal into the sample.

INSTRUMENTATION GC-equipped with halide-specific detector. (With purge-and-trap unit).

RANGE 8.0–500 µg/L. ***MDL*** 1.18 µg/L.

PRECISION 0.36X+0.94 µg/L (overall precision).

ACCURACY 0.76C–1.27 µg/L (as recovery).

SAMPLING METHOD 25-mL glass vial. Teflon®-lined septum.

STABILITY cool, 4°C, 0.008% Na2S2O3.

MHT 14 days.

QUALITY CONTROL The lab must on an ongoing basis, spike at least 10% of the samples from each sample site being monitored to assess accuracy.

REFERENCE Method 601, *Federal Register* Part VIII 40 CFR Part 136, Oct 26, 1984.

Bromomethane EPA Method 624
CAS #74-83-9

TITLE Purgeables

MATRIX Wastewater

APPLICATION Method covers 31 purgeable organics. An inert gas is bubbled through a 5-mL water sample in a specially designed purging chamber. Here, purgeables are transferred from aqueous to gaseous phase, passed onto a sorbent column, and trapped. Trap is heated and backflushed with inert gas to desorb purgeables onto a GC column, where purgeables are separated.

INTERFERENCES Impurities in the purge gas, organic compounds outgassing from the plumbing ahead of the trap, and solvent vapors in the lab. With high- and low-level samples, there can be carryover contamination.

INSTRUMENTATION GC/MS with purge-and-trap unit.

RANGE 5–600 µg/L ***MDL*** not determined

PRECISION 0.58X µg/L (overall precision).

ACCURACY 1.00C µg/L (as recovery).

SAMPLING METHOD 25-mL glass vial. Teflon®-lined septum.

STABILITY Cool, 4°C, 0.008% Na2S2O3.

MHT 14 days.

QUALITY CONTROL The lab must on an ongoing basis, spike at least 5% of the samples from each sample site being monitored to assess accuracy.

REFERENCE Method 624, *Federal Register* Part VIII 40 CFR Part 136, Oct 26, 1984.

Bromomethane
CAS #74-83-9
EPA Method 8010

TITLE Halogenated Volatile Organics

MATRIX Groundwater, soils, sludges, water miscible liquid wastes, and non-water miscible wastes.

APPLICATION This method is used for the analysis of 39 halogenated VOCs. Samples are analyzed using direct injection or purge-and-trap methods. Groundwater must be analyzed by the purge-and-trap method. The method provides an optional GC column which is used for analyte confirmation and that may help resolve analytes from interferences.

INTERFERENCES There can be carryover contamination with high- and low-level samples. Impurities may come from the purge-and-trap apparatus, organic compounds outgassing from the plumbing ahead of trap, diffusion of VOCs through the sample bottle septum during shipping or storage, or from solvent vapors in the lab.

INSTRUMENTATION GC capable of on-column injections or purge-and-trap sample introduction and a halogen specific detector. Column 1: 8 ft by 0.1 in 1%. SP-1000 on Carbopack-B. Column 2: 6 ft by 0.1 in bonded n-octane on Porasil-C.

RANGE 8 to 500 µg/L (reagent water).

MDL Not determined.

PQL FACTORS FOR MULTIPLYING × FID MDL VALUE

Matrix	Multiplication Factor
Groundwater	10
Low-level soil	10
Water miscible liquid waste	500
High-level soil and sludge	1250
Non-water miscible waste	1250

PRECISION 0.36X + 0.94 µg/L (overall precision).

ACCURACY 0.76C–1.27 µg/L (as recovery).

SAMPLING METHOD For water and liquid samples; use glass 40-mL vials with Teflon®-lined septum caps and collect two vials per sample location with no headspace. For solids and concentrated waste samples; use widemouth glass bottles with Teflon® liners.

STABILITY For concentrated wastes, soils, sediments, or sludges: cool to 4°C. For liquids: add 4 drops of concentrated hydrochloric acid and cool to 4°C.

MHT 14 days.

QUALITY CONTROL Analyze a reagent blank, matrix spike, and matrix spike duplicate/duplicate for each analytical batch (up to 20 samples). Demonstrate the purity of glassware and reagents by analyzing a reagent water method blank. Internal, surrogate, and five concentration level calibration standards are used.

REFERENCE Test Methods for Evaluating Solid Waste (SW-846), U.S. EPA Office of Solid Waste, Washington, DC, Method 8010B, Rev. 2, Nov. 1992.

4-Bromophenyl phenyl ether
CAS #101-55-3
EPA Method 1625

TITLE Semivolatile Organic Compounds by Isotope Dilution GC/MS

MATRIX The compounds may be determined in waters, soils, and municipal sludges by this method.

METHOD SUMMARY This method is used to determine 176 semivolatile toxic organic pollutants associated with the CWA (as amended 1987); the RCRA (as amended 1986); the CERCLA (as amended 1986); and other compounds amenable to extraction and analysis by capillary column gas chromatography-mass spectrometry (GC/MS).

Stable isotopically-labeled analogs of the compounds of interest are added to the sample. If the solids content is less than 1%, a 1-L sample is extracted at pH 12–13, then at pH <2 with methylene chloride using continuous extraction techniques.

If the solids content is 30% or less, the sample is diluted to 1% solids with reagent water, homogenized ultrasonically, and extracted at pH 12–13, then at pH <2 with methylene chloride using continuous extraction techniques. If the solids content is greater than 30%, the sample is extracted using ultrasonic techniques.

Each extract is dried over sodium sulfate, concentrated to a volume of 5 mL, cleaned up using GPC, if necessary, and concentrated. Extracts are concentrated to 1 mL if GPC is not performed, and to 0.5 mL if GPC is performed.

An internal standard is added to the extract, and a 1-mL aliquot of the extract is injected into the GC. The compounds are separated by GC and detected by a MS. The labeled compounds serve to correct the variability of the analytical technique.

INTERFERENCES Solvents, reagents, glassware, and other sample processing hardware may yield artifacts and/or elevated baselines causing misinterpretation of chromatograms and spectra. Materials used in the analysis must be demonstrated to be free from interferences under the conditions of analysis

by running method blanks initially and with each sample lot (sample started through the extraction process on a given 8-h shift, to a maximum of 20). Specific selection of reagents and purification of solvents by distillation in all glass systems may be required. Glassware and, where possible, reagents are cleaned by solvent rinse and baking at 450°C for 1-h minimum. Interferences coextracted from samples will vary considerably from source to source, depending on the diversity of the site being sampled.

INSTRUMENTATION Major instrumentation includes a GC with a splitless or on-column injection port for capillary column, a MS with 70 eV electron impact ionization, and a data system to collect and record MS data, and process it. A K-D apparatus is used to concentrate extracts.

GC Column: 30 m × 0.25 mm I.D. 5% phenyl, 94% methyl, 1% vinyl silicone bonded phased fused silica capillary column.

PRECISION & ACCURACY The detection limits of the method are usually dependent on the level of interferences rather than instrumental limitations. The limits typify the minimum quantities that can be detected with no interferences present.

The minimum level (in µg/mL) was 10. This is defined as a minimum level at which the analytical system shall give recognizable mass spectra (background corrected) and acceptable calibration points.

The MDL (in µg/kg) in low solids was 55 and in high solids was 17; these were determined in digested sludge (low solids) and in filter cake or compost (high solids).

The labeled and native compound initial precision as standard deviation (in µg/L) was 44.

The labeled and native compound initial accuracy as average recovery (in µg/L) was 44–140.

SAMPLE COLLECTION, PRESERVATION & HANDLING Collect samples in glass containers. Aqueous samples which flow freely are collected in refrigerated bottles using automatic sampling equipment. Solid samples are collected as grab samples using widemouth jars. Maintain samples at 0 to 4°C from the time of collection until extraction. If residual chlorine is present in aqueous samples, add 80 mg sodium thiosulfate/L of water. Begin sample extraction within 7 days of collection, and analyze all extracts within 40 days of extraction.

SAMPLE PREPARATION Samples containing 1% solids or less are extracted directly using continuous liquid-liquid extraction techniques. Samples containing 1 to 30% solids are diluted to the 1% level with reagent water and extracted using continuous liquid-liquid extraction techniques. Samples containing greater than 30% solids are extracted using ultrasonic techniques.

Base/neutral extraction — Adjust the pH of the waters in the extractors to 12–13 with 6 N NaOH. Extract with methylene chloride for 24–48 h.

Acid extraction — Adjust the pH of the waters in the extractors to 2 or less using 6 N sulfuric acid. Extract with methylene chloride for 24–48 h.

Ultrasonic extraction of high solids samples — Add anhydrous sodium sulfate to the sample and QC aliquot(s).

Add acetone:methylene chloride (1:1) to the sample and mix thoroughly

Concentrate extracts using a K-D apparatus.

QUALITY CONTROL The analyst is permitted to modify this method to improve separations or lower the costs of measurements, provided all performance specifications are met. Analyses of blanks are required to demonstrate freedom from contamination. When results of spikes indicate atypical method performance for samples, the samples are diluted to bring method performance within acceptable limits.

For low solids (aqueous samples), extract, concentrate, and analyze two sets of four 1-L aliquots (8 aliquots total) of the precision and recovery standard. For high solids samples, two sets of four 30-g aliquots of the high solids reference matrix are used.

Spike all samples with labeled compounds to assess method performance. Compute percent recovery of the labeled compounds using the internal standard method. Compare the labeled compound recovery for each compound with the corresponding labeled compound recovery.

Reagent water and high solids reference matrix blanks are analyzed to demonstrate freedom from contamination. Extract and concentrate a 1-L reagent water blank or a high solids reference matrix blank with each sample's lot (samples started through the extraction process on the same 8-h shift, to a maximum of 20 samples).

Field replicates may be collected to determine the precision of the sampling technique, and spiked samples may be required to determine the accuracy of the analysis when the internal standard method is used.

REFERENCE Semivolatile Organic Compounds by Isotope Dilution GC/MS. Office of Water Regulation and Standards, U.S. EPA Industrial Technology Division, Washington, DC, EPA Method 1625, Rev. C, June 1989 (contact W.A. Telliard, U.S. EPA, Office of Water Regulations and Standards, 401 M St., SW, Washington, DC, 20460. Phone: 202-382-7131).

4-Bromophenyl phenyl ether   **EPA Method 625**
CAS #101-55-3

TITLE Base/Neutrals and Acids, U.S. EPA Method 625

MATRIX This methods covers municipal and industrial wastewaters.

METHOD SUMMARY Approximately 1 L of sample is serially extracted with methylene chloride at a pH greater than 11 and again at a pH less than 2 using a separatory funnel or a continuous extractor. The methylene chloride extract is dried, concentrated to a volume of 1 mL, and analyzed by GC/MS. Qualitative identification of the parameters in the extract is performed using the retention time and the relative abundance of three characteristic masses (m/z). Qualitative analysis is performed using either external or internal standard techniques with a single characteristic m/z.

INTERFERENCES Method interferences may be caused by contaminants in solvents, reagents, glassware, and other sample processing hardware. Glassware must be scrupulously cleaned. Glassware should be heated in a muffle furnace at 400°C for 5 to 30 min. Some thermally stable materials, such as PCBs, may not be eliminated by this treatment. Solvent rinses with acetone and pesticide quality hexane may be substituted for the muffle furnace heating. Matrix interferences may be caused by contaminants that are coextracted from the sample. The base-neutral extraction may cause significantly reduced recovery of phenols. The packed gas chromatographic columns recommended for the basic fraction may not exhibit sufficient resolution for some analytes.

INSTRUMENTATION A GC/MS system with an injection port designed for on-column injection when using packed columns and for splitless injection when using capillary columns.

Column for base/neutrals: 1.8 m long × 2 mm I.D. glass, packed with 3% SP-2550 on Supelcoport (100/120 mesh) or equivalent.
Column for acids: 1.8 m long × 2 mm I.D. glass, packed with 1% SP-1240DA on Supelcoport (100/120 mesh) or equivalent.

PRECISION & ACCURACY The MDL concentrations were obtained using reagent water. The MDL actually achieved in a given analysis will vary depending on instrument sensitivity and matrix effects. This method was tested by 15 laboratories using reagent water, drinking water, surface water, and industrial wastewaters spiked at six concentrations over the range 5 to 100 µg/L. Single operator precision, overall precision, and method accuracy were found to be directly related to the concentration of the parameter matrix.

The MDL (in µg/L) in reagent water was 1.9.
The standard deviation (in µg/L based on 4 recovery measurements) was 23.0.
The range (in µg/L) for average recovery for 4 measurements was 64.9–114.4.
The range (in %) for percent recovery was 53–127.
Accuracy (in µg/L) as expected recovery for one or more measurements of a sample containing a true concentration of C was 0.91C–1.34.
Precision (in µg/L) as expected single analyst standard deviation of measurements at an average concentration found at X was 0.13X+0.66.
Overall precision (in µg/L) as expected interlaboratory standard deviation of measurements in an average concentration found at X was 0.16X+0.66.

C = *True value of the concentration in µg/L.*
X = *Average recovery found for measurements of samples containing a concentration at C in µg/L.*

SAMPLE PREPARATION Adjust the pH to >11 with sodium hydroxide and serially extract in a separatory funnel with methylene chloride or else in a continuous extractor. Next, adjust the pH to <2 with sulfuric acid and serially extract in a separatory funnel with methylene chloride or else in a continuous extractor. Dry the extracts separately through a column of anhydrous sodium sulfate and then concentrate each of the extracts to 1.0 mL using a K-D apparatus.

SAMPLE COLLECTION, PRESERVATION & HANDLING Grab samples must be collected in glass containers. All samples must be refrigerated at 4°C from the time of collection until extraction. If residual chlorine is present, add 80 mg of sodium thiosulfate/L of sample and mix well. All samples must be extracted within 7 days of collection and completely analyzed within 40 days of extraction.

QUALITY CONTROL Make an initial, one-time, demonstration of the ability to generate acceptable accuracy and precision with this method. Before processing any samples, the analyst must analyze a reagent water blank to demonstrate that interferences from the analytical system and glassware are under control. Each time a set of samples is extracted or reagents are changed, a reagent water blank must be processed. Spike and analyze a minimum of 5% of all samples to monitor and evaluate lab data quality. A QC check sample concentrate that contains each parameter of interest at a concentration of 100 µg/mL in acetone is required. PCBs and multicomponent pesticides may be omitted from this test.

After analysis of five spiked wastewater samples, calculate the average percent recovery and the standard deviation of the percent recovery. Spike all samples with the surrogate standard spiking solution and calculate the percent recovery of each surrogate compound.

REFERENCE *Federal Register*, Vol. 49, No. 209. Friday, Oct. 26, 1984.

4-Bromophenyl phenyl ether   **EPA Method 8270**
CAS #101-55-3

TITLE Semivolatile Organic Compounds by GC/MS

MATRIX This method is used to determine the concentration of semivolatile organic compounds in extracts prepared from all types of solid waste matrices, soils, and groundwater. Although surface waters are not specifically mentioned, this method should be applicable to water samples from rivers, lakes, etc.

METHOD SUMMARY This method covers 259 semivolatile organic compounds. In very limited applications direct injection of the sample into the GC/MS system may be appropriate, but this results in very high detection limits (approximately 10,000 µg/L). Typically, a 1-L liquid sample, containing surrogate, and matrix spiking standards, is extracted in a continuous extractor first under acid conditions and then under basic conditions. Typically 30 g of a solid sample, containing surrogate, and matrix spiking standards, is extracted ultrasonically. After concentrating the extract to 1 mL it is spiked with 10 µL of an internal standard solution just prior to analysis by GC/MS. The volume injected should contain about 100 ng of base/neutral and 200 ng of acid surrogates (for a 1-µL injection). Analysis is performed by GC/MS using a capillary GC column.

INTERFERENCES Raw GC/MS data from all blanks, samples, and spikes must be evaluated for interferences. Contamination by carryover can occur whenever high-concentration and low-concentration samples are sequentially analyzed. To reduce carryover, the sample syringe must be rinsed out between samples with solvent. Whenever an unusually concentrated sample is encountered, it should be followed by the analysis of blank solvent to check for cross-contamination.

INSTRUMENTATION A GC/MS and a data system are required. The GC column used is a 30 m × 0.25 mm I.D. (or 0.32 mm I.D.) 1um film thickness silicone-coated fused silica capillary column. A continuous liquid-liquid extractor equipped with Teflon® or glass connection joints and stopcocks requiring no lubrication, a K-D concentrating apparatus, water bath, and an ultrasonic disrupter with a minimum power of 300 W and with pulsing capability are also required.

PRECISION & ACCURACY The estimated quantitation limit (EQL) of Method 8270B for determining an individual compound is approximately 1 mg/kg (wet weight) for soil or sediment samples, 1–200 mg/kg for wastes (dependent on matrix and method of preparation), and 10 µg/L for groundwater samples. EQLs will be proportionately higher for sample extracts that require dilution to avoid saturation of the detector.

The EQL(b) for groundwater in µg/L is 10.
The EQL (a, b) for low concentrations in soil and sediment in µg/kg is 660.
Accuracy as µg/L is $0.91C - 1.34$.
Overall precision in µg/L is $0.16X + 0.66$.

(a) EQLs listed for soil/sediment are based on wet weight. Normally data is reported in a dry-weight basis; therefore, EQLs will be higher based on the % dry weight of each sample. This calculation is based on a 30-g sample and gel permeation chromatography cleanup.
(b) Sample EQLs are highly matrix-dependent. The EQLs are provided for guidance and may not always be achievable.
C = *True value for concentration, in µg/L.*
X = *Average recovery found for measurements of samples containing a concentration of C, in µg/L.*

ESTIMATED QUANTITATION LIMIT FOR OTHER MATRICES

Other Matrices	Factor (a)
High-concentration soil and sludges by sonicator	7.5
Non-water miscible waste	75

(a) EQL for other matrices = [EQL for low soil/sediment] × [Factor]. This estimated EQL is similar to an EPA "Practical Quantitation Limit."

SAMPLING METHOD
Liquid samples — Use a 1 or 2½ gallon amber glass bottle with a screw-top Teflon®-lined cover that has been prewashed with detergent and rinsed with distilled water and methanol (or isopropanol).

Soils, sediments, or sludges — Use an 8-oz. widemouth glass with a screw-top Teflon®-lined cover that has been prewashed with detergent and rinsed with distilled water and methanol (or isopropanol).

SAMPLE PRESERVATION
Liquid samples — If residual chlorine is present, add 3 mL of 10% sodium thiosulfate per gallon, cool to 4°C and store in a solvent-free refrigerator until analysis; if chlorine is not present, then eliminate the sodium thiosulfate addition.

Soils, sediments, or sludges — Cool samples to 4°C and store in a solvent-free refrigerator.

MHT Liquid samples must be extracted within 7 days and the extracts analyzed within 40 days. Soils, sediments, or sludges may be stored for a maximum of 14 days and the extracts analyzed within 40 days.

SAMPLE PREPARATION
Liquid samples — Transfer 1 L quantitatively to a continuous extractor. If high concentrations are anticipated, a smaller volume may be used and then diluted with organic-free reagent water to 1 L. Adjust pH, if necessary, to pH <2 using 1:1 (V/V) sulfuric acid. Pipette 1.0 mL of a surrogate standard spiking solution into each sample. For the sample in each analytical batch selected for spiking, add 1.0 mL of a matrix spiking standard. For base/neutral acid analysis, the amount of the surrogates and matrix spiking compounds added to the sample should result in a final concentration of 100 ng/µL of each analyte in the extract to be analyzed (assuming a 1-µL injection). Extract with methylene chloride for 18–24 h. Next, adjust the pH of the aqueous phase to pH >11 using 10 N sodium hydroxide and extract it with methylene chloride again for 18–24 h. Dry the extract through a column containing anhydrous sodium sulfate and concentrate it to 1 mL using a K-D concentrator.

Soils, sediments, or sludges — Use 30 g of sample. Nonporous or wet samples (gummy or clay type) that do not have a free-flowing sandy texture must be mixed with anhydrous sodium sulfate until the sample is free flowing. Add 1 mL of surrogate standards to all samples, spikes, standards, and blanks. For the sample in each analytical batch selected for spiking, add 1.0 mL of a matrix spiking standard. For base/neutral acid analysis, the amount added of the surrogates and matrix spiking compounds should result in a final concentration of 100 ng/µL of each base/neutral analyte and 200 ng/µL of each acid analyte in the extract to be analyzed (assuming a 1-µL injection). Immediately add a 100 mL mixture of 1:1 methylene chloride:acetone and extract the sample ultrasonically for 3 min and then decant or filter the extracts. Repeat the extraction two or more times. Dry the extract using a column with anhydrous sodium sulfate and concentrate it to 1 mL in a K-D concentrator.

QUALITY CONTROL A methylene chloride solution containing 50 ng/µL of decafluorotriphenylphosphine (DFTPP) is used for tuning the GC/MS system each 12-h shift. A system performance check also must be made during every 12-h shift. A standard containing 50 ng/µL each of 4,4'-DDT, pentachlorophenol, and benzidine is required to verify injection port inertness and GC column performance. A calibration standard at mid-concentration, containing each compound of interest, including all required surrogates, must be performed every 12 h during analysis. After the system performance check is met, calibration check compounds (CCCs) are used to check the validity of the initial calibration.

The internal standard responses and retention times in the calibration check standard must be evaluated immediately after

or during data acquisition. If the retention time for any internal standard changes by more than 30 seconds from the last check calibration (12 h), the chromatographic system must be inspected for malfunctions and corrections must be made, as required. If the electron ionization current plot (EICP) area for any of the internal standards changes by a factor of two from the last daily calibration standard check, the mass spectrometer must be inspected for malfunctions and corrections must be made, as appropriate.

Demonstrate, through the analysis of a reagent water blank, that interferences from the analytical system, glassware, and reagents are under control. The blank samples should be carried through all stages of the sample preparation and measurement steps. For each analytical batch (up to 20 samples), a reagent blank, matrix spike, and matrix spike duplicate/duplicate must be analyzed (the frequency of the spikes may be different for different monitoring programs). The blank and spiked samples must be carried through all stages of the sample preparation and measurement steps. A QC reference sample concentrate containing each analyte at a concentration of 100 mg/L in methanol is required.

REFERENCE Test Methods for Evaluating Solid Waste (SW-846). U.S. EPA 1983, Method 8270B, Rev. 2, Nov. 1990. Office of Solid Waste, Washington, DC.

Bromoxynil	EPA Method 8270
CAS #1689-84-5	

TITLE Semivolatile Organic Compounds by GC/MS

MATRIX This method is used to determine the concentration of semivolatile organic compounds in extracts prepared from all types of solid waste matrices, soils, and groundwater. Although surface waters are not specifically mentioned, this method should be applicable to water samples from rivers, lakes, etc.

METHOD SUMMARY This method covers 259 semivolatile organic compounds. In very limited applications direct injection of the sample into the GC/MS system may be appropriate, but this results in very high detection limits (approximately 10,000 μg/L). Typically, a 1-L liquid sample, containing surrogate, and matrix spiking standards, is extracted in a continuous extractor first under acid conditions and then under basic conditions. Typically 30 g of a solid sample, containing surrogate, and matrix spiking standards, is extracted ultrasonically. After concentrating the extract to 1 mL it is spiked with 10 μL of an internal standard solution just prior to analysis by GC/MS. The volume injected should contain about 100 ng of base/neutral and 200 ng of acid surrogates (for a 1-μL injection). Analysis is performed by GC/MS using a capillary GC column.

INTERFERENCES Raw GC/MS data from all blanks, samples, and spikes must be evaluated for interferences. Contamination by carryover can occur whenever high-concentration and low-concentration samples are sequentially analyzed. To reduce carryover, the sample syringe must be rinsed out between samples with solvent. Whenever an unusually concentrated sample is encountered, it should be followed by the analysis of blank solvent to check for cross-contamination.

INSTRUMENTATION A GC/MS and a data system are required. The GC column used is a 30 m × 0.25 mm I.D. (or 0.32 mm I.D.) 1um film thickness silicone-coated fused silica capillary column. A continuous liquid-liquid extractor equipped with Teflon® or glass connection joints and stopcocks requiring no lubrication, a K-D concentrating apparatus, water bath, and an ultrasonic disrupter with a minimum power of 300 W and with pulsing capability are also required.

PRECISION & ACCURACY The estimated quantitation limit (EQL) of Method 8270B for determining an individual compound is approximately 1 mg/kg (wet weight) for soil or sediment samples, 1–200 mg/kg for wastes (dependent on matrix and method of preparation), and 10 μg/L for groundwater samples. EQLs will be proportionately higher for sample extracts that require dilution to avoid saturation of the detector.

The EQL(b) for groundwater in μg/L is 10.
The EQL (a, b) for low concentrations in soil and sediment in μg/kg is not determined.
Accuracy as μg/L is not listed.
Overall precision in μg/L is not listed.

(a) *EQLs listed for soil/sediment are based on wet weight. Normally data is reported in a dry-weight basis; therefore, EQLs will be higher based on the % dry weight of each sample. This calculation is based on a 30-g sample and gel permeation chromatography cleanup.*
(b) *Sample EQLs are highly matrix-dependent. The EQLs are provided for guidance and may not always be achievable.*
$C =$ *True value for concentration, in μg/L.*
$X =$ *Average recovery found for measurements of samples containing a concentration of C, in μg/L.*

ESTIMATED QUANTITATION LIMIT FOR OTHER MATRICES

Other Matrices	Factor (a)
High-concentration soil and sludges by sonicator	7.5
Non-water miscible waste	75

(a) *EQL for other matrices = [EQL for low soil/sediment] × [Factor]. This estimated EQL is similar to an EPA "Practical Quantitation Limit."*

SAMPLING METHOD
Liquid samples — Use a 1 or 2½ gallon amber glass bottle with a screw-top Teflon®-lined cover that has been prewashed with detergent and rinsed with distilled water and methanol (or isopropanol).

Soils, sediments, or sludges — Use an 8-oz. widemouth glass with a screw-top Teflon®-lined cover that has been prewashed with detergent and rinsed with distilled water and methanol (or isopropanol).

SAMPLE PRESERVATION
Liquid samples — If residual chlorine is present, add 3 mL of 10% sodium thiosulfate per gallon, cool to 4°C and store in a solvent-free refrigerator until analysis; if chlorine is not present, then eliminate the sodium thiosulfate addition.

Soils, sediments, or sludges — Cool samples to 4°C and store in a solvent-free refrigerator.

MHT Liquid samples must be extracted within 7 days and the extracts analyzed within 40 days. Soils, sediments, or sludges may be stored for a maximum of 14 days and the extracts analyzed within 40 days.

SAMPLE PREPARATION
Liquid samples — Transfer 1 L quantitatively to a continuous extractor. If high concentrations are anticipated, a smaller volume may be used and then diluted with organic-free reagent water to 1 L. Adjust pH, if necessary, to pH <2 using 1:1 (V/V) sulfuric acid. Pipette 1.0 mL of a surrogate standard spiking solution into each sample. For the sample in each analytical batch selected for spiking, add 1.0 mL of a matrix spiking standard. For base/neutral acid analysis, the amount of the surrogates and matrix spiking compounds added to the sample should result in a final concentration of 100 ng/µL of each analyte in the extract to be analyzed (assuming a 1-µL injection). Extract with methylene chloride for 18–24 h. Next, adjust the pH of the aqueous phase to pH >11 using 10 N sodium hydroxide and extract it with methylene chloride again for 18–24 h. Dry the extract through a column containing anhydrous sodium sulfate and concentrate it to 1 mL using a K-D concentrator.

Soils, sediments, or sludges — Use 30 g of sample. Nonporous or wet samples (gummy or clay type) that do not have a free-flowing sandy texture must be mixed with anhydrous sodium sulfate until the sample is free flowing. Add 1 mL of surrogate standards to all samples, spikes, standards, and blanks. For the sample in each analytical batch selected for spiking, add 1.0 mL of a matrix spiking standard. For base/neutral acid analysis, the amount added of the surrogates and matrix spiking compounds should result in a final concentration of 100 ng/µL of each base/neutral analyte and 200 ng/µL of each acid analyte in the extract to be analyzed (assuming a 1-µL injection). Immediately add a 100 mL mixture of 1:1 methylene chloride:acetone and extract the sample ultrasonically for 3 min and then decant or filter the extracts. Repeat the extraction two or more times. Dry the extract using a column with anhydrous sodium sulfate and concentrate it to 1 mL in a K-D concentrator.

QUALITY CONTROL A methylene chloride solution containing 50 ng/µL of decafluorotriphenylphosphine (DFTPP) is used for tuning the GC/MS system each 12-h shift. A system performance check also must be made during every 12-h shift. A standard containing 50 ng/µL each of 4,4'-DDT, pentachlorophenol, and benzidine is required to verify injection port inertness and GC column performance. A calibration standard at mid-concentration, containing each compound of interest, including all required surrogates, must be performed every 12 h during analysis. After the system performance check is met, calibration check compounds (CCCs) are used to check the validity of the initial calibration.

The internal standard responses and retention times in the calibration check standard must be evaluated immediately after or during data acquisition. If the retention time for any internal standard changes by more than 30 seconds from the last check calibration (12 h), the chromatographic system must be inspected for malfunctions and corrections must be made, as required. If the electron ionization current plot (EICP) area for any of the internal standards changes by a factor of two from the last daily calibration standard check, the mass spectrometer must be inspected for malfunctions and corrections must be made, as appropriate.

Demonstrate, through the analysis of a reagent water blank, that interferences from the analytical system, glassware, and reagents are under control. The blank samples should be carried through all stages of the sample preparation and measurement steps. For each analytical batch (up to 20 samples), a reagent blank, matrix spike, and matrix spike duplicate/duplicate must be analyzed (the frequency of the spikes may be different for different monitoring programs). The blank and spiked samples must be carried through all stages of the sample preparation and measurement steps. A QC reference sample concentrate containing each analyte at a concentration of 100 mg/L in methanol is required.

REFERENCE Test Methods for Evaluating Solid Waste (SW-846). U.S. EPA 1983, Method 8270B, Rev. 2, Nov. 1990. Office of Solid Waste, Washington, DC.

Butachlor   **EPA Method 507**
CAS #23184-66-9

TITLE Determination of Nitrogen and Phosphorus-Containing Pesticides in Water by GC/NPD

MATRIX This method is applicable to the determination of certain nitrogen and phosphorus-containing pesticides in finished drinking water and groundwater.

METHOD SUMMARY Method 507 covers 46 nitrogen- and phosphorus-containing pesticides. A 1 L sample is fortified with a surrogate standard, salted, buffered, extracted with methylene chloride, and concentrated; then the solvent is exchanged with methyl tert-butyl ether (MTBE) and concentrated again, and a 2-µL aliquot of a sample extract is injected into a GC system equipped with a selective nitrogen-phosphorus detector and a capillary column for analysis.

INTERFERENCES Method interferences may be caused by contaminants in solvents, reagents, glassware, and other sample processing apparatus. Interfering contamination may occur when a sample containing low concentrations of analytes is analyzed immediately following a sample containing relatively high concentrations. One or more injections of MTBE should be made following the analysis of a sample with high concentrations of analytes to check for analyte carryover. Matrix interferences may be caused by contaminants that are coextracted from the sample. The extent of matrix interferences will vary considerably from source to source, depending upon the water sampled.

INSTRUMENTATION A gas chromatograph system (GC) equipped with a nitrogen-phosphorus detector (NPD) is needed.

Column 1: 30 m × 0.25 mm I.D. DB-5 bonded fused silica column, 0.25 μm film thickness, or equivalent.

Column 2: 30 m × 0.25 mm I.D. DB-1701 bonded fused silica column, 0.25 μm film thickness, or equivalent.

PRECISION & ACCURACY This method has been validated in a single lab and estimated detection limits (EDLs) have been determined for each analyte. Observed detection limits may vary among waters, depending upon the nature of the interferences in the sample matrix and the specific instrumentation used. Analytes that are not separated chromatographically cannot be individually identified and measured unless an alternative technique for identification and quantification exist.

The estimated detection limit (in μg/L) was 0.38. The EDL is defined as either method detection limit or a level of compound in a sample yielding a peak in the final extract with signal-to-noise ratio of approximately 5, whichever value is higher.

The concentration used for these measurements (in μg/L) was 3.8.

The accuracy (as % recovery) was 96.

The precision (% RSD) was 4.

SAMPLING METHOD Grab samples are collected in 1-L glass sample bottles (prewashed with detergent and hot tap water, rinsed with reagent water, and dried in an oven at 400°C for 1 h) with screw caps lined with PTFE-fluorocarbon.

SAMPLE PRESERVATION Add mercuric chloride to the sample bottle in amounts to produce a concentration of 10 mg/L. If residual chlorine is present, add 80 mg of sodium thiosulfate/L of sample to the sample bottle prior to collection. After collection, seal bottle and shake vigorously for 1 min, then cool the sample to 4°C immediately and store it at 4°C in the dark until extraction.

MHT Maximum holding time of the samples, and in some cases the extracts, is 14 days.

SAMPLE PREPARATION Fortify the sample with 50 μL of the surrogate standard solution, adjust to pH 7 with phosphate buffer, add 100 g NaCl to the sample, and seal and shake to dissolve the salt; then extract with methylene chloride in a separatory funnel or in a mechanical tumbler bottle. Dry the extract by pouring it through a solvent-rinsed drying column containing about 10 cm of anhydrous sodium sulfate. Collect the extract in a Kuderna-Danish (K-D) concentrator and rinse the column with 20–30 mL methylene chloride. Concentrate the extract to about 2 mL and rinse the flask and its lower joint into the concentrator tube with 1 to 2 mL of methyl t-butyl ether (MTBE). Add 5–10 mL of MTBE and concentrate the extract twice (adding more MTBE) to a final volume of 5.0 mL and store it at 4°C until analysis.

Note: If methylene chloride is not completely removed from the final extract, it may cause detector problems.

QUALITY CONTROL Minimum quality control requirements are initial demonstration of lab capability, determination of surrogate compound recoveries in each sample and blank, monitoring internal standard peak area or height in each sample and blank, analysis of lab reagent blanks, lab fortified samples, lab fortified blanks, and other QC samples. A lab reagent blank is analyzed to demonstrate that all glassware and reagent interferences are under control.

Initial demonstration of capability is fulfilled by analyzing four fortified reagent water samples with the recovery value for each analyte falling within the acceptable range (±30% average recovery). Surrogate recoveries from samples or method blanks must be 70–130%. The internal standard response for any sample chromatogram should not deviate from the daily calibration check standard's internal standard response by more than 30% or lab fortified blanks and sample matrices are used to assess lab performance and analyte recovery, respectively.

If the response for the target analyte peak exceeds the working range of the system, dilute the extract and reanalyze. Alternative techniques such as an alternate detector or second chromatography column should be used to confirm peak identification when sample components are not resolved adequately.

EPA CONTACT & HOTLINE For technical questions contact Dr. Baldev Bathija, U.S. EPA, Office of Ground Water and Drinking Water (WH-550D), 401 M St. SW, Washington, DC 20460. Tel. (202) 260-3040. For further information the EPA Safe Drinking Water Hotline may be called at: (800) 426-4791.

REFERENCE Methods for the Determination of Organic Compounds in Drinking Water, EPA/600/4-88/039 (revised July 1991). U.S. EPA Environmental Monitoring Systems Laboratory, Cincinnati, OH, 45268, U.S.A. Available from the National Technical Information Service (NTIS), 5285 Port Royal Road, Springfield, VA 22161; Tel. 800-553-6847. NTIS Order Number is PB91-231480.

2-Butanone   **EPA Method 8240**
CAS #78-93-3

TITLE Volatile Organics By GC/MS: Packed Column Technique

MATRIX Nearly all types of sample matarices, regardless of water content, can be analyzed using this method. This includes groundwater, aqueous sludges, caustic liquors, acid liquors, waste solvents, oily wastes, mousses, tars, fibrous wastes, polymetric emulsions, filter cakes, spent carbons, spent catalysts, soils, and sediments.

METHOD SUMMARY Method 8240B covers 80 volatile organic compounds that are introduced into a gas chromatograph by the purge-and-trap method or by direct injection (in limited applications). For the purge-and-trap method an inert gas (zero grade nitrogen or helium) is bubbled through a 5-mL solution at ambient temperature. Purged sample components are trapped in a tube of sorbent materials. When purging is complete, the sorbent tube is heated and backflushed with inert gas to desorb the trapped components onto a GC column.

INTERFERENCES Impurities in the purge gas and from organic compounds outgassing from the plumbing ahead of the trap account for many contamination problems. Interferences purged or coextracted from the samples will vary considerably from source to source. Cross-contamination can occur whenever high-level and low-level samples are analyzed sequentially. Whenever an unusually concentrated sample is analyzed, it should be followed by the analysis of organic-free reagent water to check for cross-contamination. Samples also can be contaminated by diffusion of volatile organics (particularly methylene chloride and fluorocarbons) through the septum seal into the sample during shipment and storage. A trip blank can serve as a check on such contamination. The lab where volatile analysis is performed and also the refrigerated storage area should be completely free of solvents.

INSTRUMENTATION A gas chromatograph/mass spectrometry/data system (GC/MS) equipped with a 6 ft × 0.1 in I.D. glass column packed with 1% SP-1000 on Carbopack-B (60/80 mesh) is required. Also needed is a 5-mL purging device, a sorbent trap, and a thermal desorption apparatus.

PRECISION & ACCURACY This method is reported to have been tested by 15 laboratories using organic-free reagent water, drinking water, surface water, and industrial wastewaters (not specified) fortified at six concentrations over the range 5–600 µg/L.

Sample estimated quantitation limits (EQLs) are highly matrix-dependent. The EQLs listed may not always be achievable. EQLs listed for soils or sediments are based on wet weight. Normally, data is reported on a dry-weight basis; therefore, EQLs will be higher, based on the percent dry weight of each sample. Note that EQLs are even more variable than MDLs and that they are highly variable depending on the matrix being analyzed.

EQL in groundwater in µg/L was 100.
EQL in low soil or sediment in µg/kg was 100.
Accuracy (a) in µg/L was not listed.
Precision (b) in µg/L was not listed.

(a) *Average recovery found for measurements of samples containing a concentration of C, in µg/L.*
(b) *Overall precision found for measurements of samples with average recovery X for samples containing a concentration of C in µg/L.*
X = *Average recovery found for measurement of samples containing a concentration of C in µg/L.*

MULTIPLICATION FACTORS FOR OTHER MATRICES

Other Matrices	Factor (a)
Waste miscible liquid waste	50
High-concentration soil and sludge	125
Non-water miscible waste	500

(a) *EQL = [EQL for low soil/sediment] × [Factor]. For non-aqueous samples, the factor is on a wet-weight basis.*

SAMPLING METHOD
Liquid samples — Use a 40-mL glass screw-cap VOA vial with a Teflon®-faced silicone septum that has been prewashed, rinsed with distilled deionized water, and oven dried. However, if residual chlorine is present, collect sample in a 40-oz. soil VOA container which has been pre-preserved with 4 drops of 10% sodium thiosulfate, mix gently, and then transfer the sample to a 40-mL VOA vial. Collect bubble-free samples in duplicate and seal them in separate plastic bags.

Soils or sediments, and sludges — Use an 8-oz. widemouth glass bottle with a Teflon®-faced silicone septum that has been prewashed with detergent, rinsed with distilled deionized water, and oven dried. Tap slightly to eliminate free air space. Collect samples in duplicate and seal them in separate plastic bags.

SAMPLE PRESERVATION
Liquid samples — Add 4 drops of concentrated HCL and immediately cool samples to 4°C and store in a solvent-free refrigerator.

Soils or sediments, and sludges — Cool samples to 4°C and store in a solvent-free refrigerator.

MHT Maximum holding time is 14 days from the date of sample collection.

SAMPLE PREPARATION
Liquid samples — Remove the plunger from a 5-mL syringe and carefully pour the sample into the syringe barrel to just short of overflowing. Replace the syringe plunger and compress the sample. Open the syringe valve and vent any residual air while adjusting the sample volume to 5.0 mL. If there is only one volatile organic analysis (VOA) vial, a second syringe should be filled at this time to protect against possible loss of sample integrity. Add 10 µL of surrogate spiking solution and 10 µL of internal standard spiking solution through the valve bore of the 5-mL syringe, then close the valve. The surrogate and internal standards may be mixed and added as a single spiking solution.

Sediments, soils, and waste samples — All samples of this type should be screened by GC analysis using a headspace method (EPA Method 3810) or the hexadecane extraction and screening method (EPA Method 3820). Use the screening data to determine whether to use the low-concentration method (0.005–1 mg/kg) or the high-concentration method (>1 mg/kg).

Low-concentration method — The low-concentration method is based on purging a heated sediment or soil sample mixed with organic-free reagent water containing the surrogate and internal standards. Analyze all reagent blanks and standards under the same conditions as the samples.

Use a 5-g sample if the expected concentration is <0.1 mg/kg or a 1-g sample for expected concentrations between 0.1 and 1 mg/kg. Mix the contents of the sample container with a narrow metal spatula. Weigh the amount of the sample into a tared purge device. Add the spiked water to the purge device, which contains the weighed amount of sample, and connect the device to the purge-and-trap system.

High-concentration method — This method is based on extracting the sediment or soil with methanol. A waste sample is either extracted or diluted, depending on its solubility in methanol. Wastes that are insoluble in methanol are diluted with reagent tetraglyme or possibly polyethylene glycol (PEG). An aliquot of the extract is added to organic-free reagent water containing surrogate and internal standards. This is purged at ambient temperature. All samples with an expected concentration of >1.0 mg/kg should be analyzed by this method.

Mix the contents of the sample container with a narrow metal spatula. For sediments or soils and solid wastes that are insoluble in methanol, weigh 4 g (wet weight) of sample into a tared 20-mL vial. For waste that is soluble in methanol, tetraglyme, or PEG, weigh 1 g (wet weight) into a tared scintillation vial or culture tube or a 10-mL volumetric flask. Quickly add 9.0 mL of appropriate solvent then add 1.0 mL of a surrogate spiking solution to the vial, cap it, and shake it for 2 min.

METHANOL EXTRACT REQUIRED FOR ANALYSIS OF HIGH-CONCENTRATION SOILS OR SEDIMENTS

Approximate Concentration Range	Volume of Methanol Extract (a)
500–10,000 µg/kg	100 µL
1,000–20,000 µg/kg	50 µL
5,000–100,000 µg/kg	10 µL
25,000–500,000 µg/kg	100 µL of 1/50 dilution (b)

Calculate appropriate dilution factor for concentrations exceeding this table.

(a) The volume of methanol added to 5 mL of water being purged should be kept constant. Therefore, add to the 5-mL syringe whatever volume of methanol is necessary to maintain a volume of 100 µL added to the syringe.
(b) Dilute an aliquot of the methanol extract and then take 100 µL for analysis.

QUALITY CONTROL Demonstrate, through the analysis of a reagent water blank, that interferences from the analytical system, glassware, and reagents are under control. Blank samples should be carried through all stages of the sample preparation and measurement steps. For each analytical batch (up to 20 samples), a reagent blank, matrix spike, and matrix spike duplicate must be analyzed (the frequency of the spikes may be different for different monitoring programs). The blank and spiked samples must be carried through all stages of the sample preparation and measurement steps. QC samples mentioned in the section on Interferences will also be needed as appropriate to those situations.

REFERENCE Test Methods for Evaluating Solid Waste (SW-846). U.S. EPA. 1983. Method 8240B, Rev. 2, Nov. 1990. Office of Solid Wastes, Washington, DC.

Butyl benzyl phthalate   **EPA Method 1625**
CAS #85-68-7

TITLE Semivolatile Organic Compounds by Isotope Dilution GC/MS

MATRIX The compounds may be determined in waters, soils, and municipal sludges by this method.

METHOD SUMMARY This method is used to determine 176 semivolatile toxic organic pollutants associated with the CWA (as amended 1987); the RCRA (as amended 1986); the CERCLA (as amended 1986); and other compounds amenable to extraction and analysis by capillary column gas chromatography-mass spectrometry (GC/MS).

Stable isotopically-labeled analogs of the compounds of interest are added to the sample. If the solids content is less than 1%, a 1-L sample is extracted at pH 12–13, then at pH <2 with methylene chloride using continuous extraction techniques.

If the solids content is 30% or less, the sample is diluted to 1% solids with reagent water, homogenized ultrasonically, and extracted at pH 12–13, then at pH <2 with methylene chloride using continuous extraction techniques. If the solids content is greater than 30%, the sample is extracted using ultrasonic techniques.

Each extract is dried over sodium sulfate, concentrated to a volume of 5 mL, cleaned up using GPC, if necessary, and concentrated. Extracts are concentrated to 1 mL if GPC is not performed, and to 0.5 mL if GPC is performed.

An internal standard is added to the extract, and a 1-mL aliquot of the extract is injected into the GC. The compounds are separated by GC and detected by a MS. The labeled compounds serve to correct the variability of the analytical technique.

INTERFERENCES Solvents, reagents, glassware, and other sample processing hardware may yield artifacts and/or elevated baselines causing misinterpretation of chromatograms and spectra. Materials used in the analysis must be demonstrated to be free from interferences under the conditions of analysis by running method blanks initially and with each sample lot (sample started through the extraction process on a given 8-h shift, to a maximum of 20). Specific selection of reagents and purification of solvents by distillation in all glass systems may be required. Glassware and, where possible, reagents are cleaned by solvent rinse and baking at 450°C for 1-h minimum. Interferences coextracted from samples will vary considerably from source to source, depending on the diversity of the site being sampled.

INSTRUMENTATION Major instrumentation includes a GC with a splitless or on-column injection port for capillary column, a MS with 70 eV electron impact ionization, and a data system to collect and record MS data, and process it. A K-D apparatus is used to concentrate extracts.

GC Column: 30 m × 0.25 mm I.D. 5% phenyl, 94% methyl, 1% vinyl silicone bonded phased fused silica capillary column.

PRECISION & ACCURACY The detection limits of the method are usually dependent on the level of interferences rather than instrumental limitations. The limits typify the minimum quantities that can be detected with no interferences present.

The minimum level (in µg/mL) was 10. This is defined as a minimum level at which the analytical system shall give recognizable mass spectra (background corrected) and acceptable calibration points.
The MDL (in µg/kg) in low solids was 60 and in high solids was 65; these were determined in digested sludge (low solids) and in filter cake or compost (high solids).
The labeled and native compound initial precision as standard deviation (in µg/L) was 31.
The labeled and native compound initial accuracy as average recovery (in µg/L) was 19–233.

SAMPLE COLLECTION, PRESERVATION & HANDLING

Collect samples in glass containers. Aqueous samples which flow freely are collected in refrigerated bottles using automatic sampling equipment. Solid samples are collected as grab samples using widemouth jars. Maintain samples at 0 to 4°C from the time of collection until extraction. If residual chlorine is present in aqueous samples, add 80 mg sodium thiosulfate/L of water. Begin sample extraction within 7 days of collection, and analyze all extracts within 40 days of extraction.

SAMPLE PREPARATION

Samples containing 1% solids or less are extracted directly using continuous liquid-liquid extraction techniques. Samples containing 1 to 30% solids are diluted to the 1% level with reagent water and extracted using continuous liquid-liquid extraction techniques. Samples containing greater than 30% solids are extracted using ultrasonic techniques.

- Base/neutral extraction — Adjust the pH of the waters in the extractors to 12–13 with 6 N NaOH. Extract with methylene chloride for 24–48 h.
- Acid extraction — Adjust the pH of the waters in the extractors to 2 or less using 6 N sulfuric acid. Extract with methylene chloride for 24–48 h.
- Ultrasonic extraction of high solids samples — Add anhydrous sodium sulfate to the sample and QC aliquot(s). Add acetone:methylene chloride (1:1) to the sample and mix thoroughly

Concentrate extracts using a K-D apparatus.

QUALITY CONTROL

The analyst is permitted to modify this method to improve separations or lower the costs of measurements, provided all performance specifications are met. Analyses of blanks are required to demonstrate freedom from contamination. When results of spikes indicate atypical method performance for samples, the samples are diluted to bring method performance within acceptable limits.

For low solids (aqueous samples), extract, concentrate, and analyze two sets of four 1-L aliquots (8 aliquots total) of the precision and recovery standard. For high solids samples, two sets of four 30-g aliquots of the high solids reference matrix are used.

Spike all samples with labeled compounds to assess method performance. Compute percent recovery of the labeled compounds using the internal standard method. Compare the labeled compound recovery for each compound with the corresponding labeled compound recovery.

Reagent water and high solids reference matrix blanks are analyzed to demonstrate freedom from contamination. Extract and concentrate a 1-L reagent water blank or a high solids reference matrix blank with each sample's lot (samples started through the extraction process on the same 8-h shift, to a maximum of 20 samples).

Field replicates may be collected to determine the precision of the sampling technique, and spiked samples may be required to determine the accuracy of the analysis when the internal standard method is used.

REFERENCE

Semivolatile Organic Compounds by Isotope Dilution GC/MS. Office of Water Regulation and Standards, U.S. EPA Industrial Technology Division, Washington, DC, EPA Method 1625, Rev. C, June 1989 (contact W.A. Telliard, U.S. EPA, Office of Water Regulations and Standards, 401 M St., SW, Washington, DC, 20460. Phone: 202-382-7131).

Butyl benzyl phthalate **EPA Method 8270**
CAS #85-68-7

TITLE
Semivolatile Organic Compounds by GC/MS

MATRIX
This method is used to determine the concentration of semivolatile organic compounds in extracts prepared from all types of solid waste matrices, soils, and groundwater. Although surface waters are not specifically mentioned, this method should be applicable to water samples from rivers, lakes, etc.

METHOD SUMMARY
This method covers 259 semivolatile organic compounds. In very limited applications direct injection of the sample into the GC/MS system may be appropriate, but this results in very high detection limits (approximately 10,000 µg/L). Typically, a 1-L liquid sample, containing surrogate, and matrix spiking standards, is extracted in a continuous extractor first under acid conditions and then under basic conditions. Typically 30 g of a solid sample, containing surrogate, and matrix spiking standards, is extracted ultrasonically. After concentrating the extract to 1 mL it is spiked with 10 µL of an internal standard solution just prior to analysis by GC/MS. The volume injected should contain about 100 ng of base/neutral and 200 ng of acid surrogates (for a 1-µL injection). Analysis is performed by GC/MS using a capillary GC column.

INTERFERENCES
Raw GC/MS data from all blanks, samples, and spikes must be evaluated for interferences. Contamination by carryover can occur whenever high-concentration and low-concentration samples are sequentially analyzed. To reduce carryover, the sample syringe must be rinsed out between samples with solvent. Whenever an unusually concentrated sample is encountered, it should be followed by the analysis of blank solvent to check for cross-contamination.

INSTRUMENTATION
A GC/MS and a data system are required. The GC column used is a 30 m × 0.25 mm I.D. (or 0.32 mm I.D.) 1um film thickness silicone-coated fused silica capillary column. A continuous liquid-liquid extractor equipped with Teflon® or glass connection joints and stopcocks requiring no lubrication, a K-D concentrating apparatus, water bath, and an ultrasonic disrupter with a minimum power of 300 W and with pulsing capability are also required.

PRECISION & ACCURACY
The estimated quantitation limit (EQL) of Method 8270B for determining an individual compound is approximately 1 mg/kg (wet weight) for soil or sediment samples, 1–200 mg/kg for wastes (dependent on matrix and method of preparation), and 10 µg/L for groundwater samples. EQLs will be proportionately higher for sample extracts that require dilution to avoid saturation of the detector.

The EQL(b) for groundwater in µg/L is 10.

The EQL (a, b) for low concentrations in soil and sediment in µg/kg is 660.
Accuracy as µg/L is 0.66C−1.68.
Overall precision in µg/L is 0.53X +0.92.

(a) *EQLs listed for soil/sediment are based on wet weight. Normally data is reported in a dry-weight basis; therefore, EQLs will be higher based on the % dry weight of each sample. This calculation is based on a 30-g sample and gel permeation chromatography cleanup.*
(b) *Sample EQLs are highly matrix-dependent. The EQLs are provided for guidance and may not always be achievable.*
C = *True value for concentration, in µg/L.*
X = *Average recovery found for measurements of samples containing a concentration of C, in µg/L.*

ESTIMATED QUANTITATION LIMIT FOR OTHER MATRICES

Other Matrices	Factor (a)
High-concentration soil and sludges by sonicator	7.5
Non-water miscible waste	75

(a) *EQL for other matrices = [EQL for low soil/sediment] × [Factor]. This estimated EQL is similar to an EPA "Practical Quantitation Limit."*

SAMPLING METHOD

Liquid samples — Use a 1 or 2½ gallon amber glass bottle with a screw-top Teflon®-lined cover that has been prewashed with detergent and rinsed with distilled water and methanol (or isopropanol).

Soils, sediments, or sludges — Use an 8-oz. widemouth glass with a screw-top Teflon®-lined cover that has been prewashed with detergent and rinsed with distilled water and methanol (or isopropanol).

SAMPLE PRESERVATION

Liquid samples — If residual chlorine is present, add 3 mL of 10% sodium thiosulfate per gallon, cool to 4°C and store in a solvent-free refrigerator until analysis; if chlorine is not present, then eliminate the sodium thiosulfate addition.

Soils, sediments, or sludges — Cool samples to 4°C and store in a solvent-free refrigerator.

MHT Liquid samples must be extracted within 7 days and the extracts analyzed within 40 days. Soils, sediments, or sludges may be stored for a maximum of 14 days and the extracts analyzed within 40 days.

SAMPLE PREPARATION

Liquid samples — Transfer 1 L quantitatively to a continuous extractor. If high concentrations are anticipated, a smaller volume may be used and then diluted with organic-free reagent water to 1 L. Adjust pH, if necessary, to pH <2 using 1:1 (V/V) sulfuric acid. Pipette 1.0 mL of a surrogate standard spiking solution into each sample. For the sample in each analytical batch selected for spiking, add 1.0 mL of a matrix spiking standard. For base/neutral acid analysis, the amount of the surrogates and matrix spiking compounds added to the sample should result in a final concentration of 100 ng/µL of each analyte in the extract to be analyzed (assuming a 1-µL injection). Extract with methylene chloride for 18–24 h. Next, adjust the pH of the aqueous phase to pH >11 using 10 N sodium hydroxide and extract it with methylene chloride again for 18–24 h. Dry the extract through a column containing anhydrous sodium sulfate and concentrate it to 1 mL using a K-D concentrator.

Soils, sediments, or sludges — Use 30 g of sample. Nonporous or wet samples (gummy or clay type) that do not have a free-flowing sandy texture must be mixed with anhydrous sodium sulfate until the sample is free flowing. Add 1 mL of surrogate standards to all samples, spikes, standards, and blanks. For the sample in each analytical batch selected for spiking, add 1.0 mL of a matrix spiking standard. For base/neutral acid analysis, the amount added of the surrogates and matrix spiking compounds should result in a final concentration of 100 ng/µL of each base/neutral analyte and 200 ng/µL of each acid analyte in the extract to be analyzed (assuming a 1-µL injection). Immediately add a 100 mL mixture of 1:1 methylene chloride:acetone and extract the sample ultrasonically for 3 min and then decant or filter the extracts. Repeat the extraction two or more times. Dry the extract using a column with anhydrous sodium sulfate and concentrate it to 1 mL in a K-D concentrator.

QUALITY CONTROL A methylene chloride solution containing 50 ng/µL of decafluorotriphenylphosphine (DFTPP) is used for tuning the GC/MS system each 12-h shift. A system performance check also must be made during every 12-h shift. A standard containing 50 ng/µL each of 4,4′-DDT, pentachlorophenol, and benzidine is required to verify injection port inertness and GC column performance. A calibration standard at mid-concentration, containing each compound of interest, including all required surrogates, must be performed every 12 h during analysis. After the system performance check is met, calibration check compounds (CCCs) are used to check the validity of the initial calibration.

The internal standard responses and retention times in the calibration check standard must be evaluated immediately after or during data acquisition. If the retention time for any internal standard changes by more than 30 seconds from the last check calibration (12 h), the chromatographic system must be inspected for malfunctions and corrections must be made, as required. If the electron ionization current plot (EICP) area for any of the internal standards changes by a factor of two from the last daily calibration standard check, the mass spectrometer must be inspected for malfunctions and corrections must be made, as appropriate.

Demonstrate, through the analysis of a reagent water blank, that interferences from the analytical system, glassware, and reagents are under control. The blank samples should be carried through all stages of the sample preparation and measurement steps. For each analytical batch (up to 20 samples), a reagent blank, matrix spike, and matrix spike duplicate/duplicate must be analyzed (the frequency of the spikes may be different for different monitoring programs). The blank and spiked samples must be carried through all stages of the sample preparation and measurement steps. A QC reference sample concentrate containing each analyte at a concentration of 100 mg/L in methanol is required.

REFERENCE Test Methods for Evaluating Solid Waste (SW-846). U.S. EPA 1983, Method 8270B, Rev. 2, Nov. 1990. Office of Solid Waste, Washington, DC.

Butyl benzyl phthalate   **EPA Method 8061**
CAS #85-68-7

TITLE Phthalate Esters by Capillary Gas Chromatography With Electron Capture Detection (GC/ECD)

MATRIX This method covers aqueous and solid matrices. This includes a wide variety such as drinking water, groundwater, industrial wastewaters, surface waters, soils, solids, and sediments.

METHOD SUMMARY This method is used to determine the identities and concentrations of phthalate esters in liquid, solid and sludge matrices. When used to analyze for any or all of the target analytes, compound identification should be supported by at least one additional qualitative technique. This method describes conditions for parallel column, dual electron capture detector analysis, which fulfills the above requirement. Alternatively, GC/MS could be used for compound confirmation.

A measured volume or weight of sample (approximately 1 L for liquids, 10 to 30 g for solids and sludges) is extracted by using the appropriate sample extraction technique specified in EPA Method 3510, EPA Method 3540, and EPA Method 3550. After cleanup, the extract is analyzed by GC/ECD.

INTERFERENCES The sensitivity of this method usually depends on the level of interferences rather than on instrumental limitations. If interferences prevent detection of the analytes, cleanup of the sample extracts is necessary. Either EPA Method 3610 or EPA Method 3620 alone or followed by EPA Method 3660, Sulfur Cleanup, may be used to eliminate interferences in the analysis. EPA Method 3640, Gel Permeation Cleanup, is applicable for samples that contain high amounts of lipids and waxes.

Interferences coextracted from the samples will vary considerably from waste to waste. Glassware must be scrupulously clean. All glassware require treatment in a muffle furnace at 400°C for 2 to 4 h, or thorough rinsing with pesticide-grade solvent, prior to use. Volumetric glassware should not be heated in a muffle furnace. Storage of glassware in the lab introduces contamination, even if the glassware is wrapped in aluminum foil. Sodium sulfate, Florisil, and alumina may be contaminated with phthalate esters and, therefore, use of these materials in sample cleanup should be employed cautiously. If these materials are used, they must be obtained packaged in glass. Heating at 400°C for sodium sulfate, 320°C for Florisil, and 210°C for alumina is recommended. Glass wool used in any step of sample preparation should be a specially treated pyrex wool, pesticide grade, and must be baked at 400°C for 4 h immediately prior to use.

Paper thimbles and filter paper must be exhaustively washed with the solvent that will be used in the sample extraction. Soxhlet extraction of paper thimbles and filter paper for 12 h with fresh solvent should be repeated for a minimum of three times. Method blanks should be obtained before any of the precleaned thimbles or filter papers are used.

INSTRUMENTATION Gas chromatograph suitable for on-column and split/splitless injections.

Column 1: 30 m × 0.53 mm ID, 5% phenyl/95% methyl silicone fused-silica open tubular column, DB-5, 1.5 µg film thickness.

Column 2: 30 m × 0.53 mm ID, 14% cyanopropyl phenyl silicone fused-silica open tubular column, DB-1701, 1.0 µg film thickness.

A dual electron capture detector (ECD) is used. A Kuderna-Danish (K-D) apparatus is required along with a vacuum manifold consisting of individually adjustable, easily accessible flow-control valves for up to 24 cartridges, sample rack, chemically resistant cover and seals, heavy-duty glass basin, removable stainless steel solvent guides, built-in vacuum gauge and valve. Also, 6-mL, 1-g solid-phase extraction cartridges, LC-Florisil or equivalent, prepackaged, ready to use will be needed.

PRECISION & ACCURACY The MDL actually achieved in a given analysis will vary, as it is dependent on instrument sensitivity and matrix effects. This method has been tested in a single lab. Single-operator precision, overall precision, and method accuracy were found to be related to the concentration of the compounds and the type of matrix.

MULTIPLICATION FACTORS FOR OTHER MATRICES (a)

Matrix	Factor (b)
Groundwater	10
Low-concentration soil by ultrasonic extraction with GPC cleanup	670
High-concentration soil and sludges by ultrasonic extraction	10,000
Non-water miscible waste	100,000

(a) Sample EQLs are highly matrix-dependent.
(b) EQL = [Method detection limit] × [Factor]. For non-aqueous samples, the factor is on a wet-weight basis.

The MDL using 7 replicate determinations and a spike concentration of 100 µg/L was 42 ng/L.

The average recovery from HPLC-grade water using 4 determinatons and a spike concentration of 100 µg/L was 97.3%.

The precision (as RSD) from HPLC-grade water using 4 determinatons and a spike concentration of 100 µg/L was 2.6%.

The average recovery from groundwater using 4 determinatons and a spike concentration of 100 µg/L was 66.0%.

The precision (as RSD) from groundwater using 4 determinatons and a spike concentration of 100 µg/L was 39.3%.

The average recovery (in %) with %RSD (in parenthesis) from 3 determinations and a spike concentration of 20 µg/L in water was 84.1 (6.4) using 3M Empore Disks and EPA Method 8061.

The average recovery (in %) with %RSD (in parenthesis) from 3 determinations and a spike concentration of 20 µg/L in leachate was 105 (20.5) using 3M Empore Disks and EPA Method 8061.

The average recovery (in %) with %RSD (in parenthesis) from 3 determinations and a spike concentration of 20 µg/L in estuarine groundwater was 89.6 (6.1) using 3M Empore Disks and EPA Method 8061.

The average recovery (in %) with %RSD (in parenthesis) from 3 determinations and a spike concentration of 1 mg/kg in estuarine sediment was 113 (12.8) after sulfur cleanup with EPA Method 3660.

The average recovery (in %) with %RSD (in parenthesis) from 3 determinations and a spike concentration of 1 mg/kg in municipal sludge was 82.8 (7.8).

The average recovery (in %) with %RSD (in parenthesis) from 3 determinations and a spike concentration of 1 mg/kg in sandy loam soil was 56.5 (5.1).

SAMPLE COLLECTION, PRESERVATION & HANDLING
Containers used to collect samples for the determination of semivolatile organic compounds should be soap and water washed followed by methanol (or isopropanol) rinsing. The sample containers should be of glass or Teflon® and have screw-top covers with Teflon® liners. Sample containers should be filled with care to prevent any portion of the collected sample coming in contact with the sampler's gloves.

No preservation is used with concentrated waste samples. With liquid samples containing no residual chlorine and with soil, sediment, and sludge samples, immediately cooling to 4°C is the only preservation used. When residual chlorine is present then 3 mL of 10% aqueous sodium sulfate is added for each gallon of sample collected, followed by cooling to 4°C.

MHT Liquid samples must be extracted within 7 days and their extracts analyzed within 40 days. Concentrated waste, soil, sediment, and sludge samples must be extracted within 14 days and their extracts analyzed within 40 days.

SAMPLE PREPARATION In general, water samples are extracted at a pH of 5 to 7 with methylene chloride in a separatory funnel (EPA Method 3510). EPA Method 3520 is not recommended for the extraction of aqueous samples because the longer chain esters tend to adsorb to the glassware and consequently, their extraction recoveries may be poor. Solid samples are extracted with hexane/acetone (1:) or methylene chloride/acetone (1:1) in a Soxhlet extractor (EPA Method 3540) or with an ultrasonic extractor (EPA Method 3550). Immediately prior to extraction, spike 500 μL of the surrogate standard spiking solution into 1-L aqueous sample or 30-g solid sample. Extraction of particulate-free aqueous samples using C–18 extraction disks is an optional method that can be used.

Prior to Florisil cleanup or GC analysis, the methylene chloride and methylene chloride/acetone extracts must be exchanged to hexane. Exchange is not required for the acetonitrile extracts. Cleanup may not be necessary for extracts from a relatively clean sample matrix. Florisil Cartridge Cleanup may be used for extract cleanup.

If PCBs and organochlorine pesticides are known to be present in the sample, and if Florisil Cartridge Cleanup is considered, then two fractions are collected: Fraction 1 is eluted with 5 mL of 20% methylene chloride in hexane and Fraction 2 is eluted with 5 mL of 10% acetone in hexane. Fraction 1 contains the organochlorine pesticides and PCBs, and can be discarded.

Fraction 2 contains the phthalate esters and is analyzed by GC/ECD.

QUALITY CONTROL Identify compounds in the sample by comparing the retention times of the peaks in the sample chromatogram with those of the peaks in standard chromatograms. The retention time window used to make identification is based upon measurements of actual retention time variations over the course of 10 consecutive injections.

Calibration standards are prepared at a minimum of five concentrations for each parameter of interest through dilution of the stock standard solutions with hexane. One of the concentrations should be at a concentration near, but above, the method detection limit. Prepare stock standard solutions in hexane. Stock standards should be checked frequently for signs of degradation or evaporation, especially just prior to preparing calibration standards from them. Stock standard solutions must be replaced after one year, or sooner if comparison with check standards indicates a problem. The suggested internal standards are: 2,5-dibromotoluene, 1,3,5-tribromobenzene, and α, α′-dibromo-m-xylene. The analyst can use any of the three compounds provided that they are resolved from matrix interferences. Recommended surrogate compounds are α-2,6-trichlorotoluene, 1,4-dichloronaphthalene, and 2,3,4,5,6-pentachlorotoluene.

Spike each sample, standard, and blank with surrogate compounds. Three surrogates are suggested for this method: diphenyl phthalate, diphenyl isophthalate, and dibenzyl phthalate.

The quality control check sample concentrate should contain the test compounds at 5 to 10 ng/μL An internal standard peak area check must be performed on all samples. The internal standard must be evaluated for acceptance by determining whether the measured area for the internal standard deviates by more than 30% from the average are for the internal standard in the calibration standards. When the internal standard peak area is outside that limit, all samples that fall outside the QC criteria must be reanalyzed. Benzyl benzoate has been tested and found appropriate as an internal standard for this method.

Any compounds confirmed by two columns may also be confirmed by GC/MS. The sample extract and associated blank should be analyzed by GC/MS. A reference standard of the compound must also be analyzed by GC/MS. Include a mid-concentration calibration standard after each group of 20 samples. The response factors for the mid-concentration calibration must be within ± 15% of the average values for the multiconcentration calibration. Demonstrate through the analyses of standards that the Florisil fractionation scheme is reproducible.

REFERENCE Test Methods for Evaluating Solid Waste, Physical/Chemical Methods, SW-846, 3rd Edition, U.S. EPA, Office of Solid Waste, Washington, DC, EPA Method 8061, Nov. 1990.

Butyl benzyl phthalate   EPA Method 8060
CAS #85-68-7

TITLE Phthalate Esters

MATRIX Groundwater, soils, sludges, water miscible liquid wastes, and non-water miscible wastes.

APPLICATION This method is used for the analysis of 6 phthalate esters.

Samples are extracted, concentrated, and analyzed using direct injection of both neat and diluted organic liquids into a gas chromatograph. Analytes are detected by a flame ionization detector (FID) or an electron capture detector (ECD). Groundwater samples should be determined by ECD. The method provides an optional GC column which is used for analyte confirmation and that may help resolve analytes from interferences.

INTERFERENCES Solvents, reagents, and glassware may introduce artifacts. Plastics, in particular, must be avoided. Other interferences may come from coextracted compounds from samples. There can be carryover contamination with high- and low-level samples.

INSTRUMENTATION GC capable of on-column injections and a flame with detector (FID) or electron capture detector (ECD). Column 1: 1.8 m by 4 mm with 1.5% SP-2250/1.95% SP-2401 on Supelcoport. Column 2: 1.8 m by 4 mm with 3% OV-1 on supelcoport.

RANGE 0.7 to 106 µg/L

MDL 15 µg/L (FID) and 0.34 µg/L (ECD)

PQL FACTORS FOR MULTIPLYING × FID MDL VALUE

Matrix	Multiplication Factor
Groundwater	10
Low-level soil by sonication with GPC cleanup	670
High-level soil and sludge by sonication	10,000
Non-water miscible waste	100,000

PRECISION 0.25X + 0.07 µg/L (overall precision using FID).

ACCURACY 0.82C + 0.13 µg/L (as recovery using FID).

SAMPLING METHOD Use 8-oz. widemouth glass bottles with Teflon®-lined caps for concentrated waste samples, soils, sediments, and sludges. Use 1 or 2½ gallon amber glass bottles with Teflon®-lined caps for liquid (water) samples.

STABILITY Cool soil, sediment, sludge, and liquid samples to 4°C. If residual chlorine is present in liquid samples add 3 mL of 10% sodium thiosulfate per gallon of sample and cool to 4°C.

MHT 14 days for concentrated waste, soil, sediment, or sludge; 7 days for liquid samples; all extracts must be analyzed within 40 days.

QUALITY CONTROL A quality control check sample concentrate containing each analyte of interest is required. The QC check sample concentrate may be prepared from pure standard materials or purchased as certified solutions. Use appropriate trip, matrix, control site, method, reagent, and solvent blanks. Internal, surrogate, and five concentration level calibration standards are used. The QC check sample concentrate should contain this compound at 10 µg/mL in acetone.

REFERENCE Method 8060, SW-846, 3rd ed., Nov.1986.

2-sec-Butyl-4,6-dinitrophenol — EPA Method 8270
CAS #88-85-7

TITLE Semivolatile Organic Compounds by GC/MS

MATRIX This method is used to determine the concentration of semivolatile organic compounds in extracts prepared from all types of solid waste matrices, soils, and groundwater. Although surface waters are not specifically mentioned, this method should be applicable to water samples from rivers, lakes, etc.

METHOD SUMMARY This method covers 259 semivolatile organic compounds. In very limited applications direct injection of the sample into the GC/MS system may be appropriate, but this results in very high detection limits (approximately 10,000 µg/L). Typically, a 1-L liquid sample, containing surrogate, and matrix spiking standards, is extracted in a continuous extractor first under acid conditions and then under basic conditions. Typically 30 g of a solid sample, containing surrogate, and matrix spiking standards, is extracted ultrasonically. After concentrating the extract to 1 mL it is spiked with 10 µL of an internal standard solution just prior to analysis by GC/MS. The volume injected should contain about 100 ng of base/neutral and 200 ng of acid surrogates (for a 1-µL injection). Analysis is performed by GC/MS using a capillary GC column.

INTERFERENCES Raw GC/MS data from all blanks, samples, and spikes must be evaluated for interferences. Contamination by carryover can occur whenever high-concentration and low-concentration samples are sequentially analyzed. To reduce carryover, the sample syringe must be rinsed out between samples with solvent. Whenever an unusually concentrated sample is encountered, it should be followed by the analysis of blank solvent to check for cross-contamination.

INSTRUMENTATION A GC/MS and a data system are required. The GC column used is a 30 m × 0.25 mm I.D. (or 0.32 mm I.D.) 1um film thickness silicone-coated fused silica capillary column. A continuous liquid-liquid extractor equipped with Teflon® or glass connection joints and stopcocks requiring no lubrication, a K-D concentrating apparatus, water bath, and an ultrasonic disrupter with a minimum power of 300 W and with pulsing capability are also required.

PRECISION & ACCURACY The estimated quantitation limit (EQL) of Method 8270B for determining an individual compound is approximately 1 mg/kg (wet weight) for soil or sediment samples, 1–200 mg/kg for wastes (dependent on matrix and method of preparation), and 10 µg/L for groundwater samples. EQLs will be proportionately higher for sample extracts that require dilution to avoid saturation of the detector.

The EQL(b) for groundwater in µg/L is not listed.
The EQL (a, b) for low concentrations in soil and sediment in µg/kg is not listed.
Accuracy as µg/L is not listed.
Overall precision in µg/L is not listed.

(a) *EQLs listed for soil/sediment are based on wet weight. Normally data is reported in a dry-weight basis; therefore, EQLs will be higher based on the % dry weight of each sample. This calculation is based on a 30-g sample and gel permeation chromatography cleanup.*

(b) *Sample EQLs are highly matrix-dependent. The EQLs are provided for guidance and may not always be achievable.*

C = *True value for concentration, in µg/L.*

X = *Average recovery found for measurements of samples containing a concentration of C, in µg/L.*

ESTIMATED QUANTITATION LIMIT FOR OTHER MATRICES

Other Matrices	Factor (a)
High-concentration soil and sludges by sonicator	7.5
Non-water miscible waste	75

(a) *EQL for other matrices = [EQL for low soil/sediment] × [Factor]. This estimated EQL is similar to an EPA "Practical Quantitation Limit."*

SAMPLING METHOD

Liquid samples — Use a 1 or 2½ gallon amber glass bottle with a screw-top Teflon®-lined cover that has been prewashed with detergent and rinsed with distilled water and methanol (or isopropanol).

Soils, sediments, or sludges — Use an 8-oz. widemouth glass with a screw-top Teflon®-lined cover that has been prewashed with detergent and rinsed with distilled water and methanol (or isopropanol).

SAMPLE PRESERVATION

Liquid samples — If residual chlorine is present, add 3 mL of 10% sodium thiosulfate per gallon, cool to 4°C and store in a solvent-free refrigerator until analysis; if chlorine is not present, then eliminate the sodium thiosulfate addition.

Soils, sediments, or sludges — Cool samples to 4°C and store in a solvent-free refrigerator.

MHT Liquid samples must be extracted within 7 days and the extracts analyzed within 40 days. Soils, sediments, or sludges may be stored for a maximum of 14 days and the extracts analyzed within 40 days.

SAMPLE PREPARATION

Liquid samples — Transfer 1 L quantitatively to a continuous extractor. If high concentrations are anticipated, a smaller volume may be used and then diluted with organic-free reagent water to 1 L. Adjust pH, if necessary, to pH <2 using 1:1 (V/V) sulfuric acid. Pipette 1.0 mL of a surrogate standard spiking solution into each sample. For the sample in each analytical batch selected for spiking, add 1.0 mL of a matrix spiking standard. For base/neutral acid analysis, the amount of the surrogates and matrix spiking compounds added to the sample should result in a final concentration of 100 ng/µL of each analyte in the extract to be analyzed (assuming a 1-µL injection). Extract with methylene chloride for 18–24 h. Next, adjust the pH of the aqueous phase to pH >11 using 10 N sodium hydroxide and extract it with methylene chloride again for 18–24 h. Dry the extract through a column containing anhydrous sodium sulfate and concentrate it to 1 mL using a K-D concentrator.

Soils, sediments, or sludges — Use 30 g of sample. Nonporous or wet samples (gummy or clay type) that do not have a free-flowing sandy texture must be mixed with anhydrous sodium sulfate until the sample is free flowing. Add 1 mL of surrogate standards to all samples, spikes, standards, and blanks. For the sample in each analytical batch selected for spiking, add 1.0 mL of a matrix spiking standard. For base/neutral acid analysis, the amount added of the surrogates and matrix spiking compounds should result in a final concentration of 100 ng/µL of each base/neutral analyte and 200 ng/µL of each acid analyte in the extract to be analyzed (assuming a 1-µL injection). Immediately add a 100 mL mixture of 1:1 methylene chloride:acetone and extract the sample ultrasonically for 3 min and then decant or filter the extracts. Repeat the extraction two or more times. Dry the extract using a column with anhydrous sodium sulfate and concentrate it to 1 mL in a K-D concentrator.

QUALITY CONTROL A methylene chloride solution containing 50 ng/µL of decafluorotriphenylphosphine (DFTPP) is used for tuning the GC/MS system each 12-h shift. A system performance check also must be made during every 12-h shift. A standard containing 50 ng/µL each of 4,4′-DDT, pentachlorophenol, and benzidine is required to verify injection port inertness and GC column performance. A calibration standard at mid-concentration, containing each compound of interest, including all required surrogates, must be performed every 12 h during analysis. After the system performance check is met, calibration check compounds (CCCs) are used to check the validity of the initial calibration.

The internal standard responses and retention times in the calibration check standard must be evaluated immediately after or during data acquisition. If the retention time for any internal standard changes by more than 30 seconds from the last check calibration (12 h), the chromatographic system must be inspected for malfunctions and corrections must be made, as required. If the electron ionization current plot (EICP) area for any of the internal standards changes by a factor of two from the last daily calibration standard check, the mass spectrometer must be inspected for malfunctions and corrections must be made, as appropriate.

Demonstrate, through the analysis of a reagent water blank, that interferences from the analytical system, glassware, and reagents are under control. The blank samples should be carried through all stages of the sample preparation and measurement steps. For each analytical batch (up to 20 samples), a reagent blank, matrix spike, and matrix spike duplicate/duplicate must be analyzed (the frequency of the spikes may be different for different monitoring programs). The blank and spiked samples must be carried through all stages of the sample preparation and measurement steps. A QC reference sample concentrate containing each analyte at a concentration of 100 mg/L in methanol is required.

REFERENCE Test Methods for Evaluating Solid Waste (SW-846). U.S. EPA 1983, Method 8270B, Rev. 2, Nov. 1990. Office of Solid Waste, Washington, DC.

Butylate
CAS #2008-41-5

EPA Method 507

TITLE Determination of Nitrogen and Phosphorus-Containing Pesticides in Water by GC/NPD

MATRIX This method is applicable to the determination of certain nitrogen and phosphorus-containing pesticides in finished drinking water and groundwater.

METHOD SUMMARY Method 507 covers 46 nitrogen- and phosphorus-containing pesticides. A 1 L sample is fortified with a surrogate standard, salted, buffered, extracted with methylene chloride, and concentrated; then the solvent is exchanged with methyl tert-butyl ether (MTBE) and concentrated again, and a 2-µL aliquot of a sample extract is injected into a GC system equipped with a selective nitrogen-phosphorus detector and a capillary column for analysis.

INTERFERENCES Method interferences may be caused by contaminants in solvents, reagents, glassware, and other sample processing apparatus. Interfering contamination may occur when a sample containing low concentrations of analytes is analyzed immediately following a sample containing relatively high concentrations. One or more injections of MTBE should be made following the analysis of a sample with high concentrations of analytes to check for analyte carryover. Matrix interferences may be caused by contaminants that are coextracted from the sample. The extent of matrix interferences will vary considerably from source to source, depending upon the water sampled.

INSTRUMENTATION A gas chromatograph system (GC) equipped with a nitrogen-phosphorus detector (NPD) is needed.

Column 1: 30 m × 0.25 mm I.D. DB-5 bonded fused silica column, 0.25 µm film thickness, or equivalent.
Column 2: 30 m × 0.25 mm I.D. DB-1701 bonded fused silica column, 0.25 µm film thickness, or equivalent

PRECISION & ACCURACY This method has been validated in a single lab and estimated detection limits (EDLs) have been determined for each analyte. Observed detection limits may vary among waters, depending upon the nature of the interferences in the sample matrix and the specific instrumentation used. Analytes that are not separated chromatographically cannot be individually identified and measured unless an alternative technique for identification and quantification exist.

The estimated detection limit (in µg/L) was 0.15. The EDL is defined as either method detection limit or a level of compound in a sample yielding a peak in the final extract with signal-to-noise ratio of approximately 5, whichever value is higher.

The concentration used for these measurements (in µg/L) was 1.5.
The accuracy (as % recovery) was 97.
The precision (% RSD) was 21.

SAMPLING METHOD Grab samples are collected in 1-L glass sample bottles (prewashed with detergent and hot tap water, rinsed with reagent water, and dried in an oven at 400°C for 1 h) with screw caps lined with PTFE-fluorocarbon.

SAMPLE PRESERVATION Add mercuric chloride to the sample bottle in amounts to produce a concentration of 10 mg/L. If residual chlorine is present, add 80 mg of sodium thiosulfate/L of sample to the sample bottle prior to collection. After collection, seal bottle and shake vigorously for 1 min, then cool the sample to 4°C immediately and store it at 4°C in the dark until extraction.

MHT Maximum holding time of the samples, and in some cases the extracts, is 14 days.

SAMPLE PREPARATION Fortify the sample with 50 µL of the surrogate standard solution, adjust to pH 7 with phosphate buffer, add 100 g NaCl to the sample, and seal and shake to dissolve the salt; then extract with methylene chloride in a separatory funnel or in a mechanical tumbler bottle. Dry the extract by pouring it through a solvent-rinsed drying column containing about 10 cm of anhydrous sodium sulfate. Collect the extract in a Kuderna-Danish (K-D) concentrator and rinse the column with 20- 30 mL methylene chloride. Concentrate the extract to about 2 mL and rinse the flask and its lower joint into the concentrator tube with 1 to 2 mL of methyl t-butyl ether (MTBE). Add 5–10 mL of MTBE and concentrate the extract twice (adding more MTBE) to a final volume of 5.0 mL and store it at 4°C until analysis.

Note: If methylene chloride is not completely removed from the final extract, it may cause detector problems.

QUALITY CONTROL Minimum quality control requirements are initial demonstration of lab capability, determination of surrogate compound recoveries in each sample and blank, monitoring internal standard peak area or height in each sample and blank, analysis of lab reagent blanks, lab fortified samples, lab fortified blanks, and other QC samples. A lab reagent blank is analyzed to demonstrate that all glassware and reagent interferences are under control.

Initial demonstration of capability is fulfilled by analyzing four fortified reagent water samples with the recovery value for each analyte falling within the acceptable range (±30% average recovery). Surrogate recoveries from samples or method blanks must be 70–130%. The internal standard response for any sample chromatogram should not deviate from the daily calibration check standard's internal standard response by more than 30% or lab fortified blanks and sample matrices are used to assess lab performance and analyte recovery, respectively.

If the response for the target analyte peak exceeds the working range of the system, dilute the extract and reanalyze. Alternative techniques such as an alternate detector or second chromatography column should be used to confirm peak identification when sample components are not resolved adequately.

EPA CONTACT & HOTLINE For technical questions contact Dr. Baldev Bathija, U.S. EPA, Office of Ground Water and Drinking Water (WH-550D), 401 M St. SW, Washington, DC 20460. Tel. (202) 260-3040. For further information the EPA Safe Drinking Water Hotline may be called at: (800) 426-4791.

REFERENCE Methods for the Determination of Organic Compounds in Drinking Water, EPA/600/4-88/039 (revised July 1991). U.S. EPA Environmental Monitoring Systems Laboratory, Cincinnati, OH, 45268, U.S.A. Available from the National Technical Information Service (NTIS), 5285 Port Royal Road, Springfield, VA 22161; Tel. 800-553-6847. NTIS Order Number is PB91-231480.

---

n-Butylbenzene   EPA Method 502
CAS #104-51-8

TITLE Volatile Organic Compounds in Water By Purge and Trap Capillary Column Gas Chromatography with Photoionization and Electrolytic Conductivity Detectors in Series. U.S. EPA Method 502.2, Rev. 2.0, 1989.

MATRIX Drinking water and raw source water. The latter should include most surface water and groundwater sources.

METHOD SUMMARY This method covers 60 volatile organic compounds that contain halogen atoms and/or that are aromatic. An inert gas (zero grade nitrogen or helium) is bubbled through a 25-mL or a 5-mL water sample (depending on the expected concentration of the analytes). Purged sample components are trapped in a tube of sorbent materials. When purging is complete, the sorbent tube is heated and backflushed with helium to desorb the trapped sample onto a capillary GC column. The column is temperature programmed to separate the method analytes which are then detected with a photoionization detector (PID) and a Hall electrolytic conductivity (HECD) placed in series. The PID is selective for aromatic compounds and the HECD is selective for halogenated compounds.

INTERFERENCES Impurities in the purge gas and from organic compounds outgassing from the plumbing ahead of the trap account for many contamination problems. Interferences purged or coextracted from the samples will vary considerably from source to source, depending upon the particular sample or extract being tested. Cross-contamination can occur whenever high-level and low-level samples are analyzed sequentially. Samples also can be contaminated by diffusion of volatile organics (particularly methylene chloride and fluorocarbons) through the septum seal into the sample during shipment and storage. The lab where volatile analysis is performed and also the refrigerated storage area should be completely free of solvents.

INSTRUMENTATION A GC containing a series configuration of a high temperature photoionization detector (PID) equipped with 10.0 eV (nominal) lamp and Hall electrolytic conductivity detector (HECD) is required. Also required is an all-glass 5-mL purging device, a sorbent trap, and a thermal desorption apparatus which is connected to the GC system.

Column 1: VOCOL glass wide-bore capillary column.
Column 2: RTX–502.2 mega-bore capillary column.
Column 3: DB-62 mega-bore capillary column.

PRECISION & ACCURACY Method detection limits are dependent upon the characteristics of the gas chromatographic system used. Analytes that are not separated chromatographically cannot be individually identified and used in the same calibration mixture or water samples unless an alternative technique for identification and quantification, such as mass spectrometry, is used.

Electrolytic conductivity detetor (c) range in µg/L (a) was 0.02–200.
Electrolytic conductivity detetor (c) MDL in µg/L (b) was not listed.
Electrolytic conductivity detetor (c) accuracy as % recovery was not listed.
Electrolytic conductivity detetor (c) precision as % RSD was not listed.
Photoionization detector (d) range in µg/L (a) was 0.02–200.
Photoionization detector (d) mDL in µg/L (b) was 0.02.
Photoionization detector (d) accuracy as % recovery was 100.
Photoionization detector (d) precision as % RSD was 4.4.

(a) *The applicable concentration range of this method is compound, instrument, and matrix-dependent. It is listed as being approximately 0.02 to 200 µg/L but no specific information is provided so caution should be observed.*
(b) *The method detection limits reports with this method are compound, instrument, and matrix-dependent. The values reported were calculated using reagent water fortified with the corresponding compounds at 10 µg/L and a GC-equipped with a 60 m × 0.75 mm VOLCOL wide bore capillary column with 1.5 µm film thickness and using helium carrier gas.*
(c) *Recoveries and relative standard deviations were determined from seven samples of reagent water fortified with 10 µg/L of each compound. 2-Bromo-1-chloropropane was used as the internal standard for calculating average recoveries.*
(d) *Recoveries and relative standard deviations were determined from seven samples of reagent water fortified with 10 µg/L of each compound. Fluorobenzene was used as the internal standard for calculating average recoveries.*

SAMPLING METHOD Collect samples using a 40- to 120-mL screw-cap vial (prewashed with detergent, rinsed with distilled water and oven dried at 105°C) with a Teflon®-faced silicone septum. Collect bubble-free samples and place the septum with the Teflon® side down on the water.

SAMPLE PRESERVATION If residual chlorine is present in the water add about 25 mg of ascorbic acid to each vial before samples are collected to remove the chlorine. Add hydrochloric acid to reduce pH to <2, immediately cool samples to 4°C, and store them in a solvent-free refrigerator at 4°C until analysis.

MHT The maximum holding time for samples is 14 days from the time they were collected.

SAMPLE PREPARATION Remove the plungers from two 5-mL syringes and attach a closed syringe valve to each. Warm the sample to room temperature, open the sample bottle, and carefully pour the sample into one of the syringe barrels to just short of overflowing. Replace the syringe plunger, invert the syringe, and compress the sample. Open the syringe valve and vent any residual air while adjusting the sample volume to

5.0 mL. Add 10 µL of the internal calibration standard to the sample through the syringe valve. Close the valve. Fill the second syringe in an identical manner from the same sample bottle. Reserve this second syringe for a reanalysis if necessary.

QUALITY CONTROL As an initial demonstration of lab accuracy and precision, analyze 4 to 7 replicates of a lab fortified blank containing analyte at 0.1–5 µg/L. Collect all samples in duplicate. Surrogate analytes (similar to those of the analytes of interest), whose concentration is known in every sample, are measured using the same internal standard calibration procedure. Duplicate field reagent water blanks (trip blanks) must be analyzed with each set of samples, lab reagent blanks (method blanks) must be analyzed with each batch of samples processed as a group within a work shift. Also, a single lab-fortified blank that contains each of the analytes of interest should be analyzed with each batch of samples processed as a group within a work shift. A 3- to 5-point calibration curve is needed depending on the calibration range factor required.

EPA CONTACT & HOTLINE For technical questions contact Dr. Baldev Bathija, U.S. EPA, Office of Ground Water and Drinking Water (WH-550D), 401 M St. SW, Washington, DC 20460. Tel. (202) 260-3040. For further information the EPA Safe Drinking Water Hotline may be called at: (800) 426-4791.

REFERENCE Methods for the Determination of Organic Compounds in Drinking Water, EPA/600/4-88/039 (revised July 1991; Final Rule for determination of compliance with the MCL for Total Trihalomethanes under 141.30, in 40 CFR Part 141, Vol. 58, No. 147, Fed. Reg., Tuesday Aug. 3, 1993). U.S. EPA Environmental Monitoring Systems Laboratory, Cincinnati, OH, 45268, U.S.A. Available from the National Technical Information Service (NTIS), 5285 Port Royal Road, Springfield, VA 22161; Tel. 800-553-6847. NTIS Order Number is PB91-231480.

---

sec-Butylbenzene   **EPA Method 502**
CAS #135-98-8

TITLE Volatile Organic Compounds in Water By Purge and Trap Capillary Column Gas Chromatography with Photoionization and Electrolytic Conductivity Detectors in Series. U.S. EPA Method 502.2, Rev. 2.0, 1989.

MATRIX Drinking water and raw source water. The latter should include most surface water and groundwater sources.

METHOD SUMMARY This method covers 60 volatile organic compounds that contain halogen atoms and/or that are aromatic. An inert gas (zero grade nitrogen or helium) is bubbled through a 25-mL or a 5-mL water sample (depending on the expected concentration of the analytes). Purged sample components are trapped in a tube of sorbent materials. When purging is complete, the sorbent tube is heated and backflushed with helium to desorb the trapped sample onto a capillary GC column. The column is temperature programmed to separate the method analytes which are then detected with a photoionization detector (PID) and a Hall electrolytic conductivity (HECD) placed in series. The PID is selective for aromatic compounds and the HECD is selective for halogenated compounds.

INTERFERENCES Impurities in the purge gas and from organic compounds outgassing from the plumbing ahead of the trap account for many contamination problems. Interferences purged or coextracted from the samples will vary considerably from source to source, depending upon the particular sample or extract being tested. Cross-contamination can occur whenever high-level and low-level samples are analyzed sequentially. Samples also can be contaminated by diffusion of volatile organics (particularly methylene chloride and fluorocarbons) through the septum seal into the sample during shipment and storage. The lab where volatile analysis is performed and also the refrigerated storage area should be completely free of solvents.

INSTRUMENTATION A GC containing a series configuration of a high temperature photoionization detector (PID) equipped with 10.0 eV (nominal) lamp and Hall electrolytic conductivity detector (HECD) is required. Also required is an all-glass 5-mL purging device, a sorbent trap, and a thermal desorption apparatus which is connected to the GC system.

Column 1: VOCOL glass wide-bore capillary column.
Column 2: RTX–502.2 mega-bore capillary column.
Column 3: DB-62 mega-bore capillary column.

PRECISION & ACCURACY Method detection limits are dependent upon the characteristics of the gas chromatographic system used. Analytes that are not separated chromatographically cannot be individually identified and used in the same calibration mixture or water samples unless an alternative technique for identification and quantification, such as mass spectrometry, is used.

Electrolytic conductivity deteter (c) range in µg/L (a) was 0.02–200.
Electrolytic conductivity deteter (c) MDL in µg/L (b) was not listed.
Electrolytic conductivity deteter (c) accuracy as % recovery was not listed.
Electrolytic conductivity deteter (c) precision as % RSD was not listed.
Photoionization detector (d) range in µg/L (a) was 0.02–200.
Photoionization detector (d) MDL in µg/L (b) was 0.02.
Photoionization detector (d) accuracy as % recovery was 97.
Photoionization detector (d) precision as % RSD was 2.7.

(a) *The applicable concentration range of this method is compound, instrument, and matrix-dependent. It is listed as being approximately 0.02 to 200 µg/L but no specific information is provided so caution should be observed.*
(b) *The method detection limits reports with this method are compound, instrument, and matrix-dependent. The values reported were calculated using reagent water fortified with the corresponding compounds at 10 µg/L and a GC-equipped with a 60 m × 0.75 mm VOLCOL wide bore capillary column with 1.5 µm film thickness and using helium carrier gas.*
(c) *Recoveries and relative standard deviations were determined from seven samples of reagent water fortified with*

10 µg/L of each compound. 2-Bromo-1-chloropropane was used as the internal standard for calculating average recoveries.

(d) *Recoveries and relative standard deviations were determined from seven samples of reagent water fortified with 10 µg/L of each compound. Fluorobenzene was used as the internal standard for calculating average recoveries.*

SAMPLING METHOD Collect samples using a 40- to 120-mL screw-cap vial (prewashed with detergent, rinsed with distilled water and oven dried at 105°C) with a Teflon®-faced silicone septum. Collect bubble-free samples and place the septum with the Teflon® side down on the water.

SAMPLE PRESERVATION If residual chlorine is present in the water add about 25 mg of ascorbic acid to each vial before samples are collected to remove the chlorine. Add hydrochloric acid to reduce pH to <2, immediately cool samples to 4°C, and store them in a solvent-free refrigerator at 4°C until analysis.

MHT The maximum holding time for samples is 14 days from the time they were collected.

SAMPLE PREPARATION Remove the plungers from two 5-mL syringes and attach a closed syringe valve to each. Warm the sample to room temperature, open the sample bottle, and carefully pour the sample into one of the syringe barrels to just short of overflowing. Replace the syringe plunger, invert the syringe, and compress the sample. Open the syringe valve and vent any residual air while adjusting the sample volume to 5.0 mL. Add 10 µL of the internal calibration standard to the sample through the syringe valve. Close the valve. Fill the second syringe in an identical manner from the same sample bottle. Reserve this second syringe for a reanalysis if necessary.

QUALITY CONTROL As an initial demonstration of lab accuracy and precision, analyze 4 to 7 replicates of a lab fortified blank containing analyte at 0.1–5 µg/L. Collect all samples in duplicate. Surrogate analytes (similar to those of the analytes of interest), whose concentration is known in every sample, are measured using the same internal standard calibration procedure. Duplicate field reagent water blanks (trip blanks) must be analyzed with each set of samples, lab reagent blanks (method blanks) must be analyzed with each batch of samples processed as a group within a work shift. Also, a single lab-fortified blank that contains each of the analytes of interest should be analyzed with each batch of samples processed as a group within a work shift. A 3- to 5-point calibration curve is needed depending on the calibration range factor required.

EPA CONTACT & HOTLINE For technical questions contact Dr. Baldev Bathija, U.S. EPA, Office of Ground Water and Drinking Water (WH-550D), 401 M St. SW, Washington, DC 20460. Tel. (202) 260-3040. For further information the EPA Safe Drinking Water Hotline may be called at: (800) 426-4791.

REFERENCE Methods for the Determination of Organic Compounds in Drinking Water, EPA/600/4-88/039 (revised July 1991; Final Rule for determination of compliance with the MCL for Total Trihalomethanes under 141.30, in 40 CFR Part 141, Vol. 58, No. 147, Fed. Reg., Tuesday Aug. 3, 1993). U.S. EPA Environmental Monitoring Systems Laboratory, Cincinnati, OH, 45268, U.S.A. Available from the National Technical Information Service (NTIS), 5285 Port Royal Road, Springfield, VA 22161; Tel. 800-553-6847. NTIS Order Number is PB91-231480.

---

tert-Butylbenzene   **EPA Method 502**
CAS #98-06-6

TITLE Volatile Organic Compounds in Water By Purge and Trap Capillary Column Gas Chromatography with Photoionization and Electrolytic Conductivity Detectors in Series. U.S. EPA Method 502.2, Rev. 2.0, 1989.

MATRIX Drinking water and raw source water. The latter should include most surface water and groundwater sources.

METHOD SUMMARY This method covers 60 volatile organic compounds that contain halogen atoms and/or that are aromatic. An inert gas (zero grade nitrogen or helium) is bubbled through a 25-mL or a 5-mL water sample (depending on the expected concentration of the analytes). Purged sample components are trapped in a tube of sorbent materials. When purging is complete, the sorbent tube is heated and backflushed with helium to desorb the trapped sample onto a capillary GC column. The column is temperature programmed to separate the method analytes which are then detected with a photoionization detector (PID) and a Hall electrolytic conductivity (HECD) placed in series. The PID is selective for aromatic compounds and the HECD is selective for halogenated compounds.

INTERFERENCES Impurities in the purge gas and from organic compounds outgassing from the plumbing ahead of the trap account for many contamination problems. Interferences purged or coextracted from the samples will vary considerably from source to source, depending upon the particular sample or extract being tested. Cross-contamination can occur whenever high-level and low-level samples are analyzed sequentially. Samples also can be contaminated by diffusion of volatile organics (particularly methylene chloride and fluorocarbons) through the septum seal into the sample during shipment and storage. The lab where volatile analysis is performed and also the refrigerated storage area should be completely free of solvents.

INSTRUMENTATION A GC containing a series configuration of a high temperature photoionization detector (PID) equipped with 10.0 eV (nominal) lamp and Hall electrolytic conductivity detector (HECD) is required. Also required is an all-glass 5-mL purging device, a sorbent trap, and a thermal desorption apparatus which is connected to the GC system.

Column 1: VOCOL glass wide-bore capillary column.
Column 2: RTX–502.2 mega-bore capillary column.
Column 3: DB-62 mega-bore capillary column.

PRECISION & ACCURACY Method detection limits are dependent upon the characteristics of the gas chromatographic system used. Analytes that are not separated chromatographically cannot be individually identified and used in the same calibration mixture or water samples unless an alternative technique

for identification and quantification, such as mass spectrometry, is used.

Electrolytic conductivity detetor (c) range in µg/L (a) was 0.02–200.

Electrolytic conductivity detetor (c) MDL in µg/L (b) was not listed.

Electrolytic conductivity detetor (c) accuracy as % recovery was not listed.

Electrolytic conductivity detetor (c) precision as % RSD was not listed.

Photoionization detector (d) range in µg/L (a) was 0.02–200.
Photoionization detector (d) MDL in µg/L (b) was 0.06.
Photoionization detector (d) accuracy as % recovery was 98.
Photoionization detector (d) precision as % RSD was 2.3.

(a) The applicable concentration range of this method is compound, instrument, and matrix-dependent. It is listed as being approximately 0.02 to 200 µg/L but no specific information is provided so caution should be observed.

(b) The method detection limits reports with this method are compound, instrument, and matrix-dependent. The values reported were calculated using reagent water fortified with the corresponding compounds at 10 µg/L and a GC-equipped with a 60 m × 0.75 mm VOLCOL wide bore capillary column with 1.5 µm film thickness and using helium carrier gas.

(c) Recoveries and relative standard deviations were determined from seven samples of reagent water fortified with 10 µg/L of each compound. 2-Bromo-1-chloropropane was used as the internal standard for calculating average recoveries.

(d) Recoveries and relative standard deviations were determined from seven samples of reagent water fortified with 10 µg/L of each compound. Fluorobenzene was used as the internal standard for calculating average recoveries.

SAMPLING METHOD Collect samples using a 40- to 120-mL screw-cap vial (prewashed with detergent, rinsed with distilled water and oven dried at 105°C) with a Teflon®-faced silicone septum. Collect bubble-free samples and place the septum with the Teflon® side down on the water.

SAMPLE PRESERVATION If residual chlorine is present in the water add about 25 mg of ascorbic acid to each vial before samples are collected to remove the chlorine. Add hydrochloric acid to reduce pH to <2, immediately cool samples to 4°C, and store them in a solvent-free refrigerator at 4°C until analysis.

MHT The maximum holding time for samples is 14 days from the time they were collected.

SAMPLE PREPARATION Remove the plungers from two 5-mL syringes and attach a closed syringe valve to each. Warm the sample to room temperature, open the sample bottle, and carefully pour the sample into one of the syringe barrels to just short of overflowing. Replace the syringe plunger, invert the syringe, and compress the sample. Open the syringe valve and vent any residual air while adjusting the sample volume to 5.0 mL. Add 10 µL of the internal calibration standard to the sample through the syringe valve. Close the valve. Fill the second syringe in an identical manner from the same sample bottle. Reserve this second syringe for a reanalysis if necessary.

QUALITY CONTROL As an initial demonstration of lab accuracy and precision, analyze 4 to 7 replicates of a lab fortified blank containing analyte at 0.1–5 µg/L. Collect all samples in duplicate. Surrogate analytes (similar to those of the analytes of interest), whose concentration is known in every sample, are measured using the same internal standard calibration procedure. Duplicate field reagent water blanks (trip blanks) must be analyzed with each set of samples, lab reagent blanks (method blanks) must be analyzed with each batch of samples processed as a group within a work shift. Also, a single lab-fortified blank that contains each of the analytes of interest should be analyzed with each batch of samples processed as a group within a work shift. A 3- to 5-point calibration curve is needed depending on the calibration range factor required.

EPA CONTACT & HOTLINE For technical questions contact Dr. Baldev Bathija, U.S. EPA, Office of Ground Water and Drinking Water (WH-550D), 401 M St. SW, Washington, DC 20460. Tel. (202) 260-3040. For further information the EPA Safe Drinking Water Hotline may be called at: (800) 426-4791.

REFERENCE Methods for the Determination of Organic Compounds in Drinking Water, EPA/600/4-88/039 (revised July 1991; Final Rule for determination of compliance with the MCL for Total Trihalomethanes under 141.30, in 40 CFR Part 141, Vol. 58, No. 147, Fed. Reg., Tuesday Aug. 3, 1993). U.S. EPA Environmental Monitoring Systems Laboratory, Cincinnati, OH, 45268, U.S.A. Available from the National Technical Information Service (NTIS), 5285 Port Royal Road, Springfield, VA 22161; Tel. 800-553-6847. NTIS Order Number is PB91-231480.

---

n-Butylbenzene   **EPA Method 524**
CAS #104-51-8

TITLE Measurement of Purgeable Organic Compounds in Water by Capillary Column GC/MS.

MATRIX Drinking water and raw source water; the latter should include most surface water and groundwater sources.

METHOD SUMMARY Method 524.2 covers 60 volatile organic compounds. An inert gas (zero grade nitrogen or helium) is bubbled through a 25-mL or a 5-mL water sample (depending on the expected concentration of the analytes). Purged sample components are trapped in a tube of sorbent materials. When purging is complete, the sorbent tube is heated and backflushed with helium to desorb the trapped sample onto a capillary GC column.

INTERFERENCES Impurities in the purge gas and from organic compounds outgassing from the plumbing ahead of the trap account for many contamination problems. Interferences purged or coextracted from the samples will vary considerably from source to source, depending upon the particular sample or extract being tested. Cross-contamination can occur whenever high-level and low-level samples are analyzed sequentially. Samples also can be contaminated by diffusion of volatile organics (particularly methylene chloride and fluorocarbons) through the septum seal into the sample during shipment and storage.

INSTRUMENTATION A GC/MS with a data system equipped with one of the following capillary GC columns:

Column 1: VOCOL glass wide bore capillary column.
Column 2: DB-624 fused silica capillary column.
Column 3: DB-5 fused silica capillary column.

Also required is an all-glass 25-mL or 5-mL purging device, a sorbent trap, and a thermal desorption apparatus which is connected to the GC/MS system.

PRECISION & ACCURACY Method detection limits are compound- and instrument-dependent, and may vary from approximately 0.02–0.35 µg/L. Note in the table below that the "true" concentration range used for accuracy and precision measurements was quite narrow. However, the applicable concentration range of this method is primarily column dependent and is approximately 0.02 to 200 µg/L for the wide-bore thick-film columns. Narrow-bore thin-film columns may have a capacity which limits the range to about 0.02 to 20 µg/L. Analytes that are inefficiently purged from water will not be detected when present at low concentrations, but they can be measured with acceptable accuracy and precision when present in sufficient amounts.

Analytes that are not separated chromatographically, but which have different mass spectra and non-interfering quantification ions, can be identified and measured in the same calibration mixture or water sample. Analytes which have very similar mass spectra cannot be individually identified and measured in the same calibration mixture or water samples unless they have different retention times. Co-eluting compounds with very similar mass spectra, typically many structural isomers, must be reported as an isomeric group or pair.

The range (in µg/L) was 0.5–10.
The method detection limit (in µg/L) was 0.11.
The accuracy (as % recovery) was 100.
The precision (in %) was 7.6.

Note: Data were obtained from 16–31 determinations using a wide-bore capillary column and a jet separator interfaced to a quadrupole mass spectrometer. All analytes were in a reagent water matrix.

SAMPLING METHOD Collect samples using a 40- to 120-mL screw-cap vial (prewashed with detergent, rinsed with distilled water and oven dried at 105°C) with a Teflon®-faced silicone septum. Collect bubble-free samples and place the septum with the Teflon® side down on the water.

SAMPLE PRESERVATION If residual chlorine is present in the water add about 25 mg of ascorbic acid to each vial before samples are collected to remove the chlorine. Add hydrochloric acid to reduce pH to <2, and immediately cool samples to 4°C, and store them in a solvent-free refrigerator at 4°C until analysis.

MHT The maximum holding time for samples is 14 days from the time they were collected.

SAMPLE PREPARATION Remove the plungers from two 25-mL (or 5-mL depending on sample size) syringes and attach a closed syringe valve to each. Warm the sample to room temperature, open the sample bottle, and carefully pour the sample into one of the syringe barrels to just short of overflowing. Replace the syringe plunger, invert the syringe, and compress the sample. Open the syringe valve and vent any residual air while adjusting the sample volume to 25.0 mL (or 5 mL). For samples and blanks, add 5 µL of the fortification solution containing the internal standard and the surrogates to the sample through the syringe valve. For calibration standards and lab fortified blanks, add 5 µL of the fortification solution containing the internal standard only. Close the valve. Fill the second syringe in an identical manner from the same sample bottle. Reserve this second syringe for a reanalysis if necessary.

QUALITY CONTROL As an initial demonstration of lab accuracy and precision, analyze 4 to 7 replicates of a lab fortified blank containing analyte at 0.2–5 µg/L. Collect all samples in duplicate. Surrogate analytes (similar to those of the analytes of interest), whose concentration is known in every sample, are measured using the same internal standard calibration procedure. Duplicate field reagent water blanks (trip blanks) must be analyzed with each set of samples, lab reagent blanks (method blanks) must be analyzed with each batch of samples processed as a group within a work shift. Also, a single lab-fortified blank that contains each of the analytes of interest should be analyzed with each batch of samples processed as a group within a work shift. A 3- to 5-point calibration curve is needed depending on the calibration range factor required.

EPA CONTACT & HOTLINE For technical questions contact Dr. Baldev Bathija, U.S. EPA, Office of Ground Water and Drinking Water (WH-550D), 401 M St. SW, Washington, DC 20460. Tel. (202) 260-3040. For further information the EPA Safe Drinking Water Hotline may be called at: (800) 426-4791.

REFERENCE Methods for the Determination of Organic Compounds in Drinking Water, EPA/600/4-88/039 (revised July 1991; Final Rule for determination of compliance with the MCL for Total Trihalomethanes under 141.30, in 40 CFR Part 141, Vol. 58, No. 147, Fed. Reg., Tuesday Aug. 3, 1993). U.S. EPA Environmental Monitoring Systems Laboratory, Cincinnati, OH, 45268, U.S.A. Available from the National Technical Information Service (NTIS), 5285 Port Royal Road, Springfield, VA 22161; Tel. 800-553-6847. NTIS Order Number is PB91-231480.

---

sec-Butylbenzene   **EPA Method 524**
CAS #135-98-8

TITLE Measurement of Purgeable Organic Compounds in Water by Capillary Column GC/MS.

MATRIX Drinking water and raw source water; the latter should include most surface water and groundwater sources.

METHOD SUMMARY Method 524.2 covers 60 volatile organic compounds. An inert gas (zero grade nitrogen or helium) is bubbled through a 25-mL or a 5-mL water sample (depending on the expected concentration of the analytes). Purged sample components are trapped in a tube of sorbent materials. When purging is complete, the sorbent tube is heated and backflushed with helium to desorb the trapped sample onto a capillary GC column.

INTERFERENCES Impurities in the purge gas and from organic compounds outgassing from the plumbing ahead of the trap account for many contamination problems. Interferences purged or coextracted from the samples will vary considerably from source to source, depending upon the particular sample or extract being tested. Cross-contamination can occur whenever high-level and low-level samples are analyzed sequentially. Samples also can be contaminated by diffusion of volatile organics (particularly methylene chloride and fluorocarbons) through the septum seal into the sample during shipment and storage.

INSTRUMENTATION A GC/MS with a data system equipped with one of the following capillary GC columns:

Column 1: VOCOL glass wide bore capillary column.
Column 2: DB-624 fused silica capillary column.
Column 3: DB-5 fused silica capillary column.

Also required is an all-glass 25-mL or 5-mL purging device, a sorbent trap, and a thermal desorption apparatus which is connected to the GC/MS system.

PRECISION & ACCURACY Method detection limits are compound- and instrument-dependent, and may vary from approximately 0.02–0.35 µg/L. Note in the table below that the "true" concentration range used for accuracy and precision measurements was quite narrow. However, the applicable concentration range of this method is primarily column dependent and is approximately 0.02 to 200 µg/L for the wide-bore thick-film columns. Narrow-bore thin-film columns may have a capacity which limits the range to about 0.02 to 20 µg/L. Analytes that are inefficiently purged from water will not be detected when present at low concentrations, but they can be measured with acceptable accuracy and precision when present in sufficient amounts.

Analytes that are not separated chromatographically, but which have different mass spectra and non-interfering quantification ions, can be identified and measured in the same calibration mixture or water sample. Analytes which have very similar mass spectra cannot be individually identified and measured in the same calibration mixture or water samples unless they have different retention times. Co-eluting compounds with very similar mass spectra, typically many structural isomers, must be reported as an isomeric group or pair.

The range (in µg/L) was 0.5–10.
The method detection limit (in µg/L) was 0.13.
The accuracy (as % recovery) was 100.
The precision (in %) was 7.6.

Note: Data were obtained from 16–31 determinations using a wide-bore capillary column and a jet separator interfaced to a quadrupole mass spectrometer. All analytes were in a reagent water matrix.

SAMPLING METHOD Collect samples using a 40- to 120-mL screw-cap vial (prewashed with detergent, rinsed with distilled water and oven dried at 105°C) with a Teflon®-faced silicone septum. Collect bubble-free samples and place the septum with the Teflon® side down on the water.

SAMPLE PRESERVATION If residual chlorine is present in the water add about 25 mg of ascorbic acid to each vial before samples are collected to remove the chlorine. Add hydrochloric acid to reduce pH to <2, and immediately cool samples to 4°C, and store them in a solvent-free refrigerator at 4°C until analysis.

MHT The maximum holding time for samples is 14 days from the time they were collected.

SAMPLE PREPARATION Remove the plungers from two 25-mL (or 5-mL depending on sample size) syringes and attach a closed syringe valve to each. Warm the sample to room temperature, open the sample bottle, and carefully pour the sample into one of the syringe barrels to just short of overflowing. Replace the syringe plunger, invert the syringe, and compress the sample. Open the syringe valve and vent any residual air while adjusting the sample volume to 25.0 mL (or 5 mL). For samples and blanks, add 5 µL of the fortification solution containing the internal standard and the surrogates to the sample through the syringe valve. For calibration standards and lab fortified blanks, add 5 µL of the fortification solution containing the internal standard only. Close the valve. Fill the second syringe in an identical manner from the same sample bottle. Reserve this second syringe for a reanalysis if necessary.

QUALITY CONTROL As an initial demonstration of lab accuracy and precision, analyze 4 to 7 replicates of a lab fortified blank containing analyte at 0.2–5 µg/L. Collect all samples in duplicate. Surrogate analytes (similar to those of the analytes of interest), whose concentration is known in every sample, are measured using the same internal standard calibration procedure. Duplicate field reagent water blanks (trip blanks) must be analyzed with each set of samples, lab reagent blanks (method blanks) must be analyzed with each batch of samples processed as a group within a work shift. Also, a single lab-fortified blank that contains each of the analytes of interest should be analyzed with each batch of samples processed as a group within a work shift. A 3- to 5-point calibration curve is needed depending on the calibration range factor required.

EPA CONTACT & HOTLINE For technical questions contact Dr. Baldev Bathija, U.S. EPA, Office of Ground Water and Drinking Water (WH-550D), 401 M St. SW, Washington, DC 20460. Tel. (202) 260-3040. For further information the EPA Safe Drinking Water Hotline may be called at: (800) 426-4791.

REFERENCE Methods for the Determination of Organic Compounds in Drinking Water, EPA/600/4-88/039 (revised July 1991; Final Rule for determination of compliance with the MCL for Total Trihalomethanes under 141.30, in 40 CFR Part 141, Vol. 58, No. 147, Fed. Reg., Tuesday Aug. 3, 1993). U.S. EPA Environmental Monitoring Systems Laboratory, Cincinnati, OH, 45268, U.S.A. Available from the National Technical Information Service (NTIS), 5285 Port Royal Road, Springfield, VA 22161; Tel. 800-553-6847. NTIS Order Number is PB91-231480.

tert-Butylbenzene   EPA Method 524
CAS #98-06-6

TITLE Measurement of Purgeable Organic Compounds in Water by Capillary Column GC/MS.

MATRIX Drinking water and raw source water; the latter should include most surface water and groundwater sources.

METHOD SUMMARY Method 524.2 covers 60 volatile organic compounds. An inert gas (zero grade nitrogen or helium) is bubbled through a 25-mL or a 5-mL water sample (depending on the expected concentration of the analytes). Purged sample components are trapped in a tube of sorbent materials. When purging is complete, the sorbent tube is heated and backflushed with helium to desorb the trapped sample onto a capillary GC column.

INTERFERENCES Impurities in the purge gas and from organic compounds outgassing from the plumbing ahead of the trap account for many contamination problems. Interferences purged or coextracted from the samples will vary considerably from source to source, depending upon the particular sample or extract being tested. Cross-contamination can occur whenever high-level and low-level samples are analyzed sequentially. Samples also can be contaminated by diffusion of volatile organics (particularly methylene chloride and fluorocarbons) through the septum seal into the sample during shipment and storage.

INSTRUMENTATION A GC/MS with a data system equipped with one of the following capillary GC columns:

Column 1: VOCOL glass wide bore capillary column.
Column 2: DB-624 fused silica capillary column.
Column 3: DB-5 fused silica capillary column.

Also required is an all-glass 25-mL or 5-mL purging device, a sorbent trap, and a thermal desorption apparatus which is connected to the GC/MS system.

PRECISION & ACCURACY Method detection limits are compound- and instrument-dependent, and may vary from approximately 0.02–0.35 µg/L. Note in the table below that the "true" concentration range used for accuracy and precision measurements was quite narrow. However, the applicable concentration range of this method is primarily column dependent and is approximately 0.02 to 200 µg/L for the wide-bore thick-film columns. Narrow-bore thin-film columns may have a capacity which limits the range to about 0.02 to 20 µg/L. Analytes that are inefficiently purged from water will not be detected when present at low concentrations, but they can be measured with acceptable accuracy and precision when present in sufficient amounts.

Analytes that are not separated chromatographically, but which have different mass spectra and non-interfering quantification ions, can be identified and measured in the same calibration mixture or water sample. Analytes which have very similar mass spectra cannot be individually identified and measured in the same calibration mixture or water samples unless they have different retention times. Co-eluting compounds with very similar mass spectra, typically many structural isomers, must be reported as an isomeric group or pair.

The range (in µg/L) was 0.5–10.
The method detection limit (in µg/L) was 0.14.
The accuracy (as % recovery) was 102.
The precision (in %) was 7.3.

Note: Data were obtained from 16–31 determinations using a wide-bore capillary column and a jet separator interfaced to a quadrupole mass spectrometer. All analytes were in a reagent water matrix.

SAMPLING METHOD Collect samples using a 40- to 120-mL screw-cap vial (prewashed with detergent, rinsed with distilled water and oven dried at 105°C) with a Teflon®-faced silicone septum. Collect bubble-free samples and place the septum with the Teflon® side down on the water.

SAMPLE PRESERVATION If residual chlorine is present in the water add about 25 mg of ascorbic acid to each vial before samples are collected to remove the chlorine. Add hydrochloric acid to reduce pH to <2, and immediately cool samples to 4°C, and store them in a solvent-free refrigerator at 4°C until analysis.

MHT The maximum holding time for samples is 14 days from the time they were collected.

SAMPLE PREPARATION Remove the plungers from two 25-mL (or 5-mL depending on sample size) syringes and attach a closed syringe valve to each. Warm the sample to room temperature, open the sample bottle, and carefully pour the sample into one of the syringe barrels to just short of overflowing. Replace the syringe plunger, invert the syringe, and compress the sample. Open the syringe valve and vent any residual air while adjusting the sample volume to 25.0 mL (or 5 mL). For samples and blanks, add 5 µL of the fortification solution containing the internal standard and the surrogates to the sample through the syringe valve. For calibration standards and lab fortified blanks, add 5 µL of the fortification solution containing the internal standard only. Close the valve. Fill the second syringe in an identical manner from the same sample bottle. Reserve this second syringe for a reanalysis if necessary.

QUALITY CONTROL As an initial demonstration of lab accuracy and precision, analyze 4 to 7 replicates of a lab fortified blank containing analyte at 0.2–5 µg/L. Collect all samples in duplicate. Surrogate analytes (similar to those of the analytes of interest), whose concentration is known in every sample, are measured using the same internal standard calibration procedure. Duplicate field reagent water blanks (trip blanks) must be analyzed with each set of samples, lab reagent blanks (method blanks) must be analyzed with each batch of samples processed as a group within a work shift. Also, a single lab-fortified blank that contains each of the analytes of interest should be analyzed with each batch of samples processed as a group within a work shift. A 3- to 5-point calibration curve is needed depending on the calibration range factor required.

EPA CONTACT & HOTLINE For technical questions contact Dr. Baldev Bathija, U.S. EPA, Office of Ground Water and Drinking Water (WH-550D), 401 M St. SW, Washington, DC

20460. Tel. (202) 260-3040. For further information the EPA Safe Drinking Water Hotline may be called at: (800) 426-4791.

REFERENCE Methods for the Determination of Organic Compounds in Drinking Water, EPA/600/4-88/039 (revised July 1991; Final Rule for determination of compliance with the MCL for Total Trihalomethanes under 141.30, in 40 CFR Part 141, Vol. 58, No. 147, Fed. Reg., Tuesday Aug. 3, 1993). U.S. EPA Environmental Monitoring Systems Laboratory, Cincinnati, OH, 45268, U.S.A. Available from the National Technical Information Service (NTIS), 5285 Port Royal Road, Springfield, VA 22161; Tel. 800-553-6847. NTIS Order Number is PB91-231480.

---

n-Butylbenzene   EPA Method 8021
CAS #104-51-8

TITLE Halogenated Volatile by Gas Chromatography Using Photoionization and Electrolytic Conductivity Detectors in Series: Capillary Column Technique

MATRIX This method is applicable to nearly all types of samples, regardless of water content, including groundwater, aqueous sludges, caustic liquors, acid liquors, waste solvents, oily wastes, mousses, tars, fibrous wastes, polymeric emulsions, filter cakes, spent carbons, spent catalysts, soils, and sediments.

METHOD SUMMARY This method is used to determine 60 volatile organic compounds in a variety of solid waste matrices. It provides GC conditions for the detection of halogenated and aromatic volatile organic compounds. Samples can be analyzed using direct injection or purge-and-trap (EPA Method 5030). Groundwater samples must be analyzed using EPA Method 5030 (where applicable). A temperature program is used with the GC. Detection is achieved by a photoionization detector (PID) and a Hall electrolytic conductivity detector (HECD) in series.

INTERFERENCES Samples can be contaminated by diffusion of volatile organics (particularly chlorofluorocarbons and methylene chloride) through the sample container septum during shipment and storage.

INSTRUMENTATION A GC-equipped with variable-constant differential flow controllers, subambient oven controller, PID and HECD detectors connected with a short piece of uncoated capillary tubing and a data system.

Column: 60 m × 0.75 mm I.D. VOCOL wide-bore capillary column with 1.5 µm film thickness.

PRECISION & ACCURACY MDLs are compound-dependent and vary with purging efficiency and concentration. The applicable concentration range of this method is compound- and instrument-dependent but is approximately 0.1 to 200 µg/L. Analytes that are inefficiently purged from water will not be detected when present at low concentrations, but they can be measured with acceptable accuracy and precision when present in sufficient amounts. The estimated quantitation limit (EQL) for an individual compound is approximately 1 µg/kg (wet weight) for soil/sediment samples, 100 µg/kg (wet weight) for wastes, and 1 µg/L for groundwater. EQLs will be proportionately higher for sample extracts and samples that require dilution or reduced sample size to avoid saturation of the detector.

MULTIPLICATION FACTORS FOR OTHER MATRICES (a)

Matrix	Factor (b)
Groundwater	10
Low-concentration soil	10
Water miscible liquid waste	500
High-concentration soil and sludge	1250
Non-water miscible waste	1250

(a) Sample EQLs are highly matrix-dependent. The EQLs listed herein are provided for guidance and may not always be achievable. (b) EQL = [Method detection limit] × [Factor]. For non-aqueous samples, the factor is on a wet-weight basis.

SINGLE LABORATORY ACCURACY & PRECISION DATA FOR VOCs IN WATER

This method was tested in a single lab using water spiked at 10 µg/L and the following data was reported:

Recoveries and standard deviations were determined from seven samples and spiked at 10 µg/L of each analyte. Recoveries were determined by the internal standard method. Internal standards were: Fluorobenzene for PID and 2-Bromo-1-chloropropane for HECD.

The average recovery (in percent) for the PID was 100.
The standard deviation of the recovery for the PID was 4.4.
The MDL (in µg/mL) for the PID was 0.02.
The average recovery (in percent) for the HECD was none (no response for this detector).
The standard deviation of the recovery for the HECD was none (no response for this detector).
The MDL (in µg/mL) for the HECD was none (no response for this detector).

SAMPLE COLLECTION, PRESERVATION & HANDLING
Volatile Organics — Standard 40-mL glass screw-cap VOA vials with Teflon®-faced silicone septum may be used for both liquid and solid matrices. When collecting samples, liquids and solids should be introduced into the vials gently to reduce agitation which might drive off volatile compounds. If there are any air bubbles present the sample must be retaken. Tap slightly as they are filled to try and eliminate as much free air space as possible. The two vials from each sampling locations should be sealed in separate plastic bags to prevent cross-contamination between samples particularly if the sampled waste is suspected of containing high levels of volatile organics.

Semivolatile organics — Containers used to collect samples for the determination of semivolatile organic compounds should be soap and water washed followed by methanol (or isopropanol) rinsing. The sample containers should be of glass or Teflon® and have screw-top covers with Teflon® liners.

Preservation for volatile organics — No preservation is used with concentrated waste samples. With liquid samples containing no residual chlorine, 4 drops of concentrated hydrochloric acid are added and the samples are immediately cooled to 4°C. When liquid samples contain residual chlorine, they are treated as above and, in addition, 4 drops of 4% aqueous sodium thiosulfate are added. Soil, sediment, and sludge samples are only cooled to 4°C.

Preservation for semivolatile organics — No preservation is used with concentrated waste samples. With liquid samples containing no residual chlorine and with soil, sediment, and sludge samples, immediately cooling to 4°C is the only preservation used. When residual chlorine is present then 3 mL of 10% aqueous sodium sulfate is added for each gallon of sample collected, followed by cooling to 4°C.

MHT The holding time for all volatile organics samples is 14 days. Liquid samples must be extracted within 7 days and their extracts analyzed within 40 days. Concentrated waste, soil, sediment, and sludge samples must be extracted within 14 days and their extracts analyzed within 40 days.

SAMPLE PREPARATION Volatile compounds are introduced into the gas chromatograph either by direct injector or purge-and-trap (EPA Method 5030). EPA Method 5030 may be used directly on groundwater samples or low-concentration contaminated soils and sediments. For medium-concentration soils or sediments, methanolic extraction, as described in EPA Method 5030, may be necessary prior to purge-and-trap analysis.

QUALITY CONTROL Calculate surrogate standard recovery on all samples, blanks, and spikes. A trip blank is recommended to check on sampling, storage, and handling contamination. Calibration standards, at a minimum of five concentration levels, are prepared in organic-free reagent water. One of the concentration levels should be at a concentration near, but above, the method detection limit.

A combination of bromochloromethane, 2-bromo-1-chloropropane, 1,4-dichlorobutane, and bromochlorobenzene are recommended as surrogate standards to encompass the range of the temperature program used in this method.

REFERENCE Test Methods for Evaluating Solid Waste, Physical/Chemical Methods, SW-846, 3rd Edition, U.S. EPA, Office of Solid Waste, Washington, DC, EPA Method 8021A, Rev. 1, Nov. 1992.

---

sec-Butylbenzene   **EPA Method 8021**
CAS #135-98-8

TITLE Halogenated Volatile by Gas Chromatography Using Photoionization and Electrolytic Conductivity Detectors in Series: Capillary Column Technique

MATRIX This method is applicable to nearly all types of samples, regardless of water content, including groundwater, aqueous sludges, caustic liquors, acid liquors, waste solvents, oily wastes, mousses, tars, fibrous wastes, polymeric emulsions, filter cakes, spent carbons, spent catalysts, soils, and sediments.

METHOD SUMMARY This method is used to determine 60 volatile organic compounds in a variety of solid waste matrices. It provides GC conditions for the detection of halogenated and aromatic volatile organic compounds. Samples can be analyzed using direct injection or purge-and-trap (EPA Method 5030). Groundwater samples must be analyzed using EPA Method 5030 (where applicable). A temperature program is used with the GC. Detection is achieved by a photoionization detector (PID) and a Hall electrolytic conductivity detector (HECD) in series.

INTERFERENCES Samples can be contaminated by diffusion of volatile organics (particularly chlorofluorocarbons and methylene chloride) through the sample container septum during shipment and storage.

INSTRUMENTATION A GC-equipped with variable-constant differential flow controllers, subambient oven controller, PID and HECD detectors connected with a short piece of uncoated capillary tubing and a data system.

Column: 60 m × 0.75 mm I.D. VOCOL wide-bore capillary column with 1.5 µm film thickness.

PRECISION & ACCURACY MDLs are compound-dependent and vary with purging efficiency and concentration. The applicable concentration range of this method is compound- and instrument-dependent but is approximately 0.1 to 200 µg/L. Analytes that are inefficiently purged from water will not be detected when present at low concentrations, but they can be measured with acceptable accuracy and precision when present in sufficient amounts. The estimated quantitation limit (EQL) for an individual compound is approximately 1 µg/kg (wet weight) for soil/sediment samples, 100 µg/kg (wet weight) for wastes, and 1 µg/L for groundwater. EQLs will be proportionately higher for sample extracts and samples that require dilution or reduced sample size to avoid saturation of the detector.

MULTIPLICATION FACTORS FOR OTHER MATRICES (a)

Matrix	Factor (b)
Groundwater	10
Low-concentration soil	10
Water miscible liquid waste	500
High-concentration soil and sludge	1250
Non-water miscible waste	1250

(a) Sample EQLs are highly matrix-dependent. The EQLs listed herein are provided for guidance and may not always be achievable. (b) EQL = [Method detection limit] × [Factor]. For non-aqueous samples, the factor is on a wet-weight basis.

SINGLE LABORATORY ACCURACY & PRECISION DATA FOR VOCs IN WATER

This method was tested in a single lab using water spiked at 10 µg/L and the following data was reported:

Recoveries and standard deviations were determined from seven samples and spiked at 10 µg/L of each analyte. Recoveries were determined by the internal standard method. Internal standards were: Fluorobenzene for PID and 2-Bromo-1-chloropropane for HECD.

The average recovery (in percent) for the PID was 97.
The standard deviation of the recovery for the PID was 2.6.
The MDL (in µg/mL) for the PID was 0.02.
The average recovery (in percent) for the HECD was none (no response for this detector).
The standard deviation of the recovery for the HECD was none (no response for this detector)-.

The MDL (in µg/mL) for the HECD was none (no response for this detector).

SAMPLE COLLECTION, PRESERVATION & HANDLING
Volatile Organics — Standard 40-mL glass screw-cap VOA vials with Teflon®-faced silicone septum may be used for both liquid and solid matrices. When collecting samples, liquids and solids should be introduced into the vials gently to reduce agitation which might drive off volatile compounds. If there are any air bubbles present the sample must be retaken. Tap slightly as they are filled to try and eliminate as much free air space as possible. The two vials from each sampling locations should be sealed in separate plastic bags to prevent cross-contamination between samples particularly if the sampled waste is suspected of containing high levels of volatile organics.

Semivolatile organics — Containers used to collect samples for the determination of semivolatile organic compounds should be soap and water washed followed by methanol (or isopropanol) rinsing. The sample containers should be of glass or Teflon® and have screw-top covers with Teflon® liners.

Preservation for volatile organics — No preservation is used with concentrated waste samples. With liquid samples containing no residual chlorine, 4 drops of concentrated hydrochloric acid are added and the samples are immediately cooled to 4°C. When liquid samples contain residual chlorine, they are treated as above and, in addition, 4 drops of 4% aqueous sodium thiosulfate are added. Soil, sediment, and sludge samples are only cooled to 4°C.

Preservation for semivolatile organics — No preservation is used with concentrated waste samples. With liquid samples containing no residual chlorine and with soil, sediment, and sludge samples, immediately cooling to 4°C is the only preservation used. When residual chlorine is present then 3 mL of 10% aqueous sodium sulfate is added for each gallon of sample collected, followed by cooling to 4°C.

MHT The holding time for all volatile organics samples is 14 days. Liquid samples must be extracted within 7 days and their extracts analyzed within 40 days. Concentrated waste, soil, sediment, and sludge samples must be extracted within 14 days and their extracts analyzed within 40 days.

SAMPLE PREPARATION Volatile compounds are introduced into the gas chromatograph either by direct injector or purge-and-trap (EPA Method 5030). EPA Method 5030 may be used directly on groundwater samples or low-concentration contaminated soils and sediments. For medium-concentration soils or sediments, methanolic extraction, as described in EPA Method 5030, may be necessary prior to purge-and-trap analysis.

QUALITY CONTROL Calculate surrogate standard recovery on all samples, blanks, and spikes. A trip blank is recommended to check on sampling, storage, and handling contamination. Calibration standards, at a minimum of five concentration levels, are prepared in organic-free reagent water. One of the concentration levels should be at a concentration near, but above, the method detection limit.

A combination of bromochloromethane, 2-bromo-1-chloropropane, 1,4-dichlorobutane, and bromochlorobenzene are recommended as surrogate standards to encompass the range of the temperature program used in this method.

REFERENCE Test Methods for Evaluating Solid Waste, Physical/Chemical Methods, SW-846, 3rd Edition, U.S. EPA, Office of Solid Waste, Washington, DC, EPA Method 8021A, Rev. 1, Nov. 1992.

---

tert-Butylbenzene   **EPA Method 8021**
CAS #98-06-6

TITLE Halogenated Volatile by Gas Chromatography Using Photoionization and Electrolytic Conductivity Detectors in Series: Capillary Column Technique

MATRIX This method is applicable to nearly all types of samples, regardless of water content, including groundwater, aqueous sludges, caustic liquors, acid liquors, waste solvents, oily wastes, mousses, tars, fibrous wastes, polymeric emulsions, filter cakes, spent carbons, spent catalysts, soils, and sediments.

METHOD SUMMARY This method is used to determine 60 volatile organic compounds in a variety of solid waste matrices. It provides GC conditions for the detection of halogenated and aromatic volatile organic compounds. Samples can be analyzed using direct injection or purge-and-trap (EPA Method 5030). Groundwater samples must be analyzed using EPA Method 5030 (where applicable). A temperature program is used with the GC. Detection is achieved by a photoionization detector (PID) and a Hall electrolytic conductivity detector (HECD) in series.

INTERFERENCES Samples can be contaminated by diffusion of volatile organics (particularly chlorofluorocarbons and methylene chloride) through the sample container septum during shipment and storage.

INSTRUMENTATION A GC-equipped with variable-constant differential flow controllers, subambient oven controller, PID and HECD detectors connected with a short piece of uncoated capillary tubing and a data system.

Column: 60 m × 0.75 mm I.D. VOCOL wide-bore capillary column with 1.5 µm film thickness.

PRECISION & ACCURACY MDLs are compound-dependent and vary with purging efficiency and concentration. The applicable concentration range of this method is compound- and instrument-dependent but is approximately 0.1 to 200 µg/L. Analytes that are inefficiently purged from water will not be detected when present at low concentrations, but they can be measured with acceptable accuracy and precision when present in sufficient amounts. The estimated quantitation limit (EQL) for an individual compound is approximately 1 µg/kg (wet weight) for soil/sediment samples, 100 µg/kg (wet weight) for wastes, and 1 µg/L for groundwater. EQLs will be proportionately higher for sample extracts and samples that require dilution or reduced sample size to avoid saturation of the detector.

MULTIPLICATION FACTORS FOR OTHER MATRICES (a)

Matrix	Factor (b)
Groundwater	10
Low-concentration soil	10
Water miscible liquid waste	500
High-concentration soil and sludge	1250
Non-water miscible waste	1250

(a) Sample EQLs are highly matrix-dependent. The EQLs listed herein are provided for guidance and may not always be achievable. (b) EQL = [Method detection limit] × [Factor]. For non-aqueous samples, the factor is on a wet-weight basis.

SINGLE LABORATORY ACCURACY & PRECISION DATA FOR VOCs IN WATER

This method was tested in a single lab using water spiked at 10 µg/L and the following data was reported:

Recoveries and standard deviations were determined from seven samples and spiked at 10 µg/L of each analyte. Recoveries were determined by the internal standard method. Internal standards were: Fluorobenzene for PID and 2-Bromo-1-chloropropane for HECD.

The average recovery (in percent) for the PID was 98.
The standard deviation of the recovery for the PID was 2.3.
The MDL (in µg/mL) for the PID was 0.06.
The average recovery (in percent) for the HECD was none (no response for this detector).
The standard deviation of the recovery for the HECD was none (no response for this detector).
The MDL (in µg/mL) for the HECD was none (no response for this detector).

SAMPLE COLLECTION, PRESERVATION & HANDLING

Volatile Organics — Standard 40-mL glass screw-cap VOA vials with Teflon®-faced silicone septum may be used for both liquid and solid matrices. When collecting samples, liquids and solids should be introduced into the vials gently to reduce agitation which might drive off volatile compounds. If there are any air bubbles present the sample must be retaken. Tap slightly as they are filled to try and eliminate as much free air space as possible. The two vials from each sampling locations should be sealed in separate plastic bags to prevent cross-contamination between samples particularly if the sampled waste is suspected of containing high levels of volatile organics.

Semivolatile organics — Containers used to collect samples for the determination of semivolatile organic compounds should be soap and water washed followed by methanol (or isopropanol) rinsing. The sample containers should be of glass or Teflon® and have screw-top covers with Teflon® liners.

Preservation for volatile organics — No preservation is used with concentrated waste samples. With liquid samples containing no residual chlorine, 4 drops of concentrated hydrochloric acid are added and the samples are immediately cooled to 4°C. When liquid samples contain residual chlorine, they are treated as above and, in addition, 4 drops of 4% aqueous sodium thiosulfate are added. Soil, sediment, and sludge samples are only cooled to 4°C.

Preservation for semivolatile organics — No preservation is used with concentrated waste samples. With liquid samples containing no residual chlorine and with soil, sediment, and sludge samples, immediately cooling to 4°C is the only preservation used. When residual chlorine is present then 3 mL of 10% aqueous sodium sulfate is added for each gallon of sample collected, followed by cooling to 4°C.

MHT The holding time for all volatile organics samples is 14 days. Liquid samples must be extracted within 7 days and their extracts analyzed within 40 days. Concentrated waste, soil, sediment, and sludge samples must be extracted within 14 days and their extracts analyzed within 40 days.

SAMPLE PREPARATION Volatile compounds are introduced into the gas chromatograph either by direct injector or purge-and-trap (EPA Method 5030). EPA Method 5030 may be used directly on groundwater samples or low-concentration contaminated soils and sediments. For medium-concentration soils or sediments, methanolic extraction, as described in EPA Method 5030, may be necessary prior to purge-and-trap analysis.

QUALITY CONTROL Calculate surrogate standard recovery on all samples, blanks, and spikes. A trip blank is recommended to check on sampling, storage, and handling contamination. Calibration standards, at a minimum of five concentration levels, are prepared in organic-free reagent water. One of the concentration levels should be at a concentration near, but above, the method detection limit.

A combination of bromochloromethane, 2-bromo-1-chloropropane, 1,4-dichlorobutane, and bromochlorobenzene are recommended as surrogate standards to encompass the range of the temperature program used in this method.

REFERENCE Test Methods for Evaluating Solid Waste, Physical/Chemical Methods, SW-846, 3rd Edition, U.S. EPA, Office of Solid Waste, Washington, DC, EPA Method 8021A, Rev. 1, Nov. 1992.

---

n-Butylbenzene   **EPA Method 8260**
CAS #104-51-8

TITLE Volatile Organic Compounds by GC/MS: Capillary Column Technique

MATRIX This method is applicable to nearly all types of samples, regardless of water content, including groundwater, soils, and sediments.

METHOD SUMMARY Method 8260A covers 58 volatile organic compounds that are introduced into a gas chromatograph by the purge-and-trap method or by direct injection (in limited applications). Zero-grade helium is bubbled through a 5-mL solution at ambient temperature. Purged sample components are trapped in a tube containing suitable sorbent materials. When purging is complete, the sorbent tube is heated and backflushed with helium to desorb trapped sample components. The analytes are desorbed directly to a large bore capillary or cryofocussed on a capillary precolumn before being flash evaporated to a narrow bore capillary for analysis.

INTERFERENCES Major contaminant sources are volatile materials in the lab and impurities in the inert purging gas and in the sorbent trap. Interfering contamination may occur when a sample containing low concentrations of volatile organic compounds is analyzed immediately after a sample containing high concentrations of volatile organic compounds. After analysis of a sample containing high concentrations of volatile organic compounds, one or more calibration blanks should be analyzed to check for cross-contamination. Screening of the samples prior to purge-and-trap GC/MS analysis is highly recommended to prevent contamination of the system. This is especially true for soil and waste samples.

Special precautions must be taken to analyze for methylene chloride. The analytical and sample storage area should be isolated from all atmospheric sources of methylene chloride. All gas chromatography carrier gas lines and purge gas plumbing should be constructed from stainless steel or copper tubing. Laboratory clothing previously exposed to methylene chloride fumes during liquid-liquid extraction procedures can contribute to sample contamination.

Samples can also be contaminated by diffusion of volatile organics (particularly methylene chloride and fluorocarbons) through the septum seal during shipment and storage. A trip blank can serve as a check on such contamination.

INSTRUMENTATION GC/MS with a temperature-programmable chromatograph suitable for splitless injection equipped with variable constant differential flow controllers, a subambient oven controller, a purging device, sorbent trap, a thermal desorption apparatus and a capillary precolumn interface when using cryogenic cooling will be needed. The following GC columns may be used:

Column 1: 60 m × 0.75 mm I.D. capillary column coated with VOCOL, 1.5 µm film thickness.
Column 2: 30 m × 0.53 mm capillary column coated with DB-624 or VOCOL, 3 µm film thickness.
Column 3: 30 m × 0.32 mm I.D. capillary column coated with DB-5 or SE-54, 1-µm film thickness.

PRECISION & ACCURACY This method has been tested in a single lab using spiked water. Using a wide-bore capillary column, water was spiked at concentrations between 0.5 and 10 µg/L. Single lab accuracy and precision data are presented. The MDL actually achieved in a given analysis will vary depending on instrument sensitivity and matrix effects.

The MDL (a) in µg/L was 0.11.
The concentration range in µg/L was 0.5–10.
The mean accuracy (% of true value) was 100.
The precision as relative standard deviation was 7.6.

Note: The MDL is based on a 25-mL sample volume instead of a 5-mL sample volume.

SAMPLING METHOD
Liquid samples — Use a 40-mL glass screw-cap VOA vial with a Teflon®-faced silicone septum that has been prewashed, rinsed with distilled deionized water, and oven dried. If residual chlorine is present, collect the sample in a 4-oz soil VOA container which has been pre-preserved with 4 drops of 10% sodium thiosulfate. Mix gently and transfer the sample to a 40-mL VOA vial. Collect bubble-free samples in duplicate and seal each sample in a separate plastic bag.

Soils, sediments, and sludges — Use an 8-oz widemouth glass bottle with Teflon®-faced silicone septum that has been prewashed, rinsed with distilled deionized water, and oven dried. **Do not** heat the septum for more than 1 h. Tap slightly to eliminate any free air space. Collect samples in duplicate and seal each one in a separate plastic bag.

SAMPLE PRESERVATION
Liquid samples — Add 4 drops of concentrated HCL, cool to 4°C and store in a solvent-free refrigerator.

Soils, sediments, and sludges — Cool samples to 4°C and store in a solvent-free refrigerator.

MHT The maximum holding time of any sample (liquids, soils, sediments, and sludges) is 14 days.

SAMPLE PREPARATION
Liquid samples — Remove the plunger from a 5-mL syringe and carefully pour the sample into the syringe barrel to just short of overflowing. Replace the syringe plunger and compress the sample. Open the syringe valve and vent any residual air while adjusting the sample volume to 5.0 mL. If there is only one volatile organic analysis (VOA) vial, a second syringe should be filled at this time to protect against possible loss of sample integrity. Add 10 µL of surrogate spiking solution and 10 µL of internal standard spiking solution through the valve bore of the 5-mL syringe, then close the valve. The surrogate and internal standards may be mixed and added as a single spiking solution.

Sediments, soils, and waste samples — All samples of this type should be screened by GC analysis using a headspace method (EPA Method 3810) or the hexadecane extraction and screening method (EPA Method 3820). Use the screening data to determine whether to use the low-concentration method (0.005–1 mg/kg) or the high-concentration method (>1 mg/kg).

Low-concentration method — The low-concentration method is based on purging a heated sediment or soil sample mixed with organic-free reagent water containing the surrogate and internal standards. Analyze all reagent blanks and standards under the same conditions as the samples.

Use a 5-g sample if the expected concentration is <0.1 mg/kg or a 1-g sample for expected concentrations between 0.1 and 1 mg/kg. Mix the contents of the sample container with a narrow metal spatula. Weigh the amount of the sample into a tared purge device. Add the spiked water to the purge device, which contains the weighed amount of sample, and connect the device to the purge-and-trap system.

High-concentration method — This method is based on extracting the sediment or soil with methanol. A waste sample is either extracted or diluted, depending on its solubility in methanol. Wastes that are insoluble in methanol are diluted with reagent tetraglyme or possibly polyethylene glycol (PEG). An aliquot of the extract is added to organic-free reagent water containing surrogate and internal standards. This is purged at

ambient temperature. All samples with an expected concentration of >1.0 mg/kg should be analyzed by this method.

Mix the contents of the sample container with a narrow metal spatula. For sediments or soils and solid wastes that are insoluble in methanol, weigh 4 g (wet weight) of sample into a tared 20-mL vial. For waste that is soluble in methanol, tetraglyme, or PEG, weigh 1 g (wet weight) into a tared scintillation vial or culture tube or a 10-mL volumetric flask. Quickly add 9.0 mL of appropriate solvent then add 1.0 mL of a surrogate spiking solution to the vial, cap it, and shake it for 2 min.

METHANOL EXTRACT REQUIRED FOR ANALYSIS OF HIGH-CONCENTRATION SOILS OR SEDIMENTS

Approximate Concentration Range	Volume of Methanol Extract (a)
500–10,000 µg/kg	100 µL
1,000–20,000 µg/kg	50 µL
5,000–100,000 µg/kg	10 µL
25,000–500,000 µg/kg	100 µL of 1/50 dilution (b)

Calculate appropriate dilution factor for concentrations exceeding this table.

(a) The volume of methanol added to 5 mL of water being purged should be kept constant. Therefore, add to the 5-mL syringe whatever volume of methanol is necessary to maintain a volume of 100 µL added to the syringe.
(b) Dilute an aliquot of the methanol extract and then take 100 µL for analysis.

QUALITY CONTROL Demonstrate, through the analysis of a reagent water blank, that interferences from the analytical system, glassware, and reagents are under control. Blank samples should be carried through all stages of the sample preparation and measurement steps. For each analytical batch (up to 20 samples), a reagent blank, matrix spike, and matrix spike duplicate must be analyzed (the frequency of the spikes may be different for different monitoring programs). The blank and spiked samples must be carried through all stages of the sample preparation and measurement steps. QC samples mentioned in the section on Interferences will also be needed as appropriate to those situations.

Matrix spiking standards should be prepared from volatile organic compounds which will be representative of the compounds being investigated. The recommended internal standards are chlorobenzene-d5, 1,4-difluorobenzene, 1,4-dichlorobenzene-d4, and pentafluorobenzene. Using stock standard solutions, prepare secondary dilution standards containing the compounds of interest, either singly or mixed together in methanol. Store them in a vial with no headspace for no more than one week. Surrogates recommended are toluene-d8, 4-bromofluorobenzene, and dibromofluoromethane. Each sample undergoing GC/MS analysis must be spiked with 10 µL of the surrogate spiking solution prior to analysis.

REFERENCE Test Methods for Evaluating Solid Waste (SW-846). U.S. EPA 1983, Method 8260A, Rev. 1, Nov. 1990. Office of Solid Waste, Washington, DC.

sec-Butylbenzene   **EPA Method 8260**
CAS #135-98-8

TITLE Volatile Organic Compounds by GC/MS: Capillary Column Technique

MATRIX This method is applicable to nearly all types of samples, regardless of water content, including groundwater, soils, and sediments.

METHOD SUMMARY Method 8260A covers 58 volatile organic compounds that are introduced into a gas chromatograph by the purge-and-trap method or by direct injection (in limited applications). Zero-grade helium is bubbled through a 5-mL solution at ambient temperature. Purged sample components are trapped in a tube containing suitable sorbent materials. When purging is complete, the sorbent tube is heated and backflushed with helium to desorb trapped sample components. The analytes are desorbed directly to a large bore capillary or cryofocussed on a capillary precolumn before being flash evaporated to a narrow bore capillary for analysis.

INTERFERENCES Major contaminant sources are volatile materials in the lab and impurities in the inert purging gas and in the sorbent trap. Interfering contamination may occur when a sample containing low concentrations of volatile organic compounds is analyzed immediately after a sample containing high concentrations of volatile organic compounds. After analysis of a sample containing high concentrations of volatile organic compounds, one or more calibration blanks should be analyzed to check for cross-contamination. Screening of the samples prior to purge-and-trap GC/MS analysis is highly recommended to prevent contamination of the system. This is especially true for soil and waste samples.

Special precautions must be taken to analyze for methylene chloride. The analytical and sample storage area should be isolated from all atmospheric sources of methylene chloride. All gas chromatography carrier gas lines and purge gas plumbing should be constructed from stainless steel or copper tubing. Laboratory clothing previously exposed to methylene chloride fumes during liquid-liquid extraction procedures can contribute to sample contamination.

Samples can also be contaminated by diffusion of volatile organics (particularly methylene chloride and fluorocarbons) through the septum seal during shipment and storage. A trip blank can serve as a check on such contamination.

INSTRUMENTATION GC/MS with a temperature-programmable chromatograph suitable for splitless injection equipped with variable constant differential flow controllers, a subambient oven controller, a purging device, sorbent trap, a thermal desorption apparatus and a capillary precolumn interface when using cryogenic cooling will be needed. The following GC columns may be used:

Column 1: 60 m × 0.75 mm I.D. capillary column coated with VOCOL, 1.5 µm film thickness.
Column 2: 30 m × 0.53 mm I.D. capillary column coated with DB-624 or VOCOL, 3 µm film thickness.

Column 3: 30 m × 0.32 mm I.D. capillary column coated with DB-5 or SE-54, 1-μm film thickness.

PRECISION & ACCURACY This method has been tested in a single lab using spiked water. Using a wide-bore capillary column, water was spiked at concentrations between 0.5 and 10 μg/L. Single lab accuracy and precision data are presented. The MDL actually achieved in a given analysis will vary depending on instrument sensitivity and matrix effects.

The MDL (a) in μg/L was 0.13.
The concentration range in μg/L was 0.5–10.
The mean accuracy (% of true value) was 100.
The precision as relative standard deviation was 7.6.

Note: The MDL is based on a 25-mL sample volume instead of a 5-mL sample volume.

SAMPLING METHOD
Liquid samples — Use a 40-mL glass screw-cap VOA vial with a Teflon®-faced silicone septum that has been prewashed, rinsed with distilled deionized water, and oven dried. If residual chlorine is present, collect the sample in a 4-oz soil VOA container which has been pre-preserved with 4 drops of 10% sodium thiosulfate. Mix gently and transfer the sample to a 40-mL VOA vial. Collect bubble-free samples in duplicate and seal each sample in a separate plastic bag.

Soils, sediments, and sludges — Use an 8-oz widemouth glass bottle with Teflon®-faced silicone septum that has been prewashed, rinsed with distilled deionized water, and oven dried. **Do not** heat the septum for more than 1 h. Tap slightly to eliminate any free air space. Collect samples in duplicate and seal each one in a separate plastic bag.

SAMPLE PRESERVATION
Liquid samples — Add 4 drops of concentrated HCL, cool to 4°C and store in a solvent-free refrigerator.

Soils, sediments, and sludges — Cool samples to 4°C and store in a solvent-free refrigerator.

MHT The maximum holding time of any sample (liquids, soils, sediments, and sludges) is 14 days.

SAMPLE PREPARATION
Liquid samples — Remove the plunger from a 5-mL syringe and carefully pour the sample into the syringe barrel to just short of overflowing. Replace the syringe plunger and compress the sample. Open the syringe valve and vent any residual air while adjusting the sample volume to 5.0 mL. If there is only one volatile organic analysis (VOA) vial, a second syringe should be filled at this time to protect against possible loss of sample integrity. Add 10 μL of surrogate spiking solution and 10 μL of internal standard spiking solution through the valve bore of the 5-mL syringe, then close the valve. The surrogate and internal standards may be mixed and added as a single spiking solution.

Sediments, soils, and waste samples — All samples of this type should be screened by GC analysis using a headspace method (EPA Method 3810) or the hexadecane extraction and screening method (EPA Method 3820). Use the screening data to determine whether to use the low-concentration method (0.005–1 mg/kg) or the high-concentration method (>1 mg/kg).

Low-concentration method — The low-concentration method is based on purging a heated sediment or soil sample mixed with organic-free reagent water containing the surrogate and internal standards. Analyze all reagent blanks and standards under the same conditions as the samples.

Use a 5-g sample if the expected concentration is <0.1 mg/kg or a 1-g sample for expected concentrations between 0.1 and 1 mg/kg. Mix the contents of the sample container with a narrow metal spatula. Weigh the amount of the sample into a tared purge device. Add the spiked water to the purge device, which contains the weighed amount of sample, and connect the device to the purge-and-trap system.

High-concentration method — This method is based on extracting the sediment or soil with methanol. A waste sample is either extracted or diluted, depending on its solubility in methanol. Wastes that are insoluble in methanol are diluted with reagent tetraglyme or possibly polyethylene glycol (PEG). An aliquot of the extract is added to organic-free reagent water containing surrogate and internal standards. This is purged at ambient temperature. All samples with an expected concentration of >1.0 mg/kg should be analyzed by this method.

Mix the contents of the sample container with a narrow metal spatula. For sediments or soils and solid wastes that are insoluble in methanol, weigh 4 g (wet weight) of sample into a tared 20-mL vial. For waste that is soluble in methanol, tetraglyme, or PEG, weigh 1 g (wet weight) into a tared scintillation vial or culture tube or a 10-mL volumetric flask. Quickly add 9.0 mL of appropriate solvent then add 1.0 mL of a surrogate spiking solution to the vial, cap it, and shake it for 2 min.

METHANOL EXTRACT REQUIRED FOR ANALYSIS OF HIGH-CONCENTRATION SOILS OR SEDIMENTS

Approximate Concentration Range	Volume of Methanol Extract (a)
500–10,000 μg/kg	100 μL
1,000–20,000 μg/kg	50 μL
5,000–100,000 μg/kg	10 μL
25,000–500,000 μg/kg	100 μL of 1/50 dilution (b)

Calculate appropriate dilution factor for concentrations exceeding this table.

(a) The volume of methanol added to 5 mL of water being purged should be kept constant. Therefore, add to the 5-mL syringe whatever volume of methanol is necessary to maintain a volume of 100 μL added to the syringe.
(b) Dilute an aliquot of the methanol extract and then take 100 μL for analysis.

QUALITY CONTROL Demonstrate, through the analysis of a reagent water blank, that interferences from the analytical system, glassware, and reagents are under control. Blank samples should be carried through all stages of the sample preparation and measurement steps. For each analytical batch (up to 20 samples), a reagent blank, matrix spike, and matrix spike duplicate must be analyzed (the frequency of the spikes may be different for different monitoring programs). The blank and

spiked samples must be carried through all stages of the sample preparation and measurement steps. QC samples mentioned in the section on Interferences will also be needed as appropriate to those situations.

Matrix spiking standards should be prepared from volatile organic compounds which will be representative of the compounds being investigated. The recommended internal standards are chlorobenzene-d5, 1,4-difluorobenzene, 1,4-dichlorobenzene-d4, and pentafluorobenzene. Using stock standard solutions, prepare secondary dilution standards containing the compounds of interest, either singly or mixed together in methanol. Store them in a vial with no headspace for no more than one week. Surrogates recommended are toluene-d8, 4-bromofluorobenzene, and dibromofluoromethane. Each sample undergoing GC/MS analysis must be spiked with 10 µL of the surrogate spiking solution prior to analysis.

REFERENCE Test Methods for Evaluating Solid Waste (SW-846). U.S. EPA 1983, Method 8260A, Rev. 1, Nov. 1990. Office of Solid Waste, Washington, DC.

---

tert-Butylbenzene   **EPA Method 8260**
CAS #98-06-6

TITLE Volatile Organic Compounds by GC/MS: Capillary Column Technique

MATRIX This method is applicable to nearly all types of samples, regardless of water content, including groundwater, soils, and sediments.

METHOD SUMMARY Method 8260A covers 58 volatile organic compounds that are introduced into a gas chromatograph by the purge-and-trap method or by direct injection (in limited applications). Zero-grade helium is bubbled through a 5-mL solution at ambient temperature. Purged sample components are trapped in a tube containing suitable sorbent materials. When purging is complete, the sorbent tube is heated and backflushed with helium to desorb trapped sample components. The analytes are desorbed directly to a large bore capillary or cryofocussed on a capillary precolumn before being flash evaporated to a narrow bore capillary for analysis.

INTERFERENCES Major contaminant sources are volatile materials in the lab and impurities in the inert purging gas and in the sorbent trap. Interfering contamination may occur when a sample containing low concentrations of volatile organic compounds is analyzed immediately after a sample containing high concentrations of volatile organic compounds. After analysis of a sample containing high concentrations of volatile organic compounds, one or more calibration blanks should be analyzed to check for cross-contamination. Screening of the samples prior to purge-and-trap GC/MS analysis is highly recommended to prevent contamination of the system. This is especially true for soil and waste samples.

Special precautions must be taken to analyze for methylene chloride. The analytical and sample storage area should be isolated from all atmospheric sources of methylene chloride. All gas chromatography carrier gas lines and purge gas plumbing should be constructed from stainless steel or copper tubing. Laboratory clothing previously exposed to methylene chloride fumes during liquid-liquid extraction procedures can contribute to sample contamination.

Samples can also be contaminated by diffusion of volatile organics (particularly methylene chloride and fluorocarbons) through the septum seal during shipment and storage. A trip blank can serve as a check on such contamination.

INSTRUMENTATION GC/MS with a temperature-programmable chromatograph suitable for splitless injection equipped with variable constant differential flow controllers, a subambient oven controller, a purging device, sorbent trap, a thermal desorption apparatus and a capillary precolumn interface when using cryogenic cooling will be needed. The following GC columns may be used:

Column 1: 60 m × 0.75 mm I.D. capillary column coated with VOCOL, 1.5 µm film thickness.
Column 2: 30 m × 0.53 mm capillary column coated with DB-624 or VOCOL, 3 µm film thickness.
Column 3: 30 m × 0.32 mm I.D. capillary column coated with DB-5 or SE-54, 1-µm film thickness.

PRECISION & ACCURACY This method has been tested in a single lab using spiked water. Using a wide-bore capillary column, water was spiked at concentrations between 0.5 and 10 µg/L. Single lab accuracy and precision data are presented. The MDL actually achieved in a given analysis will vary depending on instrument sensitivity and matrix effects.

The MDL (a) in µg/L was 0.14.
The concentration range in µg/L was 0.5–10.
The mean accuracy (% of true value) was 102.
The precision as relative standard deviation was 7.3.

Note: The MDL is based on a 25-mL sample volume instead of a 5-mL sample volume.

SAMPLING METHOD
Liquid samples — Use a 40-mL glass screw-cap VOA vial with a Teflon®-faced silicone septum that has been prewashed, rinsed with distilled deionized water, and oven dried. If residual chlorine is present, collect the sample in a 4-oz soil VOA container which has been pre-preserved with 4 drops of 10% sodium thiosulfate. Mix gently and transfer the sample to a 40-mL VOA vial. Collect bubble-free samples in duplicate and seal each sample in a separate plastic bag.

Soils, sediments, and sludges — Use an 8-oz widemouth glass bottle with Teflon®-faced silicone septum that has been prewashed, rinsed with distilled deionized water, and oven dried. **Do not** heat the septum for more than 1 h. Tap slightly to eliminate any free air space. Collect samples in duplicate and seal each one in a separate plastic bag.

SAMPLE PRESERVATION
Liquid samples — Add 4 drops of concentrated HCL, cool to 4°C and store in a solvent-free refrigerator.

Soils, sediments, and sludges — Cool samples to 4°C and store in a solvent-free refrigerator.

MHT The maximum holding time of any sample (liquids, soils, sediments, and sludges) is 14 days.

SAMPLE PREPARATION

Liquid samples — Remove the plunger from a 5-mL syringe and carefully pour the sample into the syringe barrel to just short of overflowing. Replace the syringe plunger and compress the sample. Open the syringe valve and vent any residual air while adjusting the sample volume to 5.0 mL. If there is only one volatile organic analysis (VOA) vial, a second syringe should be filled at this time to protect against possible loss of sample integrity. Add 10 µL of surrogate spiking solution and 10 µL of internal standard spiking solution through the valve bore of the 5-mL syringe, then close the valve. The surrogate and internal standards may be mixed and added as a single spiking solution.

Sediments, soils, and waste samples — All samples of this type should be screened by GC analysis using a headspace method (EPA Method 3810) or the hexadecane extraction and screening method (EPA Method 3820). Use the screening data to determine whether to use the low-concentration method (0.005–1 mg/kg) or the high-concentration method (>1 mg/kg).

Low-concentration method — The low-concentration method is based on purging a heated sediment or soil sample mixed with organic-free reagent water containing the surrogate and internal standards. Analyze all reagent blanks and standards under the same conditions as the samples.

Use a 5-g sample if the expected concentration is <0.1 mg/kg or a 1-g sample for expected concentrations between 0.1 and 1 mg/kg. Mix the contents of the sample container with a narrow metal spatula. Weigh the amount of the sample into a tared purge device. Add the spiked water to the purge device, which contains the weighed amount of sample, and connect the device to the purge-and-trap system.

High-concentration method — This method is based on extracting the sediment or soil with methanol. A waste sample is either extracted or diluted, depending on its solubility in methanol. Wastes that are insoluble in methanol are diluted with reagent tetraglyme or possibly polyethylene glycol (PEG). An aliquot of the extract is added to organic-free reagent water containing surrogate and internal standards. This is purged at ambient temperature. All samples with an expected concentration of >1.0 mg/kg should be analyzed by this method.

Mix the contents of the sample container with a narrow metal spatula. For sediments or soils and solid wastes that are insoluble in methanol, weigh 4 g (wet weight) of sample into a tared 20-mL vial. For waste that is soluble in methanol, tetraglyme, or PEG, weigh 1 g (wet weight) into a tared scintillation vial or culture tube or a 10-mL volumetric flask. Quickly add 9.0 mL of appropriate solvent then add 1.0 mL of a surrogate spiking solution to the vial, cap it, and shake it for 2 min.

METHANOL EXTRACT REQUIRED FOR ANALYSIS OF HIGH-CONCENTRATION SOILS OR SEDIMENTS

Approximate Concentration Range	Volume of Methanol Extract (a)
500–10,000 µg/kg	100 µL
1,000–20,000 µg/kg	50 µL
5,000–100,000 µg/kg	10 µL
25,000–500,000 µg/kg	100 µL of 1/50 dilution (b)

Calculate appropriate dilution factor for concentrations exceeding this table.

(a) The volume of methanol added to 5 mL of water being purged should be kept constant. Therefore, add to the 5-mL syringe whatever volume of methanol is necessary to maintain a volume of 100 µL added to the syringe.

(b) Dilute an aliquot of the methanol extract and then take 100 µL for analysis.

QUALITY CONTROL Demonstrate, through the analysis of a reagent water blank, that interferences from the analytical system, glassware, and reagents are under control. Blank samples should be carried through all stages of the sample preparation and measurement steps. For each analytical batch (up to 20 samples), a reagent blank, matrix spike, and matrix spike duplicate must be analyzed (the frequency of the spikes may be different for different monitoring programs). The blank and spiked samples must be carried through all stages of the sample preparation and measurement steps. QC samples mentioned in the section on Interferences will also be needed as appropriate to those situations.

Matrix spiking standards should be prepared from volatile organic compounds which will be representative of the compounds being investigated. The recommended internal standards are chlorobenzene-d5, 1,4-difluorobenzene, 1,4-dichlorobenzene-d4, and pentafluorobenzene. Using stock standard solutions, prepare secondary dilution standards containing the compounds of interest, either singly or mixed together in methanol. Store them in a vial with no headspace for no more than one week. Surrogates recommended are toluene-d8, 4-bromofluorobenzene, and dibromofluoromethane. Each sample undergoing GC/MS analysis must be spiked with 10 µL of the surrogate spiking solution prior to analysis.

REFERENCE Test Methods for Evaluating Solid Waste (SW-846). U.S. EPA 1983, Method 8260A, Rev. 1, Nov. 1990. Office of Solid Waste, Washington, DC.

---

n-Butylbenzene   **EPA Method 503.1**
CAS #104-51-8

TITLE Aromatic & Unsaturated VOCs

MATRIX Drinking water (finished or in Water any treatment stage) and raw source water.

APPLICATION Method covers 28 aromatic and unsaturated VOCs. An inert gas is bubbled through a 5-mL water sample. Purged sample components are trapped in tube of sorbent materials. When purging is complete, sorbent tube is heated and backflushed with inert gas to desorb trapped sample onto a packed GC column.

INTERFERENCES During analysis, major contaminant sources are volatile materials in the lab and impurities in purging gas and sorbent trap. With high and low level samples, there can be carryover contamination. Excess water causes a negative baseline deflection.

INSTRUMENTATION Purge and Trap GC w/photoionization detector. (Two GC columns are recommended); Column 1: 5% SP-1200 and 1.75% Bentone 34 on Supelcoport; Column 2: 1,2,3-tris(2-cyanoethoxy)propane on Chromosorb W.

RANGE 2.2–600 µg/L. (drinking water).

MDL 0.02 µg/L in water

PRECISION RSD = 15.7% at 0.40 µg/L conc.; 7 samples

ACCURACY Average recovery = 78% at 0.40 µg/L conc.; 7 samples

SAMPLING METHOD Use a 40- to 120-mL screw-cap vial (prewashed with detergent, rinsed with distilled water and oven dried at 105°C) with a PTFE-faced silicone septum. If residual chlorine is in the water add about 25 mg of ascorbic acid to each vial before sample collection. Collect bubble-free samples.

STABILITY Cool to 4°C; HCl to pH <2.

MHT 14 days.

QUALITY CONTROL As initial demonstration of lab accuracy and precision, analyze 4 to 7 replicates of a lab fortified blank containing the analyte at 0.1–5 µg/L. Collect all samples in duplicate.

REFERENCE Method 503.1, Volatile Aromatic and Unsaturated Organic Compounds in H2O by Purge and Trap GC, EPA 600/4-88/039.

---

sec-Butylbenzene　　　　　　　　　　EPA Method 503.1
CAS #135-98-8

TITLE Aromatic & Unsaturated VOCs

MATRIX Drinking water (finished or in Water any treatment stage) and raw source water.

APPLICATION Method covers 28 aromatic and unsaturated VOCs. An inert gas is bubbled through a 5-mL water sample. Purged sample components are trapped in tube of sorbent materials. When purging is complete, sorbent tube is heated and backflushed with inert gas to desorb trapped sample onto a packed GC column.

INTERFERENCES During analysis, major contaminant sources are volatile materials in the lab and impurities in purging gas and sorbent trap. With high and low level samples, there can be carryover contamination. Excess water causes a negative baseline deflection.

INSTRUMENTATION Purge and Trap GC w/photoionization detector. (Two GC columns are recommended); Column 1: 5% SP-1200 and 1.75% Bentone 34 on Supelcoport; Column 2: 1,2,3-tris(2-cyanoethoxy)propane on Chromosorb W.

RANGE 2.2–600 µg/L. (drinking water).

MDL 0.02 µg/L in water

PRECISION RSD = 11.0% at 0.40 µg/L conc.; 7 samples

ACCURACY Average recovery = 80% at 0.40 µg/L conc.; 7 samples

SAMPLING METHOD Use a 40- to 120-mL screw-cap vial (prewashed with detergent, rinsed with distilled water and oven dried at 105°C) with a PTFE-faced silicone septum. If residual chlorine is in the water add about 25 mg of ascorbic acid to each vial before sample collection. Collect bubble-free samples.

STABILITY Cool to 4°C; HCl to pH <2.

MHT 14 days.

QUALITY CONTROL As initial demonstration of lab accuracy and precision, analyze 4 to 7 replicates of a lab fortified blank containing the analyte at 0.1–5 µg/L. Collect all samps in duplicate.

REFERENCE Method 503.1, Volatile Aromatic & Unsaturated Organic Compounds in H2O by Purge and Trap GC, EPA 600/4-88/039.

---

tert-Butylbenzene　　　　　　　　　　EPA Method 503.1
CAS #98-06-6

TITLE Aromatic & Unsaturated VOCs

MATRIX Drinking water (finished or in Water any treatment stage) and raw source water.

APPLICATION Method covers 28 aromatic and unsaturated VOCs. An inert gas is bubbled through a 5-mL water sample. Purged sample components are trapped in tube of sorbent materials. When purging is complete, sorbent tube is heated and backflushed with inert gas to desorb trapped sample onto a packed GC column.

INTERFERENCES During analysis, major contaminant sources are volatile materials in the lab and impurities in purging gas and sorbent trap. With high and low level samples, there can be carryover contamination. Excess water causes a negative baseline deflection.

INSTRUMENTATION Purge and Trap GC w/photoionization detector. (Two GC columns are recommended); Column 1: 5% SP-1200 and 1.75% Bentone 34 on Supelcoport; Column 2: 1,2,3-tris(2-cyanoethoxy)propane on Chromosorb W.

RANGE 2.2–600 µg/L. (drinking water).

MDL 0.006 µg/L in water

PRECISION RSD = 8.7% at 0.40 µg/L conc.; 7 samples

ACCURACY Average recovery = 88% at 0.40 µg/L conc.; 7 samples

SAMPLING METHOD Use a 40- to 120-mL screw-cap vial (prewashed with detergent, rinsed with distilled water and oven dried at 105°C) with a PTFE-faced silicone septum. If residual chlorine is in the water add about 25 mg of ascorbic acid to each vial before sample collection. Collect bubble-free samples.

STABILITY Cool to 4°C; HCl to pH <2.

MHT 14 days.

QUALITY CONTROL As initial demonstration of lab accuracy and precision, analyze 4 to 7 replicates of a lab fortified blank containing the analyte at 0.1–5 µg/L. Collect all samples in duplicate.

REFERENCE Method 503.1, Volatile Aromatic & Unsaturated Organic Compounds in H2O by Purge and Trap GC, EPA 600/4-88/039.

C

Cadmium
CAS #7440-43-9

EPA Method 6010

TITLE Inductively Coupled Plasma-Atomic Emission Spectroscopy

MATRIX This method is applicable to the determination of trace elements, including metals, in groundwater, soils, sludges, sediments, and other solid wastes. All matrices require digestion prior to analysis. The method of standard addition must be used for the analysis of all sample digests unless either serial dilution or matrix spike addition demonstrates it is not required.

METHOD SUMMARY Method 6010 covers 25 elements using ICP analysis. It measures element-emitted light by optical spectrometry. Samples, following an appropriate acid digestion, are nebulized and the resulting aerosol is transported to the plasma torch. Element-specific atomic line emission spectra are produced by a radio-frequency inductively coupled plasma.

INTERFERENCES Interferences may be categorized as spectral or non-spectral. Spectral interferences are caused by overlap of a spectral line from another element, unresolved overlap of molecular band spectra, background contribution from continuous or recombination phenomenon, and stray light from the line emission of high concentration elements. Non-spectral interferences include physical and chemical interferences. Physical interferences are effects associated with the sample nebulization and transport processes. Changes in viscosity and surface tension can cause significant inaccuracies. Chemical interferences include molecular compound formation, ionization effects, and solute vaporization effects. Normally these effects are not significant and can be minimized by careful selection of operating conditions. Chemical interferences are highly dependent on matrix type and the specific analyte element.

INSTRUMENTATION An inductively coupled argon plasma emission spectrometer (ICP) capable of background correction is required.

PRECISION & ACCURACY Detection limits, sensitivity, and optimum ranges of the metals will vary with the matrices and model of the spectrometer. In a single lab evaluation, seven wastes were analyzed for 22 elements. The mean percent relative standard deviation from triplicate analyses for all elements and wastes was 9 ± 2%. The mean percent recovery of spiked elements for all wastes was 93 ± 6%. Spike levels ranged from 100 µg/L to 100 mg/L. The wastes included sludges and industrial wastewaters.

Estimated instrument detection limit in µg/L is 4.
Spiked concentration in µg/L is 2.5.
Mean reported value in µg/L is 2.9.
Precision as RSD % is 16.

SAMPLING METHOD Samples should be collected in borosilicate glass, linear polyethylene, polypropylene, or Teflon® bottles that have been prewashed with detergent and tap water, and rinsed with 1:1 nitric acid and tap water or 1:1 hydrochloric acid and tap water. Collect at least 2 g of solids and 200 mL of aqueous samples.

SAMPLE PRESERVATION Add nitric acid to make the samples pH <2.

MHT The maximum holding time for properly preserved samples is 6 months.

SAMPLE PREPARATION Preliminary treatment of most matrices is necessary because of the complexity and variability of sample matrices. Water samples that have been prefiltered and acidified will not need acid digestion. Methods for acid digestion of waters for total recoverable or dissolved metals, acid digestions of aqueous samples and extracts for total metals, and acid digestion of sediments, sludges, and soils are summarized below.

Total recoverable or dissolved metals in water — To prepare surface and groundwater samples for determination of total recoverable and dissolved metals, a 100-mL aliquot of well-mixed sample is acidified with concentrated nitric acid and concentrated hydrochloric acid, then heated until the volume is reduced to 15–20 mL. Adjust the final volume to 100 mL with reagent water.

Total metals in aqueous samples, soil and sediment extracts — To prepare aqueous samples, soil and sediment extracts, and wastes that contain suspended solids, a 100-mL aliquot is made acidic with concentrated nitric acid and the solution is evaporated to about 5 mL on a hot plate. Continue heating and adding additional acid until sample digestion is complete, which is usually indicated when the digestate is light in color or does not change in appearance. Evaporate the solution to about 3 mL and cool it and add a small quantity of 1:1 hydrochloric acid (10 mL/100 mL of final solution). Cover the beaker and reflux for 15 min. Wash down the beaker walls and filters or centrifuge the sample to remove silicates and other insoluble material. Filter the sample and adjust the final volume to 100 mL with reagent water and the final acid concentration to 10%.

Sediments, sludges, and soils — To prepare sediments, sludges and soil samples, transfer 1–2 g to a conical beaker and add 10 mL of 1:1 nitric acid, mix the slurry, and cover it with a watch glass. Heat the sample and reflux for 10–15 min without boiling. Allow it to cool, then add 5 mL of concentrated nitric acid and reflux for 30 min. Repeat last step and then allow the solution to evaporate to 5 mL without boiling. Cool and add 2 mL of water and 3 mL of 30% hydrogen peroxide. Cover and place the beaker on the hot plate. Heat and add 30% hydrogen peroxide in 1-mL aliquots with warming until the effervescence is minimal but do not add more than a total of 10 mL of 30% hydrogen peroxide. If the sample is being prepared for the analysis of Ag, Al, As, Ba, Be, Ca, Cd, Co, Cr, Cu, Fe, K, Mg, Mn, Mo, Na, Ni, Os, Pb, Se, Tl, V, and Zn, then add 5 mL of concentrated hydrochloric acid and 10 mL of water and return the covered beaker to a hot plate for 15 min of additional refluxing without boiling. Dilute the sample to a 100 mL volume with

water after cooling and filter or centrifuge to remove particulates.

QUALITY CONTROL Laboratory control samples must be analyzed for each analytical method. A method blank should be analyzed with each batch of samples. The effect of the matrix on method performance must be demonstrated: when appropriate, there should be at least one matrix spike and either one matrix duplicate or one matrix spike duplicate per analytical batch. The bias and precision of the method, as well as the method detection limit for each specific matrix type, must be measured.

Dilute and reanalyze samples that are more concentrated than the linear calibration limit. Employ a minimum of one reagent blank per sample batch to determine if contamination or any memory effects are occurring. Whenever a new or unusual sample matrix is encountered, perform either a serial dilution test or a matrix spike addition test to ensure that neither positive or negative interferences are operating on any of the analyte elements. Check the instrument standardization by verifying calibration every 10 samples using a calibration blank and a check standard.

REFERENCE Test Methods for Evaluating Solid Waste (SW-846). U.S. EPA. 1983. Method 6010, Rev. 0, Sept. 1986. Office of Solid Wastes, Washington, DC.

Cadmium   **EPA Method 200.7**
CAS #7440-43-9

TITLE Inductively Coupled Plasma (ICP)

MATRIX Dissolved, suspended, or total element in drinking and surface waters and in domestic and industrial wastewaters.

APPLICATION The method covers the determination of 25 metals. Dissolved elements are determined in filtered and acidified samples after appropriate digestion (which increases dissolved solids). Its primary advantage is that ICP instruments allow simultaneous or rapid sequential determination of many elements in a short time. Samples are first nebulized and the aerosol is transported to a plasma torch in which element specific atomic line emission spectra are produced by a radio frequency inductively coupled plasma. Background correction is required for trace element detection except in the case of line broadning.

INTERFERENCES There are spectral, physical, and chemical interferences. The primary disadvantage of ICP instruments is background radiation from other elements and the plasma gases (spectral interferences). Changes in sample viscosity and surface tension with samples containing high dissolved solids (especially those exceeding 1500 mg/L) or high acid concentrations can cause physical interferences. Ionization effects, solute vaporization and molecular compound formation can cause chemical interferences. Iron and nickel can cause interference at the 100 mg/L level.

INSTRUMENTATION Inductively coupled argon plasma emission spectroscopy. 266.502 nm Wavelength

RANGE Not listed.

MDL 4 µg/L.

PRECISION SD = 12% Mean at true value 50 µg/L.

ACCURACY Mean recovery = 93% ± 6% of spiked elements for all wastes.

SAMPLING METHOD Wash sample container with detergent and tap water, rinse with 1 + 1 nitric acid and tap water, then rinse with 1 + 1 hydrochloric acid and tap water, then rinse with deionized, distilled water in that order. Perform any filtration or acid preservation steps when the sample is collected or as soon as possible thereafter.

STABILITY Cool samples to 4°C.

MHT 24 h.

QUALITY CONTROL Mixed calibration standards, an instrument check standard, and an interference check solution are used in addition to a quality control sample. The quality control sample should be prepared in the same acid matrix as the calibration standards at 10 times the instrumental detection limits and in accordance with the instructions provided by the supplier. Furthermore, two types of blanks are required: a calibration blank and a reagent blank.

REFERENCE Method 200.7, U.S. EPA, EMSL-Cincinnati, OH, Nov. 1980

Cadmium   **EPA Method 7130**
CAS #7440-43-9

TITLE Atomic Absorption, (AA)

MATRIX Drinking, Surface and Direct Aspiration Saline Waters, Wastewater

APPLICATION Sample is aspirated and atomized in a flame. A light beam from a cadmium hollow cathode lamp is directed through the flame into a monochromator and onto a detector. Since wavelength of light beam is specific for cadmium, light energy absorbed by detector is measure of cadmium.

INTERFERENCES The most troublesome type is chemical, caused by lack of absorption of atoms bound in molecular combination in the flame. High dissolved solids in sample may result in nonatomic absorbance interference. Non specific absorption and light scattering interfere.

INSTRUMENTATION Atomic absorption spectrometer. Cadmium hollow cathode lamp (228.8 nm Wavelength).

RANGE 0.05–2 mg/L

MDL 0.005 mg/L

PRECISION Standard deviation = 21 µg/L at 71 µg/L (true value) 74 labs

ACCURACY As bias = -2.2% at 71 µg/L (true value) 74 labs

SAMPLING METHOD Use glass or plastic containers. Collect 200 g of solids and 600 mL of liquid samples.

STABILITY Cool solid samples to 4°C and analyze as soon as possible. Add nitric acid to liquid samples to pH <2.

MHT 6 months.

QUALITY CONTROL At least one duplicate and one spike sample should be run every 20 samples or with each matrix type to verify precision of the method. For 20 or more samples per day, verify working standard curve. Run an additional standard at or near mid-range every 10 samples.

REFERENCE Method 7130, SW-846, 3rd ed., Nov.1986.

Cadmium   EPA Method 7131
CAS #7440-43-9

TITLE Atomic Absorption, (AA) Furnace Technique

MATRIX Wastes, mobility procedure extracts, soils and groundwater

APPLICATION Aqueous samples, EP extracts, industrial wastes, soils, sludges, sediments, and solid wastes require digestion before analysis. An aliquot of sample is placed in the graphite tube in the furnace and slowly evaporated, charred and atomized. Absorption of lamp radiation during atomization is proportional to the cadmium concentration.

INTERFERENCES The furnace technique is subject to chemical interferences. Composition of sample matrix can effect analysis. Cd analysis can suffer severe nonspecific absorption and light scattering; background correction is required. Use cadmium-free plastic pipette tips.

INSTRUMENTATION Atomic absorption spectrometer. Cadmium hollow cathode lamp or electrodeless discharge lamp. Graphite furnace. Strip-chart recorder

RANGE 0.5–10 µg/L

MDL 0.1 µg/L (228.8 nm Wavelength)

PRECISION Standard deviation = ±0.10, 0.16, 0.33 at 2.5, 5.0, 10.0 µg Cd/L

ACCURACY Recoveries = 96, 99, 98% at 2.5, 5.0, 10.0 µg Cd/L

SAMPLING METHOD Use glass or plastic containers. Collect 200 g of solids and 600 mL of liquid samples.

STABILITY Cool solid samples to 4°C and analyze as soon as possible. Add nitric acid to liquid samples to pH <2.

MHT 6 months.

QUALITY CONTROL At least one duplicate and one spike sample should be run every 20 samples, or with each matrix type to verify method precision. If 20 or more samples are run a day, run a standard (at or near mid-range) every 10 samples.

REFERENCE Method 7131, SW-846, 3rd ed., Nov.1986.

Calcium   EPA Method 6010
CAS #7440-70-2

TITLE Inductively Coupled Plasma-Atomic Emission Spectroscopy

MATRIX This method is applicable to the determination of trace elements, including metals, in groundwater, soils, sludges, sediments, and other solid wastes. All matrices require digestion prior to analysis. The method of standard addition must be used for the analysis of all sample digests unless either serial dilution or matrix spike addition demonstrates it is not required.

METHOD SUMMARY Method 6010 covers 25 elements using ICP analysis. It measures element-emitted light by optical spectrometry. Samples, following an appropriate acid digestion, are nebulized and the resulting aerosol is transported to the plasma torch. Element-specific atomic line emission spectra are produced by a radio-frequency inductively coupled plasma.

INTERFERENCES Interferences may be categorized as spectral or non-spectral. Spectral interferences are caused by overlap of a spectral line from another element, unresolved overlap of molecular band spectra, background contribution from continuous or recombination phenomenon, and stray light from the line emission of high concentration elements. Non-spectral interferences include physical and chemical interferences. Physical interferences are effects associated with the sample nebulization and transport processes. Changes in viscosity and surface tension can cause significant inaccuracies. Chemical interferences include molecular compound formation, ionization effects, and solute vaporization effects. Normally these effects are not significant and can be minimized by careful selection of operating conditions. Chemical interferences are highly dependent on matrix type and the specific analyte element.

INSTRUMENTATION An inductively coupled argon plasma emission spectrometer (ICP) capable of background correction is required.

PRECISION & ACCURACY Detection limits, sensitivity, and optimum ranges of the metals will vary with the matrices and model of the spectrometer. In a single lab evaluation, seven wastes were analyzed for 22 elements. The mean percent relative standard deviation from triplicate analyses for all elements and wastes was 9 ± 2%. The mean percent recovery of spiked elements for all wastes was 93 ± 6%. Spike levels ranged from 100 µg/L to 100 mg/L. The wastes included sludges and industrial wastewaters.

Estimated instrument detection limit in µg/L is 10.
Spiked concentration in µg/L is not listed.
Mean reported value in µg/L is not listed.
Precision as RSD % is not listed.

SAMPLING METHOD Samples should be collected in borosilicate glass, linear polyethylene, polypropylene, or Teflon® bottles that have been prewashed with detergent and tap water, and rinsed with 1:1 nitric acid and tap water or 1:1 hydrochloric acid and tap water. Collect at least 2 g of solids and 200 mL of aqueous samples.

SAMPLE PRESERVATION Add nitric acid to make the samples pH <2.

MHT The maximum holding time for properly preserved samples is 6 months.

SAMPLE PREPARATION Preliminary treatment of most matrices is necessary because of the complexity and variability of sample matrices. Water samples that have been prefiltered and acidified will not need acid digestion. Methods for acid digestion of waters for total recoverable or dissolved metals, acid digestions of aqueous samples and extracts for total metals, and acid digestion of sediments, sludges, and soils are summarized below.

Total recoverable or dissolved metals in water — To prepare surface and groundwater samples for determination of total recoverable and dissolved metals, a 100-mL aliquot of well-mixed sample is acidified with concentrated nitric acid and concentrated hydrochloric acid, then heated until the volume is reduced to 15–20 mL. Adjust the final volume to 100 mL with reagent water.

Total metals in aqueous samples, soil and sediment extracts — To prepare aqueous samples, soil and sediment extracts, and wastes that contain suspended solids, a 100-mL aliquot is made acidic with concentrated nitric acid and the solution is evaporated to about 5 mL on a hot plate. Continue heating and adding additional acid until sample digestion is complete, which is usually indicated when the digestate is light in color or does not change in appearance. Evaporate the solution to about 3 mL and cool it and add a small quantity of 1:1 hydrochloric acid (10 mL/100 mL of final solution). Cover the beaker and reflux for 15 min. Wash down the beaker walls and filters or centrifuge the sample to remove silicates and other insoluble material. Filter the sample and adjust the final volume to 100 mL with reagent water and the final acid concentration to 10%.

Sediments, sludges, and soils — To prepare sediments, sludges and soil samples, transfer 1–2 g to a conical beaker and add 10 mL of 1:1 nitric acid, mix the slurry, and cover it with a watch glass. Heat the sample and reflux for 10–15 min without boiling. Allow it to cool, then add 5 mL of concentrated nitric acid and reflux for 30 min. Repeat last step and then allow the solution to evaporate to 5 mL without boiling. Cool and add 2 mL of water and 3 mL of 30% hydrogen peroxide. Cover and place the beaker on the hot plate. Heat and add 30% hydrogen peroxide in 1-mL aliquots with warming until the effervescence is minimal but do not add more than a total of 10 mL of 30% hydrogen peroxide. If the sample is being prepared for the analysis of Ag, Al, As, Ba, Be, Ca, Cd, Co, Cr, Cu, Fe, K, Mg, Mn, Mo, Na, Ni, Os, Pb, Se, Tl, V, and Zn, then add 5 mL of concentrated hydrochloric acid and 10 mL of water and return the covered beaker to a hot plate for 15 min of additional refluxing without boiling. Dilute the sample to a 100 mL volume with water after cooling and filter or centrifuge to remove particulates.

QUALITY CONTROL Laboratory control samples must be analyzed for each analytical method. A method blank should be analyzed with each batch of samples. The effect of the matrix on method performance must be demonstrated: when appropriate, there should be at least one matrix spike and either one matrix duplicate or one matrix spike duplicate per analytical batch. The bias and precision of the method, as well as the method detection limit for each specific matrix type, must be measured.

Dilute and reanalyze samples that are more concentrated than the linear calibration limit. Employ a minimum of one reagent blank per sample batch to determine if contamination or any memory effects are occurring. Whenever a new or unusual sample matrix is encountered, perform either a serial dilution test or a matrix spike addition test to ensure that neither positive or negative interferences are operating on any of the analyte elements. Check the instrument standardization by verifying calibration every 10 samples using a calibration blank and a check standard.

REFERENCE Test Methods for Evaluating Solid Waste (SW-846). U.S. EPA. 1983. Method 6010, Rev. 0, Sept. 1986. Office of Solid Wastes, Washington, DC.

Calcium   **EPA Method 200.7**
CAS #7440-70-2

TITLE Inductively Coupled Plasma (ICP)

MATRIX Dissolved, suspended or total element in drinking and surface waters and in domestic and industrial wastewaters.

APPLICATION The method covers the determination of 25 metals. Dissolved elements are determined in filtered and acidified samples after appropriate digestion (which increases dissolved solids). Its primary advantage is that ICP instruments allow simultaneous or rapid sequential determination of many elements in a short time. Samples are first nebulized and the aerosol is transported to a plasma torch in which element specific atomic line emission spectra are produced by a radio frequency inductively coupled plasma. Background correction is required for trace element detection except in the case of line broadning.

INTERFERENCES There are spectral, physical, and chemical interferences. The primary disadvantage of ICP instruments is background radiation from other elements and the plasma gases (spectral interferences). Changes in sample viscosity and surface tension with samples containing high dissolved solids (especially those exceeding 1500 mg/L) or high acid concentrations can cause physical interferences. Ionization effects, solute vaporization and molecular compound formation can cause chemical interferences. Chromium, iron, magnesium, manganese, thallium and vanadium can cause interference at the 100 mg/L level.

INSTRUMENTATION Inductively coupled argon plasma emission spectroscopy. 317.933 nm Wavelength

RANGE Not listed.

MDL 10 μg/L.

PRECISION Not listed.

ACCURACY Mean recovery = 93% ± 6% of spiked elements for all wastes.

SAMPLING METHOD Wash sample container with detergent and tap water, rinse with 1 + 1 nitric acid and tap water, then rinse with 1 + 1 hydrochloric acid and tap water, then rinse with deionized, distilled water in that order. Perform any filtration or acid preservation steps when the sample is collected or as soon as possible thereafter.

STABILITY Cool samples to 4°C.

MHT 24 h.

QUALITY CONTROL Mixed calibration standards, an instrument check standard, and an interference check solution are used in addition to a quality control sample. The quality control sample should be prepared in the same acid matrix as the calibration standards at 10 times the instrumental detection limits and in accordance with the instructions provided by the supplier. Furthermore, two types of blanks are required: a calibration blank and a reagent blank.

REFERENCE Method 200.7, U.S. EPA, EMSL-Cincinnati, OH, Nov. 1980

Calcium EPA Method 215.1
CAS #7440-70-2

TITLE Metals (Total and Dissolved)AAS, Direct Aspiration

MATRIX Drinking, Surface and Saline waters. Wastewater.

APPLICATION Date issued 1971. Editorial Rev. 1974. Sample is aspirated and atomized in a flame. Light beam from hollow cathode (made of Ca) lamp is directed through flame into monochromator, then to detector which measures amount absorbed light. Energy absorbed is proportional to Ca.

INTERFERENCES Phosphate, sulfate and aluminum interfere; are masked by lanthanum addition. Low Ca values, if sample pH > 7. (Prepare in dilute HCl solution). Low Ca values with magnesium conc >1000 mg/L. Control with interferences using large amounts of alkali for samples and standards.

INSTRUMENTATION AAS. Calcium (Ca) hollow cathode lamp. Burner. Pipettes. Strip chart recorder.

RANGE 0.2–7 mg/L at 422.7 nm Wavelength.

MDL 0.001 mg/L.

PRECISION SD = ±(0.3 and 0.6) at 9.0 and 36 mg Ca/L.

ACCURACY Recoveries = 99% at 9.0 and 36 mg Ca/L.

SAMPLING METHOD Plastic or glass (prewashed).

STABILITY HNO_3 to pH <2.

MHT 6 months.

QUALITY CONTROL After calibration curve composed of a minimum of a reagent blank and 3 standards has been prepared, subsequent calibration curves must be verified by use of at least a reagent blank and one standard near MCL. Must check within 10% of original curve. (For drinking water analysis)

REFERENCE EPA Methods for the Chemical Analysis of Water and Wastes, EPA-600/4-79-020, U.S. EPA, EMSL, 1979.

Calcium EPA Method 7140
CAS #7440-70-2

TITLE Atomic Absorption, (AA)

MATRIX Drinking, Surface and Direct Aspiration Saline Waters, Wastewater

APPLICATION Sample is aspirated and atomized in a flame. A light beam from a calcium hollow cathode lamp is directed through the flame into a monochromator and onto a detector. Since wavelength of light beam is specific for calcium, light energy absorbed by detector is measure of calcium.

INTERFERENCES The most troublesome type is chemical, caused by lack of absorption of atoms bound in molecular combination in the flame. High dissolved solids in sample may result in nonatomic absorbance interference. Add lanthanum to prevent complexing problems.

INSTRUMENTATION Atomic absorption spectrometer. Calcium hollow cathode lamp. (422.7 nm Wavelength).

RANGE 0.2–7 mg/L

MDL 0.01 mg/L

PRECISION Standard deviation = ±0.3 and 0.6 at 9.0 and 36 mg Ca/L

ACCURACY Recoveries = 99 and 99% at 9.0 and 36 mg Ca/L

SAMPLING METHOD Use glass or plastic containers. Collect 200 g of solids and 600 mL of liquid samples.

STABILITY Cool solid samples to 4°C and analyze as soon as possible. Add nitric acid to liquid samples to pH <2.

MHT 6 months.

QUALITY CONTROL At least one duplicate and one spike sample should be run every 20 samples or with each matrix type to verify precision of the method. For 20 or more samples per day, verify working standard curve. Run an additional standard at or near mid-range every 10 samples.

REFERENCE Method 7140, SW-846, 3rd ed., Nov. 1986.

Captafol EPA Method 8270
CAS #2425-06-1

TITLE Semivolatile Organic Compounds by GC/MS

MATRIX This method is used to determine the concentration of semivolatile organic compounds in extracts prepared from all types of solid waste matrices, soils, and groundwater.

Although surface waters are not specifically mentioned, this method should be applicable to water samples from rivers, lakes, etc.

METHOD SUMMARY This method covers 259 semivolatile organic compounds. In very limited applications direct injection of the sample into the GC/MS system may be appropriate, but this results in very high detection limits (approximately 10,000 µg/L). Typically, a 1-L liquid sample, containing surrogate, and matrix spiking standards, is extracted in a continuous extractor first under acid conditions and then under basic conditions. Typically 30 g of a solid sample, containing surrogate, and matrix spiking standards, is extracted ultrasonically. After concentrating the extract to 1 mL it is spiked with 10 µL of an internal standard solution just prior to analysis by GC/MS. The volume injected should contain about 100 ng of base/neutral and 200 ng of acid surrogates (for a 1-µL injection). Analysis is performed by GC/MS using a capillary GC column.

INTERFERENCES Raw GC/MS data from all blanks, samples, and spikes must be evaluated for interferences. Contamination by carryover can occur whenever high-concentration and low-concentration samples are sequentially analyzed. To reduce carryover, the sample syringe must be rinsed out between samples with solvent. Whenever an unusually concentrated sample is encountered, it should be followed by the analysis of blank solvent to check for cross-contamination.

INSTRUMENTATION A GC/MS and a data system are required. The GC column used is a 30 m × 0.25 mm I.D. (or 0.32 mm I.D.) 1um film thickness silicone-coated fused silica capillary column. A continuous liquid-liquid extractor equipped with Teflon® or glass connection joints and stopcocks requiring no lubrication, a K-D concentrating apparatus, water bath, and an ultrasonic disrupter with a minimum power of 300 W and with pulsing capability are also required.

PRECISION & ACCURACY The estimated quantitation limit (EQL) of Method 8270B for determining an individual compound is approximately 1 mg/kg (wet weight) for soil or sediment samples, 1–200 mg/kg for wastes (dependent on matrix and method of preparation), and 10 µg/L for groundwater samples. EQLs will be proportionately higher for sample extracts that require dilution to avoid saturation of the detector.

The EQL(b) for groundwater in µg/L is 20.
The EQL (a, b) for low concentrations in soil and sediment in µg/kg is not determined.
Accuracy as µg/L is not listed.
Overall precision in µg/L is not listed.

(a) *EQLs listed for soil/sediment are based on wet weight. Normally data is reported in a dry-weight basis; therefore, EQLs will be higher based on the % dry weight of each sample. This calculation is based on a 30 g sample and gel permeation chromatography cleanup.*
(b) *Sample EQLs are highly matrix-dependent. The EQLs are provided for guidance and may not always be achievable.*
$C =$ *True value for concentration, in µg/L.*
$X =$ *Average recovery found for measurements of samples containing a concentration of C, in µg/L.*

ESTIMATED QUANTITATION LIMIT FOR OTHER MATRICES

Other Matrices	Factor (a)
High-concentration soil and sludges by sonicator	7.5
Non-water miscible waste	75

(a) *EQL for other matrices = [EQL for low soil/sediment] × [Factor]. This estimated EQL is similar to an EPA "Practical Quantitation Limit."*

SAMPLING METHOD
Liquid samples — Use a 1 or 2½ gallon amber glass bottle with a screw-top Teflon®-lined cover that has been prewashed with detergent and rinsed with distilled water and methanol (or isopropanol).

Soils, sediments, or sludges — Use an 8-oz. widemouth glass with a screw-top Teflon®-lined cover that has been prewashed with detergent and rinsed with distilled water and methanol (or isopropanol).

SAMPLE PRESERVATION
Liquid samples — If residual chlorine is present, add 3 mL of 10% sodium thiosulfate per gallon, cool to 4°C and store in a solvent-free refrigerator until analysis; if chlorine is not present, then eliminate the sodium thiosulfate addition.

Soils, sediments, or sludges — Cool samples to 4°C and store in a solvent-free refrigerator.

MHT Liquid samples must be extracted within 7 days and the extracts analyzed within 40 days. Soils, sediments, or sludges may be stored for a maximum of 14 days and the extracts analyzed within 40 days.

SAMPLE PREPARATION
Liquid samples — Transfer 1 L quantitatively to a continuous extractor. If high concentrations are anticipated, a smaller volume may be used and then diluted with organic-free reagent water to 1 L. Adjust pH, if necessary, to pH <2 using 1:1 (V/V) sulfuric acid. Pipette 1.0 mL of a surrogate standard spiking solution into each sample. For the sample in each analytical batch selected for spiking, add 1.0 mL of a matrix spiking standard. For base/neutral acid analysis, the amount of the surrogates and matrix spiking compounds added to the sample should result in a final concentration of 100 ng/µL of each analyte in the extract to be analyzed (assuming a 1-µL injection). Extract with methylene chloride for 18–24 h. Next, adjust the pH of the aqueous phase to pH >11 using 10 N sodium hydroxide and extract it with methylene chloride again for 18–24 h. Dry the extract through a column containing anhydrous sodium sulfate and concentrate it to 1 mL using a K-D concentrator.

Soils, sediments, or sludges — Use 30 g of sample. Nonporous or wet samples (gummy or clay type) that do not have a free-flowing sandy texture must be mixed with anhydrous sodium sulfate until the sample is free flowing. Add 1 mL of surrogate standards to all samples, spikes, standards, and blanks. For the sample in each analytical batch selected for spiking, add 1.0 mL of a matrix spiking standard. For base/neutral acid analysis, the amount added of the surrogates and matrix spiking compounds should result in a final concentration of 100 ng/µL of

each base/neutral analyte and 200 ng/µL of each acid analyte in the extract to be analyzed (assuming a 1-µL injection). Immediately add a 100-mL mixture of 1:1 methylene chloride:acetone and extract the sample ultrasonically for 3 min and then decant or filter the extracts. Repeat the extraction two or more times. Dry the extract using a column with anhydrous sodium sulfate and concentrate it to 1 mL in a K-D concentrator.

QUALITY CONTROL A methylene chloride solution containing 50 ng/µL of decafluorotriphenylphosphine (DFTPP) is used for tuning the GC/MS system each 12-h shift. A system performance check also must be made during every 12-h shift. A standard containing 50 ng/µL each of 4,4'-DDT, pentachlorophenol, and benzidine is required to verify injection port inertness and GC column performance. A calibration standard at mid-concentration, containing each compound of interest, including all required surrogates, must be performed every 12 h during analysis. After the system performance check is met, calibration check compounds (CCCs) are used to check the validity of the initial calibration.

The internal standard responses and retention times in the calibration check standard must be evaluated immediately after or during data acquisition. If the retention time for any internal standard changes by more than 30 seconds from the last check calibration (12 h), the chromatographic system must be inspected for malfunctions and corrections must be made, as required. If the electron ionization current plot (EICP) area for any of the internal standards changes by a factor of two from the last daily calibration standard check, the mass spectrometer must be inspected for malfunctions and corrections must be made, as appropriate.

Demonstrate, through the analysis of a reagent water blank, that interferences from the analytical system, glassware, and reagents are under control. The blank samples should be carried through all stages of the sample preparation and measurement steps. For each analytical batch (up to 20 samples), a reagent blank, matrix spike, and matrix spike duplicate/duplicate must be analyzed (the frequency of the spikes may be different for different monitoring programs). The blank and spiked samples must be carried through all stages of the sample preparation and measurement steps. A QC reference sample concentrate containing each analyte at a concentration of 100 mg/L in methanol is required.

REFERENCE Test Methods for Evaluating Solid Waste (SW-846). U.S. EPA 1983, Method 8270B, Rev. 2, Nov. 1990. Office of Solid Waste, Washington, DC.

Captan   **EPA Method 8270**
CAS #133-06-2

TITLE Semivolatile Organic Compounds by GC/MS

MATRIX This method is used to determine the concentration of semivolatile organic compounds in extracts prepared from all types of solid waste matrices, soils, and groundwater. Although surface waters are not specifically mentioned, this method should be applicable to water samples from rivers, lakes, etc.

METHOD SUMMARY This method covers 259 semivolatile organic compounds. In very limited applications direct injection of the sample into the GC/MS system may be appropriate, but this results in very high detection limits (approximately 10,000 µg/L). Typically, a 1-L liquid sample, containing surrogate, and matrix spiking standards, is extracted in a continuous extractor first under acid conditions and then under basic conditions. Typically 30 g of a solid sample, containing surrogate, and matrix spiking standards, is extracted ultrasonically. After concentrating the extract to 1 mL it is spiked with 10 µL of an internal standard solution just prior to analysis by GC/MS. The volume injected should contain about 100 ng of base/neutral and 200 ng of acid surrogates (for a 1-µL injection). Analysis is performed by GC/MS using a capillary GC column.

INTERFERENCES Raw GC/MS data from all blanks, samples, and spikes must be evaluated for interferences. Contamination by carryover can occur whenever high-concentration and low-concentration samples are sequentially analyzed. To reduce carryover, the sample syringe must be rinsed out between samples with solvent. Whenever an unusually concentrated sample is encountered, it should be followed by the analysis of blank solvent to check for cross-contamination.

INSTRUMENTATION A GC/MS and a data system are required. The GC column used is a 30 m × 0.25 mm I.D. (or 0.32 mm I.D.) 1um film thickness silicone-coated fused silica capillary column. A continuous liquid-liquid extractor equipped with Teflon® or glass connection joints and stopcocks requiring no lubrication, a K-D concentrating apparatus, water bath, and an ultrasonic disrupter with a minimum power of 300 W and with pulsing capability are also required.

PRECISION & ACCURACY The estimated quantitation limit (EQL) of Method 8270B for determining an individual compound is approximately 1 mg/kg (wet weight) for soil or sediment samples, 1–200 mg/kg for wastes (dependent on matrix and method of preparation), and 10 µg/L for groundwater samples. EQLs will be proportionately higher for sample extracts that require dilution to avoid saturation of the detector.

The EQL(b) for groundwater in µg/L is 50.
The EQL (a, b) for low concentrations in soil and sediment in µg/kg is not determined.
Accuracy as µg/L is not listed.
Overall precision in µg/L is not listed.

(a) *EQLs listed for soil/sediment are based on wet weight. Normally data is reported in a dry-weight basis; therefore, EQLs will be higher based on the % dry weight of each sample. This calculation is based on a 30 g sample and gel permeation chromatography cleanup.*
(b) *Sample EQLs are highly matrix-dependent. The EQLs are provided for guidance and may not always be achievable.*
$C =$ *True value for concentration, in µg/L.*
$X =$ *Average recovery found for measurements of samples containing a concentration of C, in µg/L.*

ESTIMATED QUANTITATION LIMIT FOR OTHER MATRICES

Other Matrices	Factor (a)
High-concentration soil and sludges by sonicator	7.5
Non-water miscible waste	75

(a) EQL for other matrices = [EQL for low soil/sediment] × [Factor]. This estimated EQL is similar to an EPA "Practical Quantitation Limit."

SAMPLING METHOD

Liquid samples — Use a 1 or 2½ gallon amber glass bottle with a screw-top Teflon®-lined cover that has been prewashed with detergent and rinsed with distilled water and methanol (or isopropanol).

Soils, sediments, or sludges — Use an 8-oz. widemouth glass with a screw-top Teflon®-lined cover that has been prewashed with detergent and rinsed with distilled water and methanol (or isopropanol).

SAMPLE PRESERVATION

Liquid samples — If residual chlorine is present, add 3 mL of 10% sodium thiosulfate per gallon, cool to 4°C and store in a solvent-free refrigerator until analysis; if chlorine is not present, then eliminate the sodium thiosulfate addition.

Soils, sediments, or sludges — Cool samples to 4°C and store in a solvent-free refrigerator.

MHT Liquid samples must be extracted within 7 days and the extracts analyzed within 40 days. Soils, sediments, or sludges may be stored for a maximum of 14 days and the extracts analyzed within 40 days.

SAMPLE PREPARATION

Liquid samples — Transfer 1 L quantitatively to a continuous extractor. If high concentrations are anticipated, a smaller volume may be used and then diluted with organic-free reagent water to 1 L. Adjust pH, if necessary, to pH <2 using 1:1 (V/V) sulfuric acid. Pipette 1.0 mL of a surrogate standard spiking solution into each sample. For the sample in each analytical batch selected for spiking, add 1.0 mL of a matrix spiking standard. For base/neutral acid analysis, the amount of the surrogates and matrix spiking compounds added to the sample should result in a final concentration of 100 ng/µL of each analyte in the extract to be analyzed (assuming a 1-µL injection). Extract with methylene chloride for 18–24 h. Next, adjust the pH of the aqueous phase to pH >11 using 10 N sodium hydroxide and extract it with methylene chloride again for 18–24 h. Dry the extract through a column containing anhydrous sodium sulfate and concentrate it to 1 mL using a K-D concentrator.

Soils, sediments, or sludges — Use 30 g of sample. Nonporous or wet samples (gummy or clay type) that do not have a free-flowing sandy texture must be mixed with anhydrous sodium sulfate until the sample is free flowing. Add 1 mL of surrogate standards to all samples, spikes, standards, and blanks. For the sample in each analytical batch selected for spiking, add 1.0 mL of a matrix spiking standard. For base/neutral acid analysis, the amount added of the surrogates and matrix spiking compounds should result in a final concentration of 100 ng/µL of each base/neutral analyte and 200 ng/µL of each acid analyte in the extract to be analyzed (assuming a 1-µL injection). Immediately add a 100-mL mixture of 1:1 methylene chloride:acetone and extract the sample ultrasonically for 3 min and then decant or filter the extracts. Repeat the extraction two or more times. Dry the extract using a column with anhydrous sodium sulfate and concentrate it to 1 mL in a K-D concentrator.

QUALITY CONTROL A methylene chloride solution containing 50 ng/µL of decafluorotriphenylphosphine (DFTPP) is used for tuning the GC/MS system each 12-h shift. A system performance check also must be made during every 12-h shift. A standard containing 50 ng/µL each of 4,4'-DDT, pentachlorophenol, and benzidine is required to verify injection port inertness and GC column performance. A calibration standard at mid-concentration, containing each compound of interest, including all required surrogates, must be performed every 12 h during analysis. After the system performance check is met, calibration check compounds (CCCs) are used to check the validity of the initial calibration.

The internal standard responses and retention times in the calibration check standard must be evaluated immediately after or during data acquisition. If the retention time for any internal standard changes by more than 30 seconds from the last check calibration (12 h), the chromatographic system must be inspected for malfunctions and corrections must be made, as required. If the electron ionization current plot (EICP) area for any of the internal standards changes by a factor of two from the last daily calibration standard check, the mass spectrometer must be inspected for malfunctions and corrections must be made, as appropriate.

Demonstrate, through the analysis of a reagent water blank, that interferences from the analytical system, glassware, and reagents are under control. The blank samples should be carried through all stages of the sample preparation and measurement steps. For each analytical batch (up to 20 samples), a reagent blank, matrix spike, and matrix spike duplicate/duplicate must be analyzed (the frequency of the spikes may be different for different monitoring programs). The blank and spiked samples must be carried through all stages of the sample preparation and measurement steps. A QC reference sample concentrate containing each analyte at a concentration of 100 mg/L in methanol is required.

REFERENCE Test Methods for Evaluating Solid Waste (SW-846). U.S. EPA 1983, Method 8270B, Rev. 2, Nov. 1990. Office of Solid Waste, Washington, DC.

Carbaryl   **EPA Method 8270**
CAS #63-25-2

TITLE Semivolatile Organic Compounds by GC/MS

MATRIX This method is used to determine the concentration of semivolatile organic compounds in extracts prepared from all types of solid waste matrices, soils, and groundwater. Although surface waters are not specifically mentioned, this

method should be applicable to water samples from rivers, lakes, etc.

METHOD SUMMARY This method covers 259 semivolatile organic compounds. In very limited applications direct injection of the sample into the GC/MS system may be appropriate, but this results in very high detection limits (approximately 10,000 µg/L). Typically, a 1-L liquid sample, containing surrogate, and matrix spiking standards, is extracted in a continuous extractor first under acid conditions and then under basic conditions. Typically 30 g of a solid sample, containing surrogate, and matrix spiking standards, is extracted ultrasonically. After concentrating the extract to 1 mL it is spiked with 10 µL of an internal standard solution just prior to analysis by GC/MS. The volume injected should contain about 100 ng of base/neutral and 200 ng of acid surrogates (for a 1-µL injection). Analysis is performed by GC/MS using a capillary GC column.

INTERFERENCES Raw GC/MS data from all blanks, samples, and spikes must be evaluated for interferences. Contamination by carryover can occur whenever high-concentration and low-concentration samples are sequentially analyzed. To reduce carryover, the sample syringe must be rinsed out between samples with solvent. Whenever an unusually concentrated sample is encountered, it should be followed by the analysis of blank solvent to check for cross-contamination.

INSTRUMENTATION A GC/MS and a data system are required. The GC column used is a 30 m × 0.25 mm I.D. (or 0.32 mm I.D.) 1um film thickness silicone-coated fused silica capillary column. A continuous liquid-liquid extractor equipped with Teflon® or glass connection joints and stopcocks requiring no lubrication, a K-D concentrating apparatus, water bath, and an ultrasonic disrupter with a minimum power of 300 W and with pulsing capability are also required.

PRECISION & ACCURACY The estimated quantitation limit (EQL) of Method 8270B for determining an individual compound is approximately 1 mg/kg (wet weight) for soil or sediment samples, 1–200 mg/kg for wastes (dependent on matrix and method of preparation), and 10 µg/L for groundwater samples. EQLs will be proportionately higher for sample extracts that require dilution to avoid saturation of the detector.

The EQL(b) for groundwater in µg/L is 10.
The EQL (a, b) for low concentrations in soil and sediment in µg/kg is not determined.
Accuracy as µg/L is not listed.
Overall precision in µg/L is not listed.

(a) EQLs listed for soil/sediment are based on wet weight. Normally data is reported in a dry-weight basis; therefore, EQLs will be higher based on the % dry weight of each sample. This calculation is based on a 30 g sample and gel permeation chromatography cleanup.
(b) Sample EQLs are highly matrix-dependent. The EQLs are provided for guidance and may not always be achievable.
C = True value for concentration, in µg/L.
X = Average recovery found for measurements of samples containing a concentration of C, in µg/L.

ESTIMATED QUANTITATION LIMIT FOR OTHER MATRICES

Other Matrices	Factor (a)
High-concentration soil and sludges by sonicator	7.5
Non-water miscible waste	75

(a) EQL for other matrices = [EQL for low soil/sediment] × [Factor]. This estimated EQL is similar to an EPA "Practical Quantitation Limit."

SAMPLING METHOD
Liquid samples — Use a 1 or 2½ gallon amber glass bottle with a screw-top Teflon®-lined cover that has been prewashed with detergent and rinsed with distilled water and methanol (or isopropanol).

Soils, sediments, or sludges — Use an 8-oz. widemouth glass with a screw-top Teflon®-lined cover that has been prewashed with detergent and rinsed with distilled water and methanol (or isopanol).

SAMPLE PRESERVATION
Liquid samples — If residual chlorine is present, add 3 mL of 10% sodium thiosulfate per gallon, cool to 4°C and store in a solvent-free refrigerator until analysis; if chlorine is not present, then eliminate the sodium thiosulfate addition.

Soils, sediments, or sludges — Cool samples to 4°C and store in a solvent-free refrigerator.

MHT Liquid samples must be extracted within 7 days and the extracts analyzed within 40 days. Soils, sediments, or sludges may be stored for a maximum of 14 days and the extracts analyzed within 40 days.

SAMPLE PREPARATION
Liquid samples — Transfer 1 L quantitatively to a continuous extractor. If high concentrations are anticipated, a smaller volume may be used and then diluted with organic-free reagent water to 1 L. Adjust pH, if necessary, to pH <2 using 1:1 (V/V) sulfuric acid. Pipette 1.0 mL of a surrogate standard spiking solution into each sample. For the sample in each analytical batch selected for spiking, add 1.0 mL of a matrix spiking standard. For base/neutral acid analysis, the amount of the surrogates and matrix spiking compounds added to the sample should result in a final concentration of 100 ng/µL of each analyte in the extract to be analyzed (assuming a 1-µL injection). Extract with methylene chloride for 18–24 h. Next, adjust the pH of the aqueous phase to pH >11 using 10 N sodium hydroxide and extract it with methylene chloride again for 18–24 h. Dry the extract through a column containing anhydrous sodium sulfate and concentrate it to 1 mL using a K-D concentrator.

Soils, sediments, or sludges — Use 30 g of sample. Nonporous or wet samples (gummy or clay type) that do not have a free-flowing sandy texture must be mixed with anhydrous sodium sulfate until the sample is free flowing. Add 1 mL of surrogate standards to all samples, spikes, standards, and blanks. For the sample in each analytical batch selected for spiking, add 1.0 mL of a matrix spiking standard. For base/neutral acid analysis, the amount added of the surrogates and matrix spiking compounds should result in a final concentration of 100 ng/µL of

each base/neutral analyte and 200 ng/μL of each acid analyte in the extract to be analyzed (assuming a 1-μL injection). Immediately add a 100-mL mixture of 1:1 methylene chloride:acetone and extract the sample ultrasonically for 3 min and then decant or filter the extracts. Repeat the extraction two or more times. Dry the extract using a column with anhydrous sodium sulfate and concentrate it to 1 mL in a K-D concentrator.

QUALITY CONTROL A methylene chloride solution containing 50 ng/μL of decafluorotriphenylphosphine (DFTPP) is used for tuning the GC/MS system each 12-h shift. A system performance check also must be made during every 12-h shift. A standard containing 50 ng/μL each of 4,4′-DDT, pentachlorophenol, and benzidine is required to verify injection port inertness and GC column performance. A calibration standard at mid-concentration, containing each compound of interest, including all required surrogates, must be performed every 12 h during analysis. After the system performance check is met, calibration check compounds (CCCs) are used to check the validity of the initial calibration.

The internal standard responses and retention times in the calibration check standard must be evaluated immediately after or during data acquisition. If the retention time for any internal standard changes by more than 30 seconds from the last check calibration (12 h), the chromatographic system must be inspected for malfunctions and corrections must be made, as required. If the electron ionization current plot (EICP) area for any of the internal standards changes by a factor of two from the last daily calibration standard check, the mass spectrometer must be inspected for malfunctions and corrections must be made, as appropriate.

Demonstrate, through the analysis of a reagent water blank, that interferences from the analytical system, glassware, and reagents are under control. The blank samples should be carried through all stages of the sample preparation and measurement steps. For each analytical batch (up to 20 samples), a reagent blank, matrix spike, and matrix spike duplicate/duplicate must be analyzed (the frequency of the spikes may be different for different monitoring programs). The blank and spiked samples must be carried through all stages of the sample preparation and measurement steps. A QC reference sample concentrate containing each analyte at a concentration of 100 mg/L in methanol is required.

REFERENCE Test Methods for Evaluating Solid Waste (SW-846). U.S. EPA 1983, Method 8270B, Rev. 2, Nov. 1990. Office of Solid Waste, Washington, DC.

Carbazole   **EPA Method 1625**
CAS #86-74-8

TITLE Semivolatile Organic Compounds by Isotope Dilution GC/MS

MATRIX The compounds may be determined in waters, soils, and municipal sludges by this method.

METHOD SUMMARY This method is used to determine 176 semivolatile toxic organic pollutants associated with the CWA (as amended 1987); the RCRA (as amended 1986); the CERCLA (as amended 1986); and other compounds amenable to extraction and analysis by capillary column gas chromatography-mass spectrometry (GC/MS).

Stable isotopically-labeled analogs of the compounds of interest are added to the sample. If the solids content is less than 1%, a 1-L sample is extracted at pH 12–13, then at pH <2 with methylene chloride using continuous extraction techniques.

If the solids content is 30% or less, the sample is diluted to 1% solids with reagent water, homogenized ultrasonically, and extracted at pH 12–13, then at pH <2 with methylene chloride using continuous extraction techniques. If the solids content is greater than 30%, the sample is extracted using ultrasonic techniques.

Each extract is dried over sodium sulfate, concentrated to a volume of 5 mL, cleaned up using GPC, if necessary, and concentrated. Extracts are concentrated to 1 mL if GPC is not performed, and to 0.5 mL if GPC is performed.

An internal standard is added to the extract, and a 1-mL aliquot of the extract is injected into the GC. The compounds are separated by GC and detected by a MS. The labeled compounds serve to correct the variability of the analytical technique.

INTERFERENCES Solvents, reagents, glassware, and other sample processing hardware may yield artifacts and/or elevated baselines causing misinterpretation of chromatograms and spectra. Materials used in the analysis must be demonstrated to be free from interferences under the conditions of analysis by running method blanks initially and with each sample lot (sample started through the extraction process on a given 8-h shift, to a maximum of 20). Specific selection of reagents and purification of solvents by distillation in all glass systems may be required. Glassware and, where possible, reagents are cleaned by solvent rinse and baking at 450°C for 1-h minimum. Interferences coextracted from samples will vary considerably from source to source, depending on the diversity of the site being sampled.

INSTRUMENTATION Major instrumentation includes a GC with a splitless or on-column injection port for capillary column, a MS with 70 eV electron impact ionization, and a data system to collect and record MS data, and process it. A K-D apparatus is used to concentrate extracts.

GC Column: 30 m × 0.25 mm I.D. 5% phenyl, 94% methyl, 1% vinyl silicone bonded phased fused silica capillary column.

PRECISION & ACCURACY The detection limits of the method are usually dependent on the level of interferences rather than instrumental limitations. The limits typify the minimum quantities that can be detected with no interferences present.

The minimum level (in μg/mL) was 20. This is defined as a minimum level at which the analytical system shall give recognizable mass spectra (background corrected) and acceptable calibration points.
The MDL (in μg/kg) in low solids was 47 and in high solids was 24; these were determined in digested sludge (low solids) and in filter cake or compost (high solids).

The labeled and native compound initial precision as standard deviation (in µg/L) was 38.

The labeled and native compound initial accuracy as average recovery (in µg/L) was 36–165.

SAMPLE COLLECTION, PRESERVATION & HANDLING
Collect samples in glass containers. Aqueous samples which flow freely are collected in refrigerated bottles using automatic sampling equipment. Solid samples are collected as grab samples using widemouth jars. Maintain samples at 0 to 4°C from the time of collection until extraction. If residual chlorine is present in aqueous samples, add 80 mg sodium thiosulfate/L of water. Begin sample extraction within 7 days of collection, and analyze all extracts within 40 days of extraction.

SAMPLE PREPARATION Samples containing 1% solids or less are extracted directly using continuous liquid-liquid extraction techniques. Samples containing 1 to 30% solids are diluted to the 1% level with reagent water and extracted using continuous liquid-liquid extraction techniques. Samples containing greater than 30% solids are extracted using ultrasonic techniques.

Base/neutral extraction — Adjust the pH of the waters in the extractors to 12–13 with 6 N NaOH. Extract with methylene chloride for 24–48 h.

Acid extraction — Adjust the pH of the waters in the extractors to 2 or less using 6 N sulfuric acid. Extract with methylene chloride for 24–48 h.

Ultrasonic extraction of high solids samples — Add anhydrous sodium sulfate to the sample and QC aliquot(s). Add acetone:methylene chloride (1:1) to the sample and mix thoroughly

Concentrate extracts using a K-D apparatus.

QUALITY CONTROL The analyst is permitted to modify this method to improve separations or lower the costs of measurements, provided all performance specifications are met. Analyses of blanks are required to demonstrate freedom from contamination. When results of spikes indicate atypical method performance for samples, the samples are diluted to bring method performance within acceptable limits.

For low solids (aqueous samples), extract, concentrate, and analyze two sets of four 1-L aliquots (8 aliquots total) of the precision and recovery standard. For high solids samples, two sets of four 30-g aliquots of the high solids reference matrix are used.

Spike all samples with labeled compounds to assess method performance. Compute percent recovery of the labeled compounds using the internal standard method. Compare the labeled compound recovery for each compound with the corresponding labeled compound recovery.

Reagent water and high solids reference matrix blanks are analyzed to demonstrate freedom from contamination. Extract and concentrate a 1-L reagent water blank or a high solids reference matrix blank with each sample's lot (samples started through the extraction process on the same 8-h shift, to a maximum of 20 samples).

Field replicates may be collected to determine the precision of the sampling technique, and spiked samples may be required to determine the accuracy of the analysis when the internal standard method is used.

REFERENCE Semivolatile Organic Compounds by Isotope Dilution GC/MS. Office of Water Regulation and Standards, U.S. EPA Industrial Technology Division, Washington, DC, EPA Method 1625, Rev. C, June 1989 (contact W.A. Telliard, U.S. EPA, Office of Water Regulations and Standards, 401 M St., SW, Washington, DC, 20460. Phone: 202-382-7131).

Carbofuran   EPA Method 8270
CAS #1563-66-2

TITLE Semivolatile Organic Compounds by GC/MS

MATRIX This method is used to determine the concentration of semivolatile organic compounds in extracts prepared from all types of solid waste matrices, soils, and groundwater. Although surface waters are not specifically mentioned, this method should be applicable to water samples from rivers, lakes, etc.

METHOD SUMMARY This method covers 259 semivolatile organic compounds. In very limited applications direct injection of the sample into the GC/MS system may be appropriate, but this results in very high detection limits (approximately 10,000 µg/L). Typically, a 1-L liquid sample, containing surrogate, and matrix spiking standards, is extracted in a continuous extractor first under acid conditions and then under basic conditions. Typically 30 g of a solid sample, containing surrogate, and matrix spiking standards, is extracted ultrasonically. After concentrating the extract to 1 mL it is spiked with 10 µL of an internal standard solution just prior to analysis by GC/MS. The volume injected should contain about 100 ng of base/neutral and 200 ng of acid surrogates (for a 1-µL injection). Analysis is performed by GC/MS using a capillary GC column.

INTERFERENCES Raw GC/MS data from all blanks, samples, and spikes must be evaluated for interferences. Contamination by carryover can occur whenever high-concentration and low-concentration samples are sequentially analyzed. To reduce carryover, the sample syringe must be rinsed out between samples with solvent. Whenever an unusually concentrated sample is encountered, it should be followed by the analysis of blank solvent to check for cross-contamination.

INSTRUMENTATION A GC/MS and a data system are required. The GC column used is a 30 m × 0.25 mm I.D. (or 0.32 mm I.D.) 1um film thickness silicone-coated fused silica capillary column. A continuous liquid-liquid extractor equipped with Teflon® or glass connection joints and stopcocks requiring no lubrication, a K-D concentrating apparatus, water bath, and an ultrasonic disrupter with a minimum power of 300 W and with pulsing capability are also required.

PRECISION & ACCURACY The estimated quantitation limit (EQL) of Method 8270B for determining an individual compound is approximately 1 mg/kg (wet weight) for soil or sediment samples, 1–200 mg/kg for wastes (dependent on

matrix and method of preparation), and 10 µg/L for groundwater samples. EQLs will be proportionately higher for sample extracts that require dilution to avoid saturation of the detector.

The EQL(b) for groundwater in µg/L is 10.
The EQL (a, b) for low concentrations in soil and sediment in µg/kg is not determined.
Accuracy as µg/L is not listed.
Overall precision in µg/L is not listed.

(a) *EQLs listed for soil/sediment are based on wet weight. Normally data is reported in a dry-weight basis; therefore, EQLs will be higher based on the % dry weight of each sample. This calculation is based on a 30 g sample and gel permeation chromatography cleanup.*
(b) *Sample EQLs are highly matrix-dependent. The EQLs are provided for guidance and may not always be achievable.*
C = *True value for concentration, in µg/L.*
X = *Average recovery found for measurements of samples containing a concentration of C, in µg/L.*

ESTIMATED QUANTITATION LIMIT FOR OTHER MATRICES

Other Matrices	Factor (a)
High-concentration soil and sludges by sonicator	7.5
Non-water miscible waste	75

(a) *EQL for other matrices = [EQL for low soil/sediment] × [Factor]. This estimated EQL is similar to an EPA "Practical Quantitation Limit."*

SAMPLING METHOD

Liquid samples — Use a 1 or 2½ gallon amber glass bottle with a screw-top Teflon®-lined cover that has been prewashed with detergent and rinsed with distilled water and methanol (or isopropanol).

Soils, sediments, or sludges — Use an 8-oz. widemouth glass with a screw-top Teflon®-lined cover that has been prewashed with detergent and rinsed with distilled water and methanol (or isopropanol).

SAMPLE PRESERVATION

Liquid samples — If residual chlorine is present, add 3 mL of 10% sodium thiosulfate per gallon, cool to 4°C and store in a solvent-free refrigerator until analysis; if chlorine is not present, then eliminate the sodium thiosulfate addition.

Soils, sediments, or sludges — Cool samples to 4°C and store in a solvent-free refrigerator.

MHT Liquid samples must be extracted within 7 days and the extracts analyzed within 40 days. Soils, sediments, or sludges may be stored for a maximum of 14 days and the extracts analyzed within 40 days.

SAMPLE PREPARATION

Liquid samples — Transfer 1 L quantitatively to a continuous extractor. If high concentrations are anticipated, a smaller volume may be used and then diluted with organic-free reagent water to 1 L. Adjust pH, if necessary, to pH <2 using 1:1 (V/V) sulfuric acid. Pipette 1.0 mL of a surrogate standard spiking solution into each sample. For the sample in each analytical batch selected for spiking, add 1.0 mL of a matrix spiking standard. For base/neutral acid analysis, the amount of the surrogates and matrix spiking compounds added to the sample should result in a final concentration of 100 ng/µL of each analyte in the extract to be analyzed (assuming a 1-µL injection). Extract with methylene chloride for 18–24 h. Next, adjust the pH of the aqueous phase to pH >11 using 10 N sodium hydroxide and extract it with methylene chloride again for 18–24 h. Dry the extract through a column containing anhydrous sodium sulfate and concentrate it to 1 mL using a K-D concentrator.

Soils, sediments, or sludges — Use 30 g of sample. Nonporous or wet samples (gummy or clay type) that do not have a free-flowing sandy texture must be mixed with anhydrous sodium sulfate until the sample is free flowing. Add 1 mL of surrogate standards to all samples, spikes, standards, and blanks. For the sample in each analytical batch selected for spiking, add 1.0 mL of a matrix spiking standard. For base/neutral acid analysis, the amount added of the surrogates and matrix spiking compounds should result in a final concentration of 100 ng/µL of each base/neutral analyte and 200 ng/µL of each acid analyte in the extract to be analyzed (assuming a 1-µL injection). Immediately add a 100-mL mixture of 1:1 methylene chloride:acetone and extract the sample ultrasonically for 3 min and then decant or filter the extracts. Repeat the extraction two or more times. Dry the extract using a column with anhydrous sodium sulfate and concentrate it to 1 mL in a K-D concentrator.

QUALITY CONTROL A methylene chloride solution containing 50 ng/µL of decafluorotriphenylphosphine (DFTPP) is used for tuning the GC/MS system each 12-h shift. A system performance check also must be made during every 12-h shift. A standard containing 50 ng/µL each of 4,4′-DDT, pentachlorophenol, and benzidine is required to verify injection port inertness and GC column performance. A calibration standard at mid-concentration, containing each compound of interest, including all required surrogates, must be performed every 12 h during analysis. After the system performance check is met, calibration check compounds (CCCs) are used to check the validity of the initial calibration.

The internal standard responses and retention times in the calibration check standard must be evaluated immediately after or during data acquisition. If the retention time for any internal standard changes by more than 30 seconds from the last check calibration (12 h), the chromatographic system must be inspected for malfunctions and corrections must be made, as required. If the electron ionization current plot (EICP) area for any of the internal standards changes by a factor of two from the last daily calibration standard check, the mass spectrometer must be inspected for malfunctions and corrections must be made, as appropriate.

Demonstrate, through the analysis of a reagent water blank, that interferences from the analytical system, glassware, and reagents are under control. The blank samples should be carried through all stages of the sample preparation and measurement steps. For each analytical batch (up to 20 samples), a reagent blank, matrix spike, and matrix spike duplicate/duplicate must be analyzed (the frequency of the spikes may be different for different monitoring programs). The blank and spiked samples must be carried through all stages of the sample preparation and measurement steps. A QC reference sample concentrate containing each analyte at a concentration of 100 mg/L in methanol is required.

REFERENCE Test Methods for Evaluating Solid Waste (SW-846). U.S. EPA 1983, Method 8270B, Rev. 2, Nov. 1990. Office of Solid Waste, Washington, DC.

Carbon disulfide   EPA Method 1624
CAS #75-15-0

TITLE Volatile Organic Compounds by Isotope Dilution GC/MS

MATRIX Compounds may be determined in waters, soils, and municipal sludges by this method.

METHOD SUMMARY This method is used to determine 58 volatile toxic organic pollutants associated with the CWA (as amended 1987); the RCRA (as amended 1986); the CERCLA (as amended 1986); and other compounds amenable to purge-and-trap gas chromatography-mass spectrometry (GC/MS).

If the solids content is less than 1%, stable isotopically-labeled analogs of the compounds of interest are added to a 5-mL sample and the sample is purged with an inert gas at 20–25°C in a chamber designed for soil or water samples. If the solids content is greater than 1%, 5 mL of reagent water and the labeled compounds are added to a 5-g aliquot of sample and the mixture is purged at 40°C. Compounds that will not purge at 20–25°C or at 40°C are purged at 78–85°C. In the purging process, the volatile compounds are transferred from the aqueous phase into the gaseous phase where they are passed into a sorbent column, and trapped. After purging is completed, the trap is backflushed and heated rapidly to desorb the compounds into a GC. The compounds are separated by the GC and detected by a MS. The labeled compounds serve to correct the variability of the analytical technique.

INTERFERENCES Impurities in the purge gas, organic compounds outgassing from the plumbing upstream of the trap, and solvent vapors in the lab account for most problems. Samples can be contaminated by diffusion of volatile organic compounds (particularly methylene chloride) through the bottle seal during shipment and storage. Contamination by carryover can occur when high-level and low-level samples are analyzed sequentially. When an unusually concentrated sample is encountered, follow it by analysis of a reagent water blank to check for carryover.

INSTRUMENTATION Major equipment includes a GC with linear temperature programming and a glass jet separator as the MS interface, a MS with 70 eV electron impact ionization, and a data system to collect and record response factors.

Column: 2.8 m × 2 mm I.D. glass, packed with 1% SP-1000 on Carbopak B, 60/80 mesh, or equivalent.

PRECISION & ACCURACY The detection limits of the method are usually dependent on the level of interferences rather than instrumental limitations. The method detection limits were determined in digested sludge (low solids) and in filter cake or compost (high solids).

The MDL (in µg/kg) for low solids is not listed and for high solids is not listed.
Labeled and native compound precision (in µg/L) as standard deviation was not listed.
Labeled and native compound accuracy (in µg/L) as average recovery was not listed.

Acceptance criteria are at 20 µg/L for this compound.

SAMPLE COLLECTION, PRESERVATION & HANDLING Grab samples are collected in glass containers having a total volume greater than 20 mL. Fill and seal each bottle so that no air bubbles are entrapped. Samples are maintained at 0 to 4°C from the time of collection until analysis. If an aqueous sample contains residual chlorine, add sodium thiosulfate preservative (10 mg/40 mL) to the empty sample bottles just prior to shipment to the sample site. All samples must be analyzed within 14 days of collection.

SAMPLE PREPARATION Samples containing less than 1% solids are analyzed directly as aqueous samples. Samples containing 1% solids or greater are analyzed as solid samples utilizing one of two methods, depending on the levels of pollutants, in the sample. Samples containing 1% solids or greater, and low to moderate levels of pollutants are analyzed by purging a known weight of sample added to 5 mL of reagent water. Samples containing 1% solids or greater, and high levels of pollutants, are extracted with methanol, and an aliquot of the methanol extract is added to reagent water and purged.

QUALITY CONTROL A field blank prepared from reagent water and carried through the sampling and handling protocol may serve as a check on contamination from shipment and storage.

The analyst is permitted to modify this method to improve separations or lower the costs of measurements, provided all performance specifications are met. Analyses of blanks are required. When results of spikes indicate atypical method performance for samples, the samples are diluted to bring method performance within acceptable limits. Analyze two sets of four 5-mL aliquots (8 aliquots total) of the aqueous performance standard. Spike all samples with labeled compounds to assess method performance on the sample matrix. Compute the percent recovery of the labeled compounds using the internal standard method. Compare the percent recovery for each compound with the corresponding labeled compound recovery. Reagent water blanks are analyzed to demonstrate freedom from carryover contamination. Field replicates may be collected to determine the precision of the sampling technique, and spiked samples may be required to determine the accuracy of the analysis when the internal method is used.

REFERENCE Volatile Organic Compounds by Isotope Dilution GC/MS. Office of Water Regulation and Standards, U.S. EPA Industrial Technology Division, Washington, DC, EPA Method 1624, Rev. C, June 1989 (contact W.A. Telliard, U.S. EPA, Office of Water Regulations and Standards, 401 M St., SW, Washington, DC, 20460. Phone: 202-382-7131).

Carbon disulfide
CAS #75-15-0

EPA Method 8240

TITLE Volatile Organics By GC/MS: Packed Column Technique

MATRIX Nearly all types of sample matarices, regardless of water content, can be analyzed using this method. This includes groundwater, aqueous sludges, caustic liquors, acid liquors, waste solvents, oily wastes, mousses, tars, fibrous wastes, polymetric emulsions, filter cakes, spent carbons, spent catalysts, soils, and sediments.

METHOD SUMMARY Method 8240B covers 80 volatile organic compounds that are introduced into a gas chromatograph by the purge-and-trap method or by direct injection (in limited applications). For the purge-and-trap method an inert gas (zero grade nitrogen or helium) is bubbled through a 5-mL solution at ambient temperature. Purged sample components are trapped in a tube of sorbent materials. When purging is complete, the sorbent tube is heated and backflushed with inert gas to desorb the trapped components onto a GC column.

INTERFERENCES Impurities in the purge gas and from organic compounds outgassing from the plumbing ahead of the trap account for many contamination problems. Interferences purged or coextracted from the samples will vary considerably from source to source. Cross-contamination can occur whenever high-level and low-level samples are analyzed sequentially. Whenever an unusually concentrated sample is analyzed, it should be followed by the analysis of organic-free reagent water to check for cross-contamination. Samples also can be contaminated by diffusion of volatile organics (particularly methylene chloride and fluorocarbons) through the septum seal into the sample during shipment and storage. A trip blank can serve as a check on such contamination. The lab where volatile analysis is performed and also the refrigerated storage area should be completely free of solvents.

INSTRUMENTATION A gas chromatograph/mass spectrometry/data system (GC/MS) equipped with a 6 ft × 0.1 in I.D. glass column packed with 1% SP-1000 on Carbopack-B (60/80 mesh) is required. Also needed is a 5-mL purging device, a sorbent trap, and a thermal desorption apparatus.

PRECISION & ACCURACY This method is reported to have been tested by 15 laboratories using organic-free reagent water, drinking water, surface water, and industrial wastewaters (not specified) fortified at six concentrations over the range 5–600 µg/L.

Sample estimated quantitation limits (EQLs) are highly matrix-dependent. The EQLs listed may not always be achievable. EQLs listed for soils or sediments are based on wet weight. Normally, data is reported on a dry-weight basis; therefore, EQLs will be higher, based on the percent dry weight of each sample. Note that EQLs are even more variable than MDLs and that they are highly variable depending on the matrix being analyzed.

EQL in groundwater in µg/L was 100.
EQL in low soil or sediment in µg/kg was 100.
Accuracy (a) in µg/L was not listed.

Precision (b) in µg/L was not listed.

(a) *Average recovery found for measurements of samples containing a concentration of C, in µg/L.*

(b) *Overall precision found for measurements of samples with average recovery X for samples containing a concentration of C in µg/L.*

$X =$ *Average recovery found for measurement of samples containing a concentration of C in µg/L.*

MULTIPLICATION FACTORS FOR OTHER MATRICES

Other Matrices	Factor (a)
Waste miscible liquid waste	50
High-concentration soil and sludge	125
Non-water miscible waste	500

(a) EQL = [EQL for low soil/sediment] × [Factor]. For non-aqueous samples, the factor is on a wet-weight basis.

SAMPLING METHOD
Liquid samples — Use a 40-mL glass screw-cap VOA vial with a Teflon®-faced silicone septum that has been prewashed, rinsed with distilled deionized water, and oven dried. However, if residual chlorine is present, collect sample in a 40-oz. soil VOA container which has been pre-preserved with 4 drops of 10% sodium thiosulfate, mix gently, and then transfer the sample to a 40-mL VOA vial. Collect bubble-free samples in duplicate and seal them in separate plastic bags.

Soils or sediments, and sludges — Use an 8-oz. widemouth glass bottle with a Teflon®-faced silicone septum that has been prewashed with detergent, rinsed with distilled deionized water, and oven dried. Tap slightly to eliminate free air space. Collect samples in duplicate and seal them in separate plastic bags.

SAMPLE PRESERVATION
Liquid samples — Add 4 drops of concentrated HCL and immediately cool samples to 4°C and store in a solvent-free refrigerator.

Soils or sediments, and sludges — Cool samples to 4°C and store in a solvent-free refrigerator.

MHT Maximum holding time is 14 days from the date of sample collection.

SAMPLE PREPARATION
Liquid samples — Remove the plunger from a 5-mL syringe and carefully pour the sample into the syringe barrel to just short of overflowing. Replace the syringe plunger and compress the sample. Open the syringe valve and vent any residual air while adjusting the sample volume to 5.0 mL. If there is only one volatile organic analysis (VOA) vial, a second syringe should be filled at this time to protect against possible loss of sample integrity. Add 10 µL of surrogate spiking solution and 10 µL of internal standard spiking solution through the valve bore of the 5-mL syringe, then close the valve. The surrogate and internal standards may be mixed and added as a single spiking solution.

Sediments, soils, and waste samples — All samples of this type should be screened by GC analysis using a headspace method (EPA Method 3810) or the hexadecane extraction and screening method (EPA Method 3820). Use the screening data to determine whether to use the low-concentration method (0.005–1 mg/kg) or the high-concentration method (>1 mg/kg).

Low-concentration method — The low-concentration method is based on purging a heated sediment or soil sample mixed with organic-free reagent water containing the surrogate and internal standards. Analyze all reagent blanks and standards under the same conditions as the samples.

Use a 5-g sample if the expected concentration is <0.1 mg/kg or a 1-g sample for expected concentrations between 0.1 and 1 mg/kg. Mix the contents of the sample container with a narrow metal spatula. Weigh the amount of the sample into a tared purge device. Add the spiked water to the purge device, which contains the weighed amount of sample, and connect the device to the purge-and-trap system.

High-concentration method — This method is based on extracting the sediment or soil with methanol. A waste sample is either extracted or diluted, depending on its solubility in methanol. Wastes that are insoluble in methanol are diluted with reagent tetraglyme or possibly polyethylene glycol (PEG). An aliquot of the extract is added to organic-free reagent water containing surrogate and internal standards. This is purged at ambient temperature. All samples with an expected concentration of >1.0 mg/kg should be analyzed by this method.

Mix the contents of the sample container with a narrow metal spatula. For sediments or soils and solid wastes that are insoluble in methanol, weigh 4 g (wet weight) of sample into a tared 20-mL vial. For waste that is soluble in methanol, tetraglyme, or PEG, weigh 1 g (wet weight) into a tared scintillation vial or culture tube or a 10-mL volumetric flask. Quickly add 9.0 mL of appropriate solvent then add 1.0 mL of a surrogate spiking solution to the vial, cap it, and shake it for 2 min.

METHANOL EXTRACT REQUIRED FOR ANALYSIS OF HIGH-CONCENTRATION SOILS OR SEDIMENTS

Approximate Concentration Range	Volume of Methanol Extract (a)
500–10,000 µg/kg	100 µL
1,000–20,000 µg/kg	50 µL
5,000–100,000 µg/kg	10 µL
25,000–500,000 µg/kg	100 µL of 1/50 dilution (b)

Calculate appropriate dilution factor for concentrations exceeding this table.

(a) The volume of methanol added to 5 mL of water being purged should be kept constant. Therefore, add to the 5-mL syringe whatever volume of methanol is necessary to maintain a volume of 100 µL added to the syringe.
(b) Dilute an aliquot of the methanol extract and then take 100 µL for analysis.

QUALITY CONTROL Demonstrate, through the analysis of a reagent water blank, that interferences from the analytical system, glassware, and reagents are under control. Blank samples should be carried through all stages of the sample preparation and measurement steps. For each analytical batch (up to 20 samples), a reagent blank, matrix spike, and matrix spike duplicate must be analyzed (the frequency of the spikes may be different for different monitoring programs). The blank and spiked samples must be carried through all stages of the sample preparation and measurement steps. QC samples mentioned in the section on Interferences will also be needed as appropriate to those situations.

REFERENCE Test Methods for Evaluating Solid Waste (SW-846). U.S. EPA. 1983. Method 8240B, Rev. 2, Nov. 1990. Office of Solid Wastes, Washington, DC.

Carbon tetrachloride   **EPA Method 1624**
CAS #56-23-5

TITLE Volatile Organic Compounds by Isotope Dilution GC/MS

MATRIX Compounds may be determined in waters, soils, and municipal sludges by this method.

METHOD SUMMARY This method is used to determine 58 volatile toxic organic pollutants associated with the CWA (as amended 1987); the RCRA (as amended 1986); the CERCLA (as amended 1986); and other compounds amenable to purge-and-trap gas chromatography-mass spectrometry (GC/MS).

If the solids content is less than 1%, stable isotopically-labeled analogs of the compounds of interest are added to a 5-mL sample and the sample is purged with an inert gas at 20–25°C in a chamber designed for soil or water samples. If the solids content is greater than 1%, 5 mL of reagent water and the labeled compounds are added to a 5-g aliquot of sample and the mixture is purged at 40°C. Compounds that will not purge at 20–25°C or at 40°C are purged at 78–85°C. In the purging process, the volatile compounds are transferred from the aqueous phase into the gaseous phase where they are passed into a sorbent column, and trapped. After purging is completed, the trap is backflushed and heated rapidly to desorb the compounds into a GC. The compounds are separated by the GC and detected by a MS. The labeled compounds serve to correct the variability of the analytical technique.

INTERFERENCES Impurities in the purge gas, organic compounds outgassing from the plumbing upstream of the trap, and solvent vapors in the lab account for most problems. Samples can be contaminated by diffusion of volatile organic compounds (particularly methylene chloride) through the bottle seal during shipment and storage. Contamination by carryover can occur when high-level and low-level samples are analyzed sequentially. When an unusually concentrated sample is encountered, follow it by analysis of a reagent water blank to check for carryover.

INSTRUMENTATION Major equipment includes a GC with linear temperature programming and a glass jet separator as the MS interface, a MS with 70 eV electron impact ionization, and a data system to collect and record response factors.

Column: 2.8 m × 2 mm I.D. glass, packed with 1% SP-1000 on Carbopak B, 60/80 mesh, or equivalent.

PRECISION & ACCURACY The detection limits of the method are usually dependent on the level of interferences

rather than instrumental limitations. The method detection limits were determined in digested sludge (low solids) and in filter cake or compost (high solids).

The MDL (in μg/kg) for low solids is 87 and for high solids is 9.
Labeled and native compound precision (in μg/L) as standard deviation was 6.9.
Labeled and native compound accuracy (in μg/L) as average recovery was 16–25.

Acceptance criteria are at 20 μg/L for this compound.

SAMPLE COLLECTION, PRESERVATION & HANDLING Grab samples are collected in glass containers having a total volume greater than 20 mL. Fill and seal each bottle so that no air bubbles are entrapped. Samples are maintained at 0 to 4°C from the time of collection until analysis. If an aqueous sample contains residual chlorine, add sodium thiosulfate preservative (10 mg/40 mL) to the empty sample bottles just prior to shipment to the sample site. All samples must be analyzed within 14 days of collection.

SAMPLE PREPARATION Samples containing less than 1% solids are analyzed directly as aqueous samples. Samples containing 1% solids or greater are analyzed as solid samples utilizing one of two methods, depending on the levels of pollutants, in the sample. Samples containing 1% solids or greater, and low to moderate levels of pollutants are analyzed by purging a known weight of sample added to 5 mL of reagent water. Samples containing 1% solids or greater, and high levels of pollutants, are extracted with methanol, and an aliquot of the methanol extract is added to reagent water and purged.

QUALITY CONTROL A field blank prepared from reagent water and carried through the sampling and handling protocol may serve as a check on contamination from shipment and storage.

The analyst is permitted to modify this method to improve separations or lower the costs of measurements, provided all performance specifications are met. Analyses of blanks are required. When results of spikes indicate atypical method performance for samples, the samples are diluted to bring method performance within acceptable limits. Analyze two sets of four 5-mL aliquots (8 aliquots total) of the aqueous performance standard. Spike all samples with labeled compounds to assess method performance on the sample matrix. Compute the percent recovery of the labeled compounds using the internal standard method. Compare the percent recovery for each compound with the corresponding labeled compound recovery. Reagent water blanks are analyzed to demonstrate freedom from carryover contamination. Field replicates may be collected to determine the precision of the sampling technique, and spiked samples may be required to determine the accuracy of the analysis when the internal method is used.

REFERENCE Volatile Organic Compounds by Isotope Dilution GC/MS. Office of Water Regulation and Standards, U.S. EPA Industrial Technology Division, Washington, DC, EPA Method 1624, Rev. C, June 1989 (contact W.A. Telliard, U.S. EPA, Office of Water Regulations and Standards, 401 M St., SW, Washington, DC, 20460. Phone: 202-382-7131).

Carbon tetrachloride   **EPA Method 502**
CAS #56-23-5

TITLE Volatile Organic Compounds in Water By Purge and Trap Capillary Column Gas Chromatography with Photoionization and Electrolytic Conductivity Detectors in Series. U.S. EPA Method 502.2, Rev. 2.0, 1989.

MATRIX Drinking water and raw source water. The latter should include most surface water and groundwater sources.

METHOD SUMMARY This method covers 60 volatile organic compounds that contain halogen atoms and/or that are aromatic. An inert gas (zero grade nitrogen or helium) is bubbled through a 25-mL or a 5-mL water sample (depending on the expected concentration of the analytes). Purged sample components are trapped in a tube of sorbent materials. When purging is complete, the sorbent tube is heated and backflushed with helium to desorb the trapped sample onto a capillary GC column. The column is temperature programmed to separate the method analytes which are then detected with a photoionization detector (PID) and a Hall electrolytic conductivity (HECD) placed in series. The PID is selective for aromatic compounds and the HECD is selective for halogenated compounds.

INTERFERENCES Impurities in the purge gas and from organic compounds outgassing from the plumbing ahead of the trap account for many contamination problems. Interferences purged or coextracted from the samples will vary considerably from source to source, depending upon the particular sample or extract being tested. Cross-contamination can occur whenever high-level and low-level samples are analyzed sequentially. Samples also can be contaminated by diffusion of volatile organics (particularly methylene chloride and fluorocarbons) through the septum seal into the sample during shipment and storage. The lab where volatile analysis is performed and also the refrigerated storage area should be completely free of solvents.

INSTRUMENTATION A GC containing a series configuration of a high temperature photoionization detector (PID) equipped with 10.0 eV (nominal) lamp and Hall electrolytic conductivity detector (HECD) is required. Also required is an all-glass 5-mL purging device, a sorbent trap, and a thermal desorption apparatus which is connected to the GC system.

Column 1: VOCOL glass wide-bore capillary column.
Column 2: RTX–502.2 mega-bore capillary column.
Column 3: DB-62 mega-bore capillary column.

PRECISION & ACCURACY Method detection limits are dependent upon the characteristics of the gas chromatographic system used. Analytes that are not separated chromatographically cannot be individually identified and used in the same calibration mixture or water samples unless an alternative technique for identification and quantification, such as mass spectrometry, is used.

Electrolytic conductivity detetor (c) range in μg/L (a) was 0.02–200.
Electrolytic conductivity detetor (c) MDL in μg/L (b) was 0.01.

Electrolytic conductivity detetor (c) accuracy as % recovery was 92.

Electrolytic conductivity detetor (c) precision as % RSD was 3.6.

Photoionization detector (d) range in µg/L (a) was 0.02–200.

Photoionization detector (d) MDL in µg/L (b) was not listed.

Photoionization detector (d) accuracy as % recovery was not listed.

Photoionization detector (d) precision as % RSD was not listed.

(a) The applicable concentration range of this method is compound, instrument, and matrix-dependent. It is listed as being approximately 0.02 to 200 µg/L but no specific information is provided so caution should be observed.

(b) The method detection limits reports with this method are compound, instrument, and matrix-dependent. The values reported were calculated using reagent water fortified with the corresponding compounds at 10 µg/L and a GC-equipped with a 60 m × 0.75 mm VOLCOL wide bore capillary column with 1.5 µm film thickness and using helium carrier gas.

(c) Recoveries and relative standard deviations were determined from seven samples of reagent water fortified with 10 µg/L of each compound. 2-Bromo-1-chloropropane was used as the internal standard for calculating average recoveries.

(d) Recoveries and relative standard deviations were determined from seven samples of reagent water fortified with 10 µg/L of each compound. Fluorobenzene was used as the internal standard for calculating average recoveries.

SAMPLING METHOD Collect samples using a 40- to 120-mL screw-cap vial (prewashed with detergent, rinsed with distilled water and oven dried at 105°C) with a Teflon®-faced silicone septum. Collect bubble-free samples and place the septum with the Teflon® side down on the water.

SAMPLE PRESERVATION If residual chlorine is present in the water add about 25 mg of ascorbic acid to each vial before samples are collected to remove the chlorine. Add hydrochloric acid to reduce pH to <2, immediately cool samples to 4°C, and store them in a solvent-free refrigerator at 4°C until analysis.

MHT The maximum holding time for samples is 14 days from the time they were collected.

SAMPLE PREPARATION Remove the plungers from two 5-mL syringes and attach a closed syringe valve to each. Warm the sample to room temperature, open the sample bottle, and carefully pour the sample into one of the syringe barrels to just short of overflowing. Replace the syringe plunger, invert the syringe, and compress the sample. Open the syringe valve and vent any residual air while adjusting the sample volume to 5.0 mL. Add 10 µL of the internal calibration standard to the sample through the syringe valve. Close the valve. Fill the second syringe in an identical manner from the same sample bottle. Reserve this second syringe for a reanalysis if necessary.

QUALITY CONTROL As an initial demonstration of lab accuracy and precision, analyze 4 to 7 replicates of a lab fortified blank containing analyte at 0.1–5 µg/L. Collect all samples in duplicate. Surrogate analytes (similar to those of the analytes of interest), whose concentration is known in every sample, are measured using the same internal standard calibration procedure. Duplicate field reagent water blanks (trip blanks) must be analyzed with each set of samples, lab reagent blanks (method blanks) must be analyzed with each batch of samples processed as a group within a work shift. Also, a single lab-fortified blank that contains each of the analytes of interest should be analyzed with each batch of samples processed as a group within a work shift. A 3- to 5-point calibration curve is needed depending on the calibration range factor required.

EPA CONTACT & HOTLINE For technical questions contact Dr. Baldev Bathija, U.S. EPA, Office of Ground Water and Drinking Water (WH-550D), 401 M St. SW, Washington, DC 20460. Tel. (202) 260-3040. For further information the EPA Safe Drinking Water Hotline may be called at: (800) 426-4791.

REFERENCE Methods for the Determination of Organic Compounds in Drinking Water, EPA/600/4-88/039 (revised July 1991; Final Rule for determination of compliance with the MCL for Total Trihalomethanes under 141.30, in 40 CFR Part 141, Vol. 58, No. 147, Fed. Reg., Tuesday Aug. 3, 1993). U.S. EPA Environmental Monitoring Systems Laboratory, Cincinnati, OH, 45268, USA. Available from the National Technical Information Service (NTIS), 5285 Port Royal Road, Springfield, VA 22161; Tel. 800-553-6847. NTIS Order Number is PB91-231480.

Carbon tetrachloride   **EPA Method 524**
CAS #56-23-5

TITLE Measurement of Purgeable Organic Compounds in Water by Capillary Column GC/MS.

MATRIX Drinking water and raw source water; the latter should include most surface water and groundwater sources.

METHOD SUMMARY Method 524.2 covers 60 volatile organic compounds. An inert gas (zero grade nitrogen or helium) is bubbled through a 25-mL or a 5-mL water sample (depending on the expected concentration of the analytes). Purged sample components are trapped in a tube of sorbent materials. When purging is complete, the sorbent tube is heated and backflushed with helium to desorb the trapped sample onto a capillary GC column.

INTERFERENCES Impurities in the purge gas and from organic compounds outgassing from the plumbing ahead of the trap account for many contamination problems. Interferences purged or coextracted from the samples will vary considerably from source to source, depending upon the particular sample or extract being tested. Cross-contamination can occur whenever high-level and low-level samples are analyzed sequentially. Samples also can be contaminated by diffusion of volatile organics (particularly methylene chloride and fluorocarbons) through the septum seal into the sample during shipment and storage.

INSTRUMENTATION A GC/MS with a data system equipped with one of the following capillary GC columns:

Column 1: VOCOL glass wide bore capillary column.

Column 2: DB-624 fused silica capillary column.
Column 3: DB-5 fused silica capillary column.

Also required is an all-glass 25 mL or 5-mL purging device, a sorbent trap, and a thermal desorption apparatus which is connected to the GC/MS system.

PRECISION & ACCURACY Method detection limits are compound- and instrument-dependent, and may vary from approximately 0.02–0.35 µg/L. Note in the table below that the "true" concentration range used for accuracy and precision measurements was quite narrow. However, the applicable concentration range of this method is primarily column dependent and is approximately 0.02 to 200 µg/L for the wide-bore thick-film columns. Narrow-bore thin-film columns may have a capacity which limits the range to about 0.02 to 20 µg/L. Analytes that are inefficiently purged from water will not be detected when present at low concentrations, but they can be measured with acceptable accuracy and precision when present in sufficient amounts.

Analytes that are not separated chromatographically, but which have different mass spectra and non-interfering quantification ions, can be identified and measured in the same calibration mixture or water sample. Analytes which have very similar mass spectra cannot be individually identified and measured in the same calibration mixture or water samples unless they have different retention times. Co-eluting compounds with very similar mass spectra, typically many structural isomers, must be reported as an isomeric group or pair.

The range (in µg/L) was 0.5–10.
The Method Detection Limig (in µg/L) was 0.21.
The accuracy (as % recovery) was 84.
The precision (in %) was 8.8.

Note: Data were obtained from 16–31 determinations using a wide-bore capillary column and a jet separator interfaced to a quadrupole mass spectrometer. All analytes were in a reagent water matrix.

SAMPLING METHOD Collect samples using a 40- to 120-mL screw-cap vial (prewashed with detergent, rinsed with distilled water and oven dried at 105°C) with a Teflon®-faced silicone septum. Collect bubble-free samples and place the septum with the Teflon® side down on the water.

SAMPLE PRESERVATION If residual chlorine is present in the water add about 25 mg of ascorbic acid to each vial before samples are collected to remove the chlorine. Add hydrochloric acid to reduce pH to <2, and immediately cool samples to 4°C, and store them in a solvent-free refrigerator at 4°C until analysis.

MHT The maximum holding time for samples is 14 days from the time they were collected.

SAMPLE PREPARATION Remove the plungers from two 25-mL (or 5-mL depending on sample size) syringes and attach a closed syringe valve to each. Warm the sample to room temperature, open the sample bottle, and carefully pour the sample into one of the syringe barrels to just short of overflowing. Replace the syringe plunger, invert the syringe, and compress the sample. Open the syringe valve and vent any residual air while adjusting the sample volume to 25.0 mL (or 5 mL). For samples and blanks, add 5 µL of the fortification solution containing the internal standard and the surrogates to the sample through the syringe valve. For calibration standards and lab fortified blanks, add 5 µL of the fortification solution containing the internal standard only. Close the valve. Fill the second syringe in an identical manner from the same sample bottle. Reserve this second syringe for a reanalysis if necessary.

QUALITY CONTROL As an initial demonstration of lab accuracy and precision, analyze 4 to 7 replicates of a lab fortified blank containing analyte at 0.2–5 µg/L. Collect all samples in duplicate. Surrogate analytes (similar to those of the analytes of interest), whose concentration is known in every sample, are measured using the same internal standard calibration procedure. Duplicate field reagent water blanks (trip blanks) must be analyzed with each set of samples, lab reagent blanks (method blanks) must be analyzed with each batch of samples processed as a group within a work shift. Also, a single lab-fortified blank that contains each of the analytes of interest should be analyzed with each batch of samples processed as a group within a work shift. A 3- to 5-point calibration curve is needed depending on the calibration range factor required.

EPA CONTACT & HOTLINE For technical questions contact Dr. Baldev Bathija, U.S. EPA, Office of Ground Water and Drinking Water (WH-550D), 401 M St. SW, Washington, DC 20460. Tel. (202) 260-3040. For further information the EPA Safe Drinking Water Hotline may be called at: (800) 426-4791.

REFERENCE Methods for the Determination of Organic Compounds in Drinking Water, EPA/600/4-88/039 (revised July 1991; Final Rule for determination of compliance with the MCL for Total Trihalomethanes under 141.30, in 40 CFR Part 141, Vol. 58, No. 147, Fed. Reg., Tuesday Aug. 3, 1993). U.S. EPA Environmental Monitoring Systems Laboratory, Cincinnati, OH, 45268, USA. Available from the National Technical Information Service (NTIS), 5285 Port Royal Road, Springfield, VA 22161; Tel. 800-553-6847. NTIS Order Number is PB91-231480.

Carbon tetrachloride   **EPA Method 8021**
CAS #56-23-5

TITLE Halogenated Volatile by Gas Chromatography Using Photoionization and Electrolytic Conductivity Detectors in Series: Capillary Column Technique

MATRIX This method is applicable to nearly all types of samples, regardless of water content, including groundwater, aqueous sludges, caustic liquors, acid liquors, waste solvents, oily wastes, mousses, tars, fibrous wastes, polymeric emulsions, filter cakes, spent carbons, spent catalysts, soils, and sediments.

METHOD SUMMARY This method is used to determine 60 volatile organic compounds in a variety of solid waste matrices. It provides GC conditions for the detection of halogenated and aromatic volatile organic compounds. Samples can be analyzed using direct injection or purge-and-trap (EPA Method 5030). Groundwater samples must be analyzed using EPA Method 5030 (where applicable). A temperature program is used with the GC. Detection is achieved by a photoionization detector (PID) and a Hall electrolytic conductivity detector (HECD) in series.

INTERFERENCES Samples can be contaminated by diffusion of volatile organics (particularly chlorofluorocarbons and methylene chloride) through the sample container septum during shipment and storage.

INSTRUMENTATION A GC-equipped with variable-constant differential flow controllers, subambient oven controller, PID and HECD detectors connected with a short piece of uncoated capillary tubing and a data system.

Column: 60 m × 0.75 mm I.D. VOCOL wide-bore capillary column with 1.5 μm film thickness.

PRECISION & ACCURACY MDLs are compound-dependent and vary with purging efficiency and concentration. The applicable concentration range of this method is compound- and instrument-dependent but is approximately 0.1 to 200 μg/L. Analytes that are inefficiently purged from water will not be detected when present at low concentrations, but they can be measured with acceptable accuracy and precision when present in sufficient amounts. The estimated quantitation limit (EQL) for an individual compound is approximately 1 μg/kg (wet weight) for soil/sediment samples, 100 μg/kg (wet weight) for wastes, and 1 μg/L for groundwater. EQLs will be proportionately higher for sample extracts and samples that require dilution or reduced sample size to avoid saturation of the detector.

MULTIPLICATION FACTORS FOR OTHER MATRICES (a)

Matrix	Factor (b)
Groundwater	10
Low-concentration soil	10
Water miscible liquid waste	500
High-concentration soil and sludge	1250
Non-water miscible waste	1250

(a) Sample EQLs are highly matrix-dependent. The EQLs listed herein are provided for guidance and may not always be achievable.
(b) EQL = [Method detection limit] × [Factor]. For non-aqueous samples, the factor is on a wet-weight basis.

SINGLE LABORATORY ACCURACY & PRECISION DATA FOR VOCs IN WATER

This method was tested in a single lab using water spiked at 10 μg/L and the following data was reported:

Recoveries and standard deviations were determined from seven samples and spiked at 10 μg/L of each analyte. Recoveries were determined by the internal standard method. Internal standards were: Fluorobenzene for PID and 2-Bromo-1-chloropropane for HECD.
The average recovery (in percent) for the PID was none (no response for this detector).
The standard deviation of the recovery for the PID was none (no response for this detector).
The MDL (in μg/mL) for the PID was none (no response for this detector).
The average recovery (in percent) for the HECD was 92.
The standard deviation of the recovery for the HECD was 3.3.

The MDL (in μg/mL) for the HECD was 0.01.

SAMPLE COLLECTION, PRESERVATION & HANDLING
Volatile Organics — Standard 40-mL glass screw-cap VOA vials with Teflon®-faced silicone septum may be used for both liquid and solid matrices. When collecting samples, liquids and solids should be introduced into the vials gently to reduce agitation which might drive off volatile compounds. If there are any air bubbles present the sample must be retaken. Tap slightly as they are filled to try and eliminate as much free air space as possible. The two vials from each sampling locations should be sealed in separate plastic bags to prevent cross-contamination between samples particularly if the sampled waste is suspected of containing high levels of volatile organics.

Semivolatile organics — Containers used to collect samples for the determination of semivolatile organic compounds should be soap and water washed followed by methanol (or isopropanol) rinsing. The sample containers should be of glass or Teflon® and have screw-top covers with Teflon® liners.

Preservation for volatile organics — No preservation is used with concentrated waste samples. With liquid samples containing no residual chlorine, 4 drops of concentrated hydrochloric acid are added and the samples are immediately cooled to 4°C. When liquid samples contain residual chlorine, they are treated as above and, in addition, 4 drops of 4% aqueous sodium thiosulfate are added. Soil, sediment, and sludge samples are only cooled to 4°C.

Preservation for semivolatile organics — No preservation is used with concentrated waste samples. With liquid samples containing no residual chlorine and with soil, sediment, and sludge samples, immediately cooling to 4°C is the only preservation used. When residual chlorine is present then 3 mL of 10% aqueous sodium sulfate is added for each gallon of sample collected, followed by cooling to 4°C.

MHT The holding time for all volatile organics samples is 14 days. Liquid samples must be extracted within 7 days and their extracts analyzed within 40 days. Concentrated waste, soil, sediment, and sludge samples must be extracted within 14 days and their extracts analyzed within 40 days.

SAMPLE PREPARATION Volatile compounds are introduced into the gas chromatograph either by direct injector or purge-and-trap (EPA Method 5030). EPA Method 5030 may be used directly on groundwater samples or low-concentration contaminated soils and sediments. For medium-concentration soils or sediments, methanolic extraction, as described in EPA Method 5030, may be necessary prior to purge-and-trap analysis.

QUALITY CONTROL Calculate surrogate standard recovery on all samples, blanks, and spikes. A trip blank is recommended to check on sampling, storage, and handling contamination. Calibration standards, at a minimum of five concentration levels, are prepared in organic-free reagent water. One of the concentration levels should be at a concentration near, but above, the method detection limit.

A combination of bromochloromethane, 2-bromo-1-chloropropane, 1,4-dichlorobutane, and bromochlorobenzene are recommended as surrogate standards to encompass the range of the temperature program used in this method.

REFERENCE Test Methods for Evaluating Solid Waste, Physical/Chemical Methods, SW-846, 3rd Edition, U.S. EPA, Office of Solid Waste, Washington, DC, EPA Method 8021A, Rev. 1, Nov. 1992.

Carbon tetrachloride   **EPA Method 8240**
CAS #56-23-5

TITLE Volatile Organics By GC/MS: Packed Column Technique

MATRIX Nearly all types of sample matarices, regardless of water content, can be analyzed using this method. This includes groundwater, aqueous sludges, caustic liquors, acid liquors, waste solvents, oily wastes, mousses, tars, fibrous wastes, polymetric emulsions, filter cakes, spent carbons, spent catalysts, soils, and sediments.

METHOD SUMMARY Method 8240B covers 80 volatile organic compounds that are introduced into a gas chromatograph by the purge-and-trap method or by direct injection (in limited applications). For the purge-and-trap method an inert gas (zero grade nitrogen or helium) is bubbled through a 5-mL solution at ambient temperature. Purged sample components are trapped in a tube of sorbent materials. When purging is complete, the sorbent tube is heated and backflushed with inert gas to desorb the trapped components onto a GC column.

INTERFERENCES Impurities in the purge gas and from organic compounds outgassing from the plumbing ahead of the trap account for many contamination problems. Interferences purged or coextracted from the samples will vary considerably from source to source. Cross-contamination can occur whenever high-level and low-level samples are analyzed sequentially. Whenever an unusually concentrated sample is analyzed, it should be followed by the analysis of organic-free reagent water to check for cross-contamination. Samples also can be contaminated by diffusion of volatile organics (particularly methylene chloride and fluorocarbons) through the septum seal into the sample during shipment and storage. A trip blank can serve as a check on such contamination. The lab where volatile analysis is performed and also the refrigerated storage area should be completely free of solvents.

INSTRUMENTATION A gas chromatograph/mass spectrometry/data system (GC/MS) equipped with a 6 ft × 0.1 in I.D. glass column packed with 1% SP-1000 on Carbopack-B (60/80 mesh) is required. Also needed is a 5-mL purging device, a sorbent trap, and a thermal desorption apparatus.

PRECISION & ACCURACY This method is reported to have been tested by 15 laboratories using organic-free reagent water, drinking water, surface water, and industrial wastewaters (not specified) fortified at six concentrations over the range 5–600 µg/L.

Sample estimated quantitation limits (EQLs) are highly matrix-dependent. The EQLs listed may not always be achievable. EQLs listed for soils or sediments are based on wet weight. Normally, data is reported on a dry-weight basis; therefore, EQLs will be higher, based on the percent dry weight of each sample. Note that EQLs are even more variable than MDLs and that they are highly variable depending on the matrix being analyzed.

EQL in groundwater in µg/L was 5.
EQL in low soil or sediment in µg/kg was 5.
Accuracy (a) in µg/L was 1.10C–1.68.
Precision (b) in µg/L was 0.11x+0.37.

(a) Average recovery found for measurements of samples containing a concentration of C, in µg/L.
(b) Overall precision found for measurements of samples with average recovery X for samples containing a concentration of C in µg/L.
X = Average recovery found for measurement of samples containing a concentration of C in µg/L.

MULTIPLICATION FACTORS FOR OTHER MATRICES

Other Matrices	Factor (a)
Waste miscible liquid waste	50
High-concentration soil and sludge	125
Non-water miscible waste	500

(a) EQL = [EQL for low soil/sediment] × [Factor]. For non-aqueous samples, the factor is on a wet-weight basis.

SAMPLING METHOD

Liquid samples — Use a 40-mL glass screw-cap VOA vial with a Teflon®-faced silicone septum that has been prewashed, rinsed with distilled deionized water, and oven dried. However, if residual chlorine is present, collect sample in a 40-oz. soil VOA container which has been pre-preserved with 4 drops of 10% sodium thiosulfate, mix gently, and then transfer the sample to a 40-mL VOA vial. Collect bubble-free samples in duplicate and seal them in separate plastic bags.

Soils or sediments, and sludges — Use an 8-oz. widemouth glass bottle with a Teflon®-faced silicone septum that has been prewashed with detergent, rinsed with distilled deionized water, and oven dried. Tap slightly to eliminate free air space. Collect samples in duplicate and seal them in separate plastic bags.

SAMPLE PRESERVATION

Liquid samples — Add 4 drops of concentrated HCL and immediately cool samples to 4°C and store in a solvent-free refrigerator.

Soils or sediments, and sludges — Cool samples to 4°C and store in a solvent-free refrigerator.

MHT Maximum holding time is 14 days from the date of sample collection.

SAMPLE PREPARATION

Liquid samples — Remove the plunger from a 5-mL syringe and carefully pour the sample into the syringe barrel to just short of overflowing. Replace the syringe plunger and compress the sample. Open the syringe valve and vent any residual air while adjusting the sample volume to 5.0 mL. If there is only

one volatile organic analysis (VOA) vial, a second syringe should be filled at this time to protect against possible loss of sample integrity. Add 10 µL of surrogate spiking solution and 10 µL of internal standard spiking solution through the valve bore of the 5-mL syringe, then close the valve. The surrogate and internal standards may be mixed and added as a single spiking solution.

Sediments, soils, and waste samples — All samples of this type should be screened by GC analysis using a headspace method (EPA Method 3810) or the hexadecane extraction and screening method (EPA Method 3820). Use the screening data to determine whether to use the low-concentration method (0.005–1 mg/kg) or the high-concentration method (>1 mg/kg).

Low-concentration method — The low-concentration method is based on purging a heated sediment or soil sample mixed with organic-free reagent water containing the surrogate and internal standards. Analyze all reagent blanks and standards under the same conditions as the samples.

Use a 5-g sample if the expected concentration is <0.1 mg/kg or a 1-g sample for expected concentrations between 0.1 and 1 mg/kg. Mix the contents of the sample container with a narrow metal spatula. Weigh the amount of the sample into a tared purge device. Add the spiked water to the purge device, which contains the weighed amount of sample, and connect the device to the purge-and-trap system.

High-concentration method — This method is based on extracting the sediment or soil with methanol. A waste sample is either extracted or diluted, depending on its solubility in methanol. Wastes that are insoluble in methanol are diluted with reagent tetraglyme or possibly polyethylene glycol (PEG). An aliquot of the extract is added to organic-free reagent water containing surrogate and internal standards. This is purged at ambient temperature. All samples with an expected concentration of >1.0 mg/kg should be analyzed by this method.

Mix the contents of the sample container with a narrow metal spatula. For sediments or soils and solid wastes that are insoluble in methanol, weigh 4 g (wet weight) of sample into a tared 20-mL vial. For waste that is soluble in methanol, tetraglyme, or PEG, weigh 1 g (wet weight) into a tared scintillation vial or culture tube or a 10-mL volumetric flask. Quickly add 9.0 mL of appropriate solvent then add 1.0 mL of a surrogate spiking solution to the vial, cap it, and shake it for 2 min.

METHANOL EXTRACT REQUIRED FOR ANALYSIS OF HIGH-CONCENTRATION SOILS OR SEDIMENTS

Approximate Concentration Range	Volume of Methanol Extract (a)
500–10,000 µg/kg	100 µL
1,000–20,000 µg/kg	50 µL
5,000–100,000 µg/kg	10 µL
25,000–500,000 µg/kg	100 µL of 1/50 dilution (b)

Calculate appropriate dilution factor for concentrations exceeding this table.

(a) The volume of methanol added to 5 mL of water being purged should be kept constant. Therefore, add to the 5-mL syringe whatever volume of methanol is necessary to maintain a volume of 100 µL added to the syringe.
(b) Dilute an aliquot of the methanol extract and then take 100 µL for analysis.

QUALITY CONTROL Demonstrate, through the analysis of a reagent water blank, that interferences from the analytical system, glassware, and reagents are under control. Blank samples should be carried through all stages of the sample preparation and measurement steps. For each analytical batch (up to 20 samples), a reagent blank, matrix spike, and matrix spike duplicate must be analyzed (the frequency of the spikes may be different for different monitoring programs). The blank and spiked samples must be carried through all stages of the sample preparation and measurement steps. QC samples mentioned in the section on Interferences will also be needed as appropriate to those situations.

REFERENCE Test Methods for Evaluating Solid Waste (SW-846). U.S. EPA. 1983. Method 8240B, Rev. 2, Nov. 1990. Office of Solid Wastes, Washington, DC.

Carbon tetrachloride   **EPA Method 8260**
CAS #56-23-5

TITLE Volatile Organic Compounds by GC/MS: Capillary Column Technique

MATRIX This method is applicable to nearly all types of samples, regardless of water content, including groundwater, soils, and sediments.

METHOD SUMMARY Method 8260A covers 58 volatile organic compounds that are introduced into a gas chromatograph by the purge-and-trap method or by direct injection (in limited applications). Zero-grade helium is bubbled through a 5-mL solution at ambient temperature. Purged sample components are trapped in a tube containing suitable sorbent materials. When purging is complete, the sorbent tube is heated and backflushed with helium to desorb trapped sample components. The analytes are desorbed directly to a large bore capillary or cryofocussed on a capillary precolumn before being flash evaporated to a narrow bore capillary for analysis.

INTERFERENCES Major contaminant sources are volatile materials in the lab and impurities in the inert purging gas and in the sorbent trap. Interfering contamination may occur when a sample containing low concentrations of volatile organic compounds is analyzed immediately after a sample containing high concentrations of volatile organic compounds. After analysis of a sample containing high concentrations of volatile organic compounds, one or more calibration blanks should be analyzed to check for cross-contamination. Screening of the samples prior to purge-and-trap GC/MS analysis is highly recommended to prevent contamination of the system. This is especially true for soil and waste samples.

Special precautions must be taken to analyze for methylene chloride. The analytical and sample storage area should be isolated from all atmospheric sources of methylene chloride. All gas chromatography carrier gas lines and purge gas plumbing should be constructed from stainless steel or copper tubing. Laboratory clothing previously exposed to methylene chloride fumes during liquid-liquid extraction procedures can contribute to sample contamination.

Samples can also be contaminated by diffusion of volatile organics (particularly methylene chloride and fluorocarbons) through the septum seal during shipment and storage. A trip blank can serve as a check on such contamination.

INSTRUMENTATION GC/MS with a temperature-programmable chromatograph suitable for splitless injection equipped with variable constant differential flow controllers, a subambient oven controller, a purging device, sorbent trap, a thermal desorption apparatus and a capillary precolumn interface when using cryogenic cooling will be needed. The following GC columns may be used:

Column 1: 60 m × 0.75mm I.D. capillary column coated with VOCOL, 1.5 µm film thickness.
Column 2: 30 m × 0.53mm capillary column coated with DB-624 or VOCOL, 3 µm film thickness.
Column 3: 30 m × 0.32mm I.D. capillary column coated with DB-5 or SE-54, 1-µm film thickness.

PRECISION & ACCURACY This method has been tested in a single lab using spiked water. Using a wide-bore capillary column, water was spiked at concentrations between 0.5 and 10 µg/L. Single lab accuracy and precision data are presented. The MDL actually achieved in a given analysis will vary depending on instrument sensitivity and matrix effects.

The MDL (a) in µg/L was 0.21.
The concentration range in µg/L was 0.5–10.
The mean accuracy (% of true value) was 84.
The precision as relative standard deviation was 8.8.

Note: The MDL is based on a 25-mL sample volume instead of a 5-mL sample volume.

SAMPLING METHOD
Liquid samples — Use a 40-mL glass screw-cap VOA vial with a Teflon®-faced silicone septum that has been prewashed, rinsed with distilled deionized water, and oven dried. If residual chlorine is present, collect the sample in a 4-oz soil VOA container which has been pre-preserved with 4 drops of 10% sodium thiosulfate. Mix gently and transfer the sample to a 40-mL VOA vial. Collect bubble-free samples in duplicate and seal each sample in a separate plastic bag.

Soils, sediments, and sludges — Use an 8-oz widemouth glass bottle with Teflon®-faced silicone septum that has been prewashed, rinsed with distilled deionized water, and oven dried. **Do not** heat the septum for more than 1 h. Tap slightly to eliminate any free air space. Collect samples in duplicate and seal each one in a separate plastic bag.

SAMPLE PRESERVATION
Liquid samples — Add 4 drops of concentrated HCL, cool to 4°C and store in a solvent-free refrigerator.

Soils, sediments and sludges — Cool samples to 4°C and store in a solvent-free refrigerator.

MHT The maximum holding time of any sample (liquids, soils, sediments, and sludges) is 14 days.

SAMPLE PREPARATION
Liquid samples — Remove the plunger from a 5-mL syringe and carefully pour the sample into the syringe barrel to just short of overflowing. Replace the syringe plunger and compress the sample. Open the syringe valve and vent any residual air while adjusting the sample volume to 5.0 mL. If there is only one volatile organic analysis (VOA) vial, a second syringe should be filled at this time to protect against possible loss of sample integrity. Add 10 µL of surrogate spiking solution and 10 µL of internal standard spiking solution through the valve bore of the 5-mL syringe, then close the valve. The surrogate and internal standards may be mixed and added as a single spiking solution.

Sediments, soils, and waste samples — All samples of this type should be screened by GC analysis using a headspace method (EPA Method 3810) or the hexadecane extraction and screening method (EPA Method 3820). Use the screening data to determine whether to use the low-concentration method (0.005–1 mg/kg) or the high-concentration method (>1 mg/kg).

Low-concentration method — The low-concentration method is based on purging a heated sediment or soil sample mixed with organic-free reagent water containing the surrogate and internal standards. Analyze all reagent blanks and standards under the same conditions as the samples.

Use a 5-g sample if the expected concentration is <0.1 mg/kg or a 1-g sample for expected concentrations between 0.1 and 1 mg/kg. Mix the contents of the sample container with a narrow metal spatula. Weigh the amount of the sample into a tared purge device. Add the spiked water to the purge device, which contains the weighed amount of sample, and connect the device to the purge-and-trap system.

High-concentration method — This method is based on extracting the sediment or soil with methanol. A waste sample is either extracted or diluted, depending on its solubility in methanol. Wastes that are insoluble in methanol are diluted with reagent tetraglyme or possibly polyethylene glycol (PEG). An aliquot of the extract is added to organic-free reagent water containing surrogate and internal standards. This is purged at ambient temperature. All samples with an expected concentration of >1.0 mg/kg should be analyzed by this method.

Mix the contents of the sample container with a narrow metal spatula. For sediments or soils and solid wastes that are insoluble in methanol, weigh 4 g (wet weight) of sample into a tared 20-mL vial. For waste that is soluble in methanol, tetraglyme, or PEG, weigh 1 g (wet weight) into a tared scintillation vial or culture tube or a 10-mL volumetric flask. Quickly add

9.0 mL of appropriate solvent then add 1.0 mL of a surrogate spiking solution to the vial, cap it, and shake it for 2 min.

METHANOL EXTRACT REQUIRED FOR ANALYSIS OF HIGH-CONCENTRATION SOILS OR SEDIMENTS

Approximate Concentration Range	Volume of Methanol Extract (a)
500–10,000 µg/kg	100 µL
1,000–20,000 µg/kg	50 µL
5,000–100,000 µg/kg	10 µL
25,000–500,000 µg/kg	100 µL of 1/50 dilution (b)

Calculate appropriate dilution factor for concentrations exceeding this table.

(a) The volume of methanol added to 5 mL of water being purged should be kept constant. Therefore, add to the 5-mL syringe whatever volume of methanol is necessary to maintain a volume of 100 µL added to the syringe.
(b) Dilute an aliquot of the methanol extract and then take 100 µL for analysis.

QUALITY CONTROL Demonstrate, through the analysis of a reagent water blank, that interferences from the analytical system, glassware, and reagents are under control. Blank samples should be carried through all stages of the sample preparation and measurement steps. For each analytical batch (up to 20 samples), a reagent blank, matrix spike, and matrix spike duplicate must be analyzed (the frequency of the spikes may be different for different monitoring programs). The blank and spiked samples must be carried through all stages of the sample preparation and measurement steps. QC samples mentioned in the section on Interferences will also be needed as appropriate to those situations.

Matrix spiking standards should be prepared from volatile organic compounds which will be representative of the compounds being investigated. The recommended internal standards are chlorobenzene-d5, 1,4-difluorobenzene, 1,4-dichlorobenzene-d4, and pentafluorobenzene. Using stock standard solutions, prepare secondary dilution standards containing the compounds of interest, either singly or mixed together in methanol. Store them in a vial with no headspace for no more than one week. Surrogates recommended are toluene-d8, 4-bromofluorobenzene, and dibromofluoromethane. Each sample undergoing GC/MS analysis must be spiked with 10 µL of the surrogate spiking solution prior to analysis.

REFERENCE Test Methods for Evaluating Solid Waste (SW-846). U.S. EPA 1983, Method 8260A, Rev. 1, Nov. 1990. Office of Solid Waste, Washington, DC.

Carbon tetrachloride   EPA Method 601
CAS #56-23-5

TITLE Purgeable Halocarbons

MATRIX Wastewater

APPLICATION Method covers 29 purgeable halocarbons. (Method 624 provides GC/MS conditions appropriate for the qualitative and quantitative confirmation of results). Method describes conditions for a 2nd GC column to confirm measurements made with primary column.

INTERFERENCES Impurities in the purge gas and organic compounds outgassing from the plumbing ahead of the trap. With high- and low-level samples, there can be carryover contamination. Diffusion of volatile organics through the septum seal into the sample.

INSTRUMENTATION GC-equipped with halide-specific detector. (With purge-and-trap unit).

RANGE 8.0–500 µg/L.

MDL 0.12 µg/L.

PRECISION 0.20X+0.39 µg/L (overall precision).

ACCURACY 0.98C–1.04 µg/L (as recovery).

SAMPLING METHOD 25-mL glass vial. Teflon®-lined septum.

STABILITY cool, 4°C, 0.008% Sodium thiosulfate.

MHT 14 days.

QUALITY CONTROL The lab must on an ongoing basis, spike at least 10% of the samples from each sample site being monitored to assess accuracy.

REFERENCE Method 601, *Federal Register* Part VIII 40 CFR Part 136, Oct 26, 1984.

Carbon tetrachloride   EPA Method 624
CAS #56-23-5

TITLE Purgeables

MATRIX Wastewater

APPLICATION Method covers 31 purgeable organics. An inert gas is bubbled through a 5-mL water sample in a specially designed purging chamber. Here, purgeables are transferred from aqueous to gaseous phase, passed onto a sorbent column, and trapped. Trap is heated and backflushed with inert gas to desorb purgeables onto a GC column, where purgeables are separated.

INTERFERENCES Impurities in the purge gas, organic compounds outgassing from the plumbing ahead of the trap, and solvent vapors in the lab. With high- and low-level samples, there can be carryover contamination.

INSTRUMENTATION GC/MS with purge-and-trap unit.

RANGE 5–600 µg/L

MDL 2.8 µg/L

PRECISION 0.11X+0.37 µg/L (overall precision).

ACCURACY 1.10C–1.68 µg/L (as recovery).

SAMPLING METHOD 25-mL glass vial. Teflon®-lined septum.

STABILITY cool, 4°C, 0.008% Sodium thiosulfate.

MHT 14 days.

QUALITY CONTROL The lab must on an ongoing basis, spike at least 5% of the samples from each sample site being monitored to assess accuracy.

REFERENCE Method 624, *Federal Register* Part VIII 40 CFR Part 136, Oct 26, 1984.

Carbon tetrachloride   **EPA Method 8010**
CAS #56-23-5

TITLE Halogenated Volatile Organics

MATRIX Groundwater, soils, sludges water miscible liquid wastes, and non-water miscible wastes.

APPLICATION This method is used for the analysis of 39 halogenated VOCs.

Samples are analyzed using direct injection or purge-and-trap methods.

Groundwater must be analyzed by the purge-and-trap method. The method provides an optional GC column which is used for analyte confirmation and that may help resolve analytes from interferences.

INTERFERENCES There can be carryover contamination with high- and low-level samples. Impurities may come from the purge-and-trap apparatus, organic compounds outgassing from the plumbing ahead of trap, diffusion of VOCs through the sample bottle septum during shipping or storage, or from solvent vapors in the lab.

INSTRUMENTATION GC capable of on-column injections or purge-and-trap sample introduction and a halogen specific detector. Column 1: 8 ft by 0.1 in 1%. SP-1000 on Carbopack-B. Column 2: 6 ft by 0.1 in bonded n-octane on Porasil-C.

RANGE 8 to 500 µg/L (reagent water)

MDL 0.12 µg/L (reagent water).

PQL FACTORS FOR MULTIPLYING × FID MDL VALUE

Matrix	Multiplication Factor
Groundwater	10
Low-level soil	10
Water miscible liquid waste	500
High-level soil and sludge	1250
Non-water miscible waste	1250

PRECISION $0.20X + 0.39$ µg/L (overall precision).

ACCURACY $0.98C-1.04$ µg/L (as recovery).

SAMPLING METHOD For water and liquid samples; use glass 40-mL vials with Teflon®-lined septum caps and collect two vials per sample location with no headspace. For solids and concentrated waste samples; use widemouth glass bottles with Teflon® liners.

STABILITY For concentrated wastes, soils, sediments, or sludges: cool to 4°C. For liquids: add 4 drops of concentrated hydrochloric acid and cool to 4°C.

MHT 14 days.

QUALITY CONTROL Analyze a reagent blank, matrix spike, and matrix spike duplicate/duplicate for each analytical batch (up to 20 samples). Demonstrate the purity of glassware and reagents by analyzing a reagent water method blank. Internal, surrogate, and five concentration level calibration standards are used.

REFERENCE Test Methods for Evaluating Solid Waste (SW-846), U.S. EPA Office of Solid Waste, Washington, DC, Method 8010B, Rev. 2, Nov. 1992.

Carbophenothion   **EPA Method 8270**
CAS #786-19-6

TITLE Semivolatile Organic Compounds by GC/MS

MATRIX This method is used to determine the concentration of semivolatile organic compounds in extracts prepared from all types of solid waste matrices, soils, and groundwater. Although surface waters are not specifically mentioned, this method should be applicable to water samples from rivers, lakes, etc.

METHOD SUMMARY This method covers 259 semivolatile organic compounds. In very limited applications direct injection of the sample into the GC/MS system may be appropriate, but this results in very high detection limits (approximately 10,000 µg/L). Typically, a 1-L liquid sample, containing surrogate, and matrix spiking standards, is extracted in a continuous extractor first under acid conditions and then under basic conditions. Typically 30 g of a solid sample, containing surrogate, and matrix spiking standards, is extracted ultrasonically. After concentrating the extract to 1 mL it is spiked with 10 µL of an internal standard solution just prior to analysis by GC/MS. The volume injected should contain about 100 ng of base/neutral and 200 ng of acid surrogates (for a 1-µL injection). Analysis is performed by GC/MS using a capillary GC column.

INTERFERENCES Raw GC/MS data from all blanks, samples, and spikes must be evaluated for interferences. Contamination by carryover can occur whenever high-concentration and low-concentration samples are sequentially analyzed. To reduce carryover, the sample syringe must be rinsed out between samples with solvent. Whenever an unusually concentrated sample is encountered, it should be followed by the analysis of blank solvent to check for cross-contamination.

INSTRUMENTATION A GC/MS and a data system are required. The GC column used is a 30 m × 0.25 mm I.D. (or 0.32 mm I.D.) 1um film thickness silicone-coated fused silica capillary column. A continuous liquid-liquid extractor equipped with Teflon® or glass connection joints and stopcocks requiring no lubrication, a K-D concentrating apparatus, water

bath, and an ultrasonic disrupter with a minimum power of 300 W and with pulsing capability are also required.

PRECISION & ACCURACY The estimated quantitation limit (EQL) of Method 8270B for determining an individual compound is approximately 1 mg/kg (wet weight) for soil or sediment samples, 1–200 mg/kg for wastes (dependent on matrix and method of preparation), and 10 µg/L for groundwater samples. EQLs will be proportionately higher for sample extracts that require dilution to avoid saturation of the detector.

The EQL(b) for groundwater in µg/L is 10.
The EQL (a, b) for low concentrations in soil and sediment in µg/kg is not determined.
Accuracy as µg/L is not listed.
Overall precision in µg/L is not listed.

(a) EQLs listed for soil/sediment are based on wet weight. Normally data is reported in a dry-weight basis; therefore, EQLs will be higher based on the % dry weight of each sample. This calculation is based on a 30 g sample and gel permeation chromatography cleanup.
(b) Sample EQLs are highly matrix-dependent. The EQLs are provided for guidance and may not always be achievable.
C = True value for concentration, in µg/L.
X = Average recovery found for measurements of samples containing a concentration of C, in µg/L.

ESTIMATED QUANTITATION LIMIT FOR OTHER MATRICES

Other Matrices	Factor (a)
High-concentration soil and sludges by sonicator	7.5
Non-water miscible waste	75

(a) EQL for other matrices = [EQL for low soil/sediment] × [Factor]. This estimated EQL is similar to an EPA "Practical Quantitation Limit."

SAMPLING METHOD
Liquid samples — Use a 1 or 2½ gallon amber glass bottle with a screw-top Teflon®-lined cover that has been prewashed with detergent and rinsed with distilled water and methanol (or isopropanol).

Soils, sediments, or sludges — Use an 8-oz. widemouth glass with a screw-top Teflon®-lined cover that has been prewashed with detergent and rinsed with distilled water and methanol (or isopropanol).

SAMPLE PRESERVATION
Liquid samples — If residual chlorine is present, add 3 mL of 10% sodium thiosulfate per gallon, cool to 4°C and store in a solvent-free refrigerator until analysis; if chlorine is not present, then eliminate the sodium thiosulfate addition.

Soils, sediments, or sludges — Cool samples to 4°C and store in a solvent-free refrigerator.

MHT Liquid samples must be extracted within 7 days and the extracts analyzed within 40 days. Soils, sediments, or sludges may be stored for a maximum of 14 days and the extracts analyzed within 40 days.

SAMPLE PREPARATION
Liquid samples — Transfer 1 L quantitatively to a continuous extractor. If high concentrations are anticipated, a smaller volume may be used and then diluted with organic-free reagent water to 1 L. Adjust pH, if necessary, to pH <2 using 1:1 (V/V) sulfuric acid. Pipette 1.0 mL of a surrogate standard spiking solution into each sample. For the sample in each analytical batch selected for spiking, add 1.0 mL of a matrix spiking standard. For base/neutral acid analysis, the amount of the surrogates and matrix spiking compounds added to the sample should result in a final concentration of 100 ng/µL of each analyte in the extract to be analyzed (assuming a 1-µL injection). Extract with methylene chloride for 18–24 h. Next, adjust the pH of the aqueous phase to pH >11 using 10 N sodium hydroxide and extract it with methylene chloride again for 18–24 h. Dry the extract through a column containing anhydrous sodium sulfate and concentrate it to 1 mL using a K-D concentrator.

Soils, sediments, or sludges — Use 30 g of sample. Nonporous or wet samples (gummy or clay type) that do not have a free-flowing sandy texture must be mixed with anhydrous sodium sulfate until the sample is free flowing. Add 1 mL of surrogate standards to all samples, spikes, standards, and blanks. For the sample in each analytical batch selected for spiking, add 1.0 mL of a matrix spiking standard. For base/neutral acid analysis, the amount added of the surrogates and matrix spiking compounds should result in a final concentration of 100 ng/µL of each base/neutral analyte and 200 ng/µL of each acid analyte in the extract to be analyzed (assuming a 1-µL injection). Immediately add a 100-mL mixture of 1:1 methylene chloride:acetone and extract the sample ultrasonically for 3 min and then decant or filter the extracts. Repeat the extraction two or more times. Dry the extract using a column with anhydrous sodium sulfate and concentrate it to 1 mL in a K-D concentrator.

QUALITY CONTROL A methylene chloride solution containing 50 ng/µL of decafluorotriphenylphosphine (DFTPP) is used for tuning the GC/MS system each 12-h shift. A system performance check also must be made during every 12-h shift. A standard containing 50 ng/µL each of 4,4'-DDT, pentachlorophenol, and benzidine is required to verify injection port inertness and GC column performance. A calibration standard at mid-concentration, containing each compound of interest, including all required surrogates, must be performed every 12 h during analysis. After the system performance check is met, calibration check compounds (CCCs) are used to check the validity of the initial calibration.

The internal standard responses and retention times in the calibration check standard must be evaluated immediately after or during data acquisition. If the retention time for any internal standard changes by more than 30 seconds from the last check calibration (12 h), the chromatographic system must be inspected for malfunctions and corrections must be made, as required. If the electron ionization current plot (EICP) area for any of the internal standards changes by a factor of two from the last daily calibration standard check, the mass spectrometer must be inspected for malfunctions and corrections must be made, as appropriate.

Demonstrate, through the analysis of a reagent water blank, that interferences from the analytical system, glassware, and reagents are under control. The blank samples should be carried through all stages of the sample preparation and measurement steps. For each analytical batch (up to 20 samples), a reagent blank, matrix spike, and matrix spike duplicate/duplicate must be analyzed (the frequency of the spikes may be different for different monitoring programs). The blank and spiked samples must be carried through all stages of the sample preparation and measurement steps. A QC reference sample concentrate containing each analyte at a concentration of 100 mg/L in methanol is required.

REFERENCE Test Methods for Evaluating Solid Waste (SW-846). U.S. EPA 1983, Method 8270B, Rev. 2, Nov. 1990. Office of Solid Waste, Washington, DC.

Carboxin   EPA Method 507
CAS #5234-68-5

TITLE Determination of Nitrogen and Phosphorus-Containing Pesticides in Water by GC/NPD

MATRIX This method is applicable to the determination of certain nitrogen and phosphorus-containing pesticides in finished drinking water and groundwater.

METHOD SUMMARY Method 507 covers 46 nitrogen- and phosphorus-containing pesticides. A 1-L sample is fortified with a surrogate standard, salted, buffered, extracted with methylene chloride, and concentrated; then the solvent is exchanged with methyl tert-butyl ether (MTBE) and concentrated again, and a 2-µL aliquot of a sample extract is injected into a GC system equipped with a selective nitrogen-phosphorus detector and a capillary column for analysis.

INTERFERENCES Method interferences may be caused by contaminants in solvents, reagents, glassware, and other sample processing apparatus. Interfering contamination may occur when a sample containing low concentrations of analytes is analyzed immediately following a sample containing relatively high concentrations. One or more injections of MTBE should be made following the analysis of a sample with high concentrations of analytes to check for analyte carryover. Matrix interferences may be caused by contaminants that are coextracted from the sample. The extent of matrix interferences will vary considerably from source to source, depending upon the water sampled.

INSTRUMENTATION A gas chromatograph system (GC) equipped with a nitrogen-phosphorus detector (NPD) is needed.

Column 1: 30 m × 0.25 mm I.D. DB-5 bonded fused silica column, 0.25 µm film thickness, or equivalent.

Column 2: 30 m × 0.25 mm I.D. DB-1701 bonded fused silica column, 0.25 µm film thickness, or equivalent.

PRECISION & ACCURACY This method has been validated in a single lab and estimated detection limits (EDLs) have been determined for each analyte. Observed detection limits may vary among waters, depending upon the nature of the interferences in the sample matrix and the specific instrumentation used. Analytes that are not separated chromatographically cannot be individually identified and measured unless an alternative technique for identification and quantification exist.

The estimated detection limit (in µg/L) was 0.6. The EDL is defined as either method detection limit or a level of compound in a sample yielding a peak in the final extract with signal-to-noise ratio of approximately 5, whichever value is higher.

The concentration used for these measurements (in µg/L) was 6.
The accuracy (as % recovery) was 102.
The precision (% RSD) was 4.

SAMPLING METHOD Grab samples are collected in 1-L glass sample bottles (prewashed with detergent and hot tap water, rinsed with reagent water, and dried in an oven at 400°C for 1 h) with screw caps lined with PTFE-fluorocarbon.

SAMPLE PRESERVATION Add mercuric chloride to the sample bottle in amounts to produce a concentration of 10 mg/L. If residual chlorine is present, add 80 mg of sodium thiosulfate/L of sample to the sample bottle prior to collection. After collection, seal bottle and shake vigorously for 1 min, then cool the sample to 4°C immediately and store it at 4°C in the dark until extraction.

MHT Maximum holding time of the samples, and in some cases the extracts, is 14 days.

Note: Samples with this compound exhibited recoveries of less than 60% after 14 days.

SAMPLE PREPARATION Fortify the sample with 50 µL of the surrogate standard solution, adjust to pH 7 with phosphate buffer, add 100 g NaCl to the sample, and seal and shake to dissolve the salt; then extract with methylene chloride in a separatory funnel or in a mechanical tumbler bottle. Dry the extract by pouring it through a solvent-rinsed drying column containing about 10 cm of anhydrous sodium sulfate. Collect the extract in a Kuderna-Danish (K-D) concentrator and rinse the column with 20–30 mL methylene chloride. Concentrate the extract to about 2 mL and rinse the flask and its lower joint into the concentrator tube with 1 to 2 mL of methyl t-butyl ether (MTBE). Add 5–10 mL of MTBE and concentrate the extract twice (adding more MTBE) to a final volume of 5.0 mL and store it at 4°C until analysis.

Note: If methylene chloride is not completely removed from the final extract, it may cause detector problems.

QUALITY CONTROL Minimum quality control requirements are initial demonstration of lab capability, determination of surrogate compound recoveries in each sample and blank, monitoring internal standard peak area or height in each sample and blank, analysis of lab reagent blanks, lab fortified samples, lab fortified blanks, and other QC samples. A lab reagent blank is analyzed to demonstrate that all glassware and reagent interferences are under control.

Initial demonstration of capability is fulfilled by analyzing four fortified reagent water samples with the recovery value for each analyte falling within the acceptable range (±30% average recovery). Surrogate recoveries from samples or method blanks must be 70–130%. The internal standard response for any sample chromatogram should not deviate from the daily calibration check standard's internal standard response by more than 30% or lab fortified blanks and sample matrices are used to assess lab performance and analyte recovery, respectively.

If the response for the target analyte peak exceeds the working range of the system, dilute the extract and reanalyze. Alternative techniques such as an alternate detector or second chromatography column should be used to confirm peak identification when sample components are not resolved adequately.

EPA CONTACT & HOTLINE For technical questions contact Dr. Baldev Bathija, U.S. EPA, Office of Ground Water and Drinking Water (WH-550D), 401 M St. SW, Washington, DC 20460. Tel. (202) 260-3040. For further information the EPA Safe Drinking Water Hotline may be called at: (800) 426-4791.

REFERENCE Methods for the Determination of Organic Compounds in Drinking Water, EPA/600/4-88/039 (revised July 1991). U.S. EPA Environmental Monitoring Systems Laboratory, Cincinnati, OH, 45268, USA. Available from the National Technical Information Service (NTIS), 5285 Port Royal Road, Springfield, VA 22161; Tel. 800-553-6847. NTIS Order Number is PB91-231480.

Chemical Oxygen Demand (COD)   EPA Method 410.1

TITLE Organics

MATRIX Wastewater

APPLICATION Date issued 1971. Editorial Rev. 1978. (Titrimetric, mid level). EPA Method determines the quantity of oxygen required to oxidize the organic matter in a waste sample, under specific conditions of oxidizing agent, temperature and time. (Use 0.25N $K_2Cr_2O_7$).

INTERFERENCES Traces of organic material from glassware or atmosphere may cause gross positive error. Avoid inclusion of organic materials in distilled water for reagent preparation or sample dilution. Mercuric sulfate removes chloride interference. Cool flask during H_2SO_4 addition.

INSTRUMENTATION Reflux apparatus. 12 Inch allihn condenser. Ground glass joint connections.

RANGE Wastewater (with TOC >50 mg/L)

MDL Not listed.

PRECISION SD = ±17.76 mg/L COD at 270 mg/L COD.

ACCURACY As bias, -4.7% at 270 mg/L COD.

SAMPLING METHOD Plastic or glass (50 mL).

STABILITY cool, 4°C. H_2SO_4 to pH <2.

MHT 28 Days.

QUALITY CONTROL Organic and oxidizable inorganic substances are oxidized by potassium dichromate in 50% H_2SO_4 solution at reflux temperature for 2 hrs. Silver sulfate is used as a catalyst. Excess dichromate is titrated with standard ferrous ammonium sulfate (0.25N) using an indicator.

REFERENCE Methods for the Chemical Analysis of Water and Wastes, EPA-600/4-79-020, U.S. EPA, EMSL, 1979.

Chemical Oxygen Demand (COD)   EPA Method 410.2

TITLE Organics

MATRIX Surface Water and Wastewater.

APPLICATION Date issued 1971. Editorial Rev. 1974 and 1978. (Titrimetric, low level). Method determines the quantity of oxygen required to oxidize the organic matter in a waste sample, under specific conditions of oxidizing agent, temp and time. (Use 0.025N $K_2Cr_2O_7$).

INTERFERENCES Traces of organic material from glassware or atmosphere may cause gross positive error. Avoid inclusion of organic materials in distilled water for reagent preparation or sample dilution. Mercuric sulfate removes chloride interference. Cool flask during H_2SO_4 addition.

INSTRUMENTATION Reflux apparatus. 12 Inch allihn condenser. Ground glass joint connections.

RANGE 5–50 mg/L COD

MDL Not listed.

PRECISION SD = ±4.15 mg/L COD at 12.3 mg/L COD.

ACCURACY As bias, 0.3% at 12.3 mg/L COD.

SAMPLING METHOD Plastic or glass (50 mL).

STABILITY Cool, 4°C. H_2SO_4 to pH <2.

MHT 28 Days.

QUALITY CONTROL Organic and oxidizable inorganic substances are oxidized by potassium dichromate in 50% H_2SO_4 solution at reflux temperature for 2 h. Excess dichromate is titrated with standard ferrous ammonium sulfate (0.025N) using orthophenanthroline ferrous complex as indicator.

REFERENCE EPA Methods for the Chemical Analysis of Water and Wastes, EPA-600/4-79-020, U.S. EPA, EMSL, 1979.

Chemical Oxygen Demand (COD)   EPA Method 410.3

TITLE Organics

MATRIX Saline Waters

APPLICATION Date issued 1971. Editorial Rev. 1978. (Titrimetric, high level for saline waters). Method determines the quantity of oxygen required to oxidize the organic matter in a waste sample, under specific conditions of oxidizing agent, temperature and time.

INTERFERENCES Traces of organic material from glassware or atmosphere may cause gross positive error. Avoid inclusion of organic materials in distilled water for reagent preparation or sample dilution. Cool flask during H2SO4 addition. Use chloride correction procedure as outlined.

INSTRUMENTATION Reflux apparatus. 12 Inch allihn condenser. Ground glass joint connections.

RANGE >250 mg/L COD at chloride>1000 mg/L

MDL Method has special calculation

PRECISION Not listed.

ACCURACY Not listed.

SAMPLING METHOD Plastic or glass (50 mL).

STABILITY cool, 4°C. H2SO4 to pH <2.

MHT 28 Days.

QUALITY CONTROL Organic and oxidizable inorganic substances are oxidized by potassium dichromate in 50% H2SO4 solution at reflux temperature for 2 h. Excess dichromate is titrated with standard ferrous ammonium sulfate (0.25N) using an indicator. (Use smaller sample, for COD >800 mg/L).

REFERENCE Methods for the Chemical Analysis of Water and Wastes, EPA-600/4-79-020, U.S. EPA, EMSL, 1979.

Chemical Oxygen Demand (COD)   **EPA Method 410.4**

TITLE Organics

MATRIX Surface Waters and Wastewaters

APPLICATION Date issued 1978. (Colorimetric, automated, manual). Method determines the quantity of oxygen required to oxidize the organic matter in a waste sample, under specific conditions of oxidizing agent, temperature, and time.

INTERFERENCES Chlorides are quantitatively oxidized by dichromate and represent a positive interference. Mercuric sulfate is added to the digestion tubes to complex the chlorides.

INSTRUMENTATION Spectrophotometer (manually) or technicon auto analyzer (automated).

RANGE 3–900 mg/L (automated).

MDL Not listed.

PRECISION Not listed.

ACCURACY Not listed.

SAMPLING METHOD Plastic or glass. (50 mL).

STABILITY cool, 4°C. H2SO4 to pH <2.

MHT 28 Days.

QUALITY CONTROL Process standards and blanks exactly as the samples.

Samples, blanks and standards in sealed tubes are heated in an oven or block digester in presence of potassium dichromate (oxidizing agent) at 150 c for 2 h.

REFERENCE Methods for the Chemical Analysis of Water and Wastes, EPA-600/4-79-020, U.S. EPA, EMSL, 1979.

Chloramben   **EPA Method 8151**
CAS #133-90-4

TITLE Chlorinated Herbicides by GC Using Methylation or Pentafluorobenzylation Derivatization: Capillary Column Technique.

MATRIX This method covers aqueous and solid matrices. This includes a wide variety such as drinking water, groundwater, industrial wastewaters, surface waters, soils, solids, and sediments.

METHOD SUMMARY This is a GC method for determining 19 chlorinated acid herbicides in aqueous, soil, and waste matrices. Because these compounds are produced and used in various forms (i.e., acid, salt, ester, etc.) a hydrolysis step is included to convert the herbicide to the acid form prior to analysis. This method provides hydrolysis, extraction, derivatization and GC conditions for the analysis of chlorinated acid herbicides in water, soil, and waste samples. Water samples are hydrolyzed *in situ*, extracted with diethyl ether, and then esterified with either diazomethane or pentafluorobenzyl bromide. The derivatives are determined by gas chromatography with an electron capture detector (GC/ECD). The results are reported as acid equivalents. The sensitivity of this method depends on the level of interferences in addition to instrumental limitations.

INTERFERENCES Method interferences may be caused by contaminants in solvents, reagents, glassware, and other sample processing hardware. Immediately prior to use, glassware should be rinsed with the next solvent to be used. Matrix interferences may be caused by contaminants that are coextracted from the sample. Organic acids, especially chlorinated acids, cause the most direct interference with the determination by methylation. Phenols, including chlorophenols, may also interfere with this procedure. The determination using pentafluorobenzylation is more sensitive, and more prone to interferences from the presence of organic acids of phenols than by methylation. Alkaline hydrolysis and subsequent extraction of the basic solution removes many chlorinated hydrocarbons and phthalate esters that might otherwise interfere with the ECD analysis. The herbicides, being strong organic acids, react readily with alkaline substances and may be lost during analysis. Therefore, glassware must be acid-rinsed and then rinsed to constant pH with organic-free reagent water.

INSTRUMENTATION A GC suitable for Grob-type injection using capillary columns. A data system for measuring peak heights and/or peak areas is recommended. An electron capture detector (ECD) is used. Also a K-D apparatus, a diazomethane generator, a centrifuge and an ultrasonic disrupter will be required.

Narrow Bore Columns:
Primary Column 1: 30 m × 0.25 mm, 5% phenyl/95% methyl silicone (DB-5), 0.25 μm film thickness.
Primary Column 1a (GC/MS): 30 m × 0.32 mm, 5% phenyl/95% methyl silicone (DB-5), 1-μm film thickness.
Column 2: 30 m × 0.25 mm DB-608 with a 25 μm film thickness.
Confirmation Column: 30 m × 0.25 mm, 14% cyanopropyl phenyl silicone (DB-1701), 0.25 μm film thickness.

Megabore Columns:
Primary Column: 30 m × 0.53 mm DB-608 with 0.83 μm film thickness.
Confirmation Column: 30 m × 0.53 mm, 14% cyanopropyl phenyl silicone (DB-1701), 1.0 μm film thickness.

PRECISION & ACCURACY Method detection limits (MDLs) are compound-dependent and vary with derivitization efficiency, derivative recovery, the matrix sampled, and herbicide concentration.

The estimated MDL (in μg/L) was 0.93 for aqueous samples using GC/ECD.

The estimated MDL (in μg/kg) was 4.0 for soil samples using GC/ECD when corrected back to 50-g samples extracted and concentrated to 10 mL with 5-μL injections.

The estimated GC/MS identification limit (in ng) was 1.7 for soil samples using GC/MS.

Mean percent recovery, calculated from 7–8 determinations of spiked reagent water, after diazomethane derivatization, from a spike concentration (in μg/L) of 0.4 was 111 with a standard deviation of the percent recovery of 14.4.

Mean percent recovery, calculated from 10 determinations of spiked clay and clay/still bottom samples over the linear concentration range (in ng/g) of no data was none reported with a percent relative standard deviation of none. The RSD % was calculated on 10 samples high in the linear concentration range and 10 low in the range. The linear concentration range was determined using standard solutions and corrected to 50 g soil samples.

SAMPLE COLLECTION, PRESERVATION & HANDLING
Containers used to collect samples for the determination of semivolatile organic compounds should be soap and water washed followed by methanol (or isopropanol) rinsing. The sample containers should be of glass or Teflon® and have screw-top covers with Teflon® liners.

No preservation is used with concentrated waste samples. With liquid samples containing no residual chlorine and with soil, sediment, and sludge samples, immediately cooling to 4°C is the only preservation used. When residual chlorine is present then 3 mL of 10% aqueous sodium sulfate is added for each gallon of sample collected, followed by cooling to 4°C.

The holding time for all volatile organics samples is 14 days. Liquid samples must be extracted within 7 days and their extracts analyzed within 40 days. Concentrated waste, soil, sediment, and sludge samples must be extracted within 14 days and their extracts analyzed within 40 days.

SAMPLE PREPARATION
Preparation of soil, sediment, and other solid samples — Acidify 30 g (dry weight) solids with 0.1 M phosphate buffer (pH = 2.5) and thoroughly mix the contents. Spike the sample with surrogate compound(s). The ultrasonic extraction of solids must be optimized for each type of sample. In order for the ultrasonic extractor to efficiently extract solid samples, the sample must be free flowing when the solvent is added. Acidified anhydrous sodium sulfate should be added to clay-type soils, or any other solid that is not a free-flowing sandy texture, until a free flowing mixture is obtained. Add methylene chloride and perform ultrasonic extraction. Combine organic extracts from the repetitive extractings of the sample and centrifuge. Add aqueous potassium hydroxide, water, and methanol to the extract and reflux the mixture on a water bath. Extract the solution three times with methylene chloride and discard the methylene chloride phase. The basic solution contains the herbicide salts. Adjust the pH of the solution to <2 with cold sulfuric acid and extract three times with methylene chloride. Combine the extracts and pour them through a pre-rinsed drying column containing acidified anhydrous sodium sulfate. Collect the dried extracts in a K-D flask and concentrate them.

Preparation of aqueous samples — Measure 1 L of sample into a 2 L separatory funnel and spike it with surrogate compound(s). Add NaCl to the sample, then add 6 N NaOH to the sample to a pH of 12 or more and let the sample sit at room temperature for 1 h to hydrolyze esters. Extract the sample three times with methylene chloride and discard the extracts. Then add cold 12 N sulfuric acid to a pH less than or equal to 2, and extract the sample three times with ethyl ether. Collect the ether phase in a flask containing acidified anhydrous sodium sulfate and allow it to remain in contact with the sodium sulfate for a minimum of 2 h. The drying step is very critical to ensuring complete esterification; any moisture remaining in the ether will result in low herbicide recoveries.

Extract Concentration and Derivatization — The combined ether extract is concentrated to about 1 mL using a K-D apparatus followed by using a micro Snyder column or nitrogen gas blowdown. If methyl esters are to be produced, then dilute the concentrated ether extract with 1 mL of isooctane and 0.5 mL of methanol, dilute to a final volume of 4 mL, and esterify with diazomethane. If pentafluorobenzene esters are to be produced, then dilute concentrated ether extract with acetone to a final volume of 4 mL and esterify with pentafluorobenzyl bromide.

QUALITY CONTROL Select a representative spike concentration for each compound (acid or ester) to be measured. Using stock standard, prepare a quality control check sample concentrate, in acetone, that is 1000 times more concentrated than the selected concentrations. Use this quality control check sample concentrate to prepare quality control check samples. Calculate surrogate standard recovery on all standards, samples, blanks, and spikes. GC/MS techniques should be judiciously employed to support qualitative identifications made with this method. When available, chemical ionization mass spectra may be employed to aid the qualitative identification process.

REFERENCE Test Methods for Evaluating Solid Waste, Physical/Chemical Methods, SW-846, 3rd Edition, U.S. EPA, Office of Solid Waste, Washington, DC, EPA Method 8151, Nov. 1990.

Chlordane   EPA Method 505
CAS #57-74-9

TITLE Analysis of Organohalide Pesticides and Commercial Polychlorinated Biphenyl (PCB) Products in Water by Microextraction and Gas Chromatography. U.S. EPA Method 505, Rev. 2.0, 1989.

MATRIX This method is applicable to drinking water and raw source water. The latter should include most surface water and groundwater sources.

METHOD SUMMARY Method 505 covers 25 pesticides and commercial PCB products. This is a very sensitive method that is more useful for monitoring than for exploratory analyses. 5-mL of water are saturated with sodium chloride and then extracted by shaking with 2 mL of hexane. The sample extracts are transferred to an autosampler setup to inject 1–2 µL portions into a gas chromatograph (GC) for analysis. Alternatively, 1–2 µL portions of samples, blanks, and standards may be manually injected. Each extract is analyzed by capillary GC/ECD with confirmation using either a second capillary column or GC/MS. The electron capture detector is easy to use, but it is a nonselective detector. The microextraction technique also eliminates the expensive sample preparation costs of other methods, but it has the disadvantage of being less sensitive than most because the extracts are not concentrated.

INTERFERENCES Method interferences may be caused by contaminants in solvents, reagents, glassware, and other sample processing apparatus that lead to discrete artifacts or elevated baselines. Interfering contamination may occur when a sample containing low concentrations of analytes is analyzed immediately following a sample containing relatively high concentrations of the analytes. Matrix interferences also may be caused by contaminants that are coextracted from the sample; cleanup of sample extracts may be necessary in these cases. Some pesticides and commercial PCB products from aqueous solutions adhere to glass surfaces, so sample transfers and contact with glass surfaces should be minimized. Some pesticides are rapidly oxidized by chlorine so dechlorination with sodium thiosulfate at the time of sample collection is important. Also, splitless injectors may cause degradation of some pesticides.

INSTRUMENTATION A gas chromatograph/electron capture detector/data system, with temperature programming and split/splitless injector suitable for use with capillary columns is needed.

Column 1: 0.32 mm I.D. × 30 m fused silica capillary with chemically bond methyl polysiloxane phase (DB-1, 1.0 µm film, or equivalent).

Column 2: 0.32 mm I.D. × 30 m fused silica capillary with 1:1 mixed phase of dimethyl silicone and polyethylene glycol (Durawax-DX3, 0.25 µm film, or equivalent).

Column 3: 0.32 mm I.D. × 25 m fused silica capillary with chemically bonded 50:50 methyl-phenyl silicone (OV-17, 1.5 µm film, or equivalent).

Column 1 should be used as the primary analytical column. Columns 2 and 3 are recommended for use as confirmatory columns when GC/MS confirmation is not available.

PRECISION & ACCURACY Method detection limits are dependent upon the characteristics of the gas chromatographic system used. Analytes that are not separated chromatographically cannot be individually identified and used in the same calibration mixture or water samples unless an alternative technique for identification and quantification, such as mass spectrometry, is used.

The concentration(s) (in µg/L) used for these QC measurements was 0.17 and 3.4.

The MDL (in µg/L) was 0.14.

The accuracy (% recovery) for reagent water at the above concentration(s) was not available and not available and the precision (%) was 8.0 and 3.6.

The accuracy (% recovery) for groundwater at the above concentration(s) was not listed and not listed and the precision (%) was not listed and not listed.

The accuracy (% recovery) for tap water at the above concentration(s) was 105 and 95 and the precision (5) was 12.4 and 9.6.

Note: No range of concentrations is provided with this method.

SAMPLING METHOD Collect samples using a 40-mL screw-cap vial (prewashed with detergent, rinsed with distilled water and oven dried at 400°C for one h) with a Teflon®-faced silicone septum. Collect bubble-free samples and place the septum with the Teflon® side down on the water.

SAMPLE PRESERVATION If residual chlorine is present in the water add about 3 mg of sodium thiosulfate to each vial before samples are collected to remove the chlorine. Alternatively, add 75 µL of 0.04 g/mL solution of sodium thiosulfate to each vial just prior to sampling. Immediately cool samples to 4°C, and store them in a solvent-free refrigerator at 4°C until analysis.

MHT The maximum holding time is 14 days from the time the sample was collected until it must be analyzed.

SAMPLE PREPARATION Remove the sample from storage and allow it to come to room temperature. Remove a 5-mL volume from each container and weigh the container to the nearest 0.1 g. Add 6 g of sodium chloride and 2.0 mL of hexane to each sample bottle. Recap the sample and shake it vigorously for one min. Allow the water and hexane phases to separate, remove the cap, and transfer 0.5 mL of hexane into an autosampler vial using a disposable glass pipette. Transfer the remaining hexane phase into a second autosampler vial and store at 4°C for reanalysis, if necessary. Discard the remaining sample/hexane mixture and reweigh the empty container to determine net weight of sample.

QUALITY CONTROL Minimum quality control requirements are initial demonstration of lab capability, analysis of lab reagent blanks, fortified blanks, fortified sample matrix, and quality control samples. The lab must analyze at least one fortified blank per sample set, or at least one for every 20 samples. The fortifying concentration of each analyte should be 10 times the method detection limit or the maximum calibration limit (MCL), whichever is less. Calculate accuracy as percent recovery and develop control limits from the mean percent recovery and standard deviation.

The lab must add a known concentration of the analytes to a minimum of 10% of the routine samples, or one lab fortified sample matrix per sample set. Calculate the percent recovery for each analyte and compare to the control limits established from the analyses of the fortified blanks.

EPA CONTACT & HOTLINE For technical questions contact Dr. Baldev Bathija, U.S. EPA, Office of Ground Water and Drinking Water (WH-550D), 401 M St. SW, Washington, DC 20460. Tel. (202) 260-3040. For further information the EPA Safe Drinking Water Hotline may be called at: (800) 426-4791.

REFERENCE Methods for the Determination of Organic Compounds in Drinking Water, EPA/600/4-88/039 (revised July 1991). U.S. EPA Environmental Monitoring Systems Laboratory, Cincinnati, OH, 45268, USA. Available from the National Technical Information Service (NTIS), 5285 Port Royal Road, Springfield, VA 22161; Tel. 800-553-6847. NTIS Order Number is PB91-231480.

α-Chlordane   **EPA Method 505**
CAS #5103-71-9

TITLE Analysis of Organohalide Pesticides and Commercial Polychlorinated Biphenyl (PCB) Products in Water by Microextraction and Gas Chromatography. U.S. EPA Method 505, Rev. 2.0, 1989.

MATRIX This method is applicable to drinking water and raw source water. The latter should include most surface water and groundwater sources.

METHOD SUMMARY Method 505 covers 25 pesticides and commercial PCB products. This is a very sensitive method that is more useful for monitoring than for exploratory analyses. 5-mL of water are saturated with sodium chloride and then extracted by shaking with 2 mL of hexane. The sample extracts are transferred to an autosampler setup to inject 1–2 μL portions into a gas chromatograph (GC) for analysis. Alternatively, 1–2 μL portions of samples, blanks, and standards may be manually injected. Each extract is analyzed by capillary GC/ECD with confirmation using either a second capillary column or GC/MS. The electron capture detector is easy to use, but it is a nonselective detector. The microextraction technique also eliminates the expensive sample preparation costs of other methods, but it has the disadvantage of being less sensitive than most because the extracts are not concentrated.

INTERFERENCES Method interferences may be caused by contaminants in solvents, reagents, glassware, and other sample processing apparatus that lead to discrete artifacts or elevated baselines. Interfering contamination may occur when a sample containing low concentrations of analytes is analyzed immediately following a sample containing relatively high concentrations of the analytes. Matrix interferences also may be caused by contaminants that are coextracted from the sample; cleanup of sample extracts may be necessary in these cases. Some pesticides and commercial PCB products from aqueous solutions adhere to glass surfaces, so sample transfers and contact with glass surfaces should be minimized. Some pesticides are rapidly oxidized by chlorine so dechlorination with sodium thiosulfate at the time of sample collection is important. Also, splitless injectors may cause degradation of some pesticides.

INSTRUMENTATION A gas chromatograph/electron capture detector/data system, with temperature programming and split/splitless injector suitable for use with capillary columns is needed.

Column 1: 0.32 mm I.D. × 30 m fused silica capillary with chemically bond methyl polysiloxane phase (DB-1, 1.0 μm film, or equivalent).
Column 2: 0.32 mm I.D. × 30 m fused silica capillary with 1:1 mixed phase of dimethyl silicone and polyethylene glycol (Durawax-DX3, 0.25 μm film, or equivalent).
Column 3: 0.32 mm I.D. × 25 m fused silica capillary with chemically bonded 50:50 methyl-phenyl silicone (OV-17, 1.5 μm film, or equivalent).

Column 1 should be used as the primary analytical column. Columns 2 and 3 are recommended for use as confirmatory columns when GC/MS confirmation is not available.

PRECISION & ACCURACY Method detection limits are dependent upon the characteristics of the gas chromatographic system used. Analytes that are not separated chromatographically cannot be individually identified and used in the same calibration mixture or water samples unless an alternative technique for identification and quantification, such as mass spectrometry, is used.

The concentration(s) (in μg/L) used for these QC measurements was 0.06 and 0.35.
The MDL (in μg/L) was 0.006.
The accuracy (% recovery) for reagent water at the above concentration(s) was 95 and 86 and the precision (%) was 3.5 and 17.0.
The accuracy (% recovery) for groundwater at the above concentration(s) was 83 and 94 and the precision (%) was 4.4 and 10.2.
The accuracy (% recovery) for tap water at the above concentration(s) was 85 and 91 and the precision (5) was 7.1 and 2.4.

Note: No range of concentrations is provided with this method.

SAMPLING METHOD Collect samples using a 40-mL screw-cap vial (prewashed with detergent, rinsed with distilled water and oven dried at 400°C for one h) with a Teflon®-faced silicone septum. Collect bubble-free samples and place the septum with the Teflon® side down on the water.

SAMPLE PRESERVATION If residual chlorine is present in the water add about 3 mg of sodium thiosulfate to each vial before samples are collected to remove the chlorine. Alternatively, add 75 μL of 0.04 g/mL solution of sodium thiosulfate to each vial just prior to sampling. Immediately cool samples to 4°C, and store them in a solvent-free refrigerator at 4°C until analysis.

MHT The maximum holding time is 14 days from the time the sample was collected until it must be analyzed.

SAMPLE PREPARATION Remove the sample from storage and allow it to come to room temperature. Remove a 5-mL volume from each container and weigh the container to the nearest 0.1 g. Add 6 g of sodium chloride and 2.0 mL of hexane to each sample bottle. Recap the sample and shake it vigorously for one min. Allow the water and hexane phases to separate, remove the cap, and transfer 0.5 mL of hexane into an autosampler vial using a disposable glass pipette. Transfer the remaining hexane phase into a second autosampler vial and store at 4°C for reanalysis, if necessary. Discard the remaining sample/hexane mixture and reweigh the empty container to determine net weight of sample.

QUALITY CONTROL Minimum quality control requirements are initial demonstration of lab capability, analysis of lab reagent blanks, fortified blanks, fortified sample matrix, and quality control samples. The lab must analyze at least one fortified blank per sample set, or at least one for every 20 samples. The fortifying concentration of each analyte should be 10 times the method detection limit or the maximum calibration limit (MCL), whichever is less. Calculate accuracy as percent recovery and develop control limits from the mean percent recovery and standard deviation.

The lab must add a known concentration of the analytes to a minimum of 10% of the routine samples, or one lab fortified sample matrix per sample set. Calculate the percent recovery for each analyte and compare to the control limits established from the analyses of the fortified blanks.

EPA CONTACT & HOTLINE For technical questions contact Dr. Baldev Bathija, U.S. EPA, Office of Ground Water and Drinking Water (WH-550D), 401 M St. SW, Washington, DC 20460. Tel. (202) 260-3040. For further information the EPA Safe Drinking Water Hotline may be called at: (800) 426-4791.

REFERENCE Methods for the Determination of Organic Compounds in Drinking Water, EPA/600/4-88/039 (revised July 1991). U.S. EPA Environmental Monitoring Systems Laboratory, Cincinnati, OH, 45268, USA. Available from the National Technical Information Service (NTIS), 5285 Port Royal Road, Springfield, VA 22161; Tel. 800-553-6847. NTIS Order Number is PB91-231480.

γ-Chlordane   EPA Method 505
CAS #5103-74-2

TITLE Analysis of Organohalide Pesticides and Commercial Polychlorinated Biphenyl (PCB) Products in Water by Microextraction and Gas Chromatography. U.S. EPA Method 505, Rev. 2.0, 1989.

MATRIX This method is applicable to drinking water and raw source water. The latter should include most surface water and groundwater sources.

METHOD SUMMARY Method 505 covers 25 pesticides and commercial PCB products. This is a very sensitive method that is more useful for monitoring than for exploratory analyses. 5-mL of water are saturated with sodium chloride and then extracted by shaking with 2 mL of hexane. The sample extracts are transferred to an autosampler setup to inject 1–2 μL portions into a gas chromatograph (GC) for analysis. Alternatively, 1–2 μL portions of samples, blanks, and standards may be manually injected. Each extract is analyzed by capillary GC/ECD with confirmation using either a second capillary column or GC/MS. The electron capture detector is easy to use, but it is a nonselective detector. The microextraction technique also eliminates the expensive sample preparation costs of other methods, but it has the disadvantage of being less sensitive than most because the extracts are not concentrated.

INTERFERENCES Method interferences may be caused by contaminants in solvents, reagents, glassware, and other sample processing apparatus that lead to discrete artifacts or elevated baselines. Interfering contamination may occur when a sample containing low concentrations of analytes is analyzed immediately following a sample containing relatively high concentrations of the analytes. Matrix interferences also may be caused by contaminants that are coextracted from the sample; cleanup of sample extracts may be necessary in these cases. Some pesticides and commercial PCB products from aqueous solutions adhere to glass surfaces, so sample transfers and contact with glass surfaces should be minimized. Some pesticides are rapidly oxidized by chlorine so dechlorination with sodium thiosulfate at the time of sample collection is important. Also, splitless injectors may cause degradation of some pesticides.

INSTRUMENTATION A gas chromatograph/electron capture detector/data system, with temperature programming and split/splitless injector suitable for use with capillary columns is needed.

Column 1: 0.32 mm I.D. × 30 m fused silica capillary with chemically bond methyl polysiloxane phase (DB-1, 1.0 μm film, or equivalent).

Column 2: 0.32 mm I.D. × 30 m fused silica capillary with 1:1 mixed phase of dimethyl silicone and polyethylene glycol (Durawax-DX3, 0.25 μm film, or equivalent).

Column 3: 0.32 mm I.D. × 25 m fused silica capillary with chemically bonded 50:50 methyl-phenyl silicone (OV-17, 1.5 μm film, or equivalent).

Column 1 should be used as the primary analytical column. Columns 2 and 3 are recommended for use as confirmatory columns when GC/MS confirmation is not available.

PRECISION & ACCURACY Method detection limits are dependent upon the characteristics of the gas chromatographic system used. Analytes that are not separated chromatographically cannot be individually identified and used in the same calibration mixture or water samples unless an alternative technique for identification and quantification, such as mass spectrometry, is used.

The concentration(s) (in µg/L) used for these QC measurements was 0.06 and 0.35.

The MDL (in µg/L) was 0.12.

The accuracy (% recovery) for reagent water at the above concentration(s) was 95 and 86 and the precision (%) was 0.4 and 18.5.

The accuracy (% recovery) for groundwater at the above concentration(s) was 86 and 95 and the precision (%) was 5.3 and 14.5.

The accuracy (% recovery) for tap water at the above concentration(s) was 83 and 91 and the precision (5) was 14.7 and 6.0.

Note: No range of concentrations is provided with this method.

SAMPLING METHOD Collect samples using a 40-mL screw-cap vial (prewashed with detergent, rinsed with distilled water and oven dried at 400°C for one h) with a Teflon®-faced silicone septum. Collect bubble-free samples and place the septum with the Teflon® side down on the water.

SAMPLE PRESERVATION If residual chlorine is present in the water add about 3 mg of sodium thiosulfate to each vial before samples are collected to remove the chlorine. Alternatively, add 75 µL of 0.04 g/mL solution of sodium thiosulfate to each vial just prior to sampling. Immediately cool samples to 4°C, and store them in a solvent-free refrigerator at 4°C until analysis.

MHT The maximum holding time is 14 days from the time the sample was collected until it must be analyzed.

SAMPLE PREPARATION Remove the sample from storage and allow it to come to room temperature. Remove a 5-mL volume from each container and weigh the container to the nearest 0.1 g. Add 6 g of sodium chloride and 2.0 mL of hexane to each sample bottle. Recap the sample and shake it vigorously for one min. Allow the water and hexane phases to separate, remove the cap, and transfer 0.5 mL of hexane into an autosampler vial using a disposable glass pipette. Transfer the remaining hexane phase into a second autosampler vial and store at 4°C for reanalysis, if necessary. Discard the remaining sample/hexane mixture and reweigh the empty container to determine net weight of sample.

QUALITY CONTROL Minimum quality control requirements are initial demonstration of lab capability, analysis of lab reagent blanks, fortified blanks, fortified sample matrix, and quality control samples. The lab must analyze at least one fortified blank per sample set, or at least one for every 20 samples. The fortifying concentration of each analyte should be 10 times the method detection limit or the maximum calibration limit (MCL), whichever is less. Calculate accuracy as percent recovery and develop control limits from the mean percent recovery and standard deviation.

The lab must add a known concentration of the analytes to a minimum of 10% of the routine samples, or one lab fortified sample matrix per sample set. Calculate the percent recovery for each analyte and compare to the control limits established from the analyses of the fortified blanks.

EPA CONTACT & HOTLINE For technical questions contact Dr. Baldev Bathija, U.S. EPA, Office of Ground Water and Drinking Water (WH-550D), 401 M St. SW, Washington, DC 20460. Tel. (202) 260-3040. For further information the EPA Safe Drinking Water Hotline may be called at: (800) 426-4791.

REFERENCE Methods for the Determination of Organic Compounds in Drinking Water, EPA/600/4-88/039 (revised July 1991). U.S. EPA Environmental Monitoring Systems Laboratory, Cincinnati, OH, 45268, USA. Available from the National Technical Information Service (NTIS), 5285 Port Royal Road, Springfield, VA 22161; Tel. 800-553-6847. NTIS Order Number is PB91-231480.

Chlordane EPA Method 625
CAS #57-74-9

TITLE Base/Neutrals and Acids, U.S. EPA Method 625

MATRIX This methods covers municipal and industrial wastewaters.

METHOD SUMMARY Approximately 1 L of sample is serially extracted with methylene chloride at a pH greater than 11 and again at a pH less than 2 using a separatory funnel or a continuous extractor. The methylene chloride extract is dried, concentrated to a volume of 1 mL, and analyzed by GC/MS. Qualitative identification of the parameters in the extract is performed using the retention time and the relative abundance of three characteristic masses (m/z). Qualitative analysis is performed using either external or internal standard techniques with a single characteristic m/z.

INTERFERENCES Method interferences may be caused by contaminants in solvents, reagents, glassware, and other sample processing hardware. Glassware must be scrupulously cleaned. Glassware should be heated in a muffle furnace at 400°C for 5 to 30 min. Some thermally stable materials, such as PCBs, may not be eliminated by this treatment. Solvent rinses with acetone and pesticide quality hexane may be substituted for the muffle furnace heating. Matrix interferences may be caused by contaminants that are coextracted from the sample. The base-neutral extraction may cause significantly reduced recovery of phenols. The packed gas chromatographic columns recommended for the basic fraction may not exhibit sufficient resolution for some analytes.

INSTRUMENTATION A GC/MS system with an injection port designed for on-column injection when using packed columns and for splitless injection when using capillary columns.

Column for base/neutrals: 1.8 m long × 2 mm I.D. glass, packed with 3% SP-2550 on Supelcoport (100/120 mesh) or equivalent.

Column for acids: 1.8 m long × 2 mm I.D. glass, packed with 1% SP-1240DA on Supelcoport (100/120 mesh) or equivalent.

PRECISION & ACCURACY The MDL concentrations were obtained using reagent water. The MDL actually achieved in a given analysis will vary depending on instrument sensitivity and matrix effects. This method was tested by 15 laboratories

using reagent water, drinking water, surface water, and industrial wastewaters spiked at six concentrations over the range 5 to 100 µg/L. Single operator precision, overall precision, and method accuracy were found to be directly related to the concentration of the parameter matrix.

The MDL (in µg/L) in reagent water was not detected.
The standard deviation (in µg/L based on 4 recovery measurements) was not reported.
The range (in µg/L) for average recovery for 4 measurements was not reported.
The range (in %) for percent recovery was not reported.
Accuracy (in µg/L) as expected recovery for one or more measurements of a sample containing a true concentration of C was not reported.
Precision (in µg/L) as expected single analyst standard deviation of measurements at an average concentration found at X was not reported.
Overall precision (in µg/L) as expected interlaboratory standard deviation of measurements in an average concentration found at X was not reported.

C = *True value of the concentration in µg/L.*
X = *Average recovery found for measurements of samples containing a concentration at C in µg/L.*

SAMPLE PREPARATION Adjust the pH to >11 with sodium hydroxide and serially extract in a separatory funnel with methylene chloride or else in a continuous extractor. Next, adjust the pH to <2 with sulfuric acid and serially extract in a separatory funnel with methylene chloride or else in a continuous extractor. Dry the extracts separately through a column of anhydrous sodium sulfate and then concentrate each of the extracts to 1.0 mL using a K-D apparatus.

SAMPLE COLLECTION, PRESERVATION & HANDLING Grab samples must be collected in glass containers. All samples must be refrigerated at 4°C from the time of collection until extraction. If residual chlorine is present, add 80 mg of sodium thiosulfate/L of sample and mix well. All samples must be extracted within 7 days of collection and completely analyzed within 40 days of extraction.

QUALITY CONTROL Make an initial, one-time, demonstration of the ability to generate acceptable accuracy and precision with this method. Before processing any samples, the analyst must analyze a reagent water blank to demonstrate that interferences from the analytical system and glassware are under control. Each time a set of samples is extracted or reagents are changed, a reagent water blank must be processed. Spike and analyze a minimum of 5% of all samples to monitor and evaluate lab data quality. A QC check sample concentrate that contains each parameter of interest at a concentration of 100 µg/mL in acetone is required. PCBs and multicomponent pesticides may be omitted from this test.

After analysis of five spiked wastewater samples, calculate the average percent recovery and the standard deviation of the percent recovery. Spike all samples with the surrogate standard spiking solution and calculate the percent recovery of each surrogate compound.

REFERENCE *Federal Register*, Vol. 49, No. 209. Friday, Oct. 26, 1984.

Chlordane   **EPA Method 8270**
CAS #57-74-9

TITLE Semivolatile Organic Compounds by GC/MS

MATRIX This method is used to determine the concentration of semivolatile organic compounds in extracts prepared from all types of solid waste matrices, soils, and groundwater. Although surface waters are not specifically mentioned, this method should be applicable to water samples from rivers, lakes, etc.

METHOD SUMMARY This method covers 259 semivolatile organic compounds. In very limited applications direct injection of the sample into the GC/MS system may be appropriate, but this results in very high detection limits (approximately 10,000 µg/L). Typically, a 1-L liquid sample, containing surrogate, and matrix spiking standards, is extracted in a continuous extractor first under acid conditions and then under basic conditions. Typically 30 g of a solid sample, containing surrogate, and matrix spiking standards, is extracted ultrasonically. After concentrating the extract to 1 mL it is spiked with 10 µL of an internal standard solution just prior to analysis by GC/MS. The volume injected should contain about 100 ng of base/neutral and 200 ng of acid surrogates (for a 1-µL injection). Analysis is performed by GC/MS using a capillary GC column.

INTERFERENCES Raw GC/MS data from all blanks, samples, and spikes must be evaluated for interferences. Contamination by carryover can occur whenever high-concentration and low-concentration samples are sequentially analyzed. To reduce carryover, the sample syringe must be rinsed out between samples with solvent. Whenever an unusually concentrated sample is encountered, it should be followed by the analysis of blank solvent to check for cross-contamination.

INSTRUMENTATION A GC/MS and a data system are required. The GC column used is a 30 m × 0.25 mm I.D. (or 0.32 mm I.D.) 1um film thickness silicone-coated fused silica capillary column. A continuous liquid-liquid extractor equipped with Teflon® or glass connection joints and stopcocks requiring no lubrication, a K-D concentrating apparatus, water bath, and an ultrasonic disrupter with a minimum power of 300 W and with pulsing capability are also required.

PRECISION & ACCURACY The estimated quantitation limit (EQL) of Method 8270B for determining an individual compound is approximately 1 mg/kg (wet weight) for soil or sediment samples, 1–200 mg/kg for wastes (dependent on matrix and method of preparation), and 10 µg/L for groundwater samples. EQLs will be proportionately higher for sample extracts that require dilution to avoid saturation of the detector.

The EQL(b) for groundwater in µg/L is not listed.
The EQL (a, b) for low concentrations in soil and sediment in µg/kg is not listed.
Accuracy as µg/L is not listed.
Overall precision in µg/L is not listed.

(a) *EQLs listed for soil/sediment are based on wet weight. Normally data is reported in a dry-weight basis; therefore, EQLs will be higher based on the % dry weight of each sample. This calculation is based on a 30 g sample and gel permeation chromatography cleanup.*
(b) *Sample EQLs are highly matrix-dependent. The EQLs are provided for guidance and may not always be achievable.*
C = True value for concentration, in µg/L.
X = Average recovery found for measurements of samples containing a concentration of C, in µg/L.

ESTIMATED QUANTITATION LIMIT FOR OTHER MATRICES

Other Matrices	Factor (a)
High-concentration soil and sludges by sonicator	7.5
Non-water miscible waste	75

(a) EQL for other matrices = [EQL for low soil/sediment] × [Factor]. This estimated EQL is similar to an EPA "Practical Quantitation Limit."

SAMPLING METHOD

Liquid samples — Use a 1 or 2½ gallon amber glass bottle with a screw-top Teflon®-lined cover that has been prewashed with detergent and rinsed with distilled water and methanol (or isopropanol).

Soils, sediments, or sludges — Use an 8-oz. widemouth glass with a screw-top Teflon®-lined cover that has been prewashed with detergent and rinsed with distilled water and methanol (or isopropanol).

SAMPLE PRESERVATION

Liquid samples — If residual chlorine is present, add 3 mL of 10% sodium thiosulfate per gallon, cool to 4°C and store in a solvent-free refrigerator until analysis; if chlorine is not present, then eliminate the sodium thiosulfate addition.

Soils, sediments, or sludges — Cool samples to 4°C and store in a solvent-free refrigerator.

MHT Liquid samples must be extracted within 7 days and the extracts analyzed within 40 days. Soils, sediments, or sludges may be stored for a maximum of 14 days and the extracts analyzed within 40 days.

SAMPLE PREPARATION

Liquid samples — Transfer 1 L quantitatively to a continuous extractor. If high concentrations are anticipated, a smaller volume may be used and then diluted with organic-free reagent water to 1 L. Adjust pH, if necessary, to pH <2 using 1:1 (V/V) sulfuric acid. Pipette 1.0 mL of a surrogate standard spiking solution into each sample. For the sample in each analytical batch selected for spiking, add 1.0 mL of a matrix spiking standard. For base/neutral acid analysis, the amount of the surrogates and matrix spiking compounds added to the sample should result in a final concentration of 100 ng/µL of each analyte in the extract to be analyzed (assuming a 1-µL injection). Extract with methylene chloride for 18–24 h. Next, adjust the pH of the aqueous phase to pH >11 using 10 N sodium hydroxide and extract it with methylene chloride again for 18–24 h. Dry the extract through a column containing anhydrous sodium sulfate and concentrate it to 1 mL using a K-D concentrator.

Soils, sediments, or sludges — Use 30 g of sample. Nonporous or wet samples (gummy or clay type) that do not have a free-flowing sandy texture must be mixed with anhydrous sodium sulfate until the sample is free flowing. Add 1 mL of surrogate standards to all samples, spikes, standards, and blanks. For the sample in each analytical batch selected for spiking, add 1.0 mL of a matrix spiking standard. For base/neutral acid analysis, the amount added of the surrogates and matrix spiking compounds should result in a final concentration of 100 ng/µL of each base/neutral analyte and 200 ng/µL of each acid analyte in the extract to be analyzed (assuming a 1-µL injection). Immediately add a 100-mL mixture of 1:1 methylene chloride:acetone and extract the sample ultrasonically for 3 min and then decant or filter the extracts. Repeat the extraction two or more times. Dry the extract using a column with anhydrous sodium sulfate and concentrate it to 1 mL in a K-D concentrator.

QUALITY CONTROL A methylene chloride solution containing 50 ng/µL of decafluorotriphenylphosphine (DFTPP) is used for tuning the GC/MS system each 12-h shift. A system performance check also must be made during every 12-h shift. A standard containing 50 ng/µL each of 4,4'-DDT, pentachlorophenol, and benzidine is required to verify injection port inertness and GC column performance. A calibration standard at mid-concentration, containing each compound of interest, including all required surrogates, must be performed every 12 h during analysis. After the system performance check is met, calibration check compounds (CCCs) are used to check the validity of the initial calibration.

The internal standard responses and retention times in the calibration check standard must be evaluated immediately after or during data acquisition. If the retention time for any internal standard changes by more than 30 seconds from the last check calibration (12 h), the chromatographic system must be inspected for malfunctions and corrections must be made, as required. If the electron ionization current plot (EICP) area for any of the internal standards changes by a factor of two from the last daily calibration standard check, the mass spectrometer must be inspected for malfunctions and corrections must be made, as appropriate.

Demonstrate, through the analysis of a reagent water blank, that interferences from the analytical system, glassware, and reagents are under control. The blank samples should be carried through all stages of the sample preparation and measurement steps. For each analytical batch (up to 20 samples), a reagent blank, matrix spike, and matrix spike duplicate/duplicate must be analyzed (the frequency of the spikes may be different for different monitoring programs). The blank and spiked samples must be carried through all stages of the sample preparation and measurement steps. A QC reference sample concentrate containing each analyte at a concentration of 100 mg/L in methanol is required.

REFERENCE Test Methods for Evaluating Solid Waste (SW-846). U.S. EPA 1983, Method 8270B, Rev. 2, Nov. 1990. Office of Solid Waste, Washington, DC.

Chlordane (technical)   EPA Method 8080
CAS #12789-03-6

TITLE Organochlorine Pesticides and Polychlorinated Biphenyls By Gas Chromatography

MATRIX This method is used to determine the concentration of various organochlorine pesticides and polychlorinated biphenyls in extracts prepared from water, groundwater, soils, and sediments.

METHOD SUMMARY This method covers 26 pesticides and Aroclor (PCB) mixtures and it is suitable for monitoring-type analyses. After extraction, concentration and solvent exchange to hexane, a 2- to 5-μL sample aliquot is injected into a GC using the solvent flush technique, and the analytes are detected by an electron capture detector (ECD) or an electrolytic conductivity detector in the halogen mode (HECD). Both neat and diluted organic liquids may be analyzed by direct injection.

INTERFERENCES Interferences coextracted from the samples will vary considerably from source to source. Interferences by phthalate esters can pose a major problem in pesticide determinations when using the ECD. Cross-contamination of clean glassware routinely occurs when plastics are handled during extraction steps, especially when solvent-wetted surfaces are handled. The contamination from phthalate esters can be completely eliminated with a microcoulometric or electrolytic conductivity detector. Solvents, reagent, glassware, and other sample processing hardware may yield artifacts and/or interferences to sample analysis.

INSTRUMENTATION A gas chromatograph capable of on-column injections is needed. It must be equipped with an ECD or a HECD and one of the following GC columns:

Column 1: Supelcoport (100/120 mesh) coated with 1.5% SP-2250/1.95% SP-2401 packed in a 1.8 m × 4 mm I.D. glass column.
Column 2: Supelcoport (100/120 mesh) coated with 3% OV-1 in a 1.8 m × 4 mm I.D. glass column.

PRECISION & ACCURACY The method was tested by 20 laboratories using organic-free reagent water, drinking water, surface water, and three industrial wastewaters spiked at six concentrations. Concentrations used in the study ranged from 0.5 to 30 μg/L for single-component pesticides and from 8.5 to 400 μg/L for multicomponent parameters. Overall precision and method accuracy were found to be directly related to the concentration of the analyte and essentially independent of the sample matrix. The sensitivity of this method usually depends on the concentration of interferences rather than on instrumental limitations.

MDL in μg/L was 0.014.
Concentration range in μg/L was 8.5–400.
Accuracy as recovery (x*) in μg/L was 0.82C-0.04 .
Overall precision (S*) in μg/L was 0.18x +0.18.

x^* *Expected recovery for one or more measurements of a sample containing concentration C, in μg/L.*

S^* = *Expected interlaboratory standard deviation of measurements at an average concentration found of the analyte in μg/L.*
C = *True value for the concentration, in μg/L.*
X = *Average recovery found for measurements of samples containing a concentration of C, in μg/L.*

SAMPLING METHOD
Liquid samples — Use a 1 or 2½ gallon amber glass bottle with a screw-top Teflon®-lined cover. Pre-wash the bottle with detergent, rinse with distilled water and methanol (or isopropanol).

Soil, sediments, and sludges — Use an 8-oz. widemouth glass with a screw-top Teflon®-lined cover. Pre-wash the bottle with detergent, rinse with distilled water and methanol (or isopropanol).

SAMPLE PRESERVATION Cool water, soil, sediment, or sludge samples immediately to 4°C.

Water samples — If residual chlorine is present, add 3 mL of 10% sodium thiosulfate per gallon and cool to 4°C. All extracts and samples should be stored under refrigeration.

MHT Liquid samples must be extracted within 7 days and the extracts must be analyzed within 40 days. Soils, sediments, and sludges may be stored for a maximum of 14 days prior to extraction.

SAMPLE PREPARATION

Liquid samples — Extract 1-L samples in a continuous extractor at pH 5–9 with methylene chloride after adding 1.0 mL of surrogate spiking solution to each sample. Pass the extract through a column of anhydrous sodium sulfate to dry and concentrate it in a K-D apparatus to 1 mL volume.

Soils, sediments, and sludges — Rapidly weigh approximately 30 g of sample into a 400-mL beaker to avoid loss of the more volatile extractables. Nonporous or wet samples (gummy or clay type) that do not have a free-flowing sandy texture must be mixed with anhydrous sodium sulfate until the sample is free flowing. Add 1 mL of surrogate standards to all samples, spikes, standards, and blanks. Add 100 mL of 1:1 methylene chloride:acetone and extract ultrasonically. Decant and filter extracts, dry the extract by passing it through a drying column containing anhydrous sodium sulfate and concentrate to 1 mL in a K-D apparatus.

Hexane solvent exchange — Add 50 mL of hexane, a new boiling chip, and concentrate until the apparent volume of liquid reaches 1 mL. Adjust the extract volume to 10.0 mL. Stopper the concentration tube and store refrigerated at 4°C if further processing will not be performed immediately. If the extract will be stored longer than two days, transfer it to a vial with Teflon®-lined screw-cap or crimp top.

QUALITY CONTROL Demonstrate through the analysis of a reagent water blank, that all glassware and reagents are interference free. Each time a set of samples is processed, a method blank should be processed as a safeguard against chronic lab contamination. A reagent blank, a matrix spike, and a duplicate or matrix spike duplicate must be performed for each analytical batch (up to a maximum of 20 samples) analyzed.

Analytical system performance must be verified by analyzing QC check samples. The QC check sample concentration should contain each single-component analyte at the following concentrations in acetone: 4,4′-DDD, 10 µg/mL; 4,4′-DDT, 10 µg/mL; endosulfan II, 10 µg/mL; endosulfan sulfate, 10 µg/mL; and any other single-component pesticide at 2 µg/mL. If the method is only to be used to analyze PCBs, Chlordane, or Toxaphene, the QC check sample concentrate should contain the most representative multicomponent parameter at a concentration of 50 µg/mL in acetone.

REFERENCE Test Methods for Evaluating Solid Waste (SW-846). U.S. EPA. 1983. Method 8080B, Rev. 2, Nov. 1990. Office of Solid Wastes, Washington, DC.

Chlorfenvinphos   **EPA Method 8270**
CAS #470-90-6

TITLE Semivolatile Organic Compounds by GC/MS

MATRIX This method is used to determine the concentration of semivolatile organic compounds in extracts prepared from all types of solid waste matrices, soils, and groundwater. Although surface waters are not specifically mentioned, this method should be applicable to water samples from rivers, lakes, etc.

METHOD SUMMARY This method covers 259 semivolatile organic compounds. In very limited applications direct injection of the sample into the GC/MS system may be appropriate, but this results in very high detection limits (approximately 10,000 µg/L). Typically, a 1-L liquid sample, containing surrogate, and matrix spiking standards, is extracted in a continuous extractor first under acid conditions and then under basic conditions. Typically 30 g of a solid sample, containing surrogate, and matrix spiking standards, is extracted ultrasonically. After concentrating the extract to 1 mL it is spiked with 10 µL of an internal standard solution just prior to analysis by GC/MS. The volume injected should contain about 100 ng of base/neutral and 200 ng of acid surrogates (for a 1-µL injection). Analysis is performed by GC/MS using a capillary GC column.

INTERFERENCES Raw GC/MS data from all blanks, samples, and spikes must be evaluated for interferences. Contamination by carryover can occur whenever high-concentration and low-concentration samples are sequentially analyzed. To reduce carryover, the sample syringe must be rinsed out between samples with solvent. Whenever an unusually concentrated sample is encountered, it should be followed by the analysis of blank solvent to check for cross-contamination.

INSTRUMENTATION A GC/MS and a data system are required. The GC column used is a 30 m × 0.25 mm I.D. (or 0.32 mm I.D.) 1um film thickness silicone-coated fused silica capillary column. A continuous liquid-liquid extractor equipped with Teflon® or glass connection joints and stopcocks requiring no lubrication, a K-D concentrating apparatus, water bath, and an ultrasonic disrupter with a minimum power of 300 W and with pulsing capability are also required.

PRECISION & ACCURACY The estimated quantitation limit (EQL) of Method 8270B for determining an individual compound is approximately 1 mg/kg (wet weight) for soil or sediment samples, 1–200 mg/kg for wastes (dependent on matrix and method of preparation), and 10 µg/L for groundwater samples. EQLs will be proportionately higher for sample extracts that require dilution to avoid saturation of the detector.

The EQL(b) for groundwater in µg/L is 20.
The EQL (a, b) for low concentrations in soil and sediment in µg/kg is not determined.
Accuracy as µg/L is not listed.
Overall precision in µg/L is not listed.

(a) EQLs listed for soil/sediment are based on wet weight. Normally data is reported in a dry-weight basis; therefore, EQLs will be higher based on the % dry weight of each sample. This calculation is based on a 30 g sample and gel permeation chromatography cleanup.

(b) Sample EQLs are highly matrix-dependent. The EQLs are provided for guidance and may not always be achievable.

C = True value for concentration, in µg/L.
X = Average recovery found for measurements of samples containing a concentration of C, in µg/L.

ESTIMATED QUANTITATION LIMIT FOR OTHER MATRICES

Other Matrices	Factor (a)
High-concentration soil and sludges by sonicator	7.5
Non-water miscible waste	75

(a) EQL for other matrices = [EQL for low soil/sediment] × [Factor]. This estimated EQL is similar to an EPA "Practical Quantitation Limit."

SAMPLING METHOD

Liquid samples — Use a 1 or 2½ gallon amber glass bottle with a screw-top Teflon®-lined cover that has been prewashed with detergent and rinsed with distilled water and methanol (or isopropanol).

Soils, sediments, or sludges — Use an 8-oz. widemouth glass with a screw-top Teflon®-lined cover that has been prewashed with detergent and rinsed with distilled water and methanol (or isopropanol).

SAMPLE PRESERVATION

Liquid samples — If residual chlorine is present, add 3 mL of 10% sodium thiosulfate per gallon, cool to 4°C and store in a solvent-free refrigerator until analysis; if chlorine is not present, then eliminate the sodium thiosulfate addition.

Soils, sediments, or sludges — Cool samples to 4°C and store in a solvent-free refrigerator.

MHT Liquid samples must be extracted within 7 days and the extracts analyzed within 40 days. Soils, sediments, or sludges may be stored for a maximum of 14 days and the extracts analyzed within 40 days.

SAMPLE PREPARATION

Liquid samples — Transfer 1 L quantitatively to a continuous extractor. If high concentrations are anticipated, a smaller volume may be used and then diluted with organic-free reagent water to 1 L. Adjust pH, if necessary, to pH <2 using 1:1 (V/V) sulfuric acid. Pipette 1.0 mL of a surrogate standard spiking solution into each sample. For the sample in each analytical batch selected for spiking, add 1.0 mL of a matrix spiking standard. For base/neutral acid analysis, the amount of the surrogates and matrix spiking compounds added to the sample should result in a final concentration of 100 ng/µL of each analyte in the extract to be analyzed (assuming a 1-µL injection). Extract with methylene chloride for 18–24 h. Next, adjust the pH of the aqueous phase to pH >11 using 10 N sodium hydroxide and extract it with methylene chloride again for 18–24 h. Dry the extract through a column containing anhydrous sodium sulfate and concentrate it to 1 mL using a K-D concentrator.

Soils, sediments, or sludges — Use 30 g of sample. Nonporous or wet samples (gummy or clay type) that do not have a free-flowing sandy texture must be mixed with anhydrous sodium sulfate until the sample is free flowing. Add 1 mL of surrogate standards to all samples, spikes, standards, and blanks. For the sample in each analytical batch selected for spiking, add 1.0 mL of a matrix spiking standard. For base/neutral acid analysis, the amount added of the surrogates and matrix spiking compounds should result in a final concentration of 100 ng/µL of each base/neutral analyte and 200 ng/µL of each acid analyte in the extract to be analyzed (assuming a 1-µL injection). Immediately add a 100-mL mixture of 1:1 methylene chloride:acetone and extract the sample ultrasonically for 3 min and then decant or filter the extracts. Repeat the extraction two or more times. Dry the extract using a column with anhydrous sodium sulfate and concentrate it to 1 mL in a K-D concentrator.

QUALITY CONTROL A methylene chloride solution containing 50 ng/µL of decafluorotriphenylphosphine (DFTPP) is used for tuning the GC/MS system each 12-h shift. A system performance check also must be made during every 12-h shift. A standard containing 50 ng/µL each of 4,4′-DDT, pentachlorophenol, and benzidine is required to verify injection port inertness and GC column performance. A calibration standard at mid-concentration, containing each compound of interest, including all required surrogates, must be performed every 12 h during analysis. After the system performance check is met, calibration check compounds (CCCs) are used to check the validity of the initial calibration.

The internal standard responses and retention times in the calibration check standard must be evaluated immediately after or during data acquisition. If the retention time for any internal standard changes by more than 30 seconds from the last check calibration (12 h), the chromatographic system must be inspected for malfunctions and corrections must be made, as required. If the electron ionization current plot (EICP) area for any of the internal standards changes by a factor of two from the last daily calibration standard check, the mass spectrometer must be inspected for malfunctions and corrections must be made, as appropriate.

Demonstrate, through the analysis of a reagent water blank, that interferences from the analytical system, glassware, and reagents are under control. The blank samples should be carried through all stages of the sample preparation and measurement steps. For each analytical batch (up to 20 samples), a reagent blank, matrix spike, and matrix spike duplicate/duplicate must be analyzed (the frequency of the spikes may be different for different monitoring programs). The blank and spiked samples must be carried through all stages of the sample preparation and measurement steps. A QC reference sample concentrate containing each analyte at a concentration of 100 mg/L in methanol is required.

REFERENCE Test Methods for Evaluating Solid Waste (SW-846). U.S. EPA 1983, Method 8270B, Rev. 2, Nov. 1990. Office of Solid Waste, Washington, DC.

Chloride   EPA Method 325.1

TITLE Inorganics, Non-metallics

MATRIX Drinking, Surface and Saline waters. Wastewater.

APPLICATION Date issued 1971. (Colorimetric, automated ferricyanide AAI).

Thiocyanate ion (SCN) is liberated from mercuric thiocyanate through sequestration of mercury by chloride ion to form un-ionized mercuric chloride. SCN forms ferric thiocyanate with ferric ions. Ferric ammonium sulfate reagent provides ferric ion which forms highly colored ferric thyiocyanate in concentration proportional to the original chloride concentration.

INTERFERENCES No significant interferences.

INSTRUMENTATION Technicon auto analyzer, 480 nm Filters, 15 mm Tubular flow cell.

RANGE 1 to 250 mg Cl/L.

MDL Not listed.

PRECISION SD = ±0.3 At conc of 1, 100 and 250 mg Cl/L.

ACCURACY At conc. of 10 and 100 mg Cl/L, 97 and 104% recoveries.

SAMPLING METHOD Plastic or glass (50 mL).

STABILITY No preservation required.

MHT 28 Days.

QUALITY CONTROL Place working standards in sampler in order of decreasing concentration.

REFERENCE EPA Methods for the Chemical Analysis of Water and Wastes, EPA-600/4-79-020, U.S. EPA, EMSL, 1979.

Chloride   EPA Method 325.2

TITLE Inorganics, Non-metallics

MATRIX Drinking, Surface and Saline waters. Wastewater.

APPLICATION Date issued 1978. (Colorimetric, automated ferricyanide aaii). Thiocyanate ion (SCN) is liberated from mercuric thiocyanate through sequestration of mercury by chloride ion to form un-ionized mercuric chloride. SCN forms ferric thiocyanate with ferric ions. Ferric nitrate reagent provides ferric ion which forms highly colored ferric thiocyanate in concentration proportional to original chloride concentration. The range may be extended with sample dilution.

INTERFERENCES No significant interferences.

INSTRUMENTATION Technicon auto analyzer. 480 nm Filters. 15 mm Tubular flow cell.

RANGE 1 to 200 mg Cl/L.

MDL Not listed.

PRECISION Not listed.

ACCURACY Not listed.

SAMPLING METHOD Plastic or glass (50 mL).

STABILITY No preservation required.

MHT 28 Days.

QUALITY CONTROL (1) where particulate matter is present, the sample must be filtered prior to the determination. Sample may be centrifuged in place of filtration, or use technicon continuous filter. (2) Place working standards in sampler in order of decreasing concentration.

REFERENCE EPA Methods for the Chemical Analysis of Water and Wastes, EPA-600/4-79-020, U.S. EPA, EMSL, 1979.

Chloride EPA Method 325.3

TITLE Inorganics, Non-metallics

MATRIX Drinking, Surface and Saline Waters. Wastewater.

APPLICATION Date issued 1971. Editorial Rev. 1978. (Titrimetric, mercuric nitrate). An acidified sample is titrated with mercuric nitrate in the presence of mixed diphenylcarbazone-bromphenol blue indicator. The end point is a blue-violet mercury DPC complex.

INTERFERENCES Sulfite interference can be eliminated by oxidizing 50 mL sample solution with 0.5 To 1.0 mL H2O2. Special precautions are necessary when chromate and iron are present, especially ferric ion. (Automated titration may be used).

INSTRUMENTATION Standard lab titrimetric equip, including 1 or 5 mL microburet (0.01 mL Graduations)

RANGE Suitable for all Cl ranges.

MDL Not listed.

PRECISION SD = 1.32 mg Cl/L at 18 mg Cl/L increment.

ACCURACY As bias, +0.6 mg Cl/L at 18 mg Cl/L increment.

SAMPLING METHOD Plastic or glass (50 mL).

STABILITY No preservation required.

MHT 28 Days.

QUALITY CONTROL (1) check concentration of a 50 mL aliquot to determine strength of titrant and sample size to be used. (2) Make practice runs with indicator to become familiar with end point. (Alphazurine sharpens end point). (3) Store mercuric nitrate in dark bottle.

REFERENCE Methods for the Chemical Analysis of Water and Wastes, EPA 600/4-79-020, U.S. EPA, EMSL, 1979.

Chloride EPA Method 300.0

TITLE Inorganic Anions in Water

MATRIX Drinking, Surface and Mixed Wastewater

APPLICATION A small volume of sample, typically 2 to 3 mL is introduced into an ion chromatograph. The anions of interest are separated and measured using a system comprised of a guard column, separator column, suppressor column and conductivity detector.

INTERFERENCES Interferences can be caused by substances with retention times similar to and overlapping those of the ion of interest. Large amounts of an anion can interfere with peak resolution of adjacent anion. EPA Method interference can be caused by reagent or equipment contamination.

INSTRUMENTATION Ion chromatograph. Analytical balance. Guard, separator and suppressor columns.

RANGE

MDL 0.015 mg/L.

PRECISION SD = 0.289 mg/L at 10.0 mg/L chloride (drinking water).

ACCURACY Recovery = 98.2% at 10.0 mg/L chloride (drinking water)

SAMPLING METHOD Plastic or glass.

STABILITY No preservation required.

MHT 28 Days.

QUALITY CONTROL The lab should spike and analyze a minimum of 10% of all samples to monitor continuing lab performance. Field and lab duplicates should be analyzed. Measure retention times of standards.

REFERENCE Test Method-The Determination of Inorganic Anions in Water by Ion Chromatography, (EPA-600/4-84-017).

Chlorine EPA Method 330.1

TITLE Inorganics, Non-Metallics

MATRIX Waters and Wastes Total Residual

APPLICATION Date issued 1974. Editorial Rev. 1978. (Titrimetric, amperometric). Amperometric Method applies to all types waters and wastes without substantial amount of organic matter. (Chlorine and chloramines stoichiometrically liberate iodine from KI at pH 4 or less)

INTERFERENCES Stirring can lower chlorine values by volatilization. Copper and silver poison the electrode.

INSTRUMENTATION Amperometer (microammeter with necessary electrical accessories); microburet.

RANGE Not reported

MDL Not reported.

PRECISION SD = 24.8 and 12.5% at 0.64 and 1.83 mg/L total chloride.

ACCURACY Relative error = 8.5 and 8.8% at 0.64 and 1.83 mg/L total Cl

SAMPLING METHOD Plastic or glass (200 mL).

STABILITY No preservation required. Analyze immediately.

QUALITY CONTROL If dilution is necessary, it must be done with distilled water which is free of chlorine, chlorine-demand and ammonia. (Phenylarsine oxide or sodium thiosulfate is used as the standard reducing agent to titrate liberated iodine using amperometer to determine end point).

REFERENCE EPA Methods for the Chemical Analysis of Water and Wastes, EPA-600/4-79-020, U.S. EPA, EMSL, 1979.

Chlorine — EPA Method 330.2

TITLE Inorganics, Non-Metallics

MATRIX All types waters, Total Residual but especially Wastewaters.

APPLICATION Date issued 1978. (Titrimetric, back, iodometric) (starch or amperometric endpoint). Iodometric back-titration is best for wastewaters but is applicable to all types waters. (Chlorine and chloramines stoichiometrically liberate iodine from KI at pH 4 or less)

INTERFERENCES Manganese, iron and nitrite interference is minimized by buffering to pH 4 before adding KI. High concentrations of organics may cause uncertainty in the endpoint. Turbidity and color make endpoint difficult to detect. Practice runs with spikes may be necessary.

INSTRUMENTATION Standard lab glassware. Microburet 0–2 mL or 0–10 mL is used. Amperometric titrater.

RANGE Not reported

MDL Not reported.

PRECISION SD = ±0.12 mg/L Cl at 3.51 mg/L Cl (river water)

ACCURACY % recovery = 107.7% at 0.84 mg/L Cl (river water)

SAMPLING METHOD plastic or glass (200 mL).

STABILITY No preservation required. Analyze immediately.

QUALITY CONTROL Use chlorine free, chlorine-demand free distilled water for dilution. (Phenylarsine oxide is used as the standard reducing agent. Iodine quantitatively oxidizes reducing agent and excess is titrated with standard iodine titrant to starch-iodine or amperometric endpoint)

REFERENCE Methods for the Chemical Analysis of Water and Wastes, EPA-600/4-79-020, U.S. EPA, EMSL, 1979.

Chlorine — EPA Method 330.3

TITLE Inorganics, Non-Metallics

MATRIX Natural and Treated Waters. Total Residual

APPLICATION Date issued 1978. (Titrimetric, iodometric). Method applies to natural and treated waters at concentrations greater than 1 mg/L. Chlorine and chloramines liberate iodine from KI at pH 4 or less. (Iodine is titrated with a standard reducing agent using a starch indicator)

INTERFERENCES Ferric, manganic and nitrite ions interfere, the neutral titration minimizes these interferences. Acetic acid is used for acid titration — never use HCl. Turbidity and color may make endpoint difficult to detect. Practice runs with spiked samples may be necessary.

INSTRUMENTATION Standard lab glassware. Microburet 0–2 mL or 0–10 mL is used.

RANGE Concentrations >1 mg/L Cl

MDL Not reported.

PRECISION SD = 27 and 23.6% at 0.64 and 1.83 mg/L Cl.

ACCURACY Relative error = 23.6 and 16.7% at 0.64 and 1.83 mg/L Cl.

SAMPLING METHOD Plastic or glass (200 mL).

STABILITY No preservation required. Analyze immediately.

QUALITY CONTROL Phenylarsine oxide is used as the standardized reducing agent to titrate liberated iodine to starch iodine endpoint. (Titrate away from direct sunlight. Run blank titration).

REFERENCE Methods for the Chemical Analysis of Water and Wastes, EPA-600/4-79-020, U.S. EPA, EMSL, 1979.

Chlorine — EPA Method 330.4

TITLE Inorganics, Non-Metallics

MATRIX Natural and Treated Waters. Total Residual

APPLICATION Date issued 1978. (Titrimetric, DPD-FAS). The N,N-diethyl-p-phenylene diamine (DPD)-ferrous ammomium sulfate (FAS). Method applies to matrix listed at concentrations above 1 mg/L Cl. (Liberated iodine is titrated with FAS using DPD as indicator).

INTERFERENCES Bromine, bromamine and iodine are interferences normally present in insignificant amounts. Oxidized manganese and copper interfere, but can be corrected for. Turbidity and color may make endpoint difficult to detect.

INSTRUMENTATION Standard lab glassware. Microburet, 0–2 mL or 0–10 mL is used.

RANGE Concentrations above 1 mg/L Cl

MDL Not listed.

PRECISION SD = 19.4 and 9.4% at 0.64 and 1.83 mg/L Cl.

ACCURACY Relative error = 8.1 and 4.3% at 0.64 and 1.83 mg/L Cl.

SAMPLING METHOD plastic or glass (200 mL).

STABILITY No preservation required. Analyze immediately.

QUALITY CONTROL This procedure gives a convenient direct reading (ml titrant = mg/L Cl) up to 4 mg/L. An aliquot should be diluted to 100 mL if higher concentrations are present. Use chlorine free distilled water to prepare indicator.

REFERENCE EPA Methods for the Chemical Analysis of Water and Wastes, EPA-600/4-79-020, U.S. EPA, EMSL, 1979.

Chlorine   EPA Method 330.5

TITLE Inorganics, Non-Metallics

MATRIX Natural and Treated Waters. Total Residual

APPLICATION Date issued 1978. (Spectrophotometric, DPD). The N,N-diethyl-p-phenylene diamine (DPD) colorimetric method applies to matrix listed at concentrations from 0.2–4 mg/L Cl. (Liberated iodine reacts with DPD to produce a red colored solution read on spectrophotometer)

INTERFERENCES Any oxidizing agents; these are usually present at insignificant concentrations compared to the residual chlorine concentrations. Turbidity and color will essentially prevent the colorimetric analysis. (Ferrous ammonium sulfate is titrant used on permanganate stds).

INSTRUMENTATION Spectrophotometer. 515 nm. Cells of light path 1 cm or longer.

RANGE 0.2–4.0 mg/L Cl

MDL Not listed.

PRECISION SD = 27.6% at 0.66 mg/L Cl.

ACCURACY Relative error = 15.6% at 0.66 mg/L Cl.

SAMPLING METHOD Plastic or glass (200 mL).

STABILITY No preservation required. Analyze immediately.

QUALITY CONTROL The solution is spectrophotometrically compared to a series of standards, using a graph or a regression analysis calculation. (Calculation is figured using absorbance and titrated concentrations of permanganate solutions (chlorine equivalent) and absorbance of sample).

REFERENCE Methods for the Chemical Analysis of Water and Wastes, EPA-600/4-79-020, U.S. EPA, EMSL, 1979.

4-Chloro-1,2-phenylenediamine   EPA Method 8270
CAS #95-83-0

TITLE Semivolatile Organic Compounds by GC/MS

MATRIX This method is used to determine the concentration of semivolatile organic compounds in extracts prepared from all types of solid waste matrices, soils, and groundwater. Although surface waters are not specifically mentioned, this method should be applicable to water samples from rivers, lakes, etc.

METHOD SUMMARY This method covers 259 semivolatile organic compounds. In very limited applications direct injection of the sample into the GC/MS system may be appropriate, but this results in very high detection limits (approximately 10,000 µg/L). Typically, a 1-L liquid sample, containing surrogate, and matrix spiking standards, is extracted in a continuous extractor first under acid conditions and then under basic conditions. Typically 30 g of a solid sample, containing surrogate, and matrix spiking standards, is extracted ultrasonically. After concentrating the extract to 1 mL it is spiked with 10 µL of an internal standard solution just prior to analysis by GC/MS. The volume injected should contain about 100 ng of base/neutral and 200 ng of acid surrogates (for a 1-µL injection). Analysis is performed by GC/MS using a capillary GC column.

INTERFERENCES Raw GC/MS data from all blanks, samples, and spikes must be evaluated for interferences. Contamination by carryover can occur whenever high-concentration and low-concentration samples are sequentially analyzed. To reduce carryover, the sample syringe must be rinsed out between samples with solvent. Whenever an unusually concentrated sample is encountered, it should be followed by the analysis of blank solvent to check for cross-contamination.

INSTRUMENTATION A GC/MS and a data system are required. The GC column used is a 30 m × 0.25 mm I.D. (or 0.32 mm I.D.) 1um film thickness silicone-coated fused silica capillary column. A continuous liquid-liquid extractor equipped with Teflon® or glass connection joints and stopcocks requiring no lubrication, a K-D concentrating apparatus, water bath, and an ultrasonic disrupter with a minimum power of 300 W and with pulsing capability are also required.

PRECISION & ACCURACY The estimated quantitation limit (EQL) of Method 8270B for determining an individual compound is approximately 1 mg/kg (wet weight) for soil or sediment samples, 1–200 mg/kg for wastes (dependent on matrix and method of preparation), and 10 µg/L for groundwater samples. EQLs will be proportionately higher for sample extracts that require dilution to avoid saturation of the detector.

The EQL(b) for groundwater in µg/L is not listed.
The EQL (a, b) for low concentrations in soil and sediment in µg/kg is not listed.
Accuracy as µg/L is not listed.
Overall precision in µg/L is not listed.

(a) EQLs listed for soil/sediment are based on wet weight. Normally data is reported in a dry-weight basis; therefore, EQLs will be higher based on the % dry weight of each sample. This calculation is based on a 30 g sample and gel permeation chromatography cleanup.

(b) Sample EQLs are highly matrix-dependent. The EQLs are provided for guidance and may not always be achievable.

C = True value for concentration, in µg/L.

X = Average recovery found for measurements of samples containing a concentration of C, in µg/L.

ESTIMATED QUANTITATION LIMIT FOR OTHER MATRICES

Other Matrices	Factor (a)
High-concentration soil and sludges by sonicator	7.5
Non-water miscible waste	75

(a) EQL for other matrices = [EQL for low soil/sediment] × [Factor]. This estimated EQL is similar to an EPA "Practical Quantitation Limit."

SAMPLING METHOD

Liquid samples — Use a 1 or 2½ gallon amber glass bottle with a screw-top Teflon®-lined cover that has been prewashed with detergent and rinsed with distilled water and methanol (or isopropanol).

Soils, sediments, or sludges — Use an 8-oz. widemouth glass with a screw-top Teflon®-lined cover that has been prewashed with detergent and rinsed with distilled water and methanol (or isopropanol).

SAMPLE PRESERVATION

Liquid samples — If residual chlorine is present, add 3 mL of 10% sodium thiosulfate per gallon, cool to 4°C and store in a solvent-free refrigerator until analysis; if chlorine is not present, then eliminate the sodium thiosulfate addition.

Soils, sediments, or sludges — Cool samples to 4°C and store in a solvent-free refrigerator.

MHT Liquid samples must be extracted within 7 days and the extracts analyzed within 40 days. Soils, sediments, or sludges may be stored for a maximum of 14 days and the extracts analyzed within 40 days.

SAMPLE PREPARATION

Liquid samples — Transfer 1 L quantitatively to a continuous extractor. If high concentrations are anticipated, a smaller volume may be used and then diluted with organic-free reagent water to 1 L. Adjust pH, if necessary, to pH <2 using 1:1 (V/V) sulfuric acid. Pipette 1.0 mL of a surrogate standard spiking solution into each sample. For the sample in each analytical batch selected for spiking, add 1.0 mL of a matrix spiking standard. For base/neutral acid analysis, the amount of the surrogates and matrix spiking compounds added to the sample should result in a final concentration of 100 ng/µL of each analyte in the extract to be analyzed (assuming a 1-µL injection). Extract with methylene chloride for 18–24 h. Next, adjust the pH of the aqueous phase to pH >11 using 10 N sodium hydroxide and extract it with methylene chloride again for 18–24 h. Dry the extract through a column containing anhydrous sodium sulfate and concentrate it to 1 mL using a K-D concentrator.

Soils, sediments, or sludges — Use 30 g of sample. Nonporous or wet samples (gummy or clay type) that do not have a free-flowing sandy texture must be mixed with anhydrous sodium sulfate until the sample is free flowing. Add 1 mL of surrogate standards to all samples, spikes, standards, and blanks. For the sample in each analytical batch selected for spiking, add 1.0 mL of a matrix spiking standard. For base/neutral acid analysis, the amount added of the surrogates and matrix spiking compounds should result in a final concentration of 100 ng/µL of each base/neutral analyte and 200 ng/µL of each acid analyte in the extract to be analyzed (assuming a 1-µL injection). Immediately add a 100-mL mixture of 1:1 methylene chloride:acetone and extract the sample ultrasonically for 3 min and then decant or filter the extracts. Repeat the extraction two or more times. Dry the extract using a column with anhydrous sodium sulfate and concentrate it to 1 mL in a K-D concentrator.

QUALITY CONTROL A methylene chloride solution containing 50 ng/µL of decafluorotriphenylphosphine (DFTPP) is used for tuning the GC/MS system each 12-h shift. A system performance check also must be made during every 12-h shift. A standard containing 50 ng/µL each of 4,4'-DDT, pentachlorophenol, and benzidine is required to verify injection port inertness and GC column performance. A calibration standard at mid-concentration, containing each compound of interest, including all required surrogates, must be performed every 12 h during analysis. After the system performance check is met, calibration check compounds (CCCs) are used to check the validity of the initial calibration.

The internal standard responses and retention times in the calibration check standard must be evaluated immediately after or during data acquisition. If the retention time for any internal standard changes by more than 30 seconds from the last check calibration (12 h), the chromatographic system must be inspected for malfunctions and corrections must be made, as required. If the electron ionization current plot (EICP) area for any of the internal standards changes by a factor of two from the last daily calibration standard check, the mass spectrometer must be inspected for malfunctions and corrections must be made, as appropriate.

Demonstrate, through the analysis of a reagent water blank, that interferences from the analytical system, glassware, and reagents are under control. The blank samples should be carried through all stages of the sample preparation and measurement steps. For each analytical batch (up to 20 samples), a reagent blank, matrix spike, and matrix spike duplicate/duplicate must be analyzed (the frequency of the spikes may be different for different monitoring programs). The blank and spiked samples must be carried through all stages of the sample preparation and measurement steps. A QC reference sample concentrate containing each analyte at a concentration of 100 mg/L in methanol is required.

REFERENCE Test Methods for Evaluating Solid Waste (SW-846). U.S. EPA 1983, Method 8270B, Rev. 2, Nov. 1990. Office of Solid Waste, Washington, DC.

2-Chloro-1,3-butadiene
CAS #126-99-8
EPA Method 1624

TITLE Volatile Organic Compounds by Isotope Dilution GC/MS

MATRIX Compounds may be determined in waters, soils, and municipal sludges by this method.

METHOD SUMMARY This method is used to determine 58 volatile toxic organic pollutants associated with the CWA (as amended 1987); the RCRA (as amended 1986); the CERCLA (as amended 1986); and other compounds amenable to purge-and-trap gas chromatography-mass spectrometry (GC/MS).

If the solids content is less than 1%, stable isotopically-labeled analogs of the compounds of interest are added to a 5-mL sample and the sample is purged with an inert gas at 20–25°C in a chamber designed for soil or water samples. If the solids content is greater than 1%, 5 mL of reagent water and the labeled compounds are added to a 5-g aliquot of sample and the mixture is purged at 40°C. Compounds that will not purge at 20–25°C or at 40°C are purged at 78–85°C. In the purging process, the volatile compounds are transferred from the aqueous phase into the gaseous phase where they are passed into a sorbent column, and trapped. After purging is completed, the trap is backflushed and heated rapidly to desorb the compounds into a GC. The compounds are separated by the GC and detected by a MS. The labeled compounds serve to correct the variability of the analytical technique.

INTERFERENCES Impurities in the purge gas, organic compounds outgassing from the plumbing upstream of the trap, and solvent vapors in the lab account for most problems. Samples can be contaminated by diffusion of volatile organic compounds (particularly methylene chloride) through the bottle seal during shipment and storage. Contamination by carryover can occur when high-level and low-level samples are analyzed sequentially. When an unusually concentrated sample is encountered, follow it by analysis of a reagent water blank to check for carryover.

INSTRUMENTATION Major equipment includes a GC with linear temperature programming and a glass jet separator as the MS interface, a MS with 70 eV electron impact ionization, and a data system to collect and record response factors.

Column: 2.8 m × 2 mm I.D. glass, packed with 1% SP-1000 on Carbopak B, 60/80 mesh, or equivalent.

PRECISION & ACCURACY The detection limits of the method are usually dependent on the level of interferences rather than instrumental limitations. The method detection limits were determined in digested sludge (low solids) and in filter cake or compost (high solids).

The MDL (in µg/kg) for low solids is not listed and for high solids is not listed.
Labeled and native compound precision (in µg/L) as standard deviation was not listed.
Labeled and native compound accuracy (in µg/L) as average recovery was not listed.
Acceptance criteria are at 20 µg/L for this compound.

SAMPLE COLLECTION, PRESERVATION & HANDLING Grab samples are collected in glass containers having a total volume greater than 20 mL. Fill and seal each bottle so that no air bubbles are entrapped. Samples are maintained at 0 to 4°C from the time of collection until analysis. If an aqueous sample contains residual chlorine, add sodium thiosulfate preservative (10 mg/40 mL) to the empty sample bottles just prior to shipment to the sample site. All samples must be analyzed within 14 days of collection.

SAMPLE PREPARATION Samples containing less than 1% solids are analyzed directly as aqueous samples. Samples containing 1% solids or greater are analyzed as solid samples utilizing one of two methods, depending on the levels of pollutants, in the sample. Samples containing 1% solids or greater, and low to moderate levels of pollutants are analyzed by purging a known weight of sample added to 5 mL of reagent water. Samples containing 1% solids or greater, and high levels of pollutants, are extracted with methanol, and an aliquot of the methanol extract is added to reagent water and purged.

QUALITY CONTROL A field blank prepared from reagent water and carried through the sampling and handling protocol may serve as a check on contamination from shipment and storage.

The analyst is permitted to modify this method to improve separations or lower the costs of measurements, provided all performance specifications are met. Analyses of blanks are required. When results of spikes indicate atypical method performance for samples, the samples are diluted to bring method performance within acceptable limits. Analyze two sets of four 5-mL aliquots (8 aliquots total) of the aqueous performance standard. Spike all samples with labeled compounds to assess method performance on the sample matrix. Compute the percent recovery of the labeled compounds using the internal standard method. Compare the percent recovery for each compound with the corresponding labeled compound recovery. Reagent water blanks are analyzed to demonstrate freedom from carryover contamination. Field replicates may be collected to determine the precision of the sampling technique, and spiked samples may be required to determine the accuracy of the analysis when the internal method is used.

REFERENCE Volatile Organic Compounds by Isotope Dilution GC/MS. Office of Water Regulation and Standards, U.S. EPA Industrial Technology Division, Washington, DC, EPA Method 1624, Rev. C, June 1989 (contact W.A. Telliard, U.S. EPA, Office of Water Regulations and Standards, 401 M St., SW, Washington, DC, 20460. Phone: 202-382-7131).

4-Chloro-1,3-phenylenediamine
CAS #5131-60-2
EPA Method 8270

TITLE Semivolatile Organic Compounds by GC/MS

MATRIX This method is used to determine the concentration of semivolatile organic compounds in extracts prepared from all types of solid waste matrices, soils, and groundwater. Although surface waters are not specifically mentioned, this

method should be applicable to water samples from rivers, lakes, etc.

METHOD SUMMARY This method covers 259 semivolatile organic compounds. In very limited applications direct injection of the sample into the GC/MS system may be appropriate, but this results in very high detection limits (approximately 10,000 µg/L). Typically, a 1-L liquid sample, containing surrogate, and matrix spiking standards, is extracted in a continuous extractor first under acid conditions and then under basic conditions. Typically 30 g of a solid sample, containing surrogate, and matrix spiking standards, is extracted ultrasonically. After concentrating the extract to 1 mL it is spiked with 10 µL of an internal standard solution just prior to analysis by GC/MS. The volume injected should contain about 100 ng of base/neutral and 200 ng of acid surrogates (for a 1-µL injection). Analysis is performed by GC/MS using a capillary GC column.

INTERFERENCES Raw GC/MS data from all blanks, samples, and spikes must be evaluated for interferences. Contamination by carryover can occur whenever high-concentration and low-concentration samples are sequentially analyzed. To reduce carryover, the sample syringe must be rinsed out between samples with solvent. Whenever an unusually concentrated sample is encountered, it should be followed by the analysis of blank solvent to check for cross-contamination.

INSTRUMENTATION A GC/MS and a data system are required. The GC column used is a 30 m × 0.25 mm I.D. (or 0.32 mm I.D.) 1um film thickness silicone-coated fused silica capillary column. A continuous liquid-liquid extractor equipped with Teflon® or glass connection joints and stopcocks requiring no lubrication, a K-D concentrating apparatus, water bath, and an ultrasonic disrupter with a minimum power of 300 W and with pulsing capability are also required.

PRECISION & ACCURACY The estimated quantitation limit (EQL) of Method 8270B for determining an individual compound is approximately 1 mg/kg (wet weight) for soil or sediment samples, 1–200 mg/kg for wastes (dependent on matrix and method of preparation), and 10 µg/L for groundwater samples. EQLs will be proportionately higher for sample extracts that require dilution to avoid saturation of the detector.

The EQL(b) for groundwater in µg/L is not listed.
The EQL (a, b) for low concentrations in soil and sediment in µg/kg is not listed.
Accuracy as µg/L is not listed.
Overall precision in µg/L is not listed.

(a) *EQLs listed for soil/sediment are based on wet weight. Normally data is reported in a dry-weight basis; therefore, EQLs will be higher based on the % dry weight of each sample. This calculation is based on a 30 g sample and gel permeation chromatography cleanup.*
(b) *Sample EQLs are highly matrix-dependent. The EQLs are provided for guidance and may not always be achievable.*
C = *True value for concentration, in µg/L.*
X = *Average recovery found for measurements of samples containing a concentration of C, in µg/L.*

ESTIMATED QUANTITATION LIMIT FOR OTHER MATRICES

Other Matrices	Factor (a)
High-concentration soil and sludges by sonicator	7.5
Non-water miscible waste	75

(a) *EQL for other matrices = [EQL for low soil/sediment] × [Factor]. This estimated EQL is similar to an EPA "Practical Quantitation Limit."*

SAMPLING METHOD
Liquid samples — Use a 1 or 2½ gallon amber glass bottle with a screw-top Teflon®-lined cover that has been prewashed with detergent and rinsed with distilled water and methanol (or isopropanol).

Soils, sediments, or sludges — Use an 8-oz. widemouth glass with a screw-top Teflon®-lined cover that has been prewashed with detergent and rinsed with distilled water and methanol (or isopropanol).

SAMPLE PRESERVATION
Liquid samples — If residual chlorine is present, add 3 mL of 10% sodium thiosulfate per gallon, cool to 4°C and store in a solvent-free refrigerator until analysis; if chlorine is not present, then eliminate the sodium thiosulfate addition.

Soils, sediments, or sludges — Cool samples to 4°C and store in a solvent-free refrigerator.

MHT Liquid samples must be extracted within 7 days and the extracts analyzed within 40 days. Soils, sediments, or sludges may be stored for a maximum of 14 days and the extracts analyzed within 40 days.

SAMPLE PREPARATION
Liquid samples — Transfer 1 L quantitatively to a continuous extractor. If high concentrations are anticipated, a smaller volume may be used and then diluted with organic-free reagent water to 1 L. Adjust pH, if necessary, to pH <2 using 1:1 (V/V) sulfuric acid. Pipette 1.0 mL of a surrogate standard spiking solution into each sample. For the sample in each analytical batch selected for spiking, add 1.0 mL of a matrix spiking standard. For base/neutral acid analysis, the amount of the surrogates and matrix spiking compounds added to the sample should result in a final concentration of 100 ng/µL of each analyte in the extract to be analyzed (assuming a 1-µL injection). Extract with methylene chloride for 18–24 h. Next, adjust the pH of the aqueous phase to pH >11 using 10 N sodium hydroxide and extract it with methylene chloride again for 18–24 h. Dry the extract through a column containing anhydrous sodium sulfate and concentrate it to 1 mL using a K-D concentrator.

Soils, sediments, or sludges — Use 30 g of sample. Nonporous or wet samples (gummy or clay type) that do not have a free-flowing sandy texture must be mixed with anhydrous sodium sulfate until the sample is free flowing. Add 1 mL of surrogate standards to all samples, spikes, standards, and blanks. For the sample in each analytical batch selected for spiking, add 1.0 mL of a matrix spiking standard. For base/neutral acid analysis, the amount added of the surrogates and matrix spiking compounds should result in a final concentration of 100 ng/µL of

each base/neutral analyte and 200 ng/µL of each acid analyte in the extract to be analyzed (assuming a 1-µL injection). Immediately add a 100-mL mixture of 1:1 methylene chloride:acetone and extract the sample ultrasonically for 3 min and then decant or filter the extracts. Repeat the extraction two or more times. Dry the extract using a column with anhydrous sodium sulfate and concentrate it to 1 mL in a K-D concentrator.

QUALITY CONTROL A methylene chloride solution containing 50 ng/µL of decafluorotriphenylphosphine (DFTPP) is used for tuning the GC/MS system each 12-h shift. A system performance check also must be made during every 12-h shift. A standard containing 50 ng/µL each of 4,4'-DDT, pentachlorophenol, and benzidine is required to verify injection port inertness and GC column performance. A calibration standard at mid-concentration, containing each compound of interest, including all required surrogates, must be performed every 12 h during analysis. After the system performance check is met, calibration check compounds (CCCs) are used to check the validity of the initial calibration.

The internal standard responses and retention times in the calibration check standard must be evaluated immediately after or during data acquisition. If the retention time for any internal standard changes by more than 30 seconds from the last check calibration (12 h), the chromatographic system must be inspected for malfunctions and corrections must be made, as required. If the electron ionization current plot (EICP) area for any of the internal standards changes by a factor of two from the last daily calibration standard check, the mass spectrometer must be inspected for malfunctions and corrections must be made, as appropriate.

Demonstrate, through the analysis of a reagent water blank, that interferences from the analytical system, glassware, and reagents are under control. The blank samples should be carried through all stages of the sample preparation and measurement steps. For each analytical batch (up to 20 samples), a reagent blank, matrix spike, and matrix spike duplicate/duplicate must be analyzed (the frequency of the spikes may be different for different monitoring programs). The blank and spiked samples must be carried through all stages of the sample preparation and measurement steps. A QC reference sample concentrate containing each analyte at a concentration of 100 mg/L in methanol is required.

REFERENCE Test Methods for Evaluating Solid Waste (SW-846). U.S. EPA 1983, Method 8270B, Rev. 2, Nov. 1990. Office of Solid Waste, Washington, DC.

5-Chloro-2-methylaniline **EPA Method 8270**
CAS #95-79-4

TITLE Semivolatile Organic Compounds by GC/MS

MATRIX This method is used to determine the concentration of semivolatile organic compounds in extracts prepared from all types of solid waste matrices, soils, and groundwater. Although surface waters are not specifically mentioned, this method should be applicable to water samples from rivers, lakes, etc.

METHOD SUMMARY This method covers 259 semivolatile organic compounds. In very limited applications direct injection of the sample into the GC/MS system may be appropriate, but this results in very high detection limits (approximately 10,000 µg/L). Typically, a 1-L liquid sample, containing surrogate, and matrix spiking standards, is extracted in a continuous extractor first under acid conditions and then under basic conditions. Typically 30 g of a solid sample, containing surrogate, and matrix spiking standards, is extracted ultrasonically. After concentrating the extract to 1 mL it is spiked with 10 µL of an internal standard solution just prior to analysis by GC/MS. The volume injected should contain about 100 ng of base/neutral and 200 ng of acid surrogates (for a 1-µL injection). Analysis is performed by GC/MS using a capillary GC column.

INTERFERENCES Raw GC/MS data from all blanks, samples, and spikes must be evaluated for interferences. Contamination by carryover can occur whenever high-concentration and low-concentration samples are sequentially analyzed. To reduce carryover, the sample syringe must be rinsed out between samples with solvent. Whenever an unusually concentrated sample is encountered, it should be followed by the analysis of blank solvent to check for cross-contamination.

INSTRUMENTATION A GC/MS and a data system are required. The GC column used is a 30 m × 0.25 mm I.D. (or 0.32 mm I.D.) 1um film thickness silicone-coated fused silica capillary column. A continuous liquid-liquid extractor equipped with Teflon® or glass connection joints and stopcocks requiring no lubrication, a K-D concentrating apparatus, water bath, and an ultrasonic disrupter with a minimum power of 300 W and with pulsing capability are also required.

PRECISION & ACCURACY The estimated quantitation limit (EQL) of Method 8270B for determining an individual compound is approximately 1 mg/kg (wet weight) for soil or sediment samples, 1–200 mg/kg for wastes (dependent on matrix and method of preparation), and 10 µg/L for groundwater samples. EQLs will be proportionately higher for sample extracts that require dilution to avoid saturation of the detector.

The EQL(b) for groundwater in µg/L is 10.
The EQL (a, b) for low concentrations in soil and sediment in µg/kg is not determined.
Accuracy as µg/L is not listed.
Overall precision in µg/L is not listed.

(a) *EQLs listed for soil/sediment are based on wet weight. Normally data is reported in a dry-weight basis; therefore, EQLs will be higher based on the % dry weight of each sample. This calculation is based on a 30 g sample and gel permeation chromatography cleanup.*
(b) *Sample EQLs are highly matrix-dependent. The EQLs are provided for guidance and may not always be achievable.*
C = *True value for concentration, in µg/L.*
X = *Average recovery found for measurements of samples containing a concentration of C, in µg/L.*

ESTIMATED QUANTITATION LIMIT FOR OTHER MATRICES

Other Matrices	Factor (a)
High-concentration soil and sludges by sonicator	7.5
Non-water miscible waste	75

(a) EQL for other matrices = [EQL for low soil/sediment] × [Factor]. This estimated EQL is similar to an EPA "Practical Quantitation Limit."

SAMPLING METHOD

Liquid samples — Use a 1 or 2½ gallon amber glass bottle with a screw-top Teflon®-lined cover that has been prewashed with detergent and rinsed with distilled water and methanol (or isopropanol).

Soils, sediments, or sludges — Use an 8-oz. widemouth glass with a screw-top Teflon®-lined cover that has been prewashed with detergent and rinsed with distilled water and methanol (or isopropanol).

SAMPLE PRESERVATION

Liquid samples — If residual chlorine is present, add 3 mL of 10% sodium thiosulfate per gallon, cool to 4°C and store in a solvent-free refrigerator until analysis; if chlorine is not present, then eliminate the sodium thiosulfate addition.

Soils, sediments, or sludges — Cool samples to 4°C and store in a solvent-free refrigerator.

MHT Liquid samples must be extracted within 7 days and the extracts analyzed within 40 days. Soils, sediments, or sludges may be stored for a maximum of 14 days and the extracts analyzed within 40 days.

SAMPLE PREPARATION

Liquid samples — Transfer 1 L quantitatively to a continuous extractor. If high concentrations are anticipated, a smaller volume may be used and then diluted with organic-free reagent water to 1 L. Adjust pH, if necessary, to pH <2 using 1:1 (V/V) sulfuric acid. Pipette 1.0 mL of a surrogate standard spiking solution into each sample. For the sample in each analytical batch selected for spiking, add 1.0 mL of a matrix spiking standard. For base/neutral acid analysis, the amount of the surrogates and matrix spiking compounds added to the sample should result in a final concentration of 100 ng/μL of each analyte in the extract to be analyzed (assuming a 1-μL injection). Extract with methylene chloride for 18–24 h. Next, adjust the pH of the aqueous phase to pH >11 using 10 N sodium hydroxide and extract it with methylene chloride again for 18–24 h. Dry the extract through a column containing anhydrous sodium sulfate and concentrate it to 1 mL using a K-D concentrator.

Soils, sediments, or sludges — Use 30 g of sample. Nonporous or wet samples (gummy or clay type) that do not have a free-flowing sandy texture must be mixed with anhydrous sodium sulfate until the sample is free flowing. Add 1 mL of surrogate standards to all samples, spikes, standards, and blanks. For the sample in each analytical batch selected for spiking, add 1.0 mL of a matrix spiking standard. For base/neutral acid analysis, the amount added of the surrogates and matrix spiking compounds should result in a final concentration of 100 ng/μL of each base/neutral analyte and 200 ng/μL of each acid analyte in the extract to be analyzed (assuming a 1-μL injection). Immediately add a 100-mL mixture of 1:1 methylene chloride:acetone and extract the sample ultrasonically for 3 min and then decant or filter the extracts. Repeat the extraction two or more times. Dry the extract using a column with anhydrous sodium sulfate and concentrate it to 1 mL in a K-D concentrator.

QUALITY CONTROL A methylene chloride solution containing 50 ng/μL of decafluorotriphenylphosphine (DFTPP) is used for tuning the GC/MS system each 12-h shift. A system performance check also must be made during every 12-h shift. A standard containing 50 ng/μL each of 4,4'-DDT, pentachlorophenol, and benzidine is required to verify injection port inertness and GC column performance. A calibration standard at mid-concentration, containing each compound of interest, including all required surrogates, must be performed every 12 h during analysis. After the system performance check is met, calibration check compounds (CCCs) are used to check the validity of the initial calibration.

The internal standard responses and retention times in the calibration check standard must be evaluated immediately after or during data acquisition. If the retention time for any internal standard changes by more than 30 seconds from the last check calibration (12 h), the chromatographic system must be inspected for malfunctions and corrections must be made, as required. If the electron ionization current plot (EICP) area for any of the internal standards changes by a factor of two from the last daily calibration standard check, the mass spectrometer must be inspected for malfunctions and corrections must be made, as appropriate.

Demonstrate, through the analysis of a reagent water blank, that interferences from the analytical system, glassware, and reagents are under control. The blank samples should be carried through all stages of the sample preparation and measurement steps. For each analytical batch (up to 20 samples), a reagent blank, matrix spike, and matrix spike duplicate/duplicate must be analyzed (the frequency of the spikes may be different for different monitoring programs). The blank and spiked samples must be carried through all stages of the sample preparation and measurement steps. A QC reference sample concentrate containing each analyte at a concentration of 100 mg/L in methanol is required.

REFERENCE Test Methods for Evaluating Solid Waste (SW-846). U.S. EPA 1983, Method 8270B, Rev. 2, Nov. 1990. Office of Solid Waste, Washington, DC.

4-Chloro-2-nitroaniline   **EPA Method 1625**
CAS #89-63-4

TITLE Semivolatile Organic Compounds by Isotope Dilution GC/MS

MATRIX The compounds may be determined in waters, soils, and municipal sludges by this method.

METHOD SUMMARY This method is used to determine 176 semivolatile toxic organic pollutants associated with the

CWA (as amended 1987); the RCRA (as amended 1986); the CERCLA (as amended 1986); and other compounds amenable to extraction and analysis by capillary column gas chromatography-mass spectrometry (GC/MS).

Stable isotopically-labeled analogs of the compounds of interest are added to the sample. If the solids content is less than 1%, a 1-L sample is extracted at pH 12–13, then at pH <2 with methylene chloride using continuous extraction techniques.

If the solids content is 30% or less, the sample is diluted to 1% solids with reagent water, homogenized ultrasonically, and extracted at pH 12–13, then at pH <2 with methylene chloride using continuous extraction techniques. If the solids content is greater than 30%, the sample is extracted using ultrasonic techniques.

Each extract is dried over sodium sulfate, concentrated to a volume of 5 mL, cleaned up using GPC, if necessary, and concentrated. Extracts are concentrated to 1 mL if GPC is not performed, and to 0.5 mL if GPC is performed.

An internal standard is added to the extract, and a 1-mL aliquot of the extract is injected into the GC. The compounds are separated by GC and detected by a MS. The labeled compounds serve to correct the variability of the analytical technique.

INTERFERENCES Solvents, reagents, glassware, and other sample processing hardware may yield artifacts and/or elevated baselines causing misinterpretation of chromatograms and spectra. Materials used in the analysis must be demonstrated to be free from interferences under the conditions of analysis by running method blanks initially and with each sample lot (sample started through the extraction process on a given 8-h shift, to a maximum of 20). Specific selection of reagents and purification of solvents by distillation in all glass systems may be required. Glassware and, where possible, reagents are cleaned by solvent rinse and baking at 450°C for 1-h minimum. Interferences coextracted from samples will vary considerably from source to source, depending on the diversity of the site being sampled.

INSTRUMENTATION Major instrumentation includes a GC with a splitless or on-column injection port for capillary column, a MS with 70 eV electron impact ionization, and a data system to collect and record MS data, and process it. A K-D apparatus is used to concentrate extracts.

GC Column: 30 m × 0.25 mm I.D. 5% phenyl, 94% methyl, 1% vinyl silicone bonded phased fused silica capillary column.

PRECISION & ACCURACY The detection limits of the method are usually dependent on the level of interferences rather than instrumental limitations. The limits typify the minimum quantities that can be detected with no interferences present.

The minimum level (in µg/mL) was not listed. This is defined as a minimum level at which the analytical system shall give recognizable mass spectra (background corrected) and acceptable calibration points.
The MDL (in µg/kg) in low solids was not listed and in high solids was not listed; these were determined in digested sludge (low solids) and in filter cake or compost (high solids).
The labeled and native compound initial precision as standard deviation (in µg/L) was not listed.
The labeled and native compound initial accuracy as average recovery (in µg/L) was not listed.

SAMPLE COLLECTION, PRESERVATION & HANDLING Collect samples in glass containers. Aqueous samples which flow freely are collected in refrigerated bottles using automatic sampling equipment. Solid samples are collected as grab samples using widemouth jars. Maintain samples at 0 to 4°C from the time of collection until extraction. If residual chlorine is present in aqueous samples, add 80 mg sodium thiosulfate/L of water. Begin sample extraction within 7 days of collection, and analyze all extracts within 40 days of extraction.

SAMPLE PREPARATION Samples containing 1% solids or less are extracted directly using continuous liquid-liquid extraction techniques. Samples containing 1 to 30% solids are diluted to the 1% level with reagent water and extracted using continuous liquid-liquid extraction techniques. Samples containing greater than 30% solids are extracted using ultrasonic techniques.

Base/neutral extraction — Adjust the pH of the waters in the extractors to 12–13 with 6 N NaOH. Extract with methylene chloride for 24–48 h.
Acid extraction — Adjust the pH of the waters in the extractors to 2 or less using 6 N sulfuric acid. Extract with methylene chloride for 24–48 h.
Ultrasonic extraction of high solids samples — Add anhydrous sodium sulfate to the sample and QC aliquot(s). Add acetone:methylene chloride (1:1) to the sample and mix thoroughly

Concentrate extracts using a K-D apparatus.

QUALITY CONTROL The analyst is permitted to modify this method to improve separations or lower the costs of measurements, provided all performance specifications are met. Analyses of blanks are required to demonstrate freedom from contamination. When results of spikes indicate atypical method performance for samples, the samples are diluted to bring method performance within acceptable limits.

For low solids (aqueous samples), extract, concentrate, and analyze two sets of four 1-L aliquots (8 aliquots total) of the precision and recovery standard. For high solids samples, two sets of four 30-g aliquots of the high solids reference matrix are used.

Spike all samples with labeled compounds to assess method performance. Compute percent recovery of the labeled compounds using the internal standard method. Compare the labeled compound recovery for each compound with the corresponding labeled compound recovery.

Reagent water and high solids reference matrix blanks are analyzed to demonstrate freedom from contamination. Extract and concentrate a 1-L reagent water blank or a high solids reference matrix blank with each sample's lot (samples started through the extraction process on the same 8-h shift, to a maximum of 20 samples).

Field replicates may be collected to determine the precision of the sampling technique, and spiked samples may be required to determine the accuracy of the analysis when the internal standard method is used.

REFERENCE Semivolatile Organic Compounds by Isotope Dilution GC/MS. Office of Water Regulation and Standards, U.S. EPA Industrial Technology Division, Washington, DC, EPA Method 1625, Rev. C, June 1989 (contact W.A. Telliard, U.S. EPA, Office of Water Regulations and Standards, 401 M St., SW, Washington, DC, 20460. Phone: 202-382-7131).

4-Chloro-3-methylphenol **EPA Method 1625**
CAS #59-50-7

TITLE Semivolatile Organic Compounds by Isotope Dilution GC/MS

MATRIX The compounds may be determined in waters, soils, and municipal sludges by this method.

METHOD SUMMARY This method is used to determine 176 semivolatile toxic organic pollutants associated with the CWA (as amended 1987); the RCRA (as amended 1986); the CERCLA (as amended 1986); and other compounds amenable to extraction and analysis by capillary column gas chromatography-mass spectrometry (GC/MS).

Stable isotopically-labeled analogs of the compounds of interest are added to the sample. If the solids content is less than 1%, a 1-L sample is extracted at pH 12–13, then at pH <2 with methylene chloride using continuous extraction techniques.

If the solids content is 30% or less, the sample is diluted to 1% solids with reagent water, homogenized ultrasonically, and extracted at pH 12–13, then at pH <2 with methylene chloride using continuous extraction techniques. If the solids content is greater than 30%, the sample is extracted using ultrasonic techniques.

Each extract is dried over sodium sulfate, concentrated to a volume of 5 mL, cleaned up using GPC, if necessary, and concentrated. Extracts are concentrated to 1 mL if GPC is not performed, and to 0.5 mL if GPC is performed.

An internal standard is added to the extract, and a 1-mL aliquot of the extract is injected into the GC. The compounds are separated by GC and detected by a MS. The labeled compounds serve to correct the variability of the analytical technique.

INTERFERENCES Solvents, reagents, glassware, and other sample processing hardware may yield artifacts and/or elevated baselines causing misinterpretation of chromatograms and spectra. Materials used in the analysis must be demonstrated to be free from interferences under the conditions of analysis by running method blanks initially and with each sample lot (sample started through the extraction process on a given 8-h shift, to a maximum of 20). Specific selection of reagents and purification of solvents by distillation in all glass systems may be required. Glassware and, where possible, reagents are cleaned by solvent rinse and baking at 450°C for 1-h minimum. Interferences coextracted from samples will vary considerably from source to source, depending on the diversity of the site being sampled.

INSTRUMENTATION Major instrumentation includes a GC with a splitless or on-column injection port for capillary column, a MS with 70 eV electron impact ionization, and a data system to collect and record MS data, and process it. A K-D apparatus is used to concentrate extracts.

GC Column: 30 m × 0.25 mm I.D. 5% phenyl, 94% methyl, 1% vinyl silicone bonded phased fused silica capillary column.

PRECISION & ACCURACY The detection limits of the method are usually dependent on the level of interferences rather than instrumental limitations. The limits typify the minimum quantities that can be detected with no interferences present.

The minimum level (in µg/mL) was 10. This is defined as a minimum level at which the analytical system shall give recognizable mass spectra (background corrected) and acceptable calibration points.

The MDL (in µg/kg) in low solids was 41 and in high solids was 62; these were determined in digested sludge (low solids) and in filter cake or compost (high solids).

The labeled and native compound initial precision as standard deviation (in µg/L) was 37.

The labeled and native compound initial accuracy as average recovery (in µg/L) was 76–131.

SAMPLE COLLECTION, PRESERVATION & HANDLING Collect samples in glass containers. Aqueous samples which flow freely are collected in refrigerated bottles using automatic sampling equipment. Solid samples are collected as grab samples using widemouth jars. Maintain samples at 0 to 4°C from the time of collection until extraction. If residual chlorine is present in aqueous samples, add 80 mg sodium thiosulfate/L of water. Begin sample extraction within 7 days of collection, and analyze all extracts within 40 days of extraction.

SAMPLE PREPARATION Samples containing 1% solids or less are extracted directly using continuous liquid-liquid extraction techniques. Samples containing 1 to 30% solids are diluted to the 1% level with reagent water and extracted using continuous liquid-liquid extraction techniques. Samples containing greater than 30% solids are extracted using ultrasonic techniques.

Base/neutral extraction — Adjust the pH of the waters in the extractors to 12–13 with 6 N NaOH. Extract with methylene chloride for 24–48 h.

Acid extraction — Adjust the pH of the waters in the extractors to 2 or less using 6 N sulfuric acid. Extract with methylene chloride for 24–48 h.

Ultrasonic extraction of high solids samples — Add anhydrous sodium sulfate to the sample and QC aliquot(s). Add acetone:methylene chloride (1:1) to the sample and mix thoroughly

Concentrate extracts using a K-D apparatus.

QUALITY CONTROL The analyst is permitted to modify this method to improve separations or lower the costs of measurements, provided all performance specifications are met.

Analyses of blanks are required to demonstrate freedom from contamination. When results of spikes indicate atypical method performance for samples, the samples are diluted to bring method performance within acceptable limits.

For low solids (aqueous samples), extract, concentrate, and analyze two sets of four 1-L aliquots (8 aliquots total) of the precision and recovery standard. For high solids samples, two sets of four 30-g aliquots of the high solids reference matrix are used.

Spike all samples with labeled compounds to assess method performance. Compute percent recovery of the labeled compounds using the internal standard method. Compare the labeled compound recovery for each compound with the corresponding labeled compound recovery.

Reagent water and high solids reference matrix blanks are analyzed to demonstrate freedom from contamination. Extract and concentrate a 1-L reagent water blank or a high solids reference matrix blank with each sample's lot (samples started through the extraction process on the same 8-h shift, to a maximum of 20 samples).

Field replicates may be collected to determine the precision of the sampling technique, and spiked samples may be required to determine the accuracy of the analysis when the internal standard method is used.

REFERENCE Semivolatile Organic Compounds by Isotope Dilution GC/MS. Office of Water Regulation and Standards, U.S. EPA Industrial Technology Division, Washington, DC, EPA Method 1625, Rev. C, June 1989 (contact W.A. Telliard, U.S. EPA, Office of Water Regulations and Standards, 401 M St., SW, Washington, DC, 20460. Phone: 202-382-7131).

4-Chloro-3-methylphenol **EPA Method 625**
CAS #59-50-7

TITLE Base/Neutrals and Acids, U.S. EPA Method 625

MATRIX This methods covers municipal and industrial wastewaters.

METHOD SUMMARY Approximately 1 L of sample is serially extracted with methylene chloride at a pH greater than 11 and again at a pH less than 2 using a separatory funnel or a continuous extractor. The methylene chloride extract is dried, concentrated to a volume of 1 mL, and analyzed by GC/MS. Qualitative identification of the parameters in the extract is performed using the retention time and the relative abundance of three characteristic masses (m/z). Qualitative analysis is performed using either external or internal standard techniques with a single characteristic m/z.

INTERFERENCES Method interferences may be caused by contaminants in solvents, reagents, glassware, and other sample processing hardware. Glassware must be scrupulously cleaned. Glassware should be heated in a muffle furnace at 400°C for 5 to 30 min. Some thermally stable materials, such as PCBs, may not be eliminated by this treatment. Solvent rinses with acetone and pesticide quality hexane may be substituted for the muffle furnace heating. Matrix interferences may be caused by contaminants that are coextracted from the sample. The base-neutral extraction may cause significantly reduced recovery of phenols. The packed gas chromatographic columns recommended for the basic fraction may not exhibit sufficient resolution for some analytes.

INSTRUMENTATION A GC/MS system with an injection port designed for on-column injection when using packed columns and for splitless injection when using capillary columns.

Column for base/neutrals: 1.8 m long × 2 mm I.D. glass, packed with 3% SP-2550 on Supelcoport (100/120 mesh) or equivalent.

Column for acids: 1.8 m long × 2 mm I.D. glass, packed with 1% SP-1240DA on Supelcoport (100/120 mesh) or equivalent.

PRECISION & ACCURACY The MDL concentrations were obtained using reagent water. The MDL actually achieved in a given analysis will vary depending on instrument sensitivity and matrix effects. This method was tested by 15 laboratories using reagent water, drinking water, surface water, and industrial wastewaters spiked at six concentrations over the range 5 to 100 µg/L. Single operator precision, overall precision, and method accuracy were found to be directly related to the concentration of the parameter matrix.

The MDL (in µg/L) in reagent water was 3.0.

The standard deviation (in µg/L based on 4 recovery measurements) was 37.2.

The range (in µg/L) for average recovery for 4 measurements was 40.8–127.9.

The range (in %) for percent recovery was 22–147.

Accuracy (in µg/L) as expected recovery for one or more measurements of a sample containing a true concentration of C was 0.84C+0.35.

Precision (in µg/L) as expected single analyst standard deviation of measurements at an average concentration found at X was 0.23X+0.75.

Overall precision (in µg/L) as expected interlaboratory standard deviation of measurements in an average concentration found at X was 0.29X+1.31.

$C = $ *True value of the concentration in µg/L.*
$X = $ *Average recovery found for measurements of samples containing a concentration at C in µg/L.*

SAMPLE PREPARATION Adjust the pH to >11 with sodium hydroxide and serially extract in a separatory funnel with methylene chloride or else in a continuous extractor. Next, adjust the pH to <2 with sulfuric acid and serially extract in a separatory funnel with methylene chloride or else in a continuous extractor. Dry the extracts separately through a column of anhydrous sodium sulfate and then concentrate each of the extracts to 1.0 mL using a K-D apparatus.

SAMPLE COLLECTION, PRESERVATION & HANDLING Grab samples must be collected in glass containers. All samples must be refrigerated at 4°C from the time of collection until extraction. If residual chlorine is present, add 80 mg of sodium thiosulfate/L of sample and mix well. All samples must be extracted within 7 days of collection and completely analyzed within 40 days of extraction.

QUALITY CONTROL Make an initial, one-time, demonstration of the ability to generate acceptable accuracy and precision with this method. Before processing any samples, the analyst must analyze a reagent water blank to demonstrate that interferences from the analytical system and glassware are under control. Each time a set of samples is extracted or reagents are changed, a reagent water blank must be processed. Spike and analyze a minimum of 5% of all samples to monitor and evaluate lab data quality. A QC check sample concentrate that contains each parameter of interest at a concentration of 100 µg/mL in acetone is required. PCBs and multicomponent pesticides may be omitted from this test.

After analysis of five spiked wastewater samples, calculate the average percent recovery and the standard deviation of the percent recovery. Spike all samples with the surrogate standard spiking solution and calculate the percent recovery of each surrogate compound.

REFERENCE *Federal Register*, Vol. 49, No. 209. Friday, Oct. 26, 1984.

4-Chloro-3-methylphenol **EPA Method 8270**
CAS #59-50-7

TITLE Semivolatile Organic Compounds by GC/MS

MATRIX This method is used to determine the concentration of semivolatile organic compounds in extracts prepared from all types of solid waste matrices, soils, and groundwater. Although surface waters are not specifically mentioned, this method should be applicable to water samples from rivers, lakes, etc.

METHOD SUMMARY This method covers 259 semivolatile organic compounds. In very limited applications direct injection of the sample into the GC/MS system may be appropriate, but this results in very high detection limits (approximately 10,000 µg/L). Typically, a 1-L liquid sample, containing surrogate, and matrix spiking standards, is extracted in a continuous extractor first under acid conditions and then under basic conditions. Typically 30 g of a solid sample, containing surrogate, and matrix spiking standards, is extracted ultrasonically. After concentrating the extract to 1 mL it is spiked with 10 µL of an internal standard solution just prior to analysis by GC/MS. The volume injected should contain about 100 ng of base/neutral and 200 ng of acid surrogates (for a 1-µL injection). Analysis is performed by GC/MS using a capillary GC column.

INTERFERENCES Raw GC/MS data from all blanks, samples, and spikes must be evaluated for interferences. Contamination by carryover can occur whenever high-concentration and low-concentration samples are sequentially analyzed. To reduce carryover, the sample syringe must be rinsed out between samples with solvent. Whenever an unusually concentrated sample is encountered, it should be followed by the analysis of blank solvent to check for cross-contamination.

INSTRUMENTATION A GC/MS and a data system are required. The GC column used is a 30 m × 0.25 mm I.D. (or 0.32 mm I.D.) 1um film thickness silicone-coated fused silica capillary column. A continuous liquid-liquid extractor equipped with Teflon® or glass connection joints and stopcocks requiring no lubrication, a K-D concentrating apparatus, water bath, and an ultrasonic disrupter with a minimum power of 300 W and with pulsing capability are also required.

PRECISION & ACCURACY The estimated quantitation limit (EQL) of Method 8270B for determining an individual compound is approximately 1 mg/kg (wet weight) for soil or sediment samples, 1–200 mg/kg for wastes (dependent on matrix and method of preparation), and 10 µg/L for groundwater samples. EQLs will be proportionately higher for sample extracts that require dilution to avoid saturation of the detector.

The EQL(b) for groundwater in µg/L is 20.
The EQL (a, b) for low concentrations in soil and sediment in µg/kg is 1300.
Accuracy as µg/L is $0.84C + 0.35$.
Overall precision in µg/L is $0.29X + 1.31$.

(a) *EQLs listed for soil/sediment are based on wet weight. Normally data is reported in a dry-weight basis; therefore, EQLs will be higher based on the % dry weight of each sample. This calculation is based on a 30 g sample and gel permeation chromatography cleanup.*
(b) *Sample EQLs are highly matrix-dependent. The EQLs are provided for guidance and may not always be achievable.*
C = *True value for concentration, in µg/L.*
X = *Average recovery found for measurements of samples containing a concentration of C, in µg/L.*

ESTIMATED QUANTITATION LIMIT
FOR OTHER MATRICES

Other Matrices	Factor (a)
High-concentration soil and sludges by sonicator	7.5
Non-water miscible waste	75

(a) *EQL for other matrices = [EQL for low soil/sediment] × [Factor]. This estimated EQL is similar to an EPA "Practical Quantitation Limit."*

SAMPLING METHOD
Liquid samples — Use a 1 or 2½ gallon amber glass bottle with a screw-top Teflon®-lined cover that has been prewashed with detergent and rinsed with distilled water and methanol (or isopropanol).

Soils, sediments, or sludges — Use an 8-oz. widemouth glass with a screw-top Teflon®-lined cover that has been prewashed with detergent and rinsed with distilled water and methanol (or isopropanol).

SAMPLE PRESERVATION
Liquid samples — If residual chlorine is present, add 3 mL of 10% sodium thiosulfate per gallon, cool to 4°C and store in a solvent-free refrigerator until analysis; if chlorine is not present, then eliminate the sodium thiosulfate addition.

Soils, sediments, or sludges — Cool samples to 4°C and store in a solvent-free refrigerator.

MHT Liquid samples must be extracted within 7 days and the extracts analyzed within 40 days. Soils, sediments, or sludges may

be stored for a maximum of 14 days and the extracts analyzed within 40 days.

SAMPLE PREPARATION

Liquid samples — Transfer 1 L quantitatively to a continuous extractor. If high concentrations are anticipated, a smaller volume may be used and then diluted with organic-free reagent water to 1 L. Adjust pH, if necessary, to pH <2 using 1:1 (V/V) sulfuric acid. Pipette 1.0 mL of a surrogate standard spiking solution into each sample. For the sample in each analytical batch selected for spiking, add 1.0 mL of a matrix spiking standard. For base/neutral acid analysis, the amount of the surrogates and matrix spiking compounds added to the sample should result in a final concentration of 100 ng/µL of each analyte in the extract to be analyzed (assuming a 1-µL injection). Extract with methylene chloride for 18–24 h. Next, adjust the pH of the aqueous phase to pH >11 using 10 N sodium hydroxide and extract it with methylene chloride again for 18–24 h. Dry the extract through a column containing anhydrous sodium sulfate and concentrate it to 1 mL using a K-D concentrator.

Soils, sediments, or sludges — Use 30 g of sample. Nonporous or wet samples (gummy or clay type) that do not have a free-flowing sandy texture must be mixed with anhydrous sodium sulfate until the sample is free flowing. Add 1 mL of surrogate standards to all samples, spikes, standards, and blanks. For the sample in each analytical batch selected for spiking, add 1.0 mL of a matrix spiking standard. For base/neutral acid analysis, the amount added of the surrogates and matrix spiking compounds should result in a final concentration of 100 ng/µL of each base/neutral analyte and 200 ng/µL of each acid analyte in the extract to be analyzed (assuming a 1-µL injection). Immediately add a 100-mL mixture of 1:1 methylene chloride:acetone and extract the sample ultrasonically for 3 min and then decant or filter the extracts. Repeat the extraction two or more times. Dry the extract using a column with anhydrous sodium sulfate and concentrate it to 1 mL in a K-D concentrator.

Note: This compound may be exhibit erratic chromatographic behavior, especially if the GC system is contaminated with high boiling material.

QUALITY CONTROL

A methylene chloride solution containing 50 ng/µL of decafluorotriphenylphosphine (DFTPP) is used for tuning the GC/MS system each 12-h shift. A system performance check also must be made during every 12-h shift. A standard containing 50 ng/µL each of 4,4'-DDT, pentachlorophenol, and benzidine is required to verify injection port inertness and GC column performance. A calibration standard at mid-concentration, containing each compound of interest, including all required surrogates, must be performed every 12 h during analysis. After the system performance check is met, calibration check compounds (CCCs) are used to check the validity of the initial calibration.

The internal standard responses and retention times in the calibration check standard must be evaluated immediately after or during data acquisition. If the retention time for any internal standard changes by more than 30 seconds from the last check calibration (12 h), the chromatographic system must be inspected for malfunctions and corrections must be made, as required. If the electron ionization current plot (EICP) area for any of the internal standards changes by a factor of two from the last daily calibration standard check, the mass spectrometer must be inspected for malfunctions and corrections must be made, as appropriate.

Demonstrate, through the analysis of a reagent water blank, that interferences from the analytical system, glassware, and reagents are under control. The blank samples should be carried through all stages of the sample preparation and measurement steps. For each analytical batch (up to 20 samples), a reagent blank, matrix spike, and matrix spike duplicate/duplicate must be analyzed (the frequency of the spikes may be different for different monitoring programs). The blank and spiked samples must be carried through all stages of the sample preparation and measurement steps. A QC reference sample concentrate containing each analyte at a concentration of 100 mg/L in methanol is required.

REFERENCE Test Methods for Evaluating Solid Waste (SW-846). U.S. EPA 1983, Method 8270B, Rev. 2, Nov. 1990. Office of Solid Waste, Washington, DC.

4-Chloro-3-methylphenol EPA Method 8040
CAS #59-50-7

TITLE Phenols

MATRIX Groundwater, soils, sludges, water miscible liquid wastes, and non-water miscible wastes.

APPLICATION This method is used for the analysis of 17 phenols. Samples are extracted, concentrated, and analyzed using direct injection of both neat and diluted organic liquids. Pentafluorobenzylbromide (PFB) derivatives also may be made to increase sensitivity of the method.

INTERFERENCES There can be carryover contamination with high- and low-level samples. Solvents, reagents, and glassware may introduce artifacts. Other interferences may come from coextracted compounds from samples.

INSTRUMENTATION GC capable of on-column injections and a flame with detector (FID) or electron capture detector (ECD). Column for underivatized phenol: 1.8 m by 2.0 mm with 1% SP-1240DA on Supelcoport. Column for derivatized phenols: 1.8 m by 2.0 mm with 5% OV-17 on Chromosorb W-AW-DMCS.

RANGE 12 to 450 µg/L

MDL 0.36 µg/L (FID) and 1.8 µg/L (ECD)

PQL FACTORS FOR MULTIPLYING × FID MDL VALUE

Matrix	Multiplication Factor
Groundwater	10
Low-level soil by sonication with GPC cleanup	670
High-level soil and sludge by sonication	10,000
Non-water miscible waste	100,000

PRECISION $0.16X + 1.41$ µg/L (overall precision using FID)

ACCURACY 0.87C–1.97 µg/L (as recovery using FID)

SAMPLING METHOD Use 8-oz. widemouth glass bottles with Teflon®-lined caps for concentrated waste samples, soils, sediments, and sludges. Use 1 or 2½ gallon amber glass bottles with Teflon®-lined caps for liquid (water) samples.

STABILITY Cool soil, sediment, sludge, and liquid samples to 4°C. If residual chlorine is present in liquid samples add 3 mL of 10% sodium thiosulfate per gallon of sample and cool to 4°C.

MHT 14 days for concentrated waste, soil, sediment, or sludge; 7 days for liquid samples; all extracts must be analyzed within 40 days.

QUALITY CONTROL A quality control check sample concentrate containing each analyte of interest is required. The QC check sample concentrate may be prepared from pure standard materials or purchased as certified solutions Use appropriate trip, matrix, control site, method, reagent, and solvent blanks. Internal, surrogate, and five concentration level calibration standards are used. The QC check sample concentrate should contain this compound at 100 µg/mL in 2-propanol.

REFERENCE Test Methods for Evaluating Solid Waste (SW-846), U.S. EPA Office of Solid Waste, Washington, DC, Method 8040A, Rev. 1, Nov. 1990.

5-Chloro-o-toluidine   **EPA Method 1625**
CAS #95-79-4

TITLE Semivolatile Organic Compounds by Isotope Dilution GC/MS

MATRIX The compounds may be determined in waters, soils, and municipal sludges by this method.

METHOD SUMMARY This method is used to determine 176 semivolatile toxic organic pollutants associated with the CWA (as amended 1987); the RCRA (as amended 1986); the CERCLA (as amended 1986); and other compounds amenable to extraction and analysis by capillary column gas chromatography-mass spectrometry (GC/MS).

Stable isotopically-labeled analogs of the compounds of interest are added to the sample. If the solids content is less than 1%, a 1-L sample is extracted at pH 12–13, then at pH <2 with methylene chloride using continuous extraction techniques.

If the solids content is 30% or less, the sample is diluted to 1% solids with reagent water, homogenized ultrasonically, and extracted at pH 12–13, then at pH <2 with methylene chloride using continuous extraction techniques. If the solids content is greater than 30%, the sample is extracted using ultrasonic techniques.

Each extract is dried over sodium sulfate, concentrated to a volume of 5 mL, cleaned up using GPC, if necessary, and concentrated. Extracts are concentrated to 1 mL if GPC is not performed, and to 0.5 mL if GPC is performed.

An internal standard is added to the extract, and a 1-mL aliquot of the extract is injected into the GC. The compounds are separated by GC and detected by a MS. The labeled compounds serve to correct the variability of the analytical technique.

INTERFERENCES Solvents, reagents, glassware, and other sample processing hardware may yield artifacts and/or elevated baselines causing misinterpretation of chromatograms and spectra. Materials used in the analysis must be demonstrated to be free from interferences under the conditions of analysis by running method blanks initially and with each sample lot (sample started through the extraction process on a given 8-h shift, to a maximum of 20). Specific selection of reagents and purification of solvents by distillation in all glass systems may be required. Glassware and, where possible, reagents are cleaned by solvent rinse and baking at 450°C for 1-h minimum. Interferences coextracted from samples will vary considerably from source to source, depending on the diversity of the site being sampled.

INSTRUMENTATION Major instrumentation includes a GC with a splitless or on-column injection port for capillary column, a MS with 70 eV electron impact ionization, and a data system to collect and record MS data, and process it. A K-D apparatus is used to concentrate extracts.

GC Column: 30 m × 0.25 mm I.D. 5% phenyl, 94% methyl, 1% vinyl silicone bonded phased fused silica capillary column.

PRECISION & ACCURACY The detection limits of the method are usually dependent on the level of interferences rather than instrumental limitations. The limits typify the minimum quantities that can be detected with no interferences present.

The minimum level (in µg/mL) was not listed. This is defined as a minimum level at which the analytical system shall give recognizable mass spectra (background corrected) and acceptable calibration points.
The MDL (in µg/kg) in low solids was not listed and in high solids was not listed; these were determined in digested sludge (low solids) and in filter cake or compost (high solids).
The labeled and native compound initial precision as standard deviation (in µg/L) was not listed.
The labeled and native compound initial accuracy as average recovery (in µg/L) was not listed.

SAMPLE COLLECTION, PRESERVATION & HANDLING Collect samples in glass containers. Aqueous samples which flow freely are collected in refrigerated bottles using automatic sampling equipment. Solid samples are collected as grab samples using widemouth jars. Maintain samples at 0 to 4°C from the time of collection until extraction. If residual chlorine is present in aqueous samples, add 80 mg sodium thiosulfate/L of water. Begin sample extraction within 7 days of collection, and analyze all extracts within 40 days of extraction.

SAMPLE PREPARATION Samples containing 1% solids or less are extracted directly using continuous liquid-liquid extraction techniques. Samples containing 1 to 30% solids are diluted to the 1% level with reagent water and extracted using continuous liquid-liquid extraction techniques. Samples containing greater than 30% solids are extracted using ultrasonic techniques.

Base/neutral extraction — Adjust the pH of the waters in the extractors to 12–13 with 6 N NaOH. Extract with methylene chloride for 24–48 h.

Acid extraction — Adjust the pH of the waters in the extractors to 2 or less using 6 N sulfuric acid. Extract with methylene chloride for 24–48 h.

Ultrasonic extraction of high solids samples — Add anhydrous sodium sulfate to the sample and QC aliquot(s). Add acetone:methylene chloride (1:1) to the sample and mix thoroughly

Concentrate extracts using a K-D apparatus.

QUALITY CONTROL The analyst is permitted to modify this method to improve separations or lower the costs of measurements, provided all performance specifications are met. Analyses of blanks are required to demonstrate freedom from contamination. When results of spikes indicate atypical method performance for samples, the samples are diluted to bring method performance within acceptable limits.

For low solids (aqueous samples), extract, concentrate, and analyze two sets of four 1-L aliquots (8 aliquots total) of the precision and recovery standard. For high solids samples, two sets of four 30-g aliquots of the high solids reference matrix are used.

Spike all samples with labeled compounds to assess method performance. Compute percent recovery of the labeled compounds using the internal standard method. Compare the labeled compound recovery for each compound with the corresponding labeled compound recovery.

Reagent water and high solids reference matrix blanks are analyzed to demonstrate freedom from contamination. Extract and concentrate a 1-L reagent water blank or a high solids reference matrix blank with each sample's lot (samples started through the extraction process on the same 8-h shift, to a maximum of 20 samples).

Field replicates may be collected to determine the precision of the sampling technique, and spiked samples may be required to determine the accuracy of the analysis when the internal standard method is used.

REFERENCE Semivolatile Organic Compounds by Isotope Dilution GC/MS. Office of Water Regulation and Standards, U.S. EPA Industrial Technology Division, Washington, DC, EPA Method 1625, Rev. C, June 1989 (contact W.A. Telliard, U.S. EPA, Office of Water Regulations and Standards, 401 M St., SW, Washington, DC, 20460. Phone: 202-382-7131).

Chloroacetonitrile EPA Method 1624
CAS #107-14-2

TITLE Volatile Organic Compounds by Isotope Dilution GC/MS

MATRIX Compounds may be determined in waters, soils, and municipal sludges by this method.

METHOD SUMMARY This method is used to determine 58 volatile toxic organic pollutants associated with the CWA (as amended 1987); the RCRA (as amended 1986); the CERCLA (as amended 1986); and other compounds amenable to purge-and-trap gas chromatography-mass spectrometry (GC/MS).

If the solids content is less than 1%, stable isotopically-labeled analogs of the compounds of interest are added to a 5-mL sample and the sample is purged with an inert gas at 20–25°C in a chamber designed for soil or water samples. If the solids content is greater than 1%, 5 mL of reagent water and the labeled compounds are added to a 5-g aliquot of sample and the mixture is purged at 40°C. Compounds that will not purge at 20–25°C or at 40°C are purged at 78–85°C. In the purging process, the volatile compounds are transferred from the aqueous phase into the gaseous phase where they are passed into a sorbent column, and trapped. After purging is completed, the trap is backflushed and heated rapidly to desorb the compounds into a GC. The compounds are separated by the GC and detected by a MS. The labeled compounds serve to correct the variability of the analytical technique.

INTERFERENCES Impurities in the purge gas, organic compounds outgassing from the plumbing upstream of the trap, and solvent vapors in the lab account for most problems. Samples can be contaminated by diffusion of volatile organic compounds (particularly methylene chloride) through the bottle seal during shipment and storage. Contamination by carryover can occur when high-level and low-level samples are analyzed sequentially. When an unusually concentrated sample is encountered, follow it by analysis of a reagent water blank to check for carryover.

INSTRUMENTATION Major equipment includes a GC with linear temperature programming and a glass jet separator as the MS interface, a MS with 70 eV electron impact ionization, and a data system to collect and record response factors.

Column: 2.8 m × 2 mm I.D. glass, packed with 1% SP-1000 on Carbopak B, 60/80 mesh, or equivalent.

PRECISION & ACCURACY The detection limits of the method are usually dependent on the level of interferences rather than instrumental limitations. The method detection limits were determined in digested sludge (low solids) and in filter cake or compost (high solids).

The MDL (in µg/kg) for low solids is not listed and for high solids is not listed.

Labeled and native compound precision (in µg/L) as standard deviation was not listed.

Labeled and native compound accuracy (in µg/L) as average recovery was not listed.

Acceptance criteria are at 20 µg/L for this compound.

SAMPLE COLLECTION, PRESERVATION & HANDLING Grab samples are collected in glass containers having a total volume greater than 20 mL. Fill and seal each bottle so that no air bubbles are entrapped. Samples are maintained at 0 to 4°C from the time of collection until analysis. If an aqueous sample contains residual chlorine, add sodium thiosulfate preservative (10 mg/40 mL) to the empty sample bottles just prior to shipment

to the sample site. All samples must be analyzed within 14 days of collection.

SAMPLE PREPARATION Samples containing less than 1% solids are analyzed directly as aqueous samples. Samples containing 1% solids or greater are analyzed as solid samples utilizing one of two methods, depending on the levels of pollutants, in the sample. Samples containing 1% solids or greater, and low to moderate levels of pollutants are analyzed by purging a known weight of sample added to 5 mL of reagent water. Samples containing 1% solids or greater, and high levels of pollutants, are extracted with methanol, and an aliquot of the methanol extract is added to reagent water and purged.

QUALITY CONTROL A field blank prepared from reagent water and carried through the sampling and handling protocol may serve as a check on contamination from shipment and storage.

The analyst is permitted to modify this method to improve separations or lower the costs of measurements, provided all performance specifications are met. Analyses of blanks are required. When results of spikes indicate atypical method performance for samples, the samples are diluted to bring method performance within acceptable limits. Analyze two sets of four 5-mL aliquots (8 aliquots total) of the aqueous performance standard. Spike all samples with labeled compounds to assess method performance on the sample matrix. Compute the percent recovery of the labeled compounds using the internal standard method. Compare the percent recovery for each compound with the corresponding labeled compound recovery. Reagent water blanks are analyzed to demonstrate freedom from carryover contamination. Field replicates may be collected to determine the precision of the sampling technique, and spiked samples may be required to determine the accuracy of the analysis when the internal method is used.

REFERENCE Volatile Organic Compounds by Isotope Dilution GC/MS. Office of Water Regulation and Standards, U.S. EPA Industrial Technology Division, Washington, DC, EPA Method 1624, Rev. C, June 1989 (contact W.A. Telliard, U.S. EPA, Office of Water Regulations and Standards, 401 M St., SW, Washington, DC, 20460. Phone: 202-382-7131).

4-Chloroaniline   EPA Method 1625
CAS #106-47-8

TITLE Semivolatile Organic Compounds by Isotope Dilution GC/MS

MATRIX The compounds may be determined in waters, soils, and municipal sludges by this method.

METHOD SUMMARY This method is used to determine 176 semivolatile toxic organic pollutants associated with the CWA (as amended 1987); the RCRA (as amended 1986); the CERCLA (as amended 1986); and other compounds amenable to extraction and analysis by capillary column gas chromatography-mass spectrometry (GC/MS).

Stable isotopically-labeled analogs of the compounds of interest are added to the sample. If the solids content is less than 1%, a 1-L sample is extracted at pH 12–13, then at pH <2 with methylene chloride using continuous extraction techniques.

If the solids content is 30% or less, the sample is diluted to 1% solids with reagent water, homogenized ultrasonically, and extracted at pH 12–13, then at pH <2 with methylene chloride using continuous extraction techniques. If the solids content is greater than 30%, the sample is extracted using ultrasonic techniques.

Each extract is dried over sodium sulfate, concentrated to a volume of 5 mL, cleaned up using GPC, if necessary, and concentrated. Extracts are concentrated to 1 mL if GPC is not performed, and to 0.5 mL if GPC is performed.

An internal standard is added to the extract, and a 1-mL aliquot of the extract is injected into the GC. The compounds are separated by GC and detected by a MS. The labeled compounds serve to correct the variability of the analytical technique.

INTERFERENCES Solvents, reagents, glassware, and other sample processing hardware may yield artifacts and/or elevated baselines causing misinterpretation of chromatograms and spectra. Materials used in the analysis must be demonstrated to be free from interferences under the conditions of analysis by running method blanks initially and with each sample lot (sample started through the extraction process on a given 8-h shift, to a maximum of 20). Specific selection of reagents and purification of solvents by distillation in all glass systems may be required. Glassware and, where possible, reagents are cleaned by solvent rinse and baking at 450°C for 1-h minimum. Interferences coextracted from samples will vary considerably from source to source, depending on the diversity of the site being sampled.

INSTRUMENTATION Major instrumentation includes a GC with a splitless or on-column injection port for capillary column, a MS with 70 eV electron impact ionization, and a data system to collect and record MS data, and process it. A K-D apparatus is used to concentrate extracts.

GC Column: 30 m × 0.25 mm I.D. 5% phenyl, 94% methyl, 1% vinyl silicone bonded phased fused silica capillary column.

PRECISION & ACCURACY The detection limits of the method are usually dependent on the level of interferences rather than instrumental limitations. The limits typify the minimum quantities that can be detected with no interferences present.

The minimum level (in µg/mL) was not listed. This is defined as a minimum level at which the analytical system shall give recognizable mass spectra (background corrected) and acceptable calibration points.
The MDL (in µg/kg) in low solids was not listed and in high solids was not listed; these were determined in digested sludge (low solids) and in filter cake or compost (high solids).
The labeled and native compound initial precision as standard deviation (in µg/L) was not listed.
The labeled and native compound initial accuracy as average recovery (in µg/L) was not listed.

SAMPLE COLLECTION, PRESERVATION & HANDLING

Collect samples in glass containers. Aqueous samples which flow freely are collected in refrigerated bottles using automatic sampling equipment. Solid samples are collected as grab samples using widemouth jars. Maintain samples at 0 to 4°C from the time of collection until extraction. If residual chlorine is present in aqueous samples, add 80 mg sodium thiosulfate/L of water. Begin sample extraction within 7 days of collection, and analyze all extracts within 40 days of extraction.

SAMPLE PREPARATION

Samples containing 1% solids or less are extracted directly using continuous liquid-liquid extraction techniques. Samples containing 1 to 30% solids are diluted to the 1% level with reagent water and extracted using continuous liquid-liquid extraction techniques. Samples containing greater than 30% solids are extracted using ultrasonic techniques.

Base/neutral extraction — Adjust the pH of the waters in the extractors to 12–13 with 6 N NaOH. Extract with methylene chloride for 24–48 h.

Acid extraction — Adjust the pH of the waters in the extractors to 2 or less using 6 N sulfuric acid. Extract with methylene chloride for 24–48 h.

Ultrasonic extraction of high solids samples — Add anhydrous sodium sulfate to the sample and QC aliquot(s). Add acetone:methylene chloride (1:1) to the sample and mix thoroughly

Concentrate extracts using a K-D apparatus.

QUALITY CONTROL

The analyst is permitted to modify this method to improve separations or lower the costs of measurements, provided all performance specifications are met. Analyses of blanks are required to demonstrate freedom from contamination. When results of spikes indicate atypical method performance for samples, the samples are diluted to bring method performance within acceptable limits.

For low solids (aqueous samples), extract, concentrate, and analyze two sets of four 1-L aliquots (8 aliquots total) of the precision and recovery standard. For high solids samples, two sets of four 30-g aliquots of the high solids reference matrix are used.

Spike all samples with labeled compounds to assess method performance. Compute percent recovery of the labeled compounds using the internal standard method. Compare the labeled compound recovery for each compound with the corresponding labeled compound recovery.

Reagent water and high solids reference matrix blanks are analyzed to demonstrate freedom from contamination. Extract and concentrate a 1-L reagent water blank or a high solids reference matrix blank with each sample's lot (samples started through the extraction process on the same 8-h shift, to a maximum of 20 samples).

Field replicates may be collected to determine the precision of the sampling technique, and spiked samples may be required to determine the accuracy of the analysis when the internal standard method is used.

REFERENCE

Semivolatile Organic Compounds by Isotope Dilution GC/MS. Office of Water Regulation and Standards, U.S. EPA Industrial Technology Division, Washington, DC, EPA Method 1625, Rev. C, June 1989 (contact W.A. Telliard, U.S. EPA, Office of Water Regulations and Standards, 401 M St., SW, Washington, DC, 20460. Phone: 202-382-7131).

4-Chloroaniline   **EPA Method 8270**
CAS #106-47-8

TITLE
Semivolatile Organic Compounds by GC/MS

MATRIX
This method is used to determine the concentration of semivolatile organic compounds in extracts prepared from all types of solid waste matrices, soils, and groundwater. Although surface waters are not specifically mentioned, this method should be applicable to water samples from rivers, lakes, etc.

METHOD SUMMARY
This method covers 259 semivolatile organic compounds. In very limited applications direct injection of the sample into the GC/MS system may be appropriate, but this results in very high detection limits (approximately 10,000 µg/L). Typically, a 1-L liquid sample, containing surrogate, and matrix spiking standards, is extracted in a continuous extractor first under acid conditions and then under basic conditions. Typically 30 g of a solid sample, containing surrogate, and matrix spiking standards, is extracted ultrasonically. After concentrating the extract to 1 mL it is spiked with 10 µL of an internal standard solution just prior to analysis by GC/MS. The volume injected should contain about 100 ng of base/neutral and 200 ng of acid surrogates (for a 1-µL injection). Analysis is performed by GC/MS using a capillary GC column.

INTERFERENCES
Raw GC/MS data from all blanks, samples, and spikes must be evaluated for interferences. Contamination by carryover can occur whenever high-concentration and low-concentration samples are sequentially analyzed. To reduce carryover, the sample syringe must be rinsed out between samples with solvent. Whenever an unusually concentrated sample is encountered, it should be followed by the analysis of blank solvent to check for cross-contamination.

INSTRUMENTATION
A GC/MS and a data system are required. The GC column used is a 30 m × 0.25 mm I.D. (or 0.32 mm I.D.) 1um film thickness silicone-coated fused silica capillary column. A continuous liquid-liquid extractor equipped with Teflon® or glass connection joints and stopcocks requiring no lubrication, a K-D concentrating apparatus, water bath, and an ultrasonic disrupter with a minimum power of 300 W and with pulsing capability are also required.

PRECISION & ACCURACY
The estimated quantitation limit (EQL) of Method 8270B for determining an individual compound is approximately 1 mg/kg (wet weight) for soil or sediment samples, 1–200 mg/kg for wastes (dependent on matrix and method of preparation), and 10 µg/L for groundwater samples. EQLs will be proportionately higher for sample extracts that require dilution to avoid saturation of the detector.

The EQL(b) for groundwater in µg/L is 20.

The EQL (a, b) for low concentrations in soil and sediment in μg/kg is 1300.
Accuracy as μg/L is not listed.
Overall precision in μg/L is not listed.

(a) *EQLs listed for soil/sediment are based on wet weight. Normally data is reported in a dry-weight basis; therefore, EQLs will be higher based on the % dry weight of each sample. This calculation is based on a 30 g sample and gel permeation chromatography cleanup.*
(b) *Sample EQLs are highly matrix-dependent. The EQLs are provided for guidance and may not always be achievable.*
$C =$ *True value for concentration, in μg/L.*
$X =$ *Average recovery found for measurements of samples containing a concentration of C, in μg/L.*

ESTIMATED QUANTITATION LIMIT FOR OTHER MATRICES

Other Matrices	Factor (a)
High-concentration soil and sludges by sonicator	7.5
Non-water miscible waste	75

(a) *EQL for other matrices = [EQL for low soil/sediment] × [Factor]. This estimated EQL is similar to an EPA "Practical Quantitation Limit."*

SAMPLING METHOD

Liquid samples — Use a 1 or 2½ gallon amber glass bottle with a screw-top Teflon®-lined cover that has been prewashed with detergent and rinsed with distilled water and methanol (or isopropanol).

Soils, sediments, or sludges — Use an 8-oz. widemouth glass with a screw-top Teflon®-lined cover that has been prewashed with detergent and rinsed with distilled water and methanol (or isopropanol).

SAMPLE PRESERVATION

Liquid samples — If residual chlorine is present, add 3 mL of 10% sodium thiosulfate per gallon, cool to 4°C and store in a solvent-free refrigerator until analysis; if chlorine is not present, then eliminate the sodium thiosulfate addition.

Soils, sediments, or sludges — Cool samples to 4°C and store in a solvent-free refrigerator.

MHT Liquid samples must be extracted within 7 days and the extracts analyzed within 40 days. Soils, sediments, or sludges may be stored for a maximum of 14 days and the extracts analyzed within 40 days.

SAMPLE PREPARATION

Liquid samples — Transfer 1 L quantitatively to a continuous extractor. If high concentrations are anticipated, a smaller volume may be used and then diluted with organic-free reagent water to 1 L. Adjust pH, if necessary, to pH <2 using 1:1 (V/V) sulfuric acid. Pipette 1.0 mL of a surrogate standard spiking solution into each sample. For the sample in each analytical batch selected for spiking, add 1.0 mL of a matrix spiking standard. For base/neutral acid analysis, the amount of the surrogates and matrix spiking compounds added to the sample should result in a final concentration of 100 ng/μL of each analyte in the extract to be analyzed (assuming a 1-μL injection). Extract with methylene chloride for 18–24 h. Next, adjust the pH of the aqueous phase to pH >11 using 10 N sodium hydroxide and extract it with methylene chloride again for 18–24 h. Dry the extract through a column containing anhydrous sodium sulfate and concentrate it to 1 mL using a K-D concentrator.

Soils, sediments, or sludges — Use 30 g of sample. Nonporous or wet samples (gummy or clay type) that do not have a free-flowing sandy texture must be mixed with anhydrous sodium sulfate until the sample is free flowing. Add 1 mL of surrogate standards to all samples, spikes, standards, and blanks. For the sample in each analytical batch selected for spiking, add 1.0 mL of a matrix spiking standard. For base/neutral acid analysis, the amount added of the surrogates and matrix spiking compounds should result in a final concentration of 100 ng/μL of each base/neutral analyte and 200 ng/μL of each acid analyte in the extract to be analyzed (assuming a 1-μL injection). Immediately add a 100-mL mixture of 1:1 methylene chloride:acetone and extract the sample ultrasonically for 3 min and then decant or filter the extracts. Repeat the extraction two or more times. Dry the extract using a column with anhydrous sodium sulfate and concentrate it to 1 mL in a K-D concentrator.

Note: This compound may be exhibit erratic chromatographic behavior, especially if the GC system is contaminated with high boiling material.

QUALITY CONTROL A methylene chloride solution containing 50 ng/μL of decafluorotriphenylphosphine (DFTPP) is used for tuning the GC/MS system each 12-h shift. A system performance check also must be made during every 12-h shift. A standard containing 50 ng/μL each of 4,4'-DDT, pentachlorophenol, and benzidine is required to verify injection port inertness and GC column performance. A calibration standard at mid-concentration, containing each compound of interest, including all required surrogates, must be performed every 12 h during analysis. After the system performance check is met, calibration check compounds (CCCs) are used to check the validity of the initial calibration.

The internal standard responses and retention times in the calibration check standard must be evaluated immediately after or during data acquisition. If the retention time for any internal standard changes by more than 30 seconds from the last check calibration (12 h), the chromatographic system must be inspected for malfunctions and corrections must be made, as required. If the electron ionization current plot (EICP) area for any of the internal standards changes by a factor of two from the last daily calibration standard check, the mass spectrometer must be inspected for malfunctions and corrections must be made, as appropriate.

Demonstrate, through the analysis of a reagent water blank, that interferences from the analytical system, glassware, and reagents are under control. The blank samples should be carried through all stages of the sample preparation and measurement steps. For each analytical batch (up to 20 samples), a reagent blank, matrix spike, and matrix spike duplicate/duplicate must be analyzed (the frequency of the spikes may be different for different monitoring programs). The blank and spiked samples must be carried through all stages of the sample

preparation and measurement steps. A QC reference sample concentrate containing each analyte at a concentration of 100 mg/L in methanol is required.

REFERENCE Test Methods for Evaluating Solid Waste (SW-846). U.S. EPA 1983, Method 8270B, Rev. 2, Nov. 1990. Office of Solid Waste, Washington, DC.

Chlorobenzene EPA Method 1624
CAS #108-90-7

TITLE Volatile Organic Compounds by Isotope Dilution GC/MS

MATRIX Compounds may be determined in waters, soils, and municipal sludges by this method.

METHOD SUMMARY This method is used to determine 58 volatile toxic organic pollutants associated with the CWA (as amended 1987); the RCRA (as amended 1986); the CERCLA (as amended 1986); and other compounds amenable to purge-and-trap gas chromatography-mass spectrometry (GC/MS).

If the solids content is less than 1%, stable isotopically-labeled analogs of the compounds of interest are added to a 5-mL sample and the sample is purged with an inert gas at 20–25°C in a chamber designed for soil or water samples. If the solids content is greater than 1%, 5 mL of reagent water and the labeled compounds are added to a 5-g aliquot of sample and the mixture is purged at 40°C. Compounds that will not purge at 20–25°C or at 40°C are purged at 78–85°C. In the purging process, the volatile compounds are transferred from the aqueous phase into the gaseous phase where they are passed into a sorbent column, and trapped. After purging is completed, the trap is backflushed and heated rapidly to desorb the compounds into a GC. The compounds are separated by the GC and detected by a MS. The labeled compounds serve to correct the variability of the analytical technique.

INTERFERENCES Impurities in the purge gas, organic compounds outgassing from the plumbing upstream of the trap, and solvent vapors in the lab account for most problems. Samples can be contaminated by diffusion of volatile organic compounds (particularly methylene chloride) through the bottle seal during shipment and storage. Contamination by carryover can occur when high-level and low-level samples are analyzed sequentially. When an unusually concentrated sample is encountered, follow it by analysis of a reagent water blank to check for carryover.

INSTRUMENTATION Major equipment includes a GC with linear temperature programming and a glass jet separator as the MS interface, a MS with 70 eV electron impact ionization, and a data system to collect and record response factors.

Column: 2.8 m × 2 mm I.D. glass, packed with 1% SP-1000 on Carbopak B, 60/80 mesh, or equivalent.

PRECISION & ACCURACY The detection limits of the method are usually dependent on the level of interferences rather than instrumental limitations. The method detection limits were determined in digested sludge (low solids) and in filter cake or compost (high solids).

The MDL (in µg/kg) for low solids is 21 and for high solids is 58.
Background levels of this compound were present in the sludge with high solids, resulting in a higher than expected MDL.
Labeled and native compound precision (in µg/L) as standard deviation was 8.2.
Labeled and native compound accuracy (in µg/L) as average recovery was 14–30.

Acceptance criteria are at 20 µg/L for this compound.

SAMPLE COLLECTION, PRESERVATION & HANDLING Grab samples are collected in glass containers having a total volume greater than 20 mL. Fill and seal each bottle so that no air bubbles are entrapped. Samples are maintained at 0 to 4°C from the time of collection until analysis. If an aqueous sample contains residual chlorine, add sodium thiosulfate preservative (10 mg/40 mL) to the empty sample bottles just prior to shipment to the sample site. All samples must be analyzed within 14 days of collection.

SAMPLE PREPARATION Samples containing less than 1% solids are analyzed directly as aqueous samples. Samples containing 1% solids or greater are analyzed as solid samples utilizing one of two methods, depending on the levels of pollutants, in the sample. Samples containing 1% solids or greater, and low to moderate levels of pollutants are analyzed by purging a known weight of sample added to 5 mL of reagent water. Samples containing 1% solids or greater, and high levels of pollutants, are extracted with methanol, and an aliquot of the methanol extract is added to reagent water and purged.

QUALITY CONTROL A field blank prepared from reagent water and carried through the sampling and handling protocol may serve as a check on contamination from shipment and storage.

The analyst is permitted to modify this method to improve separations or lower the costs of measurements, provided all performance specifications are met. Analyses of blanks are required. When results of spikes indicate atypical method performance for samples, the samples are diluted to bring method performance within acceptable limits. Analyze two sets of four 5-mL aliquots (8 aliquots total) of the aqueous performance standard. Spike all samples with labeled compounds to assess method performance on the sample matrix. Compute the percent recovery of the labeled compounds using the internal standard method. Compare the percent recovery for each compound with the corresponding labeled compound recovery. Reagent water blanks are analyzed to demonstrate freedom from carryover contamination. Field replicates may be collected to determine the precision of the sampling technique, and spiked samples may be required to determine the accuracy of the analysis when the internal method is used.

REFERENCE Volatile Organic Compounds by Isotope Dilution GC/MS. Office of Water Regulation and Standards, U.S. EPA Industrial Technology Division, Washington, DC, EPA Method 1624, Rev. C, June 1989 (contact W.A. Telliard, U.S.

EPA, Office of Water Regulations and Standards, 401 M St., SW, Washington, DC, 20460. Phone: 202-382-7131).

Chlorobenzene	EPA Method 502
CAS #108-90-7	

TITLE Volatile Organic Compounds in Water By Purge and Trap Capillary Column Gas Chromatography with Photoionization and Electrolytic Conductivity Detectors in Series. U.S. EPA Method 502.2, Rev. 2.0, 1989.

MATRIX Drinking water and raw source water. The latter should include most surface water and groundwater sources.

METHOD SUMMARY This method covers 60 volatile organic compounds that contain halogen atoms and/or that are aromatic. An inert gas (zero grade nitrogen or helium) is bubbled through a 25-mL or a 5-mL water sample (depending on the expected concentration of the analytes). Purged sample components are trapped in a tube of sorbent materials. When purging is complete, the sorbent tube is heated and backflushed with helium to desorb the trapped sample onto a capillary GC column. The column is temperature programmed to separate the method analytes which are then detected with a photoionization detector (PID) and a Hall electrolytic conductivity (HECD) placed in series. The PID is selective for aromatic compounds and the HECD is selective for halogenated compounds.

INTERFERENCES Impurities in the purge gas and from organic compounds outgassing from the plumbing ahead of the trap account for many contamination problems. Interferences purged or coextracted from the samples will vary considerably from source to source, depending upon the particular sample or extract being tested. Cross-contamination can occur whenever high-level and low-level samples are analyzed sequentially. Samples also can be contaminated by diffusion of volatile organics (particularly methylene chloride and fluorocarbons) through the septum seal into the sample during shipment and storage. The lab where volatile analysis is performed and also the refrigerated storage area should be completely free of solvents.

INSTRUMENTATION A GC containing a series configuration of a high temperature photoionization detector (PID) equipped with 10.0 eV (nominal) lamp and Hall electrolytic conductivity detector (HECD) is required. Also required is an all-glass 5-mL purging device, a sorbent trap, and a thermal desorption apparatus which is connected to the GC system.

Column 1: VOCOL glass wide-bore capillary column.
Column 2: RTX–502.2 mega-bore capillary column.
Column 3: DB-62 mega-bore capillary column.

PRECISION & ACCURACY Method detection limits are dependent upon the characteristics of the gas chromatographic system used. Analytes that are not separated chromatographically cannot be individually identified and used in the same calibration mixture or water samples unless an alternative technique for identification and quantification, such as mass spectrometry, is used.

Electrolytic conductivity detetor (c) range in µg/L (a) was 0.02–200.
Electrolytic conductivity detetor (c) MDL in µg/L (b) was 0.01.
Electrolytic conductivity detetor (c) accuracy as % recovery was 103.
Electrolytic conductivity detetor (c) precision as % RSD was 3.6.
Photoionization detector (d) range in µg/L (a) was 0.02–200.
Photoionization detector (d) MDL in µg/L (b) was 0.01.
Photoionization detector (d) accuracy as % recovery was 100.
Photoionization detector (d) precision as % RSD was 1.0.

(a) *The applicable concentration range of this method is compound, instrument, and matrix-dependent. It is listed as being approximately 0.02 to 200 µg/L but no specific information is provided so caution should be observed.*
(b) *The method detection limits reports with this method are compound, instrument, and matrix-dependent. The values reported were calculated using reagent water fortified with the corresponding compounds at 10 µg/L and a GC-equipped with a 60 m × 0.75 mm VOLCOL wide bore capillary column with 1.5 µm film thickness and using helium carrier gas.*
(c) *Recoveries and relative standard deviations were determined from seven samples of reagent water fortified with 10 µg/L of each compound. 2-Bromo-1-chloropropane was used as the internal standard for calculating average recoveries.*
(d) *Recoveries and relative standard deviations were determined from seven samples of reagent water fortified with 10 µg/L of each compound. Fluorobenzene was used as the internal standard for calculating average recoveries.*

SAMPLING METHOD Collect samples using a 40- to 120-mL screw-cap vial (prewashed with detergent, rinsed with distilled water and oven dried at 105°C) with a Teflon®-faced silicone septum. Collect bubble-free samples and place the septum with the Teflon® side down on the water.

SAMPLE PRESERVATION If residual chlorine is present in the water add about 25 mg of ascorbic acid to each vial before samples are collected to remove the chlorine. Add hydrochloric acid to reduce pH to <2, immediately cool samples to 4°C, and store them in a solvent-free refrigerator at 4°C until analysis.

MHT The maximum holding time for samples is 14 days from the time they were collected.

SAMPLE PREPARATION Remove the plungers from two 5-mL syringes and attach a closed syringe valve to each. Warm the sample to room temperature, open the sample bottle, and carefully pour the sample into one of the syringe barrels to just short of overflowing. Replace the syringe plunger, invert the syringe, and compress the sample. Open the syringe valve and vent any residual air while adjusting the sample volume to 5.0 mL. Add 10 µL of the internal calibration standard to the sample through the syringe valve. Close the valve. Fill the second syringe in an identical manner from the same sample bottle. Reserve this second syringe for a reanalysis if necessary.

QUALITY CONTROL As an initial demonstration of lab accuracy and precision, analyze 4 to 7 replicates of a lab fortified blank containing analyte at 0.1–5 µg/L. Collect all samples in

duplicate. Surrogate analytes (similar to those of the analytes of interest), whose concentration is known in every sample, are measured using the same internal standard calibration procedure. Duplicate field reagent water blanks (trip blanks) must be analyzed with each set of samples, lab reagent blanks (method blanks) must be analyzed with each batch of samples processed as a group within a work shift. Also, a single lab-fortified blank that contains each of the analytes of interest should be analyzed with each batch of samples processed as a group within a work shift. A 3- to 5-point calibration curve is needed depending on the calibration range factor required.

EPA CONTACT & HOTLINE For technical questions contact Dr. Baldev Bathija, U.S. EPA, Office of Ground Water and Drinking Water (WH-550D), 401 M St. SW, Washington, DC 20460. Tel. (202) 260-3040. For further information the EPA Safe Drinking Water Hotline may be called at: (800) 426-4791.

REFERENCE Methods for the Determination of Organic Compounds in Drinking Water, EPA/600/4-88/039 (revised July 1991; Final Rule for determination of compliance with the MCL for Total Trihalomethanes under 141.30, in 40 CFR Part 141, Vol. 58, No. 147, Fed. Reg., Tuesday Aug. 3, 1993). U.S. EPA Environmental Monitoring Systems Laboratory, Cincinnati, OH, 45268, USA. Available from the National Technical Information Service (NTIS), 5285 Port Royal Road, Springfield, VA 22161; Tel. 800-553-6847. NTIS Order Number is PB91-231480.

Chlorobenzene   **EPA Method 524**
CAS #108-90-7

TITLE Measurement of Purgeable Organic Compounds in Water by Capillary Column GC/MS.

MATRIX Drinking water and raw source water; the latter should include most surface water and groundwater sources.

METHOD SUMMARY Method 524.2 covers 60 volatile organic compounds. An inert gas (zero grade nitrogen or helium) is bubbled through a 25-mL or a 5-mL water sample (depending on the expected concentration of the analytes). Purged sample components are trapped in a tube of sorbent materials. When purging is complete, the sorbent tube is heated and backflushed with helium to desorb the trapped sample onto a capillary GC column.

INTERFERENCES Impurities in the purge gas and from organic compounds outgassing from the plumbing ahead of the trap account for many contamination problems. Interferences purged or coextracted from the samples will vary considerably from source to source, depending upon the particular sample or extract being tested. Cross-contamination can occur whenever high-level and low-level samples are analyzed sequentially. Samples also can be contaminated by diffusion of volatile organics (particularly methylene chloride and fluorocarbons) through the septum seal into the sample during shipment and storage.

INSTRUMENTATION A GC/MS with a data system equipped with one of the following capillary GC columns:

Column 1: VOCOL glass wide bore capillary column.
Column 2: DB-624 fused silica capillary column.
Column 3: DB-5 fused silica capillary column.

Also required is an all-glass 25 mL or 5-mL purging device, a sorbent trap, and a thermal desorption apparatus which is connected to the GC/MS system.

PRECISION & ACCURACY Method detection limits are compound- and instrument-dependent, and may vary from approximately 0.02–0.35 µg/L. Note in the table below that the "true" concentration range used for accuracy and precision measurements was quite narrow. However, the applicable concentration range of this method is primarily column dependent and is approximately 0.02 to 200 µg/L for the wide-bore thick-film columns. Narrow-bore thin-film columns may have a capacity which limits the range to about 0.02 to 20 µg/L. Analytes that are inefficiently purged from water will not be detected when present at low concentrations, but they can be measured with acceptable accuracy and precision when present in sufficient amounts.

Analytes that are not separated chromatographically, but which have different mass spectra and non-interfering quantification ions, can be identified and measured in the same calibration mixture or water sample. Analytes which have very similar mass spectra cannot be individually identified and measured in the same calibration mixture or water samples unless they have different retention times. Co-eluting compounds with very similar mass spectra, typically many structural isomers, must be reported as an isomeric group or pair.

The range (in µg/L) was 0.1–10.
The Method Detection Limig (in µg/L) was 0.04.
The accuracy (as % recovery) was 98.
The precision (in %) was 5.9.

Note: Data were obtained from 16–31 determinations using a wide-bore capillary column and a jet separator interfaced to a quadrupole mass spectrometer. All analytes were in a reagent water matrix.

SAMPLING METHOD Collect samples using a 40- to 120-mL screw-cap vial (prewashed with detergent, rinsed with distilled water and oven dried at 105°C) with a Teflon®-faced silicone septum. Collect bubble-free samples and place the septum with the Teflon® side down on the water.

SAMPLE PRESERVATION If residual chlorine is present in the water add about 25 mg of ascorbic acid to each vial before samples are collected to remove the chlorine. Add hydrochloric acid to reduce pH to <2, and immediately cool samples to 4°C, and store them in a solvent-free refrigerator at 4°C until analysis.

MHT The maximum holding time for samples is 14 days from the time they were collected.

SAMPLE PREPARATION Remove the plungers from two 25-mL (or 5-mL depending on sample size) syringes and attach a closed syringe valve to each. Warm the sample to room temperature, open the sample bottle, and carefully pour the sample into one of the syringe barrels to just short of overflowing. Replace the syringe plunger, invert the syringe, and compress the sample. Open the syringe valve and vent any residual air

while adjusting the sample volume to 25.0 mL (or 5 mL). For samples and blanks, add 5 µL of the fortification solution containing the internal standard and the surrogates to the sample through the syringe valve. For calibration standards and lab fortified blanks, add 5 µL of the fortification solution containing the internal standard only. Close the valve. Fill the second syringe in an identical manner from the same sample bottle. Reserve this second syringe for a reanalysis if necessary.

QUALITY CONTROL As an initial demonstration of lab accuracy and precision, analyze 4 to 7 replicates of a lab fortified blank containing analyte at 0.2–5 µg/L. Collect all samples in duplicate. Surrogate analytes (similar to those of the analytes of interest), whose concentration is known in every sample, are measured using the same internal standard calibration procedure. Duplicate field reagent water blanks (trip blanks) must be analyzed with each set of samples, lab reagent blanks (method blanks) must be analyzed with each batch of samples processed as a group within a work shift. Also, a single lab-fortified blank that contains each of the analytes of interest should be analyzed with each batch of samples processed as a group within a work shift. A 3- to 5-point calibration curve is needed depending on the calibration range factor required.

EPA CONTACT & HOTLINE For technical questions contact Dr. Baldev Bathija, U.S. EPA, Office of Ground Water and Drinking Water (WH-550D), 401 M St. SW, Washington, DC 20460. Tel. (202) 260-3040. For further information the EPA Safe Drinking Water Hotline may be called at: (800) 426-4791.

REFERENCE Methods for the Determination of Organic Compounds in Drinking Water, EPA/600/4-88/039 (revised July 1991; Final Rule for determination of compliance with the MCL for Total Trihalomethanes under 141.30, in 40 CFR Part 141, Vol. 58, No. 147, Fed. Reg., Tuesday Aug. 3, 1993). U.S. EPA Environmental Monitoring Systems Laboratory, Cincinnati, OH, 45268, USA. Available from the National Technical Information Service (NTIS), 5285 Port Royal Road, Springfield, VA 22161; Tel. 800-553-6847. NTIS Order Number is PB91-231480.

Chlorobenzene EPA Method 8021
CAS #108-90-7

TITLE Halogenated Volatile by Gas Chromatography Using Photoionization and Electrolytic Conductivity Detectors in Series: Capillary Column Technique

MATRIX This method is applicable to nearly all types of samples, regardless of water content, including groundwater, aqueous sludges, caustic liquors, acid liquors, waste solvents, oily wastes, mousses, tars, fibrous wastes, polymeric emulsions, filter cakes, spent carbons, spent catalysts, soils, and sediments.

METHOD SUMMARY This method is used to determine 60 volatile organic compounds in a variety of solid waste matrices. It provides GC conditions for the detection of halogenated and aromatic volatile organic compounds. Samples can be analyzed using direct injection or purge-and-trap (EPA Method 5030). Groundwater samples must be analyzed using EPA Method 5030 (where applicable). A temperature program is used with the GC. Detection is achieved by a photoionization detector (PID) and a Hall electrolytic conductivity detector (HECD) in series.

INTERFERENCES Samples can be contaminated by diffusion of volatile organics (particularly chlorofluorocarbons and methylene chloride) through the sample container septum during shipment and storage.

INSTRUMENTATION A GC-equipped with variable-constant differential flow controllers, subambient oven controller, PID and HECD detectors connected with a short piece of uncoated capillary tubing and a data system.

Column: 60 m × 0.75 mm I.D. VOCOL wide-bore capillary column with 1.5 µm film thickness.

PRECISION & ACCURACY MDLs are compound-dependent and vary with purging efficiency and concentration. The applicable concentration range of this method is compound- and instrument-dependent but is approximately 0.1 to 200 µg/L. Analytes that are inefficiently purged from water will not be detected when present at low concentrations, but they can be measured with acceptable accuracy and precision when present in sufficient amounts. The estimated quantitation limit (EQL) for an individual compound is approximately 1 µg/kg (wet weight) for soil/sediment samples, 100 µg/kg (wet weight) for wastes, and 1 µg/L for groundwater. EQLs will be proportionately higher for sample extracts and samples that require dilution or reduced sample size to avoid saturation of the detector.

MULTIPLICATION FACTORS FOR OTHER MATRICES (a)

Matrix	Factor (b)
Groundwater	10
Low-concentration soil	10
Water miscible liquid waste	500
High-concentration soil and sludge	1250
Non-water miscible waste	1250

(a) Sample EQLs are highly matrix-dependent. The EQLs listed herein are provided for guidance and may not always be achievable. (b) EQL = [Method detection limit] × [Factor]. For non-aqueous samples, the factor is on a wet-weight basis.

SINGLE LABORATORY ACCURACY & PRECISION DATA FOR VOCs IN WATER

This method was tested in a single lab using water spiked at 10 µg/L and the following data was reported:

Recoveries and standard deviations were determined from seven samples and spiked at 10 µg/L of each analyte. Recoveries were determined by the internal standard method. Internal standards were: Fluorobenzene for PID and 2-Bromo-1-chloropropane for HECD.

The average recovery (in percent) for the PID was 100.
The standard deviation of the recovery for the PID was 1.0.
The MDL (in µg/mL) for the PID was 0.003.
The average recovery (in percent) for the HECD was 103.
The standard deviation of the recovery for the HECD was 3.7.
The MDL (in µg/mL) for the HECD was 0.01.

SAMPLE COLLECTION, PRESERVATION & HANDLING

Volatile Organics — Standard 40-mL glass screw-cap VOA vials with Teflon®-faced silicone septum may be used for both liquid and solid matrices. When collecting samples, liquids and solids should be introduced into the vials gently to reduce agitation which might drive off volatile compounds. If there are any air bubbles present the sample must be retaken. Tap slightly as they are filled to try and eliminate as much free air space as possible. The two vials from each sampling locations should be sealed in separate plastic bags to prevent cross-contamination between samples particularly if the sampled waste is suspected of containing high levels of volatile organics.

Semivolatile organics — Containers used to collect samples for the determination of semivolatile organic compounds should be soap and water washed followed by methanol (or isopropanol) rinsing. The sample containers should be of glass or Teflon® and have screw-top covers with Teflon® liners.

Preservation for volatile organics — No preservation is used with concentrated waste samples. With liquid samples containing no residual chlorine, 4 drops of concentrated hydrochloric acid are added and the samples are immediately cooled to 4°C. When liquid samples contain residual chlorine, they are treated as above and, in addition, 4 drops of 4% aqueous sodium thiosulfate are added. Soil, sediment, and sludge samples are only cooled to 4°C.

Preservation for semivolatile organics — No preservation is used with concentrated waste samples. With liquid samples containing no residual chlorine and with soil, sediment, and sludge samples, immediately cooling to 4°C is the only preservation used. When residual chlorine is present then 3 mL of 10% aqueous sodium sulfate is added for each gallon of sample collected, followed by cooling to 4°C.

MHT The holding time for all volatile organics samples is 14 days. Liquid samples must be extracted within 7 days and their extracts analyzed within 40 days. Concentrated waste, soil, sediment, and sludge samples must be extracted within 14 days and their extracts analyzed within 40 days.

SAMPLE PREPARATION Volatile compounds are introduced into the gas chromatograph either by direct injector or purge-and-trap (EPA Method 5030). EPA Method 5030 may be used directly on groundwater samples or low-concentration contaminated soils and sediments. For medium-concentration soils or sediments, methanolic extraction, as described in EPA Method 5030, may be necessary prior to purge-and-trap analysis.

QUALITY CONTROL Calculate surrogate standard recovery on all samples, blanks, and spikes. A trip blank is recommended to check on sampling, storage, and handling contamination. Calibration standards, at a minimum of five concentration levels, are prepared in organic-free reagent water. One of the concentration levels should be at a concentration near, but above, the method detection limit.

A combination of bromochloromethane, 2-bromo-1-chloropropane, 1,4-dichlorobutane, and bromochlorobenzene are recommended as surrogate standards to encompass the range of the temperature program used in this method.

REFERENCE Test Methods for Evaluating Solid Waste, Physical/Chemical Methods, SW-846, 3rd Edition, U.S. EPA, Office of Solid Waste, Washington, DC, EPA Method 8021A, Rev. 1, Nov. 1992.

Chlorobenzene   **EPA Method 8240**
CAS #108-90-7

TITLE Volatile Organics By GC/MS: Packed Column Technique

MATRIX Nearly all types of sample matarices, regardless of water content, can be analyzed using this method. This includes groundwater, aqueous sludges, caustic liquors, acid liquors, waste solvents, oily wastes, mousses, tars, fibrous wastes, polymetric emulsions, filter cakes, spent carbons, spent catalysts, soils, and sediments.

METHOD SUMMARY Method 8240B covers 80 volatile organic compounds that are introduced into a gas chromatograph by the purge-and-trap method or by direct injection (in limited applications). For the purge-and-trap method an inert gas (zero grade nitrogen or helium) is bubbled through a 5-mL solution at ambient temperature. Purged sample components are trapped in a tube of sorbent materials. When purging is complete, the sorbent tube is heated and backflushed with inert gas to desorb the trapped components onto a GC column.

INTERFERENCES Impurities in the purge gas and from organic compounds outgassing from the plumbing ahead of the trap account for many contamination problems. Interferences purged or coextracted from the samples will vary considerably from source to source. Cross-contamination can occur whenever high-level and low-level samples are analyzed sequentially. Whenever an unusually concentrated sample is analyzed, it should be followed by the analysis of organic-free reagent water to check for cross-contamination. Samples also can be contaminated by diffusion of volatile organics (particularly methylene chloride and fluorocarbons) through the septum seal into the sample during shipment and storage. A trip blank can serve as a check on such contamination. The lab where volatile analysis is performed and also the refrigerated storage area should be completely free of solvents.

INSTRUMENTATION A gas chromatograph/mass spectrometry/data system (GC/MS) equipped with a 6 ft × 0.1 in I.D. glass column packed with 1% SP-1000 on Carbopack-B (60/80 mesh) is required. Also needed is a 5-mL purging device, a sorbent trap, and a thermal desorption apparatus.

PRECISION & ACCURACY This method is reported to have been tested by 15 laboratories using organic-free reagent water, drinking water, surface water, and industrial wastewaters (not specified) fortified at six concentrations over the range 5–600 µg/L.

Sample estimated quantitation limits (EQLs) are highly matrix-dependent. The EQLs listed may not always be achievable. EQLs listed for soils or sediments are based on wet weight. Normally, data is reported on a dry-weight basis; therefore, EQLs will be higher, based on the percent dry weight of each sample. Note that EQLs are even more variable than MDLs and

that they are highly variable depending on the matrix being analyzed.

EQL in groundwater in µg/L was 5.
EQL in low soil or sediment in µg/kg was 5.
Accuracy (a) in µg/L was $0.98C+2.28$.
Precision (b) in µg/L was $0.26x-1.92$.

(a) *Average recovery found for measurements of samples containing a concentration of C, in µg/L.*
(b) *Overall precision found for measurements of samples with average recovery X for samples containing a concentration of C in µg/L.*
X = *Average recovery found for measurement of samples containing a concentration of C in µg/L.*

MULTIPLICATION FACTORS FOR OTHER MATRICES

Other Matrices	Factor (a)
Waste miscible liquid waste	50
High-concentration soil and sludge	125
Non-water miscible waste	500

(a) *EQL = [EQL for low soil/sediment] × [Factor]. For non-aqueous samples, the factor is on a wet-weight basis.*

SAMPLING METHOD

Liquid samples — Use a 40-mL glass screw-cap VOA vial with a Teflon®-faced silicone septum that has been prewashed, rinsed with distilled deionized water, and oven dried. However, if residual chlorine is present, collect sample in a 40-oz. soil VOA container which has been pre-preserved with 4 drops of 10% sodium thiosulfate, mix gently, and then transfer the sample to a 40-mL VOA vial. Collect bubble-free samples in duplicate and seal them in separate plastic bags.

Soils or sediments, and sludges — Use an 8-oz. widemouth glass bottle with a Teflon®-faced silicone septum that has been prewashed with detergent, rinsed with distilled deionized water, and oven dried. Tap slightly to eliminate free air space. Collect samples in duplicate and seal them in separate plastic bags.

SAMPLE PRESERVATION

Liquid samples — Add 4 drops of concentrated HCL and immediately cool samples to 4°C and store in a solvent-free refrigerator.

Soils or sediments, and sludges — Cool samples to 4°C and store in a solvent-free refrigerator.

MHT Maximum holding time is 14 days from the date of sample collection.

SAMPLE PREPARATION

Liquid samples — Remove the plunger from a 5-mL syringe and carefully pour the sample into the syringe barrel to just short of overflowing. Replace the syringe plunger and compress the sample. Open the syringe valve and vent any residual air while adjusting the sample volume to 5.0 mL. If there is only one volatile organic analysis (VOA) vial, a second syringe should be filled at this time to protect against possible loss of sample integrity. Add 10 µL of surrogate spiking solution and 10 µL of internal standard spiking solution through the valve bore of the 5-mL syringe, then close the valve. The surrogate and internal standards may be mixed and added as a single spiking solution.

Sediments, soils, and waste samples — All samples of this type should be screened by GC analysis using a headspace method (EPA Method 3810) or the hexadecane extraction and screening method (EPA Method 3820). Use the screening data to determine whether to use the low-concentration method (0.005–1 mg/kg) or the high-concentration method (>1 mg/kg).

Low-concentration method — The low-concentration method is based on purging a heated sediment or soil sample mixed with organic-free reagent water containing the surrogate and internal standards. Analyze all reagent blanks and standards under the same conditions as the samples.

Use a 5-g sample if the expected concentration is <0.1 mg/kg or a 1-g sample for expected concentrations between 0.1 and 1 mg/kg. Mix the contents of the sample container with a narrow metal spatula. Weigh the amount of the sample into a tared purge device. Add the spiked water to the purge device, which contains the weighed amount of sample, and connect the device to the purge-and-trap system.

High-concentration method — This method is based on extracting the sediment or soil with methanol. A waste sample is either extracted or diluted, depending on its solubility in methanol. Wastes that are insoluble in methanol are diluted with reagent tetraglyme or possibly polyethylene glycol (PEG). An aliquot of the extract is added to organic-free reagent water containing surrogate and internal standards. This is purged at ambient temperature. All samples with an expected concentration of >1.0 mg/kg should be analyzed by this method.

Mix the contents of the sample container with a narrow metal spatula. For sediments or soils and solid wastes that are insoluble in methanol, weigh 4 g (wet weight) of sample into a tared 20-mL vial. For waste that is soluble in methanol, tetraglyme, or PEG, weigh 1 g (wet weight) into a tared scintillation vial or culture tube or a 10-mL volumetric flask. Quickly add 9.0 mL of appropriate solvent then add 1.0 mL of a surrogate spiking solution to the vial, cap it, and shake it for 2 min.

METHANOL EXTRACT REQUIRED FOR ANALYSIS OF HIGH-CONCENTRATION SOILS OR SEDIMENTS

Approximate Concentration Range	Volume of Methanol Extract (a)
500–10,000 µg/kg	100 µL
1,000–20,000 µg/kg	50 µL
5,000–100,000 µg/kg	10 µL
25,000–500,000 µg/kg	100 µL of 1/50 dilution (b)

Calculate appropriate dilution factor for concentrations exceeding this table.

(a) *The volume of methanol added to 5 mL of water being purged should be kept constant. Therefore, add to the 5-mL syringe whatever volume of methanol is necessary to maintain a volume of 100 µL added to the syringe.*
(b) *Dilute an aliquot of the methanol extract and then take 100 µL for analysis.*

QUALITY CONTROL Demonstrate, through the analysis of a reagent water blank, that interferences from the analytical system, glassware, and reagents are under control. Blank samples should be carried through all stages of the sample preparation and measurement steps. For each analytical batch (up to 20 samples), a reagent blank, matrix spike, and matrix spike duplicate must be analyzed (the frequency of the spikes may be different for different monitoring programs). The blank and spiked samples must be carried through all stages of the sample preparation and measurement steps. QC samples mentioned in the section on Interferences will also be needed as appropriate to those situations.

REFERENCE Test Methods for Evaluating Solid Waste (SW-846). U.S. EPA. 1983. Method 8240B, Rev. 2, Nov. 1990. Office of Solid Wastes, Washington, DC.

Chlorobenzene   EPA Method 8260
CAS #108-90-7

TITLE Volatile Organic Compounds by GC/MS: Capillary Column Technique

MATRIX This method is applicable to nearly all types of samples, regardless of water content, including groundwater, soils, and sediments.

METHOD SUMMARY Method 8260A covers 58 volatile organic compounds that are introduced into a gas chromatograph by the purge-and-trap method or by direct injection (in limited applications). Zero-grade helium is bubbled through a 5-mL solution at ambient temperature. Purged sample components are trapped in a tube containing suitable sorbent materials. When purging is complete, the sorbent tube is heated and backflushed with helium to desorb trapped sample components. The analytes are desorbed directly to a large bore capillary or cryofocussed on a capillary precolumn before being flash evaporated to a narrow bore capillary for analysis.

INTERFERENCES Major contaminant sources are volatile materials in the lab and impurities in the inert purging gas and in the sorbent trap. Interfering contamination may occur when a sample containing low concentrations of volatile organic compounds is analyzed immediately after a sample containing high concentrations of volatile organic compounds. After analysis of a sample containing high concentrations of volatile organic compounds, one or more calibration blanks should be analyzed to check for cross-contamination. Screening of the samples prior to purge-and-trap GC/MS analysis is highly recommended to prevent contamination of the system. This is especially true for soil and waste samples.

Special precautions must be taken to analyze for methylene chloride. The analytical and sample storage area should be isolated from all atmospheric sources of methylene chloride. All gas chromatography carrier gas lines and purge gas plumbing should be constructed from stainless steel or copper tubing. Laboratory clothing previously exposed to methylene chloride fumes during liquid-liquid extraction procedures can contribute to sample contamination.

Samples can also be contaminated by diffusion of volatile organics (particularly methylene chloride and fluorocarbons) through the septum seal during shipment and storage. A trip blank can serve as a check on such contamination.

INSTRUMENTATION GC/MS with a temperature-programmable chromatograph suitable for splitless injection equipped with variable constant differential flow controllers, a subambient oven controller, a purging device, sorbent trap, a thermal desorption apparatus and a capillary precolumn interface when using cryogenic cooling will be needed. The following GC columns may be used:

Column 1: 60 m × 0.75mm I.D. capillary column coated with VOCOL, 1.5 µm film thickness.
Column 2: 30 m × 0.53mm capillary column coated with DB-624 or VOCOL, 3 µm film thickness.
Column 3: 30 m × 0.32mm I.D. capillary column coated with DB-5 or SE-54, 1-µm film thickness.

PRECISION & ACCURACY This method has been tested in a single lab using spiked water. Using a wide-bore capillary column, water was spiked at concentrations between 0.5 and 10 µg/L. Single lab accuracy and precision data are presented. The MDL actually achieved in a given analysis will vary depending on instrument sensitivity and matrix effects.

The MDL (a) in µg/L was 0.04.
The concentration range in µg/L was 0.1–10.
The mean accuracy (% of true value) was 98.
The precision as relative standard deviation was 5.9.

Note: The MDL is based on a 25-mL sample volume instead of a 5-mL sample volume.

SAMPLING METHOD
Liquid samples — Use a 40-mL glass screw-cap VOA vial with a Teflon®-faced silicone septum that has been prewashed, rinsed with distilled deionized water, and oven dried. If residual chlorine is present, collect the sample in a 4-oz soil VOA container which has been pre-preserved with 4 drops of 10% sodium thiosulfate. Mix gently and transfer the sample to a 40-mL VOA vial. Collect bubble-free samples in duplicate and seal each sample in a separate plastic bag.

Soils, sediments, and sludges — Use an 8-oz widemouth glass bottle with Teflon®-faced silicone septum that has been prewashed, rinsed with distilled deionized water, and oven dried. **Do not** heat the septum for more than 1 h. Tap slightly to eliminate any free air space. Collect samples in duplicate and seal each one in a separate plastic bag.

SAMPLE PRESERVATION
Liquid samples — Add 4 drops of concentrated HCL, cool to 4°C and store in a solvent-free refrigerator.

Soils, sediments and sludges — Cool samples to 4°C and store in a solvent-free refrigerator.

MHT The maximum holding time of any sample (liquids, soils, sediments, and sludges) is 14 days.

SAMPLE PREPARATION
Liquid samples — Remove the plunger from a 5-mL syringe and carefully pour the sample into the syringe barrel to just short of overflowing. Replace the syringe plunger and compress

the sample. Open the syringe valve and vent any residual air while adjusting the sample volume to 5.0 mL. If there is only one volatile organic analysis (VOA) vial, a second syringe should be filled at this time to protect against possible loss of sample integrity. Add 10 μL of surrogate spiking solution and 10 μL of internal standard spiking solution through the valve bore of the 5-mL syringe, then close the valve. The surrogate and internal standards may be mixed and added as a single spiking solution.

Sediments, soils, and waste samples — All samples of this type should be screened by GC analysis using a headspace method (EPA Method 3810) or the hexadecane extraction and screening method (EPA Method 3820). Use the screening data to determine whether to use the low-concentration method (0.005–1 mg/kg) or the high-concentration method (>1 mg/kg).

Low-concentration method — The low-concentration method is based on purging a heated sediment or soil sample mixed with organic-free reagent water containing the surrogate and internal standards. Analyze all reagent blanks and standards under the same conditions as the samples.

Use a 5-g sample if the expected concentration is <0.1 mg/kg or a 1-g sample for expected concentrations between 0.1 and 1 mg/kg. Mix the contents of the sample container with a narrow metal spatula. Weigh the amount of the sample into a tared purge device. Add the spiked water to the purge device, which contains the weighed amount of sample, and connect the device to the purge-and-trap system.

High-concentration method — This method is based on extracting the sediment or soil with methanol. A waste sample is either extracted or diluted, depending on its solubility in methanol. Wastes that are insoluble in methanol are diluted with reagent tetraglyme or possibly polyethylene glycol (PEG). An aliquot of the extract is added to organic-free reagent water containing surrogate and internal standards. This is purged at ambient temperature. All samples with an expected concentration of >1.0 mg/kg should be analyzed by this method.

Mix the contents of the sample container with a narrow metal spatula. For sediments or soils and solid wastes that are insoluble in methanol, weigh 4 g (wet weight) of sample into a tared 20-mL vial. For waste that is soluble in methanol, tetraglyme, or PEG, weigh 1 g (wet weight) into a tared scintillation vial or culture tube or a 10-mL volumetric flask. Quickly add 9.0 mL of appropriate solvent then add 1.0 mL of a surrogate spiking solution to the vial, cap it, and shake it for 2 min.

METHANOL EXTRACT REQUIRED FOR ANALYSIS OF HIGH-CONCENTRATION SOILS OR SEDIMENTS

Approximate Concentration Range	Volume of Methanol Extract (a)
500–10,000 μg/kg	100 μL
1,000–20,000 μg/kg	50 μL
5,000–100,000 μg/kg	10 μL
25,000–500,000 μg/kg	100 μL of 1/50 dilution (b)

Calculate appropriate dilution factor for concentrations exceeding this table.

(a) The volume of methanol added to 5 mL of water being purged should be kept constant. Therefore, add to the 5-mL syringe whatever volume of methanol is necessary to maintain a volume of 100 μL added to the syringe.
(b) Dilute an aliquot of the methanol extract and then take 100 μL for analysis.

QUALITY CONTROL Demonstrate, through the analysis of a reagent water blank, that interferences from the analytical system, glassware, and reagents are under control. Blank samples should be carried through all stages of the sample preparation and measurement steps. For each analytical batch (up to 20 samples), a reagent blank, matrix spike, and matrix spike duplicate must be analyzed (the frequency of the spikes may be different for different monitoring programs). The blank and spiked samples must be carried through all stages of the sample preparation and measurement steps. QC samples mentioned in the section on Interferences will also be needed as appropriate to those situations.

Matrix spiking standards should be prepared from volatile organic compounds which will be representative of the compounds being investigated. The recommended internal standards are chlorobenzene-d5, 1,4-difluorobenzene, 1,4-dichlorobenzene-d4, and pentafluorobenzene. Using stock standard solutions, prepare secondary dilution standards containing the compounds of interest, either singly or mixed together in methanol. Store them in a vial with no headspace for no more than one week. Surrogates recommended are toluene-d8, 4-bromofluorobenzene, and dibromofluoromethane. Each sample undergoing GC/MS analysis must be spiked with 10 μL of the surrogate spiking solution prior to analysis.

REFERENCE Test Methods for Evaluating Solid Waste (SW-846). U.S. EPA 1983, Method 8260A, Rev. 1, Nov. 1990. Office of Solid Waste, Washington, DC.

Chlorobenzene   **EPA Method 503.1**
CAS #108-90-7

TITLE Aromatic & Unsaturated VOCs

MATRIX Drinking water (finished or in Water any treatment stage) and raw source water.

APPLICATION Method covers 28 aromatic and unsaturated VOCs. An inert gas is bubbled through a 5-mL water sample. Purged sample components are trapped in tube of sorbent materials. When purging is complete, sorbent tube is heated and backflushed with inert gas to desorb trapped sample onto a packed GC column.

INTERFERENCES During analysis, major contaminant sources are volatile materials in the lab and impurities in purging gas and sorbent trap. With high and low level samples, there can be carryover contamination. Excess water causes a negative baseline deflection.

INSTRUMENTATION Purge and Trap GC w/photoionization detector. (Two GC columns are recommended); Column 1: 5% SP-1200 and 1.75% Bentone 34 on Supelcoport; 5% 1,2,3-tris(2-cyanoethoxy)propane on Chromosorb W.

RANGE 2.2–600 μg/L (Drinking water).

MDL 0.004 µg/L in water

PRECISION RSD = 12.1% at 27.6 µg/L; 7 labs

ACCURACY Average recovery = 98% at 27.6 µg/L; 7 labs

SAMPLING METHOD Use a 40–120-mL screw-cap vial (prewashed with detergent, rinsed with distilled water and oven dried at 105°C) with a PTFE-faced silicone septum. If residual chlorine is in the water add about 25 mg of ascorbic acid to each vial before sample collection. Collect bubble-free samples.

STABILITY Cool to 4°C; HCl to pH <2.

MHT 14 days.

QUALITY CONTROL As an initial demonstration of lab accuracy and precision, analyze 4 to 7 replicates of a lab fortified blank containing analyte at 0.1–5 µg/L. Collect all samples in duplicate.

REFERENCE Method 503.1, Volatile aromatic & unsaturated organic compounds in H2O by Purge and Trap GC, EPA 600/4-88/039.

Chlorobenzene   **EPA Method 601**
CAS #108-90-7

TITLE Purgeable Halocarbons

MATRIX Wastewater

APPLICATION Method covers 29 purgeable halocarbons. (Method 624 provides GC/MS conditions appropriate for the qualitative and quantitative confirmation of results). Method describes conditions for a 2nd GC column to confirm measurements made with primary column.

INTERFERENCES Impurities in the purge gas and organic compounds outgassing from the plumbing ahead of the trap. With high- and low-level samples, there can be carryover contamination. Diffusion of volatile organics through the septum seal into the sample.

INSTRUMENTATION GC-equipped with halide-specific detector. (With purge-and-trap unit).

RANGE 8.0–500 µg/L.

MDL 0.25 µg/L.

PRECISION 0.18X+1.21 µg/L (overall precision).

ACCURACY 1.00C–1.23 µg/L (as recovery).

SAMPLING METHOD 25-mL glass vial. Teflon®-lined septum.

STABILITY Cool, 4°C, 0.008% Na2S2O3.

MHT 14 days.

QUALITY CONTROL The lab must on an ongoing basis, spike at least 10% of the samples from each sample site being monitored to assess accuracy.

REFERENCE Method 601, *Federal Register* Part VIII 40 CFR Part 136, Oct 26, 1984.

Chlorobenzene   **EPA Method 602**
CAS #108-90-7

TITLE Purgeable Aromatics

MATRIX Wastewater

APPLICATION Method covers 7 purgeable aromatics. (Method 624 provides GC/MS conditions appropriate for the qualitative and quantitative confirmation of results). Method describes conditions for a 2nd GC column to confirm measurements made with primary column.

INTERFERENCES Impurities in the purge gas and organic compounds outgassing from the plumbing ahead of the trap. With high- and low-level samples, there can be carryover contamination. Diffusion of volatile organics through the septum seal into the sample.

INSTRUMENTATION GC-equipped with photoionization detector. (With purge-and-trap unit)

RANGE 2.1–550 µg/L.

MDL 0.2 µg/L.

PRECISION 0.17X+0.10 µg/L (overall precision).

ACCURACY 0.95C+0.02 µg/L (as recovery).

SAMPLING METHOD 25-mL glass vial. Teflon®-lined septum.

STABILITY Cool, 4°C, 0.008% Na2S2O3.

MHT 14 days.

QUALITY CONTROL The lab must on an ongoing basis, spike at least 10% of the samples from each sample site being monitored to assess accuracy.

REFERENCE Method 602, *Federal Register* Part VIII 40 CFR Part 136, Oct 26, 1984.

Chlorobenzene   **EPA Method 624**
CAS #108-90-7

TITLE Purgeables

MATRIX Wastewater

APPLICATION Method covers 31 purgeable organics. An inert gas is bubbled through a 5-mL water sample in a specially designed purging chamber. Here, purgeables are transferred from aqueous to gaseous phase, passed onto a sorbent column, and trapped. Trap is heated and backflushed with inert gas to desorb purgeables onto a GC column, where purgeables are separated.

INTERFERENCES Impurities in the purge gas, organic compounds outgassing from the plumbing ahead of the trap, and solvent vapors in the lab. With high- and low-level samples, there can be carryover contamination.

INSTRUMENTATION GC/MS with purge-and-trap unit.

RANGE 5–600 µg/L

MDL 6.0 µg/L

PRECISION 0.26X–1.92 µg/L (overall precision).

ACCURACY 0.98C+2.28 µg/L (as recovery).

SAMPLING METHOD 25-mL glass vial. Teflon®-lined septum.

STABILITY Cool, 4°C, 0.008% Na2S2O3.

MHT 14 days.

QUALITY CONTROL The lab must on an ongoing basis, spike at least 5% of the samples from each sample site being monitored to assess accuracy.

REFERENCE Method 624, *Federal Register* Part VIII 40 CFR Part 136, Oct 26, 1984.

Chlorobenzene — EPA Method 8010
CAS #108-90-7

TITLE Halogenated Volatile Organics

MATRIX Groundwater, soils, sludges, water miscible liquid wastes, and non-water miscible wastes.

APPLICATION This method is used for the analysis of 39 halogenated VOCs. Samples are analyzed using direct injection or purge-and-trap methods. Groundwater must be analyzed by the purge-and-trap method. The method provides an optional GC column which is used for analyte confirmation and that may help resolve analytes from interferences.

INTERFERENCES There can be carryover contamination with high- and low-level samples. Impurities may come from the purge-and-trap apparatus, organic compounds outgassing from the plumbing ahead of trap, diffusion of VOCs through the sample bottle septum during shipping or storage, or from solvent vapors in the lab.

INSTRUMENTATION GC capable of on-column injections or purge-and-trap sample introduction and a halogen specific detector. Column 1: 8 ft by 0.1 in 1%. SP-1000 on Carbopack-B. Column 2: 6 ft by 0.1 in bonded n-octane on Porasil-C.

RANGE 8 to 500 µg/L (reagent water)

MDL 0.25 µg/L (reagent water).

PQL FACTORS FOR MULTIPLYING × FID MDL VALUE

Matrix	Multiplication Factor
Groundwater	10
Low-level soil	10
Water miscible liquid waste	500
High-level soil and sludge	1250
Non-water miscible waste	1250

PRECISION 0.18X + 1.21 µg/L (overall precision).

ACCURACY 1.00C–1.23 µg/L (as recovery).

SAMPLING METHOD For water and liquid samples; use glass 40-mL vials with Teflon®-lined septum caps and collect two vials per sample location with no headspace. For solids and concentrated waste samples; use widemouth glass bottles with Teflon® liners.

STABILITY For concentrated wastes, soils, sediments, or sludges: cool to 4°C. For liquids: add 4 drops of concentrated hydrochloric acid and cool to 4°C.

MHT 14 days.

QUALITY CONTROL Analyze a reagent blank, matrix spike, and matrix spike duplicate/duplicate for each analytical batch (up to 20 samples). Demonstrate the purity of glassware and reagents by analyzing a reagent water method blank. Internal, surrogate, and five concentration level calibration standards are used.

REFERENCE Test Methods for Evaluating Solid Waste (SW-846), U.S. EPA Office of Solid Waste, Washington, DC, Method 8010B, Rev. 2, Nov. 1992.

Chlorobenzene — EPA Method 8020
CAS #108-90-7

TITLE Aromatic Volatile Organics

MATRIX Groundwater, soils, sludges, water miscible liquid wastes, and non-water miscible wastes.

APPLICATION This method is used to analyze for 8 aromatic VOCs. Samples are analyzed using direct injection or purge-and-trap methods. Groundwater must be analyzed by the purge-and-trap method. The method provides an optional GC column that is used for analyte confirmation and may also help resolve analytes from interferences.

INTERFERENCES There can be carryover contamination with high- and low-level samples. Impurities may come from the purge-and-trap apparatus, organic compounds outgassing from the plumbing ahead of trap, diffusion of VOCs through the sample bottle septum during shipping or storage, or from solvent vapors in the lab.

INSTRUMENTATION GC capable of on-column injections or purge-and-trap sample introduction and a photoionization detector (PID). Column 1: 6 ft by 0.082 in with 5% SP-1200 and 1.75% Bentone-34 on Supelcoport. Column 2: 8 ft by 0.1 in with 5% 1,2,3-tris(2-cyanoethoxy)propane on Chromosorb W-AW.

RANGE 2.1 to 500 µg/L

MDL 0.2 µg/L (reagent water).

PQL FACTORS FOR MULTIPLYING × FID MDL VALUE

Matrix	Multiplication Factor
Groundwater	10
Low-level soil	10
Water miscible liquid waste	500
High-level soil and sludge	1250
Non-water miscible waste	1250

PRECISION 0.17X + 0.10 µg/L (overall precision).

ACCURACY 0.95C + 0.02 µg/L (as recovery).

SAMPLING METHOD For water and liquid samples use glass 40-mL vials with Teflon®-lined septum caps and collect two vials per sample location with no headspace. For solids and concentrated waste samples use widemouth glass bottles with Teflon® liners. Cool all samples to 4°C

STABILITY For concentrated wastes, soils, sediments, or sludges cool to 4°C. For liquids, add 4 drops of concentrated hydrochloric acid and cool to 4°C.

MHT 14 days.

QUALITY CONTROL Analyze a reagent blank, matrix spike, and matrix spike duplicate/duplicate for each analytical batch (up to 20 samples). Demonstrate the purity of glassware and reagents by analyzing a reagent water method blank. Internal, surrogate, and five concentration level calibration standards are used. The QC check sample concentrate should contain this compound at 10 µg/mL in methanol.

REFERENCE Test Methods for Evaluating Solid Waste (SW-846), U.S. EPA Office of Solid Waste, Washington, DC, Method 8020A, Rev. 1, Nov. 1992.

Chlorobenzilate
CAS #510-15-6
EPA Method 8270

TITLE Semivolatile Organic Compounds by GC/MS

MATRIX This method is used to determine the concentration of semivolatile organic compounds in extracts prepared from all types of solid waste matrices, soils, and groundwater. Although surface waters are not specifically mentioned, this method should be applicable to water samples from rivers, lakes, etc.

METHOD SUMMARY This method covers 259 semivolatile organic compounds. In very limited applications direct injection of the sample into the GC/MS system may be appropriate, but this results in very high detection limits (approximately 10,000 µg/L). Typically, a 1-L liquid sample, containing surrogate, and matrix spiking standards, is extracted in a continuous extractor first under acid conditions and then under basic conditions. Typically 30 g of a solid sample, containing surrogate, and matrix spiking standards, is extracted ultrasonically. After concentrating the extract to 1 mL it is spiked with 10 µL of an internal standard solution just prior to analysis by GC/MS. The volume injected should contain about 100 ng of base/neutral and 200 ng of acid surrogates (for a 1-µL injection). Analysis is performed by GC/MS using a capillary GC column.

INTERFERENCES Raw GC/MS data from all blanks, samples, and spikes must be evaluated for interferences. Contamination by carryover can occur whenever high-concentration and low-concentration samples are sequentially analyzed. To reduce carryover, the sample syringe must be rinsed out between samples with solvent. Whenever an unusually concentrated sample is encountered, it should be followed by the analysis of blank solvent to check for cross-contamination.

INSTRUMENTATION A GC/MS and a data system are required. The GC column used is a 30 m × 0.25 mm I.D. (or 0.32 mm I.D.) 1um film thickness silicone-coated fused silica capillary column. A continuous liquid-liquid extractor equipped with Teflon® or glass connection joints and stopcocks requiring no lubrication, a K-D concentrating apparatus, water bath, and an ultrasonic disrupter with a minimum power of 300 W and with pulsing capability are also required.

PRECISION & ACCURACY The estimated quantitation limit (EQL) of Method 8270B for determining an individual compound is approximately 1 mg/kg (wet weight) for soil or sediment samples, 1–200 mg/kg for wastes (dependent on matrix and method of preparation), and 10 µg/L for groundwater samples. EQLs will be proportionately higher for sample extracts that require dilution to avoid saturation of the detector.

The EQL(b) for groundwater in µg/L is 10.
The EQL (a, b) for low concentrations in soil and sediment in µg/kg is not determined.
Accuracy as µg/L is not listed.
Overall precision in µg/L is not listed.

(a) EQLs listed for soil/sediment are based on wet weight. Normally data is reported in a dry-weight basis; therefore, EQLs will be higher based on the % dry weight of each sample. This calculation is based on a 30 g sample and gel permeation chromatography cleanup.

(b) Sample EQLs are highly matrix-dependent. The EQLs are provided for guidance and may not always be achievable.

C = True value for concentration, in µg/L.
X = Average recovery found for measurements of samples containing a concentration of C, in µg/L.

ESTIMATED QUANTITATION LIMIT FOR OTHER MATRICES

Other Matrices	Factor (a)
High-concentration soil and sludges by sonicator	7.5
Non-water miscible waste	75

(a) EQL for other matrices = [EQL for low soil/sediment] × [Factor]. This estimated EQL is similar to an EPA "Practical Quantitation Limit."

SAMPLING METHOD
Liquid samples — Use a 1 or 2½ gallon amber glass bottle with a screw-top Teflon®-lined cover that has been prewashed with detergent and rinsed with distilled water and methanol (or isopropanol).

Soils, sediments, or sludges — Use an 8-oz. widemouth glass with a screw-top Teflon®-lined cover that has been prewashed with detergent and rinsed with distilled water and methanol (or isopropanol).

SAMPLE PRESERVATION
Liquid samples — If residual chlorine is present, add 3 mL of 10% sodium thiosulfate per gallon, cool to 4°C and store in a solvent-free refrigerator until analysis; if chlorine is not present, then eliminate the sodium thiosulfate addition.

Soils, sediments, or sludges — Cool samples to 4°C and store in a solvent-free refrigerator.

MHT Liquid samples must be extracted within 7 days and the extracts analyzed within 40 days. Soils, sediments, or sludges may be stored for a maximum of 14 days and the extracts analyzed within 40 days.

SAMPLE PREPARATION
Liquid samples — Transfer 1 L quantitatively to a continuous extractor. If high concentrations are anticipated, a smaller volume may be used and then diluted with organic-free reagent water to 1 L. Adjust pH, if necessary, to pH <2 using 1:1 (V/V) sulfuric acid. Pipette 1.0 mL of a surrogate standard spiking solution into each sample. For the sample in each analytical batch selected for spiking, add 1.0 mL of a matrix spiking standard. For base/neutral acid analysis, the amount of the surrogates and matrix spiking compounds added to the sample should result in a final concentration of 100 ng/µL of each analyte in the extract to be analyzed (assuming a 1-µL injection). Extract with methylene chloride for 18–24 h. Next, adjust the pH of the aqueous phase to pH >11 using 10 N sodium hydroxide and extract it with methylene chloride again for 18–24 h. Dry the extract through a column containing anhydrous sodium sulfate and concentrate it to 1 mL using a K-D concentrator.

Soils, sediments, or sludges — Use 30 g of sample. Nonporous or wet samples (gummy or clay type) that do not have a free-flowing sandy texture must be mixed with anhydrous sodium sulfate until the sample is free flowing. Add 1 mL of surrogate standards to all samples, spikes, standards, and blanks. For the sample in each analytical batch selected for spiking, add 1.0 mL of a matrix spiking standard. For base/neutral acid analysis, the amount added of the surrogates and matrix spiking compounds should result in a final concentration of 100 ng/µL of each base/neutral analyte and 200 ng/µL of each acid analyte in the extract to be analyzed (assuming a 1-µL injection). Immediately add a 100-mL mixture of 1:1 methylene chloride:acetone and extract the sample ultrasonically for 3 min and then decant or filter the extracts. Repeat the extraction two or more times. Dry the extract using a column with anhydrous sodium sulfate and concentrate it to 1 mL in a K-D concentrator.

QUALITY CONTROL A methylene chloride solution containing 50 ng/µL of decafluorotriphenylphosphine (DFTPP) is used for tuning the GC/MS system each 12-h shift. A system performance check also must be made during every 12-h shift. A standard containing 50 ng/µL each of 4,4′-DDT, pentachlorophenol, and benzidine is required to verify injection port inertness and GC column performance. A calibration standard at mid-concentration, containing each compound of interest, including all required surrogates, must be performed every 12 h during analysis. After the system performance check is met, calibration check compounds (CCCs) are used to check the validity of the initial calibration.

The internal standard responses and retention times in the calibration check standard must be evaluated immediately after or during data acquisition. If the retention time for any internal standard changes by more than 30 seconds from the last check calibration (12 h), the chromatographic system must be inspected for malfunctions and corrections must be made, as required. If the electron ionization current plot (EICP) area for any of the internal standards changes by a factor of two from the last daily calibration standard check, the mass spectrometer must be inspected for malfunctions and corrections must be made, as appropriate.

Demonstrate, through the analysis of a reagent water blank, that interferences from the analytical system, glassware, and reagents are under control. The blank samples should be carried through all stages of the sample preparation and measurement steps. For each analytical batch (up to 20 samples), a reagent blank, matrix spike, and matrix spike duplicate/duplicate must be analyzed (the frequency of the spikes may be different for different monitoring programs). The blank and spiked samples must be carried through all stages of the sample preparation and measurement steps. A QC reference sample concentrate containing each analyte at a concentration of 100 mg/L in methanol is required.

REFERENCE Test Methods for Evaluating Solid Waste (SW-846). U.S. EPA 1983, Method 8270B, Rev. 2, Nov. 1990. Office of Solid Waste, Washington, DC.

Chlorodibromomethane EPA Method 1624
CAS #124–48–1

TITLE Volatile Organic Compounds by Isotope Dilution GC/MS

MATRIX Compounds may be determined in waters, soils, and municipal sludges by this method.

METHOD SUMMARY This method is used to determine 58 volatile toxic organic pollutants associated with the CWA (as amended 1987); the RCRA (as amended 1986); the CERCLA (as amended 1986); and other compounds amenable to purge-and-trap gas chromatography-mass spectrometry (GC/MS).

If the solids content is less than 1%, stable isotopically-labeled analogs of the compounds of interest are added to a 5-mL sample and the sample is purged with an inert gas at 20–25°C in a chamber designed for soil or water samples. If the solids content is greater than 1%, 5 mL of reagent water and the labeled compounds are added to a 5-g aliquot of sample and the mixture is purged at 40°C. Compounds that will not purge at 20–25°C or at 40°C are purged at 78–85°C. In the purging process, the volatile compounds are transferred from the aqueous phase into the gaseous phase where they are passed into a sorbent column, and trapped. After purging is completed, the trap is backflushed and heated rapidly to desorb the compounds into a GC. The compounds are separated by the GC and detected by a MS. The labeled compounds serve to correct the variability of the analytical technique.

INTERFERENCES Impurities in the purge gas, organic compounds outgassing from the plumbing upstream of the trap, and solvent vapors in the lab account for most problems. Samples can be contaminated by diffusion of volatile organic compounds (particularly methylene chloride) through the bottle seal during shipment and storage. Contamination by carryover can occur when high-level and low-level samples are analyzed sequentially. When an unusually concentrated sample is encountered, follow it by analysis of a reagent water blank to check for carryover.

INSTRUMENTATION Major equipment includes a GC with linear temperature programming and a glass jet separator as the MS interface, a MS with 70 eV electron impact ionization, and a data system to collect and record response factors.

Column: 2.8 m × 2 mm I.D. glass, packed with 1% SP-1000 on Carbopak B, 60/80 mesh, or equivalent.

PRECISION & ACCURACY The detection limits of the method are usually dependent on the level of interferences rather than instrumental limitations. The method detection limits were determined in digested sludge (low solids) and in filter cake or compost (high solids).

The MDL (in µg/kg) for low solids is 15 and for high solids is 2.
Labeled and native compound precision (in µg/L) as standard deviation was not listed.
Labeled and native compound accuracy (in µg/L) as average recovery was not listed.

Acceptance criteria are at 20 µg/L for this compound.

SAMPLE COLLECTION, PRESERVATION & HANDLING Grab samples are collected in glass containers having a total volume greater than 20 mL. Fill and seal each bottle so that no air bubbles are entrapped. Samples are maintained at 0 to 4°C from the time of collection until analysis. If an aqueous sample contains residual chlorine, add sodium thiosulfate preservative (10 mg/40 mL) to the empty sample bottles just prior to shipment to the sample site. All samples must be analyzed within 14 days of collection.

SAMPLE PREPARATION Samples containing less than 1% solids are analyzed directly as aqueous samples. Samples containing 1% solids or greater are analyzed as solid samples utilizing one of two methods, depending on the levels of pollutants, in the sample. Samples containing 1% solids or greater, and low to moderate levels of pollutants are analyzed by purging a known weight of sample added to 5 mL of reagent water. Samples containing 1% solids or greater, and high levels of pollutants, are extracted with methanol, and an aliquot of the methanol extract is added to reagent water and purged.

QUALITY CONTROL A field blank prepared from reagent water and carried through the sampling and handling protocol may serve as a check on contamination from shipment and storage.

The analyst is permitted to modify this method to improve separations or lower the costs of measurements, provided all performance specifications are met. Analyses of blanks are required. When results of spikes indicate atypical method performance for samples, the samples are diluted to bring method performance within acceptable limits. Analyze two sets of four 5-mL aliquots (8 aliquots total) of the aqueous performance standard. Spike all samples with labeled compounds to assess method performance on the sample matrix. Compute the percent recovery of the labeled compounds using the internal standard method. Compare the percent recovery for each compound with the corresponding labeled compound recovery. Reagent water blanks are analyzed to demonstrate freedom from carryover contamination. Field replicates may be collected to determine the precision of the sampling technique, and spiked samples may be required to determine the accuracy of the analysis when the internal method is used.

REFERENCE Volatile Organic Compounds by Isotope Dilution GC/MS. Office of Water Regulation and Standards, U.S. EPA Industrial Technology Division, Washington, DC, EPA Method 1624, Rev. C, June 1989 (contact W.A. Telliard, U.S. EPA, Office of Water Regulations and Standards, 401 M St., SW, Washington, DC, 20460. Phone: 202-382-7131).

Chlorodibromomethane   **EPA Method 8021**
CAS #124–48-1

TITLE Halogenated Volatile by Gas Chromatography Using Photoionization and Electrolytic Conductivity Detectors in Series: Capillary Column Technique

MATRIX This method is applicable to nearly all types of samples, regardless of water content, including groundwater, aqueous sludges, caustic liquors, acid liquors, waste solvents, oily wastes, mousses, tars, fibrous wastes, polymeric emulsions, filter cakes, spent carbons, spent catalysts, soils, and sediments.

METHOD SUMMARY This method is used to determine 60 volatile organic compounds in a variety of solid waste matrices. It provides GC conditions for the detection of halogenated and aromatic volatile organic compounds. Samples can be analyzed using direct injection or purge-and-trap (EPA Method 5030). Groundwater samples must be analyzed using EPA Method 5030 (where applicable). A temperature program is used with the GC. Detection is achieved by a photoionization detector (PID) and a Hall electrolytic conductivity detector (HECD) in series.

INTERFERENCES Samples can be contaminated by diffusion of volatile organics (particularly chlorofluorocarbons and methylene chloride) through the sample container septum during shipment and storage.

INSTRUMENTATION A GC-equipped with variable-constant differential flow controllers, subambient oven controller, PID and HECD detectors connected with a short piece of uncoated capillary tubing and a data system.

Column: 60 m × 0.75 mm I.D. VOCOL wide-bore capillary column with 1.5 µm film thickness.

PRECISION & ACCURACY MDLs are compound-dependent and vary with purging efficiency and concentration. The applicable concentration range of this method is compound- and instrument-dependent but is approximately 0.1 to 200 µg/L. Analytes that are inefficiently purged from water will not be detected when present at low concentrations, but they can be measured with acceptable accuracy and precision when present in sufficient amounts. The estimated quantitation limit (EQL) for an individual compound is approximately 1 µg/kg (wet weight) for soil/sediment samples, 100 µg/kg (wet weight) for wastes, and 1 µg/L for groundwater. EQLs will be proportionately higher for sample extracts and samples that require dilution or reduced sample size to avoid saturation of the detector.

MULTIPLICATION FACTORS FOR OTHER MATRICES (a)

Matrix	Factor (b)
Groundwater	10
Low-concentration soil	10
Water miscible liquid waste	500
High-concentration soil and sludge	1250
Non-water miscible waste	1250

(a) Sample EQLs are highly matrix-dependent. The EQLs listed herein are provided for guidance and may not always be achievable. (b) EQL = [Method detection limit] × [Factor]. For non-aqueous samples, the factor is on a wet-weight basis.

SINGLE LABORATORY ACCURACY & PRECISION DATA FOR VOCs IN WATER

This method was tested in a single lab using water spiked at 10 µg/L and the following data was reported:

Recoveries and standard deviations were determined from seven samples and spiked at 10 µg/L of each analyte. Recoveries were determined by the internal standard method. Internal standards were: Fluorobenzene for PID and 2-Bromo-1-chloropropane for HECD.

The average recovery (in percent) for the PID was none (no response for this detector).

The standard deviation of the recovery for the PID was none (no response for this detector).

The MDL (in µg/mL) for the PID was none (no response for this detector).

The average recovery (in percent) for the HECD was none (no response for this detector).

The standard deviation of the recovery for the HECD was none (no response for this detector).

The MDL (in µg/mL) for the HECD was none (no response for this detector).

SAMPLE COLLECTION, PRESERVATION & HANDLING
Volatile Organics — Standard 40-mL glass screw-cap VOA vials with Teflon®-faced silicone septum may be used for both liquid and solid matrices. When collecting samples, liquids and solids should be introduced into the vials gently to reduce agitation which might drive off volatile compounds. If there are any air bubbles present the sample must be retaken. Tap slightly as they are filled to try and eliminate as much free air space as possible. The two vials from each sampling locations should be sealed in separate plastic bags to prevent cross-contamination between samples particularly if the sampled waste is suspected of containing high levels of volatile organics.

Semivolatile organics — Containers used to collect samples for the determination of semivolatile organic compounds should be soap and water washed followed by methanol (or isopropanol) rinsing. The sample containers should be of glass or Teflon® and have screw-top covers with Teflon® liners.

Preservation for volatile organics — No preservation is used with concentrated waste samples. With liquid samples containing no residual chlorine, 4 drops of concentrated hydrochloric acid are added and the samples are immediately cooled to 4°C. When liquid samples contain residual chlorine, they are treated as above and, in addition, 4 drops of 4% aqueous sodium thiosulfate are added. Soil, sediment, and sludge samples are only cooled to 4°C.

Preservation for semivolatile organics — No preservation is used with concentrated waste samples. With liquid samples containing no residual chlorine and with soil, sediment, and sludge samples, immediately cooling to 4°C is the only preservation used. When residual chlorine is present then 3 mL of 10% aqueous sodium sulfate is added for each gallon of sample collected, followed by cooling to 4°C.

MHT The holding time for all volatile organics samples is 14 days. Liquid samples must be extracted within 7 days and their extracts analyzed within 40 days. Concentrated waste, soil, sediment, and sludge samples must be extracted within 14 days and their extracts analyzed within 40 days.

SAMPLE PREPARATION Volatile compounds are introduced into the gas chromatograph either by direct injector or purge-and-trap (EPA Method 5030). EPA Method 5030 may be used directly on groundwater samples or low-concentration contaminated soils and sediments. For medium-concentration soils or sediments, methanolic extraction, as described in EPA Method 5030, may be necessary prior to purge-and-trap analysis.

QUALITY CONTROL Calculate surrogate standard recovery on all samples, blanks, and spikes. A trip blank is recommended to check on sampling, storage, and handling contamination. Calibration standards, at a minimum of five concentration levels, are prepared in organic-free reagent water. One of the concentration levels should be at a concentration near, but above, the method detection limit.

A combination of bromochloromethane, 2-bromo-1-chloropropane, 1,4-dichlorobutane, and bromochlorobenzene are recommended as surrogate standards to encompass the range of the temperature program used in this method.

REFERENCE Test Methods for Evaluating Solid Waste, Physical/Chemical Methods, SW-846, 3rd Edition, U.S. EPA, Office of Solid Waste, Washington, DC, EPA Method 8021A, Rev. 1, Nov. 1992.

Chlorodibromomethane   **EPA Method 8240**
CAS #124–48–1

TITLE Volatile Organics By GC/MS: Packed Column Technique

MATRIX Nearly all types of sample matarices, regardless of water content, can be analyzed using this method. This includes groundwater, aqueous sludges, caustic liquors, acid liquors, waste solvents, oily wastes, mousses, tars, fibrous wastes, polymetric emulsions, filter cakes, spent carbons, spent catalysts, soils, and sediments.

METHOD SUMMARY Method 8240B covers 80 volatile organic compounds that are introduced into a gas chromatograph by the purge-and-trap method or by direct injection (in limited applications). For the purge-and-trap method an inert gas (zero grade nitrogen or helium) is bubbled through a 5-mL solution at ambient temperature. Purged sample components are trapped in a tube of sorbent materials. When purging is complete, the sorbent tube is heated and backflushed with inert gas to desorb the trapped components onto a GC column.

INTERFERENCES Impurities in the purge gas and from organic compounds outgassing from the plumbing ahead of the trap account for many contamination problems. Interferences purged or coextracted from the samples will vary considerably from source to source. Cross-contamination can occur whenever high-level and low-level samples are analyzed sequentially. Whenever an unusually concentrated sample is analyzed, it should be followed by the analysis of organic-free reagent water to check for cross-contamination. Samples also can be contaminated by diffusion of volatile organics (particularly methylene chloride and fluorocarbons) through the septum seal into the sample during shipment and storage. A trip blank can serve as a check on such contamination. The lab where volatile analysis is performed and also the refrigerated storage area should be completely free of solvents.

INSTRUMENTATION A gas chromatograph/mass spectrometry/data system (GC/MS) equipped with a 6 ft × 0.1 in I.D. glass column packed with 1% SP-1000 on Carbopack-B (60/80 mesh) is required. Also needed is a 5-mL purging device, a sorbent trap, and a thermal desorption apparatus.

PRECISION & ACCURACY This method is reported to have been tested by 15 laboratories using organic-free reagent water, drinking water, surface water, and industrial wastewaters (not specified) fortified at six concentrations over the range 5–600 µg/L.

Sample estimated quantitation limits (EQLs) are highly matrix-dependent. The EQLs listed may not always be achievable. EQLs listed for soils or sediments are based on wet weight. Normally, data is reported on a dry-weight basis; therefore, EQLs will be higher, based on the percent dry weight of each sample. Note that EQLs are even more variable than MDLs and that they are highly variable depending on the matrix being analyzed.

EQL in groundwater in µg/L was 5.
EQL in low soil or sediment in µg/kg was 5.
Accuracy (a) in µg/L was not listed.
Precision (b) in µg/L was not listed.

(a) Average recovery found for measurements of samples containing a concentration of C, in µg/L.
(b) Overall precision found for measurements of samples with average recovery X for samples containing a concentration of C in µg/L.
X = Average recovery found for measurement of samples containing a concentration of C in µg/L.

MULTIPLICATION FACTORS FOR OTHER MATRICES

Other Matrices	Factor (a)
Waste miscible liquid waste	50
High-concentration soil and sludge	125
Non-water miscible waste	500

(a) EQL = [EQL for low soil/sediment] × [Factor]. For non-aqueous samples, the factor is on a wet-weight basis.

SAMPLING METHOD
Liquid samples — Use a 40-mL glass screw-cap VOA vial with a Teflon®-faced silicone septum that has been prewashed, rinsed with distilled deionized water, and oven dried. However, if residual chlorine is present, collect sample in a 40-oz. soil VOA container which has been pre-preserved with 4 drops of 10% sodium thiosulfate, mix gently, and then transfer the sample to a 40-mL VOA vial. Collect bubble-free samples in duplicate and seal them in separate plastic bags.

Soils or sediments, and sludges — Use an 8-oz. widemouth glass bottle with a Teflon®-faced silicone septum that has been prewashed with detergent, rinsed with distilled deionized water, and oven dried. Tap slightly to eliminate free air space. Collect samples in duplicate and seal them in separate plastic bags.

SAMPLE PRESERVATION
Liquid samples — Add 4 drops of concentrated HCL and immediately cool samples to 4°C and store in a solvent-free refrigerator.

Soils or sediments, and sludges — Cool samples to 4°C and store in a solvent-free refrigerator.

MHT Maximum holding time is 14 days from the date of sample collection.

SAMPLE PREPARATION
Liquid samples — Remove the plunger from a 5-mL syringe and carefully pour the sample into the syringe barrel to just short of overflowing. Replace the syringe plunger and compress the sample. Open the syringe valve and vent any residual air while adjusting the sample volume to 5.0 mL. If there is only one volatile organic analysis (VOA) vial, a second syringe should be filled at this time to protect against possible loss of sample integrity. Add 10 µL of surrogate spiking solution and 10 µL of internal standard spiking solution through the valve bore of the 5-mL syringe, then close the valve. The surrogate and internal standards may be mixed and added as a single spiking solution.

Sediments, soils, and waste samples — All samples of this type should be screened by GC analysis using a headspace method (EPA Method 3810) or the hexadecane extraction and screening method (EPA Method 3820). Use the screening data to determine whether to use the low-concentration method (0.005–1 mg/kg) or the high-concentration method (>1 mg/kg).

Low-concentration method — The low-concentration method is based on purging a heated sediment or soil sample mixed with organic-free reagent water containing the surrogate and internal standards. Analyze all reagent blanks and standards under the same conditions as the samples.

Use a 5-g sample if the expected concentration is <0.1 mg/kg or a 1-g sample for expected concentrations between 0.1 and 1 mg/kg. Mix the contents of the sample container with a narrow metal spatula. Weigh the amount of the sample into a tared purge device. Add the spiked water to the purge device, which contains the weighed amount of sample, and connect the device to the purge-and-trap system.

High-concentration method — This method is based on extracting the sediment or soil with methanol. A waste sample is either extracted or diluted, depending on its solubility in methanol. Wastes that are insoluble in methanol are diluted with reagent tetraglyme or possibly polyethylene glycol (PEG). An aliquot of the extract is added to organic-free reagent water containing surrogate and internal standards. This is purged at ambient temperature. All samples with an expected concentration of >1.0 mg/kg should be analyzed by this method.

Mix the contents of the sample container with a narrow metal spatula. For sediments or soils and solid wastes that are insoluble in methanol, weigh 4 g (wet weight) of sample into a tared 20-mL vial. For waste that is soluble in methanol, tetraglyme, or PEG, weigh 1 g (wet weight) into a tared scintillation vial or culture tube or a 10-mL volumetric flask. Quickly add 9.0 mL of appropriate solvent then add 1.0 mL of a surrogate spiking solution to the vial, cap it, and shake it for 2 min.

METHANOL EXTRACT REQUIRED FOR ANALYSIS OF HIGH-CONCENTRATION SOILS OR SEDIMENTS

Approximate Concentration Range	Volume of Methanol Extract (a)
500–10,000 µg/kg	100 µL
1,000–20,000 µg/kg	50 µL
5,000–100,000 µg/kg	10 µL
25,000–500,000 µg/kg	100 µL of 1/50 dilution (b)

Calculate appropriate dilution factor for concentrations exceeding this table.

(a) The volume of methanol added to 5 mL of water being purged should be kept constant. Therefore, add to the 5-mL syringe whatever volume of methanol is necessary to maintain a volume of 100 µL added to the syringe.

(b) Dilute an aliquot of the methanol extract and then take 100 µL for analysis.

QUALITY CONTROL Demonstrate, through the analysis of a reagent water blank, that interferences from the analytical system, glassware, and reagents are under control. Blank samples should be carried through all stages of the sample preparation and measurement steps. For each analytical batch (up to 20 samples), a reagent blank, matrix spike, and matrix spike duplicate must be analyzed (the frequency of the spikes may be different for different monitoring programs). The blank and spiked samples must be carried through all stages of the sample preparation and measurement steps. QC samples mentioned in the section on Interferences will also be needed as appropriate to those situations.

REFERENCE Test Methods for Evaluating Solid Waste (SW-846). U.S. EPA. 1983. Method 8240B, Rev. 2, Nov. 1990. Office of Solid Wastes, Washington, DC.

Chloroethane
CAS #75-00-3
EPA Method 1624

TITLE Volatile Organic Compounds by Isotope Dilution GC/MS

MATRIX Compounds may be determined in waters, soils, and municipal sludges by this method.

METHOD SUMMARY This method is used to determine 58 volatile toxic organic pollutants associated with the CWA (as amended 1987); the RCRA (as amended 1986); the CERCLA (as amended 1986); and other compounds amenable to purge-and-trap gas chromatography-mass spectrometry (GC/MS).

If the solids content is less than 1%, stable isotopically-labeled analogs of the compounds of interest are added to a 5-mL sample and the sample is purged with an inert gas at 20–25°C in a chamber designed for soil or water samples. If the solids content is greater than 1%, 5 mL of reagent water and the labeled compounds are added to a 5-g aliquot of sample and the mixture is purged at 40°C. Compounds that will not purge at 20–25°C or at 40°C are purged at 78–85°C. In the purging process, the volatile compounds are transferred from the aqueous phase into the gaseous phase where they are passed into a sorbent column, and trapped. After purging is completed, the trap is backflushed and heated rapidly to desorb the compounds into a GC. The compounds are separated by the GC and detected by a MS. The labeled compounds serve to correct the variability of the analytical technique.

INTERFERENCES Impurities in the purge gas, organic compounds outgassing from the plumbing upstream of the trap, and solvent vapors in the lab account for most problems. Samples can be contaminated by diffusion of volatile organic compounds (particularly methylene chloride) through the bottle seal during shipment and storage. Contamination by carryover can occur when high-level and low-level samples are analyzed sequentially. When an unusually concentrated sample is encountered, follow it by analysis of a reagent water blank to check for carryover.

INSTRUMENTATION Major equipment includes a GC with linear temperature programming and a glass jet separator as the MS interface, a MS with 70 eV electron impact ionization, and a data system to collect and record response factors.

Column: 2.8 m × 2 mm I.D. glass, packed with 1% SP-1000 on Carbopak B, 60/80 mesh, or equivalent.

PRECISION & ACCURACY The detection limits of the method are usually dependent on the level of interferences rather than instrumental limitations. The method detection limits were determined in digested sludge (low solids) and in filter cake or compost (high solids).

The MDL (in µg/kg) for low solids is 789 and for high solids is 24.
Background levels of this compound were present in the sludge with low solids, resulting in a higher than expected MDL.
Labeled and native compound precision (in µg/L) as standard deviation was 15.0.
Labeled and native compound accuracy (in µg/L) as average recovery was detected to 47.

Acceptance criteria are at 20 µg/L for this compound.

SAMPLE COLLECTION, PRESERVATION & HANDLING Grab samples are collected in glass containers having a total volume greater than 20 mL. Fill and seal each bottle so that no air bubbles are entrapped. Samples are maintained at 0 to 4°C from the time of collection until analysis. If an aqueous sample contains residual chlorine, add sodium thiosulfate preservative (10 mg/40 mL) to the empty sample bottles just prior to shipment to the sample site. All samples must be analyzed within 14 days of collection.

SAMPLE PREPARATION Samples containing less than 1% solids are analyzed directly as aqueous samples. Samples containing 1% solids or greater are analyzed as solid samples utilizing one of two methods, depending on the levels of pollutants, in the sample. Samples containing 1% solids or greater, and low to moderate levels of pollutants are analyzed by purging a known weight of sample added to 5 mL of reagent water. Samples containing 1% solids or greater, and high levels of pollutants, are extracted with methanol, and an aliquot of the methanol extract is added to reagent water and purged.

QUALITY CONTROL A field blank prepared from reagent water and carried through the sampling and handling protocol may serve as a check on contamination from shipment and storage.

The analyst is permitted to modify this method to improve separations or lower the costs of measurements, provided all performance specifications are met. Analyses of blanks are required. When results of spikes indicate atypical method performance for samples, the samples are diluted to bring method performance within acceptable limits. Analyze two sets of four 5-mL aliquots (8 aliquots total) of the aqueous performance standard. Spike all samples with labeled compounds to assess method performance on the sample matrix. Compute the percent recovery of the labeled compounds using the internal standard method. Compare the percent recovery for each compound with the corresponding labeled compound recovery. Reagent water blanks are analyzed to demonstrate freedom from carryover contamination. Field replicates may be collected to determine the precision of the sampling technique, and spiked samples may be required to determine the accuracy of the analysis when the internal method is used.

REFERENCE Volatile Organic Compounds by Isotope Dilution GC/MS. Office of Water Regulation and Standards, U.S. EPA Industrial Technology Division, Washington, DC, EPA Method 1624, Rev. C, June 1989 (contact W.A. Telliard, U.S. EPA, Office of Water Regulations and Standards, 401 M St., SW, Washington, DC, 20460. Phone: 202-382-7131).

Chloroethane   EPA Method 502
CAS #75-00-3

TITLE Volatile Organic Compounds in Water By Purge and Trap Capillary Column Gas Chromatography with Photoionization and Electrolytic Conductivity Detectors in Series. U.S. EPA Method 502.2, Rev. 2.0, 1989.

MATRIX Drinking water and raw source water. The latter should include most surface water and groundwater sources.

METHOD SUMMARY This method covers 60 volatile organic compounds that contain halogen atoms and/or that are aromatic. An inert gas (zero grade nitrogen or helium) is bubbled through a 25-mL or a 5-mL water sample (depending on the expected concentration of the analytes). Purged sample components are trapped in a tube of sorbent materials. When purging is complete, the sorbent tube is heated and backflushed with helium to desorb the trapped sample onto a capillary GC column. The column is temperature programmed to separate the method analytes which are then detected with a photoionization detector (PID) and a Hall electrolytic conductivity (HECD) placed in series. The PID is selective for aromatic compounds and the HECD is selective for halogenated compounds.

INTERFERENCES Impurities in the purge gas and from organic compounds outgassing from the plumbing ahead of the trap account for many contamination problems. Interferences purged or coextracted from the samples will vary considerably from source to source, depending upon the particular sample or extract being tested. Cross-contamination can occur whenever high-level and low-level samples are analyzed sequentially. Samples also can be contaminated by diffusion of volatile organics (particularly methylene chloride and fluorocarbons) through the septum seal into the sample during shipment and storage. The lab where volatile analysis is performed and also the refrigerated storage area should be completely free of solvents.

INSTRUMENTATION A GC containing a series configuration of a high temperature photoionization detector (PID) equipped with 10.0 eV (nominal) lamp and Hall electrolytic conductivity detector (HECD) is required. Also required is an all-glass 5-mL purging device, a sorbent trap, and a thermal desorption apparatus which is connected to the GC system.

Column 1: VOCOL glass wide-bore capillary column.
Column 2: RTX–502.2 mega-bore capillary column.
Column 3: DB-62 mega-bore capillary column.

PRECISION & ACCURACY Method detection limits are dependent upon the characteristics of the gas chromatographic system used. Analytes that are not separated chromatographically cannot be individually identified and used in the same calibration mixture or water samples unless an alternative technique for identification and quantification, such as mass spectrometry, is used.

Electrolytic conductivity detetor (c) range in µg/L (a) was 0.02–200.
Electrolytic conductivity detetor (c) MDL in µg/L (b) was 0.1.
Electrolytic conductivity detetor (c) accuracy as % recovery was 96.
Electrolytic conductivity detetor (c) precision as % RSD was 3.9.
Photoionization detector (d) range in µg/L (a) was 0.02–200.
Photoionization detector (d) MDL in µg/L (b) was not listed.
Photoionization detector (d) accuracy as % recovery was not listed.
Photoionization detector (d) precision as % RSD was not listed.

(a) *The applicable concentration range of this method is compound, instrument, and matrix-dependent. It is listed as being approximately 0.02 to 200 µg/L but no specific information is provided so caution should be observed.*

(b) The method detection limits reports with this method are compound, instrument, and matrix-dependent. The values reported were calculated using reagent water fortified with the corresponding compounds at 10 µg/L and a GC-equipped with a 60 m × 0.75 mm VOLCOL wide bore capillary column with 1.5 µm film thickness and using helium carrier gas.

(c) Recoveries and relative standard deviations were determined from seven samples of reagent water fortified with 10 µg/L of each compound. 2-Bromo-1-chloropropane was used as the internal standard for calculating average recoveries.

(d) Recoveries and relative standard deviations were determined from seven samples of reagent water fortified with 10 µg/L of each compound. Fluorobenzene was used as the internal standard for calculating average recoveries.

SAMPLING METHOD Collect samples using a 40- to 120-mL screw-cap vial (prewashed with detergent, rinsed with distilled water and oven dried at 105°C) with a Teflon®-faced silicone septum. Collect bubble-free samples and place the septum with the Teflon® side down on the water.

SAMPLE PRESERVATION If residual chlorine is present in the water add about 25 mg of ascorbic acid to each vial before samples are collected to remove the chlorine. Add hydrochloric acid to reduce pH to <2, immediately cool samples to 4°C, and store them in a solvent-free refrigerator at 4°C until analysis.

MHT The maximum holding time for samples is 14 days from the time they were collected.

SAMPLE PREPARATION Remove the plungers from two 5-mL syringes and attach a closed syringe valve to each. Warm the sample to room temperature, open the sample bottle, and carefully pour the sample into one of the syringe barrels to just short of overflowing. Replace the syringe plunger, invert the syringe, and compress the sample. Open the syringe valve and vent any residual air while adjusting the sample volume to 5.0 mL. Add 10 µL of the internal calibration standard to the sample through the syringe valve. Close the valve. Fill the second syringe in an identical manner from the same sample bottle. Reserve this second syringe for a reanalysis if necessary.

QUALITY CONTROL As an initial demonstration of lab accuracy and precision, analyze 4 to 7 replicates of a lab fortified blank containing analyte at 0.1–5 µg/L. Collect all samples in duplicate. Surrogate analytes (similar to those of the analytes of interest), whose concentration is known in every sample, are measured using the same internal standard calibration procedure. Duplicate field reagent water blanks (trip blanks) must be analyzed with each set of samples, lab reagent blanks (method blanks) must be analyzed with each batch of samples processed as a group within a work shift. Also, a single lab-fortified blank that contains each of the analytes of interest should be analyzed with each batch of samples processed as a group within a work shift. A 3- to 5-point calibration curve is needed depending on the calibration range factor required.

EPA CONTACT & HOTLINE For technical questions contact Dr. Baldev Bathija, U.S. EPA, Office of Ground Water and Drinking Water (WH-550D), 401 M St. SW, Washington, DC 20460. Tel. (202) 260-3040. For further information the EPA Safe Drinking Water Hotline may be called at: (800) 426-4791.

REFERENCE Methods for the Determination of Organic Compounds in Drinking Water, EPA/600/4-88/039 (revised July 1991; Final Rule for determination of compliance with the MCL for Total Trihalomethanes under 141.30, in 40 CFR Part 141, Vol. 58, No. 147, Fed. Reg., Tuesday Aug. 3, 1993). U.S. EPA Environmental Monitoring Systems Laboratory, Cincinnati, OH, 45268, USA. Available from the National Technical Information Service (NTIS), 5285 Port Royal Road, Springfield, VA 22161; Tel. 800-553-6847. NTIS Order Number is PB91-231480.

Chloroethane   **EPA Method 524**
CAS #75-00-3

TITLE Measurement of Purgeable Organic Compounds in Water by Capillary Column GC/MS.

MATRIX Drinking water and raw source water; the latter should include most surface water and groundwater sources.

METHOD SUMMARY Method 524.2 covers 60 volatile organic compounds. An inert gas (zero grade nitrogen or helium) is bubbled through a 25-mL or a 5-mL water sample (depending on the expected concentration of the analytes). Purged sample components are trapped in a tube of sorbent materials. When purging is complete, the sorbent tube is heated and backflushed with helium to desorb the trapped sample onto a capillary GC column.

INTERFERENCES Impurities in the purge gas and from organic compounds outgassing from the plumbing ahead of the trap account for many contamination problems. Interferences purged or coextracted from the samples will vary considerably from source to source, depending upon the particular sample or extract being tested. Cross-contamination can occur whenever high-level and low-level samples are analyzed sequentially. Samples also can be contaminated by diffusion of volatile organics (particularly methylene chloride and fluorocarbons) through the septum seal into the sample during shipment and storage.

INSTRUMENTATION A GC/MS with a data system equipped with one of the following capillary GC columns:

Column 1: VOCOL glass wide bore capillary column.
Column 2: DB-624 fused silica capillary column.
Column 3: DB-5 fused silica capillary column.

Also required is an all-glass 25 mL or 5-mL purging device, a sorbent trap, and a thermal desorption apparatus which is connected to the GC/MS system.

PRECISION & ACCURACY Method detection limits are compound- and instrument-dependent, and may vary from approximately 0.02–0.35 µg/L. Note in the table below that the "true" concentration range used for accuracy and precision measurements was quite narrow. However, the applicable concentration range of this method is primarily column dependent and is approximately 0.02 to 200 µg/L for the wide-bore thick-film columns. Narrow-bore thin-film columns may have a capacity which limits the range to about 0.02 to 20 µg/L. Analytes that are inefficiently purged from water will not be detected when present at low concentrations, but they can be measured with acceptable accuracy and precision when present in sufficient amounts.

Analytes that are not separated chromatographically, but which have different mass spectra and non-interfering quantification ions, can be identified and measured in the same calibration mixture or water sample. Analytes which have very similar mass spectra cannot be individually identified and measured in the same calibration mixture or water samples unless they have different retention times. Co-eluting compounds with very similar mass spectra, typically many structural isomers, must be reported as an isomeric group or pair.

The range (in µg/L) was 0.5–10.
The Method Detection Limig (in µg/L) was 0.10.
The accuracy (as % recovery) was 89.
The precision (in %) was 9.0.

Note: Data were obtained from 16–31 determinations using a wide-bore capillary column and a jet separator interfaced to a quadrupole mass spectrometer. All analytes were in a reagent water matrix.

SAMPLING METHOD Collect samples using a 40- to 120-mL screw-cap vial (prewashed with detergent, rinsed with distilled water and oven dried at 105°C) with a Teflon®-faced silicone septum. Collect bubble-free samples and place the septum with the Teflon® side down on the water.

SAMPLE PRESERVATION If residual chlorine is present in the water add about 25 mg of ascorbic acid to each vial before samples are collected to remove the chlorine. Add hydrochloric acid to reduce pH to <2, and immediately cool samples to 4°C, and store them in a solvent-free refrigerator at 4°C until analysis.

MHT The maximum holding time for samples is 14 days from the time they were collected.

SAMPLE PREPARATION Remove the plungers from two 25-mL (or 5-mL depending on sample size) syringes and attach a closed syringe valve to each. Warm the sample to room temperature, open the sample bottle, and carefully pour the sample into one of the syringe barrels to just short of overflowing. Replace the syringe plunger, invert the syringe, and compress the sample. Open the syringe valve and vent any residual air while adjusting the sample volume to 25.0 mL (or 5 mL). For samples and blanks, add 5 µL of the fortification solution containing the internal standard and the surrogates to the sample through the syringe valve. For calibration standards and lab fortified blanks, add 5 µL of the fortification solution containing the internal standard only. Close the valve. Fill the second syringe in an identical manner from the same sample bottle. Reserve this second syringe for a reanalysis if necessary.

QUALITY CONTROL As an initial demonstration of lab accuracy and precision, analyze 4 to 7 replicates of a lab fortified blank containing analyte at 0.2–5 µg/L. Collect all samples in duplicate. Surrogate analytes (similar to those of the analytes of interest), whose concentration is known in every sample, are measured using the same internal standard calibration procedure. Duplicate field reagent water blanks (trip blanks) must be analyzed with each set of samples, lab reagent blanks (method blanks) must be analyzed with each batch of samples processed as a group within a work shift. Also, a single lab-fortified blank that contains each of the analytes of interest should be analyzed with each batch of samples processed as a group within a work shift. A 3- to 5-point calibration curve is needed depending on the calibration range factor required.

EPA CONTACT & HOTLINE For technical questions contact Dr. Baldev Bathija, U.S. EPA, Office of Ground Water and Drinking Water (WH-550D), 401 M St. SW, Washington, DC 20460. Tel. (202) 260-3040. For further information the EPA Safe Drinking Water Hotline may be called at: (800) 426-4791.

REFERENCE Methods for the Determination of Organic Compounds in Drinking Water, EPA/600/4-88/039 (revised July 1991; Final Rule for determination of compliance with the MCL for Total Trihalomethanes under 141.30, in 40 CFR Part 141, Vol. 58, No. 147, Fed. Reg., Tuesday Aug. 3, 1993). U.S. EPA Environmental Monitoring Systems Laboratory, Cincinnati, OH, 45268, USA. Available from the National Technical Information Service (NTIS), 5285 Port Royal Road, Springfield, VA 22161; Tel. 800-553-6847. NTIS Order Number is PB91-231480.

Chloroethane   EPA Method 8021
CAS #75-00-3

TITLE Halogenated Volatile by Gas Chromatography Using Photoionization and Electrolytic Conductivity Detectors in Series: Capillary Column Technique

MATRIX This method is applicable to nearly all types of samples, regardless of water content, including groundwater, aqueous sludges, caustic liquors, acid liquors, waste solvents, oily wastes, mousses, tars, fibrous wastes, polymeric emulsions, filter cakes, spent carbons, spent catalysts, soils, and sediments.

METHOD SUMMARY This method is used to determine 60 volatile organic compounds in a variety of solid waste matrices. It provides GC conditions for the detection of halogenated and aromatic volatile organic compounds. Samples can be analyzed using direct injection or purge-and-trap (EPA Method 5030). Groundwater samples must be analyzed using EPA Method 5030 (where applicable). A temperature program is used with the GC. Detection is achieved by a photoionization detector (PID) and a Hall electrolytic conductivity detector (HECD) in series.

INTERFERENCES Samples can be contaminated by diffusion of volatile organics (particularly chlorofluorocarbons and methylene chloride) through the sample container septum during shipment and storage.

INSTRUMENTATION A GC-equipped with variable-constant differential flow controllers, subambient oven controller, PID and HECD detectors connected with a short piece of uncoated capillary tubing and a data system.

Column: 60 m × 0.75 mm I.D. VOCOL wide-bore capillary column with 1.5 μm film thickness.

PRECISION & ACCURACY MDLs are compound-dependent and vary with purging efficiency and concentration. The applicable concentration range of this method is compound- and instrument-dependent but is approximately 0.1 to 200 μg/L. Analytes that are inefficiently purged from water will not be detected when present at low concentrations, but they can be measured with acceptable accuracy and precision when present in sufficient amounts. The estimated quantitation limit (EQL) for an individual compound is approximately 1 μg/kg (wet weight) for soil/sediment samples, 100 μg/kg (wet weight) for wastes, and 1 μg/L for groundwater. EQLs will be proportionately higher for sample extracts and samples that require dilution or reduced sample size to avoid saturation of the detector.

MULTIPLICATION FACTORS FOR OTHER MATRICES (a)

Matrix	Factor (b)
Groundwater	10
Low-concentration soil	10
Water miscible liquid waste	500
High-concentration soil and sludge	1250
Non-water miscible waste	1250

(a) Sample EQLs are highly matrix-dependent. The EQLs listed herein are provided for guidance and may not always be achievable. (b) EQL = [Method detection limit] × [Factor]. For non-aqueous samples, the factor is on a wet-weight basis.

SINGLE LABORATORY ACCURACY & PRECISION DATA FOR VOCs IN WATER

This method was tested in a single lab using water spiked at 10 μg/L and the following data was reported:

Recoveries and standard deviations were determined from seven samples and spiked at 10 μg/L of each analyte. Recoveries were determined by the internal standard method. Internal standards were: Fluorobenzene for PID and 2-Bromo-1-chloropropane for HECD.

The average recovery (in percent) for the PID was none (no response for this detector).
The standard deviation of the recovery for the PID was none (no response for this detector).
The MDL (in μg/mL) for the PID was none (no response for this detector).
The average recovery (in percent) for the HECD was 96.
The standard deviation of the recovery for the HECD was 3.8.
The MDL (in μg/mL) for the HECD was 0.1.

SAMPLE COLLECTION, PRESERVATION & HANDLING
Volatile Organics — Standard 40-mL glass screw-cap VOA vials with Teflon®-faced silicone septum may be used for both liquid and solid matrices. When collecting samples, liquids and solids should be introduced into the vials gently to reduce agitation which might drive off volatile compounds. If there are any air bubbles present the sample must be retaken. Tap slightly as they are filled to try and eliminate as much free air space as possible. The two vials from each sampling locations should be sealed in separate plastic bags to prevent cross-contamination between samples particularly if the sampled waste is suspected of containing high levels of volatile organics.

Semivolatile organics — Containers used to collect samples for the determination of semivolatile organic compounds should be soap and water washed followed by methanol (or isopropanol) rinsing. The sample containers should be of glass or Teflon® and have screw-top covers with Teflon® liners.

Preservation for volatile organics — No preservation is used with concentrated waste samples. With liquid samples containing no residual chlorine, 4 drops of concentrated hydrochloric acid are added and the samples are immediately cooled to 4°C. When liquid samples contain residual chlorine, they are treated as above and, in addition, 4 drops of 4% aqueous sodium thiosulfate are added. Soil, sediment, and sludge samples are only cooled to 4°C.

Preservation for semivolatile organics — No preservation is used with concentrated waste samples. With liquid samples containing no residual chlorine and with soil, sediment, and sludge samples, immediately cooling to 4°C is the only preservation used. When residual chlorine is present then 3 mL of 10% aqueous sodium sulfate is added for each gallon of sample collected, followed by cooling to 4°C.

MHT The holding time for all volatile organics samples is 14 days. Liquid samples must be extracted within 7 days and their extracts analyzed within 40 days. Concentrated waste, soil, sediment, and sludge samples must be extracted within 14 days and their extracts analyzed within 40 days.

SAMPLE PREPARATION Volatile compounds are introduced into the gas chromatograph either by direct injector or purge-and-trap (EPA Method 5030). EPA Method 5030 may be used directly on groundwater samples or low-concentration contaminated soils and sediments. For medium-concentration soils or sediments, methanolic extraction, as described in EPA Method 5030, may be necessary prior to purge-and-trap analysis.

QUALITY CONTROL Calculate surrogate standard recovery on all samples, blanks, and spikes. A trip blank is recommended to check on sampling, storage, and handling contamination. Calibration standards, at a minimum of five concentration levels, are prepared in organic-free reagent water. One of the concentration levels should be at a concentration near, but above, the method detection limit.

A combination of bromochloromethane, 2-bromo-1-chloropropane, 1,4-dichlorobutane, and bromochlorobenzene are recommended as surrogate standards to encompass the range of the temperature program used in this method.

REFERENCE Test Methods for Evaluating Solid Waste, Physical/Chemical Methods, SW-846, 3rd Edition, U.S. EPA, Office of Solid Waste, Washington, DC, EPA Method 8021A, Rev. 1, Nov. 1992.

Chloroethane   **EPA Method 8240**
CAS #75-00-3

TITLE Volatile Organics By GC/MS: Packed Column Technique

MATRIX Nearly all types of sample matarices, regardless of water content, can be analyzed using this method. This includes groundwater, aqueous sludges, caustic liquors, acid liquors, waste solvents, oily wastes, mousses, tars, fibrous wastes, polymetric emulsions, filter cakes, spent carbons, spent catalysts, soils, and sediments.

METHOD SUMMARY Method 8240B covers 80 volatile organic compounds that are introduced into a gas chromatograph by the purge-and-trap method or by direct injection (in limited applications). For the purge-and-trap method an inert gas (zero grade nitrogen or helium) is bubbled through a 5-mL solution at ambient temperature. Purged sample components are trapped in a tube of sorbent materials. When purging is complete, the sorbent tube is heated and backflushed with inert gas to desorb the trapped components onto a GC column.

INTERFERENCES Impurities in the purge gas and from organic compounds outgassing from the plumbing ahead of the trap account for many contamination problems. Interferences purged or coextracted from the samples will vary considerably from source to source. Cross-contamination can occur whenever high-level and low-level samples are analyzed sequentially. Whenever an unusually concentrated sample is analyzed, it should be followed by the analysis of organic-free reagent water to check for cross-contamination. Samples also can be contaminated by diffusion of volatile organics (particularly methylene chloride and fluorocarbons) through the septum seal into the sample during shipment and storage. A trip blank can serve as a check on such contamination. The lab where volatile analysis is performed and also the refrigerated storage area should be completely free of solvents.

INSTRUMENTATION A gas chromatograph/mass spectrometry/data system (GC/MS) equipped with a 6 ft × 0.1 in I.D. glass column packed with 1% SP-1000 on Carbopack-B (60/80 mesh) is required. Also needed is a 5-mL purging device, a sorbent trap, and a thermal desorption apparatus.

PRECISION & ACCURACY This method is reported to have been tested by 15 laboratories using organic-free reagent water, drinking water, surface water, and industrial wastewaters (not specified) fortified at six concentrations over the range 5–600 µg/L.

Sample estimated quantitation limits (EQLs) are highly matrix-dependent. The EQLs listed may not always be achievable. EQLs listed for soils or sediments are based on wet weight. Normally, data is reported on a dry-weight basis; therefore, EQLs will be higher, based on the percent dry weight of each sample. Note that EQLs are even more variable than MDLs and that they are highly variable depending on the matrix being analyzed.

EQL in groundwater in µg/L was 10.
EQL in low soil or sediment in µg/kg was 10.
Accuracy (a) in µg/L was $1.18C + 0.81$.
Precision (b) in µg/L was $0.29x + 1.75$.

(a) Average recovery found for measurements of samples containing a concentration of C, in µg/L.

(b) Overall precision found for measurements of samples with average recovery X for samples containing a concentration of C in µg/L.

X = Average recovery found for measurement of samples containing a concentration of C in µg/L.

MULTIPLICATION FACTORS FOR OTHER MATRICES

Other Matrices	Factor (a)
Waste miscible liquid waste	50
High-concentration soil and sludge	125
Non-water miscible waste	500

(a) EQL = [EQL for low soil/sediment] × [Factor]. For non-aqueous samples, the factor is on a wet-weight basis.

SAMPLING METHOD

Liquid samples — Use a 40-mL glass screw-cap VOA vial with a Teflon®-faced silicone septum that has been prewashed, rinsed with distilled deionized water, and oven dried. However, if residual chlorine is present, collect sample in a 40-oz. soil VOA container which has been pre-preserved with 4 drops of 10% sodium thiosulfate, mix gently, and then transfer the sample to a 40-mL VOA vial. Collect bubble-free samples in duplicate and seal them in separate plastic bags.

Soils or sediments, and sludges — Use an 8-oz. widemouth glass bottle with a Teflon®-faced silicone septum that has been prewashed with detergent, rinsed with distilled deionized water, and oven dried. Tap slightly to eliminate free air space. Collect samples in duplicate and seal them in separate plastic bags.

SAMPLE PRESERVATION

Liquid samples — Add 4 drops of concentrated HCL and immediately cool samples to 4°C and store in a solvent-free refrigerator.

Soils or sediments, and sludges — Cool samples to 4°C and store in a solvent-free refrigerator.

MHT Maximum holding time is 14 days from the date of sample collection.

SAMPLE PREPARATION

Liquid samples — Remove the plunger from a 5-mL syringe and carefully pour the sample into the syringe barrel to just short of overflowing. Replace the syringe plunger and compress the sample. Open the syringe valve and vent any residual air while adjusting the sample volume to 5.0 mL. If there is only one volatile organic analysis (VOA) vial, a second syringe should be filled at this time to protect against possible loss of sample integrity. Add 10 µL of surrogate spiking solution and 10 µL of internal standard spiking solution through the valve bore of the 5-mL syringe, then close the valve. The surrogate and internal standards may be mixed and added as a single spiking solution.

Sediments, soils, and waste samples — All samples of this type should be screened by GC analysis using a headspace method (EPA Method 3810) or the hexadecane extraction and screening method (EPA Method 3820). Use the screening data to determine whether to use the low-concentration method (0.005–1 mg/kg) or the high-concentration method (>1 mg/kg).

Low-concentration method — The low-concentration method is based on purging a heated sediment or soil sample mixed with organic-free reagent water containing the surrogate and internal standards. Analyze all reagent blanks and standards under the same conditions as the samples.

Use a 5-g sample if the expected concentration is <0.1 mg/kg or a 1-g sample for expected concentrations between 0.1 and 1 mg/kg. Mix the contents of the sample container with a narrow metal spatula. Weigh the amount of the sample into a tared purge device. Add the spiked water to the purge device, which contains the weighed amount of sample, and connect the device to the purge-and-trap system.

High-concentration method — This method is based on extracting the sediment or soil with methanol. A waste sample is either extracted or diluted, depending on its solubility in methanol. Wastes that are insoluble in methanol are diluted with reagent tetraglyme or possibly polyethylene glycol (PEG). An aliquot of the extract is added to organic-free reagent water containing surrogate and internal standards. This is purged at ambient temperature. All samples with an expected concentration of >1.0 mg/kg should be analyzed by this method.

Mix the contents of the sample container with a narrow metal spatula. For sediments or soils and solid wastes that are insoluble in methanol, weigh 4 g (wet weight) of sample into a tared 20-mL vial. For waste that is soluble in methanol, tetraglyme, or PEG, weigh 1 g (wet weight) into a tared scintillation vial or culture tube or a 10-mL volumetric flask. Quickly add 9.0 mL of appropriate solvent then add 1.0 mL of a surrogate spiking solution to the vial, cap it, and shake it for 2 min.

METHANOL EXTRACT REQUIRED FOR ANALYSIS OF HIGH-CONCENTRATION SOILS OR SEDIMENTS

Approximate Concentration Range	Volume of Methanol Extract (a)
500–10,000 µg/kg	100 µL
1,000–20,000 µg/kg	50 µL
5,000–100,000 µg/kg	10 µL
25,000–500,000 µg/kg	100 µL of 1/50 dilution (b)

Calculate appropriate dilution factor for concentrations exceeding this table.

(a) The volume of methanol added to 5 mL of water being purged should be kept constant. Therefore, add to the 5-mL syringe whatever volume of methanol is necessary to maintain a volume of 100 µL added to the syringe.
(b) Dilute an aliquot of the methanol extract and then take 100 µL for analysis.

QUALITY CONTROL Demonstrate, through the analysis of a reagent water blank, that interferences from the analytical system, glassware, and reagents are under control. Blank samples should be carried through all stages of the sample preparation and measurement steps. For each analytical batch (up to 20 samples), a reagent blank, matrix spike, and matrix spike duplicate must be analyzed (the frequency of the spikes may be different for different monitoring programs). The blank and spiked samples must be carried through all stages of the sample preparation and measurement steps. QC samples mentioned in the section on Interferences will also be needed as appropriate to those situations.

REFERENCE Test Methods for Evaluating Solid Waste (SW-846). U.S. EPA. 1983. Method 8240B, Rev. 2, Nov. 1990. Office of Solid Wastes, Washington, DC.

Chloroethane   **EPA Method 8260**
CAS #75-00-3

TITLE Volatile Organic Compounds by GC/MS: Capillary Column Technique

MATRIX This method is applicable to nearly all types of samples, regardless of water content, including groundwater, soils, and sediments.

METHOD SUMMARY Method 8260A covers 58 volatile organic compounds that are introduced into a gas chromatograph by the purge-and-trap method or by direct injection (in limited applications). Zero-grade helium is bubbled through a 5-mL solution at ambient temperature. Purged sample components are trapped in a tube containing suitable sorbent materials. When purging is complete, the sorbent tube is heated and backflushed with helium to desorb trapped sample components. The analytes are desorbed directly to a large bore capillary or cryofocussed on a capillary precolumn before being flash evaporated to a narrow bore capillary for analysis.

INTERFERENCES Major contaminant sources are volatile materials in the lab and impurities in the inert purging gas and in the sorbent trap. Interfering contamination may occur when a sample containing low concentrations of volatile organic compounds is analyzed immediately after a sample containing high concentrations of volatile organic compounds. After analysis of a sample containing high concentrations of volatile organic compounds, one or more calibration blanks should be analyzed to check for cross-contamination. Screening of the samples prior to purge-and-trap GC/MS analysis is highly recommended to prevent contamination of the system. This is especially true for soil and waste samples.

Special precautions must be taken to analyze for methylene chloride. The analytical and sample storage area should be isolated from all atmospheric sources of methylene chloride. All gas chromatography carrier gas lines and purge gas plumbing should be constructed from stainless steel or copper tubing. Laboratory clothing previously exposed to methylene chloride fumes during liquid-liquid extraction procedures can contribute to sample contamination.

Samples can also be contaminated by diffusion of volatile organics (particularly methylene chloride and fluorocarbons) through the septum seal during shipment and storage. A trip blank can serve as a check on such contamination.

INSTRUMENTATION GC/MS with a temperature-programmable chromatograph suitable for splitless injection equipped with variable constant differential flow controllers, a subambient oven controller, a purging device, sorbent trap, a thermal desorption apparatus and a capillary precolumn interface when using cryogenic cooling will be needed. The following GC columns may be used:

Column 1: 60 m × 0.75mm I.D. capillary column coated with VOCOL, 1.5 μm film thickness.
Column 2: 30 m × 0.53mm capillary column coated with DB-624 or VOCOL, 3 μm film thickness.
Column 3: 30 m × 0.32mm I.D. capillary column coated with DB-5 or SE-54, 1-μm film thickness.

PRECISION & ACCURACY This method has been tested in a single lab using spiked water. Using a wide-bore capillary column, water was spiked at concentrations between 0.5 and 10 μg/L. Single lab accuracy and precision data are presented. The MDL actually achieved in a given analysis will vary depending on instrument sensitivity and matrix effects.

The MDL (a) in μg/L was 0.10.
The concentration range in μg/L was 0.5–10.
The mean accuracy (% of true value) was 89.
The precision as relative standard deviation was 9.0.

Note: The MDL is based on a 25-mL sample volume instead of a 5-mL sample volume.

SAMPLING METHOD
Liquid samples — Use a 40-mL glass screw-cap VOA vial with a Teflon®-faced silicone septum that has been prewashed, rinsed with distilled deionized water, and oven dried. If residual chlorine is present, collect the sample in a 4-oz soil VOA container which has been pre-preserved with 4 drops of 10% sodium thiosulfate. Mix gently and transfer the sample to a 40-mL VOA vial. Collect bubble-free samples in duplicate and seal each sample in a separate plastic bag.

Soils, sediments, and sludges — Use an 8-oz widemouth glass bottle with Teflon®-faced silicone septum that has been prewashed, rinsed with distilled deionized water, and oven dried. **Do not** heat the septum for more than 1 h. Tap slightly to eliminate any free air space. Collect samples in duplicate and seal each one in a separate plastic bag.

SAMPLE PRESERVATION
Liquid samples — Add 4 drops of concentrated HCL, cool to 4°C and store in a solvent-free refrigerator.

Soils, sediments and sludges — Cool samples to 4°C and store in a solvent-free refrigerator.

MHT The maximum holding time of any sample (liquids, soils, sediments, and sludges) is 14 days.

SAMPLE PREPARATION
Liquid samples — Remove the plunger from a 5-mL syringe and carefully pour the sample into the syringe barrel to just short of overflowing. Replace the syringe plunger and compress the sample. Open the syringe valve and vent any residual air while adjusting the sample volume to 5.0 mL. If there is only one volatile organic analysis (VOA) vial, a second syringe should be filled at this time to protect against possible loss of sample integrity. Add 10 μL of surrogate spiking solution and 10 μL of internal standard spiking solution through the valve bore of the 5-mL syringe, then close the valve. The surrogate and internal standards may be mixed and added as a single spiking solution.

Sediments, soils, and waste samples — All samples of this type should be screened by GC analysis using a headspace method (EPA Method 3810) or the hexadecane extraction and screening method (EPA Method 3820). Use the screening data to determine whether to use the low-concentration method (0.005–1 mg/kg) or the high-concentration method (>1 mg/kg).

Low-concentration method — The low-concentration method is based on purging a heated sediment or soil sample mixed with organic-free reagent water containing the surrogate and internal standards. Analyze all reagent blanks and standards under the same conditions as the samples.

Use a 5-g sample if the expected concentration is <0.1 mg/kg or a 1-g sample for expected concentrations between 0.1 and 1 mg/kg. Mix the contents of the sample container with a narrow metal spatula. Weigh the amount of the sample into a tared purge device. Add the spiked water to the purge device, which contains the weighed amount of sample, and connect the device to the purge-and-trap system.

High-concentration method — This method is based on extracting the sediment or soil with methanol. A waste sample is either extracted or diluted, depending on its solubility in methanol. Wastes that are insoluble in methanol are diluted with reagent tetraglyme or possibly polyethylene glycol (PEG). An aliquot of the extract is added to organic-free reagent water containing surrogate and internal standards. This is purged at ambient temperature. All samples with an expected concentration of >1.0 mg/kg should be analyzed by this method.

Mix the contents of the sample container with a narrow metal spatula. For sediments or soils and solid wastes that are insoluble in methanol, weigh 4 g (wet weight) of sample into a tared 20-mL vial. For waste that is soluble in methanol, tetraglyme, or PEG, weigh 1 g (wet weight) into a tared scintillation vial or culture tube or a 10-mL volumetric flask. Quickly add 9.0 mL of appropriate solvent then add 1.0 mL of a surrogate spiking solution to the vial, cap it, and shake it for 2 min.

METHANOL EXTRACT REQUIRED FOR ANALYSIS OF HIGH-CONCENTRATION SOILS OR SEDIMENTS

Approximate Concentration Range	Volume of Methanol Extract (a)
500–10,000 μg/kg	100 μL
1,000–20,000 μg/kg	50 μL
5,000–100,000 μg/kg	10 μL
25,000–500,000 μg/kg	100 μL of 1/50 dilution (b)

Calculate appropriate dilution factor for concentrations exceeding this table.

(a) The volume of methanol added to 5 mL of water being purged should be kept constant. Therefore, add to the 5-mL syringe whatever volume of methanol is necessary to maintain a volume of 100 μL added to the syringe.
(b) Dilute an aliquot of the methanol extract and then take 100 μL for analysis.

QUALITY CONTROL Demonstrate, through the analysis of a reagent water blank, that interferences from the analytical system, glassware, and reagents are under control. Blank samples should be carried through all stages of the sample preparation and measurement steps. For each analytical batch (up to 20 samples), a reagent blank, matrix spike, and matrix spike duplicate must be analyzed (the frequency of the spikes may be different for different monitoring programs). The blank and spiked samples must be carried through all stages of the sample preparation and measurement steps. QC samples mentioned in the section on Interferences will also be needed as appropriate to those situations.

Matrix spiking standards should be prepared from volatile organic compounds which will be representative of the compounds being investigated. The recommended internal standards are chlorobenzene-d5, 1,4-difluorobenzene, 1,4-dichlorobenzene-d4, and pentafluorobenzene. Using stock standard solutions, prepare secondary dilution standards containing the compounds of interest, either singly or mixed together in methanol. Store them in a vial with no headspace for no more than one week. Surrogates recommended are toluene-d8, 4-bromofluorobenzene, and dibromofluoromethane. Each sample undergoing GC/MS analysis must be spiked with 10 µL of the surrogate spiking solution prior to analysis.

REFERENCE Test Methods for Evaluating Solid Waste (SW-846). U.S. EPA 1983, Method 8260A, Rev. 1, Nov. 1990. Office of Solid Waste, Washington, DC.

Chloroethane   **EPA Method 601**
CAS #75-00-3

TITLE Purgeable Halocarbons

MATRIX Wastewater

APPLICATION Method covers 29 purgeable halocarbons. (Method 624 provides GC/MS conditions appropriate for the qualitative and quantitative confirmation of results). Method describes conditions for a 2nd GC column to confirm measurements made with primary column.

INTERFERENCES Impurities in the purge gas and organic compounds outgassing from the plumbing ahead of the trap. With high- and low-level samples, there can be carryover contamination. Diffusion of volatile organics through the septum seal into the sample.

INSTRUMENTATION GC-equipped with halide-specific detector. (With purge-and-trap unit).

RANGE 8.0–500 µg/L.

MDL 0.52 µg/L.

PRECISION 0.17X+0.63 µg/L (overall precision).

ACCURACY 0.99C–1.53 µg/L (as recovery).

SAMPLING METHOD 25-mL glass vial. Teflon®-lined septum.

STABILITY Cool, 4°C, 0.008% Sodium thiosulfate.

MHT 14 days.

QUALITY CONTROL The lab must on an ongoing basis, spike at least 10% of the samples from each sample site being monitored to assess accuracy.

REFERENCE Method 601, *Federal Register* Part VIII 40 CFR Part 136, Oct 26, 1984.

Chloroethane   **EPA Method 624**
CAS #75-00-3

TITLE Purgeables

MATRIX Wastewater

APPLICATION Method covers 31 purgeable organics. An inert gas is bubbled through a 5-mL water sample in a specially designed purging chamber. Here, purgeables are transferred from aqueous to gaseous phase, passed onto a sorbent column, and trapped. Trap is heated and backflushed with inert gas to desorb purgeables onto a GC column, where purgeables are separated.

INTERFERENCES Impurities in the purge gas, organic compounds outgassing from the plumbing ahead of the trap, and solvent vapors in the lab. With high- and low-level samples, there can be carryover contamination.

INSTRUMENTATION GC/MS with purge-and-trap unit.

RANGE 5–600 µg/L

MDL Not determined.

PRECISION 0.29X+1.75 µg/L (overall precision).

ACCURACY 1.18C+0.81 µg/L (as recovery).

SAMPLING METHOD 25-mL glass vial. Teflon®-lined septum.

STABILITY Cool, 4°C, 0.008% Sodium thiosulfate.

MHT 14 days.

QUALITY CONTROL The lab must on an ongoing basis, spike at least 5% of the samples from each sample site being monitored to assess accuracy.

REFERENCE Method 624, *Federal Register* Part VIII 40 CFR Part 136, Oct 26, 1984.

Chloroethane   **EPA Method 8010**
CAS #75-00-3

TITLE Halogenated Volatile Organics

MATRIX Groundwater, soils, sludges, water miscible liquid wastes, and non-water miscible wastes.

APPLICATION This method is used for the analysis of 39 halogenated VOCs. Samples are analyzed using direct injection or purge-and-trap methods. Groundwater must be analyzed by the purge-and-trap method. The method provides an optional

GC column which is used for analyte confirmation and that may help resolve analytes from interferences.

INTERFERENCES There can be carryover contamination with high- and low-level samples. Impurities may come from the purge-and-trap apparatus, organic compounds outgassing from the plumbing ahead of trap, diffusion of VOCs through the sample bottle septum during shipping or storage, or from solvent vapors in the lab.

INSTRUMENTATION GC capable of on-column injections or purge-and-trap sample introduction and a halogen specific detector. Column 1: 8 ft by 0.1 in 1%. SP-1000 on Carbopack-B. Column 2: 6 ft by 0.1 in bonded n-octane on Porasil-C.

RANGE 8 to 500 µg/L (reagent water)

MDL 0.52 µg/L (reagent water).

PQL FACTORS FOR MULTIPLYING × FID MDL VALUE

Matrix	Multiplication Factor
Groundwater	10
Low-level soil	10
Water miscible liquid waste	500
High-level soil and sludge	1250
Non-water miscible waste	1250

PRECISION 0.17X + 0.63 µg/L (overall precision).

ACCURACY 0.99C–1.53 µg/L (as recovery).

SAMPLING METHOD For water and liquid samples; use glass 40-mL vials with Teflon®-lined septum caps and collect two vials per sample location with no headspace. For solids and concentrated waste samples; use widemouth glass bottles with Teflon® liners.

STABILITY For concentrated wastes, soils, sediments, or sludges: cool to 4°C. For liquids: add 4 drops of concentrated hydrochloric acid and cool to 4°C.

MHT 14 days.

QUALITY CONTROL Analyze a reagent blank, matrix spike, and matrix spike duplicate/duplicate for each analytical batch (up to 20 samples). Demonstrate the purity of glassware and reagents by analyzing a reagent water method blank. Internal, surrogate, and five concentration level calibration standards are used.

REFERENCE Test Methods for Evaluating Solid Waste (SW-846), U.S. EPA Office of Solid Waste, Washington, DC, Method 8010B, Rev. 2, Nov. 1992.

2-Chloroethanol EPA Method 8240
CAS #107-07-3

TITLE Volatile Organics By GC/MS: Packed Column Technique

MATRIX Nearly all types of sample matarices, regardless of water content, can be analyzed using this method. This includes groundwater, aqueous sludges, caustic liquors, acid liquors, waste solvents, oily wastes, mousses, tars, fibrous wastes, polymetric emulsions, filter cakes, spent carbons, spent catalysts, soils, and sediments.

METHOD SUMMARY Method 8240B covers 80 volatile organic compounds that are introduced into a gas chromatograph by the purge-and-trap method or by direct injection (in limited applications). For the purge-and-trap method an inert gas (zero grade nitrogen or helium) is bubbled through a 5-mL solution at ambient temperature. Purged sample components are trapped in a tube of sorbent materials. When purging is complete, the sorbent tube is heated and backflushed with inert gas to desorb the trapped components onto a GC column.

INTERFERENCES Impurities in the purge gas and from organic compounds outgassing from the plumbing ahead of the trap account for many contamination problems. Interferences purged or coextracted from the samples will vary considerably from source to source. Cross-contamination can occur whenever high-level and low-level samples are analyzed sequentially. Whenever an unusually concentrated sample is analyzed, it should be followed by the analysis of organic-free reagent water to check for cross-contamination. Samples also can be contaminated by diffusion of volatile organics (particularly methylene chloride and fluorocarbons) through the septum seal into the sample during shipment and storage. A trip blank can serve as a check on such contamination. The lab where volatile analysis is performed and also the refrigerated storage area should be completely free of solvents.

INSTRUMENTATION A gas chromatograph/mass spectrometry/data system (GC/MS) equipped with a 6 ft × 0.1 in I.D. glass column packed with 1% SP-1000 on Carbopack-B (60/80 mesh) is required. Also needed is a 5-mL purging device, a sorbent trap, and a thermal desorption apparatus.

PRECISION & ACCURACY This method is reported to have been tested by 15 laboratories using organic-free reagent water, drinking water, surface water, and industrial wastewaters (not specified) fortified at six concentrations over the range 5–600 µg/L.

Sample estimated quantitation limits (EQLs) are highly matrix-dependent. The EQLs listed may not always be achievable. EQLs listed for soils or sediments are based on wet weight. Normally, data is reported on a dry-weight basis; therefore, EQLs will be higher, based on the percent dry weight of each sample. Note that EQLs are even more variable than MDLs and that they are highly variable depending on the matrix being analyzed.

EQL in groundwater in µg/L was not listed.
EQL in low soil or sediment in µg/kg was not listed.
Accuracy (a) in µg/L was not listed.
Precision (b) in µg/L was not listed.

(a) *Average recovery found for measurements of samples containing a concentration of C, in µg/L.*
(b) *Overall precision found for measurements of samples with average recovery X for samples containing a concentration of C in µg/L.*
X = *Average recovery found for measurement of samples containing a concentration of C in µg/L.*

MULTIPLICATION FACTORS FOR OTHER MATRICES

Other Matrices	Factor (a)
Waste miscible liquid waste	50
High-concentration soil and sludge	125
Non-water miscible waste	500

(a) EQL = [EQL for low soil/sediment] × [Factor]. For non-aqueous samples, the factor is on a wet-weight basis.

SAMPLING METHOD

Liquid samples — Use a 40-mL glass screw-cap VOA vial with a Teflon®-faced silicone septum that has been prewashed, rinsed with distilled deionized water, and oven dried. However, if residual chlorine is present, collect sample in a 40-oz. soil VOA container which has been pre-preserved with 4 drops of 10% sodium thiosulfate, mix gently, and then transfer the sample to a 40-mL VOA vial. Collect bubble-free samples in duplicate and seal them in separate plastic bags.

Soils or sediments, and sludges — Use an 8-oz. widemouth glass bottle with a Teflon®-faced silicone septum that has been prewashed with detergent, rinsed with distilled deionized water, and oven dried. Tap slightly to eliminate free air space. Collect samples in duplicate and seal them in separate plastic bags.

SAMPLE PRESERVATION

Liquid samples — Add 4 drops of concentrated HCL and immediately cool samples to 4°C and store in a solvent-free refrigerator.

Soils or sediments, and sludges — Cool samples to 4°C and store in a solvent-free refrigerator.

MHT Maximum holding time is 14 days from the date of sample collection.

SAMPLE PREPARATION

Liquid samples — Remove the plunger from a 5-mL syringe and carefully pour the sample into the syringe barrel to just short of overflowing. Replace the syringe plunger and compress the sample. Open the syringe valve and vent any residual air while adjusting the sample volume to 5.0 mL. If there is only one volatile organic analysis (VOA) vial, a second syringe should be filled at this time to protect against possible loss of sample integrity. Add 10 µL of surrogate spiking solution and 10 µL of internal standard spiking solution through the valve bore of the 5-mL syringe, then close the valve. The surrogate and internal standards may be mixed and added as a single spiking solution.

Sediments, soils, and waste samples — All samples of this type should be screened by GC analysis using a headspace method (EPA Method 3810) or the hexadecane extraction and screening method (EPA Method 3820). Use the screening data to determine whether to use the low-concentration method (0.005–1 mg/kg) or the high-concentration method (>1 mg/kg).

Low-concentration method — The low-concentration method is based on purging a heated sediment or soil sample mixed with organic-free reagent water containing the surrogate and internal standards. Analyze all reagent blanks and standards under the same conditions as the samples.

Use a 5-g sample if the expected concentration is <0.1 mg/kg or a 1-g sample for expected concentrations between 0.1 and 1 mg/kg. Mix the contents of the sample container with a narrow metal spatula. Weigh the amount of the sample into a tared purge device. Add the spiked water to the purge device, which contains the weighed amount of sample, and connect the device to the purge-and-trap system.

High-concentration method — This method is based on extracting the sediment or soil with methanol. A waste sample is either extracted or diluted, depending on its solubility in methanol. Wastes that are insoluble in methanol are diluted with reagent tetraglyme or possibly polyethylene glycol (PEG). An aliquot of the extract is added to organic-free reagent water containing surrogate and internal standards. This is purged at ambient temperature. All samples with an expected concentration of >1.0 mg/kg should be analyzed by this method.

Mix the contents of the sample container with a narrow metal spatula. For sediments or soils and solid wastes that are insoluble in methanol, weigh 4 g (wet weight) of sample into a tared 20-mL vial. For waste that is soluble in methanol, tetraglyme, or PEG, weigh 1 g (wet weight) into a tared scintillation vial or culture tube or a 10-mL volumetric flask. Quickly add 9.0 mL of appropriate solvent then add 1.0 mL of a surrogate spiking solution to the vial, cap it, and shake it for 2 min.

METHANOL EXTRACT REQUIRED FOR ANALYSIS OF HIGH-CONCENTRATION SOILS OR SEDIMENTS

Approximate Concentration Range	Volume of Methanol Extract (a)
500–10,000 µg/kg	100 µL
1,000–20,000 µg/kg	50 µL
5,000–100,000 µg/kg	10 µL
25,000–500,000 µg/kg	100 µL of 1/50 dilution (b)

Calculate appropriate dilution factor for concentrations exceeding this table.

(a) The volume of methanol added to 5 mL of water being purged should be kept constant. Therefore, add to the 5-mL syringe whatever volume of methanol is necessary to maintain a volume of 100 µL added to the syringe.

(b) Dilute an aliquot of the methanol extract and then take 100 µL for analysis.

QUALITY CONTROL Demonstrate, through the analysis of a reagent water blank, that interferences from the analytical system, glassware, and reagents are under control. Blank samples should be carried through all stages of the sample preparation and measurement steps. For each analytical batch (up to 20 samples), a reagent blank, matrix spike, and matrix spike duplicate must be analyzed (the frequency of the spikes may be different for different monitoring programs). The blank and spiked samples must be carried through all stages of the sample preparation and measurement steps. QC samples mentioned in the section on Interferences will also be needed as appropriate to those situations.

REFERENCE Test Methods for Evaluating Solid Waste (SW-846). U.S. EPA. 1983. Method 8240B, Rev. 2, Nov. 1990. Office of Solid Wastes, Washington, DC.

2-Chloroethyl vinyl ether
CAS #110-75-8

EPA Method 1624

TITLE Volatile Organic Compounds by Isotope Dilution GC/MS

MATRIX Compounds may be determined in waters, soils, and municipal sludges by this method.

METHOD SUMMARY This method is used to determine 58 volatile toxic organic pollutants associated with the CWA (as amended 1987); the RCRA (as amended 1986); the CERCLA (as amended 1986); and other compounds amenable to purge-and-trap gas chromatography-mass spectrometry (GC/MS).

If the solids content is less than 1%, stable isotopically-labeled analogs of the compounds of interest are added to a 5-mL sample and the sample is purged with an inert gas at 20–25°C in a chamber designed for soil or water samples. If the solids content is greater than 1%, 5 mL of reagent water and the labeled compounds are added to a 5-g aliquot of sample and the mixture is purged at 40°C. Compounds that will not purge at 20–25°C or at 40°C are purged at 78–85°C. In the purging process, the volatile compounds are transferred from the aqueous phase into the gaseous phase where they are passed into a sorbent column, and trapped. After purging is completed, the trap is backflushed and heated rapidly to desorb the compounds into a GC. The compounds are separated by the GC and detected by a MS. The labeled compounds serve to correct the variability of the analytical technique.

INTERFERENCES Impurities in the purge gas, organic compounds outgassing from the plumbing upstream of the trap, and solvent vapors in the lab account for most problems. Samples can be contaminated by diffusion of volatile organic compounds (particularly methylene chloride) through the bottle seal during shipment and storage. Contamination by carryover can occur when high-level and low-level samples are analyzed sequentially. When an unusually concentrated sample is encountered, follow it by analysis of a reagent water blank to check for carryover.

INSTRUMENTATION Major equipment includes a GC with linear temperature programming and a glass jet separator as the MS interface, a MS with 70 eV electron impact ionization, and a data system to collect and record response factors.

Column: 2.8 m × 2 mm I.D. glass, packed with 1% SP-1000 on Carbopak B, 60/80 mesh, or equivalent.

PRECISION & ACCURACY The detection limits of the method are usually dependent on the level of interferences rather than instrumental limitations. The method detection limits were determined in digested sludge (low solids) and in filter cake or compost (high solids).

The MDL (in µg/kg) for low solids is 122 and for high solids is 21.
Labeled and native compound precision (in µg/L) as standard deviation was 36.0.
Labeled and native compound accuracy (in µg/L) as average recovery was detected to 70.

Acceptance criteria are at 20 µg/L for this compound.

SAMPLE COLLECTION, PRESERVATION & HANDLING Grab samples are collected in glass containers having a total volume greater than 20 mL. Fill and seal each bottle so that no air bubbles are entrapped. Samples are maintained at 0 to 4°C from the time of collection until analysis. If an aqueous sample contains residual chlorine, add sodium thiosulfate preservative (10 mg/40 mL) to the empty sample bottles just prior to shipment to the sample site. All samples must be analyzed within 14 days of collection.

SAMPLE PREPARATION Samples containing less than 1% solids are analyzed directly as aqueous samples. Samples containing 1% solids or greater are analyzed as solid samples utilizing one of two methods, depending on the levels of pollutants, in the sample. Samples containing 1% solids or greater, and low to moderate levels of pollutants are analyzed by purging a known weight of sample added to 5 mL of reagent water. Samples containing 1% solids or greater, and high levels of pollutants, are extracted with methanol, and an aliquot of the methanol extract is added to reagent water and purged.

QUALITY CONTROL A field blank prepared from reagent water and carried through the sampling and handling protocol may serve as a check on contamination from shipment and storage.

The analyst is permitted to modify this method to improve separations or lower the costs of measurements, provided all performance specifications are met. Analyses of blanks are required. When results of spikes indicate atypical method performance for samples, the samples are diluted to bring method performance within acceptable limits. Analyze two sets of four 5-mL aliquots (8 aliquots total) of the aqueous performance standard. Spike all samples with labeled compounds to assess method performance on the sample matrix. Compute the percent recovery of the labeled compounds using the internal standard method. Compare the percent recovery for each compound with the corresponding labeled compound recovery. Reagent water blanks are analyzed to demonstrate freedom from carryover contamination. Field replicates may be collected to determine the precision of the sampling technique, and spiked samples may be required to determine the accuracy of the analysis when the internal method is used.

REFERENCE Volatile Organic Compounds by Isotope Dilution GC/MS. Office of Water Regulation and Standards, U.S. EPA Industrial Technology Division, Washington, DC, EPA Method 1624, Rev. C, June 1989 (contact W.A. Telliard, U.S. EPA, Office of Water Regulations and Standards, 401 M St., SW, Washington, DC, 20460. Phone: 202-382-7131).

2-Chloroethyl vinyl ether
CAS #110-75-8

EPA Method 8240

TITLE Volatile Organics By GC/MS: Packed Column Technique

MATRIX Nearly all types of sample matarices, regardless of water content, can be analyzed using this method. This includes groundwater, aqueous sludges, caustic liquors, acid liquors, waste solvents, oily wastes, mousses, tars, fibrous wastes, polymetric emulsions, filter cakes, spent carbons, spent catalysts, soils, and sediments.

METHOD SUMMARY Method 8240B covers 80 volatile organic compounds that are introduced into a gas chromatograph by the purge-and-trap method or by direct injection (in limited applications). For the purge-and-trap method an inert gas (zero grade nitrogen or helium) is bubbled through a 5-mL solution at ambient temperature. Purged sample components are trapped in a tube of sorbent materials. When purging is complete, the sorbent tube is heated and backflushed with inert gas to desorb the trapped components onto a GC column.

INTERFERENCES Impurities in the purge gas and from organic compounds outgassing from the plumbing ahead of the trap account for many contamination problems. Interferences purged or coextracted from the samples will vary considerably from source to source. Cross-contamination can occur whenever high-level and low-level samples are analyzed sequentially. Whenever an unusually concentrated sample is analyzed, it should be followed by the analysis of organic-free reagent water to check for cross-contamination. Samples also can be contaminated by diffusion of volatile organics (particularly methylene chloride and fluorocarbons) through the septum seal into the sample during shipment and storage. A trip blank can serve as a check on such contamination. The lab where volatile analysis is performed and also the refrigerated storage area should be completely free of solvents.

INSTRUMENTATION A gas chromatograph/mass spectrometry/data system (GC/MS) equipped with a 6 ft × 0.1 in I.D. glass column packed with 1% SP-1000 on Carbopack-B (60/80 mesh) is required. Also needed is a 5-mL purging device, a sorbent trap, and a thermal desorption apparatus.

PRECISION & ACCURACY This method is reported to have been tested by 15 laboratories using organic-free reagent water, drinking water, surface water, and industrial wastewaters (not specified) fortified at six concentrations over the range 5–600 µg/L.

Sample estimated quantitation limits (EQLs) are highly matrix-dependent. The EQLs listed may not always be achievable. EQLs listed for soils or sediments are based on wet weight. Normally, data is reported on a dry-weight basis; therefore, EQLs will be higher, based on the percent dry weight of each sample. Note that EQLs are even more variable than MDLs and that they are highly variable depending on the matrix being analyzed.

EQL in groundwater in µg/L was 10.
EQL in low soil or sediment in µg/kg was 10.
Accuracy (a) in µg/L was 1.00C.
Precision (b) in µg/L was 0.84x.

(a) *Average recovery found for measurements of samples containing a concentration of C, in µg/L.*
(b) *Overall precision found for measurements of samples with average recovery X for samples containing a concentration of C in µg/L.*
X = *Average recovery found for measurement of samples containing a concentration of C in µg/L.*

MULTIPLICATION FACTORS FOR OTHER MATRICES

Other Matrices	Factor (a)
Waste miscible liquid waste	50
High-concentration soil and sludge	125
Non-water miscible waste	500

(a) *EQL = [EQL for low soil/sediment] × [Factor]. For non-aqueous samples, the factor is on a wet-weight basis.*

SAMPLING METHOD
Liquid samples — Use a 40-mL glass screw-cap VOA vial with a Teflon®-faced silicone septum that has been prewashed, rinsed with distilled deionized water, and oven dried. However, if residual chlorine is present, collect sample in a 40-oz. soil VOA container which has been pre-preserved with 4 drops of 10% sodium thiosulfate, mix gently, and then transfer the sample to a 40-mL VOA vial. Collect bubble-free samples in duplicate and seal them in separate plastic bags.

Soils or sediments, and sludges — Use an 8-oz. widemouth glass bottle with a Teflon®-faced silicone septum that has been prewashed with detergent, rinsed with distilled deionized water, and oven dried. Tap slightly to eliminate free air space. Collect samples in duplicate and seal them in separate plastic bags.

SAMPLE PRESERVATION
Liquid samples — Add 4 drops of concentrated HCL and immediately cool samples to 4°C and store in a solvent-free refrigerator.

Soils or sediments, and sludges — Cool samples to 4°C and store in a solvent-free refrigerator.

MHT Maximum holding time is 14 days from the date of sample collection.

SAMPLE PREPARATION
Liquid samples — Remove the plunger from a 5-mL syringe and carefully pour the sample into the syringe barrel to just short of overflowing. Replace the syringe plunger and compress the sample. Open the syringe valve and vent any residual air while adjusting the sample volume to 5.0 mL. If there is only one volatile organic analysis (VOA) vial, a second syringe should be filled at this time to protect against possible loss of sample integrity. Add 10 µL of surrogate spiking solution and 10 µL of internal standard spiking solution through the valve bore of the 5-mL syringe, then close the valve. The surrogate and internal standards may be mixed and added as a single spiking solution.

Sediments, soils, and waste samples — All samples of this type should be screened by GC analysis using a headspace method (EPA Method 3810) or the hexadecane extraction and screening method (EPA Method 3820). Use the screening data to determine whether to use the low-concentration method (0.005–1 mg/kg) or the high-concentration method (>1 mg/kg).

Low-concentration method — The low-concentration method is based on purging a heated sediment or soil sample mixed with organic-free reagent water containing the surrogate and internal standards. Analyze all reagent blanks and standards under the same conditions as the samples.

Use a 5-g sample if the expected concentration is <0.1 mg/kg or a 1-g sample for expected concentrations between 0.1 and 1 mg/kg. Mix the contents of the sample container with a narrow metal spatula. Weigh the amount of the sample into a tared purge device. Add the spiked water to the purge device, which contains the weighed amount of sample, and connect the device to the purge-and-trap system.

High-concentration method — This method is based on extracting the sediment or soil with methanol. A waste sample is either extracted or diluted, depending on its solubility in methanol. Wastes that are insoluble in methanol are diluted with reagent tetraglyme or possibly polyethylene glycol (PEG). An aliquot of the extract is added to organic-free reagent water containing surrogate and internal standards. This is purged at ambient temperature. All samples with an expected concentration of >1.0 mg/kg should be analyzed by this method.

Mix the contents of the sample container with a narrow metal spatula. For sediments or soils and solid wastes that are insoluble in methanol, weigh 4 g (wet weight) of sample into a tared 20-mL vial. For waste that is soluble in methanol, tetraglyme, or PEG, weigh 1 g (wet weight) into a tared scintillation vial or culture tube or a 10-mL volumetric flask. Quickly add 9.0 mL of appropriate solvent then add 1.0 mL of a surrogate spiking solution to the vial, cap it, and shake it for 2 min.

METHANOL EXTRACT REQUIRED FOR ANALYSIS OF HIGH-CONCENTRATION SOILS OR SEDIMENTS

Approximate Concentration Range	Volume of Methanol Extract (a)
500–10,000 µg/kg	100 µL
1,000–20,000 µg/kg	50 µL
5,000–100,000 µg/kg	10 µL
25,000–500,000 µg/kg	100 µL of 1/50 dilution (b)

Calculate appropriate dilution factor for concentrations exceeding this table.

(a) The volume of methanol added to 5 mL of water being purged should be kept constant. Therefore, add to the 5-mL syringe whatever volume of methanol is necessary to maintain a volume of 100 µL added to the syringe.
(b) Dilute an aliquot of the methanol extract and then take 100 µL for analysis.

QUALITY CONTROL Demonstrate, through the analysis of a reagent water blank, that interferences from the analytical system, glassware, and reagents are under control. Blank samples should be carried through all stages of the sample preparation and measurement steps. For each analytical batch (up to 20 samples), a reagent blank, matrix spike, and matrix spike duplicate must be analyzed (the frequency of the spikes may be different for different monitoring programs). The blank and spiked samples must be carried through all stages of the sample preparation and measurement steps. QC samples mentioned in the section on Interferences will also be needed as appropriate to those situations.

REFERENCE Test Methods for Evaluating Solid Waste (SW-846). U.S. EPA. 1983. Method 8240B, Rev. 2, Nov. 1990. Office of Solid Wastes, Washington, DC.

2-Chloroethyl vinyl ether   EPA Method 601
CAS #110-75-8

TITLE Purgeable Halocarbons

MATRIX Wastewater

APPLICATION Method covers 29 purgeable halocarbons. (Method 624 provides GC/MS conditions appropriate for the qualitative and quantitative confirmation of results). Method describes conditions for a 2nd GC column to confirm measurements made with primary column.

INTERFERENCES Impurities in the purge gas and organic compounds outgassing from the plumbing ahead of the trap. With high- and low-level samples, there can be carryover contamination. Diffusion of volatile organics through the septum seal into the sample.

INSTRUMENTATION GC-equipped with halide-specific detector. (With purge-and-trap unit).

RANGE 8.0–500 µg/L.

MDL 0.13 µg/L.

PRECISION 0.35X µg/L (overall precision).

ACCURACY 1.00C µg/L (as recovery).

SAMPLING METHOD 25-mL glass vial. Teflon®-lined septum.

STABILITY Cool, 4°C, 0.008% Sodium thiosulfate.

MHT 14 days.

QUALITY CONTROL The lab must on an ongoing basis, spike at least 10% of the samples from each sample site being monitored to assess accuracy.

REFERENCE Method 601, *Federal Register* Part VIII 40 CFR Part 136, Oct 26, 1984.

2-Chloroethyl vinyl ether   EPA Method 624
CAS #110-75-8

TITLE Purgeables

MATRIX Wastewater

APPLICATION Method covers 31 purgeable organics. An inert gas is bubbled through a 5-mL water sample in a specially designed purging chamber. Here, purgeables are transferred from aqueous to gaseous phase, passed onto a sorbent column, and trapped. Trap is heated and backflushed with inert gas to desorb purgeables onto a GC column, where purgeables are separated.

INTERFERENCES Impurities in the purge gas, organic compounds outgassing from the plumbing ahead of the trap, and solvent vapors in the lab. With high- and low-level samples, there can be carryover contamination.

INSTRUMENTATION GC/MS with purge-and-trap unit.

RANGE 5–600 µg/L

MDL Not determined.

PRECISION 0.84X µg/L (overall precision).

ACCURACY 1.00C µg/L (as recovery).

SAMPLING METHOD 25-mL glass vial. Teflon®-lined septum.

STABILITY Cool, 4°C, 0.008% Sodium thiosulfate.

MHT 14 days.

QUALITY CONTROL The lab must on an ongoing basis, spike at least 5% of the samples from each sample site being monitored to assess accuracy.

REFERENCE Method 624, *Federal Register* Part VIII 40 CFR Part 136, Oct 26, 1984.

2-Chloroethyl vinyl ether   EPA Method 8010
CAS #110-75-8

TITLE Halogenated Volatile Organics

MATRIX Groundwater, soils, sludges, water miscible liquid wastes, and non-water miscible wastes.

APPLICATION This method is used for the analysis of 39 halogenated VOCs. Samples are analyzed using direct injection or purge-and-trap methods. Groundwater must be analyzed by the purge-and-trap method. The method provides an optional GC column which is used for analyte confirmation and that may help resolve analytes from interferences.

INTERFERENCES There can be carryover contamination with high- and low-level samples. Impurities may come from the purge-and-trap apparatus, organic compounds outgassing from the plumbing ahead of trap, diffusion of VOCs through the sample bottle septum during shipping or storage, or from solvent vapors in the lab.

INSTRUMENTATION GC capable of on-column injections or purge-and-trap sample introduction and a halogen specific detector. Column 1: 8 ft by 0.1 in 1%. SP-1000 on Carbopack-B. Column 2: 6 ft by 0.1 in bonded n-octane on Porasil-C.

RANGE 8 to 500 µg/L (reagent water)

MDL 0.13 µg/L (reagent water).

PQL FACTORS FOR MULTIPLYING × FID MDL VALUE

Matrix	Multiplication Factor
Groundwater	10
Low-level soil	10
Water miscible liquid waste	500
High-level soil and sludge	1250
Non-water miscible waste	1250

PRECISION 0.35X µg/L (overall precision; estimate)

ACCURACY 1.00C µg/L (as recovery; estimate)

SAMPLING METHOD For water and liquid samples; use glass 40-mL vials with Teflon®-lined septum caps and collect two vials per sample location with no headspace. For solids and concentrated waste samples; use widemouth glass bottles with Teflon® liners.

STABILITY For concentrated wastes, soils, sediments, or sludges: cool to 4°C. For liquids: add 4 drops of concentrated hydrochloric acid and cool to 4°C.

MHT 14 days.

QUALITY CONTROL Analyze a reagent blank, matrix spike, and matrix spike duplicate/duplicate for each analytical batch (up to 20 samples). Demonstrate the purity of glassware and reagents by analyzing a reagent water method blank. Internal, surrogate, and five concentration level calibration standards are used.

REFERENCE Test Methods for Evaluating Solid Waste (SW-846), U.S. EPA Office of Solid Waste, Washington, DC, Method 8010B, Rev. 2, Nov. 1992.

Chloroform   EPA Method 1624
CAS #67-66-3

TITLE Volatile Organic Compounds by Isotope Dilution GC/MS

MATRIX Compounds may be determined in waters, soils, and municipal sludges by this method.

METHOD SUMMARY This method is used to determine 58 volatile toxic organic pollutants associated with the CWA (as amended 1987); the RCRA (as amended 1986); the CERCLA (as amended 1986); and other compounds amenable to purge-and-trap gas chromatography-mass spectrometry (GC/MS).

If the solids content is less than 1%, stable isotopically-labeled analogs of the compounds of interest are added to a 5-mL sample and the sample is purged with an inert gas at 20–25°C in a chamber designed for soil or water samples. If the solids content is greater than 1%, 5 mL of reagent water and the labeled compounds are added to a 5-g aliquot of sample and the mixture is purged at 40°C. Compounds that will not purge at 20–25°C or at 40°C are purged at 78–85°C. In the purging process, the volatile compounds are transferred from the aqueous phase into the gaseous phase where they are passed into a sorbent column, and trapped. After purging is completed, the trap is backflushed and heated rapidly to desorb the compounds into a GC. The compounds are separated by the GC and detected by a MS. The labeled compounds serve to correct the variability of the analytical technique.

INTERFERENCES Impurities in the purge gas, organic compounds outgassing from the plumbing upstream of the trap, and solvent vapors in the lab account for most problems. Samples can be contaminated by diffusion of volatile organic compounds (particularly methylene chloride) through the bottle seal during shipment and storage. Contamination by carryover can occur when high-level and low-level samples are analyzed sequentially. When an unusually concentrated sample is encountered, follow it by analysis of a reagent water blank to check for carryover.

INSTRUMENTATION Major equipment includes a GC with linear temperature programming and a glass jet separator as the MS interface, a MS with 70 eV electron impact ionization, and a data system to collect and record response factors.

Column: 2.8 m × 2 mm I.D. glass, packed with 1% SP-1000 on Carbopak B, 60/80 mesh, or equivalent.

PRECISION & ACCURACY The detection limits of the method are usually dependent on the level of interferences rather than instrumental limitations. The method detection limits were determined in digested sludge (low solids) and in filter cake or compost (high solids).

The MDL (in µg/kg) for low solids is 21 and for high solids is 2.
Labeled and native compound precision (in µg/L) as standard deviation was 7.9.
Labeled and native compound accuracy (in µg/L) as average recovery was 12–26.

Acceptance criteria are at 20 µg/L for this compound.

SAMPLE COLLECTION, PRESERVATION & HANDLING Grab samples are collected in glass containers having a total volume greater than 20 mL. Fill and seal each bottle so that no air bubbles are entrapped. Samples are maintained at 0 to 4°C from the time of collection until analysis. If an aqueous sample contains residual chlorine, add sodium thiosulfate preservative (10 mg/40 mL) to the empty sample bottles just prior to shipment to the sample site. All samples must be analyzed within 14 days of collection.

SAMPLE PREPARATION Samples containing less than 1% solids are analyzed directly as aqueous samples. Samples containing 1% solids or greater are analyzed as solid samples utilizing one of two methods, depending on the levels of pollutants, in the sample. Samples containing 1% solids or greater, and low to moderate levels of pollutants are analyzed by purging a known weight of sample added to 5 mL of reagent water. Samples containing 1% solids or greater, and high levels of pollutants, are extracted with methanol, and an aliquot of the methanol extract is added to reagent water and purged.

QUALITY CONTROL A field blank prepared from reagent water and carried through the sampling and handling protocol may serve as a check on contamination from shipment and storage.

The analyst is permitted to modify this method to improve separations or lower the costs of measurements, provided all performance specifications are met. Analyses of blanks are required. When results of spikes indicate atypical method performance for samples, the samples are diluted to bring method performance within acceptable limits. Analyze two sets of four 5-mL aliquots (8 aliquots total) of the aqueous performance standard. Spike all samples with labeled compounds to assess method performance on the sample matrix. Compute the percent recovery of the labeled compounds using the internal standard method. Compare the percent recovery for each compound with the corresponding labeled compound recovery. Reagent water blanks are analyzed to demonstrate freedom from carryover contamination. Field replicates may be collected to determine the precision of the sampling technique, and spiked samples may be required to determine the accuracy of the analysis when the internal method is used.

REFERENCE Volatile Organic Compounds by Isotope Dilution GC/MS. Office of Water Regulation and Standards, U.S. EPA Industrial Technology Division, Washington, DC, EPA Method 1624, Rev. C, June 1989 (contact W.A. Telliard, U.S. EPA, Office of Water Regulations and Standards, 401 M St., SW, Washington, DC, 20460. Phone: 202-382-7131).

Chloroform　　　　　　　　　　　　　EPA Method 502
CAS #67-66-3

TITLE Volatile Organic Compounds in Water By Purge and Trap Capillary Column Gas Chromatography with Photoionization and Electrolytic Conductivity Detectors in Series. U.S. EPA Method 502.2, Rev. 2.0, 1989.

MATRIX Drinking water and raw source water. The latter should include most surface water and groundwater sources.

METHOD SUMMARY This method covers 60 volatile organic compounds that contain halogen atoms and/or that are aromatic. An inert gas (zero grade nitrogen or helium) is bubbled through a 25-mL or a 5-mL water sample (depending on the expected concentration of the analytes). Purged sample components are trapped in a tube of sorbent materials. When purging is complete, the sorbent tube is heated and backflushed with helium to desorb the trapped sample onto a capillary GC column. The column is temperature programmed to separate the method analytes which are then detected with a photoionization detector (PID) and a Hall electrolytic conductivity (HECD) placed in series. The PID is selective for aromatic compounds and the HECD is selective for halogenated compounds.

INTERFERENCES Impurities in the purge gas and from organic compounds outgassing from the plumbing ahead of the trap account for many contamination problems. Interferences purged or coextracted from the samples will vary considerably from source to source, depending upon the particular sample or extract being tested. Cross-contamination can occur whenever high-level and low-level samples are analyzed sequentially. Samples also can be contaminated by diffusion of volatile organics (particularly methylene chloride and fluorocarbons) through the septum seal into the sample during shipment and storage. The lab where volatile analysis is performed and also the refrigerated storage area should be completely free of solvents.

INSTRUMENTATION A GC containing a series configuration of a high temperature photoionization detector (PID) equipped with 10.0 eV (nominal) lamp and Hall electrolytic conductivity detector (HECD) is required. Also required is an all-glass 5-mL purging device, a sorbent trap, and a thermal desorption apparatus which is connected to the GC system.

Column 1: VOCOL glass wide-bore capillary column.
Column 2: RTX–502.2 mega-bore capillary column.
Column 3: DB-62 mega-bore capillary column.

PRECISION & ACCURACY Method detection limits are dependent upon the characteristics of the gas chromatographic system used. Analytes that are not separated chromatographically cannot be individually identified and used in the same calibration mixture or water samples unless an alternative technique for identification and quantification, such as mass spectrometry, is used.

Electrolytic conductivity detetor (c) range in μg/L (a) was 0.02–200.

Electrolytic conductivity detetor (c) MDL in μg/L (b) was 0.02.

Electrolytic conductivity detetor (c) accuracy as % recovery was 98.

Electrolytic conductivity detetor (c) precision as % RSD was 2.5.

Photoionization detector (d) range in μg/L (a) was 0.02–200.

Photoionization detector (d) MDL in μg/L (b) was not listed.

Photoionization detector (d) accuracy as % recovery was not listed.

Photoionization detector (d) precision as % RSD was not listed.

(a) *The applicable concentration range of this method is compound, instrument, and matrix-dependent. It is listed as being approximately 0.02 to 200 μg/L but no specific information is provided so caution should be observed.*

(b) *The method detection limits reports with this method are compound, instrument, and matrix-dependent. The values reported were calculated using reagent water fortified with the corresponding compounds at 10 μg/L and a GC-equipped with a 60 m × 0.75 mm VOLCOL wide bore capillary column with 1.5 μm film thickness and using helium carrier gas.*

(c) *Recoveries and relative standard deviations were determined from seven samples of reagent water fortified with 10 μg/L of each compound. 2-Bromo-1-chloropropane was used as the internal standard for calculating average recoveries.*

(d) *Recoveries and relative standard deviations were determined from seven samples of reagent water fortified with 10 μg/L of each compound. Fluorobenzene was used as the internal standard for calculating average recoveries.*

SAMPLING METHOD Collect samples using a 40- to 120-mL screw-cap vial (prewashed with detergent, rinsed with distilled water and oven dried at 105°C) with a Teflon®-faced silicone septum. Collect bubble-free samples and place the septum with the Teflon® side down on the water.

SAMPLE PRESERVATION If residual chlorine is present in the water add about 25 mg of ascorbic acid to each vial before samples are collected to remove the chlorine. Add hydrochloric acid to reduce pH to <2, immediately cool samples to 4°C, and store them in a solvent-free refrigerator at 4°C until analysis.

MHT The maximum holding time for samples is 14 days from the time they were collected.

SAMPLE PREPARATION Remove the plungers from two 5-mL syringes and attach a closed syringe valve to each. Warm the sample to room temperature, open the sample bottle, and carefully pour the sample into one of the syringe barrels to just short of overflowing. Replace the syringe plunger, invert the syringe, and compress the sample. Open the syringe valve and vent any residual air while adjusting the sample volume to 5.0 mL. Add 10 μL of the internal calibration standard to the sample through the syringe valve. Close the valve. Fill the second syringe in an identical manner from the same sample bottle. Reserve this second syringe for a reanalysis if necessary.

QUALITY CONTROL As an initial demonstration of lab accuracy and precision, analyze 4 to 7 replicates of a lab fortified blank containing analyte at 0.1–5 μg/L. Collect all samples in duplicate. Surrogate analytes (similar to those of the analytes of interest), whose concentration is known in every sample, are measured using the same internal standard calibration procedure. Duplicate field reagent water blanks (trip blanks) must be analyzed with each set of samples, lab reagent blanks (method blanks) must be analyzed with each batch of samples processed as a group within a work shift. Also, a single lab-fortified blank that contains each of the analytes of interest should be analyzed with each batch of samples processed as a group within a work shift. A 3- to 5-point calibration curve is needed depending on the calibration range factor required.

EPA CONTACT & HOTLINE For technical questions contact Dr. Baldev Bathija, U.S. EPA, Office of Ground Water and Drinking Water (WH-550D), 401 M St. SW, Washington, DC 20460. Tel. (202) 260-3040. For further information the EPA Safe Drinking Water Hotline may be called at: (800) 426-4791.

REFERENCE Methods for the Determination of Organic Compounds in Drinking Water, EPA/600/4-88/039 (revised July 1991; Final Rule for determination of compliance with the MCL for Total Trihalomethanes under 141.30, in 40 CFR Part 141, Vol. 58, No. 147, Fed. Reg., Tuesday Aug. 3, 1993). U.S. EPA Environmental Monitoring Systems Laboratory, Cincinnati, OH, 45268, USA. Available from the National Technical Information Service (NTIS), 5285 Port Royal Road, Springfield, VA 22161; Tel. 800-553-6847. NTIS Order Number is PB91-231480.

Chloroform   **EPA Method 524**
CAS #67-66-3

TITLE Measurement of Purgeable Organic Compounds in Water by Capillary Column GC/MS.

MATRIX Drinking water and raw source water; the latter should include most surface water and groundwater sources.

METHOD SUMMARY Method 524.2 covers 60 volatile organic compounds. An inert gas (zero grade nitrogen or helium) is bubbled through a 25-mL or a 5-mL water sample (depending on the expected concentration of the analytes). Purged sample components are trapped in a tube of sorbent materials. When purging is complete, the sorbent tube is heated and backflushed with helium to desorb the trapped sample onto a capillary GC column.

INTERFERENCES Impurities in the purge gas and from organic compounds outgassing from the plumbing ahead of the trap account for many contamination problems. Interferences purged or coextracted from the samples will vary considerably from source to source, depending upon the particular sample or extract being tested. Cross-contamination can occur whenever high-level and low-level samples are analyzed sequentially. Samples also can be contaminated by diffusion of volatile organics (particularly methylene chloride and fluorocarbons) through the septum seal into the sample during shipment and storage.

INSTRUMENTATION A GC/MS with a data system equipped with one of the following capillary GC columns:

Column 1: VOCOL glass wide bore capillary column.
Column 2: DB-624 fused silica capillary column.
Column 3: DB-5 fused silica capillary column.

Also required is an all-glass 25 mL or 5-mL purging device, a sorbent trap, and a thermal desorption apparatus which is connected to the GC/MS system.

PRECISION & ACCURACY Method detection limits are compound- and instrument-dependent, and may vary from approximately 0.02–0.35 µg/L. Note in the table below that the "true" concentration range used for accuracy and precision measurements was quite narrow. However, the applicable concentration range of this method is primarily column dependent and is approximately 0.02 to 200 µg/L for the wide-bore thick-film columns. Narrow-bore thin-film columns may have a capacity which limits the range to about 0.02 to 20 µg/L. Analytes that are inefficiently purged from water will not be detected when present at low concentrations, but they can be measured with acceptable accuracy and precision when present in sufficient amounts.

Analytes that are not separated chromatographically, but which have different mass spectra and non-interfering quantification ions, can be identified and measured in the same calibration mixture or water sample. Analytes which have very similar mass spectra cannot be individually identified and measured in the same calibration mixture or water samples unless they have different retention times. Co-eluting compounds with very similar mass spectra, typically many structural isomers, must be reported as an isomeric group or pair.

The range (in µg/L) was 0.5–10.
The Method Detection Limig (in µg/L) was 0.03.
The accuracy (as % recovery) was 90.
The precision (in %) was 6.1.

Note: Data were obtained from 16–31 determinations using a wide-bore capillary column and a jet separator interfaced to a quadrupole mass spectrometer. All analytes were in a reagent water matrix.

SAMPLING METHOD Collect samples using a 40- to 120-mL screw-cap vial (prewashed with detergent, rinsed with distilled water and oven dried at 105°C) with a Teflon®-faced silicone septum. Collect bubble-free samples and place the septum with the Teflon® side down on the water.

SAMPLE PRESERVATION If residual chlorine is present in the water add about 25 mg of ascorbic acid to each vial before samples are collected to remove the chlorine. Add hydrochloric acid to reduce pH to <2, and immediately cool samples to 4°C, and store them in a solvent-free refrigerator at 4°C until analysis.

MHT The maximum holding time for samples is 14 days from the time they were collected.

SAMPLE PREPARATION Remove the plungers from two 25-mL (or 5-mL depending on sample size) syringes and attach a closed syringe valve to each. Warm the sample to room temperature, open the sample bottle, and carefully pour the sample into one of the syringe barrels to just short of overflowing. Replace the syringe plunger, invert the syringe, and compress the sample. Open the syringe valve and vent any residual air while adjusting the sample volume to 25.0 mL (or 5 mL). For samples and blanks, add 5 µL of the fortification solution containing the internal standard and the surrogates to the sample through the syringe valve. For calibration standards and lab fortified blanks, add 5 µL of the fortification solution containing the internal standard only. Close the valve. Fill the second syringe in an identical manner from the same sample bottle. Reserve this second syringe for a reanalysis if necessary.

QUALITY CONTROL As an initial demonstration of lab accuracy and precision, analyze 4 to 7 replicates of a lab fortified blank containing analyte at 0.2–5 µg/L. Collect all samples in duplicate. Surrogate analytes (similar to those of the analytes of interest), whose concentration is known in every sample, are measured using the same internal standard calibration procedure. Duplicate field reagent water blanks (trip blanks) must be analyzed with each set of samples, lab reagent blanks (method blanks) must be analyzed with each batch of samples processed as a group within a work shift. Also, a single lab-fortified blank that contains each of the analytes of interest should be analyzed with each batch of samples processed as a group within a work shift. A 3- to 5-point calibration curve is needed depending on the calibration range factor required.

EPA CONTACT & HOTLINE For technical questions contact Dr. Baldev Bathija, U.S. EPA, Office of Ground Water and Drinking Water (WH-550D), 401 M St. SW, Washington, DC 20460. Tel. (202) 260-3040. For further information the EPA Safe Drinking Water Hotline may be called at: (800) 426-4791.

REFERENCE Methods for the Determination of Organic Compounds in Drinking Water, EPA/600/4-88/039 (revised July 1991; Final Rule for determination of compliance with the MCL for Total Trihalomethanes under 141.30, in 40 CFR Part 141, Vol. 58, No. 147, Fed. Reg., Tuesday Aug. 3, 1993). U.S. EPA Environmental Monitoring Systems Laboratory, Cincinnati, OH, 45268, USA. Available from the National Technical Information Service (NTIS), 5285 Port Royal Road, Springfield, VA 22161; Tel. 800-553-6847. NTIS Order Number is PB91-231480.

Chloroform   **EPA Method 8021**
CAS #67-66-3

TITLE Halogenated Volatile by Gas Chromatography Using Photoionization and Electrolytic Conductivity Detectors in Series: Capillary Column Technique

MATRIX This method is applicable to nearly all types of samples, regardless of water content, including groundwater, aqueous sludges, caustic liquors, acid liquors, waste solvents, oily wastes, mousses, tars, fibrous wastes, polymeric emulsions, filter cakes, spent carbons, spent catalysts, soils, and sediments.

METHOD SUMMARY This method is used to determine 60 volatile organic compounds in a variety of solid waste matrices. It provides GC conditions for the detection of halogenated and aromatic volatile organic compounds. Samples can be analyzed using direct injection or purge-and-trap (EPA Method 5030). Groundwater samples must be analyzed using EPA Method 5030 (where applicable). A temperature program is used with the GC. Detection is achieved by a photoionization detector (PID) and a Hall electrolytic conductivity detector (HECD) in series.

INTERFERENCES Samples can be contaminated by diffusion of volatile organics (particularly chlorofluorocarbons and methylene chloride) through the sample container septum during shipment and storage.

INSTRUMENTATION A GC-equipped with variable-constant differential flow controllers, subambient oven controller, PID and HECD detectors connected with a short piece of uncoated capillary tubing and a data system.

Column: 60 m × 0.75 mm I.D. VOCOL wide-bore capillary column with 1.5 µm film thickness.

PRECISION & ACCURACY MDLs are compound-dependent and vary with purging efficiency and concentration. The applicable concentration range of this method is compound- and instrument-dependent but is approximately 0.1 to 200 µg/L. Analytes that are inefficiently purged from water will not be detected when present at low concentrations, but they can be measured with acceptable accuracy and precision when present in sufficient amounts. The estimated quantitation limit (EQL) for an individual compound is approximately 1 µg/kg (wet weight) for soil/sediment samples, 100 µg/kg (wet weight) for wastes, and 1 µg/L for groundwater. EQLs will be proportionately higher for sample extracts and samples that require dilution or reduced sample size to avoid saturation of the detector.

MULTIPLICATION FACTORS FOR OTHER MATRICES (a)

Matrix	Factor (b)
Groundwater	10
Low-concentration soil	10
Water miscible liquid waste	500
High-concentration soil and sludge	1250
Non-water miscible waste	1250

(a) Sample EQLs are highly matrix-dependent. The EQLs listed herein are provided for guidance and may not always be achievable. (b) EQL = [Method detection limit] × [Factor]. For non-aqueous samples, the factor is on a wet-weight basis.

SINGLE LABORATORY ACCURACY & PRECISION DATA FOR VOCs IN WATER

This method was tested in a single lab using water spiked at 10 µg/L and the following data was reported:

Recoveries and standard deviations were determined from seven samples and spiked at 10 µg/L of each analyte. Recoveries were determined by the internal standard method. Internal standards were: Fluorobenzene for PID and 2-Bromo-1-chloropropane for HECD.

The average recovery (in percent) for the PID was none (no response for this detector).
The standard deviation of the recovery for the PID was none (no response for this detector).
The MDL (in µg/mL) for the PID was none (no response for this detector).
The average recovery (in percent) for the HECD was 98.
The standard deviation of the recovery for the HECD was 2.5.
The MDL (in µg/mL) for the HECD was 0.02.

SAMPLE COLLECTION, PRESERVATION & HANDLING
Volatile Organics — Standard 40-mL glass screw-cap VOA vials with Teflon®-faced silicone septum may be used for both liquid and solid matrices. When collecting samples, liquids and solids should be introduced into the vials gently to reduce agitation which might drive off volatile compounds. If there are any air bubbles present the sample must be retaken. Tap slightly as they are filled to try and eliminate as much free air space as possible. The two vials from each sampling locations should be sealed in separate plastic bags to prevent cross-contamination between samples particularly if the sampled waste is suspected of containing high levels of volatile organics.

Semivolatile organics — Containers used to collect samples for the determination of semivolatile organic compounds should be soap and water washed followed by methanol (or isopropanol) rinsing. The sample containers should be of glass or Teflon® and have screw-top covers with Teflon® liners.

Preservation for volatile organics — No preservation is used with concentrated waste samples. With liquid samples containing no residual chlorine, 4 drops of concentrated hydrochloric acid are added and the samples are immediately cooled to 4°C. When liquid samples contain residual chlorine, they are treated as above and, in addition, 4 drops of 4% aqueous sodium thiosulfate are added. Soil, sediment, and sludge samples are only cooled to 4°C.

Preservation for semivolatile organics — No preservation is used with concentrated waste samples. With liquid samples containing no residual chlorine and with soil, sediment, and sludge samples, immediately cooling to 4°C is the only preservation used. When residual chlorine is present then 3 mL of 10% aqueous sodium sulfate is added for each gallon of sample collected, followed by cooling to 4°C.

MHT The holding time for all volatile organics samples is 14 days. Liquid samples must be extracted within 7 days and their extracts analyzed within 40 days. Concentrated waste, soil, sediment, and sludge samples must be extracted within 14 days and their extracts analyzed within 40 days.

SAMPLE PREPARATION Volatile compounds are introduced into the gas chromatograph either by direct injector or purge-and-trap (EPA Method 5030). EPA Method 5030 may be used directly on groundwater samples or low-concentration contaminated soils and sediments. For medium-concentration soils or sediments, methanolic extraction, as described in EPA Method 5030, may be necessary prior to purge-and-trap analysis.

QUALITY CONTROL Calculate surrogate standard recovery on all samples, blanks, and spikes. A trip blank is recommended to check on sampling, storage, and handling contamination. Calibration standards, at a minimum of five concentration levels, are prepared in organic-free reagent water. One of the concentration levels should be at a concentration near, but above, the method detection limit.

A combination of bromochloromethane, 2-bromo-1-chloropropane, 1,4-dichlorobutane, and bromochlorobenzene are recommended as surrogate standards to encompass the range of the temperature program used in this method.

REFERENCE Test Methods for Evaluating Solid Waste, Physical/Chemical Methods, SW-846, 3rd Edition, U.S. EPA, Office of Solid Waste, Washington, DC, EPA Method 8021A, Rev. 1, Nov. 1992.

Chloroform	EPA Method 8240
CAS #67-66-3	

TITLE Volatile Organics By GC/MS: Packed Column Technique

MATRIX Nearly all types of sample matarices, regardless of water content, can be analyzed using this method. This includes groundwater, aqueous sludges, caustic liquors, acid liquors, waste solvents, oily wastes, mousses, tars, fibrous wastes, polymetric emulsions, filter cakes, spent carbons, spent catalysts, soils, and sediments.

METHOD SUMMARY Method 8240B covers 80 volatile organic compounds that are introduced into a gas chromatograph by the purge-and-trap method or by direct injection (in limited applications). For the purge-and-trap method an inert gas (zero grade nitrogen or helium) is bubbled through a 5-mL solution at ambient temperature. Purged sample components are trapped in a tube of sorbent materials. When purging is complete, the sorbent tube is heated and backflushed with inert gas to desorb the trapped components onto a GC column.

INTERFERENCES Impurities in the purge gas and from organic compounds outgassing from the plumbing ahead of the trap account for many contamination problems. Interferences purged or coextracted from the samples will vary considerably from source to source. Cross-contamination can occur whenever high-level and low-level samples are analyzed sequentially. Whenever an unusually concentrated sample is analyzed, it should be followed by the analysis of organic-free reagent water to check for cross-contamination. Samples also can be contaminated by diffusion of volatile organics (particularly methylene chloride and fluorocarbons) through the septum seal into the sample during shipment and storage. A trip blank can serve as a check on such contamination. The lab where volatile analysis is performed and also the refrigerated storage area should be completely free of solvents.

INSTRUMENTATION A gas chromatograph/mass spectrometry/data system (GC/MS) equipped with a 6 ft × 0.1 in I.D. glass column packed with 1% SP-1000 on Carbopack-B (60/80 mesh) is required. Also needed is a 5-mL purging device, a sorbent trap, and a thermal desorption apparatus.

PRECISION & ACCURACY This method is reported to have been tested by 15 laboratories using organic-free reagent water, drinking water, surface water, and industrial wastewaters (not specified) fortified at six concentrations over the range 5–600 µg/L.

Sample estimated quantitation limits (EQLs) are highly matrix-dependent. The EQLs listed may not always be achievable. EQLs listed for soils or sediments are based on wet weight. Normally, data is reported on a dry-weight basis; therefore, EQLs will be higher, based on the percent dry weight of each sample. Note that EQLs are even more variable than MDLs and that they are highly variable depending on the matrix being analyzed.

EQL in groundwater in µg/L was 5.
EQL in low soil or sediment in µg/kg was 5.
Accuracy (a) in µg/L was 0.93C+0.33.
Precision (b) in µg/L was 0.18x+0.16.

(a) *Average recovery found for measurements of samples containing a concentration of C, in µg/L.*
(b) *Overall precision found for measurements of samples with average recovery X for samples containing a concentration of C in µg/L.*
X = *Average recovery found for measurement of samples containing a concentration of C in µg/L.*

MULTIPLICATION FACTORS FOR OTHER MATRICES

Other Matrices	Factor (a)
Waste miscible liquid waste	50
High-concentration soil and sludge	125
Non-water miscible waste	500

(a) *EQL = [EQL for low soil/sediment] × [Factor]. For non-aqueous samples, the factor is on a wet-weight basis.*

SAMPLING METHOD

Liquid samples — Use a 40-mL glass screw-cap VOA vial with a Teflon®-faced silicone septum that has been prewashed, rinsed with distilled deionized water, and oven dried. However, if residual chlorine is present, collect sample in a 40-oz. soil VOA container which has been pre-preserved with 4 drops of 10% sodium thiosulfate, mix gently, and then transfer the sample to a 40-mL VOA vial. Collect bubble-free samples in duplicate and seal them in separate plastic bags.

Soils or sediments, and sludges — Use an 8-oz. widemouth glass bottle with a Teflon®-faced silicone septum that has been prewashed with detergent, rinsed with distilled deionized water, and oven dried. Tap slightly to eliminate free air space. Collect samples in duplicate and seal them in separate plastic bags.

SAMPLE PRESERVATION

Liquid samples — Add 4 drops of concentrated HCL and immediately cool samples to 4°C and store in a solvent-free refrigerator.

Soils or sediments, and sludges — Cool samples to 4°C and store in a solvent-free refrigerator.

MHT Maximum holding time is 14 days from the date of sample collection.

SAMPLE PREPARATION

Liquid samples — Remove the plunger from a 5-mL syringe and carefully pour the sample into the syringe barrel to just short of overflowing. Replace the syringe plunger and compress the sample. Open the syringe valve and vent any residual air while adjusting the sample volume to 5.0 mL. If there is only one volatile organic analysis (VOA) vial, a second syringe should be filled at this time to protect against possible loss of sample integrity. Add 10 µL of surrogate spiking solution and 10 µL of internal standard spiking solution through the valve bore of the 5-mL syringe, then close the valve. The surrogate and internal standards may be mixed and added as a single spiking solution.

Sediments, soils, and waste samples — All samples of this type should be screened by GC analysis using a headspace method (EPA Method 3810) or the hexadecane extraction and screening method (EPA Method 3820). Use the screening data to determine whether to use the low-concentration method (0.005–1 mg/kg) or the high-concentration method (>1 mg/kg).

Low-concentration method — The low-concentration method is based on purging a heated sediment or soil sample mixed with organic-free reagent water containing the surrogate and internal standards. Analyze all reagent blanks and standards under the same conditions as the samples.

Use a 5-g sample if the expected concentration is <0.1 mg/kg or a 1-g sample for expected concentrations between 0.1 and 1 mg/kg. Mix the contents of the sample container with a narrow metal spatula. Weigh the amount of the sample into a tared purge device. Add the spiked water to the purge device, which contains the weighed amount of sample, and connect the device to the purge-and-trap system.

High-concentration method — This method is based on extracting the sediment or soil with methanol. A waste sample is either extracted or diluted, depending on its solubility in methanol. Wastes that are insoluble in methanol are diluted with reagent tetraglyme or possibly polyethylene glycol (PEG). An aliquot of the extract is added to organic-free reagent water containing surrogate and internal standards. This is purged at ambient temperature. All samples with an expected concentration of >1.0 mg/kg should be analyzed by this method.

Mix the contents of the sample container with a narrow metal spatula. For sediments or soils and solid wastes that are insoluble in methanol, weigh 4 g (wet weight) of sample into a tared 20-mL vial. For waste that is soluble in methanol, tetraglyme, or PEG, weigh 1 g (wet weight) into a tared scintillation vial or culture tube or a 10-mL volumetric flask. Quickly add 9.0 mL of appropriate solvent then add 1.0 mL of a surrogate spiking solution to the vial, cap it, and shake it for 2 min.

METHANOL EXTRACT REQUIRED FOR ANALYSIS OF HIGH-CONCENTRATION SOILS OR SEDIMENTS

Approximate Concentration Range	Volume of Methanol Extract (a)
500–10,000 µg/kg	100 µL
1,000–20,000 µg/kg	50 µL
5,000–100,000 µg/kg	10 µL
25,000–500,000 µg/kg	100 µL of 1/50 dilution (b)

Calculate appropriate dilution factor for concentrations exceeding this table.

(a) The volume of methanol added to 5 mL of water being purged should be kept constant. Therefore, add to the 5-mL syringe whatever volume of methanol is necessary to maintain a volume of 100 µL added to the syringe.

(b) Dilute an aliquot of the methanol extract and then take 100 µL for analysis.

QUALITY CONTROL Demonstrate, through the analysis of a reagent water blank, that interferences from the analytical system, glassware, and reagents are under control. Blank samples should be carried through all stages of the sample preparation and measurement steps. For each analytical batch (up to 20 samples), a reagent blank, matrix spike, and matrix spike duplicate must be analyzed (the frequency of the spikes may be different for different monitoring programs). The blank and spiked samples must be carried through all stages of the sample preparation and measurement steps. QC samples mentioned in the section on Interferences will also be needed as appropriate to those situations.

REFERENCE Test Methods for Evaluating Solid Waste (SW-846). U.S. EPA. 1983. Method 8240B, Rev. 2, Nov. 1990. Office of Solid Wastes, Washington, DC.

Chloroform   **EPA Method 8260**
CAS #67-66-3

TITLE Volatile Organic Compounds by GC/MS: Capillary Column Technique

MATRIX This method is applicable to nearly all types of samples, regardless of water content, including groundwater, soils, and sediments.

METHOD SUMMARY Method 8260A covers 58 volatile organic compounds that are introduced into a gas chromatograph by the purge-and-trap method or by direct injection (in limited applications). Zero-grade helium is bubbled through a 5-mL solution at ambient temperature. Purged sample components are trapped in a tube containing suitable sorbent materials. When purging is complete, the sorbent tube is heated and backflushed with helium to desorb trapped sample components. The analytes are desorbed directly to a large bore capillary or cryofocussed on a capillary precolumn before being flash evaporated to a narrow bore capillary for analysis.

INTERFERENCES Major contaminant sources are volatile materials in the lab and impurities in the inert purging gas and in the sorbent trap. Interfering contamination may occur when a sample containing low concentrations of volatile organic compounds is analyzed immediately after a sample containing high concentrations of volatile organic compounds. After analysis of a sample containing high concentrations of volatile organic compounds, one or more calibration blanks should be analyzed to check for cross-contamination. Screening of the samples prior to purge-and-trap GC/MS analysis is highly recommended to prevent contamination of the system. This is especially true for soil and waste samples.

Special precautions must be taken to analyze for methylene chloride. The analytical and sample storage area should be isolated from all atmospheric sources of methylene chloride. All gas chromatography carrier gas lines and purge gas plumbing should be constructed from stainless steel or copper tubing. Laboratory clothing previously exposed to methylene chloride fumes during liquid-liquid extraction procedures can contribute to sample contamination.

Samples can also be contaminated by diffusion of volatile organics (particularly methylene chloride and fluorocarbons) through the septum seal during shipment and storage. A trip blank can serve as a check on such contamination.

INSTRUMENTATION GC/MS with a temperature-programmable chromatograph suitable for splitless injection equipped with variable constant differential flow controllers, a subambient oven controller, a purging device, sorbent trap, a thermal desorption apparatus and a capillary precolumn interface when using cryogenic cooling will be needed. The following GC columns may be used:

Column 1: 60 m × 0.75mm I.D. capillary column coated with VOCOL, 1.5 µm film thickness.
Column 2: 30 m × 0.53mm capillary column coated with DB-624 or VOCOL, 3 µm film thickness.
Column 3: 30 m × 0.32mm I.D. capillary column coated with DB-5 or SE-54, 1-µm film thickness.

PRECISION & ACCURACY This method has been tested in a single lab using spiked water. Using a wide-bore capillary column, water was spiked at concentrations between 0.5 and 10 µg/L. Single lab accuracy and precision data are presented. The MDL actually achieved in a given analysis will vary depending on instrument sensitivity and matrix effects.

The MDL (a) in µg/L was 0.03.
The concentration range in µg/L was 0.5–10.
The mean accuracy (% of true value) was 90.
The precision as relative standard deviation was 6.1.

Note: The MDL is based on a 25-mL sample volume instead of a 5-mL sample volume.

SAMPLING METHOD
Liquid samples — Use a 40-mL glass screw-cap VOA vial with a Teflon®-faced silicone septum that has been prewashed, rinsed with distilled deionized water, and oven dried. If residual chlorine is present, collect the sample in a 4-oz soil VOA container which has been pre-preserved with 4 drops of 10% sodium thiosulfate. Mix gently and transfer the sample to a 40-mL VOA vial. Collect bubble-free samples in duplicate and seal each sample in a separate plastic bag.

Soils, sediments, and sludges — Use an 8-oz widemouth glass bottle with Teflon®-faced silicone septum that has been prewashed, rinsed with distilled deionized water, and oven dried. **Do not** heat the septum for more than 1 h. Tap slightly to eliminate any free air space. Collect samples in duplicate and seal each one in a separate plastic bag.

SAMPLE PRESERVATION
Liquid samples — Add 4 drops of concentrated HCL, cool to 4°C and store in a solvent-free refrigerator.

Soils, sediments and sludges — Cool samples to 4°C and store in a solvent-free refrigerator.

MHT The maximum holding time of any sample (liquids, soils, sediments, and sludges) is 14 days.

SAMPLE PREPARATION
Liquid samples — Remove the plunger from a 5-mL syringe and carefully pour the sample into the syringe barrel to just short of overflowing. Replace the syringe plunger and compress the sample. Open the syringe valve and vent any residual air while adjusting the sample volume to 5.0 mL. If there is only one volatile organic analysis (VOA) vial, a second syringe should be filled at this time to protect against possible loss of sample integrity. Add 10 µL of surrogate spiking solution and 10 µL of internal standard spiking solution through the valve bore of the 5-mL syringe, then close the valve. The surrogate and internal standards may be mixed and added as a single spiking solution.

Sediments, soils, and waste samples — All samples of this type should be screened by GC analysis using a headspace method (EPA Method 3810) or the hexadecane extraction and screening method (EPA Method 3820). Use the screening data to determine whether to use the low-concentration method (0.005–1 mg/kg) or the high-concentration method (>1 dg¯kg).

Low-concentration method — The low-concentration method is based on purging a heated sediment or soil sample mixed with organic-free reagent water containing the surrogate and internal standards. Analyze all reagent blanks and standards undep the same conditions as the samples.

Use a 5-g sample if the expected concentration is <0.1 mg/kg or a 1-g sample for expected concentrations between 0.1 and 1 mg/kg. Mix the contents of the sample container with a narrow metal spatula. Weigh the amount of the sample into a tared purge device. Add the spiked water to the purge device, which contains the weighed amount of sample, and connect the device to the purge-and-trap system.

High-concentration method — This method is based on extracting the sediment or soil with methanol. A waste sample is either extracted or diluted, depending on its solubility in methanol. Wastes that are insoluble in methanol are diluted with reagent tetraglyme or possibly polyethylene glycol (PEG). An aliquot of the extract is added to organic-free reagent water containing surrogate and internal standards. This is purged at ambient temperature. All samples with an expected concentration of >1.0 mg/kg should be analyzed by this method.

Mix the contents of the sample container with a narrow metal spatula. For sediments or soils and solid wastes that are insoluble in methanol, weigh 4 g (wet weight) of sample into a tared 20-mL vial. For waste that is soluble in methanol, tetraglyme, or PEG, weigh 1 g (wet weight) into a tared scintillation vial or culture tube or a 10-mL volumetric flask. Quickly add 9.0 mL of appropriate solvent then add 1.0 mL of a surrogate spiking solution to the vial, cap it, and shake it for 2 min.

METHANOL EXTRACT REQUIRED FOR ANALYSIS OF HIGH-CONCENTRATION SOILS OR SEDIMENTS

Approximate Concentration Range	Volume of Methanol Extract (a)
500–10,000 µg/kg	100 µL
1,000–20,000 µg/kg	50 µL
5,000–100,000 µg/kg	10 µL
25,000–500,000 µg/kg	100 µL of 1/50 dilution (b)

Calculate appropriate dilution factor for concentrations exceeding this table.

(a) The volume of methanol added to 5 mL of water being purged should be kept constant. Therefore, add to the 5-mL syringe whatever volume of methanol is necessary to maintain a volume of 100 µL added to the syringe.
(b) Dilute an aliquot of the methanol extract and then take 100 µL for analysis.

QUALITY CONTROL Demonstrate, through the analysis of a reagent water blank, that interferences from the analytical system, glassware, and reagents are under control. Blank samples should be carried through all stages of the sample preparation and measurement steps. For each analytical batch (up to 20 samples), a reagent blank, matrix spike, and matrix spike duplicate must be analyzed (the frequency of the spikes may be different for different monitoring programs). The blank and spiked samples must be carried through all stages of the sample preparation and measurement steps. QC samples mentioned in the section on Interferences will also be needed as appropriate to those situations.

Matrix spiking standards should be prepared from volatile organic compounds which will be representative of the compounds being investigated. The recommended internal standards are chlorobenzene-d5, 1,4-difluorobenzene, 1,4-dichlorobenzene-d4, and pentafluorobenzene. Using stock standard solutions, prepare secondary dilution standards containing the compounds of interest, either singly or mixed together in methanol. Store them in a vial with no headspace for no more than one week. Surrogates recommended are toluene-d8, 4-bromofluorobenzene, and dibromofluoromethane. Each sample undergoing GC/MS analysis must be spiked with 10 µL of the surrogate spiking solution prior to analysis.

REFERENCE Test Methods for Evaluating Solid Waste (SW-846). U.S. EPA 1983, Method 8260A, Rev. 1, Nov. 1990. Office of Solid Waste, Washington, DC.

Chloroform   **EPA Method 601**
CAS #67-66-3

TITLE Purgeable Halocarbons

MATRIX Wastewater

APPLICATION Method covers 29 purgeable halocarbons. (Method 624 provides GC/MS conditions appropriate for the qualitative and quantitative confirmation of results). Method describes conditions for a 2nd GC column to confirm measurements made with primary column.

INTERFERENCES Impurities in the purge gas and organic compounds outgassing from the plumbing ahead of the trap. With high- and low-level samples, there can be carryover contamination. Diffusion of volatile organics through the septum seal into the sample.

INSTRUMENTATION GC-equipped with halide-specific detector. (With purge-and-trap unit).

RANGE 8.0–500 µg/L.

MDL 0.05 µg/L.

PRECISION 0.19X–0.02 µg/L (overall precision).

ACCURACY 0.93C-0.39 µg/L (as recovery).

SAMPLING METHOD 25-mL glass vial. Teflon®-lined septum.

STABILITY Cool, 4°C, 0.008% Sodium thiosulfate.

MHT 14 days.

QUALITY CONTROL The lab must on an ongoing basis, spike at least 10% of the samples from each sample site being monitored to assess accuracy.

REFERENCE Method 601, *Federal Register* Part VIII 40 CFR Part 136, Oct 26, 1984.

Chloroform   **EPA Method 624**
CAS #67-66-3

TITLE Purgeables

MATRIX Wastewater

APPLICATION Method covers 31 purgeable organics. An inert gas is bubbled through a 5-mL water sample in a specially designed purging chamber. Here, purgeables are transferred from aqueous to gaseous phase, passed onto a sorbent column, and trapped. Trap is heated and backflushed with inert gas to desorb purgeables onto a GC column, where purgeables are separated.

INTERFERENCES Impurities in the purge gas, organic compounds outgassing from the plumbing ahead of the trap, and solvent vapors in the lab. With high- and low-level samples, there can be carryover contamination.

INSTRUMENTATION GC/MS with purge-and-trap unit.

RANGE 5–600 µg/L

MDL 1.6 µg/L

PRECISION 0.18X+0.16 µg/L (overall precision).

ACCURACY 0.93C+0.33 µg/L (as recovery).

SAMPLING METHOD 25-mL glass vial. Teflon®-lined septum.

STABILITY Cool, 4°C, 0.008% Sodium thiosulfate.

MHT 14 days.

QUALITY CONTROL The lab must on an ongoing basis, spike at least 5% of the samples from each sample site being monitored to assess accuracy.

REFERENCE Method 624, *Federal Register* Part VIII 40 CFR Part 136, Oct 26, 1984.

Chloroform — EPA Method 8010
CAS #67-66-3

TITLE Halogenated Volatile Organics

MATRIX Groundwater, soils, sludges, water miscible liquid wastes, and non-water miscible wastes.

APPLICATION This method is used for the analysis of 39 halogenated VOCs. Samples are analyzed using direct injection or purge-and-trap methods. Groundwater must be analyzed by the purge-and-trap method. The method provides an optional GC column which is used for analyte confirmation and that may help resolve analytes from interferences.

INTERFERENCES There can be carryover contamination with high- and low-level samples. Impurities may come from the purge-and-trap apparatus, organic compounds outgassing from the plumbing ahead of trap, diffusion of VOCs through the sample bottle septum during shipping or storage, or from solvent vapors in the lab.

INSTRUMENTATION GC capable of on-column injections or purge-and-trap sample introduction and a halogen specific detector. Column 1: 8 ft by 0.1 in 1%. SP-1000 on Carbopack-B. Column 2: 6 ft by 0.1 in bonded n-octane on Porasil-C.

RANGE 8 to 500 µg/L (reagent water)

MDL 0.05 µg/L (reagent water).

PQL FACTORS FOR MULTIPLYING × FID MDL VALUE

Matrix	Multiplication Factor
Groundwater	10
Low-level soil	10
Water miscible liquid waste	500
High-level soil and sludge	1250
Non-water miscible waste	1250

PRECISION 0.19X–0.02 µg/L (overall precision).

ACCURACY 0.93C–0.39 µg/L (as recovery).

SAMPLING METHOD For water and liquid samples; use glass 40-mL vials with Teflon®-lined septum caps and collect two vials per sample location with no headspace. For solids and concentrated waste samples; use widemouth glass bottles with Teflon® liners.

STABILITY For concentrated wastes, soils, sediments, or sludges: cool to 4°C. For liquids: add 4 drops of concentrated hydrochloric acid and cool to 4°C.

MHT 14 days.

QUALITY CONTROL Analyze a reagent blank, matrix spike, and matrix spike duplicate/duplicate for each analytical batch (up to 20 samples). Demonstrate the purity of glassware and reagents by analyzing a reagent water method blank. Internal, surrogate, and five concentration level calibration standards are used.

REFERENCE Test Methods for Evaluating Solid Waste (SW-846), U.S. EPA Office of Solid Waste, Washington, DC, Method 8010B, Rev. 2, Nov. 1992.

Chloromethane — EPA Method 1624
CAS #74-87-3

TITLE Volatile Organic Compounds by Isotope Dilution GC/MS

MATRIX Compounds may be determined in waters, soils, and municipal sludges by this method.

METHOD SUMMARY This method is used to determine 58 volatile toxic organic pollutants associated with the CWA (as amended 1987); the RCRA (as amended 1986); the CERCLA (as amended 1986); and other compounds amenable to purge-and-trap gas chromatography-mass spectrometry (GC/MS).

If the solids content is less than 1%, stable isotopically-labeled analogs of the compounds of interest are added to a 5-mL sample and the sample is purged with an inert gas at 20–25°C in a chamber designed for soil or water samples. If the solids content is greater than 1%, 5 mL of reagent water and the labeled compounds are added to a 5-g aliquot of sample and the mixture is purged at 40°C. Compounds that will not purge at 20–25°C or at 40°C are purged at 78–85°C. In the purging process, the volatile compounds are transferred from the aqueous phase into the gaseous phase where they are passed into a sorbent column, and trapped. After purging is completed, the trap is backflushed and heated rapidly to desorb the compounds into a GC. The compounds are separated by the GC and detected by a MS. The labeled compounds serve to correct the variability of the analytical technique.

INTERFERENCES Impurities in the purge gas, organic compounds outgassing from the plumbing upstream of the trap, and solvent vapors in the lab account for most problems. Samples can be contaminated by diffusion of volatile organic compounds (particularly methylene chloride) through the bottle seal during shipment and storage. Contamination by carryover can occur when high-level and low-level samples are analyzed sequentially. When an unusually concentrated sample is encountered, follow it by analysis of a reagent water blank to check for carryover.

INSTRUMENTATION Major equipment includes a GC with linear temperature programming and a glass jet separator as the MS interface, a MS with 70 eV electron impact ionization, and a data system to collect and record response factors.

Column: 2.8 m × 2 mm I.D. glass, packed with 1% SP-1000 on Carbopak B, 60/80 mesh, or equivalent.

PRECISION & ACCURACY The detection limits of the method are usually dependent on the level of interferences rather than instrumental limitations. The method detection

limits were determined in digested sludge (low solids) and in filter cake or compost (high solids).

The MDL (in µg/kg) for low solids is 207 and for high solids is 13.

Background levels of this compound were present in the sludge with low solids, resulting in a higher than expected MDL.

Labeled and native compound precision (in µg/L) as standard deviation was 26.0.

Labeled and native compound accuracy (in µg/L) as average recovery was detected to 56.

Acceptance criteria are at 20 µg/L for this compound.

SAMPLE COLLECTION, PRESERVATION & HANDLING Grab samples are collected in glass containers having a total volume greater than 20 mL. Fill and seal each bottle so that no air bubbles are entrapped. Samples are maintained at 0 to 4°C from the time of collection until analysis. If an aqueous sample contains residual chlorine, add sodium thiosulfate preservative (10 mg/40 mL) to the empty sample bottles just prior to shipment to the sample site. All samples must be analyzed within 14 days of collection.

SAMPLE PREPARATION Samples containing less than 1% solids are analyzed directly as aqueous samples. Samples containing 1% solids or greater are analyzed as solid samples utilizing one of two methods, depending on the levels of pollutants, in the sample. Samples containing 1% solids or greater, and low to moderate levels of pollutants are analyzed by purging a known weight of sample added to 5 mL of reagent water. Samples containing 1% solids or greater, and high levels of pollutants, are extracted with methanol, and an aliquot of the methanol extract is added to reagent water and purged.

QUALITY CONTROL A field blank prepared from reagent water and carried through the sampling and handling protocol may serve as a check on contamination from shipment and storage.

The analyst is permitted to modify this method to improve separations or lower the costs of measurements, provided all performance specifications are met. Analyses of blanks are required. When results of spikes indicate atypical method performance for samples, the samples are diluted to bring method performance within acceptable limits. Analyze two sets of four 5-mL aliquots (8 aliquots total) of the aqueous performance standard. Spike all samples with labeled compounds to assess method performance on the sample matrix. Compute the percent recovery of the labeled compounds using the internal standard method. Compare the percent recovery for each compound with the corresponding labeled compound recovery. Reagent water blanks are analyzed to demonstrate freedom from carryover contamination. Field replicates may be collected to determine the precision of the sampling technique, and spiked samples may be required to determine the accuracy of the analysis when the internal method is used.

REFERENCE Volatile Organic Compounds by Isotope Dilution GC/MS. Office of Water Regulation and Standards, U.S. EPA Industrial Technology Division, Washington, DC, EPA Method 1624, Rev. C, June 1989 (contact W.A. Telliard, U.S. EPA, Office of Water Regulations and Standards, 401 M St., SW, Washington, DC, 20460. Phone: 202-382-7131).

Chloromethane   **EPA Method 502**
CAS #74-87-3

TITLE Volatile Organic Compounds in Water By Purge and Trap Capillary Column Gas Chromatography with Photoionization and Electrolytic Conductivity Detectors in Series. U.S. EPA Method 502.2, Rev. 2.0, 1989.

MATRIX Drinking water and raw source water. The latter should include most surface water and groundwater sources.

METHOD SUMMARY This method covers 60 volatile organic compounds that contain halogen atoms and/or that are aromatic. An inert gas (zero grade nitrogen or helium) is bubbled through a 25-mL or a 5-mL water sample (depending on the expected concentration of the analytes). Purged sample components are trapped in a tube of sorbent materials. When purging is complete, the sorbent tube is heated and backflushed with helium to desorb the trapped sample onto a capillary GC column. The column is temperature programmed to separate the method analytes which are then detected with a photoionization detector (PID) and a Hall electrolytic conductivity (HECD) placed in series. The PID is selective for aromatic compounds and the HECD is selective for halogenated compounds.

INTERFERENCES Impurities in the purge gas and from organic compounds outgassing from the plumbing ahead of the trap account for many contamination problems. Interferences purged or coextracted from the samples will vary considerably from source to source, depending upon the particular sample or extract being tested. Cross-contamination can occur whenever high-level and low-level samples are analyzed sequentially. Samples also can be contaminated by diffusion of volatile organics (particularly methylene chloride and fluorocarbons) through the septum seal into the sample during shipment and storage. The lab where volatile analysis is performed and also the refrigerated storage area should be completely free of solvents.

INSTRUMENTATION A GC containing a series configuration of a high temperature photoionization detector (PID) equipped with 10.0 eV (nominal) lamp and Hall electrolytic conductivity detector (HECD) is required. Also required is an all-glass 5-mL purging device, a sorbent trap, and a thermal desorption apparatus which is connected to the GC system.

Column 1: VOCOL glass wide-bore capillary column.
Column 2: RTX–502.2 mega-bore capillary column.
Column 3: DB-62 mega-bore capillary column.

PRECISION & ACCURACY Method detection limits are dependent upon the characteristics of the gas chromatographic system used. Analytes that are not separated chromatographically cannot be individually identified and used in the same calibration mixture or water samples unless an alternative technique for identification and quantification, such as mass spectrometry, is used.

Electrolytic conductivity detetor (c) range in µg/L (a) was 0.02–200.
Electrolytic conductivity detetor (c) MDL in µg/L (b) was 0.03.
Electrolytic conductivity detetor (c) accuracy as % recovery was 96.
Electrolytic conductivity detetor (c) precision as % RSD was 9.2.
Photoionization detector (d) range in µg/L (a) was 0.02–200.
Photoionization detector (d) MDL in µg/L (b) was not listed.
Photoionization detector (d) accuracy as % recovery was not listed.
Photoionization detector (d) precision as % RSD was not listed.

(a) *The applicable concentration range of this method is compound, instrument, and matrix-dependent. It is listed as being approximately 0.02 to 200 µg/L but no specific information is provided so caution should be observed.*
(b) *The method detection limits reports with this method are compound, instrument, and matrix-dependent. The values reported were calculated using reagent water fortified with the corresponding compounds at 10 µg/L and a GC-equipped with a 60 m × 0.75 mm VOLCOL wide bore capillary column with 1.5 µm film thickness and using helium carrier gas.*
(c) *Recoveries and relative standard deviations were determined from seven samples of reagent water fortified with 10 µg/L of each compound. 2-Bromo-1-chloropropane was used as the internal standard for calculating average recoveries.*
(d) *Recoveries and relative standard deviations were determined from seven samples of reagent water fortified with 10 µg/L of each compound. Fluorobenzene was used as the internal standard for calculating average recoveries.*

SAMPLING METHOD Collect samples using a 40- to 120-mL screw-cap vial (prewashed with detergent, rinsed with distilled water and oven dried at 105°C) with a Teflon®-faced silicone septum. Collect bubble-free samples and place the septum with the Teflon® side down on the water.

SAMPLE PRESERVATION If residual chlorine is present in the water add about 25 mg of ascorbic acid to each vial before samples are collected to remove the chlorine. Add hydrochloric acid to reduce pH to <2, immediately cool samples to 4°C, and store them in a solvent-free refrigerator at 4°C until analysis.

MHT The maximum holding time for samples is 14 days from the time they were collected.

SAMPLE PREPARATION Remove the plungers from two 5-mL syringes and attach a closed syringe valve to each. Warm the sample to room temperature, open the sample bottle, and carefully pour the sample into one of the syringe barrels to just short of overflowing. Replace the syringe plunger, invert the syringe, and compress the sample. Open the syringe valve and vent any residual air while adjusting the sample volume to 5.0 mL. Add 10 µL of the internal calibration standard to the sample through the syringe valve. Close the valve. Fill the second syringe in an identical manner from the same sample bottle. Reserve this second syringe for a reanalysis if necessary.

QUALITY CONTROL As an initial demonstration of lab accuracy and precision, analyze 4 to 7 replicates of a lab fortified blank containing analyte at 0.1–5 µg/L. Collect all samples in duplicate. Surrogate analytes (similar to those of the analytes of interest), whose concentration is known in every sample, are measured using the same internal standard calibration procedure. Duplicate field reagent water blanks (trip blanks) must be analyzed with each set of samples, lab reagent blanks (method blanks) must be analyzed with each batch of samples processed as a group within a work shift. Also, a single lab-fortified blank that contains each of the analytes of interest should be analyzed with each batch of samples processed as a group within a work shift. A 3- to 5-point calibration curve is needed depending on the calibration range factor required.

EPA CONTACT & HOTLINE For technical questions contact Dr. Baldev Bathija, U.S. EPA, Office of Ground Water and Drinking Water (WH-550D), 401 M St. SW, Washington, DC 20460. Tel. (202) 260-3040. For further information the EPA Safe Drinking Water Hotline may be called at: (800) 426-4791.

REFERENCE Methods for the Determination of Organic Compounds in Drinking Water, EPA/600/4-88/039 (revised July 1991; Final Rule for determination of compliance with the MCL for Total Trihalomethanes under 141.30, in 40 CFR Part 141, Vol. 58, No. 147, Fed. Reg., Tuesday Aug. 3, 1993). U.S. EPA Environmental Monitoring Systems Laboratory, Cincinnati, OH, 45268, USA. Available from the National Technical Information Service (NTIS), 5285 Port Royal Road, Springfield, VA 22161; Tel. 800-553-6847. NTIS Order Number is PB91-231480.

Chloromethane   **EPA Method 524**
CAS #74-87-3

TITLE Measurement of Purgeable Organic Compounds in Water by Capillary Column GC/MS.

MATRIX Drinking water and raw source water; the latter should include most surface water and groundwater sources.

METHOD SUMMARY Method 524.2 covers 60 volatile organic compounds. An inert gas (zero grade nitrogen or helium) is bubbled through a 25-mL or a 5-mL water sample (depending on the expected concentration of the analytes). Purged sample components are trapped in a tube of sorbent materials. When purging is complete, the sorbent tube is heated and backflushed with helium to desorb the trapped sample onto a capillary GC column.

INTERFERENCES Impurities in the purge gas and from organic compounds outgassing from the plumbing ahead of the trap account for many contamination problems. Interferences purged or coextracted from the samples will vary considerably from source to source, depending upon the particular sample or extract being tested. Cross-contamination can occur whenever high-level and low-level samples are analyzed sequentially. Samples also can be contaminated by diffusion of volatile organics (particularly methylene chloride and fluorocarbons) through the septum seal into the sample during shipment and storage.

INSTRUMENTATION A GC/MS with a data system equipped with one of the following capillary GC columns:

Column 1: VOCOL glass wide bore capillary column.
Column 2: DB-624 fused silica capillary column.
Column 3: DB-5 fused silica capillary column.

Also required is an all-glass 25 mL or 5-mL purging device, a sorbent trap, and a thermal desorption apparatus which is connected to the GC/MS system.

PRECISION & ACCURACY Method detection limits are compound- and instrument-dependent, and may vary from approximately 0.02–0.35 µg/L. Note in the table below that the "true" concentration range used for accuracy and precision measurements was quite narrow. However, the applicable concentration range of this method is primarily column dependent and is approximately 0.02 to 200 µg/L for the wide-bore thick-film columns. Narrow-bore thin-film columns may have a capacity which limits the range to about 0.02 to 20 µg/L. Analytes that are inefficiently purged from water will not be detected when present at low concentrations, but they can be measured with acceptable accuracy and precision when present in sufficient amounts.

Analytes that are not separated chromatographically, but which have different mass spectra and non-interfering quantification ions, can be identified and measured in the same calibration mixture or water sample. Analytes which have very similar mass spectra cannot be individually identified and measured in the same calibration mixture or water samples unless they have different retention times. Co-eluting compounds with very similar mass spectra, typically many structural isomers, must be reported as an isomeric group or pair.

The range (in µg/L) was 0.5–10.
The Method Detection Limig (in µg/L) was 0.13.
The accuracy (as % recovery) was 93.
The precision (in %) was 8.9.

Note: Data were obtained from 16–31 determinations using a wide-bore capillary column and a jet separator interfaced to a quadrupole mass spectrometer. All analytes were in a reagent water matrix.

SAMPLING METHOD Collect samples using a 40- to 120-mL screw-cap vial (prewashed with detergent, rinsed with distilled water and oven dried at 105°C) with a Teflon®-faced silicone septum. Collect bubble-free samples and place the septum with the Teflon® side down on the water.

SAMPLE PRESERVATION If residual chlorine is present in the water add about 25 mg of ascorbic acid to each vial before samples are collected to remove the chlorine. Add hydrochloric acid to reduce pH to <2, and immediately cool samples to 4°C, and store them in a solvent-free refrigerator at 4°C until analysis.

MHT The maximum holding time for samples is 14 days from the time they were collected.

SAMPLE PREPARATION Remove the plungers from two 25-mL (or 5-mL depending on sample size) syringes and attach a closed syringe valve to each. Warm the sample to room temperature, open the sample bottle, and carefully pour the sample into one of the syringe barrels to just short of overflowing. Replace the syringe plunger, invert the syringe, and compress the sample. Open the syringe valve and vent any residual air while adjusting the sample volume to 25.0 mL (or 5 mL). For samples and blanks, add 5 µL of the fortification solution containing the internal standard and the surrogates to the sample through the syringe valve. For calibration standards and lab fortified blanks, add 5 µL of the fortification solution containing the internal standard only. Close the valve. Fill the second syringe in an identical manner from the same sample bottle. Reserve this second syringe for a reanalysis if necessary.

QUALITY CONTROL As an initial demonstration of lab accuracy and precision, analyze 4 to 7 replicates of a lab fortified blank containing analyte at 0.2–5 µg/L. Collect all samples in duplicate. Surrogate analytes (similar to those of the analytes of interest), whose concentration is known in every sample, are measured using the same internal standard calibration procedure. Duplicate field reagent water blanks (trip blanks) must be analyzed with each set of samples, lab reagent blanks (method blanks) must be analyzed with each batch of samples processed as a group within a work shift. Also, a single lab-fortified blank that contains each of the analytes of interest should be analyzed with each batch of samples processed as a group within a work shift. A 3- to 5-point calibration curve is needed depending on the calibration range factor required.

EPA CONTACT & HOTLINE For technical questions contact Dr. Baldev Bathija, U.S. EPA, Office of Ground Water and Drinking Water (WH-550D), 401 M St. SW, Washington, DC 20460. Tel. (202) 260-3040. For further information the EPA Safe Drinking Water Hotline may be called at: (800) 426-4791.

REFERENCE Methods for the Determination of Organic Compounds in Drinking Water, EPA/600/4-88/039 (revised July 1991; Final Rule for determination of compliance with the MCL for Total Trihalomethanes under 141.30, in 40 CFR Part 141, Vol. 58, No. 147, Fed. Reg., Tuesday Aug. 3, 1993). U.S. EPA Environmental Monitoring Systems Laboratory, Cincinnati, OH, 45268, USA. Available from the National Technical Information Service (NTIS), 5285 Port Royal Road, Springfield, VA 22161; Tel. 800-553-6847. NTIS Order Number is PB91-231480.

Chloromethane   **EPA Method 8021**
CAS #74-87-3

TITLE Halogenated Volatile by Gas Chromatography Using Photoionization and Electrolytic Conductivity Detectors in Series: Capillary Column Technique

MATRIX This method is applicable to nearly all types of samples, regardless of water content, including groundwater, aqueous sludges, caustic liquors, acid liquors, waste solvents, oily wastes, mousses, tars, fibrous wastes, polymeric emulsions, filter cakes, spent carbons, spent catalysts, soils, and sediments.

METHOD SUMMARY This method is used to determine 60 volatile organic compounds in a variety of solid waste matrices. It provides GC conditions for the detection of halogenated and aromatic volatile organic compounds. Samples can be analyzed using direct injection or purge-and-trap (EPA Method 5030). Groundwater samples must be analyzed using EPA Method 5030 (where applicable). A temperature program is used with the GC. Detection is achieved by a photoionization detector (PID) and a Hall electrolytic conductivity detector (HECD) in series.

INTERFERENCES Samples can be contaminated by diffusion of volatile organics (particularly chlorofluorocarbons and methylene chloride) through the sample container septum during shipment and storage.

INSTRUMENTATION A GC-equipped with variable-constant differential flow controllers, subambient oven controller, PID and HECD detectors connected with a short piece of uncoated capillary tubing and a data system.

Column: 60 m × 0.75 mm I.D. VOCOL wide-bore capillary column with 1.5 µm film thickness.

PRECISION & ACCURACY MDLs are compound-dependent and vary with purging efficiency and concentration. The applicable concentration range of this method is compound- and instrument-dependent but is approximately 0.1 to 200 µg/L. Analytes that are inefficiently purged from water will not be detected when present at low concentrations, but they can be measured with acceptable accuracy and precision when present in sufficient amounts. The estimated quantitation limit (EQL) for an individual compound is approximately 1 µg/kg (wet weight) for soil/sediment samples, 100 µg/kg (wet weight) for wastes, and 1 µg/L for groundwater. EQLs will be proportionately higher for sample extracts and samples that require dilution or reduced sample size to avoid saturation of the detector.

MULTIPLICATION FACTORS FOR OTHER MATRICES (a)

Matrix	Factor (b)
Groundwater	10
Low-concentration soil	10
Water miscible liquid waste	500
High-concentration soil and sludge	1250
Non-water miscible waste	1250

(a) Sample EQLs are highly matrix-dependent. The EQLs listed herein are provided for guidance and may not always be achievable.
(b) EQL = [Method detection limit] × [Factor]. For non-aqueous samples, the factor is on a wet-weight basis.

SINGLE LABORATORY ACCURACY & PRECISION DATA FOR VOCs IN WATER

This method was tested in a single lab using water spiked at 10 µg/L and the following data was reported:

Recoveries and standard deviations were determined from seven samples and spiked at 10 µg/L of each analyte. Recoveries were determined by the internal standard method. Internal standards were: Fluorobenzene for PID and 2-Bromo-1-chloropropane for HECD.

The average recovery (in percent) for the PID was none (no response for this detector).

The standard deviation of the recovery for the PID was none (no response for this detector).

The MDL (in µg/mL) for the PID was none (no response for this detector).

The average recovery (in percent) for the HECD was 96.

The standard deviation of the recovery for the HECD was 8.9.

The MDL (in µg/mL) for the HECD was 0.03.

SAMPLE COLLECTION, PRESERVATION & HANDLING
Volatile Organics — Standard 40-mL glass screw-cap VOA vials with Teflon®-faced silicone septum may be used for both liquid and solid matrices. When collecting samples, liquids and solids should be introduced into the vials gently to reduce agitation which might drive off volatile compounds. If there are any air bubbles present the sample must be retaken. Tap slightly as they are filled to try and eliminate as much free air space as possible. The two vials from each sampling locations should be sealed in separate plastic bags to prevent cross-contamination between samples particularly if the sampled waste is suspected of containing high levels of volatile organics.

Semivolatile organics — Containers used to collect samples for the determination of semivolatile organic compounds should be soap and water washed followed by methanol (or isopropanol) rinsing. The sample containers should be of glass or Teflon® and have screw-top covers with Teflon® liners.

Preservation for volatile organics — No preservation is used with concentrated waste samples. With liquid samples containing no residual chlorine, 4 drops of concentrated hydrochloric acid are added and the samples are immediately cooled to 4°C. When liquid samples contain residual chlorine, they are treated as above and, in addition, 4 drops of 4% aqueous sodium thiosulfate are added. Soil, sediment, and sludge samples are only cooled to 4°C.

Preservation for semivolatile organics — No preservation is used with concentrated waste samples. With liquid samples containing no residual chlorine and with soil, sediment, and sludge samples, immediately cooling to 4°C is the only preservation used. When residual chlorine is present then 3 mL of 10% aqueous sodium sulfate is added for each gallon of sample collected, followed by cooling to 4°C.

MHT The holding time for all volatile organics samples is 14 days. Liquid samples must be extracted within 7 days and their extracts analyzed within 40 days. Concentrated waste, soil, sediment, and sludge samples must be extracted within 14 days and their extracts analyzed within 40 days.

SAMPLE PREPARATION Volatile compounds are introduced into the gas chromatograph either by direct injector or purge-and-trap (EPA Method 5030). EPA Method 5030 may be used directly on groundwater samples or low-concentration contaminated soils and sediments. For medium-concentration soils or sediments, methanolic extraction, as described in EPA Method 5030, may be necessary prior to purge-and-trap analysis.

QUALITY CONTROL Calculate surrogate standard recovery on all samples, blanks, and spikes. A trip blank is recommended to check on sampling, storage, and handling contamination. Calibration standards, at a minimum of five concentration levels, are prepared in organic-free reagent water. One of the concentration levels should be at a concentration near, but above, the method detection limit.

A combination of bromochloromethane, 2-bromo-1-chloropropane, 1,4-dichlorobutane, and bromochlorobenzene are recommended as surrogate standards to encompass the range of the temperature program used in this method.

REFERENCE Test Methods for Evaluating Solid Waste, Physical/Chemical Methods, SW-846, 3rd Edition, U.S. EPA, Office of Solid Waste, Washington, DC, EPA Method 8021A, Rev. 1, Nov. 1992.

Chloromethane   **EPA Method 8240**
CAS #74-87-3

TITLE Volatile Organics By GC/MS: Packed Column Technique

MATRIX Nearly all types of sample matarices, regardless of water content, can be analyzed using this method. This includes groundwater, aqueous sludges, caustic liquors, acid liquors, waste solvents, oily wastes, mousses, tars, fibrous wastes, polymetric emulsions, filter cakes, spent carbons, spent catalysts, soils, and sediments.

METHOD SUMMARY Method 8240B covers 80 volatile organic compounds that are introduced into a gas chromatograph by the purge-and-trap method or by direct injection (in limited applications). For the purge-and-trap method an inert gas (zero grade nitrogen or helium) is bubbled through a 5-mL solution at ambient temperature. Purged sample components are trapped in a tube of sorbent materials. When purging is complete, the sorbent tube is heated and backflushed with inert gas to desorb the trapped components onto a GC column.

INTERFERENCES Impurities in the purge gas and from organic compounds outgassing from the plumbing ahead of the trap account for many contamination problems. Interferences purged or coextracted from the samples will vary considerably from source to source. Cross-contamination can occur whenever high-level and low-level samples are analyzed sequentially. Whenever an unusually concentrated sample is analyzed, it should be followed by the analysis of organic-free reagent water to check for cross-contamination. Samples also can be contaminated by diffusion of volatile organics (particularly methylene chloride and fluorocarbons) through the septum seal into the sample during shipment and storage. A trip blank can serve as a check on such contamination. The lab where volatile analysis is performed and also the refrigerated storage area should be completely free of solvents.

INSTRUMENTATION A gas chromatograph/mass spectrometry/data system (GC/MS) equipped with a 6 ft × 0.1 in I.D. glass column packed with 1% SP-1000 on Carbopack-B (60/80 mesh) is required. Also needed is a 5-mL purging device, a sorbent trap, and a thermal desorption apparatus.

PRECISION & ACCURACY This method is reported to have been tested by 15 laboratories using organic-free reagent water, drinking water, surface water, and industrial wastewaters (not specified) fortified at six concentrations over the range 5–600 µg/L.

Sample estimated quantitation limits (EQLs) are highly matrix-dependent. The EQLs listed may not always be achievable. EQLs listed for soils or sediments are based on wet weight. Normally, data is reported on a dry-weight basis; therefore, EQLs will be higher, based on the percent dry weight of each sample. Note that EQLs are even more variable than MDLs and that they are highly variable depending on the matrix being analyzed.

EQL in groundwater in µg/L was 10.
EQL in low soil or sediment in µg/kg was 10.
Accuracy (a) in µg/L was 1.03C–1.81.
Precision (b) in µg/L was 0.58x+0.43.

(a) *Average recovery found for measurements of samples containing a concentration of C, in µg/L.*
(b) *Overall precision found for samples with average recovery X for samples containing a concentration of C in µg/L.*
X = *Average recovery found for measurement of samples containing a concentration of C in µg/L.*

MULTIPLICATION FACTORS FOR OTHER MATRICES

Other Matrices	Factor (a)
Waste miscible liquid waste	50
High-concentration soil and sludge	125
Non-water miscible waste	500

(a) EQL = [EQL for low soil/sediment] × [Factor]. For non-aqueous samples, the factor is on a wet-weight basis.

SAMPLING METHOD
Liquid samples — Use a 40-mL glass screw-cap VOA vial with a Teflon®-faced silicone septum that has been prewashed, rinsed with distilled deionized water, and oven dried. However, if residual chlorine is present, collect sample in a 40-oz. soil VOA container which has been pre-preserved with 4 drops of 10% sodium thiosulfate, mix gently, and then transfer the sample to a 40-mL VOA vial. Collect bubble-free samples in duplicate and seal them in separate plastic bags.

Soils or sediments, and sludges — Use an 8-oz. widemouth glass bottle with a Teflon®-faced silicone septum that has been prewashed with detergent, rinsed with distilled deionized water, and oven dried. Tap slightly to eliminate free air space. Collect samples in duplicate and seal them in separate plastic bags.

SAMPLE PRESERVATION
Liquid samples — Add 4 drops of concentrated HCL and immediately cool samples to 4°C and store in a solvent-free refrigerator.

Soils or sediments, and sludges — Cool samples to 4°C and store in a solvent-free refrigerator.

MHT Maximum holding time is 14 days from the date of sample collection.

SAMPLE PREPARATION
Liquid samples — Remove the plunger from a 5-mL syringe and carefully pour the sample into the syringe barrel to just

short of overflowing. Replace the syringe plunger and compress the sample. Open the syringe valve and vent any residual air while adjusting the sample volume to 5.0 mL. If there is only one volatile organic analysis (VOA) vial, a second syringe should be filled at this time to protect against possible loss of sample integrity. Add 10 μL of surrogate spiking solution and 10 μL of internal standard spiking solution through the valve bore of the 5-mL syringe, then close the valve. The surrogate and internal standards may be mixed and added as a single spiking solution.

Sediments, soils, and waste samples — All samples of this type should be screened by GC analysis using a headspace method (EPA Method 3810) or the hexadecane extraction and screening method (EPA Method 3820). Use the screening data to determine whether to use the low-concentration method (0.005–1 mg/kg) or the high-concentration method (>1 mg/kg).

Low-concentration method — The low-concentration method is based on purging a heated sediment or soil sample mixed with organic-free reagent water containing the surrogate and internal standards. Analyze all reagent blanks and standards under the same conditions as the samples.

Use a 5-g sample if the expected concentration is <0.1 mg/kg or a 1-g sample for expected concentrations between 0.1 and 1 mg/kg. Mix the contents of the sample container with a narrow metal spatula. Weigh the amount of the sample into a tared purge device. Add the spiked water to the purge device, which contains the weighed amount of sample, and connect the device to the purge-and-trap system.

High-concentration method — This method is based on extracting the sediment or soil with methanol. A waste sample is either extracted or diluted, depending on its solubility in methanol. Wastes that are insoluble in methanol are diluted with reagent tetraglyme or possibly polyethylene glycol (PEG). An aliquot of the extract is added to organic-free reagent water containing surrogate and internal standards. This is purged at ambient temperature. All samples with an expected concentration of >1.0 mg/kg should be analyzed by this method.

Mix the contents of the sample container with a narrow metal spatula. For sediments or soils and solid wastes that are insoluble in methanol, weigh 4 g (wet weight) of sample into a tared 20-mL vial. For waste that is soluble in methanol, tetraglyme, or PEG, weigh 1 g (wet weight) into a tared scintillation vial or culture tube or a 10-mL volumetric flask. Quickly add 9.0 mL of appropriate solvent then add 1.0 mL of a surrogate spiking solution to the vial, cap it, and shake it for 2 min.

METHANOL EXTRACT REQUIRED FOR ANALYSIS OF HIGH-CONCENTRATION SOILS OR SEDIMENTS

Approximate Concentration Range	Volume of Methanol Extract (a)
500–10,000 μg/kg	100 μL
1,000–20,000 μg/kg	50 μL
5,000–100,000 μg/kg	10 μL
25,000–500,000 μg/kg	100 μL of 1/50 dilution (b)

Calculate appropriate dilution factor for concentrations exceeding this table.

(a) The volume of methanol added to 5 mL of water being purged should be kept constant. Therefore, add to the 5-mL syringe whatever volume of methanol is necessary to maintain a volume of 100 μL added to the syringe.
(b) Dilute an aliquot of the methanol extract and then take 100 μL for analysis.

QUALITY CONTROL Demonstrate, through the analysis of a reagent water blank, that interferences from the analytical system, glassware, and reagents are under control. Blank samples should be carried through all stages of the sample preparation and measurement steps. For each analytical batch (up to 20 samples), a reagent blank, matrix spike, and matrix spike duplicate must be analyzed (the frequency of the spikes may be different for different monitoring programs). The blank and spiked samples must be carried through all stages of the sample preparation and measurement steps. QC samples mentioned in the section on Interferences will also be needed as appropriate to those situations.

REFERENCE Test Methods for Evaluating Solid Waste (SW-846). U.S. EPA. 1983. Method 8240B, Rev. 2, Nov. 1990. Office of Solid Wastes, Washington, DC.

Chloromethane   **EPA Method 8260**
CAS #74-87-3

TITLE Volatile Organic Compounds by GC/MS: Capillary Column Technique

MATRIX This method is applicable to nearly all types of samples, regardless of water content, including groundwater, soils, and sediments.

METHOD SUMMARY Method 8260A covers 58 volatile organic compounds that are introduced into a gas chromatograph by the purge-and-trap method or by direct injection (in limited applications). Zero-grade helium is bubbled through a 5-mL solution at ambient temperature. Purged sample components are trapped in a tube containing suitable sorbent materials. When purging is complete, the sorbent tube is heated and backflushed with helium to desorb trapped sample components. The analytes are desorbed directly to a large bore capillary or cryofocussed on a capillary precolumn before being flash evaporated to a narrow bore capillary for analysis.

INTERFERENCES Major contaminant sources are volatile materials in the lab and impurities in the inert purging gas and in the sorbent trap. Interfering contamination may occur when a sample containing low concentrations of volatile organic compounds is analyzed immediately after a sample containing high concentrations of volatile organic compounds. After analysis of a sample containing high concentrations of volatile organic compounds, one or more calibration blanks should be analyzed to check for cross-contamination. Screening of the samples prior to purge-and-trap GC/MS analysis is highly recommended to prevent contamination of the system. This is especially true for soil and waste samples.

Special precautions must be taken to analyze for methylene chloride. The analytical and sample storage area should be isolated from all atmospheric sources of methylene chloride. All gas chromatography carrier gas lines and purge gas plumbing should be constructed from stainless steel or copper tubing. Laboratory clothing previously exposed to methylene chloride fumes during liquid-liquid extraction procedures can contribute to sample contamination.

Samples can also be contaminated by diffusion of volatile organics (particularly methylene chloride and fluorocarbons) through the septum seal during shipment and storage. A trip blank can serve as a check on such contamination.

INSTRUMENTATION GC/MS with a temperature-programmable chromatograph suitable for splitless injection equipped with variable constant differential flow controllers, a subambient oven controller, a purging device, sorbent trap, a thermal desorption apparatus and a capillary precolumn interface when using cryogenic cooling will be needed. The following GC columns may be used:

Column 1: 60 m × 0.75mm I.D. capillary column coated with VOCOL, 1.5 µm film thickness.
Column 2: 30 m × 0.53mm capillary column coated with DB-624 or VOCOL, 3 µm film thickness.
Column 3: 30 m × 0.32mm I.D. capillary column coated with DB-5 or SE-54, 1-µm film thickness.

PRECISION & ACCURACY This method has been tested in a single lab using spiked water. Using a wide-bore capillary column, water was spiked at concentrations between 0.5 and 10 µg/L. Single lab accuracy and precision data are presented. The MDL actually achieved in a given analysis will vary depending on instrument sensitivity and matrix effects.

The MDL (a) in µg/L was 0.13.
The concentration range in µg/L was 0.5–10.
The mean accuracy (% of true value) was 93.
The precision as relative standard deviation was 8.9.

Note: The MDL is based on a 25-mL sample volume instead of a 5-mL sample volume.

SAMPLING METHOD
Liquid samples — Use a 40-mL glass screw-cap VOA vial with a Teflon®-faced silicone septum that has been prewashed, rinsed with distilled deionized water, and oven dried. If residual chlorine is present, collect the sample in a 4-oz soil VOA container which has been pre-preserved with 4 drops of 10% sodium thiosulfate. Mix gently and transfer the sample to a 40-mL VOA vial. Collect bubble-free samples in duplicate and seal each sample in a separate plastic bag.

Soils, sediments, and sludges — Use an 8-oz widemouth glass bottle with Teflon®-faced silicone septum that has been prewashed, rinsed with distilled deionized water, and oven dried. **Do not** heat the septum for more than 1 h. Tap slightly to eliminate any free air space. Collect samples in duplicate and seal each one in a separate plastic bag.

SAMPLE PRESERVATION
Liquid samples — Add 4 drops of concentrated HCL, cool to 4°C and store in a solvent-free refrigerator.

Soils, sediments and sludges — Cool samples to 4°C and store in a solvent-free refrigerator.

MHT The maximum holding time of any sample (liquids, soils, sediments, and sludges) is 14 days.

SAMPLE PREPARATION
Liquid samples — Remove the plunger from a 5-mL syringe and carefully pour the sample into the syringe barrel to just short of overflowing. Replace the syringe plunger and compress the sample. Open the syringe valve and vent any residual air while adjusting the sample volume to 5.0 mL. If there is only one volatile organic analysis (VOA) vial, a second syringe should be filled at this time to protect against possible loss of sample integrity. Add 10 µL of surrogate spiking solution and 10 µL of internal standard spiking solution through the valve bore of the 5-mL syringe, then close the valve. The surrogate and internal standards may be mixed and added as a single spiking solution.

Sediments, soils, and waste samples — All samples of this type should be screened by GC analysis using a headspace method (EPA Method 3810) or the hexadecane extraction and screening method (EPA Method 3820). Use the screening data to determine whether to use the low-concentration method (0.005–1 mg/kg) or the high-concentration method (>1 mg/kg).

Low-concentration method — The low-concentration method is based on purging a heated sediment or soil sample mixed with organic-free reagent water containing the surrogate and internal standards. Analyze all reagent blanks and standards under the same conditions as the samples.

Use a 5-g sample if the expected concentration is <0.1 mg/kg or a 1-g sample for expected concentrations between 0.1 and 1 mg/kg. Mix the contents of the sample container with a narrow metal spatula. Weigh the amount of the sample into a tared purge device. Add the spiked water to the purge device, which contains the weighed amount of sample, and connect the device to the purge-and-trap system.

High-concentration method — This method is based on extracting the sediment or soil with methanol. A waste sample is either extracted or diluted, depending on its solubility in methanol. Wastes that are insoluble in methanol are diluted with reagent tetraglyme or possibly polyethylene glycol (PEG). An aliquot of the extract is added to organic-free reagent water containing surrogate and internal standards. This is purged at ambient temperature. All samples with an expected concentration of >1.0 mg/kg should be analyzed by this method.

Mix the contents of the sample container with a narrow metal spatula. For sediments or soils and solid wastes that are insoluble in methanol, weigh 4 g (wet weight) of sample into a tared 20-mL vial. For waste that is soluble in methanol, tetraglyme, or PEG, weigh 1 g (wet weight) into a tared scintillation vial or culture tube or a 10-mL volumetric flask. Quickly add

9.0 mL of appropriate solvent then add 1.0 mL of a surrogate spiking solution to the vial, cap it, and shake it for 2 min.

METHANOL EXTRACT REQUIRED FOR ANALYSIS OF HIGH-CONCENTRATION SOILS OR SEDIMENTS

Approximate Concentration Range	Volume of Methanol Extract (a)
500–10,000 µg/kg	100 µL
1,000–20,000 µg/kg	50 µL
5,000–100,000 µg/kg	10 µL
25,000–500,000 µg/kg	100 µL of 1/50 dilution (b)

Calculate appropriate dilution factor for concentrations exceeding this table.

(a) The volume of methanol added to 5 mL of water being purged should be kept constant. Therefore, add to the 5-mL syringe whatever volume of methanol is necessary to maintain a volume of 100 µL added to the syringe.
(b) Dilute an aliquot of the methanol extract and then take 100 µL for analysis.

QUALITY CONTROL Demonstrate, through the analysis of a reagent water blank, that interferences from the analytical system, glassware, and reagents are under control. Blank samples should be carried through all stages of the sample preparation and measurement steps. For each analytical batch (up to 20 samples), a reagent blank, matrix spike, and matrix spike duplicate must be analyzed (the frequency of the spikes may be different for different monitoring programs). The blank and spiked samples must be carried through all stages of the sample preparation and measurement steps. QC samples mentioned in the section on Interferences will also be needed as appropriate to those situations.

Matrix spiking standards should be prepared from volatile organic compounds which will be representative of the compounds being investigated. The recommended internal standards are chlorobenzene-d5, 1,4-difluorobenzene, 1,4-dichlorobenzene-d4, and pentafluorobenzene. Using stock standard solutions, prepare secondary dilution standards containing the compounds of interest, either singly or mixed together in methanol. Store them in a vial with no headspace for no more than one week. Surrogates recommended are toluene-d8, 4-bromofluorobenzene, and dibromofluoromethane. Each sample undergoing GC/MS analysis must be spiked with 10 µL of the surrogate spiking solution prior to analysis.

REFERENCE Test Methods for Evaluating Solid Waste (SW-846). U.S. EPA 1983, Method 8260A, Rev. 1, Nov. 1990. Office of Solid Waste, Washington, DC.

Chloromethane   **EPA Method 601**
CAS #74-87-3

TITLE Purgeable Halocarbons

MATRIX Wastewater

APPLICATION Method covers 29 purgeable halocarbons. (Method 624 provides GC/MS conditions appropriate for the qualitative and quantitative confirmation of results). Method describes conditions for a 2nd GC column to confirm measurements made with primary column.

INTERFERENCES Impurities in the purge gas and organic compounds outgassing from the plumbing ahead of the trap. With high- and low-level samples, there can be carryover contamination. Diffusion of volatile organics through the septum seal into the sample.

INSTRUMENTATION GC-equipped with halide-specific detector. (With purge-and-trap unit).

RANGE 8.0–500 µg/L.

MDL 0.08 µg/L.

PRECISION 0.52X+1.31 µg/L (overall precision).

ACCURACY 0.77C+0.18 µg/L (as recovery).

SAMPLING METHOD 25-mL glass vial. Teflon®-lined septum.

STABILITY Cool, 4°C, 0.008% Sodium thiosulfate.

MHT 14 days.

QUALITY CONTROL The lab must on an ongoing basis, spike at least 10% of the samples from each sample site being monitored to assess accuracy.

REFERENCE Method 601, *Federal Register* Part VIII 40 CFR Part 136, Oct 26, 1984.

Chloromethane   **EPA Method 624**
CAS #74-87-3

TITLE Purgeables

MATRIX Wastewater

APPLICATION Method covers 31 purgeable organics. An inert gas is bubbled through a 5-mL water sample in a specially designed purging chamber. Here, purgeables are transferred from aqueous to gaseous phase, passed onto a sorbent column, and trapped. Trap is heated and backflushed with inert gas to desorb purgeables onto a GC column, where purgeables are separated.

INTERFERENCES Impurities in the purge gas, organic compounds outgassing from the plumbing ahead of the trap, and solvent vapors in the lab. With high- and low-level samples, there can be carryover contamination.

INSTRUMENTATION GC/MS with purge-and-trap unit.

RANGE 5–600 µg/L

MDL Not determined.

PRECISION 0.58X+0.43 µg/L (overall precision).

ACCURACY 1.03C–1.81 µg/L (as recovery).

SAMPLING METHOD 25-mL glass vial. Teflon®-lined septum.

STABILITY Cool, 4°C, 0.008% Sodium thiosulfate.

MHT 14 days.

QUALITY CONTROL The lab must on an ongoing basis, spike at least 5% of the samples from each sample site being monitored to assess accuracy.

REFERENCE Method 624, *Federal Register* Part VIII 40 CFR Part 136, Oct 26, 1984.

Chloromethane EPA Method 8010
CAS #74-87-3

TITLE Halogenated Volatile Organics

MATRIX Groundwater, soils, sludges, water miscible liquid wastes, and non-water miscible wastes.

APPLICATION This method is used for the analysis of 39 halogenated VOCs. Samples are analyzed using direct injection or purge-and-trap methods. Groundwater must be analyzed by the purge-and-trap method. The method provides an optional GC column which is used for analyte confirmation and that may help resolve analytes from interferences.

INTERFERENCES There can be carryover contamination with high- and low-level samples. Impurities may come from the purge-and-trap apparatus, organic compounds outgassing from the plumbing ahead of trap, diffusion of VOCs through the sample bottle septum during shipping or storage, or from solvent vapors in the lab.

INSTRUMENTATION GC capable of on-column injections or purge-and-trap sample introduction and a halogen specific detector. Column 1: 8 ft by 0.1 in 1%. SP-1000 on Carbopack-B. Column 2: 6 ft by 0.1 in bonded n-octane on Porasil-C.

RANGE 8 to 500 µg/L (reagent water)

MDL 0.08 µg/L (reagent water).

PQL FACTORS FOR MULTIPLYING × FID MDL VALUE

Matrix	Multiplication Factor
Groundwater	10
Low-level soil	10
Water miscible liquid waste	500
High-level soil and sludge	1250
Non-water miscible waste	1250

PRECISION $0.52X + 1.31$ µg/L (overall precision).

ACCURACY $0.77C + 0.18$ µg/L (as recovery).

SAMPLING METHOD For water and liquid samples; use glass 40-mL vials with Teflon®-lined septum caps and collect two vials per sample location with no headspace. For solids and concentrated waste samples; use widemouth glass bottles with Teflon® liners.

STABILITY For concentrated wastes, soils, sediments, or sludges: cool to 4°C. For liquids: add 4 drops of concentrated hydrochloric acid and cool to 4°C.

MHT 14 days.

QUALITY CONTROL Analyze a reagent blank, matrix spike, and matrix spike duplicate/duplicate for each analytical batch (up to 20 samples). Demonstrate the purity of glassware and reagents by analyzing a reagent water method blank. Internal, surrogate, and five concentration level calibration standards are used.

REFERENCE Test Methods for Evaluating Solid Waste (SW-846), U.S. EPA Office of Solid Waste, Washington, DC, Method 8010B, Rev. 2, Nov. 1992.

3-(Chloromethyl)pyridine hydrochloride EPA Method 8270
CAS #6959-48-4

TITLE Semivolatile Organic Compounds by GC/MS

MATRIX This method is used to determine the concentration of semivolatile organic compounds in extracts prepared from all types of solid waste matrices, soils, and groundwater. Although surface waters are not specifically mentioned, this method should be applicable to water samples from rivers, lakes, etc.

METHOD SUMMARY This method covers 259 semivolatile organic compounds. In very limited applications direct injection of the sample into the GC/MS system may be appropriate, but this results in very high detection limits (approximately 10,000 µg/L). Typically, a 1-L liquid sample, containing surrogate, and matrix spiking standards, is extracted in a continuous extractor first under acid conditions and then under basic conditions. Typically 30 g of a solid sample, containing surrogate, and matrix spiking standards, is extracted ultrasonically. After concentrating the extract to 1 mL it is spiked with 10 µL of an internal standard solution just prior to analysis by GC/MS. The volume injected should contain about 100 ng of base/neutral and 200 ng of acid surrogates (for a 1-µL injection). Analysis is performed by GC/MS using a capillary GC column.

INTERFERENCES Raw GC/MS data from all blanks, samples, and spikes must be evaluated for interferences. Contamination by carryover can occur whenever high-concentration and low-concentration samples are sequentially analyzed. To reduce carryover, the sample syringe must be rinsed out between samples with solvent. Whenever an unusually concentrated sample is encountered, it should be followed by the analysis of blank solvent to check for cross-contamination.

INSTRUMENTATION A GC/MS and a data system are required. The GC column used is a 30 m × 0.25 mm I.D. (or 0.32 mm I.D.) 1um film thickness silicone-coated fused silica capillary column. A continuous liquid-liquid extractor equipped with Teflon® or glass connection joints and stopcocks requiring no lubrication, a K-D concentrating apparatus, water bath, and an ultrasonic disrupter with a minimum power of 300 W and with pulsing capability are also required.

PRECISION & ACCURACY The estimated quantitation limit (EQL) of Method 8270B for determining an individual compound is approximately 1 mg/kg (wet weight) for soil or sediment samples, 1–200 mg/kg for wastes (dependent on matrix and method of preparation), and 10 µg/L for groundwater

samples. EQLs will be proportionately higher for sample extracts that require dilution to avoid saturation of the detector.

The EQL(b) for groundwater in µg/L is 100.
The EQL (a, b) for low concentrations in soil and sediment in µg/kg is not determined.
Accuracy as µg/L is not listed.
Overall precision in µg/L is not listed.

(a) EQLs listed for soil/sediment are based on wet weight. Normally data is reported in a dry-weight basis; therefore, EQLs will be higher based on the % dry weight of each sample. This calculation is based on a 30 g sample and gel permeation chromatography cleanup.
(b) Sample EQLs are highly matrix-dependent. The EQLs are provided for guidance and may not always be achievable.
C = *True value for concentration, in µg/L.*
X = *Average recovery found for measurements of samples containing a concentration of C, in µg/L.*

ESTIMATED QUANTITATION LIMIT FOR OTHER MATRICES

Other Matrices	Factor (a)
High-concentration soil and sludges by sonicator	7.5
Non-water miscible waste	75

(a) EQL for other matrices = [EQL for low soil/sediment] × [Factor]. This estimated EQL is similar to an EPA "Practical Quantitation Limit."

SAMPLING METHOD

Liquid samples — Use a 1 or 2½ gallon amber glass bottle with a screw-top Teflon®-lined cover that has been prewashed with detergent and rinsed with distilled water and methanol (or isopropanol).

Soils, sediments, or sludges — Use an 8-oz. widemouth glass with a screw-top Teflon®-lined cover that has been prewashed with detergent and rinsed with distilled water and methanol (or isopropanol).

SAMPLE PRESERVATION

Liquid samples — If residual chlorine is present, add 3 mL of 10% sodium thiosulfate per gallon, cool to 4°C and store in a solvent-free refrigerator until analysis; if chlorine is not present, then eliminate the sodium thiosulfate addition.

Soils, sediments, or sludges — Cool samples to 4°C and store in a solvent-free refrigerator.

MHT Liquid samples must be extracted within 7 days and the extracts analyzed within 40 days. Soils, sediments, or sludges may be stored for a maximum of 14 days and the extracts analyzed within 40 days.

SAMPLE PREPARATION

Liquid samples — Transfer 1 L quantitatively to a continuous extractor. If high concentrations are anticipated, a smaller volume may be used and then diluted with organic-free reagent water to 1 L. Adjust pH, if necessary, to pH <2 using 1:1 (V/V) sulfuric acid. Pipette 1.0 mL of a surrogate standard spiking solution into each sample. For the sample in each analytical batch selected for spiking, add 1.0 mL of a matrix spiking standard. For base/neutral acid analysis, the amount of the surrogates and matrix spiking compounds added to the sample should result in a final concentration of 100 ng/µL of each analyte in the extract to be analyzed (assuming a 1-µL injection). Extract with methylene chloride for 18–24 h. Next, adjust the pH of the aqueous phase to pH >11 using 10 N sodium hydroxide and extract it with methylene chloride again for 18–24 h. Dry the extract through a column containing anhydrous sodium sulfate and concentrate it to 1 mL using a K-D concentrator.

Soils, sediments, or sludges — Use 30 g of sample. Nonporous or wet samples (gummy or clay type) that do not have a free-flowing sandy texture must be mixed with anhydrous sodium sulfate until the sample is free flowing. Add 1 mL of surrogate standards to all samples, spikes, standards, and blanks. For the sample in each analytical batch selected for spiking, add 1.0 mL of a matrix spiking standard. For base/neutral acid analysis, the amount added of the surrogates and matrix spiking compounds should result in a final concentration of 100 ng/µL of each base/neutral analyte and 200 ng/µL of each acid analyte in the extract to be analyzed (assuming a 1-µL injection). Immediately add a 100-mL mixture of 1:1 methylene chloride:acetone and extract the sample ultrasonically for 3 min and then decant or filter the extracts. Repeat the extraction two or more times. Dry the extract using a column with anhydrous sodium sulfate and concentrate it to 1 mL in a K-D concentrator.

QUALITY CONTROL A methylene chloride solution containing 50 ng/µL of decafluorotriphenylphosphine (DFTPP) is used for tuning the GC/MS system each 12-h shift. A system performance check also must be made during every 12-h shift. A standard containing 50 ng/µL each of 4,4'-DDT, pentachlorophenol, and benzidine is required to verify injection port inertness and GC column performance. A calibration standard at mid-concentration, containing each compound of interest, including all required surrogates, must be performed every 12 h during analysis. After the system performance check is met, calibration check compounds (CCCs) are used to check the validity of the initial calibration.

The internal standard responses and retention times in the calibration check standard must be evaluated immediately after or during data acquisition. If the retention time for any internal standard changes by more than 30 seconds from the last check calibration (12 h), the chromatographic system must be inspected for malfunctions and corrections must be made, as required. If the electron ionization current plot (EICP) area for any of the internal standards changes by a factor of two from the last daily calibration standard check, the mass spectrometer must be inspected for malfunctions and corrections must be made, as appropriate.

Demonstrate, through the analysis of a reagent water blank, that interferences from the analytical system, glassware, and reagents are under control. The blank samples should be carried through all stages of the sample preparation and measurement steps. For each analytical batch (up to 20 samples), a reagent blank, matrix spike, and matrix spike duplicate/duplicate must be analyzed (the frequency of the spikes may be different for different monitoring programs). The blank and

spiked samples must be carried through all stages of the sample preparation and measurement steps. A QC reference sample concentrate containing each analyte at a concentration of 100 mg/L in methanol is required.

REFERENCE Test Methods for Evaluating Solid Waste (SW-846). U.S. EPA 1983, Method 8270B, Rev. 2, Nov. 1990. Office of Solid Waste, Washington, DC.

1-Chloronaphthalene   **EPA Method 8270**
CAS #90-13-1

TITLE Semivolatile Organic Compounds by GC/MS

MATRIX This method is used to determine the concentration of semivolatile organic compounds in extracts prepared from all types of solid waste matrices, soils, and groundwater. Although surface waters are not specifically mentioned, this method should be applicable to water samples from rivers, lakes, etc.

METHOD SUMMARY This method covers 259 semivolatile organic compounds. In very limited applications direct injection of the sample into the GC/MS system may be appropriate, but this results in very high detection limits (approximately 10,000 µg/L). Typically, a 1-L liquid sample, containing surrogate, and matrix spiking standards, is extracted in a continuous extractor first under acid conditions and then under basic conditions. Typically 30 g of a solid sample, containing surrogate, and matrix spiking standards, is extracted ultrasonically. After concentrating the extract to 1 mL it is spiked with 10 µL of an internal standard solution just prior to analysis by GC/MS. The volume injected should contain about 100 ng of base/neutral and 200 ng of acid surrogates (for a 1-µL injection). Analysis is performed by GC/MS using a capillary GC column.

INTERFERENCES Raw GC/MS data from all blanks, samples, and spikes must be evaluated for interferences. Contamination by carryover can occur whenever high-concentration and low-concentration samples are sequentially analyzed. To reduce carryover, the sample syringe must be rinsed out between samples with solvent. Whenever an unusually concentrated sample is encountered, it should be followed by the analysis of blank solvent to check for cross-contamination.

INSTRUMENTATION A GC/MS and a data system are required. The GC column used is a 30 m × 0.25 mm I.D. (or 0.32 mm I.D.) 1um film thickness silicone-coated fused silica capillary column. A continuous liquid-liquid extractor equipped with Teflon® or glass connection joints and stopcocks requiring no lubrication, a K-D concentrating apparatus, water bath, and an ultrasonic disrupter with a minimum power of 300 W and with pulsing capability are also required.

PRECISION & ACCURACY The estimated quantitation limit (EQL) of Method 8270B for determining an individual compound is approximately 1 mg/kg (wet weight) for soil or sediment samples, 1–200 mg/kg for wastes (dependent on matrix and method of preparation), and 10 µg/L for groundwater samples. EQLs will be proportionately higher for sample extracts that require dilution to avoid saturation of the detector.

The EQL(b) for groundwater in µg/L is not listed.
The EQL (a, b) for low concentrations in soil and sediment in µg/kg is not listed.
Accuracy as µg/L is 0.89C +0.01.
Overall precision in µg/L is 0.13X +0.34.

(a) *EQLs listed for soil/sediment are based on wet weight. Normally data is reported in a dry-weight basis; therefore, EQLs will be higher based on the % dry weight of each sample. This calculation is based on a 30 g sample and gel permeation chromatography cleanup.*

(b) *Sample EQLs are highly matrix-dependent. The EQLs are provided for guidance and may not always be achievable.*

$C =$ *True value for concentration, in µg/L.*
$X =$ *Average recovery found for measurements of samples containing a concentration of C, in µg/L.*

ESTIMATED QUANTITATION LIMIT FOR OTHER MATRICES

Other Matrices	Factor (a)
High-concentration soil and sludges by sonicator	7.5
Non-water miscible waste	75

(a) *EQL for other matrices = [EQL for low soil/sediment] × [Factor]. This estimated EQL is similar to an EPA "Practical Quantitation Limit."*

SAMPLING METHOD
Liquid samples — Use a 1 or 2½ gallon amber glass bottle with a screw-top Teflon®-lined cover that has been prewashed with detergent and rinsed with distilled water and methanol (or isopropanol).

Soils, sediments, or sludges — Use an 8-oz. widemouth glass with a screw-top Teflon®-lined cover that has been prewashed with detergent and rinsed with distilled water and methanol (or isopropanol).

SAMPLE PRESERVATION
Liquid samples — If residual chlorine is present, add 3 mL of 10% sodium thiosulfate per gallon, cool to 4°C and store in a solvent-free refrigerator until analysis; if chlorine is not present, then eliminate the sodium thiosulfate addition.

Soils, sediments, or sludges — Cool samples to 4°C and store in a solvent-free refrigerator.

MHT Liquid samples must be extracted within 7 days and the extracts analyzed within 40 days. Soils, sediments, or sludges may be stored for a maximum of 14 days and the extracts analyzed within 40 days.

SAMPLE PREPARATION
Liquid samples — Transfer 1 L quantitatively to a continuous extractor. If high concentrations are anticipated, a smaller volume may be used and then diluted with organic-free reagent water to 1 L. Adjust pH, if necessary, to pH <2 using 1:1 (V/V) sulfuric acid. Pipette 1.0 mL of a surrogate standard spiking solution into each sample. For the sample in each analytical batch selected for spiking, add 1.0 mL of a matrix spiking standard. For base/neutral acid analysis, the amount of the surrogates and matrix spiking compounds added to the sample should result in a final concentration of 100 ng/µL of each

analyte in the extract to be analyzed (assuming a 1-µL injection). Extract with methylene chloride for 18–24 h. Next, adjust the pH of the aqueous phase to pH >11 using 10 N sodium hydroxide and extract it with methylene chloride again for 18–24 h. Dry the extract through a column containing anhydrous sodium sulfate and concentrate it to 1 mL using a K-D concentrator.

Soils, sediments, or sludges — Use 30 g of sample. Nonporous or wet samples (gummy or clay type) that do not have a free-flowing sandy texture must be mixed with anhydrous sodium sulfate until the sample is free flowing. Add 1 mL of surrogate standards to all samples, spikes, standards, and blanks. For the sample in each analytical batch selected for spiking, add 1.0 mL of a matrix spiking standard. For base/neutral acid analysis, the amount added of the surrogates and matrix spiking compounds should result in a final concentration of 100 ng/µL of each base/neutral analyte and 200 ng/µL of each acid analyte in the extract to be analyzed (assuming a 1-µL injection). Immediately add a 100-mL mixture of 1:1 methylene chloride:acetone and extract the sample ultrasonically for 3 min and then decant or filter the extracts. Repeat the extraction two or more times. Dry the extract using a column with anhydrous sodium sulfate and concentrate it to 1 mL in a K-D concentrator.

QUALITY CONTROL A methylene chloride solution containing 50 ng/µL of decafluorotriphenylphosphine (DFTPP) is used for tuning the GC/MS system each 12-h shift. A system performance check also must be made during every 12-h shift. A standard containing 50 ng/µL each of 4,4′-DDT, pentachlorophenol, and benzidine is required to verify injection port inertness and GC column performance. A calibration standard at mid-concentration, containing each compound of interest, including all required surrogates, must be performed every 12 h during analysis. After the system performance check is met, calibration check compounds (CCCs) are used to check the validity of the initial calibration.

The internal standard responses and retention times in the calibration check standard must be evaluated immediately after or during data acquisition. If the retention time for any internal standard changes by more than 30 seconds from the last check calibration (12 h), the chromatographic system must be inspected for malfunctions and corrections must be made, as required. If the electron ionization current plot (EICP) area for any of the internal standards changes by a factor of two from the last daily calibration standard check, the mass spectrometer must be inspected for malfunctions and corrections must be made, as appropriate.

Demonstrate, through the analysis of a reagent water blank, that interferences from the analytical system, glassware, and reagents are under control. The blank samples should be carried through all stages of the sample preparation and measurement steps. For each analytical batch (up to 20 samples), a reagent blank, matrix spike, and matrix spike duplicate/duplicate must be analyzed (the frequency of the spikes may be different for different monitoring programs). The blank and spiked samples must be carried through all stages of the sample preparation and measurement steps. A QC reference sample concentrate containing each analyte at a concentration of 100 mg/L in methanol is required.

REFERENCE Test Methods for Evaluating Solid Waste (SW-846). U.S. EPA 1983, Method 8270B, Rev. 2, Nov. 1990. Office of Solid Waste, Washington, DC.

2-Chloronaphthalene   EPA Method 1625
CAS #91-58-7

TITLE Semivolatile Organic Compounds by Isotope Dilution GC/MS

MATRIX The compounds may be determined in waters, soils, and municipal sludges by this method.

METHOD SUMMARY This method is used to determine 176 semivolatile toxic organic pollutants associated with the CWA (as amended 1987); the RCRA (as amended 1986); the CERCLA (as amended 1986); and other compounds amenable to extraction and analysis by capillary column gas chromatography-mass spectrometry (GC/MS).

Stable isotopically-labeled analogs of the compounds of interest are added to the sample. If the solids content is less than 1%, a 1-L sample is extracted at pH 12–13, then at pH <2 with methylene chloride using continuous extraction techniques.

If the solids content is 30% or less, the sample is diluted to 1% solids with reagent water, homogenized ultrasonically, and extracted at pH 12–13, then at pH <2 with methylene chloride using continuous extraction techniques. If the solids content is greater than 30%, the sample is extracted using ultrasonic techniques.

Each extract is dried over sodium sulfate, concentrated to a volume of 5 mL, cleaned up using GPC, if necessary, and concentrated. Extracts are concentrated to 1 mL if GPC is not performed, and to 0.5 mL if GPC is performed.

An internal standard is added to the extract, and a 1-mL aliquot of the extract is injected into the GC. The compounds are separated by GC and detected by a MS. The labeled compounds serve to correct the variability of the analytical technique.

INTERFERENCES Solvents, reagents, glassware, and other sample processing hardware may yield artifacts and/or elevated baselines causing misinterpretation of chromatograms and spectra. Materials used in the analysis must be demonstrated to be free from interferences under the conditions of analysis by running method blanks initially and with each sample lot (sample started through the extraction process on a given 8-h shift, to a maximum of 20). Specific selection of reagents and purification of solvents by distillation in all glass systems may be required. Glassware and, where possible, reagents are cleaned by solvent rinse and baking at 450°C for 1-h minimum. Interferences coextracted from samples will vary considerably from source to source, depending on the diversity of the site being sampled.

INSTRUMENTATION Major instrumentation includes a GC with a splitless or on-column injection port for capillary column,

a MS with 70 eV electron impact ionization, and a data system to collect and record MS data, and process it. A K-D apparatus is used to concentrate extracts.

GC Column: 30 m × 0.25 mm I.D. 5% phenyl, 94% methyl, 1% vinyl silicone bonded phased fused silica capillary column.

PRECISION & ACCURACY The detection limits of the method are usually dependent on the level of interferences rather than instrumental limitations. The limits typify the minimum quantities that can be detected with no interferences present.

The minimum level (in µg/mL) was 10. This is defined as a minimum level at which the analytical system shall give recognizable mass spectra (background corrected) and acceptable calibration points.

The MDL (in µg/kg) in low solids was 80 and in high solids was 59; these were determined in digested sludge (low solids) and in filter cake or compost (high solids).

The labeled and native compound initial precision as standard deviation (in µg/L) was 100.

The labeled and native compound initial accuracy as average recovery (in µg/L) was 46–357.

SAMPLE COLLECTION, PRESERVATION & HANDLING Collect samples in glass containers. Aqueous samples which flow freely are collected in refrigerated bottles using automatic sampling equipment. Solid samples are collected as grab samples using widemouth jars. Maintain samples at 0 to 4°C from the time of collection until extraction. If residual chlorine is present in aqueous samples, add 80 mg sodium thiosulfate/L of water. Begin sample extraction within 7 days of collection, and analyze all extracts within 40 days of extraction.

SAMPLE PREPARATION Samples containing 1% solids or less are extracted directly using continuous liquid-liquid extraction techniques. Samples containing 1 to 30% solids are diluted to the 1% level with reagent water and extracted using continuous liquid-liquid extraction techniques. Samples containing greater than 30% solids are extracted using ultrasonic techniques.

Base/neutral extraction — Adjust the pH of the waters in the extractors to 12–13 with 6 N NaOH. Extract with methylene chloride for 24–48 h.
Acid extraction — Adjust the pH of the waters in the extractors to 2 or less using 6 N sulfuric acid. Extract with methylene chloride for 24–48 h.
Ultrasonic extraction of high solids samples — Add anhydrous sodium sulfate to the sample and QC aliquot(s). Add acetone:methylene chloride (1:1) to the sample and mix thoroughly

Concentrate extracts using a K-D apparatus.

QUALITY CONTROL The analyst is permitted to modify this method to improve separations or lower the costs of measurements, provided all performance specifications are met. Analyses of blanks are required to demonstrate freedom from contamination. When results of spikes indicate atypical method performance for samples, the samples are diluted to bring method performance within acceptable limits.

For low solids (aqueous samples), extract, concentrate, and analyze two sets of four 1-L aliquots (8 aliquots total) of the precision and recovery standard. For high solids samples, two sets of four 30-g aliquots of the high solids reference matrix are used.

Spike all samples with labeled compounds to assess method performance. Compute percent recovery of the labeled compounds using the internal standard method. Compare the labeled compound recovery for each compound with the corresponding labeled compound recovery.

Reagent water and high solids reference matrix blanks are analyzed to demonstrate freedom from contamination. Extract and concentrate a 1-L reagent water blank or a high solids reference matrix blank with each sample's lot (samples started through the extraction process on the same 8-h shift, to a maximum of 20 samples).

Field replicates may be collected to determine the precision of the sampling technique, and spiked samples may be required to determine the accuracy of the analysis when the internal standard method is used.

REFERENCE Semivolatile Organic Compounds by Isotope Dilution GC/MS. Office of Water Regulation and Standards, U.S. EPA Industrial Technology Division, Washington, DC, EPA Method 1625, Rev. C, June 1989 (contact W.A. Telliard, U.S. EPA, Office of Water Regulations and Standards, 401 M St., SW, Washington, DC, 20460. Phone: 202-382-7131).

2-Chloronaphthalene   **EPA Method 625**
CAS #91-58-7

TITLE Base/Neutrals and Acids, U.S. EPA Method 625

MATRIX This methods covers municipal and industrial wastewaters.

METHOD SUMMARY Approximately 1 L of sample is serially extracted with methylene chloride at a pH greater than 11 and again at a pH less than 2 using a separatory funnel or a continuous extractor. The methylene chloride extract is dried, concentrated to a volume of 1 mL, and analyzed by GC/MS. Qualitative identification of the parameters in the extract is performed using the retention time and the relative abundance of three characteristic masses (m/z). Qualitative analysis is performed using either external or internal standard techniques with a single characteristic m/z.

INTERFERENCES Method interferences may be caused by contaminants in solvents, reagents, glassware, and other sample processing hardware. Glassware must be scrupulously cleaned. Glassware should be heated in a muffle furnace at 400°C for 5 to 30 min. Some thermally stable materials, such as PCBs, may not be eliminated by this treatment. Solvent rinses with acetone and pesticide quality hexane may be substituted for the muffle furnace heating. Matrix interferences may be caused by contaminants that are coextracted from the sample. The base-neutral extraction may cause significantly reduced recovery of phenols. The packed gas chromatographic columns recommended for the basic fraction may not exhibit sufficient resolution for some analytes.

INSTRUMENTATION A GC/MS system with an injection port designed for on-column injection when using packed columns and for splitless injection when using capillary columns.

Column for base/neutrals: 1.8 m long × 2 mm I.D. glass, packed with 3% SP-2550 on Supelcoport (100/120 mesh) or equivalent.

Column for acids: 1.8 m long × 2 mm I.D. glass, packed with 1% SP-1240DA on Supelcoport (100/120 mesh) or equivalent.

PRECISION & ACCURACY The MDL concentrations were obtained using reagent water. The MDL actually achieved in a given analysis will vary depending on instrument sensitivity and matrix effects. This method was tested by 15 laboratories using reagent water, drinking water, surface water, and industrial wastewaters spiked at six concentrations over the range 5 to 100 µg/L. Single operator precision, overall precision, and method accuracy were found to be directly related to the concentration of the parameter matrix.

The MDL (in µg/L) in reagent water was not reported.
The standard deviation (in µg/L based on 4 recovery measurements) was 13.0.
The range (in µg/L) for average recovery for 4 measurements was 64.5–113.5.
The range (in %) for percent recovery was 60–118.
Accuracy (in µg/L) as expected recovery for one or more measurements of a sample containing a true concentration of C was 0.89C+0.01.
Precision (in µg/L) as expected single analyst standard deviation of measurements at an average concentration found at X was 0.07X+0.52.
Overall precision (in µg/L) as expected interlaboratory standard deviation of measurements in an average concentration found at X was 0.13X+0.34.

C = *True value of the concentration in µg/L.*
X = *Average recovery found for measurements of samples containing a concentration at C in µg/L.*

SAMPLE PREPARATION Adjust the pH to >11 with sodium hydroxide and serially extract in a separatory funnel with methylene chloride or else in a continuous extractor. Next, adjust the pH to <2 with sulfuric acid and serially extract in a separatory funnel with methylene chloride or else in a continuous extractor. Dry the extracts separately through a column of anhydrous sodium sulfate and then concentrate each of the extracts to 1.0 mL using a K-D apparatus.

SAMPLE COLLECTION, PRESERVATION & HANDLING Grab samples must be collected in glass containers. All samples must be refrigerated at 4°C from the time of collection until extraction. If residual chlorine is present, add 80 mg of sodium thiosulfate/L of sample and mix well. All samples must be extracted within 7 days of collection and completely analyzed within 40 days of extraction.

QUALITY CONTROL Make an initial, one-time, demonstration of the ability to generate acceptable accuracy and precision with this method. Before processing any samples, the analyst must analyze a reagent water blank to demonstrate that interferences from the analytical system and glassware are under control. Each time a set of samples is extracted or reagents are changed, a reagent water blank must be processed. Spike and analyze a minimum of 5% of all samples to monitor and evaluate lab data quality. A QC check sample concentrate that contains each parameter of interest at a concentration of 100 µg/mL in acetone is required. PCBs and multicomponent pesticides may be omitted from this test.

After analysis of five spiked wastewater samples, calculate the average percent recovery and the standard deviation of the percent recovery. Spike all samples with the surrogate standard spiking solution and calculate the percent recovery of each surrogate compound.

REFERENCE *Federal Register*, Vol. 49, No. 209. Friday, Oct. 26, 1984.

2-Chloronaphthalene   **EPA Method 8120**
CAS #91-58-7

TITLE Chlorinated Hydrocarbons by Gas Chromatography

MATRIX This method covers aqueous and solid matrices. This includes a wide variety such as drinking water, groundwater, industrial wastewaters, surface waters, soils, solids, and sediments.

METHOD SUMMARY This method is used to determine the concentration of 14 chlorinated hydrocarbons. It provides gas chromatographic conditions for the detection of ppb concentrations of certain chlorinated hydrocarbons. Prior to use of this method, appropriate sample extraction techniques must be used. Both neat and diluted organic liquids (EPA Method 3580, Waste Dilution) may be analyzed by direct injection. A 2 to 5 µg/mL aliquot of the extract is injected into a gas chromatograph (GC) using the solvent flush technique, and compounds in the GC effluent are detected by an electron capture detector (ECD).

INTERFERENCES Solvents, reagents, glassware, and other sample processing hardware may yield discrete artifacts and/or elevated baselines causing misinterpretation of gas chromatograms. Interferences coextracted from samples will vary considerably from source to source, depending upon the waste being sampled.

INSTRUMENTATION An analytical system complete with GC suitable for on-column injections and accessories, including detectors, column supplies, recorder, gases and syringes is required. A data system for measuring peak areas and/or peak heights is recommended. The GC is equipped with an electron capture detector (ECD). A K-D apparatus is needed for sample preparation.

Column 1: 1.8 m × 2 mm I.D. glass column packed with 1% SP-1000 on Supelcoport (100/120 mesh) or equivalent.

Column 2: 1.8 m × 2 mm I.D. glass column packed with 1.5% OV-1/2.4% OV-225 on Supelcoport (80/100 mesh) or equivalent.

PRECISION & ACCURACY The method was tested by 20 laboratories using organic-free reagent water, drinking water, surface water, and three industrial wastewaters spiked at six concentrations over the range 1.0 to 356 µg/L. Single operator precision, overall precision, and method accuracy were found to be directly related to the concentration of the parameter and essentially independent of the sample matrix.

MULTIPLICATION FACTORS FOR OTHER MATRICES (a)

Matrix	Factor (b)
Groundwater	10
Low-concentration soil by ultrasonic extraction with GPC cleanup	670
High-concentration soil and sludges by ultrasonic extraction	10,000
Non-water miscible waste	100,000

(a) Sample EQLs are highly matrix-dependent. The EQLs listed are provided for guidance and may not always be achievable.
(b) EQL = [Method detection limit] × [Factor]. For non-aqueous samples, the factor is on a wet-weight basis.

PRECISION & ACCURACY The estimates below are based upon the performance in a single lab.

The accuracy (in µg/L) as expected recovery for one or more measurements of a sample containing a concentration of C was 0.75C+3.21.

The precision (in µg/L) as expected single analyst standard deviation of measurements at an average concentration of x" was 0.28x"−1.17.

The precision (in µg/L) as expected interlaboratory standard deviation measurements at an average concentration found of x" was 0.38x"−1.39.

C = *True value for the concentration, in µg/L.*
$x"$ = *Average recovery found for measurements of samples containing a concentration of C, in µg/L.*

SAMPLE COLLECTION, PRESERVATION & HANDLING Extracts must be stored under refrigeration at 4°C and analyzed within 40 days of extraction.

SAMPLE PREPARATION In general, water samples are extracted at a neutral, or as is, pH with methylene chloride using either EPA Method 3510 or EPA Method 3520. Solid samples are extracted using either EPA Method 3540 or EPA Method 3550. Prior to gas chromatographic analysis, the extraction solvent must be exchanged to hexane.

QUALITY CONTROL The quality control check concentrate (EPA Method 8000) should contain each parameter of interest in acetone at the following concentrations: hexachloro-substituted hydrocarbon, 10 µg/mL; and any other chlorinated hydrocarbon, 100 µg/mL. Calculate surrogate standard recovery on all samples, blanks, and spikes.

Prepare stock standard solutions in isooctane or hexane. Calibration standards at a minimum of five concentrations should be prepared through dilution of the stock standards with isooctane or hexane. Internal standards and surrogate standards are also needed.

REFERENCE Test Methods for Evaluating Solid Waste, Physical/Chemical Methods, SW-846, 3rd Edition, U.S. EPA, Office of Solid Waste, Washington, DC, 1990. EPA Method 8120 A Rev. 1, Nov. 1990.

2-Chloronaphthalene   EPA Method 8121
CAS #91-58-7

TITLE Chlorinated Hyrrocarbons by GC: Capillary Column Technique

MATRIX This method covers aqueous and solid matrices. This includes a wide variety such as drinking water, groundwater, industrial wastewaters, surface waters, soils, solids, and sediments.

METHOD SUMMARY This method provides procedures for the determination of 22 chlorinated hydrocarbons in water, soil/sediment, and waste matrices. A measured volume or weight of sample is extracted by using one of the appropriate sample extraction techniques specified in EPA Method 3510, EPA Method 3520, EPA Method 3540, or EPA Method 3550, or diluted using EPA Method 3580. Aqueous samples are extracted at neutral pH with methylene chloride by using either a separatory funnel (EPA Method 3510) or a continuous liquid-liquid extractor (EPA Method 3520). Solid samples are extracted with hexane/acetone (1:1) by using a Soxhlet extractor (EPA Method 3540) or with methylene chloride/acetone (1:1) by using an ultrasonic extractor (EPA Method 3550). After cleanup, the extract or diluted sample is analyzed by gas chromatography with electron capture detection (GC/ECD).

The sensitivity level of this method usually depends on the level of interferences rather than on instrumental limitations. This method may be used in conjunction with EPA Method 3620, Florisil Column Cleanup, EPA Method 3660, Sulfur Cleanup, and EPA Method 3640, Gel Permeation Chromatography, to aid in the elimination of interferences.

INTERFERENCES Solvents, reagents, glassware, and other hardware used in sample processing may introduce artifacts which may result in elevated baselines, causing misinterpretation of gas chromatograms. Interferants coextracted from the samples will vary considerably from waste to waste. Glassware must be scrupulously clean. Phthalate esters, if present in a sample, will interfere only with the BHC isomers. The presence of elemental sulfur will result in large peaks, and can often mask the region of compounds eluting after 1,2,4,5-tetrachlorobenzene. The tetrabutylammonium (TBA)-sulfite procedure (EPA Method 3660) works well for the removal of elemental sulfur. Waxes and lipids can be removed by gel permeation chromatography (EPA Method 3640).

INSTRUMENTATION A GC suitable for on-column injections and all required accessories, including and electron capture detector (ECD), analytical columns, recorder, gases, and syringes are needed. A data system for measuring peak heights and/or peak areas is recommended. A Kuderna-Danish (K-D) apparatus will also be needed to concentrate extracts.

Column 1: 30 m × 0.53 mm I.D. fused-silica capillary column chemically bonded with trifluoropropyl methyl silicone (DB-210 or equivalent).

Column 2: 30 m × 0.53 mm I.D. fused-silica capillary column chemically bonded with polyethylene glycol (DB-WAX or equivalent).

PRECISION & ACCURACY This method has been tested in a single lab by using organic-free reagent water, sandy loam samples, and extracts which were spiked with the test compounds at one concentration. Single-operator precision and method accuracy were found to be related to the concentration of compound and the type of matrix. The accuracy and precision technique will be determined by the sample matrix, sample preparation technique, optional cleanup techniques, and calibration procedures used.

ESTIMATED QUANTITATION LIMIT (EQL) FACTORS FOR VARIOUS MATRICES (a)

Matrix	Factor (b)
Groundwater	10
Low-concentration soil by ultrasonic extraction with GPC cleanup	670
High-concentration soil and sludges by ultrasonic extraction	10,000
Non-water miscible waste	100,000

(a) Sample EQLs are highly matrix-dependent. The EQLs listed herein are provided for guidance and may not always be achievable. (b) EQL = [Method detection limit] × [Factor]. For nonaqueous samples, the factor is on a wet-weight basis.

PRECISION & ACCURACY MDL is the method detection limit for organic-free reagent water. MDL was determined from the analysis of eight replicate aliquots processed through the entire analytical method (extraction, Florisil cartridge cleanup, and GC/ECD analysis).

The MDL (in ng/L) was 1,300.

The accuracy (as average % recovery using 5 determinations and no Florisil cleanup) from a spike concentration of 200 µg/L and separatory funnel extraction was 91% with a final volume of 10 mL.

The precision (as RSD% using 5 determinations and no Florisil cleanup) from a spike concentration of 200 µg/L and separatory funnel extraction was 6.5% with a final volume of 10 mL.

The accuracy (as average % recovery using 5 determinations and no Florisil cleanup), from a spike concentration of 66,000µg/L and ultrasonic extraction of solid samples using 1:1 methylene chloride and acetone, was 100% with a final volume of 10 mL.

The precision (as RSD% using 5 determinations and no Florisil cleanup), from a spike concentration of 66,000µg/L and ultrasonic extraction of solid samples using 1:1 methylene chloride and acetone, was 6.4% with a final volume of 10 mL.

SAMPLE COLLECTION, PRESERVATION & HANDLING
Volatile organics — Standard 40-mL glass screw-cap VOA vials with Teflon®-faced silicone septum may be used for both liquid and solid matrices. When collecting samples, liquids and solids should be introduced into the vials gently to reduce agitation which might drive off volatile compounds. If there are any air bubbles present the sample must be retaken. The vials with solids should be tapped slightly as they are filled to try and eliminate as much free air space as possible. Two vials from each sampling location should be sealed in separate plastic bags to prevent cross-contamination between samples.

Semivolatile organics — Containers used to collect samples for the determination of semivolatile organic compounds should be soap and water washed followed by methanol (or isopropanol) rinsing. The sample containers should be of glass or Teflon® and have screw-top covers with Teflon® liners.

Preservation for volatile organics — No preservation is used with concentrated waste samples. With liquid samples containing no residual chlorine, 4 drops of concentrated hydrochloric acid are added and the samples are immediately cooled to 4°C. When liquid samples contain residual chlorine, they are treated as above and, in addition, 4 drops of 4% aqueous sodium thiosulfate are added to remove the residual chlorine. Soil, sediment, and sludge samples are only cooled to 4°C.

Preservation for semivolatile organics — No preservation is used with concentrated waste samples. With liquid samples containing no residual chlorine and with soil, sediment, and sludge samples, immediately cooling to 4°C is the only preservation used. When residual chlorine is present then 3 mL of 10% aqueous sodium sulfate is added for each gallon of sample collected, followed by cooling to 4°C.

Holding times — The holding time for all volatile organics samples is 14 days. Liquid samples must be extracted within 7 days and their extracts analyzed within 40 days. Concentrated waste, soil, sediment, and sludge samples must be extracted within 14 days and their extracts analyzed within 40 days.

SAMPLE PREPARATION Prepare stock standard solutions in hexane. Calibration standards at a minimum of five concentrations should be prepared through dilution of the stock standards with hexane. The suggested internal standards are: 2,5-dibromotoluene, 1,3,5-tribromobenzene, and α, α-dibromo-m-xylene. The analyst can use any of the three compounds provided that they are resolved from matrix interferences. Recommended surrogate compounds are α-2,6-trichlorotoluene, 1,4-dichloronaphthalene, and 2,3,4,5,6-pentachlorotoluene.

In general, water samples are extracted at a neutral pH with methylene chloride using a separatory funnel (EPA Method 3510) or a continuous liquid-liquid extractor (EPA Method 3520). Solid samples are extracted with hexane/acetone (1:1 v:v) using a Soxhlet extractor (EPA Method 3540) or with methylene chloride/acetone (1:1 v:v) using an ultrasonic extractor (EPA Method 3550). Non-aqueous waste samples may be diluted using EPA Method 3580. Prior to Florisil cleanup or gas chromatographic analysis, the extraction solvent must be exchanged to hexane. Sample extracts that will be subjected to gel permeation chromatography do not need solvent exchange.

Cleanup procedures may not be necessary for a relatively clean matrix. If removal of interferences such as chlorinated phenols, phthalate esters, etc., is required, proceed with the procedure outlined in EPA Method 3620.

QUALITY CONTROL Analyze a quality control check standard to demonstrate that the operation of the GC is in control. The frequency of the check standard analysis is equivalent to 10% of the samples analyzed. If the recovery of any compound found in the check standard is less than 80% of the certified value, the problem must be corrected and a new set of calibration standards must be prepared and analyzed. Calculate surrogate standard recoveries for all samples, blanks, and spikes. An internal standard peak area check must be performed on all samples. The internal standard must be evaluated for acceptance by determining whether the measured area for the internal standard deviates by more than 30% from the average area for the internal standard in the calibration standards. When the internal standard peak area is outside that limit, all samples that fall outside the QC criteria must be reanalyzed. Any compound confirmed by two columns may also be confirmed by GC/MS (EPA Method 8270). The GC/MS would normally require a minimum concentration of 1 ng/µL in the final extract for each compound. Include a mid-concentration calibration standard after each group of 20 samples in the analysis sequence. The response factors for the mid-concentration calibration must be within 15% of the average values for the multiconcentration calibration.

REFERENCE Test Methods for Evaluating Solid Waste, Physical/Chemical Methods, SW-846, 3rd Edition, U.S. EPA, Office of Solid Waste, Washington, DC, 1990. EPA Method 8121, Rev. 0, Nov. 1990.

2-Chloronaphthalene	EPA Method 8270
CAS #91-58-7	

TITLE Semivolatile Organic Compounds by GC/MS

MATRIX This method is used to determine the concentration of semivolatile organic compounds in extracts prepared from all types of solid waste matrices, soils, and groundwater. Although surface waters are not specifically mentioned, this method should be applicable to water samples from rivers, lakes, etc.

METHOD SUMMARY This method covers 259 semivolatile organic compounds. In very limited applications direct injection of the sample into the GC/MS system may be appropriate, but this results in very high detection limits (approximately 10,000 µg/L). Typically, a 1-L liquid sample, containing surrogate, and matrix spiking standards, is extracted in a continuous extractor first under acid conditions and then under basic conditions. Typically 30 g of a solid sample, containing surrogate, and matrix spiking standards, is extracted ultrasonically. After concentrating the extract to 1 mL it is spiked with 10 µL of an internal standard solution just prior to analysis by GC/MS. The volume injected should contain about 100 ng of base/neutral and 200 ng of acid surrogates (for a 1-µL injection). Analysis is performed by GC/MS using a capillary GC column.

INTERFERENCES Raw GC/MS data from all blanks, samples, and spikes must be evaluated for interferences. Contamination by carryover can occur whenever high-concentration and low-concentration samples are sequentially analyzed. To reduce carryover, the sample syringe must be rinsed out between samples with solvent. Whenever an unusually concentrated sample is encountered, it should be followed by the analysis of blank solvent to check for cross-contamination.

INSTRUMENTATION A GC/MS and a data system are required. The GC column used is a 30 m × 0.25 mm I.D. (or 0.32 mm I.D.) 1um film thickness silicone-coated fused silica capillary column. A continuous liquid-liquid extractor equipped with Teflon® or glass connection joints and stopcocks requiring no lubrication, a K-D concentrating apparatus, water bath, and an ultrasonic disrupter with a minimum power of 300 W and with pulsing capability are also required.

PRECISION & ACCURACY The estimated quantitation limit (EQL) of Method 8270B for determining an individual compound is approximately 1 mg/kg (wet weight) for soil or sediment samples, 1–200 mg/kg for wastes (dependent on matrix and method of preparation), and 10 µg/L for groundwater samples. EQLs will be proportionately higher for sample extracts that require dilution to avoid saturation of the detector.

The EQL(b) for groundwater in µg/L is 10.
The EQL (a, b) for low concentrations in soil and sediment in µg/kg is 660.
Accuracy as µg/L is not listed.
Overall precision in µg/L is not listed.

(a) *EQLs listed for soil/sediment are based on wet weight. Normally data is reported in a dry-weight basis; therefore, EQLs will be higher based on the % dry weight of each sample. This calculation is based on a 30 g sample and gel permeation chromatography cleanup.*
(b) *Sample EQLs are highly matrix-dependent. The EQLs are provided for guidance and may not always be achievable.*
C = *True value for concentration, in µg/L.*
X = *Average recovery found for measurements of samples containing a concentration of C, in µg/L.*

ESTIMATED QUANTITATION LIMIT FOR OTHER MATRICES

Other Matrices	Factor (a)
High-concentration soil and sludges by sonicator	7.5
Non-water miscible waste	75

(a) *EQL for other matrices = [EQL for low soil/sediment] × [Factor]. This estimated EQL is similar to an EPA "Practical Quantitation Limit."*

SAMPLING METHOD

Liquid samples — Use a 1 or 2½ gallon amber glass bottle with a screw-top Teflon®-lined cover that has been prewashed with detergent and rinsed with distilled water and methanol (or isopropanol).

Soils, sediments, or sludges — Use an 8-oz. widemouth glass with a screw-top Teflon®-lined cover that has been prewashed with detergent and rinsed with distilled water and methanol (or isopropanol).

SAMPLE PRESERVATION

Liquid samples — If residual chlorine is present, add 3 mL of 10% sodium thiosulfate per gallon, cool to 4°C and store in a solvent-free refrigerator until analysis; if chlorine is not present, then eliminate the sodium thiosulfate addition.

Soils, sediments, or sludges — Cool samples to 4°C and store in a solvent-free refrigerator.

MHT Liquid samples must be extracted within 7 days and the extracts analyzed within 40 days. Soils, sediments, or sludges may be stored for a maximum of 14 days and the extracts analyzed within 40 days.

SAMPLE PREPARATION

Liquid samples — Transfer 1 L quantitatively to a continuous extractor. If high concentrations are anticipated, a smaller volume may be used and then diluted with organic-free reagent water to 1 L. Adjust pH, if necessary, to pH <2 using 1:1 (V/V) sulfuric acid. Pipette 1.0 mL of a surrogate standard spiking solution into each sample. For the sample in each analytical batch selected for spiking, add 1.0 mL of a matrix spiking standard. For base/neutral acid analysis, the amount of the surrogates and matrix spiking compounds added to the sample should result in a final concentration of 100 ng/µL of each analyte in the extract to be analyzed (assuming a 1-µL injection). Extract with methylene chloride for 18–24 h. Next, adjust the pH of the aqueous phase to pH >11 using 10 N sodium hydroxide and extract it with methylene chloride again for 18–24 h. Dry the extract through a column containing anhydrous sodium sulfate and concentrate it to 1 mL using a K-D concentrator.

Soils, sediments, or sludges — Use 30 g of sample. Nonporous or wet samples (gummy or clay type) that do not have a free-flowing sandy texture must be mixed with anhydrous sodium sulfate until the sample is free flowing. Add 1 mL of surrogate standards to all samples, spikes, standards, and blanks. For the sample in each analytical batch selected for spiking, add 1.0 mL of a matrix spiking standard. For base/neutral acid analysis, the amount added of the surrogates and matrix spiking compounds should result in a final concentration of 100 ng/µL of each base/neutral analyte and 200 ng/µL of each acid analyte in the extract to be analyzed (assuming a 1-µL injection). Immediately add a 100-mL mixture of 1:1 methylene chloride:acetone and extract the sample ultrasonically for 3 min and then decant or filter the extracts. Repeat the extraction two or more times. Dry the extract using a column with anhydrous sodium sulfate and concentrate it to 1 mL in a K-D concentrator.

QUALITY CONTROL

A methylene chloride solution containing 50 ng/µL of decafluorotriphenylphosphine (DFTPP) is used for tuning the GC/MS system each 12-h shift. A system performance check also must be made during every 12-h shift. A standard containing 50 ng/µL each of 4,4′-DDT, pentachlorophenol, and benzidine is required to verify injection port inertness and GC column performance. A calibration standard at mid-concentration, containing each compound of interest, including all required surrogates, must be performed every 12 h during analysis. After the system performance check is met, calibration check compounds (CCCs) are used to check the validity of the initial calibration.

The internal standard responses and retention times in the calibration check standard must be evaluated immediately after or during data acquisition. If the retention time for any internal standard changes by more than 30 seconds from the last check calibration (12 h), the chromatographic system must be inspected for malfunctions and corrections must be made, as required. If the electron ionization current plot (EICP) area for any of the internal standards changes by a factor of two from the last daily calibration standard check, the mass spectrometer must be inspected for malfunctions and corrections must be made, as appropriate.

Demonstrate, through the analysis of a reagent water blank, that interferences from the analytical system, glassware, and reagents are under control. The blank samples should be carried through all stages of the sample preparation and measurement steps. For each analytical batch (up to 20 samples), a reagent blank, matrix spike, and matrix spike duplicate/duplicate must be analyzed (the frequency of the spikes may be different for different monitoring programs). The blank and spiked samples must be carried through all stages of the sample preparation and measurement steps. A QC reference sample concentrate containing each analyte at a concentration of 100 mg/L in methanol is required.

REFERENCE Test Methods for Evaluating Solid Waste (SW-846). U.S. EPA 1983, Method 8270B, Rev. 2, Nov. 1990. Office of Solid Waste, Washington, DC.

3-Chloronitrobenzene   **EPA Method 1625**
CAS #121-73-3

TITLE Semivolatile Organic Compounds by Isotope Dilution GC/MS

MATRIX The compounds may be determined in waters, soils, and municipal sludges by this method.

METHOD SUMMARY This method is used to determine 176 semivolatile toxic organic pollutants associated with the CWA (as amended 1987); the RCRA (as amended 1986); the CERCLA (as amended 1986); and other compounds amenable to extraction and analysis by capillary column gas chromatography-mass spectrometry (GC/MS).

Stable isotopically-labeled analogs of the compounds of interest are added to the sample. If the solids content is less than 1%, a 1-L sample is extracted at pH 12–13, then at pH <2 with methylene chloride using continuous extraction techniques.

If the solids content is 30% or less, the sample is diluted to 1% solids with reagent water, homogenized ultrasonically, and extracted at pH 12–13, then at pH <2 with methylene chloride using continuous extraction techniques. If the solids content is

greater than 30%, the sample is extracted using ultrasonic techniques.

Each extract is dried over sodium sulfate, concentrated to a volume of 5 mL, cleaned up using GPC, if necessary, and concentrated. Extracts are concentrated to 1 mL if GPC is not performed, and to 0.5 mL if GPC is performed.

An internal standard is added to the extract, and a 1-mL aliquot of the extract is injected into the GC. The compounds are separated by GC and detected by a MS. The labeled compounds serve to correct the variability of the analytical technique.

INTERFERENCES Solvents, reagents, glassware, and other sample processing hardware may yield artifacts and/or elevated baselines causing misinterpretation of chromatograms and spectra. Materials used in the analysis must be demonstrated to be free from interferences under the conditions of analysis by running method blanks initially and with each sample lot (sample started through the extraction process on a given 8-h shift, to a maximum of 20). Specific selection of reagents and purification of solvents by distillation in all glass systems may be required. Glassware and, where possible, reagents are cleaned by solvent rinse and baking at 450°C for 1-h minimum. Interferences coextracted from samples will vary considerably from source to source, depending on the diversity of the site being sampled.

INSTRUMENTATION Major instrumentation includes a GC with a splitless or on-column injection port for capillary column, a MS with 70 eV electron impact ionization, and a data system to collect and record MS data, and process it. A K-D apparatus is used to concentrate extracts.

GC Column: 30 m × 0.25 mm I.D. 5% phenyl, 94% methyl, 1% vinyl silicone bonded phased fused silica capillary column.

PRECISION & ACCURACY The detection limits of the method are usually dependent on the level of interferences rather than instrumental limitations. The limits typify the minimum quantities that can be detected with no interferences present.

The minimum level (in µg/mL) was not listed. This is defined as a minimum level at which the analytical system shall give recognizable mass spectra (background corrected) and acceptable calibration points.

The MDL (in µg/kg) in low solids was not listed and in high solids was not listed; these were determined in digested sludge (low solids) and in filter cake or compost (high solids).

The labeled and native compound initial precision as standard deviation (in µg/L) was not listed.

The labeled and native compound initial accuracy as average recovery (in µg/L) was not listed.

SAMPLE COLLECTION, PRESERVATION & HANDLING Collect samples in glass containers. Aqueous samples which flow freely are collected in refrigerated bottles using automatic sampling equipment. Solid samples are collected as grab samples using widemouth jars. Maintain samples at 0 to 4°C from the time of collection until extraction. If residual chlorine is present in aqueous samples, add 80 mg sodium thiosulfate/L of water. Begin sample extraction within 7 days of collection, and analyze all extracts within 40 days of extraction.

SAMPLE PREPARATION Samples containing 1% solids or less are extracted directly using continuous liquid-liquid extraction techniques. Samples containing 1 to 30% solids are diluted to the 1% level with reagent water and extracted using continuous liquid-liquid extraction techniques. Samples containing greater than 30% solids are extracted using ultrasonic techniques.

Base/neutral extraction — Adjust the pH of the waters in the extractors to 12–13 with 6 N NaOH. Extract with methylene chloride for 24–48 h.

Acid extraction — Adjust the pH of the waters in the extractors to 2 or less using 6 N sulfuric acid. Extract with methylene chloride for 24–48 h.

Ultrasonic extraction of high solids samples — Add anhydrous sodium sulfate to the sample and QC aliquot(s). Add acetone:methylene chloride (1:1) to the sample and mix thoroughly

Concentrate extracts using a K-D apparatus.

QUALITY CONTROL The analyst is permitted to modify this method to improve separations or lower the costs of measurements, provided all performance specifications are met. Analyses of blanks are required to demonstrate freedom from contamination. When results of spikes indicate atypical method performance for samples, the samples are diluted to bring method performance within acceptable limits.

For low solids (aqueous samples), extract, concentrate, and analyze two sets of four 1-L aliquots (8 aliquots total) of the precision and recovery standard. For high solids samples, two sets of four 30-g aliquots of the high solids reference matrix are used.

Spike all samples with labeled compounds to assess method performance. Compute percent recovery of the labeled compounds using the internal standard method. Compare the labeled compound recovery for each compound with the corresponding labeled compound recovery.

Reagent water and high solids reference matrix blanks are analyzed to demonstrate freedom from contamination. Extract and concentrate a 1-L reagent water blank or a high solids reference matrix blank with each sample's lot (samples started through the extraction process on the same 8-h shift, to a maximum of 20 samples).

Field replicates may be collected to determine the precision of the sampling technique, and spiked samples may be required to determine the accuracy of the analysis when the internal standard method is used.

REFERENCE Semivolatile Organic Compounds by Isotope Dilution GC/MS. Office of Water Regulation and Standards, U.S. EPA Industrial Technology Division, Washington, DC, EPA Method 1625, Rev. C, June 1989 (contact W.A. Telliard, U.S. EPA, Office of Water Regulations and Standards, 401 M St., SW, Washington, DC, 20460. Phone: 202-382-7131).

2-Chlorophenol
CAS #95-57-8

EPA Method 1625

TITLE Semivolatile Organic Compounds by Isotope Dilution GC/MS

MATRIX The compounds may be determined in waters, soils, and municipal sludges by this method.

METHOD SUMMARY This method is used to determine 176 semivolatile toxic organic pollutants associated with the CWA (as amended 1987); the RCRA (as amended 1986); the CERCLA (as amended 1986); and other compounds amenable to extraction and analysis by capillary column gas chromatography-mass spectrometry (GC/MS).

Stable isotopically-labeled analogs of the compounds of interest are added to the sample. If the solids content is less than 1%, a 1-L sample is extracted at pH 12–13, then at pH <2 with methylene chloride using continuous extraction techniques.

If the solids content is 30% or less, the sample is diluted to 1% solids with reagent water, homogenized ultrasonically, and extracted at pH 12–13, then at pH <2 with methylene chloride using continuous extraction techniques. If the solids content is greater than 30%, the sample is extracted using ultrasonic techniques.

Each extract is dried over sodium sulfate, concentrated to a volume of 5 mL, cleaned up using GPC, if necessary, and concentrated. Extracts are concentrated to 1 mL if GPC is not performed, and to 0.5 mL if GPC is performed.

An internal standard is added to the extract, and a 1-mL aliquot of the extract is injected into the GC. The compounds are separated by GC and detected by a MS. The labeled compounds serve to correct the variability of the analytical technique.

INTERFERENCES Solvents, reagents, glassware, and other sample processing hardware may yield artifacts and/or elevated baselines causing misinterpretation of chromatograms and spectra. Materials used in the analysis must be demonstrated to be free from interferences under the conditions of analysis by running method blanks initially and with each sample lot (sample started through the extraction process on a given 8-h shift, to a maximum of 20). Specific selection of reagents and purification of solvents by distillation in all glass systems may be required. Glassware and, where possible, reagents are cleaned by solvent rinse and baking at 450°C for 1-h minimum. Interferences coextracted from samples will vary considerably from source to source, depending on the diversity of the site being sampled.

INSTRUMENTATION Major instrumentation includes a GC with a splitless or on-column injection port for capillary column, a MS with 70 eV electron impact ionization, and a data system to collect and record MS data, and process it. A K-D apparatus is used to concentrate extracts.

GC Column: 30 m × 0.25 mm I.D. 5% phenyl, 94% methyl, 1% vinyl silicone bonded phased fused silica capillary column.

PRECISION & ACCURACY The detection limits of the method are usually dependent on the level of interferences rather than instrumental limitations. The limits typify the minimum quantities that can be detected with no interferences present.

The minimum level (in µg/mL) was 10. This is defined as a minimum level at which the analytical system shall give recognizable mass spectra (background corrected) and acceptable calibration points.

The MDL (in µg/kg) in low solids was 18 and in high solids was 10; these were determined in digested sludge (low solids) and in filter cake or compost (high solids).

The labeled and native compound initial precision as standard deviation (in µg/L) was 13.

The labeled and native compound initial accuracy as average recovery (in µg/L) was 79–135.

SAMPLE COLLECTION, PRESERVATION & HANDLING
Collect samples in glass containers. Aqueous samples which flow freely are collected in refrigerated bottles using automatic sampling equipment. Solid samples are collected as grab samples using widemouth jars. Maintain samples at 0 to 4°C from the time of collection until extraction. If residual chlorine is present in aqueous samples, add 80 mg sodium thiosulfate/L of water. Begin sample extraction within 7 days of collection, and analyze all extracts within 40 days of extraction.

SAMPLE PREPARATION Samples containing 1% solids or less are extracted directly using continuous liquid-liquid extraction techniques. Samples containing 1 to 30% solids are diluted to the 1% level with reagent water and extracted using continuous liquid-liquid extraction techniques. Samples containing greater than 30% solids are extracted using ultrasonic techniques.

Base/neutral extraction — Adjust the pH of the waters in the extractors to 12–13 with 6 N NaOH. Extract with methylene chloride for 24–48 h.

Acid extraction — Adjust the pH of the waters in the extractors to 2 or less using 6 N sulfuric acid. Extract with methylene chloride for 24–48 h.

Ultrasonic extraction of high solids samples — Add anhydrous sodium sulfate to the sample and QC aliquot(s). Add acetone:methylene chloride (1:1) to the sample and mix thoroughly

Concentrate extracts using a K-D apparatus.

QUALITY CONTROL The analyst is permitted to modify this method to improve separations or lower the costs of measurements, provided all performance specifications are met. Analyses of blanks are required to demonstrate freedom from contamination. When results of spikes indicate atypical method performance for samples, the samples are diluted to bring method performance within acceptable limits.

For low solids (aqueous samples), extract, concentrate, and analyze two sets of four 1-L aliquots (8 aliquots total) of the precision and recovery standard. For high solids samples, two sets of four 30-g aliquots of the high solids reference matrix are used.

Spike all samples with labeled compounds to assess method performance. Compute percent recovery of the labeled compounds

using the internal standard method. Compare the labeled compound recovery for each compound with the corresponding labeled compound recovery.

Reagent water and high solids reference matrix blanks are analyzed to demonstrate freedom from contamination. Extract and concentrate a 1-L reagent water blank or a high solids reference matrix blank with each sample's lot (samples started through the extraction process on the same 8-h shift, to a maximum of 20 samples).

Field replicates may be collected to determine the precision of the sampling technique, and spiked samples may be required to determine the accuracy of the analysis when the internal standard method is used.

REFERENCE Semivolatile Organic Compounds by Isotope Dilution GC/MS. Office of Water Regulation and Standards, U.S. EPA Industrial Technology Division, Washington, DC, EPA Method 1625, Rev. C, June 1989 (contact W.A. Telliard, U.S. EPA, Office of Water Regulations and Standards, 401 M St., SW, Washington, DC, 20460. Phone: 202-382-7131).

2-Chlorophenol   **EPA Method 625**
CAS #95-57-8

TITLE Base/Neutrals and Acids, U.S. EPA Method 625

MATRIX This methods covers municipal and industrial wastewaters.

METHOD SUMMARY Approximately 1 L of sample is serially extracted with methylene chloride at a pH greater than 11 and again at a pH less than 2 using a separatory funnel or a continuous extractor. The methylene chloride extract is dried, concentrated to a volume of 1 mL, and analyzed by GC/MS. Qualitative identification of the parameters in the extract is performed using the retention time and the relative abundance of three characteristic masses (m/z). Qualitative analysis is performed using either external or internal standard techniques with a single characteristic m/z.

INTERFERENCES Method interferences may be caused by contaminants in solvents, reagents, glassware, and other sample processing hardware. Glassware must be scrupulously cleaned. Glassware should be heated in a muffle furnace at 400°C for 5 to 30 min. Some thermally stable materials, such as PCBs, may not be eliminated by this treatment. Solvent rinses with acetone and pesticide quality hexane may be substituted for the muffle furnace heating. Matrix interferences may be caused by contaminants that are coextracted from the sample. The base-neutral extraction may cause significantly reduced recovery of phenols. The packed gas chromatographic columns recommended for the basic fraction may not exhibit sufficient resolution for some analytes.

INSTRUMENTATION A GC/MS system with an injection port designed for on-column injection when using packed columns and for splitless injection when using capillary columns. Column for base/neutrals: 1.8 m long × 2 mm I.D. glass, packed with 3% SP-2550 on Supelcoport (100/120 mesh) or equivalent.
Column for acids: 1.8 m long × 2 mm I.D. glass, packed with 1% SP-1240DA on Supelcoport (100/120 mesh) or equivalent.

PRECISION & ACCURACY The MDL concentrations were obtained using reagent water. The MDL actually achieved in a given analysis will vary depending on instrument sensitivity and matrix effects. This method was tested by 15 laboratories using reagent water, drinking water, surface water, and industrial wastewaters spiked at six concentrations over the range 5 to 100 µg/L. Single operator precision, overall precision, and method accuracy were found to be directly related to the concentration of the parameter matrix.

The MDL (in µg/L) in reagent water was 3.3.
The standard deviation (in µg/L based on 4 recovery measurements) was 28.7.
The range (in µg/L) for average recovery for 4 measurements was 36.2–120.4.
The range (in %) for percent recovery was 23–134.
Accuracy (in µg/L) as expected recovery for one or more measurements of a sample containing a true concentration of C was 0.78C+0.29.
Precision (in µg/L) as expected single analyst standard deviation of measurements at an average concentration found at X was 0.18X+1.46.
Overall precision (in µg/L) as expected interlaboratory standard deviation of measurements in an average concentration found at X was 0.28X+0.97.

$C =$ *True value of the concentration in µg/L.*
$X =$ *Average recovery found for measurements of samples containing a concentration at C in µg/L.*

SAMPLE PREPARATION Adjust the pH to >11 with sodium hydroxide and serially extract in a separatory funnel with methylene chloride or else in a continuous extractor. Next, adjust the pH to <2 with sulfuric acid and serially extract in a separatory funnel with methylene chloride or else in a continuous extractor. Dry the extracts separately through a column of anhydrous sodium sulfate and then concentrate each of the extracts to 1.0 mL using a K-D apparatus.

SAMPLE COLLECTION, PRESERVATION & HANDLING Grab samples must be collected in glass containers. All samples must be refrigerated at 4°C from the time of collection until extraction. If residual chlorine is present, add 80 mg of sodium thiosulfate/L of sample and mix well. All samples must be extracted within 7 days of collection and completely analyzed within 40 days of extraction.

QUALITY CONTROL Make an initial, one-time, demonstration of the ability to generate acceptable accuracy and precision with this method. Before processing any samples, the analyst must analyze a reagent water blank to demonstrate that interferences from the analytical system and glassware are under control. Each time a set of samples is extracted or reagents are changed, a reagent water blank must be processed. Spike and analyze a minimum of 5% of all samples to monitor and evaluate lab data quality. A QC check sample concentrate that

contains each parameter of interest at a concentration of 100 µg/mL in acetone is required. PCBs and multicomponent pesticides may be omitted from this test.

After analysis of five spiked wastewater samples, calculate the average percent recovery and the standard deviation of the percent recovery. Spike all samples with the surrogate standard spiking solution and calculate the percent recovery of each surrogate compound.

REFERENCE *Federal Register*, Vol. 49, No. 209. Friday, Oct. 26, 1984.

2-Chlorophenol **EPA Method 8270**
CAS #95-57-8

TITLE Semivolatile Organic Compounds by GC/MS

MATRIX This method is used to determine the concentration of semivolatile organic compounds in extracts prepared from all types of solid waste matrices, soils, and groundwater. Although surface waters are not specifically mentioned, this method should be applicable to water samples from rivers, lakes, etc.

METHOD SUMMARY This method covers 259 semivolatile organic compounds. In very limited applications direct injection of the sample into the GC/MS system may be appropriate, but this results in very high detection limits (approximately 10,000 µg/L). Typically, a 1-L liquid sample, containing surrogate, and matrix spiking standards, is extracted in a continuous extractor first under acid conditions and then under basic conditions. Typically 30 g of a solid sample, containing surrogate, and matrix spiking standards, is extracted ultrasonically. After concentrating the extract to 1 mL it is spiked with 10 µL of an internal standard solution just prior to analysis by GC/MS. The volume injected should contain about 100 ng of base/neutral and 200 ng of acid surrogates (for a 1-µL injection). Analysis is performed by GC/MS using a capillary GC column.

INTERFERENCES Raw GC/MS data from all blanks, samples, and spikes must be evaluated for interferences. Contamination by carryover can occur whenever high-concentration and low-concentration samples are sequentially analyzed. To reduce carryover, the sample syringe must be rinsed out between samples with solvent. Whenever an unusually concentrated sample is encountered, it should be followed by the analysis of blank solvent to check for cross-contamination.

INSTRUMENTATION A GC/MS and a data system are required. The GC column used is a 30 m × 0.25 mm I.D. (or 0.32 mm I.D.) 1um film thickness silicone-coated fused silica capillary column. A continuous liquid-liquid extractor equipped with Teflon® or glass connection joints and stopcocks requiring no lubrication, a K-D concentrating apparatus, water bath, and an ultrasonic disrupter with a minimum power of 300 W and with pulsing capability are also required.

PRECISION & ACCURACY The estimated quantitation limit (EQL) of Method 8270B for determining an individual compound is approximately 1 mg/kg (wet weight) for soil or sediment samples, 1–200 mg/kg for wastes (dependent on matrix and method of preparation), and 10 µg/L for groundwater samples. EQLs will be proportionately higher for sample extracts that require dilution to avoid saturation of the detector.

The EQL(b) for groundwater in µg/L is 10.
The EQL (a, b) for low concentrations in soil and sediment in µg/kg is 660.
Accuracy as µg/L is 0.78C +0.29.
Overall precision in µg/L is 0.28X +0.97.

(a) *EQLs listed for soil/sediment are based on wet weight. Normally data is reported in a dry-weight basis; therefore, EQLs will be higher based on the % dry weight of each sample. This calculation is based on a 30 g sample and gel permeation chromatography cleanup.*
(b) *Sample EQLs are highly matrix-dependent. The EQLs are provided for guidance and may not always be achievable.*
$C =$ *True value for concentration, in µg/L.*
$X =$ *Average recovery found for measurements of samples containing a concentration of C, in µg/L.*

ESTIMATED QUANTITATION LIMIT
FOR OTHER MATRICES

Other Matrices	Factor (a)
High-concentration soil and sludges by sonicator	7.5
Non-water miscible waste	75

(a) *EQL for other matrices = [EQL for low soil/sediment] × [Factor]. This estimated EQL is similar to an EPA "Practical Quantitation Limit."*

SAMPLING METHOD
Liquid samples — Use a 1 or 2½ gallon amber glass bottle with a screw-top Teflon®-lined cover that has been prewashed with detergent and rinsed with distilled water and methanol (or isopropanol).

Soils, sediments, or sludges — Use an 8-oz. widemouth glass with a screw-top Teflon®-lined cover that has been prewashed with detergent and rinsed with distilled water and methanol (or isopropanol).

SAMPLE PRESERVATION
Liquid samples — If residual chlorine is present, add 3 mL of 10% sodium thiosulfate per gallon, cool to 4°C and store in a solvent-free refrigerator until analysis; if chlorine is not present, then eliminate the sodium thiosulfate addition.

Soils, sediments, or sludges — Cool samples to 4°C and store in a solvent-free refrigerator.

MHT Liquid samples must be extracted within 7 days and the extracts analyzed within 40 days. Soils, sediments, or sludges may be stored for a maximum of 14 days and the extracts analyzed within 40 days.

SAMPLE PREPARATION
Liquid samples — Transfer 1 L quantitatively to a continuous extractor. If high concentrations are anticipated, a smaller volume may be used and then diluted with organic-free reagent water to 1 L. Adjust pH, if necessary, to pH <2 using 1:1 (V/V) sulfuric acid. Pipette 1.0 mL of a surrogate standard spiking

solution into each sample. For the sample in each analytical batch selected for spiking, add 1.0 mL of a matrix spiking standard. For base/neutral acid analysis, the amount of the surrogates and matrix spiking compounds added to the sample should result in a final concentration of 100 ng/µL of each analyte in the extract to be analyzed (assuming a 1-µL injection). Extract with methylene chloride for 18–24 h. Next, adjust the pH of the aqueous phase to pH >11 using 10 N sodium hydroxide and extract it with methylene chloride again for 18–24 h. Dry the extract through a column containing anhydrous sodium sulfate and concentrate it to 1 mL using a K-D concentrator.

Soils, sediments, or sludges — Use 30 g of sample. Nonporous or wet samples (gummy or clay type) that do not have a free-flowing sandy texture must be mixed with anhydrous sodium sulfate until the sample is free flowing. Add 1 mL of surrogate standards to all samples, spikes, standards, and blanks. For the sample in each analytical batch selected for spiking, add 1.0 mL of a matrix spiking standard. For base/neutral acid analysis, the amount added of the surrogates and matrix spiking compounds should result in a final concentration of 100 ng/µL of each base/neutral analyte and 200 ng/µL of each acid analyte in the extract to be analyzed (assuming a 1-µL injection). Immediately add a 100-mL mixture of 1:1 methylene chloride:acetone and extract the sample ultrasonically for 3 min and then decant or filter the extracts. Repeat the extraction two or more times. Dry the extract using a column with anhydrous sodium sulfate and concentrate it to 1 mL in a K-D concentrator.

QUALITY CONTROL A methylene chloride solution containing 50 ng/µL of decafluorotriphenylphosphine (DFTPP) is used for tuning the GC/MS system each 12-h shift. A system performance check also must be made during every 12-h shift. A standard containing 50 ng/µL each of 4,4′-DDT, pentachlorophenol, and benzidine is required to verify injection port inertness and GC column performance. A calibration standard at mid-concentration, containing each compound of interest, including all required surrogates, must be performed every 12 h during analysis. After the system performance check is met, calibration check compounds (CCCs) are used to check the validity of the initial calibration.

The internal standard responses and retention times in the calibration check standard must be evaluated immediately after or during data acquisition. If the retention time for any internal standard changes by more than 30 seconds from the last check calibration (12 h), the chromatographic system must be inspected for malfunctions and corrections must be made, as required. If the electron ionization current plot (EICP) area for any of the internal standards changes by a factor of two from the last daily calibration standard check, the mass spectrometer must be inspected for malfunctions and corrections must be made, as appropriate.

Demonstrate, through the analysis of a reagent water blank, that interferences from the analytical system, glassware, and reagents are under control. The blank samples should be carried through all stages of the sample preparation and measurement steps. For each analytical batch (up to 20 samples), a reagent blank, matrix spike, and matrix spike duplicate/duplicate must be analyzed (the frequency of the spikes may be different for different monitoring programs). The blank and spiked samples must be carried through all stages of the sample preparation and measurement steps. A QC reference sample concentrate containing each analyte at a concentration of 100 mg/L in methanol is required.

REFERENCE Test Methods for Evaluating Solid Waste (SW-846). U.S. EPA 1983, Method 8270B, Rev. 2, Nov. 1990. Office of Solid Waste, Washington, DC.

2-Chlorophenol   EPA Method 8040
CAS #95-57-8

TITLE Phenols

MATRIX Groundwater, soils, sludges, water miscible liquid wastes, and non-water miscible wastes.

APPLICATION This method is used for the analysis of 17 phenols. Samples are extracted, concentrated, and analyzed using direct injection of both neat and diluted organic liquids. Pentafluorobenzylbromide (PFB) derivatives also may be made to increase sensitivity of the method.

INTERFERENCES There can be carryover contamination with high- and low-level samples. Solvents, reagents, and glassware may introduce artifacts. Other interferences may come from coextracted compounds from samples.

INSTRUMENTATION GC capable of on-column injections and a flame with detector (FID) or electron capture detector (ECD). Column for underivatized phenol: 1.8 m by 2.0 mm with 1% SP-1240DA on Supelcoport. Column for derivatized phenols: 1.8 m by 2.0 mm with 5% OV-17 on Chromosorb W-AW-DMCS.

RANGE 12 to 450 µg/L

MDL 0.31 µg/L (FID) and 0.58 µg/L (ECD)

PQL FACTORS FOR MULTIPLYING × FID MDL VALUE

Matrix	Multiplication Factor
Groundwater	10
Low-level soil by sonication with GPC cleanup	670
High-level soil and sludge by sonication	10,000
Non-water miscible waste	100,000

PRECISION 0.21X + 0.75 µg/L (overall precision using FID)

ACCURACY 0.83C–0.84 µg/L (as recovery using FID)

SAMPLING METHOD Use 8-oz. widemouth glass bottles with Teflon®-lined caps for concentrated waste samples, soils, sediments, and sludges. Use 1 or 2½ gallon amber glass bottles with Teflon®-lined caps for liquid (water) samples.

STABILITY Cool soil, sediment, sludge, and liquid samples to 4°C. If residual chlorine is present in liquid samples add 3 mL of 10% sodium thiosulfate per gallon of sample and cool to 4°C.

MHT 14 days for concentrated waste, soil, sediment, or sludge; 7 days for liquid samples; all extracts must be analyzed within 40 days.

QUALITY CONTROL A quality control check sample concentrate containing each analyte of interest is required. The QC check sample concentrate may be prepared from pure standard materials or purchased as certified solutions Use appropriate trip, matrix, control site, method, reagent, and solvent blanks. Internal, surrogate, and five concentration level calibration standards are used. The QC check sample concentrate should contain this compound at 100 µg/mL in 2-propanol.

REFERENCE Test Methods for Evaluating Solid Waste (SW-846), U.S. EPA Office of Solid Waste, Washington, DC, Method 8040A, Rev. 1, Nov. 1990.

4-Chlorophenyl phenyl ether   **EPA Method 1625**
CAS #7005-72-3

TITLE Semivolatile Organic Compounds by Isotope Dilution GC/MS

MATRIX The compounds may be determined in waters, soils, and municipal sludges by this method.

METHOD SUMMARY This method is used to determine 176 semivolatile toxic organic pollutants associated with the CWA (as amended 1987); the RCRA (as amended 1986); the CERCLA (as amended 1986); and other compounds amenable to extraction and analysis by capillary column gas chromatography-mass spectrometry (GC/MS).

Stable isotopically-labeled analogs of the compounds of interest are added to the sample. If the solids content is less than 1%, a 1-L sample is extracted at pH 12–13, then at pH <2 with methylene chloride using continuous extraction techniques.

If the solids content is 30% or less, the sample is diluted to 1% solids with reagent water, homogenized ultrasonically, and extracted at pH 12–13, then at pH <2 with methylene chloride using continuous extraction techniques. If the solids content is greater than 30%, the sample is extracted using ultrasonic techniques.

Each extract is dried over sodium sulfate, concentrated to a volume of 5 mL, cleaned up using GPC, if necessary, and concentrated. Extracts are concentrated to 1 mL if GPC is not performed, and to 0.5 mL if GPC is performed.

An internal standard is added to the extract, and a 1-mL aliquot of the extract is injected into the GC. The compounds are separated by GC and detected by a MS. The labeled compounds serve to correct the variability of the analytical technique.

INTERFERENCES Solvents, reagents, glassware, and other sample processing hardware may yield artifacts and/or elevated baselines causing misinterpretation of chromatograms and spectra. Materials used in the analysis must be demonstrated to be free from interferences under the conditions of analysis by running method blanks initially and with each sample lot (sample started through the extraction process on a given 8-h shift, to a maximum of 20). Specific selection of reagents and purification of solvents by distillation in all glass systems may be required. Glassware and, where possible, reagents are cleaned by solvent rinse and baking at 450°C for 1-h minimum. Interferences coextracted from samples will vary considerably from source to source, depending on the diversity of the site being sampled.

INSTRUMENTATION Major instrumentation includes a GC with a splitless or on-column injection port for capillary column, a MS with 70 eV electron impact ionization, and a data system to collect and record MS data, and process it. A K-D apparatus is used to concentrate extracts.

GC Column: 30 m × 0.25 mm I.D. 5% phenyl, 94% methyl, 1% vinyl silicone bonded phased fused silica capillary column.

PRECISION & ACCURACY The detection limits of the method are usually dependent on the level of interferences rather than instrumental limitations. The limits typify the minimum quantities that can be detected with no interferences present.

The minimum level (in µg/mL) was 10. This is defined as a minimum level at which the analytical system shall give recognizable mass spectra (background corrected) and acceptable calibration points.

The MDL (in µg/kg) in low solids was 73 and in high solids was 59; these were determined in digested sludge (low solids) and in filter cake or compost (high solids).

The labeled and native compound initial precision as standard deviation (in µg/L) was 42.

The labeled and native compound initial accuracy as average recovery (in µg/L) was 75–166.

SAMPLE COLLECTION, PRESERVATION & HANDLING Collect samples in glass containers. Aqueous samples which flow freely are collected in refrigerated bottles using automatic sampling equipment. Solid samples are collected as grab samples using widemouth jars. Maintain samples at 0 to 4°C from the time of collection until extraction. If residual chlorine is present in aqueous samples, add 80 mg sodium thiosulfate/L of water. Begin sample extraction within 7 days of collection, and analyze all extracts within 40 days of extraction.

SAMPLE PREPARATION Samples containing 1% solids or less are extracted directly using continuous liquid-liquid extraction techniques. Samples containing 1 to 30% solids are diluted to the 1% level with reagent water and extracted using continuous liquid-liquid extraction techniques. Samples containing greater than 30% solids are extracted using ultrasonic techniques.

Base/neutral extraction — Adjust the pH of the waters in the extractors to 12–13 with 6 N NaOH. Extract with methylene chloride for 24–48 h.

Acid extraction — Adjust the pH of the waters in the extractors to 2 or less using 6 N sulfuric acid. Extract with methylene chloride for 24–48 h.

Ultrasonic extraction of high solids samples — Add anhydrous sodium sulfate to the sample and QC aliquot(s). Add acetone:methylene chloride (1:1) to the sample and mix thoroughly

Concentrate extracts using a K-D apparatus.

QUALITY CONTROL The analyst is permitted to modify this method to improve separations or lower the costs of measurements, provided all performance specifications are met. Analyses of blanks are required to demonstrate freedom from contamination. When results of spikes indicate atypical method performance for samples, the samples are diluted to bring method performance within acceptable limits.

For low solids (aqueous samples), extract, concentrate, and analyze two sets of four 1-L aliquots (8 aliquots total) of the precision and recovery standard. For high solids samples, two sets of four 30-g aliquots of the high solids reference matrix are used.

Spike all samples with labeled compounds to assess method performance. Compute percent recovery of the labeled compounds using the internal standard method. Compare the labeled compound recovery for each compound with the corresponding labeled compound recovery.

Reagent water and high solids reference matrix blanks are analyzed to demonstrate freedom from contamination. Extract and concentrate a 1-L reagent water blank or a high solids reference matrix blank with each sample's lot (samples started through the extraction process on the same 8-h shift, to a maximum of 20 samples).

Field replicates may be collected to determine the precision of the sampling technique, and spiked samples may be required to determine the accuracy of the analysis when the internal standard method is used.

REFERENCE Semivolatile Organic Compounds by Isotope Dilution GC/MS. Office of Water Regulation and Standards, U.S. EPA Industrial Technology Division, Washington, DC, EPA Method 1625, Rev. C, June 1989 (contact W.A. Telliard, U.S. EPA, Office of Water Regulations and Standards, 401 M St., SW, Washington, DC, 20460. Phone: 202-382-7131).

4-Chlorophenyl phenyl ether   **EPA Method 625**
CAS #7005-72-3

TITLE Base/Neutrals and Acids, U.S. EPA Method 625

MATRIX This methods covers municipal and industrial wastewaters.

METHOD SUMMARY Approximately 1 L of sample is serially extracted with methylene chloride at a pH greater than 11 and again at a pH less than 2 using a separatory funnel or a continuous extractor. The methylene chloride extract is dried, concentrated to a volume of 1 mL, and analyzed by GC/MS. Qualitative identification of the parameters in the extract is performed using the retention time and the relative abundance of three characteristic masses (m/z). Qualitative analysis is performed using either external or internal standard techniques with a single characteristic m/z.

INTERFERENCES Method interferences may be caused by contaminants in solvents, reagents, glassware, and other sample processing hardware. Glassware must be scrupulously cleaned. Glassware should be heated in a muffle furnace at 400°C for 5 to 30 min. Some thermally stable materials, such as PCBs, may not be eliminated by this treatment. Solvent rinses with acetone and pesticide quality hexane may be substituted for the muffle furnace heating. Matrix interferences may be caused by contaminants that are coextracted from the sample. The base-neutral extraction may cause significantly reduced recovery of phenols. The packed gas chromatographic columns recommended for the basic fraction may not exhibit sufficient resolution for some analytes.

INSTRUMENTATION A GC/MS system with an injection port designed for on-column injection when using packed columns and for splitless injection when using capillary columns.

Column for base/neutrals: 1.8 m long × 2 mm I.D. glass, packed with 3% SP-2550 on Supelcoport (100/120 mesh) or equivalent.

Column for acids: 1.8 m long × 2 mm I.D. glass, packed with 1% SP-1240DA on Supelcoport (100/120 mesh) or equivalent.

PRECISION & ACCURACY The MDL concentrations were obtained using reagent water. The MDL actually achieved in a given analysis will vary depending on instrument sensitivity and matrix effects. This method was tested by 15 laboratories using reagent water, drinking water, surface water, and industrial wastewaters spiked at six concentrations over the range 5 to 100 µg/L. Single operator precision, overall precision, and method accuracy were found to be directly related to the concentration of the parameter matrix.

The MDL (in µg/L) in reagent water was 4.2.

The standard deviation (in µg/L based on 4 recovery measurements) was 33.4.

The range (in µg/L) for average recovery for 4 measurements was 38.4–144.7.

The range (in %) for percent recovery was 25–158.

Accuracy (in µg/L) as expected recovery for one or more measurements of a sample containing a true concentration of C was $0.91C+0.53$.

Precision (in µg/L) as expected single analyst standard deviation of measurements at an average concentration found at X was $0.20X-0.94$.

Overall precision (in µg/L) as expected interlaboratory standard deviation of measurements in an average concentration found at X was $0.30X-0.46$.

$C =$ *True value of the concentration in µg/L.*
$X =$ *Average recovery found for measurements of samples containing a concentration at C in µg/L.*

SAMPLE PREPARATION Adjust the pH to >11 with sodium hydroxide and serially extract in a separatory funnel with methylene chloride or else in a continuous extractor. Next, adjust the pH to <2 with sulfuric acid and serially extract in a separatory funnel with methylene chloride or else in a continuous extractor. Dry the extracts separately through a column of anhydrous sodium sulfate and then concentrate each of the extracts to 1.0 mL using a K-D apparatus.

SAMPLE COLLECTION, PRESERVATION & HANDLING
Grab samples must be collected in glass containers. All samples must be refrigerated at 4°C from the time of collection until extraction. If residual chlorine is present, add 80 mg of sodium thiosulfate/L of sample and mix well. All samples must be extracted within 7 days of collection and completely analyzed within 40 days of extraction.

QUALITY CONTROL Make an initial, one-time, demonstration of the ability to generate acceptable accuracy and precision with this method. Before processing any samples, the analyst must analyze a reagent water blank to demonstrate that interferences from the analytical system and glassware are under control. Each time a set of samples is extracted or reagents are changed, a reagent water blank must be processed. Spike and analyze a minimum of 5% of all samples to monitor and evaluate lab data quality. A QC check sample concentrate that contains each parameter of interest at a concentration of 100 µg/mL in acetone is required. PCBs and multicomponent pesticides may be omitted from this test.

After analysis of five spiked wastewater samples, calculate the average percent recovery and the standard deviation of the percent recovery. Spike all samples with the surrogate standard spiking solution and calculate the percent recovery of each surrogate compound.

REFERENCE *Federal Register*, Vol. 49, No. 209. Friday, Oct. 26, 1984.

4-Chlorophenyl phenyl ether   **EPA Method 8270**
CAS #7005-72-3

TITLE Semivolatile Organic Compounds by GC/MS

MATRIX This method is used to determine the concentration of semivolatile organic compounds in extracts prepared from all types of solid waste matrices, soils, and groundwater. Although surface waters are not specifically mentioned, this method should be applicable to water samples from rivers, lakes, etc.

METHOD SUMMARY This method covers 259 semivolatile organic compounds. In very limited applications direct injection of the sample into the GC/MS system may be appropriate, but this results in very high detection limits (approximately 10,000 µg/L). Typically, a 1-L liquid sample, containing surrogate, and matrix spiking standards, is extracted in a continuous extractor first under acid conditions and then under basic conditions. Typically 30 g of a solid sample, containing surrogate, and matrix spiking standards, is extracted ultrasonically. After concentrating the extract to 1 mL it is spiked with 10 µL of an internal standard solution just prior to analysis by GC/MS. The volume injected should contain about 100 ng of base/neutral and 200 ng of acid surrogates (for a 1-µL injection). Analysis is performed by GC/MS using a capillary GC column.

INTERFERENCES Raw GC/MS data from all blanks, samples, and spikes must be evaluated for interferences. Contamination by carryover can occur whenever high-concentration and low-concentration samples are sequentially analyzed. To reduce carryover, the sample syringe must be rinsed out between samples with solvent. Whenever an unusually concentrated sample is encountered, it should be followed by the analysis of blank solvent to check for cross-contamination.

INSTRUMENTATION A GC/MS and a data system are required. The GC column used is a 30 m × 0.25 mm I.D. (or 0.32 mm I.D.) 1um film thickness silicone-coated fused silica capillary column. A continuous liquid-liquid extractor equipped with Teflon® or glass connection joints and stopcocks requiring no lubrication, a K-D concentrating apparatus, water bath, and an ultrasonic disrupter with a minimum power of 300 W and with pulsing capability are also required.

PRECISION & ACCURACY The estimated quantitation limit (EQL) of Method 8270B for determining an individual compound is approximately 1 mg/kg (wet weight) for soil or sediment samples, 1–200 mg/kg for wastes (dependent on matrix and method of preparation), and 10 µg/L for groundwater samples. EQLs will be proportionately higher for sample extracts that require dilution to avoid saturation of the detector.

The EQL(b) for groundwater in µg/L is 10.
The EQL (a, b) for low concentrations in soil and sediment in µg/kg is 660.
Accuracy as µg/L is 0.91C +0.53.
Overall precision in µg/L is 0.30X–0.46.

(a) *EQLs listed for soil/sediment are based on wet weight. Normally data is reported in a dry-weight basis; therefore, EQLs will be higher based on the % dry weight of each sample. This calculation is based on a 30 g sample and gel permeation chromatography cleanup.*
(b) *Sample EQLs are highly matrix-dependent. The EQLs are provided for guidance and may not always be achievable.*
C = *True value for concentration, in µg/L.*
X = *Average recovery found for measurements of samples containing a concentration of C, in µg/L.*

ESTIMATED QUANTITATION LIMIT
FOR OTHER MATRICES

Other Matrices	Factor (a)
High-concentration soil and sludges by sonicator	7.5
Non-water miscible waste	75

(a) *EQL for other matrices = [EQL for low soil/sediment] × [Factor]. This estimated EQL is similar to an EPA "Practical Quantitation Limit."*

SAMPLING METHOD
Liquid samples — Use a 1 or 2½ gallon amber glass bottle with a screw-top Teflon®-lined cover that has been prewashed with detergent and rinsed with distilled water and methanol (or isopropanol).

Soils, sediments, or sludges — Use an 8-oz. widemouth glass with a screw-top Teflon®-lined cover that has been prewashed with detergent and rinsed with distilled water and methanol (or isopropanol).

SAMPLE PRESERVATION
Liquid samples — If residual chlorine is present, add 3 mL of 10% sodium thiosulfate per gallon, cool to 4°C and store in a

solvent-free refrigerator until analysis; if chlorine is not present, then eliminate the sodium thiosulfate addition.

Soils, sediments, or sludges — Cool samples to 4°C and store in a solvent-free refrigerator.

MHT Liquid samples must be extracted within 7 days and the extracts analyzed within 40 days. Soils, sediments, or sludges may be stored for a maximum of 14 days and the extracts analyzed within 40 days.

SAMPLE PREPARATION

Liquid samples — Transfer 1 L quantitatively to a continuous extractor. If high concentrations are anticipated, a smaller volume may be used and then diluted with organic-free reagent water to 1 L. Adjust pH, if necessary, to pH <2 using 1:1 (V/V) sulfuric acid. Pipette 1.0 mL of a surrogate standard spiking solution into each sample. For the sample in each analytical batch selected for spiking, add 1.0 mL of a matrix spiking standard. For base/neutral acid analysis, the amount of the surrogates and matrix spiking compounds added to the sample should result in a final concentration of 100 ng/µL of each analyte in the extract to be analyzed (assuming a 1-µL injection). Extract with methylene chloride for 18–24 h. Next, adjust the pH of the aqueous phase to pH >11 using 10 N sodium hydroxide and extract it with methylene chloride again for 18–24 h. Dry the extract through a column containing anhydrous sodium sulfate and concentrate it to 1 mL using a K-D concentrator.

Soils, sediments, or sludges — Use 30 g of sample. Nonporous or wet samples (gummy or clay type) that do not have a free-flowing sandy texture must be mixed with anhydrous sodium sulfate until the sample is free flowing. Add 1 mL of surrogate standards to all samples, spikes, standards, and blanks. For the sample in each analytical batch selected for spiking, add 1.0 mL of a matrix spiking standard. For base/neutral acid analysis, the amount added of the surrogates and matrix spiking compounds should result in a final concentration of 100 ng/µL of each base/neutral analyte and 200 ng/µL of each acid analyte in the extract to be analyzed (assuming a 1-µL injection). Immediately add a 100-mL mixture of 1:1 methylene chloride:acetone and extract the sample ultrasonically for 3 min and then decant or filter the extracts. Repeat the extraction two or more times. Dry the extract using a column with anhydrous sodium sulfate and concentrate it to 1 mL in a K-D concentrator.

QUALITY CONTROL A methylene chloride solution containing 50 ng/µL of decafluorotriphenylphosphine (DFTPP) is used for tuning the GC/MS system each 12-h shift. A system performance check also must be made during every 12-h shift. A standard containing 50 ng/µL each of 4,4′-DDT, pentachlorophenol, and benzidine is required to verify injection port inertness and GC column performance. A calibration standard at mid-concentration, containing each compound of interest, including all required surrogates, must be performed every 12 h during analysis. After the system performance check is met, calibration check compounds (CCCs) are used to check the validity of the initial calibration.

The internal standard responses and retention times in the calibration check standard must be evaluated immediately after or during data acquisition. If the retention time for any internal standard changes by more than 30 seconds from the last check calibration (12 h), the chromatographic system must be inspected for malfunctions and corrections must be made, as required. If the electron ionization current plot (EICP) area for any of the internal standards changes by a factor of two from the last daily calibration standard check, the mass spectrometer must be inspected for malfunctions and corrections must be made, as appropriate.

Demonstrate, through the analysis of a reagent water blank, that interferences from the analytical system, glassware, and reagents are under control. The blank samples should be carried through all stages of the sample preparation and measurement steps. For each analytical batch (up to 20 samples), a reagent blank, matrix spike, and matrix spike duplicate/duplicate must be analyzed (the frequency of the spikes may be different for different monitoring programs). The blank and spiked samples must be carried through all stages of the sample preparation and measurement steps. A QC reference sample concentrate containing each analyte at a concentration of 100 mg/L in methanol is required.

REFERENCE Test Methods for Evaluating Solid Waste (SW-846). U.S. EPA 1983, Method 8270B, Rev. 2, Nov. 1990. Office of Solid Waste, Washington, DC.

Chloroprene EPA Method 8240
CAS #126-99-8

TITLE Volatile Organics By GC/MS: Packed Column Technique

MATRIX Nearly all types of sample matarices, regardless of water content, can be analyzed using this method. This includes groundwater, aqueous sludges, caustic liquors, acid liquors, waste solvents, oily wastes, mousses, tars, fibrous wastes, polymetric emulsions, filter cakes, spent carbons, spent catalysts, soils, and sediments.

METHOD SUMMARY Method 8240B covers 80 volatile organic compounds that are introduced into a gas chromatograph by the purge-and-trap method or by direct injection (in limited applications). For the purge-and-trap method an inert gas (zero grade nitrogen or helium) is bubbled through a 5-mL solution at ambient temperature. Purged sample components are trapped in a tube of sorbent materials. When purging is complete, the sorbent tube is heated and backflushed with inert gas to desorb the trapped components onto a GC column.

INTERFERENCES Impurities in the purge gas and from organic compounds outgassing from the plumbing ahead of the trap account for many contamination problems. Interferences purged or coextracted from the samples will vary considerably from source to source. Cross-contamination can occur whenever high-level and low-level samples are analyzed sequentially. Whenever an unusually concentrated sample is analyzed, it should be followed by the analysis of organic-free reagent water to check for cross-contamination. Samples also can be contaminated by diffusion of volatile organics (particularly methylene chloride and fluorocarbons) through the septum seal into the sample during shipment and storage. A trip blank can serve as a check on such contamination. The lab where volatile analysis is performed and also the refrigerated storage area should be completely free of solvents.

INSTRUMENTATION A gas chromatograph/mass spectrometry/data system (GC/MS) equipped with a 6 ft × 0.1 in I.D. glass column packed with 1% SP-1000 on Carbopack-B (60/80 mesh) is required. Also needed is a 5-mL purging device, a sorbent trap, and a thermal desorption apparatus.

PRECISION & ACCURACY This method is reported to have been tested by 15 laboratories using organic-free reagent water, drinking water, surface water, and industrial wastewaters (not specified) fortified at six concentrations over the range 5–600 µg/L.

Sample estimated quantitation limits (EQLs) are highly matrix-dependent. The EQLs listed may not always be achievable. EQLs listed for soils or sediments are based on wet weight. Normally, data is reported on a dry-weight basis; therefore, EQLs will be higher, based on the percent dry weight of each sample. Note that EQLs are even more variable than MDLs and that they are highly variable depending on the matrix being analyzed.

EQL in groundwater in µg/L was 5.
EQL in low soil or sediment in µg/kg was 5.
Accuracy (a) in µg/L was not listed.
Precision (b) in µg/L was not listed.

(a) *Average recovery found for measurements of samples containing a concentration of C, in µg/L.*
(b) *Overall precision found for measurements of samples with average recovery X for samples containing a concentration of C in µg/L.*
X = *Average recovery found for measurement of samples containing a concentration of C in µg/L.*

MULTIPLICATION FACTORS FOR OTHER MATRICES

Other Matrices	Factor (a)
Waste miscible liquid waste	50
High-concentration soil and sludge	125
Non-water miscible waste	500

(a) *EQL = [EQL for low soil/sediment] × [Factor]. For non-aqueous samples, the factor is on a wet-weight basis.*

SAMPLING METHOD

Liquid samples — Use a 40-mL glass screw-cap VOA vial with a Teflon®-faced silicone septum that has been prewashed, rinsed with distilled deionized water, and oven dried. However, if residual chlorine is present, collect sample in a 40-oz. soil VOA container which has been pre-preserved with 4 drops of 10% sodium thiosulfate, mix gently, and then transfer the sample to a 40-mL VOA vial. Collect bubble-free samples in duplicate and seal them in separate plastic bags.

Soils or sediments, and sludges — Use an 8-oz. widemouth glass bottle with a Teflon®-faced silicone septum that has been prewashed with detergent, rinsed with distilled deionized water, and oven dried. Tap slightly to eliminate free air space. Collect samples in duplicate and seal them in separate plastic bags.

SAMPLE PRESERVATION

Liquid samples — Add 4 drops of concentrated HCL and immediately cool samples to 4°C and store in a solvent-free refrigerator.

Soils or sediments, and sludges — Cool samples to 4°C and store in a solvent-free refrigerator.

MHT Maximum holding time is 14 days from the date of sample collection.

SAMPLE PREPARATION

Liquid samples — Remove the plunger from a 5-mL syringe and carefully pour the sample into the syringe barrel to just short of overflowing. Replace the syringe plunger and compress the sample. Open the syringe valve and vent any residual air while adjusting the sample volume to 5.0 mL. If there is only one volatile organic analysis (VOA) vial, a second syringe should be filled at this time to protect against possible loss of sample integrity. Add 10 µL of surrogate spiking solution and 10 µL of internal standard spiking solution through the valve bore of the 5-mL syringe, then close the valve. The surrogate and internal standards may be mixed and added as a single spiking solution.

Sediments, soils, and waste samples — All samples of this type should be screened by GC analysis using a headspace method (EPA Method 3810) or the hexadecane extraction and screening method (EPA Method 3820). Use the screening data to determine whether to use the low-concentration method (0.005–1 mg/kg) or the high-concentration method (>1 mg/kg).

Low-concentration method — The low-concentration method is based on purging a heated sediment or soil sample mixed with organic-free reagent water containing the surrogate and internal standards. Analyze all reagent blanks and standards under the same conditions as the samples.

Use a 5-g sample if the expected concentration is <0.1 mg/kg or a 1-g sample for expected concentrations between 0.1 and 1 mg/kg. Mix the contents of the sample container with a narrow metal spatula. Weigh the amount of the sample into a tared purge device. Add the spiked water to the purge device, which contains the weighed amount of sample, and connect the device to the purge-and-trap system.

High-concentration method — This method is based on extracting the sediment or soil with methanol. A waste sample is either extracted or diluted, depending on its solubility in methanol. Wastes that are insoluble in methanol are diluted with reagent tetraglyme or possibly polyethylene glycol (PEG). An aliquot of the extract is added to organic-free reagent water containing surrogate and internal standards. This is purged at ambient temperature. All samples with an expected concentration of >1.0 mg/kg should be analyzed by this method.

Mix the contents of the sample container with a narrow metal spatula. For sediments or soils and solid wastes that are insoluble in methanol, weigh 4 g (wet weight) of sample into a tared 20-mL vial. For waste that is soluble in methanol, tetraglyme, or PEG, weigh 1 g (wet weight) into a tared scintillation vial or culture tube or a 10-mL volumetric flask. Quickly add

9.0 mL of appropriate solvent then add 1.0 mL of a surrogate spiking solution to the vial, cap it, and shake it for 2 min.

METHANOL EXTRACT REQUIRED FOR ANALYSIS OF HIGH-CONCENTRATION SOILS OR SEDIMENTS

Approximate Concentration Range	Volume of Methanol Extract (a)
500–10,000 µg/kg	100 µL
1,000–20,000 µg/kg	50 µL
5,000–100,000 µg/kg	10 µL
25,000–500,000 µg/kg	100 µL of 1/50 dilution (b)

Calculate appropriate dilution factor for concentrations exceeding this table.

(a) The volume of methanol added to 5 mL of water being purged should be kept constant. Therefore, add to the 5-mL syringe whatever volume of methanol is necessary to maintain a volume of 100 µL added to the syringe.

(b) Dilute an aliquot of the methanol extract and then take 100 µL for analysis.

QUALITY CONTROL Demonstrate, through the analysis of a reagent water blank, that interferences from the analytical system, glassware, and reagents are under control. Blank samples should be carried through all stages of the sample preparation and measurement steps. For each analytical batch (up to 20 samples), a reagent blank, matrix spike, and matrix spike duplicate must be analyzed (the frequency of the spikes may be different for different monitoring programs). The blank and spiked samples must be carried through all stages of the sample preparation and measurement steps. QC samples mentioned in the section on Interferences will also be needed as appropriate to those situations.

REFERENCE Test Methods for Evaluating Solid Waste (SW-846). U.S. EPA. 1983. Method 8240B, Rev. 2, Nov. 1990. Office of Solid Wastes, Washington, DC.

3-Chloropropene   **EPA Method 1624**
CAS #107-05-1

TITLE Volatile Organic Compounds by Isotope Dilution GC/MS

MATRIX Compounds may be determined in waters, soils, and municipal sludges by this method.

METHOD SUMMARY This method is used to determine 58 volatile toxic organic pollutants associated with the CWA (as amended 1987); the RCRA (as amended 1986); the CERCLA (as amended 1986); and other compounds amenable to purge-and-trap gas chromatography-mass spectrometry (GC/MS).

If the solids content is less than 1%, stable isotopically-labeled analogs of the compounds of interest are added to a 5-mL sample and the sample is purged with an inert gas at 20–25°C in a chamber designed for soil or water samples. If the solids content is greater than 1%, 5 mL of reagent water and the labeled compounds are added to a 5-g aliquot of sample and the mixture is purged at 40°C. Compounds that will not purge at 20–25°C or at 40°C are purged at 78–85°C. In the purging process, the volatile compounds are transferred from the aqueous phase into the gaseous phase where they are passed into a sorbent column, and trapped. After purging is completed, the trap is backflushed and heated rapidly to desorb the compounds into a GC. The compounds are separated by the GC and detected by a MS. The labeled compounds serve to correct the variability of the analytical technique.

INTERFERENCES Impurities in the purge gas, organic compounds outgassing from the plumbing upstream of the trap, and solvent vapors in the lab account for most problems. Samples can be contaminated by diffusion of volatile organic compounds (particularly methylene chloride) through the bottle seal during shipment and storage. Contamination by carryover can occur when high-level and low-level samples are analyzed sequentially. When an unusually concentrated sample is encountered, follow it by analysis of a reagent water blank to check for carryover.

INSTRUMENTATION Major equipment includes a GC with linear temperature programming and a glass jet separator as the MS interface, a MS with 70 eV electron impact ionization, and a data system to collect and record response factors.

Column: 2.8 m × 2 mm I.D. glass, packed with 1% SP-1000 on Carbopak B, 60/80 mesh, or equivalent.

PRECISION & ACCURACY The detection limits of the method are usually dependent on the level of interferences rather than instrumental limitations. The method detection limits were determined in digested sludge (low solids) and in filter cake or compost (high solids).

The MDL (in µg/kg) for low solids is not listed and for high solids is not listed.

Labeled and native compound precision (in µg/L) as standard deviation was not listed.

Labeled and native compound accuracy (in µg/L) as average recovery was not listed.

Acceptance criteria are at 20 µg/L for this compound.

SAMPLE COLLECTION, PRESERVATION & HANDLING Grab samples are collected in glass containers having a total volume greater than 20 mL. Fill and seal each bottle so that no air bubbles are entrapped. Samples are maintained at 0 to 4°C from the time of collection until analysis. If an aqueous sample contains residual chlorine, add sodium thiosulfate preservative (10 mg/40 mL) to the empty sample bottles just prior to shipment to the sample site. All samples must be analyzed within 14 days of collection.

SAMPLE PREPARATION Samples containing less than 1% solids are analyzed directly as aqueous samples. Samples containing 1% solids or greater are analyzed as solid samples utilizing one of two methods, depending on the levels of pollutants, in the sample. Samples containing 1% solids or greater, and low to moderate levels of pollutants are analyzed by purging a known weight of sample added to 5 mL of reagent water. Samples containing 1% solids or greater, and high levels of pollutants, are extracted with methanol, and an aliquot of the methanol extract is added to reagent water and purged.

QUALITY CONTROL A field blank prepared from reagent water and carried through the sampling and handling protocol may serve as a check on contamination from shipment and storage.

The analyst is permitted to modify this method to improve separations or lower the costs of measurements, provided all performance specifications are met. Analyses of blanks are required. When results of spikes indicate atypical method performance for samples, the samples are diluted to bring method performance within acceptable limits. Analyze two sets of four 5-mL aliquots (8 aliquots total) of the aqueous performance standard. Spike all samples with labeled compounds to assess method performance on the sample matrix. Compute the percent recovery of the labeled compounds using the internal standard method. Compare the percent recovery for each compound with the corresponding labeled compound recovery. Reagent water blanks are analyzed to demonstrate freedom from carryover contamination. Field replicates may be collected to determine the precision of the sampling technique, and spiked samples may be required to determine the accuracy of the analysis when the internal method is used.

REFERENCE Volatile Organic Compounds by Isotope Dilution GC/MS. Office of Water Regulation and Standards, U.S. EPA Industrial Technology Division, Washington, DC, EPA Method 1624, Rev. C, June 1989 (contact W.A. Telliard, U.S. EPA, Office of Water Regulations and Standards, 401 M St., SW, Washington, DC, 20460. Phone: 202-382-7131).

3-Chloropropionitrile **EPA Method 8240**
CAS #542-76-7

TITLE Volatile Organics By GC/MS: Packed Column Technique

MATRIX Nearly all types of sample matarices, regardless of water content, can be analyzed using this method. This includes groundwater, aqueous sludges, caustic liquors, acid liquors, waste solvents, oily wastes, mousses, tars, fibrous wastes, polymetric emulsions, filter cakes, spent carbons, spent catalysts, soils, and sediments.

METHOD SUMMARY Method 8240B covers 80 volatile organic compounds that are introduced into a gas chromatograph by the purge-and-trap method or by direct injection (in limited applications). For the purge-and-trap method an inert gas (zero grade nitrogen or helium) is bubbled through a 5-mL solution at ambient temperature. Purged sample components are trapped in a tube of sorbent materials. When purging is complete, the sorbent tube is heated and backflushed with inert gas to desorb the trapped components onto a GC column.

INTERFERENCES Impurities in the purge gas and from organic compounds outgassing from the plumbing ahead of the trap account for many contamination problems. Interferences purged or coextracted from the samples will vary considerably from source to source. Cross-contamination can occur whenever high-level and low-level samples are analyzed sequentially. Whenever an unusually concentrated sample is analyzed, it should be followed by the analysis of organic-free reagent water to check for cross-contamination. Samples also can be contaminated by diffusion of volatile organics (particularly methylene chloride and fluorocarbons) through the septum seal into the sample during shipment and storage. A trip blank can serve as a check on such contamination. The lab where volatile analysis is performed and also the refrigerated storage area should be completely free of solvents.

INSTRUMENTATION A gas chromatograph/mass spectrometry/data system (GC/MS) equipped with a 6 ft × 0.1 in I.D. glass column packed with 1% SP-1000 on Carbopack-B (60/80 mesh) is required. Also needed is a 5-mL purging device, a sorbent trap, and a thermal desorption apparatus.

PRECISION & ACCURACY This method is reported to have been tested by 15 laboratories using organic-free reagent water, drinking water, surface water, and industrial wastewaters (not specified) fortified at six concentrations over the range 5–600 µg/L.

Sample estimated quantitation limits (EQLs) are highly matrix-dependent. The EQLs listed may not always be achievable. EQLs listed for soils or sediments are based on wet weight. Normally, data is reported on a dry-weight basis; therefore, EQLs will be higher, based on the percent dry weight of each sample. Note that EQLs are even more variable than MDLs and that they are highly variable depending on the matrix being analyzed.

EQL in groundwater in µg/L was not listed.
EQL in low soil or sediment in µg/kg was not listed.
Accuracy (a) in µg/L was not listed.
Precision (b) in µg/L was not listed.

(a) *Average recovery found for measurements of samples containing a concentration of C, in µg/L.*
(b) *Overall precision found for measurements of samples with average recovery X for samples containing a concentration of C in µg/L.*
X = *Average recovery found for measurement of samples containing a concentration of C in µg/L.*

MULTIPLICATION FACTORS FOR OTHER MATRICES

Other Matrices	Factor (a)
Waste miscible liquid waste	50
High-concentration soil and sludge	125
Non-water miscible waste	500

(a) EQL = [EQL for low soil/sediment] × [Factor]. For non-aqueous samples, the factor is on a wet-weight basis.

SAMPLING METHOD

Liquid samples — Use a 40-mL glass screw-cap VOA vial with a Teflon®-faced silicone septum that has been prewashed, rinsed with distilled deionized water, and oven dried. However, if residual chlorine is present, collect sample in a 40-oz. soil VOA container which has been pre-preserved with 4 drops of 10% sodium thiosulfate, mix gently, and then transfer the sample to a 40-mL VOA vial. Collect bubble-free samples in duplicate and seal them in separate plastic bags.

Soils or sediments, and sludges — Use an 8-oz. widemouth glass bottle with a Teflon®-faced silicone septum that has been prewashed with detergent, rinsed with distilled deionized water, and oven dried. Tap slightly to eliminate free air space. Collect samples in duplicate and seal them in separate plastic bags.

SAMPLE PRESERVATION
Liquid samples — Add 4 drops of concentrated HCL and immediately cool samples to 4°C and store in a solvent-free refrigerator.

Soils or sediments, and sludges — Cool samples to 4°C and store in a solvent-free refrigerator.

MHT Maximum holding time is 14 days from the date of sample collection.

SAMPLE PREPARATION
Liquid samples — Remove the plunger from a 5-mL syringe and carefully pour the sample into the syringe barrel to just short of overflowing. Replace the syringe plunger and compress the sample. Open the syringe valve and vent any residual air while adjusting the sample volume to 5.0 mL. If there is only one volatile organic analysis (VOA) vial, a second syringe should be filled at this time to protect against possible loss of sample integrity. Add 10 µL of surrogate spiking solution and 10 µL of internal standard spiking solution through the valve bore of the 5-mL syringe, then close the valve. The surrogate and internal standards may be mixed and added as a single spiking solution.

Sediments, soils, and waste samples — All samples of this type should be screened by GC analysis using a headspace method (EPA Method 3810) or the hexadecane extraction and screening method (EPA Method 3820). Use the screening data to determine whether to use the low-concentration method (0.005–1 mg/kg) or the high-concentration method (>1 mg/kg).

Low-concentration method — The low-concentration method is based on purging a heated sediment or soil sample mixed with organic-free reagent water containing the surrogate and internal standards. Analyze all reagent blanks and standards under the same conditions as the samples.

Use a 5-g sample if the expected concentration is <0.1 mg/kg or a 1-g sample for expected concentrations between 0.1 and 1 mg/kg. Mix the contents of the sample container with a narrow metal spatula. Weigh the amount of the sample into a tared purge device. Add the spiked water to the purge device, which contains the weighed amount of sample, and connect the device to the purge-and-trap system.

High-concentration method — This method is based on extracting the sediment or soil with methanol. A waste sample is either extracted or diluted, depending on its solubility in methanol. Wastes that are insoluble in methanol are diluted with reagent tetraglyme or possibly polyethylene glycol (PEG). An aliquot of the extract is added to organic-free reagent water containing surrogate and internal standards. This is purged at ambient temperature. All samples with an expected concentration of >1.0 mg/kg should be analyzed by this method.

Mix the contents of the sample container with a narrow metal spatula. For sediments or soils and solid wastes that are insoluble in methanol, weigh 4 g (wet weight) of sample into a tared 20-mL vial. For waste that is soluble in methanol, tetraglyme, or PEG, weigh 1 g (wet weight) into a tared scintillation vial or culture tube or a 10-mL volumetric flask. Quickly add 9.0 mL of appropriate solvent then add 1.0 mL of a surrogate spiking solution to the vial, cap it, and shake it for 2 min.

METHANOL EXTRACT REQUIRED FOR ANALYSIS OF HIGH-CONCENTRATION SOILS OR SEDIMENTS

Approximate Concentration Range	Volume of Methanol Extract (a)
500–10,000 µg/kg	100 µL
1,000–20,000 µg/kg	50 µL
5,000–100,000 µg/kg	10 µL
25,000–500,000 µg/kg	100 µL of 1/50 dilution (b)

Calculate appropriate dilution factor for concentrations exceeding this table.

(a) The volume of methanol added to 5 mL of water being purged should be kept constant. Therefore, add to the 5-mL syringe whatever volume of methanol is necessary to maintain a volume of 100 µL added to the syringe.
(b) Dilute an aliquot of the methanol extract and then take 100 µL for analysis.

QUALITY CONTROL Demonstrate, through the analysis of a reagent water blank, that interferences from the analytical system, glassware, and reagents are under control. Blank samples should be carried through all stages of the sample preparation and measurement steps. For each analytical batch (up to 20 samples), a reagent blank, matrix spike, and matrix spike duplicate must be analyzed (the frequency of the spikes may be different for different monitoring programs). The blank and spiked samples must be carried through all stages of the sample preparation and measurement steps. QC samples mentioned in the section on Interferences will also be needed as appropriate to those situations.

REFERENCE Test Methods for Evaluating Solid Waste (SW-846). U.S. EPA. 1983. Method 8240B, Rev. 2, Nov. 1990. Office of Solid Wastes, Washington, DC.

2-Chlorotoluene   **EPA Method 502**
CAS #95-49-8

TITLE Volatile Organic Compounds in Water By Purge and Trap Capillary Column Gas Chromatography with Photoionization and Electrolytic Conductivity Detectors in Series. U.S. EPA Method 502.2, Rev. 2.0, 1989.

MATRIX Drinking water and raw source water. The latter should include most surface water and groundwater sources.

METHOD SUMMARY This method covers 60 volatile organic compounds that contain halogen atoms and/or that are aromatic. An inert gas (zero grade nitrogen or helium) is bubbled through a 25-mL or a 5-mL water sample (depending on the expected concentration of the analytes). Purged sample components are trapped in a tube of sorbent materials. When purging is complete, the sorbent tube is heated and backflushed with helium to desorb the trapped sample onto a capillary GC column. The column is temperature programmed to separate the method analytes which are then detected with a photoionization detector (PID) and a Hall electrolytic conductivity (HECD) placed in series. The PID is selective for aromatic compounds and the HECD is selective for halogenated compounds.

INTERFERENCES Impurities in the purge gas and from organic compounds outgassing from the plumbing ahead of the trap account for many contamination problems. Interferences purged or coextracted from the samples will vary considerably from source to source, depending upon the particular sample or extract being tested. Cross-contamination can occur whenever high-level and low-level samples are analyzed sequentially. Samples also can be contaminated by diffusion of volatile organics (particularly methylene chloride and fluorocarbons) through the septum seal into the sample during shipment and storage. The lab where volatile analysis is performed and also the refrigerated storage area should be completely free of solvents.

INSTRUMENTATION A GC containing a series configuration of a high temperature photoionization detector (PID) equipped with 10.0 eV (nominal) lamp and Hall electrolytic conductivity detector (HECD) is required. Also required is an all-glass 5-mL purging device, a sorbent trap, and a thermal desorption apparatus which is connected to the GC system.

Column 1: VOCOL glass wide-bore capillary column.
Column 2: RTX–502.2 mega-bore capillary column.
Column 3: DB-62 mega-bore capillary column.

PRECISION & ACCURACY Method detection limits are dependent upon the characteristics of the gas chromatographic system used. Analytes that are not separated chromatographically cannot be individually identified and used in the same calibration mixture or water samples unless an alternative technique for identification and quantification, such as mass spectrometry, is used.

Electrolytic conductivity detetor (c) range in µg/L (a) was 0.02–200.
Electrolytic conductivity detetor (c) MDL in µg/L (b) was 0.01.
Electrolytic conductivity detetor (c) accuracy as % recovery was 97.
Electrolytic conductivity detetor (c) precision as % RSD was 2.7.
Photoionization detector (d) range in µg/L (a) was 0.02–200.
Photoionization detector (d) MDL in µg/L (b) was not detected.
Photoionization detector (d) accuracy as % recovery was not detected.
Photoionization detector (d) precision as % RSD was not detected.

(a) The applicable concentration range of this method is compound, instrument, and matrix-dependent. It is listed as being approximately 0.02 to 200 µg/L but no specific information is provided so caution should be observed.
(b) The method detection limits reports with this method are compound, instrument, and matrix-dependent. The values reported were calculated using reagent water fortified with the corresponding compounds at 10 µg/L and a GC-equipped with a 60 m × 0.75 mm VOLCOL wide bore capillary column with 1.5 µm film thickness and using helium carrier gas.
(c) Recoveries and relative standard deviations were determined from seven samples of reagent water fortified with 10 µg/L of each compound. 2-Bromo-1-chloropropane was used as the internal standard for calculating average recoveries.
(d) Recoveries and relative standard deviations were determined from seven samples of reagent water fortified with 10 µg/L of each compound. Fluorobenzene was used as the internal standard for calculating average recoveries.

SAMPLING METHOD Collect samples using a 40- to 120-mL screw-cap vial (prewashed with detergent, rinsed with distilled water and oven dried at 105°C) with a Teflon®-faced silicone septum. Collect bubble-free samples and place the septum with the Teflon® side down on the water.

SAMPLE PRESERVATION If residual chlorine is present in the water add about 25 mg of ascorbic acid to each vial before samples are collected to remove the chlorine. Add hydrochloric acid to reduce pH to <2, immediately cool samples to 4°C, and store them in a solvent-free refrigerator at 4°C until analysis.

MHT The maximum holding time for samples is 14 days from the time they were collected.

SAMPLE PREPARATION Remove the plungers from two 5-mL syringes and attach a closed syringe valve to each. Warm the sample to room temperature, open the sample bottle, and carefully pour the sample into one of the syringe barrels to just short of overflowing. Replace the syringe plunger, invert the syringe, and compress the sample. Open the syringe valve and vent any residual air while adjusting the sample volume to 5.0 mL. Add 10 µL of the internal calibration standard to the sample through the syringe valve. Close the valve. Fill the second syringe in an identical manner from the same sample bottle. Reserve this second syringe for a reanalysis if necessary.

QUALITY CONTROL As an initial demonstration of lab accuracy and precision, analyze 4 to 7 replicates of a lab fortified blank containing analyte at 0.1–5 µg/L. Collect all samples in duplicate. Surrogate analytes (similar to those of the analytes of interest), whose concentration is known in every sample, are measured using the same internal standard calibration procedure. Duplicate field reagent water blanks (trip blanks) must be analyzed with each set of samples, lab reagent blanks (method blanks) must be analyzed with each batch of samples processed as a group within a work shift. Also, a single lab-fortified blank that contains each of the analytes of interest should be analyzed with each batch of samples processed as a group within a work shift. A 3- to 5-point calibration curve is needed depending on the calibration range factor required.

EPA CONTACT & HOTLINE For technical questions contact Dr. Baldev Bathija, U.S. EPA, Office of Ground Water and Drinking Water (WH-550D), 401 M St. SW, Washington, DC 20460. Tel. (202) 260-3040. For further information the EPA Safe Drinking Water Hotline may be called at: (800) 426-4791.

REFERENCE Methods for the Determination of Organic Compounds in Drinking Water, EPA/600/4-88/039 (revised

July 1991; Final Rule for determination of compliance with the MCL for Total Trihalomethanes under 141.30, in 40 CFR Part 141, Vol. 58, No. 147, Fed. Reg., Tuesday Aug. 3, 1993). U.S. EPA Environmental Monitoring Systems Laboratory, Cincinnati, OH, 45268, USA. Available from the National Technical Information Service (NTIS), 5285 Port Royal Road, Springfield, VA 22161; Tel. 800-553-6847. NTIS Order Number is PB91-231480.

4-Chlorotoluene   **EPA Method 502**
CAS #106-43-4

TITLE Volatile Organic Compounds in Water By Purge and Trap Capillary Column Gas Chromatography with Photoionization and Electrolytic Conductivity Detectors in Series. U.S. EPA Method 502.2, Rev. 2.0, 1989.

MATRIX Drinking water and raw source water. The latter should include most surface water and groundwater sources.

METHOD SUMMARY This method covers 60 volatile organic compounds that contain halogen atoms and/or that are aromatic. An inert gas (zero grade nitrogen or helium) is bubbled through a 25-mL or a 5-mL water sample (depending on the expected concentration of the analytes). Purged sample components are trapped in a tube of sorbent materials. When purging is complete, the sorbent tube is heated and backflushed with helium to desorb the trapped sample onto a capillary GC column. The column is temperature programmed to separate the method analytes which are then detected with a photoionization detector (PID) and a Hall electrolytic conductivity (HECD) placed in series. The PID is selective for aromatic compounds and the HECD is selective for halogenated compounds.

INTERFERENCES Impurities in the purge gas and from organic compounds outgassing from the plumbing ahead of the trap account for many contamination problems. Interferences purged or coextracted from the samples will vary considerably from source to source, depending upon the particular sample or extract being tested. Cross-contamination can occur whenever high-level and low-level samples are analyzed sequentially. Samples also can be contaminated by diffusion of volatile organics (particularly methylene chloride and fluorocarbons) through the septum seal into the sample during shipment and storage. The lab where volatile analysis is performed and also the refrigerated storage area should be completely free of solvents.

INSTRUMENTATION A GC containing a series configuration of a high temperature photoionization detector (PID) equipped with 10.0 eV (nominal) lamp and Hall electrolytic conductivity detector (HECD) is required. Also required is an all-glass 5-mL purging device, a sorbent trap, and a thermal desorption apparatus which is connected to the GC system.

Column 1: VOCOL glass wide-bore capillary column.
Column 2: RTX–502.2 mega-bore capillary column.
Column 3: DB-62 mega-bore capillary column.

PRECISION & ACCURACY Method detection limits are dependent upon the characteristics of the gas chromatographic system used. Analytes that are not separated chromatographically cannot be individually identified and used in the same calibration mixture or water samples unless an alternative technique for identification and quantification, such as mass spectrometry, is used.

Electrolytic conductivity detetor (c) range in µg/L (a) was 0.02–200.
Electrolytic conductivity detetor (c) MDL in µg/L (b) was 0.01.
Electrolytic conductivity detetor (c) accuracy as % recovery was 97.
Electrolytic conductivity detetor (c) precision as % RSD was 3.2.
Photoionization detector (d) range in µg/L (a) was 0.02–200.
Photoionization detector (d) MDL in µg/L (b) was 0.02.
Photoionization detector (d) accuracy as % recovery was 101.
Photoionization detector (d) precision as % RSD was 1.0.

(a) *The applicable concentration range of this method is compound, instrument, and matrix-dependent. It is listed as being approximately 0.02 to 200 µg/L but no specific information is provided so caution should be observed.*
(b) *The method detection limits reports with this method are compound, instrument, and matrix-dependent. The values reported were calculated using reagent water fortified with the corresponding compounds at 10 µg/L and a GC-equipped with a 60 m × 0.75 mm VOLCOL wide bore capillary column with 1.5 µm film thickness and using helium carrier gas.*
(c) *Recoveries and relative standard deviations were determined from seven samples of reagent water fortified with 10 µg/L of each compound. 2-Bromo-1-chloropropane was used as the internal standard for calculating average recoveries.*
(d) *Recoveries and relative standard deviations were determined from seven samples of reagent water fortified with 10 µg/L of each compound. Fluorobenzene was used as the internal standard for calculating average recoveries.*

SAMPLING METHOD Collect samples using a 40- to 120-mL screw-cap vial (prewashed with detergent, rinsed with distilled water and oven dried at 105°C) with a Teflon®-faced silicone septum. Collect bubble-free samples and place the septum with the Teflon® side down on the water.

SAMPLE PRESERVATION If residual chlorine is present in the water add about 25 mg of ascorbic acid to each vial before samples are collected to remove the chlorine. Add hydrochloric acid to reduce pH to <2, immediately cool samples to 4°C, and store them in a solvent-free refrigerator at 4°C until analysis.

MHT The maximum holding time for samples is 14 days from the time they were collected.

SAMPLE PREPARATION Remove the plungers from two 5-mL syringes and attach a closed syringe valve to each. Warm the sample to room temperature, open the sample bottle, and carefully pour the sample into one of the syringe barrels to just short of overflowing. Replace the syringe plunger, invert the syringe, and compress the sample. Open the syringe valve and vent any residual air while adjusting the sample volume to 5.0 mL. Add 10 µL of the internal calibration standard to the sample through the syringe valve. Close the valve. Fill the second syringe in an identical manner from the same sample bottle. Reserve this second syringe for a reanalysis if necessary.

QUALITY CONTROL As an initial demonstration of lab accuracy and precision, analyze 4 to 7 replicates of a lab fortified blank containing analyte at 0.1–5 µg/L. Collect all samples in duplicate. Surrogate analytes (similar to those of the analytes of interest), whose concentration is known in every sample, are measured using the same internal standard calibration procedure. Duplicate field reagent water blanks (trip blanks) must be analyzed with each set of samples, lab reagent blanks (method blanks) must be analyzed with each batch of samples processed as a group within a work shift. Also, a single lab-fortified blank that contains each of the analytes of interest should be analyzed with each batch of samples processed as a group within a work shift. A 3- to 5-point calibration curve is needed depending on the calibration range factor required.

EPA CONTACT & HOTLINE For technical questions contact Dr. Baldev Bathija, U.S. EPA, Office of Ground Water and Drinking Water (WH-550D), 401 M St. SW, Washington, DC 20460. Tel. (202) 260-3040. For further information the EPA Safe Drinking Water Hotline may be called at: (800) 426-4791.

REFERENCE Methods for the Determination of Organic Compounds in Drinking Water, EPA/600/4-88/039 (revised July 1991; Final Rule for determination of compliance with the MCL for Total Trihalomethanes under 141.30, in 40 CFR Part 141, Vol. 58, No. 147, Fed. Reg., Tuesday Aug. 3, 1993). U.S. EPA Environmental Monitoring Systems Laboratory, Cincinnati, OH, 45268, USA. Available from the National Technical Information Service (NTIS), 5285 Port Royal Road, Springfield, VA 22161; Tel. 800-553-6847. NTIS Order Number is PB91-231480.

2-Chlorotoluene   EPA Method 524
CAS #95-49-8

TITLE Measurement of Purgeable Organic Compounds in Water by Capillary Column GC/MS.

MATRIX Drinking water and raw source water; the latter should include most surface water and groundwater sources.

METHOD SUMMARY Method 524.2 covers 60 volatile organic compounds. An inert gas (zero grade nitrogen or helium) is bubbled through a 25-mL or a 5-mL water sample (depending on the expected concentration of the analytes). Purged sample components are trapped in a tube of sorbent materials. When purging is complete, the sorbent tube is heated and backflushed with helium to desorb the trapped sample onto a capillary GC column.

INTERFERENCES Impurities in the purge gas and from organic compounds outgassing from the plumbing ahead of the trap account for many contamination problems. Interferences purged or coextracted from the samples will vary considerably from source to source, depending upon the particular sample or extract being tested. Cross-contamination can occur whenever high-level and low-level samples are analyzed sequentially. Samples also can be contaminated by diffusion of volatile organics (particularly methylene chloride and fluorocarbons) through the septum seal into the sample during shipment and storage.

INSTRUMENTATION A GC/MS with a data system equipped with one of the following capillary GC columns:

Column 1: VOCOL glass wide bore capillary column.
Column 2: DB-624 fused silica capillary column.
Column 3: DB-5 fused silica capillary column.

Also required is an all-glass 25 mL or 5-mL purging device, a sorbent trap, and a thermal desorption apparatus which is connected to the GC/MS system.

PRECISION & ACCURACY Method detection limits are compound- and instrument-dependent, and may vary from approximately 0.02–0.35 µg/L. Note in the table below that the "true" concentration range used for accuracy and precision measurements was quite narrow. However, the applicable concentration range of this method is primarily column dependent and is approximately 0.02 to 200 µg/L for the wide-bore thick-film columns. Narrow-bore thin-film columns may have a capacity which limits the range to about 0.02 to 20 µg/L. Analytes that are inefficiently purged from water will not be detected when present at low concentrations, but they can be measured with acceptable accuracy and precision when present in sufficient amounts.

Analytes that are not separated chromatographically, but which have different mass spectra and non-interfering quantification ions, can be identified and measured in the same calibration mixture or water sample. Analytes which have very similar mass spectra cannot be individually identified and measured in the same calibration mixture or water samples unless they have different retention times. Co-eluting compounds with very similar mass spectra, typically many structural isomers, must be reported as an isomeric group or pair.

The range (in µg/L) was 0.1–10.
The Method Detection Limig (in µg/L) was 0.04.
The accuracy (as % recovery) was 90.
The precision (in %) was 6.2.

Note: Data were obtained from 16–31 determinations using a wide-bore capillary column and a jet separator interfaced to a quadrupole mass spectrometer. All analytes were in a reagent water matrix.

SAMPLING METHOD Collect samples using a 40- to 120-mL screw-cap vial (prewashed with detergent, rinsed with distilled water and oven dried at 105°C) with a Teflon®-faced silicone septum. Collect bubble-free samples and place the septum with the Teflon® side down on the water.

SAMPLE PRESERVATION If residual chlorine is present in the water add about 25 mg of ascorbic acid to each vial before samples are collected to remove the chlorine. Add hydrochloric

acid to reduce pH to <2, and immediately cool samples to 4°C, and store them in a solvent-free refrigerator at 4°C until analysis.

MHT The maximum holding time for samples is 14 days from the time they were collected.

SAMPLE PREPARATION Remove the plungers from two 25-mL (or 5-mL depending on sample size) syringes and attach a closed syringe valve to each. Warm the sample to room temperature, open the sample bottle, and carefully pour the sample into one of the syringe barrels to just short of overflowing. Replace the syringe plunger, invert the syringe, and compress the sample. Open the syringe valve and vent any residual air while adjusting the sample volume to 25.0 mL (or 5 mL). For samples and blanks, add 5 µL of the fortification solution containing the internal standard and the surrogates to the sample through the syringe valve. For calibration standards and lab fortified blanks, add 5 µL of the fortification solution containing the internal standard only. Close the valve. Fill the second syringe in an identical manner from the same sample bottle. Reserve this second syringe for a reanalysis if necessary.

QUALITY CONTROL As an initial demonstration of lab accuracy and precision, analyze 4 to 7 replicates of a lab fortified blank containing analyte at 0.2–5 µg/L. Collect all samples in duplicate. Surrogate analytes (similar to those of the analytes of interest), whose concentration is known in every sample, are measured using the same internal standard calibration procedure. Duplicate field reagent water blanks (trip blanks) must be analyzed with each set of samples, lab reagent blanks (method blanks) must be analyzed with each batch of samples processed as a group within a work shift. Also, a single lab-fortified blank that contains each of the analytes of interest should be analyzed with each batch of samples processed as a group within a work shift. A 3- to 5-point calibration curve is needed depending on the calibration range factor required.

EPA CONTACT & HOTLINE For technical questions contact Dr. Baldev Bathija, U.S. EPA, Office of Ground Water and Drinking Water (WH-550D), 401 M St. SW, Washington, DC 20460. Tel. (202) 260-3040. For further information the EPA Safe Drinking Water Hotline may be called at: (800) 426-4791.

REFERENCE Methods for the Determination of Organic Compounds in Drinking Water, EPA/600/4-88/039 (revised July 1991; Final Rule for determination of compliance with the MCL for Total Trihalomethanes under 141.30, in 40 CFR Part 141, Vol. 58, No. 147, Fed. Reg., Tuesday Aug. 3, 1993). U.S. EPA Environmental Monitoring Systems Laboratory, Cincinnati, OH, 45268, USA. Available from the National Technical Information Service (NTIS), 5285 Port Royal Road, Springfield, VA 22161; Tel. 800-553-6847. NTIS Order Number is PB91-231480.

4-Chlorotoluene   EPA Method 524
CAS #106-43-4

TITLE Measurement of Purgeable Organic Compounds in Water by Capillary Column GC/MS.

MATRIX Drinking water and raw source water; the latter should include most surface water and groundwater sources.

METHOD SUMMARY Method 524.2 covers 60 volatile organic compounds. An inert gas (zero grade nitrogen or helium) is bubbled through a 25-mL or a 5-mL water sample (depending on the expected concentration of the analytes). Purged sample components are trapped in a tube of sorbent materials. When purging is complete, the sorbent tube is heated and backflushed with helium to desorb the trapped sample onto a capillary GC column.

INTERFERENCES Impurities in the purge gas and from organic compounds outgassing from the plumbing ahead of the trap account for many contamination problems. Interferences purged or coextracted from the samples will vary considerably from source to source, depending upon the particular sample or extract being tested. Cross-contamination can occur whenever high-level and low-level samples are analyzed sequentially. Samples also can be contaminated by diffusion of volatile organics (particularly methylene chloride and fluorocarbons) through the septum seal into the sample during shipment and storage.

INSTRUMENTATION A GC/MS with a data system equipped with one of the following capillary GC columns:

Column 1: VOCOL glass wide bore capillary column.
Column 2: DB-624 fused silica capillary column.
Column 3: DB-5 fused silica capillary column.

Also required is an all-glass 25 mL or 5-mL purging device, a sorbent trap, and a thermal desorption apparatus which is connected to the GC/MS system.

PRECISION & ACCURACY Method detection limits are compound- and instrument-dependent, and may vary from approximately 0.02–0.35 µg/L. Note in the table below that the "true" concentration range used for accuracy and precision measurements was quite narrow. However, the applicable concentration range of this method is primarily column dependent and is approximately 0.02 to 200 µg/L for the wide-bore thick-film columns. Narrow-bore thin-film columns may have a capacity which limits the range to about 0.02 to 20 µg/L. Analytes that are inefficiently purged from water will not be detected when present at low concentrations, but they can be measured with acceptable accuracy and precision when present in sufficient amounts.

Analytes that are not separated chromatographically, but which have different mass spectra and non-interfering quantification ions, can be identified and measured in the same calibration mixture or water sample. Analytes which have very similar mass spectra cannot be individually identified and measured in the same calibration mixture or water samples unless they have different retention times. Co-eluting compounds with very similar mass spectra, typically many structural isomers, must be reported as an isomeric group or pair.

The range (in µg/L) was 0.1–10.
The Method Detection Limig (in µg/L) was 0.06.
The accuracy (as % recovery) was 99.

The precision (in %) was 8.3.

Note: Data were obtained from 16–31 determinations using a wide-bore capillary column and a jet separator interfaced to a quadrupole mass spectrometer. All analytes were in a reagent water matrix.

SAMPLING METHOD Collect samples using a 40- to 120-mL screw-cap vial (prewashed with detergent, rinsed with distilled water and oven dried at 105°C) with a Teflon®-faced silicone septum. Collect bubble-free samples and place the septum with the Teflon® side down on the water.

SAMPLE PRESERVATION If residual chlorine is present in the water add about 25 mg of ascorbic acid to each vial before samples are collected to remove the chlorine. Add hydrochloric acid to reduce pH to <2, and immediately cool samples to 4°C, and store them in a solvent-free refrigerator at 4°C until analysis.

MHT The maximum holding time for samples is 14 days from the time they were collected.

SAMPLE PREPARATION Remove the plungers from two 25-mL (or 5-mL depending on sample size) syringes and attach a closed syringe valve to each. Warm the sample to room temperature, open the sample bottle, and carefully pour the sample into one of the syringe barrels to just short of overflowing. Replace the syringe plunger, invert the syringe, and compress the sample. Open the syringe valve and vent any residual air while adjusting the sample volume to 25.0 mL (or 5 mL). For samples and blanks, add 5 μL of the fortification solution containing the internal standard and the surrogates to the sample through the syringe valve. For calibration standards and lab fortified blanks, add 5 μL of the fortification solution containing the internal standard only. Close the valve. Fill the second syringe in an identical manner from the same sample bottle. Reserve this second syringe for a reanalysis if necessary.

QUALITY CONTROL As an initial demonstration of lab accuracy and precision, analyze 4 to 7 replicates of a lab fortified blank containing analyte at 0.2–5 μg/L. Collect all samples in duplicate. Surrogate analytes (similar to those of the analytes of interest), whose concentration is known in every sample, are measured using the same internal standard calibration procedure. Duplicate field reagent water blanks (trip blanks) must be analyzed with each set of samples, lab reagent blanks (method blanks) must be analyzed with each batch of samples processed as a group within a work shift. Also, a single lab-fortified blank that contains each of the analytes of interest should be analyzed with each batch of samples processed as a group within a work shift. A 3- to 5-point calibration curve is needed depending on the calibration range factor required.

EPA CONTACT & HOTLINE For technical questions contact Dr. Baldev Bathija, U.S. EPA, Office of Ground Water and Drinking Water (WH-550D), 401 M St. SW, Washington, DC 20460. Tel. (202) 260-3040. For further information the EPA Safe Drinking Water Hotline may be called at: (800) 426-4791.

REFERENCE Methods for the Determination of Organic Compounds in Drinking Water, EPA/600/4-88/039 (revised July 1991; Final Rule for determination of compliance with the MCL for Total Trihalomethanes under 141.30, in 40 CFR Part 141, Vol. 58, No. 147, Fed. Reg., Tuesday Aug. 3, 1993). U.S. EPA Environmental Monitoring Systems Laboratory, Cincinnati, OH, 45268, USA. Available from the National Technical Information Service (NTIS), 5285 Port Royal Road, Springfield, VA 22161; Tel. 800-553-6847. NTIS Order Number is PB91-231480.

2-Chlorotoluene — EPA Method 8021
CAS #95-49-8

TITLE Halogenated Volatile by Gas Chromatography Using Photoionization and Electrolytic Conductivity Detectors in Series: Capillary Column Technique

MATRIX This method is applicable to nearly all types of samples, regardless of water content, including groundwater, aqueous sludges, caustic liquors, acid liquors, waste solvents, oily wastes, mousses, tars, fibrous wastes, polymeric emulsions, filter cakes, spent carbons, spent catalysts, soils, and sediments.

METHOD SUMMARY This method is used to determine 60 volatile organic compounds in a variety of solid waste matrices. It provides GC conditions for the detection of halogenated and aromatic volatile organic compounds. Samples can be analyzed using direct injection or purge-and-trap (EPA Method 5030). Groundwater samples must be analyzed using EPA Method 5030 (where applicable). A temperature program is used with the GC. Detection is achieved by a photoionization detector (PID) and a Hall electrolytic conductivity detector (HECD) in series.

INTERFERENCES Samples can be contaminated by diffusion of volatile organics (particularly chlorofluorocarbons and methylene chloride) through the sample container septum during shipment and storage.

INSTRUMENTATION A GC-equipped with variable-constant differential flow controllers, subambient oven controller, PID and HECD detectors connected with a short piece of uncoated capillary tubing and a data system.

Column: 60 m × 0.75 mm I.D. VOCOL wide-bore capillary column with 1.5 μm film thickness.

PRECISION & ACCURACY MDLs are compound-dependent and vary with purging efficiency and concentration. The applicable concentration range of this method is compound- and instrument-dependent but is approximately 0.1 to 200 μg/L. Analytes that are inefficiently purged from water will not be detected when present at low concentrations, but they can be measured with acceptable accuracy and precision when present in sufficient amounts. The estimated quantitation limit (EQL) for an individual compound is approximately 1 μg/kg (wet weight) for soil/sediment samples, 100 μg/kg (wet weight) for wastes, and 1 μg/L for groundwater. EQLs will be proportionately higher for sample extracts and samples that require dilution or reduced sample size to avoid saturation of the detector.

MULTIPLICATION FACTORS FOR OTHER MATRICES (a)

Matrix	Factor (b)
Groundwater	10
Low-concentration soil	10
Water miscible liquid waste	500
High-concentration soil and sludge	1250
Non-water miscible waste	1250

(a) Sample EQLs are highly matrix-dependent. The EQLs listed herein are provided for guidance and may not always be achievable.
(b) EQL = [Method detection limit] × [Factor]. For non-aqueous samples, the factor is on a wet-weight basis.

SINGLE LABORATORY ACCURACY & PRECISION DATA FOR VOCs IN WATER

This method was tested in a single lab using water spiked at 10 µg/L and the following data was reported:

Recoveries and standard deviations were determined from seven samples and spiked at 10 µg/L of each analyte. Recoveries were determined by the internal standard method. Internal standards were: Fluorobenzene for PID and 2-Bromo-1-chloropropane for HECD.

The average recovery (in percent) for the PID was not determined.
The standard deviation of the recovery for the PID was not determined.
The MDL (in µg/mL) for the PID was not determined.
The average recovery (in percent) for the HECD was 97.
The standard deviation of the recovery for the HECD was 2.6.
The MDL (in µg/mL) for the HECD was 0.01.

SAMPLE COLLECTION, PRESERVATION & HANDLING

Volatile Organics — Standard 40-mL glass screw-cap VOA vials with Teflon®-faced silicone septum may be used for both liquid and solid matrices. When collecting samples, liquids and solids should be introduced into the vials gently to reduce agitation which might drive off volatile compounds. If there are any air bubbles present the sample must be retaken. Tap slightly as they are filled to try and eliminate as much free air space as possible. The two vials from each sampling locations should be sealed in separate plastic bags to prevent cross-contamination between samples particularly if the sampled waste is suspected of containing high levels of volatile organics.

Semivolatile organics — Containers used to collect samples for the determination of semivolatile organic compounds should be soap and water washed followed by methanol (or isopropanol) rinsing. The sample containers should be of glass or Teflon® and have screw-top covers with Teflon® liners.

Preservation for volatile organics — No preservation is used with concentrated waste samples. With liquid samples containing no residual chlorine, 4 drops of concentrated hydrochloric acid are added and the samples are immediately cooled to 4°C. When liquid samples contain residual chlorine, they are treated as above and, in addition, 4 drops of 4% aqueous sodium thiosulfate are added. Soil, sediment, and sludge samples are only cooled to 4°C.

Preservation for semivolatile organics — No preservation is used with concentrated waste samples. With liquid samples containing no residual chlorine and with soil, sediment, and sludge samples, immediately cooling to 4°C is the only preservation used. When residual chlorine is present then 3 mL of 10% aqueous sodium sulfate is added for each gallon of sample collected, followed by cooling to 4°C.

MHT The holding time for all volatile organics samples is 14 days. Liquid samples must be extracted within 7 days and their extracts analyzed within 40 days. Concentrated waste, soil, sediment, and sludge samples must be extracted within 14 days and their extracts analyzed within 40 days.

SAMPLE PREPARATION Volatile compounds are introduced into the gas chromatograph either by direct injector or purge-and-trap (EPA Method 5030). EPA Method 5030 may be used directly on groundwater samples or low-concentration contaminated soils and sediments. For medium-concentration soils or sediments, methanolic extraction, as described in EPA Method 5030, may be necessary prior to purge-and-trap analysis.

QUALITY CONTROL Calculate surrogate standard recovery on all samples, blanks, and spikes. A trip blank is recommended to check on sampling, storage, and handling contamination. Calibration standards, at a minimum of five concentration levels, are prepared in organic-free reagent water. One of the concentration levels should be at a concentration near, but above, the method detection limit.

A combination of bromochloromethane, 2-bromo-1-chloropropane, 1,4-dichlorobutane, and bromochlorobenzene are recommended as surrogate standards to encompass the range of the temperature program used in this method.

REFERENCE Test Methods for Evaluating Solid Waste, Physical/Chemical Methods, SW-846, 3rd Edition, U.S. EPA, Office of Solid Waste, Washington, DC, EPA Method 8021A, Rev. 1, Nov. 1992.

4-Chlorotoluene   **EPA Method 8021**
CAS #106-43-4

TITLE Halogenated Volatile by Gas Chromatography Using Photoionization and Electrolytic Conductivity Detectors in Series: Capillary Column Technique

MATRIX This method is applicable to nearly all types of samples, regardless of water content, including groundwater, aqueous sludges, caustic liquors, acid liquors, waste solvents, oily wastes, mousses, tars, fibrous wastes, polymeric emulsions, filter cakes, spent carbons, spent catalysts, soils, and sediments.

METHOD SUMMARY This method is used to determine 60 volatile organic compounds in a variety of solid waste matrices. It provides GC conditions for the detection of halogenated and aromatic volatile organic compounds. Samples can be analyzed using direct injection or purge-and-trap (EPA Method 5030). Groundwater samples must be analyzed using EPA Method 5030 (where applicable). A temperature program is used with the GC. Detection is achieved by a photoionization detector

(PID) and a Hall electrolytic conductivity detector (HECD) in series.

INTERFERENCES Samples can be contaminated by diffusion of volatile organics (particularly chlorofluorocarbons and methylene chloride) through the sample container septum during shipment and storage.

INSTRUMENTATION A GC-equipped with variable-constant differential flow controllers, subambient oven controller, PID and HECD detectors connected with a short piece of uncoated capillary tubing and a data system.

Column: 60 m × 0.75 mm I.D. VOCOL wide-bore capillary column with 1.5 µm film thickness.

PRECISION & ACCURACY MDLs are compound-dependent and vary with purging efficiency and concentration. The applicable concentration range of this method is compound- and instrument-dependent but is approximately 0.1 to 200 µg/L. Analytes that are inefficiently purged from water will not be detected when present at low concentrations, but they can be measured with acceptable accuracy and precision when present in sufficient amounts. The estimated quantitation limit (EQL) for an individual compound is approximately 1 µg/kg (wet weight) for soil/sediment samples, 100 µg/kg (wet weight) for wastes, and 1 µg/L for groundwater. EQLs will be proportionately higher for sample extracts and samples that require dilution or reduced sample size to avoid saturation of the detector.

MULTIPLICATION FACTORS FOR OTHER MATRICES (a)

Matrix	Factor (b)
Groundwater	10
Low-concentration soil	10
Water miscible liquid waste	500
High-concentration soil and sludge	1250
Non-water miscible waste	1250

(a) Sample EQLs are highly matrix-dependent. The EQLs listed herein are provided for guidance and may not always be achievable.
(b) EQL = [Method detection limit] × [Factor]. For non-aqueous samples, the factor is on a wet-weight basis.

SINGLE LABORATORY ACCURACY & PRECISION DATA FOR VOCs IN WATER

This method was tested in a single lab using water spiked at 10 µg/L and the following data was reported:

Recoveries and standard deviations were determined from seven samples and spiked at 10 µg/L of each analyte. Recoveries were determined by the internal standard method. Internal standards were: Fluorobenzene for PID and 2-Bromo-1-chloropropane for HECD.

The average recovery (in percent) for the PID was 101.
The standard deviation of the recovery for the PID was 1.0.
The MDL (in µg/mL) for the PID was 0.02.
The average recovery (in percent) for the HECD was 97.
The standard deviation of the recovery for the HECD was 3.1.
The MDL (in µg/mL) for the HECD was 0.01.

SAMPLE COLLECTION, PRESERVATION & HANDLING
Volatile Organics — Standard 40-mL glass screw-cap VOA vials with Teflon®-faced silicone septum may be used for both liquid and solid matrices. When collecting samples, liquids and solids should be introduced into the vials gently to reduce agitation which might drive off volatile compounds. If there are any air bubbles present the sample must be retaken. Tap slightly as they are filled to try and eliminate as much free air space as possible. The two vials from each sampling locations should be sealed in separate plastic bags to prevent cross-contamination between samples particularly if the sampled waste is suspected of containing high levels of volatile organics.

Semivolatile organics — Containers used to collect samples for the determination of semivolatile organic compounds should be soap and water washed followed by methanol (or isopropanol) rinsing. The sample containers should be of glass or Teflon® and have screw-top covers with Teflon® liners.

Preservation for volatile organics — No preservation is used with concentrated waste samples. With liquid samples containing no residual chlorine, 4 drops of concentrated hydrochloric acid are added and the samples are immediately cooled to 4°C. When liquid samples contain residual chlorine, they are treated as above and, in addition, 4 drops of 4% aqueous sodium thiosulfate are added. Soil, sediment, and sludge samples are only cooled to 4°C.

Preservation for semivolatile organics — No preservation is used with concentrated waste samples. With liquid samples containing no residual chlorine and with soil, sediment, and sludge samples, immediately cooling to 4°C is the only preservation used. When residual chlorine is present then 3 mL of 10% aqueous sodium sulfate is added for each gallon of sample collected, followed by cooling to 4°C.

MHT The holding time for all volatile organics samples is 14 days. Liquid samples must be extracted within 7 days and their extracts analyzed within 40 days. Concentrated waste, soil, sediment, and sludge samples must be extracted within 14 days and their extracts analyzed within 40 days.

SAMPLE PREPARATION Volatile compounds are introduced into the gas chromatograph either by direct injector or purge-and-trap (EPA Method 5030). EPA Method 5030 may be used directly on groundwater samples or low-concentration contaminated soils and sediments. For medium-concentration soils or sediments, methanolic extraction, as described in EPA Method 5030, may be necessary prior to purge-and-trap analysis.

QUALITY CONTROL Calculate surrogate standard recovery on all samples, blanks, and spikes. A trip blank is recommended to check on sampling, storage, and handling contamination. Calibration standards, at a minimum of five concentration levels, are prepared in organic-free reagent water. One of the concentration levels should be at a concentration near, but above, the method detection limit.

A combination of bromochloromethane, 2-bromo-1-chloropropane, 1,4-dichlorobutane, and bromochlorobenzene are recommended as surrogate standards to encompass the range of the temperature program used in this method.

REFERENCE Test Methods for Evaluating Solid Waste, Physical/Chemical Methods, SW-846, 3rd Edition, U.S. EPA, Office of Solid Waste, Washington, DC, EPA Method 8021A, Rev. 1, Nov. 1992.

2-Chlorotoluene
CAS #95-49-8
EPA Method 8260

TITLE Volatile Organic Compounds by GC/MS: Capillary Column Technique

MATRIX This method is applicable to nearly all types of samples, regardless of water content, including groundwater, soils, and sediments.

METHOD SUMMARY Method 8260A covers 58 volatile organic compounds that are introduced into a gas chromatograph by the purge-and-trap method or by direct injection (in limited applications). Zero-grade helium is bubbled through a 5-mL solution at ambient temperature. Purged sample components are trapped in a tube containing suitable sorbent materials. When purging is complete, the sorbent tube is heated and backflushed with helium to desorb trapped sample components. The analytes are desorbed directly to a large bore capillary or cryofocussed on a capillary precolumn before being flash evaporated to a narrow bore capillary for analysis.

INTERFERENCES Major contaminant sources are volatile materials in the lab and impurities in the inert purging gas and in the sorbent trap. Interfering contamination may occur when a sample containing low concentrations of volatile organic compounds is analyzed immediately after a sample containing high concentrations of volatile organic compounds. After analysis of a sample containing high concentrations of volatile organic compounds, one or more calibration blanks should be analyzed to check for cross-contamination. Screening of the samples prior to purge-and-trap GC/MS analysis is highly recommended to prevent contamination of the system. This is especially true for soil and waste samples.

Special precautions must be taken to analyze for methylene chloride. The analytical and sample storage area should be isolated from all atmospheric sources of methylene chloride. All gas chromatography carrier gas lines and purge gas plumbing should be constructed from stainless steel or copper tubing. Laboratory clothing previously exposed to methylene chloride fumes during liquid-liquid extraction procedures can contribute to sample contamination.

Samples can also be contaminated by diffusion of volatile organics (particularly methylene chloride and fluorocarbons) through the septum seal during shipment and storage. A trip blank can serve as a check on such contamination.

INSTRUMENTATION GC/MS with a temperature-programmable chromatograph suitable for splitless injection equipped with variable constant differential flow controllers, a subambient oven controller, a purging device, sorbent trap, a thermal desorption apparatus and a capillary precolumn interface when using cryogenic cooling will be needed. The following GC columns may be used:

Column 1: 60 m × 0.75mm I.D. capillary column coated with VOCOL, 1.5 μm film thickness.
Column 2: 30 m × 0.53mm capillary column coated with DB-624 or VOCOL, 3 μm film thickness.
Column 3: 30 m × 0.32mm I.D. capillary column coated with DB-5 or SE-54, 1-μm film thickness.

PRECISION & ACCURACY This method has been tested in a single lab using spiked water. Using a wide-bore capillary column, water was spiked at concentrations between 0.5 and 10 μg/L. Single lab accuracy and precision data are presented. The MDL actually achieved in a given analysis will vary depending on instrument sensitivity and matrix effects.

The MDL (a) in μg/L was 0.04.
The concentration range in μg/L was 0.1–10.
The mean accuracy (% of true value) was 90.
The precision as relative standard deviation was 6.2.

Note: The MDL is based on a 25-mL sample volume instead of a 5-mL sample volume.

SAMPLING METHOD
Liquid samples — Use a 40-mL glass screw-cap VOA vial with a Teflon®-faced silicone septum that has been prewashed, rinsed with distilled deionized water, and oven dried. If residual chlorine is present, collect the sample in a 4-oz soil VOA container which has been pre-preserved with 4 drops of 10% sodium thiosulfate. Mix gently and transfer the sample to a 40-mL VOA vial. Collect bubble-free samples in duplicate and seal each sample in a separate plastic bag.

Soils, sediments, and sludges — Use an 8-oz widemouth glass bottle with Teflon®-faced silicone septum that has been prewashed, rinsed with distilled deionized water, and oven dried. **Do not** heat the septum for more than 1 h. Tap slightly to eliminate any free air space. Collect samples in duplicate and seal each one in a separate plastic bag.

SAMPLE PRESERVATION
Liquid samples — Add 4 drops of concentrated HCL, cool to 4°C and store in a solvent-free refrigerator.

Soils, sediments and sludges — Cool samples to 4°C and store in a solvent-free refrigerator.

MHT The maximum holding time of any sample (liquids, soils, sediments, and sludges) is 14 days.

SAMPLE PREPARATION
Liquid samples — Remove the plunger from a 5-mL syringe and carefully pour the sample into the syringe barrel to just short of overflowing. Replace the syringe plunger and compress the sample. Open the syringe valve and vent any residual air while adjusting the sample volume to 5.0 mL. If there is only one volatile organic analysis (VOA) vial, a second syringe should be filled at this time to protect against possible loss of sample integrity. Add 10 μL of surrogate spiking solution and 10 μL of internal standard spiking solution through the valve bore of the 5-mL syringe, then close the valve. The surrogate and internal standards may be mixed and added as a single spiking solution.

Sediments, soils, and waste samples — All samples of this type should be screened by GC analysis using a headspace method (EPA Method 3810) or the hexadecane extraction and screening method (EPA Method 3820). Use the screening data to determine whether to use the low-concentration method (0.005–1 mg/kg) or the high-concentration method (>1 mg/kg).

Low-concentration method — The low-concentration method is based on purging a heated sediment or soil sample mixed with organic-free reagent water containing the surrogate and internal standards. Analyze all reagent blanks and standards under the same conditions as the samples.

Use a 5-g sample if the expected concentration is <0.1 mg/kg or a 1-g sample for expected concentrations between 0.1 and 1 mg/kg. Mix the contents of the sample container with a narrow metal spatula. Weigh the amount of the sample into a tared purge device. Add the spiked water to the purge device, which contains the weighed amount of sample, and connect the device to the purge-and-trap system.

High-concentration method — This method is based on extracting the sediment or soil with methanol. A waste sample is either extracted or diluted, depending on its solubility in methanol. Wastes that are insoluble in methanol are diluted with reagent tetraglyme or possibly polyethylene glycol (PEG). An aliquot of the extract is added to organic-free reagent water containing surrogate and internal standards. This is purged at ambient temperature. All samples with an expected concentration of >1.0 mg/kg should be analyzed by this method.

Mix the contents of the sample container with a narrow metal spatula. For sediments or soils and solid wastes that are insoluble in methanol, weigh 4 g (wet weight) of sample into a tared 20-mL vial. For waste that is soluble in methanol, tetraglyme, or PEG, weigh 1 g (wet weight) into a tared scintillation vial or culture tube or a 10-mL volumetric flask. Quickly add 9.0 mL of appropriate solvent then add 1.0 mL of a surrogate spiking solution to the vial, cap it, and shake it for 2 min.

METHANOL EXTRACT REQUIRED FOR ANALYSIS OF HIGH-CONCENTRATION SOILS OR SEDIMENTS

Approximate Concentration Range	Volume of Methanol Extract (a)
500–10,000 µg/kg	100 µL
1,000–20,000 µg/kg	50 µL
5,000–100,000 µg/kg	10 µL
25,000–500,000 µg/kg	100 µL of 1/50 dilution (b)

Calculate appropriate dilution factor for concentrations exceeding this table.

(a) The volume of methanol added to 5 mL of water being purged should be kept constant. Therefore, add to the 5-mL syringe whatever volume of methanol is necessary to maintain a volume of 100 µL added to the syringe.
(b) Dilute an aliquot of the methanol extract and then take 100 µL for analysis.

QUALITY CONTROL Demonstrate, through the analysis of a reagent water blank, that interferences from the analytical system, glassware, and reagents are under control. Blank samples should be carried through all stages of the sample preparation and measurement steps. For each analytical batch (up to 20 samples), a reagent blank, matrix spike, and matrix spike duplicate must be analyzed (the frequency of the spikes may be different for different monitoring programs). The blank and spiked samples must be carried through all stages of the sample preparation and measurement steps. QC samples mentioned in the section on Interferences will also be needed as appropriate to those situations.

Matrix spiking standards should be prepared from volatile organic compounds which will be representative of the compounds being investigated. The recommended internal standards are chlorobenzene-d5, 1,4-difluorobenzene, 1,4-dichlorobenzene-d4, and pentafluorobenzene. Using stock standard solutions, prepare secondary dilution standards containing the compounds of interest, either singly or mixed together in methanol. Store them in a vial with no headspace for no more than one week. Surrogates recommended are toluene-d8, 4-bromofluorobenzene, and dibromofluoromethane. Each sample undergoing GC/MS analysis must be spiked with 10 µL of the surrogate spiking solution prior to analysis.

REFERENCE Test Methods for Evaluating Solid Waste (SW-846). U.S. EPA 1983, Method 8260A, Rev. 1, Nov. 1990. Office of Solid Waste, Washington, DC.

4-Chlorotoluene   EPA Method 8260
CAS #106-43-4

TITLE Volatile Organic Compounds by GC/MS: Capillary Column Technique

MATRIX This method is applicable to nearly all types of samples, regardless of water content, including groundwater, soils, and sediments.

METHOD SUMMARY Method 8260A covers 58 volatile organic compounds that are introduced into a gas chromatograph by the purge-and-trap method or by direct injection (in limited applications). Zero-grade helium is bubbled through a 5-mL solution at ambient temperature. Purged sample components are trapped in a tube containing suitable sorbent materials. When purging is complete, the sorbent tube is heated and backflushed with helium to desorb trapped sample components. The analytes are desorbed directly to a large bore capillary or cryofocussed on a capillary precolumn before being flash evaporated to a narrow bore capillary for analysis.

INTERFERENCES Major contaminant sources are volatile materials in the lab and impurities in the inert purging gas and in the sorbent trap. Interfering contamination may occur when a sample containing low concentrations of volatile organic compounds is analyzed immediately after a sample containing high concentrations of volatile organic compounds. After analysis of a sample containing high concentrations of volatile organic compounds, one or more calibration blanks should be analyzed to check for cross-contamination. Screening of the samples prior to purge-and-trap GC/MS analysis is highly recommended to prevent contamination of the system. This is especially true for soil and waste samples.

Special precautions must be taken to analyze for methylene chloride. The analytical and sample storage area should be isolated from all atmospheric sources of methylene chloride. All gas chromatography carrier gas lines and purge gas plumbing should be constructed from stainless steel or copper tubing. Laboratory clothing previously exposed to methylene chloride fumes during liquid-liquid extraction procedures can contribute to sample contamination.

Samples can also be contaminated by diffusion of volatile organics (particularly methylene chloride and fluorocarbons) through the septum seal during shipment and storage. A trip blank can serve as a check on such contamination.

INSTRUMENTATION GC/MS with a temperature-programmable chromatograph suitable for splitless injection equipped with variable constant differential flow controllers, a subambient oven controller, a purging device, sorbent trap, a thermal desorption apparatus and a capillary precolumn interface when using cryogenic cooling will be needed. The following GC columns may be used:

Column 1: 60 m × 0.75mm I.D. capillary column coated with VOCOL, 1.5 µm film thickness.
Column 2: 30 m × 0.53mm capillary column coated with DB-624 or VOCOL, 3 µm film thickness.
Column 3: 30 m × 0.32mm I.D. capillary column coated with DB-5 or SE-54, 1-µm film thickness.

PRECISION & ACCURACY This method has been tested in a single lab using spiked water. Using a wide-bore capillary column, water was spiked at concentrations between 0.5 and 10 µg/L. Single lab accuracy and precision data are presented. The MDL actually achieved in a given analysis will vary depending on instrument sensitivity and matrix effects.

The MDL (a) in µg/L was 0.06.
The concentration range in µg/L was 0.1–10.
The mean accuracy (% of true value) was 99.
The precision as relative standard deviation was 8.3.

Note: The MDL is based on a 25-mL sample volume instead of a 5-mL sample volume.

SAMPLING METHOD
Liquid samples — Use a 40-mL glass screw-cap VOA vial with a Teflon®-faced silicone septum that has been prewashed, rinsed with distilled deionized water, and oven dried. If residual chlorine is present, collect the sample in a 4-oz soil VOA container which has been pre-preserved with 4 drops of 10% sodium thiosulfate. Mix gently and transfer the sample to a 40-mL VOA vial. Collect bubble-free samples in duplicate and seal each sample in a separate plastic bag.

Soils, sediments, and sludges — Use an 8-oz widemouth glass bottle with Teflon®-faced silicone septum that has been prewashed, rinsed with distilled deionized water, and oven dried. **Do not** heat the septum for more than 1 h. Tap slightly to eliminate any free air space. Collect samples in duplicate and seal each one in a separate plastic bag.

SAMPLE PRESERVATION
Liquid samples — Add 4 drops of concentrated HCL, cool to 4°C and store in a solvent-free refrigerator.

Soils, sediments and sludges — Cool samples to 4°C and store in a solvent-free refrigerator.

MHT The maximum holding time of any sample (liquids, soils, sediments, and sludges) is 14 days.

SAMPLE PREPARATION
Liquid samples — Remove the plunger from a 5-mL syringe and carefully pour the sample into the syringe barrel to just short of overflowing. Replace the syringe plunger and compress the sample. Open the syringe valve and vent any residual air while adjusting the sample volume to 5.0 mL. If there is only one volatile organic analysis (VOA) vial, a second syringe should be filled at this time to protect against possible loss of sample integrity. Add 10 µL of surrogate spiking solution and 10 µL of internal standard spiking solution through the valve bore of the 5-mL syringe, then close the valve. The surrogate and internal standards may be mixed and added as a single spiking solution.

Sediments, soils, and waste samples — All samples of this type should be screened by GC analysis using a headspace method (EPA Method 3810) or the hexadecane extraction and screening method (EPA Method 3820). Use the screening data to determine whether to use the low-concentration method (0.005–1 mg/kg) or the high-concentration method (>1 mg/kg).

Low-concentration method — The low-concentration method is based on purging a heated sediment or soil sample mixed with organic-free reagent water containing the surrogate and internal standards. Analyze all reagent blanks and standards under the same conditions as the samples.

Use a 5-g sample if the expected concentration is <0.1 mg/kg or a 1-g sample for expected concentrations between 0.1 and 1 mg/kg. Mix the contents of the sample container with a narrow metal spatula. Weigh the amount of the sample into a tared purge device. Add the spiked water to the purge device, which contains the weighed amount of sample, and connect the device to the purge-and-trap system.

High-concentration method — This method is based on extracting the sediment or soil with methanol. A waste sample is either extracted or diluted, depending on its solubility in methanol. Wastes that are insoluble in methanol are diluted with reagent tetraglyme or possibly polyethylene glycol (PEG). An aliquot of the extract is added to organic-free reagent water containing surrogate and internal standards. This is purged at ambient temperature. All samples with an expected concentration of >1.0 mg/kg should be analyzed by this method.

Mix the contents of the sample container with a narrow metal spatula. For sediments or soils and solid wastes that are insoluble in methanol, weigh 4 g (wet weight) of sample into a tared 20-mL vial. For waste that is soluble in methanol, tetraglyme, or PEG, weigh 1 g (wet weight) into a tared scintillation vial or culture tube or a 10-mL volumetric flask. Quickly add

9.0 mL of appropriate solvent then add 1.0 mL of a surrogate spiking solution to the vial, cap it, and shake it for 2 min.

METHANOL EXTRACT REQUIRED FOR ANALYSIS OF HIGH-CONCENTRATION SOILS OR SEDIMENTS

Approximate Concentration Range	Volume of Methanol Extract (a)
500–10,000 µg/kg	100 µL
1,000–20,000 µg/kg	50 µL
5,000–100,000 µg/kg	10 µL
25,000–500,000 µg/kg	100 µL of 1/50 dilution (b)

Calculate appropriate dilution factor for concentrations exceeding this table.

(a) The volume of methanol added to 5 mL of water being purged should be kept constant. Therefore, add to the 5-mL syringe whatever volume of methanol is necessary to maintain a volume of 100 µL added to the syringe.

(b) Dilute an aliquot of the methanol extract and then take 100 µL for analysis.

QUALITY CONTROL Demonstrate, through the analysis of a reagent water blank, that interferences from the analytical system, glassware, and reagents are under control. Blank samples should be carried through all stages of the sample preparation and measurement steps. For each analytical batch (up to 20 samples), a reagent blank, matrix spike, and matrix spike duplicate must be analyzed (the frequency of the spikes may be different for different monitoring programs). The blank and spiked samples must be carried through all stages of the sample preparation and measurement steps. QC samples mentioned in the section on Interferences will also be needed as appropriate to those situations.

Matrix spiking standards should be prepared from volatile organic compounds which will be representative of the compounds being investigated. The recommended internal standards are chlorobenzene-d5, 1,4-difluorobenzene, 1,4-dichlorobenzene-d4, and pentafluorobenzene. Using stock standard solutions, prepare secondary dilution standards containing the compounds of interest, either singly or mixed together in methanol. Store them in a vial with no headspace for no more than one week. Surrogates recommended are toluene-d8, 4-bromofluorobenzene, and dibromofluoromethane. Each sample undergoing GC/MS analysis must be spiked with 10 µL of the surrogate spiking solution prior to analysis.

REFERENCE Test Methods for Evaluating Solid Waste (SW-846). U.S. EPA 1983, Method 8260A, Rev. 1, Nov. 1990. Office of Solid Waste, Washington, DC.

2-Chlorotoluene   **EPA Method 503.1**
CAS #95-49-8

TITLE Aromatic & Unsaturated VOCs

MATRIX Drinking water (finished or in Water any treatment stage) and raw source water.

APPLICATION Method covers 28 aromatic and unsaturated VOCs. An inert gas is bubbled through a 5-mL water sample. Purged sample components are trapped in tube of sorbent materials. When purging is complete, sorbent tube is heated and backflushed with inert gas to desorb trapped sample onto a packed GC column.

INTERFERENCES During analysis, major contaminant sources are volatile materials in the lab and impurities in purging gas and sorbent trap. With high and low level samples, there can be carryover contamination. Excess water causes a negative baseline deflection.

INSTRUMENTATION Purge and Trap GC w/photoionization detector. (Two GC columns are recommended); Column 1: 5% SP-1200 and 1.75% Bentone 34 on Supelcoport; 5% 1,2,3-tris(2-cyanoethoxy)propane on Chromosorb W.

RANGE 2.2–600 µg/L (Drinking water).

MDL 0.008 µg/L in water.

PRECISION Not listed.

ACCURACY Not listed.

SAMPLING METHOD Use a 40–120-mL screw-cap vial (prewashed with detergent, rinsed with distilled water and oven dried at 105°C) with a PTFE-faced silicone septum. If residual chlorine is in the water add about 25 mg of ascorbic acid to each vial before sample collection. Collect bubble-free samples.

STABILITY Cool to 4°C; HCl to pH <2.

MHT 14 days.

QUALITY CONTROL As an initial demonstration of lab accuracy and precision, analyze 4 to 7 replicates of a lab fortified blank containing analyte at 0.1–5 µg/L. Collect all samples in duplicate.

REFERENCE Method 503.1, Volatile Aromatic & Unsaturated Organic Compounds in H2O by Purge and Trap GC, EPA 600/4-88/039.

4-Chlorotoluene   **EPA Method 503.1**
CAS #106-43-4

TITLE Aromatic & Unsaturated VOCs

MATRIX Drinking water (finished or in Water any treatment stage) and raw source water.

APPLICATION Method covers 28 aromatic and unsaturated VOCs. An inert gas is bubbled through a 5-mL water sample. Purged sample components are trapped in tube of sorbent materials. When purging is complete, sorbent tube is heated and backflushed with inert gas to desorb trapped sample onto a packed GC column.

INTERFERENCES During analysis, major contaminant sources are volatile materials in the lab and impurities in purging gas and sorbent trap. With high and low level samples, there can be carryover contamination. Excess water causes a negative baseline deflection.

INSTRUMENTATION Purge and Trap GC w/photoionization detector. (Two GC columns are recommended); Column 1: 5% SP-1200 and 1.75% Bentone 34 on Supelcoport; 5% 1,2,3-tris(2-cyanoethoxy)propane on Chromosorb W.

RANGE 2.2–600 µg/L (Drinking water).

MDL Not listed.

PRECISION RSD = 5.0% at 0.50 µg/L; 17 samples

ACCURACY Average recovery = 91% at 0.50 µg/L; 17 saples

SAMPLING METHOD Use a 40–120-mL screw-cap vial (prewashed with detergent, rinsed with distilled water and oven dried at 105°C) with a PTFE-faced silicone septum. If residual chlorine is in the water add about 25 mg of ascorbic acid to each vial before sample collection. Collect bubble-free samples.

STABILITY Cool to 4°C; HCl to pH <2.

MHT 14 days.

QUALITY CONTROL As an initial demonstration of lab accuracy and precision, analyze 4 to 7 replicates of a lab fortified blank containing analyte at 0.1–5 µg/L. Collect all samples in duplicate.

REFERENCE Method 503.1, Volatile Aromatic & Unsaturated Organic Compounds in H2O by Purge and Trap GC, EPA 600/4-88/039.

Chlorpropham
CAS #101-21-3

EPA Method 507

TITLE Determination of Nitrogen and Phosphorus-Containing Pesticides in Water by GC/NPD

MATRIX This method is applicable to the determination of certain nitrogen and phosphorus-containing pesticides in finished drinking water and groundwater.

METHOD SUMMARY Method 507 covers 46 nitrogen- and phosphorus-containing pesticides. A 1-L sample is fortified with a surrogate standard, salted, buffered, extracted with methylene chloride, and concentrated; then the solvent is exchanged with methyl tert-butyl ether (MTBE) and concentrated again, and a 2-µL aliquot of a sample extract is injected into a GC system equipped with a selective nitrogen-phosphorus detector and a capillary column for analysis.

INTERFERENCES Method interferences may be caused by contaminants in solvents, reagents, glassware, and other sample processing apparatus. Interfering contamination may occur when a sample containing low concentrations of analytes is analyzed immediately following a sample containing relatively high concentrations. One or more injections of MTBE should be made following the analysis of a sample with high concentrations of analytes to check for analyte carryover. Matrix interferences may be caused by contaminants that are coextracted from the sample. The extent of matrix interferences will vary considerably from source to source, depending upon the water sampled.

INSTRUMENTATION A gas chromatograph system (GC) equipped with a nitrogen-phosphorus detector (NPD) is needed.

Column 1: 30 m × 0.25 mm I.D. DB-5 bonded fused silica column, 0.25 µm film thickness, or equivalent.

Column 2: 30 m × 0.25 mm I.D. DB-1701 bonded fused silica column, 0.25 µm film thickness, or equivalent.

PRECISION & ACCURACY This method has been validated in a single lab and estimated detection limits (EDLs) have been determined for each analyte. Observed detection limits may vary among waters, depending upon the nature of the interferences in the sample matrix and the specific instrumentation used. Analytes that are not separated chromatographically cannot be individually identified and measured unless an alternative technique for identification and quantification exist.

The estimated detection limit (in µg/L) was 0.5. The EDL is defined as either method detection limit or a level of compound in a sample yielding a peak in the final extract with signal-to-noise ratio of approximately 5, whichever value is higher.

The concentration used for these measurements (in µg/L) was 5.
The accuracy (as % recovery) was 93.
The precision (% RSD) was 11.

SAMPLING METHOD Grab samples are collected in 1-L glass sample bottles (prewashed with detergent and hot tap water, rinsed with reagent water, and dried in an oven at 400°C for 1 h) with screw caps lined with PTFE-fluorocarbon.

SAMPLE PRESERVATION Add mercuric chloride to the sample bottle in amounts to produce a concentration of 10 mg/L. If residual chlorine is present, add 80 mg of sodium thiosulfate/L of sample to the sample bottle prior to collection. After collection, seal bottle and shake vigorously for 1 min, then cool the sample to 4°C immediately and store it at 4°C in the dark until extraction.

MHT Maximum holding time of the samples, and in some cases the extracts, is 14 days.

SAMPLE PREPARATION Fortify the sample with 50 µL of the surrogate standard solution, adjust to pH 7 with phosphate buffer, add 100 g NaCl to the sample, and seal and shake to dissolve the salt; then extract with methylene chloride in a separatory funnel or in a mechanical tumbler bottle. Dry the extract by pouring it through a solvent-rinsed drying column containing about 10 cm of anhydrous sodium sulfate. Collect the extract in a Kuderna-Danish (K-D) concentrator and rinse the column with 20–30 mL methylene chloride. Concentrate the extract to about 2 mL and rinse the flask and its lower joint into the concentrator tube with 1 to 2 mL of methyl t-butyl ether (MTBE). Add 5–10 mL of MTBE and concentrate the extract twice (adding more MTBE) to a final volume of 5.0 mL and store it at 4°C until analysis.

Note: If methylene chloride is not completely removed from the final extract, it may cause detector problems.

QUALITY CONTROL Minimum quality control requirements are initial demonstration of lab capability, determination of surrogate compound recoveries in each sample and blank, monitoring internal standard peak area or height in each sample and blank, analysis of lab reagent blanks, lab fortified samples, lab fortified blanks, and other QC samples. A lab reagent blank is analyzed to demonstrate that all glassware and reagent interferences are under control.

Initial demonstration of capability is fulfilled by analyzing four fortified reagent water samples with the recovery value for each analyte falling within the acceptable range (±30% average recovery). Surrogate recoveries from samples or method blanks must be 70–130%. The internal standard response for any sample chromatogram should not deviate from the daily calibration check standard's internal standard response by more than 30% or lab fortified blanks and sample matrices are used to assess lab performance and analyte recovery, respectively.

If the response for the target analyte peak exceeds the working range of the system, dilute the extract and reanalyze. Alternative techniques such as an alternate detector or second chromatography column should be used to confirm peak identification when sample components are not resolved adequately.

EPA CONTACT & HOTLINE For technical questions contact Dr. Baldev Bathija, U.S. EPA, Office of Ground Water and Drinking Water (WH-550D), 401 M St. SW, Washington, DC 20460. Tel. (202) 260-3040. For further information the EPA Safe Drinking Water Hotline may be called at: (800) 426-4791.

REFERENCE Methods for the Determination of Organic Compounds in Drinking Water, EPA/600/4-88/039 (revised July 1991). U.S. EPA Environmental Monitoring Systems Laboratory, Cincinnati, OH, 45268, USA. Available from the National Technical Information Service (NTIS), 5285 Port Royal Road, Springfield, VA 22161; Tel. 800-553-6847. NTIS Order Number is PB91-231480.

Chlorpyrifos   **EPA Method 8141**
CAS #2921-88-2

TITLE Organophosphorus Compounds by Gas Chromatography: Capillary Column Technique

MATRIX This method covers aqueous and solid matrices. This includes a wide variety such as drinking water, groundwater, industrial wastewaters, surface waters, soils, solids, and sediments.

METHOD SUMMARY This is a GC method used to determine the concentration of 28 organophosphorus pesticides.

The use of Gel Permeation Cleanup (EPA Method 3640) for sample cleanup has been demonstrated to yield recoveries of less than 85% for many method analytes and is therefore not recommended for use with this method.

This method provides GC conditions for the detection of ppb concentrations of organophosphorus compounds. Prior to the use of this method, appropriate sample preparation techniques must be used. Water samples are extracted at a neutral pH with methylene chloride as a solvent by using a separatory funnel (EPA Method 3510) or a continuous liquid-liquid extractor (EPA Method 3520). Soxhlet extraction (EPA Method 3540) or ultrasonic extraction (EPA Method 3550) using methylene chloride/acetone (1:1) are used for solid samples. Both neat and diluted organic liquids (EPA Method 3580) may be analyzed by direct injection. Spiked samples are used to verify the applicability of the chosen extraction technique to each new sample type. A GC with a flame photometric (FPD) or nitrogen-phosphorus detector (NPD) is used for this multiresidue procedure.

INTERFERENCES The use of Florisil cleanup materials (EPA Method 3620) for some of the compounds in this method has been demonstrated to yield recoveries less than 85% and is therefore not recommended for all compounds. Use of phosphorus or halogen specific detectors, however, often obviates the necessity for cleanup for relatively clean sample matrices. If particular circumstances demand the use of an alternative cleanup procedure, the analyst must determine the elution profile and demonstrate that the recovery of each analyte is no less than 85%.

Use of a flame photometric detector (FPD) in the phosphorus mode will minimize interferences from materials that do not contain phosphorus. Elemental sulfur, however, may interfere with the determination of certain organophosphorus compounds by flame photometric gas chromatography. Sulfur cleanup using EPA Method 3660 may alleviate this interference. A nitrogen phosphorus detector (NPD) is also recommended.

A few analytes coelute on certain columns. Therefore, select a second column for confirmation where coelution of the analytes of interest does not occur.

Method interferences may be caused by contaminants in solvents, reagents, glassware, and other sample processing hardware that lead to discrete artifacts or elevated baselines in gas chromatograms. All these materials must be routinely demonstrated to be free from interferences under the conditions of the analysis by analyzing reagent blanks.

INSTRUMENTATION A GC with a NPD or a FPD will be needed. A data system or integrator is recommended for measuring peak areas and/or peak heights. A Kuderna-Danish (K-D) apparatus will be needed for extract concentration.

Column 1: 15 m × 0.53 mm megabore capillary column, 1.0 µm film thickness, DB-210.
Column 2: 15 m × 0.53 mm megabore capillary column, 1.5 µm film thickness, SPB-608.
Column 3: 15 m × 0.53 mm megabore capillary column, 1.0 µm film thickness, DB-5.

Three megabore capillary columns are included for analysis of organophosphates by this method. Column 1 (DB-210 or equivalent) and Column 2 (SPB-608 or equivalent) are recommended if a large number of organophosphorus analytes are to be determined. If the superior resolution offered by Column 1 and Column 2 is not required, Column 3 (DB-5 or equivalent) may be used. For megabore capillary columns, automatic injections of 1 µL are recommended.

PRECISION & ACCURACY The MDL actually achieved in a given analysis will vary, as it is dependent on instrument sensitivity and matrix effects. Single operator accuracy and precision studies have been conducted with spiked water and soil samples.

MULTIPLICATION FACTORS FOR OTHER MATRICES (a)

Matrix	Factor (b)
Groundwater (EPA Method 3510 or EPA Method 3520)	10
Low-concentration soil by Soxhlet and no cleanup	10 (c)
Low-concentration soil by ultrasonic extraction with GPC cleanup	6.7 (c)
High-concentration soil and sludges by ultrasonic extraction	500 (c)
Non-Water miscible waste (EPA Method 3580)	1000 (c)

(a) Sample EQLs are highly matrix-dependent. The EQLs listed here are provided for guidance and may not always be achievable.
(b) EQL = [Method detection limit] × [Factor]. For non-aqueous samples the factor is on a wet-weight basis.
(c) Multiply this factory times the soil MDL.

The MDL (in µg/L) when reagent water was extracted using a separatory funnel was 0.07.

The MDL (in µg/kg) when soil was extracted using Soxhlet extraction (EPA Method 3540) was 5.0.

Accuracy (as % recovery) with separatory funnel extraction ranged from 7 (with low spikes) to 86 (with high spikes).

Accuracy (as % recovery) with continuous liquid-liquid extraction ranged from 13 (with low spikes) to 88 (with high spikes).

Accuracy (as % recovery) with Soxhlet extraction of soils ranged from not recovered (with low spikes to 79 (with high spikes).

Accuracy (as % recovery) with ultrasonic extraction of soils ranged from not recovered (with low spikes) to 77 (with high spikes).

SAMPLE COLLECTION, PRESERVATION & HANDLING
Containers used to collect samples for the determination of semivolatile organic compounds should be soap and water washed followed by methanol (or isopropanol) rinsing. The sample containers should be of glass or Teflon® and have screw-top covers with Teflon® liners.

No preservation is used with concentrated waste samples. With liquid samples containing no residual chlorine and with soil, sediment, and sludge samples, immediately cooling to 4°C is the only preservation used. When residual chlorine is present then 3 mL of 10% aqueous sodium sulfate is added for each gallon of sample collected, followed by cooling to 4°C.

Liquid samples must be extracted within 7 days and their extracts analyzed within 40 days. Concentrated waste, soil, sediment, and sludge samples must be extracted within 14 days and their extracts analyzed within 40 days.

SAMPLE PREPARATION In general, water samples are extracted at a neutral pH with methylene chloride, using either EPA Method 3510 or EPA Method 3520. Solid samples are extracted using either EPA Method 3540 or EPA Method 3550 with methylene chloride/acetone (1:1) as the extraction solvent.

Prior to GC analysis, the extraction solvent may be exchanged to hexane. Single lab data indicates that samples should not be transferred with 100% hexane during sample workup as the more water soluble organophosphorus compounds may be lost.

If cleanup is performed on the samples, the analyst should analyze the samples by GC. This will confirm elution patterns and the absence of interferences from the reagents. If peak detection and identification is prevented by the presence of interferences, further cleanup is required.

QUALITY CONTROL The analyst should monitor the performance of the extraction, cleanup (when used), and analytical system and the effectiveness of the method in dealing with each sample matrix by spiking each sample, standard, and blank with one or two surrogates (e.g., organophosphorus compounds not expected to be present in the sample). Deuterated analogs of analytes should not be used as surrogates for gas chromatographic analysis due to coelution problems.

A minimum of five concentrations for each analyte of interest should be prepared through dilution of the stock standards with isooctane. One of the concentrations should be at a concentration near, but above, the MDL.

Include a mid-level check standard after each group of 10 samples in the analysis sequence. GC/MS techniques should be judiciously employed to support qualitative identifications made with this method. Follow the GC/MS operating requirements specified in EPA Method 8270.

When available, chemical ionization mass spectra may be employed to aid in the qualitative identification process. To confirm an identification of a compound, the background-corrected mass spectrum of the compound must be obtained from the sample extract and must be compared with a mass spectrum from a stock or calibration standard analyzed under the same chromatographic conditions. The molecular ion and all other ions present above 20% relative abundance in the mass spectrum of the standard must be present in the mass spectrum of the sample with agreement to ± 20%. The retention time of the compound in the sample must be within six seconds of the retention time for the same compound in the standard solution.

Should the MS procedure fail to provide satisfactory results, additional steps may be taken before reanalysis. These steps may include the use of alternate packed or capillary GC columns or additional sample cleanup.

REFERENCE Test Methods for Evaluating Solid Waste, Physical/Chemical Methods, SW-846, 3rd Edition, U.S. EPA, Office of Solid Waste, Washington, DC, EPA Method 8141 July 1992.

Chlorpyrifos   EPA Method 8140
CAS #2921-88-2

TITLE Organophosphorus Pesticides

MATRIX Groundwater, soils, sludges, water miscible liquid wastes, and non-water miscible wastes.

APPLICATION This method is used for the analysis of 21 organophosphorus pesticides. Samples are extracted, concentrated, and analyzed using direct injection of both neat and diluted organic liquid into a gas chromatograph (GC).

INTERFERENCES Solvents, reagents, and glassware may introduce artifacts. Other interferences may come from coextracted compounds from samples. The use of Florisil cleanup materials may produce low recoveries. Elemental sulfur may interfere with some compounds when using a flame photometric detector. Sulfur cleanup (Method 3660) may alleviate sulfur interference.

INSTRUMENTATION GC capable of on-column injections and a flame photometric detector (FPD) or a thermionic detector. Column 1: 1.8 m by 2 mm with 5% SP-2401 on Supelcoport. Column 2: 1.8 m by 2 mm with 3% SP-2401 on Supelcoport. Column 3: 50 cm by ⅛ in Teflon® with 15% SE-54 on Gas Chrom Q. The preferred column is Column Number 2.

RANGE 1.0 to 50.5 µg/L

MDL 0.3 µg/L (in reagent water).

PQL FACTORS FOR MULTIPLYING × FID MDL VALUE

Matrix	Multiplication Factor
Groundwater	10
Low-level soil by sonication with GPC cleanup	670
High-level soil and sludge by sonication	10,000
Non-water miscible waste	100,000

PRECISION 5.5% (single operator standard deviation)

ACCURACY 98.3% (single operator average recovery)

SAMPLING METHOD Use 8-oz. widemouth glass bottles with Teflon®-lined caps for concentrated waste samples, soils, sediments, and sludges. Use 1 or 2½ gallon amber glass bottles with Teflon®-lined caps for liquid (water) samples.

STABILITY Cool soil, sediment, sludge, and liquid samples to 4°C. If residual chlorine is present in liquid samples add 3 mL of 10% sodium thiosulfate per gallon of sample and cool to 4°C.

MHT 14 days for concentrated waste, soil, sediment, or sludge; 7 days for liquid samples; all extracts must be analyzed within 40 days.

QUALITY CONTROL A quality control check sample concentrate containing this compound in acetone at a concentration 1,000 times more concentrated than the selected spike concentration is required. The QC check sample concentrate may be prepared from pure standard materials or purchased as certified solutions. Use appropriate trip, matrix, control site, method, reagent, and solvent blanks. Internal, surrogate, and five concentration level calibration standards are used.

REFERENCE Method 8140, SW-846, 3rd ed., Sept. 1986.

Chromium EPA Method 200.7
CAS #7440-47-3

TITLE Inductively Coupled Plasma

MATRIX Dissolved, suspended or (ICP) total element in drinking and surface waters and in domestic and industrial wastewaters.

APPLICATION The method covers the determination of 25 metals. Dissolved elements are determined in filtered and acidified samples after appropriate digestion (which increases dissolved solids). Its primary advantage is that ICP instruments allow simultaneous or rapid sequential determination of many elements in a short time. Samples are first nebulized and the aerosol is transported to a plasma torch in which element specific atomic line emission spectra are produced by a radio frequency inductively coupled plasma. Background correction is required for trace element detection except in the case of line broadning.

INTERFERENCES There are spectral, physical, and chemical interferences. The primary disadvantage of ICP instruments is background radiation from other elements and the plasma gases (spectral interferences). Changes in sample viscosity and surface tension with samples containing high dissolved solids (especially those exceeding 1500 mg/L) or high acid concentrations can cause physical interferences. Ionization effects, solute vaporization and molecular compound formation can cause chemical interferences. Iron, manganese and vanadium can cause interference at the 100 mg/L level.

INSTRUMENTATION Inductively coupled argon plasma emission spectroscopy. 267.716 nm Wavelength.

RANGE Not listed.

MDL 7 µg/L.

PRECISION SD = 3.8% Mean at true value 150 µg/L.

ACCURACY Mean recovery = 93% ± 6% of spiked elements for all wastes.

SAMPLING METHOD
Wash sample container with detergent and tap water, rinse with 1 + 1 nitric acid and tap water, then rinse with 1 + 1 hydrochloric acid and tap water, then rinse with deionized, distilled water in that order. Perform any filtration or acid preservation steps when the sample is collected or as soon as possible thereafter.

STABILITY Cool samples to 4°C.

MHT 24 h.

QUALITY CONTROL Mixed calibration standards, an instrument check standard, and an interference check solution are used in addition to a quality control sample. The quality control sample should be prepared in the same acid matrix as the calibration standards at 10 times the instrumental detection limits and in accordance with the instructions provided by the supplier. Furthermore, two types of blanks are required: a calibration blank and a reagent blank.

REFERENCE Method 200.7, U.S. EPA, EMSL-Cincinnati, OH, Nov. 1980

Chromium   **EPA Method 7190**
CAS #7440-47-3

TITLE Atomic Absorption, (AA)

MATRIX Drinking, Surface and Direct Aspiration Saline Waters, Wastewater

APPLICATION Sample is aspirated and atomized in a flame. A light beam from a chromium hollow cathode lamp is directed through the flame into a monochromator and onto a detector. Since wavelength of light beam is specific for chromium, light energy absorbed by detector is measure of chromium.

INTERFERENCES The most troublesome type is chemical, caused by lack of absorption of atoms bound in molecular combination in the flame. High dissolved solids in sample may result in nonatomic absorbance interference. Very high alkali metal contents can interfere.

INSTRUMENTATION Atomic absorption spectrometer. Chromium hollow cathode lamp. (357.9 nm Wavelength).

RANGE 0.5–10 mg/L

MDL 0.05 mg/L

PRECISION Standard deviation = 105 µg/L at 370 µg/L (true value) 74 labs

ACCURACY As bias = -4.5% at 370 µg/L (true value) 74 labs

SAMPLING METHOD Use glass or plastic containers. Collect 200 g of solids and 600 mL of liquid samples.

STABILITY Cool solid samples to 4°C and analyze as soon as possible. Add nitric acid to liquid samples to pH <2.

MHT 6 months.

QUALITY CONTROL At least one duplicate and one spike sample should be run every 20 samples or with each matrix type to verify precision of the method. For 20 or more samples per day, verify working standard curve. Run an additional standard at or near mid-range every 10 samples.

REFERENCE Method 7190, SW-846, 3rd ed., Nov.1986.

Chromium   **EPA Method 7191**
CAS #7440-47-3

TITLE Atomic Absorption, (AA)

MATRIX Wastes, mobility procedure. Furnace Technique extracts, soils and groundwater.

APPLICATION Aqueous samples, EP e–tracts, industrial wastes, soils, sludges, sediments, and solid wastes require digestion before analysis. An aliquot of sample is placed in the graphite tube in the furnace and slowly evaporated, charred and atomized. Absorption of lamp radiation during atomization is proportional to chromium concentration.

INTERFERENCES The furnace technique is subject to chemical interferences. Composition of sample matrix can have major effect on analysis. Low concentrations of calcium and/or phosphate may interfere. Background correction may be required. Don't use nitrogen as purge gas.

INSTRUMENTATION Atomic absorption spectrometer. Chromium hollow cathode lamp or electrodeless discharge lamp. Graphite furnace. Strip-chart recorder

RANGE 5–100 µg/L

MDL 1 µg/L (357.9 nm Wavelength)

PRECISION Standard deviation = ±0.10, 0.20, 0.80 at 19, 48, 77 µg Cr/L

ACCURACY Recoveries = 97, 101, 102% at 19, 48, 77 µg Cr/L

SAMPLING METHOD Use glass or plastic containers. Collect 200 g of solids and 600 mL of liquid samples.

STABILITY Cool solid samples to 4°C and analyze as soon as possible. Add nitric acid to liquid samples to pH <2.

MHT 6 months.

QUALITY CONTROL At least one duplicate and one spike sample should be run every 20 samples, or with each matrix type to verify method precision. If 20 or more samples are run a day, run a standard (at or near mid-range) every 10 samples.

REFERENCE Method 7191, SW-846, 3rd ed., Nov.1986.

Chromium (total)   **EPA Method 6010**
CAS #7440-47-3

TITLE Inductively Coupled Plasma-Atomic Emission Spectroscopy

MATRIX This method is applicable to the determination of trace elements, including metals, in groundwater, soils, sludges, sediments, and other solid wastes. All matrices require digestion prior to analysis. The method of standard addition must be used for the analysis of all sample digests unless either serial dilution or matrix spike addition demonstrates it is not required.

METHOD SUMMARY Method 6010 covers 25 elements using ICP analysis. It measures element-emitted light by optical spectrometry. Samples, following an appropriate acid digestion, are nebulized and the resulting aerosol is transported to the plasma torch. Element-specific atomic line emission spectra are produced by a radio-frequency inductively coupled plasma.

INTERFERENCES Interferences may be categorized as spectral or non-spectral. Spectral interferences are caused by overlap of a spectral line from another element, unresolved overlap of molecular band spectra, background contribution from continuous or recombination phenomena, and stray light from the line emission of high concentration elements. Non-spectral interferences include physical and chemical interferences. Physical interferences are effects associated with the sample nebulization and transport processes. Changes in viscosity and surface tension can cause significant inaccuracies. Chemical interferences include molecular compound formation, ionization effects, and solute vaporization effects. Normally these effects are not significant and can be minimized by careful selection of operating conditions. Chemical interferences are highly dependent on matrix type and the specific analyte element.

INSTRUMENTATION An inductively coupled argon plasma emission spectrometer (ICP) capable of background correction is required.

PRECISION & ACCURACY Detection limits, sensitivity, and optimum ranges of the metals will vary with the matrices and model of the spectrometer. In a single lab evaluation, seven wastes were analyzed for 22 elements. The mean percent relative standard deviation from triplicate analyses for all elements and wastes was 9 ± 2%. The mean percent recovery of spiked elements for all wastes was 93 ± 6%. Spike levels ranged from 100 µg/L to 100 mg/L. The wastes included sludges and industrial wastewaters.

Estimated instrument detection limit in µg/L is 7.
Spiked concentration in µg/L is 10.
Mean reported value in µg/L is 10.
Precision as RSD % is 18.

SAMPLING METHOD Samples should be collected in borosilicate glass, linear polyethylene, polypropylene, or Teflon® bottles that have been prewashed with detergent and tap water, and rinsed with 1:1 nitric acid and tap water or 1:1 hydrochloric acid and tap water. Collect at least 2 g of solids and 200 mL of aqueous samples.

SAMPLE PRESERVATION Add nitric acid to make the samples pH <2.

MHT The maximum holding time for properly preserved samples is 6 months.

SAMPLE PREPARATION Preliminary treatment of most matrices is necessary because of the complexity and variability of sample matrices. Water samples that have been prefiltered and acidified will not need acid digestion. Methods for acid digestion of waters for total recoverable or dissolved metals, acid digestions of aqueous samples and extracts for total metals, and acid digestion of sediments, sludges, and soils are summarized below.

Total recoverable or dissolved metals in water — To prepare surface and groundwater samples for determination of total recoverable and dissolved metals, a 100-mL aliquot of well-mixed sample is acidified with concentrated nitric acid and concentrated hydrochloric acid, then heated until the volume is reduced to 15–20 mL. Adjust the final volume to 100 mL with reagent water.

Total metals in aqueous samples, soil and sediment extracts — To prepare aqueous samples, soil and sediment extracts, and wastes that contain suspended solids, a 100-mL aliquot is made acidic with concentrated nitric acid and the solution is evaporated to about 5 mL on a hot plate. Continue heating and adding additional acid until sample digestion is complete, which is usually indicated when the digestate is light in color or does not change in appearance. Evaporate the solution to about 3 mL and cool it and add a small quantity of 1:1 hydrochloric acid (10 mL/100 mL of final solution). Cover the beaker and reflux for 15 min. Wash down the beaker walls and filters or centrifuge the sample to remove silicates and other insoluble material. Filter the sample and adjust the final volume to 100 mL with reagent water and the final acid concentration to 10%.

Sediments, sludges, and soils — To prepare sediments, sludges and soil samples, transfer 1–2 g to a conical beaker and add 10 mL of 1:1 nitric acid, mix the slurry, and cover it with a watch glass. Heat the sample and reflux for 10–15 min without boiling. Allow it to cool, then add 5 mL of concentrated nitric acid and reflux for 30 min. Repeat last step and then allow the solution to evaporate to 5 mL without boiling. Cool and add 2 mL of water and 3 mL of 30% hydrogen peroxide. Cover and place the beaker on the hot plate. Heat and add 30% hydrogen peroxide in 1-mL aliquots with warming until the effervescence is minimal but do not add more than a total of 10 mL of 30% hydrogen peroxide. If the sample is being prepared for the analysis of Ag, Al, As, Ba, Be, Ca, Cd, Co, Cr, Cu, Fe, K, Mg, Mn, Mo, Na, Ni, Os, Pb, Se, Tl, V, and Zn, then add 5 mL of concentrated hydrochloric acid and 10 mL of water and return the covered beaker to a hot plate for 15 min of additional refluxing without boiling. Dilute the sample to a 100 mL volume with water after cooling and filter or centrifuge to remove particulates.

QUALITY CONTROL Laboratory control samples must be analyzed for each analytical method. A method blank should be analyzed with each batch of samples. The effect of the matrix on method performance must be demonstrated: when appropriate, there should be at least one matrix spike and either one matrix duplicate or one matrix spike duplicate per analytical batch. The bias and precision of the method, as well as the method detection limit for each specific matrix type, must be measured.

Dilute and reanalyze samples that are more concentrated than the linear calibration limit. Employ a minimum of one reagent blank per sample batch to determine if contamination or any memory effects are occurring. Whenever a new or unusual sample matrix is encountered, perform either a serial dilution test or a matrix spike addition test to ensure that neither positive or negative interferences are operating on any of the analyte elements. Check the instrument standardization by verifying calibration every 10 samples using a calibration blank and a check standard.

REFERENCE Test Methods for Evaluating Solid Waste (SW-846). U.S. EPA. 1983. Method 6010, Rev. 0, Sept. 1986. Office of Solid Wastes, Washington, DC.

Chromium VI, Hexavalent   EPA Method 7196
CAS #«cas»

TITLE Chromium, Hexavalent (Colorimetric).

MATRIX This method is used to determine the concentration of dissolved hexavalent chromium (Cr VI) in Extraction Procedure (EP) toxicity characteristic extracts and groundwaters. It may also be applicable to certain domestic and industrial wastes, provided that no interfering substances are present.

METHOD SUMMARY This method is specific for hexavalent chromium (Cr VI). A sample is reacted with diphenylcarbazide under strongly acidic conditions and the hexavalent chromium is measured in a 1 cm cell using its absorbance at 540 nm vs. Type II water as a reference. The reaction is very sensitive with the addition of an excess of diphenylcarbazide yielding a red-violet product with its absorbance measured photometrically at 540 nm.

INTERFERENCES The chromium reaction with diphenylcarbazide is usually free from interferences. However, certain substances may interfere if the chromium concentration is relatively low. Hexavalent molybdenum and mercury salts also react to form color with the reagent; however, the red-violet intensities produced are much lower than those for chromium at the specified pH. Concentrations of up to 200 mg/L of molybdenum and mercury can be tolerated. Vanadium interferes strongly, but concentrations up to 10 times that of chromium will not cause trouble.

INSTRUMENTATION Either a spectrophotometer for use at 540 nm, providing a light path of 1 cm or longer; or a filter photometer, providing a light path of 1 cm or longer and equipped with a greenish-yellow filter having maximum transmittance near 540 nm.

PRECISION & ACCURACY This method may be used to analyze samples containing from 0.5 to 50 mg of hexavalent chromium/L.

SAMPLING METHOD Collect samples in 500 mL or 1 L plastic bottles previously washed with detergent, rinsed with tap water, 1:1 hydrochloric acid, tap water and Type II water.

SAMPLE PRESERVATION Immediately cool the sample to 4°C. To retard the chemical activity of Cr VI, the samples and extracts should be stored at 4°C.

MHT The maximum holding time prior to analysis is 24 h.

SAMPLE PREPARATION Transfer 95 mL of the sample to be tested to a 100 mL volumetric flask. Add 2 mL diphenylcarbazide solution and mix. Add sulfuric acid solution to give a pH of 2 ± 0.5 pH units, dilute to 100 mL with Type II water, and let stand 5 to 10 min for full color development.

QUALITY CONTROL Dilute samples if they are more concentrated than the highest standard or if they fall on the plateau of a calibration curve. Employ a minimum of one blank per sample batch to determine if contamination or any memory effects are occurring. Verify calibration with an independently prepared check standard every 15 samples. Run one spike duplicate for every 10 samples. The method of standard addition must be used for the analysis of all EP extracts and whenever a new sample matrix is being analyzed.

REFERENCE Test Methods for Evaluating Solid Waste (SW-846). U.S. EPA. 1983. Method 6010, Rev. 0, Sept. 1986. Office of Solid Wastes, Washington, DC.

Chromium VI, Hexavalent   **EPA Method 7195**
CAS #7440-47-3

TITLE Coprecipitation

MATRIX Ground Waters and Certain Wastewaters.

APPLICATION Method is used to determine concentration of dissolved Cr(VI) in extraction procedure toxicity characteristic extracts and ground waters. Method may also apply to certain wastewaters, if no interfering substances are present. Cr(VI) is separated from solution by coprecipitation.

INTERFERENCES Extracts containing either sulfate or chloride in concentrations above 1,000 mg/L should be diluted prior to analysis.

INSTRUMENTATION Flame or furnace atomic absorption spectroscopy.

RANGE Samples with >5 µg Cr(VI)/L.

MDL Not listed.

PRECISION Not listed.

ACCURACY Not listed.

SAMPLING METHOD Use plastic or glass containers.

STABILITY Cool to 4°C.

MHT 24 h.

QUALITY CONTROL Dilute samples if they are more concentrated than the highest standard or if they fall on the plateau of a calibration curve. run one spike duplicate sample for every 10 samples.

REFERENCE Method 7195, SW-846, 3rd ed., Nov.1986.

Chromium VI, Hexavalent   **EPA Method 7198**
CAS #7440-47-3

TITLE Differential Pulse Polarography

MATRIX EP Extracts, Natural and Wastewaters

APPLICATION The method measures the peak current produced from the reduction of Cr(VI) to Cr(III) at a dropping mercury electrode during a differential pulse voltage ramp. EPA Method uses 0.125M ammonium hydroxide and 0.125M ammonium chloride as the supporting electrolyte. Cr(VI) reduction results in the peak current occurring at a peak potential of -0.25V vs. a silver/silver chloride reference electrode.

INTERFERENCES Copper ion at concentrations higher than Cr(VI) concentration is a potential interference due to peak overlap when using the 0.125M ammoniacal electrolyte. Reductants such as ferrous iron, sulfite, and sulfide will reduce

Cr(VI) to Cr(III); thus it is imperative to analyze samples as soon as possible.

INSTRUMENTATION Polarographic instrumentation with a dropping mercury electrode assembly. Reference electrode and strip chart recorder.

RANGE 1.0–5.0 mg/L (higher by dilution)

MDL 10 µg/L

PRECISION relative standard deviation = 0.69% at 1.87 (Average value) 3 samples (leachate)

ACCURACY Average recovery = 92.8% at 5.0 mg/L spike. (8 samples; EP extracts)

SAMPLING METHOD Use plastic or glass containers. Sample as per chapter nine.

STABILITY Cool to 4°C and run as soon as possible.

MHT 24 Hrs.

QUALITY CONTROL Quantitation must be performed by the method of standard additions. Verify calibration with an independently prepared check standard every 15 samples.

REFERENCE Method 7198, SW-846, 3rd ed., Nov.1986.

Chromium VI, Hexavalent   **EPA Method 7197**
CAS #7440-47-3

TITLE Atomic Absorption, (AA)

MATRIX ep (toxicity characteristic) Chelation Extraction extracts and groundwater

APPLICATION Method also applies to certain wastewaters, provided that no interfering substances are present. Method is based on chelation of Cr(VI) with ammonium pyrolidine dithiocarbamate and extraction with methyl isobutyl ketone. The extract is aspirated and atomized in a flame of an atomic absorption spectrometer. A light beam from a chromium hollow cathode lamp is directed through the flame into a monochromator and onto a detector. Since wavelength is specific for chromium, the light energy absorbed is a measure of chromium.

INTERFERENCES High concentrations of other metals, as may be found in wastewaters, may interfere.

INSTRUMENTATION Atomic absorption spectrometer. Chromium hollow cathode lamp. Strip-chart recorder. (357.9 nm Wavelength)

RANGE 1.0 to 25 µg Cr(VI)/L

MDL Not listed.

PRECISION Standard deviation = ±2.6 at 50 µg Cr(VI)/L

ACCURACY Recovery = 96% at 50 µg Cr(VI)/L

SAMPLING METHOD Use plastic or glass containers. Sample as per chapter nine.

STABILITY Cool to 4°C. Run as soon as possible.

MHT 24 Hrs

QUALITY CONTROL Run one spike duplicate sample for every 10 samples. Verify calibration with an independently prepared check standard every 15 samples.

REFERENCE Method 7197, SW-846, 3rd ed., Nov.1986.

Chrysene   **EPA Method 1625**
CAS #218-01-9

TITLE Semivolatile Organic Compounds by Isotope Dilution GC/MS

MATRIX The compounds may be determined in waters, soils, and municipal sludges by this method.

METHOD SUMMARY This method is used to determine 176 semivolatile toxic organic pollutants associated with the CWA (as amended 1987); the RCRA (as amended 1986); the CERCLA (as amended 1986); and other compounds amenable to extraction and analysis by capillary column gas chromatography-mass spectrometry (GC/MS).

Stable isotopically-labeled analogs of the compounds of interest are added to the sample. If the solids content is less than 1%, a 1-L sample is extracted at pH 12–13, then at pH <2 with methylene chloride using continuous extraction techniques.

If the solids content is 30% or less, the sample is diluted to 1% solids with reagent water, homogenized ultrasonically, and extracted at pH 12–13, then at pH <2 with methylene chloride using continuous extraction techniques. If the solids content is greater than 30%, the sample is extracted using ultrasonic techniques.

Each extract is dried over sodium sulfate, concentrated to a volume of 5 mL, cleaned up using GPC, if necessary, and concentrated. Extracts are concentrated to 1 mL if GPC is not performed, and to 0.5 mL if GPC is performed.

An internal standard is added to the extract, and a 1-mL aliquot of the extract is injected into the GC. The compounds are separated by GC and detected by a MS. The labeled compounds serve to correct the variability of the analytical technique.

INTERFERENCES Solvents, reagents, glassware, and other sample processing hardware may yield artifacts and/or elevated baselines causing misinterpretation of chromatograms and spectra. Materials used in the analysis must be demonstrated to be free from interferences under the conditions of analysis by running method blanks initially and with each sample lot (sample started through the extraction process on a given 8-h shift, to a maximum of 20). Specific selection of reagents and purification of solvents by distillation in all glass systems may be required. Glassware and, where possible, reagents are cleaned by solvent rinse and baking at 450°C for 1-h minimum. Interferences coextracted from samples will vary considerably from source to source, depending on the diversity of the site being sampled.

INSTRUMENTATION Major instrumentation includes a GC with a splitless or on-column injection port for capillary column,

a MS with 70 eV electron impact ionization, and a data system to collect and record MS data, and process it. A K-D apparatus is used to concentrate extracts.

GC Column: 30 m × 0.25 mm I.D. 5% phenyl, 94% methyl, 1% vinyl silicone bonded phased fused silica capillary column.

PRECISION & ACCURACY The detection limits of the method are usually dependent on the level of interferences rather than instrumental limitations. The limits typify the minimum quantities that can be detected with no interferences present.

- The minimum level (in µg/mL) was 10. This is defined as a minimum level at which the analytical system shall give recognizable mass spectra (background corrected) and acceptable calibration points.
- The MDL (in µg/kg) in low solids was 51 and in high solids was 48; these were determined in digested sludge (low solids) and in filter cake or compost (high solids).
- The labeled and native compound initial precision as standard deviation (in µg/L) was 51.
- The labeled and native compound initial accuracy as average recovery (in µg/L) was 59–186.

SAMPLE COLLECTION, PRESERVATION & HANDLING Collect samples in glass containers. Aqueous samples which flow freely are collected in refrigerated bottles using automatic sampling equipment. Solid samples are collected as grab samples using widemouth jars. Maintain samples at 0 to 4°C from the time of collection until extraction. If residual chlorine is present in aqueous samples, add 80 mg sodium thiosulfate/L of water. Begin sample extraction within 7 days of collection, and analyze all extracts within 40 days of extraction.

SAMPLE PREPARATION Samples containing 1% solids or less are extracted directly using continuous liquid-liquid extraction techniques. Samples containing 1 to 30% solids are diluted to the 1% level with reagent water and extracted using continuous liquid-liquid extraction techniques. Samples containing greater than 30% solids are extracted using ultrasonic techniques.

- Base/neutral extraction — Adjust the pH of the waters in the extractors to 12–13 with 6 N NaOH. Extract with methylene chloride for 24–48 h.
- Acid extraction — Adjust the pH of the waters in the extractors to 2 or less using 6 N sulfuric acid. Extract with methylene chloride for 24–48 h.
- Ultrasonic extraction of high solids samples — Add anhydrous sodium sulfate to the sample and QC aliquot(s). Add acetone:methylene chloride (1:1) to the sample and mix thoroughly

Concentrate extracts using a K-D apparatus.

QUALITY CONTROL The analyst is permitted to modify this method to improve separations or lower the costs of measurements, provided all performance specifications are met. Analyses of blanks are required to demonstrate freedom from contamination. When results of spikes indicate atypical method performance for samples, the samples are diluted to bring method performance within acceptable limits.

For low solids (aqueous samples), extract, concentrate, and analyze two sets of four 1-L aliquots (8 aliquots total) of the precision and recovery standard. For high solids samples, two sets of four 30-g aliquots of the high solids reference matrix are used.

Spike all samples with labeled compounds to assess method performance. Compute percent recovery of the labeled compounds using the internal standard method. Compare the labeled compound recovery for each compound with the corresponding labeled compound recovery.

Reagent water and high solids reference matrix blanks are analyzed to demonstrate freedom from contamination. Extract and concentrate a 1-L reagent water blank or a high solids reference matrix blank with each sample's lot (samples started through the extraction process on the same 8-h shift, to a maximum of 20 samples).

Field replicates may be collected to determine the precision of the sampling technique, and spiked samples may be required to determine the accuracy of the analysis when the internal standard method is used.

REFERENCE Semivolatile Organic Compounds by Isotope Dilution GC/MS. Office of Water Regulation and Standards, U.S. EPA Industrial Technology Division, Washington, DC, EPA Method 1625, Rev. C, June 1989 (contact W.A. Telliard, U.S. EPA, Office of Water Regulations and Standards, 401 M St., SW, Washington, DC, 20460. Phone: 202-382-7131).

Chrysene   **EPA Method 625**
CAS #218-01-9

TITLE Base/Neutrals and Acids, U.S. EPA Method 625

MATRIX This methods covers municipal and industrial wastewaters.

METHOD SUMMARY Approximately 1 L of sample is serially extracted with methylene chloride at a pH greater than 11 and again at a pH less than 2 using a separatory funnel or a continuous extractor. The methylene chloride extract is dried, concentrated to a volume of 1 mL, and analyzed by GC/MS. Qualitative identification of the parameters in the extract is performed using the retention time and the relative abundance of three characteristic masses (m/z). Qualitative analysis is performed using either external or internal standard techniques with a single characteristic m/z.

INTERFERENCES Method interferences may be caused by contaminants in solvents, reagents, glassware, and other sample processing hardware. Glassware must be scrupulously cleaned. Glassware should be heated in a muffle furnace at 400°C for 5 to 30 min. Some thermally stable materials, such as PCBs, may not be eliminated by this treatment. Solvent rinses with acetone and pesticide quality hexane may be substituted for the muffle furnace heating. Matrix interferences may be caused by contaminants that are coextracted from the sample. The base-neutral extraction may cause significantly reduced recovery of phenols. The packed gas chromatographic columns recommended for the basic fraction may not exhibit sufficient resolution for some analytes.

INSTRUMENTATION A GC/MS system with an injection port designed for on-column injection when using packed columns and for splitless injection when using capillary columns.

Column for base/neutrals: 1.8 m long × 2 mm I.D. glass, packed with 3% SP-2550 on Supelcoport (100/120 mesh) or equivalent.

Column for acids: 1.8 m long × 2 mm I.D. glass, packed with 1% SP-1240DA on Supelcoport (100/120 mesh) or equivalent.

PRECISION & ACCURACY The MDL concentrations were obtained using reagent water. The MDL actually achieved in a given analysis will vary depending on instrument sensitivity and matrix effects. This method was tested by 15 laboratories using reagent water, drinking water, surface water, and industrial wastewaters spiked at six concentrations over the range 5 to 100 µg/L. Single operator precision, overall precision, and method accuracy were found to be directly related to the concentration of the parameter matrix.

The MDL (in µg/L) in reagent water was 2.5.

The standard deviation (in µg/L based on 4 recovery measurements) was 48.3.

The range (in µg/L) for average recovery for 4 measurements was 44.1–139.9.

The range (in %) for percent recovery was 17–168.

Accuracy (in µg/L) as expected recovery for one or more measurements of a sample containing a true concentration of C was 0.93C–1.00.

Precision (in µg/L) as expected single analyst standard deviation of measurements at an average concentration found at X was 0.28X+0.13.

Overall precision (in µg/L) as expected interlaboratory standard deviation of measurements in an average concentration found at X was 0.33X–0.09.

C = *True value of the concentration in µg/L.*
X = *Average recovery found for measurements of samples containing a concentration at C in µg/L.*

SAMPLE PREPARATION Adjust the pH to >11 with sodium hydroxide and serially extract in a separatory funnel with methylene chloride or else in a continuous extractor. Next, adjust the pH to <2 with sulfuric acid and serially extract in a separatory funnel with methylene chloride or else in a continuous extractor. Dry the extracts separately through a column of anhydrous sodium sulfate and then concentrate each of the extracts to 1.0 mL using a K-D apparatus.

SAMPLE COLLECTION, PRESERVATION & HANDLING Grab samples must be collected in glass containers. All samples must be refrigerated at 4°C from the time of collection until extraction. If residual chlorine is present, add 80 mg of sodium thiosulfate/L of sample and mix well. All samples must be extracted within 7 days of collection and completely analyzed within 40 days of extraction.

QUALITY CONTROL Make an initial, one-time, demonstration of the ability to generate acceptable accuracy and precision with this method. Before processing any samples, the analyst must analyze a reagent water blank to demonstrate that interferences from the analytical system and glassware are under control. Each time a set of samples is extracted or reagents are changed, a reagent water blank must be processed. Spike and analyze a minimum of 5% of all samples to monitor and evaluate lab data quality. A QC check sample concentrate that contains each parameter of interest at a concentration of 100 µg/mL in acetone is required. PCBs and multicomponent pesticides may be omitted from this test.

After analysis of five spiked wastewater samples, calculate the average percent recovery and the standard deviation of the percent recovery. Spike all samples with the surrogate standard spiking solution and calculate the percent recovery of each surrogate compound.

REFERENCE Federal Register, Vol. 49, No. 209. Friday, Oct. 26, 1984.

Chrysene   **EPA Method 8270**
CAS #218-01-9

TITLE Semivolatile Organic Compounds by GC/MS

MATRIX This method is used to determine the concentration of semivolatile organic compounds in extracts prepared from all types of solid waste matrices, soils, and groundwater. Although surface waters are not specifically mentioned, this method should be applicable to water samples from rivers, lakes, etc.

METHOD SUMMARY This method covers 259 semivolatile organic compounds. In very limited applications direct injection of the sample into the GC/MS system may be appropriate, but this results in very high detection limits (approximately 10,000 µg/L). Typically, a 1-L liquid sample, containing surrogate, and matrix spiking standards, is extracted in a continuous extractor first under acid conditions and then under basic conditions. Typically 30 g of a solid sample, containing surrogate, and matrix spiking standards, is extracted ultrasonically. After concentrating the extract to 1 mL it is spiked with 10 µL of an internal standard solution just prior to analysis by GC/MS. The volume injected should contain about 100 ng of base/neutral and 200 ng of acid surrogates (for a 1-µL injection). Analysis is performed by GC/MS using a capillary GC column.

INTERFERENCES Raw GC/MS data from all blanks, samples, and spikes must be evaluated for interferences. Contamination by carryover can occur whenever high-concentration and low-concentration samples are sequentially analyzed. To reduce carryover, the sample syringe must be rinsed out between samples with solvent. Whenever an unusually concentrated sample is encountered, it should be followed by the analysis of blank solvent to check for cross-contamination.

INSTRUMENTATION A GC/MS and a data system are required. The GC column used is a 30 m × 0.25 mm I.D. (or 0.32 mm I.D.) 1um film thickness silicone-coated fused silica capillary column. A continuous liquid-liquid extractor equipped with Teflon® or glass connection joints and stopcocks requiring no lubrication, a K-D concentrating apparatus, water

bath, and an ultrasonic disrupter with a minimum power of 300 W and with pulsing capability are also required.

PRECISION & ACCURACY The estimated quantitation limit (EQL) of Method 8270B for determining an individual compound is approximately 1 mg/kg (wet weight) for soil or sediment samples, 1–200 mg/kg for wastes (dependent on matrix and method of preparation), and 10 µg/L for groundwater samples. EQLs will be proportionately higher for sample extracts that require dilution to avoid saturation of the detector.

The EQL(b) for groundwater in µg/L is 10.
The EQL (a, b) for low concentrations in soil and sediment in µg/kg is 660.
Accuracy as µg/L is 0.93C–1.00.
Overall precision in µg/L is 0.33X–0.09.

(a) *EQLs listed for soil/sediment are based on wet weight. Normally data is reported in a dry-weight basis; therefore, EQLs will be higher based on the % dry weight of each sample. This calculation is based on a 30 g sample and gel permeation chromatography cleanup.*
(b) *Sample EQLs are highly matrix-dependent. The EQLs are provided for guidance and may not always be achievable.*
C = *True value for concentration, in µg/L.*
X = *Average recovery found for measurements of samples containing a concentration of C, in µg/L.*

ESTIMATED QUANTITATION LIMIT FOR OTHER MATRICES

Other Matrices	Factor (a)
High-concentration soil and sludges by sonicator	7.5
Non-water miscible waste	75

(a) *EQL for other matrices = [EQL for low soil/sediment] × [Factor]. This estimated EQL is similar to an EPA "Practical Quantitation Limit."*

SAMPLING METHOD

Liquid samples — Use a 1 or 2½ gallon amber glass bottle with a screw-top Teflon®-lined cover that has been prewashed with detergent and rinsed with distilled water and methanol (or isopropanol).

Soils, sediments, or sludges — Use an 8-oz. widemouth glass with a screw-top Teflon®-lined cover that has been prewashed with detergent and rinsed with distilled water and methanol (or isopropanol).

SAMPLE PRESERVATION

Liquid samples — If residual chlorine is present, add 3 mL of 10% sodium thiosulfate per gallon, cool to 4°C and store in a solvent-free refrigerator until analysis; if chlorine is not present, then eliminate the sodium thiosulfate addition.

Soils, sediments, or sludges — Cool samples to 4°C and store in a solvent-free refrigerator.

MHT Liquid samples must be extracted within 7 days and the extracts analyzed within 40 days. Soils, sediments, or sludges may be stored for a maximum of 14 days and the extracts analyzed within 40 days.

SAMPLE PREPARATION

Liquid samples — Transfer 1 L quantitatively to a continuous extractor. If high concentrations are anticipated, a smaller volume may be used and then diluted with organic-free reagent water to 1 L. Adjust pH, if necessary, to pH <2 using 1:1 (V/V) sulfuric acid. Pipette 1.0 mL of a surrogate standard spiking solution into each sample. For the sample in each analytical batch selected for spiking, add 1.0 mL of a matrix spiking standard. For base/neutral acid analysis, the amount of the surrogates and matrix spiking compounds added to the sample should result in a final concentration of 100 ng/µL of each analyte in the extract to be analyzed (assuming a 1-µL injection). Extract with methylene chloride for 18–24 h. Next, adjust the pH of the aqueous phase to pH >11 using 10 N sodium hydroxide and extract it with methylene chloride again for 18–24 h. Dry the extract through a column containing anhydrous sodium sulfate and concentrate it to 1 mL using a K-D concentrator.

Soils, sediments, or sludges — Use 30 g of sample. Nonporous or wet samples (gummy or clay type) that do not have a free-flowing sandy texture must be mixed with anhydrous sodium sulfate until the sample is free flowing. Add 1 mL of surrogate standards to all samples, spikes, standards, and blanks. For the sample in each analytical batch selected for spiking, add 1.0 mL of a matrix spiking standard. For base/neutral acid analysis, the amount added of the surrogates and matrix spiking compounds should result in a final concentration of 100 ng/µL of each base/neutral analyte and 200 ng/µL of each acid analyte in the extract to be analyzed (assuming a 1-µL injection). Immediately add a 100-mL mixture of 1:1 methylene chloride:acetone and extract the sample ultrasonically for 3 min and then decant or filter the extracts. Repeat the extraction two or more times. Dry the extract using a column with anhydrous sodium sulfate and concentrate it to 1 mL in a K-D concentrator.

QUALITY CONTROL A methylene chloride solution containing 50 ng/µL of decafluorotriphenylphosphine (DFTPP) is used for tuning the GC/MS system each 12-h shift. A system performance check also must be made during every 12-h shift. A standard containing 50 ng/µL each of 4,4'-DDT, pentachlorophenol, and benzidine is required to verify injection port inertness and GC column performance. A calibration standard at mid-concentration, containing each compound of interest, including all required surrogates, must be performed every 12 h during analysis. After the system performance check is met, calibration check compounds (CCCs) are used to check the validity of the initial calibration.

The internal standard responses and retention times in the calibration check standard must be evaluated immediately after or during data acquisition. If the retention time for any internal standard changes by more than 30 seconds from the last check calibration (12 h), the chromatographic system must be inspected for malfunctions and corrections must be made, as required. If the electron ionization current plot (EICP) area for any of the internal standards changes by a factor of two from the last daily calibration standard check, the mass spectrometer must be inspected for malfunctions and corrections must be made, as appropriate.

Demonstrate, through the analysis of a reagent water blank, that interferences from the analytical system, glassware, and reagents are under control. The blank samples should be carried through all stages of the sample preparation and measurement steps. For each analytical batch (up to 20 samples), a reagent blank, matrix spike, and matrix spike duplicate/duplicate must be analyzed (the frequency of the spikes may be different for different monitoring programs). The blank and spiked samples must be carried through all stages of the sample preparation and measurement steps. A QC reference sample concentrate containing each analyte at a concentration of 100 mg/L in methanol is required.

REFERENCE Test Methods for Evaluating Solid Waste (SW-846). U.S. EPA 1983, Method 8270B, Rev. 2, Nov. 1990. Office of Solid Waste, Washington, DC.

Chrysene — EPA Method 8100
CAS #218-01-9

TITLE Polynuclear Aromatic Hydrocarbons

MATRIX Groundwater, soils, sludges, water miscible liquid wastes, and non-water miscible wastes.

APPLICATION This method is used for the analysis of various PAHs. Samples are extracted, concentrated, and analyzed using direct injection of both neat and diluted organic liquids. The method provides two optional GC columns that are better than Column 1 and that may help resolve analytes from interferences.

INTERFERENCES Solvents, reagents, and glassware may introduce artifacts. Other interferences may come from coextracted compounds from samples.

INSTRUMENTATION GC capable of on-column injections and a flame with detector (FID). Column 1: a 1.8 m by 2 mm 3% OV-17 on Chromosorb W-AW-DCMS column. Column 2: a 30 m by 0.25 mm SE-54 fused silica capillary colunm. Column 3: a 30 m by 0.32 mm SE-54 fused silica capillary column.

RANGE 0.1–425 µg/L

MDL Not reported.

PQL FACTORS FOR MULTIPLYING × FID MDL VALUE
Not available.

PRECISION 0.66X–0.22 µg/L (overall precision).

ACCURACY 0.77C–0.18 µg/L (as recovery).

SAMPLING METHOD Use 8-oz. widemouth glass bottles with Teflon®-lined caps for concentrated waste samples, soils, sediments, and sludges. Use 1 or 2½ gallon amber glass bottles with Teflon®-lined caps for liquid (water) samples.

STABILITY Cool soil, sediment, sludge, and liquid samples to 4°C. If residual chlorine is present in liquid samples add 3 mL of 10% sodium thiosulfate per gallon of sample and cool to 4°C.

MHT 14 days for concentrated waste, soil, sediment, or sludge; 7 days for liquid samples. All extracts must be analyzed within 40 days.

QUALITY CONTROL A quality control check sample concentrate containing each analyte of interest is required. The QC check sample concentrate may be prepared from pure standard materials or purchased as certified solutions Use appropriate trip, matrix, control site, method, reagent, and solvent blanks. Internal, surrogate, and five concentration level calibration standards are used. The quality control check sample concentrate should contain chrysene at 10 µg/mL in acetonitrile.

REFERENCE Test Methods for Evaluating Solid Waste (SW-846), U.S. EPA Office of Solid Waste, Washington, DC, Method 8100, Nov. 1986.

Chrysene — EPA Method 8310
CAS #218-01-9

TITLE Polynuclear Aromatic Hydrocarbons

MATRIX Groundwater, soils, sludges, water miscible liquid wastes, and non-water miscible wastes.

APPLICATION This method is used for the analysis of 16 polynuclear aromatic hydrocarbons (PAHs). Samples are extracted, concentrated, and analyzed using HPLC with detection by UV and fluorescence detectors.

INTERFERENCES Solvents, reagents, and glassware may introduce artifacts. Other interferences may come from coextracted compounds from samples.

INSTRUMENTATION HPLC with a gradient pumping system and a 250 mm by 2.6 mm reverse phase HC-ODS Sil-X 5-micron particle-size column. The fluorescence detector uses an excitation wavelength of 280 nm and emission greater than 389 nm cutoff with dispersive optics.

RANGE 0.1–425 µg/L

MDL 0.15 µg/L (fluorescence; reagent water).

PQL FACTORS FOR MULTIPLYING × FID MDL VALUE

Matrix	Multiplication Factor
Groundwater	10
Low-level soil by sonication with GPC cleanup	670
High-level soil and sludge by sonication	10,000
Non-water miscible waste	100,000

PRECISION 0.66X–0.22 µg/L (overall precision).

ACCURACY 0.77C–0.18 µg/L (as recovery).

SAMPLING METHOD Use 8-oz. widemouth glass bottles with Teflon®-lined caps for concentrated waste samples, soils, sediments, and sludges. Use 1 or 2½ gallon amber glass bottles with Teflon®-lined caps for liquid (water) samples.

STABILITY Cool soil, sediment, sludge, and liquid samples to 4°C. If residual chlorine is present in liquid samples add

3 mL of 10% sodium thiosulfate per gallon of sample and cool to 4°C.

MHT 14 days for concentrated waste, soil, sediment, or sludge; 7 days for liquid samples. All extracts must be analyzed within 40 days.

QUALITY CONTROL Internal, surrogate, and five concentration level calibration standards are used. The calibration standards must be used with the analytical method blank. A quality control check sample concentrate containing chrysene at 10 µg/mL is required. The QC check sample concentrate may be prepared from pure standard materials or purchased as certified solutions. Use appropriate trip, matrix, control site, method, reagent, and solvent blanks.

REFERENCE Test Methods for Evaluating Solid Waste (SW-846), U.S. EPA Office of Solid Waste, Washington, DC, Method 8310, Rev. 0, Nov. 1986.

Cobalt EPA Method 6010
CAS #7440-48-4

TITLE Inductively Coupled Plasma-Atomic Emission Spectroscopy

MATRIX This method is applicable to the determination of trace elements, including metals, in groundwater, soils, sludges, sediments, and other solid wastes. All matrices require digestion prior to analysis. The method of standard addition must be used for the analysis of all sample digests unless either serial dilution or matrix spike addition demonstrates it is not required.

METHOD SUMMARY Method 6010 covers 25 elements using ICP analysis. It measures element-emitted light by optical spectrometry. Samples, following an appropriate acid digestion, are nebulized and the resulting aerosol is transported to the plasma torch. Element-specific atomic line emission spectra are produced by a radio-frequency inductively coupled plasma.

INTERFERENCES Interferences may be categorized as spectral or non-spectral. Spectral interferences are caused by overlap of a spectral line from another element, unresolved overlap of molecular band spectra, background contribution from continuous or recombination phenomenon, and stray light from the line emission of high concentration elements. Non-spectral interferences include physical and chemical interferences. Physical interferences are effects associated with the sample nebulization and transport processes. Changes in viscosity and surface tension can cause significant inaccuracies. Chemical interferences include molecular compound formation, ionization effects, and solute vaporization effects. Normally these effects are not significant and can be minimized by careful selection of operating conditions. Chemical interferences are highly dependent on matrix type and the specific analyte element.

INSTRUMENTATION An inductively coupled argon plasma emission spectrometer (ICP) capable of background correction is required.

PRECISION & ACCURACY Detection limits, sensitivity, and optimum ranges of the metals will vary with the matrices and model of the spectrometer. In a single lab evaluation, seven wastes were analyzed for 22 elements. The mean percent relative standard deviation from triplicate analyses for all elements and wastes was 9 ± 2%. The mean percent recovery of spiked elements for all wastes was 93 ± 6%. Spike levels ranged from 100 µg/L to 100 mg/L. The wastes included sludges and industrial wastewaters.

Estimated instrument detection limit in µg/L is 7.
Spiked concentration in µg/L is 20.
Mean reported value in µg/L is 20.
Precision as RSD % is 4.1.

SAMPLING METHOD Samples should be collected in borosilicate glass, linear polyethylene, polypropylene, or Teflon® bottles that have been prewashed with detergent and tap water, and rinsed with 1:1 nitric acid and tap water or 1:1 hydrochloric acid and tap water. Collect at least 2 g of solids and 200 mL of aqueous samples.

SAMPLE PRESERVATION Add nitric acid to make the samples pH <2.

MHT The maximum holding time for properly preserved samples is 6 months.

SAMPLE PREPARATION Preliminary treatment of most matrices is necessary because of the complexity and variability of sample matrices. Water samples that have been prefiltered and acidified will not need acid digestion. Methods for acid digestion of waters for total recoverable or dissolved metals, acid digestions of aqueous samples and extracts for total metals, and acid digestion of sediments, sludges, and soils are summarized below.

Total recoverable or dissolved metals in water — To prepare surface and groundwater samples for determination of total recoverable and dissolved metals, a 100-mL aliquot of well-mixed sample is acidified with concentrated nitric acid and concentrated hydrochloric acid, then heated until the volume is reduced to 15–20 mL. Adjust the final volume to 100 mL with reagent water.

Total metals in aqueous samples, soil and sediment extracts — To prepare aqueous samples, soil and sediment extracts, and wastes that contain suspended solids, a 100-mL aliquot is made acidic with concentrated nitric acid and the solution is evaporated to about 5 mL on a hot plate. Continue heating and adding additional acid until sample digestion is complete, which is usually indicated when the digestate is light in color or does not change in appearance. Evaporate the solution to about 3 mL and cool it and add a small quantity of 1:1 hydrochloric acid (10 mL/100 mL of final solution). Cover the beaker and reflux for 15 min. Wash down the beaker walls and filters or centrifuge the sample to remove silicates and other insoluble material. Filter the sample and adjust the final volume to 100 mL with reagent water and the final acid concentration to 10%.

Sediments, sludges, and soils — To prepare sediments, sludges and soil samples, transfer 1–2 g to a conical beaker and add 10 mL of 1:1 nitric acid, mix the slurry, and cover it with a watch glass. Heat the sample and reflux for 10–15 min without boiling. Allow it to cool, then add 5 mL of concentrated nitric acid and reflux for 30 min. Repeat last step and then allow the solution to evaporate to 5 mL without boiling. Cool and add 2 mL of water and 3 mL of 30% hydrogen peroxide. Cover and place the beaker on the hot plate. Heat and add 30% hydrogen peroxide in 1-mL aliquots with warming until the effervescence is minimal but do not add more than a total of 10 mL of 30% hydrogen peroxide. If the sample is being prepared for the analysis of Ag, Al, As, Ba, Be, Ca, Cd, Co, Cr, Cu, Fe, K, Mg, Mn, Mo, Na, Ni, Os, Pb, Se, Tl, V, and Zn, then add 5 mL of concentrated hydrochloric acid and 10 mL of water and return the covered beaker to a hot plate for 15 min of additional refluxing without boiling. Dilute the sample to a 100 mL volume with water after cooling and filter or centrifuge to remove particulates.

QUALITY CONTROL Laboratory control samples must be analyzed for each analytical method. A method blank should be analyzed with each batch of samples. The effect of the matrix on method performance must be demonstrated: when appropriate, there should be at least one matrix spike and either one matrix duplicate or one matrix spike duplicate per analytical batch. The bias and precision of the method, as well as the method detection limit for each specific matrix type, must be measured.

Dilute and reanalyze samples that are more concentrated than the linear calibration limit. Employ a minimum of one reagent blank per sample batch to determine if contamination or any memory effects are occurring. Whenever a new or unusual sample matrix is encountered, perform either a serial dilution test or a matrix spike addition test to ensure that neither positive or negative interferences are operating on any of the analyte elements. Check the instrument standardization by verifying calibration every 10 samples using a calibration blank and a check standard.

REFERENCE Test Methods for Evaluating Solid Waste (SW-846). U.S. EPA. 1983. Method 6010, Rev. 0, Sept. 1986. Office of Solid Wastes, Washington, DC.

Cobalt   EPA Method 200.7
CAS #7440-48-4

TITLE Inductively Coupled Plasma

MATRIX Dissolved, suspended or (ICP) total element in drinking and surface waters and in domestic and industrial wastewaters.

APPLICATION The method covers the determination of 25 metals. Dissolved elements are determined in filtered and acidified samples after appropriate digestion (which increases dissolved solids). Its primary advantage is that ICP instruments allow simultaneous or rapid sequential determination of many elements in a short time. Samples are first nebulized and the aerosol is transported to a plasma torch in which element specific atomic line emission spectra are produced by a radio frequency inductively coupled plasma. Background correction is required for trace element detection except in the case of line broadning.

INTERFERENCES There are spectral, physical, and chemical interferences. The primary disadvantage of ICP instruments is background radiation from other elements and the plasma gases (spectral interferences). Changes in sample viscosity and surface tension with samples containing high dissolved solids (especially those exceeding 1500 mg/L) or high acid concentrations can cause physical interferences. Ionization effects, solute vaporization and molecular compound formation can cause chemical interferences. Chromium, iron, nickel and thallium can cause interference at the 100 mg/L level.

INSTRUMENTATION Inductively coupled argon plasma emission spectroscopy. 228.616 nm Wavelength.

RANGE Not listed.

MDL 7 µg/L.

PRECISION SD = 10% Mean at true value 700 µg/L.

ACCURACY Mean recovery = 93% ± 6% of spiked elements for all wastes.

SAMPLING METHOD Wash sample container with detergent and tap water, rinse with 1 + 1 nitric acid and tap water, then rinse with 1 + 1 hydrochloric acid and tap water, then rinse with deionized, distilled water in that order. Perform any filtration or acid preservation steps when the sample is collected or as soon as possible thereafter.

STABILITY Cool samples to 4°C.

MHT 24 h.

QUALITY CONTROL Mixed calibration standards, an instrument check standard, and an interference check solution are used in addition to a quality control sample. The quality control sample should be prepared in the same acid matrix as the calibration standards at 10 times the instrumental detection limits and in accordance with the instructions provided by the supplier. Furthermore, two types of blanks are required: a calibration blank and a reagent blank.

REFERENCE Method 200.7, U.S. EPA, EMSL-Cincinnati, OH, Nov. 1980

Cobalt   EPA Method 7200
CAS #7440-48-4

TITLE Atomic Absorption, (AA)

MATRIX Drinking, Surface and Direct Aspiration Saline Waters, Wastewater

APPLICATION Sample is aspirated and atomized in a flame. A light beam from a Co hollow cathode lamp is directed through the flame into a monochromator and onto a detector. Since wavelength of light beam is specific for Co, light energy absorbed by detector is measure of cobalt.

INTERFERENCES The most troublesome type is chemical, caused by lack of absorption of atoms bound in molecular combination in the flame. High dissolved solids in sample may result in nonatomic absorbance interference. Excess of other transition metals may interfere.

INSTRUMENTATION Atomic absorption spectrometer. Cobalt hollow cathode lamp. (240.7 nm Wavelength).

RANGE 0.5–5 mg/L

MDL 0.05 mg/L

PRECISION Standard deviation = ±0.013, 0.01, 0.05 at 0.20, 1.0, 5.0 mg Co/L

ACCURACY Recoveries = 98, 98, 97% at 0.20, 1.0, 5.0 mg Co/L

SAMPLING METHOD Use glass or plastic containers. Collect 200 g of solids and 600 mL of liquid samples.

STABILITY Cool solid samples to 4°C and analyze as soon as possible. Add nitric acid to liquid samples to pH <2.

MHT 6 months.

QUALITY CONTROL At least one duplicate and one spike sample should be run every 20 samples or with each matrix type to verify precision of the method. For 20 or more samples per day, verify working standard curve. Run an additional standard at or near mid-range every 10 samples.

REFERENCE Method 7200, SW-846, 3rd ed., Nov.1986.

Cobalt — EPA Method 7201
CAS #7440-48-4

TITLE Atomic Absorption, (AA)

MATRIX Wastes, mobility procedure. Furnace Technique extracts, soils and groundwater.

APPLICATION Aqueous samples, EP extracts, industrial wastes, soils, sludges, sediments, and solid wastes require digestion before analysis. An aliquot of sample is placed in the graphite tube in the furnace and slowly evaporated, charred and atomized. Absorption of lamp radiation during atomization is proportional to cobalt concentration.

INTERFERENCES The furnace technique is subject to chemical interferences.

Composition of sample matrix can effect analysis. Modify matrix to remove interferences or to stabilize the analyte. Background correction is required. Excess chloride may interfere.

INSTRUMENTATION Atomic absorption spectrometer. Cobalt hollow cathode lamp or electrodeless discharge lamp. Graphite furnace. Strip-chart recorder

RANGE 5–100 µg/L

MDL 1 µg/L (240.7 nm Wavelength)

PRECISION Not listed.

ACCURACY Not listed.

SAMPLING METHOD Use glass or plastic containers. Collect 200 g of solids and 600 mL of liquid samples.

STABILITY Cool solid samples to 4°C and analyze as soon as possible. Add nitric acid to liquid samples to pH <2.

MHT 6 months.

QUALITY CONTROL At least one duplicate and one spike sample should be run every 20 samples, or with each matrix type to verify method precision. If 20 or more samples are run a day, run a standard (at or near mid-range) every 10 samples.

REFERENCE Method 7201, SW-846, 3rd ed., Nov.1986.

Conductance — EPA Method 120.1

TITLE Physical Properties

MATRIX Drinking, Surface and Saline Waters. Wastewater.

APPLICATION Date issued 1971. (Specific conductance, umhos at 25 c). The specific conductance of a sample is measured by use of a self-contained conductivity meter. Field measurements with comparable instruments are reliable.

INTERFERENCES NA

INSTRUMENTATION Wheatstone type-bridge or equivalent.

RANGE Not listed.

MDL Not listed.

PRECISION SD = 7.55 at Specific conductance increment of 100.

ACCURACY As bias, -2.0 umhos/cm at Specific conductance 100.

SAMPLING METHOD plastic or glass (100 mL).

STABILITY Cool, 4°C.

MHT 28 Days.

QUALITY CONTROL Instrument must be standardized with KCl solution before daily use. Conductivity cell must be kept clean. Make temperature corrections, and report results at 25 C, if sample not analyzed at 25 C.

REFERENCE Methods for the Chemical Analysis of Water and Wastes, EPA-600/4-79-020, U.S. EPA, EMSL, 1979.

Copper — EPA Method 6010
CAS #7440-50-8

TITLE Inductively Coupled Plasma-Atomic Emission Spectroscopy

MATRIX This method is applicable to the determination of trace elements, including metals, in groundwater, soils, sludges, sediments, and other solid wastes. All matrices require digestion prior to analysis. The method of standard addition must be used for the analysis of all sample digests unless either serial dilution or matrix spike addition demonstrates it is not required.

METHOD SUMMARY Method 6010 covers 25 elements using ICP analysis. It measures element-emitted light by optical spectrometry. Samples, following an appropriate acid digestion, are nebulized and the resulting aerosol is transported to the plasma torch. Element-specific atomic line emission spectra are produced by a radio-frequency inductively coupled plasma.

INTERFERENCES Interferences may be categorized as spectral or non-spectral. Spectral interferences are caused by overlap of a spectral line from another element, unresolved overlap of molecular band spectra, background contribution from continuous or recombination phenomenon, and stray light from the line emission of high concentration elements. Non-spectral interferences include physical and chemical interferences. Physical interferences are effects associated with the sample nebulization and transport processes. Changes in viscosity and surface tension can cause significant inaccuracies. Chemical interferences include molecular compound formation, ionization effects, and solute vaporization effects. Normally these effects are not significant and can be minimized by careful selection of operating conditions. Chemical interferences are highly dependent on matrix type and the specific analyte element.

INSTRUMENTATION An inductively coupled argon plasma emission spectrometer (ICP) capable of background correction is required.

PRECISION & ACCURACY Detection limits, sensitivity, and optimum ranges of the metals will vary with the matrices and model of the spectrometer. In a single lab evaluation, seven wastes were analyzed for 22 elements. The mean percent relative standard deviation from triplicate analyses for all elements and wastes was 9 ± 2%. The mean percent recovery of spiked elements for all wastes was 93 ± 6%. Spike levels ranged from 100 µg/L to 100 mg/L. The wastes included sludges and industrial wastewaters.

Estimated instrument detection limit in µg/L is 6.
Spiked concentration in µg/L is 11.
Mean reported value in µg/L is 11.
Precision as RSD % is 40.

SAMPLING METHOD Samples should be collected in borosilicate glass, linear polyethylene, polypropylene, or Teflon® bottles that have been prewashed with detergent and tap water, and rinsed with 1:1 nitric acid and tap water or 1:1 hydrochloric acid and tap water. Collect at least 2 g of solids and 200 mL of aqueous samples.

SAMPLE PRESERVATION Add nitric acid to make the samples pH <2.

MHT The maximum holding time for properly preserved samples is 6 months.

SAMPLE PREPARATION Preliminary treatment of most matrices is necessary because of the complexity and variability of sample matrices. Water samples that have been prefiltered and acidified will not need acid digestion. Methods for acid digestion of waters for total recoverable or dissolved metals, acid digestions of aqueous samples and extracts for total metals, and acid digestion of sediments, sludges, and soils are summarized below.

Total recoverable or dissolved metals in water — To prepare surface and groundwater samples for determination of total recoverable and dissolved metals, a 100-mL aliquot of well-mixed sample is acidified with concentrated nitric acid and concentrated hydrochloric acid, then heated until the volume is reduced to 15–20 mL. Adjust the final volume to 100 mL with reagent water.

Total metals in aqueous samples, soil and sediment extracts — To prepare aqueous samples, soil and sediment extracts, and wastes that contain suspended solids, a 100-mL aliquot is made acidic with concentrated nitric acid and the solution is evaporated to about 5 mL on a hot plate. Continue heating and adding additional acid until sample digestion is complete, which is usually indicated when the digestate is light in color or does not change in appearance. Evaporate the solution to about 3 mL and cool it and add a small quantity of 1:1 hydrochloric acid (10 mL/100 mL of final solution). Cover the beaker and reflux for 15 min. Wash down the beaker walls and filters or centrifuge the sample to remove silicates and other insoluble material. Filter the sample and adjust the final volume to 100 mL with reagent water and the final acid concentration to 10%.

Sediments, sludges, and soils — To prepare sediments, sludges and soil samples, transfer 1–2 g to a conical beaker and add 10 mL of 1:1 nitric acid, mix the slurry, and cover it with a watch glass. Heat the sample and reflux for 10–15 min without boiling. Allow it to cool, then add 5 mL of concentrated nitric acid and reflux for 30 min. Repeat last step and then allow the solution to evaporate to 5 mL without boiling. Cool and add 2 mL of water and 3 mL of 30% hydrogen peroxide. Cover and place the beaker on the hot plate. Heat and add 30% hydrogen peroxide in 1-mL aliquots with warming until the effervescence is minimal but do not add more than a total of 10 mL of 30% hydrogen peroxide. If the sample is being prepared for the analysis of Ag, Al, As, Ba, Be, Ca, Cd, Co, Cr, Cu, Fe, K, Mg, Mn, Mo, Na, Ni, Os, Pb, Se, Tl, V, and Zn, then add 5 mL of concentrated hydrochloric acid and 10 mL of water and return the covered beaker to a hot plate for 15 min of additional refluxing without boiling. Dilute the sample to a 100 mL volume with water after cooling and filter or centrifuge to remove particulates.

QUALITY CONTROL Laboratory control samples must be analyzed for each analytical method. A method blank should be analyzed with each batch of samples. The effect of the matrix on method performance must be demonstrated: when appropriate, there should be at least one matrix spike and either one matrix duplicate or one matrix spike duplicate per analytical batch. The bias and precision of the method, as well as the method detection limit for each specific matrix type, must be measured.

Dilute and reanalyze samples that are more concentrated than the linear calibration limit. Employ a minimum of one reagent blank per sample batch to determine if contamination or any memory effects are occurring. Whenever a new or unusual sample matrix is encountered, perform either a serial dilution test or a matrix spike addition test to ensure that neither positive or negative interferences are operating on any of the analyte elements. Check the instrument standardization by verifying calibration every 10 samples using a calibration blank and a check standard.

REFERENCE Test Methods for Evaluating Solid Waste (SW-846). U.S. EPA. 1983. Method 6010, Rev. 0, Sept. 1986. Office of Solid Wastes, Washington, DC.

Copper   **EPA Method 200.7**
CAS #7440-50-8

TITLE Inductively Coupled Plasma

MATRIX Dissolved, suspended or (ICP) total element in drinking and surface waters and in domestic and industrial wastewaters.

APPLICATION The method covers the determination of 25 metals. Dissolved elements are determined in filtered and acidified samples after appropriate digestion (which increases dissolved solids). Its primary advantage is that ICP instruments allow simultaneous or rapid sequential determination of many elements in a short time. Samples are first nebulized and the aerosol is transported to a plasma torch in which element specific atomic line emission spectra are produced by a radio frequency inductively coupled plasma. Background correction is required for trace element detection except in the case of line broadning.

INTERFERENCES There are spectral, physical, and chemical interferences. The primary disadvantage of ICP instruments is background radiation from other elements and the plasma gases (spectral interferences). Changes in sample viscosity and surface tension with samples containing high dissolved solids (especially those exceeding 1500 mg/L) or high acid concentrations can cause physical interferences. Ionization effects, solute vaporization and molecular compound formation can cause chemical interferences. Iron, thallium and vanadium can cause interference at the 100 mg/L level.

INSTRUMENTATION Inductively coupled argon plasma emission spectroscopy. 324.754 nm Wavelength

RANGE Not listed.

MDL 6 µg/L.

PRECISION SD = 5.1% Mean at true value 250 µg/L.

ACCURACY Mean recovery = 93% ± 6% of spiked elements for all wastes.

SAMPLING METHOD
Wash sample container with detergent and tap water, rinse with 1 + 1 nitric acid and tap water, then rinse with 1 + 1 hydrochloric acid and tap water, then rinse with deionized, distilled water in that order. Perform any filtration or acid preservation steps when the sample is collected or as soon as possible thereafter.

STABILITY Cool samples to 4°C.

MHT 24 h.

QUALITY CONTROL Mixed calibration standards, an instrument check standard, and an interference check solution are used in addition to a quality control sample. The quality control sample should be prepared in the same acid matrix as the calibration standards at 10 times the instrumental detection limits and in accordance with the instructions provided by the supplier. Furthermore, two types of blanks are required: a calibration blank and a reagent blank.

REFERENCE Method 200.7, U.S. EPA, EMSL-Cincinnati, OH, Nov. 1980

Copper   **EPA Method 7210**
CAS #7440-50-8

TITLE Atomic Absorption, (AA)

MATRIX Drinking, Surface and Direct Aspiration Saline Waters, Wastewater

APPLICATION Sample is aspirated and atomized in a flame. A light beam from a Cu hollow cathode lamp is directed through the flame into a monochromator and onto a detector. Since wavelength of light beam is specific for Cu, light energy absorbed by detector is measure of copper.

INTERFERENCES The most troublesomee type is chemical, caused by lack of absorption of atoms bound in molecular combination in the flame. High dissolved solids in sample may result in nonatomic absorbance interference. Non specific absorption and scattering may interfere.

INSTRUMENTATION Atomic absorption spectrometer. Copper hollow cathode lamp. (324.7 nm Wavelength).

RANGE 0.2–5 mg/L

MDL 0.02 mg/L

PRECISION Standard deviation = 56 µg/L at 332 µg/L (true value) 92 labs

ACCURACY As bias = -2.4% at 332 µg/L (true value) 92 labs

SAMPLING METHOD Use glass or plastic containers. Collect 200 g of solids and 600 mL of liquid samples.

STABILITY Cool solid samples to 4°C and analyze as soon as possible. Add nitric acid to liquid samples to pH <2.

MHT 6 months.

QUALITY CONTROL At least one duplicate and one spike sample should be run every 20 samples or with each matrix type to verify precision of the method. For 20 or more samples per day, verify working standard curve. Run an additional standard at or near mid-range every 10 samples.

REFERENCE Method 7210, SW-846, 3rd ed., Nov.1986.

Copper
CAS #7440-50-8
EPA Method 7211

TITLE Atomic Absorption, (AA)

MATRIX Wastes, mobility procedure. Furnace Technique extracts, soils and groundwater.

APPLICATION Aqueous samples, EP extracts, industrial wastes, soils, sludges, sediments, and solid wastes require digestion before analysis. An aliquot of sample is placed in the graphite tube in the furnace and slowly evaporated, charred and atomized. Absorption of lamp radiation during atomization is proportional to copper concentration.

INTERFERENCES The furnace technique is subject to chemical interferences. Composition of sample matrix can have major effect on analysis. Modify matrix to remove interferences. Background correction may be required. Nonspecific absorption and scattering may be significant

INSTRUMENTATION Atomic absorption spectrometer. Copper hollow cathode lamp or electrodeless discharge lamp. Graphite furnace. Strip-chart recorder

RANGE 5–100 µg/L

MDL 1 µg/L (324.7 nm Wavelength)

PRECISION Not listed.

ACCURACY Not listed.

SAMPLING METHOD Use glass or plastic containers. Collect 200 g of solids and 600 mL of liquid samples.

STABILITY Cool solid samples to 4°C and analyze as soon as possible. Add nitric acid to liquid samples to pH <2.

MHT 6 months.

QUALITY CONTROL At least one duplicate and one spike sample should be run every 20 samples, or with each matrix type to verify method precision. If 20 or more samples are run a day, run a standard (at or near mid-range) every 10 samples.

REFERENCE Method 7211, SW-846, 3rd ed., (Included as Rev. 0, Dec. 1987)

Coumaphos
CAS #56-72-4
EPA Method 8141

TITLE Organophosphorus Compounds by Gas Chromatography: Capillary Column Technique

MATRIX This method covers aqueous and solid matrices. This includes a wide variety such as drinking water, groundwater, industrial wastewaters, surface waters, soils, solids, and sediments.

METHOD SUMMARY This is a GC method used to determine the concentration of 28 organophosphorus pesticides. The use of Gel Permeation Cleanup (EPA Method 3640) for sample cleanup has been demonstrated to yield recoveries of less than 85% for many method analytes and is therefore not recommended for use with this method.

This method provides GC conditions for the detection of ppb concentrations of organophosphorus compounds. Prior to the use of this method, appropriate sample preparation techniques must be used. Water samples are extracted at a neutral pH with methylene chloride as a solvent by using a separatory funnel (EPA Method 3510) or a continuous liquid-liquid extractor (EPA Method 3520). Soxhlet extraction (EPA Method 3540) or ultrasonic extraction (EPA Method 3550) using methylene chloride/acetone (1:1) are used for solid samples. Both neat and diluted organic liquids (EPA Method 3580) may be analyzed by direct injection. Spiked samples are used to verify the applicability of the chosen extraction technique to each new sample type. A GC with a flame photometric (FPD) or nitrogen-phosphorus detector (NPD) is used for this multiresidue procedure.

INTERFERENCES The use of Florisil cleanup materials (EPA Method 3620) for some of the compounds in this method has been demonstrated to yield recoveries less than 85% and is therefore not recommended for all compounds. Use of phosphorus or halogen specific detectors, however, often obviates the necessity for cleanup for relatively clean sample matrices. If particular circumstances demand the use of an alternative cleanup procedure, the analyst must determine the elution profile and demonstrate that the recovery of each analyte is no less than 85%.

Use of a flame photometric detector (FPD) in the phosphorus mode will minimize interferences from materials that do not contain phosphorus. Elemental sulfur, however, may interfere with the determination of certain organophosphorus compounds by flame photometric gas chromatography. Sulfur cleanup using EPA Method 3660 may alleviate this interference. A nitrogen phosphorus detector (NPD) is also recommended.

A few analytes coelute on certain columns. Therefore, select a second column for confirmation where coelution of the analytes of interest does not occur.

Method interferences may be caused by contaminants in solvents, reagents, glassware, and other sample processing hardware that lead to discrete artifacts or elevated baselines in gas chromatograms. All these materials must be routinely demonstrated to be free from interferences under the conditions of the analysis by analyzing reagent blanks.

INSTRUMENTATION A GC with a NPD or a FPD will be needed. A data system or integrator is recommended for measuring peak areas and/or peak heights. A Kuderna-Danish (K-D) apparatus will be needed for extract concentration.

Column 1: 15 m × 0.53 mm megabore capillary column, 1.0 µm film thickness, DB-210.
Column 2: 15 m × 0.53 mm megabore capillary column, 1.5 µm film thickness, SPB-608.
Column 3: 15 m × 0.53 mm megabore capillary column, 1.0 µm film thickness, DB-5.

Three megabore capillary columns are included for analysis of organophosphates by this method. Column 1 (DB-210 or equivalent) and Column 2 (SPB-608 or equivalent) are recommended if a large number of organophosphorus analytes are to be determined. If the superior resolution offered by Column 1 and Column 2 is not required, Column 3 (DB-5 or equivalent) may be used. For megabore capillary columns, automatic injections of 1 µL are recommended.

PRECISION & ACCURACY The MDL actually achieved in a given analysis will vary, as it is dependent on instrument sensitivity and matrix effects. Single operator accuracy and precision studies have been conducted with spiked water and soil samples.

MULTIPLICATION FACTORS FOR OTHER MATRICES (a)

Matrix	Factor (b)
Groundwater (EPA Method 3510 or EPA Method 3520)	10
Low-concentration soil by Soxhlet and no cleanup	10 (c)
Low-concentration soil by ultrasonic extraction with GPC cleanup	6.7 (c)
High-concentration soil and sludges by ultrasonic extraction	500 (c)
Non-Water miscible waste (EPA Method 3580)	1000 (c)

(a) Sample EQLs are highly matrix-dependent. The EQLs listed here are provided for guidance and may not always be achievable.
(b) EQL = [Method detection limit] × [Factor]. For non-aqueous samples the factor is on a wet-weight basis.
(c) Multiply this factory times the soil MDL.

The MDL (in µg/L) when reagent water was extracted using a separatory funnel was 0.20.
The MDL (in µg/kg) when soil was extracted using Soxhlet extraction (EPA Method 3540) was 10.0.
Accuracy (as % recovery) with separatory funnel extraction ranged from 103 (with low spikes) to 96 (with high spikes).
Accuracy (as % recovery) with continuous liquid-liquid extraction ranged from 94 (with low spikes) to 89 (with high spikes).
Accuracy (as % recovery) with Soxhlet extraction of soils ranged from 93 (with low spikes to 90 (with high spikes).
Accuracy (as % recovery) with ultrasonic extraction of soils ranged from not recovered (with low spikes) to 15 (with high spikes).

SAMPLE COLLECTION, PRESERVATION & HANDLING Containers used to collect samples for the determination of semivolatile organic compounds should be soap and water washed followed by methanol (or isopropanol) rinsing. The sample containers should be of glass or Teflon® and have screw-top covers with Teflon® liners.

No preservation is used with concentrated waste samples. With liquid samples containing no residual chlorine and with soil, sediment, and sludge samples, immediately cooling to 4°C is the only preservation used. When residual chlorine is present then 3 mL of 10% aqueous sodium sulfate is added for each gallon of sample collected, followed by cooling to 4°C.

Liquid samples must be extracted within 7 days and their extracts analyzed within 40 days. Concentrated waste, soil, sediment, and sludge samples must be extracted within 14 days and their extracts analyzed within 40 days.

SAMPLE PREPARATION In general, water samples are extracted at a neutral pH with methylene chloride, using either EPA Method 3510 or EPA Method 3520. Solid samples are extracted using either EPA Method 3540 or EPA Method 3550 with methylene chloride/acetone (1:1) as the extraction solvent.

Prior to GC analysis, the extraction solvent may be exchanged to hexane. Single lab data indicates that samples should not be transferred with 100% hexane during sample workup as the more water soluble organophosphorus compounds may be lost.

If cleanup is performed on the samples, the analyst should analyze the samples by GC. This will confirm elution patterns and the absence of interferences from the reagents. If peak detection and identification is prevented by the presence of interferences, further cleanup is required.

QUALITY CONTROL The analyst should monitor the performance of the extraction, cleanup (when used), and analytical system and the effectiveness of the method in dealing with each sample matrix by spiking each sample, standard, and blank with one or two surrogates (e.g., organophosphorus compounds not expected to be present in the sample). Deuterated analogs of analytes should not be used as surrogates for gas chromatographic analysis due to coelution problems.

A minimum of five concentrations for each analyte of interest should be prepared through dilution of the stock standards with isooctane. One of the concentrations should be at a concentration near, but above, the MDL.

Include a mid-level check standard after each group of 10 samples in the analysis sequence. GC/MS techniques should be judiciously employed to support qualitative identifications made with this method. Follow the GC/MS operating requirements specified in EPA Method 8270.

When available, chemical ionization mass spectra may be employed to aid in the qualitative identification process. To confirm an identification of a compound, the background-corrected mass spectrum of the compound must be obtained from the sample extract and must be compared with a mass spectrum from a stock or calibration standard analyzed under the same chromatographic conditions. The molecular ion and all other ions present above 20% relative abundance in the mass spectrum of the standard must be present in the mass spectrum of the sample with agreement to ± 20%. The retention time of the compound in the sample must be within six seconds of the retention time for the same compound in the standard solution.

Should the MS procedure fail to provide satisfactory results, additional steps may be taken before reanalysis. These steps may include the use of alternate packed or capillary GC columns or additional sample cleanup.

REFERENCE Test Methods for Evaluating Solid Waste, Physical/Chemical Methods, SW-846, 3rd Edition, U.S. EPA, Office of Solid Waste, Washington, DC, EPA Method 8141 July 1992.

Coumaphos
CAS #56-72-4

EPA Method 8270

TITLE Semivolatile Organic Compounds by GC/MS

MATRIX This method is used to determine the concentration of semivolatile organic compounds in extracts prepared from all types of solid waste matrices, soils, and groundwater. Although surface waters are not specifically mentioned, this method should be applicable to water samples from rivers, lakes, etc.

METHOD SUMMARY This method covers 259 semivolatile organic compounds. In very limited applications direct injection of the sample into the GC/MS system may be appropriate, but this results in very high detection limits (approximately 10,000 µg/L). Typically, a 1-L liquid sample, containing surrogate, and matrix spiking standards, is extracted in a continuous extractor first under acid conditions and then under basic conditions. Typically 30 g of a solid sample, containing surrogate, and matrix spiking standards, is extracted ultrasonically. After concentrating the extract to 1 mL it is spiked with 10 µL of an internal standard solution just prior to analysis by GC/MS. The volume injected should contain about 100 ng of base/neutral and 200 ng of acid surrogates (for a 1-µL injection). Analysis is performed by GC/MS using a capillary GC column.

INTERFERENCES Raw GC/MS data from all blanks, samples, and spikes must be evaluated for interferences. Contamination by carryover can occur whenever high-concentration and low-concentration samples are sequentially analyzed. To reduce carryover, the sample syringe must be rinsed out between samples with solvent. Whenever an unusually concentrated sample is encountered, it should be followed by the analysis of blank solvent to check for cross-contamination.

INSTRUMENTATION A GC/MS and a data system are required. The GC column used is a 30 m × 0.25 mm I.D. (or 0.32 mm I.D.) 1um film thickness silicone-coated fused silica capillary column. A continuous liquid-liquid extractor equipped with Teflon® or glass connection joints and stopcocks requiring no lubrication, a K-D concentrating apparatus, water bath, and an ultrasonic disrupter with a minimum power of 300 W and with pulsing capability are also required.

PRECISION & ACCURACY The estimated quantitation limit (EQL) of Method 8270B for determining an individual compound is approximately 1 mg/kg (wet weight) for soil or sediment samples, 1–200 mg/kg for wastes (dependent on matrix and method of preparation), and 10 µg/L for groundwater samples. EQLs will be proportionately higher for sample extracts that require dilution to avoid saturation of the detector.

The EQL(b) for groundwater in µg/L is 40.
The EQL (a, b) for low concentrations in soil and sediment in µg/kg is not determined.
Accuracy as µg/L is not listed.
Overall precision in µg/L is not listed.

(a) *EQLs listed for soil/sediment are based on wet weight. Normally data is reported in a dry-weight basis; therefore, EQLs will be higher based on the % dry weight of each sample. This calculation is based on a 30 g sample and gel permeation chromatography cleanup.*
(b) *Sample EQLs are highly matrix-dependent. The EQLs are provided for guidance and may not always be achievable.*
C = *True value for concentration, in µg/L.*
X = *Average recovery found for measurements of samples containing a concentration of C, in µg/L.*

ESTIMATED QUANTITATION LIMIT FOR OTHER MATRICES

Other Matrices	Factor (a)
High-concentration soil and sludges by sonicator	7.5
Non-water miscible waste	75

(a) EQL for other matrices = [EQL for low soil/sediment] × [Factor]. This estimated EQL is similar to an EPA "Practical Quantitation Limit."

SAMPLING METHOD

Liquid samples — Use a 1 or 2½ gallon amber glass bottle with a screw-top Teflon®-lined cover that has been prewashed with detergent and rinsed with distilled water and methanol (or isopropanol).

Soils, sediments, or sludges — Use an 8-oz. widemouth glass with a screw-top Teflon®-lined cover that has been prewashed with detergent and rinsed with distilled water and methanol (or isopropanol).

SAMPLE PRESERVATION

Liquid samples — If residual chlorine is present, add 3 mL of 10% sodium thiosulfate per gallon, cool to 4°C and store in a solvent-free refrigerator until analysis; if chlorine is not present, then eliminate the sodium thiosulfate addition.

Soils, sediments, or sludges — Cool samples to 4°C and store in a solvent-free refrigerator.

MHT Liquid samples must be extracted within 7 days and the extracts analyzed within 40 days. Soils, sediments, or sludges may be stored for a maximum of 14 days and the extracts analyzed within 40 days.

SAMPLE PREPARATION

Liquid samples — Transfer 1 L quantitatively to a continuous extractor. If high concentrations are anticipated, a smaller volume may be used and then diluted with organic-free reagent water to 1 L. Adjust pH, if necessary, to pH <2 using 1:1 (V/V) sulfuric acid. Pipette 1.0 mL of a surrogate standard spiking solution into each sample. For the sample in each analytical batch selected for spiking, add 1.0 mL of a matrix spiking standard. For base/neutral acid analysis, the amount of the surrogates and matrix spiking compounds added to the sample should result in a final concentration of 100 ng/µL of each analyte in the extract to be analyzed (assuming a 1-µL injection). Extract with methylene chloride for 18–24 h. Next, adjust the pH of the aqueous phase to pH >11 using 10 N sodium hydroxide and extract it with methylene chloride again for 18–24 h. Dry the extract through a column containing anhydrous sodium sulfate and concentrate it to 1 mL using a K-D concentrator.

Soils, sediments, or sludges — Use 30 g of sample. Nonporous or wet samples (gummy or clay type) that do not have a free-flowing sandy texture must be mixed with anhydrous sodium sulfate until the sample is free flowing. Add 1 mL of surrogate standards to all samples, spikes, standards, and blanks. For the sample in each analytical batch selected for spiking, add 1.0 mL of a matrix spiking standard. For base/neutral acid analysis, the amount added of the surrogates and matrix spiking compounds should result in a final concentration of 100 ng/µL of each base/neutral analyte and 200 ng/µL of each acid analyte in the extract to be analyzed (assuming a 1-µL injection). Immediately add a 100-mL mixture of 1:1 methylene chloride:acetone and extract the sample ultrasonically for 3 min and then decant or filter the extracts. Repeat the extraction two or more times. Dry the extract using a column with anhydrous sodium sulfate and concentrate it to 1 mL in a K-D concentrator.

QUALITY CONTROL A methylene chloride solution containing 50 ng/µL of decafluorotriphenylphosphine (DFTPP) is used for tuning the GC/MS system each 12-h shift. A system performance check also must be made during every 12-h shift. A standard containing 50 ng/µL each of 4,4'-DDT, pentachlorophenol, and benzidine is required to verify injection port inertness and GC column performance. A calibration standard at mid-concentration, containing each compound of interest, including all required surrogates, must be performed every 12 h during analysis. After the system performance check is met, calibration check compounds (CCCs) are used to check the validity of the initial calibration.

The internal standard responses and retention times in the calibration check standard must be evaluated immediately after or during data acquisition. If the retention time for any internal standard changes by more than 30 seconds from the last check calibration (12 h), the chromatographic system must be inspected for malfunctions and corrections must be made, as required. If the electron ionization current plot (EICP) area for any of the internal standards changes by a factor of two from the last daily calibration standard check, the mass spectrometer must be inspected for malfunctions and corrections must be made, as appropriate.

Demonstrate, through the analysis of a reagent water blank, that interferences from the analytical system, glassware, and reagents are under control. The blank samples should be carried through all stages of the sample preparation and measurement steps. For each analytical batch (up to 20 samples), a reagent blank, matrix spike, and matrix spike duplicate/duplicate must be analyzed (the frequency of the spikes may be different for different monitoring programs). The blank and spiked samples must be carried through all stages of the sample preparation and measurement steps. A QC reference sample concentrate containing each analyte at a concentration of 100 mg/L in methanol is required.

REFERENCE Test Methods for Evaluating Solid Waste (SW-846). U.S. EPA 1983, Method 8270B, Rev. 2, Nov. 1990. Office of Solid Waste, Washington, DC.

Coumaphos   EPA Method 8140
CAS #56-72-4

TITLE Organophosphorus Pesticides

MATRIX Groundwater, soils, sludges, water miscible liquid wastes, and non-water miscible wastes.

APPLICATION This method is used for the analysis of 21 organophosphorus pesticides. Samples are extracted, concentrated, and analyzed using direct injection of both neat and diluted organic liquid into a gas chromatograph (GC).

INTERFERENCES Solvents, reagents, and glassware may introduce artifacts. Other interferences may come from coextracted compounds from samples. The use of Florisil cleanup materials may produce low recoveries. Elemental sulfur may interfere with some compounds when using a flame photometric detector. Sulfur cleanup (Method 3660) may alleviate sulfur interference.

INSTRUMENTATION GC capable of on-column injections and a flame photometric detector (FPD) or a thermionic detector. Column 1: 1.8 m by 2 mm with 5% SP-2401 on Supelcoport. Column 2: 1.8 m by 2 mm with 3% SP-2401 on Supelcoport. Column 3: 50 cm by ⅛ in Teflon® with 15% SE-54 on Gas Chrom Q. The preferred column is Column Number 1.

RANGE 25 to 225 µg/L

MDL 1.5 µg/L (in reagent water).

PQL FACTORS FOR MULTIPLYING × FID MDL VALUE

Matrix	Multiplication Factor
Groundwater	10
Low-level soil by sonication with GPC cleanup	670
High-level soil and sludge by sonication	10,000
Non-water miscible waste	100,000

PRECISION 12.7% (single operator standard deviation)

ACCURACY 109% (single operator average recovery)

SAMPLING METHOD Use 8-oz. widemouth glass bottles with Teflon®-lined caps for concentrated waste samples, soils, sediments, and sludges. Use 1 or 2½ gallon amber glass bottles with Teflon®-lined caps for liquid (water) samples.

STABILITY Cool soil, sediment, sludge, and liquid samples to 4°C. If residual chlorine is present in liquid samples add 3 mL of 10% sodium thiosulfate per gallon of sample and cool to 4°C.

MHT 14 days for concentrated waste, soil, sediment, or sludge; 7 days for liquid samples. All extracts must be analyzed within 40 days.

QUALITY CONTROL A quality control check sample concentrate containing this compound in acetone at a concentration 1,000 times more concentrated than the selected spike concentration is required. The QC check sample concentrate may be prepared from pure standard materials or purchased as certified solutions. Use appropriate trip, matrix, control site, method, reagent, and solvent blanks. Internal, surrogate, and five concentration level calibration standards are used.

REFERENCE   Method 8140, SW-846, 3rd ed., Sept. 1986.

---

p-Cresidine **EPA Method 8270**
CAS #120-71-8

TITLE   Semivolatile Organic Compounds by GC/MS

MATRIX   This method is used to determine the concentration of semivolatile organic compounds in extracts prepared from all types of solid waste matrices, soils, and groundwater. Although surface waters are not specifically mentioned, this method should be applicable to water samples from rivers, lakes, etc.

METHOD SUMMARY   This method covers 259 semivolatile organic compounds. In very limited applications direct injection of the sample into the GC/MS system may be appropriate, but this results in very high detection limits (approximately 10,000 µg/L). Typically, a 1-L liquid sample, containing surrogate, and matrix spiking standards, is extracted in a continuous extractor first under acid conditions and then under basic conditions. Typically 30 g of a solid sample, containing surrogate, and matrix spiking standards, is extracted ultrasonically. After concentrating the extract to 1 mL it is spiked with 10 µL of an internal standard solution just prior to analysis by GC/MS. The volume injected should contain about 100 ng of base/neutral and 200 ng of acid surrogates (for a 1-µL injection). Analysis is performed by GC/MS using a capillary GC column.

INTERFERENCES   Raw GC/MS data from all blanks, samples, and spikes must be evaluated for interferences. Contamination by carryover can occur whenever high-concentration and low-concentration samples are sequentially analyzed. To reduce carryover, the sample syringe must be rinsed out between samples with solvent. Whenever an unusually concentrated sample is encountered, it should be followed by the analysis of blank solvent to check for cross-contamination.

INSTRUMENTATION   A GC/MS and a data system are required. The GC column used is a 30 m × 0.25 mm I.D. (or 0.32 mm I.D.) 1um film thickness silicone-coated fused silica capillary column. A continuous liquid-liquid extractor equipped with Teflon® or glass connection joints and stopcocks requiring no lubrication, a K-D concentrating apparatus, water bath, and an ultrasonic disrupter with a minimum power of 300 W and with pulsing capability are also required.

PRECISION & ACCURACY   The estimated quantitation limit (EQL) of Method 8270B for determining an individual compound is approximately 1 mg/kg (wet weight) for soil or sediment samples, 1–200 mg/kg for wastes (dependent on matrix and method of preparation), and 10 µg/L for groundwater samples. EQLs will be proportionately higher for sample extracts that require dilution to avoid saturation of the detector.

The EQL(b) for groundwater in µg/L is 10.
The EQL (a, b) for low concentrations in soil and sediment in µg/kg is not determined.

Accuracy as µg/L is not listed.
Overall precision in µg/L is not listed.

(a)   EQLs listed for soil/sediment are based on wet weight. Normally data is reported in a dry-weight basis; therefore, EQLs will be higher based on the % dry weight of each sample. This calculation is based on a 30 g sample and gel permeation chromatography cleanup.
(b)   Sample EQLs are highly matrix-dependent. The EQLs are provided for guidance and may not always be achievable.
$C =$ True value for concentration, in µg/L.
$X =$ Average recovery found for measurements of samples containing a concentration of C, in µg/L.

ESTIMATED QUANTITATION LIMIT FOR OTHER MATRICES

Other Matrices	Factor (a)
High-concentration soil and sludges by sonicator	7.5
Non-water miscible waste	75

(a) EQL for other matrices = [EQL for low soil/sediment] × [Factor]. This estimated EQL is similar to an EPA "Practical Quantitation Limit."

SAMPLING METHOD
Liquid samples — Use a 1 or 2½ gallon amber glass bottle with a screw-top Teflon®-lined cover that has been prewashed with detergent and rinsed with distilled water and methanol (or isopropanol).

Soils, sediments, or sludges — Use an 8-oz. widemouth glass with a screw-top Teflon®-lined cover that has been prewashed with detergent and rinsed with distilled water and methanol (or isopropanol).

SAMPLE PRESERVATION
Liquid samples — If residual chlorine is present, add 3 mL of 10% sodium thiosulfate per gallon, cool to 4°C and store in a solvent-free refrigerator until analysis; if chlorine is not present, then eliminate the sodium thiosulfate addition.

Soils, sediments, or sludges — Cool samples to 4°C and store in a solvent-free refrigerator.

MHT   Liquid samples must be extracted within 7 days and the extracts analyzed within 40 days. Soils, sediments, or sludges may be stored for a maximum of 14 days and the extracts analyzed within 40 days.

SAMPLE PREPARATION
Liquid samples — Transfer 1 L quantitatively to a continuous extractor. If high concentrations are anticipated, a smaller volume may be used and then diluted with organic-free reagent water to 1 L. Adjust pH, if necessary, to pH <2 using 1:1 (V/V) sulfuric acid. Pipette 1.0 mL of a surrogate standard spiking solution into each sample. For the sample in each analytical batch selected for spiking, add 1.0 mL of a matrix spiking standard. For base/neutral acid analysis, the amount of the surrogates and matrix spiking compounds added to the sample should result in a final concentration of 100 ng/µL of each analyte in the extract to be analyzed (assuming a 1-µL injection). Extract with methylene chloride for 18–24 h. Next, adjust the pH of the aqueous phase to pH >11 using 10 N sodium

hydroxide and extract it with methylene chloride again for 18–24 h. Dry the extract through a column containing anhydrous sodium sulfate and concentrate it to 1 mL using a K-D concentrator.

Soils, sediments, or sludges — Use 30 g of sample. Nonporous or wet samples (gummy or clay type) that do not have a free-flowing sandy texture must be mixed with anhydrous sodium sulfate until the sample is free flowing. Add 1 mL of surrogate standards to all samples, spikes, standards, and blanks. For the sample in each analytical batch selected for spiking, add 1.0 mL of a matrix spiking standard. For base/neutral acid analysis, the amount added of the surrogates and matrix spiking compounds should result in a final concentration of 100 ng/µL of each base/neutral analyte and 200 ng/µL of each acid analyte in the extract to be analyzed (assuming a 1-µL injection). Immediately add a 100-mL mixture of 1:1 methylene chloride:acetone and extract the sample ultrasonically for 3 min and then decant or filter the extracts. Repeat the extraction two or more times. Dry the extract using a column with anhydrous sodium sulfate and concentrate it to 1 mL in a K-D concentrator.

QUALITY CONTROL A methylene chloride solution containing 50 ng/µL of decafluorotriphenylphosphine (DFTPP) is used for tuning the GC/MS system each 12-h shift. A system performance check also must be made during every 12-h shift. A standard containing 50 ng/µL each of 4,4′-DDT, pentachlorophenol, and benzidine is required to verify injection port inertness and GC column performance. A calibration standard at mid-concentration, containing each compound of interest, including all required surrogates, must be performed every 12 h during analysis. After the system performance check is met, calibration check compounds (CCCs) are used to check the validity of the initial calibration.

The internal standard responses and retention times in the calibration check standard must be evaluated immediately after or during data acquisition. If the retention time for any internal standard changes by more than 30 seconds from the last check calibration (12 h), the chromatographic system must be inspected for malfunctions and corrections must be made, as required. If the electron ionization current plot (EICP) area for any of the internal standards changes by a factor of two from the last daily calibration standard check, the mass spectrometer must be inspected for malfunctions and corrections must be made, as appropriate.

Demonstrate, through the analysis of a reagent water blank, that interferences from the analytical system, glassware, and reagents are under control. The blank samples should be carried through all stages of the sample preparation and measurement steps. For each analytical batch (up to 20 samples), a reagent blank, matrix spike, and matrix spike duplicate/duplicate must be analyzed (the frequency of the spikes may be different for different monitoring programs). The blank and spiked samples must be carried through all stages of the sample preparation and measurement steps. A QC reference sample concentrate containing each analyte at a concentration of 100 mg/L in methanol is required.

REFERENCE Test Methods for Evaluating Solid Waste (SW-846). U.S. EPA 1983, Method 8270B, Rev. 2, Nov. 1990. Office of Solid Waste, Washington, DC.

---

o-Cresol EPA Method 1625
CAS #95-48-7

TITLE Semivolatile Organic Compounds by Isotope Dilution GC/MS

MATRIX The compounds may be determined in waters, soils, and municipal sludges by this method.

METHOD SUMMARY This method is used to determine 176 semivolatile toxic organic pollutants associated with the CWA (as amended 1987); the RCRA (as amended 1986); the CERCLA (as amended 1986); and other compounds amenable to extraction and analysis by capillary column gas chromatography-mass spectrometry (GC/MS).

Stable isotopically-labeled analogs of the compounds of interest are added to the sample. If the solids content is less than 1%, a 1-L sample is extracted at pH 12–13, then at pH <2 with methylene chloride using continuous extraction techniques.

If the solids content is 30% or less, the sample is diluted to 1% solids with reagent water, homogenized ultrasonically, and extracted at pH 12–13, then at pH <2 with methylene chloride using continuous extraction techniques. If the solids content is greater than 30%, the sample is extracted using ultrasonic techniques.

Each extract is dried over sodium sulfate, concentrated to a volume of 5 mL, cleaned up using GPC, if necessary, and concentrated. Extracts are concentrated to 1 mL if GPC is not performed, and to 0.5 mL if GPC is performed.

An internal standard is added to the extract, and a 1-mL aliquot of the extract is injected into the GC. The compounds are separated by GC and detected by a MS. The labeled compounds serve to correct the variability of the analytical technique.

INTERFERENCES Solvents, reagents, glassware, and other sample processing hardware may yield artifacts and/or elevated baselines causing misinterpretation of chromatograms and spectra. Materials used in the analysis must be demonstrated to be free from interferences under the conditions of analysis by running method blanks initially and with each sample lot (sample started through the extraction process on a given 8-h shift, to a maximum of 20). Specific selection of reagents and purification of solvents by distillation in all glass systems may be required. Glassware and, where possible, reagents are cleaned by solvent rinse and baking at 450°C for 1-h minimum. Interferences coextracted from samples will vary considerably from source to source, depending on the diversity of the site being sampled.

INSTRUMENTATION Major instrumentation includes a GC with a splitless or on-column injection port for capillary column, a MS with 70 eV electron impact ionization, and a data system to collect and record MS data, and process it. A K-D apparatus is used to concentrate extracts.

GC Column: 30 m × 0.25 mm I.D. 5% phenyl, 94% methyl, 1% vinyl silicone bonded phased fused silica capillary column.

PRECISION & ACCURACY The detection limits of the method are usually dependent on the level of interferences rather than instrumental limitations. The limits typify the minimum quantities that can be detected with no interferences present.

The minimum level (in µg/mL) was not listed. This is defined as a minimum level at which the analytical system shall give recognizable mass spectra (background corrected) and acceptable calibration points.

The MDL (in µg/kg) in low solids was not listed and in high solids was not listed; these were determined in digested sludge (low solids) and in filter cake or compost (high solids).

The labeled and native compound initial precision as standard deviation (in µg/L) was not listed.

The labeled and native compound initial accuracy as average recovery (in µg/L) was not listed.

SAMPLE COLLECTION, PRESERVATION & HANDLING Collect samples in glass containers. Aqueous samples which flow freely are collected in refrigerated bottles using automatic sampling equipment. Solid samples are collected as grab samples using widemouth jars. Maintain samples at 0 to 4°C from the time of collection until extraction. If residual chlorine is present in aqueous samples, add 80 mg sodium thiosulfate/L of water. Begin sample extraction within 7 days of collection, and analyze all extracts within 40 days of extraction.

SAMPLE PREPARATION Samples containing 1% solids or less are extracted directly using continuous liquid-liquid extraction techniques. Samples containing 1 to 30% solids are diluted to the 1% level with reagent water and extracted using continuous liquid-liquid extraction techniques. Samples containing greater than 30% solids are extracted using ultrasonic techniques.

Base/neutral extraction — Adjust the pH of the waters in the extractors to 12–13 with 6 N NaOH. Extract with methylene chloride for 24–48 h.

Acid extraction — Adjust the pH of the waters in the extractors to 2 or less using 6 N sulfuric acid. Extract with methylene chloride for 24–48 h.

Ultrasonic extraction of high solids samples — Add anhydrous sodium sulfate to the sample and QC aliquot(s). Add acetone:methylene chloride (1:1) to the sample and mix thoroughly

Concentrate extracts using a K-D apparatus.

QUALITY CONTROL The analyst is permitted to modify this method to improve separations or lower the costs of measurements, provided all performance specifications are met. Analyses of blanks are required to demonstrate freedom from contamination. When results of spikes indicate atypical method performance for samples, the samples are diluted to bring method performance within acceptable limits.

For low solids (aqueous samples), extract, concentrate, and analyze two sets of four 1-L aliquots (8 aliquots total) of the precision and recovery standard. For high solids samples, two sets of four 30-g aliquots of the high solids reference matrix are used.

Spike all samples with labeled compounds to assess method performance. Compute percent recovery of the labeled compounds using the internal standard method. Compare the labeled compound recovery for each compound with the corresponding labeled compound recovery.

Reagent water and high solids reference matrix blanks are analyzed to demonstrate freedom from contamination. Extract and concentrate a 1-L reagent water blank or a high solids reference matrix blank with each sample's lot (samples started through the extraction process on the same 8-h shift, to a maximum of 20 samples).

Field replicates may be collected to determine the precision of the sampling technique, and spiked samples may be required to determine the accuracy of the analysis when the internal standard method is used.

REFERENCE Semivolatile Organic Compounds by Isotope Dilution GC/MS. Office of Water Regulation and Standards, U.S. EPA Industrial Technology Division, Washington, DC, EPA Method 1625, Rev. C, June 1989 (contact W.A. Telliard, U.S. EPA, Office of Water Regulations and Standards, 401 M St., SW, Washington, DC, 20460. Phone: 202-382-7131).

---

p-Cresol **EPA Method 1625**
CAS #106-44-5

TITLE Semivolatile Organic Compounds by Isotope Dilution GC/MS

MATRIX The compounds may be determined in waters, soils, and municipal sludges by this method.

METHOD SUMMARY This method is used to determine 176 semivolatile toxic organic pollutants associated with the CWA (as amended 1987); the RCRA (as amended 1986); the CERCLA (as amended 1986); and other compounds amenable to extraction and analysis by capillary column gas chromatography-mass spectrometry (GC/MS).

Stable isotopically-labeled analogs of the compounds of interest are added to the sample. If the solids content is less than 1%, a 1-L sample is extracted at pH 12–13, then at pH <2 with methylene chloride using continuous extraction techniques.

If the solids content is 30% or less, the sample is diluted to 1% solids with reagent water, homogenized ultrasonically, and extracted at pH 12–13, then at pH <2 with methylene chloride using continuous extraction techniques. If the solids content is greater than 30%, the sample is extracted using ultrasonic techniques.

Each extract is dried over sodium sulfate, concentrated to a volume of 5 mL, cleaned up using GPC, if necessary, and concentrated. Extracts are concentrated to 1 mL if GPC is not performed, and to 0.5 mL if GPC is performed.

An internal standard is added to the extract, and a 1-mL aliquot of the extract is injected into the GC. The compounds are

separated by GC and detected by a MS. The labeled compounds serve to correct the variability of the analytical technique.

INTERFERENCES Solvents, reagents, glassware, and other sample processing hardware may yield artifacts and/or elevated baselines causing misinterpretation of chromatograms and spectra. Materials used in the analysis must be demonstrated to be free from interferences under the conditions of analysis by running method blanks initially and with each sample lot (sample started through the extraction process on a given 8-h shift, to a maximum of 20). Specific selection of reagents and purification of solvents by distillation in all glass systems may be required. Glassware and, where possible, reagents are cleaned by solvent rinse and baking at 450°C for 1-h minimum. Interferences coextracted from samples will vary considerably from source to source, depending on the diversity of the site being sampled.

INSTRUMENTATION Major instrumentation includes a GC with a splitless or on-column injection port for capillary column, a MS with 70 eV electron impact ionization, and a data system to collect and record MS data, and process it. A K-D apparatus is used to concentrate extracts.

GC Column: 30 m × 0.25 mm I.D. 5% phenyl, 94% methyl, 1% vinyl silicone bonded phased fused silica capillary column.

PRECISION & ACCURACY The detection limits of the method are usually dependent on the level of interferences rather than instrumental limitations. The limits typify the minimum quantities that can be detected with no interferences present.

The minimum level (in µg/mL) was not listed. This is defined as a minimum level at which the analytical system shall give recognizable mass spectra (background corrected) and acceptable calibration points.

The MDL (in µg/kg) in low solids was not listed and in high solids was not listed; these were determined in digested sludge (low solids) and in filter cake or compost (high solids).

The labeled and native compound initial precision as standard deviation (in µg/L) was not listed.

The labeled and native compound initial accuracy as average recovery (in µg/L) was not listed.

SAMPLE COLLECTION, PRESERVATION & HANDLING Collect samples in glass containers. Aqueous samples which flow freely are collected in refrigerated bottles using automatic sampling equipment. Solid samples are collected as grab samples using widemouth jars. Maintain samples at 0 to 4°C from the time of collection until extraction. If residual chlorine is present in aqueous samples, add 80 mg sodium thiosulfate/L of water. Begin sample extraction within 7 days of collection, and analyze all extracts within 40 days of extraction.

SAMPLE PREPARATION Samples containing 1% solids or less are extracted directly using continuous liquid-liquid extraction techniques. Samples containing 1 to 30% solids are diluted to the 1% level with reagent water and extracted using continuous liquid-liquid extraction techniques. Samples containing greater than 30% solids are extracted using ultrasonic techniques.

- Base/neutral extraction — Adjust the pH of the waters in the extractors to 12–13 with 6 N NaOH. Extract with methylene chloride for 24–48 h.
- Acid extraction — Adjust the pH of the waters in the extractors to 2 or less using 6 N sulfuric acid. Extract with methylene chloride for 24–48 h.
- Ultrasonic extraction of high solids samples — Add anhydrous sodium sulfate to the sample and QC aliquot(s). Add acetone:methylene chloride (1:1) to the sample and mix thoroughly

Concentrate extracts using a K-D apparatus.

QUALITY CONTROL The analyst is permitted to modify this method to improve separations or lower the costs of measurements, provided all performance specifications are met. Analyses of blanks are required to demonstrate freedom from contamination. When results of spikes indicate atypical method performance for samples, the samples are diluted to bring method performance within acceptable limits.

For low solids (aqueous samples), extract, concentrate, and analyze two sets of four 1-L aliquots (8 aliquots total) of the precision and recovery standard. For high solids samples, two sets of four 30-g aliquots of the high solids reference matrix are used.

Spike all samples with labeled compounds to assess method performance. Compute percent recovery of the labeled compounds using the internal standard method. Compare the labeled compound recovery for each compound with the corresponding labeled compound recovery.

Reagent water and high solids reference matrix blanks are analyzed to demonstrate freedom from contamination. Extract and concentrate a 1-L reagent water blank or a high solids reference matrix blank with each sample's lot (samples started through the extraction process on the same 8-h shift, to a maximum of 20 samples).

Field replicates may be collected to determine the precision of the sampling technique, and spiked samples may be required to determine the accuracy of the analysis when the internal standard method is used.

REFERENCE Semivolatile Organic Compounds by Isotope Dilution GC/MS. Office of Water Regulation and Standards, U.S. EPA Industrial Technology Division, Washington, DC, EPA Method 1625, Rev. C, June 1989 (contact W.A. Telliard, U.S. EPA, Office of Water Regulations and Standards, 401 M St., SW, Washington, DC, 20460. Phone: 202-382-7131).

Crotonaldehyde   EPA Method 1624
CAS #123-73-9

TITLE Volatile Organic Compounds by Isotope Dilution GC/MS

MATRIX Compounds may be determined in waters, soils, and municipal sludges by this method.

METHOD SUMMARY This method is used to determine 58 volatile toxic organic pollutants associated with the CWA (as amended 1987); the RCRA (as amended 1986); the CERCLA (as amended 1986); and other compounds amenable to purge-and-trap gas chromatography-mass spectrometry (GC/MS).

If the solids content is less than 1%, stable isotopically-labeled analogs of the compounds of interest are added to a 5-mL sample and the sample is purged with an inert gas at 20–25°C in a chamber designed for soil or water samples. If the solids content is greater than 1%, 5 mL of reagent water and the labeled compounds are added to a 5-g aliquot of sample and the mixture is purged at 40°C. Compounds that will not purge at 20–25°C or at 40°C are purged at 78–85°C. In the purging process, the volatile compounds are transferred from the aqueous phase into the gaseous phase where they are passed into a sorbent column, and trapped. After purging is completed, the trap is backflushed and heated rapidly to desorb the compounds into a GC. The compounds are separated by the GC and detected by a MS. The labeled compounds serve to correct the variability of the analytical technique.

INTERFERENCES Impurities in the purge gas, organic compounds outgassing from the plumbing upstream of the trap, and solvent vapors in the lab account for most problems. Samples can be contaminated by diffusion of volatile organic compounds (particularly methylene chloride) through the bottle seal during shipment and storage. Contamination by carryover can occur when high-level and low-level samples are analyzed sequentially. When an unusually concentrated sample is encountered, follow it by analysis of a reagent water blank to check for carryover.

INSTRUMENTATION Major equipment includes a GC with linear temperature programming and a glass jet separator as the MS interface, a MS with 70 eV electron impact ionization, and a data system to collect and record response factors.

Column: 2.8 m × 2 mm I.D. glass, packed with 1% SP-1000 on Carbopak B, 60/80 mesh, or equivalent.

PRECISION & ACCURACY The detection limits of the method are usually dependent on the level of interferences rather than instrumental limitations. The method detection limits were determined in digested sludge (low solids) and in filter cake or compost (high solids).

The MDL (in µg/kg) for low solids is not listed and for high solids is not listed.
Labeled and native compound precision (in µg/L) as standard deviation was not listed.
Labeled and native compound accuracy (in µg/L) as average recovery was not listed.
Acceptance criteria are at 20 µg/L for this compound.

SAMPLE COLLECTION, PRESERVATION & HANDLING
Grab samples are collected in glass containers having a total volume greater than 20 mL. Fill and seal each bottle so that no air bubbles are entrapped. Samples are maintained at 0 to 4°C from the time of collection until analysis. If an aqueous sample contains residual chlorine, add sodium thiosulfate preservative (10 mg/40 mL) to the empty sample bottles just prior to shipment to the sample site. All samples must be analyzed within 14 days of collection.

SAMPLE PREPARATION Samples containing less than 1% solids are analyzed directly as aqueous samples. Samples containing 1% solids or greater are analyzed as solid samples utilizing one of two methods, depending on the levels of pollutants, in the sample. Samples containing 1% solids or greater, and low to moderate levels of pollutants are analyzed by purging a known weight of sample added to 5 mL of reagent water. Samples containing 1% solids or greater, and high levels of pollutants, are extracted with methanol, and an aliquot of the methanol extract is added to reagent water and purged.

QUALITY CONTROL A field blank prepared from reagent water and carried through the sampling and handling protocol may serve as a check on contamination from shipment and storage.

The analyst is permitted to modify this method to improve separations or lower the costs of measurements, provided all performance specifications are met. Analyses of blanks are required. When results of spikes indicate atypical method performance for samples, the samples are diluted to bring method performance within acceptable limits. Analyze two sets of four 5-mL aliquots (8 aliquots total) of the aqueous performance standard. Spike all samples with labeled compounds to assess method performance on the sample matrix. Compute the percent recovery of the labeled compounds using the internal standard method. Compare the percent recovery for each compound with the corresponding labeled compound recovery. Reagent water blanks are analyzed to demonstrate freedom from carryover contamination. Field replicates may be collected to determine the precision of the sampling technique, and spiked samples may be required to determine the accuracy of the analysis when the internal method is used.

REFERENCE Volatile Organic Compounds by Isotope Dilution GC/MS. Office of Water Regulation and Standards, U.S. EPA Industrial Technology Division, Washington, DC, EPA Method 1624, Rev. C, June 1989 (contact W.A. Telliard, U.S. EPA, Office of Water Regulations and Standards, 401 M St., SW, Washington, DC, 20460. Phone: 202-382-7131).

Crotoxyphos   **EPA Method 1625**
CAS #7700-17-6

TITLE Semivolatile Organic Compounds by Isotope Dilution GC/MS

MATRIX The compounds may be determined in waters, soils, and municipal sludges by this method.

METHOD SUMMARY This method is used to determine 176 semivolatile toxic organic pollutants associated with the CWA (as amended 1987); the RCRA (as amended 1986); the CERCLA (as amended 1986); and other compounds amenable to extraction and analysis by capillary column gas chromatography-mass spectrometry (GC/MS).

Stable isotopically-labeled analogs of the compounds of interest are added to the sample. If the solids content is less than 1%, a 1-L sample is extracted at pH 12–13, then at pH <2 with methylene chloride using continuous extraction techniques.

If the solids content is 30% or less, the sample is diluted to 1% solids with reagent water, homogenized ultrasonically, and extracted at pH 12–13, then at pH <2 with methylene chloride using continuous extraction techniques. If the solids content is greater than 30%, the sample is extracted using ultrasonic techniques.

Each extract is dried over sodium sulfate, concentrated to a volume of 5 mL, cleaned up using GPC, if necessary, and concentrated. Extracts are concentrated to 1 mL if GPC is not performed, and to 0.5 mL if GPC is performed.

An internal standard is added to the extract, and a 1-mL aliquot of the extract is injected into the GC. The compounds are separated by GC and detected by a MS. The labeled compounds serve to correct the variability of the analytical technique.

INTERFERENCES Solvents, reagents, glassware, and other sample processing hardware may yield artifacts and/or elevated baselines causing misinterpretation of chromatograms and spectra. Materials used in the analysis must be demonstrated to be free from interferences under the conditions of analysis by running method blanks initially and with each sample lot (sample started through the extraction process on a given 8-h shift, to a maximum of 20). Specific selection of reagents and purification of solvents by distillation in all glass systems may be required. Glassware and, where possible, reagents are cleaned by solvent rinse and baking at 450°C for 1-h minimum. Interferences coextracted from samples will vary considerably from source to source, depending on the diversity of the site being sampled.

INSTRUMENTATION Major instrumentation includes a GC with a splitless or on-column injection port for capillary column, a MS with 70 eV electron impact ionization, and a data system to collect and record MS data, and process it. A K-D apparatus is used to concentrate extracts.

- GC Column: 30 m × 0.25 mm I.D. 5% phenyl, 94% methyl, 1% vinyl silicone bonded phased fused silica capillary column.

PRECISION & ACCURACY The detection limits of the method are usually dependent on the level of interferences rather than instrumental limitations. The limits typify the minimum quantities that can be detected with no interferences present.

- The minimum level (in µg/mL) was not listed. This is defined as a minimum level at which the analytical system shall give recognizable mass spectra (background corrected) and acceptable calibration points.
- The MDL (in µg/kg) in low solids was not listed and in high solids was not listed; these were determined in digested sludge (low solids) and in filter cake or compost (high solids).
- The labeled and native compound initial precision as standard deviation (in µg/L) was not listed.
- The labeled and native compound initial accuracy as average recovery (in µg/L) was not listed.

SAMPLE COLLECTION, PRESERVATION & HANDLING Collect samples in glass containers. Aqueous samples which flow freely are collected in refrigerated bottles using automatic sampling equipment. Solid samples are collected as grab samples using widemouth jars. Maintain samples at 0 to 4°C from the time of collection until extraction. If residual chlorine is present in aqueous samples, add 80 mg sodium thiosulfate/L of water. Begin sample extraction within 7 days of collection, and analyze all extracts within 40 days of extraction.

SAMPLE PREPARATION Samples containing 1% solids or less are extracted directly using continuous liquid-liquid extraction techniques. Samples containing 1 to 30% solids are diluted to the 1% level with reagent water and extracted using continuous liquid-liquid extraction techniques. Samples containing greater than 30% solids are extracted using ultrasonic techniques.

- Base/neutral extraction — Adjust the pH of the waters in the extractors to 12–13 with 6 N NaOH. Extract with methylene chloride for 24–48 h.
- Acid extraction — Adjust the pH of the waters in the extractors to 2 or less using 6 N sulfuric acid. Extract with methylene chloride for 24–48 h.
- Ultrasonic extraction of high solids samples — Add anhydrous sodium sulfate to the sample and QC aliquot(s). Add acetone:methylene chloride (1:1) to the sample and mix thoroughly

Concentrate extracts using a K-D apparatus.

QUALITY CONTROL The analyst is permitted to modify this method to improve separations or lower the costs of measurements, provided all performance specifications are met. Analyses of blanks are required to demonstrate freedom from contamination. When results of spikes indicate atypical method performance for samples, the samples are diluted to bring method performance within acceptable limits.

For low solids (aqueous samples), extract, concentrate, and analyze two sets of four 1-L aliquots (8 aliquots total) of the precision and recovery standard. For high solids samples, two sets of four 30-g aliquots of the high solids reference matrix are used.

Spike all samples with labeled compounds to assess method performance. Compute percent recovery of the labeled compounds using the internal standard method. Compare the labeled compound recovery for each compound with the corresponding labeled compound recovery.

Reagent water and high solids reference matrix blanks are analyzed to demonstrate freedom from contamination. Extract and concentrate a 1-L reagent water blank or a high solids reference matrix blank with each sample's lot (samples started through the extraction process on the same 8-h shift, to a maximum of 20 samples).

Field replicates may be collected to determine the precision of the sampling technique, and spiked samples may be required to determine the accuracy of the analysis when the internal standard method is used.

REFERENCE Semivolatile Organic Compounds by Isotope Dilution GC/MS. Office of Water Regulation and Standards, U.S. EPA Industrial Technology Division, Washington, DC, EPA Method 1625, Rev. C, June 1989 (contact W.A. Telliard, U.S. EPA, Office of Water Regulations and Standards, 401 M St., SW, Washington, DC, 20460. Phone: 202-382-7131).

Crotoxyphos   **EPA Method 8270**
CAS #7700-17-6

TITLE Semivolatile Organic Compounds by GC/MS

MATRIX This method is used to determine the concentration of semivolatile organic compounds in extracts prepared from all types of solid waste matrices, soils, and groundwater. Although surface waters are not specifically mentioned, this method should be applicable to water samples from rivers, lakes, etc.

METHOD SUMMARY This method covers 259 semivolatile organic compounds. In very limited applications direct injection of the sample into the GC/MS system may be appropriate, but this results in very high detection limits (approximately 10,000 µg/L). Typically, a 1-L liquid sample, containing surrogate, and matrix spiking standards, is extracted in a continuous extractor first under acid conditions and then under basic conditions. Typically 30 g of a solid sample, containing surrogate, and matrix spiking standards, is extracted ultrasonically. After concentrating the extract to 1 mL it is spiked with 10 µL of an internal standard solution just prior to analysis by GC/MS. The volume injected should contain about 100 ng of base/neutral and 200 ng of acid surrogates (for a 1-µL injection). Analysis is performed by GC/MS using a capillary GC column.

INTERFERENCES Raw GC/MS data from all blanks, samples, and spikes must be evaluated for interferences. Contamination by carryover can occur whenever high-concentration and low-concentration samples are sequentially analyzed. To reduce carryover, the sample syringe must be rinsed out between samples with solvent. Whenever an unusually concentrated sample is encountered, it should be followed by the analysis of blank solvent to check for cross-contamination.

INSTRUMENTATION A GC/MS and a data system are required. The GC column used is a 30 m × 0.25 mm I.D. (or 0.32 mm I.D.) 1um film thickness silicone-coated fused silica capillary column. A continuous liquid-liquid extractor equipped with Teflon® or glass connection joints and stopcocks requiring no lubrication, a K-D concentrating apparatus, water bath, and an ultrasonic disrupter with a minimum power of 300 W and with pulsing capability are also required.

PRECISION & ACCURACY The estimated quantitation limit (EQL) of Method 8270B for determining an individual compound is approximately 1 mg/kg (wet weight) for soil or sediment samples, 1–200 mg/kg for wastes (dependent on matrix and method of preparation), and 10 µg/L for groundwater samples. EQLs will be proportionately higher for sample extracts that require dilution to avoid saturation of the detector.

The EQL(b) for groundwater in µg/L is 20.

The EQL (a, b) for low concentrations in soil and sediment in µg/kg is not determined.
Accuracy as µg/L is not listed.
Overall precision in µg/L is not listed.

(a) EQLs listed for soil/sediment are based on wet weight. Normally data is reported in a dry-weight basis; therefore, EQLs will be higher based on the % dry weight of each sample. This calculation is based on a 30 g sample and gel permeation chromatography cleanup.
(b) Sample EQLs are highly matrix-dependent. The EQLs are provided for guidance and may not always be achievable.
C = True value for concentration, in µg/L.
X = Average recovery found for measurements of samples containing a concentration of C, in µg/L.

ESTIMATED QUANTITATION LIMIT FOR OTHER MATRICES

Other Matrices	Factor (a)
High-concentration soil and sludges by sonicator	7.5
Non-water miscible waste	75

(a) EQL for other matrices = [EQL for low soil/sediment] × [Factor]. This estimated EQL is similar to an EPA "Practical Quantitation Limit."

SAMPLING METHOD
Liquid samples — Use a 1 or 2½ gallon amber glass bottle with a screw-top Teflon®-lined cover that has been prewashed with detergent and rinsed with distilled water and methanol (or isopropanol).

Soils, sediments, or sludges — Use an 8-oz. widemouth glass with a screw-top Teflon®-lined cover that has been prewashed with detergent and rinsed with distilled water and methanol (or isopropanol).

SAMPLE PRESERVATION
Liquid samples — If residual chlorine is present, add 3 mL of 10% sodium thiosulfate per gallon, cool to 4°C and store in a solvent-free refrigerator until analysis; if chlorine is not present, then eliminate the sodium thiosulfate addition.

Soils, sediments, or sludges — Cool samples to 4°C and store in a solvent-free refrigerator.

MHT Liquid samples must be extracted within 7 days and the extracts analyzed within 40 days. Soils, sediments, or sludges may be stored for a maximum of 14 days and the extracts analyzed within 40 days.

SAMPLE PREPARATION
Liquid samples — Transfer 1 L quantitatively to a continuous extractor. If high concentrations are anticipated, a smaller volume may be used and then diluted with organic-free reagent water to 1 L. Adjust pH, if necessary, to pH <2 using 1:1 (V/V) sulfuric acid. Pipette 1.0 mL of a surrogate standard spiking solution into each sample. For the sample in each analytical batch selected for spiking, add 1.0 mL of a matrix spiking standard. For base/neutral acid analysis, the amount of the surrogates and matrix spiking compounds added to the sample should result in a final concentration of 100 ng/µL of each analyte in the extract to be analyzed (assuming a 1-µL injection). Extract

with methylene chloride for 18–24 h. Next, adjust the pH of the aqueous phase to pH >11 using 10 N sodium hydroxide and extract it with methylene chloride again for 18–24 h. Dry the extract through a column containing anhydrous sodium sulfate and concentrate it to 1 mL using a K-D concentrator.

Soils, sediments, or sludges — Use 30 g of sample. Nonporous or wet samples (gummy or clay type) that do not have a free-flowing sandy texture must be mixed with anhydrous sodium sulfate until the sample is free flowing. Add 1 mL of surrogate standards to all samples, spikes, standards, and blanks. For the sample in each analytical batch selected for spiking, add 1.0 mL of a matrix spiking standard. For base/neutral acid analysis, the amount added of the surrogates and matrix spiking compounds should result in a final concentration of 100 ng/µL of each base/neutral analyte and 200 ng/µL of each acid analyte in the extract to be analyzed (assuming a 1-µL injection). Immediately add a 100-mL mixture of 1:1 methylene chloride:acetone and extract the sample ultrasonically for 3 min and then decant or filter the extracts. Repeat the extraction two or more times. Dry the extract using a column with anhydrous sodium sulfate and concentrate it to 1 mL in a K-D concentrator.

QUALITY CONTROL A methylene chloride solution containing 50 ng/µL of decafluorotriphenylphosphine (DFTPP) is used for tuning the GC/MS system each 12-h shift. A system performance check also must be made during every 12-h shift. A standard containing 50 ng/µL each of 4,4′-DDT, pentachlorophenol, and benzidine is required to verify injection port inertness and GC column performance. A calibration standard at mid-concentration, containing each compound of interest, including all required surrogates, must be performed every 12 h during analysis. After the system performance check is met, calibration check compounds (CCCs) are used to check the validity of the initial calibration.

The internal standard responses and retention times in the calibration check standard must be evaluated immediately after or during data acquisition. If the retention time for any internal standard changes by more than 30 seconds from the last check calibration (12 h), the chromatographic system must be inspected for malfunctions and corrections must be made, as required. If the electron ionization current plot (EICP) area for any of the internal standards changes by a factor of two from the last daily calibration standard check, the mass spectrometer must be inspected for malfunctions and corrections must be made, as appropriate.

Demonstrate, through the analysis of a reagent water blank, that interferences from the analytical system, glassware, and reagents are under control. The blank samples should be carried through all stages of the sample preparation and measurement steps. For each analytical batch (up to 20 samples), a reagent blank, matrix spike, and matrix spike duplicate/duplicate must be analyzed (the frequency of the spikes may be different for different monitoring programs). The blank and spiked samples must be carried through all stages of the sample preparation and measurement steps. A QC reference sample concentrate containing each analyte at a concentration of 100 mg/L in methanol is required.

REFERENCE Test Methods for Evaluating Solid Waste (SW-846). U.S. EPA 1983, Method 8270B, Rev. 2, Nov. 1990. Office of Solid Waste, Washington, DC.

Cyanide   EPA Method 9012
CAS #None

TITLE Total and Amenable Cyanide (Colorimetric, Automated UV

MATRIX This method is used to determine the concentration of inorganic cyanide in an aqueous waste or leachate. It addresses total and amenable cyanides. Cyanide compounds that are amenable to chlorination include free cyanide as well as the complex cyanides.

METHOD SUMMARY Method 9012A covers the determination of total cyanide and amenable cyanide which includes free cyanide. Using a colorimetric measurement, the cyanide is converted to cyanogen chloride (CNCl) by reaction with Chloramine-T reagent at a pH less than 9 without hydrolyzing the cyanate. After the reaction is complete, color is formed upon the addition of pyridine-barbituric acid reagent. The concentration of NaOH must be the same in the standards, the scrubber solutions, and any dilution of the original scrubber solution to obtain color of comparable intensity.

Note: This method is essentially the same as EPA Method 335 and either may be used with comparable results. The matrices listed for EPA-335.3 are drinking waters, surface waters, domestic and industrial wastes.

INTERFERENCES Sulfides adversely affect the colorimetric procedures. Samples that contain hydrogen sulfide, metal sulfides, or other compounds that may produce hydrogen sulfide during the distillation should be treated by addition of bismuth nitrate prior to distillation.

High results may be obtained for samples that contain nitrate and/or nitrite ions. During the distillation, nitrate and nitrite will form nitrous acid, which will react with some organic compounds to form oximes. These compounds will decompose under test conditions to generate hydrogen cyanide. The possible interference of nitrate and nitrite is eliminated by pretreatment with sulfamic acid.

INSTRUMENTATION The reflux distillation apparatus consists of a 1 L boiling flask with an inlet tube and provision for condenser. The gas absorber is a Fisher-Milligan scrubber. An automated continuous-flow analytical instrument includes a sampler, manifold with UV digestor, proportioning pump, heating bath with distillation coil, distillation head, colorimeter equipped with at 15 mm flow cell and 570 nm filter and a recorder.

PRECISION & ACCURACY No precision or accuracy data are available for this method.

SAMPLING METHOD Collect samples in 1 L or larger, plastic or glass bottles. All bottles must be thoroughly cleaned and rinsed to remove soluble materials.

SAMPLE PRESERVATION Oxidizing agents such as chlorine decompose most cyanides. To determine whether oxidizing agents are present, test a drop of the sample with acidified potassium iodide (KI) — starch test paper at the time of collection; a blue color indicates the need for treatment. Add ascorbic acid a few crystals at a time until a drop of sample produces no color on the indicator. Then, add an additional 0.6 g of ascorbic acid for each liter of water. Ascorbic acid does not need to be added if chlorine and other oxidizing agents are absent. Samples are preserved by adding 10 N sodium hydroxide until sample pH is greater than or equal to 12 at time of collection. Samples should be stored at 4°C and analyzed as soon as possible.

MHT Samples must be analyzed within 14 days of their collection.

SAMPLE PREPARATION

Pretreatment for cyanides amenable to chlorination:
Two sample aliquots are required to determine cyanides amenable to chlorination. To one 500 mL aliquot, or to a volume diluted to 500 mL, add calcium hypochlorite solution dropwise while agitating and maintaining the pH between 11 and 12 with sodium hydroxide. This should be done in a fume hood. Test for residual chlorine with potassium iodide-starch paper and maintain this excess for one h, continuing agitation. A distinct blue color on the test paper indicates a sufficient chlorine level. If necessary, add additional hypochlorite solution. After 1 h add 0.5 g portions of ascorbic acid until potassium iodide-starch paper shows no residual chlorine. Add an additional 0.5 g of ascorbic acid to ensure the presence of excess reducing agent. Test for total cyanide in both the chlorinated and unchlorinated aliquots. The difference of total cyanide in the chlorinated and unchlorinated aliquots is the cyanide amenable to chlorination.

Distillation Procedure:
Place 500 mL of sample, or an aliquot diluted to 500 mL, in a 1 L boiling flask and pipette 50 mL of sodium hydroxide into the absorbing tube. Add reagent water until the spiral is covered. Connect the boiling flask, condenser, absorber, and trap in the train. By adjusting the vacuum source, start a slow stream of air entering the boiling flask so that approximately two bubbles of air per second enter the flask through the air inlet tube. Use lead acetate paper to check the sample for the presence of sulfide. A positive test is indicated by a black color on the paper. If positive, treat the sample by adding 50 mL of bismuth nitrate solution through the air inlet tube after the air rate is set. Mix for 3 min prior to addition of sulfuric acid. If samples are suspected to contain nitrate and/or nitrite ions, add 50 mL of sulfamic acid solution after the air rate is set through the air inlet tube. Mix for 3 min prior to addition of sulfuric acid.

Slowly add 50 mL 1:1 water and sulfuric acid mixture through the air inlet tube. Rinse the tube with reagent water and allow the airflow to mix the flask contents for 3 min. Pour 20 mL of magnesium chloride into the air inlet and wash down with a stream of water. Heat the solution to boiling. Reflux for 1 h. Turn off heat and continue the airflow for at least 15 min. After cooling the boiling flask, disconnect absorber and close off the vacuum source. Drain the solution from the absorber into a 250 mL volumetric flask. Wash the absorber with reagent water and add the washings to the flask. Dilute to the mark with reagent water.

QUALITY CONTROL Employ a minimum of one blank per sample batch to determine if contamination or any memory effects are occurring. Verify calibration with an independently prepared check standard every 15 samples. Run one spike duplicate sample for every 10 samples. A duplicate sample is a sample brought through the whole sample preparation process. The method of standard additions should be used for the analysis of all samples that suffer from matrix interferences.

REFERENCE Test Methods for Evaluating Solid Waste (SW-846). U.S. EPA. 1983. Method 9012, Rev. 0, Sept. 1986. Office of Solid Wastes, Washington, DC.

Cycloate	EPA Method 507
CAS #1134-23-2	

TITLE Determination of Nitrogen and Phosphorus-Containing Pesticides in Water by GC/NPD

MATRIX This method is applicable to the determination of certain nitrogen and phosphorus-containing pesticides in finished drinking water and groundwater.

METHOD SUMMARY Method 507 covers 46 nitrogen- and phosphorus-containing pesticides. A 1-L sample is fortified with a surrogate standard, salted, buffered, extracted with methylene chloride, and concentrated; then the solvent is exchanged with methyl tert-butyl ether (MTBE) and concentrated again, and a 2-µL aliquot of a sample extract is injected into a GC system equipped with a selective nitrogen-phosphorus detector and a capillary column for analysis.

INTERFERENCES Method interferences may be caused by contaminants in solvents, reagents, glassware, and other sample processing apparatus. Interfering contamination may occur when a sample containing low concentrations of analytes is analyzed immediately following a sample containing relatively high concentrations. One or more injections of MTBE should be made following the analysis of a sample with high concentrations of analytes to check for analyte carryover. Matrix interferences may be caused by contaminants that are coextracted from the sample. The extent of matrix interferences will vary considerably from source to source, depending upon the water sampled.

INSTRUMENTATION A gas chromatograph system (GC) equipped with a nitrogen-phosphorus detector (NPD) is needed.

Column 1: 30 m × 0.25 mm I.D. DB-5 bonded fused silica column, 0.25 µm film thickness, or equivalent.
Column 2: 30 m × 0.25 mm I.D. DB-1701 bonded fused silica column, 0.25 µm film thickness, or equivalent.

PRECISION & ACCURACY This method has been validated in a single lab and estimated detection limits (EDLs) have been determined for each analyte. Observed detection limits may vary among waters, depending upon the nature of the interferences in the sample matrix and the specific instrumentation used. Analytes that are not separated chromatographically cannot be individually identified and measured unless an alternative technique for identification and quantification exist.

The estimated detection limit (in µg/L) was 0.25. The EDL is defined as either method detection limit or a level of compound in a sample yielding a peak in the final extract with signal-to-noise ratio of approximately 5, whichever value is higher.

The concentration used for these measurements (in µg/L) was 2.5.

The accuracy (as % recovery) was 89.

The precision (% RSD) was 9.

SAMPLING METHOD Grab samples are collected in 1-L glass sample bottles (prewashed with detergent and hot tap water, rinsed with reagent water, and dried in an oven at 400°C for 1 h) with screw caps lined with PTFE-fluorocarbon.

SAMPLE PRESERVATION Add mercuric chloride to the sample bottle in amounts to produce a concentration of 10 mg/L. If residual chlorine is present, add 80 mg of sodium thiosulfate/L of sample to the sample bottle prior to collection. After collection, seal bottle and shake vigorously for 1 min, then cool the sample to 4°C immediately and store it at 4°C in the dark until extraction.

MHT Maximum holding time of the samples, and in some cases the extracts, is 14 days.

SAMPLE PREPARATION Fortify the sample with 50 µL of the surrogate standard solution, adjust to pH 7 with phosphate buffer, add 100 g NaCl to the sample, and seal and shake to dissolve the salt; then extract with methylene chloride in a separatory funnel or in a mechanical tumbler bottle. Dry the extract by pouring it through a solvent-rinsed drying column containing about 10 cm of anhydrous sodium sulfate. Collect the extract in a Kuderna-Danish (K-D) concentrator and rinse the column with 20–30 mL methylene chloride. Concentrate the extract to about 2 mL and rinse the flask and its lower joint into the concentrator tube with 1 to 2 mL of methyl t-butyl ether (MTBE). Add 5–10 mL of MTBE and concentrate the extract twice (adding more MTBE) to a final volume of 5.0 mL and store it at 4°C until analysis.

Note: If methylene chloride is not completely removed from the final extract, it may cause detector problems.

QUALITY CONTROL Minimum quality control requirements are initial demonstration of lab capability, determination of surrogate compound recoveries in each sample and blank, monitoring internal standard peak area or height in each sample and blank, analysis of lab reagent blanks, lab fortified samples, lab fortified blanks, and other QC samples. A lab reagent blank is analyzed to demonstrate that all glassware and reagent interferences are under control.

Initial demonstration of capability is fulfilled by analyzing four fortified reagent water samples with the recovery value for each analyte falling within the acceptable range (±30% average recovery). Surrogate recoveries from samples or method blanks must be 70–130%. The internal standard response for any sample chromatogram should not deviate from the daily calibration check standard's internal standard response by more than 30% or lab fortified blanks and sample matrices are used to assess lab performance and analyte recovery, respectively.

If the response for the target analyte peak exceeds the working range of the system, dilute the extract and reanalyze. Alternative techniques such as an alternate detector or second chromatography column should be used to confirm peak identification when sample components are not resolved adequately.

EPA CONTACT & HOTLINE For technical questions contact Dr. Baldev Bathija, U.S. EPA, Office of Ground Water and Drinking Water (WH-550D), 401 M St. SW, Washington, DC 20460. Tel. (202) 260-3040. For further information the EPA Safe Drinking Water Hotline may be called at: (800) 426-4791.

REFERENCE Methods for the Determination of Organic Compounds in Drinking Water, EPA/600/4-88/039 (revised July 1991). U.S. EPA Environmental Monitoring Systems Laboratory, Cincinnati, OH, 45268, USA. Available from the National Technical Information Service (NTIS), 5285 Port Royal Road, Springfield, VA 22161; Tel. 800-553-6847. NTIS Order Number is PB91-231480.

2-Cyclohexyl-4,6-dinitrophenol   EPA Method 8270
CAS #131-89-5

TITLE Semivolatile Organic Compounds by GC/MS

MATRIX This method is used to determine the concentration of semivolatile organic compounds in extracts prepared from all types of solid waste matrices, soils, and groundwater. Although surface waters are not specifically mentioned, this method should be applicable to water samples from rivers, lakes, etc.

METHOD SUMMARY This method covers 259 semivolatile organic compounds. In very limited applications direct injection of the sample into the GC/MS system may be appropriate, but this results in very high detection limits (approximately 10,000 µg/L). Typically, a 1-L liquid sample, containing surrogate, and matrix spiking standards, is extracted in a continuous extractor first under acid conditions and then under basic conditions. Typically 30 g of a solid sample, containing surrogate, and matrix spiking standards, is extracted ultrasonically. After concentrating the extract to 1 mL it is spiked with 10 µL of an internal standard solution just prior to analysis by GC/MS. The volume injected should contain about 100 ng of base/neutral and 200 ng of acid surrogates (for a 1-µL injection). Analysis is performed by GC/MS using a capillary GC column.

INTERFERENCES Raw GC/MS data from all blanks, samples, and spikes must be evaluated for interferences. Contamination by carryover can occur whenever high-concentration and low-concentration samples are sequentially analyzed. To reduce carryover, the sample syringe must be rinsed out between samples with solvent. Whenever an unusually concentrated sample is encountered, it should be followed by the analysis of blank solvent to check for cross-contamination.

INSTRUMENTATION A GC/MS and a data system are required. The GC column used is a 30 m × 0.25 mm I.D. (or 0.32 mm I.D.) 1um film thickness silicone-coated fused silica capillary column. A continuous liquid-liquid extractor equipped with Teflon® or glass connection joints and stopcocks requiring no lubrication, a K-D concentrating apparatus, water bath, and an ultrasonic disrupter with a minimum power of 300 W and with pulsing capability are also required.

PRECISION & ACCURACY The estimated quantitation limit (EQL) of Method 8270B for determining an individual compound is approximately 1 mg/kg (wet weight) for soil or sediment samples, 1–200 mg/kg for wastes (dependent on matrix and method of preparation), and 10 µg/L for groundwater samples. EQLs will be proportionately higher for sample extracts that require dilution to avoid saturation of the detector.

The EQL(b) for groundwater in µg/L is 100.
The EQL (a, b) for low concentrations in soil and sediment in µg/kg is not determined.
Accuracy as µg/L is not listed.
Overall precision in µg/L is not listed.

(a) *EQLs listed for soil/sediment are based on wet weight. Normally data is reported in a dry-weight basis; therefore, EQLs will be higher based on the % dry weight of each sample. This calculation is based on a 30 g sample and gel permeation chromatography cleanup.*
(b) *Sample EQLs are highly matrix-dependent. The EQLs are provided for guidance and may not always be achievable.*
C = *True value for concentration, in µg/L.*
X = *Average recovery found for measurements of samples containing a concentration of C, in µg/L.*

ESTIMATED QUANTITATION LIMIT FOR OTHER MATRICES

Other Matrices	Factor (a)
High-concentration soil and sludges by sonicator	7.5
Non-water miscible waste	75

(a) *EQL for other matrices = [EQL for low soil/sediment] × [Factor]. This estimated EQL is similar to an EPA "Practical Quantitation Limit."*

SAMPLING METHOD

Liquid samples — Use a 1 or 2½ gallon amber glass bottle with a screw-top Teflon®-lined cover that has been prewashed with detergent and rinsed with distilled water and methanol (or isopropanol).

Soils, sediments, or sludges — Use an 8-oz. widemouth glass with a screw-top Teflon®-lined cover that has been prewashed with detergent and rinsed with distilled water and methanol (or isopropanol).

SAMPLE PRESERVATION

Liquid samples — If residual chlorine is present, add 3 mL of 10% sodium thiosulfate per gallon, cool to 4°C and store in a solvent-free refrigerator until analysis; if chlorine is not present, then eliminate the sodium thiosulfate addition.

Soils, sediments, or sludges — Cool samples to 4°C and store in a solvent-free refrigerator.

MHT Liquid samples must be extracted within 7 days and the extracts analyzed within 40 days. Soils, sediments, or sludges may be stored for a maximum of 14 days and the extracts analyzed within 40 days.

SAMPLE PREPARATION

Liquid samples — Transfer 1 L quantitatively to a continuous extractor. If high concentrations are anticipated, a smaller volume may be used and then diluted with organic-free reagent water to 1 L. Adjust pH, if necessary, to pH <2 using 1:1 (V/V) sulfuric acid. Pipette 1.0 mL of a surrogate standard spiking solution into each sample. For the sample in each analytical batch selected for spiking, add 1.0 mL of a matrix spiking standard. For base/neutral acid analysis, the amount of the surrogates and matrix spiking compounds added to the sample should result in a final concentration of 100 ng/µL of each analyte in the extract to be analyzed (assuming a 1-µL injection). Extract with methylene chloride for 18–24 h. Next, adjust the pH of the aqueous phase to pH >11 using 10 N sodium hydroxide and extract it with methylene chloride again for 18–24 h. Dry the extract through a column containing anhydrous sodium sulfate and concentrate it to 1 mL using a K-D concentrator.

Soils, sediments, or sludges — Use 30 g of sample. Nonporous or wet samples (gummy or clay type) that do not have a free-flowing sandy texture must be mixed with anhydrous sodium sulfate until the sample is free flowing. Add 1 mL of surrogate standards to all samples, spikes, standards, and blanks. For the sample in each analytical batch selected for spiking, add 1.0 mL of a matrix spiking standard. For base/neutral acid analysis, the amount added of the surrogates and matrix spiking compounds should result in a final concentration of 100 ng/µL of each base/neutral analyte and 200 ng/µL of each acid analyte in the extract to be analyzed (assuming a 1-µL injection). Immediately add a 100-mL mixture of 1:1 methylene chloride:acetone and extract the sample ultrasonically for 3 min and then decant or filter the extracts. Repeat the extraction two or more times. Dry the extract using a column with anhydrous sodium sulfate and concentrate it to 1 mL in a K-D concentrator.

QUALITY CONTROL A methylene chloride solution containing 50 ng/µL of decafluorotriphenylphosphine (DFTPP) is used for tuning the GC/MS system each 12-h shift. A system performance check also must be made during every 12-h shift. A standard containing 50 ng/µL each of 4,4'-DDT, pentachlorophenol, and benzidine is required to verify injection port inertness and GC column performance. A calibration standard at mid-concentration, containing each compound of interest, including all required surrogates, must be performed every 12 h during analysis. After the system performance check is met, calibration check compounds (CCCs) are used to check the validity of the initial calibration.

The internal standard responses and retention times in the calibration check standard must be evaluated immediately after or during data acquisition. If the retention time for any internal standard changes by more than 30 seconds from the last check calibration (12 h), the chromatographic system must be inspected for malfunctions and corrections must be made, as required. If the electron ionization current plot (EICP) area for

any of the internal standards changes by a factor of two from the last daily calibration standard check, the mass spectrometer must be inspected for malfunctions and corrections must be made, as appropriate.

Demonstrate, through the analysis of a reagent water blank, that interferences from the analytical system, glassware, and reagents are under control. The blank samples should be carried through all stages of the sample preparation and measurement steps. For each analytical batch (up to 20 samples), a reagent blank, matrix spike, and matrix spike duplicate/duplicate must be analyzed (the frequency of the spikes may be different for different monitoring programs). The blank and spiked samples must be carried through all stages of the sample preparation and measurement steps. A QC reference sample concentrate containing each analyte at a concentration of 100 mg/L in methanol is required.

REFERENCE Test Methods for Evaluating Solid Waste (SW-846). U.S. EPA 1983, Method 8270B, Rev. 2, Nov. 1990. Office of Solid Waste, Washington, DC.

---

p-Cymene   **EPA Method 1625**
CAS #99-87-6

TITLE Semivolatile Organic Compounds by Isotope Dilution GC/MS

MATRIX The compounds may be determined in waters, soils, and municipal sludges by this method.

METHOD SUMMARY This method is used to determine 176 semivolatile toxic organic pollutants associated with the CWA (as amended 1987); the RCRA (as amended 1986); the CERCLA (as amended 1986); and other compounds amenable to extraction and analysis by capillary column gas chromatography-mass spectrometry (GC/MS).

Stable isotopically-labeled analogs of the compounds of interest are added to the sample. If the solids content is less than 1%, a 1-L sample is extracted at pH 12–13, then at pH <2 with methylene chloride using continuous extraction techniques.

If the solids content is 30% or less, the sample is diluted to 1% solids with reagent water, homogenized ultrasonically, and extracted at pH 12–13, then at pH <2 with methylene chloride using continuous extraction techniques. If the solids content is greater than 30%, the sample is extracted using ultrasonic techniques.

Each extract is dried over sodium sulfate, concentrated to a volume of 5 mL, cleaned up using GPC, if necessary, and concentrated. Extracts are concentrated to 1 mL if GPC is not performed, and to 0.5 mL if GPC is performed.

An internal standard is added to the extract, and a 1-mL aliquot of the extract is injected into the GC. The compounds are separated by GC and detected by a MS. The labeled compounds serve to correct the variability of the analytical technique.

INTERFERENCES Solvents, reagents, glassware, and other sample processing hardware may yield artifacts and/or elevated baselines causing misinterpretation of chromatograms and spectra. Materials used in the analysis must be demonstrated to be free from interferences under the conditions of analysis by running method blanks initially and with each sample lot (sample started through the extraction process on a given 8-h shift, to a maximum of 20). Specific selection of reagents and purification of solvents by distillation in all glass systems may be required. Glassware and, where possible, reagents are cleaned by solvent rinse and baking at 450°C for 1-h minimum. Interferences coextracted from samples will vary considerably from source to source, depending on the diversity of the site being sampled.

INSTRUMENTATION Major instrumentation includes a GC with a splitless or on-column injection port for capillary column, a MS with 70 eV electron impact ionization, and a data system to collect and record MS data, and process it. A K-D apparatus is used to concentrate extracts.

GC Column: 30 m × 0.25 mm I.D. 5% phenyl, 94% methyl, 1% vinyl silicone bonded phased fused silica capillary column.

PRECISION & ACCURACY The detection limits of the method are usually dependent on the level of interferences rather than instrumental limitations. The limits typify the minimum quantities that can be detected with no interferences present.

The minimum level (in µg/mL) was 10. This is defined as a minimum level at which the analytical system shall give recognizable mass spectra (background corrected) and acceptable calibration points.

The MDL (in µg/kg) in low solids was 426 and in high solids was 912; these were determined in digested sludge (low solids) and in filter cake or compost (high solids).

Note: Background levels of this compound were present in the sludge tested, resulting in higher than expected MDLs. The MDL for this compound is expected to be approximately 50 µg/kg with no interferences present.

The labeled and native compound initial precision as standard deviation (in µg/L) was 18.
The labeled and native compound initial accuracy as average recovery (in µg/L) was 76–140.

SAMPLE COLLECTION, PRESERVATION & HANDLING
Collect samples in glass containers. Aqueous samples which flow freely are collected in refrigerated bottles using automatic sampling equipment. Solid samples are collected as grab samples using widemouth jars. Maintain samples at 0 to 4°C from the time of collection until extraction. If residual chlorine is present in aqueous samples, add 80 mg sodium thiosulfate/L of water. Begin sample extraction within 7 days of collection, and analyze all extracts within 40 days of extraction.

SAMPLE PREPARATION Samples containing 1% solids or less are extracted directly using continuous liquid-liquid extraction techniques. Samples containing 1 to 30% solids are diluted to the 1% level with reagent water and extracted using continuous liquid-liquid extraction techniques. Samples containing greater than 30% solids are extracted using ultrasonic techniques.

- Base/neutral extraction — Adjust the pH of the waters in the extractors to 12–13 with 6 N NaOH. Extract with methylene chloride for 24–48 h.
- Acid extraction — Adjust the pH of the waters in the extractors to 2 or less using 6 N sulfuric acid. Extract with methylene chloride for 24–48 h.
- Ultrasonic extraction of high solids samples — Add anhydrous sodium sulfate to the sample and QC aliquot(s). Add acetone:methylene chloride (1:1) to the sample and mix thoroughly

Concentrate extracts using a K-D apparatus.

QUALITY CONTROL The analyst is permitted to modify this method to improve separations or lower the costs of measurements, provided all performance specifications are met. Analyses of blanks are required to demonstrate freedom from contamination. When results of spikes indicate atypical method performance for samples, the samples are diluted to bring method performance within acceptable limits.

For low solids (aqueous samples), extract, concentrate, and analyze two sets of four 1-L aliquots (8 aliquots total) of the precision and recovery standard. For high solids samples, two sets of four 30-g aliquots of the high solids reference matrix are used.

Spike all samples with labeled compounds to assess method performance. Compute percent recovery of the labeled compounds using the internal standard method. Compare the labeled compound recovery for each compound with the corresponding labeled compound recovery.

Reagent water and high solids reference matrix blanks are analyzed to demonstrate freedom from contamination. Extract and concentrate a 1-L reagent water blank or a high solids reference matrix blank with each sample's lot (samples started through the extraction process on the same 8-h shift, to a maximum of 20 samples).

Field replicates may be collected to determine the precision of the sampling technique, and spiked samples may be required to determine the accuracy of the analysis when the internal standard method is used.

REFERENCE Semivolatile Organic Compounds by Isotope Dilution GC/MS. Office of Water Regulation and Standards, U.S. EPA Industrial Technology Division, Washington, DC, EPA Method 1625, Rev. C, June 1989 (contact W.A. Telliard, U.S. EPA, Office of Water Regulations and Standards, 401 M St., SW, Washington, DC, 20460. Phone: 202-382-7131).

D

2,4-D
CAS #94-75-7

EPA Method 8151

TITLE Chlorinated Herbicides by GC Using Methylation or Pentafluorobenzylation Derivatization: Capillary Column Technique.

MATRIX This method covers aqueous and soslid matrices. This includes a wide variety such as drinking water, groundwater, industrial wastewaters, surface waters, soils, solids, and sediments.

METHOD SUMMARY This is a GC method for determining 19 chlorinated acid herbicides in aqueous, soil, and waste matrices. Because these compounds are produced and used in various forms (i.e., acid, salt, ester, etc.) a hydrolysis step is included to convert the herbicide to the acid form prior to analysis. This method provides hydrolysis, extraction, derivatization and GC conditions for the analysis of chlorinated acid herbicides in water, soil, and waste samples. Water samples are hydrolyzed *in situ*, extracted with diethyl ether, and then esterified with either diazomethane or pentafluorobenzyl bromide. The derivatives are determined by gas chromatography with an electron capture detector (GC/ECD). The results are reported as acid equivalents. The sensitivity of this method depends on the level of interferences in addition to instrumental limitations.

INTERFERENCES Method interferences may be caused by contaminants in solvents, reagents, glassware, and other sample processing hardware. Immediately prior to use, glassware should be rinsed with the next solvent to be used. Matrix interferences may be caused by contaminants that are coextracted from the sample. Organic acids, especially chlorinated acids, cause the most direct interference with the determination by methylation. Phenols, including chlorophenols, may also interfere with this procedure. The determination using pentafluorobenzylation is more sensitive, and more prone to interferences from the presence of organic acids of phenols than by methylation. Alkaline hydrolysis and subsequent extraction of the basic solution removes many chlorinated hydrocarbons and phthalate esters that might otherwise interfere with the ECD analysis. The herbicides, being strong organic acids, react readily with alkaline substances and may be lost during analysis. Therefore, glassware must be acid-rinsed and then rinsed to constant pH with organic-free reagent water.

INSTRUMENTATION A GC suitable for Grob-type injection using capillary columns. A data system for measuring peak heights and/or peak areas is recommended. An electron capture detector (ECD) is used. Also a K-D apparatus, a diazomethane generator, a centrifuge and an ultrasonic disrupter will be required.

Narrow Bore Columns:
Primary Column 1: 30 m × 0.25 mm, 5% phenyl/95% methyl silicone (DB-5), 0.25 µm film thickness.
Primary Column 1a (GC/MS): 30 m × 0.32 mm, 5% phenyl/95% methyl silicone (DB-5), 1-µm film thickness.
Column 2: 30 m × 0.25 mm DB-608 with a 25 µm film thickness.
Confirmation Column: 30 m × 0.25 mm, 14% cyanopropyl phenyl silicone (DB-1701), 0.25 µm film thickness.

Megabore Columns:
Primary Column: 30 m × 0.53 mm DB-608 with 0.83 µm film thickness.
Confirmation Column: 30 m × 0.53 mm, 14% cyanopropyl phenyl silicone (DB-1701), 1.0 µm film thickness.

PRECISION & ACCURACY Method detection limits (MDLs) are compound-dependent and vary with derivitization efficiency, derivative recovery, the matrix sampled, and herbicide concentration.

The estimated MDL (in µg/L) was 0.2 for aqueous samples using GC/ECD.

The estimated MDL (in µg/kg) was 0.11 for soil samples using GC/ECD when corrected back to 50-g samples extracted and concentrated to 10 mL with 5-µL injections.

The estimated GC/MS identification limit (in ng) was 1.25 for soil samples using GC/MS.

Mean percent recovery, calculated from 7–8 determinations of spiked reagent water, after diazomethane derivatization, from a spike concentration (in µg/L) of 1 was 131 with a standard deviation of the percent recovery of 27.5.

Mean percent recovery, calculated from 10 determinations of spiked clay and clay/still bottom samples over the linear concentration range (in ng/g) of 1.2–2,440 was 84.3 with a percent relative standard deviation of 5.3. The RSD % was calculated on 10 samples high in the linear concentration range and 10 low in the range. The linear concentration range was determined using standard solutions and corrected to 50 g soil samples.

SAMPLE COLLECTION, PRESERVATION & HANDLING Containers used to collect samples for the determination of semivolatile organic compounds should be soap and water washed followed by methanol (or isopropanol) rinsing. The sample containers should be of glass or Teflon® and have screw-top covers with Teflon® liners.

No preservation is used with concentrated waste samples. With liquid samples containing no residual chlorine and with soil, sediment, and sludge samples, immediately cooling to 4°C is the only preservation used. When residual chlorine is present then 3 mL of 10% aqueous sodium sulfate is added for each gallon of sample collected, followed by cooling to 4°C.

The holding time for all volatile organics samples is 14 days. Liquid samples must be extracted within 7 days and their extracts analyzed within 40 days. Concentrated waste, soil, sediment, and sludge samples must be extracted within 14 days and their extracts analyzed within 40 days.

SAMPLE PREPARATION
Preparation of soil, sediment, and other solid samples — Acidify 30 g (dry weight) solids with 0.1 M phosphate buffer (pH = 2.5) and thoroughly mix the contents. Spike the sample with surrogate compound(s). The ultrasonic extraction of solids

must be optimized for each type of sample. In order for the ultrasonic extractor to efficiently extract solid samples, the sample must be free flowing when the solvent is added. Acidified anhydrous sodium sulfate should be added to clay-type soils, or any other solid that is not a free-flowing sandy texture, until a free flowing mixture is obtained. Add methylene chloride and perform ultrasonic extraction. Combine organic extracts from the repetitive extractings of the sample and centrifuge. Add aqueous potassium hydroxide, water, and methanol to the extract and reflux the mixture on a water bath. Extract the solution three times with methylene chloride and discard the methylene chloride phase. The basic solution contains the herbicide salts. Adjust the pH of the solution to <2 with cold sulfuric acid and extract three times with methylene chloride. Combine the extracts and pour them through a pre-rinsed drying column containing acidified anhydrous sodium sulfate. Collect the dried extracts in a K-D flask and concentrate them.

Preparation of aqueous samples — Measure 1 L of sample into a 2 L separatory funnel and spike it with surrogate compound(s). Add NaCl to the sample, then add 6 N NaOH to the sample to a pH of 12 or more and let the sample sit at room temperature for 1 h to hydrolyze esters. Extract the sample three times with methylene chloride and discard the extracts. Then add cold 12 N sulfuric acid to a pH less than or equal to 2, and extract the sample three times with ethyl ether. Collect the ether phase in a flask containing acidified anhydrous sodium sulfate and allow it to remain in contact with the sodium sulfate for a minimum of 2 h. The drying step is very critical to ensuring complete esterification; any moisture remaining in the ether will result in low herbicide recoveries.

Extract concentration and derivatization — The combined ether extract is concentrated to about 1 mL using a K-D apparatus followed by using a micro Snyder column or nitrogen gas blowdown. If methyl esters are to be produced, then dilute the concentrated ether extract with 1 mL of isooctane and 0.5 mL of methanol, dilute to a final volume of 4 mL, and esterify with diazomethane. If pentafluorobenzene esters are to be produced, then dilute concentrated ether extract with acetone to a final volume of 4 mL and esterify with pentafluorobenzyl bromide.

QUALITY CONTROL Select a representative spike concentration for each compound (acid or ester) to be measured. Using stock standard, prepare a quality control check sample concentrate, in acetone, that is 1000 times more concentrated than the selected concentrations. Use this quality control check sample concentrate to prepare quality control check samples. Calculate surrogate standard recovery on all standards, samples, blanks, and spikes. GC/MS techniques should be judiciously employed to support qualitative identifications made with this method. When available, chemical ionization mass spectra may be employed to aid the qualitative identification process.

REFERENCE Test Methods for Evaluating Solid Waste, Physical/Chemical Methods, SW-846, 3rd Edition, U.S. EPA, Office of Solid Waste, Washington, DC, EPA Method 8151, Nov. 1990.

2,4-D   EPA Method 8150
CAS #94-75-7

TITLE Chlorinated Herbicides

MATRIX Groundwater, soils, sludges, water miscible liquid wastes, and non-water miscible wastes.

APPLICATION This method is used for the analysis of 10 chlorinated herbicides. Samples are extracted, hydrolyzed with potassium hydroxide, and extraneous organics are removed by a solvent wash. After acidification, the acids are extracted, concentrated and converted to their methyl esters using diazomethane. They are then analyzed using direct injection into a gas chromatograph (GC). Be very careful because diazomethane can explode under certain conditions and it is also a carcinogen.

INTERFERENCES Organic acids and phenols (especially chlorinated acids and phenols) may cause interferences. Phthalate esters are not as significant an interference as with other GC-ECD methods if an electron capture detector is used. The herbicides may react readily with alkaline substances and be lost during analysis so all glassware and glass wool must be acid rinsed and sodium sulfate must be acidified with sulfuric acid prior to use. Sensitivity usually depends on the level of interferences rather than on instrumentation.

INSTRUMENTATION GC capable of on-column injections and an electron capture detector (ECD) or a halogen specific detector. Column 1: 1.8 m by 4 mm with 1.5% SP-2250/1.95% SP-2401 on Supelcoport. Column 2: 1.8 m by 4 mm with 5% OV-210 on Gas Chrom Q. Column 3: 1.98 m by 2 mm with 0.1%. SP-1000 on Carbopack C. The preferred column is Column Number 1 or 2.

RANGE Not listed.

MDL 1.2 µg/L (in reagent water; ECD).

PQL FACTORS FOR MULTIPLYING × FID MDL VALUE

Matrix	Multiplication Factor
Groundwater	10
Low-level soil by sonication with GPC cleanup	670
High-level soil and sludge by sonication	10,000
Non-water miscible waste	100,000

PRECISION (as standard deviation) 4% with 10.9 µg/L spike in drinking water. 4% with 10.1 µg/L in municipal water.

ACCURACY (as mean recovery) 75% with 10.9 µg/L spike in drinking water. 77% with 10.1 µg/L in municipal water.

SAMPLING METHOD Use 8-oz. widemouth glass bottles with Teflon®-lined caps for concentrated waste samples, soils, sediments, and sludges. Use 1 or 2½ gallon amber glass bottles with Teflon®-lined caps for liquid (water) samples.

STABILITY Cool soil, sediment, sludge, and liquid samples to 4°C. If residual chlorine is present in liquid samples add 3 mL of 10% sodium thiosulfate per gallon of sample and cool to 4°C.

MHT 14 days for concentrated waste, soil, sediment, or sludge; 7 days for liquid samples; all extracts must be analyzed within 40 days.

QUALITY CONTROL A quality control check sample concentrate containing this compound in acetone at a concentration 1,000 times more concentrated than the selected spike concentration is required. The QC check sample concentrate may be prepared from pure standard materials or purchased as certified solutions. Use appropriate trip, matrix, control site, method, reagent, and solvent blanks. Internal, surrogate, and five concentration level calibration standards are used.

REFERENCE Method 8150, SW-846, 3rd ed., Sept. 1986.

Dalapon EPA Method 8151
CAS #75-99-0

TITLE Chlorinated Herbicides by GC Using Methylation or Pentafluorobenzylation Derivatization: Capillary Column Technique.

MATRIX This method covers aqueous and solid matrices. This includes a wide variety such as drinking water, groundwater, industrial wastewaters, surface waters, soils, solids, and sediments.

METHOD SUMMARY This is a GC method for determining 19 chlorinated acid herbicides in aqueous, soil, and waste matrices. Because these compounds are produced and used in various forms (i.e., acid, salt, ester, etc.) a hydrolysis step is included to convert the herbicide to the acid form prior to analysis. This method provides hydrolysis, extraction, derivatization and GC conditions for the analysis of chlorinated acid herbicides in water, soil, and waste samples. Water samples are hydrolyzed *in situ*, extracted with diethyl ether, and then esterified with either diazomethane or pentafluorobenzyl bromide. The derivatives are determined by gas chromatography with an electron capture detector (GC/ECD). The results are reported as acid equivalents. The sensitivity of this method depends on the level of interferences in addition to instrumental limitations.

INTERFERENCES Method interferences may be caused by contaminants in solvents, reagents, glassware, and other sample processing hardware. Immediately prior to use, glassware should be rinsed with the next solvent to be used. Matrix interferences may be caused by contaminants that are coextracted from the sample. Organic acids, especially chlorinated acids, cause the most direct interference with the determination by methylation. Phenols, including chlorophenols, may also interfere with this procedure. The determination using pentafluorobenzylation is more sensitive, and more prone to interferences from the presence of organic acids of phenols than by methylation. Alkaline hydrolysis and subsequent extraction of the basic solution removes many chlorinated hydrocarbons and phthalate esters that might otherwise interfere with the ECD analysis. The herbicides, being strong organic acids, react readily with alkaline substances and may be lost during analysis. Therefore, glassware must be acid-rinsed and then rinsed to constant pH with organic-free reagent water.

INSTRUMENTATION A GC suitable for Grob-type injection using capillary columns. A data system for measuring peak heights and/or peak areas is recommended. An electron capture detector (ECD) is used. Also a K-D apparatus, a diazomethane generator, a centrifuge and an ultrasonic disrupter will be required.

Narrow Bore Columns:
Primary Column 1: 30 m × 0.25 mm, 5% phenyl/95% methyl silicone (DB-5), 0.25 µm film thickness.
Primary Column 1a (GC/MS): 30 m × 0.32 mm, 5% phenyl/95% methyl silicone (DB-5), 1-µm film thickness.
Column 2: 30 m × 0.25 mm DB-608 with a 25 µm film thickness.
Confirmation Column: 30 m × 0.25 mm, 14% cyanopropyl phenyl silicone (DB-1701), 0.25 µm film thickness.

Megabore Columns:
Primary Column: 30 m × 0.53 mm DB-608 with 0.83 µm film thickness.
Confirmation Column: 30 m × 0.53 mm, 14% cyanopropyl phenyl silicone (DB-1701), 1.0 µm film thickness.

PRECISION & ACCURACY Method detection limits (MDLs) are compound-dependent and vary with derivitization efficiency, derivative recovery, the matrix sampled, and herbicide concentration.

The estimated MDL (in µg/L) was 1.3 for aqueous samples using GC/ECD.

The estimated MDL (in µg/kg) was 0.12 for soil samples using GC/ECD when corrected back to 50-g samples extracted and concentrated to 10 mL with 5-µL injections.

The estimated GC/MS identification limit (in ng) was 0.5 for soil samples using GC/MS.

Mean percent recovery, calculated from 7–8 determinations of spiked reagent water, after diazomethane derivatization, from a spike concentration (in µg/L) of 10 was 100 with a standard deviation of the percent recovery of 20.0.

Mean percent recovery, calculated from 10 determinations of spiked clay and clay/still bottom samples over the linear concentration range (in ng/g) of no data was none reported with a percent relative standard deviation of none. The RSD % was calculated on 10 samples high in the linear concentration range and 10 low in the range. The linear concentration range was determined using standard solutions and corrected to 50 g soil samples.

SAMPLE COLLECTION, PRESERVATION & HANDLING Containers used to collect samples for the determination of semivolatile organic compounds should be soap and water washed followed by methanol (or isopropanol) rinsing. The sample containers should be of glass or Teflon® and have screw-top covers with Teflon® liners.

No preservation is used with concentrated waste samples. With liquid samples containing no residual chlorine and with soil, sediment, and sludge samples, immediately cooling to 4°C is the only preservation used. When residual chlorine is present then 3 mL of 10% aqueous sodium sulfate is added for each gallon of sample collected, followed by cooling to 4°C.

The holding time for all volatile organics samples is 14 days. Liquid samples must be extracted within 7 days and their extracts analyzed within 40 days. Concentrated waste, soil, sediment, and sludge samples must be extracted within 14 days and their extracts analyzed within 40 days.

SAMPLE PREPARATION
Preparation of soil, sediment, and other solid samples — Acidify 30 g (dry weight) solids with 0.1 M phosphate buffer (pH = 2.5) and thoroughly mix the contents. Spike the sample with surrogate compound(s). The ultrasonic extraction of solids must be optimized for each type of sample. In order for the ultrasonic extractor to efficiently extract solid samples, the sample must be free flowing when the solvent is added. Acidified anhydrous sodium sulfate should be added to clay-type soils, or any other solid that is not a free-flowing sandy texture, until a free flowing mixture is obtained. Add methylene chloride and perform ultrasonic extraction. Combine organic extracts from the repetitive extractings of the sample and centrifuge. Add aqueous potassium hydroxide, water, and methanol to the extract and reflux the mixture on a water bath. Extract the solution three times with methylene chloride and discard the methylene chloride phase. The basic solution contains the herbicide salts. Adjust the pH of the solution to <2 with cold sulfuric acid and extract three times with methylene chloride. Combine the extracts and pour them through a pre-rinsed drying column containing acidified anhydrous sodium sulfate. Collect the dried extracts in a K-D flask and concentrate them.

Preparation of aqueous samples — Measure 1 L of sample into a 2 L separatory funnel and spike it with surrogate compound(s). Add NaCl to the sample, then add 6 N NaOH to the sample to a pH of 12 or more and let the sample sit at room temperature for 1 h to hydrolyze esters. Extract the sample three times with methylene chloride and discard the extracts. Then add cold 12 N sulfuric acid to a pH less than or equal to 2, and extract the sample three times with ethyl ether. Collect the ether phase in a flask containing acidified anhydrous sodium sulfate and allow it to remain in contact with the sodium sulfate for a minimum of 2 h. The drying step is very critical to ensuring complete esterification; any moisture remaining in the ether will result in low herbicide recoveries.

Extract concentration and derivatization — The combined ether extract is concentrated to about 1 mL using a K-D apparatus followed by using a micro Snyder column or nitrogen gas blowdown. If methyl esters are to be produced, then dilute the concentrated ether extract with 1 mL of isooctane and 0.5 mL of methanol, dilute to a final volume of 4 mL, and esterify with diazomethane. If pentafluorobenzene esters are to be produced, then dilute concentrated ether extract with acetone to a final volume of 4 mL and esterify with pentafluorobenzyl bromide.

QUALITY CONTROL Select a representative spike concentration for each compound (acid or ester) to be measured. Using stock standard, prepare a quality control check sample concentrate, in acetone, that is 1000 times more concentrated than the selected concentrations. Use this quality control check sample concentrate to prepare quality control check samples. Calculate surrogate standard recovery on all standards, samples, blanks, and spikes. GC/MS techniques should be judiciously employed to support qualitative identifications made with this method. When available, chemical ionization mass spectra may be employed to aid the qualitative identification process.

REFERENCE Test Methods for Evaluating Solid Waste, Physical/Chemical Methods, SW-846, 3rd Edition, U.S. EPA, Office of Solid Waste, Washington, DC, EPA Method 8151, Nov. 1990.

Dalapon — EPA Method 8150
CAS #75-99-0

TITLE Chlorinated Herbicides

MATRIX Groundwater, soils, sludges, water miscible liquid wastes, and non-water miscible wastes.

APPLICATION This method is used for the analysis of 10 chlorinated herbicides. Samples are extracted, hydrolyzed with potassium hydroxide, and extraneous organics are removed by a solvent wash. After acidification, the acids are extracted, concentrated and converted to their methyl esters using diazomethane. They are then analyzed using direct injection into a gas chromatograph (GC). Be very careful because diazomethane can explode under certain conditions and it is also a carcinogen.

INTERFERENCES Organic acids and phenols (especially chlorinated acids and phenols) may cause interferences. Phthalate esters are not as significant an interference as with other GC-ECD methods if an electron capture detector is used. The herbicides may react readily with alkaline substances and be lost during analysis so all glassware and glass wool must be acid rinsed and sodium sulfate must be acidified with sulfuric acid prior to use. Sensitivity usually depends on the level of interferences rather than on instrumentation.

INSTRUMENTATION GC capable of on-column injections and an electron capture detector (ECD) or a halogen specific detector. Column 1: 1.8 m by 4 mm with 1.5% SP-2250/1.95% SP-2401 on Supelcoport. Column 2: 1.8 m by 4 mm with 5% OV-210 on Gas Chrom Q. Column 3: 1.98 m by 2 mm with 0.1%. SP-1000 on Carbopack C. The preferred column is Column Number 3.

RANGE Not listed.

MDL 5.8 µg/L (in reagent water; ECD).

PQL FACTORS FOR MULTIPLYING × FID MDL VALUE

Matrix	Multiplication Factor
Groundwater	10
Low-level soil by sonication with GPC cleanup	670
High-level soil and sludge by sonication	10,000
Non-water miscible waste	100,000

PRECISION (as standard deviation) 8% with 23.4 µg/L spike in drinking water. 5% with 200 µg/L in municipal water.

ACCURACY (as mean recovery) 66% with 23.4 µg/L spike in drinking water. 65% with 200 µg/L in municipal water.

SAMPLING METHOD Use 8-oz. widemouth glass bottles with Teflon®-lined caps for concentrated waste samples, soils, sediments, and sludges. Use 1 or 2½ gallon amber glass bottles with Teflon®-lined caps for liquid (water) samples.

STABILITY Cool soil, sediment, sludge, and liquid samples to 4°C. If residual chlorine is present in liquid samples add 3 mL of 10% sodium thiosulfate per gallon of sample and cool to 4°C.

MHT 14 days for concentrated waste, soil, sediment, or sludge; 7 days for liquid samples; all extracts must be analyzed within 40 days.

QUALITY CONTROL A quality control check sample concentrate containing this compound in acetone at a concentration 1,000 times more concentrated than the selected spike concentration is required. The QC check sample concentrate may be prepared from pure standard materials or purchased as certified solutions. Use appropriate trip, matrix, control site, method, reagent, and solvent blanks. Internal, surrogate, and five concentration level calibration standards are used.

REFERENCE Method 8150, SW-846, 3rd ed., Sept. 1986.

2,4-DB   EPA Method 8151
CAS #94-82-6

TITLE Chlorinated Herbicides by GC Using Methylation or Pentafluorobenzylation Derivatization: Capillary Column Technique.

MATRIX This method covers aqueous and solid matrices. This includes a wide variety such as drinking water, groundwater, industrial wastewaters, surface waters, soils, solids, and sediments.

METHOD SUMMARY This is a GC method for determining 19 chlorinated acid herbicides in aqueous, soil, and waste matrices. Because these compounds are produced and used in various forms (i.e., acid, salt, ester, etc.) a hydrolysis step is included to convert the herbicide to the acid form prior to analysis. This method provides hydrolysis, extraction, derivatization and GC conditions for the analysis of chlorinated acid herbicides in water, soil, and waste samples. Water samples are hydrolyzed *in situ*, extracted with diethyl ether, and then esterified with either diazomethane or pentafluorobenzyl bromide. The derivatives are determined by gas chromatography with an electron capture detector (GC/ECD). The results are reported as acid equivalents. The sensitivity of this method depends on the level of interferences in addition to instrumental limitations.

INTERFERENCES Method interferences may be caused by contaminants in solvents, reagents, glassware, and other sample processing hardware. Immediately prior to use, glassware should be rinsed with the next solvent to be used. Matrix interferences may be caused by contaminants that are coextracted from the sample. Organic acids, especially chlorinated acids, cause the most direct interference with the determination by methylation. Phenols, including chlorophenols, may also interfere with this procedure. The determination using pentafluorobenzylation is more sensitive, and more prone to interferences from the presence of organic acids of phenols than by methylation. Alkaline hydrolysis and subsequent extraction of the basic solution removes many chlorinated hydrocarbons and phthalate esters that might otherwise interfere with the ECD analysis. The herbicides, being strong organic acids, react readily with alkaline substances and may be lost during analysis. Therefore, glassware must be acid-rinsed and then rinsed to constant pH with organic-free reagent water.

INSTRUMENTATION A GC suitable for Grob-type injection using capillary columns. A data system for measuring peak heights and/or peak areas is recommended. An electron capture detector (ECD) is used. Also a K-D apparatus, a diazomethane generator, a centrifuge and an ultrasonic disrupter will be required.

Narrow Bore Columns:
Primary Column 1: 30 m × 0.25 mm, 5% phenyl/95% methyl silicone (DB-5), 0.25 µm film thickness.
Primary Column 1a (GC/MS): 30 m × 0.32 mm, 5% phenyl/95% methyl silicone (DB-5), 1-µm film thickness.
Column 2: 30 m × 0.25 mm DB-608 with a 25 µm film thickness.
Confirmation Column: 30 m × 0.25 mm, 14% cyanopropyl phenyl silicone (DB-1701), 0.25 µm film thickness.

Megabore Columns:
Primary Column: 30 m × 0.53 mm DB-608 with 0.83 µm film thickness.
Confirmation Column: 30 m × 0.53 mm, 14% cyanopropyl phenyl silicone (DB-1701), 1.0 µm film thickness.

PRECISION & ACCURACY Method detection limits (MDLs) are compound-dependent and vary with derivitization efficiency, derivative recovery, the matrix sampled, and herbicide concentration.

The estimated MDL (in µg/L) was 0.8 for aqueous samples using GC/ECD.

The estimated MDL (in µg/kg) was not reported for soil samples using GC/ECD when corrected back to 50-g samples extracted and concentrated to 10 mL with 5-µL injections.

The estimated GC/MS identification limit (in ng) was not reported for soil samples using GC/MS.

Mean percent recovery, calculated from 7–8 determinations of spiked reagent water, after diazomethane derivatization, from a spike concentration (in µg/L) of 4 was 87 with a standard deviation of the percent recovery of 13.1.

Mean percent recovery, calculated from 10 determinations of spiked clay and clay/still bottom samples over the linear concentration range (in ng/g) of 4.0–8,060 was 90.7 with a percent relative standard deviation of 7.6. The RSD % was calculated on 10 samples high in the linear concentration range and 10 low in the range. The linear concentration range was determined using standard solutions and corrected to 50 g soil samples.

SAMPLE COLLECTION, PRESERVATION & HANDLING Containers used to collect samples for the determination of semivolatile organic compounds should be soap and water washed followed by methanol (or isopropanol) rinsing. The sample containers should be of glass or Teflon® and have screw-top covers with Teflon® liners.

No preservation is used with concentrated waste samples. With liquid samples containing no residual chlorine and with soil, sediment, and sludge samples, immediately cooling to 4°C is the only preservation used. When residual chlorine is present then 3 mL of 10% aqueous sodium sulfate is added for each gallon of sample collected, followed by cooling to 4°C.

The holding time for all volatile organics samples is 14 days. Liquid samples must be extracted within 7 days and their extracts analyzed within 40 days. Concentrated waste, soil, sediment, and sludge samples must be extracted within 14 days and their extracts analyzed within 40 days.

SAMPLE PREPARATION

Preparation of soil, sediment, and other solid samples — Acidify 30 g (dry weight) solids with 0.1 M phosphate buffer (pH = 2.5) and thoroughly mix the contents. Spike the sample with surrogate compound(s). The ultrasonic extraction of solids must be optimized for each type of sample. In order for the ultrasonic extractor to efficiently extract solid samples, the sample must be free flowing when the solvent is added. Acidified anhydrous sodium sulfate should be added to clay-type soils, or any other solid that is not a free-flowing sandy texture, until a free flowing mixture is obtained. Add methylene chloride and perform ultrasonic extraction. Combine organic extracts from the repetitive extractings of the sample and centrifuge. Add aqueous potassium hydroxide, water, and methanol to the extract and reflux the mixture on a water bath. Extract the solution three times with methylene chloride and discard the methylene chloride phase. The basic solution contains the herbicide salts. Adjust the pH of the solution to <2 with cold sulfuric acid and extract three times with methylene chloride. Combine the extracts and pour them through a pre-rinsed drying column containing acidified anhydrous sodium sulfate. Collect the dried extracts in a K-D flask and concentrate them.

Preparation of aqueous samples — Measure 1 L of sample into a 2 L separatory funnel and spike it with surrogate compound(s). Add NaCl to the sample, then add 6 N NaOH to the sample to a pH of 12 or more and let the sample sit at room temperature for 1 h to hydrolyze esters. Extract the sample three times with methylene chloride and discard the extracts. Then add cold 12 N sulfuric acid to a pH less than or equal to 2, and extract the sample three times with ethyl ether. Collect the ether phase in a flask containing acidified anhydrous sodium sulfate and allow it to remain in contact with the sodium sulfate for a minimum of 2 h. The drying step is very critical to ensuring complete esterification; any moisture remaining in the ether will result in low herbicide recoveries.

Extract concentration and derivatization — The combined ether extract is concentrated to about 1 mL using a K-D apparatus followed by using a micro Snyder column or nitrogen gas blowdown. If methyl esters are to be produced, then dilute the concentrated ether extract with 1 mL of isooctane and 0.5 mL of methanol, dilute to a final volume of 4 mL, and esterify with diazomethane. If pentafluorobenzene esters are to be produced, then dilute concentrated ether extract with acetone to a final volume of 4 mL and esterify with pentafluorobenzyl bromide.

QUALITY CONTROL Select a representative spike concentration for each compound (acid or ester) to be measured. Using stock standard, prepare a quality control check sample concentrate, in acetone, that is 1000 times more concentrated than the selected concentrations. Use this quality control check sample concentrate to prepare quality control check samples. Calculate surrogate standard recovery on all standards, samples, blanks, and spikes. GC/MS techniques should be judiciously employed to support qualitative identifications made with this method. When available, chemical ionization mass spectra may be employed to aid the qualitative identification process.

REFERENCE Test Methods for Evaluating Solid Waste, Physical/Chemical Methods, SW-846, 3rd Edition, U.S. EPA, Office of Solid Waste, Washington, DC, EPA Method 8151, Nov. 1990.

2,4-DB   **EPA Method 8150**
CAS #94-82-6

TITLE Chlorinated Herbicides

MATRIX Groundwater, soils, sludges, water miscible liquid wastes, and non-water miscible wastes.

APPLICATION This method is used for the analysis of 10 chlorinated herbicides. Samples are extracted, hydrolyzed with potassium hydroxide, and extraneous organics are removed by a solvent wash. After acidification, the acids are extracted, concentrated and converted to their methyl esters using diazomethane. They are then analyzed using direct injection into a gas chromatograph (GC). Be very careful because diazomethane can explode under certain conditions and it is also a carcinogen.

INTERFERENCES Organic acids and phenols (especially chlorinated acids and phenols) may cause interferences. Phthalate esters are not as significant an interference as with other GC-ECD methods if an electron capture detector is used. The herbicides may react readily with alkaline substances and be lost during analysis so all glassware and glass wool must be acid rinsed and sodium sulfate must be acidified with sulfuric acid prior to use. Sensitivity usually depends on the level of interferences rather than on instrumentation.

INSTRUMENTATION GC capable of on-column injections and an electron capture detector (ECD) or a halogen specific detector. Column 1: 1.8 m by 4 mm with 1.5% SP-2250/1.95% SP-2401 on Supelcoport. Column 2: 1.8 m by 4 mm with 5% OV-210 on Gas Chrom Q. Column 3: 1.98 m by 2 mm with 0.1%. SP-1000 on Carbopack C. The preferred column is Column Number 1.

RANGE Not listed.

MDL 0.91 µg/L (in reagent water; ECD).

PQL FACTORS FOR MULTIPLYING × FID MDL VALUE

Matrix	Multiplication Factor
Groundwater	10
Low-level soil by sonication with GPC cleanup	670
High-level soil and sludge by sonication	10,000
Non-water miscible waste	100,000

PRECISION (as standard deviation) 3% with 10.3 µg/L spike in drinking water. 3% with 10.4 µg/L in municipal water.

ACCURACY (as mean recovery) 93% with 10.3 µg/L spike in drinking water. 93% with 10.4 µg/L in municipal water.

SAMPLING METHOD Use 8-oz. widemouth glass bottles with Teflon®-lined caps for concentrated waste samples, soils, sediments, and sludges. Use 1 or 2½ gallon amber glass bottles with Teflon®-lined caps for liquid (water) samples.

STABILITY Cool soil, sediment, sludge, and liquid samples to 4°C. If residual chlorine is present in liquid samples add 3 mL of 10% sodium thiosulfate per gallon of sample and cool to 4°C.

MHT 14 days for concentrated waste, soil, sediment, or sludge; 7 days for liquid samples; all extracts must be analyzed within 40 days.

QUALITY CONTROL A quality control check sample concentrate containing this compound in acetone at a concentration 1,000 times more concentrated than the selected spike concentration is required. The QC check sample concentrate may be prepared from pure standard materials or purchased as certified solutions. Use appropriate trip, matrix, control site, method, reagent, and solvent blanks. Internal, surrogate, and five concentration level calibration standards are used.

REFERENCE Method 8150, SW-846, 3rd ed., Sept. 1986.

DCPA diacid   **EPA Method 8151**
CAS #2136-79-0

TITLE Chlorinated Herbicides by GC Using Methylation or Pentafluorobenzylation Derivatization: Capillary Column Technique.

MATRIX This method covers aqueous and solid matrices. This includes a wide variety such as drinking water, groundwater, industrial wastewaters, surface waters, soils, solids, and sediments.

METHOD SUMMARY This is a GC method for determining 19 chlorinated acid herbicides in aqueous, soil, and waste matrices. Because these compounds are produced and used in various forms (i.e., acid, salt, ester, etc.) a hydrolysis step is included to convert the herbicide to the acid form prior to analysis. This method provides hydrolysis, extraction, derivatization and GC conditions for the analysis of chlorinated acid herbicides in water, soil, and waste samples. Water samples are hydrolyzed *in situ*, extracted with diethyl ether, and then esterified with either diazomethane or pentafluorobenzyl bromide. The derivatives are determined by gas chromatography with an electron capture detector (GC/ECD). The results are reported as acid equivalents. The sensitivity of this method depends on the level of interferences in addition to instrumental limitations.

INTERFERENCES Method interferences may be caused by contaminants in solvents, reagents, glassware, and other sample processing hardware. Immediately prior to use, glassware should be rinsed with the next solvent to be used. Matrix interferences may be caused by contaminants that are coextracted from the sample. Organic acids, especially chlorinated acids, cause the most direct interference with the determination by methylation. Phenols, including chlorophenols, may also interfere with this procedure. The determination using pentafluorobenzylation is more sensitive, and more prone to interferences from the presence of organic acids of phenols than by methylation. Alkaline hydrolysis and subsequent extraction of the basic solution removes many chlorinated hydrocarbons and phthalate esters that might otherwise interfere with the ECD analysis. The herbicides, being strong organic acids, react readily with alkaline substances and may be lost during analysis. Therefore, glassware must be acid-rinsed and then rinsed to constant pH with organic-free reagent water.

INSTRUMENTATION A GC suitable for Grob-type injection using capillary columns. A data system for measuring peak heights and/or peak areas is recommended. An electron capture detector (ECD) is used. Also a K-D apparatus, a diazomethane generator, a centrifuge and an ultrasonic disrupter will be required.

Narrow Bore Columns:
Primary Column 1: 30 m × 0.25 mm, 5% phenyl/95% methyl silicone (DB-5), 0.25 µm film thickness.
Primary Column 1a (GC/MS): 30 m × 0.32 mm, 5% phenyl/95% methyl silicone (DB-5), 1-µm film thickness.
Column 2: 30 m × 0.25 mm DB-608 with a 25 µm film thickness.
Confirmation Column: 30 m × 0.25 mm, 14% cyanopropyl phenyl silicone (DB-1701), 0.25 µm film thickness.

Megabore Columns:
Primary Column: 30 m × 0.53 mm DB-608 with 0.83 µm film thickness.
Confirmation Column: 30 m × 0.53 mm, 14% cyanopropyl phenyl silicone (DB-1701), 1.0 µm film thickness.

PRECISION & ACCURACY Method detection limits (MDLs) are compound-dependent and vary with derivitization efficiency, derivative recovery, the matrix sampled, and herbicide concentration.

The estimated MDL (in µg/L) was 0.02 for aqueous samples using GC/ECD.

The estimated MDL (in µg/kg) was not reported for soil samples using GC/ECD when corrected back to 50-g samples extracted and concentrated to 10 mL with 5-µL injections.

The estimated GC/MS identification limit (in ng) was not reported for soil samples using GC/MS.

Mean percent recovery, calculated from 7–8 determinations of spiked reagent water, after diazomethane derivatization, from a spike concentration (in µg/L) of 0.2 was 74 with a standard deviation of the percent recovery of 9.7.

Mean percent recovery, calculated from 10 determinations of spiked clay and clay/still bottom samples over the linear concentration range (in ng/g) of no data was none reported with a percent relative standard deviation of none. The RSD % was calculated on 10 samples high in the linear concentration range and 10 low in the range. The linear concentration range was determined using standard solutions and corrected to 50 g soil samples.

SAMPLE COLLECTION, PRESERVATION & HANDLING
Containers used to collect samples for the determination of semivolatile organic compounds should be soap and water washed followed by methanol (or isopropanol) rinsing. The sample containers should be of glass or Teflon® and have screw-top covers with Teflon® liners.

No preservation is used with concentrated waste samples. With liquid samples containing no residual chlorine and with soil, sediment, and sludge samples, immediately cooling to 4°C is the only preservation used. When residual chlorine is present then 3 mL of 10% aqueous sodium sulfate is added for each gallon of sample collected, followed by cooling to 4°C.

The holding time for all volatile organics samples is 14 days. Liquid samples must be extracted within 7 days and their extracts analyzed within 40 days. Concentrated waste, soil, sediment, and sludge samples must be extracted within 14 days and their extracts analyzed within 40 days.

SAMPLE PREPARATION
Preparation of soil, sediment, and other solid samples — Acidify 30 g (dry weight) solids with 0.1 M phosphate buffer (pH = 2.5) and thoroughly mix the contents. Spike the sample with surrogate compound(s). The ultrasonic extraction of solids must be optimized for each type of sample. In order for the ultrasonic extractor to efficiently extract solid samples, the sample must be free flowing when the solvent is added. Acidified anhydrous sodium sulfate should be added to clay-type soils, or any other solid that is not a free-flowing sandy texture, until a free flowing mixture is obtained. Add methylene chloride and perform ultrasonic extraction. Combine organic extracts from the repetitive extractings of the sample and centrifuge. Add aqueous potassium hydroxide, water, and methanol to the extract and reflux the mixture on a water bath. Extract the solution three times with methylene chloride and discard the methylene chloride phase. The basic solution contains the herbicide salts. Adjust the pH of the solution to <2 with cold sulfuric acid and extract three times with methylene chloride. Combine the extracts and pour them through a pre-rinsed drying column containing acidified anhydrous sodium sulfate. Collect the dried extracts in a K-D flask and concentrate them.

Preparation of aqueous samples — Measure 1 L of sample into a 2 L separatory funnel and spike it with surrogate compound(s). Add NaCl to the sample, then add 6 N NaOH to the sample to a pH of 12 or more and let the sample sit at room temperature for 1 h to hydrolyze esters. Extract the sample three times with methylene chloride and discard the extracts. Then add cold 12 N sulfuric acid to a pH less than or equal to 2, and extract the sample three times with ethyl ether. Collect the ether phase in a flask containing acidified anhydrous sodium sulfate and allow it to remain in contact with the sodium sulfate for a minimum of 2 h. The drying step is very critical to ensuring complete esterification; any moisture remaining in the ether will result in low herbicide recoveries.

Extract concentration and derivatization — The combined ether extract is concentrated to about 1 mL using a K-D apparatus followed by using a micro Snyder column or nitrogen gas blowdown. If methyl esters are to be produced, then dilute the concentrated ether extract with 1 mL of isooctane and 0.5 mL of methanol, dilute to a final volume of 4 mL, and esterify with diazomethane. If pentafluorobenzene esters are to be produced, then dilute concentrated ether extract with acetone to a final volume of 4 mL and esterify with pentafluorobenzyl bromide.

QUALITY CONTROL   Select a representative spike concentration for each compound (acid or ester) to be measured. Using stock standard, prepare a quality control check sample concentrate, in acetone, that is 1000 times more concentrated than the selected concentrations. Use this quality control check sample concentrate to prepare quality control check samples. Calculate surrogate standard recovery on all standards, samples, blanks, and spikes. GC/MS techniques should be judiciously employed to support qualitative identifications made with this method. When available, chemical ionization mass spectra may be employed to aid the qualitative identification process.

REFERENCE   Test Methods for Evaluating Solid Waste, Physical/Chemical Methods, SW-846, 3rd Edition, U.S. EPA, Office of Solid Waste, Washington, DC, EPA Method 8151, Nov. 1990.

4,4'-DDD   **EPA Method 625**
CAS #72-54-8

TITLE   Base/Neutrals and Acids, U.S. EPA Method 625

MATRIX   This methods covers municipal and industrial wastewaters.

METHOD SUMMARY   Approximately 1 L of sample is serially extracted with methylene chloride at a pH greater than 11 and again at a pH less than 2 using a separatory funnel or a continuous extractor. The methylene chloride extract is dried, concentrated to a volume of 1 mL, and analyzed by GC/MS. Qualitative identification of the parameters in the extract is performed using the retention time and the relative abundance of three characteristic masses (m/z). Qualitative analysis is performed using either external or internal standard techniques with a single characteristic m/z.

INTERFERENCES   Method interferences may be caused by contaminants in solvents, reagents, glassware, and other sample processing hardware. Glassware must be scrupulously cleaned. Glassware should be heated in a muffle furnace at 400°C for 5 to 30 min. Some thermally stable materials, such as PCBs, may not be eliminated by this treatment. Solvent rinses with acetone and pesticide quality hexane may be substituted for the muffle furnace heating. Matrix interferences may be caused by contaminants that are coextracted from the sample. The base-neutral extraction may cause significantly reduced recovery of phenols. The packed gas chromatographic columns recommended for the basic fraction may not exhibit sufficient resolution for some analytes.

INSTRUMENTATION A GC/MS system with an injection port designed for on-column injection when using packed columns and for splitless injection when using capillary columns.

Column for base/neutrals: 1.8 m long × 2 mm I.D. glass, packed with 3% SP-2550 on Supelcoport (100/120 mesh) or equivalent.

Column for acids: 1.8 m long × 2 mm I.D. glass, packed with 1% SP-1240DA on Supelcoport (100/120 mesh) or equivalent.

PRECISION & ACCURACY The MDL concentrations were obtained using reagent water. The MDL actually achieved in a given analysis will vary depending on instrument sensitivity and matrix effects. This method was tested by 15 laboratories using reagent water, drinking water, surface water, and industrial wastewaters spiked at six concentrations over the range 5 to 100 µg/L. Single operator precision, overall precision, and method accuracy were found to be directly related to the concentration of the parameter matrix.

The MDL (in µg/L) in reagent water was 2.8.

The standard deviation (in µg/L based on 4 recovery measurements) was 31.0.

The range (in µg/L) for average recovery for 4 measurements was D-134.5.

The range (in %) for percent recovery was D-145.

Accuracy (in µg/L) as expected recovery for one or more measurements of a sample containing a true concentration of C was 0.56C-0.40.

Precision (in µg/L) as expected single analyst standard deviation of measurements at an average concentration found at X was 0.29X–0.32.

Overall precision (in µg/L) as expected interlaboratory standard deviation of measurements in an average concentration found at X was 0.66X–0.96.

C = *True value of the concentration in µg/L.*

X = *Average recovery found for measurements of samples containing a concentration at C in µg/L.*

SAMPLE PREPARATION Adjust the pH to >11 with sodium hydroxide and serially extract in a separatory funnel with methylene chloride or else in a continuous extractor. Next, adjust the pH to <2 with sulfuric acid and serially extract in a separatory funnel with methylene chloride or else in a continuous extractor. Dry the extracts separately through a column of anhydrous sodium sulfate and then concentrate each of the extracts to 1.0 mL using a K-D apparatus.

SAMPLE COLLECTION, PRESERVATION & HANDLING Grab samples must be collected in glass containers. All samples must be refrigerated at 4°C from the time of collection until extraction. If residual chlorine is present, add 80 mg of sodium thiosulfate/L of sample and mix well. All samples must be extracted within 7 days of collection and completely analyzed within 40 days of extraction.

QUALITY CONTROL Make an initial, one-time, demonstration of the ability to generate acceptable accuracy and precision with this method. Before processing any samples, the analyst must analyze a reagent water blank to demonstrate that interferences from the analytical system and glassware are under control. Each time a set of samples is extracted or reagents are changed, a reagent water blank must be processed. Spike and analyze a minimum of 5% of all samples to monitor and evaluate lab data quality. A QC check sample concentrate that contains each parameter of interest at a concentration of 100 µg/mL in acetone is required. PCBs and multicomponent pesticides may be omitted from this test.

After analysis of five spiked wastewater samples, calculate the average percent recovery and the standard deviation of the percent recovery. Spike all samples with the surrogate standard spiking solution and calculate the percent recovery of each surrogate compound.

REFERENCE *Federal Register*, Vol. 49, No. 209. Friday, Oct. 26, 1984.

4,4'-DDD   **EPA Method 8080**
CAS #72-54-8

TITLE Organochlorine Pesticides and Polychlorinated Biphenyls By Gas Chromatography

MATRIX This method is used to determine the concentration of various organochlorine pesticides and polychlorinated biphenyls in extracts prepared from water, groundwater, soils, and sediments.

METHOD SUMMARY This method covers 26 pesticides and Aroclor (PCB) mixtures and it is suitable for monitoring-type analyses. After extraction, concentration and solvent exchange to hexane, a 2- to 5-µL sample aliquot is injected into a GC using the solvent flush technique, and the analytes are detected by an electron capture detector (ECD) or an electrolytic conductivity detector in the halogen mode (HECD). Both neat and diluted organic liquids may be analyzed by direct injection.

INTERFERENCES Interferences coextracted from the samples will vary considerably from source to source. Interferences by phthalate esters can pose a major problem in pesticide determinations when using the ECD. Cross-contamination of clean glassware routinely occurs when plastics are handled during extraction steps, especially when solvent-wetted surfaces are handled. The contamination from phthalate esters can be completely eliminated with a microcoulometric or electrolytic conductivity detector. Solvents, reagent, glassware, and other sample processing hardware may yield artifacts and/or interferences to sample analysis.

INSTRUMENTATION A gas chromatograph capable of on-column injections is needed. It must be equipped with an ECD or a HECD and one of the following GC columns:

Column 1: Supelcoport (100/120 mesh) coated with 1.5% SP-2250/1.95% SP-2401 packed in a 1.8 m × 4 mm I.D. glass column.

Column 2: Supelcoport (100/120 mesh) coated with 3% OV-1 in a 1.8 m × 4 mm I.D. glass column.

PRECISION & ACCURACY The method was tested by 20 laboratories using organic-free reagent water, drinking water, surface water, and three industrial wastewaters spiked at six concentrations. Concentrations used in the study ranged from 0.5 to 30 µg/L for single-component pesticides and from 8.5 to 400 µg/L for multicomponent parameters. Overall precision and method accuracy were found to be directly related to the concentration of the analyte and essentially independent of the sample matrix. The sensitivity of this method usually depends on the concentration of interferences rather than on instrumental limitations.

MDL in µg/L was 0.011.
Concentration range in µg/L was 0.5–30.
Accuracy as recovery (x^*) in µg/L was $0.84C+0.30$.
Overall precision (S^*) in µg/L was $0.27x-0.14$.

x^* *Expected recovery for one or more measurements of a sample containing concentration C, in µg/L.*
$S^* =$ *Expected interlaboratory standard deviation of measurements at an average concentration found of the analyte in µg/L.*
$C =$ *True value for the concentration, in µg/L.*
$X =$ *Average recovery found for measurements of samples containing a concentration of C, in µg/L.*

SAMPLING METHOD
Liquid samples — Use a 1 or 2½ gallon amber glass bottle with a screw-top Teflon®-lined cover. Pre-wash the bottle with detergent, rinse with distilled water and methanol (or isopropanol).

Soil, sediments, and sludges — Use an 8-oz. widemouth glass with a screw-top Teflon®-lined cover. Pre-wash the bottle with detergent, rinse with distilled water and methanol (or isopropanol).

SAMPLE PRESERVATION Cool water, soil, sediment, or sludge samples immediately to 4°C.

Water samples — If residual chlorine is present, add 3 mL of 10% sodium thiosulfate per gallon and cool to 4°C. All extracts and samples should be stored under refrigeration.

MHT Liquid samples must be extracted within 7 days and the extracts must be analyzed within 40 days. Soils, sediments, and sludges may be stored for a maximum of 14 days prior to extraction.

SAMPLE PREPARATION
Liquid samples — Extract 1 L samples in a continuous extractor at pH 5–9 with methylene chloride after adding 1.0 mL of surrogate spiking solution to each sample. Pass the extract through a column of anhydrous sodium sulfate to dry and concentrate it in a K-D apparatus to 1 mL volume.

Soils, sediments, and sludges — Rapidly weigh approximately 30 g of sample into a 400-mL beaker to avoid loss of the more volatile extractables. Nonporous or wet samples (gummy or clay type) that do not have a free-flowing sandy texture must be mixed with anhydrous sodium sulfate until the sample is free flowing. Add 1 mL of surrogate standards to all samples, spikes, standards, and blanks. Add 100 mL of 1:1 methylene chloride:acetone and extract ultrasonically. Decant and filter extracts, dry the extract by passing it through a drying column containing anhydrous sodium sulfate and concentrate to 1 mL in a K-D apparatus.

Hexane solvent exchange — Add 50 mL of hexane, a new boiling chip, and concentrate until the apparent volume of liquid reaches 1 mL. Adjust the extract volume to 10.0 mL. Stopper the concentration tube and store refrigerated at 4°C if further processing will not be performed immediately. If the extract will be stored longer than two days, transfer it to a vial with Teflon®-lined screw-cap or crimp top.

QUALITY CONTROL Demonstrate through the analysis of a reagent water blank, that all glassware and reagents are interference free. Each time a set of samples is processed, a method blank should be processed as a safeguard against chronic lab contamination. A reagent blank, a matrix spike, and a duplicate or matrix spike duplicate must be performed for each analytical batch (up to a maximum of 20 samples) analyzed.

Analytical system performance must be verified by analyzing QC check samples. The QC check sample concentration should contain each single-component analyte at the following concentrations in acetone: 4,4'-DDD, 10 µg/mL; 4,4'-DDT, 10 µg/mL; endosulfan II, 10 µg/mL; endosulfan sulfate, 10 µg/mL; and any other single-component pesticide at 2 µg/mL. If the method is only to be used to analyze PCBs, Chlordane, or Toxaphene, the QC check sample concentrate should contain the most representative multicomponent parameter at a concentration of 50 µg/mL in acetone.

REFERENCE Test Methods for Evaluating Solid Waste (SW-846). U.S. EPA. 1983. Method 8080B, Rev. 2, Nov. 1990. Office of Solid Wastes, Washington, DC.

4,4'-DDD   **EPA Method 8270**
CAS #72-54-8

TITLE Semivolatile Organic Compounds by GC/MS

MATRIX This method is used to determine the concentration of semivolatile organic compounds in extracts prepared from all types of solid waste matrices, soils, and groundwater. Although surface waters are not specifically mentioned, this method should be applicable to water samples from rivers, lakes, etc.

METHOD SUMMARY This method covers 259 semivolatile organic compounds. In very limited applications direct injection of the sample into the GC/MS system may be appropriate, but this results in very high detection limits (approximately 10,000 µg/L). Typically, a 1-L liquid sample, containing surrogate, and matrix spiking standards, is extracted in a continuous extractor first under acid conditions and then under basic conditions. Typically 30 g of a solid sample, containing surrogate, and matrix spiking standards, is extracted ultrasonically. After concentrating the extract to 1 mL it is spiked with 10 µL of an internal standard solution just prior to analysis by GC/MS. The volume injected should contain about 100 ng of base/neutral and 200 ng of acid surrogates (for a 1-µL injection). Analysis is performed by GC/MS using a capillary GC column.

INTERFERENCES Raw GC/MS data from all blanks, samples, and spikes must be evaluated for interferences. Contamination by carryover can occur whenever high-concentration and low-concentration samples are sequentially analyzed. To reduce carryover, the sample syringe must be rinsed out between samples with solvent. Whenever an unusually concentrated sample is encountered, it should be followed by the analysis of blank solvent to check for cross-contamination.

INSTRUMENTATION A GC/MS and a data system are required. The GC column used is a 30 m × 0.25 mm I.D. (or 0.32 mm I.D.) 1um film thickness silicone-coated fused silica capillary column. A continuous liquid-liquid extractor equipped with Teflon® or glass connection joints and stopcocks requiring no lubrication, a K-D concentrating apparatus, water bath, and an ultrasonic disrupter with a minimum power of 300 W and with pulsing capability are also required.

PRECISION & ACCURACY The estimated quantitation limit (EQL) of Method 8270B for determining an individual compound is approximately 1 mg/kg (wet weight) for soil or sediment samples, 1–200 mg/kg for wastes (dependent on matrix and method of preparation), and 10 µg/L for groundwater samples. EQLs will be proportionately higher for sample extracts that require dilution to avoid saturation of the detector.

The EQL(b) for groundwater in µg/L is not listed.
The EQL (a, b) for low concentrations in soil and sediment in µg/kg is not listed.
Accuracy as µg/L is 0.56C–0.40.
Overall precision in µg/L is 0.66X–0.96.

(a) EQLs listed for soil/sediment are based on wet weight. Normally data is reported in a dry-weight basis; therefore, EQLs will be higher based on the % dry weight of each sample. This calculation is based on a 30 g sample and gel permeation chromatography cleanup.
(b) Sample EQLs are highly matrix-dependent. The EQLs are provided for guidance and may not always be achievable.
C = *True value for concentration, in µg/L.*
X = *Average recovery found for measurements of samples containing a concentration of C, in µg/L.*

ESTIMATED QUANTITATION LIMIT

Other Matrices	Factor (a)
High-concentration soil and sludges by sonicator	7.5
Non-water miscible waste	75

(a) EQL for other matrices = [EQL for low soil/sediment] × [Factor]. This estimated EQL is similar to an EPA "Practical Quantitation Limit."

SAMPLING METHOD

Liquid samples — Use a 1 or 2½ gallon amber glass bottle with a screw-top Teflon®-lined cover that has been prewashed with detergent and rinsed with distilled water and methanol (or isopropanol).

Soils, sediments, or sludges — Use an 8-oz. widemouth glass with a screw-top Teflon®-lined cover that has been prewashed with detergent and rinsed with distilled water and methanol (or isopropanol).

SAMPLE PRESERVATION
Liquid samples — If residual chlorine is present, add 3 mL of 10% sodium thiosulfate per gallon, cool to 4°C and store in a solvent-free refrigerator until analysis; if chlorine is not present, then eliminate the sodium thiosulfate addition.

Soils, sediments, or sludges — Cool samples to 4°C and store in a solvent-free refrigerator.

MHT Liquid samples must be extracted within 7 days and the extracts analyzed within 40 days. Soils, sediments, or sludges may be stored for a maximum of 14 days and the extracts analyzed within 40 days.

SAMPLE PREPARATION
Liquid samples — Transfer 1 L quantitatively to a continuous extractor. If high concentrations are anticipated, a smaller volume may be used and then diluted with organic-free reagent water to 1 L. Adjust pH, if necessary, to pH <2 using 1:1 (V/V) sulfuric acid. Pipette 1.0 mL of a surrogate standard spiking solution into each sample. For the sample in each analytical batch selected for spiking, add 1.0 mL of a matrix spiking standard. For base/neutral acid analysis, the amount of the surrogates and matrix spiking compounds added to the sample should result in a final concentration of 100 ng/µL of each analyte in the extract to be analyzed (assuming a 1-µL injection). Extract with methylene chloride for 18–24 h. Next, adjust the pH of the aqueous phase to pH >11 using 10 N sodium hydroxide and extract it with methylene chloride again for 18–24 h. Dry the extract through a column containing anhydrous sodium sulfate and concentrate it to 1 mL using a K-D concentrator.

Soils, sediments, or sludges — Use 30 g of sample. Nonporous or wet samples (gummy or clay type) that do not have a free-flowing sandy texture must be mixed with anhydrous sodium sulfate until the sample is free flowing. Add 1 mL of surrogate standards to all samples, spikes, standards, and blanks. For the sample in each analytical batch selected for spiking, add 1.0 mL of a matrix spiking standard. For base/neutral acid analysis, the amount added of the surrogates and matrix spiking compounds should result in a final concentration of 100 ng/µL of each base/neutral analyte and 200 ng/µL of each acid analyte in the extract to be analyzed (assuming a 1-µL injection). Immediately add a 100-mL mixture of 1:1 methylene chloride:acetone and extract the sample ultrasonically for 3 min and then decant or filter the extracts. Repeat the extraction two or more times. Dry the extract using a column with anhydrous sodium sulfate and concentrate it to 1 mL in a K-D concentrator.

QUALITY CONTROL A methylene chloride solution containing 50 ng/µL of decafluorotriphenylphosphine (DFTPP) is used for tuning the GC/MS system each 12-h shift. A system performance check also must be made during every 12-h shift. A standard containing 50 ng/µL each of 4,4'-DDT, pentachlorophenol, and benzidine is required to verify injection port inertness and GC column performance. A calibration standard at mid-concentration, containing each compound of interest, including all required surrogates, must be performed every 12 h during analysis. After the system performance check is met, calibration check compounds (CCCs) are used to check the validity of the initial calibration.

The internal standard responses and retention times in the calibration check standard must be evaluated immediately after or during data acquisition. If the retention time for any internal standard changes by more than 30 seconds from the last check calibration (12 h), the chromatographic system must be inspected for malfunctions and corrections must be made, as required. If the electron ionization current plot (EICP) area for any of the internal standards changes by a factor of two from the last daily calibration standard check, the mass spectrometer must be inspected for malfunctions and corrections must be made, as appropriate.

Demonstrate, through the analysis of a reagent water blank, that interferences from the analytical system, glassware, and reagents are under control. The blank samples should be carried through all stages of the sample preparation and measurement steps. For each analytical batch (up to 20 samples), a reagent blank, matrix spike, and matrix spike duplicate/duplicate must be analyzed (the frequency of the spikes may be different for different monitoring programs). The blank and spiked samples must be carried through all stages of the sample preparation and measurement steps. A QC reference sample concentrate containing each analyte at a concentration of 100 mg/L in methanol is required.

REFERENCE Test Methods for Evaluating Solid Waste (SW-846). U.S. EPA 1983, Method 8270B, Rev. 2, Nov. 1990. Office of Solid Waste, Washington, DC.

4,4'-DDE   **EPA Method 625**
CAS #72-55-9

TITLE Base/Neutrals and Acids, U.S. EPA Method 625

MATRIX This methods covers municipal and industrial wastewaters.

METHOD SUMMARY Approximately 1 L of sample is serially extracted with methylene chloride at a pH greater than 11 and again at a pH less than 2 using a separatory funnel or a continuous extractor. The methylene chloride extract is dried, concentrated to a volume of 1 mL, and analyzed by GC/MS. Qualitative identification of the parameters in the extract is performed using the retention time and the relative abundance of three characteristic masses (m/z). Qualitative analysis is performed using either external or internal standard techniques with a single characteristic m/z.

INTERFERENCES Method interferences may be caused by contaminants in solvents, reagents, glassware, and other sample processing hardware. Glassware must be scrupulously cleaned. Glassware should be heated in a muffle furnace at 400°C for 5 to 30 min. Some thermally stable materials, such as PCBs, may not be eliminated by this treatment. Solvent rinses with acetone and pesticide quality hexane may be substituted for the muffle furnace heating. Matrix interferences may be caused by contaminants that are coextracted from the sample. The base-neutral extraction may cause significantly reduced recovery of phenols. The packed gas chromatographic columns recommended for the basic fraction may not exhibit sufficient resolution for some analytes.

INSTRUMENTATION A GC/MS system with an injection port designed for on-column injection when using packed columns and for splitless injection when using capillary columns.

Column for base/neutrals: 1.8 m long × 2 mm I.D. glass, packed with 3% SP-2550 on Supelcoport (100/120 mesh) or equivalent.

Column for acids: 1.8 m long × 2 mm I.D. glass, packed with 1% SP-1240DA on Supelcoport (100/120 mesh) or equivalent.

PRECISION & ACCURACY The MDL concentrations were obtained using reagent water. The MDL actually achieved in a given analysis will vary depending on instrument sensitivity and matrix effects. This method was tested by 15 laboratories using reagent water, drinking water, surface water, and industrial wastewaters spiked at six concentrations over the range 5 to 100 µg/L. Single operator precision, overall precision, and method accuracy were found to be directly related to the concentration of the parameter matrix.

The MDL (in µg/L) in reagent water was 5.6.

The standard deviation (in µg/L based on 4 recovery measurements) was 32.0.

The range (in µg/L) for average recovery for 4 measurements was 19.2–119.7.

The range (in %) for percent recovery was 4–136.

Accuracy (in µg/L) as expected recovery for one or more measurements of a sample containing a true concentration of C was 0.70C-0.54.

Precision (in µg/L) as expected single analyst standard deviation of measurements at an average concentration found at X was 0.26X–1.17.

Overall precision (in µg/L) as expected interlaboratory standard deviation of measurements in an average concentration found at X was 0.39X–1.04.

$C =$ *True value of the concentration in µg/L.*
$X =$ *Average recovery found for measurements of samples containing a concentration at C in µg/L.*

SAMPLE PREPARATION Adjust the pH to >11 with sodium hydroxide and serially extract in a separatory funnel with methylene chloride or else in a continuous extractor. Next, adjust the pH to <2 with sulfuric acid and serially extract in a separatory funnel with methylene chloride or else in a continuous extractor. Dry the extracts separately through a column of anhydrous sodium sulfate and then concentrate each of the extracts to 1.0 mL using a K-D apparatus.

SAMPLE COLLECTION, PRESERVATION & HANDLING Grab samples must be collected in glass containers. All samples must be refrigerated at 4°C from the time of collection until extraction. If residual chlorine is present, add 80 mg of sodium thiosulfate/L of sample and mix well. All samples must be extracted within 7 days of collection and completely analyzed within 40 days of extraction.

QUALITY CONTROL Make an initial, one-time, demonstration of the ability to generate acceptable accuracy and precision with this method. Before processing any samples, the analyst must analyze a reagent water blank to demonstrate that interferences from the analytical system and glassware are under control. Each time a set of samples is extracted or reagents are changed, a reagent water blank must be processed. Spike and analyze a minimum of 5% of all samples to monitor and evaluate lab data quality. A QC check sample concentrate that contains each parameter of interest at a concentration of 100 µg/mL in acetone is required. PCBs and multicomponent pesticides may be omitted from this test.

After analysis of five spiked wastewater samples, calculate the average percent recovery and the standard deviation of the percent recovery. Spike all samples with the surrogate standard spiking solution and calculate the percent recovery of each surrogate compound.

REFERENCE *Federal Register*, Vol. 49, No. 209. Friday, Oct. 26, 1984.

4,4'-DDE **EPA Method 8080**
CAS #72-55-9

TITLE Organochlorine Pesticides and Polychlorinated Biphenyls By Gas Chromatography

MATRIX This method is used to determine the concentration of various organochlorine pesticides and polychlorinated biphenyls in extracts prepared from water, groundwater, soils, and sediments.

METHOD SUMMARY This method covers 26 pesticides and Aroclor (PCB) mixtures and it is suitable for monitoring-type analyses. After extraction, concentration and solvent exchange to hexane, a 2- to 5-µL sample aliquot is injected into a GC using the solvent flush technique, and the analytes are detected by an electron capture detector (ECD) or an electrolytic conductivity detector in the halogen mode (HECD). Both neat and diluted organic liquids may be analyzed by direct injection.

INTERFERENCES Interferences coextracted from the samples will vary considerably from source to source. Interferences by phthalate esters can pose a major problem in pesticide determinations when using the ECD. Cross-contamination of clean glassware routinely occurs when plastics are handled during extraction steps, especially when solvent-wetted surfaces are handled. The contamination from phthalate esters can be completely eliminated with a microcoulometric or electrolytic conductivity detector. Solvents, reagent, glassware, and other sample processing hardware may yield artifacts and/or interferences to sample analysis.

INSTRUMENTATION A gas chromatograph capable of on-column injections is needed. It must be equipped with an ECD or a HECD and one of the following GC columns:

Column 1: Supelcoport (100/120 mesh) coated with 1.5% SP-2250/1.95% SP-2401 packed in a 1.8 m × 4 mm I.D. glass column.

Column 2: Supelcoport (100/120 mesh) coated with 3% OV-1 in a 1.8 m × 4 mm I.D. glass column.

PRECISION & ACCURACY The method was tested by 20 laboratories using organic-free reagent water, drinking water, surface water, and three industrial wastewaters spiked at six concentrations. Concentrations used in the study ranged from 0.5 to 30 µg/L for single-component pesticides and from 8.5 to 400 µg/L for multicomponent parameters. Overall precision and method accuracy were found to be directly related to the concentration of the analyte and essentially independent of the sample matrix. The sensitivity of this method usually depends on the concentration of interferences rather than on instrumental limitations.

MDL in µg/L was 0.004.
Concentration range in µg/L was 0.5–30.
Accuracy as recovery (x^*) in µg/L was $0.85C+0.14$.
Overall precision (S^*) in µg/L was $0.28x-0.09$.

x^* Expected recovery for one or more measurements of a sample containing concentration C, in µg/L.
S^* = Expected interlaboratory standard deviation of measurements at an average concentration found of the analyte in µg/L.
C = True value for the concentration, in µg/L.
X = Average recovery found for measurements of samples containing a concentration of C, in µg/L.

SAMPLING METHOD

Liquid samples — Use a 1 or 2½ gallon amber glass bottle with a screw-top Teflon®-lined cover. Pre-wash the bottle with detergent, rinse with distilled water and methanol (or isopropanol).

Soil, sediments, and sludges — Use an 8-oz. widemouth glass with a screw-top Teflon®-lined cover. Pre-wash the bottle with detergent, rinse with distilled water and methanol (or isopropanol).

SAMPLE PRESERVATION Cool water, soil, sediment, or sludge samples immediately to 4°C.

Water samples — If residual chlorine is present, add 3 mL of 10% sodium thiosulfate per gallon and cool to 4°C. All extracts and samples should be stored under refrigeration.

MHT Liquid samples must be extracted within 7 days and the extracts must be analyzed within 40 days. Soils, sediments, and sludges may be stored for a maximum of 14 days prior to extraction.

SAMPLE PREPARATION

Liquid samples — Extract 1 L samples in a continuous extractor at pH 5–9 with methylene chloride after adding 1.0 mL of surrogate spiking solution to each sample. Pass the extract through a column of anhydrous sodium sulfate to dry and concentrate it in a K-D apparatus to 1 mL volume.

Soils, sediments, and sludges — Rapidly weigh approximately 30 g of sample into a 400-mL beaker to avoid loss of the more volatile extractables. Nonporous or wet samples (gummy or clay type) that do not have a free-flowing sandy texture must be mixed with anhydrous sodium sulfate until the sample is free flowing. Add 1 mL of surrogate standards to all samples, spikes, standards, and blanks. Add 100 mL of 1:1 methylene chloride:acetone and extract ultrasonically. Decant and filter extracts, dry the extract by passing it through a drying column containing anhydrous sodium sulfate and concentrate to 1 mL in a K-D apparatus.

Hexane solvent exchange — Add 50 mL of hexane, a new boiling chip, and concentrate until the apparent volume of liquid reaches 1 mL. Adjust the extract volume to 10.0 mL. Stopper the concentration tube and store refrigerated at 4°C if further processing will not be performed immediately. If the extract will be stored longer than two days, transfer it to a vial with Teflon®-lined screw-cap or crimp top.

QUALITY CONTROL Demonstrate through the analysis of a reagent water blank, that all glassware and reagents are interference free. Each time a set of samples is processed, a method blank should be processed as a safeguard against chronic lab contamination. A reagent blank, a matrix spike, and a duplicate or matrix spike duplicate must be performed for each analytical batch (up to a maximum of 20 samples) analyzed.

Analytical system performance must be verified by analyzing QC check samples. The QC check sample concentration should contain each single-component analyte at the following concentrations in acetone: 4,4'-DDD, 10 µg/mL; 4,4'-DDT, 10 µg/mL; endosulfan II, 10 µg/mL; endosulfan sulfate, 10 µg/mL; and any other single-component pesticide at 2 µg/mL. If the method is only to be used to analyze PCBs, Chlordane, or Toxaphene, the QC check sample concentrate should contain the most representative multicomponent parameter at a concentration of 50 µg/mL in acetone.

REFERENCE Test Methods for Evaluating Solid Waste (SW-846). U.S. EPA. 1983. Method 8080B, Rev. 2, Nov. 1990. Office of Solid Wastes, Washington, DC.

4,4'-DDE   **EPA Method 8270**
CAS #72-55-9

TITLE Semivolatile Organic Compounds by GC/MS

MATRIX This method is used to determine the concentration of semivolatile organic compounds in extracts prepared from all types of solid waste matrices, soils, and groundwater. Although surface waters are not specifically mentioned, this method should be applicable to water samples from rivers, lakes, etc.

METHOD SUMMARY This method covers 259 semivolatile organic compounds. In very limited applications direct injection of the sample into the GC/MS system may be appropriate, but this results in very high detection limits (approximately 10,000 µg/L). Typically, a 1-L liquid sample, containing surrogate, and matrix spiking standards, is extracted in a continuous extractor first under acid conditions and then under basic conditions. Typically 30 g of a solid sample, containing surrogate, and matrix spiking standards, is extracted ultrasonically. After concentrating the extract to 1 mL it is spiked with 10 µL of an internal standard solution just prior to analysis by GC/MS. The volume injected should contain about 100 ng of base/neutral and 200 ng of acid surrogates (for a 1-µL injection). Analysis is performed by GC/MS using a capillary GC column.

INTERFERENCES Raw GC/MS data from all blanks, samples, and spikes must be evaluated for interferences. Contamination by carryover can occur whenever high-concentration and low-concentration samples are sequentially analyzed. To reduce carryover, the sample syringe must be rinsed out between samples with solvent. Whenever an unusually concentrated sample is encountered, it should be followed by the analysis of blank solvent to check for cross-contamination.

INSTRUMENTATION A GC/MS and a data system are required. The GC column used is a 30 m × 0.25 mm I.D. (or 0.32 mm I.D.) 1um film thickness silicone-coated fused silica capillary column. A continuous liquid-liquid extractor equipped with Teflon® or glass connection joints and stopcocks requiring no lubrication, a K-D concentrating apparatus, water bath, and an ultrasonic disrupter with a minimum power of 300 W and with pulsing capability are also required.

PRECISION & ACCURACY The estimated quantitation limit (EQL) of Method 8270B for determining an individual compound is approximately 1 mg/kg (wet weight) for soil or sediment samples, 1–200 mg/kg for wastes (dependent on matrix and method of preparation), and 10 µg/L for groundwater samples. EQLs will be proportionately higher for sample extracts that require dilution to avoid saturation of the detector.

The EQL(b) for groundwater in µg/L is not listed.
The EQL (a, b) for low concentrations in soil and sediment in µg/kg is not listed.
Accuracy as µg/L is 0.70C–0.54.
Overall precision in µg/L is 0.39X–1.04.

(a) *EQLs listed for soil/sediment are based on wet weight. Normally data is reported in a dry-weight basis; therefore, EQLs will be higher based on the % dry weight of each sample. This calculation is based on a 30 g sample and gel permeation chromatography cleanup.*
(b) *Sample EQLs are highly matrix-dependent. The EQLs are provided for guidance and may not always be achievable.*
$C =$ *True value for concentration, in µg/L.*
$X =$ *Average recovery found for measurements of samples containing a concentration of C, in µg/L.*

ESTIMATED QUANTITATION LIMIT

Other Matrices	Factor (a)
High-concentration soil and sludges by sonicator	7.5
Non-water miscible waste	75

(a) EQL for other matrices = [EQL for low soil/sediment] × [Factor]. This estimated EQL is similar to an EPA "Practical Quantitation Limit."

SAMPLING METHOD

Liquid samples — Use a 1 or 2½ gallon amber glass bottle with a screw-top Teflon®-lined cover that has been prewashed with detergent and rinsed with distilled water and methanol (or isopropanol).

Soils, sediments, or sludges — Use an 8-oz. widemouth glass with a screw-top Teflon®-lined cover that has been prewashed with detergent and rinsed with distilled water and methanol (or isopropanol).

SAMPLE PRESERVATION

Liquid samples — If residual chlorine is present, add 3 mL of 10% sodium thiosulfate per gallon, cool to 4°C and store in a solvent-free refrigerator until analysis; if chlorine is not present, then eliminate the sodium thiosulfate addition.

Soils, sediments, or sludges — Cool samples to 4°C and store in a solvent-free refrigerator.

MHT Liquid samples must be extracted within 7 days and the extracts analyzed within 40 days. Soils, sediments, or sludges may be stored for a maximum of 14 days and the extracts analyzed within 40 days.

SAMPLE PREPARATION

Liquid samples — Transfer 1 L quantitatively to a continuous extractor. If high concentrations are anticipated, a smaller volume may be used and then diluted with organic-free reagent water to 1 L. Adjust pH, if necessary, to pH <2 using 1:1 (V/V) sulfuric acid. Pipette 1.0 mL of a surrogate standard spiking solution into each sample. For the sample in each analytical batch selected for spiking, add 1.0 mL of a matrix spiking standard. For base/neutral acid analysis, the amount of the surrogates and matrix spiking compounds added to the sample should result in a final concentration of 100 ng/µL of each analyte in the extract to be analyzed (assuming a 1-µL injection). Extract with methylene chloride for 18–24 h. Next, adjust the pH of the aqueous phase to pH >11 using 10 N sodium hydroxide and extract it with methylene chloride again for 18–24 h. Dry the extract through a column containing anhydrous sodium sulfate and concentrate it to 1 mL using a K-D concentrator.

Soils, sediments, or sludges — Use 30 g of sample. Nonporous or wet samples (gummy or clay type) that do not have a free-flowing sandy texture must be mixed with anhydrous sodium sulfate until the sample is free flowing. Add 1 mL of surrogate standards to all samples, spikes, standards, and blanks. For the sample in each analytical batch selected for spiking, add 1.0 mL of a matrix spiking standard. For base/neutral acid analysis, the amount added of the surrogates and matrix spiking compounds should result in a final concentration of 100 ng/µL of each base/neutral analyte and 200 ng/µL of each acid analyte in the extract to be analyzed (assuming a 1-µL injection). Immediately add a 100-mL mixture of 1:1 methylene chloride:acetone and extract the sample ultrasonically for 3 min and then decant or filter the extracts. Repeat the extraction two or more times. Dry the extract using a column with anhydrous sodium sulfate and concentrate it to 1 mL in a K-D concentrator.

QUALITY CONTROL

A methylene chloride solution containing 50 ng/µL of decafluorotriphenylphosphine (DFTPP) is used for tuning the GC/MS system each 12-h shift. A system performance check also must be made during every 12-h shift. A standard containing 50 ng/µL each of 4,4'-DDT, pentachlorophenol, and benzidine is required to verify injection port inertness and GC column performance. A calibration standard at mid-concentration, containing each compound of interest, including all required surrogates, must be performed every 12 h during analysis. After the system performance check is met, calibration check compounds (CCCs) are used to check the validity of the initial calibration.

The internal standard responses and retention times in the calibration check standard must be evaluated immediately after or during data acquisition. If the retention time for any internal standard changes by more than 30 seconds from the last check calibration (12 h), the chromatographic system must be inspected for malfunctions and corrections must be made, as required. If the electron ionization current plot (EICP) area for any of the internal standards changes by a factor of two from the last daily calibration standard check, the mass spectrometer must be inspected for malfunctions and corrections must be made, as appropriate.

Demonstrate, through the analysis of a reagent water blank, that interferences from the analytical system, glassware, and reagents are under control. The blank samples should be carried through all stages of the sample preparation and measurement steps. For each analytical batch (up to 20 samples), a reagent blank, matrix spike, and matrix spike duplicate/duplicate must be analyzed (the frequency of the spikes may be different for different monitoring programs). The blank and spiked samples must be carried through all stages of the sample preparation and measurement steps. A QC reference sample concentrate containing each analyte at a concentration of 100 mg/L in methanol is required.

REFERENCE Test Methods for Evaluating Solid Waste (SW-846). U.S. EPA 1983, Method 8270B, Rev. 2, Nov. 1990. Office of Solid Waste, Washington, DC.

4,4'-DDT   EPA Method 625
CAS #50-29-3

TITLE Base/Neutrals and Acids, U.S. EPA Method 625

MATRIX This methods covers municipal and industrial wastewaters.

METHOD SUMMARY Approximately 1 L of sample is serially extracted with methylene chloride at a pH greater than 11 and again at a pH less than 2 using a separatory funnel or a continuous extractor. The methylene chloride extract is dried, concentrated to a volume of 1 mL, and analyzed by GC/MS. Qualitative identification of the parameters in the extract is performed using the retention time and the relative abundance of three characteristic masses (m/z). Qualitative analysis is performed using either external or internal standard techniques with a single characteristic m/z.

INTERFERENCES Method interferences may be caused by contaminants in solvents, reagents, glassware, and other sample processing hardware. Glassware must be scrupulously cleaned. Glassware should be heated in a muffle furnace at 400°C for 5 to 30 min. Some thermally stable materials, such as PCBs, may not be eliminated by this treatment. Solvent rinses with acetone and pesticide quality hexane may be substituted for the muffle furnace heating. Matrix interferences may be caused by contaminants that are coextracted from the sample. The base-neutral extraction may cause significantly reduced recovery of phenols. The packed gas chromatographic columns recommended for the basic fraction may not exhibit sufficient resolution for some analytes.

INSTRUMENTATION A GC/MS system with an injection port designed for on-column injection when using packed columns and for splitless injection when using capillary columns.

Column for base/neutrals: 1.8 m long × 2 mm I.D. glass, packed with 3% SP-2550 on Supelcoport (100/120 mesh) or equivalent.

Column for acids: 1.8 m long × 2 mm I.D. glass, packed with 1% SP-1240DA on Supelcoport (100/120 mesh) or equivalent.

PRECISION & ACCURACY The MDL concentrations were obtained using reagent water. The MDL actually achieved in a given analysis will vary depending on instrument sensitivity and matrix effects. This method was tested by 15 laboratories using reagent water, drinking water, surface water, and industrial wastewaters spiked at six concentrations over the range 5 to 100 µg/L. Single operator precision, overall precision, and method accuracy were found to be directly related to the concentration of the parameter matrix.

The MDL (in µg/L) in reagent water was 4.7.

The standard deviation (in µg/L based on 4 recovery measurements) was 61.6.

The range (in µg/L) for average recovery for 4 measurements was D-170.6.

The range (in %) for percent recovery was D-203.

Accuracy (in µg/L) as expected recovery for one or more measurements of a sample containing a true concentration of C was 0.79C-3.28.

Precision (in µg/L) as expected single analyst standard deviation of measurements at an average concentration found at X was $0.42X+0.19$.

Overall precision (in µg/L) as expected interlaboratory standard deviation of measurements in an average concentration found at X was $0.65X-0.58$.

C = *True value of the concentration in µg/L.*

X = *Average recovery found for measurements of samples containing a concentration at C in µg/L.*

SAMPLE PREPARATION Adjust the pH to >11 with sodium hydroxide and serially extract in a separatory funnel with methylene chloride or else in a continuous extractor. Next, adjust the pH to <2 with sulfuric acid and serially extract in a separatory funnel with methylene chloride or else in a continuous extractor. Dry the extracts separately through a column of anhydrous sodium sulfate and then concentrate each of the extracts to 1.0 mL using a K-D apparatus.

SAMPLE COLLECTION, PRESERVATION & HANDLING Grab samples must be collected in glass containers. All samples must be refrigerated at 4°C from the time of collection until extraction. If residual chlorine is present, add 80 mg of sodium thiosulfate/L of sample and mix well. All samples must be extracted within 7 days of collection and completely analyzed within 40 days of extraction.

QUALITY CONTROL Make an initial, one-time, demonstration of the ability to generate acceptable accuracy and precision with this method. Before processing any samples, the analyst must analyze a reagent water blank to demonstrate that interferences from the analytical system and glassware are under control. Each time a set of samples is extracted or reagents are changed, a reagent water blank must be processed. Spike and analyze a minimum of 5% of all samples to monitor and evaluate lab data quality. A QC check sample concentrate that contains each parameter of interest at a concentration of 100 µg/mL in acetone is required. PCBs and multicomponent pesticides may be omitted from this test.

After analysis of five spiked wastewater samples, calculate the average percent recovery and the standard deviation of the percent recovery. Spike all samples with the surrogate standard spiking solution and calculate the percent recovery of each surrogate compound.

REFERENCE *Federal Register,* Vol. 49, No. 209. Friday, Oct. 26, 1984.

4,4'-DDT   **EPA Method 8080**
CAS #50-29-3

TITLE Organochlorine Pesticides and Polychlorinated Biphenyls By Gas Chromatography

MATRIX This method is used to determine the concentration of various organochlorine pesticides and polychlorinated biphenyls in extracts prepared from water, groundwater, soils, and sediments.

METHOD SUMMARY This method covers 26 pesticides and Aroclor (PCB) mixtures and it is suitable for monitoring-type analyses. After extraction, concentration and solvent exchange to hexane, a 2- to 5-µL sample aliquot is injected into a GC using the solvent flush technique, and the analytes are detected by an electron capture detector (ECD) or an electrolytic conductivity detector in the halogen mode (HECD). Both neat and diluted organic liquids may be analyzed by direct injection.

INTERFERENCES Interferences coextracted from the samples will vary considerably from source to source. Interferences by phthalate esters can pose a major problem in pesticide determinations when using the ECD. Cross-contamination of clean glassware routinely occurs when plastics are handled during extraction steps, especially when solvent-wetted surfaces are handled. The contamination from phthalate esters can be completely eliminated with a microcoulometric or electrolytic conductivity detector. Solvents, reagent, glassware, and other sample processing hardware may yield artifacts and/or interferences to sample analysis.

INSTRUMENTATION A gas chromatograph capable of on-column injections is needed. It must be equipped with an ECD or a HECD and one of the following GC columns:

Column 1: Supelcoport (100/120 mesh) coated with 1.5% SP-2250/1.95% SP-2401 packed in a 1.8 m × 4 mm I.D. glass column.

Column 2: Supelcoport (100/120 mesh) coated with 3% OV-1 in a 1.8 m × 4 mm I.D. glass column.

PRECISION & ACCURACY The method was tested by 20 laboratories using organic-free reagent water, drinking water, surface water, and three industrial wastewaters spiked at six concentrations. Concentrations used in the study ranged from 0.5 to 30 µg/L for single-component pesticides and from 8.5 to 400 µg/L for multicomponent parameters. Overall precision and method accuracy were found to be directly related to the concentration of the analyte and essentially independent of the sample matrix. The sensitivity of this method usually depends on the concentration of interferences rather than on instrumental limitations.

MDL in µg/L was 0.012.
Concentration range in µg/L was 0.5–30.
Accuracy as recovery (x*) in µg/L was 0.93C–0.13 .
Overall precision (S*) in µg/L was 0.31x–0.21.

- x^* Expected recovery for one or more measurements of a sample containing concentration C, in µg/L.
- S^* = Expected interlaboratory standard deviation of measurements at an average concentration found of the analyte in µg/L.
- C = True value for the concentration, in µg/L.
- X = Average recovery found for measurements of samples containing a concentration of C, in µg/L.

SAMPLING METHOD
Liquid samples — Use a 1 or 2½ gallon amber glass bottle with a screw-top Teflon®-lined cover. Pre-wash the bottle with detergent, rinse with distilled water and methanol (or isopropanol).

Soil, sediments, and sludges — Use an 8-oz. widemouth glass with a screw-top Teflon®-lined cover. Pre-wash the bottle with detergent, rinse with distilled water and methanol (or isopropanol).

SAMPLE PRESERVATION Cool water, soil, sediment, or sludge samples immediately to 4°C.

Water samples — If residual chlorine is present, add 3 mL of 10% sodium thiosulfate per gallon and cool to 4°C. All extracts and samples should be stored under refrigeration.

MHT Liquid samples must be extracted within 7 days and the extracts must be analyzed within 40 days. Soils, sediments, and sludges may be stored for a maximum of 14 days prior to extraction.

SAMPLE PREPARATION
Liquid samples — Extract 1 L samples in a continuous extractor at pH 5–9 with methylene chloride after adding 1.0 mL of surrogate spiking solution to each sample. Pass the extract through a column of anhydrous sodium sulfate to dry and concentrate it in a K-D apparatus to 1 mL volume.

Soils, sediments, and sludges — Rapidly weigh approximately 30 g of sample into a 400-mL beaker to avoid loss of the more volatile extractables. Nonporous or wet samples (gummy or clay type) that do not have a free-flowing sandy texture must be mixed with anhydrous sodium sulfate until the sample is free flowing. Add 1 mL of surrogate standards to all samples, spikes, standards, and blanks. Add 100 mL of 1:1 methylene chloride:acetone and extract ultrasonically. Decant and filter extracts, dry the extract by passing it through a drying column containing anhydrous sodium sulfate and concentrate to 1 mL in a K-D apparatus.

Hexane solvent exchange — Add 50 mL of hexane, a new boiling chip, and concentrate until the apparent volume of liquid reaches 1 mL. Adjust the extract volume to 10.0 mL. Stopper the concentration tube and store refrigerated at 4°C if further processing will not be performed immediately. If the extract will be stored longer than two days, transfer it to a vial with Teflon®-lined screw-cap or crimp top.

QUALITY CONTROL Demonstrate through the analysis of a reagent water blank, that all glassware and reagents are interference free. Each time a set of samples is processed, a method blank should be processed as a safeguard against chronic lab contamination. A reagent blank, a matrix spike, and a duplicate or matrix spike duplicate must be performed for each analytical batch (up to a maximum of 20 samples) analyzed.

Analytical system performance must be verified by analyzing QC check samples. The QC check sample concentration should contain each single-component analyte at the following concentrations in acetone: 4,4'-DDD, 10 µg/mL; 4,4'-DDT, 10 µg/mL; endosulfan II, 10 µg/mL; endosulfan sulfate, 10 µg/mL; and any other single-component pesticide at 2 µg/mL. If the method is only to be used to analyze PCBs, Chlordane, or Toxaphene, the QC check sample concentrate should contain the most representative multicomponent parameter at a concentration of 50 µg/mL in acetone.

REFERENCE Test Methods for Evaluating Solid Waste (SW-846). U.S. EPA. 1983. Method 8080B, Rev. 2, Nov. 1990. Office of Solid Wastes, Washington, DC.

4,4'-DDT **EPA Method 8270**
CAS #50-29-3

TITLE Semivolatile Organic Compounds by GC/MS

MATRIX This method is used to determine the concentration of semivolatile organic compounds in extracts prepared from all types of solid waste matrices, soils, and groundwater. Although surface waters are not specifically mentioned, this method should be applicable to water samples from rivers, lakes, etc.

METHOD SUMMARY This method covers 259 semivolatile organic compounds. In very limited applications direct injection of the sample into the GC/MS system may be appropriate, but this results in very high detection limits (approximately 10,000 µg/L). Typically, a 1-L liquid sample, containing surrogate, and matrix spiking standards, is extracted in a continuous extractor first under acid conditions and then under basic conditions. Typically 30 g of a solid sample, containing surrogate, and matrix spiking standards, is extracted ultrasonically. After concentrating the extract to 1 mL it is spiked with 10 µL of an internal standard solution just prior to analysis by GC/MS. The volume injected should contain about 100 ng of base/neutral and 200 ng of acid surrogates (for a 1-µL injection). Analysis is performed by GC/MS using a capillary GC column.

INTERFERENCES Raw GC/MS data from all blanks, samples, and spikes must be evaluated for interferences. Contamination by carryover can occur whenever high-concentration and low-concentration samples are sequentially analyzed. To reduce carryover, the sample syringe must be rinsed out between samples with solvent. Whenever an unusually concentrated sample is encountered, it should be followed by the analysis of blank solvent to check for cross-contamination.

INSTRUMENTATION A GC/MS and a data system are required. The GC column used is a 30 m × 0.25 mm I.D. (or 0.32 mm I.D.) 1um film thickness silicone-coated fused silica capillary column. A continuous liquid-liquid extractor equipped with Teflon® or glass connection joints and stopcocks requiring no lubrication, a K-D concentrating apparatus, water bath, and an ultrasonic disrupter with a minimum power of 300 W and with pulsing capability are also required.

PRECISION & ACCURACY The estimated quantitation limit (EQL) of Method 8270B for determining an individual compound is approximately 1 mg/kg (wet weight) for soil or sediment samples, 1–200 mg/kg for wastes (dependent on matrix and method of preparation), and 10 µg/L for groundwater samples. EQLs will be proportionately higher for sample extracts that require dilution to avoid saturation of the detector.

The EQL(b) for groundwater in µg/L is not listed.
The EQL (a, b) for low concentrations in soil and sediment in µg/kg is not listed.
Accuracy as µg/L is 0.79C–3.28.
Overall precision in µg/L is 0.65X–0.58.

(a) EQLs listed for soil/sediment are based on wet weight. Normally data is reported in a dry-weight basis; therefore, EQLs will be higher based on the % dry weight of each sample. This calculation is based on a 30 g sample and gel permeation chromatography cleanup.

(b) Sample EQLs are highly matrix-dependent. The EQLs are provided for guidance and may not always be achievable.

C = True value for concentration, in µg/L.

X = Average recovery found for measurements of samples containing a concentration of C, in µg/L.

ESTIMATED QUANTITATION LIMIT

Other Matrices	Factor (a)
High-concentration soil and sludges by sonicator	7.5
Non-water miscible waste	75

(a) EQL for other matrices = [EQL for low soil/sediment] × [Factor]. This estimated EQL is similar to an EPA "Practical Quantitation Limit."

SAMPLING METHOD

Liquid samples — Use a 1 or 2½ gallon amber glass bottle with a screw-top Teflon®-lined cover that has been prewashed with detergent and rinsed with distilled water and methanol (or isopropanol).

Soils, sediments, or sludges — Use an 8-oz. widemouth glass with a screw-top Teflon®-lined cover that has been prewashed with detergent and rinsed with distilled water and methanol (or isopropanol).

SAMPLE PRESERVATION
Liquid samples — If residual chlorine is present, add 3 mL of 10% sodium thiosulfate per gallon, cool to 4°C and store in a solvent-free refrigerator until analysis; if chlorine is not present, then eliminate the sodium thiosulfate addition.

Soils, sediments, or sludges — Cool samples to 4°C and store in a solvent-free refrigerator.

MHT Liquid samples must be extracted within 7 days and the extracts analyzed within 40 days. Soils, sediments, or sludges may be stored for a maximum of 14 days and the extracts analyzed within 40 days.

SAMPLE PREPARATION
Liquid samples — Transfer 1 L quantitatively to a continuous extractor. If high concentrations are anticipated, a smaller volume may be used and then diluted with organic-free reagent water to 1 L. Adjust pH, if necessary, to pH <2 using 1:1 (V/V) sulfuric acid. Pipette 1.0 mL of a surrogate standard spiking solution into each sample. For the sample in each analytical batch selected for spiking, add 1.0 mL of a matrix spiking standard. For base/neutral acid analysis, the amount of the surrogates and matrix spiking compounds added to the sample should result in a final concentration of 100 ng/µL of each analyte in the extract to be analyzed (assuming a 1-µL injection). Extract with methylene chloride for 18–24 h. Next, adjust the pH of the aqueous phase to pH >11 using 10 N sodium hydroxide and extract it with methylene chloride again for 18–24 h. Dry the extract through a column containing anhydrous sodium sulfate and concentrate it to 1 mL using a K-D concentrator.

Soils, sediments, or sludges — Use 30 g of sample. Nonporous or wet samples (gummy or clay type) that do not have a free-flowing sandy texture must be mixed with anhydrous sodium sulfate until the sample is free flowing. Add 1 mL of surrogate standards to all samples, spikes, standards, and blanks. For the sample in each analytical batch selected for spiking, add 1.0 mL of a matrix spiking standard. For base/neutral acid analysis, the amount added of the surrogates and matrix spiking compounds should result in a final concentration of 100 ng/µL of each base/neutral analyte and 200 ng/µL of each acid analyte in the extract to be analyzed (assuming a 1-µL injection). Immediately add a 100-mL mixture of 1:1 methylene chloride:acetone and extract the sample ultrasonically for 3 min and then decant or filter the extracts. Repeat the extraction two or more times. Dry the extract using a column with anhydrous sodium sulfate and concentrate it to 1 mL in a K-D concentrator.

QUALITY CONTROL A methylene chloride solution containing 50 ng/µL of decafluorotriphenylphosphine (DFTPP) is used for tuning the GC/MS system each 12-h shift. A system performance check also must be made during every 12-h shift. A standard containing 50 ng/µL each of 4,4'-DDT, pentachlorophenol, and benzidine is required to verify injection port inertness and GC column performance. A calibration standard at mid-concentration, containing each compound of interest, including all required surrogates, must be performed every 12 h during analysis. After the system performance check is met, calibration check compounds (CCCs) are used to check the validity of the initial calibration.

The internal standard responses and retention times in the calibration check standard must be evaluated immediately after or during data acquisition. If the retention time for any internal standard changes by more than 30 seconds from the last check calibration (12 h), the chromatographic system must be inspected for malfunctions and corrections must be made, as required. If the electron ionization current plot (EICP) area for any of the internal standards changes by a factor of two from the last daily calibration standard check, the mass spectrometer must be inspected for malfunctions and corrections must be made, as appropriate.

Demonstrate, through the analysis of a reagent water blank, that interferences from the analytical system, glassware, and reagents are under control. The blank samples should be carried through all stages of the sample preparation and measurement steps. For each analytical batch (up to 20 samples), a reagent blank, matrix spike, and matrix spike duplicate/duplicate must be analyzed (the frequency of the spikes may be different for different monitoring programs). The blank and spiked samples must be carried through all stages of the sample preparation and measurement steps. A QC reference sample concentrate containing each analyte at a concentration of 100 mg/L in methanol is required.

REFERENCE Test Methods for Evaluating Solid Waste (SW-846). U.S. EPA 1983, Method 8270B, Rev. 2, Nov. 1990. Office of Solid Waste, Washington, DC.

---

n-Decane **EPA Method 1625**
CAS #124-18-5

TITLE Semivolatile Organic Compounds by Isotope Dilution GC/MS

MATRIX The compounds may be determined in waters, soils, and municipal sludges by this method.

METHOD SUMMARY This method is used to determine 176 semivolatile toxic organic pollutants associated with the CWA (as amended 1987); the RCRA (as amended 1986); the CERCLA (as amended 1986); and other compounds amenable to extraction and analysis by capillary column gas chromatography-mass spectrometry (GC/MS).

Stable isotopically-labeled analogs of the compounds of interest are added to the sample. If the solids content is less than 1%, a 1-L sample is extracted at pH 12–13, then at pH <2 with methylene chloride using continuous extraction techniques.

If the solids content is 30% or less, the sample is diluted to 1% solids with reagent water, homogenized ultrasonically, and extracted at pH 12–13, then at pH <2 with methylene chloride using continuous extraction techniques. If the solids content is greater than 30%, the sample is extracted using ultrasonic techniques.

Each extract is dried over sodium sulfate, concentrated to a volume of 5 mL, cleaned up using GPC, if necessary, and concentrated. Extracts are concentrated to 1 mL if GPC is not performed, and to 0.5 mL if GPC is performed.

An internal standard is added to the extract, and a 1-mL aliquot of the extract is injected into the GC. The compounds are separated by GC and detected by a MS. The labeled compounds serve to correct the variability of the analytical technique.

INTERFERENCES Solvents, reagents, glassware, and other sample processing hardware may yield artifacts and/or elevated baselines causing misinterpretation of chromatograms and spectra. Materials used in the analysis must be demonstrated to be free from interferences under the conditions of analysis by running method blanks initially and with each sample lot (sample started through the extraction process on a given 8-h shift, to a maximum of 20). Specific selection of reagents and purification of solvents by distillation in all glass systems may be required. Glassware and, where possible, reagents are cleaned by solvent rinse and baking at 450°C for 1-h minimum. Interferences coextracted from samples will vary considerably from source to source, depending on the diversity of the site being sampled.

INSTRUMENTATION Major instrumentation includes a GC with a splitless or on-column injection port for capillary column, a MS with 70 eV electron impact ionization, and a data system to collect and record MS data, and process it. A K-D apparatus is used to concentrate extracts.

GC Column: 30 m × 0.25 mm I.D. 5% phenyl, 94% methyl, 1% vinyl silicone bonded phased fused silica capillary column.

PRECISION & ACCURACY The detection limits of the method are usually dependent on the level of interferences rather than instrumental limitations. The limits typify the minimum quantities that can be detected with no interferences present.

The minimum level (in µg/mL) was 10. This is defined as a minimum level at which the analytical system shall give recognizable mass spectra (background corrected) and acceptable calibration points.

The MDL (in µg/kg) in low solids was 299 and in high solids was 1188; these were determined in digested sludge (low solids) and in filter cake or compost (high solids).

Note: Background levels of this compound were present in the sludge tested, resulting in higher than expected MDLs. The MDL for this compound is expected to be approximately 50 µg/kg with no interferences present.

The labeled and native compound initial precision as standard deviation (in µg/L) was 51.
The labeled and native compound initial accuracy as average recovery (in µg/L) was 24–195.

SAMPLE COLLECTION, PRESERVATION & HANDLING
Collect samples in glass containers. Aqueous samples which flow freely are collected in refrigerated bottles using automatic sampling equipment. Solid samples are collected as grab samples using widemouth jars. Maintain samples at 0 to 4°C from the time of collection until extraction. If residual chlorine is present in aqueous samples, add 80 mg sodium thiosulfate/L of water. Begin sample extraction within 7 days of collection, and analyze all extracts within 40 days of extraction.

SAMPLE PREPARATION Samples containing 1% solids or less are extracted directly using continuous liquid-liquid extraction techniques. Samples containing 1 to 30% solids are diluted to the 1% level with reagent water and extracted using continuous liquid-liquid extraction techniques. Samples containing greater than 30% solids are extracted using ultrasonic techniques.

Base/neutral extraction — Adjust the pH of the waters in the extractors to 12–13 with 6 N NaOH. Extract with methylene chloride for 24–48 h.

Acid extraction — Adjust the pH of the waters in the extractors to 2 or less using 6 N sulfuric acid. Extract with methylene chloride for 24–48 h.

Ultrasonic extraction of high solids samples — Add anhydrous sodium sulfate to the sample and QC aliquot(s). Add acetone:methylene chloride (1:1) to the sample and mix thoroughly

Concentrate extracts using a K-D apparatus.

QUALITY CONTROL The analyst is permitted to modify this method to improve separations or lower the costs of measurements, provided all performance specifications are met. Analyses of blanks are required to demonstrate freedom from contamination. When results of spikes indicate atypical method performance for samples, the samples are diluted to bring method performance within acceptable limits.

For low solids (aqueous samples), extract, concentrate, and analyze two sets of four 1-L aliquots (8 aliquots total) of the precision and recovery standard. For high solids samples, two sets of four 30-g aliquots of the high solids reference matrix are used.

Spike all samples with labeled compounds to assess method performance. Compute percent recovery of the labeled compounds using the internal standard method. Compare the labeled compound recovery for each compound with the corresponding labeled compound recovery.

Reagent water and high solids reference matrix blanks are analyzed to demonstrate freedom from contamination. Extract and concentrate a 1-L reagent water blank or a high solids reference matrix blank with each sample's lot (samples started through the extraction process on the same 8-h shift, to a maximum of 20 samples).

Field replicates may be collected to determine the precision of the sampling technique, and spiked samples may be required to determine the accuracy of the analysis when the internal standard method is used.

REFERENCE Semivolatile Organic Compounds by Isotope Dilution GC/MS. Office of Water Regulation and Standards, U.S. EPA Industrial Technology Division, Washington, DC, EPA Method 1625, Rev. C, June 1989 (contact W.A. Telliard, U.S. EPA, Office of Water Regulations and Standards, 401 M St., SW, Washington, DC, 20460. Phone: 202-382-7131).

Demeton, -O and -S **EPA Method 8141**
CAS #8065-48-3

TITLE Organophosphorus Compounds by Gas Chromatography: Capillary Column Technique

MATRIX This method covers aqueous and solid matrices. This includes a wide variety such as drinking water, groundwater, industrial wastewaters, surface waters, soils, solids, and sediments.

METHOD SUMMARY This is a GC method used to determine the concentration of 28 organophosphorus pesticides.

The use of Gel Permeation Cleanup (EPA Method 3640) for sample cleanup has been demonstrated to yield recoveries of less than 85% for many method analytes and is therefore not recommended for use with this method.

This method provides GC conditions for the detection of ppb concentrations of organophosphorus compounds. Prior to the use of this method, appropriate sample preparation techniques must be used. Water samples are extracted at a neutral pH with methylene chloride as a solvent by using a separatory funnel (EPA Method 3510) or a continuous liquid-liquid extractor (EPA Method 3520). Soxhlet extraction (EPA Method 3540) or ultrasonic extraction (EPA Method 3550) using methylene chloride/acetone (1:1) are used for solid samples. Both neat and diluted organic liquids (EPA Method 3580) may be analyzed by direct injection. Spiked samples are used to verify the applicability of the chosen extraction technique to each new sample type. A GC with a flame photometric (FPD) or nitrogen-phosphorus detector (NPD) is used for this multiresidue procedure.

INTERFERENCES The use of Florisil cleanup materials (EPA Method 3620) for some of the compounds in this method has been demonstrated to yield recoveries less than 85% and is therefore not recommended for all compounds. Use of phosphorus or halogen specific detectors, however, often obviates the necessity for cleanup for relatively clean sample matrices. If particular circumstances demand the use of an alternative cleanup procedure, the analyst must determine the elution profile and demonstrate that the recovery of each analyte is no less than 85%.

Use of a flame photometric detector (FPD) in the phosphorus mode will minimize interferences from materials that do not contain phosphorus. Elemental sulfur, however, may interfere with the determination of certain organophosphorus compounds by flame photometric gas chromatography. Sulfur cleanup using EPA Method 3660 may alleviate this interference. A nitrogen phosphorus detector (NPD) is also recommended.

A few analytes coelute on certain columns. Therefore, select a second column for confirmation where coelution of the analytes of interest does not occur.

Method interferences may be caused by contaminants in solvents, reagents, glassware, and other sample processing hardware that lead to discrete artifacts or elevated baselines in gas chromatograms. All these materials must be routinely demonstrated to be free from interferences under the conditions of the analysis by analyzing reagent blanks.

INSTRUMENTATION A GC with a NPD or a FPD will be needed. A data system or integrator is recommended for measuring peak areas and/or peak heights. A Kuderna-Danish (K-D) apparatus will be needed for extract concentration.

Column 1: 15 m × 0.53 mm megabore capillary column, 1.0 μm film thickness, DB-210.
Column 2: 15 m × 0.53 mm megabore capillary column, 1.5 μm film thickness, SPB-608.
Column 3: 15 m × 0.53 mm megabore capillary column, 1.0 μm film thickness, DB-5.

Three megabore capillary columns are included for analysis of organophosphates by this method. Column 1 (DB-210 or equivalent) and Column 2 (SPB-608 or equivalent) are recommended if a large number of organophosphorus analytes are to be determined. If the superior resolution offered by Column 1 and Column 2 is not required, Column 3 (DB-5 or equivalent) may be used. For megabore capillary columns, automatic injections of 1 μL are recommended.

PRECISION & ACCURACY The MDL actually achieved in a given analysis will vary, as it is dependent on instrument sensitivity and matrix effects. Single operator accuracy and precision studies have been conducted with spiked water and soil samples.

MULTIPLICATION FACTORS FOR OTHER MATRICES (a)

Matrix	Factor (b)
Groundwater (EPA Method 3510 or EPA Method 3520)	10
Low-concentration soil by Soxhlet and no cleanup	10 (c)
Low-concentration soil by ultrasonic extraction with GPC cleanup	6.7 (c)
High-concentration soil and sludges by ultrasonic extraction	500 (c)
Non-water miscible waste (EPA Method 3580)	1000 (c)

(a) Sample EQLs are highly matrix-dependent. The EQLs listed here are provided for guidance and may not always be achievable.
(b) EQL = [Method detection limit] × [Factor]. For non-aqueous samples the factor is on a wet-weight basis.
(c) Multiply this factory times the soil MDL.

The MDL (in μg/L) when reagent water was extracted using a separatory funnel was 0.12.
The MDL (in μg/kg) when soil was extracted using Soxhlet extraction (EPA Method 3540) was 6.0.
Accuracy (as % recovery) with separatory funnel extraction ranged from 33 (with low spikes) to 74 (with high spikes).
Accuracy (as % recovery) with continuous liquid-liquid extraction ranged from 38 (with low spikes) to 41 (with high spikes).
Accuracy (as % recovery) with Soxhlet extraction of soils ranged from 169 (with low spikes to 75 (with high spikes).
Accuracy (as % recovery) with ultrasonic extraction of soils ranged from not recovered (with low spikes) to 16 (with high spikes).

SAMPLE COLLECTION, PRESERVATION & HANDLING
Containers used to collect samples for the determination of semivolatile organic compounds should be soap and water washed followed by methanol (or isopropanol) rinsing. The sample containers should be of glass or Teflon® and have screw-top covers with Teflon® liners.

No preservation is used with concentrated waste samples. With liquid samples containing no residual chlorine and with soil, sediment, and sludge samples, immediately cooling to 4°C is the only preservation used. When residual chlorine is present then 3 mL of 10% aqueous sodium sulfate is added for each gallon of sample collected, followed by cooling to 4°C.

Liquid samples must be extracted within 7 days and their extracts analyzed within 40 days. Concentrated waste, soil, sediment, and sludge samples must be extracted within 14 days and their extracts analyzed within 40 days.

SAMPLE PREPARATION In general, water samples are extracted at a neutral pH with methylene chloride, using either EPA Method 3510 or EPA Method 3520. Solid samples are extracted using either EPA Method 3540 or EPA Method 3550 with methylene chloride/acetone (1:1) as the extraction solvent.

Prior to GC analysis, the extraction solvent may be exchanged to hexane. Single lab data indicates that samples should not be transferred with 100% hexane during sample workup as the more water soluble organophosphorus compounds may be lost.

If cleanup is performed on the samples, the analyst should analyze the samples by GC. This will confirm elution patterns and the absence of interferences from the reagents. If peak detection and identification is prevented by the presence of interferences, further cleanup is required.

QUALITY CONTROL The analyst should monitor the performance of the extraction, cleanup (when used), and analytical system and the effectiveness of the method in dealing with each sample matrix by spiking each sample, standard, and blank with one or two surrogates (e.g., organophosphorus compounds not expected to be present in the sample). Deuterated analogs of analytes should not be used as surrogates for gas chromatographic analysis due to coelution problems.

A minimum of five concentrations for each analyte of interest should be prepared through dilution of the stock standards with isooctane. One of the concentrations should be at a concentration near, but above, the MDL.

Include a mid-level check standard after each group of 10 samples in the analysis sequence. GC/MS techniques should be judiciously employed to support qualitative identifications made with this method. Follow the GC/MS operating requirements specified in EPA Method 8270.

When available, chemical ionization mass spectra may be employed to aid in the qualitative identification process. To confirm an identification of a compound, the background-corrected mass spectrum of the compound must be obtained from the sample extract and must be compared with a mass spectrum from a stock or calibration standard analyzed under the same chromatographic conditions. The molecular ion and all other ions present above 20% relative abundance in the mass spectrum of the standard must be present in the mass spectrum of the sample with agreement to ±20%. The retention time of the compound in the sample must be within six seconds of the retention time for the same compound in the standard solution.

Should the MS procedure fail to provide satisfactory results, additional steps may be taken before reanalysis. These steps may include the use of alternate packed or capillary GC columns or additional sample cleanup.

REFERENCE Test Methods for Evaluating Solid Waste, Physical/Chemical Methods, SW-846, 3rd Edition, U.S. EPA, Office of Solid Waste, Washington, DC, EPA Method 8141 July 1992.

Demeton-O	EPA Method 8270
CAS #298-03-3	

TITLE Semivolatile Organic Compounds by GC/MS

MATRIX This method is used to determine the concentration of semivolatile organic compounds in extracts prepared from all types of solid waste matrices, soils, and groundwater. Although surface waters are not specifically mentioned, this method should be applicable to water samples from rivers, lakes, etc.

METHOD SUMMARY This method covers 259 semivolatile organic compounds. In very limited applications direct injection of the sample into the GC/MS system may be appropriate, but this results in very high detection limits (approximately 10,000 µg/L). Typically, a 1-L liquid sample, containing surrogate, and matrix spiking standards, is extracted in a continuous extractor first under acid conditions and then under basic conditions. Typically 30 g of a solid sample, containing surrogate, and matrix spiking standards, is extracted ultrasonically. After concentrating the extract to 1 mL it is spiked with 10 µL of an internal standard solution just prior to analysis by GC/MS. The volume injected should contain about 100 ng of base/neutral and 200 ng of acid surrogates (for a 1-µL injection). Analysis is performed by GC/MS using a capillary GC column.

INTERFERENCES Raw GC/MS data from all blanks, samples, and spikes must be evaluated for interferences. Contamination by carryover can occur whenever high-concentration and low-concentration samples are sequentially analyzed. To reduce carryover, the sample syringe must be rinsed out between samples with solvent. Whenever an unusually concentrated sample is encountered, it should be followed by the analysis of blank solvent to check for cross-contamination.

INSTRUMENTATION A GC/MS and a data system are required. The GC column used is a 30 m × 0.25 mm I.D. (or 0.32 mm I.D.) 1um film thickness silicone-coated fused silica capillary column. A continuous liquid-liquid extractor equipped with Teflon® or glass connection joints and stopcocks requiring no lubrication, a K-D concentrating apparatus, water bath, and an ultrasonic disrupter with a minimum power of 300 W and with pulsing capability are also required.

PRECISION & ACCURACY The estimated quantitation limit (EQL) of Method 8270B for determining an individual compound is approximately 1 mg/kg (wet weight) for soil or sediment samples, 1–200 mg/kg for wastes (dependent on matrix and method of preparation), and 10 µg/L for groundwater samples. EQLs will be proportionately higher for sample extracts that require dilution to avoid saturation of the detector.

The EQL(b) for groundwater in µg/L is 10.
The EQL (a, b) for low concentrations in soil and sediment in µg/kg is not determined.
Accuracy as µg/L is not listed.
Overall precision in µg/L is not listed.

(a) *EQLs listed for soil/sediment are based on wet weight. Normally data is reported in a dry-weight basis; therefore, EQLs will be higher based on the % dry weight of each sample. This calculation is based on a 30 g sample and gel permeation chromatography cleanup.*
(b) *Sample EQLs are highly matrix-dependent. The EQLs are provided for guidance and may not always be achievable.*
C = *True value for concentration, in µg/L.*

X = Average recovery found for measurements of samples containing a concentration of C, in µg/L.

ESTIMATED QUANTITATION LIMIT

Other Matrices	Factor (a)
High-concentration soil and sludges by sonicator	7.5
Non-water miscible waste	75

(a) *EQL for other matrices = [EQL for low soil/sediment] × [Factor]. This estimated EQL is similar to an EPA "Practical Quantitation Limit."*

SAMPLING METHOD
Liquid samples — Use a 1 or 2½ gallon amber glass bottle with a screw-top Teflon®-lined cover that has been prewashed with detergent and rinsed with distilled water and methanol (or isopropanol).

Soils, sediments, or sludges — Use an 8-oz. widemouth glass with a screw-top Teflon®-lined cover that has been prewashed with detergent and rinsed with distilled water and methanol (or isopropanol).

SAMPLE PRESERVATION
Liquid samples — If residual chlorine is present, add 3 mL of 10% sodium thiosulfate per gallon, cool to 4°C and store in a solvent-free refrigerator until analysis; if chlorine is not present, then eliminate the sodium thiosulfate addition.

Soils, sediments, or sludges — Cool samples to 4°C and store in a solvent-free refrigerator.

MHT Liquid samples must be extracted within 7 days and the extracts analyzed within 40 days. Soils, sediments, or sludges may be stored for a maximum of 14 days and the extracts analyzed within 40 days.

SAMPLE PREPARATION
Liquid samples — Transfer 1 L quantitatively to a continuous extractor. If high concentrations are anticipated, a smaller volume may be used and then diluted with organic-free reagent water to 1 L. Adjust pH, if necessary, to pH <2 using 1:1 (V/V) sulfuric acid. Pipette 1.0 mL of a surrogate standard spiking

solution into each sample. For the sample in each analytical batch selected for spiking, add 1.0 mL of a matrix spiking standard. For base/neutral acid analysis, the amount of the surrogates and matrix spiking compounds added to the sample should result in a final concentration of 100 ng/µL of each analyte in the extract to be analyzed (assuming a 1-µL injection). Extract with methylene chloride for 18–24 h. Next, adjust the pH of the aqueous phase to pH >11 using 10 N sodium hydroxide and extract it with methylene chloride again for 18–24 h. Dry the extract through a column containing anhydrous sodium sulfate and concentrate it to 1 mL using a K-D concentrator.

Soils, sediments, or sludges — Use 30 g of sample. Nonporous or wet samples (gummy or clay type) that do not have a free-flowing sandy texture must be mixed with anhydrous sodium sulfate until the sample is free flowing. Add 1 mL of surrogate standards to all samples, spikes, standards, and blanks. For the sample in each analytical batch selected for spiking, add 1.0 mL of a matrix spiking standard. For base/neutral acid analysis, the amount added of the surrogates and matrix spiking compounds should result in a final concentration of 100 ng/µL of each base/neutral analyte and 200 ng/µL of each acid analyte in the extract to be analyzed (assuming a 1-µL injection). Immediately add a 100-mL mixture of 1:1 methylene chloride:acetone and extract the sample ultrasonically for 3 min and then decant or filter the extracts. Repeat the extraction two or more times. Dry the extract using a column with anhydrous sodium sulfate and concentrate it to 1 mL in a K-D concentrator.

QUALITY CONTROL A methylene chloride solution containing 50 ng/µL of decafluorotriphenylphosphine (DFTPP) is used for tuning the GC/MS system each 12-h shift. A system performance check also must be made during every 12-h shift. A standard containing 50 ng/µL each of 4,4′-DDT, pentachlorophenol, and benzidine is required to verify injection port inertness and GC column performance. A calibration standard at mid-concentration, containing each compound of interest, including all required surrogates, must be performed every 12 h during analysis. After the system performance check is met, calibration check compounds (CCCs) are used to check the validity of the initial calibration.

The internal standard responses and retention times in the calibration check standard must be evaluated immediately after or during data acquisition. If the retention time for any internal standard changes by more than 30 seconds from the last check calibration (12 h), the chromatographic system must be inspected for malfunctions and corrections must be made, as required. If the electron ionization current plot (EICP) area for any of the internal standards changes by a factor of two from the last daily calibration standard check, the mass spectrometer must be inspected for malfunctions and corrections must be made, as appropriate.

Demonstrate, through the analysis of a reagent water blank, that interferences from the analytical system, glassware, and reagents are under control. The blank samples should be carried through all stages of the sample preparation and measurement steps. For each analytical batch (up to 20 samples), a reagent blank, matrix spike, and matrix spike duplicate/duplicate must be analyzed (the frequency of the spikes may be different for different monitoring programs). The blank and spiked samples must be carried through all stages of the sample preparation and measurement steps. A QC reference sample concentrate containing each analyte at a concentration of 100 mg/L in methanol is required.

REFERENCE Test Methods for Evaluating Solid Waste (SW-846). U.S. EPA 1983, Method 8270B, Rev. 2, Nov. 1990. Office of Solid Waste, Washington, DC.

Demeton-O EPA Method 8140
CAS #298-03-3

TITLE Organophosphorus Pesticides

MATRIX Groundwater, soils, sludges, water miscible liquid wastes, and non-water miscible wastes.

APPLICATION This method is used for the analysis of 21 organophosphorus pesticides. Samples are extracted, concentrated, and analyzed using direct injection of both neat and diluted organic liquid into a gas chromatograph (GC).

INTERFERENCES Solvents, reagents, and glassware may introduce artifacts. Other interferences may come from coextracted compounds from samples. The use of Florisil cleanup materials may produce low recoveries. Elemental sulfur may interfere with some compounds when using a flame photometric detector. Sulfur cleanup (Method 3660) may alleviate sulfur interference.

INSTRUMENTATION GC capable of on-column injections and a flame photometric detector (FPD) or a thermionic detector. Column 1: 1.8 m by 2 mm with 5% SP-2401 on Supelcoport. Column 2: 1.8 m by 2 mm with 3% SP-2401 on Supelcoport. Column 3: 50 cm by ⅛ IN EFLON® with 15% SE-54 on Gas Chrom Q. The preferred column is Column Number 1.

RANGE 11.9 to 314 µg/L.

MDL 0.25 µg/L (in reagent water).

PQL FACTORS FOR MULTIPLYING × FID MDL VALUE

Matrix	Multiplication Factor
Groundwater	10
Low-level soil by sonication with GPC cleanup	670
High-level soil and sludge by sonication	10,000
Non-water miscible waste	100,000

PRECISION 10.5% (single operator standard deviation).

ACCURACY 67.4% (single operator average recovery).

SAMPLING METHOD Use 8-oz. widemouth glass bottles with Teflon®-lined caps for concentrated waste samples, soils, sediments, and sludges. Use 1 or 2½ gallon amber glass bottles with Teflon®-lined caps for liquid (water) samples.

STABILITY Cool soil, sediment, sludge, and liquid samples to 4°C. If residual chlorine is present in liquid samples add 3 mL of 10% sodium thiosulfate per gallon of sample and cool to 4°C.

MHT 14 days for concentrated waste, soil, sediment, or sludge; 7 days for liquid samples; all extracts must be analyzed within 40 days.

QUALITY CONTROL A quality control check sample concentrate containing this compound in acetone at a concentration 1,000 times more concentrated than the selected spike concentration is required. The QC check sample concentrate may be prepared from pure standard materials or purchased as certified solutions. Use appropriate trip, matrix, control site, method, reagent, and solvent blanks. Internal, surrogate, and five concentration level calibration standards are used.

REFERENCE Method 8140, SW-846, 3rd ed., Sept. 1986.

Demeton-S	EPA Method 8270
CAS #126-75-0	

TITLE Semivolatile Organic Compounds by GC/MS

MATRIX This method is used to determine the concentration of semivolatile organic compounds in extracts prepared from all types of solid waste matrices, soils, and groundwater. Although surface waters are not specifically mentioned, this method should be applicable to water samples from rivers, lakes, etc.

METHOD SUMMARY This method covers 259 semivolatile organic compounds. In very limited applications direct injection of the sample into the GC/MS system may be appropriate, but this results in very high detection limits (approximately 10,000 µg/L). Typically, a 1-L liquid sample, containing surrogate, and matrix spiking standards, is extracted in a continuous extractor first under acid conditions and then under basic conditions. Typically 30 g of a solid sample, containing surrogate, and matrix spiking standards, is extracted ultrasonically. After concentrating the extract to 1 mL it is spiked with 10 µL of an internal standard solution just prior to analysis by GC/MS. The volume injected should contain about 100 ng of base/neutral and 200 ng of acid surrogates (for a 1-µL injection). Analysis is performed by GC/MS using a capillary GC column.

INTERFERENCES Raw GC/MS data from all blanks, samples, and spikes must be evaluated for interferences. Contamination by carryover can occur whenever high-concentration and low-concentration samples are sequentially analyzed. To reduce carryover, the sample syringe must be rinsed out between samples with solvent. Whenever an unusually concentrated sample is encountered, it should be followed by the analysis of blank solvent to check for cross-contamination.

INSTRUMENTATION A GC/MS and a data system are required. The GC column used is a 30 m × 0.25 mm I.D. (or 0.32 mm I.D.) 1um film thickness silicone-coated fused silica capillary column. A continuous liquid-liquid extractor equipped with Teflon® or glass connection joints and stopcocks requiring no lubrication, a K-D concentrating apparatus, water bath, and an ultrasonic disrupter with a minimum power of 300 W and with pulsing capability are also required.

PRECISION & ACCURACY The estimated quantitation limit (EQL) of Method 8270B for determining an individual compound is approximately 1 mg/kg (wet weight) for soil or sediment samples, 1–200 mg/kg for wastes (dependent on matrix and method of preparation), and 10 µg/L for groundwater samples. EQLs will be proportionately higher for sample extracts that require dilution to avoid saturation of the detector.

The EQL(b) for groundwater in µg/L is 10.
The EQL (a, b) for low concentrations in soil and sediment in µg/kg is not determined.
Accuracy as µg/L is not listed.
Overall precision in µg/L is not listed.

(a) *EQLs listed for soil/sediment are based on wet weight. Normally data is reported in a dry-weight basis; therefore, EQLs will be higher based on the % dry weight of each sample. This calculation is based on a 30 g sample and gel permeation chromatography cleanup.*
(b) *Sample EQLs are highly matrix-dependent. The EQLs are provided for guidance and may not always be achievable.*
C = *True value for concentration, in µg/L.*
X = *Average recovery found for measurements of samples containing a concentration of C, in µg/L.*

ESTIMATED QUANTITATION LIMIT

Other Matrices	Factor (a)
High-concentration soil and sludges by sonicator	7.5
Non-water miscible waste	75

(a) *EQL for other matrices = [EQL for low soil/sediment] × [Factor]. This estimated EQL is similar to an EPA "Practical Quantitation Limit."*

SAMPLING METHOD
Liquid samples — Use a 1 or 2½ gallon amber glass bottle with a screw-top Teflon®-lined cover that has been prewashed with detergent and rinsed with distilled water and methanol (or isopropanol).

Soils, sediments, or sludges — Use an 8-oz. widemouth glass with a screw-top Teflon®-lined cover that has been prewashed with detergent and rinsed with distilled water and methanol (or isopropanol).

SAMPLE PRESERVATION
Liquid samples — If residual chlorine is present, add 3 mL of 10% sodium thiosulfate per gallon, cool to 4°C and store in a solvent-free refrigerator until analysis; if chlorine is not present, then eliminate the sodium thiosulfate addition.

Soils, sediments, or sludges — Cool samples to 4°C and store in a solvent-free refrigerator.

MHT Liquid samples must be extracted within 7 days and the extracts analyzed within 40 days. Soils, sediments, or sludges may be stored for a maximum of 14 days and the extracts analyzed within 40 days.

SAMPLE PREPARATION
Liquid samples — Transfer 1 L quantitatively to a continuous extractor. If high concentrations are anticipated, a smaller volume may be used and then diluted with organic-free reagent

water to 1 L. Adjust pH, if necessary, to pH <2 using 1:1 (V/V) sulfuric acid. Pipette 1.0 mL of a surrogate standard spiking solution into each sample. For the sample in each analytical batch selected for spiking, add 1.0 mL of a matrix spiking standard. For base/neutral acid analysis, the amount of the surrogates and matrix spiking compounds added to the sample should result in a final concentration of 100 ng/μL of each analyte in the extract to be analyzed (assuming a 1-μL injection). Extract with methylene chloride for 18–24 h. Next, adjust the pH of the aqueous phase to pH >11 using 10 N sodium hydroxide and extract it with methylene chloride again for 18–24 h. Dry the extract through a column containing anhydrous sodium sulfate and concentrate it to 1 mL using a K-D concentrator.

Soils, sediments, or sludges — Use 30 g of sample. Nonporous or wet samples (gummy or clay type) that do not have a free-flowing sandy texture must be mixed with anhydrous sodium sulfate until the sample is free flowing. Add 1 mL of surrogate standards to all samples, spikes, standards, and blanks. For the sample in each analytical batch selected for spiking, add 1.0 mL of a matrix spiking standard. For base/neutral acid analysis, the amount added of the surrogates and matrix spiking compounds should result in a final concentration of 100 ng/μL of each base/neutral analyte and 200 ng/μL of each acid analyte in the extract to be analyzed (assuming a 1-μL injection). Immediately add a 100-mL mixture of 1:1 methylene chloride:acetone and extract the sample ultrasonically for 3 min and then decant or filter the extracts. Repeat the extraction two or more times. Dry the extract using a column with anhydrous sodium sulfate and concentrate it to 1 mL in a K-D concentrator.

QUALITY CONTROL A methylene chloride solution containing 50 ng/μL of decafluorotriphenylphosphine (DFTPP) is used for tuning the GC/MS system each 12-h shift. A system performance check also must be made during every 12-h shift. A standard containing 50 ng/μL each of 4,4′-DDT, pentachlorophenol, and benzidine is required to verify injection port inertness and GC column performance. A calibration standard at mid-concentration, containing each compound of interest, including all required surrogates, must be performed every 12 h during analysis. After the system performance check is met, calibration check compounds (CCCs) are used to check the validity of the initial calibration.

The internal standard responses and retention times in the calibration check standard must be evaluated immediately after or during data acquisition. If the retention time for any internal standard changes by more than 30 seconds from the last check calibration (12 h), the chromatographic system must be inspected for malfunctions and corrections must be made, as required. If the electron ionization current plot (EICP) area for any of the internal standards changes by a factor of two from the last daily calibration standard check, the mass spectrometer must be inspected for malfunctions and corrections must be made, as appropriate.

Demonstrate, through the analysis of a reagent water blank, that interferences from the analytical system, glassware, and reagents are under control. The blank samples should be carried through all stages of the sample preparation and measurement steps. For each analytical batch (up to 20 samples), a reagent blank, matrix spike, and matrix spike duplicate/duplicate must be analyzed (the frequency of the spikes may be different for different monitoring programs). The blank and spiked samples must be carried through all stages of the sample preparation and measurement steps. A QC reference sample concentrate containing each analyte at a concentration of 100 mg/L in methanol is required.

REFERENCE Test Methods for Evaluating Solid Waste (SW-846). U.S. EPA 1983, Method 8270B, Rev. 2, Nov. 1990. Office of Solid Waste, Washington, DC.

Demeton-S   EPA Method 8140
CAS #126-75-0

TITLE Organophosphorus Pesticides

MATRIX Groundwater, soils, sludges, water miscible liquid wastes, and non-water miscible wastes.

APPLICATION This method is used for the analysis of 21 organophosphorus pesticides. Samples are extracted, concentrated, and analyzed using direct injection of both neat and diluted organic liquid into a gas chromatograph (GC).

INTERFERENCES Solvents, reagents, and glassware may introduce artifacts. Other interferences may come from coextracted compounds from samples. The use of Florisil cleanup materials may produce low recoveries. Elemental sulfur may interfere with some compounds when using a flame photometric detector. Sulfur cleanup (Method 3660) may alleviate sulfur interference.

INSTRUMENTATION GC capable of on-column injections and a flame photometric detector (FPD) or a thermionic detector. Column 1: 1.8 m by 2 mm with 5% SP-2401 on Supelcoport. Column 2: 1.8 m by 2 mm with 3% SP-2401 on Supelcoport. Column 3: 50 cm by ⅛ IN EFLON® with 15% SE-54 on Gas Chrom Q. The preferred column is Column Number 1.

RANGE 11.9 to 314 μg/L.

MDL 0.25 μg/L (in reagent water).

PQL FACTORS FOR MULTIPLYING × FID MDL VALUE

Matrix	Multiplication Factor
Groundwater	10
Low-level soil by sonication with GPC cleanup	670
High-level soil and sludge by sonication	10,000
Non-water miscible waste	100,000

PRECISION 10.5% (single operator standard deviation).

ACCURACY 67.4% (single operator average recovery).

SAMPLING METHOD Use 8-oz. widemouth glass bottles with Teflon®-lined caps for concentrated waste samples, soils, sediments, and sludges. Use 1 or 2½ gallon amber glass bottles with Teflon®-lined caps for liquid (water) samples.

STABILITY Cool soil, sediment, sludge, and liquid samples to 4°C. If residual chlorine is present in liquid samples add

3 mL of 10% sodium thiosulfate per gallon of sample and cool to 4°C.

MHT 14 days for concentrated waste, soil, sediment, or sludge; 7 days for liquid samples; all extracts must be analyzed within 40 days.

QUALITY CONTROL A quality control check sample concentrate containing this compound in acetone at a concentration 1,000 times more concentrated than the selected spike concentration is required. The QC check sample concentrate may be prepared from pure standard materials or purchased as certified solutions. Use appropriate trip, matrix, control site, method, reagent, and solvent blanks. Internal, surrogate, and five concentration level calibration standards are used.

REFERENCE Method 8140, SW-846, 3rd ed., Sept. 1986.

Di-n-butyl phthalate   EPA Method 1625
CAS #84-74-2

TITLE Semivolatile Organic Compounds by Isotope Dilution GC/MS

MATRIX The compounds may be determined in waters, soils, and municipal sludges by this method.

METHOD SUMMARY This method is used to determine 176 semivolatile toxic organic pollutants associated with the CWA (as amended 1987); the RCRA (as amended 1986); the CERCLA (as amended 1986); and other compounds amenable to extraction and analysis by capillary column gas chromatography-mass spectrometry (GC/MS).

Stable isotopically-labeled analogs of the compounds of interest are added to the sample. If the solids content is less than 1%, a 1-L sample is extracted at pH 12–13, then at pH <2 with methylene chloride using continuous extraction techniques.

If the solids content is 30% or less, the sample is diluted to 1% solids with reagent water, homogenized ultrasonically, and extracted at pH 12–13, then at pH <2 with methylene chloride using continuous extraction techniques. If the solids content is greater than 30%, the sample is extracted using ultrasonic techniques.

Each extract is dried over sodium sulfate, concentrated to a volume of 5 mL, cleaned up using GPC, if necessary, and concentrated. Extracts are concentrated to 1 mL if GPC is not performed, and to 0.5 mL if GPC is performed.

An internal standard is added to the extract, and a 1-mL aliquot of the extract is injected into the GC. The compounds are separated by GC and detected by a MS. The labeled compounds serve to correct the variability of the analytical technique.

INTERFERENCES Solvents, reagents, glassware, and other sample processing hardware may yield artifacts and/or elevated baselines causing misinterpretation of chromatograms and spectra. Materials used in the analysis must be demonstrated to be free from interferences under the conditions of analysis by running method blanks initially and with each sample lot (sample started through the extraction process on a given 8-h shift, to a maximum of 20). Specific selection of reagents and purification of solvents by distillation in all glass systems may be required. Glassware and, where possible, reagents are cleaned by solvent rinse and baking at 450°C for 1-h minimum. Interferences coextracted from samples will vary considerably from source to source, depending on the diversity of the site being sampled.

INSTRUMENTATION Major instrumentation includes a GC with a splitless or on-column injection port for capillary column, a MS with 70 eV electron impact ionization, and a data system to collect and record MS data, and process it. A K-D apparatus is used to concentrate extracts.

GC Column: 30 m × 0.25 mm I.D. 5% phenyl, 94% methyl, 1% vinyl silicone bonded phased fused silica capillary column.

PRECISION & ACCURACY The detection limits of the method are usually dependent on the level of interferences rather than instrumental limitations. The limits typify the minimum quantities that can be detected with no interferences present.

The minimum level (in µg/mL) was 10. This is defined as a minimum level at which the analytical system shall give recognizable mass spectra (background corrected) and acceptable calibration points.

The MDL (in µg/kg) in low solids was 64 and in high solids was 80; these were determined in digested sludge (low solids) and in filter cake or compost (high solids).

The labeled and native compound initial precision as standard deviation (in µg/L) was 15.

The labeled and native compound initial accuracy as average recovery (in µg/L) was 76–165.

SAMPLE COLLECTION, PRESERVATION & HANDLING Collect samples in glass containers. Aqueous samples which flow freely are collected in refrigerated bottles using automatic sampling equipment. Solid samples are collected as grab samples using widemouth jars. Maintain samples at 0 to 4°C from the time of collection until extraction. If residual chlorine is present in aqueous samples, add 80 mg sodium thiosulfate/L of water. Begin sample extraction within 7 days of collection, and analyze all extracts within 40 days of extraction.

SAMPLE PREPARATION Samples containing 1% solids or less are extracted directly using continuous liquid-liquid extraction techniques. Samples containing 1 to 30% solids are diluted to the 1% level with reagent water and extracted using continuous liquid-liquid extraction techniques. Samples containing greater than 30% solids are extracted using ultrasonic techniques.

Base/neutral extraction — Adjust the pH of the waters in the extractors to 12–13 with 6 N NaOH. Extract with methylene chloride for 24–48 h.

Acid extraction — Adjust the pH of the waters in the extractors to 2 or less using 6 N sulfuric acid. Extract with methylene chloride for 24–48 h.

Ultrasonic extraction of high solids samples — Add anhydrous sodium sulfate to the sample and QC aliquot(s).

Add acetone:methylene chloride (1:1) to the sample and mix thoroughly

Concentrate extracts using a K-D apparatus.

QUALITY CONTROL The analyst is permitted to modify this method to improve separations or lower the costs of measurements, provided all performance specifications are met. Analyses of blanks are required to demonstrate freedom from contamination. When results of spikes indicate atypical method performance for samples, the samples are diluted to bring method performance within acceptable limits.

For low solids (aqueous samples), extract, concentrate, and analyze two sets of four 1-L aliquots (8 aliquots total) of the precision and recovery standard. For high solids samples, two sets of four 30-g aliquots of the high solids reference matrix are used.

Spike all samples with labeled compounds to assess method performance. Compute percent recovery of the labeled compounds using the internal standard method. Compare the labeled compound recovery for each compound with the corresponding labeled compound recovery.

Reagent water and high solids reference matrix blanks are analyzed to demonstrate freedom from contamination. Extract and concentrate a 1-L reagent water blank or a high solids reference matrix blank with each sample's lot (samples started through the extraction process on the same 8-h shift, to a maximum of 20 samples).

Field replicates may be collected to determine the precision of the sampling technique, and spiked samples may be required to determine the accuracy of the analysis when the internal standard method is used.

REFERENCE Semivolatile Organic Compounds by Isotope Dilution GC/MS. Office of Water Regulation and Standards, U.S. EPA Industrial Technology Division, Washington, DC, EPA Method 1625, Rev. C, June 1989 (contact W.A. Telliard, U.S. EPA, Office of Water Regulations and Standards, 401 M St., SW, Washington, DC, 20460. Phone: 202-382-7131).

Di-n-butyl phthalate   **EPA Method 625**
CAS #84-74-2

TITLE Base/Neutrals and Acids, U.S. EPA Method 625

MATRIX This methods covers municipal and industrial wastewaters.

METHOD SUMMARY Approximately 1 L of sample is serially extracted with methylene chloride at a pH greater than 11 and again at a pH less than 2 using a separatory funnel or a continuous extractor. The methylene chloride extract is dried, concentrated to a volume of 1 mL, and analyzed by GC/MS. Qualitative identification of the parameters in the extract is performed using the retention time and the relative abundance of three characteristic masses (m/z). Qualitative analysis is performed using either external or internal standard techniques with a single characteristic m/z.

INTERFERENCES Method interferences may be caused by contaminants in solvents, reagents, glassware, and other sample processing hardware. Glassware must be scrupulously cleaned. Glassware should be heated in a muffle furnace at 400°C for 5 to 30 min. Some thermally stable materials, such as PCBs, may not be eliminated by this treatment. Solvent rinses with acetone and pesticide quality hexane may be substituted for the muffle furnace heating. Matrix interferences may be caused by contaminants that are coextracted from the sample. The base-neutral extraction may cause significantly reduced recovery of phenols. The packed gas chromatographic columns recommended for the basic fraction may not exhibit sufficient resolution for some analytes.

INSTRUMENTATION A GC/MS system with an injection port designed for on-column injection when using packed columns and for splitless injection when using capillary columns.

Column for base/neutrals: 1.8 m long × 2 mm I.D. glass, packed with 3% SP-2550 on Supelcoport (100/120 mesh) or equivalent.

Column for acids: 1.8 m long × 2 mm I.D. glass, packed with 1% SP-1240DA on Supelcoport (100/120 mesh) or equivalent.

PRECISION & ACCURACY The MDL concentrations were obtained using reagent water. The MDL actually achieved in a given analysis will vary depending on instrument sensitivity and matrix effects. This method was tested by 15 laboratories using reagent water, drinking water, surface water, and industrial wastewaters spiked at six concentrations over the range 5 to 100 µg/L. Single operator precision, overall precision, and method accuracy were found to be directly related to the concentration of the parameter matrix.

The MDL (in µg/L) in reagent water was not reported.

The standard deviation (in µg/L based on 4 recovery measurements) was 16.7.

The range (in µg/L) for average recovery for 4 measurements was 8.4–111.0.

The range (in %) for percent recovery was 1–118.

Accuracy (in µg/L) as expected recovery for one or more measurements of a sample containing a true concentration of C was $0.59C+0.71$.

Precision (in µg/L) as expected single analyst standard deviation of measurements at an average concentration found at X was $0.13X+1.16$.

Overall precision (in µg/L) as expected interlaboratory standard deviation of measurements in an average concentration found at X was $0.39X+0.60$.

C = *True value of the concentration in µg/L.*
X = *Average recovery found for measurements of samples containing a concentration at C in µg/L.*

SAMPLE PREPARATION Adjust the pH to >11 with sodium hydroxide and serially extract in a separatory funnel with methylene chloride or else in a continuous extractor. Next, adjust the pH to <2 with sulfuric acid and serially extract in a separatory funnel with methylene chloride or else in a continuous extractor. Dry the extracts separately through a column of anhydrous sodium sulfate and then concentrate each of the extracts to 1.0 mL using a K-D apparatus.

SAMPLE COLLECTION, PRESERVATION & HANDLING
Grab samples must be collected in glass containers. All samples must be refrigerated at 4°C from the time of collection until extraction. If residual chlorine is present, add 80 mg of sodium thiosulfate/L of sample and mix well. All samples must be extracted within 7 days of collection and completely analyzed within 40 days of extraction.

QUALITY CONTROL Make an initial, one-time, demonstration of the ability to generate acceptable accuracy and precision with this method. Before processing any samples, the analyst must analyze a reagent water blank to demonstrate that interferences from the analytical system and glassware are under control. Each time a set of samples is extracted or reagents are changed, a reagent water blank must be processed. Spike and analyze a minimum of 5% of all samples to monitor and evaluate lab data quality. A QC check sample concentrate that contains each parameter of interest at a concentration of 100 µg/mL in acetone is required. PCBs and multicomponent pesticides may be omitted from this test.

After analysis of five spiked wastewater samples, calculate the average percent recovery and the standard deviation of the percent recovery. Spike all samples with the surrogate standard spiking solution and calculate the percent recovery of each surrogate compound.

REFERENCE *Federal Register*, Vol. 49, No. 209. Friday, Oct. 26, 1984.

Di-n-butyl phthalate   **EPA Method 8270**
CAS #84-74-2

TITLE Semivolatile Organic Compounds by GC/MS

MATRIX This method is used to determine the concentration of semivolatile organic compounds in extracts prepared from all types of solid waste matrices, soils, and groundwater. Although surface waters are not specifically mentioned, this method should be applicable to water samples from rivers, lakes, etc.

METHOD SUMMARY This method covers 259 semivolatile organic compounds. In very limited applications direct injection of the sample into the GC/MS system may be appropriate, but this results in very high detection limits (approximately 10,000 µg/L). Typically, a 1-L liquid sample, containing surrogate, and matrix spiking standards, is extracted in a continuous extractor first under acid conditions and then under basic conditions. Typically 30 g of a solid sample, containing surrogate, and matrix spiking standards, is extracted ultrasonically. After concentrating the extract to 1 mL it is spiked with 10 µL of an internal standard solution just prior to analysis by GC/MS. The volume injected should contain about 100 ng of base/neutral and 200 ng of acid surrogates (for a 1-µL injection). Analysis is performed by GC/MS using a capillary GC column.

INTERFERENCES Raw GC/MS data from all blanks, samples, and spikes must be evaluated for interferences. Contamination by carryover can occur whenever high-concentration and low-concentration samples are sequentially analyzed. To reduce carryover, the sample syringe must be rinsed out between samples with solvent. Whenever an unusually concentrated sample is encountered, it should be followed by the analysis of blank solvent to check for cross-contamination.

INSTRUMENTATION A GC/MS and a data system are required. The GC column used is a 30 m × 0.25 mm I.D. (or 0.32 mm I.D.) 1um film thickness silicone-coated fused silica capillary column. A continuous liquid-liquid extractor equipped with Teflon® or glass connection joints and stopcocks requiring no lubrication, a K-D concentrating apparatus, water bath, and an ultrasonic disrupter with a minimum power of 300 W and with pulsing capability are also required.

PRECISION & ACCURACY The estimated quantitation limit (EQL) of Method 8270B for determining an individual compound is approximately 1 mg/kg (wet weight) for soil or sediment samples, 1–200 mg/kg for wastes (dependent on matrix and method of preparation), and 10 µg/L for groundwater samples. EQLs will be proportionately higher for sample extracts that require dilution to avoid saturation of the detector.

The EQL(b) for groundwater in µg/L is 10.
The EQL (a, b) for low concentrations in soil and sediment in µg/kg is not determined.
Accuracy as µg/L is 0.59C +0.71.
Overall precision in µg/L is 0.39X +0.60.

(a) *EQLs listed for soil/sediment are based on wet weight. Normally data is reported in a dry-weight basis; therefore, EQLs will be higher based on the % dry weight of each sample. This calculation is based on a 30 g sample and gel permeation chromatography cleanup.*
(b) *Sample EQLs are highly matrix-dependent. The EQLs are provided for guidance and may not always be achievable.*
C = *True value for concentration, in µg/L.*
X = *Average recovery found for measurements of samples containing a concentration of C, in µg/L.*

ESTIMATED QUANTITATION LIMIT

Other Matrices	Factor (a)
High-concentration soil and sludges by sonicator	7.5
Non-water miscible waste	75

(a) *EQL for other matrices = [EQL for low soil/sediment] × [Factor]. This estimated EQL is similar to an EPA "Practical Quantitation Limit."*

SAMPLING METHOD

Liquid samples — Use a 1 or 2½ gallon amber glass bottle with a screw-top Teflon®-lined cover that has been prewashed with detergent and rinsed with distilled water and methanol (or isopropanol).

Soils, sediments, or sludges — Use an 8-oz. widemouth glass with a screw-top Teflon®-lined cover that has been prewashed

with detergent and rinsed with distilled water and methanol (or isopropanol).

SAMPLE PRESERVATION

Liquid samples — If residual chlorine is present, add 3 mL of 10% sodium thiosulfate per gallon, cool to 4°C and store in a solvent-free refrigerator until analysis; if chlorine is not present, then eliminate the sodium thiosulfate addition.

Soils, sediments, or sludges — Cool samples to 4°C and store in a solvent-free refrigerator.

MHT Liquid samples must be extracted within 7 days and the extracts analyzed within 40 days. Soils, sediments, or sludges may be stored for a maximum of 14 days and the extracts analyzed within 40 days.

SAMPLE PREPARATION

Liquid samples — Transfer 1 L quantitatively to a continuous extractor. If high concentrations are anticipated, a smaller volume may be used and then diluted with organic-free reagent water to 1 L. Adjust pH, if necessary, to pH <2 using 1:1 (V/V) sulfuric acid. Pipette 1.0 mL of a surrogate standard spiking solution into each sample. For the sample in each analytical batch selected for spiking, add 1.0 mL of a matrix spiking standard. For base/neutral acid analysis, the amount of the surrogates and matrix spiking compounds added to the sample should result in a final concentration of 100 ng/µL of each analyte in the extract to be analyzed (assuming a 1-µL injection). Extract with methylene chloride for 18–24 h. Next, adjust the pH of the aqueous phase to pH >11 using 10 *N* sodium hydroxide and extract it with methylene chloride again for 18–24 h. Dry the extract through a column containing anhydrous sodium sulfate and concentrate it to 1 mL using a K-D concentrator.

Soils, sediments, or sludges — Use 30 g of sample. Nonporous or wet samples (gummy or clay type) that do not have a free-flowing sandy texture must be mixed with anhydrous sodium sulfate until the sample is free flowing. Add 1 mL of surrogate standards to all samples, spikes, standards, and blanks. For the sample in each analytical batch selected for spiking, add 1.0 mL of a matrix spiking standard. For base/neutral acid analysis, the amount added of the surrogates and matrix spiking compounds should result in a final concentration of 100 ng/µL of each base/neutral analyte and 200 ng/µL of each acid analyte in the extract to be analyzed (assuming a 1-µL injection). Immediately add a 100-mL mixture of 1:1 methylene chloride:acetone and extract the sample ultrasonically for 3 min and then decant or filter the extracts. Repeat the extraction two or more times. Dry the extract using a column with anhydrous sodium sulfate and concentrate it to 1 mL in a K-D concentrator.

QUALITY CONTROL

A methylene chloride solution containing 50 ng/µL of decafluorotriphenylphosphine (DFTPP) is used for tuning the GC/MS system each 12-h shift. A system performance check also must be made during every 12-h shift. A standard containing 50 ng/µL each of 4,4'-DDT, pentachlorophenol, and benzidine is required to verify injection port inertness and GC column performance. A calibration standard at mid-concentration, containing each compound of interest, including all required surrogates, must be performed every 12 h during analysis. After the system performance check is met, calibration check compounds (CCCs) are used to check the validity of the initial calibration.

The internal standard responses and retention times in the calibration check standard must be evaluated immediately after or during data acquisition. If the retention time for any internal standard changes by more than 30 seconds from the last check calibration (12 h), the chromatographic system must be inspected for malfunctions and corrections must be made, as required. If the electron ionization current plot (EICP) area for any of the internal standards changes by a factor of two from the last daily calibration standard check, the mass spectrometer must be inspected for malfunctions and corrections must be made, as appropriate.

Demonstrate, through the analysis of a reagent water blank, that interferences from the analytical system, glassware, and reagents are under control. The blank samples should be carried through all stages of the sample preparation and measurement steps. For each analytical batch (up to 20 samples), a reagent blank, matrix spike, and matrix spike duplicate/duplicate must be analyzed (the frequency of the spikes may be different for different monitoring programs). The blank and spiked samples must be carried through all stages of the sample preparation and measurement steps. A QC reference sample concentrate containing each analyte at a concentration of 100 mg/L in methanol is required.

REFERENCE Test Methods for Evaluating Solid Waste (SW-846). U.S. EPA 1983, Method 8270B, Rev. 2, Nov. 1990. Office of Solid Waste, Washington, DC.

Di-n-butyl phthalate   **EPA Method 8061**
CAS #84-74-2

TITLE Phthalate Esters by Capillary Gas Chromatography With Electron Capture Detection (GC/ECD)

MATRIX This method covers aqueous and solid matrices. This includes a wide variety such as drinking water, groundwater, industrial wastewaters, surface waters, soils, solids, and sediments.

METHOD SUMMARY This method is used to determine the identities and concentrations of phthalate esters in liquid, solid and sludge matrices. When used to analyze for any or all of the target analytes, compound identification should be supported by at least one additional qualitative technique. This method describes conditions for parallel column, dual electron capture detector analysis, which fulfills the above requirement. Alternatively, GC/MS could be used for compound confirmation.

A measured volume or weight of sample (approximately 1 L for liquids, 10 to 30 g for solids and sludges) is extracted by using the appropriate sample extraction technique specified in EPA Method 3510, EPA Method 3540, and EPA Method 3550. After cleanup, the extract is analyzed by GC/ECD.

INTERFERENCES The sensitivity of this method usually depends on the level of interferences rather than on instrumental limitations. If interferences prevent detection of the analytes, cleanup of the sample extracts is necessary. Either EPA Method 3610 or EPA Method 3620 alone or followed by EPA Method 3660, Sulfur Cleanup, may be used to eliminate interferences in the analysis. EPA Method 3640, Gel Permeation Cleanup, is applicable for samples that contain high amounts of lipids and waxes.

Interferences coextracted from the samples will vary considerably from waste to waste. Glassware must be scrupulously clean. All glassware require treatment in a muffle furnace at 400°C for 2 to 4 h, or thorough rinsing with pesticide-grade solvent, prior to use. Volumetric glassware should not be heated in a muffle furnace. Storage of glassware in the lab introduces contamination, even if the glassware is wrapped in aluminum foil. Sodium sulfate, Florisil, and alumina may be contaminated with phthalate esters and, therefore, use of these materials in sample cleanup should be employed cautiously. If these materials are used, they must be obtained packaged in glass. Heating at 400°C for sodium sulfate, 320°C for Florisil, and 210°C for alumina is recommended. Glass wool used in any step of sample preparation should be a specially treated pyrex wool, pesticide grade, and must be baked at 400°C for 4 h immediately prior to use.

Paper thimbles and filter paper must be exhaustively washed with the solvent that will be used in the sample extraction. Soxhlet extraction of paper thimbles and filter paper for 12 h with fresh solvent should be repeated for a minimum of three times. Method blanks should be obtained before any of the precleaned thimbles or filter papers are used.

INSTRUMENTATION Gas chromatograph suitable for on-column and split/splitless injections.

Column 1: 30 m × 0.53 mm ID, 5% phenyl/95% methyl silicone fused-silica open tubular column, DB-5, 1.5 µg film thickness.

Column 2: 30 m × 0.53 mm ID, 14% cyanopropyl phenyl silicone fused-silica open tubular column, DB-1701, 1.0 µg film thickness.

A dual electron capture detector (ECD) is used. A Kuderna-Danish (K-D) apparatus is required along with a vacuum manifold consisting of individually adjustable, easily accessible flow-control valves for up to 24 cartridges, sample rack, chemically resistant cover and seals, heavy-duty glass basin, removable stainless steel solvent guides, built-in vacuum gauge and valve. Also, 6-mL, 1-g solid-phase extraction cartridges, LC-Florisil or equivalent, prepackaged, ready to use will be needed.

PRECISION & ACCURACY The MDL actually achieved in a given analysis will vary, as it is dependent on instrument sensitivity and matrix effects. This method has been tested in a single lab. Single-operator precision, overall precision, and method accuracy were found to be related to the concentration of the compounds and the type of matrix.

MULTIPLICATION FACTORS FOR OTHER MATRICES (a)

Matrix	Factor (b)
Groundwater	10
Low-concentration soil by ultrasonic extraction with GPC cleanup	670
High-concentration soil and sludges by ultrasonic extraction	10,000
Non-water miscible waste	100,000

(a) Sample EQLs are highly matrix-dependent.
(b) EQL = [Method detection limit] × [Factor]. For non-aqueous samples, the factor is on a wet-weight basis.

The MDL using 7 replicate determinations and a spike concentration of 100 µg/L was 330 ng/L.

The average recovery from HPLC-grade water using 4 determinatons and a spike concentration of 100 µg/L was 90.3%.

The precision (as RSD) from HPLC-grade water using 4 determinatons and a spike concentration of 100 µg/L was 13.2%.

The average recovery from groundwater using 4 determinatons and a spike concentration of 100 µg/L was 95.0%.

The precision (as RSD) from groundwater using 4 determinatons and a spike concentration of 100 µg/L was 1.5%.

The average recovery (in %) with %RSD (in parenthesis) from 3 determinations and a spike concentration of 20 µg/L in water was 83.2 (6.5) using 3M Empore Disks and EPA Method 8061.

The average recovery (in %) with %RSD (in parenthesis) from 3 determinations and a spike concentration of 20 µg/L in leachate was 97.5 (22.3) using 3M Empore Disks and EPA Method 8061.

The average recovery (in %) with %RSD (in parenthesis) from 3 determinations and a spike concentration of 20 µg/L in estuarine groundwater was 91.0 (10.7) using 3M Empore Disks and EPA Method 8061.

The average recovery (in %) with %RSD (in parenthesis) from 3 determinations and a spike concentration of 1 mg/kg in estuarine sediment was 121 (25.8) after sulfur cleanup with EPA Method 3660.

The average recovery (in %) with %RSD (in parenthesis) from 3 determinations and a spike concentration of 1 mg/kg in municipal sludge was 86.3 (17.7).

The average recovery (in %) with %RSD (in parenthesis) from 3 determinations and a spike concentration of 1 mg/kg in sandy loam soil was 72.6 (3.7).

SAMPLE COLLECTION, PRESERVATION & HANDLING
Containers used to collect samples for the determination of semivolatile organic compounds should be soap and water washed followed by methanol (or isopropanol) rinsing. The sample containers should be of glass or Teflon® and have screw-top covers with Teflon® liners. Sample containers should be filled with care to prevent any portion of the collected sample coming in contact with the sampler's gloves.

No preservation is used with concentrated waste samples. With liquid samples containing no residual chlorine and with soil, sediment, and sludge samples, immediately cooling to 4°C is the only preservation used. When residual chlorine is present then 3 mL of 10% aqueous sodium sulfate is added for each gallon of sample collected, followed by cooling to 4°C.

MHT Liquid samples must be extracted within 7 days and their extracts analyzed within 40 days. Concentrated waste, soil, sediment, and sludge samples must be extracted within 14 days and their extracts analyzed within 40 days.

SAMPLE PREPARATION In general, water samples are extracted at a pH of 5 to 7 with methylene chloride in a separatory funnel (EPA Method 3510). EPA Method 3520 is not recommended for the extraction of aqueous samples because the longer chain esters tend to adsorb to the glassware and consequently, their extraction recoveries may be poor. Solid samples are extracted with hexane/acetone (1:) or methylene chloride/acetone (1:1) in a Soxhlet extractor (EPA Method 3540) or with an ultrasonic extractor (EPA Method 3550). Immediately prior to extraction, spike 500 µL of the surrogate standard spiking solution into 1-L aqueous sample or 30-g solid sample. Extraction of particulate-free aqueous samples using C-18 extraction disks is an optional method that can be used.

Prior to Florisil cleanup or GC analysis, the methylene chloride and methylene chloride/acetone extracts must be exchanged to hexane. Exchange is not required for the acetonitrile extracts. Cleanup may not be necessary for extracts from a relatively clean sample matrix. Florisil Cartridge Cleanup may be used for extract cleanup.

If PCBs and organochlorine pesticides are known to be present in the sample, and if Florisil Cartridge Cleanup is considered, then two fractions are collected: Fraction 1 is eluted with 5 mL of 20% methylene chloride in hexane and Fraction 2 is eluted with 5 mL of 10% acetone in hexane. Fraction 1 contains the organochlorine pesticides and PCBs, and can be discarded. Fraction 2 contains the phthalate esters and is analyzed by GC/ECD.

QUALITY CONTROL Identify compounds in the sample by comparing the retention times of the peaks in the sample chromatogram with those of the peaks in standard chromatograms. The retention time window used to make identification is based upon measurements of actual retention time variations over the course of 10 consecutive injections.

Calibration standards are prepared at a minimum of five concentrations for each parameter of interest through dilution of the stock standard solutions with hexane. One of the concentrations should be at a concentration near, but above, the method detection limit. Prepare stock standard solutions in hexane. Stock standards should be checked frequently for signs of degradation or evaporation, especially just prior to preparing calibration standards from them. Stock standard solutions must be replaced after one year, or sooner if comparison with check standards indicates a problem. The suggested internal standards are: 2,5-dibromotoluene, 1,3,5-tribromobenzene, and α, α'-dibromo-m-xylene. The analyst can use any of the three compounds provided that they are resolved from matrix interferences. Recommended surrogate compounds are α-2,6-trichlorotoluene, 1,4-dichloronaphthalene, and 2,3,4,5,6-pentachlorotoluene.

Spike each sample, standard, and blank with surrogate compounds. Three surrogates are suggested for this method: diphenyl phthalate, diphenyl isophthalate, and dibenzyl phthalate.

The quality control check sample concentrate should contain the test compounds at 5 to 10 ng/µL An internal standard peak area check must be performed on all samples. The internal standard must be evaluated for acceptance by determining whether the measured area for the internal standard deviates by more than 30% from the average are for the internal standard in the calibration standards. When the internal standard peak area is outside that limit, all samples that fall outside the QC criteria must be reanalyzed. Benzyl benzoate has been tested and found appropriate as an internal standard for this method.

Any compounds confirmed by two columns may also be confirmed by GC/MS. The sample extract and associated blank should be analyzed by GC/MS. A reference standard of the compound must also be analyzed by GC/MS. Include a mid-concentration calibration standard after each group of 20 samples. The response factors for the mid-concentration calibration must be within ±15% of the average values for the multiconcentration calibration. Demonstrate through the analyses of standards that the Florisil fractionation scheme is reproducible.

REFERENCE Test Methods for Evaluating Solid Waste, Physical/Chemical Methods, SW-846, 3rd Edition, U.S. EPA, Office of Solid Waste, Washington, DC, EPA Method 8061, Nov. 1990.

Di-n-butyl phthalate   **EPA Method 8060**
CAS #84-74-2

TITLE Phthalate Esters

MATRIX Groundwater, soils, sludges, water miscible liquid wastes, and non-water miscible wastes.

APPLICATION This method is used for the analysis of 6 phthalate esters. Samples are extracted, concentrated, and analyzed using direct injection of both neat and diluted organic liquids into a gas chromatograph. Analytes are detected by a flame ionization detector (FID) or an electron capture detector (ECD). Groundwater samples should be determined by ECD. The method provides an optional GC column which is used for analyte confirmation and that may help resolve analytes from interferences.

INTERFERENCES Solvents, reagents, and glassware may introduce artifacts. Plastics, in particular, must be avoided. Other interferences may come from coextracted compounds from samples. There can be carryover contamination with high- and low-level samples.

INSTRUMENTATION GC capable of on-column injections and a flame with detector (FID) or electron capture detector (ECD). Column 1: 1.8 m by 4 mm with 1.5% SP-2250/1.95% SP-2401 on Supelcoport. Column 2: 1.8 m by 4 mm with 3% OV-1 on supelcoport.

RANGE 0.7 to 106 µg/L.

MDL 14 µg/L (FID) and 0.36 µg/L (ECD).

PQL FACTORS FOR MULTIPLYING × FID MDL VALUE

Matrix	Multiplication Factor
Groundwater	10
Low-level soil by sonication with GPC cleanup	670
High-level soil and sludge by sonication	10,000
Non-water miscible waste	100,000

PRECISION 0.29X + 0.06 µg/L (overall precision using FID).

ACCURACY 0.79C + 0.17 µg/L (as recovery using FID).

SAMPLING METHOD Use 8-oz. widemouth glass bottles with Teflon®-lined caps for concentrated waste samples, soils, sediments, and sludges. Use 1 or 2½ gallon amber glass bottles with Teflon®-lined caps for liquid (water) samples.

STABILITY Cool soil, sediment, sludge, and liquid samples to 4°C. If residual chlorine is present in liquid samples add 3 mL of 10% sodium thiosulfate per gallon of sample and cool to 4°C.

MHT 14 days for concentrated waste, soil, sediment, or sludge; 7 days for liquid samples; all extracts must be analyzed within 40 days.

QUALITY CONTROL A quality control check sample concentrate containing each analyte of interest is required. The QC check sample concentrate may be prepared from pure standard materials or purchased as certified solutions Use appropriate trip, matrix, control site, method, reagent, and solvent blanks. Internal, surrogate, and five concentration level calibration standards are used. The QC check sample concentrate should contain this compound at 25 µg/mL in acetone.

REFERENCE Method 8060, SW-846, 3rd ed., Nov.1986.

Di-n-octyl phthalate **EPA Method 1625**
CAS #117-84-0

TITLE Semivolatile Organic Compounds by Isotope Dilution GC/MS

MATRIX The compounds may be determined in waters, soils, and municipal sludges by this method.

METHOD SUMMARY This method is used to determine 176 semivolatile toxic organic pollutants associated with the CWA (as amended 1987); the RCRA (as amended 1986); the CERCLA (as amended 1986); and other compounds amenable to extraction and analysis by capillary column gas chromatography-mass spectrometry (GC/MS).

Stable isotopically-labeled analogs of the compounds of interest are added to the sample. If the solids content is less than 1%, a 1-L sample is extracted at pH 12–13, then at pH <2 with methylene chloride using continuous extraction techniques.

If the solids content is 30% or less, the sample is diluted to 1% solids with reagent water, homogenized ultrasonically, and extracted at pH 12–13, then at pH <2 with methylene chloride using continuous extraction techniques. If the solids content is greater than 30%, the sample is extracted using ultrasonic techniques.

Each extract is dried over sodium sulfate, concentrated to a volume of 5 mL, cleaned up using GPC, if necessary, and concentrated. Extracts are concentrated to 1 mL if GPC is not performed, and to 0.5 mL if GPC is performed.

An internal standard is added to the extract, and a 1-mL aliquot of the extract is injected into the GC. The compounds are separated by GC and detected by a MS. The labeled compounds serve to correct the variability of the analytical technique.

INTERFERENCES Solvents, reagents, glassware, and other sample processing hardware may yield artifacts and/or elevated baselines causing misinterpretation of chromatograms and spectra. Materials used in the analysis must be demonstrated to be free from interferences under the conditions of analysis by running method blanks initially and with each sample lot (sample started through the extraction process on a given 8-h shift, to a maximum of 20). Specific selection of reagents and purification of solvents by distillation in all glass systems may be required. Glassware and, where possible, reagents are cleaned by solvent rinse and baking at 450°C for 1-h minimum. Interferences coextracted from samples will vary considerably from source to source, depending on the diversity of the site being sampled.

INSTRUMENTATION Major instrumentation includes a GC with a splitless or on-column injection port for capillary column, a MS with 70 eV electron impact ionization, and a data system to collect and record MS data, and process it. A K-D apparatus is used to concentrate extracts.

GC Column: 30 m × 0.25 mm I.D. 5% phenyl, 94% methyl, 1% vinyl silicone bonded phased fused silica capillary column.

PRECISION & ACCURACY The detection limits of the method are usually dependent on the level of interferences rather than instrumental limitations. The limits typify the minimum quantities that can be detected with no interferences present.

The minimum level (in µg/mL) was 10. This is defined as a minimum level at which the analytical system shall give recognizable mass spectra (background corrected) and acceptable calibration points.

The MDL (in µg/kg) in low solids was 72 and in high solids was 62; these were determined in digested sludge (low solids) and in filter cake or compost (high solids).

The labeled and native compound initial precision as standard deviation (in µg/L) was 16.
The labeled and native compound initial accuracy as average recovery (in µg/L) was 77–161.

SAMPLE COLLECTION, PRESERVATION & HANDLING
Collect samples in glass containers. Aqueous samples which flow freely are collected in refrigerated bottles using automatic sampling equipment. Solid samples are collected as grab samples using widemouth jars. Maintain samples at 0 to 4°C from the time of collection until extraction. If residual chlorine is present in aqueous samples, add 80 mg sodium thiosulfate/L

of water. Begin sample extraction within 7 days of collection, and analyze all extracts within 40 days of extraction.

SAMPLE PREPARATION Samples containing 1% solids or less are extracted directly using continuous liquid-liquid extraction techniques. Samples containing 1 to 30% solids are diluted to the 1% level with reagent water and extracted using continuous liquid-liquid extraction techniques. Samples containing greater than 30% solids are extracted using ultrasonic techniques.

Base/neutral extraction — Adjust the pH of the waters in the extractors to 12–13 with 6 N NaOH. Extract with methylene chloride for 24–48 h.
Acid extraction — Adjust the pH of the waters in the extractors to 2 or less using 6 N sulfuric acid. Extract with methylene chloride for 24–48 h.
Ultrasonic extraction of high solids samples — Add anhydrous sodium sulfate to the sample and QC aliquot(s). Add acetone:methylene chloride (1:1) to the sample and mix thoroughly

Concentrate extracts using a K-D apparatus.

QUALITY CONTROL The analyst is permitted to modify this method to improve separations or lower the costs of measurements, provided all performance specifications are met. Analyses of blanks are required to demonstrate freedom from contamination. When results of spikes indicate atypical method performance for samples, the samples are diluted to bring method performance within acceptable limits.

For low solids (aqueous samples), extract, concentrate, and analyze two sets of four 1-L aliquots (8 aliquots total) of the precision and recovery standard. For high solids samples, two sets of four 30-g aliquots of the high solids reference matrix are used.

Spike all samples with labeled compounds to assess method performance. Compute percent recovery of the labeled compounds using the internal standard method. Compare the labeled compound recovery for each compound with the corresponding labeled compound recovery.

Reagent water and high solids reference matrix blanks are analyzed to demonstrate freedom from contamination. Extract and concentrate a 1-L reagent water blank or a high solids reference matrix blank with each sample's lot (samples started through the extraction process on the same 8-h shift, to a maximum of 20 samples).

Field replicates may be collected to determine the precision of the sampling technique, and spiked samples may be required to determine the accuracy of the analysis when the internal standard method is used.

REFERENCE Semivolatile Organic Compounds by Isotope Dilution GC/MS. Office of Water Regulation and Standards, U.S. EPA Industrial Technology Division, Washington, DC, EPA Method 1625, Rev. C, June 1989 (contact W.A. Telliard, U.S. EPA, Office of Water Regulations and Standards, 401 M St., SW, Washington, DC, 20460. Phone: 202-382-7131).

Di-n-octyl phthalate **EPA Method 625**
CAS #117-84-0

TITLE Base/Neutrals and Acids, U.S. EPA Method 625

MATRIX This methods covers municipal and industrial wastewaters.

METHOD SUMMARY Approximately 1 L of sample is serially extracted with methylene chloride at a pH greater than 11 and again at a pH less than 2 using a separatory funnel or a continuous extractor. The methylene chloride extract is dried, concentrated to a volume of 1 mL, and analyzed by GC/MS. Qualitative identification of the parameters in the extract is performed using the retention time and the relative abundance of three characteristic masses (m/z). Qualitative analysis is performed using either external or internal standard techniques with a single characteristic m/z.

INTERFERENCES Method interferences may be caused by contaminants in solvents, reagents, glassware, and other sample processing hardware. Glassware must be scrupulously cleaned. Glassware should be heated in a muffle furnace at 400°C for 5 to 30 min. Some thermally stable materials, such as PCBs, may not be eliminated by this treatment. Solvent rinses with acetone and pesticide quality hexane may be substituted for the muffle furnace heating. Matrix interferences may be caused by contaminants that are coextracted from the sample. The base-neutral extraction may cause significantly reduced recovery of phenols. The packed gas chromatographic columns recommended for the basic fraction may not exhibit sufficient resolution for some analytes.

INSTRUMENTATION A GC/MS system with an injection port designed for on-column injection when using packed columns and for splitless injection when using capillary columns.

Column for base/neutrals: 1.8 m long × 2 mm I.D. glass, packed with 3% SP-2550 on Supelcoport (100/120 mesh) or equivalent.
Column for acids: 1.8 m long × 2 mm I.D. glass, packed with 1% SP-1240DA on Supelcoport (100/120 mesh) or equivalent.

PRECISION & ACCURACY The MDL concentrations were obtained using reagent water. The MDL actually achieved in a given analysis will vary depending on instrument sensitivity and matrix effects. This method was tested by 15 laboratories using reagent water, drinking water, surface water, and industrial wastewaters spiked at six concentrations over the range 5 to 100 µg/L. Single operator precision, overall precision, and method accuracy were found to be directly related to the concentration of the parameter matrix.

The MDL (in µg/L) in reagent water was 2.5.
The standard deviation (in µg/L based on 4 recovery measurements) was 31.4.
The range (in µg/L) for average recovery for 4 measurements was 18.6–131.8.
The range (in %) for percent recovery was 4–146.

Accuracy (in µg/L) as expected recovery for one or more measurements of a sample containing a true concentration of C was 0.76C-0.79.

Precision (in µg/L) as expected single analyst standard deviation of measurements at an average concentration found at X was 0.21X+1.19.

Overall precision (in µg/L) as expected interlaboratory standard deviation of measurements in an average concentration found at X was 0.37X+1.19.

C = *True value of the concentration in µg/L.*
X = *Average recovery found for measurements of samples containing a concentration at C in µg/L.*

SAMPLE PREPARATION Adjust the pH to >11 with sodium hydroxide and serially extract in a separatory funnel with methylene chloride or else in a continuous extractor. Next, adjust the pH to <2 with sulfuric acid and serially extract in a separatory funnel with methylene chloride or else in a continuous extractor. Dry the extracts separately through a column of anhydrous sodium sulfate and then concentrate each of the extracts to 1.0 mL using a K-D apparatus.

SAMPLE COLLECTION, PRESERVATION & HANDLING Grab samples must be collected in glass containers. All samples must be refrigerated at 4°C from the time of collection until extraction. If residual chlorine is present, add 80 mg of sodium thiosulfate/L of sample and mix well. All samples must be extracted within 7 days of collection and completely analyzed within 40 days of extraction.

QUALITY CONTROL Make an initial, one-time, demonstration of the ability to generate acceptable accuracy and precision with this method. Before processing any samples, the analyst must analyze a reagent water blank to demonstrate that interferences from the analytical system and glassware are under control. Each time a set of samples is extracted or reagents are changed, a reagent water blank must be processed. Spike and analyze a minimum of 5% of all samples to monitor and evaluate lab data quality. A QC check sample concentrate that contains each parameter of interest at a concentration of 100 µg/mL in acetone is required. PCBs and multicomponent pesticides may be omitted from this test.

After analysis of five spiked wastewater samples, calculate the average percent recovery and the standard deviation of the percent recovery. Spike all samples with the surrogate standard spiking solution and calculate the percent recovery of each surrogate compound.

REFERENCE *Federal Register*, Vol. 49, No. 209. Friday, Oct. 26, 1984.

Di-n-octyl phthalate **EPA Method 8270**
CAS #117-84-0

TITLE Semivolatile Organic Compounds by GC/MS

MATRIX This method is used to determine the concentration of semivolatile organic compounds in extracts prepared from all types of solid waste matrices, soils, and groundwater. Although surface waters are not specifically mentioned, this method should be applicable to water samples from rivers, lakes, etc.

METHOD SUMMARY This method covers 259 semivolatile organic compounds. In very limited applications direct injection of the sample into the GC/MS system may be appropriate, but this results in very high detection limits (approximately 10,000 µg/L). Typically, a 1-L liquid sample, containing surrogate, and matrix spiking standards, is extracted in a continuous extractor first under acid conditions and then under basic conditions. Typically 30 g of a solid sample, containing surrogate, and matrix spiking standards, is extracted ultrasonically. After concentrating the extract to 1 mL it is spiked with 10 µL of an internal standard solution just prior to analysis by GC/MS. The volume injected should contain about 100 ng of base/neutral and 200 ng of acid surrogates (for a 1-µL injection). Analysis is performed by GC/MS using a capillary GC column.

INTERFERENCES Raw GC/MS data from all blanks, samples, and spikes must be evaluated for interferences. Contamination by carryover can occur whenever high-concentration and low-concentration samples are sequentially analyzed. To reduce carryover, the sample syringe must be rinsed out between samples with solvent. Whenever an unusually concentrated sample is encountered, it should be followed by the analysis of blank solvent to check for cross-contamination.

INSTRUMENTATION A GC/MS and a data system are required. The GC column used is a 30 m × 0.25 mm I.D. (or 0.32 mm I.D.) 1um film thickness silicone-coated fused silica capillary column. A continuous liquid-liquid extractor equipped with Teflon® or glass connection joints and stopcocks requiring no lubrication, a K-D concentrating apparatus, water bath, and an ultrasonic disrupter with a minimum power of 300 W and with pulsing capability are also required.

PRECISION & ACCURACY The estimated quantitation limit (EQL) of Method 8270B for determining an individual compound is approximately 1 mg/kg (wet weight) for soil or sediment samples, 1–200 mg/kg for wastes (dependent on matrix and method of preparation), and 10 µg/L for groundwater samples. EQLs will be proportionately higher for sample extracts that require dilution to avoid saturation of the detector.

The EQL(b) for groundwater in µg/L is 10.
The EQL (a, b) for low concentrations in soil and sediment in µg/kg is 660.
Accuracy as µg/L is 0.76C−0.79.
Overall precision in µg/L is 0.37X +1.19.

(a) *EQLs listed for soil/sediment are based on wet weight. Normally data is reported in a dry-weight basis; therefore, EQLs will be higher based on the % dry weight of each sample. This calculation is based on a 30 g sample and gel permeation chromatography cleanup.*
(b) *Sample EQLs are highly matrix-dependent. The EQLs are provided for guidance and may not always be achievable.*
C = *True value for concentration, in µg/L.*
X = *Average recovery found for measurements of samples containing a concentration of C, in µg/L.*

ESTIMATED QUANTITATION LIMIT

Other Matrices	Factor (a)
High-concentration soil and sludges by sonicator	7.5
Non-water miscible waste	75

(a) EQL for other matrices = [EQL for low soil/sediment] × [Factor]. This estimated EQL is similar to an EPA "Practical Quantitation Limit."

SAMPLING METHOD
Liquid samples — Use a 1 or 2½ gallon amber glass bottle with a screw-top Teflon®-lined cover that has been prewashed with detergent and rinsed with distilled water and methanol (or isopropanol).

Soils, sediments, or sludges — Use an 8-oz. widemouth glass with a screw-top Teflon®-lined cover that has been prewashed with detergent and rinsed with distilled water and methanol (or isopropanol).

SAMPLE PRESERVATION
Liquid samples — If residual chlorine is present, add 3 mL of 10% sodium thiosulfate per gallon, cool to 4°C and store in a solvent-free refrigerator until analysis; if chlorine is not present, then eliminate the sodium thiosulfate addition.

Soils, sediments, or sludges — Cool samples to 4°C and store in a solvent-free refrigerator.

MHT Liquid samples must be extracted within 7 days and the extracts analyzed within 40 days. Soils, sediments, or sludges may be stored for a maximum of 14 days and the extracts analyzed within 40 days.

SAMPLE PREPARATION
Liquid samples — Transfer 1 L quantitatively to a continuous extractor. If high concentrations are anticipated, a smaller volume may be used and then diluted with organic-free reagent water to 1 L. Adjust pH, if necessary, to pH <2 using 1:1 (V/V) sulfuric acid. Pipette 1.0 mL of a surrogate standard spiking solution into each sample. For the sample in each analytical batch selected for spiking, add 1.0 mL of a matrix spiking standard. For base/neutral acid analysis, the amount of the surrogates and matrix spiking compounds added to the sample should result in a final concentration of 100 ng/µL of each analyte in the extract to be analyzed (assuming a 1-µL injection). Extract with methylene chloride for 18–24 h. Next, adjust the pH of the aqueous phase to pH >11 using 10 N sodium hydroxide and extract it with methylene chloride again for 18–24 h. Dry the extract through a column containing anhydrous sodium sulfate and concentrate it to 1 mL using a K-D concentrator.

Soils, sediments, or sludges — Use 30 g of sample. Nonporous or wet samples (gummy or clay type) that do not have a free-flowing sandy texture must be mixed with anhydrous sodium sulfate until the sample is free flowing. Add 1 mL of surrogate standards to all samples, spikes, standards, and blanks. For the sample in each analytical batch selected for spiking, add 1.0 mL of a matrix spiking standard. For base/neutral acid analysis, the amount added of the surrogates and matrix spiking compounds should result in a final concentration of 100 ng/µL of each base/neutral analyte and 200 ng/µL of each acid analyte in the extract to be analyzed (assuming a 1-µL injection). Immediately add a 100-mL mixture of 1:1 methylene chloride:acetone and extract the sample ultrasonically for 3 min and then decant or filter the extracts. Repeat the extraction two or more times. Dry the extract using a column with anhydrous sodium sulfate and concentrate it to 1 mL in a K-D concentrator.

QUALITY CONTROL A methylene chloride solution containing 50 ng/µL of decafluorotriphenylphosphine (DFTPP) is used for tuning the GC/MS system each 12-h shift. A system performance check also must be made during every 12-h shift. A standard containing 50 ng/µL each of 4,4'-DDT, pentachlorophenol, and benzidine is required to verify injection port inertness and GC column performance. A calibration standard at mid-concentration, containing each compound of interest, including all required surrogates, must be performed every 12 h during analysis. After the system performance check is met, calibration check compounds (CCCs) are used to check the validity of the initial calibration.

The internal standard responses and retention times in the calibration check standard must be evaluated immediately after or during data acquisition. If the retention time for any internal standard changes by more than 30 seconds from the last check calibration (12 h), the chromatographic system must be inspected for malfunctions and corrections must be made, as required. If the electron ionization current plot (EICP) area for any of the internal standards changes by a factor of two from the last daily calibration standard check, the mass spectrometer must be inspected for malfunctions and corrections must be made, as appropriate.

Demonstrate, through the analysis of a reagent water blank, that interferences from the analytical system, glassware, and reagents are under control. The blank samples should be carried through all stages of the sample preparation and measurement steps. For each analytical batch (up to 20 samples), a reagent blank, matrix spike, and matrix spike duplicate/duplicate must be analyzed (the frequency of the spikes may be different for different monitoring programs). The blank and spiked samples must be carried through all stages of the sample preparation and measurement steps. A QC reference sample concentrate containing each analyte at a concentration of 100 mg/L in methanol is required.

REFERENCE Test Methods for Evaluating Solid Waste (SW-846). U.S. EPA 1983, Method 8270B, Rev. 2, Nov. 1990. Office of Solid Waste, Washington, DC.

Di-n-octyl phthalate   EPA Method 8061
CAS #117-84-0

TITLE Phthalate Esters by Capillary Gas Chromatography With Electron Capture Detection (GC/ECD)

MATRIX This method covers aqueous and solid matrices. This includes a wide variety such as drinking water, groundwater, industrial wastewaters, surface waters, soils, solids, and sediments.

METHOD SUMMARY This method is used to determine the identities and concentrations of phthalate esters in liquid, solid and sludge matrices. When used to analyze for any or all of the target analytes, compound identification should be supported by at least one additional qualitative technique. This method describes conditions for parallel column, dual electron capture detector analysis, which fulfills the above requirement. Alternatively, GC/MS could be used for compound confirmation.

A measured volume or weight of sample (approximately 1 L for liquids, 10 to 30 g for solids and sludges) is extracted by using the appropriate sample extraction technique specified in EPA Method 3510, EPA Method 3540, and EPA Method 3550. After cleanup, the extract is analyzed by GC/ECD.

INTERFERENCES The sensitivity of this method usually depends on the level of interferences rather than on instrumental limitations. If interferences prevent detection of the analytes, cleanup of the sample extracts is necessary. Either EPA Method 3610 or EPA Method 3620 alone or followed by EPA Method 3660, Sulfur Cleanup, may be used to eliminate interferences in the analysis. EPA Method 3640, Gel Permeation Cleanup, is applicable for samples that contain high amounts of lipids and waxes.

Interferences coextracted from the samples will vary considerably from waste to waste. Glassware must be scrupulously clean. All glassware require treatment in a muffle furnace at 400°C for 2 to 4 h, or thorough rinsing with pesticide-grade solvent, prior to use. Volumetric glassware should not be heated in a muffle furnace. Storage of glassware in the lab introduces contamination, even if the glassware is wrapped in aluminum foil. Sodium sulfate, Florisil, and alumina may be contaminated with phthalate esters and, therefore, use of these materials in sample cleanup should be employed cautiously. If these materials are used, they must be obtained packaged in glass. Heating at 400°C for sodium sulfate, 320°C for Florisil, and 210°C for alumina is recommended. Glass wool used in any step of sample preparation should be a specially treated pyrex wool, pesticide grade, and must be baked at 400°C for 4 h immediately prior to use.

Paper thimbles and filter paper must be exhaustively washed with the solvent that will be used in the sample extraction. Soxhlet extraction of paper thimbles and filter paper for 12 h with fresh solvent should be repeated for a minimum of three times. Method blanks should be obtained before any of the precleaned thimbles or filter papers are used.

INSTRUMENTATION Gas chromatograph suitable for on-column and split/splitless injections.

Column 1: 30 m × 0.53 mm ID, 5% phenyl/95% methyl silicone fused-silica open tubular column, DB-5, 1.5 µg film thickness.
Column 2: 30 m × 0.53 mm ID, 14% cyanopropyl phenyl silicone fused-silica open tubular column, DB-1701, 1.0 µg film thickness.

A dual electron capture detector (ECD) is used. A Kuderna-Danish (K-D) apparatus is required along with a vacuum manifold consisting of individually adjustable, easily accessible flow-control valves for up to 24 cartridges, sample rack, chemically resistant cover and seals, heavy-duty glass basin, removable stainless steel solvent guides, built-in vacuum gauge and valve. Also, 6-mL, 1-g solid-phase extraction cartridges, LC-Florisil or equivalent, prepackaged, ready to use will be needed.

PRECISION & ACCURACY The MDL actually achieved in a given analysis will vary, as it is dependent on instrument sensitivity and matrix effects. This method has been tested in a single lab. Single-operator precision, overall precision, and method accuracy were found to be related to the concentration of the compounds and the type of matrix.

MULTIPLICATION FACTORS FOR OTHER MATRICES (a)

Matrix	Factor (b)
Groundwater	10
Low-concentration soil by ultrasonic extraction with GPC cleanup	670
High-concentration soil and sludges by ultrasonic extraction	10,000
Non-water miscible waste	100,000

(a) Sample EQLs are highly matrix-dependent.
(b) EQL = [Method detection limit] × [Factor]. For non-aqueous samples, the factor is on a wet-weight basis.

The MDL using 7 replicate determinations and a spike concentration of 100 µg/L was 49 ng/L.
The average recovery from HPLC-grade water using 4 determinatons and a spike concentration of 100 µg/L was 84.9%.
The precision (as RSD) from HPLC-grade water using 4 determinatons and a spike concentration of 100 µg/L was 3.8%.
The average recovery from groundwater using 4 determinatons and a spike concentration of 100 µg/L was 90.1%.
The precision (as RSD) from groundwater using 4 determinatons and a spike concentration of 100 µg/L was 6.1%.
The average recovery (in %) with %RSD (in parenthesis) from 3 determinations and a spike concentration of 20 µg/L in water was 74.9 (4.9) using 3M Empore Disks and EPA Method 8061.
The average recovery (in %) with %RSD (in parenthesis) from 3 determinations and a spike concentration of 20 µg/L in leachate was 87.5 (18.7) using 3M Empore Disks and EPA Method 8061.
The average recovery (in %) with %RSD (in parenthesis) from 3 determinations and a spike concentration of 20 µg/L in estuarine groundwater was 87.2 (3.7) using 3M Empore Disks and EPA Method 8061.
The average recovery (in %) with %RSD (in parenthesis) from 3 determinations and a spike concentration of 1 mg/kg in estuarine sediment was not determined (matrix interferant) after sulfur cleanup with EPA Method 3660.
The average recovery (in %) with %RSD (in parenthesis) from 3 determinations and a spike concentration of 1 mg/kg in municipal sludge was 93.3 (14.6).
The average recovery (in %) with %RSD (in parenthesis) from 3 determinations and a spike concentration of 1 mg/kg in sandy loam soil was 84.7 (9.3).

SAMPLE COLLECTION, PRESERVATION & HANDLING
Containers used to collect samples for the determination of semivolatile organic compounds should be soap and water

washed followed by methanol (or isopropanol) rinsing. The sample containers should be of glass or Teflon® and have screw-top covers with Teflon® liners. Sample containers should be filled with care to prevent any portion of the collected sample coming in contact with the sampler's gloves.

No preservation is used with concentrated waste samples. With liquid samples containing no residual chlorine and with soil, sediment, and sludge samples, immediately cooling to 4°C is the only preservation used. When residual chlorine is present then 3 mL of 10% aqueous sodium sulfate is added for each gallon of sample collected, followed by cooling to 4°C.

MHT Liquid samples must be extracted within 7 days and their extracts analyzed within 40 days. Concentrated waste, soil, sediment, and sludge samples must be extracted within 14 days and their extracts analyzed within 40 days.

SAMPLE PREPARATION In general, water samples are extracted at a pH of 5 to 7 with methylene chloride in a separatory funnel (EPA Method 3510). EPA Method 3520 is not recommended for the extraction of aqueous samples because the longer chain esters tend to adsorb to the glassware and consequently, their extraction recoveries may be poor. Solid samples are extracted with hexane/acetone (1:) or methylene chloride/acetone (1:1) in a Soxhlet extractor (EPA Method 3540) or with an ultrasonic extractor (EPA Method 3550). Immediately prior to extraction, spike 500 µL of the surrogate standard spiking solution into 1-L aqueous sample or 30-g solid sample. Extraction of particulate-free aqueous samples using C-18 extraction disks is an optional method that can be used.

Prior to Florisil cleanup or GC analysis, the methylene chloride and methylene chloride/acetone extracts must be exchanged to hexane. Exchange is not required for the acetonitrile extracts. Cleanup may not be necessary for extracts from a relatively clean sample matrix. Florisil Cartridge Cleanup may be used for extract cleanup.

If PCBs and organochlorine pesticides are known to be present in the sample, and if Florisil Cartridge Cleanup is considered, then two fractions are collected: Fraction 1 is eluted with 5 mL of 20% methylene chloride in hexane and Fraction 2 is eluted with 5 mL of 10% acetone in hexane. Fraction 1 contains the organochlorine pesticides and PCBs, and can be discarded. Fraction 2 contains the phthalate esters and is analyzed by GC/ECD.

QUALITY CONTROL Identify compounds in the sample by comparing the retention times of the peaks in the sample chromatogram with those of the peaks in standard chromatograms. The retention time window used to make identification is based upon measurements of actual retention time variations over the course of 10 consecutive injections.

Calibration standards are prepared at a minimum of five concentrations for each parameter of interest through dilution of the stock standard solutions with hexane. One of the concentrations should be at a concentration near, but above, the method detection limit. Prepare stock standard solutions in hexane. Stock standards should be checked frequently for signs of degradation or evaporation, especially just prior to preparing calibration standards from them. Stock standard solutions must be replaced after one year, or sooner if comparison with check standards indicates a problem. The suggested internal standards are: 2,5-dibromotoluene, 1,3,5-tribromobenzene, and α, α'-dibromo-m-xylene. The analyst can use any of the three compounds provided that they are resolved from matrix interferences. Recommended surrogate compounds are α-2,6-trichlorotoluene, 1,4-dichloronaphthalene, and 2,3,4,5,6-pentachlorotoluene.

Spike each sample, standard, and blank with surrogate compounds. Three surrogates are suggested for this method: diphenyl phthalate, diphenyl isophthalate, and dibenzyl phthalate.

The quality control check sample concentrate should contain the test compounds at 5 to 10 ng/µL An internal standard peak area check must be performed on all samples. The internal standard must be evaluated for acceptance by determining whether the measured area for the internal standard deviates by more than 30% from the average are for the internal standard in the calibration standards. When the internal standard peak area is outside that limit, all samples that fall outside the QC criteria must be reanalyzed. Benzyl benzoate has been tested and found appropriate as an internal standard for this method.

Any compounds confirmed by two columns may also be confirmed by GC/MS. The sample extract and associated blank should be analyzed by GC/MS. A reference standard of the compound must also be analyzed by GC/MS. Include a mid-concentration calibration standard after each group of 20 samples. The response factors for the mid-concentration calibration must be within ±15% of the average values for the multiconcentration calibration. Demonstrate through the analyses of standards that the Florisil fractionation scheme is reproducible.

REFERENCE Test Methods for Evaluating Solid Waste, Physical/Chemical Methods, SW-846, 3rd Edition, U.S. EPA, Office of Solid Waste, Washington, DC, EPA Method 8061, Nov. 1990.

Di-n-octyl phthalate EPA Method 8060
CAS #117-84-0

TITLE Phthalate Esters

MATRIX Groundwater, soils, sludges, water miscible liquid wastes, and non-water miscible wastes.

APPLICATION This method is used for the analysis of 6 phthalate esters.

Samples are extracted, concentrated, and analyzed using direct injection of both neat and diluted organic liquids into a gas chromatograph. Analytes are detected by a flame ionization detector (FID) or an electron capture detector (ECD). Groundwater samples should be determined by ECD. The method provides an optional GC column which is used for analyte confirmation and that may help resolve analytes from interferences.

INTERFERENCES Solvents, reagents, and glassware may introduce artifacts. Plastics, in particular, must be avoided. Other interferences may come from coextracted compounds from samples. There can be carryover contamination with high- and low-level samples.

INSTRUMENTATION GC capable of on-column injections and a flame with detector (FID) or electron capture detector (ECD). Column 1: 1.8 m by 4 mm with 1.5% SP-2250/1.95% SP-2401 on Supelcoport. Column 2: 1.8 m by 4 mm with 3% OV-1 on supelcoport.

RANGE 0.7 to 106 µg/L.

MDL 31 µg/L (FID) and 3.0 µg/L (ECD).

PQL FACTORS FOR MULTIPLYING × FID MDL VALUE

Matrix	Multiplication Factor
Groundwater	10
Low-level soil by sonication with GPC cleanup	670
High-level soil and sludge by sonication	10,000
Non-water miscible waste	100,000

PRECISION 0.62X + 0.34 µg/L (overall precision using FID).

ACCURACY 0.35C–0.71 µg/L (as recovery using FID).

SAMPLING METHOD Use 8-oz. widemouth glass bottles with Teflon®-lined caps for concentrated waste samples, soils, sediments, and sludges. Use 1 or 2½ gallon amber glass bottles with Teflon®-lined caps for liquid (water) samples.

STABILITY Cool soil, sediment, sludge, and liquid samples to 4°C. If residual chlorine is present in liquid samples add 3 mL of 10% sodium thiosulfate per gallon of sample and cool to 4°C.

MHT 14 days for concentrated waste, soil, sediment, or sludge; 7 days for liquid samples; all extracts must be analyzed within 40 days.

QUALITY CONTROL A quality control check sample concentrate containing each analyte of interest is required. The QC check sample concentrate may be prepared from pure standard materials or purchased as certified solutions Use appropriate trip, matrix, control site, method, reagent, and solvent blanks. Internal, surrogate, and five concentration level calibration standards are used. The QC check sample concentrate should contain this compound at 50 µg/mL in acetone.

REFERENCE Method 8060, SW-846, 3rd ed., Nov.1986.

2,6-Di-tert-butyl-p-benzoquinone   **EPA Method 1625**
CAS #719-22-2

TITLE Semivolatile Organic Compounds by Isotope Dilution GC/MS

MATRIX The compounds may be determined in waters, soils, and municipal sludges by this method.

METHOD SUMMARY This method is used to determine 176 semivolatile toxic organic pollutants associated with the CWA (as amended 1987); the RCRA (as amended 1986); the CERCLA (as amended 1986); and other compounds amenable to extraction and analysis by capillary column gas chromatography-mass spectrometry (GC/MS).

Stable isotopically-labeled analogs of the compounds of interest are added to the sample. If the solids content is less than 1%, a 1-L sample is extracted at pH 12–13, then at pH <2 with methylene chloride using continuous extraction techniques.

If the solids content is 30% or less, the sample is diluted to 1% solids with reagent water, homogenized ultrasonically, and extracted at pH 12–13, then at pH <2 with methylene chloride using continuous extraction techniques. If the solids content is greater than 30%, the sample is extracted using ultrasonic techniques.

Each extract is dried over sodium sulfate, concentrated to a volume of 5 mL, cleaned up using GPC, if necessary, and concentrated. Extracts are concentrated to 1 mL if GPC is not performed, and to 0.5 mL if GPC is performed.

An internal standard is added to the extract, and a 1-mL aliquot of the extract is injected into the GC. The compounds are separated by GC and detected by a MS. The labeled compounds serve to correct the variability of the analytical technique.

INTERFERENCES Solvents, reagents, glassware, and other sample processing hardware may yield artifacts and/or elevated baselines causing misinterpretation of chromatograms and spectra. Materials used in the analysis must be demonstrated to be free from interferences under the conditions of analysis by running method blanks initially and with each sample lot (sample started through the extraction process on a given 8-h shift, to a maximum of 20). Specific selection of reagents and purification of solvents by distillation in all glass systems may be required. Glassware and, where possible, reagents are cleaned by solvent rinse and baking at 450°C for 1-h minimum. Interferences coextracted from samples will vary considerably from source to source, depending on the diversity of the site being sampled.

INSTRUMENTATION Major instrumentation includes a GC with a splitless or on-column injection port for capillary column, a MS with 70 eV electron impact ionization, and a data system to collect and record MS data, and process it. A K-D apparatus is used to concentrate extracts.

GC Column: 30 m × 0.25 mm I.D. 5% phenyl, 94% methyl, 1% vinyl silicone bonded phased fused silica capillary column.

PRECISION & ACCURACY The detection limits of the method are usually dependent on the level of interferences rather than instrumental limitations. The limits typify the minimum quantities that can be detected with no interferences present.

The minimum level (in µg/mL) was not listed. This is defined as a minimum level at which the analytical system shall give recognizable mass spectra (background corrected) and acceptable calibration points.

The MDL (in µg/kg) in low solids was not listed and in high solids was not listed; these were determined in digested sludge (low solids) and in filter cake or compost (high solids).

The labeled and native compound initial precision as standard deviation (in µg/L) was not listed.

The labeled and native compound initial accuracy as average recovery (in µg/L) was not listed.

SAMPLE COLLECTION, PRESERVATION & HANDLING Collect samples in glass containers. Aqueous samples which flow freely are collected in refrigerated bottles using automatic sampling equipment. Solid samples are collected as grab samples using widemouth jars. Maintain samples at 0 to 4°C from the time of collection until extraction. If residual chlorine is present in aqueous samples, add 80 mg sodium thiosulfate/L of water. Begin sample extraction within 7 days of collection, and analyze all extracts within 40 days of extraction.

SAMPLE PREPARATION Samples containing 1% solids or less are extracted directly using continuous liquid-liquid extraction techniques. Samples containing 1 to 30% solids are diluted to the 1% level with reagent water and extracted using continuous liquid-liquid extraction techniques. Samples containing greater than 30% solids are extracted using ultrasonic techniques.

- Base/neutral extraction — Adjust the pH of the waters in the extractors to 12–13 with 6 N NaOH. Extract with methylene chloride for 24–48 h.
- Acid extraction — Adjust the pH of the waters in the extractors to 2 or less using 6 N sulfuric acid. Extract with methylene chloride for 24–48 h.
- Ultrasonic extraction of high solids samples — Add anhydrous sodium sulfate to the sample and QC aliquot(s). Add acetone:methylene chloride (1:1) to the sample and mix thoroughly

Concentrate extracts using a K-D apparatus.

QUALITY CONTROL The analyst is permitted to modify this method to improve separations or lower the costs of measurements, provided all performance specifications are met. Analyses of blanks are required to demonstrate freedom from contamination. When results of spikes indicate atypical method performance for samples, the samples are diluted to bring method performance within acceptable limits.

For low solids (aqueous samples), extract, concentrate, and analyze two sets of four 1-L aliquots (8 aliquots total) of the precision and recovery standard. For high solids samples, two sets of four 30-g aliquots of the high solids reference matrix are used.

Spike all samples with labeled compounds to assess method performance. Compute percent recovery of the labeled compounds using the internal standard method. Compare the labeled compound recovery for each compound with the corresponding labeled compound recovery.

Reagent water and high solids reference matrix blanks are analyzed to demonstrate freedom from contamination. Extract and concentrate a 1-L reagent water blank or a high solids reference matrix blank with each sample's lot (samples started through the extraction process on the same 8-h shift, to a maximum of 20 samples).

Field replicates may be collected to determine the precision of the sampling technique, and spiked samples may be required to determine the accuracy of the analysis when the internal standard method is used.

REFERENCE Semivolatile Organic Compounds by Isotope Dilution GC/MS. Office of Water Regulation and Standards, U.S. EPA Industrial Technology Division, Washington, DC, EPA Method 1625, Rev. C, June 1989 (contact W.A. Telliard, U.S. EPA, Office of Water Regulations and Standards, 401 M St., SW, Washington, DC, 20460. Phone: 202-382-7131).

Diallate (*cis* or *trans*)   EPA Method 8270
CAS #2303-16-4

TITLE Semivolatile Organic Compounds by GC/MS

MATRIX This method is used to determine the concentration of semivolatile organic compounds in extracts prepared from all types of solid waste matrices, soils, and groundwater. Although surface waters are not specifically mentioned, this method should be applicable to water samples from rivers, lakes, etc.

METHOD SUMMARY This method covers 259 semivolatile organic compounds. In very limited applications direct injection of the sample into the GC/MS system may be appropriate, but this results in very high detection limits (approximately 10,000 µg/L). Typically, a 1-L liquid sample, containing surrogate, and matrix spiking standards, is extracted in a continuous extractor first under acid conditions and then under basic conditions. Typically 30 g of a solid sample, containing surrogate, and matrix spiking standards, is extracted ultrasonically. After concentrating the extract to 1 mL it is spiked with 10 µL of an internal standard solution just prior to analysis by GC/MS. The volume injected should contain about 100 ng of base/neutral and 200 ng of acid surrogates (for a 1-µL injection). Analysis is performed by GC/MS using a capillary GC column.

INTERFERENCES Raw GC/MS data from all blanks, samples, and spikes must be evaluated for interferences. Contamination by carryover can occur whenever high-concentration and low-concentration samples are sequentially analyzed. To reduce carryover, the sample syringe must be rinsed out between samples with solvent. Whenever an unusually concentrated sample is encountered, it should be followed by the analysis of blank solvent to check for cross-contamination.

INSTRUMENTATION A GC/MS and a data system are required. The GC column used is a 30 m × 0.25 mm I.D. (or 0.32 mm I.D.) 1um film thickness silicone-coated fused silica capillary column. A continuous liquid-liquid extractor equipped with Teflon® or glass connection joints and stopcocks requiring no lubrication, a K-D concentrating apparatus, water bath, and an ultrasonic disrupter with a minimum power of 300 W and with pulsing capability are also required.

PRECISION & ACCURACY The estimated quantitation limit (EQL) of Method 8270B for determining an individual compound is approximately 1 mg/kg (wet weight) for soil or sediment samples, 1–200 mg/kg for wastes (dependent on

matrix and method of preparation), and 10 µg/L for groundwater samples. EQLs will be proportionately higher for sample extracts that require dilution to avoid saturation of the detector.

The EQL(b) for groundwater in µg/L is 10.
The EQL (a, b) for low concentrations in soil and sediment in µg/kg is not determined.
Accuracy as µg/L is not listed.
Overall precision in µg/L is not listed.

(a) *EQLs listed for soil/sediment are based on wet weight. Normally data is reported in a dry-weight basis; therefore, EQLs will be higher based on the % dry weight of each sample. This calculation is based on a 30 g sample and gel permeation chromatography cleanup.*
(b) *Sample EQLs are highly matrix-dependent. The EQLs are provided for guidance and may not always be achievable.*
$C =$ *True value for concentration, in µg/L.*
$X =$ *Average recovery found for measurements of samples containing a concentration of C, in µg/L.*

ESTIMATED QUANTITATION LIMIT

Other Matrices	Factor (a)
High-concentration soil and sludges by sonicator	7.5
Non-water miscible waste	75

(a) *EQL for other matrices = [EQL for low soil/sediment] × [Factor]. This estimated EQL is similar to an EPA "Practical Quantitation Limit."*

SAMPLING METHOD
Liquid samples — Use a 1 or 2½ gallon amber glass bottle with a screw-top Teflon®-lined cover that has been prewashed with detergent and rinsed with distilled water and methanol (or isopropanol).

Soils, sediments, or sludges — Use an 8-oz. widemouth glass with a screw-top Teflon®-lined cover that has been prewashed with detergent and rinsed with distilled water and methanol (or isopropanol).

SAMPLE PRESERVATION
Liquid samples — If residual chlorine is present, add 3 mL of 10% sodium thiosulfate per gallon, cool to 4°C and store in a solvent-free refrigerator until analysis; if chlorine is not present, then eliminate the sodium thiosulfate addition.

Soils, sediments, or sludges — Cool samples to 4°C and store in a solvent-free refrigerator.

MHT Liquid samples must be extracted within 7 days and the extracts analyzed within 40 days. Soils, sediments, or sludges may be stored for a maximum of 14 days and the extracts analyzed within 40 days.

SAMPLE PREPARATION
Liquid samples — Transfer 1 L quantitatively to a continuous extractor. If high concentrations are anticipated, a smaller volume may be used and then diluted with organic-free reagent water to 1 L. Adjust pH, if necessary, to pH <2 using 1:1 (V/V) sulfuric acid. Pipette 1.0 mL of a surrogate standard spiking solution into each sample. For the sample in each analytical batch selected for spiking, add 1.0 mL of a matrix spiking standard. For base/neutral acid analysis, the amount of the surrogates and matrix spiking compounds added to the sample should result in a final concentration of 100 ng/µL of each analyte in the extract to be analyzed (assuming a 1-µL injection). Extract with methylene chloride for 18–24 h. Next, adjust the pH of the aqueous phase to pH >11 using 10 N sodium hydroxide and extract it with methylene chloride again for 18–24 h. Dry the extract through a column containing anhydrous sodium sulfate and concentrate it to 1 mL using a K-D concentrator.

Soils, sediments, or sludges — Use 30 g of sample. Nonporous or wet samples (gummy or clay type) that do not have a free-flowing sandy texture must be mixed with anhydrous sodium sulfate until the sample is free flowing. Add 1 mL of surrogate standards to all samples, spikes, standards, and blanks. For the sample in each analytical batch selected for spiking, add 1.0 mL of a matrix spiking standard. For base/neutral acid analysis, the amount added of the surrogates and matrix spiking compounds should result in a final concentration of 100 ng/µL of each base/neutral analyte and 200 ng/µL of each acid analyte in the extract to be analyzed (assuming a 1-µL injection). Immediately add a 100-mL mixture of 1:1 methylene chloride:acetone and extract the sample ultrasonically for 3 min and then decant or filter the extracts. Repeat the extraction two or more times. Dry the extract using a column with anhydrous sodium sulfate and concentrate it to 1 mL in a K-D concentrator.

QUALITY CONTROL A methylene chloride solution containing 50 ng/µL of decafluorotriphenylphosphine (DFTPP) is used for tuning the GC/MS system each 12-h shift. A system performance check also must be made during every 12-h shift. A standard containing 50 ng/µL each of 4,4′-DDT, pentachlorophenol, and benzidine is required to verify injection port inertness and GC column performance. A calibration standard at mid-concentration, containing each compound of interest, including all required surrogates, must be performed every 12 h during analysis. After the system performance check is met, calibration check compounds (CCCs) are used to check the validity of the initial calibration.

The internal standard responses and retention times in the calibration check standard must be evaluated immediately after or during data acquisition. If the retention time for any internal standard changes by more than 30 seconds from the last check calibration (12 h), the chromatographic system must be inspected for malfunctions and corrections must be made, as required. If the electron ionization current plot (EICP) area for any of the internal standards changes by a factor of two from the last daily calibration standard check, the mass spectrometer must be inspected for malfunctions and corrections must be made, as appropriate.

Demonstrate, through the analysis of a reagent water blank, that interferences from the analytical system, glassware, and reagents are under control. The blank samples should be carried through all stages of the sample preparation and measurement steps. For each analytical batch (up to 20 samples), a reagent blank, matrix spike, and matrix spike duplicate/duplicate must be analyzed (the frequency of the spikes may be different for different monitoring programs). The blank and

spiked samples must be carried through all stages of the sample preparation and measurement steps. A QC reference sample concentrate containing each analyte at a concentration of 100 mg/L in methanol is required.

REFERENCE Test Methods for Evaluating Solid Waste (SW-846). U.S. EPA 1983, Method 8270B, Rev. 2, Nov. 1990. Office of Solid Waste, Washington, DC.

2,4-Diaminotoluene **EPA Method 1625**
CAS #95-80-7

TITLE Semivolatile Organic Compounds by Isotope Dilution GC/MS

MATRIX The compounds may be determined in waters, soils, and municipal sludges by this method.

METHOD SUMMARY This method is used to determine 176 semivolatile toxic organic pollutants associated with the CWA (as amended 1987); the RCRA (as amended 1986); the CERCLA (as amended 1986); and other compounds amenable to extraction and analysis by capillary column gas chromatography-mass spectrometry (GC/MS).

Stable isotopically-labeled analogs of the compounds of interest are added to the sample. If the solids content is less than 1%, a 1-L sample is extracted at pH 12–13, then at pH <2 with methylene chloride using continuous extraction techniques.

If the solids content is 30% or less, the sample is diluted to 1% solids with reagent water, homogenized ultrasonically, and extracted at pH 12–13, then at pH <2 with methylene chloride using continuous extraction techniques. If the solids content is greater than 30%, the sample is extracted using ultrasonic techniques.

Each extract is dried over sodium sulfate, concentrated to a volume of 5 mL, cleaned up using GPC, if necessary, and concentrated. Extracts are concentrated to 1 mL if GPC is not performed, and to 0.5 mL if GPC is performed.

An internal standard is added to the extract, and a 1-mL aliquot of the extract is injected into the GC. The compounds are separated by GC and detected by a MS. The labeled compounds serve to correct the variability of the analytical technique.

INTERFERENCES Solvents, reagents, glassware, and other sample processing hardware may yield artifacts and/or elevated baselines causing misinterpretation of chromatograms and spectra. Materials used in the analysis must be demonstrated to be free from interferences under the conditions of analysis by running method blanks initially and with each sample lot (sample started through the extraction process on a given 8-h shift, to a maximum of 20). Specific selection of reagents and purification of solvents by distillation in all glass systems may be required. Glassware and, where possible, reagents are cleaned by solvent rinse and baking at 450°C for 1-h minimum. Interferences coextracted from samples will vary considerably from source to source, depending on the diversity of the site being sampled.

INSTRUMENTATION Major instrumentation includes a GC with a splitless or on-column injection port for capillary column, a MS with 70 eV electron impact ionization, and a data system to collect and record MS data, and process it. A K-D apparatus is used to concentrate extracts.

GC Column: 30 m × 0.25 mm I.D. 5% phenyl, 94% methyl, 1% vinyl silicone bonded phased fused silica capillary column.

PRECISION & ACCURACY The detection limits of the method are usually dependent on the level of interferences rather than instrumental limitations. The limits typify the minimum quantities that can be detected with no interferences present.

The minimum level (in µg/mL) was not listed. This is defined as a minimum level at which the analytical system shall give recognizable mass spectra (background corrected) and acceptable calibration points.

The MDL (in µg/kg) in low solids was not listed and in high solids was not listed; these were determined in digested sludge (low solids) and in filter cake or compost (high solids).

The labeled and native compound initial precision as standard deviation (in µg/L) was not listed.
The labeled and native compound initial accuracy as average recovery (in µg/L) was not listed.

SAMPLE COLLECTION, PRESERVATION & HANDLING Collect samples in glass containers. Aqueous samples which flow freely are collected in refrigerated bottles using automatic sampling equipment. Solid samples are collected as grab samples using widemouth jars. Maintain samples at 0 to 4°C from the time of collection until extraction. If residual chlorine is present in aqueous samples, add 80 mg sodium thiosulfate/L of water. Begin sample extraction within 7 days of collection, and analyze all extracts within 40 days of extraction.

SAMPLE PREPARATION Samples containing 1% solids or less are extracted directly using continuous liquid-liquid extraction techniques. Samples containing 1 to 30% solids are diluted to the 1% level with reagent water and extracted using continuous liquid-liquid extraction techniques. Samples containing greater than 30% solids are extracted using ultrasonic techniques.

Base/neutral extraction — Adjust the pH of the waters in the extractors to 12–13 with 6 *N* NaOH. Extract with methylene chloride for 24–48 h.
Acid extraction — Adjust the pH of the waters in the extractors to 2 or less using 6 *N* sulfuric acid. Extract with methylene chloride for 24–48 h.
Ultrasonic extraction of high solids samples — Add anhydrous sodium sulfate to the sample and QC aliquot(s). Add acetone:methylene chloride (1:1) to the sample and mix thoroughly

Concentrate extracts using a K-D apparatus.

QUALITY CONTROL The analyst is permitted to modify this method to improve separations or lower the costs of measurements, provided all performance specifications are met. Analyses of blanks are required to demonstrate freedom from contamination. When results of spikes indicate atypical method performance for samples, the samples are diluted to bring method performance within acceptable limits.

For low solids (aqueous samples), extract, concentrate, and analyze two sets of four 1-L aliquots (8 aliquots total) of the precision and recovery standard. For high solids samples, two sets of four 30-g aliquots of the high solids reference matrix are used.

Spike all samples with labeled compounds to assess method performance. Compute percent recovery of the labeled compounds using the internal standard method. Compare the labeled compound recovery for each compound with the corresponding labeled compound recovery.

Reagent water and high solids reference matrix blanks are analyzed to demonstrate freedom from contamination. Extract and concentrate a 1-L reagent water blank or a high solids reference matrix blank with each sample's lot (samples started through the extraction process on the same 8-h shift, to a maximum of 20 samples).

Field replicates may be collected to determine the precision of the sampling technique, and spiked samples may be required to determine the accuracy of the analysis when the internal standard method is used.

REFERENCE Semivolatile Organic Compounds by Isotope Dilution GC/MS. Office of Water Regulation and Standards, U.S. EPA Industrial Technology Division, Washington, DC, EPA Method 1625, Rev. C, June 1989 (contact W.A. Telliard, U.S. EPA, Office of Water Regulations and Standards, 401 M St., SW, Washington, DC, 20460. Phone: 202-382-7131).

2,4-Diaminotoluene   **EPA Method 8270**
CAS #95-80-7

TITLE Semivolatile Organic Compounds by GC/MS

MATRIX This method is used to determine the concentration of semivolatile organic compounds in extracts prepared from all types of solid waste matrices, soils, and groundwater. Although surface waters are not specifically mentioned, this method should be applicable to water samples from rivers, lakes, etc.

METHOD SUMMARY This method covers 259 semivolatile organic compounds. In very limited applications direct injection of the sample into the GC/MS system may be appropriate, but this results in very high detection limits (approximately 10,000 µg/L). Typically, a 1-L liquid sample, containing surrogate, and matrix spiking standards, is extracted in a continuous extractor first under acid conditions and then under basic conditions. Typically 30 g of a solid sample, containing surrogate, and matrix spiking standards, is extracted ultrasonically. After concentrating the extract to 1 mL it is spiked with 10 µL of an internal standard solution just prior to analysis by GC/MS. The volume injected should contain about 100 ng of base/neutral and 200 ng of acid surrogates (for a 1-µL injection). Analysis is performed by GC/MS using a capillary GC column.

INTERFERENCES Raw GC/MS data from all blanks, samples, and spikes must be evaluated for interferences. Contamination by carryover can occur whenever high-concentration and low-concentration samples are sequentially analyzed. To reduce carryover, the sample syringe must be rinsed out between samples with solvent. Whenever an unusually concentrated sample is encountered, it should be followed by the analysis of blank solvent to check for cross-contamination.

INSTRUMENTATION A GC/MS and a data system are required. The GC column used is a 30 m × 0.25 mm I.D. (or 0.32 mm I.D.) 1um film thickness silicone-coated fused silica capillary column. A continuous liquid-liquid extractor equipped with Teflon® or glass connection joints and stopcocks requiring no lubrication, a K-D concentrating apparatus, water bath, and an ultrasonic disrupter with a minimum power of 300 W and with pulsing capability are also required.

PRECISION & ACCURACY The estimated quantitation limit (EQL) of Method 8270B for determining an individual compound is approximately 1 mg/kg (wet weight) for soil or sediment samples, 1–200 mg/kg for wastes (dependent on matrix and method of preparation), and 10 µg/L for groundwater samples. EQLs will be proportionately higher for sample extracts that require dilution to avoid saturation of the detector.

The EQL(b) for groundwater in µg/L is 20.
The EQL (a, b) for low concentrations in soil and sediment in µg/kg is not determined.
Accuracy as µg/L is not listed.
Overall precision in µg/L is not listed.

(a) *EQLs listed for soil/sediment are based on wet weight. Normally data is reported in a dry-weight basis; therefore, EQLs will be higher based on the % dry weight of each sample. This calculation is based on a 30 g sample and gel permeation chromatography cleanup.*
(b) *Sample EQLs are highly matrix-dependent. The EQLs are provided for guidance and may not always be achievable.*
$C =$ *True value for concentration, in µg/L.*
$X =$ *Average recovery found for measurements of samples containing a concentration of C, in µg/L.*

ESTIMATED QUANTITATION LIMIT

Other Matrices	Factor (a)
High-concentration soil and sludges by sonicator	7.5
Non-water miscible waste	75

(a) *EQL for other matrices = [EQL for low soil/sediment] × [Factor]. This estimated EQL is similar to an EPA "Practical Quantitation Limit."*

SAMPLING METHOD

Liquid samples — Use a 1 or 2½ gallon amber glass bottle with a screw-top Teflon®-lined cover that has been prewashed with detergent and rinsed with distilled water and methanol (or isopropanol).

Soils, sediments, or sludges — Use an 8-oz. widemouth glass with a screw-top Teflon®-lined cover that has been prewashed with detergent and rinsed with distilled water and methanol (or isopropanol).

SAMPLE PRESERVATION

Liquid samples — If residual chlorine is present, add 3 mL of 10% sodium thiosulfate per gallon, cool to 4°C and store in a solvent-free refrigerator until analysis; if chlorine is not present, then eliminate the sodium thiosulfate addition.

Soils, sediments, or sludges — Cool samples to 4°C and store in a solvent-free refrigerator.

MHT Liquid samples must be extracted within 7 days and the extracts analyzed within 40 days. Soils, sediments, or sludges may be stored for a maximum of 14 days and the extracts analyzed within 40 days.

SAMPLE PREPARATION

Liquid samples — Transfer 1 L quantitatively to a continuous extractor. If high concentrations are anticipated, a smaller volume may be used and then diluted with organic-free reagent water to 1 L. Adjust pH, if necessary, to pH <2 using 1:1 (V/V) sulfuric acid. Pipette 1.0 mL of a surrogate standard spiking solution into each sample. For the sample in each analytical batch selected for spiking, add 1.0 mL of a matrix spiking standard. For base/neutral acid analysis, the amount of the surrogates and matrix spiking compounds added to the sample should result in a final concentration of 100 ng/μL of each analyte in the extract to be analyzed (assuming a 1-μL injection). Extract with methylene chloride for 18–24 h. Next, adjust the pH of the aqueous phase to pH >11 using 10 N sodium hydroxide and extract it with methylene chloride again for 18–24 h. Dry the extract through a column containing anhydrous sodium sulfate and concentrate it to 1 mL using a K-D concentrator.

Soils, sediments, or sludges — Use 30 g of sample. Nonporous or wet samples (gummy or clay type) that do not have a free-flowing sandy texture must be mixed with anhydrous sodium sulfate until the sample is free flowing. Add 1 mL of surrogate standards to all samples, spikes, standards, and blanks. For the sample in each analytical batch selected for spiking, add 1.0 mL of a matrix spiking standard. For base/neutral acid analysis, the amount added of the surrogates and matrix spiking compounds should result in a final concentration of 100 ng/μL of each base/neutral analyte and 200 ng/μL of each acid analyte in the extract to be analyzed (assuming a 1-μL injection). Immediately add a 100-mL mixture of 1:1 methylene chloride:acetone and extract the sample ultrasonically for 3 min and then decant or filter the extracts. Repeat the extraction two or more times. Dry the extract using a column with anhydrous sodium sulfate and concentrate it to 1 mL in a K-D concentrator.

QUALITY CONTROL A methylene chloride solution containing 50 ng/μL of decafluorotriphenylphosphine (DFTPP) is used for tuning the GC/MS system each 12-h shift. A system performance check also must be made during every 12-h shift. A standard containing 50 ng/μL each of 4,4′-DDT, pentachlorophenol, and benzidine is required to verify injection port inertness and GC column performance. A calibration standard at mid-concentration, containing each compound of interest, including all required surrogates, must be performed every 12 h during analysis. After the system performance check is met, calibration check compounds (CCCs) are used to check the validity of the initial calibration.

The internal standard responses and retention times in the calibration check standard must be evaluated immediately after or during data acquisition. If the retention time for any internal standard changes by more than 30 seconds from the last check calibration (12 h), the chromatographic system must be inspected for malfunctions and corrections must be made, as required. If the electron ionization current plot (EICP) area for any of the internal standards changes by a factor of two from the last daily calibration standard check, the mass spectrometer must be inspected for malfunctions and corrections must be made, as appropriate.

Demonstrate, through the analysis of a reagent water blank, that interferences from the analytical system, glassware, and reagents are under control. The blank samples should be carried through all stages of the sample preparation and measurement steps. For each analytical batch (up to 20 samples), a reagent blank, matrix spike, and matrix spike duplicate/duplicate must be analyzed (the frequency of the spikes may be different for different monitoring programs). The blank and spiked samples must be carried through all stages of the sample preparation and measurement steps. A QC reference sample concentrate containing each analyte at a concentration of 100 mg/L in methanol is required.

REFERENCE Test Methods for Evaluating Solid Waste (SW-846). U.S. EPA 1983, Method 8270B, Rev. 2, Nov. 1990. Office of Solid Waste, Washington, DC.

Diamyl phthalate EPA Method 8061
CAS #131-18-0

TITLE Phthalate Esters by Capillary Gas Chromatography With Electron Capture Detection (GC/ECD)

MATRIX This method covers aqueous and solid matrices. This includes a wide variety such as drinking water, groundwater, industrial wastewaters, surface waters, soils, solids, and sediments.

METHOD SUMMARY This method is used to determine the identities and concentrations of phthalate esters in liquid, solid and sludge matrices. When used to analyze for any or all of the target analytes, compound identification should be supported by at least one additional qualitative technique. This method describes conditions for parallel column, dual electron capture detector analysis, which fulfills the above requirement. Alternatively, GC/MS could be used for compound confirmation.

A measured volume or weight of sample (approximately 1 L for liquids, 10 to 30 g for solids and sludges) is extracted by using the appropriate sample extraction technique specified in

EPA Method 3510, EPA Method 3540, and EPA Method 3550. After cleanup, the extract is analyzed by GC/ECD.

INTERFERENCES The sensitivity of this method usually depends on the level of interferences rather than on instrumental limitations. If interferences prevent detection of the analytes, cleanup of the sample extracts is necessary. Either EPA Method 3610 or EPA Method 3620 alone or followed by EPA Method 3660, Sulfur Cleanup, may be used to eliminate interferences in the analysis. EPA Method 3640, Gel Permeation Cleanup, is applicable for samples that contain high amounts of lipids and waxes.

Interferences coextracted from the samples will vary considerably from waste to waste. Glassware must be scrupulously clean. All glassware require treatment in a muffle furnace at 400°C for 2 to 4 h, or thorough rinsing with pesticide-grade solvent, prior to use. Volumetric glassware should not be heated in a muffle furnace. Storage of glassware in the lab introduces contamination, even if the glassware is wrapped in aluminum foil. Sodium sulfate, Florisil, and alumina may be contaminated with phthalate esters and, therefore, use of these materials in sample cleanup should be employed cautiously. If these materials are used, they must be obtained packaged in glass. Heating at 400°C for sodium sulfate, 320°C for Florisil, and 210°C for alumina is recommended. Glass wool used in any step of sample preparation should be a specially treated pyrex wool, pesticide grade, and must be baked at 400°C for 4 h immediately prior to use.

Paper thimbles and filter paper must be exhaustively washed with the solvent that will be used in the sample extraction. Soxhlet extraction of paper thimbles and filter paper for 12 h with fresh solvent should be repeated for a minimum of three times. Method blanks should be obtained before any of the precleaned thimbles or filter papers are used.

INSTRUMENTATION Gas chromatograph suitable for on-column and split/splitless injections.

Column 1: 30 m × 0.53 mm ID, 5% phenyl/95% methyl silicone fused-silica open tubular column, DB-5, 1.5 µg film thickness.
Column 2: 30 m × 0.53 mm ID, 14% cyanopropyl phenyl silicone fused-silica open tubular column, DB-1701, 1.0 µg film thickness.

A dual electron capture detector (ECD) is used. A Kuderna-Danish (K-D) apparatus is required along with a vacuum manifold consisting of individually adjustable, easily accessible flow-control valves for up to 24 cartridges, sample rack, chemically resistant cover and seals, heavy-duty glass basin, removable stainless steel solvent guides, built-in vacuum gauge and valve. Also, 6-mL, 1-g solid-phase extraction cartridges, LC-Florisil or equivalent, prepackaged, ready to use will be needed.

PRECISION & ACCURACY The MDL actually achieved in a given analysis will vary, as it is dependent on instrument sensitivity and matrix effects. This method has been tested in a single lab. Single-operator precision, overall precision, and method accuracy were found to be related to the concentration of the compounds and the type of matrix.

MULTIPLICATION FACTORS FOR OTHER MATRICES (a)

Matrix	Factor (b)
Groundwater	10
Low-concentration soil by ultrasonic extraction with GPC cleanup	670
High-concentration soil and sludges by ultrasonic extraction	10,000
Non-water miscible waste	100,000

(a) Sample EQLs are highly matrix-dependent.
(b) EQL = [Method detection limit] × [Factor]. For non-aqueous samples, the factor is on a wet-weight basis.

The MDL using 7 replicate determinations and a spike concentration of 100 µg/L was 110 ng/L.
The average recovery from HPLC-grade water using 4 determinatons and a spike concentration of 100 µg/L was 93.6%.
The precision (as RSD) from HPLC-grade water using 4 determinatons and a spike concentration of 100 µg/L was 21.0%.
The average recovery from groundwater using 4 determinatons and a spike concentration of 100 µg/L was 78.9%.
The precision (as RSD) from groundwater using 4 determinatons and a spike concentration of 100 µg/L was 5.8%.
The average recovery (in %) with %RSD (in parenthesis) from 3 determinations and a spike concentration of 20 µg/L in water was 78.2 (7.3) using 3M Empore Disks and EPA Method 8061.
The average recovery (in %) with %RSD (in parenthesis) from 3 determinations and a spike concentration of 20 µg/L in leachate was 92.1 (21.5) using 3M Empore Disks and EPA Method 8061.
The average recovery (in %) with %RSD (in parenthesis) from 3 determinations and a spike concentration of 20 µg/L in estuarine groundwater was 88.8 (7.5) using 3M Empore Disks and EPA Method 8061.
The average recovery (in %) with %RSD (in parenthesis) from 3 determinations and a spike concentration of 1 mg/kg in estuarine sediment was 95.0 (10.2) after sulfur cleanup with EPA Method 3660.
The average recovery (in %) with %RSD (in parenthesis) from 3 determinations and a spike concentration of 1 mg/kg in municipal sludge was 81.9 (7.1).
The average recovery (in %) with %RSD (in parenthesis) from 3 determinations and a spike concentration of 1 mg/kg in sandy loam soil was 81.9 (15.9).

SAMPLE COLLECTION, PRESERVATION & HANDLING
Containers used to collect samples for the determination of semivolatile organic compounds should be soap and water washed followed by methanol (or isopropanol) rinsing. The sample containers should be of glass or Teflon® and have screw-top covers with Teflon® liners. Sample containers should be filled with care to prevent any portion of the collected sample coming in contact with the sampler's gloves.

No preservation is used with concentrated waste samples. With liquid samples containing no residual chlorine and with soil, sediment, and sludge samples, immediately cooling to 4°C is the only preservation used. When residual chlorine is present then 3 mL of 10% aqueous sodium sulfate is added for each gallon of sample collected, followed by cooling to 4°C.

MHT Liquid samples must be extracted within 7 days and their extracts analyzed within 40 days. Concentrated waste, soil, sediment, and sludge samples must be extracted within 14 days and their extracts analyzed within 40 days.

SAMPLE PREPARATION In general, water samples are extracted at a pH of 5 to 7 with methylene chloride in a separatory funnel (EPA Method 3510). EPA Method 3520 is not recommended for the extraction of aqueous samples because the longer chain esters tend to adsorb to the glassware and consequently, their extraction recoveries may be poor. Solid samples are extracted with hexane/acetone (1:) or methylene chloride/acetone (1:1) in a Soxhlet extractor (EPA Method 3540) or with an ultrasonic extractor (EPA Method 3550). Immediately prior to extraction, spike 500 µL of the surrogate standard spiking solution into 1-L aqueous sample or 30-g solid sample. Extraction of particulate-free aqueous samples using C-18 extraction disks is an optional method that can be used.

Prior to Florisil cleanup or GC analysis, the methylene chloride and methylene chloride/acetone extracts must be exchanged to hexane. Exchange is not required for the acetonitrile extracts. Cleanup may not be necessary for extracts from a relatively clean sample matrix. Florisil Cartridge Cleanup may be used for extract cleanup.

If PCBs and organochlorine pesticides are known to be present in the sample, and if Florisil Cartridge Cleanup is considered, then two fractions are collected: Fraction 1 is eluted with 5 mL of 20% methylene chloride in hexane and Fraction 2 is eluted with 5 mL of 10% acetone in hexane. Fraction 1 contains the organochlorine pesticides and PCBs, and can be discarded. Fraction 2 contains the phthalate esters and is analyzed by GC/ECD.

QUALITY CONTROL Identify compounds in the sample by comparing the retention times of the peaks in the sample chromatogram with those of the peaks in standard chromatograms. The retention time window used to make identification is based upon measurements of actual retention time variations over the course of 10 consecutive injections.

Calibration standards are prepared at a minimum of five concentrations for each parameter of interest through dilution of the stock standard solutions with hexane. One of the concentrations should be at a concentration near, but above, the method detection limit. Prepare stock standard solutions in hexane. Stock standards should be checked frequently for signs of degradation or evaporation, especially just prior to preparing calibration standards from them. Stock standard solutions must be replaced after one year, or sooner if comparison with check standards indicates a problem. The suggested internal standards are: 2,5-dibromotoluene, 1,3,5-tribromobenzene, and α, α'-dibromo-m-xylene. The analyst can use any of the three compounds provided that they are resolved from matrix interferences. Recommended surrogate compounds are α-2,6-trichlorotoluene, 1,4-dichloronaphthalene, and 2,3,4,5,6-pentachlorotoluene.

Spike each sample, standard, and blank with surrogate compounds. Three surrogates are suggested for this method: diphenyl phthalate, diphenyl isophthalate, and dibenzyl phthalate.

The quality control check sample concentrate should contain the test compounds at 5 to 10 ng/µL An internal standard peak area check must be performed on all samples. The internal standard must be evaluated for acceptance by determining whether the measured area for the internal standard deviates by more than 30% from the average are for the internal standard in the calibration standards. When the internal standard peak area is outside that limit, all samples that fall outside the QC criteria must be reanalyzed. Benzyl benzoate has been tested and found appropriate as an internal standard for this method.

Any compounds confirmed by two columns may also be confirmed by GC/MS. The sample extract and associated blank should be analyzed by GC/MS. A reference standard of the compound must also be analyzed by GC/MS. Include a mid-concentration calibration standard after each group of 20 samples. The response factors for the mid-concentration calibration must be within ±15% of the average values for the multiconcentration calibration. Demonstrate through the analyses of standards that the Florisil fractionation scheme is reproducible.

REFERENCE Test Methods for Evaluating Solid Waste, Physical/Chemical Methods, SW-846, 3rd Edition, U.S. EPA, Office of Solid Waste, Washington, DC, EPA Method 8061, Nov. 1990.

Diazinon   **EPA Method 507**
CAS #333-41-5

TITLE Determination of Nitrogen and Phosphorus-Containing Pesticides in Water by GC/NPD

MATRIX This method is applicable to the determination of certain nitrogen and phosphorus-containing pesticides in finished drinking water and groundwater.

METHOD SUMMARY Method 507 covers 46 nitrogen- and phosphorus-containing pesticides. A 1-L sample is fortified with a surrogate standard, salted, buffered, extracted with methylene chloride, and concentrated; then the solvent is exchanged with methyl tert-butyl ether (MTBE) and concentrated again, and a 2-µL aliquot of a sample extract is injected into a GC system equipped with a selective nitrogen-phosphorus detector and a capillary column for analysis.

INTERFERENCES Method interferences may be caused by contaminants in solvents, reagents, glassware, and other sample processing apparatus. Interfering contamination may occur when a sample containing low concentrations of analytes is analyzed immediately following a sample containing relatively high concentrations. One or more injections of MTBE should be made following the analysis of a sample with high concentrations of analytes to check for analyte carryover. Matrix interferences may be caused by contaminants that are coextracted from the sample. The extent of matrix interferences will vary considerably from source to source, depending upon the water sampled.

INSTRUMENTATION A gas chromatograph system (GC) equipped with a nitrogen-phosphorus detector (NPD) is needed.

Column 1: 30 m × 0.25 mm I.D. DB-5 bonded fused silica column, 0.25 μm film thickness, or equivalent.

Column 2: 30 m × 0.25 mm I.D. DB-1701 bonded fused silica column, 0.25 μm film thickness, or equivalent.

PRECISION & ACCURACY This method has been validated in a single lab and estimated detection limits (EDLs) have been determined for each analyte. Observed detection limits may vary among waters, depending upon the nature of the interferences in the sample matrix and the specific instrumentation used. Analytes that are not separated chromatographically cannot be individually identified and measured unless an alternative technique for identification and quantification exist.

The estimated detection limit (in μg/L) was 0.25. The EDL is defined as either method detection limit or a level of compound in a sample yielding a peak in the final extract with signal-to-noise ratio of approximately 5, whichever value is higher.

The concentration used for these measurements (in μg/L) was 2.5.
The accuracy (as % recovery) was 115.
The precision (% RSD) was 7.

SAMPLING METHOD Grab samples are collected in 1-L glass sample bottles (prewashed with detergent and hot tap water, rinsed with reagent water, and dried in an oven at 400°C for 1 h) with screw caps lined with PTFE-fluorocarbon.

SAMPLE PRESERVATION Add mercuric chloride to the sample bottle in amounts to produce a concentration of 10 mg/L. If residual chlorine is present, add 80 mg of sodium thiosulfate/L of sample to the sample bottle prior to collection. After collection, seal bottle and shake vigorously for 1 min, then cool the sample to 4°C immediately and store it at 4°C in the dark until extraction.

MHT Maximum holding time of the samples, and in some cases the extracts, is 14 days.

Note: Samples with this compound must be extracted immediately.

SAMPLE PREPARATION Fortify the sample with 50 μL of the surrogate standard solution, adjust to pH 7 with phosphate buffer, add 100 g NaCl to the sample, and seal and shake to dissolve the salt; then extract with methylene chloride in a separatory funnel or in a mechanical tumbler bottle. Dry the extract by pouring it through a solvent-rinsed drying column containing about 10 cm of anhydrous sodium sulfate. Collect the extract in a Kuderna-Danish (K-D) concentrator and rinse the column with 20–30 mL methylene chloride. Concentrate the extract to about 2 mL and rinse the flask and its lower joint into the concentrator tube with 1 to 2 mL of methyl t-butyl ether (MTBE). Add 5–10 mL of MTBE and concentrate the extract twice (adding more MTBE) to a final volume of 5.0 mL and store it at 4°C until analysis.

Note: If methylene chloride is not completely removed from the final extract, it may cause detector problems.

QUALITY CONTROL Minimum quality control requirements are initial demonstration of lab capability, determination of surrogate compound recoveries in each sample and blank, monitoring internal standard peak area or height in each sample and blank, analysis of lab reagent blanks, lab fortified samples, lab fortified blanks, and other QC samples. A lab reagent blank is analyzed to demonstrate that all glassware and reagent interferences are under control.

Initial demonstration of capability is fulfilled by analyzing four fortified reagent water samples with the recovery value for each analyte falling within the acceptable range (±30% average recovery). Surrogate recoveries from samples or method blanks must be 70–130%. The internal standard response for any sample chromatogram should not deviate from the daily calibration check standard's internal standard response by more than 30% or lab fortified blanks and sample matrices are used to assess lab performance and analyte recovery, respectively.

If the response for the target analyte peak exceeds the working range of the system, dilute the extract and reanalyze. Alternative techniques such as an alternate detector or second chromatography column should be used to confirm peak identification when sample components are not resolved adequately.

EPA CONTACT & HOTLINE For technical questions contact Dr. Baldev Bathija, U.S. EPA, Office of Ground Water and Drinking Water (WH-550D), 401 M St. SW, Washington, DC 20460. Tel. (202) 260-3040. For further information the EPA Safe Drinking Water Hotline may be called at: (800) 426-4791.

REFERENCE Methods for the Determination of Organic Compounds in Drinking Water, EPA/600/4-88/039 (revised July 1991). U.S. EPA Environmental Monitoring Systems Laboratory, Cincinnati, OH, 45268, U.S.A. Available from the National Technical Information Service (NTIS), 5285 Port Royal Road, Springfield, VA 22161; Tel. 800-553-6847. NTIS Order Number is PB91-231480.

Diazinon   **EPA Method 8141**
CAS #333-41-5

TITLE Organophosphorus Compounds by Gas Chromatography: Capillary Column Technique

MATRIX This method covers aqueous and solid matrices. This includes a wide variety such as drinking water, groundwater, industrial wastewaters, surface waters, soils, solids, and sediments.

METHOD SUMMARY This is a GC method used to determine the concentration of 28 organophosphorus pesticides.

The use of Gel Permeation Cleanup (EPA Method 3640) for sample cleanup has been demonstrated to yield recoveries of less than 85% for many method analytes and is therefore not recommended for use with this method.

This method provides GC conditions for the detection of ppb concentrations of organophosphorus compounds. Prior to the use of this method, appropriate sample preparation techniques must be used. Water samples are extracted at a neutral pH with methylene chloride as a solvent by using a separatory funnel (EPA Method 3510) or a continuous liquid-liquid extractor (EPA Method 3520). Soxhlet extraction (EPA Method 3540) or ultrasonic extraction (EPA Method 3550) using methylene chloride/acetone (1:1) are used for solid samples. Both neat and diluted organic liquids (EPA Method 3580) may be analyzed by direct injection. Spiked samples are used to verify the applicability of the chosen extraction technique to each new sample type. A GC with a flame photometric (FPD) or nitrogen-phosphorus detector (NPD) is used for this multiresidue procedure.

INTERFERENCES The use of Florisil cleanup materials (EPA Method 3620) for some of the compounds in this method has been demonstrated to yield recoveries less than 85% and is therefore not recommended for all compounds. Use of phosphorus or halogen specific detectors, however, often obviates the necessity for cleanup for relatively clean sample matrices. If particular circumstances demand the use of an alternative cleanup procedure, the analyst must determine the elution profile and demonstrate that the recovery of each analyte is no less than 85%.

Use of a flame photometric detector (FPD) in the phosphorus mode will minimize interferences from materials that do not contain phosphorus. Elemental sulfur, however, may interfere with the determination of certain organophosphorus compounds by flame photometric gas chromatography. Sulfur cleanup using EPA Method 3660 may alleviate this interference. A nitrogen phosphorus detector (NPD) is also recommended.

A few analytes coelute on certain columns. Therefore, select a second column for confirmation where coelution of the analytes of interest does not occur.

Method interferences may be caused by contaminants in solvents, reagents, glassware, and other sample processing hardware that lead to discrete artifacts or elevated baselines in gas chromatograms. All these materials must be routinely demonstrated to be free from interferences under the conditions of the analysis by analyzing reagent blanks.

INSTRUMENTATION A GC with a NPD or a FPD will be needed. A data system or integrator is recommended for measuring peak areas and/or peak heights. A Kuderna-Danish (K-D) apparatus will be needed for extract concentration.

Column 1: 15 m × 0.53 mm megabore capillary column, 1.0 µm film thickness, DB-210.
Column 2: 15 m × 0.53 mm megabore capillary column, 1.5 µm film thickness, SPB-608.
Column 3: 15 m × 0.53 mm megabore capillary column, 1.0 µm film thickness, DB-5.

Three megabore capillary columns are included for analysis of organophosphates by this method. Column 1 (DB-210 or equivalent) and Column 2 (SPB-608 or equivalent) are recommended if a large number of organophosphorus analytes are to be determined. If the superior resolution offered by Column 1 and Column 2 is not required, Column 3 (DB-5 or equivalent) may be used. For megabore capillary columns, automatic injections of 1 µL are recommended.

PRECISION & ACCURACY The MDL actually achieved in a given analysis will vary, as it is dependent on instrument sensitivity and matrix effects. Single operator accuracy and precision studies have been conducted with spiked water and soil samples.

MULTIPLICATION FACTORS FOR OTHER MATRICES (a)

Matrix	Factor (b)
Groundwater (EPA Method 3510 or EPA Method 3520)	10
Low-concentration soil by Soxhlet and no cleanup	10 (c)
Low-concentration soil by ultrasonic extraction with GPC cleanup	6.7 (c)
High-concentration soil and sludges by ultrasonic extraction	500 (c)
Non-water miscible waste (EPA Method 3580)	1000 (c)

(a) Sample EQLs are highly matrix-dependent. The EQLs listed here are provided for guidance and may not always be achievable.
(b) EQL = [Method detection limit] × [Factor]. For non-aqueous samples the factor is on a wet-weight basis.
(c) Multiply this factory times the soil MDL.

The MDL (in µg/L) when reagent water was extracted using a separatory funnel was 0.20.
The MDL (in µg/kg) when soil was extracted using Soxhlet extraction (EPA Method 3540) was 10.0.
Accuracy (as % recovery) with separatory funnel extraction ranged from 136 (with low spikes) to 82 (with high spikes).
Accuracy (as % recovery) with continuous liquid-liquid extraction ranged from not recovered (with low spikes) to 118 (with high spikes).
Accuracy (as % recovery) with Soxhlet extraction of soils ranged from 87 (with low spikes to 75 (with high spikes).
Accuracy (as % recovery) with ultrasonic extraction of soils ranged from not recovered (with low spikes) to 78 (with high spikes).

SAMPLE COLLECTION, PRESERVATION & HANDLING
Containers used to collect samples for the determination of semivolatile organic compounds should be soap and water washed followed by methanol (or isopropanol) rinsing. The sample containers should be of glass or Teflon® and have screw-top covers with Teflon® liners.

No preservation is used with concentrated waste samples. With liquid samples containing no residual chlorine and with soil, sediment, and sludge samples, immediately cooling to 4°C is the only preservation used. When residual chlorine is present then 3 mL of 10% aqueous sodium sulfate is added for each gallon of sample collected, followed by cooling to 4°C.

Liquid samples must be extracted within 7 days and their extracts analyzed within 40 days. Concentrated waste, soil, sediment, and sludge samples must be extracted within 14 days and their extracts analyzed within 40 days.

SAMPLE PREPARATION In general, water samples are extracted at a neutral pH with methylene chloride, using either EPA Method 3510 or EPA Method 3520. Solid samples are

extracted using either EPA Method 3540 or EPA Method 3550 with methylene chloride/acetone (1:1) as the extraction solvent.

Prior to GC analysis, the extraction solvent may be exchanged to hexane. Single lab data indicates that samples should not be transferred with 100% hexane during sample workup as the more water soluble organophosphorus compounds may be lost.

If cleanup is performed on the samples, the analyst should analyze the samples by GC. This will confirm elution patterns and the absence of interferences from the reagents. If peak detection and identification is prevented by the presence of interferences, further cleanup is required.

QUALITY CONTROL The analyst should monitor the performance of the extraction, cleanup (when used), and analytical system and the effectiveness of the method in dealing with each sample matrix by spiking each sample, standard, and blank with one or two surrogates (e.g., organophosphorus compounds not expected to be present in the sample). Deuterated analogs of analytes should not be used as surrogates for gas chromatographic analysis due to coelution problems.

A minimum of five concentrations for each analyte of interest should be prepared through dilution of the stock standards with isooctane. One of the concentrations should be at a concentration near, but above, the MDL.

Include a mid-level check standard after each group of 10 samples in the analysis sequence. GC/MS techniques should be judiciously employed to support qualitative identifications made with this method. Follow the GC/MS operating requirements specified in EPA Method 8270.

When available, chemical ionization mass spectra may be employed to aid in the qualitative identification process. To confirm an identification of a compound, the background-corrected mass spectrum of the compound must be obtained from the sample extract and must be compared with a mass spectrum from a stock or calibration standard analyzed under the same chromatographic conditions. The molecular ion and all other ions present above 20% relative abundance in the mass spectrum of the standard must be present in the mass spectrum of the sample with agreement to ±20%. The retention time of the compound in the sample must be within six seconds of the retention time for the same compound in the standard solution.

Should the MS procedure fail to provide satisfactory results, additional steps may be taken before reanalysis. These steps may include the use of alternate packed or capillary GC columns or additional sample cleanup.

REFERENCE Test Methods for Evaluating Solid Waste, Physical/Chemical Methods, SW-846, 3rd Edition, U.S. EPA, Office of Solid Waste, Washington, DC, EPA Method 8141 July 1992.

Diazinon   **EPA Method 8140**
CAS #333-41-5

TITLE Organophosphorus Pesticides

MATRIX Groundwater, soils, sludges, water miscible liquid wastes, and non-water miscible wastes.

APPLICATION This method is used for the analysis of 21 organophosphorus pesticides. Samples are extracted, concentrated, and analyzed using direct injection of both neat and diluted organic liquid into a gas chromatograph (GC).

INTERFERENCES Solvents, reagents, and glassware may introduce artifacts. Other interferences may come from coextracted compounds from samples. The use of Florisil cleanup materials may produce low recoveries. Elemental sulfur may interfere with some compounds when using a flame photometric detector. Sulfur cleanup (Method 3660) may alleviate sulfur interference.

INSTRUMENTATION GC capable of on-column injections and a flame photometric detector (FPD) or a thermionic detector. Column 1: 1.8 m by 2 mm with 5% SP-2401 on Supelcoport. Column 2: 1.8 m by 2 mm with 3% SP-2401 on Supelcoport. Column 3: 50 cm by ⅛ IN EFLON® with 15% SE-54 on Gas Chrom Q. The preferred column is Column Number 2.

RANGE 5.6 µg/L only.

MDL 0.6 µg/L (in reagent water).

PQL FACTORS FOR MULTIPLYING × FID MDL VALUE

Matrix	Multiplication Factor
Groundwater	10
Low-level soil by sonication with GPC cleanup	670
High-level soil and sludge by sonication	10,000
Non-water miscible waste	100,000

PRECISION 6.0% (single operator standard deviation).

ACCURACY 67.0% (single operator average recovery).

SAMPLING METHOD Use 8-oz. widemouth glass bottles with Teflon®-lined caps for concentrated waste samples, soils, sediments, and sludges. Use 1 or 2½ gallon amber glass bottles with Teflon®-lined caps for liquid (water) samples.

STABILITY Cool soil, sediment, sludge, and liquid samples to 4°C. If residual chlorine is present in liquid samples add 3 mL of 10% sodium thiosulfate per gallon of sample and cool to 4°C.

MHT 14 days for concentrated waste, soil, sediment, or sludge; 7 days for liquid samples; all extracts must be analyzed within 40 days.

QUALITY CONTROL A quality control check sample concentrate containing this compound in acetone at a concentration 1,000 times more concentrated than the selected spike concentration is required. The QC check sample concentrate may be prepared from pure standard materials or purchased as certified solutions. Use appropriate trip, matrix, control site, method, reagent, and solvent blanks. Internal, surrogate, and five concentration level calibration standards are used.

REFERENCE Method 8140, SW-846, 3rd ed., Sept. 1986.

Dibenz(a,h)anthracene
CAS #53-70-3

EPA Method 1625

TITLE Semivolatile Organic Compounds by Isotope Dilution GC/MS

MATRIX The compounds may be determined in waters, soils, and municipal sludges by this method.

METHOD SUMMARY This method is used to determine 176 semivolatile toxic organic pollutants associated with the CWA (as amended 1987); the RCRA (as amended 1986); the CERCLA (as amended 1986); and other compounds amenable to extraction and analysis by capillary column gas chromatography-mass spectrometry (GC/MS).

Stable isotopically-labeled analogs of the compounds of interest are added to the sample. If the solids content is less than 1%, a 1-L sample is extracted at pH 12–13, then at pH <2 with methylene chloride using continuous extraction techniques.

If the solids content is 30% or less, the sample is diluted to 1% solids with reagent water, homogenized ultrasonically, and extracted at pH 12–13, then at pH <2 with methylene chloride using continuous extraction techniques. If the solids content is greater than 30%, the sample is extracted using ultrasonic techniques.

Each extract is dried over sodium sulfate, concentrated to a volume of 5 mL, cleaned up using GPC, if necessary, and concentrated. Extracts are concentrated to 1 mL if GPC is not performed, and to 0.5 mL if GPC is performed.

An internal standard is added to the extract, and a 1-mL aliquot of the extract is injected into the GC. The compounds are separated by GC and detected by a MS. The labeled compounds serve to correct the variability of the analytical technique.

INTERFERENCES Solvents, reagents, glassware, and other sample processing hardware may yield artifacts and/or elevated baselines causing misinterpretation of chromatograms and spectra. Materials used in the analysis must be demonstrated to be free from interferences under the conditions of analysis by running method blanks initially and with each sample lot (sample started through the extraction process on a given 8-h shift, to a maximum of 20). Specific selection of reagents and purification of solvents by distillation in all glass systems may be required. Glassware and, where possible, reagents are cleaned by solvent rinse and baking at 450°C for 1-h minimum. Interferences coextracted from samples will vary considerably from source to source, depending on the diversity of the site being sampled.

INSTRUMENTATION Major instrumentation includes a GC with a splitless or on-column injection port for capillary column, a MS with 70 eV electron impact ionization, and a data system to collect and record MS data, and process it. A K-D apparatus is used to concentrate extracts.

GC Column: 30 m × 0.25 mm I.D. 5% phenyl, 94% methyl, 1% vinyl silicone bonded phased fused silica capillary column.

PRECISION & ACCURACY The detection limits of the method are usually dependent on the level of interferences rather than instrumental limitations. The limits typify the minimum quantities that can be detected with no interferences present.

The minimum level (in µg/mL) was 20. This is defined as a minimum level at which the analytical system shall give recognizable mass spectra (background corrected) and acceptable calibration points.

The MDL (in µg/kg) in low solids was 49 and in high solids was 125; these were determined in digested sludge (low solids) and in filter cake or compost (high solids).

The labeled and native compound initial precision as standard deviation (in µg/L) was 55.

The labeled and native compound initial accuracy as average recovery (in µg/L) was 23–299.

SAMPLE COLLECTION, PRESERVATION & HANDLING Collect samples in glass containers. Aqueous samples which flow freely are collected in refrigerated bottles using automatic sampling equipment. Solid samples are collected as grab samples using widemouth jars. Maintain samples at 0 to 4°C from the time of collection until extraction. If residual chlorine is present in aqueous samples, add 80 mg sodium thiosulfate/L of water. Begin sample extraction within 7 days of collection, and analyze all extracts within 40 days of extraction.

SAMPLE PREPARATION Samples containing 1% solids or less are extracted directly using continuous liquid-liquid extraction techniques. Samples containing 1 to 30% solids are diluted to the 1% level with reagent water and extracted using continuous liquid-liquid extraction techniques. Samples containing greater than 30% solids are extracted using ultrasonic techniques.

Base/neutral extraction — Adjust the pH of the waters in the extractors to 12–13 with 6 N NaOH. Extract with methylene chloride for 24–48 h.
Acid extraction — Adjust the pH of the waters in the extractors to 2 or less using 6 N sulfuric acid. Extract with methylene chloride for 24–48 h.
Ultrasonic extraction of high solids samples — Add anhydrous sodium sulfate to the sample and QC aliquot(s). Add acetone:methylene chloride (1:1) to the sample and mix thoroughly

Concentrate extracts using a K-D apparatus.

QUALITY CONTROL The analyst is permitted to modify this method to improve separations or lower the costs of measurements, provided all performance specifications are met. Analyses of blanks are required to demonstrate freedom from contamination. When results of spikes indicate atypical method performance for samples, the samples are diluted to bring method performance within acceptable limits.

For low solids (aqueous samples), extract, concentrate, and analyze two sets of four 1-L aliquots (8 aliquots total) of the precision and recovery standard. For high solids samples, two sets of four 30-g aliquots of the high solids reference matrix are used.

Spike all samples with labeled compounds to assess method performance. Compute percent recovery of the labeled compounds using the internal standard method. Compare the labeled compound recovery for each compound with the corresponding labeled compound recovery.

Reagent water and high solids reference matrix blanks are analyzed to demonstrate freedom from contamination. Extract and concentrate a 1-L reagent water blank or a high solids reference matrix blank with each sample's lot (samples started through the extraction process on the same 8-h shift, to a maximum of 20 samples).

Field replicates may be collected to determine the precision of the sampling technique, and spiked samples may be required to determine the accuracy of the analysis when the internal standard method is used.

REFERENCE Semivolatile Organic Compounds by Isotope Dilution GC/MS. Office of Water Regulation and Standards, U.S. EPA Industrial Technology Division, Washington, DC, EPA Method 1625, Rev. C, June 1989 (contact W.A. Telliard, U.S. EPA, Office of Water Regulations and Standards, 401 M St., SW, Washington, DC, 20460. Phone: 202-382-7131).

Dibenz(a,h)anthracene   **EPA Method 8270**
CAS #53-70-3

TITLE Semivolatile Organic Compounds by GC/MS

MATRIX This method is used to determine the concentration of semivolatile organic compounds in extracts prepared from all types of solid waste matrices, soils, and groundwater. Although surface waters are not specifically mentioned, this method should be applicable to water samples from rivers, lakes, etc.

METHOD SUMMARY This method covers 259 semivolatile organic compounds. In very limited applications direct injection of the sample into the GC/MS system may be appropriate, but this results in very high detection limits (approximately 10,000 µg/L). Typically, a 1-L liquid sample, containing surrogate, and matrix spiking standards, is extracted in a continuous extractor first under acid conditions and then under basic conditions. Typically 30 g of a solid sample, containing surrogate, and matrix spiking standards, is extracted ultrasonically. After concentrating the extract to 1 mL it is spiked with 10 µL of an internal standard solution just prior to analysis by GC/MS. The volume injected should contain about 100 ng of base/neutral and 200 ng of acid surrogates (for a 1-µL injection). Analysis is performed by GC/MS using a capillary GC column.

INTERFERENCES Raw GC/MS data from all blanks, samples, and spikes must be evaluated for interferences. Contamination by carryover can occur whenever high-concentration and low-concentration samples are sequentially analyzed. To reduce carryover, the sample syringe must be rinsed out between samples with solvent. Whenever an unusually concentrated sample is encountered, it should be followed by the analysis of blank solvent to check for cross-contamination.

INSTRUMENTATION A GC/MS and a data system are required. The GC column used is a 30 m × 0.25 mm I.D. (or 0.32 mm I.D.) 1um film thickness silicone-coated fused silica capillary column. A continuous liquid-liquid extractor equipped with Teflon® or glass connection joints and stopcocks requiring no lubrication, a K-D concentrating apparatus, water bath, and an ultrasonic disrupter with a minimum power of 300 W and with pulsing capability are also required.

PRECISION & ACCURACY The estimated quantitation limit (EQL) of Method 8270B for determining an individual compound is approximately 1 mg/kg (wet weight) for soil or sediment samples, 1–200 mg/kg for wastes (dependent on matrix and method of preparation), and 10 µg/L for groundwater samples. EQLs will be proportionately higher for sample extracts that require dilution to avoid saturation of the detector.

The EQL(b) for groundwater in µg/L is 10.
The EQL (a, b) for low concentrations in soil and sediment in µg/kg is 660.
Accuracy as µg/L is 0.88C +4.72.
Overall precision in µg/L is 0.59X +0.25.

(a) EQLs listed for soil/sediment are based on wet weight. Normally data is reported in a dry-weight basis; therefore, EQLs will be higher based on the % dry weight of each sample. This calculation is based on a 30 g sample and gel permeation chromatography cleanup.
(b) Sample EQLs are highly matrix-dependent. The EQLs are provided for guidance and may not always be achievable.
C = True value for concentration, in µg/L.
X = Average recovery found for measurements of samples containing a concentration of C, in µg/L.

ESTIMATED QUANTITATION LIMIT

Other Matrices	Factor (a)
High-concentration soil and sludges by sonicator	7.5
Non-water miscible waste	75

(a) EQL for other matrices = [EQL for low soil/sediment] × [Factor]. This estimated EQL is similar to an EPA "Practical Quantitation Limit."

SAMPLING METHOD
Liquid samples — Use a 1 or 2½ gallon amber glass bottle with a screw-top Teflon®-lined cover that has been prewashed with detergent and rinsed with distilled water and methanol (or isopropanol).

Soils, sediments, or sludges — Use an 8-oz. widemouth glass with a screw-top Teflon®-lined cover that has been prewashed with detergent and rinsed with distilled water and methanol (or isopropanol).

SAMPLE PRESERVATION
Liquid samples — If residual chlorine is present, add 3 mL of 10% sodium thiosulfate per gallon, cool to 4°C and store in a solvent-free refrigerator until analysis; if chlorine is not present, then eliminate the sodium thiosulfate addition.

Soils, sediments, or sludges — Cool samples to 4°C and store in a solvent-free refrigerator.

MHT Liquid samples must be extracted within 7 days and the extracts analyzed within 40 days. Soils, sediments, or sludges may be stored for a maximum of 14 days and the extracts analyzed within 40 days.

SAMPLE PREPARATION

Liquid samples — Transfer 1 L quantitatively to a continuous extractor. If high concentrations are anticipated, a smaller volume may be used and then diluted with organic-free reagent water to 1 L. Adjust pH, if necessary, to pH <2 using 1:1 (V/V) sulfuric acid. Pipette 1.0 mL of a surrogate standard spiking solution into each sample. For the sample in each analytical batch selected for spiking, add 1.0 mL of a matrix spiking standard. For base/neutral acid analysis, the amount of the surrogates and matrix spiking compounds added to the sample should result in a final concentration of 100 ng/µL of each analyte in the extract to be analyzed (assuming a 1-µL injection). Extract with methylene chloride for 18–24 h. Next, adjust the pH of the aqueous phase to pH >11 using 10 N sodium hydroxide and extract it with methylene chloride again for 18–24 h. Dry the extract through a column containing anhydrous sodium sulfate and concentrate it to 1 mL using a K-D concentrator.

Soils, sediments, or sludges — Use 30 g of sample. Nonporous or wet samples (gummy or clay type) that do not have a free-flowing sandy texture must be mixed with anhydrous sodium sulfate until the sample is free flowing. Add 1 mL of surrogate standards to all samples, spikes, standards, and blanks. For the sample in each analytical batch selected for spiking, add 1.0 mL of a matrix spiking standard. For base/neutral acid analysis, the amount added of the surrogates and matrix spiking compounds should result in a final concentration of 100 ng/µL of each base/neutral analyte and 200 ng/µL of each acid analyte in the extract to be analyzed (assuming a 1-µL injection). Immediately add a 100-mL mixture of 1:1 methylene chloride:acetone and extract the sample ultrasonically for 3 min and then decant or filter the extracts. Repeat the extraction two or more times. Dry the extract using a column with anhydrous sodium sulfate and concentrate it to 1 mL in a K-D concentrator.

QUALITY CONTROL

A methylene chloride solution containing 50 ng/µL of decafluorotriphenylphosphine (DFTPP) is used for tuning the GC/MS system each 12-h shift. A system performance check also must be made during every 12-h shift. A standard containing 50 ng/µL each of 4,4′-DDT, pentachlorophenol, and benzidine is required to verify injection port inertness and GC column performance. A calibration standard at mid-concentration, containing each compound of interest, including all required surrogates, must be performed every 12 h during analysis. After the system performance check is met, calibration check compounds (CCCs) are used to check the validity of the initial calibration.

The internal standard responses and retention times in the calibration check standard must be evaluated immediately after or during data acquisition. If the retention time for any internal standard changes by more than 30 seconds from the last check calibration (12 h), the chromatographic system must be inspected for malfunctions and corrections must be made, as required. If the electron ionization current plot (EICP) area for any of the internal standards changes by a factor of two from the last daily calibration standard check, the mass spectrometer must be inspected for malfunctions and corrections must be made, as appropriate.

Demonstrate, through the analysis of a reagent water blank, that interferences from the analytical system, glassware, and reagents are under control. The blank samples should be carried through all stages of the sample preparation and measurement steps. For each analytical batch (up to 20 samples), a reagent blank, matrix spike, and matrix spike duplicate/duplicate must be analyzed (the frequency of the spikes may be different for different monitoring programs). The blank and spiked samples must be carried through all stages of the sample preparation and measurement steps. A QC reference sample concentrate containing each analyte at a concentration of 100 mg/L in methanol is required.

REFERENCE Test Methods for Evaluating Solid Waste (SW-846). U.S. EPA 1983, Method 8270B, Rev. 2, Nov. 1990. Office of Solid Waste, Washington, DC.

Dibenz(a,j)acridine   EPA Method 8270
CAS #224-42-0

TITLE Semivolatile Organic Compounds by GC/MS

MATRIX This method is used to determine the concentration of semivolatile organic compounds in extracts prepared from all types of solid waste matrices, soils, and groundwater. Although surface waters are not specifically mentioned, this method should be applicable to water samples from rivers, lakes, etc.

METHOD SUMMARY This method covers 259 semivolatile organic compounds. In very limited applications direct injection of the sample into the GC/MS system may be appropriate, but this results in very high detection limits (approximately 10,000 µg/L). Typically, a 1-L liquid sample, containing surrogate, and matrix spiking standards, is extracted in a continuous extractor first under acid conditions and then under basic conditions. Typically 30 g of a solid sample, containing surrogate, and matrix spiking standards, is extracted ultrasonically. After concentrating the extract to 1 mL it is spiked with 10 µL of an internal standard solution just prior to analysis by GC/MS. The volume injected should contain about 100 ng of base/neutral and 200 ng of acid surrogates (for a 1-µL injection). Analysis is performed by GC/MS using a capillary GC column.

INTERFERENCES Raw GC/MS data from all blanks, samples, and spikes must be evaluated for interferences. Contamination by carryover can occur whenever high-concentration and low-concentration samples are sequentially analyzed. To reduce carryover, the sample syringe must be rinsed out between samples with solvent. Whenever an unusually concentrated sample is encountered, it should be followed by the analysis of blank solvent to check for cross-contamination.

INSTRUMENTATION A GC/MS and a data system are required. The GC column used is a 30 m × 0.25 mm I.D. (or 0.32 mm I.D.) 1um film thickness silicone-coated fused silica

capillary column. A continuous liquid-liquid extractor equipped with Teflon® or glass connection joints and stopcocks requiring no lubrication, a K-D concentrating apparatus, water bath, and an ultrasonic disrupter with a minimum power of 300 W and with pulsing capability are also required.

PRECISION & ACCURACY The estimated quantitation limit (EQL) of Method 8270B for determining an individual compound is approximately 1 mg/kg (wet weight) for soil or sediment samples, 1–200 mg/kg for wastes (dependent on matrix and method of preparation), and 10 µg/L for groundwater samples. EQLs will be proportionately higher for sample extracts that require dilution to avoid saturation of the detector.

The EQL(b) for groundwater in µg/L is 10.
The EQL (a, b) for low concentrations in soil and sediment in µg/kg is not determined.
Accuracy as µg/L is not listed.
Overall precision in µg/L is not listed.

(a) *EQLs listed for soil/sediment are based on wet weight. Normally data is reported in a dry-weight basis; therefore, EQLs will be higher based on the % dry weight of each sample. This calculation is based on a 30 g sample and gel permeation chromatography cleanup.*
(b) *Sample EQLs are highly matrix-dependent. The EQLs are provided for guidance and may not always be achievable.*
C = *True value for concentration, in µg/L.*
X = *Average recovery found for measurements of samples containing a concentration of C, in µg/L.*

ESTIMATED QUANTITATION LIMIT

Other Matrices	Factor (a)
High-concentration soil and sludges by sonicator	7.5
Non-water miscible waste	75

(a) *EQL for other matrices = [EQL for low soil/sediment] × [Factor]. This estimated EQL is similar to an EPA "Practical Quantitation Limit."*

SAMPLING METHOD
Liquid samples — Use a 1 or 2½ gallon amber glass bottle with a screw-top Teflon®-lined cover that has been prewashed with detergent and rinsed with distilled water and methanol (or isopropanol).

Soils, sediments, or sludges — Use an 8-oz. widemouth glass with a screw-top Teflon®-lined cover that has been prewashed with detergent and rinsed with distilled water and methanol (or isopropanol).

SAMPLE PRESERVATION
Liquid samples — If residual chlorine is present, add 3 mL of 10% sodium thiosulfate per gallon, cool to 4°C and store in a solvent-free refrigerator until analysis; if chlorine is not present, then eliminate the sodium thiosulfate addition.

Soils, sediments, or sludges — Cool samples to 4°C and store in a solvent-free refrigerator.

MHT Liquid samples must be extracted within 7 days and the extracts analyzed within 40 days. Soils, sediments, or sludges may be stored for a maximum of 14 days and the extracts analyzed within 40 days.

SAMPLE PREPARATION
Liquid samples — Transfer 1 L quantitatively to a continuous extractor. If high concentrations are anticipated, a smaller volume may be used and then diluted with organic-free reagent water to 1 L. Adjust pH, if necessary, to pH <2 using 1:1 (V/V) sulfuric acid. Pipette 1.0 mL of a surrogate standard spiking solution into each sample. For the sample in each analytical batch selected for spiking, add 1.0 mL of a matrix spiking standard. For base/neutral acid analysis, the amount of the surrogates and matrix spiking compounds added to the sample should result in a final concentration of 100 ng/µL of each analyte in the extract to be analyzed (assuming a 1-µL injection). Extract with methylene chloride for 18–24 h. Next, adjust the pH of the aqueous phase to pH >11 using 10 N sodium hydroxide and extract it with methylene chloride again for 18–24 h. Dry the extract through a column containing anhydrous sodium sulfate and concentrate it to 1 mL using a K-D concentrator.

Soils, sediments, or sludges — Use 30 g of sample. Nonporous or wet samples (gummy or clay type) that do not have a free-flowing sandy texture must be mixed with anhydrous sodium sulfate until the sample is free flowing. Add 1 mL of surrogate standards to all samples, spikes, standards, and blanks. For the sample in each analytical batch selected for spiking, add 1.0 mL of a matrix spiking standard. For base/neutral acid analysis, the amount added of the surrogates and matrix spiking compounds should result in a final concentration of 100 ng/µL of each base/neutral analyte and 200 ng/µL of each acid analyte in the extract to be analyzed (assuming a 1-µL injection). Immediately add a 100-mL mixture of 1:1 methylene chloride:acetone and extract the sample ultrasonically for 3 min and then decant or filter the extracts. Repeat the extraction two or more times. Dry the extract using a column with anhydrous sodium sulfate and concentrate it to 1 mL in a K-D concentrator.

QUALITY CONTROL A methylene chloride solution containing 50 ng/µL of decafluorotriphenylphosphine (DFTPP) is used for tuning the GC/MS system each 12-h shift. A system performance check also must be made during every 12-h shift. A standard containing 50 ng/µL each of 4,4'-DDT, pentachlorophenol, and benzidine is required to verify injection port inertness and GC column performance. A calibration standard at mid-concentration, containing each compound of interest, including all required surrogates, must be performed every 12 h during analysis. After the system performance check is met, calibration check compounds (CCCs) are used to check the validity of the initial calibration.

The internal standard responses and retention times in the calibration check standard must be evaluated immediately after or during data acquisition. If the retention time for any internal standard changes by more than 30 seconds from the last check calibration (12 h), the chromatographic system must be inspected for malfunctions and corrections must be made, as required. If the electron ionization current plot (EICP) area for any of the internal standards changes by a factor of two from the last daily calibration standard check, the mass spectrometer

must be inspected for malfunctions and corrections must be made, as appropriate.

Demonstrate, through the analysis of a reagent water blank, that interferences from the analytical system, glassware, and reagents are under control. The blank samples should be carried through all stages of the sample preparation and measurement steps. For each analytical batch (up to 20 samples), a reagent blank, matrix spike, and matrix spike duplicate/duplicate must be analyzed (the frequency of the spikes may be different for different monitoring programs). The blank and spiked samples must be carried through all stages of the sample preparation and measurement steps. A QC reference sample concentrate containing each analyte at a concentration of 100 mg/L in methanol is required.

REFERENCE Test Methods for Evaluating Solid Waste (SW-846). U.S. EPA 1983, Method 8270B, Rev. 2, Nov. 1990. Office of Solid Waste, Washington, DC.

Dibenzo(a,e)pyrene   **EPA Method 8270**
CAS #192-65-4

TITLE Semivolatile Organic Compounds by GC/MS

MATRIX This method is used to determine the concentration of semivolatile organic compounds in extracts prepared from all types of solid waste matrices, soils, and groundwater. Although surface waters are not specifically mentioned, this method should be applicable to water samples from rivers, lakes, etc.

METHOD SUMMARY This method covers 259 semivolatile organic compounds. In very limited applications direct injection of the sample into the GC/MS system may be appropriate, but this results in very high detection limits (approximately 10,000 µg/L). Typically, a 1-L liquid sample, containing surrogate, and matrix spiking standards, is extracted in a continuous extractor first under acid conditions and then under basic conditions. Typically 30 g of a solid sample, containing surrogate, and matrix spiking standards, is extracted ultrasonically. After concentrating the extract to 1 mL it is spiked with 10 µL of an internal standard solution just prior to analysis by GC/MS. The volume injected should contain about 100 ng of base/neutral and 200 ng of acid surrogates (for a 1-µL injection). Analysis is performed by GC/MS using a capillary GC column.

INTERFERENCES Raw GC/MS data from all blanks, samples, and spikes must be evaluated for interferences. Contamination by carryover can occur whenever high-concentration and low-concentration samples are sequentially analyzed. To reduce carryover, the sample syringe must be rinsed out between samples with solvent. Whenever an unusually concentrated sample is encountered, it should be followed by the analysis of blank solvent to check for cross-contamination.

INSTRUMENTATION A GC/MS and a data system are required. The GC column used is a 30 m × 0.25 mm I.D. (or 0.32 mm I.D.) 1um film thickness silicone-coated fused silica capillary column. A continuous liquid-liquid extractor equipped with Teflon® or glass connection joints and stopcocks requiring no lubrication, a K-D concentrating apparatus, water bath, and an ultrasonic disrupter with a minimum power of 300 W and with pulsing capability are also required.

PRECISION & ACCURACY The estimated quantitation limit (EQL) of Method 8270B for determining an individual compound is approximately 1 mg/kg (wet weight) for soil or sediment samples, 1–200 mg/kg for wastes (dependent on matrix and method of preparation), and 10 µg/L for groundwater samples. EQLs will be proportionately higher for sample extracts that require dilution to avoid saturation of the detector.

The EQL(b) for groundwater in µg/L is 10.
The EQL (a, b) for low concentrations in soil and sediment in µg/kg is not determined.
Accuracy as µg/L is not listed.
Overall precision in µg/L is not listed.

(a) EQLs listed for soil/sediment are based on wet weight. Normally data is reported in a dry-weight basis; therefore, EQLs will be higher based on the % dry weight of each sample. This calculation is based on a 30 g sample and gel permeation chromatography cleanup.

(b) Sample EQLs are highly matrix-dependent. The EQLs are provided for guidance and may not always be achievable.

C = True value for concentration, in µg/L.
X = Average recovery found for measurements of samples containing a concentration of C, in µg/L.

ESTIMATED QUANTITATION LIMIT

Other Matrices	Factor (a)
High-concentration soil and sludges by sonicator	7.5
Non-water miscible waste	75

(a) EQL for other matrices = [EQL for low soil/sediment] × [Factor]. This estimated EQL is similar to an EPA "Practical Quantitation Limit."

SAMPLING METHOD
Liquid samples — Use a 1 or 2½ gallon amber glass bottle with a screw-top Teflon®-lined cover that has been prewashed with detergent and rinsed with distilled water and methanol (or isopropanol).

Soils, sediments, or sludges — Use an 8-oz. widemouth glass with a screw-top Teflon®-lined cover that has been prewashed with detergent and rinsed with distilled water and methanol (or isopropanol).

SAMPLE PRESERVATION
Liquid samples — If residual chlorine is present, add 3 mL of 10% sodium thiosulfate per gallon, cool to 4°C and store in a solvent-free refrigerator until analysis; if chlorine is not present, then eliminate the sodium thiosulfate addition.

Soils, sediments, or sludges — Cool samples to 4°C and store in a solvent-free refrigerator.

MHT Liquid samples must be extracted within 7 days and the extracts analyzed within 40 days. Soils, sediments, or sludges may be stored for a maximum of 14 days and the extracts analyzed within 40 days.

SAMPLE PREPARATION

Liquid samples — Transfer 1 L quantitatively to a continuous extractor. If high concentrations are anticipated, a smaller volume may be used and then diluted with organic-free reagent water to 1 L. Adjust pH, if necessary, to pH <2 using 1:1 (V/V) sulfuric acid. Pipette 1.0 mL of a surrogate standard spiking solution into each sample. For the sample in each analytical batch selected for spiking, add 1.0 mL of a matrix spiking standard. For base/neutral acid analysis, the amount of the surrogates and matrix spiking compounds added to the sample should result in a final concentration of 100 ng/µL of each analyte in the extract to be analyzed (assuming a 1-µL injection). Extract with methylene chloride for 18–24 h. Next, adjust the pH of the aqueous phase to pH >11 using 10 N sodium hydroxide and extract it with methylene chloride again for 18–24 h. Dry the extract through a column containing anhydrous sodium sulfate and concentrate it to 1 mL using a K-D concentrator.

Soils, sediments, or sludges — Use 30 g of sample. Nonporous or wet samples (gummy or clay type) that do not have a free-flowing sandy texture must be mixed with anhydrous sodium sulfate until the sample is free flowing. Add 1 mL of surrogate standards to all samples, spikes, standards, and blanks. For the sample in each analytical batch selected for spiking, add 1.0 mL of a matrix spiking standard. For base/neutral acid analysis, the amount added of the surrogates and matrix spiking compounds should result in a final concentration of 100 ng/µL of each base/neutral analyte and 200 ng/µL of each acid analyte in the extract to be analyzed (assuming a 1-µL injection). Immediately add a 100-mL mixture of 1:1 methylene chloride:acetone and extract the sample ultrasonically for 3 min and then decant or filter the extracts. Repeat the extraction two or more times. Dry the extract using a column with anhydrous sodium sulfate and concentrate it to 1 mL in a K-D concentrator.

QUALITY CONTROL
A methylene chloride solution containing 50 ng/µL of decafluorotriphenylphosphine (DFTPP) is used for tuning the GC/MS system each 12-h shift. A system performance check also must be made during every 12-h shift. A standard containing 50 ng/µL each of 4,4′-DDT, pentachlorophenol, and benzidine is required to verify injection port inertness and GC column performance. A calibration standard at mid-concentration, containing each compound of interest, including all required surrogates, must be performed every 12 h during analysis. After the system performance check is met, calibration check compounds (CCCs) are used to check the validity of the initial calibration.

The internal standard responses and retention times in the calibration check standard must be evaluated immediately after or during data acquisition. If the retention time for any internal standard changes by more than 30 seconds from the last check calibration (12 h), the chromatographic system must be inspected for malfunctions and corrections must be made, as required. If the electron ionization current plot (EICP) area for any of the internal standards changes by a factor of two from the last daily calibration standard check, the mass spectrometer must be inspected for malfunctions and corrections must be made, as appropriate.

Demonstrate, through the analysis of a reagent water blank, that interferences from the analytical system, glassware, and reagents are under control. The blank samples should be carried through all stages of the sample preparation and measurement steps. For each analytical batch (up to 20 samples), a reagent blank, matrix spike, and matrix spike duplicate/duplicate must be analyzed (the frequency of the spikes may be different for different monitoring programs). The blank and spiked samples must be carried through all stages of the sample preparation and measurement steps. A QC reference sample concentrate containing each analyte at a concentration of 100 mg/L in methanol is required.

REFERENCE Test Methods for Evaluating Solid Waste (SW-846). U.S. EPA 1983, Method 8270B, Rev. 2, Nov. 1990. Office of Solid Waste, Washington, DC.

Dibenz(a,h)anthracene EPA Method 625
CAS #53-70-3

TITLE Base/Neutrals and Acids, U.S. EPA Method 625

MATRIX This methods covers municipal and industrial wastewaters.

METHOD SUMMARY Approximately 1 L of sample is serially extracted with methylene chloride at a pH greater than 11 and again at a pH less than 2 using a separatory funnel or a continuous extractor. The methylene chloride extract is dried, concentrated to a volume of 1 mL, and analyzed by GC/MS. Qualitative identification of the parameters in the extract is performed using the retention time and the relative abundance of three characteristic masses (m/z). Qualitative analysis is performed using either external or internal standard techniques with a single characteristic m/z.

INTERFERENCES Method interferences may be caused by contaminants in solvents, reagents, glassware, and other sample processing hardware. Glassware must be scrupulously cleaned. Glassware should be heated in a muffle furnace at 400°C for 5 to 30 min. Some thermally stable materials, such as PCBs, may not be eliminated by this treatment. Solvent rinses with acetone and pesticide quality hexane may be substituted for the muffle furnace heating. Matrix interferences may be caused by contaminants that are coextracted from the sample. The base-neutral extraction may cause significantly reduced recovery of phenols. The packed gas chromatographic columns recommended for the basic fraction may not exhibit sufficient resolution for some analytes.

INSTRUMENTATION A GC/MS system with an injection port designed for on-column injection when using packed columns and for splitless injection when using capillary columns.

Column for base/neutrals: 1.8 m long × 2 mm I.D. glass, packed with 3% SP-2550 on Supelcoport (100/120 mesh) or equivalent.

Column for acids: 1.8 m long × 2 mm I.D. glass, packed with 1% SP-1240DA on Supelcoport (100/120 mesh) or equivalent.

PRECISION & ACCURACY The MDL concentrations were obtained using reagent water. The MDL actually achieved in a given analysis will vary depending on instrument sensitivity and matrix effects. This method was tested by 15 laboratories using reagent water, drinking water, surface water, and industrial wastewaters spiked at six concentrations over the range 5 to 100 µg/L. Single operator precision, overall precision, and method accuracy were found to be directly related to the concentration of the parameter matrix.

The MDL (in µg/L) in reagent water was 2.5.

The standard deviation (in µg/L based on 4 recovery measurements) was 70.0.

The range (in µg/L) for average recovery for 4 measurements was D-199.7.

The range (in %) for percent recovery was D-227.

Accuracy (in µg/L) as expected recovery for one or more measurements of a sample containing a true concentration of C was $0.88C + 4.72$.

Precision (in µg/L) as expected single analyst standard deviation of measurements at an average concentration found at X was $0.30X + 8.51$.

Overall precision (in µg/L) as expected interlaboratory standard deviation of measurements in an average concentration found at X was $0.59X + 0.25$.

C = *True value of the concentration in µg/L.*
X = *Average recovery found for measurements of samples containing a concentration at C in µg/L.*

SAMPLE PREPARATION Adjust the pH to >11 with sodium hydroxide and serially extract in a separatory funnel with methylene chloride or else in a continuous extractor. Next, adjust the pH to <2 with sulfuric acid and serially extract in a separatory funnel with methylene chloride or else in a continuous extractor. Dry the extracts separately through a column of anhydrous sodium sulfate and then concentrate each of the extracts to 1.0 mL using a K-D apparatus.

SAMPLE COLLECTION, PRESERVATION & HANDLING Grab samples must be collected in glass containers. All samples must be refrigerated at 4°C from the time of collection until extraction. If residual chlorine is present, add 80 mg of sodium thiosulfate/L of sample and mix well. All samples must be extracted within 7 days of collection and completely analyzed within 40 days of extraction.

QUALITY CONTROL Make an initial, one-time, demonstration of the ability to generate acceptable accuracy and precision with this method. Before processing any samples, the analyst must analyze a reagent water blank to demonstrate that interferences from the analytical system and glassware are under control. Each time a set of samples is extracted or reagents are changed, a reagent water blank must be processed. Spike and analyze a minimum of 5% of all samples to monitor and evaluate lab data quality. A QC check sample concentrate that contains each parameter of interest at a concentration of 100 µg/mL in acetone is required. PCBs and multicomponent pesticides may be omitted from this test.

After analysis of five spiked wastewater samples, calculate the average percent recovery and the standard deviation of the percent recovery. Spike all samples with the surrogate standard spiking solution and calculate the percent recovery of each surrogate compound.

REFERENCE *Federal Register*, Vol. 49, No. 209. Friday, Oct. 26, 1984.

Dibenz(a,h)anthracene   EPA Method 8100
CAS #53-70-3

TITLE Polynuclear Aromatic Hydrocarbons

MATRIX Groundwater, soils, sludges, water miscible liquid wastes, and non-water miscible wastes.

APPLICATION This method is used for the analysis of various PAHs. Samples are extracted, concentrated, and analyzed using direct injection of both neat and diluted organic liquids. The method provides two optional GC columns that are better than Column 1 and that may help resolve analytes from interferences.

INTERFERENCES Solvents, reagents, and glassware may introduce artifacts. Other interferences may come from coextracted compounds from samples.

INSTRUMENTATION GC capable of on-column injections and a flame with detector (FID). Column 1: a 1.8 m by 2 mm 3% OV-17 on Chromosorb W-AW-DCMS column. Column 2: a 30 m by 0.25 mm SE-54 fused silica capillary colunm. Column 3: a 30 m by 0.32 mm SE-54 fused silica capillary column.

RANGE 0.1–425 µg/L.

MDL Not reported.

PQL FACTORS FOR MULTIPLYING × FID MDL VALUE Not available.

PRECISION $0.45X + 0.03$ µg/L (overall precision).

ACCURACY $0.41C - 0.11$ µg/L (as recovery).

SAMPLING METHOD Use 8-oz. widemouth glass bottles with Teflon®-lined caps for concentrated waste samples, soils, sediments, and sludges. Use 1 or 2½ gallon amber glass bottles with Teflon®-lined caps for liquid (water) samples.

STABILITY Cool soil, sediment, sludge, and liquid samples to 4°C. If residual chlorine is present in liquid samples add 3 mL of 10% sodium thiosulfate per gallon of sample and cool to 4°C.

MHT 14 days for concentrated waste, soil, sediment, or sludge; 7 days for liquid samples; all extracts must be analyzed within 40 days.

QUALITY CONTROL A quality control check sample concentrate containing each analyte of interest is required. The QC check sample concentrate may be prepared from pure standard materials or purchased as certified solutions Use appropriate trip, matrix, control site, method, reagent, and solvent blanks. Internal, surrogate, and five concentration level calibration standards are used. The quality control check sample concentrate should contain Dibenz(a,h)anthracene at 10 µg/mL in acetonitrile.

REFERENCE Test Methods for Evaluating Solid Waste (SW-846), U.S. EPA Office of Solid Waste, Washington, DC, Method 8100, Nov. 1986.

Dibenz(a,h)anthracene   **EPA Method 8310**
CAS #53-70-3

TITLE Polynuclear Aromatic Hydrocarbons

MATRIX Groundwater, soils, sludges, water miscible liquid wastes, and non-water miscible wastes.

APPLICATION This method is used for the analysis of 16 polynuclear aromatic hydrocarbons (PAHs). Samples are extracted, concentrated, and analyzed using HPLC with detection by UV and fluorescence detectors.

INTERFERENCES Solvents, reagents, and glassware may introduce artifacts. Other interferences may come from coextracted compounds from samples.

INSTRUMENTATION HPLC with a gradient pumping system and a 250 mm by 2.6 mm reverse phase HC-ODS Sil-X 5-micron particle-size column. The fluorescence detector uses an excitation wavelength of 280 nm and emission greater than 389 nm cutoff with dispersive optics.

RANGE 0.1–425 µg/L.

MDL 0.030 µg/L (fluorescence; reagent water).

PQL FACTORS FOR MULTIPLYING × FID MDL VALUE

Matrix	Multiplication Factor
Groundwater	10
Low-level soil by sonication with GPC cleanup	670
High-level soil and sludge by sonication	10,000
Non-water miscible waste	100,000

PRECISION 0.45X + 0.03 µg/L (overall precision).

ACCURACY 0.41C–0.11 µg/L (as recovery).

SAMPLING METHOD Use 8-oz. widemouth glass bottles with Teflon®-lined caps for concentrated waste samples, soils, sediments, and sludges. Use 1 or 2½ gallon amber glass bottles with Teflon®-lined caps for liquid (water) samples.

STABILITY Cool soil, sediment, sludge, and liquid samples to 4°C. If residual chlorine is present in liquid samples add 3 mL of 10% sodium thiosulfate per gallon of sample and cool to 4°C.

MHT 14 days for concentrated waste, soil, sediment, or sludge; 7 days for liquid samples; all extracts must be analyzed within 40 days.

QUALITY CONTROL Internal, surrogate, and five concentration level calibration standards are used. The calibration standards must be used with the analytical method blank. A quality control check sample concentrate containing dibenz(a,h)anthracene at 10 µg/mL is required. The QC check sample concentrate may be prepared from pure standard materials or purchased as certified solutions. Use appropriate trip, matrix, control site, method, reagent, and solvent blanks.

REFERENCE Test Methods for Evaluating Solid Waste (SW-846), U.S. EPA Office of Solid Waste, Washington, DC, Method 8310, Rev. 0, Nov. 1986.

Dibenzofuran   **EPA Method 1625**
CAS #132-64-9

TITLE Semivolatile Organic Compounds by Isotope Dilution GC/MS

MATRIX The compounds may be determined in waters, soils, and municipal sludges by this method.

METHOD SUMMARY This method is used to determine 176 semivolatile toxic organic pollutants associated with the CWA (as amended 1987); the RCRA (as amended 1986); the CERCLA (as amended 1986); and other compounds amenable to extraction and analysis by capillary column gas chromatography-mass spectrometry (GC/MS).

Stable isotopically-labeled analogs of the compounds of interest are added to the sample. If the solids content is less than 1%, a 1-L sample is extracted at pH 12–13, then at pH <2 with methylene chloride using continuous extraction techniques.

If the solids content is 30% or less, the sample is diluted to 1% solids with reagent water, homogenized ultrasonically, and extracted at pH 12–13, then at pH <2 with methylene chloride using continuous extraction techniques. If the solids content is greater than 30%, the sample is extracted using ultrasonic techniques.

Each extract is dried over sodium sulfate, concentrated to a volume of 5 mL, cleaned up using GPC, if necessary, and concentrated. Extracts are concentrated to 1 mL if GPC is not performed, and to 0.5 mL if GPC is performed.

An internal standard is added to the extract, and a 1-mL aliquot of the extract is injected into the GC. The compounds are separated by GC and detected by a MS. The labeled compounds serve to correct the variability of the analytical technique.

INTERFERENCES Solvents, reagents, glassware, and other sample processing hardware may yield artifacts and/or elevated baselines causing misinterpretation of chromatograms and spectra. Materials used in the analysis must be demonstrated to be free from interferences under the conditions of analysis by running method blanks initially and with each sample lot (sample started through the extraction process on a given 8-h shift, to a maximum of 20). Specific selection of reagents and purification of solvents by distillation in all glass systems may be required. Glassware and, where possible, reagents are cleaned by solvent rinse and baking at 450°C for 1-h minimum. Interferences coextracted from samples will vary considerably from source to source, depending on the diversity of the site being sampled.

INSTRUMENTATION Major instrumentation includes a GC with a splitless or on-column injection port for capillary column, a MS with 70 eV electron impact ionization, and a data system to collect and record MS data, and process it. A K-D apparatus is used to concentrate extracts.

GC Column: 30 m × 0.25 mm I.D. 5% phenyl, 94% methyl, 1% vinyl silicone bonded phased fused silica capillary column.

PRECISION & ACCURACY The detection limits of the method are usually dependent on the level of interferences rather than instrumental limitations. The limits typify the minimum quantities that can be detected with no interferences present.

The minimum level (in μg/mL) was 10. This is defined as a minimum level at which the analytical system shall give recognizable mass spectra (background corrected) and acceptable calibration points.

The MDL (in μg/kg) in low solids was 77 and in high solids was 210; these were determined in digested sludge (low solids) and in filter cake or compost (high solids).

Note: Background levels of this compound were present in the sludge tested, resulting in higher than expected MDLs. The MDL for this compound is expected to be approximately 50 μg/kg with no interferences present.

The labeled and native compound initial precision as standard deviation (in μg/L) was 20.

The labeled and native compound initial accuracy as average recovery (in μg/L) was 85–136.

SAMPLE COLLECTION, PRESERVATION & HANDLING Collect samples in glass containers. Aqueous samples which flow freely are collected in refrigerated bottles using automatic sampling equipment. Solid samples are collected as grab samples using widemouth jars. Maintain samples at 0 to 4°C from the time of collection until extraction. If residual chlorine is present in aqueous samples, add 80 mg sodium thiosulfate/L of water. Begin sample extraction within 7 days of collection, and analyze all extracts within 40 days of extraction.

SAMPLE PREPARATION Samples containing 1% solids or less are extracted directly using continuous liquid-liquid extraction techniques. Samples containing 1 to 30% solids are diluted to the 1% level with reagent water and extracted using continuous liquid-liquid extraction techniques. Samples containing greater than 30% solids are extracted using ultrasonic techniques.

Base/neutral extraction — Adjust the pH of the waters in the extractors to 12–13 with 6 N NaOH. Extract with methylene chloride for 24–48 h.

Acid extraction — Adjust the pH of the waters in the extractors to 2 or less using 6 N sulfuric acid. Extract with methylene chloride for 24–48 h.

Ultrasonic extraction of high solids samples — Add anhydrous sodium sulfate to the sample and QC aliquot(s). Add acetone:methylene chloride (1:1) to the sample and mix thoroughly

Concentrate extracts using a K-D apparatus.

QUALITY CONTROL The analyst is permitted to modify this method to improve separations or lower the costs of measurements, provided all performance specifications are met. Analyses of blanks are required to demonstrate freedom from contamination. When results of spikes indicate atypical method performance for samples, the samples are diluted to bring method performance within acceptable limits.

For low solids (aqueous samples), extract, concentrate, and analyze two sets of four 1-L aliquots (8 aliquots total) of the precision and recovery standard. For high solids samples, two sets of four 30-g aliquots of the high solids reference matrix are used.

Spike all samples with labeled compounds to assess method performance. Compute percent recovery of the labeled compounds using the internal standard method. Compare the labeled compound recovery for each compound with the corresponding labeled compound recovery.

Reagent water and high solids reference matrix blanks are analyzed to demonstrate freedom from contamination. Extract and concentrate a 1-L reagent water blank or a high solids reference matrix blank with each sample's lot (samples started through the extraction process on the same 8-h shift, to a maximum of 20 samples).

Field replicates may be collected to determine the precision of the sampling technique, and spiked samples may be required to determine the accuracy of the analysis when the internal standard method is used.

REFERENCE Semivolatile Organic Compounds by Isotope Dilution GC/MS. Office of Water Regulation and Standards, U.S. EPA Industrial Technology Division, Washington, DC, EPA Method 1625, Rev. C, June 1989 (contact W.A. Telliard, U.S. EPA, Office of Water Regulations and Standards, 401 M St., SW, Washington, DC, 20460. Phone: 202-382-7131).

Dibenzofuran   **EPA Method 8270**
CAS #132-64-9

TITLE Semivolatile Organic Compounds by GC/MS

MATRIX This method is used to determine the concentration of semivolatile organic compounds in extracts prepared from all types of solid waste matrices, soils, and groundwater. Although surface waters are not specifically mentioned, this method should be applicable to water samples from rivers, lakes, etc.

METHOD SUMMARY This method covers 259 semivolatile organic compounds. In very limited applications direct injection of the sample into the GC/MS system may be appropriate, but this results in very high detection limits (approximately 10,000 μg/L). Typically, a 1-L liquid sample, containing surrogate, and matrix spiking standards, is extracted in a continuous extractor first under acid conditions and then under basic conditions. Typically 30 g of a solid sample, containing surrogate, and matrix spiking standards, is extracted ultrasonically. After concentrating the extract to 1 mL it is spiked with 10 μL of an

internal standard solution just prior to analysis by GC/MS. The volume injected should contain about 100 ng of base/neutral and 200 ng of acid surrogates (for a 1-μL injection). Analysis is performed by GC/MS using a capillary GC column.

INTERFERENCES Raw GC/MS data from all blanks, samples, and spikes must be evaluated for interferences. Contamination by carryover can occur whenever high-concentration and low-concentration samples are sequentially analyzed. To reduce carryover, the sample syringe must be rinsed out between samples with solvent. Whenever an unusually concentrated sample is encountered, it should be followed by the analysis of blank solvent to check for cross-contamination.

INSTRUMENTATION A GC/MS and a data system are required. The GC column used is a 30 m × 0.25 mm I.D. (or 0.32 mm I.D.) 1um film thickness silicone-coated fused silica capillary column. A continuous liquid-liquid extractor equipped with Teflon® or glass connection joints and stopcocks requiring no lubrication, a K-D concentrating apparatus, water bath, and an ultrasonic disrupter with a minimum power of 300 W and with pulsing capability are also required.

PRECISION & ACCURACY The estimated quantitation limit (EQL) of Method 8270B for determining an individual compound is approximately 1 mg/kg (wet weight) for soil or sediment samples, 1–200 mg/kg for wastes (dependent on matrix and method of preparation), and 10 μg/L for groundwater samples. EQLs will be proportionately higher for sample extracts that require dilution to avoid saturation of the detector.

The EQL(b) for groundwater in μg/L is 10.
The EQL (a, b) for low concentrations in soil and sediment in μg/kg is 660.
Accuracy as μg/L is not listed.
Overall precision in μg/L is not listed.

(a) *EQLs listed for soil/sediment are based on wet weight. Normally data is reported in a dry-weight basis; therefore, EQLs will be higher based on the % dry weight of each sample. This calculation is based on a 30 g sample and gel permeation chromatography cleanup.*
(b) *Sample EQLs are highly matrix-dependent. The EQLs are provided for guidance and may not always be achievable.*
$C =$ *True value for concentration, in μg/L.*
$X =$ *Average recovery found for measurements of samples containing a concentration of C, in μg/L.*

ESTIMATED QUANTITATION LIMIT

Other Matrices	Factor (a)
High-concentration soil and sludges by sonicator	7.5
Non-water miscible waste	75

(a) EQL for other matrices = [EQL for low soil/sediment] × [Factor]. This estimated EQL is similar to an EPA "Practical Quantitation Limit."

SAMPLING METHOD

Liquid samples — Use a 1 or 2½ gallon amber glass bottle with a screw-top Teflon®-lined cover that has been prewashed with detergent and rinsed with distilled water and methanol (or isopropanol).

Soils, sediments, or sludges — Use an 8-oz. widemouth glass with a screw-top Teflon®-lined cover that has been prewashed with detergent and rinsed with distilled water and methanol (or isopropanol).

SAMPLE PRESERVATION

Liquid samples — If residual chlorine is present, add 3 mL of 10% sodium thiosulfate per gallon, cool to 4°C and store in a solvent-free refrigerator until analysis; if chlorine is not present, then eliminate the sodium thiosulfate addition.

Soils, sediments, or sludges — Cool samples to 4°C and store in a solvent-free refrigerator.

MHT Liquid samples must be extracted within 7 days and the extracts analyzed within 40 days. Soils, sediments, or sludges may be stored for a maximum of 14 days and the extracts analyzed within 40 days.

SAMPLE PREPARATION

Liquid samples — Transfer 1 L quantitatively to a continuous extractor. If high concentrations are anticipated, a smaller volume may be used and then diluted with organic-free reagent water to 1 L. Adjust pH, if necessary, to pH <2 using 1:1 (V/V) sulfuric acid. Pipette 1.0 mL of a surrogate standard spiking solution into each sample. For the sample in each analytical batch selected for spiking, add 1.0 mL of a matrix spiking standard. For base/neutral acid analysis, the amount of the surrogates and matrix spiking compounds added to the sample should result in a final concentration of 100 ng/μL of each analyte in the extract to be analyzed (assuming a 1-μL injection). Extract with methylene chloride for 18–24 h. Next, adjust the pH of the aqueous phase to pH >11 using 10 N sodium hydroxide and extract it with methylene chloride again for 18–24 h. Dry the extract through a column containing anhydrous sodium sulfate and concentrate it to 1 mL using a K-D concentrator.

Soils, sediments, or sludges — Use 30 g of sample. Nonporous or wet samples (gummy or clay type) that do not have a free-flowing sandy texture must be mixed with anhydrous sodium sulfate until the sample is free flowing. Add 1 mL of surrogate standards to all samples, spikes, standards, and blanks. For the sample in each analytical batch selected for spiking, add 1.0 mL of a matrix spiking standard. For base/neutral acid analysis, the amount added of the surrogates and matrix spiking compounds should result in a final concentration of 100 ng/μL of each base/neutral analyte and 200 ng/μL of each acid analyte in the extract to be analyzed (assuming a 1-μL injection). Immediately add a 100-mL mixture of 1:1 methylene chloride:acetone and extract the sample ultrasonically for 3 min and then decant or filter the extracts. Repeat the extraction two or more times. Dry the extract using a column with anhydrous sodium sulfate and concentrate it to 1 mL in a K-D concentrator.

QUALITY CONTROL A methylene chloride solution containing 50 ng/μL of decafluorotriphenylphosphine (DFTPP) is used for tuning the GC/MS system each 12-h shift. A system performance check also must be made during every 12-h shift. A standard containing 50 ng/μL each of 4,4'-DDT, pentachlorophenol, and benzidine is required to verify injection port inertness and GC column performance. A calibration standard

at mid-concentration, containing each compound of interest, including all required surrogates, must be performed every 12 h during analysis. After the system performance check is met, calibration check compounds (CCCs) are used to check the validity of the initial calibration.

The internal standard responses and retention times in the calibration check standard must be evaluated immediately after or during data acquisition. If the retention time for any internal standard changes by more than 30 seconds from the last check calibration (12 h), the chromatographic system must be inspected for malfunctions and corrections must be made, as required. If the electron ionization current plot (EICP) area for any of the internal standards changes by a factor of two from the last daily calibration standard check, the mass spectrometer must be inspected for malfunctions and corrections must be made, as appropriate.

Demonstrate, through the analysis of a reagent water blank, that interferences from the analytical system, glassware, and reagents are under control. The blank samples should be carried through all stages of the sample preparation and measurement steps. For each analytical batch (up to 20 samples), a reagent blank, matrix spike, and matrix spike duplicate/duplicate must be analyzed (the frequency of the spikes may be different for different monitoring programs). The blank and spiked samples must be carried through all stages of the sample preparation and measurement steps. A QC reference sample concentrate containing each analyte at a concentration of 100 mg/L in methanol is required.

REFERENCE Test Methods for Evaluating Solid Waste (SW-846). U.S. EPA 1983, Method 8270B, Rev. 2, Nov. 1990. Office of Solid Waste, Washington, DC.

Dibenzothiophene **EPA Method 1625**
CAS #132-65-0

TITLE Semivolatile Organic Compounds by Isotope Dilution GC/MS

MATRIX The compounds may be determined in waters, soils, and municipal sludges by this method.

METHOD SUMMARY This method is used to determine 176 semivolatile toxic organic pollutants associated with the CWA (as amended 1987); the RCRA (as amended 1986); the CERCLA (as amended 1986); and other compounds amenable to extraction and analysis by capillary column gas chromatography-mass spectrometry (GC/MS).

Stable isotopically-labeled analogs of the compounds of interest are added to the sample. If the solids content is less than 1%, a 1-L sample is extracted at pH 12–13, then at pH <2 with methylene chloride using continuous extraction techniques.

If the solids content is 30% or less, the sample is diluted to 1% solids with reagent water, homogenized ultrasonically, and extracted at pH 12–13, then at pH <2 with methylene chloride using continuous extraction techniques. If the solids content is greater than 30%, the sample is extracted using ultrasonic techniques.

Each extract is dried over sodium sulfate, concentrated to a volume of 5 mL, cleaned up using GPC, if necessary, and concentrated. Extracts are concentrated to 1 mL if GPC is not performed, and to 0.5 mL if GPC is performed.

An internal standard is added to the extract, and a 1-mL aliquot of the extract is injected into the GC. The compounds are separated by GC and detected by a MS. The labeled compounds serve to correct the variability of the analytical technique.

INTERFERENCES Solvents, reagents, glassware, and other sample processing hardware may yield artifacts and/or elevated baselines causing misinterpretation of chromatograms and spectra. Materials used in the analysis must be demonstrated to be free from interferences under the conditions of analysis by running method blanks initially and with each sample lot (sample started through the extraction process on a given 8-h shift, to a maximum of 20). Specific selection of reagents and purification of solvents by distillation in all glass systems may be required. Glassware and, where possible, reagents are cleaned by solvent rinse and baking at 450°C for 1-h minimum. Interferences coextracted from samples will vary considerably from source to source, depending on the diversity of the site being sampled.

INSTRUMENTATION Major instrumentation includes a GC with a splitless or on-column injection port for capillary column, a MS with 70 eV electron impact ionization, and a data system to collect and record MS data, and process it. A K-D apparatus is used to concentrate extracts.

GC Column: 30 m × 0.25 mm I.D. 5% phenyl, 94% methyl, 1% vinyl silicone bonded phased fused silica capillary column.

PRECISION & ACCURACY The detection limits of the method are usually dependent on the level of interferences rather than instrumental limitations. The limits typify the minimum quantities that can be detected with no interferences present.

The minimum level (in µg/mL) was 10. This is defined as a minimum level at which the analytical system shall give recognizable mass spectra (background corrected) and acceptable calibration points.

The MDL (in µg/kg) in low solids was 72 and in high solids was 71; these were determined in digested sludge (low solids) and in filter cake or compost (high solids).

The labeled and native compound initial precision as standard deviation (in µg/L) was 31.
The labeled and native compound initial accuracy as average recovery (in µg/L) was 79–150.

SAMPLE COLLECTION, PRESERVATION & HANDLING Collect samples in glass containers. Aqueous samples which flow freely are collected in refrigerated bottles using automatic sampling equipment. Solid samples are collected as grab samples using widemouth jars. Maintain samples at 0 to 4°C from the time of collection until extraction. If residual chlorine is present in aqueous samples, add 80 mg sodium thiosulfate/L

of water. Begin sample extraction within 7 days of collection, and analyze all extracts within 40 days of extraction.

SAMPLE PREPARATION Samples containing 1% solids or less are extracted directly using continuous liquid-liquid extraction techniques. Samples containing 1 to 30% solids are diluted to the 1% level with reagent water and extracted using continuous liquid-liquid extraction techniques. Samples containing greater than 30% solids are extracted using ultrasonic techniques.

Base/neutral extraction — Adjust the pH of the waters in the extractors to 12–13 with 6 N NaOH. Extract with methylene chloride for 24–48 h.

Acid extraction — Adjust the pH of the waters in the extractors to 2 or less using 6 N sulfuric acid. Extract with methylene chloride for 24–48 h.

Ultrasonic extraction of high solids samples — Add anhydrous sodium sulfate to the sample and QC aliquot(s). Add acetone:methylene chloride (1:1) to the sample and mix thoroughly

Concentrate extracts using a K-D apparatus.

QUALITY CONTROL The analyst is permitted to modify this method to improve separations or lower the costs of measurements, provided all performance specifications are met. Analyses of blanks are required to demonstrate freedom from contamination. When results of spikes indicate atypical method performance for samples, the samples are diluted to bring method performance within acceptable limits.

For low solids (aqueous samples), extract, concentrate, and analyze two sets of four 1-L aliquots (8 aliquots total) of the precision and recovery standard. For high solids samples, two sets of four 30-g aliquots of the high solids reference matrix are used.

Spike all samples with labeled compounds to assess method performance. Compute percent recovery of the labeled compounds using the internal standard method. Compare the labeled compound recovery for each compound with the corresponding labeled compound recovery.

Reagent water and high solids reference matrix blanks are analyzed to demonstrate freedom from contamination. Extract and concentrate a 1-L reagent water blank or a high solids reference matrix blank with each sample's lot (samples started through the extraction process on the same 8-h shift, to a maximum of 20 samples).

Field replicates may be collected to determine the precision of the sampling technique, and spiked samples may be required to determine the accuracy of the analysis when the internal standard method is used.

REFERENCE Semivolatile Organic Compounds by Isotope Dilution GC/MS. Office of Water Regulation and Standards, U.S. EPA Industrial Technology Division, Washington, DC, EPA Method 1625, Rev. C, June 1989 (contact W.A. Telliard, U.S. EPA, Office of Water Regulations and Standards, 401 M St., SW, Washington, DC, 20460. Phone: 202-382-7131).

1,2-Dibromo-3-chloropropane   **EPA Method 1625**
CAS #96-12-8

TITLE Semivolatile Organic Compounds by Isotope Dilution GC/MS

MATRIX The compounds may be determined in waters, soils, and municipal sludges by this method.

METHOD SUMMARY This method is used to determine 176 semivolatile toxic organic pollutants associated with the CWA (as amended 1987); the RCRA (as amended 1986); the CERCLA (as amended 1986); and other compounds amenable to extraction and analysis by capillary column gas chromatography-mass spectrometry (GC/MS).

Stable isotopically-labeled analogs of the compounds of interest are added to the sample. If the solids content is less than 1%, a 1-L sample is extracted at pH 12–13, then at pH <2 with methylene chloride using continuous extraction techniques.

If the solids content is 30% or less, the sample is diluted to 1% solids with reagent water, homogenized ultrasonically, and extracted at pH 12–13, then at pH <2 with methylene chloride using continuous extraction techniques. If the solids content is greater than 30%, the sample is extracted using ultrasonic techniques.

Each extract is dried over sodium sulfate, concentrated to a volume of 5 mL, cleaned up using GPC, if necessary, and concentrated. Extracts are concentrated to 1 mL if GPC is not performed, and to 0.5 mL if GPC is performed.

An internal standard is added to the extract, and a 1-mL aliquot of the extract is injected into the GC. The compounds are separated by GC and detected by a MS. The labeled compounds serve to correct the variability of the analytical technique.

INTERFERENCES Solvents, reagents, glassware, and other sample processing hardware may yield artifacts and/or elevated baselines causing misinterpretation of chromatograms and spectra. Materials used in the analysis must be demonstrated to be free from interferences under the conditions of analysis by running method blanks initially and with each sample lot (sample started through the extraction process on a given 8-h shift, to a maximum of 20). Specific selection of reagents and purification of solvents by distillation in all glass systems may be required. Glassware and, where possible, reagents are cleaned by solvent rinse and baking at 450°C for 1-h minimum. Interferences coextracted from samples will vary considerably from source to source, depending on the diversity of the site being sampled.

INSTRUMENTATION Major instrumentation includes a GC with a splitless or on-column injection port for capillary column, a MS with 70 eV electron impact ionization, and a data system to collect and record MS data, and process it. A K-D apparatus is used to concentrate extracts.

GC Column: 30 m × 0.25 mm I.D. 5% phenyl, 94% methyl, 1% vinyl silicone bonded phased fused silica capillary column.

PRECISION & ACCURACY The detection limits of the method are usually dependent on the level of interferences

rather than instrumental limitations. The limits typify the minimum quantities that can be detected with no interferences present.

The minimum level (in µg/mL) was not listed. This is defined as a minimum level at which the analytical system shall give recognizable mass spectra (background corrected) and acceptable calibration points.

The MDL (in µg/kg) in low solids was not listed and in high solids was not listed; these were determined in digested sludge (low solids) and in filter cake or compost (high solids).

The labeled and native compound initial precision as standard deviation (in µg/L) was not listed.
The labeled and native compound initial accuracy as average recovery (in µg/L) was not listed.

SAMPLE COLLECTION, PRESERVATION & HANDLING
Collect samples in glass containers. Aqueous samples which flow freely are collected in refrigerated bottles using automatic sampling equipment. Solid samples are collected as grab samples using widemouth jars. Maintain samples at 0 to 4°C from the time of collection until extraction. If residual chlorine is present in aqueous samples, add 80 mg sodium thiosulfate/L of water. Begin sample extraction within 7 days of collection, and analyze all extracts within 40 days of extraction.

SAMPLE PREPARATION Samples containing 1% solids or less are extracted directly using continuous liquid-liquid extraction techniques. Samples containing 1 to 30% solids are diluted to the 1% level with reagent water and extracted using continuous liquid-liquid extraction techniques. Samples containing greater than 30% solids are extracted using ultrasonic techniques.

Base/neutral extraction — Adjust the pH of the waters in the extractors to 12–13 with 6 N NaOH. Extract with methylene chloride for 24–48 h.
Acid extraction — Adjust the pH of the waters in the extractors to 2 or less using 6 N sulfuric acid. Extract with methylene chloride for 24–48 h.
Ultrasonic extraction of high solids samples — Add anhydrous sodium sulfate to the sample and QC aliquot(s). Add acetone:methylene chloride (1:1) to the sample and mix thoroughly

Concentrate extracts using a K-D apparatus.

QUALITY CONTROL The analyst is permitted to modify this method to improve separations or lower the costs of measurements, provided all performance specifications are met. Analyses of blanks are required to demonstrate freedom from contamination. When results of spikes indicate atypical method performance for samples, the samples are diluted to bring method performance within acceptable limits.

For low solids (aqueous samples), extract, concentrate, and analyze two sets of four 1-L aliquots (8 aliquots total) of the precision and recovery standard. For high solids samples, two sets of four 30-g aliquots of the high solids reference matrix are used.

Spike all samples with labeled compounds to assess method performance. Compute percent recovery of the labeled compounds using the internal standard method. Compare the labeled compound recovery for each compound with the corresponding labeled compound recovery.

Reagent water and high solids reference matrix blanks are analyzed to demonstrate freedom from contamination. Extract and concentrate a 1-L reagent water blank or a high solids reference matrix blank with each sample's lot (samples started through the extraction process on the same 8-h shift, to a maximum of 20 samples).

Field replicates may be collected to determine the precision of the sampling technique, and spiked samples may be required to determine the accuracy of the analysis when the internal standard method is used.

REFERENCE Semivolatile Organic Compounds by Isotope Dilution GC/MS. Office of Water Regulation and Standards, U.S. EPA Industrial Technology Division, Washington, DC, EPA Method 1625, Rev. C, June 1989 (contact W.A. Telliard, U.S. EPA, Office of Water Regulations and Standards, 401 M St., SW, Washington, DC, 20460. Phone: 202-382-7131).

1,2-Dibromo-3-chloropropane   **EPA Method 502**
CAS #96-12-8

TITLE Volatile Organic Compounds in Water By Purge and Trap Capillary Column Gas Chromatography with Photoionization and Electrolytic Conductivity Detectors in Series. U.S. EPA Method 502.2, Rev. 2.0, 1989.

MATRIX Drinking water and raw source water. The latter should include most surface water and groundwater sources.

METHOD SUMMARY This method covers 60 volatile organic compounds that contain halogen atoms and/or that are aromatic. An inert gas (zero grade nitrogen or helium) is bubbled through a 25-mL or a 5-mL water sample (depending on the expected concentration of the analytes). Purged sample components are trapped in a tube of sorbent materials. When purging is complete, the sorbent tube is heated and backflushed with helium to desorb the trapped sample onto a capillary GC column. The column is temperature programmed to separate the method analytes which are then detected with a photoionization detector (PID) and a Hall electrolytic conductivity (HECD) placed in series. The PID is selective for aromatic compounds and the HECD is selective for halogenated compounds.

INTERFERENCES Impurities in the purge gas and from organic compounds outgassing from the plumbing ahead of the trap account for many contamination problems. Interferences purged or coextracted from the samples will vary considerably from source to source, depending upon the particular sample or extract being tested. Cross-contamination can occur whenever high-level and low-level samples are analyzed sequentially. Samples also can be contaminated by diffusion of volatile organics (particularly methylene chloride and fluorocarbons) through the septum seal into the sample during shipment and storage. The lab where volatile analysis is performed and also the refrigerated storage area should be completely free of solvents.

INSTRUMENTATION A GC containing a series configuration of a high temperature photoionization detector (PID) equipped with 10.0 eV (nominal) lamp and Hall electrolytic conductivity detector (HECD) is required. Also required is an all-glass 5-mL purging device, a sorbent trap, and a thermal desorption apparatus which is connected to the GC system.

Column 1: VOCOL glass wide-bore capillary column.
Column 2: RTX–502.2 mega-bore capillary column.
Column 3: DB-62 mega-bore capillary column.

PRECISION & ACCURACY Method detection limits are dependent upon the characteristics of the gas chromatographic system used. Analytes that are not separated chromatographically cannot be individually identified and used in the same calibration mixture or water samples unless an alternative technique for identification and quantification, such as mass spectrometry, is used.

Electrolytic conductivity detetor (c) range in µg/L (a) was 0.02–200.
Electrolytic conductivity detetor (c) MDL in µg/L (b) was 3.0.
Electrolytic conductivity detetor (c) accuracy as % recovery was 86.
Electrolytic conductivity detetor (c) precision as % RSD was 11.3.
Photoionization detector (d) range in µg/L (a) was 0.02–200.
Photoionization detector (d) MDL in µg/L (b) was not listed.
Photoionization detector (d) accuracy as % recovery was not listed.
Photoionization detector (d) precision as % RSD was not listed.

(a) *The applicable concentration range of this method is compound, instrument, and matrix-dependent. It is listed as being approximately 0.02 to 200 µg/L but no specific information is provided so caution should be observed.*
(b) *The method detection limits reports with this method are compound, instrument, and matrix-dependent. The values reported were calculated using reagent water fortified with the corresponding compounds at 10 µg/L and a GC-equipped with a 60 m × 0.75 mm VOLCOL wide bore capillary column with 1.5 µm film thickness and using helium carrier gas.*
(c) *Recoveries and relative standard deviations were determined from seven samples of reagent water fortified with 10 µg/L of each compound. 2-Bromo-1-chloropropane was used as the internal standard for calculating average recoveries.*
(d) *Recoveries and relative standard deviations were determined from seven samples of reagent water fortified with 10 µg/L of each compound. Fluorobenzene was used as the internal standard for calculating average recoveries.*

SAMPLING METHOD Collect samples using a 40- to 120-mL screw-cap vial (prewashed with detergent, rinsed with distilled water and oven dried at 105°C) with a Teflon®-faced silicone septum. Collect bubble-free samples and place the septum with the Teflon® side down on the water.

SAMPLE PRESERVATION If residual chlorine is present in the water add about 25 mg of ascorbic acid to each vial before samples are collected to remove the chlorine. Add hydrochloric acid to reduce pH to <2, immediately cool samples to 4°C, and store them in a solvent-free refrigerator at 4°C until analysis.

MHT The maximum holding time for samples is 14 days from the time they were collected.

SAMPLE PREPARATION Remove the plungers from two 5-mL syringes and attach a closed syringe valve to each. Warm the sample to room temperature, open the sample bottle, and carefully pour the sample into one of the syringe barrels to just short of overflowing. Replace the syringe plunger, invert the syringe, and compress the sample. Open the syringe valve and vent any residual air while adjusting the sample volume to 5.0 mL. Add 10 µL of the internal calibration standard to the sample through the syringe valve. Close the valve. Fill the second syringe in an identical manner from the same sample bottle. Reserve this second syringe for a reanalysis if necessary.

QUALITY CONTROL As an initial demonstration of lab accuracy and precision, analyze 4 to 7 replicates of a lab fortified blank containing analyte at 0.1–5 µg/L. Collect all samples in duplicate. Surrogate analytes (similar to those of the analytes of interest), whose concentration is known in every sample, are measured using the same internal standard calibration procedure. Duplicate field reagent water blanks (trip blanks) must be analyzed with each set of samples, lab reagent blanks (method blanks) must be analyzed with each batch of samples processed as a group within a work shift. Also, a single lab-fortified blank that contains each of the analytes of interest should be analyzed with each batch of samples processed as a group within a work shift. A 3- to 5-point calibration curve is needed depending on the calibration range factor required.

EPA CONTACT & HOTLINE For technical questions contact Dr. Baldev Bathija, U.S. EPA, Office of Ground Water and Drinking Water (WH-550D), 401 M St. SW, Washington, DC 20460. Tel. (202) 260-3040. For further information the EPA Safe Drinking Water Hotline may be called at: (800) 426-4791.

REFERENCE Methods for the Determination of Organic Compounds in Drinking Water, EPA/600/4-88/039 (revised July 1991; Final Rule for determination of compliance with the MCL for Total Trihalomethanes under 141.30, in 40 CFR Part 141, Vol. 58, No. 147, Fed. Reg., Tuesday Aug. 3, 1993). U.S. EPA Environmental Monitoring Systems Laboratory, Cincinnati, OH, 45268, U.S.A. Available from the National Technical Information Service (NTIS), 5285 Port Royal Road, Springfield, VA 22161; Tel. 800-553-6847. NTIS Order Number is PB91-231480.

1,2-Dibromo-3-chloropropane **EPA Method 524**
CAS #96-12-8

TITLE Measurement of Purgeable Organic Compounds in Water by Capillary Column GC/MS.

MATRIX Drinking water and raw source water; the latter should include most surface water and groundwater sources.

METHOD SUMMARY Method 524.2 covers 60 volatile organic compounds. An inert gas (zero grade nitrogen or helium) is bubbled through a 25-mL or a 5-mL water sample (depending on the expected concentration of the analytes). Purged sample components are trapped in a tube of sorbent materials. When purging is complete, the sorbent tube is heated and backflushed with helium to desorb the trapped sample onto a capillary GC column.

INTERFERENCES Impurities in the purge gas and from organic compounds outgassing from the plumbing ahead of the trap account for many contamination problems. Interferences purged or coextracted from the samples will vary considerably from source to source, depending upon the particular sample or extract being tested. Cross-contamination can occur whenever high-level and low-level samples are analyzed sequentially. Samples also can be contaminated by diffusion of volatile organics (particularly methylene chloride and fluorocarbons) through the septum seal into the sample during shipment and storage.

INSTRUMENTATION A GC/MS with a data system equipped with one of the following capillary GC columns:

Column 1: VOCOL glass wide bore capillary column.
Column 2: DB-624 fused silica capillary column.
Column 3: DB-5 fused silica capillary column.

Also required is an all-glass 25 mL or 5-mL purging device, a sorbent trap, and a thermal desorption apparatus which is connected to the GC/MS system.

PRECISION & ACCURACY Method detection limits are compound- and instrument-dependent, and may vary from approximately 0.02–0.35 µg/L. Note in the table below that the "true" concentration range used for accuracy and precision measurements was quite narrow. However, the applicable concentration range of this method is primarily column dependent and is approximately 0.02 to 200 µg/L for the wide-bore thick-film columns. Narrow-bore thin-film columns may have a capacity which limits the range to about 0.02 to 20 µg/L. Analytes that are inefficiently purged from water will not be detected when present at low concentrations, but they can be measured with acceptable accuracy and precision when present in sufficient amounts.

Analytes that are not separated chromatographically, but which have different mass spectra and non-interfering quantification ions, can be identified and measured in the same calibration mixture or water sample. Analytes which have very similar mass spectra cannot be individually identified and measured in the same calibration mixture or water samples unless they have different retention times. Co-eluting compounds with very similar mass spectra, typically many structural isomers, must be reported as an isomeric group or pair.

The range (in µg/L) was 0.5–10.
The Method Detection Limig (in µg/L) was 0.26.
The accuracy (as % recovery) was 83.

The precision (in %) was 19.9.

Note: Data were obtained from 16–31 determinations using a wide-bore capillary column and a jet separator interfaced to a quadrupole mass spectrometer. All analytes were in a reagent water matrix.

SAMPLING METHOD Collect samples using a 40- to 120-mL screw-cap vial (prewashed with detergent, rinsed with distilled water and oven dried at 105°C) with a Teflon®-faced silicone septum. Collect bubble-free samples and place the septum with the Teflon® side down on the water.

SAMPLE PRESERVATION If residual chlorine is present in the water add about 25 mg of ascorbic acid to each vial before samples are collected to remove the chlorine. Add hydrochloric acid to reduce pH to <2, and immediately cool samples to 4°C, and store them in a solvent-free refrigerator at 4°C until analysis.

MHT The maximum holding time for samples is 14 days from the time they were collected.

SAMPLE PREPARATION Remove the plungers from two 25-mL (or 5-mL depending on sample size) syringes and attach a closed syringe valve to each. Warm the sample to room temperature, open the sample bottle, and carefully pour the sample into one of the syringe barrels to just short of overflowing. Replace the syringe plunger, invert the syringe, and compress the sample. Open the syringe valve and vent any residual air while adjusting the sample volume to 25.0 mL (or 5 mL). For samples and blanks, add 5 µL of the fortification solution containing the internal standard and the surrogates to the sample through the syringe valve. For calibration standards and lab fortified blanks, add 5 µL of the fortification solution containing the internal standard only. Close the valve. Fill the second syringe in an identical manner from the same sample bottle. Reserve this second syringe for a reanalysis if necessary.

QUALITY CONTROL As an initial demonstration of lab accuracy and precision, analyze 4 to 7 replicates of a lab fortified blank containing analyte at 0.2–5 µg/L. Collect all samples in duplicate. Surrogate analytes (similar to those of the analytes of interest), whose concentration is known in every sample, are measured using the same internal standard calibration procedure. Duplicate field reagent water blanks (trip blanks) must be analyzed with each set of samples, lab reagent blanks (method blanks) must be analyzed with each batch of samples processed as a group within a work shift. Also, a single lab-fortified blank that contains each of the analytes of interest should be analyzed with each batch of samples processed as a group within a work shift. A 3- to 5-point calibration curve is needed depending on the calibration range factor required.

EPA CONTACT & HOTLINE For technical questions contact Dr. Baldev Bathija, U.S. EPA, Office of Ground Water and Drinking Water (WH-550D), 401 M St. SW, Washington, DC 20460. Tel. (202) 260-3040. For further information the EPA Safe Drinking Water Hotline may be called at: (800) 426-4791.

REFERENCE Methods for the Determination of Organic Compounds in Drinking Water, EPA/600/4-88/039 (revised July 1991; Final Rule for determination of compliance with the

MCL for Total Trihalomethanes under 141.30, in 40 CFR Part 141, Vol. 58, No. 147, Fed. Reg., Tuesday Aug. 3, 1993). U.S. EPA Environmental Monitoring Systems Laboratory, Cincinnati, OH, 45268, U.S.A. Available from the National Technical Information Service (NTIS), 5285 Port Royal Road, Springfield, VA 22161; Tel. 800-553-6847. NTIS Order Number is PB91-231480.

1,2-Dibromo-3-chloropropane   **EPA Method 8021**
CAS #96-12-8

TITLE Halogenated Volatile by Gas Chromatography Using Photoionization and Electrolytic Conductivity Detectors in Series: Capillary Column Technique

MATRIX This method is applicable to nearly all types of samples, regardless of water content, including groundwater, aqueous sludges, caustic liquors, acid liquors, waste solvents, oily wastes, mousses, tars, fibrous wastes, polymeric emulsions, filter cakes, spent carbons, spent catalysts, soils, and sediments.

METHOD SUMMARY This method is used to determine 60 volatile organic compounds in a variety of solid waste matrices. It provides GC conditions for the detection of halogenated and aromatic volatile organic compounds. Samples can be analyzed using direct injection or purge-and-trap (EPA Method 5030). Groundwater samples must be analyzed using EPA Method 5030 (where applicable). A temperature program is used with the GC. Detection is achieved by a photoionization detector (PID) and a Hall electrolytic conductivity detector (HECD) in series.

INTERFERENCES Samples can be contaminated by diffusion of volatile organics (particularly chlorofluorocarbons and methylene chloride) through the sample container septum during shipment and storage.

INSTRUMENTATION A GC-equipped with variable-constant differential flow controllers, subambient oven controller, PID and HECD detectors connected with a short piece of uncoated capillary tubing and a data system.

Column: 60 m × 0.75 mm I.D. VOCOL wide-bore capillary column with 1.5 µm film thickness.

PRECISION & ACCURACY MDLs are compound-dependent and vary with purging efficiency and concentration. The applicable concentration range of this method is compound- and instrument-dependent but is approximately 0.1 to 200 µg/L. Analytes that are inefficiently purged from water will not be detected when present at low concentrations, but they can be measured with acceptable accuracy and precision when present in sufficient amounts. The estimated quantitation limit (EQL) for an individual compound is approximately 1 µg/kg (wet weight) for soil/sediment samples, 100 µg/kg (wet weight) for wastes, and 1 µg/L for groundwater. EQLs will be proportionately higher for sample extracts and samples that require dilution or reduced sample size to avoid saturation of the detector.

MULTIPLICATION FACTORS FOR OTHER MATRICES (a)

Matrix	Factor (b)
Groundwater	10
Low-concentration soil	10
Water miscible liquid waste	500
High-concentration soil and sludge	1250
Non-water miscible waste	1250

(a) Sample EQLs are highly matrix-dependent. The EQLs listed herein are provided for guidance and may not always be achievable.
(b) EQL = [Method detection limit] × [Factor]. For non-aqueous samples, the factor is on a wet-weight basis.

SINGLE LABORATORY ACCURACY & PRECISION DATA FOR VOCs IN WATER

This method was tested in a single lab using water spiked at 10 µg/L and the following data was reported:

Recoveries and standard deviations were determined from seven samples and spiked at 10 µg/L of each analyte. Recoveries were determined by the internal standard method. Internal standards were: Fluorobenzene for PID and 2-Bromo-1-chloropropane for HECD.

The average recovery (in percent) for the PID was none (no response for this detector).
The standard deviation of the recovery for the PID was none (no response for this detector).
The MDL (in µg/mL) for the PID was none (no response for this detector).
The average recovery (in percent) for the HECD was 86.
The standard deviation of the recovery for the HECD was 9.9.
The MDL (in µg/mL) for the HECD was 3.0.

SAMPLE COLLECTION, PRESERVATION & HANDLING

Volatile Organics — Standard 40-mL glass screw-cap VOA vials with Teflon®-faced silicone septum may be used for both liquid and solid matrices. When collecting samples, liquids and solids should be introduced into the vials gently to reduce agitation which might drive off volatile compounds. If there are any air bubbles present the sample must be retaken. Tap slightly as they are filled to try and eliminate as much free air space as possible. The two vials from each sampling locations should be sealed in separate plastic bags to prevent cross-contamination between samples particularly if the sampled waste is suspected of containing high levels of volatile organics.

Semivolatile organics — Containers used to collect samples for the determination of semivolatile organic compounds should be soap and water washed followed by methanol (or isopropanol) rinsing. The sample containers should be of glass or Teflon® and have screw-top covers with Teflon® liners.

Preservation for volatile organics — No preservation is used with concentrated waste samples. With liquid samples containing no residual chlorine, 4 drops of concentrated hydrochloric acid are added and the samples are immediately cooled to 4°C. When liquid samples contain residual chlorine, they are treated as above and, in addition, 4 drops of 4% aqueous sodium thiosulfate are added. Soil, sediment, and sludge samples are only cooled to 4°C.

Preservation for semivolatile organics — No preservation is used with concentrated waste samples. With liquid samples containing no residual chlorine and with soil, sediment, and sludge samples, immediately cooling to 4°C is the only preservation used. When residual chlorine is present then 3 mL of 10% aqueous sodium sulfate is added for each gallon of sample collected, followed by cooling to 4°C.

MHT The holding time for all volatile organics samples is 14 days. Liquid samples must be extracted within 7 days and their extracts analyzed within 40 days. Concentrated waste, soil, sediment, and sludge samples must be extracted within 14 days and their extracts analyzed within 40 days.

SAMPLE PREPARATION Volatile compounds are introduced into the gas chromatograph either by direct injector or purge-and-trap (EPA Method 5030). EPA Method 5030 may be used directly on groundwater samples or low-concentration contaminated soils and sediments. For medium-concentration soils or sediments, methanolic extraction, as described in EPA Method 5030, may be necessary prior to purge-and-trap analysis.

QUALITY CONTROL Calculate surrogate standard recovery on all samples, blanks, and spikes. A trip blank is recommended to check on sampling, storage, and handling contamination. Calibration standards, at a minimum of five concentration levels, are prepared in organic-free reagent water. One of the concentration levels should be at a concentration near, but above, the method detection limit.

A combination of bromochloromethane, 2-bromo-1-chloropropane, 1,4-dichlorobutane, and bromochlorobenzene are recommended as surrogate standards to encompass the range of the temperature program used in this method.

REFERENCE Test Methods for Evaluating Solid Waste, Physical/Chemical Methods, SW-846, 3rd Edition, U.S. EPA, Office of Solid Waste, Washington, DC, EPA Method 8021A, Rev. 1, Nov. 1992.

1,2-Dibromo-3-chloropropane   **EPA Method 8240**
CAS #96-12-8

TITLE Volatile Organics By GC/MS: Packed Column Technique

MATRIX Nearly all types of sample matarices, regardless of water content, can be analyzed using this method. This includes groundwater, aqueous sludges, caustic liquors, acid liquors, waste solvents, oily wastes, mousses, tars, fibrous wastes, polymetric emulsions, filter cakes, spent carbons, spent catalysts, soils, and sediments.

METHOD SUMMARY Method 8240B covers 80 volatile organic compounds that are introduced into a gas chromatograph by the purge-and-trap method or by direct injection (in limited applications). For the purge-and-trap method an inert gas (zero grade nitrogen or helium) is bubbled through a 5-mL solution at ambient temperature. Purged sample components are trapped in a tube of sorbent materials. When purging is complete, the sorbent tube is heated and backflushed with inert gas to desorb the trapped components onto a GC column.

INTERFERENCES Impurities in the purge gas and from organic compounds outgassing from the plumbing ahead of the trap account for many contamination problems. Interferences purged or coextracted from the samples will vary considerably from source to source. Cross-contamination can occur whenever high-level and low-level samples are analyzed sequentially. Whenever an unusually concentrated sample is analyzed, it should be followed by the analysis of organic-free reagent water to check for cross-contamination. Samples also can be contaminated by diffusion of volatile organics (particularly methylene chloride and fluorocarbons) through the septum seal into the sample during shipment and storage. A trip blank can serve as a check on such contamination. The lab where volatile analysis is performed and also the refrigerated storage area should be completely free of solvents.

INSTRUMENTATION A gas chromatograph/mass spectrometry/data system (GC/MS) equipped with a 6 ft × 0.1 in I.D. glass column packed with 1% SP-1000 on Carbopack-B (60/80 mesh) is required. Also needed is a 5-mL purging device, a sorbent trap, and a thermal desorption apparatus.

PRECISION & ACCURACY This method is reported to have been tested by 15 laboratories using organic-free reagent water, drinking water, surface water, and industrial wastewaters (not specified) fortified at six concentrations over the range 5–600 µg/L.

Sample estimated quantitation limits (EQLs) are highly matrix-dependent. The EQLs listed may not always be achievable. EQLs listed for soils or sediments are based on wet weight. Normally, data is reported on a dry-weight basis; therefore, EQLs will be higher, based on the percent dry weight of each sample. Note that EQLs are even more variable than MDLs and that they are highly variable depending on the matrix being analyzed.

EQL in groundwater in µg/L was 100.
EQL in low soil or sediment in µg/kg was 100.
Accuracy (a) in µg/L was not listed.
Precision (b) in µg/L was not listed.

(a) *Average recovery found for measurements of samples containing a concentration of C, in µg/L.*
(b) *Overall precision found for measurements of samples with average recovery X for samples containing a concentration of C in µg/L.*
X = *Average recovery found for measurement of samples containing a concentration of C in µg/L.*

MULTIPLICATION FACTORS FOR OTHER MATRICES

Other Matrices	Factor (a)
Waste miscible liquid waste	50
High-concentration soil and sludge	125
Non-water miscible waste	500

(a) *EQL = [EQL for low soil sediment] × [Factor]. For non-aqueous samples, the factor is on a wet-weight basis.*

SAMPLING METHOD

Liquid samples — Use a 40-mL glass screw-cap VOA vial with a Teflon®-faced silicone septum that has been prewashed, rinsed with distilled deionized water, and oven dried. However,

if residual chlorine is present, collect sample in a 40-oz. soil VOA container which has been pre-preserved with 4 drops of 10% sodium thiosulfate, mix gently, and then transfer the sample to a 40-mL VOA vial. Collect bubble-free samples in duplicate and seal them in separate plastic bags.

Soils or sediments, and sludges — Use an 8-oz. widemouth glass bottle with a Teflon®-faced silicone septum that has been prewashed with detergent, rinsed with distilled deionized water, and oven dried. Tap slightly to eliminate free air space. Collect samples in duplicate and seal them in separate plastic bags.

SAMPLE PRESERVATION
Liquid samples — Add 4 drops of concentrated HCL and immediately cool samples to 4°C and store in a solvent-free refrigerator.

Soils or sediments, and sludges — Cool samples to 4°C and store in a solvent-free refrigerator.

MHT Maximum holding time is 14 days from the date of sample collection.

SAMPLE PREPARATION
Liquid samples — Remove the plunger from a 5-mL syringe and carefully pour the sample into the syringe barrel to just short of overflowing. Replace the syringe plunger and compress the sample. Open the syringe valve and vent any residual air while adjusting the sample volume to 5.0 mL. If there is only one volatile organic analysis (VOA) vial, a second syringe should be filled at this time to protect against possible loss of sample integrity. Add 10 µL of surrogate spiking solution and 10 µL of internal standard spiking solution through the valve bore of the 5-mL syringe, then close the valve. The surrogate and internal standards may be mixed and added as a single spiking solution.

Sediments, soils, and waste samples — All samples of this type should be screened by GC analysis using a headspace method (EPA Method 3810) or the hexadecane extraction and screening method (EPA Method 3820). Use the screening data to determine whether to use the low-concentration method (0.005–1 mg/kg) or the high-concentration method (>1 mg/kg).

Low-concentration method — The low-concentration method is based on purging a heated sediment or soil sample mixed with organic-free reagent water containing the surrogate and internal standards. Analyze all reagent blanks and standards under the same conditions as the samples.

Use a 5-g sample if the expected concentration is <0.1 mg/kg or a 1-g sample for expected concentrations between 0.1 and 1 mg/kg. Mix the contents of the sample container with a narrow metal spatula. Weigh the amount of the sample into a tared purge device. Add the spiked water to the purge device, which contains the weighed amount of sample, and connect the device to the purge-and-trap system.

High-concentration method — This method is based on extracting the sediment or soil with methanol. A waste sample is either extracted or diluted, depending on its solubility in methanol. Wastes that are insoluble in methanol are diluted with reagent tetraglyme or possibly polyethylene glycol (PEG).

An aliquot of the extract is added to organic-free reagent water containing surrogate and internal standards. This is purged at ambient temperature. All samples with an expected concentration of >1.0 mg/kg should be analyzed by this method.

Mix the contents of the sample container with a narrow metal spatula. For sediments or soils and solid wastes that are insoluble in methanol, weigh 4 g (wet weight) of sample into a tared 20-mL vial. For waste that is soluble in methanol, tetraglyme, or PEG, weigh 1 g (wet weight) into a tared scintillation vial or culture tube or a 10-mL volumetric flask. Quickly add 9.0 mL of appropriate solvent then add 1.0 mL of a surrogate spiking solution to the vial, cap it, and shake it for 2 min.

METHANOL EXTRACT REQUIRED FOR ANALYSIS OF HIGH-CONCENTRATION SOILS OR SEDIMENTS

Approximate Concentration Range	Volume of Methanol Extract (a)
500–10,000 µg/kg	100 µL
1,000–20,000 µg/kg	50 µL
5,000–100,000 µg/kg	10 µL
25,000–500,000 µg/kg	100 µL of 1/50 dilution (b)

Calculate appropriate dilution factor for concentrations exceeding this table.

(a) The volume of methanol added to 5 mL of water being purged should be kept constant. Therefore, add to the 5-mL syringe whatever volume of methanol is necessary to maintain a volume of 100 µL added to the syringe.

(b) Dilute an aliquot of the methanol extract and then take 100 µL for analysis.

QUALITY CONTROL Demonstrate, through the analysis of a reagent water blank, that interferences from the analytical system, glassware, and reagents are under control. Blank samples should be carried through all stages of the sample preparation and measurement steps. For each analytical batch (up to 20 samples), a reagent blank, matrix spike, and matrix spike duplicate must be analyzed (the frequency of the spikes may be different for different monitoring programs). The blank and spiked samples must be carried through all stages of the sample preparation and measurement steps. QC samples mentioned in the section on Interferences will also be needed as appropriate to those situations.

REFERENCE Test Methods for Evaluating Solid Waste (SW-846). U.S. EPA. 1983. Method 8240B, Rev. 2, Nov. 1990. Office of Solid Wastes, Washington, DC.

1,2-Dibromo-3-chloropropane EPA Method 8260
CAS #96-12-8

TITLE Volatile Organic Compounds by GC/MS: Capillary Column Technique

MATRIX This method is applicable to nearly all types of samples, regardless of water content, including groundwater, soils, and sediments.

METHOD SUMMARY Method 8260A covers 58 volatile organic compounds that are introduced into a gas chromatograph by the purge-and-trap method or by direct injection (in limited applications). Zero-grade helium is bubbled through a 5-mL solution at ambient temperature. Purged sample components are trapped in a tube containing suitable sorbent materials. When purging is complete, the sorbent tube is heated and backflushed with helium to desorb trapped sample components. The analytes are desorbed directly to a large bore capillary or cryofocussed on a capillary precolumn before being flash evaporated to a narrow bore capillary for analysis.

INTERFERENCES Major contaminant sources are volatile materials in the lab and impurities in the inert purging gas and in the sorbent trap. Interfering contamination may occur when a sample containing low concentrations of volatile organic compounds is analyzed immediately after a sample containing high concentrations of volatile organic compounds. After analysis of a sample containing high concentrations of volatile organic compounds, one or more calibration blanks should be analyzed to check for cross-contamination. Screening of the samples prior to purge-and-trap GC/MS analysis is highly recommended to prevent contamination of the system. This is especially true for soil and waste samples.

Special precautions must be taken to analyze for methylene chloride. The analytical and sample storage area should be isolated from all atmospheric sources of methylene chloride. All gas chromatography carrier gas lines and purge gas plumbing should be constructed from stainless steel or copper tubing. Laboratory clothing previously exposed to methylene chloride fumes during liquid-liquid extraction procedures can contribute to sample contamination.

Samples can also be contaminated by diffusion of volatile organics (particularly methylene chloride and fluorocarbons) through the septum seal during shipment and storage. A trip blank can serve as a check on such contamination.

INSTRUMENTATION GC/MS with a temperature-programmable chromatograph suitable for splitless injection equipped with variable constant differential flow controllers, a subambient oven controller, a purging device, sorbent trap, a thermal desorption apparatus and a capillary precolumn interface when using cryogenic cooling will be needed. The following GC columns may be used:

Column 1: 60 m × 0.75mm I.D. capillary column coated with VOCOL, 1.5 µm film thickness.
Column 2: 30 m × 0.53mm capillary column coated with DB-624 or VOCOL, 3 µm film thickness.
Column 3: 30 m × 0.32mm I.D. capillary column coated with DB-5 or SE-54, 1-µm film thickness.

PRECISION & ACCURACY This method has been tested in a single lab using spiked water. Using a wide-bore capillary column, water was spiked at concentrations between 0.5 and 10 µg/L. Single lab accuracy and precision data are presented. The MDL actually achieved in a given analysis will vary depending on instrument sensitivity and matrix effects.

The MDL (a) in µg/L was 0.26.
The concentration range in µg/L was 0.5–10.

The mean accuracy (% of true value) was 92.
The precision as relative standard deviation was 19.9.

Note: The MDL is based on a 25-mL sample volume instead of a 5-mL sample volume.

SAMPLING METHOD
Liquid samples — Use a 40-mL glass screw-cap VOA vial with a Teflon®-faced silicone septum that has been prewashed, rinsed with distilled deionized water, and oven dried. If residual chlorine is present, collect the sample in a 4-oz soil VOA container which has been pre-preserved with 4 drops of 10% sodium thiosulfate. Mix gently and transfer the sample to a 40-mL VOA vial. Collect bubble-free samples in duplicate and seal each sample in a separate plastic bag.

Soils, sediments, and sludges — Use an 8-oz widemouth glass bottle with Teflon®-faced silicone septum that has been prewashed, rinsed with distilled deionized water, and oven dried. **Do not** heat the septum for more than 1 h. Tap slightly to eliminate any free air space. Collect samples in duplicate and seal each one in a separate plastic bag.

SAMPLE PRESERVATION
Liquid samples — Add 4 drops of concentrated HCL, cool to 4°C and store in a solvent-free refrigerator.

Soils, sediments and sludges — Cool samples to 4°C and store in a solvent-free refrigerator.

MHT The maximum holding time of any sample (liquids, soils, sediments, and sludges) is 14 days.

SAMPLE PREPARATION
Liquid samples — Remove the plunger from a 5-mL syringe and carefully pour the sample into the syringe barrel to just short of overflowing. Replace the syringe plunger and compress the sample. Open the syringe valve and vent any residual air while adjusting the sample volume to 5.0 mL. If there is only one volatile organic analysis (VOA) vial, a second syringe should be filled at this time to protect against possible loss of sample integrity. Add 10 µL of surrogate spiking solution and 10 µL of internal standard spiking solution through the valve bore of the 5-mL syringe, then close the valve. The surrogate and internal standards may be mixed and added as a single spiking solution.

Sediments, soils, and waste samples — All samples of this type should be screened by GC analysis using a headspace method (EPA Method 3810) or the hexadecane extraction and screening method (EPA Method 3820). Use the screening data to determine whether to use the low-concentration method (0.005–1 mg/kg) or the high-concentration method (>1 mg/kg).

Low-concentration method — The low-concentration method is based on purging a heated sediment or soil sample mixed with organic-free reagent water containing the surrogate and internal standards. Analyze all reagent blanks and standards under the same conditions as the samples.

Use a 5-g sample if the expected concentration is <0.1 mg/kg or a 1-g sample for expected concentrations between 0.1 and 1 mg/kg. Mix the contents of the sample container with a narrow metal spatula. Weigh the amount of the sample into a tared

purge device. Add the spiked water to the purge device, which contains the weighed amount of sample, and connect the device to the purge-and-trap system.

High-concentration method — This method is based on extracting the sediment or soil with methanol. A waste sample is either extracted or diluted, depending on its solubility in methanol. Wastes that are insoluble in methanol are diluted with reagent tetraglyme or possibly polyethylene glycol (PEG). An aliquot of the extract is added to organic-free reagent water containing surrogate and internal standards. This is purged at ambient temperature. All samples with an expected concentration of >1.0 mg/kg should be analyzed by this method.

Mix the contents of the sample container with a narrow metal spatula. For sediments or soils and solid wastes that are insoluble in methanol, weigh 4 g (wet weight) of sample into a tared 20-mL vial. For waste that is soluble in methanol, tetraglyme, or PEG, weigh 1 g (wet weight) into a tared scintillation vial or culture tube or a 10-mL volumetric flask. Quickly add 9.0 mL of appropriate solvent then add 1.0 mL of a surrogate spiking solution to the vial, cap it, and shake it for 2 min.

METHANOL EXTRACT REQUIRED FOR ANALYSIS OF HIGH-CONCENTRATION SOILS OR SEDIMENTS

Approximate Concentration Range	Volume of Methanol Extract (a)
500–10,000 µg/kg	100 µL
1,000–20,000 µg/kg	50 µL
5,000–100,000 µg/kg	10 µL
25,000–500,000 µg/kg	100 µL of 1/50 dilution (b)

Calculate appropriate dilution factor for concentrations exceeding this table.

(a) The volume of methanol added to 5 mL of water being purged should be kept constant. Therefore, add to the 5-mL syringe whatever volume of methanol is necessary to maintain a volume of 100 µL added to the syringe.
(b) Dilute an aliquot of the methanol extract and then take 100 µL for analysis.

QUALITY CONTROL Demonstrate, through the analysis of a reagent water blank, that interferences from the analytical system, glassware, and reagents are under control. Blank samples should be carried through all stages of the sample preparation and measurement steps. For each analytical batch (up to 20 samples), a reagent blank, matrix spike, and matrix spike duplicate must be analyzed (the frequency of the spikes may be different for different monitoring programs). The blank and spiked samples must be carried through all stages of the sample preparation and measurement steps. QC samples mentioned in the section on Interferences will also be needed as appropriate to those situations.

Matrix spiking standards should be prepared from volatile organic compounds which will be representative of the compounds being investigated. The recommended internal standards are chlorobenzene-d5, 1,4-difluorobenzene, 1,4-dichlorobenzene-d4, and pentafluorobenzene. Using stock standard solutions, prepare secondary dilution standards containing the compounds of interest, either singly or mixed together in methanol. Store them in a vial with no headspace for no more than one week. Surrogates recommended are toluene-d8, 4-bromofluorobenzene, and dibromofluoromethane. Each sample undergoing GC/MS analysis must be spiked with 10 µL of the surrogate spiking solution prior to analysis.

REFERENCE Test Methods for Evaluating Solid Waste (SW-846). U.S. EPA 1983, Method 8260A, Rev. 1, Nov. 1990. Office of Solid Waste, Washington, DC.

1,2-Dibromo-3-chloropropane   **EPA Method 8270**
CAS #96-12-8

TITLE Semivolatile Organic Compounds by GC/MS

MATRIX This method is used to determine the concentration of semivolatile organic compounds in extracts prepared from all types of solid waste matrices, soils, and groundwater. Although surface waters are not specifically mentioned, this method should be applicable to water samples from rivers, lakes, etc.

METHOD SUMMARY This method covers 259 semivolatile organic compounds. In very limited applications direct injection of the sample into the GC/MS system may be appropriate, but this results in very high detection limits (approximately 10,000 µg/L). Typically, a 1-L liquid sample, containing surrogate, and matrix spiking standards, is extracted in a continuous extractor first under acid conditions and then under basic conditions. Typically 30 g of a solid sample, containing surrogate, and matrix spiking standards, is extracted ultrasonically. After concentrating the extract to 1 mL it is spiked with 10 µL of an internal standard solution just prior to analysis by GC/MS. The volume injected should contain about 100 ng of base/neutral and 200 ng of acid surrogates (for a 1-µL injection). Analysis is performed by GC/MS using a capillary GC column.

INTERFERENCES Raw GC/MS data from all blanks, samples, and spikes must be evaluated for interferences. Contamination by carryover can occur whenever high-concentration and low-concentration samples are sequentially analyzed. To reduce carryover, the sample syringe must be rinsed out between samples with solvent. Whenever an unusually concentrated sample is encountered, it should be followed by the analysis of blank solvent to check for cross-contamination.

INSTRUMENTATION A GC/MS and a data system are required. The GC column used is a 30 m × 0.25 mm I.D. (or 0.32 mm I.D.) 1um film thickness silicone-coated fused silica capillary column. A continuous liquid-liquid extractor equipped with Teflon® or glass connection joints and stopcocks requiring no lubrication, a K-D concentrating apparatus, water bath, and an ultrasonic disrupter with a minimum power of 300 W and with pulsing capability are also required.

PRECISION & ACCURACY The estimated quantitation limit (EQL) of Method 8270B for determining an individual compound is approximately 1 mg/kg (wet weight) for soil or sediment samples, 1–200 mg/kg for wastes (dependent on matrix and method of preparation), and 10 µg/L for groundwater

samples. EQLs will be proportionately higher for sample extracts that require dilution to avoid saturation of the detector.

The EQL(b) for groundwater in μg/L is not listed.
The EQL (a, b) for low concentrations in soil and sediment in μg/kg is not listed.
Accuracy as μg/L is not listed.
Overall precision in μg/L is not listed.

(a) EQLs listed for soil/sediment are based on wet weight. Normally data is reported in a dry-weight basis; therefore, EQLs will be higher based on the % dry weight of each sample. This calculation is based on a 30 g sample and gel permeation chromatography cleanup.
(b) Sample EQLs are highly matrix-dependent. The EQLs are provided for guidance and may not always be achievable.
C = *True value for concentration, in μg/L.*
X = *Average recovery found for measurements of samples containing a concentration of C, in μg/L.*

ESTIMATED QUANTITATION LIMIT

Other Matrices	Factor (a)
High-concentration soil and sludges by sonicator	7.5
Non-water miscible waste	75

(a) EQL for other matrices = [EQL for low soil/sediment] × [Factor]. This estimated EQL is similar to an EPA "Practical Quantitation Limit."

SAMPLING METHOD

Liquid samples — Use a 1 or 2½ gallon amber glass bottle with a screw-top Teflon®-lined cover that has been prewashed with detergent and rinsed with distilled water and methanol (or isopropanol).

Soils, sediments, or sludges — Use an 8-oz. widemouth glass with a screw-top Teflon®-lined cover that has been prewashed with detergent and rinsed with distilled water and methanol (or isopropanol).

SAMPLE PRESERVATION

Liquid samples — If residual chlorine is present, add 3 mL of 10% sodium thiosulfate per gallon, cool to 4°C and store in a solvent-free refrigerator until analysis; if chlorine is not present, then eliminate the sodium thiosulfate addition.

Soils, sediments, or sludges — Cool samples to 4°C and store in a solvent-free refrigerator.

MHT Liquid samples must be extracted within 7 days and the extracts analyzed within 40 days. Soils, sediments, or sludges may be stored for a maximum of 14 days and the extracts analyzed within 40 days.

SAMPLE PREPARATION

Liquid samples — Transfer 1 L quantitatively to a continuous extractor. If high concentrations are anticipated, a smaller volume may be used and then diluted with organic-free reagent water to 1 L. Adjust pH, if necessary, to pH <2 using 1:1 (V/V) sulfuric acid. Pipette 1.0 mL of a surrogate standard spiking solution into each sample. For the sample in each analytical batch selected for spiking, add 1.0 mL of a matrix spiking standard. For base/neutral acid analysis, the amount of the surrogates and matrix spiking compounds added to the sample should result in a final concentration of 100 ng/μL of each analyte in the extract to be analyzed (assuming a 1-μL injection). Extract with methylene chloride for 18–24 h. Next, adjust the pH of the aqueous phase to pH >11 using 10 N sodium hydroxide and extract it with methylene chloride again for 18–24 h. Dry the extract through a column containing anhydrous sodium sulfate and concentrate it to 1 mL using a K-D concentrator.

Soils, sediments, or sludges — Use 30 g of sample. Nonporous or wet samples (gummy or clay type) that do not have a free-flowing sandy texture must be mixed with anhydrous sodium sulfate until the sample is free flowing. Add 1 mL of surrogate standards to all samples, spikes, standards, and blanks. For the sample in each analytical batch selected for spiking, add 1.0 mL of a matrix spiking standard. For base/neutral acid analysis, the amount added of the surrogates and matrix spiking compounds should result in a final concentration of 100 ng/μL of each base/neutral analyte and 200 ng/μL of each acid analyte in the extract to be analyzed (assuming a 1-μL injection). Immediately add a 100-mL mixture of 1:1 methylene chloride:acetone and extract the sample ultrasonically for 3 min and then decant or filter the extracts. Repeat the extraction two or more times. Dry the extract using a column with anhydrous sodium sulfate and concentrate it to 1 mL in a K-D concentrator.

QUALITY CONTROL A methylene chloride solution containing 50 ng/μL of decafluorotriphenylphosphine (DFTPP) is used for tuning the GC/MS system each 12-h shift. A system performance check also must be made during every 12-h shift. A standard containing 50 ng/μL each of 4,4'-DDT, pentachlorophenol, and benzidine is required to verify injection port inertness and GC column performance. A calibration standard at mid-concentration, containing each compound of interest, including all required surrogates, must be performed every 12 h during analysis. After the system performance check is met, calibration check compounds (CCCs) are used to check the validity of the initial calibration.

The internal standard responses and retention times in the calibration check standard must be evaluated immediately after or during data acquisition. If the retention time for any internal standard changes by more than 30 seconds from the last check calibration (12 h), the chromatographic system must be inspected for malfunctions and corrections must be made, as required. If the electron ionization current plot (EICP) area for any of the internal standards changes by a factor of two from the last daily calibration standard check, the mass spectrometer must be inspected for malfunctions and corrections must be made, as appropriate.

Demonstrate, through the analysis of a reagent water blank, that interferences from the analytical system, glassware, and reagents are under control. The blank samples should be carried through all stages of the sample preparation and measurement steps. For each analytical batch (up to 20 samples), a reagent blank, matrix spike, and matrix spike duplicate/duplicate must be analyzed (the frequency of the spikes may be different for different monitoring programs). The blank and spiked samples must be carried through all stages of the sample

preparation and measurement steps. A QC reference sample concentrate containing each analyte at a concentration of 100 mg/L in methanol is required.

REFERENCE Test Methods for Evaluating Solid Waste (SW-846). U.S. EPA 1983, Method 8270B, Rev. 2, Nov. 1990. Office of Solid Waste, Washington, DC.

1,2-Dibromo-3-chloropropane   **EPA Method 504**
CAS #96-12-8

TITLE (EDB)&(DBCP) in H2O by

MATRIX Finished drinking water and Microextraction unfinished groundwater.

APPLICATION 35 mL of sample are extracted with 2ml of hexane. 2ul Of the extract are injected into a GC with a linearized electron capture detector for separation & analysis. Aqueous calibration standards are run in same manner to compensate for possible extraction losses.

INTERFERENCES Impurities contained in the extracting solvent usually account for the majority of analytical problems. (Run solvent blanks as checks). Store interference-free solvents in area free of chlorinated solvents. Extraction technique extracts polars with non-polars.

INSTRUMENTATION GC with EC detector & capillary column splitless injector. Use two GC columns. Column 1: Durawax-DX3 fused silica; Column 2: DB-1 fused silica; Column 3: RTX–Volatiles wide bore.

RANGE 0.03–200 µg/L.

MDL 0.01 µg/L

PRECISION (overall) $S = 0.143X - 0.00$ (RW); $S = 0.160X + 0.006$ (groundwater).

ACCURACY (recovery) $X = 0.987C - 0.00$ (RW); $X = 0.972C + 0.007$ (groundwater).

SAMPLING METHOD Use a 40–120-mL screw-cap vial (prewashed with detergent, rinsed with distilled water and oven dried at 105°C) with a PTFE-faced silicone septum. If residual chlorine is in the water add about 25 mg of ascorbic acid to each vial before sample collection. Collect bubble-free samples.

STABILITY Cool to 4°C; add 4 drops of 10% sodium thiosulfate and 4 drops of hydrochloric acid.

MHT 28 Days.

QUALITY CONTROL As an initial demonstration of lab accuracy and precision, analyze 4 to 7 replicates of a lab fortified blank containing analyte at 0.1–5 µg/L. control. The frequency of the qc check standard analyses is equivalent to 5% of all samples analyzed. Collect all samps in duplicate.

REFERENCE Method 504, (EDB) & (DBCP) in Water by Microextraction & GC, EPA 600/4-88/039.

3,5-Dibromo-4-hydroxybenzonitrile   **EPA Method 1625**
CAS #1689-84-5

TITLE Semivolatile Organic Compounds by Isotope Dilution GC/MS

MATRIX The compounds may be determined in waters, soils, and municipal sludges by this method.

METHOD SUMMARY This method is used to determine 176 semivolatile toxic organic pollutants associated with the CWA (as amended 1987); the RCRA (as amended 1986); the CERCLA (as amended 1986); and other compounds amenable to extraction and analysis by capillary column gas chromatography-mass spectrometry (GC/MS).

Stable isotopically-labeled analogs of the compounds of interest are added to the sample. If the solids content is less than 1%, a 1-L sample is extracted at pH 12–13, then at pH <2 with methylene chloride using continuous extraction techniques.

If the solids content is 30% or less, the sample is diluted to 1% solids with reagent water, homogenized ultrasonically, and extracted at pH 12–13, then at pH <2 with methylene chloride using continuous extraction techniques. If the solids content is greater than 30%, the sample is extracted using ultrasonic techniques.

Each extract is dried over sodium sulfate, concentrated to a volume of 5 mL, cleaned up using GPC, if necessary, and concentrated. Extracts are concentrated to 1 mL if GPC is not performed, and to 0.5 mL if GPC is performed.

An internal standard is added to the extract, and a 1-mL aliquot of the extract is injected into the GC. The compounds are separated by GC and detected by a MS. The labeled compounds serve to correct the variability of the analytical technique.

INTERFERENCES Solvents, reagents, glassware, and other sample processing hardware may yield artifacts and/or elevated baselines causing misinterpretation of chromatograms and spectra. Materials used in the analysis must be demonstrated to be free from interferences under the conditions of analysis by running method blanks initially and with each sample lot (sample started through the extraction process on a given 8-h shift, to a maximum of 20). Specific selection of reagents and purification of solvents by distillation in all glass systems may be required. Glassware and, where possible, reagents are cleaned by solvent rinse and baking at 450°C for 1-h minimum. Interferences coextracted from samples will vary considerably from source to source, depending on the diversity of the site being sampled.

INSTRUMENTATION Major instrumentation includes a GC with a splitless or on-column injection port for capillary column, a MS with 70 eV electron impact ionization, and a data system to collect and record MS data, and process it. A K-D apparatus is used to concentrate extracts.

GC Column: 30 m × 0.25 mm I.D. 5% phenyl, 94% methyl, 1% vinyl silicone bonded phased fused silica capillary column.

PRECISION & ACCURACY The detection limits of the method are usually dependent on the level of interferences

rather than instrumental limitations. The limits typify the minimum quantities that can be detected with no interferences present.

The minimum level (in µg/mL) was not listed. This is defined as a minimum level at which the analytical system shall give recognizable mass spectra (background corrected) and acceptable calibration points.

The MDL (in µg/kg) in low solids was not listed and in high solids was not listed; these were determined in digested sludge (low solids) and in filter cake or compost (high solids).

The labeled and native compound initial precision as standard deviation (in µg/L) was not listed.
The labeled and native compound initial accuracy as average recovery (in µg/L) was not listed.

SAMPLE COLLECTION, PRESERVATION & HANDLING
Collect samples in glass containers. Aqueous samples which flow freely are collected in refrigerated bottles using automatic sampling equipment. Solid samples are collected as grab samples using widemouth jars. Maintain samples at 0 to 4°C from the time of collection until extraction. If residual chlorine is present in aqueous samples, add 80 mg sodium thiosulfate/L of water. Begin sample extraction within 7 days of collection, and analyze all extracts within 40 days of extraction.

SAMPLE PREPARATION Samples containing 1% solids or less are extracted directly using continuous liquid-liquid extraction techniques. Samples containing 1 to 30% solids are diluted to the 1% level with reagent water and extracted using continuous liquid-liquid extraction techniques. Samples containing greater than 30% solids are extracted using ultrasonic techniques.

Base/neutral extraction — Adjust the pH of the waters in the extractors to 12–13 with 6 N NaOH. Extract with methylene chloride for 24–48 h.
Acid extraction — Adjust the pH of the waters in the extractors to 2 or less using 6 N sulfuric acid. Extract with methylene chloride for 24–48 h.
Ultrasonic extraction of high solids samples — Add anhydrous sodium sulfate to the sample and QC aliquot(s). Add acetone:methylene chloride (1:1) to the sample and mix thoroughly

Concentrate extracts using a K-D apparatus.

QUALITY CONTROL The analyst is permitted to modify this method to improve separations or lower the costs of measurements, provided all performance specifications are met. Analyses of blanks are required to demonstrate freedom from contamination. When results of spikes indicate atypical method performance for samples, the samples are diluted to bring method performance within acceptable limits.

For low solids (aqueous samples), extract, concentrate, and analyze two sets of four 1-L aliquots (8 aliquots total) of the precision and recovery standard. For high solids samples, two sets of four 30-g aliquots of the high solids reference matrix are used.

Spike all samples with labeled compounds to assess method performance. Compute percent recovery of the labeled compounds using the internal standard method. Compare the labeled compound recovery for each compound with the corresponding labeled compound recovery.

Reagent water and high solids reference matrix blanks are analyzed to demonstrate freedom from contamination. Extract and concentrate a 1-L reagent water blank or a high solids reference matrix blank with each sample's lot (samples started through the extraction process on the same 8-h shift, to a maximum of 20 samples).

Field replicates may be collected to determine the precision of the sampling technique, and spiked samples may be required to determine the accuracy of the analysis when the internal standard method is used.

REFERENCE Semivolatile Organic Compounds by Isotope Dilution GC/MS. Office of Water Regulation and Standards, U.S. EPA Industrial Technology Division, Washington, DC, EPA Method 1625, Rev. C, June 1989 (contact W.A. Telliard, U.S. EPA, Office of Water Regulations and Standards, 401 M St., SW, Washington, DC, 20460. Phone: 202-382-7131).

Dibromochloromethane   **EPA Method 1624**
CAS #124–48-1

TITLE Volatile Organic Compounds by Isotope Dilution GC/MS

MATRIX Compounds may be determined in waters, soils, and municipal sludges by this method.

METHOD SUMMARY This method is used to determine 58 volatile toxic organic pollutants associated with the CWA (as amended 1987); the RCRA (as amended 1986); the CERCLA (as amended 1986); and other compounds amenable to purge-and-trap gas chromatography-mass spectrometry (GC/MS).

If the solids content is less than 1%, stable isotopically-labeled analogs of the compounds of interest are added to a 5-mL sample and the sample is purged with an inert gas at 20–25°C in a chamber designed for soil or water samples. If the solids content is greater than 1%, 5 mL of reagent water and the labeled compounds are added to a 5-g aliquot of sample and the mixture is purged at 40°C. Compounds that will not purge at 20–25°C or at 40°C are purged at 78–85°C. In the purging process, the volatile compounds are transferred from the aqueous phase into the gaseous phase where they are passed into a sorbent column, and trapped. After purging is completed, the trap is backflushed and heated rapidly to desorb the compounds into a GC. The compounds are separated by the GC and detected by a MS. The labeled compounds serve to correct the variability of the analytical technique.

INTERFERENCES Impurities in the purge gas, organic compounds outgassing from the plumbing upstream of the trap, and solvent vapors in the lab account for most problems. Samples can be contaminated by diffusion of volatile organic compounds (particularly methylene chloride) through the bottle seal during shipment and storage. Contamination by carryover can occur when high-level and low-level samples are analyzed sequentially. When an unusually concentrated sample is encountered, follow it by analysis of a reagent water blank to check for carryover.

INSTRUMENTATION Major equipment includes a GC with linear temperature programming and a glass jet separator as the MS interface, a MS with 70 eV electron impact ionization, and a data system to collect and record response factors.

Column: 2.8 m × 2 mm I.D. glass, packed with 1% SP-1000 on Carbopak B, 60/80 mesh, or equivalent.

PRECISION & ACCURACY The detection limits of the method are usually dependent on the level of interferences rather than instrumental limitations. The method detection limits were determined in digested sludge (low solids) and in filter cake or compost (high solids).

The MDL (in µg/kg) for low solids is not listed and for high solids is not listed.

Labeled and native compound precision (in µg/L) as standard deviation was 7.9.

Labeled and native compound accuracy (in µg/L) as average recovery was 11–29.

Acceptance criteria are at 20 µg/L for this compound.

SAMPLE COLLECTION, PRESERVATION & HANDLING Grab samples are collected in glass containers having a total volume greater than 20 mL. Fill and seal each bottle so that no air bubbles are entrapped. Samples are maintained at 0 to 4°C from the time of collection until analysis. If an aqueous sample contains residual chlorine, add sodium thiosulfate preservative (10 mg/40 mL) to the empty sample bottles just prior to shipment to the sample site. All samples must be analyzed within 14 days of collection.

SAMPLE PREPARATION Samples containing less than 1% solids are analyzed directly as aqueous samples. Samples containing 1% solids or greater are analyzed as solid samples utilizing one of two methods, depending on the levels of pollutants, in the sample. Samples containing 1% solids or greater, and low to moderate levels of pollutants are analyzed by purging a known weight of sample added to 5 mL of reagent water. Samples containing 1% solids or greater, and high levels of pollutants, are extracted with methanol, and an aliquot of the methanol extract is added to reagent water and purged.

QUALITY CONTROL A field blank prepared from reagent water and carried through the sampling and handling protocol may serve as a check on contamination from shipment and storage.

The analyst is permitted to modify this method to improve separations or lower the costs of measurements, provided all performance specifications are met. Analyses of blanks are required. When results of spikes indicate atypical method performance for samples, the samples are diluted to bring method performance within acceptable limits. Analyze two sets of four 5-mL aliquots (8 aliquots total) of the aqueous performance standard. Spike all samples with labeled compounds to assess method performance on the sample matrix. Compute the percent recovery of the labeled compounds using the internal standard method. Compare the percent recovery for each compound with the corresponding labeled compound recovery. Reagent water blanks are analyzed to demonstrate freedom from carryover contamination. Field replicates may be collected to determine the precision of the sampling technique, and spiked samples may be required to determine the accuracy of the analysis when the internal method is used.

REFERENCE Volatile Organic Compounds by Isotope Dilution GC/MS. Office of Water Regulation and Standards, U.S. EPA Industrial Technology Division, Washington, DC, EPA Method 1624, Rev. C, June 1989 (contact W.A. Telliard, U.S. EPA, Office of Water Regulations and Standards, 401 M St., SW, Washington, DC, 20460. Phone: 202-382-7131).

Dibromochloromethane   **EPA Method 502**
CAS #124–48–1

TITLE Volatile Organic Compounds in Water By Purge and Trap Capillary Column Gas Chromatography with Photoionization and Electrolytic Conductivity Detectors in Series. U.S. EPA Method 502.2, Rev. 2.0, 1989.

MATRIX Drinking water and raw source water. The latter should include most surface water and groundwater sources.

METHOD SUMMARY This method covers 60 volatile organic compounds that contain halogen atoms and/or that are aromatic. An inert gas (zero grade nitrogen or helium) is bubbled through a 25-mL or a 5-mL water sample (depending on the expected concentration of the analytes). Purged sample components are trapped in a tube of sorbent materials. When purging is complete, the sorbent tube is heated and backflushed with helium to desorb the trapped sample onto a capillary GC column. The column is temperature programmed to separate the method analytes which are then detected with a photoionization detector (PID) and a Hall electrolytic conductivity (HECD) placed in series. The PID is selective for aromatic compounds and the HECD is selective for halogenated compounds.

INTERFERENCES Impurities in the purge gas and from organic compounds outgassing from the plumbing ahead of the trap account for many contamination problems. Interferences purged or coextracted from the samples will vary considerably from source to source, depending upon the particular sample or extract being tested. Cross-contamination can occur whenever high-level and low-level samples are analyzed sequentially. Samples also can be contaminated by diffusion of volatile organics (particularly methylene chloride and fluorocarbons) through the septum seal into the sample during shipment and storage. The lab where volatile analysis is performed and also the refrigerated storage area should be completely free of solvents.

INSTRUMENTATION A GC containing a series configuration of a high temperature photoionization detector (PID) equipped with 10.0 eV (nominal) lamp and Hall electrolytic conductivity detector (HECD) is required. Also required is an

all-glass 5-mL purging device, a sorbent trap, and a thermal desorption apparatus which is connected to the GC system.

Column 1: VOCOL glass wide-bore capillary column.
Column 2: RTX–502.2 mega-bore capillary column.
Column 3: DB-62 mega-bore capillary column.

PRECISION & ACCURACY Method detection limits are dependent upon the characteristics of the gas chromatographic system used. Analytes that are not separated chromatographically cannot be individually identified and used in the same calibration mixture or water samples unless an alternative technique for identification and quantification, such as mass spectrometry, is used.

Electrolytic conductivity detetor (c) range in µg/L (a) was 0.02–200.
Electrolytic conductivity detetor (c) MDL in µg/L (b) was 0.3.
Electrolytic conductivity detetor (c) accuracy as % recovery was 102.
Electrolytic conductivity detetor (c) precision as % RSD was 3.3.
Photoionization detector (d) range in µg/L (a) was 0.02–200.
Photoionization detector (d) MDL in µg/L (b) was not listed.
Photoionization detector (d) accuracy as % recovery was not listed.
Photoionization detector (d) precision as % RSD was not listed.

(a) The applicable concentration range of this method is compound, instrument, and matrix-dependent. It is listed as being approximately 0.02 to 200 µg/L but no specific information is provided so caution should be observed.

(b) The method detection limits reports with this method are compound, instrument, and matrix-dependent. The values reported were calculated using reagent water fortified with the corresponding compounds at 10 µg/L and a GC-equipped with a 60 m × 0.75 mm VOLCOL wide bore capillary column with 1.5 µm film thickness and using helium carrier gas.

(c) Recoveries and relative standard deviations were determined from seven samples of reagent water fortified with 10 µg/L of each compound. 2-Bromo-1-chloropropane was used as the internal standard for calculating average recoveries.

(d) Recoveries and relative standard deviations were determined from seven samples of reagent water fortified with 10 µg/L of each compound. Fluorobenzene was used as the internal standard for calculating average recoveries.

SAMPLING METHOD Collect samples using a 40- to 120-mL screw-cap vial (prewashed with detergent, rinsed with distilled water and oven dried at 105°C) with a Teflon®-faced silicone septum. Collect bubble-free samples and place the septum with the Teflon® side down on the water.

SAMPLE PRESERVATION If residual chlorine is present in the water add about 25 mg of ascorbic acid to each vial before samples are collected to remove the chlorine. Add hydrochloric acid to reduce pH to <2, immediately cool samples to 4°C, and store them in a solvent-free refrigerator at 4°C until analysis.

MHT The maximum holding time for samples is 14 days from the time they were collected.

SAMPLE PREPARATION Remove the plungers from two 5-mL syringes and attach a closed syringe valve to each. Warm the sample to room temperature, open the sample bottle, and carefully pour the sample into one of the syringe barrels to just short of overflowing. Replace the syringe plunger, invert the syringe, and compress the sample. Open the syringe valve and vent any residual air while adjusting the sample volume to 5.0 mL. Add 10 µL of the internal calibration standard to the sample through the syringe valve. Close the valve. Fill the second syringe in an identical manner from the same sample bottle. Reserve this second syringe for a reanalysis if necessary.

QUALITY CONTROL As an initial demonstration of lab accuracy and precision, analyze 4 to 7 replicates of a lab fortified blank containing analyte at 0.1–5 µg/L. Collect all samples in duplicate. Surrogate analytes (similar to those of the analytes of interest), whose concentration is known in every sample, are measured using the same internal standard calibration procedure. Duplicate field reagent water blanks (trip blanks) must be analyzed with each set of samples, lab reagent blanks (method blanks) must be analyzed with each batch of samples processed as a group within a work shift. Also, a single lab-fortified blank that contains each of the analytes of interest should be analyzed with each batch of samples processed as a group within a work shift. A 3- to 5-point calibration curve is needed depending on the calibration range factor required.

EPA CONTACT & HOTLINE For technical questions contact Dr. Baldev Bathija, U.S. EPA, Office of Ground Water and Drinking Water (WH-550D), 401 M St. SW, Washington, DC 20460. Tel. (202) 260-3040. For further information the EPA Safe Drinking Water Hotline may be called at: (800) 426-4791.

REFERENCE Methods for the Determination of Organic Compounds in Drinking Water, EPA/600/4-88/039 (revised July 1991; Final Rule for determination of compliance with the MCL for Total Trihalomethanes under 141.30, in 40 CFR Part 141, Vol. 58, No. 147, Fed. Reg., Tuesday Aug. 3, 1993). U.S. EPA Environmental Monitoring Systems Laboratory, Cincinnati, OH, 45268, U.S.A. Available from the National Technical Information Service (NTIS), 5285 Port Royal Road, Springfield, VA 22161; Tel. 800-553-6847. NTIS Order Number is PB91-231480.

Dibromochloromethane — EPA Method 524
CAS #124–48-1

TITLE Measurement of Purgeable Organic Compounds in Water by Capillary Column GC/MS.

MATRIX Drinking water and raw source water; the latter should include most surface water and groundwater sources.

METHOD SUMMARY Method 524.2 covers 60 volatile organic compounds. An inert gas (zero grade nitrogen or helium) is bubbled through a 25-mL or a 5-mL water sample (depending on the expected concentration of the analytes). Purged sample components are trapped in a tube of sorbent materials. When purging is complete, the sorbent tube is heated and backflushed with helium to desorb the trapped sample onto a capillary GC column.

INTERFERENCES Impurities in the purge gas and from organic compounds outgassing from the plumbing ahead of the trap account for many contamination problems. Interferences purged or coextracted from the samples will vary considerably from source to source, depending upon the particular sample or extract being tested. Cross-contamination can occur whenever high-level and low-level samples are analyzed sequentially. Samples also can be contaminated by diffusion of volatile organics (particularly methylene chloride and fluorocarbons) through the septum seal into the sample during shipment and storage.

INSTRUMENTATION A GC/MS with a data system equipped with one of the following capillary GC columns:

Column 1: VOCOL glass wide bore capillary column.
Column 2: DB-624 fused silica capillary column.
Column 3: DB-5 fused silica capillary column.

Also required is an all-glass 25 mL or 5-mL purging device, a sorbent trap, and a thermal desorption apparatus which is connected to the GC/MS system.

PRECISION & ACCURACY Method detection limits are compound- and instrument-dependent, and may vary from approximately 0.02–0.35 µg/L. Note in the table below that the "true" concentration range used for accuracy and precision measurements was quite narrow. However, the applicable concentration range of this method is primarily column dependent and is approximately 0.02 to 200 µg/L for the wide-bore thick-film columns. Narrow-bore thin-film columns may have a capacity which limits the range to about 0.02 to 20 µg/L. Analytes that are inefficiently purged from water will not be detected when present at low concentrations, but they can be measured with acceptable accuracy and precision when present in sufficient amounts.

Analytes that are not separated chromatographically, but which have different mass spectra and non-interfering quantification ions, can be identified and measured in the same calibration mixture or water sample. Analytes which have very similar mass spectra cannot be individually identified and measured in the same calibration mixture or water samples unless they have different retention times. Co-eluting compounds with very similar mass spectra, typically many structural isomers, must be reported as an isomeric group or pair.

The range (in µg/L) was 0.1–10.
The Method Detection Limig (in µg/L) was 0.05.
The accuracy (as % recovery) was 92.
The precision (in %) was 7.0.

Note: Data were obtained from 16–31 determinations using a wide-bore capillary column and a jet separator interfaced to a quadrupole mass spectrometer. All analytes were in a reagent water matrix.

SAMPLING METHOD Collect samples using a 40- to 120-mL screw-cap vial (prewashed with detergent, rinsed with distilled water and oven dried at 105°C) with a Teflon®-faced silicone septum. Collect bubble-free samples and place the septum with the Teflon® side down on the water.

SAMPLE PRESERVATION If residual chlorine is present in the water add about 25 mg of ascorbic acid to each vial before samples are collected to remove the chlorine. Add hydrochloric acid to reduce pH to <2, and immediately cool samples to 4°C, and store them in a solvent-free refrigerator at 4°C until analysis.

MHT The maximum holding time for samples is 14 days from the time they were collected.

SAMPLE PREPARATION Remove the plungers from two 25-mL (or 5-mL depending on sample size) syringes and attach a closed syringe valve to each. Warm the sample to room temperature, open the sample bottle, and carefully pour the sample into one of the syringe barrels to just short of overflowing. Replace the syringe plunger, invert the syringe, and compress the sample. Open the syringe valve and vent any residual air while adjusting the sample volume to 25.0 mL (or 5 mL). For samples and blanks, add 5 µL of the fortification solution containing the internal standard and the surrogates to the sample through the syringe valve. For calibration standards and lab fortified blanks, add 5 µL of the fortification solution containing the internal standard only. Close the valve. Fill the second syringe in an identical manner from the same sample bottle. Reserve this second syringe for a reanalysis if necessary.

QUALITY CONTROL As an initial demonstration of lab accuracy and precision, analyze 4 to 7 replicates of a lab fortified blank containing analyte at 0.2–5 µg/L. Collect all samples in duplicate. Surrogate analytes (similar to those of the analytes of interest), whose concentration is known in every sample, are measured using the same internal standard calibration procedure. Duplicate field reagent water blanks (trip blanks) must be analyzed with each set of samples, lab reagent blanks (method blanks) must be analyzed with each batch of samples processed as a group within a work shift. Also, a single lab-fortified blank that contains each of the analytes of interest should be analyzed with each batch of samples processed as a group within a work shift. A 3- to 5-point calibration curve is needed depending on the calibration range factor required.

EPA CONTACT & HOTLINE For technical questions contact Dr. Baldev Bathija, U.S. EPA, Office of Ground Water and Drinking Water (WH-550D), 401 M St. SW, Washington, DC 20460. Tel. (202) 260-3040. For further information the EPA Safe Drinking Water Hotline may be called at: (800) 426-4791.

REFERENCE Methods for the Determination of Organic Compounds in Drinking Water, EPA/600/4-88/039 (revised July 1991; Final Rule for determination of compliance with the MCL for Total Trihalomethanes under 141.30, in 40 CFR Part 141, Vol. 58, No. 147, Fed. Reg., Tuesday Aug. 3, 1993). U.S. EPA Environmental Monitoring Systems Laboratory, Cincinnati, OH, 45268, U.S.A. Available from the National Technical Information Service (NTIS), 5285 Port Royal Road, Springfield, VA 22161; Tel. 800-553-6847. NTIS Order Number is PB91-231480.

Dibromochloromethane
CAS #124-48-1

EPA Method 8021

TITLE Halogenated Volatile by Gas Chromatography Using Photoionization and Electrolytic Conductivity Detectors in Series: Capillary Column Technique

MATRIX This method is applicable to nearly all types of samples, regardless of water content, including groundwater, aqueous sludges, caustic liquors, acid liquors, waste solvents, oily wastes, mousses, tars, fibrous wastes, polymeric emulsions, filter cakes, spent carbons, spent catalysts, soils, and sediments.

METHOD SUMMARY This method is used to determine 60 volatile organic compounds in a variety of solid waste matrices. It provides GC conditions for the detection of halogenated and aromatic volatile organic compounds. Samples can be analyzed using direct injection or purge-and-trap (EPA Method 5030). Groundwater samples must be analyzed using EPA Method 5030 (where applicable). A temperature program is used with the GC. Detection is achieved by a photoionization detector (PID) and a Hall electrolytic conductivity detector (HECD) in series.

INTERFERENCES Samples can be contaminated by diffusion of volatile organics (particularly chlorofluorocarbons and methylene chloride) through the sample container septum during shipment and storage.

INSTRUMENTATION A GC-equipped with variable-constant differential flow controllers, subambient oven controller, PID and HECD detectors connected with a short piece of uncoated capillary tubing and a data system.

Column: 60 m × 0.75 mm I.D. VOCOL wide-bore capillary column with 1.5 µm film thickness.

PRECISION & ACCURACY MDLs are compound-dependent and vary with purging efficiency and concentration. The applicable concentration range of this method is compound- and instrument-dependent but is approximately 0.1 to 200 µg/L. Analytes that are inefficiently purged from water will not be detected when present at low concentrations, but they can be measured with acceptable accuracy and precision when present in sufficient amounts. The estimated quantitation limit (EQL) for an individual compound is approximately 1 µg/kg (wet weight) for soil/sediment samples, 100 µg/kg (wet weight) for wastes, and 1 µg/L for groundwater. EQLs will be proportionately higher for sample extracts and samples that require dilution or reduced sample size to avoid saturation of the detector.

MULTIPLICATION FACTORS FOR OTHER MATRICES (a)

Matrix	Factor (b)
Groundwater	10
Low-concentration soil	10
Water miscible liquid waste	500
High-concentration soil and sludge	1250
Non-water miscible waste	1250

(a) Sample EQLs are highly matrix-dependent. The EQLs listed herein are provided for guidance and may not always be achievable. (b) EQL = [Method detection limit] × [Factor]. For non-aqueous samples, the factor is on a wet-weight basis.

SINGLE LABORATORY ACCURACY & PRECISION DATA FOR VOCs IN WATER

This method was tested in a single lab using water spiked at 10 µg/L and the following data was reported:

Recoveries and standard deviations were determined from seven samples and spiked at 10 µg/L of each analyte. Recoveries were determined by the internal standard method. Internal standards were: Fluorobenzene for PID and 2-Bromo-1-chloropropane for HECD.

The average recovery (in percent) for the PID was none (no response for this detector).
The standard deviation of the recovery for the PID was none (no response for this detector).
The MDL (in µg/mL) for the PID was none (no response for this detector).
The average recovery (in percent) for the HECD was 102.
The standard deviation of the recovery for the HECD was 3.3.
The MDL (in µg/mL) for the HECD was 0.03.

SAMPLE COLLECTION, PRESERVATION & HANDLING
Volatile Organics — Standard 40-mL glass screw-cap VOA vials with Teflon®-faced silicone septum may be used for both liquid and solid matrices. When collecting samples, liquids and solids should be introduced into the vials gently to reduce agitation which might drive off volatile compounds. If there are any air bubbles present the sample must be retaken. Tap slightly as they are filled to try and eliminate as much free air space as possible. The two vials from each sampling locations should be sealed in separate plastic bags to prevent cross-contamination between samples particularly if the sampled waste is suspected of containing high levels of volatile organics.

Semivolatile organics — Containers used to collect samples for the determination of semivolatile organic compounds should be soap and water washed followed by methanol (or isopropanol) rinsing. The sample containers should be of glass or Teflon® and have screw-top covers with Teflon® liners.

Preservation for volatile organics — No preservation is used with concentrated waste samples. With liquid samples containing no residual chlorine, 4 drops of concentrated hydrochloric acid are added and the samples are immediately cooled to 4°C. When liquid samples contain residual chlorine, they are treated as above and, in addition, 4 drops of 4% aqueous sodium thiosulfate are added. Soil, sediment, and sludge samples are only cooled to 4°C.

Preservation for semivolatile organics — No preservation is used with concentrated waste samples. With liquid samples containing no residual chlorine and with soil, sediment, and sludge samples, immediately cooling to 4°C is the only preservation used. When residual chlorine is present then 3 mL of 10% aqueous sodium sulfate is added for each gallon of sample collected, followed by cooling to 4°C.

MHT The holding time for all volatile organics samples is 14 days. Liquid samples must be extracted within 7 days and their extracts analyzed within 40 days. Concentrated waste, soil, sediment, and sludge samples must be extracted within 14 days and their extracts analyzed within 40 days.

SAMPLE PREPARATION Volatile compounds are introduced into the gas chromatograph either by direct injector or purge-and-trap (EPA Method 5030). EPA Method 5030 may be used directly on groundwater samples or low-concentration contaminated soils and sediments. For medium-concentration soils or sediments, methanolic extraction, as described in EPA Method 5030, may be necessary prior to purge-and-trap analysis.

QUALITY CONTROL Calculate surrogate standard recovery on all samples, blanks, and spikes. A trip blank is recommended to check on sampling, storage, and handling contamination. Calibration standards, at a minimum of five concentration levels, are prepared in organic-free reagent water. One of the concentration levels should be at a concentration near, but above, the method detection limit.

A combination of bromochloromethane, 2-bromo-1-chloropropane, 1,4-dichlorobutane, and bromochlorobenzene are recommended as surrogate standards to encompass the range of the temperature program used in this method.

REFERENCE Test Methods for Evaluating Solid Waste, Physical/Chemical Methods, SW-846, 3rd Edition, U.S. EPA, Office of Solid Waste, Washington, DC, EPA Method 8021A, Rev. 1, Nov. 1992.

Dibromochloromethane   **EPA Method 8240**
CAS #124–48-1

TITLE Volatile Organics By GC/MS: Packed Column Technique

MATRIX Nearly all types of sample matarices, regardless of water content, can be analyzed using this method. This includes groundwater, aqueous sludges, caustic liquors, acid liquors, waste solvents, oily wastes, mousses, tars, fibrous wastes, polymetric emulsions, filter cakes, spent carbons, spent catalysts, soils, and sediments.

METHOD SUMMARY Method 8240B covers 80 volatile organic compounds that are introduced into a gas chromatograph by the purge-and-trap method or by direct injection (in limited applications). For the purge-and-trap method an inert gas (zero grade nitrogen or helium) is bubbled through a 5-mL solution at ambient temperature. Purged sample components are trapped in a tube of sorbent materials. When purging is complete, the sorbent tube is heated and backflushed with inert gas to desorb the trapped components onto a GC column.

INTERFERENCES Impurities in the purge gas and from organic compounds outgassing from the plumbing ahead of the trap account for many contamination problems. Interferences purged or coextracted from the samples will vary considerably from source to source. Cross-contamination can occur whenever high-level and low-level samples are analyzed sequentially. Whenever an unusually concentrated sample is analyzed, it should be followed by the analysis of organic-free reagent water to check for cross-contamination. Samples also can be contaminated by diffusion of volatile organics (particularly methylene chloride and fluorocarbons) through the septum seal into the sample during shipment and storage. A trip blank can serve as a check on such contamination. The lab where volatile analysis is performed and also the refrigerated storage area should be completely free of solvents.

INSTRUMENTATION A gas chromatograph/mass spectrometry/data system (GC/MS) equipped with a 6 ft × 0.1 in I.D. glass column packed with 1% SP-1000 on Carbopack-B (60/80 mesh) is required. Also needed is a 5-mL purging device, a sorbent trap, and a thermal desorption apparatus.

PRECISION & ACCURACY This method is reported to have been tested by 15 laboratories using organic-free reagent water, drinking water, surface water, and industrial wastewaters (not specified) fortified at six concentrations over the range 5–600 µg/L.

Sample estimated quantitation limits (EQLs) are highly matrix-dependent. The EQLs listed may not always be achievable. EQLs listed for soils or sediments are based on wet weight. Normally, data is reported on a dry-weight basis; therefore, EQLs will be higher, based on the percent dry weight of each sample. Note that EQLs are even more variable than MDLs and that they are highly variable depending on the matrix being analyzed.

EQL in groundwater in µg/L was not listed.
EQL in low soil or sediment in µg/kg was not listed.
Accuracy (a) in µg/L was 1.01C-0.03.
Precision (b) in µg/L was 0.17x+0.49.

(a) *Average recovery found for measurements of samples containing a concentration of C, in µg/L.*
(b) *Overall precision found for measurements of samples with average recovery X for samples containing a concentration of C in µg/L.*
X = *Average recovery found for measurement of samples containing a concentration of C in µg/L.*

MULTIPLICATION FACTORS FOR OTHER MATRICES

Other Matrices	Factor (a)
Waste miscible liquid waste	50
High-concentration soil and sludge	125
Non-water miscible waste	500

(a) *EQL = [EQL for low soil sediment] × [Factor]. For non-aqueous samples, the factor is on a wet-weight basis.*

SAMPLING METHOD

Liquid samples — Use a 40-mL glass screw-cap VOA vial with a Teflon®-faced silicone septum that has been prewashed, rinsed with distilled deionized water, and oven dried. However, if residual chlorine is present, collect sample in a 40-oz. soil VOA container which has been pre-preserved with 4 drops of 10% sodium thiosulfate, mix gently, and then transfer the sample to a 40-mL VOA vial. Collect bubble-free samples in duplicate and seal them in separate plastic bags.

Soils or sediments, and sludges — Use an 8-oz. widemouth glass bottle with a Teflon®-faced silicone septum that has been prewashed with detergent, rinsed with distilled deionized water, and oven dried. Tap slightly to eliminate free air space. Collect samples in duplicate and seal them in separate plastic bags.

SAMPLE PRESERVATION

Liquid samples — Add 4 drops of concentrated HCL and immediately cool samples to 4°C and store in a solvent-free refrigerator.

Soils or sediments, and sludges — Cool samples to 4°C and store in a solvent-free refrigerator.

MHT Maximum holding time is 14 days from the date of sample collection.

SAMPLE PREPARATION

Liquid samples — Remove the plunger from a 5-mL syringe and carefully pour the sample into the syringe barrel to just short of overflowing. Replace the syringe plunger and compress the sample. Open the syringe valve and vent any residual air while adjusting the sample volume to 5.0 mL. If there is only one volatile organic analysis (VOA) vial, a second syringe should be filled at this time to protect against possible loss of sample integrity. Add 10 µL of surrogate spiking solution and 10 µL of internal standard spiking solution through the valve bore of the 5-mL syringe, then close the valve. The surrogate and internal standards may be mixed and added as a single spiking solution.

Sediments, soils, and waste samples — All samples of this type should be screened by GC analysis using a headspace method (EPA Method 3810) or the hexadecane extraction and screening method (EPA Method 3820). Use the screening data to determine whether to use the low-concentration method (0.005–1 mg/kg) or the high-concentration method (>1 mg/kg).

Low-concentration method — The low-concentration method is based on purging a heated sediment or soil sample mixed with organic-free reagent water containing the surrogate and internal standards. Analyze all reagent blanks and standards under the same conditions as the samples.

Use a 5-g sample if the expected concentration is <0.1 mg/kg or a 1-g sample for expected concentrations between 0.1 and 1 mg/kg. Mix the contents of the sample container with a narrow metal spatula. Weigh the amount of the sample into a tared purge device. Add the spiked water to the purge device, which contains the weighed amount of sample, and connect the device to the purge-and-trap system.

High-concentration method — This method is based on extracting the sediment or soil with methanol. A waste sample is either extracted or diluted, depending on its solubility in methanol. Wastes that are insoluble in methanol are diluted with reagent tetraglyme or possibly polyethylene glycol (PEG). An aliquot of the extract is added to organic-free reagent water containing surrogate and internal standards. This is purged at ambient temperature. All samples with an expected concentration of >1.0 mg/kg should be analyzed by this method.

Mix the contents of the sample container with a narrow metal spatula. For sediments or soils and solid wastes that are insoluble in methanol, weigh 4 g (wet weight) of sample into a tared 20-mL vial. For waste that is soluble in methanol, tetraglyme, or PEG, weigh 1 g (wet weight) into a tared scintillation vial or culture tube or a 10-mL volumetric flask. Quickly add 9.0 mL of appropriate solvent then add 1.0 mL of a surrogate spiking solution to the vial, cap it, and shake it for 2 min.

METHANOL EXTRACT REQUIRED FOR ANALYSIS OF HIGH-CONCENTRATION SOILS OR SEDIMENTS

Approximate Concentration Range	Volume of Methanol Extract (a)
500–10,000 µg/kg	100 µL
1,000–20,000 µg/kg	50 µL
5,000–100,000 µg/kg	10 µL
25,000–500,000 µg/kg	100 µL of 1/50 dilution (b)

Calculate appropriate dilution factor for concentrations exceeding this table.

(a) The volume of methanol added to 5 mL of water being purged should be kept constant. Therefore, add to the 5-mL syringe whatever volume of methanol is necessary to maintain a volume of 100 µL added to the syringe.
(b) Dilute an aliquot of the methanol extract and then take 100 µL for analysis.

QUALITY CONTROL Demonstrate, through the analysis of a reagent water blank, that interferences from the analytical system, glassware, and reagents are under control. Blank samples should be carried through all stages of the sample preparation and measurement steps. For each analytical batch (up to 20 samples), a reagent blank, matrix spike, and matrix spike duplicate must be analyzed (the frequency of the spikes may be different for different monitoring programs). The blank and spiked samples must be carried through all stages of the sample preparation and measurement steps. QC samples mentioned in the section on Interferences will also be needed as appropriate to those situations.

REFERENCE Test Methods for Evaluating Solid Waste (SW-846). U.S. EPA. 1983. Method 8240B, Rev. 2, Nov. 1990. Office of Solid Wastes, Washington, DC.

Dibromochloromethane   **EPA Method 8260**
CAS #124–48-1

TITLE Volatile Organic Compounds by GC/MS: Capillary Column Technique

MATRIX This method is applicable to nearly all types of samples, regardless of water content, including groundwater, soils, and sediments.

METHOD SUMMARY Method 8260A covers 58 volatile organic compounds that are introduced into a gas chromatograph by the purge-and-trap method or by direct injection (in limited applications). Zero-grade helium is bubbled through a 5-mL solution at ambient temperature. Purged sample components are trapped in a tube containing suitable sorbent materials. When purging is complete, the sorbent tube is heated and backflushed with helium to desorb trapped sample components. The analytes are desorbed directly to a large bore capillary or cryofocussed on a capillary precolumn before being flash evaporated to a narrow bore capillary for analysis.

INTERFERENCES Major contaminant sources are volatile materials in the lab and impurities in the inert purging gas and in the sorbent trap. Interfering contamination may occur when a sample containing low concentrations of volatile organic compounds is analyzed immediately after a sample containing high concentrations of volatile organic compounds. After analysis of a sample containing high concentrations of volatile organic compounds, one or more calibration blanks should be analyzed to check for cross-contamination. Screening of the samples prior to purge-and-trap GC/MS analysis is highly recommended to prevent contamination of the system. This is especially true for soil and waste samples.

Special precautions must be taken to analyze for methylene chloride. The analytical and sample storage area should be isolated from all atmospheric sources of methylene chloride. All gas chromatography carrier gas lines and purge gas plumbing should be constructed from stainless steel or copper tubing. Laboratory clothing previously exposed to methylene chloride fumes during liquid-liquid extraction procedures can contribute to sample contamination.

Samples can also be contaminated by diffusion of volatile organics (particularly methylene chloride and fluorocarbons) through the septum seal during shipment and storage. A trip blank can serve as a check on such contamination.

INSTRUMENTATION GC/MS with a temperature-programmable chromatograph suitable for splitless injection equipped with variable constant differential flow controllers, a subambient oven controller, a purging device, sorbent trap, a thermal desorption apparatus and a capillary precolumn interface when using cryogenic cooling will be needed. The following GC columns may be used:

Column 1: 60 m × 0.75mm I.D. capillary column coated with VOCOL, 1.5 µm film thickness.
Column 2: 30 m × 0.53mm capillary column coated with DB-624 or VOCOL, 3 µm film thickness.
Column 3: 30 m × 0.32mm I.D. capillary column coated with DB-5 or SE-54, 1-µm film thickness.

PRECISION & ACCURACY This method has been tested in a single lab using spiked water. Using a wide-bore capillary column, water was spiked at concentrations between 0.5 and 10 µg/L. Single lab accuracy and precision data are presented. The MDL actually achieved in a given analysis will vary depending on instrument sensitivity and matrix effects.

The MDL (a) in µg/L was 0.05.
The concentration range in µg/L was 0.1–10.
The mean accuracy (% of true value) was 83.
The precision as relative standard deviation was 7.0.

Note: The MDL is based on a 25-mL sample volume instead of a 5-mL sample volume.

SAMPLING METHOD
Liquid samples — Use a 40-mL glass screw-cap VOA vial with a Teflon®-faced silicone septum that has been prewashed, rinsed with distilled deionized water, and oven dried. If residual chlorine is present, collect the sample in a 4-oz soil VOA container which has been pre-preserved with 4 drops of 10% sodium thiosulfate. Mix gently and transfer the sample to a 40-mL VOA vial. Collect bubble-free samples in duplicate and seal each sample in a separate plastic bag.

Soils, sediments, and sludges — Use an 8-oz widemouth glass bottle with Teflon®-faced silicone septum that has been prewashed, rinsed with distilled deionized water, and oven dried. **Do not** heat the septum for more than 1 h. Tap slightly to eliminate any free air space. Collect samples in duplicate and seal each one in a separate plastic bag.

SAMPLE PRESERVATION
Liquid samples — Add 4 drops of concentrated HCL, cool to 4°C and store in a solvent-free refrigerator.

Soils, sediments and sludges — Cool samples to 4°C and store in a solvent-free refrigerator.

MHT The maximum holding time of any sample (liquids, soils, sediments, and sludges) is 14 days.

SAMPLE PREPARATION
Liquid samples — Remove the plunger from a 5-mL syringe and carefully pour the sample into the syringe barrel to just short of overflowing. Replace the syringe plunger and compress the sample. Open the syringe valve and vent any residual air while adjusting the sample volume to 5.0 mL. If there is only one volatile organic analysis (VOA) vial, a second syringe should be filled at this time to protect against possible loss of sample integrity. Add 10 µL of surrogate spiking solution and 10 µL of internal standard spiking solution through the valve bore of the 5-mL syringe, then close the valve. The surrogate and internal standards may be mixed and added as a single spiking solution.

Sediments, soils, and waste samples — All samples of this type should be screened by GC analysis using a headspace method (EPA Method 3810) or the hexadecane extraction and screening method (EPA Method 3820). Use the screening data to determine whether to use the low-concentration method (0.005–1 mg/kg) or the high-concentration method (>1 mg/kg).

Low-concentration method — The low-concentration method is based on purging a heated sediment or soil sample mixed with organic-free reagent water containing the surrogate and internal standards. Analyze all reagent blanks and standards under the same conditions as the samples.

Use a 5-g sample if the expected concentration is <0.1 mg/kg or a 1-g sample for expected concentrations between 0.1 and 1 mg/kg. Mix the contents of the sample container with a narrow metal spatula. Weigh the amount of the sample into a tared purge device. Add the spiked water to the purge device, which contains the weighed amount of sample, and connect the device to the purge-and-trap system.

High-concentration method — This method is based on extracting the sediment or soil with methanol. A waste sample is either extracted or diluted, depending on its solubility in methanol. Wastes that are insoluble in methanol are diluted with reagent tetraglyme or possibly polyethylene glycol (PEG). An aliquot of the extract is added to organic-free reagent water containing surrogate and internal standards. This is purged at

ambient temperature. All samples with an expected concentration of >1.0 mg/kg should be analyzed by this method.

Mix the contents of the sample container with a narrow metal spatula. For sediments or soils and solid wastes that are insoluble in methanol, weigh 4 g (wet weight) of sample into a tared 20-mL vial. For waste that is soluble in methanol, tetraglyme, or PEG, weigh 1 g (wet weight) into a tared scintillation vial or culture tube or a 10-mL volumetric flask. Quickly add 9.0 mL of appropriate solvent then add 1.0 mL of a surrogate spiking solution to the vial, cap it, and shake it for 2 min.

METHANOL EXTRACT REQUIRED FOR ANALYSIS OF HIGH-CONCENTRATION SOILS OR SEDIMENTS

Approximate Concentration Range	Volume of Methanol Extract (a)
500–10,000 µg/kg	100 µL
1,000–20,000 µg/kg	50 µL
5,000–100,000 µg/kg	10 µL
25,000–500,000 µg/kg	100 µL of 1/50 dilution (b)

Calculate appropriate dilution factor for concentrations exceeding this table.

(a) The volume of methanol added to 5 mL of water being purged should be kept constant. Therefore, add to the 5-mL syringe whatever volume of methanol is necessary to maintain a volume of 100 µL added to the syringe.

(b) Dilute an aliquot of the methanol extract and then take 100 µL for analysis.

QUALITY CONTROL Demonstrate, through the analysis of a reagent water blank, that interferences from the analytical system, glassware, and reagents are under control. Blank samples should be carried through all stages of the sample preparation and measurement steps. For each analytical batch (up to 20 samples), a reagent blank, matrix spike, and matrix spike duplicate must be analyzed (the frequency of the spikes may be different for different monitoring programs). The blank and spiked samples must be carried through all stages of the sample preparation and measurement steps. QC samples mentioned in the section on Interferences will also be needed as appropriate to those situations.

Matrix spiking standards should be prepared from volatile organic compounds which will be representative of the compounds being investigated. The recommended internal standards are chlorobenzene-d5, 1,4-difluorobenzene, 1,4-dichlorobenzene-d4, and pentafluorobenzene. Using stock standard solutions, prepare secondary dilution standards containing the compounds of interest, either singly or mixed together in methanol. Store them in a vial with no headspace for no more than one week. Surrogates recommended are toluene-d8, 4-bromofluorobenzene, and dibromofluoromethane. Each sample undergoing GC/MS analysis must be spiked with 10 µL of the surrogate spiking solution prior to analysis.

REFERENCE Test Methods for Evaluating Solid Waste (SW-846). U.S. EPA 1983, Method 8260A, Rev. 1, Nov. 1990. Office of Solid Waste, Washington, DC.

Dibromochloromethane   **EPA Method 601**
CAS #124–48-1

TITLE Purgeable Halocarbons

MATRIX Wastewater

APPLICATION Method covers 29 purgeable halocarbons. (Method 624 provides GC/MS conditions appropriate for the qualitative and quantitative confirmation of results). Method describes conditions for a 2nd GC column to confirm measurements made with primary column.

INTERFERENCES Impurities in the purge gas and organic compounds outgassing from the plumbing ahead of the trap. With high- and low-level samples, there can be carryover contamination. Diffusion of volatile organics through the septum seal into the sample.

INSTRUMENTATION GC-equipped with halide-specific detector. (With purge-and-trap unit).

RANGE 8.0–500 µg/L.

MDL 0.09 µg/L.

PRECISION 0.24X+1.68 µg/L (overall precision).

ACCURACY 0.94C+2.72 µg/L (as recovery).

SAMPLING METHOD 25-mL glass vial. Teflon®-lined septum.

STABILITY cool, 4°C, 0.008% $Na_2S_2O_3$.

MHT 14 days.

QUALITY CONTROL The lab must on an ongoing basis, spike at least 10% of the samples from each sample site being monitored to assess accuracy.

REFERENCE Method 601, *Federal Register* Part VIII 40 CFR Part 136, Oct 26, 1984.

Dibromochloromethane   **EPA Method 624**
CAS #124–48-1

TITLE Purgeables

MATRIX Wastewater

APPLICATION Method covers 31 purgeable organics. An inert gas is bubbled through a 5-mL water sample in a specially designed purging chamber. Here, purgeables are transferred from aqueous to gaseous phase, passed onto a sorbent column, and trapped. Trap is heated and backflushed with inert gas to desorb purgeables onto a GC column, where purgeables are separated.

INTERFERENCES Impurities in the purge gas, organic compounds outgassing from the plumbing ahead of the trap, and solvent vapors in the lab. With high- and low-level samples, there can be carryover contamination.

INSTRUMENTATION GC/MS with purge-and-trap unit.

RANGE 5–600 µg/L.

MDL 3.1 µg/L

PRECISION 0.17X+0.49 µg/L (overall precision).

ACCURACY 1.01C-0.03 µg/L (as recovery).

SAMPLING METHOD 25-mL glass vial. Teflon®-lined septum.

STABILITY cool, 4°C, 0.008% Na2S2O3.

MHT 14 days.

QUALITY CONTROL The lab must on an ongoing basis, spike at least 5% of the samples from each sample site being monitored to assess accuracy.

REFERENCE Method 624, *Federal Register* Part VIII 40 CFR Part 136, Oct 26, 1984.

Dibromochloromethane   **EPA Method 8010**
CAS #124–48-1

TITLE Halogenated Volatile Organics

MATRIX Groundwater, soils, sludges, water miscible liquid wastes, and non-water miscible wastes.

APPLICATION This method is used for the analysis of 39 halogenated VOCs. Samples are analyzed using direct injection or purge-and-trap methods. Groundwater must be analyzed by the purge-and-trap method. The method provides an optional GC column which is used for analyte confirmation and that may help resolve analytes from interferences.

INTERFERENCES There can be carryover contamination with high- and low-level samples. Impurities may come from the purge-and-trap apparatus, organic compounds outgassing from the plumbing ahead of trap, diffusion of VOCs through the sample bottle septum during shipping or storage, or from solvent vapors in the lab.

INSTRUMENTATION GC capable of on-column injections or purge-and-trap sample introduction and a halogen specific detector. Column 1: 8 ft by 0.1 in 1%. SP-1000 on Carbopack-B. Column 2: 6 ft by 0.1 in bonded n-octane on Porasil-C.

RANGE 8 to 500 µg/L (reagent water).

MDL 0.09 µg/L (reagent water).

PQL FACTORS FOR MULTIPLYING × FID MDL VALUE

Matrix	Multiplication Factor
Groundwater	10
Low-level soil	10
Water miscible liquid waste	500
High-level soil and sludge	1250
Non-water miscible waste	1250

PRECISION 0.24X + 1.68 µg/L (overall precision).

ACCURACY 0.94C + 2.72 µg/L (as recovery).

SAMPLING METHOD For water and liquid samples; use glass 40-mL vials with Teflon®-lined septum caps and collect two vials per sample location with no headspace. For solids and concentrated waste samples; use widemouth glass bottles with Teflon® liners.

STABILITY For concentrated wastes, soils, sediments, or sludges: cool to 4°C. For liquids: add 4 drops of concentrated hydrochloric acid and cool to 4°C.

MHT 14 days.

QUALITY CONTROL Analyze a reagent blank, matrix spike, and matrix spike duplicate/duplicate for each analytical batch (up to 20 samples). Demonstrate the purity of glassware and reagents by analyzing a reagent water method blank. Internal, surrogate, and five concentration level calibration standards are used.

REFERENCE Test Methods for Evaluating Solid Waste (SW-846), U.S. EPA Office of Solid Waste, Washington, DC, Method 8010B, Rev. 2, Nov. 1992.

1,2-Dibromoethane   **EPA Method 502**
CAS #106-93-4

TITLE Volatile Organic Compounds in Water By Purge and Trap Capillary Column Gas Chromatography with Photoionization and Electrolytic Conductivity Detectors in Series. U.S. EPA Method 502.2, Rev. 2.0, 1989.

MATRIX Drinking water and raw source water. The latter should include most surface water and groundwater sources.

METHOD SUMMARY This method covers 60 volatile organic compounds that contain halogen atoms and/or that are aromatic. An inert gas (zero grade nitrogen or helium) is bubbled through a 25-mL or a 5-mL water sample (depending on the expected concentration of the analytes). Purged sample components are trapped in a tube of sorbent materials. When purging is complete, the sorbent tube is heated and backflushed with helium to desorb the trapped sample onto a capillary GC column. The column is temperature programmed to separate the method analytes which are then detected with a photoionization detector (PID) and a Hall electrolytic conductivity (HECD) placed in series. The PID is selective for aromatic compounds and the HECD is selective for halogenated compounds.

INTERFERENCES Impurities in the purge gas and from organic compounds outgassing from the plumbing ahead of the trap account for many contamination problems. Interferences purged or coextracted from the samples will vary considerably from source to source, depending upon the particular sample or extract being tested. Cross-contamination can occur whenever high-level and low-level samples are analyzed sequentially. Samples also can be contaminated by diffusion of volatile organics (particularly methylene chloride and fluorocarbons) through the septum seal into the sample during shipment and storage. The lab where volatile analysis is performed and also the refrigerated storage area should be completely free of solvents.

INSTRUMENTATION A GC containing a series configuration of a high temperature photoionization detector (PID) equipped with 10.0 eV (nominal) lamp and Hall electrolytic conductivity detector (HECD) is required. Also required is an all-glass 5-mL purging device, a sorbent trap, and a thermal desorption apparatus which is connected to the GC system.

Column 1: VOCOL glass wide-bore capillary column.
Column 2: RTX–502.2 mega-bore capillary column.
Column 3: DB-62 mega-bore capillary column.

PRECISION & ACCURACY Method detection limits are dependent upon the characteristics of the gas chromatographic system used. Analytes that are not separated chromatographically cannot be individually identified and used in the same calibration mixture or water samples unless an alternative technique for identification and quantification, such as mass spectrometry, is used.

Electrolytic conductivity detetor (c) range in µg/L (a) was 0.02–200.
Electrolytic conductivity detetor (c) MDL in µg/L (b) was 0.8.
Electrolytic conductivity detetor (c) accuracy as % recovery was 97.
Electrolytic conductivity detetor (c) precision as % RSD was 2.8.
Photoionization detector (d) range in µg/L (a) was 0.02–200.
Photoionization detector (d) MDL in µg/L (b) was not listed.
Photoionization detector (d) accuracy as % recovery was not listed.
Photoionization detector (d) precision as % RSD was not listed.

(a) The applicable concentration range of this method is compound, instrument, and matrix-dependent. It is listed as being approximately 0.02 to 200 µg/L but no specific information is provided so caution should be observed.

(b) The method detection limits reports with this method are compound, instrument, and matrix-dependent. The values reported were calculated using reagent water fortified with the corresponding compounds at 10 µg/L and a GC-equipped with a 60 m × 0.75 mm VOLCOL wide bore capillary column with 1.5 µm film thickness and using helium carrier gas.

(c) Recoveries and relative standard deviations were determined from seven samples of reagent water fortified with 10 µg/L of each compound. 2-Bromo-1-chloropropane was used as the internal standard for calculating average recoveries.

(d) Recoveries and relative standard deviations were determined from seven samples of reagent water fortified with 10 µg/L of each compound. Fluorobenzene was used as the internal standard for calculating average recoveries.

SAMPLING METHOD Collect samples using a 40- to 120-mL screw-cap vial (prewashed with detergent, rinsed with distilled water and oven dried at 105°C) with a Teflon®-faced silicone septum. Collect bubble-free samples and place the septum with the Teflon® side down on the water.

SAMPLE PRESERVATION If residual chlorine is present in the water add about 25 mg of ascorbic acid to each vial before samples are collected to remove the chlorine. Add hydrochloric acid to reduce pH to <2, immediately cool samples to 4°C, and store them in a solvent-free refrigerator at 4°C until analysis.

MHT The maximum holding time for samples is 14 days from the time they were collected.

SAMPLE PREPARATION Remove the plungers from two 5-mL syringes and attach a closed syringe valve to each. Warm the sample to room temperature, open the sample bottle, and carefully pour the sample into one of the syringe barrels to just short of overflowing. Replace the syringe plunger, invert the syringe, and compress the sample. Open the syringe valve and vent any residual air while adjusting the sample volume to 5.0 mL. Add 10 µL of the internal calibration standard to the sample through the syringe valve. Close the valve. Fill the second syringe in an identical manner from the same sample bottle. Reserve this second syringe for a reanalysis if necessary.

QUALITY CONTROL As an initial demonstration of lab accuracy and precision, analyze 4 to 7 replicates of a lab fortified blank containing analyte at 0.1–5 µg/L. Collect all samples in duplicate. Surrogate analytes (similar to those of the analytes of interest), whose concentration is known in every sample, are measured using the same internal standard calibration procedure. Duplicate field reagent water blanks (trip blanks) must be analyzed with each set of samples, lab reagent blanks (method blanks) must be analyzed with each batch of samples processed as a group within a work shift. Also, a single lab-fortified blank that contains each of the analytes of interest should be analyzed with each batch of samples processed as a group within a work shift. A 3- to 5-point calibration curve is needed depending on the calibration range factor required.

EPA CONTACT & HOTLINE For technical questions contact Dr. Baldev Bathija, U.S. EPA, Office of Ground Water and Drinking Water (WH-550D), 401 M St. SW, Washington, DC 20460. Tel. (202) 260-3040. For further information the EPA Safe Drinking Water Hotline may be called at: (800) 426-4791.

REFERENCE Methods for the Determination of Organic Compounds in Drinking Water, EPA/600/4-88/039 (revised July 1991; Final Rule for determination of compliance with the MCL for Total Trihalomethanes under 141.30, in 40 CFR Part 141, Vol. 58, No. 147, Fed. Reg., Tuesday Aug. 3, 1993). U.S. EPA Environmental Monitoring Systems Laboratory, Cincinnati, OH, 45268, U.S.A. Available from the National Technical Information Service (NTIS), 5285 Port Royal Road, Springfield, VA 22161; Tel. 800-553-6847. NTIS Order Number is PB91-231480.

1,2-Dibromoethane   **EPA Method 524**
CAS #106-93-4

TITLE Measurement of Purgeable Organic Compounds in Water by Capillary Column GC/MS.

MATRIX Drinking water and raw source water; the latter should include most surface water and groundwater sources.

METHOD SUMMARY Method 524.2 covers 60 volatile organic compounds. An inert gas (zero grade nitrogen or helium) is bubbled through a 25-mL or a 5-mL water sample (depending on the expected concentration of the analytes). Purged sample components are trapped in a tube of sorbent materials. When purging is complete, the sorbent tube is heated and backflushed with helium to desorb the trapped sample onto a capillary GC column.

INTERFERENCES Impurities in the purge gas and from organic compounds outgassing from the plumbing ahead of the trap account for many contamination problems. Interferences purged or coextracted from the samples will vary considerably from source to source, depending upon the particular sample or extract being tested. Cross-contamination can occur whenever high-level and low-level samples are analyzed sequentially. Samples also can be contaminated by diffusion of volatile organics (particularly methylene chloride and fluorocarbons) through the septum seal into the sample during shipment and storage.

INSTRUMENTATION A GC/MS with a data system equipped with one of the following capillary GC columns:

Column 1: VOCOL glass wide bore capillary column.
Column 2: DB-624 fused silica capillary column.
Column 3: DB-5 fused silica capillary column.

Also required is an all-glass 25 mL or 5-mL purging device, a sorbent trap, and a thermal desorption apparatus which is connected to the GC/MS system.

PRECISION & ACCURACY Method detection limits are compound- and instrument-dependent, and may vary from approximately 0.02–0.35 µg/L. Note in the table below that the "true" concentration range used for accuracy and precision measurements was quite narrow. However, the applicable concentration range of this method is primarily column dependent and is approximately 0.02 to 200 µg/L for the wide-bore thick-film columns. Narrow-bore thin-film columns may have a capacity which limits the range to about 0.02 to 20 µg/L. Analytes that are inefficiently purged from water will not be detected when present at low concentrations, but they can be measured with acceptable accuracy and precision when present in sufficient amounts.

Analytes that are not separated chromatographically, but which have different mass spectra and non-interfering quantification ions, can be identified and measured in the same calibration mixture or water sample. Analytes which have very similar mass spectra cannot be individually identified and measured in the same calibration mixture or water samples unless they have different retention times. Co-eluting compounds with very similar mass spectra, typically many structural isomers, must be reported as an isomeric group or pair.

The range (in µg/L) was 0.5–10.
The Method Detection Limig (in µg/L) was 0.06.
The accuracy (as % recovery) was 102.
The precision (in %) was 3.9.

Note: Data were obtained from 16–31 determinations using a wide-bore capillary column and a jet separator interfaced to a quadrupole mass spectrometer. All analytes were in a reagent water matrix.

SAMPLING METHOD Collect samples using a 40- to 120-mL screw-cap vial (prewashed with detergent, rinsed with distilled water and oven dried at 105°C) with a Teflon®-faced silicone septum. Collect bubble-free samples and place the septum with the Teflon® side down on the water.

SAMPLE PRESERVATION If residual chlorine is present in the water add about 25 mg of ascorbic acid to each vial before samples are collected to remove the chlorine. Add hydrochloric acid to reduce pH to <2, and immediately cool samples to 4°C, and store them in a solvent-free refrigerator at 4°C until analysis.

MHT The maximum holding time for samples is 14 days from the time they were collected.

SAMPLE PREPARATION Remove the plungers from two 25-mL (or 5-mL depending on sample size) syringes and attach a closed syringe valve to each. Warm the sample to room temperature, open the sample bottle, and carefully pour the sample into one of the syringe barrels to just short of overflowing. Replace the syringe plunger, invert the syringe, and compress the sample. Open the syringe valve and vent any residual air while adjusting the sample volume to 25.0 mL (or 5 mL). For samples and blanks, add 5 µL of the fortification solution containing the internal standard and the surrogates to the sample through the syringe valve. For calibration standards and lab fortified blanks, add 5 µL of the fortification solution containing the internal standard only. Close the valve. Fill the second syringe in an identical manner from the same sample bottle. Reserve this second syringe for a reanalysis if necessary.

QUALITY CONTROL As an initial demonstration of lab accuracy and precision, analyze 4 to 7 replicates of a lab fortified blank containing analyte at 0.2–5 µg/L. Collect all samples in duplicate. Surrogate analytes (similar to those of the analytes of interest), whose concentration is known in every sample, are measured using the same internal standard calibration procedure. Duplicate field reagent water blanks (trip blanks) must be analyzed with each set of samples, lab reagent blanks (method blanks) must be analyzed with each batch of samples processed as a group within a work shift. Also, a single lab-fortified blank that contains each of the analytes of interest should be analyzed with each batch of samples processed as a group within a work shift. A 3- to 5-point calibration curve is needed depending on the calibration range factor required.

EPA CONTACT & HOTLINE For technical questions contact Dr. Baldev Bathija, U.S. EPA, Office of Ground Water and Drinking Water (WH-550D), 401 M St. SW, Washington, DC 20460. Tel. (202) 260-3040. For further information the EPA Safe Drinking Water Hotline may be called at: (800) 426-4791.

REFERENCE Methods for the Determination of Organic Compounds in Drinking Water, EPA/600/4-88/039 (revised July 1991; Final Rule for determination of compliance with the MCL for Total Trihalomethanes under 141.30, in 40 CFR Part 141, Vol. 58, No. 147, Fed. Reg., Tuesday Aug. 3, 1993). U.S. EPA Environmental Monitoring Systems Laboratory, Cincinnati, OH, 45268, U.S.A. Available from the National Technical Information Service (NTIS), 5285 Port Royal Road, Springfield, VA 22161; Tel. 800-553-6847. NTIS Order Number is PB91-231480.

1,2-Dibromoethane
CAS #106-93-4

EPA Method 8021

TITLE Halogenated Volatile by Gas Chromatography Using Photoionization and Electrolytic Conductivity Detectors in Series: Capillary Column Technique

MATRIX This method is applicable to nearly all types of samples, regardless of water content, including groundwater, aqueous sludges, caustic liquors, acid liquors, waste solvents, oily wastes, mousses, tars, fibrous wastes, polymeric emulsions, filter cakes, spent carbons, spent catalysts, soils, and sediments.

METHOD SUMMARY This method is used to determine 60 volatile organic compounds in a variety of solid waste matrices. It provides GC conditions for the detection of halogenated and aromatic volatile organic compounds. Samples can be analyzed using direct injection or purge-and-trap (EPA Method 5030). Groundwater samples must be analyzed using EPA Method 5030 (where applicable). A temperature program is used with the GC. Detection is achieved by a photoionization detector (PID) and a Hall electrolytic conductivity detector (HECD) in series.

INTERFERENCES Samples can be contaminated by diffusion of volatile organics (particularly chlorofluorocarbons and methylene chloride) through the sample container septum during shipment and storage.

INSTRUMENTATION A GC-equipped with variable-constant differential flow controllers, subambient oven controller, PID and HECD detectors connected with a short piece of uncoated capillary tubing and a data system.

Column: 60 m × 0.75 mm I.D. VOCOL wide-bore capillary column with 1.5 µm film thickness.

PRECISION & ACCURACY MDLs are compound-dependent and vary with purging efficiency and concentration. The applicable concentration range of this method is compound- and instrument-dependent but is approximately 0.1 to 200 µg/L. Analytes that are inefficiently purged from water will not be detected when present at low concentrations, but they can be measured with acceptable accuracy and precision when present in sufficient amounts. The estimated quantitation limit (EQL) for an individual compound is approximately 1 µg/kg (wet weight) for soil/sediment samples, 100 µg/kg (wet weight) for wastes, and 1 µg/L for groundwater. EQLs will be proportionately higher for sample extracts and samples that require dilution or reduced sample size to avoid saturation of the detector.

MULTIPLICATION FACTORS FOR OTHER MATRICES (a)

Matrix	Factor (b)
Groundwater	10
Low-concentration soil	10
Water miscible liquid waste	500
High-concentration soil and sludge	1250
Non-water miscible waste	1250

(a) Sample EQLs are highly matrix-dependent. The EQLs listed herein are provided for guidance and may not always be achievable. (b) EQL = [Method detection limit] × [Factor]. For non-aqueous samples, the factor is on a wet-weight basis.

SINGLE LABORATORY ACCURACY & PRECISION DATA FOR VOCs IN WATER

This method was tested in a single lab using water spiked at 10 µg/L and the following data was reported:

Recoveries and standard deviations were determined from seven samples and spiked at 10 µg/L of each analyte. Recoveries were determined by the internal standard method. Internal standards were: Fluorobenzene for PID and 2-Bromo-1-chloropropane for HECD.

The average recovery (in percent) for the PID was none (no response for this detector).
The standard deviation of the recovery for the PID was none (no response for this detector).
The MDL (in µg/mL) for the PID was none (no response for this detector).
The average recovery (in percent) for the HECD was 97.
The standard deviation of the recovery for the HECD was 2.7.
The MDL (in µg/mL) for the HECD was 0.8.

SAMPLE COLLECTION, PRESERVATION & HANDLING

Volatile Organics — Standard 40-mL glass screw-cap VOA vials with Teflon®-faced silicone septum may be used for both liquid and solid matrices. When collecting samples, liquids and solids should be introduced into the vials gently to reduce agitation which might drive off volatile compounds. If there are any air bubbles present the sample must be retaken. Tap slightly as they are filled to try and eliminate as much free air space as possible. The two vials from each sampling locations should be sealed in separate plastic bags to prevent cross-contamination between samples particularly if the sampled waste is suspected of containing high levels of volatile organics.

Semivolatile organics — Containers used to collect samples for the determination of semivolatile organic compounds should be soap and water washed followed by methanol (or isopropanol) rinsing. The sample containers should be of glass or Teflon® and have screw-top covers with Teflon® liners.

Preservation for volatile organics — No preservation is used with concentrated waste samples. With liquid samples containing no residual chlorine, 4 drops of concentrated hydrochloric acid are added and the samples are immediately cooled to 4°C. When liquid samples contain residual chlorine, they are treated as above and, in addition, 4 drops of 4% aqueous sodium thiosulfate are added. Soil, sediment, and sludge samples are only cooled to 4°C.

Preservation for semivolatile organics — No preservation is used with concentrated waste samples. With liquid samples containing no residual chlorine and with soil, sediment, and sludge samples, immediately cooling to 4°C is the only preservation used. When residual chlorine is present then 3 mL of 10% aqueous sodium sulfate is added for each gallon of sample collected, followed by cooling to 4°C.

MHT The holding time for all volatile organics samples is 14 days. Liquid samples must be extracted within 7 days and their extracts analyzed within 40 days. Concentrated waste, soil, sediment, and sludge samples must be extracted within 14 days and their extracts analyzed within 40 days.

SAMPLE PREPARATION Volatile compounds are introduced into the gas chromatograph either by direct injector or purge-and-trap (EPA Method 5030). EPA Method 5030 may be used directly on groundwater samples or low-concentration contaminated soils and sediments. For medium-concentration soils or sediments, methanolic extraction, as described in EPA Method 5030, may be necessary prior to purge-and-trap analysis.

QUALITY CONTROL Calculate surrogate standard recovery on all samples, blanks, and spikes. A trip blank is recommended to check on sampling, storage, and handling contamination. Calibration standards, at a minimum of five concentration levels, are prepared in organic-free reagent water. One of the concentration levels should be at a concentration near, but above, the method detection limit.

A combination of bromochloromethane, 2-bromo-1-chloropropane, 1,4-dichlorobutane, and bromochlorobenzene are recommended as surrogate standards to encompass the range of the temperature program used in this method.

REFERENCE Test Methods for Evaluating Solid Waste, Physical/Chemical Methods, SW-846, 3rd Edition, U.S. EPA, Office of Solid Waste, Washington, DC, EPA Method 8021A, Rev. 1, Nov. 1992.

1,2-Dibromoethane	EPA Method 8240
CAS #106-93-4	

TITLE Volatile Organics By GC/MS: Packed Column Technique

MATRIX Nearly all types of sample matarices, regardless of water content, can be analyzed using this method. This includes groundwater, aqueous sludges, caustic liquors, acid liquors, waste solvents, oily wastes, mousses, tars, fibrous wastes, polymetric emulsions, filter cakes, spent carbons, spent catalysts, soils, and sediments.

METHOD SUMMARY Method 8240B covers 80 volatile organic compounds that are introduced into a gas chromatograph by the purge-and-trap method or by direct injection (in limited applications). For the purge-and-trap method an inert gas (zero grade nitrogen or helium) is bubbled through a 5-mL solution at ambient temperature. Purged sample components are trapped in a tube of sorbent materials. When purging is complete, the sorbent tube is heated and backflushed with inert gas to desorb the trapped components onto a GC column.

INTERFERENCES Impurities in the purge gas and from organic compounds outgassing from the plumbing ahead of the trap account for many contamination problems. Interferences purged or coextracted from the samples will vary considerably from source to source. Cross-contamination can occur whenever high-level and low-level samples are analyzed sequentially. Whenever an unusually concentrated sample is analyzed, it should be followed by the analysis of organic-free reagent water to check for cross-contamination. Samples also can be contaminated by diffusion of volatile organics (particularly methylene chloride and fluorocarbons) through the septum seal into the sample during shipment and storage. A trip blank can serve as a check on such contamination. The lab where volatile analysis is performed and also the refrigerated storage area should be completely free of solvents.

INSTRUMENTATION A gas chromatograph/mass spectrometry/data system (GC/MS) equipped with a 6 ft × 0.1 in I.D. glass column packed with 1% SP-1000 on Carbopack-B (60/80 mesh) is required. Also needed is a 5-mL purging device, a sorbent trap, and a thermal desorption apparatus.

PRECISION & ACCURACY This method is reported to have been tested by 15 laboratories using organic-free reagent water, drinking water, surface water, and industrial wastewaters (not specified) fortified at six concentrations over the range 5–600 µg/L.

Sample estimated quantitation limits (EQLs) are highly matrix-dependent. The EQLs listed may not always be achievable. EQLs listed for soils or sediments are based on wet weight. Normally, data is reported on a dry-weight basis; therefore, EQLs will be higher, based on the percent dry weight of each sample. Note that EQLs are even more variable than MDLs and that they are highly variable depending on the matrix being analyzed.

EQL in groundwater in µg/L was 5.
EQL in low soil or sediment in µg/kg was 5.
Accuracy (a) in µg/L was not listed.
Precision (b) in µg/L was not listed.

(a) Average recovery found for measurements of samples containing a concentration of C, in µg/L.
(b) Overall precision found for measurements of samples with average recovery X for samples containing a concentration of C in µg/L.
X = Average recovery found for measurement of samples containing a concentration of C in µg/L.

MULTIPLICATION FACTORS FOR OTHER MATRICES

Other Matrices	Factor (a)
Waste miscible liquid waste	50
High-concentration soil and sludge	125
Non-water miscible waste	500

(a) EQL = [EQL for low soil sediment] × [Factor]. For non-aqueous samples, the factor is on a wet-weight basis.

SAMPLING METHOD

Liquid samples — Use a 40-mL glass screw-cap VOA vial with a Teflon®-faced silicone septum that has been prewashed, rinsed with distilled deionized water, and oven dried. However, if residual chlorine is present, collect sample in a 40-oz. soil VOA container which has been pre-preserved with 4 drops of 10% sodium thiosulfate, mix gently, and then transfer the sample to a 40-mL VOA vial. Collect bubble-free samples in duplicate and seal them in separate plastic bags.

Soils or sediments, and sludges — Use an 8-oz. widemouth glass bottle with a Teflon®-faced silicone septum that has been prewashed with detergent, rinsed with distilled deionized water, and oven dried. Tap slightly to eliminate free air space. Collect samples in duplicate and seal them in separate plastic bags.

SAMPLE PRESERVATION

Liquid samples — Add 4 drops of concentrated HCL and immediately cool samples to 4°C and store in a solvent-free refrigerator.

Soils or sediments, and sludges — Cool samples to 4°C and store in a solvent-free refrigerator.

MHT Maximum holding time is 14 days from the date of sample collection.

SAMPLE PREPARATION

Liquid samples — Remove the plunger from a 5-mL syringe and carefully pour the sample into the syringe barrel to just short of overflowing. Replace the syringe plunger and compress the sample. Open the syringe valve and vent any residual air while adjusting the sample volume to 5.0 mL. If there is only one volatile organic analysis (VOA) vial, a second syringe should be filled at this time to protect against possible loss of sample integrity. Add 10 µL of surrogate spiking solution and 10 µL of internal standard spiking solution through the valve bore of the 5-mL syringe, then close the valve. The surrogate and internal standards may be mixed and added as a single spiking solution.

Sediments, soils, and waste samples — All samples of this type should be screened by GC analysis using a headspace method (EPA Method 3810) or the hexadecane extraction and screening method (EPA Method 3820). Use the screening data to determine whether to use the low-concentration method (0.005–1 mg/kg) or the high-concentration method (>1 mg/kg).

Low-concentration method — The low-concentration method is based on purging a heated sediment or soil sample mixed with organic-free reagent water containing the surrogate and internal standards. Analyze all reagent blanks and standards under the same conditions as the samples.

Use a 5-g sample if the expected concentration is <0.1 mg/kg or a 1-g sample for expected concentrations between 0.1 and 1 mg/kg. Mix the contents of the sample container with a narrow metal spatula. Weigh the amount of the sample into a tared purge device. Add the spiked water to the purge device, which contains the weighed amount of sample, and connect the device to the purge-and-trap system.

High-concentration method — This method is based on extracting the sediment or soil with methanol. A waste sample is either extracted or diluted, depending on its solubility in methanol. Wastes that are insoluble in methanol are diluted with reagent tetraglyme or possibly polyethylene glycol (PEG). An aliquot of the extract is added to organic-free reagent water containing surrogate and internal standards. This is purged at ambient temperature. All samples with an expected concentration of >1.0 mg/kg should be analyzed by this method.

Mix the contents of the sample container with a narrow metal spatula. For sediments or soils and solid wastes that are insoluble in methanol, weigh 4 g (wet weight) of sample into a tared 20-mL vial. For waste that is soluble in methanol, tetraglyme, or PEG, weigh 1 g (wet weight) into a tared scintillation vial or culture tube or a 10-mL volumetric flask. Quickly add 9.0 mL of appropriate solvent then add 1.0 mL of a surrogate spiking solution to the vial, cap it, and shake it for 2 min.

METHANOL EXTRACT REQUIRED FOR ANALYSIS OF HIGH-CONCENTRATION SOILS OR SEDIMENTS

Approximate Concentration Range	Volume of Methanol Extract (a)
500–10,000 µg/kg	100 µL
1,000–20,000 µg/kg	50 µL
5,000–100,000 µg/kg	10 µL
25,000–500,000 µg/kg	100 µL of 1/50 dilution (b)

Calculate appropriate dilution factor for concentrations exceeding this table.

(a) The volume of methanol added to 5 mL of water being purged should be kept constant. Therefore, add to the 5-mL syringe whatever volume of methanol is necessary to maintain a volume of 100 µL added to the syringe.
(b) Dilute an aliquot of the methanol extract and then take 100 µL for analysis.

QUALITY CONTROL Demonstrate, through the analysis of a reagent water blank, that interferences from the analytical system, glassware, and reagents are under control. Blank samples should be carried through all stages of the sample preparation and measurement steps. For each analytical batch (up to 20 samples), a reagent blank, matrix spike, and matrix spike duplicate must be analyzed (the frequency of the spikes may be different for different monitoring programs). The blank and spiked samples must be carried through all stages of the sample preparation and measurement steps. QC samples mentioned in the section on Interferences will also be needed as appropriate to those situations.

REFERENCE Test Methods for Evaluating Solid Waste (SW-846). U.S. EPA. 1983. Method 8240B, Rev. 2, Nov. 1990. Office of Solid Wastes, Washington, DC.

1,2-Dibromoethane   **EPA Method 8260**
CAS #106-93-4

TITLE Volatile Organic Compounds by GC/MS: Capillary Column Technique

MATRIX This method is applicable to nearly all types of samples, regardless of water content, including groundwater, soils, and sediments.

METHOD SUMMARY Method 8260A covers 58 volatile organic compounds that are introduced into a gas chromatograph by the purge-and-trap method or by direct injection (in limited applications). Zero-grade helium is bubbled through a 5-mL solution at ambient temperature. Purged sample components are trapped in a tube containing suitable sorbent materials. When purging is complete, the sorbent tube is heated and backflushed with helium to desorb trapped sample components. The analytes are desorbed directly to a large bore capillary or cryofocussed on a capillary precolumn before being flash evaporated to a narrow bore capillary for analysis.

INTERFERENCES Major contaminant sources are volatile materials in the lab and impurities in the inert purging gas and in the sorbent trap. Interfering contamination may occur when a sample containing low concentrations of volatile organic compounds is analyzed immediately after a sample containing high concentrations of volatile organic compounds. After analysis of a sample containing high concentrations of volatile organic compounds, one or more calibration blanks should be analyzed to check for cross-contamination. Screening of the samples prior to purge-and-trap GC/MS analysis is highly recommended to prevent contamination of the system. This is especially true for soil and waste samples.

Special precautions must be taken to analyze for methylene chloride. The analytical and sample storage area should be isolated from all atmospheric sources of methylene chloride. All gas chromatography carrier gas lines and purge gas plumbing should be constructed from stainless steel or copper tubing. Laboratory clothing previously exposed to methylene chloride fumes during liquid-liquid extraction procedures can contribute to sample contamination.

Samples can also be contaminated by diffusion of volatile organics (particularly methylene chloride and fluorocarbons) through the septum seal during shipment and storage. A trip blank can serve as a check on such contamination.

INSTRUMENTATION GC/MS with a temperature-programmable chromatograph suitable for splitless injection equipped with variable constant differential flow controllers, a subambient oven controller, a purging device, sorbent trap, a thermal desorption apparatus and a capillary precolumn interface when using cryogenic cooling will be needed. The following GC columns may be used:

Column 1: 60 m × 0.75mm I.D. capillary column coated with VOCOL, 1.5 µm film thickness.
Column 2: 30 m × 0.53mm capillary column coated with DB-624 or VOCOL, 3 µm film thickness.
Column 3: 30 m × 0.32mm I.D. capillary column coated with DB-5 or SE-54, 1-µm film thickness.

PRECISION & ACCURACY This method has been tested in a single lab using spiked water. Using a wide-bore capillary column, water was spiked at concentrations between 0.5 and 10 µg/L. Single lab accuracy and precision data are presented. The MDL actually achieved in a given analysis will vary depending on instrument sensitivity and matrix effects.

The MDL (a) in µg/L was 0.06.
The concentration range in µg/L was 0.5–10.
The mean accuracy (% of true value) was 102.
The precision as relative standard deviation was 3.9.

Note: The MDL is based on a 25-mL sample volume instead of a 5-mL sample volume.

SAMPLING METHOD
Liquid samples — Use a 40-mL glass screw-cap VOA vial with a Teflon®-faced silicone septum that has been prewashed, rinsed with distilled deionized water, and oven dried. If residual chlorine is present, collect the sample in a 4-oz soil VOA container which has been pre-preserved with 4 drops of 10% sodium thiosulfate. Mix gently and transfer the sample to a 40-mL VOA vial. Collect bubble-free samples in duplicate and seal each sample in a separate plastic bag.

Soils, sediments, and sludges — Use an 8-oz widemouth glass bottle with Teflon®-faced silicone septum that has been prewashed, rinsed with distilled deionized water, and oven dried. **Do not** heat the septum for more than 1 h. Tap slightly to eliminate any free air space. Collect samples in duplicate and seal each one in a separate plastic bag.

SAMPLE PRESERVATION
Liquid samples — Add 4 drops of concentrated HCL, cool to 4°C and store in a solvent-free refrigerator.

Soils, sediments and sludges — Cool samples to 4°C and store in a solvent-free refrigerator.

MHT The maximum holding time of any sample (liquids, soils, sediments, and sludges) is 14 days.

SAMPLE PREPARATION
Liquid samples — Remove the plunger from a 5-mL syringe and carefully pour the sample into the syringe barrel to just short of overflowing. Replace the syringe plunger and compress the sample. Open the syringe valve and vent any residual air while adjusting the sample volume to 5.0 mL. If there is only one volatile organic analysis (VOA) vial, a second syringe should be filled at this time to protect against possible loss of sample integrity. Add 10 µL of surrogate spiking solution and 10 µL of internal standard spiking solution through the valve bore of the 5-mL syringe, then close the valve. The surrogate and internal standards may be mixed and added as a single spiking solution.

Sediments, soils, and waste samples — All samples of this type should be screened by GC analysis using a headspace method (EPA Method 3810) or the hexadecane extraction and screening method (EPA Method 3820). Use the screening data to determine whether to use the low-concentration method (0.005–1 mg/kg) or the high-concentration method (>1 mg/kg).

Low-concentration method — The low-concentration method is based on purging a heated sediment or soil sample mixed with organic-free reagent water containing the surrogate and internal standards. Analyze all reagent blanks and standards under the same conditions as the samples.

Use a 5-g sample if the expected concentration is <0.1 mg/kg or a 1-g sample for expected concentrations between 0.1 and 1 mg/kg. Mix the contents of the sample container with a narrow metal spatula. Weigh the amount of the sample into a tared purge device. Add the spiked water to the purge device, which contains the weighed amount of sample, and connect the device to the purge-and-trap system.

High-concentration method — This method is based on extracting the sediment or soil with methanol. A waste sample is either extracted or diluted, depending on its solubility in methanol. Wastes that are insoluble in methanol are diluted with reagent tetraglyme or possibly polyethylene glycol (PEG). An aliquot of the extract is added to organic-free reagent water

containing surrogate and internal standards. This is purged at ambient temperature. All samples with an expected concentration of >1.0 mg/kg should be analyzed by this method.

Mix the contents of the sample container with a narrow metal spatula. For sediments or soils and solid wastes that are insoluble in methanol, weigh 4 g (wet weight) of sample into a tared 20-mL vial. For waste that is soluble in methanol, tetraglyme, or PEG, weigh 1 g (wet weight) into a tared scintillation vial or culture tube or a 10-mL volumetric flask. Quickly add 9.0 mL of appropriate solvent then add 1.0 mL of a surrogate spiking solution to the vial, cap it, and shake it for 2 min.

METHANOL EXTRACT REQUIRED FOR ANALYSIS OF HIGH-CONCENTRATION SOILS OR SEDIMENTS

Approximate Concentration Range	Volume of Methanol Extract (a)
500–10,000 µg/kg	100 µL
1,000–20,000 µg/kg	50 µL
5,000–100,000 µg/kg	10 µL
25,000–500,000 µg/kg	100 µL of 1/50 dilution (b)

Calculate appropriate dilution factor for concentrations exceeding this table.

(a) The volume of methanol added to 5 mL of water being purged should be kept constant. Therefore, add to the 5-mL syringe whatever volume of methanol is necessary to maintain a volume of 100 µL added to the syringe.
(b) Dilute an aliquot of the methanol extract and then take 100 µL for analysis.

QUALITY CONTROL Demonstrate, through the analysis of a reagent water blank, that interferences from the analytical system, glassware, and reagents are under control. Blank samples should be carried through all stages of the sample preparation and measurement steps. For each analytical batch (up to 20 samples), a reagent blank, matrix spike, and matrix spike duplicate must be analyzed (the frequency of the spikes may be different for different monitoring programs). The blank and spiked samples must be carried through all stages of the sample preparation and measurement steps. QC samples mentioned in the section on Interferences will also be needed as appropriate to those situations.

Matrix spiking standards should be prepared from volatile organic compounds which will be representative of the compounds being investigated. The recommended internal standards are chlorobenzene-d5, 1,4-difluorobenzene, 1,4-dichlorobenzene-d4, and pentafluorobenzene. Using stock standard solutions, prepare secondary dilution standards containing the compounds of interest, either singly or mixed together in methanol. Store them in a vial with no headspace for no more than one week. Surrogates recommended are toluene-d8, 4-bromofluorobenzene, and dibromofluoromethane. Each sample undergoing GC/MS analysis must be spiked with 10 µL of the surrogate spiking solution prior to analysis.

REFERENCE Test Methods for Evaluating Solid Waste (SW-846). U.S. EPA 1983, Method 8260A, Rev. 1, Nov. 1990. Office of Solid Waste, Washington, DC.

1,2-Dibromoethane   **EPA Method 1624**
CAS #106-93-4

TITLE Volatile Organic Compounds by Isotope Dilution GC/MS

MATRIX Compounds may be determined in waters, soils, and municipal sludges by this method.

METHOD SUMMARY This method is used to determine 58 volatile toxic organic pollutants associated with the CWA (as amended 1987); the RCRA (as amended 1986); the CERCLA (as amended 1986); and other compounds amenable to purge-and-trap gas chromatography-mass spectrometry (GC/MS).

If the solids content is less than 1%, stable isotopically-labeled analogs of the compounds of interest are added to a 5-mL sample and the sample is purged with an inert gas at 20–25°C in a chamber designed for soil or water samples. If the solids content is greater than 1%, 5 mL of reagent water and the labeled compounds are added to a 5-g aliquot of sample and the mixture is purged at 40°C. Compounds that will not purge at 20–25°C or at 40°C are purged at 78–85°C. In the purging process, the volatile compounds are transferred from the aqueous phase into the gaseous phase where they are passed into a sorbent column, and trapped. After purging is completed, the trap is backflushed and heated rapidly to desorb the compounds into a GC. The compounds are separated by the GC and detected by a MS. The labeled compounds serve to correct the variability of the analytical technique.

INTERFERENCES Impurities in the purge gas, organic compounds outgassing from the plumbing upstream of the trap, and solvent vapors in the lab account for most problems. Samples can be contaminated by diffusion of volatile organic compounds (particularly methylene chloride) through the bottle seal during shipment and storage. Contamination by carryover can occur when high-level and low-level samples are analyzed sequentially. When an unusually concentrated sample is encountered, follow it by analysis of a reagent water blank to check for carryover.

INSTRUMENTATION Major equipment includes a GC with linear temperature programming and a glass jet separator as the MS interface, a MS with 70 eV electron impact ionization, and a data system to collect and record response factors.

Column: 2.8 m × 2 mm I.D. glass, packed with 1% SP-1000 on Carbopak B, 60/80 mesh, or equivalent.

PRECISION & ACCURACY The detection limits of the method are usually dependent on the level of interferences rather than instrumental limitations. The method detection limits were determined in digested sludge (low solids) and in filter cake or compost (high solids).

The MDL (in µg/kg) for low solids is not listed and for high solids is not listed.
Labeled and native compound precision (in µg/L) as standard deviation was not listed.
Labeled and native compound accuracy (in µg/L) as average recovery was not listed.

Acceptance criteria are at 20 µg/L for this compound.

SAMPLE COLLECTION, PRESERVATION & HANDLING
Grab samples are collected in glass containers having a total volume greater than 20 mL. Fill and seal each bottle so that no air bubbles are entrapped. Samples are maintained at 0 to 4°C from the time of collection until analysis. If an aqueous sample contains residual chlorine, add sodium thiosulfate preservative (10 mg/40 mL) to the empty sample bottles just prior to shipment to the sample site. All samples must be analyzed within 14 days of collection.

SAMPLE PREPARATION Samples containing less than 1% solids are analyzed directly as aqueous samples. Samples containing 1% solids or greater are analyzed as solid samples utilizing one of two methods, depending on the levels of pollutants, in the sample. Samples containing 1% solids or greater, and low to moderate levels of pollutants are analyzed by purging a known weight of sample added to 5 mL of reagent water. Samples containing 1% solids or greater, and high levels of pollutants, are extracted with methanol, and an aliquot of the methanol extract is added to reagent water and purged.

QUALITY CONTROL A field blank prepared from reagent water and carried through the sampling and handling protocol may serve as a check on contamination from shipment and storage.

The analyst is permitted to modify this method to improve separations or lower the costs of measurements, provided all performance specifications are met. Analyses of blanks are required. When results of spikes indicate atypical method performance for samples, the samples are diluted to bring method performance within acceptable limits. Analyze two sets of four 5-mL aliquots (8 aliquots total) of the aqueous performance standard. Spike all samples with labeled compounds to assess method performance on the sample matrix. Compute the percent recovery of the labeled compounds using the internal standard method. Compare the percent recovery for each compound with the corresponding labeled compound recovery. Reagent water blanks are analyzed to demonstrate freedom from carryover contamination. Field replicates may be collected to determine the precision of the sampling technique, and spiked samples may be required to determine the accuracy of the analysis when the internal method is used.

REFERENCE Volatile Organic Compounds by Isotope Dilution GC/MS. Office of Water Regulation and Standards, U.S. EPA Industrial Technology Division, Washington, DC, EPA Method 1624, Rev. C, June 1989 (contact W.A. Telliard, U.S. EPA, Office of Water Regulations and Standards, 401 M St., SW, Washington, DC, 20460. Phone: 202-382-7131).

1,2-Dibromoethane   **EPA Method 504**
CAS #106-93-4

TITLE (EDB)&(DBCP) in H2O by

MATRIX Finished drinking water and Microextraction groundwater.

APPLICATION 35 mL Of sample are extracted with 2ml of hexane. 2ul Of the extract are injected into a GC with a linearized electron capture detector for separation & analysis. Aqueous calibration standards are run in same manner to compensate for possible extraction losses.

INTERFERENCES Impurities contained in the extracting solvent usually account for the majority of analytical problems. (Run solvent blanks as checks). EDB at low concentrations may be masked by high levels of dibromochloromethane. Extraction technique extracts polars with non-polars.

INSTRUMENTATION GC with EC detector & capillary column splitless injector. Use two GC columns. Column 1: Durawax-DX3 fused silica; Column 2: DB-1 fused silica; Column 3: RTX–Volatiles wide bore.

RANGE 0.03–200 µg/L.

MDL 0.01 µg/L

PRECISION (overall) $S = 0.075X + 0.008$ (RW); $S = 0.102X + 0.006$ (groundwater).

ACCURACY (recovery) $X = 1.072C - 0.006$ (RW); $X = 1.077C - 0.001$ (groundwater).

SAMPLING METHOD 40-mL screw-cap vials, PTFE-faced silicon septum.

STABILITY Cool to 4°C; add 4 drops of 10% sodium thiosulfate and 4 drops of hydrochloric acid.

MHT 28 Days.

QUALITY CONTROL Laboratory must, on an ongoing basis, demonstrate thrugh analyses of lab fortified blanks that the operation of the measurement system is in control. The frequency of the lab fortified blank analyses must be equivalent to 10% of all samples analyzed.

REFERENCE Method 504, (EDB) & (DBCP) in Water by Microextraction & GC, EPA

600/4-88/039.

Dibromomethane   **EPA Method 1624**
CAS #74-95-3

TITLE Volatile Organic Compounds by Isotope Dilution GC/MS

MATRIX Compounds may be determined in waters, soils, and municipal sludges by this method.

METHOD SUMMARY This method is used to determine 58 volatile toxic organic pollutants associated with the CWA (as amended 1987); the RCRA (as amended 1986); the CERCLA (as amended 1986); and other compounds amenable to purge-and-trap gas chromatography-mass spectrometry (GC/MS).

If the solids content is less than 1%, stable isotopically-labeled analogs of the compounds of interest are added to a 5-mL sample and the sample is purged with an inert gas at 20–25°C in a chamber designed for soil or water samples. If the solids content is greater than 1%, 5 mL of reagent water and the labeled compounds are added to a 5-g aliquot of sample and the mixture is purged at 40°C. Compounds that will not purge at 20–25°C or at 40°C are purged at 78–85°C. In the purging process, the volatile compounds are transferred from the aqueous phase into the gaseous phase where they are passed into a sorbent column, and trapped. After purging is completed, the trap is backflushed and heated rapidly to desorb the compounds into a GC. The compounds are separated by the GC and detected by a MS. The labeled compounds serve to correct the variability of the analytical technique.

INTERFERENCES Impurities in the purge gas, organic compounds outgassing from the plumbing upstream of the trap, and solvent vapors in the lab account for most problems. Samples can be contaminated by diffusion of volatile organic compounds (particularly methylene chloride) through the bottle seal during shipment and storage. Contamination by carryover can occur when high-level and low-level samples are analyzed sequentially. When an unusually concentrated sample is encountered, follow it by analysis of a reagent water blank to check for carryover.

INSTRUMENTATION Major equipment includes a GC with linear temperature programming and a glass jet separator as the MS interface, a MS with 70 eV electron impact ionization, and a data system to collect and record response factors.

Column: 2.8 m × 2 mm I.D. glass, packed with 1% SP-1000 on Carbopak B, 60/80 mesh, or equivalent.

PRECISION & ACCURACY The detection limits of the method are usually dependent on the level of interferences rather than instrumental limitations. The method detection limits were determined in digested sludge (low solids) and in filter cake or compost (high solids).

The MDL (in µg/kg) for low solids is not listed and for high solids is not listed.
Labeled and native compound precision (in µg/L) as standard deviation was not listed.
Labeled and native compound accuracy (in µg/L) as average recovery was not listed.

Acceptance criteria are at 20 µg/L for this compound.

SAMPLE COLLECTION, PRESERVATION & HANDLING Grab samples are collected in glass containers having a total volume greater than 20 mL. Fill and seal each bottle so that no air bubbles are entrapped. Samples are maintained at 0 to 4°C from the time of collection until analysis. If an aqueous sample contains residual chlorine, add sodium thiosulfate preservative (10 mg/40 mL) to the empty sample bottles just prior to shipment to the sample site. All samples must be analyzed within 14 days of collection.

SAMPLE PREPARATION Samples containing less than 1% solids are analyzed directly as aqueous samples. Samples containing 1% solids or greater are analyzed as solid samples utilizing one of two methods, depending on the levels of pollutants, in the sample. Samples containing 1% solids or greater, and low to moderate levels of pollutants are analyzed by purging a known weight of sample added to 5 mL of reagent water. Samples containing 1% solids or greater, and high levels of pollutants, are extracted with methanol, and an aliquot of the methanol extract is added to reagent water and purged.

QUALITY CONTROL A field blank prepared from reagent water and carried through the sampling and handling protocol may serve as a check on contamination from shipment and storage.

The analyst is permitted to modify this method to improve separations or lower the costs of measurements, provided all performance specifications are met. Analyses of blanks are required. When results of spikes indicate atypical method performance for samples, the samples are diluted to bring method performance within acceptable limits. Analyze two sets of four 5-mL aliquots (8 aliquots total) of the aqueous performance standard. Spike all samples with labeled compounds to assess method performance on the sample matrix. Compute the percent recovery of the labeled compounds using the internal standard method. Compare the percent recovery for each compound with the corresponding labeled compound recovery. Reagent water blanks are analyzed to demonstrate freedom from carryover contamination. Field replicates may be collected to determine the precision of the sampling technique, and spiked samples may be required to determine the accuracy of the analysis when the internal method is used.

REFERENCE Volatile Organic Compounds by Isotope Dilution GC/MS. Office of Water Regulation and Standards, U.S. EPA Industrial Technology Division, Washington, DC, EPA Method 1624, Rev. C, June 1989 (contact W.A. Telliard, U.S. EPA, Office of Water Regulations and Standards, 401 M St., SW, Washington, DC, 20460. Phone: 202-382-7131).

Dibromomethane   EPA Method 502
CAS #74-95-3

TITLE Volatile Organic Compounds in Water By Purge and Trap Capillary Column Gas Chromatography with Photoionization and Electrolytic Conductivity Detectors in Series. U.S. EPA Method 502.2, Rev. 2.0, 1989.

MATRIX Drinking water and raw source water. The latter should include most surface water and groundwater sources.

METHOD SUMMARY This method covers 60 volatile organic compounds that contain halogen atoms and/or that are aromatic. An inert gas (zero grade nitrogen or helium) is bubbled through a 25-mL or a 5-mL water sample (depending on the expected concentration of the analytes). Purged sample components are trapped in a tube of sorbent materials. When purging is complete, the sorbent tube is heated and backflushed with helium to desorb the trapped sample onto a capillary GC column. The column is temperature programmed to separate the method analytes which are then detected with a photoionization detector (PID) and a Hall electrolytic conductivity (HECD) placed in series. The PID is selective for aromatic compounds and the HECD is selective for halogenated compounds.

INTERFERENCES Impurities in the purge gas and from organic compounds outgassing from the plumbing ahead of the trap account for many contamination problems. Interferences purged or coextracted from the samples will vary considerably from source to source, depending upon the particular sample or extract being tested. Cross-contamination can occur whenever high-level and low-level samples are analyzed sequentially. Samples also can be contaminated by diffusion of volatile organics (particularly methylene chloride and fluorocarbons) through the septum seal into the sample during shipment and storage. The lab where volatile analysis is performed and also the refrigerated storage area should be completely free of solvents.

INSTRUMENTATION A GC containing a series configuration of a high temperature photoionization detector (PID) equipped with 10.0 eV (nominal) lamp and Hall electrolytic conductivity detector (HECD) is required. Also required is an all-glass 5-mL purging device, a sorbent trap, and a thermal desorption apparatus which is connected to the GC system.

Column 1: VOCOL glass wide-bore capillary column.
Column 2: RTX–502.2 mega-bore capillary column.
Column 3: DB-62 mega-bore capillary column.

PRECISION & ACCURACY Method detection limits are dependent upon the characteristics of the gas chromatographic system used. Analytes that are not separated chromatographically cannot be individually identified and used in the same calibration mixture or water samples unless an alternative technique for identification and quantification, such as mass spectrometry, is used.

Electrolytic conductivity detetor (c) range in µg/L (a) was 0.02–200.
Electrolytic conductivity detetor (c) MDL in µg/L (b) was 2.2.
Electrolytic conductivity detetor (c) accuracy as % recovery was 109.
Electrolytic conductivity detetor (c) precision as % RSD was 6.7.
Photoionization detector (d) range in µg/L (a) was 0.02–200.
Photoionization detector (d) MDL in µg/L (b) was not listed.
Photoionization detector (d) accuracy as % recovery was not listed.
Photoionization detector (d) precision as % RSD was not listed.

(a) The applicable concentration range of this method is compound, instrument, and matrix-dependent. It is listed as being approximately 0.02 to 200 µg/L but no specific information is provided so caution should be observed.
(b) *The method detection limits reports with this method are compound, instrument, and matrix-dependent. The values reported were calculated using reagent water fortified with the corresponding compounds at 10 µg/L and a GC-equipped with a 60 m × 0.75 mm VOLCOL wide bore capillary column with 1.5 µm film thickness and using helium carrier gas.*
(c) *Recoveries and relative standard deviations were determined from seven samples of reagent water fortified with 10 µg/L of each compound. 2-Bromo-1-chloropropane was used as the internal standard for calculating average recoveries.*
(d) *Recoveries and relative standard deviations were determined from seven samples of reagent water fortified with 10 µg/L of each compound. Fluorobenzene was used as the internal standard for calculating average recoveries.*

SAMPLING METHOD Collect samples using a 40- to 120-mL screw-cap vial (prewashed with detergent, rinsed with distilled water and oven dried at 105°C) with a Teflon®-faced silicone septum. Collect bubble-free samples and place the septum with the Teflon® side down on the water.

SAMPLE PRESERVATION If residual chlorine is present in the water add about 25 mg of ascorbic acid to each vial before samples are collected to remove the chlorine. Add hydrochloric acid to reduce pH to <2, immediately cool samples to 4°C, and store them in a solvent-free refrigerator at 4°C until analysis.

MHT The maximum holding time for samples is 14 days from the time they were collected.

SAMPLE PREPARATION Remove the plungers from two 5-mL syringes and attach a closed syringe valve to each. Warm the sample to room temperature, open the sample bottle, and carefully pour the sample into one of the syringe barrels to just short of overflowing. Replace the syringe plunger, invert the syringe, and compress the sample. Open the syringe valve and vent any residual air while adjusting the sample volume to 5.0 mL. Add 10 µL of the internal calibration standard to the sample through the syringe valve. Close the valve. Fill the second syringe in an identical manner from the same sample bottle. Reserve this second syringe for a reanalysis if necessary.

QUALITY CONTROL As an initial demonstration of lab accuracy and precision, analyze 4 to 7 replicates of a lab fortified blank containing analyte at 0.1–5 µg/L. Collect all samples in duplicate. Surrogate analytes (similar to those of the analytes of interest), whose concentration is known in every sample, are measured using the same internal standard calibration procedure. Duplicate field reagent water blanks (trip blanks) must be analyzed with each set of samples, lab reagent blanks (method blanks) must be analyzed with each batch of samples processed as a group within a work shift. Also, a single lab-fortified blank that contains each of the analytes of interest should be analyzed with each batch of samples processed as a group within a work shift. A 3- to 5-point calibration curve is needed depending on the calibration range factor required.

EPA CONTACT & HOTLINE For technical questions contact Dr. Baldev Bathija, U.S. EPA, Office of Ground Water and Drinking Water (WH-550D), 401 M St. SW, Washington, DC 20460. Tel. (202) 260-3040. For further information the EPA Safe Drinking Water Hotline may be called at: (800) 426-4791.

REFERENCE Methods for the Determination of Organic Compounds in Drinking Water, EPA/600/4-88/039 (revised July 1991; Final Rule for determination of compliance with the MCL for Total Trihalomethanes under 141.30, in 40 CFR Part 141, Vol. 58, No. 147, Fed. Reg., Tuesday Aug. 3, 1993). U.S. EPA Environmental Monitoring Systems Laboratory, Cincinnati, OH, 45268, U.S.A. Available from the National Technical Information Service (NTIS), 5285 Port Royal Road, Springfield, VA 22161; Tel. 800-553-6847. NTIS Order Number is PB91-231480.

Dibromomethane
CAS #74-95-3

EPA Method 524

TITLE Measurement of Purgeable Organic Compounds in Water by Capillary Column GC/MS.

MATRIX Drinking water and raw source water; the latter should include most surface water and groundwater sources.

METHOD SUMMARY Method 524.2 covers 60 volatile organic compounds. An inert gas (zero grade nitrogen or helium) is bubbled through a 25-mL or a 5-mL water sample (depending on the expected concentration of the analytes). Purged sample components are trapped in a tube of sorbent materials. When purging is complete, the sorbent tube is heated and backflushed with helium to desorb the trapped sample onto a capillary GC column.

INTERFERENCES Impurities in the purge gas and from organic compounds outgassing from the plumbing ahead of the trap account for many contamination problems. Interferences purged or coextracted from the samples will vary considerably from source to source, depending upon the particular sample or extract being tested. Cross-contamination can occur whenever high-level and low-level samples are analyzed sequentially. Samples also can be contaminated by diffusion of volatile organics (particularly methylene chloride and fluorocarbons) through the septum seal into the sample during shipment and storage.

INSTRUMENTATION A GC/MS with a data system equipped with one of the following capillary GC columns:

Column 1: VOCOL glass wide bore capillary column.
Column 2: DB-624 fused silica capillary column.
Column 3: DB-5 fused silica capillary column.

Also required is an all-glass 25 mL or 5-mL purging device, a sorbent trap, and a thermal desorption apparatus which is connected to the GC/MS system.

PRECISION & ACCURACY Method detection limits are compound- and instrument-dependent, and may vary from approximately 0.02–0.35 µg/L. Note in the table below that the "true" concentration range used for accuracy and precision measurements was quite narrow. However, the applicable concentration range of this method is primarily column dependent and is approximately 0.02 to 200 µg/L for the wide-bore thick-film columns. Narrow-bore thin-film columns may have a capacity which limits the range to about 0.02 to 20 µg/L. Analytes that are inefficiently purged from water will not be detected when present at low concentrations, but they can be measured with acceptable accuracy and precision when present in sufficient amounts.

Analytes that are not separated chromatographically, but which have different mass spectra and non-interfering quantification ions, can be identified and measured in the same calibration mixture or water sample. Analytes which have very similar mass spectra cannot be individually identified and measured in the same calibration mixture or water samples unless they have different retention times. Co-eluting compounds with very similar mass spectra, typically many structural isomers, must be reported as an isomeric group or pair.

The range (in µg/L) was 0.5–10.
The Method Detection Limig (in µg/L) was 0.24.
The accuracy (as % recovery) was 100.
The precision (in %) was 5.6.

Note: Data were obtained from 16–31 determinations using a wide-bore capillary column and a jet separator interfaced to a quadrupole mass spectrometer. All analytes were in a reagent water matrix.

SAMPLING METHOD Collect samples using a 40- to 120-mL screw-cap vial (prewashed with detergent, rinsed with distilled water and oven dried at 105°C) with a Teflon®-faced silicone septum. Collect bubble-free samples and place the septum with the Teflon® side down on the water.

SAMPLE PRESERVATION If residual chlorine is present in the water add about 25 mg of ascorbic acid to each vial before samples are collected to remove the chlorine. Add hydrochloric acid to reduce pH to <2, and immediately cool samples to 4°C, and store them in a solvent-free refrigerator at 4°C until analysis.

MHT The maximum holding time for samples is 14 days from the time they were collected.

SAMPLE PREPARATION Remove the plungers from two 25-mL (or 5-mL depending on sample size) syringes and attach a closed syringe valve to each. Warm the sample to room temperature, open the sample bottle, and carefully pour the sample into one of the syringe barrels to just short of overflowing. Replace the syringe plunger, invert the syringe, and compress the sample. Open the syringe valve and vent any residual air while adjusting the sample volume to 25.0 mL (or 5 mL). For samples and blanks, add 5 µL of the fortification solution containing the internal standard and the surrogates to the sample through the syringe valve. For calibration standards and lab fortified blanks, add 5 µL of the fortification solution containing the internal standard only. Close the valve. Fill the second syringe in an identical manner from the same sample bottle. Reserve this second syringe for a reanalysis if necessary.

QUALITY CONTROL As an initial demonstration of lab accuracy and precision, analyze 4 to 7 replicates of a lab fortified blank containing analyte at 0.2–5 µg/L. Collect all samples in duplicate. Surrogate analytes (similar to those of the analytes of interest), whose concentration is known in every sample, are measured using the same internal standard calibration procedure. Duplicate field reagent water blanks (trip blanks) must be analyzed with each set of samples, lab reagent blanks (method blanks) must be analyzed with each batch of samples processed as a group within a work shift. Also, a single lab-fortified blank that contains each of the analytes of interest should be analyzed with each batch of samples processed as a group within a work shift. A 3- to 5-point calibration curve is needed depending on the calibration range factor required.

EPA CONTACT & HOTLINE For technical questions contact Dr. Baldev Bathija, U.S. EPA, Office of Ground Water and Drinking Water (WH-550D), 401 M St. SW, Washington, DC

20460. Tel. (202) 260-3040. For further information the EPA Safe Drinking Water Hotline may be called at: (800) 426-4791.

REFERENCE Methods for the Determination of Organic Compounds in Drinking Water, EPA/600/4-88/039 (revised July 1991; Final Rule for determination of compliance with the MCL for Total Trihalomethanes under 141.30, in 40 CFR Part 141, Vol. 58, No. 147, Fed. Reg., Tuesday Aug. 3, 1993). U.S. EPA Environmental Monitoring Systems Laboratory, Cincinnati, OH, 45268, U.S.A. Available from the National Technical Information Service (NTIS), 5285 Port Royal Road, Springfield, VA 22161; Tel. 800-553-6847. NTIS Order Number is PB91-231480.

Dibromomethane **EPA Method 8021**
CAS #74-95-3

TITLE Halogenated Volatile by Gas Chromatography Using Photoionization and Electrolytic Conductivity Detectors in Series: Capillary Column Technique

MATRIX This method is applicable to nearly all types of samples, regardless of water content, including groundwater, aqueous sludges, caustic liquors, acid liquors, waste solvents, oily wastes, mousses, tars, fibrous wastes, polymeric emulsions, filter cakes, spent carbons, spent catalysts, soils, and sediments.

METHOD SUMMARY This method is used to determine 60 volatile organic compounds in a variety of solid waste matrices. It provides GC conditions for the detection of halogenated and aromatic volatile organic compounds. Samples can be analyzed using direct injection or purge-and-trap (EPA Method 5030). Groundwater samples must be analyzed using EPA Method 5030 (where applicable). A temperature program is used with the GC. Detection is achieved by a photoionization detector (PID) and a Hall electrolytic conductivity detector (HECD) in series.

INTERFERENCES Samples can be contaminated by diffusion of volatile organics (particularly chlorofluorocarbons and methylene chloride) through the sample container septum during shipment and storage.

INSTRUMENTATION A GC-equipped with variable-constant differential flow controllers, subambient oven controller, PID and HECD detectors connected with a short piece of uncoated capillary tubing and a data system.

Column: 60 m × 0.75 mm I.D. VOCOL wide-bore capillary column with 1.5 µm film thickness.

PRECISION & ACCURACY MDLs are compound-dependent and vary with purging efficiency and concentration. The applicable concentration range of this method is compound- and instrument-dependent but is approximately 0.1 to 200 µg/L. Analytes that are inefficiently purged from water will not be detected when present at low concentrations, but they can be measured with acceptable accuracy and precision when present in sufficient amounts. The estimated quantitation limit (EQL) for an individual compound is approximately 1 µg/kg (wet weight) for soil/sediment samples, 100 µg/kg (wet weight) for wastes, and 1 µg/L for groundwater. EQLs will be proportionately higher for sample extracts and samples that require dilution or reduced sample size to avoid saturation of the detector.

MULTIPLICATION FACTORS FOR OTHER MATRICES (a)

Matrix	Factor (b)
Groundwater	10
Low-concentration soil	10
Water miscible liquid waste	500
High-concentration soil and sludge	1250
Non-water miscible waste	1250

(a) Sample EQLs are highly matrix-dependent. The EQLs listed herein are provided for guidance and may not always be achievable. (b) EQL = [Method detection limit] × [Factor]. For non-aqueous samples, the factor is on a wet-weight basis.

SINGLE LABORATORY ACCURACY & PRECISION DATA FOR VOCs IN WATER

This method was tested in a single lab using water spiked at 10 µg/L and the following data was reported:

Recoveries and standard deviations were determined from seven samples and spiked at 10 µg/L of each analyte. Recoveries were determined by the internal standard method. Internal standards were: Fluorobenzene for PID and 2-Bromo-1-chloropropane for HECD.

The average recovery (in percent) for the PID was none (no response for this detector).
The standard deviation of the recovery for the PID was none (no response for this detector).
The MDL (in µg/mL) for the PID was none (no response for this detector).
The average recovery (in percent) for the HECD was 109.
The standard deviation of the recovery for the HECD was 7.4.
The MDL (in µg/mL) for the HECD was 2.2.

SAMPLE COLLECTION, PRESERVATION & HANDLING
Volatile Organics — Standard 40-mL glass screw-cap VOA vials with Teflon®-faced silicone septum may be used for both liquid and solid matrices. When collecting samples, liquids and solids should be introduced into the vials gently to reduce agitation which might drive off volatile compounds. If there are any air bubbles present the sample must be retaken. Tap slightly as they are filled to try and eliminate as much free air space as possible. The two vials from each sampling locations should be sealed in separate plastic bags to prevent cross-contamination between samples particularly if the sampled waste is suspected of containing high levels of volatile organics.

Semivolatile organics — Containers used to collect samples for the determination of semivolatile organic compounds should be soap and water washed followed by methanol (or isopropanol) rinsing. The sample containers should be of glass or Teflon® and have screw-top covers with Teflon® liners.

Preservation for volatile organics — No preservation is used with concentrated waste samples. With liquid samples containing no residual chlorine, 4 drops of concentrated hydrochloric acid are added and the samples are immediately cooled to 4°C. When liquid samples contain residual chlorine, they are treated as above and, in addition, 4 drops of 4% aqueous sodium thiosulfate are added. Soil, sediment, and sludge samples are only cooled to 4°C.

Preservation for semivolatile organics — No preservation is used with concentrated waste samples. With liquid samples containing no residual chlorine and with soil, sediment, and sludge samples, immediately cooling to 4°C is the only preservation used. When residual chlorine is present then 3 mL of 10% aqueous sodium sulfate is added for each gallon of sample collected, followed by cooling to 4°C.

MHT The holding time for all volatile organics samples is 14 days. Liquid samples must be extracted within 7 days and their extracts analyzed within 40 days. Concentrated waste, soil, sediment, and sludge samples must be extracted within 14 days and their extracts analyzed within 40 days.

SAMPLE PREPARATION Volatile compounds are introduced into the gas chromatograph either by direct injector or purge-and-trap (EPA Method 5030). EPA Method 5030 may be used directly on groundwater samples or low-concentration contaminated soils and sediments. For medium-concentration soils or sediments, methanolic extraction, as described in EPA Method 5030, may be necessary prior to purge-and-trap analysis.

QUALITY CONTROL Calculate surrogate standard recovery on all samples, blanks, and spikes. A trip blank is recommended to check on sampling, storage, and handling contamination. Calibration standards, at a minimum of five concentration levels, are prepared in organic-free reagent water. One of the concentration levels should be at a concentration near, but above, the method detection limit.

A combination of bromochloromethane, 2-bromo-1-chloropropane, 1,4-dichlorobutane, and bromochlorobenzene are recommended as surrogate standards to encompass the range of the temperature program used in this method.

REFERENCE Test Methods for Evaluating Solid Waste, Physical/Chemical Methods, SW-846, 3rd Edition, U.S. EPA, Office of Solid Waste, Washington, DC, EPA Method 8021A, Rev. 1, Nov. 1992.

Dibromomethane	EPA Method 8240
CAS #74-95-3	

TITLE Volatile Organics By GC/MS: Packed Column Technique

MATRIX Nearly all types of sample matarices, regardless of water content, can be analyzed using this method. This includes groundwater, aqueous sludges, caustic liquors, acid liquors, waste solvents, oily wastes, mousses, tars, fibrous wastes, polymetric emulsions, filter cakes, spent carbons, spent catalysts, soils, and sediments.

METHOD SUMMARY Method 8240B covers 80 volatile organic compounds that are introduced into a gas chromatograph by the purge-and-trap method or by direct injection (in limited applications). For the purge-and-trap method an inert gas (zero grade nitrogen or helium) is bubbled through a 5-mL solution at ambient temperature. Purged sample components are trapped in a tube of sorbent materials. When purging is complete, the sorbent tube is heated and backflushed with inert gas to desorb the trapped components onto a GC column.

INTERFERENCES Impurities in the purge gas and from organic compounds outgassing from the plumbing ahead of the trap account for many contamination problems. Interferences purged or coextracted from the samples will vary considerably from source to source. Cross-contamination can occur whenever high-level and low-level samples are analyzed sequentially. Whenever an unusually concentrated sample is analyzed, it should be followed by the analysis of organic-free reagent water to check for cross-contamination. Samples also can be contaminated by diffusion of volatile organics (particularly methylene chloride and fluorocarbons) through the septum seal into the sample during shipment and storage. A trip blank can serve as a check on such contamination. The lab where volatile analysis is performed and also the refrigerated storage area should be completely free of solvents.

INSTRUMENTATION A gas chromatograph/mass spectrometry/data system (GC/MS) equipped with a 6 ft × 0.1 in I.D. glass column packed with 1% SP-1000 on Carbopack-B (60/80 mesh) is required. Also needed is a 5-mL purging device, a sorbent trap, and a thermal desorption apparatus.

PRECISION & ACCURACY This method is reported to have been tested by 15 laboratories using organic-free reagent water, drinking water, surface water, and industrial wastewaters (not specified) fortified at six concentrations over the range 5–600 µg/L.

Sample estimated quantitation limits (EQLs) are highly matrix-dependent. The EQLs listed may not always be achievable. EQLs listed for soils or sediments are based on wet weight. Normally, data is reported on a dry-weight basis; therefore, EQLs will be higher, based on the percent dry weight of each sample. Note that EQLs are even more variable than MDLs and that they are highly variable depending on the matrix being analyzed.

EQL in groundwater in µg/L was 5.
EQL in low soil or sediment in µg/kg was 5.
Accuracy (a) in µg/L was not listed.
Precision (b) in µg/L was not listed.

(a) *Average recovery found for measurements of samples containing a concentration of C, in µg/L.*
(b) *Overall precision found for measurements of samples with average recovery X for samples containing a concentration of C in µg/L.*
X = *Average recovery found for measurement of samples containing a concentration of C in µg/L.*

MULTIPLICATION FACTORS FOR OTHER MATRICES

Other Matrices	Factor (a)
Waste miscible liquid waste	50
High-concentration soil and sludge	125
Non-water miscible waste	500

(a) *EQL = [EQL for low soil sediment] × [Factor]. For non-aqueous samples, the factor is on a wet-weight basis.*

SAMPLING METHOD

Liquid samples — Use a 40-mL glass screw-cap VOA vial with a Teflon®-faced silicone septum that has been prewashed, rinsed with distilled deionized water, and oven dried. However, if residual chlorine is present, collect sample in a 40-oz. soil VOA container which has been pre-preserved with 4 drops of 10% sodium thiosulfate, mix gently, and then transfer the sample to a 40-mL VOA vial. Collect bubble-free samples in duplicate and seal them in separate plastic bags.

Soils or sediments, and sludges — Use an 8-oz. widemouth glass bottle with a Teflon®-faced silicone septum that has been prewashed with detergent, rinsed with distilled deionized water, and oven dried. Tap slightly to eliminate free air space. Collect samples in duplicate and seal them in separate plastic bags.

SAMPLE PRESERVATION

Liquid samples — Add 4 drops of concentrated HCL and immediately cool samples to 4°C and store in a solvent-free refrigerator.

Soils or sediments, and sludges — Cool samples to 4°C and store in a solvent-free refrigerator.

MHT Maximum holding time is 14 days from the date of sample collection.

SAMPLE PREPARATION

Liquid samples — Remove the plunger from a 5-mL syringe and carefully pour the sample into the syringe barrel to just short of overflowing. Replace the syringe plunger and compress the sample. Open the syringe valve and vent any residual air while adjusting the sample volume to 5.0 mL. If there is only one volatile organic analysis (VOA) vial, a second syringe should be filled at this time to protect against possible loss of sample integrity. Add 10 µL of surrogate spiking solution and 10 µL of internal standard spiking solution through the valve bore of the 5-mL syringe, then close the valve. The surrogate and internal standards may be mixed and added as a single spiking solution.

Sediments, soils, and waste samples — All samples of this type should be screened by GC analysis using a headspace method (EPA Method 3810) or the hexadecane extraction and screening method (EPA Method 3820). Use the screening data to determine whether to use the low-concentration method (0.005–1 mg/kg) or the high-concentration method (>1 mg/kg).

Low-concentration method — The low-concentration method is based on purging a heated sediment or soil sample mixed with organic-free reagent water containing the surrogate and internal standards. Analyze all reagent blanks and standards under the same conditions as the samples.

Use a 5-g sample if the expected concentration is <0.1 mg/kg or a 1-g sample for expected concentrations between 0.1 and 1 mg/kg. Mix the contents of the sample container with a narrow metal spatula. Weigh the amount of the sample into a tared purge device. Add the spiked water to the purge device, which contains the weighed amount of sample, and connect the device to the purge-and-trap system.

High-concentration method — This method is based on extracting the sediment or soil with methanol. A waste sample is either extracted or diluted, depending on its solubility in methanol. Wastes that are insoluble in methanol are diluted with reagent tetraglyme or possibly polyethylene glycol (PEG). An aliquot of the extract is added to organic-free reagent water containing surrogate and internal standards. This is purged at ambient temperature. All samples with an expected concentration of >1.0 mg/kg should be analyzed by this method.

Mix the contents of the sample container with a narrow metal spatula. For sediments or soils and solid wastes that are insoluble in methanol, weigh 4 g (wet weight) of sample into a tared 20-mL vial. For waste that is soluble in methanol, tetraglyme, or PEG, weigh 1 g (wet weight) into a tared scintillation vial or culture tube or a 10-mL volumetric flask. Quickly add 9.0 mL of appropriate solvent then add 1.0 mL of a surrogate spiking solution to the vial, cap it, and shake it for 2 min.

METHANOL EXTRACT REQUIRED FOR ANALYSIS OF HIGH-CONCENTRATION SOILS OR SEDIMENTS

Approximate Concentration Range	Volume of Methanol Extract (a)
500–10,000 µg/kg	100 µL
1,000–20,000 µg/kg	50 µL
5,000–100,000 µg/kg	10 µL
25,000–500,000 µg/kg	100 µL of 1/50 dilution (b)

Calculate appropriate dilution factor for concentrations exceeding this table.

(a) The volume of methanol added to 5 mL of water being purged should be kept constant. Therefore, add to the 5-mL syringe whatever volume of methanol is necessary to maintain a volume of 100 µL added to the syringe.

(b) Dilute an aliquot of the methanol extract and then take 100 µL for analysis.

QUALITY CONTROL Demonstrate, through the analysis of a reagent water blank, that interferences from the analytical system, glassware, and reagents are under control. Blank samples should be carried through all stages of the sample preparation and measurement steps. For each analytical batch (up to 20 samples), a reagent blank, matrix spike, and matrix spike duplicate must be analyzed (the frequency of the spikes may be different for different monitoring programs). The blank and spiked samples must be carried through all stages of the sample preparation and measurement steps. QC samples mentioned in the section on Interferences will also be needed as appropriate to those situations.

REFERENCE Test Methods for Evaluating Solid Waste (SW-846). U.S. EPA. 1983. Method 8240B, Rev. 2, Nov. 1990. Office of Solid Wastes, Washington, DC.

Dibromomethane   EPA Method 8260
CAS #74-95-3

TITLE Volatile Organic Compounds by GC/MS: Capillary Column Technique

MATRIX This method is applicable to nearly all types of samples, regardless of water content, including groundwater, soils, and sediments.

METHOD SUMMARY Method 8260A covers 58 volatile organic compounds that are introduced into a gas chromatograph by the purge-and-trap method or by direct injection (in limited applications). Zero-grade helium is bubbled through a 5-mL solution at ambient temperature. Purged sample components are trapped in a tube containing suitable sorbent materials. When purging is complete, the sorbent tube is heated and backflushed with helium to desorb trapped sample components. The analytes are desorbed directly to a large bore capillary or cryofocussed on a capillary precolumn before being flash evaporated to a narrow bore capillary for analysis.

INTERFERENCES Major contaminant sources are volatile materials in the lab and impurities in the inert purging gas and in the sorbent trap. Interfering contamination may occur when a sample containing low concentrations of volatile organic compounds is analyzed immediately after a sample containing high concentrations of volatile organic compounds. After analysis of a sample containing high concentrations of volatile organic compounds, one or more calibration blanks should be analyzed to check for cross-contamination. Screening of the samples prior to purge-and-trap GC/MS analysis is highly recommended to prevent contamination of the system. This is especially true for soil and waste samples.

Special precautions must be taken to analyze for methylene chloride. The analytical and sample storage area should be isolated from all atmospheric sources of methylene chloride. All gas chromatography carrier gas lines and purge gas plumbing should be constructed from stainless steel or copper tubing. Laboratory clothing previously exposed to methylene chloride fumes during liquid-liquid extraction procedures can contribute to sample contamination.

Samples can also be contaminated by diffusion of volatile organics (particularly methylene chloride and fluorocarbons) through the septum seal during shipment and storage. A trip blank can serve as a check on such contamination.

INSTRUMENTATION GC/MS with a temperature-programmable chromatograph suitable for splitless injection equipped with variable constant differential flow controllers, a subambient oven controller, a purging device, sorbent trap, a thermal desorption apparatus and a capillary precolumn interface when using cryogenic cooling will be needed. The following GC columns may be used:

Column 1: 60 m × 0.75mm I.D. capillary column coated with VOCOL, 1.5 µm film thickness.
Column 2: 30 m × 0.53mm capillary column coated with DB-624 or VOCOL, 3 µm film thickness.
Column 3: 30 m × 0.32mm I.D. capillary column coated with DB-5 or SE-54, 1-µm film thickness.

PRECISION & ACCURACY This method has been tested in a single lab using spiked water. Using a wide-bore capillary column, water was spiked at concentrations between 0.5 and 10 µg/L. Single lab accuracy and precision data are presented. The MDL actually achieved in a given analysis will vary depending on instrument sensitivity and matrix effects.

The MDL (a) in µg/L was 0.24.
The concentration range in µg/L was 0.5–10.
The mean accuracy (% of true value) was 100.
The precision as relative standard deviation was 5.6.

Note: The MDL is based on a 25-mL sample volume instead of a 5-mL sample volume.

SAMPLING METHOD
Liquid samples — Use a 40-mL glass screw-cap VOA vial with a Teflon®-faced silicone septum that has been prewashed, rinsed with distilled deionized water, and oven dried. If residual chlorine is present, collect the sample in a 4-oz soil VOA container which has been pre-preserved with 4 drops of 10% sodium thiosulfate. Mix gently and transfer the sample to a 40-mL VOA vial. Collect bubble-free samples in duplicate and seal each sample in a separate plastic bag.

Soils, sediments, and sludges — Use an 8-oz widemouth glass bottle with Teflon®-faced silicone septum that has been prewashed, rinsed with distilled deionized water, and oven dried. **Do not** heat the septum for more than 1 h. Tap slightly to eliminate any free air space. Collect samples in duplicate and seal each one in a separate plastic bag.

SAMPLE PRESERVATION
Liquid samples — Add 4 drops of concentrated HCL, cool to 4°C and store in a solvent-free refrigerator.

Soils, sediments and sludges — Cool samples to 4°C and store in a solvent-free refrigerator.

MHT The maximum holding time of any sample (liquids, soils, sediments, and sludges) is 14 days.

SAMPLE PREPARATION
Liquid samples — Remove the plunger from a 5-mL syringe and carefully pour the sample into the syringe barrel to just short of overflowing. Replace the syringe plunger and compress the sample. Open the syringe valve and vent any residual air while adjusting the sample volume to 5.0 mL. If there is only one volatile organic analysis (VOA) vial, a second syringe should be filled at this time to protect against possible loss of sample integrity. Add 10 µL of surrogate spiking solution and 10 µL of internal standard spiking solution through the valve bore of the 5-mL syringe, then close the valve. The surrogate and internal standards may be mixed and added as a single spiking solution.

Sediments, soils, and waste samples — All samples of this type should be screened by GC analysis using a headspace method (EPA Method 3810) or the hexadecane extraction and screening method (EPA Method 3820). Use the screening data to determine whether to use the low-concentration method (0.005–1 mg/kg) or the high-concentration method (>1 mg/kg).

Low-concentration method — The low-concentration method is based on purging a heated sediment or soil sample mixed with organic-free reagent water containing the surrogate and internal standards. Analyze all reagent blanks and standards under the same conditions as the samples.

Use a 5-g sample if the expected concentration is <0.1 mg/kg or a 1-g sample for expected concentrations between 0.1 and 1 mg/kg. Mix the contents of the sample container with a narrow metal spatula. Weigh the amount of the sample into a tared purge device. Add the spiked water to the purge device, which contains the weighed amount of sample, and connect the device to the purge-and-trap system.

High-concentration method — This method is based on extracting the sediment or soil with methanol. A waste sample is either extracted or diluted, depending on its solubility in methanol. Wastes that are insoluble in methanol are diluted with reagent tetraglyme or possibly polyethylene glycol (PEG). An aliquot of the extract is added to organic-free reagent water containing surrogate and internal standards. This is purged at ambient temperature. All samples with an expected concentration of >1.0 mg/kg should be analyzed by this method.

Mix the contents of the sample container with a narrow metal spatula. For sediments or soils and solid wastes that are insoluble in methanol, weigh 4 g (wet weight) of sample into a tared 20-mL vial. For waste that is soluble in methanol, tetraglyme, or PEG, weigh 1 g (wet weight) into a tared scintillation vial or culture tube or a 10-mL volumetric flask. Quickly add 9.0 mL of appropriate solvent then add 1.0 mL of a surrogate spiking solution to the vial, cap it, and shake it for 2 min.

METHANOL EXTRACT REQUIRED FOR ANALYSIS OF HIGH-CONCENTRATION SOILS OR SEDIMENTS

Approximate Concentration Range	Volume of Methanol Extract (a)
500–10,000 µg/kg	100 µL
1,000–20,000 µg/kg	50 µL
5,000–100,000 µg/kg	10 µL
25,000–500,000 µg/kg	100 µL of 1/50 dilution (b)

Calculate appropriate dilution factor for concentrations exceeding this table.

(a) The volume of methanol added to 5 mL of water being purged should be kept constant. Therefore, add to the 5-mL syringe whatever volume of methanol is necessary to maintain a volume of 100 µL added to the syringe.

(b) Dilute an aliquot of the methanol extract and then take 100 µL for analysis.

QUALITY CONTROL Demonstrate, through the analysis of a reagent water blank, that interferences from the analytical system, glassware, and reagents are under control. Blank samples should be carried through all stages of the sample preparation and measurement steps. For each analytical batch (up to 20 samples), a reagent blank, matrix spike, and matrix spike duplicate must be analyzed (the frequency of the spikes may be different for different monitoring programs). The blank and spiked samples must be carried through all stages of the sample preparation and measurement steps. QC samples mentioned in the section on Interferences will also be needed as appropriate to those situations.

Matrix spiking standards should be prepared from volatile organic compounds which will be representative of the compounds being investigated. The recommended internal standards are chlorobenzene-d5, 1,4-difluorobenzene, 1,4-dichlorobenzene-d4, and pentafluorobenzene. Using stock standard solutions, prepare secondary dilution standards containing the compounds of interest, either singly or mixed together in methanol. Store them in a vial with no headspace for no more than one week. Surrogates recommended are toluene-d8, 4-bromofluorobenzene, and dibromofluoromethane. Each sample undergoing GC/MS analysis must be spiked with 10 µL of the surrogate spiking solution prior to analysis.

REFERENCE Test Methods for Evaluating Solid Waste (SW-846). U.S. EPA 1983, Method 8260A, Rev. 1, Nov. 1990. Office of Solid Waste, Washington, DC.

Dicamba   **EPA Method 8151**
CAS #1918-00-9

TITLE Chlorinated Herbicides by GC Using Methylation or Pentafluorobenzylation Derivatization: Capillary Column Technique.

MATRIX This method covers aqueous and solid matrices. This includes a wide variety such as drinking water, groundwater, industrial wastewaters, surface waters, soils, solids, and sediments.

METHOD SUMMARY This is a GC method for determining 19 chlorinated acid herbicides in aqueous, soil, and waste matrices. Because these compounds are produced and used in various forms (i.e., acid, salt, ester, etc.) a hydrolysis step is included to convert the herbicide to the acid form prior to analysis. This method provides hydrolysis, extraction, derivatization and GC conditions for the analysis of chlorinated acid herbicides in water, soil, and waste samples. Water samples are hydrolyzed *in situ*, extracted with diethyl ether, and then esterified with either diazomethane or pentafluorobenzyl bromide. The derivatives are determined by gas chromatography with an electron capture detector (GC/ECD). The results are reported as acid equivalents. The sensitivity of this method depends on the level of interferences in addition to instrumental limitations.

INTERFERENCES Method interferences may be caused by contaminants in solvents, reagents, glassware, and other sample processing hardware. Immediately prior to use, glassware should be rinsed with the next solvent to be used. Matrix interferences may be caused by contaminants that are coextracted from the sample. Organic acids, especially chlorinated acids, cause the most direct interference with the determination by methylation. Phenols, including chlorophenols, may also interfere with this procedure. The determination using pentafluorobenzylation is more sensitive, and more prone to interferences from the presence of organic acids of phenols than by methylation. Alkaline hydrolysis and subsequent extraction of the basic solution removes many chlorinated hydrocarbons and phthalate esters that might otherwise interfere with the ECD analysis. The herbicides, being strong organic acids, react readily with alkaline substances and may be lost during analysis. Therefore, glassware must be acid-rinsed and then rinsed to constant pH with organic-free reagent water.

INSTRUMENTATION A GC suitable for Grob-type injection using capillary columns. A data system for measuring peak heights and/or peak areas is recommended. An electron capture detector (ECD) is used. Also a K-D apparatus, a diazomethane generator, a centrifuge and an ultrasonic disrupter will be required.

Narrow Bore Columns:
Primary Column 1: 30 m × 0.25 mm, 5% phenyl/95% methyl silicone (DB-5), 0.25 μm film thickness.
Primary Column 1a (GC/MS): 30 m × 0.32 mm, 5% phenyl/95% methyl silicone (DB-5), 1-μm film thickness.
Column 2: 30 m × 0.25 mm DB-608 with a 25 μm film thickness.
Confirmation Column: 30 m × 0.25 mm, 14% cyanopropyl phenyl silicone (DB-1701), 0.25 μm film thickness.

Megabore Columns:
Primary Column: 30 m × 0.53 mm DB-608 with 0.83 μm film thickness.
Confirmation Column: 30 m × 0.53 mm, 14% cyanopropyl phenyl silicone (DB-1701), 1.0 μm film thickness.

PRECISION & ACCURACY Method detection limits (MDLs) are compound-dependent and vary with derivitization efficiency, derivative recovery, the matrix sampled, and herbicide concentration.

The estimated MDL (in μg/L) was 0.081 for aqueous samples using GC/ECD.

The estimated MDL (in μg/kg) was not reported for soil samples using GC/ECD when corrected back to 50-g samples extracted and concentrated to 10 mL with 5-μL injections.

The estimated GC/MS identification limit (in ng) was not reported for soil samples using GC/MS.

Mean percent recovery, calculated from 7–8 determinations of spiked reagent water, after diazomethane derivatization, from a spike concentration (in μg/L) of 0.4 was 135 with a standard deviation of the percent recovery of 32.4.

Mean percent recovery, calculated from 10 determinations of spiked clay and clay/still bottom samples over the linear concentration range (in ng/g) of 0.52–104 was 95.7 with a percent relative standard deviation of 7.5. The RSD % was calculated on 10 samples high in the linear concentration range and 10 low in the range. The linear concentration range was determined using standard solutions and corrected to 50 g soil samples.

SAMPLE COLLECTION, PRESERVATION & HANDLING
Containers used to collect samples for the determination of semivolatile organic compounds should be soap and water washed followed by methanol (or isopropanol) rinsing. The sample containers should be of glass or Teflon® and have screw-top covers with Teflon® liners.

No preservation is used with concentrated waste samples. With liquid samples containing no residual chlorine and with soil, sediment, and sludge samples, immediately cooling to 4°C is the only preservation used. When residual chlorine is present then 3 mL of 10% aqueous sodium sulfate is added for each gallon of sample collected, followed by cooling to 4°C.

The holding time for all volatile organics samples is 14 days. Liquid samples must be extracted within 7 days and their extracts analyzed within 40 days. Concentrated waste, soil, sediment, and sludge samples must be extracted within 14 days and their extracts analyzed within 40 days.

SAMPLE PREPARATION
Preparation of soil, sediment, and other solid samples — Acidify 30 g (dry weight) solids with 0.1 M phosphate buffer (pH = 2.5) and thoroughly mix the contents. Spike the sample with surrogate compound(s). The ultrasonic extraction of solids must be optimized for each type of sample. In order for the ultrasonic extractor to efficiently extract solid samples, the sample must be free flowing when the solvent is added. Acidified anhydrous sodium sulfate should be added to clay-type soils, or any other solid that is not a free-flowing sandy texture, until a free flowing mixture is obtained. Add methylene chloride and perform ultrasonic extraction. Combine organic extracts from the repetitive extractings of the sample and centrifuge. Add aqueous potassium hydroxide, water, and methanol to the extract and reflux the mixture on a water bath. Extract the solution three times with methylene chloride and discard the methylene chloride phase. The basic solution contains the herbicide salts. Adjust the pH of the solution to <2 with cold sulfuric acid and extract three times with methylene chloride. Combine the extracts and pour them through a prerinsed drying column containing acidified anhydrous sodium sulfate. Collect the dried extracts in a K-D flask and concentrate them.

Preparation of aqueous samples — Measure 1 L of sample into a 2 L separatory funnel and spike it with surrogate compound(s). Add NaCl to the sample, then add 6 N NaOH to the sample to a pH of 12 or more and let the sample sit at room temperature for 1 h to hydrolyze esters. Extract the sample three times with methylene chloride and discard the extracts. Then add cold 12 N sulfuric acid to a pH less than or equal to 2, and extract the sample three times with ethyl ether. Collect the ether phase in a flask containing acidified anhydrous sodium sulfate and allow it to remain in contact with the sodium sulfate for a minimum of 2 h. The drying step is very critical to ensuring complete esterification; any moisture remaining in the ether will result in low herbicide recoveries.

Extract concentration and derivatization — The combined ether extract is concentrated to about 1 mL using a K-D apparatus followed by using a micro Snyder column or nitrogen gas blowdown. If methyl esters are to be produced, then dilute the concentrated ether extract with 1 mL of isooctane and 0.5 mL of methanol, dilute to a final volume of 4 mL, and esterify with diazomethane. If pentafluorobenzene esters are to be produced, then dilute concentrated ether extract with acetone to a final volume of 4 mL and esterify with pentafluorobenzyl bromide.

QUALITY CONTROL Select a representative spike concentration for each compound (acid or ester) to be measured. Using stock standard, prepare a quality control check sample concentrate, in acetone, that is 1000 times more concentrated than the selected concentrations. Use this quality control check sample concentrate to prepare quality control check samples. Calculate surrogate standard recovery on all standards, samples, blanks, and spikes. GC/MS techniques should be judiciously employed to support qualitative identifications made with this method. When available, chemical ionization mass spectra may be employed to aid the qualitative identification process.

REFERENCE Test Methods for Evaluating Solid Waste, Physical/Chemical Methods, SW-846, 3rd Edition, U.S. EPA, Office of Solid Waste, Washington, DC, EPA Method 8151, Nov. 1990.

Dicamba   **EPA Method 8150**
CAS #1918-00-9

TITLE Chlorinated Herbicides

MATRIX Groundwater, soils, sludges, water miscible liquid wastes, and non-water miscible wastes.

APPLICATION This method is used for the analysis of 10 chlorinated herbicides. Samples are extracted, hydrolyzed with potassium hydroxide, and extraneous organics are removed by a solvent wash. After acidification, the acids are extracted, concentrated and converted to their methyl esters using diazomethane. They are then analyzed using direct injection into a gas chromatograph (GC). Be very careful because diazomethane can explode under certain conditions and it is also a carcinogen.

INTERFERENCES Organic acids and phenols (especially chlorinated acids and phenols) may cause interferences. Phthalate esters are not as significant an interference as with other GC-ECD methods if an electron capture detector is used. The herbicides may react readily with alkaline substances and be lost during analysis so all glassware and glass wool must be acid rinsed and sodium sulfate must be acidified with sulfuric acid prior to use. Sensitivity usually depends on the level of interferences rather than on instrumentation.

INSTRUMENTATION GC capable of on-column injections and an electron capture detector (ECD)or a halogen specific detector. Column 1: 1.8 m by 4 mm with 1.5% SP-2250/1.95% SP-2401 on Supelcoport. Column 2: 1.8 m by 4 mm with 5% OV-210 on Gas Chrom Q. Column 3: 1.98 m by 2 mm with 0.1%. SP-1000 on Carbopack C. The preferred column is Column Number 1 or 2.

RANGE Not listed.

MDL 0.27 µg/L (in reagent water; ECD).

PQL FACTORS FOR MULTIPLYING × FID MDL VALUE

Matrix	Multiplication Factor
Groundwater	10
Low-level soil by sonication with GPC cleanup	670
High-level soil and sludge by sonication	10,000
Non-water miscible waste	100,000

PRECISION (as standard deviation) 7% with 1.2 µg/L spike in drinking water. 9% with 1.1 µg/L in municipal water.

ACCURACY (as mean recovery) 79% with 1.2 µg/L spike in drinking water. 86% with 1.1 µg/L in municipal water.

SAMPLING METHOD Use 8-oz. widemouth glass bottles with Teflon®-lined caps for concentrated waste samples, soils, sediments, and sludges. Use 1 or 2½ gallon amber glass bottles with Teflon®-lined caps for liquid (water) samples.

STABILITY Cool soil, sediment, sludge, and liquid samples to 4°C. If residual chlorine is present in liquid samples add 3 mL of 10% sodium thiosulfate per gallon of sample and cool to 4°C.

MHT 14 days for concentrated waste, soil, sediment, or sludge; 7 days for liquid samples; all extracts must be analyzed within 40 days.

QUALITY CONTROL A quality control check sample concentrate containing this compound in acetone at a concentration 1,000 times more concentrated than the selected spike concentration is required. The QC check sample concentrate may be prepared from pure standard materials or purchased as certified solutions. Use appropriate trip, matrix, control site, method, reagent, and solvent blanks. Internal, surrogate, and five concentration level calibration standards are used.

REFERENCE Method 8150, SW-846, 3rd ed., Sept. 1986.

Dichlone   **EPA Method 8270**
CAS #117-80-6

TITLE Semivolatile Organic Compounds by GC/MS

MATRIX This method is used to determine the concentration of semivolatile organic compounds in extracts prepared from all types of solid waste matrices, soils, and groundwater. Although surface waters are not specifically mentioned, this method should be applicable to water samples from rivers, lakes, etc.

METHOD SUMMARY This method covers 259 semivolatile organic compounds. In very limited applications direct injection of the sample into the GC/MS system may be appropriate, but this results in very high detection limits (approximately 10,000 µg/L). Typically, a 1-L liquid sample, containing surrogate, and matrix spiking standards, is extracted in a continuous extractor first under acid conditions and then under basic conditions. Typically 30 g of a solid sample, containing surrogate, and matrix spiking standards, is extracted ultrasonically. After concentrating the extract to 1 mL it is spiked with 10 µL of an internal standard solution just prior to analysis by GC/MS. The volume injected should contain about 100 ng of base/neutral and 200 ng of acid surrogates (for a 1-µL injection). Analysis is performed by GC/MS using a capillary GC column.

INTERFERENCES Raw GC/MS data from all blanks, samples, and spikes must be evaluated for interferences. Contamination by carryover can occur whenever high-concentration and low-concentration samples are sequentially analyzed. To reduce carryover, the sample syringe must be rinsed out between samples with solvent. Whenever an unusually concentrated sample is encountered, it should be followed by the analysis of blank solvent to check for cross-contamination.

INSTRUMENTATION A GC/MS and a data system are required. The GC column used is a 30 m × 0.25 mm I.D. (or 0.32 mm I.D.) 1um film thickness silicone-coated fused silica capillary column. A continuous liquid-liquid extractor equipped with Teflon® or glass connection joints and stopcocks requiring no lubrication, a K-D concentrating apparatus, water bath, and an ultrasonic disrupter with a minimum power of 300 W and with pulsing capability are also required.

PRECISION & ACCURACY The estimated quantitation limit (EQL) of Method 8270B for determining an individual compound is approximately 1 mg/kg (wet weight) for soil or sediment samples, 1–200 mg/kg for wastes (dependent on matrix and method of preparation), and 10 µg/L for groundwater samples. EQLs will be proportionally higher for sample extracts that require dilution to avoid saturation of the detector.

The EQL(b) for groundwater in µg/L is not Applicable.
The EQL (a, b) for low concentrations in soil and sediment in µg/kg is not determined.
Accuracy as µg/L is not listed.
Overall precision in µg/L is not listed.

(a) EQLs listed for soil/sediment are based on wet weight. Normally data is reported in a dry-weight basis; therefore, EQLs will be higher based on the % dry weight of each sample. This calculation is based on a 30 g sample and gel permeation chromatography cleanup.
(b) Sample EQLs are highly matrix-dependent. The EQLs are provided for guidance and may not always be achievable.
C = True value for concentration, in µg/L.
X = Average recovery found for measurements of samples containing a concentration of C, in µg/L.

ESTIMATED QUANTITATION LIMIT

Other Matrices	Factor (a)
High-concentration soil and sludges by sonicator	7.5
Non-water miscible waste	75

(a) EQL for other matrices = [EQL for low soil/sediment] × [Factor]. This estimated EQL is similar to an EPA "Practical Quantitation Limit."

SAMPLING METHOD

Liquid samples — Use a 1 or 2½ gallon amber glass bottle with a screw-top Teflon®-lined cover that has been prewashed with detergent and rinsed with distilled water and methanol (or isopropanol).

Soils, sediments, or sludges — Use an 8-oz. widemouth glass with a screw-top Teflon®-lined cover that has been prewashed with detergent and rinsed with distilled water and methanol (or isopropanol).

SAMPLE PRESERVATION

Liquid samples — If residual chlorine is present, add 3 mL of 10% sodium thiosulfate per gallon, cool to 4°C and store in a solvent-free refrigerator until analysis; if chlorine is not present, then eliminate the sodium thiosulfate addition.

Soils, sediments, or sludges — Cool samples to 4°C and store in a solvent-free refrigerator.

MHT Liquid samples must be extracted within 7 days and the extracts analyzed within 40 days. Soils, sediments, or sludges may be stored for a maximum of 14 days and the extracts analyzed within 40 days.

SAMPLE PREPARATION

Liquid samples — Transfer 1 L quantitatively to a continuous extractor. If high concentrations are anticipated, a smaller volume may be used and then diluted with organic-free reagent water to 1 L. Adjust pH, if necessary, to pH <2 using 1:1 (V/V) sulfuric acid. Pipette 1.0 mL of a surrogate standard spiking solution into each sample. For the sample in each analytical batch selected for spiking, add 1.0 mL of a matrix spiking standard. For base/neutral acid analysis, the amount of the surrogates and matrix spiking compounds added to the sample should result in a final concentration of 100 ng/µL of each analyte in the extract to be analyzed (assuming a 1-µL injection). Extract with methylene chloride for 18–24 h. Next, adjust the pH of the aqueous phase to pH >11 using 10 N sodium hydroxide and extract it with methylene chloride again for 18–24 h. Dry the extract through a column containing anhydrous sodium sulfate and concentrate it to 1 mL using a K-D concentrator.

Soils, sediments, or sludges — Use 30 g of sample. Nonporous or wet samples (gummy or clay type) that do not have a free-flowing sandy texture must be mixed with anhydrous sodium sulfate until the sample is free flowing. Add 1 mL of surrogate standards to all samples, spikes, standards, and blanks. For the sample in each analytical batch selected for spiking, add 1.0 mL of a matrix spiking standard. For base/neutral acid analysis, the amount added of the surrogates and matrix spiking compounds should result in a final concentration of 100 ng/µL of each base/neutral analyte and 200 ng/µL of each acid analyte in the extract to be analyzed (assuming a 1-µL injection). Immediately add a 100-mL mixture of 1:1 methylene chloride:acetone and extract the sample ultrasonically for 3 min and then decant or filter the extracts. Repeat the extraction two or more times. Dry the extract using a column with anhydrous sodium sulfate and concentrate it to 1 mL in a K-D concentrator.

QUALITY CONTROL A methylene chloride solution containing 50 ng/µL of decafluorotriphenylphosphine (DFTPP) is used for tuning the GC/MS system each 12-h shift. A system performance check also must be made during every 12-h shift. A standard containing 50 ng/µL each of 4,4'-DDT, pentachlorophenol, and benzidine is required to verify injection port inertness and GC column performance. A calibration standard at mid-concentration, containing each compound of interest, including all required surrogates, must be performed every 12 h during analysis. After the system performance check is met, calibration check compounds (CCCs) are used to check the validity of the initial calibration.

The internal standard responses and retention times in the calibration check standard must be evaluated immediately after or during data acquisition. If the retention time for any internal standard changes by more than 30 seconds from the last check calibration (12 h), the chromatographic system must be inspected for malfunctions and corrections must be made, as required. If the electron ionization current plot (EICP) area for

any of the internal standards changes by a factor of two from the last daily calibration standard check, the mass spectrometer must be inspected for malfunctions and corrections must be made, as appropriate.

Demonstrate, through the analysis of a reagent water blank, that interferences from the analytical system, glassware, and reagents are under control. The blank samples should be carried through all stages of the sample preparation and measurement steps. For each analytical batch (up to 20 samples), a reagent blank, matrix spike, and matrix spike duplicate/duplicate must be analyzed (the frequency of the spikes may be different for different monitoring programs). The blank and spiked samples must be carried through all stages of the sample preparation and measurement steps. A QC reference sample concentrate containing each analyte at a concentration of 100 mg/L in methanol is required.

REFERENCE Test Methods for Evaluating Solid Waste (SW-846). U.S. EPA 1983, Method 8270B, Rev. 2, Nov. 1990. Office of Solid Waste, Washington, DC.

trans-1,4-Dichloro-2-butene   EPA Method 1624
CAS #110-57-6

TITLE Volatile Organic Compounds by Isotope Dilution GC/MS

MATRIX Compounds may be determined in waters, soils, and municipal sludges by this method.

METHOD SUMMARY This method is used to determine 58 volatile toxic organic pollutants associated with the CWA (as amended 1987); the RCRA (as amended 1986); the CERCLA (as amended 1986); and other compounds amenable to purge-and-trap gas chromatography-mass spectrometry (GC/MS).

If the solids content is less than 1%, stable isotopically-labeled analogs of the compounds of interest are added to a 5-mL sample and the sample is purged with an inert gas at 20–25°C in a chamber designed for soil or water samples. If the solids content is greater than 1%, 5 mL of reagent water and the labeled compounds are added to a 5-g aliquot of sample and the mixture is purged at 40°C. Compounds that will not purge at 20–25°C or at 40°C are purged at 78–85°C. In the purging process, the volatile compounds are transferred from the aqueous phase into the gaseous phase where they are passed into a sorbent column, and trapped. After purging is completed, the trap is backflushed and heated rapidly to desorb the compounds into a GC. The compounds are separated by the GC and detected by a MS. The labeled compounds serve to correct the variability of the analytical technique.

INTERFERENCES Impurities in the purge gas, organic compounds outgassing from the plumbing upstream of the trap, and solvent vapors in the lab account for most problems. Samples can be contaminated by diffusion of volatile organic compounds (particularly methylene chloride) through the bottle seal during shipment and storage. Contamination by carryover can occur when high-level and low-level samples are analyzed sequentially. When an unusually concentrated sample is encountered, follow it by analysis of a reagent water blank to check for carryover.

INSTRUMENTATION Major equipment includes a GC with linear temperature programming and a glass jet separator as the MS interface, a MS with 70 eV electron impact ionization, and a data system to collect and record response factors.

Column: 2.8 m × 2 mm I.D. glass, packed with 1% SP-1000 on Carbopak B, 60/80 mesh, or equivalent.

PRECISION & ACCURACY The detection limits of the method are usually dependent on the level of interferences rather than instrumental limitations. The method detection limits were determined in digested sludge (low solids) and in filter cake or compost (high solids).

The MDL (in µg/kg) for low solids is not listed and for high solids is not listed.

Labeled and native compound precision (in µg/L) as standard deviation was not listed.

Labeled and native compound accuracy (in µg/L) as average recovery was not listed.

Acceptance criteria are at 20 µg/L for this compound.

SAMPLE COLLECTION, PRESERVATION & HANDLING Grab samples are collected in glass containers having a total volume greater than 20 mL. Fill and seal each bottle so that no air bubbles are entrapped. Samples are maintained at 0 to 4°C from the time of collection until analysis. If an aqueous sample contains residual chlorine, add sodium thiosulfate preservative (10 mg/40 mL) to the empty sample bottles just prior to shipment to the sample site. All samples must be analyzed within 14 days of collection.

SAMPLE PREPARATION Samples containing less than 1% solids are analyzed directly as aqueous samples. Samples containing 1% solids or greater are analyzed as solid samples utilizing one of two methods, depending on the levels of pollutants, in the sample. Samples containing 1% solids or greater, and low to moderate levels of pollutants are analyzed by purging a known weight of sample added to 5 mL of reagent water. Samples containing 1% solids or greater, and high levels of pollutants, are extracted with methanol, and an aliquot of the methanol extract is added to reagent water and purged.

QUALITY CONTROL A field blank prepared from reagent water and carried through the sampling and handling protocol may serve as a check on contamination from shipment and storage.

The analyst is permitted to modify this method to improve separations or lower the costs of measurements, provided all performance specifications are met. Analyses of blanks are required. When results of spikes indicate atypical method performance for samples, the samples are diluted to bring method performance within acceptable limits. Analyze two sets of four 5-mL aliquots (8 aliquots total) of the aqueous performance standard. Spike all samples with labeled compounds to assess method performance on the sample matrix. Compute the percent recovery of the labeled compounds using the internal standard method. Compare the percent recovery for each compound with the corresponding labeled compound recovery. Reagent water blanks are analyzed to demonstrate freedom from carryover contamination. Field replicates may be collected to determine the precision of the sampling technique, and spiked samples may be required to determine the accuracy of the analysis when the internal method is used.

REFERENCE Volatile Organic Compounds by Isotope Dilution GC/MS. Office of Water Regulation and Standards, U.S. EPA Industrial Technology Division, Washington, DC, EPA Method 1624, Rev. C, June 1989 (contact W.A. Telliard, U.S. EPA, Office of Water Regulations and Standards, 401 M St., SW, Washington, DC, 20460. Phone: 202-382-7131).

1,4-Dichloro-2-butene   **EPA Method 8240**
CAS #764-41-0

TITLE Volatile Organics By GC/MS: Packed Column Technique

MATRIX Nearly all types of sample matarices, regardless of water content, can be analyzed using this method. This includes groundwater, aqueous sludges, caustic liquors, acid liquors, waste solvents, oily wastes, mousses, tars, fibrous wastes, polymetric emulsions, filter cakes, spent carbons, spent catalysts, soils, and sediments.

METHOD SUMMARY Method 8240B covers 80 volatile organic compounds that are introduced into a gas chromatograph by the purge-and-trap method or by direct injection (in limited applications). For the purge-and-trap method an inert gas (zero grade nitrogen or helium) is bubbled through a 5-mL solution at ambient temperature. Purged sample components are trapped in a tube of sorbent materials. When purging is complete, the sorbent tube is heated and backflushed with inert gas to desorb the trapped components onto a GC column.

INTERFERENCES Impurities in the purge gas and from organic compounds outgassing from the plumbing ahead of the trap account for many contamination problems. Interferences purged or coextracted from the samples will vary considerably from source to source. Cross-contamination can occur whenever high-level and low-level samples are analyzed sequentially. Whenever an unusually concentrated sample is analyzed, it should be followed by the analysis of organic-free reagent water to check for cross-contamination. Samples also can be contaminated by diffusion of volatile organics (particularly methylene chloride and fluorocarbons) through the septum seal into the sample during shipment and storage. A trip blank can serve as a check on such contamination. The lab where volatile analysis is performed and also the refrigerated storage area should be completely free of solvents.

INSTRUMENTATION A gas chromatograph/mass spectrometry/data system (GC/MS) equipped with a 6 ft × 0.1 in I.D. glass column packed with 1% SP-1000 on Carbopack-B (60/80 mesh) is required. Also needed is a 5-mL purging device, a sorbent trap, and a thermal desorption apparatus.

PRECISION & ACCURACY This method is reported to have been tested by 15 laboratories using organic-free reagent water, drinking water, surface water, and industrial wastewaters (not specified) fortified at six concentrations over the range 5–600 µg/L.

Sample estimated quantitation limits (EQLs) are highly matrix-dependent. The EQLs listed may not always be achievable. EQLs listed for soils or sediments are based on wet weight. Normally, data is reported on a dry-weight basis; therefore, EQLs will be higher, based on the percent dry weight of each sample. Note that EQLs are even more variable than MDLs and that they are highly variable depending on the matrix being analyzed.

EQL in groundwater in µg/L was 100.
EQL in low soil or sediment in µg/kg was 100.
Accuracy (a) in µg/L was not listed.
Precision (b) in µg/L was not listed.

(a) *Average recovery found for measurements of samples containing a concentration of C, in µg/L.*
(b) *Overall precision found for measurements of samples with average recovery X for samples containing a concentration of C in µg/L.*
X = *Average recovery found for measurement of samples containing a concentration of C in µg/L.*

MULTIPLICATION FACTORS FOR OTHER MATRICES

Other Matrices	Factor (a)
Waste miscible liquid waste	50
High-concentration soil and sludge	125
Non-water miscible waste	500

(a) EQL = [EQL for low soil sediment] × [Factor]. For non-aqueous samples, the factor is on a wet-weight basis.

SAMPLING METHOD

Liquid samples — Use a 40-mL glass screw-cap VOA vial with a Teflon®-faced silicone septum that has been prewashed, rinsed with distilled deionized water, and oven dried. However, if residual chlorine is present, collect sample in a 40-oz. soil VOA container which has been pre-preserved with 4 drops of 10% sodium thiosulfate, mix gently, and then transfer the sample to a 40-mL VOA vial. Collect bubble-free samples in duplicate and seal them in separate plastic bags.

Soils or sediments, and sludges — Use an 8-oz. widemouth glass bottle with a Teflon®-faced silicone septum that has been prewashed with detergent, rinsed with distilled deionized water, and oven dried. Tap slightly to eliminate free air space. Collect samples in duplicate and seal them in separate plastic bags.

SAMPLE PRESERVATION

Liquid samples — Add 4 drops of concentrated HCL and immediately cool samples to 4°C and store in a solvent-free refrigerator.

Soils or sediments, and sludges — Cool samples to 4°C and store in a solvent-free refrigerator.

MHT Maximum holding time is 14 days from the date of sample collection.

SAMPLE PREPARATION

Liquid samples — Remove the plunger from a 5-mL syringe and carefully pour the sample into the syringe barrel to just short of overflowing. Replace the syringe plunger and compress the sample. Open the syringe valve and vent any residual air while adjusting the sample volume to 5.0 mL. If there is only one volatile organic analysis (VOA) vial, a second syringe should be filled at this time to protect against possible loss of sample integrity. Add 10 µL of surrogate spiking solution and 10 µL of internal standard spiking solution through the valve bore of the 5-mL syringe, then close the valve. The surrogate and internal standards may be mixed and added as a single spiking solution.

Sediments, soils, and waste samples — All samples of this type should be screened by GC analysis using a headspace method (EPA Method 3810) or the hexadecane extraction and screening method (EPA Method 3820). Use the screening data to determine whether to use the low-concentration method (0.005–1 mg/kg) or the high-concentration method (>1 mg/kg).

Low-concentration method — The low-concentration method is based on purging a heated sediment or soil sample mixed with organic-free reagent water containing the surrogate and internal standards. Analyze all reagent blanks and standards under the same conditions as the samples.

Use a 5-g sample if the expected concentration is <0.1 mg/kg or a 1-g sample for expected concentrations between 0.1 and 1 mg/kg. Mix the contents of the sample container with a narrow metal spatula. Weigh the amount of the sample into a tared purge device. Add the spiked water to the purge device, which contains the weighed amount of sample, and connect the device to the purge-and-trap system.

High-concentration method — This method is based on extracting the sediment or soil with methanol. A waste sample is either extracted or diluted, depending on its solubility in methanol. Wastes that are insoluble in methanol are diluted with reagent tetraglyme or possibly polyethylene glycol (PEG). An aliquot of the extract is added to organic-free reagent water containing surrogate and internal standards. This is purged at ambient temperature. All samples with an expected concentration of >1.0 mg/kg should be analyzed by this method.

Mix the contents of the sample container with a narrow metal spatula. For sediments or soils and solid wastes that are insoluble in methanol, weigh 4 g (wet weight) of sample into a tared 20-mL vial. For waste that is soluble in methanol, tetraglyme, or PEG, weigh 1 g (wet weight) into a tared scintillation vial or culture tube or a 10-mL volumetric flask. Quickly add 9.0 mL of appropriate solvent then add 1.0 mL of a surrogate spiking solution to the vial, cap it, and shake it for 2 min.

METHANOL EXTRACT REQUIRED FOR ANALYSIS OF HIGH-CONCENTRATION SOILS OR SEDIMENTS

Approximate Concentration Range	Volume of Methanol Extract (a)
500–10,000 µg/kg	100 µL
1,000–20,000 µg/kg	50 µL
5,000–100,000 µg/kg	10 µL
25,000–500,000 µg/kg	100 µL of 1/50 dilution (b)

Calculate appropriate dilution factor for concentrations exceeding this table.

(a) The volume of methanol added to 5 mL of water being purged should be kept constant. Therefore, add to the 5-mL syringe whatever volume of methanol is necessary to maintain a volume of 100 µL added to the syringe.

(b) Dilute an aliquot of the methanol extract and then take 100 µL for analysis.

QUALITY CONTROL Demonstrate, through the analysis of a reagent water blank, that interferences from the analytical system, glassware, and reagents are under control. Blank samples should be carried through all stages of the sample preparation and measurement steps. For each analytical batch (up to 20 samples), a reagent blank, matrix spike, and matrix spike duplicate must be analyzed (the frequency of the spikes may be different for different monitoring programs). The blank and spiked samples must be carried through all stages of the sample preparation and measurement steps. QC samples mentioned in the section on Interferences will also be needed as appropriate to those situations.

REFERENCE Test Methods for Evaluating Solid Waste (SW-846). U.S. EPA. 1983. Method 8240B, Rev. 2, Nov. 1990. Office of Solid Wastes, Washington, DC.

1,3-Dichloro-2-propanol **EPA Method 1625**
CAS #96-23-1

TITLE Semivolatile Organic Compounds by Isotope Dilution GC/MS

MATRIX The compounds may be determined in waters, soils, and municipal sludges by this method.

METHOD SUMMARY This method is used to determine 176 semivolatile toxic organic pollutants associated with the CWA (as amended 1987); the RCRA (as amended 1986); the CERCLA (as amended 1986); and other compounds amenable to extraction and analysis by capillary column gas chromatography-mass spectrometry (GC/MS).

Stable isotopically-labeled analogs of the compounds of interest are added to the sample. If the solids content is less than 1%, a 1-L sample is extracted at pH 12–13, then at pH <2 with methylene chloride using continuous extraction techniques.

If the solids content is 30% or less, the sample is diluted to 1% solids with reagent water, homogenized ultrasonically, and extracted at pH 12–13, then at pH <2 with methylene chloride using continuous extraction techniques. If the solids content is greater than 30%, the sample is extracted using ultrasonic techniques.

Each extract is dried over sodium sulfate, concentrated to a volume of 5 mL, cleaned up using GPC, if necessary, and concentrated. Extracts are concentrated to 1 mL if GPC is not performed, and to 0.5 mL if GPC is performed.

An internal standard is added to the extract, and a 1-mL aliquot of the extract is injected into the GC. The compounds are

separated by GC and detected by a MS. The labeled compounds serve to correct the variability of the analytical technique.

INTERFERENCES Solvents, reagents, glassware, and other sample processing hardware may yield artifacts and/or elevated baselines causing misinterpretation of chromatograms and spectra. Materials used in the analysis must be demonstrated to be free from interferences under the conditions of analysis by running method blanks initially and with each sample lot (sample started through the extraction process on a given 8-h shift, to a maximum of 20). Specific selection of reagents and purification of solvents by distillation in all glass systems may be required. Glassware and, where possible, reagents are cleaned by solvent rinse and baking at 450°C for 1-h minimum. Interferences coextracted from samples will vary considerably from source to source, depending on the diversity of the site being sampled.

INSTRUMENTATION Major instrumentation includes a GC with a splitless or on-column injection port for capillary column, a MS with 70 eV electron impact ionization, and a data system to collect and record MS data, and process it. A K-D apparatus is used to concentrate extracts.

GC Column: 30 m × 0.25 mm I.D. 5% phenyl, 94% methyl, 1% vinyl silicone bonded phased fused silica capillary column.

PRECISION & ACCURACY The detection limits of the method are usually dependent on the level of interferences rather than instrumental limitations. The limits typify the minimum quantities that can be detected with no interferences present.

The minimum level (in μg/mL) was not listed. This is defined as a minimum level at which the analytical system shall give recognizable mass spectra (background corrected) and acceptable calibration points.

The MDL (in μg/kg) in low solids was not listed and in high solids was not listed; these were determined in digested sludge (low solids) and in filter cake or compost (high solids).

The labeled and native compound initial precision as standard deviation (in μg/L) was not listed.
The labeled and native compound initial accuracy as average recovery (in μg/L) was not listed.

SAMPLE COLLECTION, PRESERVATION & HANDLING Collect samples in glass containers. Aqueous samples which flow freely are collected in refrigerated bottles using automatic sampling equipment. Solid samples are collected as grab samples using widemouth jars. Maintain samples at 0 to 4°C from the time of collection until extraction. If residual chlorine is present in aqueous samples, add 80 mg sodium thiosulfate/L of water. Begin sample extraction within 7 days of collection, and analyze all extracts within 40 days of extraction.

SAMPLE PREPARATION Samples containing 1% solids or less are extracted directly using continuous liquid-liquid extraction techniques. Samples containing 1 to 30% solids are diluted to the 1% level with reagent water and extracted using continuous liquid-liquid extraction techniques. Samples containing greater than 30% solids are extracted using ultrasonic techniques.

Base/neutral extraction — Adjust the pH of the waters in the extractors to 12–13 with 6 N NaOH. Extract with methylene chloride for 24–48 h.

Acid extraction — Adjust the pH of the waters in the extractors to 2 or less using 6 N sulfuric acid. Extract with methylene chloride for 24–48 h.

Ultrasonic extraction of high solids samples — Add anhydrous sodium sulfate to the sample and QC aliquot(s). Add acetone:methylene chloride (1:1) to the sample and mix thoroughly

Concentrate extracts using a K-D apparatus.

QUALITY CONTROL The analyst is permitted to modify this method to improve separations or lower the costs of measurements, provided all performance specifications are met. Analyses of blanks are required to demonstrate freedom from contamination. When results of spikes indicate atypical method performance for samples, the samples are diluted to bring method performance within acceptable limits.

For low solids (aqueous samples), extract, concentrate, and analyze two sets of four 1-L aliquots (8 aliquots total) of the precision and recovery standard. For high solids samples, two sets of four 30-g aliquots of the high solids reference matrix are used.

Spike all samples with labeled compounds to assess method performance. Compute percent recovery of the labeled compounds using the internal standard method. Compare the labeled compound recovery for each compound with the corresponding labeled compound recovery.

Reagent water and high solids reference matrix blanks are analyzed to demonstrate freedom from contamination. Extract and concentrate a 1-L reagent water blank or a high solids reference matrix blank with each sample's lot (samples started through the extraction process on the same 8-h shift, to a maximum of 20 samples).

Field replicates may be collected to determine the precision of the sampling technique, and spiked samples may be required to determine the accuracy of the analysis when the internal standard method is used.

REFERENCE Semivolatile Organic Compounds by Isotope Dilution GC/MS. Office of Water Regulation and Standards, U.S. EPA Industrial Technology Division, Washington, DC, EPA Method 1625, Rev. C, June 1989 (contact W.A. Telliard, U.S. EPA, Office of Water Regulations and Standards, 401 M St., SW, Washington, DC, 20460. Phone: 202-382-7131).

1,3-Dichloro-2-propanol   **EPA Method 8240**
CAS #96-23-1

TITLE Volatile Organics By GC/MS: Packed Column Technique

MATRIX Nearly all types of sample matarices, regardless of water content, can be analyzed using this method. This includes groundwater, aqueous sludges, caustic liquors, acid liquors, waste solvents, oily wastes, mousses, tars, fibrous wastes, polymetric

emulsions, filter cakes, spent carbons, spent catalysts, soils, and sediments.

METHOD SUMMARY Method 8240B covers 80 volatile organic compounds that are introduced into a gas chromatograph by the purge-and-trap method or by direct injection (in limited applications). For the purge-and-trap method an inert gas (zero grade nitrogen or helium) is bubbled through a 5-mL solution at ambient temperature. Purged sample components are trapped in a tube of sorbent materials. When purging is complete, the sorbent tube is heated and backflushed with inert gas to desorb the trapped components onto a GC column.

INTERFERENCES Impurities in the purge gas and from organic compounds outgassing from the plumbing ahead of the trap account for many contamination problems. Interferences purged or coextracted from the samples will vary considerably from source to source. Cross-contamination can occur whenever high-level and low-level samples are analyzed sequentially. Whenever an unusually concentrated sample is analyzed, it should be followed by the analysis of organic-free reagent water to check for cross-contamination. Samples also can be contaminated by diffusion of volatile organics (particularly methylene chloride and fluorocarbons) through the septum seal into the sample during shipment and storage. A trip blank can serve as a check on such contamination. The lab where volatile analysis is performed and also the refrigerated storage area should be completely free of solvents.

INSTRUMENTATION A gas chromatograph/mass spectrometry/data system (GC/MS) equipped with a 6 ft × 0.1 in I.D. glass column packed with 1% SP-1000 on Carbopack-B (60/80 mesh) is required. Also needed is a 5-mL purging device, a sorbent trap, and a thermal desorption apparatus.

PRECISION & ACCURACY This method is reported to have been tested by 15 laboratories using organic-free reagent water, drinking water, surface water, and industrial wastewaters (not specified) fortified at six concentrations over the range 5–600 µg/L.

Sample estimated quantitation limits (EQLs) are highly matrix-dependent. The EQLs listed may not always be achievable. EQLs listed for soils or sediments are based on wet weight. Normally, data is reported on a dry-weight basis; therefore, EQLs will be higher, based on the percent dry weight of each sample. Note that EQLs are even more variable than MDLs and that they are highly variable depending on the matrix being analyzed.

EQL in groundwater in µg/L was not listed.
EQL in low soil or sediment in µg/kg was not listed.
Accuracy (a) in µg/L was not listed.
Precision (b) in µg/L was not listed.

(a) *Average recovery found for measurements of samples containing a concentration of C, in µg/L.*
(b) *Overall precision found for measurements of samples with average recovery X for samples containing a concentration of C in µg/L.*
X = *Average recovery found for measurement of samples containing a concentration of C in µg/L.*

MULTIPLICATION FACTORS FOR OTHER MATRICES

Other Matrices	Factor (a)
Waste miscible liquid waste	50
High-concentration soil and sludge	125
Non-water miscible waste	500

(a) EQL = [EQL for low soil sediment] × [Factor]. For non-aqueous samples, the factor is on a wet-weight basis.

SAMPLING METHOD
Liquid samples — Use a 40-mL glass screw-cap VOA vial with a Teflon®-faced silicone septum that has been prewashed, rinsed with distilled deionized water, and oven dried. However, if residual chlorine is present, collect sample in a 40-oz. soil VOA container which has been pre-preserved with 4 drops of 10% sodium thiosulfate, mix gently, and then transfer the sample to a 40-mL VOA vial. Collect bubble-free samples in duplicate and seal them in separate plastic bags.

Soils or sediments, and sludges — Use an 8-oz. widemouth glass bottle with a Teflon®-faced silicone septum that has been prewashed with detergent, rinsed with distilled deionized water, and oven dried. Tap slightly to eliminate free air space. Collect samples in duplicate and seal them in separate plastic bags.

SAMPLE PRESERVATION
Liquid samples — Add 4 drops of concentrated HCL and immediately cool samples to 4°C and store in a solvent-free refrigerator.

Soils or sediments, and sludges — Cool samples to 4°C and store in a solvent-free refrigerator.

MHT Maximum holding time is 14 days from the date of sample collection.

SAMPLE PREPARATION
Liquid samples — Remove the plunger from a 5-mL syringe and carefully pour the sample into the syringe barrel to just short of overflowing. Replace the syringe plunger and compress the sample. Open the syringe valve and vent any residual air while adjusting the sample volume to 5.0 mL. If there is only one volatile organic analysis (VOA) vial, a second syringe should be filled at this time to protect against possible loss of sample integrity. Add 10 µL of surrogate spiking solution and 10 µL of internal standard spiking solution through the valve bore of the 5-mL syringe, then close the valve. The surrogate and internal standards may be mixed and added as a single spiking solution.

Sediments, soils, and waste samples — All samples of this type should be screened by GC analysis using a headspace method (EPA Method 3810) or the hexadecane extraction and screening method (EPA Method 3820). Use the screening data to determine whether to use the low-concentration method (0.005–1 mg/kg) or the high-concentration method (>1 mg/kg).

Low-concentration method — The low-concentration method is based on purging a heated sediment or soil sample mixed with organic-free reagent water containing the surrogate and

internal standards. Analyze all reagent blanks and standards under the same conditions as the samples.

Use a 5-g sample if the expected concentration is <0.1 mg/kg or a 1-g sample for expected concentrations between 0.1 and 1 mg/kg. Mix the contents of the sample container with a narrow metal spatula. Weigh the amount of the sample into a tared purge device. Add the spiked water to the purge device, which contains the weighed amount of sample, and connect the device to the purge-and-trap system.

High-concentration method — This method is based on extracting the sediment or soil with methanol. A waste sample is either extracted or diluted, depending on its solubility in methanol. Wastes that are insoluble in methanol are diluted with reagent tetraglyme or possibly polyethylene glycol (PEG). An aliquot of the extract is added to organic-free reagent water containing surrogate and internal standards. This is purged at ambient temperature. All samples with an expected concentration of >1.0 mg/kg should be analyzed by this method.

Mix the contents of the sample container with a narrow metal spatula. For sediments or soils and solid wastes that are insoluble in methanol, weigh 4 g (wet weight) of sample into a tared 20-mL vial. For waste that is soluble in methanol, tetraglyme, or PEG, weigh 1 g (wet weight) into a tared scintillation vial or culture tube or a 10-mL volumetric flask. Quickly add 9.0 mL of appropriate solvent then add 1.0 mL of a surrogate spiking solution to the vial, cap it, and shake it for 2 min.

METHANOL EXTRACT REQUIRED FOR ANALYSIS OF HIGH-CONCENTRATION SOILS OR SEDIMENTS

Approximate Concentration Range	Volume of Methanol Extract (a)
500–10,000 μg/kg	100 μL
1,000–20,000 μg/kg	50 μL
5,000–100,000 μg/kg	10 μL
25,000–500,000 μg/kg	100 μL of 1/50 dilution (b)

Calculate appropriate dilution factor for concentrations exceeding this table.

(a) The volume of methanol added to 5 mL of water being purged should be kept constant. Therefore, add to the 5-mL syringe whatever volume of methanol is necessary to maintain a volume of 100 μL added to the syringe.
(b) Dilute an aliquot of the methanol extract and then take 100 μL for analysis.

QUALITY CONTROL Demonstrate, through the analysis of a reagent water blank, that interferences from the analytical system, glassware, and reagents are under control. Blank samples should be carried through all stages of the sample preparation and measurement steps. For each analytical batch (up to 20 samples), a reagent blank, matrix spike, and matrix spike duplicate must be analyzed (the frequency of the spikes may be different for different monitoring programs). The blank and spiked samples must be carried through all stages of the sample preparation and measurement steps. QC samples mentioned in the section on Interferences will also be needed as appropriate to those situations.

REFERENCE Test Methods for Evaluating Solid Waste (SW-846). U.S. EPA. 1983. Method 8240B, Rev. 2, Nov. 1990. Office of Solid Wastes, Washington, DC.

2,6-Dichloro-4-nitroaniline   EPA Method 1625
CAS #99-30-9

TITLE Semivolatile Organic Compounds by Isotope Dilution GC/MS

MATRIX The compounds may be determined in waters, soils, and municipal sludges by this method.

METHOD SUMMARY This method is used to determine 176 semivolatile toxic organic pollutants associated with the CWA (as amended 1987); the RCRA (as amended 1986); the CERCLA (as amended 1986); and other compounds amenable to extraction and analysis by capillary column gas chromatography-mass spectrometry (GC/MS).

Stable isotopically-labeled analogs of the compounds of interest are added to the sample. If the solids content is less than 1%, a 1-L sample is extracted at pH 12–13, then at pH <2 with methylene chloride using continuous extraction techniques.

If the solids content is 30% or less, the sample is diluted to 1% solids with reagent water, homogenized ultrasonically, and extracted at pH 12–13, then at pH <2 with methylene chloride using continuous extraction techniques. If the solids content is greater than 30%, the sample is extracted using ultrasonic techniques.

Each extract is dried over sodium sulfate, concentrated to a volume of 5 mL, cleaned up using GPC, if necessary, and concentrated. Extracts are concentrated to 1 mL if GPC is not performed, and to 0.5 mL if GPC is performed.

An internal standard is added to the extract, and a 1-mL aliquot of the extract is injected into the GC. The compounds are separated by GC and detected by a MS. The labeled compounds serve to correct the variability of the analytical technique.

INTERFERENCES Solvents, reagents, glassware, and other sample processing hardware may yield artifacts and/or elevated baselines causing misinterpretation of chromatograms and spectra. Materials used in the analysis must be demonstrated to be free from interferences under the conditions of analysis by running method blanks initially and with each sample lot (sample started through the extraction process on a given 8-h shift, to a maximum of 20). Specific selection of reagents and purification of solvents by distillation in all glass systems may be required. Glassware and, where possible, reagents are cleaned by solvent rinse and baking at 450°C for 1-h minimum. Interferences coextracted from samples will vary considerably from source to source, depending on the diversity of the site being sampled.

INSTRUMENTATION Major instrumentation includes a GC with a splitless or on-column injection port for capillary column, a MS with 70 eV electron impact ionization, and a data system to collect and record MS data, and process it. A K-D apparatus is used to concentrate extracts.

GC Column: 30 m × 0.25 mm I.D. 5% phenyl, 94% methyl, 1% vinyl silicone bonded phased fused silica capillary column.

PRECISION & ACCURACY The detection limits of the method are usually dependent on the level of interferences rather than instrumental limitations. The limits typify the minimum quantities that can be detected with no interferences present.

The minimum level (in μg/mL) was not listed. This is defined as a minimum level at which the analytical system shall give recognizable mass spectra (background corrected) and acceptable calibration points.

The MDL (in μg/kg) in low solids was not listed and in high solids was not listed; these were determined in digested sludge (low solids) and in filter cake or compost (high solids).

The labeled and native compound initial precision as standard deviation (in μg/L) was not listed.
The labeled and native compound initial accuracy as average recovery (in μg/L) was not listed.

SAMPLE COLLECTION, PRESERVATION & HANDLING
Collect samples in glass containers. Aqueous samples which flow freely are collected in refrigerated bottles using automatic sampling equipment. Solid samples are collected as grab samples using widemouth jars. Maintain samples at 0 to 4°C from the time of collection until extraction. If residual chlorine is present in aqueous samples, add 80 mg sodium thiosulfate/L of water. Begin sample extraction within 7 days of collection, and analyze all extracts within 40 days of extraction.

SAMPLE PREPARATION Samples containing 1% solids or less are extracted directly using continuous liquid-liquid extraction techniques. Samples containing 1 to 30% solids are diluted to the 1% level with reagent water and extracted using continuous liquid-liquid extraction techniques. Samples containing greater than 30% solids are extracted using ultrasonic techniques.

Base/neutral extraction — Adjust the pH of the waters in the extractors to 12–13 with 6 *N* NaOH. Extract with methylene chloride for 24–48 h.

Acid extraction — Adjust the pH of the waters in the extractors to 2 or less using 6 *N* sulfuric acid. Extract with methylene chloride for 24–48 h.

Ultrasonic extraction of high solids samples — Add anhydrous sodium sulfate to the sample and QC aliquot(s). Add acetone:methylene chloride (1:1) to the sample and mix thoroughly

Concentrate extracts using a K-D apparatus.

QUALITY CONTROL The analyst is permitted to modify this method to improve separations or lower the costs of measurements, provided all performance specifications are met. Analyses of blanks are required to demonstrate freedom from contamination. When results of spikes indicate atypical method performance for samples, the samples are diluted to bring method performance within acceptable limits.

For low solids (aqueous samples), extract, concentrate, and analyze two sets of four 1-L aliquots (8 aliquots total) of the precision and recovery standard. For high solids samples, two sets of four 30-g aliquots of the high solids reference matrix are used.

Spike all samples with labeled compounds to assess method performance. Compute percent recovery of the labeled compounds using the internal standard method. Compare the labeled compound recovery for each compound with the corresponding labeled compound recovery.

Reagent water and high solids reference matrix blanks are analyzed to demonstrate freedom from contamination. Extract and concentrate a 1-L reagent water blank or a high solids reference matrix blank with each sample's lot (samples started through the extraction process on the same 8-h shift, to a maximum of 20 samples).

Field replicates may be collected to determine the precision of the sampling technique, and spiked samples may be required to determine the accuracy of the analysis when the internal standard method is used.

REFERENCE Semivolatile Organic Compounds by Isotope Dilution GC/MS. Office of Water Regulation and Standards, U.S. EPA Industrial Technology Division, Washington, DC, EPA Method 1625, Rev. C, June 1989 (contact W.A. Telliard, U.S. EPA, Office of Water Regulations and Standards, 401 M St., SW, Washington, DC, 20460. Phone: 202-382-7131).

2,3-Dichloroaniline   **EPA Method 1625**
CAS #608-27-5

TITLE Semivolatile Organic Compounds by Isotope Dilution GC/MS

MATRIX The compounds may be determined in waters, soils, and municipal sludges by this method.

METHOD SUMMARY This method is used to determine 176 semivolatile toxic organic pollutants associated with the CWA (as amended 1987); the RCRA (as amended 1986); the CERCLA (as amended 1986); and other compounds amenable to extraction and analysis by capillary column gas chromatography-mass spectrometry (GC/MS).

Stable isotopically-labeled analogs of the compounds of interest are added to the sample. If the solids content is less than 1%, a 1-L sample is extracted at pH 12–13, then at pH <2 with methylene chloride using continuous extraction techniques.

If the solids content is 30% or less, the sample is diluted to 1% solids with reagent water, homogenized ultrasonically, and extracted at pH 12–13, then at pH <2 with methylene chloride using continuous extraction techniques. If the solids content is greater than 30%, the sample is extracted using ultrasonic techniques.

Each extract is dried over sodium sulfate, concentrated to a volume of 5 mL, cleaned up using GPC, if necessary, and concentrated. Extracts are concentrated to 1 mL if GPC is not performed, and to 0.5 mL if GPC is performed.

An internal standard is added to the extract, and a 1-mL aliquot of the extract is injected into the GC. The compounds are separated by GC and detected by a MS. The labeled compounds serve to correct the variability of the analytical technique.

INTERFERENCES Solvents, reagents, glassware, and other sample processing hardware may yield artifacts and/or elevated baselines causing misinterpretation of chromatograms and spectra. Materials used in the analysis must be demonstrated to be free from interferences under the conditions of analysis by running method blanks initially and with each sample lot (sample started through the extraction process on a given 8-h shift, to a maximum of 20). Specific selection of reagents and purification of solvents by distillation in all glass systems may be required. Glassware and, where possible, reagents are cleaned by solvent rinse and baking at 450°C for 1-h minimum. Interferences coextracted from samples will vary considerably from source to source, depending on the diversity of the site being sampled.

INSTRUMENTATION Major instrumentation includes a GC with a splitless or on-column injection port for capillary column, a MS with 70 eV electron impact ionization, and a data system to collect and record MS data, and process it. A K-D apparatus is used to concentrate extracts.

GC Column: 30 m × 0.25 mm I.D. 5% phenyl, 94% methyl, 1% vinyl silicone bonded phased fused silica capillary column.

PRECISION & ACCURACY The detection limits of the method are usually dependent on the level of interferences rather than instrumental limitations. The limits typify the minimum quantities that can be detected with no interferences present.

The minimum level (in µg/mL) was not listed. This is defined as a minimum level at which the analytical system shall give recognizable mass spectra (background corrected) and acceptable calibration points.

The MDL (in µg/kg) in low solids was not listed and in high solids was not listed; these were determined in digested sludge (low solids) and in filter cake or compost (high solids).

The labeled and native compound initial precision as standard deviation (in µg/L) was not listed.
The labeled and native compound initial accuracy as average recovery (in µg/L) was not listed.

SAMPLE COLLECTION, PRESERVATION & HANDLING Collect samples in glass containers. Aqueous samples which flow freely are collected in refrigerated bottles using automatic sampling equipment. Solid samples are collected as grab samples using widemouth jars. Maintain samples at 0 to 4°C from the time of collection until extraction. If residual chlorine is present in aqueous samples, add 80 mg sodium thiosulfate/L of water. Begin sample extraction within 7 days of collection, and analyze all extracts within 40 days of extraction.

SAMPLE PREPARATION Samples containing 1% solids or less are extracted directly using continuous liquid-liquid extraction techniques. Samples containing 1 to 30% solids are diluted to the 1% level with reagent water and extracted using continuous liquid-liquid extraction techniques. Samples containing greater than 30% solids are extracted using ultrasonic techniques.

Base/neutral extraction — Adjust the pH of the waters in the extractors to 12–13 with 6 N NaOH. Extract with methylene chloride for 24–48 h.

Acid extraction — Adjust the pH of the waters in the extractors to 2 or less using 6 N sulfuric acid. Extract with methylene chloride for 24–48 h.

Ultrasonic extraction of high solids samples — Add anhydrous sodium sulfate to the sample and QC aliquot(s). Add acetone:methylene chloride (1:1) to the sample and mix thoroughly

Concentrate extracts using a K-D apparatus.

QUALITY CONTROL The analyst is permitted to modify this method to improve separations or lower the costs of measurements, provided all performance specifications are met. Analyses of blanks are required to demonstrate freedom from contamination. When results of spikes indicate atypical method performance for samples, the samples are diluted to bring method performance within acceptable limits.

For low solids (aqueous samples), extract, concentrate, and analyze two sets of four 1-L aliquots (8 aliquots total) of the precision and recovery standard. For high solids samples, two sets of four 30-g aliquots of the high solids reference matrix are used.

Spike all samples with labeled compounds to assess method performance. Compute percent recovery of the labeled compounds using the internal standard method. Compare the labeled compound recovery for each compound with the corresponding labeled compound recovery.

Reagent water and high solids reference matrix blanks are analyzed to demonstrate freedom from contamination. Extract and concentrate a 1-L reagent water blank or a high solids reference matrix blank with each sample's lot (samples started through the extraction process on the same 8-h shift, to a maximum of 20 samples).

Field replicates may be collected to determine the precision of the sampling technique, and spiked samples may be required to determine the accuracy of the analysis when the internal standard method is used.

REFERENCE Semivolatile Organic Compounds by Isotope Dilution GC/MS. Office of Water Regulation and Standards, U.S. EPA Industrial Technology Division, Washington, DC, EPA Method 1625, Rev. C, June 1989 (contact W.A. Telliard, U.S. EPA, Office of Water Regulations and Standards, 401 M St., SW, Washington, DC, 20460. Phone: 202-382-7131).

1,2-Dichlorobenzene   **EPA Method 1625**
CAS #95-50-1

TITLE Semivolatile Organic Compounds by Isotope Dilution GC/MS

MATRIX The compounds may be determined in waters, soils, and municipal sludges by this method.

METHOD SUMMARY This method is used to determine 176 semivolatile toxic organic pollutants associated with the CWA (as amended 1987); the RCRA (as amended 1986); the CERCLA (as amended 1986); and other compounds amenable to extraction and analysis by capillary column gas chromatography-mass spectrometry (GC/MS).

Stable isotopically-labeled analogs of the compounds of interest are added to the sample. If the solids content is less than 1%, a 1-L sample is extracted at pH 12–13, then at pH <2 with methylene chloride using continuous extraction techniques.

If the solids content is 30% or less, the sample is diluted to 1% solids with reagent water, homogenized ultrasonically, and extracted at pH 12–13, then at pH <2 with methylene chloride using continuous extraction techniques. If the solids content is greater than 30%, the sample is extracted using ultrasonic techniques.

Each extract is dried over sodium sulfate, concentrated to a volume of 5 mL, cleaned up using GPC, if necessary, and concentrated. Extracts are concentrated to 1 mL if GPC is not performed, and to 0.5 mL if GPC is performed.

An internal standard is added to the extract, and a 1-mL aliquot of the extract is injected into the GC. The compounds are separated by GC and detected by a MS. The labeled compounds serve to correct the variability of the analytical technique.

INTERFERENCES Solvents, reagents, glassware, and other sample processing hardware may yield artifacts and/or elevated baselines causing misinterpretation of chromatograms and spectra. Materials used in the analysis must be demonstrated to be free from interferences under the conditions of analysis by running method blanks initially and with each sample lot (sample started through the extraction process on a given 8-h shift, to a maximum of 20). Specific selection of reagents and purification of solvents by distillation in all glass systems may be required. Glassware and, where possible, reagents are cleaned by solvent rinse and baking at 450°C for 1-h minimum. Interferences coextracted from samples will vary considerably from source to source, depending on the diversity of the site being sampled.

INSTRUMENTATION Major instrumentation includes a GC with a splitless or on-column injection port for capillary column, a MS with 70 eV electron impact ionization, and a data system to collect and record MS data, and process it. A K-D apparatus is used to concentrate extracts.

GC Column: 30 m × 0.25 mm I.D. 5% phenyl, 94% methyl, 1% vinyl silicone bonded phased fused silica capillary column.

PRECISION & ACCURACY The detection limits of the method are usually dependent on the level of interferences rather than instrumental limitations. The limits typify the minimum quantities that can be detected with no interferences present.

The minimum level (in µg/mL) was 10. This is defined as a minimum level at which the analytical system shall give recognizable mass spectra (background corrected) and acceptable calibration points.

The MDL (in µg/kg) in low solids was 63 and in high solids was 16; these were determined in digested sludge (low solids) and in filter cake or compost (high solids).

The labeled and native compound initial precision as standard deviation (in µg/L) was 17.

The labeled and native compound initial accuracy as average recovery (in µg/L) was 73–146.

SAMPLE COLLECTION, PRESERVATION & HANDLING Collect samples in glass containers. Aqueous samples which flow freely are collected in refrigerated bottles using automatic sampling equipment. Solid samples are collected as grab samples using widemouth jars. Maintain samples at 0 to 4°C from the time of collection until extraction. If residual chlorine is present in aqueous samples, add 80 mg sodium thiosulfate/L of water. Begin sample extraction within 7 days of collection, and analyze all extracts within 40 days of extraction.

SAMPLE PREPARATION Samples containing 1% solids or less are extracted directly using continuous liquid-liquid extraction techniques. Samples containing 1 to 30% solids are diluted to the 1% level with reagent water and extracted using continuous liquid-liquid extraction techniques. Samples containing greater than 30% solids are extracted using ultrasonic techniques.

Base/neutral extraction — Adjust the pH of the waters in the extractors to 12–13 with 6 N NaOH. Extract with methylene chloride for 24–48 h.

Acid extraction — Adjust the pH of the waters in the extractors to 2 or less using 6 N sulfuric acid. Extract with methylene chloride for 24–48 h.

Ultrasonic extraction of high solids samples — Add anhydrous sodium sulfate to the sample and QC aliquot(s). Add acetone:methylene chloride (1:1) to the sample and mix thoroughly

Concentrate extracts using a K-D apparatus.

QUALITY CONTROL The analyst is permitted to modify this method to improve separations or lower the costs of measurements, provided all performance specifications are met. Analyses of blanks are required to demonstrate freedom from contamination. When results of spikes indicate atypical method performance for samples, the samples are diluted to bring method performance within acceptable limits.

For low solids (aqueous samples), extract, concentrate, and analyze two sets of four 1-L aliquots (8 aliquots total) of the precision and recovery standard. For high solids samples, two sets of four 30-g aliquots of the high solids reference matrix are used.

Spike all samples with labeled compounds to assess method performance. Compute percent recovery of the labeled compounds using the internal standard method. Compare the labeled compound recovery for each compound with the corresponding labeled compound recovery.

Reagent water and high solids reference matrix blanks are analyzed to demonstrate freedom from contamination. Extract and concentrate a 1-L reagent water blank or a high solids reference matrix blank with each sample's lot (samples started through the extraction process on the same 8-h shift, to a maximum of 20 samples).

Field replicates may be collected to determine the precision of the sampling technique, and spiked samples may be required to determine the accuracy of the analysis when the internal standard method is used.

REFERENCE Semivolatile Organic Compounds by Isotope Dilution GC/MS. Office of Water Regulation and Standards, U.S. EPA Industrial Technology Division, Washington, DC, EPA Method 1625, Rev. C, June 1989 (contact W.A. Telliard, U.S. EPA, Office of Water Regulations and Standards, 401 M St., SW, Washington, DC, 20460. Phone: 202-382-7131).

1,3-Dichlorobenzene   **EPA Method 1625**
CAS #541-73-1

TITLE Semivolatile Organic Compounds by Isotope Dilution GC/MS

MATRIX The compounds may be determined in waters, soils, and municipal sludges by this method.

METHOD SUMMARY This method is used to determine 176 semivolatile toxic organic pollutants associated with the CWA (as amended 1987); the RCRA (as amended 1986); the CERCLA (as amended 1986); and other compounds amenable to extraction and analysis by capillary column gas chromatography-mass spectrometry (GC/MS).

Stable isotopically-labeled analogs of the compounds of interest are added to the sample. If the solids content is less than 1%, a 1-L sample is extracted at pH 12–13, then at pH <2 with methylene chloride using continuous extraction techniques.

If the solids content is 30% or less, the sample is diluted to 1% solids with reagent water, homogenized ultrasonically, and extracted at pH 12–13, then at pH <2 with methylene chloride using continuous extraction techniques. If the solids content is greater than 30%, the sample is extracted using ultrasonic techniques.

Each extract is dried over sodium sulfate, concentrated to a volume of 5 mL, cleaned up using GPC, if necessary, and concentrated. Extracts are concentrated to 1 mL if GPC is not performed, and to 0.5 mL if GPC is performed.

An internal standard is added to the extract, and a 1-mL aliquot of the extract is injected into the GC. The compounds are separated by GC and detected by a MS. The labeled compounds serve to correct the variability of the analytical technique.

INTERFERENCES Solvents, reagents, glassware, and other sample processing hardware may yield artifacts and/or elevated baselines causing misinterpretation of chromatograms and spectra. Materials used in the analysis must be demonstrated to be free from interferences under the conditions of analysis by running method blanks initially and with each sample lot (sample started through the extraction process on a given 8-h shift, to a maximum of 20). Specific selection of reagents and purification of solvents by distillation in all glass systems may be required. Glassware and, where possible, reagents are cleaned by solvent rinse and baking at 450°C for 1-h minimum. Interferences coextracted from samples will vary considerably from source to source, depending on the diversity of the site being sampled.

INSTRUMENTATION Major instrumentation includes a GC with a splitless or on-column injection port for capillary column, a MS with 70 eV electron impact ionization, and a data system to collect and record MS data, and process it. A K-D apparatus is used to concentrate extracts.

GC Column: 30 m × 0.25 mm I.D. 5% phenyl, 94% methyl, 1% vinyl silicone bonded phased fused silica capillary column.

PRECISION & ACCURACY The detection limits of the method are usually dependent on the level of interferences rather than instrumental limitations. The limits typify the minimum quantities that can be detected with no interferences present.

The minimum level (in µg/mL) was 10. This is defined as a minimum level at which the analytical system shall give recognizable mass spectra (background corrected) and acceptable calibration points.

The MDL (in µg/kg) in low solids was 46 and in high solids was 26; these were determined in digested sludge (low solids) and in filter cake or compost (high solids).

The labeled and native compound initial precision as standard deviation (in µg/L) was 43.
The labeled and native compound initial accuracy as average recovery (in µg/L) was 63–201.

SAMPLE COLLECTION, PRESERVATION & HANDLING Collect samples in glass containers. Aqueous samples which flow freely are collected in refrigerated bottles using automatic sampling equipment. Solid samples are collected as grab samples using widemouth jars. Maintain samples at 0 to 4°C from the time of collection until extraction. If residual chlorine is present in aqueous samples, add 80 mg sodium thiosulfate/L of water. Begin sample extraction within 7 days of collection, and analyze all extracts within 40 days of extraction.

SAMPLE PREPARATION Samples containing 1% solids or less are extracted directly using continuous liquid-liquid extraction techniques. Samples containing 1 to 30% solids are diluted to the 1% level with reagent water and extracted using continuous liquid-liquid extraction techniques. Samples containing greater than 30% solids are extracted using ultrasonic techniques.

Base/neutral extraction — Adjust the pH of the waters in the extractors to 12–13 with 6 N NaOH. Extract with methylene chloride for 24–48 h.

Acid extraction — Adjust the pH of the waters in the extractors to 2 or less using 6 N sulfuric acid. Extract with methylene chloride for 24–48 h.

Ultrasonic extraction of high solids samples — Add anhydrous sodium sulfate to the sample and QC aliquot(s). Add acetone:methylene chloride (1:1) to the sample and mix thoroughly

Concentrate extracts using a K-D apparatus.

QUALITY CONTROL The analyst is permitted to modify this method to improve separations or lower the costs of measurements, provided all performance specifications are met. Analyses of blanks are required to demonstrate freedom from contamination. When results of spikes indicate atypical method performance for samples, the samples are diluted to bring method performance within acceptable limits.

For low solids (aqueous samples), extract, concentrate, and analyze two sets of four 1-L aliquots (8 aliquots total) of the precision and recovery standard. For high solids samples, two sets of four 30-g aliquots of the high solids reference matrix are used.

Spike all samples with labeled compounds to assess method performance. Compute percent recovery of the labeled compounds using the internal standard method. Compare the labeled compound recovery for each compound with the corresponding labeled compound recovery.

Reagent water and high solids reference matrix blanks are analyzed to demonstrate freedom from contamination. Extract and concentrate a 1-L reagent water blank or a high solids reference matrix blank with each sample's lot (samples started through the extraction process on the same 8-h shift, to a maximum of 20 samples).

Field replicates may be collected to determine the precision of the sampling technique, and spiked samples may be required to determine the accuracy of the analysis when the internal standard method is used.

REFERENCE Semivolatile Organic Compounds by Isotope Dilution GC/MS. Office of Water Regulation and Standards, U.S. EPA Industrial Technology Division, Washington, DC, EPA Method 1625, Rev. C, June 1989 (contact W.A. Telliard, U.S. EPA, Office of Water Regulations and Standards, 401 M St., SW, Washington, DC, 20460. Phone: 202-382-7131).

1,4-Dichlorobenzene   **EPA Method 1625**
CAS #106-46-7

TITLE Semivolatile Organic Compounds by Isotope Dilution GC/MS

MATRIX The compounds may be determined in waters, soils, and municipal sludges by this method.

METHOD SUMMARY This method is used to determine 176 semivolatile toxic organic pollutants associated with the CWA (as amended 1987); the RCRA (as amended 1986); the CERCLA (as amended 1986); and other compounds amenable to extraction and analysis by capillary column gas chromatography-mass spectrometry (GC/MS).

Stable isotopically-labeled analogs of the compounds of interest are added to the sample. If the solids content is less than 1%, a 1-L sample is extracted at pH 12–13, then at pH <2 with methylene chloride using continuous extraction techniques.

If the solids content is 30% or less, the sample is diluted to 1% solids with reagent water, homogenized ultrasonically, and extracted at pH 12–13, then at pH <2 with methylene chloride using continuous extraction techniques. If the solids content is greater than 30%, the sample is extracted using ultrasonic techniques.

Each extract is dried over sodium sulfate, concentrated to a volume of 5 mL, cleaned up using GPC, if necessary, and concentrated. Extracts are concentrated to 1 mL if GPC is not performed, and to 0.5 mL if GPC is performed.

An internal standard is added to the extract, and a 1-mL aliquot of the extract is injected into the GC. The compounds are separated by GC and detected by a MS. The labeled compounds serve to correct the variability of the analytical technique.

INTERFERENCES Solvents, reagents, glassware, and other sample processing hardware may yield artifacts and/or elevated baselines causing misinterpretation of chromatograms and spectra. Materials used in the analysis must be demonstrated to be free from interferences under the conditions of analysis by running method blanks initially and with each sample lot (sample started through the extraction process on a given 8-h shift, to a maximum of 20). Specific selection of reagents and purification of solvents by distillation in all glass systems may be required. Glassware and, where possible, reagents are cleaned by solvent rinse and baking at 450°C for 1-h minimum. Interferences coextracted from samples will vary considerably from source to source, depending on the diversity of the site being sampled.

INSTRUMENTATION Major instrumentation includes a GC with a splitless or on-column injection port for capillary column, a MS with 70 eV electron impact ionization, and a data system to collect and record MS data, and process it. A K-D apparatus is used to concentrate extracts.

GC Column: 30 m × 0.25 mm I.D. 5% phenyl, 94% methyl, 1% vinyl silicone bonded phased fused silica capillary column.

PRECISION & ACCURACY The detection limits of the method are usually dependent on the level of interferences rather than instrumental limitations. The limits typify the minimum quantities that can be detected with no interferences present.

The minimum level (in μg/mL) was 10. This is defined as a minimum level at which the analytical system shall give recognizable mass spectra (background corrected) and acceptable calibration points.

The MDL (in μg/kg) in low solids was 35 and in high solids was 20; these were determined in digested sludge (low solids) and in filter cake or compost (high solids).

The labeled and native compound initial precision as standard deviation (in μg/L) was 42.

The labeled and native compound initial accuracy as average recovery (in µg/L) was 61–194.

SAMPLE COLLECTION, PRESERVATION & HANDLING Collect samples in glass containers. Aqueous samples which flow freely are collected in refrigerated bottles using automatic sampling equipment. Solid samples are collected as grab samples using widemouth jars. Maintain samples at 0 to 4°C from the time of collection until extraction. If residual chlorine is present in aqueous samples, add 80 mg sodium thiosulfate/L of water. Begin sample extraction within 7 days of collection, and analyze all extracts within 40 days of extraction.

SAMPLE PREPARATION Samples containing 1% solids or less are extracted directly using continuous liquid-liquid extraction techniques. Samples containing 1 to 30% solids are diluted to the 1% level with reagent water and extracted using continuous liquid-liquid extraction techniques. Samples containing greater than 30% solids are extracted using ultrasonic techniques.

- Base/neutral extraction — Adjust the pH of the waters in the extractors to 12–13 with 6 N NaOH. Extract with methylene chloride for 24–48 h.
- Acid extraction — Adjust the pH of the waters in the extractors to 2 or less using 6 N sulfuric acid. Extract with methylene chloride for 24–48 h.
- Ultrasonic extraction of high solids samples — Add anhydrous sodium sulfate to the sample and QC aliquot(s). Add acetone:methylene chloride (1:1) to the sample and mix thoroughly

Concentrate extracts using a K-D apparatus.

QUALITY CONTROL The analyst is permitted to modify this method to improve separations or lower the costs of measurements, provided all performance specifications are met. Analyses of blanks are required to demonstrate freedom from contamination. When results of spikes indicate atypical method performance for samples, the samples are diluted to bring method performance within acceptable limits.

For low solids (aqueous samples), extract, concentrate, and analyze two sets of four 1-L aliquots (8 aliquots total) of the precision and recovery standard. For high solids samples, two sets of four 30-g aliquots of the high solids reference matrix are used.

Spike all samples with labeled compounds to assess method performance. Compute percent recovery of the labeled compounds using the internal standard method. Compare the labeled compound recovery for each compound with the corresponding labeled compound recovery.

Reagent water and high solids reference matrix blanks are analyzed to demonstrate freedom from contamination. Extract and concentrate a 1-L reagent water blank or a high solids reference matrix blank with each sample's lot (samples started through the extraction process on the same 8-h shift, to a maximum of 20 samples).

Field replicates may be collected to determine the precision of the sampling technique, and spiked samples may be required to determine the accuracy of the analysis when the internal standard method is used.

REFERENCE Semivolatile Organic Compounds by Isotope Dilution GC/MS. Office of Water Regulation and Standards, U.S. EPA Industrial Technology Division, Washington, DC, EPA Method 1625, Rev. C, June 1989 (contact W.A. Telliard, U.S. EPA, Office of Water Regulations and Standards, 401 M St., SW, Washington, DC, 20460. Phone: 202-382-7131).

1,2-Dichlorobenzene   EPA Method 502
CAS #95-50-1

TITLE Volatile Organic Compounds in Water By Purge and Trap Capillary Column Gas Chromatography with Photoionization and Electrolytic Conductivity Detectors in Series. U.S. EPA Method 502.2, Rev. 2.0, 1989.

MATRIX Drinking water and raw source water. The latter should include most surface water and groundwater sources.

METHOD SUMMARY This method covers 60 volatile organic compounds that contain halogen atoms and/or that are aromatic. An inert gas (zero grade nitrogen or helium) is bubbled through a 25-mL or a 5-mL water sample (depending on the expected concentration of the analytes). Purged sample components are trapped in a tube of sorbent materials. When purging is complete, the sorbent tube is heated and backflushed with helium to desorb the trapped sample onto a capillary GC column. The column is temperature programmed to separate the method analytes which are then detected with a photoionization detector (PID) and a Hall electrolytic conductivity (HECD) placed in series. The PID is selective for aromatic compounds and the HECD is selective for halogenated compounds.

INTERFERENCES Impurities in the purge gas and from organic compounds outgassing from the plumbing ahead of the trap account for many contamination problems. Interferences purged or coextracted from the samples will vary considerably from source to source, depending upon the particular sample or extract being tested. Cross-contamination can occur whenever high-level and low-level samples are analyzed sequentially. Samples also can be contaminated by diffusion of volatile organics (particularly methylene chloride and fluorocarbons) through the septum seal into the sample during shipment and storage. The lab where volatile analysis is performed and also the refrigerated storage area should be completely free of solvents.

INSTRUMENTATION A GC containing a series configuration of a high temperature photoionization detector (PID) equipped with 10.0 eV (nominal) lamp and Hall electrolytic conductivity detector (HECD) is required. Also required is an all-glass 5-mL purging device, a sorbent trap, and a thermal desorption apparatus which is connected to the GC system.

Column 1: VOCOL glass wide-bore capillary column.
Column 2: RTX–502.2 mega-bore capillary column.
Column 3: DB-62 mega-bore capillary column.

PRECISION & ACCURACY Method detection limits are dependent upon the characteristics of the gas chromatographic system used. Analytes that are not separated chromatographically cannot be individually identified and used in the same calibration mixture or water samples unless an alternative technique for identification and quantification, such as mass spectrometry, is used.

Electrolytic conductivity detetor (c) range in µg/L (a) was 0.02–200.
Electrolytic conductivity detetor (c) MDL in µg/L (b) was 0.02.
Electrolytic conductivity detetor (c) accuracy as % recovery was 100.
Electrolytic conductivity detetor (c) precision as % RSD was 1.5.
Photoionization detector (d) range in µg/L (a) was 0.02–200.
Photoionization detector (d) MDL in µg/L (b) was 0.05.
Photoionization detector (d) accuracy as % recovery was 102.
Photoionization detector (d) precision as % RSD was 2.1.

(a) *The applicable concentration range of this method is compound, instrument, and matrix-dependent. It is listed as being approximately 0.02 to 200 µg/L but no specific information is provided so caution should be observed.*

(b) *The method detection limits reports with this method are compound, instrument, and matrix-dependent. The values reported were calculated using reagent water fortified with the corresponding compounds at 10 µg/L and a GC-equipped with a 60 m × 0.75 mm VOLCOL wide bore capillary column with 1.5 µm film thickness and using helium carrier gas.*

(c) *Recoveries and relative standard deviations were determined from seven samples of reagent water fortified with 10 µg/L of each compound. 2-Bromo-1-chloropropane was used as the internal standard for calculating average recoveries.*

(d) *Recoveries and relative standard deviations were determined from seven samples of reagent water fortified with 10 µg/L of each compound. Fluorobenzene was used as the internal standard for calculating average recoveries.*

SAMPLING METHOD Collect samples using a 40- to 120-mL screw-cap vial (prewashed with detergent, rinsed with distilled water and oven dried at 105°C) with a Teflon®-faced silicone septum. Collect bubble-free samples and place the septum with the Teflon® side down on the water.

SAMPLE PRESERVATION If residual chlorine is present in the water add about 25 mg of ascorbic acid to each vial before samples are collected to remove the chlorine. Add hydrochloric acid to reduce pH to <2, immediately cool samples to 4°C, and store them in a solvent-free refrigerator at 4°C until analysis.

MHT The maximum holding time for samples is 14 days from the time they were collected.

SAMPLE PREPARATION Remove the plungers from two 5-mL syringes and attach a closed syringe valve to each. Warm the sample to room temperature, open the sample bottle, and carefully pour the sample into one of the syringe barrels to just short of overflowing. Replace the syringe plunger, invert the syringe, and compress the sample. Open the syringe valve and vent any residual air while adjusting the sample volume to 5.0 mL. Add 10 µL of the internal calibration standard to the sample through the syringe valve. Close the valve. Fill the second syringe in an identical manner from the same sample bottle. Reserve this second syringe for a reanalysis if necessary.

QUALITY CONTROL As an initial demonstration of lab accuracy and precision, analyze 4 to 7 replicates of a lab fortified blank containing analyte at 0.1–5 µg/L. Collect all samples in duplicate. Surrogate analytes (similar to those of the analytes of interest), whose concentration is known in every sample, are measured using the same internal standard calibration procedure. Duplicate field reagent water blanks (trip blanks) must be analyzed with each set of samples, lab reagent blanks (method blanks) must be analyzed with each batch of samples processed as a group within a work shift. Also, a single lab-fortified blank that contains each of the analytes of interest should be analyzed with each batch of samples processed as a group within a work shift. A 3- to 5-point calibration curve is needed depending on the calibration range factor required.

EPA CONTACT & HOTLINE For technical questions contact Dr. Baldev Bathija, U.S. EPA, Office of Ground Water and Drinking Water (WH-550D), 401 M St. SW, Washington, DC 20460. Tel. (202) 260-3040. For further information the EPA Safe Drinking Water Hotline may be called at: (800) 426-4791.

REFERENCE Methods for the Determination of Organic Compounds in Drinking Water, EPA/600/4-88/039 (revised July 1991; Final Rule for determination of compliance with the MCL for Total Trihalomethanes under 141.30, in 40 CFR Part 141, Vol. 58, No. 147, Fed. Reg., Tuesday Aug. 3, 1993). U.S. EPA Environmental Monitoring Systems Laboratory, Cincinnati, OH, 45268, U.S.A. Available from the National Technical Information Service (NTIS), 5285 Port Royal Road, Springfield, VA 22161; Tel. 800-553-6847. NTIS Order Number is PB91-231480.

1,3-Dichlorobenzene   **EPA Method 502**
CAS #541-73-1

TITLE Volatile Organic Compounds in Water By Purge and Trap Capillary Column Gas Chromatography with Photoionization and Electrolytic Conductivity Detectors in Series. U.S. EPA Method 502.2, Rev. 2.0, 1989.

MATRIX Drinking water and raw source water. The latter should include most surface water and groundwater sources.

METHOD SUMMARY This method covers 60 volatile organic compounds that contain halogen atoms and/or that are aromatic. An inert gas (zero grade nitrogen or helium) is bubbled through a 25-mL or a 5-mL water sample (depending on the expected concentration of the analytes). Purged sample components are trapped in a tube of sorbent materials. When purging is complete, the sorbent tube is heated and backflushed with helium to desorb the trapped sample onto a capillary GC column. The column is temperature programmed to separate the method analytes which are then detected with a photoionization detector (PID) and a Hall electrolytic conductivity (HECD) placed in series. The PID is selective for aromatic compounds and the HECD is selective for halogenated compounds.

INTERFERENCES Impurities in the purge gas and from organic compounds outgassing from the plumbing ahead of the trap account for many contamination problems. Interferences purged or coextracted from the samples will vary considerably from source to source, depending upon the particular sample or extract being tested. Cross-contamination can occur whenever high-level and low-level samples are analyzed sequentially. Samples also can be contaminated by diffusion of volatile organics (particularly methylene chloride and fluorocarbons) through the septum seal into the sample during shipment and storage. The lab where volatile analysis is performed and also the refrigerated storage area should be completely free of solvents.

INSTRUMENTATION A GC containing a series configuration of a high temperature photoionization detector (PID) equipped with 10.0 eV (nominal) lamp and Hall electrolytic conductivity detector (HECD) is required. Also required is an all-glass 5-mL purging device, a sorbent trap, and a thermal desorption apparatus which is connected to the GC system.

Column 1: VOCOL glass wide-bore capillary column.
Column 2: RTX–502.2 mega-bore capillary column.
Column 3: DB-62 mega-bore capillary column.

PRECISION & ACCURACY Method detection limits are dependent upon the characteristics of the gas chromatographic system used. Analytes that are not separated chromatographically cannot be individually identified and used in the same calibration mixture or water samples unless an alternative technique for identification and quantification, such as mass spectrometry, is used.

Electrolytic conductivity detetor (c) range in µg/L (a) was 0.02–200.
Electrolytic conductivity detetor (c) MDL in µg/L (b) was 0.02.
Electrolytic conductivity detetor (c) accuracy as % recovery was 106.
Electrolytic conductivity detetor (c) precision as % RSD was 4.0.
Photoionization detector (d) range in µg/L (a) was 0.02–200.
Photoionization detector (d) MDL in µg/L (b) was 0.02.
Photoionization detector (d) accuracy as % recovery was 104.
Photoionization detector (d) precision as % RSD was 1.6.

(a) *The applicable concentration range of this method is compound, instrument, and matrix-dependent. It is listed as being approximately 0.02 to 200 µg/L but no specific information is provided so caution should be observed.*
(b) *The method detection limits reports with this method are compound, instrument, and matrix-dependent. The values reported were calculated using reagent water fortified with the corresponding compounds at 10 µg/L and a GC-equipped with a 60 m × 0.75 mm VOLCOL wide bore capillary column with 1.5 µm film thickness and using helium carrier gas.*
(c) *Recoveries and relative standard deviations were determined from seven samples of reagent water fortified with 10 µg/L of each compound. 2-Bromo-1-chloropropane was used as the internal standard for calculating average recoveries.*
(d) *Recoveries and relative standard deviations were determined from seven samples of reagent water fortified with 10 µg/L of each compound. Fluorobenzene was used as the internal standard for calculating average recoveries.*

SAMPLING METHOD Collect samples using a 40- to 120-mL screw-cap vial (prewashed with detergent, rinsed with distilled water and oven dried at 105°C) with a Teflon®-faced silicone septum. Collect bubble-free samples and place the septum with the Teflon® side down on the water.

SAMPLE PRESERVATION If residual chlorine is present in the water add about 25 mg of ascorbic acid to each vial before samples are collected to remove the chlorine. Add hydrochloric acid to reduce pH to <2, immediately cool samples to 4°C, and store them in a solvent-free refrigerator at 4°C until analysis.

MHT The maximum holding time for samples is 14 days from the time they were collected.

SAMPLE PREPARATION Remove the plungers from two 5-mL syringes and attach a closed syringe valve to each. Warm the sample to room temperature, open the sample bottle, and carefully pour the sample into one of the syringe barrels to just short of overflowing. Replace the syringe plunger, invert the syringe, and compress the sample. Open the syringe valve and vent any residual air while adjusting the sample volume to 5.0 mL. Add 10 µL of the internal calibration standard to the sample through the syringe valve. Close the valve. Fill the second syringe in an identical manner from the same sample bottle. Reserve this second syringe for a reanalysis if necessary.

QUALITY CONTROL As an initial demonstration of lab accuracy and precision, analyze 4 to 7 replicates of a lab fortified blank containing analyte at 0.1–5 µg/L. Collect all samples in duplicate. Surrogate analytes (similar to those of the analytes of interest), whose concentration is known in every sample, are measured using the same internal standard calibration procedure. Duplicate field reagent water blanks (trip blanks) must be analyzed with each set of samples, lab reagent blanks (method blanks) must be analyzed with each batch of samples processed as a group within a work shift. Also, a single lab-fortified blank that contains each of the analytes of interest should be analyzed with each batch of samples processed as a group within a work shift. A 3- to 5-point calibration curve is needed depending on the calibration range factor required.

EPA CONTACT & HOTLINE For technical questions contact Dr. Baldev Bathija, U.S. EPA, Office of Ground Water and Drinking Water (WH-550D), 401 M St. SW, Washington, DC 20460. Tel. (202) 260-3040. For further information the EPA Safe Drinking Water Hotline may be called at: (800) 426-4791.

REFERENCE Methods for the Determination of Organic Compounds in Drinking Water, EPA/600/4-88/039 (revised July 1991; Final Rule for determination of compliance with the MCL for Total Trihalomethanes under 141.30, in 40 CFR Part 141, Vol. 58, No. 147, Fed. Reg., Tuesday Aug. 3, 1993). U.S. EPA Environmental Monitoring Systems Laboratory, Cincinnati, OH, 45268, U.S.A. Available from the National Technical Information Service (NTIS), 5285 Port Royal Road, Springfield,

VA 22161; Tel. 800-553-6847. NTIS Order Number is PB91-231480.

1,4-Dichlorobenzene   **EPA Method 502**
CAS #106-46-7

TITLE Volatile Organic Compounds in Water By Purge and Trap Capillary Column Gas Chromatography with Photoionization and Electrolytic Conductivity Detectors in Series. U.S. EPA Method 502.2, Rev. 2.0, 1989.

MATRIX Drinking water and raw source water. The latter should include most surface water and groundwater sources.

METHOD SUMMARY This method covers 60 volatile organic compounds that contain halogen atoms and/or that are aromatic. An inert gas (zero grade nitrogen or helium) is bubbled through a 25-mL or a 5-mL water sample (depending on the expected concentration of the analytes). Purged sample components are trapped in a tube of sorbent materials. When purging is complete, the sorbent tube is heated and backflushed with helium to desorb the trapped sample onto a capillary GC column. The column is temperature programmed to separate the method analytes which are then detected with a photoionization detector (PID) and a Hall electrolytic conductivity (HECD) placed in series. The PID is selective for aromatic compounds and the HECD is selective for halogenated compounds.

INTERFERENCES Impurities in the purge gas and from organic compounds outgassing from the plumbing ahead of the trap account for many contamination problems. Interferences purged or coextracted from the samples will vary considerably from source to source, depending upon the particular sample or extract being tested. Cross-contamination can occur whenever high-level and low-level samples are analyzed sequentially. Samples also can be contaminated by diffusion of volatile organics (particularly methylene chloride and fluorocarbons) through the septum seal into the sample during shipment and storage. The lab where volatile analysis is performed and also the refrigerated storage area should be completely free of solvents.

INSTRUMENTATION A GC containing a series configuration of a high temperature photoionization detector (PID) equipped with 10.0 eV (nominal) lamp and Hall electrolytic conductivity detector (HECD) is required. Also required is an all-glass 5-mL purging device, a sorbent trap, and a thermal desorption apparatus which is connected to the GC system.

Column 1: VOCOL glass wide-bore capillary column.
Column 2: RTX–502.2 mega-bore capillary column.
Column 3: DB-62 mega-bore capillary column.

PRECISION & ACCURACY Method detection limits are dependent upon the characteristics of the gas chromatographic system used. Analytes that are not separated chromatographically cannot be individually identified and used in the same calibration mixture or water samples unless an alternative technique for identification and quantification, such as mass spectrometry, is used.

Electrolytic conductivity detetor (c) range in µg/L (a) was 0.2–20.
Electrolytic conductivity detetor (c) MDL in µg/L (b) was 0.01.
Electrolytic conductivity detetor (c) accuracy as % recovery was 98.
Electrolytic conductivity detetor (c) precision as % RSD was 2.3.
Photoionization detector (d) range in µg/L (a) was 0.2–20.
Photoionization detector (d) MDL in µg/L (b) was 0.02.
Photoionization detector (d) accuracy as % recovery was 103.
Photoionization detector (d) precision as % RSD was 2.1.

(a) *The applicable concentration range of this method is compound, instrument, and matrix-dependent. It is listed as being approximately 0.02 to 200 µg/L but no specific information is provided so caution should be observed.*

(b) *The method detection limits reports with this method are compound, instrument, and matrix-dependent. The values reported were calculated using reagent water fortified with the corresponding compounds at 10 µg/L and a GC-equipped with a 60 m × 0.75 mm VOLCOL wide bore capillary column with 1.5 µm film thickness and using helium carrier gas.*

(c) *Recoveries and relative standard deviations were determined from seven samples of reagent water fortified with 10 µg/L of each compound. 2-Bromo-1-chloropropane was used as the internal standard for calculating average recoveries.*

(d) *Recoveries and relative standard deviations were determined from seven samples of reagent water fortified with 10 µg/L of each compound. Fluorobenzene was used as the internal standard for calculating average recoveries.*

SAMPLING METHOD Collect samples using a 40- to 120-mL screw-cap vial (prewashed with detergent, rinsed with distilled water and oven dried at 105°C) with a Teflon®-faced silicone septum. Collect bubble-free samples and place the septum with the Teflon® side down on the water.

SAMPLE PRESERVATION If residual chlorine is present in the water add about 25 mg of ascorbic acid to each vial before samples are collected to remove the chlorine. Add hydrochloric acid to reduce pH to <2, immediately cool samples to 4°C, and store them in a solvent-free refrigerator at 4°C until analysis.

MHT The maximum holding time for samples is 14 days from the time they were collected.

SAMPLE PREPARATION Remove the plungers from two 5-mL syringes and attach a closed syringe valve to each. Warm the sample to room temperature, open the sample bottle, and carefully pour the sample into one of the syringe barrels to just short of overflowing. Replace the syringe plunger, invert the syringe, and compress the sample. Open the syringe valve and vent any residual air while adjusting the sample volume to 5.0 mL. Add 10 µL of the internal calibration standard to the sample through the syringe valve. Close the valve. Fill the second syringe in an identical manner from the same sample bottle. Reserve this second syringe for a reanalysis if necessary.

QUALITY CONTROL As an initial demonstration of lab accuracy and precision, analyze 4 to 7 replicates of a lab fortified blank containing analyte at 0.1–5 µg/L. Collect all samples in

duplicate. Surrogate analytes (similar to those of the analytes of interest), whose concentration is known in every sample, are measured using the same internal standard calibration procedure. Duplicate field reagent water blanks (trip blanks) must be analyzed with each set of samples, lab reagent blanks (method blanks) must be analyzed with each batch of samples processed as a group within a work shift. Also, a single lab-fortified blank that contains each of the analytes of interest should be analyzed with each batch of samples processed as a group within a work shift. A 3- to 5-point calibration curve is needed depending on the calibration range factor required.

EPA CONTACT & HOTLINE For technical questions contact Dr. Baldev Bathija, U.S. EPA, Office of Ground Water and Drinking Water (WH-550D), 401 M St. SW, Washington, DC 20460. Tel. (202) 260-3040. For further information the EPA Safe Drinking Water Hotline may be called at: (800) 426-4791.

REFERENCE Methods for the Determination of Organic Compounds in Drinking Water, EPA/600/4-88/039 (revised July 1991; Final Rule for determination of compliance with the MCL for Total Trihalomethanes under 141.30, in 40 CFR Part 141, Vol. 58, No. 147, Fed. Reg., Tuesday Aug. 3, 1993). U.S. EPA Environmental Monitoring Systems Laboratory, Cincinnati, OH, 45268, U.S.A. Available from the National Technical Information Service (NTIS), 5285 Port Royal Road, Springfield, VA 22161; Tel. 800-553-6847. NTIS Order Number is PB91-231480.

1,2-Dichlorobenzene   **EPA Method 524**
CAS #95-50-1

TITLE Measurement of Purgeable Organic Compounds in Water by Capillary Column GC/MS.

MATRIX Drinking water and raw source water; the latter should include most surface water and groundwater sources.

METHOD SUMMARY Method 524.2 covers 60 volatile organic compounds. An inert gas (zero grade nitrogen or helium) is bubbled through a 25-mL or a 5-mL water sample (depending on the expected concentration of the analytes). Purged sample components are trapped in a tube of sorbent materials. When purging is complete, the sorbent tube is heated and backflushed with helium to desorb the trapped sample onto a capillary GC column.

INTERFERENCES Impurities in the purge gas and from organic compounds outgassing from the plumbing ahead of the trap account for many contamination problems. Interferences purged or coextracted from the samples will vary considerably from source to source, depending upon the particular sample or extract being tested. Cross-contamination can occur whenever high-level and low-level samples are analyzed sequentially. Samples also can be contaminated by diffusion of volatile organics (particularly methylene chloride and fluorocarbons) through the septum seal into the sample during shipment and storage.

INSTRUMENTATION A GC/MS with a data system equipped with one of the following capillary GC columns:

Column 1: VOCOL glass wide bore capillary column.
Column 2: DB-624 fused silica capillary column.
Column 3: DB-5 fused silica capillary column.

Also required is an all-glass 25 mL or 5-mL purging device, a sorbent trap, and a thermal desorption apparatus which is connected to the GC/MS system.

PRECISION & ACCURACY Method detection limits are compound- and instrument-dependent, and may vary from approximately 0.02–0.35 µg/L. Note in the table below that the "true" concentration range used for accuracy and precision measurements was quite narrow. However, the applicable concentration range of this method is primarily column dependent and is approximately 0.02 to 200 µg/L for the wide-bore thick-film columns. Narrow-bore thin-film columns may have a capacity which limits the range to about 0.02 to 20 µg/L. Analytes that are inefficiently purged from water will not be detected when present at low concentrations, but they can be measured with acceptable accuracy and precision when present in sufficient amounts.

Analytes that are not separated chromatographically, but which have different mass spectra and non-interfering quantification ions, can be identified and measured in the same calibration mixture or water sample. Analytes which have very similar mass spectra cannot be individually identified and measured in the same calibration mixture or water samples unless they have different retention times. Co-eluting compounds with very similar mass spectra, typically many structural isomers, must be reported as an isomeric group or pair.

The range (in µg/L) was 0.1–10.
The Method Detection Limig (in µg/L) was 0.03.
The accuracy (as % recovery) was 93.
The precision (in %) was 6.2.

Note: Data were obtained from 16–31 determinations using a wide-bore capillary column and a jet separator interfaced to a quadrupole mass spectrometer. All analytes were in a reagent water matrix.

SAMPLING METHOD Collect samples using a 40- to 120-mL screw-cap vial (prewashed with detergent, rinsed with distilled water and oven dried at 105°C) with a Teflon®-faced silicone septum. Collect bubble-free samples and place the septum with the Teflon® side down on the water.

SAMPLE PRESERVATION If residual chlorine is present in the water add about 25 mg of ascorbic acid to each vial before samples are collected to remove the chlorine. Add hydrochloric acid to reduce pH to <2, and immediately cool samples to 4°C, and store them in a solvent-free refrigerator at 4°C until analysis.

MHT The maximum holding time for samples is 14 days from the time they were collected.

SAMPLE PREPARATION Remove the plungers from two 25-mL (or 5-mL depending on sample size) syringes and attach a closed syringe valve to each. Warm the sample to room temperature, open the sample bottle, and carefully pour the sample into one of the syringe barrels to just short of overflowing. Replace the syringe plunger, invert the syringe, and compress the sample. Open the syringe valve and vent any residual air

while adjusting the sample volume to 25.0 mL (or 5 mL). For samples and blanks, add 5 µL of the fortification solution containing the internal standard and the surrogates to the sample through the syringe valve. For calibration standards and lab fortified blanks, add 5 µL of the fortification solution containing the internal standard only. Close the valve. Fill the second syringe in an identical manner from the same sample bottle. Reserve this second syringe for a reanalysis if necessary.

QUALITY CONTROL As an initial demonstration of lab accuracy and precision, analyze 4 to 7 replicates of a lab fortified blank containing analyte at 0.2–5 µg/L. Collect all samples in duplicate. Surrogate analytes (similar to those of the analytes of interest), whose concentration is known in every sample, are measured using the same internal standard calibration procedure. Duplicate field reagent water blanks (trip blanks) must be analyzed with each set of samples, lab reagent blanks (method blanks) must be analyzed with each batch of samples processed as a group within a work shift. Also, a single lab-fortified blank that contains each of the analytes of interest should be analyzed with each batch of samples processed as a group within a work shift. A 3- to 5-point calibration curve is needed depending on the calibration range factor required.

EPA CONTACT & HOTLINE For technical questions contact Dr. Baldev Bathija, U.S. EPA, Office of Ground Water and Drinking Water (WH-550D), 401 M St. SW, Washington, DC 20460. Tel. (202) 260-3040. For further information the EPA Safe Drinking Water Hotline may be called at: (800) 426-4791.

REFERENCE Methods for the Determination of Organic Compounds in Drinking Water, EPA/600/4-88/039 (revised July 1991; Final Rule for determination of compliance with the MCL for Total Trihalomethanes under 141.30, in 40 CFR Part 141, Vol. 58, No. 147, Fed. Reg., Tuesday Aug. 3, 1993). U.S. EPA Environmental Monitoring Systems Laboratory, Cincinnati, OH, 45268, U.S.A. Available from the National Technical Information Service (NTIS), 5285 Port Royal Road, Springfield, VA 22161; Tel. 800-553-6847. NTIS Order Number is PB91-231480.

1,3-Dichlorobenzene   **EPA Method 524**
CAS #541-73-1

TITLE Measurement of Purgeable Organic Compounds in Water by Capillary Column GC/MS.

MATRIX Drinking water and raw source water; the latter should include most surface water and groundwater sources.

METHOD SUMMARY Method 524.2 covers 60 volatile organic compounds. An inert gas (zero grade nitrogen or helium) is bubbled through a 25-mL or a 5-mL water sample (depending on the expected concentration of the analytes). Purged sample components are trapped in a tube of sorbent materials. When purging is complete, the sorbent tube is heated and backflushed with helium to desorb the trapped sample onto a capillary GC column.

INTERFERENCES Impurities in the purge gas and from organic compounds outgassing from the plumbing ahead of the trap account for many contamination problems. Interferences purged or coextracted from the samples will vary considerably from source to source, depending upon the particular sample or extract being tested. Cross-contamination can occur whenever high-level and low-level samples are analyzed sequentially. Samples also can be contaminated by diffusion of volatile organics (particularly methylene chloride and fluorocarbons) through the septum seal into the sample during shipment and storage.

INSTRUMENTATION A GC/MS with a data system equipped with one of the following capillary GC columns:

Column 1: VOCOL glass wide bore capillary column.
Column 2: DB-624 fused silica capillary column.
Column 3: DB-5 fused silica capillary column.

Also required is an all-glass 25 mL or 5-mL purging device, a sorbent trap, and a thermal desorption apparatus which is connected to the GC/MS system.

PRECISION & ACCURACY Method detection limits are compound- and instrument-dependent, and may vary from approximately 0.02–0.35 µg/L. Note in the table below that the "true" concentration range used for accuracy and precision measurements was quite narrow. However, the applicable concentration range of this method is primarily column dependent and is approximately 0.02 to 200 µg/L for the wide-bore thick-film columns. Narrow-bore thin-film columns may have a capacity which limits the range to about 0.02 to 20 µg/L. Analytes that are inefficiently purged from water will not be detected when present at low concentrations, but they can be measured with acceptable accuracy and precision when present in sufficient amounts.

Analytes that are not separated chromatographically, but which have different mass spectra and non-interfering quantification ions, can be identified and measured in the same calibration mixture or water sample. Analytes which have very similar mass spectra cannot be individually identified and measured in the same calibration mixture or water samples unless they have different retention times. Co-eluting compounds with very similar mass spectra, typically many structural isomers, must be reported as an isomeric group or pair.

The range (in µg/L) was 0.5–10.
The Method Detection Limig (in µg/L) was 0.12.
The accuracy (as % recovery) was 99.
The precision (in %) was 6.9.

Note: Data were obtained from 16–31 determinations using a wide-bore capillary column and a jet separator interfaced to a quadrupole mass spectrometer. All analytes were in a reagent water matrix.

SAMPLING METHOD Collect samples using a 40- to 120-mL screw-cap vial (prewashed with detergent, rinsed with distilled water and oven dried at 105°C) with a Teflon®-faced silicone septum. Collect bubble-free samples and place the septum with the Teflon® side down on the water.

SAMPLE PRESERVATION If residual chlorine is present in the water add about 25 mg of ascorbic acid to each vial before samples are collected to remove the chlorine. Add hydrochloric

acid to reduce pH to <2, and immediately cool samples to 4°C, and store them in a solvent-free refrigerator at 4°C until analysis.

MHT The maximum holding time for samples is 14 days from the time they were collected.

SAMPLE PREPARATION Remove the plungers from two 25-mL (or 5-mL depending on sample size) syringes and attach a closed syringe valve to each. Warm the sample to room temperature, open the sample bottle, and carefully pour the sample into one of the syringe barrels to just short of overflowing. Replace the syringe plunger, invert the syringe, and compress the sample. Open the syringe valve and vent any residual air while adjusting the sample volume to 25.0 mL (or 5 mL). For samples and blanks, add 5 µL of the fortification solution containing the internal standard and the surrogates to the sample through the syringe valve. For calibration standards and lab fortified blanks, add 5 µL of the fortification solution containing the internal standard only. Close the valve. Fill the second syringe in an identical manner from the same sample bottle. Reserve this second syringe for a reanalysis if necessary.

QUALITY CONTROL As an initial demonstration of lab accuracy and precision, analyze 4 to 7 replicates of a lab fortified blank containing analyte at 0.2–5 µg/L. Collect all samples in duplicate. Surrogate analytes (similar to those of the analytes of interest), whose concentration is known in every sample, are measured using the same internal standard calibration procedure. Duplicate field reagent water blanks (trip blanks) must be analyzed with each set of samples, lab reagent blanks (method blanks) must be analyzed with each batch of samples processed as a group within a work shift. Also, a single lab-fortified blank that contains each of the analytes of interest should be analyzed with each batch of samples processed as a group within a work shift. A 3- to 5-point calibration curve is needed depending on the calibration range factor required.

EPA CONTACT & HOTLINE For technical questions contact Dr. Baldev Bathija, U.S. EPA, Office of Ground Water and Drinking Water (WH-550D), 401 M St. SW, Washington, DC 20460. Tel. (202) 260-3040. For further information the EPA Safe Drinking Water Hotline may be called at: (800) 426-4791.

REFERENCE Methods for the Determination of Organic Compounds in Drinking Water, EPA/600/4-88/039 (revised July 1991; Final Rule for determination of compliance with the MCL for Total Trihalomethanes under 141.30, in 40 CFR Part 141, Vol. 58, No. 147, Fed. Reg., Tuesday Aug. 3, 1993). U.S. EPA Environmental Monitoring Systems Laboratory, Cincinnati, OH, 45268, U.S.A. Available from the National Technical Information Service (NTIS), 5285 Port Royal Road, Springfield, VA 22161; Tel. 800-553-6847. NTIS Order Number is PB91-231480.

1,4-Dichlorobenzene   EPA Method 524
CAS #106-46-7

TITLE Measurement of Purgeable Organic Compounds in Water by Capillary Column GC/MS.

MATRIX Drinking water and raw source water; the latter should include most surface water and groundwater sources.

METHOD SUMMARY Method 524.2 covers 60 volatile organic compounds. An inert gas (zero grade nitrogen or helium) is bubbled through a 25-mL or a 5-mL water sample (depending on the expected concentration of the analytes). Purged sample components are trapped in a tube of sorbent materials. When purging is complete, the sorbent tube is heated and backflushed with helium to desorb the trapped sample onto a capillary GC column.

INTERFERENCES Impurities in the purge gas and from organic compounds outgassing from the plumbing ahead of the trap account for many contamination problems. Interferences purged or coextracted from the samples will vary considerably from source to source, depending upon the particular sample or extract being tested. Cross-contamination can occur whenever high-level and low-level samples are analyzed sequentially. Samples also can be contaminated by diffusion of volatile organics (particularly methylene chloride and fluorocarbons) through the septum seal into the sample during shipment and storage.

INSTRUMENTATION A GC/MS with a data system equipped with one of the following capillary GC columns:

Column 1: VOCOL glass wide bore capillary column.
Column 2: DB-624 fused silica capillary column.
Column 3: DB-5 fused silica capillary column.

Also required is an all-glass 25 mL or 5-mL purging device, a sorbent trap, and a thermal desorption apparatus which is connected to the GC/MS system.

PRECISION & ACCURACY Method detection limits are compound- and instrument-dependent, and may vary from approximately 0.02–0.35 µg/L. Note in the table below that the "true" concentration range used for accuracy and precision measurements was quite narrow. However, the applicable concentration range of this method is primarily column dependent and is approximately 0.02 to 200 µg/L for the wide-bore thick-film columns. Narrow-bore thin-film columns may have a capacity which limits the range to about 0.02 to 20 µg/L. Analytes that are inefficiently purged from water will not be detected when present at low concentrations, but they can be measured with acceptable accuracy and precision when present in sufficient amounts.

Analytes that are not separated chromatographically, but which have different mass spectra and non-interfering quantification ions, can be identified and measured in the same calibration mixture or water sample. Analytes which have very similar mass spectra cannot be individually identified and measured in the same calibration mixture or water samples unless they have different retention times. Co-eluting compounds with very similar mass spectra, typically many structural isomers, must be reported as an isomeric group or pair.

The range (in µg/L) was 0.2–20.
The Method Detection Limig (in µg/L) was 0.03.
The accuracy (as % recovery) was 103.
The precision (in %) was 6.4.

Note: Data were obtained from 16–31 determinations using a wide-bore capillary column and a jet separator interfaced to a quadrupole mass spectrometer. All analytes were in a reagent water matrix.

SAMPLING METHOD Collect samples using a 40- to 120-mL screw-cap vial (prewashed with detergent, rinsed with distilled water and oven dried at 105°C) with a Teflon®-faced silicone septum. Collect bubble-free samples and place the septum with the Teflon® side down on the water.

SAMPLE PRESERVATION If residual chlorine is present in the water add about 25 mg of ascorbic acid to each vial before samples are collected to remove the chlorine. Add hydrochloric acid to reduce pH to <2, and immediately cool samples to 4°C, and store them in a solvent-free refrigerator at 4°C until analysis.

MHT The maximum holding time for samples is 14 days from the time they were collected.

SAMPLE PREPARATION Remove the plungers from two 25-mL (or 5-mL depending on sample size) syringes and attach a closed syringe valve to each. Warm the sample to room temperature, open the sample bottle, and carefully pour the sample into one of the syringe barrels to just short of overflowing. Replace the syringe plunger, invert the syringe, and compress the sample. Open the syringe valve and vent any residual air while adjusting the sample volume to 25.0 mL (or 5 mL). For samples and blanks, add 5 µL of the fortification solution containing the internal standard and the surrogates to the sample through the syringe valve. For calibration standards and lab fortified blanks, add 5 µL of the fortification solution containing the internal standard only. Close the valve. Fill the second syringe in an identical manner from the same sample bottle. Reserve this second syringe for a reanalysis if necessary.

QUALITY CONTROL As an initial demonstration of lab accuracy and precision, analyze 4 to 7 replicates of a lab fortified blank containing analyte at 0.2–5 µg/L. Collect all samples in duplicate. Surrogate analytes (similar to those of the analytes of interest), whose concentration is known in every sample, are measured using the same internal standard calibration procedure. Duplicate field reagent water blanks (trip blanks) must be analyzed with each set of samples, lab reagent blanks (method blanks) must be analyzed with each batch of samples processed as a group within a work shift. Also, a single lab-fortified blank that contains each of the analytes of interest should be analyzed with each batch of samples processed as a group within a work shift. A 3- to 5-point calibration curve is needed depending on the calibration range factor required.

EPA CONTACT & HOTLINE For technical questions contact Dr. Baldev Bathija, U.S. EPA, Office of Ground Water and Drinking Water (WH-550D), 401 M St. SW, Washington, DC 20460. Tel. (202) 260-3040. For further information the EPA Safe Drinking Water Hotline may be called at: (800) 426-4791.

REFERENCE Methods for the Determination of Organic Compounds in Drinking Water, EPA/600/4-88/039 (revised July 1991; Final Rule for determination of compliance with the MCL for Total Trihalomethanes under 141.30, in 40 CFR Part 141, Vol. 58, No. 147, Fed. Reg., Tuesday Aug. 3, 1993). U.S. EPA Environmental Monitoring Systems Laboratory, Cincinnati, OH, 45268, U.S.A. Available from the National Technical Information Service (NTIS), 5285 Port Royal Road, Springfield, VA 22161; Tel. 800-553-6847. NTIS Order Number is PB91-231480.

1,2-Dichlorobenzene   **EPA Method 625**
CAS #95-50-1

TITLE Base/Neutrals and Acids, U.S. EPA Method 625

MATRIX This methods covers municipal and industrial wastewaters.

METHOD SUMMARY Approximately 1 L of sample is serially extracted with methylene chloride at a pH greater than 11 and again at a pH less than 2 using a separatory funnel or a continuous extractor. The methylene chloride extract is dried, concentrated to a volume of 1 mL, and analyzed by GC/MS. Qualitative identification of the parameters in the extract is performed using the retention time and the relative abundance of three characteristic masses (m/z). Qualitative analysis is performed using either external or internal standard techniques with a single characteristic m/z.

INTERFERENCES Method interferences may be caused by contaminants in solvents, reagents, glassware, and other sample processing hardware. Glassware must be scrupulously cleaned. Glassware should be heated in a muffle furnace at 400°C for 5 to 30 min. Some thermally stable materials, such as PCBs, may not be eliminated by this treatment. Solvent rinses with acetone and pesticide quality hexane may be substituted for the muffle furnace heating. Matrix interferences may be caused by contaminants that are coextracted from the sample. The base-neutral extraction may cause significantly reduced recovery of phenols. The packed gas chromatographic columns recommended for the basic fraction may not exhibit sufficient resolution for some analytes.

INSTRUMENTATION A GC/MS system with an injection port designed for on-column injection when using packed columns and for splitless injection when using capillary columns.

Column for base/neutrals: 1.8 m long × 2 mm I.D. glass, packed with 3% SP-2550 on Supelcoport (100/120 mesh) or equivalent.
Column for acids: 1.8 m long × 2 mm I.D. glass, packed with 1% SP-1240DA on Supelcoport (100/120 mesh) or equivalent.

PRECISION & ACCURACY The MDL concentrations were obtained using reagent water. The MDL actually achieved in a given analysis will vary depending on instrument sensitivity and matrix effects. This method was tested by 15 laboratories using reagent water, drinking water, surface water, and industrial wastewaters spiked at six concentrations over the range 5 to 100 µg/L. Single operator precision, overall precision, and method accuracy were found to be directly related to the concentration of the parameter matrix.

The MDL (in µg/L) in reagent water was not reported.
The standard deviation (in µg/L based on 4 recovery measurements) was 30.9.

The range (in μg/L) for average recovery for 4 measurements was 48.6–112.0.

The range (in %) for percent recovery was 32–129.

Accuracy (in μg/L) as expected recovery for one or more measurements of a sample containing a true concentration of C was 0.80C+0.28.

Precision (in μg/L) as expected single analyst standard deviation of measurements at an average concentration found at X was 0.20X+0.47.

Overall precision (in μg/L) as expected interlaboratory standard deviation of measurements in an average concentration found at X was 0.24X+0.39.

$C =$ *True value of the concentration in μg/L.*
$X =$ *Average recovery found for measurements of samples containing a concentration at C in μg/L.*

SAMPLE PREPARATION Adjust the pH to >11 with sodium hydroxide and serially extract in a separatory funnel with methylene chloride or else in a continuous extractor. Next, adjust the pH to <2 with sulfuric acid and serially extract in a separatory funnel with methylene chloride or else in a continuous extractor. Dry the extracts separately through a column of anhydrous sodium sulfate and then concentrate each of the extracts to 1.0 mL using a K-D apparatus.

SAMPLE COLLECTION, PRESERVATION & HANDLING Grab samples must be collected in glass containers. All samples must be refrigerated at 4°C from the time of collection until extraction. If residual chlorine is present, add 80 mg of sodium thiosulfate/L of sample and mix well. All samples must be extracted within 7 days of collection and completely analyzed within 40 days of extraction.

QUALITY CONTROL Make an initial, one-time, demonstration of the ability to generate acceptable accuracy and precision with this method. Before processing any samples, the analyst must analyze a reagent water blank to demonstrate that interferences from the analytical system and glassware are under control. Each time a set of samples is extracted or reagents are changed, a reagent water blank must be processed. Spike and analyze a minimum of 5% of all samples to monitor and evaluate lab data quality. A QC check sample concentrate that contains each parameter of interest at a concentration of 100 μg/mL in acetone is required. PCBs and multicomponent pesticides may be omitted from this test.

After analysis of five spiked wastewater samples, calculate the average percent recovery and the standard deviation of the percent recovery. Spike all samples with the surrogate standard spiking solution and calculate the percent recovery of each surrogate compound.

REFERENCE *Federal Register,* Vol. 49, No. 209. Friday, Oct. 26, 1984.

1,3-Dichlorobenzene   **EPA Method 625**
CAS #541-73-1

татLE Base/Neutrals and Acids, U.S. EPA Method 625

MATRIX This methods covers municipal and industrial wastewaters.

METHOD SUMMARY Approximately 1 L of sample is serially extracted with methylene chloride at a pH greater than 11 and again at a pH less than 2 using a separatory funnel or a continuous extractor. The methylene chloride extract is dried, concentrated to a volume of 1 mL, and analyzed by GC/MS. Qualitative identification of the parameters in the extract is performed using the retention time and the relative abundance of three characteristic masses (m/z). Qualitative analysis is performed using either external or internal standard techniques with a single characteristic m/z.

INTERFERENCES Method interferences may be caused by contaminants in solvents, reagents, glassware, and other sample processing hardware. Glassware must be scrupulously cleaned. Glassware should be heated in a muffle furnace at 400°C for 5 to 30 min. Some thermally stable materials, such as PCBs, may not be eliminated by this treatment. Solvent rinses with acetone and pesticide quality hexane may be substituted for the muffle furnace heating. Matrix interferences may be caused by contaminants that are coextracted from the sample. The base-neutral extraction may cause significantly reduced recovery of phenols. The packed gas chromatographic columns recommended for the basic fraction may not exhibit sufficient resolution for some analytes.

INSTRUMENTATION A GC/MS system with an injection port designed for on-column injection when using packed columns and for splitless injection when using capillary columns.

Column for base/neutrals: 1.8 m long × 2 mm I.D. glass, packed with 3% SP-2550 on Supelcoport (100/120 mesh) or equivalent.

Column for acids: 1.8 m long × 2 mm I.D. glass, packed with 1% SP-1240DA on Supelcoport (100/120 mesh) or equivalent.

PRECISION & ACCURACY The MDL concentrations were obtained using reagent water. The MDL actually achieved in a given analysis will vary depending on instrument sensitivity and matrix effects. This method was tested by 15 laboratories using reagent water, drinking water, surface water, and industrial wastewaters spiked at six concentrations over the range 5 to 100 μg/L. Single operator precision, overall precision, and method accuracy were found to be directly related to the concentration of the parameter matrix.

The MDL (in μg/L) in reagent water was not reported.

The standard deviation (in μg/L based on 4 recovery measurements) was 41.7.

The range (in μg/L) for average recovery for 4 measurements was 16.7–153.9.

The range (in %) for percent recovery was D-17.

Accuracy (in μg/L) as expected recovery for one or more measurements of a sample containing a true concentration of C was 0.86C−0.70.

Precision (in μg/L) as expected single analyst standard deviation of measurements at an average concentration found at X was 0.25X+0.68.

Overall precision (in µg/L) as expected interlaboratory standard deviation of measurements in an average concentration found at X was 0.41X+0.11.

C = *True value of the concentration in µg/L.*
X = *Average recovery found for measurements of samples containing a concentration at C in µg/L.*

SAMPLE PREPARATION Adjust the pH to >11 with sodium hydroxide and serially extract in a separatory funnel with methylene chloride or else in a continuous extractor. Next, adjust the pH to <2 with sulfuric acid and serially extract in a separatory funnel with methylene chloride or else in a continuous extractor. Dry the extracts separately through a column of anhydrous sodium sulfate and then concentrate each of the extracts to 1.0 mL using a K-D apparatus.

SAMPLE COLLECTION, PRESERVATION & HANDLING Grab samples must be collected in glass containers. All samples must be refrigerated at 4°C from the time of collection until extraction. If residual chlorine is present, add 80 mg of sodium thiosulfate/L of sample and mix well. All samples must be extracted within 7 days of collection and completely analyzed within 40 days of extraction.

QUALITY CONTROL Make an initial, one-time, demonstration of the ability to generate acceptable accuracy and precision with this method. Before processing any samples, the analyst must analyze a reagent water blank to demonstrate that interferences from the analytical system and glassware are under control. Each time a set of samples is extracted or reagents are changed, a reagent water blank must be processed. Spike and analyze a minimum of 5% of all samples to monitor and evaluate lab data quality. A QC check sample concentrate that contains each parameter of interest at a concentration of 100 µg/mL in acetone is required. PCBs and multicomponent pesticides may be omitted from this test.

After analysis of five spiked wastewater samples, calculate the average percent recovery and the standard deviation of the percent recovery. Spike all samples with the surrogate standard spiking solution and calculate the percent recovery of each surrogate compound.

REFERENCE *Federal Register*, Vol. 49, No. 209. Friday, Oct. 26, 1984.

1,4-Dichlorobenzene   **EPA Method 625**
CAS #106-46-7

TITLE Base/Neutrals and Acids, U.S. EPA Method 625

MATRIX This methods covers municipal and industrial wastewaters.

METHOD SUMMARY Approximately 1 L of sample is serially extracted with methylene chloride at a pH greater than 11 and again at a pH less than 2 using a separatory funnel or a continuous extractor. The methylene chloride extract is dried, concentrated to a volume of 1 mL, and analyzed by GC/MS. Qualitative identification of the parameters in the extract is performed using the retention time and the relative abundance of three characteristic masses (m/z). Qualitative analysis is performed using either external or internal standard techniques with a single characteristic m/z.

INTERFERENCES Method interferences may be caused by contaminants in solvents, reagents, glassware, and other sample processing hardware. Glassware must be scrupulously cleaned. Glassware should be heated in a muffle furnace at 400°C for 5 to 30 min. Some thermally stable materials, such as PCBs, may not be eliminated by this treatment. Solvent rinses with acetone and pesticide quality hexane may be substituted for the muffle furnace heating. Matrix interferences may be caused by contaminants that are coextracted from the sample. The base-neutral extraction may cause significantly reduced recovery of phenols. The packed gas chromatographic columns recommended for the basic fraction may not exhibit sufficient resolution for some analytes.

INSTRUMENTATION A GC/MS system with an injection port designed for on-column injection when using packed columns and for splitless injection when using capillary columns.

Column for base/neutrals: 1.8 m long × 2 mm I.D. glass, packed with 3% SP-2550 on Supelcoport (100/120 mesh) or equivalent.
Column for acids: 1.8 m long × 2 mm I.D. glass, packed with 1% SP-1240DA on Supelcoport (100/120 mesh) or equivalent.

PRECISION & ACCURACY The MDL concentrations were obtained using reagent water. The MDL actually achieved in a given analysis will vary depending on instrument sensitivity and matrix effects. This method was tested by 15 laboratories using reagent water, drinking water, surface water, and industrial wastewaters spiked at six concentrations over the range 5 to 100 µg/L. Single operator precision, overall precision, and method accuracy were found to be directly related to the concentration of the parameter matrix.

The MDL (in µg/L) in reagent water was not reported.
The standard deviation (in µg/L based on 4 recovery measurements) was 32.1.
The range (in µg/L) for average recovery for 4 measurements was 37.3–105.7.
The range (in %) for percent recovery was 20–124.
Accuracy (in µg/L) as expected recovery for one or more measurements of a sample containing a true concentration of C was 0.73C–1.47.
Precision (in µg/L) as expected single analyst standard deviation of measurements at an average concentration found at X was 0.24X+0.23.
Overall precision (in µg/L) as expected interlaboratory standard deviation of measurements in an average concentration found at X was 0.29X+0.36.

C = *True value of the concentration in µg/L.*
X = *Average recovery found for measurements of samples containing a concentration at C in µg/L.*

SAMPLE PREPARATION Adjust the pH to >11 with sodium hydroxide and serially extract in a separatory funnel with methylene chloride or else in a continuous extractor. Next, adjust the pH to <2 with sulfuric acid and serially extract in a separatory funnel with methylene chloride or else in a continuous extractor. Dry the extracts separately through a column of anhydrous sodium sulfate and then concentrate each of the extracts to 1.0 mL using a K-D apparatus.

SAMPLE COLLECTION, PRESERVATION & HANDLING Grab samples must be collected in glass containers. All samples must be refrigerated at 4°C from the time of collection until extraction. If residual chlorine is present, add 80 mg of sodium thiosulfate/L of sample and mix well. All samples must be extracted within 7 days of collection and completely analyzed within 40 days of extraction.

QUALITY CONTROL Make an initial, one-time, demonstration of the ability to generate acceptable accuracy and precision with this method. Before processing any samples, the analyst must analyze a reagent water blank to demonstrate that interferences from the analytical system and glassware are under control. Each time a set of samples is extracted or reagents are changed, a reagent water blank must be processed. Spike and analyze a minimum of 5% of all samples to monitor and evaluate lab data quality. A QC check sample concentrate that contains each parameter of interest at a concentration of 100 µg/mL in acetone is required. PCBs and multicomponent pesticides may be omitted from this test.

After analysis of five spiked wastewater samples, calculate the average percent recovery and the standard deviation of the percent recovery. Spike all samples with the surrogate standard spiking solution and calculate the percent recovery of each surrogate compound.

REFERENCE Federal Register, Vol. 49, No. 209. Friday, Oct. 26, 1984.

1,2-Dichlorobenzene   EPA Method 8021
CAS #95-50-1

TITLE Halogenated Volatile by Gas Chromatography Using Photoionization and Electrolytic Conductivity Detectors in Series: Capillary Column Technique

MATRIX This method is applicable to nearly all types of samples, regardless of water content, including groundwater, aqueous sludges, caustic liquors, acid liquors, waste solvents, oily wastes, mousses, tars, fibrous wastes, polymeric emulsions, filter cakes, spent carbons, spent catalysts, soils, and sediments.

METHOD SUMMARY This method is used to determine 60 volatile organic compounds in a variety of solid waste matrices. It provides GC conditions for the detection of halogenated and aromatic volatile organic compounds. Samples can be analyzed using direct injection or purge-and-trap (EPA Method 5030). Groundwater samples must be analyzed using EPA Method 5030 (where applicable). A temperature program is used with the GC. Detection is achieved by a photoionization detector (PID) and a Hall electrolytic conductivity detector (HECD) in series.

INTERFERENCES Samples can be contaminated by diffusion of volatile organics (particularly chlorofluorocarbons and methylene chloride) through the sample container septum during shipment and storage.

INSTRUMENTATION A GC-equipped with variable-constant differential flow controllers, subambient oven controller, PID and HECD detectors connected with a short piece of uncoated capillary tubing and a data system.

Column: 60 m × 0.75 mm I.D. VOCOL wide-bore capillary column with 1.5 µm film thickness.

PRECISION & ACCURACY MDLs are compound-dependent and vary with purging efficiency and concentration. The applicable concentration range of this method is compound- and instrument-dependent but is approximately 0.1 to 200 µg/L. Analytes that are inefficiently purged from water will not be detected when present at low concentrations, but they can be measured with acceptable accuracy and precision when present in sufficient amounts. The estimated quantitation limit (EQL) for an individual compound is approximately 1 µg/kg (wet weight) for soil/sediment samples, 100 µg/kg (wet weight) for wastes, and 1 µg/L for groundwater. EQLs will be proportionately higher for sample extracts and samples that require dilution or reduced sample size to avoid saturation of the detector.

MULTIPLICATION FACTORS FOR OTHER MATRICES (a)

Matrix	Factor (b)
Groundwater	10
Low-concentration soil	10
Water miscible liquid waste	500
High-concentration soil and sludge	1250
Non-water miscible waste	1250

(a) Sample EQLs are highly matrix-dependent. The EQLs listed herein are provided for guidance and may not always be achievable. (b) EQL = [Method detection limit] × [Factor]. For non-aqueous samples, the factor is on a wet-weight basis.

SINGLE LABORATORY ACCURACY & PRECISION DATA FOR VOCs IN WATER

This method was tested in a single lab using water spiked at 10 µg/L and the following data was reported:

Recoveries and standard deviations were determined from seven samples and spiked at 10 µg/L of each analyte. Recoveries were determined by the internal standard method. Internal standards were: Fluorobenzene for PID and 2-Bromo-1-chloropropane for HECD.

The average recovery (in percent) for the PID was 102.
The standard deviation of the recovery for the PID was 2.1.
The MDL (in µg/mL) for the PID was 0.05.
The average recovery (in percent) for the HECD was 100.
The standard deviation of the recovery for the HECD was 1.5.
The MDL (in µg/mL) for the HECD was 0.02.

SAMPLE COLLECTION, PRESERVATION & HANDLING
Volatile Organics — Standard 40-mL glass screw-cap VOA vials with Teflon®-faced silicone septum may be used for both liquid and solid matrices. When collecting samples, liquids and solids should be introduced into the vials gently to reduce agitation which might drive off volatile compounds. If there are any air

bubbles present the sample must be retaken. Tap slightly as they are filled to try and eliminate as much free air space as possible. The two vials from each sampling locations should be sealed in separate plastic bags to prevent cross-contamination between samples particularly if the sampled waste is suspected of containing high levels of volatile organics.

Semivolatile organics — Containers used to collect samples for the determination of semivolatile organic compounds should be soap and water washed followed by methanol (or isopropanol) rinsing. The sample containers should be of glass or Teflon® and have screw-top covers with Teflon® liners.

Preservation for volatile organics — No preservation is used with concentrated waste samples. With liquid samples containing no residual chlorine, 4 drops of concentrated hydrochloric acid are added and the samples are immediately cooled to 4°C. When liquid samples contain residual chlorine, they are treated as above and, in addition, 4 drops of 4% aqueous sodium thiosulfate are added. Soil, sediment, and sludge samples are only cooled to 4°C.

Preservation for semivolatile organics — No preservation is used with concentrated waste samples. With liquid samples containing no residual chlorine and with soil, sediment, and sludge samples, immediately cooling to 4°C is the only preservation used. When residual chlorine is present then 3 mL of 10% aqueous sodium sulfate is added for each gallon of sample collected, followed by cooling to 4°C.

MHT The holding time for all volatile organics samples is 14 days. Liquid samples must be extracted within 7 days and their extracts analyzed within 40 days. Concentrated waste, soil, sediment, and sludge samples must be extracted within 14 days and their extracts analyzed within 40 days.

SAMPLE PREPARATION Volatile compounds are introduced into the gas chromatograph either by direct injector or purge-and-trap (EPA Method 5030). EPA Method 5030 may be used directly on groundwater samples or low-concentration contaminated soils and sediments. For medium-concentration soils or sediments, methanolic extraction, as described in EPA Method 5030, may be necessary prior to purge-and-trap analysis.

QUALITY CONTROL Calculate surrogate standard recovery on all samples, blanks, and spikes. A trip blank is recommended to check on sampling, storage, and handling contamination. Calibration standards, at a minimum of five concentration levels, are prepared in organic-free reagent water. One of the concentration levels should be at a concentration near, but above, the method detection limit.

A combination of bromochloromethane, 2-bromo-1-chloropropane, 1,4-dichlorobutane, and bromochlorobenzene are recommended as surrogate standards to encompass the range of the temperature program used in this method.

REFERENCE Test Methods for Evaluating Solid Waste, Physical/Chemical Methods, SW-846, 3rd Edition, U.S. EPA, Office of Solid Waste, Washington, DC, EPA Method 8021A, Rev. 1, Nov. 1992.

1,3-Dichlorobenzene   EPA Method 8021
CAS #541-73-1

TITLE Halogenated Volatile by Gas Chromatography Using Photoionization and Electrolytic Conductivity Detectors in Series: Capillary Column Technique

MATRIX This method is applicable to nearly all types of samples, regardless of water content, including groundwater, aqueous sludges, caustic liquors, acid liquors, waste solvents, oily wastes, mousses, tars, fibrous wastes, polymeric emulsions, filter cakes, spent carbons, spent catalysts, soils, and sediments.

METHOD SUMMARY This method is used to determine 60 volatile organic compounds in a variety of solid waste matrices. It provides GC conditions for the detection of halogenated and aromatic volatile organic compounds. Samples can be analyzed using direct injection or purge-and-trap (EPA Method 5030). Groundwater samples must be analyzed using EPA Method 5030 (where applicable). A temperature program is used with the GC. Detection is achieved by a photoionization detector (PID) and a Hall electrolytic conductivity detector (HECD) in series.

INTERFERENCES Samples can be contaminated by diffusion of volatile organics (particularly chlorofluorocarbons and methylene chloride) through the sample container septum during shipment and storage.

INSTRUMENTATION A GC-equipped with variable-constant differential flow controllers, subambient oven controller, PID and HECD detectors connected with a short piece of uncoated capillary tubing and a data system.

Column: 60 m × 0.75 mm I.D. VOCOL wide-bore capillary column with 1.5 µm film thickness.

PRECISION & ACCURACY MDLs are compound-dependent and vary with purging efficiency and concentration. The applicable concentration range of this method is compound- and instrument-dependent but is approximately 0.1 to 200 µg/L. Analytes that are inefficiently purged from water will not be detected when present at low concentrations, but they can be measured with acceptable accuracy and precision when present in sufficient amounts. The estimated quantitation limit (EQL) for an individual compound is approximately 1 µg/kg (wet weight) for soil/sediment samples, 100 µg/kg (wet weight) for wastes, and 1 µg/L for groundwater. EQLs will be proportionately higher for sample extracts and samples that require dilution or reduced sample size to avoid saturation of the detector.

MULTIPLICATION FACTORS FOR OTHER MATRICES (a)

Matrix	Factor (b)
Groundwater	10
Low-concentration soil	10
Water miscible liquid waste	500
High-concentration soil and sludge	1250
Non-water miscible waste	1250

(a) Sample EQLs are highly matrix-dependent. The EQLs listed herein are provided for guidance and may not always be achievable. (b) EQL = [Method detection limit] × [Factor]. For non-aqueous samples, the factor is on a wet-weight basis.

SINGLE LABORATORY ACCURACY & PRECISION DATA FOR VOCs IN WATER

This method was tested in a single lab using water spiked at 10 µg/L and the following data was reported:

Recoveries and standard deviations were determined from seven samples and spiked at 10 µg/L of each analyte. Recoveries were determined by the internal standard method. Internal standards were: Fluorobenzene for PID and 2-Bromo-1-chloropropane for HECD.

The average recovery (in percent) for the PID was 104.
The standard deviation of the recovery for the PID was 1.7.
The MDL (in µg/mL) for the PID was 0.02.
The average recovery (in percent) for the HECD was 106.
The standard deviation of the recovery for the HECD was 4.3.
The MDL (in µg/mL) for the HECD was 0.02.

SAMPLE COLLECTION, PRESERVATION & HANDLING
Volatile Organics — Standard 40-mL glass screw-cap VOA vials with Teflon®-faced silicone septum may be used for both liquid and solid matrices. When collecting samples, liquids and solids should be introduced into the vials gently to reduce agitation which might drive off volatile compounds. If there are any air bubbles present the sample must be retaken. Tap slightly as they are filled to try and eliminate as much free air space as possible. The two vials from each sampling locations should be sealed in separate plastic bags to prevent cross-contamination between samples particularly if the sampled waste is suspected of containing high levels of volatile organics.

Semivolatile organics — Containers used to collect samples for the determination of semivolatile organic compounds should be soap and water washed followed by methanol (or isopropanol) rinsing. The sample containers should be of glass or Teflon® and have screw-top covers with Teflon® liners.

Preservation for volatile organics — No preservation is used with concentrated waste samples. With liquid samples containing no residual chlorine, 4 drops of concentrated hydrochloric acid are added and the samples are immediately cooled to 4°C. When liquid samples contain residual chlorine, they are treated as above and, in addition, 4 drops of 4% aqueous sodium thiosulfate are added. Soil, sediment, and sludge samples are only cooled to 4°C.

Preservation for semivolatile organics — No preservation is used with concentrated waste samples. With liquid samples containing no residual chlorine and with soil, sediment, and sludge samples, immediately cooling to 4°C is the only preservation used. When residual chlorine is present then 3 mL of 10% aqueous sodium sulfate is added for each gallon of sample collected, followed by cooling to 4°C.

MHT The holding time for all volatile organics samples is 14 days. Liquid samples must be extracted within 7 days and their extracts analyzed within 40 days. Concentrated waste, soil, sediment, and sludge samples must be extracted within 14 days and their extracts analyzed within 40 days.

SAMPLE PREPARATION Volatile compounds are introduced into the gas chromatograph either by direct injector or purge-and-trap (EPA Method 5030). EPA Method 5030 may be used directly on groundwater samples or low-concentration contaminated soils and sediments. For medium-concentration soils or sediments, methanolic extraction, as described in EPA Method 5030, may be necessary prior to purge-and-trap analysis.

QUALITY CONTROL Calculate surrogate standard recovery on all samples, blanks, and spikes. A trip blank is recommended to check on sampling, storage, and handling contamination. Calibration standards, at a minimum of five concentration levels, are prepared in organic-free reagent water. One of the concentration levels should be at a concentration near, but above, the method detection limit.

A combination of bromochloromethane, 2-bromo-1-chloropropane, 1,4-dichlorobutane, and bromochlorobenzene are recommended as surrogate standards to encompass the range of the temperature program used in this method.

REFERENCE Test Methods for Evaluating Solid Waste, Physical/Chemical Methods, SW-846, 3rd Edition, U.S. EPA, Office of Solid Waste, Washington, DC, EPA Method 8021A, Rev. 1, Nov. 1992.

1,4-Dichlorobenzene   **EPA Method 8021**
CAS #106-46-7

TITLE Halogenated Volatile by Gas Chromatography Using Photoionization and Electrolytic Conductivity Detectors in Series: Capillary Column Technique

MATRIX This method is applicable to nearly all types of samples, regardless of water content, including groundwater, aqueous sludges, caustic liquors, acid liquors, waste solvents, oily wastes, mousses, tars, fibrous wastes, polymeric emulsions, filter cakes, spent carbons, spent catalysts, soils, and sediments.

METHOD SUMMARY This method is used to determine 60 volatile organic compounds in a variety of solid waste matrices. It provides GC conditions for the detection of halogenated and aromatic volatile organic compounds. Samples can be analyzed using direct injection or purge-and-trap (EPA Method 5030). Groundwater samples must be analyzed using EPA Method 5030 (where applicable). A temperature program is used with the GC. Detection is achieved by a photoionization detector (PID) and a Hall electrolytic conductivity detector (HECD) in series.

INTERFERENCES Samples can be contaminated by diffusion of volatile organics (particularly chlorofluorocarbons and methylene chloride) through the sample container septum during shipment and storage.

INSTRUMENTATION A GC-equipped with variable-constant differential flow controllers, subambient oven controller, PID and HECD detectors connected with a short piece of uncoated capillary tubing and a data system.

Column: 60 m × 0.75 mm I.D. VOCOL wide-bore capillary column with 1.5 µm film thickness.

PRECISION & ACCURACY MDLs are compound-dependent and vary with purging efficiency and concentration. The applicable concentration range of this method is compound- and instrument-dependent but is approximately 0.1 to 200 µg/L. Analytes that are inefficiently purged from water will not be detected when present at low concentrations, but they can be measured with acceptable accuracy and precision when present in sufficient amounts. The estimated quantitation limit (EQL) for an individual compound is approximately 1 µg/kg (wet weight) for soil/sediment samples, 100 µg/kg (wet weight) for wastes, and 1 µg/L for groundwater. EQLs will be proportionately higher for sample extracts and samples that require dilution or reduced sample size to avoid saturation of the detector.

MULTIPLICATION FACTORS FOR OTHER MATRICES (a)

Matrix	Factor (b)
Groundwater	10
Low-concentration soil	10
Water miscible liquid waste	500
High-concentration soil and sludge	1250
Non-water miscible waste	1250

(a) Sample EQLs are highly matrix-dependent. The EQLs listed herein are provided for guidance and may not always be achievable.
(b) EQL = [Method detection limit] × [Factor]. For non-aqueous samples, the factor is on a wet-weight basis.

SINGLE LABORATORY ACCURACY & PRECISION DATA FOR VOCs IN WATER

This method was tested in a single lab using water spiked at 10 µg/L and the following data was reported:

Recoveries and standard deviations were determined from seven samples and spiked at 10 µg/L of each analyte. Recoveries were determined by the internal standard method. Internal standards were: Fluorobenzene for PID and 2-Bromo-1-chloropropane for HECD.

The average recovery (in percent) for the PID was 103.
The standard deviation of the recovery for the PID was 2.2.
The MDL (in µg/mL) for the PID was 0.007.
The average recovery (in percent) for the HECD was 98.
The standard deviation of the recovery for the HECD was 2.3.
The MDL (in µg/mL) for the HECD was 0.01.

SAMPLE COLLECTION, PRESERVATION & HANDLING

Volatile Organics — Standard 40-mL glass screw-cap VOA vials with Teflon®-faced silicone septum may be used for both liquid and solid matrices. When collecting samples, liquids and solids should be introduced into the vials gently to reduce agitation which might drive off volatile compounds. If there are any air bubbles present the sample must be retaken. Tap slightly as they are filled to try and eliminate as much free air space as possible. The two vials from each sampling locations should be sealed in separate plastic bags to prevent cross-contamination between samples particularly if the sampled waste is suspected of containing high levels of volatile organics.

Semivolatile organics — Containers used to collect samples for the determination of semivolatile organic compounds should be soap and water washed followed by methanol (or isopropanol) rinsing. The sample containers should be of glass or Teflon® and have screw-top covers with Teflon® liners.

Preservation for volatile organics — No preservation is used with concentrated waste samples. With liquid samples containing no residual chlorine, 4 drops of concentrated hydrochloric acid are added and the samples are immediately cooled to 4°C. When liquid samples contain residual chlorine, they are treated as above and, in addition, 4 drops of 4% aqueous sodium thiosulfate are added. Soil, sediment, and sludge samples are only cooled to 4°C.

Preservation for semivolatile organics — No preservation is used with concentrated waste samples. With liquid samples containing no residual chlorine and with soil, sediment, and sludge samples, immediately cooling to 4°C is the only preservation used. When residual chlorine is present then 3 mL of 10% aqueous sodium sulfate is added for each gallon of sample collected, followed by cooling to 4°C.

MHT The holding time for all volatile organics samples is 14 days. Liquid samples must be extracted within 7 days and their extracts analyzed within 40 days. Concentrated waste, soil, sediment, and sludge samples must be extracted within 14 days and their extracts analyzed within 40 days.

SAMPLE PREPARATION Volatile compounds are introduced into the gas chromatograph either by direct injector or purge-and-trap (EPA Method 5030). EPA Method 5030 may be used directly on groundwater samples or low-concentration contaminated soils and sediments. For medium-concentration soils or sediments, methanolic extraction, as described in EPA Method 5030, may be necessary prior to purge-and-trap analysis.

QUALITY CONTROL Calculate surrogate standard recovery on all samples, blanks, and spikes. A trip blank is recommended to check on sampling, storage, and handling contamination. Calibration standards, at a minimum of five concentration levels, are prepared in organic-free reagent water. One of the concentration levels should be at a concentration near, but above, the method detection limit.

A combination of bromochloromethane, 2-bromo-1-chloropropane, 1,4-dichlorobutane, and bromochlorobenzene are recommended as surrogate standards to encompass the range of the temperature program used in this method.

REFERENCE Test Methods for Evaluating Solid Waste, Physical/Chemical Methods, SW-846, 3rd Edition, U.S. EPA, Office of Solid Waste, Washington, DC, EPA Method 8021A, Rev. 1, Nov. 1992.

1,2-Dichlorobenzene   **EPA Method 8120**
CAS #95-50-1

TITLE Chlorinated Hydrocarbons by Gas Chromatography

MATRIX This method covers aqueous and solid matrices. This includes a wide variety such as drinking water, groundwater, industrial wastewaters, surface waters, soils, solids, and sediments.

METHOD SUMMARY This method is used to determine the concentration of 14 chlorinated hydrocarbons. It provides gas chromatographic conditions for the detection of ppb concentrations of certain chlorinated hydrocarbons. Prior to use of this method, appropriate sample extraction techniques must be used. Both neat and diluted organic liquids (EPA Method 3580, Waste Dilution) may be analyzed by direct injection. A 2 to 5 µg/mL aliquot of the extract is injected into a gas chromatograph (GC) using the solvent flush technique, and compounds in the GC effluent are detected by an electron capture detector (ECD).

INTERFERENCES Solvents, reagents, glassware, and other sample processing hardware may yield discrete artifacts and/or elevated baselines causing misinterpretation of gas chromatograms. Interferences coextracted from samples will vary considerably from source to source, depending upon the waste being sampled.

INSTRUMENTATION An analytical system complete with GC suitable for on-column injections and accessories, including detectors, column supplies, recorder, gases and syringes is required. A data system for measuring peak areas and/or peak heights is recommended. The GC is equipped with an electron capture detector (ECD). A K-D apparatus is needed for sample preparation.

Column 1: 1.8 m × 2 mm I.D. glass column packed with 1% SP-1000 on Supelcoport (100/120 mesh) or equivalent.

Column 2: 1.8 m × 2 mm I.D. glass column packed with 1.5% OV-1/2.4% OV-225 on Supelcoport (80/100 mesh) or equivalent.

PRECISION & ACCURACY The method was tested by 20 laboratories using organic-free reagent water, drinking water, surface water, and three industrial wastewaters spiked at six concentrations over the range 1.0 to 356 µg/L. Single operator precision, overall precision, and method accuracy were found to be directly related to the concentration of the parameter and essentially independent of the sample matrix.

MULTIPLICATION FACTORS FOR OTHER MATRICES (a)

Matrix	Factor (b)
Groundwater	10
Low-concentration soil by ultrasonic extraction with GPC cleanup	670
High-concentration soil and sludges by ultrasonic extraction	10,000
Non-water miscible waste	100,000

(a) Sample EQLs are highly matrix-dependent. The EQLs listed are provided for guidance and may not always be achievable.
(b) EQL = [Method detection limit] × [Factor]. For non-aqueous samples, the factor is on a wet-weight basis.

PRECISION & ACCURACY The estimates below are based upon the performance in a single lab.

The accuracy (in µg/L) as expected recovery for one or more measurements of a sample containing a concentration of C was 0.85C-0.70.

The precision (in µg/L) as expected single analyst standard deviation of measurements at an average concentration of x" was 0.22x"-2.95.

The precision (in µg/L) as expected interlaboratory standard deviation measurements at an average concentration found of x" was 0.41x"-3.92.

C = *True value for the concentration, in µg/L.*
$x"$ = *Average recovery found for measurements of samples containing a concentration of C, in µg/L.*

SAMPLE COLLECTION, PRESERVATION & HANDLING
Extracts must be stored under refrigeration at 4°C and analyzed within 40 days of extraction.

SAMPLE PREPARATION In general, water samples are extracted at a neutral, or as is, pH with methylene chloride using either EPA Method 3510 or EPA Method 3520. Solid samples are extracted using either EPA Method 3540 or EPA Method 3550. Prior to gas chromatographic analysis, the extraction solvent must be exchanged to hexane.

QUALITY CONTROL The quality control check concentrate (EPA Method 8000) should contain each parameter of interest in acetone at the following concentrations: hexachloro-substituted hydrocarbon, 10 µg/mL; and any other chlorinated hydrocarbon, 100 µg/mL. Calculate surrogate standard recovery on all samples, blanks, and spikes.

Prepare stock standard solutions in isooctane or hexane. Calibration standards at a minimum of five concentrations should be prepared through dilution of the stock standards with isooctane or hexane. Internal standards and surrogate standards are also needed.

REFERENCE Test Methods for Evaluating Solid Waste, Physical/Chemical Methods, SW-846, 3rd Edition, U.S. EPA, Office of Solid Waste, Washington, DC, 1990. EPA Method 8120 A Rev. 1, Nov. 1990.

1,3-Dichlorobenzene   **EPA Method 8120**
CAS #541-73-1

TITLE Chlorinated Hydrocarbons by Gas Chromatography

MATRIX This method covers aqueous and solid matrices. This includes a wide variety such as drinking water, groundwater, industrial wastewaters, surface waters, soils, solids, and sediments.

METHOD SUMMARY This method is used to determine the concentration of 14 chlorinated hydrocarbons. It provides gas chromatographic conditions for the detection of ppb concentrations of certain chlorinated hydrocarbons. Prior to use of this method, appropriate sample extraction techniques must be used. Both neat and diluted organic liquids (EPA Method 3580, Waste Dilution) may be analyzed by direct injection. A 2 to 5 µg/mL aliquot of the extract is injected into a gas chromatograph (GC) using the solvent flush technique, and compounds in the GC effluent are detected by an electron capture detector (ECD).

INTERFERENCES Solvents, reagents, glassware, and other sample processing hardware may yield discrete artifacts and/or elevated baselines causing misinterpretation of gas chromatograms. Interferences coextracted from samples will vary considerably from source to source, depending upon the waste being sampled.

INSTRUMENTATION An analytical system complete with GC suitable for on-column injections and accessories, including detectors, column supplies, recorder, gases and syringes is required. A data system for measuring peak areas and/or peak heights is recommended. The GC is equipped with an electron capture detector (ECD). A K-D apparatus is needed for sample preparation.

Column 1: 1.8 m × 2 mm I.D. glass column packed with 1% SP-1000 on Supelcoport (100/120 mesh) or equivalent.

Column 2: 1.8 m × 2 mm I.D. glass column packed with 1.5% OV-1/2.4% OV-225 on Supelcoport (80/100 mesh) or equivalent.

PRECISION & ACCURACY The method was tested by 20 laboratories using organic-free reagent water, drinking water, surface water, and three industrial wastewaters spiked at six concentrations over the range 1.0 to 356 µg/L. Single operator precision, overall precision, and method accuracy were found to be directly related to the concentration of the parameter and essentially independent of the sample matrix.

MULTIPLICATION FACTORS FOR OTHER MATRICES (a)

Matrix	Factor (b)
Groundwater	10
Low-concentration soil by ultrasonic extraction with GPC cleanup	670
High-concentration soil and sludges by ultrasonic extraction	10,000
Non-water miscible waste	100,000

(a) Sample EQLs are highly matrix-dependent. The EQLs listed are provided for guidance and may not always be achievable.
(b) EQL = [Method detection limit] × [Factor]. For non-aqueous samples, the factor is on a wet-weight basis.

PRECISION & ACCURACY The estimates below are based upon the performance in a single lab.

The accuracy (in µg/L) as expected recovery for one or more measurements of a sample containing a concentration of C was $0.72C + 0.87$.

The precision (in µg/L) as expected single analyst standard deviation of measurements at an average concentration of x'' was $0.21x'' - 1.03$.

The precision (in µg/L) as expected interlaboratory standard deviation measurements at an average concentration found of x'' was $0.49x'' - 3.98$.

C = True value for the concentration, in µg/L.
x'' = Average recovery found for measurements of samples containing a concentration of C, in µg/L.

SAMPLE COLLECTION, PRESERVATION & HANDLING Extracts must be stored under refrigeration at 4°C and analyzed within 40 days of extraction.

SAMPLE PREPARATION In general, water samples are extracted at a neutral, or as is, pH with methylene chloride using either EPA Method 3510 or EPA Method 3520. Solid samples are extracted using either EPA Method 3540 or EPA Method 3550. Prior to gas chromatographic analysis, the extraction solvent must be exchanged to hexane.

QUALITY CONTROL The quality control check concentrate (EPA Method 8000) should contain each parameter of interest in acetone at the following concentrations: hexachloro-substituted hydrocarbon, 10 µg/mL; and any other chlorinated hydrocarbon, 100 µg/mL. Calculate surrogate standard recovery on all samples, blanks, and spikes.

Prepare stock standard solutions in isooctane or hexane. Calibration standards at a minimum of five concentrations should be prepared through dilution of the stock standards with isooctane or hexane. Internal standards and surrogate standards are also needed.

REFERENCE Test Methods for Evaluating Solid Waste, Physical/Chemical Methods, SW-846, 3rd Edition, U.S. EPA, Office of Solid Waste, Washington, DC, 1990. EPA Method 8120 A Rev. 1, Nov. 1990.

1,4-Dichlorobenzene   **EPA Method 8120**
CAS #106-46-7

TITLE Chlorinated Hydrocarbons by Gas Chromatography

MATRIX This method covers aqueous and solid matrices. This includes a wide variety such as drinking water, groundwater, industrial wastewaters, surface waters, soils, solids, and sediments.

METHOD SUMMARY This method is used to determine the concentration of 14 chlorinated hydrocarbons. It provides gas chromatographic conditions for the detection of ppb concentrations of certain chlorinated hydrocarbons. Prior to use of this method, appropriate sample extraction techniques must be used. Both neat and diluted organic liquids (EPA Method 3580, Waste Dilution) may be analyzed by direct injection. A 2 to 5 µg/mL aliquot of the extract is injected into a gas chromatograph (GC) using the solvent flush technique, and compounds in the GC effluent are detected by an electron capture detector (ECD).

INTERFERENCES Solvents, reagents, glassware, and other sample processing hardware may yield discrete artifacts and/or elevated baselines causing misinterpretation of gas chromatograms. Interferences coextracted from samples will vary considerably from source to source, depending upon the waste being sampled.

INSTRUMENTATION An analytical system complete with GC suitable for on-column injections and accessories, including detectors, column supplies, recorder, gases and syringes is required. A data system for measuring peak areas and/or peak heights is recommended. The GC is equipped with an electron capture detector (ECD). A K-D apparatus is needed for sample preparation.

Column 1: 1.8 m × 2 mm I.D. glass column packed with 1% SP-1000 on Supelcoport (100/120 mesh) or equivalent.

Column 2: 1.8 m × 2 mm I.D. glass column packed with 1.5% OV-1/2.4% OV-225 on Supelcoport (80/100 mesh) or equivalent.

PRECISION & ACCURACY The method was tested by 20 laboratories using organic-free reagent water, drinking water, surface water, and three industrial wastewaters spiked at six concentrations over the range 1.0 to 356 µg/L. Single operator precision, overall precision, and method accuracy were found to be directly related to the concentration of the parameter and essentially independent of the sample matrix.

MULTIPLICATION FACTORS FOR OTHER MATRICES (a)

Matrix	Factor (b)
Groundwater	10
Low-concentration soil by ultrasonic extraction with GPC cleanup	670
High-concentration soil and sludges by ultrasonic extraction	10,000
Non-water miscible waste	100,000

(a) Sample EQLs are highly matrix-dependent. The EQLs listed are provided for guidance and may not always be achievable.
(b) EQL = [Method detection limit] × [Factor]. For non-aqueous samples, the factor is on a wet-weight basis.

PRECISION & ACCURACY The estimates below are based upon the performance in a single lab.

The accuracy (in µg/L) as expected recovery for one or more measurements of a sample containing a concentration of C was $0.72C+2.80$.

The precision (in µg/L) as expected single analyst standard deviation of measurements at an average concentration of x" was $0.16x"-0.48$.

The precision (in µg/L) as expected interlaboratory standard deviation measurements at an average concentration found of x" was $0.35x"-0.57$.

C = *True value for the concentration, in µg/L.*
x''' = *Average recovery found for measurements of samples containing a concentration of C, in µg/L.*

SAMPLE COLLECTION, PRESERVATION & HANDLING Extracts must be stored under refrigeration at 4°C and analyzed within 40 days of extraction.

SAMPLE PREPARATION In general, water samples are extracted at a neutral, or as is, pH with methylene chloride using either EPA Method 3510 or EPA Method 3520. Solid samples are extracted using either EPA Method 3540 or EPA Method 3550. Prior to gas chromatographic analysis, the extraction solvent must be exchanged to hexane.

QUALITY CONTROL The quality control check concentrate (EPA Method 8000) should contain each parameter of interest in acetone at the following concentrations: hexachloro-substituted hydrocarbon, 10 µg/mL; and any other chlorinated hydrocarbon, 100 µg/mL. Calculate surrogate standard recovery on all samples, blanks, and spikes.

Prepare stock standard solutions in isooctane or hexane. Calibration standards at a minimum of five concentrations should be prepared through dilution of the stock standards with isooctane or hexane. Internal standards and surrogate standards are also needed.

REFERENCE Test Methods for Evaluating Solid Waste, Physical/Chemical Methods, SW-846, 3rd Edition, U.S. EPA, Office of Solid Waste, Washington, DC, 1990. EPA Method 8120 A Rev. 1, Nov. 1990.

1,2-Dichlorobenzene   **EPA Method 8121**
CAS #95-50-1

TITLE Chlorinated Hydrocarbons by GC: Capillary Column Technique

MATRIX This method covers aqueous and solid matrices. This includes a wide variety such as drinking water, groundwater, industrial wastewaters, surface waters, soils, solids, and sediments.

METHOD SUMMARY This method provides procedures for the determination of 22 chlorinated hydrocarbons in water, soil/sediment, and waste matrices. A measured volume or weight of sample is extracted by using one of the appropriate sample extraction techniques specified in EPA Method 3510, EPA Method 3520, EPA Method 3540, or EPA Method 3550, or diluted using EPA Method 3580. Aqueous samples are extracted at neutral pH with methylene chloride by using either a separatory funnel (EPA Method 3510) or a continuous liquid-liquid extractor (EPA Method 3520). Solid samples are extracted with hexane/acetone (1:1) by using a Soxhlet extractor (EPA Method 3540) or with methylene chloride/acetone (1:1) by using an ultrasonic extractor (EPA Method 3550). After cleanup, the extract or diluted sample is analyzed by gas chromatography with electron capture detection (GC/ECD).

The sensitivity level of this method usually depends on the level of interferences rather than on instrumental limitations. This method may be used in conjunction with EPA Method 3620, Florisil Column Cleanup, EPA Method 3660, Sulfur Cleanup, and EPA Method 3640, Gel Permeation Chromatography, to aid in the elimination of interferences.

INTERFERENCES Solvents, reagents, glassware, and other hardware used in sample processing may introduce artifacts which may result in elevated baselines, causing misinterpretation of gas chromatograms. Interferants coextracted from the samples will vary considerably from waste to waste. Glassware must be scrupulously clean. Phthalate esters, if present in a sample, will interfere only with the BHC isomers. The presence of elemental sulfur will result in large peaks, and can often mask the region of compounds eluting after 1,2,4,5-tetrachlorobenzene. The tetrabutylammonium (TBA)-sulfite procedure (EPA Method 3660) works well for the removal of elemental sulfur. Waxes and lipids can be removed by gel permeation chromatography (EPA Method 3640).

INSTRUMENTATION A GC suitable for on-column injections and all required accessories, including and electron capture detector (ECD), analytical columns, recorder, gases, and syringes are needed. A data system for measuring peak heights and/or peak areas is recommended. A Kuderna-Danish (K-D) apparatus will also be needed to concentrate extracts.

Column 1: 30 m × 0.53 mm I.D. fused-silica capillary column chemically bonded with trifluoropropyl methyl silicone (DB-210 or equivalent).

Column 2: 30 m × 0.53 mm I.D. fused-silica capillary column chemically bonded with polyethylene glycol (DB-WAX or equivalent).

PRECISION & ACCURACY This method has been tested in a single lab by using organic-free reagent water, sandy loam samples, and extracts which were spiked with the test compounds at one concentration. Single-operator precision and method accuracy were found to be related to the concentration of compound and the type of matrix. The accuracy and precision technique will be determined by the sample matrix, sample preparation technique, optional cleanup techniques, and calibration procedures used.

MULTIPLICATION FACTORS FOR OTHER MATRICES (a)

Matrix	Factor (b)
Groundwater	10
Low-concentration soil by ultrasonic extraction with GPC cleanup	670
High-concentration soil and sludges by ultrasonic extraction	10,000
Non-water miscible waste	100,000

(a) Sample EQLs are highly matrix-dependent. The EQLs listed herein are provided for guidance and may not always be achievable. (b) EQL = [Method detection limit] × [Factor]. For nonaqueous samples, the factor is on a wet-weight basis.

PRECISION & ACCURACY MDL is the method detection limit for organic-free reagent water. MDL was determined from the analysis of eight replicate aliquots processed through the entire analytical method (extraction, Florisil cartridge cleanup, and GC/ECD analysis).

The MDL (in ng/L) was 270.

The accuracy (as average % recovery using 5 determinations and no Florisil cleanup) from a spike concentration of 100 µg/L and separatory funnel extraction was 92% with a final volume of 10 mL.

The precision (as RSD% using 5 determinations and no Florisil cleanup) from a spike concentration of 100 µg/L and separatory funnel extraction was 5.7% with a final volume of 10 mL.

The accuracy (as average % recovery using 5 determinations and no Florisil cleanup), from a spike concentration of 33,000µg/L and ultrasonic extraction of solid samples using 1:1 methylene chloride and acetone, was 84% with a final volume of 10 mL.

The precision (as RSD% using 5 determinations and no Florisil cleanup), from a spike concentration of 33,000µg/L and ultrasonic extraction of solid samples using 1:1 methylene chloride and acetone, was 7.1% with a final volume of 10 mL.

SAMPLE COLLECTION, PRESERVATION & HANDLING
Volatile Organics — Standard 40-mL glass screw-cap VOA vials with Teflon®-faced silicone septum may be used for both liquid and solid matrices. When collecting samples, liquids and solids should be introduced into the vials gently to reduce agitation which might drive off volatile compounds. If there are any air bubbles present the sample must be retaken. The vials with solids should be tapped slightly as they are filled to try and eliminate as much free air space as possible. Two vials from each sampling location should be sealed in separate plastic bags to prevent cross-contamination between samples.

Semivolatile organics — Containers used to collect samples for the determination of semivolatile organic compounds should be soap and water washed followed by methanol (or isopropanol) rinsing. The sample containers should be of glass or Teflon® and have screw-top covers with Teflon® liners.

Preservation for volatile organics — No preservation is used with concentrated waste samples. With liquid samples containing no residual chlorine, 4 drops of concentrated hydrochloric acid are added and the samples are immediately cooled to 4°C. When liquid samples contain residual chlorine, they are treated as above and, in addition, 4 drops of 4% aqueous sodium thiosulfate are added to remove the residual chlorine. Soil, sediment, and sludge samples are only cooled to 4°C.

Preservation for semivolatile organics — No preservation is used with concentrated waste samples. With liquid samples containing no residual chlorine and with soil, sediment, and sludge samples, immediately cooling to 4°C is the only preservation used. When residual chlorine is present then 3 mL of 10% aqueous sodium sulfate is added for each gallon of sample collected, followed by cooling to 4°C.

Holding times — The holding time for all volatile organics samples is 14 days. Liquid samples must be extracted within 7 days and their extracts analyzed within 40 days. Concentrated waste, soil, sediment, and sludge samples must be extracted within 14 days and their extracts analyzed within 40 days.

SAMPLE PREPARATION Prepare stock standard solutions in hexane. Calibration standards at a minimum of five concentrations should be prepared through dilution of the stock standards with hexane. The suggested internal standards are: 2,5-dibromotoluene, 1,3,5-tribromobenzene, and α, α-dibromo-m-xylene. The analyst can use any of the three compounds provided that they are resolved from matrix interferences. Recommended surrogate compounds are α-2,6-trichlorotoluene, 1,4-dichloronaphthalene, and 2,3,4,5,6-pentachlorotoluene.

In general, water samples are extracted at a neutral pH with methylene chloride using a separatory funnel (EPA Method 3510) or a continuous liquid-liquid extractor (EPA Method 3520). Solid samples are extracted with hexane/acetone (1:1 v:v) using a Soxhlet extractor (EPA Method 3540) or with methylene chloride/acetone (1:1 v:v) using an ultrasonic extractor (EPA Method 3550). Non-aqueous waste samples may be diluted using EPA Method 3580. Prior to Florisil cleanup or gas chromatographic analysis, the extraction solvent must be exchanged to hexane. Sample extracts that will be subjected to gel permeation chromatography do not need solvent exchange.

Cleanup procedures may not be necessary for a relatively clean matrix. If removal of interferences such as chlorinated phenols, phthalate esters, etc., is required, proceed with the procedure outlined in EPA Method 3620.

QUALITY CONTROL Analyze a quality control check standard to demonstrate that the operation of the GC is in control. The frequency of the check standard analysis is equivalent to 10% of the samples analyzed. If the recovery of any compound found in the check standard is less than 80% of the certified value, the problem must be corrected and a new set of calibration standards must be prepared and analyzed. Calculate surrogate standard recoveries for all samples, blanks, and spikes. An internal standard peak area check must be performed on all samples. The internal standard must be evaluated for acceptance by determining whether the measured area for the internal standard deviates by more than 30% from the average area for the internal standard in the calibration standards. When the internal standard peak area is outside that limit, all samples that fall outside the QC criteria must be reanalyzed. Any compound confirmed by two columns may also be confirmed by GC/MS (EPA Method 8270). The GC/MS would normally require a minimum concentration of 1 ng/µL in the final extract for each compound. Include a mid-concentration calibration standard after each group of 20 samples in the analysis sequence. The response factors for the mid-concentration calibration must be within 15% of the average values for the multiconcentration calibration.

REFERENCE Test Methods for Evaluating Solid Waste, Physical/Chemical Methods, SW-846, 3rd Edition, U.S. EPA, Office of Solid Waste, Washington, DC, 1990. EPA Method 8121, Rev. 0, Nov. 1990.

1,3-Dichlorobenzene   **EPA Method 8121**
CAS #541-73-1

TITLE Chlorinated Hydrocarbons by GC: Capillary Column Technique

MATRIX This method covers aqueous and solid matrices. This includes a wide variety such as drinking water, groundwater, industrial wastewaters, surface waters, soils, solids, and sediments.

METHOD SUMMARY This method provides procedures for the determination of 22 chlorinated hydrocarbons in water, soil/sediment, and waste matrices. A measured volume or weight of sample is extracted by using one of the appropriate sample extraction techniques specified in EPA Method 3510, EPA Method 3520, EPA Method 3540, or EPA Method 3550, or diluted using EPA Method 3580. Aqueous samples are extracted at neutral pH with methylene chloride by using either a separatory funnel (EPA Method 3510) or a continuous liquid-liquid extractor (EPA Method 3520). Solid samples are extracted with hexane/acetone (1:1) by using a Soxhlet extractor (EPA Method 3540) or with methylene chloride/acetone (1:1) by using an ultrasonic extractor (EPA Method 3550). After cleanup, the extract or diluted sample is analyzed by gas chromatography with electron capture detection (GC/ECD).

The sensitivity level of this method usually depends on the level of interferences rather than on instrumental limitations. This method may be used in conjunction with EPA Method 3620, Florisil Column Cleanup, EPA Method 3660, Sulfur Cleanup, and EPA Method 3640, Gel Permeation Chromatography, to aid in the elimination of interferences.

INTERFERENCES Solvents, reagents, glassware, and other hardware used in sample processing may introduce artifacts which may result in elevated baselines, causing misinterpretation of gas chromatograms. Interferants coextracted from the samples will vary considerably from waste to waste. Glassware must be scrupulously clean. Phthalate esters, if present in a sample, will interfere only with the BHC isomers. The presence of elemental sulfur will result in large peaks, and can often mask the region of compounds eluting after 1,2,4,5-tetrachlorobenzene. The tetrabutylammonium (TBA)-sulfite procedure (EPA Method 3660) works well for the removal of elemental sulfur. Waxes and lipids can be removed by gel permeation chromatography (EPA Method 3640).

INSTRUMENTATION A GC suitable for on-column injections and all required accessories, including and electron capture detector (ECD), analytical columns, recorder, gases, and syringes are needed. A data system for measuring peak heights and/or peak areas is recommended. A Kuderna-Danish (K-D) apparatus will also be needed to concentrate extracts.

Column 1: 30 m × 0.53 mm I.D. fused-silica capillary column chemically bonded with trifluoropropyl methyl silicone (DB-210 or equivalent).
Column 2: 30 m × 0.53 mm I.D. fused-silica capillary column chemically bonded with polyethylene glycol (DB-WAX or equivalent).

PRECISION & ACCURACY This method has been tested in a single lab by using organic-free reagent water, sandy loam samples, and extracts which were spiked with the test compounds at one concentration. Single-operator precision and method accuracy were found to be related to the concentration of compound and the type of matrix. The accuracy and precision technique will be determined by the sample matrix, sample preparation technique, optional cleanup techniques, and calibration procedures used.

MULTIPLICATION FACTORS FOR OTHER MATRICES (a)

Matrix	Factor (b)
Groundwater	10
Low-concentration soil by ultrasonic extraction with GPC cleanup	670
High-concentration soil and sludges by ultrasonic extraction	10,000
Non-water miscible waste	100,000

(a) Sample EQLs are highly matrix-dependent. The EQLs listed herein are provided for guidance and may not always be achievable. (b) EQL = [Method detection limit] × [Factor]. For nonaqueous samples, the factor is on a wet-weight basis.

PRECISION & ACCURACY MDL is the method detection limit for organic-free reagent water. MDL was determined from the analysis of eight replicate aliquots processed through the entire analytical method (extraction, Florisil cartridge cleanup, and GC/ECD analysis).

The MDL (in ng/L) was 250.

The accuracy (as average % recovery using 5 determinations and no Florisil cleanup) from a spike concentration of 100 µg/L and separatory funnel extraction was 87% with a final volume of 10 mL.

The precision (as RSD% using 5 determinations and no Florisil cleanup) from a spike concentration of 100 µg/L and separatory funnel extraction was 8.7% with a final volume of 10 mL.

The accuracy (as average % recovery using 5 determinations and no Florisil cleanup), from a spike concentration of 33,000µg/L and ultrasonic extraction of solid samples using 1:1 methylene chloride and acetone, was 81% with a final volume of 10 mL.

The precision (as RSD% using 5 determinations and no Florisil cleanup), from a spike concentration of 33,000µg/L and ultrasonic extraction of solid samples using 1:1 methylene chloride and acetone, was 12.6% with a final volume of 10 mL.

SAMPLE COLLECTION, PRESERVATION & HANDLING
Volatile Organics — Standard 40-mL glass screw-cap VOA vials with Teflon®-faced silicone septum may be used for both liquid and solid matrices. When collecting samples, liquids and solids should be introduced into the vials gently to reduce agitation which might drive off volatile compounds. If there are any air bubbles present the sample must be retaken. The vials with solids should be tapped slightly as they are filled to try and eliminate as much free air space as possible. Two vials from each sampling location should be sealed in separate plastic bags to prevent cross-contamination between samples.

Semivolatile organics — Containers used to collect samples for the determination of semivolatile organic compounds should be soap and water washed followed by methanol (or isopropanol) rinsing. The sample containers should be of glass or Teflon® and have screw-top covers with Teflon® liners.

Preservation for volatile organics — No preservation is used with concentrated waste samples. With liquid samples containing no residual chlorine, 4 drops of concentrated hydrochloric acid are added and the samples are immediately cooled to 4°C. When liquid samples contain residual chlorine, they are treated as above and, in addition, 4 drops of 4% aqueous sodium thiosulfate are added to remove the residual chlorine. Soil, sediment, and sludge samples are only cooled to 4°C.

Preservation for semivolatile organics — No preservation is used with concentrated waste samples. With liquid samples containing no residual chlorine and with soil, sediment, and sludge samples, immediately cooling to 4°C is the only preservation used. When residual chlorine is present then 3 mL of 10% aqueous sodium sulfate is added for each gallon of sample collected, followed by cooling to 4°C.

Holding times — The holding time for all volatile organics samples is 14 days. Liquid samples must be extracted within 7 days and their extracts analyzed within 40 days. Concentrated waste, soil, sediment, and sludge samples must be extracted within 14 days and their extracts analyzed within 40 days.

SAMPLE PREPARATION Prepare stock standard solutions in hexane. Calibration standards at a minimum of five concentrations should be prepared through dilution of the stock standards with hexane. The suggested internal standards are: 2,5-dibromotoluene, 1,3,5-tribromobenzene, and α, α-dibromo-m-xylene. The analyst can use any of the three compounds provided that they are resolved from matrix interferences. Recommended surrogate compounds are α-2,6-trichlorotoluene, 1,4-dichloronaphthalene, and 2,3,4,5,6-pentachlorotoluene.

In general, water samples are extracted at a neutral pH with methylene chloride using a separatory funnel (EPA Method 3510) or a continuous liquid-liquid extractor (EPA Method 3520). Solid samples are extracted with hexane/acetone (1:1 v:v) using a Soxhlet extractor (EPA Method 3540) or with methylene chloride/acetone (1:1 v:v) using an ultrasonic extractor (EPA Method 3550). Non-aqueous waste samples may be diluted using EPA Method 3580. Prior to Florisil cleanup or gas chromatographic analysis, the extraction solvent must be exchanged to hexane. Sample extracts that will be subjected to gel permeation chromatography do not need solvent exchange.

Cleanup procedures may not be necessary for a relatively clean matrix. If removal of interferences such as chlorinated phenols, phthalate esters, etc., is required, proceed with the procedure outlined in EPA Method 3620.

QUALITY CONTROL Analyze a quality control check standard to demonstrate that the operation of the GC is in control. The frequency of the check standard analysis is equivalent to 10% of the samples analyzed. If the recovery of any compound found in the check standard is less than 80% of the certified value, the problem must be corrected and a new set of calibration standards must be prepared and analyzed. Calculate surrogate standard recoveries for all samples, blanks, and spikes. An internal standard peak area check must be performed on all samples. The internal standard must be evaluated for acceptance by determining whether the measured area for the internal standard deviates by more than 30% from the average area for the internal standard in the calibration standards. When the internal standard peak area is outside that limit, all samples that fall outside the QC criteria must be reanalyzed. Any compound confirmed by two columns may also be confirmed by GC/MS (EPA Method 8270). The GC/MS would normally require a minimum concentration of 1 ng/µL in the final extract for each compound. Include a mid-concentration calibration standard after each group of 20 samples in the analysis sequence. The response factors for the mid-concentration calibration must be within 15% of the average values for the multiconcentration calibration.

REFERENCE Test Methods for Evaluating Solid Waste, Physical/Chemical Methods, SW-846, 3rd Edition, U.S. EPA, Office of Solid Waste, Washington, DC, 1990. EPA Method 8121, Rev. 0, Nov. 1990.

1,4-Dichlorobenzene EPA Method 8121
CAS #106-46-7

TITLE Chlorinated Hydrocarbons by GC: Capillary Column Technique

MATRIX This method covers aqueous and solid matrices. This includes a wide variety such as drinking water, groundwater, industrial wastewaters, surface waters, soils, solids, and sediments.

METHOD SUMMARY This method provides procedures for the determination of 22 chlorinated hydrocarbons in water, soil/sediment, and waste matrices. A measured volume or weight of sample is extracted by using one of the appropriate sample extraction techniques specified in EPA Method 3510, EPA Method 3520, EPA Method 3540, or EPA Method 3550, or diluted using EPA Method 3580. Aqueous samples are extracted at neutral pH with methylene chloride by using either a separatory funnel (EPA Method 3510) or a continuous liquid-liquid extractor (EPA Method 3520). Solid samples are extracted with hexane/acetone (1:1) by using a Soxhlet extractor (EPA Method 3540) or with methylene chloride/acetone (1:1) by using an ultrasonic extractor (EPA Method 3550). After cleanup, the extract or diluted sample is analyzed by gas chromatography with electron capture detection (GC/ECD).

The sensitivity level of this method usually depends on the level of interferences rather than on instrumental limitations. This method may be used in conjunction with EPA Method 3620, Florisil Column Cleanup, EPA Method 3660, Sulfur Cleanup, and EPA Method 3640, Gel Permeation Chromatography, to aid in the elimination of interferences.

INTERFERENCES Solvents, reagents, glassware, and other hardware used in sample processing may introduce artifacts which may result in elevated baselines, causing misinterpretation of gas chromatograms. Interferants coextracted from the samples will vary considerably from waste to waste. Glassware must be scrupulously clean. Phthalate esters, if present in a sample, will interfere only with the BHC isomers. The presence of elemental sulfur will result in large peaks, and can often mask the region of compounds eluting after 1,2,4,5-tetrachlorobenzene. The tetrabutylammonium (TBA)-sulfite procedure (EPA Method 3660) works well for the removal of elemental sulfur. Waxes and lipids can be removed by gel permeation chromatography (EPA Method 3640).

INSTRUMENTATION A GC suitable for on-column injections and all required accessories, including and electron capture detector (ECD), analytical columns, recorder, gases, and syringes are needed. A data system for measuring peak heights and/or peak areas is recommended. A Kuderna-Danish (K-D) apparatus will also be needed to concentrate extracts.

Column 1: 30 m × 0.53 mm I.D. fused-silica capillary column chemically bonded with trifluoropropyl methyl silicone (DB-210 or equivalent).

Column 2: 30 m × 0.53 mm I.D. fused-silica capillary column chemically bonded with polyethylene glycol (DB-WAX or equivalent).

PRECISION & ACCURACY This method has been tested in a single lab by using organic-free reagent water, sandy loam samples, and extracts which were spiked with the test compounds at one concentration. Single-operator precision and method accuracy were found to be related to the concentration of compound and the type of matrix. The accuracy and precision technique will be determined by the sample matrix, sample preparation technique, optional cleanup techniques, and calibration procedures used.

MULTIPLICATION FACTORS FOR OTHER MATRICES (a)

Matrix	Factor (b)
Groundwater	10
Low-concentration soil by ultrasonic extraction with GPC cleanup	670
High-concentration soil and sludges by ultrasonic extraction	10,000
Non-water miscible waste	100,000

(a) Sample EQLs are highly matrix-dependent. The EQLs listed herein are provided for guidance and may not always be achievable. (b) EQL = [Method detection limit] × [Factor]. For nonaqueous samples, the factor is on a wet-weight basis.

PRECISION & ACCURACY MDL is the method detection limit for organic-free reagent water. MDL was determined from the analysis of eight replicate aliquots processed through the entire analytical method (extraction, Florisil cartridge cleanup, and GC/ECD analysis).

The MDL (in ng/L) was 890.

The accuracy (as average % recovery using 5 determinations and no Florisil cleanup) from a spike concentration of 100 µg/L and separatory funnel extraction was 89% with a final volume of 10 mL.

The precision (as RSD% using 5 determinations and no Florisil cleanup) from a spike concentration of 100 µg/L and separatory funnel extraction was 8.9% with a final volume of 10 mL.

The accuracy (as average % recovery using 5 determinations and no Florisil cleanup), from a spike concentration of 33,000µg/L and ultrasonic extraction of solid samples using 1:1 methylene chloride and acetone, was 89% with a final volume of 10 mL.

The precision (as RSD% using 5 determinations and no Florisil cleanup), from a spike concentration of 33,000µg/L and ultrasonic extraction of solid samples using 1:1 methylene chloride and acetone, was 11.0% with a final volume of 10 mL.

SAMPLE COLLECTION, PRESERVATION & HANDLING
Volatile Organics — Standard 40-mL glass screw-cap VOA vials with Teflon®-faced silicone septum may be used for both liquid and solid matrices. When collecting samples, liquids and solids should be introduced into the vials gently to reduce agitation which might drive off volatile compounds. If there are any air bubbles present the sample must be retaken. The vials with solids should be tapped slightly as they are filled to try and eliminate as much free air space as possible. Two vials from each sampling location should be sealed in separate plastic bags to prevent cross-contamination between samples.

Semivolatile organics — Containers used to collect samples for the determination of semivolatile organic compounds should be soap and water washed followed by methanol (or isopropanol) rinsing. The sample containers should be of glass or Teflon® and have screw-top covers with Teflon® liners.

Preservation for volatile organics — No preservation is used with concentrated waste samples. With liquid samples containing no residual chlorine, 4 drops of concentrated hydrochloric acid are added and the samples are immediately cooled to 4°C. When liquid samples contain residual chlorine, they are treated as above and, in addition, 4 drops of 4% aqueous sodium thiosulfate are added to remove the residual chlorine. Soil, sediment, and sludge samples are only cooled to 4°C.

Preservation for semivolatile organics — No preservation is used with concentrated waste samples. With liquid samples containing no residual chlorine and with soil, sediment, and sludge samples, immediately cooling to 4°C is the only preservation used. When residual chlorine is present then 3 mL of 10% aqueous sodium sulfate is added for each gallon of sample collected, followed by cooling to 4°C.

Holding times — The holding time for all volatile organics samples is 14 days. Liquid samples must be extracted within 7 days and their extracts analyzed within 40 days. Concentrated waste, soil, sediment, and sludge samples must be extracted within 14 days and their extracts analyzed within 40 days.

SAMPLE PREPARATION Prepare stock standard solutions in hexane. Calibration standards at a minimum of five concentrations should be prepared through dilution of the stock standards with hexane. The suggested internal standards are: 2,5-dibromotoluene, 1,3,5-tribromobenzene, and α,α-dibromo-m-xylene. The analyst can use any of the three compounds provided that they are resolved from matrix interferences. Recommended surrogate compounds are α-2,6-trichlorotoluene, 1,4-dichloronaphthalene, and 2,3,4,5,6-pentachlorotoluene.

In general, water samples are extracted at a neutral pH with methylene chloride using a separatory funnel (EPA Method 3510) or a continuous liquid-liquid extractor (EPA Method 3520). Solid samples are extracted with hexane/acetone (1:1 v:v) using a Soxhlet extractor (EPA Method 3540) or with methylene chloride/acetone (1:1 v:v) using an ultrasonic extractor (EPA Method 3550). Non-aqueous waste samples may be diluted using EPA Method 3580. Prior to Florisil cleanup or gas chromatographic analysis, the extraction solvent must be exchanged to hexane. Sample extracts that will be subjected to gel permeation chromatography do not need solvent exchange.

Cleanup procedures may not be necessary for a relatively clean matrix. If removal of interferences such as chlorinated phenols, phthalate esters, etc., is required, proceed with the procedure outlined in EPA Method 3620.

QUALITY CONTROL Analyze a quality control check standard to demonstrate that the operation of the GC is in control. The frequency of the check standard analysis is equivalent to 10% of the samples analyzed. If the recovery of any compound found in the check standard is less than 80% of the certified value, the problem must be corrected and a new set of calibration standards must be prepared and analyzed. Calculate surrogate standard recoveries for all samples, blanks, and spikes. An internal standard peak area check must be performed on all samples. The internal standard must be evaluated for acceptance by determining whether the measured area for the internal standard deviates by more than 30% from the average area for the internal standard in the calibration standards. When the internal standard peak area is outside that limit, all samples that fall outside the QC criteria must be reanalyzed. Any compound confirmed by two columns may also be confirmed by GC/MS (EPA Method 8270). The GC/MS would normally require a minimum concentration of 1 ng/µL in the final extract for each compound. Include a mid-concentration calibration standard after each group of 20 samples in the analysis sequence. The response factors for the mid-concentration calibration must be within 15% of the average values for the multiconcentration calibration.

REFERENCE Test Methods for Evaluating Solid Waste, Physical/Chemical Methods, SW-846, 3rd Edition, U.S. EPA, Office of Solid Waste, Washington, DC, 1990. EPA Method 8121, Rev. 0, Nov. 1990.

1,2-Dichlorobenzene   **EPA Method 8240**
CAS #95-50-1

TITLE Volatile Organics By GC/MS: Packed Column Technique

MATRIX Nearly all types of sample matarices, regardless of water content, can be analyzed using this method. This includes groundwater, aqueous sludges, caustic liquors, acid liquors, waste solvents, oily wastes, mousses, tars, fibrous wastes, polymetric emulsions, filter cakes, spent carbons, spent catalysts, soils, and sediments.

METHOD SUMMARY Method 8240B covers 80 volatile organic compounds that are introduced into a gas chromatograph by the purge-and-trap method or by direct injection (in limited applications). For the purge-and-trap method an inert gas (zero grade nitrogen or helium) is bubbled through a 5-mL solution at ambient temperature. Purged sample components are trapped in a tube of sorbent materials. When purging is complete, the sorbent tube is heated and backflushed with inert gas to desorb the trapped components onto a GC column.

INTERFERENCES Impurities in the purge gas and from organic compounds outgassing from the plumbing ahead of the trap account for many contamination problems. Interferences purged or coextracted from the samples will vary considerably from source to source. Cross-contamination can occur whenever high-level and low-level samples are analyzed sequentially. Whenever an unusually concentrated sample is analyzed, it should be followed by the analysis of organic-free reagent water to check for cross-contamination. Samples also can be contaminated by diffusion of volatile organics (particularly methylene chloride and fluorocarbons) through the septum seal into the sample during shipment and storage. A trip blank can serve as a check on such contamination. The lab where volatile analysis is performed and also the refrigerated storage area should be completely free of solvents.

INSTRUMENTATION A gas chromatograph/mass spectrometry/data system (GC/MS) equipped with a 6 ft × 0.1 in I.D. glass column packed with 1% SP-1000 on Carbopack-B (60/80 mesh) is required. Also needed is a 5-mL purging device, a sorbent trap, and a thermal desorption apparatus.

PRECISION & ACCURACY This method is reported to have been tested by 15 laboratories using organic-free reagent water, drinking water, surface water, and industrial wastewaters (not specified) fortified at six concentrations over the range 5–600 µg/L.

Sample estimated quantitation limits (EQLs) are highly matrix-dependent. The EQLs listed may not always be achievable. EQLs listed for soils or sediments are based on wet weight. Normally, data is reported on a dry-weight basis; therefore, EQLs will be higher, based on the percent dry weight of each sample. Note that EQLs are even more variable than MDLs and that they are highly variable depending on the matrix being analyzed.

EQL in groundwater in µg/L was not listed.
EQL in low soil or sediment in µg/kg was not listed.
Accuracy (a) in µg/L was 0.94C+4.47.
Precision (b) in µg/L was 0.30x-1.20.

(a) *Average recovery found for measurements of samples containing a concentration of C, in µg/L.*
(b) *Overall precision found for measurements of samples with average recovery X for samples containing a concentration of C in µg/L.*

X = Average recovery found for measurement of samples containing a concentration of C in µg/L.

MULTIPLICATION FACTORS FOR OTHER MATRICES

Other Matrices	Factor (a)
Waste miscible liquid waste	50
High-concentration soil and sludge	125
Non-water miscible waste	500

(a) *EQL = [EQL for low soil sediment] × [Factor]. For non-aqueous samples, the factor is on a wet-weight basis.*

SAMPLING METHOD
Liquid samples — Use a 40-mL glass screw-cap VOA vial with a Teflon®-faced silicone septum that has been prewashed, rinsed with distilled deionized water, and oven dried. However, if residual chlorine is present, collect sample in a 40-oz. soil VOA container which has been pre-preserved with 4 drops of 10% sodium thiosulfate, mix gently, and then transfer the sample to a 40-mL VOA vial. Collect bubble-free samples in duplicate and seal them in separate plastic bags.

Soils or sediments, and sludges — Use an 8-oz. widemouth glass bottle with a Teflon®-faced silicone septum that has been prewashed with detergent, rinsed with distilled deionized water, and oven dried. Tap slightly to eliminate free air space. Collect samples in duplicate and seal them in separate plastic bags.

SAMPLE PRESERVATION
Liquid samples — Add 4 drops of concentrated HCL and immediately cool samples to 4°C and store in a solvent-free refrigerator.

Soils or sediments, and sludges — Cool samples to 4°C and store in a solvent-free refrigerator.

MHT Maximum holding time is 14 days from the date of sample collection.

SAMPLE PREPARATION
Liquid samples — Remove the plunger from a 5-mL syringe and carefully pour the sample into the syringe barrel to just short of overflowing. Replace the syringe plunger and compress the sample. Open the syringe valve and vent any residual air while adjusting the sample volume to 5.0 mL. If there is only one volatile organic analysis (VOA) vial, a second syringe should be filled at this time to protect against possible loss of sample integrity. Add 10 µL of surrogate spiking solution and 10 µL of internal standard spiking solution through the valve bore of the 5-mL syringe, then close the valve. The surrogate and internal standards may be mixed and added as a single spiking solution.

Sediments, soils, and waste samples — All samples of this type should be screened by GC analysis using a headspace method (EPA Method 3810) or the hexadecane extraction and screening method (EPA Method 3820). Use the screening data to determine whether to use the low-concentration method (0.005–1 mg/kg) or the high-concentration method (>1 mg/kg).

Low-concentration method — The low-concentration method is based on purging a heated sediment or soil sample mixed with organic-free reagent water containing the surrogate and internal standards. Analyze all reagent blanks and standards under the same conditions as the samples.

Use a 5-g sample if the expected concentration is <0.1 mg/kg or a 1-g sample for expected concentrations between 0.1 and 1 mg/kg. Mix the contents of the sample container with a narrow metal spatula. Weigh the amount of the sample into a tared purge device. Add the spiked water to the purge device, which contains the weighed amount of sample, and connect the device to the purge-and-trap system.

High-concentration method — This method is based on extracting the sediment or soil with methanol. A waste sample is either extracted or diluted, depending on its solubility in methanol. Wastes that are insoluble in methanol are diluted with reagent tetraglyme or possibly polyethylene glycol (PEG). An aliquot of the extract is added to organic-free reagent water containing surrogate and internal standards. This is purged at ambient temperature. All samples with an expected concentration of >1.0 mg/kg should be analyzed by this method.

Mix the contents of the sample container with a narrow metal spatula. For sediments or soils and solid wastes that are insoluble in methanol, weigh 4 g (wet weight) of sample into a tared 20-mL vial. For waste that is soluble in methanol, tetraglyme, or PEG, weigh 1 g (wet weight) into a tared scintillation vial or culture tube or a 10-mL volumetric flask. Quickly add 9.0 mL of appropriate solvent then add 1.0 mL of a surrogate spiking solution to the vial, cap it, and shake it for 2 min.

METHANOL EXTRACT REQUIRED FOR ANALYSIS OF HIGH-CONCENTRATION SOILS OR SEDIMENTS

Approximate Concentration Range	Volume of Methanol Extract (a)
500–10,000 µg/kg	100 µL
1,000–20,000 µg/kg	50 µL
5,000–100,000 µg/kg	10 µL
25,000–500,000 µg/kg	100 µL of 1/50 dilution (b)

Calculate appropriate dilution factor for concentrations exceeding this table.

(a) The volume of methanol added to 5 mL of water being purged should be kept constant. Therefore, add to the 5-mL syringe whatever volume of methanol is necessary to maintain a volume of 100 µL added to the syringe.
(b) Dilute an aliquot of the methanol extract and then take 100 µL for analysis.

QUALITY CONTROL Demonstrate, through the analysis of a reagent water blank, that interferences from the analytical system, glassware, and reagents are under control. Blank samples should be carried through all stages of the sample preparation and measurement steps. For each analytical batch (up to 20 samples), a reagent blank, matrix spike, and matrix spike duplicate must be analyzed (the frequency of the spikes may be different for different monitoring programs). The blank and spiked samples must be carried through all stages of the sample preparation and measurement steps. QC samples mentioned in the section on Interferences will also be needed as appropriate to those situations.

REFERENCE Test Methods for Evaluating Solid Waste (SW-846). U.S. EPA. 1983. Method 8240B, Rev. 2, Nov. 1990. Office of Solid Wastes, Washington, DC.

1,3-Dichlorobenzene   **EPA Method 8240**
CAS #541-73-1

TITLE Volatile Organics By GC/MS: Packed Column Technique

MATRIX Nearly all types of sample matarices, regardless of water content, can be analyzed using this method. This includes groundwater, aqueous sludges, caustic liquors, acid liquors, waste solvents, oily wastes, mousses, tars, fibrous wastes, polymetric emulsions, filter cakes, spent carbons, spent catalysts, soils, and sediments.

METHOD SUMMARY Method 8240B covers 80 volatile organic compounds that are introduced into a gas chromatograph by the purge-and-trap method or by direct injection (in limited applications). For the purge-and-trap method an inert gas (zero grade nitrogen or helium) is bubbled through a 5-mL solution at ambient temperature. Purged sample components are trapped in a tube of sorbent materials. When purging is complete, the sorbent tube is heated and backflushed with inert gas to desorb the trapped components onto a GC column.

INTERFERENCES Impurities in the purge gas and from organic compounds outgassing from the plumbing ahead of the trap account for many contamination problems. Interferences purged or coextracted from the samples will vary considerably from source to source. Cross-contamination can occur whenever high-level and low-level samples are analyzed sequentially. Whenever an unusually concentrated sample is analyzed, it should be followed by the analysis of organic-free reagent water to check for cross-contamination. Samples also can be contaminated by diffusion of volatile organics (particularly methylene chloride and fluorocarbons) through the septum seal into the sample during shipment and storage. A trip blank can serve as a check on such contamination. The lab where volatile analysis is performed and also the refrigerated storage area should be completely free of solvents.

INSTRUMENTATION A gas chromatograph/mass spectrometry/data system (GC/MS) equipped with a 6 ft × 0.1 in I.D. glass column packed with 1% SP-1000 on Carbopack-B (60/80 mesh) is required. Also needed is a 5-mL purging device, a sorbent trap, and a thermal desorption apparatus.

PRECISION & ACCURACY This method is reported to have been tested by 15 laboratories using organic-free reagent water, drinking water, surface water, and industrial wastewaters (not specified) fortified at six concentrations over the range 5–600 µg/L.

Sample estimated quantitation limits (EQLs) are highly matrix-dependent. The EQLs listed may not always be achievable. EQLs listed for soils or sediments are based on wet weight. Normally, data is reported on a dry-weight basis; therefore, EQLs will be higher, based on the percent dry weight of each sample. Note that EQLs are even more variable than MDLs and that they are highly variable depending on the matrix being analyzed.

EQL in groundwater in µg/L was not listed.
EQL in low soil or sediment in µg/kg was not listed.
Accuracy (a) in µg/L was 1.06C+1.68.
Precision (b) in µg/L was 0.18x-0.82.

(a) Average recovery found for measurements of samples containing a concentration of C, in µg/L.

(b) *Overall precision found for measurements of samples with average recovery X for samples containing a concentration of C in µg/L.*

X = *Average recovery found for measurement of samples containing a concentration of C in µg/L.*

MULTIPLICATION FACTORS FOR OTHER MATRICES

Other Matrices	Factor (a)
Waste miscible liquid waste	50
High-concentration soil and sludge	125
Non-water miscible waste	500

(a) EQL = [EQL for low soil sediment] × [Factor]. For non-aqueous samples, the factor is on a wet-weight basis.

SAMPLING METHOD

Liquid samples — Use a 40-mL glass screw-cap VOA vial with a Teflon®-faced silicone septum that has been prewashed, rinsed with distilled deionized water, and oven dried. However, if residual chlorine is present, collect sample in a 40-oz. soil VOA container which has been pre-preserved with 4 drops of 10% sodium thiosulfate, mix gently, and then transfer the sample to a 40-mL VOA vial. Collect bubble-free samples in duplicate and seal them in separate plastic bags.

Soils or sediments, and sludges — Use an 8-oz. widemouth glass bottle with a Teflon®-faced silicone septum that has been prewashed with detergent, rinsed with distilled deionized water, and oven dried. Tap slightly to eliminate free air space. Collect samples in duplicate and seal them in separate plastic bags.

SAMPLE PRESERVATION

Liquid samples — Add 4 drops of concentrated HCL and immediately cool samples to 4°C and store in a solvent-free refrigerator.

Soils or sediments, and sludges — Cool samples to 4°C and store in a solvent-free refrigerator.

MHT Maximum holding time is 14 days from the date of sample collection.

SAMPLE PREPARATION

Liquid samples — Remove the plunger from a 5-mL syringe and carefully pour the sample into the syringe barrel to just short of overflowing. Replace the syringe plunger and compress the sample. Open the syringe valve and vent any residual air while adjusting the sample volume to 5.0 mL. If there is only one volatile organic analysis (VOA) vial, a second syringe should be filled at this time to protect against possible loss of sample integrity. Add 10 µL of surrogate spiking solution and 10 µL of internal standard spiking solution through the valve bore of the 5-mL syringe, then close the valve. The surrogate and internal standards may be mixed and added as a single spiking solution.

Sediments, soils, and waste samples — All samples of this type should be screened by GC analysis using a headspace method (EPA Method 3810) or the hexadecane extraction and screening method (EPA Method 3820). Use the screening data to determine whether to use the low-concentration method (0.005–1 mg/kg) or the high-concentration method (>1 mg/kg).

Low-concentration method — The low-concentration method is based on purging a heated sediment or soil sample mixed with organic-free reagent water containing the surrogate and internal standards. Analyze all reagent blanks and standards under the same conditions as the samples.

Use a 5-g sample if the expected concentration is <0.1 mg/kg or a 1-g sample for expected concentrations between 0.1 and 1 mg/kg. Mix the contents of the sample container with a narrow metal spatula. Weigh the amount of the sample into a tared purge device. Add the spiked water to the purge device, which contains the weighed amount of sample, and connect the device to the purge-and-trap system.

High-concentration method — This method is based on extracting the sediment or soil with methanol. A waste sample is either extracted or diluted, depending on its solubility in methanol. Wastes that are insoluble in methanol are diluted with reagent tetraglyme or possibly polyethylene glycol (PEG). An aliquot of the extract is added to organic-free reagent water containing surrogate and internal standards. This is purged at ambient temperature. All samples with an expected concentration of >1.0 mg/kg should be analyzed by this method.

Mix the contents of the sample container with a narrow metal spatula. For sediments or soils and solid wastes that are insoluble in methanol, weigh 4 g (wet weight) of sample into a tared 20-mL vial. For waste that is soluble in methanol, tetraglyme, or PEG, weigh 1 g (wet weight) into a tared scintillation vial or culture tube or a 10-mL volumetric flask. Quickly add 9.0 mL of appropriate solvent then add 1.0 mL of a surrogate spiking solution to the vial, cap it, and shake it for 2 min.

METHANOL EXTRACT REQUIRED FOR ANALYSIS OF HIGH-CONCENTRATION SOILS OR SEDIMENTS

Approximate Concentration Range	Volume of Methanol Extract (a)
500–10,000 µg/kg	100 µL
1,000–20,000 µg/kg	50 µL
5,000–100,000 µg/kg	10 µL
25,000–500,000 µg/kg	100 µL of 1/50 dilution (b)

Calculate appropriate dilution factor for concentrations exceeding this table.

(a) The volume of methanol added to 5 mL of water being purged should be kept constant. Therefore, add to the 5-mL syringe whatever volume of methanol is necessary to maintain a volume of 100 µL added to the syringe.

(b) Dilute an aliquot of the methanol extract and then take 100 µL for analysis.

QUALITY CONTROL Demonstrate, through the analysis of a reagent water blank, that interferences from the analytical system, glassware, and reagents are under control. Blank samples should be carried through all stages of the sample preparation and measurement steps. For each analytical batch (up to 20 samples), a reagent blank, matrix spike, and matrix spike duplicate must be analyzed (the frequency of the spikes may be different for different monitoring programs). The blank and spiked samples must be carried through all stages of the sample preparation and measurement steps. QC samples mentioned in the section on Interferences will also be needed as appropriate to those situations.

REFERENCE Test Methods for Evaluating Solid Waste (SW-846). U.S. EPA. 1983. Method 8240B, Rev. 2, Nov. 1990. Office of Solid Wastes, Washington, DC.

1,4-Dichlorobenzene   **EPA Method 8240**
CAS #106-46-7

TITLE Volatile Organics By GC/MS: Packed Column Technique

MATRIX Nearly all types of sample matarices, regardless of water content, can be analyzed using this method. This includes groundwater, aqueous sludges, caustic liquors, acid liquors, waste solvents, oily wastes, mousses, tars, fibrous wastes, polymetric emulsions, filter cakes, spent carbons, spent catalysts, soils, and sediments.

METHOD SUMMARY Method 8240B covers 80 volatile organic compounds that are introduced into a gas chromatograph by the purge-and-trap method or by direct injection (in limited applications). For the purge-and-trap method an inert gas (zero grade nitrogen or helium) is bubbled through a 5-mL solution at ambient temperature. Purged sample components are trapped in a tube of sorbent materials. When purging is complete, the sorbent tube is heated and backflushed with inert gas to desorb the trapped components onto a GC column.

INTERFERENCES Impurities in the purge gas and from organic compounds outgassing from the plumbing ahead of the trap account for many contamination problems. Interferences purged or coextracted from the samples will vary considerably from source to source. Cross-contamination can occur whenever high-level and low-level samples are analyzed sequentially. Whenever an unusually concentrated sample is analyzed, it should be followed by the analysis of organic-free reagent water to check for cross-contamination. Samples also can be contaminated by diffusion of volatile organics (particularly methylene chloride and fluorocarbons) through the septum seal into the sample during shipment and storage. A trip blank can serve as a check on such contamination. The lab where volatile analysis is performed and also the refrigerated storage area should be completely free of solvents.

INSTRUMENTATION A gas chromatograph/mass spectrometry/data system (GC/MS) equipped with a 6 ft × 0.1 in I.D. glass column packed with 1% SP-1000 on Carbopack-B (60/80 mesh) is required. Also needed is a 5-mL purging device, a sorbent trap, and a thermal desorption apparatus.

PRECISION & ACCURACY This method is reported to have been tested by 15 laboratories using organic-free reagent water, drinking water, surface water, and industrial wastewaters (not specified) fortified at six concentrations over the range 5–600 µg/L.

Sample estimated quantitation limits (EQLs) are highly matrix-dependent. The EQLs listed may not always be achievable. EQLs listed for soils or sediments are based on wet weight. Normally, data is reported on a dry-weight basis; therefore, EQLs will be higher, based on the percent dry weight of each sample. Note that EQLs are even more variable than MDLs and that they are highly variable depending on the matrix being analyzed.

EQL in groundwater in µg/L was not listed.
EQL in low soil or sediment in µg/kg was not listed.
Accuracy (a) in µg/L was 0.94C+4.47.
Precision (b) in µg/L was 0.30x-1.20.

(a) *Average recovery found for measurements of samples containing a concentration of C, in µg/L.*
(b) *Overall precision found for measurements of samples with average recovery X for samples containing a concentration of C in µg/L.*
X = *Average recovery found for measurement of samples containing a concentration of C in µg/L.*

MULTIPLICATION FACTORS FOR OTHER MATRICES

Other Matrices	Factor (a)
Waste miscible liquid waste	50
High-concentration soil and sludge	125
Non-water miscible waste	500

(a) EQL = [EQL for low soil sediment] × [Factor]. For non-aqueous samples, the factor is on a wet-weight basis.

SAMPLING METHOD
Liquid samples — Use a 40-mL glass screw-cap VOA vial with a Teflon®-faced silicone septum that has been prewashed, rinsed with distilled deionized water, and oven dried. However, if residual chlorine is present, collect sample in a 40-oz. soil VOA container which has been pre-preserved with 4 drops of 10% sodium thiosulfate, mix gently, and then transfer the sample to a 40-mL VOA vial. Collect bubble-free samples in duplicate and seal them in separate plastic bags.

Soils or sediments, and sludges — Use an 8-oz. widemouth glass bottle with a Teflon®-faced silicone septum that has been prewashed with detergent, rinsed with distilled deionized water, and oven dried. Tap slightly to eliminate free air space. Collect samples in duplicate and seal them in separate plastic bags.

SAMPLE PRESERVATION
Liquid samples — Add 4 drops of concentrated HCL and immediately cool samples to 4°C and store in a solvent-free refrigerator.

Soils or sediments, and sludges — Cool samples to 4°C and store in a solvent-free refrigerator.

MHT Maximum holding time is 14 days from the date of sample collection.

SAMPLE PREPARATION
Liquid samples — Remove the plunger from a 5-mL syringe and carefully pour the sample into the syringe barrel to just short of overflowing. Replace the syringe plunger and compress the sample. Open the syringe valve and vent any residual air while adjusting the sample volume to 5.0 mL. If there is only

one volatile organic analysis (VOA) vial, a second syringe should be filled at this time to protect against possible loss of sample integrity. Add 10 µL of surrogate spiking solution and 10 µL of internal standard spiking solution through the valve bore of the 5-mL syringe, then close the valve. The surrogate and internal standards may be mixed and added as a single spiking solution.

Sediments, soils, and waste samples — All samples of this type should be screened by GC analysis using a headspace method (EPA Method 3810) or the hexadecane extraction and screening method (EPA Method 3820). Use the screening data to determine whether to use the low-concentration method (0.005–1 mg/kg) or the high-concentration method (>1 mg/kg).

Low-concentration method — The low-concentration method is based on purging a heated sediment or soil sample mixed with organic-free reagent water containing the surrogate and internal standards. Analyze all reagent blanks and standards under the same conditions as the samples.

Use a 5-g sample if the expected concentration is <0.1 mg/kg or a 1-g sample for expected concentrations between 0.1 and 1 mg/kg. Mix the contents of the sample container with a narrow metal spatula. Weigh the amount of the sample into a tared purge device. Add the spiked water to the purge device, which contains the weighed amount of sample, and connect the device to the purge-and-trap system.

High-concentration method — This method is based on extracting the sediment or soil with methanol. A waste sample is either extracted or diluted, depending on its solubility in methanol. Wastes that are insoluble in methanol are diluted with reagent tetraglyme or possibly polyethylene glycol (PEG). An aliquot of the extract is added to organic-free reagent water containing surrogate and internal standards. This is purged at ambient temperature. All samples with an expected concentration of >1.0 mg/kg should be analyzed by this method.

Mix the contents of the sample container with a narrow metal spatula. For sediments or soils and solid wastes that are insoluble in methanol, weigh 4 g (wet weight) of sample into a tared 20-mL vial. For waste that is soluble in methanol, tetraglyme, or PEG, weigh 1 g (wet weight) into a tared scintillation vial or culture tube or a 10-mL volumetric flask. Quickly add 9.0 mL of appropriate solvent then add 1.0 mL of a surrogate spiking solution to the vial, cap it, and shake it for 2 min.

METHANOL EXTRACT REQUIRED FOR ANALYSIS OF HIGH-CONCENTRATION SOILS OR SEDIMENTS

Approximate Concentration Range	Volume of Methanol Extract (a)
500–10,000 µg/kg	100 µL
1,000–20,000 µg/kg	50 µL
5,000–100,000 µg/kg	10 µL
25,000–500,000 µg/kg	100 µL of 1/50 dilution (b)

Calculate appropriate dilution factor for concentrations exceeding this table.

(a) The volume of methanol added to 5 mL of water being purged should be kept constant. Therefore, add to the 5-mL syringe whatever volume of methanol is necessary to maintain a volume of 100 µL added to the syringe.
(b) Dilute an aliquot of the methanol extract and then take 100 µL for analysis.

QUALITY CONTROL Demonstrate, through the analysis of a reagent water blank, that interferences from the analytical system, glassware, and reagents are under control. Blank samples should be carried through all stages of the sample preparation and measurement steps. For each analytical batch (up to 20 samples), a reagent blank, matrix spike, and matrix spike duplicate must be analyzed (the frequency of the spikes may be different for different monitoring programs). The blank and spiked samples must be carried through all stages of the sample preparation and measurement steps. QC samples mentioned in the section on Interferences will also be needed as appropriate to those situations.

REFERENCE Test Methods for Evaluating Solid Waste (SW-846). U.S. EPA. 1983. Method 8240B, Rev. 2, Nov. 1990. Office of Solid Wastes, Washington, DC.

1,2-Dichlorobenzene   **EPA Method 8260**
CAS #95-50-1

TITLE Volatile Organic Compounds by GC/MS: Capillary Column Technique

MATRIX This method is applicable to nearly all types of samples, regardless of water content, including groundwater, soils, and sediments.

METHOD SUMMARY Method 8260A covers 58 volatile organic compounds that are introduced into a gas chromatograph by the purge-and-trap method or by direct injection (in limited applications). Zero-grade helium is bubbled through a 5-mL solution at ambient temperature. Purged sample components are trapped in a tube containing suitable sorbent materials. When purging is complete, the sorbent tube is heated and backflushed with helium to desorb trapped sample components. The analytes are desorbed directly to a large bore capillary or cryofocussed on a capillary precolumn before being flash evaporated to a narrow bore capillary for analysis.

INTERFERENCES Major contaminant sources are volatile materials in the lab and impurities in the inert purging gas and in the sorbent trap. Interfering contamination may occur when a sample containing low concentrations of volatile organic compounds is analyzed immediately after a sample containing high concentrations of volatile organic compounds. After analysis of a sample containing high concentrations of volatile organic compounds, one or more calibration blanks should be analyzed to check for cross-contamination. Screening of the samples prior to purge-and-trap GC/MS analysis is highly recommended to prevent contamination of the system. This is especially true for soil and waste samples.

Special precautions must be taken to analyze for methylene chloride. The analytical and sample storage area should be isolated from all atmospheric sources of methylene chloride. All gas chromatography carrier gas lines and purge gas plumbing should be constructed from stainless steel or copper tubing. Laboratory clothing previously exposed to methylene chloride fumes during liquid-liquid extraction procedures can contribute to sample contamination.

Samples can also be contaminated by diffusion of volatile organics (particularly methylene chloride and fluorocarbons) through the septum seal during shipment and storage. A trip blank can serve as a check on such contamination.

INSTRUMENTATION GC/MS with a temperature-programmable chromatograph suitable for splitless injection equipped with variable constant differential flow controllers, a subambient oven controller, a purging device, sorbent trap, a thermal desorption apparatus and a capillary precolumn interface when using cryogenic cooling will be needed. The following GC columns may be used:

Column 1: 60 m × 0.75mm I.D. capillary column coated with VOCOL, 1.5 µm film thickness.
Column 2: 30 m × 0.53mm capillary column coated with DB-624 or VOCOL, 3 µm film thickness.
Column 3: 30 m × 0.32mm I.D. capillary column coated with DB-5 or SE-54, 1-µm film thickness.

PRECISION & ACCURACY This method has been tested in a single lab using spiked water. Using a wide-bore capillary column, water was spiked at concentrations between 0.5 and 10 µg/L. Single lab accuracy and precision data are presented. The MDL actually achieved in a given analysis will vary depending on instrument sensitivity and matrix effects.

The MDL (a) in µg/L was 0.03.
The concentration range in µg/L was 0.1–10.
The mean accuracy (% of true value) was 93.
The precision as relative standard deviation was 6.2.

Note: The MDL is based on a 25-mL sample volume instead of a 5-mL sample volume.

SAMPLING METHOD
Liquid samples — Use a 40-mL glass screw-cap VOA vial with a Teflon®-faced silicone septum that has been prewashed, rinsed with distilled deionized water, and oven dried. If residual chlorine is present, collect the sample in a 4-oz soil VOA container which has been pre-preserved with 4 drops of 10% sodium thiosulfate. Mix gently and transfer the sample to a 40-mL VOA vial. Collect bubble-free samples in duplicate and seal each sample in a separate plastic bag.

Soils, sediments, and sludges — Use an 8-oz widemouth glass bottle with Teflon®-faced silicone septum that has been prewashed, rinsed with distilled deionized water, and oven dried. **Do not** heat the septum for more than 1 h. Tap slightly to eliminate any free air space. Collect samples in duplicate and seal each one in a separate plastic bag.

SAMPLE PRESERVATION
Liquid samples — Add 4 drops of concentrated HCL, cool to 4°C and store in a solvent-free refrigerator.

Soils, sediments and sludges — Cool samples to 4°C and store in a solvent-free refrigerator.

MHT The maximum holding time of any sample (liquids, soils, sediments, and sludges) is 14 days.

SAMPLE PREPARATION
Liquid samples — Remove the plunger from a 5-mL syringe and carefully pour the sample into the syringe barrel to just short of overflowing. Replace the syringe plunger and compress the sample. Open the syringe valve and vent any residual air while adjusting the sample volume to 5.0 mL. If there is only one volatile organic analysis (VOA) vial, a second syringe should be filled at this time to protect against possible loss of sample integrity. Add 10 µL of surrogate spiking solution and 10 µL of internal standard spiking solution through the valve bore of the 5-mL syringe, then close the valve. The surrogate and internal standards may be mixed and added as a single spiking solution.

Sediments, soils, and waste samples — All samples of this type should be screened by GC analysis using a headspace method (EPA Method 3810) or the hexadecane extraction and screening method (EPA Method 3820). Use the screening data to determine whether to use the low-concentration method (0.005–1 mg/kg) or the high-concentration method (>1 mg/kg).

Low-concentration method — The low-concentration method is based on purging a heated sediment or soil sample mixed with organic-free reagent water containing the surrogate and internal standards. Analyze all reagent blanks and standards under the same conditions as the samples.

Use a 5-g sample if the expected concentration is <0.1 mg/kg or a 1-g sample for expected concentrations between 0.1 and 1 mg/kg. Mix the contents of the sample container with a narrow metal spatula. Weigh the amount of the sample into a tared purge device. Add the spiked water to the purge device, which contains the weighed amount of sample, and connect the device to the purge-and-trap system.

High-concentration method — This method is based on extracting the sediment or soil with methanol. A waste sample is either extracted or diluted, depending on its solubility in methanol. Wastes that are insoluble in methanol are diluted with reagent tetraglyme or possibly polyethylene glycol (PEG). An aliquot of the extract is added to organic-free reagent water containing surrogate and internal standards. This is purged at ambient temperature. All samples with an expected concentration of >1.0 mg/kg should be analyzed by this method.

Mix the contents of the sample container with a narrow metal spatula. For sediments or soils and solid wastes that are insoluble in methanol, weigh 4 g (wet weight) of sample into a tared 20-mL vial. For waste that is soluble in methanol, tetraglyme, or PEG, weigh 1 g (wet weight) into a tared scintillation vial or culture tube or a 10-mL volumetric flask. Quickly add

9.0 mL of appropriate solvent then add 1.0 mL of a surrogate spiking solution to the vial, cap it, and shake it for 2 min.

METHANOL EXTRACT REQUIRED FOR ANALYSIS OF HIGH-CONCENTRATION SOILS OR SEDIMENTS

Approximate Concentration Range	Volume of Methanol Extract (a)
500–10,000 µg/kg	100 µL
1,000–20,000 µg/kg	50 µL
5,000–100,000 µg/kg	10 µL
25,000–500,000 µg/kg	100 µL of 1/50 dilution (b)

Calculate appropriate dilution factor for concentrations exceeding this table.

(a) The volume of methanol added to 5 mL of water being purged should be kept constant. Therefore, add to the 5-mL syringe whatever volume of methanol is necessary to maintain a volume of 100 µL added to the syringe.

(b) Dilute an aliquot of the methanol extract and then take 100 µL for analysis.

QUALITY CONTROL Demonstrate, through the analysis of a reagent water blank, that interferences from the analytical system, glassware, and reagents are under control. Blank samples should be carried through all stages of the sample preparation and measurement steps. For each analytical batch (up to 20 samples), a reagent blank, matrix spike, and matrix spike duplicate must be analyzed (the frequency of the spikes may be different for different monitoring programs). The blank and spiked samples must be carried through all stages of the sample preparation and measurement steps. QC samples mentioned in the section on Interferences will also be needed as appropriate to those situations.

Matrix spiking standards should be prepared from volatile organic compounds which will be representative of the compounds being investigated. The recommended internal standards are chlorobenzene-d5, 1,4-difluorobenzene, 1,4-dichlorobenzene-d4, and pentafluorobenzene. Using stock standard solutions, prepare secondary dilution standards containing the compounds of interest, either singly or mixed together in methanol. Store them in a vial with no headspace for no more than one week. Surrogates recommended are toluene-d8, 4-bromofluorobenzene, and dibromofluoromethane. Each sample undergoing GC/MS analysis must be spiked with 10 µL of the surrogate spiking solution prior to analysis.

REFERENCE Test Methods for Evaluating Solid Waste (SW-846). U.S. EPA 1983, Method 8260A, Rev. 1, Nov. 1990. Office of Solid Waste, Washington, DC.

1,3-Dichlorobenzene   **EPA Method 8260**
CAS #541-73-1

TITLE Volatile Organic Compounds by GC/MS: Capillary Column Technique

MATRIX This method is applicable to nearly all types of samples, regardless of water content, including groundwater, soils, and sediments.

METHOD SUMMARY Method 8260A covers 58 volatile organic compounds that are introduced into a gas chromatograph by the purge-and-trap method or by direct injection (in limited applications). Zero-grade helium is bubbled through a 5-mL solution at ambient temperature. Purged sample components are trapped in a tube containing suitable sorbent materials. When purging is complete, the sorbent tube is heated and backflushed with helium to desorb trapped sample components. The analytes are desorbed directly to a large bore capillary or cryofocussed on a capillary precolumn before being flash evaporated to a narrow bore capillary for analysis.

INTERFERENCES Major contaminant sources are volatile materials in the lab and impurities in the inert purging gas and in the sorbent trap. Interfering contamination may occur when a sample containing low concentrations of volatile organic compounds is analyzed immediately after a sample containing high concentrations of volatile organic compounds. After analysis of a sample containing high concentrations of volatile organic compounds, one or more calibration blanks should be analyzed to check for cross-contamination. Screening of the samples prior to purge-and-trap GC/MS analysis is highly recommended to prevent contamination of the system. This is especially true for soil and waste samples.

Special precautions must be taken to analyze for methylene chloride. The analytical and sample storage area should be isolated from all atmospheric sources of methylene chloride. All gas chromatography carrier gas lines and purge gas plumbing should be constructed from stainless steel or copper tubing. Laboratory clothing previously exposed to methylene chloride fumes during liquid-liquid extraction procedures can contribute to sample contamination.

Samples can also be contaminated by diffusion of volatile organics (particularly methylene chloride and fluorocarbons) through the septum seal during shipment and storage. A trip blank can serve as a check on such contamination.

INSTRUMENTATION GC/MS with a temperature-programmable chromatograph suitable for splitless injection equipped with variable constant differential flow controllers, a subambient oven controller, a purging device, sorbent trap, a thermal desorption apparatus and a capillary precolumn interface when using cryogenic cooling will be needed. The following GC columns may be used:

Column 1: 60 m × 0.75mm I.D. capillary column coated with VOCOL, 1.5 µm film thickness.
Column 2: 30 m × 0.53mm capillary column coated with DB-624 or VOCOL, 3 µm film thickness.
Column 3: 30 m × 0.32mm I.D. capillary column coated with DB-5 or SE-54, 1-µm film thickness.

PRECISION & ACCURACY This method has been tested in a single lab using spiked water. Using a wide-bore capillary column, water was spiked at concentrations between 0.5 and 10 µg/L. Single lab accuracy and precision data are presented.

The MDL actually achieved in a given analysis will vary depending on instrument sensitivity and matrix effects.

The MDL (a) in μg/L was 0.12.
The concentration range in μg/L was 0.5–10.
The mean accuracy (% of true value) was 99.
The precision as relative standard deviation was 6.9.

Note: The MDL is based on a 25-mL sample volume instead of a 5-mL sample volume.

SAMPLING METHOD
Liquid samples — Use a 40-mL glass screw-cap VOA vial with a Teflon®-faced silicone septum that has been prewashed, rinsed with distilled deionized water, and oven dried. If residual chlorine is present, collect the sample in a 4-oz soil VOA container which has been pre-preserved with 4 drops of 10% sodium thiosulfate. Mix gently and transfer the sample to a 40-mL VOA vial. Collect bubble-free samples in duplicate and seal each sample in a separate plastic bag.

Soils, sediments, and sludges — Use an 8-oz widemouth glass bottle with Teflon®-faced silicone septum that has been prewashed, rinsed with distilled deionized water, and oven dried. **Do not** heat the septum for more than 1 h. Tap slightly to eliminate any free air space. Collect samples in duplicate and seal each one in a separate plastic bag.

SAMPLE PRESERVATION
Liquid samples — Add 4 drops of concentrated HCL, cool to 4°C and store in a solvent-free refrigerator.

Soils, sediments and sludges — Cool samples to 4°C and store in a solvent-free refrigerator.

MHT The maximum holding time of any sample (liquids, soils, sediments, and sludges) is 14 days.

SAMPLE PREPARATION
Liquid samples — Remove the plunger from a 5-mL syringe and carefully pour the sample into the syringe barrel to just short of overflowing. Replace the syringe plunger and compress the sample. Open the syringe valve and vent any residual air while adjusting the sample volume to 5.0 mL. If there is only one volatile organic analysis (VOA) vial, a second syringe should be filled at this time to protect against possible loss of sample integrity. Add 10 μL of surrogate spiking solution and 10 μL of internal standard spiking solution through the valve bore of the 5-mL syringe, then close the valve. The surrogate and internal standards may be mixed and added as a single spiking solution.

Sediments, soils, and waste samples — All samples of this type should be screened by GC analysis using a headspace method (EPA Method 3810) or the hexadecane extraction and screening method (EPA Method 3820). Use the screening data to determine whether to use the low-concentration method (0.005–1 mg/kg) or the high-concentration method (>1 mg/kg).

Low-concentration method — The low-concentration method is based on purging a heated sediment or soil sample mixed with organic-free reagent water containing the surrogate and internal standards. Analyze all reagent blanks and standards under the same conditions as the samples.

Use a 5-g sample if the expected concentration is <0.1 mg/kg or a 1-g sample for expected concentrations between 0.1 and 1 mg/kg. Mix the contents of the sample container with a narrow metal spatula. Weigh the amount of the sample into a tared purge device. Add the spiked water to the purge device, which contains the weighed amount of sample, and connect the device to the purge-and-trap system.

High-concentration method — This method is based on extracting the sediment or soil with methanol. A waste sample is either extracted or diluted, depending on its solubility in methanol. Wastes that are insoluble in methanol are diluted with reagent tetraglyme or possibly polyethylene glycol (PEG). An aliquot of the extract is added to organic-free reagent water containing surrogate and internal standards. This is purged at ambient temperature. All samples with an expected concentration of >1.0 mg/kg should be analyzed by this method.

Mix the contents of the sample container with a narrow metal spatula. For sediments or soils and solid wastes that are insoluble in methanol, weigh 4 g (wet weight) of sample into a tared 20-mL vial. For waste that is soluble in methanol, tetraglyme, or PEG, weigh 1 g (wet weight) into a tared scintillation vial or culture tube or a 10-mL volumetric flask. Quickly add 9.0 mL of appropriate solvent then add 1.0 mL of a surrogate spiking solution to the vial, cap it, and shake it for 2 min.

METHANOL EXTRACT REQUIRED FOR ANALYSIS OF HIGH-CONCENTRATION SOILS OR SEDIMENTS

Approximate Concentration Range	Volume of Methanol Extract (a)
500–10,000 μg/kg	100 μL
1,000–20,000 μg/kg	50 μL
5,000–100,000 μg/kg	10 μL
25,000–500,000 μg/kg	100 μL of 1/50 dilution (b)

Calculate appropriate dilution factor for concentrations exceeding this table.

(a) The volume of methanol added to 5 mL of water being purged should be kept constant. Therefore, add to the 5-mL syringe whatever volume of methanol is necessary to maintain a volume of 100 μL added to the syringe.
(b) Dilute an aliquot of the methanol extract and then take 100 μL for analysis.

QUALITY CONTROL Demonstrate, through the analysis of a reagent water blank, that interferences from the analytical system, glassware, and reagents are under control. Blank samples should be carried through all stages of the sample preparation and measurement steps. For each analytical batch (up to 20 samples), a reagent blank, matrix spike, and matrix spike duplicate must be analyzed (the frequency of the spikes may be different for different monitoring programs). The blank and spiked samples must be carried through all stages of the sample preparation and measurement steps. QC samples mentioned in the section on Interferences will also be needed as appropriate to those situations.

Matrix spiking standards should be prepared from volatile organic compounds which will be representative of the compounds being investigated. The recommended internal standards

are chlorobenzene-d5, 1,4-difluorobenzene, 1,4-dichlorobenzene-d4, and pentafluorobenzene. Using stock standard solutions, prepare secondary dilution standards containing the compounds of interest, either singly or mixed together in methanol. Store them in a vial with no headspace for no more than one week. Surrogates recommended are toluene-d8, 4-bromofluorobenzene, and dibromofluoromethane. Each sample undergoing GC/MS analysis must be spiked with 10 µL of the surrogate spiking solution prior to analysis.

REFERENCE Test Methods for Evaluating Solid Waste (SW-846). U.S. EPA 1983, Method 8260A, Rev. 1, Nov. 1990. Office of Solid Waste, Washington, DC.

1,4-Dichlorobenzene   **EPA Method 8260**
CAS #106-46-7

TITLE Volatile Organic Compounds by GC/MS: Capillary Column Technique

MATRIX This method is applicable to nearly all types of samples, regardless of water content, including groundwater, soils, and sediments.

METHOD SUMMARY Method 8260A covers 58 volatile organic compounds that are introduced into a gas chromatograph by the purge-and-trap method or by direct injection (in limited applications). Zero-grade helium is bubbled through a 5-mL solution at ambient temperature. Purged sample components are trapped in a tube containing suitable sorbent materials. When purging is complete, the sorbent tube is heated and backflushed with helium to desorb trapped sample components. The analytes are desorbed directly to a large bore capillary or cryofocussed on a capillary precolumn before being flash evaporated to a narrow bore capillary for analysis.

INTERFERENCES Major contaminant sources are volatile materials in the lab and impurities in the inert purging gas and in the sorbent trap. Interfering contamination may occur when a sample containing low concentrations of volatile organic compounds is analyzed immediately after a sample containing high concentrations of volatile organic compounds. After analysis of a sample containing high concentrations of volatile organic compounds, one or more calibration blanks should be analyzed to check for cross-contamination. Screening of the samples prior to purge-and-trap GC/MS analysis is highly recommended to prevent contamination of the system. This is especially true for soil and waste samples.

Special precautions must be taken to analyze for methylene chloride. The analytical and sample storage area should be isolated from all atmospheric sources of methylene chloride. All gas chromatography carrier gas lines and purge gas plumbing should be constructed from stainless steel or copper tubing. Laboratory clothing previously exposed to methylene chloride fumes during liquid-liquid extraction procedures can contribute to sample contamination.

Samples can also be contaminated by diffusion of volatile organics (particularly methylene chloride and fluorocarbons) through the septum seal during shipment and storage. A trip blank can serve as a check on such contamination.

INSTRUMENTATION GC/MS with a temperature-programmable chromatograph suitable for splitless injection equipped with variable constant differential flow controllers, a subambient oven controller, a purging device, sorbent trap, a thermal desorption apparatus and a capillary precolumn interface when using cryogenic cooling will be needed. The following GC columns may be used:

Column 1: 60 m × 0.75mm I.D. capillary column coated with VOCOL, 1.5 µm film thickness.
Column 2: 30 m × 0.53mm capillary column coated with DB-624 or VOCOL, 3 µm film thickness.
Column 3: 30 m × 0.32mm I.D. capillary column coated with DB-5 or SE-54, 1-µm film thickness.

PRECISION & ACCURACY This method has been tested in a single lab using spiked water. Using a wide-bore capillary column, water was spiked at concentrations between 0.5 and 10 µg/L. Single lab accuracy and precision data are presented. The MDL actually achieved in a given analysis will vary depending on instrument sensitivity and matrix effects.

The MDL (a) in µg/L was 0.03.
The concentration range in µg/L was 0.2–20.
The mean accuracy (% of true value) was 103.
The precision as relative standard deviation was 6.4.

Note: The MDL is based on a 25-mL sample volume instead of a 5-mL sample volume.

SAMPLING METHOD
Liquid samples — Use a 40-mL glass screw-cap VOA vial with a Teflon®-faced silicone septum that has been prewashed, rinsed with distilled deionized water, and oven dried. If residual chlorine is present, collect the sample in a 4-oz soil VOA container which has been pre-preserved with 4 drops of 10% sodium thiosulfate. Mix gently and transfer the sample to a 40-mL VOA vial. Collect bubble-free samples in duplicate and seal each sample in a separate plastic bag.

Soils, sediments, and sludges — Use an 8-oz widemouth glass bottle with Teflon®-faced silicone septum that has been prewashed, rinsed with distilled deionized water, and oven dried. **Do not** heat the septum for more than 1 h. Tap slightly to eliminate any free air space. Collect samples in duplicate and seal each one in a separate plastic bag.

SAMPLE PRESERVATION
Liquid samples — Add 4 drops of concentrated HCL, cool to 4°C and store in a solvent-free refrigerator.

Soils, sediments and sludges — Cool samples to 4°C and store in a solvent-free refrigerator.

MHT The maximum holding time of any sample (liquids, soils, sediments, and sludges) is 14 days.

SAMPLE PREPARATION
Liquid samples — Remove the plunger from a 5-mL syringe and carefully pour the sample into the syringe barrel to just short of overflowing. Replace the syringe plunger and compress the sample. Open the syringe valve and vent any residual air

while adjusting the sample volume to 5.0 mL. If there is only one volatile organic analysis (VOA) vial, a second syringe should be filled at this time to protect against possible loss of sample integrity. Add 10 μL of surrogate spiking solution and 10 μL of internal standard spiking solution through the valve bore of the 5-mL syringe, then close the valve. The surrogate and internal standards may be mixed and added as a single spiking solution.

Sediments, soils, and waste samples — All samples of this type should be screened by GC analysis using a headspace method (EPA Method 3810) or the hexadecane extraction and screening method (EPA Method 3820). Use the screening data to determine whether to use the low-concentration method (0.005–1 mg/kg) or the high-concentration method (>1 mg/kg).

Low-concentration method — The low-concentration method is based on purging a heated sediment or soil sample mixed with organic-free reagent water containing the surrogate and internal standards. Analyze all reagent blanks and standards under the same conditions as the samples.

Use a 5-g sample if the expected concentration is <0.1 mg/kg or a 1-g sample for expected concentrations between 0.1 and 1 mg/kg. Mix the contents of the sample container with a narrow metal spatula. Weigh the amount of the sample into a tared purge device. Add the spiked water to the purge device, which contains the weighed amount of sample, and connect the device to the purge-and-trap system.

High-concentration method — This method is based on extracting the sediment or soil with methanol. A waste sample is either extracted or diluted, depending on its solubility in methanol. Wastes that are insoluble in methanol are diluted with reagent tetraglyme or possibly polyethylene glycol (PEG). An aliquot of the extract is added to organic-free reagent water containing surrogate and internal standards. This is purged at ambient temperature. All samples with an expected concentration of >1.0 mg/kg should be analyzed by this method.

Mix the contents of the sample container with a narrow metal spatula. For sediments or soils and solid wastes that are insoluble in methanol, weigh 4 g (wet weight) of sample into a tared 20-mL vial. For waste that is soluble in methanol, tetraglyme, or PEG, weigh 1 g (wet weight) into a tared scintillation vial or culture tube or a 10-mL volumetric flask. Quickly add 9.0 mL of appropriate solvent then add 1.0 mL of a surrogate spiking solution to the vial, cap it, and shake it for 2 min.

METHANOL EXTRACT REQUIRED FOR ANALYSIS OF HIGH-CONCENTRATION SOILS OR SEDIMENTS

Approximate Concentration Range	Volume of Methanol Extract (a)
500–10,000 μg/kg	100 μL
1,000–20,000 μg/kg	50 μL
5,000–100,000 μg/kg	10 μL
25,000–500,000 μg/kg	100 μL of 1/50 dilution (b)

Calculate appropriate dilution factor for concentrations exceeding this table.

(a) The volume of methanol added to 5 mL of water being purged should be kept constant. Therefore, add to the 5-mL syringe whatever volume of methanol is necessary to maintain a volume of 100 μL added to the syringe.
(b) Dilute an aliquot of the methanol extract and then take 100 μL for analysis.

QUALITY CONTROL Demonstrate, through the analysis of a reagent water blank, that interferences from the analytical system, glassware, and reagents are under control. Blank samples should be carried through all stages of the sample preparation and measurement steps. For each analytical batch (up to 20 samples), a reagent blank, matrix spike, and matrix spike duplicate must be analyzed (the frequency of the spikes may be different for different monitoring programs). The blank and spiked samples must be carried through all stages of the sample preparation and measurement steps. QC samples mentioned in the section on Interferences will also be needed as appropriate to those situations.

Matrix spiking standards should be prepared from volatile organic compounds which will be representative of the compounds being investigated. The recommended internal standards are chlorobenzene-d5, 1,4-difluorobenzene, 1,4-dichlorobenzene-d4, and pentafluorobenzene. Using stock standard solutions, prepare secondary dilution standards containing the compounds of interest, either singly or mixed together in methanol. Store them in a vial with no headspace for no more than one week. Surrogates recommended are toluene-d8, 4-bromofluorobenzene, and dibromofluoromethane. Each sample undergoing GC/MS analysis must be spiked with 10 μL of the surrogate spiking solution prior to analysis.

REFERENCE Test Methods for Evaluating Solid Waste (SW-846). U.S. EPA 1983, Method 8260A, Rev. 1, Nov. 1990. Office of Solid Waste, Washington, DC.

1,2-Dichlorobenzene   **EPA Method 8270**
CAS #95-50-1

TITLE Semivolatile Organic Compounds by GC/MS

MATRIX This method is used to determine the concentration of semivolatile organic compounds in extracts prepared from all types of solid waste matrices, soils, and groundwater. Although surface waters are not specifically mentioned, this method should be applicable to water samples from rivers, lakes, etc.

METHOD SUMMARY This method covers 259 semivolatile organic compounds. In very limited applications direct injection of the sample into the GC/MS system may be appropriate, but this results in very high detection limits (approximately 10,000 μg/L). Typically, a 1-L liquid sample, containing surrogate, and matrix spiking standards, is extracted in a continuous extractor first under acid conditions and then under basic conditions. Typically 30 g of a solid sample, containing surrogate, and matrix spiking standards, is extracted ultrasonically. After concentrating the extract to 1 mL it is spiked with 10 μL of an internal standard solution just prior to analysis by GC/MS. The volume injected should contain about 100 ng of base/neutral

and 200 ng of acid surrogates (for a 1-μL injection). Analysis is performed by GC/MS using a capillary GC column.

INTERFERENCES Raw GC/MS data from all blanks, samples, and spikes must be evaluated for interferences. Contamination by carryover can occur whenever high-concentration and low-concentration samples are sequentially analyzed. To reduce carryover, the sample syringe must be rinsed out between samples with solvent. Whenever an unusually concentrated sample is encountered, it should be followed by the analysis of blank solvent to check for cross-contamination.

INSTRUMENTATION A GC/MS and a data system are required. The GC column used is a 30 m × 0.25 mm I.D. (or 0.32 mm I.D.) 1um film thickness silicone-coated fused silica capillary column. A continuous liquid-liquid extractor equipped with Teflon® or glass connection joints and stopcocks requiring no lubrication, a K-D concentrating apparatus, water bath, and an ultrasonic disrupter with a minimum power of 300 W and with pulsing capability are also required.

PRECISION & ACCURACY The estimated quantitation limit (EQL) of Method 8270B for determining an individual compound is approximately 1 mg/kg (wet weight) for soil or sediment samples, 1–200 mg/kg for wastes (dependent on matrix and method of preparation), and 10 μg/L for groundwater samples. EQLs will be proportionately higher for sample extracts that require dilution to avoid saturation of the detector.

The EQL(b) for groundwater in μg/L is 10.
The EQL (a, b) for low concentrations in soil and sediment in μg/kg is 660.
Accuracy as μg/L is $0.80C + 0.28$.
Overall precision in μg/L is $0.24X + 0.39$.

(a) EQLs listed for soil/sediment are based on wet weight. Normally data is reported in a dry-weight basis; therefore, EQLs will be higher based on the % dry weight of each sample. This calculation is based on a 30 g sample and gel permeation chromatography cleanup.
(b) Sample EQLs are highly matrix-dependent. The EQLs are provided for guidance and may not always be achievable.
C = True value for concentration, in μg/L.
X = Average recovery found for measurements of samples containing a concentration of C, in μg/L.

ESTIMATED QUANTITATION LIMIT

Other Matrices	Factor (a)
High-concentration soil and sludges by sonicator	7.5
Non-water miscible waste	75

(a) EQL for other matrices = [EQL for low soil/sediment] × [Factor]. This estimated EQL is similar to an EPA "Practical Quantitation Limit."

SAMPLING METHOD
Liquid samples — Use a 1 or 2½ gallon amber glass bottle with a screw-top Teflon®-lined cover that has been prewashed with detergent and rinsed with distilled water and methanol (or isopropanol).

Soils, sediments, or sludges — Use an 8-oz. widemouth glass with a screw-top Teflon®-lined cover that has been prewashed with detergent and rinsed with distilled water and methanol (or isopropanol).

SAMPLE PRESERVATION
Liquid samples — If residual chlorine is present, add 3 mL of 10% sodium thiosulfate per gallon, cool to 4°C and store in a solvent-free refrigerator until analysis; if chlorine is not present, then eliminate the sodium thiosulfate addition.

Soils, sediments, or sludges — Cool samples to 4°C and store in a solvent-free refrigerator.

MHT Liquid samples must be extracted within 7 days and the extracts analyzed within 40 days. Soils, sediments, or sludges may be stored for a maximum of 14 days and the extracts analyzed within 40 days.

SAMPLE PREPARATION
Liquid samples — Transfer 1 L quantitatively to a continuous extractor. If high concentrations are anticipated, a smaller volume may be used and then diluted with organic-free reagent water to 1 L. Adjust pH, if necessary, to pH <2 using 1:1 (V/V) sulfuric acid. Pipette 1.0 mL of a surrogate standard spiking solution into each sample. For the sample in each analytical batch selected for spiking, add 1.0 mL of a matrix spiking standard. For base/neutral acid analysis, the amount of the surrogates and matrix spiking compounds added to the sample should result in a final concentration of 100 ng/μL of each analyte in the extract to be analyzed (assuming a 1-μL injection). Extract with methylene chloride for 18–24 h. Next, adjust the pH of the aqueous phase to pH >11 using 10 N sodium hydroxide and extract it with methylene chloride again for 18–24 h. Dry the extract through a column containing anhydrous sodium sulfate and concentrate it to 1 mL using a K-D concentrator.

Soils, sediments, or sludges — Use 30 g of sample. Nonporous or wet samples (gummy or clay type) that do not have a free-flowing sandy texture must be mixed with anhydrous sodium sulfate until the sample is free flowing. Add 1 mL of surrogate standards to all samples, spikes, standards, and blanks. For the sample in each analytical batch selected for spiking, add 1.0 mL of a matrix spiking standard. For base/neutral acid analysis, the amount added of the surrogates and matrix spiking compounds should result in a final concentration of 100 ng/μL of each base/neutral analyte and 200 ng/μL of each acid analyte in the extract to be analyzed (assuming a 1-μL injection). Immediately add a 100-mL mixture of 1:1 methylene chloride:acetone and extract the sample ultrasonically for 3 min and then decant or filter the extracts. Repeat the extraction two or more times. Dry the extract using a column with anhydrous sodium sulfate and concentrate it to 1 mL in a K-D concentrator.

QUALITY CONTROL A methylene chloride solution containing 50 ng/μL of decafluorotriphenylphosphine (DFTPP) is used for tuning the GC/MS system each 12-h shift. A system performance check also must be made during every 12-h shift. A standard containing 50 ng/μL each of 4,4'-DDT, pentachlorophenol, and benzidine is required to verify injection port inertness and GC column performance. A calibration standard at mid-concentration, containing each compound of interest, including all required surrogates, must be performed every 12 h

during analysis. After the system performance check is met, calibration check compounds (CCCs) are used to check the validity of the initial calibration.

The internal standard responses and retention times in the calibration check standard must be evaluated immediately after or during data acquisition. If the retention time for any internal standard changes by more than 30 seconds from the last check calibration (12 h), the chromatographic system must be inspected for malfunctions and corrections must be made, as required. If the electron ionization current plot (EICP) area for any of the internal standards changes by a factor of two from the last daily calibration standard check, the mass spectrometer must be inspected for malfunctions and corrections must be made, as appropriate.

Demonstrate, through the analysis of a reagent water blank, that interferences from the analytical system, glassware, and reagents are under control. The blank samples should be carried through all stages of the sample preparation and measurement steps. For each analytical batch (up to 20 samples), a reagent blank, matrix spike, and matrix spike duplicate/duplicate must be analyzed (the frequency of the spikes may be different for different monitoring programs). The blank and spiked samples must be carried through all stages of the sample preparation and measurement steps. A QC reference sample concentrate containing each analyte at a concentration of 100 mg/L in methanol is required.

REFERENCE Test Methods for Evaluating Solid Waste (SW-846). U.S. EPA 1983, Method 8270B, Rev. 2, Nov. 1990. Office of Solid Waste, Washington, DC.

1,3-Dichlorobenzene   **EPA Method 8270**
CAS #541-73-1

TITLE Semivolatile Organic Compounds by GC/MS

MATRIX This method is used to determine the concentration of semivolatile organic compounds in extracts prepared from all types of solid waste matrices, soils, and groundwater. Although surface waters are not specifically mentioned, this method should be applicable to water samples from rivers, lakes, etc.

METHOD SUMMARY This method covers 259 semivolatile organic compounds. In very limited applications direct injection of the sample into the GC/MS system may be appropriate, but this results in very high detection limits (approximately 10,000 µg/L). Typically, a 1-L liquid sample, containing surrogate, and matrix spiking standards, is extracted in a continuous extractor first under acid conditions and then under basic conditions. Typically 30 g of a solid sample, containing surrogate, and matrix spiking standards, is extracted ultrasonically. After concentrating the extract to 1 mL it is spiked with 10 µL of an internal standard solution just prior to analysis by GC/MS. The volume injected should contain about 100 ng of base/neutral and 200 ng of acid surrogates (for a 1-µL injection). Analysis is performed by GC/MS using a capillary GC column.

INTERFERENCES Raw GC/MS data from all blanks, samples, and spikes must be evaluated for interferences. Contamination by carryover can occur whenever high-concentration and low-concentration samples are sequentially analyzed. To reduce carryover, the sample syringe must be rinsed out between samples with solvent. Whenever an unusually concentrated sample is encountered, it should be followed by the analysis of blank solvent to check for cross-contamination.

INSTRUMENTATION A GC/MS and a data system are required. The GC column used is a 30 m × 0.25 mm I.D. (or 0.32 mm I.D.) 1um film thickness silicone-coated fused silica capillary column. A continuous liquid-liquid extractor equipped with Teflon® or glass connection joints and stopcocks requiring no lubrication, a K-D concentrating apparatus, water bath, and an ultrasonic disrupter with a minimum power of 300 W and with pulsing capability are also required.

PRECISION & ACCURACY The estimated quantitation limit (EQL) of Method 8270B for determining an individual compound is approximately 1 mg/kg (wet weight) for soil or sediment samples, 1–200 mg/kg for wastes (dependent on matrix and method of preparation), and 10 µg/L for groundwater samples. EQLs will be proportionately higher for sample extracts that require dilution to avoid saturation of the detector.

The EQL(b) for groundwater in µg/L is 10.
The EQL (a, b) for low concentrations in soil and sediment in µg/kg is 660.
Accuracy as µg/L is 0.86C–0.70.
Overall precision in µg/L is 0.41X +0.11.

(a) *EQLs listed for soil/sediment are based on wet weight. Normally data is reported in a dry-weight basis; therefore, EQLs will be higher based on the % dry weight of each sample. This calculation is based on a 30 g sample and gel permeation chromatography cleanup.*
(b) *Sample EQLs are highly matrix-dependent. The EQLs are provided for guidance and may not always be achievable.*
$C =$ *True value for concentration, in µg/L.*
$X =$ *Average recovery found for measurements of samples containing a concentration of C, in µg/L.*

ESTIMATED QUANTITATION LIMIT

Other Matrices	Factor (a)
High-concentration soil and sludges by sonicator	7.5
Non-water miscible waste	75

(a) *EQL for other matrices = [EQL for low soil/sediment] × [Factor]. This estimated EQL is similar to an EPA "Practical Quantitation Limit."*

SAMPLING METHOD

Liquid samples — Use a 1 or 2½ gallon amber glass bottle with a screw-top Teflon®-lined cover that has been prewashed with detergent and rinsed with distilled water and methanol (or isopropanol).

Soils, sediments, or sludges — Use an 8-oz. widemouth glass with a screw-top Teflon®-lined cover that has been prewashed with detergent and rinsed with distilled water and methanol (or isopropanol).

SAMPLE PRESERVATION

Liquid samples — If residual chlorine is present, add 3 mL of 10% sodium thiosulfate per gallon, cool to 4°C and store in a solvent-free refrigerator until analysis; if chlorine is not present, then eliminate the sodium thiosulfate addition.

Soils, sediments, or sludges — Cool samples to 4°C and store in a solvent-free refrigerator.

MHT Liquid samples must be extracted within 7 days and the extracts analyzed within 40 days. Soils, sediments, or sludges may be stored for a maximum of 14 days and the extracts analyzed within 40 days.

SAMPLE PREPARATION

Liquid samples — Transfer 1 L quantitatively to a continuous extractor. If high concentrations are anticipated, a smaller volume may be used and then diluted with organic-free reagent water to 1 L. Adjust pH, if necessary, to pH <2 using 1:1 (V/V) sulfuric acid. Pipette 1.0 mL of a surrogate standard spiking solution into each sample. For the sample in each analytical batch selected for spiking, add 1.0 mL of a matrix spiking standard. For base/neutral acid analysis, the amount of the surrogates and matrix spiking compounds added to the sample should result in a final concentration of 100 ng/µL of each analyte in the extract to be analyzed (assuming a 1-µL injection). Extract with methylene chloride for 18–24 h. Next, adjust the pH of the aqueous phase to pH >11 using 10 N sodium hydroxide and extract it with methylene chloride again for 18–24 h. Dry the extract through a column containing anhydrous sodium sulfate and concentrate it to 1 mL using a K-D concentrator.

Soils, sediments, or sludges — Use 30 g of sample. Nonporous or wet samples (gummy or clay type) that do not have a free-flowing sandy texture must be mixed with anhydrous sodium sulfate until the sample is free flowing. Add 1 mL of surrogate standards to all samples, spikes, standards, and blanks. For the sample in each analytical batch selected for spiking, add 1.0 mL of a matrix spiking standard. For base/neutral acid analysis, the amount added of the surrogates and matrix spiking compounds should result in a final concentration of 100 ng/µL of each base/neutral analyte and 200 ng/µL of each acid analyte in the extract to be analyzed (assuming a 1-µL injection). Immediately add a 100-mL mixture of 1:1 methylene chloride:acetone and extract the sample ultrasonically for 3 min and then decant or filter the extracts. Repeat the extraction two or more times. Dry the extract using a column with anhydrous sodium sulfate and concentrate it to 1 mL in a K-D concentrator.

QUALITY CONTROL A methylene chloride solution containing 50 ng/µL of decafluorotriphenylphosphine (DFTPP) is used for tuning the GC/MS system each 12-h shift. A system performance check also must be made during every 12-h shift. A standard containing 50 ng/µL each of 4,4'-DDT, pentachlorophenol, and benzidine is required to verify injection port inertness and GC column performance. A calibration standard at mid-concentration, containing each compound of interest, including all required surrogates, must be performed every 12 h during analysis. After the system performance check is met, calibration check compounds (CCCs) are used to check the validity of the initial calibration.

The internal standard responses and retention times in the calibration check standard must be evaluated immediately after or during data acquisition. If the retention time for any internal standard changes by more than 30 seconds from the last check calibration (12 h), the chromatographic system must be inspected for malfunctions and corrections must be made, as required. If the electron ionization current plot (EICP) area for any of the internal standards changes by a factor of two from the last daily calibration standard check, the mass spectrometer must be inspected for malfunctions and corrections must be made, as appropriate.

Demonstrate, through the analysis of a reagent water blank, that interferences from the analytical system, glassware, and reagents are under control. The blank samples should be carried through all stages of the sample preparation and measurement steps. For each analytical batch (up to 20 samples), a reagent blank, matrix spike, and matrix spike duplicate/duplicate must be analyzed (the frequency of the spikes may be different for different monitoring programs). The blank and spiked samples must be carried through all stages of the sample preparation and measurement steps. A QC reference sample concentrate containing each analyte at a concentration of 100 mg/L in methanol is required.

REFERENCE Test Methods for Evaluating Solid Waste (SW-846). U.S. EPA 1983, Method 8270B, Rev. 2, Nov. 1990. Office of Solid Waste, Washington, DC.

1,4-Dichlorobenzene   **EPA Method 8270**
CAS #106-46-7

TITLE Semivolatile Organic Compounds by GC/MS

MATRIX This method is used to determine the concentration of semivolatile organic compounds in extracts prepared from all types of solid waste matrices, soils, and groundwater. Although surface waters are not specifically mentioned, this method should be applicable to water samples from rivers, lakes, etc.

METHOD SUMMARY This method covers 259 semivolatile organic compounds. In very limited applications direct injection of the sample into the GC/MS system may be appropriate, but this results in very high detection limits (approximately 10,000 µg/L). Typically, a 1-L liquid sample, containing surrogate, and matrix spiking standards, is extracted in a continuous extractor first under acid conditions and then under basic conditions. Typically 30 g of a solid sample, containing surrogate, and matrix spiking standards, is extracted ultrasonically. After concentrating the extract to 1 mL it is spiked with 10 µL of an internal standard solution just prior to analysis by GC/MS. The volume injected should contain about 100 ng of base/neutral and 200 ng of acid surrogates (for a 1-µL injection). Analysis is performed by GC/MS using a capillary GC column.

INTERFERENCES Raw GC/MS data from all blanks, samples, and spikes must be evaluated for interferences. Contamination by carryover can occur whenever high-concentration and low-concentration samples are sequentially analyzed. To reduce carryover, the sample syringe must be rinsed out between samples with solvent. Whenever an unusually concentrated sample is encountered, it should be followed by the analysis of blank solvent to check for cross-contamination.

INSTRUMENTATION A GC/MS and a data system are required. The GC column used is a 30 m × 0.25 mm I.D. (or 0.32 mm I.D.) 1um film thickness silicone-coated fused silica capillary column. A continuous liquid-liquid extractor equipped with Teflon® or glass connection joints and stopcocks requiring no lubrication, a K-D concentrating apparatus, water bath, and an ultrasonic disrupter with a minimum power of 300 W and with pulsing capability are also required.

PRECISION & ACCURACY The estimated quantitation limit (EQL) of Method 8270B for determining an individual compound is approximately 1 mg/kg (wet weight) for soil or sediment samples, 1–200 mg/kg for wastes (dependent on matrix and method of preparation), and 10 µg/L for groundwater samples. EQLs will be proportionately higher for sample extracts that require dilution to avoid saturation of the detector.

The EQL(b) for groundwater in µg/L is 10.
The EQL (a, b) for low concentrations in soil and sediment in µg/kg is 660.
Accuracy as µg/L is $0.73C-1.47$.
Overall precision in µg/L is $0.29X +0.36$.

(a) EQLs listed for soil/sediment are based on wet weight. Normally data is reported in a dry-weight basis; therefore, EQLs will be higher based on the % dry weight of each sample. This calculation is based on a 30 g sample and gel permeation chromatography cleanup.

(b) Sample EQLs are highly matrix-dependent. The EQLs are provided for guidance and may not always be achievable.

$C =$ *True value for concentration, in µg/L.*
$X =$ *Average recovery found for measurements of samples containing a concentration of C, in µg/L.*

ESTIMATED QUANTITATION LIMIT

Other Matrices	Factor (a)
High-concentration soil and sludges by sonicator	7.5
Non-water miscible waste	75

(a) EQL for other matrices = [EQL for low soil/sediment] × [Factor]. This estimated EQL is similar to an EPA "Practical Quantitation Limit."

SAMPLING METHOD

Liquid samples — Use a 1 or 2½ gallon amber glass bottle with a screw-top Teflon®-lined cover that has been prewashed with detergent and rinsed with distilled water and methanol (or isopropanol).

Soils, sediments, or sludges — Use an 8-oz. widemouth glass with a screw-top Teflon®-lined cover that has been prewashed with detergent and rinsed with distilled water and methanol (or isopropanol).

SAMPLE PRESERVATION

Liquid samples — If residual chlorine is present, add 3 mL of 10% sodium thiosulfate per gallon, cool to 4°C and store in a solvent-free refrigerator until analysis; if chlorine is not present, then eliminate the sodium thiosulfate addition.

Soils, sediments, or sludges — Cool samples to 4°C and store in a solvent-free refrigerator.

MHT Liquid samples must be extracted within 7 days and the extracts analyzed within 40 days. Soils, sediments, or sludges may be stored for a maximum of 14 days and the extracts analyzed within 40 days.

SAMPLE PREPARATION

Liquid samples — Transfer 1 L quantitatively to a continuous extractor. If high concentrations are anticipated, a smaller volume may be used and then diluted with organic-free reagent water to 1 L. Adjust pH, if necessary, to pH <2 using 1:1 (V/V) sulfuric acid. Pipette 1.0 mL of a surrogate standard spiking solution into each sample. For the sample in each analytical batch selected for spiking, add 1.0 mL of a matrix spiking standard. For base/neutral acid analysis, the amount of the surrogates and matrix spiking compounds added to the sample should result in a final concentration of 100 ng/µL of each analyte in the extract to be analyzed (assuming a 1-µL injection). Extract with methylene chloride for 18–24 h. Next, adjust the pH of the aqueous phase to pH >11 using 10 N sodium hydroxide and extract it with methylene chloride again for 18–24 h. Dry the extract through a column containing anhydrous sodium sulfate and concentrate it to 1 mL using a K-D concentrator.

Soils, sediments, or sludges — Use 30 g of sample. Nonporous or wet samples (gummy or clay type) that do not have a free-flowing sandy texture must be mixed with anhydrous sodium sulfate until the sample is free flowing. Add 1 mL of surrogate standards to all samples, spikes, standards, and blanks. For the sample in each analytical batch selected for spiking, add 1.0 mL of a matrix spiking standard. For base/neutral acid analysis, the amount added of the surrogates and matrix spiking compounds should result in a final concentration of 100 ng/µL of each base/neutral analyte and 200 ng/µL of each acid analyte in the extract to be analyzed (assuming a 1-µL injection). Immediately add a 100-mL mixture of 1:1 methylene chloride:acetone and extract the sample ultrasonically for 3 min and then decant or filter the extracts. Repeat the extraction two or more times. Dry the extract using a column with anhydrous sodium sulfate and concentrate it to 1 mL in a K-D concentrator.

QUALITY CONTROL A methylene chloride solution containing 50 ng/µL of decafluorotriphenylphosphine (DFTPP) is used for tuning the GC/MS system each 12-h shift. A system performance check also must be made during every 12-h shift. A standard containing 50 ng/µL each of 4,4'-DDT, pentachlorophenol, and benzidine is required to verify injection port inertness and GC column performance. A calibration standard at mid-concentration, containing each compound of interest, including all required surrogates, must be performed every 12 h during analysis. After the system performance check is met, calibration check compounds (CCCs) are used to check the validity of the initial calibration.

The internal standard responses and retention times in the calibration check standard must be evaluated immediately after

or during data acquisition. If the retention time for any internal standard changes by more than 30 seconds from the last check calibration (12 h), the chromatographic system must be inspected for malfunctions and corrections must be made, as required. If the electron ionization current plot (EICP) area for any of the internal standards changes by a factor of two from the last daily calibration standard check, the mass spectrometer must be inspected for malfunctions and corrections must be made, as appropriate.

Demonstrate, through the analysis of a reagent water blank, that interferences from the analytical system, glassware, and reagents are under control. The blank samples should be carried through all stages of the sample preparation and measurement steps. For each analytical batch (up to 20 samples), a reagent blank, matrix spike, and matrix spike duplicate/duplicate must be analyzed (the frequency of the spikes may be different for different monitoring programs). The blank and spiked samples must be carried through all stages of the sample preparation and measurement steps. A QC reference sample concentrate containing each analyte at a concentration of 100 mg/L in methanol is required.

REFERENCE Test Methods for Evaluating Solid Waste (SW-846). U.S. EPA 1983, Method 8270B, Rev. 2, Nov. 1990. Office of Solid Waste, Washington, DC.

1,2-Dichlorobenzene   **EPA Method 503.1**
CAS #95-50-1

TITLE Aromatic & Unsaturated VOCs

MATRIX Drinking water (finished or in Water any treatment stage) and raw source water.

APPLICATION Method covers 28 aromatic and unsaturated VOCs. An inert gas is bubbled through a 5-mL water sample. Purged sample components are trapped in tube of sorbent materials. When purging is complete, sorbent tube is heated and backflushed with inert gas to desorb trapped sample onto a packed GC column.

INTERFERENCES During analysis, major contaminant sources are volatile materials in the lab and impurities in purging gas and sorbent trap. With high and low level samples, there can be carryover contamination. Excess water causes a negative baseline deflection.

INSTRUMENTATION Purge and Trap GC w/photoionization detector. (Two GC columns are recommended); Column 1: 5% SP-1200 and 1.75% Bentone 34 on Supelcoport; 5% 1,2,3-tris(2-cyanoethoxy)propane on Chromosorb W.

RANGE 2.2–600 µg/L (Drinking water).

MDL 0.02 µg/L in water

PRECISION RSD = 18.8% at 19.4 µg/L; 4 labs

ACCURACY Average recovery = 85% at 19.4 µg/L; 4 labs

SAMPLING METHOD Use a 40–120-mL screw-cap vial (prewashed with detergent, rinsed with distilled water and oven dried at 105°C) with a PTFE-faced silicone septum. If residual chlorine is in the water add about 25 mg of ascorbic acid to each vial before sample collection. Collect bubble-free samples.

STABILITY Cool to 4°C; HCl to pH <2.

MHT 14 days.

QUALITY CONTROL As an initial demonstration of lab accuracy and precision, analyze 4 to 7 replicates of a lab fortified blank containing analyte at 0.1–5 µg/L. Collect all samples in duplicate.

REFERENCE Method 503.1, Volatile Aromatic & Unsaturated Organic Compounds in H2O by Purge and Trap GC, EPA 600/4-88/039.

1,3-Dichlorobenzene   **EPA Method 503.1**
CAS #541-73-1

TITLE Aromatic & Unsaturated VOCs

MATRIX Drinking water (finished or in Water any treatment stage) and raw source water.

APPLICATION Method covers 28 aromatic and unsaturated VOCs. An inert gas is bubbled through a 5-mL water sample. Purged sample components are trapped in tube of sorbent materials. When purging is complete, sorbent tube is heated and backflushed with inert gas to desorb trapped sample onto a packed GC column.

INTERFERENCES During analysis, major contaminant sources are volatile materials in the lab and impurities in purging gas and sorbent trap. With high and low level samples, there can be carryover contamination. Excess water causes a negative baseline deflection.

INSTRUMENTATION Purge and Trap GC w/photoionization detector. (Two GC columns are recommended); Column 1: 5% SP-1200 and 1.75% Bentone 34 on Supelcoport; 5% 1,2,3-tris(2-cyanoethoxy)propane on Chromosorb W.

RANGE 2.2–600 µg/L (Drinking water).

MDL 0.006 µg/L in water

PRECISION RSD = 8.5% at 0.50 µg/L; 19 samples

ACCURACY Average recovery = 91% at 0.50 µg/L; 19 samples

SAMPLING METHOD Use a 40–120-mL screw-cap vial (prewashed with detergent, rinsed with distilled water and oven dried at 105°C) with a PTFE-faced silicone septum. If residual chlorine is in the water add about 25 mg of ascorbic acid to each vial before sample collection. Collect bubble-free samples.

STABILITY Cool to 4°C; HCl to pH <2.

MHT 14 days.

QUALITY CONTROL As an initial demonstration of lab accuracy and precision, analyze 4 to 7 replicates of a lab fortified blank containing analyte at 0.1–5 µg/L. Collect all samples in duplicate.

REFERENCE Method 503.1, Volatile Aromatic & Unsaturated Organic Compounds in H2O by Purge and Trap GC, EPA 600/4-88/039.

1,4-Dichlorobenzene   **EPA Method 503.1**
CAS #106-46-7

TITLE Aromatic & Unsaturated VOCs

MATRIX Drinking water (finished or in Water any treatment stage) and raw source water.

APPLICATION Method covers 28 aromatic and unsaturated VOCs. An inert gas is bubbled through a 5-mL water sample. Purged sample components are trapped in tube of sorbent materials. When purging is complete, sorbent tube is heated and backflushed with inert gas to desorb trapped sample onto a packed GC column.

INTERFERENCES During analysis, major contaminant sources are volatile materials in the lab and impurities in purging gas and sorbent trap. With high and low level samples, there can be carryover contamination. Excess water causes a negative baseline deflection.

INSTRUMENTATION Purge and Trap GC w/photoionization detector. (Two GC columns are recommended); Column 1: 5% SP-1200 and 1.75% Bentone 34 on Supelcoport; 5% 1,2,3-tris(2-cyanoethoxy)propane on Chromosorb W.

RANGE 2.2–600 µg/L (Drinking water).

MDL 0.006 µg/L in water

PRECISION RSD = 22.8% at 68.6 µg/L; 5 labs

ACCURACY Average recovery = 91% at 68.6 µg/L; 5 labs

SAMPLING METHOD Use a 40–120-mL screw-cap vial (prewashed with detergent, rinsed with distilled water and oven dried at 105°C) with a PTFE-faced silicone septum. If residual chlorine is in the water add about 25 mg of ascorbic acid to each vial before sample collection. Collect bubble-free samples.

STABILITY Cool to 4°C; HCl to pH <2.

MHT 14 days.

QUALITY CONTROL As an initial demonstration of lab accuracy and precision, analyze 4 to 7 replicates of a lab fortified blank containing analyte at 0.1–5 µg/L. Collect all samples in duplicate.

REFERENCE Method 503.1, Volatile aromatic & unsaturated organic compounds in H2O by Purge and Trap GC, EPA 600/4-88/039.

1,2-Dichlorobenzene   **EPA Method 601**
CAS #95-50-1

TITLE Purgeable Halocarbons

MATRIX Wastewater

APPLICATION Method covers 29 purgeable halocarbons. (Method 624 provides GC/MS conditions appropriate for the qualitative and quantitative confirmation of results). Method describes conditions for a 2nd GC column to confirm measurements made with primary column.

INTERFERENCES Impurities in the purge gas and organic compounds outgassing from the plumbing ahead of the trap. With high- and low-level samples, there can be carryover contamination. Diffusion of volatile organics through the septum seal into the sample.

INSTRUMENTATION GC-equipped with halide-specific detector. (With purge-and-trap unit).

RANGE 8.0–500 µg/L.

MDL 0.15 µg/L.

PRECISION 0.13X+6.13 µg/L (overall precision).

ACCURACY 0.93C+1.70 µg/L (as recovery).

SAMPLING METHOD 25-mL glass vial. Teflon®-lined septum.

STABILITY cool, 4°C, 0.008% Na2S2O3.

MHT 14 days.

QUALITY CONTROL The lab must on an ongoing basis, spike at least 10% of the samples from each sample site being monitored to assess accuracy.

REFERENCE Method 601, *Federal Register* Part VIII 40 CFR Part 136, Oct 26, 1984.

1,3-Dichlorobenzene   **EPA Method 601**
CAS #541-73-1

TITLE Purgeable Halocarbons

MATRIX Wastewater

APPLICATION Method covers 29 purgeable halocarbons. (Method 624 provides GC/MS conditions appropriate for the qualitative and quantitative confirmation of results). Method describes conditions for a 2nd GC column to confirm measurements made with primary column.

INTERFERENCES Impurities in the purge gas and organic compounds outgassing from the plumbing ahead of the trap. With high- and low-level samples, there can be carryover contamination. Diffusion of volatile organics through the septum seal into the sample.

INSTRUMENTATION GC-equipped with halide-specific detector. (With purge-and-trap unit).

RANGE 8.0–500 µg/L.

MDL 0.32 µg/L.

PRECISION 0.26X+2.34 µg/L (overall precision).

ACCURACY 0.95C+0.43 µg/L (as recovery).

SAMPLING METHOD 25-mL glass vial. Teflon®-lined septum.

STABILITY cool, 4°C, 0.008% Na2S2O3.

MHT 14 days.

QUALITY CONTROL The lab must on an ongoing basis, spike at least 10% of the samples from each sample site being monitored to assess accuracy.

REFERENCE Method 601, *Federal Register* Part VIII 40 CFR Part 136, Oct 26, 1984.

1,4-Dichlorobenzene   EPA Method 601
CAS #106-46-7

TITLE Purgeable Halocarbons

MATRIX Wastewater

APPLICATION Method covers 29 purgeable halocarbons. (Method 624 provides GC/MS conditions appropriate for the qualitative and quantitative confirmation of results). Method describes conditions for a 2nd GC column to confirm measurements made with primary column.

INTERFERENCES Impurities in the purge gas and organic compounds outgassing from the plumbing ahead of the trap. With high- and low-level samples, there can be carryover contamination. Diffusion of volatile organics through the septum seal into the sample.

INSTRUMENTATION GC-equipped with halide-specific detector. (With purge-and-trap unit).

RANGE 8.0–500 µg/L.

MDL 0.24 µg/L.

PRECISION 0.20X+0.41 µg/L (overall precision).

ACCURACY 0.93C-0.09 µg/L (as recovery).

SAMPLING METHOD 25-mL glass vial. Teflon®-lined septum.

STABILITY cool, 4°C, 0.008% Na2S2O3.

MHT 14 days.

QUALITY CONTROL The lab must on an ongoing basis, spike at least 10% of the samples from each sample site being monitored to assess accuracy.

REFERENCE Method 601, *Federal Register* Part VIII 40 CFR Part 136, Oct 26, 1984.

1,2-Dichlorobenzene   EPA Method 602
CAS #95-50-1

TITLE Purgeable Aromatics

MATRIX Wastewater

APPLICATION Method covers 7 purgeable aromatics. (Method 624 provides GC/MS conditions appropriate for the qualitative and quantitative confirmation of results). Method describes conditions for a 2nd GC column to confirm measurements made with primary column.

INTERFERENCES Impurities in the purge gas and organic compounds outgassing from the plumbing ahead of the trap. With high- and low-level samples, there can be carryover contamination. Diffusion of volatile organics through the septum seal into the sample.

INSTRUMENTATION GC-equipped with photoionization detector. (With purge-and-trap unit).

RANGE 2.1 To 550 µg/L.

MDL 0.4 µg/L.

PRECISION 0.22X+0.53 µg/L (overall precision).

ACCURACY 0.93C+0.52 µg/L (as recovery).

SAMPLING METHOD 25-mL glass vial. Teflon®-lined septum.

STABILITY cool, 4°C, 0.008% Na2S2O3.

MHT 14 days.

QUALITY CONTROL The lab must on an ongoing basis, spike at least 10% of the samples from each sample site being monitored to assess accuracy.

REFERENCE Method 602, *Federal Register* Part VIII 40 CFR Part 136, Oct 26, 1984.

1,3-Dichlorobenzene   EPA Method 602
CAS #541-73-1

TITLE Purgeable Aromatics

MATRIX Wastewater

APPLICATION Method covers 7 purgeable aromatics. (Method 624 provides GC/MS conditions appropriate for the qualitative and quantitative confirmation of results). EPA Method describes conditions for a 2nd GC column to confirm measurements made with primary column.

INTERFERENCES Impurities in the purge gas and organic compounds outgassing from the plumbing ahead of the trap. With high- and low-level samples, there can be carryover contamination. Diffusion of volatile organics through the septum seal into the sample.

INSTRUMENTATION GC-equipped with photoionization detector. (With purge-and-trap unit).

RANGE 2.1 To 550 µg/L.

MDL 0.4 µg/L.

PRECISION 0.19X+0.09 µg/L (overall precision).

ACCURACY 0.96C-0.04 µg/L (as recovery).

SAMPLING METHOD 25-mL glass vial. Teflon®-lined septum.

STABILITY cool, 4°C, 0.008% Na2S2O3.

MHT 14 days.

QUALITY CONTROL The lab must on an ongoing basis, spike at least 10% of the samples from each sample site being monitored to assess accuracy.

REFERENCE Method 602, *Federal Register* Part VIII 40 CFR Part 136, Oct 26, 1984.

1,4-Dichlorobenzene **EPA Method 602**
CAS #106-46-7

TITLE Purgeable Aromatics

MATRIX Wastewater

APPLICATION Method covers 7 purgeable aromatics. (Method 624 provides GC/MS conditions appropriate for the qualitative and quantitative confirmation of results). EPA Method describes conditions for a 2nd GC column to confirm measurements made with primary column.

INTERFERENCES Impurities in the purge gas and organic compounds outgassing from the plumbing ahead of the trap. With high- and low-level samples, there can be carryover contamination. Diffusion of volatile organics through the septum seal into the sample.

INSTRUMENTATION GC-equipped with photoionization detector. (With purge-and-trap unit).

RANGE 2.1 To 550 µg/L.

MDL 0.3 µg/L.

PRECISION 0.20X+0.41 µg/L (overall precision).

ACCURACY 0.93C+0.09 µg/L (as recovery).

SAMPLING METHOD 25-mL glass vial. Teflon®-lined septum.

STABILITY cool, 4°C, 0.008% Na2S2O3.

MHT 14 days.

QUALITY CONTROL The lab must on an ongoing basis, spike at least 10% of the samples from each sample site being monitored to assess accuracy.

REFERENCE Method 602, *Federal Register* Part VIII 40 CFR Part 136, Oct 26, 1984.

1,2-Dichlorobenzene **EPA Method 624**
CAS #95-50-1

TITLE Purgeables

MATRIX Wastewater

APPLICATION Method covers 31 purgeable organics. An inert gas is bubbled through a 5-mL water sample in a specially designed purging chamber. Here, purgeables are transferred from aqueous to gaseous phase, passed onto a sorbent column, and trapped. Trap is heated and backflushed with inert gas to desorb purgeables onto a GC column, where purgeables are separated.

INTERFERENCES Impurities in the purge gas, organic compounds outgassing from the plumbing ahead of the trap, and solvent vapors in the lab. With high- and low-level samples, there can be carryover contamination.

INSTRUMENTATION GC/MS with purge-and-trap unit.

RANGE 5–600 µg/L.

MDL Not determined

PRECISION 0.30X–1.20 µg/L (overall precision).

ACCURACY 0.94C+4.47 µg/L (as recovery).

SAMPLING METHOD 25-mL glass vial. Teflon®-lined septum.

STABILITY cool, 4°C, 0.008% Na2S2O3.

MHT 14 days.

QUALITY CONTROL The lab must on an ongoing basis, spike at least 5% of the samples from each sample site being monitored to assess accuracy.

REFERENCE Method 624, *Federal Register* Part VIII 40 CFR Part 136, Oct 26, 1984.

1,3-Dichlorobenzene **EPA Method 624**
CAS #541-73-1

TITLE Purgeables

MATRIX Wastewater

APPLICATION Method covers 31 purgeable organics. An inert gas is bubbled through a 5-mL water sample in a specially designed purging chamber. Here, purgeables are transferred from aqueous to gaseous phase, passed onto a sorbent column, and trapped. Trap is heated and backflushed with inert gas to desorb purgeables onto a GC column, where purgeables are separated.

INTERFERENCES Impurities in the purge gas, organic compounds outgassing from the plumbing ahead of the trap, and solvent vapors in the lab. With high- and low-level samples, there can be carryover contamination.

INSTRUMENTATION GC/MS with purge-and-trap unit.

RANGE 5–600 µg/L.

MDL Not determined

PRECISION 0.18X–0.82 µg/L (overall precision).

ACCURACY 1.06C+1.68 µg/L (as recovery).

SAMPLING METHOD 25-mL glass vial. Teflon®-lined septum.

STABILITY cool, 4°C, 0.008% Na2S2O3.

MHT 14 days.

QUALITY CONTROL The lab must on an ongoing basis, spike at least 5% of the samples from each sample site being monitored to assess accuracy.

REFERENCE Method 624, *Federal Register* Part VIII 40 CFR Part 136, Oct 26, 1984.

1,4-Dichlorobenzene   **EPA Method 624**
CAS #106-46-7

TITLE Purgeables

MATRIX Wastewater

APPLICATION Method covers 31 purgeable organics. An inert gas is bubbled through a 5-mL water sample in a specially designed purging chamber. Here, purgeables are transferred from aqueous to gaseous phase, passed onto a sorbent column, and trapped. Trap is heated and backflushed with inert gas to desorb purgeables onto a GC column, where purgeables are separated.

INTERFERENCES Impurities in the purge gas, organic compounds outgassing from the plumbing ahead of the trap, and solvent vapors in the lab. With high- and low-level samples, there can be carryover contamination.

INSTRUMENTATION GC/MS with purge-and-trap unit.

RANGE 5–600 µg/L.

MDL Not determined

PRECISION 0.30X–1.20 µg/L (overall precision).

ACCURACY 0.94C+4.47 µg/L (as recovery).

SAMPLING METHOD 25-mL glass vial. Teflon®-lined septum.

STABILITY cool, 4°C, 0.008% Na2S2O3.

MHT 14 days.

QUALITY CONTROL The lab must on an ongoing basis, spike at least 5% of the samples from each sample site being monitored to assess accuracy.

REFERENCE Method 624, *Federal Register* Part VIII 40 CFR Part 136, Oct 26, 1984.

1,2-Dichlorobenzene   **EPA Method 8010**
CAS #95-50-1

TITLE Halogenated Volatile Organics

MATRIX Groundwater, soils, sludges, water miscible liquid wastes, and non-water miscible wastes.

APPLICATION This method is used for the analysis of 39 halogenated VOCs. Samples are analyzed using direct injection or purge-and-trap methods. Groundwater must be analyzed by the purge-and-trap method. The method provides an optional GC column which is used for analyte confirmation and that may help resolve analytes from interferences.

INTERFERENCES There can be carryover contamination with high- and low-level samples. Impurities may come from the purge-and-trap apparatus, organic compounds outgassing from the plumbing ahead of trap, diffusion of VOCs through the sample bottle septum during shipping or storage, or from solvent vapors in the lab.

INSTRUMENTATION GC capable of on-column injections or purge-and-trap sample introduction and a halogen specific detector. Column 1: 8 ft by 0.1 in 1%. SP-1000 on Carbopack-B. Column 2: 6 ft by 0.1 in bonded n-octane on Porasil-C.

RANGE 8 to 500 µg/L (reagent water).

MDL 0.15 µg/L (reagent water).

PQL FACTORS FOR MULTIPLYING × FID MDL VALUE

Matrix	Multiplication Factor
Groundwater	10
Low-level soil	10
Water miscible liquid waste	500
High-level soil and sludge	1250
Non-water miscible waste	1250

PRECISION 0.13X + 6.13 µg/L (overall precision).

ACCURACY 0.93C + 1.70 µg/L (as recovery).

SAMPLING METHOD For water and liquid samples; use glass 40-mL vials with Teflon®-lined septum caps and collect two vials per sample location with no headspace. For solids and concentrated waste samples; use widemouth glass bottles with Teflon® liners.

STABILITY For concentrated wastes, soils, sediments, or sludges: cool to 4°C. For liquids: add 4 drops of concentrated hydrochloric acid and cool to 4°C.

MHT 14 days.

QUALITY CONTROL Analyze a reagent blank, matrix spike, and matrix spike duplicate/duplicate for each analytical batch (up to 20 samples). Demonstrate the purity of glassware and reagents by analyzing a reagent water method blank. Internal, surrogate, and five concentration level calibration standards are used.

REFERENCE Test Methods for Evaluating Solid Waste (SW-846), U.S. EPA Office of Solid Waste, Washington, DC, Method 8010B, Rev. 2, Nov. 1992.

1,3-Dichlorobenzene   **EPA Method 8010**
CAS #541-73-1

TITLE Halogenated Volatile Organics

MATRIX Groundwater, soils, sludges, water miscible liquid wastes, and non-water miscible wastes.

APPLICATION This method is used for the analysis of 39 halogenated VOCs. Samples are analyzed using direct injection or purge-and-trap methods. Groundwater must be analyzed by the purge-and-trap method. The method provides an optional GC column which is used for analyte confirmation and that may help resolve analytes from interferences.

INTERFERENCES There can be carryover contamination with high- and low-level samples. Impurities may come from the purge-and-trap apparatus, organic compounds outgassing from the plumbing ahead of trap, diffusion of VOCs through the sample bottle septum during shipping or storage, or from solvent vapors in the lab.

INSTRUMENTATION GC capable of on-column injections or purge-and-trap sample introduction and a halogen specific detector. Column 1: 8 ft by 0.1 in 1%. SP-1000 on Carbopack-B. Column 2: 6 ft by 0.1 in bonded n-octane on Porasil-C.

RANGE 8 to 500 µg/L (reagent water).

MDL 0.32 µg/L (reagent water).

PQL FACTORS FOR MULTIPLYING × FID MDL VALUE

Matrix	Multiplication Factor
Groundwater	10
Low-level soil	10
Water miscible liquid waste	500
High-level soil and sludge	1250
Non-water miscible waste	1250

PRECISION 0.26X + 2.34 µg/L (overall precision).

ACCURACY 0.95C + 0.43 µg/L (as recovery).

SAMPLING METHOD For water and liquid samples; use glass 40-mL vials with Teflon®-lined septum caps and collect two vials per sample location with no headspace. For solids and concentrated waste samples; use widemouth glass bottles with Teflon® liners.

STABILITY For concentrated wastes, soils, sediments, or sludges: cool to 4°C. For liquids: add 4 drops of concentrated hydrochloric acid and cool to 4°C.

MHT 14 days.

QUALITY CONTROL Analyze a reagent blank, matrix spike, and matrix spike duplicate/duplicate for each analytical batch (up to 20 samples). Demonstrate the purity of glassware and reagents by analyzing a reagent water method blank. Internal, surrogate, and five concentration level calibration standards are used.

REFERENCE Test Methods for Evaluating Solid Waste (SW-846), U.S. EPA Office of Solid Waste, Washington, DC, Method 8010B, Rev. 2, Nov. 1992.

1,4-Dichlorobenzene   **EPA Method 8010**
CAS #106-46-7

TITLE Halogenated Volatile Organics

MATRIX Groundwater, soils, sludges, water miscible liquid wastes, and non-water miscible wastes.

APPLICATION This method is used for the analysis of 39 halogenated VOCs. Samples are analyzed using direct injection or purge-and-trap methods. Groundwater must be analyzed by the purge-and-trap method. The method provides an optional GC column which is used for analyte confirmation and that may help resolve analytes from interferences.

INTERFERENCES There can be carryover contamination with high- and low-level samples. Impurities may come from the purge-and-trap apparatus, organic compounds outgassing from the plumbing ahead of trap, diffusion of VOCs through the sample bottle septum during shipping or storage, or from solvent vapors in the lab.

INSTRUMENTATION GC capable of on-column injections or purge-and-trap sample introduction and a halogen specific detector. Column 1: 8 ft by 0.1 in 1%. SP-1000 on Carbopack-B. Column 2: 6 ft by 0.1 in bonded n-octane on Porasil-C.

RANGE 8 to 500 µg/L (reagent water).

MDL 0.24 µg/L (reagent water).

PQL FACTORS FOR MULTIPLYING × FID MDL VALUE

Matrix	Multiplication Factor
Groundwater	10
Low-level soil	10
Water miscible liquid waste	500
High-level soil and sludge	1250
Non-water miscible waste	1250

PRECISION 0.20X + 0.41 µg/L (overall precision).

ACCURACY 0.93C–0.09 µg/L (as recovery).

SAMPLING METHOD For water and liquid samples; use glass 40-mL vials with Teflon®-lined septum caps and collect two vials per sample location with no headspace. For solids and concentrated waste samples; use widemouth glass bottles with Teflon® liners.

STABILITY For concentrated wastes, soils, sediments, or sludges: cool to 4°C. For liquids: add 4 drops of concentrated hydrochloric acid and cool to 4°C.

MHT 14 days.

QUALITY CONTROL Analyze a reagent blank, matrix spike, and matrix spike duplicate/duplicate for each analytical batch (up to 20 samples). Demonstrate the purity of glassware and reagents by analyzing a reagent water method blank. Internal, surrogate, and five concentration level calibration standards are used.

REFERENCE Test Methods for Evaluating Solid Waste (SW-846), U.S. EPA Office of Solid Waste, Washington, DC, Method 8010B, Rev. 2, Nov. 1992.

1,2-Dichlorobenzene   **EPA Method 8020**
CAS #95-50-1

TITLE Aromatic Volatile Organics

MATRIX Groundwater, soils, sludges, water miscible liquid wastes, and non-water miscible wastes.

APPLICATION This method is used to analyze for 8 aromatic VOCs. Samples are analyzed using direct injection or purge-and-trap methods. Groundwater must be analyzed by the purge-and-trap method. The method provides an optional GC column that is used for analyte confirmation and may also help resolve analytes from interferences.

INTERFERENCES There can be carryover contamination with high- and low-level samples. Impurities may come from the purge-and-trap apparatus, organic compounds outgassing from the plumbing ahead of trap, diffusion of VOCs through the sample bottle septum during shipping or storage, or from solvent vapors in the lab.

INSTRUMENTATION GC capable of on-column injections or purge-and-trap sample introduction and a photoionization detector (PID). Column 1: 6 ft by 0.082 in with 5% SP-1200 and 1.75% Bentone-34 on Supelcoport. Column 2: 8 ft by 0.1 in with 5% 1,2,3-tris(2-cyanoethoxy)propane on Chromosorb W-AW.

RANGE 2.1 to 500 µg/L.

MDL 0.4 µg/L (reagent water).

PQL FACTORS FOR MULTIPLYING × FID MDL VALUE

Matrix	Multiplication Factor
Groundwater	10
Low-level soil	10
Water miscible liquid waste	500
High-level soil and sludge	1250
Non-water miscible waste	1250

PRECISION 0.22X + 0.53 µg/L (overall precision).

ACCURACY 0.93C + 0.52 µg/L (as recovery).

SAMPLING METHOD For water and liquid samples use glass 40-mL vials with Teflon®-lined septum caps and collect two vials per sample location with no headspace. For solids and concentrated waste samples use widemouth glass bottles with Teflon® liners. Cool all samples to 4°C

STABILITY For concentrated wastes, soils, sediments, or sludges cool to 4°C. For liquids, add 4 drops of concentrated hydrochloric acid and cool to 4°C.

MHT 14 days.

QUALITY CONTROL Analyze a reagent blank, matrix spike, and matrix spike duplicate/duplicate for each analytical batch (up to 20 samples). Demonstrate the purity of glassware and reagents by analyzing a reagent water method blank. Internal, surrogate, and five concentration level calibration standards are used. The QC check sample concentrate should contain this compound at 10 µg/mL in methanol.

REFERENCE Test Methods for Evaluating Solid Waste (SW-846), U.S. EPA Office of Solid Waste, Washington, DC, Method 8020A, Rev. 1, Nov. 1992.

1,3-Dichlorobenzene   **EPA Method 8020**
CAS #541-73-1

TITLE Aromatic Volatile Organics

MATRIX Groundwater, soils, sludges, water miscible liquid wastes, and non-water miscible wastes.

APPLICATION This method is used to analyze for 8 aromatic VOCs. Samples are analyzed using direct injection or purge-and-trap methods. Groundwater must be analyzed by the purge-and-trap method. The method provides an optional GC column that is used for analyte confirmation and may also help resolve analytes from interferences.

INTERFERENCES There can be carryover contamination with high- and low-level samples. Impurities may come from the purge-and-trap apparatus, organic compounds outgassing from the plumbing ahead of trap, diffusion of VOCs through the sample bottle septum during shipping or storage, or from solvent vapors in the lab.

INSTRUMENTATION GC capable of on-column injections or purge-and-trap sample introduction and a photoionization detector (PID). Column 1: 6 ft by 0.082 in with 5% SP-1200 and 1.75% Bentone-34 on Supelcoport. Column 2: 8 ft by 0.1 in with 5% 1,2,3-tris(2-cyanoethoxy)propane on Chromosorb W-AW.

RANGE 2.1 to 500 µg/L.

MDL 0.4 µg/L (reagent water).

PQL FACTORS FOR MULTIPLYING × FID MDL VALUE

Matrix	Multiplication Factor
Groundwater	10
Low-level soil	10
Water miscible liquid waste	500
High-level soil and sludge	1250
Non-water miscible waste	1250

PRECISION 0.19X + 0.09 µg/L (overall precision).

ACCURACY 0.96C + 0.04 µg/L (as recovery).

SAMPLING METHOD For water and liquid samples use glass 40-mL vials with Teflon®-lined septum caps and collect two vials per sample location with no headspace. For solids and concentrated waste samples use widemouth glass bottles with Teflon® liners. Cool all samples to 4°C

STABILITY For concentrated wastes, soils, sediments, or sludges cool to 4°C. For liquids, add 4 drops of concentrated hydrochloric acid and cool to 4°C.

MHT 14 days.

QUALITY CONTROL Analyze a reagent blank, matrix spike, and matrix spike duplicate/duplicate for each analytical batch (up to 20 samples). Demonstrate the purity of glassware and reagents by analyzing a reagent water method blank. Internal, surrogate, and five concentration level calibration standards are used. The QC check sample concentrate should contain this compound at 10 µg/mL in methanol.

REFERENCE Test Methods for Evaluating Solid Waste (SW-846), U.S. EPA Office of Solid Waste, Washington, DC, Method 8020A, Rev. 1, Nov. 1992.

1,4-Dichlorobenzene
CAS #106-46-7
EPA Method 8020

TITLE Aromatic Volatile Organics

MATRIX Groundwater, soils, sludges, water miscible liquid wastes, and non-water miscible wastes.

APPLICATION This method is used to analyze for 8 aromatic VOCs. Samples are analyzed using direct injection or purge-and-trap methods. Groundwater must be analyzed by the purge-and-trap method. The method provides an optional GC column that is used for analyte confirmation and may also help resolve analytes from interferences.

INTERFERENCES There can be carryover contamination with high- and low-level samples. Impurities may come from the purge-and-trap apparatus, organic compounds outgassing from the plumbing ahead of trap, diffusion of VOCs through the sample bottle septum during shipping or storage, or from solvent vapors in the lab.

INSTRUMENTATION GC capable of on-column injections or purge-and-trap sample introduction and a photoionization detector (PID). Column 1: 6 ft by 0.082 in with 5% SP-1200 and 1.75% Bentone-34 on Supelcoport. Column 2: 8 ft by 0.1 in with 5% 1,2,3-tris(2-cyanoethoxy)propane on Chromosorb W-AW.

RANGE 2.1 to 500 µg/L.

MDL 0.3 µg/L (reagent water).

PQL FACTORS FOR MULTIPLYING × FID MDL VALUE

Matrix	Multiplication Factor
Groundwater	10
Low-level soil	10
Water miscible liquid waste	500
High-level soil and sludge	1250
Non-water miscible waste	1250

PRECISION 0.20X + 0.41 µg/L (overall precision).

ACCURACY 0.93C + 0.09 µg/L (as recovery).

SAMPLING METHOD For water and liquid samples use glass 40-mL vials with Teflon®-lined septum caps and collect two vials per sample location with no headspace. For solids and concentrated waste samples use widemouth glass bottles with Teflon® liners. Cool all samples to 4°C

STABILITY For concentrated wastes, soils, sediments, or sludges cool to 4°C. For liquids, add 4 drops of concentrated hydrochloric acid and cool to 4°C.

MHT 14 days.

QUALITY CONTROL Analyze a reagent blank, matrix spike, and matrix spike duplicate/duplicate for each analytical batch (up to 20 samples). Demonstrate the purity of glassware and reagents by analyzing a reagent water method blank. Internal, surrogate, and five concentration level calibration standards are used. The QC check sample concentrate should contain this compound at 10 µg/mL in methanol.

REFERENCE Test Methods for Evaluating Solid Waste (SW-846), U.S. EPA Office of Solid Waste, Washington, DC, Method 8020A, Rev. 1, Nov. 1992.

3,3'-Dichlorobenzidine
CAS #91-94-1
EPA Method 1625

TITLE Semivolatile Organic Compounds by Isotope Dilution GC/MS

MATRIX The compounds may be determined in waters, soils, and municipal sludges by this method.

METHOD SUMMARY This method is used to determine 176 semivolatile toxic organic pollutants associated with the CWA (as amended 1987); the RCRA (as amended 1986); the CERCLA (as amended 1986); and other compounds amenable to extraction and analysis by capillary column gas chromatography-mass spectrometry (GC/MS).

Stable isotopically-labeled analogs of the compounds of interest are added to the sample. If the solids content is less than 1%, a 1-L sample is extracted at pH 12–13, then at pH <2 with methylene chloride using continuous extraction techniques.

If the solids content is 30% or less, the sample is diluted to 1% solids with reagent water, homogenized ultrasonically, and extracted at pH 12–13, then at pH <2 with methylene chloride using continuous extraction techniques. If the solids content is greater than 30%, the sample is extracted using ultrasonic techniques.

Each extract is dried over sodium sulfate, concentrated to a volume of 5 mL, cleaned up using GPC, if necessary, and concentrated. Extracts are concentrated to 1 mL if GPC is not performed, and to 0.5 mL if GPC is performed.

An internal standard is added to the extract, and a 1-mL aliquot of the extract is injected into the GC. The compounds are separated by GC and detected by a MS. The labeled compounds serve to correct the variability of the analytical technique.

INTERFERENCES Solvents, reagents, glassware, and other sample processing hardware may yield artifacts and/or elevated baselines causing misinterpretation of chromatograms and spectra. Materials used in the analysis must be demonstrated to be free from interferences under the conditions of analysis by running method blanks initially and with each sample lot (sample started through the extraction process on a given 8-h shift, to a maximum of 20). Specific selection of reagents and purification of solvents by distillation in all glass systems may be required. Glassware and, where possible, reagents are cleaned by solvent rinse and baking at 450°C for 1-h minimum. Interferences coextracted from samples will vary considerably from source to source, depending on the diversity of the site being sampled.

INSTRUMENTATION Major instrumentation includes a GC with a splitless or on-column injection port for capillary column, a MS with 70 eV electron impact ionization, and a data system to collect and record MS data, and process it. A K-D apparatus is used to concentrate extracts.

GC Column: 30 m × 0.25 mm I.D. 5% phenyl, 94% methyl, 1% vinyl silicone bonded phased fused silica capillary column.

PRECISION & ACCURACY The detection limits of the method are usually dependent on the level of interferences rather than instrumental limitations. The limits typify the minimum quantities that can be detected with no interferences present.

The minimum level (in µg/mL) was 50. This is defined as a minimum level at which the analytical system shall give recognizable mass spectra (background corrected) and acceptable calibration points.

The MDL (in µg/kg) in low solids was 62 and in high solids was 111; these were determined in digested sludge (low solids) and in filter cake or compost (high solids).

The labeled and native compound initial precision as standard deviation (in µg/L) was 26.
The labeled and native compound initial accuracy as average recovery (in µg/L) was 68–174.

SAMPLE COLLECTION, PRESERVATION & HANDLING
Collect samples in glass containers. Aqueous samples which flow freely are collected in refrigerated bottles using automatic sampling equipment. Solid samples are collected as grab samples using widemouth jars. Maintain samples at 0 to 4°C from the time of collection until extraction. If residual chlorine is present in aqueous samples, add 80 mg sodium thiosulfate/L of water. Begin sample extraction within 7 days of collection, and analyze all extracts within 40 days of extraction.

SAMPLE PREPARATION Samples containing 1% solids or less are extracted directly using continuous liquid-liquid extraction techniques. Samples containing 1 to 30% solids are diluted to the 1% level with reagent water and extracted using continuous liquid-liquid extraction techniques. Samples containing greater than 30% solids are extracted using ultrasonic techniques.

Base/neutral extraction — Adjust the pH of the waters in the extractors to 12–13 with 6 N NaOH. Extract with methylene chloride for 24–48 h.
Acid extraction — Adjust the pH of the waters in the extractors to 2 or less using 6 N sulfuric acid. Extract with methylene chloride for 24–48 h.
Ultrasonic extraction of high solids samples — Add anhydrous sodium sulfate to the sample and QC aliquot(s). Add acetone:methylene chloride (1:1) to the sample and mix thoroughly

Concentrate extracts using a K-D apparatus.

QUALITY CONTROL The analyst is permitted to modify this method to improve separations or lower the costs of measurements, provided all performance specifications are met. Analyses of blanks are required to demonstrate freedom from contamination. When results of spikes indicate atypical method performance for samples, the samples are diluted to bring method performance within acceptable limits.

For low solids (aqueous samples), extract, concentrate, and analyze two sets of four 1-L aliquots (8 aliquots total) of the precision and recovery standard. For high solids samples, two sets of four 30-g aliquots of the high solids reference matrix are used.

Spike all samples with labeled compounds to assess method performance. Compute percent recovery of the labeled compounds using the internal standard method. Compare the labeled compound recovery for each compound with the corresponding labeled compound recovery.

Reagent water and high solids reference matrix blanks are analyzed to demonstrate freedom from contamination. Extract and concentrate a 1-L reagent water blank or a high solids reference matrix blank with each sample's lot (samples started through the extraction process on the same 8-h shift, to a maximum of 20 samples).

Field replicates may be collected to determine the precision of the sampling technique, and spiked samples may be required to determine the accuracy of the analysis when the internal standard method is used.

REFERENCE Semivolatile Organic Compounds by Isotope Dilution GC/MS. Office of Water Regulation and Standards, U.S. EPA Industrial Technology Division, Washington, DC, EPA Method 1625, Rev. C, June 1989 (contact W.A. Telliard, U.S. EPA, Office of Water Regulations and Standards, 401 M St., SW, Washington, DC, 20460. Phone: 202-382-7131).

3,3′-Dichlorobenzidine   **EPA Method 625**
CAS #91-94-1

TITLE Base/Neutrals and Acids, U.S. EPA Method 625

MATRIX This methods covers municipal and industrial wastewaters.

METHOD SUMMARY Approximately 1 L of sample is serially extracted with methylene chloride at a pH greater than 11 and again at a pH less than 2 using a separatory funnel or a continuous extractor. The methylene chloride extract is dried, concentrated to a volume of 1 mL, and analyzed by GC/MS. Qualitative identification of the parameters in the extract is performed using the retention time and the relative abundance of three characteristic masses (m/z). Qualitative analysis is performed using either external or internal standard techniques with a single characteristic m/z.

INTERFERENCES Method interferences may be caused by contaminants in solvents, reagents, glassware, and other sample processing hardware. Glassware must be scrupulously cleaned. Glassware should be heated in a muffle furnace at 400°C for 5 to 30 min. Some thermally stable materials, such as PCBs, may not be eliminated by this treatment. Solvent rinses with acetone and pesticide quality hexane may be substituted for the muffle furnace heating. Matrix interferences may be caused by contaminants that are coextracted from the sample. The base-neutral extraction may cause significantly reduced recovery of phenols. The packed gas chromatographic columns recommended for the basic fraction may not exhibit sufficient resolution for some analytes.

INSTRUMENTATION A GC/MS system with an injection port designed for on-column injection when using packed columns and for splitless injection when using capillary columns.

Column for base/neutrals: 1.8 m long × 2 mm I.D. glass, packed with 3% SP-2550 on Supelcoport (100/120 mesh) or equivalent.

Column for acids: 1.8 m long × 2 mm I.D. glass, packed with 1% SP-1240DA on Supelcoport (100/120 mesh) or equivalent.

PRECISION & ACCURACY The MDL concentrations were obtained using reagent water. The MDL actually achieved in a given analysis will vary depending on instrument sensitivity and matrix effects. This method was tested by 15 laboratories using reagent water, drinking water, surface water, and industrial wastewaters spiked at six concentrations over the range 5 to 100 µg/L. Single operator precision, overall precision, and method accuracy were found to be directly related to the concentration of the parameter matrix.

The MDL (in µg/L) in reagent water was 16.5.

The standard deviation (in µg/L based on 4 recovery measurements) was 71.4.

The range (in µg/L) for average recovery for 4 measurements was 8.2–212.5.

The range (in %) for percent recovery was D-262.

Accuracy (in µg/L) as expected recovery for one or more measurements of a sample containing a true concentration of C was $1.23C-12.65$.

Precision (in µg/L) as expected single analyst standard deviation of measurements at an average concentration found at X was $0.28X+7.33$.

Overall precision (in µg/L) as expected interlaboratory standard deviation of measurements in an average concentration found at X was $0.47X+3.45$.

C = *True value of the concentration in µg/L.*
X = *Average recovery found for measurements of samples containing a concentration at C in µg/L.*

SAMPLE PREPARATION Adjust the pH to >11 with sodium hydroxide and serially extract in a separatory funnel with methylene chloride or else in a continuous extractor. Next, adjust the pH to <2 with sulfuric acid and serially extract in a separatory funnel with methylene chloride or else in a continuous extractor. Dry the extracts separately through a column of anhydrous sodium sulfate and then concentrate each of the extracts to 1.0 mL using a K-D apparatus.

SAMPLE COLLECTION, PRESERVATION & HANDLING Grab samples must be collected in glass containers. All samples must be refrigerated at 4°C from the time of collection until extraction. If residual chlorine is present, add 80 mg of sodium thiosulfate/L of sample and mix well. All samples must be extracted within 7 days of collection and completely analyzed within 40 days of extraction.

QUALITY CONTROL Make an initial, one-time, demonstration of the ability to generate acceptable accuracy and precision with this method. Before processing any samples, the analyst must analyze a reagent water blank to demonstrate that interferences from the analytical system and glassware are under control. Each time a set of samples is extracted or reagents are changed, a reagent water blank must be processed. Spike and analyze a minimum of 5% of all samples to monitor and evaluate lab data quality. A QC check sample concentrate that contains each parameter of interest at a concentration of 100 µg/mL in acetone is required. PCBs and multicomponent pesticides may be omitted from this test.

After analysis of five spiked wastewater samples, calculate the average percent recovery and the standard deviation of the percent recovery. Spike all samples with the surrogate standard spiking solution and calculate the percent recovery of each surrogate compound.

REFERENCE *Federal Register*, Vol. 49, No. 209. Friday, Oct. 26, 1984.

3,3'-Dichlorobenzidine   EPA Method 8270
CAS #91-94-1

TITLE Semivolatile Organic Compounds by GC/MS

MATRIX This method is used to determine the concentration of semivolatile organic compounds in extracts prepared from all types of solid waste matrices, soils, and groundwater. Although surface waters are not specifically mentioned, this method should be applicable to water samples from rivers, lakes, etc.

METHOD SUMMARY This method covers 259 semivolatile organic compounds. In very limited applications direct injection of the sample into the GC/MS system may be appropriate, but this results in very high detection limits (approximately 10,000 µg/L). Typically, a 1-L liquid sample, containing surrogate, and matrix spiking standards, is extracted in a continuous extractor first under acid conditions and then under basic conditions. Typically 30 g of a solid sample, containing surrogate, and matrix spiking standards, is extracted ultrasonically. After concentrating the extract to 1 mL it is spiked with 10 µL of an internal standard solution just prior to analysis by GC/MS. The volume injected should contain about 100 ng of base/neutral and 200 ng of acid surrogates (for a 1-µL injection). Analysis is performed by GC/MS using a capillary GC column.

INTERFERENCES Raw GC/MS data from all blanks, samples, and spikes must be evaluated for interferences. Contamination by carryover can occur whenever high-concentration and low-concentration samples are sequentially analyzed. To reduce carryover, the sample syringe must be rinsed out between samples with solvent. Whenever an unusually concentrated sample is encountered, it should be followed by the analysis of blank solvent to check for cross-contamination.

INSTRUMENTATION A GC/MS and a data system are required. The GC column used is a 30 m × 0.25 mm I.D. (or 0.32 mm I.D.) 1um film thickness silicone-coated fused silica capillary column. A continuous liquid-liquid extractor equipped with Teflon® or glass connection joints and stopcocks requiring no lubrication, a K-D concentrating apparatus, water bath, and an ultrasonic disrupter with a minimum power of 300 W and with pulsing capability are also required.

PRECISION & ACCURACY The estimated quantitation limit (EQL) of Method 8270B for determining an individual compound is approximately 1 mg/kg (wet weight) for soil or sediment samples, 1–200 mg/kg for wastes (dependent on matrix and method of preparation), and 10 μg/L for groundwater samples. EQLs will be proportionately higher for sample extracts that require dilution to avoid saturation of the detector.

The EQL(b) for groundwater in μg/L is 20.
The EQL (a, b) for low concentrations in soil and sediment in μg/kg is 1300.
Accuracy as μg/L is $1.23C - 12.65$.
Overall precision in μg/L is $0.47X + 3.45$.

(a) *EQLs listed for soil/sediment are based on wet weight. Normally data is reported in a dry-weight basis; therefore, EQLs will be higher based on the % dry weight of each sample. This calculation is based on a 30 g sample and gel permeation chromatography cleanup.*
(b) *Sample EQLs are highly matrix-dependent. The EQLs are provided for guidance and may not always be achievable.*
$C =$ *True value for concentration, in μg/L.*
$X =$ *Average recovery found for measurements of samples containing a concentration of C, in μg/L.*

ESTIMATED QUANTITATION LIMIT

Other Matrices	Factor (a)
High-concentration soil and sludges by sonicator	7.5
Non-water miscible waste	75

(a) EQL for other matrices = [EQL for low soil/sediment] × [Factor]. This estimated EQL is similar to an EPA "Practical Quantitation Limit."

SAMPLING METHOD

Liquid samples — Use a 1 or 2½ gallon amber glass bottle with a screw-top Teflon®-lined cover that has been prewashed with detergent and rinsed with distilled water and methanol (or isopropanol).

Soils, sediments, or sludges — Use an 8-oz. widemouth glass with a screw-top Teflon®-lined cover that has been prewashed with detergent and rinsed with distilled water and methanol (or isopropanol).

SAMPLE PRESERVATION

Liquid samples — If residual chlorine is present, add 3 mL of 10% sodium thiosulfate per gallon, cool to 4°C and store in a solvent-free refrigerator until analysis; if chlorine is not present, then eliminate the sodium thiosulfate addition.

Soils, sediments, or sludges — Cool samples to 4°C and store in a solvent-free refrigerator.

MHT Liquid samples must be extracted within 7 days and the extracts analyzed within 40 days. Soils, sediments, or sludges may be stored for a maximum of 14 days and the extracts analyzed within 40 days.

SAMPLE PREPARATION

Liquid samples — Transfer 1 L quantitatively to a continuous extractor. If high concentrations are anticipated, a smaller volume may be used and then diluted with organic-free reagent water to 1 L. Adjust pH, if necessary, to pH <2 using 1:1 (V/V) sulfuric acid. Pipette 1.0 mL of a surrogate standard spiking solution into each sample. For the sample in each analytical batch selected for spiking, add 1.0 mL of a matrix spiking standard. For base/neutral acid analysis, the amount of the surrogates and matrix spiking compounds added to the sample should result in a final concentration of 100 ng/μL of each analyte in the extract to be analyzed (assuming a 1-μL injection). Extract with methylene chloride for 18–24 h. Next, adjust the pH of the aqueous phase to pH >11 using 10 N sodium hydroxide and extract it with methylene chloride again for 18–24 h. Dry the extract through a column containing anhydrous sodium sulfate and concentrate it to 1 mL using a K-D concentrator.

Soils, sediments, or sludges — Use 30 g of sample. Nonporous or wet samples (gummy or clay type) that do not have a free-flowing sandy texture must be mixed with anhydrous sodium sulfate until the sample is free flowing. Add 1 mL of surrogate standards to all samples, spikes, standards, and blanks. For the sample in each analytical batch selected for spiking, add 1.0 mL of a matrix spiking standard. For base/neutral acid analysis, the amount added of the surrogates and matrix spiking compounds should result in a final concentration of 100 ng/μL of each base/neutral analyte and 200 ng/μL of each acid analyte in the extract to be analyzed (assuming a 1-μL injection). Immediately add a 100-mL mixture of 1:1 methylene chloride:acetone and extract the sample ultrasonically for 3 min and then decant or filter the extracts. Repeat the extraction two or more times. Dry the extract using a column with anhydrous sodium sulfate and concentrate it to 1 mL in a K-D concentrator.

QUALITY CONTROL A methylene chloride solution containing 50 ng/μL of decafluorotriphenylphosphine (DFTPP) is used for tuning the GC/MS system each 12-h shift. A system performance check also must be made during every 12-h shift. A standard containing 50 ng/μL each of 4,4'-DDT, pentachlorophenol, and benzidine is required to verify injection port inertness and GC column performance. A calibration standard at mid-concentration, containing each compound of interest, including all required surrogates, must be performed every 12 h during analysis. After the system performance check is met, calibration check compounds (CCCs) are used to check the validity of the initial calibration.

The internal standard responses and retention times in the calibration check standard must be evaluated immediately after or during data acquisition. If the retention time for any internal standard changes by more than 30 seconds from the last check calibration (12 h), the chromatographic system must be inspected for malfunctions and corrections must be made, as required. If the electron ionization current plot (EICP) area for any of the internal standards changes by a factor of two from the last daily calibration standard check, the mass spectrometer must be inspected for malfunctions and corrections must be made, as appropriate.

Demonstrate, through the analysis of a reagent water blank, that interferences from the analytical system, glassware, and reagents are under control. The blank samples should be carried through all stages of the sample preparation and measurement

steps. For each analytical batch (up to 20 samples), a reagent blank, matrix spike, and matrix spike duplicate/duplicate must be analyzed (the frequency of the spikes may be different for different monitoring programs). The blank and spiked samples must be carried through all stages of the sample preparation and measurement steps. A QC reference sample concentrate containing each analyte at a concentration of 100 mg/L in methanol is required.

REFERENCE Test Methods for Evaluating Solid Waste (SW-846). U.S. EPA 1983, Method 8270B, Rev. 2, Nov. 1990. Office of Solid Waste, Washington, DC.

3,5-Dichlorobenzoic acid   **EPA Method 8151**
CAS #51-36-5

TITLE Chlorinated Herbicides by GC Using Methylation or Pentafluorobenzylation Derivatization: Capillary Column Technique.

MATRIX This method covers aqueous and solid matrices. This includes a wide variety such as drinking water, groundwater, industrial wastewaters, surface waters, soils, solids, and sediments.

METHOD SUMMARY This is a GC method for determining 19 chlorinated acid herbicides in aqueous, soil, and waste matrices. Because these compounds are produced and used in various forms (i.e., acid, salt, ester, etc.) a hydrolysis step is included to convert the herbicide to the acid form prior to analysis. This method provides hydrolysis, extraction, derivatization and GC conditions for the analysis of chlorinated acid herbicides in water, soil, and waste samples. Water samples are hydrolyzed *in situ*, extracted with diethyl ether, and then esterified with either diazomethane or pentafluorobenzyl bromide. The derivatives are determined by gas chromatography with an electron capture detector (GC/ECD). The results are reported as acid equivalents. The sensitivity of this method depends on the level of interferences in addition to instrumental limitations.

INTERFERENCES Method interferences may be caused by contaminants in solvents, reagents, glassware, and other sample processing hardware. Immediately prior to use, glassware should be rinsed with the next solvent to be used. Matrix interferences may be caused by contaminants that are coextracted from the sample. Organic acids, especially chlorinated acids, cause the most direct interference with the determination by methylation. Phenols, including chlorophenols, may also interfere with this procedure. The determination using pentafluorobenzylation is more sensitive, and more prone to interferences from the presence of organic acids of phenols than by methylation. Alkaline hydrolysis and subsequent extraction of the basic solution removes many chlorinated hydrocarbons and phthalate esters that might otherwise interfere with the ECD analysis. The herbicides, being strong organic acids, react readily with alkaline substances and may be lost during analysis. Therefore, glassware must be acid-rinsed and then rinsed to constant pH with organic-free reagent water.

INSTRUMENTATION A GC suitable for Grob-type injection using capillary columns. A data system for measuring peak heights and/or peak areas is recommended. An electron capture detector (ECD) is used. Also a K-D apparatus, a diazomethane generator, a centrifuge and an ultrasonic disrupter will be required.

Narrow Bore Columns:
Primary Column 1: 30 m × 0.25 mm, 5% phenyl/95% methyl silicone (DB-5), 0.25 µm film thickness.
Primary Column 1a (GC/MS): 30 m × 0.32 mm, 5% phenyl/95% methyl silicone (DB-5), 1-µm film thickness.
Column 2: 30 m × 0.25 mm DB-608 with a 25 µm film thickness.
Confirmation Column: 30 m × 0.25 mm, 14% cyanopropyl phenyl silicone (DB-1701), 0.25 µm film thickness.

Megabore Columns:
Primary Column: 30 m × 0.53 mm DB-608 with 0.83 µm film thickness.
Confirmation Column: 30 m × 0.53 mm, 14% cyanopropyl phenyl silicone (DB-1701), 1.0 µm film thickness.

PRECISION & ACCURACY Method detection limits (MDLs) are compound-dependent and vary with derivitization efficiency, derivative recovery, the matrix sampled, and herbicide concentration.

The estimated MDL (in µg/L) was 0.061 for aqueous samples using GC/ECD.

The estimated MDL (in µg/kg) was 0.38 for soil samples using GC/ECD when corrected back to 50-g samples extracted and concentrated to 10 mL with 5-µL injections.

The estimated GC/MS identification limit (in ng) was 0.65 for soil samples using GC/MS.

Mean percent recovery, calculated from 7–8 determinations of spiked reagent water, after diazomethane derivatization, from a spike concentration (in µg/L) of 0.6 was 102 with a standard deviation of the percent recovery of 16.3.

Mean percent recovery, calculated from 10 determinations of spiked clay and clay/still bottom samples over the linear concentration range (in ng/g) of no data was none reported with a percent relative standard deviation of none. The RSD % was calculated on 10 samples high in the linear concentration range and 10 low in the range. The linear concentration range was determined using standard solutions and corrected to 50 g soil samples.

SAMPLE COLLECTION, PRESERVATION & HANDLING Containers used to collect samples for the determination of semivolatile organic compounds should be soap and water washed followed by methanol (or isopropanol) rinsing. The sample containers should be of glass or Teflon® and have screw-top covers with Teflon® liners.

No preservation is used with concentrated waste samples. With liquid samples containing no residual chlorine and with soil, sediment, and sludge samples, immediately cooling to 4°C is the only preservation used. When residual chlorine is present then 3 mL of 10% aqueous sodium sulfate is added for each gallon of sample collected, followed by cooling to 4°C.

The holding time for all volatile organics samples is 14 days. Liquid samples must be extracted within 7 days and their extracts analyzed within 40 days. Concentrated waste, soil, sediment, and sludge samples must be extracted within 14 days and their extracts analyzed within 40 days.

SAMPLE PREPARATION

Preparation of soil, sediment, and other solid samples — Acidify 30 g (dry weight) solids with 0.1 M phosphate buffer (pH = 2.5) and thoroughly mix the contents. Spike the sample with surrogate compound(s). The ultrasonic extraction of solids must be optimized for each type of sample. In order for the ultrasonic extractor to efficiently extract solid samples, the sample must be free flowing when the solvent is added. Acidified anhydrous sodium sulfate should be added to clay-type soils, or any other solid that is not a free-flowing sandy texture, until a free flowing mixture is obtained. Add methylene chloride and perform ultrasonic extraction. Combine organic extracts from the repetitive extractings of the sample and centrifuge. Add aqueous potassium hydroxide, water, and methanol to the extract and reflux the mixture on a water bath. Extract the solution three times with methylene chloride and discard the methylene chloride phase. The basic solution contains the herbicide salts. Adjust the pH of the solution to <2 with cold sulfuric acid and extract three times with methylene chloride. Combine the extracts and pour them through a pre-rinsed drying column containing acidified anhydrous sodium sulfate. Collect the dried extracts in a K-D flask and concentrate them.

Preparation of aqueous samples — Measure 1 L of sample into a 2 L separatory funnel and spike it with surrogate compound(s). Add NaCl to the sample, then add 6 N NaOH to the sample to a pH of 12 or more and let the sample sit at room temperature for 1 h to hydrolyze esters. Extract the sample three times with methylene chloride and discard the extracts. Then add cold 12 N sulfuric acid to a pH less than or equal to 2, and extract the sample three times with ethyl ether. Collect the ether phase in a flask containing acidified anhydrous sodium sulfate and allow it to remain in contact with the sodium sulfate for a minimum of 2 h. The drying step is very critical to ensuring complete esterification; any moisture remaining in the ether will result in low herbicide recoveries.

Extract concentration and derivatization — The combined ether extract is concentrated to about 1 mL using a K-D apparatus followed by using a micro Snyder column or nitrogen gas blowdown. If methyl esters are to be produced, then dilute the concentrated ether extract with 1 mL of isooctane and 0.5 mL of methanol, dilute to a final volume of 4 mL, and esterify with diazomethane. If pentafluorobenzene esters are to be produced, then dilute concentrated ether extract with acetone to a final volume of 4 mL and esterify with pentafluorobenzyl bromide.

QUALITY CONTROL
Select a representative spike concentration for each compound (acid or ester) to be measured. Using stock standard, prepare a quality control check sample concentrate, in acetone, that is 1000 times more concentrated than the selected concentrations. Use this quality control check sample concentrate to prepare quality control check samples. Calculate surrogate standard recovery on all standards, samples, blanks, and spikes. GC/MS techniques should be judiciously employed to support qualitative identifications made with this method. When available, chemical ionization mass spectra may be employed to aid the qualitative identification process.

REFERENCE Test Methods for Evaluating Solid Waste, Physical/Chemical Methods, SW-846, 3rd Edition, U.S. EPA, Office of Solid Waste, Washington, DC, EPA Method 8151, Nov. 1990.

Dichlorodifluoromethane   **EPA Method 502**
CAS #75-71-8

TITLE Volatile Organic Compounds in Water By Purge and Trap Capillary Column Gas Chromatography with Photoionization and Electrolytic Conductivity Detectors in Series. U.S. EPA Method 502.2, Rev. 2.0, 1989.

MATRIX Drinking water and raw source water. The latter should include most surface water and groundwater sources.

METHOD SUMMARY This method covers 60 volatile organic compounds that contain halogen atoms and/or that are aromatic. An inert gas (zero grade nitrogen or helium) is bubbled through a 25-mL or a 5-mL water sample (depending on the expected concentration of the analytes). Purged sample components are trapped in a tube of sorbent materials. When purging is complete, the sorbent tube is heated and backflushed with helium to desorb the trapped sample onto a capillary GC column. The column is temperature programmed to separate the method analytes which are then detected with a photoionization detector (PID) and a Hall electrolytic conductivity (HECD) placed in series. The PID is selective for aromatic compounds and the HECD is selective for halogenated compounds.

INTERFERENCES Impurities in the purge gas and from organic compounds outgassing from the plumbing ahead of the trap account for many contamination problems. Interferences purged or coextracted from the samples will vary considerably from source to source, depending upon the particular sample or extract being tested. Cross-contamination can occur whenever high-level and low-level samples are analyzed sequentially. Samples also can be contaminated by diffusion of volatile organics (particularly methylene chloride and fluorocarbons) through the septum seal into the sample during shipment and storage. The lab where volatile analysis is performed and also the refrigerated storage area should be completely free of solvents.

INSTRUMENTATION A GC containing a series configuration of a high temperature photoionization detector (PID) equipped with 10.0 eV (nominal) lamp and Hall electrolytic conductivity detector (HECD) is required. Also required is an all-glass 5-mL purging device, a sorbent trap, and a thermal desorption apparatus which is connected to the GC system.

Column 1: VOCOL glass wide-bore capillary column.
Column 2: RTX–502.2 mega-bore capillary column.
Column 3: DB-62 mega-bore capillary column.

PRECISION & ACCURACY Method detection limits are dependent upon the characteristics of the gas chromatographic system used. Analytes that are not separated chromatographically cannot be individually identified and used in the same calibration mixture or water samples unless an alternative technique for identification and quantification, such as mass spectrometry, is used.

Electrolytic conductivity detetor (c) range in µg/L (a) was 0.02–200.
Electrolytic conductivity detetor (c) MDL in µg/L (b) was 0.05.
Electrolytic conductivity detetor (c) accuracy as % recovery was 89.
Electrolytic conductivity detetor (c) precision as % RSD was 6.6.
Photoionization detector (d) range in µg/L (a) was 0.02–200.
Photoionization detector (d) MDL in µg/L (b) was not listed.
Photoionization detector (d) accuracy as % recovery was not listed.
Photoionization detector (d) precision as % RSD was not listed.

(a) *The applicable concentration range of this method is compound, instrument, and matrix-dependent. It is listed as being approximately 0.02 to 200 µg/L but no specific information is provided so caution should be observed.*
(b) *The method detection limits reports with this method are compound, instrument, and matrix-dependent. The values reported were calculated using reagent water fortified with the corresponding compounds at 10 µg/L and a GC-equipped with a 60 m × 0.75 mm VOLCOL wide bore capillary column with 1.5 µm film thickness and using helium carrier gas.*
(c) *Recoveries and relative standard deviations were determined from seven samples of reagent water fortified with 10 µg/L of each compound. 2-Bromo-1-chloropropane was used as the internal standard for calculating average recoveries.*
(d) *Recoveries and relative standard deviations were determined from seven samples of reagent water fortified with 10 µg/L of each compound. Fluorobenzene was used as the internal standard for calculating average recoveries.*

SAMPLING METHOD Collect samples using a 40- to 120-mL screw-cap vial (prewashed with detergent, rinsed with distilled water and oven dried at 105°C) with a Teflon®-faced silicone septum. Collect bubble-free samples and place the septum with the Teflon® side down on the water.

SAMPLE PRESERVATION If residual chlorine is present in the water add about 25 mg of ascorbic acid to each vial before samples are collected to remove the chlorine. Add hydrochloric acid to reduce pH to <2, immediately cool samples to 4°C, and store them in a solvent-free refrigerator at 4°C until analysis.

MHT The maximum holding time for samples is 14 days from the time they were collected.

SAMPLE PREPARATION Remove the plungers from two 5-mL syringes and attach a closed syringe valve to each. Warm the sample to room temperature, open the sample bottle, and carefully pour the sample into one of the syringe barrels to just short of overflowing. Replace the syringe plunger, invert the syringe, and compress the sample. Open the syringe valve and vent any residual air while adjusting the sample volume to 5.0 mL. Add 10 µL of the internal calibration standard to the sample through the syringe valve. Close the valve. Fill the second syringe in an identical manner from the same sample bottle. Reserve this second syringe for a reanalysis if necessary.

QUALITY CONTROL As an initial demonstration of lab accuracy and precision, analyze 4 to 7 replicates of a lab fortified blank containing analyte at 0.1–5 µg/L. Collect all samples in duplicate. Surrogate analytes (similar to those of the analytes of interest), whose concentration is known in every sample, are measured using the same internal standard calibration procedure. Duplicate field reagent water blanks (trip blanks) must be analyzed with each set of samples, lab reagent blanks (method blanks) must be analyzed with each batch of samples processed as a group within a work shift. Also, a single lab-fortified blank that contains each of the analytes of interest should be analyzed with each batch of samples processed as a group within a work shift. A 3- to 5-point calibration curve is needed depending on the calibration range factor required.

EPA CONTACT & HOTLINE For technical questions contact Dr. Baldev Bathija, U.S. EPA, Office of Ground Water and Drinking Water (WH-550D), 401 M St. SW, Washington, DC 20460. Tel. (202) 260-3040. For further information the EPA Safe Drinking Water Hotline may be called at: (800) 426-4791.

REFERENCE Methods for the Determination of Organic Compounds in Drinking Water, EPA/600/4-88/039 (revised July 1991; Final Rule for determination of compliance with the MCL for Total Trihalomethanes under 141.30, in 40 CFR Part 141, Vol. 58, No. 147, Fed. Reg., Tuesday Aug. 3, 1993). U.S. EPA Environmental Monitoring Systems Laboratory, Cincinnati, OH, 45268, U.S.A. Available from the National Technical Information Service (NTIS), 5285 Port Royal Road, Springfield, VA 22161; Tel. 800-553-6847. NTIS Order Number is PB91-231480.

Dichlorodifluoromethane   **EPA Method 524**
CAS #75-71-8

TITLE Measurement of Purgeable Organic Compounds in Water by Capillary Column GC/MS.

MATRIX Drinking water and raw source water; the latter should include most surface water and groundwater sources.

METHOD SUMMARY Method 524.2 covers 60 volatile organic compounds. An inert gas (zero grade nitrogen or helium) is bubbled through a 25-mL or a 5-mL water sample (depending on the expected concentration of the analytes). Purged sample components are trapped in a tube of sorbent materials. When purging is complete, the sorbent tube is heated and backflushed with helium to desorb the trapped sample onto a capillary GC column.

INTERFERENCES Impurities in the purge gas and from organic compounds outgassing from the plumbing ahead of the trap account for many contamination problems. Interferences purged or coextracted from the samples will vary considerably from source to source, depending upon the particular sample or extract being tested. Cross-contamination can occur whenever high-level and low-level samples are analyzed sequentially. Samples also can be contaminated by diffusion of volatile organics (particularly methylene chloride and fluorocarbons) through the septum seal into the sample during shipment and storage.

INSTRUMENTATION A GC/MS with a data system equipped with one of the following capillary GC columns:

Column 1: VOCOL glass wide bore capillary column.
Column 2: DB-624 fused silica capillary column.
Column 3: DB-5 fused silica capillary column.

Also required is an all-glass 25 mL or 5-mL purging device, a sorbent trap, and a thermal desorption apparatus which is connected to the GC/MS system.

PRECISION & ACCURACY Method detection limits are compound- and instrument-dependent, and may vary from approximately 0.02–0.35 µg/L. Note in the table below that the "true" concentration range used for accuracy and precision measurements was quite narrow. However, the applicable concentration range of this method is primarily column dependent and is approximately 0.02 to 200 µg/L for the wide-bore thick-film columns. Narrow-bore thin-film columns may have a capacity which limits the range to about 0.02 to 20 µg/L. Analytes that are inefficiently purged from water will not be detected when present at low concentrations, but they can be measured with acceptable accuracy and precision when present in sufficient amounts.

Analytes that are not separated chromatographically, but which have different mass spectra and non-interfering quantification ions, can be identified and measured in the same calibration mixture or water sample. Analytes which have very similar mass spectra cannot be individually identified and measured in the same calibration mixture or water samples unless they have different retention times. Co-eluting compounds with very similar mass spectra, typically many structural isomers, must be reported as an isomeric group or pair.

The range (in µg/L) was 0.5–10.
The Method Detection Limig (in µg/L) was 0.10.
The accuracy (as % recovery) was 90.
The precision (in %) was 7.7.

Note: Data were obtained from 16–31 determinations using a wide-bore capillary column and a jet separator interfaced to a quadrupole mass spectrometer. All analytes were in a reagent water matrix.

SAMPLING METHOD Collect samples using a 40- to 120-mL screw-cap vial (prewashed with detergent, rinsed with distilled water and oven dried at 105°C) with a Teflon®-faced silicone septum. Collect bubble-free samples and place the septum with the Teflon® side down on the water.

SAMPLE PRESERVATION If residual chlorine is present in the water add about 25 mg of ascorbic acid to each vial before samples are collected to remove the chlorine. Add hydrochloric acid to reduce pH to <2, and immediately cool samples to 4°C, and store them in a solvent-free refrigerator at 4°C until analysis.

MHT The maximum holding time for samples is 14 days from the time they were collected.

SAMPLE PREPARATION Remove the plungers from two 25-mL (or 5-mL depending on sample size) syringes and attach a closed syringe valve to each. Warm the sample to room temperature, open the sample bottle, and carefully pour the sample into one of the syringe barrels to just short of overflowing. Replace the syringe plunger, invert the syringe, and compress the sample. Open the syringe valve and vent any residual air while adjusting the sample volume to 25.0 mL (or 5 mL). For samples and blanks, add 5 µL of the fortification solution containing the internal standard and the surrogates to the sample through the syringe valve. For calibration standards and lab fortified blanks, add 5 µL of the fortification solution containing the internal standard only. Close the valve. Fill the second syringe in an identical manner from the same sample bottle. Reserve this second syringe for a reanalysis if necessary.

QUALITY CONTROL As an initial demonstration of lab accuracy and precision, analyze 4 to 7 replicates of a lab fortified blank containing analyte at 0.2–5 µg/L. Collect all samples in duplicate. Surrogate analytes (similar to those of the analytes of interest), whose concentration is known in every sample, are measured using the same internal standard calibration procedure. Duplicate field reagent water blanks (trip blanks) must be analyzed with each set of samples, lab reagent blanks (method blanks) must be analyzed with each batch of samples processed as a group within a work shift. Also, a single lab-fortified blank that contains each of the analytes of interest should be analyzed with each batch of samples processed as a group within a work shift. A 3- to 5-point calibration curve is needed depending on the calibration range factor required.

EPA CONTACT & HOTLINE For technical questions contact Dr. Baldev Bathija, U.S. EPA, Office of Ground Water and Drinking Water (WH-550D), 401 M St. SW, Washington, DC 20460. Tel. (202) 260-3040. For further information the EPA Safe Drinking Water Hotline may be called at: (800) 426-4791.

REFERENCE Methods for the Determination of Organic Compounds in Drinking Water, EPA/600/4-88/039 (revised July 1991; Final Rule for determination of compliance with the MCL for Total Trihalomethanes under 141.30, in 40 CFR Part 141, Vol. 58, No. 147, Fed. Reg., Tuesday Aug. 3, 1993). U.S. EPA Environmental Monitoring Systems Laboratory, Cincinnati, OH, 45268, U.S.A. Available from the National Technical Information Service (NTIS), 5285 Port Royal Road, Springfield, VA 22161; Tel. 800-553-6847. NTIS Order Number is PB91-231480.

Dichlorodifluoromethane   **EPA Method 8021**
CAS #75-71-8

TITLE Halogenated Volatile by Gas Chromatography Using Photoionization and Electrolytic Conductivity Detectors in Series: Capillary Column Technique

MATRIX This method is applicable to nearly all types of samples, regardless of water content, including groundwater, aqueous sludges, caustic liquors, acid liquors, waste solvents, oily wastes, mousses, tars, fibrous wastes, polymeric emulsions, filter cakes, spent carbons, spent catalysts, soils, and sediments.

METHOD SUMMARY This method is used to determine 60 volatile organic compounds in a variety of solid waste matrices. It provides GC conditions for the detection of halogenated and aromatic volatile organic compounds. Samples can be analyzed using direct injection or purge-and-trap (EPA Method 5030). Groundwater samples must be analyzed using EPA Method 5030 (where applicable). A temperature program is used with the GC. Detection is achieved by a photoionization detector (PID) and a Hall electrolytic conductivity detector (HECD) in series.

INTERFERENCES Samples can be contaminated by diffusion of volatile organics (particularly chlorofluorocarbons and methylene chloride) through the sample container septum during shipment and storage.

INSTRUMENTATION A GC-equipped with variable-constant differential flow controllers, subambient oven controller, PID and HECD detectors connected with a short piece of uncoated capillary tubing and a data system.

Column: 60 m × 0.75 mm I.D. VOCOL wide-bore capillary column with 1.5 µm film thickness.

PRECISION & ACCURACY MDLs are compound-dependent and vary with purging efficiency and concentration. The applicable concentration range of this method is compound- and instrument-dependent but is approximately 0.1 to 200 µg/L. Analytes that are inefficiently purged from water will not be detected when present at low concentrations, but they can be measured with acceptable accuracy and precision when present in sufficient amounts. The estimated quantitation limit (EQL) for an individual compound is approximately 1 µg/kg (wet weight) for soil/sediment samples, 100 µg/kg (wet weight) for wastes, and 1 µg/L for groundwater. EQLs will be proportionately higher for sample extracts and samples that require dilution or reduced sample size to avoid saturation of the detector.

MULTIPLICATION FACTORS FOR OTHER MATRICES (a)

Matrix	Factor (b)
Groundwater	10
Low-concentration soil	10
Water miscible liquid waste	500
High-concentration soil and sludge	1250
Non-water miscible waste	1250

(a) Sample EQLs are highly matrix-dependent. The EQLs listed herein are provided for guidance and may not always be achievable.
(b) EQL = [Method detection limit] × [Factor]. For non-aqueous samples, the factor is on a wet-weight basis.

SINGLE LABORATORY ACCURACY & PRECISION DATA FOR VOCs IN WATER

This method was tested in a single lab using water spiked at 10 µg/L and the following data was reported:

Recoveries and standard deviations were determined from seven samples and spiked at 10 µg/L of each analyte. Recoveries were determined by the internal standard method. Internal standards were: Fluorobenzene for PID and 2-Bromo-1-chloropropane for HECD.

The average recovery (in percent) for the PID was none (no response for this detector).
The standard deviation of the recovery for the PID was none (no response for this detector).
The MDL (in µg/mL) for the PID was none (no response for this detector).
The average recovery (in percent) for the HECD was 89.
The standard deviation of the recovery for the HECD was 5.9.
The MDL (in µg/mL) for the HECD was 0.05.

SAMPLE COLLECTION, PRESERVATION & HANDLING
Volatile Organics — Standard 40-mL glass screw-cap VOA vials with Teflon®-faced silicone septum may be used for both liquid and solid matrices. When collecting samples, liquids and solids should be introduced into the vials gently to reduce agitation which might drive off volatile compounds. If there are any air bubbles present the sample must be retaken. Tap slightly as they are filled to try and eliminate as much free air space as possible. The two vials from each sampling locations should be sealed in separate plastic bags to prevent cross-contamination between samples particularly if the sampled waste is suspected of containing high levels of volatile organics.

Semivolatile organics — Containers used to collect samples for the determination of semivolatile organic compounds should be soap and water washed followed by methanol (or isopropanol) rinsing. The sample containers should be of glass or Teflon® and have screw-top covers with Teflon® liners.

Preservation for volatile organics — No preservation is used with concentrated waste samples. With liquid samples containing no residual chlorine, 4 drops of concentrated hydrochloric acid are added and the samples are immediately cooled to 4°C. When liquid samples contain residual chlorine, they are treated as above and, in addition, 4 drops of 4% aqueous sodium thiosulfate are added. Soil, sediment, and sludge samples are only cooled to 4°C.

Preservation for semivolatile organics — No preservation is used with concentrated waste samples. With liquid samples containing no residual chlorine and with soil, sediment, and sludge samples, immediately cooling to 4°C is the only preservation used. When residual chlorine is present then 3 mL of 10% aqueous sodium sulfate is added for each gallon of sample collected, followed by cooling to 4°C.

MHT The holding time for all volatile organics samples is 14 days. Liquid samples must be extracted within 7 days and their extracts analyzed within 40 days. Concentrated waste, soil, sediment, and sludge samples must be extracted within 14 days and their extracts analyzed within 40 days.

SAMPLE PREPARATION Volatile compounds are introduced into the gas chromatograph either by direct injector or purge-and-trap (EPA Method 5030). EPA Method 5030 may be used directly on groundwater samples or low-concentration contaminated soils and sediments. For medium-concentration soils or sediments, methanolic extraction, as described in EPA Method 5030, may be necessary prior to purge-and-trap analysis.

QUALITY CONTROL Calculate surrogate standard recovery on all samples, blanks, and spikes. A trip blank is recommended to check on sampling, storage, and handling contamination. Calibration standards, at a minimum of five concentration levels, are prepared in organic-free reagent water. One of the concentration levels should be at a concentration near, but above, the method detection limit.

A combination of bromochloromethane, 2-bromo-1-chloropropane, 1,4-dichlorobutane, and bromochlorobenzene are recommended as surrogate standards to encompass the range of the temperature program used in this method.

REFERENCE Test Methods for Evaluating Solid Waste, Physical/Chemical Methods, SW-846, 3rd Edition, U.S. EPA, Office of Solid Waste, Washington, DC, EPA Method 8021A, Rev. 1, Nov. 1992.

Dichlorodifluoromethane   EPA Method 8240
CAS #75-71-8

TITLE Volatile Organics By GC/MS: Packed Column Technique

MATRIX Nearly all types of sample matarices, regardless of water content, can be analyzed using this method. This includes groundwater, aqueous sludges, caustic liquors, acid liquors, waste solvents, oily wastes, mousses, tars, fibrous wastes, polymetric emulsions, filter cakes, spent carbons, spent catalysts, soils, and sediments.

METHOD SUMMARY Method 8240B covers 80 volatile organic compounds that are introduced into a gas chromatograph by the purge-and-trap method or by direct injection (in limited applications). For the purge-and-trap method an inert gas (zero grade nitrogen or helium) is bubbled through a 5-mL solution at ambient temperature. Purged sample components are trapped in a tube of sorbent materials. When purging is complete, the sorbent tube is heated and backflushed with inert gas to desorb the trapped components onto a GC column.

INTERFERENCES Impurities in the purge gas and from organic compounds outgassing from the plumbing ahead of the trap account for many contamination problems. Interferences purged or coextracted from the samples will vary considerably from source to source. Cross-contamination can occur whenever high-level and low-level samples are analyzed sequentially. Whenever an unusually concentrated sample is analyzed, it should be followed by the analysis of organic-free reagent water to check for cross-contamination. Samples also can be contaminated by diffusion of volatile organics (particularly methylene chloride and fluorocarbons) through the septum seal into the sample during shipment and storage. A trip blank can serve as a check on such contamination. The lab where volatile analysis is performed and also the refrigerated storage area should be completely free of solvents.

INSTRUMENTATION A gas chromatograph/mass spectrometry/data system (GC/MS) equipped with a 6 ft × 0.1 in I.D. glass column packed with 1% SP-1000 on Carbopack-B (60/80 mesh) is required. Also needed is a 5-mL purging device, a sorbent trap, and a thermal desorption apparatus.

PRECISION & ACCURACY This method is reported to have been tested by 15 laboratories using organic-free reagent water, drinking water, surface water, and industrial wastewaters (not specified) fortified at six concentrations over the range 5–600 µg/L.

Sample estimated quantitation limits (EQLs) are highly matrix-dependent. The EQLs listed may not always be achievable. EQLs listed for soils or sediments are based on wet weight. Normally, data is reported on a dry-weight basis; therefore, EQLs will be higher, based on the percent dry weight of each sample. Note that EQLs are even more variable than MDLs and that they are highly variable depending on the matrix being analyzed.

EQL in groundwater in µg/L was 5.
EQL in low soil or sediment in µg/kg was 5.
Accuracy (a) in µg/L was not listed.
Precision (b) in µg/L was not listed.

(a) *Average recovery found for measurements of samples containing a concentration of C, in µg/L.*
(b) *Overall precision found for measurements of samples with average recovery X for samples containing a concentration of C in µg/L.*
$X =$ *Average recovery found for measurement of samples containing a concentration of C in µg/L.*

MULTIPLICATION FACTORS FOR OTHER MATRICES

Other Matrices	Factor (a)
Waste miscible liquid waste	50
High-concentration soil and sludge	125
Non-water miscible waste	500

(a) *EQL = [EQL for low soil sediment] × [Factor]. For non-aqueous samples, the factor is on a wet-weight basis.*

SAMPLING METHOD

Liquid samples — Use a 40-mL glass screw-cap VOA vial with a Teflon®-faced silicone septum that has been prewashed, rinsed with distilled deionized water, and oven dried. However, if residual chlorine is present, collect sample in a 40-oz. soil VOA container which has been pre-preserved with 4 drops of 10% sodium thiosulfate, mix gently, and then transfer the sample to a 40-mL VOA vial. Collect bubble-free samples in duplicate and seal them in separate plastic bags.

Soils or sediments, and sludges — Use an 8-oz. widemouth glass bottle with a Teflon®-faced silicone septum that has been prewashed with detergent, rinsed with distilled deionized water, and oven dried. Tap slightly to eliminate free air space. Collect samples in duplicate and seal them in separate plastic bags.

SAMPLE PRESERVATION

Liquid samples — Add 4 drops of concentrated HCL and immediately cool samples to 4°C and store in a solvent-free refrigerator.

Soils or sediments, and sludges — Cool samples to 4°C and store in a solvent-free refrigerator.

MHT Maximum holding time is 14 days from the date of sample collection.

SAMPLE PREPARATION

Liquid samples — Remove the plunger from a 5-mL syringe and carefully pour the sample into the syringe barrel to just short of overflowing. Replace the syringe plunger and compress the sample. Open the syringe valve and vent any residual air while adjusting the sample volume to 5.0 mL. If there is only one volatile organic analysis (VOA) vial, a second syringe should be filled at this time to protect against possible loss of sample integrity. Add 10 µL of surrogate spiking solution and 10 µL of internal standard spiking solution through the valve bore of the 5-mL syringe, then close the valve. The surrogate and internal standards may be mixed and added as a single spiking solution.

Sediments, soils, and waste samples — All samples of this type should be screened by GC analysis using a headspace method (EPA Method 3810) or the hexadecane extraction and screening method (EPA Method 3820). Use the screening data to determine whether to use the low-concentration method (0.005–1 mg/kg) or the high-concentration method (>1 mg/kg).

Low-concentration method — The low-concentration method is based on purging a heated sediment or soil sample mixed with organic-free reagent water containing the surrogate and internal standards. Analyze all reagent blanks and standards under the same conditions as the samples.

Use a 5-g sample if the expected concentration is <0.1 mg/kg or a 1-g sample for expected concentrations between 0.1 and 1 mg/kg. Mix the contents of the sample container with a narrow metal spatula. Weigh the amount of the sample into a tared purge device. Add the spiked water to the purge device, which contains the weighed amount of sample, and connect the device to the purge-and-trap system.

High-concentration method — This method is based on extracting the sediment or soil with methanol. A waste sample is either extracted or diluted, depending on its solubility in methanol. Wastes that are insoluble in methanol are diluted with reagent tetraglyme or possibly polyethylene glycol (PEG). An aliquot of the extract is added to organic-free reagent water containing surrogate and internal standards. This is purged at ambient temperature. All samples with an expected concentration of >1.0 mg/kg should be analyzed by this method.

Mix the contents of the sample container with a narrow metal spatula. For sediments or soils and solid wastes that are insoluble in methanol, weigh 4 g (wet weight) of sample into a tared 20-mL vial. For waste that is soluble in methanol, tetraglyme, or PEG, weigh 1 g (wet weight) into a tared scintillation vial or culture tube or a 10-mL volumetric flask. Quickly add 9.0 mL of appropriate solvent then add 1.0 mL of a surrogate spiking solution to the vial, cap it, and shake it for 2 min.

METHANOL EXTRACT REQUIRED FOR ANALYSIS OF HIGH-CONCENTRATION SOILS OR SEDIMENTS

Approximate Concentration Range	Volume of Methanol Extract (a)
500–10,000 µg/kg	100 µL
1,000–20,000 µg/kg	50 µL
5,000–100,000 µg/kg	10 µL
25,000–500,000 µg/kg	100 µL of 1/50 dilution (b)

Calculate appropriate dilution factor for concentrations exceeding this table.

(a) The volume of methanol added to 5 mL of water being purged should be kept constant. Therefore, add to the 5-mL syringe whatever volume of methanol is necessary to maintain a volume of 100 µL added to the syringe.

(b) Dilute an aliquot of the methanol extract and then take 100 µL for analysis.

QUALITY CONTROL Demonstrate, through the analysis of a reagent water blank, that interferences from the analytical system, glassware, and reagents are under control. Blank samples should be carried through all stages of the sample preparation and measurement steps. For each analytical batch (up to 20 samples), a reagent blank, matrix spike, and matrix spike duplicate must be analyzed (the frequency of the spikes may be different for different monitoring programs). The blank and spiked samples must be carried through all stages of the sample preparation and measurement steps. QC samples mentioned in the section on Interferences will also be needed as appropriate to those situations.

REFERENCE Test Methods for Evaluating Solid Waste (SW-846). U.S. EPA. 1983. Method 8240B, Rev. 2, Nov. 1990. Office of Solid Wastes, Washington, DC.

Dichlorodifluoromethane   **EPA Method 8260**
CAS #75-71-8

TITLE Volatile Organic Compounds by GC/MS: Capillary Column Technique

MATRIX This method is applicable to nearly all types of samples, regardless of water content, including groundwater, soils, and sediments.

METHOD SUMMARY Method 8260A covers 58 volatile organic compounds that are introduced into a gas chromatograph by the purge-and-trap method or by direct injection (in limited applications). Zero-grade helium is bubbled through a 5-mL solution at ambient temperature. Purged sample components are trapped in a tube containing suitable sorbent materials. When purging is complete, the sorbent tube is heated and backflushed with helium to desorb trapped sample components. The analytes are desorbed directly to a large bore capillary or cryofocussed on a capillary precolumn before being flash evaporated to a narrow bore capillary for analysis.

INTERFERENCES Major contaminant sources are volatile materials in the lab and impurities in the inert purging gas and in the sorbent trap. Interfering contamination may occur when a sample containing low concentrations of volatile organic compounds is analyzed immediately after a sample containing high concentrations of volatile organic compounds. After analysis of a sample containing high concentrations of volatile organic compounds, one or more calibration blanks should be analyzed to check for cross-contamination. Screening of the samples prior to purge-and-trap GC/MS analysis is highly recommended to prevent contamination of the system. This is especially true for soil and waste samples.

Special precautions must be taken to analyze for methylene chloride. The analytical and sample storage area should be isolated from all atmospheric sources of methylene chloride. All gas chromatography carrier gas lines and purge gas plumbing should be constructed from stainless steel or copper tubing. Laboratory clothing previously exposed to methylene chloride fumes during liquid-liquid extraction procedures can contribute to sample contamination.

Samples can also be contaminated by diffusion of volatile organics (particularly methylene chloride and fluorocarbons) through the septum seal during shipment and storage. A trip blank can serve as a check on such contamination.

INSTRUMENTATION GC/MS with a temperature-programmable chromatograph suitable for splitless injection equipped with variable constant differential flow controllers, a subambient oven controller, a purging device, sorbent trap, a thermal desorption apparatus and a capillary precolumn interface when using cryogenic cooling will be needed. The following GC columns may be used:

Column 1: 60 m × 0.75mm I.D. capillary column coated with VOCOL, 1.5 µm film thickness.
Column 2: 30 m × 0.53mm capillary column coated with DB-624 or VOCOL, 3 µm film thickness.
Column 3: 30 m × 0.32mm I.D. capillary column coated with DB-5 or SE-54, 1-µm film thickness.

PRECISION & ACCURACY This method has been tested in a single lab using spiked water. Using a wide-bore capillary column, water was spiked at concentrations between 0.5 and 10 µg/L. Single lab accuracy and precision data are presented. The MDL actually achieved in a given analysis will vary depending on instrument sensitivity and matrix effects.

The MDL (a) in µg/L was 0.10.
The concentration range in µg/L was 0.5–10.
The mean accuracy (% of true value) was 90.
The precision as relative standard deviation was 7.7.

Note: The MDL is based on a 25-mL sample volume instead of a 5-mL sample volume.

SAMPLING METHOD
Liquid samples — Use a 40-mL glass screw-cap VOA vial with a Teflon®-faced silicone septum that has been prewashed, rinsed with distilled deionized water, and oven dried. If residual chlorine is present, collect the sample in a 4-oz soil VOA container which has been pre-preserved with 4 drops of 10% sodium thiosulfate. Mix gently and transfer the sample to a 40-mL VOA vial. Collect bubble-free samples in duplicate and seal each sample in a separate plastic bag.

Soils, sediments, and sludges — Use an 8-oz widemouth glass bottle with Teflon®-faced silicone septum that has been prewashed, rinsed with distilled deionized water, and oven dried. **Do not** heat the septum for more than 1 h. Tap slightly to eliminate any free air space. Collect samples in duplicate and seal each one in a separate plastic bag.

SAMPLE PRESERVATION
Liquid samples — Add 4 drops of concentrated HCL, cool to 4°C and store in a solvent-free refrigerator.

Soils, sediments and sludges — Cool samples to 4°C and store in a solvent-free refrigerator.

MHT The maximum holding time of any sample (liquids, soils, sediments, and sludges) is 14 days.

SAMPLE PREPARATION
Liquid samples — Remove the plunger from a 5-mL syringe and carefully pour the sample into the syringe barrel to just short of overflowing. Replace the syringe plunger and compress the sample. Open the syringe valve and vent any residual air while adjusting the sample volume to 5.0 mL. If there is only one volatile organic analysis (VOA) vial, a second syringe should be filled at this time to protect against possible loss of sample integrity. Add 10 µL of surrogate spiking solution and 10 µL of internal standard spiking solution through the valve bore of the 5-mL syringe, then close the valve. The surrogate and internal standards may be mixed and added as a single spiking solution.

Sediments, soils, and waste samples — All samples of this type should be screened by GC analysis using a headspace method (EPA Method 3810) or the hexadecane extraction and screening method (EPA Method 3820). Use the screening data to determine whether to use the low-concentration method (0.005–1 mg/kg) or the high-concentration method (>1 mg/kg).

Low-concentration method — The low-concentration method is based on purging a heated sediment or soil sample mixed with organic-free reagent water containing the surrogate and internal standards. Analyze all reagent blanks and standards under the same conditions as the samples.

Use a 5-g sample if the expected concentration is <0.1 mg/kg or a 1-g sample for expected concentrations between 0.1 and 1 mg/kg. Mix the contents of the sample container with a narrow metal spatula. Weigh the amount of the sample into a tared purge device. Add the spiked water to the purge device, which contains the weighed amount of sample, and connect the device to the purge-and-trap system.

High-concentration method — This method is based on extracting the sediment or soil with methanol. A waste sample is either extracted or diluted, depending on its solubility in methanol. Wastes that are insoluble in methanol are diluted with reagent tetraglyme or possibly polyethylene glycol (PEG). An aliquot of the extract is added to organic-free reagent water containing surrogate and internal standards. This is purged at ambient temperature. All samples with an expected concentration of >1.0 mg/kg should be analyzed by this method.

Mix the contents of the sample container with a narrow metal spatula. For sediments or soils and solid wastes that are insoluble in methanol, weigh 4 g (wet weight) of sample into a tared 20-mL vial. For waste that is soluble in methanol, tetraglyme, or PEG, weigh 1 g (wet weight) into a tared scintillation vial or culture tube or a 10-mL volumetric flask. Quickly add

9.0 mL of appropriate solvent then add 1.0 mL of a surrogate spiking solution to the vial, cap it, and shake it for 2 min.

METHANOL EXTRACT REQUIRED FOR ANALYSIS OF HIGH-CONCENTRATION SOILS OR SEDIMENTS

Approximate Concentration Range	Volume of Methanol Extract (a)
500–10,000 µg/kg	100 µL
1,000–20,000 µg/kg	50 µL
5,000–100,000 µg/kg	10 µL
25,000–500,000 µg/kg	100 µL of 1/50 dilution (b)

Calculate appropriate dilution factor for concentrations exceeding this table.

(a) The volume of methanol added to 5 mL of water being purged should be kept constant. Therefore, add to the 5-mL syringe whatever volume of methanol is necessary to maintain a volume of 100 µL added to the syringe.

(b) Dilute an aliquot of the methanol extract and then take 100 µL for analysis.

QUALITY CONTROL Demonstrate, through the analysis of a reagent water blank, that interferences from the analytical system, glassware, and reagents are under control. Blank samples should be carried through all stages of the sample preparation and measurement steps. For each analytical batch (up to 20 samples), a reagent blank, matrix spike, and matrix spike duplicate must be analyzed (the frequency of the spikes may be different for different monitoring programs). The blank and spiked samples must be carried through all stages of the sample preparation and measurement steps. QC samples mentioned in the section on Interferences will also be needed as appropriate to those situations.

Matrix spiking standards should be prepared from volatile organic compounds which will be representative of the compounds being investigated. The recommended internal standards are chlorobenzene-d5, 1,4-difluorobenzene, 1,4-dichlorobenzene-d4, and pentafluorobenzene. Using stock standard solutions, prepare secondary dilution standards containing the compounds of interest, either singly or mixed together in methanol. Store them in a vial with no headspace for no more than one week. Surrogates recommended are toluene-d8, 4-bromofluorobenzene, and dibromofluoromethane. Each sample undergoing GC/MS analysis must be spiked with 10 µL of the surrogate spiking solution prior to analysis.

REFERENCE Test Methods for Evaluating Solid Waste (SW-846). U.S. EPA 1983, Method 8260A, Rev. 1, Nov. 1990. Office of Solid Waste, Washington, DC.

Dichlorodifluoromethane   **EPA Method 601**
CAS #75-71-8

TITLE Purgeable Halocarbons

MATRIX Wastewater

APPLICATION Method covers 29 purgeable halocarbons. EPA Method describes conditions for a 2nd GC column to confirm measurements made with primary column.

INTERFERENCES impurities in the purge gas and organic compounds outgassing from the plumbing ahead of the trap. With high- and low-level samples, there can be carryover contamination. Diffusion of volatile organics through the septum seal into the sample.

INSTRUMENTATION GC-equipped with halide-specific detector. (With purge-and-trap unit).

RANGE 8.0–500 µg/L.

MDL 1.81 µg/L

PRECISION not listed

ACCURACY not listed

SAMPLING METHOD 25-mL glass vial. Teflon®-lined septum.

STABILITY cool, 4°C, 0.008% $Na_2S_2O_3$.

MHT 14 days.

QUALITY CONTROL The lab must on an ongoing basis, spike at least 10% of the samples from each sample site being monitored to assess accuracy.

REFERENCE Method 601, *Federal Register* Part VIII 40 CFR Part 136, Oct 26, 1984.

1,1-Dichloroethane   **EPA Method 1624**
CAS #75-34-3

TITLE Volatile Organic Compounds by Isotope Dilution GC/MS

MATRIX Compounds may be determined in waters, soils, and municipal sludges by this method.

METHOD SUMMARY This method is used to determine 58 volatile toxic organic pollutants associated with the CWA (as amended 1987); the RCRA (as amended 1986); the CERCLA (as amended 1986); and other compounds amenable to purge-and-trap gas chromatography-mass spectrometry (GC/MS).

If the solids content is less than 1%, stable isotopically-labeled analogs of the compounds of interest are added to a 5-mL sample and the sample is purged with an inert gas at 20–25°C in a chamber designed for soil or water samples. If the solids content is greater than 1%, 5 mL of reagent water and the labeled compounds are added to a 5-g aliquot of sample and the mixture is purged at 40°C. Compounds that will not purge at 20–25°C or at 40°C are purged at 78–85°C. In the purging process, the volatile compounds are transferred from the aqueous phase into the gaseous phase where they are passed into a sorbent column, and trapped. After purging is completed, the trap is backflushed and heated rapidly to desorb the compounds into a GC. The compounds are separated by the GC and detected by a MS. The labeled compounds serve to correct the variability of the analytical technique.

INTERFERENCES Impurities in the purge gas, organic compounds outgassing from the plumbing upstream of the trap, and solvent vapors in the lab account for most problems. Samples can be contaminated by diffusion of volatile organic compounds (particularly methylene chloride) through the bottle seal during shipment and storage. Contamination by carryover can occur when high-level and low-level samples are analyzed sequentially. When an unusually concentrated sample is encountered, follow it by analysis of a reagent water blank to check for carryover.

INSTRUMENTATION Major equipment includes a GC with linear temperature programming and a glass jet separator as the MS interface, a MS with 70 eV electron impact ionization, and a data system to collect and record response factors.

Column: 2.8 m × 2 mm I.D. glass, packed with 1% SP-1000 on Carbopak B, 60/80 mesh, or equivalent.

PRECISION & ACCURACY The detection limits of the method are usually dependent on the level of interferences rather than instrumental limitations. The method detection limits were determined in digested sludge (low solids) and in filter cake or compost (high solids).

The MDL (in µg/kg) for low solids is 16 and for high solids is 1.
Labeled and native compound precision (in µg/L) as standard deviation was 6.7.
Labeled and native compound accuracy (in µg/L) as average recovery was 11–31.
Acceptance criteria are at 20 µg/L for this compound.

SAMPLE COLLECTION, PRESERVATION & HANDLING Grab samples are collected in glass containers having a total volume greater than 20 mL. Fill and seal each bottle so that no air bubbles are entrapped. Samples are maintained at 0 to 4°C from the time of collection until analysis. If an aqueous sample contains residual chlorine, add sodium thiosulfate preservative (10 mg/40 mL) to the empty sample bottles just prior to shipment to the sample site. All samples must be analyzed within 14 days of collection.

SAMPLE PREPARATION Samples containing less than 1% solids are analyzed directly as aqueous samples. Samples containing 1% solids or greater are analyzed as solid samples utilizing one of two methods, depending on the levels of pollutants, in the sample. Samples containing 1% solids or greater, and low to moderate levels of pollutants are analyzed by purging a known weight of sample added to 5 mL of reagent water. Samples containing 1% solids or greater, and high levels of pollutants, are extracted with methanol, and an aliquot of the methanol extract is added to reagent water and purged.

QUALITY CONTROL A field blank prepared from reagent water and carried through the sampling and handling protocol may serve as a check on contamination from shipment and storage.

The analyst is permitted to modify this method to improve separations or lower the costs of measurements, provided all performance specifications are met. Analyses of blanks are required. When results of spikes indicate atypical method performance for samples, the samples are diluted to bring method performance within acceptable limits. Analyze two sets of four 5-mL aliquots (8 aliquots total) of the aqueous performance standard. Spike all samples with labeled compounds to assess method performance on the sample matrix. Compute the percent recovery of the labeled compounds using the internal standard method. Compare the percent recovery for each compound with the corresponding labeled compound recovery. Reagent water blanks are analyzed to demonstrate freedom from carryover contamination. Field replicates may be collected to determine the precision of the sampling technique, and spiked samples may be required to determine the accuracy of the analysis when the internal method is used.

REFERENCE Volatile Organic Compounds by Isotope Dilution GC/MS. Office of Water Regulation and Standards, U.S. EPA Industrial Technology Division, Washington, DC, EPA Method 1624, Rev. C, June 1989 (contact W.A. Telliard, U.S. EPA, Office of Water Regulations and Standards, 401 M St., SW, Washington, DC, 20460. Phone: 202-382-7131).

1,2-Dichloroethane   **EPA Method 1624**
CAS #107-06-2

TITLE Volatile Organic Compounds by Isotope Dilution GC/MS

MATRIX Compounds may be determined in waters, soils, and municipal sludges by this method.

METHOD SUMMARY This method is used to determine 58 volatile toxic organic pollutants associated with the CWA (as amended 1987); the RCRA (as amended 1986); the CERCLA (as amended 1986); and other compounds amenable to purge-and-trap gas chromatography-mass spectrometry (GC/MS).

If the solids content is less than 1%, stable isotopically-labeled analogs of the compounds of interest are added to a 5-mL sample and the sample is purged with an inert gas at 20–25°C in a chamber designed for soil or water samples. If the solids content is greater than 1%, 5 mL of reagent water and the labeled compounds are added to a 5-g aliquot of sample and the mixture is purged at 40°C. Compounds that will not purge at 20–25°C or at 40°C are purged at 78–85°C. In the purging process, the volatile compounds are transferred from the aqueous phase into the gaseous phase where they are passed into a sorbent column, and trapped. After purging is completed, the trap is backflushed and heated rapidly to desorb the compounds into a GC. The compounds are separated by the GC and detected by a MS. The labeled compounds serve to correct the variability of the analytical technique.

INTERFERENCES Impurities in the purge gas, organic compounds outgassing from the plumbing upstream of the trap, and solvent vapors in the lab account for most problems. Samples can be contaminated by diffusion of volatile organic compounds (particularly methylene chloride) through the bottle seal during shipment and storage. Contamination by carryover can occur when high-level and low-level samples are analyzed sequentially. When an unusually concentrated sample is encountered, follow it by analysis of a reagent water blank to check for carryover.

INSTRUMENTATION Major equipment includes a GC with linear temperature programming and a glass jet separator as the MS interface, a MS with 70 eV electron impact ionization, and a data system to collect and record response factors.

Column: 2.8 m × 2 mm I.D. glass, packed with 1% SP-1000 on Carbopak B, 60/80 mesh, or equivalent.

PRECISION & ACCURACY The detection limits of the method are usually dependent on the level of interferences rather than instrumental limitations. The method detection limits were determined in digested sludge (low solids) and in filter cake or compost (high solids).

The MDL (in µg/kg) for low solids is 23 and for high solids is 3.
Labeled and native compound precision (in µg/L) as standard deviation was 7.7.
Labeled and native compound accuracy (in µg/L) as average recovery was 12–30.

Acceptance criteria are at 20 µg/L for this compound.

SAMPLE COLLECTION, PRESERVATION & HANDLING Grab samples are collected in glass containers having a total volume greater than 20 mL. Fill and seal each bottle so that no air bubbles are entrapped. Samples are maintained at 0 to 4°C from the time of collection until analysis. If an aqueous sample contains residual chlorine, add sodium thiosulfate preservative (10 mg/40 mL) to the empty sample bottles just prior to shipment to the sample site. All samples must be analyzed within 14 days of collection.

SAMPLE PREPARATION Samples containing less than 1% solids are analyzed directly as aqueous samples. Samples containing 1% solids or greater are analyzed as solid samples utilizing one of two methods, depending on the levels of pollutants, in the sample. Samples containing 1% solids or greater, and low to moderate levels of pollutants are analyzed by purging a known weight of sample added to 5 mL of reagent water. Samples containing 1% solids or greater, and high levels of pollutants, are extracted with methanol, and an aliquot of the methanol extract is added to reagent water and purged.

QUALITY CONTROL A field blank prepared from reagent water and carried through the sampling and handling protocol may serve as a check on contamination from shipment and storage.

The analyst is permitted to modify this method to improve separations or lower the costs of measurements, provided all performance specifications are met. Analyses of blanks are required. When results of spikes indicate atypical method performance for samples, the samples are diluted to bring method performance within acceptable limits. Analyze two sets of four 5-mL aliquots (8 aliquots total) of the aqueous performance standard. Spike all samples with labeled compounds to assess method performance on the sample matrix. Compute the percent recovery of the labeled compounds using the internal standard method. Compare the percent recovery for each compound with the corresponding labeled compound recovery. Reagent water blanks are analyzed to demonstrate freedom from carryover contamination. Field replicates may be collected to determine the precision of the sampling technique, and spiked samples may be required to determine the accuracy of the analysis when the internal method is used.

REFERENCE Volatile Organic Compounds by Isotope Dilution GC/MS. Office of Water Regulation and Standards, U.S. EPA Industrial Technology Division, Washington, DC, EPA Method 1624, Rev. C, June 1989 (contact W.A. Telliard, U.S. EPA, Office of Water Regulations and Standards, 401 M St., SW, Washington, DC, 20460. Phone: 202-382-7131).

1,1-Dichloroethane   EPA Method 502
CAS #75-34-3

TITLE Volatile Organic Compounds in Water By Purge and Trap Capillary Column Gas Chromatography with Photoionization and Electrolytic Conductivity Detectors in Series. U.S. EPA Method 502.2, Rev. 2.0, 1989.

MATRIX Drinking water and raw source water. The latter should include most surface water and groundwater sources.

METHOD SUMMARY This method covers 60 volatile organic compounds that contain halogen atoms and/or that are aromatic. An inert gas (zero grade nitrogen or helium) is bubbled through a 25-mL or a 5-mL water sample (depending on the expected concentration of the analytes). Purged sample components are trapped in a tube of sorbent materials. When purging is complete, the sorbent tube is heated and backflushed with helium to desorb the trapped sample onto a capillary GC column. The column is temperature programmed to separate the method analytes which are then detected with a photoionization detector (PID) and a Hall electrolytic conductivity (HECD) placed in series. The PID is selective for aromatic compounds and the HECD is selective for halogenated compounds.

INTERFERENCES Impurities in the purge gas and from organic compounds outgassing from the plumbing ahead of the trap account for many contamination problems. Interferences purged or coextracted from the samples will vary considerably from source to source, depending upon the particular sample or extract being tested. Cross-contamination can occur whenever high-level and low-level samples are analyzed sequentially. Samples also can be contaminated by diffusion of volatile organics (particularly methylene chloride and fluorocarbons) through the septum seal into the sample during shipment and storage. The lab where volatile analysis is performed and also the refrigerated storage area should be completely free of solvents.

INSTRUMENTATION A GC containing a series configuration of a high temperature photoionization detector (PID) equipped with 10.0 eV (nominal) lamp and Hall electrolytic conductivity detector (HECD) is required. Also required is an all-glass 5-mL purging device, a sorbent trap, and a thermal desorption apparatus which is connected to the GC system.

Column 1: VOCOL glass wide-bore capillary column.
Column 2: RTX–502.2 mega-bore capillary column.

Column 3: DB-62 mega-bore capillary column.

PRECISION & ACCURACY Method detection limits are dependent upon the characteristics of the gas chromatographic system used. Analytes that are not separated chromatographically cannot be individually identified and used in the same calibration mixture or water samples unless an alternative technique for identification and quantification, such as mass spectrometry, is used.

Electrolytic conductivity detetor (c) range in µg/L (a) was 0.02–200.

Electrolytic conductivity detetor (c) MDL in µg/L (b) was 0.07.

Electrolytic conductivity detetor (c) accuracy as % recovery was 100.

Electrolytic conductivity detetor (c) precision as % RSD was 5.7.

Photoionization detector (d) range in µg/L (a) was 0.02–200.

Photoionization detector (d) MDL in µg/L (b) was not listed.

Photoionization detector (d) accuracy as % recovery was not listed.

Photoionization detector (d) precision as % RSD was not listed.

(a) *The applicable concentration range of this method is compound, instrument, and matrix-dependent. It is listed as being approximately 0.02 to 200 µg/L but no specific information is provided so caution should be observed.*

(b) *The method detection limits reports with this method are compound, instrument, and matrix-dependent. The values reported were calculated using reagent water fortified with the corresponding compounds at 10 µg/L and a GC-equipped with a 60 m × 0.75 mm VOLCOL wide bore capillary column with 1.5 µm film thickness and using helium carrier gas.*

(c) *Recoveries and relative standard deviations were determined from seven samples of reagent water fortified with 10 µg/L of each compound. 2-Bromo-1-chloropropane was used as the internal standard for calculating average recoveries.*

(d) *Recoveries and relative standard deviations were determined from seven samples of reagent water fortified with 10 µg/L of each compound. Fluorobenzene was used as the internal standard for calculating average recoveries.*

SAMPLING METHOD Collect samples using a 40- to 120-mL screw-cap vial (prewashed with detergent, rinsed with distilled water and oven dried at 105°C) with a Teflon®-faced silicone septum. Collect bubble-free samples and place the septum with the Teflon® side down on the water.

SAMPLE PRESERVATION If residual chlorine is present in the water add about 25 mg of ascorbic acid to each vial before samples are collected to remove the chlorine. Add hydrochloric acid to reduce pH to <2, immediately cool samples to 4°C, and store them in a solvent-free refrigerator at 4°C until analysis.

MHT The maximum holding time for samples is 14 days from the time they were collected.

SAMPLE PREPARATION Remove the plungers from two 5-mL syringes and attach a closed syringe valve to each. Warm the sample to room temperature, open the sample bottle, and carefully pour the sample into one of the syringe barrels to just short of overflowing. Replace the syringe plunger, invert the syringe, and compress the sample. Open the syringe valve and vent any residual air while adjusting the sample volume to 5.0 mL. Add 10 µL of the internal calibration standard to the sample through the syringe valve. Close the valve. Fill the second syringe in an identical manner from the same sample bottle. Reserve this second syringe for a reanalysis if necessary.

QUALITY CONTROL As an initial demonstration of lab accuracy and precision, analyze 4 to 7 replicates of a lab fortified blank containing analyte at 0.1–5 µg/L. Collect all samples in duplicate. Surrogate analytes (similar to those of the analytes of interest), whose concentration is known in every sample, are measured using the same internal standard calibration procedure. Duplicate field reagent water blanks (trip blanks) must be analyzed with each set of samples, lab reagent blanks (method blanks) must be analyzed with each batch of samples processed as a group within a work shift. Also, a single lab-fortified blank that contains each of the analytes of interest should be analyzed with each batch of samples processed as a group within a work shift. A 3- to 5-point calibration curve is needed depending on the calibration range factor required.

EPA CONTACT & HOTLINE For technical questions contact Dr. Baldev Bathija, U.S. EPA, Office of Ground Water and Drinking Water (WH-550D), 401 M St. SW, Washington, DC 20460. Tel. (202) 260-3040. For further information the EPA Safe Drinking Water Hotline may be called at: (800) 426-4791.

REFERENCE Methods for the Determination of Organic Compounds in Drinking Water, EPA/600/4-88/039 (revised July 1991; Final Rule for determination of compliance with the MCL for Total Trihalomethanes under 141.30, in 40 CFR Part 141, Vol. 58, No. 147, Fed. Reg., Tuesday Aug. 3, 1993). U.S. EPA Environmental Monitoring Systems Laboratory, Cincinnati, OH, 45268, U.S.A. Available from the National Technical Information Service (NTIS), 5285 Port Royal Road, Springfield, VA 22161; Tel. 800-553-6847. NTIS Order Number is PB91-231480.

1,2-Dichloroethane   **EPA Method 502**
CAS #107-06-2

TITLE Volatile Organic Compounds in Water By Purge and Trap Capillary Column Gas Chromatography with Photoionization and Electrolytic Conductivity Detectors in Series. U.S. EPA Method 502.2, Rev. 2.0, 1989.

MATRIX Drinking water and raw source water. The latter should include most surface water and groundwater sources.

METHOD SUMMARY This method covers 60 volatile organic compounds that contain halogen atoms and/or that are aromatic. An inert gas (zero grade nitrogen or helium) is bubbled through a 25-mL or a 5-mL water sample (depending on the expected concentration of the analytes). Purged sample components are trapped in a tube of sorbent materials. When purging is complete, the sorbent tube is heated and backflushed with helium to desorb the trapped sample onto a capillary GC column. The column is temperature programmed to separate the method analytes which are then detected with a photoionization detector (PID) and a Hall electrolytic conductivity (HECD) placed in series. The PID is selective for aromatic compounds and the HECD is selective for halogenated compounds.

INTERFERENCES Impurities in the purge gas and from organic compounds outgassing from the plumbing ahead of the trap account for many contamination problems. Interferences purged or coextracted from the samples will vary considerably from source to source, depending upon the particular sample or extract being tested. Cross-contamination can occur whenever high-level and low-level samples are analyzed sequentially. Samples also can be contaminated by diffusion of volatile organics (particularly methylene chloride and fluorocarbons) through the septum seal into the sample during shipment and storage. The lab where volatile analysis is performed and also the refrigerated storage area should be completely free of solvents.

INSTRUMENTATION A GC containing a series configuration of a high temperature photoionization detector (PID) equipped with 10.0 eV (nominal) lamp and Hall electrolytic conductivity detector (HECD) is required. Also required is an all-glass 5-mL purging device, a sorbent trap, and a thermal desorption apparatus which is connected to the GC system.

Column 1: VOCOL glass wide-bore capillary column.
Column 2: RTX–502.2 mega-bore capillary column.
Column 3: DB-62 mega-bore capillary column.

PRECISION & ACCURACY Method detection limits are dependent upon the characteristics of the gas chromatographic system used. Analytes that are not separated chromatographically cannot be individually identified and used in the same calibration mixture or water samples unless an alternative technique for identification and quantification, such as mass spectrometry, is used.

Electrolytic conductivity detector (c) range in µg/L (a) was 0.02–200.
Electrolytic conductivity detetor (c) MDL in µg/L (b) was 0.03.
Electrolytic conductivity detetor (c) accuracy as % recovery was 100.
Electrolytic conductivity detetor (c) precision as % RSD was 3.8.
Photoionization detector (d) range in µg/L (a) was 0.02–200.
Photoionization detector (d) MDL in µg/L (b) was not listed.
Photoionization detector (d) accuracy as % recovery was not listed.
Photoionization detector (d) precision as % RSD was not listed.

(a) *The applicable concentration range of this method is compound, instrument, and matrix-dependent. It is listed as being approximately 0.02 to 200 µg/L but no specific information is provided so caution should be observed.*
(b) *The method detection limits reports with this method are compound, instrument, and matrix-dependent. The values reported were calculated using reagent water fortified with the corresponding compounds at 10 µg/L and a GC-equipped with a 60 m × 0.75 mm VOLCOL wide bore capillary column with 1.5 µm film thickness and using helium carrier gas.*
(c) *Recoveries and relative standard deviations were determined from seven samples of reagent water fortified with 10 µg/L of each compound. 2-Bromo-1-chloropropane was used as the internal standard for calculating average recoveries.*
(d) *Recoveries and relative standard deviations were determined from seven samples of reagent water fortified with 10 µg/L of each compound. Fluorobenzene was used as the internal standard for calculating average recoveries.*

SAMPLING METHOD Collect samples using a 40- to 120-mL screw-cap vial (prewashed with detergent, rinsed with distilled water and oven dried at 105°C) with a Teflon®-faced silicone septum. Collect bubble-free samples and place the septum with the Teflon® side down on the water.

SAMPLE PRESERVATION If residual chlorine is present in the water add about 25 mg of ascorbic acid to each vial before samples are collected to remove the chlorine. Add hydrochloric acid to reduce pH to <2, immediately cool samples to 4°C, and store them in a solvent-free refrigerator at 4°C until analysis.

MHT The maximum holding time for samples is 14 days from the time they were collected.

SAMPLE PREPARATION Remove the plungers from two 5-mL syringes and attach a closed syringe valve to each. Warm the sample to room temperature, open the sample bottle, and carefully pour the sample into one of the syringe barrels to just short of overflowing. Replace the syringe plunger, invert the syringe, and compress the sample. Open the syringe valve and vent any residual air while adjusting the sample volume to 5.0 mL. Add 10 µL of the internal calibration standard to the sample through the syringe valve. Close the valve. Fill the second syringe in an identical manner from the same sample bottle. Reserve this second syringe for a reanalysis if necessary.

QUALITY CONTROL As an initial demonstration of lab accuracy and precision, analyze 4 to 7 replicates of a lab fortified blank containing analyte at 0.1–5 µg/L. Collect all samples in duplicate. Surrogate analytes (similar to those of the analytes of interest), whose concentration is known in every sample, are measured using the same internal standard calibration procedure. Duplicate field reagent water blanks (trip blanks) must be analyzed with each set of samples, lab reagent blanks (method blanks) must be analyzed with each batch of samples processed as a group within a work shift. Also, a single lab-fortified blank that contains each of the analytes of interest should be analyzed with each batch of samples processed as a group within a work shift. A 3- to 5-point calibration curve is needed depending on the calibration range factor required.

EPA CONTACT & HOTLINE For technical questions contact Dr. Baldev Bathija, U.S. EPA, Office of Ground Water and Drinking Water (WH-550D), 401 M St. SW, Washington, DC 20460. Tel. (202) 260-3040. For further information the EPA Safe Drinking Water Hotline may be called at: (800) 426-4791.

REFERENCE Methods for the Determination of Organic Compounds in Drinking Water, EPA/600/4-88/039 (revised July 1991; Final Rule for determination of compliance with the MCL for Total Trihalomethanes under 141.30, in 40 CFR Part 141, Vol. 58, No. 147, Fed. Reg., Tuesday Aug. 3, 1993). U.S. EPA Environmental Monitoring Systems Laboratory, Cincinnati, OH,

45268, U.S.A. Available from the National Technical Information Service (NTIS), 5285 Port Royal Road, Springfield, VA 22161; Tel. 800-553-6847. NTIS Order Number is PB91-231480.

1,1-Dichloroethane   EPA Method 524
CAS #75-34-3

TITLE Measurement of Purgeable Organic Compounds in Water by Capillary Column GC/MS.

MATRIX Drinking water and raw source water; the latter should include most surface water and groundwater sources.

METHOD SUMMARY Method 524.2 covers 60 volatile organic compounds. An inert gas (zero grade nitrogen or helium) is bubbled through a 25-mL or a 5-mL water sample (depending on the expected concentration of the analytes). Purged sample components are trapped in a tube of sorbent materials. When purging is complete, the sorbent tube is heated and backflushed with helium to desorb the trapped sample onto a capillary GC column.

INTERFERENCES Impurities in the purge gas and from organic compounds outgassing from the plumbing ahead of the trap account for many contamination problems. Interferences purged or coextracted from the samples will vary considerably from source to source, depending upon the particular sample or extract being tested. Cross-contamination can occur whenever high-level and low-level samples are analyzed sequentially. Samples also can be contaminated by diffusion of volatile organics (particularly methylene chloride and fluorocarbons) through the septum seal into the sample during shipment and storage.

INSTRUMENTATION A GC/MS with a data system equipped with one of the following capillary GC columns:

Column 1: VOCOL glass wide bore capillary column.
Column 2: DB-624 fused silica capillary column.
Column 3: DB-5 fused silica capillary column.

Also required is an all-glass 25 mL or 5-mL purging device, a sorbent trap, and a thermal desorption apparatus which is connected to the GC/MS system.

PRECISION & ACCURACY Method detection limits are compound- and instrument-dependent, and may vary from approximately 0.02–0.35 µg/L. Note in the table below that the "true" concentration range used for accuracy and precision measurements was quite narrow. However, the applicable concentration range of this method is primarily column dependent and is approximately 0.02 to 200 µg/L for the wide-bore thick-film columns. Narrow-bore thin-film columns may have a capacity which limits the range to about 0.02 to 20 µg/L. Analytes that are inefficiently purged from water will not be detected when present at low concentrations, but they can be measured with acceptable accuracy and precision when present in sufficient amounts.

Analytes that are not separated chromatographically, but which have different mass spectra and non-interfering quantification ions, can be identified and measured in the same calibration mixture or water sample. Analytes which have very similar mass spectra cannot be individually identified and measured in the same calibration mixture or water samples unless they have different retention times. Co-eluting compounds with very similar mass spectra, typically many structural isomers, must be reported as an isomeric group or pair.

The range (in µg/L) was 0.5–10.
The Method Detection Limig (in µg/L) was 0.04.
The accuracy (as % recovery) was 96.
The precision (in %) was 5.3.

Note: Data were obtained from 16–31 determinations using a wide-bore capillary column and a jet separator interfaced to a quadrupole mass spectrometer. All analytes were in a reagent water matrix.

SAMPLING METHOD Collect samples using a 40- to 120-mL screw-cap vial (prewashed with detergent, rinsed with distilled water and oven dried at 105°C) with a Teflon®-faced silicone septum. Collect bubble-free samples and place the septum with the Teflon® side down on the water.

SAMPLE PRESERVATION If residual chlorine is present in the water add about 25 mg of ascorbic acid to each vial before samples are collected to remove the chlorine. Add hydrochloric acid to reduce pH to <2, and immediately cool samples to 4°C, and store them in a solvent-free refrigerator at 4°C until analysis.

MHT The maximum holding time for samples is 14 days from the time they were collected.

SAMPLE PREPARATION Remove the plungers from two 25-mL (or 5-mL depending on sample size) syringes and attach a closed syringe valve to each. Warm the sample to room temperature, open the sample bottle, and carefully pour the sample into one of the syringe barrels to just short of overflowing. Replace the syringe plunger, invert the syringe, and compress the sample. Open the syringe valve and vent any residual air while adjusting the sample volume to 25.0 mL (or 5 mL). For samples and blanks, add 5 µL of the fortification solution containing the internal standard and the surrogates to the sample through the syringe valve. For calibration standards and lab fortified blanks, add 5 µL of the fortification solution containing the internal standard only. Close the valve. Fill the second syringe in an identical manner from the same sample bottle. Reserve this second syringe for a reanalysis if necessary.

QUALITY CONTROL As an initial demonstration of lab accuracy and precision, analyze 4 to 7 replicates of a lab fortified blank containing analyte at 0.2–5 µg/L. Collect all samples in duplicate. Surrogate analytes (similar to those of the analytes of interest), whose concentration is known in every sample, are measured using the same internal standard calibration procedure. Duplicate field reagent water blanks (trip blanks) must be analyzed with each set of samples, lab reagent blanks (method blanks) must be analyzed with each batch of samples processed as a group within a work shift. Also, a single lab-fortified blank that contains each of the analytes of interest should be analyzed with each batch of samples processed as a group within a work shift. A 3- to 5-point calibration curve is needed depending on the calibration range factor required.

EPA CONTACT & HOTLINE For technical questions contact Dr. Baldev Bathija, U.S. EPA, Office of Ground Water and Drinking Water (WH-550D), 401 M St. SW, Washington, DC 20460. Tel. (202) 260-3040. For further information the EPA Safe Drinking Water Hotline may be called at: (800) 426-4791.

REFERENCE Methods for the Determination of Organic Compounds in Drinking Water, EPA/600/4-88/039 (revised July 1991; Final Rule for determination of compliance with the MCL for Total Trihalomethanes under 141.30, in 40 CFR Part 141, Vol. 58, No. 147, Fed. Reg., Tuesday Aug. 3, 1993). U.S. EPA Environmental Monitoring Systems Laboratory, Cincinnati, OH, 45268, U.S.A. Available from the National Technical Information Service (NTIS), 5285 Port Royal Road, Springfield, VA 22161; Tel. 800-553-6847. NTIS Order Number is PB91-231480.

1,2-Dichloroethane   EPA Method 524
CAS #107-06-2

TITLE Measurement of Purgeable Organic Compounds in Water by Capillary Column GC/MS.

MATRIX Drinking water and raw source water; the latter should include most surface water and groundwater sources.

METHOD SUMMARY Method 524.2 covers 60 volatile organic compounds. An inert gas (zero grade nitrogen or helium) is bubbled through a 25-mL or a 5-mL water sample (depending on the expected concentration of the analytes). Purged sample components are trapped in a tube of sorbent materials. When purging is complete, the sorbent tube is heated and backflushed with helium to desorb the trapped sample onto a capillary GC column.

INTERFERENCES Impurities in the purge gas and from organic compounds outgassing from the plumbing ahead of the trap account for many contamination problems. Interferences purged or coextracted from the samples will vary considerably from source to source, depending upon the particular sample or extract being tested. Cross-contamination can occur whenever high-level and low-level samples are analyzed sequentially. Samples also can be contaminated by diffusion of volatile organics (particularly methylene chloride and fluorocarbons) through the septum seal into the sample during shipment and storage.

INSTRUMENTATION A GC/MS with a data system equipped with one of the following capillary GC columns:

Column 1: VOCOL glass wide bore capillary column.
Column 2: DB-624 fused silica capillary column.
Column 3: DB-5 fused silica capillary column.

Also required is an all-glass 25 mL or 5-mL purging device, a sorbent trap, and a thermal desorption apparatus which is connected to the GC/MS system.

PRECISION & ACCURACY Method detection limits are compound- and instrument-dependent, and may vary from approximately 0.02–0.35 µg/L. Note in the table below that the "true" concentration range used for accuracy and precision measurements was quite narrow. However, the applicable concentration range of this method is primarily column dependent and is approximately 0.02 to 200 µg/L for the wide-bore thick-film columns. Narrow-bore thin-film columns may have a capacity which limits the range to about 0.02 to 20 µg/L. Analytes that are inefficiently purged from water will not be detected when present at low concentrations, but they can be measured with acceptable accuracy and precision when present in sufficient amounts.

Analytes that are not separated chromatographically, but which have different mass spectra and non-interfering quantification ions, can be identified and measured in the same calibration mixture or water sample. Analytes which have very similar mass spectra cannot be individually identified and measured in the same calibration mixture or water samples unless they have different retention times. Co-eluting compounds with very similar mass spectra, typically many structural isomers, must be reported as an isomeric group or pair.

The range (in µg/L) was 0.1–10.
The Method Detection Limig (in µg/L) was 0.06.
The accuracy (as % recovery) was 95.
The precision (in %) was 5.4.

Note: Data were obtained from 16–31 determinations using a wide-bore capillary column and a jet separator interfaced to a quadrupole mass spectrometer. All analytes were in a reagent water matrix.

SAMPLING METHOD Collect samples using a 40- to 120-mL screw-cap vial (prewashed with detergent, rinsed with distilled water and oven dried at 105°C) with a Teflon®-faced silicone septum. Collect bubble-free samples and place the septum with the Teflon® side down on the water.

SAMPLE PRESERVATION If residual chlorine is present in the water add about 25 mg of ascorbic acid to each vial before samples are collected to remove the chlorine. Add hydrochloric acid to reduce pH to <2, and immediately cool samples to 4°C, and store them in a solvent-free refrigerator at 4°C until analysis.

MHT The maximum holding time for samples is 14 days from the time they were collected.

SAMPLE PREPARATION Remove the plungers from two 25-mL (or 5-mL depending on sample size) syringes and attach a closed syringe valve to each. Warm the sample to room temperature, open the sample bottle, and carefully pour the sample into one of the syringe barrels to just short of overflowing. Replace the syringe plunger, invert the syringe, and compress the sample. Open the syringe valve and vent any residual air while adjusting the sample volume to 25.0 mL (or 5 mL). For samples and blanks, add 5 µL of the fortification solution containing the internal standard and the surrogates to the sample through the syringe valve. For calibration standards and lab fortified blanks, add 5 µL of the fortification solution containing the internal standard only. Close the valve. Fill the second syringe in an identical manner from the same sample bottle. Reserve this second syringe for a reanalysis if necessary.

QUALITY CONTROL As an initial demonstration of lab accuracy and precision, analyze 4 to 7 replicates of a lab fortified blank containing analyte at 0.2–5 µg/L. Collect all samples in duplicate. Surrogate analytes (similar to those of the analytes

of interest), whose concentration is known in every sample, are measured using the same internal standard calibration procedure. Duplicate field reagent water blanks (trip blanks) must be analyzed with each set of samples, lab reagent blanks (method blanks) must be analyzed with each batch of samples processed as a group within a work shift. Also, a single lab-fortified blank that contains each of the analytes of interest should be analyzed with each batch of samples processed as a group within a work shift. A 3- to 5-point calibration curve is needed depending on the calibration range factor required.

EPA CONTACT & HOTLINE For technical questions contact Dr. Baldev Bathija, U.S. EPA, Office of Ground Water and Drinking Water (WH-550D), 401 M St. SW, Washington, DC 20460. Tel. (202) 260-3040. For further information the EPA Safe Drinking Water Hotline may be called at: (800) 426-4791.

REFERENCE Methods for the Determination of Organic Compounds in Drinking Water, EPA/600/4-88/039 (revised July 1991; Final Rule for determination of compliance with the MCL for Total Trihalomethanes under 141.30, in 40 CFR Part 141, Vol. 58, No. 147, Fed. Reg., Tuesday Aug. 3, 1993). U.S. EPA Environmental Monitoring Systems Laboratory, Cincinnati, OH, 45268, U.S.A. Available from the National Technical Information Service (NTIS), 5285 Port Royal Road, Springfield, VA 22161; Tel. 800-553-6847. NTIS Order Number is PB91-231480.

1,1-Dichloroethane   **EPA Method 8021**
CAS #75-34-3

TITLE Halogenated Volatile by Gas Chromatography Using Photoionization and Electrolytic Conductivity Detectors in Series: Capillary Column Technique

MATRIX This method is applicable to nearly all types of samples, regardless of water content, including groundwater, aqueous sludges, caustic liquors, acid liquors, waste solvents, oily wastes, mousses, tars, fibrous wastes, polymeric emulsions, filter cakes, spent carbons, spent catalysts, soils, and sediments.

METHOD SUMMARY This method is used to determine 60 volatile organic compounds in a variety of solid waste matrices. It provides GC conditions for the detection of halogenated and aromatic volatile organic compounds. Samples can be analyzed using direct injection or purge-and-trap (EPA Method 5030). Groundwater samples must be analyzed using EPA Method 5030 (where applicable). A temperature program is used with the GC. Detection is achieved by a photoionization detector (PID) and a Hall electrolytic conductivity detector (HECD) in series.

INTERFERENCES Samples can be contaminated by diffusion of volatile organics (particularly chlorofluorocarbons and methylene chloride) through the sample container septum during shipment and storage.

INSTRUMENTATION A GC-equipped with variable-constant differential flow controllers, subambient oven controller, PID and HECD detectors connected with a short piece of uncoated capillary tubing and a data system.

Column: 60 m × 0.75 mm I.D. VOCOL wide-bore capillary column with 1.5 µm film thickness.

PRECISION & ACCURACY MDLs are compound-dependent and vary with purging efficiency and concentration. The applicable concentration range of this method is compound- and instrument-dependent but is approximately 0.1 to 200 µg/L. Analytes that are inefficiently purged from water will not be detected when present at low concentrations, but they can be measured with acceptable accuracy and precision when present in sufficient amounts. The estimated quantitation limit (EQL) for an individual compound is approximately 1 µg/kg (wet weight) for soil/sediment samples, 100 µg/kg (wet weight) for wastes, and 1 µg/L for groundwater. EQLs will be proportionately higher for sample extracts and samples that require dilution or reduced sample size to avoid saturation of the detector.

MULTIPLICATION FACTORS FOR OTHER MATRICES (a)

Matrix	Factor (b)
Groundwater	10
Low-concentration soil	10
Water miscible liquid waste	500
High-concentration soil and sludge	1250
Non-water miscible waste	1250

(a) Sample EQLs are highly matrix-dependent. The EQLs listed herein are provided for guidance and may not always be achievable. (b) EQL = [Method detection limit] × [Factor]. For non-aqueous samples, the factor is on a wet-weight basis.

SINGLE LABORATORY ACCURACY & PRECISION DATA FOR VOCs IN WATER

This method was tested in a single lab using water spiked at 10 µg/L and the following data was reported:

Recoveries and standard deviations were determined from seven samples and spiked at 10 µg/L of each analyte. Recoveries were determined by the internal standard method. Internal standards were: Fluorobenzene for PID and 2-Bromo-1-chloropropane for HECD.

The average recovery (in percent) for the PID was none (no response for this detector).
The standard deviation of the recovery for the PID was none (no response for this detector).
The MDL (in µg/mL) for the PID was none (no response for this detector).
The average recovery (in percent) for the HECD was 100.
The standard deviation of the recovery for the HECD was 5.7.
The MDL (in µg/mL) for the HECD was 0.07.

SAMPLE COLLECTION, PRESERVATION & HANDLING
Volatile Organics — Standard 40-mL glass screw-cap VOA vials with Teflon®-faced silicone septum may be used for both liquid and solid matrices. When collecting samples, liquids and solids should be introduced into the vials gently to reduce agitation which might drive off volatile compounds. If there are any air bubbles present the sample must be retaken. Tap slightly as they are filled to try and eliminate as much free air space as possible. The two vials from each sampling locations should be sealed in separate plastic bags to prevent cross-contamination between samples particularly if the sampled waste is suspected of containing high levels of volatile organics.

Semivolatile organics — Containers used to collect samples for the determination of semivolatile organic compounds should be soap and water washed followed by methanol (or isopropanol) rinsing. The sample containers should be of glass or Teflon® and have screw-top covers with Teflon® liners.

Preservation for volatile organics — No preservation is used with concentrated waste samples. With liquid samples containing no residual chlorine, 4 drops of concentrated hydrochloric acid are added and the samples are immediately cooled to 4°C. When liquid samples contain residual chlorine, they are treated as above and, in addition, 4 drops of 4% aqueous sodium thiosulfate are added. Soil, sediment, and sludge samples are only cooled to 4°C.

Preservation for semivolatile organics — No preservation is used with concentrated waste samples. With liquid samples containing no residual chlorine and with soil, sediment, and sludge samples, immediately cooling to 4°C is the only preservation used. When residual chlorine is present then 3 mL of 10% aqueous sodium sulfate is added for each gallon of sample collected, followed by cooling to 4°C.

MHT The holding time for all volatile organics samples is 14 days. Liquid samples must be extracted within 7 days and their extracts analyzed within 40 days. Concentrated waste, soil, sediment, and sludge samples must be extracted within 14 days and their extracts analyzed within 40 days.

SAMPLE PREPARATION Volatile compounds are introduced into the gas chromatograph either by direct injector or purge-and-trap (EPA Method 5030). EPA Method 5030 may be used directly on groundwater samples or low-concentration contaminated soils and sediments. For medium-concentration soils or sediments, methanolic extraction, as described in EPA Method 5030, may be necessary prior to purge-and-trap analysis.

QUALITY CONTROL Calculate surrogate standard recovery on all samples, blanks, and spikes. A trip blank is recommended to check on sampling, storage, and handling contamination. Calibration standards, at a minimum of five concentration levels, are prepared in organic-free reagent water. One of the concentration levels should be at a concentration near, but above, the method detection limit.

A combination of bromochloromethane, 2-bromo-1-chloropropane, 1,4-dichlorobutane, and bromochlorobenzene are recommended as surrogate standards to encompass the range of the temperature program used in this method.

REFERENCE Test Methods for Evaluating Solid Waste, Physical/Chemical Methods, SW-846, 3rd Edition, U.S. EPA, Office of Solid Waste, Washington, DC, EPA Method 8021A, Rev. 1, Nov. 1992.

1,2-Dichloroethane   **EPA Method 8021**
CAS #107-06-2

TITLE Halogenated Volatile by Gas Chromatography Using Photoionization and Electrolytic Conductivity Detectors in Series: Capillary Column Technique

MATRIX This method is applicable to nearly all types of samples, regardless of water content, including groundwater, aqueous sludges, caustic liquors, acid liquors, waste solvents, oily wastes, mousses, tars, fibrous wastes, polymeric emulsions, filter cakes, spent carbons, spent catalysts, soils, and sediments.

METHOD SUMMARY This method is used to determine 60 volatile organic compounds in a variety of solid waste matrices. It provides GC conditions for the detection of halogenated and aromatic volatile organic compounds. Samples can be analyzed using direct injection or purge-and-trap (EPA Method 5030). Groundwater samples must be analyzed using EPA Method 5030 (where applicable). A temperature program is used with the GC. Detection is achieved by a photoionization detector (PID) and a Hall electrolytic conductivity detector (HECD) in series.

INTERFERENCES Samples can be contaminated by diffusion of volatile organics (particularly chlorofluorocarbons and methylene chloride) through the sample container septum during shipment and storage.

INSTRUMENTATION A GC-equipped with variable-constant differential flow controllers, subambient oven controller, PID and HECD detectors connected with a short piece of uncoated capillary tubing and a data system.

Column: 60 m × 0.75 mm I.D. VOCOL wide-bore capillary column with 1.5 µm film thickness.

PRECISION & ACCURACY MDLs are compound-dependent and vary with purging efficiency and concentration. The applicable concentration range of this method is compound- and instrument-dependent but is approximately 0.1 to 200 µg/L. Analytes that are inefficiently purged from water will not be detected when present at low concentrations, but they can be measured with acceptable accuracy and precision when present in sufficient amounts. The estimated quantitation limit (EQL) for an individual compound is approximately 1 µg/kg (wet weight) for soil/sediment samples, 100 µg/kg (wet weight) for wastes, and 1 µg/L for groundwater. EQLs will be proportionately higher for sample extracts and samples that require dilution or reduced sample size to avoid saturation of the detector.

MULTIPLICATION FACTORS FOR OTHER MATRICES (a)

Matrix	Factor (b)
Groundwater	10
Low-concentration soil	10
Water miscible liquid waste	500
High-concentration soil and sludge	1250
Non-water miscible waste	1250

(a) Sample EQLs are highly matrix-dependent. The EQLs listed herein are provided for guidance and may not always be achievable. (b) EQL = [Method detection limit] × [Factor]. For non-aqueous samples, the factor is on a wet-weight basis.

SINGLE LABORATORY ACCURACY & PRECISION DATA FOR VOCs IN WATER

This method was tested in a single lab using water spiked at 10 µg/L and the following data was reported:

Recoveries and standard deviations were determined from seven samples and spiked at 10 µg/L of each analyte. Recoveries were determined by the internal standard method. Internal standards were: Fluorobenzene for PID and 2-Bromo-1-chloropropane for HECD.

The average recovery (in percent) for the PID was none (no response for this detector).
The standard deviation of the recovery for the PID was none (no response for this detector).
The MDL (in µg/mL) for the PID was none (no response for this detector).
The average recovery (in percent) for the HECD was 100.
The standard deviation of the recovery for the HECD was 3.8.
The MDL (in µg/mL) for the HECD was 0.03.

SAMPLE COLLECTION, PRESERVATION & HANDLING
Volatile Organics — Standard 40-mL glass screw-cap VOA vials with Teflon®-faced silicone septum may be used for both liquid and solid matrices. When collecting samples, liquids and solids should be introduced into the vials gently to reduce agitation which might drive off volatile compounds. If there are any air bubbles present the sample must be retaken. Tap slightly as they are filled to try and eliminate as much free air space as possible. The two vials from each sampling locations should be sealed in separate plastic bags to prevent cross-contamination between samples particularly if the sampled waste is suspected of containing high levels of volatile organics.

Semivolatile organics — Containers used to collect samples for the determination of semivolatile organic compounds should be soap and water washed followed by methanol (or isopropanol) rinsing. The sample containers should be of glass or Teflon® and have screw-top covers with Teflon® liners.

Preservation for volatile organics — No preservation is used with concentrated waste samples. With liquid samples containing no residual chlorine, 4 drops of concentrated hydrochloric acid are added and the samples are immediately cooled to 4°C. When liquid samples contain residual chlorine, they are treated as above and, in addition, 4 drops of 4% aqueous sodium thiosulfate are added. Soil, sediment, and sludge samples are only cooled to 4°C.

Preservation for semivolatile organics — No preservation is used with concentrated waste samples. With liquid samples containing no residual chlorine and with soil, sediment, and sludge samples, immediately cooling to 4°C is the only preservation used. When residual chlorine is present then 3 mL of 10% aqueous sodium sulfate is added for each gallon of sample collected, followed by cooling to 4°C.

MHT The holding time for all volatile organics samples is 14 days. Liquid samples must be extracted within 7 days and their extracts analyzed within 40 days. Concentrated waste, soil, sediment, and sludge samples must be extracted within 14 days and their extracts analyzed within 40 days.

SAMPLE PREPARATION Volatile compounds are introduced into the gas chromatograph either by direct injector or purge-and-trap (EPA Method 5030). EPA Method 5030 may be used directly on groundwater samples or low-concentration contaminated soils and sediments. For medium-concentration soils or sediments, methanolic extraction, as described in EPA Method 5030, may be necessary prior to purge-and-trap analysis.

QUALITY CONTROL Calculate surrogate standard recovery on all samples, blanks, and spikes. A trip blank is recommended to check on sampling, storage, and handling contamination. Calibration standards, at a minimum of five concentration levels, are prepared in organic-free reagent water. One of the concentration levels should be at a concentration near, but above, the method detection limit.

A combination of bromochloromethane, 2-bromo-1-chloropropane, 1,4-dichlorobutane, and bromochlorobenzene are recommended as surrogate standards to encompass the range of the temperature program used in this method.

REFERENCE Test Methods for Evaluating Solid Waste, Physical/Chemical Methods, SW-846, 3rd Edition, U.S. EPA, Office of Solid Waste, Washington, DC, EPA Method 8021A, Rev. 1, Nov. 1992.

1,1-Dichloroethane   **EPA Method 8240**
CAS #75-34-3

TITLE Volatile Organics By GC/MS: Packed Column Technique

MATRIX Nearly all types of sample matarices, regardless of water content, can be analyzed using this method. This includes groundwater, aqueous sludges, caustic liquors, acid liquors, waste solvents, oily wastes, mousses, tars, fibrous wastes, polymetric emulsions, filter cakes, spent carbons, spent catalysts, soils, and sediments.

METHOD SUMMARY Method 8240B covers 80 volatile organic compounds that are introduced into a gas chromatograph by the purge-and-trap method or by direct injection (in limited applications). For the purge-and-trap method an inert gas (zero grade nitrogen or helium) is bubbled through a 5-mL solution at ambient temperature. Purged sample components are trapped in a tube of sorbent materials. When purging is complete, the sorbent tube is heated and backflushed with inert gas to desorb the trapped components onto a GC column.

INTERFERENCES Impurities in the purge gas and from organic compounds outgassing from the plumbing ahead of the trap account for many contamination problems. Interferences purged or coextracted from the samples will vary considerably from source to source. Cross-contamination can occur whenever high-level and low-level samples are analyzed sequentially. Whenever an unusually concentrated sample is analyzed, it should be followed by the analysis of organic-free reagent water to check for cross-contamination. Samples also can be contaminated by diffusion of volatile organics (particularly methylene chloride and fluorocarbons) through the septum seal into the sample during shipment and storage. A trip blank can serve as a check on such contamination. The lab where volatile analysis is performed and also the refrigerated storage area should be completely free of solvents.

INSTRUMENTATION A gas chromatograph/mass spectrometry/data system (GC/MS) equipped with a 6 ft × 0.1 in I.D. glass column packed with 1% SP-1000 on Carbopack-B (60/80 mesh) is required. Also needed is a 5-mL purging device, a sorbent trap, and a thermal desorption apparatus.

PRECISION & ACCURACY This method is reported to have been tested by 15 laboratories using organic-free reagent water, drinking water, surface water, and industrial wastewaters (not specified) fortified at six concentrations over the range 5–600 µg/L.

Sample estimated quantitation limits (EQLs) are highly matrix-dependent. The EQLs listed may not always be achievable. EQLs listed for soils or sediments are based on wet weight. Normally, data is reported on a dry-weight basis; therefore, EQLs will be higher, based on the percent dry weight of each sample. Note that EQLs are even more variable than MDLs and that they are highly variable depending on the matrix being analyzed.

EQL in groundwater in µg/L was 5.
EQL in low soil or sediment in µg/kg was 5.
Accuracy (a) in µg/L was 1.05C+.036.
Precision (b) in µg/L was 0.16x+0.47.

(a) *Average recovery found for measurements of samples containing a concentration of C, in µg/L.*
(b) *Overall precision found for measurements of samples with average recovery X for samples containing a concentration of C in µg/L.*
X = *Average recovery found for measurement of samples containing a concentration of C in µg/L.*

MULTIPLICATION FACTORS FOR OTHER MATRICES

Other Matrices	Factor (a)
Waste miscible liquid waste	50
High-concentration soil and sludge	125
Non-water miscible waste	500

(a) *EQL = [EQL for low soil sediment] × [Factor]. For non-aqueous samples, the factor is on a wet-weight basis.*

SAMPLING METHOD

Liquid samples — Use a 40-mL glass screw-cap VOA vial with a Teflon®-faced silicone septum that has been prewashed, rinsed with distilled deionized water, and oven dried. However, if residual chlorine is present, collect sample in a 40-oz. soil VOA container which has been pre-preserved with 4 drops of 10% sodium thiosulfate, mix gently, and then transfer the sample to a 40-mL VOA vial. Collect bubble-free samples in duplicate and seal them in separate plastic bags.

Soils or sediments, and sludges — Use an 8-oz. widemouth glass bottle with a Teflon®-faced silicone septum that has been prewashed with detergent, rinsed with distilled deionized water, and oven dried. Tap slightly to eliminate free air space. Collect samples in duplicate and seal them in separate plastic bags.

SAMPLE PRESERVATION
Liquid samples — Add 4 drops of concentrated HCL and immediately cool samples to 4°C and store in a solvent-free refrigerator.

Soils or sediments, and sludges — Cool samples to 4°C and store in a solvent-free refrigerator.

MHT Maximum holding time is 14 days from the date of sample collection.

SAMPLE PREPARATION
Liquid samples — Remove the plunger from a 5-mL syringe and carefully pour the sample into the syringe barrel to just short of overflowing. Replace the syringe plunger and compress the sample. Open the syringe valve and vent any residual air while adjusting the sample volume to 5.0 mL. If there is only one volatile organic analysis (VOA) vial, a second syringe should be filled at this time to protect against possible loss of sample integrity. Add 10 µL of surrogate spiking solution and 10 µL of internal standard spiking solution through the valve bore of the 5-mL syringe, then close the valve. The surrogate and internal standards may be mixed and added as a single spiking solution.

Sediments, soils, and waste samples — All samples of this type should be screened by GC analysis using a headspace method (EPA Method 3810) or the hexadecane extraction and screening method (EPA Method 3820). Use the screening data to determine whether to use the low-concentration method (0.005–1 mg/kg) or the high-concentration method (>1 mg/kg).

Low-concentration method — The low-concentration method is based on purging a heated sediment or soil sample mixed with organic-free reagent water containing the surrogate and internal standards. Analyze all reagent blanks and standards under the same conditions as the samples.

Use a 5-g sample if the expected concentration is <0.1 mg/kg or a 1-g sample for expected concentrations between 0.1 and 1 mg/kg. Mix the contents of the sample container with a narrow metal spatula. Weigh the amount of the sample into a tared purge device. Add the spiked water to the purge device, which contains the weighed amount of sample, and connect the device to the purge-and-trap system.

High-concentration method — This method is based on extracting the sediment or soil with methanol. A waste sample is either extracted or diluted, depending on its solubility in methanol. Wastes that are insoluble in methanol are diluted with reagent tetraglyme or possibly polyethylene glycol (PEG). An aliquot of the extract is added to organic-free reagent water containing surrogate and internal standards. This is purged at ambient temperature. All samples with an expected concentration of >1.0 mg/kg should be analyzed by this method.

Mix the contents of the sample container with a narrow metal spatula. For sediments or soils and solid wastes that are insoluble in methanol, weigh 4 g (wet weight) of sample into a tared 20-mL vial. For waste that is soluble in methanol, tetraglyme, or PEG, weigh 1 g (wet weight) into a tared scintillation vial or culture tube or a 10-mL volumetric flask. Quickly add

9.0 mL of appropriate solvent then add 1.0 mL of a surrogate spiking solution to the vial, cap it, and shake it for 2 min.

METHANOL EXTRACT REQUIRED FOR ANALYSIS OF HIGH-CONCENTRATION SOILS OR SEDIMENTS

Approximate Concentration Range	Volume of Methanol Extract (a)
500–10,000 µg/kg	100 µL
1,000–20,000 µg/kg	50 µL
5,000–100,000 µg/kg	10 µL
25,000–500,000 µg/kg	100 µL of 1/50 dilution (b)

Calculate appropriate dilution factor for concentrations exceeding this table.

(a) The volume of methanol added to 5 mL of water being purged should be kept constant. Therefore, add to the 5-mL syringe whatever volume of methanol is necessary to maintain a volume of 100 µL added to the syringe.

(b) Dilute an aliquot of the methanol extract and then take 100 µL for analysis.

QUALITY CONTROL Demonstrate, through the analysis of a reagent water blank, that interferences from the analytical system, glassware, and reagents are under control. Blank samples should be carried through all stages of the sample preparation and measurement steps. For each analytical batch (up to 20 samples), a reagent blank, matrix spike, and matrix spike duplicate must be analyzed (the frequency of the spikes may be different for different monitoring programs). The blank and spiked samples must be carried through all stages of the sample preparation and measurement steps. QC samples mentioned in the section on Interferences will also be needed as appropriate to those situations.

REFERENCE Test Methods for Evaluating Solid Waste (SW-846). U.S. EPA. 1983. Method 8240B, Rev. 2, Nov. 1990. Office of Solid Wastes, Washington, DC.

1,2-Dichloroethane   **EPA Method 8240**
CAS #107-06-2

TITLE Volatile Organics By GC/MS: Packed Column Technique

MATRIX Nearly all types of sample matarices, regardless of water content, can be analyzed using this method. This includes groundwater, aqueous sludges, caustic liquors, acid liquors, waste solvents, oily wastes, mousses, tars, fibrous wastes, polymetric emulsions, filter cakes, spent carbons, spent catalysts, soils, and sediments.

METHOD SUMMARY Method 8240B covers 80 volatile organic compounds that are introduced into a gas chromatograph by the purge-and-trap method or by direct injection (in limited applications). For the purge-and-trap method an inert gas (zero grade nitrogen or helium) is bubbled through a 5-mL solution at ambient temperature. Purged sample components are trapped in a tube of sorbent materials. When purging is complete, the sorbent tube is heated and backflushed with inert gas to desorb the trapped components onto a GC column.

INTERFERENCES Impurities in the purge gas and from organic compounds outgassing from the plumbing ahead of the trap account for many contamination problems. Interferences purged or coextracted from the samples will vary considerably from source to source. Cross-contamination can occur whenever high-level and low-level samples are analyzed sequentially. Whenever an unusually concentrated sample is analyzed, it should be followed by the analysis of organic-free reagent water to check for cross-contamination. Samples also can be contaminated by diffusion of volatile organics (particularly methylene chloride and fluorocarbons) through the septum seal into the sample during shipment and storage. A trip blank can serve as a check on such contamination. The lab where volatile analysis is performed and also the refrigerated storage area should be completely free of solvents.

INSTRUMENTATION A gas chromatograph/mass spectrometry/data system (GC/MS) equipped with a 6 ft × 0.1 in I.D. glass column packed with 1% SP-1000 on Carbopack-B (60/80 mesh) is required. Also needed is a 5-mL purging device, a sorbent trap, and a thermal desorption apparatus.

PRECISION & ACCURACY This method is reported to have been tested by 15 laboratories using organic-free reagent water, drinking water, surface water, and industrial wastewaters (not specified) fortified at six concentrations over the range 5–600 µg/L.

Sample estimated quantitation limits (EQLs) are highly matrix-dependent. The EQLs listed may not always be achievable. EQLs listed for soils or sediments are based on wet weight. Normally, data is reported on a dry-weight basis; therefore, EQLs will be higher, based on the percent dry weight of each sample. Note that EQLs are even more variable than MDLs and that they are highly variable depending on the matrix being analyzed.

EQL in groundwater in µg/L was 5.
EQL in low soil or sediment in µg/kg was 5.
Accuracy (a) in µg/L was $1.02C + 0.45$.
Precision (b) in µg/L was $0.21x - 0.38$.

(a) Average recovery found for measurements of samples containing a concentration of C, in µg/L.
(b) Overall precision found for measurements of samples with average recovery X for samples containing a concentration of C in µg/L.
X = Average recovery found for measurement of samples containing a concentration of C in µg/L.

MULTIPLICATION FACTORS FOR OTHER MATRICES

Other Matrices	Factor (a)
Waste miscible liquid waste	50
High-concentration soil and sludge	125
Non-water miscible waste	500

(a) EQL = [EQL for low soil sediment] × [Factor]. For non-aqueous samples, the factor is on a wet-weight basis.

SAMPLING METHOD

Liquid samples — Use a 40-mL glass screw-cap VOA vial with a Teflon®-faced silicone septum that has been prewashed, rinsed with distilled deionized water, and oven dried. However,

if residual chlorine is present, collect sample in a 40-oz. soil VOA container which has been pre-preserved with 4 drops of 10% sodium thiosulfate, mix gently, and then transfer the sample to a 40-mL VOA vial. Collect bubble-free samples in duplicate and seal them in separate plastic bags.

Soils or sediments, and sludges — Use an 8-oz. widemouth glass bottle with a Teflon®-faced silicone septum that has been prewashed with detergent, rinsed with distilled deionized water, and oven dried. Tap slightly to eliminate free air space. Collect samples in duplicate and seal them in separate plastic bags.

SAMPLE PRESERVATION
Liquid samples — Add 4 drops of concentrated HCL and immediately cool samples to 4°C and store in a solvent-free refrigerator.

Soils or sediments, and sludges — Cool samples to 4°C and store in a solvent-free refrigerator.

MHT Maximum holding time is 14 days from the date of sample collection.

SAMPLE PREPARATION
Liquid samples — Remove the plunger from a 5-mL syringe and carefully pour the sample into the syringe barrel to just short of overflowing. Replace the syringe plunger and compress the sample. Open the syringe valve and vent any residual air while adjusting the sample volume to 5.0 mL. If there is only one volatile organic analysis (VOA) vial, a second syringe should be filled at this time to protect against possible loss of sample integrity. Add 10 µL of surrogate spiking solution and 10 µL of internal standard spiking solution through the valve bore of the 5-mL syringe, then close the valve. The surrogate and internal standards may be mixed and added as a single spiking solution.

Sediments, soils, and waste samples — All samples of this type should be screened by GC analysis using a headspace method (EPA Method 3810) or the hexadecane extraction and screening method (EPA Method 3820). Use the screening data to determine whether to use the low-concentration method (0.005–1 mg/kg) or the high-concentration method (>1 mg/kg).

Low-concentration method — The low-concentration method is based on purging a heated sediment or soil sample mixed with organic-free reagent water containing the surrogate and internal standards. Analyze all reagent blanks and standards under the same conditions as the samples.

Use a 5-g sample if the expected concentration is <0.1 mg/kg or a 1-g sample for expected concentrations between 0.1 and 1 mg/kg. Mix the contents of the sample container with a narrow metal spatula. Weigh the amount of the sample into a tared purge device. Add the spiked water to the purge device, which contains the weighed amount of sample, and connect the device to the purge-and-trap system.

High-concentration method — This method is based on extracting the sediment or soil with methanol. A waste sample is either extracted or diluted, depending on its solubility in methanol. Wastes that are insoluble in methanol are diluted with reagent tetraglyme or possibly polyethylene glycol (PEG).

An aliquot of the extract is added to organic-free reagent water containing surrogate and internal standards. This is purged at ambient temperature. All samples with an expected concentration of >1.0 mg/kg should be analyzed by this method.

Mix the contents of the sample container with a narrow metal spatula. For sediments or soils and solid wastes that are insoluble in methanol, weigh 4 g (wet weight) of sample into a tared 20-mL vial. For waste that is soluble in methanol, tetraglyme, or PEG, weigh 1 g (wet weight) into a tared scintillation vial or culture tube or a 10-mL volumetric flask. Quickly add 9.0 mL of appropriate solvent then add 1.0 mL of a surrogate spiking solution to the vial, cap it, and shake it for 2 min.

METHANOL EXTRACT REQUIRED FOR ANALYSIS OF HIGH-CONCENTRATION SOILS OR SEDIMENTS

Approximate Concentration Range	Volume of Methanol Extract (a)
500–10,000 µg/kg	100 µL
1,000–20,000 µg/kg	50 µL
5,000–100,000 µg/kg	10 µL
25,000–500,000 µg/kg	100 µL of 1/50 dilution (b)

Calculate appropriate dilution factor for concentrations exceeding this table.

(a) The volume of methanol added to 5 mL of water being purged should be kept constant. Therefore, add to the 5-mL syringe whatever volume of methanol is necessary to maintain a volume of 100 µL added to the syringe.
(b) Dilute an aliquot of the methanol extract and then take 100 µL for analysis.

QUALITY CONTROL Demonstrate, through the analysis of a reagent water blank, that interferences from the analytical system, glassware, and reagents are under control. Blank samples should be carried through all stages of the sample preparation and measurement steps. For each analytical batch (up to 20 samples), a reagent blank, matrix spike, and matrix spike duplicate must be analyzed (the frequency of the spikes may be different for different monitoring programs). The blank and spiked samples must be carried through all stages of the sample preparation and measurement steps. QC samples mentioned in the section on Interferences will also be needed as appropriate to those situations.

REFERENCE Test Methods for Evaluating Solid Waste (SW-846). U.S. EPA. 1983. Method 8240B, Rev. 2, Nov. 1990. Office of Solid Wastes, Washington, DC.

1,1-Dichloroethane **EPA Method 8260**
CAS #75-34-3

TITLE Volatile Organic Compounds by GC/MS: Capillary Column Technique

MATRIX This method is applicable to nearly all types of samples, regardless of water content, including groundwater, soils, and sediments.

METHOD SUMMARY Method 8260A covers 58 volatile organic compounds that are introduced into a gas chromatograph by the purge-and-trap method or by direct injection (in limited applications). Zero-grade helium is bubbled through a 5-mL solution at ambient temperature. Purged sample components are trapped in a tube containing suitable sorbent materials. When purging is complete, the sorbent tube is heated and backflushed with helium to desorb trapped sample components. The analytes are desorbed directly to a large bore capillary or cryofocussed on a capillary precolumn before being flash evaporated to a narrow bore capillary for analysis.

INTERFERENCES Major contaminant sources are volatile materials in the lab and impurities in the inert purging gas and in the sorbent trap. Interfering contamination may occur when a sample containing low concentrations of volatile organic compounds is analyzed immediately after a sample containing high concentrations of volatile organic compounds. After analysis of a sample containing high concentrations of volatile organic compounds, one or more calibration blanks should be analyzed to check for cross-contamination. Screening of the samples prior to purge-and-trap GC/MS analysis is highly recommended to prevent contamination of the system. This is especially true for soil and waste samples.

Special precautions must be taken to analyze for methylene chloride. The analytical and sample storage area should be isolated from all atmospheric sources of methylene chloride. All gas chromatography carrier gas lines and purge gas plumbing should be constructed from stainless steel or copper tubing. Laboratory clothing previously exposed to methylene chloride fumes during liquid-liquid extraction procedures can contribute to sample contamination.

Samples can also be contaminated by diffusion of volatile organics (particularly methylene chloride and fluorocarbons) through the septum seal during shipment and storage. A trip blank can serve as a check on such contamination.

INSTRUMENTATION GC/MS with a temperature-programmable chromatograph suitable for splitless injection equipped with variable constant differential flow controllers, a subambient oven controller, a purging device, sorbent trap, a thermal desorption apparatus and a capillary precolumn interface when using cryogenic cooling will be needed. The following GC columns may be used:

Column 1: 60 m × 0.75mm I.D. capillary column coated with VOCOL, 1.5 µm film thickness.
Column 2: 30 m × 0.53mm capillary column coated with DB-624 or VOCOL, 3 µm film thickness.
Column 3: 30 m × 0.32mm I.D. capillary column coated with DB-5 or SE-54, 1-µm film thickness.

PRECISION & ACCURACY This method has been tested in a single lab using spiked water. Using a wide-bore capillary column, water was spiked at concentrations between 0.5 and 10 µg/L. Single lab accuracy and precision data are presented. The MDL actually achieved in a given analysis will vary depending on instrument sensitivity and matrix effects.

The MDL (a) in µg/L was 0.04.
The concentration range in µg/L was 0.5–10.
The mean accuracy (% of true value) was 96.
The precision as relative standard deviation was 5.3.

Note: The MDL is based on a 25-mL sample volume instead of a 5-mL sample volume.

SAMPLING METHOD
Liquid samples — Use a 40-mL glass screw-cap VOA vial with a Teflon®-faced silicone septum that has been prewashed, rinsed with distilled deionized water, and oven dried. If residual chlorine is present, collect the sample in a 4-oz soil VOA container which has been pre-preserved with 4 drops of 10% sodium thiosulfate. Mix gently and transfer the sample to a 40-mL VOA vial. Collect bubble-free samples in duplicate and seal each sample in a separate plastic bag.

Soils, sediments, and sludges — Use an 8-oz widemouth glass bottle with Teflon®-faced silicone septum that has been prewashed, rinsed with distilled deionized water, and oven dried. **Do not** heat the septum for more than 1 h. Tap slightly to eliminate any free air space. Collect samples in duplicate and seal each one in a separate plastic bag.

SAMPLE PRESERVATION
Liquid samples — Add 4 drops of concentrated HCL, cool to 4°C and store in a solvent-free refrigerator.

Soils, sediments and sludges — Cool samples to 4°C and store in a solvent-free refrigerator.

MHT The maximum holding time of any sample (liquids, soils, sediments, and sludges) is 14 days.

SAMPLE PREPARATION
Liquid samples — Remove the plunger from a 5-mL syringe and carefully pour the sample into the syringe barrel to just short of overflowing. Replace the syringe plunger and compress the sample. Open the syringe valve and vent any residual air while adjusting the sample volume to 5.0 mL. If there is only one volatile organic analysis (VOA) vial, a second syringe should be filled at this time to protect against possible loss of sample integrity. Add 10 µL of surrogate spiking solution and 10 µL of internal standard spiking solution through the valve bore of the 5-mL syringe, then close the valve. The surrogate and internal standards may be mixed and added as a single spiking solution.

Sediments, soils, and waste samples — All samples of this type should be screened by GC analysis using a headspace method (EPA Method 3810) or the hexadecane extraction and screening method (EPA Method 3820). Use the screening data to determine whether to use the low-concentration method (0.005–1 mg/kg) or the high-concentration method (>1 mg/kg).

Low-concentration method — The low-concentration method is based on purging a heated sediment or soil sample mixed with organic-free reagent water containing the surrogate and internal standards. Analyze all reagent blanks and standards under the same conditions as the samples.

Use a 5-g sample if the expected concentration is <0.1 mg/kg or a 1-g sample for expected concentrations between 0.1 and 1 mg/kg. Mix the contents of the sample container with a narrow metal spatula. Weigh the amount of the sample into a tared

purge device. Add the spiked water to the purge device, which contains the weighed amount of sample, and connect the device to the purge-and-trap system.

High-concentration method — This method is based on extracting the sediment or soil with methanol. A waste sample is either extracted or diluted, depending on its solubility in methanol. Wastes that are insoluble in methanol are diluted with reagent tetraglyme or possibly polyethylene glycol (PEG). An aliquot of the extract is added to organic-free reagent water containing surrogate and internal standards. This is purged at ambient temperature. All samples with an expected concentration of >1.0 mg/kg should be analyzed by this method.

Mix the contents of the sample container with a narrow metal spatula. For sediments or soils and solid wastes that are insoluble in methanol, weigh 4 g (wet weight) of sample into a tared 20-mL vial. For waste that is soluble in methanol, tetraglyme, or PEG, weigh 1 g (wet weight) into a tared scintillation vial or culture tube or a 10-mL volumetric flask. Quickly add 9.0 mL of appropriate solvent then add 1.0 mL of a surrogate spiking solution to the vial, cap it, and shake it for 2 min.

METHANOL EXTRACT REQUIRED FOR ANALYSIS OF HIGH-CONCENTRATION SOILS OR SEDIMENTS

Approximate Concentration Range	Volume of Methanol Extract (a)
500–10,000 µg/kg	100 µL
1,000–20,000 µg/kg	50 µL
5,000–100,000 µg/kg	10 µL
25,000–500,000 µg/kg	100 µL of 1/50 dilution (b)

Calculate appropriate dilution factor for concentrations exceeding this table.

(a) The volume of methanol added to 5 mL of water being purged should be kept constant. Therefore, add to the 5-mL syringe whatever volume of methanol is necessary to maintain a volume of 100 µL added to the syringe.
(b) Dilute an aliquot of the methanol extract and then take 100 µL for analysis.

QUALITY CONTROL Demonstrate, through the analysis of a reagent water blank, that interferences from the analytical system, glassware, and reagents are under control. Blank samples should be carried through all stages of the sample preparation and measurement steps. For each analytical batch (up to 20 samples), a reagent blank, matrix spike, and matrix spike duplicate must be analyzed (the frequency of the spikes may be different for different monitoring programs). The blank and spiked samples must be carried through all stages of the sample preparation and measurement steps. QC samples mentioned in the section on Interferences will also be needed as appropriate to those situations.

Matrix spiking standards should be prepared from volatile organic compounds which will be representative of the compounds being investigated. The recommended internal standards are chlorobenzene-d5, 1,4-difluorobenzene, 1,4-dichlorobenzene-d4, and pentafluorobenzene. Using stock standard solutions, prepare secondary dilution standards containing the compounds of interest, either singly or mixed together in methanol. Store them in a vial with no headspace for no more than one week. Surrogates recommended are toluene-d8, 4-bromofluorobenzene, and dibromofluoromethane. Each sample undergoing GC/MS analysis must be spiked with 10 µL of the surrogate spiking solution prior to analysis.

REFERENCE Test Methods for Evaluating Solid Waste (SW-846). U.S. EPA 1983, Method 8260A, Rev. 1, Nov. 1990. Office of Solid Waste, Washington, DC.

1,2-Dichloroethane   **EPA Method 8260**
CAS #107-06-2

TITLE Volatile Organic Compounds by GC/MS: Capillary Column Technique

MATRIX This method is applicable to nearly all types of samples, regardless of water content, including groundwater, soils, and sediments.

METHOD SUMMARY Method 8260A covers 58 volatile organic compounds that are introduced into a gas chromatograph by the purge-and-trap method or by direct injection (in limited applications). Zero-grade helium is bubbled through a 5-mL solution at ambient temperature. Purged sample components are trapped in a tube containing suitable sorbent materials. When purging is complete, the sorbent tube is heated and backflushed with helium to desorb trapped sample components. The analytes are desorbed directly to a large bore capillary or cryofocussed on a capillary precolumn before being flash evaporated to a narrow bore capillary for analysis.

INTERFERENCES Major contaminant sources are volatile materials in the lab and impurities in the inert purging gas and in the sorbent trap. Interfering contamination may occur when a sample containing low concentrations of volatile organic compounds is analyzed immediately after a sample containing high concentrations of volatile organic compounds. After analysis of a sample containing high concentrations of volatile organic compounds, one or more calibration blanks should be analyzed to check for cross-contamination. Screening of the samples prior to purge-and-trap GC/MS analysis is highly recommended to prevent contamination of the system. This is especially true for soil and waste samples.

Special precautions must be taken to analyze for methylene chloride. The analytical and sample storage area should be isolated from all atmospheric sources of methylene chloride. All gas chromatography carrier gas lines and purge gas plumbing should be constructed from stainless steel or copper tubing. Laboratory clothing previously exposed to methylene chloride fumes during liquid-liquid extraction procedures can contribute to sample contamination.

Samples can also be contaminated by diffusion of volatile organics (particularly methylene chloride and fluorocarbons) through the septum seal during shipment and storage. A trip blank can serve as a check on such contamination.

INSTRUMENTATION GC/MS with a temperature-programmable chromatograph suitable for splitless injection equipped with variable constant differential flow controllers, a subambient oven controller, a purging device, sorbent trap, a thermal desorption apparatus and a capillary precolumn interface when using cryogenic cooling will be needed. The following GC columns may be used:

Column 1: 60 m × 0.75mm I.D. capillary column coated with VOCOL, 1.5 µm film thickness.
Column 2: 30 m × 0.53mm capillary column coated with DB-624 or VOCOL, 3 µm film thickness.
Column 3: 30 m × 0.32mm I.D. capillary column coated with DB-5 or SE-54, 1-µm film thickness.

PRECISION & ACCURACY This method has been tested in a single lab using spiked water. Using a wide-bore capillary column, water was spiked at concentrations between 0.5 and 10 µg/L. Single lab accuracy and precision data are presented. The MDL actually achieved in a given analysis will vary depending on instrument sensitivity and matrix effects.

The MDL (a) in µg/L was 0.06.
The concentration range in µg/L was 0.1–10.
The mean accuracy (% of true value) was 95.
The precision as relative standard deviation was 5.4.

Note: The MDL is based on a 25-mL sample volume instead of a 5-mL sample volume.

SAMPLING METHOD
Liquid samples — Use a 40-mL glass screw-cap VOA vial with a Teflon®-faced silicone septum that has been prewashed, rinsed with distilled deionized water, and oven dried. If residual chlorine is present, collect the sample in a 4-oz soil VOA container which has been pre-preserved with 4 drops of 10% sodium thiosulfate. Mix gently and transfer the sample to a 40-mL VOA vial. Collect bubble-free samples in duplicate and seal each sample in a separate plastic bag.

Soils, sediments, and sludges — Use an 8-oz widemouth glass bottle with Teflon®-faced silicone septum that has been prewashed, rinsed with distilled deionized water, and oven dried. **Do not** heat the septum for more than 1 h. Tap slightly to eliminate any free air space. Collect samples in duplicate and seal each one in a separate plastic bag.

SAMPLE PRESERVATION
Liquid samples — Add 4 drops of concentrated HCL, cool to 4°C and store in a solvent-free refrigerator.

Soils, sediments and sludges — Cool samples to 4°C and store in a solvent-free refrigerator.

MHT The maximum holding time of any sample (liquids, soils, sediments, and sludges) is 14 days.

SAMPLE PREPARATION
Liquid samples — Remove the plunger from a 5-mL syringe and carefully pour the sample into the syringe barrel to just short of overflowing. Replace the syringe plunger and compress the sample. Open the syringe valve and vent any residual air while adjusting the sample volume to 5.0 mL. If there is only one volatile organic analysis (VOA) vial, a second syringe should be filled at this time to protect against possible loss of sample integrity. Add 10 µL of surrogate spiking solution and 10 µL of internal standard spiking solution through the valve bore of the 5-mL syringe, then close the valve. The surrogate and internal standards may be mixed and added as a single spiking solution.

Sediments, soils, and waste samples — All samples of this type should be screened by GC analysis using a headspace method (EPA Method 3810) or the hexadecane extraction and screening method (EPA Method 3820). Use the screening data to determine whether to use the low-concentration method (0.005–1 mg/kg) or the high-concentration method (>1 mg/kg).

Low-concentration method — The low-concentration method is based on purging a heated sediment or soil sample mixed with organic-free reagent water containing the surrogate and internal standards. Analyze all reagent blanks and standards under the same conditions as the samples.

Use a 5-g sample if the expected concentration is <0.1 mg/kg or a 1-g sample for expected concentrations between 0.1 and 1 mg/kg. Mix the contents of the sample container with a narrow metal spatula. Weigh the amount of the sample into a tared purge device. Add the spiked water to the purge device, which contains the weighed amount of sample, and connect the device to the purge-and-trap system.

High-concentration method — This method is based on extracting the sediment or soil with methanol. A waste sample is either extracted or diluted, depending on its solubility in methanol. Wastes that are insoluble in methanol are diluted with reagent tetraglyme or possibly polyethylene glycol (PEG). An aliquot of the extract is added to organic-free reagent water containing surrogate and internal standards. This is purged at ambient temperature. All samples with an expected concentration of >1.0 mg/kg should be analyzed by this method.

Mix the contents of the sample container with a narrow metal spatula. For sediments or soils and solid wastes that are insoluble in methanol, weigh 4 g (wet weight) of sample into a tared 20-mL vial. For waste that is soluble in methanol, tetraglyme, or PEG, weigh 1 g (wet weight) into a tared scintillation vial or culture tube or a 10-mL volumetric flask. Quickly add 9.0 mL of appropriate solvent then add 1.0 mL of a surrogate spiking solution to the vial, cap it, and shake it for 2 min.

METHANOL EXTRACT REQUIRED FOR ANALYSIS OF HIGH-CONCENTRATION SOILS OR SEDIMENTS

Approximate Concentration Range	Volume of Methanol Extract (a)
500–10,000 µg/kg	100 µL
1,000–20,000 µg/kg	50 µL
5,000–100,000 µg/kg	10 µL
25,000–500,000 µg/kg	100 µL of 1/50 dilution (b)

Calculate appropriate dilution factor for concentrations exceeding this table.

(a) The volume of methanol added to 5 mL of water being purged should be kept constant. Therefore, add to the 5-mL syringe whatever volume of methanol is necessary to maintain a volume of 100 µL added to the syringe.

(b) Dilute an aliquot of the methanol extract and then take 100 µL for analysis.

QUALITY CONTROL Demonstrate, through the analysis of a reagent water blank, that interferences from the analytical system, glassware, and reagents are under control. Blank samples should be carried through all stages of the sample preparation and measurement steps. For each analytical batch (up to 20 samples), a reagent blank, matrix spike, and matrix spike duplicate must be analyzed (the frequency of the spikes may be different for different monitoring programs). The blank and spiked samples must be carried through all stages of the sample preparation and measurement steps. QC samples mentioned in the section on Interferences will also be needed as appropriate to those situations.

Matrix spiking standards should be prepared from volatile organic compounds which will be representative of the compounds being investigated. The recommended internal standards are chlorobenzene-d5, 1,4-difluorobenzene, 1,4-dichlorobenzene-d4, and pentafluorobenzene. Using stock standard solutions, prepare secondary dilution standards containing the compounds of interest, either singly or mixed together in methanol. Store them in a vial with no headspace for no more than one week. Surrogates recommended are toluene-d8, 4-bromofluorobenzene, and dibromofluoromethane. Each sample undergoing GC/MS analysis must be spiked with 10 μL of the surrogate spiking solution prior to analysis.

REFERENCE Test Methods for Evaluating Solid Waste (SW-846). U.S. EPA 1983, Method 8260A, Rev. 1, Nov. 1990. Office of Solid Waste, Washington, DC.

1,1-Dichloroethane   **EPA Method 601**
CAS #75-34-3

TITLE Purgeable Halocarbons

MATRIX Wastewater

APPLICATION Method covers 29 purgeable halocarbons. (Method 624 provides GC/MS conditions appropriate for the qualitative and quantitative confirmation of results). Method describes conditions for a 2nd GC column to confirm measurements made with primary column.

INTERFERENCES Impurities in the purge gas and organic compounds outgassing from the plumbing ahead of the trap. With high- and low-level samples, there can be carryover contamination. Diffusion of volatile organics through the septum seal into the sample.

INSTRUMENTATION GC-equipped with halide-specific detector. (With purge-and-trap unit).

RANGE 8.0–500 μg/L.

MDL 0.07 μg/L.

PRECISION 0.14X+0.94 μg/L (overall precision).

ACCURACY 0.95C–1.08 μg/L (as recovery).

SAMPLING METHOD 25-mL glass vial Teflon®-lined septum.

STABILITY cool, 4°C, 0.008% Sodium thiosulfate.

MHT 14 days.

QUALITY CONTROL The lab must on an ongoing basis, spike at least 10% of the samples from each sample site being monitored to assess accuracy.

REFERENCE Method 601, *Federal Register* Part VIII 40 CFR Part 136, Oct 26, 1984.

1,2-Dichloroethane   **EPA Method 601**
CAS #107-06-2

TITLE Purgeable Halocarbons

MATRIX Wastewater

APPLICATION Method covers 29 purgeable halocarbons. (Method 624 provides GC/MS conditions appropriate for the qualitative and quantitative confirmation of results). Method describes conditions for a 2nd GC column to confirm measurements made with primary column.

INTERFERENCES Impurities in the purge gas and organic compounds outgassing from the plumbing ahead of the trap. With high- and low-level samples, there can be carryover contamination. Diffusion of volatile organics through the septum seal into the sample.

INSTRUMENTATION GC-equipped with halide-specific detector. (With purge-and-trap unit).

RANGE 8.0–500 μg/L.

MDL 0.03 μg/L.

PRECISION 0.15X+0.94 μg/L (overall precision).

ACCURACY 1.04C–1.06 μg/L (as recovery).

SAMPLING METHOD 25-mL glass vial. Teflon®-lined septum.

STABILITY cool, 4°C, 0.008% Sodium thiosulfate.

MHT 14 days.

QUALITY CONTROL The lab must on an ongoing basis, spike at least 10% of the samples from each sample site being monitored to assess accuracy.

REFERENCE Method 601, *Federal Register* Part VIII 40 CFR Part 136, Oct 26, 1984.

1,1-Dichloroethane   **EPA Method 624**
CAS #75-34-3

TITLE Purgeables

MATRIX Wastewater

APPLICATION Method covers 31 purgeable organics. An inert gas is bubbled through a 5-mL water sample in a specially designed purging chamber. Here, purgeables are transferred from aqueous to gaseous phase, passed onto a sorbent column, and trapped. Trap is heated and backflushed with inert gas to desorb purgeables onto a GC column, where purgeables are separated.

INTERFERENCES Impurities in the purge gas, organic compounds outgassing from the plumbing ahead of the trap, and solvent vapors in the lab. With high- and low-level samples, there can be carryover contamination.

INSTRUMENTATION GC/MS with purge-and-trap unit.

RANGE 5–600 µg/L.

MDL 4.7 µg/L

PRECISION 0.16X+0.47 µg/L (overall precision).

ACCURACY 1.05C+0.36 µg/L (as recovery).

SAMPLING METHOD 25-mL glass vial. Teflon®-lined septum.

STABILITY cool, 4°C, 0.008% Sodium thiosulfate.

MHT 14 days.

QUALITY CONTROL The lab must on an ongoing basis, spike at least 5% of the samples from each sample site being monitored to assess accuracy.

REFERENCE Method 624, *Federal Register* Part VIII 40 CFR Part 136, Oct 26, 1984.

1,2-Dichloroethane   **EPA Method 624**
CAS #107-06-2

TITLE Purgeables

MATRIX Wastewater

APPLICATION Method covers 31 purgeable organics. An inert gas is bubbled through a 5-mL water sample in a specially designed purging chamber. Here, purgeables are transferred from aqueous to gaseous phase, passed onto a sorbent column, and trapped. Trap is heated and backflushed with inert gas to desorb purgeables onto a GC column, where purgeables are separated.

INTERFERENCES Impurities in the purge gas, organic compounds outgassing from the plumbing ahead of the trap, and solvent vapors in the lab. With high- and low-level samples, there can be carryover contamination.

INSTRUMENTATION GC/MS with purge-and-trap unit.

RANGE 5–600 µg/L.

MDL 2.8 µg/L

PRECISION 0.21X–0.38 µg/L (overall precision).

ACCURACY 1.02C+0.45 µg/L (as recovery).

SAMPLING METHOD 25-mL glass vial. Teflon®-lined septum.

STABILITY cool, 4°C, 0.008% Sodium thiosulfate.

MHT 14 days.

QUALITY CONTROL The lab must on an ongoing basis, spike at least 5% of the samples from each sample site being monitored to assess accuracy.

REFERENCE Method 624, *Federal Register* Part VIII 40 CFR Part 136, Oct 26, 1984.

1,1-Dichloroethane   **EPA Method 8010**
CAS #75-34-3

TITLE Halogenated Volatile Organics

MATRIX Groundwater, soils, sludges, water miscible liquid wastes, and non-water miscible wastes.

APPLICATION This method is used for the analysis of 39 halogenated VOCs. Samples are analyzed using direct injection or purge-and-trap methods. Groundwater must be analyzed by the purge-and-trap method. The method provides an optional GC column which is used for analyte confirmation and that may help resolve analytes from interferences.

INTERFERENCES There can be carryover contamination with high- and low-level samples. Impurities may come from the purge-and-trap apparatus, organic compounds outgassing from the plumbing ahead of trap, diffusion of VOCs through the sample bottle septum during shipping or storage, or from solvent vapors in the lab.

INSTRUMENTATION GC capable of on-column injections or purge-and-trap sample introduction and a halogen specific detector. Column 1: 8 ft by 0.1 in 1%. SP-1000 on Carbopack-B. Column 2: 6 ft by 0.1 in bonded n-octane on Porasil-C.

RANGE 8 to 500 µg/L (reagent water).

MDL 0.07 µg/L (reagent water).

PQL FACTORS FOR MULTIPLYING × FID MDL VALUE

Matrix	Multiplication Factor
Groundwater	10
Low-level soil	10
Water miscible liquid waste	500
High-level soil and sludge	1250
Non-water miscible waste	1250

PRECISION 0.14X + 0.94 µg/L (overall precision).

ACCURACY 0.95C–1.08 µg/L (as recovery).

SAMPLING METHOD For water and liquid samples; use glass 40-mL vials with Teflon®-lined septum caps and collect two vials per sample location with no headspace. For solids and concentrated waste samples; use widemouth glass bottles with Teflon® liners.

STABILITY For concentrated wastes, soils, sediments, or sludges: cool to 4°C. For liquids: add 4 drops of concentrated hydrochloric acid and cool to 4°C.

MHT 14 days.

QUALITY CONTROL Analyze a reagent blank, matrix spike, and matrix spike duplicate/duplicate for each analytical batch (up to 20 samples). Demonstrate the purity of glassware and reagents by analyzing a reagent water method blank. Internal, surrogate, and five concentration level calibration standards are used.

REFERENCE Test Methods for Evaluating Solid Waste (SW-846), U.S. EPA Office of Solid Waste, Washington, DC, Method 8010B, Rev. 2, Nov. 1992.

1,2-Dichloroethane   EPA Method 8010
CAS #107-06-2

TITLE Halogenated Volatile Organics

MATRIX Groundwater, soils, sludges water miscible liquid wastes, and non-water miscible wastes.

APPLICATION This method is used for the analysis of 39 halogenated VOCs. Samples are analyzed using direct injection or purge-and-trap methods. Groundwater must be analyzed by the purge-and-trap method. The method provides an optional GC column which is used for analyte confirmation and that may help resolve analytes from interferences.

INTERFERENCES There can be carryover contamination with high- and low-level samples. Impurities may come from the purge-and-trap apparatus, organic compounds outgassing from the plumbing ahead of trap, diffusion of VOCs through the sample bottle septum during shipping or storage, or from solvent vapors in the lab.

INSTRUMENTATION GC capable of on-column injections or purge-and-trap sample introduction and a halogen specific detector. Column 1: 8 ft by 0.1 in 1%. SP-1000 on Carbopack-B. Column 2: 6 ft by 0.1 in bonded n-octane on Porasil-C.

RANGE 8 to 500 µg/L (reagent water).

MDL 0.03 µg/L (reagent water).

PQL FACTORS FOR MULTIPLYING × FID MDL VALUE

Matrix	Multiplication Factor
Groundwater	10
Low-level soil	10
Water miscible liquid waste	500
High-level soil and sludge	1250
Non-water miscible waste	1250

PRECISION $0.15X + 0.94$ µg/L (overall precision).

ACCURACY $1.04C - 1.06$ µg/L (as recovery).

SAMPLING METHOD For water and liquid samples; use glass 40-mL vials with Teflon®-lined septum caps and collect two vials per sample location with no headspace. For solids and concentrated waste samples; use widemouth glass bottles with Teflon® liners.

STABILITY For concentrated wastes, soils, sediments, or sludges: cool to 4°C. For liquids: add 4 drops of concentrated hydrochloric acid and cool to 4°C.

MHT 14 days.

QUALITY CONTROL Analyze a reagent blank, matrix spike, and matrix spike duplicate/duplicate for each analytical batch (up to 20 samples). Demonstrate the purity of glassware and reagents by analyzing a reagent water method blank. Internal, surrogate, and five concentration level calibration standards are used.

REFERENCE Test Methods for Evaluating Solid Waste (SW-846), U.S. EPA Office of Solid Waste, Washington, DC, Method 8010B, Rev. 2, Nov. 1992.

1,1-Dichloroethene   EPA Method 1624
CAS #75-35-4

TITLE Volatile Organic Compounds by Isotope Dilution GC/MS

MATRIX Compounds may be determined in waters, soils, and municipal sludges by this method.

METHOD SUMMARY This method is used to determine 58 volatile toxic organic pollutants associated with the CWA (as amended 1987); the RCRA (as amended 1986); the CERCLA (as amended 1986); and other compounds amenable to purge-and-trap gas chromatography-mass spectrometry (GC/MS).

If the solids content is less than 1%, stable isotopically-labeled analogs of the compounds of interest are added to a 5-mL sample and the sample is purged with an inert gas at 20–25°C in a chamber designed for soil or water samples. If the solids content is greater than 1%, 5 mL of reagent water and the labeled compounds are added to a 5-g aliquot of sample and the mixture is purged at 40°C. Compounds that will not purge at 20–25°C or at 40°C are purged at 78–85°C. In the purging process, the volatile compounds are transferred from the aqueous phase into the gaseous phase where they are passed into a sorbent column, and trapped. After purging is completed, the trap is backflushed and heated rapidly to desorb the compounds into a GC. The compounds are separated by the GC and detected by a MS. The labeled compounds serve to correct the variability of the analytical technique.

INTERFERENCES Impurities in the purge gas, organic compounds outgassing from the plumbing upstream of the trap, and solvent vapors in the lab account for most problems. Samples can be contaminated by diffusion of volatile organic compounds (particularly methylene chloride) through the bottle seal during shipment and storage. Contamination by carryover can occur when high-level and low-level samples are analyzed sequentially. When an unusually concentrated sample is encountered, follow it by analysis of a reagent water blank to check for carryover.

INSTRUMENTATION Major equipment includes a GC with linear temperature programming and a glass jet separator as the MS interface, a MS with 70 eV electron impact ionization, and a data system to collect and record response factors.

Column: 2.8 m × 2 mm I.D. glass, packed with 1% SP-1000 on Carbopak B, 60/80 mesh, or equivalent.

PRECISION & ACCURACY The detection limits of the method are usually dependent on the level of interferences

rather than instrumental limitations. The method detection limits were determined in digested sludge (low solids) and in filter cake or compost (high solids).

The MDL (in µg/kg) for low solids is 31 and for high solids is 5.
Labeled and native compound precision (in µg/L) as standard deviation was 12.0.
Labeled and native compound accuracy (in µg/L) as average recovery was detected to 50.

Acceptance criteria are at 20 µg/L for this compound.

SAMPLE COLLECTION, PRESERVATION & HANDLING Grab samples are collected in glass containers having a total volume greater than 20 mL. Fill and seal each bottle so that no air bubbles are entrapped. Samples are maintained at 0 to 4°C from the time of collection until analysis. If an aqueous sample contains residual chlorine, add sodium thiosulfate preservative (10 mg/40 mL) to the empty sample bottles just prior to shipment to the sample site. All samples must be analyzed within 14 days of collection.

SAMPLE PREPARATION Samples containing less than 1% solids are analyzed directly as aqueous samples. Samples containing 1% solids or greater are analyzed as solid samples utilizing one of two methods, depending on the levels of pollutants, in the sample. Samples containing 1% solids or greater, and low to moderate levels of pollutants are analyzed by purging a known weight of sample added to 5 mL of reagent water. Samples containing 1% solids or greater, and high levels of pollutants, are extracted with methanol, and an aliquot of the methanol extract is added to reagent water and purged.

QUALITY CONTROL A field blank prepared from reagent water and carried through the sampling and handling protocol may serve as a check on contamination from shipment and storage.

The analyst is permitted to modify this method to improve separations or lower the costs of measurements, provided all performance specifications are met. Analyses of blanks are required. When results of spikes indicate atypical method performance for samples, the samples are diluted to bring method performance within acceptable limits. Analyze two sets of four 5-mL aliquots (8 aliquots total) of the aqueous performance standard. Spike all samples with labeled compounds to assess method performance on the sample matrix. Compute the percent recovery of the labeled compounds using the internal standard method. Compare the percent recovery for each compound with the corresponding labeled compound recovery. Reagent water blanks are analyzed to demonstrate freedom from carryover contamination. Field replicates may be collected to determine the precision of the sampling technique, and spiked samples may be required to determine the accuracy of the analysis when the internal method is used.

REFERENCE Volatile Organic Compounds by Isotope Dilution GC/MS. Office of Water Regulation and Standards, U.S. EPA Industrial Technology Division, Washington, DC, EPA Method 1624, Rev. C, June 1989 (contact W.A. Telliard, U.S. EPA, Office of Water Regulations and Standards, 401 M St., SW, Washington, DC, 20460. Phone: 202-382-7131).

---

*trans-*1,2-Dichloroethene   **EPA Method 1624**
CAS #156-60-5

TITLE Volatile Organic Compounds by Isotope Dilution GC/MS

MATRIX Compounds may be determined in waters, soils, and municipal sludges by this method.

METHOD SUMMARY This method is used to determine 58 volatile toxic organic pollutants associated with the CWA (as amended 1987); the RCRA (as amended 1986); the CERCLA (as amended 1986); and other compounds amenable to purge-and-trap gas chromatography-mass spectrometry (GC/MS).

If the solids content is less than 1%, stable isotopically-labeled analogs of the compounds of interest are added to a 5-mL sample and the sample is purged with an inert gas at 20–25°C in a chamber designed for soil or water samples. If the solids content is greater than 1%, 5 mL of reagent water and the labeled compounds are added to a 5-g aliquot of sample and the mixture is purged at 40°C. Compounds that will not purge at 20–25°C or at 40°C are purged at 78–85°C. In the purging process, the volatile compounds are transferred from the aqueous phase into the gaseous phase where they are passed into a sorbent column, and trapped. After purging is completed, the trap is backflushed and heated rapidly to desorb the compounds into a GC. The compounds are separated by the GC and detected by a MS. The labeled compounds serve to correct the variability of the analytical technique.

INTERFERENCES Impurities in the purge gas, organic compounds outgassing from the plumbing upstream of the trap, and solvent vapors in the lab account for most problems. Samples can be contaminated by diffusion of volatile organic compounds (particularly methylene chloride) through the bottle seal during shipment and storage. Contamination by carryover can occur when high-level and low-level samples are analyzed sequentially. When an unusually concentrated sample is encountered, follow it by analysis of a reagent water blank to check for carryover.

INSTRUMENTATION Major equipment includes a GC with linear temperature programming and a glass jet separator as the MS interface, a MS with 70 eV electron impact ionization, and a data system to collect and record response factors.

Column: 2.8 m × 2 mm I.D. glass, packed with 1% SP-1000 on Carbopak B, 60/80 mesh, or equivalent.

PRECISION & ACCURACY The detection limits of the method are usually dependent on the level of interferences rather than instrumental limitations. The method detection limits were determined in digested sludge (low solids) and in filter cake or compost (high solids).

The MDL (in µg/kg) for low solids is 41 and for high solids is 3.
Labeled and native compound precision (in µg/L) as standard deviation was 7.4.
Labeled and native compound accuracy (in µg/L) as average recovery was 11–32.

Acceptance criteria are at 20 µg/L for this compound.

SAMPLE COLLECTION, PRESERVATION & HANDLING
Grab samples are collected in glass containers having a total volume greater than 20 mL. Fill and seal each bottle so that no air bubbles are entrapped. Samples are maintained at 0 to 4°C from the time of collection until analysis. If an aqueous sample contains residual chlorine, add sodium thiosulfate preservative (10 mg/40 mL) to the empty sample bottles just prior to shipment to the sample site. All samples must be analyzed within 14 days of collection.

SAMPLE PREPARATION Samples containing less than 1% solids are analyzed directly as aqueous samples. Samples containing 1% solids or greater are analyzed as solid samples utilizing one of two methods, depending on the levels of pollutants, in the sample. Samples containing 1% solids or greater, and low to moderate levels of pollutants are analyzed by purging a known weight of sample added to 5 mL of reagent water. Samples containing 1% solids or greater, and high levels of pollutants, are extracted with methanol, and an aliquot of the methanol extract is added to reagent water and purged.

QUALITY CONTROL A field blank prepared from reagent water and carried through the sampling and handling protocol may serve as a check on contamination from shipment and storage.

The analyst is permitted to modify this method to improve separations or lower the costs of measurements, provided all performance specifications are met. Analyses of blanks are required. When results of spikes indicate atypical method performance for samples, the samples are diluted to bring method performance within acceptable limits. Analyze two sets of four 5-mL aliquots (8 aliquots total) of the aqueous performance standard. Spike all samples with labeled compounds to assess method performance on the sample matrix. Compute the percent recovery of the labeled compounds using the internal standard method. Compare the percent recovery for each compound with the corresponding labeled compound recovery. Reagent water blanks are analyzed to demonstrate freedom from carryover contamination. Field replicates may be collected to determine the precision of the sampling technique, and spiked samples may be required to determine the accuracy of the analysis when the internal method is used.

REFERENCE Volatile Organic Compounds by Isotope Dilution GC/MS. Office of Water Regulation and Standards, U.S. EPA Industrial Technology Division, Washington, DC, EPA Method 1624, Rev. C, June 1989 (contact W.A. Telliard, U.S. EPA, Office of Water Regulations and Standards, 401 M St., SW, Washington, DC, 20460. Phone: 202-382-7131).

1,1-Dichloroethene EPA Method 502
CAS #75-35-4

TITLE Volatile Organic Compounds in Water By Purge and Trap Capillary Column Gas Chromatography with Photoionization and Electrolytic Conductivity Detectors in Series. U.S. EPA Method 502.2, Rev. 2.0, 1989.

MATRIX Drinking water and raw source water. The latter should include most surface water and groundwater sources.

METHOD SUMMARY This method covers 60 volatile organic compounds that contain halogen atoms and/or that are aromatic. An inert gas (zero grade nitrogen or helium) is bubbled through a 25-mL or a 5-mL water sample (depending on the expected concentration of the analytes). Purged sample components are trapped in a tube of sorbent materials. When purging is complete, the sorbent tube is heated and backflushed with helium to desorb the trapped sample onto a capillary GC column. The column is temperature programmed to separate the method analytes which are then detected with a photoionization detector (PID) and a Hall electrolytic conductivity (HECD) placed in series. The PID is selective for aromatic compounds and the HECD is selective for halogenated compounds.

INTERFERENCES Impurities in the purge gas and from organic compounds outgassing from the plumbing ahead of the trap account for many contamination problems. Interferences purged or coextracted from the samples will vary considerably from source to source, depending upon the particular sample or extract being tested. Cross-contamination can occur whenever high-level and low-level samples are analyzed sequentially. Samples also can be contaminated by diffusion of volatile organics (particularly methylene chloride and fluorocarbons) through the septum seal into the sample during shipment and storage. The lab where volatile analysis is performed and also the refrigerated storage area should be completely free of solvents.

INSTRUMENTATION A GC containing a series configuration of a high temperature photoionization detector (PID) equipped with 10.0 eV (nominal) lamp and Hall electrolytic conductivity detector (HECD) is required. Also required is an all-glass 5-mL purging device, a sorbent trap, and a thermal desorption apparatus which is connected to the GC system.

Column 1: VOCOL glass wide-bore capillary column.
Column 2: RTX–502.2 mega-bore capillary column.
Column 3: DB-62 mega-bore capillary column.

PRECISION & ACCURACY Method detection limits are dependent upon the characteristics of the gas chromatographic system used. Analytes that are not separated chromatographically cannot be individually identified and used in the same calibration mixture or water samples unless an alternative technique for identification and quantification, such as mass spectrometry, is used.

Electrolytic conductivity detector (c) range in µg/L (a) was 0.02–200.
Electrolytic conductivity detector (c) MDL in µg/L (b) was 0.07.
Electrolytic conductivity detetor (c) accuracy as % recovery was 103.
Electrolytic conductivity detetor (c) precision as % RSD was 2.8.
Photoionization detector (d) range in µg/L (a) was 0.02–200.
Photoionization detector (d) MDL in µg/L (b) was not detected.
Photoionization detector (d) accuracy as % recovery was 100.

Photoionization detector (d) precision as % RSD was 2.4.

(a) The applicable concentration range of this method is compound, instrument, and matrix-dependent. It is listed as being approximately 0.02 to 200 µg/L but no specific information is provided so caution should be observed.

(b) The method detection limits reports with this method are compound, instrument, and matrix-dependent. The values reported were calculated using reagent water fortified with the corresponding compounds at 10 µg/L and a GC-equipped with a 60 m × 0.75 mm VOLCOL wide bore capillary column with 1.5 µm film thickness and using helium carrier gas.

(c) Recoveries and relative standard deviations were determined from seven samples of reagent water fortified with 10 µg/L of each compound. 2-Bromo-1-chloropropane was used as the internal standard for calculating average recoveries.

(d) Recoveries and relative standard deviations were determined from seven samples of reagent water fortified with 10 µg/L of each compound. Fluorobenzene was used as the internal standard for calculating average recoveries.

SAMPLING METHOD Collect samples using a 40- to 120-mL screw-cap vial (prewashed with detergent, rinsed with distilled water and oven dried at 105°C) with a Teflon®-faced silicone septum. Collect bubble-free samples and place the septum with the Teflon® side down on the water.

SAMPLE PRESERVATION If residual chlorine is present in the water add about 25 mg of ascorbic acid to each vial before samples are collected to remove the chlorine. Add hydrochloric acid to reduce pH to <2, immediately cool samples to 4°C, and store them in a solvent-free refrigerator at 4°C until analysis.

MHT The maximum holding time for samples is 14 days from the time they were collected.

SAMPLE PREPARATION Remove the plungers from two 5-mL syringes and attach a closed syringe valve to each. Warm the sample to room temperature, open the sample bottle, and carefully pour the sample into one of the syringe barrels to just short of overflowing. Replace the syringe plunger, invert the syringe, and compress the sample. Open the syringe valve and vent any residual air while adjusting the sample volume to 5.0 mL. Add 10 µL of the internal calibration standard to the sample through the syringe valve. Close the valve. Fill the second syringe in an identical manner from the same sample bottle. Reserve this second syringe for a reanalysis if necessary.

QUALITY CONTROL As an initial demonstration of lab accuracy and precision, analyze 4 to 7 replicates of a lab fortified blank containing analyte at 0.1–5 µg/L. Collect all samples in duplicate. Surrogate analytes (similar to those of the analytes of interest), whose concentration is known in every sample, are measured using the same internal standard calibration procedure. Duplicate field reagent water blanks (trip blanks) must be analyzed with each set of samples, lab reagent blanks (method blanks) must be analyzed with each batch of samples processed as a group within a work shift. Also, a single lab-fortified blank that contains each of the analytes of interest should be analyzed with each batch of samples processed as a group within a work shift. A 3- to 5-point calibration curve is needed depending on the calibration range factor required.

EPA CONTACT & HOTLINE For technical questions contact Dr. Baldev Bathija, U.S. EPA, Office of Ground Water and Drinking Water (WH-550D), 401 M St. SW, Washington, DC 20460. Tel. (202) 260-3040. For further information the EPA Safe Drinking Water Hotline may be called at: (800) 426-4791.

REFERENCE Methods for the Determination of Organic Compounds in Drinking Water, EPA/600/4-88/039 (revised July 1991; Final Rule for determination of compliance with the MCL for Total Trihalomethanes under 141.30, in 40 CFR Part 141, Vol. 58, No. 147, Fed. Reg., Tuesday Aug. 3, 1993). U.S. EPA Environmental Monitoring Systems Laboratory, Cincinnati, OH, 45268, U.S.A. Available from the National Technical Information Service (NTIS), 5285 Port Royal Road, Springfield, VA 22161; Tel. 800-553-6847. NTIS Order Number is PB91-231480.

***cis*-1,2-Dichloroethene**   **EPA Method 502**
CAS #156-59-4

TITLE Volatile Organic Compounds in Water By Purge and Trap Capillary Column Gas Chromatography with Photoionization and Electrolytic Conductivity Detectors in Series. U.S. EPA Method 502.2, Rev. 2.0, 1989.

MATRIX Drinking water and raw source water. The latter should include most surface water and groundwater sources.

METHOD SUMMARY This method covers 60 volatile organic compounds that contain halogen atoms and/or that are aromatic. An inert gas (zero grade nitrogen or helium) is bubbled through a 25-mL or a 5-mL water sample (depending on the expected concentration of the analytes). Purged sample components are trapped in a tube of sorbent materials. When purging is complete, the sorbent tube is heated and backflushed with helium to desorb the trapped sample onto a capillary GC column. The column is temperature programmed to separate the method analytes which are then detected with a photoionization detector (PID) and a Hall electrolytic conductivity (HECD) placed in series. The PID is selective for aromatic compounds and the HECD is selective for halogenated compounds.

INTERFERENCES Impurities in the purge gas and from organic compounds outgassing from the plumbing ahead of the trap account for many contamination problems. Interferences purged or coextracted from the samples will vary considerably from source to source, depending upon the particular sample or extract being tested. Cross-contamination can occur whenever high-level and low-level samples are analyzed sequentially. Samples also can be contaminated by diffusion of volatile organics (particularly methylene chloride and fluorocarbons) through the septum seal into the sample during shipment and storage. The lab where volatile analysis is performed and also the refrigerated storage area should be completely free of solvents.

INSTRUMENTATION A GC containing a series configuration of a high temperature photoionization detector (PID) equipped with 10.0 eV (nominal) lamp and Hall electrolytic conductivity detector (HECD) is required. Also required is an all-glass 5-mL purging device, a sorbent trap, and a thermal desorption apparatus which is connected to the GC system.

Column 1: VOCOL glass wide-bore capillary column.
Column 2: RTX–502.2 mega-bore capillary column.
Column 3: DB-62 mega-bore capillary column.

PRECISION & ACCURACY Method detection limits are dependent upon the characteristics of the gas chromatographic system used. Analytes that are not separated chromatographically cannot be individually identified and used in the same calibration mixture or water samples unless an alternative technique for identification and quantification, such as mass spectrometry, is used.

Electrolytic conductivity detetor (c) range in µg/L (a) was 0.02–200.
Electrolytic conductivity detetor (c) MDL in µg/L (b) was 0.01.
Electrolytic conductivity detetor (c) accuracy as % recovery was 105.
Electrolytic conductivity detetor (c) precision as % RSD was 3.3.
Photoionization detector (d) range in µg/L (a) was 0.02–200.
Photoionization detector (d) MDL in µg/L (b) was 0.02.
Photoionization detector (d) accuracy as % recovery was not detected.
Photoionization detector (d) precision as % RSD was not detected.

(a) *The applicable concentration range of this method is compound, instrument, and matrix-dependent. It is listed as being approximately 0.02 to 200 µg/L but no specific information is provided so caution should be observed.*
(b) *The method detection limits reports with this method are compound, instrument, and matrix-dependent. The values reported were calculated using reagent water fortified with the corresponding compounds at 10 µg/L and a GC-equipped with a 60 m × 0.75 mm VOLCOL wide bore capillary column with 1.5 µm film thickness and using helium carrier gas.*
(c) *Recoveries and relative standard deviations were determined from seven samples of reagent water fortified with 10 µg/L of each compound. 2-Bromo-1-chloropropane was used as the internal standard for calculating average recoveries.*
(d) *Recoveries and relative standard deviations were determined from seven samples of reagent water fortified with 10 µg/L of each compound. Fluorobenzene was used as the internal standard for calculating average recoveries.*

SAMPLING METHOD Collect samples using a 40- to 120-mL screw-cap vial (prewashed with detergent, rinsed with distilled water and oven dried at 105°C) with a Teflon®-faced silicone septum. Collect bubble-free samples and place the septum with the Teflon® side down on the water.

SAMPLE PRESERVATION If residual chlorine is present in the water add about 25 mg of ascorbic acid to each vial before samples are collected to remove the chlorine. Add hydrochloric acid to reduce pH to <2, immediately cool samples to 4°C, and store them in a solvent-free refrigerator at 4°C until analysis.

MHT The maximum holding time for samples is 14 days from the time they were collected.

SAMPLE PREPARATION Remove the plungers from two 5-mL syringes and attach a closed syringe valve to each. Warm the sample to room temperature, open the sample bottle, and carefully pour the sample into one of the syringe barrels to just short of overflowing. Replace the syringe plunger, invert the syringe, and compress the sample. Open the syringe valve and vent any residual air while adjusting the sample volume to 5.0 mL. Add 10 µL of the internal calibration standard to the sample through the syringe valve. Close the valve. Fill the second syringe in an identical manner from the same sample bottle. Reserve this second syringe for a reanalysis if necessary.

QUALITY CONTROL As an initial demonstration of lab accuracy and precision, analyze 4 to 7 replicates of a lab fortified blank containing analyte at 0.1–5 µg/L. Collect all samples in duplicate. Surrogate analytes (similar to those of the analytes of interest), whose concentration is known in every sample, are measured using the same internal standard calibration procedure. Duplicate field reagent water blanks (trip blanks) must be analyzed with each set of samples, lab reagent blanks (method blanks) must be analyzed with each batch of samples processed as a group within a work shift. Also, a single lab-fortified blank that contains each of the analytes of interest should be analyzed with each batch of samples processed as a group within a work shift. A 3- to 5-point calibration curve is needed depending on the calibration range factor required.

EPA CONTACT & HOTLINE For technical questions contact Dr. Baldev Bathija, U.S. EPA, Office of Ground Water and Drinking Water (WH-550D), 401 M St. SW, Washington, DC 20460. Tel. (202) 260-3040. For further information the EPA Safe Drinking Water Hotline may be called at: (800) 426-4791.

REFERENCE Methods for the Determination of Organic Compounds in Drinking Water, EPA/600/4-88/039 (revised July 1991; Final Rule for determination of compliance with the MCL for Total Trihalomethanes under 141.30, in 40 CFR Part 141, Vol. 58, No. 147, Fed. Reg., Tuesday Aug. 3, 1993). U.S. EPA Environmental Monitoring Systems Laboratory, Cincinnati, OH, 45268, U.S.A. Available from the National Technical Information Service (NTIS), 5285 Port Royal Road, Springfield, VA 22161; Tel. 800-553-6847. NTIS Order Number is PB91-231480.

***trans*-1,2-Dichloroethene**   **EPA Method 502**
CAS #156-60-5

TITLE Volatile Organic Compounds in Water By Purge and Trap Capillary Column Gas Chromatography with Photoionization and Electrolytic Conductivity Detectors in Series. U.S. EPA Method 502.2, Rev. 2.0, 1989.

MATRIX Drinking water and raw source water. The latter should include most surface water and groundwater sources.

METHOD SUMMARY This method covers 60 volatile organic compounds that contain halogen atoms and/or that are aromatic. An inert gas (zero grade nitrogen or helium) is bubbled through a 25-mL or a 5-mL water sample (depending on the expected concentration of the analytes). Purged sample components are trapped in a tube of sorbent materials. When purging is complete, the sorbent tube is heated and backflushed with helium to desorb the trapped sample onto a capillary GC column. The column is temperature programmed to separate the method analytes which are then detected with a photoionization detector (PID) and a Hall electrolytic conductivity (HECD) placed in series. The PID is selective for aromatic compounds and the HECD is selective for halogenated compounds.

INTERFERENCES Impurities in the purge gas and from organic compounds outgassing from the plumbing ahead of the trap account for many contamination problems. Interferences purged or coextracted from the samples will vary considerably from source to source, depending upon the particular sample or extract being tested. Cross-contamination can occur whenever high-level and low-level samples are analyzed sequentially. Samples also can be contaminated by diffusion of volatile organics (particularly methylene chloride and fluorocarbons) through the septum seal into the sample during shipment and storage. The lab where volatile analysis is performed and also the refrigerated storage area should be completely free of solvents.

INSTRUMENTATION A GC containing a series configuration of a high temperature photoionization detector (PID) equipped with 10.0 eV (nominal) lamp and Hall electrolytic conductivity detector (HECD) is required. Also required is an all-glass 5-mL purging device, a sorbent trap, and a thermal desorption apparatus which is connected to the GC system.

Column 1: VOCOL glass wide-bore capillary column.
Column 2: RTX–502.2 mega-bore capillary column.
Column 3: DB-62 mega-bore capillary column.

PRECISION & ACCURACY Method detection limits are dependent upon the characteristics of the gas chromatographic system used. Analytes that are not separated chromatographically cannot be individually identified and used in the same calibration mixture or water samples unless an alternative technique for identification and quantification, such as mass spectrometry, is used.

Electrolytic conductivity detetor (c) range in µg/L (a) was 0.02–200.
Electrolytic conductivity detector (c) MDL in µg/L (b) was 0.06.
Electrolytic conductivity detetor (c) accuracy as % recovery was 99.
Electrolytic conductivity detetor (c) precision as % RSD was 3.7.
Photoionization detector (d) range in µg/L (a) was 0.02–200.
Photoionization detector (d) MDL in µg/L (b) was 0.05.
Photoionization detector (d) accuracy as % recovery was 93.
Photoionization detector (d) precision as % RSD was 4.0.

(a) *The applicable concentration range of this method is compound, instrument, and matrix-dependent. It is listed as being approximately 0.02 to 200 µg/L but no specific information is provided so caution should be observed.*
(b) *The method detection limits reports with this method are compound, instrument, and matrix-dependent. The values reported were calculated using reagent water fortified with the corresponding compounds at 10 µg/L and a GC-equipped with a 60 m × 0.75 mm VOLCOL wide bore capillary column with 1.5 µm film thickness and using helium carrier gas.*
(c) *Recoveries and relative standard deviations were determined from seven samples of reagent water fortified with 10 µg/L of each compound. 2-Bromo-1-chloropropane was used as the internal standard for calculating average recoveries.*
(d) *Recoveries and relative standard deviations were determined from seven samples of reagent water fortified with 10 µg/L of each compound. Fluorobenzene was used as the internal standard for calculating average recoveries.*

SAMPLING METHOD Collect samples using a 40- to 120-mL screw-cap vial (prewashed with detergent, rinsed with distilled water and oven dried at 105°C) with a Teflon®-faced silicone septum. Collect bubble-free samples and place the septum with the Teflon® side down on the water.

SAMPLE PRESERVATION If residual chlorine is present in the water add about 25 mg of ascorbic acid to each vial before samples are collected to remove the chlorine. Add hydrochloric acid to reduce pH to <2, immediately cool samples to 4°C, and store them in a solvent-free refrigerator at 4°C until analysis.

MHT The maximum holding time for samples is 14 days from the time they were collected.

SAMPLE PREPARATION Remove the plungers from two 5-mL syringes and attach a closed syringe valve to each. Warm the sample to room temperature, open the sample bottle, and carefully pour the sample into one of the syringe barrels to just short of overflowing. Replace the syringe plunger, invert the syringe, and compress the sample. Open the syringe valve and vent any residual air while adjusting the sample volume to 5.0 mL. Add 10 µL of the internal calibration standard to the sample through the syringe valve. Close the valve. Fill the second syringe in an identical manner from the same sample bottle. Reserve this second syringe for a reanalysis if necessary.

QUALITY CONTROL As an initial demonstration of lab accuracy and precision, analyze 4 to 7 replicates of a lab fortified blank containing analyte at 0.1–5 µg/L. Collect all samples in duplicate. Surrogate analytes (similar to those of the analytes of interest), whose concentration is known in every sample, are measured using the same internal standard calibration procedure. Duplicate field reagent water blanks (trip blanks) must be analyzed with each set of samples, lab reagent blanks (method blanks) must be analyzed with each batch of samples processed as a group within a work shift. Also, a single lab-fortified blank that contains each of the analytes of interest should be analyzed with each batch of samples processed as a group within a work shift. A 3- to 5-point calibration curve is needed depending on the calibration range factor required.

EPA CONTACT & HOTLINE For technical questions contact Dr. Baldev Bathija, U.S. EPA, Office of Ground Water and

Drinking Water (WH-550D), 401 M St. SW, Washington, DC 20460. Tel. (202) 260-3040. For further information the EPA Safe Drinking Water Hotline may be called at: (800) 426-4791.

REFERENCE Methods for the Determination of Organic Compounds in Drinking Water, EPA/600/4-88/039 (revised July 1991; Final Rule for determination of compliance with the MCL for Total Trihalomethanes under 141.30, in 40 CFR Part 141, Vol. 58, No. 147, Fed. Reg., Tuesday Aug. 3, 1993). U.S. EPA Environmental Monitoring Systems Laboratory, Cincinnati, OH, 45268, U.S.A. Available from the National Technical Information Service (NTIS), 5285 Port Royal Road, Springfield, VA 22161; Tel. 800-553-6847. NTIS Order Number is PB91-231480.

1,1-Dichloroethene	EPA Method 524
CAS #75-35-4	

TITLE Measurement of Purgeable Organic Compounds in Water by Capillary Column GC/MS.

MATRIX Drinking water and raw source water; the latter should include most surface water and groundwater sources.

METHOD SUMMARY Method 524.2 covers 60 volatile organic compounds. An inert gas (zero grade nitrogen or helium) is bubbled through a 25-mL or a 5-mL water sample (depending on the expected concentration of the analytes). Purged sample components are trapped in a tube of sorbent materials. When purging is complete, the sorbent tube is heated and backflushed with helium to desorb the trapped sample onto a capillary GC column.

INTERFERENCES Impurities in the purge gas and from organic compounds outgassing from the plumbing ahead of the trap account for many contamination problems. Interferences purged or coextracted from the samples will vary considerably from source to source, depending upon the particular sample or extract being tested. Cross-contamination can occur whenever high-level and low-level samples are analyzed sequentially. Samples also can be contaminated by diffusion of volatile organics (particularly methylene chloride and fluorocarbons) through the septum seal into the sample during shipment and storage.

INSTRUMENTATION A GC/MS with a data system equipped with one of the following capillary GC columns:

Column 1: VOCOL glass wide bore capillary column.
Column 2: DB-624 fused silica capillary column.
Column 3: DB-5 fused silica capillary column.

Also required is an all-glass 25 mL or 5-mL purging device, a sorbent trap, and a thermal desorption apparatus which is connected to the GC/MS system.

PRECISION & ACCURACY Method detection limits are compound- and instrument-dependent, and may vary from approximately 0.02–0.35 µg/L. Note in the table below that the "true" concentration range used for accuracy and precision measurements was quite narrow. However, the applicable concentration range of this method is primarily column dependent and is approximately 0.02 to 200 µg/L for the wide-bore thick-film columns. Narrow-bore thin-film columns may have a capacity which limits the range to about 0.02 to 20 µg/L. Analytes that are inefficiently purged from water will not be detected when present at low concentrations, but they can be measured with acceptable accuracy and precision when present in sufficient amounts.

Analytes that are not separated chromatographically, but which have different mass spectra and non-interfering quantification ions, can be identified and measured in the same calibration mixture or water sample. Analytes which have very similar mass spectra cannot be individually identified and measured in the same calibration mixture or water samples unless they have different retention times. Co-eluting compounds with very similar mass spectra, typically many structural isomers, must be reported as an isomeric group or pair.

The range (in µg/L) was 0.1–10.
The Method Detection Limig (in µg/L) was 0.12.
The accuracy (as % recovery) was 94.
The precision (in %) was 6.7.

Note: Data were obtained from 16–31 determinations using a wide-bore capillary column and a jet separator interfaced to a quadrupole mass spectrometer. All analytes were in a reagent water matrix.

SAMPLING METHOD Collect samples using a 40- to 120-mL screw-cap vial (prewashed with detergent, rinsed with distilled water and oven dried at 105°C) with a Teflon®-faced silicone septum. Collect bubble-free samples and place the septum with the Teflon® side down on the water.

SAMPLE PRESERVATION If residual chlorine is present in the water add about 25 mg of ascorbic acid to each vial before samples are collected to remove the chlorine. Add hydrochloric acid to reduce pH to <2, and immediately cool samples to 4°C, and store them in a solvent-free refrigerator at 4°C until analysis.

MHT The maximum holding time for samples is 14 days from the time they were collected.

SAMPLE PREPARATION Remove the plungers from two 25-mL (or 5-mL depending on sample size) syringes and attach a closed syringe valve to each. Warm the sample to room temperature, open the sample bottle, and carefully pour the sample into one of the syringe barrels to just short of overflowing. Replace the syringe plunger, invert the syringe, and compress the sample. Open the syringe valve and vent any residual air while adjusting the sample volume to 25.0 mL (or 5 mL). For samples and blanks, add 5 µL of the fortification solution containing the internal standard and the surrogates to the sample through the syringe valve. For calibration standards and lab fortified blanks, add 5 µL of the fortification solution containing the internal standard only. Close the valve. Fill the second syringe in an identical manner from the same sample bottle. Reserve this second syringe for a reanalysis if necessary.

QUALITY CONTROL As an initial demonstration of lab accuracy and precision, analyze 4 to 7 replicates of a lab fortified blank containing analyte at 0.2–5 µg/L. Collect all samples in duplicate. Surrogate analytes (similar to those of the analytes

of interest), whose concentration is known in every sample, are measured using the same internal standard calibration procedure. Duplicate field reagent water blanks (trip blanks) must be analyzed with each set of samples, lab reagent blanks (method blanks) must be analyzed with each batch of samples processed as a group within a work shift. Also, a single lab-fortified blank that contains each of the analytes of interest should be analyzed with each batch of samples processed as a group within a work shift. A 3- to 5-point calibration curve is needed depending on the calibration range factor required.

EPA CONTACT & HOTLINE For technical questions contact Dr. Baldev Bathija, U.S. EPA, Office of Ground Water and Drinking Water (WH-550D), 401 M St. SW, Washington, DC 20460. Tel. (202) 260-3040. For further information the EPA Safe Drinking Water Hotline may be called at: (800) 426-4791.

REFERENCE Methods for the Determination of Organic Compounds in Drinking Water, EPA/600/4-88/039 (revised July 1991; Final Rule for determination of compliance with the MCL for Total Trihalomethanes under 141.30, in 40 CFR Part 141, Vol. 58, No. 147, Fed. Reg., Tuesday Aug. 3, 1993). U.S. EPA Environmental Monitoring Systems Laboratory, Cincinnati, OH, 45268, U.S.A. Available from the National Technical Information Service (NTIS), 5285 Port Royal Road, Springfield, VA 22161; Tel. 800-553-6847. NTIS Order Number is PB91-231480.

---

cis-1,2-Dichloroethene   EPA Method 524
CAS #156-59-4

TITLE Measurement of Purgeable Organic Compounds in Water by Capillary Column GC/MS.

MATRIX Drinking water and raw source water; the latter should include most surface water and groundwater sources.

METHOD SUMMARY Method 524.2 covers 60 volatile organic compounds. An inert gas (zero grade nitrogen or helium) is bubbled through a 25-mL or a 5-mL water sample (depending on the expected concentration of the analytes). Purged sample components are trapped in a tube of sorbent materials. When purging is complete, the sorbent tube is heated and backflushed with helium to desorb the trapped sample onto a capillary GC column.

INTERFERENCES Impurities in the purge gas and from organic compounds outgassing from the plumbing ahead of the trap account for many contamination problems. Interferences purged or coextracted from the samples will vary considerably from source to source, depending upon the particular sample or extract being tested. Cross-contamination can occur whenever high-level and low-level samples are analyzed sequentially. Samples also can be contaminated by diffusion of volatile organics (particularly methylene chloride and fluorocarbons) through the septum seal into the sample during shipment and storage.

INSTRUMENTATION A GC/MS with a data system equipped with one of the following capillary GC columns:

Column 1: VOCOL glass wide bore capillary column.
Column 2: DB-624 fused silica capillary column.
Column 3: DB-5 fused silica capillary column.

Also required is an all-glass 25 mL or 5-mL purging device, a sorbent trap, and a thermal desorption apparatus which is connected to the GC/MS system.

PRECISION & ACCURACY Method detection limits are compound- and instrument-dependent, and may vary from approximately 0.02–0.35 µg/L. Note in the table below that the "true" concentration range used for accuracy and precision measurements was quite narrow. However, the applicable concentration range of this method is primarily column dependent and is approximately 0.02 to 200 µg/L for the wide-bore thick-film columns. Narrow-bore thin-film columns may have a capacity which limits the range to about 0.02 to 20 µg/L. Analytes that are inefficiently purged from water will not be detected when present at low concentrations, but they can be measured with acceptable accuracy and precision when present in sufficient amounts.

Analytes that are not separated chromatographically, but which have different mass spectra and non-interfering quantification ions, can be identified and measured in the same calibration mixture or water sample. Analytes which have very similar mass spectra cannot be individually identified and measured in the same calibration mixture or water samples unless they have different retention times. Co-eluting compounds with very similar mass spectra, typically many structural isomers, must be reported as an isomeric group or pair.

The range (in µg/L) was 0.5–10.
The Method Detection Limig (in µg/L) was 0.12.
The accuracy (as % recovery) was 101.
The precision (in %) was 6.7.

Note: Data were obtained from 16–31 determinations using a wide-bore capillary column and a jet separator interfaced to a quadrupole mass spectrometer. All analytes were in a reagent water matrix.

SAMPLING METHOD Collect samples using a 40- to 120-mL screw-cap vial (prewashed with detergent, rinsed with distilled water and oven dried at 105°C) with a Teflon®-faced silicone septum. Collect bubble-free samples and place the septum with the Teflon® side down on the water.

SAMPLE PRESERVATION If residual chlorine is present in the water add about 25 mg of ascorbic acid to each vial before samples are collected to remove the chlorine. Add hydrochloric acid to reduce pH to <2, and immediately cool samples to 4°C, and store them in a solvent-free refrigerator at 4°C until analysis.

MHT The maximum holding time for samples is 14 days from the time they were collected.

SAMPLE PREPARATION Remove the plungers from two 25-mL (or 5-mL depending on sample size) syringes and attach a closed syringe valve to each. Warm the sample to room temperature, open the sample bottle, and carefully pour the sample into one of the syringe barrels to just short of overflowing. Replace the syringe plunger, invert the syringe, and compress the sample. Open the syringe valve and vent any residual air

while adjusting the sample volume to 25.0 mL (or 5 mL). For samples and blanks, add 5 µL of the fortification solution containing the internal standard and the surrogates to the sample through the syringe valve. For calibration standards and lab fortified blanks, add 5 µL of the fortification solution containing the internal standard only. Close the valve. Fill the second syringe in an identical manner from the same sample bottle. Reserve this second syringe for a reanalysis if necessary.

QUALITY CONTROL As an initial demonstration of lab accuracy and precision, analyze 4 to 7 replicates of a lab fortified blank containing analyte at 0.2–5 µg/L. Collect all samples in duplicate. Surrogate analytes (similar to those of the analytes of interest), whose concentration is known in every sample, are measured using the same internal standard calibration procedure. Duplicate field reagent water blanks (trip blanks) must be analyzed with each set of samples, lab reagent blanks (method blanks) must be analyzed with each batch of samples processed as a group within a work shift. Also, a single lab-fortified blank that contains each of the analytes of interest should be analyzed with each batch of samples processed as a group within a work shift. A 3- to 5-point calibration curve is needed depending on the calibration range factor required.

EPA CONTACT & HOTLINE For technical questions contact Dr. Baldev Bathija, U.S. EPA, Office of Ground Water and Drinking Water (WH-550D), 401 M St. SW, Washington, DC 20460. Tel. (202) 260-3040. For further information the EPA Safe Drinking Water Hotline may be called at: (800) 426-4791.

REFERENCE Methods for the Determination of Organic Compounds in Drinking Water, EPA/600/4-88/039 (revised July 1991; Final Rule for determination of compliance with the MCL for Total Trihalomethanes under 141.30, in 40 CFR Part 141, Vol. 58, No. 147, Fed. Reg., Tuesday Aug. 3, 1993). U.S. EPA Environmental Monitoring Systems Laboratory, Cincinnati, OH, 45268, U.S.A. Available from the National Technical Information Service (NTIS), 5285 Port Royal Road, Springfield, VA 22161; Tel. 800-553-6847. NTIS Order Number is PB91-231480.

---

trans-1,2-Dichloroethene   EPA Method 524
CAS #156-60-5

TITLE Measurement of Purgeable Organic Compounds in Water by Capillary Column GC/MS.

MATRIX Drinking water and raw source water; the latter should include most surface water and groundwater sources.

METHOD SUMMARY Method 524.2 covers 60 volatile organic compounds. An inert gas (zero grade nitrogen or helium) is bubbled through a 25-mL or a 5-mL water sample (depending on the expected concentration of the analytes). Purged sample components are trapped in a tube of sorbent materials. When purging is complete, the sorbent tube is heated and backflushed with helium to desorb the trapped sample onto a capillary GC column.

INTERFERENCES Impurities in the purge gas and from organic compounds outgassing from the plumbing ahead of the trap account for many contamination problems. Interferences purged or coextracted from the samples will vary considerably from source to source, depending upon the particular sample or extract being tested. Cross-contamination can occur whenever high-level and low-level samples are analyzed sequentially. Samples also can be contaminated by diffusion of volatile organics (particularly methylene chloride and fluorocarbons) through the septum seal into the sample during shipment and storage.

INSTRUMENTATION A GC/MS with a data system equipped with one of the following capillary GC columns:

Column 1: VOCOL glass wide bore capillary column.
Column 2: DB-624 fused silica capillary column.
Column 3: DB-5 fused silica capillary column.

Also required is an all-glass 25 mL or 5-mL purging device, a sorbent trap, and a thermal desorption apparatus which is connected to the GC/MS system.

PRECISION & ACCURACY Method detection limits are compound- and instrument-dependent, and may vary from approximately 0.02–0.35 µg/L. Note in the table below that the "true" concentration range used for accuracy and precision measurements was quite narrow. However, the applicable concentration range of this method is primarily column dependent and is approximately 0.02 to 200 µg/L for the wide-bore thick-film columns. Narrow-bore thin-film columns may have a capacity which limits the range to about 0.02 to 20 µg/L. Analytes that are inefficiently purged from water will not be detected when present at low concentrations, but they can be measured with acceptable accuracy and precision when present in sufficient amounts.

Analytes that are not separated chromatographically, but which have different mass spectra and non-interfering quantification ions, can be identified and measured in the same calibration mixture or water sample. Analytes which have very similar mass spectra cannot be individually identified and measured in the same calibration mixture or water samples unless they have different retention times. Co-eluting compounds with very similar mass spectra, typically many structural isomers, must be reported as an isomeric group or pair.

The range (in µg/L) was 0.1–10.
The Method Detection Limig (in µg/L) was 0.06.
The accuracy (as % recovery) was 93.
The precision (in %) was 5.6.

Note: Data were obtained from 16–31 determinations using a wide-bore capillary column and a jet separator interfaced to a quadrupole mass spectrometer. All analytes were in a reagent water matrix.

SAMPLING METHOD Collect samples using a 40- to 120-mL screw-cap vial (prewashed with detergent, rinsed with distilled water and oven dried at 105°C) with a Teflon®-faced silicone septum. Collect bubble-free samples and place the septum with the Teflon® side down on the water.

SAMPLE PRESERVATION If residual chlorine is present in the water add about 25 mg of ascorbic acid to each vial before samples are collected to remove the chlorine. Add hydrochloric

acid to reduce pH to <2, and immediately cool samples to 4°C, and store them in a solvent-free refrigerator at 4°C until analysis.

MHT The maximum holding time for samples is 14 days from the time they were collected.

SAMPLE PREPARATION Remove the plungers from two 25-mL (or 5-mL depending on sample size) syringes and attach a closed syringe valve to each. Warm the sample to room temperature, open the sample bottle, and carefully pour the sample into one of the syringe barrels to just short of overflowing. Replace the syringe plunger, invert the syringe, and compress the sample. Open the syringe valve and vent any residual air while adjusting the sample volume to 25.0 mL (or 5 mL). For samples and blanks, add 5 µL of the fortification solution containing the internal standard and the surrogates to the sample through the syringe valve. For calibration standards and lab fortified blanks, add 5 µL of the fortification solution containing the internal standard only. Close the valve. Fill the second syringe in an identical manner from the same sample bottle. Reserve this second syringe for a reanalysis if necessary.

QUALITY CONTROL As an initial demonstration of lab accuracy and precision, analyze 4 to 7 replicates of a lab fortified blank containing analyte at 0.2–5 µg/L. Collect all samples in duplicate. Surrogate analytes (similar to those of the analytes of interest), whose concentration is known in every sample, are measured using the same internal standard calibration procedure. Duplicate field reagent water blanks (trip blanks) must be analyzed with each set of samples, lab reagent blanks (method blanks) must be analyzed with each batch of samples processed as a group within a work shift. Also, a single lab-fortified blank that contains each of the analytes of interest should be analyzed with each batch of samples processed as a group within a work shift. A 3- to 5-point calibration curve is needed depending on the calibration range factor required.

EPA CONTACT & HOTLINE For technical questions contact Dr. Baldev Bathija, U.S. EPA, Office of Ground Water and Drinking Water (WH-550D), 401 M St. SW, Washington, DC 20460. Tel. (202) 260-3040. For further information the EPA Safe Drinking Water Hotline may be called at: (800) 426-4791.

REFERENCE Methods for the Determination of Organic Compounds in Drinking Water, EPA/600/4-88/039 (revised July 1991; Final Rule for determination of compliance with the MCL for Total Trihalomethanes under 141.30, in 40 CFR Part 141, Vol. 58, No. 147, Fed. Reg., Tuesday Aug. 3, 1993). U.S. EPA Environmental Monitoring Systems Laboratory, Cincinnati, OH, 45268, U.S.A. Available from the National Technical Information Service (NTIS), 5285 Port Royal Road, Springfield, VA 22161; Tel. 800-553-6847. NTIS Order Number is PB91-231480.

1,1-Dichloroethene   **EPA Method 8021**
CAS #75-35-4

TITLE Halogenated Volatile by Gas Chromatography Using Photoionization and Electrolytic Conductivity Detectors in Series: Capillary Column Technique

MATRIX This method is applicable to nearly all types of samples, regardless of water content, including groundwater, aqueous sludges, caustic liquors, acid liquors, waste solvents, oily wastes, mousses, tars, fibrous wastes, polymeric emulsions, filter cakes, spent carbons, spent catalysts, soils, and sediments.

METHOD SUMMARY This method is used to determine 60 volatile organic compounds in a variety of solid waste matrices. It provides GC conditions for the detection of halogenated and aromatic volatile organic compounds. Samples can be analyzed using direct injection or purge-and-trap (EPA Method 5030). Groundwater samples must be analyzed using EPA Method 5030 (where applicable). A temperature program is used with the GC. Detection is achieved by a photoionization detector (PID) and a Hall electrolytic conductivity detector (HECD) in series.

INTERFERENCES Samples can be contaminated by diffusion of volatile organics (particularly chlorofluorocarbons and methylene chloride) through the sample container septum during shipment and storage.

INSTRUMENTATION A GC-equipped with variable-constant differential flow controllers, subambient oven controller, PID and HECD detectors connected with a short piece of uncoated capillary tubing and a data system.

Column: 60 m × 0.75 mm I.D. VOCOL wide-bore capillary column with 1.5 µm film thickness.

PRECISION & ACCURACY MDLs are compound-dependent and vary with purging efficiency and concentration. The applicable concentration range of this method is compound- and instrument-dependent but is approximately 0.1 to 200 µg/L. Analytes that are inefficiently purged from water will not be detected when present at low concentrations, but they can be measured with acceptable accuracy and precision when present in sufficient amounts. The estimated quantitation limit (EQL) for an individual compound is approximately 1 µg/kg (wet weight) for soil/sediment samples, 100 µg/kg (wet weight) for wastes, and 1 µg/L for groundwater. EQLs will be proportionately higher for sample extracts and samples that require dilution or reduced sample size to avoid saturation of the detector.

MULTIPLICATION FACTORS FOR OTHER MATRICES (a)

Matrix	Factor (b)
Groundwater	10
Low-concentration soil	10
Water miscible liquid waste	500
High-concentration soil and sludge	1250
Non-water miscible waste	1250

(a) Sample EQLs are highly matrix-dependent. The EQLs listed herein are provided for guidance and may not always be achievable. (b) EQL = [Method detection limit] × [Factor]. For non-aqueous samples, the factor is on a wet-weight basis.

SINGLE LABORATORY ACCURACY & PRECISION DATA FOR VOCs IN WATER

This method was tested in a single lab using water spiked at 10 µg/L and the following data was reported:

Recoveries and standard deviations were determined from seven samples and spiked at 10 µg/L of each analyte. Recoveries were determined by the internal standard method. Internal standards were: Fluorobenzene for PID and 2-Bromo-1-chloropropane for HECD.

The average recovery (in percent) for the PID was 100.
The standard deviation of the recovery for the PID was 2.4.
The MDL (in µg/mL) for the PID was not determined.
The average recovery (in percent) for the HECD was 103.
The standard deviation of the recovery for the HECD was 2.9.
The MDL (in µg/mL) for the HECD was 0.07.

SAMPLE COLLECTION, PRESERVATION & HANDLING
Volatile Organics — Standard 40-mL glass screw-cap VOA vials with Teflon®-faced silicone septum may be used for both liquid and solid matrices. When collecting samples, liquids and solids should be introduced into the vials gently to reduce agitation which might drive off volatile compounds. If there are any air bubbles present the sample must be retaken. Tap slightly as they are filled to try and eliminate as much free air space as possible. The two vials from each sampling locations should be sealed in separate plastic bags to prevent cross-contamination between samples particularly if the sampled waste is suspected of containing high levels of volatile organics.

Semivolatile organics — Containers used to collect samples for the determination of semivolatile organic compounds should be soap and water washed followed by methanol (or isopropanol) rinsing. The sample containers should be of glass or Teflon® and have screw-top covers with Teflon® liners.

Preservation for volatile organics — No preservation is used with concentrated waste samples. With liquid samples containing no residual chlorine, 4 drops of concentrated hydrochloric acid are added and the samples are immediately cooled to 4°C. When liquid samples contain residual chlorine, they are treated as above and, in addition, 4 drops of 4% aqueous sodium thiosulfate are added. Soil, sediment, and sludge samples are only cooled to 4°C.

Preservation for semivolatile organics — No preservation is used with concentrated waste samples. With liquid samples containing no residual chlorine and with soil, sediment, and sludge samples, immediately cooling to 4°C is the only preservation used. When residual chlorine is present then 3 mL of 10% aqueous sodium sulfate is added for each gallon of sample collected, followed by cooling to 4°C.

MHT The holding time for all volatile organics samples is 14 days. Liquid samples must be extracted within 7 days and their extracts analyzed within 40 days. Concentrated waste, soil, sediment, and sludge samples must be extracted within 14 days and their extracts analyzed within 40 days.

SAMPLE PREPARATION Volatile compounds are introduced into the gas chromatograph either by direct injector or purge-and-trap (EPA Method 5030). EPA Method 5030 may be used directly on groundwater samples or low-concentration contaminated soils and sediments. For medium-concentration soils or sediments, methanolic extraction, as described in EPA Method 5030, may be necessary prior to purge-and-trap analysis.

QUALITY CONTROL Calculate surrogate standard recovery on all samples, blanks, and spikes. A trip blank is recommended to check on sampling, storage, and handling contamination. Calibration standards, at a minimum of five concentration levels, are prepared in organic-free reagent water. One of the concentration levels should be at a concentration near, but above, the method detection limit.

A combination of bromochloromethane, 2-bromo-1-chloropropane, 1,4-dichlorobutane, and bromochlorobenzene are recommended as surrogate standards to encompass the range of the temperature program used in this method.

REFERENCE Test Methods for Evaluating Solid Waste, Physical/Chemical Methods, SW-846, 3rd Edition, U.S. EPA, Office of Solid Waste, Washington, DC, EPA Method 8021A, Rev. 1, Nov. 1992.

---

cis-1,2-Dichloroethene   **EPA Method 8021**
CAS #156-59-4

TITLE Halogenated Volatile by Gas Chromatography Using Photoionization and Electrolytic Conductivity Detectors in Series: Capillary Column Technique

MATRIX This method is applicable to nearly all types of samples, regardless of water content, including groundwater, aqueous sludges, caustic liquors, acid liquors, waste solvents, oily wastes, mousses, tars, fibrous wastes, polymeric emulsions, filter cakes, spent carbons, spent catalysts, soils, and sediments.

METHOD SUMMARY This method is used to determine 60 volatile organic compounds in a variety of solid waste matrices. It provides GC conditions for the detection of halogenated and aromatic volatile organic compounds. Samples can be analyzed using direct injection or purge-and-trap (EPA Method 5030). Groundwater samples must be analyzed using EPA Method 5030 (where applicable). A temperature program is used with the GC. Detection is achieved by a photoionization detector (PID) and a Hall electrolytic conductivity detector (HECD) in series.

INTERFERENCES Samples can be contaminated by diffusion of volatile organics (particularly chlorofluorocarbons and methylene chloride) through the sample container septum during shipment and storage.

INSTRUMENTATION A GC-equipped with variable-constant differential flow controllers, subambient oven controller, PID and HECD detectors connected with a short piece of uncoated capillary tubing and a data system.

Column: 60 m × 0.75 mm I.D. VOCOL wide-bore capillary column with 1.5 µm film thickness.

PRECISION & ACCURACY MDLs are compound-dependent and vary with purging efficiency and concentration. The applicable concentration range of this method is compound- and instrument-dependent but is approximately 0.1 to 200 µg/L. Analytes that are inefficiently purged from water will not be detected when present at low concentrations, but they can be measured with acceptable accuracy and precision when present in sufficient amounts. The estimated quantitation limit (EQL) for an individual compound is approximately 1 µg/kg (wet weight) for soil/sediment samples, 100 µg/kg (wet weight) for wastes, and 1 µg/L for groundwater. EQLs will be proportionately higher for sample extracts and samples that require dilution or reduced sample size to avoid saturation of the detector.

MULTIPLICATION FACTORS FOR OTHER MATRICES (a)

Matrix	Factor (b)
Groundwater	10
Low-concentration soil	10
Water miscible liquid waste	500
High-concentration soil and sludge	1250
Non-water miscible waste	1250

(a) Sample EQLs are highly matrix-dependent. The EQLs listed herein are provided for guidance and may not always be achievable.
(b) EQL = [Method detection limit] × [Factor]. For non-aqueous samples, the factor is on a wet-weight basis.

SINGLE LABORATORY ACCURACY & PRECISION DATA FOR VOCs IN WATER

This method was tested in a single lab using water spiked at 10 µg/L and the following data was reported:

Recoveries and standard deviations were determined from seven samples and spiked at 10 µg/L of each analyte. Recoveries were determined by the internal standard method. Internal standards were: Fluorobenzene for PID and 2-Bromo-1-chloropropane for HECD.

The average recovery (in percent) for the PID was not determined.
The standard deviation of the recovery for the PID was not determined.
The MDL (in µg/mL) for the PID was none (no response for this detector).
The average recovery (in percent) for the HECD was 105.
The standard deviation of the recovery for the HECD was 3.5.
The MDL (in µg/mL) for the HECD was none (no response for this detector).

SAMPLE COLLECTION, PRESERVATION & HANDLING

Volatile Organics — Standard 40-mL glass screw-cap VOA vials with Teflon®-faced silicone septum may be used for both liquid and solid matrices. When collecting samples, liquids and solids should be introduced into the vials gently to reduce agitation which might drive off volatile compounds. If there are any air bubbles present the sample must be retaken. Tap slightly as they are filled to try and eliminate as much free air space as possible. The two vials from each sampling locations should be sealed in separate plastic bags to prevent cross-contamination between samples particularly if the sampled waste is suspected of containing high levels of volatile organics.

Semivolatile organics — Containers used to collect samples for the determination of semivolatile organic compounds should be soap and water washed followed by methanol (or isopropanol) rinsing. The sample containers should be of glass or Teflon® and have screw-top covers with Teflon® liners.

Preservation for volatile organics — No preservation is used with concentrated waste samples. With liquid samples containing no residual chlorine, 4 drops of concentrated hydrochloric acid are added and the samples are immediately cooled to 4°C. When liquid samples contain residual chlorine, they are treated as above and, in addition, 4 drops of 4% aqueous sodium thiosulfate are added. Soil, sediment, and sludge samples are only cooled to 4°C.

Preservation for semivolatile organics — No preservation is used with concentrated waste samples. With liquid samples containing no residual chlorine and with soil, sediment, and sludge samples, immediately cooling to 4°C is the only preservation used. When residual chlorine is present then 3 mL of 10% aqueous sodium sulfate is added for each gallon of sample collected, followed by cooling to 4°C.

MHT The holding time for all volatile organics samples is 14 days. Liquid samples must be extracted within 7 days and their extracts analyzed within 40 days. Concentrated waste, soil, sediment, and sludge samples must be extracted within 14 days and their extracts analyzed within 40 days.

SAMPLE PREPARATION Volatile compounds are introduced into the gas chromatograph either by direct injector or purge-and-trap (EPA Method 5030). EPA Method 5030 may be used directly on groundwater samples or low-concentration contaminated soils and sediments. For medium-concentration soils or sediments, methanolic extraction, as described in EPA Method 5030, may be necessary prior to purge-and-trap analysis.

QUALITY CONTROL Calculate surrogate standard recovery on all samples, blanks, and spikes. A trip blank is recommended to check on sampling, storage, and handling contamination. Calibration standards, at a minimum of five concentration levels, are prepared in organic-free reagent water. One of the concentration levels should be at a concentration near, but above, the method detection limit.

A combination of bromochloromethane, 2-bromo-1-chloropropane, 1,4-dichlorobutane, and bromochlorobenzene are recommended as surrogate standards to encompass the range of the temperature program used in this method.

REFERENCE Test Methods for Evaluating Solid Waste, Physical/Chemical Methods, SW-846, 3rd Edition, U.S. EPA, Office of Solid Waste, Washington, DC, EPA Method 8021A, Rev. 1, Nov. 1992.

---

trans-1,2-Dichloroethene   EPA Method 8021
CAS #156-60-5

TITLE Halogenated Volatile by Gas Chromatography Using Photoionization and Electrolytic Conductivity Detectors in Series: Capillary Column Technique

MATRIX This method is applicable to nearly all types of samples, regardless of water content, including groundwater, aqueous sludges, caustic liquors, acid liquors, waste solvents, oily wastes, mousses, tars, fibrous wastes, polymeric emulsions, filter cakes, spent carbons, spent catalysts, soils, and sediments.

METHOD SUMMARY This method is used to determine 60 volatile organic compounds in a variety of solid waste matrices. It provides GC conditions for the detection of halogenated and aromatic volatile organic compounds. Samples can be analyzed using direct injection or purge-and-trap (EPA Method 5030). Groundwater samples must be analyzed using EPA Method 5030 (where applicable). A temperature program is used with the GC. Detection is achieved by a photoionization detector (PID) and a Hall electrolytic conductivity detector (HECD) in series.

INTERFERENCES Samples can be contaminated by diffusion of volatile organics (particularly chlorofluorocarbons and methylene chloride) through the sample container septum during shipment and storage.

INSTRUMENTATION A GC-equipped with variable-constant differential flow controllers, subambient oven controller, PID and HECD detectors connected with a short piece of uncoated capillary tubing and a data system.

Column: 60 m × 0.75 mm I.D. VOCOL wide-bore capillary column with 1.5 μm film thickness.

PRECISION & ACCURACY MDLs are compound-dependent and vary with purging efficiency and concentration. The applicable concentration range of this method is compound- and instrument-dependent but is approximately 0.1 to 200 μg/L. Analytes that are inefficiently purged from water will not be detected when present at low concentrations, but they can be measured with acceptable accuracy and precision when present in sufficient amounts. The estimated quantitation limit (EQL) for an individual compound is approximately 1 μg/kg (wet weight) for soil/sediment samples, 100 μg/kg (wet weight) for wastes, and 1 μg/L for groundwater. EQLs will be proportionately higher for sample extracts and samples that require dilution or reduced sample size to avoid saturation of the detector.

MULTIPLICATION FACTORS FOR OTHER MATRICES (a)

Matrix	Factor (b)
Groundwater	10
Low-concentration soil	10
Water miscible liquid waste	500
High-concentration soil and sludge	1250
Non-water miscible waste	1250

(a) Sample EQLs are highly matrix-dependent. The EQLs listed herein are provided for guidance and may not always be achievable.
(b) EQL = [Method detection limit] × [Factor]. For non-aqueous samples, the factor is on a wet-weight basis.

SINGLE LABORATORY ACCURACY & PRECISION DATA FOR VOCs IN WATER

This method was tested in a single lab using water spiked at 10 μg/L and the following data was reported:

Recoveries and standard deviations were determined from seven samples and spiked at 10 μg/L of each analyte. Recoveries were determined by the internal standard method. Internal standards were: Fluorobenzene for PID and 2-Bromo-1-chloropropane for HECD.

The average recovery (in percent) for the PID was 93.
The standard deviation of the recovery for the PID was 3.7.
The MDL (in μg/mL) for the PID was 0.05.
The average recovery (in percent) for the HECD was 99.
The standard deviation of the recovery for the HECD was 3.7.
The MDL (in μg/mL) for the HECD was 0.06.

SAMPLE COLLECTION, PRESERVATION & HANDLING
Volatile Organics — Standard 40-mL glass screw-cap VOA vials with Teflon®-faced silicone septum may be used for both liquid and solid matrices. When collecting samples, liquids and solids should be introduced into the vials gently to reduce agitation which might drive off volatile compounds. If there are any air bubbles present the sample must be retaken. Tap slightly as they are filled to try and eliminate as much free air space as possible. The two vials from each sampling locations should be sealed in separate plastic bags to prevent cross-contamination between samples particularly if the sampled waste is suspected of containing high levels of volatile organics.

Semivolatile organics — Containers used to collect samples for the determination of semivolatile organic compounds should be soap and water washed followed by methanol (or isopropanol) rinsing. The sample containers should be of glass or Teflon® and have screw-top covers with Teflon® liners.

Preservation for volatile organics — No preservation is used with concentrated waste samples. With liquid samples containing no residual chlorine, 4 drops of concentrated hydrochloric acid are added and the samples are immediately cooled to 4°C. When liquid samples contain residual chlorine, they are treated as above and, in addition, 4 drops of 4% aqueous sodium thiosulfate are added. Soil, sediment, and sludge samples are only cooled to 4°C.

Preservation for semivolatile organics — No preservation is used with concentrated waste samples. With liquid samples containing no residual chlorine and with soil, sediment, and sludge samples, immediately cooling to 4°C is the only preservation used. When residual chlorine is present then 3 mL of 10% aqueous sodium sulfate is added for each gallon of sample collected, followed by cooling to 4°C.

MHT The holding time for all volatile organics samples is 14 days. Liquid samples must be extracted within 7 days and their extracts analyzed within 40 days. Concentrated waste, soil, sediment, and sludge samples must be extracted within 14 days and their extracts analyzed within 40 days.

SAMPLE PREPARATION Volatile compounds are introduced into the gas chromatograph either by direct injector or purge-and-trap (EPA Method 5030). EPA Method 5030 may be used directly on groundwater samples or low-concentration contaminated soils and sediments. For medium-concentration soils or sediments, methanolic extraction, as described in EPA Method 5030, may be necessary prior to purge-and-trap analysis.

QUALITY CONTROL Calculate surrogate standard recovery on all samples, blanks, and spikes. A trip blank is recommended to check on sampling, storage, and handling contamination. Calibration standards, at a minimum of five concentration levels, are prepared in organic-free reagent water. One of the concentration levels should be at a concentration near, but above, the method detection limit.

A combination of bromochloromethane, 2-bromo-1-chloropropane, 1,4-dichlorobutane, and bromochlorobenzene are recommended as surrogate standards to encompass the range of the temperature program used in this method.

REFERENCE Test Methods for Evaluating Solid Waste, Physical/Chemical Methods, SW-846, 3rd Edition, U.S. EPA, Office of Solid Waste, Washington, DC, EPA Method 8021A, Rev. 1, Nov. 1992.

1,1-Dichloroethene   **EPA Method 8240**
CAS #75-35-4

TITLE Volatile Organics By GC/MS: Packed Column Technique

MATRIX Nearly all types of sample matarices, regardless of water content, can be analyzed using this method. This includes groundwater, aqueous sludges, caustic liquors, acid liquors, waste solvents, oily wastes, mousses, tars, fibrous wastes, polymetric emulsions, filter cakes, spent carbons, spent catalysts, soils, and sediments.

METHOD SUMMARY Method 8240B covers 80 volatile organic compounds that are introduced into a gas chromatograph by the purge-and-trap method or by direct injection (in limited applications). For the purge-and-trap method an inert gas (zero grade nitrogen or helium) is bubbled through a 5-mL solution at ambient temperature. Purged sample components are trapped in a tube of sorbent materials. When purging is complete, the sorbent tube is heated and backflushed with inert gas to desorb the trapped components onto a GC column.

INTERFERENCES Impurities in the purge gas and from organic compounds outgassing from the plumbing ahead of the trap account for many contamination problems. Interferences purged or coextracted from the samples will vary considerably from source to source. Cross-contamination can occur whenever high-level and low-level samples are analyzed sequentially. Whenever an unusually concentrated sample is analyzed, it should be followed by the analysis of organic-free reagent water to check for cross-contamination. Samples also can be contaminated by diffusion of volatile organics (particularly methylene chloride and fluorocarbons) through the septum seal into the sample during shipment and storage. A trip blank can serve as a check on such contamination. The lab where volatile analysis is performed and also the refrigerated storage area should be completely free of solvents.

INSTRUMENTATION A gas chromatograph/mass spectrometry/data system (GC/MS) equipped with a 6 ft × 0.1 in I.D. glass column packed with 1% SP-1000 on Carbopack-B (60/80 mesh) is required. Also needed is a 5-mL purging device, a sorbent trap, and a thermal desorption apparatus.

PRECISION & ACCURACY This method is reported to have been tested by 15 laboratories using organic-free reagent water, drinking water, surface water, and industrial wastewaters (not specified) fortified at six concentrations over the range 5–600 µg/L.

Sample estimated quantitation limits (EQLs) are highly matrix-dependent. The EQLs listed may not always be achievable. EQLs listed for soils or sediments are based on wet weight. Normally, data is reported on a dry-weight basis; therefore, EQLs will be higher, based on the percent dry weight of each sample. Note that EQLs are even more variable than MDLs and that they are highly variable depending on the matrix being analyzed.

EQL in groundwater in µg/L was 5.
EQL in low soil or sediment in µg/kg was 5.
Accuracy (a) in µg/L was $1.12C + 0.61$.
Precision (b) in µg/L was $0.43x - 0.22$.

(a) Average recovery found for measurements of samples containing a concentration of C, in µg/L.

(b) Overall precision found for measurements of samples with average recovery X for samples containing a concentration of C in µg/L.

X = Average recovery found for measurement of samples containing a concentration of C in µg/L.

MULTIPLICATION FACTORS FOR OTHER MATRICES

Other Matrices	Factor (a)
Waste miscible liquid waste	50
High-concentration soil and sludge	125
Non-water miscible waste	500

(a) EQL = [EQL for low soil sediment] × [Factor]. For non-aqueous samples, the factor is on a wet-weight basis.

SAMPLING METHOD

Liquid samples — Use a 40-mL glass screw-cap VOA vial with a Teflon®-faced silicone septum that has been prewashed, rinsed with distilled deionized water, and oven dried. However, if residual chlorine is present, collect sample in a 40-oz. soil VOA container which has been pre-preserved with 4 drops of 10% sodium thiosulfate, mix gently, and then transfer the sample to a 40-mL VOA vial. Collect bubble-free samples in duplicate and seal them in separate plastic bags.

Soils or sediments, and sludges — Use an 8-oz. widemouth glass bottle with a Teflon®-faced silicone septum that has been prewashed with detergent, rinsed with distilled deionized water, and oven dried. Tap slightly to eliminate free air space. Collect samples in duplicate and seal them in separate plastic bags.

SAMPLE PRESERVATION

Liquid samples — Add 4 drops of concentrated HCL and immediately cool samples to 4°C and store in a solvent-free refrigerator.

Soils or sediments, and sludges — Cool samples to 4°C and store in a solvent-free refrigerator.

MHT Maximum holding time is 14 days from the date of sample collection.

SAMPLE PREPARATION

Liquid samples — Remove the plunger from a 5-mL syringe and carefully pour the sample into the syringe barrel to just short of overflowing. Replace the syringe plunger and compress the sample. Open the syringe valve and vent any residual air while adjusting the sample volume to 5.0 mL. If there is only one volatile organic analysis (VOA) vial, a second syringe should be filled at this time to protect against possible loss of sample integrity. Add 10 μL of surrogate spiking solution and 10 μL of internal standard spiking solution through the valve bore of the 5-mL syringe, then close the valve. The surrogate and internal standards may be mixed and added as a single spiking solution.

Sediments, soils, and waste samples — All samples of this type should be screened by GC analysis using a headspace method (EPA Method 3810) or the hexadecane extraction and screening method (EPA Method 3820). Use the screening data to determine whether to use the low-concentration method (0.005–1 mg/kg) or the high-concentration method (>1 mg/kg).

Low-concentration method — The low-concentration method is based on purging a heated sediment or soil sample mixed with organic-free reagent water containing the surrogate and internal standards. Analyze all reagent blanks and standards under the same conditions as the samples.

Use a 5-g sample if the expected concentration is <0.1 mg/kg or a 1-g sample for expected concentrations between 0.1 and 1 mg/kg. Mix the contents of the sample container with a narrow metal spatula. Weigh the amount of the sample into a tared purge device. Add the spiked water to the purge device, which contains the weighed amount of sample, and connect the device to the purge-and-trap system.

High-concentration method — This method is based on extracting the sediment or soil with methanol. A waste sample is either extracted or diluted, depending on its solubility in methanol. Wastes that are insoluble in methanol are diluted with reagent tetraglyme or possibly polyethylene glycol (PEG). An aliquot of the extract is added to organic-free reagent water containing surrogate and internal standards. This is purged at ambient temperature. All samples with an expected concentration of >1.0 mg/kg should be analyzed by this method.

Mix the contents of the sample container with a narrow metal spatula. For sediments or soils and solid wastes that are insoluble in methanol, weigh 4 g (wet weight) of sample into a tared 20-mL vial. For waste that is soluble in methanol, tetraglyme, or PEG, weigh 1 g (wet weight) into a tared scintillation vial or culture tube or a 10-mL volumetric flask. Quickly add 9.0 mL of appropriate solvent then add 1.0 mL of a surrogate spiking solution to the vial, cap it, and shake it for 2 min.

METHANOL EXTRACT REQUIRED FOR ANALYSIS OF HIGH-CONCENTRATION SOILS OR SEDIMENTS

Approximate Concentration Range	Volume of Methanol Extract (a)
500–10,000 μg/kg	100 μL
1,000–20,000 μg/kg	50 μL
5,000–100,000 μg/kg	10 μL
25,000–500,000 μg/kg	100 μL of 1/50 dilution (b)

Calculate appropriate dilution factor for concentrations exceeding this table.

(a) The volume of methanol added to 5 mL of water being purged should be kept constant. Therefore, add to the 5-mL syringe whatever volume of methanol is necessary to maintain a volume of 100 μL added to the syringe.

(b) Dilute an aliquot of the methanol extract and then take 100 μL for analysis.

QUALITY CONTROL Demonstrate, through the analysis of a reagent water blank, that interferences from the analytical system, glassware, and reagents are under control. Blank samples should be carried through all stages of the sample preparation and measurement steps. For each analytical batch (up to 20 samples), a reagent blank, matrix spike, and matrix spike duplicate must be analyzed (the frequency of the spikes may be different for different monitoring programs). The blank and spiked samples must be carried through all stages of the sample preparation and measurement steps. QC samples mentioned in the section on Interferences will also be needed as appropriate to those situations.

REFERENCE Test Methods for Evaluating Solid Waste (SW-846). U.S. EPA. 1983. Method 8240B, Rev. 2, Nov. 1990. Office of Solid Wastes, Washington, DC.

---

trans-1,2-Dichloroethene　　　　　　　EPA Method 8240
CAS #156-60-5

TITLE Volatile Organics By GC/MS: Packed Column Technique

MATRIX Nearly all types of sample matarices, regardless of water content, can be analyzed using this method. This includes groundwater, aqueous sludges, caustic liquors, acid liquors, waste solvents, oily wastes, mousses, tars, fibrous wastes, polymetric emulsions, filter cakes, spent carbons, spent catalysts, soils, and sediments.

METHOD SUMMARY Method 8240B covers 80 volatile organic compounds that are introduced into a gas chromatograph by the purge-and-trap method or by direct injection (in limited applications). For the purge-and-trap method an inert gas (zero grade nitrogen or helium) is bubbled through a 5-mL solution at ambient temperature. Purged sample components are trapped in a tube of sorbent materials. When purging is complete, the sorbent tube is heated and backflushed with inert gas to desorb the trapped components onto a GC column.

INTERFERENCES Impurities in the purge gas and from organic compounds outgassing from the plumbing ahead of the trap account for many contamination problems. Interferences purged or coextracted from the samples will vary considerably from source to source. Cross-contamination can occur whenever high-level and low-level samples are analyzed sequentially. Whenever an unusually concentrated sample is analyzed, it should be followed by the analysis of organic-free reagent water to check for cross-contamination. Samples also can be contaminated by diffusion of volatile organics (particularly methylene chloride and fluorocarbons) through the septum seal into the sample during shipment and storage. A trip blank can serve as a check on such contamination. The lab where volatile analysis is performed and also the refrigerated storage area should be completely free of solvents.

INSTRUMENTATION A gas chromatograph/mass spectrometry/data system (GC/MS) equipped with a 6 ft × 0.1 in I.D. glass column packed with 1% SP-1000 on Carbopack-B (60/80 mesh) is required. Also needed is a 5-mL purging device, a sorbent trap, and a thermal desorption apparatus.

PRECISION & ACCURACY This method is reported to have been tested by 15 laboratories using organic-free reagent water, drinking water, surface water, and industrial wastewaters (not specified) fortified at six concentrations over the range 5–600 µg/L.

Sample estimated quantitation limits (EQLs) are highly matrix-dependent. The EQLs listed may not always be achievable. EQLs listed for soils or sediments are based on wet weight. Normally, data is reported on a dry-weight basis; therefore, EQLs will be higher, based on the percent dry weight of each sample. Note that EQLs are even more variable than MDLs and that they are highly variable depending on the matrix being analyzed.

EQL in groundwater in µg/L was 5.
EQL in low soil or sediment in µg/kg was 5.
Accuracy (a) in µg/L was 1.05C+0.03.
Precision (b) in µg/L was 0.19x+0.17.

(a) *Average recovery found for measurements of samples containing a concentration of C, in µg/L.*
(b) *Overall precision found for measurements of samples with average recovery X for samples containing a concentration of C in µg/L.*
X = *Average recovery found for measurement of samples containing a concentration of C in µg/L.*

MULTIPLICATION FACTORS FOR OTHER MATRICES

Other Matrices	Factor (a)
Waste miscible liquid waste	50
High-concentration soil and sludge	125
Non-water miscible waste	500

(a) *EQL = [EQL for low soil sediment] × [Factor]. For non-aqueous samples, the factor is on a wet-weight basis.*

SAMPLING METHOD

Liquid samples — Use a 40-mL glass screw-cap VOA vial with a Teflon®-faced silicone septum that has been prewashed, rinsed with distilled deionized water, and oven dried. However, if residual chlorine is present, collect sample in a 40-oz. soil VOA container which has been pre-preserved with 4 drops of 10% sodium thiosulfate, mix gently, and then transfer the sample to a 40-mL VOA vial. Collect bubble-free samples in duplicate and seal them in separate plastic bags.

Soils or sediments, and sludges — Use an 8-oz. widemouth glass bottle with a Teflon®-faced silicone septum that has been prewashed with detergent, rinsed with distilled deionized water, and oven dried. Tap slightly to eliminate free air space. Collect samples in duplicate and seal them in separate plastic bags.

SAMPLE PRESERVATION

Liquid samples — Add 4 drops of concentrated HCL and immediately cool samples to 4°C and store in a solvent-free refrigerator.

Soils or sediments, and sludges — Cool samples to 4°C and store in a solvent-free refrigerator.

MHT Maximum holding time is 14 days from the date of sample collection.

SAMPLE PREPARATION

Liquid samples — Remove the plunger from a 5-mL syringe and carefully pour the sample into the syringe barrel to just short of overflowing. Replace the syringe plunger and compress the sample. Open the syringe valve and vent any residual air while adjusting the sample volume to 5.0 mL. If there is only one volatile organic analysis (VOA) vial, a second syringe should be filled at this time to protect against possible loss of sample integrity. Add 10 µL of surrogate spiking solution and 10 µL of internal standard spiking solution through the valve bore of the 5-mL syringe, then close the valve. The surrogate and internal standards may be mixed and added as a single spiking solution.

Sediments, soils, and waste samples — All samples of this type should be screened by GC analysis using a headspace method (EPA Method 3810) or the hexadecane extraction and screening method (EPA Method 3820). Use the screening data to determine whether to use the low-concentration method (0.005–1 mg/kg) or the high-concentration method (>1 mg/kg).

Low-concentration method — The low-concentration method is based on purging a heated sediment or soil sample mixed with organic-free reagent water containing the surrogate and internal standards. Analyze all reagent blanks and standards under the same conditions as the samples.

Use a 5-g sample if the expected concentration is <0.1 mg/kg or a 1-g sample for expected concentrations between 0.1 and 1 mg/kg. Mix the contents of the sample container with a narrow metal spatula. Weigh the amount of the sample into a tared purge device. Add the spiked water to the purge device, which contains the weighed amount of sample, and connect the device to the purge-and-trap system.

High-concentration method — This method is based on extracting the sediment or soil with methanol. A waste sample is either extracted or diluted, depending on its solubility in methanol. Wastes that are insoluble in methanol are diluted with reagent tetraglyme or possibly polyethylene glycol (PEG). An aliquot of the extract is added to organic-free reagent water containing surrogate and internal standards. This is purged at ambient temperature. All samples with an expected concentration of >1.0 mg/kg should be analyzed by this method.

Mix the contents of the sample container with a narrow metal spatula. For sediments or soils and solid wastes that are insoluble in methanol, weigh 4 g (wet weight) of sample into a tared 20-mL vial. For waste that is soluble in methanol, tetraglyme, or PEG, weigh 1 g (wet weight) into a tared scintillation vial or culture tube or a 10-mL volumetric flask. Quickly add

9.0 mL of appropriate solvent then add 1.0 mL of a surrogate spiking solution to the vial, cap it, and shake it for 2 min.

METHANOL EXTRACT REQUIRED FOR ANALYSIS OF HIGH-CONCENTRATION SOILS OR SEDIMENTS

Approximate Concentration Range	Volume of Methanol Extract (a)
500–10,000 μg/kg	100 μL
1,000–20,000 μg/kg	50 μL
5,000–100,000 μg/kg	10 μL
25,000–500,000 μg/kg	100 μL of 1/50 dilution (b)

Calculate appropriate dilution factor for concentrations exceeding this table.

(a) The volume of methanol added to 5 mL of water being purged should be kept constant. Therefore, add to the 5-mL syringe whatever volume of methanol is necessary to maintain a volume of 100 μL added to the syringe.
(b) Dilute an aliquot of the methanol extract and then take 100 μL for analysis.

QUALITY CONTROL Demonstrate, through the analysis of a reagent water blank, that interferences from the analytical system, glassware, and reagents are under control. Blank samples should be carried through all stages of the sample preparation and measurement steps. For each analytical batch (up to 20 samples), a reagent blank, matrix spike, and matrix spike duplicate must be analyzed (the frequency of the spikes may be different for different monitoring programs). The blank and spiked samples must be carried through all stages of the sample preparation and measurement steps. QC samples mentioned in the section on Interferences will also be needed as appropriate to those situations.

REFERENCE Test Methods for Evaluating Solid Waste (SW-846). U.S. EPA. 1983. Method 8240B, Rev. 2, Nov. 1990. Office of Solid Wastes, Washington, DC.

1,1-Dichloroethene   **EPA Method 8260**
CAS #75-35-4

TITLE Volatile Organic Compounds by GC/MS: Capillary Column Technique

MATRIX This method is applicable to nearly all types of samples, regardless of water content, including groundwater, soils, and sediments.

METHOD SUMMARY Method 8260A covers 58 volatile organic compounds that are introduced into a gas chromatograph by the purge-and-trap method or by direct injection (in limited applications). Zero-grade helium is bubbled through a 5-mL solution at ambient temperature. Purged sample components are trapped in a tube containing suitable sorbent materials. When purging is complete, the sorbent tube is heated and backflushed with helium to desorb trapped sample components. The analytes are desorbed directly to a large bore capillary or cryofocussed on a capillary precolumn before being flash evaporated to a narrow bore capillary for analysis.

INTERFERENCES Major contaminant sources are volatile materials in the lab and impurities in the inert purging gas and in the sorbent trap. Interfering contamination may occur when a sample containing low concentrations of volatile organic compounds is analyzed immediately after a sample containing high concentrations of volatile organic compounds. After analysis of a sample containing high concentrations of volatile organic compounds, one or more calibration blanks should be analyzed to check for cross-contamination. Screening of the samples prior to purge-and-trap GC/MS analysis is highly recommended to prevent contamination of the system. This is especially true for soil and waste samples.

Special precautions must be taken to analyze for methylene chloride. The analytical and sample storage area should be isolated from all atmospheric sources of methylene chloride. All gas chromatography carrier gas lines and purge gas plumbing should be constructed from stainless steel or copper tubing. Laboratory clothing previously exposed to methylene chloride fumes during liquid-liquid extraction procedures can contribute to sample contamination.

Samples can also be contaminated by diffusion of volatile organics (particularly methylene chloride and fluorocarbons) through the septum seal during shipment and storage. A trip blank can serve as a check on such contamination.

INSTRUMENTATION GC/MS with a temperature-programmable chromatograph suitable for splitless injection equipped with variable constant differential flow controllers, a subambient oven controller, a purging device, sorbent trap, a thermal desorption apparatus and a capillary precolumn interface when using cryogenic cooling will be needed. The following GC columns may be used:

Column 1: 60 m × 0.75mm I.D. capillary column coated with VOCOL, 1.5 μm film thickness.
Column 2: 30 m × 0.53mm capillary column coated with DB-624 or VOCOL, 3 μm film thickness.
Column 3: 30 m × 0.32mm I.D. capillary column coated with DB-5 or SE-54, 1-μm film thickness.

PRECISION & ACCURACY This method has been tested in a single lab using spiked water. Using a wide-bore capillary column, water was spiked at concentrations between 0.5 and 10 μg/L. Single lab accuracy and precision data are presented. The MDL actually achieved in a given analysis will vary depending on instrument sensitivity and matrix effects.

The MDL (a) in μg/L was 0.12.
The concentration range in μg/L was 0.1–10.
The mean accuracy (% of true value) was 94.
The precision as relative standard deviation was 6.7.

Note: The MDL is based on a 25-mL sample volume instead of a 5-mL sample volume.

SAMPLING METHOD
Liquid samples — Use a 40-mL glass screw-cap VOA vial with a Teflon®-faced silicone septum that has been prewashed, rinsed with distilled deionized water, and oven dried. If residual chlorine is present, collect the sample in a 4-oz soil VOA container which has been pre-preserved with 4 drops of 10%

sodium thiosulfate. Mix gently and transfer the sample to a 40-mL VOA vial. Collect bubble-free samples in duplicate and seal each sample in a separate plastic bag.

Soils, sediments, and sludges — Use an 8-oz widemouth glass bottle with Teflon®-faced silicone septum that has been prewashed, rinsed with distilled deionized water, and oven dried. **Do not** heat the septum for more than 1 h. Tap slightly to eliminate any free air space. Collect samples in duplicate and seal each one in a separate plastic bag.

SAMPLE PRESERVATION
Liquid samples — Add 4 drops of concentrated HCL, cool to 4°C and store in a solvent-free refrigerator.

Soils, sediments and sludges — Cool samples to 4°C and store in a solvent-free refrigerator.

MHT The maximum holding time of any sample (liquids, soils, sediments, and sludges) is 14 days.

SAMPLE PREPARATION
Liquid samples — Remove the plunger from a 5-mL syringe and carefully pour the sample into the syringe barrel to just short of overflowing. Replace the syringe plunger and compress the sample. Open the syringe valve and vent any residual air while adjusting the sample volume to 5.0 mL. If there is only one volatile organic analysis (VOA) vial, a second syringe should be filled at this time to protect against possible loss of sample integrity. Add 10 µL of surrogate spiking solution and 10 µL of internal standard spiking solution through the valve bore of the 5-mL syringe, then close the valve. The surrogate and internal standards may be mixed and added as a single spiking solution.

Sediments, soils, and waste samples — All samples of this type should be screened by GC analysis using a headspace method (EPA Method 3810) or the hexadecane extraction and screening method (EPA Method 3820). Use the screening data to determine whether to use the low-concentration method (0.005–1 mg/kg) or the high-concentration method (>1 mg/kg).

Low-concentration method — The low-concentration method is based on purging a heated sediment or soil sample mixed with organic-free reagent water containing the surrogate and internal standards. Analyze all reagent blanks and standards under the same conditions as the samples.

Use a 5-g sample if the expected concentration is <0.1 mg/kg or a 1-g sample for expected concentrations between 0.1 and 1 mg/kg. Mix the contents of the sample container with a narrow metal spatula. Weigh the amount of the sample into a tared purge device. Add the spiked water to the purge device, which contains the weighed amount of sample, and connect the device to the purge-and-trap system.

High-concentration method — This method is based on extracting the sediment or soil with methanol. A waste sample is either extracted or diluted, depending on its solubility in methanol. Wastes that are insoluble in methanol are diluted with reagent tetraglyme or possibly polyethylene glycol (PEG). An aliquot of the extract is added to organic-free reagent water containing surrogate and internal standards. This is purged at ambient temperature. All samples with an expected concentration of >1.0 mg/kg should be analyzed by this method.

Mix the contents of the sample container with a narrow metal spatula. For sediments or soils and solid wastes that are insoluble in methanol, weigh 4 g (wet weight) of sample into a tared 20-mL vial. For waste that is soluble in methanol, tetraglyme, or PEG, weigh 1 g (wet weight) into a tared scintillation vial or culture tube or a 10-mL volumetric flask. Quickly add 9.0 mL of appropriate solvent then add 1.0 mL of a surrogate spiking solution to the vial, cap it, and shake it for 2 min.

METHANOL EXTRACT REQUIRED FOR ANALYSIS OF HIGH-CONCENTRATION SOILS OR SEDIMENTS

Approximate Concentration Range	Volume of Methanol Extract (a)
500–10,000 µg/kg	100 µL
1,000–20,000 µg/kg	50 µL
5,000–100,000 µg/kg	10 µL
25,000–500,000 µg/kg	100 µL of 1/50 dilution (b)

Calculate appropriate dilution factor for concentrations exceeding this table.

(a) The volume of methanol added to 5 mL of water being purged should be kept constant. Therefore, add to the 5-mL syringe whatever volume of methanol is necessary to maintain a volume of 100 µL added to the syringe.
(b) Dilute an aliquot of the methanol extract and then take 100 µL for analysis.

QUALITY CONTROL Demonstrate, through the analysis of a reagent water blank, that interferences from the analytical system, glassware, and reagents are under control. Blank samples should be carried through all stages of the sample preparation and measurement steps. For each analytical batch (up to 20 samples), a reagent blank, matrix spike, and matrix spike duplicate must be analyzed (the frequency of the spikes may be different for different monitoring programs). The blank and spiked samples must be carried through all stages of the sample preparation and measurement steps. QC samples mentioned in the section on Interferences will also be needed as appropriate to those situations.

Matrix spiking standards should be prepared from volatile organic compounds which will be representative of the compounds being investigated. The recommended internal standards are chlorobenzene-d5, 1,4-difluorobenzene, 1,4-dichlorobenzene-d4, and pentafluorobenzene. Using stock standard solutions, prepare secondary dilution standards containing the compounds of interest, either singly or mixed together in methanol. Store them in a vial with no headspace for no more than one week. Surrogates recommended are toluene-d8, 4-bromofluorobenzene, and dibromofluoromethane. Each sample undergoing GC/MS analysis must be spiked with 10 µL of the surrogate spiking solution prior to analysis.

REFERENCE Test Methods for Evaluating Solid Waste (SW-846). U.S. EPA 1983, Method 8260A, Rev. 1, Nov. 1990. Office of Solid Waste, Washington, DC.

cis-1,2-Dichloroethene
CAS #156-59-4

EPA Method 8260

TITLE Volatile Organic Compounds by GC/MS: Capillary Column Technique

MATRIX This method is applicable to nearly all types of samples, regardless of water content, including groundwater, soils, and sediments.

METHOD SUMMARY Method 8260A covers 58 volatile organic compounds that are introduced into a gas chromatograph by the purge-and-trap method or by direct injection (in limited applications). Zero-grade helium is bubbled through a 5-mL solution at ambient temperature. Purged sample components are trapped in a tube containing suitable sorbent materials. When purging is complete, the sorbent tube is heated and backflushed with helium to desorb trapped sample components. The analytes are desorbed directly to a large bore capillary or cryofocussed on a capillary precolumn before being flash evaporated to a narrow bore capillary for analysis.

INTERFERENCES Major contaminant sources are volatile materials in the lab and impurities in the inert purging gas and in the sorbent trap. Interfering contamination may occur when a sample containing low concentrations of volatile organic compounds is analyzed immediately after a sample containing high concentrations of volatile organic compounds. After analysis of a sample containing high concentrations of volatile organic compounds, one or more calibration blanks should be analyzed to check for cross-contamination. Screening of the samples prior to purge-and-trap GC/MS analysis is highly recommended to prevent contamination of the system. This is especially true for soil and waste samples.

Special precautions must be taken to analyze for methylene chloride. The analytical and sample storage area should be isolated from all atmospheric sources of methylene chloride. All gas chromatography carrier gas lines and purge gas plumbing should be constructed from stainless steel or copper tubing. Laboratory clothing previously exposed to methylene chloride fumes during liquid-liquid extraction procedures can contribute to sample contamination.

Samples can also be contaminated by diffusion of volatile organics (particularly methylene chloride and fluorocarbons) through the septum seal during shipment and storage. A trip blank can serve as a check on such contamination.

INSTRUMENTATION GC/MS with a temperature-programmable chromatograph suitable for splitless injection equipped with variable constant differential flow controllers, a subambient oven controller, a purging device, sorbent trap, a thermal desorption apparatus and a capillary precolumn interface when using cryogenic cooling will be needed. The following GC columns may be used:

Column 1: 60 m × 0.75mm I.D. capillary column coated with VOCOL, 1.5 μm film thickness.

Column 2: 30 m × 0.53mm capillary column coated with DB-624 or VOCOL, 3 μm film thickness.

Column 3: 30 m × 0.32mm I.D. capillary column coated with DB-5 or SE-54, 1-μm film thickness.

PRECISION & ACCURACY This method has been tested in a single lab using spiked water. Using a wide-bore capillary column, water was spiked at concentrations between 0.5 and 10 μg/L. Single lab accuracy and precision data are presented. The MDL actually achieved in a given analysis will vary depending on instrument sensitivity and matrix effects.

The MDL (a) in μg/L was 0.12.
The concentration range in μg/L was 0.5–10.
The mean accuracy (% of true value) was 101.
The precision as relative standard deviation was 6.7.

Note: The MDL is based on a 25-mL sample volume instead of a 5-mL sample volume.

SAMPLING METHOD
Liquid samples — Use a 40-mL glass screw-cap VOA vial with a Teflon®-faced silicone septum that has been prewashed, rinsed with distilled deionized water, and oven dried. If residual chlorine is present, collect the sample in a 4-oz soil VOA container which has been pre-preserved with 4 drops of 10% sodium thiosulfate. Mix gently and transfer the sample to a 40-mL VOA vial. Collect bubble-free samples in duplicate and seal each sample in a separate plastic bag.

Soils, sediments, and sludges — Use an 8-oz widemouth glass bottle with Teflon®-faced silicone septum that has been prewashed, rinsed with distilled deionized water, and oven dried. **Do not** heat the septum for more than 1 h. Tap slightly to eliminate any free air space. Collect samples in duplicate and seal each one in a separate plastic bag.

SAMPLE PRESERVATION
Liquid samples — Add 4 drops of concentrated HCL, cool to 4°C and store in a solvent-free refrigerator.

Soils, sediments and sludges — Cool samples to 4°C and store in a solvent-free refrigerator.

MHT The maximum holding time of any sample (liquids, soils, sediments, and sludges) is 14 days.

SAMPLE PREPARATION
Liquid samples — Remove the plunger from a 5-mL syringe and carefully pour the sample into the syringe barrel to just short of overflowing. Replace the syringe plunger and compress the sample. Open the syringe valve and vent any residual air while adjusting the sample volume to 5.0 mL. If there is only one volatile organic analysis (VOA) vial, a second syringe should be filled at this time to protect against possible loss of sample integrity. Add 10 μL of surrogate spiking solution and 10 μL of internal standard spiking solution through the valve bore of the 5-mL syringe, then close the valve. The surrogate and internal standards may be mixed and added as a single spiking solution.

Sediments, soils, and waste samples — All samples of this type should be screened by GC analysis using a headspace method (EPA Method 3810) or the hexadecane extraction and screening method (EPA Method 3820). Use the screening data to determine whether to use the low-concentration method (0.005–1 mg/kg) or the high-concentration method (>1 mg/kg).

Low-concentration method — The low-concentration method is based on purging a heated sediment or soil sample mixed with organic-free reagent water containing the surrogate and internal standards. Analyze all reagent blanks and standards under the same conditions as the samples.

Use a 5-g sample if the expected concentration is <0.1 mg/kg or a 1-g sample for expected concentrations between 0.1 and 1 mg/kg. Mix the contents of the sample container with a narrow metal spatula. Weigh the amount of the sample into a tared purge device. Add the spiked water to the purge device, which contains the weighed amount of sample, and connect the device to the purge-and-trap system.

High-concentration method — This method is based on extracting the sediment or soil with methanol. A waste sample is either extracted or diluted, depending on its solubility in methanol. Wastes that are insoluble in methanol are diluted with reagent tetraglyme or possibly polyethylene glycol (PEG). An aliquot of the extract is added to organic-free reagent water containing surrogate and internal standards. This is purged at ambient temperature. All samples with an expected concentration of >1.0 mg/kg should be analyzed by this method.

Mix the contents of the sample container with a narrow metal spatula. For sediments or soils and solid wastes that are insoluble in methanol, weigh 4 g (wet weight) of sample into a tared 20-mL vial. For waste that is soluble in methanol, tetraglyme, or PEG, weigh 1 g (wet weight) into a tared scintillation vial or culture tube or a 10-mL volumetric flask. Quickly add 9.0 mL of appropriate solvent then add 1.0 mL of a surrogate spiking solution to the vial, cap it, and shake it for 2 min.

METHANOL EXTRACT REQUIRED FOR ANALYSIS OF HIGH-CONCENTRATION SOILS OR SEDIMENTS

Approximate Concentration Range	Volume of Methanol Extract (a)
500–10,000 µg/kg	100 µL
1,000–20,000 µg/kg	50 µL
5,000–100,000 µg/kg	10 µL
25,000–500,000 µg/kg	100 µL of 1/50 dilution (b)

Calculate appropriate dilution factor for concentrations exceeding this table.

(a) The volume of methanol added to 5 mL of water being purged should be kept constant. Therefore, add to the 5-mL syringe whatever volume of methanol is necessary to maintain a volume of 100 µL added to the syringe.
(b) Dilute an aliquot of the methanol extract and then take 100 µL for analysis.

QUALITY CONTROL Demonstrate, through the analysis of a reagent water blank, that interferences from the analytical system, glassware, and reagents are under control. Blank samples should be carried through all stages of the sample preparation and measurement steps. For each analytical batch (up to 20 samples), a reagent blank, matrix spike, and matrix spike duplicate must be analyzed (the frequency of the spikes may be different for different monitoring programs). The blank and spiked samples must be carried through all stages of the sample preparation and measurement steps. QC samples mentioned in the section on Interferences will also be needed as appropriate to those situations.

Matrix spiking standards should be prepared from volatile organic compounds which will be representative of the compounds being investigated. The recommended internal standards are chlorobenzene-d5, 1,4-difluorobenzene, 1,4-dichlorobenzene-d4, and pentafluorobenzene. Using stock standard solutions, prepare secondary dilution standards containing the compounds of interest, either singly or mixed together in methanol. Store them in a vial with no headspace for no more than one week. Surrogates recommended are toluene-d8, 4-bromofluorobenzene, and dibromofluoromethane. Each sample undergoing GC/MS analysis must be spiked with 10 µL of the surrogate spiking solution prior to analysis.

REFERENCE Test Methods for Evaluating Solid Waste (SW-846). U.S. EPA 1983, Method 8260A, Rev. 1, Nov. 1990. Office of Solid Waste, Washington, DC.

---

trans-1,2-Dichloroethene   **EPA Method 8260**
CAS #156-60-5

TITLE Volatile Organic Compounds by GC/MS: Capillary Column Technique

MATRIX This method is applicable to nearly all types of samples, regardless of water content, including groundwater, soils, and sediments.

METHOD SUMMARY Method 8260A covers 58 volatile organic compounds that are introduced into a gas chromatograph by the purge-and-trap method or by direct injection (in limited applications). Zero-grade helium is bubbled through a 5-mL solution at ambient temperature. Purged sample components are trapped in a tube containing suitable sorbent materials. When purging is complete, the sorbent tube is heated and backflushed with helium to desorb trapped sample components. The analytes are desorbed directly to a large bore capillary or cryofocussed on a capillary precolumn before being flash evaporated to a narrow bore capillary for analysis.

INTERFERENCES Major contaminant sources are volatile materials in the lab and impurities in the inert purging gas and in the sorbent trap. Interfering contamination may occur when a sample containing low concentrations of volatile organic compounds is analyzed immediately after a sample containing high concentrations of volatile organic compounds. After analysis of a sample containing high concentrations of volatile organic compounds, one or more calibration blanks should be analyzed to check for cross-contamination. Screening of the samples prior to purge-and-trap GC/MS analysis is highly recommended to prevent contamination of the system. This is especially true for soil and waste samples.

Special precautions must be taken to analyze for methylene chloride. The analytical and sample storage area should be isolated from all atmospheric sources of methylene chloride. All gas chromatography carrier gas lines and purge gas plumbing should be constructed from stainless steel or copper tubing. Laboratory clothing previously exposed to methylene chloride fumes during liquid-liquid extraction procedures can contribute to sample contamination.

Samples can also be contaminated by diffusion of volatile organics (particularly methylene chloride and fluorocarbons) through the septum seal during shipment and storage. A trip blank can serve as a check on such contamination.

INSTRUMENTATION GC/MS with a temperature-programmable chromatograph suitable for splitless injection equipped with variable constant differential flow controllers, a subambient oven controller, a purging device, sorbent trap, a thermal desorption apparatus and a capillary precolumn interface when using cryogenic cooling will be needed. The following GC columns may be used:

Column 1: 60 m × 0.75mm I.D. capillary column coated with VOCOL, 1.5 µm film thickness.
Column 2: 30 m × 0.53mm capillary column coated with DB-624 or VOCOL, 3 µm film thickness.
Column 3: 30 m × 0.32mm I.D. capillary column coated with DB-5 or SE-54, 1-µm film thickness.

PRECISION & ACCURACY This method has been tested in a single lab using spiked water. Using a wide-bore capillary column, water was spiked at concentrations between 0.5 and 10 µg/L. Single lab accuracy and precision data are presented. The MDL actually achieved in a given analysis will vary depending on instrument sensitivity and matrix effects.

The MDL (a) in µg/L was 0.06.
The concentration range in µg/L was 0.1–10.
The mean accuracy (% of true value) was 93.
The precision as relative standard deviation was 5.6.

Note: The MDL is based on a 25-mL sample volume instead of a 5-mL sample volume.

SAMPLING METHOD
Liquid samples — Use a 40-mL glass screw-cap VOA vial with a Teflon®-faced silicone septum that has been prewashed, rinsed with distilled deionized water, and oven dried. If residual chlorine is present, collect the sample in a 4-oz soil VOA container which has been pre-preserved with 4 drops of 10% sodium thiosulfate. Mix gently and transfer the sample to a 40-mL VOA vial. Collect bubble-free samples in duplicate and seal each sample in a separate plastic bag.

Soils, sediments, and sludges — Use an 8-oz widemouth glass bottle with Teflon®-faced silicone septum that has been prewashed, rinsed with distilled deionized water, and oven dried. **Do not** heat the septum for more than 1 h. Tap slightly to eliminate any free air space. Collect samples in duplicate and seal each one in a separate plastic bag.

SAMPLE PRESERVATION
Liquid samples — Add 4 drops of concentrated HCL, cool to 4°C and store in a solvent-free refrigerator.

Soils, sediments and sludges — Cool samples to 4°C and store in a solvent-free refrigerator.

MHT The maximum holding time of any sample (liquids, soils, sediments, and sludges) is 14 days.

SAMPLE PREPARATION
Liquid samples — Remove the plunger from a 5-mL syringe and carefully pour the sample into the syringe barrel to just short of overflowing. Replace the syringe plunger and compress the sample. Open the syringe valve and vent any residual air while adjusting the sample volume to 5.0 mL. If there is only one volatile organic analysis (VOA) vial, a second syringe should be filled at this time to protect against possible loss of sample integrity. Add 10 µL of surrogate spiking solution and 10 µL of internal standard spiking solution through the valve bore of the 5-mL syringe, then close the valve. The surrogate and internal standards may be mixed and added as a single spiking solution.

Sediments, soils, and waste samples — All samples of this type should be screened by GC analysis using a headspace method (EPA Method 3810) or the hexadecane extraction and screening method (EPA Method 3820). Use the screening data to determine whether to use the low-concentration method (0.005–1 mg/kg) or the high-concentration method (>1 mg/kg).

Low-concentration method — The low-concentration method is based on purging a heated sediment or soil sample mixed with organic-free reagent water containing the surrogate and internal standards. Analyze all reagent blanks and standards under the same conditions as the samples.

Use a 5-g sample if the expected concentration is <0.1 mg/kg or a 1-g sample for expected concentrations between 0.1 and 1 mg/kg. Mix the contents of the sample container with a narrow metal spatula. Weigh the amount of the sample into a tared purge device. Add the spiked water to the purge device, which contains the weighed amount of sample, and connect the device to the purge-and-trap system.

High-concentration method — This method is based on extracting the sediment or soil with methanol. A waste sample is either extracted or diluted, depending on its solubility in methanol. Wastes that are insoluble in methanol are diluted with reagent tetraglyme or possibly polyethylene glycol (PEG). An aliquot of the extract is added to organic-free reagent water containing surrogate and internal standards. This is purged at ambient temperature. All samples with an expected concentration of >1.0 mg/kg should be analyzed by this method.

Mix the contents of the sample container with a narrow metal spatula. For sediments or soils and solid wastes that are insoluble in methanol, weigh 4 g (wet weight) of sample into a tared 20-mL vial. For waste that is soluble in methanol, tetraglyme, or PEG, weigh 1 g (wet weight) into a tared scintillation vial or culture tube or a 10-mL volumetric flask. Quickly add

9.0 mL of appropriate solvent then add 1.0 mL of a surrogate spiking solution to the vial, cap it, and shake it for 2 min.

METHANOL EXTRACT REQUIRED FOR ANALYSIS OF HIGH-CONCENTRATION SOILS OR SEDIMENTS

Approximate Concentration Range	Volume of Methanol Extract (a)
500–10,000 µg/kg	100 µL
1,000–20,000 µg/kg	50 µL
5,000–100,000 µg/kg	10 µL
25,000–500,000 µg/kg	100 µL of 1/50 dilution (b)

Calculate appropriate dilution factor for concentrations exceeding this table.

(a) The volume of methanol added to 5 mL of water being purged should be kept constant. Therefore, add to the 5-mL syringe whatever volume of methanol is necessary to maintain a volume of 100 µL added to the syringe.

(b) Dilute an aliquot of the methanol extract and then take 100 µL for analysis.

QUALITY CONTROL Demonstrate, through the analysis of a reagent water blank, that interferences from the analytical system, glassware, and reagents are under control. Blank samples should be carried through all stages of the sample preparation and measurement steps. For each analytical batch (up to 20 samples), a reagent blank, matrix spike, and matrix spike duplicate must be analyzed (the frequency of the spikes may be different for different monitoring programs). The blank and spiked samples must be carried through all stages of the sample preparation and measurement steps. QC samples mentioned in the section on Interferences will also be needed as appropriate to those situations.

Matrix spiking standards should be prepared from volatile organic compounds which will be representative of the compounds being investigated. The recommended internal standards are chlorobenzene-d5, 1,4-difluorobenzene, 1,4-dichlorobenzene-d4, and pentafluorobenzene. Using stock standard solutions, prepare secondary dilution standards containing the compounds of interest, either singly or mixed together in methanol. Store them in a vial with no headspace for no more than one week. Surrogates recommended are toluene-d8, 4-bromofluorobenzene, and dibromofluoromethane. Each sample undergoing GC/MS analysis must be spiked with 10 µL of the surrogate spiking solution prior to analysis.

REFERENCE Test Methods for Evaluating Solid Waste (SW-846). U.S. EPA 1983, Method 8260A, Rev. 1, Nov. 1990. Office of Solid Waste, Washington, DC.

1,1-Dichloroethene   **EPA Method 601**
CAS #75-35-4

TITLE Purgeable Halocarbons

MATRIX Wastewater

APPLICATION Method covers 29 purgeable halocarbons. (Method 624 provides GC/MS conditions appropriate for the qualitative and quantitative confirmation of results). Method describes conditions for a 2nd GC column to confirm measurements made with primary column.

INTERFERENCES Impurities in the purge gas and organic compounds outgassing from the plumbing ahead of the trap. With high- and low-level samples, there can be carryover contamination. Diffusion of volatile organics through the septum seal into the sample.

INSTRUMENTATION GC-equipped with halide-specific detector. (With purge-and-trap unit).

RANGE 8.0–500 µg/L.

MDL 0.13 µg/L.

PRECISION 0.29X–0.40 µg/L (overall precision).

ACCURACY 0.98C-0.87 µg/L (as recovery).

SAMPLING METHOD 25-mL glass vial. Teflon®-lined septum.

STABILITY cool, 4°C, 0.008% Sodium thiosulfate.

MHT 14 days.

QUALITY CONTROL The lab must on an ongoing basis, spike at least 10% of the samples from each sample site being monitored to assess accuracy.

REFERENCE Method 601, *Federal Register* Part VIII 40 CFR Part 136, Oct 26, 1984.

1,1-Dichloroethene   **EPA Method 624**
CAS #75-35-4

TITLE Purgeables

MATRIX Wastewater

APPLICATION Method covers 31 purgeable organics. An inert gas is bubbled through a 5-mL water sample in a specially designed purging chamber. Here, purgeables are transferred from aqueous to gaseous phase, passed onto a sorbent column, and trapped. Trap is heated and backflushed with inert gas to desorb purgeables onto a GC column, where purgeables are separated.

INTERFERENCES Impurities in the purge gas, organic compounds outgassing from the plumbing ahead of the trap, and solvent vapors in the lab. With high- and low-level samples, there can be carryover contamination.

INSTRUMENTATION GC/MS with purge-and-trap unit.

RANGE 5–600 µg/L.

MDL 2.8 µg/L

PRECISION 0.43X–0.22 µg/L (overall precision).

ACCURACY 1.12C+0.61 µg/L (as recovery).

SAMPLING METHOD 25-mL glass vial. Teflon®-lined septum.

STABILITY cool, 4°C, 0.008% Sodium thiosulfate.

MHT 14 days.

QUALITY CONTROL The lab must on an ongoing basis, spike at least 5% of the samples from each sample site being monitored to assess accuracy.

REFERENCE Method 624, *Federal Register* Part VIII 40 CFR Part 136, Oct 26, 1984.

1,1-Dichloroethene   **EPA Method 8010**
CAS #75-35-4

TITLE Halogenated Volatile Organics

MATRIX Groundwater, soils, sludges, water miscible liquid wastes, and non-water miscible wastes.

APPLICATION This method is used for the analysis of 39 halogenated VOCs. Samples are analyzed using direct injection or purge-and-trap methods. Groundwater must be analyzed by the purge-and-trap method. The method provides an optional GC column which is used for analyte confirmation and that may help resolve analytes from interferences.

INTERFERENCES There can be carryover contamination with high- and low-level samples. Impurities may come from the purge-and-trap apparatus, organic compounds outgassing from the plumbing ahead of trap, diffusion of VOCs through the sample bottle septum during shipping or storage, or from solvent vapors in the lab.

INSTRUMENTATION GC capable of on-column injections or purge-and-trap sample introduction and a halogen specific detector. Column 1: 8 ft by 0.1 in 1%. SP-1000 on Carbopack-B. Column 2: 6 ft by 0.1 in bonded n-octane on Porasil-C.

RANGE 8 to 500 µg/L (reagent water).

MDL 0.13 µg/L (reagent water).

PQL FACTORS FOR MULTIPLYING × FID MDL VALUE

Matrix	Multiplication Factor
Groundwater	10
Low-level soil	10
Water miscible liquid waste	500
High-level soil and sludge	1250
Non-water miscible waste	1250

PRECISION 0.29X–0.04 µg/L (overall precision).

ACCURACY 0.98C–0.87 µg/L (as recovery).

SAMPLING METHOD For water and liquid samples; use glass 40-mL vials with Teflon®-lined septum caps and collect two vials per sample location with no headspace. For solids and concentrated waste samples; use widemouth glass bottles with Teflon® liners.

STABILITY For concentrated wastes, soils, sediments, or sludges: cool to 4°C. For liquids: add 4 drops of concentrated hydrochloric acid and cool to 4°C.

MHT 14 days.

QUALITY CONTROL Analyze a reagent blank, matrix spike, and matrix spike duplicate/duplicate for each analytical batch (up to 20 samples). Demonstrate the purity of glassware and reagents by analyzing a reagent water method blank. Internal, surrogate, and five concentration level calibration standards are used.

REFERENCE Test Methods for Evaluating Solid Waste (SW-846), U.S. EPA Office of Solid Waste, Washington, DC, Method 8010B, Rev. 2, Nov. 1992.

***trans*-1,2-Dichloroethene**   **EPA Method 601**
CAS #156-60-5

TITLE Purgeable Halocarbons

MATRIX Wastewater

APPLICATION Method covers 29 purgeable halocarbons. (Method 624 provides GC/MS conditions appropriate for the qualitative and quantitative confirmation of results). Method describes conditions for a 2nd GC column to confirm measurements made with primary column.

INTERFERENCES Impurities in the purge gas and organic compounds outgassing from the plumbing ahead of the trap. With high- and low-level samples, there can be carryover contamination. Diffusion of volatile organics through the septum seal into the sample.

INSTRUMENTATION GC-equipped with halide-specific detector. (With purge-and-trap unit).

RANGE 8.0–500 µg/L.

MDL 0.10 µg/L.

PRECISION 0.17X+1.46 µg/L (overall precision).

ACCURACY 0.97C–0.16 µg/L (as recovery).

SAMPLING METHOD 25-mL glass vial. Teflon®-lined septum.

STABILITY cool, 4°C, 0.008% Sodium thiosulfate.

MHT 14 days.

QUALITY CONTROL The lab must on an ongoing basis, spike at least 10% of the samples from each sample site being monitored to assess accuracy.

REFERENCE Method 601, *Federal Register* Part VIII 40 CFR Part 136, Oct 26, 1984.

***trans*-1,2-Dichloroethene**   **EPA Method 624**
CAS #156-60-5

TITLE Purgeables

MATRIX Wastewater

APPLICATION Method covers 31 purgeable organics. An inert gas is bubbled through a 5-mL water sample in a specially designed purging chamber. Here, purgeables are transferred from aqueous to gaseous phase, passed onto a sorbent column, and trapped. Trap is heated and backflushed with inert gas to desorb purgeables onto a GC column, where purgeables are separated.

INTERFERENCES Impurities in the purge gas, organic compounds outgassing from the plumbing ahead of the trap, and solvent vapors in the lab. With high- and low-level samples, there can be carryover contamination.

INSTRUMENTATION GC/MS with purge-and-trap unit.

RANGE 5–600 µg/L.

MDL 1.6 µg/L

PRECISION 0.19X+0.17 µg/L (overall precision).

ACCURACY 1.05C+0.03 µg/L (as recovery).

SAMPLING METHOD 25-mL glass vial. Teflon®-lined septum.

STABILITY cool, 4°C, 0.008% Sodium thiosulfate.

MHT 14 days.

QUALITY CONTROL The lab must on an ongoing basis, spike at least 5% of the samples from each sample site being monitored to assess accuracy.

REFERENCE Method 624, *Federal Register* Part VIII 40 CFR Part 136, Oct 26, 1984.

---

trans-1,2-Dichloroethene   EPA Method 8010
CAS #156-60-5

TITLE Halogenated Volatile Organics

MATRIX Groundwater, soils, sludges, water miscible liquid wastes, and non-water miscible wastes.

APPLICATION This method is used for the analysis of 39 halogenated VOCs. Samples are analyzed using direct injection or purge-and-trap methods. Groundwater must be analyzed by the purge-and-trap method. The method provides an optional GC column which is used for analyte confirmation and that may help resolve analytes from interferences.

INTERFERENCES There can be carryover contamination with high- and low-level samples. Impurities may come from the purge-and-trap apparatus, organic compounds outgassing from the plumbing ahead of trap, diffusion of VOCs through the sample bottle septum during shipping or storage, or from solvent vapors in the lab.

INSTRUMENTATION GC capable of on-column injections or purge-and-trap sample introduction and a halogen specific detector. Column 1: 8 ft by 0.1 in 1%. SP-1000 on Carbopack-B. Column 2: 6 ft by 0.1 in bonded n-octane on Porasil-C.

RANGE 8 to 500 µg/L (reagent water).

MDL 0.10 µg/L (reagent water).

PQL FACTORS FOR MULTIPLYING × FID MDL VALUE

Matrix	Multiplication Factor
Groundwater	10
Low-level soil	10
Water miscible liquid waste	500
High-level soil and sludge	1250
Non-water miscible waste	1250

PRECISION 0.17X + 1.46 µg/L (overall precision).

ACCURACY 0.97C–0.16 µg/L (as recovery).

SAMPLING METHOD For water and liquid samples; use glass 40-mL vials with Teflon®-lined septum caps and collect two vials per sample location with no headspace. For solids and concentrated waste samples; use widemouth glass bottles with Teflon® liners.

STABILITY For concentrated wastes, soils, sediments, or sludges: cool to 4°C. For liquids: add 4 drops of concentrated hydrochloric acid and cool to 4°C.

MHT 14 days.

QUALITY CONTROL Analyze a reagent blank, matrix spike, and matrix spike duplicate/duplicate for each analytical batch (up to 20 samples). Demonstrate the purity of glassware and reagents by analyzing a reagent water method blank. Internal, surrogate, and five concentration level calibration standards are used.

REFERENCE Test Methods for Evaluating Solid Waste (SW-846), U.S. EPA Office of Solid Waste, Washington, DC, Method 8010B, Rev. 2, Nov. 1992.

2,3-Dichloronitrobenzene   EPA Method 1625
CAS #3209-22-1

TITLE Semivolatile Organic Compounds by Isotope Dilution GC/MS

MATRIX The compounds may be determined in waters, soils, and municipal sludges by this method.

METHOD SUMMARY This method is used to determine 176 semivolatile toxic organic pollutants associated with the CWA (as amended 1987); the RCRA (as amended 1986); the CERCLA (as amended 1986); and other compounds amenable to extraction and analysis by capillary column gas chromatography-mass spectrometry (GC/MS).

Stable isotopically-labeled analogs of the compounds of interest are added to the sample. If the solids content is less than 1%, a 1-L sample is extracted at pH 12–13, then at pH <2 with methylene chloride using continuous extraction techniques.

If the solids content is 30% or less, the sample is diluted to 1% solids with reagent water, homogenized ultrasonically, and extracted at pH 12–13, then at pH <2 with methylene chloride using continuous extraction techniques. If the solids content is greater than 30%, the sample is extracted using ultrasonic techniques.

Each extract is dried over sodium sulfate, concentrated to a volume of 5 mL, cleaned up using GPC, if necessary, and concentrated. Extracts are concentrated to 1 mL if GPC is not performed, and to 0.5 mL if GPC is performed.

An internal standard is added to the extract, and a 1-mL aliquot of the extract is injected into the GC. The compounds are separated by GC and detected by a MS. The labeled compounds serve to correct the variability of the analytical technique.

INTERFERENCES Solvents, reagents, glassware, and other sample processing hardware may yield artifacts and/or elevated baselines causing misinterpretation of chromatograms and spectra. Materials used in the analysis must be demonstrated to be free from interferences under the conditions of analysis by running method blanks initially and with each sample lot (sample started through the extraction process on a given 8-h shift, to a maximum of 20). Specific selection of reagents and purification of solvents by distillation in all glass systems may be required. Glassware and, where possible, reagents are cleaned by solvent rinse and baking at 450°C for 1-h minimum. Interferences coextracted from samples will vary considerably from source to source, depending on the diversity of the site being sampled.

INSTRUMENTATION Major instrumentation includes a GC with a splitless or on-column injection port for capillary column, a MS with 70 eV electron impact ionization, and a data system to collect and record MS data, and process it. A K-D apparatus is used to concentrate extracts.

GC Column: 30 m × 0.25 mm I.D. 5% phenyl, 94% methyl, 1% vinyl silicone bonded phased fused silica capillary column.

PRECISION & ACCURACY The detection limits of the method are usually dependent on the level of interferences rather than instrumental limitations. The limits typify the minimum quantities that can be detected with no interferences present.

The minimum level (in µg/mL) was not listed. This is defined as a minimum level at which the analytical system shall give recognizable mass spectra (background corrected) and acceptable calibration points.

The MDL (in µg/kg) in low solids was not listed and in high solids was not listed; these were determined in digested sludge (low solids) and in filter cake or compost (high solids).

The labeled and native compound initial precision as standard deviation (in µg/L) was not listed.
The labeled and native compound initial accuracy as average recovery (in µg/L) was not listed.

SAMPLE COLLECTION, PRESERVATION & HANDLING
Collect samples in glass containers. Aqueous samples which flow freely are collected in refrigerated bottles using automatic sampling equipment. Solid samples are collected as grab samples using widemouth jars. Maintain samples at 0 to 4°C from the time of collection until extraction. If residual chlorine is present in aqueous samples, add 80 mg sodium thiosulfate/L of water. Begin sample extraction within 7 days of collection, and analyze all extracts within 40 days of extraction.

SAMPLE PREPARATION Samples containing 1% solids or less are extracted directly using continuous liquid-liquid extraction techniques. Samples containing 1 to 30% solids are diluted to the 1% level with reagent water and extracted using continuous liquid-liquid extraction techniques. Samples containing greater than 30% solids are extracted using ultrasonic techniques.

Base/neutral extraction — Adjust the pH of the waters in the extractors to 12–13 with 6 N NaOH. Extract with methylene chloride for 24–48 h.
Acid extraction — Adjust the pH of the waters in the extractors to 2 or less using 6 N sulfuric acid. Extract with methylene chloride for 24–48 h.
Ultrasonic extraction of high solids samples — Add anhydrous sodium sulfate to the sample and QC aliquot(s). Add acetone:methylene chloride (1:1) to the sample and mix thoroughly

Concentrate extracts using a K-D apparatus.

QUALITY CONTROL The analyst is permitted to modify this method to improve separations or lower the costs of measurements, provided all performance specifications are met. Analyses of blanks are required to demonstrate freedom from contamination. When results of spikes indicate atypical method performance for samples, the samples are diluted to bring method performance within acceptable limits.

For low solids (aqueous samples), extract, concentrate, and analyze two sets of four 1-L aliquots (8 aliquots total) of the precision and recovery standard. For high solids samples, two sets of four 30-g aliquots of the high solids reference matrix are used.

Spike all samples with labeled compounds to assess method performance. Compute percent recovery of the labeled compounds using the internal standard method. Compare the labeled compound recovery for each compound with the corresponding labeled compound recovery.

Reagent water and high solids reference matrix blanks are analyzed to demonstrate freedom from contamination. Extract and concentrate a 1-L reagent water blank or a high solids reference matrix blank with each sample's lot (samples started through the extraction process on the same 8-h shift, to a maximum of 20 samples).

Field replicates may be collected to determine the precision of the sampling technique, and spiked samples may be required to determine the accuracy of the analysis when the internal standard method is used.

REFERENCE Semivolatile Organic Compounds by Isotope Dilution GC/MS. Office of Water Regulation and Standards, U.S. EPA Industrial Technology Division, Washington, DC, EPA Method 1625, Rev. C, June 1989 (contact W.A. Telliard, U.S. EPA, Office of Water Regulations and Standards, 401 M St., SW, Washington, DC, 20460. Phone: 202-382-7131).

2,4-Dichlorophenol
CAS #120-83-2

EPA Method 1625

TITLE Semivolatile Organic Compounds by Isotope Dilution GC/MS

MATRIX The compounds may be determined in waters, soils, and municipal sludges by this method.

METHOD SUMMARY This method is used to determine 176 semivolatile toxic organic pollutants associated with the CWA (as amended 1987); the RCRA (as amended 1986); the CERCLA (as amended 1986); and other compounds amenable to extraction and analysis by capillary column gas chromatography-mass spectrometry (GC/MS).

Stable isotopically-labeled analogs of the compounds of interest are added to the sample. If the solids content is less than 1%, a 1-L sample is extracted at pH 12–13, then at pH <2 with methylene chloride using continuous extraction techniques.

If the solids content is 30% or less, the sample is diluted to 1% solids with reagent water, homogenized ultrasonically, and extracted at pH 12–13, then at pH <2 with methylene chloride using continuous extraction techniques. If the solids content is greater than 30%, the sample is extracted using ultrasonic techniques.

Each extract is dried over sodium sulfate, concentrated to a volume of 5 mL, cleaned up using GPC, if necessary, and concentrated. Extracts are concentrated to 1 mL if GPC is not performed, and to 0.5 mL if GPC is performed.

An internal standard is added to the extract, and a 1-mL aliquot of the extract is injected into the GC. The compounds are separated by GC and detected by a MS. The labeled compounds serve to correct the variability of the analytical technique.

INTERFERENCES Solvents, reagents, glassware, and other sample processing hardware may yield artifacts and/or elevated baselines causing misinterpretation of chromatograms and spectra. Materials used in the analysis must be demonstrated to be free from interferences under the conditions of analysis by running method blanks initially and with each sample lot (sample started through the extraction process on a given 8-h shift, to a maximum of 20). Specific selection of reagents and purification of solvents by distillation in all glass systems may be required. Glassware and, where possible, reagents are cleaned by solvent rinse and baking at 450°C for 1-h minimum. Interferences coextracted from samples will vary considerably from source to source, depending on the diversity of the site being sampled.

INSTRUMENTATION Major instrumentation includes a GC with a splitless or on-column injection port for capillary column, a MS with 70 eV electron impact ionization, and a data system to collect and record MS data, and process it. A K-D apparatus is used to concentrate extracts.

GC Column: 30 m × 0.25 mm I.D. 5% phenyl, 94% methyl, 1% vinyl silicone bonded phased fused silica capillary column.

PRECISION & ACCURACY The detection limits of the method are usually dependent on the level of interferences rather than instrumental limitations. The limits typify the minimum quantities that can be detected with no interferences present.

The minimum level (in µg/mL) was 10. This is defined as a minimum level at which the analytical system shall give recognizable mass spectra (background corrected) and acceptable calibration points.

The MDL (in µg/kg) in low solids was 24 and in high solids was 116; these were determined in digested sludge (low solids) and in filter cake or compost (high solids).

The labeled and native compound initial precision as standard deviation (in µg/L) was 12.

The labeled and native compound initial accuracy as average recovery (in µg/L) was 85–131.

SAMPLE COLLECTION, PRESERVATION & HANDLING
Collect samples in glass containers. Aqueous samples which flow freely are collected in refrigerated bottles using automatic sampling equipment. Solid samples are collected as grab samples using widemouth jars. Maintain samples at 0 to 4°C from the time of collection until extraction. If residual chlorine is present in aqueous samples, add 80 mg sodium thiosulfate/L of water. Begin sample extraction within 7 days of collection, and analyze all extracts within 40 days of extraction.

SAMPLE PREPARATION Samples containing 1% solids or less are extracted directly using continuous liquid-liquid extraction techniques. Samples containing 1 to 30% solids are diluted to the 1% level with reagent water and extracted using continuous liquid-liquid extraction techniques. Samples containing greater than 30% solids are extracted using ultrasonic techniques.

Base/neutral extraction — Adjust the pH of the waters in the extractors to 12–13 with 6 N NaOH. Extract with methylene chloride for 24–48 h.

Acid extraction — Adjust the pH of the waters in the extractors to 2 or less using 6 N sulfuric acid. Extract with methylene chloride for 24–48 h.

Ultrasonic extraction of high solids samples — Add anhydrous sodium sulfate to the sample and QC aliquot(s). Add acetone:methylene chloride (1:1) to the sample and mix thoroughly

Concentrate extracts using a K-D apparatus.

QUALITY CONTROL The analyst is permitted to modify this method to improve separations or lower the costs of measurements, provided all performance specifications are met. Analyses of blanks are required to demonstrate freedom from contamination. When results of spikes indicate atypical method performance for samples, the samples are diluted to bring method performance within acceptable limits.

For low solids (aqueous samples), extract, concentrate, and analyze two sets of four 1-L aliquots (8 aliquots total) of the precision and recovery standard. For high solids samples, two sets of four 30-g aliquots of the high solids reference matrix are used.

Spike all samples with labeled compounds to assess method performance. Compute percent recovery of the labeled compounds using the internal standard method. Compare the labeled compound recovery for each compound with the corresponding labeled compound recovery.

Reagent water and high solids reference matrix blanks are analyzed to demonstrate freedom from contamination. Extract and concentrate a 1-L reagent water blank or a high solids reference matrix blank with each sample's lot (samples started through the extraction process on the same 8-h shift, to a maximum of 20 samples).

Field replicates may be collected to determine the precision of the sampling technique, and spiked samples may be required to determine the accuracy of the analysis when the internal standard method is used.

REFERENCE Semivolatile Organic Compounds by Isotope Dilution GC/MS. Office of Water Regulation and Standards, U.S. EPA Industrial Technology Division, Washington, DC, EPA Method 1625, Rev. C, June 1989 (contact W.A. Telliard, U.S. EPA, Office of Water Regulations and Standards, 401 M St., SW, Washington, DC, 20460. Phone: 202-382-7131).

2,6-Dichlorophenol   **EPA Method 1625**
CAS #87-65-0

TITLE Semivolatile Organic Compounds by Isotope Dilution GC/MS

MATRIX The compounds may be determined in waters, soils, and municipal sludges by this method.

METHOD SUMMARY This method is used to determine 176 semivolatile toxic organic pollutants associated with the CWA (as amended 1987); the RCRA (as amended 1986); the CERCLA (as amended 1986); and other compounds amenable to extraction and analysis by capillary column gas chromatography-mass spectrometry (GC/MS).

Stable isotopically-labeled analogs of the compounds of interest are added to the sample. If the solids content is less than 1%, a 1-L sample is extracted at pH 12–13, then at pH <2 with methylene chloride using continuous extraction techniques.

If the solids content is 30% or less, the sample is diluted to 1% solids with reagent water, homogenized ultrasonically, and extracted at pH 12–13, then at pH <2 with methylene chloride using continuous extraction techniques. If the solids content is greater than 30%, the sample is extracted using ultrasonic techniques.

Each extract is dried over sodium sulfate, concentrated to a volume of 5 mL, cleaned up using GPC, if necessary, and concentrated. Extracts are concentrated to 1 mL if GPC is not performed, and to 0.5 mL if GPC is performed.

An internal standard is added to the extract, and a 1-mL aliquot of the extract is injected into the GC. The compounds are separated by GC and detected by a MS. The labeled compounds serve to correct the variability of the analytical technique.

INTERFERENCES Solvents, reagents, glassware, and other sample processing hardware may yield artifacts and/or elevated baselines causing misinterpretation of chromatograms and spectra. Materials used in the analysis must be demonstrated to be free from interferences under the conditions of analysis by running method blanks initially and with each sample lot (sample started through the extraction process on a given 8-h shift, to a maximum of 20). Specific selection of reagents and purification of solvents by distillation in all glass systems may be required. Glassware and, where possible, reagents are cleaned by solvent rinse and baking at 450°C for 1-h minimum. Interferences coextracted from samples will vary considerably from source to source, depending on the diversity of the site being sampled.

INSTRUMENTATION Major instrumentation includes a GC with a splitless or on-column injection port for capillary column, a MS with 70 eV electron impact ionization, and a data system to collect and record MS data, and process it. A K-D apparatus is used to concentrate extracts.

GC Column: 30 m × 0.25 mm I.D. 5% phenyl, 94% methyl, 1% vinyl silicone bonded phased fused silica capillary column.

PRECISION & ACCURACY The detection limits of the method are usually dependent on the level of interferences rather than instrumental limitations. The limits typify the minimum quantities that can be detected with no interferences present.

The minimum level (in µg/mL) was not listed. This is defined as a minimum level at which the analytical system shall give recognizable mass spectra (background corrected) and acceptable calibration points.

The MDL (in µg/kg) in low solids was not listed and in high solids was not listed; these were determined in digested sludge (low solids) and in filter cake or compost (high solids).

The labeled and native compound initial precision as standard deviation (in µg/L) was not listed.
The labeled and native compound initial accuracy as average recovery (in µg/L) was not listed.

SAMPLE COLLECTION, PRESERVATION & HANDLING Collect samples in glass containers. Aqueous samples which flow freely are collected in refrigerated bottles using automatic sampling equipment. Solid samples are collected as grab samples using widemouth jars. Maintain samples at 0 to 4°C from the time of collection until extraction. If residual chlorine is present in aqueous samples, add 80 mg sodium thiosulfate/L of water. Begin sample extraction within 7 days of collection, and analyze all extracts within 40 days of extraction.

SAMPLE PREPARATION Samples containing 1% solids or less are extracted directly using continuous liquid-liquid extraction techniques. Samples containing 1 to 30% solids are diluted to the 1% level with reagent water and extracted using continuous liquid-liquid extraction techniques. Samples containing greater than 30% solids are extracted using ultrasonic techniques.

Base/neutral extraction — Adjust the pH of the waters in the extractors to 12–13 with 6 N NaOH. Extract with methylene chloride for 24–48 h.

Acid extraction — Adjust the pH of the waters in the extractors to 2 or less using 6 N sulfuric acid. Extract with methylene chloride for 24–48 h.

Ultrasonic extraction of high solids samples — Add anhydrous sodium sulfate to the sample and QC aliquot(s). Add acetone:methylene chloride (1:1) to the sample and mix thoroughly

Concentrate extracts using a K-D apparatus.

QUALITY CONTROL The analyst is permitted to modify this method to improve separations or lower the costs of measurements, provided all performance specifications are met. Analyses of blanks are required to demonstrate freedom from contamination. When results of spikes indicate atypical method performance for samples, the samples are diluted to bring method performance within acceptable limits.

For low solids (aqueous samples), extract, concentrate, and analyze two sets of four 1-L aliquots (8 aliquots total) of the precision and recovery standard. For high solids samples, two sets of four 30-g aliquots of the high solids reference matrix are used.

Spike all samples with labeled compounds to assess method performance. Compute percent recovery of the labeled compounds using the internal standard method. Compare the labeled compound recovery for each compound with the corresponding labeled compound recovery.

Reagent water and high solids reference matrix blanks are analyzed to demonstrate freedom from contamination. Extract and concentrate a 1-L reagent water blank or a high solids reference matrix blank with each sample's lot (samples started through the extraction process on the same 8-h shift, to a maximum of 20 samples).

Field replicates may be collected to determine the precision of the sampling technique, and spiked samples may be required to determine the accuracy of the analysis when the internal standard method is used.

REFERENCE Semivolatile Organic Compounds by Isotope Dilution GC/MS. Office of Water Regulation and Standards, U.S. EPA Industrial Technology Division, Washington, DC, EPA Method 1625, Rev. C, June 1989 (contact W.A. Telliard, U.S. EPA, Office of Water Regulations and Standards, 401 M St., SW, Washington, DC, 20460. Phone: 202-382-7131).

2,4-Dichlorophenol   **EPA Method 625**
CAS #120-83-2

TITLE Base/Neutrals and Acids, U.S. EPA Method 625

MATRIX This methods covers municipal and industrial wastewaters.

METHOD SUMMARY Approximately 1 L of sample is serially extracted with methylene chloride at a pH greater than 11 and again at a pH less than 2 using a separatory funnel or a continuous extractor. The methylene chloride extract is dried, concentrated to a volume of 1 mL, and analyzed by GC/MS. Qualitative identification of the parameters in the extract is performed using the retention time and the relative abundance of three characteristic masses (m/z). Qualitative analysis is performed using either external or internal standard techniques with a single characteristic m/z.

INTERFERENCES Method interferences may be caused by contaminants in solvents, reagents, glassware, and other sample processing hardware. Glassware must be scrupulously cleaned. Glassware should be heated in a muffle furnace at 400°C for 5 to 30 min. Some thermally stable materials, such as PCBs, may not be eliminated by this treatment. Solvent rinses with acetone and pesticide quality hexane may be substituted for the muffle furnace heating. Matrix interferences may be caused by contaminants that are coextracted from the sample. The base-neutral extraction may cause significantly reduced recovery of phenols. The packed gas chromatographic columns recommended for the basic fraction may not exhibit sufficient resolution for some analytes.

INSTRUMENTATION A GC/MS system with an injection port designed for on-column injection when using packed columns and for splitless injection when using capillary columns.

Column for base/neutrals: 1.8 m long × 2 mm I.D. glass, packed with 3% SP-2550 on Supelcoport (100/120 mesh) or equivalent.

Column for acids: 1.8 m long × 2 mm I.D. glass, packed with 1% SP-1240DA on Supelcoport (100/120 mesh) or equivalent.

PRECISION & ACCURACY The MDL concentrations were obtained using reagent water. The MDL actually achieved in a given analysis will vary depending on instrument sensitivity and matrix effects. This method was tested by 15 laboratories using reagent water, drinking water, surface water, and industrial wastewaters spiked at six concentrations over the range 5 to 100 µg/L. Single operator precision, overall precision, and method accuracy were found to be directly related to the concentration of the parameter matrix.

The MDL (in µg/L) in reagent water was 2.7.

The standard deviation (in µg/L based on 4 recovery measurements) was 26.4.

The range (in µg/L) for average recovery for 4 measurements was 52.5–121.7.

The range (in %) for percent recovery was 39–135.

Accuracy (in µg/L) as expected recovery for one or more measurements of a sample containing a true concentration of C was 0.87C+0.13.

Precision (in µg/L) as expected single analyst standard deviation of measurements at an average concentration found at X was 0.15X+1.25.

Overall precision (in µg/L) as expected interlaboratory standard deviation of measurements in an average concentration found at X was 0.21X+1.28.

$C =$ *True value of the concentration in µg/L.*
$X =$ *Average recovery found for measurements of samples containing a concentration at C in µg/L.*

SAMPLE PREPARATION Adjust the pH to >11 with sodium hydroxide and serially extract in a separatory funnel with methylene chloride or else in a continuous extractor. Next, adjust the pH to <2 with sulfuric acid and serially extract in a separatory funnel with methylene chloride or else in a continuous extractor. Dry the extracts separately through a column of anhydrous sodium sulfate and then concentrate each of the extracts to 1.0 mL using a K-D apparatus.

SAMPLE COLLECTION, PRESERVATION & HANDLING Grab samples must be collected in glass containers. All samples must be refrigerated at 4°C from the time of collection until extraction. If residual chlorine is present, add 80 mg of sodium thiosulfate/L of sample and mix well. All samples must be extracted within 7 days of collection and completely analyzed within 40 days of extraction.

QUALITY CONTROL Make an initial, one-time, demonstration of the ability to generate acceptable accuracy and precision with this method. Before processing any samples, the analyst must analyze a reagent water blank to demonstrate that interferences from the analytical system and glassware are under control. Each time a set of samples is extracted or reagents are changed, a reagent water blank must be processed. Spike and analyze a minimum of 5% of all samples to monitor and evaluate lab data quality. A QC check sample concentrate that contains each parameter of interest at a concentration of 100 µg/mL in acetone is required. PCBs and multicomponent pesticides may be omitted from this test.

After analysis of five spiked wastewater samples, calculate the average percent recovery and the standard deviation of the percent recovery. Spike all samples with the surrogate standard spiking solution and calculate the percent recovery of each surrogate compound.

REFERENCE *Federal Register*, Vol. 49, No. 209. Friday, Oct. 26, 1984.

2,4-Dichlorophenol   **EPA Method 8270**
CAS #120-83-2

TITLE Semivolatile Organic Compounds by GC/MS

MATRIX This method is used to determine the concentration of semivolatile organic compounds in extracts prepared from all types of solid waste matrices, soils, and groundwater. Although surface waters are not specifically mentioned, this method should be applicable to water samples from rivers, lakes, etc.

METHOD SUMMARY This method covers 259 semivolatile organic compounds. In very limited applications direct injection of the sample into the GC/MS system may be appropriate, but this results in very high detection limits (approximately 10,000 µg/L). Typically, a 1-L liquid sample, containing surrogate, and matrix spiking standards, is extracted in a continuous extractor first under acid conditions and then under basic conditions. Typically 30 g of a solid sample, containing surrogate, and matrix spiking standards, is extracted ultrasonically. After concentrating the extract to 1 mL it is spiked with 10 µL of an internal standard solution just prior to analysis by GC/MS. The volume injected should contain about 100 ng of base/neutral and 200 ng of acid surrogates (for a 1-µL injection). Analysis is performed by GC/MS using a capillary GC column.

INTERFERENCES Raw GC/MS data from all blanks, samples, and spikes must be evaluated for interferences. Contamination by carryover can occur whenever high-concentration and low-concentration samples are sequentially analyzed. To reduce carryover, the sample syringe must be rinsed out between samples with solvent. Whenever an unusually concentrated sample is encountered, it should be followed by the analysis of blank solvent to check for cross-contamination.

INSTRUMENTATION A GC/MS and a data system are required. The GC column used is a 30 m × 0.25 mm I.D. (or 0.32 mm I.D.) 1um film thickness silicone-coated fused silica capillary column. A continuous liquid-liquid extractor equipped with Teflon® or glass connection joints and stopcocks requiring no lubrication, a K-D concentrating apparatus, water bath, and an ultrasonic disrupter with a minimum power of 300 W and with pulsing capability are also required.

PRECISION & ACCURACY The estimated quantitation limit (EQL) of Method 8270B for determining an individual compound is approximately 1 mg/kg (wet weight) for soil or sediment samples, 1–200 mg/kg for wastes (dependent on matrix and method of preparation), and 10 µg/L for groundwater samples. EQLs will be proportionately higher for sample extracts that require dilution to avoid saturation of the detector.

The EQL(b) for groundwater in µg/L is 10.
The EQL (a, b) for low concentrations in soil and sediment in µg/kg is 660.
Accuracy as µg/L is $0.87C - 0.13$.
Overall precision in µg/L is $0.21X + 1.28$.

(a) *EQLs listed for soil/sediment are based on wet weight. Normally data is reported in a dry-weight basis; therefore, EQLs will be higher based on the % dry weight of each sample. This calculation is based on a 30 g sample and gel permeation chromatography cleanup.*
(b) *Sample EQLs are highly matrix-dependent. The EQLs are provided for guidance and may not always be achievable.*
C = *True value for concentration, in µg/L.*
X = *Average recovery found for measurements of samples containing a concentration of C, in µg/L.*

ESTIMATED QUANTITATION LIMIT

Other Matrices	Factor (a)
High-concentration soil and sludges by sonicator	7.5
Non-water miscible waste	75

(a) *EQL for other matrices = [EQL for low soil/sediment] × [Factor]. This estimated EQL is similar to an EPA "Practical Quantitation Limit."*

SAMPLING METHOD
Liquid samples — Use a 1 or 2½ gallon amber glass bottle with a screw-top Teflon®-lined cover that has been prewashed with detergent and rinsed with distilled water and methanol (or isopropanol).

Soils, sediments, or sludges — Use an 8-oz. widemouth glass with a screw-top Teflon®-lined cover that has been prewashed with detergent and rinsed with distilled water and methanol (or isopropanol).

SAMPLE PRESERVATION
Liquid samples — If residual chlorine is present, add 3 mL of 10% sodium thiosulfate per gallon, cool to 4°C and store in a solvent-free refrigerator until analysis; if chlorine is not present, then eliminate the sodium thiosulfate addition.

Soils, sediments, or sludges — Cool samples to 4°C and store in a solvent-free refrigerator.

MHT Liquid samples must be extracted within 7 days and the extracts analyzed within 40 days. Soils, sediments, or sludges may be stored for a maximum of 14 days and the extracts analyzed within 40 days.

SAMPLE PREPARATION
Liquid samples — Transfer 1 L quantitatively to a continuous extractor. If high concentrations are anticipated, a smaller volume may be used and then diluted with organic-free reagent water to 1 L. Adjust pH, if necessary, to pH <2 using 1:1 (V/V) sulfuric acid. Pipette 1.0 mL of a surrogate standard spiking solution into each sample. For the sample in each analytical batch selected for spiking, add 1.0 mL of a matrix spiking standard. For base/neutral acid analysis, the amount of the surrogates and matrix spiking compounds added to the sample should result in a final concentration of 100 ng/µL of each analyte in the extract to be analyzed (assuming a 1-µL injection). Extract with methylene chloride for 18–24 h. Next, adjust the pH of the aqueous phase to pH >11 using 10 N sodium hydroxide and extract it with methylene chloride again for 18–24 h. Dry the extract through a column containing anhydrous sodium sulfate and concentrate it to 1 mL using a K-D concentrator.

Soils, sediments, or sludges — Use 30 g of sample. Nonporous or wet samples (gummy or clay type) that do not have a free-flowing sandy texture must be mixed with anhydrous sodium sulfate until the sample is free flowing. Add 1 mL of surrogate standards to all samples, spikes, standards, and blanks. For the sample in each analytical batch selected for spiking, add 1.0 mL of a matrix spiking standard. For base/neutral acid analysis, the amount added of the surrogates and matrix spiking compounds should result in a final concentration of 100 ng/µL of each base/neutral analyte and 200 ng/µL of each acid analyte in the extract to be analyzed (assuming a 1-µL injection). Immediately add a 100-mL mixture of 1:1 methylene chloride:acetone and extract the sample ultrasonically for 3 min and then decant or filter the extracts. Repeat the extraction two or more times. Dry the extract using a column with anhydrous sodium sulfate and concentrate it to 1 mL in a K-D concentrator.

QUALITY CONTROL A methylene chloride solution containing 50 ng/µL of decafluorotriphenylphosphine (DFTPP) is used for tuning the GC/MS system each 12-h shift. A system performance check also must be made during every 12-h shift. A standard containing 50 ng/µL each of 4,4'-DDT, pentachlorophenol, and benzidine is required to verify injection port inertness and GC column performance. A calibration standard at mid-concentration, containing each compound of interest, including all required surrogates, must be performed every 12 h during analysis. After the system performance check is met, calibration check compounds (CCCs) are used to check the validity of the initial calibration.

The internal standard responses and retention times in the calibration check standard must be evaluated immediately after or during data acquisition. If the retention time for any internal standard changes by more than 30 seconds from the last check calibration (12 h), the chromatographic system must be inspected for malfunctions and corrections must be made, as required. If the electron ionization current plot (EICP) area for any of the internal standards changes by a factor of two from the last daily calibration standard check, the mass spectrometer must be inspected for malfunctions and corrections must be made, as appropriate.

Demonstrate, through the analysis of a reagent water blank, that interferences from the analytical system, glassware, and reagents are under control. The blank samples should be carried through all stages of the sample preparation and measurement steps. For each analytical batch (up to 20 samples), a reagent blank, matrix spike, and matrix spike duplicate/duplicate must be analyzed (the frequency of the spikes may be different for different monitoring programs). The blank and spiked samples must be carried through all stages of the sample preparation and measurement steps. A QC reference sample concentrate containing each analyte at a concentration of 100 mg/L in methanol is required.

REFERENCE Test Methods for Evaluating Solid Waste (SW-846). U.S. EPA 1983, Method 8270B, Rev. 2, Nov. 1990. Office of Solid Waste, Washington, DC.

2,6-Dichlorophenol　　　　　　　　　　**EPA Method 8270**
CAS #87-65-0

TITLE Semivolatile Organic Compounds by GC/MS

MATRIX This method is used to determine the concentration of semivolatile organic compounds in extracts prepared from all types of solid waste matrices, soils, and groundwater. Although surface waters are not specifically mentioned, this method should be applicable to water samples from rivers, lakes, etc.

METHOD SUMMARY This method covers 259 semivolatile organic compounds. In very limited applications direct injection of the sample into the GC/MS system may be appropriate, but this results in very high detection limits (approximately 10,000 µg/L). Typically, a 1-L liquid sample, containing surrogate, and matrix spiking standards, is extracted in a continuous extractor first under acid conditions and then under basic conditions. Typically 30 g of a solid sample, containing surrogate, and matrix spiking standards, is extracted ultrasonically. After concentrating the extract to 1 mL it is spiked with 10 µL of an internal standard solution just prior to analysis by GC/MS. The volume injected should contain about 100 ng of base/neutral

and 200 ng of acid surrogates (for a 1-µL injection). Analysis is performed by GC/MS using a capillary GC column.

INTERFERENCES Raw GC/MS data from all blanks, samples, and spikes must be evaluated for interferences. Contamination by carryover can occur whenever high-concentration and low-concentration samples are sequentially analyzed. To reduce carryover, the sample syringe must be rinsed out between samples with solvent. Whenever an unusually concentrated sample is encountered, it should be followed by the analysis of blank solvent to check for cross-contamination.

INSTRUMENTATION A GC/MS and a data system are required. The GC column used is a 30 m × 0.25 mm I.D. (or 0.32 mm I.D.) 1um film thickness silicone-coated fused silica capillary column. A continuous liquid-liquid extractor equipped with Teflon® or glass connection joints and stopcocks requiring no lubrication, a K-D concentrating apparatus, water bath, and an ultrasonic disrupter with a minimum power of 300 W and with pulsing capability are also required.

PRECISION & ACCURACY The estimated quantitation limit (EQL) of Method 8270B for determining an individual compound is approximately 1 mg/kg (wet weight) for soil or sediment samples, 1–200 mg/kg for wastes (dependent on matrix and method of preparation), and 10 µg/L for groundwater samples. EQLs will be proportionately higher for sample extracts that require dilution to avoid saturation of the detector.

The EQL(b) for groundwater in µg/L is 10.
The EQL (a, b) for low concentrations in soil and sediment in µg/kg is not determined.
Accuracy as µg/L is not listed.
Overall precision in µg/L is not listed.

(a) EQLs listed for soil/sediment are based on wet weight. Normally data is reported in a dry-weight basis; therefore, EQLs will be higher based on the % dry weight of each sample. This calculation is based on a 30 g sample and gel permeation chromatography cleanup.
(b) Sample EQLs are highly matrix-dependent. The EQLs are provided for guidance and may not always be achievable.
C = True value for concentration, in µg/L.
X = Average recovery found for measurements of samples containing a concentration of C, in µg/L.

ESTIMATED QUANTITATION LIMIT

Other Matrices	Factor (a)
High-concentration soil and sludges by sonicator	7.5
Non-water miscible waste	75

(a) EQL for other matrices = [EQL for low soil/sediment] × [Factor]. This estimated EQL is similar to an EPA "Practical Quantitation Limit."

SAMPLING METHOD
Liquid samples — Use a 1 or 2½ gallon amber glass bottle with a screw-top Teflon®-lined cover that has been prewashed with detergent and rinsed with distilled water and methanol (or isopropanol).

Soils, sediments, or sludges — Use an 8-oz. widemouth glass with a screw-top Teflon®-lined cover that has been prewashed with detergent and rinsed with distilled water and methanol (or isopropanol).

SAMPLE PRESERVATION
Liquid samples — If residual chlorine is present, add 3 mL of 10% sodium thiosulfate per gallon, cool to 4°C and store in a solvent-free refrigerator until analysis; if chlorine is not present, then eliminate the sodium thiosulfate addition.

Soils, sediments, or sludges — Cool samples to 4°C and store in a solvent-free refrigerator.

MHT Liquid samples must be extracted within 7 days and the extracts analyzed within 40 days. Soils, sediments, or sludges may be stored for a maximum of 14 days and the extracts analyzed within 40 days.

SAMPLE PREPARATION
Liquid samples — Transfer 1 L quantitatively to a continuous extractor. If high concentrations are anticipated, a smaller volume may be used and then diluted with organic-free reagent water to 1 L. Adjust pH, if necessary, to pH <2 using 1:1 (V/V) sulfuric acid. Pipette 1.0 mL of a surrogate standard spiking solution into each sample. For the sample in each analytical batch selected for spiking, add 1.0 mL of a matrix spiking standard. For base/neutral acid analysis, the amount of the surrogates and matrix spiking compounds added to the sample should result in a final concentration of 100 ng/µL of each analyte in the extract to be analyzed (assuming a 1-µL injection). Extract with methylene chloride for 18–24 h. Next, adjust the pH of the aqueous phase to pH >11 using 10 N sodium hydroxide and extract it with methylene chloride again for 18–24 h. Dry the extract through a column containing anhydrous sodium sulfate and concentrate it to 1 mL using a K-D concentrator.

Soils, sediments, or sludges — Use 30 g of sample. Nonporous or wet samples (gummy or clay type) that do not have a free-flowing sandy texture must be mixed with anhydrous sodium sulfate until the sample is free flowing. Add 1 mL of surrogate standards to all samples, spikes, standards, and blanks. For the sample in each analytical batch selected for spiking, add 1.0 mL of a matrix spiking standard. For base/neutral acid analysis, the amount added of the surrogates and matrix spiking compounds should result in a final concentration of 100 ng/µL of each base/neutral analyte and 200 ng/µL of each acid analyte in the extract to be analyzed (assuming a 1-µL injection). Immediately add a 100-mL mixture of 1:1 methylene chloride:acetone and extract the sample ultrasonically for 3 min and then decant or filter the extracts. Repeat the extraction two or more times. Dry the extract using a column with anhydrous sodium sulfate and concentrate it to 1 mL in a K-D concentrator.

QUALITY CONTROL A methylene chloride solution containing 50 ng/µL of decafluorotriphenylphosphine (DFTPP) is used for tuning the GC/MS system each 12-h shift. A system performance check also must be made during every 12-h shift. A standard containing 50 ng/µL each of 4,4'-DDT, pentachlorophenol, and benzidine is required to verify injection port inertness and GC column performance. A calibration standard at mid-concentration, containing each compound of interest, including all required surrogates, must be performed every 12 h

during analysis. After the system performance check is met, calibration check compounds (CCCs) are used to check the validity of the initial calibration.

The internal standard responses and retention times in the calibration check standard must be evaluated immediately after or during data acquisition. If the retention time for any internal standard changes by more than 30 seconds from the last check calibration (12 h), the chromatographic system must be inspected for malfunctions and corrections must be made, as required. If the electron ionization current plot (EICP) area for any of the internal standards changes by a factor of two from the last daily calibration standard check, the mass spectrometer must be inspected for malfunctions and corrections must be made, as appropriate.

Demonstrate, through the analysis of a reagent water blank, that interferences from the analytical system, glassware, and reagents are under control. The blank samples should be carried through all stages of the sample preparation and measurement steps. For each analytical batch (up to 20 samples), a reagent blank, matrix spike, and matrix spike duplicate/duplicate must be analyzed (the frequency of the spikes may be different for different monitoring programs). The blank and spiked samples must be carried through all stages of the sample preparation and measurement steps. A QC reference sample concentrate containing each analyte at a concentration of 100 mg/L in methanol is required.

REFERENCE Test Methods for Evaluating Solid Waste (SW-846). U.S. EPA 1983, Method 8270B, Rev. 2, Nov. 1990. Office of Solid Waste, Washington, DC.

2,4-Dichlorophenol   EPA Method 8040
CAS #120-83-2

TITLE Phenols

MATRIX Groundwater, soils, sludges, water miscible liquid wastes, and non-water miscible wastes.

APPLICATION This method is used for the analysis of 17 phenols. Samples are extracted, concentrated, and analyzed using direct injection of both neat and diluted organic liquids. Pentafluorobenzylbromide (PFB) derivatives also may be made to increase sensitivity of the method.

INTERFERENCES There can be carryover contamination with high- and low-level samples. Solvents, reagents, and glassware may introduce artifacts. Other interferences may come from coextracted compounds from samples.

INSTRUMENTATION GC capable of on-column injections and a flame with detector (FID) or electron capture detector (ECD). Column for underivatized phenol: 1.8 m by 2.0 mm with 1% SP-1240DA on Supelcoport. Column for derivatized phenols: 1.8 m by 2.0 mm with 5% OV-17 on Chromosorb W-AW-DMCS.

RANGE 12 to 450 µg/L.

MDL 0.39 µg/L (FID) and 0.68 µg/L (ECD).

PQL FACTORS FOR MULTIPLYING × FID MDL VALUE

Matrix	Multiplication Factor
Groundwater	10
Low-level soil by sonication with GPC cleanup	670
High-level soil and sludge by sonication	10,000
Non-water miscible waste	100,000

PRECISION 0.18X + 0.62 µg/L (overall precision using FID).

ACCURACY 0.81C + 0.48 µg/L (as recovery using FID).

SAMPLING METHOD Use 8-oz. widemouth glass bottles with Teflon®-lined caps for concentrated waste samples, soils, sediments, and sludges. Use 1 or 2½ gallon amber glass bottles with Teflon®-lined caps for liquid (water) samples.

STABILITY Cool soil, sediment, sludge, and liquid samples to 4°C. If residual chlorine is present in liquid samples add 3 mL of 10% sodium thiosulfate per gallon of sample and cool to 4°C.

MHT 14 days for concentrated waste, soil, sediment, or sludge; 7 days for liquid samples; all extracts must be analyzed within 40 days.

QUALITY CONTROL A quality control check sample concentrate containing each analyte of interest is required. The QC check sample concentrate may be prepared from pure standard materials or purchased as certified solutions Use appropriate trip, matrix, control site, method, reagent, and solvent blanks. Internal, surrogate, and five concentration level calibration standards are used. The QC check sample concentrate should contain this compound at 100 µg/mL in 2-propanol.

REFERENCE Test Methods for Evaluating Solid Waste (SW-846), U.S. EPA Office of Solid Waste, Washington, DC, Method 8040A, Rev. 1, Nov. 1990.

Dichloroprop   EPA Method 8150
CAS #120-36-5

TITLE Chlorinated Herbicides

MATRIX Groundwater, soils, sludges, water miscible liquid wastes, and non-water miscible wastes.

APPLICATION This method is used for the analysis of 10 chlorinated herbicides. Samples are extracted, hydrolyzed with potassium hydroxide, and extraneous organics are removed by a solvent wash. After acidification, the acids are extracted, concentrated and converted to their methyl esters using diazomethane. They are then analyzed using direct injection into a gas chromatograph (GC). Be very careful because diazomethane can explode under certain conditions and it is also a carcinogen.

INTERFERENCES Organic acids and phenols (especially chlorinated acids and phenols) may cause interferences. Phthalate esters are not as significant an interference as with other GC-ECD methods if an electron capture detector is used. The herbicides may react readily with alkaline substances and be lost during analysis so all glassware and glass wool must be acid rinsed and sodium sulfate must be acidified with sulfuric acid prior to use. Sensitivity usually depends on the level of interferences rather than on instrumentation.

INSTRUMENTATION GC capable of on-column injections and an electron capture detector (ECD) or a halogen specific detector. Column 1: 1.8 m by 4 mm with 1.5% SP-2250/1.95% SP-2401 on Supelcoport. Column 2: 1.8 m by 4 mm with 5% OV-210 on Gas Chrom Q. Column 3: 1.98 m by 2 mm with 0.1%. SP-1000 on Carbopack C. The preferred column is Column Number 1.

RANGE Not listed.

MDL 0.65 µg/L (in reagent water; ECD).

PQL FACTORS FOR MULTIPLYING × FID MDL VALUE

Matrix	Multiplication Factor
Groundwater	10
Low-level soil by sonication with GPC cleanup	670
High-level soil and sludge by sonication	10,000
Non-water miscible waste	100,000

PRECISION (as standard deviation) 2% with 10.7 µg/L spike in drinking water. 3% with 10.7 µg/L in municipal water.

ACCURACY (as mean recovery) 97% with 10.7 µg/L spike in drinking water. 72% with 10.7 µg/L in municipal water.

SAMPLING METHOD Use 8-oz. widemouth glass bottles with Teflon®-lined caps for concentrated waste samples, soils, sediments, and sludges. Use 1 or 2½ gallon amber glass bottles with Teflon®-lined caps for liquid (water) samples.

STABILITY Cool soil, sediment, sludge, and liquid samples to 4°C. If residual chlorine is present in liquid samples add 3 mL of 10% sodium thiosulfate per gallon of sample and cool to 4°C.

MHT 14 days for concentrated waste, soil, sediment, or sludge; 7 days for liquid samples; all extracts must be analyzed within 40 days.

QUALITY CONTROL A quality control check sample concentrate containing this compound in acetone at a concentration 1,000 times more concentrated than the selected spike concentration is required. The QC check sample concentrate may be prepared from pure standard materials or purchased as certified solutions. Use appropriate trip, matrix, control site, method, reagent, and solvent blanks. Internal, surrogate, and five concentration level calibration standards are used.

REFERENCE Method 8150, SW-846, 3rd ed., Sept. 1986.

1,2-Dichloropropane   **EPA Method 1624**
CAS #78-87-5

TITLE Volatile Organic Compounds by Isotope Dilution GC/MS

MATRIX Compounds may be determined in waters, soils, and municipal sludges by this method.

METHOD SUMMARY This method is used to determine 58 volatile toxic organic pollutants associated with the CWA (as amended 1987); the RCRA (as amended 1986); the CERCLA (as amended 1986); and other compounds amenable to purge-and-trap gas chromatography-mass spectrometry (GC/MS).

If the solids content is less than 1%, stable isotopically-labeled analogs of the compounds of interest are added to a 5-mL sample and the sample is purged with an inert gas at 20–25°C in a chamber designed for soil or water samples. If the solids content is greater than 1%, 5 mL of reagent water and the labeled compounds are added to a 5-g aliquot of sample and the mixture is purged at 40°C. Compounds that will not purge at 20–25°C or at 40°C are purged at 78–85°C. In the purging process, the volatile compounds are transferred from the aqueous phase into the gaseous phase where they are passed into a sorbent column, and trapped. After purging is completed, the trap is backflushed and heated rapidly to desorb the compounds into a GC. The compounds are separated by the GC and detected by a MS. The labeled compounds serve to correct the variability of the analytical technique.

INTERFERENCES Impurities in the purge gas, organic compounds outgassing from the plumbing upstream of the trap, and solvent vapors in the lab account for most problems. Samples can be contaminated by diffusion of volatile organic compounds (particularly methylene chloride) through the bottle seal during shipment and storage. Contamination by carryover can occur when high-level and low-level samples are analyzed sequentially. When an unusually concentrated sample is encountered, follow it by analysis of a reagent water blank to check for carryover.

INSTRUMENTATION Major equipment includes a GC with linear temperature programming and a glass jet separator as the MS interface, a MS with 70 eV electron impact ionization, and a data system to collect and record response factors.

Column: 2.8 m × 2 mm I.D. glass, packed with 1% SP-1000 on Carbopak B, 60/80 mesh, or equivalent.

PRECISION & ACCURACY The detection limits of the method are usually dependent on the level of interferences rather than instrumental limitations. The method detection limits were determined in digested sludge (low solids) and in filter cake or compost (high solids).

The MDL (in µg/kg) for low solids is 29 and for high solids is 55.

Labeled and native compound precision (in µg/L) as standard deviation was 19.0.

Labeled and native compound accuracy (in µg/L) as average recovery was detected to 47.

Acceptance criteria are at 20 µg/L for this compound.

SAMPLE COLLECTION, PRESERVATION & HANDLING Grab samples are collected in glass containers having a total volume greater than 20 mL. Fill and seal each bottle so that no air bubbles are entrapped. Samples are maintained at 0 to 4°C from the time of collection until analysis. If an aqueous sample contains residual chlorine, add sodium thiosulfate preservative (10 mg/40 mL) to the empty sample bottles just prior to shipment

to the sample site. All samples must be analyzed within 14 days of collection.

SAMPLE PREPARATION Samples containing less than 1% solids are analyzed directly as aqueous samples. Samples containing 1% solids or greater are analyzed as solid samples utilizing one of two methods, depending on the levels of pollutants, in the sample. Samples containing 1% solids or greater, and low to moderate levels of pollutants are analyzed by purging a known weight of sample added to 5 mL of reagent water. Samples containing 1% solids or greater, and high levels of pollutants, are extracted with methanol, and an aliquot of the methanol extract is added to reagent water and purged.

QUALITY CONTROL A field blank prepared from reagent water and carried through the sampling and handling protocol may serve as a check on contamination from shipment and storage.

The analyst is permitted to modify this method to improve separations or lower the costs of measurements, provided all performance specifications are met. Analyses of blanks are required. When results of spikes indicate atypical method performance for samples, the samples are diluted to bring method performance within acceptable limits. Analyze two sets of four 5-mL aliquots (8 aliquots total) of the aqueous performance standard. Spike all samples with labeled compounds to assess method performance on the sample matrix. Compute the percent recovery of the labeled compounds using the internal standard method. Compare the percent recovery for each compound with the corresponding labeled compound recovery. Reagent water blanks are analyzed to demonstrate freedom from carryover contamination. Field replicates may be collected to determine the precision of the sampling technique, and spiked samples may be required to determine the accuracy of the analysis when the internal method is used.

REFERENCE Volatile Organic Compounds by Isotope Dilution GC/MS. Office of Water Regulation and Standards, U.S. EPA Industrial Technology Division, Washington, DC, EPA Method 1624, Rev. C, June 1989 (contact W.A. Telliard, U.S. EPA, Office of Water Regulations and Standards, 401 M St., SW, Washington, DC, 20460. Phone: 202-382-7131).

1,3-Dichloropropane   **EPA Method 1624**
CAS #142-28-9

TITLE Volatile Organic Compounds by Isotope Dilution GC/MS

MATRIX Compounds may be determined in waters, soils, and municipal sludges by this method.

METHOD SUMMARY This method is used to determine 58 volatile toxic organic pollutants associated with the CWA (as amended 1987); the RCRA (as amended 1986); the CERCLA (as amended 1986); and other compounds amenable to purge-and-trap gas chromatography-mass spectrometry (GC/MS).

If the solids content is less than 1%, stable isotopically-labeled analogs of the compounds of interest are added to a 5-mL sample and the sample is purged with an inert gas at 20–25°C in a chamber designed for soil or water samples. If the solids content is greater than 1%, 5 mL of reagent water and the labeled compounds are added to a 5-g aliquot of sample and the mixture is purged at 40°C. Compounds that will not purge at 20–25°C or at 40°C are purged at 78–85°C. In the purging process, the volatile compounds are transferred from the aqueous phase into the gaseous phase where they are passed into a sorbent column, and trapped. After purging is completed, the trap is backflushed and heated rapidly to desorb the compounds into a GC. The compounds are separated by the GC and detected by a MS. The labeled compounds serve to correct the variability of the analytical technique.

INTERFERENCES Impurities in the purge gas, organic compounds outgassing from the plumbing upstream of the trap, and solvent vapors in the lab account for most problems. Samples can be contaminated by diffusion of volatile organic compounds (particularly methylene chloride) through the bottle seal during shipment and storage. Contamination by carryover can occur when high-level and low-level samples are analyzed sequentially. When an unusually concentrated sample is encountered, follow it by analysis of a reagent water blank to check for carryover.

INSTRUMENTATION Major equipment includes a GC with linear temperature programming and a glass jet separator as the MS interface, a MS with 70 eV electron impact ionization, and a data system to collect and record response factors.

Column: 2.8 m × 2 mm I.D. glass, packed with 1% SP-1000 on Carbopak B, 60/80 mesh, or equivalent.

PRECISION & ACCURACY The detection limits of the method are usually dependent on the level of interferences rather than instrumental limitations. The method detection limits were determined in digested sludge (low solids) and in filter cake or compost (high solids).

The MDL (in µg/kg) for low solids is 41 and for high solids is 2.
Labeled and native compound precision (in µg/L) as standard deviation was not listed.
Labeled and native compound accuracy (in µg/L) as average recovery was not listed.

Acceptance criteria are at 20 µg/L for this compound.

SAMPLE COLLECTION, PRESERVATION & HANDLING Grab samples are collected in glass containers having a total volume greater than 20 mL. Fill and seal each bottle so that no air bubbles are entrapped. Samples are maintained at 0 to 4°C from the time of collection until analysis. If an aqueous sample contains residual chlorine, add sodium thiosulfate preservative (10 mg/40 mL) to the empty sample bottles just prior to shipment to the sample site. All samples must be analyzed within 14 days of collection.

SAMPLE PREPARATION Samples containing less than 1% solids are analyzed directly as aqueous samples. Samples containing 1% solids or greater are analyzed as solid samples utilizing one of two methods, depending on the levels of pollutants, in the sample. Samples containing 1% solids or greater, and low to moderate levels of pollutants are analyzed by purging a known weight of sample added to 5 mL of reagent water. Samples containing 1% solids or greater, and high levels of pollutants, are extracted with methanol, and an aliquot of the methanol extract is added to reagent water and purged.

QUALITY CONTROL A field blank prepared from reagent water and carried through the sampling and handling protocol may serve as a check on contamination from shipment and storage.

The analyst is permitted to modify this method to improve separations or lower the costs of measurements, provided all performance specifications are met. Analyses of blanks are required. When results of spikes indicate atypical method performance for samples, the samples are diluted to bring method performance within acceptable limits. Analyze two sets of four 5-mL aliquots (8 aliquots total) of the aqueous performance standard. Spike all samples with labeled compounds to assess method performance on the sample matrix. Compute the percent recovery of the labeled compounds using the internal standard method. Compare the percent recovery for each compound with the corresponding labeled compound recovery. Reagent water blanks are analyzed to demonstrate freedom from carryover contamination. Field replicates may be collected to determine the precision of the sampling technique, and spiked samples may be required to determine the accuracy of the analysis when the internal method is used.

REFERENCE Volatile Organic Compounds by Isotope Dilution GC/MS. Office of Water Regulation and Standards, U.S. EPA Industrial Technology Division, Washington, DC, EPA Method 1624, Rev. C, June 1989 (contact W.A. Telliard, U.S. EPA, Office of Water Regulations and Standards, 401 M St., SW, Washington, DC, 20460. Phone: 202-382-7131).

1,2-Dichloropropane   **EPA Method 502**
CAS #78-87-5

TITLE Volatile Organic Compounds in Water By Purge and Trap Capillary Column Gas Chromatography with Photoionization and Electrolytic Conductivity Detectors in Series. U.S. EPA Method 502.2, Rev. 2.0, 1989.

MATRIX Drinking water and raw source water. The latter should include most surface water and groundwater sources.

METHOD SUMMARY This method covers 60 volatile organic compounds that contain halogen atoms and/or that are aromatic. An inert gas (zero grade nitrogen or helium) is bubbled through a 25-mL or a 5-mL water sample (depending on the expected concentration of the analytes). Purged sample components are trapped in a tube of sorbent materials. When purging is complete, the sorbent tube is heated and backflushed with helium to desorb the trapped sample onto a capillary GC column. The column is temperature programmed to separate the method analytes which are then detected with a photoionization detector (PID) and a Hall electrolytic conductivity (HECD) placed in series. The PID is selective for aromatic compounds and the HECD is selective for halogenated compounds.

INTERFERENCES Impurities in the purge gas and from organic compounds outgassing from the plumbing ahead of the trap account for many contamination problems. Interferences purged or coextracted from the samples will vary considerably from source to source, depending upon the particular sample or extract being tested. Cross-contamination can occur whenever high-level and low-level samples are analyzed sequentially. Samples also can be contaminated by diffusion of volatile organics (particularly methylene chloride and fluorocarbons) through the septum seal into the sample during shipment and storage. The lab where volatile analysis is performed and also the refrigerated storage area should be completely free of solvents.

INSTRUMENTATION A GC containing a series configuration of a high temperature photoionization detector (PID) equipped with 10.0 eV (nominal) lamp and Hall electrolytic conductivity detector (HECD) is required. Also required is an all-glass 5-mL purging device, a sorbent trap, and a thermal desorption apparatus which is connected to the GC system.

Column 1: VOCOL glass wide-bore capillary column.
Column 2: RTX–502.2 mega-bore capillary column.
Column 3: DB-62 mega-bore capillary column.

PRECISION & ACCURACY Method detection limits are dependent upon the characteristics of the gas chromatographic system used. Analytes that are not separated chromatographically cannot be individually identified and used in the same calibration mixture or water samples unless an alternative technique for identification and quantification, such as mass spectrometry, is used.

Electrolytic conductivity detetor (c) range in µg/L (a) was 0.02–200.
Electrolytic conductivity detetor (c) MDL in µg/L (b) was 0.01.
Electrolytic conductivity detetor (c) accuracy as % recovery was 103.
Electrolytic conductivity detetor (c) precision as % RSD was 3.7.
Photoionization detector (d) range in µg/L (a) was 0.02–200.
Photoionization detector (d) MDL in µg/L (b) was not listed.
Photoionization detector (d) accuracy as % recovery was not listed.
Photoionization detector (d) precision as % RSD was not listed.

(a) *The applicable concentration range of this method is compound, instrument, and matrix-dependent. It is listed as being approximately 0.02 to 200 µg/L but no specific information is provided so caution should be observed.*
(b) *The method detection limits reports with this method are compound, instrument, and matrix-dependent. The values reported were calculated using reagent water fortified with the corresponding compounds at 10 µg/L and a GC-equipped with a 60 m × 0.75 mm VOLCOL wide bore capillary column with 1.5 µm film thickness and using helium carrier gas.*
(c) *Recoveries and relative standard deviations were determined from seven samples of reagent water fortified with 10 µg/L of each compound. 2-Bromo-1-chloropropane was used as the internal standard for calculating average recoveries.*

(d) Recoveries and relative standard deviations were determined from seven samples of reagent water fortified with 10 µg/L of each compound. Fluorobenzene was used as the internal standard for calculating average recoveries.

SAMPLING METHOD Collect samples using a 40- to 120-mL screw-cap vial (prewashed with detergent, rinsed with distilled water and oven dried at 105°C) with a Teflon®-faced silicone septum. Collect bubble-free samples and place the septum with the Teflon® side down on the water.

SAMPLE PRESERVATION If residual chlorine is present in the water add about 25 mg of ascorbic acid to each vial before samples are collected to remove the chlorine. Add hydrochloric acid to reduce pH to <2, immediately cool samples to 4°C, and store them in a solvent-free refrigerator at 4°C until analysis.

MHT The maximum holding time for samples is 14 days from the time they were collected.

SAMPLE PREPARATION Remove the plungers from two 5-mL syringes and attach a closed syringe valve to each. Warm the sample to room temperature, open the sample bottle, and carefully pour the sample into one of the syringe barrels to just short of overflowing. Replace the syringe plunger, invert the syringe, and compress the sample. Open the syringe valve and vent any residual air while adjusting the sample volume to 5.0 mL. Add 10 µL of the internal calibration standard to the sample through the syringe valve. Close the valve. Fill the second syringe in an identical manner from the same sample bottle. Reserve this second syringe for a reanalysis if necessary.

QUALITY CONTROL As an initial demonstration of lab accuracy and precision, analyze 4 to 7 replicates of a lab fortified blank containing analyte at 0.1–5 µg/L. Collect all samples in duplicate. Surrogate analytes (similar to those of the analytes of interest), whose concentration is known in every sample, are measured using the same internal standard calibration procedure. Duplicate field reagent water blanks (trip blanks) must be analyzed with each set of samples, lab reagent blanks (method blanks) must be analyzed with each batch of samples processed as a group within a work shift. Also, a single lab-fortified blank that contains each of the analytes of interest should be analyzed with each batch of samples processed as a group within a work shift. A 3- to 5-point calibration curve is needed depending on the calibration range factor required.

EPA CONTACT & HOTLINE For technical questions contact Dr. Baldev Bathija, U.S. EPA, Office of Ground Water and Drinking Water (WH-550D), 401 M St. SW, Washington, DC 20460. Tel. (202) 260-3040. For further information the EPA Safe Drinking Water Hotline may be called at: (800) 426-4791.

REFERENCE Methods for the Determination of Organic Compounds in Drinking Water, EPA/600/4-88/039 (revised July 1991; Final Rule for determination of compliance with the MCL for Total Trihalomethanes under 141.30, in 40 CFR Part 141, Vol. 58, No. 147, Fed. Reg., Tuesday Aug. 3, 1993). U.S. EPA Environmental Monitoring Systems Laboratory, Cincinnati, OH, 45268, U.S.A. Available from the National Technical Information Service (NTIS), 5285 Port Royal Road, Springfield, VA 22161; Tel. 800-553-6847. NTIS Order Number is PB91-231480.

1,3-Dichloropropane   EPA Method 502
CAS #142-28-9

TITLE Volatile Organic Compounds in Water By Purge and Trap Capillary Column Gas Chromatography with Photoionization and Electrolytic Conductivity Detectors in Series. U.S. EPA Method 502.2, Rev. 2.0, 1989.

MATRIX Drinking water and raw source water. The latter should include most surface water and groundwater sources.

METHOD SUMMARY This method covers 60 volatile organic compounds that contain halogen atoms and/or that are aromatic. An inert gas (zero grade nitrogen or helium) is bubbled through a 25-mL or a 5-mL water sample (depending on the expected concentration of the analytes). Purged sample components are trapped in a tube of sorbent materials. When purging is complete, the sorbent tube is heated and backflushed with helium to desorb the trapped sample onto a capillary GC column. The column is temperature programmed to separate the method analytes which are then detected with a photoionization detector (PID) and a Hall electrolytic conductivity (HECD) placed in series. The PID is selective for aromatic compounds and the HECD is selective for halogenated compounds.

INTERFERENCES Impurities in the purge gas and from organic compounds outgassing from the plumbing ahead of the trap account for many contamination problems. Interferences purged or coextracted from the samples will vary considerably from source to source, depending upon the particular sample or extract being tested. Cross-contamination can occur whenever high-level and low-level samples are analyzed sequentially. Samples also can be contaminated by diffusion of volatile organics (particularly methylene chloride and fluorocarbons) through the septum seal into the sample during shipment and storage. The lab where volatile analysis is performed and also the refrigerated storage area should be completely free of solvents.

INSTRUMENTATION A GC containing a series configuration of a high temperature photoionization detector (PID) equipped with 10.0 eV (nominal) lamp and Hall electrolytic conductivity detector (HECD) is required. Also required is an all-glass 5-mL purging device, a sorbent trap, and a thermal desorption apparatus which is connected to the GC system.

Column 1: VOCOL glass wide-bore capillary column.
Column 2: RTX–502.2 mega-bore capillary column.
Column 3: DB-62 mega-bore capillary column.

PRECISION & ACCURACY Method detection limits are dependent upon the characteristics of the gas chromatographic system used. Analytes that are not separated chromatographically cannot be individually identified and used in the same calibration mixture or water samples unless an alternative technique for identification and quantification, such as mass spectrometry, is used.

Electrolytic conductivity detetor (c) range in µg/L (a) was 0.02–200.

Electrolytic conductivity detetor (c) MDL in µg/L (b) was 0.03.

Electrolytic conductivity detetor (c) accuracy as % recovery was 100.

Electrolytic conductivity detetor (c) precision as % RSD was 3.4.

Photoionization detector (d) range in µg/L (a) was 0.02–200.

Photoionization detector (d) MDL in µg/L (b) was not listed.

Photoionization detector (d) accuracy as % recovery was not listed.

Photoionization detector (d) precision as % RSD was not listed.

(a) The applicable concentration range of this method is compound, instrument, and matrix-dependent. It is listed as being approximately 0.02 to 200 µg/L but no specific information is provided so caution should be observed.

(b) The method detection limits reports with this method are compound, instrument, and matrix-dependent. The values reported were calculated using reagent water fortified with the corresponding compounds at 10 µg/L and a GC-equipped with a 60 m × 0.75 mm VOLCOL wide bore capillary column with 1.5 µm film thickness and using helium carrier gas.

(c) Recoveries and relative standard deviations were determined from seven samples of reagent water fortified with 10 µg/L of each compound. 2-Bromo-1-chloropropane was used as the internal standard for calculating average recoveries.

(d) Recoveries and relative standard deviations were determined from seven samples of reagent water fortified with 10 µg/L of each compound. Fluorobenzene was used as the internal standard for calculating average recoveries.

SAMPLING METHOD Collect samples using a 40- to 120-mL screw-cap vial (prewashed with detergent, rinsed with distilled water and oven dried at 105°C) with a Teflon®-faced silicone septum. Collect bubble-free samples and place the septum with the Teflon® side down on the water.

SAMPLE PRESERVATION If residual chlorine is present in the water add about 25 mg of ascorbic acid to each vial before samples are collected to remove the chlorine. Add hydrochloric acid to reduce pH to <2, immediately cool samples to 4°C, and store them in a solvent-free refrigerator at 4°C until analysis.

MHT The maximum holding time for samples is 14 days from the time they were collected.

SAMPLE PREPARATION Remove the plungers from two 5-mL syringes and attach a closed syringe valve to each. Warm the sample to room temperature, open the sample bottle, and carefully pour the sample into one of the syringe barrels to just short of overflowing. Replace the syringe plunger, invert the syringe, and compress the sample. Open the syringe valve and vent any residual air while adjusting the sample volume to 5.0 mL. Add 10 µL of the internal calibration standard to the sample through the syringe valve. Close the valve. Fill the second syringe in an identical manner from the same sample bottle. Reserve this second syringe for a reanalysis if necessary.

QUALITY CONTROL As an initial demonstration of lab accuracy and precision, analyze 4 to 7 replicates of a lab fortified blank containing analyte at 0.1–5 µg/L. Collect all samples in duplicate. Surrogate analytes (similar to those of the analytes of interest), whose concentration is known in every sample, are measured using the same internal standard calibration procedure. Duplicate field reagent water blanks (trip blanks) must be analyzed with each set of samples, lab reagent blanks (method blanks) must be analyzed with each batch of samples processed as a group within a work shift. Also, a single lab-fortified blank that contains each of the analytes of interest should be analyzed with each batch of samples processed as a group within a work shift. A 3- to 5-point calibration curve is needed depending on the calibration range factor required.

EPA CONTACT & HOTLINE For technical questions contact Dr. Baldev Bathija, U.S. EPA, Office of Ground Water and Drinking Water (WH-550D), 401 M St. SW, Washington, DC 20460. Tel. (202) 260-3040. For further information the EPA Safe Drinking Water Hotline may be called at: (800) 426-4791.

REFERENCE Methods for the Determination of Organic Compounds in Drinking Water, EPA/600/4-88/039 (revised July 1991; Final Rule for determination of compliance with the MCL for Total Trihalomethanes under 141.30, in 40 CFR Part 141, Vol. 58, No. 147, Fed. Reg., Tuesday Aug. 3, 1993). U.S. EPA Environmental Monitoring Systems Laboratory, Cincinnati, OH, 45268, U.S.A. Available from the National Technical Information Service (NTIS), 5285 Port Royal Road, Springfield, VA 22161; Tel. 800-553-6847. NTIS Order Number is PB91-231480.

2,2-Dichloropropane   **EPA Method 502**
CAS #590-20-7

TITLE Volatile Organic Compounds in Water By Purge and Trap Capillary Column Gas Chromatography with Photoionization and Electrolytic Conductivity Detectors in Series. U.S. EPA Method 502.2, Rev. 2.0, 1989.

MATRIX Drinking water and raw source water. The latter should include most surface water and groundwater sources.

METHOD SUMMARY This method covers 60 volatile organic compounds that contain halogen atoms and/or that are aromatic. An inert gas (zero grade nitrogen or helium) is bubbled through a 25-mL or a 5-mL water sample (depending on the expected concentration of the analytes). Purged sample components are trapped in a tube of sorbent materials. When purging is complete, the sorbent tube is heated and backflushed with helium to desorb the trapped sample onto a capillary GC column. The column is temperature programmed to separate the method analytes which are then detected with a photoionization detector (PID) and a Hall electrolytic conductivity (HECD) placed in series. The PID is selective for aromatic compounds and the HECD is selective for halogenated compounds.

INTERFERENCES Impurities in the purge gas and from organic compounds outgassing from the plumbing ahead of the trap account for many contamination problems. Interferences purged or coextracted from the samples will vary considerably from source to source, depending upon the particular sample or extract being tested. Cross-contamination can occur whenever high-level and low-level samples are analyzed sequentially. Samples also can be contaminated by diffusion of volatile organics (particularly methylene chloride and fluorocarbons) through the septum seal into the sample during shipment and storage. The lab where volatile analysis is performed and also the refrigerated storage area should be completely free of solvents.

INSTRUMENTATION A GC containing a series configuration of a high temperature photoionization detector (PID) equipped with 10.0 eV (nominal) lamp and Hall electrolytic conductivity detector (HECD) is required. Also required is an all-glass 5-mL purging device, a sorbent trap, and a thermal desorption apparatus which is connected to the GC system.

Column 1: VOCOL glass wide-bore capillary column.
Column 2: RTX-502.2 mega-bore capillary column.
Column 3: DB-62 mega-bore capillary column.

PRECISION & ACCURACY Method detection limits are dependent upon the characteristics of the gas chromatographic system used. Analytes that are not separated chromatographically cannot be individually identified and used in the same calibration mixture or water samples unless an alternative technique for identification and quantification, such as mass spectrometry, is used.

Electrolytic conductivity detetor (c) range in µg/L (a) was 0.02–200.
Electrolytic conductivity detetor (c) MDL in µg/L (b) was 0.05.
Electrolytic conductivity detetor (c) accuracy as % recovery was 105.
Electrolytic conductivity detetor (c) precision as % RSD was 3.4.
Photoionization detector (d) range in µg/L (a) was 0.02–200.
Photoionization detector (d) MDL in µg/L (b) was not listed.
Photoionization detector (d) accuracy as % recovery was not listed.
Photoionization detector (d) precision as % RSD was not listed.

(a) The applicable concentration range of this method is compound, instrument, and matrix-dependent. It is listed as being approximately 0.02 to 200 µg/L but no specific information is provided so caution should be observed.
(b) The method detection limits reports with this method are compound, instrument, and matrix-dependent. The values reported were calculated using reagent water fortified with the corresponding compounds at 10 µg/L and a GC-equipped with a 60 m × 0.75 mm VOLCOL wide bore capillary column with 1.5 µm film thickness and using helium carrier gas.
(c) Recoveries and relative standard deviations were determined from seven samples of reagent water fortified with 10 µg/L of each compound. 2-Bromo-1-chloropropane was used as the internal standard for calculating average recoveries.
(d) Recoveries and relative standard deviations were determined from seven samples of reagent water fortified with 10 µg/L of each compound. Fluorobenzene was used as the internal standard for calculating average recoveries.

SAMPLING METHOD Collect samples using a 40- to 120-mL screw-cap vial (prewashed with detergent, rinsed with distilled water and oven dried at 105°C) with a Teflon®-faced silicone septum. Collect bubble-free samples and place the septum with the Teflon® side down on the water.

SAMPLE PRESERVATION If residual chlorine is present in the water add about 25 mg of ascorbic acid to each vial before samples are collected to remove the chlorine. Add hydrochloric acid to reduce pH to <2, immediately cool samples to 4°C, and store them in a solvent-free refrigerator at 4°C until analysis.

MHT The maximum holding time for samples is 14 days from the time they were collected.

SAMPLE PREPARATION Remove the plungers from two 5-mL syringes and attach a closed syringe valve to each. Warm the sample to room temperature, open the sample bottle, and carefully pour the sample into one of the syringe barrels to just short of overflowing. Replace the syringe plunger, invert the syringe, and compress the sample. Open the syringe valve and vent any residual air while adjusting the sample volume to 5.0 mL. Add 10 µL of the internal calibration standard to the sample through the syringe valve. Close the valve. Fill the second syringe in an identical manner from the same sample bottle. Reserve this second syringe for a reanalysis if necessary.

QUALITY CONTROL As an initial demonstration of lab accuracy and precision, analyze 4 to 7 replicates of a lab fortified blank containing analyte at 0.1–5 µg/L. Collect all samples in duplicate. Surrogate analytes (similar to those of the analytes of interest), whose concentration is known in every sample, are measured using the same internal standard calibration procedure. Duplicate field reagent water blanks (trip blanks) must be analyzed with each set of samples, lab reagent blanks (method blanks) must be analyzed with each batch of samples processed as a group within a work shift. Also, a single lab-fortified blank that contains each of the analytes of interest should be analyzed with each batch of samples processed as a group within a work shift. A 3- to 5-point calibration curve is needed depending on the calibration range factor required.

EPA CONTACT & HOTLINE For technical questions contact Dr. Baldev Bathija, U.S. EPA, Office of Ground Water and Drinking Water (WH-550D), 401 M St. SW, Washington, DC 20460. Tel. (202) 260-3040. For further information the EPA Safe Drinking Water Hotline may be called at: (800) 426-4791.

REFERENCE Methods for the Determination of Organic Compounds in Drinking Water, EPA/600/4-88/039 (revised July 1991; Final Rule for determination of compliance with the MCL for Total Trihalomethanes under 141.30, in 40 CFR Part 141, Vol. 58, No. 147, Fed. Reg., Tuesday Aug. 3, 1993). U.S. EPA Environmental Monitoring Systems Laboratory, Cincinnati, OH, 45268, U.S.A. Available from the National Technical Information Service (NTIS), 5285 Port Royal Road, Springfield, VA 22161; Tel. 800-553-6847. NTIS Order Number is PB91-231480.

1,2-Dichloropropane EPA Method 524
CAS #78-87-5

TITLE Measurement of Purgeable Organic Compounds in Water by Capillary Column GC/MS.

MATRIX Drinking water and raw source water; the latter should include most surface water and groundwater sources.

METHOD SUMMARY Method 524.2 covers 60 volatile organic compounds. An inert gas (zero grade nitrogen or helium) is bubbled through a 25-mL or a 5-mL water sample (depending on the expected concentration of the analytes). Purged sample components are trapped in a tube of sorbent materials. When purging is complete, the sorbent tube is heated and backflushed with helium to desorb the trapped sample onto a capillary GC column.

INTERFERENCES Impurities in the purge gas and from organic compounds outgassing from the plumbing ahead of the trap account for many contamination problems. Interferences purged or coextracted from the samples will vary considerably from source to source, depending upon the particular sample or extract being tested. Cross-contamination can occur whenever high-level and low-level samples are analyzed sequentially. Samples also can be contaminated by diffusion of volatile organics (particularly methylene chloride and fluorocarbons) through the septum seal into the sample during shipment and storage.

INSTRUMENTATION A GC/MS with a data system equipped with one of the following capillary GC columns:

Column 1: VOCOL glass wide bore capillary column.
Column 2: DB-624 fused silica capillary column.
Column 3: DB-5 fused silica capillary column.

Also required is an all-glass 25 mL or 5-mL purging device, a sorbent trap, and a thermal desorption apparatus which is connected to the GC/MS system.

PRECISION & ACCURACY Method detection limits are compound- and instrument-dependent, and may vary from approximately 0.02–0.35 µg/L. Note in the table below that the "true" concentration range used for accuracy and precision measurements was quite narrow. However, the applicable concentration range of this method is primarily column dependent and is approximately 0.02 to 200 µg/L for the wide-bore thick-film columns. Narrow-bore thin-film columns may have a capacity which limits the range to about 0.02 to 20 µg/L. Analytes that are inefficiently purged from water will not be detected when present at low concentrations, but they can be measured with acceptable accuracy and precision when present in sufficient amounts.

Analytes that are not separated chromatographically, but which have different mass spectra and non-interfering quantification ions, can be identified and measured in the same calibration mixture or water sample. Analytes which have very similar mass spectra cannot be individually identified and measured in the same calibration mixture or water samples unless they have different retention times. Co-eluting compounds with very similar mass spectra, typically many structural isomers, must be reported as an isomeric group or pair.

The range (in µg/L) was 0.1–10.
The Method Detection Limig (in µg/L) was 0.04.
The accuracy (as % recovery) was 97.
The precision (in %) was 6.1.

Note: Data were obtained from 16–31 determinations using a wide-bore capillary column and a jet separator interfaced to a quadrupole mass spectrometer. All analytes were in a reagent water matrix.

SAMPLING METHOD Collect samples using a 40- to 120-mL screw-cap vial (prewashed with detergent, rinsed with distilled water and oven dried at 105°C) with a Teflon®-faced silicone septum. Collect bubble-free samples and place the septum with the Teflon® side down on the water.

SAMPLE PRESERVATION If residual chlorine is present in the water add about 25 mg of ascorbic acid to each vial before samples are collected to remove the chlorine. Add hydrochloric acid to reduce pH to <2, and immediately cool samples to 4°C, and store them in a solvent-free refrigerator at 4°C until analysis.

MHT The maximum holding time for samples is 14 days from the time they were collected.

SAMPLE PREPARATION Remove the plungers from two 25-mL (or 5-mL depending on sample size) syringes and attach a closed syringe valve to each. Warm the sample to room temperature, open the sample bottle, and carefully pour the sample into one of the syringe barrels to just short of overflowing. Replace the syringe plunger, invert the syringe, and compress the sample. Open the syringe valve and vent any residual air while adjusting the sample volume to 25.0 mL (or 5 mL). For samples and blanks, add 5 µL of the fortification solution containing the internal standard and the surrogates to the sample through the syringe valve. For calibration standards and lab fortified blanks, add 5 µL of the fortification solution containing the internal standard only. Close the valve. Fill the second syringe in an identical manner from the same sample bottle. Reserve this second syringe for a reanalysis if necessary.

QUALITY CONTROL As an initial demonstration of lab accuracy and precision, analyze 4 to 7 replicates of a lab fortified blank containing analyte at 0.2–5 µg/L. Collect all samples in duplicate. Surrogate analytes (similar to those of the analytes of interest), whose concentration is known in every sample, are measured using the same internal standard calibration procedure. Duplicate field reagent water blanks (trip blanks) must be analyzed with each set of samples, lab reagent blanks (method blanks) must be analyzed with each batch of samples processed as a group within a work shift. Also, a single lab-fortified blank that contains each of the analytes of interest should be analyzed with each batch of samples processed as a group within a work shift. A 3- to 5-point calibration curve is needed depending on the calibration range factor required.

EPA CONTACT & HOTLINE For technical questions contact Dr. Baldev Bathija, U.S. EPA, Office of Ground Water and Drinking Water (WH-550D), 401 M St. SW, Washington, DC 20460. Tel. (202) 260-3040. For further information the EPA Safe Drinking Water Hotline may be called at: (800) 426-4791.

REFERENCE Methods for the Determination of Organic Compounds in Drinking Water, EPA/600/4-88/039 (revised July 1991; Final Rule for determination of compliance with the MCL for Total Trihalomethanes under 141.30, in 40 CFR Part 141, Vol. 58, No. 147, Fed. Reg., Tuesday Aug. 3, 1993). U.S. EPA Environmental Monitoring Systems Laboratory, Cincinnati, OH,

45268, U.S.A. Available from the National Technical Information Service (NTIS), 5285 Port Royal Road, Springfield, VA 22161; Tel. 800-553-6847. NTIS Order Number is PB91-231480.

1,3-Dichloropropane   EPA Method 524
CAS #142-28-9

TITLE Measurement of Purgeable Organic Compounds in Water by Capillary Column GC/MS.

MATRIX Drinking water and raw source water; the latter should include most surface water and groundwater sources.

METHOD SUMMARY Method 524.2 covers 60 volatile organic compounds. An inert gas (zero grade nitrogen or helium) is bubbled through a 25-mL or a 5-mL water sample (depending on the expected concentration of the analytes). Purged sample components are trapped in a tube of sorbent materials. When purging is complete, the sorbent tube is heated and backflushed with helium to desorb the trapped sample onto a capillary GC column.

INTERFERENCES Impurities in the purge gas and from organic compounds outgassing from the plumbing ahead of the trap account for many contamination problems. Interferences purged or coextracted from the samples will vary considerably from source to source, depending upon the particular sample or extract being tested. Cross-contamination can occur whenever high-level and low-level samples are analyzed sequentially. Samples also can be contaminated by diffusion of volatile organics (particularly methylene chloride and fluorocarbons) through the septum seal into the sample during shipment and storage.

INSTRUMENTATION A GC/MS with a data system equipped with one of the following capillary GC columns:

Column 1: VOCOL glass wide bore capillary column.
Column 2: DB-624 fused silica capillary column.
Column 3: DB-5 fused silica capillary column.

Also required is an all-glass 25 mL or 5-mL purging device, a sorbent trap, and a thermal desorption apparatus which is connected to the GC/MS system.

PRECISION & ACCURACY Method detection limits are compound- and instrument-dependent, and may vary from approximately 0.02–0.35 µg/L. Note in the table below that the "true" concentration range used for accuracy and precision measurements was quite narrow. However, the applicable concentration range of this method is primarily column dependent and is approximately 0.02 to 200 µg/L for the wide-bore thick-film columns. Narrow-bore thin-film columns may have a capacity which limits the range to about 0.02 to 20 µg/L. Analytes that are inefficiently purged from water will not be detected when present at low concentrations, but they can be measured with acceptable accuracy and precision when present in sufficient amounts.

Analytes that are not separated chromatographically, but which have different mass spectra and non-interfering quantification ions, can be identified and measured in the same calibration mixture or water sample. Analytes which have very similar mass spectra cannot be individually identified and measured in the same calibration mixture or water samples unless they have different retention times. Co-eluting compounds with very similar mass spectra, typically many structural isomers, must be reported as an isomeric group or pair.

The range (in µg/L) was 0.1–10.
The Method Detection Limig (in µg/L) was 0.04.
The accuracy (as % recovery) was 96.
The precision (in %) was 6.0.

Note: Data were obtained from 16–31 determinations using a wide-bore capillary column and a jet separator interfaced to a quadrupole mass spectrometer. All analytes were in a reagent water matrix.

SAMPLING METHOD Collect samples using a 40- to 120-mL screw-cap vial (prewashed with detergent, rinsed with distilled water and oven dried at 105°C) with a Teflon®-faced silicone septum. Collect bubble-free samples and place the septum with the Teflon® side down on the water.

SAMPLE PRESERVATION If residual chlorine is present in the water add about 25 mg of ascorbic acid to each vial before samples are collected to remove the chlorine. Add hydrochloric acid to reduce pH to <2, and immediately cool samples to 4°C, and store them in a solvent-free refrigerator at 4°C until analysis.

MHT The maximum holding time for samples is 14 days from the time they were collected.

SAMPLE PREPARATION Remove the plungers from two 25-mL (or 5-mL depending on sample size) syringes and attach a closed syringe valve to each. Warm the sample to room temperature, open the sample bottle, and carefully pour the sample into one of the syringe barrels to just short of overflowing. Replace the syringe plunger, invert the syringe, and compress the sample. Open the syringe valve and vent any residual air while adjusting the sample volume to 25.0 mL (or 5 mL). For samples and blanks, add 5 µL of the fortification solution containing the internal standard and the surrogates to the sample through the syringe valve. For calibration standards and lab fortified blanks, add 5 µL of the fortification solution containing the internal standard only. Close the valve. Fill the second syringe in an identical manner from the same sample bottle. Reserve this second syringe for a reanalysis if necessary.

QUALITY CONTROL As an initial demonstration of lab accuracy and precision, analyze 4 to 7 replicates of a lab fortified blank containing analyte at 0.2–5 µg/L. Collect all samples in duplicate. Surrogate analytes (similar to those of the analytes of interest), whose concentration is known in every sample, are measured using the same internal standard calibration procedure. Duplicate field reagent water blanks (trip blanks) must be analyzed with each set of samples, lab reagent blanks (method blanks) must be analyzed with each batch of samples processed as a group within a work shift. Also, a single lab-fortified blank that contains each of the analytes of interest should be analyzed with each batch of samples processed as a group within a work shift. A 3- to 5-point calibration curve is needed depending on the calibration range factor required.

EPA CONTACT & HOTLINE For technical questions contact Dr. Baldev Bathija, U.S. EPA, Office of Ground Water and Drinking Water (WH-550D), 401 M St. SW, Washington, DC 20460. Tel. (202) 260-3040. For further information the EPA Safe Drinking Water Hotline may be called at: (800) 426-4791.

REFERENCE Methods for the Determination of Organic Compounds in Drinking Water, EPA/600/4-88/039 (revised July 1991; Final Rule for determination of compliance with the MCL for Total Trihalomethanes under 141.30, in 40 CFR Part 141, Vol. 58, No. 147, Fed. Reg., Tuesday Aug. 3, 1993). U.S. EPA Environmental Monitoring Systems Laboratory, Cincinnati, OH, 45268, U.S.A. Available from the National Technical Information Service (NTIS), 5285 Port Royal Road, Springfield, VA 22161; Tel. 800-553-6847. NTIS Order Number is PB91-231480.

2,2-Dichloropropane   **EPA Method 524**
CAS #590-20-7

TITLE Measurement of Purgeable Organic Compounds in Water by Capillary Column GC/MS.

MATRIX Drinking water and raw source water; the latter should include most surface water and groundwater sources.

METHOD SUMMARY Method 524.2 covers 60 volatile organic compounds. An inert gas (zero grade nitrogen or helium) is bubbled through a 25-mL or a 5-mL water sample (depending on the expected concentration of the analytes). Purged sample components are trapped in a tube of sorbent materials. When purging is complete, the sorbent tube is heated and backflushed with helium to desorb the trapped sample onto a capillary GC column.

INTERFERENCES Impurities in the purge gas and from organic compounds outgassing from the plumbing ahead of the trap account for many contamination problems. Interferences purged or coextracted from the samples will vary considerably from source to source, depending upon the particular sample or extract being tested. Cross-contamination can occur whenever high-level and low-level samples are analyzed sequentially. Samples also can be contaminated by diffusion of volatile organics (particularly methylene chloride and fluorocarbons) through the septum seal into the sample during shipment and storage.

INSTRUMENTATION A GC/MS with a data system equipped with one of the following capillary GC columns:

Column 1: VOCOL glass wide bore capillary column.
Column 2: DB-624 fused silica capillary column.
Column 3: DB-5 fused silica capillary column.

Also required is an all-glass 25 mL or 5-mL purging device, a sorbent trap, and a thermal desorption apparatus which is connected to the GC/MS system.

PRECISION & ACCURACY Method detection limits are compound- and instrument-dependent, and may vary from approximately 0.02–0.35 µg/L. Note in the table below that the "true" concentration range used for accuracy and precision measurements was quite narrow. However, the applicable concentration range of this method is primarily column dependent and is approximately 0.02 to 200 µg/L for the wide-bore thick-film columns. Narrow-bore thin-film columns may have a capacity which limits the range to about 0.02 to 20 µg/L. Analytes that are inefficiently purged from water will not be detected when present at low concentrations, but they can be measured with acceptable accuracy and precision when present in sufficient amounts.

Analytes that are not separated chromatographically, but which have different mass spectra and non-interfering quantification ions, can be identified and measured in the same calibration mixture or water sample. Analytes which have very similar mass spectra cannot be individually identified and measured in the same calibration mixture or water samples unless they have different retention times. Co-eluting compounds with very similar mass spectra, typically many structural isomers, must be reported as an isomeric group or pair.

The range (in µg/L) was 0.5–10.
The Method Detection Limig (in µg/L) was 0.35.
The accuracy (as % recovery) was 86.
The precision (in %) was 16.9.

Note: Data were obtained from 16–31 determinations using a wide-bore capillary column and a jet separator interfaced to a quadrupole mass spectrometer. All analytes were in a reagent water matrix.

SAMPLING METHOD Collect samples using a 40- to 120-mL screw-cap vial (prewashed with detergent, rinsed with distilled water and oven dried at 105°C) with a Teflon®-faced silicone septum. Collect bubble-free samples and place the septum with the Teflon® side down on the water.

SAMPLE PRESERVATION If residual chlorine is present in the water add about 25 mg of ascorbic acid to each vial before samples are collected to remove the chlorine. Add hydrochloric acid to reduce pH to <2, and immediately cool samples to 4°C, and store them in a solvent-free refrigerator at 4°C until analysis.

MHT The maximum holding time for samples is 14 days from the time they were collected.

SAMPLE PREPARATION Remove the plungers from two 25-mL (or 5-mL depending on sample size) syringes and attach a closed syringe valve to each. Warm the sample to room temperature, open the sample bottle, and carefully pour the sample into one of the syringe barrels to just short of overflowing. Replace the syringe plunger, invert the syringe, and compress the sample. Open the syringe valve and vent any residual air while adjusting the sample volume to 25.0 mL (or 5 mL). For samples and blanks, add 5 µL of the fortification solution containing the internal standard and the surrogates to the sample through the syringe valve. For calibration standards and lab fortified blanks, add 5 µL of the fortification solution containing the internal standard only. Close the valve. Fill the second syringe in an identical manner from the same sample bottle. Reserve this second syringe for a reanalysis if necessary.

QUALITY CONTROL As an initial demonstration of lab accuracy and precision, analyze 4 to 7 replicates of a lab fortified blank containing analyte at 0.2–5 µg/L. Collect all samples in duplicate. Surrogate analytes (similar to those of the analytes

of interest), whose concentration is known in every sample, are measured using the same internal standard calibration procedure. Duplicate field reagent water blanks (trip blanks) must be analyzed with each set of samples, lab reagent blanks (method blanks) must be analyzed with each batch of samples processed as a group within a work shift. Also, a single lab-fortified blank that contains each of the analytes of interest should be analyzed with each batch of samples processed as a group within a work shift. A 3- to 5-point calibration curve is needed depending on the calibration range factor required.

EPA CONTACT & HOTLINE For technical questions contact Dr. Baldev Bathija, U.S. EPA, Office of Ground Water and Drinking Water (WH-550D), 401 M St. SW, Washington, DC 20460. Tel. (202) 260-3040. For further information the EPA Safe Drinking Water Hotline may be called at: (800) 426-4791.

REFERENCE Methods for the Determination of Organic Compounds in Drinking Water, EPA/600/4-88/039 (revised July 1991; Final Rule for determination of compliance with the MCL for Total Trihalomethanes under 141.30, in 40 CFR Part 141, Vol. 58, No. 147, Fed. Reg., Tuesday Aug. 3, 1993). U.S. EPA Environmental Monitoring Systems Laboratory, Cincinnati, OH, 45268, U.S.A. Available from the National Technical Information Service (NTIS), 5285 Port Royal Road, Springfield, VA 22161; Tel. 800-553-6847. NTIS Order Number is PB91-231480.

1,2-Dichloropropane   **EPA Method 8021**
CAS #78-87-5

TITLE Halogenated Volatile by Gas Chromatography Using Photoionization and Electrolytic Conductivity Detectors in Series: Capillary Column Technique

MATRIX This method is applicable to nearly all types of samples, regardless of water content, including groundwater, aqueous sludges, caustic liquors, acid liquors, waste solvents, oily wastes, mousses, tars, fibrous wastes, polymeric emulsions, filter cakes, spent carbons, spent catalysts, soils, and sediments.

METHOD SUMMARY This method is used to determine 60 volatile organic compounds in a variety of solid waste matrices. It provides GC conditions for the detection of halogenated and aromatic volatile organic compounds. Samples can be analyzed using direct injection or purge-and-trap (EPA Method 5030). Groundwater samples must be analyzed using EPA Method 5030 (where applicable). A temperature program is used with the GC. Detection is achieved by a photoionization detector (PID) and a Hall electrolytic conductivity detector (HECD) in series.

INTERFERENCES Samples can be contaminated by diffusion of volatile organics (particularly chlorofluorocarbons and methylene chloride) through the sample container septum during shipment and storage.

INSTRUMENTATION A GC-equipped with variable-constant differential flow controllers, subambient oven controller, PID and HECD detectors connected with a short piece of uncoated capillary tubing and a data system.

Column: 60 m × 0.75 mm I.D. VOCOL wide-bore capillary column with 1.5 μm film thickness.

PRECISION & ACCURACY MDLs are compound-dependent and vary with purging efficiency and concentration. The applicable concentration range of this method is compound- and instrument-dependent but is approximately 0.1 to 200 μg/L. Analytes that are inefficiently purged from water will not be detected when present at low concentrations, but they can be measured with acceptable accuracy and precision when present in sufficient amounts. The estimated quantitation limit (EQL) for an individual compound is approximately 1 μg/kg (wet weight) for soil/sediment samples, 100 μg/kg (wet weight) for wastes, and 1 μg/L for groundwater. EQLs will be proportionately higher for sample extracts and samples that require dilution or reduced sample size to avoid saturation of the detector.

MULTIPLICATION FACTORS FOR OTHER MATRICES (a)

Matrix	Factor (b)
Groundwater	10
Low-concentration soil	10
Water miscible liquid waste	500
High-concentration soil and sludge	1250
Non-water miscible waste	1250

(a) Sample EQLs are highly matrix-dependent. The EQLs listed herein are provided for guidance and may not always be achievable.
(b) EQL = [Method detection limit] × [Factor]. For non-aqueous samples, the factor is on a wet-weight basis.

SINGLE LABORATORY ACCURACY & PRECISION DATA FOR VOCs IN WATER

This method was tested in a single lab using water spiked at 10 μg/L and the following data was reported:

Recoveries and standard deviations were determined from seven samples and spiked at 10 μg/L of each analyte. Recoveries were determined by the internal standard method. Internal standards were: Fluorobenzene for PID and 2-Bromo-1-chloropropane for HECD.

The average recovery (in percent) for the PID was none (no response for this detector).
The standard deviation of the recovery for the PID was none (no response for this detector).
The MDL (in μg/mL) for the PID was none (no response for this detector).
The average recovery (in percent) for the HECD was 103.
The standard deviation of the recovery for the HECD was 3.8.
The MDL (in μg/mL) for the HECD was 0.006.

SAMPLE COLLECTION, PRESERVATION & HANDLING
Volatile Organics — Standard 40-mL glass screw-cap VOA vials with Teflon®-faced silicone septum may be used for both liquid and solid matrices. When collecting samples, liquids and solids should be introduced into the vials gently to reduce agitation which might drive off volatile compounds. If there are any air bubbles present the sample must be retaken. Tap slightly as they are filled to try and eliminate as much free air space as possible. The two vials from each sampling locations should be sealed in separate plastic bags to prevent cross-contamination between samples particularly if the sampled waste is suspected of containing high levels of volatile organics.

Semivolatile organics — Containers used to collect samples for the determination of semivolatile organic compounds should be soap and water washed followed by methanol (or isopropanol) rinsing. The sample containers should be of glass or Teflon® and have screw-top covers with Teflon® liners.

Preservation for volatile organics — No preservation is used with concentrated waste samples. With liquid samples containing no residual chlorine, 4 drops of concentrated hydrochloric acid are added and the samples are immediately cooled to 4°C. When liquid samples contain residual chlorine, they are treated as above and, in addition, 4 drops of 4% aqueous sodium thiosulfate are added. Soil, sediment, and sludge samples are only cooled to 4°C.

Preservation for semivolatile organics — No preservation is used with concentrated waste samples. With liquid samples containing no residual chlorine and with soil, sediment, and sludge samples, immediately cooling to 4°C is the only preservation used. When residual chlorine is present then 3 mL of 10% aqueous sodium sulfate is added for each gallon of sample collected, followed by cooling to 4°C.

MHT The holding time for all volatile organics samples is 14 days. Liquid samples must be extracted within 7 days and their extracts analyzed within 40 days. Concentrated waste, soil, sediment, and sludge samples must be extracted within 14 days and their extracts analyzed within 40 days.

SAMPLE PREPARATION Volatile compounds are introduced into the gas chromatograph either by direct injector or purge-and-trap (EPA Method 5030). EPA Method 5030 may be used directly on groundwater samples or low-concentration contaminated soils and sediments. For medium-concentration soils or sediments, methanolic extraction, as described in EPA Method 5030, may be necessary prior to purge-and-trap analysis.

QUALITY CONTROL Calculate surrogate standard recovery on all samples, blanks, and spikes. A trip blank is recommended to check on sampling, storage, and handling contamination. Calibration standards, at a minimum of five concentration levels, are prepared in organic-free reagent water. One of the concentration levels should be at a concentration near, but above, the method detection limit.

A combination of bromochloromethane, 2-bromo-1-chloropropane, 1,4-dichlorobutane, and bromochlorobenzene are recommended as surrogate standards to encompass the range of the temperature program used in this method.

REFERENCE Test Methods for Evaluating Solid Waste, Physical/Chemical Methods, SW-846, 3rd Edition, U.S. EPA, Office of Solid Waste, Washington, DC, EPA Method 8021A, Rev. 1, Nov. 1992.

1,3-Dichloropropane EPA Method 8021
CAS #142-28-9

TITLE Halogenated Volatile by Gas Chromatography Using Photoionization and Electrolytic Conductivity Detectors in Series: Capillary Column Technique

MATRIX This method is applicable to nearly all types of samples, regardless of water content, including groundwater, aqueous sludges, caustic liquors, acid liquors, waste solvents, oily wastes, mousses, tars, fibrous wastes, polymeric emulsions, filter cakes, spent carbons, spent catalysts, soils, and sediments.

METHOD SUMMARY This method is used to determine 60 volatile organic compounds in a variety of solid waste matrices. It provides GC conditions for the detection of halogenated and aromatic volatile organic compounds. Samples can be analyzed using direct injection or purge-and-trap (EPA Method 5030). Groundwater samples must be analyzed using EPA Method 5030 (where applicable). A temperature program is used with the GC. Detection is achieved by a photoionization detector (PID) and a Hall electrolytic conductivity detector (HECD) in series.

INTERFERENCES Samples can be contaminated by diffusion of volatile organics (particularly chlorofluorocarbons and methylene chloride) through the sample container septum during shipment and storage.

INSTRUMENTATION A GC-equipped with variable-constant differential flow controllers, subambient oven controller, PID and HECD detectors connected with a short piece of uncoated capillary tubing and a data system.

Column: 60 m × 0.75 mm I.D. VOCOL wide-bore capillary column with 1.5 µm film thickness.

PRECISION & ACCURACY MDLs are compound-dependent and vary with purging efficiency and concentration. The applicable concentration range of this method is compound- and instrument-dependent but is approximately 0.1 to 200 µg/L. Analytes that are inefficiently purged from water will not be detected when present at low concentrations, but they can be measured with acceptable accuracy and precision when present in sufficient amounts. The estimated quantitation limit (EQL) for an individual compound is approximately 1 µg/kg (wet weight) for soil/sediment samples, 100 µg/kg (wet weight) for wastes, and 1 µg/L for groundwater. EQLs will be proportionately higher for sample extracts and samples that require dilution or reduced sample size to avoid saturation of the detector.

MULTIPLICATION FACTORS FOR OTHER MATRICES (a)

Matrix	Factor (b)
Groundwater	10
Low-concentration soil	10
Water miscible liquid waste	500
High-concentration soil and sludge	1250
Non-water miscible waste	1250

(a) Sample EQLs are highly matrix-dependent. The EQLs listed herein are provided for guidance and may not always be achievable. (b) EQL = [Method detection limit] × [Factor]. For non-aqueous samples, the factor is on a wet-weight basis.

SINGLE LABORATORY ACCURACY & PRECISION DATA FOR VOCs IN WATER

This method was tested in a single lab using water spiked at 10 µg/L and the following data was reported:

Recoveries and standard deviations were determined from seven samples and spiked at 10 µg/L of each analyte. Recoveries were determined by the internal standard method. Internal standards were: Fluorobenzene for PID and 2-Bromo-1-chloropropane for HECD.

The average recovery (in percent) for the PID was none (no response for this detector).

The standard deviation of the recovery for the PID was none (no response for this detector).

The MDL (in µg/mL) for the PID was none (no response for this detector).

The average recovery (in percent) for the HECD was 100.

The standard deviation of the recovery for the HECD was 3.4.

The MDL (in µg/mL) for the HECD was 0.03.

SAMPLE COLLECTION, PRESERVATION & HANDLING
Volatile Organics — Standard 40-mL glass screw-cap VOA vials with Teflon®-faced silicone septum may be used for both liquid and solid matrices. When collecting samples, liquids and solids should be introduced into the vials gently to reduce agitation which might drive off volatile compounds. If there are any air bubbles present the sample must be retaken. Tap slightly as they are filled to try and eliminate as much free air space as possible. The two vials from each sampling locations should be sealed in separate plastic bags to prevent cross-contamination between samples particularly if the sampled waste is suspected of containing high levels of volatile organics.

Semivolatile organics — Containers used to collect samples for the determination of semivolatile organic compounds should be soap and water washed followed by methanol (or isopropanol) rinsing. The sample containers should be of glass or Teflon® and have screw-top covers with Teflon® liners.

Preservation for volatile organics — No preservation is used with concentrated waste samples. With liquid samples containing no residual chlorine, 4 drops of concentrated hydrochloric acid are added and the samples are immediately cooled to 4°C. When liquid samples contain residual chlorine, they are treated as above and, in addition, 4 drops of 4% aqueous sodium thiosulfate are added. Soil, sediment, and sludge samples are only cooled to 4°C.

Preservation for semivolatile organics — No preservation is used with concentrated waste samples. With liquid samples containing no residual chlorine and with soil, sediment, and sludge samples, immediately cooling to 4°C is the only preservation used. When residual chlorine is present then 3 mL of 10% aqueous sodium sulfate is added for each gallon of sample collected, followed by cooling to 4°C.

MHT The holding time for all volatile organics samples is 14 days. Liquid samples must be extracted within 7 days and their extracts analyzed within 40 days. Concentrated waste, soil, sediment, and sludge samples must be extracted within 14 days and their extracts analyzed within 40 days.

SAMPLE PREPARATION Volatile compounds are introduced into the gas chromatograph either by direct injector or purge-and-trap (EPA Method 5030). EPA Method 5030 may be used directly on groundwater samples or low-concentration contaminated soils and sediments. For medium-concentration soils or sediments, methanolic extraction, as described in EPA Method 5030, may be necessary prior to purge-and-trap analysis.

QUALITY CONTROL Calculate surrogate standard recovery on all samples, blanks, and spikes. A trip blank is recommended to check on sampling, storage, and handling contamination. Calibration standards, at a minimum of five concentration levels, are prepared in organic-free reagent water. One of the concentration levels should be at a concentration near, but above, the method detection limit.

A combination of bromochloromethane, 2-bromo-1-chloropropane, 1,4-dichlorobutane, and bromochlorobenzene are recommended as surrogate standards to encompass the range of the temperature program used in this method.

REFERENCE Test Methods for Evaluating Solid Waste, Physical/Chemical Methods, SW-846, 3rd Edition, U.S. EPA, Office of Solid Waste, Washington, DC, EPA Method 8021A, Rev. 1, Nov. 1992.

2,2-Dichloropropane   **EPA Method 8021**
CAS #590-20-7

TITLE Halogenated Volatile by Gas Chromatography Using Photoionization and Electrolytic Conductivity Detectors in Series: Capillary Column Technique

MATRIX This method is applicable to nearly all types of samples, regardless of water content, including groundwater, aqueous sludges, caustic liquors, acid liquors, waste solvents, oily wastes, mousses, tars, fibrous wastes, polymeric emulsions, filter cakes, spent carbons, spent catalysts, soils, and sediments.

METHOD SUMMARY This method is used to determine 60 volatile organic compounds in a variety of solid waste matrices. It provides GC conditions for the detection of halogenated and aromatic volatile organic compounds. Samples can be analyzed using direct injection or purge-and-trap (EPA Method 5030). Groundwater samples must be analyzed using EPA Method 5030 (where applicable). A temperature program is used with the GC. Detection is achieved by a photoionization detector (PID) and a Hall electrolytic conductivity detector (HECD) in series.

INTERFERENCES Samples can be contaminated by diffusion of volatile organics (particularly chlorofluorocarbons and methylene chloride) through the sample container septum during shipment and storage.

INSTRUMENTATION A GC-equipped with variable-constant differential flow controllers, subambient oven controller, PID and HECD detectors connected with a short piece of uncoated capillary tubing and a data system.

Column: 60 m × 0.75 mm I.D. VOCOL wide-bore capillary column with 1.5 µm film thickness.

PRECISION & ACCURACY MDLs are compound-dependent and vary with purging efficiency and concentration. The applicable concentration range of this method is compound- and instrument-dependent but is approximately 0.1 to 200 µg/L. Analytes that are inefficiently purged from water will not be detected when present at low concentrations, but they can be measured with acceptable accuracy and precision when present in sufficient amounts. The estimated quantitation limit (EQL) for an individual compound is approximately 1 µg/kg (wet weight) for soil/sediment samples, 100 µg/kg (wet weight) for wastes, and 1 µg/L for groundwater. EQLs will be proportionately higher for sample extracts and samples that require dilution or reduced sample size to avoid saturation of the detector.

MULTIPLICATION FACTORS FOR OTHER MATRICES (a)

Matrix	Factor (b)
Groundwater	10
Low-concentration soil	10
Water miscible liquid waste	500
High-concentration soil and sludge	1250
Non-water miscible waste	1250

(a) Sample EQLs are highly matrix-dependent. The EQLs listed herein are provided for guidance and may not always be achievable.
(b) EQL = [Method detection limit] × [Factor]. For non-aqueous samples, the factor is on a wet-weight basis.

SINGLE LABORATORY ACCURACY & PRECISION DATA FOR VOCs IN WATER

This method was tested in a single lab using water spiked at 10 µg/L and the following data was reported:

Recoveries and standard deviations were determined from seven samples and spiked at 10 µg/L of each analyte. Recoveries were determined by the internal standard method. Internal standards were: Fluorobenzene for PID and 2-Bromo-1-chloropropane for HECD.

The average recovery (in percent) for the PID was none (no response for this detector).
The standard deviation of the recovery for the PID was none (no response for this detector).
The MDL (in µg/mL) for the PID was none (no response for this detector).
The average recovery (in percent) for the HECD was 105.
The standard deviation of the recovery for the HECD was 3.6.
The MDL (in µg/mL) for the HECD was 0.05.

SAMPLE COLLECTION, PRESERVATION & HANDLING
Volatile Organics — Standard 40-mL glass screw-cap VOA vials with Teflon®-faced silicone septum may be used for both liquid and solid matrices. When collecting samples, liquids and solids should be introduced into the vials gently to reduce agitation which might drive off volatile compounds. If there are any air bubbles present the sample must be retaken. Tap slightly as they are filled to try and eliminate as much free air space as possible. The two vials from each sampling locations should be sealed in separate plastic bags to prevent cross-contamination between samples particularly if the sampled waste is suspected of containing high levels of volatile organics.

Semivolatile organics — Containers used to collect samples for the determination of semivolatile organic compounds should be soap and water washed followed by methanol (or isopropanol) rinsing. The sample containers should be of glass or Teflon® and have screw-top covers with Teflon® liners.

Preservation for volatile organics — No preservation is used with concentrated waste samples. With liquid samples containing no residual chlorine, 4 drops of concentrated hydrochloric acid are added and the samples are immediately cooled to 4°C. When liquid samples contain residual chlorine, they are treated as above and, in addition, 4 drops of 4% aqueous sodium thiosulfate are added. Soil, sediment, and sludge samples are only cooled to 4°C.

Preservation for semivolatile organics — No preservation is used with concentrated waste samples. With liquid samples containing no residual chlorine and with soil, sediment, and sludge samples, immediately cooling to 4°C is the only preservation used. When residual chlorine is present then 3 mL of 10% aqueous sodium sulfate is added for each gallon of sample collected, followed by cooling to 4°C.

MHT The holding time for all volatile organics samples is 14 days. Liquid samples must be extracted within 7 days and their extracts analyzed within 40 days. Concentrated waste, soil, sediment, and sludge samples must be extracted within 14 days and their extracts analyzed within 40 days.

SAMPLE PREPARATION Volatile compounds are introduced into the gas chromatograph either by direct injector or purge-and-trap (EPA Method 5030). EPA Method 5030 may be used directly on groundwater samples or low-concentration contaminated soils and sediments. For medium-concentration soils or sediments, methanolic extraction, as described in EPA Method 5030, may be necessary prior to purge-and-trap analysis.

QUALITY CONTROL Calculate surrogate standard recovery on all samples, blanks, and spikes. A trip blank is recommended to check on sampling, storage, and handling contamination. Calibration standards, at a minimum of five concentration levels, are prepared in organic-free reagent water. One of the concentration levels should be at a concentration near, but above, the method detection limit.

A combination of bromochloromethane, 2-bromo-1-chloropropane, 1,4-dichlorobutane, and bromochlorobenzene are recommended as surrogate standards to encompass the range of the temperature program used in this method.

REFERENCE Test Methods for Evaluating Solid Waste, Physical/Chemical Methods, SW-846, 3rd Edition, U.S. EPA, Office of Solid Waste, Washington, DC, EPA Method 8021A, Rev. 1, Nov. 1992.

1,2-Dichloropropane   **EPA Method 8240**
CAS #78-87-5

TITLE Volatile Organics By GC/MS: Packed Column Technique

MATRIX Nearly all types of sample matarices, regardless of water content, can be analyzed using this method. This includes groundwater, aqueous sludges, caustic liquors, acid liquors, waste solvents, oily wastes, mousses, tars, fibrous wastes, polymetric emulsions, filter cakes, spent carbons, spent catalysts, soils, and sediments.

METHOD SUMMARY Method 8240B covers 80 volatile organic compounds that are introduced into a gas chromatograph by the purge-and-trap method or by direct injection (in limited applications). For the purge-and-trap method an inert gas (zero grade nitrogen or helium) is bubbled through a 5-mL solution at ambient temperature. Purged sample components are trapped in a tube of sorbent materials. When purging is complete, the sorbent tube is heated and backflushed with inert gas to desorb the trapped components onto a GC column.

INTERFERENCES Impurities in the purge gas and from organic compounds outgassing from the plumbing ahead of the trap account for many contamination problems. Interferences purged or coextracted from the samples will vary considerably from source to source. Cross-contamination can occur whenever high-level and low-level samples are analyzed sequentially. Whenever an unusually concentrated sample is analyzed, it should be followed by the analysis of organic-free reagent water to check for cross-contamination. Samples also can be contaminated by diffusion of volatile organics (particularly methylene chloride and fluorocarbons) through the septum seal into the sample during shipment and storage. A trip blank can serve as a check on such contamination. The lab where volatile analysis is performed and also the refrigerated storage area should be completely free of solvents.

INSTRUMENTATION A gas chromatograph/mass spectrometry/data system (GC/MS) equipped with a 6 ft × 0.1 in I.D. glass column packed with 1% SP-1000 on Carbopack-B (60/80 mesh) is required. Also needed is a 5-mL purging device, a sorbent trap, and a thermal desorption apparatus.

PRECISION & ACCURACY This method is reported to have been tested by 15 laboratories using organic-free reagent water, drinking water, surface water, and industrial wastewaters (not specified) fortified at six concentrations over the range 5–600 µg/L.

Sample estimated quantitation limits (EQLs) are highly matrix-dependent. The EQLs listed may not always be achievable. EQLs listed for soils or sediments are based on wet weight. Normally, data is reported on a dry-weight basis; therefore, EQLs will be higher, based on the percent dry weight of each sample. Note that EQLs are even more variable than MDLs and that they are highly variable depending on the matrix being analyzed.

EQL in groundwater in µg/L was 5.
EQL in low soil or sediment in µg/kg was 5.
Accuracy (a) in µg/L was 1.00C.
Precision (b) in µg/L was 0.45x.

(a) *Average recovery found for measurements of samples containing a concentration of C, in µg/L.*
(b) *Overall precision found for measurements of samples with average recovery X for samples containing a concentration of C in µg/L.*
X = *Average recovery found for measurement of samples containing a concentration of C in µg/L.*

MULTIPLICATION FACTORS FOR OTHER MATRICES

Other Matrices	Factor (a)
Waste miscible liquid waste	50
High-concentration soil and sludge	125
Non-water miscible waste	500

(a) *EQL = [EQL for low soil sediment] × [Factor]. For non-aqueous samples, the factor is on a wet-weight basis.*

SAMPLING METHOD

Liquid samples — Use a 40-mL glass screw-cap VOA vial with a Teflon®-faced silicone septum that has been prewashed, rinsed with distilled deionized water, and oven dried. However, if residual chlorine is present, collect sample in a 40-oz. soil VOA container which has been pre-preserved with 4 drops of 10% sodium thiosulfate, mix gently, and then transfer the sample to a 40-mL VOA vial. Collect bubble-free samples in duplicate and seal them in separate plastic bags.

Soils or sediments, and sludges — Use an 8-oz. widemouth glass bottle with a Teflon®-faced silicone septum that has been prewashed with detergent, rinsed with distilled deionized water, and oven dried. Tap slightly to eliminate free air space. Collect samples in duplicate and seal them in separate plastic bags.

SAMPLE PRESERVATION

Liquid samples — Add 4 drops of concentrated HCL and immediately cool samples to 4°C and store in a solvent-free refrigerator.

Soils or sediments, and sludges — Cool samples to 4°C and store in a solvent-free refrigerator.

MHT Maximum holding time is 14 days from the date of sample collection.

SAMPLE PREPARATION

Liquid samples — Remove the plunger from a 5-mL syringe and carefully pour the sample into the syringe barrel to just short of overflowing. Replace the syringe plunger and compress the sample. Open the syringe valve and vent any residual air while adjusting the sample volume to 5.0 mL. If there is only one volatile organic analysis (VOA) vial, a second syringe should be filled at this time to protect against possible loss of sample integrity. Add 10 µL of surrogate spiking solution and 10 µL of internal standard spiking solution through the valve bore of the 5-mL syringe, then close the valve. The surrogate and internal standards may be mixed and added as a single spiking solution.

Sediments, soils, and waste samples — All samples of this type should be screened by GC analysis using a headspace method (EPA Method 3810) or the hexadecane extraction and screening method (EPA Method 3820). Use the screening data to determine whether to use the low-concentration method (0.005–1 mg/kg) or the high-concentration method (>1 mg/kg).

Low-concentration method — The low-concentration method is based on purging a heated sediment or soil sample mixed with organic-free reagent water containing the surrogate and

internal standards. Analyze all reagent blanks and standards under the same conditions as the samples.

Use a 5-g sample if the expected concentration is <0.1 mg/kg or a 1-g sample for expected concentrations between 0.1 and 1 mg/kg. Mix the contents of the sample container with a narrow metal spatula. Weigh the amount of the sample into a tared purge device. Add the spiked water to the purge device, which contains the weighed amount of sample, and connect the device to the purge-and-trap system.

High-concentration method — This method is based on extracting the sediment or soil with methanol. A waste sample is either extracted or diluted, depending on its solubility in methanol. Wastes that are insoluble in methanol are diluted with reagent tetraglyme or possibly polyethylene glycol (PEG). An aliquot of the extract is added to organic-free reagent water containing surrogate and internal standards. This is purged at ambient temperature. All samples with an expected concentration of >1.0 mg/kg should be analyzed by this method.

Mix the contents of the sample container with a narrow metal spatula. For sediments or soils and solid wastes that are insoluble in methanol, weigh 4 g (wet weight) of sample into a tared 20-mL vial. For waste that is soluble in methanol, tetraglyme, or PEG, weigh 1 g (wet weight) into a tared scintillation vial or culture tube or a 10-mL volumetric flask. Quickly add 9.0 mL of appropriate solvent then add 1.0 mL of a surrogate spiking solution to the vial, cap it, and shake it for 2 min.

METHANOL EXTRACT REQUIRED FOR ANALYSIS OF HIGH-CONCENTRATION SOILS OR SEDIMENTS

Approximate Concentration Range	Volume of Methanol Extract (a)
500–10,000 μg/kg	100 μL
1,000–20,000 μg/kg	50 μL
5,000–100,000 μg/kg	10 μL
25,000–500,000 μg/kg	100 μL of 1/50 dilution (b)

Calculate appropriate dilution factor for concentrations exceeding this table.

(a) The volume of methanol added to 5 mL of water being purged should be kept constant. Therefore, add to the 5-mL syringe whatever volume of methanol is necessary to maintain a volume of 100 μL added to the syringe.
(b) Dilute an aliquot of the methanol extract and then take 100 μL for analysis.

QUALITY CONTROL Demonstrate, through the analysis of a reagent water blank, that interferences from the analytical system, glassware, and reagents are under control. Blank samples should be carried through all stages of the sample preparation and measurement steps. For each analytical batch (up to 20 samples), a reagent blank, matrix spike, and matrix spike duplicate must be analyzed (the frequency of the spikes may be different for different monitoring programs). The blank and spiked samples must be carried through all stages of the sample preparation and measurement steps. QC samples mentioned in the section on Interferences will also be needed as appropriate to those situations.

REFERENCE Test Methods for Evaluating Solid Waste (SW-846). U.S. EPA. 1983. Method 8240B, Rev. 2, Nov. 1990. Office of Solid Wastes, Washington, DC.

1,2-Dichloropropane   **EPA Method 8260**
CAS #78-87-5

TITLE Volatile Organic Compounds by GC/MS: Capillary Column Technique

MATRIX This method is applicable to nearly all types of samples, regardless of water content, including groundwater, soils, and sediments.

METHOD SUMMARY Method 8260A covers 58 volatile organic compounds that are introduced into a gas chromatograph by the purge-and-trap method or by direct injection (in limited applications). Zero-grade helium is bubbled through a 5-mL solution at ambient temperature. Purged sample components are trapped in a tube containing suitable sorbent materials. When purging is complete, the sorbent tube is heated and backflushed with helium to desorb trapped sample components. The analytes are desorbed directly to a large bore capillary or cryofocussed on a capillary precolumn before being flash evaporated to a narrow bore capillary for analysis.

INTERFERENCES Major contaminant sources are volatile materials in the lab and impurities in the inert purging gas and in the sorbent trap. Interfering contamination may occur when a sample containing low concentrations of volatile organic compounds is analyzed immediately after a sample containing high concentrations of volatile organic compounds. After analysis of a sample containing high concentrations of volatile organic compounds, one or more calibration blanks should be analyzed to check for cross-contamination. Screening of the samples prior to purge-and-trap GC/MS analysis is highly recommended to prevent contamination of the system. This is especially true for soil and waste samples.

Special precautions must be taken to analyze for methylene chloride. The analytical and sample storage area should be isolated from all atmospheric sources of methylene chloride. All gas chromatography carrier gas lines and purge gas plumbing should be constructed of stainless steel or copper tubing. Laboratory clothing previously exposed to methylene chloride fumes during liquid-liquid extraction procedures can contribute to sample contamination.

Samples can also be contaminated by diffusion of volatile organics (particularly methylene chloride and fluorocarbons) through the septum seal during shipment and storage. A trip blank can serve as a check on such contamination.

INSTRUMENTATION GC/MS with a temperature-programmable chromatograph suitable for splitless injection equipped with variable constant differential flow controllers, a subambient oven controller, a purging device, sorbent trap, a thermal desorption apparatus and a capillary precolumn interface when using cryogenic cooling will be needed. The following GC columns may be used:

Column 1: 60 m × 0.75mm I.D. capillary column coated with VOCOL, 1.5 µm film thickness.
Column 2: 30 m × 0.53mm capillary column coated with DB-624 or VOCOL, 3 µm film thickness.
Column 3: 30 m × 0.32mm I.D. capillary column coated with DB-5 or SE-54, 1-µm film thickness.

PRECISION & ACCURACY This method has been tested in a single lab using spiked water. Using a wide-bore capillary column, water was spiked at concentrations between 0.5 and 10 µg/L. Single lab accuracy and precision data are presented. The MDL actually achieved in a given analysis will vary depending on instrument sensitivity and matrix effects.

The MDL (a) in µg/L was 0.04.
The concentration range in µg/L was 0.1–10.
The mean accuracy (% of true value) was 97.
The precision as relative standard deviation was 6.1.

Note: The MDL is based on a 25-mL sample volume instead of a 5-mL sample volume.

SAMPLING METHOD
Liquid samples — Use a 40-mL glass screw-cap VOA vial with a Teflon®-faced silicone septum that has been prewashed, rinsed with distilled deionized water, and oven dried. If residual chlorine is present, collect the sample in a 4-oz soil VOA container which has been pre-preserved with 4 drops of 10% sodium thiosulfate. Mix gently and transfer the sample to a 40-mL VOA vial. Collect bubble-free samples in duplicate and seal each sample in a separate plastic bag.

Soils, sediments, and sludges — Use an 8-oz widemouth glass bottle with Teflon®-faced silicone septum that has been prewashed, rinsed with distilled deionized water, and oven dried. **Do not** heat the septum for more than 1 h. Tap slightly to eliminate any free air space. Collect samples in duplicate and seal each one in a separate plastic bag.

SAMPLE PRESERVATION
Liquid samples — Add 4 drops of concentrated HCL, cool to 4°C and store in a solvent-free refrigerator.

Soils, sediments and sludges — Cool samples to 4°C and store in a solvent-free refrigerator.

MHT The maximum holding time of any sample (liquids, soils, sediments, and sludges) is 14 days.

SAMPLE PREPARATION
Liquid samples — Remove the plunger from a 5-mL syringe and carefully pour the sample into the syringe barrel to just short of overflowing. Replace the syringe plunger and compress the sample. Open the syringe valve and vent any residual air while adjusting the sample volume to 5.0 mL. If there is only one volatile organic analysis (VOA) vial, a second syringe should be filled at this time to protect against possible loss of sample integrity. Add 10 µL of surrogate spiking solution and 10 µL of internal standard spiking solution through the valve bore of the 5-mL syringe, then close the valve. The surrogate and internal standards may be mixed and added as a single spiking solution.

Sediments, soils, and waste samples — All samples of this type should be screened by GC analysis using a headspace method (EPA Method 3810) or the hexadecane extraction and screening method (EPA Method 3820). Use the screening data to determine whether to use the low-concentration method (0.005–1 mg/kg) or the high-concentration method (>1 mg/kg).

Low-concentration method — The low-concentration method is based on purging a heated sediment or soil sample mixed with organic-free reagent water containing the surrogate and internal standards. Analyze all reagent blanks and standards under the same conditions as the samples.

Use a 5-g sample if the expected concentration is <0.1 mg/kg or a 1-g sample for expected concentrations between 0.1 and 1 mg/kg. Mix the contents of the sample container with a narrow metal spatula. Weigh the amount of the sample into a tared purge device. Add the spiked water to the purge device, which contains the weighed amount of sample, and connect the device to the purge-and-trap system.

High-concentration method — This method is based on extracting the sediment or soil with methanol. A waste sample is either extracted or diluted, depending on its solubility in methanol. Wastes that are insoluble in methanol are diluted with reagent tetraglyme or possibly polyethylene glycol (PEG). An aliquot of the extract is added to organic-free reagent water containing surrogate and internal standards. This is purged at ambient temperature. All samples with an expected concentration of >1.0 mg/kg should be analyzed by this method.

Mix the contents of the sample container with a narrow metal spatula. For sediments or soils and solid wastes that are insoluble in methanol, weigh 4 g (wet weight) of sample into a tared 20-mL vial. For waste that is soluble in methanol, tetraglyme, or PEG, weigh 1 g (wet weight) into a tared scintillation vial or culture tube or a 10-mL volumetric flask. Quickly add 9.0 mL of appropriate solvent then add 1.0 mL of a surrogate spiking solution to the vial, cap it, and shake it for 2 min.

METHANOL EXTRACT REQUIRED FOR ANALYSIS OF HIGH-CONCENTRATION SOILS OR SEDIMENTS

Approximate Concentration Range	Volume of Methanol Extract (a)
500–10,000 µg/kg	100 µL
1,000–20,000 µg/kg	50 µL
5,000–100,000 µg/kg	10 µL
25,000–500,000 µg/kg	100 µL of 1/50 dilution (b)

Calculate appropriate dilution factor for concentrations exceeding this table.

(a) The volume of methanol added to 5 mL of water being purged should be kept constant. Therefore, add to the 5-mL syringe whatever volume of methanol is necessary to maintain a volume of 100 µL added to the syringe.
(b) Dilute an aliquot of the methanol extract and then take 100 µL for analysis.

QUALITY CONTROL Demonstrate, through the analysis of a reagent water blank, that interferences from the analytical system, glassware, and reagents are under control. Blank samples should be carried through all stages of the sample preparation and measurement steps. For each analytical batch (up to 20 samples), a reagent blank, matrix spike, and matrix spike duplicate must be analyzed (the frequency of the spikes may be different for different monitoring programs). The blank and spiked samples must be carried through all stages of the sample preparation and measurement steps. QC samples mentioned in the section on Interferences will also be needed as appropriate to those situations.

Matrix spiking standards should be prepared from volatile organic compounds which will be representative of the compounds being investigated. The recommended internal standards are chlorobenzene-d5, 1,4-difluorobenzene, 1,4-dichlorobenzene-d4, and pentafluorobenzene. Using stock standard solutions, prepare secondary dilution standards containing the compounds of interest, either singly or mixed together in methanol. Store them in a vial with no headspace for no more than one week. Surrogates recommended are toluene-d8, 4-bromofluorobenzene, and dibromofluoromethane. Each sample undergoing GC/MS analysis must be spiked with 10 µL of the surrogate spiking solution prior to analysis.

REFERENCE Test Methods for Evaluating Solid Waste (SW-846). U.S. EPA 1983, Method 8260A, Rev. 1, Nov. 1990. Office of Solid Waste, Washington, DC.

1,3-Dichloropropane   **EPA Method 8260**
CAS #142-28-9

TITLE Volatile Organic Compounds by GC/MS: Capillary Column Technique

MATRIX This method is applicable to nearly all types of samples, regardless of water content, including groundwater, soils, and sediments.

METHOD SUMMARY Method 8260A covers 58 volatile organic compounds that are introduced into a gas chromatograph by the purge-and-trap method or by direct injection (in limited applications). Zero-grade helium is bubbled through a 5-mL solution at ambient temperature. Purged sample components are trapped in a tube containing suitable sorbent materials. When purging is complete, the sorbent tube is heated and backflushed with helium to desorb trapped sample components. The analytes are desorbed directly to a large bore capillary or cryofocussed on a capillary precolumn before being flash evaporated to a narrow bore capillary for analysis.

INTERFERENCES Major contaminant sources are volatile materials in the lab and impurities in the inert purging gas and in the sorbent trap. Interfering contamination may occur when a sample containing low concentrations of volatile organic compounds is analyzed immediately after a sample containing high concentrations of volatile organic compounds. After analysis of a sample containing high concentrations of volatile organic compounds, one or more calibration blanks should be analyzed to check for cross-contamination. Screening of the samples prior to purge-and-trap GC/MS analysis is highly recommended to prevent contamination of the system. This is especially true for soil and waste samples.

Special precautions must be taken to analyze for methylene chloride. The analytical and sample storage area should be isolated from all atmospheric sources of methylene chloride. All gas chromatography carrier gas lines and purge gas plumbing should be constructed from stainless steel or copper tubing. Laboratory clothing previously exposed to methylene chloride fumes during liquid-liquid extraction procedures can contribute to sample contamination.

Samples can also be contaminated by diffusion of volatile organics (particularly methylene chloride and fluorocarbons) through the septum seal during shipment and storage. A trip blank can serve as a check on such contamination.

INSTRUMENTATION GC/MS with a temperature-programmable chromatograph suitable for splitless injection equipped with variable constant differential flow controllers, a subambient oven controller, a purging device, sorbent trap, a thermal desorption apparatus and a capillary precolumn interface when using cryogenic cooling will be needed. The following GC columns may be used:

Column 1: 60 m × 0.75mm I.D. capillary column coated with VOCOL, 1.5 µm film thickness.
Column 2: 30 m × 0.53mm capillary column coated with DB-624 or VOCOL, 3 µm film thickness.
Column 3: 30 m × 0.32mm I.D. capillary column coated with DB-5 or SE-54, 1-µm film thickness.

PRECISION & ACCURACY This method has been tested in a single lab using spiked water. Using a wide-bore capillary column, water was spiked at concentrations between 0.5 and 10 µg/L. Single lab accuracy and precision data are presented. The MDL actually achieved in a given analysis will vary depending on instrument sensitivity and matrix effects.

The MDL (a) in µg/L was 0.04.
The concentration range in µg/L was 0.1–10.
The mean accuracy (% of true value) was 96.
The precision as relative standard deviation was 6.0.

Note: The MDL is based on a 25-mL sample volume instead of a 5-mL sample volume.

SAMPLING METHOD

Liquid samples — Use a 40-mL glass screw-cap VOA vial with a Teflon®-faced silicone septum that has been prewashed, rinsed with distilled deionized water, and oven dried. If residual chlorine is present, collect the sample in a 4-oz soil VOA container which has been pre-preserved with 4 drops of 10% sodium thiosulfate. Mix gently and transfer the sample to a 40-mL VOA vial. Collect bubble-free samples in duplicate and seal each sample in a separate plastic bag.

Soils, sediments, and sludges — Use an 8-oz widemouth glass bottle with Teflon®-faced silicone septum that has been prewashed, rinsed with distilled deionized water, and oven dried. **Do not** heat the septum for more than 1 h. Tap slightly to eliminate any free air space. Collect samples in duplicate and seal each one in a separate plastic bag.

SAMPLE PRESERVATION

Liquid samples — Add 4 drops of concentrated HCL, cool to 4°C and store in a solvent-free refrigerator.

Soils, sediments and sludges — Cool samples to 4°C and store in a solvent-free refrigerator.

MHT The maximum holding time of any sample (liquids, soils, sediments, and sludges) is 14 days.

SAMPLE PREPARATION
Liquid samples — Remove the plunger from a 5-mL syringe and carefully pour the sample into the syringe barrel to just short of overflowing. Replace the syringe plunger and compress the sample. Open the syringe valve and vent any residual air while adjusting the sample volume to 5.0 mL. If there is only one volatile organic analysis (VOA) vial, a second syringe should be filled at this time to protect against possible loss of sample integrity. Add 10 µL of surrogate spiking solution and 10 µL of internal standard spiking solution through the valve bore of the 5-mL syringe, then close the valve. The surrogate and internal standards may be mixed and added as a single spiking solution.

Sediments, soils, and waste samples — All samples of this type should be screened by GC analysis using a headspace method (EPA Method 3810) or the hexadecane extraction and screening method (EPA Method 3820). Use the screening data to determine whether to use the low-concentration method (0.005–1 mg/kg) or the high-concentration method (>1 mg/kg).

Low-concentration method — The low-concentration method is based on purging a heated sediment or soil sample mixed with organic-free reagent water containing the surrogate and internal standards. Analyze all reagent blanks and standards under the same conditions as the samples.

Use a 5-g sample if the expected concentration is <0.1 mg/kg or a 1-g sample for expected concentrations between 0.1 and 1 mg/kg. Mix the contents of the sample container with a narrow metal spatula. Weigh the amount of the sample into a tared purge device. Add the spiked water to the purge device, which contains the weighed amount of sample, and connect the device to the purge-and-trap system.

High-concentration method — This method is based on extracting the sediment or soil with methanol. A waste sample is either extracted or diluted, depending on its solubility in methanol. Wastes that are insoluble in methanol are diluted with reagent tetraglyme or possibly polyethylene glycol (PEG). An aliquot of the extract is added to organic-free reagent water containing surrogate and internal standards. This is purged at ambient temperature. All samples with an expected concentration of >1.0 mg/kg should be analyzed by this method.

Mix the contents of the sample container with a narrow metal spatula. For sediments or soils and solid wastes that are insoluble in methanol, weigh 4 g (wet weight) of sample into a tared 20-mL vial. For waste that is soluble in methanol, tetraglyme, or PEG, weigh 1 g (wet weight) into a tared scintillation vial or culture tube or a 10-mL volumetric flask. Quickly add 9.0 mL of appropriate solvent then add 1.0 mL of a surrogate spiking solution to the vial, cap it, and shake it for 2 min.

METHANOL EXTRACT REQUIRED FOR ANALYSIS OF HIGH-CONCENTRATION SOILS OR SEDIMENTS

Approximate Concentration Range	Volume of Methanol Extract (a)
500–10,000 µg/kg	100 µL
1,000–20,000 µg/kg	50 µL
5,000–100,000 µg/kg	10 µL
25,000–500,000 µg/kg	100 µL of 1/50 dilution (b)

Calculate appropriate dilution factor for concentrations exceeding this table.

(a) The volume of methanol added to 5 mL of water being purged should be kept constant. Therefore, add to the 5-mL syringe whatever volume of methanol is necessary to maintain a volume of 100 µL added to the syringe.

(b) Dilute an aliquot of the methanol extract and then take 100 µL for analysis.

QUALITY CONTROL Demonstrate, through the analysis of a reagent water blank, that interferences from the analytical system, glassware, and reagents are under control. Blank samples should be carried through all stages of the sample preparation and measurement steps. For each analytical batch (up to 20 samples), a reagent blank, matrix spike, and matrix spike duplicate must be analyzed (the frequency of the spikes may be different for different monitoring programs). The blank and spiked samples must be carried through all stages of the sample preparation and measurement steps. QC samples mentioned in the section on Interferences will also be needed as appropriate to those situations.

Matrix spiking standards should be prepared from volatile organic compounds which will be representative of the compounds being investigated. The recommended internal standards are chlorobenzene-d5, 1,4-difluorobenzene, 1,4-dichlorobenzene-d4, and pentafluorobenzene. Using stock standard solutions, prepare secondary dilution standards containing the compounds of interest, either singly or mixed together in methanol. Store them in a vial with no headspace for no more than one week. Surrogates recommended are toluene-d8, 4-bromofluorobenzene, and dibromofluoromethane. Each sample undergoing GC/MS analysis must be spiked with 10 µL of the surrogate spiking solution prior to analysis.

REFERENCE Test Methods for Evaluating Solid Waste (SW-846). U.S. EPA 1983, Method 8260A, Rev. 1, Nov. 1990. Office of Solid Waste, Washington, DC.

2,2-Dichloropropane   **EPA Method 8260**
CAS #590-20-7

TITLE Volatile Organic Compounds by GC/MS: Capillary Column Technique

MATRIX This method is applicable to nearly all types of samples, regardless of water content, including groundwater, soils, and sediments.

METHOD SUMMARY Method 8260A covers 58 volatile organic compounds that are introduced into a gas chromatograph by the purge-and-trap method or by direct injection (in limited applications). Zero-grade helium is bubbled through a 5-mL solution at ambient temperature. Purged sample components are trapped in a tube containing suitable sorbent materials. When purging is complete, the sorbent tube is heated and backflushed with helium to desorb trapped sample components. The analytes are desorbed directly to a large bore capillary or cryofocussed on a capillary precolumn before being flash evaporated to a narrow bore capillary for analysis.

INTERFERENCES Major contaminant sources are volatile materials in the lab and impurities in the inert purging gas and in the sorbent trap. Interfering contamination may occur when a sample containing low concentrations of volatile organic compounds is analyzed immediately after a sample containing high concentrations of volatile organic compounds. After analysis of a sample containing high concentrations of volatile organic compounds, one or more calibration blanks should be analyzed to check for cross-contamination. Screening of the samples prior to purge-and-trap GC/MS analysis is highly recommended to prevent contamination of the system. This is especially true for soil and waste samples.

Special precautions must be taken to analyze for methylene chloride. The analytical and sample storage area should be isolated from all atmospheric sources of methylene chloride. All gas chromatography carrier gas lines and purge gas plumbing should be constructed from stainless steel or copper tubing. Laboratory clothing previously exposed to methylene chloride fumes during liquid-liquid extraction procedures can contribute to sample contamination.

Samples can also be contaminated by diffusion of volatile organics (particularly methylene chloride and fluorocarbons) through the septum seal during shipment and storage. A trip blank can serve as a check on such contamination.

INSTRUMENTATION GC/MS with a temperature-programmable chromatograph suitable for splitless injection equipped with variable constant differential flow controllers, a subambient oven controller, a purging device, sorbent trap, a thermal desorption apparatus and a capillary precolumn interface when using cryogenic cooling will be needed. The following GC columns may be used:

Column 1: 60 m × 0.75mm I.D. capillary column coated with VOCOL, 1.5 µm film thickness.
Column 2: 30 m × 0.53mm capillary column coated with DB-624 or VOCOL, 3 µm film thickness.
Column 3: 30 m × 0.32mm I.D. capillary column coated with DB-5 or SE-54, 1-µm film thickness.

PRECISION & ACCURACY This method has been tested in a single lab using spiked water. Using a wide-bore capillary column, water was spiked at concentrations between 0.5 and 10 µg/L. Single lab accuracy and precision data are presented. The MDL actually achieved in a given analysis will vary depending on instrument sensitivity and matrix effects.

The MDL (a) in µg/L was 0.12.
The concentration range in µg/L was 0.5–10.

The mean accuracy (% of true value) was 86.
The precision as relative standard deviation was 16.9.

Note: The MDL is based on a 25-mL sample volume instead of a 5-mL sample volume.

SAMPLING METHOD
Liquid samples — Use a 40-mL glass screw-cap VOA vial with a Teflon®-faced silicone septum that has been prewashed, rinsed with distilled deionized water, and oven dried. If residual chlorine is present, collect the sample in a 4-oz soil VOA container which has been pre-preserved with 4 drops of 10% sodium thiosulfate. Mix gently and transfer the sample to a 40-mL VOA vial. Collect bubble-free samples in duplicate and seal each sample in a separate plastic bag.

Soils, sediments, and sludges — Use an 8-oz widemouth glass bottle with Teflon®-faced silicone septum that has been prewashed, rinsed with distilled deionized water, and oven dried. **Do not** heat the septum for more than 1 h. Tap slightly to eliminate any free air space. Collect samples in duplicate and seal each one in a separate plastic bag.

SAMPLE PRESERVATION
Liquid samples — Add 4 drops of concentrated HCL, cool to 4°C and store in a solvent-free refrigerator.

Soils, sediments and sludges — Cool samples to 4°C and store in a solvent-free refrigerator.

MHT The maximum holding time of any sample (liquids, soils, sediments, and sludges) is 14 days.

SAMPLE PREPARATION
Liquid samples — Remove the plunger from a 5-mL syringe and carefully pour the sample into the syringe barrel to just short of overflowing. Replace the syringe plunger and compress the sample. Open the syringe valve and vent any residual air while adjusting the sample volume to 5.0 mL. If there is only one volatile organic analysis (VOA) vial, a second syringe should be filled at this time to protect against possible loss of sample integrity. Add 10 µL of surrogate spiking solution and 10 µL of internal standard spiking solution through the valve bore of the 5-mL syringe, then close the valve. The surrogate and internal standards may be mixed and added as a single spiking solution.

Sediments, soils, and waste samples — All samples of this type should be screened by GC analysis using a headspace method (EPA Method 3810) or the hexadecane extraction and screening method (EPA Method 3820). Use the screening data to determine whether to use the low-concentration method (0.005–1 mg/kg) or the high-concentration method (>1 mg/kg).

Low-concentration method — The low-concentration method is based on purging a heated sediment or soil sample mixed with organic-free reagent water containing the surrogate and internal standards. Analyze all reagent blanks and standards under the same conditions as the samples.

Use a 5-g sample if the expected concentration is <0.1 mg/kg or a 1-g sample for expected concentrations between 0.1 and 1 mg/kg. Mix the contents of the sample container with a narrow metal spatula. Weigh the amount of the sample into a tared

purge device. Add the spiked water to the purge device, which contains the weighed amount of sample, and connect the device to the purge-and-trap system.

High-concentration method — This method is based on extracting the sediment or soil with methanol. A waste sample is either extracted or diluted, depending on its solubility in methanol. Wastes that are insoluble in methanol are diluted with reagent tetraglyme or possibly polyethylene glycol (PEG). An aliquot of the extract is added to organic-free reagent water containing surrogate and internal standards. This is purged at ambient temperature. All samples with an expected concentration of >1.0 mg/kg should be analyzed by this method.

Mix the contents of the sample container with a narrow metal spatula. For sediments or soils and solid wastes that are insoluble in methanol, weigh 4 g (wet weight) of sample into a tared 20-mL vial. For waste that is soluble in methanol, tetraglyme, or PEG, weigh 1 g (wet weight) into a tared scintillation vial or culture tube or a 10-mL volumetric flask. Quickly add 9.0 mL of appropriate solvent then add 1.0 mL of a surrogate spiking solution to the vial, cap it, and shake it for 2 min.

METHANOL EXTRACT REQUIRED FOR ANALYSIS OF HIGH-CONCENTRATION SOILS OR SEDIMENTS

Approximate Concentration Range	Volume of Methanol Extract (a)
500–10,000 µg/kg	100 µL
1,000–20,000 µg/kg	50 µL
5,000–100,000 µg/kg	10 µL
25,000–500,000 µg/kg	100 µL of 1/50 dilution (b)

Calculate appropriate dilution factor for concentrations exceeding this table.

(a) The volume of methanol added to 5 mL of water being purged should be kept constant. Therefore, add to the 5-mL syringe whatever volume of methanol is necessary to maintain a volume of 100 µL added to the syringe.

(b) Dilute an aliquot of the methanol extract and then take 100 µL for analysis.

QUALITY CONTROL Demonstrate, through the analysis of a reagent water blank, that interferences from the analytical system, glassware, and reagents are under control. Blank samples should be carried through all stages of the sample preparation and measurement steps. For each analytical batch (up to 20 samples), a reagent blank, matrix spike, and matrix spike duplicate must be analyzed (the frequency of the spikes may be different for different monitoring programs). The blank and spiked samples must be carried through all stages of the sample preparation and measurement steps. QC samples mentioned in the section on Interferences will also be needed as appropriate to those situations.

Matrix spiking standards should be prepared from volatile organic compounds which will be representative of the compounds being investigated. The recommended internal standards are chlorobenzene-d5, 1,4-difluorobenzene, 1,4-dichlorobenzene-d4, and pentafluorobenzene. Using stock standard solutions, prepare secondary dilution standards containing the compounds of interest, either singly or mixed together in methanol. Store them in a vial with no headspace for no more than one week. Surrogates recommended are toluene-d8, 4-bromofluorobenzene, and dibromofluoromethane. Each sample undergoing GC/MS analysis must be spiked with 10 µL of the surrogate spiking solution prior to analysis.

REFERENCE Test Methods for Evaluating Solid Waste (SW-846). U.S. EPA 1983, Method 8260A, Rev. 1, Nov. 1990. Office of Solid Waste, Washington, DC.

1,2-Dichloropropane   **EPA Method 601**
CAS #78-87-5

TITLE Purgeable Halocarbons

MATRIX Wastewater

APPLICATION Method covers 29 purgeable halocarbons. (Method 624 provides GC/MS conditions appropriate for the qualitative and quantitative confirmation of results). Method describes conditions for a 2nd GC column to confirm measurements made with primary column.

INTERFERENCES Impurities in the purge gas and organic compounds outgassing from the plumbing ahead of the trap. With high- and low-level samples, there can be carryover contamination. Diffusion of volatile organics through the septum seal into the sample.

INSTRUMENTATION GC-equipped with halide-specific detector. (With purge-and-trap unit).

RANGE 8.0–500 µg/L.

MDL 0.04 µg/L.

PRECISION 0.23X µg/L (overall precision).

ACCURACY 1.00C µg/L (as recovery).

SAMPLING METHOD 25-mL glass vial. Teflon®-lined septum.

STABILITY cool, 4°C, 0.008% Sodium thiosulfate.

MHT 14 days.

QUALITY CONTROL The lab must on an ongoing basis, spike at least 10% of the samples from each sample site being monitored to assess accuracy.

REFERENCE Method 601, *Federal Register* Part VIII 40 CFR Part 136, Oct 26, 1984.

1,2-Dichloropropane   **EPA Method 624**
CAS #78-87-5

TITLE Purgeables

MATRIX Wastewater

APPLICATION Method covers 31 purgeable organics. An inert gas is bubbled through a 5-mL water sample in a specially designed purging chamber. Here, purgeables are transferred from aqueous to gaseous phase, passed onto a sorbent column, and trapped. Trap is heated and backflushed with inert gas to desorb purgeables onto a GC column, where purgeables are separated.

INTERFERENCES Impurities in the purge gas, organic compounds outgassing from the plumbing ahead of the trap, and solvent vapors in the lab. With high- and low-level samples, there can be carryover contamination.

INSTRUMENTATION GC/MS with purge-and-trap unit.

RANGE 5–600 µg/L.

MDL 6.0 µg/L

PRECISION 0.45X µg/L (overall precision).

ACCURACY 1.00C µg/L (as recovery).

SAMPLING METHOD 25-mL glass vial. Teflon®-lined septum.

STABILITY cool, 4°C, 0.008% Sodium thiosulfate.

MHT 14 days.

QUALITY CONTROL The lab must on an ongoing basis, spike at least 5% of the samples from each sample site being monitored to assess accuracy.

REFERENCE Method 624, *Federal Register* Part VIII 40 CFR Part 136, Oct 26, 1984.

1,2-Dichloropropane   **EPA Method 8010**
CAS #78-87-5

TITLE Halogenated Volatile Organics

MATRIX Groundwater, soils, sludges, water miscible liquid wastes, and non-water miscible wastes.

APPLICATION This method is used for the analysis of 39 halogenated VOCs. Samples are analyzed using direct injection or purge-and-trap methods. Groundwater must be analyzed by the purge-and-trap method. The method provides an optional GC column which is used for analyte confirmation and that may help resolve analytes from interferences.

INTERFERENCES There can be carryover contamination with high- and low-level samples. Impurities may come from the purge-and-trap apparatus, organic compounds outgassing from the plumbing ahead of trap, diffusion of VOCs through the sample bottle septum during shipping or storage, or from solvent vapors in the lab.

INSTRUMENTATION GC capable of on-column injections or purge-and-trap sample introduction and a halogen specific detector. Column 1: 8 ft by 0.1 in 1%. SP-1000 on Carbopack-B. Column 2: 6 ft by 0.1 in bonded n-octane on Porasil-C.

RANGE 8 to 500 µg/L (reagent water).

MDL 0.04 µg/L (reagent water).

PQL FACTORS FOR MULTIPLYING × FID MDL VALUE

Matrix	Multiplication Factor
Groundwater	10
Low-level soil	10
Water miscible liquid waste	500
High-level soil and sludge	1250
Non-water miscible waste	1250

PRECISION 0.23X µg/L (overall precision; estimated).

ACCURACY 1.00C µg/L (as recovery; estimated).

SAMPLING METHOD For water and liquid samples; use glass 40-mL vials with Teflon®-lined septum caps and collect two vials per sample location with no headspace. For solids and concentrated waste samples; use widemouth glass bottles with Teflon® liners.

STABILITY For concentrated wastes, soils, sediments, or sludges: cool to 4°C. For liquids: add 4 drops of concentrated hydrochloric acid and cool to 4°C.

MHT 14 days.

QUALITY CONTROL Analyze a reagent blank, matrix spike, and matrix spike duplicate/duplicate for each analytical batch (up to 20 samples). Demonstrate the purity of glassware and reagents by analyzing a reagent water method blank. Internal, surrogate, and five concentration level calibration standards are used.

REFERENCE Test Methods for Evaluating Solid Waste (SW-846), U.S. EPA Office of Solid Waste, Washington, DC, Method 8010B, Rev. 2, Nov. 1992.

***cis*-1,3-Dichloropropene**   **EPA Method 1624**
CAS #10061-01-5

TITLE Volatile Organic Compounds by Isotope Dilution GC/MS

MATRIX Compounds may be determined in waters, soils, and municipal sludges by this method.

METHOD SUMMARY This method is used to determine 58 volatile toxic organic pollutants associated with the CWA (as amended 1987); the RCRA (as amended 1986); the CERCLA (as amended 1986); and other compounds amenable to purge-and-trap gas chromatography-mass spectrometry (GC/MS).

If the solids content is less than 1%, stable isotopically-labeled analogs of the compounds of interest are added to a 5-mL sample and the sample is purged with an inert gas at 20–25°C in a chamber designed for soil or water samples. If the solids content is greater than 1%, 5 mL of reagent water and the labeled compounds are added to a 5-g aliquot of sample and the mixture is purged at 40°C. Compounds that will not purge at 20–25°C or at 40°C are purged at 78–85°C. In the purging process, the volatile compounds are transferred from the aqueous phase into the gaseous phase where they are passed into a sorbent column, and trapped. After purging is completed, the trap is backflushed and heated rapidly to desorb the compounds into a GC. The compounds are separated by the GC and detected by a MS. The labeled compounds serve to correct the variability of the analytical technique.

INTERFERENCES Impurities in the purge gas, organic compounds outgassing from the plumbing upstream of the trap, and solvent vapors in the lab account for most problems. Samples can be contaminated by diffusion of volatile organic compounds (particularly methylene chloride) through the bottle seal during shipment and storage. Contamination by carryover can occur when high-level and low-level samples are analyzed sequentially. When an unusually concentrated sample is encountered, follow it by analysis of a reagent water blank to check for carryover.

INSTRUMENTATION Major equipment includes a GC with linear temperature programming and a glass jet separator as the MS interface, a MS with 70 eV electron impact ionization, and a data system to collect and record response factors.

Column: 2.8 m × 2 mm I.D. glass, packed with 1% SP-1000 on Carbopak B, 60/80 mesh, or equivalent.

PRECISION & ACCURACY The detection limits of the method are usually dependent on the level of interferences rather than instrumental limitations. The method detection limits were determined in digested sludge (low solids) and in filter cake or compost (high solids).

The MDL (in µg/kg) for low solids is not listed and for high solids is not listed.

Labeled and native compound precision (in µg/L) as standard deviation was not listed.

Labeled and native compound accuracy (in µg/L) as average recovery was not listed.

Acceptance criteria are at 20 µg/L for this compound.

SAMPLE COLLECTION, PRESERVATION & HANDLING Grab samples are collected in glass containers having a total volume greater than 20 mL. Fill and seal each bottle so that no air bubbles are entrapped. Samples are maintained at 0 to 4°C from the time of collection until analysis. If an aqueous sample contains residual chlorine, add sodium thiosulfate preservative (10 mg/40 mL) to the empty sample bottles just prior to shipment to the sample site. All samples must be analyzed within 14 days of collection.

SAMPLE PREPARATION Samples containing less than 1% solids are analyzed directly as aqueous samples. Samples containing 1% solids or greater are analyzed as solid samples utilizing one of two methods, depending on the levels of pollutants, in the sample. Samples containing 1% solids or greater, and low to moderate levels of pollutants are analyzed by purging a known weight of sample added to 5 mL of reagent water. Samples containing 1% solids or greater, and high levels of pollutants, are extracted with methanol, and an aliquot of the methanol extract is added to reagent water and purged.

QUALITY CONTROL A field blank prepared from reagent water and carried through the sampling and handling protocol may serve as a check on contamination from shipment and storage.

The analyst is permitted to modify this method to improve separations or lower the costs of measurements, provided all performance specifications are met. Analyses of blanks are required. When results of spikes indicate atypical method performance for samples, the samples are diluted to bring method performance within acceptable limits. Analyze two sets of four 5-mL aliquots (8 aliquots total) of the aqueous performance standard. Spike all samples with labeled compounds to assess method performance on the sample matrix. Compute the percent recovery of the labeled compounds using the internal standard method. Compare the percent recovery for each compound with the corresponding labeled compound recovery. Reagent water blanks are analyzed to demonstrate freedom from carryover contamination. Field replicates may be collected to determine the precision of the sampling technique, and spiked samples may be required to determine the accuracy of the analysis when the internal method is used.

REFERENCE Volatile Organic Compounds by Isotope Dilution GC/MS. Office of Water Regulation and Standards, U.S. EPA Industrial Technology Division, Washington, DC, EPA Method 1624, Rev. C, June 1989 (contact W.A. Telliard, U.S. EPA, Office of Water Regulations and Standards, 401 M St., SW, Washington, DC, 20460. Phone: 202-382-7131).

---

trans-1,3-Dichloropropene　　　　　　　EPA Method 1624
CAS #10061-02-6

TITLE Volatile Organic Compounds by Isotope Dilution GC/MS

MATRIX Compounds may be determined in waters, soils, and municipal sludges by this method.

METHOD SUMMARY This method is used to determine 58 volatile toxic organic pollutants associated with the CWA (as amended 1987); the RCRA (as amended 1986); the CERCLA (as amended 1986); and other compounds amenable to purge-and-trap gas chromatography-mass spectrometry (GC/MS).

If the solids content is less than 1%, stable isotopically-labeled analogs of the compounds of interest are added to a 5-mL sample and the sample is purged with an inert gas at 20–25°C in a chamber designed for soil or water samples. If the solids content is greater than 1%, 5 mL of reagent water and the labeled compounds are added to a 5-g aliquot of sample and the mixture is purged at 40°C. Compounds that will not purge at 20–25°C or at 40°C are purged at 78–85°C. In the purging process, the volatile compounds are transferred from the aqueous phase into the gaseous phase where they are passed into a sorbent column, and trapped. After purging is completed, the trap is backflushed and heated rapidly to desorb the compounds into a GC. The compounds are separated by the GC and detected by a MS. The labeled compounds serve to correct the variability of the analytical technique.

INTERFERENCES Impurities in the purge gas, organic compounds outgassing from the plumbing upstream of the trap, and solvent vapors in the lab account for most problems. Samples can be contaminated by diffusion of volatile organic compounds (particularly methylene chloride) through the bottle seal during shipment and storage. Contamination by carryover can occur when high-level and low-level samples are analyzed sequentially. When an unusually concentrated sample is encountered, follow it by analysis of a reagent water blank to check for carryover.

INSTRUMENTATION Major equipment includes a GC with linear temperature programming and a glass jet separator as the MS interface, a MS with 70 eV electron impact ionization, and a data system to collect and record response factors.

Column: 2.8 m × 2 mm I.D. glass, packed with 1% SP-1000 on Carbopak B, 60/80 mesh, or equivalent.

PRECISION & ACCURACY The detection limits of the method are usually dependent on the level of interferences rather than instrumental limitations. The method detection limits were determined in digested sludge (low solids) and in filter cake or compost (high solids).

The MDL (in µg/kg) for low solids is not listed and for high solids is not listed.

Background levels of this compound were present in the sludge with low solids, resulting in a higher than expected MDL. Background levels of this compound were present in the sludge with high solids, resulting in a higher than expected MDL.

Labeled and native compound precision (in µg/L) as standard deviation was 15.0.
Labeled and native compound accuracy (in µg/L) as average recovery was detected to 40.

Acceptance criteria are at 20 µg/L for this compound.

SAMPLE COLLECTION, PRESERVATION & HANDLING Grab samples are collected in glass containers having a total volume greater than 20 mL. Fill and seal each bottle so that no air bubbles are entrapped. Samples are maintained at 0 to 4°C from the time of collection until analysis. If an aqueous sample contains residual chlorine, add sodium thiosulfate preservative (10 mg/40 mL) to the empty sample bottles just prior to shipment to the sample site. All samples must be analyzed within 14 days of collection.

SAMPLE PREPARATION Samples containing less than 1% solids are analyzed directly as aqueous samples. Samples containing 1% solids or greater are analyzed as solid samples utilizing one of two methods, depending on the levels of pollutants, in the sample. Samples containing 1% solids or greater, and low to moderate levels of pollutants are analyzed by purging a known weight of sample added to 5 mL of reagent water. Samples containing 1% solids or greater, and high levels of pollutants, are extracted with methanol, and an aliquot of the methanol extract is added to reagent water and purged.

QUALITY CONTROL A field blank prepared from reagent water and carried through the sampling and handling protocol may serve as a check on contamination from shipment and storage.

The analyst is permitted to modify this method to improve separations or lower the costs of measurements, provided all performance specifications are met. Analyses of blanks are required. When results of spikes indicate atypical method performance for samples, the samples are diluted to bring method performance within acceptable limits. Analyze two sets of four 5-mL aliquots (8 aliquots total) of the aqueous performance standard. Spike all samples with labeled compounds to assess method performance on the sample matrix. Compute the percent recovery of the labeled compounds using the internal standard method. Compare the percent recovery for each compound with the corresponding labeled compound recovery. Reagent water blanks are analyzed to demonstrate freedom from carryover contamination. Field replicates may be collected to determine the precision of the sampling technique, and spiked samples may be required to determine the accuracy of the analysis when the internal method is used.

REFERENCE Volatile Organic Compounds by Isotope Dilution GC/MS. Office of Water Regulation and Standards, U.S. EPA Industrial Technology Division, Washington, DC, EPA Method 1624, Rev. C, June 1989 (contact W.A. Telliard, U.S. EPA, Office of Water Regulations and Standards, 401 M St., SW, Washington, DC, 20460. Phone: 202-382-7131).

1,1-Dichloropropene   **EPA Method 502**
CAS #563-58-6

TITLE Volatile Organic Compounds in Water By Purge and Trap Capillary Column Gas Chromatography with Photoionization and Electrolytic Conductivity Detectors in Series. U.S. EPA Method 502.2, Rev. 2.0, 1989.

MATRIX Drinking water and raw source water. The latter should include most surface water and groundwater sources.

METHOD SUMMARY This method covers 60 volatile organic compounds that contain halogen atoms and/or that are aromatic. An inert gas (zero grade nitrogen or helium) is bubbled through a 25-mL or a 5-mL water sample (depending on the expected concentration of the analytes). Purged sample components are trapped in a tube of sorbent materials. When purging is complete, the sorbent tube is heated and backflushed with helium to desorb the trapped sample onto a capillary GC column. The column is temperature programmed to separate the method analytes which are then detected with a photoionization detector (PID) and a Hall electrolytic conductivity (HECD) placed in series. The PID is selective for aromatic compounds and the HECD is selective for halogenated compounds.

INTERFERENCES Impurities in the purge gas and from organic compounds outgassing from the plumbing ahead of the trap account for many contamination problems. Interferences purged or coextracted from the samples will vary considerably from source to source, depending upon the particular sample or extract being tested. Cross-contamination can occur whenever high-level and low-level samples are analyzed sequentially. Samples also can be contaminated by diffusion of volatile organics (particularly methylene chloride and fluorocarbons) through the septum seal into the sample during shipment and storage. The lab where volatile analysis is performed and also the refrigerated storage area should be completely free of solvents.

INSTRUMENTATION A GC containing a series configuration of a high temperature photoionization detector (PID) equipped with 10.0 eV (nominal) lamp and Hall electrolytic conductivity detector (HECD) is required. Also required is an all-glass 5-mL purging device, a sorbent trap, and a thermal desorption apparatus which is connected to the GC system.

Column 1: VOCOL glass wide-bore capillary column.
Column 2: RTX–502.2 mega-bore capillary column.
Column 3: DB-62 mega-bore capillary column.

PRECISION & ACCURACY Method detection limits are dependent upon the characteristics of the gas chromatographic system used. Analytes that are not separated chromatographically cannot be individually identified and used in the same calibration mixture or water samples unless an alternative technique for identification and quantification, such as mass spectrometry, is used.

Electrolytic conductivity detetor (c) range in µg/L (a) was 0.02–200.
Electrolytic conductivity detetor (c) MDL in µg/L (b) was 0.02.
Electrolytic conductivity detetor (c) accuracy as % recovery was 103.
Electrolytic conductivity detetor (c) precision as % RSD was 3.3.
Photoionization detector (d) range in µg/L (a) was 0.02–200.
Photoionization detector (d) MDL in µg/L (b) was 0.02.
Photoionization detector (d) accuracy as % recovery was 103.
Photoionization detector (d) precision as % RSD was 3.5.

(a) *The applicable concentration range of this method is compound, instrument, and matrix-dependent. It is listed as being approximately 0.02 to 200 µg/L but no specific information is provided so caution should be observed.*
(b) *The method detection limits reports with this method are compound, instrument, and matrix-dependent. The values reported were calculated using reagent water fortified with the corresponding compounds at 10 µg/L and a GC-equipped with a 60 m × 0.75 mm VOLCOL wide bore capillary column with 1.5 µm film thickness and using helium carrier gas.*
(c) *Recoveries and relative standard deviations were determined from seven samples of reagent water fortified with 10 µg/L of each compound. 2-Bromo-1-chloropropane was used as the internal standard for calculating average recoveries.*
(d) *Recoveries and relative standard deviations were determined from seven samples of reagent water fortified with 10 µg/L of each compound. Fluorobenzene was used as the internal standard for calculating average recoveries.*

SAMPLING METHOD Collect samples using a 40- to 120-mL screw-cap vial (prewashed with detergent, rinsed with distilled water and oven dried at 105°C) with a Teflon®-faced silicone septum. Collect bubble-free samples and place the septum with the Teflon® side down on the water.

SAMPLE PRESERVATION If residual chlorine is present in the water add about 25 mg of ascorbic acid to each vial before samples are collected to remove the chlorine. Add hydrochloric acid to reduce pH to <2, immediately cool samples to 4°C, and store them in a solvent-free refrigerator at 4°C until analysis.

MHT The maximum holding time for samples is 14 days from the time they were collected.

SAMPLE PREPARATION Remove the plungers from two 5-mL syringes and attach a closed syringe valve to each. Warm the sample to room temperature, open the sample bottle, and carefully pour the sample into one of the syringe barrels to just short of overflowing. Replace the syringe plunger, invert the syringe, and compress the sample. Open the syringe valve and vent any residual air while adjusting the sample volume to 5.0 mL. Add 10 µL of the internal calibration standard to the sample through the syringe valve. Close the valve. Fill the second syringe in an identical manner from the same sample bottle. Reserve this second syringe for a reanalysis if necessary.

QUALITY CONTROL As an initial demonstration of lab accuracy and precision, analyze 4 to 7 replicates of a lab fortified blank containing analyte at 0.1–5 µg/L. Collect all samples in duplicate. Surrogate analytes (similar to those of the analytes of interest), whose concentration is known in every sample, are measured using the same internal standard calibration procedure. Duplicate field reagent water blanks (trip blanks) must be analyzed with each set of samples, lab reagent blanks (method blanks) must be analyzed with each batch of samples processed as a group within a work shift. Also, a single lab-fortified blank that contains each of the analytes of interest should be analyzed with each batch of samples processed as a group within a work shift. A 3- to 5-point calibration curve is needed depending on the calibration range factor required.

EPA CONTACT & HOTLINE For technical questions contact Dr. Baldev Bathija, U.S. EPA, Office of Ground Water and Drinking Water (WH-550D), 401 M St. SW, Washington, DC 20460. Tel. (202) 260-3040. For further information the EPA Safe Drinking Water Hotline may be called at: (800) 426-4791.

REFERENCE Methods for the Determination of Organic Compounds in Drinking Water, EPA/600/4-88/039 (revised July 1991; Final Rule for determination of compliance with the MCL for Total Trihalomethanes under 141.30, in 40 CFR Part 141, Vol. 58, No. 147, Fed. Reg., Tuesday Aug. 3, 1993). U.S. EPA Environmental Monitoring Systems Laboratory, Cincinnati, OH, 45268, U.S.A. Available from the National Technical Information Service (NTIS), 5285 Port Royal Road, Springfield, VA 22161; Tel. 800-553-6847. NTIS Order Number is PB91-231480.

---

cis-1,3-Dichloropropene EPA Method 502
CAS #10061-01-5

TITLE Volatile Organic Compounds in Water By Purge and Trap Capillary Column Gas Chromatography with Photoionization and Electrolytic Conductivity Detectors in Series. U.S. EPA Method 502.2, Rev. 2.0, 1989.

MATRIX Drinking water and raw source water. The latter should include most surface water and groundwater sources.

METHOD SUMMARY This method covers 60 volatile organic compounds that contain halogen atoms and/or that are aromatic. An inert gas (zero grade nitrogen or helium) is bubbled through a 25-mL or a 5-mL water sample (depending on the expected concentration of the analytes). Purged sample components are trapped in a tube of sorbent materials. When purging is complete, the sorbent tube is heated and backflushed with helium to desorb the trapped sample onto a capillary GC column. The column is temperature programmed to separate the method analytes which are then detected with a photoionization detector (PID) and a Hall electrolytic conductivity (HECD) placed in series. The PID is selective for aromatic compounds and the HECD is selective for halogenated compounds.

INTERFERENCES Impurities in the purge gas and from organic compounds outgassing from the plumbing ahead of the trap account for many contamination problems. Interferences purged or coextracted from the samples will vary considerably from source to source, depending upon the particular sample or extract being tested. Cross-contamination can occur whenever high-level and low-level samples are analyzed sequentially. Samples also can be contaminated by diffusion of volatile organics (particularly methylene chloride and fluorocarbons) through the septum seal into the sample during shipment and storage. The lab where volatile analysis is performed and also the refrigerated storage area should be completely free of solvents.

INSTRUMENTATION A GC containing a series configuration of a high temperature photoionization detector (PID) equipped with 10.0 eV (nominal) lamp and Hall electrolytic conductivity detector (HECD) is required. Also required is an all-glass 5-mL purging device, a sorbent trap, and a thermal desorption apparatus which is connected to the GC system.

Column 1: VOCOL glass wide-bore capillary column.
Column 2: RTX–502.2 mega-bore capillary column.
Column 3: DB-62 mega-bore capillary column.

PRECISION & ACCURACY Method detection limits are dependent upon the characteristics of the gas chromatographic system used. Analytes that are not separated chromatographically cannot be individually identified and used in the same calibration mixture or water samples unless an alternative technique for identification and quantification, such as mass spectrometry, is used.

Electrolytic conductivity detetor (c) range in μg/L (a) was not listed.
Electrolytic conductivity detetor (c) MDL in μg/L (b) was not listed.
Electrolytic conductivity detetor (c) accuracy as % recovery was not listed.
Electrolytic conductivity detetor (c) precision as % RSD was not listed.
Photoionization detector (d) range in μg/L (a) was not listed.
Photoionization detector (d) MDL in μg/L (b) was not listed.
Photoionization detector (d) accuracy as % recovery was not listed.
Photoionization detector (d) precision as % RSD was not listed.

(a) *The applicable concentration range of this method is compound, instrument, and matrix-dependent. It is listed as being approximately 0.02 to 200 μg/L but no specific information is provided so caution should be observed.*
(b) *The method detection limits reports with this method are compound, instrument, and matrix-dependent. The values reported were calculated using reagent water fortified with the corresponding compounds at 10 μg/L and a GC-equipped with a 60 m × 0.75 mm VOLCOL wide bore capillary column with 1.5 μm film thickness and using helium carrier gas.*
(c) *Recoveries and relative standard deviations were determined from seven samples of reagent water fortified with 10 μg/L of each compound. 2-Bromo-1-chloropropane was used as the internal standard for calculating average recoveries.*
(d) *Recoveries and relative standard deviations were determined from seven samples of reagent water fortified with 10 μg/L of each compound. Fluorobenzene was used as the internal standard for calculating average recoveries.*

SAMPLING METHOD Collect samples using a 40- to 120-mL screw-cap vial (prewashed with detergent, rinsed with distilled water and oven dried at 105°C) with a Teflon®-faced silicone septum. Collect bubble-free samples and place the septum with the Teflon® side down on the water.

SAMPLE PRESERVATION If residual chlorine is present in the water add about 25 mg of ascorbic acid to each vial before samples are collected to remove the chlorine. Add hydrochloric acid to reduce pH to <2, immediately cool samples to 4°C, and store them in a solvent-free refrigerator at 4°C until analysis.

MHT The maximum holding time for samples is 14 days from the time they were collected.

SAMPLE PREPARATION Remove the plungers from two 5-mL syringes and attach a closed syringe valve to each. Warm the sample to room temperature, open the sample bottle, and carefully pour the sample into one of the syringe barrels to just short of overflowing. Replace the syringe plunger, invert the syringe, and compress the sample. Open the syringe valve and vent any residual air while adjusting the sample volume to 5.0 mL. Add 10 μL of the internal calibration standard to the sample through the syringe valve. Close the valve. Fill the second syringe in an identical manner from the same sample bottle. Reserve this second syringe for a reanalysis if necessary.

QUALITY CONTROL As an initial demonstration of lab accuracy and precision, analyze 4 to 7 replicates of a lab fortified blank containing analyte at 0.1–5 μg/L. Collect all samples in duplicate. Surrogate analytes (similar to those of the analytes of interest), whose concentration is known in every sample, are measured using the same internal standard calibration procedure. Duplicate field reagent water blanks (trip blanks) must be analyzed with each set of samples, lab reagent blanks (method blanks) must be analyzed with each batch of samples processed as a group within a work shift. Also, a single lab-fortified blank that contains each of the analytes of interest should be analyzed with each batch of samples processed as a group within a work shift. A 3- to 5-point calibration curve is needed depending on the calibration range factor required.

EPA CONTACT & HOTLINE For technical questions contact Dr. Baldev Bathija, U.S. EPA, Office of Ground Water and Drinking Water (WH-550D), 401 M St. SW, Washington, DC 20460. Tel. (202) 260-3040. For further information the EPA Safe Drinking Water Hotline may be called at: (800) 426-4791.

REFERENCE Methods for the Determination of Organic Compounds in Drinking Water, EPA/600/4-88/039 (revised July 1991; Final Rule for determination of compliance with the MCL for Total Trihalomethanes under 141.30, in 40 CFR Part 141, Vol. 58, No. 147, Fed. Reg., Tuesday Aug. 3, 1993). U.S. EPA Environmental Monitoring Systems Laboratory, Cincinnati, OH, 45268, U.S.A. Available from the National Technical Information Service (NTIS), 5285 Port Royal Road, Springfield, VA 22161; Tel. 800-553-6847. NTIS Order Number is PB91-231480.

---

trans-1,3-Dichloropropene **EPA Method 502**
CAS #10061-02-6

TITLE Volatile Organic Compounds in Water By Purge and Trap Capillary Column Gas Chromatography with Photoionization and Electrolytic Conductivity Detectors in Series. U.S. EPA Method 502.2, Rev. 2.0, 1989.

MATRIX Drinking water and raw source water. The latter should include most surface water and groundwater sources.

METHOD SUMMARY This method covers 60 volatile organic compounds that contain halogen atoms and/or that are aromatic. An inert gas (zero grade nitrogen or helium) is bubbled through a 25-mL or a 5-mL water sample (depending on the expected concentration of the analytes). Purged sample components are trapped in a tube of sorbent materials. When purging is complete, the sorbent tube is heated and backflushed with helium to desorb the trapped sample onto a capillary GC column. The column is temperature programmed to separate the method analytes which are then detected with a photoionization detector (PID) and a Hall electrolytic conductivity (HECD) placed in series. The PID is selective for aromatic compounds and the HECD is selective for halogenated compounds.

INTERFERENCES Impurities in the purge gas and from organic compounds outgassing from the plumbing ahead of the trap account for many contamination problems. Interferences purged or coextracted from the samples will vary considerably from source to source, depending upon the particular sample or extract being tested. Cross-contamination can occur whenever high-level and low-level samples are analyzed sequentially. Samples also can be contaminated by diffusion of volatile organics (particularly methylene chloride and fluorocarbons) through the septum seal into the sample during shipment and storage. The lab where volatile analysis is performed and also the refrigerated storage area should be completely free of solvents.

INSTRUMENTATION A GC containing a series configuration of a high temperature photoionization detector (PID) equipped with 10.0 eV (nominal) lamp and Hall electrolytic conductivity detector (HECD) is required. Also required is an all-glass 5-mL purging device, a sorbent trap, and a thermal desorption apparatus which is connected to the GC system.

Column 1: VOCOL glass wide-bore capillary column.
Column 2: RTX–502.2 mega-bore capillary column.
Column 3: DB-62 mega-bore capillary column.

PRECISION & ACCURACY Method detection limits are dependent upon the characteristics of the gas chromatographic system used. Analytes that are not separated chromatographically cannot be individually identified and used in the same calibration mixture or water samples unless an alternative technique for identification and quantification, such as mass spectrometry, is used.

Electrolytic conductivity detector (c) range in µg/L (a) was not listed.
Electrolytic conductivity detector (c) MDL in µg/L (b) was not listed.
Electrolytic conductivity detetor (c) accuracy as % recovery was not listed.
Electrolytic conductivity detetor (c) precision as % RSD was not listed.
Photoionization detector (d) range in µg/L (a) was not listed.
Photoionization detector (d) MDL in µg/L (b) was not listed.
Photoionization detector (d) accuracy as % recovery was not listed.
Photoionization detector (d) precision as % RSD was not listed.

(a) The applicable concentration range of this method is compound, instrument, and matrix-dependent. It is listed as being approximately 0.02 to 200 µg/L but no specific information is provided so caution should be observed.

(b) The method detection limits reports with this method are compound, instrument, and matrix-dependent. The values reported were calculated using reagent water fortified with the corresponding compounds at 10 µg/L and a GC-equipped with a 60 m × 0.75 mm VOLCOL wide bore capillary column with 1.5 µm film thickness and using helium carrier gas.

(c) Recoveries and relative standard deviations were determined from seven samples of reagent water fortified with 10 µg/L of each compound. 2-Bromo-1-chloropropane was used as the internal standard for calculating average recoveries.

(d) Recoveries and relative standard deviations were determined from seven samples of reagent water fortified with 10 µg/L of each compound. Fluorobenzene was used as the internal standard for calculating average recoveries.

SAMPLING METHOD Collect samples using a 40- to 120-mL screw-cap vial (prewashed with detergent, rinsed with distilled water and oven dried at 105°C) with a Teflon®-faced silicone septum. Collect bubble-free samples and place the septum with the Teflon® side down on the water.

SAMPLE PRESERVATION If residual chlorine is present in the water add about 25 mg of ascorbic acid to each vial before samples are collected to remove the chlorine. Add hydrochloric acid to reduce pH to <2, immediately cool samples to 4°C, and store them in a solvent-free refrigerator at 4°C until analysis.

MHT The maximum holding time for samples is 14 days from the time they were collected.

SAMPLE PREPARATION Remove the plungers from two 5-mL syringes and attach a closed syringe valve to each. Warm the sample to room temperature, open the sample bottle, and carefully pour the sample into one of the syringe barrels to just short of overflowing. Replace the syringe plunger, invert the syringe, and compress the sample. Open the syringe valve and vent any residual air while adjusting the sample volume to 5.0 mL. Add 10 µL of the internal calibration standard to the sample through the syringe valve. Close the valve. Fill the second syringe in an identical manner from the same sample bottle. Reserve this second syringe for a reanalysis if necessary.

QUALITY CONTROL As an initial demonstration of lab accuracy and precision, analyze 4 to 7 replicates of a lab fortified blank containing analyte at 0.1–5 µg/L. Collect all samples in duplicate. Surrogate analytes (similar to those of the analytes of interest), whose concentration is known in every sample, are measured using the same internal standard calibration procedure. Duplicate field reagent water blanks (trip blanks) must be analyzed with each set of samples, lab reagent blanks (method blanks) must be analyzed with each batch of samples processed as a group within a work shift. Also, a single lab-fortified blank that contains each of the analytes of interest should be analyzed with each batch of samples processed as a group within a work shift. A 3- to 5-point calibration curve is needed depending on the calibration range factor required.

EPA CONTACT & HOTLINE For technical questions contact Dr. Baldev Bathija, U.S. EPA, Office of Ground Water and Drinking Water (WH-550D), 401 M St. SW, Washington, DC 20460. Tel. (202) 260-3040. For further information the EPA Safe Drinking Water Hotline may be called at: (800) 426-4791.

REFERENCE Methods for the Determination of Organic Compounds in Drinking Water, EPA/600/4-88/039 (revised July 1991; Final Rule for determination of compliance with the MCL for Total Trihalomethanes under 141.30, in 40 CFR Part 141, Vol. 58, No. 147, Fed. Reg., Tuesday Aug. 3, 1993). U.S. EPA Environmental Monitoring Systems Laboratory, Cincinnati, OH, 45268, U.S.A. Available from the National Technical Information Service (NTIS), 5285 Port Royal Road, Springfield, VA 22161; Tel. 800-553-6847. NTIS Order Number is PB91-231480.

1,1-Dichloropropene   **EPA Method 524**
CAS #563-58-6

TITLE Measurement of Purgeable Organic Compounds in Water by Capillary Column GC/MS.

MATRIX Drinking water and raw source water; the latter should include most surface water and groundwater sources.

METHOD SUMMARY Method 524.2 covers 60 volatile organic compounds. An inert gas (zero grade nitrogen or helium) is bubbled through a 25-mL or a 5-mL water sample (depending on the expected concentration of the analytes). Purged sample components are trapped in a tube of sorbent materials. When purging is complete, the sorbent tube is heated and backflushed with helium to desorb the trapped sample onto a capillary GC column.

INTERFERENCES Impurities in the purge gas and from organic compounds outgassing from the plumbing ahead of the trap account for many contamination problems. Interferences purged or coextracted from the samples will vary considerably from source to source, depending upon the particular sample or extract being tested. Cross-contamination can occur whenever high-level and low-level samples are analyzed sequentially. Samples also can be contaminated by diffusion of volatile organics (particularly methylene chloride and fluorocarbons) through the septum seal into the sample during shipment and storage.

INSTRUMENTATION A GC/MS with a data system equipped with one of the following capillary GC columns:

Column 1: VOCOL glass wide bore capillary column.
Column 2: DB-624 fused silica capillary column.
Column 3: DB-5 fused silica capillary column.

Also required is an all-glass 25 mL or 5-mL purging device, a sorbent trap, and a thermal desorption apparatus which is connected to the GC/MS system.

PRECISION & ACCURACY Method detection limits are compound- and instrument-dependent, and may vary from approximately 0.02–0.35 µg/L. Note in the table below that the "true" concentration range used for accuracy and precision measurements was quite narrow. However, the applicable concentration range of this method is primarily column dependent and is approximately 0.02 to 200 µg/L for the wide-bore thick-film columns. Narrow-bore thin-film columns may have a capacity which limits the range to about 0.02 to 20 µg/L. Analytes that are inefficiently purged from water will not be detected when present at low concentrations, but they can be measured with acceptable accuracy and precision when present in sufficient amounts.

Analytes that are not separated chromatographically, but which have different mass spectra and non-interfering quantification ions, can be identified and measured in the same calibration mixture or water sample. Analytes which have very similar mass spectra cannot be individually identified and measured in the same calibration mixture or water samples unless they have different retention times. Co-eluting compounds with very similar mass spectra, typically many structural isomers, must be reported as an isomeric group or pair.

The range (in µg/L) was 0.5–10.
The Method Detection Limig (in µg/L) was 0.10.
The accuracy (as % recovery) was 98.
The precision (in %) was 8.9.

Note: Data were obtained from 16–31 determinations using a wide-bore capillary column and a jet separator interfaced to a quadrupole mass spectrometer. All analytes were in a reagent water matrix.

SAMPLING METHOD Collect samples using a 40- to 120-mL screw-cap vial (prewashed with detergent, rinsed with distilled water and oven dried at 105°C) with a Teflon®-faced silicone septum. Collect bubble-free samples and place the septum with the Teflon® side down on the water.

SAMPLE PRESERVATION If residual chlorine is present in the water add about 25 mg of ascorbic acid to each vial before samples are collected to remove the chlorine. Add hydrochloric acid to reduce pH to <2, and immediately cool samples to 4°C, and store them in a solvent-free refrigerator at 4°C until analysis.

MHT The maximum holding time for samples is 14 days from the time they were collected.

SAMPLE PREPARATION Remove the plungers from two 25-mL (or 5-mL depending on sample size) syringes and attach a closed syringe valve to each. Warm the sample to room temperature, open the sample bottle, and carefully pour the sample into one of the syringe barrels to just short of overflowing. Replace the syringe plunger, invert the syringe, and compress the sample. Open the syringe valve and vent any residual air while adjusting the sample volume to 25.0 mL (or 5 mL). For samples and blanks, add 5 µL of the fortification solution containing the internal standard and the surrogates to the sample through the syringe valve. For calibration standards and lab fortified blanks, add 5 µL of the fortification solution containing the internal standard only. Close the valve. Fill the second syringe in an identical manner from the same sample bottle. Reserve this second syringe for a reanalysis if necessary.

QUALITY CONTROL As an initial demonstration of lab accuracy and precision, analyze 4 to 7 replicates of a lab fortified blank containing analyte at 0.2–5 µg/L. Collect all samples in duplicate. Surrogate analytes (similar to those of the analytes of interest), whose concentration is known in every sample, are measured using the same internal standard calibration procedure. Duplicate field reagent water blanks (trip blanks) must be analyzed with each set of samples, lab reagent blanks (method blanks) must be analyzed with each batch of samples processed as a group within a work shift. Also, a single lab-fortified blank that contains each of the analytes of interest should be analyzed with each batch of samples processed as a group within a work shift. A 3- to 5-point calibration curve is needed depending on the calibration range factor required.

EPA CONTACT & HOTLINE For technical questions contact Dr. Baldev Bathija, U.S. EPA, Office of Ground Water and Drinking Water (WH-550D), 401 M St. SW, Washington, DC 20460. Tel. (202) 260-3040. For further information the EPA Safe Drinking Water Hotline may be called at: (800) 426-4791.

REFERENCE Methods for the Determination of Organic Compounds in Drinking Water, EPA/600/4-88/039 (revised July 1991; Final Rule for determination of compliance with the MCL for Total Trihalomethanes under 141.30, in 40 CFR Part 141, Vol. 58, No. 147, Fed. Reg., Tuesday Aug. 3, 1993). U.S. EPA Environmental Monitoring Systems Laboratory, Cincinnati, OH, 45268, U.S.A. Available from the National Technical Information Service (NTIS), 5285 Port Royal Road, Springfield, VA 22161; Tel. 800-553-6847. NTIS Order Number is PB91-231480.

cis-1,3-Dichloropropene   EPA Method 524
CAS #10061-01-5

TITLE Measurement of Purgeable Organic Compounds in Water by Capillary Column GC/MS.

MATRIX Drinking water and raw source water; the latter should include most surface water and groundwater sources.

METHOD SUMMARY Method 524.2 covers 60 volatile organic compounds. An inert gas (zero grade nitrogen or helium) is bubbled through a 25-mL or a 5-mL water sample (depending on the expected concentration of the analytes). Purged sample components are trapped in a tube of sorbent materials. When purging is complete, the sorbent tube is heated and backflushed with helium to desorb the trapped sample onto a capillary GC column.

INTERFERENCES Impurities in the purge gas and from organic compounds outgassing from the plumbing ahead of the trap account for many contamination problems. Interferences purged or coextracted from the samples will vary considerably from source to source, depending upon the particular sample or extract being tested. Cross-contamination can occur whenever high-level and low-level samples are analyzed sequentially. Samples also can be contaminated by diffusion of volatile organics (particularly methylene chloride and fluorocarbons) through the septum seal into the sample during shipment and storage.

INSTRUMENTATION A GC/MS with a data system equipped with one of the following capillary GC columns:

Column 1: VOCOL glass wide bore capillary column.
Column 2: DB-624 fused silica capillary column.
Column 3: DB-5 fused silica capillary column.

Also required is an all-glass 25 mL or 5-mL purging device, a sorbent trap, and a thermal desorption apparatus which is connected to the GC/MS system.

PRECISION & ACCURACY Method detection limits are compound- and instrument-dependent, and may vary from approximately 0.02–0.35 µg/L. Note in the table below that the "true" concentration range used for accuracy and precision measurements was quite narrow. However, the applicable concentration range of this method is primarily column dependent and is approximately 0.02 to 200 µg/L for the wide-bore thick-film columns. Narrow-bore thin-film columns may have a capacity which limits the range to about 0.02 to 20 µg/L. Analytes that are inefficiently purged from water will not be detected when present at low concentrations, but they can be measured with acceptable accuracy and precision when present in sufficient amounts.

Analytes that are not separated chromatographically, but which have different mass spectra and non-interfering quantification ions, can be identified and measured in the same calibration mixture or water sample. Analytes which have very similar mass spectra cannot be individually identified and measured in the same calibration mixture or water samples unless they have different retention times. Co-eluting compounds with very

similar mass spectra, typically many structural isomers, must be reported as an isomeric group or pair.

The range (in μg/L) was not listed.
The Method Detection Limig (in μg/L) was not listed.
The accuracy (as % recovery) was not listed.
The precision (in %) was not listed.

Note: Data were obtained from 16–31 determinations using a wide-bore capillary column and a jet separator interfaced to a quadrupole mass spectrometer. All analytes were in a reagent water matrix.

SAMPLING METHOD Collect samples using a 40- to 120-mL screw-cap vial (prewashed with detergent, rinsed with distilled water and oven dried at 105°C) with a Teflon®-faced silicone septum. Collect bubble-free samples and place the septum with the Teflon® side down on the water.

SAMPLE PRESERVATION If residual chlorine is present in the water add about 25 mg of ascorbic acid to each vial before samples are collected to remove the chlorine. Add hydrochloric acid to reduce pH to <2, and immediately cool samples to 4°C, and store them in a solvent-free refrigerator at 4°C until analysis.

MHT The maximum holding time for samples is 14 days from the time they were collected.

SAMPLE PREPARATION Remove the plungers from two 25-mL (or 5-mL depending on sample size) syringes and attach a closed syringe valve to each. Warm the sample to room temperature, open the sample bottle, and carefully pour the sample into one of the syringe barrels to just short of overflowing. Replace the syringe plunger, invert the syringe, and compress the sample. Open the syringe valve and vent any residual air while adjusting the sample volume to 25.0 mL (or 5 mL). For samples and blanks, add 5 μL of the fortification solution containing the internal standard and the surrogates to the sample through the syringe valve. For calibration standards and lab fortified blanks, add 5 μL of the fortification solution containing the internal standard only. Close the valve. Fill the second syringe in an identical manner from the same sample bottle. Reserve this second syringe for a reanalysis if necessary.

QUALITY CONTROL As an initial demonstration of lab accuracy and precision, analyze 4 to 7 replicates of a lab fortified blank containing analyte at 0.2–5 μg/L. Collect all samples in duplicate. Surrogate analytes (similar to those of the analytes of interest), whose concentration is known in every sample, are measured using the same internal standard calibration procedure. Duplicate field reagent water blanks (trip blanks) must be analyzed with each set of samples, lab reagent blanks (method blanks) must be analyzed with each batch of samples processed as a group within a work shift. Also, a single lab-fortified blank that contains each of the analytes of interest should be analyzed with each batch of samples processed as a group within a work shift. A 3- to 5-point calibration curve is needed depending on the calibration range factor required.

EPA CONTACT & HOTLINE For technical questions contact Dr. Baldev Bathija, U.S. EPA, Office of Ground Water and Drinking Water (WH-550D), 401 M St. SW, Washington, DC 20460. Tel. (202) 260-3040. For further information the EPA Safe Drinking Water Hotline may be called at: (800) 426-4791.

REFERENCE Methods for the Determination of Organic Compounds in Drinking Water, EPA/600/4-88/039 (revised July 1991; Final Rule for determination of compliance with the MCL for Total Trihalomethanes under 141.30, in 40 CFR Part 141, Vol. 58, No. 147, Fed. Reg., Tuesday Aug. 3, 1993). U.S. EPA Environmental Monitoring Systems Laboratory, Cincinnati, OH, 45268, U.S.A. Available from the National Technical Information Service (NTIS), 5285 Port Royal Road, Springfield, VA 22161; Tel. 800-553-6847. NTIS Order Number is PB91-231480.

***trans*-1,3-Dichloropropene**   **EPA Method 524**
CAS #10061-02-6

TITLE Measurement of Purgeable Organic Compounds in Water by Capillary Column GC/MS.

MATRIX Drinking water and raw source water; the latter should include most surface water and groundwater sources.

METHOD SUMMARY Method 524.2 covers 60 volatile organic compounds. An inert gas (zero grade nitrogen or helium) is bubbled through a 25-mL or a 5-mL water sample (depending on the expected concentration of the analytes). Purged sample components are trapped in a tube of sorbent materials. When purging is complete, the sorbent tube is heated and backflushed with helium to desorb the trapped sample onto a capillary GC column.

INTERFERENCES Impurities in the purge gas and from organic compounds outgassing from the plumbing ahead of the trap account for many contamination problems. Interferences purged or coextracted from the samples will vary considerably from source to source, depending upon the particular sample or extract being tested. Cross-contamination can occur whenever high-level and low-level samples are analyzed sequentially. Samples also can be contaminated by diffusion of volatile organics (particularly methylene chloride and fluorocarbons) through the septum seal into the sample during shipment and storage.

INSTRUMENTATION A GC/MS with a data system equipped with one of the following capillary GC columns:

Column 1: VOCOL glass wide bore capillary column.
Column 2: DB-624 fused silica capillary column.
Column 3: DB-5 fused silica capillary column.

Also required is an all-glass 25 mL or 5-mL purging device, a sorbent trap, and a thermal desorption apparatus which is connected to the GC/MS system.

PRECISION & ACCURACY Method detection limits are compound- and instrument-dependent, and may vary from approximately 0.02–0.35 μg/L. Note in the table below that the "true" concentration range used for accuracy and precision measurements was quite narrow. However, the applicable concentration range of this method is primarily column dependent and is approximately 0.02 to 200 μg/L for the wide-bore thick-film

columns. Narrow-bore thin-film columns may have a capacity which limits the range to about 0.02 to 20 µg/L. Analytes that are inefficiently purged from water will not be detected when present at low concentrations, but they can be measured with acceptable accuracy and precision when present in sufficient amounts.

Analytes that are not separated chromatographically, but which have different mass spectra and non-interfering quantification ions, can be identified and measured in the same calibration mixture or water sample. Analytes which have very similar mass spectra cannot be individually identified and measured in the same calibration mixture or water samples unless they have different retention times. Co-eluting compounds with very similar mass spectra, typically many structural isomers, must be reported as an isomeric group or pair.

The range (in µg/L) was not listed.
The Method Detection Limig (in µg/L) was not listed.
The accuracy (as % recovery) was not listed.
The precision (in %) was not listed.

Note: Data were obtained from 16–31 determinations using a wide-bore capillary column and a jet separator interfaced to a quadrupole mass spectrometer. All analytes were in a reagent water matrix.

SAMPLING METHOD Collect samples using a 40- to 120-mL screw-cap vial (prewashed with detergent, rinsed with distilled water and oven dried at 105°C) with a Teflon®-faced silicone septum. Collect bubble-free samples and place the septum with the Teflon® side down on the water.

SAMPLE PRESERVATION If residual chlorine is present in the water add about 25 mg of ascorbic acid to each vial before samples are collected to remove the chlorine. Add hydrochloric acid to reduce pH to <2, and immediately cool samples to 4°C, and store them in a solvent-free refrigerator at 4°C until analysis.

MHT The maximum holding time for samples is 14 days from the time they were collected.

SAMPLE PREPARATION Remove the plungers from two 25-mL (or 5-mL depending on sample size) syringes and attach a closed syringe valve to each. Warm the sample to room temperature, open the sample bottle, and carefully pour the sample into one of the syringe barrels to just short of overflowing. Replace the syringe plunger, invert the syringe, and compress the sample. Open the syringe valve and vent any residual air while adjusting the sample volume to 25.0 mL (or 5 mL). For samples and blanks, add 5 µL of the fortification solution containing the internal standard and the surrogates to the sample through the syringe valve. For calibration standards and lab fortified blanks, add 5 µL of the fortification solution containing the internal standard only. Close the valve. Fill the second syringe in an identical manner from the same sample bottle. Reserve this second syringe for a reanalysis if necessary.

QUALITY CONTROL As an initial demonstration of lab accuracy and precision, analyze 4 to 7 replicates of a lab fortified blank containing analyte at 0.2–5 µg/L. Collect all samples in duplicate. Surrogate analytes (similar to those of the analytes of interest), whose concentration is known in every sample, are measured using the same internal standard calibration procedure. Duplicate field reagent water blanks (trip blanks) must be analyzed with each set of samples, lab reagent blanks (method blanks) must be analyzed with each batch of samples processed as a group within a work shift. Also, a single lab-fortified blank that contains each of the analytes of interest should be analyzed with each batch of samples processed as a group within a work shift. A 3- to 5-point calibration curve is needed depending on the calibration range factor required.

EPA CONTACT & HOTLINE For technical questions contact Dr. Baldev Bathija, U.S. EPA, Office of Ground Water and Drinking Water (WH-550D), 401 M St. SW, Washington, DC 20460. Tel. (202) 260-3040. For further information the EPA Safe Drinking Water Hotline may be called at: (800) 426-4791.

REFERENCE Methods for the Determination of Organic Compounds in Drinking Water, EPA/600/4-88/039 (revised July 1991; Final Rule for determination of compliance with the MCL for Total Trihalomethanes under 141.30, in 40 CFR Part 141, Vol. 58, No. 147, Fed. Reg., Tuesday Aug. 3, 1993). U.S. EPA Environmental Monitoring Systems Laboratory, Cincinnati, OH, 45268, U.S.A. Available from the National Technical Information Service (NTIS), 5285 Port Royal Road, Springfield, VA 22161; Tel. 800-553-6847. NTIS Order Number is PB91-231480.

1,1-Dichloropropene   **EPA Method 8021**
CAS #563-58-6

TITLE Halogenated Volatile by Gas Chromatography Using Photoionization and Electrolytic Conductivity Detectors in Series: Capillary Column Technique

MATRIX This method is applicable to nearly all types of samples, regardless of water content, including groundwater, aqueous sludges, caustic liquors, acid liquors, waste solvents, oily wastes, mousses, tars, fibrous wastes, polymeric emulsions, filter cakes, spent carbons, spent catalysts, soils, and sediments.

METHOD SUMMARY This method is used to determine 60 volatile organic compounds in a variety of solid waste matrices. It provides GC conditions for the detection of halogenated and aromatic volatile organic compounds. Samples can be analyzed using direct injection or purge-and-trap (EPA Method 5030). Groundwater samples must be analyzed using EPA Method 5030 (where applicable). A temperature program is used with the GC. Detection is achieved by a photoionization detector (PID) and a Hall electrolytic conductivity detector (HECD) in series.

INTERFERENCES Samples can be contaminated by diffusion of volatile organics (particularly chlorofluorocarbons and methylene chloride) through the sample container septum during shipment and storage.

INSTRUMENTATION A GC-equipped with variable-constant differential flow controllers, subambient oven controller, PID and HECD detectors connected with a short piece of uncoated capillary tubing and a data system.

Column: 60 m × 0.75 mm I.D. VOCOL wide-bore capillary column with 1.5 μm film thickness.

PRECISION & ACCURACY MDLs are compound-dependent and vary with purging efficiency and concentration. The applicable concentration range of this method is compound- and instrument-dependent but is approximately 0.1 to 200 μg/L. Analytes that are inefficiently purged from water will not be detected when present at low concentrations, but they can be measured with acceptable accuracy and precision when present in sufficient amounts. The estimated quantitation limit (EQL) for an individual compound is approximately 1 μg/kg (wet weight) for soil/sediment samples, 100 μg/kg (wet weight) for wastes, and 1 μg/L for groundwater. EQLs will be proportionately higher for sample extracts and samples that require dilution or reduced sample size to avoid saturation of the detector.

MULTIPLICATION FACTORS FOR OTHER MATRICES (a)

Matrix	Factor (b)
Groundwater	10
Low-concentration soil	10
Water miscible liquid waste	500
High-concentration soil and sludge	1250
Non-water miscible waste	1250

(a) Sample EQLs are highly matrix-dependent. The EQLs listed herein are provided for guidance and may not always be achievable. (b) EQL = [Method detection limit] × [Factor]. For non-aqueous samples, the factor is on a wet-weight basis.

SINGLE LABORATORY ACCURACY & PRECISION DATA FOR VOCs IN WATER

This method was tested in a single lab using water spiked at 10 μg/L and the following data was reported:

Recoveries and standard deviations were determined from seven samples and spiked at 10 μg/L of each analyte. Recoveries were determined by the internal standard method. Internal standards were: Fluorobenzene for PID and 2-Bromo-1-chloropropane for HECD.

The average recovery (in percent) for the PID was 103.
The standard deviation of the recovery for the PID was 3.6.
The MDL (in μg/mL) for the PID was 0.02.
The average recovery (in percent) for the HECD was 103.
The standard deviation of the recovery for the HECD was 3.4.
The MDL (in μg/mL) for the HECD was 0.02.

SAMPLE COLLECTION, PRESERVATION & HANDLING
Volatile Organics — Standard 40-mL glass screw-cap VOA vials with Teflon®-faced silicone septum may be used for both liquid and solid matrices. When collecting samples, liquids and solids should be introduced into the vials gently to reduce agitation which might drive off volatile compounds. If there are any air bubbles present the sample must be retaken. Tap slightly as they are filled to try and eliminate as much free air space as possible. The two vials from each sampling locations should be sealed in separate plastic bags to prevent cross-contamination between samples particularly if the sampled waste is suspected of containing high levels of volatile organics.

Semivolatile organics — Containers used to collect samples for the determination of semivolatile organic compounds should be soap and water washed followed by methanol (or isopropanol) rinsing. The sample containers should be of glass or Teflon® and have screw-top covers with Teflon® liners.

Preservation for volatile organics — No preservation is used with concentrated waste samples. With liquid samples containing no residual chlorine, 4 drops of concentrated hydrochloric acid are added and the samples are immediately cooled to 4°C. When liquid samples contain residual chlorine, they are treated as above and, in addition, 4 drops of 4% aqueous sodium thiosulfate are added. Soil, sediment, and sludge samples are only cooled to 4°C.

Preservation for semivolatile organics — No preservation is used with concentrated waste samples. With liquid samples containing no residual chlorine and with soil, sediment, and sludge samples, immediately cooling to 4°C is the only preservation used. When residual chlorine is present then 3 mL of 10% aqueous sodium sulfate is added for each gallon of sample collected, followed by cooling to 4°C.

MHT The holding time for all volatile organics samples is 14 days. Liquid samples must be extracted within 7 days and their extracts analyzed within 40 days. Concentrated waste, soil, sediment, and sludge samples must be extracted within 14 days and their extracts analyzed within 40 days.

SAMPLE PREPARATION Volatile compounds are introduced into the gas chromatograph either by direct injector or purge-and-trap (EPA Method 5030). EPA Method 5030 may be used directly on groundwater samples or low-concentration contaminated soils and sediments. For medium-concentration soils or sediments, methanolic extraction, as described in EPA Method 5030, may be necessary prior to purge-and-trap analysis.

QUALITY CONTROL Calculate surrogate standard recovery on all samples, blanks, and spikes. A trip blank is recommended to check on sampling, storage, and handling contamination. Calibration standards, at a minimum of five concentration levels, are prepared in organic-free reagent water. One of the concentration levels should be at a concentration near, but above, the method detection limit.

A combination of bromochloromethane, 2-bromo-1-chloropropane, 1,4-dichlorobutane, and bromochlorobenzene are recommended as surrogate standards to encompass the range of the temperature program used in this method.

REFERENCE Test Methods for Evaluating Solid Waste, Physical/Chemical Methods, SW-846, 3rd Edition, U.S. EPA, Office of Solid Waste, Washington, DC, EPA Method 8021A, Rev. 1, Nov. 1992.

---

cis-1,3-Dichloropropene   EPA Method 8021
CAS #10061-01-5

TITLE Halogenated Volatile by Gas Chromatography Using Photoionization and Electrolytic Conductivity Detectors in Series: Capillary Column Technique

MATRIX This method is applicable to nearly all types of samples, regardless of water content, including groundwater, aqueous sludges, caustic liquors, acid liquors, waste solvents, oily wastes, mousses, tars, fibrous wastes, polymeric emulsions, filter cakes, spent carbons, spent catalysts, soils, and sediments.

METHOD SUMMARY This method is used to determine 60 volatile organic compounds in a variety of solid waste matrices. It provides GC conditions for the detection of halogenated and aromatic volatile organic compounds. Samples can be analyzed using direct injection or purge-and-trap (EPA Method 5030). Groundwater samples must be analyzed using EPA Method 5030 (where applicable). A temperature program is used with the GC. Detection is achieved by a photoionization detector (PID) and a Hall electrolytic conductivity detector (HECD) in series.

INTERFERENCES Samples can be contaminated by diffusion of volatile organics (particularly chlorofluorocarbons and methylene chloride) through the sample container septum during shipment and storage.

INSTRUMENTATION A GC-equipped with variable-constant differential flow controllers, subambient oven controller, PID and HECD detectors connected with a short piece of uncoated capillary tubing and a data system.

Column: 60 m × 0.75 mm I.D. VOCOL wide-bore capillary column with 1.5 µm film thickness.

PRECISION & ACCURACY MDLs are compound-dependent and vary with purging efficiency and concentration. The applicable concentration range of this method is compound- and instrument-dependent but is approximately 0.1 to 200 µg/L. Analytes that are inefficiently purged from water will not be detected when present at low concentrations, but they can be measured with acceptable accuracy and precision when present in sufficient amounts. The estimated quantitation limit (EQL) for an individual compound is approximately 1 µg/kg (wet weight) for soil/sediment samples, 100 µg/kg (wet weight) for wastes, and 1 µg/L for groundwater. EQLs will be proportionately higher for sample extracts and samples that require dilution or reduced sample size to avoid saturation of the detector.

MULTIPLICATION FACTORS FOR OTHER MATRICES (a)

Matrix	Factor (b)
Groundwater	10
Low-concentration soil	10
Water miscible liquid waste	500
High-concentration soil and sludge	1250
Non-water miscible waste	1250

(a) Sample EQLs are highly matrix-dependent. The EQLs listed herein are provided for guidance and may not always be achievable. (b) EQL = [Method detection limit] × [Factor]. For non-aqueous samples, the factor is on a wet-weight basis.

SINGLE LABORATORY ACCURACY & PRECISION DATA FOR VOCs IN WATER

This method was tested in a single lab using water spiked at 10 µg/L and the following data was reported:

Recoveries and standard deviations were determined from seven samples and spiked at 10 µg/L of each analyte. Recoveries were determined by the internal standard method. Internal standards were: Fluorobenzene for PID and 2-Bromo-1-chloropropane for HECD.

The average recovery (in percent) for the PID was none (no response for this detector).
The standard deviation of the recovery for the PID was none (no response for this detector).
The MDL (in µg/mL) for the PID was none (no response for this detector).
The average recovery (in percent) for the HECD was none (no response for this detector).
The standard deviation of the recovery for the HECD was none (no response for this detector)-.
The MDL (in µg/mL) for the HECD was none (no response for this detector).

SAMPLE COLLECTION, PRESERVATION & HANDLING
Volatile Organics — Standard 40-mL glass screw-cap VOA vials with Teflon®-faced silicone septum may be used for both liquid and solid matrices. When collecting samples, liquids and solids should be introduced into the vials gently to reduce agitation which might drive off volatile compounds. If there are any air bubbles present the sample must be retaken. Tap slightly as they are filled to try and eliminate as much free air space as possible. The two vials from each sampling locations should be sealed in separate plastic bags to prevent cross-contamination between samples particularly if the sampled waste is suspected of containing high levels of volatile organics.

Semivolatile organics — Containers used to collect samples for the determination of semivolatile organic compounds should be soap and water washed followed by methanol (or isopropanol) rinsing. The sample containers should be of glass or Teflon® and have screw-top covers with Teflon® liners.

Preservation for volatile organics — No preservation is used with concentrated waste samples. With liquid samples containing no residual chlorine, 4 drops of concentrated hydrochloric acid are added and the samples are immediately cooled to 4°C. When liquid samples contain residual chlorine, they are treated as above and, in addition, 4 drops of 4% aqueous sodium thiosulfate are added. Soil, sediment, and sludge samples are only cooled to 4°C.

Preservation for semivolatile organics — No preservation is used with concentrated waste samples. With liquid samples containing no residual chlorine and with soil, sediment, and sludge samples, immediately cooling to 4°C is the only preservation used. When residual chlorine is present then 3 mL of 10% aqueous sodium sulfate is added for each gallon of sample collected, followed by cooling to 4°C.

MHT The holding time for all volatile organics samples is 14 days. Liquid samples must be extracted within 7 days and their extracts analyzed within 40 days. Concentrated waste, soil, sediment, and sludge samples must be extracted within 14 days and their extracts analyzed within 40 days.

SAMPLE PREPARATION Volatile compounds are introduced into the gas chromatograph either by direct injector or purge-and-trap (EPA Method 5030). EPA Method 5030 may be used directly on groundwater samples or low-concentration contaminated soils and sediments. For medium-concentration soils or sediments, methanolic extraction, as described in EPA Method 5030, may be necessary prior to purge-and-trap analysis.

QUALITY CONTROL Calculate surrogate standard recovery on all samples, blanks, and spikes. A trip blank is recommended to check on sampling, storage, and handling contamination. Calibration standards, at a minimum of five concentration levels, are prepared in organic-free reagent water. One of the concentration levels should be at a concentration near, but above, the method detection limit.

A combination of bromochloromethane, 2-bromo-1-chloropropane, 1,4-dichlorobutane, and bromochlorobenzene are recommended as surrogate standards to encompass the range of the temperature program used in this method.

REFERENCE Test Methods for Evaluating Solid Waste, Physical/Chemical Methods, SW-846, 3rd Edition, U.S. EPA, Office of Solid Waste, Washington, DC, EPA Method 8021A, Rev. 1, Nov. 1992.

---

trans-1,3-Dichloropropene   **EPA Method 8021**
CAS #10061-02-6

TITLE Halogenated Volatile by Gas Chromatography Using Photoionization and Electrolytic Conductivity Detectors in Series: Capillary Column Technique

MATRIX This method is applicable to nearly all types of samples, regardless of water content, including groundwater, aqueous sludges, caustic liquors, acid liquors, waste solvents, oily wastes, mousses, tars, fibrous wastes, polymeric emulsions, filter cakes, spent carbons, spent catalysts, soils, and sediments.

METHOD SUMMARY This method is used to determine 60 volatile organic compounds in a variety of solid waste matrices. It provides GC conditions for the detection of halogenated and aromatic volatile organic compounds. Samples can be analyzed using direct injection or purge-and-trap (EPA Method 5030). Groundwater samples must be analyzed using EPA Method 5030 (where applicable). A temperature program is used with the GC. Detection is achieved by a photoionization detector (PID) and a Hall electrolytic conductivity detector (HECD) in series.

INTERFERENCES Samples can be contaminated by diffusion of volatile organics (particularly chlorofluorocarbons and methylene chloride) through the sample container septum during shipment and storage.

INSTRUMENTATION A GC-equipped with variable-constant differential flow controllers, subambient oven controller, PID and HECD detectors connected with a short piece of uncoated capillary tubing and a data system.

Column: 60 m × 0.75 mm I.D. VOCOL wide-bore capillary column with 1.5 µm film thickness.

PRECISION & ACCURACY MDLs are compound-dependent and vary with purging efficiency and concentration. The applicable concentration range of this method is compound- and instrument-dependent but is approximately 0.1 to 200 µg/L. Analytes that are inefficiently purged from water will not be detected when present at low concentrations, but they can be measured with acceptable accuracy and precision when present in sufficient amounts. The estimated quantitation limit (EQL) for an individual compound is approximately 1 µg/kg (wet weight) for soil/sediment samples, 100 µg/kg (wet weight) for wastes, and 1 µg/L for groundwater. EQLs will be proportionately higher for sample extracts and samples that require dilution or reduced sample size to avoid saturation of the detector.

MULTIPLICATION FACTORS FOR OTHER MATRICES (a)

Matrix	Factor (b)
Groundwater	10
Low-concentration soil	10
Water miscible liquid waste	500
High-concentration soil and sludge	1250
Non-water miscible waste	1250

(a) Sample EQLs are highly matrix-dependent. The EQLs listed herein are provided for guidance and may not always be achievable. (b) EQL = [Method detection limit] × [Factor]. For non-aqueous samples, the factor is on a wet-weight basis.

SINGLE LABORATORY ACCURACY & PRECISION DATA FOR VOCs IN WATER

This method was tested in a single lab using water spiked at 10 µg/L and the following data was reported:

Recoveries and standard deviations were determined from seven samples and spiked at 10 µg/L of each analyte. Recoveries were determined by the internal standard method. Internal standards were: Fluorobenzene for PID and 2-Bromo-1-chloropropane for HECD.

The average recovery (in percent) for the PID was none (no response for this detector).
The standard deviation of the recovery for the PID was none (no response for this detector).
The MDL (in µg/mL) for the PID was none (no response for this detector).
The average recovery (in percent) for the HECD was none (no response for this detector).
The standard deviation of the recovery for the HECD was none (no response for this detector).
The MDL (in µg/mL) for the HECD was none (no response for this detector).

SAMPLE COLLECTION, PRESERVATION & HANDLING
Volatile Organics — Standard 40-mL glass screw-cap VOA vials with Teflon®-faced silicone septum may be used for both liquid and solid matrices. When collecting samples, liquids and solids should be introduced into the vials gently to reduce agitation which might drive off volatile compounds. If there are any air bubbles present the sample must be retaken. Tap slightly as they are filled to try and eliminate as much free air space as possible. The two vials from each sampling locations should be sealed in separate plastic bags to prevent cross-contamination

between samples particularly if the sampled waste is suspected of containing high levels of volatile organics.

Semivolatile organics — Containers used to collect samples for the determination of semivolatile organic compounds should be soap and water washed followed by methanol (or isopropanol) rinsing. The sample containers should be of glass or Teflon® and have screw-top covers with Teflon® liners.

Preservation for volatile organics — No preservation is used with concentrated waste samples. With liquid samples containing no residual chlorine, 4 drops of concentrated hydrochloric acid are added and the samples are immediately cooled to 4°C. When liquid samples contain residual chlorine, they are treated as above and, in addition, 4 drops of 4% aqueous sodium thiosulfate are added. Soil, sediment, and sludge samples are only cooled to 4°C.

Preservation for semivolatile organics — No preservation is used with concentrated waste samples. With liquid samples containing no residual chlorine and with soil, sediment, and sludge samples, immediately cooling to 4°C is the only preservation used. When residual chlorine is present then 3 mL of 10% aqueous sodium sulfate is added for each gallon of sample collected, followed by cooling to 4°C.

MHT The holding time for all volatile organics samples is 14 days. Liquid samples must be extracted within 7 days and their extracts analyzed within 40 days. Concentrated waste, soil, sediment, and sludge samples must be extracted within 14 days and their extracts analyzed within 40 days.

SAMPLE PREPARATION Volatile compounds are introduced into the gas chromatograph either by direct injector or purge-and-trap (EPA Method 5030). EPA Method 5030 may be used directly on groundwater samples or low-concentration contaminated soils and sediments. For medium-concentration soils or sediments, methanolic extraction, as described in EPA Method 5030, may be necessary prior to purge-and-trap analysis.

QUALITY CONTROL Calculate surrogate standard recovery on all samples, blanks, and spikes. A trip blank is recommended to check on sampling, storage, and handling contamination. Calibration standards, at a minimum of five concentration levels, are prepared in organic-free reagent water. One of the concentration levels should be at a concentration near, but above, the method detection limit.

A combination of bromochloromethane, 2-bromo-1-chloropropane, 1,4-dichlorobutane, and bromochlorobenzene are recommended as surrogate standards to encompass the range of the temperature program used in this method.

REFERENCE Test Methods for Evaluating Solid Waste, Physical/Chemical Methods, SW-846, 3rd Edition, U.S. EPA, Office of Solid Waste, Washington, DC, EPA Method 8021A, Rev. 1, Nov. 1992.

---

cis-1,3-Dichloropropene **EPA Method 8240**
CAS #10061-01-5

TITLE Volatile Organics By GC/MS: Packed Column Technique

MATRIX Nearly all types of sample matarices, regardless of water content, can be analyzed using this method. This includes groundwater, aqueous sludges, caustic liquors, acid liquors, waste solvents, oily wastes, mousses, tars, fibrous wastes, polymetric emulsions, filter cakes, spent carbons, spent catalysts, soils, and sediments.

METHOD SUMMARY Method 8240B covers 80 volatile organic compounds that are introduced into a gas chromatograph by the purge-and-trap method or by direct injection (in limited applications). For the purge-and-trap method an inert gas (zero grade nitrogen or helium) is bubbled through a 5-mL solution at ambient temperature. Purged sample components are trapped in a tube of sorbent materials. When purging is complete, the sorbent tube is heated and backflushed with inert gas to desorb the trapped components onto a GC column.

INTERFERENCES Impurities in the purge gas and from organic compounds outgassing from the plumbing ahead of the trap account for many contamination problems. Interferences purged or coextracted from the samples will vary considerably from source to source. Cross-contamination can occur whenever high-level and low-level samples are analyzed sequentially. Whenever an unusually concentrated sample is analyzed, it should be followed by the analysis of organic-free reagent water to check for cross-contamination. Samples also can be contaminated by diffusion of volatile organics (particularly methylene chloride and fluorocarbons) through the septum seal into the sample during shipment and storage. A trip blank can serve as a check on such contamination. The lab where volatile analysis is performed and also the refrigerated storage area should be completely free of solvents.

INSTRUMENTATION A gas chromatograph/mass spectrometry/data system (GC/MS) equipped with a 6 ft × 0.1 in I.D. glass column packed with 1% SP-1000 on Carbopack-B (60/80 mesh) is required. Also needed is a 5-mL purging device, a sorbent trap, and a thermal desorption apparatus.

PRECISION & ACCURACY This method is reported to have been tested by 15 laboratories using organic-free reagent water, drinking water, surface water, and industrial wastewaters (not specified) fortified at six concentrations over the range 5–600 µg/L.

Sample estimated quantitation limits (EQLs) are highly matrix-dependent. The EQLs listed may not always be achievable. EQLs listed for soils or sediments are based on wet weight. Normally, data is reported on a dry-weight basis; therefore, EQLs will be higher, based on the percent dry weight of each sample. Note that EQLs are even more variable than MDLs and that they are highly variable depending on the matrix being analyzed.

EQL in groundwater in µg/L was 5.
EQL in low soil or sediment in µg/kg was 5.
Accuracy (a) in µg/L was 1.00C.
Precision (b) in µg/L was 0.52x.

(a) Average recovery found for measurements of samples containing a concentration of C, in µg/L.

(b) Overall precision found for measurements of samples with average recovery X for samples containing a concentration of C in µg/L.

X = *Average recovery found for measurement of samples containing a concentration of C in µg/L.*

MULTIPLICATION FACTORS FOR OTHER MATRICES

Other Matrices	Factor (a)
Waste miscible liquid waste	50
High-concentration soil and sludge	125
Non-water miscible waste	500

(a) EQL = [EQL for low soil sediment] × [Factor]. For non-aqueous samples, the factor is on a wet-weight basis.

SAMPLING METHOD

Liquid samples — Use a 40-mL glass screw-cap VOA vial with a Teflon®-faced silicone septum that has been prewashed, rinsed with distilled deionized water, and oven dried. However, if residual chlorine is present, collect sample in a 40-oz. soil VOA container which has been pre-preserved with 4 drops of 10% sodium thiosulfate, mix gently, and then transfer the sample to a 40-mL VOA vial. Collect bubble-free samples in duplicate and seal them in separate plastic bags.

Soils or sediments, and sludges — Use an 8-oz. widemouth glass bottle with a Teflon®-faced silicone septum that has been prewashed with detergent, rinsed with distilled deionized water, and oven dried. Tap slightly to eliminate free air space. Collect samples in duplicate and seal them in separate plastic bags.

SAMPLE PRESERVATION

Liquid samples — Add 4 drops of concentrated HCL and immediately cool samples to 4°C and store in a solvent-free refrigerator.

Soils or sediments, and sludges — Cool samples to 4°C and store in a solvent-free refrigerator.

MHT Maximum holding time is 14 days from the date of sample collection.

SAMPLE PREPARATION

Liquid samples — Remove the plunger from a 5-mL syringe and carefully pour the sample into the syringe barrel to just short of overflowing. Replace the syringe plunger and compress the sample. Open the syringe valve and vent any residual air while adjusting the sample volume to 5.0 mL. If there is only one volatile organic analysis (VOA) vial, a second syringe should be filled at this time to protect against possible loss of sample integrity. Add 10 µL of surrogate spiking solution and 10 µL of internal standard spiking solution through the valve bore of the 5-mL syringe, then close the valve. The surrogate and internal standards may be mixed and added as a single spiking solution.

Sediments, soils, and waste samples — All samples of this type should be screened by GC analysis using a headspace method (EPA Method 3810) or the hexadecane extraction and screening method (EPA Method 3820). Use the screening data to determine whether to use the low-concentration method (0.005–1 mg/kg) or the high-concentration method (>1 mg/kg).

Low-concentration method — The low-concentration method is based on purging a heated sediment or soil sample mixed with organic-free reagent water containing the surrogate and internal standards. Analyze all reagent blanks and standards under the same conditions as the samples.

Use a 5-g sample if the expected concentration is <0.1 mg/kg or a 1-g sample for expected concentrations between 0.1 and 1 mg/kg. Mix the contents of the sample container with a narrow metal spatula. Weigh the amount of the sample into a tared purge device. Add the spiked water to the purge device, which contains the weighed amount of sample, and connect the device to the purge-and-trap system.

High-concentration method — This method is based on extracting the sediment or soil with methanol. A waste sample is either extracted or diluted, depending on its solubility in methanol. Wastes that are insoluble in methanol are diluted with reagent tetraglyme or possibly polyethylene glycol (PEG). An aliquot of the extract is added to organic-free reagent water containing surrogate and internal standards. This is purged at ambient temperature. All samples with an expected concentration of >1.0 mg/kg should be analyzed by this method.

Mix the contents of the sample container with a narrow metal spatula. For sediments or soils and solid wastes that are insoluble in methanol, weigh 4 g (wet weight) of sample into a tared 20-mL vial. For waste that is soluble in methanol, tetraglyme, or PEG, weigh 1 g (wet weight) into a tared scintillation vial or culture tube or a 10-mL volumetric flask. Quickly add 9.0 mL of appropriate solvent then add 1.0 mL of a surrogate spiking solution to the vial, cap it, and shake it for 2 min.

METHANOL EXTRACT REQUIRED FOR ANALYSIS OF HIGH-CONCENTRATION SOILS OR SEDIMENTS

Approximate Concentration Range	Volume of Methanol Extract (a)
500–10,000 µg/kg	100 µL
1,000–20,000 µg/kg	50 µL
5,000–100,000 µg/kg	10 µL
25,000–500,000 µg/kg	100 µL of 1/50 dilution (b)

Calculate appropriate dilution factor for concentrations exceeding this table.

(a) The volume of methanol added to 5 mL of water being purged should be kept constant. Therefore, add to the 5-mL syringe whatever volume of methanol is necessary to maintain a volume of 100 µL added to the syringe.

(b) Dilute an aliquot of the methanol extract and then take 100 µL for analysis.

QUALITY CONTROL Demonstrate, through the analysis of a reagent water blank, that interferences from the analytical system, glassware, and reagents are under control. Blank samples should be carried through all stages of the sample preparation and measurement steps. For each analytical batch (up to 20 samples), a reagent blank, matrix spike, and matrix spike duplicate must be analyzed (the frequency of the spikes may be different for different monitoring programs). The blank and spiked samples must be carried through all stages of the sample preparation and measurement steps. QC samples mentioned in the section on Interferences will also be needed as appropriate to those situations.

REFERENCE Test Methods for Evaluating Solid Waste (SW-846). U.S. EPA. 1983. Method 8240B, Rev. 2, Nov. 1990. Office of Solid Wastes, Washington, DC.

---

trans-1,3-Dichloropropene
CAS #10061-02-6
EPA Method 8240

TITLE Volatile Organics By GC/MS: Packed Column Technique

MATRIX Nearly all types of sample matarices, regardless of water content, can be analyzed using this method. This includes groundwater, aqueous sludges, caustic liquors, acid liquors, waste solvents, oily wastes, mousses, tars, fibrous wastes, polymetric emulsions, filter cakes, spent carbons, spent catalysts, soils, and sediments.

METHOD SUMMARY Method 8240B covers 80 volatile organic compounds that are introduced into a gas chromatograph by the purge-and-trap method or by direct injection (in limited applications). For the purge-and-trap method an inert gas (zero grade nitrogen or helium) is bubbled through a 5-mL solution at ambient temperature. Purged sample components are trapped in a tube of sorbent materials. When purging is complete, the sorbent tube is heated and backflushed with inert gas to desorb the trapped components onto a GC column.

INTERFERENCES Impurities in the purge gas and from organic compounds outgassing from the plumbing ahead of the trap account for many contamination problems. Interferences purged or coextracted from the samples will vary considerably from source to source. Cross-contamination can occur whenever high-level and low-level samples are analyzed sequentially. Whenever an unusually concentrated sample is analyzed, it should be followed by the analysis of organic-free reagent water to check for cross-contamination. Samples also can be contaminated by diffusion of volatile organics (particularly methylene chloride and fluorocarbons) through the septum seal into the sample during shipment and storage. A trip blank can serve as a check on such contamination. The lab where volatile analysis is performed and also the refrigerated storage area should be completely free of solvents.

INSTRUMENTATION A gas chromatograph/mass spectrometry/data system (GC/MS) equipped with a 6 ft × 0.1 in I.D. glass column packed with 1% SP-1000 on Carbopack-B (60/80 mesh) is required. Also needed is a 5-mL purging device, a sorbent trap, and a thermal desorption apparatus.

PRECISION & ACCURACY This method is reported to have been tested by 15 laboratories using organic-free reagent water, drinking water, surface water, and industrial wastewaters (not specified) fortified at six concentrations over the range 5–600 µg/L.

Sample estimated quantitation limits (EQLs) are highly matrix-dependent. The EQLs listed may not always be achievable. EQLs listed for soils or sediments are based on wet weight. Normally, data is reported on a dry-weight basis; therefore, EQLs will be higher, based on the percent dry weight of each sample. Note that EQLs are even more variable than MDLs and that they are highly variable depending on the matrix being analyzed.

EQL in groundwater in µg/L was 5.
EQL in low soil or sediment in µg/kg was 5.
Accuracy (a) in µg/L was 1.00C.
Precision (b) in µg/L was 0.34x.

(a) *Average recovery found for measurements of samples containing a concentration of C, in µg/L.*
(b) *Overall precision found for measurements of samples with average recovery X for samples containing a concentration of C in µg/L.*
X = *Average recovery found for measurement of samples containing a concentration of C in µg/L.*

MULTIPLICATION FACTORS FOR OTHER MATRICES

Other Matrices	Factor (a)
Waste miscible liquid waste	50
High-concentration soil and sludge	125
Non-water miscible waste	500

(a) *EQL = [EQL for low soil sediment] × [Factor]. For non-aqueous samples, the factor is on a wet-weight basis.*

SAMPLING METHOD
Liquid samples — Use a 40-mL glass screw-cap VOA vial with a Teflon®-faced silicone septum that has been prewashed, rinsed with distilled deionized water, and oven dried. However, if residual chlorine is present, collect sample in a 40-oz. soil VOA container which has been pre-preserved with 4 drops of 10% sodium thiosulfate, mix gently, and then transfer the sample to a 40-mL VOA vial. Collect bubble-free samples in duplicate and seal them in separate plastic bags.

Soils or sediments, and sludges — Use an 8-oz. widemouth glass bottle with a Teflon®-faced silicone septum that has been prewashed with detergent, rinsed with distilled deionized water, and oven dried. Tap slightly to eliminate free air space. Collect samples in duplicate and seal them in separate plastic bags.

SAMPLE PRESERVATION
Liquid samples — Add 4 drops of concentrated HCL and immediately cool samples to 4°C and store in a solvent-free refrigerator.

Soils or sediments, and sludges — Cool samples to 4°C and store in a solvent-free refrigerator.

MHT Maximum holding time is 14 days from the date of sample collection.

SAMPLE PREPARATION
Liquid samples — Remove the plunger from a 5-mL syringe and carefully pour the sample into the syringe barrel to just short of overflowing. Replace the syringe plunger and compress the sample. Open the syringe valve and vent any residual air while adjusting the sample volume to 5.0 mL. If there is only

one volatile organic analysis (VOA) vial, a second syringe should be filled at this time to protect against possible loss of sample integrity. Add 10 μL of surrogate spiking solution and 10 μL of internal standard spiking solution through the valve bore of the 5-mL syringe, then close the valve. The surrogate and internal standards may be mixed and added as a single spiking solution.

Sediments, soils, and waste samples — All samples of this type should be screened by GC analysis using a headspace method (EPA Method 3810) or the hexadecane extraction and screening method (EPA Method 3820). Use the screening data to determine whether to use the low-concentration method (0.005–1 mg/kg) or the high-concentration method (>1 mg/kg).

Low-concentration method — The low-concentration method is based on purging a heated sediment or soil sample mixed with organic-free reagent water containing the surrogate and internal standards. Analyze all reagent blanks and standards under the same conditions as the samples.

Use a 5-g sample if the expected concentration is <0.1 mg/kg or a 1-g sample for expected concentrations between 0.1 and 1 mg/kg. Mix the contents of the sample container with a narrow metal spatula. Weigh the amount of the sample into a tared purge device. Add the spiked water to the purge device, which contains the weighed amount of sample, and connect the device to the purge-and-trap system.

High-concentration method — This method is based on extracting the sediment or soil with methanol. A waste sample is either extracted or diluted, depending on its solubility in methanol. Wastes that are insoluble in methanol are diluted with reagent tetraglyme or possibly polyethylene glycol (PEG). An aliquot of the extract is added to organic-free reagent water containing surrogate and internal standards. This is purged at ambient temperature. All samples with an expected concentration of >1.0 mg/kg should be analyzed by this method.

Mix the contents of the sample container with a narrow metal spatula. For sediments or soils and solid wastes that are insoluble in methanol, weigh 4 g (wet weight) of sample into a tared 20-mL vial. For waste that is soluble in methanol, tetraglyme, or PEG, weigh 1 g (wet weight) into a tared scintillation vial or culture tube or a 10-mL volumetric flask. Quickly add 9.0 mL of appropriate solvent then add 1.0 mL of a surrogate spiking solution to the vial, cap it, and shake it for 2 min.

METHANOL EXTRACT REQUIRED FOR ANALYSIS OF HIGH-CONCENTRATION SOILS OR SEDIMENTS

Approximate Concentration Range	Volume of Methanol Extract (a)
500–10,000 μg/kg	100 μL
1,000–20,000 μg/kg	50 μL
5,000–100,000 μg/kg	10 μL
25,000–500,000 μg/kg	100 μL of 1/50 dilution (b)

Calculate appropriate dilution factor for concentrations exceeding this table.

(a) The volume of methanol added to 5 mL of water being purged should be kept constant. Therefore, add to the 5-mL syringe whatever volume of methanol is necessary to maintain a volume of 100 μL added to the syringe.
(b) Dilute an aliquot of the methanol extract and then take 100 μL for analysis.

QUALITY CONTROL Demonstrate, through the analysis of a reagent water blank, that interferences from the analytical system, glassware, and reagents are under control. Blank samples should be carried through all stages of the sample preparation and measurement steps. For each analytical batch (up to 20 samples), a reagent blank, matrix spike, and matrix spike duplicate must be analyzed (the frequency of the spikes may be different for different monitoring programs). The blank and spiked samples must be carried through all stages of the sample preparation and measurement steps. QC samples mentioned in the section on Interferences will also be needed as appropriate to those situations.

REFERENCE Test Methods for Evaluating Solid Waste (SW-846). U.S. EPA. 1983. Method 8240B, Rev. 2, Nov. 1990. Office of Solid Wastes, Washington, DC.

1,1-Dichloropropene   **EPA Method 8260**
CAS #563-58-6

TITLE Volatile Organic Compounds by GC/MS: Capillary Column Technique

MATRIX This method is applicable to nearly all types of samples, regardless of water content, including groundwater, soils, and sediments.

METHOD SUMMARY Method 8260A covers 58 volatile organic compounds that are introduced into a gas chromatograph by the purge-and-trap method or by direct injection (in limited applications). Zero-grade helium is bubbled through a 5-mL solution at ambient temperature. Purged sample components are trapped in a tube containing suitable sorbent materials. When purging is complete, the sorbent tube is heated and backflushed with helium to desorb trapped sample components. The analytes are desorbed directly to a large bore capillary or cryofocussed on a capillary precolumn before being flash evaporated to a narrow bore capillary for analysis.

INTERFERENCES Major contaminant sources are volatile materials in the lab and impurities in the inert purging gas and in the sorbent trap. Interfering contamination may occur when a sample containing low concentrations of volatile organic compounds is analyzed immediately after a sample containing high concentrations of volatile organic compounds. After analysis of a sample containing high concentrations of volatile organic compounds, one or more calibration blanks should be analyzed to check for cross-contamination. Screening of the samples prior to purge-and-trap GC/MS analysis is highly recommended to prevent contamination of the system. This is especially true for soil and waste samples.

Special precautions must be taken to analyze for methylene chloride. The analytical and sample storage area should be isolated from all atmospheric sources of methylene chloride. All gas chromatography carrier gas lines and purge gas plumbing should be constructed from stainless steel or copper tubing. Laboratory clothing previously exposed to methylene chloride fumes during liquid-liquid extraction procedures can contribute to sample contamination.

Samples can also be contaminated by diffusion of volatile organics (particularly methylene chloride and fluorocarbons) through the septum seal during shipment and storage. A trip blank can serve as a check on such contamination.

INSTRUMENTATION GC/MS with a temperature-programmable chromatograph suitable for splitless injection equipped with variable constant differential flow controllers, a subambient oven controller, a purging device, sorbent trap, a thermal desorption apparatus and a capillary precolumn interface when using cryogenic cooling will be needed. The following GC columns may be used:

Column 1: 60 m × 0.75mm I.D. capillary column coated with VOCOL, 1.5 µm film thickness.
Column 2: 30 m × 0.53mm capillary column coated with DB-624 or VOCOL, 3 µm film thickness.
Column 3: 30 m × 0.32mm I.D. capillary column coated with DB-5 or SE-54, 1-µm film thickness.

PRECISION & ACCURACY This method has been tested in a single lab using spiked water. Using a wide-bore capillary column, water was spiked at concentrations between 0.5 and 10 µg/L. Single lab accuracy and precision data are presented. The MDL actually achieved in a given analysis will vary depending on instrument sensitivity and matrix effects.

The MDL (a) in µg/L was 0.10.
The concentration range in µg/L was 0.5–10.
The mean accuracy (% of true value) was 98.
The precision as relative standard deviation was 8.9.

Note: The MDL is based on a 25-mL sample volume instead of a 5-mL sample volume.

SAMPLING METHOD
Liquid samples — Use a 40-mL glass screw-cap VOA vial with a Teflon®-faced silicone septum that has been prewashed, rinsed with distilled deionized water, and oven dried. If residual chlorine is present, collect the sample in a 4-oz soil VOA container which has been pre-preserved with 4 drops of 10% sodium thiosulfate. Mix gently and transfer the sample to a 40-mL VOA vial. Collect bubble-free samples in duplicate and seal each sample in a separate plastic bag.

Soils, sediments, and sludges — Use an 8-oz widemouth glass bottle with Teflon®-faced silicone septum that has been prewashed, rinsed with distilled deionized water, and oven dried. **Do not** heat the septum for more than 1 h. Tap slightly to eliminate any free air space. Collect samples in duplicate and seal each one in a separate plastic bag.

SAMPLE PRESERVATION
Liquid samples — Add 4 drops of concentrated HCL, cool to 4°C and store in a solvent-free refrigerator.

Soils, sediments and sludges — Cool samples to 4°C and store in a solvent-free refrigerator.

MHT The maximum holding time of any sample (liquids, soils, sediments, and sludges) is 14 days.

SAMPLE PREPARATION
Liquid samples — Remove the plunger from a 5-mL syringe and carefully pour the sample into the syringe barrel to just short of overflowing. Replace the syringe plunger and compress the sample. Open the syringe valve and vent any residual air while adjusting the sample volume to 5.0 mL. If there is only one volatile organic analysis (VOA) vial, a second syringe should be filled at this time to protect against possible loss of sample integrity. Add 10 µL of surrogate spiking solution and 10 µL of internal standard spiking solution through the valve bore of the 5-mL syringe, then close the valve. The surrogate and internal standards may be mixed and added as a single spiking solution.

Sediments, soils, and waste samples — All samples of this type should be screened by GC analysis using a headspace method (EPA Method 3810) or the hexadecane extraction and screening method (EPA Method 3820). Use the screening data to determine whether to use the low-concentration method (0.005–1 mg/kg) or the high-concentration method (>1 mg/kg).

Low-concentration method — The low-concentration method is based on purging a heated sediment or soil sample mixed with organic-free reagent water containing the surrogate and internal standards. Analyze all reagent blanks and standards under the same conditions as the samples.

Use a 5-g sample if the expected concentration is <0.1 mg/kg or a 1-g sample for expected concentrations between 0.1 and 1 mg/kg. Mix the contents of the sample container with a narrow metal spatula. Weigh the amount of the sample into a tared purge device. Add the spiked water to the purge device, which contains the weighed amount of sample, and connect the device to the purge-and-trap system.

High-concentration method — This method is based on extracting the sediment or soil with methanol. A waste sample is either extracted or diluted, depending on its solubility in methanol. Wastes that are insoluble in methanol are diluted with reagent tetraglyme or possibly polyethylene glycol (PEG). An aliquot of the extract is added to organic-free reagent water containing surrogate and internal standards. This is purged at ambient temperature. All samples with an expected concentration of >1.0 mg/kg should be analyzed by this method.

Mix the contents of the sample container with a narrow metal spatula. For sediments or soils and solid wastes that are insoluble in methanol, weigh 4 g (wet weight) of sample into a tared 20-mL vial. For waste that is soluble in methanol, tetraglyme, or PEG, weigh 1 g (wet weight) into a tared scintillation vial or culture tube or a 10-mL volumetric flask. Quickly add

9.0 mL of appropriate solvent then add 1.0 mL of a surrogate spiking solution to the vial, cap it, and shake it for 2 min.

METHANOL EXTRACT REQUIRED FOR ANALYSIS OF HIGH-CONCENTRATION SOILS OR SEDIMENTS

Approximate Concentration Range	Volume of Methanol Extract (a)
500–10,000 µg/kg	100 µL
1,000–20,000 µg/kg	50 µL
5,000–100,000 µg/kg	10 µL
25,000–500,000 µg/kg	100 µL of 1/50 dilution (b)

Calculate appropriate dilution factor for concentrations exceeding this table.

(a) The volume of methanol added to 5 mL of water being purged should be kept constant. Therefore, add to the 5-mL syringe whatever volume of methanol is necessary to maintain a volume of 100 µL added to the syringe.
(b) Dilute an aliquot of the methanol extract and then take 100 µL for analysis.

QUALITY CONTROL Demonstrate, through the analysis of a reagent water blank, that interferences from the analytical system, glassware, and reagents are under control. Blank samples should be carried through all stages of the sample preparation and measurement steps. For each analytical batch (up to 20 samples), a reagent blank, matrix spike, and matrix spike duplicate must be analyzed (the frequency of the spikes may be different for different monitoring programs). The blank and spiked samples must be carried through all stages of the sample preparation and measurement steps. QC samples mentioned in the section on Interferences will also be needed as appropriate to those situations.

Matrix spiking standards should be prepared from volatile organic compounds which will be representative of the compounds being investigated. The recommended internal standards are chlorobenzene-d5, 1,4-difluorobenzene, 1,4-dichlorobenzene-d4, and pentafluorobenzene. Using stock standard solutions, prepare secondary dilution standards containing the compounds of interest, either singly or mixed together in methanol. Store them in a vial with no headspace for no more than one week. Surrogates recommended are toluene-d8, 4-bromofluorobenzene, and dibromofluoromethane. Each sample undergoing GC/MS analysis must be spiked with 10 µL of the surrogate spiking solution prior to analysis.

REFERENCE Test Methods for Evaluating Solid Waste (SW-846). U.S. EPA 1983, Method 8260A, Rev. 1, Nov. 1990. Office of Solid Waste, Washington, DC.

---

cis-1,3-Dichloropropene   EPA Method 601
CAS #10061-01-5

TITLE Purgeable Halocarbons

MATRIX Wastewater

APPLICATION Method covers 29 purgeable halocarbons. (Method 624 provides GC/MS conditions appropriate for the qualitative and quantitative confirmation of results). Method describes conditions for a 2nd GC column to confirm measurements made with primary column.

INTERFERENCES Impurities in the purge gas and organic compounds outgassing from the plumbing ahead of the trap. With high- and low-level samples, there can be carryover contamination. Diffusion of volatile organics through the septum seal into the sample.

INSTRUMENTATION GC-equipped with halide-specific detector. (With purge-and-trap unit).

RANGE 8.0–500 µg/L.

MDL 0.34 µg/L.

PRECISION 0.32X µg/L (overall precision).

ACCURACY 1.00C µg/L (as recovery).

SAMPLING METHOD 25-mL glass vial. Teflon®-lined septum.

STABILITY cool, 4°C, 0.008% Sodium thiosulfate.

MHT 14 days.

QUALITY CONTROL The lab must on an ongoing basis, spike at least 10% of the samples from each sample site being monitored to assess accuracy.

REFERENCE Method 601, *Federal Register* Part VIII 40 CFR Part 136, Oct 26, 1984.

---

cis-1,3-Dichloropropene   EPA Method 624
CAS #10061-01-5

TITLE Purgeables

MATRIX Wastewater

APPLICATION Method covers 31 purgeable organics. An inert gas is bubbled through a 5-mL water sample in a specially designed purging chamber. Here, purgeables are transferred from aqueous to gaseous phase, passed onto a sorbent column, and trapped. Trap is heated and backflushed with inert gas to desorb purgeables onto a GC column, where purgeables are separated.

INTERFERENCES Impurities in the purge gas, organic compounds outgassing from the plumbing ahead of the trap, and solvent vapors in the lab. With high- and low-level samples, there can be carryover contamination.

INSTRUMENTATION GC/MS with purge-and-trap unit.

RANGE 5–600 µg/L.

MDL 5.0 µg/L

PRECISION 0.52X µg/L (overall precision).

ACCURACY 1.00C µg/L (as recovery).

SAMPLING METHOD 25-mL glass vial. Teflon®-lined septum.

STABILITY cool, 4°C, 0.008% Sodium thiosulfate.

MHT 14 days.

QUALITY CONTROL The lab must on an ongoing basis, spike at least 5% of the samples from each sample site being monitored to assess accuracy.

REFERENCE Method 624, *Federal Register* Part VIII 40 CFR Part 136, Oct 26, 1984.

---

cis-1,3-Dichloropropene   **EPA Method 8010**
CAS #10061-01-5

TITLE Halogenated Volatile Organics

MATRIX Groundwater, soils, sludges, water miscible liquid wastes, and non-water miscible wastes.

APPLICATION This method is used for the analysis of 39 halogenated VOCs. Samples are analyzed using direct injection or purge-and-trap methods. Groundwater must be analyzed by the purge-and-trap method. The method provides an optional GC column which is used for analyte confirmation and that may help resolve analytes from interferences.

INTERFERENCES There can be carryover contamination with high- and low-level samples. Impurities may come from the purge-and-trap apparatus, organic compounds outgassing from the plumbing ahead of trap, diffusion of VOCs through the sample bottle septum during shipping or storage, or from solvent vapors in the lab.

INSTRUMENTATION GC capable of on-column injections or purge-and-trap sample introduction and a halogen specific detector. Column 1: 8 ft by 0.1 in 1%. SP-1000 on Carbopack-B. Column 2: 6 ft by 0.1 in bonded n-octane on Porasil-C.

RANGE 8 to 500 µg/L (reagent water).

MDL Not determined.

PQL FACTORS FOR MULTIPLYING × FID MDL VALUE

Matrix	Multiplication Factor
Groundwater	10
Low-level soil	10
Water miscible liquid waste	500
High-level soil and sludge	1250
Non-water miscible waste	1250

PRECISION 0.32X µg/L (overall precision; estimated).

ACCURACY 1.00C µg/L (as recovery; estimated).

SAMPLING METHOD For water and liquid samples; use glass 40-mL vials with Teflon®-lined septum caps and collect two vials per sample location with no headspace. For solids and concentrated waste samples; use widemouth glass bottles with Teflon® liners.

STABILITY For concentrated wastes, soils, sediments, or sludges: cool to 4°C. For liquids: add 4 drops of concentrated hydrochloric acid and cool to 4°C.

MHT 14 days.

QUALITY CONTROL Analyze a reagent blank, matrix spike, and matrix spike duplicate/duplicate for each analytical batch (up to 20 samples). Demonstrate the purity of glassware and reagents by analyzing a reagent water method blank. Internal, surrogate, and five concentration level calibration standards are used.

REFERENCE Test Methods for Evaluating Solid Waste (SW-846), U.S. EPA Office of Solid Waste, Washington, DC, Method 8010B, Rev. 2, Nov. 1992.

---

trans-1,3-Dichloropropene   **EPA Method 601**
CAS #10061-02-6

TITLE Purgeable Halocarbons

MATRIX Wastewater

APPLICATION Method covers 29 purgeable halocarbons. (Method 624 provides GC/MS conditions appropriate for the qualitative and quantitative confirmation of results). Method describes conditions for a 2nd GC column to confirm measurements made with primary column.

INTERFERENCES Impurities in the purge gas and organic compounds outgassing from the plumbing ahead of the trap. With high- and low-level samples, there can be carryover contamination. Diffusion of volatile organics through the septum seal into the sample.

INSTRUMENTATION GC-equipped with halide-specific detector. (With purge-and-trap unit).

RANGE 8.0–500 µg/L.

MDL 0.20 µg/L.

PRECISION 0.32X µg/L (overall precision).

ACCURACY 1.00C µg/L (as recovery).

SAMPLING METHOD 25-mL glass vial. Teflon®-lined septum.

STABILITY cool, 4°C, 0.008% Sodium thiosulfate.

MHT 14 days.

QUALITY CONTROL The lab must on an ongoing basis, spike at least 10% of the samples from each sample site being monitored to assess accuracy.

REFERENCE Method 601, *Federal Register* Part VIII 40 CFR Part 136, Oct 26, 1984.

---

trans-1,3-Dichloropropene   **EPA Method 624**
CAS #10061-02-6

TITLE Purgeables

MATRIX Wastewater

APPLICATION Method covers 31 purgeable organics. An inert gas is bubbled through a 5-mL water sample in a specially designed purging chamber. Here, purgeables are transferred from aqueous to gaseous phase, passed onto a sorbent column, and trapped. Trap is heated and backflushed with inert gas to desorb purgeables onto a GC column, where purgeables are separated.

INTERFERENCES Impurities in the purge gas, organic compounds outgassing from the plumbing ahead of the trap, and solvent vapors in the lab. With high- and low-level samples, there can be carryover contamination.

INSTRUMENTATION GC/MS with purge-and-trap unit.

RANGE 5–600 µg/L.

MDL Not determined.

PRECISION 0.34X µg/L (overall precision).

ACCURACY 1.00C µg/L (as recovery).

SAMPLING METHOD 25-mL glass vial. Teflon®-lined septum.

STABILITY cool, 4°C, 0.008% Sodium thiosulfate.

MHT 14 days.

QUALITY CONTROL The lab must on an ongoing basis, spike at least 5% of the samples from each sample site being monitored to assess accuracy.

REFERENCE Method 624, *Federal Register* Part VIII 40 CFR Part 136, Oct 26, 1984.

---

trans-1,3-Dichloropropene   **EPA Method 8010**
CAS #10061-02-6

TITLE Halogenated Volatile Organics

MATRIX Groundwater, soils, sludges, water miscible liquid wastes, and non-water miscible wastes.

APPLICATION This method is used for the analysis of 39 halogenated VOCs. Samples are analyzed using direct injection or purge-and-trap methods. Groundwater must be analyzed by the purge-and-trap method. The method provides an optional GC column which is used for analyte confirmation and that may help resolve analytes from interferences.

INTERFERENCES There can be carryover contamination with high- and low-level samples. Impurities may come from the purge-and-trap apparatus, organic compounds outgassing from the plumbing ahead of trap, diffusion of VOCs through the sample bottle septum during shipping or storage, or from solvent vapors in the lab.

INSTRUMENTATION GC capable of on-column injections or purge-and-trap sample introduction and a halogen specific detector. Column 1: 8 ft by 0.1 in 1%. SP-1000 on Carbopack-B. Column 2: 6 ft by 0.1 in bonded n-octane on Porasil-C.

RANGE 8 to 500 µg/L (reagent water).

MDL 0.34 µg/L (reagent water).

PQL FACTORS FOR MULTIPLYING × FID MDL VALUE

Matrix	Multiplication Factor
Groundwater	10
Low-level soil	10
Water miscible liquid waste	500
High-level soil and sludge	1250
Non-water miscible waste	1250

PRECISION 0.32X µg/L (overall precision; estimated).

ACCURACY 1.00C µg/L (as recovery; estimated).

SAMPLING METHOD For water and liquid samples; use glass 40-mL vials with Teflon®-lined septum caps and collect two vials per sample location with no headspace. For solids and concentrated waste samples; use widemouth glass bottles with Teflon® liners.

STABILITY For concentrated wastes, soils, sediments, or sludges: cool to 4°C. For liquids: add 4 drops of concentrated hydrochloric acid and cool to 4°C.

MHT 14 days.

QUALITY CONTROL Analyze a reagent blank, matrix spike, and matrix spike duplicate/duplicate for each analytical batch (up to 20 samples). Demonstrate the purity of glassware and reagents by analyzing a reagent water method blank. Internal, surrogate, and five concentration level calibration standards are used.

REFERENCE Test Methods for Evaluating Solid Waste (SW-846), U.S. EPA Office of Solid Waste, Washington, DC, Method 8010B, Rev. 2, Nov. 1992.

Dichlorovos   **EPA Method 507**
CAS #62-73-7

TITLE Determination of Nitrogen and Phosphorus-Containing Pesticides in Water by GC/NPD

MATRIX This method is applicable to the determination of certain nitrogen and phosphorus-containing pesticides in finished drinking water and groundwater.

METHOD SUMMARY Method 507 covers 46 nitrogen- and phosphorus-containing pesticides. A 1-L sample is fortified with a surrogate standard, salted, buffered, extracted with methylene chloride, and concentrated; then the solvent is exchanged with methyl tert-butyl ether (MTBE) and concentrated again, and a 2-µL aliquot of a sample extract is injected into a GC system equipped with a selective nitrogen-phosphorus detector and a capillary column for analysis.

INTERFERENCES Method interferences may be caused by contaminants in solvents, reagents, glassware, and other sample processing apparatus. Interfering contamination may occur when a sample containing low concentrations of analytes is analyzed immediately following a sample containing relatively high concentrations. One or more injections of MTBE should be made following the analysis of a sample with high concentrations of analytes to check for analyte carryover. Matrix interferences may be caused by contaminants that are coextracted from the sample. The extent of matrix interferences will vary considerably from source to source, depending upon the water sampled.

INSTRUMENTATION A gas chromatograph system (GC) equipped with a nitrogen-phosphorus detector (NPD) is needed.

Column 1: 30 m × 0.25 mm I.D. DB-5 bonded fused silica column, 0.25 μm film thickness, or equivalent.

Column 2: 30 m × 0.25 mm I.D. DB-1701 bonded fused silica column, 0.25 μm film thickness, or equivalent.

PRECISION & ACCURACY This method has been validated in a single lab and estimated detection limits (EDLs) have been determined for each analyte. Observed detection limits may vary among waters, depending upon the nature of the interferences in the sample matrix and the specific instrumentation used. Analytes that are not separated chromatographically cannot be individually identified and measured unless an alternative technique for identification and quantification exist.

The estimated detection limit (in μg/L) was 2.5. The EDL is defined as either method detection limit or a level of compound in a sample yielding a peak in the final extract with signal-to-noise ratio of approximately 5, whichever value is higher.

The concentration used for these measurements (in μg/L) was 25.
The accuracy (as % recovery) was 97.
The precision (% RSD) was 6.

SAMPLING METHOD Grab samples are collected in 1-L glass sample bottles (prewashed with detergent and hot tap water, rinsed with reagent water, and dried in an oven at 400°C for 1 h) with screw caps lined with PTFE-fluorocarbon.

SAMPLE PRESERVATION Add mercuric chloride to the sample bottle in amounts to produce a concentration of 10 mg/L. If residual chlorine is present, add 80 mg of sodium thiosulfate/L of sample to the sample bottle prior to collection. After collection, seal bottle and shake vigorously for 1 min, then cool the sample to 4°C immediately and store it at 4°C in the dark until extraction.

MHT Maximum holding time of the samples, and in some cases the extracts, is 14 days.

SAMPLE PREPARATION Fortify the sample with 50 μL of the surrogate standard solution, adjust to pH 7 with phosphate buffer, add 100 g NaCl to the sample, and seal and shake to dissolve the salt; then extract with methylene chloride in a separatory funnel or in a mechanical tumbler bottle. Dry the extract by pouring it through a solvent-rinsed drying column containing about 10 cm of anhydrous sodium sulfate. Collect the extract in a Kuderna-Danish (K-D) concentrator and rinse the column with 20–30 mL methylene chloride. Concentrate the extract to about 2 mL and rinse the flask and its lower joint into the concentrator tube with 1 to 2 mL of methyl t-butyl ether (MTBE). Add 5–10 mL of MTBE and concentrate the extract twice (adding more MTBE) to a final volume of 5.0 mL and store it at 4°C until analysis.

Note: If methylene chloride is not completely removed from the final extract, it may cause detector problems.

QUALITY CONTROL Minimum quality control requirements are initial demonstration of lab capability, determination of surrogate compound recoveries in each sample and blank, monitoring internal standard peak area or height in each sample and blank, analysis of lab reagent blanks, lab fortified samples, lab fortified blanks, and other QC samples. A lab reagent blank is analyzed to demonstrate that all glassware and reagent interferences are under control.

Initial demonstration of capability is fulfilled by analyzing four fortified reagent water samples with the recovery value for each analyte falling within the acceptable range (±30% average recovery). Surrogate recoveries from samples or method blanks must be 70–130%. The internal standard response for any sample chromatogram should not deviate from the daily calibration check standard's internal standard response by more than 30% or lab fortified blanks and sample matrices are used to assess lab performance and analyte recovery, respectively.

If the response for the target analyte peak exceeds the working range of the system, dilute the extract and reanalyze. Alternative techniques such as an alternate detector or second chromatography column should be used to confirm peak identification when sample components are not resolved adequately.

EPA CONTACT & HOTLINE For technical questions contact Dr. Baldev Bathija, U.S. EPA, Office of Ground Water and Drinking Water (WH-550D), 401 M St. SW, Washington, DC 20460. Tel. (202) 260-3040. For further information the EPA Safe Drinking Water Hotline may be called at: (800) 426-4791.

REFERENCE Methods for the Determination of Organic Compounds in Drinking Water, EPA/600/4-88/039 (revised July 1991). U.S. EPA Environmental Monitoring Systems Laboratory, Cincinnati, OH, 45268, U.S.A. Available from the National Technical Information Service (NTIS), 5285 Port Royal Road, Springfield, VA 22161; Tel. 800-553-6847. NTIS Order Number is PB91-231480.

Dichlorovos   **EPA Method 8141**
CAS #62-73-7

TITLE Organophosphorus Compounds by Gas Chromatography: Capillary Column Technique

MATRIX This method covers aqueous and solid matrices. This includes a wide variety such as drinking water, groundwater, industrial wastewaters, surface waters, soils, solids, and sediments.

METHOD SUMMARY This is a GC method used to determine the concentration of 28 organophosphorus pesticides.

The use of Gel Permeation Cleanup (EPA Method 3640) for sample cleanup has been demonstrated to yield recoveries of less than 85% for many method analytes and is therefore not recommended for use with this method.

This method provides GC conditions for the detection of ppb concentrations of organophosphorus compounds. Prior to the use of this method, appropriate sample preparation techniques must be used. Water samples are extracted at a neutral pH with methylene chloride as a solvent by using a separatory funnel (EPA Method 3510) or a continuous liquid-liquid extractor (EPA Method 3520). Soxhlet extraction (EPA Method 3540) or ultrasonic extraction (EPA Method 3550) using methylene chloride/acetone (1:1) are used for solid samples. Both neat and diluted organic liquids (EPA Method 3580) may be analyzed by direct injection. Spiked samples are used to verify the applicability of the chosen extraction technique to each new sample type. A GC with a flame photometric (FPD) or nitrogen-phosphorus detector (NPD) is used for this multiresidue procedure.

INTERFERENCES The use of Florisil cleanup materials (EPA Method 3620) for some of the compounds in this method has been demonstrated to yield recoveries less than 85% and is therefore not recommended for all compounds. Use of phosphorus or halogen specific detectors, however, often obviates the necessity for cleanup for relatively clean sample matrices. If particular circumstances demand the use of an alternative cleanup procedure, the analyst must determine the elution profile and demonstrate that the recovery of each analyte is no less than 85%.

Use of a flame photometric detector (FPD) in the phosphorus mode will minimize interferences from materials that do not contain phosphorus. Elemental sulfur, however, may interfere with the determination of certain organophosphorus compounds by flame photometric gas chromatography. Sulfur cleanup using EPA Method 3660 may alleviate this interference. A nitrogen phosphorus detector (NPD) is also recommended.

A few analytes coelute on certain columns. Therefore, select a second column for confirmation where coelution of the analytes of interest does not occur.

Method interferences may be caused by contaminants in solvents, reagents, glassware, and other sample processing hardware that lead to discrete artifacts or elevated baselines in gas chromatograms. All these materials must be routinely demonstrated to be free from interferences under the conditions of the analysis by analyzing reagent blanks.

INSTRUMENTATION A GC with a NPD or a FPD will be needed. A data system or integrator is recommended for measuring peak areas and/or peak heights. A Kuderna-Danish (K-D) apparatus will be needed for extract concentration.

Column 1: 15 m × 0.53 mm megabore capillary column, 1.0 μm film thickness, DB-210.
Column 2: 15 m × 0.53 mm megabore capillary column, 1.5 μm film thickness, SPB-608.
Column 3: 15 m × 0.53 mm megabore capillary column, 1.0 μm film thickness, DB-5.

Three megabore capillary columns are included for analysis of organophosphates by this method. Column 1 (DB-210 or equivalent) and Column 2 (SPB-608 or equivalent) are recommended if a large number of organophosphorus analytes are to be determined. If the superior resolution offered by Column 1 and Column 2 is not required, Column 3 (DB-5 or equivalent) may be used. For megabore capillary columns, automatic injections of 1 μL are recommended.

PRECISION & ACCURACY The MDL actually achieved in a given analysis will vary, as it is dependent on instrument sensitivity and matrix effects. Single operator accuracy and precision studies have been conducted with spiked water and soil samples.

MULTIPLICATION FACTORS FOR OTHER MATRICES (a)

Matrix	Factor (b)
Groundwater (EPA Method 3510 or EPA Method 3520)	10
Low-concentration soil by Soxhlet and no cleanup	10 (c)
Low-concentration soil by ultrasonic extraction with GPC cleanup	6.7 (c)
High-concentration soil and sludges by ultrasonic extraction	500 (c)
Non-water miscible waste (EPA Method 3580)	1000 (c)

(a) Sample EQLs are highly matrix-dependent. The EQLs listed here are provided for guidance and may not always be achievable.
(b) EQL = [Method detection limit] × [Factor]. For non-aqueous samples the factor is on a wet-weight basis.
(c) Multiply this factory times the soil MDL.

The MDL (in μg/L) when reagent water was extracted using a separatory funnel was 0.80.
The MDL (in μg/kg) when soil was extracted using Soxhlet extraction (EPA Method 3540) was 40.0.
Accuracy (as % recovery) with separatory funnel extraction ranged from 80 (with low spikes) to 72 (with high spikes).
Accuracy (as % recovery) with continuous liquid-liquid extraction ranged from 81 (with low spikes) to 74 (with high spikes).
Accuracy (as % recovery) with Soxhlet extraction of soils ranged from 84 (with low spikes to 71 (with high spikes).
Accuracy (as % recovery) with ultrasonic extraction of soils ranged from 41 (with low spikes) to 27 (with high spikes).

SAMPLE COLLECTION, PRESERVATION & HANDLING
Containers used to collect samples for the determination of semivolatile organic compounds should be soap and water washed followed by methanol (or isopropanol) rinsing. The sample containers should be of glass or Teflon® and have screw-top covers with Teflon® liners.

No preservation is used with concentrated waste samples. With liquid samples containing no residual chlorine and with soil, sediment, and sludge samples, immediately cooling to 4°C is the only preservation used. When residual chlorine is present then 3 mL of 10% aqueous sodium sulfate is added for each gallon of sample collected, followed by cooling to 4°C.

Liquid samples must be extracted within 7 days and their extracts analyzed within 40 days. Concentrated waste, soil, sediment, and sludge samples must be extracted within 14 days and their extracts analyzed within 40 days.

SAMPLE PREPARATION In general, water samples are extracted at a neutral pH with methylene chloride, using either EPA Method 3510 or EPA Method 3520. Solid samples are

extracted using either EPA Method 3540 or EPA Method 3550 with methylene chloride/acetone (1:1) as the extraction solvent.

Prior to GC analysis, the extraction solvent may be exchanged to hexane. Single lab data indicates that samples should not be transferred with 100% hexane during sample workup as the more water soluble organophosphorus compounds may be lost.

If cleanup is performed on the samples, the analyst should analyze the samples by GC. This will confirm elution patterns and the absence of interferences from the reagents. If peak detection and identification is prevented by the presence of interferences, further cleanup is required.

QUALITY CONTROL The analyst should monitor the performance of the extraction, cleanup (when used), and analytical system and the effectiveness of the method in dealing with each sample matrix by spiking each sample, standard, and blank with one or two surrogates (e.g., organophosphorus compounds not expected to be present in the sample). Deuterated analogs of analytes should not be used as surrogates for gas chromatographic analysis due to coelution problems.

A minimum of five concentrations for each analyte of interest should be prepared through dilution of the stock standards with isooctane. One of the concentrations should be at a concentration near, but above, the MDL.

Include a mid-level check standard after each group of 10 samples in the analysis sequence. GC/MS techniques should be judiciously employed to support qualitative identifications made with this method. Follow the GC/MS operating requirements specified in EPA Method 8270.

When available, chemical ionization mass spectra may be employed to aid in the qualitative identification process. To confirm an identification of a compound, the background-corrected mass spectrum of the compound must be obtained from the sample extract and must be compared with a mass spectrum from a stock or calibration standard analyzed under the same chromatographic conditions. The molecular ion and all other ions present above 20% relative abundance in the mass spectrum of the standard must be present in the mass spectrum of the sample with agreement to ±20%. The retention time of the compound in the sample must be within six seconds of the retention time for the same compound in the standard solution.

Should the MS procedure fail to provide satisfactory results, additional steps may be taken before reanalysis. These steps may include the use of alternate packed or capillary GC columns or additional sample cleanup.

REFERENCE Test Methods for Evaluating Solid Waste, Physical/Chemical Methods, SW-846, 3rd Edition, U.S. EPA, Office of Solid Waste, Washington, DC, EPA Method 8141 July 1992.

Dichlorovos   **EPA Method 8270**
CAS #62-73-7

TITLE Semivolatile Organic Compounds by GC/MS

MATRIX This method is used to determine the concentration of semivolatile organic compounds in extracts prepared from all types of solid waste matrices, soils, and groundwater. Although surface waters are not specifically mentioned, this method should be applicable to water samples from rivers, lakes, etc.

METHOD SUMMARY This method covers 259 semivolatile organic compounds. In very limited applications direct injection of the sample into the GC/MS system may be appropriate, but this results in very high detection limits (approximately 10,000 µg/L). Typically, a 1-L liquid sample, containing surrogate, and matrix spiking standards, is extracted in a continuous extractor first under acid conditions and then under basic conditions. Typically 30 g of a solid sample, containing surrogate, and matrix spiking standards, is extracted ultrasonically. After concentrating the extract to 1 mL it is spiked with 10 µL of an internal standard solution just prior to analysis by GC/MS. The volume injected should contain about 100 ng of base/neutral and 200 ng of acid surrogates (for a 1-µL injection). Analysis is performed by GC/MS using a capillary GC column.

INTERFERENCES Raw GC/MS data from all blanks, samples, and spikes must be evaluated for interferences. Contamination by carryover can occur whenever high-concentration and low-concentration samples are sequentially analyzed. To reduce carryover, the sample syringe must be rinsed out between samples with solvent. Whenever an unusually concentrated sample is encountered, it should be followed by the analysis of blank solvent to check for cross-contamination.

INSTRUMENTATION A GC/MS and a data system are required. The GC column used is a 30 m × 0.25 mm I.D. (or 0.32 mm I.D.) 1um film thickness silicone-coated fused silica capillary column. A continuous liquid-liquid extractor equipped with Teflon® or glass connection joints and stopcocks requiring no lubrication, a K-D concentrating apparatus, water bath, and an ultrasonic disrupter with a minimum power of 300 W and with pulsing capability are also required.

PRECISION & ACCURACY The estimated quantitation limit (EQL) of Method 8270B for determining an individual compound is approximately 1 mg/kg (wet weight) for soil or sediment samples, 1–200 mg/kg for wastes (dependent on matrix and method of preparation), and 10 µg/L for groundwater samples. EQLs will be proportionately higher for sample extracts that require dilution to avoid saturation of the detector.

The EQL(b) for groundwater in µg/L is 10.
The EQL (a, b) for low concentrations in soil and sediment in µg/kg is not determined.
Accuracy as µg/L is not listed.
Overall precision in µg/L is not listed.

(a) *EQLs listed for soil/sediment are based on wet weight. Normally data is reported in a dry-weight basis; therefore, EQLs will be higher based on the % dry weight of each sample. This calculation is based on a 30 g sample and gel permeation chromatography cleanup.*

(b) *Sample EQLs are highly matrix-dependent. The EQLs are provided for guidance and may not always be achievable.*

C = *True value for concentration, in µg/L.*

X = Average recovery found for measurements of samples containing a concentration of C, in µg/L.

ESTIMATED QUANTITATION LIMIT

Other Matrices	Factor (a)
High-concentration soil and sludges by sonicator	7.5
Non-water miscible waste	75

(a) EQL for other matrices = [EQL for low soil/sediment] × [Factor]. This estimated EQL is similar to an EPA "Practical Quantitation Limit."

SAMPLING METHOD

Liquid samples — Use a 1 or 2½ gallon amber glass bottle with a screw-top Teflon®-lined cover that has been prewashed with detergent and rinsed with distilled water and methanol (or isopropanol).

Soils, sediments, or sludges — Use an 8-oz. widemouth glass with a screw-top Teflon®-lined cover that has been prewashed with detergent and rinsed with distilled water and methanol (or isopropanol).

SAMPLE PRESERVATION

Liquid samples — If residual chlorine is present, add 3 mL of 10% sodium thiosulfate per gallon, cool to 4°C and store in a solvent-free refrigerator until analysis; if chlorine is not present, then eliminate the sodium thiosulfate addition.

Soils, sediments, or sludges — Cool samples to 4°C and store in a solvent-free refrigerator.

MHT Liquid samples must be extracted within 7 days and the extracts analyzed within 40 days. Soils, sediments, or sludges may be stored for a maximum of 14 days and the extracts analyzed within 40 days.

SAMPLE PREPARATION

Liquid samples — Transfer 1 L quantitatively to a continuous extractor. If high concentrations are anticipated, a smaller volume may be used and then diluted with organic-free reagent water to 1 L. Adjust pH, if necessary, to pH <2 using 1:1 (V/V) sulfuric acid. Pipette 1.0 mL of a surrogate standard spiking solution into each sample. For the sample in each analytical batch selected for spiking, add 1.0 mL of a matrix spiking standard. For base/neutral acid analysis, the amount of the surrogates and matrix spiking compounds added to the sample should result in a final concentration of 100 ng/µL of each analyte in the extract to be analyzed (assuming a 1-µL injection). Extract with methylene chloride for 18–24 h. Next, adjust the pH of the aqueous phase to pH >11 using 10 N sodium hydroxide and extract it with methylene chloride again for 18–24 h. Dry the extract through a column containing anhydrous sodium sulfate and concentrate it to 1 mL using a K-D concentrator.

Soils, sediments, or sludges — Use 30 g of sample. Nonporous or wet samples (gummy or clay type) that do not have a free-flowing sandy texture must be mixed with anhydrous sodium sulfate until the sample is free flowing. Add 1 mL of surrogate standards to all samples, spikes, standards, and blanks. For the sample in each analytical batch selected for spiking, add 1.0 mL of a matrix spiking standard. For base/neutral acid analysis, the amount added of the surrogates and matrix spiking compounds should result in a final concentration of 100 ng/µL of each base/neutral analyte and 200 ng/µL of each acid analyte in the extract to be analyzed (assuming a 1-µL injection). Immediately add a 100-mL mixture of 1:1 methylene chloride:acetone and extract the sample ultrasonically for 3 min and then decant or filter the extracts. Repeat the extraction two or more times. Dry the extract using a column with anhydrous sodium sulfate and concentrate it to 1 mL in a K-D concentrator.

QUALITY CONTROL A methylene chloride solution containing 50 ng/µL of decafluorotriphenylphosphine (DFTPP) is used for tuning the GC/MS system each 12-h shift. A system performance check also must be made during every 12-h shift. A standard containing 50 ng/µL each of 4,4′-DDT, pentachlorophenol, and benzidine is required to verify injection port inertness and GC column performance. A calibration standard at mid-concentration, containing each compound of interest, including all required surrogates, must be performed every 12 h during analysis. After the system performance check is met, calibration check compounds (CCCs) are used to check the validity of the initial calibration.

The internal standard responses and retention times in the calibration check standard must be evaluated immediately after or during data acquisition. If the retention time for any internal standard changes by more than 30 seconds from the last check calibration (12 h), the chromatographic system must be inspected for malfunctions and corrections must be made, as required. If the electron ionization current plot (EICP) area for any of the internal standards changes by a factor of two from the last daily calibration standard check, the mass spectrometer must be inspected for malfunctions and corrections must be made, as appropriate.

Demonstrate, through the analysis of a reagent water blank, that interferences from the analytical system, glassware, and reagents are under control. The blank samples should be carried through all stages of the sample preparation and measurement steps. For each analytical batch (up to 20 samples), a reagent blank, matrix spike, and matrix spike duplicate/duplicate must be analyzed (the frequency of the spikes may be different for different monitoring programs). The blank and spiked samples must be carried through all stages of the sample preparation and measurement steps. A QC reference sample concentrate containing each analyte at a concentration of 100 mg/L in methanol is required.

REFERENCE Test Methods for Evaluating Solid Waste (SW-846). U.S. EPA 1983, Method 8270B, Rev. 2, Nov. 1990. Office of Solid Waste, Washington, DC.

Dichlorovos   EPA Method 8140
CAS #62-73-7

TITLE Organophosphorus Pesticides

MATRIX Groundwater, soils, sludges, water miscible liquid wastes, and non-water miscible wastes.

APPLICATION This method is used for the analysis of 21 organophosphorus pesticides. Samples are extracted, concentrated, and analyzed using direct injection of both neat and diluted organic liquid into a gas chromatograph (GC).

INTERFERENCES Solvents, reagents, and glassware may introduce artifacts. Other interferences may come from coextracted compounds from samples. The use of Florisil cleanup materials may produce low recoveries. Elemental sulfur may interfere with some compounds when using a flame photometric detector. Sulfur cleanup (Method 3660) may alleviate sulfur interference.

INSTRUMENTATION GC capable of on-column injections and a flame photometric detector (FPD) or a thermionic detector. A halogen specific detector may also be used and may have the advantage of fewer interferences. Column 1: 1.8 meter by 2 mm with 5% SP-2401 on Supelcoport. Column 2: 1.8 m by 2 mm with 3% SP-2401 on Supelcoport. Column 3: 50 cm by 1/8 IN EFLON® with 15% SE-54 on Gas Chrom Q. The preferred column is Column Number 1 or 3.

RANGE 15.6 to 517 µg/L.

MDL 0.1 µg/L (in reagent water).

PQL FACTORS FOR MULTIPLYING × FID MDL VALUE

Matrix	Multiplication Factor
Groundwater	10
Low-level soil by sonication with GPC cleanup	670
High-level soil and sludge by sonication	10,000
Non-water miscible waste	100,000

PRECISION 7.7% (single operator standard deviation).

ACCURACY 72.1% (single operator average recovery).

SAMPLING METHOD Use 8-oz. widemouth glass bottles with Teflon®-lined caps for concentrated waste samples, soils, sediments, and sludges. Use 1 or 2½ gallon amber glass bottles with Teflon®-lined caps for liquid (water) samples.

STABILITY Cool soil, sediment, sludge, and liquid samples to 4°C. If residual chlorine is present in liquid samples add 3 mL of 10% sodium thiosulfate per gallon of sample and cool to 4°C.

MHT 14 days for concentrated waste, soil, sediment, or sludge; 7 days for liquid samples; all extracts must be analyzed within 40 days.

QUALITY CONTROL A quality control check sample concentrate containing this compound in acetone at a concentration 1,000 times more concentrated than the selected spike concentration is required. The QC check sample concentrate may be prepared from pure standard materials or purchased as certified solutions. Use appropriate trip, matrix, control site, method, reagent, and solvent blanks. Internal, surrogate, and five concentration level calibration standards are used.

REFERENCE Method 8140, SW-846, 3rd ed., Sept. 1986.

Dichlorprop EPA Method 8151
CAS #120-36-5

TITLE Chlorinated Herbicides by GC Using Methylation or Pentafluorobenzylation Derivatization: Capillary Column Technique.

MATRIX This method covers aqueous and solid matrices. This includes a wide variety such as drinking water, groundwater, industrial wastewaters, surface waters, soils, solids, and sediments.

METHOD SUMMARY This is a GC method for determining 19 chlorinated acid herbicides in aqueous, soil, and waste matrices. Because these compounds are produced and used in various forms (i.e., acid, salt, ester, etc.) a hydrolysis step is included to convert the herbicide to the acid form prior to analysis. This method provides hydrolysis, extraction, derivatization and GC conditions for the analysis of chlorinated acid herbicides in water, soil, and waste samples. Water samples are hydrolyzed *in situ*, extracted with diethyl ether, and then esterified with either diazomethane or pentafluorobenzyl bromide. The derivatives are determined by gas chromatography with an electron capture detector (GC/ECD). The results are reported as acid equivalents. The sensitivity of this method depends on the level of interferences in addition to instrumental limitations.

INTERFERENCES Method interferences may be caused by contaminants in solvents, reagents, glassware, and other sample processing hardware. Immediately prior to use, glassware should be rinsed with the next solvent to be used. Matrix interferences may be caused by contaminants that are coextracted from the sample. Organic acids, especially chlorinated acids, cause the most direct interference with the determination by methylation. Phenols, including chlorophenols, may also interfere with this procedure. The determination using pentafluorobenzylation is more sensitive, and more prone to interferences from the presence of organic acids of phenols than by methylation. Alkaline hydrolysis and subsequent extraction of the basic solution removes many chlorinated hydrocarbons and phthalate esters that might otherwise interfere with the ECD analysis. The herbicides, being strong organic acids, react readily with alkaline substances and may be lost during analysis. Therefore, glassware must be acid-rinsed and then rinsed to constant pH with organic-free reagent water.

INSTRUMENTATION A GC suitable for Grob-type injection using capillary columns. A data system for measuring peak heights and/or peak areas is recommended. An electron capture detector (ECD) is used. Also a K-D apparatus, a diazomethane generator, a centrifuge and an ultrasonic disrupter will be required.

Narrow Bore Columns:
Primary Column 1: 30 m × 0.25 mm, 5% phenyl/95% methyl silicone (DB-5), 0.25 µm film thickness.
Primary Column 1a (GC/MS): 30 m × 0.32 mm, 5% phenyl/95% methyl silicone (DB-5), 1-µm film thickness.
Column 2: 30 m × 0.25 mm DB-608 with a 25 µm film thickness.
Confirmation Column: 30 m × 0.25 mm, 14% cyanopropyl phenyl silicone (DB-1701), 0.25 µm film thickness.

Megabore Columns:
Primary Column: 30 m × 0.53 mm DB-608 with 0.83 μm film thickness.
Confirmation Column: 30 m × 0.53 mm, 14% cyanopropyl phenyl silicone (DB-1701), 1.0 μm film thickness.

PRECISION & ACCURACY Method detection limits (MDLs) are compound-dependent and vary with derivitization efficiency, derivative recovery, the matrix sampled, and herbicide concentration.

The estimated MDL (in μg/L) was 0.26 for aqueous samples using GC/ECD.

The estimated MDL (in μg/kg) was not reported for soil samples using GC/ECD when corrected back to 50-g samples extracted and concentrated to 10 mL with 5-μL injections.

The estimated GC/MS identification limit (in ng) was not reported for soil samples using GC/MS.

Mean percent recovery, calculated from 7–8 determinations of spiked reagent water, after diazomethane derivatization, from a spike concentration (in μg/L) of 2 was 107 with a standard deviation of the percent recovery of 20.3.

Mean percent recovery, calculated from 10 determinations of spiked clay and clay/still bottom samples over the linear concentration range (in ng/g) of 1.5–3,000 was 97.3 with a percent relative standard deviation of 5.0. The RSD % was calculated on 10 samples high in the linear concentration range and 10 low in the range. The linear concentration range was determined using standard solutions and corrected to 50 g soil samples.

SAMPLE COLLECTION, PRESERVATION & HANDLING
Containers used to collect samples for the determination of semivolatile organic compounds should be soap and water washed followed by methanol (or isopropanol) rinsing. The sample containers should be of glass or Teflon® and have screw-top covers with Teflon® liners.

No preservation is used with concentrated waste samples. With liquid samples containing no residual chlorine and with soil, sediment, and sludge samples, immediately cooling to 4°C is the only preservation used. When residual chlorine is present then 3 mL of 10% aqueous sodium sulfate is added for each gallon of sample collected, followed by cooling to 4°C.

The holding time for all volatile organics samples is 14 days. Liquid samples must be extracted within 7 days and their extracts analyzed within 40 days. Concentrated waste, soil, sediment, and sludge samples must be extracted within 14 days and their extracts analyzed within 40 days.

SAMPLE PREPARATION
Preparation of soil, sediment, and other solid samples — Acidify 30 g (dry weight) solids with 0.1 M phosphate buffer (pH = 2.5) and thoroughly mix the contents. Spike the sample with surrogate compound(s). The ultrasonic extraction of solids must be optimized for each type of sample. In order for the ultrasonic extractor to efficiently extract solid samples, the sample must be free flowing when the solvent is added. Acidified anhydrous sodium sulfate should be added to clay-type soils, or any other solid that is not a free-flowing sandy texture, until a free flowing mixture is obtained. Add methylene chloride and perform ultrasonic extraction. Combine organic extracts from the repetitive extractings of the sample and centrifuge. Add aqueous potassium hydroxide, water, and methanol to the extract and reflux the mixture on a water bath. Extract the solution three times with methylene chloride and discard the methylene chloride phase. The basic solution contains the herbicide salts. Adjust the pH of the solution to <2 with cold sulfuric acid and extract three times with methylene chloride. Combine the extracts and pour them through a pre-rinsed drying column containing acidified anhydrous sodium sulfate. Collect the dried extracts in a K-D flask and concentrate them.

Preparation of aqueous samples — Measure 1 L of sample into a 2 L separatory funnel and spike it with surrogate compound(s). Add NaCl to the sample, then add 6 N NaOH to the sample to a pH of 12 or more and let the sample sit at room temperature for 1 h to hydrolyze esters. Extract the sample three times with methylene chloride and discard the extracts. Then add cold 12 N sulfuric acid to a pH less than or equal to 2, and extract the sample three times with ethyl ether. Collect the ether phase in a flask containing acidified anhydrous sodium sulfate and allow it to remain in contact with the sodium sulfate for a minimum of 2 h. The drying step is very critical to ensuring complete esterification; any moisture remaining in the ether will result in low herbicide recoveries.

Extract concentration and derivatization — The combined ether extract is concentrated to about 1 mL using a K-D apparatus followed by using a micro Snyder column or nitrogen gas blowdown. If methyl esters are to be produced, then dilute the concentrated ether extract with 1 mL of isooctane and 0.5 mL of methanol, dilute to a final volume of 4 mL, and esterify with diazomethane. If pentafluorobenzene esters are to be produced, then dilute concentrated ether extract with acetone to a final volume of 4 mL and esterify with pentafluorobenzyl bromide.

QUALITY CONTROL Select a representative spike concentration for each compound (acid or ester) to be measured. Using stock standard, prepare a quality control check sample concentrate, in acetone, that is 1000 times more concentrated than the selected concentrations. Use this quality control check sample concentrate to prepare quality control check samples. Calculate surrogate standard recovery on all standards, samples, blanks, and spikes. GC/MS techniques should be judiciously employed to support qualitative identifications made with this method. When available, chemical ionization mass spectra may be employed to aid the qualitative identification process.

REFERENCE Test Methods for Evaluating Solid Waste, Physical/Chemical Methods, SW-846, 3rd Edition, U.S. EPA, Office of Solid Waste, Washington, DC, EPA Method 8151, Nov. 1990.

Dicrotophos   **EPA Method 8270**
CAS #141-66-2

TITLE Semivolatile Organic Compounds by GC/MS

MATRIX This method is used to determine the concentration of semivolatile organic compounds in extracts prepared from all types of solid waste matrices, soils, and groundwater. Although surface waters are not specifically mentioned, this method should be applicable to water samples from rivers, lakes, etc.

METHOD SUMMARY This method covers 259 semivolatile organic compounds. In very limited applications direct injection of the sample into the GC/MS system may be appropriate, but this results in very high detection limits (approximately 10,000 µg/L). Typically, a 1-L liquid sample, containing surrogate, and matrix spiking standards, is extracted in a continuous extractor first under acid conditions and then under basic conditions. Typically 30 g of a solid sample, containing surrogate, and matrix spiking standards, is extracted ultrasonically. After concentrating the extract to 1 mL it is spiked with 10 µL of an internal standard solution just prior to analysis by GC/MS. The volume injected should contain about 100 ng of base/neutral and 200 ng of acid surrogates (for a 1-µL injection). Analysis is performed by GC/MS using a capillary GC column.

INTERFERENCES Raw GC/MS data from all blanks, samples, and spikes must be evaluated for interferences. Contamination by carryover can occur whenever high-concentration and low-concentration samples are sequentially analyzed. To reduce carryover, the sample syringe must be rinsed out between samples with solvent. Whenever an unusually concentrated sample is encountered, it should be followed by the analysis of blank solvent to check for cross-contamination.

INSTRUMENTATION A GC/MS and a data system are required. The GC column used is a 30 m × 0.25 mm I.D. (or 0.32 mm I.D.) 1um film thickness silicone-coated fused silica capillary column. A continuous liquid-liquid extractor equipped with Teflon® or glass connection joints and stopcocks requiring no lubrication, a K-D concentrating apparatus, water bath, and an ultrasonic disrupter with a minimum power of 300 W and with pulsing capability are also required.

PRECISION & ACCURACY The estimated quantitation limit (EQL) of Method 8270B for determining an individual compound is approximately 1 mg/kg (wet weight) for soil or sediment samples, 1–200 mg/kg for wastes (dependent on matrix and method of preparation), and 10 µg/L for groundwater samples. EQLs will be proportionately higher for sample extracts that require dilution to avoid saturation of the detector.

The EQL(b) for groundwater in µg/L is 10.
The EQL (a, b) for low concentrations in soil and sediment in µg/kg is not determined.
Accuracy as µg/L is not listed.
Overall precision in µg/L is not listed.

(a) *EQLs listed for soil/sediment are based on wet weight. Normally data is reported in a dry-weight basis; therefore, EQLs will be higher based on the % dry weight of each sample. This calculation is based on a 30 g sample and gel permeation chromatography cleanup.*
(b) *Sample EQLs are highly matrix-dependent. The EQLs are provided for guidance and may not always be achievable.*

$C =$ *True value for concentration, in µg/L.*

$X =$ *Average recovery found for measurements of samples containing a concentration of C, in µg/L.*

ESTIMATED QUANTITATION LIMIT

Other Matrices	Factor (a)
High-concentration soil and sludges by sonicator	7.5
Non-water miscible waste	75

(a) *EQL for other matrices = [EQL for low soil/sediment] × [Factor]. This estimated EQL is similar to an EPA "Practical Quantitation Limit."*

SAMPLING METHOD
Liquid samples — Use a 1 or 2½ gallon amber glass bottle with a screw-top Teflon®-lined cover that has been prewashed with detergent and rinsed with distilled water and methanol (or isopropanol).

Soils, sediments, or sludges — Use an 8-oz. widemouth glass with a screw-top Teflon®-lined cover that has been prewashed with detergent and rinsed with distilled water and methanol (or isopropanol).

SAMPLE PRESERVATION
Liquid samples — If residual chlorine is present, add 3 mL of 10% sodium thiosulfate per gallon, cool to 4°C and store in a solvent-free refrigerator until analysis; if chlorine is not present, then eliminate the sodium thiosulfate addition.

Soils, sediments, or sludges — Cool samples to 4°C and store in a solvent-free refrigerator.

MHT Liquid samples must be extracted within 7 days and the extracts analyzed within 40 days. Soils, sediments, or sludges may be stored for a maximum of 14 days and the extracts analyzed within 40 days.

SAMPLE PREPARATION
Liquid samples — Transfer 1 L quantitatively to a continuous extractor. If high concentrations are anticipated, a smaller volume may be used and then diluted with organic-free reagent water to 1 L. Adjust pH, if necessary, to pH <2 using 1:1 (V/V) sulfuric acid. Pipette 1.0 mL of a surrogate standard spiking solution into each sample. For the sample in each analytical batch selected for spiking, add 1.0 mL of a matrix spiking standard. For base/neutral acid analysis, the amount of the surrogates and matrix spiking compounds added to the sample should result in a final concentration of 100 ng/µL of each analyte in the extract to be analyzed (assuming a 1-µL injection). Extract with methylene chloride for 18–24 h. Next, adjust the pH of the aqueous phase to pH >11 using 10 *N* sodium hydroxide and extract it with methylene chloride again for 18–24 h. Dry the extract through a column containing anhydrous sodium sulfate and concentrate it to 1 mL using a K-D concentrator.

Soils, sediments, or sludges — Use 30 g of sample. Nonporous or wet samples (gummy or clay type) that do not have a free-flowing sandy texture must be mixed with anhydrous sodium sulfate until the sample is free flowing. Add 1 mL of surrogate standards to all samples, spikes, standards, and blanks. For the sample in each analytical batch selected for spiking, add 1.0 mL of a matrix spiking standard. For base/neutral acid analysis, the amount added of the surrogates and matrix spiking compounds should result in a final concentration of 100 ng/µL of each base/neutral analyte and 200 ng/µL of each acid analyte in the extract to be analyzed (assuming a 1-µL injection). Immediately add a 100-mL mixture of 1:1 methylene chloride:acetone and extract the sample ultrasonically for 3 min and then decant or filter the extracts. Repeat the extraction two or more times. Dry the extract using a column with anhydrous sodium sulfate and concentrate it to 1 mL in a K-D concentrator.

QUALITY CONTROL A methylene chloride solution containing 50 ng/µL of decafluorotriphenylphosphine (DFTPP) is used for tuning the GC/MS system each 12-h shift. A system performance check also must be made during every 12-h shift. A standard containing 50 ng/µL each of 4,4'-DDT, pentachlorophenol, and benzidine is required to verify injection port inertness and GC column performance. A calibration standard at mid-concentration, containing each compound of interest, including all required surrogates, must be performed every 12 h during analysis. After the system performance check is met, calibration check compounds (CCCs) are used to check the validity of the initial calibration.

The internal standard responses and retention times in the calibration check standard must be evaluated immediately after or during data acquisition. If the retention time for any internal standard changes by more than 30 seconds from the last check calibration (12 h), the chromatographic system must be inspected for malfunctions and corrections must be made, as required. If the electron ionization current plot (EICP) area for any of the internal standards changes by a factor of two from the last daily calibration standard check, the mass spectrometer must be inspected for malfunctions and corrections must be made, as appropriate.

Demonstrate, through the analysis of a reagent water blank, that interferences from the analytical system, glassware, and reagents are under control. The blank samples should be carried through all stages of the sample preparation and measurement steps. For each analytical batch (up to 20 samples), a reagent blank, matrix spike, and matrix spike duplicate/duplicate must be analyzed (the frequency of the spikes may be different for different monitoring programs). The blank and spiked samples must be carried through all stages of the sample preparation and measurement steps. A QC reference sample concentrate containing each analyte at a concentration of 100 mg/L in methanol is required.

REFERENCE Test Methods for Evaluating Solid Waste (SW-846). U.S. EPA 1983, Method 8270B, Rev. 2, Nov. 1990. Office of Solid Waste, Washington, DC.

Dicyclohexyl phthalate   **EPA Method 8061**
CAS #84-61-7

TITLE Phthalate Esters by Capillary Gas Chromatography With Electron Capture Detection (GC/ECD)

MATRIX This method covers aqueous and solid matrices. This includes a wide variety such as drinking water, groundwater, industrial wastewaters, surface waters, soils, solids, and sediments.

METHOD SUMMARY This method is used to determine the identities and concentrations of phthalate esters in liquid, solid and sludge matrices. When used to analyze for any or all of the target analytes, compound identification should be supported by at least one additional qualitative technique. This method describes conditions for parallel column, dual electron capture detector analysis, which fulfills the above requirement. Alternatively, GC/MS could be used for compound confirmation.

A measured volume or weight of sample (approximately 1 L for liquids, 10 to 30 g for solids and sludges) is extracted by using the appropriate sample extraction technique specified in EPA Method 3510, EPA Method 3540, and EPA Method 3550. After cleanup, the extract is analyzed by GC/ECD.

INTERFERENCES The sensitivity of this method usually depends on the level of interferences rather than on instrumental limitations. If interferences prevent detection of the analytes, cleanup of the sample extracts is necessary. Either EPA Method 3610 or EPA Method 3620 alone or followed by EPA Method 3660, Sulfur Cleanup, may be used to eliminate interferences in the analysis. EPA Method 3640, Gel Permeation Cleanup, is applicable for samples that contain high amounts of lipids and waxes.

Interferences coextracted from the samples will vary considerably from waste to waste. Glassware must be scrupulously clean. All glassware require treatment in a muffle furnace at 400°C for 2 to 4 h, or thorough rinsing with pesticide-grade solvent, prior to use. Volumetric glassware should not be heated in a muffle furnace. Storage of glassware in the lab introduces contamination, even if the glassware is wrapped in aluminum foil. Sodium sulfate, Florisil, and alumina may be contaminated with phthalate esters and, therefore, use of these materials in sample cleanup should be employed cautiously. If these materials are used, they must be obtained packaged in glass. Heating at 400°C for sodium sulfate, 320°C for Florisil, and 210°C for alumina is recommended. Glass wool used in any step of sample preparation should be a specially treated pyrex wool, pesticide grade, and must be baked at 400°C for 4 h immediately prior to use.

Paper thimbles and filter paper must be exhaustively washed with the solvent that will be used in the sample extraction. Soxhlet extraction of paper thimbles and filter paper for 12 h with fresh solvent should be repeated for a minimum of three times. Method blanks should be obtained before any of the precleaned thimbles or filter papers are used.

INSTRUMENTATION Gas chromatograph suitable for on-column and split/splitless injections.

Column 1: 30 m × 0.53 mm ID, 5% phenyl/95% methyl silicone fused-silica open tubular column, DB-5, 1.5 µg film thickness.
Column 2: 30 m × 0.53 mm ID, 14% cyanopropyl phenyl silicone fused-silica open tubular column, DB-1701, 1.0 µg film thickness.

A dual electron capture detector (ECD) is used. A Kuderna-Danish (K-D) apparatus is required along with a vacuum manifold consisting of individually adjustable, easily accessible flow-control valves for up to 24 cartridges, sample rack, chemically resistant cover and seals, heavy-duty glass basin, removable stainless steel solvent guides, built-in vacuum gauge and valve. Also, 6-mL, 1-g solid-phase extraction cartridges, LC-Florisil or equivalent, prepackaged, ready to use will be needed.

PRECISION & ACCURACY The MDL actually achieved in a given analysis will vary, as it is dependent on instrument sensitivity and matrix effects. This method has been tested in a single lab. Single-operator precision, overall precision, and method accuracy were found to be related to the concentration of the compounds and the type of matrix.

MULTIPLICATION FACTORS FOR OTHER MATRICES (a)

Matrix	Factor (b)
Groundwater	10
Low-concentration soil by ultrasonic extraction with GPC cleanup	670
High-concentration soil and sludges by ultrasonic extraction	10,000
Non-water miscible waste	100,000

(a) Sample EQLs are highly matrix-dependent.
(b) EQL = [Method detection limit] × [Factor]. For non-aqueous samples, the factor is on a wet-weight basis.

The MDL using 7 replicate determinations and a spike concentration of 100 µg/L was 22 ng/L.
The average recovery from HPLC-grade water using 4 determinatons and a spike concentration of 100 µg/L was 106%.
The precision (as RSD) from HPLC-grade water using 4 determinatons and a spike concentration of 100 µg/L was 19.9%.
The average recovery from groundwater using 4 determinatons and a spike concentration of 100 µg/L was 108%.
The precision (as RSD) from groundwater using 4 determinatons and a spike concentration of 100 µg/L was 13.3%.
The average recovery (in %) with %RSD (in parenthesis) from 3 determinations and a spike concentration of 20 µg/L in water was 77.4 (6.5) using 3M Empore Disks and EPA Method 8061.
The average recovery (in %) with %RSD (in parenthesis) from 3 determinations and a spike concentration of 20 µg/L in leachate was 88.2 (13.2) using 3M Empore Disks and EPA Method 8061.
The average recovery (in %) with %RSD (in parenthesis) from 3 determinations and a spike concentration of 20 µg/L in estuarine groundwater was 91.7 (15.2) using 3M Empore Disks and EPA Method 8061.
The average recovery (in %) with %RSD (in parenthesis) from 3 determinations and a spike concentration of 1 mg/kg in estuarine sediment was 36.6 (48.8) after sulfur cleanup with EPA Method 3660.
The average recovery (in %) with %RSD (in parenthesis) from 3 determinations and a spike concentration of 1 mg/kg in municipal sludge was 65.8 (15.7).
The average recovery (in %) with %RSD (in parenthesis) from 3 determinations and a spike concentration of 1 mg/kg in sandy loam soil was 92.8 (35.9).

SAMPLE COLLECTION, PRESERVATION & HANDLING
Containers used to collect samples for the determination of semivolatile organic compounds should be soap and water washed followed by methanol (or isopropanol) rinsing. The sample containers should be of glass or Teflon® and have screw-top covers with Teflon® liners. Sample containers should be filled with care to prevent any portion of the collected sample coming in contact with the sampler's gloves.

No preservation is used with concentrated waste samples. With liquid samples containing no residual chlorine and with soil, sediment, and sludge samples, immediately cooling to 4°C is the only preservation used. When residual chlorine is present then 3 mL of 10% aqueous sodium sulfate is added for each gallon of sample collected, followed by cooling to 4°C.

MHT Liquid samples must be extracted within 7 days and their extracts analyzed within 40 days. Concentrated waste, soil, sediment, and sludge samples must be extracted within 14 days and their extracts analyzed within 40 days.

SAMPLE PREPARATION In general, water samples are extracted at a pH of 5 to 7 with methylene chloride in a separatory funnel (EPA Method 3510). EPA Method 3520 is not recommended for the extraction of aqueous samples because the longer chain esters tend to adsorb to the glassware and consequently, their extraction recoveries may be poor. Solid samples are extracted with hexane/acetone (1:) or methylene chloride/acetone (1:1) in a Soxhlet extractor (EPA Method 3540) or with an ultrasonic extractor (EPA Method 3550). Immediately prior to extraction, spike 500 µL of the surrogate standard spiking solution into 1-L aqueous sample or 30-g solid sample. Extraction of particulate-free aqueous samples using C-18 extraction disks is an optional method that can be used.

Prior to Florisil cleanup or GC analysis, the methylene chloride and methylene chloride/acetone extracts must be exchanged to hexane. Exchange is not required for the acetonitrile extracts. Cleanup may not be necessary for extracts from a relatively clean sample matrix. Florisil Cartridge Cleanup may be used for extract cleanup.

If PCBs and organochlorine pesticides are known to be present in the sample, and if Florisil Cartridge Cleanup is considered, then two fractions are collected: Fraction 1 is eluted with 5 mL of 20% methylene chloride in hexane and Fraction 2 is eluted with 5 mL of 10% acetone in hexane. Fraction 1 contains the organochlorine pesticides and PCBs, and can be discarded. Fraction 2 contains the phthalate esters and is analyzed by GC/ECD.

QUALITY CONTROL Identify compounds in the sample by comparing the retention times of the peaks in the sample chromatogram with those of the peaks in standard chromatograms. The retention time window used to make identification is based upon measurements of actual retention time variations over the course of 10 consecutive injections.

Calibration standards are prepared at a minimum of five concentrations for each parameter of interest through dilution of the stock standard solutions with hexane. One of the concentrations should be at a concentration near, but above, the method detection limit. Prepare stock standard solutions in hexane. Stock standards should be checked frequently for signs of degradation or evaporation, especially just prior to preparing calibration standards from them. Stock standard solutions must be replaced after one year, or sooner if comparison with check standards indicates a problem. The suggested internal standards are: 2,5-dibromotoluene, 1,3,5-tribromobenzene, and α, α'-dibromo-m-xylene. The analyst can use any of the three compounds provided that they are resolved from matrix interferences. Recommended surrogate compounds are α-2,6-trichlorotoluene, 1,4-dichloronaphthalene, and 2,3,4,5,6-pentachlorotoluene.

Spike each sample, standard, and blank with surrogate compounds. Three surrogates are suggested for this method: diphenyl phthalate, diphenyl isophthalate, and dibenzyl phthalate.

The quality control check sample concentrate should contain the test compounds at 5 to 10 ng/μL An internal standard peak area check must be performed on all samples. The internal standard must be evaluated for acceptance by determining whether the measured area for the internal standard deviates by more than 30% from the average are for the internal standard in the calibration standards. When the internal standard peak area is outside that limit, all samples that fall outside the QC criteria must be reanalyzed. Benzyl benzoate has been tested and found appropriate as an internal standard for this method.

Any compounds confirmed by two columns may also be confirmed by GC/MS. The sample extract and associated blank should be analyzed by GC/MS. A reference standard of the compound must also be analyzed by GC/MS. Include a mid-concentration calibration standard after each group of 20 samples. The response factors for the mid-concentration calibration must be within ±15% of the average values for the multiconcentration calibration. Demonstrate through the analyses of standards that the Florisil fractionation scheme is reproducible.

REFERENCE Test Methods for Evaluating Solid Waste, Physical/Chemical Methods, SW-846, 3rd Edition, U.S. EPA, Office of Solid Waste, Washington, DC, EPA Method 8061, Nov. 1990.

Dieldrin   EPA Method 505
CAS #60-57-1

TITLE Analysis of Organohalide Pesticides and Commercial Polychlorinated Biphenyl (PCB) Products in Water by Microextraction and Gas Chromatography. U.S. EPA Method 505, Rev. 2.0, 1989.

MATRIX This method is applicable to drinking water and raw source water. The latter should include most surface water and groundwater sources.

METHOD SUMMARY Method 505 covers 25 pesticides and commercial PCB products. This is a very sensitive method that is more useful for monitoring than for exploratory analyses. 5-mL of water are saturated with sodium chloride and then extracted by shaking with 2 mL of hexane. The sample extracts are transferred to an autosampler setup to inject 1–2 μL portions into a gas chromatograph (GC) for analysis. Alternatively, 1–2 μL portions of samples, blanks, and standards may be manually injected. Each extract is analyzed by capillary GC/ECD with confirmation using either a second capillary column or GC/MS. The electron capture detector is easy to use, but it is a nonselective detector. The microextraction technique also eliminates the expensive sample preparation costs of other methods, but it has the disadvantage of being less sensitive than most because the extracts are not concentrated.

INTERFERENCES Method interferences may be caused by contaminants in solvents, reagents, glassware, and other sample processing apparatus that lead to discrete artifacts or elevated baselines. Interfering contamination may occur when a sample containing low concentrations of analytes is analyzed immediately following a sample containing relatively high concentrations of the analytes. Matrix interferences also may be caused by contaminants that are coextracted from the sample; cleanup of sample extracts may be necessary in these cases. Some pesticides and commercial PCB products from aqueous solutions adhere to glass surfaces, so sample transfers and contact with glass surfaces should be minimized. Some pesticides are rapidly oxidized by chlorine so dechlorination with sodium thiosulfate at the time of sample collection is important. Also, splitless injectors may cause degradation of some pesticides.

INSTRUMENTATION A gas chromatograph/electron capture detector/data system, with temperature programming and split/splitless injector suitable for use with capillary columns is needed.

Column 1: 0.32 mm I.D. × 30 m fused silica capillary with chemically bond methyl polysiloxane phase (DB-1, 1.0 μm film, or equivalent).
Column 2: 0.32 mm I.D. × 30 m fused silica capillary with 1:1 mixed phase of dimethyl silicone and polyethylene glycol (Durawax-DX3, 0.25 μm film, or equivalent).
Column 3: 0.32 mm I.D. × 25 m fused silica capillary with chemically bonded 50:50 methyl-phenyl silicone (OV-17, 1.5 μm film, or equivalent).

Column 1 should be used as the primary analytical column. Columns 2 and 3 are recommended for use as confirmatory columns when GC/MS confirmation is not available.

PRECISION & ACCURACY Method detection limits are dependent upon the characteristics of the gas chromatographic system used. Analytes that are not separated chromatographically cannot be individually identified and used in the same calibration mixture or water samples unless an alternative technique for identification and quantification, such as mass spectrometry, is used.

The concentration(s) (in μg/L) used for these QC measurements was 0.10 and 3.6.
The MDL (in μg/L) was 0.012.
The accuracy (% recovery) for reagent water at the above concentration(s) was 87 and 114 and the precision (%) was 17.1 and 9.1.
The accuracy (% recovery) for groundwater at the above concentration(s) was 67 and 94 and the precision (%) was 10.1 and 8.6.

The accuracy (% recovery) for tap water at the above concentration(s) was 92 and 81 and the precision (5) was 15.7 and 14.0.

Note: No range of concentrations is provided with this method.

SAMPLING METHOD Collect samples using a 40-mL screw-cap vial (prewashed with detergent, rinsed with distilled water and oven dried at 400°C for one h) with a Teflon®-faced silicone septum. Collect bubble-free samples and place the septum with the Teflon® side down on the water.

SAMPLE PRESERVATION If residual chlorine is present in the water add about 3 mg of sodium thiosulfate to each vial before samples are collected to remove the chlorine. Alternatively, add 75 µL of 0.04 g/mL solution of sodium thiosulfate to each vial just prior to sampling. Immediately cool samples to 4°C, and store them in a solvent-free refrigerator at 4°C until analysis.

MHT The maximum holding time is 14 days from the time the sample was collected until it must be analyzed.

SAMPLE PREPARATION Remove the sample from storage and allow it to come to room temperature. Remove a 5-mL volume from each container and weigh the container to the nearest 0.1 g. Add 6 g of sodium chloride and 2.0 mL of hexane to each sample bottle. Recap the sample and shake it vigorously for one min. Allow the water and hexane phases to separate, remove the cap, and transfer 0.5 mL of hexane into an autosampler vial using a disposable glass pipette. Transfer the remaining hexane phase into a second autosampler vial and store at 4°C for reanalysis, if necessary. Discard the remaining sample/hexane mixture and reweigh the empty container to determine net weight of sample.

QUALITY CONTROL Minimum quality control requirements are initial demonstration of lab capability, analysis of lab reagent blanks, fortified blanks, fortified sample matrix, and quality control samples. The lab must analyze at least one fortified blank per sample set, or at least one for every 20 samples. The fortifying concentration of each analyte should be 10 times the method detection limit or the maximum calibration limit (MCL), whichever is less. Calculate accuracy as percent recovery and develop control limits from the mean percent recovery and standard deviation.

The lab must add a known concentration of the analytes to a minimum of 10% of the routine samples, or one lab fortified sample matrix per sample set. Calculate the percent recovery for each analyte and compare to the control limits established from the analyses of the fortified blanks.

EPA CONTACT & HOTLINE For technical questions contact Dr. Baldev Bathija, U.S. EPA, Office of Ground Water and Drinking Water (WH-550D), 401 M St. SW, Washington, DC 20460. Tel. (202) 260-3040. For further information the EPA Safe Drinking Water Hotline may be called at: (800) 426-4791.

REFERENCE Methods for the Determination of Organic Compounds in Drinking Water, EPA/600/4-88/039 (revised July 1991). U.S. EPA Environmental Monitoring Systems Laboratory, Cincinnati, OH, 45268, U.S.A. Available from the National Technical Information Service (NTIS), 5285 Port Royal Road, Springfield, VA 22161; Tel. 800-553-6847. NTIS Order Number is PB91-231480.

Dieldrin EPA Method 625
CAS #60-57-1

TITLE Base/Neutrals and Acids, U.S. EPA Method 625

MATRIX This methods covers municipal and industrial wastewaters.

METHOD SUMMARY Approximately 1 L of sample is serially extracted with methylene chloride at a pH greater than 11 and again at a pH less than 2 using a separatory funnel or a continuous extractor. The methylene chloride extract is dried, concentrated to a volume of 1 mL, and analyzed by GC/MS. Qualitative identification of the parameters in the extract is performed using the retention time and the relative abundance of three characteristic masses (m/z). Qualitative analysis is performed using either external or internal standard techniques with a single characteristic m/z.

INTERFERENCES Method interferences may be caused by contaminants in solvents, reagents, glassware, and other sample processing hardware. Glassware must be scrupulously cleaned. Glassware should be heated in a muffle furnace at 400°C for 5 to 30 min. Some thermally stable materials, such as PCBs, may not be eliminated by this treatment. Solvent rinses with acetone and pesticide quality hexane may be substituted for the muffle furnace heating. Matrix interferences may be caused by contaminants that are coextracted from the sample. The base-neutral extraction may cause significantly reduced recovery of phenols. The packed gas chromatographic columns recommended for the basic fraction may not exhibit sufficient resolution for some analytes.

INSTRUMENTATION A GC/MS system with an injection port designed for on-column injection when using packed columns and for splitless injection when using capillary columns.

Column for base/neutrals: 1.8 m long × 2 mm I.D. glass, packed with 3% SP-2550 on Supelcoport (100/120 mesh) or equivalent.
Column for acids: 1.8 m long × 2 mm I.D. glass, packed with 1% SP-1240DA on Supelcoport (100/120 mesh) or equivalent.

PRECISION & ACCURACY The MDL concentrations were obtained using reagent water. The MDL actually achieved in a given analysis will vary depending on instrument sensitivity and matrix effects. This method was tested by 15 laboratories using reagent water, drinking water, surface water, and industrial wastewaters spiked at six concentrations over the range 5 to 100 µg/L. Single operator precision, overall precision, and method accuracy were found to be directly related to the concentration of the parameter matrix.

The MDL (in µg/L) in reagent water was 2.5.
The standard deviation (in µg/L based on 4 recovery measurements) was 30.7.
The range (in µg/L) for average recovery for 4 measurements was 44.3–119.3.

The range (in %) for percent recovery was 29–136.

Accuracy (in µg/L) as expected recovery for one or more measurements of a sample containing a true concentration of C was 0.82C−0.16.

Precision (in µg/L) as expected single analyst standard deviation of measurements at an average concentration found at X was 0.20X−0.16.

Overall precision (in µg/L) as expected interlaboratory standard deviation of measurements in an average concentration found at X was 0.26X−0.07.

C = *True value of the concentration in µg/L.*
X = *Average recovery found for measurements of samples containing a concentration at C in µg/L.*

SAMPLE PREPARATION Adjust the pH to >11 with sodium hydroxide and serially extract in a separatory funnel with methylene chloride or else in a continuous extractor. Next, adjust the pH to <2 with sulfuric acid and serially extract in a separatory funnel with methylene chloride or else in a continuous extractor. Dry the extracts separately through a column of anhydrous sodium sulfate and then concentrate each of the extracts to 1.0 mL using a K-D apparatus.

SAMPLE COLLECTION, PRESERVATION & HANDLING Grab samples must be collected in glass containers. All samples must be refrigerated at 4°C from the time of collection until extraction. If residual chlorine is present, add 80 mg of sodium thiosulfate/L of sample and mix well. All samples must be extracted within 7 days of collection and completely analyzed within 40 days of extraction.

QUALITY CONTROL Make an initial, one-time, demonstration of the ability to generate acceptable accuracy and precision with this method. Before processing any samples, the analyst must analyze a reagent water blank to demonstrate that interferences from the analytical system and glassware are under control. Each time a set of samples is extracted or reagents are changed, a reagent water blank must be processed. Spike and analyze a minimum of 5% of all samples to monitor and evaluate lab data quality. A QC check sample concentrate that contains each parameter of interest at a concentration of 100 µg/mL in acetone is required. PCBs and multicomponent pesticides may be omitted from this test.

After analysis of five spiked wastewater samples, calculate the average percent recovery and the standard deviation of the percent recovery. Spike all samples with the surrogate standard spiking solution and calculate the percent recovery of each surrogate compound.

REFERENCE *Federal Register*, Vol. 49, No. 209. Friday, Oct. 26, 1984.

Dieldrin **EPA Method 8080**
CAS #60-57-1

TITLE Organochlorine Pesticides and Polychlorinated Biphenyls By Gas Chromatography

MATRIX This method is used to determine the concentration of various organochlorine pesticides and polychlorinated biphenyls in extracts prepared from water, groundwater, soils, and sediments.

METHOD SUMMARY This method covers 26 pesticides and Aroclor (PCB) mixtures and it is suitable for monitoring-type analyses. After extraction, concentration and solvent exchange to hexane, a 2- to 5-µL sample aliquot is injected into a GC using the solvent flush technique, and the analytes are detected by an electron capture detector (ECD) or an electrolytic conductivity detector in the halogen mode (HECD). Both neat and diluted organic liquids may be analyzed by direct injection.

INTERFERENCES Interferences coextracted from the samples will vary considerably from source to source. Interferences by phthalate esters can pose a major problem in pesticide determinations when using the ECD. Cross-contamination of clean glassware routinely occurs when plastics are handled during extraction steps, especially when solvent-wetted surfaces are handled. The contamination from phthalate esters can be completely eliminated with a microcoulometric or electrolytic conductivity detector. Solvents, reagent, glassware, and other sample processing hardware may yield artifacts and/or interferences to sample analysis.

INSTRUMENTATION A gas chromatograph capable of on-column injections is needed. It must be equipped with an ECD or a HECD and one of the following GC columns:

Column 1: Supelcoport (100/120 mesh) coated with 1.5% SP-2250/1.95% SP-2401 packed in a 1.8 m × 4 mm I.D. glass column.

Column 2: Supelcoport (100/120 mesh) coated with 3% OV-1 in a 1.8 m × 4 mm I.D. glass column.

PRECISION & ACCURACY The method was tested by 20 laboratories using organic-free reagent water, drinking water, surface water, and three industrial wastewaters spiked at six concentrations. Concentrations used in the study ranged from 0.5 to 30 µg/L for single-component pesticides and from 8.5 to 400 µg/L for multicomponent parameters. Overall precision and method accuracy were found to be directly related to the concentration of the analyte and essentially independent of the sample matrix. The sensitivity of this method usually depends on the concentration of interferences rather than on instrumental limitations.

MDL in µg/L was 0.002.
Concentration range in µg/L was 0.5–30.
Accuracy as recovery (x^*) in µg/L was 0.90C+0.02.
Overall precision (S^*) in µg/L was 0.16x +0.16.

x^* *Expected recovery for one or more measurements of a sample containing concentration C, in µg/L.*
S^* = *Expected interlaboratory standard deviation of measurements at an average concentration found of the analyte in µg/L.*
C = *True value for the concentration, in µg/L.*
X = *Average recovery found for measurements of samples containing a concentration of C, in µg/L.*

SAMPLING METHOD

Liquid samples — Use a 1 or 2½ gallon amber glass bottle with a screw-top Teflon®-lined cover. Pre-wash the bottle with detergent, rinse with distilled water and methanol (or isopropanol).

Soil, sediments, and sludges — Use an 8-oz. widemouth glass with a screw-top Teflon®-lined cover. Pre-wash the bottle with detergent, rinse with distilled water and methanol (or isopropanol).

SAMPLE PRESERVATION Cool water, soil, sediment, or sludge samples immediately to 4°C.

Water samples — If residual chlorine is present, add 3 mL of 10% sodium thiosulfate per gallon and cool to 4°C. All extracts and samples should be stored under refrigeration.

MHT Liquid samples must be extracted within 7 days and the extracts must be analyzed within 40 days. Soils, sediments, and sludges may be stored for a maximum of 14 days prior to extraction.

SAMPLE PREPARATION

Liquid samples — Extract 1 L samples in a continuous extractor at pH 5–9 with methylene chloride after adding 1.0 mL of surrogate spiking solution to each sample. Pass the extract through a column of anhydrous sodium sulfate to dry and concentrate it in a K-D apparatus to 1 mL volume.

Soils, sediments, and sludges — Rapidly weigh approximately 30 g of sample into a 400-mL beaker to avoid loss of the more volatile extractables. Nonporous or wet samples (gummy or clay type) that do not have a free-flowing sandy texture must be mixed with anhydrous sodium sulfate until the sample is free flowing. Add 1 mL of surrogate standards to all samples, spikes, standards, and blanks. Add 100 mL of 1:1 methylene chloride:acetone and extract ultrasonically. Decant and filter extracts, dry the extract by passing it through a drying column containing anhydrous sodium sulfate and concentrate to 1 mL in a K-D apparatus.

Hexane solvent exchange — Add 50 mL of hexane, a new boiling chip, and concentrate until the apparent volume of liquid reaches 1 mL. Adjust the extract volume to 10.0 mL. Stopper the concentration tube and store refrigerated at 4°C if further processing will not be performed immediately. If the extract will be stored longer than two days, transfer it to a vial with Teflon®-lined screw-cap or crimp top.

QUALITY CONTROL Demonstrate through the analysis of a reagent water blank, that all glassware and reagents are interference free. Each time a set of samples is processed, a method blank should be processed as a safeguard against chronic lab contamination. A reagent blank, a matrix spike, and a duplicate or matrix spike duplicate must be performed for each analytical batch (up to a maximum of 20 samples) analyzed.

Analytical system performance must be verified by analyzing QC check samples. The QC check sample concentration should contain each single-component analyte at the following concentrations in acetone: 4,4'-DDD, 10 µg/mL; 4,4'-DDT, 10 µg/mL; endosulfan II, 10 µg/mL; endosulfan sulfate, 10 µg/mL; and any other single-component pesticide at 2 µg/mL. If the method is only to be used to analyze PCBs, Chlordane, or Toxaphene, the QC check sample concentrate should contain the most representative multicomponent parameter at a concentration of 50 µg/mL in acetone.

REFERENCE Test Methods for Evaluating Solid Waste (SW-846). U.S. EPA. 1983. Method 8080B, Rev. 2, Nov. 1990. Office of Solid Wastes, Washington, DC.

Dieldrin — EPA Method 8270
CAS #60-57-1

TITLE Semivolatile Organic Compounds by GC/MS

MATRIX This method is used to determine the concentration of semivolatile organic compounds in extracts prepared from all types of solid waste matrices, soils, and groundwater. Although surface waters are not specifically mentioned, this method should be applicable to water samples from rivers, lakes, etc.

METHOD SUMMARY This method covers 259 semivolatile organic compounds. In very limited applications direct injection of the sample into the GC/MS system may be appropriate, but this results in very high detection limits (approximately 10,000 µg/L). Typically, a 1-L liquid sample, containing surrogate, and matrix spiking standards, is extracted in a continuous extractor first under acid conditions and then under basic conditions. Typically 30 g of a solid sample, containing surrogate, and matrix spiking standards, is extracted ultrasonically. After concentrating the extract to 1 mL it is spiked with 10 µL of an internal standard solution just prior to analysis by GC/MS. The volume injected should contain about 100 ng of base/neutral and 200 ng of acid surrogates (for a 1-µL injection). Analysis is performed by GC/MS using a capillary GC column.

INTERFERENCES Raw GC/MS data from all blanks, samples, and spikes must be evaluated for interferences. Contamination by carryover can occur whenever high-concentration and low-concentration samples are sequentially analyzed. To reduce carryover, the sample syringe must be rinsed out between samples with solvent. Whenever an unusually concentrated sample is encountered, it should be followed by the analysis of blank solvent to check for cross-contamination.

INSTRUMENTATION A GC/MS and a data system are required. The GC column used is a 30 m × 0.25 mm I.D. (or 0.32 mm I.D.) 1um film thickness silicone-coated fused silica capillary column. A continuous liquid-liquid extractor equipped with Teflon® or glass connection joints and stopcocks requiring no lubrication, a K-D concentrating apparatus, water bath, and an ultrasonic disrupter with a minimum power of 300 W and with pulsing capability are also required.

PRECISION & ACCURACY The estimated quantitation limit (EQL) of Method 8270B for determining an individual compound is approximately 1 mg/kg (wet weight) for soil or sediment samples, 1–200 mg/kg for wastes (dependent on matrix and method of preparation), and 10 µg/L for groundwater samples. EQLs will be proportionately higher for sample extracts that require dilution to avoid saturation of the detector.

The EQL(b) for groundwater in μg/L is not listed.
The EQL (a, b) for low concentrations in soil and sediment in μg/kg is not listed.
Accuracy as μg/L is 0.82C–0.16.
Overall precision in μg/L is 0.26X–0.07.

(a) *EQLs listed for soil/sediment are based on wet weight. Normally data is reported in a dry-weight basis; therefore, EQLs will be higher based on the % dry weight of each sample. This calculation is based on a 30 g sample and gel permeation chromatography cleanup.*
(b) *Sample EQLs are highly matrix-dependent. The EQLs are provided for guidance and may not always be achievable.*
C = *True value for concentration, in μg/L.*
X = *Average recovery found for measurements of samples containing a concentration of C, in μg/L.*

ESTIMATED QUANTITATION LIMIT

Other Matrices	Factor (a)
High-concentration soil and sludges by sonicator	7.5
Non-water miscible waste	75

(a) EQL for other matrices = [EQL for low soil/sediment] × [Factor]. This estimated EQL is similar to an EPA "Practical Quantitation Limit."

SAMPLING METHOD

Liquid samples — Use a 1 or 2½ gallon amber glass bottle with a screw-top Teflon®-lined cover that has been prewashed with detergent and rinsed with distilled water and methanol (or isopropanol).

Soils, sediments, or sludges — Use an 8-oz. widemouth glass with a screw-top Teflon®-lined cover that has been prewashed with detergent and rinsed with distilled water and methanol (or isopropanol).

SAMPLE PRESERVATION

Liquid samples — If residual chlorine is present, add 3 mL of 10% sodium thiosulfate per gallon, cool to 4°C and store in a solvent-free refrigerator until analysis; if chlorine is not present, then eliminate the sodium thiosulfate addition.

Soils, sediments, or sludges — Cool samples to 4°C and store in a solvent-free refrigerator.

MHT Liquid samples must be extracted within 7 days and the extracts analyzed within 40 days. Soils, sediments, or sludges may be stored for a maximum of 14 days and the extracts analyzed within 40 days.

SAMPLE PREPARATION

Liquid samples — Transfer 1 L quantitatively to a continuous extractor. If high concentrations are anticipated, a smaller volume may be used and then diluted with organic-free reagent water to 1 L. Adjust pH, if necessary, to pH <2 using 1:1 (V/V) sulfuric acid. Pipette 1.0 mL of a surrogate standard spiking solution into each sample. For the sample in each analytical batch selected for spiking, add 1.0 mL of a matrix spiking standard. For base/neutral acid analysis, the amount of the surrogates and matrix spiking compounds added to the sample should result in a final concentration of 100 ng/μL of each analyte in the extract to be analyzed (assuming a 1-μL injection). Extract with methylene chloride for 18–24 h. Next, adjust the pH of the aqueous phase to pH >11 using 10 N sodium hydroxide and extract it with methylene chloride again for 18–24 h. Dry the extract through a column containing anhydrous sodium sulfate and concentrate it to 1 mL using a K-D concentrator.

Soils, sediments, or sludges — Use 30 g of sample. Nonporous or wet samples (gummy or clay type) that do not have a free-flowing sandy texture must be mixed with anhydrous sodium sulfate until the sample is free flowing. Add 1 mL of surrogate standards to all samples, spikes, standards, and blanks. For the sample in each analytical batch selected for spiking, add 1.0 mL of a matrix spiking standard. For base/neutral acid analysis, the amount added of the surrogates and matrix spiking compounds should result in a final concentration of 100 ng/μL of each base/neutral analyte and 200 ng/μL of each acid analyte in the extract to be analyzed (assuming a 1-μL injection). Immediately add a 100-mL mixture of 1:1 methylene chloride:acetone and extract the sample ultrasonically for 3 min and then decant or filter the extracts. Repeat the extraction two or more times. Dry the extract using a column with anhydrous sodium sulfate and concentrate it to 1 mL in a K-D concentrator.

QUALITY CONTROL A methylene chloride solution containing 50 ng/μL of decafluorotriphenylphosphine (DFTPP) is used for tuning the GC/MS system each 12-h shift. A system performance check also must be made during every 12-h shift. A standard containing 50 ng/μL each of 4,4'-DDT, pentachlorophenol, and benzidine is required to verify injection port inertness and GC column performance. A calibration standard at mid-concentration, containing each compound of interest, including all required surrogates, must be performed every 12 h during analysis. After the system performance check is met, calibration check compounds (CCCs) are used to check the validity of the initial calibration.

The internal standard responses and retention times in the calibration check standard must be evaluated immediately after or during data acquisition. If the retention time for any internal standard changes by more than 30 seconds from the last check calibration (12 h), the chromatographic system must be inspected for malfunctions and corrections must be made, as required. If the electron ionization current plot (EICP) area for any of the internal standards changes by a factor of two from the last daily calibration standard check, the mass spectrometer must be inspected for malfunctions and corrections must be made, as appropriate.

Demonstrate, through the analysis of a reagent water blank, that interferences from the analytical system, glassware, and reagents are under control. The blank samples should be carried through all stages of the sample preparation and measurement steps. For each analytical batch (up to 20 samples), a reagent blank, matrix spike, and matrix spike duplicate/duplicate must be analyzed (the frequency of the spikes may be different for different monitoring programs). The blank and spiked samples must be carried through all stages of the sample preparation and measurement steps. A QC reference sample concentrate containing each analyte at a concentration of 100 mg/L in methanol is required.

REFERENCE Test Methods for Evaluating Solid Waste (SW-846). U.S. EPA 1983, Method 8270B, Rev. 2, Nov. 1990. Office of Solid Waste, Washington, DC.

1,2:3,4-Diepoxybutane   **EPA Method 1625**
CAS #1464-53-5

TITLE Semivolatile Organic Compounds by Isotope Dilution GC/MS

MATRIX The compounds may be determined in waters, soils, and municipal sludges by this method.

METHOD SUMMARY This method is used to determine 176 semivolatile toxic organic pollutants associated with the CWA (as amended 1987); the RCRA (as amended 1986); the CERCLA (as amended 1986); and other compounds amenable to extraction and analysis by capillary column gas chromatography-mass spectrometry (GC/MS).

Stable isotopically-labeled analogs of the compounds of interest are added to the sample. If the solids content is less than 1%, a 1-L sample is extracted at pH 12–13, then at pH <2 with methylene chloride using continuous extraction techniques.

If the solids content is 30% or less, the sample is diluted to 1% solids with reagent water, homogenized ultrasonically, and extracted at pH 12–13, then at pH <2 with methylene chloride using continuous extraction techniques. If the solids content is greater than 30%, the sample is extracted using ultrasonic techniques.

Each extract is dried over sodium sulfate, concentrated to a volume of 5 mL, cleaned up using GPC, if necessary, and concentrated. Extracts are concentrated to 1 mL if GPC is not performed, and to 0.5 mL if GPC is performed.

An internal standard is added to the extract, and a 1-mL aliquot of the extract is injected into the GC. The compounds are separated by GC and detected by a MS. The labeled compounds serve to correct the variability of the analytical technique.

INTERFERENCES Solvents, reagents, glassware, and other sample processing hardware may yield artifacts and/or elevated baselines causing misinterpretation of chromatograms and spectra. Materials used in the analysis must be demonstrated to be free from interferences under the conditions of analysis by running method blanks initially and with each sample lot (sample started through the extraction process on a given 8-h shift, to a maximum of 20). Specific selection of reagents and purification of solvents by distillation in all glass systems may be required. Glassware and, where possible, reagents are cleaned by solvent rinse and baking at 450°C for 1-h minimum. Interferences coextracted from samples will vary considerably from source to source, depending on the diversity of the site being sampled.

INSTRUMENTATION Major instrumentation includes a GC with a splitless or on-column injection port for capillary column, a MS with 70 eV electron impact ionization, and a data system to collect and record MS data, and process it. A K-D apparatus is used to concentrate extracts.

GC Column: 30 m × 0.25 mm I.D. 5% phenyl, 94% methyl, 1% vinyl silicone bonded phased fused silica capillary column.

PRECISION & ACCURACY The detection limits of the method are usually dependent on the level of interferences rather than instrumental limitations. The limits typify the minimum quantities that can be detected with no interferences present.

The minimum level (in µg/mL) was not listed. This is defined as a minimum level at which the analytical system shall give recognizable mass spectra (background corrected) and acceptable calibration points.

The MDL (in µg/kg) in low solids was not listed and in high solids was not listed; these were determined in digested sludge (low solids) and in filter cake or compost (high solids).

The labeled and native compound initial precision as standard deviation (in µg/L) was not listed.
The labeled and native compound initial accuracy as average recovery (in µg/L) was not listed.

SAMPLE COLLECTION, PRESERVATION & HANDLING Collect samples in glass containers. Aqueous samples which flow freely are collected in refrigerated bottles using automatic sampling equipment. Solid samples are collected as grab samples using widemouth jars. Maintain samples at 0 to 4°C from the time of collection until extraction. If residual chlorine is present in aqueous samples, add 80 mg sodium thiosulfate/L of water. Begin sample extraction within 7 days of collection, and analyze all extracts within 40 days of extraction.

SAMPLE PREPARATION Samples containing 1% solids or less are extracted directly using continuous liquid-liquid extraction techniques. Samples containing 1 to 30% solids are diluted to the 1% level with reagent water and extracted using continuous liquid-liquid extraction techniques. Samples containing greater than 30% solids are extracted using ultrasonic techniques.

Base/neutral extraction — Adjust the pH of the waters in the extractors to 12–13 with 6 *N* NaOH. Extract with methylene chloride for 24–48 h.
Acid extraction — Adjust the pH of the waters in the extractors to 2 or less using 6 *N* sulfuric acid. Extract with methylene chloride for 24–48 h.
Ultrasonic extraction of high solids samples — Add anhydrous sodium sulfate to the sample and QC aliquot(s). Add acetone:methylene chloride (1:1) to the sample and mix thoroughly

Concentrate extracts using a K-D apparatus.

QUALITY CONTROL The analyst is permitted to modify this method to improve separations or lower the costs of measurements, provided all performance specifications are met. Analyses of blanks are required to demonstrate freedom from contamination. When results of spikes indicate atypical method performance for samples, the samples are diluted to bring method performance within acceptable limits.

For low solids (aqueous samples), extract, concentrate, and analyze two sets of four 1-L aliquots (8 aliquots total) of the

precision and recovery standard. For high solids samples, two sets of four 30-g aliquots of the high solids reference matrix are used.

Spike all samples with labeled compounds to assess method performance. Compute percent recovery of the labeled compounds using the internal standard method. Compare the labeled compound recovery for each compound with the corresponding labeled compound recovery.

Reagent water and high solids reference matrix blanks are analyzed to demonstrate freedom from contamination. Extract and concentrate a 1-L reagent water blank or a high solids reference matrix blank with each sample's lot (samples started through the extraction process on the same 8-h shift, to a maximum of 20 samples).

Field replicates may be collected to determine the precision of the sampling technique, and spiked samples may be required to determine the accuracy of the analysis when the internal standard method is used.

REFERENCE Semivolatile Organic Compounds by Isotope Dilution GC/MS. Office of Water Regulation and Standards, U.S. EPA Industrial Technology Division, Washington, DC, EPA Method 1625, Rev. C, June 1989 (contact W.A. Telliard, U.S. EPA, Office of Water Regulations and Standards, 401 M St., SW, Washington, DC, 20460. Phone: 202-382-7131).

1,2:3,4-Diepoxybutane   **EPA Method 8240**
CAS #1464-53-5

TITLE Volatile Organics By GC/MS: Packed Column Technique

MATRIX Nearly all types of sample matarices, regardless of water content, can be analyzed using this method. This includes groundwater, aqueous sludges, caustic liquors, acid liquors, waste solvents, oily wastes, mousses, tars, fibrous wastes, polymetric emulsions, filter cakes, spent carbons, spent catalysts, soils, and sediments.

METHOD SUMMARY Method 8240B covers 80 volatile organic compounds that are introduced into a gas chromatograph by the purge-and-trap method or by direct injection (in limited applications). For the purge-and-trap method an inert gas (zero grade nitrogen or helium) is bubbled through a 5-mL solution at ambient temperature. Purged sample components are trapped in a tube of sorbent materials. When purging is complete, the sorbent tube is heated and backflushed with inert gas to desorb the trapped components onto a GC column.

INTERFERENCES Impurities in the purge gas and from organic compounds outgassing from the plumbing ahead of the trap account for many contamination problems. Interferences purged or coextracted from the samples will vary considerably from source to source. Cross-contamination can occur whenever high-level and low-level samples are analyzed sequentially. Whenever an unusually concentrated sample is analyzed, it should be followed by the analysis of organic-free reagent water to check for cross-contamination. Samples also can be contaminated by diffusion of volatile organics (particularly methylene chloride and fluorocarbons) through the septum seal into the sample during shipment and storage. A trip blank can serve as a check on such contamination. The lab where volatile analysis is performed and also the refrigerated storage area should be completely free of solvents.

INSTRUMENTATION A gas chromatograph/mass spectrometry/data system (GC/MS) equipped with a 6 ft × 0.1 in I.D. glass column packed with 1% SP-1000 on Carbopack-B (60/80 mesh) is required. Also needed is a 5-mL purging device, a sorbent trap, and a thermal desorption apparatus.

PRECISION & ACCURACY This method is reported to have been tested by 15 laboratories using organic-free reagent water, drinking water, surface water, and industrial wastewaters (not specified) fortified at six concentrations over the range 5–600 µg/L.

Sample estimated quantitation limits (EQLs) are highly matrix-dependent. The EQLs listed may not always be achievable. EQLs listed for soils or sediments are based on wet weight. Normally, data is reported on a dry-weight basis; therefore, EQLs will be higher, based on the percent dry weight of each sample. Note that EQLs are even more variable than MDLs and that they are highly variable depending on the matrix being analyzed.

EQL in groundwater in µg/L was not listed.
EQL in low soil or sediment in µg/kg was not listed.
Accuracy (a) in µg/L was not listed.
Precision (b) in µg/L was not listed.

(a) *Average recovery found for measurements of samples containing a concentration of C, in µg/L.*
(b) *Overall precision found for measurements of samples with average recovery X for samples containing a concentration of C in µg/L.*
X = *Average recovery found for measurement of samples containing a concentration of C in µg/L.*

MULTIPLICATION FACTORS FOR OTHER MATRICES

Other Matrices	Factor (a)
Waste miscible liquid waste	50
High-concentration soil and sludge	125
Non-water miscible waste	500

(a) EQL = [EQL for low soil sediment] × [Factor]. For non-aqueous samples, the factor is on a wet-weight basis.

SAMPLING METHOD

Liquid samples — Use a 40-mL glass screw-cap VOA vial with a Teflon®-faced silicone septum that has been prewashed, rinsed with distilled deionized water, and oven dried. However, if residual chlorine is present, collect sample in a 40-oz. soil VOA container which has been pre-preserved with 4 drops of 10% sodium thiosulfate, mix gently, and then transfer the sample to a 40-mL VOA vial. Collect bubble-free samples in duplicate and seal them in separate plastic bags.

Soils or sediments, and sludges — Use an 8-oz. widemouth glass bottle with a Teflon®-faced silicone septum that has been prewashed with detergent, rinsed with distilled deionized water, and oven dried. Tap slightly to eliminate free air space.

Collect samples in duplicate and seal them in separate plastic bags.

SAMPLE PRESERVATION

Liquid samples — Add 4 drops of concentrated HCL and immediately cool samples to 4°C and store in a solvent-free refrigerator.

Soils or sediments, and sludges — Cool samples to 4°C and store in a solvent-free refrigerator.

MHT Maximum holding time is 14 days from the date of sample collection.

SAMPLE PREPARATION

Liquid samples — Remove the plunger from a 5-mL syringe and carefully pour the sample into the syringe barrel to just short of overflowing. Replace the syringe plunger and compress the sample. Open the syringe valve and vent any residual air while adjusting the sample volume to 5.0 mL. If there is only one volatile organic analysis (VOA) vial, a second syringe should be filled at this time to protect against possible loss of sample integrity. Add 10 µL of surrogate spiking solution and 10 µL of internal standard spiking solution through the valve bore of the 5-mL syringe, then close the valve. The surrogate and internal standards may be mixed and added as a single spiking solution.

Sediments, soils, and waste samples — All samples of this type should be screened by GC analysis using a headspace method (EPA Method 3810) or the hexadecane extraction and screening method (EPA Method 3820). Use the screening data to determine whether to use the low-concentration method (0.005–1 mg/kg) or the high-concentration method (>1 mg/kg).

Low-concentration method — The low-concentration method is based on purging a heated sediment or soil sample mixed with organic-free reagent water containing the surrogate and internal standards. Analyze all reagent blanks and standards under the same conditions as the samples.

Use a 5-g sample if the expected concentration is <0.1 mg/kg or a 1-g sample for expected concentrations between 0.1 and 1 mg/kg. Mix the contents of the sample container with a narrow metal spatula. Weigh the amount of the sample into a tared purge device. Add the spiked water to the purge device, which contains the weighed amount of sample, and connect the device to the purge-and-trap system.

High-concentration method — This method is based on extracting the sediment or soil with methanol. A waste sample is either extracted or diluted, depending on its solubility in methanol. Wastes that are insoluble in methanol are diluted with reagent tetraglyme or possibly polyethylene glycol (PEG). An aliquot of the extract is added to organic-free reagent water containing surrogate and internal standards. This is purged at ambient temperature. All samples with an expected concentration of >1.0 mg/kg should be analyzed by this method.

Mix the contents of the sample container with a narrow metal spatula. For sediments or soils and solid wastes that are insoluble in methanol, weigh 4 g (wet weight) of sample into a tared 20-mL vial. For waste that is soluble in methanol, tetraglyme, or PEG, weigh 1 g (wet weight) into a tared scintillation vial or culture tube or a 10-mL volumetric flask. Quickly add 9.0 mL of appropriate solvent then add 1.0 mL of a surrogate spiking solution to the vial, cap it, and shake it for 2 min.

METHANOL EXTRACT REQUIRED FOR ANALYSIS OF HIGH-CONCENTRATION SOILS OR SEDIMENTS

Approximate Concentration Range	Volume of Methanol Extract (a)
500–10,000 µg/kg	100 µL
1,000–20,000 µg/kg	50 µL
5,000–100,000 µg/kg	10 µL
25,000–500,000 µg/kg	100 µL of 1/50 dilution (b)

Calculate appropriate dilution factor for concentrations exceeding this table.

(a) The volume of methanol added to 5 mL of water being purged should be kept constant. Therefore, add to the 5-mL syringe whatever volume of methanol is necessary to maintain a volume of 100 µL added to the syringe.

(b) Dilute an aliquot of the methanol extract and then take 100 µL for analysis.

QUALITY CONTROL Demonstrate, through the analysis of a reagent water blank, that interferences from the analytical system, glassware, and reagents are under control. Blank samples should be carried through all stages of the sample preparation and measurement steps. For each analytical batch (up to 20 samples), a reagent blank, matrix spike, and matrix spike duplicate must be analyzed (the frequency of the spikes may be different for different monitoring programs). The blank and spiked samples must be carried through all stages of the sample preparation and measurement steps. QC samples mentioned in the section on Interferences will also be needed as appropriate to those situations.

REFERENCE Test Methods for Evaluating Solid Waste (SW-846). U.S. EPA. 1983. Method 8240B, Rev. 2, Nov. 1990. Office of Solid Wastes, Washington, DC.

Diethyl ether   **EPA Method 1624**
CAS #60-29-7

TITLE Volatile Organic Compounds by Isotope Dilution GC/MS

MATRIX Compounds may be determined in waters, soils, and municipal sludges by this method.

METHOD SUMMARY This method is used to determine 58 volatile toxic organic pollutants associated with the CWA (as amended 1987); the RCRA (as amended 1986); the CERCLA (as amended 1986); and other compounds amenable to purge-and-trap gas chromatography-mass spectrometry (GC/MS).

If the solids content is less than 1%, stable isotopically-labeled analogs of the compounds of interest are added to a 5-mL sample and the sample is purged with an inert gas at 20–25°C in a chamber designed for soil or water samples. If the solids content is greater than 1%, 5 mL of reagent water and the labeled compounds are added to a 5-g aliquot of sample and the mixture is purged at 40°C. Compounds that will not purge at 20–25°C or at 40°C are purged at 78–85°C. In the purging process, the volatile compounds are transferred from the aqueous phase into the gaseous phase where they are passed into a sorbent column, and trapped. After purging is completed, the trap is backflushed and heated rapidly to desorb the compounds into a GC. The compounds are separated by the GC and detected by a MS. The labeled compounds serve to correct the variability of the analytical technique.

INTERFERENCES Impurities in the purge gas, organic compounds outgassing from the plumbing upstream of the trap, and solvent vapors in the lab account for most problems. Samples can be contaminated by diffusion of volatile organic compounds (particularly methylene chloride) through the bottle seal during shipment and storage. Contamination by carryover can occur when high-level and low-level samples are analyzed sequentially. When an unusually concentrated sample is encountered, follow it by analysis of a reagent water blank to check for carryover.

INSTRUMENTATION Major equipment includes a GC with linear temperature programming and a glass jet separator as the MS interface, a MS with 70 eV electron impact ionization, and a data system to collect and record response factors.

Column: 2.8 m × 2 mm I.D. glass, packed with 1% SP-1000 on Carbopak B, 60/80 mesh, or equivalent.

PRECISION & ACCURACY The detection limits of the method are usually dependent on the level of interferences rather than instrumental limitations. The method detection limits were determined in digested sludge (low solids) and in filter cake or compost (high solids).

The MDL (in µg/kg) for low solids is 63 and for high solids is 12.

Labeled and native compound precision (in µg/L) as standard deviation was 44.0.

Labeled and native compound accuracy (in µg/L) as average recovery was 75–146.

Acceptance criteria are at 100 µg/L for this compound.

SAMPLE COLLECTION, PRESERVATION & HANDLING Grab samples are collected in glass containers having a total volume greater than 20 mL. Fill and seal each bottle so that no air bubbles are entrapped. Samples are maintained at 0 to 4°C from the time of collection until analysis. If an aqueous sample contains residual chlorine, add sodium thiosulfate preservative (10 mg/40 mL) to the empty sample bottles just prior to shipment to the sample site. All samples must be analyzed within 14 days of collection.

SAMPLE PREPARATION Samples containing less than 1% solids are analyzed directly as aqueous samples. Samples containing 1% solids or greater are analyzed as solid samples utilizing one of two methods, depending on the levels of pollutants, in the sample. Samples containing 1% solids or greater, and low to moderate levels of pollutants are analyzed by purging a known weight of sample added to 5 mL of reagent water. Samples containing 1% solids or greater, and high levels of pollutants, are extracted with methanol, and an aliquot of the methanol extract is added to reagent water and purged.

QUALITY CONTROL A field blank prepared from reagent water and carried through the sampling and handling protocol may serve as a check on contamination from shipment and storage.

The analyst is permitted to modify this method to improve separations or lower the costs of measurements, provided all performance specifications are met. Analyses of blanks are required. When results of spikes indicate atypical method performance for samples, the samples are diluted to bring method performance within acceptable limits. Analyze two sets of four 5-mL aliquots (8 aliquots total) of the aqueous performance standard. Spike all samples with labeled compounds to assess method performance on the sample matrix. Compute the percent recovery of the labeled compounds using the internal standard method. Compare the percent recovery for each compound with the corresponding labeled compound recovery. Reagent water blanks are analyzed to demonstrate freedom from carryover contamination. Field replicates may be collected to determine the precision of the sampling technique, and spiked samples may be required to determine the accuracy of the analysis when the internal method is used.

REFERENCE Volatile Organic Compounds by Isotope Dilution GC/MS. Office of Water Regulation and Standards, U.S. EPA Industrial Technology Division, Washington, DC, EPA Method 1624, Rev. C, June 1989 (contact W.A. Telliard, U.S. EPA, Office of Water Regulations and Standards, 401 M St., SW, Washington, DC, 20460. Phone: 202-382-7131).

Diethyl ether   **EPA Method 8015**
CAS #60-29-7

TITLE Nonhalogenated Volatile Organics

MATRIX Groundwater, soils, sludges, water miscible liquid wastes, and non-water miscible wastes.

APPLICATION This method is used for the analysis of 6 nonhalogenated VOCs. Samples are analyzed using direct injection or purge-and-trap methods. Groundwater must be analyzed by the purge-and-trap method. The method provides an optional GC column that may help resolve analytes from interferences and which is also used for analyte confirmation.

INTERFERENCES There can be carryover contamination with high- and low-level samples. Impurities may come from the purge-and-trap apparatus, organic compounds outgassing from the plumbing ahead of trap, diffusion of VOCs through the sample bottle septum during shipping or storage, or from solvent vapors in the lab.

INSTRUMENTATION GC capable of on-column injections or purge-and-trap sample introduction and a flame ionization detector (FID). Column 1: an 8 ft by 0.1 in 1% SP-1000 on Carbopack-B column. Column 2: a 6 ft by 0.1 in bonded n-octane on Porasil-C.

RANGE Not available.

MDL Not available.

PRECISION Not available.

ACCURACY Not available.

SAMPLING METHOD For water and liquid samples; use glass 40-mL vials with Teflon®-lined septum caps and collect two vials per sample location with no headspace. For solids and concentrated waste samples; use widemouth glass bottles with Teflon® liners. Cool all samples to 4°C

STABILITY For concentrated wastes, soils, sediments, or sludges: cool to 4°C. For liquids: add 4 drops of concentrated hydrochloric acid, cool to 4°C. MHT = 14 days.

QUALITY CONTROL Analyze a reagent blank, matrix spike, and matrix spike duplicate/duplicate for each analytical batch (up to 20 samples). Demonstrate the purity of glassware and reagents by analyzing a reagent water method blank. Internal, surrogate, and five concentration level calibration standards are used.

REFERENCE Method 8015, SW-846, 3rd ed., Nov.1986.

Diethyl phthalate   **EPA Method 1625**
CAS #84-66-2

TITLE Semivolatile Organic Compounds by Isotope Dilution GC/MS

MATRIX The compounds may be determined in waters, soils, and municipal sludges by this method.

METHOD SUMMARY This method is used to determine 176 semivolatile toxic organic pollutants associated with the CWA (as amended 1987); the RCRA (as amended 1986); the CERCLA (as amended 1986); and other compounds amenable to extraction and analysis by capillary column gas chromatography-mass spectrometry (GC/MS).

Stable isotopically-labeled analogs of the compounds of interest are added to the sample. If the solids content is less than 1%, a 1-L sample is extracted at pH 12–13, then at pH <2 with methylene chloride using continuous extraction techniques.

If the solids content is 30% or less, the sample is diluted to 1% solids with reagent water, homogenized ultrasonically, and extracted at pH 12–13, then at pH <2 with methylene chloride using continuous extraction techniques. If the solids content is greater than 30%, the sample is extracted using ultrasonic techniques.

Each extract is dried over sodium sulfate, concentrated to a volume of 5 mL, cleaned up using GPC, if necessary, and concentrated. Extracts are concentrated to 1 mL if GPC is not performed, and to 0.5 mL if GPC is performed.

An internal standard is added to the extract, and a 1-mL aliquot of the extract is injected into the GC. The compounds are separated by GC and detected by a MS. The labeled compounds serve to correct the variability of the analytical technique.

INTERFERENCES Solvents, reagents, glassware, and other sample processing hardware may yield artifacts and/or elevated baselines causing misinterpretation of chromatograms and spectra. Materials used in the analysis must be demonstrated to be free from interferences under the conditions of analysis by running method blanks initially and with each sample lot (sample started through the extraction process on a given 8-h shift, to a maximum of 20). Specific selection of reagents and purification of solvents by distillation in all glass systems may be required. Glassware and, where possible, reagents are cleaned by solvent rinse and baking at 450°C for 1-h minimum. Interferences coextracted from samples will vary considerably from source to source, depending on the diversity of the site being sampled.

INSTRUMENTATION Major instrumentation includes a GC with a splitless or on-column injection port for capillary column, a MS with 70 eV electron impact ionization, and a data system to collect and record MS data, and process it. A K-D apparatus is used to concentrate extracts.

GC Column: 30 m × 0.25 mm I.D. 5% phenyl, 94% methyl, 1% vinyl silicone bonded phased fused silica capillary column.

PRECISION & ACCURACY The detection limits of the method are usually dependent on the level of interferences rather than instrumental limitations. The limits typify the minimum quantities that can be detected with no interferences present.

The minimum level (in µg/mL) was 10. This is defined as a minimum level at which the analytical system shall give recognizable mass spectra (background corrected) and acceptable calibration points.

The MDL (in µg/kg) in low solids was 52 and in high solids was 16; these were determined in digested sludge (low solids) and in filter cake or compost (high solids).

The labeled and native compound initial precision as standard deviation (in µg/L) was 44.

The labeled and native compound initial accuracy as average recovery (in µg/L) was 75–196.

SAMPLE COLLECTION, PRESERVATION & HANDLING Collect samples in glass containers. Aqueous samples which flow freely are collected in refrigerated bottles using automatic sampling equipment. Solid samples are collected as grab samples using widemouth jars. Maintain samples at 0 to 4°C from the time of collection until extraction. If residual chlorine is present in aqueous samples, add 80 mg sodium thiosulfate/L of water. Begin sample extraction within 7 days of collection, and analyze all extracts within 40 days of extraction.

SAMPLE PREPARATION Samples containing 1% solids or less are extracted directly using continuous liquid-liquid extraction techniques. Samples containing 1 to 30% solids are diluted to the 1% level with reagent water and extracted using continuous liquid-liquid extraction techniques. Samples containing greater than 30% solids are extracted using ultrasonic techniques.

Base/neutral extraction — Adjust the pH of the waters in the extractors to 12–13 with 6 N NaOH. Extract with methylene chloride for 24–48 h.

Acid extraction — Adjust the pH of the waters in the extractors to 2 or less using 6 N sulfuric acid. Extract with methylene chloride for 24–48 h.

Ultrasonic extraction of high solids samples — Add anhydrous sodium sulfate to the sample and QC aliquot(s). Add acetone:methylene chloride (1:1) to the sample and mix thoroughly

Concentrate extracts using a K-D apparatus.

QUALITY CONTROL The analyst is permitted to modify this method to improve separations or lower the costs of measurements, provided all performance specifications are met. Analyses of blanks are required to demonstrate freedom from contamination. When results of spikes indicate atypical method performance for samples, the samples are diluted to bring method performance within acceptable limits.

For low solids (aqueous samples), extract, concentrate, and analyze two sets of four 1-L aliquots (8 aliquots total) of the precision and recovery standard. For high solids samples, two sets of four 30-g aliquots of the high solids reference matrix are used.

Spike all samples with labeled compounds to assess method performance. Compute percent recovery of the labeled compounds using the internal standard method. Compare the labeled compound recovery for each compound with the corresponding labeled compound recovery.

Reagent water and high solids reference matrix blanks are analyzed to demonstrate freedom from contamination. Extract and concentrate a 1-L reagent water blank or a high solids reference matrix blank with each sample's lot (samples started through the extraction process on the same 8-h shift, to a maximum of 20 samples).

Field replicates may be collected to determine the precision of the sampling technique, and spiked samples may be required to determine the accuracy of the analysis when the internal standard method is used.

REFERENCE Semivolatile Organic Compounds by Isotope Dilution GC/MS. Office of Water Regulation and Standards, U.S. EPA Industrial Technology Division, Washington, DC, EPA Method 1625, Rev. C, June 1989 (contact W.A. Telliard, U.S. EPA, Office of Water Regulations and Standards, 401 M St., SW, Washington, DC, 20460. Phone: 202-382-7131).

Diethyl phthalate **EPA Method 625**
CAS #84-66-2

TITLE Base/Neutrals and Acids, U.S. EPA Method 625

MATRIX This methods covers municipal and industrial wastewaters.

METHOD SUMMARY Approximately 1 L of sample is serially extracted with methylene chloride at a pH greater than 11 and again at a pH less than 2 using a separatory funnel or a continuous extractor. The methylene chloride extract is dried, concentrated to a volume of 1 mL, and analyzed by GC/MS. Qualitative identification of the parameters in the extract is performed using the retention time and the relative abundance of three characteristic masses (m/z). Qualitative analysis is performed using either external or internal standard techniques with a single characteristic m/z.

INTERFERENCES Method interferences may be caused by contaminants in solvents, reagents, glassware, and other sample processing hardware. Glassware must be scrupulously cleaned. Glassware should be heated in a muffle furnace at 400°C for 5 to 30 min. Some thermally stable materials, such as PCBs, may not be eliminated by this treatment. Solvent rinses with acetone and pesticide quality hexane may be substituted for the muffle furnace heating. Matrix interferences may be caused by contaminants that are coextracted from the sample. The base-neutral extraction may cause significantly reduced recovery of phenols. The packed gas chromatographic columns recommended for the basic fraction may not exhibit sufficient resolution for some analytes.

INSTRUMENTATION A GC/MS system with an injection port designed for on-column injection when using packed columns and for splitless injection when using capillary columns.

Column for base/neutrals: 1.8 m long × 2 mm I.D. glass, packed with 3% SP-2550 on Supelcoport (100/120 mesh) or equivalent.

Column for acids: 1.8 m long × 2 mm I.D. glass, packed with 1% SP-1240DA on Supelcoport (100/120 mesh) or equivalent.

PRECISION & ACCURACY The MDL concentrations were obtained using reagent water. The MDL actually achieved in a given analysis will vary depending on instrument sensitivity and matrix effects. This method was tested by 15 laboratories using reagent water, drinking water, surface water, and industrial wastewaters spiked at six concentrations over the range 5 to 100 µg/L. Single operator precision, overall precision, and method accuracy were found to be directly related to the concentration of the parameter matrix.

The MDL (in µg/L) in reagent water was 1.9.

The standard deviation (in µg/L based on 4 recovery measurements) was 26.5.

The range (in µg/L) for average recovery for 4 measurements was D-100.0.

The range (in %) for percent recovery was D-114.

Accuracy (in µg/L) as expected recovery for one or more measurements of a sample containing a true concentration of C was $0.43C + 1.00$.

Precision (in µg/L) as expected single analyst standard deviation of measurements at an average concentration found at X was $0.28X + 1.44$.

Overall precision (in µg/L) as expected interlaboratory standard deviation of measurements in an average concentration found at X was $0.52X + 0.22$.

C = *True value of the concentration in µg/L.*

X = *Average recovery found for measurements of samples containing a concentration at C in µg/L.*

SAMPLE PREPARATION Adjust the pH to >11 with sodium hydroxide and serially extract in a separatory funnel with methylene chloride or else in a continuous extractor. Next, adjust the pH to <2 with sulfuric acid and serially extract in a separatory funnel with methylene chloride or else in a continuous extractor. Dry the extracts separately through a column of anhydrous sodium sulfate and then concentrate each of the extracts to 1.0 mL using a K-D apparatus.

SAMPLE COLLECTION, PRESERVATION & HANDLING Grab samples must be collected in glass containers. All samples must be refrigerated at 4°C from the time of collection until extraction. If residual chlorine is present, add 80 mg of sodium thiosulfate/L of sample and mix well. All samples must be extracted within 7 days of collection and completely analyzed within 40 days of extraction.

QUALITY CONTROL Make an initial, one-time, demonstration of the ability to generate acceptable accuracy and precision with this method. Before processing any samples, the analyst must analyze a reagent water blank to demonstrate that interferences from the analytical system and glassware are under control. Each time a set of samples is extracted or reagents are changed, a reagent water blank must be processed. Spike and analyze a minimum of 5% of all samples to monitor and evaluate lab data quality. A QC check sample concentrate that contains each parameter of interest at a concentration of 100 µg/mL in acetone is required. PCBs and multicomponent pesticides may be omitted from this test.

After analysis of five spiked wastewater samples, calculate the average percent recovery and the standard deviation of the percent recovery. Spike all samples with the surrogate standard spiking solution and calculate the percent recovery of each surrogate compound.

REFERENCE Federal Register, Vol. 49, No. 209. Friday, Oct. 26, 1984.

Diethyl phthalate   **EPA Method 8061**
CAS #84-66-2

TITLE Phthalate Esters by Capillary Gas Chromatography With Electron Capture Detection (GC/ECD)

MATRIX This method covers aqueous and solid matrices. This includes a wide variety such as drinking water, groundwater, industrial wastewaters, surface waters, soils, solids, and sediments.

METHOD SUMMARY This method is used to determine the identities and concentrations of phthalate esters in liquid, solid and sludge matrices. When used to analyze for any or all of the target analytes, compound identification should be supported by at least one additional qualitative technique. This method describes conditions for parallel column, dual electron capture detector analysis, which fulfills the above requirement. Alternatively, GC/MS could be used for compound confirmation.

A measured volume or weight of sample (approximately 1 L for liquids, 10 to 30 g for solids and sludges) is extracted by using the appropriate sample extraction technique specified in EPA Method 3510, EPA Method 3540, and EPA Method 3550. After cleanup, the extract is analyzed by GC/ECD.

INTERFERENCES The sensitivity of this method usually depends on the level of interferences rather than on instrumental limitations. If interferences prevent detection of the analytes, cleanup of the sample extracts is necessary. Either EPA Method 3610 or EPA Method 3620 alone or followed by EPA Method 3660, Sulfur Cleanup, may be used to eliminate interferences in the analysis. EPA Method 3640, Gel Permeation Cleanup, is applicable for samples that contain high amounts of lipids and waxes.

Interferences coextracted from the samples will vary considerably from waste to waste. Glassware must be scrupulously clean. All glassware require treatment in a muffle furnace at 400°C for 2 to 4 h, or thorough rinsing with pesticide-grade solvent, prior to use. Volumetric glassware should not be heated in a muffle furnace. Storage of glassware in the lab introduces contamination, even if the glassware is wrapped in aluminum foil. Sodium sulfate, Florisil, and alumina may be contaminated with phthalate esters and, therefore, use of these materials in sample cleanup should be employed cautiously. If these materials are used, they must be obtained packaged in glass. Heating at 400°C for sodium sulfate, 320°C for Florisil, and 210°C for alumina is recommended. Glass wool used in any step of sample preparation should be a specially treated pyrex wool, pesticide grade, and must be baked at 400°C for 4 h immediately prior to use.

Paper thimbles and filter paper must be exhaustively washed with the solvent that will be used in the sample extraction. Soxhlet extraction of paper thimbles and filter paper for 12 h with fresh solvent should be repeated for a minimum of three times. Method blanks should be obtained before any of the precleaned thimbles or filter papers are used.

INSTRUMENTATION Gas chromatograph suitable for on-column and split/splitless injections.

Column 1: 30 m × 0.53 mm ID, 5% phenyl/95% methyl silicone fused-silica open tubular column, DB-5, 1.5 µg film thickness.

Column 2: 30 m × 0.53 mm ID, 14% cyanopropyl phenyl silicone fused-silica open tubular column, DB-1701, 1.0 µg film thickness.

A dual electron capture detector (ECD) is used. A Kuderna-Danish (K-D) apparatus is required along with a vacuum manifold consisting of individually adjustable, easily accessible flow-control valves for up to 24 cartridges, sample rack, chemically resistant cover and seals, heavy-duty glass basin, removable stainless steel solvent guides, built-in vacuum gauge and valve. Also, 6-mL, 1-g solid-phase extraction cartridges, LC-Florisil or equivalent, prepackaged, ready to use will be needed.

PRECISION & ACCURACY The MDL actually achieved in a given analysis will vary, as it is dependent on instrument sensitivity and matrix effects. This method has been tested in a single lab. Single-operator precision, overall precision, and method accuracy were found to be related to the concentration of the compounds and the type of matrix.

MULTIPLICATION FACTORS FOR OTHER MATRICES (a)

Matrix	Factor (b)
Groundwater	10
Low-concentration soil by ultrasonic extraction with GPC cleanup	670
High-concentration soil and sludges by ultrasonic extraction	10,000
Non-water miscible waste	100,000

(a) Sample EQLs are highly matrix-dependent.
(b) EQL = [Method detection limit] × [Factor]. For non-aqueous samples, the factor is on a wet-weight basis.

The MDL using 7 replicate determinations and a spike concentration of 100 µg/L was 250 ng/L.

The average recovery from HPLC-grade water using 4 determinatons and a spike concentration of 100 µg/L was 92.3%.

The precision (as RSD) from HPLC-grade water using 4 determinatons and a spike concentration of 100 µg/L was 10.3%.

The average recovery from groundwater using 4 determinatons and a spike concentration of 100 µg/L was 92.6%.

The precision (as RSD) from groundwater using 4 determinatons and a spike concentration of 100 µg/L was 7.2%.

The average recovery (in %) with %RSD (in parenthesis) from 3 determinations and a spike concentration of 20 µg/L in water was 71.2 (3.8) using 3M Empore Disks and EPA Method 8061.

The average recovery (in %) with %RSD (in parenthesis) from 3 determinations and a spike concentration of 20 µg/L in leachate was 82.8 (19.3) using 3M Empore Disks and EPA Method 8061.

The average recovery (in %) with %RSD (in parenthesis) from 3 determinations and a spike concentration of 20 µg/L in estuarine groundwater was 88.5 (15.3) using 3M Empore Disks and EPA Method 8061.

The average recovery (in %) with %RSD (in parenthesis) from 3 determinations and a spike concentration of 1 mg/kg in estuarine sediment was 68.4 (1.7) after sulfur cleanup with EPA Method 3660.

The average recovery (in %) with %RSD (in parenthesis) from 3 determinations and a spike concentration of 1 mg/kg in municipal sludge was 68.6 (9.1).

The average recovery (in %) with %RSD (in parenthesis) from 3 determinations and a spike concentration of 1 mg/kg in sandy loam soil was 54.7 (6.2).

SAMPLE COLLECTION, PRESERVATION & HANDLING
Containers used to collect samples for the determination of semivolatile organic compounds should be soap and water washed followed by methanol (or isopropanol) rinsing. The sample containers should be of glass or Teflon® and have screw-top covers with Teflon® liners. Sample containers should be filled with care to prevent any portion of the collected sample coming in contact with the sampler's gloves.

No preservation is used with concentrated waste samples. With liquid samples containing no residual chlorine and with soil, sediment, and sludge samples, immediately cooling to 4°C is the only preservation used. When residual chlorine is present then 3 mL of 10% aqueous sodium sulfate is added for each gallon of sample collected, followed by cooling to 4°C.

MHT Liquid samples must be extracted within 7 days and their extracts analyzed within 40 days. Concentrated waste, soil, sediment, and sludge samples must be extracted within 14 days and their extracts analyzed within 40 days.

SAMPLE PREPARATION In general, water samples are extracted at a pH of 5 to 7 with methylene chloride in a separatory funnel (EPA Method 3510). EPA Method 3520 is not recommended for the extraction of aqueous samples because the longer chain esters tend to adsorb to the glassware and consequently, their extraction recoveries may be poor. Solid samples are extracted with hexane/acetone (1:) or methylene chloride/acetone (1:1) in a Soxhlet extractor (EPA Method 3540) or with an ultrasonic extractor (EPA Method 3550). Immediately prior to extraction, spike 500 µL of the surrogate standard spiking solution into 1-L aqueous sample or 30-g solid sample. Extraction of particulate-free aqueous samples using C-18 extraction disks is an optional method that can be used.

Prior to Florisil cleanup or GC analysis, the methylene chloride and methylene chloride/acetone extracts must be exchanged to hexane. Exchange is not required for the acetonitrile extracts. Cleanup may not be necessary for extracts from a relatively clean sample matrix. Florisil Cartridge Cleanup may be used for extract cleanup.

If PCBs and organochlorine pesticides are known to be present in the sample, and if Florisil Cartridge Cleanup is considered, then two fractions are collected: Fraction 1 is eluted with 5 mL of 20% methylene chloride in hexane and Fraction 2 is eluted with 5 mL of 10% acetone in hexane. Fraction 1 contains the organochlorine pesticides and PCBs, and can be discarded. Fraction 2 contains the phthalate esters and is analyzed by GC/ECD.

QUALITY CONTROL Identify compounds in the sample by comparing the retention times of the peaks in the sample chromatogram with those of the peaks in standard chromatograms. The retention time window used to make identification is based upon measurements of actual retention time variations over the course of 10 consecutive injections.

Calibration standards are prepared at a minimum of five concentrations for each parameter of interest through dilution of the stock standard solutions with hexane. One of the concentrations should be at a concentration near, but above, the method detection limit. Prepare stock standard solutions in hexane. Stock standards should be checked frequently for signs of degradation or evaporation, especially just prior to preparing calibration standards from them. Stock standard solutions must be replaced after one year, or sooner if comparison with check standards indicates a problem. The suggested internal standards are: 2,5-dibromotoluene, 1,3,5-tribromobenzene, and α, α'-dibromo-m-xylene. The analyst can use any of the three compounds provided that they are resolved from matrix interferences. Recommended surrogate compounds are α-2,6-trichlorotoluene, 1,4-dichloronaphthalene, and 2,3,4,5,6-pentachlorotoluene.

Spike each sample, standard, and blank with surrogate compounds. Three surrogates are suggested for this method: diphenyl phthalate, diphenyl isophthalate, and dibenzyl phthalate.

The quality control check sample concentrate should contain the test compounds at 5 to 10 ng/μL An internal standard peak area check must be performed on all samples. The internal standard must be evaluated for acceptance by determining whether the measured area for the internal standard deviates by more than 30% from the average are for the internal standard in the calibration standards. When the internal standard peak area is outside that limit, all samples that fall outside the QC criteria must be reanalyzed. Benzyl benzoate has been tested and found appropriate as an internal standard for this method.

Any compounds confirmed by two columns may also be confirmed by GC/MS. The sample extract and associated blank should be analyzed by GC/MS. A reference standard of the compound must also be analyzed by GC/MS. Include a mid-concentration calibration standard after each group of 20 samples. The response factors for the mid-concentration calibration must be within ±15% of the average values for the multiconcentration calibration. Demonstrate through the analyses of standards that the Florisil fractionation scheme is reproducible.

REFERENCE Test Methods for Evaluating Solid Waste, Physical/Chemical Methods, SW-846, 3rd Edition, U.S. EPA, Office of Solid Waste, Washington, DC, EPA Method 8061, Nov. 1990.

Diethyl phthalate   **EPA Method 8270**
CAS #84-66-2

TITLE Semivolatile Organic Compounds by GC/MS

MATRIX This method is used to determine the concentration of semivolatile organic compounds in extracts prepared from all types of solid waste matrices, soils, and groundwater. Although surface waters are not specifically mentioned, this method should be applicable to water samples from rivers, lakes, etc.

METHOD SUMMARY This method covers 259 semivolatile organic compounds. In very limited applications direct injection of the sample into the GC/MS system may be appropriate, but this results in very high detection limits (approximately 10,000 μg/L). Typically, a 1-L liquid sample, containing surrogate, and matrix spiking standards, is extracted in a continuous extractor first under acid conditions and then under basic conditions. Typically 30 g of a solid sample, containing surrogate, and matrix spiking standards, is extracted ultrasonically. After concentrating the extract to 1 mL it is spiked with 10 μL of an internal standard solution just prior to analysis by GC/MS. The volume injected should contain about 100 ng of base/neutral and 200 ng of acid surrogates (for a 1-μL injection). Analysis is performed by GC/MS using a capillary GC column.

INTERFERENCES Raw GC/MS data from all blanks, samples, and spikes must be evaluated for interferences. Contamination by carryover can occur whenever high-concentration and low-concentration samples are sequentially analyzed. To reduce carryover, the sample syringe must be rinsed out between samples with solvent. Whenever an unusually concentrated sample is encountered, it should be followed by the analysis of blank solvent to check for cross-contamination.

INSTRUMENTATION A GC/MS and a data system are required. The GC column used is a 30 m × 0.25 mm I.D. (or 0.32 mm I.D.) 1um film thickness silicone-coated fused silica capillary column. A continuous liquid-liquid extractor equipped with Teflon® or glass connection joints and stopcocks requiring no lubrication, a K-D concentrating apparatus, water bath, and an ultrasonic disrupter with a minimum power of 300 W and with pulsing capability are also required.

PRECISION & ACCURACY The estimated quantitation limit (EQL) of Method 8270B for determining an individual compound is approximately 1 mg/kg (wet weight) for soil or sediment samples, 1–200 mg/kg for wastes (dependent on matrix and method of preparation), and 10 μg/L for groundwater samples. EQLs will be proportionately higher for sample extracts that require dilution to avoid saturation of the detector.

The EQL(b) for groundwater in μg/L is 10.
The EQL (a, b) for low concentrations in soil and sediment in μg/kg is 660.
Accuracy as μg/L is 0.43C +1.00.
Overall precision in μg/L is 0.52X +0.22.

(a) *EQLs listed for soil/sediment are based on wet weight. Normally data is reported in a dry-weight basis; therefore, EQLs will be higher based on the % dry weight of each sample. This calculation is based on a 30 g sample and gel permeation chromatography cleanup.*

(b) *Sample EQLs are highly matrix-dependent. The EQLs are provided for guidance and may not always be achievable.*

C = *True value for concentration, in μg/L.*
X = *Average recovery found for measurements of samples containing a concentration of C, in μg/L.*

ESTIMATED QUANTITATION LIMIT

Other Matrices	Factor (a)
High-concentration soil and sludges by sonicator	7.5
Non-water miscible waste	75

(a) *EQL for other matrices = [EQL for low soil/sediment] × [Factor]. This estimated EQL is similar to an EPA "Practical Quantitation Limit."*

SAMPLING METHOD
Liquid samples — Use a 1 or 2½ gallon amber glass bottle with a screw-top Teflon®-lined cover that has been prewashed with detergent and rinsed with distilled water and methanol (or isopropanol).

Soils, sediments, or sludges — Use an 8-oz. widemouth glass with a screw-top Teflon®-lined cover that has been prewashed with detergent and rinsed with distilled water and methanol (or isopropanol).

SAMPLE PRESERVATION
Liquid samples — If residual chlorine is present, add 3 mL of 10% sodium thiosulfate per gallon, cool to 4°C and store in a

solvent-free refrigerator until analysis; if chlorine is not present, then eliminate the sodium thiosulfate addition.

Soils, sediments, or sludges — Cool samples to 4°C and store in a solvent-free refrigerator.

MHT Liquid samples must be extracted within 7 days and the extracts analyzed within 40 days. Soils, sediments, or sludges may be stored for a maximum of 14 days and the extracts analyzed within 40 days.

SAMPLE PREPARATION
Liquid samples — Transfer 1 L quantitatively to a continuous extractor. If high concentrations are anticipated, a smaller volume may be used and then diluted with organic-free reagent water to 1 L. Adjust pH, if necessary, to pH <2 using 1:1 (V/V) sulfuric acid. Pipette 1.0 mL of a surrogate standard spiking solution into each sample. For the sample in each analytical batch selected for spiking, add 1.0 mL of a matrix spiking standard. For base/neutral acid analysis, the amount of the surrogates and matrix spiking compounds added to the sample should result in a final concentration of 100 ng/µL of each analyte in the extract to be analyzed (assuming a 1-µL injection). Extract with methylene chloride for 18–24 h. Next, adjust the pH of the aqueous phase to pH >11 using 10 *N* sodium hydroxide and extract it with methylene chloride again for 18–24 h. Dry the extract through a column containing anhydrous sodium sulfate and concentrate it to 1 mL using a K-D concentrator.

Soils, sediments, or sludges — Use 30 g of sample. Nonporous or wet samples (gummy or clay type) that do not have a free-flowing sandy texture must be mixed with anhydrous sodium sulfate until the sample is free flowing. Add 1 mL of surrogate standards to all samples, spikes, standards, and blanks. For the sample in each analytical batch selected for spiking, add 1.0 mL of a matrix spiking standard. For base/neutral acid analysis, the amount added of the surrogates and matrix spiking compounds should result in a final concentration of 100 ng/µL of each base/neutral analyte and 200 ng/µL of each acid analyte in the extract to be analyzed (assuming a 1-µL injection). Immediately add a 100-mL mixture of 1:1 methylene chloride:acetone and extract the sample ultrasonically for 3 min and then decant or filter the extracts. Repeat the extraction two or more times. Dry the extract using a column with anhydrous sodium sulfate and concentrate it to 1 mL in a K-D concentrator.

QUALITY CONTROL A methylene chloride solution containing 50 ng/µL of decafluorotriphenylphosphine (DFTPP) is used for tuning the GC/MS system each 12-h shift. A system performance check also must be made during every 12-h shift. A standard containing 50 ng/µL each of 4,4'-DDT, pentachlorophenol, and benzidine is required to verify injection port inertness and GC column performance. A calibration standard at mid-concentration, containing each compound of interest, including all required surrogates, must be performed every 12 h during analysis. After the system performance check is met, calibration check compounds (CCCs) are used to check the validity of the initial calibration.

The internal standard responses and retention times in the calibration check standard must be evaluated immediately after or during data acquisition. If the retention time for any internal standard changes by more than 30 seconds from the last check calibration (12 h), the chromatographic system must be inspected for malfunctions and corrections must be made, as required. If the electron ionization current plot (EICP) area for any of the internal standards changes by a factor of two from the last daily calibration standard check, the mass spectrometer must be inspected for malfunctions and corrections must be made, as appropriate.

Demonstrate, through the analysis of a reagent water blank, that interferences from the analytical system, glassware, and reagents are under control. The blank samples should be carried through all stages of the sample preparation and measurement steps. For each analytical batch (up to 20 samples), a reagent blank, matrix spike, and matrix spike duplicate/duplicate must be analyzed (the frequency of the spikes may be different for different monitoring programs). The blank and spiked samples must be carried through all stages of the sample preparation and measurement steps. A QC reference sample concentrate containing each analyte at a concentration of 100 mg/L in methanol is required.

REFERENCE Test Methods for Evaluating Solid Waste (SW-846). U.S. EPA 1983, Method 8270B, Rev. 2, Nov. 1990. Office of Solid Waste, Washington, DC.

Diethyl phthalate   **EPA Method 8060**
CAS #84-66-2

TITLE Phthalate Esters

MATRIX Groundwater, soils, sludges, water miscible liquid wastes, and non-water miscible wastes.

APPLICATION This method is used for the analysis of 6 phthalate esters. Samples are extracted, concentrated, and analyzed using direct injection of both neat and diluted organic liquids into a gas chromatograph. Analytes are detected by a flame ionization detector (FID) or an electron capture detector (ECD). Groundwater samples should be determined by ECD. The method provides an optional GC column which is used for analyte confirmation and that may help resolve analytes from interferences.

INTERFERENCES Solvents, reagents, and glassware may introduce artifacts. Plastics, in particular, must be avoided. Other interferences may come from coextracted compounds from samples. There can be carryover contamination with high- and low-level samples.

INSTRUMENTATION GC capable of on-column injections and a flame with detector (FID) or electron capture detector (ECD). Column 1: 1.8 m by 4 mm with 1.5% SP-2250/1.95% SP-2401 on Supelcoport. Column 2: 1.8 m by 4 mm with 3% OV-1 on supelcoport.

RANGE 0.7 to 106 µg/L.

MDL 31 µg/L (FID) and 0.49 µg/L (ECD).

PQL FACTORS FOR MULTIPLYING × FID MDL VALUE

Matrix	Multiplication Factor
Groundwater	10
Low-level soil by sonication with GPC cleanup	670
High-level soil and sludge by sonication	10,000
Non-water miscible waste	100,000

PRECISION $0.45X + 0.11$ µg/L (overall precision using FID).

ACCURACY $0.70C + 0.13$ µg/L (as recovery using FID).

SAMPLING METHOD Use 8-oz. widemouth glass bottles with Teflon®-lined caps for concentrated waste samples, soils, sediments, and sludges. Use 1 or 2½ gallon amber glass bottles with Teflon®-lined caps for liquid (water) samples.

STABILITY Cool soil, sediment, sludge, and liquid samples to 4°C. If residual chlorine is present in liquid samples add 3 mL of 10% sodium thiosulfate per gallon of sample and cool to 4°C.

MHT 14 days for concentrated waste, soil, sediment, or sludge; 7 days for liquid samples; all extracts must be analyzed within 40 days.

QUALITY CONTROL A quality control check sample concentrate containing each analyte of interest is required. The QC check sample concentrate may be prepared from pure standard materials or purchased as certified solutions Use appropriate trip, matrix, control site, method, reagent, and solvent blanks. Internal, surrogate, and five concentration level calibration standards are used. The QC check sample concentrate should contain this compound at 25 µg/mL in acetone.

REFERENCE Method 8060, SW-846, 3rd ed., Nov.1986.

Diethyl sulfate EPA Method 8270
CAS #64-67-5

TITLE Semivolatile Organic Compounds by GC/MS

MATRIX This method is used to determine the concentration of semivolatile organic compounds in extracts prepared from all types of solid waste matrices, soils, and groundwater. Although surface waters are not specifically mentioned, this method should be applicable to water samples from rivers, lakes, etc.

METHOD SUMMARY This method covers 259 semivolatile organic compounds. In very limited applications direct injection of the sample into the GC/MS system may be appropriate, but this results in very high detection limits (approximately 10,000 µg/L). Typically, a 1-L liquid sample, containing surrogate, and matrix spiking standards, is extracted in a continuous extractor first under acid conditions and then under basic conditions. Typically 30 g of a solid sample, containing surrogate, and matrix spiking standards, is extracted ultrasonically. After concentrating the extract to 1 mL it is spiked with 10 µL of an internal standard solution just prior to analysis by GC/MS. The volume injected should contain about 100 ng of base/neutral and 200 ng of acid surrogates (for a 1-µL injection). Analysis is performed by GC/MS using a capillary GC column.

INTERFERENCES Raw GC/MS data from all blanks, samples, and spikes must be evaluated for interferences. Contamination by carryover can occur whenever high-concentration and low-concentration samples are sequentially analyzed. To reduce carryover, the sample syringe must be rinsed out between samples with solvent. Whenever an unusually concentrated sample is encountered, it should be followed by the analysis of blank solvent to check for cross-contamination.

INSTRUMENTATION A GC/MS and a data system are required. The GC column used is a 30 m × 0.25 mm I.D. (or 0.32 mm I.D.) 1um film thickness silicone-coated fused silica capillary column. A continuous liquid-liquid extractor equipped with Teflon® or glass connection joints and stopcocks requiring no lubrication, a K-D concentrating apparatus, water bath, and an ultrasonic disrupter with a minimum power of 300 W and with pulsing capability are also required.

PRECISION & ACCURACY The estimated quantitation limit (EQL) of Method 8270B for determining an individual compound is approximately 1 mg/kg (wet weight) for soil or sediment samples, 1–200 mg/kg for wastes (dependent on matrix and method of preparation), and 10 µg/L for groundwater samples. EQLs will be proportionately higher for sample extracts that require dilution to avoid saturation of the detector.

The EQL(b) for groundwater in µg/L is 100.
The EQL (a, b) for low concentrations in soil and sediment in µg/kg is not determined.
Accuracy as µg/L is not listed.
Overall precision in µg/L is not listed.

(a) *EQLs listed for soil/sediment are based on wet weight. Normally data is reported in a dry-weight basis; therefore, EQLs will be higher based on the % dry weight of each sample. This calculation is based on a 30 g sample and gel permeation chromatography cleanup.*

(b) *Sample EQLs are highly matrix-dependent. The EQLs are provided for guidance and may not always be achievable.*

C = *True value for concentration, in µg/L.*
X = *Average recovery found for measurements of samples containing a concentration of C, in µg/L.*

ESTIMATED QUANTITATION LIMIT

Other Matrices	Factor (a)
High-concentration soil and sludges by sonicator	7.5
Non-water miscible waste	75

(a) *EQL for other matrices = [EQL for low soil/sediment] × [Factor]. This estimated EQL is similar to an EPA "Practical Quantitation Limit."*

SAMPLING METHOD

Liquid samples — Use a 1 or 2½ gallon amber glass bottle with a screw-top Teflon®-lined cover that has been prewashed with detergent and rinsed with distilled water and methanol (or isopropanol).

Soils, sediments, or sludges — Use an 8-oz. widemouth glass with a screw-top Teflon®-lined cover that has been prewashed

with detergent and rinsed with distilled water and methanol (or isopropanol).

SAMPLE PRESERVATION
Liquid samples — If residual chlorine is present, add 3 mL of 10% sodium thiosulfate per gallon, cool to 4°C and store in a solvent-free refrigerator until analysis; if chlorine is not present, then eliminate the sodium thiosulfate addition.

Soils, sediments, or sludges — Cool samples to 4°C and store in a solvent-free refrigerator.

MHT Liquid samples must be extracted within 7 days and the extracts analyzed within 40 days. Soils, sediments, or sludges may be stored for a maximum of 14 days and the extracts analyzed within 40 days.

SAMPLE PREPARATION
Liquid samples — Transfer 1 L quantitatively to a continuous extractor. If high concentrations are anticipated, a smaller volume may be used and then diluted with organic-free reagent water to 1 L. Adjust pH, if necessary, to pH <2 using 1:1 (V/V) sulfuric acid. Pipette 1.0 mL of a surrogate standard spiking solution into each sample. For the sample in each analytical batch selected for spiking, add 1.0 mL of a matrix spiking standard. For base/neutral acid analysis, the amount of the surrogates and matrix spiking compounds added to the sample should result in a final concentration of 100 ng/µL of each analyte in the extract to be analyzed (assuming a 1-µL injection). Extract with methylene chloride for 18–24 h. Next, adjust the pH of the aqueous phase to pH >11 using 10 N sodium hydroxide and extract it with methylene chloride again for 18–24 h. Dry the extract through a column containing anhydrous sodium sulfate and concentrate it to 1 mL using a K-D concentrator.

Soils, sediments, or sludges — Use 30 g of sample. Nonporous or wet samples (gummy or clay type) that do not have a free-flowing sandy texture must be mixed with anhydrous sodium sulfate until the sample is free flowing. Add 1 mL of surrogate standards to all samples, spikes, standards, and blanks. For the sample in each analytical batch selected for spiking, add 1.0 mL of a matrix spiking standard. For base/neutral acid analysis, the amount added of the surrogates and matrix spiking compounds should result in a final concentration of 100 ng/µL of each base/neutral analyte and 200 ng/µL of each acid analyte in the extract to be analyzed (assuming a 1-µL injection). Immediately add a 100-mL mixture of 1:1 methylene chloride:acetone and extract the sample ultrasonically for 3 min and then decant or filter the extracts. Repeat the extraction two or more times. Dry the extract using a column with anhydrous sodium sulfate and concentrate it to 1 mL in a K-D concentrator.

QUALITY CONTROL A methylene chloride solution containing 50 ng/µL of decafluorotriphenylphosphine (DFTPP) is used for tuning the GC/MS system each 12-h shift. A system performance check also must be made during every 12-h shift. A standard containing 50 ng/µL each of 4,4′-DDT, pentachlorophenol, and benzidine is required to verify injection port inertness and GC column performance. A calibration standard at mid-concentration, containing each compound of interest, including all required surrogates, must be performed every 12 h during analysis. After the system performance check is met, calibration check compounds (CCCs) are used to check the validity of the initial calibration.

The internal standard responses and retention times in the calibration check standard must be evaluated immediately after or during data acquisition. If the retention time for any internal standard changes by more than 30 seconds from the last check calibration (12 h), the chromatographic system must be inspected for malfunctions and corrections must be made, as required. If the electron ionization current plot (EICP) area for any of the internal standards changes by a factor of two from the last daily calibration standard check, the mass spectrometer must be inspected for malfunctions and corrections must be made, as appropriate.

Demonstrate, through the analysis of a reagent water blank, that interferences from the analytical system, glassware, and reagents are under control. The blank samples should be carried through all stages of the sample preparation and measurement steps. For each analytical batch (up to 20 samples), a reagent blank, matrix spike, and matrix spike duplicate/duplicate must be analyzed (the frequency of the spikes may be different for different monitoring programs). The blank and spiked samples must be carried through all stages of the sample preparation and measurement steps. A QC reference sample concentrate containing each analyte at a concentration of 100 mg/L in methanol is required.

REFERENCE Test Methods for Evaluating Solid Waste (SW-846). U.S. EPA 1983, Method 8270B, Rev. 2, Nov. 1990. Office of Solid Waste, Washington, DC.

Diethylstilbestrol   EPA Method 8270
CAS #56-53-1

TITLE Semivolatile Organic Compounds by GC/MS

MATRIX This method is used to determine the concentration of semivolatile organic compounds in extracts prepared from all types of solid waste matrices, soils, and groundwater. Although surface waters are not specifically mentioned, this method should be applicable to water samples from rivers, lakes, etc.

METHOD SUMMARY This method covers 259 semivolatile organic compounds. In very limited applications direct injection of the sample into the GC/MS system may be appropriate, but this results in very high detection limits (approximately 10,000 µg/L). Typically, a 1-L liquid sample, containing surrogate, and matrix spiking standards, is extracted in a continuous extractor first under acid conditions and then under basic conditions. Typically 30 g of a solid sample, containing surrogate, and matrix spiking standards, is extracted ultrasonically. After concentrating the extract to 1 mL it is spiked with 10 µL of an internal standard solution just prior to analysis by GC/MS. The volume injected should contain about 100 ng of base/neutral and 200 ng of acid surrogates (for a 1-µL injection). Analysis is performed by GC/MS using a capillary GC column.

INTERFERENCES Raw GC/MS data from all blanks, samples, and spikes must be evaluated for interferences. Contamination by carryover can occur whenever high-concentration and low-concentration samples are sequentially analyzed. To reduce carryover, the sample syringe must be rinsed out between samples with solvent. Whenever an unusually concentrated sample is encountered, it should be followed by the analysis of blank solvent to check for cross-contamination.

INSTRUMENTATION A GC/MS and a data system are required. The GC column used is a 30 m × 0.25 mm I.D. (or 0.32 mm I.D.) 1um film thickness silicone-coated fused silica capillary column. A continuous liquid-liquid extractor equipped with Teflon® or glass connection joints and stopcocks requiring no lubrication, a K-D concentrating apparatus, water bath, and an ultrasonic disrupter with a minimum power of 300 W and with pulsing capability are also required.

PRECISION & ACCURACY The estimated quantitation limit (EQL) of Method 8270B for determining an individual compound is approximately 1 mg/kg (wet weight) for soil or sediment samples, 1–200 mg/kg for wastes (dependent on matrix and method of preparation), and 10 µg/L for groundwater samples. EQLs will be proportionately higher for sample extracts that require dilution to avoid saturation of the detector.

The EQL(b) for groundwater in µg/L is 20.
The EQL (a, b) for low concentrations in soil and sediment in µg/kg is not determined.
Accuracy as µg/L is not listed.
Overall precision in µg/L is not listed.

(a) *EQLs listed for soil/sediment are based on wet weight. Normally data is reported in a dry-weight basis; therefore, EQLs will be higher based on the % dry weight of each sample. This calculation is based on a 30 g sample and gel permeation chromatography cleanup.*
(b) *Sample EQLs are highly matrix-dependent. The EQLs are provided for guidance and may not always be achievable.*
C = *True value for concentration, in µg/L.*
X = *Average recovery found for measurements of samples containing a concentration of C, in µg/L.*

ESTIMATED QUANTITATION LIMIT

Other Matrices	Factor (a)
High-concentration soil and sludges by sonicator	7.5
Non-water miscible waste	75

(a) *EQL for other matrices = [EQL for low soil/sediment] × [Factor]. This estimated EQL is similar to an EPA "Practical Quantitation Limit."*

SAMPLING METHOD
Liquid samples — Use a 1 or 2½ gallon amber glass bottle with a screw-top Teflon®-lined cover that has been prewashed with detergent and rinsed with distilled water and methanol (or isopropanol).

Soils, sediments, or sludges — Use an 8-oz. widemouth glass with a screw-top Teflon®-lined cover that has been prewashed with detergent and rinsed with distilled water and methanol (or isopropanol).

SAMPLE PRESERVATION
Liquid samples — If residual chlorine is present, add 3 mL of 10% sodium thiosulfate per gallon, cool to 4°C and store in a solvent-free refrigerator until analysis; if chlorine is not present, then eliminate the sodium thiosulfate addition.

Soils, sediments, or sludges — Cool samples to 4°C and store in a solvent-free refrigerator.

MHT Liquid samples must be extracted within 7 days and the extracts analyzed within 40 days. Soils, sediments, or sludges may be stored for a maximum of 14 days and the extracts analyzed within 40 days.

SAMPLE PREPARATION
Liquid samples — Transfer 1 L quantitatively to a continuous extractor. If high concentrations are anticipated, a smaller volume may be used and then diluted with organic-free reagent water to 1 L. Adjust pH, if necessary, to pH <2 using 1:1 (V/V) sulfuric acid. Pipette 1.0 mL of a surrogate standard spiking solution into each sample. For the sample in each analytical batch selected for spiking, add 1.0 mL of a matrix spiking standard. For base/neutral acid analysis, the amount of the surrogates and matrix spiking compounds added to the sample should result in a final concentration of 100 ng/µL of each analyte in the extract to be analyzed (assuming a 1-µL injection). Extract with methylene chloride for 18–24 h. Next, adjust the pH of the aqueous phase to pH >11 using 10 N sodium hydroxide and extract it with methylene chloride again for 18–24 h. Dry the extract through a column containing anhydrous sodium sulfate and concentrate it to 1 mL using a K-D concentrator.

Soils, sediments, or sludges — Use 30 g of sample. Nonporous or wet samples (gummy or clay type) that do not have a free-flowing sandy texture must be mixed with anhydrous sodium sulfate until the sample is free flowing. Add 1 mL of surrogate standards to all samples, spikes, standards, and blanks. For the sample in each analytical batch selected for spiking, add 1.0 mL of a matrix spiking standard. For base/neutral acid analysis, the amount added of the surrogates and matrix spiking compounds should result in a final concentration of 100 ng/µL of each base/neutral analyte and 200 ng/µL of each acid analyte in the extract to be analyzed (assuming a 1-µL injection). Immediately add a 100-mL mixture of 1:1 methylene chloride:acetone and extract the sample ultrasonically for 3 min and then decant or filter the extracts. Repeat the extraction two or more times. Dry the extract using a column with anhydrous sodium sulfate and concentrate it to 1 mL in a K-D concentrator.

QUALITY CONTROL A methylene chloride solution containing 50 ng/µL of decafluorotriphenylphosphine (DFTPP) is used for tuning the GC/MS system each 12-h shift. A system performance check also must be made during every 12-h shift. A standard containing 50 ng/µL each of 4,4'-DDT, pentachlorophenol, and benzidine is required to verify injection port inertness and GC column performance. A calibration standard at mid-concentration, containing each compound of interest,

including all required surrogates, must be performed every 12 h during analysis. After the system performance check is met, calibration check compounds (CCCs) are used to check the validity of the initial calibration.

The internal standard responses and retention times in the calibration check standard must be evaluated immediately after or during data acquisition. If the retention time for any internal standard changes by more than 30 seconds from the last check calibration (12 h), the chromatographic system must be inspected for malfunctions and corrections must be made, as required. If the electron ionization current plot (EICP) area for any of the internal standards changes by a factor of two from the last daily calibration standard check, the mass spectrometer must be inspected for malfunctions and corrections must be made, as appropriate.

Demonstrate, through the analysis of a reagent water blank, that interferences from the analytical system, glassware, and reagents are under control. The blank samples should be carried through all stages of the sample preparation and measurement steps. For each analytical batch (up to 20 samples), a reagent blank, matrix spike, and matrix spike duplicate/duplicate must be analyzed (the frequency of the spikes may be different for different monitoring programs). The blank and spiked samples must be carried through all stages of the sample preparation and measurement steps. A QC reference sample concentrate containing each analyte at a concentration of 100 mg/L in methanol is required.

REFERENCE Test Methods for Evaluating Solid Waste (SW-846). U.S. EPA 1983, Method 8270B, Rev. 2, Nov. 1990. Office of Solid Waste, Washington, DC.

Dihexyl phthalate   **EPA Method 8061**
CAS #84-75-3

TITLE Phthalate Esters by Capillary Gas Chromatography With Electron Capture Detection (GC/ECD)

MATRIX This method covers aqueous and solid matrices. This includes a wide variety such as drinking water, groundwater, industrial wastewaters, surface waters, soils, solids, and sediments.

METHOD SUMMARY This method is used to determine the identities and concentrations of phthalate esters in liquid, solid and sludge matrices. When used to analyze for any or all of the target analytes, compound identification should be supported by at least one additional qualitative technique. This method describes conditions for parallel column, dual electron capture detector analysis, which fulfills the above requirement. Alternatively, GC/MS could be used for compound confirmation.

A measured volume or weight of sample (approximately 1 L for liquids, 10 to 30 g for solids and sludges) is extracted by using the appropriate sample extraction technique specified in EPA Method 3510, EPA Method 3540, and EPA Method 3550. After cleanup, the extract is analyzed by GC/ECD.

INTERFERENCES The sensitivity of this method usually depends on the level of interferences rather than on instrumental limitations. If interferences prevent detection of the analytes, cleanup of the sample extracts is necessary. Either EPA Method 3610 or EPA Method 3620 alone or followed by EPA Method 3660, Sulfur Cleanup, may be used to eliminate interferences in the analysis. EPA Method 3640, Gel Permeation Cleanup, is applicable for samples that contain high amounts of lipids and waxes.

Interferences coextracted from the samples will vary considerably from waste to waste. Glassware must be scrupulously clean. All glassware require treatment in a muffle furnace at 400°C for 2 to 4 h, or thorough rinsing with pesticide-grade solvent, prior to use. Volumetric glassware should not be heated in a muffle furnace. Storage of glassware in the lab introduces contamination, even if the glassware is wrapped in aluminum foil. Sodium sulfate, Florisil, and alumina may be contaminated with phthalate esters and, therefore, use of these materials in sample cleanup should be employed cautiously. If these materials are used, they must be obtained packaged in glass. Heating at 400°C for sodium sulfate, 320°C for Florisil, and 210°C for alumina is recommended. Glass wool used in any step of sample preparation should be a specially treated pyrex wool, pesticide grade, and must be baked at 400°C for 4 h immediately prior to use.

Paper thimbles and filter paper must be exhaustively washed with the solvent that will be used in the sample extraction. Soxhlet extraction of paper thimbles and filter paper for 12 h with fresh solvent should be repeated for a minimum of three times. Method blanks should be obtained before any of the precleaned thimbles or filter papers are used.

INSTRUMENTATION Gas chromatograph suitable for on-column and split/splitless injections.

Column 1: 30 m × 0.53 mm ID, 5% phenyl/95% methyl silicone fused-silica open tubular column, DB-5, 1.5 µg film thickness.

Column 2: 30 m × 0.53 mm ID, 14% cyanopropyl phenyl silicone fused-silica open tubular column, DB-1701, 1.0 µg film thickness.

A dual electron capture detector (ECD) is used. A Kuderna-Danish (K-D) apparatus is required along with a vacuum manifold consisting of individually adjustable, easily accessible flow-control valves for up to 24 cartridges, sample rack, chemically resistant cover and seals, heavy-duty glass basin, removable stainless steel solvent guides, built-in vacuum gauge and valve. Also, 6-mL, 1-g solid-phase extraction cartridges, LC-Florisil or equivalent, prepackaged, ready to use will be needed.

PRECISION & ACCURACY The MDL actually achieved in a given analysis will vary, as it is dependent on instrument sensitivity and matrix effects. This method has been tested in a single lab. Single-operator precision, overall precision, and method accuracy were found to be related to the concentration of the compounds and the type of matrix.

MULTIPLICATION FACTORS FOR OTHER MATRICES (a)

Matrix	Factor (b)
Groundwater	10
Low-concentration soil by ultrasonic extraction with GPC cleanup	670
High-concentration soil and sludges by ultrasonic extraction	10,000
Non-water miscible waste	100,000

(a) Sample EQLs are highly matrix-dependent.
(b) EQL = [Method detection limit] × [Factor]. For non-aqueous samples, the factor is on a wet-weight basis.

The MDL using 7 replicate determinations and a spike concentration of 100 µg/L was 68 ng/L.

The average recovery from HPLC-grade water using 4 determinatons and a spike concentration of 100 µg/L was 98.4%.

The precision (as RSD) from HPLC-grade water using 4 determinatons and a spike concentration of 100 µg/L was 5.0%.

The average recovery from groundwater using 4 determinatons and a spike concentration of 100 µg/L was 97.7%.

The precision (as RSD) from groundwater using 4 determinatons and a spike concentration of 100 µg/L was 14.8%.

The average recovery (in %) with %RSD (in parenthesis) from 3 determinations and a spike concentration of 20 µg/L in water was 79.8 (7.2) using 3M Empore Disks and EPA Method 8061.

The average recovery (in %) with %RSD (in parenthesis) from 3 determinations and a spike concentration of 20 µg/L in leachate was (21.5) using 3M Empore Disks and EPA Method 8061.

The average recovery (in %) with %RSD (in parenthesis) from 3 determinations and a spike concentration of 20 µg/L in estuarine groundwater was 90.9 (7.6) using 3M Empore Disks and EPA Method 8061.

The average recovery (in %) with %RSD (in parenthesis) from 3 determinations and a spike concentration of 1 mg/kg in estuarine sediment was 103 (3.6) after sulfur cleanup with EPA Method 3660.

The average recovery (in %) with %RSD (in parenthesis) from 3 determinations and a spike concentration of 1 mg/kg in municipal sludge was 96.4 (10.7).

The average recovery (in %) with %RSD (in parenthesis) from 3 determinations and a spike concentration of 1 mg/kg in sandy loam soil was 77.9 (2.4).

SAMPLE COLLECTION, PRESERVATION & HANDLING
Containers used to collect samples for the determination of semivolatile organic compounds should be soap and water washed followed by methanol (or isopropanol) rinsing. The sample containers should be of glass or Teflon® and have screw-top covers with Teflon® liners. Sample containers should be filled with care to prevent any portion of the collected sample coming in contact with the sampler's gloves.

No preservation is used with concentrated waste samples. With liquid samples containing no residual chlorine and with soil, sediment, and sludge samples, immediately cooling to 4°C is the only preservation used. When residual chlorine is present then 3 mL of 10% aqueous sodium sulfate is added for each gallon of sample collected, followed by cooling to 4°C.

MHT Liquid samples must be extracted within 7 days and their extracts analyzed within 40 days. Concentrated waste, soil, sediment, and sludge samples must be extracted within 14 days and their extracts analyzed within 40 days.

SAMPLE PREPARATION In general, water samples are extracted at a pH of 5 to 7 with methylene chloride in a separatory funnel (EPA Method 3510). EPA Method 3520 is not recommended for the extraction of aqueous samples because the longer chain esters tend to adsorb to the glassware and consequently, their extraction recoveries may be poor. Solid samples are extracted with hexane/acetone (1:) or methylene chloride/acetone (1:1) in a Soxhlet extractor (EPA Method 3540) or with an ultrasonic extractor (EPA Method 3550). Immediately prior to extraction, spike 500 µL of the surrogate standard spiking solution into 1-L aqueous sample or 30-g solid sample. Extraction of particulate-free aqueous samples using C-18 extraction disks is an optional method that can be used.

Prior to Florisil cleanup or GC analysis, the methylene chloride and methylene chloride/acetone extracts must be exchanged to hexane. Exchange is not required for the acetonitrile extracts. Cleanup may not be necessary for extracts from a relatively clean sample matrix. Florisil Cartridge Cleanup may be used for extract cleanup.

If PCBs and organochlorine pesticides are known to be present in the sample, and if Florisil Cartridge Cleanup is considered, then two fractions are collected: Fraction 1 is eluted with 5 mL of 20% methylene chloride in hexane and Fraction 2 is eluted with 5 mL of 10% acetone in hexane. Fraction 1 contains the organochlorine pesticides and PCBs, and can be discarded. Fraction 2 contains the phthalate esters and is analyzed by GC/ECD.

QUALITY CONTROL Identify compounds in the sample by comparing the retention times of the peaks in the sample chromatogram with those of the peaks in standard chromatograms. The retention time window used to make identification is based upon measurements of actual retention time variations over the course of 10 consecutive injections.

Calibration standards are prepared at a minimum of five concentrations for each parameter of interest through dilution of the stock standard solutions with hexane. One of the concentrations should be at a concentration near, but above, the method detection limit. Prepare stock standard solutions in hexane. Stock standards should be checked frequently for signs of degradation or evaporation, especially just prior to preparing calibration standards from them. Stock standard solutions must be replaced after one year, or sooner if comparison with check standards indicates a problem. The suggested internal standards are: 2,5-dibromotoluene, 1,3,5-tribromobenzene, and α, α'-dibromo-m-xylene. The analyst can use any of the three compounds provided that they are resolved from matrix interferences. Recommended surrogate compounds are α-2,6-trichlorotoluene, 1,4-dichloronaphthalene, and 2,3,4,5,6-pentachlorotoluene.

Spike each sample, standard, and blank with surrogate compounds. Three surrogates are suggested for this method: diphenyl phthalate, diphenyl isophthalate, and dibenzyl phthalate.

The quality control check sample concentrate should contain the test compounds at 5 to 10 ng/μL An internal standard peak area check must be performed on all samples. The internal standard must be evaluated for acceptance by determining whether the measured area for the internal standard deviates by more than 30% from the average are for the internal standard in the calibration standards. When the internal standard peak area is outside that limit, all samples that fall outside the QC criteria must be reanalyzed. Benzyl benzoate has been tested and found appropriate as an internal standard for this method.

Any compounds confirmed by two columns may also be confirmed by GC/MS. The sample extract and associated blank should be analyzed by GC/MS. A reference standard of the compound must also be analyzed by GC/MS. Include a mid-concentration calibration standard after each group of 20 samples. The response factors for the mid-concentration calibration must be within ±15% of the average values for the multiconcentration calibration. Demonstrate through the analyses of standards that the Florisil fractionation scheme is reproducible.

REFERENCE Test Methods for Evaluating Solid Waste, Physical/Chemical Methods, SW-846, 3rd Edition, U.S. EPA, Office of Solid Waste, Washington, DC, EPA Method 8061, Nov. 1990.

Dihydrosaffrole   EPA Method 8270
CAS #56312-13-1

TITLE Semivolatile Organic Compounds by GC/MS

MATRIX This method is used to determine the concentration of semivolatile organic compounds in extracts prepared from all types of solid waste matrices, soils, and groundwater. Although surface waters are not specifically mentioned, this method should be applicable to water samples from rivers, lakes, etc.

METHOD SUMMARY This method covers 259 semivolatile organic compounds. In very limited applications direct injection of the sample into the GC/MS system may be appropriate, but this results in very high detection limits (approximately 10,000 μg/L). Typically, a 1-L liquid sample, containing surrogate, and matrix spiking standards, is extracted in a continuous extractor first under acid conditions and then under basic conditions. Typically 30 g of a solid sample, containing surrogate, and matrix spiking standards, is extracted ultrasonically. After concentrating the extract to 1 mL it is spiked with 10 μL of an internal standard solution just prior to analysis by GC/MS. The volume injected should contain about 100 ng of base/neutral and 200 ng of acid surrogates (for a 1-μL injection). Analysis is performed by GC/MS using a capillary GC column.

INTERFERENCES Raw GC/MS data from all blanks, samples, and spikes must be evaluated for interferences. Contamination by carryover can occur whenever high-concentration and low-concentration samples are sequentially analyzed. To reduce carryover, the sample syringe must be rinsed out between samples with solvent. Whenever an unusually concentrated sample is encountered, it should be followed by the analysis of blank solvent to check for cross-contamination.

INSTRUMENTATION A GC/MS and a data system are required. The GC column used is a 30 m × 0.25 mm I.D. (or 0.32 mm I.D.) 1um film thickness silicone-coated fused silica capillary column. A continuous liquid-liquid extractor equipped with Teflon® or glass connection joints and stopcocks requiring no lubrication, a K-D concentrating apparatus, water bath, and an ultrasonic disrupter with a minimum power of 300 W and with pulsing capability are also required.

PRECISION & ACCURACY The estimated quantitation limit (EQL) of Method 8270B for determining an individual compound is approximately 1 mg/kg (wet weight) for soil or sediment samples, 1–200 mg/kg for wastes (dependent on matrix and method of preparation), and 10 μg/L for groundwater samples. EQLs will be proportionately higher for sample extracts that require dilution to avoid saturation of the detector.

The EQL(b) for groundwater in μg/L is not listed.
The EQL (a, b) for low concentrations in soil and sediment in μg/kg is not listed.
Accuracy as μg/L is not listed.
Overall precision in μg/L is not listed.

(a) *EQLs listed for soil/sediment are based on wet weight. Normally data is reported in a dry-weight basis; therefore, EQLs will be higher based on the % dry weight of each sample. This calculation is based on a 30 g sample and gel permeation chromatography cleanup.*
(b) *Sample EQLs are highly matrix-dependent. The EQLs are provided for guidance and may not always be achievable.*
C = *True value for concentration, in μg/L.*
X = *Average recovery found for measurements of samples containing a concentration of C, in μg/L.*

ESTIMATED QUANTITATION LIMIT

Other Matrices	Factor (a)
High-concentration soil and sludges by sonicator	7.5
Non-water miscible waste	75

(a) *EQL for other matrices = [EQL for low soil/sediment] × [Factor]. This estimated EQL is similar to an EPA "Practical Quantitation Limit."*

SAMPLING METHOD
Liquid samples — Use a 1 or 2½ gallon amber glass bottle with a screw-top Teflon®-lined cover that has been prewashed with detergent and rinsed with distilled water and methanol (or isopropanol).

Soils, sediments, or sludges — Use an 8-oz. widemouth glass with a screw-top Teflon®-lined cover that has been prewashed with detergent and rinsed with distilled water and methanol (or isopropanol).

SAMPLE PRESERVATION
Liquid samples — If residual chlorine is present, add 3 mL of 10% sodium thiosulfate per gallon, cool to 4°C and store in a solvent-free refrigerator until analysis; if chlorine is not present, then eliminate the sodium thiosulfate addition.

Soils, sediments, or sludges — Cool samples to 4°C and store in a solvent-free refrigerator.

MHT Liquid samples must be extracted within 7 days and the extracts analyzed within 40 days. Soils, sediments, or sludges may be stored for a maximum of 14 days and the extracts analyzed within 40 days.

SAMPLE PREPARATION
Liquid samples — Transfer 1 L quantitatively to a continuous extractor. If high concentrations are anticipated, a smaller volume may be used and then diluted with organic-free reagent water to 1 L. Adjust pH, if necessary, to pH <2 using 1:1 (V/V) sulfuric acid. Pipette 1.0 mL of a surrogate standard spiking solution into each sample. For the sample in each analytical batch selected for spiking, add 1.0 mL of a matrix spiking standard. For base/neutral acid analysis, the amount of the surrogates and matrix spiking compounds added to the sample should result in a final concentration of 100 ng/µL of each analyte in the extract to be analyzed (assuming a 1-µL injection). Extract with methylene chloride for 18–24 h. Next, adjust the pH of the aqueous phase to pH >11 using 10 *N* sodium hydroxide and extract it with methylene chloride again for 18–24 h. Dry the extract through a column containing anhydrous sodium sulfate and concentrate it to 1 mL using a K-D concentrator.

Soils, sediments, or sludges — Use 30 g of sample. Nonporous or wet samples (gummy or clay type) that do not have a free-flowing sandy texture must be mixed with anhydrous sodium sulfate until the sample is free flowing. Add 1 mL of surrogate standards to all samples, spikes, standards, and blanks. For the sample in each analytical batch selected for spiking, add 1.0 mL of a matrix spiking standard. For base/neutral acid analysis, the amount added of the surrogates and matrix spiking compounds should result in a final concentration of 100 ng/µL of each base/neutral analyte and 200 ng/µL of each acid analyte in the extract to be analyzed (assuming a 1-µL injection). Immediately add a 100-mL mixture of 1:1 methylene chloride:acetone and extract the sample ultrasonically for 3 min and then decant or filter the extracts. Repeat the extraction two or more times. Dry the extract using a column with anhydrous sodium sulfate and concentrate it to 1 mL in a K-D concentrator.

QUALITY CONTROL A methylene chloride solution containing 50 ng/µL of decafluorotriphenylphosphine (DFTPP) is used for tuning the GC/MS system each 12-h shift. A system performance check also must be made during every 12-h shift. A standard containing 50 ng/µL each of 4,4'-DDT, pentachlorophenol, and benzidine is required to verify injection port inertness and GC column performance. A calibration standard at mid-concentration, containing each compound of interest, including all required surrogates, must be performed every 12 h during analysis. After the system performance check is met, calibration check compounds (CCCs) are used to check the validity of the initial calibration.

The internal standard responses and retention times in the calibration check standard must be evaluated immediately after or during data acquisition. If the retention time for any internal standard changes by more than 30 seconds from the last check calibration (12 h), the chromatographic system must be inspected for malfunctions and corrections must be made, as required. If the electron ionization current plot (EICP) area for any of the internal standards changes by a factor of two from the last daily calibration standard check, the mass spectrometer must be inspected for malfunctions and corrections must be made, as appropriate.

Demonstrate, through the analysis of a reagent water blank, that interferences from the analytical system, glassware, and reagents are under control. The blank samples should be carried through all stages of the sample preparation and measurement steps. For each analytical batch (up to 20 samples), a reagent blank, matrix spike, and matrix spike duplicate/duplicate must be analyzed (the frequency of the spikes may be different for different monitoring programs). The blank and spiked samples must be carried through all stages of the sample preparation and measurement steps. A QC reference sample concentrate containing each analyte at a concentration of 100 mg/L in methanol is required.

REFERENCE Test Methods for Evaluating Solid Waste (SW-846). U.S. EPA 1983, Method 8270B, Rev. 2, Nov. 1990. Office of Solid Waste, Washington, DC.

Diisobutyl phthalate   **EPA Method 8061**
CAS #84-69-5

TITLE Phthalate Esters by Capillary Gas Chromatography With Electron Capture Detection (GC/ECD)

MATRIX This method covers aqueous and solid matrices. This includes a wide variety such as drinking water, groundwater, industrial wastewaters, surface waters, soils, solids, and sediments.

METHOD SUMMARY This method is used to determine the identities and concentrations of phthalate esters in liquid, solid and sludge matrices. When used to analyze for any or all of the target analytes, compound identification should be supported by at least one additional qualitative technique. This method describes conditions for parallel column, dual electron capture detector analysis, which fulfills the above requirement. Alternatively, GC/MS could be used for compound confirmation.

A measured volume or weight of sample (approximately 1 L for liquids, 10 to 30 g for solids and sludges) is extracted by using the appropriate sample extraction technique specified in EPA Method 3510, EPA Method 3540, and EPA Method 3550. After cleanup, the extract is analyzed by GC/ECD.

INTERFERENCES The sensitivity of this method usually depends on the level of interferences rather than on instrumental limitations. If interferences prevent detection of the analytes, cleanup of the sample extracts is necessary. Either EPA Method 3610 or EPA Method 3620 alone or followed by EPA Method 3660, Sulfur Cleanup, may be used to eliminate interferences in the analysis. EPA Method 3640, Gel Permeation Cleanup, is applicable for samples that contain high amounts of lipids and waxes.

Interferences coextracted from the samples will vary considerably from waste to waste. Glassware must be scrupulously clean. All glassware require treatment in a muffle furnace at 400°C for 2 to 4 h, or thorough rinsing with pesticide-grade solvent, prior to use. Volumetric glassware should not be heated in a muffle furnace. Storage of glassware in the lab introduces contamination, even if the glassware is wrapped in aluminum foil. Sodium sulfate, Florisil, and alumina may be contaminated with phthalate esters and, therefore, use of these materials in sample cleanup should be employed cautiously. If these materials are used, they must be obtained packaged in glass. Heating at 400°C for sodium sulfate, 320°C for Florisil, and 210°C for alumina is recommended. Glass wool used in any step of sample preparation should be a specially treated pyrex wool, pesticide grade, and must be baked at 400°C for 4 h immediately prior to use.

Paper thimbles and filter paper must be exhaustively washed with the solvent that will be used in the sample extraction. Soxhlet extraction of paper thimbles and filter paper for 12 h with fresh solvent should be repeated for a minimum of three times. Method blanks should be obtained before any of the precleaned thimbles or filter papers are used.

INSTRUMENTATION Gas chromatograph suitable for on-column and split/splitless injections.

Column 1: 30 m × 0.53 mm ID, 5% phenyl/95% methyl silicone fused-silica open tubular column, DB-5, 1.5 µg film thickness.

Column 2: 30 m × 0.53 mm ID, 14% cyanopropyl phenyl silicone fused-silica open tubular column, DB-1701, 1.0 µg film thickness.

A dual electron capture detector (ECD) is used. A Kuderna-Danish (K-D) apparatus is required along with a vacuum manifold consisting of individually adjustable, easily accessible flow-control valves for up to 24 cartridges, sample rack, chemically resistant cover and seals, heavy-duty glass basin, removable stainless steel solvent guides, built-in vacuum gauge and valve. Also, 6-mL, 1-g solid-phase extraction cartridges, LC-Florisil or equivalent, prepackaged, ready to use will be needed.

PRECISION & ACCURACY The MDL actually achieved in a given analysis will vary, as it is dependent on instrument sensitivity and matrix effects. This method has been tested in a single lab. Single-operator precision, overall precision, and method accuracy were found to be related to the concentration of the compounds and the type of matrix.

MULTIPLICATION FACTORS FOR OTHER MATRICES (a)

Matrix	Factor (b)
Groundwater	10
Low-concentration soil by ultrasonic extraction with GPC cleanup	670
High-concentration soil and sludges by ultrasonic extraction	10,000
Non-water miscible waste	100,000

(a) Sample EQLs are highly matrix-dependent.
(b) EQL = [Method detection limit] × [Factor]. For non-aqueous samples, the factor is on a wet-weight basis.

The MDL using 7 replicate determinations and a spike concentration of 100 µg/L was 120 ng/L.

The average recovery from HPLC-grade water using 4 determinatons and a spike concentration of 100 µg/L was 87.6%.

The precision (as RSD) from HPLC-grade water using 4 determinatons and a spike concentration of 100 µg/L was 16.2%.

The average recovery from groundwater using 4 determinatons and a spike concentration of 100 µg/L was 89.3%.

The precision (as RSD) from groundwater using 4 determinatons and a spike concentration of 100 µg/L was 1.6%.

The average recovery (in %) with %RSD (in parenthesis) from 3 determinations and a spike concentration of 20 µg/L in water was 76.0 (6.5) using 3M Empore Disks and EPA Method 8061.

The average recovery (in %) with %RSD (in parenthesis) from 3 determinations and a spike concentration of 20 µg/L in leachate was 95.3 (16.9) using 3M Empore Disks and EPA Method 8061.

The average recovery (in %) with %RSD (in parenthesis) from 3 determinations and a spike concentration of 20 µg/L in estuarine groundwater was 92.7 (17.1) using 3M Empore Disks and EPA Method 8061.

The average recovery (in %) with %RSD (in parenthesis) from 3 determinations and a spike concentration of 1 mg/kg in estuarine sediment was 103 (31.) after sulfur cleanup with EPA Method 3660.

The average recovery (in %) with %RSD (in parenthesis) from 3 determinations and a spike concentration of 1 mg/kg in municipal sludge was 106 (5.3).

The average recovery (in %) with %RSD (in parenthesis) from 3 determinations and a spike concentration of 1 mg/kg in sandy loam soil was 70.3 (3.7).

SAMPLE COLLECTION, PRESERVATION & HANDLING
Containers used to collect samples for the determination of semivolatile organic compounds should be soap and water washed followed by methanol (or isopropanol) rinsing. The sample containers should be of glass or Teflon® and have screw-top covers with Teflon® liners. Sample containers should be filled with care to prevent any portion of the collected sample coming in contact with the sampler's gloves.

No preservation is used with concentrated waste samples. With liquid samples containing no residual chlorine and with soil, sediment, and sludge samples, immediately cooling to 4°C is the only preservation used. When residual chlorine is present then 3 mL of 10% aqueous sodium sulfate is added for each gallon of sample collected, followed by cooling to 4°C.

MHT Liquid samples must be extracted within 7 days and their extracts analyzed within 40 days. Concentrated waste, soil, sediment, and sludge samples must be extracted within 14 days and their extracts analyzed within 40 days.

SAMPLE PREPARATION In general, water samples are extracted at a pH of 5 to 7 with methylene chloride in a separatory funnel (EPA Method 3510). EPA Method 3520 is not recommended for the extraction of aqueous samples because the longer chain esters tend to adsorb to the glassware and consequently, their extraction recoveries may be poor. Solid samples are extracted with hexane/acetone (1:) or methylene chloride/acetone (1:1) in a Soxhlet extractor (EPA Method 3540) or with an ultrasonic extractor (EPA Method 3550). Immediately prior to extraction, spike 500 µL of the surrogate standard spiking solution into 1-L aqueous sample or 30-g solid sample. Extraction of particulate-free aqueous samples using C-18 extraction disks is an optional method that can be used.

Prior to Florisil cleanup or GC analysis, the methylene chloride and methylene chloride/acetone extracts must be exchanged to hexane. Exchange is not required for the acetonitrile extracts. Cleanup may not be necessary for extracts from a relatively clean sample matrix. Florisil Cartridge Cleanup may be used for extract cleanup.

If PCBs and organochlorine pesticides are known to be present in the sample, and if Florisil Cartridge Cleanup is considered, then two fractions are collected: Fraction 1 is eluted with 5 mL of 20% methylene chloride in hexane and Fraction 2 is eluted with 5 mL of 10% acetone in hexane. Fraction 1 contains the organochlorine pesticides and PCBs, and can be discarded. Fraction 2 contains the phthalate esters and is analyzed by GC/ECD.

QUALITY CONTROL Identify compounds in the sample by comparing the retention times of the peaks in the sample chromatogram with those of the peaks in standard chromatograms. The retention time window used to make identification is based upon measurements of actual retention time variations over the course of 10 consecutive injections.

Calibration standards are prepared at a minimum of five concentrations for each parameter of interest through dilution of the stock standard solutions with hexane. One of the concentrations should be at a concentration near, but above, the method detection limit. Prepare stock standard solutions in hexane. Stock standards should be checked frequently for signs of degradation or evaporation, especially just prior to preparing calibration standards from them. Stock standard solutions must be replaced after one year, or sooner if comparison with check standards indicates a problem. The suggested internal standards are: 2,5-dibromotoluene, 1,3,5-tribromobenzene, and α, α'-dibromo-m-xylene. The analyst can use any of the three compounds provided that they are resolved from matrix interferences. Recommended surrogate compounds are α-2,6-trichlorotoluene, 1,4-dichloronaphthalene, and 2,3,4,5,6-pentachlorotoluene.

Spike each sample, standard, and blank with surrogate compounds. Three surrogates are suggested for this method: diphenyl phthalate, diphenyl isophthalate, and dibenzyl phthalate.

The quality control check sample concentrate should contain the test compounds at 5 to 10 ng/µL An internal standard peak area check must be performed on all samples. The internal standard must be evaluated for acceptance by determining whether the measured area for the internal standard deviates by more than 30% from the average are for the internal standard in the calibration standards. When the internal standard peak area is outside that limit, all samples that fall outside the QC criteria must be reanalyzed. Benzyl benzoate has been tested and found appropriate as an internal standard for this method.

Any compounds confirmed by two columns may also be confirmed by GC/MS. The sample extract and associated blank should be analyzed by GC/MS. A reference standard of the compound must also be analyzed by GC/MS. Include a mid-concentration calibration standard after each group of 20 samples. The response factors for the mid-concentration calibration must be within ±15% of the average values for the multiconcentration calibration. Demonstrate through the analyses of standards that the Florisil fractionation scheme is reproducible.

REFERENCE Test Methods for Evaluating Solid Waste, Physical/Chemical Methods, SW-846, 3rd Edition, U.S. EPA, Office of Solid Waste, Washington, DC, EPA Method 8061, Nov. 1990.

Dimethoate   **EPA Method 8141**
CAS #60-51-5

TITLE Organophosphorus Compounds by Gas Chromatography: Capillary Column Technique

MATRIX This method covers aqueous and solid matrices. This includes a wide variety such as drinking water, groundwater, industrial wastewaters, surface waters, soils, solids, and sediments.

METHOD SUMMARY This is a GC method used to determine the concentration of 28 organophosphorus pesticides.

The use of Gel Permeation Cleanup (EPA Method 3640) for sample cleanup has been demonstrated to yield recoveries of less than 85% for many method analytes and is therefore not recommended for use with this method.

This method provides GC conditions for the detection of ppb concentrations of organophosphorus compounds. Prior to the use of this method, appropriate sample preparation techniques must be used. Water samples are extracted at a neutral pH with methylene chloride as a solvent by using a separatory funnel (EPA Method 3510) or a continuous liquid-liquid extractor (EPA Method 3520). Soxhlet extraction (EPA Method 3540) or ultrasonic extraction (EPA Method 3550) using methylene chloride/acetone (1:1) are used for solid samples. Both neat and diluted organic liquids (EPA Method 3580) may be analyzed by direct injection. Spiked samples are used to verify the applicability of the chosen extraction technique to each new sample type. A GC with a flame photometric (FPD) or nitrogen-phosphorus detector (NPD) is used for this multiresidue procedure.

INTERFERENCES The use of Florisil cleanup materials (EPA Method 3620) for some of the compounds in this method has been demonstrated to yield recoveries less than 85% and is therefore not recommended for all compounds. Use of phosphorus or halogen specific detectors, however, often obviates the necessity for cleanup for relatively clean sample matrices. If particular circumstances demand the use of an alternative cleanup procedure, the analyst must determine the elution profile and demonstrate that the recovery of each analyte is no less than 85%.

Use of a flame photometric detector (FPD) in the phosphorus mode will minimize interferences from materials that do not contain phosphorus. Elemental sulfur, however, may interfere with the determination of certain organophosphorus compounds by flame photometric gas chromatography. Sulfur cleanup using EPA Method 3660 may alleviate this interference. A nitrogen phosphorus detector (NPD) is also recommended.

A few analytes coelute on certain columns. Therefore, select a second column for confirmation where coelution of the analytes of interest does not occur.

Method interferences may be caused by contaminants in solvents, reagents, glassware, and other sample processing hardware that lead to discrete artifacts or elevated baselines in gas chromatograms. All these materials must be routinely demonstrated to be free from interferences under the conditions of the analysis by analyzing reagent blanks.

INSTRUMENTATION A GC with a NPD or a FPD will be needed. A data system or integrator is recommended for measuring peak areas and/or peak heights. A Kuderna-Danish (K-D) apparatus will be needed for extract concentration.

Column 1: 15 m × 0.53 mm megabore capillary column, 1.0 µm film thickness, DB-210.
Column 2: 15 m × 0.53 mm megabore capillary column, 1.5 µm film thickness, SPB-608.
Column 3: 15 m × 0.53 mm megabore capillary column, 1.0 µm film thickness, DB-5.

Three megabore capillary columns are included for analysis of organophosphates by this method. Column 1 (DB-210 or equivalent) and Column 2 (SPB-608 or equivalent) are recommended if a large number of organophosphorus analytes are to be determined. If the superior resolution offered by Column 1 and Column 2 is not required, Column 3 (DB-5 or equivalent) may be used. For megabore capillary columns, automatic injections of 1 µL are recommended.

PRECISION & ACCURACY The MDL actually achieved in a given analysis will vary, as it is dependent on instrument sensitivity and matrix effects. Single operator accuracy and precision studies have been conducted with spiked water and soil samples.

MULTIPLICATION FACTORS FOR OTHER MATRICES (a)

Matrix	Factor (b)
Groundwater (EPA Method 3510 or EPA Method 3520)	10
Low-concentration soil by Soxhlet and no cleanup	10 (c)
Low-concentration soil by ultrasonic extraction with GPC cleanup	6.7 (c)
High-concentration soil and sludges by ultrasonic extraction	500 (c)
Non-water miscible waste (EPA Method 3580)	1000 (c)

(a) Sample EQLs are highly matrix-dependent. The EQLs listed here are provided for guidance and may not always be achievable.
(b) EQL = [Method detection limit] × [Factor]. For non-aqueous samples the factor is on a wet-weight basis.
(c) Multiply this factory times the soil MDL.

The MDL (in µg/L) when reagent water was extracted using a separatory funnel was 0.26.
The MDL (in µg/kg) when soil was extracted using Soxhlet extraction (EPA Method 3540) was 13.0.
Accuracy (as % recovery) with separatory funnel extraction ranged from not recovered (with low spikes) to 101 (with high spikes).
Accuracy (as % recovery) with continuous liquid-liquid extraction ranged from not recovered (with low spikes) to 102 (with high spikes).
Accuracy (as % recovery) with Soxhlet extraction of soils ranged from not recovered (with low spikes to 98 (with high spikes).
Accuracy (as % recovery) with ultrasonic extraction of soils ranged from not recovered (with low spikes) to not recovered (with high spikes).

SAMPLE COLLECTION, PRESERVATION & HANDLING
Containers used to collect samples for the determination of semivolatile organic compounds should be soap and water washed followed by methanol (or isopropanol) rinsing. The sample containers should be of glass or Teflon® and have screw-top covers with Teflon® liners.

No preservation is used with concentrated waste samples. With liquid samples containing no residual chlorine and with soil, sediment, and sludge samples, immediately cooling to 4°C is the only preservation used. When residual chlorine is present then 3 mL of 10% aqueous sodium sulfate is added for each gallon of sample collected, followed by cooling to 4°C.

Liquid samples must be extracted within 7 days and their extracts analyzed within 40 days. Concentrated waste, soil, sediment, and sludge samples must be extracted within 14 days and their extracts analyzed within 40 days.

SAMPLE PREPARATION In general, water samples are extracted at a neutral pH with methylene chloride, using either EPA Method 3510 or EPA Method 3520. Solid samples are extracted using either EPA Method 3540 or EPA Method 3550 with methylene chloride/acetone (1:1) as the extraction solvent.

Prior to GC analysis, the extraction solvent may be exchanged to hexane. Single lab data indicates that samples should not be transferred with 100% hexane during sample workup as the more water soluble organophosphorus compounds may be lost.

If cleanup is performed on the samples, the analyst should analyze the samples by GC. This will confirm elution patterns and the absence of interferences from the reagents. If peak detection and identification is prevented by the presence of interferences, further cleanup is required.

QUALITY CONTROL The analyst should monitor the performance of the extraction, cleanup (when used), and analytical system and the effectiveness of the method in dealing with each sample matrix by spiking each sample, standard, and blank with one or two surrogates (e.g., organophosphorus compounds not expected to be present in the sample). Deuterated analogs of analytes should not be used as surrogates for gas chromatographic analysis due to coelution problems.

A minimum of five concentrations for each analyte of interest should be prepared through dilution of the stock standards with isooctane. One of the concentrations should be at a concentration near, but above, the MDL.

Include a mid-level check standard after each group of 10 samples in the analysis sequence. GC/MS techniques should be judiciously employed to support qualitative identifications made with this method. Follow the GC/MS operating requirements specified in EPA Method 8270.

When available, chemical ionization mass spectra may be employed to aid in the qualitative identification process. To confirm an identification of a compound, the background-corrected mass spectrum of the compound must be obtained from the sample extract and must be compared with a mass spectrum from a stock or calibration standard analyzed under the same chromatographic conditions. The molecular ion and all other ions present above 20% relative abundance in the mass spectrum of the standard must be present in the mass spectrum of the sample with agreement to ±20%. The retention time of the compound in the sample must be within six seconds of the retention time for the same compound in the standard solution.

Should the MS procedure fail to provide satisfactory results, additional steps may be taken before reanalysis. These steps may include the use of alternate packed or capillary GC columns or additional sample cleanup.

REFERENCE Test Methods for Evaluating Solid Waste, Physical/Chemical Methods, SW-846, 3rd Edition, U.S. EPA, Office of Solid Waste, Washington, DC, EPA Method 8141 July 1992.

Dimethoate   **EPA Method 8270**
CAS #60-51-5

TITLE Semivolatile Organic Compounds by GC/MS

MATRIX This method is used to determine the concentration of semivolatile organic compounds in extracts prepared from all types of solid waste matrices, soils, and groundwater. Although surface waters are not specifically mentioned, this method should be applicable to water samples from rivers, lakes, etc.

METHOD SUMMARY This method covers 259 semivolatile organic compounds. In very limited applications direct injection of the sample into the GC/MS system may be appropriate, but this results in very high detection limits (approximately 10,000 µg/L). Typically, a 1-L liquid sample, containing surrogate, and matrix spiking standards, is extracted in a continuous extractor first under acid conditions and then under basic conditions. Typically 30 g of a solid sample, containing surrogate, and matrix spiking standards, is extracted ultrasonically. After concentrating the extract to 1 mL it is spiked with 10 µL of an internal standard solution just prior to analysis by GC/MS. The volume injected should contain about 100 ng of base/neutral and 200 ng of acid surrogates (for a 1-µL injection). Analysis is performed by GC/MS using a capillary GC column.

INTERFERENCES Raw GC/MS data from all blanks, samples, and spikes must be evaluated for interferences. Contamination by carryover can occur whenever high-concentration and low-concentration samples are sequentially analyzed. To reduce carryover, the sample syringe must be rinsed out between samples with solvent. Whenever an unusually concentrated sample is encountered, it should be followed by the analysis of blank solvent to check for cross-contamination.

INSTRUMENTATION A GC/MS and a data system are required. The GC column used is a 30 m × 0.25 mm I.D. (or 0.32 mm I.D.) 1um film thickness silicone-coated fused silica capillary column. A continuous liquid-liquid extractor equipped with Teflon® or glass connection joints and stopcocks requiring no lubrication, a K-D concentrating apparatus, water bath, and an ultrasonic disrupter with a minimum power of 300 W and with pulsing capability are also required.

PRECISION & ACCURACY The estimated quantitation limit (EQL) of Method 8270B for determining an individual compound is approximately 1 mg/kg (wet weight) for soil or sediment samples, 1–200 mg/kg for wastes (dependent on matrix and method of preparation), and 10 µg/L for groundwater samples. EQLs will be proportionately higher for sample extracts that require dilution to avoid saturation of the detector.

The EQL(b) for groundwater in µg/L is 20.
The EQL (a, b) for low concentrations in soil and sediment in µg/kg is not determined.
Accuracy as µg/L is not listed.
Overall precision in µg/L is not listed.

(a) *EQLs listed for soil/sediment are based on wet weight. Normally data is reported in a dry-weight basis; therefore, EQLs will be higher based on the % dry weight of each sample. This calculation is based on a 30 g sample and gel permeation chromatography cleanup.*
(b) *Sample EQLs are highly matrix-dependent. The EQLs are provided for guidance and may not always be achievable.*
C = *True value for concentration, in µg/L.*
X = *Average recovery found for measurements of samples containing a concentration of C, in µg/L.*

ESTIMATED QUANTITATION LIMIT

Other Matrices	Factor (a)
High-concentration soil and sludges by sonicator	7.5
Non-water miscible waste	75

(a) *EQL for other matrices = [EQL for low soil/sediment] × [Factor]. This estimated EQL is similar to an EPA "Practical Quantitation Limit."*

SAMPLING METHOD

Liquid samples — Use a 1 or 2½ gallon amber glass bottle with a screw-top Teflon®-lined cover that has been prewashed with detergent and rinsed with distilled water and methanol (or isopropanol).

Soils, sediments, or sludges — Use an 8-oz. widemouth glass with a screw-top Teflon®-lined cover that has been prewashed with detergent and rinsed with distilled water and methanol (or isopropanol).

SAMPLE PRESERVATION

Liquid samples — If residual chlorine is present, add 3 mL of 10% sodium thiosulfate per gallon, cool to 4°C and store in a solvent-free refrigerator until analysis; if chlorine is not present, then eliminate the sodium thiosulfate addition.

Soils, sediments, or sludges — Cool samples to 4°C and store in a solvent-free refrigerator.

MHT Liquid samples must be extracted within 7 days and the extracts analyzed within 40 days. Soils, sediments, or sludges may be stored for a maximum of 14 days and the extracts analyzed within 40 days.

SAMPLE PREPARATION

Liquid samples — Transfer 1 L quantitatively to a continuous extractor. If high concentrations are anticipated, a smaller volume may be used and then diluted with organic-free reagent water to 1 L. Adjust pH, if necessary, to pH <2 using 1:1 (V/V) sulfuric acid. Pipette 1.0 mL of a surrogate standard spiking solution into each sample. For the sample in each analytical batch selected for spiking, add 1.0 mL of a matrix spiking standard. For base/neutral acid analysis, the amount of the surrogates and matrix spiking compounds added to the sample should result in a final concentration of 100 ng/µL of each analyte in the extract to be analyzed (assuming a 1-µL injection). Extract with methylene chloride for 18–24 h. Next, adjust the pH of the aqueous phase to pH >11 using 10 N sodium hydroxide and extract it with methylene chloride again for 18–24 h. Dry the extract through a column containing anhydrous sodium sulfate and concentrate it to 1 mL using a K-D concentrator.

Soils, sediments, or sludges — Use 30 g of sample. Nonporous or wet samples (gummy or clay type) that do not have a free-flowing sandy texture must be mixed with anhydrous sodium sulfate until the sample is free flowing. Add 1 mL of surrogate standards to all samples, spikes, standards, and blanks. For the sample in each analytical batch selected for spiking, add 1.0 mL of a matrix spiking standard. For base/neutral acid analysis, the amount added of the surrogates and matrix spiking compounds should result in a final concentration of 100 ng/µL of each base/neutral analyte and 200 ng/µL of each acid analyte in the extract to be analyzed (assuming a 1-µL injection). Immediately add a 100-mL mixture of 1:1 methylene chloride:acetone and extract the sample ultrasonically for 3 min and then decant or filter the extracts. Repeat the extraction two or more times. Dry the extract using a column with anhydrous sodium sulfate and concentrate it to 1 mL in a K-D concentrator.

QUALITY CONTROL

A methylene chloride solution containing 50 ng/µL of decafluorotriphenylphosphine (DFTPP) is used for tuning the GC/MS system each 12-h shift. A system performance check also must be made during every 12-h shift. A standard containing 50 ng/µL each of 4,4'-DDT, pentachlorophenol, and benzidine is required to verify injection port inertness and GC column performance. A calibration standard at mid-concentration, containing each compound of interest, including all required surrogates, must be performed every 12 h during analysis. After the system performance check is met, calibration check compounds (CCCs) are used to check the validity of the initial calibration.

The internal standard responses and retention times in the calibration check standard must be evaluated immediately after or during data acquisition. If the retention time for any internal standard changes by more than 30 seconds from the last check calibration (12 h), the chromatographic system must be inspected for malfunctions and corrections must be made, as required. If the electron ionization current plot (EICP) area for any of the internal standards changes by a factor of two from the last daily calibration standard check, the mass spectrometer must be inspected for malfunctions and corrections must be made, as appropriate.

Demonstrate, through the analysis of a reagent water blank, that interferences from the analytical system, glassware, and reagents are under control. The blank samples should be carried through all stages of the sample preparation and measurement steps. For each analytical batch (up to 20 samples), a reagent blank, matrix spike, and matrix spike duplicate/duplicate must be analyzed (the frequency of the spikes may be different for different monitoring programs). The blank and spiked samples must be carried through all stages of the sample preparation and measurement steps. A QC reference sample concentrate containing each analyte at a concentration of 100 mg/L in methanol is required.

REFERENCE Test Methods for Evaluating Solid Waste (SW-846). U.S. EPA 1983, Method 8270B, Rev. 2, Nov. 1990. Office of Solid Waste, Washington, DC.

3,3'-Dimethoxybenzidine   EPA Method 1625
CAS #119-90-4

TITLE Semivolatile Organic Compounds by Isotope Dilution GC/MS

MATRIX The compounds may be determined in waters, soils, and municipal sludges by this method.

METHOD SUMMARY This method is used to determine 176 semivolatile toxic organic pollutants associated with the CWA (as amended 1987); the RCRA (as amended 1986); the CERCLA (as amended 1986); and other compounds amenable to extraction and analysis by capillary column gas chromatography-mass spectrometry (GC/MS).

Stable isotopically-labeled analogs of the compounds of interest are added to the sample. If the solids content is less than 1%, a 1-L sample is extracted at pH 12–13, then at pH <2 with methylene chloride using continuous extraction techniques.

If the solids content is 30% or less, the sample is diluted to 1% solids with reagent water, homogenized ultrasonically, and extracted at pH 12–13, then at pH <2 with methylene chloride using continuous extraction techniques. If the solids content is greater than 30%, the sample is extracted using ultrasonic techniques.

Each extract is dried over sodium sulfate, concentrated to a volume of 5 mL, cleaned up using GPC, if necessary, and concentrated. Extracts are concentrated to 1 mL if GPC is not performed, and to 0.5 mL if GPC is performed.

An internal standard is added to the extract, and a 1-mL aliquot of the extract is injected into the GC. The compounds are separated by GC and detected by a MS. The labeled compounds serve to correct the variability of the analytical technique.

INTERFERENCES Solvents, reagents, glassware, and other sample processing hardware may yield artifacts and/or elevated baselines causing misinterpretation of chromatograms and spectra. Materials used in the analysis must be demonstrated to be free from interferences under the conditions of analysis by running method blanks initially and with each sample lot (sample started through the extraction process on a given 8-h shift, to a maximum of 20). Specific selection of reagents and purification of solvents by distillation in all glass systems may be required. Glassware and, where possible, reagents are cleaned by solvent rinse and baking at 450°C for 1-h minimum. Interferences coextracted from samples will vary considerably from source to source, depending on the diversity of the site being sampled.

INSTRUMENTATION Major instrumentation includes a GC with a splitless or on-column injection port for capillary column, a MS with 70 eV electron impact ionization, and a data system to collect and record MS data, and process it. A K-D apparatus is used to concentrate extracts.

GC Column: 30 m × 0.25 mm I.D. 5% phenyl, 94% methyl, 1% vinyl silicone bonded phased fused silica capillary column.

PRECISION & ACCURACY The detection limits of the method are usually dependent on the level of interferences rather than instrumental limitations. The limits typify the minimum quantities that can be detected with no interferences present.

The minimum level (in µg/mL) was not listed. This is defined as a minimum level at which the analytical system shall give recognizable mass spectra (background corrected) and acceptable calibration points.

The MDL (in µg/kg) in low solids was not listed and in high solids was not listed; these were determined in digested sludge (low solids) and in filter cake or compost (high solids).

The labeled and native compound initial precision as standard deviation (in µg/L) was not listed.
The labeled and native compound initial accuracy as average recovery (in µg/L) was not listed.

SAMPLE COLLECTION, PRESERVATION & HANDLING Collect samples in glass containers. Aqueous samples which flow freely are collected in refrigerated bottles using automatic sampling equipment. Solid samples are collected as grab samples using widemouth jars. Maintain samples at 0 to 4°C from the time of collection until extraction. If residual chlorine is present in aqueous samples, add 80 mg sodium thiosulfate/L of water. Begin sample extraction within 7 days of collection, and analyze all extracts within 40 days of extraction.

SAMPLE PREPARATION Samples containing 1% solids or less are extracted directly using continuous liquid-liquid extraction techniques. Samples containing 1 to 30% solids are diluted to the 1% level with reagent water and extracted using continuous liquid-liquid extraction techniques. Samples containing greater than 30% solids are extracted using ultrasonic techniques.

Base/neutral extraction — Adjust the pH of the waters in the extractors to 12–13 with 6 N NaOH. Extract with methylene chloride for 24–48 h.

Acid extraction — Adjust the pH of the waters in the extractors to 2 or less using 6 N sulfuric acid. Extract with methylene chloride for 24–48 h.

Ultrasonic extraction of high solids samples — Add anhydrous sodium sulfate to the sample and QC aliquot(s). Add acetone:methylene chloride (1:1) to the sample and mix thoroughly

Concentrate extracts using a K-D apparatus.

QUALITY CONTROL The analyst is permitted to modify this method to improve separations or lower the costs of measurements, provided all performance specifications are met. Analyses of blanks are required to demonstrate freedom from contamination. When results of spikes indicate atypical method performance for samples, the samples are diluted to bring method performance within acceptable limits.

For low solids (aqueous samples), extract, concentrate, and analyze two sets of four 1-L aliquots (8 aliquots total) of the precision and recovery standard. For high solids samples, two sets of four 30-g aliquots of the high solids reference matrix are used.

Spike all samples with labeled compounds to assess method performance. Compute percent recovery of the labeled compounds using the internal standard method. Compare the labeled compound recovery for each compound with the corresponding labeled compound recovery.

Reagent water and high solids reference matrix blanks are analyzed to demonstrate freedom from contamination. Extract and concentrate a 1-L reagent water blank or a high solids reference matrix blank with each sample's lot (samples started through the extraction process on the same 8-h shift, to a maximum of 20 samples).

Field replicates may be collected to determine the precision of the sampling technique, and spiked samples may be required to determine the accuracy of the analysis when the internal standard method is used.

REFERENCE Semivolatile Organic Compounds by Isotope Dilution GC/MS. Office of Water Regulation and Standards, U.S. EPA Industrial Technology Division, Washington, DC, EPA Method 1625, Rev. C, June 1989 (contact W.A. Telliard, U.S. EPA, Office of Water Regulations and Standards, 401 M St., SW, Washington, DC, 20460. Phone: 202-382-7131).

3,3'-Dimethoxybenzidine   EPA Method 8270
CAS #119-90-4

TITLE Semivolatile Organic Compounds by GC/MS

MATRIX This method is used to determine the concentration of semivolatile organic compounds in extracts prepared

from all types of solid waste matrices, soils, and groundwater. Although surface waters are not specifically mentioned, this method should be applicable to water samples from rivers, lakes, etc.

METHOD SUMMARY This method covers 259 semivolatile organic compounds. In very limited applications direct injection of the sample into the GC/MS system may be appropriate, but this results in very high detection limits (approximately 10,000 μg/L). Typically, a 1-L liquid sample, containing surrogate, and matrix spiking standards, is extracted in a continuous extractor first under acid conditions and then under basic conditions. Typically 30 g of a solid sample, containing surrogate, and matrix spiking standards, is extracted ultrasonically. After concentrating the extract to 1 mL it is spiked with 10 μL of an internal standard solution just prior to analysis by GC/MS. The volume injected should contain about 100 ng of base/neutral and 200 ng of acid surrogates (for a 1-μL injection). Analysis is performed by GC/MS using a capillary GC column.

INTERFERENCES Raw GC/MS data from all blanks, samples, and spikes must be evaluated for interferences. Contamination by carryover can occur whenever high-concentration and low-concentration samples are sequentially analyzed. To reduce carryover, the sample syringe must be rinsed out between samples with solvent. Whenever an unusually concentrated sample is encountered, it should be followed by the analysis of blank solvent to check for cross-contamination.

INSTRUMENTATION A GC/MS and a data system are required. The GC column used is a 30 m × 0.25 mm I.D. (or 0.32 mm I.D.) 1um film thickness silicone-coated fused silica capillary column. A continuous liquid-liquid extractor equipped with Teflon® or glass connection joints and stopcocks requiring no lubrication, a K-D concentrating apparatus, water bath, and an ultrasonic disrupter with a minimum power of 300 W and with pulsing capability are also required.

PRECISION & ACCURACY The estimated quantitation limit (EQL) of Method 8270B for determining an individual compound is approximately 1 mg/kg (wet weight) for soil or sediment samples, 1–200 mg/kg for wastes (dependent on matrix and method of preparation), and 10 μg/L for groundwater samples. EQLs will be proportionately higher for sample extracts that require dilution to avoid saturation of the detector.

The EQL(b) for groundwater in μg/L is 100.
The EQL (a, b) for low concentrations in soil and sediment in μg/kg is not determined.
Accuracy as μg/L is not listed.
Overall precision in μg/L is not listed.

(a) EQLs listed for soil/sediment are based on wet weight. Normally data is reported in a dry-weight basis; therefore, EQLs will be higher based on the % dry weight of each sample. This calculation is based on a 30 g sample and gel permeation chromatography cleanup.
(b) Sample EQLs are highly matrix-dependent. The EQLs are provided for guidance and may not always be achievable.
C = True value for concentration, in μg/L.
X = Average recovery found for measurements of samples containing a concentration of C, in μg/L.

ESTIMATED QUANTITATION LIMIT

Other Matrices	Factor (a)
High-concentration soil and sludges by sonicator	7.5
Non-water miscible waste	75

(a) EQL for other matrices = [EQL for low soil/sediment] × [Factor]. This estimated EQL is similar to an EPA "Practical Quantitation Limit."

SAMPLING METHOD
Liquid samples — Use a 1 or 2½ gallon amber glass bottle with a screw-top Teflon®-lined cover that has been prewashed with detergent and rinsed with distilled water and methanol (or isopropanol).

Soils, sediments, or sludges — Use an 8-oz. widemouth glass with a screw-top Teflon®-lined cover that has been prewashed with detergent and rinsed with distilled water and methanol (or isopropanol).

SAMPLE PRESERVATION
Liquid samples — If residual chlorine is present, add 3 mL of 10% sodium thiosulfate per gallon, cool to 4°C and store in a solvent-free refrigerator until analysis; if chlorine is not present, then eliminate the sodium thiosulfate addition.

Soils, sediments, or sludges — Cool samples to 4°C and store in a solvent-free refrigerator.

MHT Liquid samples must be extracted within 7 days and the extracts analyzed within 40 days. Soils, sediments, or sludges may be stored for a maximum of 14 days and the extracts analyzed within 40 days.

SAMPLE PREPARATION
Liquid samples — Transfer 1 L quantitatively to a continuous extractor. If high concentrations are anticipated, a smaller volume may be used and then diluted with organic-free reagent water to 1 L. Adjust pH, if necessary, to pH <2 using 1:1 (V/V) sulfuric acid. Pipette 1.0 mL of a surrogate standard spiking solution into each sample. For the sample in each analytical batch selected for spiking, add 1.0 mL of a matrix spiking standard. For base/neutral acid analysis, the amount of the surrogates and matrix spiking compounds added to the sample should result in a final concentration of 100 ng/μL of each analyte in the extract to be analyzed (assuming a 1-μL injection). Extract with methylene chloride for 18–24 h. Next, adjust the pH of the aqueous phase to pH >11 using 10 N sodium hydroxide and extract it with methylene chloride again for 18–24 h. Dry the extract through a column containing anhydrous sodium sulfate and concentrate it to 1 mL using a K-D concentrator.

Soils, sediments, or sludges — Use 30 g of sample. Nonporous or wet samples (gummy or clay type) that do not have a free-flowing sandy texture must be mixed with anhydrous sodium sulfate until the sample is free flowing. Add 1 mL of surrogate standards to all samples, spikes, standards, and blanks. For the sample in each analytical batch selected for spiking, add 1.0 mL of a matrix spiking standard. For base/neutral acid analysis, the amount added of the surrogates and matrix spiking compounds should result in a final concentration of 100 ng/μL of each base/neutral analyte and 200 ng/μL of each acid analyte

in the extract to be analyzed (assuming a 1-µL injection). Immediately add a 100-mL mixture of 1:1 methylene chloride:acetone and extract the sample ultrasonically for 3 min and then decant or filter the extracts. Repeat the extraction two or more times. Dry the extract using a column with anhydrous sodium sulfate and concentrate it to 1 mL in a K-D concentrator.

QUALITY CONTROL A methylene chloride solution containing 50 ng/µL of decafluorotriphenylphosphine (DFTPP) is used for tuning the GC/MS system each 12-h shift. A system performance check also must be made during every 12-h shift. A standard containing 50 ng/µL each of 4,4'-DDT, pentachlorophenol, and benzidine is required to verify injection port inertness and GC column performance. A calibration standard at mid-concentration, containing each compound of interest, including all required surrogates, must be performed every 12 h during analysis. After the system performance check is met, calibration check compounds (CCCs) are used to check the validity of the initial calibration.

The internal standard responses and retention times in the calibration check standard must be evaluated immediately after or during data acquisition. If the retention time for any internal standard changes by more than 30 seconds from the last check calibration (12 h), the chromatographic system must be inspected for malfunctions and corrections must be made, as required. If the electron ionization current plot (EICP) area for any of the internal standards changes by a factor of two from the last daily calibration standard check, the mass spectrometer must be inspected for malfunctions and corrections must be made, as appropriate.

Demonstrate, through the analysis of a reagent water blank, that interferences from the analytical system, glassware, and reagents are under control. The blank samples should be carried through all stages of the sample preparation and measurement steps. For each analytical batch (up to 20 samples), a reagent blank, matrix spike, and matrix spike duplicate/duplicate must be analyzed (the frequency of the spikes may be different for different monitoring programs). The blank and spiked samples must be carried through all stages of the sample preparation and measurement steps. A QC reference sample concentrate containing each analyte at a concentration of 100 mg/L in methanol is required.

REFERENCE Test Methods for Evaluating Solid Waste (SW-846). U.S. EPA 1983, Method 8270B, Rev. 2, Nov. 1990. Office of Solid Waste, Washington, DC.

Dimethyl phthalate   **EPA Method 1625**
CAS #131-11-3

TITLE Semivolatile Organic Compounds by Isotope Dilution GC/MS

MATRIX The compounds may be determined in waters, soils, and municipal sludges by this method.

METHOD SUMMARY This method is used to determine 176 semivolatile toxic organic pollutants associated with the CWA (as amended 1987); the RCRA (as amended 1986); the CERCLA (as amended 1986); and other compounds amenable to extraction and analysis by capillary column gas chromatography-mass spectrometry (GC/MS).

Stable isotopically-labeled analogs of the compounds of interest are added to the sample. If the solids content is less than 1%, a 1-L sample is extracted at pH 12–13, then at pH <2 with methylene chloride using continuous extraction techniques.

If the solids content is 30% or less, the sample is diluted to 1% solids with reagent water, homogenized ultrasonically, and extracted at pH 12–13, then at pH <2 with methylene chloride using continuous extraction techniques. If the solids content is greater than 30%, the sample is extracted using ultrasonic techniques.

Each extract is dried over sodium sulfate, concentrated to a volume of 5 mL, cleaned up using GPC, if necessary, and concentrated. Extracts are concentrated to 1 mL if GPC is not performed, and to 0.5 mL if GPC is performed.

An internal standard is added to the extract, and a 1-mL aliquot of the extract is injected into the GC. The compounds are separated by GC and detected by a MS. The labeled compounds serve to correct the variability of the analytical technique.

INTERFERENCES Solvents, reagents, glassware, and other sample processing hardware may yield artifacts and/or elevated baselines causing misinterpretation of chromatograms and spectra. Materials used in the analysis must be demonstrated to be free from interferences under the conditions of analysis by running method blanks initially and with each sample lot (sample started through the extraction process on a given 8-h shift, to a maximum of 20). Specific selection of reagents and purification of solvents by distillation in all glass systems may be required. Glassware and, where possible, reagents are cleaned by solvent rinse and baking at 450°C for 1-h minimum. Interferences coextracted from samples will vary considerably from source to source, depending on the diversity of the site being sampled.

INSTRUMENTATION Major instrumentation includes a GC with a splitless or on-column injection port for capillary column, a MS with 70 eV electron impact ionization, and a data system to collect and record MS data, and process it. A K-D apparatus is used to concentrate extracts.

GC Column: 30 m × 0.25 mm I.D. 5% phenyl, 94% methyl, 1% vinyl silicone bonded phased fused silica capillary column.

PRECISION & ACCURACY The detection limits of the method are usually dependent on the level of interferences rather than instrumental limitations. The limits typify the minimum quantities that can be detected with no interferences present.

The minimum level (in µg/mL) was 10. This is defined as a minimum level at which the analytical system shall give recognizable mass spectra (background corrected) and acceptable calibration points.

The MDL (in µg/kg) in low solids was 62 and in high solids was 21; these were determined in digested sludge (low solids) and in filter cake or compost (high solids).

The labeled and native compound initial precision as standard deviation (in μg/L) was 44.

The labeled and native compound initial accuracy as average recovery (in μg/L) was 75–196.

SAMPLE COLLECTION, PRESERVATION & HANDLING
Collect samples in glass containers. Aqueous samples which flow freely are collected in refrigerated bottles using automatic sampling equipment. Solid samples are collected as grab samples using widemouth jars. Maintain samples at 0 to 4°C from the time of collection until extraction. If residual chlorine is present in aqueous samples, add 80 mg sodium thiosulfate/L of water. Begin sample extraction within 7 days of collection, and analyze all extracts within 40 days of extraction.

SAMPLE PREPARATION Samples containing 1% solids or less are extracted directly using continuous liquid-liquid extraction techniques. Samples containing 1 to 30% solids are diluted to the 1% level with reagent water and extracted using continuous liquid-liquid extraction techniques. Samples containing greater than 30% solids are extracted using ultrasonic techniques.

Base/neutral extraction — Adjust the pH of the waters in the extractors to 12–13 with 6 N NaOH. Extract with methylene chloride for 24–48 h.

Acid extraction — Adjust the pH of the waters in the extractors to 2 or less using 6 N sulfuric acid. Extract with methylene chloride for 24–48 h.

Ultrasonic extraction of high solids samples — Add anhydrous sodium sulfate to the sample and QC aliquot(s). Add acetone:methylene chloride (1:1) to the sample and mix thoroughly

Concentrate extracts using a K-D apparatus.

QUALITY CONTROL The analyst is permitted to modify this method to improve separations or lower the costs of measurements, provided all performance specifications are met. Analyses of blanks are required to demonstrate freedom from contamination. When results of spikes indicate atypical method performance for samples, the samples are diluted to bring method performance within acceptable limits.

For low solids (aqueous samples), extract, concentrate, and analyze two sets of four 1-L aliquots (8 aliquots total) of the precision and recovery standard. For high solids samples, two sets of four 30-g aliquots of the high solids reference matrix are used.

Spike all samples with labeled compounds to assess method performance. Compute percent recovery of the labeled compounds using the internal standard method. Compare the labeled compound recovery for each compound with the corresponding labeled compound recovery.

Reagent water and high solids reference matrix blanks are analyzed to demonstrate freedom from contamination. Extract and concentrate a 1-L reagent water blank or a high solids reference matrix blank with each sample's lot (samples started through the extraction process on the same 8-h shift, to a maximum of 20 samples).

Field replicates may be collected to determine the precision of the sampling technique, and spiked samples may be required to determine the accuracy of the analysis when the internal standard method is used.

REFERENCE Semivolatile Organic Compounds by Isotope Dilution GC/MS. Office of Water Regulation and Standards, U.S. EPA Industrial Technology Division, Washington, DC, EPA Method 1625, Rev. C, June 1989 (contact W.A. Telliard, U.S. EPA, Office of Water Regulations and Standards, 401 M St., SW, Washington, DC, 20460. Phone: 202-382-7131).

Dimethyl phthalate EPA Method 625
CAS #131-11-3

TITLE Base/Neutrals and Acids, U.S. EPA Method 625

MATRIX This methods covers municipal and industrial wastewaters.

METHOD SUMMARY Approximately 1 L of sample is serially extracted with methylene chloride at a pH greater than 11 and again at a pH less than 2 using a separatory funnel or a continuous extractor. The methylene chloride extract is dried, concentrated to a volume of 1 mL, and analyzed by GC/MS. Qualitative identification of the parameters in the extract is performed using the retention time and the relative abundance of three characteristic masses (m/z). Qualitative analysis is performed using either external or internal standard techniques with a single characteristic m/z.

INTERFERENCES Method interferences may be caused by contaminants in solvents, reagents, glassware, and other sample processing hardware. Glassware must be scrupulously cleaned. Glassware should be heated in a muffle furnace at 400°C for 5 to 30 min. Some thermally stable materials, such as PCBs, may not be eliminated by this treatment. Solvent rinses with acetone and pesticide quality hexane may be substituted for the muffle furnace heating. Matrix interferences may be caused by contaminants that are coextracted from the sample. The base-neutral extraction may cause significantly reduced recovery of phenols. The packed gas chromatographic columns recommended for the basic fraction may not exhibit sufficient resolution for some analytes.

INSTRUMENTATION A GC/MS system with an injection port designed for on-column injection when using packed columns and for splitless injection when using capillary columns.

Column for base/neutrals: 1.8 m long × 2 mm I.D. glass, packed with 3% SP-2550 on Supelcoport (100/120 mesh) or equivalent.

Column for acids: 1.8 m long × 2 mm I.D. glass, packed with 1% SP-1240DA on Supelcoport (100/120 mesh) or equivalent.

PRECISION & ACCURACY The MDL concentrations were obtained using reagent water. The MDL actually achieved in a given analysis will vary depending on instrument sensitivity and matrix effects. This method was tested by 15 laboratories using reagent water, drinking water, surface water, and industrial wastewaters spiked at six concentrations over the range 5

to 100 µg/L. Single operator precision, overall precision, and method accuracy were found to be directly related to the concentration of the parameter matrix.

The MDL (in µg/L) in reagent water was 1.6.

The standard deviation (in µg/L based on 4 recovery measurements) was 23.2.

The range (in µg/L) for average recovery for 4 measurements was D-100.0.

The range (in %) for percent recovery was D-112.

Accuracy (in µg/L) as expected recovery for one or more measurements of a sample containing a true concentration of C was $0.20C+1.03$.

Precision (in µg/L) as expected single analyst standard deviation of measurements at an average concentration found at X was $0.54X+0.19$.

Overall precision (in µg/L) as expected interlaboratory standard deviation of measurements in an average concentration found at X was $1.05X-0.92$.

$C =$ *True value of the concentration in µg/L.*

$X =$ *Average recovery found for measurements of samples containing a concentration at C in µg/L.*

SAMPLE PREPARATION Adjust the pH to >11 with sodium hydroxide and serially extract in a separatory funnel with methylene chloride or else in a continuous extractor. Next, adjust the pH to <2 with sulfuric acid and serially extract in a separatory funnel with methylene chloride or else in a continuous extractor. Dry the extracts separately through a column of anhydrous sodium sulfate and then concentrate each of the extracts to 1.0 mL using a K-D apparatus.

SAMPLE COLLECTION, PRESERVATION & HANDLING Grab samples must be collected in glass containers. All samples must be refrigerated at 4°C from the time of collection until extraction. If residual chlorine is present, add 80 mg of sodium thiosulfate/L of sample and mix well. All samples must be extracted within 7 days of collection and completely analyzed within 40 days of extraction.

QUALITY CONTROL Make an initial, one-time, demonstration of the ability to generate acceptable accuracy and precision with this method. Before processing any samples, the analyst must analyze a reagent water blank to demonstrate that interferences from the analytical system and glassware are under control. Each time a set of samples is extracted or reagents are changed, a reagent water blank must be processed. Spike and analyze a minimum of 5% of all samples to monitor and evaluate lab data quality. A QC check sample concentrate that contains each parameter of interest at a concentration of 100 µg/mL in acetone is required. PCBs and multicomponent pesticides may be omitted from this test.

After analysis of five spiked wastewater samples, calculate the average percent recovery and the standard deviation of the percent recovery. Spike all samples with the surrogate standard spiking solution and calculate the percent recovery of each surrogate compound.

REFERENCE Federal Register, Vol. 49, No. 209. Friday, Oct. 26, 1984.

Dimethyl phthalate
CAS #131-11-3

EPA Method 8061

TITLE Phthalate Esters by Capillary Gas Chromatography With Electron Capture Detection (GC/ECD)

MATRIX This method covers aqueous and solid matrices. This includes a wide variety such as drinking water, groundwater, industrial wastewaters, surface waters, soils, solids, and sediments.

METHOD SUMMARY This method is used to determine the identities and concentrations of phthalate esters in liquid, solid and sludge matrices. When used to analyze for any or all of the target analytes, compound identification should be supported by at least one additional qualitative technique. This method describes conditions for parallel column, dual electron capture detector analysis, which fulfills the above requirement. Alternatively, GC/MS could be used for compound confirmation.

A measured volume or weight of sample (approximately 1 L for liquids, 10 to 30 g for solids and sludges) is extracted by using the appropriate sample extraction technique specified in EPA Method 3510, EPA Method 3540, and EPA Method 3550. After cleanup, the extract is analyzed by GC/ECD.

INTERFERENCES The sensitivity of this method usually depends on the level of interferences rather than on instrumental limitations. If interferences prevent detection of the analytes, cleanup of the sample extracts is necessary. Either EPA Method 3610 or EPA Method 3620 alone or followed by EPA Method 3660, Sulfur Cleanup, may be used to eliminate interferences in the analysis. EPA Method 3640, Gel Permeation Cleanup, is applicable for samples that contain high amounts of lipids and waxes.

Interferences coextracted from the samples will vary considerably from waste to waste. Glassware must be scrupulously clean. All glassware require treatment in a muffle furnace at 400°C for 2 to 4 h, or thorough rinsing with pesticide-grade solvent, prior to use. Volumetric glassware should not be heated in a muffle furnace. Storage of glassware in the lab introduces contamination, even if the glassware is wrapped in aluminum foil. Sodium sulfate, Florisil, and alumina may be contaminated with phthalate esters and, therefore, use of these materials in sample cleanup should be employed cautiously. If these materials are used, they must be obtained packaged in glass. Heating at 400°C for sodium sulfate, 320°C for Florisil, and 210°C for alumina is recommended. Glass wool used in any step of sample preparation should be a specially treated pyrex wool, pesticide grade, and must be baked at 400°C for 4 h immediately prior to use.

Paper thimbles and filter paper must be exhaustively washed with the solvent that will be used in the sample extraction. Soxhlet extraction of paper thimbles and filter paper for 12 h with fresh solvent should be repeated for a minimum of three times. Method blanks should be obtained before any of the precleaned thimbles or filter papers are used.

INSTRUMENTATION Gas chromatograph suitable for on-column and split/splitless injections.

Column 1: 30 m × 0.53 mm ID, 5% phenyl/95% methyl silicone fused-silica open tubular column, DB-5, 1.5 µg film thickness.

Column 2: 30 m × 0.53 mm ID, 14% cyanopropyl phenyl silicone fused-silica open tubular column, DB-1701, 1.0 µg film thickness.

A dual electron capture detector (ECD) is used. A Kuderna-Danish (K-D) apparatus is required along with a vacuum manifold consisting of individually adjustable, easily accessible flow-control valves for up to 24 cartridges, sample rack, chemically resistant cover and seals, heavy-duty glass basin, removable stainless steel solvent guides, built-in vacuum gauge and valve. Also, 6-mL, 1-g solid-phase extraction cartridges, LC-Florisil or equivalent, prepackaged, ready to use will be needed.

PRECISION & ACCURACY The MDL actually achieved in a given analysis will vary, as it is dependent on instrument sensitivity and matrix effects. This method has been tested in a single lab. Single-operator precision, overall precision, and method accuracy were found to be related to the concentration of the compounds and the type of matrix.

MULTIPLICATION FACTORS FOR OTHER MATRICES (a)

Matrix	Factor (b)
Groundwater	10
Low-concentration soil by ultrasonic extraction with GPC cleanup	670
High-concentration soil and sludges by ultrasonic extraction	10,000
Non-water miscible waste	100,000

(a) Sample EQLs are highly matrix-dependent.
(b) EQL = [Method detection limit] × [Factor]. For non-aqueous samples, the factor is on a wet-weight basis.

The MDL using 7 replicate determinations and a spike concentration of 100 µg/L was 640 ng/L.

The average recovery from HPLC-grade water using 4 determinatons and a spike concentration of 100 µg/L was 88.6%.

The precision (as RSD) from HPLC-grade water using 4 determinatons and a spike concentration of 100 µg/L was 17.7%.

The average recovery from groundwater using 4 determinatons and a spike concentration of 100 µg/L was 86.8%.

The precision (as RSD) from groundwater using 4 determinatons and a spike concentration of 100 µg/L was 14.3%.

The average recovery (in %) with %RSD (in parenthesis) from 3 determinations and a spike concentration of 20 µg/L in water was 84.0 (4.1) using 3M Empore Disks and EPA Method 8061.

The average recovery (in %) with %RSD (in parenthesis) from 3 determinations and a spike concentration of 20 µg/L in leachate was 98.9 (19.6) using 3M Empore Disks and EPA Method 8061.

The average recovery (in %) with %RSD (in parenthesis) from 3 determinations and a spike concentration of 20 µg/L in estuarine groundwater was 87.1 (8.1) using 3M Empore Disks and EPA Method 8061.

The average recovery (in %) with %RSD (in parenthesis) from 3 determinations and a spike concentration of 1 mg/kg in estuarine sediment was 77.9 (42.8) after sulfur cleanup with EPA Method 3660.

The average recovery (in %) with %RSD (in parenthesis) from 3 determinations and a spike concentration of 1 mg/kg in municipal sludge was 52.1 (35.5).

The average recovery (in %) with %RSD (in parenthesis) from 3 determinations and a spike concentration of 1 mg/kg in sandy loam soil was not determined (matrix interferant).

SAMPLE COLLECTION, PRESERVATION & HANDLING
Containers used to collect samples for the determination of semivolatile organic compounds should be soap and water washed followed by methanol (or isopropanol) rinsing. The sample containers should be of glass or Teflon® and have screw-top covers with Teflon® liners. Sample containers should be filled with care to prevent any portion of the collected sample coming in contact with the sampler's gloves.

No preservation is used with concentrated waste samples. With liquid samples containing no residual chlorine and with soil, sediment, and sludge samples, immediately cooling to 4°C is the only preservation used. When residual chlorine is present then 3 mL of 10% aqueous sodium sulfate is added for each gallon of sample collected, followed by cooling to 4°C.

MHT Liquid samples must be extracted within 7 days and their extracts analyzed within 40 days. Concentrated waste, soil, sediment, and sludge samples must be extracted within 14 days and their extracts analyzed within 40 days.

SAMPLE PREPARATION In general, water samples are extracted at a pH of 5 to 7 with methylene chloride in a separatory funnel (EPA Method 3510). EPA Method 3520 is not recommended for the extraction of aqueous samples because the longer chain esters tend to adsorb to the glassware and consequently, their extraction recoveries may be poor. Solid samples are extracted with hexane/acetone (1:) or methylene chloride/acetone (1:1) in a Soxhlet extractor (EPA Method 3540) or with an ultrasonic extractor (EPA Method 3550). Immediately prior to extraction, spike 500 µL of the surrogate standard spiking solution into 1-L aqueous sample or 30-g solid sample. Extraction of particulate-free aqueous samples using C-18 extraction disks is an optional method that can be used.

Prior to Florisil cleanup or GC analysis, the methylene chloride and methylene chloride/acetone extracts must be exchanged to hexane. Exchange is not required for the acetonitrile extracts. Cleanup may not be necessary for extracts from a relatively clean sample matrix. Florisil Cartridge Cleanup may be used for extract cleanup.

If PCBs and organochlorine pesticides are known to be present in the sample, and if Florisil Cartridge Cleanup is considered, then two fractions are collected: Fraction 1 is eluted with 5 mL of 20% methylene chloride in hexane and Fraction 2 is eluted with 5 mL of 10% acetone in hexane. Fraction 1 contains the organochlorine pesticides and PCBs, and can be discarded. Fraction 2 contains the phthalate esters and is analyzed by GC/ECD.

QUALITY CONTROL Identify compounds in the sample by comparing the retention times of the peaks in the sample chromatogram with those of the peaks in standard chromatograms.

The retention time window used to make identification is based upon measurements of actual retention time variations over the course of 10 consecutive injections.

Calibration standards are prepared at a minimum of five concentrations for each parameter of interest through dilution of the stock standard solutions with hexane. One of the concentrations should be at a concentration near, but above, the method detection limit. Prepare stock standard solutions in hexane. Stock standards should be checked frequently for signs of degradation or evaporation, especially just prior to preparing calibration standards from them. Stock standard solutions must be replaced after one year, or sooner if comparison with check standards indicates a problem. The suggested internal standards are: 2,5-dibromotoluene, 1,3,5-tribromobenzene, and α, α'-dibromo-m-xylene. The analyst can use any of the three compounds provided that they are resolved from matrix interferences. Recommended surrogate compounds are α-2,6-trichlorotoluene, 1,4-dichloronaphthalene, and 2,3,4,5,6-pentachlorotoluene.

Spike each sample, standard, and blank with surrogate compounds. Three surrogates are suggested for this method: diphenyl phthalate, diphenyl isophthalate, and dibenzyl phthalate.

The quality control check sample concentrate should contain the test compounds at 5 to 10 ng/µL An internal standard peak area check must be performed on all samples. The internal standard must be evaluated for acceptance by determining whether the measured area for the internal standard deviates by more than 30% from the average are for the internal standard in the calibration standards. When the internal standard peak area is outside that limit, all samples that fall outside the QC criteria must be reanalyzed. Benzyl benzoate has been tested and found appropriate as an internal standard for this method.

Any compounds confirmed by two columns may also be confirmed by GC/MS. The sample extract and associated blank should be analyzed by GC/MS. A reference standard of the compound must also be analyzed by GC/MS. Include a mid-concentration calibration standard after each group of 20 samples. The response factors for the mid-concentration calibration must be within ±15% of the average values for the multiconcentration calibration. Demonstrate through the analyses of standards that the Florisil fractionation scheme is reproducible.

REFERENCE Test Methods for Evaluating Solid Waste, Physical/Chemical Methods, SW-846, 3rd Edition, U.S. EPA, Office of Solid Waste, Washington, DC, EPA Method 8061, Nov. 1990.

Dimethyl phthalate   **EPA Method 8270**
CAS #131-11-3

TITLE Semivolatile Organic Compounds by GC/MS

MATRIX This method is used to determine the concentration of semivolatile organic compounds in extracts prepared from all types of solid waste matrices, soils, and groundwater. Although surface waters are not specifically mentioned, this method should be applicable to water samples from rivers, lakes, etc.

METHOD SUMMARY This method covers 259 semivolatile organic compounds. In very limited applications direct injection of the sample into the GC/MS system may be appropriate, but this results in very high detection limits (approximately 10,000 µg/L). Typically, a 1-L liquid sample, containing surrogate, and matrix spiking standards, is extracted in a continuous extractor first under acid conditions and then under basic conditions. Typically 30 g of a solid sample, containing surrogate, and matrix spiking standards, is extracted ultrasonically. After concentrating the extract to 1 mL it is spiked with 10 µL of an internal standard solution just prior to analysis by GC/MS. The volume injected should contain about 100 ng of base/neutral and 200 ng of acid surrogates (for a 1-µL injection). Analysis is performed by GC/MS using a capillary GC column.

INTERFERENCES Raw GC/MS data from all blanks, samples, and spikes must be evaluated for interferences. Contamination by carryover can occur whenever high-concentration and low-concentration samples are sequentially analyzed. To reduce carryover, the sample syringe must be rinsed out between samples with solvent. Whenever an unusually concentrated sample is encountered, it should be followed by the analysis of blank solvent to check for cross-contamination.

INSTRUMENTATION A GC/MS and a data system are required. The GC column used is a 30 m × 0.25 mm I.D. (or 0.32 mm I.D.) 1um film thickness silicone-coated fused silica capillary column. A continuous liquid-liquid extractor equipped with Teflon® or glass connection joints and stopcocks requiring no lubrication, a K-D concentrating apparatus, water bath, and an ultrasonic disrupter with a minimum power of 300 W and with pulsing capability are also required.

PRECISION & ACCURACY The estimated quantitation limit (EQL) of Method 8270B for determining an individual compound is approximately 1 mg/kg (wet weight) for soil or sediment samples, 1–200 mg/kg for wastes (dependent on matrix and method of preparation), and 10 µg/L for groundwater samples. EQLs will be proportionately higher for sample extracts that require dilution to avoid saturation of the detector.

The EQL(b) for groundwater in µg/L is 10.
The EQL (a, b) for low concentrations in soil and sediment in µg/kg is 660.
Accuracy as µg/L is $0.20C + 1.03$.
Overall precision in µg/L is $1.05X - 0.92$.

(a) *EQLs listed for soil/sediment are based on wet weight. Normally data is reported in a dry-weight basis; therefore, EQLs will be higher based on the % dry weight of each sample. This calculation is based on a 30 g sample and gel permeation chromatography cleanup.*
(b) *Sample EQLs are highly matrix-dependent. The EQLs are provided for guidance and may not always be achievable.*
$C =$ *True value for concentration, in µg/L.*
$X =$ *Average recovery found for measurements of samples containing a concentration of C, in µg/L.*

ESTIMATED QUANTITATION LIMIT

Other Matrices	Factor (a)
High-concentration soil and sludges by sonicator	7.5
Non-water miscible waste	75

(a) EQL for other matrices = [EQL for low soil/sediment] × [Factor]. This estimated EQL is similar to an EPA "Practical Quantitation Limit."

SAMPLING METHOD

Liquid samples — Use a 1 or 2½ gallon amber glass bottle with a screw-top Teflon®-lined cover that has been prewashed with detergent and rinsed with distilled water and methanol (or isopropanol).

Soils, sediments, or sludges — Use an 8-oz. widemouth glass with a screw-top Teflon®-lined cover that has been prewashed with detergent and rinsed with distilled water and methanol (or isopropanol).

SAMPLE PRESERVATION

Liquid samples — If residual chlorine is present, add 3 mL of 10% sodium thiosulfate per gallon, cool to 4°C and store in a solvent-free refrigerator until analysis; if chlorine is not present, then eliminate the sodium thiosulfate addition.

Soils, sediments, or sludges — Cool samples to 4°C and store in a solvent-free refrigerator.

MHT Liquid samples must be extracted within 7 days and the extracts analyzed within 40 days. Soils, sediments, or sludges may be stored for a maximum of 14 days and the extracts analyzed within 40 days.

SAMPLE PREPARATION

Liquid samples — Transfer 1 L quantitatively to a continuous extractor. If high concentrations are anticipated, a smaller volume may be used and then diluted with organic-free reagent water to 1 L. Adjust pH, if necessary, to pH <2 using 1:1 (V/V) sulfuric acid. Pipette 1.0 mL of a surrogate standard spiking solution into each sample. For the sample in each analytical batch selected for spiking, add 1.0 mL of a matrix spiking standard. For base/neutral acid analysis, the amount of the surrogates and matrix spiking compounds added to the sample should result in a final concentration of 100 ng/µL of each analyte in the extract to be analyzed (assuming a 1-µL injection). Extract with methylene chloride for 18–24 h. Next, adjust the pH of the aqueous phase to pH >11 using 10 N sodium hydroxide and extract it with methylene chloride again for 18–24 h. Dry the extract through a column containing anhydrous sodium sulfate and concentrate it to 1 mL using a K-D concentrator.

Soils, sediments, or sludges — Use 30 g of sample. Nonporous or wet samples (gummy or clay type) that do not have a free-flowing sandy texture must be mixed with anhydrous sodium sulfate until the sample is free flowing. Add 1 mL of surrogate standards to all samples, spikes, standards, and blanks. For the sample in each analytical batch selected for spiking, add 1.0 mL of a matrix spiking standard. For base/neutral acid analysis, the amount added of the surrogates and matrix spiking compounds should result in a final concentration of 100 ng/µL of each base/neutral analyte and 200 ng/µL of each acid analyte in the extract to be analyzed (assuming a 1-µL injection). Immediately add a 100-mL mixture of 1:1 methylene chloride:acetone and extract the sample ultrasonically for 3 min and then decant or filter the extracts. Repeat the extraction two or more times. Dry the extract using a column with anhydrous sodium sulfate and concentrate it to 1 mL in a K-D concentrator.

QUALITY CONTROL A methylene chloride solution containing 50 ng/µL of decafluorotriphenylphosphine (DFTPP) is used for tuning the GC/MS system each 12-h shift. A system performance check also must be made during every 12-h shift. A standard containing 50 ng/µL each of 4,4'-DDT, pentachlorophenol, and benzidine is required to verify injection port inertness and GC column performance. A calibration standard at mid-concentration, containing each compound of interest, including all required surrogates, must be performed every 12 h during analysis. After the system performance check is met, calibration check compounds (CCCs) are used to check the validity of the initial calibration.

The internal standard responses and retention times in the calibration check standard must be evaluated immediately after or during data acquisition. If the retention time for any internal standard changes by more than 30 seconds from the last check calibration (12 h), the chromatographic system must be inspected for malfunctions and corrections must be made, as required. If the electron ionization current plot (EICP) area for any of the internal standards changes by a factor of two from the last daily calibration standard check, the mass spectrometer must be inspected for malfunctions and corrections must be made, as appropriate.

Demonstrate, through the analysis of a reagent water blank, that interferences from the analytical system, glassware, and reagents are under control. The blank samples should be carried through all stages of the sample preparation and measurement steps. For each analytical batch (up to 20 samples), a reagent blank, matrix spike, and matrix spike duplicate/duplicate must be analyzed (the frequency of the spikes may be different for different monitoring programs). The blank and spiked samples must be carried through all stages of the sample preparation and measurement steps. A QC reference sample concentrate containing each analyte at a concentration of 100 mg/L in methanol is required.

REFERENCE Test Methods for Evaluating Solid Waste (SW-846). U.S. EPA 1983, Method 8270B, Rev. 2, Nov. 1990. Office of Solid Waste, Washington, DC.

Dimethyl phthalate   **EPA Method 8060**
CAS #131-11-13

TITLE Phthalate Esters

MATRIX Groundwater, soils, sludges, water miscible liquid wastes, and non-water miscible wastes.

APPLICATION This method is used for the analysis of 6 phthalate esters. Samples are extracted, concentrated, and analyzed using direct injection of both neat and diluted organic liquids into a gas chromatograph. Analytes are detected by a flame ionization detector (FID) or an electron capture detector (ECD). Groundwater samples should be determined by ECD. The method provides an optional GC column which is used for analyte confirmation and that may help resolve analytes from interferences.

INTERFERENCES Solvents, reagents, and glassware may introduce artifacts. Plastics, in particular, must be avoided. Other interferences may come from coextracted compounds from samples. There can be carryover contamination with high- and low-level samples.

INSTRUMENTATION GC capable of on-column injections and a flame with detector (FID) or electron capture detector (ECD). Column 1: 1.8 m by 4 mm with 1.5% SP-2250/1.95% SP-2401 on Supelcoport. Column 2: 1.8 m by 4 mm with 3% OV-1 on supelcoport.

RANGE 0.7 to 106 µg/L.

MDL 19 µg/L (FID) and 0.29 µg/L (ECD).

PQL FACTORS FOR MULTIPLYING × FID MDL VALUE

Matrix	Multiplication Factor
Groundwater	10
Low-level soil by sonication with GPC cleanup	670
High-level soil and sludge by sonication	10,000
Non-water miscible waste	100,000

PRECISION 0.44X + 0.31 µg/L (overall precision using FID).

ACCURACY 0.73C + 0.17 µg/L (as recovery using FID).

SAMPLING METHOD Use 8-oz. widemouth glass bottles with Teflon®-lined caps for concentrated waste samples, soils, sediments, and sludges. Use 1 or 2½ gallon amber glass bottles with Teflon®-lined caps for liquid (water) samples.

STABILITY Cool soil, sediment, sludge, and liquid samples to 4°C. If residual chlorine is present in liquid samples add 3 mL of 10% sodium thiosulfate per gallon of sample and cool to 4°C.

MHT 14 days for concentrated waste, soil, sediment, or sludge; 7 days for liquid samples; all extracts must be analyzed within 40 days.

QUALITY CONTROL A quality control check sample concentrate containing each analyte of interest is required. The QC check sample concentrate may be prepared from pure standard materials or purchased as certified solutions Use appropriate trip, matrix, control site, method, reagent, and solvent blanks. Internal, surrogate, and five concentration level calibration standards are used. The QC check sample concentrate should contain this compound at 25 µg/mL in acetone.

REFERENCE Method 8060, SW-846, 3rd ed., Nov.1986.

Dimethyl sulfone   EPA Method 1625
CAS #67-71-0

TITLE Semivolatile Organic Compounds by Isotope Dilution GC/MS

MATRIX The compounds may be determined in waters, soils, and municipal sludges by this method.

METHOD SUMMARY This method is used to determine 176 semivolatile toxic organic pollutants associated with the CWA (as amended 1987); the RCRA (as amended 1986); the CERCLA (as amended 1986); and other compounds amenable to extraction and analysis by capillary column gas chromatography-mass spectrometry (GC/MS).

Stable isotopically-labeled analogs of the compounds of interest are added to the sample. If the solids content is less than 1%, a 1-L sample is extracted at pH 12–13, then at pH <2 with methylene chloride using continuous extraction techniques.

If the solids content is 30% or less, the sample is diluted to 1% solids with reagent water, homogenized ultrasonically, and extracted at pH 12–13, then at pH <2 with methylene chloride using continuous extraction techniques. If the solids content is greater than 30%, the sample is extracted using ultrasonic techniques.

Each extract is dried over sodium sulfate, concentrated to a volume of 5 mL, cleaned up using GPC, if necessary, and concentrated. Extracts are concentrated to 1 mL if GPC is not performed, and to 0.5 mL if GPC is performed.

An internal standard is added to the extract, and a 1-mL aliquot of the extract is injected into the GC. The compounds are separated by GC and detected by a MS. The labeled compounds serve to correct the variability of the analytical technique.

INTERFERENCES Solvents, reagents, glassware, and other sample processing hardware may yield artifacts and/or elevated baselines causing misinterpretation of chromatograms and spectra. Materials used in the analysis must be demonstrated to be free from interferences under the conditions of analysis by running method blanks initially and with each sample lot (sample started through the extraction process on a given 8-h shift, to a maximum of 20). Specific selection of reagents and purification of solvents by distillation in all glass systems may be required. Glassware and, where possible, reagents are cleaned by solvent rinse and baking at 450°C for 1-h minimum. Interferences coextracted from samples will vary considerably from source to source, depending on the diversity of the site being sampled.

INSTRUMENTATION Major instrumentation includes a GC with a splitless or on-column injection port for capillary column, a MS with 70 eV electron impact ionization, and a data system to collect and record MS data, and process it. A K-D apparatus is used to concentrate extracts.

GC Column: 30 m × 0.25 mm I.D. 5% phenyl, 94% methyl, 1% vinyl silicone bonded phased fused silica capillary column.

PRECISION & ACCURACY The detection limits of the method are usually dependent on the level of interferences rather than instrumental limitations. The limits typify the minimum quantities that can be detected with no interferences present.

The minimum level (in µg/mL) was not listed. This is defined as a minimum level at which the analytical system shall give

recognizable mass spectra (background corrected) and acceptable calibration points.

The MDL (in µg/kg) in low solids was not listed and in high solids was not listed; these were determined in digested sludge (low solids) and in filter cake or compost (high solids).

The labeled and native compound initial precision as standard deviation (in µg/L) was not listed.
The labeled and native compound initial accuracy as average recovery (in µg/L) was not listed.

SAMPLE COLLECTION, PRESERVATION & HANDLING
Collect samples in glass containers. Aqueous samples which flow freely are collected in refrigerated bottles using automatic sampling equipment. Solid samples are collected as grab samples using widemouth jars. Maintain samples at 0 to 4°C from the time of collection until extraction. If residual chlorine is present in aqueous samples, add 80 mg sodium thiosulfate/L of water. Begin sample extraction within 7 days of collection, and analyze all extracts within 40 days of extraction.

SAMPLE PREPARATION Samples containing 1% solids or less are extracted directly using continuous liquid-liquid extraction techniques. Samples containing 1 to 30% solids are diluted to the 1% level with reagent water and extracted using continuous liquid-liquid extraction techniques. Samples containing greater than 30% solids are extracted using ultrasonic techniques.

Base/neutral extraction — Adjust the pH of the waters in the extractors to 12–13 with 6 N NaOH. Extract with methylene chloride for 24–48 h.
Acid extraction — Adjust the pH of the waters in the extractors to 2 or less using 6 N sulfuric acid. Extract with methylene chloride for 24–48 h.
Ultrasonic extraction of high solids samples — Add anhydrous sodium sulfate to the sample and QC aliquot(s). Add acetone:methylene chloride (1:1) to the sample and mix thoroughly

Concentrate extracts using a K-D apparatus.

QUALITY CONTROL The analyst is permitted to modify this method to improve separations or lower the costs of measurements, provided all performance specifications are met. Analyses of blanks are required to demonstrate freedom from contamination. When results of spikes indicate atypical method performance for samples, the samples are diluted to bring method performance within acceptable limits.

For low solids (aqueous samples), extract, concentrate, and analyze two sets of four 1-L aliquots (8 aliquots total) of the precision and recovery standard. For high solids samples, two sets of four 30-g aliquots of the high solids reference matrix are used.

Spike all samples with labeled compounds to assess method performance. Compute percent recovery of the labeled compounds using the internal standard method. Compare the labeled compound recovery for each compound with the corresponding labeled compound recovery.

Reagent water and high solids reference matrix blanks are analyzed to demonstrate freedom from contamination. Extract and concentrate a 1-L reagent water blank or a high solids reference matrix blank with each sample's lot (samples started through the extraction process on the same 8-h shift, to a maximum of 20 samples).

Field replicates may be collected to determine the precision of the sampling technique, and spiked samples may be required to determine the accuracy of the analysis when the internal standard method is used.

REFERENCE Semivolatile Organic Compounds by Isotope Dilution GC/MS. Office of Water Regulation and Standards, U.S. EPA Industrial Technology Division, Washington, DC, EPA Method 1625, Rev. C, June 1989 (contact W.A. Telliard, U.S. EPA, Office of Water Regulations and Standards, 401 M St., SW, Washington, DC, 20460. Phone: 202-382-7131).

p-Dimethylaminoazobenzene   EPA Method 1625
CAS #60-11-7

TITLE Semivolatile Organic Compounds by Isotope Dilution GC/MS

MATRIX The compounds may be determined in waters, soils, and municipal sludges by this method.

METHOD SUMMARY This method is used to determine 176 semivolatile toxic organic pollutants associated with the CWA (as amended 1987); the RCRA (as amended 1986); the CERCLA (as amended 1986); and other compounds amenable to extraction and analysis by capillary column gas chromatography-mass spectrometry (GC/MS).

Stable isotopically-labeled analogs of the compounds of interest are added to the sample. If the solids content is less than 1%, a 1-L sample is extracted at pH 12–13, then at pH <2 with methylene chloride using continuous extraction techniques.

If the solids content is 30% or less, the sample is diluted to 1% solids with reagent water, homogenized ultrasonically, and extracted at pH 12–13, then at pH <2 with methylene chloride using continuous extraction techniques. If the solids content is greater than 30%, the sample is extracted using ultrasonic techniques.

Each extract is dried over sodium sulfate, concentrated to a volume of 5 mL, cleaned up using GPC, if necessary, and concentrated. Extracts are concentrated to 1 mL if GPC is not performed, and to 0.5 mL if GPC is performed.

An internal standard is added to the extract, and a 1-mL aliquot of the extract is injected into the GC. The compounds are separated by GC and detected by a MS. The labeled compounds serve to correct the variability of the analytical technique.

INTERFERENCES Solvents, reagents, glassware, and other sample processing hardware may yield artifacts and/or elevated baselines causing misinterpretation of chromatograms and spectra. Materials used in the analysis must be demonstrated to be free from interferences under the conditions of analysis

by running method blanks initially and with each sample lot (sample started through the extraction process on a given 8-h shift, to a maximum of 20). Specific selection of reagents and purification of solvents by distillation in all glass systems may be required. Glassware and, where possible, reagents are cleaned by solvent rinse and baking at 450°C for 1-h minimum. Interferences coextracted from samples will vary considerably from source to source, depending on the diversity of the site being sampled.

INSTRUMENTATION Major instrumentation includes a GC with a splitless or on-column injection port for capillary column, a MS with 70 eV electron impact ionization, and a data system to collect and record MS data, and process it. A K-D apparatus is used to concentrate extracts.

GC Column: 30 m × 0.25 mm I.D. 5% phenyl, 94% methyl, 1% vinyl silicone bonded phased fused silica capillary column.

PRECISION & ACCURACY The detection limits of the method are usually dependent on the level of interferences rather than instrumental limitations. The limits typify the minimum quantities that can be detected with no interferences present.

The minimum level (in µg/mL) was not listed. This is defined as a minimum level at which the analytical system shall give recognizable mass spectra (background corrected) and acceptable calibration points.

The MDL (in µg/kg) in low solids was not listed and in high solids was not listed; these were determined in digested sludge (low solids) and in filter cake or compost (high solids).

The labeled and native compound initial precision as standard deviation (in µg/L) was not listed.
The labeled and native compound initial accuracy as average recovery (in µg/L) was not listed.

SAMPLE COLLECTION, PRESERVATION & HANDLING Collect samples in glass containers. Aqueous samples which flow freely are collected in refrigerated bottles using automatic sampling equipment. Solid samples are collected as grab samples using widemouth jars. Maintain samples at 0 to 4°C from the time of collection until extraction. If residual chlorine is present in aqueous samples, add 80 mg sodium thiosulfate/L of water. Begin sample extraction within 7 days of collection, and analyze all extracts within 40 days of extraction.

SAMPLE PREPARATION Samples containing 1% solids or less are extracted directly using continuous liquid-liquid extraction techniques. Samples containing 1 to 30% solids are diluted to the 1% level with reagent water and extracted using continuous liquid-liquid extraction techniques. Samples containing greater than 30% solids are extracted using ultrasonic techniques.

Base/neutral extraction — Adjust the pH of the waters in the extractors to 12–13 with 6 N NaOH. Extract with methylene chloride for 24–48 h.
Acid extraction — Adjust the pH of the waters in the extractors to 2 or less using 6 N sulfuric acid. Extract with methylene chloride for 24–48 h.
Ultrasonic extraction of high solids samples — Add anhydrous sodium sulfate to the sample and QC aliquot(s). Add acetone:methylene chloride (1:1) to the sample and mix thoroughly

Concentrate extracts using a K-D apparatus.

QUALITY CONTROL The analyst is permitted to modify this method to improve separations or lower the costs of measurements, provided all performance specifications are met. Analyses of blanks are required to demonstrate freedom from contamination. When results of spikes indicate atypical method performance for samples, the samples are diluted to bring method performance within acceptable limits.

For low solids (aqueous samples), extract, concentrate, and analyze two sets of four 1-L aliquots (8 aliquots total) of the precision and recovery standard. For high solids samples, two sets of four 30-g aliquots of the high solids reference matrix are used.

Spike all samples with labeled compounds to assess method performance. Compute percent recovery of the labeled compounds using the internal standard method. Compare the labeled compound recovery for each compound with the corresponding labeled compound recovery.

Reagent water and high solids reference matrix blanks are analyzed to demonstrate freedom from contamination. Extract and concentrate a 1-L reagent water blank or a high solids reference matrix blank with each sample's lot (samples started through the extraction process on the same 8-h shift, to a maximum of 20 samples).

Field replicates may be collected to determine the precision of the sampling technique, and spiked samples may be required to determine the accuracy of the analysis when the internal standard method is used.

REFERENCE Semivolatile Organic Compounds by Isotope Dilution GC/MS. Office of Water Regulation and Standards, U.S. EPA Industrial Technology Division, Washington, DC, EPA Method 1625, Rev. C, June 1989 (contact W.A. Telliard, U.S. EPA, Office of Water Regulations and Standards, 401 M St., SW, Washington, DC, 20460. Phone: 202-382-7131).

Dimethylaminoazobenzene   **EPA Method 8270**
CAS #60-11-7

TITLE Semivolatile Organic Compounds by GC/MS

MATRIX This method is used to determine the concentration of semivolatile organic compounds in extracts prepared from all types of solid waste matrices, soils, and groundwater. Although surface waters are not specifically mentioned, this method should be applicable to water samples from rivers, lakes, etc.

METHOD SUMMARY This method covers 259 semivolatile organic compounds. In very limited applications direct injection of the sample into the GC/MS system may be appropriate, but this results in very high detection limits (approximately

10,000 μg/L). Typically, a 1-L liquid sample, containing surrogate, and matrix spiking standards, is extracted in a continuous extractor first under acid conditions and then under basic conditions. Typically 30 g of a solid sample, containing surrogate, and matrix spiking standards, is extracted ultrasonically. After concentrating the extract to 1 mL it is spiked with 10 μL of an internal standard solution just prior to analysis by GC/MS. The volume injected should contain about 100 ng of base/neutral and 200 ng of acid surrogates (for a 1-μL injection). Analysis is performed by GC/MS using a capillary GC column.

INTERFERENCES Raw GC/MS data from all blanks, samples, and spikes must be evaluated for interferences. Contamination by carryover can occur whenever high-concentration and low-concentration samples are sequentially analyzed. To reduce carryover, the sample syringe must be rinsed out between samples with solvent. Whenever an unusually concentrated sample is encountered, it should be followed by the analysis of blank solvent to check for cross-contamination.

INSTRUMENTATION A GC/MS and a data system are required. The GC column used is a 30 m × 0.25 mm I.D. (or 0.32 mm I.D.) 1um film thickness silicone-coated fused silica capillary column. A continuous liquid-liquid extractor equipped with Teflon® or glass connection joints and stopcocks requiring no lubrication, a K-D concentrating apparatus, water bath, and an ultrasonic disrupter with a minimum power of 300 W and with pulsing capability are also required.

PRECISION & ACCURACY The estimated quantitation limit (EQL) of Method 8270B for determining an individual compound is approximately 1 mg/kg (wet weight) for soil or sediment samples, 1–200 mg/kg for wastes (dependent on matrix and method of preparation), and 10 μg/L for groundwater samples. EQLs will be proportionately higher for sample extracts that require dilution to avoid saturation of the detector.

The EQL(b) for groundwater in μg/L is 10.
The EQL (a, b) for low concentrations in soil and sediment in μg/kg is not determined.
Accuracy as μg/L is not listed.
Overall precision in μg/L is not listed.

(a) EQLs listed for soil/sediment are based on wet weight. Normally data is reported in a dry-weight basis; therefore, EQLs will be higher based on the % dry weight of each sample. This calculation is based on a 30 g sample and gel permeation chromatography cleanup.
(b) Sample EQLs are highly matrix-dependent. The EQLs are provided for guidance and may not always be achievable.
$C =$ *True value for concentration, in μg/L.*
$X =$ *Average recovery found for measurements of samples containing a concentration of C, in μg/L.*

ESTIMATED QUANTITATION LIMIT

Other Matrices	Factor (a)
High-concentration soil and sludges by sonicator	7.5
Non-water miscible waste	75

(a) EQL for other matrices = [EQL for low soil/sediment] × [Factor]. This estimated EQL is similar to an EPA "Practical Quantitation Limit."

SAMPLING METHOD
Liquid samples — Use a 1 or 2½ gallon amber glass bottle with a screw-top Teflon®-lined cover that has been prewashed with detergent and rinsed with distilled water and methanol (or isopropanol).

Soils, sediments, or sludges — Use an 8-oz. widemouth glass with a screw-top Teflon®-lined cover that has been prewashed with detergent and rinsed with distilled water and methanol (or isopropanol).

SAMPLE PRESERVATION
Liquid samples — If residual chlorine is present, add 3 mL of 10% sodium thiosulfate per gallon, cool to 4°C and store in a solvent-free refrigerator until analysis; if chlorine is not present, then eliminate the sodium thiosulfate addition.

Soils, sediments, or sludges — Cool samples to 4°C and store in a solvent-free refrigerator.

MHT Liquid samples must be extracted within 7 days and the extracts analyzed within 40 days. Soils, sediments, or sludges may be stored for a maximum of 14 days and the extracts analyzed within 40 days.

SAMPLE PREPARATION
Liquid samples — Transfer 1 L quantitatively to a continuous extractor. If high concentrations are anticipated, a smaller volume may be used and then diluted with organic-free reagent water to 1 L. Adjust pH, if necessary, to pH <2 using 1:1 (V/V) sulfuric acid. Pipette 1.0 mL of a surrogate standard spiking solution into each sample. For the sample in each analytical batch selected for spiking, add 1.0 mL of a matrix spiking standard. For base/neutral acid analysis, the amount of the surrogates and matrix spiking compounds added to the sample should result in a final concentration of 100 ng/μL of each analyte in the extract to be analyzed (assuming a 1-μL injection). Extract with methylene chloride for 18–24 h. Next, adjust the pH of the aqueous phase to pH >11 using 10 N sodium hydroxide and extract it with methylene chloride again for 18–24 h. Dry the extract through a column containing anhydrous sodium sulfate and concentrate it to 1 mL using a K-D concentrator.

Soils, sediments, or sludges — Use 30 g of sample. Nonporous or wet samples (gummy or clay type) that do not have a free-flowing sandy texture must be mixed with anhydrous sodium sulfate until the sample is free flowing. Add 1 mL of surrogate standards to all samples, spikes, standards, and blanks. For the sample in each analytical batch selected for spiking, add 1.0 mL of a matrix spiking standard. For base/neutral acid analysis, the amount added of the surrogates and matrix spiking compounds should result in a final concentration of 100 ng/μL of each base/neutral analyte and 200 ng/μL of each acid analyte in the extract to be analyzed (assuming a 1-μL injection). Immediately add a 100-mL mixture of 1:1 methylene chloride:acetone and extract the sample ultrasonically for 3 min and then decant or filter the extracts. Repeat the extraction two or more times. Dry the extract using a column with anhydrous sodium sulfate and concentrate it to 1 mL in a K-D concentrator.

QUALITY CONTROL A methylene chloride solution containing 50 ng/μL of decafluorotriphenylphosphine (DFTPP) is

used for tuning the GC/MS system each 12-h shift. A system performance check also must be made during every 12-h shift. A standard containing 50 ng/μL each of 4,4'-DDT, pentachlorophenol, and benzidine is required to verify injection port inertness and GC column performance. A calibration standard at mid-concentration, containing each compound of interest, including all required surrogates, must be performed every 12 h during analysis. After the system performance check is met, calibration check compounds (CCCs) are used to check the validity of the initial calibration.

The internal standard responses and retention times in the calibration check standard must be evaluated immediately after or during data acquisition. If the retention time for any internal standard changes by more than 30 seconds from the last check calibration (12 h), the chromatographic system must be inspected for malfunctions and corrections must be made, as required. If the electron ionization current plot (EICP) area for any of the internal standards changes by a factor of two from the last daily calibration standard check, the mass spectrometer must be inspected for malfunctions and corrections must be made, as appropriate.

Demonstrate, through the analysis of a reagent water blank, that interferences from the analytical system, glassware, and reagents are under control. The blank samples should be carried through all stages of the sample preparation and measurement steps. For each analytical batch (up to 20 samples), a reagent blank, matrix spike, and matrix spike duplicate/duplicate must be analyzed (the frequency of the spikes may be different for different monitoring programs). The blank and spiked samples must be carried through all stages of the sample preparation and measurement steps. A QC reference sample concentrate containing each analyte at a concentration of 100 mg/L in methanol is required.

REFERENCE Test Methods for Evaluating Solid Waste (SW-846). U.S. EPA 1983, Method 8270B, Rev. 2, Nov. 1990. Office of Solid Waste, Washington, DC.

7,12-Dimethylbenz(a)anthracene   EPA Method 1625
CAS #57-97-6

TITLE Semivolatile Organic Compounds by Isotope Dilution GC/MS

MATRIX The compounds may be determined in waters, soils, and municipal sludges by this method.

METHOD SUMMARY This method is used to determine 176 semivolatile toxic organic pollutants associated with the CWA (as amended 1987); the RCRA (as amended 1986); the CERCLA (as amended 1986); and other compounds amenable to extraction and analysis by capillary column gas chromatography-mass spectrometry (GC/MS).

Stable isotopically-labeled analogs of the compounds of interest are added to the sample. If the solids content is less than 1%, a 1-L sample is extracted at pH 12–13, then at pH <2 with methylene chloride using continuous extraction techniques.

If the solids content is 30% or less, the sample is diluted to 1% solids with reagent water, homogenized ultrasonically, and extracted at pH 12–13, then at pH <2 with methylene chloride using continuous extraction techniques. If the solids content is greater than 30%, the sample is extracted using ultrasonic techniques.

Each extract is dried over sodium sulfate, concentrated to a volume of 5 mL, cleaned up using GPC, if necessary, and concentrated. Extracts are concentrated to 1 mL if GPC is not performed, and to 0.5 mL if GPC is performed.

An internal standard is added to the extract, and a 1-mL aliquot of the extract is injected into the GC. The compounds are separated by GC and detected by a MS. The labeled compounds serve to correct the variability of the analytical technique.

INTERFERENCES Solvents, reagents, glassware, and other sample processing hardware may yield artifacts and/or elevated baselines causing misinterpretation of chromatograms and spectra. Materials used in the analysis must be demonstrated to be free from interferences under the conditions of analysis by running method blanks initially and with each sample lot (sample started through the extraction process on a given 8-h shift, to a maximum of 20). Specific selection of reagents and purification of solvents by distillation in all glass systems may be required. Glassware and, where possible, reagents are cleaned by solvent rinse and baking at 450°C for 1-h minimum. Interferences coextracted from samples will vary considerably from source to source, depending on the diversity of the site being sampled.

INSTRUMENTATION Major instrumentation includes a GC with a splitless or on-column injection port for capillary column, a MS with 70 eV electron impact ionization, and a data system to collect and record MS data, and process it. A K-D apparatus is used to concentrate extracts.

GC Column: 30 m × 0.25 mm I.D. 5% phenyl, 94% methyl, 1% vinyl silicone bonded phased fused silica capillary column.

PRECISION & ACCURACY The detection limits of the method are usually dependent on the level of interferences rather than instrumental limitations. The limits typify the minimum quantities that can be detected with no interferences present.

The minimum level (in μg/mL) was not listed. This is defined as a minimum level at which the analytical system shall give recognizable mass spectra (background corrected) and acceptable calibration points.

The MDL (in μg/kg) in low solids was not listed and in high solids was not listed; these were determined in digested sludge (low solids) and in filter cake or compost (high solids).

The labeled and native compound initial precision as standard deviation (in μg/L) was not listed.
The labeled and native compound initial accuracy as average recovery (in μg/L) was not listed.

SAMPLE COLLECTION, PRESERVATION & HANDLING
Collect samples in glass containers. Aqueous samples which flow freely are collected in refrigerated bottles using automatic

sampling equipment. Solid samples are collected as grab samples using widemouth jars. Maintain samples at 0 to 4°C from the time of collection until extraction. If residual chlorine is present in aqueous samples, add 80 mg sodium thiosulfate/L of water. Begin sample extraction within 7 days of collection, and analyze all extracts within 40 days of extraction.

SAMPLE PREPARATION Samples containing 1% solids or less are extracted directly using continuous liquid-liquid extraction techniques. Samples containing 1 to 30% solids are diluted to the 1% level with reagent water and extracted using continuous liquid-liquid extraction techniques. Samples containing greater than 30% solids are extracted using ultrasonic techniques.

Base/neutral extraction — Adjust the pH of the waters in the extractors to 12–13 with 6 N NaOH. Extract with methylene chloride for 24–48 h.

Acid extraction — Adjust the pH of the waters in the extractors to 2 or less using 6 N sulfuric acid. Extract with methylene chloride for 24–48 h.

Ultrasonic extraction of high solids samples — Add anhydrous sodium sulfate to the sample and QC aliquot(s). Add acetone:methylene chloride (1:1) to the sample and mix thoroughly

Concentrate extracts using a K-D apparatus.

QUALITY CONTROL The analyst is permitted to modify this method to improve separations or lower the costs of measurements, provided all performance specifications are met. Analyses of blanks are required to demonstrate freedom from contamination. When results of spikes indicate atypical method performance for samples, the samples are diluted to bring method performance within acceptable limits.

For low solids (aqueous samples), extract, concentrate, and analyze two sets of four 1-L aliquots (8 aliquots total) of the precision and recovery standard. For high solids samples, two sets of four 30-g aliquots of the high solids reference matrix are used.

Spike all samples with labeled compounds to assess method performance. Compute percent recovery of the labeled compounds using the internal standard method. Compare the labeled compound recovery for each compound with the corresponding labeled compound recovery.

Reagent water and high solids reference matrix blanks are analyzed to demonstrate freedom from contamination. Extract and concentrate a 1-L reagent water blank or a high solids reference matrix blank with each sample's lot (samples started through the extraction process on the same 8-h shift, to a maximum of 20 samples).

Field replicates may be collected to determine the precision of the sampling technique, and spiked samples may be required to determine the accuracy of the analysis when the internal standard method is used.

REFERENCE Semivolatile Organic Compounds by Isotope Dilution GC/MS. Office of Water Regulation and Standards, U.S. EPA Industrial Technology Division, Washington, DC, EPA Method 1625, Rev. C, June 1989 (contact W.A. Telliard, U.S. EPA, Office of Water Regulations and Standards, 401 M St., SW, Washington, DC, 20460. Phone: 202-382-7131).

7,12-Dimethylbenz(a)anthracene   EPA Method 8270
CAS #57-97-6

TITLE Semivolatile Organic Compounds by GC/MS

MATRIX This method is used to determine the concentration of semivolatile organic compounds in extracts prepared from all types of solid waste matrices, soils, and groundwater. Although surface waters are not specifically mentioned, this method should be applicable to water samples from rivers, lakes, etc.

METHOD SUMMARY This method covers 259 semivolatile organic compounds. In very limited applications direct injection of the sample into the GC/MS system may be appropriate, but this results in very high detection limits (approximately 10,000 µg/L). Typically, a 1-L liquid sample, containing surrogate, and matrix spiking standards, is extracted in a continuous extractor first under acid conditions and then under basic conditions. Typically 30 g of a solid sample, containing surrogate, and matrix spiking standards, is extracted ultrasonically. After concentrating the extract to 1 mL it is spiked with 10 µL of an internal standard solution just prior to analysis by GC/MS. The volume injected should contain about 100 ng of base/neutral and 200 ng of acid surrogates (for a 1-µL injection). Analysis is performed by GC/MS using a capillary GC column.

INTERFERENCES Raw GC/MS data from all blanks, samples, and spikes must be evaluated for interferences. Contamination by carryover can occur whenever high-concentration and low-concentration samples are sequentially analyzed. To reduce carryover, the sample syringe must be rinsed out between samples with solvent. Whenever an unusually concentrated sample is encountered, it should be followed by the analysis of blank solvent to check for cross-contamination.

INSTRUMENTATION A GC/MS and a data system are required. The GC column used is a 30 m × 0.25 mm I.D. (or 0.32 mm I.D.) 1um film thickness silicone-coated fused silica capillary column. A continuous liquid-liquid extractor equipped with Teflon® or glass connection joints and stopcocks requiring no lubrication, a K-D concentrating apparatus, water bath, and an ultrasonic disrupter with a minimum power of 300 W and with pulsing capability are also required.

PRECISION & ACCURACY The estimated quantitation limit (EQL) of Method 8270B for determining an individual compound is approximately 1 mg/kg (wet weight) for soil or sediment samples, 1–200 mg/kg for wastes (dependent on matrix and method of preparation), and 10 µg/L for groundwater samples. EQLs will be proportionately higher for sample extracts that require dilution to avoid saturation of the detector.

The EQL(b) for groundwater in µg/L is 10.
The EQL (a, b) for low concentrations in soil and sediment in µg/kg is not determined.
Accuracy as µg/L is not listed.
Overall precision in µg/L is not listed.

(a) *EQLs listed for soil/sediment are based on wet weight. Normally data is reported in a dry-weight basis; therefore, EQLs will be higher based on the % dry weight of each sample. This calculation is based on a 30 g sample and gel permeation chromatography cleanup.*
(b) *Sample EQLs are highly matrix-dependent. The EQLs are provided for guidance and may not always be achievable.*
C = *True value for concentration, in μg/L.*
X = *Average recovery found for measurements of samples containing a concentration of C, in μg/L.*

ESTIMATED QUANTITATION LIMIT

Other Matrices	Factor (a)
High-concentration soil and sludges by sonicator	7.5
Non-water miscible waste	75

(a) EQL for other matrices = [EQL for low soil/sediment] × [Factor]. This estimated EQL is similar to an EPA "Practical Quantitation Limit."

SAMPLING METHOD

Liquid samples — Use a 1 or 2½ gallon amber glass bottle with a screw-top Teflon®-lined cover that has been prewashed with detergent and rinsed with distilled water and methanol (or isopropanol).

Soils, sediments, or sludges — Use an 8-oz. widemouth glass with a screw-top Teflon®-lined cover that has been prewashed with detergent and rinsed with distilled water and methanol (or isopropanol).

SAMPLE PRESERVATION

Liquid samples — If residual chlorine is present, add 3 mL of 10% sodium thiosulfate per gallon, cool to 4°C and store in a solvent-free refrigerator until analysis; if chlorine is not present, then eliminate the sodium thiosulfate addition.

Soils, sediments, or sludges — Cool samples to 4°C and store in a solvent-free refrigerator.

MHT Liquid samples must be extracted within 7 days and the extracts analyzed within 40 days. Soils, sediments, or sludges may be stored for a maximum of 14 days and the extracts analyzed within 40 days.

SAMPLE PREPARATION

Liquid samples — Transfer 1 L quantitatively to a continuous extractor. If high concentrations are anticipated, a smaller volume may be used and then diluted with organic-free reagent water to 1 L. Adjust pH, if necessary, to pH <2 using 1:1 (V/V) sulfuric acid. Pipette 1.0 mL of a surrogate standard spiking solution into each sample. For the sample in each analytical batch selected for spiking, add 1.0 mL of a matrix spiking standard. For base/neutral acid analysis, the amount of the surrogates and matrix spiking compounds added to the sample should result in a final concentration of 100 ng/μL of each analyte in the extract to be analyzed (assuming a 1-μL injection). Extract with methylene chloride for 18–24 h. Next, adjust the pH of the aqueous phase to pH >11 using 10 N sodium hydroxide and extract it with methylene chloride again for 18–24 h. Dry the extract through a column containing anhydrous sodium sulfate and concentrate it to 1 mL using a K-D concentrator.

Soils, sediments, or sludges — Use 30 g of sample. Nonporous or wet samples (gummy or clay type) that do not have a free-flowing sandy texture must be mixed with anhydrous sodium sulfate until the sample is free flowing. Add 1 mL of surrogate standards to all samples, spikes, standards, and blanks. For the sample in each analytical batch selected for spiking, add 1.0 mL of a matrix spiking standard. For base/neutral acid analysis, the amount added of the surrogates and matrix spiking compounds should result in a final concentration of 100 ng/μL of each base/neutral analyte and 200 ng/μL of each acid analyte in the extract to be analyzed (assuming a 1-μL injection). Immediately add a 100-mL mixture of 1:1 methylene chloride:acetone and extract the sample ultrasonically for 3 min and then decant or filter the extracts. Repeat the extraction two or more times. Dry the extract using a column with anhydrous sodium sulfate and concentrate it to 1 mL in a K-D concentrator.

QUALITY CONTROL A methylene chloride solution containing 50 ng/μL of decafluorotriphenylphosphine (DFTPP) is used for tuning the GC/MS system each 12-h shift. A system performance check also must be made during every 12-h shift. A standard containing 50 ng/μL each of 4,4'-DDT, pentachlorophenol, and benzidine is required to verify injection port inertness and GC column performance. A calibration standard at mid-concentration, containing each compound of interest, including all required surrogates, must be performed every 12 h during analysis. After the system performance check is met, calibration check compounds (CCCs) are used to check the validity of the initial calibration.

The internal standard responses and retention times in the calibration check standard must be evaluated immediately after or during data acquisition. If the retention time for any internal standard changes by more than 30 seconds from the last check calibration (12 h), the chromatographic system must be inspected for malfunctions and corrections must be made, as required. If the electron ionization current plot (EICP) area for any of the internal standards changes by a factor of two from the last daily calibration standard check, the mass spectrometer must be inspected for malfunctions and corrections must be made, as appropriate.

Demonstrate, through the analysis of a reagent water blank, that interferences from the analytical system, glassware, and reagents are under control. The blank samples should be carried through all stages of the sample preparation and measurement steps. For each analytical batch (up to 20 samples), a reagent blank, matrix spike, and matrix spike duplicate/duplicate must be analyzed (the frequency of the spikes may be different for different monitoring programs). The blank and spiked samples must be carried through all stages of the sample preparation and measurement steps. A QC reference sample concentrate containing each analyte at a concentration of 100 mg/L in methanol is required.

REFERENCE Test Methods for Evaluating Solid Waste (SW-846). U.S. EPA 1983, Method 8270B, Rev. 2, Nov. 1990. Office of Solid Waste, Washington, DC.

3,3'-Dimethylbenzidine
CAS #119-93-7
EPA Method 8270

TITLE Semivolatile Organic Compounds by GC/MS

MATRIX This method is used to determine the concentration of semivolatile organic compounds in extracts prepared from all types of solid waste matrices, soils, and groundwater. Although surface waters are not specifically mentioned, this method should be applicable to water samples from rivers, lakes, etc.

METHOD SUMMARY This method covers 259 semivolatile organic compounds. In very limited applications direct injection of the sample into the GC/MS system may be appropriate, but this results in very high detection limits (approximately 10,000 µg/L). Typically, a 1-L liquid sample, containing surrogate, and matrix spiking standards, is extracted in a continuous extractor first under acid conditions and then under basic conditions. Typically 30 g of a solid sample, containing surrogate, and matrix spiking standards, is extracted ultrasonically. After concentrating the extract to 1 mL it is spiked with 10 µL of an internal standard solution just prior to analysis by GC/MS. The volume injected should contain about 100 ng of base/neutral and 200 ng of acid surrogates (for a 1-µL injection). Analysis is performed by GC/MS using a capillary GC column.

INTERFERENCES Raw GC/MS data from all blanks, samples, and spikes must be evaluated for interferences. Contamination by carryover can occur whenever high-concentration and low-concentration samples are sequentially analyzed. To reduce carryover, the sample syringe must be rinsed out between samples with solvent. Whenever an unusually concentrated sample is encountered, it should be followed by the analysis of blank solvent to check for cross-contamination.

INSTRUMENTATION A GC/MS and a data system are required. The GC column used is a 30 m × 0.25 mm I.D. (or 0.32 mm I.D.) 1um film thickness silicone-coated fused silica capillary column. A continuous liquid-liquid extractor equipped with Teflon® or glass connection joints and stopcocks requiring no lubrication, a K-D concentrating apparatus, water bath, and an ultrasonic disrupter with a minimum power of 300 W and with pulsing capability are also required.

PRECISION & ACCURACY The estimated quantitation limit (EQL) of Method 8270B for determining an individual compound is approximately 1 mg/kg (wet weight) for soil or sediment samples, 1–200 mg/kg for wastes (dependent on matrix and method of preparation), and 10 µg/L for groundwater samples. EQLs will be proportionately higher for sample extracts that require dilution to avoid saturation of the detector.

The EQL(b) for groundwater in µg/L is 10.
The EQL (a, b) for low concentrations in soil and sediment in µg/kg is not determined.
Accuracy as µg/L is not listed.
Overall precision in µg/L is not listed.

(a) *EQLs listed for soil/sediment are based on wet weight. Normally data is reported in a dry-weight basis; therefore, EQLs will be higher based on the % dry weight of each sample.*

This calculation is based on a 30 g sample and gel permeation chromatography cleanup.

(b) *Sample EQLs are highly matrix-dependent. The EQLs are provided for guidance and may not always be achievable.*

$C =$ *True value for concentration, in µg/L.*
$X =$ *Average recovery found for measurements of samples containing a concentration of C, in µg/L.*

ESTIMATED QUANTITATION LIMIT

Other Matrices	Factor (a)
High-concentration soil and sludges by sonicator	7.5
Non-water miscible waste	75

(a) *EQL for other matrices = [EQL for low soil/sediment] × [Factor]. This estimated EQL is similar to an EPA "Practical Quantitation Limit."*

SAMPLING METHOD
Liquid samples — Use a 1 or 2½ gallon amber glass bottle with a screw-top Teflon®-lined cover that has been prewashed with detergent and rinsed with distilled water and methanol (or isopropanol).

Soils, sediments, or sludges — Use an 8-oz. widemouth glass with a screw-top Teflon®-lined cover that has been prewashed with detergent and rinsed with distilled water and methanol (or isopropanol).

SAMPLE PRESERVATION
Liquid samples — If residual chlorine is present, add 3 mL of 10% sodium thiosulfate per gallon, cool to 4°C and store in a solvent-free refrigerator until analysis; if chlorine is not present, then eliminate the sodium thiosulfate addition.

Soils, sediments, or sludges — Cool samples to 4°C and store in a solvent-free refrigerator.

MHT Liquid samples must be extracted within 7 days and the extracts analyzed within 40 days. Soils, sediments, or sludges may be stored for a maximum of 14 days and the extracts analyzed within 40 days.

SAMPLE PREPARATION
Liquid samples — Transfer 1 L quantitatively to a continuous extractor. If high concentrations are anticipated, a smaller volume may be used and then diluted with organic-free reagent water to 1 L. Adjust pH, if necessary, to pH <2 using 1:1 (V/V) sulfuric acid. Pipette 1.0 mL of a surrogate standard spiking solution into each sample. For the sample in each analytical batch selected for spiking, add 1.0 mL of a matrix spiking standard. For base/neutral acid analysis, the amount of the surrogates and matrix spiking compounds added to the sample should result in a final concentration of 100 ng/µL of each analyte in the extract to be analyzed (assuming a 1-µL injection). Extract with methylene chloride for 18–24 h. Next, adjust the pH of the aqueous phase to pH >11 using 10 N sodium hydroxide and extract it with methylene chloride again for 18–24 h. Dry the extract through a column containing anhydrous sodium sulfate and concentrate it to 1 mL using a K-D concentrator.

Soils, sediments, or sludges — Use 30 g of sample. Nonporous or wet samples (gummy or clay type) that do not have a free-flowing

sandy texture must be mixed with anhydrous sodium sulfate until the sample is free flowing. Add 1 mL of surrogate standards to all samples, spikes, standards, and blanks. For the sample in each analytical batch selected for spiking, add 1.0 mL of a matrix spiking standard. For base/neutral acid analysis, the amount added of the surrogates and matrix spiking compounds should result in a final concentration of 100 ng/µL of each base/neutral analyte and 200 ng/µL of each acid analyte in the extract to be analyzed (assuming a 1-µL injection). Immediately add a 100-mL mixture of 1:1 methylene chloride:acetone and extract the sample ultrasonically for 3 min and then decant or filter the extracts. Repeat the extraction two or more times. Dry the extract using a column with anhydrous sodium sulfate and concentrate it to 1 mL in a K-D concentrator.

QUALITY CONTROL A methylene chloride solution containing 50 ng/µL of decafluorotriphenylphosphine (DFTPP) is used for tuning the GC/MS system each 12-h shift. A system performance check also must be made during every 12-h shift. A standard containing 50 ng/µL each of 4,4'-DDT, pentachlorophenol, and benzidine is required to verify injection port inertness and GC column performance. A calibration standard at mid-concentration, containing each compound of interest, including all required surrogates, must be performed every 12 h during analysis. After the system performance check is met, calibration check compounds (CCCs) are used to check the validity of the initial calibration.

The internal standard responses and retention times in the calibration check standard must be evaluated immediately after or during data acquisition. If the retention time for any internal standard changes by more than 30 seconds from the last check calibration (12 h), the chromatographic system must be inspected for malfunctions and corrections must be made, as required. If the electron ionization current plot (EICP) area for any of the internal standards changes by a factor of two from the last daily calibration standard check, the mass spectrometer must be inspected for malfunctions and corrections must be made, as appropriate.

Demonstrate, through the analysis of a reagent water blank, that interferences from the analytical system, glassware, and reagents are under control. The blank samples should be carried through all stages of the sample preparation and measurement steps. For each analytical batch (up to 20 samples), a reagent blank, matrix spike, and matrix spike duplicate/duplicate must be analyzed (the frequency of the spikes may be different for different monitoring programs). The blank and spiked samples must be carried through all stages of the sample preparation and measurement steps. A QC reference sample concentrate containing each analyte at a concentration of 100 mg/L in methanol is required.

REFERENCE Test Methods for Evaluating Solid Waste (SW-846). U.S. EPA 1983, Method 8270B, Rev. 2, Nov. 1990. Office of Solid Waste, Washington, DC.

N,N-Dimethylformamide   **EPA Method 1625**
CAS #68-12-2

TITLE Semivolatile Organic Compounds by Isotope Dilution GC/MS

MATRIX The compounds may be determined in waters, soils, and municipal sludges by this method.

METHOD SUMMARY This method is used to determine 176 semivolatile toxic organic pollutants associated with the CWA (as amended 1987); the RCRA (as amended 1986); the CERCLA (as amended 1986); and other compounds amenable to extraction and analysis by capillary column gas chromatography-mass spectrometry (GC/MS).

Stable isotopically-labeled analogs of the compounds of interest are added to the sample. If the solids content is less than 1%, a 1-L sample is extracted at pH 12–13, then at pH <2 with methylene chloride using continuous extraction techniques.

If the solids content is 30% or less, the sample is diluted to 1% solids with reagent water, homogenized ultrasonically, and extracted at pH 12–13, then at pH <2 with methylene chloride using continuous extraction techniques. If the solids content is greater than 30%, the sample is extracted using ultrasonic techniques.

Each extract is dried over sodium sulfate, concentrated to a volume of 5 mL, cleaned up using GPC, if necessary, and concentrated. Extracts are concentrated to 1 mL if GPC is not performed, and to 0.5 mL if GPC is performed.

An internal standard is added to the extract, and a 1-mL aliquot of the extract is injected into the GC. The compounds are separated by GC and detected by a MS. The labeled compounds serve to correct the variability of the analytical technique.

INTERFERENCES Solvents, reagents, glassware, and other sample processing hardware may yield artifacts and/or elevated baselines causing misinterpretation of chromatograms and spectra. Materials used in the analysis must be demonstrated to be free from interferences under the conditions of analysis by running method blanks initially and with each sample lot (sample started through the extraction process on a given 8-h shift, to a maximum of 20). Specific selection of reagents and purification of solvents by distillation in all glass systems may be required. Glassware and, where possible, reagents are cleaned by solvent rinse and baking at 450°C for 1-h minimum. Interferences coextracted from samples will vary considerably from source to source, depending on the diversity of the site being sampled.

INSTRUMENTATION Major instrumentation includes a GC with a splitless or on-column injection port for capillary column, a MS with 70 eV electron impact ionization, and a data system to collect and record MS data, and process it. A K-D apparatus is used to concentrate extracts.

GC Column: 30 m × 0.25 mm I.D. 5% phenyl, 94% methyl, 1% vinyl silicone bonded phased fused silica capillary column.

PRECISION & ACCURACY The detection limits of the method are usually dependent on the level of interferences rather than instrumental limitations. The limits typify the minimum quantities that can be detected with no interferences present.

The minimum level (in μg/mL) was not listed. This is defined as a minimum level at which the analytical system shall give recognizable mass spectra (background corrected) and acceptable calibration points.

The MDL (in μg/kg) in low solids was not listed and in high solids was not listed; these were determined in digested sludge (low solids) and in filter cake or compost (high solids).

The labeled and native compound initial precision as standard deviation (in μg/L) was not listed.
The labeled and native compound initial accuracy as average recovery (in μg/L) was not listed.

SAMPLE COLLECTION, PRESERVATION & HANDLING Collect samples in glass containers. Aqueous samples which flow freely are collected in refrigerated bottles using automatic sampling equipment. Solid samples are collected as grab samples using widemouth jars. Maintain samples at 0 to 4°C from the time of collection until extraction. If residual chlorine is present in aqueous samples, add 80 mg sodium thiosulfate/L of water. Begin sample extraction within 7 days of collection, and analyze all extracts within 40 days of extraction.

SAMPLE PREPARATION Samples containing 1% solids or less are extracted directly using continuous liquid-liquid extraction techniques. Samples containing 1 to 30% solids are diluted to the 1% level with reagent water and extracted using continuous liquid-liquid extraction techniques. Samples containing greater than 30% solids are extracted using ultrasonic techniques.

Base/neutral extraction — Adjust the pH of the waters in the extractors to 12–13 with 6 N NaOH. Extract with methylene chloride for 24–48 h.
Acid extraction — Adjust the pH of the waters in the extractors to 2 or less using 6 N sulfuric acid. Extract with methylene chloride for 24–48 h.
Ultrasonic extraction of high solids samples — Add anhydrous sodium sulfate to the sample and QC aliquot(s). Add acetone:methylene chloride (1:1) to the sample and mix thoroughly

Concentrate extracts using a K-D apparatus.

QUALITY CONTROL The analyst is permitted to modify this method to improve separations or lower the costs of measurements, provided all performance specifications are met. Analyses of blanks are required to demonstrate freedom from contamination. When results of spikes indicate atypical method performance for samples, the samples are diluted to bring method performance within acceptable limits.

For low solids (aqueous samples), extract, concentrate, and analyze two sets of four 1-L aliquots (8 aliquots total) of the precision and recovery standard. For high solids samples, two sets of four 30-g aliquots of the high solids reference matrix are used.

Spike all samples with labeled compounds to assess method performance. Compute percent recovery of the labeled compounds using the internal standard method. Compare the labeled compound recovery for each compound with the corresponding labeled compound recovery.

Reagent water and high solids reference matrix blanks are analyzed to demonstrate freedom from contamination. Extract and concentrate a 1-L reagent water blank or a high solids reference matrix blank with each sample's lot (samples started through the extraction process on the same 8-h shift, to a maximum of 20 samples).

Field replicates may be collected to determine the precision of the sampling technique, and spiked samples may be required to determine the accuracy of the analysis when the internal standard method is used.

REFERENCE Semivolatile Organic Compounds by Isotope Dilution GC/MS. Office of Water Regulation and Standards, U.S. EPA Industrial Technology Division, Washington, DC, EPA Method 1625, Rev. C, June 1989 (contact W.A. Telliard, U.S. EPA, Office of Water Regulations and Standards, 401 M St., SW, Washington, DC, 20460. Phone: 202-382-7131).

3,6-Dimethylphenanthrene   **EPA Method 1625**
CAS #1576-67-6

TITLE Semivolatile Organic Compounds by Isotope Dilution GC/MS

MATRIX The compounds may be determined in waters, soils, and municipal sludges by this method.

METHOD SUMMARY This method is used to determine 176 semivolatile toxic organic pollutants associated with the CWA (as amended 1987); the RCRA (as amended 1986); the CERCLA (as amended 1986); and other compounds amenable to extraction and analysis by capillary column gas chromatography-mass spectrometry (GC/MS).

Stable isotopically-labeled analogs of the compounds of interest are added to the sample. If the solids content is less than 1%, a 1-L sample is extracted at pH 12–13, then at pH <2 with methylene chloride using continuous extraction techniques.

If the solids content is 30% or less, the sample is diluted to 1% solids with reagent water, homogenized ultrasonically, and extracted at pH 12–13, then at pH <2 with methylene chloride using continuous extraction techniques. If the solids content is greater than 30%, the sample is extracted using ultrasonic techniques.

Each extract is dried over sodium sulfate, concentrated to a volume of 5 mL, cleaned up using GPC, if necessary, and concentrated. Extracts are concentrated to 1 mL if GPC is not performed, and to 0.5 mL if GPC is performed.

An internal standard is added to the extract, and a 1-mL aliquot of the extract is injected into the GC. The compounds are separated by GC and detected by a MS. The labeled compounds serve to correct the variability of the analytical technique.

INTERFERENCES Solvents, reagents, glassware, and other sample processing hardware may yield artifacts and/or elevated baselines causing misinterpretation of chromatograms and spectra. Materials used in the analysis must be demonstrated to be free from interferences under the conditions of analysis by running method blanks initially and with each sample lot (sample started through the extraction process on a given 8-h shift, to a maximum of 20). Specific selection of reagents and purification of solvents by distillation in all glass systems may be required. Glassware and, where possible, reagents are cleaned by solvent rinse and baking at 450°C for 1-h minimum. Interferences coextracted from samples will vary considerably from source to source, depending on the diversity of the site being sampled.

INSTRUMENTATION Major instrumentation includes a GC with a splitless or on-column injection port for capillary column, a MS with 70 eV electron impact ionization, and a data system to collect and record MS data, and process it. A K-D apparatus is used to concentrate extracts.

GC Column: 30 m × 0.25 mm I.D. 5% phenyl, 94% methyl, 1% vinyl silicone bonded phased fused silica capillary column.

PRECISION & ACCURACY The detection limits of the method are usually dependent on the level of interferences rather than instrumental limitations. The limits typify the minimum quantities that can be detected with no interferences present.

The minimum level (in µg/mL) was not listed. This is defined as a minimum level at which the analytical system shall give recognizable mass spectra (background corrected) and acceptable calibration points.

The MDL (in µg/kg) in low solids was not listed and in high solids was not listed; these were determined in digested sludge (low solids) and in filter cake or compost (high solids).

The labeled and native compound initial precision as standard deviation (in µg/L) was not listed.

The labeled and native compound initial accuracy as average recovery (in µg/L) was not listed.

SAMPLE COLLECTION, PRESERVATION & HANDLING Collect samples in glass containers. Aqueous samples which flow freely are collected in refrigerated bottles using automatic sampling equipment. Solid samples are collected as grab samples using widemouth jars. Maintain samples at 0 to 4°C from the time of collection until extraction. If residual chlorine is present in aqueous samples, add 80 mg sodium thiosulfate/L of water. Begin sample extraction within 7 days of collection, and analyze all extracts within 40 days of extraction.

SAMPLE PREPARATION Samples containing 1% solids or less are extracted directly using continuous liquid-liquid extraction techniques. Samples containing 1 to 30% solids are diluted to the 1% level with reagent water and extracted using continuous liquid-liquid extraction techniques. Samples containing greater than 30% solids are extracted using ultrasonic techniques.

Base/neutral extraction — Adjust the pH of the waters in the extractors to 12–13 with 6 N NaOH. Extract with methylene chloride for 24–48 h.

Acid extraction — Adjust the pH of the waters in the extractors to 2 or less using 6 N sulfuric acid. Extract with methylene chloride for 24–48 h.

Ultrasonic extraction of high solids samples — Add anhydrous sodium sulfate to the sample and QC aliquot(s). Add acetone:methylene chloride (1:1) to the sample and mix thoroughly

Concentrate extracts using a K-D apparatus.

QUALITY CONTROL The analyst is permitted to modify this method to improve separations or lower the costs of measurements, provided all performance specifications are met. Analyses of blanks are required to demonstrate freedom from contamination. When results of spikes indicate atypical method performance for samples, the samples are diluted to bring method performance within acceptable limits.

For low solids (aqueous samples), extract, concentrate, and analyze two sets of four 1-L aliquots (8 aliquots total) of the precision and recovery standard. For high solids samples, two sets of four 30-g aliquots of the high solids reference matrix are used.

Spike all samples with labeled compounds to assess method performance. Compute percent recovery of the labeled compounds using the internal standard method. Compare the labeled compound recovery for each compound with the corresponding labeled compound recovery.

Reagent water and high solids reference matrix blanks are analyzed to demonstrate freedom from contamination. Extract and concentrate a 1-L reagent water blank or a high solids reference matrix blank with each sample's lot (samples started through the extraction process on the same 8-h shift, to a maximum of 20 samples).

Field replicates may be collected to determine the precision of the sampling technique, and spiked samples may be required to determine the accuracy of the analysis when the internal standard method is used.

REFERENCE Semivolatile Organic Compounds by Isotope Dilution GC/MS. Office of Water Regulation and Standards, U.S. EPA Industrial Technology Division, Washington, DC, EPA Method 1625, Rev. C, June 1989 (contact W.A. Telliard, U.S. EPA, Office of Water Regulations and Standards, 401 M St., SW, Washington, DC, 20460. Phone: 202-382-7131).

α,α-Dimethylphenethylamine   **EPA Method 8270**
CAS #122-09-8

TITLE Semivolatile Organic Compounds by GC/MS

MATRIX This method is used to determine the concentration of semivolatile organic compounds in extracts prepared from all types of solid waste matrices, soils, and groundwater. Although surface waters are not specifically mentioned, this

method should be applicable to water samples from rivers, lakes, etc.

METHOD SUMMARY This method covers 259 semivolatile organic compounds. In very limited applications direct injection of the sample into the GC/MS system may be appropriate, but this results in very high detection limits (approximately 10,000 µg/L). Typically, a 1-L liquid sample, containing surrogate, and matrix spiking standards, is extracted in a continuous extractor first under acid conditions and then under basic conditions. Typically 30 g of a solid sample, containing surrogate, and matrix spiking standards, is extracted ultrasonically. After concentrating the extract to 1 mL it is spiked with 10 µL of an internal standard solution just prior to analysis by GC/MS. The volume injected should contain about 100 ng of base/neutral and 200 ng of acid surrogates (for a 1-µL injection). Analysis is performed by GC/MS using a capillary GC column.

INTERFERENCES Raw GC/MS data from all blanks, samples, and spikes must be evaluated for interferences. Contamination by carryover can occur whenever high-concentration and low-concentration samples are sequentially analyzed. To reduce carryover, the sample syringe must be rinsed out between samples with solvent. Whenever an unusually concentrated sample is encountered, it should be followed by the analysis of blank solvent to check for cross-contamination.

INSTRUMENTATION A GC/MS and a data system are required. The GC column used is a 30 m × 0.25 mm I.D. (or 0.32 mm I.D.) 1um film thickness silicone-coated fused silica capillary column. A continuous liquid-liquid extractor equipped with Teflon® or glass connection joints and stopcocks requiring no lubrication, a K-D concentrating apparatus, water bath, and an ultrasonic disrupter with a minimum power of 300 W and with pulsing capability are also required.

PRECISION & ACCURACY The estimated quantitation limit (EQL) of Method 8270B for determining an individual compound is approximately 1 mg/kg (wet weight) for soil or sediment samples, 1–200 mg/kg for wastes (dependent on matrix and method of preparation), and 10 µg/L for groundwater samples. EQLs will be proportionately higher for sample extracts that require dilution to avoid saturation of the detector.

The EQL(b) for groundwater in µg/L is not determined.
The EQL (a, b) for low concentrations in soil and sediment in µg/kg is not determined.
Accuracy as µg/L is not listed.
Overall precision in µg/L is not listed.

(a) *EQLs listed for soil/sediment are based on wet weight. Normally data is reported in a dry-weight basis; therefore, EQLs will be higher based on the % dry weight of each sample. This calculation is based on a 30 g sample and gel permeation chromatography cleanup.*
(b) *Sample EQLs are highly matrix-dependent. The EQLs are provided for guidance and may not always be achievable.*
C = *True value for concentration, in µg/L.*
X = *Average recovery found for measurements of samples containing a concentration of C, in µg/L.*

ESTIMATED QUANTITATION LIMIT

Other Matrices	Factor (a)
High-concentration soil and sludges by sonicator	7.5
Non-water miscible waste	75

(a) EQL for other matrices = [EQL for low soil/sediment] × [Factor]. This estimated EQL is similar to an EPA "Practical Quantitation Limit."

SAMPLING METHOD
Liquid samples — Use a 1 or 2½ gallon amber glass bottle with a screw-top Teflon®-lined cover that has been prewashed with detergent and rinsed with distilled water and methanol (or isopropanol).

Soils, sediments, or sludges — Use an 8-oz. widemouth glass with a screw-top Teflon®-lined cover that has been prewashed with detergent and rinsed with distilled water and methanol (or isopropanol).

SAMPLE PRESERVATION
Liquid samples — If residual chlorine is present, add 3 mL of 10% sodium thiosulfate per gallon, cool to 4°C and store in a solvent-free refrigerator until analysis; if chlorine is not present, then eliminate the sodium thiosulfate addition.

Soils, sediments, or sludges — Cool samples to 4°C and store in a solvent-free refrigerator.

MHT Liquid samples must be extracted within 7 days and the extracts analyzed within 40 days. Soils, sediments, or sludges may be stored for a maximum of 14 days and the extracts analyzed within 40 days.

SAMPLE PREPARATION
Liquid samples — Transfer 1 L quantitatively to a continuous extractor. If high concentrations are anticipated, a smaller volume may be used and then diluted with organic-free reagent water to 1 L. Adjust pH, if necessary, to pH <2 using 1:1 (V/V) sulfuric acid. Pipette 1.0 mL of a surrogate standard spiking solution into each sample. For the sample in each analytical batch selected for spiking, add 1.0 mL of a matrix spiking standard. For base/neutral acid analysis, the amount of the surrogates and matrix spiking compounds added to the sample should result in a final concentration of 100 ng/µL of each analyte in the extract to be analyzed (assuming a 1-µL injection). Extract with methylene chloride for 18–24 h. Next, adjust the pH of the aqueous phase to pH >11 using 10 N sodium hydroxide and extract it with methylene chloride again for 18–24 h. Dry the extract through a column containing anhydrous sodium sulfate and concentrate it to 1 mL using a K-D concentrator.

Soils, sediments, or sludges — Use 30 g of sample. Nonporous or wet samples (gummy or clay type) that do not have a free-flowing sandy texture must be mixed with anhydrous sodium sulfate until the sample is free flowing. Add 1 mL of surrogate standards to all samples, spikes, standards, and blanks. For the sample in each analytical batch selected for spiking, add 1.0 mL of a matrix spiking standard. For base/neutral acid analysis, the amount added of the surrogates and matrix spiking compounds should result in a final concentration of 100 ng/µL of each base/neutral analyte and 200 ng/µL of each acid analyte

in the extract to be analyzed (assuming a 1-µL injection). Immediately add a 100-mL mixture of 1:1 methylene chloride:acetone and extract the sample ultrasonically for 3 min and then decant or filter the extracts. Repeat the extraction two or more times. Dry the extract using a column with anhydrous sodium sulfate and concentrate it to 1 mL in a K-D concentrator.

QUALITY CONTROL A methylene chloride solution containing 50 ng/µL of decafluorotriphenylphosphine (DFTPP) is used for tuning the GC/MS system each 12-h shift. A system performance check also must be made during every 12-h shift. A standard containing 50 ng/µL each of 4,4'-DDT, pentachlorophenol, and benzidine is required to verify injection port inertness and GC column performance. A calibration standard at mid-concentration, containing each compound of interest, including all required surrogates, must be performed every 12 h during analysis. After the system performance check is met, calibration check compounds (CCCs) are used to check the validity of the initial calibration.

The internal standard responses and retention times in the calibration check standard must be evaluated immediately after or during data acquisition. If the retention time for any internal standard changes by more than 30 seconds from the last check calibration (12 h), the chromatographic system must be inspected for malfunctions and corrections must be made, as required. If the electron ionization current plot (EICP) area for any of the internal standards changes by a factor of two from the last daily calibration standard check, the mass spectrometer must be inspected for malfunctions and corrections must be made, as appropriate.

Demonstrate, through the analysis of a reagent water blank, that interferences from the analytical system, glassware, and reagents are under control. The blank samples should be carried through all stages of the sample preparation and measurement steps. For each analytical batch (up to 20 samples), a reagent blank, matrix spike, and matrix spike duplicate/duplicate must be analyzed (the frequency of the spikes may be different for different monitoring programs). The blank and spiked samples must be carried through all stages of the sample preparation and measurement steps. A QC reference sample concentrate containing each analyte at a concentration of 100 mg/L in methanol is required.

REFERENCE Test Methods for Evaluating Solid Waste (SW-846). U.S. EPA 1983, Method 8270B, Rev. 2, Nov. 1990. Office of Solid Waste, Washington, DC.

2,4-Dimethylphenol **EPA Method 1625**
CAS #105-67-9

TITLE Semivolatile Organic Compounds by Isotope Dilution GC/MS

MATRIX The compounds may be determined in waters, soils, and municipal sludges by this method.

METHOD SUMMARY This method is used to determine 176 semivolatile toxic organic pollutants associated with the CWA (as amended 1987); the RCRA (as amended 1986); the CERCLA (as amended 1986); and other compounds amenable to extraction and analysis by capillary column gas chromatography-mass spectrometry (GC/MS).

Stable isotopically-labeled analogs of the compounds of interest are added to the sample. If the solids content is less than 1%, a 1-L sample is extracted at pH 12–13, then at pH <2 with methylene chloride using continuous extraction techniques.

If the solids content is 30% or less, the sample is diluted to 1% solids with reagent water, homogenized ultrasonically, and extracted at pH 12–13, then at pH <2 with methylene chloride using continuous extraction techniques. If the solids content is greater than 30%, the sample is extracted using ultrasonic techniques.

Each extract is dried over sodium sulfate, concentrated to a volume of 5 mL, cleaned up using GPC, if necessary, and concentrated. Extracts are concentrated to 1 mL if GPC is not performed, and to 0.5 mL if GPC is performed.

An internal standard is added to the extract, and a 1-mL aliquot of the extract is injected into the GC. The compounds are separated by GC and detected by a MS. The labeled compounds serve to correct the variability of the analytical technique.

INTERFERENCES Solvents, reagents, glassware, and other sample processing hardware may yield artifacts and/or elevated baselines causing misinterpretation of chromatograms and spectra. Materials used in the analysis must be demonstrated to be free from interferences under the conditions of analysis by running method blanks initially and with each sample lot (sample started through the extraction process on a given 8-h shift, to a maximum of 20). Specific selection of reagents and purification of solvents by distillation in all glass systems may be required. Glassware and, where possible, reagents are cleaned by solvent rinse and baking at 450°C for 1-h minimum. Interferences coextracted from samples will vary considerably from source to source, depending on the diversity of the site being sampled.

INSTRUMENTATION Major instrumentation includes a GC with a splitless or on-column injection port for capillary column, a MS with 70 eV electron impact ionization, and a data system to collect and record MS data, and process it. A K-D apparatus is used to concentrate extracts.

GC Column: 30 m × 0.25 mm I.D. 5% phenyl, 94% methyl, 1% vinyl silicone bonded phased fused silica capillary column.

PRECISION & ACCURACY The detection limits of the method are usually dependent on the level of interferences rather than instrumental limitations. The limits typify the minimum quantities that can be detected with no interferences present.

The minimum level (in µg/mL) was 10. This is defined as a minimum level at which the analytical system shall give recognizable mass spectra (background corrected) and acceptable calibration points.

The MDL (in µg/kg) in low solids was 26 and in high solids was 13; these were determined in digested sludge (low solids) and in filter cake or compost (high solids).

The labeled and native compound initial precision as standard deviation (in μg/L) was 13.

The labeled and native compound initial accuracy as average recovery (in μg/L) was 62–153.

SAMPLE COLLECTION, PRESERVATION & HANDLING
Collect samples in glass containers. Aqueous samples which flow freely are collected in refrigerated bottles using automatic sampling equipment. Solid samples are collected as grab samples using widemouth jars. Maintain samples at 0 to 4°C from the time of collection until extraction. If residual chlorine is present in aqueous samples, add 80 mg sodium thiosulfate/L of water. Begin sample extraction within 7 days of collection, and analyze all extracts within 40 days of extraction.

SAMPLE PREPARATION Samples containing 1% solids or less are extracted directly using continuous liquid-liquid extraction techniques. Samples containing 1 to 30% solids are diluted to the 1% level with reagent water and extracted using continuous liquid-liquid extraction techniques. Samples containing greater than 30% solids are extracted using ultrasonic techniques.

Base/neutral extraction — Adjust the pH of the waters in the extractors to 12–13 with 6 N NaOH. Extract with methylene chloride for 24–48 h.

Acid extraction — Adjust the pH of the waters in the extractors to 2 or less using 6 N sulfuric acid. Extract with methylene chloride for 24–48 h.

Ultrasonic extraction of high solids samples — Add anhydrous sodium sulfate to the sample and QC aliquot(s). Add acetone:methylene chloride (1:1) to the sample and mix thoroughly

Concentrate extracts using a K-D apparatus.

QUALITY CONTROL The analyst is permitted to modify this method to improve separations or lower the costs of measurements, provided all performance specifications are met. Analyses of blanks are required to demonstrate freedom from contamination. When results of spikes indicate atypical method performance for samples, the samples are diluted to bring method performance within acceptable limits.

For low solids (aqueous samples), extract, concentrate, and analyze two sets of four 1-L aliquots (8 aliquots total) of the precision and recovery standard. For high solids samples, two sets of four 30-g aliquots of the high solids reference matrix are used.

Spike all samples with labeled compounds to assess method performance. Compute percent recovery of the labeled compounds using the internal standard method. Compare the labeled compound recovery for each compound with the corresponding labeled compound recovery.

Reagent water and high solids reference matrix blanks are analyzed to demonstrate freedom from contamination. Extract and concentrate a 1-L reagent water blank or a high solids reference matrix blank with each sample's lot (samples started through the extraction process on the same 8-h shift, to a maximum of 20 samples).

Field replicates may be collected to determine the precision of the sampling technique, and spiked samples may be required to determine the accuracy of the analysis when the internal standard method is used.

REFERENCE Semivolatile Organic Compounds by Isotope Dilution GC/MS. Office of Water Regulation and Standards, U.S. EPA Industrial Technology Division, Washington, DC, EPA Method 1625, Rev. C, June 1989 (contact W.A. Telliard, U.S. EPA, Office of Water Regulations and Standards, 401 M St., SW, Washington, DC, 20460. Phone: 202-382-7131).

2,4-Dimethylphenol   EPA Method 625
CAS #105-67-9

TITLE Base/Neutrals and Acids, U.S. EPA Method 625

MATRIX This methods covers municipal and industrial wastewaters.

METHOD SUMMARY Approximately 1 L of sample is serially extracted with methylene chloride at a pH greater than 11 and again at a pH less than 2 using a separatory funnel or a continuous extractor. The methylene chloride extract is dried, concentrated to a volume of 1 mL, and analyzed by GC/MS. Qualitative identification of the parameters in the extract is performed using the retention time and the relative abundance of three characteristic masses (m/z). Qualitative analysis is performed using either external or internal standard techniques with a single characteristic m/z.

INTERFERENCES Method interferences may be caused by contaminants in solvents, reagents, glassware, and other sample processing hardware. Glassware must be scrupulously cleaned. Glassware should be heated in a muffle furnace at 400°C for 5 to 30 min. Some thermally stable materials, such as PCBs, may not be eliminated by this treatment. Solvent rinses with acetone and pesticide quality hexane may be substituted for the muffle furnace heating. Matrix interferences may be caused by contaminants that are coextracted from the sample. The base-neutral extraction may cause significantly reduced recovery of phenols. The packed gas chromatographic columns recommended for the basic fraction may not exhibit sufficient resolution for some analytes.

INSTRUMENTATION A GC/MS system with an injection port designed for on-column injection when using packed columns and for splitless injection when using capillary columns.

Column for base/neutrals: 1.8 m long × 2 mm I.D. glass, packed with 3% SP-2550 on Supelcoport (100/120 mesh) or equivalent.

Column for acids: 1.8 m long × 2 mm I.D. glass, packed with 1% SP-1240DA on Supelcoport (100/120 mesh) or equivalent.

PRECISION & ACCURACY The MDL concentrations were obtained using reagent water. The MDL actually achieved in a given analysis will vary depending on instrument sensitivity and matrix effects. This method was tested by 15 laboratories using reagent water, drinking water, surface water, and industrial wastewaters spiked at six concentrations over the range 5

to 100 µg/L. Single operator precision, overall precision, and method accuracy were found to be directly related to the concentration of the parameter matrix.

The MDL (in µg/L) in reagent water was 2.7.

The standard deviation (in µg/L based on 4 recovery measurements) was 26.1.

The range (in µg/L) for average recovery for 4 measurements was 41.8–109.0.

The range (in %) for percent recovery was 32–118.

Accuracy (in µg/L) as expected recovery for one or more measurements of a sample containing a true concentration of C was 0.71C+4.41.

Precision (in µg/L) as expected single analyst standard deviation of measurements at an average concentration found at X was 0.16X+1.21.

Overall precision (in µg/L) as expected interlaboratory standard deviation of measurements in an average concentration found at X was 0.22X+1.31.

C = *True value of the concentration in µg/L.*
X = *Average recovery found for measurements of samples containing a concentration at C in µg/L.*

SAMPLE PREPARATION Adjust the pH to >11 with sodium hydroxide and serially extract in a separatory funnel with methylene chloride or else in a continuous extractor. Next, adjust the pH to <2 with sulfuric acid and serially extract in a separatory funnel with methylene chloride or else in a continuous extractor. Dry the extracts separately through a column of anhydrous sodium sulfate and then concentrate each of the extracts to 1.0 mL using a K-D apparatus.

SAMPLE COLLECTION, PRESERVATION & HANDLING Grab samples must be collected in glass containers. All samples must be refrigerated at 4°C from the time of collection until extraction. If residual chlorine is present, add 80 mg of sodium thiosulfate/L of sample and mix well. All samples must be extracted within 7 days of collection and completely analyzed within 40 days of extraction.

QUALITY CONTROL Make an initial, one-time, demonstration of the ability to generate acceptable accuracy and precision with this method. Before processing any samples, the analyst must analyze a reagent water blank to demonstrate that interferences from the analytical system and glassware are under control. Each time a set of samples is extracted or reagents are changed, a reagent water blank must be processed. Spike and analyze a minimum of 5% of all samples to monitor and evaluate lab data quality. A QC check sample concentrate that contains each parameter of interest at a concentration of 100 µg/mL in acetone is required. PCBs and multicomponent pesticides may be omitted from this test.

After analysis of five spiked wastewater samples, calculate the average percent recovery and the standard deviation of the percent recovery. Spike all samples with the surrogate standard spiking solution and calculate the percent recovery of each surrogate compound.

REFERENCE *Federal Register*, Vol. 49, No. 209. Friday, Oct. 26, 1984.

2,4-Dimethylphenol   EPA Method 8270
CAS #105-67-9

TITLE Semivolatile Organic Compounds by GC/MS

MATRIX This method is used to determine the concentration of semivolatile organic compounds in extracts prepared from all types of solid waste matrices, soils, and groundwater. Although surface waters are not specifically mentioned, this method should be applicable to water samples from rivers, lakes, etc.

METHOD SUMMARY This method covers 259 semivolatile organic compounds. In very limited applications direct injection of the sample into the GC/MS system may be appropriate, but this results in very high detection limits (approximately 10,000 µg/L). Typically, a 1-L liquid sample, containing surrogate, and matrix spiking standards, is extracted in a continuous extractor first under acid conditions and then under basic conditions. Typically 30 g of a solid sample, containing surrogate, and matrix spiking standards, is extracted ultrasonically. After concentrating the extract to 1 mL it is spiked with 10 µL of an internal standard solution just prior to analysis by GC/MS. The volume injected should contain about 100 ng of base/neutral and 200 ng of acid surrogates (for a 1-µL injection). Analysis is performed by GC/MS using a capillary GC column.

INTERFERENCES Raw GC/MS data from all blanks, samples, and spikes must be evaluated for interferences. Contamination by carryover can occur whenever high-concentration and low-concentration samples are sequentially analyzed. To reduce carryover, the sample syringe must be rinsed out between samples with solvent. Whenever an unusually concentrated sample is encountered, it should be followed by the analysis of blank solvent to check for cross-contamination.

INSTRUMENTATION A GC/MS and a data system are required. The GC column used is a 30 m × 0.25 mm I.D. (or 0.32 mm I.D.) 1um film thickness silicone-coated fused silica capillary column. A continuous liquid-liquid extractor equipped with Teflon® or glass connection joints and stopcocks requiring no lubrication, a K-D concentrating apparatus, water bath, and an ultrasonic disrupter with a minimum power of 300 W and with pulsing capability are also required.

PRECISION & ACCURACY The estimated quantitation limit (EQL) of Method 8270B for determining an individual compound is approximately 1 mg/kg (wet weight) for soil or sediment samples, 1–200 mg/kg for wastes (dependent on matrix and method of preparation), and 10 µg/L for groundwater samples. EQLs will be proportionately higher for sample extracts that require dilution to avoid saturation of the detector.

The EQL(b) for groundwater in µg/L is 10.

The EQL (a, b) for low concentrations in soil and sediment in µg/kg is 660.

Accuracy as µg/L is 0.71C +4.41.

Overall precision in µg/L is 0.22X +1.31.

(a) *EQLs listed for soil/sediment are based on wet weight. Normally data is reported in a dry-weight basis; therefore, EQLs will be higher based on the % dry weight of each sample.*

This calculation is based on a 30 g sample and gel permeation chromatography cleanup.

(b) *Sample EQLs are highly matrix-dependent. The EQLs are provided for guidance and may not always be achievable.*

C = *True value for concentration, in µg/L.*

X = *Average recovery found for measurements of samples containing a concentration of C, in µg/L.*

ESTIMATED QUANTITATION LIMIT

Other Matrices	Factor (a)
High-concentration soil and sludges by sonicator	7.5
Non-water miscible waste	75

(a) *EQL for other matrices = [EQL for low soil/sediment] × [Factor]. This estimated EQL is similar to an EPA "Practical Quantitation Limit."*

SAMPLING METHOD

Liquid samples — Use a 1 or 2½ gallon amber glass bottle with a screw-top Teflon®-lined cover that has been prewashed with detergent and rinsed with distilled water and methanol (or isopropanol).

Soils, sediments, or sludges — Use an 8-oz. widemouth glass with a screw-top Teflon®-lined cover that has been prewashed with detergent and rinsed with distilled water and methanol (or isopropanol).

SAMPLE PRESERVATION

Liquid samples — If residual chlorine is present, add 3 mL of 10% sodium thiosulfate per gallon, cool to 4°C and store in a solvent-free refrigerator until analysis; if chlorine is not present, then eliminate the sodium thiosulfate addition.

Soils, sediments, or sludges — Cool samples to 4°C and store in a solvent-free refrigerator.

MHT Liquid samples must be extracted within 7 days and the extracts analyzed within 40 days. Soils, sediments, or sludges may be stored for a maximum of 14 days and the extracts analyzed within 40 days.

SAMPLE PREPARATION

Liquid samples — Transfer 1 L quantitatively to a continuous extractor. If high concentrations are anticipated, a smaller volume may be used and then diluted with organic-free reagent water to 1 L. Adjust pH, if necessary, to pH <2 using 1:1 (V/V) sulfuric acid. Pipette 1.0 mL of a surrogate standard spiking solution into each sample. For the sample in each analytical batch selected for spiking, add 1.0 mL of a matrix spiking standard. For base/neutral acid analysis, the amount of the surrogates and matrix spiking compounds added to the sample should result in a final concentration of 100 ng/µL of each analyte in the extract to be analyzed (assuming a 1-µL injection). Extract with methylene chloride for 18–24 h. Next, adjust the pH of the aqueous phase to pH >11 using 10 N sodium hydroxide and extract it with methylene chloride again for 18–24 h. Dry the extract through a column containing anhydrous sodium sulfate and concentrate it to 1 mL using a K-D concentrator.

Soils, sediments, or sludges — Use 30 g of sample. Nonporous or wet samples (gummy or clay type) that do not have a free-flowing sandy texture must be mixed with anhydrous sodium sulfate until the sample is free flowing. Add 1 mL of surrogate standards to all samples, spikes, standards, and blanks. For the sample in each analytical batch selected for spiking, add 1.0 mL of a matrix spiking standard. For base/neutral acid analysis, the amount added of the surrogates and matrix spiking compounds should result in a final concentration of 100 ng/µL of each base/neutral analyte and 200 ng/µL of each acid analyte in the extract to be analyzed (assuming a 1-µL injection). Immediately add a 100-mL mixture of 1:1 methylene chloride:acetone and extract the sample ultrasonically for 3 min and then decant or filter the extracts. Repeat the extraction two or more times. Dry the extract using a column with anhydrous sodium sulfate and concentrate it to 1 mL in a K-D concentrator.

QUALITY CONTROL A methylene chloride solution containing 50 ng/µL of decafluorotriphenylphosphine (DFTPP) is used for tuning the GC/MS system each 12-h shift. A system performance check also must be made during every 12-h shift. A standard containing 50 ng/µL each of 4,4'-DDT, pentachlorophenol, and benzidine is required to verify injection port inertness and GC column performance. A calibration standard at mid-concentration, containing each compound of interest, including all required surrogates, must be performed every 12 h during analysis. After the system performance check is met, calibration check compounds (CCCs) are used to check the validity of the initial calibration.

The internal standard responses and retention times in the calibration check standard must be evaluated immediately after or during data acquisition. If the retention time for any internal standard changes by more than 30 seconds from the last check calibration (12 h), the chromatographic system must be inspected for malfunctions and corrections must be made, as required. If the electron ionization current plot (EICP) area for any of the internal standards changes by a factor of two from the last daily calibration standard check, the mass spectrometer must be inspected for malfunctions and corrections must be made, as appropriate.

Demonstrate, through the analysis of a reagent water blank, that interferences from the analytical system, glassware, and reagents are under control. The blank samples should be carried through all stages of the sample preparation and measurement steps. For each analytical batch (up to 20 samples), a reagent blank, matrix spike, and matrix spike duplicate/duplicate must be analyzed (the frequency of the spikes may be different for different monitoring programs). The blank and spiked samples must be carried through all stages of the sample preparation and measurement steps. A QC reference sample concentrate containing each analyte at a concentration of 100 mg/L in methanol is required.

REFERENCE Test Methods for Evaluating Solid Waste (SW-846). U.S. EPA 1983, Method 8270B, Rev. 2, Nov. 1990. Office of Solid Waste, Washington, DC.

2,4-Dimethylphenol
CAS #105-67-9

EPA Method 8040

TITLE Phenols

MATRIX Groundwater, soils, sludges, water miscible liquid wastes, and non-water miscible wastes.

APPLICATION This method is used for the analysis of 17 phenols. Samples are extracted, concentrated, and analyzed using direct injection of both neat and diluted organic liquids. Pentafluorobenzylbromide (PFB) derivatives also may be made to increase sensitivity of the method.

INTERFERENCES There can be carryover contamination with high- and low-level samples. Solvents, reagents, and glassware may introduce artifacts. Other interferences may come from coextracted compounds from samples.

INSTRUMENTATION GC capable of on-column injections and a flame with detector (FID) or electron capture detector (ECD). Column for underivatized phenol: 1.8 m by 2.0 mm with 1% SP-1240DA on Supelcoport. Column for derivatized phenols: 1.8 m by 2.0 mm with 5% OV-17 on Chromosorb W-AW-DMCS.

RANGE 12 to 450 µg/L.

MDL 0.32 µg/L (FID) and 0.63 µg/L (ECD).

PQL FACTORS FOR MULTIPLYING × FID MDL VALUE

Matrix	Multiplication Factor
Groundwater	10
Low-level soil by sonication with GPC cleanup	670
High-level soil and sludge by sonication	10,000
Non-water miscible waste	100,000

PRECISION $0.25X + 0.48$ µg/L (overall precision using FID).

ACCURACY $0.62C - 1.64$ µg/L (as recovery using FID).

SAMPLING METHOD Use 8-oz. widemouth glass bottles with Teflon®-lined caps for concentrated waste samples, soils, sediments, and sludges. Use 1 or 2½ gallon amber glass bottles with Teflon®-lined caps for liquid (water) samples.

STABILITY Cool soil, sediment, sludge, and liquid samples to 4°C. If residual chlorine is present in liquid samples add 3 mL of 10% sodium thiosulfate per gallon of sample and cool to 4°C.

MHT 14 days for concentrated waste, soil, sediment, or sludge; 7 days for liquid samples; all extracts must be analyzed within 40 days.

QUALITY CONTROL A quality control check sample concentrate containing each analyte of interest is required. The QC check sample concentrate may be prepared from pure standard materials or purchased as certified solutions Use appropriate trip, matrix, control site, method, reagent, and solvent blanks. Internal, surrogate, and five concentration level calibration standards are used. The QC check sample concentrate should contain this compound at 100 µg/mL in 2-propanol.

REFERENCE Test Methods for Evaluating Solid Waste (SW-846), U.S. EPA Office of Solid Waste, Washington, DC, Method 8040A, Rev. 1, Nov. 1990.

4,6-Dinitro-2-methylphenol
CAS #534-52-1

EPA Method 8270

TITLE Semivolatile Organic Compounds by GC/MS

MATRIX This method is used to determine the concentration of semivolatile organic compounds in extracts prepared from all types of solid waste matrices, soils, and groundwater. Although surface waters are not specifically mentioned, this method should be applicable to water samples from rivers, lakes, etc.

METHOD SUMMARY This method covers 259 semivolatile organic compounds. In very limited applications direct injection of the sample into the GC/MS system may be appropriate, but this results in very high detection limits (approximately 10,000 µg/L). Typically, a 1-L liquid sample, containing surrogate, and matrix spiking standards, is extracted in a continuous extractor first under acid conditions and then under basic conditions. Typically 30 g of a solid sample, containing surrogate, and matrix spiking standards, is extracted ultrasonically. After concentrating the extract to 1 mL it is spiked with 10 µL of an internal standard solution just prior to analysis by GC/MS. The volume injected should contain about 100 ng of base/neutral and 200 ng of acid surrogates (for a 1-µL injection). Analysis is performed by GC/MS using a capillary GC column.

INTERFERENCES Raw GC/MS data from all blanks, samples, and spikes must be evaluated for interferences. Contamination by carryover can occur whenever high-concentration and low-concentration samples are sequentially analyzed. To reduce carryover, the sample syringe must be rinsed out between samples with solvent. Whenever an unusually concentrated sample is encountered, it should be followed by the analysis of blank solvent to check for cross-contamination.

INSTRUMENTATION A GC/MS and a data system are required. The GC column used is a 30 m × 0.25 mm I.D. (or 0.32 mm I.D.) 1um film thickness silicone-coated fused silica capillary column. A continuous liquid-liquid extractor equipped with Teflon® or glass connection joints and stopcocks requiring no lubrication, a K-D concentrating apparatus, water bath, and an ultrasonic disrupter with a minimum power of 300 W and with pulsing capability are also required.

PRECISION & ACCURACY The estimated quantitation limit (EQL) of Method 8270B for determining an individual compound is approximately 1 mg/kg (wet weight) for soil or sediment samples, 1–200 mg/kg for wastes (dependent on matrix and method of preparation), and 10 µg/L for groundwater samples. EQLs will be proportionately higher for sample extracts that require dilution to avoid saturation of the detector.

The EQL(b) for groundwater in µg/L is 50.
The EQL (a, b) for low concentrations in soil and sediment in µg/kg is 3300.
Accuracy as µg/L is not listed.

Overall precision in µg/L is not listed.

(a) *EQLs listed for soil/sediment are based on wet weight. Normally data is reported in a dry-weight basis; therefore, EQLs will be higher based on the % dry weight of each sample. This calculation is based on a 30 g sample and gel permeation chromatography cleanup.*

(b) *Sample EQLs are highly matrix-dependent. The EQLs are provided for guidance and may not always be achievable.*

C = *True value for concentration, in µg/L.*

X = *Average recovery found for measurements of samples containing a concentration of C, in µg/L.*

ESTIMATED QUANTITATION LIMIT

Other Matrices	Factor (a)
High-concentration soil and sludges by sonicator	7.5
Non-water miscible waste	75

(a) *EQL for other matrices = [EQL for low soil/sediment] × [Factor]. This estimated EQL is similar to an EPA "Practical Quantitation Limit."*

SAMPLING METHOD

Liquid samples — Use a 1 or 2½ gallon amber glass bottle with a screw-top Teflon®-lined cover that has been prewashed with detergent and rinsed with distilled water and methanol (or isopropanol).

Soils, sediments, or sludges — Use an 8-oz. widemouth glass with a screw-top Teflon®-lined cover that has been prewashed with detergent and rinsed with distilled water and methanol (or isopropanol).

SAMPLE PRESERVATION

Liquid samples — If residual chlorine is present, add 3 mL of 10% sodium thiosulfate per gallon, cool to 4°C and store in a solvent-free refrigerator until analysis; if chlorine is not present, then eliminate the sodium thiosulfate addition.

Soils, sediments, or sludges — Cool samples to 4°C and store in a solvent-free refrigerator.

MHT Liquid samples must be extracted within 7 days and the extracts analyzed within 40 days. Soils, sediments, or sludges may be stored for a maximum of 14 days and the extracts analyzed within 40 days.

SAMPLE PREPARATION

Liquid samples — Transfer 1 L quantitatively to a continuous extractor. If high concentrations are anticipated, a smaller volume may be used and then diluted with organic-free reagent water to 1 L. Adjust pH, if necessary, to pH <2 using 1:1 (V/V) sulfuric acid. Pipette 1.0 mL of a surrogate standard spiking solution into each sample. For the sample in each analytical batch selected for spiking, add 1.0 mL of a matrix spiking standard. For base/neutral acid analysis, the amount of the surrogates and matrix spiking compounds added to the sample should result in a final concentration of 100 ng/µL of each analyte in the extract to be analyzed (assuming a 1-µL injection). Extract with methylene chloride for 18–24 h. Next, adjust the pH of the aqueous phase to pH >11 using 10 N sodium hydroxide and extract it with methylene chloride again for 18–24 h. Dry the extract through a column containing anhydrous sodium sulfate and concentrate it to 1 mL using a K-D concentrator.

Soils, sediments, or sludges — Use 30 g of sample. Nonporous or wet samples (gummy or clay type) that do not have a free-flowing sandy texture must be mixed with anhydrous sodium sulfate until the sample is free flowing. Add 1 mL of surrogate standards to all samples, spikes, standards, and blanks. For the sample in each analytical batch selected for spiking, add 1.0 mL of a matrix spiking standard. For base/neutral acid analysis, the amount added of the surrogates and matrix spiking compounds should result in a final concentration of 100 ng/µL of each base/neutral analyte and 200 ng/µL of each acid analyte in the extract to be analyzed (assuming a 1-µL injection). Immediately add a 100-mL mixture of 1:1 methylene chloride:acetone and extract the sample ultrasonically for 3 min and then decant or filter the extracts. Repeat the extraction two or more times. Dry the extract using a column with anhydrous sodium sulfate and concentrate it to 1 mL in a K-D concentrator.

Note: N-nitrosodiphenylamine decomposes in the gas chromatographic inlet and cannot be separated from diphenylamine.

QUALITY CONTROL A methylene chloride solution containing 50 ng/µL of decafluorotriphenylphosphine (DFTPP) is used for tuning the GC/MS system each 12-h shift. A system performance check also must be made during every 12-h shift. A standard containing 50 ng/µL each of 4,4'-DDT, pentachlorophenol, and benzidine is required to verify injection port inertness and GC column performance. A calibration standard at mid-concentration, containing each compound of interest, including all required surrogates, must be performed every 12 h during analysis. After the system performance check is met, calibration check compounds (CCCs) are used to check the validity of the initial calibration.

The internal standard responses and retention times in the calibration check standard must be evaluated immediately after or during data acquisition. If the retention time for any internal standard changes by more than 30 seconds from the last check calibration (12 h), the chromatographic system must be inspected for malfunctions and corrections must be made, as required. If the electron ionization current plot (EICP) area for any of the internal standards changes by a factor of two from the last daily calibration standard check, the mass spectrometer must be inspected for malfunctions and corrections must be made, as appropriate.

Demonstrate, through the analysis of a reagent water blank, that interferences from the analytical system, glassware, and reagents are under control. The blank samples should be carried through all stages of the sample preparation and measurement steps. For each analytical batch (up to 20 samples), a reagent blank, matrix spike, and matrix spike duplicate/duplicate must be analyzed (the frequency of the spikes may be different for different monitoring programs). The blank and spiked samples must be carried through all stages of the sample preparation and measurement steps. A QC reference sample concentrate containing each analyte at a concentration of 100 mg/L in methanol is required.

REFERENCE Test Methods for Evaluating Solid Waste (SW-846). U.S. EPA 1983, Method 8270B, Rev. 2, Nov. 1990. Office of Solid Waste, Washington, DC.

1,4-Dinitrobenzene   **EPA Method 1625**
CAS #100-25-4

TITLE Semivolatile Organic Compounds by Isotope Dilution GC/MS

MATRIX The compounds may be determined in waters, soils, and municipal sludges by this method.

METHOD SUMMARY This method is used to determine 176 semivolatile toxic organic pollutants associated with the CWA (as amended 1987); the RCRA (as amended 1986); the CERCLA (as amended 1986); and other compounds amenable to extraction and analysis by capillary column gas chromatography-mass spectrometry (GC/MS).

Stable isotopically-labeled analogs of the compounds of interest are added to the sample. If the solids content is less than 1%, a 1-L sample is extracted at pH 12–13, then at pH <2 with methylene chloride using continuous extraction techniques.

If the solids content is 30% or less, the sample is diluted to 1% solids with reagent water, homogenized ultrasonically, and extracted at pH 12–13, then at pH <2 with methylene chloride using continuous extraction techniques. If the solids content is greater than 30%, the sample is extracted using ultrasonic techniques.

Each extract is dried over sodium sulfate, concentrated to a volume of 5 mL, cleaned up using GPC, if necessary, and concentrated. Extracts are concentrated to 1 mL if GPC is not performed, and to 0.5 mL if GPC is performed.

An internal standard is added to the extract, and a 1-mL aliquot of the extract is injected into the GC. The compounds are separated by GC and detected by a MS. The labeled compounds serve to correct the variability of the analytical technique.

INTERFERENCES Solvents, reagents, glassware, and other sample processing hardware may yield artifacts and/or elevated baselines causing misinterpretation of chromatograms and spectra. Materials used in the analysis must be demonstrated to be free from interferences under the conditions of analysis by running method blanks initially and with each sample lot (sample started through the extraction process on a given 8-h shift, to a maximum of 20). Specific selection of reagents and purification of solvents by distillation in all glass systems may be required. Glassware and, where possible, reagents are cleaned by solvent rinse and baking at 450°C for 1-h minimum. Interferences coextracted from samples will vary considerably from source to source, depending on the diversity of the site being sampled.

INSTRUMENTATION Major instrumentation includes a GC with a splitless or on-column injection port for capillary column, a MS with 70 eV electron impact ionization, and a data system to collect and record MS data, and process it. A K-D apparatus is used to concentrate extracts.

GC Column: 30 m × 0.25 mm I.D. 5% phenyl, 94% methyl, 1% vinyl silicone bonded phased fused silica capillary column.

PRECISION & ACCURACY The detection limits of the method are usually dependent on the level of interferences rather than instrumental limitations. The limits typify the minimum quantities that can be detected with no interferences present.

The minimum level (in µg/mL) was not listed. This is defined as a minimum level at which the analytical system shall give recognizable mass spectra (background corrected) and acceptable calibration points.

The MDL (in µg/kg) in low solids was not listed and in high solids was not listed; these were determined in digested sludge (low solids) and in filter cake or compost (high solids).

The labeled and native compound initial precision as standard deviation (in µg/L) was not listed.
The labeled and native compound initial accuracy as average recovery (in µg/L) was not listed.

SAMPLE COLLECTION, PRESERVATION & HANDLING Collect samples in glass containers. Aqueous samples which flow freely are collected in refrigerated bottles using automatic sampling equipment. Solid samples are collected as grab samples using widemouth jars. Maintain samples at 0 to 4°C from the time of collection until extraction. If residual chlorine is present in aqueous samples, add 80 mg sodium thiosulfate/L of water. Begin sample extraction within 7 days of collection, and analyze all extracts within 40 days of extraction.

SAMPLE PREPARATION Samples containing 1% solids or less are extracted directly using continuous liquid-liquid extraction techniques. Samples containing 1 to 30% solids are diluted to the 1% level with reagent water and extracted using continuous liquid-liquid extraction techniques. Samples containing greater than 30% solids are extracted using ultrasonic techniques.

Base/neutral extraction — Adjust the pH of the waters in the extractors to 12–13 with 6 *N* NaOH. Extract with methylene chloride for 24–48 h.
Acid extraction — Adjust the pH of the waters in the extractors to 2 or less using 6 *N* sulfuric acid. Extract with methylene chloride for 24–48 h.
Ultrasonic extraction of high solids samples — Add anhydrous sodium sulfate to the sample and QC aliquot(s). Add acetone:methylene chloride (1:1) to the sample and mix thoroughly

Concentrate extracts using a K-D apparatus.

QUALITY CONTROL The analyst is permitted to modify this method to improve separations or lower the costs of measurements, provided all performance specifications are met. Analyses of blanks are required to demonstrate freedom from contamination. When results of spikes indicate atypical method performance for samples, the samples are diluted to bring method performance within acceptable limits.

For low solids (aqueous samples), extract, concentrate, and analyze two sets of four 1-L aliquots (8 aliquots total) of the

precision and recovery standard. For high solids samples, two sets of four 30-g aliquots of the high solids reference matrix are used.

Spike all samples with labeled compounds to assess method performance. Compute percent recovery of the labeled compounds using the internal standard method. Compare the labeled compound recovery for each compound with the corresponding labeled compound recovery.

Reagent water and high solids reference matrix blanks are analyzed to demonstrate freedom from contamination. Extract and concentrate a 1-L reagent water blank or a high solids reference matrix blank with each sample's lot (samples started through the extraction process on the same 8-h shift, to a maximum of 20 samples).

Field replicates may be collected to determine the precision of the sampling technique, and spiked samples may be required to determine the accuracy of the analysis when the internal standard method is used.

REFERENCE Semivolatile Organic Compounds by Isotope Dilution GC/MS. Office of Water Regulation and Standards, U.S. EPA Industrial Technology Division, Washington, DC, EPA Method 1625, Rev. C, June 1989 (contact W.A. Telliard, U.S. EPA, Office of Water Regulations and Standards, 401 M St., SW, Washington, DC, 20460. Phone: 202-382-7131).

1,2-Dinitrobenzene   **EPA Method 8270**
CAS #528-29-0

TITLE Semivolatile Organic Compounds by GC/MS

MATRIX This method is used to determine the concentration of semivolatile organic compounds in extracts prepared from all types of solid waste matrices, soils, and groundwater. Although surface waters are not specifically mentioned, this method should be applicable to water samples from rivers, lakes, etc.

METHOD SUMMARY This method covers 259 semivolatile organic compounds. In very limited applications direct injection of the sample into the GC/MS system may be appropriate, but this results in very high detection limits (approximately 10,000 µg/L). Typically, a 1-L liquid sample, containing surrogate, and matrix spiking standards, is extracted in a continuous extractor first under acid conditions and then under basic conditions. Typically 30 g of a solid sample, containing surrogate, and matrix spiking standards, is extracted ultrasonically. After concentrating the extract to 1 mL it is spiked with 10 µL of an internal standard solution just prior to analysis by GC/MS. The volume injected should contain about 100 ng of base/neutral and 200 ng of acid surrogates (for a 1-µL injection). Analysis is performed by GC/MS using a capillary GC column.

INTERFERENCES Raw GC/MS data from all blanks, samples, and spikes must be evaluated for interferences. Contamination by carryover can occur whenever high-concentration and low-concentration samples are sequentially analyzed. To reduce carryover, the sample syringe must be rinsed out between samples with solvent. Whenever an unusually concentrated sample is encountered, it should be followed by the analysis of blank solvent to check for cross-contamination.

INSTRUMENTATION A GC/MS and a data system are required. The GC column used is a 30 m × 0.25 mm I.D. (or 0.32 mm I.D.) 1um film thickness silicone-coated fused silica capillary column. A continuous liquid-liquid extractor equipped with Teflon® or glass connection joints and stopcocks requiring no lubrication, a K-D concentrating apparatus, water bath, and an ultrasonic disrupter with a minimum power of 300 W and with pulsing capability are also required.

PRECISION & ACCURACY The estimated quantitation limit (EQL) of Method 8270B for determining an individual compound is approximately 1 mg/kg (wet weight) for soil or sediment samples, 1–200 mg/kg for wastes (dependent on matrix and method of preparation), and 10 µg/L for groundwater samples. EQLs will be proportionately higher for sample extracts that require dilution to avoid saturation of the detector.

The EQL(b) for groundwater in µg/L is 40.
The EQL (a, b) for low concentrations in soil and sediment in µg/kg is not determined.
Accuracy as µg/L is not listed.
Overall precision in µg/L is not listed.

(a) *EQLs listed for soil/sediment are based on wet weight. Normally data is reported in a dry-weight basis; therefore, EQLs will be higher based on the % dry weight of each sample. This calculation is based on a 30 g sample and gel permeation chromatography cleanup.*

(b) *Sample EQLs are highly matrix-dependent. The EQLs are provided for guidance and may not always be achievable.*

$C =$ *True value for concentration, in µg/L.*
$X =$ *Average recovery found for measurements of samples containing a concentration of C, in µg/L.*

ESTIMATED QUANTITATION LIMIT

Other Matrices	Factor (a)
High-concentration soil and sludges by sonicator	7.5
Non-water miscible waste	75

(a) EQL for other matrices = [EQL for low soil/sediment] × [Factor]. This estimated EQL is similar to an EPA "Practical Quantitation Limit."

SAMPLING METHOD
Liquid samples — Use a 1 or 2½ gallon amber glass bottle with a screw-top Teflon®-lined cover that has been prewashed with detergent and rinsed with distilled water and methanol (or isopropanol).

Soils, sediments, or sludges — Use an 8-oz. widemouth glass with a screw-top Teflon®-lined cover that has been prewashed with detergent and rinsed with distilled water and methanol (or isopropanol).

SAMPLE PRESERVATION
Liquid samples — If residual chlorine is present, add 3 mL of 10% sodium thiosulfate per gallon, cool to 4°C and store in a solvent-free refrigerator until analysis; if chlorine is not present, then eliminate the sodium thiosulfate addition.

Soils, sediments, or sludges — Cool samples to 4°C and store in a solvent-free refrigerator.

MHT Liquid samples must be extracted within 7 days and the extracts analyzed within 40 days. Soils, sediments, or sludges may be stored for a maximum of 14 days and the extracts analyzed within 40 days.

SAMPLE PREPARATION
Liquid samples — Transfer 1 L quantitatively to a continuous extractor. If high concentrations are anticipated, a smaller volume may be used and then diluted with organic-free reagent water to 1 L. Adjust pH, if necessary, to pH <2 using 1:1 (V/V) sulfuric acid. Pipette 1.0 mL of a surrogate standard spiking solution into each sample. For the sample in each analytical batch selected for spiking, add 1.0 mL of a matrix spiking standard. For base/neutral acid analysis, the amount of the surrogates and matrix spiking compounds added to the sample should result in a final concentration of 100 ng/μL of each analyte in the extract to be analyzed (assuming a 1-μL injection). Extract with methylene chloride for 18–24 h. Next, adjust the pH of the aqueous phase to pH >11 using 10 N sodium hydroxide and extract it with methylene chloride again for 18–24 h. Dry the extract through a column containing anhydrous sodium sulfate and concentrate it to 1 mL using a K-D concentrator.

Soils, sediments, or sludges — Use 30 g of sample. Nonporous or wet samples (gummy or clay type) that do not have a free-flowing sandy texture must be mixed with anhydrous sodium sulfate until the sample is free flowing. Add 1 mL of surrogate standards to all samples, spikes, standards, and blanks. For the sample in each analytical batch selected for spiking, add 1.0 mL of a matrix spiking standard. For base/neutral acid analysis, the amount added of the surrogates and matrix spiking compounds should result in a final concentration of 100 ng/μL of each base/neutral analyte and 200 ng/μL of each acid analyte in the extract to be analyzed (assuming a 1-μL injection). Immediately add a 100-mL mixture of 1:1 methylene chloride:acetone and extract the sample ultrasonically for 3 min and then decant or filter the extracts. Repeat the extraction two or more times. Dry the extract using a column with anhydrous sodium sulfate and concentrate it to 1 mL in a K-D concentrator.

QUALITY CONTROL A methylene chloride solution containing 50 ng/μL of decafluorotriphenylphosphine (DFTPP) is used for tuning the GC/MS system each 12-h shift. A system performance check also must be made during every 12-h shift. A standard containing 50 ng/μL each of 4,4'-DDT, pentachlorophenol, and benzidine is required to verify injection port inertness and GC column performance. A calibration standard at mid-concentration, containing each compound of interest, including all required surrogates, must be performed every 12 h during analysis. After the system performance check is met, calibration check compounds (CCCs) are used to check the validity of the initial calibration.

The internal standard responses and retention times in the calibration check standard must be evaluated immediately after or during data acquisition. If the retention time for any internal standard changes by more than 30 seconds from the last check calibration (12 h), the chromatographic system must be inspected for malfunctions and corrections must be made, as required. If the electron ionization current plot (EICP) area for any of the internal standards changes by a factor of two from the last daily calibration standard check, the mass spectrometer must be inspected for malfunctions and corrections must be made, as appropriate.

Demonstrate, through the analysis of a reagent water blank, that interferences from the analytical system, glassware, and reagents are under control. The blank samples should be carried through all stages of the sample preparation and measurement steps. For each analytical batch (up to 20 samples), a reagent blank, matrix spike, and matrix spike duplicate/duplicate must be analyzed (the frequency of the spikes may be different for different monitoring programs). The blank and spiked samples must be carried through all stages of the sample preparation and measurement steps. A QC reference sample concentrate containing each analyte at a concentration of 100 mg/L in methanol is required.

REFERENCE Test Methods for Evaluating Solid Waste (SW-846). U.S. EPA 1983, Method 8270B, Rev. 2, Nov. 1990. Office of Solid Waste, Washington, DC.

1,3-Dinitrobenzene   **EPA Method 8270**
CAS #99-65-0

TITLE Semivolatile Organic Compounds by GC/MS

MATRIX This method is used to determine the concentration of semivolatile organic compounds in extracts prepared from all types of solid waste matrices, soils, and groundwater. Although surface waters are not specifically mentioned, this method should be applicable to water samples from rivers, lakes, etc.

METHOD SUMMARY This method covers 259 semivolatile organic compounds. In very limited applications direct injection of the sample into the GC/MS system may be appropriate, but this results in very high detection limits (approximately 10,000 μg/L). Typically, a 1-L liquid sample, containing surrogate, and matrix spiking standards, is extracted in a continuous extractor first under acid conditions and then under basic conditions. Typically 30 g of a solid sample, containing surrogate, and matrix spiking standards, is extracted ultrasonically. After concentrating the extract to 1 mL it is spiked with 10 μL of an internal standard solution just prior to analysis by GC/MS. The volume injected should contain about 100 ng of base/neutral and 200 ng of acid surrogates (for a 1-μL injection). Analysis is performed by GC/MS using a capillary GC column.

INTERFERENCES Raw GC/MS data from all blanks, samples, and spikes must be evaluated for interferences. Contamination by carryover can occur whenever high-concentration and low-concentration samples are sequentially analyzed. To reduce carryover, the sample syringe must be rinsed out between samples with solvent. Whenever an unusually concentrated sample is encountered, it should be followed by the analysis of blank solvent to check for cross-contamination.

INSTRUMENTATION A GC/MS and a data system are required. The GC column used is a 30 m × 0.25 mm I.D. (or 0.32 mm I.D.) 1um film thickness silicone-coated fused silica capillary column. A continuous liquid-liquid extractor equipped with Teflon® or glass connection joints and stopcocks requiring no lubrication, a K-D concentrating apparatus, water bath, and an ultrasonic disrupter with a minimum power of 300 W and with pulsing capability are also required.

PRECISION & ACCURACY The estimated quantitation limit (EQL) of Method 8270B for determining an individual compound is approximately 1 mg/kg (wet weight) for soil or sediment samples, 1–200 mg/kg for wastes (dependent on matrix and method of preparation), and 10 µg/L for groundwater samples. EQLs will be proportionately higher for sample extracts that require dilution to avoid saturation of the detector.

The EQL(b) for groundwater in µg/L is 20.
The EQL (a, b) for low concentrations in soil and sediment in µg/kg is not determined.
Accuracy as µg/L is not listed.
Overall precision in µg/L is not listed.

(a) EQLs listed for soil/sediment are based on wet weight. Normally data is reported in a dry-weight basis; therefore, EQLs will be higher based on the % dry weight of each sample. This calculation is based on a 30 g sample and gel permeation chromatography cleanup.
(b) Sample EQLs are highly matrix-dependent. The EQLs are provided for guidance and may not always be achievable.
C = *True value for concentration, in µg/L.*
X = *Average recovery found for measurements of samples containing a concentration of C, in µg/L.*

ESTIMATED QUANTITATION LIMIT

Other Matrices	Factor (a)
High-concentration soil and sludges by sonicator	7.5
Non-water miscible waste	75

(a) EQL for other matrices = [EQL for low soil/sediment] × [Factor]. This estimated EQL is similar to an EPA "Practical Quantitation Limit."

SAMPLING METHOD

Liquid samples — Use a 1 or 2½ gallon amber glass bottle with a screw-top Teflon®-lined cover that has been prewashed with detergent and rinsed with distilled water and methanol (or isopropanol).

Soils, sediments, or sludges — Use an 8-oz. widemouth glass with a screw-top Teflon®-lined cover that has been prewashed with detergent and rinsed with distilled water and methanol (or isopropanol).

SAMPLE PRESERVATION

Liquid samples — If residual chlorine is present, add 3 mL of 10% sodium thiosulfate per gallon, cool to 4°C and store in a solvent-free refrigerator until analysis; if chlorine is not present, then eliminate the sodium thiosulfate addition.

Soils, sediments, or sludges — Cool samples to 4°C and store in a solvent-free refrigerator.

MHT Liquid samples must be extracted within 7 days and the extracts analyzed within 40 days. Soils, sediments, or sludges may be stored for a maximum of 14 days and the extracts analyzed within 40 days.

SAMPLE PREPARATION

Liquid samples — Transfer 1 L quantitatively to a continuous extractor. If high concentrations are anticipated, a smaller volume may be used and then diluted with organic-free reagent water to 1 L. Adjust pH, if necessary, to pH <2 using 1:1 (V/V) sulfuric acid. Pipette 1.0 mL of a surrogate standard spiking solution into each sample. For the sample in each analytical batch selected for spiking, add 1.0 mL of a matrix spiking standard. For base/neutral acid analysis, the amount of the surrogates and matrix spiking compounds added to the sample should result in a final concentration of 100 ng/µL of each analyte in the extract to be analyzed (assuming a 1-µL injection). Extract with methylene chloride for 18–24 h. Next, adjust the pH of the aqueous phase to pH >11 using 10 N sodium hydroxide and extract it with methylene chloride again for 18–24 h. Dry the extract through a column containing anhydrous sodium sulfate and concentrate it to 1 mL using a K-D concentrator.

Soils, sediments, or sludges — Use 30 g of sample. Nonporous or wet samples (gummy or clay type) that do not have a free-flowing sandy texture must be mixed with anhydrous sodium sulfate until the sample is free flowing. Add 1 mL of surrogate standards to all samples, spikes, standards, and blanks. For the sample in each analytical batch selected for spiking, add 1.0 mL of a matrix spiking standard. For base/neutral acid analysis, the amount added of the surrogates and matrix spiking compounds should result in a final concentration of 100 ng/µL of each base/neutral analyte and 200 ng/µL of each acid analyte in the extract to be analyzed (assuming a 1-µL injection). Immediately add a 100-mL mixture of 1:1 methylene chloride:acetone and extract the sample ultrasonically for 3 min and then decant or filter the extracts. Repeat the extraction two or more times. Dry the extract using a column with anhydrous sodium sulfate and concentrate it to 1 mL in a K-D concentrator.

QUALITY CONTROL A methylene chloride solution containing 50 ng/µL of decafluorotriphenylphosphine (DFTPP) is used for tuning the GC/MS system each 12-h shift. A system performance check also must be made during every 12-h shift. A standard containing 50 ng/µL each of 4,4'-DDT, pentachlorophenol, and benzidine is required to verify injection port inertness and GC column performance. A calibration standard at mid-concentration, containing each compound of interest, including all required surrogates, must be performed every 12 h during analysis. After the system performance check is met, calibration check compounds (CCCs) are used to check the validity of the initial calibration.

The internal standard responses and retention times in the calibration check standard must be evaluated immediately after or during data acquisition. If the retention time for any internal standard changes by more than 30 seconds from the last check calibration (12 h), the chromatographic system must be inspected for malfunctions and corrections must be made, as required. If the electron ionization current plot (EICP) area for

any of the internal standards changes by a factor of two from the last daily calibration standard check, the mass spectrometer must be inspected for malfunctions and corrections must be made, as appropriate.

Demonstrate, through the analysis of a reagent water blank, that interferences from the analytical system, glassware, and reagents are under control. The blank samples should be carried through all stages of the sample preparation and measurement steps. For each analytical batch (up to 20 samples), a reagent blank, matrix spike, and matrix spike duplicate/duplicate must be analyzed (the frequency of the spikes may be different for different monitoring programs). The blank and spiked samples must be carried through all stages of the sample preparation and measurement steps. A QC reference sample concentrate containing each analyte at a concentration of 100 mg/L in methanol is required.

REFERENCE Test Methods for Evaluating Solid Waste (SW-846). U.S. EPA 1983, Method 8270B, Rev. 2, Nov. 1990. Office of Solid Waste, Washington, DC.

1,4-Dinitrobenzene   EPA Method 8270
CAS #100-25-4

TITLE Semivolatile Organic Compounds by GC/MS

MATRIX This method is used to determine the concentration of semivolatile organic compounds in extracts prepared from all types of solid waste matrices, soils, and groundwater. Although surface waters are not specifically mentioned, this method should be applicable to water samples from rivers, lakes, etc.

METHOD SUMMARY This method covers 259 semivolatile organic compounds. In very limited applications direct injection of the sample into the GC/MS system may be appropriate, but this results in very high detection limits (approximately 10,000 µg/L). Typically, a 1-L liquid sample, containing surrogate, and matrix spiking standards, is extracted in a continuous extractor first under acid conditions and then under basic conditions. Typically 30 g of a solid sample, containing surrogate, and matrix spiking standards, is extracted ultrasonically. After concentrating the extract to 1 mL it is spiked with 10 µL of an internal standard solution just prior to analysis by GC/MS. The volume injected should contain about 100 ng of base/neutral and 200 ng of acid surrogates (for a 1-µL injection). Analysis is performed by GC/MS using a capillary GC column.

INTERFERENCES Raw GC/MS data from all blanks, samples, and spikes must be evaluated for interferences. Contamination by carryover can occur whenever high-concentration and low-concentration samples are sequentially analyzed. To reduce carryover, the sample syringe must be rinsed out between samples with solvent. Whenever an unusually concentrated sample is encountered, it should be followed by the analysis of blank solvent to check for cross-contamination.

INSTRUMENTATION A GC/MS and a data system are required. The GC column used is a 30 m × 0.25 mm I.D. (or 0.32 mm I.D.) 1um film thickness silicone-coated fused silica capillary column. A continuous liquid-liquid extractor equipped with Teflon® or glass connection joints and stopcocks requiring no lubrication, a K-D concentrating apparatus, water bath, and an ultrasonic disrupter with a minimum power of 300 W and with pulsing capability are also required.

PRECISION & ACCURACY The estimated quantitation limit (EQL) of Method 8270B for determining an individual compound is approximately 1 mg/kg (wet weight) for soil or sediment samples, 1–200 mg/kg for wastes (dependent on matrix and method of preparation), and 10 µg/L for groundwater samples. EQLs will be proportionately higher for sample extracts that require dilution to avoid saturation of the detector.

The EQL(b) for groundwater in µg/L is 40.
The EQL (a, b) for low concentrations in soil and sediment in µg/kg is not determined.
Accuracy as µg/L is not listed.
Overall precision in µg/L is not listed.

(a) *EQLs listed for soil/sediment are based on wet weight. Normally data is reported in a dry-weight basis; therefore, EQLs will be higher based on the % dry weight of each sample. This calculation is based on a 30 g sample and gel permeation chromatography cleanup.*
(b) *Sample EQLs are highly matrix-dependent. The EQLs are provided for guidance and may not always be achievable.*
C = *True value for concentration, in µg/L.*
X = *Average recovery found for measurements of samples containing a concentration of C, in µg/L.*

ESTIMATED QUANTITATION LIMIT

Other Matrices	Factor (a)
High-concentration soil and sludges by sonicator	7.5
Non-water miscible waste	75

(a) *EQL for other matrices = [EQL for low soil/sediment] × [Factor]. This estimated EQL is similar to an EPA "Practical Quantitation Limit."*

SAMPLING METHOD

Liquid samples — Use a 1 or 2½ gallon amber glass bottle with a screw-top Teflon®-lined cover that has been prewashed with detergent and rinsed with distilled water and methanol (or isopropanol).

Soils, sediments, or sludges — Use an 8-oz. widemouth glass with a screw-top Teflon®-lined cover that has been prewashed with detergent and rinsed with distilled water and methanol (or isopropanol).

SAMPLE PRESERVATION

Liquid samples — If residual chlorine is present, add 3 mL of 10% sodium thiosulfate per gallon, cool to 4°C and store in a solvent-free refrigerator until analysis; if chlorine is not present, then eliminate the sodium thiosulfate addition.

Soils, sediments, or sludges — Cool samples to 4°C and store in a solvent-free refrigerator.

MHT Liquid samples must be extracted within 7 days and the extracts analyzed within 40 days. Soils, sediments, or sludges may

be stored for a maximum of 14 days and the extracts analyzed within 40 days.

SAMPLE PREPARATION
Liquid samples — Transfer 1 L quantitatively to a continuous extractor. If high concentrations are anticipated, a smaller volume may be used and then diluted with organic-free reagent water to 1 L. Adjust pH, if necessary, to pH <2 using 1:1 (V/V) sulfuric acid. Pipette 1.0 mL of a surrogate standard spiking solution into each sample. For the sample in each analytical batch selected for spiking, add 1.0 mL of a matrix spiking standard. For base/neutral acid analysis, the amount of the surrogates and matrix spiking compounds added to the sample should result in a final concentration of 100 ng/µL of each analyte in the extract to be analyzed (assuming a 1-µL injection). Extract with methylene chloride for 18–24 h. Next, adjust the pH of the aqueous phase to pH >11 using 10 N sodium hydroxide and extract it with methylene chloride again for 18–24 h. Dry the extract through a column containing anhydrous sodium sulfate and concentrate it to 1 mL using a K-D concentrator.

Soils, sediments, or sludges — Use 30 g of sample. Nonporous or wet samples (gummy or clay type) that do not have a free-flowing sandy texture must be mixed with anhydrous sodium sulfate until the sample is free flowing. Add 1 mL of surrogate standards to all samples, spikes, standards, and blanks. For the sample in each analytical batch selected for spiking, add 1.0 mL of a matrix spiking standard. For base/neutral acid analysis, the amount added of the surrogates and matrix spiking compounds should result in a final concentration of 100 ng/µL of each base/neutral analyte and 200 ng/µL of each acid analyte in the extract to be analyzed (assuming a 1-µL injection). Immediately add a 100-mL mixture of 1:1 methylene chloride:acetone and extract the sample ultrasonically for 3 min and then decant or filter the extracts. Repeat the extraction two or more times. Dry the extract using a column with anhydrous sodium sulfate and concentrate it to 1 mL in a K-D concentrator.

QUALITY CONTROL A methylene chloride solution containing 50 ng/µL of decafluorotriphenylphosphine (DFTPP) is used for tuning the GC/MS system each 12-h shift. A system performance check also must be made during every 12-h shift. A standard containing 50 ng/µL each of 4,4'-DDT, pentachlorophenol, and benzidine is required to verify injection port inertness and GC column performance. A calibration standard at mid-concentration, containing each compound of interest, including all required surrogates, must be performed every 12 h during analysis. After the system performance check is met, calibration check compounds (CCCs) are used to check the validity of the initial calibration.

The internal standard responses and retention times in the calibration check standard must be evaluated immediately after or during data acquisition. If the retention time for any internal standard changes by more than 30 seconds from the last check calibration (12 h), the chromatographic system must be inspected for malfunctions and corrections must be made, as required. If the electron ionization current plot (EICP) area for any of the internal standards changes by a factor of two from the last daily calibration standard check, the mass spectrometer must be inspected for malfunctions and corrections must be made, as appropriate.

Demonstrate, through the analysis of a reagent water blank, that interferences from the analytical system, glassware, and reagents are under control. The blank samples should be carried through all stages of the sample preparation and measurement steps. For each analytical batch (up to 20 samples), a reagent blank, matrix spike, and matrix spike duplicate/duplicate must be analyzed (the frequency of the spikes may be different for different monitoring programs). The blank and spiked samples must be carried through all stages of the sample preparation and measurement steps. A QC reference sample concentrate containing each analyte at a concentration of 100 mg/L in methanol is required.

REFERENCE Test Methods for Evaluating Solid Waste (SW-846). U.S. EPA 1983, Method 8270B, Rev. 2, Nov. 1990. Office of Solid Waste, Washington, DC.

2,4-Dinitrophenol   **EPA Method 1625**
CAS #51-28-5

TITLE Semivolatile Organic Compounds by Isotope Dilution GC/MS

MATRIX The compounds may be determined in waters, soils, and municipal sludges by this method.

METHOD SUMMARY This method is used to determine 176 semivolatile toxic organic pollutants associated with the CWA (as amended 1987); the RCRA (as amended 1986); the CERCLA (as amended 1986); and other compounds amenable to extraction and analysis by capillary column gas chromatography-mass spectrometry (GC/MS).

Stable isotopically-labeled analogs of the compounds of interest are added to the sample. If the solids content is less than 1%, a 1-L sample is extracted at pH 12–13, then at pH <2 with methylene chloride using continuous extraction techniques.

If the solids content is 30% or less, the sample is diluted to 1% solids with reagent water, homogenized ultrasonically, and extracted at pH 12–13, then at pH <2 with methylene chloride using continuous extraction techniques. If the solids content is greater than 30%, the sample is extracted using ultrasonic techniques.

Each extract is dried over sodium sulfate, concentrated to a volume of 5 mL, cleaned up using GPC, if necessary, and concentrated. Extracts are concentrated to 1 mL if GPC is not performed, and to 0.5 mL if GPC is performed.

An internal standard is added to the extract, and a 1-mL aliquot of the extract is injected into the GC. The compounds are separated by GC and detected by a MS. The labeled compounds serve to correct the variability of the analytical technique.

INTERFERENCES Solvents, reagents, glassware, and other sample processing hardware may yield artifacts and/or elevated baselines causing misinterpretation of chromatograms and spectra. Materials used in the analysis must be demonstrated

to be free from interferences under the conditions of analysis by running method blanks initially and with each sample lot (sample started through the extraction process on a given 8-h shift, to a maximum of 20). Specific selection of reagents and purification of solvents by distillation in all glass systems may be required. Glassware and, where possible, reagents are cleaned by solvent rinse and baking at 450°C for 1-h minimum. Interferences coextracted from samples will vary considerably from source to source, depending on the diversity of the site being sampled.

INSTRUMENTATION Major instrumentation includes a GC with a splitless or on-column injection port for capillary column, a MS with 70 eV electron impact ionization, and a data system to collect and record MS data, and process it. A K-D apparatus is used to concentrate extracts.

GC Column: 30 m × 0.25 mm I.D. 5% phenyl, 94% methyl, 1% vinyl silicone bonded phased fused silica capillary column.

PRECISION & ACCURACY The detection limits of the method are usually dependent on the level of interferences rather than instrumental limitations. The limits typify the minimum quantities that can be detected with no interferences present.

The minimum level (in µg/mL) was 50. This is defined as a minimum level at which the analytical system shall give recognizable mass spectra (background corrected) and acceptable calibration points.

The MDL (in µg/kg) in low solids was 565 and in high solids was 642; these were determined in digested sludge (low solids) and in filter cake or compost (high solids).

The labeled and native compound initial precision as standard deviation (in µg/L) was 18.
The labeled and native compound initial accuracy as average recovery (in µg/L) was 72–134.

SAMPLE COLLECTION, PRESERVATION & HANDLING
Collect samples in glass containers. Aqueous samples which flow freely are collected in refrigerated bottles using automatic sampling equipment. Solid samples are collected as grab samples using widemouth jars. Maintain samples at 0 to 4°C from the time of collection until extraction. If residual chlorine is present in aqueous samples, add 80 mg sodium thiosulfate/L of water. Begin sample extraction within 7 days of collection, and analyze all extracts within 40 days of extraction.

SAMPLE PREPARATION Samples containing 1% solids or less are extracted directly using continuous liquid-liquid extraction techniques. Samples containing 1 to 30% solids are diluted to the 1% level with reagent water and extracted using continuous liquid-liquid extraction techniques. Samples containing greater than 30% solids are extracted using ultrasonic techniques.

Base/neutral extraction — Adjust the pH of the waters in the extractors to 12–13 with 6 N NaOH. Extract with methylene chloride for 24–48 h.
Acid extraction — Adjust the pH of the waters in the extractors to 2 or less using 6 N sulfuric acid. Extract with methylene chloride for 24–48 h.
Ultrasonic extraction of high solids samples — Add anhydrous sodium sulfate to the sample and QC aliquot(s). Add acetone:methylene chloride (1:1) to the sample and mix thoroughly

Concentrate extracts using a K-D apparatus.

QUALITY CONTROL The analyst is permitted to modify this method to improve separations or lower the costs of measurements, provided all performance specifications are met. Analyses of blanks are required to demonstrate freedom from contamination. When results of spikes indicate atypical method performance for samples, the samples are diluted to bring method performance within acceptable limits.

For low solids (aqueous samples), extract, concentrate, and analyze two sets of four 1-L aliquots (8 aliquots total) of the precision and recovery standard. For high solids samples, two sets of four 30-g aliquots of the high solids reference matrix are used.

Spike all samples with labeled compounds to assess method performance. Compute percent recovery of the labeled compounds using the internal standard method. Compare the labeled compound recovery for each compound with the corresponding labeled compound recovery.

Reagent water and high solids reference matrix blanks are analyzed to demonstrate freedom from contamination. Extract and concentrate a 1-L reagent water blank or a high solids reference matrix blank with each sample's lot (samples started through the extraction process on the same 8-h shift, to a maximum of 20 samples).

Field replicates may be collected to determine the precision of the sampling technique, and spiked samples may be required to determine the accuracy of the analysis when the internal standard method is used.

REFERENCE Semivolatile Organic Compounds by Isotope Dilution GC/MS. Office of Water Regulation and Standards, U.S. EPA Industrial Technology Division, Washington, DC, EPA Method 1625, Rev. C, June 1989 (contact W.A. Telliard, U.S. EPA, Office of Water Regulations and Standards, 401 M St., SW, Washington, DC, 20460. Phone: 202-382-7131).

2,4-Dinitrophenol   EPA Method 625
CAS #51-28-5

TITLE Base/Neutrals and Acids, U.S. EPA Method 625

MATRIX This methods covers municipal and industrial wastewaters.

METHOD SUMMARY Approximately 1 L of sample is serially extracted with methylene chloride at a pH greater than 11 and again at a pH less than 2 using a separatory funnel or a continuous extractor. The methylene chloride extract is dried, concentrated to a volume of 1 mL, and analyzed by GC/MS. Qualitative identification of the parameters in the extract is performed using the retention time and the relative abundance of three characteristic masses (m/z). Qualitative analysis is performed using either external or internal standard techniques with a single characteristic m/z.

INTERFERENCES Method interferences may be caused by contaminants in solvents, reagents, glassware, and other sample processing hardware. Glassware must be scrupulously cleaned. Glassware should be heated in a muffle furnace at 400°C for 5 to 30 min. Some thermally stable materials, such as PCBs, may not be eliminated by this treatment. Solvent rinses with acetone and pesticide quality hexane may be substituted for the muffle furnace heating. Matrix interferences may be caused by contaminants that are coextracted from the sample. The base-neutral extraction may cause significantly reduced recovery of phenols. The packed gas chromatographic columns recommended for the basic fraction may not exhibit sufficient resolution for some analytes.

INSTRUMENTATION A GC/MS system with an injection port designed for on-column injection when using packed columns and for splitless injection when using capillary columns.

Column for base/neutrals: 1.8 m long × 2 mm I.D. glass, packed with 3% SP-2550 on Supelcoport (100/120 mesh) or equivalent.

Column for acids: 1.8 m long × 2 mm I.D. glass, packed with 1% SP-1240DA on Supelcoport (100/120 mesh) or equivalent.

PRECISION & ACCURACY The MDL concentrations were obtained using reagent water. The MDL actually achieved in a given analysis will vary depending on instrument sensitivity and matrix effects. This method was tested by 15 laboratories using reagent water, drinking water, surface water, and industrial wastewaters spiked at six concentrations over the range 5 to 100 µg/L. Single operator precision, overall precision, and method accuracy were found to be directly related to the concentration of the parameter matrix.

The MDL (in µg/L) in reagent water was 42.

The standard deviation (in µg/L based on 4 recovery measurements) was 49.8.

The range (in µg/L) for average recovery for 4 measurements was D-172.9.

The range (in %) for percent recovery was D-191.

Accuracy (in µg/L) as expected recovery for one or more measurements of a sample containing a true concentration of C was $0.61C-18.04$.

Precision (in µg/L) as expected single analyst standard deviation of measurements at an average concentration found at X was $0.38X+2.36$.

Overall precision (in µg/L) as expected interlaboratory standard deviation of measurements in an average concentration found at X was $0.42X+26.29$.

$C = $ True value of the concentration in µg/L.
$X = $ Average recovery found for measurements of samples containing a concentration at C in µg/L.

SAMPLE PREPARATION Adjust the pH to >11 with sodium hydroxide and serially extract in a separatory funnel with methylene chloride or else in a continuous extractor. Next, adjust the pH to <2 with sulfuric acid and serially extract in a separatory funnel with methylene chloride or else in a continuous extractor. Dry the extracts separately through a column of anhydrous sodium sulfate and then concentrate each of the extracts to 1.0 mL using a K-D apparatus.

SAMPLE COLLECTION, PRESERVATION & HANDLING Grab samples must be collected in glass containers. All samples must be refrigerated at 4°C from the time of collection until extraction. If residual chlorine is present, add 80 mg of sodium thiosulfate/L of sample and mix well. All samples must be extracted within 7 days of collection and completely analyzed within 40 days of extraction.

QUALITY CONTROL Make an initial, one-time, demonstration of the ability to generate acceptable accuracy and precision with this method. Before processing any samples, the analyst must analyze a reagent water blank to demonstrate that interferences from the analytical system and glassware are under control. Each time a set of samples is extracted or reagents are changed, a reagent water blank must be processed. Spike and analyze a minimum of 5% of all samples to monitor and evaluate lab data quality. A QC check sample concentrate that contains each parameter of interest at a concentration of 100 µg/mL in acetone is required. PCBs and multicomponent pesticides may be omitted from this test.

After analysis of five spiked wastewater samples, calculate the average percent recovery and the standard deviation of the percent recovery. Spike all samples with the surrogate standard spiking solution and calculate the percent recovery of each surrogate compound.

REFERENCE Federal Register, Vol. 49, No. 209. Friday, Oct. 26, 1984.

2,4-Dinitrophenol   **EPA Method 8270**
CAS #51-28-5

TITLE Semivolatile Organic Compounds by GC/MS

MATRIX This method is used to determine the concentration of semivolatile organic compounds in extracts prepared from all types of solid waste matrices, soils, and groundwater. Although surface waters are not specifically mentioned, this method should be applicable to water samples from rivers, lakes, etc.

METHOD SUMMARY This method covers 259 semivolatile organic compounds. In very limited applications direct injection of the sample into the GC/MS system may be appropriate, but this results in very high detection limits (approximately 10,000 µg/L). Typically, a 1-L liquid sample, containing surrogate, and matrix spiking standards, is extracted in a continuous extractor first under acid conditions and then under basic conditions. Typically 30 g of a solid sample, containing surrogate, and matrix spiking standards, is extracted ultrasonically. After concentrating the extract to 1 mL it is spiked with 10 µL of an internal standard solution just prior to analysis by GC/MS. The volume injected should contain about 100 ng of base/neutral and 200 ng of acid surrogates (for a 1-µL injection). Analysis is performed by GC/MS using a capillary GC column.

INTERFERENCES Raw GC/MS data from all blanks, samples, and spikes must be evaluated for interferences. Contamination by carryover can occur whenever high-concentration and low-concentration samples are sequentially analyzed. To reduce carryover, the sample syringe must be rinsed out between samples with solvent. Whenever an unusually concentrated sample is encountered, it should be followed by the analysis of blank solvent to check for cross-contamination.

INSTRUMENTATION A GC/MS and a data system are required. The GC column used is a 30 m × 0.25 mm I.D. (or 0.32 mm I.D.) 1um film thickness silicone-coated fused silica capillary column. A continuous liquid-liquid extractor equipped with Teflon® or glass connection joints and stopcocks requiring no lubrication, a K-D concentrating apparatus, water bath, and an ultrasonic disrupter with a minimum power of 300 W and with pulsing capability are also required.

PRECISION & ACCURACY The estimated quantitation limit (EQL) of Method 8270B for determining an individual compound is approximately 1 mg/kg (wet weight) for soil or sediment samples, 1–200 mg/kg for wastes (dependent on matrix and method of preparation), and 10 µg/L for groundwater samples. EQLs will be proportionately higher for sample extracts that require dilution to avoid saturation of the detector.

The EQL(b) for groundwater in µg/L is 50.
The EQL (a, b) for low concentrations in soil and sediment in µg/kg is 3300.
Accuracy as µg/L is 0.81C–18.04.
Overall precision in µg/L is 0.42X +26.29.

(a) EQLs listed for soil/sediment are based on wet weight. Normally data is reported in a dry-weight basis; therefore, EQLs will be higher based on the % dry weight of each sample. This calculation is based on a 30 g sample and gel permeation chromatography cleanup.
(b) Sample EQLs are highly matrix-dependent. The EQLs are provided for guidance and may not always be achievable.
C = *True value for concentration, in µg/L.*
X = *Average recovery found for measurements of samples containing a concentration of C, in µg/L.*

ESTIMATED QUANTITATION LIMIT

Other Matrices	Factor (a)
High-concentration soil and sludges by sonicator	7.5
Non-water miscible waste	75

(a) EQL for other matrices = [EQL for low soil/sediment] × [Factor]. This estimated EQL is similar to an EPA "Practical Quantitation Limit."

SAMPLING METHOD
Liquid samples — Use a 1 or 2½ gallon amber glass bottle with a screw-top Teflon®-lined cover that has been prewashed with detergent and rinsed with distilled water and methanol (or isopropanol).

Soils, sediments, or sludges — Use an 8-oz. widemouth glass with a screw-top Teflon®-lined cover that has been prewashed with detergent and rinsed with distilled water and methanol (or isopropanol).

SAMPLE PRESERVATION
Liquid samples — If residual chlorine is present, add 3 mL of 10% sodium thiosulfate per gallon, cool to 4°C and store in a solvent-free refrigerator until analysis; if chlorine is not present, then eliminate the sodium thiosulfate addition.

Soils, sediments, or sludges — Cool samples to 4°C and store in a solvent-free refrigerator.

MHT Liquid samples must be extracted within 7 days and the extracts analyzed within 40 days. Soils, sediments, or sludges may be stored for a maximum of 14 days and the extracts analyzed within 40 days.

SAMPLE PREPARATION
Liquid samples — Transfer 1 L quantitatively to a continuous extractor. If high concentrations are anticipated, a smaller volume may be used and then diluted with organic-free reagent water to 1 L. Adjust pH, if necessary, to pH <2 using 1:1 (V/V) sulfuric acid. Pipette 1.0 mL of a surrogate standard spiking solution into each sample. For the sample in each analytical batch selected for spiking, add 1.0 mL of a matrix spiking standard. For base/neutral acid analysis, the amount of the surrogates and matrix spiking compounds added to the sample should result in a final concentration of 100 ng/µL of each analyte in the extract to be analyzed (assuming a 1-µL injection). Extract with methylene chloride for 18–24 h. Next, adjust the pH of the aqueous phase to pH >11 using 10 N sodium hydroxide and extract it with methylene chloride again for 18–24 h. Dry the extract through a column containing anhydrous sodium sulfate and concentrate it to 1 mL using a K-D concentrator.

Soils, sediments, or sludges — Use 30 g of sample. Nonporous or wet samples (gummy or clay type) that do not have a free-flowing sandy texture must be mixed with anhydrous sodium sulfate until the sample is free flowing. Add 1 mL of surrogate standards to all samples, spikes, standards, and blanks. For the sample in each analytical batch selected for spiking, add 1.0 mL of a matrix spiking standard. For base/neutral acid analysis, the amount added of the surrogates and matrix spiking compounds should result in a final concentration of 100 ng/µL of each base/neutral analyte and 200 ng/µL of each acid analyte in the extract to be analyzed (assuming a 1-µL injection). Immediately add a 100-mL mixture of 1:1 methylene chloride:acetone and extract the sample ultrasonically for 3 min and then decant or filter the extracts. Repeat the extraction two or more times. Dry the extract using a column with anhydrous sodium sulfate and concentrate it to 1 mL in a K-D concentrator.

Note: N-nitrosodiphenylamine decomposes in the gas chromatographic inlet and cannot be separated from diphenylamine.

QUALITY CONTROL A methylene chloride solution containing 50 ng/µL of decafluorotriphenylphosphine (DFTPP) is used for tuning the GC/MS system each 12-h shift. A system performance check also must be made during every 12-h shift. A standard containing 50 ng/µL each of 4,4'-DDT, pentachlorophenol, and benzidine is required to verify injection port inertness and GC column performance. A calibration standard at mid-concentration, containing each compound of interest,

including all required surrogates, must be performed every 12 h during analysis. After the system performance check is met, calibration check compounds (CCCs) are used to check the validity of the initial calibration.

The internal standard responses and retention times in the calibration check standard must be evaluated immediately after or during data acquisition. If the retention time for any internal standard changes by more than 30 seconds from the last check calibration (12 h), the chromatographic system must be inspected for malfunctions and corrections must be made, as required. If the electron ionization current plot (EICP) area for any of the internal standards changes by a factor of two from the last daily calibration standard check, the mass spectrometer must be inspected for malfunctions and corrections must be made, as appropriate.

Demonstrate, through the analysis of a reagent water blank, that interferences from the analytical system, glassware, and reagents are under control. The blank samples should be carried through all stages of the sample preparation and measurement steps. For each analytical batch (up to 20 samples), a reagent blank, matrix spike, and matrix spike duplicate/duplicate must be analyzed (the frequency of the spikes may be different for different monitoring programs). The blank and spiked samples must be carried through all stages of the sample preparation and measurement steps. A QC reference sample concentrate containing each analyte at a concentration of 100 mg/L in methanol is required.

REFERENCE Test Methods for Evaluating Solid Waste (SW-846). U.S. EPA 1983, Method 8270B, Rev. 2, Nov. 1990. Office of Solid Waste, Washington, DC.

2,4-Dinitrophenol   EPA Method 8040
CAS #51-28-5

TITLE Phenols

MATRIX Groundwater, soils, sludges, water miscible liquid wastes, and non-water miscible wastes.

APPLICATION This method is used for the analysis of 17 phenols. Samples are extracted, concentrated, and analyzed using direct injection of both neat and diluted organic liquids. Pentafluorobenzylbromide (PFB) derivatives also may be made to increase sensitivity of the method.

INTERFERENCES There can be carryover contamination with high- and low-level samples. Solvents, reagents, and glassware may introduce artifacts. Other interferences may come from coextracted compounds from samples.

INSTRUMENTATION GC capable of on-column injections and a flame with detector (FID) or electron capture detector (ECD). Column for underivatized phenol: 1.8 m by 2.0 mm with 1% SP-1240DA on Supelcoport. Column for derivatized phenols: 1.8 m by 2.0 mm with 5% OV-17 on Chromosorb W-AW-DMCS.

RANGE 12 to 450 µg/L.

MDL 13.0 µg/L (FID).

PQL FACTORS FOR MULTIPLYING × FID MDL VALUE

Matrix	Multiplication Factor
Groundwater	10
Low-level soil by sonication with GPC cleanup	670
High-level soil and sludge by sonication	10,000
Non-water miscible waste	100,000

PRECISION 0.29X + 4.51 µg/L (overall precision using FID).

ACCURACY 0.80C–1.58 µg/L (as recovery using FID).

SAMPLING METHOD Use 8-oz. widemouth glass bottles with Teflon®-lined caps for concentrated waste samples, soils, sediments, and sludges. Use 1 or 2½ gallon amber glass bottles with Teflon®-lined caps for liquid (water) samples.

STABILITY Cool soil, sediment, sludge, and liquid samples to 4°C. If residual chlorine is present in liquid samples add 3 mL of 10% sodium thiosulfate per gallon of sample and cool to 4°C.

MHT 14 days for concentrated waste, soil, sediment, or sludge; 7 days for liquid samples; all extracts must be analyzed within 40 days.

QUALITY CONTROL A quality control check sample concentrate containing each analyte of interest is required. The QC check sample concentrate may be prepared from pure standard materials or purchased as certified solutions Use appropriate trip, matrix, control site, method, reagent, and solvent blanks. Internal, surrogate, and five concentration level calibration standards are used. The QC check sample concentrate should contain this compound at 100 µg/mL in 2-propanol.

REFERENCE Test Methods for Evaluating Solid Waste (SW-846), U.S. EPA Office of Solid Waste, Washington, DC, Method 8040A, Rev. 1, Nov. 1990.

2,4-Dinitrotoluene   EPA Method 1625
CAS #121-14-2

TITLE Semivolatile Organic Compounds by Isotope Dilution GC/MS

MATRIX The compounds may be determined in waters, soils, and municipal sludges by this method.

METHOD SUMMARY This method is used to determine 176 semivolatile toxic organic pollutants associated with the CWA (as amended 1987); the RCRA (as amended 1986); the CERCLA (as amended 1986); and other compounds amenable to extraction and analysis by capillary column gas chromatography-mass spectrometry (GC/MS).

Stable isotopically-labeled analogs of the compounds of interest are added to the sample. If the solids content is less than 1%, a 1-L sample is extracted at pH 12–13, then at pH <2 with methylene chloride using continuous extraction techniques.

If the solids content is 30% or less, the sample is diluted to 1% solids with reagent water, homogenized ultrasonically, and extracted at pH 12–13, then at pH <2 with methylene chloride using continuous extraction techniques. If the solids content is greater than 30%, the sample is extracted using ultrasonic techniques.

Each extract is dried over sodium sulfate, concentrated to a volume of 5 mL, cleaned up using GPC, if necessary, and concentrated. Extracts are concentrated to 1 mL if GPC is not performed, and to 0.5 mL if GPC is performed.

An internal standard is added to the extract, and a 1-mL aliquot of the extract is injected into the GC. The compounds are separated by GC and detected by a MS. The labeled compounds serve to correct the variability of the analytical technique.

INTERFERENCES Solvents, reagents, glassware, and other sample processing hardware may yield artifacts and/or elevated baselines causing misinterpretation of chromatograms and spectra. Materials used in the analysis must be demonstrated to be free from interferences under the conditions of analysis by running method blanks initially and with each sample lot (sample started through the extraction process on a given 8-h shift, to a maximum of 20). Specific selection of reagents and purification of solvents by distillation in all glass systems may be required. Glassware and, where possible, reagents are cleaned by solvent rinse and baking at 450°C for 1-h minimum. Interferences coextracted from samples will vary considerably from source to source, depending on the diversity of the site being sampled.

INSTRUMENTATION Major instrumentation includes a GC with a splitless or on-column injection port for capillary column, a MS with 70 eV electron impact ionization, and a data system to collect and record MS data, and process it. A K-D apparatus is used to concentrate extracts.

GC Column: 30 m × 0.25 mm I.D. 5% phenyl, 94% methyl, 1% vinyl silicone bonded phased fused silica capillary column.

PRECISION & ACCURACY The detection limits of the method are usually dependent on the level of interferences rather than instrumental limitations. The limits typify the minimum quantities that can be detected with no interferences present.

The minimum level (in µg/mL) was 10. This is defined as a minimum level at which the analytical system shall give recognizable mass spectra (background corrected) and acceptable calibration points.

The MDL (in µg/kg) in low solids was 65 and in high solids was 209; these were determined in digested sludge (low solids) and in filter cake or compost (high solids).

Note: Background levels of this compound were present in the sludge tested, resulting in higher than expected MDLs. The MDL for this compound is expected to be approximately 50 µg/kg with no interferences present.

The labeled and native compound initial precision as standard deviation (in µg/L) was 18.
The labeled and native compound initial accuracy as average recovery (in µg/L) was 75–158.

SAMPLE COLLECTION, PRESERVATION & HANDLING
Collect samples in glass containers. Aqueous samples which flow freely are collected in refrigerated bottles using automatic sampling equipment. Solid samples are collected as grab samples using widemouth jars. Maintain samples at 0 to 4°C from the time of collection until extraction. If residual chlorine is present in aqueous samples, add 80 mg sodium thiosulfate/L of water. Begin sample extraction within 7 days of collection, and analyze all extracts within 40 days of extraction.

SAMPLE PREPARATION Samples containing 1% solids or less are extracted directly using continuous liquid-liquid extraction techniques. Samples containing 1 to 30% solids are diluted to the 1% level with reagent water and extracted using continuous liquid-liquid extraction techniques. Samples containing greater than 30% solids are extracted using ultrasonic techniques.

Base/neutral extraction — Adjust the pH of the waters in the extractors to 12–13 with 6 N NaOH. Extract with methylene chloride for 24–48 h.
Acid extraction — Adjust the pH of the waters in the extractors to 2 or less using 6 N sulfuric acid. Extract with methylene chloride for 24–48 h.
Ultrasonic extraction of high solids samples — Add anhydrous sodium sulfate to the sample and QC aliquot(s). Add acetone:methylene chloride (1:1) to the sample and mix thoroughly

Concentrate extracts using a K-D apparatus.

QUALITY CONTROL The analyst is permitted to modify this method to improve separations or lower the costs of measurements, provided all performance specifications are met. Analyses of blanks are required to demonstrate freedom from contamination. When results of spikes indicate atypical method performance for samples, the samples are diluted to bring method performance within acceptable limits.

For low solids (aqueous samples), extract, concentrate, and analyze two sets of four 1-L aliquots (8 aliquots total) of the precision and recovery standard. For high solids samples, two sets of four 30-g aliquots of the high solids reference matrix are used.

Spike all samples with labeled compounds to assess method performance. Compute percent recovery of the labeled compounds using the internal standard method. Compare the labeled compound recovery for each compound with the corresponding labeled compound recovery.

Reagent water and high solids reference matrix blanks are analyzed to demonstrate freedom from contamination. Extract and concentrate a 1-L reagent water blank or a high solids reference matrix blank with each sample's lot (samples started through the extraction process on the same 8-h shift, to a maximum of 20 samples).

Field replicates may be collected to determine the precision of the sampling technique, and spiked samples may be required to determine the accuracy of the analysis when the internal standard method is used.

REFERENCE Semivolatile Organic Compounds by Isotope Dilution GC/MS. Office of Water Regulation and Standards, U.S. EPA Industrial Technology Division, Washington, DC, EPA Method 1625, Rev. C, June 1989 (contact W.A. Telliard, U.S. EPA, Office of Water Regulations and Standards, 401 M St., SW, Washington, DC, 20460. Phone: 202-382-7131).

2,6-Dinitrotoluene   **EPA Method 1625**
CAS #606-20-2

TITLE Semivolatile Organic Compounds by Isotope Dilution GC/MS

MATRIX The compounds may be determined in waters, soils, and municipal sludges by this method.

METHOD SUMMARY This method is used to determine 176 semivolatile toxic organic pollutants associated with the CWA (as amended 1987); the RCRA (as amended 1986); the CERCLA (as amended 1986); and other compounds amenable to extraction and analysis by capillary column gas chromatography-mass spectrometry (GC/MS).

Stable isotopically-labeled analogs of the compounds of interest are added to the sample. If the solids content is less than 1%, a 1-L sample is extracted at pH 12–13, then at pH <2 with methylene chloride using continuous extraction techniques.

If the solids content is 30% or less, the sample is diluted to 1% solids with reagent water, homogenized ultrasonically, and extracted at pH 12–13, then at pH <2 with methylene chloride using continuous extraction techniques. If the solids content is greater than 30%, the sample is extracted using ultrasonic techniques.

Each extract is dried over sodium sulfate, concentrated to a volume of 5 mL, cleaned up using GPC, if necessary, and concentrated. Extracts are concentrated to 1 mL if GPC is not performed, and to 0.5 mL if GPC is performed.

An internal standard is added to the extract, and a 1-mL aliquot of the extract is injected into the GC. The compounds are separated by GC and detected by a MS. The labeled compounds serve to correct the variability of the analytical technique.

INTERFERENCES Solvents, reagents, glassware, and other sample processing hardware may yield artifacts and/or elevated baselines causing misinterpretation of chromatograms and spectra. Materials used in the analysis must be demonstrated to be free from interferences under the conditions of analysis by running method blanks initially and with each sample lot (sample started through the extraction process on a given 8-h shift, to a maximum of 20). Specific selection of reagents and purification of solvents by distillation in all glass systems may be required. Glassware and, where possible, reagents are cleaned by solvent rinse and baking at 450°C for 1-h minimum. Interferences coextracted from samples will vary considerably from source to source, depending on the diversity of the site being sampled.

INSTRUMENTATION Major instrumentation includes a GC with a splitless or on-column injection port for capillary column, a MS with 70 eV electron impact ionization, and a data system to collect and record MS data, and process it. A K-D apparatus is used to concentrate extracts.

GC Column: 30 m × 0.25 mm I.D. 5% phenyl, 94% methyl, 1% vinyl silicone bonded phased fused silica capillary column.

PRECISION & ACCURACY The detection limits of the method are usually dependent on the level of interferences rather than instrumental limitations. The limits typify the minimum quantities that can be detected with no interferences present.

The minimum level (in µg/mL) was 10. This is defined as a minimum level at which the analytical system shall give recognizable mass spectra (background corrected) and acceptable calibration points.

The MDL (in µg/kg) in low solids was 55 and in high solids was 47; these were determined in digested sludge (low solids) and in filter cake or compost (high solids).

The labeled and native compound initial precision as standard deviation (in µg/L) was 30.
The labeled and native compound initial accuracy as average recovery (in µg/L) was 80–141.

SAMPLE COLLECTION, PRESERVATION & HANDLING Collect samples in glass containers. Aqueous samples which flow freely are collected in refrigerated bottles using automatic sampling equipment. Solid samples are collected as grab samples using widemouth jars. Maintain samples at 0 to 4°C from the time of collection until extraction. If residual chlorine is present in aqueous samples, add 80 mg sodium thiosulfate/L of water. Begin sample extraction within 7 days of collection, and analyze all extracts within 40 days of extraction.

SAMPLE PREPARATION Samples containing 1% solids or less are extracted directly using continuous liquid-liquid extraction techniques. Samples containing 1 to 30% solids are diluted to the 1% level with reagent water and extracted using continuous liquid-liquid extraction techniques. Samples containing greater than 30% solids are extracted using ultrasonic techniques.

Base/neutral extraction — Adjust the pH of the waters in the extractors to 12–13 with 6 N NaOH. Extract with methylene chloride for 24–48 h.
Acid extraction — Adjust the pH of the waters in the extractors to 2 or less using 6 N sulfuric acid. Extract with methylene chloride for 24–48 h.
Ultrasonic extraction of high solids samples — Add anhydrous sodium sulfate to the sample and QC aliquot(s). Add acetone:methylene chloride (1:1) to the sample and mix thoroughly

Concentrate extracts using a K-D apparatus.

QUALITY CONTROL The analyst is permitted to modify this method to improve separations or lower the costs of measurements, provided all performance specifications are met. Analyses of blanks are required to demonstrate freedom from contamination. When results of spikes indicate atypical

method performance for samples, the samples are diluted to bring method performance within acceptable limits.

For low solids (aqueous samples), extract, concentrate, and analyze two sets of four 1-L aliquots (8 aliquots total) of the precision and recovery standard. For high solids samples, two sets of four 30-g aliquots of the high solids reference matrix are used.

Spike all samples with labeled compounds to assess method performance. Compute percent recovery of the labeled compounds using the internal standard method. Compare the labeled compound recovery for each compound with the corresponding labeled compound recovery.

Reagent water and high solids reference matrix blanks are analyzed to demonstrate freedom from contamination. Extract and concentrate a 1-L reagent water blank or a high solids reference matrix blank with each sample's lot (samples started through the extraction process on the same 8-h shift, to a maximum of 20 samples).

Field replicates may be collected to determine the precision of the sampling technique, and spiked samples may be required to determine the accuracy of the analysis when the internal standard method is used.

REFERENCE Semivolatile Organic Compounds by Isotope Dilution GC/MS. Office of Water Regulation and Standards, U.S. EPA Industrial Technology Division, Washington, DC, EPA Method 1625, Rev. C, June 1989 (contact W.A. Telliard, U.S. EPA, Office of Water Regulations and Standards, 401 M St., SW, Washington, DC, 20460. Phone: 202-382-7131).

2,4-Dinitrotoluene **EPA Method 625**
CAS #121-14-2

TITLE Base/Neutrals and Acids, U.S. EPA Method 625

MATRIX This methods covers municipal and industrial wastewaters.

METHOD SUMMARY Approximately 1 L of sample is serially extracted with methylene chloride at a pH greater than 11 and again at a pH less than 2 using a separatory funnel or a continuous extractor. The methylene chloride extract is dried, concentrated to a volume of 1 mL, and analyzed by GC/MS. Qualitative identification of the parameters in the extract is performed using the retention time and the relative abundance of three characteristic masses (m/z). Qualitative analysis is performed using either external or internal standard techniques with a single characteristic m/z.

INTERFERENCES Method interferences may be caused by contaminants in solvents, reagents, glassware, and other sample processing hardware. Glassware must be scrupulously cleaned. Glassware should be heated in a muffle furnace at 400°C for 5 to 30 min. Some thermally stable materials, such as PCBs, may not be eliminated by this treatment. Solvent rinses with acetone and pesticide quality hexane may be substituted for the muffle furnace heating. Matrix interferences may be caused by contaminants that are coextracted from the sample. The base-neutral extraction may cause significantly reduced recovery of phenols. The packed gas chromatographic columns recommended for the basic fraction may not exhibit sufficient resolution for some analytes.

INSTRUMENTATION A GC/MS system with an injection port designed for on-column injection when using packed columns and for splitless injection when using capillary columns.

Column for base/neutrals: 1.8 m long × 2 mm I.D. glass, packed with 3% SP-2550 on Supelcoport (100/120 mesh) or equivalent.

Column for acids: 1.8 m long × 2 mm I.D. glass, packed with 1% SP-1240DA on Supelcoport (100/120 mesh) or equivalent.

PRECISION & ACCURACY The MDL concentrations were obtained using reagent water. The MDL actually achieved in a given analysis will vary depending on instrument sensitivity and matrix effects. This method was tested by 15 laboratories using reagent water, drinking water, surface water, and industrial wastewaters spiked at six concentrations over the range 5 to 100 µg/L. Single operator precision, overall precision, and method accuracy were found to be directly related to the concentration of the parameter matrix.

The MDL (in µg/L) in reagent water was 5.7.
The standard deviation (in µg/L based on 4 recovery measurements) was 21.8.
The range (in µg/L) for average recovery for 4 measurements was 47.5–126.9.
The range (in %) for percent recovery was 39–139.
Accuracy (in µg/L) as expected recovery for one or more measurements of a sample containing a true concentration of C was 0.92C-4.81.
Precision (in µg/L) as expected single analyst standard deviation of measurements at an average concentration found at X was 0.12X+1.06.
Overall precision (in µg/L) as expected interlaboratory standard deviation of measurements in an average concentration found at X was 0.21X+1.50.

$C =$ *True value of the concentration in µg/L.*
$X =$ *Average recovery found for measurements of samples containing a concentration at C in µg/L.*

SAMPLE PREPARATION Adjust the pH to >11 with sodium hydroxide and serially extract in a separatory funnel with methylene chloride or else in a continuous extractor. Next, adjust the pH to <2 with sulfuric acid and serially extract in a separatory funnel with methylene chloride or else in a continuous extractor. Dry the extracts separately through a column of anhydrous sodium sulfate and then concentrate each of the extracts to 1.0 mL using a K-D apparatus.

SAMPLE COLLECTION, PRESERVATION & HANDLING Grab samples must be collected in glass containers. All samples must be refrigerated at 4°C from the time of collection until extraction. If residual chlorine is present, add 80 mg of sodium thiosulfate/L of sample and mix well. All samples must be extracted within 7 days of collection and completely analyzed within 40 days of extraction.

QUALITY CONTROL Make an initial, one-time, demonstration of the ability to generate acceptable accuracy and precision with this method. Before processing any samples, the analyst must analyze a reagent water blank to demonstrate that interferences from the analytical system and glassware are under control. Each time a set of samples is extracted or reagents are changed, a reagent water blank must be processed. Spike and analyze a minimum of 5% of all samples to monitor and evaluate lab data quality. A QC check sample concentrate that contains each parameter of interest at a concentration of 100 µg/mL in acetone is required. PCBs and multicomponent pesticides may be omitted from this test.

After analysis of five spiked wastewater samples, calculate the average percent recovery and the standard deviation of the percent recovery. Spike all samples with the surrogate standard spiking solution and calculate the percent recovery of each surrogate compound.

REFERENCE *Federal Register*, Vol. 49, No. 209. Friday, Oct. 26, 1984.

2,6-Dinitrotoluene   EPA Method 625
CAS #606-20-2

TITLE Base/Neutrals and Acids, U.S. EPA Method 625

MATRIX This methods covers municipal and industrial wastewaters.

METHOD SUMMARY Approximately 1 L of sample is serially extracted with methylene chloride at a pH greater than 11 and again at a pH less than 2 using a separatory funnel or a continuous extractor. The methylene chloride extract is dried, concentrated to a volume of 1 mL, and analyzed by GC/MS. Qualitative identification of the parameters in the extract is performed using the retention time and the relative abundance of three characteristic masses (m/z). Qualitative analysis is performed using either external or internal standard techniques with a single characteristic m/z.

INTERFERENCES Method interferences may be caused by contaminants in solvents, reagents, glassware, and other sample processing hardware. Glassware must be scrupulously cleaned. Glassware should be heated in a muffle furnace at 400°C for 5 to 30 min. Some thermally stable materials, such as PCBs, may not be eliminated by this treatment. Solvent rinses with acetone and pesticide quality hexane may be substituted for the muffle furnace heating. Matrix interferences may be caused by contaminants that are coextracted from the sample. The base-neutral extraction may cause significantly reduced recovery of phenols. The packed gas chromatographic columns recommended for the basic fraction may not exhibit sufficient resolution for some analytes.

INSTRUMENTATION A GC/MS system with an injection port designed for on-column injection when using packed columns and for splitless injection when using capillary columns.

Column for base/neutrals: 1.8 m long × 2 mm I.D. glass, packed with 3% SP-2550 on Supelcoport (100/120 mesh) or equivalent.

Column for acids: 1.8 m long × 2 mm I.D. glass, packed with 1% SP-1240DA on Supelcoport (100/120 mesh) or equivalent.

PRECISION & ACCURACY The MDL concentrations were obtained using reagent water. The MDL actually achieved in a given analysis will vary depending on instrument sensitivity and matrix effects. This method was tested by 15 laboratories using reagent water, drinking water, surface water, and industrial wastewaters spiked at six concentrations over the range 5 to 100 µg/L. Single operator precision, overall precision, and method accuracy were found to be directly related to the concentration of the parameter matrix.

The MDL (in µg/L) in reagent water was 1.9.

The standard deviation (in µg/L based on 4 recovery measurements) was 29.6.

The range (in µg/L) for average recovery for 4 measurements was 68.1–136.7.

The range (in %) for percent recovery was 50–158.

Accuracy (in µg/L) as expected recovery for one or more measurements of a sample containing a true concentration of C was 1.06C-3.60.

Precision (in µg/L) as expected single analyst standard deviation of measurements at an average concentration found at X was 0.14X+1.26.

Overall precision (in µg/L) as expected interlaboratory standard deviation of measurements in an average concentration found at X was 0.19X+0.35.

$C =$ *True value of the concentration in µg/L.*
$X =$ *Average recovery found for measurements of samples containing a concentration at C in µg/L.*

SAMPLE PREPARATION Adjust the pH to >11 with sodium hydroxide and serially extract in a separatory funnel with methylene chloride or else in a continuous extractor. Next, adjust the pH to <2 with sulfuric acid and serially extract in a separatory funnel with methylene chloride or else in a continuous extractor. Dry the extracts separately through a column of anhydrous sodium sulfate and then concentrate each of the extracts to 1.0 mL using a K-D apparatus.

SAMPLE COLLECTION, PRESERVATION & HANDLING
Grab samples must be collected in glass containers. All samples must be refrigerated at 4°C from the time of collection until extraction. If residual chlorine is present, add 80 mg of sodium thiosulfate/L of sample and mix well. All samples must be extracted within 7 days of collection and completely analyzed within 40 days of extraction.

QUALITY CONTROL Make an initial, one-time, demonstration of the ability to generate acceptable accuracy and precision with this method. Before processing any samples, the analyst must analyze a reagent water blank to demonstrate that interferences from the analytical system and glassware are under control. Each time a set of samples is extracted or reagents are changed, a reagent water blank must be processed. Spike and analyze a minimum of 5% of all samples to monitor and evaluate lab data quality. A QC check sample concentrate that

contains each parameter of interest at a concentration of 100 µg/mL in acetone is required. PCBs and multicomponent pesticides may be omitted from this test.

After analysis of five spiked wastewater samples, calculate the average percent recovery and the standard deviation of the percent recovery. Spike all samples with the surrogate standard spiking solution and calculate the percent recovery of each surrogate compound.

REFERENCE *Federal Register*, Vol. 49, No. 209. Friday, Oct. 26, 1984.

2,4-Dinitrotoluene   **EPA Method 8270**
CAS #121-14-2

TITLE Semivolatile Organic Compounds by GC/MS

MATRIX This method is used to determine the concentration of semivolatile organic compounds in extracts prepared from all types of solid waste matrices, soils, and groundwater. Although surface waters are not specifically mentioned, this method should be applicable to water samples from rivers, lakes, etc.

METHOD SUMMARY This method covers 259 semivolatile organic compounds. In very limited applications direct injection of the sample into the GC/MS system may be appropriate, but this results in very high detection limits (approximately 10,000 µg/L). Typically, a 1-L liquid sample, containing surrogate, and matrix spiking standards, is extracted in a continuous extractor first under acid conditions and then under basic conditions. Typically 30 g of a solid sample, containing surrogate, and matrix spiking standards, is extracted ultrasonically. After concentrating the extract to 1 mL it is spiked with 10 µL of an internal standard solution just prior to analysis by GC/MS. The volume injected should contain about 100 ng of base/neutral and 200 ng of acid surrogates (for a 1-µL injection). Analysis is performed by GC/MS using a capillary GC column.

INTERFERENCES Raw GC/MS data from all blanks, samples, and spikes must be evaluated for interferences. Contamination by carryover can occur whenever high-concentration and low-concentration samples are sequentially analyzed. To reduce carryover, the sample syringe must be rinsed out between samples with solvent. Whenever an unusually concentrated sample is encountered, it should be followed by the analysis of blank solvent to check for cross-contamination.

INSTRUMENTATION A GC/MS and a data system are required. The GC column used is a 30 m × 0.25 mm I.D. (or 0.32 mm I.D.) 1um film thickness silicone-coated fused silica capillary column. A continuous liquid-liquid extractor equipped with Teflon® or glass connection joints and stopcocks requiring no lubrication, a K-D concentrating apparatus, water bath, and an ultrasonic disrupter with a minimum power of 300 W and with pulsing capability are also required.

PRECISION & ACCURACY The estimated quantitation limit (EQL) of Method 8270B for determining an individual compound is approximately 1 mg/kg (wet weight) for soil or sediment samples, 1–200 mg/kg for wastes (dependent on matrix and method of preparation), and 10 µg/L for groundwater samples. EQLs will be proportionately higher for sample extracts that require dilution to avoid saturation of the detector.

The EQL(b) for groundwater in µg/L is 10.
The EQL (a, b) for low concentrations in soil and sediment in µg/kg is 660.
Accuracy as µg/L is 0.92C–4.81.
Overall precision in µg/L is 0.21X +1.50.

(a) *EQLs listed for soil/sediment are based on wet weight. Normally data is reported in a dry-weight basis; therefore, EQLs will be higher based on the % dry weight of each sample. This calculation is based on a 30 g sample and gel permeation chromatography cleanup.*
(b) *Sample EQLs are highly matrix-dependent. The EQLs are provided for guidance and may not always be achievable.*
C = *True value for concentration, in µg/L.*
X = *Average recovery found for measurements of samples containing a concentration of C, in µg/L.*

ESTIMATED QUANTITATION LIMIT

Other Matrices	Factor (a)
High-concentration soil and sludges by sonicator	7.5
Non-water miscible waste	75

(a) *EQL for other matrices = [EQL for low soil/sediment] × [Factor]. This estimated EQL is similar to an EPA "Practical Quantitation Limit."*

SAMPLING METHOD
Liquid samples — Use a 1 or 2½ gallon amber glass bottle with a screw-top Teflon®-lined cover that has been prewashed with detergent and rinsed with distilled water and methanol (or isopropanol).

Soils, sediments, or sludges — Use an 8-oz. widemouth glass with a screw-top Teflon®-lined cover that has been prewashed with detergent and rinsed with distilled water and methanol (or isopropanol).

SAMPLE PRESERVATION
Liquid samples — If residual chlorine is present, add 3 mL of 10% sodium thiosulfate per gallon, cool to 4°C and store in a solvent-free refrigerator until analysis; if chlorine is not present, then eliminate the sodium thiosulfate addition.

Soils, sediments, or sludges — Cool samples to 4°C and store in a solvent-free refrigerator.

MHT Liquid samples must be extracted within 7 days and the extracts analyzed within 40 days. Soils, sediments, or sludges may be stored for a maximum of 14 days and the extracts analyzed within 40 days.

SAMPLE PREPARATION
Liquid samples — Transfer 1 L quantitatively to a continuous extractor. If high concentrations are anticipated, a smaller volume may be used and then diluted with organic-free reagent water to 1 L. Adjust pH, if necessary, to pH <2 using 1:1 (V/V) sulfuric acid. Pipette 1.0 mL of a surrogate standard spiking solution into each sample. For the sample in each analytical

batch selected for spiking, add 1.0 mL of a matrix spiking standard. For base/neutral acid analysis, the amount of the surrogates and matrix spiking compounds added to the sample should result in a final concentration of 100 ng/µL of each analyte in the extract to be analyzed (assuming a 1-µL injection). Extract with methylene chloride for 18–24 h. Next, adjust the pH of the aqueous phase to pH >11 using 10 N sodium hydroxide and extract it with methylene chloride again for 18–24 h. Dry the extract through a column containing anhydrous sodium sulfate and concentrate it to 1 mL using a K-D concentrator.

Soils, sediments, or sludges — Use 30 g of sample. Nonporous or wet samples (gummy or clay type) that do not have a free-flowing sandy texture must be mixed with anhydrous sodium sulfate until the sample is free flowing. Add 1 mL of surrogate standards to all samples, spikes, standards, and blanks. For the sample in each analytical batch selected for spiking, add 1.0 mL of a matrix spiking standard. For base/neutral acid analysis, the amount added of the surrogates and matrix spiking compounds should result in a final concentration of 100 ng/µL of each base/neutral analyte and 200 ng/µL of each acid analyte in the extract to be analyzed (assuming a 1-µL injection). Immediately add a 100-mL mixture of 1:1 methylene chloride:acetone and extract the sample ultrasonically for 3 min and then decant or filter the extracts. Repeat the extraction two or more times. Dry the extract using a column with anhydrous sodium sulfate and concentrate it to 1 mL in a K-D concentrator.

QUALITY CONTROL   A methylene chloride solution containing 50 ng/µL of decafluorotriphenylphosphine (DFTPP) is used for tuning the GC/MS system each 12-h shift. A system performance check also must be made during every 12-h shift. A standard containing 50 ng/µL each of 4,4′-DDT, pentachlorophenol, and benzidine is required to verify injection port inertness and GC column performance. A calibration standard at mid-concentration, containing each compound of interest, including all required surrogates, must be performed every 12 h during analysis. After the system performance check is met, calibration check compounds (CCCs) are used to check the validity of the initial calibration.

The internal standard responses and retention times in the calibration check standard must be evaluated immediately after or during data acquisition. If the retention time for any internal standard changes by more than 30 seconds from the last check calibration (12 h), the chromatographic system must be inspected for malfunctions and corrections must be made, as required. If the electron ionization current plot (EICP) area for any of the internal standards changes by a factor of two from the last daily calibration standard check, the mass spectrometer must be inspected for malfunctions and corrections must be made, as appropriate.

Demonstrate, through the analysis of a reagent water blank, that interferences from the analytical system, glassware, and reagents are under control. The blank samples should be carried through all stages of the sample preparation and measurement steps. For each analytical batch (up to 20 samples), a reagent blank, matrix spike, and matrix spike duplicate/duplicate must be analyzed (the frequency of the spikes may be different for different monitoring programs). The blank and spiked samples must be carried through all stages of the sample preparation and measurement steps. A QC reference sample concentrate containing each analyte at a concentration of 100 mg/L in methanol is required.

REFERENCE   Test Methods for Evaluating Solid Waste (SW-846). U.S. EPA 1983, Method 8270B, Rev. 2, Nov. 1990. Office of Solid Waste, Washington, DC.

2,6-Dinitrotoluene   **EPA Method 8270**
CAS #606-20-2

TITLE   Semivolatile Organic Compounds by GC/MS

MATRIX   This method is used to determine the concentration of semivolatile organic compounds in extracts prepared from all types of solid waste matrices, soils, and groundwater. Although surface waters are not specifically mentioned, this method should be applicable to water samples from rivers, lakes, etc.

METHOD SUMMARY   This method covers 259 semivolatile organic compounds. In very limited applications direct injection of the sample into the GC/MS system may be appropriate, but this results in very high detection limits (approximately 10,000 µg/L). Typically, a 1-L liquid sample, containing surrogate, and matrix spiking standards, is extracted in a continuous extractor first under acid conditions and then under basic conditions. Typically 30 g of a solid sample, containing surrogate, and matrix spiking standards, is extracted ultrasonically. After concentrating the extract to 1 mL it is spiked with 10 µL of an internal standard solution just prior to analysis by GC/MS. The volume injected should contain about 100 ng of base/neutral and 200 ng of acid surrogates (for a 1-µL injection). Analysis is performed by GC/MS using a capillary GC column.

INTERFERENCES   Raw GC/MS data from all blanks, samples, and spikes must be evaluated for interferences. Contamination by carryover can occur whenever high-concentration and low-concentration samples are sequentially analyzed. To reduce carryover, the sample syringe must be rinsed out between samples with solvent. Whenever an unusually concentrated sample is encountered, it should be followed by the analysis of blank solvent to check for cross-contamination.

INSTRUMENTATION   A GC/MS and a data system are required. The GC column used is a 30 m × 0.25 mm I.D. (or 0.32 mm I.D.) 1um film thickness silicone-coated fused silica capillary column. A continuous liquid-liquid extractor equipped with Teflon® or glass connection joints and stopcocks requiring no lubrication, a K-D concentrating apparatus, water bath, and an ultrasonic disrupter with a minimum power of 300 W and with pulsing capability are also required.

PRECISION & ACCURACY   The estimated quantitation limit (EQL) of Method 8270B for determining an individual compound is approximately 1 mg/kg (wet weight) for soil or sediment samples, 1–200 mg/kg for wastes (dependent on matrix and method of preparation), and 10 µg/L for groundwater samples.

EQLs will be proportionately higher for sample extracts that require dilution to avoid saturation of the detector.

The EQL(b) for groundwater in µg/L is 10.
The EQL (a, b) for low concentrations in soil and sediment in µg/kg is 660.
Accuracy as µg/L is 1.06C–3.60.
Overall precision in µg/L is 0.19X +0.35.

(a) *EQLs listed for soil/sediment are based on wet weight. Normally data is reported in a dry-weight basis; therefore, EQLs will be higher based on the % dry weight of each sample. This calculation is based on a 30 g sample and gel permeation chromatography cleanup.*
(b) *Sample EQLs are highly matrix-dependent. The EQLs are provided for guidance and may not always be achievable.*
$C =$ *True value for concentration, in µg/L.*
$X =$ *Average recovery found for measurements of samples containing a concentration of C, in µg/L.*

ESTIMATED QUANTITATION LIMIT

Other Matrices	Factor (a)
High-concentration soil and sludges by sonicator	7.5
Non-water miscible waste	75

(a) *EQL for other matrices = [EQL for low soil/sediment] × [Factor]. This estimated EQL is similar to an EPA "Practical Quantitation Limit."*

SAMPLING METHOD
Liquid samples — Use a 1 or 2½ gallon amber glass bottle with a screw-top Teflon®-lined cover that has been prewashed with detergent and rinsed with distilled water and methanol (or isopropanol).

Soils, sediments, or sludges — Use an 8-oz. widemouth glass with a screw-top Teflon®-lined cover that has been prewashed with detergent and rinsed with distilled water and methanol (or isopropanol).

SAMPLE PRESERVATION
Liquid samples — If residual chlorine is present, add 3 mL of 10% sodium thiosulfate per gallon, cool to 4°C and store in a solvent-free refrigerator until analysis; if chlorine is not present, then eliminate the sodium thiosulfate addition.

Soils, sediments, or sludges — Cool samples to 4°C and store in a solvent-free refrigerator.

MHT Liquid samples must be extracted within 7 days and the extracts analyzed within 40 days. Soils, sediments, or sludges may be stored for a maximum of 14 days and the extracts analyzed within 40 days.

SAMPLE PREPARATION
Liquid samples — Transfer 1 L quantitatively to a continuous extractor. If high concentrations are anticipated, a smaller volume may be used and then diluted with organic-free reagent water to 1 L. Adjust pH, if necessary, to pH <2 using 1:1 (V/V) sulfuric acid. Pipette 1.0 mL of a surrogate standard spiking solution into each sample. For the sample in each analytical batch selected for spiking, add 1.0 mL of a matrix spiking standard. For base/neutral acid analysis, the amount of the surrogates and matrix spiking compounds added to the sample should result in a final concentration of 100 ng/µL of each analyte in the extract to be analyzed (assuming a 1-µL injection). Extract with methylene chloride for 18–24 h. Next, adjust the pH of the aqueous phase to pH >11 using 10 N sodium hydroxide and extract it with methylene chloride again for 18–24 h. Dry the extract through a column containing anhydrous sodium sulfate and concentrate it to 1 mL using a K-D concentrator.

Soils, sediments, or sludges — Use 30 g of sample. Nonporous or wet samples (gummy or clay type) that do not have a free-flowing sandy texture must be mixed with anhydrous sodium sulfate until the sample is free flowing. Add 1 mL of surrogate standards to all samples, spikes, standards, and blanks. For the sample in each analytical batch selected for spiking, add 1.0 mL of a matrix spiking standard. For base/neutral acid analysis, the amount added of the surrogates and matrix spiking compounds should result in a final concentration of 100 ng/µL of each base/neutral analyte and 200 ng/µL of each acid analyte in the extract to be analyzed (assuming a 1-µL injection). Immediately add a 100-mL mixture of 1:1 methylene chloride:acetone and extract the sample ultrasonically for 3 min and then decant or filter the extracts. Repeat the extraction two or more times. Dry the extract using a column with anhydrous sodium sulfate and concentrate it to 1 mL in a K-D concentrator.

QUALITY CONTROL A methylene chloride solution containing 50 ng/µL of decafluorotriphenylphosphine (DFTPP) is used for tuning the GC/MS system each 12-h shift. A system performance check also must be made during every 12-h shift. A standard containing 50 ng/µL each of 4,4'-DDT, pentachlorophenol, and benzidine is required to verify injection port inertness and GC column performance. A calibration standard at mid-concentration, containing each compound of interest, including all required surrogates, must be performed every 12 h during analysis. After the system performance check is met, calibration check compounds (CCCs) are used to check the validity of the initial calibration.

The internal standard responses and retention times in the calibration check standard must be evaluated immediately after or during data acquisition. If the retention time for any internal standard changes by more than 30 seconds from the last check calibration (12 h), the chromatographic system must be inspected for malfunctions and corrections must be made, as required. If the electron ionization current plot (EICP) area for any of the internal standards changes by a factor of two from the last daily calibration standard check, the mass spectrometer must be inspected for malfunctions and corrections must be made, as appropriate.

Demonstrate, through the analysis of a reagent water blank, that interferences from the analytical system, glassware, and reagents are under control. The blank samples should be carried through all stages of the sample preparation and measurement steps. For each analytical batch (up to 20 samples), a reagent blank, matrix spike, and matrix spike duplicate/duplicate must be analyzed (the frequency of the spikes may be different for different monitoring programs). The blank and spiked samples must be carried through all stages of the sample

preparation and measurement steps. A QC reference sample concentrate containing each analyte at a concentration of 100 mg/L in methanol is required.

REFERENCE Test Methods for Evaluating Solid Waste (SW-846). U.S. EPA 1983, Method 8270B, Rev. 2, Nov. 1990. Office of Solid Waste, Washington, DC.

2,4-Dinitrotoluene **EPA Method 8090**
CAS #121-14-2

TITLE Nitroaromatics & Cyclic Ketones

MATRIX Groundwater, soils, sludges, water miscible liquid wastes, and non-water miscible wastes.

APPLICATION This method is used for the analysis of various nitroaromatic and cyclic ketone compounds. Samples are extracted, concentrated, and analyzed using direct injection of both neat and diluted organic liquids.

Dinitrotoluenes are determined using ECD and the other compounds amenable to this method are determined using FID. The method provides an optional GC column which is used for analyte confirmation and that may help resolve analytes from interferences.

INTERFERENCES Solvents, reagents, and glassware may introduce artifacts. Other interferences may come from coextracted compounds from samples.

INSTRUMENTATION GC capable of on-column injections and a flame with detector (FID) or electron capture detector (ECD). Column 1: a 1.2 m by 2 mm or 4 mm with 1.95% QF-1/1.5% OV-17 on Gas-Chrom Q. Column 2: a 3 meter by 2 mm or 4 mm with 3% OV-101 on Gas-Chrom Q.

RANGE 1 to 515 µg/L.

MDL 0.02 µg/L (ECD).

PQL FACTORS FOR MULTIPLYING × FID MDL VALUE

Matrix	Multiplication Factor
Groundwater	10
Low-level soil by sonication with GPC cleanup	670
High-level soil and sludge by sonication	10,000
Non-water miscible waste	100,000

PRECISION 0.37X–0.07 µg/L (overall precision).

ACCURACY 0.65C + 0.22 µg/L (as recovery).

SAMPLING METHOD Use 8-oz. widemouth glass bottles with Teflon®-lined caps for concentrated waste samples, soils, sediments, and sludges. Use 1 or 2½ gallon amber glass bottles with Teflon®-lined caps for liquid (water) samples.

STABILITY Cool soil, sediment, sludge, and liquid samples to 4°C. If residual chlorine is present in liquid samples add 3 mL of 10% sodium thiosulfate per gallon of sample and cool to 4°C.

MHT 14 days for concentrated waste, soil, sediment, or sludge; 7 days for liquid samples; all extracts must be analyzed within 40 days.

QUALITY CONTROL A quality control check sample concentrate containing each analyte of interest is required. The QC check sample concentrate may be prepared from pure standard materials or purchased as certified solutions Use appropriate trip, matrix, control site, method, reagent, and solvent blanks. Internal, surrogate, and five concentration level calibration standards are used. The QC check sample concentrate should contain this compound at 20 µg/mL in acetone.

REFERENCE Method 8090, SW-846, 3rd ed., Nov.1986.

2,6-Dinitrotoluene **EPA Method 8090**
CAS #606-20-2

TITLE Nitroaromatics & Cyclic Ketones

MATRIX Groundwater, soils, sludges, water miscible liquid wastes, and non-water miscible wastes.

APPLICATION This method is used for the analysis of various nitroaromatic and cyclic ketone compounds. Samples are extracted, concentrated, and analyzed using direct injection of both neat and diluted organic liquids.

Dinitrotoluenes are determined using ECD and the other compounds amenable to this method are determined using FID. The method provides an optional GC column which is used for analyte confirmation and that may help resolve analytes from interferences.

INTERFERENCES Solvents, reagents, and glassware may introduce artifacts. Other interferences may come from coextracted compounds from samples.

INSTRUMENTATION GC capable of on-column injections and a flame with detector (FID) or electron capture detector (ECD). Column 1: a 1.2 m by 2 mm or 4 mm with 1.95% QF-1/1.5% OV-17 on Gas-Chrom Q. Column 2: a 3 meter by 2 mm or 4 mm with 3% OV-101 on Gas-Chrom Q.

RANGE 1 to 515 µg/L.

MDL 0.01 µg/L (ECD).

PQL FACTORS FOR MULTIPLYING × FID MDL VALUE

Matrix	Multiplication Factor
Groundwater	10
Low-level soil by sonication with GPC cleanup	670
High-level soil and sludge by sonication	10,000
Non-water miscible waste	100,000

PRECISION 0.36X–0.00 µg/L (overall precision).

ACCURACY 0.66C + 0.20 µg/L (as recovery).

SAMPLING METHOD Use 8-oz. widemouth glass bottles with Teflon®-lined caps for concentrated waste samples, soils, sediments, and sludges. Use 1 or 2½ gallon amber glass bottles with Teflon®-lined caps for liquid (water) samples.

STABILITY Cool soil, sediment, sludge, and liquid samples to 4°C. If residual chlorine is present in liquid samples add 3 mL of 10% sodium thiosulfate per gallon of sample and cool to 4°C.

MHT 14 days for concentrated waste, soil, sediment, or sludge; 7 days for liquid samples; all extracts must be analyzed within 40 days.

QUALITY CONTROL A quality control check sample concentrate containing each analyte of interest is required. The QC check sample concentrate may be prepared from pure standard materials or purchased as certified solutions Use appropriate trip, matrix, control site, method, reagent, and solvent blanks. Internal, surrogate, and five concentration level calibration standards are used. The QC check sample concentrate should contain this compound at 20 µg/mL in acetone.

REFERENCE Method 8090, SW-846, 3rd ed., Nov.1986.

Dinocap   **EPA Method 8270**
CAS #39300-45-3

TITLE Semivolatile Organic Compounds by GC/MS

MATRIX This method is used to determine the concentration of semivolatile organic compounds in extracts prepared from all types of solid waste matrices, soils, and groundwater. Although surface waters are not specifically mentioned, this method should be applicable to water samples from rivers, lakes, etc.

METHOD SUMMARY This method covers 259 semivolatile organic compounds. In very limited applications direct injection of the sample into the GC/MS system may be appropriate, but this results in very high detection limits (approximately 10,000 µg/L). Typically, a 1-L liquid sample, containing surrogate, and matrix spiking standards, is extracted in a continuous extractor first under acid conditions and then under basic conditions. Typically 30 g of a solid sample, containing surrogate, and matrix spiking standards, is extracted ultrasonically. After concentrating the extract to 1 mL it is spiked with 10 µL of an internal standard solution just prior to analysis by GC/MS. The volume injected should contain about 100 ng of base/neutral and 200 ng of acid surrogates (for a 1-µL injection). Analysis is performed by GC/MS using a capillary GC column.

INTERFERENCES Raw GC/MS data from all blanks, samples, and spikes must be evaluated for interferences. Contamination by carryover can occur whenever high-concentration and low-concentration samples are sequentially analyzed. To reduce carryover, the sample syringe must be rinsed out between samples with solvent. Whenever an unusually concentrated sample is encountered, it should be followed by the analysis of blank solvent to check for cross-contamination.

INSTRUMENTATION A GC/MS and a data system are required. The GC column used is a 30 m × 0.25 mm I.D. (or 0.32 mm I.D.) 1um film thickness silicone-coated fused silica capillary column. A continuous liquid-liquid extractor equipped with Teflon® or glass connection joints and stopcocks requiring no lubrication, a K-D concentrating apparatus, water bath, and an ultrasonic disrupter with a minimum power of 300 W and with pulsing capability are also required.

PRECISION & ACCURACY The estimated quantitation limit (EQL) of Method 8270B for determining an individual compound is approximately 1 mg/kg (wet weight) for soil or sediment samples, 1–200 mg/kg for wastes (dependent on matrix and method of preparation), and 10 µg/L for groundwater samples. EQLs will be proportionately higher for sample extracts that require dilution to avoid saturation of the detector.

The EQL(b) for groundwater in µg/L is 100.
The EQL (a, b) for low concentrations in soil and sediment in µg/kg is not determined.
Accuracy as µg/L is not listed.
Overall precision in µg/L is not listed.

(a) *EQLs listed for soil/sediment are based on wet weight. Normally data is reported in a dry-weight basis; therefore, EQLs will be higher based on the % dry weight of each sample. This calculation is based on a 30 g sample and gel permeation chromatography cleanup.*
(b) *Sample EQLs are highly matrix-dependent. The EQLs are provided for guidance and may not always be achievable.*
C = *True value for concentration, in µg/L.*
X = *Average recovery found for measurements of samples containing a concentration of C, in µg/L.*

ESTIMATED QUANTITATION LIMIT

Other Matrices	Factor (a)
High-concentration soil and sludges by sonicator	7.5
Non-water miscible waste	75

(a) *EQL for other matrices = [EQL for low soil/sediment] × [Factor]. This estimated EQL is similar to an EPA "Practical Quantitation Limit."*

SAMPLING METHOD
Liquid samples — Use a 1 or 2½ gallon amber glass bottle with a screw-top Teflon®-lined cover that has been prewashed with detergent and rinsed with distilled water and methanol (or isopropanol).

Soils, sediments, or sludges — Use an 8-oz. widemouth glass with a screw-top Teflon®-lined cover that has been prewashed with detergent and rinsed with distilled water and methanol (or isopropanol).

SAMPLE PRESERVATION
Liquid samples — If residual chlorine is present, add 3 mL of 10% sodium thiosulfate per gallon, cool to 4°C and store in a solvent-free refrigerator until analysis; if chlorine is not present, then eliminate the sodium thiosulfate addition.

Soils, sediments, or sludges — Cool samples to 4°C and store in a solvent-free refrigerator.

MHT Liquid samples must be extracted within 7 days and the extracts analyzed within 40 days. Soils, sediments, or sludges may be stored for a maximum of 14 days and the extracts analyzed within 40 days.

SAMPLE PREPARATION
Liquid samples — Transfer 1 L quantitatively to a continuous extractor. If high concentrations are anticipated, a smaller volume may be used and then diluted with organic-free reagent water to 1 L. Adjust pH, if necessary, to pH <2 using 1:1 (V/V) sulfuric acid. Pipette 1.0 mL of a surrogate standard spiking solution into each sample. For the sample in each analytical batch selected for spiking, add 1.0 mL of a matrix spiking standard. For base/neutral acid analysis, the amount of the surrogates and matrix spiking compounds added to the sample should result in a final concentration of 100 ng/µL of each analyte in the extract to be analyzed (assuming a 1-µL injection). Extract with methylene chloride for 18–24 h. Next, adjust the pH of the aqueous phase to pH >11 using 10 N sodium hydroxide and extract it with methylene chloride again for 18–24 h. Dry the extract through a column containing anhydrous sodium sulfate and concentrate it to 1 mL using a K-D concentrator.

Soils, sediments, or sludges — Use 30 g of sample. Nonporous or wet samples (gummy or clay type) that do not have a free-flowing sandy texture must be mixed with anhydrous sodium sulfate until the sample is free flowing. Add 1 mL of surrogate standards to all samples, spikes, standards, and blanks. For the sample in each analytical batch selected for spiking, add 1.0 mL of a matrix spiking standard. For base/neutral acid analysis, the amount added of the surrogates and matrix spiking compounds should result in a final concentration of 100 ng/µL of each base/neutral analyte and 200 ng/µL of each acid analyte in the extract to be analyzed (assuming a 1-µL injection). Immediately add a 100-mL mixture of 1:1 methylene chloride:acetone and extract the sample ultrasonically for 3 min and then decant or filter the extracts. Repeat the extraction two or more times. Dry the extract using a column with anhydrous sodium sulfate and concentrate it to 1 mL in a K-D concentrator.

QUALITY CONTROL A methylene chloride solution containing 50 ng/µL of decafluorotriphenylphosphine (DFTPP) is used for tuning the GC/MS system each 12-h shift. A system performance check also must be made during every 12-h shift. A standard containing 50 ng/µL each of 4,4′-DDT, pentachlorophenol, and benzidine is required to verify injection port inertness and GC column performance. A calibration standard at mid-concentration, containing each compound of interest, including all required surrogates, must be performed every 12 h during analysis. After the system performance check is met, calibration check compounds (CCCs) are used to check the validity of the initial calibration.

The internal standard responses and retention times in the calibration check standard must be evaluated immediately after or during data acquisition. If the retention time for any internal standard changes by more than 30 seconds from the last check calibration (12 h), the chromatographic system must be inspected for malfunctions and corrections must be made, as required. If the electron ionization current plot (EICP) area for any of the internal standards changes by a factor of two from the last daily calibration standard check, the mass spectrometer must be inspected for malfunctions and corrections must be made, as appropriate.

Demonstrate, through the analysis of a reagent water blank, that interferences from the analytical system, glassware, and reagents are under control. The blank samples should be carried through all stages of the sample preparation and measurement steps. For each analytical batch (up to 20 samples), a reagent blank, matrix spike, and matrix spike duplicate/duplicate must be analyzed (the frequency of the spikes may be different for different monitoring programs). The blank and spiked samples must be carried through all stages of the sample preparation and measurement steps. A QC reference sample concentrate containing each analyte at a concentration of 100 mg/L in methanol is required.

REFERENCE Test Methods for Evaluating Solid Waste (SW-846). U.S. EPA 1983, Method 8270B, Rev. 2, Nov. 1990. Office of Solid Waste, Washington, DC.

Dinonyl phthalate   **EPA Method 8061**
CAS #84-76-4

TITLE Phthalate Esters by Capillary Gas Chromatography With Electron Capture Detection (GC/ECD)

MATRIX This method covers aqueous and solid matrices. This includes a wide variety such as drinking water, groundwater, industrial wastewaters, surface waters, soils, solids, and sediments.

METHOD SUMMARY This method is used to determine the identities and concentrations of phthalate esters in liquid, solid and sludge matrices. When used to analyze for any or all of the target analytes, compound identification should be supported by at least one additional qualitative technique. This method describes conditions for parallel column, dual electron capture detector analysis, which fulfills the above requirement. Alternatively, GC/MS could be used for compound confirmation.

A measured volume or weight of sample (approximately 1 L for liquids, 10 to 30 g for solids and sludges) is extracted by using the appropriate sample extraction technique specified in EPA Method 3510, EPA Method 3540, and EPA Method 3550. After cleanup, the extract is analyzed by GC/ECD.

INTERFERENCES The sensitivity of this method usually depends on the level of interferences rather than on instrumental limitations. If interferences prevent detection of the analytes, cleanup of the sample extracts is necessary. Either EPA Method 3610 or EPA Method 3620 alone or followed by EPA Method 3660, Sulfur Cleanup, may be used to eliminate interferences in the analysis. EPA Method 3640, Gel Permeation Cleanup, is applicable for samples that contain high amounts of lipids and waxes.

Interferences coextracted from the samples will vary considerably from waste to waste. Glassware must be scrupulously clean. All glassware require treatment in a muffle furnace at 400°C for 2 to 4 h, or thorough rinsing with pesticide-grade solvent, prior to use. Volumetric glassware should not be heated in a muffle furnace. Storage of glassware in the lab introduces contamination, even if the glassware is wrapped in aluminum foil. Sodium sulfate, Florisil, and alumina may be contaminated with phthalate esters and, therefore, use of these materials in sample cleanup should be employed cautiously. If these materials are used, they must be obtained packaged in glass. Heating at 400°C for sodium sulfate, 320°C for Florisil, and 210°C for alumina is recommended. Glass wool used in any step of sample preparation should be a specially treated pyrex wool, pesticide grade, and must be baked at 400°C for 4 h immediately prior to use.

Paper thimbles and filter paper must be exhaustively washed with the solvent that will be used in the sample extraction. Soxhlet extraction of paper thimbles and filter paper for 12 h with fresh solvent should be repeated for a minimum of three times. Method blanks should be obtained before any of the precleaned thimbles or filter papers are used.

INSTRUMENTATION   Gas chromatograph suitable for on-column and split/splitless injections.

Column 1: 30 m × 0.53 mm ID, 5% phenyl/95% methyl silicone fused-silica open tubular column, DB-5, 1.5 µg film thickness.

Column 2: 30 m × 0.53 mm ID, 14% cyanopropyl phenyl silicone fused-silica open tubular column, DB-1701, 1.0 µg film thickness.

A dual electron capture detector (ECD) is used. A Kuderna-Danish (K-D) apparatus is required along with a vacuum manifold consisting of individually adjustable, easily accessible flow-control valves for up to 24 cartridges, sample rack, chemically resistant cover and seals, heavy-duty glass basin, removable stainless steel solvent guides, built-in vacuum gauge and valve. Also, 6-mL, 1-g solid-phase extraction cartridges, LC-Florisil or equivalent, prepackaged, ready to use will be needed.

PRECISION & ACCURACY   The MDL actually achieved in a given analysis will vary, as it is dependent on instrument sensitivity and matrix effects. This method has been tested in a single lab. Single-operator precision, overall precision, and method accuracy were found to be related to the concentration of the compounds and the type of matrix.

MULTIPLICATION FACTORS FOR OTHER MATRICES (a)

Matrix	Factor (b)
Groundwater	10
Low-concentration soil by ultrasonic extraction with GPC cleanup	670
High-concentration soil and sludges by ultrasonic extraction	10,000
Non-water miscible waste	100,000

(a) Sample EQLs are highly matrix-dependent.
(b) EQL = [Method detection limit] × [Factor]. For non-aqueous samples, the factor is on a wet-weight basis.

The MDL using 7 replicate determinations and a spike concentration of 100 µg/L was 22 ng/L.

The average recovery from HPLC-grade water using 4 determinatons and a spike concentration of 100 µg/L was 96.9%.

The precision (as RSD) from HPLC-grade water using 4 determinatons and a spike concentration of 100 µg/L was 11.1%.

The average recovery from groundwater using 4 determinatons and a spike concentration of 100 µg/L was 95.2%.

The precision (as RSD) from groundwater using 4 determinatons and a spike concentration of 100 µg/L was 12.7%.

The average recovery (in %) with %RSD (in parenthesis) from 3 determinations and a spike concentration of 20 µg/L in water was 59.5 (6.1) using 3M Empore Disks and EPA Method 8061.

The average recovery (in %) with %RSD (in parenthesis) from 3 determinations and a spike concentration of 20 µg/L in leachate was 77.3 (4.2) using 3M Empore Disks and EPA Method 8061.

The average recovery (in %) with %RSD (in parenthesis) from 3 determinations and a spike concentration of 20 µg/L in estuarine groundwater was 67.2 (8.0) using 3M Empore Disks and EPA Method 8061.

The average recovery (in %) with %RSD (in parenthesis) from 3 determinations and a spike concentration of 1 mg/kg in estuarine sediment was not determined (matrix interferant) after sulfur cleanup with EPA Method 3660.

The average recovery (in %) with %RSD (in parenthesis) from 3 determinations and a spike concentration of 1 mg/kg in municipal sludge was 80.0 (41.1).

The average recovery (in %) with %RSD (in parenthesis) from 3 determinations and a spike concentration of 1 mg/kg in sandy loam soil was 64.2 (17.2).

SAMPLE COLLECTION, PRESERVATION & HANDLING
Containers used to collect samples for the determination of semivolatile organic compounds should be soap and water washed followed by methanol (or isopropanol) rinsing. The sample containers should be of glass or Teflon® and have screw-top covers with Teflon® liners. Sample containers should be filled with care to prevent any portion of the collected sample coming in contact with the sampler's gloves.

No preservation is used with concentrated waste samples. With liquid samples containing no residual chlorine and with soil, sediment, and sludge samples, immediately cooling to 4°C is the only preservation used. When residual chlorine is present then 3 mL of 10% aqueous sodium sulfate is added for each gallon of sample collected, followed by cooling to 4°C.

MHT   Liquid samples must be extracted within 7 days and their extracts analyzed within 40 days. Concentrated waste, soil, sediment, and sludge samples must be extracted within 14 days and their extracts analyzed within 40 days.

SAMPLE PREPARATION   In general, water samples are extracted at a pH of 5 to 7 with methylene chloride in a separatory funnel (EPA Method 3510). EPA Method 3520 is not recommended for the extraction of aqueous samples because the longer chain esters tend to adsorb to the glassware and consequently, their extraction recoveries may be poor. Solid samples are extracted with hexane/acetone (1:) or methylene chloride/acetone (1:1) in a Soxhlet extractor (EPA Method 3540) or with an ultrasonic extractor (EPA Method 3550). Immediately prior to extraction, spike 500 µL of the surrogate standard spiking solution into 1-L aqueous sample or 30-g solid sample. Extraction of particulate-free aqueous samples using C-18 extraction disks is an optional method that can be used.

Prior to Florisil cleanup or GC analysis, the methylene chloride and methylene chloride/acetone extracts must be exchanged to hexane. Exchange is not required for the acetonitrile extracts. Cleanup may not be necessary for extracts from a relatively clean sample matrix. Florisil Cartridge Cleanup may be used for extract cleanup.

If PCBs and organochlorine pesticides are known to be present in the sample, and if Florisil Cartridge Cleanup is considered, then two fractions are collected: Fraction 1 is eluted with 5 mL of 20% methylene chloride in hexane and Fraction 2 is eluted with 5 mL of 10% acetone in hexane. Fraction 1 contains the organochlorine pesticides and PCBs, and can be discarded. Fraction 2 contains the phthalate esters and is analyzed by GC/ECD.

QUALITY CONTROL Identify compounds in the sample by comparing the retention times of the peaks in the sample chromatogram with those of the peaks in standard chromatograms. The retention time window used to make identification is based upon measurements of actual retention time variations over the course of 10 consecutive injections.

Calibration standards are prepared at a minimum of five concentrations for each parameter of interest through dilution of the stock standard solutions with hexane. One of the concentrations should be at a concentration near, but above, the method detection limit. Prepare stock standard solutions in hexane. Stock standards should be checked frequently for signs of degradation or evaporation, especially just prior to preparing calibration standards from them. Stock standard solutions must be replaced after one year, or sooner if comparison with check standards indicates a problem. The suggested internal standards are: 2,5-dibromotoluene, 1,3,5-tribromobenzene, and α, α'-dibromo-m-xylene. The analyst can use any of the three compounds provided that they are resolved from matrix interferences. Recommended surrogate compounds are α-2,6-trichlorotoluene, 1,4-dichloronaphthalene, and 2,3,4,5,6-pentachlorotoluene.

Spike each sample, standard, and blank with surrogate compounds. Three surrogates are suggested for this method: diphenyl phthalate, diphenyl isophthalate, and dibenzyl phthalate.

The quality control check sample concentrate should contain the test compounds at 5 to 10 ng/μL An internal standard peak area check must be performed on all samples. The internal standard must be evaluated for acceptance by determining whether the measured area for the internal standard deviates by more than 30% from the average are for the internal standard in the calibration standards. When the internal standard peak area is outside that limit, all samples that fall outside the QC criteria must be reanalyzed. Benzyl benzoate has been tested and found appropriate as an internal standard for this method.

Any compounds confirmed by two columns may also be confirmed by GC/MS. The sample extract and associated blank should be analyzed by GC/MS. A reference standard of the compound must also be analyzed by GC/MS. Include a mid-concentration calibration standard after each group of 20 samples. The response factors for the mid-concentration calibration must be within ±15% of the average values for the multiconcentration calibration. Demonstrate through the analyses of standards that the Florisil fractionation scheme is reproducible.

REFERENCE Test Methods for Evaluating Solid Waste, Physical/Chemical Methods, SW-846, 3rd Edition, U.S. EPA, Office of Solid Waste, Washington, DC, EPA Method 8061, Nov. 1990.

Dinoseb   EPA Method 8151
CAS #88-85-7

TITLE Chlorinated Herbicides by GC Using Methylation or Pentafluorobenzylation Derivatization: Capillary Column Technique.

MATRIX This method covers aqueous and solid matrices. This includes a wide variety such as drinking water, groundwater, industrial wastewaters, surface waters, soils, solids, and sediments.

METHOD SUMMARY This is a GC method for determining 19 chlorinated acid herbicides in aqueous, soil, and waste matrices. Because these compounds are produced and used in various forms (i.e., acid, salt, ester, etc.) a hydrolysis step is included to convert the herbicide to the acid form prior to analysis. This method provides hydrolysis, extraction, derivatization and GC conditions for the analysis of chlorinated acid herbicides in water, soil, and waste samples. Water samples are hydrolyzed *in situ,* extracted with diethyl ether, and then esterified with either diazomethane or pentafluorobenzyl bromide. The derivatives are determined by gas chromatography with an electron capture detector (GC/ECD). The results are reported as acid equivalents. The sensitivity of this method depends on the level of interferences in addition to instrumental limitations.

INTERFERENCES Method interferences may be caused by contaminants in solvents, reagents, glassware, and other sample processing hardware. Immediately prior to use, glassware should be rinsed with the next solvent to be used. Matrix interferences may be caused by contaminants that are coextracted from the sample. Organic acids, especially chlorinated acids, cause the most direct interference with the determination by methylation. Phenols, including chlorophenols, may also interfere with this procedure. The determination using pentafluorobenzylation is more sensitive, and more prone to interferences from the presence of organic acids of phenols than by methylation. Alkaline hydrolysis and subsequent extraction of the basic solution removes many chlorinated hydrocarbons and phthalate esters that might otherwise interfere with the ECD analysis. The herbicides, being strong organic acids, react readily with alkaline substances and may be lost during analysis. Therefore, glassware must be acid-rinsed and then rinsed to constant pH with organic-free reagent water.

INSTRUMENTATION A GC suitable for Grob-type injection using capillary columns. A data system for measuring peak heights and/or peak areas is recommended. An electron capture detector (ECD) is used. Also a K-D apparatus, a diazomethane generator, a centrifuge and an ultrasonic disrupter will be required.

Narrow Bore Columns:
Primary Column 1: 30 m × 0.25 mm, 5% phenyl/95% methyl silicone (DB-5), 0.25 μm film thickness.
Primary Column 1a (GC/MS): 30 m × 0.32 mm, 5% phenyl/95% methyl silicone (DB-5), 1-μm film thickness.
Column 2: 30 m × 0.25 mm DB-608 with a 25 μm film thickness.
Confirmation Column: 30 m × 0.25 mm, 14% cyanopropyl phenyl silicone (DB-1701), 0.25 μm film thickness.

Megabore Columns:
Primary Column: 30 m × 0.53 mm DB-608 with 0.83 μm film thickness.
Confirmation Column: 30 m × 0.53 mm, 14% cyanopropyl phenyl silicone (DB-1701), 1.0 μm film thickness.

PRECISION & ACCURACY Method detection limits (MDLs) are compound-dependent and vary with derivitization efficiency, derivative recovery, the matrix sampled, and herbicide concentration.

The estimated MDL (in μg/L) was 0.19 for aqueous samples using GC/ECD.

The estimated MDL (in μg/kg) was not reported for soil samples using GC/ECD when corrected back to 50-g samples extracted and concentrated to 10 mL with 5-μL injections.

The estimated GC/MS identification limit (in ng) was not reported for soil samples using GC/MS.

Mean percent recovery, calculated from 7–8 determinations of spiked reagent water, after diazomethane derivatization, from a spike concentration (in μg/L) of 0.4 was 42 with a standard deviation of the percent recovery of 14.3.

Mean percent recovery, calculated from 10 determinations of spiked clay and clay/still bottom samples over the linear concentration range (in ng/g) of 0.82–1,620 was 93.7 with a percent relative standard deviation of 8.7. The RSD % was calculated on 10 samples high in the linear concentration range and 10 low in the range. The linear concentration range was determined using standard solutions and corrected to 50 g soil samples.

SAMPLE COLLECTION, PRESERVATION & HANDLING
Containers used to collect samples for the determination of semivolatile organic compounds should be soap and water washed followed by methanol (or isopropanol) rinsing. The sample containers should be of glass or Teflon® and have screw-top covers with Teflon® liners.

No preservation is used with concentrated waste samples. With liquid samples containing no residual chlorine and with soil, sediment, and sludge samples, immediately cooling to 4°C is the only preservation used. When residual chlorine is present then 3 mL of 10% aqueous sodium sulfate is added for each gallon of sample collected, followed by cooling to 4°C.

The holding time for all volatile organics samples is 14 days. Liquid samples must be extracted within 7 days and their extracts analyzed within 40 days. Concentrated waste, soil, sediment, and sludge samples must be extracted within 14 days and their extracts analyzed within 40 days.

SAMPLE PREPARATION
Preparation of soil, sediment, and other solid samples — Acidify 30 g (dry weight) solids with 0.1 M phosphate buffer (pH = 2.5) and thoroughly mix the contents. Spike the sample with surrogate compound(s). The ultrasonic extraction of solids must be optimized for each type of sample. In order for the ultrasonic extractor to efficiently extract solid samples, the sample must be free flowing when the solvent is added. Acidified anhydrous sodium sulfate should be added to clay-type soils, or any other solid that is not a free-flowing sandy texture, until a free flowing mixture is obtained. Add methylene chloride and perform ultrasonic extraction. Combine organic extracts from the repetitive extractings of the sample and centrifuge. Add aqueous potassium hydroxide, water, and methanol to the extract and reflux the mixture on a water bath. Extract the solution three times with methylene chloride and discard the methylene chloride phase. The basic solution contains the herbicide salts. Adjust the pH of the solution to <2 with cold sulfuric acid and extract three times with methylene chloride. Combine the extracts and pour them through a prerinsed drying column containing acidified anhydrous sodium sulfate. Collect the dried extracts in a K-D flask and concentrate them.

Preparation of aqueous samples — Measure 1 L of sample into a 2 L separatory funnel and spike it with surrogate compound(s). Add NaCl to the sample, then add 6 N NaOH to the sample to a pH of 12 or more and let the sample sit at room temperature for 1 h to hydrolyze esters. Extract the sample three times with methylene chloride and discard the extracts. Then add cold 12 N sulfuric acid to a pH less than or equal to 2, and extract the sample three times with ethyl ether. Collect the ether phase in a flask containing acidified anhydrous sodium sulfate and allow it to remain in contact with the sodium sulfate for a minimum of 2 h. The drying step is very critical to ensuring complete esterification; any moisture remaining in the ether will result in low herbicide recoveries.

Extract concentration and derivatization — The combined ether extract is concentrated to about 1 mL using a K-D apparatus followed by using a micro Snyder column or nitrogen gas blowdown. If methyl esters are to be produced, then dilute the concentrated ether extract with 1 mL of isooctane and 0.5 mL of methanol, dilute to a final volume of 4 mL, and esterify with diazomethane. If pentafluorobenzene esters are to be produced, then dilute concentrated ether extract with acetone to a final volume of 4 mL and esterify with pentafluorobenzyl bromide.

QUALITY CONTROL Select a representative spike concentration for each compound (acid or ester) to be measured. Using stock standard, prepare a quality control check sample concentrate, in acetone, that is 1000 times more concentrated than the selected concentrations. Use this quality control check sample concentrate to prepare quality control check samples. Calculate surrogate standard recovery on all standards, samples, blanks, and spikes. GC/MS techniques should be judiciously employed to support qualitative identifications made with this method. When available, chemical ionization mass spectra may be employed to aid the qualitative identification process.

REFERENCE Test Methods for Evaluating Solid Waste, Physical/Chemical Methods, SW-846, 3rd Edition, U.S. EPA, Office of Solid Waste, Washington, DC, EPA Method 8151, Nov. 1990.

Dinoseb   **EPA Method 8270**
CAS #88-85-7

TITLE Semivolatile Organic Compounds by GC/MS

MATRIX This method is used to determine the concentration of semivolatile organic compounds in extracts prepared from all types of solid waste matrices, soils, and groundwater. Although surface waters are not specifically mentioned, this method should be applicable to water samples from rivers, lakes, etc.

METHOD SUMMARY This method covers 259 semivolatile organic compounds. In very limited applications direct injection of the sample into the GC/MS system may be appropriate, but this results in very high detection limits (approximately 10,000 µg/L). Typically, a 1-L liquid sample, containing surrogate, and matrix spiking standards, is extracted in a continuous extractor first under acid conditions and then under basic conditions. Typically 30 g of a solid sample, containing surrogate, and matrix spiking standards, is extracted ultrasonically. After concentrating the extract to 1 mL it is spiked with 10 µL of an internal standard solution just prior to analysis by GC/MS. The volume injected should contain about 100 ng of base/neutral and 200 ng of acid surrogates (for a 1-µL injection). Analysis is performed by GC/MS using a capillary GC column.

INTERFERENCES Raw GC/MS data from all blanks, samples, and spikes must be evaluated for interferences. Contamination by carryover can occur whenever high-concentration and low-concentration samples are sequentially analyzed. To reduce carryover, the sample syringe must be rinsed out between samples with solvent. Whenever an unusually concentrated sample is encountered, it should be followed by the analysis of blank solvent to check for cross-contamination.

INSTRUMENTATION A GC/MS and a data system are required. The GC column used is a 30 m × 0.25 mm I.D. (or 0.32 mm I.D.) 1um film thickness silicone-coated fused silica capillary column. A continuous liquid-liquid extractor equipped with Teflon® or glass connection joints and stopcocks requiring no lubrication, a K-D concentrating apparatus, water bath, and an ultrasonic disrupter with a minimum power of 300 W and with pulsing capability are also required.

PRECISION & ACCURACY The estimated quantitation limit (EQL) of Method 8270B for determining an individual compound is approximately 1 mg/kg (wet weight) for soil or sediment samples, 1–200 mg/kg for wastes (dependent on matrix and method of preparation), and 10 µg/L for groundwater samples. EQLs will be proportionately higher for sample extracts that require dilution to avoid saturation of the detector.

The EQL(b) for groundwater in µg/L is 20.
The EQL (a, b) for low concentrations in soil and sediment in µg/kg is not determined.
Accuracy as µg/L is not listed.

Overall precision in µg/L is not listed.

(a) EQLs listed for soil/sediment are based on wet weight. Normally data is reported in a dry-weight basis; therefore, EQLs will be higher based on the % dry weight of each sample. This calculation is based on a 30 g sample and gel permeation chromatography cleanup.
(b) Sample EQLs are highly matrix-dependent. The EQLs are provided for guidance and may not always be achievable.
$C =$ True value for concentration, in µg/L.
$X =$ Average recovery found for measurements of samples containing a concentration of C, in µg/L.

ESTIMATED QUANTITATION LIMIT

Other Matrices	Factor (a)
High-concentration soil and sludges by sonicator	7.5
Non-water miscible waste	75

(a) EQL for other matrices = [EQL for low soil/sediment] × [Factor]. This estimated EQL is similar to an EPA "Practical Quantitation Limit."

SAMPLING METHOD
Liquid samples — Use a 1 or 2½ gallon amber glass bottle with a screw-top Teflon®-lined cover that has been prewashed with detergent and rinsed with distilled water and methanol (or isopropanol).

Soils, sediments, or sludges — Use an 8-oz. widemouth glass with a screw-top Teflon®-lined cover that has been prewashed with detergent and rinsed with distilled water and methanol (or isopropanol).

SAMPLE PRESERVATION
Liquid samples — If residual chlorine is present, add 3 mL of 10% sodium thiosulfate per gallon, cool to 4°C and store in a solvent-free refrigerator until analysis; if chlorine is not present, then eliminate the sodium thiosulfate addition.

Soils, sediments, or sludges — Cool samples to 4°C and store in a solvent-free refrigerator.

MHT Liquid samples must be extracted within 7 days and the extracts analyzed within 40 days. Soils, sediments, or sludges may be stored for a maximum of 14 days and the extracts analyzed within 40 days.

SAMPLE PREPARATION
Liquid samples — Transfer 1 L quantitatively to a continuous extractor. If high concentrations are anticipated, a smaller volume may be used and then diluted with organic-free reagent water to 1 L. Adjust pH, if necessary, to pH <2 using 1:1 (V/V) sulfuric acid. Pipette 1.0 mL of a surrogate standard spiking solution into each sample. For the sample in each analytical batch selected for spiking, add 1.0 mL of a matrix spiking standard. For base/neutral acid analysis, the amount of the surrogates and matrix spiking compounds added to the sample should result in a final concentration of 100 ng/µL of each analyte in the extract to be analyzed (assuming a 1-µL injection). Extract with methylene chloride for 18–24 h. Next, adjust the pH of the aqueous phase to pH >11 using 10 N sodium hydroxide and extract it with methylene chloride again for 18–24 h. Dry the extract through a column containing anhydrous

sodium sulfate and concentrate it to 1 mL using a K-D concentrator.

Soils, sediments, or sludges — Use 30 g of sample. Nonporous or wet samples (gummy or clay type) that do not have a free-flowing sandy texture must be mixed with anhydrous sodium sulfate until the sample is free flowing. Add 1 mL of surrogate standards to all samples, spikes, standards, and blanks. For the sample in each analytical batch selected for spiking, add 1.0 mL of a matrix spiking standard. For base/neutral acid analysis, the amount added of the surrogates and matrix spiking compounds should result in a final concentration of 100 ng/µL of each base/neutral analyte and 200 ng/µL of each acid analyte in the extract to be analyzed (assuming a 1-µL injection). Immediately add a 100-mL mixture of 1:1 methylene chloride:acetone and extract the sample ultrasonically for 3 min and then decant or filter the extracts. Repeat the extraction two or more times. Dry the extract using a column with anhydrous sodium sulfate and concentrate it to 1 mL in a K-D concentrator.

QUALITY CONTROL A methylene chloride solution containing 50 ng/µL of decafluorotriphenylphosphine (DFTPP) is used for tuning the GC/MS system each 12-h shift. A system performance check also must be made during every 12-h shift. A standard containing 50 ng/µL each of 4,4'-DDT, pentachlorophenol, and benzidine is required to verify injection port inertness and GC column performance. A calibration standard at mid-concentration, containing each compound of interest, including all required surrogates, must be performed every 12 h during analysis. After the system performance check is met, calibration check compounds (CCCs) are used to check the validity of the initial calibration.

The internal standard responses and retention times in the calibration check standard must be evaluated immediately after or during data acquisition. If the retention time for any internal standard changes by more than 30 seconds from the last check calibration (12 h), the chromatographic system must be inspected for malfunctions and corrections must be made, as required. If the electron ionization current plot (EICP) area for any of the internal standards changes by a factor of two from the last daily calibration standard check, the mass spectrometer must be inspected for malfunctions and corrections must be made, as appropriate.

Demonstrate, through the analysis of a reagent water blank, that interferences from the analytical system, glassware, and reagents are under control. The blank samples should be carried through all stages of the sample preparation and measurement steps. For each analytical batch (up to 20 samples), a reagent blank, matrix spike, and matrix spike duplicate/duplicate must be analyzed (the frequency of the spikes may be different for different monitoring programs). The blank and spiked samples must be carried through all stages of the sample preparation and measurement steps. A QC reference sample concentrate containing each analyte at a concentration of 100 mg/L in methanol is required.

REFERENCE Test Methods for Evaluating Solid Waste (SW-846). U.S. EPA 1983, Method 8270B, Rev. 2, Nov. 1990. Office of Solid Waste, Washington, DC.

Dinoseb EPA Method 8150
CAS #88-85-7

TITLE Chlorinated Herbicides

MATRIX Groundwater, soils, sludges, water miscible liquid wastes, and non-water miscible wastes.

APPLICATION This method is used for the analysis of 10 chlorinated herbicides. Samples are extracted, hydrolyzed with potassium hydroxide, and extraneous organics are removed by a solvent wash. After acidification, the acids are extracted, concentrated and converted to their methyl esters using diazomethane. They are then analyzed using direct injection into a gas chromatograph (GC). Be very careful because diazomethane can explode under certain conditions and it is also a carcinogen.

INTERFERENCES Organic acids and phenols (especially chlorinated acids and phenols) may cause interferences. Phthalate esters are not as significant an interference as with other GC-ECD methods if an electron capture detector is used. The herbicides may react readily with alkaline substances and be lost during analysis so all glassware and glass wool must be acid rinsed and sodium sulfate must be acidified with sulfuric acid prior to use. Sensitivity usually depends on the level of interferences rather than on instrumentation.

INSTRUMENTATION GC capable of on-column injections and an electron capture detector (ECD) or a halogen specific detector. Column 1: 1.8 m by 4 mm with 1.5% SP-2250/1.95% SP-2401 on Supelcoport. Column 2: 1.8 m by 4 mm with 5% OV-210 on Gas Chrom Q. Column 3: 1.98 m by 2 mm with 0.1%. SP-1000 on Carbopack C. The preferred column is Column Number 1.

RANGE Not listed.

MDL 0.07 µg/L (in reagent water; ECD).

PQL FACTORS FOR MULTIPLYING × FID MDL VALUE

Matrix	Multiplication Factor
Groundwater	10
Low-level soil by sonication with GPC cleanup	670
High-level soil and sludge by sonication	10,000
Non-water miscible waste	100,000

PRECISION (as standard deviation) 4% with 0.5 µg/L spike in municipal water. 3% with 102 µg/L in municipal water.

ACCURACY (as mean recovery) 86% with 0.5 µg/L spike in municipal water. 81% with 102 µg/L in municipal water.

SAMPLING METHOD Use 8-oz. widemouth glass bottles with Teflon®-lined caps for concentrated waste samples, soils, sediments, and sludges. Use 1 or 2½ gallon amber glass bottles with Teflon®-lined caps for liquid (water) samples.

STABILITY Cool soil, sediment, sludge, and liquid samples to 4°C. If residual chlorine is present in liquid samples add 3 mL of 10% sodium thiosulfate per gallon of sample and cool to 4°C.

MHT 14 days for concentrated waste, soil, sediment, or sludge; 7 days for liquid samples; all extracts must be analyzed within 40 days.

QUALITY CONTROL A quality control check sample concentrate containing this compound in acetone at a concentration 1,000 times more concentrated than the selected spike concentration is required. The QC check sample concentrate may be prepared from pure standard materials or purchased as certified solutions. Use appropriate trip, matrix, control site, method, reagent, and solvent blanks. Internal, surrogate, and five concentration level calibration standards are used.

REFERENCE Method 8150, SW-846, 3rd ed., Sept. 1986.

1,4-Dioxane (p-Dioxane)	EPA Method 1624
CAS #123-91-1	

TITLE Volatile Organic Compounds by Isotope Dilution GC/MS

MATRIX Compounds may be determined in waters, soils, and municipal sludges by this method.

METHOD SUMMARY This method is used to determine 58 volatile toxic organic pollutants associated with the CWA (as amended 1987); the RCRA (as amended 1986); the CERCLA (as amended 1986); and other compounds amenable to purge-and-trap gas chromatography-mass spectrometry (GC/MS).

If the solids content is less than 1%, stable isotopically-labeled analogs of the compounds of interest are added to a 5-mL sample and the sample is purged with an inert gas at 20–25°C in a chamber designed for soil or water samples. If the solids content is greater than 1%, 5 mL of reagent water and the labeled compounds are added to a 5-g aliquot of sample and the mixture is purged at 40°C. Compounds that will not purge at 20–25°C or at 40°C are purged at 78–85°C. In the purging process, the volatile compounds are transferred from the aqueous phase into the gaseous phase where they are passed into a sorbent column, and trapped. After purging is completed, the trap is backflushed and heated rapidly to desorb the compounds into a GC. The compounds are separated by the GC and detected by a MS. The labeled compounds serve to correct the variability of the analytical technique.

INTERFERENCES Impurities in the purge gas, organic compounds outgassing from the plumbing upstream of the trap, and solvent vapors in the lab account for most problems. Samples can be contaminated by diffusion of volatile organic compounds (particularly methylene chloride) through the bottle seal during shipment and storage. Contamination by carryover can occur when high-level and low-level samples are analyzed sequentially. When an unusually concentrated sample is encountered, follow it by analysis of a reagent water blank to check for carryover.

INSTRUMENTATION Major equipment includes a GC with linear temperature programming and a glass jet separator as the MS interface, a MS with 70 eV electron impact ionization, and a data system to collect and record response factors.

Column: 2.8 m × 2 mm I.D. glass, packed with 1% SP-1000 on Carbopak B, 60/80 mesh, or equivalent.

PRECISION & ACCURACY The detection limits of the method are usually dependent on the level of interferences rather than instrumental limitations. The method detection limits were determined in digested sludge (low solids) and in filter cake or compost (high solids).

The MDL (in µg/kg) for low solids is not listed and for high solids is 140.

Background levels of this compound were present in the sludge with high solids, resulting in a higher than expected MDL.

Labeled and native compound precision (in µg/L) as standard deviation was 7.2.

Labeled and native compound accuracy (in µg/L) as average recovery was 13–27.

Acceptance criteria are at 100 µg/L for this compound.

SAMPLE COLLECTION, PRESERVATION & HANDLING Grab samples are collected in glass containers having a total volume greater than 20 mL. Fill and seal each bottle so that no air bubbles are entrapped. Samples are maintained at 0 to 4°C from the time of collection until analysis. If an aqueous sample contains residual chlorine, add sodium thiosulfate preservative (10 mg/40 mL) to the empty sample bottles just prior to shipment to the sample site. All samples must be analyzed within 14 days of collection.

SAMPLE PREPARATION Samples containing less than 1% solids are analyzed directly as aqueous samples. Samples containing 1% solids or greater are analyzed as solid samples utilizing one of two methods, depending on the levels of pollutants, in the sample. Samples containing 1% solids or greater, and low to moderate levels of pollutants are analyzed by purging a known weight of sample added to 5 mL of reagent water. Samples containing 1% solids or greater, and high levels of pollutants, are extracted with methanol, and an aliquot of the methanol extract is added to reagent water and purged.

QUALITY CONTROL A field blank prepared from reagent water and carried through the sampling and handling protocol may serve as a check on contamination from shipment and storage.

The analyst is permitted to modify this method to improve separations or lower the costs of measurements, provided all performance specifications are met. Analyses of blanks are required. When results of spikes indicate atypical method performance for samples, the samples are diluted to bring method performance within acceptable limits. Analyze two sets of four 5-mL aliquots (8 aliquots total) of the aqueous performance standard. Spike all samples with labeled compounds to assess method performance on the sample matrix. Compute the percent recovery of the labeled compounds using the internal standard method. Compare the percent recovery for each compound with the corresponding labeled compound recovery. Reagent water blanks are analyzed to demonstrate freedom from carryover contamination. Field replicates may be collected to determine the precision of the sampling technique, and spiked samples may be required to determine the accuracy of the analysis when the internal method is used.

REFERENCE Volatile Organic Compounds by Isotope Dilution GC/MS. Office of Water Regulation and Standards, U.S. EPA Industrial Technology Division, Washington, DC, EPA Method 1624, Rev. C, June 1989 (contact W.A. Telliard, U.S. EPA, Office of Water Regulations and Standards, 401 M St., SW, Washington, DC, 20460. Phone: 202-382-7131).

1,4-Dioxane (p-Dioxane)   **EPA Method 8240**
CAS #123-91-1

TITLE Volatile Organics By GC/MS: Packed Column Technique

MATRIX Nearly all types of sample matarices, regardless of water content, can be analyzed using this method. This includes groundwater, aqueous sludges, caustic liquors, acid liquors, waste solvents, oily wastes, mousses, tars, fibrous wastes, polymetric emulsions, filter cakes, spent carbons, spent catalysts, soils, and sediments.

METHOD SUMMARY Method 8240B covers 80 volatile organic compounds that are introduced into a gas chromatograph by the purge-and-trap method or by direct injection (in limited applications). For the purge-and-trap method an inert gas (zero grade nitrogen or helium) is bubbled through a 5-mL solution at ambient temperature. Purged sample components are trapped in a tube of sorbent materials. When purging is complete, the sorbent tube is heated and backflushed with inert gas to desorb the trapped components onto a GC column.

INTERFERENCES Impurities in the purge gas and from organic compounds outgassing from the plumbing ahead of the trap account for many contamination problems. Interferences purged or coextracted from the samples will vary considerably from source to source. Cross-contamination can occur whenever high-level and low-level samples are analyzed sequentially. Whenever an unusually concentrated sample is analyzed, it should be followed by the analysis of organic-free reagent water to check for cross-contamination. Samples also can be contaminated by diffusion of volatile organics (particularly methylene chloride and fluorocarbons) through the septum seal into the sample during shipment and storage. A trip blank can serve as a check on such contamination. The lab where volatile analysis is performed and also the refrigerated storage area should be completely free of solvents.

INSTRUMENTATION A gas chromatograph/mass spectrometry/data system (GC/MS) equipped with a 6 ft × 0.1 in I.D. glass column packed with 1% SP-1000 on Carbopack-B (60/80 mesh) is required. Also needed is a 5-mL purging device, a sorbent trap, and a thermal desorption apparatus.

PRECISION & ACCURACY This method is reported to have been tested by 15 laboratories using organic-free reagent water, drinking water, surface water, and industrial wastewaters (not specified) fortified at six concentrations over the range 5–600 µg/L.

Sample estimated quantitation limits (EQLs) are highly matrix-dependent. The EQLs listed may not always be achievable. EQLs listed for soils or sediments are based on wet weight. Normally, data is reported on a dry-weight basis; therefore, EQLs will be higher, based on the percent dry weight of each sample. Note that EQLs are even more variable than MDLs and that they are highly variable depending on the matrix being analyzed.

EQL in groundwater in µg/L was not listed.
EQL in low soil or sediment in µg/kg was not listed.
Accuracy (a) in µg/L was not listed.
Precision (b) in µg/L was not listed.

(a) *Average recovery found for measurements of samples containing a concentration of C, in µg/L.*
(b) *Overall precision found for measurements of samples with average recovery X for samples containing a concentration of C in µg/L.*
X = *Average recovery found for measurement of samples containing a concentration of C in µg/L.*

MULTIPLICATION FACTORS FOR OTHER MATRICES

Other Matrices	Factor (a)
Waste miscible liquid waste	50
High-concentration soil and sludge	125
Non-water miscible waste	500

(a) *EQL = [EQL for low soil sediment] × [Factor]. For non-aqueous samples, the factor is on a wet-weight basis.*

SAMPLING METHOD

Liquid samples — Use a 40-mL glass screw-cap VOA vial with a Teflon®-faced silicone septum that has been prewashed, rinsed with distilled deionized water, and oven dried. However, if residual chlorine is present, collect sample in a 40-oz. soil VOA container which has been pre-preserved with 4 drops of 10% sodium thiosulfate, mix gently, and then transfer the sample to a 40-mL VOA vial. Collect bubble-free samples in duplicate and seal them in separate plastic bags.

Soils or sediments, and sludges — Use an 8-oz. widemouth glass bottle with a Teflon®-faced silicone septum that has been prewashed with detergent, rinsed with distilled deionized water, and oven dried. Tap slightly to eliminate free air space. Collect samples in duplicate and seal them in separate plastic bags.

SAMPLE PRESERVATION

Liquid samples — Add 4 drops of concentrated HCL and immediately cool samples to 4°C and store in a solvent-free refrigerator.

Soils or sediments, and sludges — Cool samples to 4°C and store in a solvent-free refrigerator.

MHT Maximum holding time is 14 days from the date of sample collection.

SAMPLE PREPARATION

Liquid samples — Remove the plunger from a 5-mL syringe and carefully pour the sample into the syringe barrel to just short of overflowing. Replace the syringe plunger and compress the sample. Open the syringe valve and vent any residual air while adjusting the sample volume to 5.0 mL. If there is only one volatile organic analysis (VOA) vial, a second syringe should be filled at this time to protect against possible loss of sample integrity. Add 10 µL of surrogate spiking solution and 10 µL of internal standard spiking solution through the valve bore of the 5-mL syringe, then close the valve. The surrogate and internal standards may be mixed and added as a single spiking solution.

Sediments, soils, and waste samples — All samples of this type should be screened by GC analysis using a headspace method (EPA Method 3810) or the hexadecane extraction and screening method (EPA Method 3820). Use the screening data to determine whether to use the low-concentration method (0.005–1 mg/kg) or the high-concentration method (>1 mg/kg).

Low-concentration method — The low-concentration method is based on purging a heated sediment or soil sample mixed with organic-free reagent water containing the surrogate and internal standards. Analyze all reagent blanks and standards under the same conditions as the samples.

Use a 5-g sample if the expected concentration is <0.1 mg/kg or a 1-g sample for expected concentrations between 0.1 and 1 mg/kg. Mix the contents of the sample container with a narrow metal spatula. Weigh the amount of the sample into a tared purge device. Add the spiked water to the purge device, which contains the weighed amount of sample, and connect the device to the purge-and-trap system.

High-concentration method — This method is based on extracting the sediment or soil with methanol. A waste sample is either extracted or diluted, depending on its solubility in methanol. Wastes that are insoluble in methanol are diluted with reagent tetraglyme or possibly polyethylene glycol (PEG). An aliquot of the extract is added to organic-free reagent water containing surrogate and internal standards. This is purged at ambient temperature. All samples with an expected concentration of >1.0 mg/kg should be analyzed by this method.

Mix the contents of the sample container with a narrow metal spatula. For sediments or soils and solid wastes that are insoluble in methanol, weigh 4 g (wet weight) of sample into a tared 20-mL vial. For waste that is soluble in methanol, tetraglyme, or PEG, weigh 1 g (wet weight) into a tared scintillation vial or culture tube or a 10-mL volumetric flask. Quickly add 9.0 mL of appropriate solvent then add 1.0 mL of a surrogate spiking solution to the vial, cap it, and shake it for 2 min.

METHANOL EXTRACT REQUIRED FOR ANALYSIS OF HIGH-CONCENTRATION SOILS OR SEDIMENTS

Approximate Concentration Range	Volume of Methanol Extract (a)
500–10,000 µg/kg	100 µL
1,000–20,000 µg/kg	50 µL
5,000–100,000 µg/kg	10 µL
25,000–500,000 µg/kg	100 µL of 1/50 dilution (b)

Calculate appropriate dilution factor for concentrations exceeding this table.

(a) The volume of methanol added to 5 mL of water being purged should be kept constant. Therefore, add to the 5-mL syringe whatever volume of methanol is necessary to maintain a volume of 100 µL added to the syringe.

(b) Dilute an aliquot of the methanol extract and then take 100 µL for analysis.

QUALITY CONTROL Demonstrate, through the analysis of a reagent water blank, that interferences from the analytical system, glassware, and reagents are under control. Blank samples should be carried through all stages of the sample preparation and measurement steps. For each analytical batch (up to 20 samples), a reagent blank, matrix spike, and matrix spike duplicate must be analyzed (the frequency of the spikes may be different for different monitoring programs). The blank and spiked samples must be carried through all stages of the sample preparation and measurement steps. QC samples mentioned in the section on Interferences will also be needed as appropriate to those situations.

REFERENCE Test Methods for Evaluating Solid Waste (SW-846). U.S. EPA. 1983. Method 8240B, Rev. 2, Nov. 1990. Office of Solid Wastes, Washington, DC.

Dioxathion   **EPA Method 8270**
CAS #78-34-2

TITLE Semivolatile Organic Compounds by GC/MS

MATRIX This method is used to determine the concentration of semivolatile organic compounds in extracts prepared from all types of solid waste matrices, soils, and groundwater. Although surface waters are not specifically mentioned, this method should be applicable to water samples from rivers, lakes, etc.

METHOD SUMMARY This method covers 259 semivolatile organic compounds. In very limited applications direct injection of the sample into the GC/MS system may be appropriate, but this results in very high detection limits (approximately 10,000 µg/L). Typically, a 1-L liquid sample, containing surrogate, and matrix spiking standards, is extracted in a continuous extractor first under acid conditions and then under basic conditions. Typically 30 g of a solid sample, containing surrogate, and matrix spiking standards, is extracted ultrasonically. After concentrating the extract to 1 mL it is spiked with 10 µL of an internal standard solution just prior to analysis by GC/MS. The volume injected should contain about 100 ng of base/neutral and 200 ng of acid surrogates (for a 1-µL injection). Analysis is performed by GC/MS using a capillary GC column.

INTERFERENCES Raw GC/MS data from all blanks, samples, and spikes must be evaluated for interferences. Contamination by carryover can occur whenever high-concentration and low-concentration samples are sequentially analyzed. To reduce carryover, the sample syringe must be rinsed out between samples with solvent. Whenever an unusually concentrated

sample is encountered, it should be followed by the analysis of blank solvent to check for cross-contamination.

INSTRUMENTATION A GC/MS and a data system are required. The GC column used is a 30 m × 0.25 mm I.D. (or 0.32 mm I.D.) 1um film thickness silicone-coated fused silica capillary column. A continuous liquid-liquid extractor equipped with Teflon® or glass connection joints and stopcocks requiring no lubrication, a K-D concentrating apparatus, water bath, and an ultrasonic disrupter with a minimum power of 300 W and with pulsing capability are also required.

PRECISION & ACCURACY The estimated quantitation limit (EQL) of Method 8270B for determining an individual compound is approximately 1 mg/kg (wet weight) for soil or sediment samples, 1–200 mg/kg for wastes (dependent on matrix and method of preparation), and 10 µg/L for groundwater samples. EQLs will be proportionately higher for sample extracts that require dilution to avoid saturation of the detector.

The EQL(b) for groundwater in µg/L is not listed.
The EQL (a, b) for low concentrations in soil and sediment in µg/kg is not listed.
Accuracy as µg/L is not listed.
Overall precision in µg/L is not listed.

(a) EQLs listed for soil/sediment are based on wet weight. Normally data is reported in a dry-weight basis; therefore, EQLs will be higher based on the % dry weight of each sample. This calculation is based on a 30 g sample and gel permeation chromatography cleanup.
(b) Sample EQLs are highly matrix-dependent. The EQLs are provided for guidance and may not always be achievable.
C = True value for concentration, in µg/L.
X = Average recovery found for measurements of samples containing a concentration of C, in µg/L.

ESTIMATED QUANTITATION LIMIT

Other Matrices	Factor (a)
High-concentration soil and sludges by sonicator	7.5
Non-water miscible waste	75

(a) EQL for other matrices = [EQL for low soil/sediment] × [Factor]. This estimated EQL is similar to an EPA "Practical Quantitation Limit."

SAMPLING METHOD

Liquid samples — Use a 1 or 2½ gallon amber glass bottle with a screw-top Teflon®-lined cover that has been prewashed with detergent and rinsed with distilled water and methanol (or isopropanol).

Soils, sediments, or sludges — Use an 8-oz. widemouth glass with a screw-top Teflon®-lined cover that has been prewashed with detergent and rinsed with distilled water and methanol (or isopropanol).

SAMPLE PRESERVATION

Liquid samples — If residual chlorine is present, add 3 mL of 10% sodium thiosulfate per gallon, cool to 4°C and store in a solvent-free refrigerator until analysis; if chlorine is not present, then eliminate the sodium thiosulfate addition.

Soils, sediments, or sludges — Cool samples to 4°C and store in a solvent-free refrigerator.

MHT Liquid samples must be extracted within 7 days and the extracts analyzed within 40 days. Soils, sediments, or sludges may be stored for a maximum of 14 days and the extracts analyzed within 40 days.

SAMPLE PREPARATION

Liquid samples — Transfer 1 L quantitatively to a continuous extractor. If high concentrations are anticipated, a smaller volume may be used and then diluted with organic-free reagent water to 1 L. Adjust pH, if necessary, to pH <2 using 1:1 (V/V) sulfuric acid. Pipette 1.0 mL of a surrogate standard spiking solution into each sample. For the sample in each analytical batch selected for spiking, add 1.0 mL of a matrix spiking standard. For base/neutral acid analysis, the amount of the surrogates and matrix spiking compounds added to the sample should result in a final concentration of 100 ng/µL of each analyte in the extract to be analyzed (assuming a 1-µL injection). Extract with methylene chloride for 18–24 h. Next, adjust the pH of the aqueous phase to pH >11 using 10 N sodium hydroxide and extract it with methylene chloride again for 18–24 h. Dry the extract through a column containing anhydrous sodium sulfate and concentrate it to 1 mL using a K-D concentrator.

Soils, sediments, or sludges — Use 30 g of sample. Nonporous or wet samples (gummy or clay type) that do not have a free-flowing sandy texture must be mixed with anhydrous sodium sulfate until the sample is free flowing. Add 1 mL of surrogate standards to all samples, spikes, standards, and blanks. For the sample in each analytical batch selected for spiking, add 1.0 mL of a matrix spiking standard. For base/neutral acid analysis, the amount added of the surrogates and matrix spiking compounds should result in a final concentration of 100 ng/µL of each base/neutral analyte and 200 ng/µL of each acid analyte in the extract to be analyzed (assuming a 1-µL injection). Immediately add a 100-mL mixture of 1:1 methylene chloride:acetone and extract the sample ultrasonically for 3 min and then decant or filter the extracts. Repeat the extraction two or more times. Dry the extract using a column with anhydrous sodium sulfate and concentrate it to 1 mL in a K-D concentrator.

QUALITY CONTROL A methylene chloride solution containing 50 ng/µL of decafluorotriphenylphosphine (DFTPP) is used for tuning the GC/MS system each 12-h shift. A system performance check also must be made during every 12-h shift. A standard containing 50 ng/µL each of 4,4'-DDT, pentachlorophenol, and benzidine is required to verify injection port inertness and GC column performance. A calibration standard at mid-concentration, containing each compound of interest, including all required surrogates, must be performed every 12 h during analysis. After the system performance check is met, calibration check compounds (CCCs) are used to check the validity of the initial calibration.

The internal standard responses and retention times in the calibration check standard must be evaluated immediately after or during data acquisition. If the retention time for any internal standard changes by more than 30 seconds from the last check calibration (12 h), the chromatographic system must be

inspected for malfunctions and corrections must be made, as required. If the electron ionization current plot (EICP) area for any of the internal standards changes by a factor of two from the last daily calibration standard check, the mass spectrometer must be inspected for malfunctions and corrections must be made, as appropriate.

Demonstrate, through the analysis of a reagent water blank, that interferences from the analytical system, glassware, and reagents are under control. The blank samples should be carried through all stages of the sample preparation and measurement steps. For each analytical batch (up to 20 samples), a reagent blank, matrix spike, and matrix spike duplicate/duplicate must be analyzed (the frequency of the spikes may be different for different monitoring programs). The blank and spiked samples must be carried through all stages of the sample preparation and measurement steps. A QC reference sample concentrate containing each analyte at a concentration of 100 mg/L in methanol is required.

REFERENCE Test Methods for Evaluating Solid Waste (SW-846). U.S. EPA 1983, Method 8270B, Rev. 2, Nov. 1990. Office of Solid Waste, Washington, DC.

Diphenamid   **EPA Method 507**
CAS #957-51-7

TITLE Determination of Nitrogen and Phosphorus-Containing Pesticides in Water by GC/NPD

MATRIX This method is applicable to the determination of certain nitrogen and phosphorus-containing pesticides in finished drinking water and groundwater.

METHOD SUMMARY Method 507 covers 46 nitrogen- and phosphorus-containing pesticides. A 1-L sample is fortified with a surrogate standard, salted, buffered, extracted with methylene chloride, and concentrated; then the solvent is exchanged with methyl tert-butyl ether (MTBE) and concentrated again, and a 2-µL aliquot of a sample extract is injected into a GC system equipped with a selective nitrogen-phosphorus detector and a capillary column for analysis.

INTERFERENCES Method interferences may be caused by contaminants in solvents, reagents, glassware, and other sample processing apparatus. Interfering contamination may occur when a sample containing low concentrations of analytes is analyzed immediately following a sample containing relatively high concentrations. One or more injections of MTBE should be made following the analysis of a sample with high concentrations of analytes to check for analyte carryover. Matrix interferences may be caused by contaminants that are coextracted from the sample. The extent of matrix interferences will vary considerably from source to source, depending upon the water sampled.

INSTRUMENTATION A gas chromatograph system (GC) equipped with a nitrogen-phosphorus detector (NPD) is needed.

Column 1: 30 m × 0.25 mm I.D. DB-5 bonded fused silica column, 0.25 µm film thickness, or equivalent.
Column 2: 30 m × 0.25 mm I.D. DB-1701 bonded fused silica column, 0.25 µm film thickness, or equivalent.

PRECISION & ACCURACY This method has been validated in a single lab and estimated detection limits (EDLs) have been determined for each analyte. Observed detection limits may vary among waters, depending upon the nature of the interferences in the sample matrix and the specific instrumentation used. Analytes that are not separated chromatographically cannot be individually identified and measured unless an alternative technique for identification and quantification exist.

The estimated detection limit (in µg/L) was 0.6. The EDL is defined as either method detection limit or a level of compound in a sample yielding a peak in the final extract with signal-to-noise ratio of approximately 5, whichever value is higher.

The concentration used for these measurements (in µg/L) was 6.
The accuracy (as % recovery) was 93.
The precision (% RSD) was 8.

SAMPLING METHOD Grab samples are collected in 1-L glass sample bottles (prewashed with detergent and hot tap water, rinsed with reagent water, and dried in an oven at 400°C for 1 h) with screw caps lined with PTFE-fluorocarbon.

SAMPLE PRESERVATION Add mercuric chloride to the sample bottle in amounts to produce a concentration of 10 mg/L. If residual chlorine is present, add 80 mg of sodium thiosulfate/L of sample to the sample bottle prior to collection. After collection, seal bottle and shake vigorously for 1 min, then cool the sample to 4°C immediately and store it at 4°C in the dark until extraction.

MHT Maximum holding time of the samples, and in some cases the extracts, is 14 days.

SAMPLE PREPARATION Fortify the sample with 50 µL of the surrogate standard solution, adjust to pH 7 with phosphate buffer, add 100 g NaCl to the sample, and seal and shake to dissolve the salt; then extract with methylene chloride in a separatory funnel or in a mechanical tumbler bottle. Dry the extract by pouring it through a solvent-rinsed drying column containing about 10 cm of anhydrous sodium sulfate. Collect the extract in a Kuderna-Danish (K-D) concentrator and rinse the column with 20–30 mL methylene chloride. Concentrate the extract to about 2 mL and rinse the flask and its lower joint into the concentrator tube with 1 to 2 mL of methyl t-butyl ether (MTBE). Add 5–10 mL of MTBE and concentrate the extract twice (adding more MTBE) to a final volume of 5.0 mL and store it at 4°C until analysis.

Note: If methylene chloride is not completely removed from the final extract, it may cause detector problems.

QUALITY CONTROL Minimum quality control requirements are initial demonstration of lab capability, determination of surrogate compound recoveries in each sample and blank, monitoring internal standard peak area or height in each sample and blank, analysis of lab reagent blanks, lab fortified samples, lab fortified blanks, and other QC samples. A lab reagent blank is analyzed to demonstrate that all glassware and reagent interferences are under control.

Initial demonstration of capability is fulfilled by analyzing four fortified reagent water samples with the recovery value for each analyte falling within the acceptable range (±30% average recovery). Surrogate recoveries from samples or method blanks must be 70–130%. The internal standard response for any sample chromatogram should not deviate from the daily calibration check standard's internal standard response by more than 30% or lab fortified blanks and sample matrices are used to assess lab performance and analyte recovery, respectively.

If the response for the target analyte peak exceeds the working range of the system, dilute the extract and reanalyze. Alternative techniques such as an alternate detector or second chromatography column should be used to confirm peak identification when sample components are not resolved adequately.

EPA CONTACT & HOTLINE For technical questions contact Dr. Baldev Bathija, U.S. EPA, Office of Ground Water and Drinking Water (WH-550D), 401 M St. SW, Washington, DC 20460. Tel. (202) 260-3040. For further information the EPA Safe Drinking Water Hotline may be called at: (800) 426-4791.

REFERENCE Methods for the Determination of Organic Compounds in Drinking Water, EPA/600/4-88/039 (revised July 1991). U.S. EPA Environmental Monitoring Systems Laboratory, Cincinnati, OH, 45268, U.S.A. Available from the National Technical Information Service (NTIS), 5285 Port Royal Road, Springfield, VA 22161; Tel. 800-553-6847. NTIS Order Number is PB91-231480.

Diphenyl ether   EPA Method 1625
CAS #101-84-8

TITLE Semivolatile Organic Compounds by Isotope Dilution GC/MS

MATRIX The compounds may be determined in waters, soils, and municipal sludges by this method.

METHOD SUMMARY This method is used to determine 176 semivolatile toxic organic pollutants associated with the CWA (as amended 1987); the RCRA (as amended 1986); the CERCLA (as amended 1986); and other compounds amenable to extraction and analysis by capillary column gas chromatography-mass spectrometry (GC/MS).

Stable isotopically-labeled analogs of the compounds of interest are added to the sample. If the solids content is less than 1%, a 1-L sample is extracted at pH 12–13, then at pH <2 with methylene chloride using continuous extraction techniques.

If the solids content is 30% or less, the sample is diluted to 1% solids with reagent water, homogenized ultrasonically, and extracted at pH 12–13, then at pH <2 with methylene chloride using continuous extraction techniques. If the solids content is greater than 30%, the sample is extracted using ultrasonic techniques.

Each extract is dried over sodium sulfate, concentrated to a volume of 5 mL, cleaned up using GPC, if necessary, and concentrated. Extracts are concentrated to 1 mL if GPC is not performed, and to 0.5 mL if GPC is performed.

An internal standard is added to the extract, and a 1-mL aliquot of the extract is injected into the GC. The compounds are separated by GC and detected by a MS. The labeled compounds serve to correct the variability of the analytical technique.

INTERFERENCES Solvents, reagents, glassware, and other sample processing hardware may yield artifacts and/or elevated baselines causing misinterpretation of chromatograms and spectra. Materials used in the analysis must be demonstrated to be free from interferences under the conditions of analysis by running method blanks initially and with each sample lot (sample started through the extraction process on a given 8-h shift, to a maximum of 20). Specific selection of reagents and purification of solvents by distillation in all glass systems may be required. Glassware and, where possible, reagents are cleaned by solvent rinse and baking at 450°C for 1-h minimum. Interferences coextracted from samples will vary considerably from source to source, depending on the diversity of the site being sampled.

INSTRUMENTATION Major instrumentation includes a GC with a splitless or on-column injection port for capillary column, a MS with 70 eV electron impact ionization, and a data system to collect and record MS data, and process it. A K-D apparatus is used to concentrate extracts.

GC Column: 30 m × 0.25 mm I.D. 5% phenyl, 94% methyl, 1% vinyl silicone bonded phased fused silica capillary column.

PRECISION & ACCURACY The detection limits of the method are usually dependent on the level of interferences rather than instrumental limitations. The limits typify the minimum quantities that can be detected with no interferences present.

The minimum level (in µg/mL) was 10. This is defined as a minimum level at which the analytical system shall give recognizable mass spectra (background corrected) and acceptable calibration points.

The MDL (in µg/kg) in low solids was 44 and in high solids was 12; these were determined in digested sludge (low solids) and in filter cake or compost (high solids).

The labeled and native compound initial precision as standard deviation (in µg/L) was 19.
The labeled and native compound initial accuracy as average recovery (in µg/L) was 82–136.

SAMPLE COLLECTION, PRESERVATION & HANDLING Collect samples in glass containers. Aqueous samples which flow freely are collected in refrigerated bottles using automatic sampling equipment. Solid samples are collected as grab samples using widemouth jars. Maintain samples at 0 to 4°C from the time of collection until extraction. If residual chlorine is present in aqueous samples, add 80 mg sodium thiosulfate/L of water. Begin sample extraction within 7 days of collection, and analyze all extracts within 40 days of extraction.

SAMPLE PREPARATION Samples containing 1% solids or less are extracted directly using continuous liquid-liquid extraction techniques. Samples containing 1 to 30% solids are diluted to the 1% level with reagent water and extracted using continuous liquid-liquid extraction techniques. Samples containing greater than 30% solids are extracted using ultrasonic techniques.

Base/neutral extraction — Adjust the pH of the waters in the extractors to 12–13 with 6 N NaOH. Extract with methylene chloride for 24–48 h.

Acid extraction — Adjust the pH of the waters in the extractors to 2 or less using 6 N sulfuric acid. Extract with methylene chloride for 24–48 h.

Ultrasonic extraction of high solids samples — Add anhydrous sodium sulfate to the sample and QC aliquot(s). Add acetone:methylene chloride (1:1) to the sample and mix thoroughly

Concentrate extracts using a K-D apparatus.

QUALITY CONTROL The analyst is permitted to modify this method to improve separations or lower the costs of measurements, provided all performance specifications are met. Analyses of blanks are required to demonstrate freedom from contamination. When results of spikes indicate atypical method performance for samples, the samples are diluted to bring method performance within acceptable limits.

For low solids (aqueous samples), extract, concentrate, and analyze two sets of four 1-L aliquots (8 aliquots total) of the precision and recovery standard. For high solids samples, two sets of four 30-g aliquots of the high solids reference matrix are used.

Spike all samples with labeled compounds to assess method performance. Compute percent recovery of the labeled compounds using the internal standard method. Compare the labeled compound recovery for each compound with the corresponding labeled compound recovery.

Reagent water and high solids reference matrix blanks are analyzed to demonstrate freedom from contamination. Extract and concentrate a 1-L reagent water blank or a high solids reference matrix blank with each sample's lot (samples started through the extraction process on the same 8-h shift, to a maximum of 20 samples).

Field replicates may be collected to determine the precision of the sampling technique, and spiked samples may be required to determine the accuracy of the analysis when the internal standard method is used.

REFERENCE Semivolatile Organic Compounds by Isotope Dilution GC/MS. Office of Water Regulation and Standards, U.S. EPA Industrial Technology Division, Washington, DC, EPA Method 1625, Rev. C, June 1989 (contact W.A. Telliard, U.S. EPA, Office of Water Regulations and Standards, 401 M St., SW, Washington, DC, 20460. Phone: 202-382-7131).

Diphenylamine   **EPA Method 1625**
CAS #122-39-4

TITLE Semivolatile Organic Compounds by Isotope Dilution GC/MS

MATRIX The compounds may be determined in waters, soils, and municipal sludges by this method.

METHOD SUMMARY This method is used to determine 176 semivolatile toxic organic pollutants associated with the CWA (as amended 1987); the RCRA (as amended 1986); the CERCLA (as amended 1986); and other compounds amenable to extraction and analysis by capillary column gas chromatography-mass spectrometry (GC/MS).

Stable isotopically-labeled analogs of the compounds of interest are added to the sample. If the solids content is less than 1%, a 1-L sample is extracted at pH 12–13, then at pH <2 with methylene chloride using continuous extraction techniques.

If the solids content is 30% or less, the sample is diluted to 1% solids with reagent water, homogenized ultrasonically, and extracted at pH 12–13, then at pH <2 with methylene chloride using continuous extraction techniques. If the solids content is greater than 30%, the sample is extracted using ultrasonic techniques.

Each extract is dried over sodium sulfate, concentrated to a volume of 5 mL, cleaned up using GPC, if necessary, and concentrated. Extracts are concentrated to 1 mL if GPC is not performed, and to 0.5 mL if GPC is performed.

An internal standard is added to the extract, and a 1-mL aliquot of the extract is injected into the GC. The compounds are separated by GC and detected by a MS. The labeled compounds serve to correct the variability of the analytical technique.

INTERFERENCES Solvents, reagents, glassware, and other sample processing hardware may yield artifacts and/or elevated baselines causing misinterpretation of chromatograms and spectra. Materials used in the analysis must be demonstrated to be free from interferences under the conditions of analysis by running method blanks initially and with each sample lot (sample started through the extraction process on a given 8-h shift, to a maximum of 20). Specific selection of reagents and purification of solvents by distillation in all glass systems may be required. Glassware and, where possible, reagents are cleaned by solvent rinse and baking at 450°C for 1-h minimum. Interferences coextracted from samples will vary considerably from source to source, depending on the diversity of the site being sampled.

INSTRUMENTATION Major instrumentation includes a GC with a splitless or on-column injection port for capillary column, a MS with 70 eV electron impact ionization, and a data system to collect and record MS data, and process it. A K-D apparatus is used to concentrate extracts.

GC Column: 30 m × 0.25 mm I.D. 5% phenyl, 94% methyl, 1% vinyl silicone bonded phased fused silica capillary column.

PRECISION & ACCURACY The detection limits of the method are usually dependent on the level of interferences rather than instrumental limitations. The limits typify the minimum quantities that can be detected with no interferences present.

The minimum level (in µg/mL) was 20. This is defined as a minimum level at which the analytical system shall give recognizable mass spectra (background corrected) and acceptable calibration points.

The MDL (in µg/kg) in low solids was 58 and in high solids was 54; these were determined in digested sludge (low solids) and in filter cake or compost (high solids).

The labeled and native compound initial precision as standard deviation (in µg/L) was 45.

The labeled and native compound initial accuracy as average recovery (in µg/L) was 58–205.

SAMPLE COLLECTION, PRESERVATION & HANDLING Collect samples in glass containers. Aqueous samples which flow freely are collected in refrigerated bottles using automatic sampling equipment. Solid samples are collected as grab samples using widemouth jars. Maintain samples at 0 to 4°C from the time of collection until extraction. If residual chlorine is present in aqueous samples, add 80 mg sodium thiosulfate/L of water. Begin sample extraction within 7 days of collection, and analyze all extracts within 40 days of extraction.

SAMPLE PREPARATION Samples containing 1% solids or less are extracted directly using continuous liquid-liquid extraction techniques. Samples containing 1 to 30% solids are diluted to the 1% level with reagent water and extracted using continuous liquid-liquid extraction techniques. Samples containing greater than 30% solids are extracted using ultrasonic techniques.

- Base/neutral extraction — Adjust the pH of the waters in the extractors to 12–13 with 6 N NaOH. Extract with methylene chloride for 24–48 h.
- Acid extraction — Adjust the pH of the waters in the extractors to 2 or less using 6 N sulfuric acid. Extract with methylene chloride for 24–48 h.
- Ultrasonic extraction of high solids samples — Add anhydrous sodium sulfate to the sample and QC aliquot(s). Add acetone:methylene chloride (1:1) to the sample and mix thoroughly

Concentrate extracts using a K-D apparatus.

QUALITY CONTROL The analyst is permitted to modify this method to improve separations or lower the costs of measurements, provided all performance specifications are met. Analyses of blanks are required to demonstrate freedom from contamination. When results of spikes indicate atypical method performance for samples, the samples are diluted to bring method performance within acceptable limits.

For low solids (aqueous samples), extract, concentrate, and analyze two sets of four 1-L aliquots (8 aliquots total) of the precision and recovery standard. For high solids samples, two sets of four 30-g aliquots of the high solids reference matrix are used.

Spike all samples with labeled compounds to assess method performance. Compute percent recovery of the labeled compounds using the internal standard method. Compare the labeled compound recovery for each compound with the corresponding labeled compound recovery.

Reagent water and high solids reference matrix blanks are analyzed to demonstrate freedom from contamination. Extract and concentrate a 1-L reagent water blank or a high solids reference matrix blank with each sample's lot (samples started through the extraction process on the same 8-h shift, to a maximum of 20 samples).

Field replicates may be collected to determine the precision of the sampling technique, and spiked samples may be required to determine the accuracy of the analysis when the internal standard method is used.

REFERENCE Semivolatile Organic Compounds by Isotope Dilution GC/MS. Office of Water Regulation and Standards, U.S. EPA Industrial Technology Division, Washington, DC, EPA Method 1625, Rev. C, June 1989 (contact W.A. Telliard, U.S. EPA, Office of Water Regulations and Standards, 401 M St., SW, Washington, DC, 20460. Phone: 202-382-7131).

Diphenylamine   EPA Method 8270
CAS #122-39-4

TITLE Semivolatile Organic Compounds by GC/MS

MATRIX This method is used to determine the concentration of semivolatile organic compounds in extracts prepared from all types of solid waste matrices, soils, and groundwater. Although surface waters are not specifically mentioned, this method should be applicable to water samples from rivers, lakes, etc.

METHOD SUMMARY This method covers 259 semivolatile organic compounds. In very limited applications direct injection of the sample into the GC/MS system may be appropriate, but this results in very high detection limits (approximately 10,000 µg/L). Typically, a 1-L liquid sample, containing surrogate, and matrix spiking standards, is extracted in a continuous extractor first under acid conditions and then under basic conditions. Typically 30 g of a solid sample, containing surrogate, and matrix spiking standards, is extracted ultrasonically. After concentrating the extract to 1 mL it is spiked with 10 µL of an internal standard solution just prior to analysis by GC/MS. The volume injected should contain about 100 ng of base/neutral and 200 ng of acid surrogates (for a 1-µL injection). Analysis is performed by GC/MS using a capillary GC column.

INTERFERENCES Raw GC/MS data from all blanks, samples, and spikes must be evaluated for interferences. Contamination by carryover can occur whenever high-concentration and low-concentration samples are sequentially analyzed. To reduce carryover, the sample syringe must be rinsed out between samples with solvent. Whenever an unusually concentrated sample is encountered, it should be followed by the analysis of blank solvent to check for cross-contamination.

INSTRUMENTATION A GC/MS and a data system are required. The GC column used is a 30 m × 0.25 mm I.D. (or 0.32 mm I.D.) 1um film thickness silicone-coated fused silica capillary column. A continuous liquid-liquid extractor equipped with Teflon® or glass connection joints and stopcocks requiring no lubrication, a K-D concentrating apparatus, water bath, and an ultrasonic disrupter with a minimum power of 300 W and with pulsing capability are also required.

PRECISION & ACCURACY The estimated quantitation limit (EQL) of Method 8270B for determining an individual compound is approximately 1 mg/kg (wet weight) for soil or sediment samples, 1–200 mg/kg for wastes (dependent on matrix and method of preparation), and 10 µg/L for groundwater samples. EQLs will be proportionately higher for sample extracts that require dilution to avoid saturation of the detector.

The EQL (b) for groundwater in µg/L is not listed.
The EQL (a, b) for low concentrations in soil and sediment in µg/kg is not listed.
Accuracy as µg/L is not listed.
Overall precision in µg/L is not listed.

(a) *EQLs listed for soil/sediment are based on wet weight. Normally data is reported in a dry-weight basis; therefore, EQLs will be higher based on the % dry weight of each sample. This calculation is based on a 30 g sample and gel permeation chromatography cleanup.*
(b) *Sample EQLs are highly matrix-dependent. The EQLs are provided for guidance and may not always be achievable.*
C = *True value for concentration, in µg/L.*
X = *Average recovery found for measurements of samples containing a concentration of C, in µg/L.*

ESTIMATED QUANTITATION LIMIT

Other Matrices	Factor (a)
High-concentration soil and sludges by sonicator	7.5
Non-water miscible waste	75

(a) *EQL for other matrices = [EQL for low soil/sediment] × [Factor]. This estimated EQL is similar to an EPA "Practical Quantitation Limit."*

SAMPLING METHOD
Liquid samples — Use a 1 or 2½ gallon amber glass bottle with a screw-top Teflon®-lined cover that has been prewashed with detergent and rinsed with distilled water and methanol (or isopropanol).

Soils, sediments, or sludges — Use an 8-oz. widemouth glass with a screw-top Teflon®-lined cover that has been prewashed with detergent and rinsed with distilled water and methanol (or isopropanol).

SAMPLE PRESERVATION
Liquid samples — If residual chlorine is present, add 3 mL of 10% sodium thiosulfate per gallon, cool to 4°C and store in a solvent-free refrigerator until analysis; if chlorine is not present, then eliminate the sodium thiosulfate addition.

Soils, sediments, or sludges — Cool samples to 4°C and store in a solvent-free refrigerator.

MHT Liquid samples must be extracted within 7 days and the extracts analyzed within 40 days. Soils, sediments, or sludges may be stored for a maximum of 14 days and the extracts analyzed within 40 days.

SAMPLE PREPARATION
Liquid samples — Transfer 1 L quantitatively to a continuous extractor. If high concentrations are anticipated, a smaller volume may be used and then diluted with organic-free reagent water to 1 L. Adjust pH, if necessary, to pH <2 using 1:1 (V/V) sulfuric acid. Pipette 1.0 mL of a surrogate standard spiking solution into each sample. For the sample in each analytical batch selected for spiking, add 1.0 mL of a matrix spiking standard. For base/neutral acid analysis, the amount of the surrogates and matrix spiking compounds added to the sample should result in a final concentration of 100 ng/µL of each analyte in the extract to be analyzed (assuming a 1-µL injection). Extract with methylene chloride for 18–24 h. Next, adjust the pH of the aqueous phase to pH >11 using 10 N sodium hydroxide and extract it with methylene chloride again for 18–24 h. Dry the extract through a column containing anhydrous sodium sulfate and concentrate it to 1 mL using a K-D concentrator.

Soils, sediments, or sludges — Use 30 g of sample. Nonporous or wet samples (gummy or clay type) that do not have a free-flowing sandy texture must be mixed with anhydrous sodium sulfate until the sample is free flowing. Add 1 mL of surrogate standards to all samples, spikes, standards, and blanks. For the sample in each analytical batch selected for spiking, add 1.0 mL of a matrix spiking standard. For base/neutral acid analysis, the amount added of the surrogates and matrix spiking compounds should result in a final concentration of 100 ng/µL of each base/neutral analyte and 200 ng/µL of each acid analyte in the extract to be analyzed (assuming a 1-µL injection). Immediately add a 100-mL mixture of 1:1 methylene chloride:acetone and extract the sample ultrasonically for 3 min and then decant or filter the extracts. Repeat the extraction two or more times. Dry the extract using a column with anhydrous sodium sulfate and concentrate it to 1 mL in a K-D concentrator.

Note: N-nitrosodiphenylamine decomposes in the gas chromatographic inlet and cannot be separated from diphenylamine.

QUALITY CONTROL A methylene chloride solution containing 50 ng/µL of decafluorotriphenylphosphine (DFTPP) is used for tuning the GC/MS system each 12-h shift. A system performance check also must be made during every 12-h shift. A standard containing 50 ng/µL each of 4,4'-DDT, pentachlorophenol, and benzidine is required to verify injection port inertness and GC column performance. A calibration standard at mid-concentration, containing each compound of interest, including all required surrogates, must be performed every 12 h during analysis. After the system performance check is met, calibration check compounds (CCCs) are used to check the validity of the initial calibration.

The internal standard responses and retention times in the calibration check standard must be evaluated immediately after or during data acquisition. If the retention time for any internal standard changes by more than 30 seconds from the last check calibration (12 h), the chromatographic system must be inspected for malfunctions and corrections must be made, as required. If the electron ionization current plot (EICP) area for any of the internal standards changes by a factor of two from the last daily calibration standard check, the mass spectrometer must be inspected for malfunctions and corrections must be made, as appropriate.

Demonstrate, through the analysis of a reagent water blank, that interferences from the analytical system, glassware, and reagents are under control. The blank samples should be carried through all stages of the sample preparation and measurement steps. For each analytical batch (up to 20 samples), a reagent blank, matrix spike, and matrix spike duplicate/duplicate must be analyzed (the frequency of the spikes may be different for different monitoring programs). The blank and spiked samples must be carried through all stages of the sample preparation and measurement steps. A QC reference sample concentrate containing each analyte at a concentration of 100 mg/L in methanol is required.

REFERENCE Test Methods for Evaluating Solid Waste (SW-846). U.S. EPA 1983, Method 8270B, Rev. 2, Nov. 1990. Office of Solid Waste. Washington, DC.

Diphenyldisulfide   **EPA Method 1625**
CAS #882-33-7

TITLE Semivolatile Organic Compounds by Isotope Dilution GC/MS

MATRIX The compounds may be determined in waters, soils, and municipal sludges by this method.

METHOD SUMMARY This method is used to determine 176 semivolatile toxic organic pollutants associated with the CWA (as amended 1987); the RCRA (as amended 1986); the CERCLA (as amended 1986); and other compounds amenable to extraction and analysis by capillary column gas chromatography-mass spectrometry (GC/MS).

Stable isotopically-labeled analogs of the compounds of interest are added to the sample. If the solids content is less than 1%, a 1-L sample is extracted at pH 12–13, then at pH <2 with methylene chloride using continuous extraction techniques.

If the solids content is 30% or less, the sample is diluted to 1% solids with reagent water, homogenized ultrasonically, and extracted at pH 12–13, then at pH <2 with methylene chloride using continuous extraction techniques. If the solids content is greater than 30%, the sample is extracted using ultrasonic techniques.

Each extract is dried over sodium sulfate, concentrated to a volume of 5 mL, cleaned up using GPC, if necessary, and concentrated. Extracts are concentrated to 1 mL if GPC is not performed, and to 0.5 mL if GPC is performed.

An internal standard is added to the extract, and a 1-mL aliquot of the extract is injected into the GC. The compounds are separated by GC and detected by a MS. The labeled compounds serve to correct the variability of the analytical technique.

INTERFERENCES Solvents, reagents, glassware, and other sample processing hardware may yield artifacts and/or elevated baselines causing misinterpretation of chromatograms and spectra. Materials used in the analysis must be demonstrated to be free from interferences under the conditions of analysis by running method blanks initially and with each sample lot (sample started through the extraction process on a given 8-h shift, to a maximum of 20). Specific selection of reagents and purification of solvents by distillation in all glass systems may be required. Glassware and, where possible, reagents are cleaned by solvent rinse and baking at 450°C for 1-h minimum. Interferences coextracted from samples will vary considerably from source to source, depending on the diversity of the site being sampled.

INSTRUMENTATION Major instrumentation includes a GC with a splitless or on-column injection port for capillary column, a MS with 70 eV electron impact ionization, and a data system to collect and record MS data, and process it. A K-D apparatus is used to concentrate extracts.

GC Column: 30 m × 0.25 mm I.D. 5% phenyl, 94% methyl, 1% vinyl silicone bonded phased fused silica capillary column.

PRECISION & ACCURACY The detection limits of the method are usually dependent on the level of interferences rather than instrumental limitations. The limits typify the minimum quantities that can be detected with no interferences present.

The minimum level (in μg/mL) was not listed. This is defined as a minimum level at which the analytical system shall give recognizable mass spectra (background corrected) and acceptable calibration points.

The MDL (in μg/kg) in low solids was not listed and in high solids was not listed; these were determined in digested sludge (low solids) and in filter cake or compost (high solids).

The labeled and native compound initial precision as standard deviation (in μg/L) was not listed.

The labeled and native compound initial accuracy as average recovery (in μg/L) was not listed.

SAMPLE COLLECTION, PRESERVATION & HANDLING Collect samples in glass containers. Aqueous samples which flow freely are collected in refrigerated bottles using automatic sampling equipment. Solid samples are collected as grab samples using widemouth jars. Maintain samples at 0 to 4°C from the time of collection until extraction. If residual chlorine is present in aqueous samples, add 80 mg sodium thiosulfate/L of water. Begin sample extraction within 7 days of collection, and analyze all extracts within 40 days of extraction.

SAMPLE PREPARATION Samples containing 1% solids or less are extracted directly using continuous liquid-liquid extraction techniques. Samples containing 1 to 30% solids are diluted to the 1% level with reagent water and extracted using continuous liquid-liquid extraction techniques. Samples containing greater than 30% solids are extracted using ultrasonic techniques.

Base/neutral extraction — Adjust the pH of the waters in the extractors to 12–13 with 6 N NaOH. Extract with methylene chloride for 24–48 h.

Acid extraction — Adjust the pH of the waters in the extractors to 2 or less using 6 N sulfuric acid. Extract with methylene chloride for 24–48 h.

Ultrasonic extraction of high solids samples — Add anhydrous sodium sulfate to the sample and QC aliquot(s). Add acetone:methylene chloride (1:1) to the sample and mix thoroughly

Concentrate extracts using a K-D apparatus.

QUALITY CONTROL The analyst is permitted to modify this method to improve separations or lower the costs of measurements, provided all performance specifications are met. Analyses of blanks are required to demonstrate freedom from contamination. When results of spikes indicate atypical method performance for samples, the samples are diluted to bring method performance within acceptable limits.

For low solids (aqueous samples), extract, concentrate, and analyze two sets of four 1-L aliquots (8 aliquots total) of the precision and recovery standard. For high solids samples, two sets of four 30-g aliquots of the high solids reference matrix are used.

Spike all samples with labeled compounds to assess method performance. Compute percent recovery of the labeled compounds using the internal standard method. Compare the labeled compound recovery for each compound with the corresponding labeled compound recovery.

Reagent water and high solids reference matrix blanks are analyzed to demonstrate freedom from contamination. Extract and concentrate a 1-L reagent water blank or a high solids reference matrix blank with each sample's lot (samples started through the extraction process on the same 8-h shift, to a maximum of 20 samples).

Field replicates may be collected to determine the precision of the sampling technique, and spiked samples may be required to determine the accuracy of the analysis when the internal standard method is used.

REFERENCE Semivolatile Organic Compounds by Isotope Dilution GC/MS. Office of Water Regulation and Standards, U.S. EPA Industrial Technology Division, Washington, DC, EPA Method 1625, Rev. C, June 1989 (contact W.A. Telliard, U.S. EPA, Office of Water Regulations and Standards, 401 M St., SW, Washington, DC, 20460. Phone: 202-382-7131).

5,5-Diphenylhydantoin   **EPA Method 8270**
CAS #57-41-0

TITLE Semivolatile Organic Compounds by GC/MS

MATRIX This method is used to determine the concentration of semivolatile organic compounds in extracts prepared from all types of solid waste matrices, soils, and groundwater. Although surface waters are not specifically mentioned, this method should be applicable to water samples from rivers, lakes, etc.

METHOD SUMMARY This method covers 259 semivolatile organic compounds. In very limited applications direct injection of the sample into the GC/MS system may be appropriate, but this results in very high detection limits (approximately 10,000 μg/L). Typically, a 1-L liquid sample, containing surrogate, and matrix spiking standards, is extracted in a continuous extractor first under acid conditions and then under basic conditions. Typically 30 g of a solid sample, containing surrogate, and matrix spiking standards, is extracted ultrasonically. After concentrating the extract to 1 mL it is spiked with 10 μL of an internal standard solution just prior to analysis by GC/MS. The volume injected should contain about 100 ng of base/neutral and 200 ng of acid surrogates (for a 1-μL injection). Analysis is performed by GC/MS using a capillary GC column.

INTERFERENCES Raw GC/MS data from all blanks, samples, and spikes must be evaluated for interferences. Contamination by carryover can occur whenever high-concentration and low-concentration samples are sequentially analyzed. To reduce carryover, the sample syringe must be rinsed out between samples with solvent. Whenever an unusually concentrated sample is encountered, it should be followed by the analysis of blank solvent to check for cross-contamination.

INSTRUMENTATION A GC/MS and a data system are required. The GC column used is a 30 m × 0.25 mm I.D. (or 0.32 mm I.D.) 1um film thickness silicone-coated fused silica capillary column. A continuous liquid-liquid extractor equipped with Teflon® or glass connection joints and stopcocks requiring no lubrication, a K-D concentrating apparatus, water bath, and an ultrasonic disrupter with a minimum power of 300 W and with pulsing capability are also required.

PRECISION & ACCURACY The estimated quantitation limit (EQL) of Method 8270B for determining an individual compound is approximately 1 mg/kg (wet weight) for soil or sediment samples, 1–200 mg/kg for wastes (dependent on matrix and method of preparation), and 10 μg/L for groundwater samples. EQLs will be proportionately higher for sample extracts that require dilution to avoid saturation of the detector.

The EQL(b) for groundwater in μg/L is 20.
The EQL (a, b) for low concentrations in soil and sediment in μg/kg is not determined.
Accuracy as μg/L is not listed.
Overall precision in μg/L is not listed.

(a) *EQLs listed for soil/sediment are based on wet weight. Normally data is reported in a dry-weight basis; therefore, EQLs will be higher based on the % dry weight of each sample. This calculation is based on a 30 g sample and gel permeation chromatography cleanup.*
(b) *Sample EQLs are highly matrix-dependent. The EQLs are provided for guidance and may not always be achievable.*
$C =$ *True value for concentration, in μg/L.*
$X =$ *Average recovery found for measurements of samples containing a concentration of C, in μg/L.*

ESTIMATED QUANTITATION LIMIT

Other Matrices	Factor (a)
High-concentration soil and sludges by sonicator	7.5
Non-water miscible waste	75

(a) EQL for other matrices = [EQL for low soil/sediment] × [Factor]. This estimated EQL is similar to an EPA "Practical Quantitation Limit."

SAMPLING METHOD

Liquid samples — Use a 1 or 2½ gallon amber glass bottle with a screw-top Teflon®-lined cover that has been prewashed with detergent and rinsed with distilled water and methanol (or isopropanol).

Soils, sediments, or sludges — Use an 8-oz. widemouth glass with a screw-top Teflon®-lined cover that has been prewashed with detergent and rinsed with distilled water and methanol (or isopropanol).

SAMPLE PRESERVATION

Liquid samples — If residual chlorine is present, add 3 mL of 10% sodium thiosulfate per gallon, cool to 4°C and store in a solvent-free refrigerator until analysis; if chlorine is not present, then eliminate the sodium thiosulfate addition.

Soils, sediments, or sludges — Cool samples to 4°C and store in a solvent-free refrigerator.

MHT Liquid samples must be extracted within 7 days and the extracts analyzed within 40 days. Soils, sediments, or sludges may be stored for a maximum of 14 days and the extracts analyzed within 40 days.

SAMPLE PREPARATION

Liquid samples — Transfer 1 L quantitatively to a continuous extractor. If high concentrations are anticipated, a smaller volume may be used and then diluted with organic-free reagent water to 1 L. Adjust pH, if necessary, to pH <2 using 1:1 (V/V) sulfuric acid. Pipette 1.0 mL of a surrogate standard spiking solution into each sample. For the sample in each analytical batch selected for spiking, add 1.0 mL of a matrix spiking standard. For base/neutral acid analysis, the amount of the surrogates and matrix spiking compounds added to the sample should result in a final concentration of 100 ng/µL of each analyte in the extract to be analyzed (assuming a 1-µL injection). Extract with methylene chloride for 18–24 h. Next, adjust the pH of the aqueous phase to pH >11 using 10 N sodium hydroxide and extract it with methylene chloride again for 18–24 h. Dry the extract through a column containing anhydrous sodium sulfate and concentrate it to 1 mL using a K-D concentrator.

Soils, sediments, or sludges — Use 30 g of sample. Nonporous or wet samples (gummy or clay type) that do not have a free-flowing sandy texture must be mixed with anhydrous sodium sulfate until the sample is free flowing. Add 1 mL of surrogate standards to all samples, spikes, standards, and blanks. For the sample in each analytical batch selected for spiking, add 1.0 mL of a matrix spiking standard. For base/neutral acid analysis, the amount added of the surrogates and matrix spiking compounds should result in a final concentration of 100 ng/µL of each base/neutral analyte and 200 ng/µL of each acid analyte in the extract to be analyzed (assuming a 1-µL injection). Immediately add a 100-mL mixture of 1:1 methylene chloride:acetone and extract the sample ultrasonically for 3 min and then decant or filter the extracts. Repeat the extraction two or more times. Dry the extract using a column with anhydrous sodium sulfate and concentrate it to 1 mL in a K-D concentrator.

QUALITY CONTROL A methylene chloride solution containing 50 ng/µL of decafluorotriphenylphosphine (DFTPP) is used for tuning the GC/MS system each 12-h shift. A system performance check also must be made during every 12-h shift. A standard containing 50 ng/µL each of 4,4'-DDT, pentachlorophenol, and benzidine is required to verify injection port inertness and GC column performance. A calibration standard at mid-concentration, containing each compound of interest, including all required surrogates, must be performed every 12 h during analysis. After the system performance check is met, calibration check compounds (CCCs) are used to check the validity of the initial calibration.

The internal standard responses and retention times in the calibration check standard must be evaluated immediately after or during data acquisition. If the retention time for any internal standard changes by more than 30 seconds from the last check calibration (12 h), the chromatographic system must be inspected for malfunctions and corrections must be made, as required. If the electron ionization current plot (EICP) area for any of the internal standards changes by a factor of two from the last daily calibration standard check, the mass spectrometer must be inspected for malfunctions and corrections must be made, as appropriate.

Demonstrate, through the analysis of a reagent water blank, that interferences from the analytical system, glassware, and reagents are under control. The blank samples should be carried through all stages of the sample preparation and measurement steps. For each analytical batch (up to 20 samples), a reagent blank, matrix spike, and matrix spike duplicate/duplicate must be analyzed (the frequency of the spikes may be different for different monitoring programs). The blank and spiked samples must be carried through all stages of the sample preparation and measurement steps. A QC reference sample concentrate containing each analyte at a concentration of 100 mg/L in methanol is required.

REFERENCE Test Methods for Evaluating Solid Waste (SW-846). U.S. EPA 1983, Method 8270B, Rev. 2, Nov. 1990. Office of Solid Waste, Washington, DC.

1,2-Diphenylhydrazine   **EPA Method 1625**
CAS #122-66-7

TITLE Semivolatile Organic Compounds by Isotope Dilution GC/MS

MATRIX The compounds may be determined in waters, soils, and municipal sludges by this method.

METHOD SUMMARY This method is used to determine 176 semivolatile toxic organic pollutants associated with the CWA (as amended 1987); the RCRA (as amended 1986); the

CERCLA (as amended 1986); and other compounds amenable to extraction and analysis by capillary column gas chromatography-mass spectrometry (GC/MS).

Stable isotopically-labeled analogs of the compounds of interest are added to the sample. If the solids content is less than 1%, a 1-L sample is extracted at pH 12–13, then at pH <2 with methylene chloride using continuous extraction techniques.

If the solids content is 30% or less, the sample is diluted to 1% solids with reagent water, homogenized ultrasonically, and extracted at pH 12–13, then at pH <2 with methylene chloride using continuous extraction techniques. If the solids content is greater than 30%, the sample is extracted using ultrasonic techniques.

Each extract is dried over sodium sulfate, concentrated to a volume of 5 mL, cleaned up using GPC, if necessary, and concentrated. Extracts are concentrated to 1 mL if GPC is not performed, and to 0.5 mL if GPC is performed.

An internal standard is added to the extract, and a 1-mL aliquot of the extract is injected into the GC. The compounds are separated by GC and detected by a MS. The labeled compounds serve to correct the variability of the analytical technique.

INTERFERENCES Solvents, reagents, glassware, and other sample processing hardware may yield artifacts and/or elevated baselines causing misinterpretation of chromatograms and spectra. Materials used in the analysis must be demonstrated to be free from interferences under the conditions of analysis by running method blanks initially and with each sample lot (sample started through the extraction process on a given 8-h shift, to a maximum of 20). Specific selection of reagents and purification of solvents by distillation in all glass systems may be required. Glassware and, where possible, reagents are cleaned by solvent rinse and baking at 450°C for 1-h minimum. Interferences coextracted from samples will vary considerably from source to source, depending on the diversity of the site being sampled.

INSTRUMENTATION Major instrumentation includes a GC with a splitless or on-column injection port for capillary column, a MS with 70 eV electron impact ionization, and a data system to collect and record MS data, and process it. A K-D apparatus is used to concentrate extracts.

GC Column: 30 m × 0.25 mm I.D. 5% phenyl, 94% methyl, 1% vinyl silicone bonded phased fused silica capillary column.

PRECISION & ACCURACY The detection limits of the method are usually dependent on the level of interferences rather than instrumental limitations. The limits typify the minimum quantities that can be detected with no interferences present.

The minimum level (in μg/mL) was 20. This is defined as a minimum level at which the analytical system shall give recognizable mass spectra (background corrected) and acceptable calibration points.

The MDL (in μg/kg) in low solids was 48 and in high solids was 27; these were determined in digested sludge (low solids) and in filter cake or compost (high solids).

The labeled and native compound initial precision as standard deviation (in μg/L) was 73.

The labeled and native compound initial accuracy as average recovery (in μg/L) was 49–308.

SAMPLE COLLECTION, PRESERVATION & HANDLING
Collect samples in glass containers. Aqueous samples which flow freely are collected in refrigerated bottles using automatic sampling equipment. Solid samples are collected as grab samples using widemouth jars. Maintain samples at 0 to 4°C from the time of collection until extraction. If residual chlorine is present in aqueous samples, add 80 mg sodium thiosulfate/L of water. Begin sample extraction within 7 days of collection, and analyze all extracts within 40 days of extraction.

SAMPLE PREPARATION Samples containing 1% solids or less are extracted directly using continuous liquid-liquid extraction techniques. Samples containing 1 to 30% solids are diluted to the 1% level with reagent water and extracted using continuous liquid-liquid extraction techniques. Samples containing greater than 30% solids are extracted using ultrasonic techniques.

Base/neutral extraction — Adjust the pH of the waters in the extractors to 12–13 with 6 N NaOH. Extract with methylene chloride for 24–48 h.

Acid extraction — Adjust the pH of the waters in the extractors to 2 or less using 6 N sulfuric acid. Extract with methylene chloride for 24–48 h.

Ultrasonic extraction of high solids samples — Add anhydrous sodium sulfate to the sample and QC aliquot(s). Add acetone:methylene chloride (1:1) to the sample and mix thoroughly

Concentrate extracts using a K-D apparatus.

QUALITY CONTROL The analyst is permitted to modify this method to improve separations or lower the costs of measurements, provided all performance specifications are met. Analyses of blanks are required to demonstrate freedom from contamination. When results of spikes indicate atypical method performance for samples, the samples are diluted to bring method performance within acceptable limits.

For low solids (aqueous samples), extract, concentrate, and analyze two sets of four 1-L aliquots (8 aliquots total) of the precision and recovery standard. For high solids samples, two sets of four 30-g aliquots of the high solids reference matrix are used.

Spike all samples with labeled compounds to assess method performance. Compute percent recovery of the labeled compounds using the internal standard method. Compare the labeled compound recovery for each compound with the corresponding labeled compound recovery.

Reagent water and high solids reference matrix blanks are analyzed to demonstrate freedom from contamination. Extract and concentrate a 1-L reagent water blank or a high solids reference matrix blank with each sample's lot (samples started through the extraction process on the same 8-h shift, to a maximum of 20 samples).

Field replicates may be collected to determine the precision of the sampling technique, and spiked samples may be required to determine the accuracy of the analysis when the internal standard method is used.

REFERENCE Semivolatile Organic Compounds by Isotope Dilution GC/MS. Office of Water Regulation and Standards, U.S. EPA Industrial Technology Division, Washington, DC, EPA Method 1625, Rev. C, June 1989 (contact W.A. Telliard, U.S. EPA, Office of Water Regulations and Standards, 401 M St., SW, Washington, DC, 20460. Phone: 202-382-7131).

1,2-Diphenylhydrazine **EPA Method 8270**
CAS #122-66-7

TITLE Semivolatile Organic Compounds by GC/MS

MATRIX This method is used to determine the concentration of semivolatile organic compounds in extracts prepared from all types of solid waste matrices, soils, and groundwater. Although surface waters are not specifically mentioned, this method should be applicable to water samples from rivers, lakes, etc.

METHOD SUMMARY This method covers 259 semivolatile organic compounds. In very limited applications direct injection of the sample into the GC/MS system may be appropriate, but this results in very high detection limits (approximately 10,000 µg/L). Typically, a 1-L liquid sample, containing surrogate, and matrix spiking standards, is extracted in a continuous extractor first under acid conditions and then under basic conditions. Typically 30 g of a solid sample, containing surrogate, and matrix spiking standards, is extracted ultrasonically. After concentrating the extract to 1 mL it is spiked with 10 µL of an internal standard solution just prior to analysis by GC/MS. The volume injected should contain about 100 ng of base/neutral and 200 ng of acid surrogates (for a 1-µL injection). Analysis is performed by GC/MS using a capillary GC column.

INTERFERENCES Raw GC/MS data from all blanks, samples, and spikes must be evaluated for interferences. Contamination by carryover can occur whenever high-concentration and low-concentration samples are sequentially analyzed. To reduce carryover, the sample syringe must be rinsed out between samples with solvent. Whenever an unusually concentrated sample is encountered, it should be followed by the analysis of blank solvent to check for cross-contamination.

INSTRUMENTATION A GC/MS and a data system are required. The GC column used is a 30 m × 0.25 mm I.D. (or 0.32 mm I.D.) 1um film thickness silicone-coated fused silica capillary column. A continuous liquid-liquid extractor equipped with Teflon® or glass connection joints and stopcocks requiring no lubrication, a K-D concentrating apparatus, water bath, and an ultrasonic disrupter with a minimum power of 300 W and with pulsing capability are also required.

PRECISION & ACCURACY The estimated quantitation limit (EQL) of Method 8270B for determining an individual compound is approximately 1 mg/kg (wet weight) for soil or sediment samples, 1–200 mg/kg for wastes (dependent on matrix and method of preparation), and 10 µg/L for groundwater samples. EQLs will be proportionately higher for sample extracts that require dilution to avoid saturation of the detector.

The EQL(b) for groundwater in µg/L is not listed.
The EQL (a, b) for low concentrations in soil and sediment in µg/kg is not listed.
Accuracy as µg/L is not listed.
Overall precision in µg/L is not listed.

(a) *EQLs listed for soil/sediment are based on wet weight. Normally data is reported in a dry-weight basis; therefore, EQLs will be higher based on the % dry weight of each sample. This calculation is based on a 30 g sample and gel permeation chromatography cleanup.*
(b) *Sample EQLs are highly matrix-dependent. The EQLs are provided for guidance and may not always be achievable.*
C = *True value for concentration, in µg/L.*
X = *Average recovery found for measurements of samples containing a concentration of C, in µg/L.*

ESTIMATED QUANTITATION LIMIT

Other Matrices	Factor (a)
High-concentration soil and sludges by sonicator	7.5
Non-water miscible waste	75

(a) *EQL for other matrices = [EQL for low soil/sediment] × [Factor]. This estimated EQL is similar to an EPA "Practical Quantitation Limit."*

SAMPLING METHOD
Liquid samples — Use a 1 or 2½ gallon amber glass bottle with a screw-top Teflon®-lined cover that has been prewashed with detergent and rinsed with distilled water and methanol (or isopropanol).

Soils, sediments, or sludges — Use an 8-oz. widemouth glass with a screw-top Teflon®-lined cover that has been prewashed with detergent and rinsed with distilled water and methanol (or isopropanol).

SAMPLE PRESERVATION
Liquid samples — If residual chlorine is present, add 3 mL of 10% sodium thiosulfate per gallon, cool to 4°C and store in a solvent-free refrigerator until analysis; if chlorine is not present, then eliminate the sodium thiosulfate addition.

Soils, sediments, or sludges — Cool samples to 4°C and store in a solvent-free refrigerator.

MHT Liquid samples must be extracted within 7 days and the extracts analyzed within 40 days. Soils, sediments, or sludges may be stored for a maximum of 14 days and the extracts analyzed within 40 days.

SAMPLE PREPARATION
Liquid samples — Transfer 1 L quantitatively to a continuous extractor. If high concentrations are anticipated, a smaller volume may be used and then diluted with organic-free reagent water to 1 L. Adjust pH, if necessary, to pH <2 using 1:1 (V/V) sulfuric acid. Pipette 1.0 mL of a surrogate standard spiking solution into each sample. For the sample in each analytical batch selected for spiking, add 1.0 mL of a matrix spiking standard. For

base/neutral acid analysis, the amount of the surrogates and matrix spiking compounds added to the sample should result in a final concentration of 100 ng/μL of each analyte in the extract to be analyzed (assuming a 1-μL injection). Extract with methylene chloride for 18–24 h. Next, adjust the pH of the aqueous phase to pH >11 using 10 N sodium hydroxide and extract it with methylene chloride again for 18–24 h. Dry the extract through a column containing anhydrous sodium sulfate and concentrate it to 1 mL using a K-D concentrator.

Soils, sediments, or sludges — Use 30 g of sample. Nonporous or wet samples (gummy or clay type) that do not have a free-flowing sandy texture must be mixed with anhydrous sodium sulfate until the sample is free flowing. Add 1 mL of surrogate standards to all samples, spikes, standards, and blanks. For the sample in each analytical batch selected for spiking, add 1.0 mL of a matrix spiking standard. For base/neutral acid analysis, the amount added of the surrogates and matrix spiking compounds should result in a final concentration of 100 ng/μL of each base/neutral analyte and 200 ng/μL of each acid analyte in the extract to be analyzed (assuming a 1-μL injection). Immediately add a 100-mL mixture of 1:1 methylene chloride:acetone and extract the sample ultrasonically for 3 min and then decant or filter the extracts. Repeat the extraction two or more times. Dry the extract using a column with anhydrous sodium sulfate and concentrate it to 1 mL in a K-D concentrator.

QUALITY CONTROL A methylene chloride solution containing 50 ng/μL of decafluorotriphenylphosphine (DFTPP) is used for tuning the GC/MS system each 12-h shift. A system performance check also must be made during every 12-h shift. A standard containing 50 ng/μL each of 4,4′-DDT, pentachlorophenol, and benzidine is required to verify injection port inertness and GC column performance. A calibration standard at mid-concentration, containing each compound of interest, including all required surrogates, must be performed every 12 h during analysis. After the system performance check is met, calibration check compounds (CCCs) are used to check the validity of the initial calibration.

The internal standard responses and retention times in the calibration check standard must be evaluated immediately after or during data acquisition. If the retention time for any internal standard changes by more than 30 seconds from the last check calibration (12 h), the chromatographic system must be inspected for malfunctions and corrections must be made, as required. If the electron ionization current plot (EICP) area for any of the internal standards changes by a factor of two from the last daily calibration standard check, the mass spectrometer must be inspected for malfunctions and corrections must be made, as appropriate.

Demonstrate, through the analysis of a reagent water blank, that interferences from the analytical system, glassware, and reagents are under control. The blank samples should be carried through all stages of the sample preparation and measurement steps. For each analytical batch (up to 20 samples), a reagent blank, matrix spike, and matrix spike duplicate/duplicate must be analyzed (the frequency of the spikes may be different for different monitoring programs). The blank and spiked samples must be carried through all stages of the sample preparation and measurement steps. A QC reference sample concentrate containing each analyte at a concentration of 100 mg/L in methanol is required.

REFERENCE Test Methods for Evaluating Solid Waste (SW-846). U.S. EPA 1983, Method 8270B, Rev. 2, Nov. 1990. Office of Solid Waste, Washington, DC.

Dissolved Oxygen Uptake Rate   **EPA Method 213 A**

TITLE Inorganics, Non-Metallics

MATRIX Activated Sludge

APPLICATION (DOUR). (Also referred to as oxygen consumption rate). this test is used to determine the oxygen consumption rate of a biological suspension sample. A routine plant operation test, it will indicate changes in operating conditions at an early stage.

INTERFERENCES Determination results are quite sensitive to temperature variations. The determination is sensitive to the time lag between sample collection and test initiation. Because test and sample site conditions can vary, observed measurements and actual DOUR may vary.

INSTRUMENTATION Oxygen probe method and m or manometric or respirometric device

RANGE Not listed.

MDL Not listed.

PRECISION Precision has not been determined.

ACCURACY Accuracy is not applicable.

SAMPLING METHOD Glass container only (300 mL). Bottle and top.

STABILITY No preservation required. Analyze immediately.

QUALITY CONTROL Record appropriate DO data at time intervals of less than 1 min, depending on rate of consumption. Record data over 15 min period or until DO becomes limiting, whichever occurs first.

REFERENCE Standard Methods for the Examination of Water and Wastewater, 16th ed., Page 127, 1985.

Disulfoton   **EPA Method 507**
CAS #298-04-4

TITLE Determination of Nitrogen and Phosphorus-Containing Pesticides in Water by GC/NPD

MATRIX This method is applicable to the determination of certain nitrogen and phosphorus-containing pesticides in finished drinking water and groundwater.

METHOD SUMMARY Method 507 covers 46 nitrogen- and phosphorus-containing pesticides. A 1-L sample is fortified with a surrogate standard, salted, buffered, extracted with methylene chloride, and concentrated; then the solvent is exchanged with methyl tert-butyl ether (MTBE) and concentrated again, and a 2-µL aliquot of a sample extract is injected into a GC system equipped with a selective nitrogen-phosphorus detector and a capillary column for analysis.

INTERFERENCES Method interferences may be caused by contaminants in solvents, reagents, glassware, and other sample processing apparatus. Interfering contamination may occur when a sample containing low concentrations of analytes is analyzed immediately following a sample containing relatively high concentrations. One or more injections of MTBE should be made following the analysis of a sample with high concentrations of analytes to check for analyte carryover. Matrix interferences may be caused by contaminants that are coextracted from the sample. The extent of matrix interferences will vary considerably from source to source, depending upon the water sampled.

INSTRUMENTATION A gas chromatograph system (GC) equipped with a nitrogen-phosphorus detector (NPD) is needed.

Column 1: 30 m × 0.25 mm I.D. DB-5 bonded fused silica column, 0.25 µm film thickness, or equivalent.
Column 2: 30 m × 0.25 mm I.D. DB-1701 bonded fused silica column, 0.25 µm film thickness, or equivalent.

PRECISION & ACCURACY This method has been validated in a single lab and estimated detection limits (EDLs) have been determined for each analyte. Observed detection limits may vary among waters, depending upon the nature of the interferences in the sample matrix and the specific instrumentation used. Analytes that are not separated chromatographically cannot be individually identified and measured unless an alternative technique for identification and quantification exist.

The estimated detection limit (in µg/L) was 0.3. The EDL is defined as either method detection limit or a level of compound in a sample yielding a peak in the final extract with signal-to-noise ratio of approximately 5, whichever value is higher.

The concentration used for these measurements (in µg/L) was 3.
The accuracy (as % recovery) was 89.
The precision (% RSD) was 10.

SAMPLING METHOD Grab samples are collected in 1-L glass sample bottles (prewashed with detergent and hot tap water, rinsed with reagent water, and dried in an oven at 400°C for 1 h) with screw caps lined with PTFE-fluorocarbon.

SAMPLE PRESERVATION Add mercuric chloride to the sample bottle in amounts to produce a concentration of 10 mg/L. If residual chlorine is present, add 80 mg of sodium thiosulfate/L of sample to the sample bottle prior to collection. After collection, seal bottle and shake vigorously for 1 min, then cool the sample to 4°C immediately and store it at 4°C in the dark until extraction.

MHT Maximum holding time of the samples, and in some cases the extracts, is 14 days.

SAMPLE PREPARATION Fortify the sample with 50 µL of the surrogate standard solution, adjust to pH 7 with phosphate buffer, add 100 g NaCl to the sample, and seal and shake to dissolve the salt; then extract with methylene chloride in a separatory funnel or in a mechanical tumbler bottle. Dry the extract by pouring it through a solvent-rinsed drying column containing about 10 cm of anhydrous sodium sulfate. Collect the extract in a Kuderna-Danish (K-D) concentrator and rinse the column with 20–30 mL methylene chloride. Concentrate the extract to about 2 mL and rinse the flask and its lower joint into the concentrator tube with 1 to 2 mL of methyl t-butyl ether (MTBE). Add 5–10 mL of MTBE and concentrate the extract twice (adding more MTBE) to a final volume of 5.0 mL and store it at 4°C until analysis.

Note: If methylene chloride is not completely removed from the final extract, it may cause detector problems.

QUALITY CONTROL Minimum quality control requirements are initial demonstration of lab capability, determination of surrogate compound recoveries in each sample and blank, monitoring internal standard peak area or height in each sample and blank, analysis of lab reagent blanks, lab fortified samples, lab fortified blanks, and other QC samples. A lab reagent blank is analyzed to demonstrate that all glassware and reagent interferences are under control.

Initial demonstration of capability is fulfilled by analyzing four fortified reagent water samples with the recovery value for each analyte falling within the acceptable range (±30% average recovery). Surrogate recoveries from samples or method blanks must be 70–130%. The internal standard response for any sample chromatogram should not deviate from the daily calibration check standard's internal standard response by more than 30% or lab fortified blanks and sample matrices are used to assess lab performance and analyte recovery, respectively.

If the response for the target analyte peak exceeds the working range of the system, dilute the extract and reanalyze. Alternative techniques such as an alternate detector or second chromatography column should be used to confirm peak identification when sample components are not resolved adequately.

EPA CONTACT & HOTLINE For technical questions contact Dr. Baldev Bathija, U.S. EPA, Office of Ground Water and Drinking Water (WH-550D), 401 M St. SW, Washington, DC 20460. Tel. (202) 260-3040. For further information the EPA Safe Drinking Water Hotline may be called at: (800) 426-4791.

REFERENCE Methods for the Determination of Organic Compounds in Drinking Water, EPA/600/4-88/039 (revised July 1991). U.S. EPA Environmental Monitoring Systems Laboratory, Cincinnati, OH, 45268, U.S.A. Available from the National Technical Information Service (NTIS), 5285 Port Royal Road, Springfield, VA 22161; Tel. 800-553-6847. NTIS Order Number is PB91-231480.

Disulfoton   **EPA Method 8141**
CAS #298-04-4

TITLE Organophosphorus Compounds by Gas Chromatography: Capillary Column Technique

MATRIX This method covers aqueous and solid matrices. This includes a wide variety such as drinking water, groundwater, industrial wastewaters, surface waters, soils, solids, and sediments.

METHOD SUMMARY This is a GC method used to determine the concentration of 28 organophosphorus pesticides.

The use of Gel Permeation Cleanup (EPA Method 3640) for sample cleanup has been demonstrated to yield recoveries of less than 85% for many method analytes and is therefore not recommended for use with this method.

This method provides GC conditions for the detection of ppb concentrations of organophosphorus compounds. Prior to the use of this method, appropriate sample preparation techniques must be used. Water samples are extracted at a neutral pH with methylene chloride as a solvent by using a separatory funnel (EPA Method 3510) or a continuous liquid-liquid extractor (EPA Method 3520). Soxhlet extraction (EPA Method 3540) or ultrasonic extraction (EPA Method 3550) using methylene chloride/acetone (1:1) are used for solid samples. Both neat and diluted organic liquids (EPA Method 3580) may be analyzed by direct injection. Spiked samples are used to verify the applicability of the chosen extraction technique to each new sample type. A GC with a flame photometric (FPD) or nitrogen-phosphorus detector (NPD) is used for this multiresidue procedure.

INTERFERENCES The use of Florisil cleanup materials (EPA Method 3620) for some of the compounds in this method has been demonstrated to yield recoveries less than 85% and is therefore not recommended for all compounds. Use of phosphorus or halogen specific detectors, however, often obviates the necessity for cleanup for relatively clean sample matrices. If particular circumstances demand the use of an alternative cleanup procedure, the analyst must determine the elution profile and demonstrate that the recovery of each analyte is no less than 85%.

Use of a flame photometric detector (FPD) in the phosphorus mode will minimize interferences from materials that do not contain phosphorus. Elemental sulfur, however, may interfere with the determination of certain organophosphorus compounds by flame photometric gas chromatography. Sulfur cleanup using EPA Method 3660 may alleviate this interference. A nitrogen phosphorus detector (NPD) is also recommended.

A few analytes coelute on certain columns. Therefore, select a second column for confirmation where coelution of the analytes of interest does not occur.

Method interferences may be caused by contaminants in solvents, reagents, glassware, and other sample processing hardware that lead to discrete artifacts or elevated baselines in gas chromatograms. All these materials must be routinely demonstrated to be free from interferences under the conditions of the analysis by analyzing reagent blanks.

INSTRUMENTATION A GC with a NPD or a FPD will be needed. A data system or integrator is recommended for measuring peak areas and/or peak heights. A Kuderna-Danish (K-D) apparatus will be needed for extract concentration.

Column 1: 15 m × 0.53 mm megabore capillary column, 1.0 μm film thickness, DB-210.
Column 2: 15 m × 0.53 mm megabore capillary column, 1.5 μm film thickness, SPB-608.
Column 3: 15 m × 0.53 mm megabore capillary column, 1.0 μm film thickness, DB-5.

Three megabore capillary columns are included for analysis of organophosphates by this method. Column 1 (DB-210 or equivalent) and Column 2 (SPB-608 or equivalent) are recommended if a large number of organophosphorus analytes are to be determined. If the superior resolution offered by Column 1 and Column 2 is not required, Column 3 (DB-5 or equivalent) may be used. For megabore capillary columns, automatic injections of 1 μL are recommended.

PRECISION & ACCURACY The MDL actually achieved in a given analysis will vary, as it is dependent on instrument sensitivity and matrix effects. Single operator accuracy and precision studies have been conducted with spiked water and soil samples.

MULTIPLICATION FACTORS FOR OTHER MATRICES (a)

Matrix	Factor (b)
Groundwater	10
(EPA Method 3510 or EPA Method 3520)	
Low-concentration soil by Soxhlet and no cleanup	10 (c)
Low-concentration soil by ultrasonic extraction with GPC cleanup	6.7 (c)
High-concentration soil and sludges by ultrasonic extraction	500 (c)
Non-water miscible waste (EPA Method 3580)	1000 (c)

(a) Sample EQLs are highly matrix-dependent. The EQLs listed here are provided for guidance and may not always be achievable.
(b) EQL = [Method detection limit] × [Factor]. For non-aqueous samples the factor is on a wet-weight basis.
(c) Multiply this factory times the soil MDL.

The MDL (in μg/L) when reagent water was extracted using a separatory funnel was 0.07.
The MDL (in μg/kg) when soil was extracted using Soxhlet extraction (EPA Method 3540) was 3.5.
Accuracy (as % recovery) with separatory funnel extraction ranged from 48 (with low spikes) to 84 (with high spikes).
Accuracy (as % recovery) with continuous liquid-liquid extraction ranged from 94 (with low spikes) to 81 (with high spikes).
Accuracy (as % recovery) with Soxhlet extraction of soils ranged from 78 (with low spikes to 76 (with high spikes).
Accuracy (as % recovery) with ultrasonic extraction of soils ranged from 30 (with low spikes) to 69 (with high spikes).

SAMPLE COLLECTION, PRESERVATION & HANDLING
Containers used to collect samples for the determination of semivolatile organic compounds should be soap and water washed followed by methanol (or isopropanol) rinsing. The sample containers should be of glass or Teflon® and have screw-top covers with Teflon® liners.

No preservation is used with concentrated waste samples. With liquid samples containing no residual chlorine and with soil, sediment, and sludge samples, immediately cooling to 4°C is the only preservation used. When residual chlorine is present then 3 mL of 10% aqueous sodium sulfate is added for each gallon of sample collected, followed by cooling to 4°C.

Liquid samples must be extracted within 7 days and their extracts analyzed within 40 days. Concentrated waste, soil, sediment, and sludge samples must be extracted within 14 days and their extracts analyzed within 40 days.

SAMPLE PREPARATION In general, water samples are extracted at a neutral pH with methylene chloride, using either EPA Method 3510 or EPA Method 3520. Solid samples are extracted using either EPA Method 3540 or EPA Method 3550 with methylene chloride/acetone (1:1) as the extraction solvent.

Prior to GC analysis, the extraction solvent may be exchanged to hexane. Single lab data indicates that samples should not be transferred with 100% hexane during sample workup as the more water soluble organophosphorus compounds may be lost.

If cleanup is performed on the samples, the analyst should analyze the samples by GC. This will confirm elution patterns and the absence of interferences from the reagents. If peak detection and identification is prevented by the presence of interferences, further cleanup is required.

QUALITY CONTROL The analyst should monitor the performance of the extraction, cleanup (when used), and analytical system and the effectiveness of the method in dealing with each sample matrix by spiking each sample, standard, and blank with one or two surrogates (e.g., organophosphorus compounds not expected to be present in the sample). Deuterated analogs of analytes should not be used as surrogates for gas chromatographic analysis due to coelution problems.

A minimum of five concentrations for each analyte of interest should be prepared through dilution of the stock standards with isooctane. One of the concentrations should be at a concentration near, but above, the MDL.

Include a mid-level check standard after each group of 10 samples in the analysis sequence. GC/MS techniques should be judiciously employed to support qualitative identifications made with this method. Follow the GC/MS operating requirements specified in EPA Method 8270.

When available, chemical ionization mass spectra may be employed to aid in the qualitative identification process. To confirm an identification of a compound, the background-corrected mass spectrum of the compound must be obtained from the sample extract and must be compared with a mass spectrum from a stock or calibration standard analyzed under the same chromatographic conditions. The molecular ion and all other ions present above 20% relative abundance in the mass spectrum of the standard must be present in the mass spectrum of the sample with agreement to ±20%. The retention time of the compound in the sample must be within six seconds of the retention time for the same compound in the standard solution.

Should the MS procedure fail to provide satisfactory results, additional steps may be taken before reanalysis. These steps may include the use of alternate packed or capillary GC columns or additional sample cleanup.

REFERENCE Test Methods for Evaluating Solid Waste, Physical/Chemical Methods, SW-846, 3rd Edition, U.S. EPA, Office of Solid Waste, Washington, DC, EPA Method 8141 July 1992.

Disulfoton   EPA Method 8270
CAS #298-04-4

TITLE Semivolatile Organic Compounds by GC/MS

MATRIX This method is used to determine the concentration of semivolatile organic compounds in extracts prepared from all types of solid waste matrices, soils, and groundwater. Although surface waters are not specifically mentioned, this method should be applicable to water samples from rivers, lakes, etc.

METHOD SUMMARY This method covers 259 semivolatile organic compounds. In very limited applications direct injection of the sample into the GC/MS system may be appropriate, but this results in very high detection limits (approximately 10,000 µg/L). Typically, a 1-L liquid sample, containing surrogate, and matrix spiking standards, is extracted in a continuous extractor first under acid conditions and then under basic conditions. Typically 30 g of a solid sample, containing surrogate, and matrix spiking standards, is extracted ultrasonically. After concentrating the extract to 1 mL it is spiked with 10 µL of an internal standard solution just prior to analysis by GC/MS. The volume injected should contain about 100 ng of base/neutral and 200 ng of acid surrogates (for a 1-µL injection). Analysis is performed by GC/MS using a capillary GC column.

INTERFERENCES Raw GC/MS data from all blanks, samples, and spikes must be evaluated for interferences. Contamination by carryover can occur whenever high-concentration and low-concentration samples are sequentially analyzed. To reduce carryover, the sample syringe must be rinsed out between samples with solvent. Whenever an unusually concentrated sample is encountered, it should be followed by the analysis of blank solvent to check for cross-contamination.

INSTRUMENTATION A GC/MS and a data system are required. The GC column used is a 30 m × 0.25 mm I.D. (or 0.32 mm I.D.) 1um film thickness silicone-coated fused silica capillary column. A continuous liquid-liquid extractor equipped with Teflon® or glass connection joints and stopcocks requiring no lubrication, a K-D concentrating apparatus, water bath, and an ultrasonic disrupter with a minimum power of 300 W and with pulsing capability are also required.

PRECISION & ACCURACY The estimated quantitation limit (EQL) of Method 8270B for determining an individual compound is approximately 1 mg/kg (wet weight) for soil or sediment samples, 1–200 mg/kg for wastes (dependent on matrix and method of preparation), and 10 µg/L for groundwater samples. EQLs will be proportionately higher for sample extracts that require dilution to avoid saturation of the detector.

The EQL(b) for groundwater in µg/L is 10.
The EQL (a, b) for low concentrations in soil and sediment in µg/kg is not determined.
Accuracy as µg/L is not listed.

Overall precision in µg/L is not listed.

(a) *EQLs listed for soil/sediment are based on wet weight. Normally data is reported in a dry-weight basis; therefore, EQLs will be higher based on the % dry weight of each sample. This calculation is based on a 30 g sample and gel permeation chromatography cleanup.*
(b) *Sample EQLs are highly matrix-dependent. The EQLs are provided for guidance and may not always be achievable.*
C = *True value for concentration, in µg/L.*
X = *Average recovery found for measurements of samples containing a concentration of C, in µg/L.*

ESTIMATED QUANTITATION LIMIT

Other Matrices	Factor (a)
High-concentration soil and sludges by sonicator	7.5
Non-water miscible waste	75

(a) *EQL for other matrices = [EQL for low soil/sediment] × [Factor]. This estimated EQL is similar to an EPA "Practical Quantitation Limit."*

SAMPLING METHOD

Liquid samples — Use a 1 or 2½ gallon amber glass bottle with a screw-top Teflon®-lined cover that has been prewashed with detergent and rinsed with distilled water and methanol (or isopropanol).

Soils, sediments, or sludges — Use an 8-oz. widemouth glass with a screw-top Teflon®-lined cover that has been prewashed with detergent and rinsed with distilled water and methanol (or isopropanol).

SAMPLE PRESERVATION

Liquid samples — If residual chlorine is present, add 3 mL of 10% sodium thiosulfate per gallon, cool to 4°C and store in a solvent-free refrigerator until analysis; if chlorine is not present, then eliminate the sodium thiosulfate addition.

Soils, sediments, or sludges — Cool samples to 4°C and store in a solvent-free refrigerator.

MHT Liquid samples must be extracted within 7 days and the extracts analyzed within 40 days. Soils, sediments, or sludges may be stored for a maximum of 14 days and the extracts analyzed within 40 days.

SAMPLE PREPARATION

Liquid samples — Transfer 1 L quantitatively to a continuous extractor. If high concentrations are anticipated, a smaller volume may be used and then diluted with organic-free reagent water to 1 L. Adjust pH, if necessary, to pH <2 using 1:1 (V/V) sulfuric acid. Pipette 1.0 mL of a surrogate standard spiking solution into each sample. For the sample in each analytical batch selected for spiking, add 1.0 mL of a matrix spiking standard. For base/neutral acid analysis, the amount of the surrogates and matrix spiking compounds added to the sample should result in a final concentration of 100 ng/µL of each analyte in the extract to be analyzed (assuming a 1-µL injection). Extract with methylene chloride for 18–24 h. Next, adjust the pH of the aqueous phase to pH >11 using 10 N sodium hydroxide and extract it with methylene chloride again for 18–24 h. Dry the extract through a column containing anhydrous sodium sulfate and concentrate it to 1 mL using a K-D concentrator.

Soils, sediments, or sludges — Use 30 g of sample. Nonporous or wet samples (gummy or clay type) that do not have a free-flowing sandy texture must be mixed with anhydrous sodium sulfate until the sample is free flowing. Add 1 mL of surrogate standards to all samples, spikes, standards, and blanks. For the sample in each analytical batch selected for spiking, add 1.0 mL of a matrix spiking standard. For base/neutral acid analysis, the amount added of the surrogates and matrix spiking compounds should result in a final concentration of 100 ng/µL of each base/neutral analyte and 200 ng/µL of each acid analyte in the extract to be analyzed (assuming a 1-µL injection). Immediately add a 100-mL mixture of 1:1 methylene chloride:acetone and extract the sample ultrasonically for 3 min and then decant or filter the extracts. Repeat the extraction two or more times. Dry the extract using a column with anhydrous sodium sulfate and concentrate it to 1 mL in a K-D concentrator.

QUALITY CONTROL A methylene chloride solution containing 50 ng/µL of decafluorotriphenylphosphine (DFTPP) is used for tuning the GC/MS system each 12-h shift. A system performance check also must be made during every 12-h shift. A standard containing 50 ng/µL each of 4,4'-DDT, pentachlorophenol, and benzidine is required to verify injection port inertness and GC column performance. A calibration standard at mid-concentration, containing each compound of interest, including all required surrogates, must be performed every 12 h during analysis. After the system performance check is met, calibration check compounds (CCCs) are used to check the validity of the initial calibration.

The internal standard responses and retention times in the calibration check standard must be evaluated immediately after or during data acquisition. If the retention time for any internal standard changes by more than 30 seconds from the last check calibration (12 h), the chromatographic system must be inspected for malfunctions and corrections must be made, as required. If the electron ionization current plot (EICP) area for any of the internal standards changes by a factor of two from the last daily calibration standard check, the mass spectrometer must be inspected for malfunctions and corrections must be made, as appropriate.

Demonstrate, through the analysis of a reagent water blank, that interferences from the analytical system, glassware, and reagents are under control. The blank samples should be carried through all stages of the sample preparation and measurement steps. For each analytical batch (up to 20 samples), a reagent blank, matrix spike, and matrix spike duplicate/duplicate must be analyzed (the frequency of the spikes may be different for different monitoring programs). The blank and spiked samples must be carried through all stages of the sample preparation and measurement steps. A QC reference sample concentrate containing each analyte at a concentration of 100 mg/L in methanol is required.

REFERENCE Test Methods for Evaluating Solid Waste (SW-846). U.S. EPA 1983, Method 8270B, Rev. 2, Nov. 1990. Office of Solid Waste, Washington, DC.

Disulfoton
CAS #298-04-4

EPA Method 8140

TITLE Organophosphorus Pesticides

MATRIX Groundwater, soils, sludges, water miscible liquid wastes, and non-water miscible wastes.

APPLICATION This method is used for the analysis of 21 organophosphorus pesticides. Samples are extracted, concentrated, and analyzed using direct injection of both neat and diluted organic liquid into a gas chromatograph (GC).

INTERFERENCES Solvents, reagents, and glassware may introduce artifacts. Other interferences may come from coextracted compounds from samples. The use of Florisil cleanup materials may produce low recoveries. Elemental sulfur may interfere with some compounds when using a flame photometric detector. Sulfur cleanup (Method 3660) may alleviate sulfur interference.

INSTRUMENTATION GC capable of on-column injections and a flame photometric detector (FPD) or a thermionic detector. Column 1: 1.8 m by 2 mm with 5% SP-2401 on Supelcoport. Column 2: 1.8 m by 2 mm with 3% SP-2401 on Supelcoport. Column 3: 50 cm by ⅛ IN EFLON® with 15% SE-54 on Gas Chrom Q. The preferred column is Column Number 1.

RANGE 5.2 to 92 µg/L.

MDL 0.20 µg/L (in reagent water).

PQL FACTORS FOR MULTIPLYING × FID MDL VALUE

Matrix	Multiplication Factor
Groundwater	10
Low-level soil by sonication with GPC cleanup	670
High-level soil and sludge by sonication	10,000
Non-water miscible waste	100,000

PRECISION 9.0% (single operator standard deviation).

ACCURACY 81.9% (single operator average recovery).

SAMPLING METHOD Use 8-oz. widemouth glass bottles with Teflon®-lined caps for concentrated waste samples, soils, sediments, and sludges. Use 1 or 2½ gallon amber glass bottles with Teflon®-lined caps for liquid (water) samples.

STABILITY Cool soil, sediment, sludge, and liquid samples to 4°C. If residual chlorine is present in liquid samples add 3 mL of 10% sodium thiosulfate per gallon of sample and cool to 4°C.

MHT 14 days for concentrated waste, soil, sediment, or sludge; 7 days for liquid samples; all extracts must be analyzed within 40 days.

QUALITY CONTROL A quality control check sample concentrate containing this compound in acetone at a concentration 1,000 times more concentrated than the selected spike concentration is required. The QC check sample concentrate may be prepared from pure standard materials or purchased as certified solutions. Use appropriate trip, matrix, control site, method, reagent, and solvent blanks. Internal, surrogate, and five concentration level calibration standards are used.

REFERENCE Method 8140, SW-846, 3rd ed., Sept. 1986.

Disulfoton sulfone
CAS #2497-06-5

EPA Method 507

TITLE Determination of Nitrogen and Phosphorus-Containing Pesticides in Water by GC/NPD

MATRIX This method is applicable to the determination of certain nitrogen and phosphorus-containing pesticides in finished drinking water and groundwater.

METHOD SUMMARY Method 507 covers 46 nitrogen- and phosphorus-containing pesticides. A 1-L sample is fortified with a surrogate standard, salted, buffered, extracted with methylene chloride, and concentrated; then the solvent is exchanged with methyl tert-butyl ether (MTBE) and concentrated again, and a 2-µL aliquot of a sample extract is injected into a GC system equipped with a selective nitrogen-phosphorus detector and a capillary column for analysis.

INTERFERENCES Method interferences may be caused by contaminants in solvents, reagents, glassware, and other sample processing apparatus. Interfering contamination may occur when a sample containing low concentrations of analytes is analyzed immediately following a sample containing relatively high concentrations. One or more injections of MTBE should be made following the analysis of a sample with high concentrations of analytes to check for analyte carryover. Matrix interferences may be caused by contaminants that are coextracted from the sample. The extent of matrix interferences will vary considerably from source to source, depending upon the water sampled.

INSTRUMENTATION A gas chromatograph system (GC) equipped with a nitrogen-phosphorus detector (NPD) is needed.

Column 1: 30 m × 0.25 mm I.D. DB-5 bonded fused silica column, 0.25 µm film thickness, or equivalent.
Column 2: 30 m × 0.25 mm I.D. DB-1701 bonded fused silica column, 0.25 µm film thickness, or equivalent.

PRECISION & ACCURACY This method has been validated in a single lab and estimated detection limits (EDLs) have been determined for each analyte. Observed detection limits may vary among waters, depending upon the nature of the interferences in the sample matrix and the specific instrumentation used. Analytes that are not separated chromatographically cannot be individually identified and measured unless an alternative technique for identification and quantification exist.

The estimated detection limit (in µg/L) was 3.8. The EDL is defined as either method detection limit or a level of compound in a sample yielding a peak in the final extract with signal-to-noise ratio of approximately 5, whichever value is higher.

The concentration used for these measurements (in µg/L) was 7.5.

The accuracy (as % recovery) was 98.
The precision (% RSD) was 10.

SAMPLING METHOD Grab samples are collected in 1-L glass sample bottles (prewashed with detergent and hot tap water, rinsed with reagent water, and dried in an oven at 400°C for 1 h) with screw caps lined with PTFE-fluorocarbon.

SAMPLE PRESERVATION Add mercuric chloride to the sample bottle in amounts to produce a concentration of 10 mg/L. If residual chlorine is present, add 80 mg of sodium thiosulfate/L of sample to the sample bottle prior to collection. After collection, seal bottle and shake vigorously for 1 min, then cool the sample to 4°C immediately and store it at 4°C in the dark until extraction.

MHT Maximum holding time of the samples, and in some cases the extracts, is 14 days.

SAMPLE PREPARATION Fortify the sample with 50 μL of the surrogate standard solution, adjust to pH 7 with phosphate buffer, add 100 g NaCl to the sample, and seal and shake to dissolve the salt; then extract with methylene chloride in a separatory funnel or in a mechanical tumbler bottle. Dry the extract by pouring it through a solvent-rinsed drying column containing about 10 cm of anhydrous sodium sulfate. Collect the extract in a Kuderna-Danish (K-D) concentrator and rinse the column with 20–30 mL methylene chloride. Concentrate the extract to about 2 mL and rinse the flask and its lower joint into the concentrator tube with 1 to 2 mL of methyl t-butyl ether (MTBE). Add 5–10 mL of MTBE and concentrate the extract twice (adding more MTBE) to a final volume of 5.0 mL and store it at 4°C until analysis.

Note: If methylene chloride is not completely removed from the final extract, it may cause detector problems.

QUALITY CONTROL Minimum quality control requirements are initial demonstration of lab capability, determination of surrogate compound recoveries in each sample and blank, monitoring internal standard peak area or height in each sample and blank, analysis of lab reagent blanks, lab fortified samples, lab fortified blanks, and other QC samples. A lab reagent blank is analyzed to demonstrate that all glassware and reagent interferences are under control.

Initial demonstration of capability is fulfilled by analyzing four fortified reagent water samples with the recovery value for each analyte falling within the acceptable range (±30% average recovery). Surrogate recoveries from samples or method blanks must be 70–130%. The internal standard response for any sample chromatogram should not deviate from the daily calibration check standard's internal standard response by more than 30% or lab fortified blanks and sample matrices are used to assess lab performance and analyte recovery, respectively.

If the response for the target analyte peak exceeds the working range of the system, dilute the extract and reanalyze. Alternative techniques such as an alternate detector or second chromatography column should be used to confirm peak identification when sample components are not resolved adequately.

EPA CONTACT & HOTLINE For technical questions contact Dr. Baldev Bathija, U.S. EPA, Office of Ground Water and Drinking Water (WH-550D), 401 M St. SW, Washington, DC 20460. Tel. (202) 260-3040. For further information the EPA Safe Drinking Water Hotline may be called at: (800) 426-4791.

REFERENCE Methods for the Determination of Organic Compounds in Drinking Water, EPA/600/4-88/039 (revised July 1991). U.S. EPA Environmental Monitoring Systems Laboratory, Cincinnati, OH, 45268, U.S.A. Available from the National Technical Information Service (NTIS), 5285 Port Royal Road, Springfield, VA 22161; Tel. 800-553-6847. NTIS Order Number is PB91-231480.

Disulfoton sulfoxide   **EPA Method 507**
CAS #2497-07-6

TITLE Determination of Nitrogen and Phosphorus-Containing Pesticides in Water by GC/NPD

MATRIX This method is applicable to the determination of certain nitrogen and phosphorus-containing pesticides in finished drinking water and groundwater.

METHOD SUMMARY Method 507 covers 46 nitrogen- and phosphorus-containing pesticides. A 1-L sample is fortified with a surrogate standard, salted, buffered, extracted with methylene chloride, and concentrated; then the solvent is exchanged with methyl tert-butyl ether (MTBE) and concentrated again, and a 2-μL aliquot of a sample extract is injected into a GC system equipped with a selective nitrogen-phosphorus detector and a capillary column for analysis.

INTERFERENCES Method interferences may be caused by contaminants in solvents, reagents, glassware, and other sample processing apparatus. Interfering contamination may occur when a sample containing low concentrations of analytes is analyzed immediately following a sample containing relatively high concentrations. One or more injections of MTBE should be made following the analysis of a sample with high concentrations of analytes to check for analyte carryover. Matrix interferences may be caused by contaminants that are coextracted from the sample. The extent of matrix interferences will vary considerably from source to source, depending upon the water sampled.

INSTRUMENTATION A gas chromatograph system (GC) equipped with a nitrogen-phosphorus detector (NPD) is needed.

Column 1: 30 m × 0.25 mm I.D. DB-5 bonded fused silica column, 0.25 μm film thickness, or equivalent.

Column 2: 30 m × 0.25 mm I.D. DB-1701 bonded fused silica column, 0.25 μm film thickness, or equivalent.

PRECISION & ACCURACY This method has been validated in a single lab and estimated detection limits (EDLs) have been determined for each analyte. Observed detection limits may vary among waters, depending upon the nature of the interferences in the sample matrix and the specific instrumentation used. Analytes that are not separated chromatographically cannot be individually identified and measured unless an alternative technique for identification and quantification exist.

The estimated detection limit (in μg/L) was 0.38. The EDL is defined as either method detection limit or a level of compound in a sample yielding a peak in the final extract with signal-to-noise ratio of approximately 5, whichever value is higher.

The concentration used for these measurements (in μg/L) was 3.8.
The accuracy (as % recovery) was 87.
The precision (% RSD) was 11.

SAMPLING METHOD Grab samples are collected in 1-L glass sample bottles (prewashed with detergent and hot tap water, rinsed with reagent water, and dried in an oven at 400°C for 1 h) with screw caps lined with PTFE-fluorocarbon.

SAMPLE PRESERVATION Add mercuric chloride to the sample bottle in amounts to produce a concentration of 10 mg/L. If residual chlorine is present, add 80 mg of sodium thiosulfate/L of sample to the sample bottle prior to collection. After collection, seal bottle and shake vigorously for 1 min, then cool the sample to 4°C immediately and store it at 4°C in the dark until extraction.

MHT Maximum holding time of the samples, and in some cases the extracts, is 14 days.

Note: Samples with this compound must be extracted immediately.

SAMPLE PREPARATION Fortify the sample with 50 μL of the surrogate standard solution, adjust to pH 7 with phosphate buffer, add 100 g NaCl to the sample, and seal and shake to dissolve the salt; then extract with methylene chloride in a separatory funnel or in a mechanical tumbler bottle. Dry the extract by pouring it through a solvent-rinsed drying column containing about 10 cm of anhydrous sodium sulfate. Collect the extract in a Kuderna-Danish (K-D) concentrator and rinse the column with 20–30 mL methylene chloride. Concentrate the extract to about 2 mL and rinse the flask and its lower joint into the concentrator tube with 1 to 2 mL of methyl t-butyl ether (MTBE). Add 5–10 mL of MTBE and concentrate the extract twice (adding more MTBE) to a final volume of 5.0 mL and store it at 4°C until analysis.

Note: If methylene chloride is not completely removed from the final extract, it may cause detector problems.

QUALITY CONTROL Minimum quality control requirements are initial demonstration of lab capability, determination of surrogate compound recoveries in each sample and blank, monitoring internal standard peak area or height in each sample and blank, analysis of lab reagent blanks, lab fortified samples, lab fortified blanks, and other QC samples. A lab reagent blank is analyzed to demonstrate that all glassware and reagent interferences are under control.

Initial demonstration of capability is fulfilled by analyzing four fortified reagent water samples with the recovery value for each analyte falling within the acceptable range (±30% average recovery). Surrogate recoveries from samples or method blanks must be 70–130%. The internal standard response for any sample chromatogram should not deviate from the daily calibration check standard's internal standard response by more than 30% or lab fortified blanks and sample matrices are used to assess lab performance and analyte recovery, respectively.

If the response for the target analyte peak exceeds the working range of the system, dilute the extract and reanalyze. Alternative techniques such as an alternate detector or second chromatography column should be used to confirm peak identification when sample components are not resolved adequately.

EPA CONTACT & HOTLINE For technical questions contact Dr. Baldev Bathija, U.S. EPA, Office of Ground Water and Drinking Water (WH-550D), 401 M St. SW, Washington, DC 20460. Tel. (202) 260-3040. For further information the EPA Safe Drinking Water Hotline may be called at: (800) 426-4791.

REFERENCE Methods for the Determination of Organic Compounds in Drinking Water, EPA/600/4-88/039 (revised July 1991). U.S. EPA Environmental Monitoring Systems Laboratory, Cincinnati, OH, 45268, U.S.A. Available from the National Technical Information Service (NTIS), 5285 Port Royal Road, Springfield, VA 22161; Tel. 800-553-6847. NTIS Order Number is PB91-231480.

---

n-Docosane **EPA Method 1625**
CAS #629-97-0

TITLE Semivolatile Organic Compounds by Isotope Dilution GC/MS

MATRIX The compounds may be determined in waters, soils, and municipal sludges by this method.

METHOD SUMMARY This method is used to determine 176 semivolatile toxic organic pollutants associated with the CWA (as amended 1987); the RCRA (as amended 1986); the CERCLA (as amended 1986); and other compounds amenable to extraction and analysis by capillary column gas chromatography-mass spectrometry (GC/MS).

Stable isotopically-labeled analogs of the compounds of interest are added to the sample. If the solids content is less than 1%, a 1-L sample is extracted at pH 12–13, then at pH <2 with methylene chloride using continuous extraction techniques.

If the solids content is 30% or less, the sample is diluted to 1% solids with reagent water, homogenized ultrasonically, and extracted at pH 12–13, then at pH <2 with methylene chloride using continuous extraction techniques. If the solids content is greater than 30%, the sample is extracted using ultrasonic techniques.

Each extract is dried over sodium sulfate, concentrated to a volume of 5 mL, cleaned up using GPC, if necessary, and concentrated. Extracts are concentrated to 1 mL if GPC is not performed, and to 0.5 mL if GPC is performed.

An internal standard is added to the extract, and a 1-mL aliquot of the extract is injected into the GC. The compounds are separated by GC and detected by a MS. The labeled compounds serve to correct the variability of the analytical technique.

INTERFERENCES Solvents, reagents, glassware, and other sample processing hardware may yield artifacts and/or elevated

baselines causing misinterpretation of chromatograms and spectra. Materials used in the analysis must be demonstrated to be free from interferences under the conditions of analysis by running method blanks initially and with each sample lot (sample started through the extraction process on a given 8-h shift, to a maximum of 20). Specific selection of reagents and purification of solvents by distillation in all glass systems may be required. Glassware and, where possible, reagents are cleaned by solvent rinse and baking at 450°C for 1-h minimum. Interferences coextracted from samples will vary considerably from source to source, depending on the diversity of the site being sampled.

INSTRUMENTATION Major instrumentation includes a GC with a splitless or on-column injection port for capillary column, a MS with 70 eV electron impact ionization, and a data system to collect and record MS data, and process it. A K-D apparatus is used to concentrate extracts.

GC Column: 30 m × 0.25 mm I.D. 5% phenyl, 94% methyl, 1% vinyl silicone bonded phased fused silica capillary column.

PRECISION & ACCURACY The detection limits of the method are usually dependent on the level of interferences rather than instrumental limitations. The limits typify the minimum quantities that can be detected with no interferences present.

The minimum level (in µg/mL) was 10. This is defined as a minimum level at which the analytical system shall give recognizable mass spectra (background corrected) and acceptable calibration points.

The MDL (in µg/kg) in low solids was 432 and in high solids was 447; these were determined in digested sludge (low solids) and in filter cake or compost (high solids).

Note: Background levels of this compound were present in the sludge tested, resulting in higher than expected MDLs. The MDL for this compound is expected to be approximately 50 µg/kg with no interferences present.

The labeled and native compound initial precision as standard deviation (in µg/L) was 31.
The labeled and native compound initial accuracy as average recovery (in µg/L) was 45–152.

SAMPLE COLLECTION, PRESERVATION & HANDLING Collect samples in glass containers. Aqueous samples which flow freely are collected in refrigerated bottles using automatic sampling equipment. Solid samples are collected as grab samples using widemouth jars. Maintain samples at 0 to 4°C from the time of collection until extraction. If residual chlorine is present in aqueous samples, add 80 mg sodium thiosulfate/L of water. Begin sample extraction within 7 days of collection, and analyze all extracts within 40 days of extraction.

SAMPLE PREPARATION Samples containing 1% solids or less are extracted directly using continuous liquid-liquid extraction techniques. Samples containing 1 to 30% solids are diluted to the 1% level with reagent water and extracted using continuous liquid-liquid extraction techniques. Samples containing greater than 30% solids are extracted using ultrasonic techniques.

Base/neutral extraction — Adjust the pH of the waters in the extractors to 12–13 with 6 N NaOH. Extract with methylene chloride for 24–48 h.
Acid extraction — Adjust the pH of the waters in the extractors to 2 or less using 6 N sulfuric acid. Extract with methylene chloride for 24–48 h.
Ultrasonic extraction of high solids samples — Add anhydrous sodium sulfate to the sample and QC aliquot(s). Add acetone:methylene chloride (1:1) to the sample and mix thoroughly

Concentrate extracts using a K-D apparatus.

QUALITY CONTROL The analyst is permitted to modify this method to improve separations or lower the costs of measurements, provided all performance specifications are met. Analyses of blanks are required to demonstrate freedom from contamination. When results of spikes indicate atypical method performance for samples, the samples are diluted to bring method performance within acceptable limits.

For low solids (aqueous samples), extract, concentrate, and analyze two sets of four 1-L aliquots (8 aliquots total) of the precision and recovery standard. For high solids samples, two sets of four 30-g aliquots of the high solids reference matrix are used.

Spike all samples with labeled compounds to assess method performance. Compute percent recovery of the labeled compounds using the internal standard method. Compare the labeled compound recovery for each compound with the corresponding labeled compound recovery.

Reagent water and high solids reference matrix blanks are analyzed to demonstrate freedom from contamination. Extract and concentrate a 1-L reagent water blank or a high solids reference matrix blank with each sample's lot (samples started through the extraction process on the same 8-h shift, to a maximum of 20 samples).

Field replicates may be collected to determine the precision of the sampling technique, and spiked samples may be required to determine the accuracy of the analysis when the internal standard method is used.

REFERENCE Semivolatile Organic Compounds by Isotope Dilution GC/MS. Office of Water Regulation and Standards, U.S. EPA Industrial Technology Division, Washington, DC, EPA Method 1625, Rev. C, June 1989 (contact W.A. Telliard, U.S. EPA, Office of Water Regulations and Standards, 401 M St., SW, Washington, DC, 20460. Phone: 202-382-7131).

---

n-Dodecane   **EPA Method 1625**
CAS #112-40-3

TITLE Semivolatile Organic Compounds by Isotope Dilution GC/MS

MATRIX The compounds may be determined in waters, soils, and municipal sludges by this method.

METHOD SUMMARY This method is used to determine 176 semivolatile toxic organic pollutants associated with the CWA (as amended 1987); the RCRA (as amended 1986); the CERCLA (as amended 1986); and other compounds amenable to extraction and analysis by capillary column gas chromatography-mass spectrometry (GC/MS).

Stable isotopically-labeled analogs of the compounds of interest are added to the sample. If the solids content is less than 1%, a 1-L sample is extracted at pH 12–13, then at pH <2 with methylene chloride using continuous extraction techniques.

If the solids content is 30% or less, the sample is diluted to 1% solids with reagent water, homogenized ultrasonically, and extracted at pH 12–13, then at pH <2 with methylene chloride using continuous extraction techniques. If the solids content is greater than 30%, the sample is extracted using ultrasonic techniques.

Each extract is dried over sodium sulfate, concentrated to a volume of 5 mL, cleaned up using GPC, if necessary, and concentrated. Extracts are concentrated to 1 mL if GPC is not performed, and to 0.5 mL if GPC is performed.

An internal standard is added to the extract, and a 1-mL aliquot of the extract is injected into the GC. The compounds are separated by GC and detected by a MS. The labeled compounds serve to correct the variability of the analytical technique.

INTERFERENCES Solvents, reagents, glassware, and other sample processing hardware may yield artifacts and/or elevated baselines causing misinterpretation of chromatograms and spectra. Materials used in the analysis must be demonstrated to be free from interferences under the conditions of analysis by running method blanks initially and with each sample lot (sample started through the extraction process on a given 8-h shift, to a maximum of 20). Specific selection of reagents and purification of solvents by distillation in all glass systems may be required. Glassware and, where possible, reagents are cleaned by solvent rinse and baking at 450°C for 1-h minimum. Interferences coextracted from samples will vary considerably from source to source, depending on the diversity of the site being sampled.

INSTRUMENTATION Major instrumentation includes a GC with a splitless or on-column injection port for capillary column, a MS with 70 eV electron impact ionization, and a data system to collect and record MS data, and process it. A K-D apparatus is used to concentrate extracts.

GC Column: 30 m × 0.25 mm I.D. 5% phenyl, 94% methyl, 1% vinyl silicone bonded phased fused silica capillary column.

PRECISION & ACCURACY The detection limits of the method are usually dependent on the level of interferences rather than instrumental limitations. The limits typify the minimum quantities that can be detected with no interferences present.

The minimum level (in µg/mL) was 10. This is defined as a minimum level at which the analytical system shall give recognizable mass spectra (background corrected) and acceptable calibration points.

The MDL (in µg/kg) in low solids was 860 and in high solids was 3885; these were determined in digested sludge (low solids) and in filter cake or compost (high solids).

Note: Background levels of this compound were present in the sludge tested, resulting in higher than expected MDLs. The MDL for this compound is expected to be approximately 50 µg/kg with no interferences present.

The labeled and native compound initial precision as standard deviation (in µg/L) was 74.

The labeled and native compound initial accuracy as average recovery (in µg/L) was 35–369.

SAMPLE COLLECTION, PRESERVATION & HANDLING
Collect samples in glass containers. Aqueous samples which flow freely are collected in refrigerated bottles using automatic sampling equipment. Solid samples are collected as grab samples using widemouth jars. Maintain samples at 0 to 4°C from the time of collection until extraction. If residual chlorine is present in aqueous samples, add 80 mg sodium thiosulfate/L of water. Begin sample extraction within 7 days of collection, and analyze all extracts within 40 days of extraction.

SAMPLE PREPARATION Samples containing 1% solids or less are extracted directly using continuous liquid-liquid extraction techniques. Samples containing 1 to 30% solids are diluted to the 1% level with reagent water and extracted using continuous liquid-liquid extraction techniques. Samples containing greater than 30% solids are extracted using ultrasonic techniques.

Base/neutral extraction — Adjust the pH of the waters in the extractors to 12–13 with 6 N NaOH. Extract with methylene chloride for 24–48 h.

Acid extraction — Adjust the pH of the waters in the extractors to 2 or less using 6 N sulfuric acid. Extract with methylene chloride for 24–48 h.

Ultrasonic extraction of high solids samples — Add anhydrous sodium sulfate to the sample and QC aliquot(s). Add acetone:methylene chloride (1:1) to the sample and mix thoroughly

Concentrate extracts using a K-D apparatus.

QUALITY CONTROL The analyst is permitted to modify this method to improve separations or lower the costs of measurements, provided all performance specifications are met. Analyses of blanks are required to demonstrate freedom from contamination. When results of spikes indicate atypical method performance for samples, the samples are diluted to bring method performance within acceptable limits.

For low solids (aqueous samples), extract, concentrate, and analyze two sets of four 1-L aliquots (8 aliquots total) of the precision and recovery standard. For high solids samples, two sets of four 30-g aliquots of the high solids reference matrix are used.

Spike all samples with labeled compounds to assess method performance. Compute percent recovery of the labeled compounds using the internal standard method. Compare the labeled compound recovery for each compound with the corresponding labeled compound recovery.

Reagent water and high solids reference matrix blanks are analyzed to demonstrate freedom from contamination. Extract and concentrate a 1-L reagent water blank or a high solids reference matrix blank with each sample's lot (samples started through the extraction process on the same 8-h shift, to a maximum of 20 samples).

Field replicates may be collected to determine the precision of the sampling technique, and spiked samples may be required to determine the accuracy of the analysis when the internal standard method is used.

REFERENCE Semivolatile Organic Compounds by Isotope Dilution GC/MS. Office of Water Regulation and Standards, U.S. EPA Industrial Technology Division, Washington, DC, EPA Method 1625, Rev. C, June 1989 (contact W.A. Telliard, U.S. EPA, Office of Water Regulations and Standards, 401 M St., SW, Washington, DC, 20460. Phone: 202-382-7131).

E

n-Eicosane
CAS #112-95-8

EPA Method 1625

TITLE Semivolatile Organic Compounds by Isotope Dilution GC/MS

MATRIX The compounds may be determined in waters, soils, and municipal sludges by this method.

METHOD SUMMARY This method is used to determine 176 semivolatile toxic organic pollutants associated with the CWA (as amended 1987); the RCRA (as amended 1986); the CERCLA (as amended 1986); and other compounds amenable to extraction and analysis by capillary column gas chromatography-mass spectrometry (GC/MS).

Stable isotopically-labeled analogs of the compounds of interest are added to the sample. If the solids content is less than 1%, a 1-L sample is extracted at pH 12–13, then at pH <2 with methylene chloride using continuous extraction techniques.

If the solids content is 30% or less, the sample is diluted to 1% solids with reagent water, homogenized ultrasonically, and extracted at pH 12–13, then at pH <2 with methylene chloride using continuous extraction techniques. If the solids content is greater than 30%, the sample is extracted using ultrasonic techniques.

Each extract is dried over sodium sulfate, concentrated to a volume of 5 mL, cleaned up using GPC, if necessary, and concentrated. Extracts are concentrated to 1 mL if GPC is not performed, and to 0.5 mL if GPC is performed.

An internal standard is added to the extract, and a 1-mL aliquot of the extract is injected into the GC. The compounds are separated by GC and detected by a MS. The labeled compounds serve to correct the variability of the analytical technique.

INTERFERENCES Solvents, reagents, glassware, and other sample processing hardware may yield artifacts and/or elevated baselines causing misinterpretation of chromatograms and spectra. Materials used in the analysis must be demonstrated to be free from interferences under the conditions of analysis by running method blanks initially and with each sample lot (sample started through the extraction process on a given 8-h shift, to a maximum of 20). Specific selection of reagents and purification of solvents by distillation in all glass systems may be required. Glassware and, where possible, reagents are cleaned by solvent rinse and baking at 450°C for 1-h minimum. Interferences coextracted from samples will vary considerably from source to source, depending on the diversity of the site being sampled.

INSTRUMENTATION Major instrumentation includes a GC with a splitless or on-column injection port for capillary column, a MS with 70 eV electron impact ionization, and a data system to collect and record MS data, and process it. A K-D apparatus is used to concentrate extracts.

GC Column: 30 m × 0.25 mm I.D. 5% phenyl, 94% methyl, 1% vinyl silicone bonded phased fused silica capillary column.

PRECISION & ACCURACY The detection limits of the method are usually dependent on the level of interferences rather than instrumental limitations. The limits typify the minimum quantities that can be detected with no interferences present.

The minimum level (in µg/mL) was 10. This is defined as a minimum level at which the analytical system shall give recognizable mass spectra (background corrected) and acceptable calibration points.

The MDL (in µg/kg) in low solids was 83 and in high solids was 229; these were determined in digested sludge (low solids) and in filter cake or compost (high solids).

Note: Background levels of this compound were present in the sludge tested, resulting in higher than expected MDLs. The MDL for this compound is expected to be approximately 50 µg/kg with no interferences present.

The labeled and native compound initial precision as standard deviation (in µg/L) was 59.
The labeled and native compound initial accuracy as average recovery (in µg/L) was 53–263.

SAMPLE COLLECTION, PRESERVATION & HANDLING Collect samples in glass containers. Aqueous samples which flow freely are collected in refrigerated bottles using automatic sampling equipment. Solid samples are collected as grab samples using widemouth jars. Maintain samples at 0 to 4°C from the time of collection until extraction. If residual chlorine is present in aqueous samples, add 80 mg sodium thiosulfate/L of water. Begin sample extraction within 7 days of collection, and analyze all extracts within 40 days of extraction.

SAMPLE PREPARATION Samples containing 1% solids or less are extracted directly using continuous liquid-liquid extraction techniques. Samples containing 1 to 30% solids are diluted to the 1% level with reagent water and extracted using continuous liquid-liquid extraction techniques. Samples containing greater than 30% solids are extracted using ultrasonic techniques.

Base/neutral extraction — Adjust the pH of the waters in the extractors to 12–13 with 6 N NaOH. Extract with methylene chloride for 24–48 h.
Acid extraction — Adjust the pH of the waters in the extractors to 2 or less using 6 N sulfuric acid. Extract with methylene chloride for 24–48 h.
Ultrasonic extraction of high solids samples — Add anhydrous sodium sulfate to the sample and QC aliquot(s). Add acetone:methylene chloride (1:1) to the sample and mix thoroughly

Concentrate extracts using a K-D apparatus.

QUALITY CONTROL The analyst is permitted to modify this method to improve separations or lower the costs of measurements, provided all performance specifications are met. Analyses of blanks are required to demonstrate freedom from contamination. When results of spikes indicate atypical

method performance for samples, the samples are diluted to bring method performance within acceptable limits.

For low solids (aqueous samples), extract, concentrate, and analyze two sets of four 1-L aliquots (8 aliquots total) of the precision and recovery standard. For high solids samples, two sets of four 30-g aliquots of the high solids reference matrix are used.

Spike all samples with labeled compounds to assess method performance. Compute percent recovery of the labeled compounds using the internal standard method. Compare the labeled compound recovery for each compound with the corresponding labeled compound recovery.

Reagent water and high solids reference matrix blanks are analyzed to demonstrate freedom from contamination. Extract and concentrate a 1-L reagent water blank or a high solids reference matrix blank with each sample's lot (samples started through the extraction process on the same 8-h shift, to a maximum of 20 samples).

Field replicates may be collected to determine the precision of the sampling technique, and spiked samples may be required to determine the accuracy of the analysis when the internal standard method is used.

REFERENCE Semivolatile Organic Compounds by Isotope Dilution GC/MS. Office of Water Regulation and Standards, U.S. EPA Industrial Technology Division, Washington, DC, EPA Method 1625, Rev. C, June 1989 (contact W.A. Telliard, U.S. EPA, Office of Water Regulations and Standards, 401 M St., SW, Washington, DC, 20460. Phone: 202-382-7131).

Endosulfan I   **EPA Method 625**
CAS #959-98-8

TITLE Base/Neutrals and Acids, U.S. EPA Method 625

MATRIX This methods covers municipal and industrial wastewaters.

METHOD SUMMARY Approximately 1 L of sample is serially extracted with methylene chloride at a pH greater than 11 and again at a pH less than 2 using a separatory funnel or a continuous extractor. The methylene chloride extract is dried, concentrated to a volume of 1 mL, and analyzed by GC/MS. Qualitative identification of the parameters in the extract is performed using the retention time and the relative abundance of three characteristic masses (m/z). Qualitative analysis is performed using either external or internal standard techniques with a single characteristic m/z.

INTERFERENCES Method interferences may be caused by contaminants in solvents, reagents, glassware, and other sample processing hardware. Glassware must be scrupulously cleaned. Glassware should be heated in a muffle furnace at 400°C for 5 to 30 min. Some thermally stable materials, such as PCBs, may not be eliminated by this treatment. Solvent rinses with acetone and pesticide quality hexane may be substituted for the muffle furnace heating. Matrix interferences may be caused by contaminants that are coextracted from the sample. The base-neutral extraction may cause significantly reduced recovery of phenols. The packed gas chromatographic columns recommended for the basic fraction may not exhibit sufficient resolution for some analytes.

INSTRUMENTATION A GC/MS system with an injection port designed for on-column injection when using packed columns and for splitless injection when using capillary columns.

Column for base/neutrals: 1.8 m long × 2 mm I.D. glass, packed with 3% SP-2550 on Supelcoport (100/120 mesh) or equivalent.

Column for acids: 1.8 m long × 2 mm I.D. glass, packed with 1% SP-1240DA on Supelcoport (100/120 mesh) or equivalent.

PRECISION & ACCURACY The MDL concentrations were obtained using reagent water. The MDL actually achieved in a given analysis will vary depending on instrument sensitivity and matrix effects. This method was tested by 15 laboratories using reagent water, drinking water, surface water, and industrial wastewaters spiked at six concentrations over the range 5 to 100 µg/L. Single operator precision, overall precision, and method accuracy were found to be directly related to the concentration of the parameter matrix.

The MDL (in µg/L) in reagent water was not detected.
The standard deviation (in µg/L based on 4 recovery measurements) was not reported.
The range (in µg/L) for average recovery for 4 measurements was not reported.
The range (in %) for percent recovery was not reported.
Accuracy (in µg/L) as expected recovery for one or more measurements of a sample containing a true concentration of C was not reported.
Precision (in µg/L) as expected single analyst standard deviation of measurements at an average concentration found at X was not reported.
Overall precision (in µg/L) as expected interlaboratory standard deviation of measurements in an average concentration found at X was not reported.

$C =$ *True value of the concentration in µg/L.*
$X =$ *Average recovery found for measurements of samples containing a concentration at C in µg/L.*

SAMPLE PREPARATION Adjust the pH to >11 with sodium hydroxide and serially extract in a separatory funnel with methylene chloride or else in a continuous extractor. Next, adjust the pH to <2 with sulfuric acid and serially extract in a separatory funnel with methylene chloride or else in a continuous extractor. Dry the extracts separately through a column of anhydrous sodium sulfate and then concentrate each of the extracts to 1.0 mL using a K-D apparatus.

SAMPLE COLLECTION, PRESERVATION & HANDLING
Grab samples must be collected in glass containers. All samples must be refrigerated at 4°C from the time of collection until extraction. If residual chlorine is present, add 80 mg of sodium thiosulfate/L of sample and mix well. All samples must be extracted within 7 days of collection and completely analyzed within 40 days of extraction.

QUALITY CONTROL Make an initial, one-time, demonstration of the ability to generate acceptable accuracy and precision with this method. Before processing any samples, the analyst must analyze a reagent water blank to demonstrate that interferences from the analytical system and glassware are under control. Each time a set of samples is extracted or reagents are changed, a reagent water blank must be processed. Spike and analyze a minimum of 5% of all samples to monitor and evaluate lab data quality. A QC check sample concentrate that contains each parameter of interest at a concentration of 100 µg/mL in acetone is required. PCBs and multicomponent pesticides may be omitted from this test.

After analysis of five spiked wastewater samples, calculate the average percent recovery and the standard deviation of the percent recovery. Spike all samples with the surrogate standard spiking solution and calculate the percent recovery of each surrogate compound.

REFERENCE Federal Register, Vol. 49, No. 209. Friday, Oct. 26, 1984.

Endosulfan I   **EPA Method 8080**
CAS #959-98-8

TITLE Organochlorine Pesticides and Polychlorinated Biphenyls By Gas Chromatography

MATRIX This method is used to determine the concentration of various organochlorine pesticides and polychlorinated biphenyls in extracts prepared from water, groundwater, soils, and sediments.

METHOD SUMMARY This method covers 26 pesticides and Aroclor (PCB) mixtures and it is suitable for monitoring-type analyses. After extraction, concentration and solvent exchange to hexane, a 2- to 5-µL sample aliquot is injected into a GC using the solvent flush technique, and the analytes are detected by an electron capture detector (ECD) or an electrolytic conductivity detector in the halogen mode (HECD). Both neat and diluted organic liquids may be analyzed by direct injection.

INTERFERENCES Interferences coextracted from the samples will vary considerably from source to source. Interferences by phthalate esters can pose a major problem in pesticide determinations when using the ECD. Cross-contamination of clean glassware routinely occurs when plastics are handled during extraction steps, especially when solvent-wetted surfaces are handled. The contamination from phthalate esters can be completely eliminated with a microcoulometric or electrolytic conductivity detector. Solvents, reagent, glassware, and other sample processing hardware may yield artifacts and/or interferences to sample analysis.

INSTRUMENTATION A gas chromatograph capable of on-column injections is needed. It must be equipped with an ECD or a HECD and one of the following GC columns:

Column 1: Supelcoport (100/120 mesh) coated with 1.5% SP-2250/1.95% SP-2401 packed in a 1.8 m × 4 mm I.D. glass column.

Column 2: Supelcoport (100/120 mesh) coated with 3% OV-1 in a 1.8 m × 4 mm I.D. glass column.

PRECISION & ACCURACY The method was tested by 20 laboratories using organic-free reagent water, drinking water, surface water, and three industrial wastewaters spiked at six concentrations. Concentrations used in the study ranged from 0.5 to 30 µg/L for single-component pesticides and from 8.5 to 400 µg/L for multicomponent parameters. Overall precision and method accuracy were found to be directly related to the concentration of the analyte and essentially independent of the sample matrix. The sensitivity of this method usually depends on the concentration of interferences rather than on instrumental limitations.

MDL in µg/L was 0.014.
Concentration range in µg/L was 0.5–30.
Accuracy as recovery (x^*) in µg/L was 0.97C + 0.04 .
Overall precision (S^*) in µg/L was 0.18x + 0.08.

x^* = *Expected recovery for one or more measurements of a sample containing concentration C, in µg/L.*
S^* = *Expected interlaboratory standard deviation of measurements at an average concentration found of the analyte in µg/L.*
C = *True value for the concentration, in µg/L.*
X = *Average recovery found for measurements of samples containing a concentration of C, in µg/L.*

SAMPLING METHOD
Liquid samples — Use a 1 or 2½ gallon amber glass bottle with a screw-top Teflon®-lined cover. Pre-wash the bottle with detergent, rinse with distilled water and methanol (or isopropanol).

Soil, sediments, and sludges — Use an 8-oz. widemouth glass with a screw-top Teflon®-lined cover. Pre-wash the bottle with detergent, rinse with distilled water and methanol (or isopropanol).

SAMPLE PRESERVATION Cool water, soil, sediment, or sludge samples immediately to 4°C.

Water samples — If residual chlorine is present, add 3 mL of 10% sodium thiosulfate per gallon and cool to 4°C. All extracts and samples should be stored under refrigeration.

MHT Liquid samples must be extracted within 7 days and the extracts must be analyzed within 40 days. Soils, sediments, and sludges may be stored for a maximum of 14 days prior to extraction.

SAMPLE PREPARATION
Liquid samples — Extract 1 L samples in a continuous extractor at pH 5–9 with methylene chloride after adding 1.0 mL of surrogate spiking solution to each sample. Pass the extract through a column of anhydrous sodium sulfate to dry and concentrate it in a K-D apparatus to 1 mL volume.

Soils, sediments and sludges — Rapidly weigh approximately 30 g of sample into a 400-mL beaker to avoid loss of the more volatile extractables. Nonporous or wet samples (gummy or clay type) that do not have a free-flowing sandy texture must be mixed with anhydrous sodium sulfate until the sample is free flowing. Add 1 mL of surrogate standards to all samples, spikes, standards, and blanks. Add 100 mL of 1:1 methylene chloride:acetone and extract ultrasonically. Decant and filter extracts, dry the extract by passing it through a drying column containing anhydrous sodium sulfate and concentrate to 1 mL in a K-D apparatus.

Hexane solvent exchange — Add 50 mL of hexane, a new boiling chip, and concentrate until the apparent volume of liquid reaches 1 mL. Adjust the extract volume to 10.0 mL. Stopper the concentration tube and store refrigerated at 4°C if further processing will not be performed immediately. If the extract will be stored longer than two days, transfer it to a vial with Teflon®-lined screw-cap or crimp top.

QUALITY CONTROL Demonstrate through the analysis of a reagent water blank, that all glassware and reagents are interference free. Each time a set of samples is processed, a method blank should be processed as a safeguard against chronic lab contamination. A reagent blank, a matrix spike, and a duplicate or matrix spike duplicate must be performed for each analytical batch (up to a maximum of 20 samples) analyzed.

Analytical system performance must be verified by analyzing QC check samples. The QC check sample concentration should contain each single-component analyte at the following concentrations in acetone: 4,4'-DDD, 10 µg/mL; 4,4'-DDT, 10 µg/mL; endosulfan II, 10 µg/mL; endosulfan sulfate, 10 µg/mL; and any other single-component pesticide at 2 µg/mL. If the method is only to be used to analyze PCBs, Chlordane, or Toxaphene, the QC check sample concentrate should contain the most representative multicomponent parameter at a concentration of 50 µg/mL in acetone.

REFERENCE Test Methods for Evaluating Solid Waste (SW-846). U.S. EPA. 1983. Method 8080B, Rev. 2, Nov. 1990. Office of Solid Wastes, Washington, DC.

Endosulfan I   EPA Method 8270
CAS #959-98-8

TITLE Semivolatile Organic Compounds by GC/MS

MATRIX This method is used to determine the concentration of semivolatile organic compounds in extracts prepared from all types of solid waste matrices, soils, and groundwater. Although surface waters are not specifically mentioned, this method should be applicable to water samples from rivers, lakes, etc.

METHOD SUMMARY This method covers 259 semivolatile organic compounds. In very limited applications direct injection of the sample into the GC/MS system may be appropriate, but this results in very high detection limits (approximately 10,000 µg/L). Typically, a 1-L liquid sample, containing surrogate, and matrix spiking standards, is extracted in a continuous extractor first under acid conditions and then under basic conditions. Typically 30 g of a solid sample, containing surrogate, and matrix spiking standards, is extracted ultrasonically. After concentrating the extract to 1 mL it is spiked with 10 µL of an internal standard solution just prior to analysis by GC/MS. The volume injected should contain about 100 ng of base/neutral and 200 ng of acid surrogates (for a 1-µL injection). Analysis is performed by GC/MS using a capillary GC column.

INTERFERENCES Raw GC/MS data from all blanks, samples, and spikes must be evaluated for interferences. Contamination by carryover can occur whenever high-concentration and low-concentration samples are sequentially analyzed. To reduce carryover, the sample syringe must be rinsed out between samples with solvent. Whenever an unusually concentrated sample is encountered, it should be followed by the analysis of blank solvent to check for cross-contamination.

INSTRUMENTATION A GC/MS and a data system are required. The GC column used is a 30 m × 0.25 mm I.D. (or 0.32 mm I.D.) 1um film thickness silicone-coated fused silica capillary column. A continuous liquid-liquid extractor equipped with Teflon® or glass connection joints and stopcocks requiring no lubrication, a K-D concentrating apparatus, water bath, and an ultrasonic disrupter with a minimum power of 300 W and with pulsing capability are also required.

PRECISION & ACCURACY The estimated quantitation limit (EQL) of Method 8270B for determining an individual compound is approximately 1 mg/kg (wet weight) for soil or sediment samples, 1–200 mg/kg for wastes (dependent on matrix and method of preparation), and 10 µg/L for groundwater samples. EQLs will be proportionately higher for sample extracts that require dilution to avoid saturation of the detector.

The EQL(b) for groundwater in µg/L is not listed.
The EQL (a, b) for low concentrations in soil and sediment in µg/kg is not listed.
Accuracy as µg/L is not listed.
Overall precision in µg/L is not listed.

(a) *EQLs listed for soil/sediment are based on wet weight. Normally data is reported in a dry-weight basis; therefore, EQLs will be higher based on the % dry weight of each sample. This calculation is based on a 30-g sample and gel permeation chromatography cleanup.*
(b) *Sample EQLs are highly matrix-dependent. The EQLs are provided for guidance and may not always be achievable.*
$C =$ *True value for concentration, in µg/L.*
$X =$ *Average recovery found for measurements of samples containing a concentration of C, in µg/L.*

ESTIMATED QUANTITATION LIMIT

Other Matrices	Factor (a)
High-concentration soil and sludges by sonicator	7.5
Non-water miscible waste	75

(a) *EQL for other matrices = [EQL for low soil/sediment] × [Factor]. This estimated EQL is similar to an EPA "Practical Quantitation Limit."*

SAMPLING METHOD
Liquid samples — Use a 1 or 2½ gallon amber glass bottle with a screw-top Teflon®-lined cover that has been prewashed with detergent and rinsed with distilled water and methanol (or isopropanol).

Soils, sediments, or sludges — Use an 8-oz. widemouth glass with a screw-top Teflon®-lined cover that has been prewashed

with detergent and rinsed with distilled water and methanol (or isopropanol).

SAMPLE PRESERVATION

Liquid samples — If residual chlorine is present, add 3 mL of 10% sodium thiosulfate per gallon, cool to 4°C and store in a solvent-free refrigerator until analysis; if chlorine is not present, then eliminate the sodium thiosulfate addition.

Soils, sediments, or sludges — Cool samples to 4°C and store in a solvent-free refrigerator.

MHT Liquid samples must be extracted within 7 days and the extracts analyzed within 40 days. Soils, sediments, or sludges may be stored for a maximum of 14 days and the extracts analyzed within 40 days.

SAMPLE PREPARATION

Liquid samples — Transfer 1 L quantitatively to a continuous extractor. If high concentrations are anticipated, a smaller volume may be used and then diluted with organic-free reagent water to 1 L. Adjust pH, if necessary, to pH <2 using 1:1 (V/V) sulfuric acid. Pipette 1.0 mL of a surrogate standard spiking solution into each sample. For the sample in each analytical batch selected for spiking, add 1.0 mL of a matrix spiking standard. For base/neutral acid analysis, the amount of the surrogates and matrix spiking compounds added to the sample should result in a final concentration of 100 ng/µL of each analyte in the extract to be analyzed (assuming a 1-µL injection). Extract with methylene chloride for 18–24 h. Next, adjust the pH of the aqueous phase to pH >11 using 10 N sodium hydroxide and extract it with methylene chloride again for 18–24 h. Dry the extract through a column containing anhydrous sodium sulfate and concentrate it to 1 mL using a K-D concentrator.

Soils, sediments, or sludges — Use 30 g of sample. Nonporous or wet samples (gummy or clay type) that do not have a free-flowing sandy texture must be mixed with anhydrous sodium sulfate until the sample is free flowing. Add 1 mL of surrogate standards to all samples, spikes, standards, and blanks. For the sample in each analytical batch selected for spiking, add 1.0 mL of a matrix spiking standard. For base/neutral acid analysis, the amount added of the surrogates and matrix spiking compounds should result in a final concentration of 100 ng/µL of each base/neutral analyte and 200 ng/µL of each acid analyte in the extract to be analyzed (assuming a 1-µL injection). Immediately add a 100-mL mixture of 1:1 methylene chloride:acetone and extract the sample ultrasonically for 3 min and then decant or filter the extracts. Repeat the extraction two or more times. Dry the extract using a column with anhydrous sodium sulfate and concentrate it to 1 mL in a K-D concentrator.

Note: Under the alkaline conditions of the extraction step endosulfan I is subject to decomposition so neutral extraction should be performed if this compound is expected.

QUALITY CONTROL A methylene chloride solution containing 50 ng/µL of decafluorotriphenylphosphine (DFTPP) is used for tuning the GC/MS system each 12-h shift. A system performance check also must be made during every 12-h shift. A standard containing 50 ng/µL each of 4,4′-DDT, pentachlorophenol, and benzidine is required to verify injection port inertness and GC column performance. A calibration standard at mid-concentration, containing each compound of interest, including all required surrogates, must be performed every 12 h during analysis. After the system performance check is met, calibration check compounds (CCCs) are used to check the validity of the initial calibration.

The internal standard responses and retention times in the calibration check standard must be evaluated immediately after or during data acquisition. If the retention time for any internal standard changes by more than 30 seconds from the last check calibration (12 h), the chromatographic system must be inspected for malfunctions and corrections must be made, as required. If the electron ionization current plot (EICP) area for any of the internal standards changes by a factor of two from the last daily calibration standard check, the mass spectrometer must be inspected for malfunctions and corrections must be made, as appropriate.

Demonstrate, through the analysis of a reagent water blank, that interferences from the analytical system, glassware, and reagents are under control. The blank samples should be carried through all stages of the sample preparation and measurement steps. For each analytical batch (up to 20 samples), a reagent blank, matrix spike, and matrix spike duplicate/duplicate must be analyzed (the frequency of the spikes may be different for different monitoring programs). The blank and spiked samples must be carried through all stages of the sample preparation and measurement steps. A QC reference sample concentrate containing each analyte at a concentration of 100 mg/L in methanol is required.

REFERENCE Test Methods for Evaluating Solid Waste (SW-846). U.S. EPA 1983, Method 8270B, Rev. 2, Nov. 1990. Office of Solid Waste, Washington, DC.

Endosulfan II
CAS #33213-65-9

EPA Method 625

TITLE Base/Neutrals and Acids, U.S. EPA Method 625

MATRIX This methods covers municipal and industrial wastewaters.

METHOD SUMMARY Approximately 1 L of sample is serially extracted with methylene chloride at a pH greater than 11 and again at a pH less than 2 using a separatory funnel or a continuous extractor. The methylene chloride extract is dried, concentrated to a volume of 1 mL, and analyzed by GC/MS. Qualitative identification of the parameters in the extract is performed using the retention time and the relative abundance of three characteristic masses (m/z). Qualitative analysis is performed using either external or internal standard techniques with a single characteristic m/z.

INTERFERENCES Method interferences may be caused by contaminants in solvents, reagents, glassware, and other sample processing hardware. Glassware must be scrupulously cleaned. Glassware should be heated in a muffle furnace at 400°C for 5 to 30 min. Some thermally stable materials, such as PCBs, may not be eliminated by this treatment. Solvent rinses with acetone and pesticide quality hexane may be substituted for the muffle furnace heating. Matrix interferences may be caused by contaminants that are coextracted from the sample. The base-neutral extraction may cause significantly reduced recovery of phenols. The packed gas chromatographic columns recommended for the basic fraction may not exhibit sufficient resolution for some analytes.

INSTRUMENTATION A GC/MS system with an injection port designed for on-column injection when using packed columns and for splitless injection when using capillary columns.

Column for base/neutrals: 1.8 m long × 2 mm I.D. glass, packed with 3% SP-2550 on Supelcoport (100/120 mesh) or equivalent.

Column for acids: 1.8 m long × 2 mm I.D. glass, packed with 1% SP-1240DA on Supelcoport (100/120 mesh) or equivalent.

PRECISION & ACCURACY The MDL concentrations were obtained using reagent water. The MDL actually achieved in a given analysis will vary depending on instrument sensitivity and matrix effects. This method was tested by 15 laboratories using reagent water, drinking water, surface water, and industrial wastewaters spiked at six concentrations over the range 5 to 100 µg/L. Single operator precision, overall precision, and method accuracy were found to be directly related to the concentration of the parameter matrix.

The MDL (in µg/L) in reagent water was not reported.
The standard deviation (in µg/L based on 4 recovery measurements) was not reported.
The range (in µg/L) for average recovery for 4 measurements was not reported.
The range (in %) for percent recovery was not reported.
Accuracy (in µg/L) as expected recovery for one or more measurements of a sample containing a true concentration of C was not reported.
Precision (in µg/L) as expected single analyst standard deviation of measurements at an average concentration found at X was not reported.
Overall precision (in µg/L) as expected interlaboratory standard deviation of measurements in an average concentration found at X was not reported.

C = *True value of the concentration in µg/L.*
X = *Average recovery found for measurements of samples containing a concentration at C in µg/L.*

SAMPLE PREPARATION Adjust the pH to >11 with sodium hydroxide and serially extract in a separatory funnel with methylene chloride or else in a continuous extractor. Next, adjust the pH to <2 with sulfuric acid and serially extract in a separatory funnel with methylene chloride or else in a continuous extractor. Dry the extracts separately through a column of anhydrous sodium sulfate and then concentrate each of the extracts to 1.0 mL using a K-D apparatus.

SAMPLE COLLECTION, PRESERVATION & HANDLING Grab samples must be collected in glass containers. All samples must be refrigerated at 4°C from the time of collection until extraction. If residual chlorine is present, add 80 mg of sodium thiosulfate/L of sample and mix well. All samples must be extracted within 7 days of collection and completely analyzed within 40 days of extraction.

QUALITY CONTROL Make an initial, one-time, demonstration of the ability to generate acceptable accuracy and precision with this method. Before processing any samples, the analyst must analyze a reagent water blank to demonstrate that interferences from the analytical system and glassware are under control. Each time a set of samples is extracted or reagents are changed, a reagent water blank must be processed. Spike and analyze a minimum of 5% of all samples to monitor and evaluate lab data quality. A QC check sample concentrate that contains each parameter of interest at a concentration of 100 µg/mL in acetone is required. PCBs and multicomponent pesticides may be omitted from this test.

After analysis of five spiked wastewater samples, calculate the average percent recovery and the standard deviation of the percent recovery. Spike all samples with the surrogate standard spiking solution and calculate the percent recovery of each surrogate compound.

REFERENCE *Federal Register,* Vol. 49, No. 209. Friday, Oct. 26, 1984.

Endosulfan II   EPA Method 8080
CAS #33213-65-9

TITLE Organochlorine Pesticides and Polychlorinated Biphenyls By Gas Chromatography

MATRIX This method is used to determine the concentration of various organochlorine pesticides and polychlorinated biphenyls in extracts prepared from water, groundwater, soils, and sediments.

METHOD SUMMARY This method covers 26 pesticides and Aroclor (PCB) mixtures and it is suitable for monitoring-type analyses. After extraction, concentration and solvent exchange to hexane, a 2- to 5-µL sample aliquot is injected into a GC using the solvent flush technique, and the analytes are detected by an electron capture detector (ECD) or an electrolytic conductivity detector in the halogen mode (HECD). Both neat and diluted organic liquids may be analyzed by direct injection.

INTERFERENCES Interferences coextracted from the samples will vary considerably from source to source. Interferences by phthalate esters can pose a major problem in pesticide determinations when using the ECD. Cross-contamination of clean glassware routinely occurs when plastics are handled during extraction steps, especially when solvent-wetted surfaces are handled. The contamination from phthalate esters can be completely eliminated with a microcoulometric or electrolytic conductivity detector. Solvents, reagent, glassware, and other sample processing hardware may yield artifacts and/or interferences to sample analysis.

INSTRUMENTATION A gas chromatograph capable of on-column injections is needed. It must be equipped with an ECD or a HECD and one of the following GC columns:

Column 1: Supelcoport (100/120 mesh) coated with 1.5% SP-2250/1.95% SP-2401 packed in a 1.8 m × 4 mm I.D. glass column.

Column 2: Supelcoport (100/120 mesh) coated with 3% OV-1 in a 1.8 m × 4 mm I.D. glass column.

PRECISION & ACCURACY The method was tested by 20 laboratories using organic-free reagent water, drinking water, surface water, and three industrial wastewaters spiked at six concentrations. Concentrations used in the study ranged from 0.5 to 30 µg/L for single-component pesticides and from 8.5 to 400 µg/L for multicomponent parameters. Overall precision and method accuracy were found to be directly related to the concentration of the analyte and essentially independent of the sample matrix. The sensitivity of this method usually depends on the concentration of interferences rather than on instrumental limitations.

MDL in µg/L was 0.004.
Concentration range in µg/L was 0.5–30.
Accuracy as recovery (x^*) in µg/L was $0.93C + 0.34$.
Overall precision (S^*) in µg/L was $0.47x - 0.20$.

x^* = *Expected recovery for one or more measurements of a sample containing concentration C, in µg/L.*
S^* = *Expected interlaboratory standard deviation of measurements at an average concentration found of the analyte in µg/L.*
C = *True value for the concentration, in µg/L.*
X = *Average recovery found for measurements of samples containing a concentration of C, in µg/L.*

SAMPLING METHOD
Liquid samples — Use a 1 or 2½ gallon amber glass bottle with a screw-top Teflon®-lined cover. Pre-wash the bottle with detergent, rinse with distilled water and methanol (or isopropanol).

Soil, sediments, and sludges — Use an 8-oz. widemouth glass with a screw-top Teflon®-lined cover. Pre-wash the bottle with detergent, rinse with distilled water and methanol (or isopropanol).

SAMPLE PRESERVATION Cool water, soil, sediment, or sludge samples immediately to 4°C.

Water samples — If residual chlorine is present, add 3 mL of 10% sodium thiosulfate per gallon and cool to 4°C. All extracts and samples should be stored under refrigeration.

MHT Liquid samples must be extracted within 7 days and the extracts must be analyzed within 40 days. Soils, sediments, and sludges may be stored for a maximum of 14 days prior to extraction.

SAMPLE PREPARATION
Liquid samples — Extract 1 L samples in a continuous extractor at pH 5–9 with methylene chloride after adding 1.0 mL of surrogate spiking solution to each sample. Pass the extract through a column of anhydrous sodium sulfate to dry and concentrate it in a K-D apparatus to 1 mL volume.

Soils, sediments and sludges — Rapidly weigh approximately 30 g of sample into a 400-mL beaker to avoid loss of the more volatile extractables. Nonporous or wet samples (gummy or clay type) that do not have a free-flowing sandy texture must be mixed with anhydrous sodium sulfate until the sample is free flowing. Add 1 mL of surrogate standards to all samples, spikes, standards, and blanks. Add 100 mL of 1:1 methylene chloride:acetone and extract ultrasonically. Decant and filter extracts, dry the extract by passing it through a drying column containing anhydrous sodium sulfate and concentrate to 1 mL in a K-D apparatus.

Hexane solvent exchange — Add 50 mL of hexane, a new boiling chip, and concentrate until the apparent volume of liquid reaches 1 mL. Adjust the extract volume to 10.0 mL. Stopper the concentration tube and store refrigerated at 4°C if further processing will not be performed immediately. If the extract will be stored longer than two days, transfer it to a vial with Teflon®-lined screw-cap or crimp top.

QUALITY CONTROL Demonstrate through the analysis of a reagent water blank, that all glassware and reagents are interference free. Each time a set of samples is processed, a method blank should be processed as a safeguard against chronic lab contamination. A reagent blank, a matrix spike, and a duplicate or matrix spike duplicate must be performed for each analytical batch (up to a maximum of 20 samples) analyzed.

Analytical system performance must be verified by analyzing QC check samples. The QC check sample concentration should contain each single-component analyte at the following concentrations in acetone: 4,4'-DDD, 10 µg/mL; 4,4'-DDT, 10 µg/mL; endosulfan II, 10 µg/mL; endosulfan sulfate, 10 µg/mL; and any other single-component pesticide at 2 µg/mL. If the method is only to be used to analyze PCBs, Chlordane, or Toxaphene, the QC check sample concentrate should contain the most representative multicomponent parameter at a concentration of 50 µg/mL in acetone.

REFERENCE Test Methods for Evaluating Solid Waste (SW-846). U.S. EPA. 1983. Method 8080B, Rev. 2, Nov. 1990. Office of Solid Wastes, Washington, DC.

Endosulfan II   **EPA Method 8270**
CAS #33213-65-9

TITLE Semivolatile Organic Compounds by GC/MS

MATRIX This method is used to determine the concentration of semivolatile organic compounds in extracts prepared from all types of solid waste matrices, soils, and groundwater. Although surface waters are not specifically mentioned, this method should be applicable to water samples from rivers, lakes, etc.

METHOD SUMMARY This method covers 259 semivolatile organic compounds. In very limited applications direct injection of the sample into the GC/MS system may be appropriate, but this results in very high detection limits (approximately 10,000 µg/L). Typically, a 1-L liquid sample, containing surrogate, and matrix spiking standards, is extracted in a continuous extractor first under acid conditions and then under basic conditions. Typically 30 g of a solid sample, containing surrogate, and matrix spiking standards, is extracted ultrasonically. After concentrating the extract to 1 mL it is spiked with 10 μL of an internal standard solution just prior to analysis by GC/MS. The volume injected should contain about 100 ng of base/neutral and 200 ng of acid surrogates (for a 1-μL injection). Analysis is performed by GC/MS using a capillary GC column.

INTERFERENCES Raw GC/MS data from all blanks, samples, and spikes must be evaluated for interferences. Contamination by carryover can occur whenever high-concentration and low-concentration samples are sequentially analyzed. To reduce carryover, the sample syringe must be rinsed out between samples with solvent. Whenever an unusually concentrated sample is encountered, it should be followed by the analysis of blank solvent to check for cross-contamination.

INSTRUMENTATION A GC/MS and a data system are required. The GC column used is a 30 m × 0.25 mm I.D. (or 0.32 mm I.D.) 1um film thickness silicone-coated fused silica capillary column. A continuous liquid-liquid extractor equipped with Teflon® or glass connection joints and stopcocks requiring no lubrication, a K-D concentrating apparatus, water bath, and an ultrasonic disrupter with a minimum power of 300 W and with pulsing capability are also required.

PRECISION & ACCURACY The estimated quantitation limit (EQL) of Method 8270B for determining an individual compound is approximately 1 mg/kg (wet weight) for soil or sediment samples, 1–200 mg/kg for wastes (dependent on matrix and method of preparation), and 10 μg/L for groundwater samples. EQLs will be proportionately higher for sample extracts that require dilution to avoid saturation of the detector.

The EQL(b) for groundwater in μg/L is not listed.
The EQL (a, b) for low concentrations in soil and sediment in μg/kg is not listed.
Accuracy as μg/L is not listed.
Overall precision in μg/L is not listed.

(a) EQLs listed for soil/sediment are based on wet weight. Normally data is reported in a dry-weight basis; therefore, EQLs will be higher based on the % dry weight of each sample. This calculation is based on a 30-g sample and gel permeation chromatography cleanup.
(b) Sample EQLs are highly matrix-dependent. The EQLs are provided for guidance and may not always be achievable.
C = True value for concentration, in μg/L.
X = Average recovery found for measurements of samples containing a concentration of C, in μg/L.

ESTIMATED QUANTITATION LIMIT

Other Matrices	Factor (a)
High-concentration soil and sludges by sonicator	7.5
Non-water miscible waste	75

(a) EQL for other matrices = [EQL for low soil/sediment] × [Factor]. This estimated EQL is similar to an EPA "Practical Quantitation Limit."

SAMPLING METHOD
Liquid samples — Use a 1 or 2½ gallon amber glass bottle with a screw-top Teflon®-lined cover that has been prewashed with detergent and rinsed with distilled water and methanol (or isopropanol).

Soils, sediments, or sludges — Use an 8-oz. widemouth glass with a screw-top Teflon®-lined cover that has been prewashed with detergent and rinsed with distilled water and methanol (or isopropanol).

SAMPLE PRESERVATION
Liquid samples — If residual chlorine is present, add 3 mL of 10% sodium thiosulfate per gallon, cool to 4°C and store in a solvent-free refrigerator until analysis; if chlorine is not present, then eliminate the sodium thiosulfate addition.

Soils, sediments, or sludges — Cool samples to 4°C and store in a solvent-free refrigerator.

MHT Liquid samples must be extracted within 7 days and the extracts analyzed within 40 days. Soils, sediments, or sludges may be stored for a maximum of 14 days and the extracts analyzed within 40 days.

SAMPLE PREPARATION
Liquid samples — Transfer 1 L quantitatively to a continuous extractor. If high concentrations are anticipated, a smaller volume may be used and then diluted with organic-free reagent water to 1 L. Adjust pH, if necessary, to pH <2 using 1:1 (V/V) sulfuric acid. Pipette 1.0 mL of a surrogate standard spiking solution into each sample. For the sample in each analytical batch selected for spiking, add 1.0 mL of a matrix spiking standard. For base/neutral acid analysis, the amount of the surrogates and matrix spiking compounds added to the sample should result in a final concentration of 100 ng/μL of each analyte in the extract to be analyzed (assuming a 1-μL injection). Extract with methylene chloride for 18–24 h. Next, adjust the pH of the aqueous phase to pH >11 using 10 N sodium hydroxide and extract it with methylene chloride again for 18–24 h. Dry the extract through a column containing anhydrous sodium sulfate and concentrate it to 1 mL using a K-D concentrator.

Soils, sediments, or sludges — Use 30 g of sample. Nonporous or wet samples (gummy or clay type) that do not have a free-flowing sandy texture must be mixed with anhydrous sodium sulfate until the sample is free flowing. Add 1 mL of surrogate standards to all samples, spikes, standards, and blanks. For the sample in each analytical batch selected for spiking, add 1.0 mL of a matrix spiking standard. For base/neutral acid analysis, the amount added of the surrogates and matrix spiking compounds should result in a final concentration of 100 ng/μL of each base/neutral analyte and 200 ng/μL of each acid analyte in the extract to be analyzed (assuming a 1-μL injection). Immediately add a 100-mL mixture of 1:1 methylene chloride:acetone and extract the sample ultrasonically for 3 min and then decant or filter the extracts. Repeat the extraction two or more times. Dry the extract using a column with anhydrous sodium sulfate and concentrate it to 1 mL in a K-D concentrator.

Note: Under the alkaline conditions of the extraction step endosulfan II is subject to decomposition so neutral extraction should be performed if this compound is expected.

QUALITY CONTROL A methylene chloride solution containing 50 ng/μL of decafluorotriphenylphosphine (DFTPP) is used for tuning the GC/MS system each 12-h shift. A system performance check also must be made during every 12-h shift.

A standard containing 50 ng/μL each of 4,4′-DDT, pentachlorophenol, and benzidine is required to verify injection port inertness and GC column performance. A calibration standard at mid-concentration, containing each compound of interest, including all required surrogates, must be performed every 12 h during analysis. After the system performance check is met, calibration check compounds (CCCs) are used to check the validity of the initial calibration.

The internal standard responses and retention times in the calibration check standard must be evaluated immediately after or during data acquisition. If the retention time for any internal standard changes by more than 30 seconds from the last check calibration (12 h), the chromatographic system must be inspected for malfunctions and corrections must be made, as required. If the electron ionization current plot (EICP) area for any of the internal standards changes by a factor of two from the last daily calibration standard check, the mass spectrometer must be inspected for malfunctions and corrections must be made, as appropriate.

Demonstrate, through the analysis of a reagent water blank, that interferences from the analytical system, glassware, and reagents are under control. The blank samples should be carried through all stages of the sample preparation and measurement steps. For each analytical batch (up to 20 samples), a reagent blank, matrix spike, and matrix spike duplicate/duplicate must be analyzed (the frequency of the spikes may be different for different monitoring programs). The blank and spiked samples must be carried through all stages of the sample preparation and measurement steps. A QC reference sample concentrate containing each analyte at a concentration of 100 mg/L in methanol is required.

REFERENCE Test Methods for Evaluating Solid Waste (SW-846). U.S. EPA 1983, Method 8270B, Rev. 2, Nov. 1990. Office of Solid Waste, Washington, DC.

Endosulfan sulfate   **EPA Method 625**
CAS #1031-07-8

TITLE Base/Neutrals and Acids, U.S. EPA Method 625

MATRIX This methods covers municipal and industrial wastewaters.

METHOD SUMMARY Approximately 1 L of sample is serially extracted with methylene chloride at a pH greater than 11 and again at a pH less than 2 using a separatory funnel or a continuous extractor. The methylene chloride extract is dried, concentrated to a volume of 1 mL, and analyzed by GC/MS. Qualitative identification of the parameters in the extract is performed using the retention time and the relative abundance of three characteristic masses (m/z). Qualitative analysis is performed using either external or internal standard techniques with a single characteristic m/z.

INTERFERENCES Method interferences may be caused by contaminants in solvents, reagents, glassware, and other sample processing hardware. Glassware must be scrupulously cleaned. Glassware should be heated in a muffle furnace at 400°C for 5 to 30 min. Some thermally stable materials, such as PCBs, may not be eliminated by this treatment. Solvent rinses with acetone and pesticide quality hexane may be substituted for the muffle furnace heating. Matrix interferences may be caused by contaminants that are coextracted from the sample. The base-neutral extraction may cause significantly reduced recovery of phenols. The packed gas chromatographic columns recommended for the basic fraction may not exhibit sufficient resolution for some analytes.

INSTRUMENTATION A GC/MS system with an injection port designed for on-column injection when using packed columns and for splitless injection when using capillary columns.

Column for base/neutrals: 1.8 m long × 2 mm I.D. glass, packed with 3% SP-2550 on Supelcoport (100/120 mesh) or equivalent.

Column for acids: 1.8 m long × 2 mm I.D. glass, packed with 1% SP-1240DA on Supelcoport (100/120 mesh) or equivalent.

PRECISION & ACCURACY The MDL concentrations were obtained using reagent water. The MDL actually achieved in a given analysis will vary depending on instrument sensitivity and matrix effects. This method was tested by 15 laboratories using reagent water, drinking water, surface water, and industrial wastewaters spiked at six concentrations over the range 5 to 100 μg/L. Single operator precision, overall precision, and method accuracy were found to be directly related to the concentration of the parameter matrix.

The MDL (in μg/L) in reagent water was 5.6.

The standard deviation (in μg/L based on 4 recovery measurements) was 16.7.

The range (in μg/L) for average recovery for 4 measurements was D-103.5.

The range (in %) for percent recovery was D-107.

Accuracy (in μg/L) as expected recovery for one or more measurements of a sample containing a true concentration of C was $0.39C + 0.41$.

Precision (in μg/L) as expected single analyst standard deviation of measurements at an average concentration found at X was $0.12X + 2.47$.

Overall precision (in μg/L) as expected interlaboratory standard deviation of measurements in an average concentration found at X was $0.63X - 1.03$.

$C =$ *True value of the concentration in μg/L.*
$X =$ *Average recovery found for measurements of samples containing a concentration at C in μg/L.*

SAMPLE PREPARATION Adjust the pH to >11 with sodium hydroxide and serially extract in a separatory funnel with methylene chloride or else in a continuous extractor. Next, adjust the pH to <2 with sulfuric acid and serially extract in a separatory funnel with methylene chloride or else in a continuous extractor. Dry the extracts separately through a column of anhydrous sodium sulfate and then concentrate each of the extracts to 1.0 mL using a K-D apparatus.

SAMPLE COLLECTION, PRESERVATION & HANDLING Grab samples must be collected in glass containers. All samples must be refrigerated at 4°C from the time of collection until extraction. If residual chlorine is present, add 80 mg of sodium thiosulfate/L of sample and mix well. All samples must be extracted within 7 days of collection and completely analyzed within 40 days of extraction.

QUALITY CONTROL Make an initial, one-time, demonstration of the ability to generate acceptable accuracy and precision with this method. Before processing any samples, the analyst must analyze a reagent water blank to demonstrate that interferences from the analytical system and glassware are under control. Each time a set of samples is extracted or reagents are changed, a reagent water blank must be processed. Spike and analyze a minimum of 5% of all samples to monitor and evaluate lab data quality. A QC check sample concentrate that contains each parameter of interest at a concentration of 100 µg/mL in acetone is required. PCBs and multicomponent pesticides may be omitted from this test.

After analysis of five spiked wastewater samples, calculate the average percent recovery and the standard deviation of the percent recovery. Spike all samples with the surrogate standard spiking solution and calculate the percent recovery of each surrogate compound.

REFERENCE Federal Register, Vol. 49, No. 209. Friday, Oct. 26, 1984.

Endosulfan Sulfate
CAS #1031-07-8

EPA Method 8080

TITLE Organochlorine Pesticides and Polychlorinated Biphenyls By Gas Chromatography

MATRIX This method is used to determine the concentration of various organochlorine pesticides and polychlorinated biphenyls in extracts prepared from water, groundwater, soils, and sediments.

METHOD SUMMARY This method covers 26 pesticides and Aroclor (PCB) mixtures and it is suitable for monitoring-type analyses. After extraction, concentration and solvent exchange to hexane, a 2- to 5-µL sample aliquot is injected into a GC using the solvent flush technique, and the analytes are detected by an electron capture detector (ECD) or an electrolytic conductivity detector in the halogen mode (HECD). Both neat and diluted organic liquids may be analyzed by direct injection.

INTERFERENCES Interferences coextracted from the samples will vary considerably from source to source. Interferences by phthalate esters can pose a major problem in pesticide determinations when using the ECD. Cross-contamination of clean glassware routinely occurs when plastics are handled during extraction steps, especially when solvent-wetted surfaces are handled. The contamination from phthalate esters can be completely eliminated with a microcoulometric or electrolytic conductivity detector. Solvents, reagent, glassware, and other sample processing hardware may yield artifacts and/or interferences to sample analysis.

INSTRUMENTATION A gas chromatograph capable of on-column injections is needed. It must be equipped with an ECD or a HECD and one of the following GC columns:

Column 1: Supelcoport (100/120 mesh) coated with 1.5% SP-2250/1.95% SP-2401 packed in a 1.8 m × 4 mm I.D. glass column.

Column 2: Supelcoport (100/120 mesh) coated with 3% OV-1 in a 1.8 m × 4 mm I.D. glass column.

PRECISION & ACCURACY The method was tested by 20 laboratories using organic-free reagent water, drinking water, surface water, and three industrial wastewaters spiked at six concentrations. Concentrations used in the study ranged from 0.5 to 30 µg/L for single-component pesticides and from 8.5 to 400 µg/L for multicomponent parameters. Overall precision and method accuracy were found to be directly related to the concentration of the analyte and essentially independent of the sample matrix. The sensitivity of this method usually depends on the concentration of interferences rather than on instrumental limitations.

MDL in µg/L was 0.066.
Concentration range in µg/L was 0.5–30.
Accuracy as recovery (x^*) in µg/L was 0.89C-0.37.
Overall precision (S^*) in µg/L was 0.24x + 0.35.

x^* = *Expected recovery for one or more measurements of a sample containing concentration C, in µg/L.*
S^* = *Expected interlaboratory standard deviation of measurements at an average concentration found of the analyte in µg/L.*
C = *True value for the concentration, in µg/L.*
X = *Average recovery found for measurements of samples containing a concentration of C, in µg/L.*

SAMPLING METHOD
Liquid samples — Use a 1 or 2½ gallon amber glass bottle with a screw-top Teflon®-lined cover. Pre-wash the bottle with detergent, rinse with distilled water and methanol (or isopropanol).

Soil, sediments, and sludges — Use an 8-oz. widemouth glass with a screw-top Teflon®-lined cover. Pre-wash the bottle with detergent, rinse with distilled water and methanol (or isopropanol).

SAMPLE PRESERVATION Cool water, soil, sediment, or sludge samples immediately to 4°C.

Water samples — If residual chlorine is present, add 3 mL of 10% sodium thiosulfate per gallon and cool to 4°C. All extracts and samples should be stored under refrigeration.

MHT Liquid samples must be extracted within 7 days and the extracts must be analyzed within 40 days. Soils, sediments, and sludges may be stored for a maximum of 14 days prior to extraction.

SAMPLE PREPARATION
Liquid samples — Extract 1 L samples in a continuous extractor at pH 5–9 with methylene chloride after adding 1.0 mL of surrogate spiking solution to each sample. Pass the extract through a column of anhydrous sodium sulfate to dry and concentrate it in a K-D apparatus to 1 mL volume.

Soils, sediments and sludges — Rapidly weigh approximately 30 g of sample into a 400-mL beaker to avoid loss of the more volatile extractables. Nonporous or wet samples (gummy or clay type) that do not have a free-flowing sandy texture must be mixed with anhydrous sodium sulfate until the sample is free flowing. Add 1 mL of surrogate standards to all samples, spikes, standards, and blanks. Add 100 mL of 1:1 methylene chloride:acetone and extract ultrasonically. Decant and filter extracts, dry the extract by passing it through a drying column containing anhydrous sodium sulfate and concentrate to 1 mL in a K-D apparatus.

Hexane solvent exchange — Add 50 mL of hexane, a new boiling chip, and concentrate until the apparent volume of liquid reaches 1 mL. Adjust the extract volume to 10.0 mL. Stopper the concentration tube and store refrigerated at 4°C if further processing will not be performed immediately. If the extract will be stored longer than two days, transfer it to a vial with Teflon®-lined screw-cap or crimp top.

QUALITY CONTROL Demonstrate through the analysis of a reagent water blank, that all glassware and reagents are interference free. Each time a set of samples is processed, a method blank should be processed as a safeguard against chronic lab contamination. A reagent blank, a matrix spike, and a duplicate or matrix spike duplicate must be performed for each analytical batch (up to a maximum of 20 samples) analyzed.

Analytical system performance must be verified by analyzing QC check samples. The QC check sample concentration should contain each single-component analyte at the following concentrations in acetone: 4,4′-DDD, 10 µg/mL; 4,4′-DDT, 10 µg/mL; endosulfan II, 10 µg/mL; endosulfan sulfate, 10 µg/mL; and any other single-component pesticide at 2 µg/mL. If the method is only to be used to analyze PCBs, Chlordane, or Toxaphene, the QC check sample concentrate should contain the most representative multicomponent parameter at a concentration of 50 µg/mL in acetone.

REFERENCE Test Methods for Evaluating Solid Waste (SW-846). U.S. EPA. 1983. Method 8080B, Rev. 2, Nov. 1990. Office of Solid Wastes, Washington, DC.

Endosulfan sulfate   **EPA Method 8270**
CAS #1031-07-8

TITLE Semivolatile Organic Compounds by GC/MS

MATRIX This method is used to determine the concentration of semivolatile organic compounds in extracts prepared from all types of solid waste matrices, soils, and groundwater. Although surface waters are not specifically mentioned, this method should be applicable to water samples from rivers, lakes, etc.

METHOD SUMMARY This method covers 259 semivolatile organic compounds. In very limited applications direct injection of the sample into the GC/MS system may be appropriate, but this results in very high detection limits (approximately 10,000 µg/L). Typically, a 1-L liquid sample, containing surrogate, and matrix spiking standards, is extracted in a continuous extractor first under acid conditions and then under basic conditions. Typically 30 g of a solid sample, containing surrogate, and matrix spiking standards, is extracted ultrasonically. After concentrating the extract to 1 mL it is spiked with 10 µL of an internal standard solution just prior to analysis by GC/MS. The volume injected should contain about 100 ng of base/neutral and 200 ng of acid surrogates (for a 1-µL injection). Analysis is performed by GC/MS using a capillary GC column.

INTERFERENCES Raw GC/MS data from all blanks, samples, and spikes must be evaluated for interferences. Contamination by carryover can occur whenever high-concentration and low-concentration samples are sequentially analyzed. To reduce carryover, the sample syringe must be rinsed out between samples with solvent. Whenever an unusually concentrated sample is encountered, it should be followed by the analysis of blank solvent to check for cross-contamination.

INSTRUMENTATION A GC/MS and a data system are required. The GC column used is a 30 m × 0.25 mm I.D. (or 0.32 mm I.D.) 1um film thickness silicone-coated fused silica capillary column. A continuous liquid-liquid extractor equipped with Teflon® or glass connection joints and stopcocks requiring no lubrication, a K-D concentrating apparatus, water bath, and an ultrasonic disrupter with a minimum power of 300 W and with pulsing capability are also required.

PRECISION & ACCURACY The estimated quantitation limit (EQL) of Method 8270B for determining an individual compound is approximately 1 mg/kg (wet weight) for soil or sediment samples, 1–200 mg/kg for wastes (dependent on matrix and method of preparation), and 10 µg/L for groundwater samples. EQLs will be proportionately higher for sample extracts that require dilution to avoid saturation of the detector.

The EQL(b) for groundwater in µg/L is not listed.
The EQL (a, b) for low concentrations in soil and sediment in µg/kg is not listed.
Accuracy as µg/L is 0.39C + 0.41.
Overall precision in µg/L is 0.63X–1.03.

(a) *EQLs listed for soil/sediment are based on wet weight. Normally data is reported in a dry-weight basis; therefore, EQLs will be higher based on the % dry weight of each sample. This calculation is based on a 30-g sample and gel permeation chromatography cleanup.*
(b) *Sample EQLs are highly matrix-dependent. The EQLs are provided for guidance and may not always be achievable.*
$C =$ *True value for concentration, in µg/L.*
$X =$ *Average recovery found for measurements of samples containing a concentration of C, in µg/L.*

ESTIMATED QUANTITATION LIMIT

Other Matrices	Factor (a)
High-concentration soil and sludges by sonicator	7.5
Non-water miscible waste	75

(a) EQL for other matrices = [EQL for low soil/sediment] ×[Factor]. This estimated EQL is similar to an EPA "Practical Quantitation Limit."

SAMPLING METHOD

Liquid samples — Use a 1 or 2½ gallon amber glass bottle with a screw-top Teflon®-lined cover that has been prewashed with detergent and rinsed with distilled water and methanol (or isopropanol).

Soils, sediments, or sludges — Use an 8-oz. widemouth glass with a screw-top Teflon®-lined cover that has been prewashed with detergent and rinsed with distilled water and methanol (or isopropanol).

SAMPLE PRESERVATION
Liquid samples — If residual chlorine is present, add 3 mL of 10% sodium thiosulfate per gallon, cool to 4°C and store in a solvent-free refrigerator until analysis; if chlorine is not present, then eliminate the sodium thiosulfate addition.

Soils, sediments, or sludges — Cool samples to 4°C and store in a solvent-free refrigerator.

MHT Liquid samples must be extracted within 7 days and the extracts analyzed within 40 days. Soils, sediments, or sludges may be stored for a maximum of 14 days and the extracts analyzed within 40 days.

SAMPLE PREPARATION
Liquid samples — Transfer 1 L quantitatively to a continuous extractor. If high concentrations are anticipated, a smaller volume may be used and then diluted with organic-free reagent water to 1 L. Adjust pH, if necessary, to pH <2 using 1:1 (V/V) sulfuric acid. Pipette 1.0 mL of a surrogate standard spiking solution into each sample. For the sample in each analytical batch selected for spiking, add 1.0 mL of a matrix spiking standard. For base/neutral acid analysis, the amount of the surrogates and matrix spiking compounds added to the sample should result in a final concentration of 100 ng/µL of each analyte in the extract to be analyzed (assuming a 1-µL injection). Extract with methylene chloride for 18–24 h. Next, adjust the pH of the aqueous phase to pH >11 using 10 N sodium hydroxide and extract it with methylene chloride again for 18–24 h. Dry the extract through a column containing anhydrous sodium sulfate and concentrate it to 1 mL using a K-D concentrator.

Soils, sediments, or sludges — Use 30 g of sample. Nonporous or wet samples (gummy or clay type) that do not have a free-flowing sandy texture must be mixed with anhydrous sodium sulfate until the sample is free flowing. Add 1 mL of surrogate standards to all samples, spikes, standards, and blanks. For the sample in each analytical batch selected for spiking, add 1.0 mL of a matrix spiking standard. For base/neutral acid analysis, the amount added of the surrogates and matrix spiking compounds should result in a final concentration of 100 ng/µL of each base/neutral analyte and 200 ng/µL of each acid analyte in the extract to be analyzed (assuming a 1-µL injection). Immediately add a 100-mL mixture of 1:1 methylene chloride:acetone and extract the sample ultrasonically for 3 min and then decant or filter the extracts. Repeat the extraction two or more times. Dry the extract using a column with anhydrous sodium sulfate and concentrate it to 1 mL in a K-D concentrator.

QUALITY CONTROL A methylene chloride solution containing 50 ng/µL of decafluorotriphenylphosphine (DFTPP) is used for tuning the GC/MS system each 12-h shift. A system performance check also must be made during every 12-h shift. A standard containing 50 ng/µL each of 4,4'-DDT, pentachlorophenol, and benzidine is required to verify injection port inertness and GC column performance. A calibration standard at mid-concentration, containing each compound of interest, including all required surrogates, must be performed every 12 h during analysis. After the system performance check is met, calibration check compounds (CCCs) are used to check the validity of the initial calibration.

The internal standard responses and retention times in the calibration check standard must be evaluated immediately after or during data acquisition. If the retention time for any internal standard changes by more than 30 seconds from the last check calibration (12 h), the chromatographic system must be inspected for malfunctions and corrections must be made, as required. If the electron ionization current plot (EICP) area for any of the internal standards changes by a factor of two from the last daily calibration standard check, the mass spectrometer must be inspected for malfunctions and corrections must be made, as appropriate.

Demonstrate, through the analysis of a reagent water blank, that interferences from the analytical system, glassware, and reagents are under control. The blank samples should be carried through all stages of the sample preparation and measurement steps. For each analytical batch (up to 20 samples), a reagent blank, matrix spike, and matrix spike duplicate/duplicate must be analyzed (the frequency of the spikes may be different for different monitoring programs). The blank and spiked samples must be carried through all stages of the sample preparation and measurement steps. A QC reference sample concentrate containing each analyte at a concentration of 100 mg/L in methanol is required.

REFERENCE Test Methods for Evaluating Solid Waste (SW-846). U.S. EPA 1983, Method 8270B, Rev. 2, Nov. 1990. Office of Solid Waste, Washington, DC.

Endrin   **EPA Method 505**
CAS #72-20-8

TITLE Analysis of Organohalide Pesticides and Commercial Polychlorinated Biphenyl (PCB) Products in Water by Microextraction and Gas Chromatography. U.S. EPA Method 505, Rev. 2.0, 1989.

MATRIX This method is applicable to drinking water and raw source water. The latter should include most surface water and groundwater sources.

METHOD SUMMARY Method 505 covers 25 pesticides and commercial PCB products. This is a very sensitive method that is more useful for monitoring than for exploratory analyses. 5-mL of water are saturated with sodium chloride and then extracted by shaking with 2 mL of hexane. The sample extracts are transferred to an autosampler setup to inject 1–2 µL portions into a gas chromatograph (GC) for analysis. Alternatively, 1–2 µL portions of samples, blanks, and standards may be manually injected. Each extract is analyzed by capillary GC/ECD with confirmation using either a second capillary column or GC/MS. The electron capture detector is easy to use, but it is a nonselective detector. The microextraction technique also eliminates the expensive sample preparation costs of other methods, but it has the disadvantage of being less sensitive than most because the extracts are not concentrated.

INTERFERENCES Method interferences may be caused by contaminants in solvents, reagents, glassware, and other sample processing apparatus that lead to discrete artifacts or elevated baselines. Interfering contamination may occur when a sample containing low concentrations of analytes is analyzed immediately following a sample containing relatively high concentrations of the analytes. Matrix interferences also may be caused by contaminants that are coextracted from the sample; cleanup of sample extracts may be necessary in these cases. Some pesticides and commercial PCB products from aqueous solutions adhere to glass surfaces, so sample transfers and contact with glass surfaces should be minimized. Some pesticides are rapidly oxidized by chlorine so dechlorination with sodium thiosulfate at the time of sample collection is important. Also, splitless injectors may cause degradation of some pesticides.

INSTRUMENTATION A gas chromatograph/electron capture detector/data system, with temperature programming and split/splitless injector suitable for use with capillary columns is needed.

Column 1: 0.32 mm I.D. × 30 m fused silica capillary with chemically bond methyl polysiloxane phase (DB-1, 1.0 μm film, or equivalent).
Column 2: 0.32 mm I.D. × 30 m fused silica capillary with 1:1 mixed phase of dimethyl silicone and polyethylene glycol (Durawax-DX3, 0.25 μm film, or equivalent).
Column 3: 0.32 mm I.D. × 25 m fused silica capillary with chemically bonded 50:50 methyl-phenyl silicone (OV-17, 1.5 μm film, or equivalent).

Column 1 should be used as the primary analytical column. Columns 2 and 3 are recommended for use as confirmatory columns when GC/MS confirmation is not available.

PRECISION & ACCURACY Method detection limits are dependent upon the characteristics of the gas chromatographic system used. Analytes that are not separated chromatographically cannot be individually identified and used in the same calibration mixture or water samples unless an alternative technique for identification and quantification, such as mass spectrometry, is used.

The concentration(s) (in μg/L) used for these QC measurements was 0.10 and 3.6.
The MDL (in μg/L) was 0.063.
The accuracy (% recovery) for reagent water at the above concentration(s) was 119 and 99 and the precision (%) was 29.8 and 6.5.
The accuracy (% recovery) for groundwater at the above concentration(s) was 94 and 100 and the precision (%) was 20.2 and 11.3.
The accuracy (% recovery) for tap water at the above concentration(s) was 106 and 85 and the precision (5) was 14.7 and 12.4.

Note: No range of concentrations is provided with this method.

SAMPLING METHOD Collect samples using a 40-mL screw-cap vial (prewashed with detergent, rinsed with distilled water and oven dried at 400°C for one h) with a Teflon®-faced silicone septum. Collect bubble-free samples and place the septum with the Teflon® side down on the water.

SAMPLE PRESERVATION If residual chlorine is present in the water add about 3 mg of sodium thiosulfate to each vial before samples are collected to remove the chlorine. Alternatively, add 75 μL of 0.04 g/mL solution of sodium thiosulfate to each vial just prior to sampling. Immediately cool samples to 4°C, and store them in a solvent-free refrigerator at 4°C until analysis.

MHT The maximum holding time is 14 days from the time the sample was collected until it must be analyzed.

SAMPLE PREPARATION Remove the sample from storage and allow it to come to room temperature. Remove a 5-mL volume from each container and weigh the container to the nearest 0.1 g. Add 6 g of sodium chloride and 2.0 mL of hexane to each sample bottle. Recap the sample and shake it vigorously for one min. Allow the water and hexane phases to separate, remove the cap, and transfer 0.5 mL of hexane into an autosampler vial using a disposable glass pipette. Transfer the remaining hexane phase into a second autosampler vial and store at 4°C for reanalysis, if necessary. Discard the remaining sample/hexane mixture and reweigh the empty container to determine net weight of sample.

QUALITY CONTROL Minimum quality control requirements are initial demonstration of lab capability, analysis of lab reagent blanks, fortified blanks, fortified sample matrix, and quality control samples. The lab must analyze at least one fortified blank per sample set, or at least one for every 20 samples. The fortifying concentration of each analyte should be 10 times the method detection limit or the maximum calibration limit (MCL), whichever is less. Calculate accuracy as percent recovery and develop control limits from the mean percent recovery and standard deviation.

The lab must add a known concentration of the analytes to a minimum of 10% of the routine samples, or one lab fortified sample matrix per sample set. Calculate the percent recovery for each analyte and compare to the control limits established from the analyses of the fortified blanks.

EPA CONTACT & HOTLINE For technical questions contact Dr. Baldev Bathija, U.S. EPA, Office of Ground Water and Drinking Water (WH-550D), 401 M St. SW, Washington, DC 20460. Tel. (202) 260-3040. For further information the EPA Safe Drinking Water Hotline may be called at: (800) 426-4791.

REFERENCE Methods for the Determination of Organic Compounds in Drinking Water, EPA/600/4-88/039 (revised July 1991). U.S. EPA Environmental Monitoring Systems Laboratory, Cincinnati, OH, 45268, U.S.A. Available from the National Technical Information Service (NTIS), 5285 Port Royal Road, Springfield, VA 22161; Tel. 800-553-6847. NTIS Order Number is PB91-231480.

Endrin
CAS #72-20-8

EPA Method 625

TITLE Base/Neutrals and Acids, U.S. EPA Method 625

MATRIX This methods covers municipal and industrial wastewaters.

METHOD SUMMARY Approximately 1 L of sample is serially extracted with methylene chloride at a pH greater than 11 and again at a pH less than 2 using a separatory funnel or a continuous extractor. The methylene chloride extract is dried, concentrated to a volume of 1 mL, and analyzed by GC/MS. Qualitative identification of the parameters in the extract is performed using the retention time and the relative abundance of three characteristic masses (m/z). Qualitative analysis is performed using either external or internal standard techniques with a single characteristic m/z.

INTERFERENCES Method interferences may be caused by contaminants in solvents, reagents, glassware, and other sample processing hardware. Glassware must be scrupulously cleaned. Glassware should be heated in a muffle furnace at 400°C for 5 to 30 min. Some thermally stable materials, such as PCBs, may not be eliminated by this treatment. Solvent rinses with acetone and pesticide quality hexane may be substituted for the muffle furnace heating. Matrix interferences may be caused by contaminants that are coextracted from the sample. The base-neutral extraction may cause significantly reduced recovery of phenols. The packed gas chromatographic columns recommended for the basic fraction may not exhibit sufficient resolution for some analytes.

INSTRUMENTATION A GC/MS system with an injection port designed for on-column injection when using packed columns and for splitless injection when using capillary columns.

Column for base/neutrals: 1.8 m long × 2 mm I.D. glass, packed with 3% SP-2550 on Supelcoport (100/120 mesh) or equivalent.

Column for acids: 1.8 m long × 2 mm I.D. glass, packed with 1% SP-1240DA on Supelcoport (100/120 mesh) or equivalent.

PRECISION & ACCURACY The MDL concentrations were obtained using reagent water. The MDL actually achieved in a given analysis will vary depending on instrument sensitivity and matrix effects. This method was tested by 15 laboratories using reagent water, drinking water, surface water, and industrial wastewaters spiked at six concentrations over the range 5 to 100 µg/L. Single operator precision, overall precision, and method accuracy were found to be directly related to the concentration of the parameter matrix.

The MDL (in µg/L) in reagent water was not reported.
The standard deviation (in µg/L based on 4 recovery measurements) was not reported.
The range (in µg/L) for average recovery for 4 measurements was not reported.
The range (in %) for percent recovery was not reported.
Accuracy (in µg/L) as expected recovery for one or more measurements of a sample containing a true concentration of C was not reported.
Precision (in µg/L) as expected single analyst standard deviation of measurements at an average concentration found at X was not reported.
Overall precision (in µg/L) as expected interlaboratory standard deviation of measurements in an average concentration found at X was not reported.

$C =$ True value of the concentration in µg/L.
$X =$ Average recovery found for measurements of samples containing a concentration at C in µg/L.

SAMPLE PREPARATION Adjust the pH to >11 with sodium hydroxide and serially extract in a separatory funnel with methylene chloride or else in a continuous extractor. Next, adjust the pH to <2 with sulfuric acid and serially extract in a separatory funnel with methylene chloride or else in a continuous extractor. Dry the extracts separately through a column of anhydrous sodium sulfate and then concentrate each of the extracts to 1.0 mL using a K-D apparatus.

SAMPLE COLLECTION, PRESERVATION & HANDLING Grab samples must be collected in glass containers. All samples must be refrigerated at 4°C from the time of collection until extraction. If residual chlorine is present, add 80 mg of sodium thiosulfate/L of sample and mix well. All samples must be extracted within 7 days of collection and completely analyzed within 40 days of extraction.

QUALITY CONTROL Make an initial, one-time, demonstration of the ability to generate acceptable accuracy and precision with this method. Before processing any samples, the analyst must analyze a reagent water blank to demonstrate that interferences from the analytical system and glassware are under control. Each time a set of samples is extracted or reagents are changed, a reagent water blank must be processed. Spike and analyze a minimum of 5% of all samples to monitor and evaluate lab data quality. A QC check sample concentrate that contains each parameter of interest at a concentration of 100 µg/mL in acetone is required. PCBs and multicomponent pesticides may be omitted from this test.

After analysis of five spiked wastewater samples, calculate the average percent recovery and the standard deviation of the percent recovery. Spike all samples with the surrogate standard spiking solution and calculate the percent recovery of each surrogate compound.

REFERENCE Federal Register, Vol. 49, No. 209. Friday, Oct. 26, 1984.

Endrin
CAS #72-20-8

EPA Method 8080

TITLE Organochlorine Pesticides and Polychlorinated Biphenyls By Gas Chromatography

MATRIX This method is used to determine the concentration of various organochlorine pesticides and polychlorinated biphenyls in extracts prepared from water, groundwater, soils, and sediments.

METHOD SUMMARY This method covers 26 pesticides and Aroclor (PCB) mixtures and it is suitable for monitoring-type analyses. After extraction, concentration and solvent exchange to hexane, a 2- to 5-μL sample aliquot is injected into a GC using the solvent flush technique, and the analytes are detected by an electron capture detector (ECD) or an electrolytic conductivity detector in the halogen mode (HECD). Both neat and diluted organic liquids may be analyzed by direct injection.

INTERFERENCES Interferences coextracted from the samples will vary considerably from source to source. Interferences by phthalate esters can pose a major problem in pesticide determinations when using the ECD. Cross-contamination of clean glassware routinely occurs when plastics are handled during extraction steps, especially when solvent-wetted surfaces are handled. The contamination from phthalate esters can be completely eliminated with a microcoulometric or electrolytic conductivity detector. Solvents, reagent, glassware, and other sample processing hardware may yield artifacts and/or interferences to sample analysis.

INSTRUMENTATION A gas chromatograph capable of on-column injections is needed. It must be equipped with an ECD or a HECD and one of the following GC columns:

Column 1: Supelcoport (100/120 mesh) coated with 1.5% SP-2250/1.95% SP-2401 packed in a 1.8 m × 4 mm I.D. glass column.
Column 2: Supelcoport (100/120 mesh) coated with 3% OV-1 in a 1.8 m × 4 mm I.D. glass column.

PRECISION & ACCURACY The method was tested by 20 laboratories using organic-free reagent water, drinking water, surface water, and three industrial wastewaters spiked at six concentrations. Concentrations used in the study ranged from 0.5 to 30 μg/L for single-component pesticides and from 8.5 to 400 μg/L for multicomponent parameters. Overall precision and method accuracy were found to be directly related to the concentration of the analyte and essentially independent of the sample matrix. The sensitivity of this method usually depends on the concentration of interferences rather than on instrumental limitations.

MDL in μg/L was 0.006.
Concentration range in μg/L was 0.5–30.
Accuracy as recovery (x*) in μg/L was 0.89C-0.04.
Overall precision (S*) in μg/L was 0.24x + 0.25.

x^* = *Expected recovery for one or more measurements of a sample containing concentration C, in μg/L.*
S^* = *Expected interlaboratory standard deviation of measurements at an average concentration found of the analyte in μg/L.*
C = *True value for the concentration, in μg/L.*
X = *Average recovery found for measurements of samples containing a concentration of C, in μg/L.*

SAMPLING METHOD
Liquid samples — Use a 1 or 2½ gallon amber glass bottle with a screw-top Teflon®-lined cover. Pre-wash the bottle with detergent, rinse with distilled water and methanol (or isopropanol).

Soil, sediments, and sludges — Use an 8-oz. widemouth glass with a screw-top Teflon®-lined cover. Pre-wash the bottle with detergent, rinse with distilled water and methanol (or isopropanol).

SAMPLE PRESERVATION Cool water, soil, sediment, or sludge samples immediately to 4°C.

Water samples — If residual chlorine is present, add 3 mL of 10% sodium thiosulfate per gallon and cool to 4°C. All extracts and samples should be stored under refrigeration.

MHT Liquid samples must be extracted within 7 days and the extracts must be analyzed within 40 days. Soils, sediments, and sludges may be stored for a maximum of 14 days prior to extraction.

SAMPLE PREPARATION
Liquid samples — Extract 1 L samples in a continuous extractor at pH 5–9 with methylene chloride after adding 1.0 mL of surrogate spiking solution to each sample. Pass the extract through a column of anhydrous sodium sulfate to dry and concentrate it in a K-D apparatus to 1 mL volume.

Soils, sediments and sludges — Rapidly weigh approximately 30 g of sample into a 400-mL beaker to avoid loss of the more volatile extractables. Nonporous or wet samples (gummy or clay type) that do not have a free-flowing sandy texture must be mixed with anhydrous sodium sulfate until the sample is free flowing. Add 1 mL of surrogate standards to all samples, spikes, standards, and blanks. Add 100 mL of 1:1 methylene chloride:acetone and extract ultrasonically. Decant and filter extracts, dry the extract by passing it through a drying column containing anhydrous sodium sulfate and concentrate to 1 mL in a K-D apparatus.

Hexane solvent exchange — Add 50 mL of hexane, a new boiling chip, and concentrate until the apparent volume of liquid reaches 1 mL. Adjust the extract volume to 10.0 mL. Stopper the concentration tube and store refrigerated at 4°C if further processing will not be performed immediately. If the extract will be stored longer than two days, transfer it to a vial with Teflon®-lined screw-cap or crimp top.

QUALITY CONTROL Demonstrate through the analysis of a reagent water blank, that all glassware and reagents are interference free. Each time a set of samples is processed, a method blank should be processed as a safeguard against chronic lab contamination. A reagent blank, a matrix spike, and a duplicate or matrix spike duplicate must be performed for each analytical batch (up to a maximum of 20 samples) analyzed.

Analytical system performance must be verified by analyzing QC check samples. The QC check sample concentration should contain each single-component analyte at the following concentrations in acetone: 4,4'-DDD, 10 μg/mL; 4,4'-DDT, 10 μg/mL; endosulfan II, 10 μg/mL; endosulfan sulfate, 10 μg/mL; and any other single-component pesticide at 2 μg/mL. If the method is only to be used to analyze PCBs, Chlordane, or Toxaphene, the QC check sample concentrate should contain the most representative multicomponent parameter at a concentration of 50 μg/mL in acetone.

REFERENCE Test Methods for Evaluating Solid Waste (SW-846). U.S. EPA. 1983. Method 8080B, Rev. 2, Nov. 1990. Office of Solid Wastes, Washington, DC.

Endrin   **EPA Method 8270**
CAS #72-20-8

TITLE Semivolatile Organic Compounds by GC/MS

MATRIX This method is used to determine the concentration of semivolatile organic compounds in extracts prepared from all types of solid waste matrices, soils, and groundwater. Although surface waters are not specifically mentioned, this method should be applicable to water samples from rivers, lakes, etc.

METHOD SUMMARY This method covers 259 semivolatile organic compounds. In very limited applications direct injection of the sample into the GC/MS system may be appropriate, but this results in very high detection limits (approximately 10,000 µg/L). Typically, a 1-L liquid sample, containing surrogate, and matrix spiking standards, is extracted in a continuous extractor first under acid conditions and then under basic conditions. Typically 30 g of a solid sample, containing surrogate, and matrix spiking standards, is extracted ultrasonically. After concentrating the extract to 1 mL it is spiked with 10 µL of an internal standard solution just prior to analysis by GC/MS. The volume injected should contain about 100 ng of base/neutral and 200 ng of acid surrogates (for a 1-µL injection). Analysis is performed by GC/MS using a capillary GC column.

INTERFERENCES Raw GC/MS data from all blanks, samples, and spikes must be evaluated for interferences. Contamination by carryover can occur whenever high-concentration and low-concentration samples are sequentially analyzed. To reduce carryover, the sample syringe must be rinsed out between samples with solvent. Whenever an unusually concentrated sample is encountered, it should be followed by the analysis of blank solvent to check for cross-contamination.

INSTRUMENTATION A GC/MS and a data system are required. The GC column used is a 30 m × 0.25 mm I.D. (or 0.32 mm I.D.) 1um film thickness silicone-coated fused silica capillary column. A continuous liquid-liquid extractor equipped with Teflon® or glass connection joints and stopcocks requiring no lubrication, a K-D concentrating apparatus, water bath, and an ultrasonic disrupter with a minimum power of 300 W and with pulsing capability are also required.

PRECISION & ACCURACY The estimated quantitation limit (EQL) of Method 8270B for determining an individual compound is approximately 1 mg/kg (wet weight) for soil or sediment samples, 1–200 mg/kg for wastes (dependent on matrix and method of preparation), and 10 µg/L for groundwater samples. EQLs will be proportionately higher for sample extracts that require dilution to avoid saturation of the detector.

The EQL(b) for groundwater in µg/L is not listed.
The EQL (a, b) for low concentrations in soil and sediment in µg/kg is not listed.
Accuracy as µg/L is not listed.

Overall precision in µg/L is not listed.

(a) *EQLs listed for soil/sediment are based on wet weight. Normally data is reported in a dry-weight basis; therefore, EQLs will be higher based on the % dry weight of each sample. This calculation is based on a 30-g sample and gel permeation chromatography cleanup.*
(b) *Sample EQLs are highly matrix-dependent. The EQLs are provided for guidance and may not always be achievable.*
C = *True value for concentration, in µg/L.*
X = *Average recovery found for measurements of samples containing a concentration of C, in µg/L.*

ESTIMATED QUANTITATION LIMIT

Other Matrices	Factor (a)
High-concentration soil and sludges by sonicator	7.5
Non-water miscible waste	75

(a) *EQL for other matrices = [EQL for low soil/sediment] × [Factor]. This estimated EQL is similar to an EPA "Practical Quantitation Limit."*

SAMPLING METHOD
Liquid samples — Use a 1 or 2½ gallon amber glass bottle with a screw-top Teflon®-lined cover that has been prewashed with detergent and rinsed with distilled water and methanol (or isopropanol).

Soils, sediments, or sludges — Use an 8-oz. widemouth glass with a screw-top Teflon®-lined cover that has been prewashed with detergent and rinsed with distilled water and methanol (or isopropanol).

SAMPLE PRESERVATION
Liquid samples — If residual chlorine is present, add 3 mL of 10% sodium thiosulfate per gallon, cool to 4°C and store in a solvent-free refrigerator until analysis; if chlorine is not present, then eliminate the sodium thiosulfate addition.

Soils, sediments, or sludges — Cool samples to 4°C and store in a solvent-free refrigerator.

MHT Liquid samples must be extracted within 7 days and the extracts analyzed within 40 days. Soils, sediments, or sludges may be stored for a maximum of 14 days and the extracts analyzed within 40 days.

SAMPLE PREPARATION
Liquid samples — Transfer 1 L quantitatively to a continuous extractor. If high concentrations are anticipated, a smaller volume may be used and then diluted with organic-free reagent water to 1 L. Adjust pH, if necessary, to pH <2 using 1:1 (V/V) sulfuric acid. Pipette 1.0 mL of a surrogate standard spiking solution into each sample. For the sample in each analytical batch selected for spiking, add 1.0 mL of a matrix spiking standard. For base/neutral acid analysis, the amount of the surrogates and matrix spiking compounds added to the sample should result in a final concentration of 100 ng/µL of each analyte in the extract to be analyzed (assuming a 1-µL injection). Extract with methylene chloride for 18–24 h. Next, adjust the pH of the aqueous phase to pH >11 using 10 N sodium hydroxide and extract it with methylene chloride again for 18–24 h. Dry the extract through a column containing anhydrous

sodium sulfate and concentrate it to 1 mL using a K-D concentrator.

Soils, sediments, or sludges — Use 30 g of sample. Nonporous or wet samples (gummy or clay type) that do not have a free-flowing sandy texture must be mixed with anhydrous sodium sulfate until the sample is free flowing. Add 1 mL of surrogate standards to all samples, spikes, standards, and blanks. For the sample in each analytical batch selected for spiking, add 1.0 mL of a matrix spiking standard. For base/neutral acid analysis, the amount added of the surrogates and matrix spiking compounds should result in a final concentration of 100 ng/µL of each base/neutral analyte and 200 ng/µL of each acid analyte in the extract to be analyzed (assuming a 1-µL injection). Immediately add a 100-mL mixture of 1:1 methylene chloride:acetone and extract the sample ultrasonically for 3 min and then decant or filter the extracts. Repeat the extraction two or more times. Dry the extract using a column with anhydrous sodium sulfate and concentrate it to 1 mL in a K-D concentrator.

QUALITY CONTROL A methylene chloride solution containing 50 ng/µL of decafluorotriphenylphosphine (DFTPP) is used for tuning the GC/MS system each 12-h shift. A system performance check also must be made during every 12-h shift. A standard containing 50 ng/µL each of 4,4'-DDT, pentachlorophenol, and benzidine is required to verify injection port inertness and GC column performance. A calibration standard at mid-concentration, containing each compound of interest, including all required surrogates, must be performed every 12 h during analysis. After the system performance check is met, calibration check compounds (CCCs) are used to check the validity of the initial calibration.

The internal standard responses and retention times in the calibration check standard must be evaluated immediately after or during data acquisition. If the retention time for any internal standard changes by more than 30 seconds from the last check calibration (12 h), the chromatographic system must be inspected for malfunctions and corrections must be made, as required. If the electron ionization current plot (EICP) area for any of the internal standards changes by a factor of two from the last daily calibration standard check, the mass spectrometer must be inspected for malfunctions and corrections must be made, as appropriate.

Demonstrate, through the analysis of a reagent water blank, that interferences from the analytical system, glassware, and reagents are under control. The blank samples should be carried through all stages of the sample preparation and measurement steps. For each analytical batch (up to 20 samples), a reagent blank, matrix spike, and matrix spike duplicate/duplicate must be analyzed (the frequency of the spikes may be different for different monitoring programs). The blank and spiked samples must be carried through all stages of the sample preparation and measurement steps. A QC reference sample concentrate containing each analyte at a concentration of 100 mg/L in methanol is required.

REFERENCE Test Methods for Evaluating Solid Waste (SW-846). U.S. EPA 1983, Method 8270B, Rev. 2, Nov. 1990. Office of Solid Waste, Washington, DC.

Endrin aldehyde   EPA Method 625
CAS #7421-93-4

TITLE Base/Neutrals and Acids, U.S. EPA Method 625

MATRIX This methods covers municipal and industrial wastewaters.

METHOD SUMMARY Approximately 1 L of sample is serially extracted with methylene chloride at a pH greater than 11 and again at a pH less than 2 using a separatory funnel or a continuous extractor. The methylene chloride extract is dried, concentrated to a volume of 1 mL, and analyzed by GC/MS. Qualitative identification of the parameters in the extract is performed using the retention time and the relative abundance of three characteristic masses (m/z). Qualitative analysis is performed using either external or internal standard techniques with a single characteristic m/z.

INTERFERENCES Method interferences may be caused by contaminants in solvents, reagents, glassware, and other sample processing hardware. Glassware must be scrupulously cleaned. Glassware should be heated in a muffle furnace at 400°C for 5 to 30 min. Some thermally stable materials, such as PCBs, may not be eliminated by this treatment. Solvent rinses with acetone and pesticide quality hexane may be substituted for the muffle furnace heating. Matrix interferences may be caused by contaminants that are coextracted from the sample. The base-neutral extraction may cause significantly reduced recovery of phenols. The packed gas chromatographic columns recommended for the basic fraction may not exhibit sufficient resolution for some analytes.

INSTRUMENTATION A GC/MS system with an injection port designed for on-column injection when using packed columns and for splitless injection when using capillary columns.

Column for base/neutrals: 1.8 m long × 2 mm I.D. glass, packed with 3% SP-2550 on Supelcoport (100/120 mesh) or equivalent.
Column for acids: 1.8 m long × 2 mm I.D. glass, packed with 1% SP-1240DA on Supelcoport (100/120 mesh) or equivalent.

PRECISION & ACCURACY The MDL concentrations were obtained using reagent water. The MDL actually achieved in a given analysis will vary depending on instrument sensitivity and matrix effects. This method was tested by 15 laboratories using reagent water, drinking water, surface water, and industrial wastewaters spiked at six concentrations over the range 5 to 100 µg/L. Single operator precision, overall precision, and method accuracy were found to be directly related to the concentration of the parameter matrix.

The MDL (in µg/L) in reagent water was not detected.
The standard deviation (in µg/L based on 4 recovery measurements) was 32.5.
The range (in µg/L) for average recovery for 4 measurements was D-188.8.
The range (in %) for percent recovery was D-209.
Accuracy (in µg/L) as expected recovery for one or more measurements of a sample containing a true concentration of C was 0.76C-3.86.

Precision (in μg/L) as expected single analyst standard deviation of measurements at an average concentration found at X was 0.18X + 3.91.

Overall precision (in μg/L) as expected interlaboratory standard deviation of measurements in an average concentration found at X was 0.73X–0.62.

C = *True value of the concentration in μg/L.*
X = *Average recovery found for measurements of samples containing a concentration at C in μg/L.*

SAMPLE PREPARATION Adjust the pH to >11 with sodium hydroxide and serially extract in a separatory funnel with methylene chloride or else in a continuous extractor. Next, adjust the pH to <2 with sulfuric acid and serially extract in a separatory funnel with methylene chloride or else in a continuous extractor. Dry the extracts separately through a column of anhydrous sodium sulfate and then concentrate each of the extracts to 1.0 mL using a K-D apparatus.

SAMPLE COLLECTION, PRESERVATION & HANDLING
Grab samples must be collected in glass containers. All samples must be refrigerated at 4°C from the time of collection until extraction. If residual chlorine is present, add 80 mg of sodium thiosulfate/L of sample and mix well. All samples must be extracted within 7 days of collection and completely analyzed within 40 days of extraction.

QUALITY CONTROL Make an initial, one-time, demonstration of the ability to generate acceptable accuracy and precision with this method. Before processing any samples, the analyst must analyze a reagent water blank to demonstrate that interferences from the analytical system and glassware are under control. Each time a set of samples is extracted or reagents are changed, a reagent water blank must be processed. Spike and analyze a minimum of 5% of all samples to monitor and evaluate lab data quality. A QC check sample concentrate that contains each parameter of interest at a concentration of 100 μg/mL in acetone is required. PCBs and multicomponent pesticides may be omitted from this test.

After analysis of five spiked wastewater samples, calculate the average percent recovery and the standard deviation of the percent recovery. Spike all samples with the surrogate standard spiking solution and calculate the percent recovery of each surrogate compound.

REFERENCE *Federal Register*, Vol. 49, No. 209. Friday, Oct. 26, 1984.

Endrin aldehyde **EPA Method 8080**
CAS #7421-93-4

TITLE Organochlorine Pesticides and Polychlorinated Biphenyls By Gas Chromatography

MATRIX This method is used to determine the concentration of various organochlorine pesticides and polychlorinated biphenyls in extracts prepared from water, groundwater, soils, and sediments.

METHOD SUMMARY This method covers 26 pesticides and Aroclor (PCB) mixtures and it is suitable for monitoring-type analyses. After extraction, concentration and solvent exchange to hexane, a 2- to 5-μL sample aliquot is injected into a GC using the solvent flush technique, and the analytes are detected by an electron capture detector (ECD) or an electrolytic conductivity detector in the halogen mode (HECD). Both neat and diluted organic liquids may be analyzed by direct injection.

INTERFERENCES Interferences coextracted from the samples will vary considerably from source to source. Interferences by phthalate esters can pose a major problem in pesticide determinations when using the ECD. Cross-contamination of clean glassware routinely occurs when plastics are handled during extraction steps, especially when solvent-wetted surfaces are handled. The contamination from phthalate esters can be completely eliminated with a microcoulometric or electrolytic conductivity detector. Solvents, reagent, glassware, and other sample processing hardware may yield artifacts and/or interferences to sample analysis.

INSTRUMENTATION A gas chromatograph capable of on-column injections is needed. It must be equipped with an ECD or a HECD and one of the following GC columns:

Column 1: Supelcoport (100/120 mesh) coated with 1.5% SP-2250/1.95% SP-2401 packed in a 1.8 m × 4 mm I.D. glass column.

Column 2: Supelcoport (100/120 mesh) coated with 3% OV-1 in a 1.8 m × 4 mm I.D. glass column.

PRECISION & ACCURACY The method was tested by 20 laboratories using organic-free reagent water, drinking water, surface water, and three industrial wastewaters spiked at six concentrations. Concentrations used in the study ranged from 0.5 to 30 μg/L for single-component pesticides and from 8.5 to 400 μg/L for multicomponent parameters. Overall precision and method accuracy were found to be directly related to the concentration of the analyte and essentially independent of the sample matrix. The sensitivity of this method usually depends on the concentration of interferences rather than on instrumental limitations.

MDL in μg/L was 0.023.
Concentration range in μg/L was 0.5–30.
Accuracy as recovery (x*) in μg/L was not listed.
Overall precision (S*) in μg/L was not listed.

x* = *Expected recovery for one or more measurements of a sample containing concentration C, in μg/L.*
S* = *Expected interlaboratory standard deviation of measurements at an average concentration found of the analyte in μg/L.*
C = *True value for the concentration, in μg/L.*
X = *Average recovery found for measurements of samples containing a concentration of C, in μg/L.*

SAMPLING METHOD

Liquid samples — Use a 1 or 2½ gallon amber glass bottle with a screw-top Teflon®-lined cover. Pre-wash the bottle with detergent, rinse with distilled water and methanol (or isopropanol).

Soil, sediments, and sludges — Use an 8-oz. widemouth glass with a screw-top Teflon®-lined cover. Pre-wash the bottle with detergent, rinse with distilled water and methanol (or isopropanol).

SAMPLE PRESERVATION Cool water, soil, sediment, or sludge samples immediately to 4°C.

Water samples — If residual chlorine is present, add 3 mL of 10% sodium thiosulfate per gallon and cool to 4°C. All extracts and samples should be stored under refrigeration.

MHT Liquid samples must be extracted within 7 days and the extracts must be analyzed within 40 days. Soils, sediments, and sludges may be stored for a maximum of 14 days prior to extraction.

SAMPLE PREPARATION
Liquid samples — Extract 1 L samples in a continuous extractor at pH 5–9 with methylene chloride after adding 1.0 mL of surrogate spiking solution to each sample. Pass the extract through a column of anhydrous sodium sulfate to dry and concentrate it in a K-D apparatus to 1 mL volume.

Soils, sediments and sludges — Rapidly weigh approximately 30 g of sample into a 400-mL beaker to avoid loss of the more volatile extractables. Nonporous or wet samples (gummy or clay type) that do not have a free-flowing sandy texture must be mixed with anhydrous sodium sulfate until the sample is free flowing. Add 1 mL of surrogate standards to all samples, spikes, standards, and blanks. Add 100 mL of 1:1 methylene chloride:acetone and extract ultrasonically. Decant and filter extracts, dry the extract by passing it through a drying column containing anhydrous sodium sulfate and concentrate to 1 mL in a K-D apparatus.

Hexane solvent exchange — Add 50 mL of hexane, a new boiling chip, and concentrate until the apparent volume of liquid reaches 1 mL. Adjust the extract volume to 10.0 mL. Stopper the concentration tube and store refrigerated at 4°C if further processing will not be performed immediately. If the extract will be stored longer than two days, transfer it to a vial with Teflon®-lined screw-cap or crimp top.

QUALITY CONTROL Demonstrate through the analysis of a reagent water blank, that all glassware and reagents are interference free. Each time a set of samples is processed, a method blank should be processed as a safeguard against chronic lab contamination. A reagent blank, a matrix spike, and a duplicate or matrix spike duplicate must be performed for each analytical batch (up to a maximum of 20 samples) analyzed.

Analytical system performance must be verified by analyzing QC check samples. The QC check sample concentration should contain each single-component analyte at the following concentrations in acetone: 4,4′-DDD, 10 µg/mL; 4,4′-DDT, 10 µg/mL; endosulfan II, 10 µg/mL; endosulfan sulfate, 10 µg/mL; and any other single-component pesticide at 2 µg/mL. If the method is only to be used to analyze PCBs, Chlordane, or Toxaphene, the QC check sample concentrate should contain the most representative multicomponent parameter at a concentration of 50 µg/mL in acetone.

REFERENCE Test Methods for Evaluating Solid Waste (SW-846). U.S. EPA. 1983. Method 8080B, Rev. 2, Nov. 1990. Office of Solid Wastes, Washington, DC.

Endrin aldehyde EPA Method 8270
CAS #7421-93-4

TITLE Semivolatile Organic Compounds by GC/MS

MATRIX This method is used to determine the concentration of semivolatile organic compounds in extracts prepared from all types of solid waste matrices, soils, and groundwater. Although surface waters are not specifically mentioned, this method should be applicable to water samples from rivers, lakes, etc.

METHOD SUMMARY This method covers 259 semivolatile organic compounds. In very limited applications direct injection of the sample into the GC/MS system may be appropriate, but this results in very high detection limits (approximately 10,000 µg/L). Typically, a 1-L liquid sample, containing surrogate, and matrix spiking standards, is extracted in a continuous extractor first under acid conditions and then under basic conditions. Typically 30 g of a solid sample, containing surrogate, and matrix spiking standards, is extracted ultrasonically. After concentrating the extract to 1 mL it is spiked with 10 µL of an internal standard solution just prior to analysis by GC/MS. The volume injected should contain about 100 ng of base/neutral and 200 ng of acid surrogates (for a 1-µL injection). Analysis is performed by GC/MS using a capillary GC column.

INTERFERENCES Raw GC/MS data from all blanks, samples, and spikes must be evaluated for interferences. Contamination by carryover can occur whenever high-concentration and low-concentration samples are sequentially analyzed. To reduce carryover, the sample syringe must be rinsed out between samples with solvent. Whenever an unusually concentrated sample is encountered, it should be followed by the analysis of blank solvent to check for cross-contamination.

INSTRUMENTATION A GC/MS and a data system are required. The GC column used is a 30 m × 0.25 mm I.D. (or 0.32 mm I.D.) 1um film thickness silicone-coated fused silica capillary column. A continuous liquid-liquid extractor equipped with Teflon® or glass connection joints and stopcocks requiring no lubrication, a K-D concentrating apparatus, water bath, and an ultrasonic disrupter with a minimum power of 300 W and with pulsing capability are also required.

PRECISION & ACCURACY The estimated quantitation limit (EQL) of Method 8270B for determining an individual compound is approximately 1 mg/kg (wet weight) for soil or sediment samples, 1–200 mg/kg for wastes (dependent on matrix and method of preparation), and 10 µg/L for groundwater samples. EQLs will be proportionately higher for sample extracts that require dilution to avoid saturation of the detector.

The EQL(b) for groundwater in µg/L is not listed.
The EQL (a, b) for low concentrations in soil and sediment in µg/kg is not listed.
Accuracy as µg/L is 0.76C–3.86.

Overall precision in µg/L is 0.73X–0.62.

(a) *EQLs listed for soil/sediment are based on wet weight. Normally data is reported in a dry-weight basis; therefore, EQLs will be higher based on the % dry weight of each sample. This calculation is based on a 30-g sample and gel permeation chromatography cleanup.*

(b) *Sample EQLs are highly matrix-dependent. The EQLs are provided for guidance and may not always be achievable.*

C = *True value for concentration, in µg/L.*

X = *Average recovery found for measurements of samples containing a concentration of C, in µg/L.*

ESTIMATED QUANTITATION LIMIT

Other Matrices	Factor (a)
High-concentration soil and sludges by sonicator	7.5
Non-water miscible waste	75

(a) EQL for other matrices = [EQL for low soil/sediment] × [Factor]. This estimated EQL is similar to an EPA "Practical Quantitation Limit."

SAMPLING METHOD

Liquid samples — Use a 1 or 2½ gallon amber glass bottle with a screw-top Teflon®-lined cover that has been prewashed with detergent and rinsed with distilled water and methanol (or isopropanol).

Soils, sediments, or sludges — Use an 8-oz. widemouth glass with a screw-top Teflon®-lined cover that has been prewashed with detergent and rinsed with distilled water and methanol (or isopropanol).

SAMPLE PRESERVATION

Liquid samples — If residual chlorine is present, add 3 mL of 10% sodium thiosulfate per gallon, cool to 4°C and store in a solvent-free refrigerator until analysis; if chlorine is not present, then eliminate the sodium thiosulfate addition.

Soils, sediments, or sludges — Cool samples to 4°C and store in a solvent-free refrigerator.

MHT Liquid samples must be extracted within 7 days and the extracts analyzed within 40 days. Soils, sediments, or sludges may be stored for a maximum of 14 days and the extracts analyzed within 40 days.

SAMPLE PREPARATION

Liquid samples — Transfer 1 L quantitatively to a continuous extractor. If high concentrations are anticipated, a smaller volume may be used and then diluted with organic-free reagent water to 1 L. Adjust pH, if necessary, to pH <2 using 1:1 (V/V) sulfuric acid. Pipette 1.0 mL of a surrogate standard spiking solution into each sample. For the sample in each analytical batch selected for spiking, add 1.0 mL of a matrix spiking standard. For base/neutral acid analysis, the amount of the surrogates and matrix spiking compounds added to the sample should result in a final concentration of 100 ng/µL of each analyte in the extract to be analyzed (assuming a 1-µL injection). Extract with methylene chloride for 18–24 h. Next, adjust the pH of the aqueous phase to pH >11 using 10 N sodium hydroxide and extract it with methylene chloride again for 18–24 h. Dry the extract through a column containing anhydrous sodium sulfate and concentrate it to 1 mL using a K-D concentrator.

Soils, sediments, or sludges — Use 30 g of sample. Nonporous or wet samples (gummy or clay type) that do not have a free-flowing sandy texture must be mixed with anhydrous sodium sulfate until the sample is free flowing. Add 1 mL of surrogate standards to all samples, spikes, standards, and blanks. For the sample in each analytical batch selected for spiking, add 1.0 mL of a matrix spiking standard. For base/neutral acid analysis, the amount added of the surrogates and matrix spiking compounds should result in a final concentration of 100 ng/µL of each base/neutral analyte and 200 ng/µL of each acid analyte in the extract to be analyzed (assuming a 1-µL injection). Immediately add a 100-mL mixture of 1:1 methylene chloride:acetone and extract the sample ultrasonically for 3 min and then decant or filter the extracts. Repeat the extraction two or more times. Dry the extract using a column with anhydrous sodium sulfate and concentrate it to 1 mL in a K-D concentrator.

QUALITY CONTROL A methylene chloride solution containing 50 ng/µL of decafluorotriphenylphosphine (DFTPP) is used for tuning the GC/MS system each 12-h shift. A system performance check also must be made during every 12-h shift. A standard containing 50 ng/µL each of 4,4'-DDT, pentachlorophenol, and benzidine is required to verify injection port inertness and GC column performance. A calibration standard at mid-concentration, containing each compound of interest, including all required surrogates, must be performed every 12 h during analysis. After the system performance check is met, calibration check compounds (CCCs) are used to check the validity of the initial calibration.

The internal standard responses and retention times in the calibration check standard must be evaluated immediately after or during data acquisition. If the retention time for any internal standard changes by more than 30 seconds from the last check calibration (12 h), the chromatographic system must be inspected for malfunctions and corrections must be made, as required. If the electron ionization current plot (EICP) area for any of the internal standards changes by a factor of two from the last daily calibration standard check, the mass spectrometer must be inspected for malfunctions and corrections must be made, as appropriate.

Demonstrate, through the analysis of a reagent water blank, that interferences from the analytical system, glassware, and reagents are under control. The blank samples should be carried through all stages of the sample preparation and measurement steps. For each analytical batch (up to 20 samples), a reagent blank, matrix spike, and matrix spike duplicate/duplicate must be analyzed (the frequency of the spikes may be different for different monitoring programs). The blank and spiked samples must be carried through all stages of the sample preparation and measurement steps. A QC reference sample concentrate containing each analyte at a concentration of 100 mg/L in methanol is required.

REFERENCE Test Methods for Evaluating Solid Waste (SW-846). U.S. EPA 1983, Method 8270B, Rev. 2, Nov. 1990. Office of Solid Waste, Washington, DC.

Endrin ketone
CAS #53494-70-5

EPA Method 8270

TITLE Semivolatile Organic Compounds by GC/MS

MATRIX This method is used to determine the concentration of semivolatile organic compounds in extracts prepared from all types of solid waste matrices, soils, and groundwater. Although surface waters are not specifically mentioned, this method should be applicable to water samples from rivers, lakes, etc.

METHOD SUMMARY This method covers 259 semivolatile organic compounds. In very limited applications direct injection of the sample into the GC/MS system may be appropriate, but this results in very high detection limits (approximately 10,000 µg/L). Typically, a 1-L liquid sample, containing surrogate, and matrix spiking standards, is extracted in a continuous extractor first under acid conditions and then under basic conditions. Typically 30 g of a solid sample, containing surrogate, and matrix spiking standards, is extracted ultrasonically. After concentrating the extract to 1 mL it is spiked with 10 µL of an internal standard solution just prior to analysis by GC/MS. The volume injected should contain about 100 ng of base/neutral and 200 ng of acid surrogates (for a 1-µL injection). Analysis is performed by GC/MS using a capillary GC column.

INTERFERENCES Raw GC/MS data from all blanks, samples, and spikes must be evaluated for interferences. Contamination by carryover can occur whenever high-concentration and low-concentration samples are sequentially analyzed. To reduce carryover, the sample syringe must be rinsed out between samples with solvent. Whenever an unusually concentrated sample is encountered, it should be followed by the analysis of blank solvent to check for cross-contamination.

INSTRUMENTATION A GC/MS and a data system are required. The GC column used is a 30 m × 0.25 mm I.D. (or 0.32 mm I.D.) 1um film thickness silicone-coated fused silica capillary column. A continuous liquid-liquid extractor equipped with Teflon® or glass connection joints and stopcocks requiring no lubrication, a K-D concentrating apparatus, water bath, and an ultrasonic disrupter with a minimum power of 300 W and with pulsing capability are also required.

PRECISION & ACCURACY The estimated quantitation limit (EQL) of Method 8270B for determining an individual compound is approximately 1 mg/kg (wet weight) for soil or sediment samples, 1–200 mg/kg for wastes (dependent on matrix and method of preparation), and 10 µg/L for groundwater samples. EQLs will be proportionately higher for sample extracts that require dilution to avoid saturation of the detector.

The EQL(b) for groundwater in µg/L is not listed.
The EQL (a, b) for low concentrations in soil and sediment in µg/kg is not listed.
Accuracy as µg/L is not listed.
Overall precision in µg/L is not listed.

(a) *EQLs listed for soil/sediment are based on wet weight. Normally data is reported in a dry-weight basis; therefore, EQLs will be higher based on the % dry weight of each sample.*

This calculation is based on a 30-g sample and gel permeation chromatography cleanup.

(b) *Sample EQLs are highly matrix-dependent. The EQLs are provided for guidance and may not always be achievable.*

$C =$ *True value for concentration, in µg/L.*
$X =$ *Average recovery found for measurements of samples containing a concentration of C, in µg/L.*

ESTIMATED QUANTITATION LIMIT

Other Matrices	Factor (a)
High-concentration soil and sludges by sonicator	7.5
Non-water miscible waste	75

(a) *EQL for other matrices = [EQL for low soil/sediment] × [Factor]. This estimated EQL is similar to an EPA "Practical Quantitation Limit."*

SAMPLING METHOD
Liquid samples — Use a 1 or 2½ gallon amber glass bottle with a screw-top Teflon®-lined cover that has been prewashed with detergent and rinsed with distilled water and methanol (or isopropanol).

Soils, sediments, or sludges — Use an 8-oz. widemouth glass with a screw-top Teflon®-lined cover that has been prewashed with detergent and rinsed with distilled water and methanol (or isopropanol).

SAMPLE PRESERVATION
Liquid samples — If residual chlorine is present, add 3 mL of 10% sodium thiosulfate per gallon, cool to 4°C and store in a solvent-free refrigerator until analysis; if chlorine is not present, then eliminate the sodium thiosulfate addition.

Soils, sediments, or sludges — Cool samples to 4°C and store in a solvent-free refrigerator.

MHT Liquid samples must be extracted within 7 days and the extracts analyzed within 40 days. Soils, sediments, or sludges may be stored for a maximum of 14 days and the extracts analyzed within 40 days.

SAMPLE PREPARATION
Liquid samples — Transfer 1 L quantitatively to a continuous extractor. If high concentrations are anticipated, a smaller volume may be used and then diluted with organic-free reagent water to 1 L. Adjust pH, if necessary, to pH <2 using 1:1 (V/V) sulfuric acid. Pipette 1.0 mL of a surrogate standard spiking solution into each sample. For the sample in each analytical batch selected for spiking, add 1.0 mL of a matrix spiking standard. For base/neutral acid analysis, the amount of the surrogates and matrix spiking compounds added to the sample should result in a final concentration of 100 ng/µL of each analyte in the extract to be analyzed (assuming a 1-µL injection). Extract with methylene chloride for 18–24 h. Next, adjust the pH of the aqueous phase to pH >11 using 10 N sodium hydroxide and extract it with methylene chloride again for 18–24 h. Dry the extract through a column containing anhydrous sodium sulfate and concentrate it to 1 mL using a K-D concentrator.

Soils, sediments, or sludges — Use 30 g of sample. Nonporous or wet samples (gummy or clay type) that do not have a free-flowing

sandy texture must be mixed with anhydrous sodium sulfate until the sample is free flowing. Add 1 mL of surrogate standards to all samples, spikes, standards, and blanks. For the sample in each analytical batch selected for spiking, add 1.0 mL of a matrix spiking standard. For base/neutral acid analysis, the amount added of the surrogates and matrix spiking compounds should result in a final concentration of 100 ng/µL of each base/neutral analyte and 200 ng/µL of each acid analyte in the extract to be analyzed (assuming a 1-µL injection). Immediately add a 100-mL mixture of 1:1 methylene chloride:acetone and extract the sample ultrasonically for 3 min and then decant or filter the extracts. Repeat the extraction two or more times. Dry the extract using a column with anhydrous sodium sulfate and concentrate it to 1 mL in a K-D concentrator.

QUALITY CONTROL A methylene chloride solution containing 50 ng/µL of decafluorotriphenylphosphine (DFTPP) is used for tuning the GC/MS system each 12-h shift. A system performance check also must be made during every 12-h shift. A standard containing 50 ng/µL each of 4,4′-DDT, pentachlorophenol, and benzidine is required to verify injection port inertness and GC column performance. A calibration standard at mid-concentration, containing each compound of interest, including all required surrogates, must be performed every 12 h during analysis. After the system performance check is met, calibration check compounds (CCCs) are used to check the validity of the initial calibration.

The internal standard responses and retention times in the calibration check standard must be evaluated immediately after or during data acquisition. If the retention time for any internal standard changes by more than 30 seconds from the last check calibration (12 h), the chromatographic system must be inspected for malfunctions and corrections must be made, as required. If the electron ionization current plot (EICP) area for any of the internal standards changes by a factor of two from the last daily calibration standard check, the mass spectrometer must be inspected for malfunctions and corrections must be made, as appropriate.

Demonstrate, through the analysis of a reagent water blank, that interferences from the analytical system, glassware, and reagents are under control. The blank samples should be carried through all stages of the sample preparation and measurement steps. For each analytical batch (up to 20 samples), a reagent blank, matrix spike, and matrix spike duplicate/duplicate must be analyzed (the frequency of the spikes may be different for different monitoring programs). The blank and spiked samples must be carried through all stages of the sample preparation and measurement steps. A QC reference sample concentrate containing each analyte at a concentration of 100 mg/L in methanol is required.

REFERENCE Test Methods for Evaluating Solid Waste (SW-846). U.S. EPA 1983, Method 8270B, Rev. 2, Nov. 1990. Office of Solid Waste, Washington, DC.

Epichlorohydrin   EPA Method 8240
CAS #106-89-8

TITLE Volatile Organics By GC/MS: Packed Column Technique

MATRIX Nearly all types of sample matarices, regardless of water content, can be analyzed using this method. This includes groundwater, aqueous sludges, caustic liquors, acid liquors, waste solvents, oily wastes, mousses, tars, fibrous wastes, polymetric emulsions, filter cakes, spent carbons, spent catalysts, soils, and sediments.

METHOD SUMMARY Method 8240B covers 80 volatile organic compounds that are introduced into a gas chromatograph by the purge-and-trap method or by direct injection (in limited applications). For the purge-and-trap method an inert gas (zero grade nitrogen or helium) is bubbled through a 5-mL solution at ambient temperature. Purged sample components are trapped in a tube of sorbent materials. When purging is complete, the sorbent tube is heated and backflushed with inert gas to desorb the trapped components onto a GC column.

INTERFERENCES Impurities in the purge gas and from organic compounds outgassing from the plumbing ahead of the trap account for many contamination problems. Interferences purged or coextracted from the samples will vary considerably from source to source. Cross-contamination can occur whenever high-level and low-level samples are analyzed sequentially. Whenever an unusually concentrated sample is analyzed, it should be followed by the analysis of organic-free reagent water to check for cross-contamination. Samples also can be contaminated by diffusion of volatile organics (particularly methylene chloride and fluorocarbons) through the septum seal into the sample during shipment and storage. A trip blank can serve as a check on such contamination. The lab where volatile analysis is performed and also the refrigerated storage area should be completely free of solvents.

INSTRUMENTATION A gas chromatograph/mass spectrometry/data system (GC/MS) equipped with a 6 ft × 0.1 in I.D. glass column packed with 1% SP-1000 on Carbopack-B (60/80 mesh) is required. Also needed is a 5-mL purging device, a sorbent trap, and a thermal desorption apparatus.

PRECISION & ACCURACY This method is reported to have been tested by 15 laboratories using organic-free reagent water, drinking water, surface water, and industrial wastewaters (not specified) fortified at six concentrations over the range 5–600 µg/L.

Sample estimated quantitation limits (EQLs) are highly matrix-dependent. The EQLs listed may not always be achievable. EQLs listed for soils or sediments are based on wet weight. Normally, data is reported on a dry-weight basis; therefore, EQLs will be higher, based on the percent dry weight of each sample. Note that EQLs are even more variable than MDLs and that they are highly variable depending on the matrix being analyzed.

EQL in groundwater in µg/L was not listed.
EQL in low soil or sediment in µg/kg was not listed.

Accuracy (a) in μg/L was not listed.
Precision (b) in μg/L was not listed.

(a) *Average recovery found for measurements of samples containing a concentration of C, in μg/L.*
(b) *Overall precision found for measurements of samples with average recovery X for samples containing a concentration of C in μg/L.*
X = *Average recovery found for measurement of samples containing a concentration of C in μg/L.*

MULTIPLICATION FACTORS FOR OTHER MATRICES

Other Matrices	Factor (a)
Waste miscible liquid waste	50
High-concentration soil and sludge	125
Non-water miscible waste	500

(a) *EQL = [EQL for low soil/sediment] × [Factor]. For non-aqueous samples, the factor is on a wet-weight basis.*

SAMPLING METHOD

Liquid samples — Use a 40-mL glass screw-cap VOA vial with a Teflon®-faced silicone septum that has been prewashed, rinsed with distilled deionized water, and oven dried. However, if residual chlorine is present, collect sample in a 40-oz. soil VOA container which has been pre-preserved with 4 drops of 10% sodium thiosulfate, mix gently, and then transfer the sample to a 40-mL VOA vial. Collect bubble-free samples in duplicate and seal them in separate plastic bags.

Soils or sediments, and sludges — Use an 8-oz. widemouth glass bottle with a Teflon®-faced silicone septum that has been prewashed with detergent, rinsed with distilled deionized water, and oven dried. Tap slightly to eliminate free air space. Collect samples in duplicate and seal them in separate plastic bags.

SAMPLE PRESERVATION

Liquid samples — Add 4 drops of concentrated HCL and immediately cool samples to 4°C and store in a solvent-free refrigerator.

Soils or sediments, and sludges — Cool samples to 4°C and store in a solvent-free refrigerator.

MHT Maximum holding time is 14 days from the date of sample collection.

SAMPLE PREPARATION

Liquid samples — Remove the plunger from a 5-mL syringe and carefully pour the sample into the syringe barrel to just short of overflowing. Replace the syringe plunger and compress the sample. Open the syringe valve and vent any residual air while adjusting the sample volume to 5.0 mL. If there is only one volatile organic analysis (VOA) vial, a second syringe should be filled at this time to protect against possible loss of sample integrity. Add 10 μL of surrogate spiking solution and 10 μL of internal standard spiking solution through the valve bore of the 5-mL syringe, then close the valve. The surrogate and internal standards may be mixed and added as a single spiking solution.

Sediments, soils, and waste samples — All samples of this type should be screened by GC analysis using a headspace method (EPA Method 3810) or the hexadecane extraction and screening method (EPA Method 3820). Use the screening data to determine whether to use the low-concentration method (0.005–1 mg/kg) or the high-concentration method (>1 mg/kg).

Low-concentration method — The low-concentration method is based on purging a heated sediment or soil sample mixed with organic-free reagent water containing the surrogate and internal standards. Analyze all reagent blanks and standards under the same conditions as the samples.

Use a 5-g sample if the expected concentration is <0.1 mg/kg or a 1-g sample for expected concentrations between 0.1 and 1 mg/kg. Mix the contents of the sample container with a narrow metal spatula. Weigh the amount of the sample into a tared purge device. Add the spiked water to the purge device, which contains the weighed amount of sample, and connect the device to the purge-and-trap system.

High-concentration method — This method is based on extracting the sediment or soil with methanol. A waste sample is either extracted or diluted, depending on its solubility in methanol. Wastes that are insoluble in methanol are diluted with reagent tetraglyme or possibly polyethylene glycol (PEG). An aliquot of the extract is added to organic-free reagent water containing surrogate and internal standards. This is purged at ambient temperature. All samples with an expected concentration of >1.0 mg/kg should be analyzed by this method.

Mix the contents of the sample container with a narrow metal spatula. For sediments or soils and solid wastes that are insoluble in methanol, weigh 4 g (wet weight) of sample into a tared 20-mL vial. For waste that is soluble in methanol, tetraglyme, or PEG, weigh 1 g (wet weight) into a tared scintillation vial or culture tube or a 10-mL volumetric flask. Quickly add 9.0 mL of appropriate solvent then add 1.0 mL of a surrogate spiking solution to the vial, cap it, and shake it for 2 min.

METHANOL EXTRACT REQUIRED FOR ANALYSIS OF HIGH-CONCENTRATION SOILS OR SEDIMENTS

Approximate Concentration Range	Volume of Methanol Extract (a)
500–10,000 μg/kg	100 μL
1,000–20,000 μg/kg	50 μL
5,000–100,000 μg/kg	10 μL
25,000–500,000 μg/kg	100 μL of 1/50 dilution (b)

Calculate appropriate dilution factor for concentrations exceeding this table.

(a) *The volume of methanol added to 5 mL of water being purged should be kept constant. Therefore, add to the 5-mL syringe whatever volume of methanol is necessary to maintain a volume of 100 μL added to the syringe.*
(b) *Dilute an aliquot of the methanol extract and then take 100 μL for analysis.*

QUALITY CONTROL Demonstrate, through the analysis of a reagent water blank, that interferences from the analytical system, glassware, and reagents are under control. Blank samples should be carried through all stages of the sample preparation and measurement steps. For each analytical batch (up to 20 samples), a reagent blank, matrix spike, and matrix spike duplicate must be analyzed (the frequency of the spikes may be different for different monitoring programs). The blank and spiked samples must be carried through all stages of the sample preparation and measurement steps. QC samples mentioned in the section on Interferences will also be needed as appropriate to those situations.

REFERENCE Test Methods for Evaluating Solid Waste (SW-846). U.S. EPA. 1983. Method 8240B, Rev. 2, Nov. 1990. Office of Solid Wastes, Washington, DC.

EPN   **EPA Method 8141**
CAS #2104-64-5

TITLE Organophosphorus Compounds by Gas Chromatography: Capillary Column Technique

MATRIX This method covers aqueous and solid matrices. This includes a wide variety such as drinking water, groundwater, industrial wastewaters, surface waters, soils, solids, and sediments.

METHOD SUMMARY This is a GC method used to determine the concentration of 28 organophosphorus pesticides.

The use of Gel Permeation Cleanup (EPA Method 3640) for sample cleanup has been demonstrated to yield recoveries of less than 85% for many method analytes and is therefore not recommended for use with this method.

This method provides GC conditions for the detection of ppb concentrations of organophosphorus compounds. Prior to the use of this method, appropriate sample preparation techniques must be used. Water samples are extracted at a neutral pH with methylene chloride as a solvent by using a separatory funnel (EPA Method 3510) or a continuous liquid-liquid extractor (EPA Method 3520). Soxhlet extraction (EPA Method 3540) or ultrasonic extraction (EPA Method 3550) using methylene chloride/acetone (1:1) are used for solid samples. Both neat and diluted organic liquids (EPA Method 3580) may be analyzed by direct injection. Spiked samples are used to verify the applicability of the chosen extraction technique to each new sample type. A GC with a flame photometric (FPD) or nitrogen-phosphorus detector (NPD) is used for this multiresidue procedure.

INTERFERENCES The use of Florisil cleanup materials (EPA Method 3620) for some of the compounds in this method has been demonstrated to yield recoveries less than 85% and is therefore not recommended for all compounds. Use of phosphorus or halogen specific detectors, however, often obviates the necessity for cleanup for relatively clean sample matrices. If particular circumstances demand the use of an alternative cleanup procedure, the analyst must determine the elution profile and demonstrate that the recovery of each analyte is no less than 85%.

Use of a flame photometric detector (FPD) in the phosphorus mode will minimize interferences from materials that do not contain phosphorus. Elemental sulfur, however, may interfere with the determination of certain organophosphorus compounds by flame photometric gas chromatography. Sulfur cleanup using EPA Method 3660 may alleviate this interference. A nitrogen phosphorus detector (NPD) is also recommended.

A few analytes coelute on certain columns. Therefore, select a second column for confirmation where coelution of the analytes of interest does not occur.

Method interferences may be caused by contaminants in solvents, reagents, glassware, and other sample processing hardware that lead to discrete artifacts or elevated baselines in gas chromatograms. All these materials must be routinely demonstrated to be free from interferences under the conditions of the analysis by analyzing reagent blanks.

INSTRUMENTATION A GC with a NPD or a FPD will be needed. A data system or integrator is recommended for measuring peak areas and/or peak heights. A Kuderna-Danish (K-D) apparatus will be needed for extract concentration.

Column 1: 15 m × 0.53 mm megabore capillary column, 1.0 μm film thickness, DB-210.
Column 2: 15 m × 0.53 mm megabore capillary column, 1.5 μm film thickness, SPB-608.
Column 3: 15 m × 0.53 mm megabore capillary column, 1.0 μm film thickness, DB-5.

Three megabore capillary columns are included for analysis of organophosphates by this method. Column 1 (DB-210 or equivalent) and Column 2 (SPB-608 or equivalent) are recommended if a large number of organophosphorus analytes are to be determined. If the superior resolution offered by Column 1 and Column 2 is not required, Column 3 (DB-5 or equivalent) may be used. For megabore capillary columns, automatic injections of 1 μL are recommended.

PRECISION & ACCURACY The MDL actually achieved in a given analysis will vary, as it is dependent on instrument sensitivity and matrix effects. Single operator accuracy and precision studies have been conducted with spiked water and soil samples.

MULTIPLICATION FACTORS FOR OTHER MATRICES (a)

Matrix	Factor (b)
Groundwater	10
(EPA Method 3510 or EPA Method 3520)	
Low-concentration soil by Soxhlet and no cleanup	10 (c)
Low-concentration soil by ultrasonic extraction with GPC cleanup	6.7 (c)
High-concentration soil and sludges by ultrasonic extraction	500 (c)
Non-water miscible waste (EPA Method 3580)	1000 (c)

(a) Sample EQLs are highly matrix-dependent. The EQLs listed here are provided for guidance and may not always be achievable.
(b) EQL = [Method detection limit] × [Factor]. For non-aqueous samples the factor is on a wet-weight basis.
(c) Multiply this factory times the soil MDL.

The MDL (in μg/L) when reagent water was extracted using a separatory funnel was 0.04.
The MDL (in μg/kg) when soil was extracted using Soxhlet extraction (EPA Method 3540) was 2.0.

Accuracy (as % recovery) with separatory funnel extraction ranged from 113 (with low spikes) to 97 (with high spikes).

Accuracy (as % recovery) with continuous liquid-liquid extraction ranged from not recovered (with low spikes) to 119 (with high spikes).

Accuracy (as % recovery) with Soxhlet extraction of soils ranged from 114 (with low spikes to 82 (with high spikes).

Accuracy (as % recovery) with ultrasonic extraction of soils ranged from 14 (with low spikes) to 105 (with high spikes).

SAMPLE COLLECTION, PRESERVATION & HANDLING
Containers used to collect samples for the determination of semivolatile organic compounds should be soap and water washed followed by methanol (or isopropanol) rinsing. The sample containers should be of glass or Teflon® and have screw-top covers with Teflon® liners.

No preservation is used with concentrated waste samples. With liquid samples containing no residual chlorine and with soil, sediment, and sludge samples, immediately cooling to 4°C is the only preservation used. When residual chlorine is present then 3 mL of 10% aqueous sodium sulfate is added for each gallon of sample collected, followed by cooling to 4°C.

Liquid samples must be extracted within 7 days and their extracts analyzed within 40 days. Concentrated waste, soil, sediment, and sludge samples must be extracted within 14 days and their extracts analyzed within 40 days.

SAMPLE PREPARATION In general, water samples are extracted at a neutral pH with methylene chloride, using either EPA Method 3510 or EPA Method 3520. Solid samples are extracted using either EPA Method 3540 or EPA Method 3550 with methylene chloride/acetone (1:1) as the extraction solvent.

Prior to GC analysis, the extraction solvent may be exchanged to hexane. Single lab data indicates that samples should not be transferred with 100% hexane during sample workup as the more water soluble organophosphorus compounds may be lost.

If cleanup is performed on the samples, the analyst should analyze the samples by GC. This will confirm elution patterns and the absence of interferences from the reagents. If peak detection and identification is prevented by the presence of interferences, further cleanup is required.

QUALITY CONTROL The analyst should monitor the performance of the extraction, cleanup (when used), and analytical system and the effectiveness of the method in dealing with each sample matrix by spiking each sample, standard, and blank with one or two surrogates (e.g., organophosphorus compounds not expected to be present in the sample). Deuterated analogs of analytes should not be used as surrogates for gas chromatographic analysis due to coelution problems.

A minimum of five concentrations for each analyte of interest should be prepared through dilution of the stock standards with isooctane. One of the concentrations should be at a concentration near, but above, the MDL.

Include a mid-level check standard after each group of 10 samples in the analysis sequence. GC/MS techniques should be judiciously employed to support qualitative identifications made with this method. Follow the GC/MS operating requirements specified in EPA Method 8270.

When available, chemical ionization mass spectra may be employed to aid in the qualitative identification process. To confirm an identification of a compound, the background-corrected mass spectrum of the compound must be obtained from the sample extract and must be compared with a mass spectrum from a stock or calibration standard analyzed under the same chromatographic conditions. The molecular ion and all other ions present above 20% relative abundance in the mass spectrum of the standard must be present in the mass spectrum of the sample with agreement to ± 20%. The retention time of the compound in the sample must be within six seconds of the retention time for the same compound in the standard solution.

Should the MS procedure fail to provide satisfactory results, additional steps may be taken before reanalysis. These steps may include the use of alternate packed or capillary GC columns or additional sample cleanup.

REFERENCE Test Methods for Evaluating Solid Waste, Physical/Chemical Methods, SW-846, 3rd Edition, U.S. EPA, Office of Solid Waste, Washington, DC, EPA Method 8141 July 1992.

EPN	EPA Method 8270
CAS #2104-64-5	

TITLE Semivolatile Organic Compounds by GC/MS

MATRIX This method is used to determine the concentration of semivolatile organic compounds in extracts prepared from all types of solid waste matrices, soils, and groundwater. Although surface waters are not specifically mentioned, this method should be applicable to water samples from rivers, lakes, etc.

METHOD SUMMARY This method covers 259 semivolatile organic compounds. In very limited applications direct injection of the sample into the GC/MS system may be appropriate, but this results in very high detection limits (approximately 10,000 µg/L). Typically, a 1-L liquid sample, containing surrogate, and matrix spiking standards, is extracted in a continuous extractor first under acid conditions and then under basic conditions. Typically 30 g of a solid sample, containing surrogate, and matrix spiking standards, is extracted ultrasonically. After concentrating the extract to 1 mL it is spiked with 10 µL of an internal standard solution just prior to analysis by GC/MS. The volume injected should contain about 100 ng of base/neutral and 200 ng of acid surrogates (for a 1-µL injection). Analysis is performed by GC/MS using a capillary GC column.

INTERFERENCES Raw GC/MS data from all blanks, samples, and spikes must be evaluated for interferences. Contamination by carryover can occur whenever high-concentration and low-concentration samples are sequentially analyzed. To reduce carryover, the sample syringe must be rinsed out between samples with solvent. Whenever an unusually concentrated sample is encountered, it should be followed by the analysis of blank solvent to check for cross-contamination.

INSTRUMENTATION A GC/MS and a data system are required. The GC column used is a 30 m × 0.25 mm I.D. (or 0.32 mm I.D.) 1um film thickness silicone-coated fused silica capillary column. A continuous liquid-liquid extractor equipped with Teflon® or glass connection joints and stopcocks requiring no lubrication, a K-D concentrating apparatus, water bath, and an ultrasonic disrupter with a minimum power of 300 W and with pulsing capability are also required.

PRECISION & ACCURACY The estimated quantitation limit (EQL) of Method 8270B for determining an individual compound is approximately 1 mg/kg (wet weight) for soil or sediment samples, 1–200 mg/kg for wastes (dependent on matrix and method of preparation), and 10 µg/L for groundwater samples. EQLs will be proportionately higher for sample extracts that require dilution to avoid saturation of the detector.

The EQL(b) for groundwater in µg/L is 10.
The EQL (a, b) for low concentrations in soil and sediment in µg/kg is not determined.
Accuracy as µg/L is not listed.
Overall precision in µg/L is not listed.

(a) *EQLs listed for soil/sediment are based on wet weight. Normally data is reported in a dry-weight basis; therefore, EQLs will be higher based on the % dry weight of each sample. This calculation is based on a 30-g sample and gel permeation chromatography cleanup.*
(b) *Sample EQLs are highly matrix-dependent. The EQLs are provided for guidance and may not always be achievable.*
C = *True value for concentration, in µg/L.*
X = *Average recovery found for measurements of samples containing a concentration of C, in µg/L.*

ESTIMATED QUANTITATION LIMIT

Other Matrices	Factor (a)
High-concentration soil and sludges by sonicator	7.5
Non-water miscible waste	75

(a) *EQL for other matrices = [EQL for low soil/sediment] × [Factor]. This estimated EQL is similar to an EPA "Practical Quantitation Limit."*

SAMPLING METHOD

Liquid samples — Use a 1 or 2½ gallon amber glass bottle with a screw-top Teflon®-lined cover that has been prewashed with detergent and rinsed with distilled water and methanol (or isopropanol).

Soils, sediments, or sludges — Use an 8-oz. widemouth glass with a screw-top Teflon®-lined cover that has been prewashed with detergent and rinsed with distilled water and methanol (or isopropanol).

SAMPLE PRESERVATION

Liquid samples — If residual chlorine is present, add 3 mL of 10% sodium thiosulfate per gallon, cool to 4°C and store in a solvent-free refrigerator until analysis; if chlorine is not present, then eliminate the sodium thiosulfate addition.

Soils, sediments, or sludges — Cool samples to 4°C and store in a solvent-free refrigerator.

MHT Liquid samples must be extracted within 7 days and the extracts analyzed within 40 days. Soils, sediments, or sludges may be stored for a maximum of 14 days and the extracts analyzed within 40 days.

SAMPLE PREPARATION

Liquid samples — Transfer 1 L quantitatively to a continuous extractor. If high concentrations are anticipated, a smaller volume may be used and then diluted with organic-free reagent water to 1 L. Adjust pH, if necessary, to pH <2 using 1:1 (V/V) sulfuric acid. Pipette 1.0 mL of a surrogate standard spiking solution into each sample. For the sample in each analytical batch selected for spiking, add 1.0 mL of a matrix spiking standard. For base/neutral acid analysis, the amount of the surrogates and matrix spiking compounds added to the sample should result in a final concentration of 100 ng/µL of each analyte in the extract to be analyzed (assuming a 1-µL injection). Extract with methylene chloride for 18–24 h. Next, adjust the pH of the aqueous phase to pH >11 using 10 N sodium hydroxide and extract it with methylene chloride again for 18–24 h. Dry the extract through a column containing anhydrous sodium sulfate and concentrate it to 1 mL using a K-D concentrator.

Soils, sediments, or sludges — Use 30 g of sample. Nonporous or wet samples (gummy or clay type) that do not have a free-flowing sandy texture must be mixed with anhydrous sodium sulfate until the sample is free flowing. Add 1 mL of surrogate standards to all samples, spikes, standards, and blanks. For the sample in each analytical batch selected for spiking, add 1.0 mL of a matrix spiking standard. For base/neutral acid analysis, the amount added of the surrogates and matrix spiking compounds should result in a final concentration of 100 ng/µL of each base/neutral analyte and 200 ng/µL of each acid analyte in the extract to be analyzed (assuming a 1-µL injection). Immediately add a 100-mL mixture of 1:1 methylene chloride:acetone and extract the sample ultrasonically for 3 min and then decant or filter the extracts. Repeat the extraction two or more times. Dry the extract using a column with anhydrous sodium sulfate and concentrate it to 1 mL in a K-D concentrator.

QUALITY CONTROL A methylene chloride solution containing 50 ng/µL of decafluorotriphenylphosphine (DFTPP) is used for tuning the GC/MS system each 12-h shift. A system performance check also must be made during every 12-h shift. A standard containing 50 ng/µL each of 4,4'-DDT, pentachlorophenol, and benzidine is required to verify injection port inertness and GC column performance. A calibration standard at mid-concentration, containing each compound of interest, including all required surrogates, must be performed every 12 h during analysis. After the system performance check is met, calibration check compounds (CCCs) are used to check the validity of the initial calibration.

The internal standard responses and retention times in the calibration check standard must be evaluated immediately after or during data acquisition. If the retention time for any internal standard changes by more than 30 seconds from the last check calibration (12 h), the chromatographic system must be inspected for malfunctions and corrections must be made, as required. If the electron ionization current plot (EICP) area for

any of the internal standards changes by a factor of two from the last daily calibration standard check, the mass spectrometer must be inspected for malfunctions and corrections must be made, as appropriate.

Demonstrate, through the analysis of a reagent water blank, that interferences from the analytical system, glassware, and reagents are under control. The blank samples should be carried through all stages of the sample preparation and measurement steps. For each analytical batch (up to 20 samples), a reagent blank, matrix spike, and matrix spike duplicate/duplicate must be analyzed (the frequency of the spikes may be different for different monitoring programs). The blank and spiked samples must be carried through all stages of the sample preparation and measurement steps. A QC reference sample concentrate containing each analyte at a concentration of 100 mg/L in methanol is required.

REFERENCE Test Methods for Evaluating Solid Waste (SW-846). U.S. EPA 1983, Method 8270B, Rev. 2, Nov. 1990. Office of Solid Waste, Washington, DC.

EPTC   **EPA Method 507**
CAS #759-94-4

TITLE Determination of Nitrogen and Phosphorus-Containing Pesticides in Water by GC/NPD

MATRIX This method is applicable to the determination of certain nitrogen and phosphorus-containing pesticides in finished drinking water and groundwater.

METHOD SUMMARY Method 507 covers 46 nitrogen- and phosphorus-containing pesticides. A 1-L sample is fortified with a surrogate standard, salted, buffered, extracted with methylene chloride, and concentrated; then the solvent is exchanged with methyl tert-butyl ether (MTBE) and concentrated again, and a 2-μL aliquot of a sample extract is injected into a GC system equipped with a selective nitrogen-phosphorus detector and a capillary column for analysis.

INTERFERENCES Method interferences may be caused by contaminants in solvents, reagents, glassware, and other sample processing apparatus. Interfering contamination may occur when a sample containing low concentrations of analytes is analyzed immediately following a sample containing relatively high concentrations. One or more injections of MTBE should be made following the analysis of a sample with high concentrations of analytes to check for analyte carryover. Matrix interferences may be caused by contaminants that are coextracted from the sample. The extent of matrix interferences will vary considerably from source to source, depending upon the water sampled.

INSTRUMENTATION A gas chromatograph system (GC) equipped with a nitrogen-phosphorus detector (NPD) is needed.

Column 1: 30 m × 0.25 mm I.D. DB-5 bonded fused silica column, 0.25 μm film thickness, or equivalent.

Column 2: 30 m × 0.25 mm I.D. DB-1701 bonded fused silica column, 0.25 μm film thickness, or equivalent.

PRECISION & ACCURACY This method has been validated in a single lab and estimated detection limits (EDLs) have been determined for each analyte. Observed detection limits may vary among waters, depending upon the nature of the interferences in the sample matrix and the specific instrumentation used. Analytes that are not separated chromatographically cannot be individually identified and measured unless an alternative technique for identification and quantification exist.

The estimated detection limit (in μg/L) was 0.25. The EDL is defined as either method detection limit or a level of compound in a sample yielding a peak in the final extract with signal-to-noise ratio of approximately 5, whichever value is higher.

The concentration used for these measurements (in μg/L) was 2.5.
The accuracy (as % recovery) was 85.
The precision (% RSD) was 9.

SAMPLING METHOD Grab samples are collected in 1-L glass sample bottles (prewashed with detergent and hot tap water, rinsed with reagent water, and dried in an oven at 400°C for 1 h) with screw caps lined with PTFE-fluorocarbon.

SAMPLE PRESERVATION Add mercuric chloride to the sample bottle in amounts to produce a concentration of 10 mg/L. If residual chlorine is present, add 80 mg of sodium thiosulfate/L of sample to the sample bottle prior to collection. After collection, seal bottle and shake vigorously for 1 min, then cool the sample to 4°C immediately and store it at 4°C in the dark until extraction.

MHT Maximum holding time of the samples, and in some cases the extracts, is 14 days.

Note: Samples with this compound exhibited recoveries of less than 60% after 14 days.

SAMPLE PREPARATION Fortify the sample with 50 μL of the surrogate standard solution, adjust to pH 7 with phosphate buffer, add 100 g NaCl to the sample, and seal and shake to dissolve the salt; then extract with methylene chloride in a separatory funnel or in a mechanical tumbler bottle. Dry the extract by pouring it through a solvent-rinsed drying column containing about 10 cm of anhydrous sodium sulfate. Collect the extract in a Kuderna-Danish (K-D) concentrator and rinse the column with 20–30 mL methylene chloride. Concentrate the extract to about 2 mL and rinse the flask and its lower joint into the concentrator tube with 1 to 2 mL of methyl t-butyl ether (MTBE). Add 5–10 mL of MTBE and concentrate the extract twice (adding more MTBE) to a final volume of 5.0 mL and store it at 4°C until analysis.

Note: If methylene chloride is not completely removed from the final extract, it may cause detector problems.

QUALITY CONTROL Minimum quality control requirements are initial demonstration of lab capability, determination of surrogate compound recoveries in each sample and blank, monitoring internal standard peak area or height in each sample and blank, analysis of lab reagent blanks, lab fortified samples, lab fortified blanks, and other QC samples. A lab reagent blank is analyzed to demonstrate that all glassware and reagent interferences are under control.

Initial demonstration of capability is fulfilled by analyzing four fortified reagent water samples with the recovery value for each analyte falling within the acceptable range (±30% average recovery). Surrogate recoveries from samples or method blanks must be 70–130%. The internal standard response for any sample chromatogram should not deviate from the daily calibration check standard's internal standard response by more than 30% or lab fortified blanks and sample matrices are used to assess lab performance and analyte recovery, respectively.

If the response for the target analyte peak exceeds the working range of the system, dilute the extract and reanalyze. Alternative techniques such as an alternate detector or second chromatography column should be used to confirm peak identification when sample components are not resolved adequately.

EPA CONTACT & HOTLINE For technical questions contact Dr. Baldev Bathija, U.S. EPA, Office of Ground Water and Drinking Water (WH-550D), 401 M St. SW, Washington, DC 20460. Tel. (202) 260-3040. For further information the EPA Safe Drinking Water Hotline may be called at: (800) 426-4791.

REFERENCE Methods for the Determination of Organic Compounds in Drinking Water, EPA/600/4-88/039 (revised July 1991). U.S. EPA Environmental Monitoring Systems Laboratory, Cincinnati, OH, 45268, U.S.A. Available from the National Technical Information Service (NTIS), 5285 Port Royal Road, Springfield, VA 22161; Tel. 800-553-6847. NTIS Order Number is PB91-231480.

Ethanol	EPA Method 8240
CAS #64-17-5	

TITLE Volatile Organics By GC/MS: Packed Column Technique

MATRIX Nearly all types of sample matarices, regardless of water content, can be analyzed using this method. This includes groundwater, aqueous sludges, caustic liquors, acid liquors, waste solvents, oily wastes, mousses, tars, fibrous wastes, polymetric emulsions, filter cakes, spent carbons, spent catalysts, soils, and sediments.

METHOD SUMMARY Method 8240B covers 80 volatile organic compounds that are introduced into a gas chromatograph by the purge-and-trap method or by direct injection (in limited applications). For the purge-and-trap method an inert gas (zero grade nitrogen or helium) is bubbled through a 5-mL solution at ambient temperature. Purged sample components are trapped in a tube of sorbent materials. When purging is complete, the sorbent tube is heated and backflushed with inert gas to desorb the trapped components onto a GC column.

INTERFERENCES Impurities in the purge gas and from organic compounds outgassing from the plumbing ahead of the trap account for many contamination problems. Interferences purged or coextracted from the samples will vary considerably from source to source. Cross-contamination can occur whenever high-level and low-level samples are analyzed sequentially. Whenever an unusually concentrated sample is analyzed, it should be followed by the analysis of organic-free reagent water to check for cross-contamination. Samples also can be contaminated by diffusion of volatile organics (particularly methylene chloride and fluorocarbons) through the septum seal into the sample during shipment and storage. A trip blank can serve as a check on such contamination. The lab where volatile analysis is performed and also the refrigerated storage area should be completely free of solvents.

INSTRUMENTATION A gas chromatograph/mass spectrometry/data system (GC/MS) equipped with a 6 ft × 0.1 in I.D. glass column packed with 1% SP-1000 on Carbopack-B (60/80 mesh) is required. Also needed is a 5-mL purging device, a sorbent trap, and a thermal desorption apparatus.

PRECISION & ACCURACY This method is reported to have been tested by 15 laboratories using organic-free reagent water, drinking water, surface water, and industrial wastewaters (not specified) fortified at six concentrations over the range 5–600 µg/L.

Sample estimated quantitation limits (EQLs) are highly matrix-dependent. The EQLs listed may not always be achievable. EQLs listed for soils or sediments are based on wet weight. Normally, data is reported on a dry-weight basis; therefore, EQLs will be higher, based on the percent dry weight of each sample. Note that EQLs are even more variable than MDLs and that they are highly variable depending on the matrix being analyzed.

EQL in groundwater in µg/L was not listed.
EQL in low soil or sediment in µg/kg was not listed.
Accuracy (a) in µg/L was not listed.
Precision (b) in µg/L was not listed.

(a) *Average recovery found for measurements of samples containing a concentration of C, in µg/L.*
(b) *Overall precision found for measurements of samples with average recovery X for samples containing a concentration of C in µg/L.*
X = *Average recovery found for measurement of samples containing a concentration of C in µg/L.*

MULTIPLICATION FACTORS FOR OTHER MATRICES

Other Matrices	Factor (a)
Waste miscible liquid waste	50
High-concentration soil and sludge	125
Non-water miscible waste	500

(a) *EQL = [EQL for low soil/sediment] × [Factor]. For non-aqueous samples, the factor is on a wet-weight basis.*

SAMPLING METHOD

Liquid samples — Use a 40-mL glass screw-cap VOA vial with a Teflon®-faced silicone septum that has been prewashed, rinsed with distilled deionized water, and oven dried. However, if residual chlorine is present, collect sample in a 40-oz. soil VOA container which has been pre-preserved with 4 drops of 10% sodium thiosulfate, mix gently, and then transfer the sample to a 40-mL VOA vial. Collect bubble-free samples in duplicate and seal them in separate plastic bags.

Soils or sediments, and sludges — Use an 8-oz. widemouth glass bottle with a Teflon®-faced silicone septum that has been prewashed with detergent, rinsed with distilled deionized water, and oven dried. Tap slightly to eliminate free air space. Collect samples in duplicate and seal them in separate plastic bags.

SAMPLE PRESERVATION

Liquid samples — Add 4 drops of concentrated HCL and immediately cool samples to 4°C and store in a solvent-free refrigerator.

Soils or sediments, and sludges — Cool samples to 4°C and store in a solvent-free refrigerator.

MHT Maximum holding time is 14 days from the date of sample collection.

SAMPLE PREPARATION

Liquid samples — Remove the plunger from a 5-mL syringe and carefully pour the sample into the syringe barrel to just short of overflowing. Replace the syringe plunger and compress the sample. Open the syringe valve and vent any residual air while adjusting the sample volume to 5.0 mL. If there is only one volatile organic analysis (VOA) vial, a second syringe should be filled at this time to protect against possible loss of sample integrity. Add 10 µL of surrogate spiking solution and 10 µL of internal standard spiking solution through the valve bore of the 5-mL syringe, then close the valve. The surrogate and internal standards may be mixed and added as a single spiking solution.

Sediments, soils, and waste samples — All samples of this type should be screened by GC analysis using a headspace method (EPA Method 3810) or the hexadecane extraction and screening method (EPA Method 3820). Use the screening data to determine whether to use the low-concentration method (0.005–1 mg/kg) or the high-concentration method (>1 mg/kg).

Low-concentration method — The low-concentration method is based on purging a heated sediment or soil sample mixed with organic-free reagent water containing the surrogate and internal standards. Analyze all reagent blanks and standards under the same conditions as the samples.

Use a 5-g sample if the expected concentration is <0.1 mg/kg or a 1-g sample for expected concentrations between 0.1 and 1 mg/kg. Mix the contents of the sample container with a narrow metal spatula. Weigh the amount of the sample into a tared purge device. Add the spiked water to the purge device, which contains the weighed amount of sample, and connect the device to the purge-and-trap system.

High-concentration method — This method is based on extracting the sediment or soil with methanol. A waste sample is either extracted or diluted, depending on its solubility in methanol. Wastes that are insoluble in methanol are diluted with reagent tetraglyme or possibly polyethylene glycol (PEG). An aliquot of the extract is added to organic-free reagent water containing surrogate and internal standards. This is purged at ambient temperature. All samples with an expected concentration of >1.0 mg/kg should be analyzed by this method.

Mix the contents of the sample container with a narrow metal spatula. For sediments or soils and solid wastes that are insoluble in methanol, weigh 4 g (wet weight) of sample into a tared 20-mL vial. For waste that is soluble in methanol, tetraglyme, or PEG, weigh 1 g (wet weight) into a tared scintillation vial or culture tube or a 10-mL volumetric flask. Quickly add 9.0 mL of appropriate solvent then add 1.0 mL of a surrogate spiking solution to the vial, cap it, and shake it for 2 min.

METHANOL EXTRACT REQUIRED FOR ANALYSIS OF HIGH-CONCENTRATION SOILS OR SEDIMENTS

Approximate Concentration Range	Volume of Methanol Extract (a)
500–10,000 µg/kg	100 µL
1,000–20,000 µg/kg	50 µL
5,000–100,000 µg/kg	10 µL
25,000–500,000 µg/kg	100 µL of 1/50 dilution (b)

Calculate appropriate dilution factor for concentrations exceeding this table.

(a) The volume of methanol added to 5 mL of water being purged should be kept constant. Therefore, add to the 5-mL syringe whatever volume of methanol is necessary to maintain a volume of 100 µL added to the syringe.

(b) Dilute an aliquot of the methanol extract and then take 100 µL for analysis.

QUALITY CONTROL Demonstrate, through the analysis of a reagent water blank, that interferences from the analytical system, glassware, and reagents are under control. Blank samples should be carried through all stages of the sample preparation and measurement steps. For each analytical batch (up to 20 samples), a reagent blank, matrix spike, and matrix spike duplicate must be analyzed (the frequency of the spikes may be different for different monitoring programs). The blank and spiked samples must be carried through all stages of the sample preparation and measurement steps. QC samples mentioned in the section on Interferences will also be needed as appropriate to those situations.

REFERENCE Test Methods for Evaluating Solid Waste (SW-846). U.S. EPA. 1983. Method 8240B, Rev. 2, Nov. 1990. Office of Solid Wastes, Washington, DC.

Ethanol **EPA Method 8015**
CAS #64-17-5

TITLE Nonhalogenated Volatile Organics

MATRIX Groundwater, soils, sludges, water miscible liquid wastes, and non-water miscible wastes.

APPLICATION This method is used for the analysis of 6 nonhalogenated VOCs. Samples are analyzed using direct injection or purge-and-trap methods. Groundwater must be analyzed by the purge-and-trap method. The method provides an optional GC column that may help resolve analytes from interferences and which is also used for analyte confirmation.

INTERFERENCES There can be carryover contamination with high- and low-level samples. Impurities may come from the purge-and-trap apparatus, organic compounds outgassing from the plumbing ahead of trap, diffusion of VOCs through the sample bottle septum during shipping or storage, or from solvent vapors in the lab.

INSTRUMENTATION GC capable of on-column injections or purge-and-trap sample introduction and a flame ionization detector (FID). Column 1: an 8 ft by 0.1 in 1% SP-1000 on Carbopack-B column. Column 2: a 6 ft by 0.1 in bonded n-octane on Porasil-C.

RANGE Not available.

MDL Not available.

PRECISION Not available.

ACCURACY Not available.

SAMPLING METHOD For water and liquid samples; use glass 40-mL vials with Teflon®-lined septum caps and collect two vials per sample location with no headspace. For solids and concentrated waste samples; use widemouth glass bottles with Teflon® liners. Cool all samples to 4°C

STABILITY For concentrated wastes, soils, sediments, or sludges: cool to 4°C. For liquids: add 4 drops of concentrated hydrochloric acid, cool to 4°C.

MHT 14 days.

QUALITY CONTROL Analyze a reagent blank, matrix spike, and matrix spike duplicate/duplicate for each analytical batch (up to 20 samples). Demonstrate the purity of glassware and reagents by analyzing a reagent water method blank. Internal, surrogate, and five concentration level calibration standards are used.

REFERENCE Method 8015, SW-846, 3rd ed., Nov.1986.

Ethion   **EPA Method 8270**
CAS #563-12-2

TITLE Semivolatile Organic Compounds by GC/MS

MATRIX This method is used to determine the concentration of semivolatile organic compounds in extracts prepared from all types of solid waste matrices, soils, and groundwater. Although surface waters are not specifically mentioned, this method should be applicable to water samples from rivers, lakes, etc.

METHOD SUMMARY This method covers 259 semivolatile organic compounds. In very limited applications direct injection of the sample into the GC/MS system may be appropriate, but this results in very high detection limits (approximately 10,000 µg/L). Typically, a 1-L liquid sample, containing surrogate, and matrix spiking standards, is extracted in a continuous extractor first under acid conditions and then under basic conditions. Typically 30 g of a solid sample, containing surrogate, and matrix spiking standards, is extracted ultrasonically. After concentrating the extract to 1 mL it is spiked with 10 µL of an internal standard solution just prior to analysis by GC/MS. The volume injected should contain about 100 ng of base/neutral and 200 ng of acid surrogates (for a 1-µL injection). Analysis is performed by GC/MS using a capillary GC column.

INTERFERENCES Raw GC/MS data from all blanks, samples, and spikes must be evaluated for interferences. Contamination by carryover can occur whenever high-concentration and low-concentration samples are sequentially analyzed. To reduce carryover, the sample syringe must be rinsed out between samples with solvent. Whenever an unusually concentrated sample is encountered, it should be followed by the analysis of blank solvent to check for cross-contamination.

INSTRUMENTATION A GC/MS and a data system are required. The GC column used is a 30 m × 0.25 mm I.D. (or 0.32 mm I.D.) 1um film thickness silicone-coated fused silica capillary column. A continuous liquid-liquid extractor equipped with Teflon® or glass connection joints and stopcocks requiring no lubrication, a K-D concentrating apparatus, water bath, and an ultrasonic disrupter with a minimum power of 300 W and with pulsing capability are also required.

PRECISION & ACCURACY The estimated quantitation limit (EQL) of Method 8270B for determining an individual compound is approximately 1 mg/kg (wet weight) for soil or sediment samples, 1–200 mg/kg for wastes (dependent on matrix and method of preparation), and 10 µg/L for groundwater samples. EQLs will be proportionately higher for sample extracts that require dilution to avoid saturation of the detector.

The EQL(b) for groundwater in µg/L is 10.
The EQL (a, b) for low concentrations in soil and sediment in µg/kg is not determined.
Accuracy as µg/L is not listed.
Overall precision in µg/L is not listed.

(a)   *EQLs listed for soil/sediment are based on wet weight. Normally data is reported in a dry-weight basis; therefore, EQLs will be higher based on the % dry weight of each sample. This calculation is based on a 30-g sample and gel permeation chromatography cleanup.*
(b)   *Sample EQLs are highly matrix-dependent. The EQLs are provided for guidance and may not always be achievable.*
C =   *True value for concentration, in µg/L.*
X =   *Average recovery found for measurements of samples containing a concentration of C, in µg/L.*

ESTIMATED QUANTITATION LIMIT

Other Matrices	Factor (a)
High-concentration soil and sludges by sonicator	7.5
Non-water miscible waste	75

(a) EQL for other matrices = [EQL for low soil/sediment] × [Factor]. This estimated EQL is similar to an EPA "Practical Quantitation Limit."

SAMPLING METHOD

Liquid samples — Use a 1 or 2½ gallon amber glass bottle with a screw-top Teflon®-lined cover that has been prewashed with detergent and rinsed with distilled water and methanol (or isopropanol).

Soils, sediments, or sludges — Use an 8-oz. widemouth glass with a screw-top Teflon®-lined cover that has been prewashed

with detergent and rinsed with distilled water and methanol (or isopropanol).

SAMPLE PRESERVATION
Liquid samples — If residual chlorine is present, add 3 mL of 10% sodium thiosulfate per gallon, cool to 4°C and store in a solvent-free refrigerator until analysis; if chlorine is not present, then eliminate the sodium thiosulfate addition.

Soils, sediments, or sludges — Cool samples to 4°C and store in a solvent-free refrigerator.

MHT Liquid samples must be extracted within 7 days and the extracts analyzed within 40 days. Soils, sediments, or sludges may be stored for a maximum of 14 days and the extracts analyzed within 40 days.

SAMPLE PREPARATION
Liquid samples — Transfer 1 L quantitatively to a continuous extractor. If high concentrations are anticipated, a smaller volume may be used and then diluted with organic-free reagent water to 1 L. Adjust pH, if necessary, to pH <2 using 1:1 (V/V) sulfuric acid. Pipette 1.0 mL of a surrogate standard spiking solution into each sample. For the sample in each analytical batch selected for spiking, add 1.0 mL of a matrix spiking standard. For base/neutral acid analysis, the amount of the surrogates and matrix spiking compounds added to the sample should result in a final concentration of 100 ng/µL of each analyte in the extract to be analyzed (assuming a 1-µL injection). Extract with methylene chloride for 18–24 h. Next, adjust the pH of the aqueous phase to pH >11 using 10 N sodium hydroxide and extract it with methylene chloride again for 18–24 h. Dry the extract through a column containing anhydrous sodium sulfate and concentrate it to 1 mL using a K-D concentrator.

Soils, sediments, or sludges — Use 30 g of sample. Nonporous or wet samples (gummy or clay type) that do not have a free-flowing sandy texture must be mixed with anhydrous sodium sulfate until the sample is free flowing. Add 1 mL of surrogate standards to all samples, spikes, standards, and blanks. For the sample in each analytical batch selected for spiking, add 1.0 mL of a matrix spiking standard. For base/neutral acid analysis, the amount added of the surrogates and matrix spiking compounds should result in a final concentration of 100 ng/µL of each base/neutral analyte and 200 ng/µL of each acid analyte in the extract to be analyzed (assuming a 1-µL injection). Immediately add a 100-mL mixture of 1:1 methylene chloride:acetone and extract the sample ultrasonically for 3 min and then decant or filter the extracts. Repeat the extraction two or more times. Dry the extract using a column with anhydrous sodium sulfate and concentrate it to 1 mL in a K-D concentrator.

QUALITY CONTROL A methylene chloride solution containing 50 ng/µL of decafluorotriphenylphosphine (DFTPP) is used for tuning the GC/MS system each 12-h shift. A system performance check also must be made during every 12-h shift. A standard containing 50 ng/µL each of 4,4'-DDT, pentachlorophenol, and benzidine is required to verify injection port inertness and GC column performance. A calibration standard at mid-concentration, containing each compound of interest, including all required surrogates, must be performed every 12 h during analysis. After the system performance check is met, calibration check compounds (CCCs) are used to check the validity of the initial calibration.

The internal standard responses and retention times in the calibration check standard must be evaluated immediately after or during data acquisition. If the retention time for any internal standard changes by more than 30 seconds from the last check calibration (12 h), the chromatographic system must be inspected for malfunctions and corrections must be made, as required. If the electron ionization current plot (EICP) area for any of the internal standards changes by a factor of two from the last daily calibration standard check, the mass spectrometer must be inspected for malfunctions and corrections must be made, as appropriate.

Demonstrate, through the analysis of a reagent water blank, that interferences from the analytical system, glassware, and reagents are under control. The blank samples should be carried through all stages of the sample preparation and measurement steps. For each analytical batch (up to 20 samples), a reagent blank, matrix spike, and matrix spike duplicate/duplicate must be analyzed (the frequency of the spikes may be different for different monitoring programs). The blank and spiked samples must be carried through all stages of the sample preparation and measurement steps. A QC reference sample concentrate containing each analyte at a concentration of 100 mg/L in methanol is required.

REFERENCE Test Methods for Evaluating Solid Waste (SW-846). U.S. EPA 1983, Method 8270B, Rev. 2, Nov. 1990. Office of Solid Waste, Washington, DC.

Ethoprop EPA Method 507
CAS #13194-48-4

TITLE Determination of Nitrogen and Phosphorus-Containing Pesticides in Water by GC/NPD

MATRIX This method is applicable to the determination of certain nitrogen and phosphorus-containing pesticides in finished drinking water and groundwater.

METHOD SUMMARY Method 507 covers 46 nitrogen- and phosphorus-containing pesticides. A 1-L sample is fortified with a surrogate standard, salted, buffered, extracted with methylene chloride, and concentrated; then the solvent is exchanged with methyl tert-butyl ether (MTBE) and concentrated again, and a 2-µL aliquot of a sample extract is injected into a GC system equipped with a selective nitrogen-phosphorus detector and a capillary column for analysis.

INTERFERENCES Method interferences may be caused by contaminants in solvents, reagents, glassware, and other sample processing apparatus. Interfering contamination may occur when a sample containing low concentrations of analytes is analyzed immediately following a sample containing relatively high concentrations. One or more injections of MTBE should be made following the analysis of a sample with high concentrations of analytes to check for analyte carryover. Matrix interferences may be caused by contaminants that are coextracted from the sample. The extent of matrix interferences will vary considerably from source to source, depending upon the water sampled.

INSTRUMENTATION A gas chromatograph system (GC) equipped with a nitrogen-phosphorus detector (NPD) is needed.

Column 1: 30 m × 0.25 mm I.D. DB-5 bonded fused silica column, 0.25 μm film thickness, or equivalent.
Column 2: 30 m × 0.25 mm I.D. DB-1701 bonded fused silica column, 0.25 μm film thickness, or equivalent.

PRECISION & ACCURACY This method has been validated in a single lab and estimated detection limits (EDLs) have been determined for each analyte. Observed detection limits may vary among waters, depending upon the nature of the interferences in the sample matrix and the specific instrumentation used. Analytes that are not separated chromatographically cannot be individually identified and measured unless an alternative technique for identification and quantification exist.

The estimated detection limit (in μg/L) was 0.19. The EDL is defined as either method detection limit or a level of compound in a sample yielding a peak in the final extract with signal-to-noise ratio of approximately 5, whichever value is higher.

The concentration used for these measurements (in μg/L) was 1.9.
The accuracy (as % recovery) was 103.
The precision (% RSD) was 5.

SAMPLING METHOD Grab samples are collected in 1-L glass sample bottles (prewashed with detergent and hot tap water, rinsed with reagent water, and dried in an oven at 400°C for 1 h) with screw caps lined with PTFE-fluorocarbon.

SAMPLE PRESERVATION Add mercuric chloride to the sample bottle in amounts to produce a concentration of 10 mg/L. If residual chlorine is present, add 80 mg of sodium thiosulfate/L of sample to the sample bottle prior to collection. After collection, seal bottle and shake vigorously for 1 min, then cool the sample to 4°C immediately and store it at 4°C in the dark until extraction.

MHT Maximum holding time of the samples, and in some cases the extracts, is 14 days.

SAMPLE PREPARATION Fortify the sample with 50 μL of the surrogate standard solution, adjust to pH 7 with phosphate buffer, add 100 g NaCl to the sample, and seal and shake to dissolve the salt; then extract with methylene chloride in a separatory funnel or in a mechanical tumbler bottle. Dry the extract by pouring it through a solvent-rinsed drying column containing about 10 cm of anhydrous sodium sulfate. Collect the extract in a Kuderna-Danish (K-D) concentrator and rinse the column with 20–30 mL methylene chloride. Concentrate the extract to about 2 mL and rinse the flask and its lower joint into the concentrator tube with 1 to 2 mL of methyl t-butyl ether (MTBE). Add 5–10 mL of MTBE and concentrate the extract twice (adding more MTBE) to a final volume of 5.0 mL and store it at 4°C until analysis.

Note: If methylene chloride is not completely removed from the final extract, it may cause detector problems.

QUALITY CONTROL Minimum quality control requirements are initial demonstration of lab capability, determination of surrogate compound recoveries in each sample and blank, monitoring internal standard peak area or height in each sample and blank, analysis of lab reagent blanks, lab fortified samples, lab fortified blanks, and other QC samples. A lab reagent blank is analyzed to demonstrate that all glassware and reagent interferences are under control.

Initial demonstration of capability is fulfilled by analyzing four fortified reagent water samples with the recovery value for each analyte falling within the acceptable range (±30% average recovery). Surrogate recoveries from samples or method blanks must be 70–130%. The internal standard response for any sample chromatogram should not deviate from the daily calibration check standard's internal standard response by more than 30% or lab fortified blanks and sample matrices are used to assess lab performance and analyte recovery, respectively.

If the response for the target analyte peak exceeds the working range of the system, dilute the extract and reanalyze. Alternative techniques such as an alternate detector or second chromatography column should be used to confirm peak identification when sample components are not resolved adequately.

EPA CONTACT & HOTLINE For technical questions contact Dr. Baldev Bathija, U.S. EPA, Office of Ground Water and Drinking Water (WH-550D), 401 M St. SW, Washington, DC 20460. Tel. (202) 260-3040. For further information the EPA Safe Drinking Water Hotline may be called at: (800) 426-4791.

REFERENCE Methods for the Determination of Organic Compounds in Drinking Water, EPA/600/4-88/039 (revised July 1991). U.S. EPA Environmental Monitoring Systems Laboratory, Cincinnati, OH, 45268, U.S.A. Available from the National Technical Information Service (NTIS), 5285 Port Royal Road, Springfield, VA 22161; Tel. 800-553-6847. NTIS Order Number is PB91-231480.

Ethoprop — EPA Method 8141
CAS #13194-48-4

TITLE Organophosphorus Compounds by Gas Chromatography: Capillary Column Technique

MATRIX This method covers aqueous and solid matrices. This includes a wide variety such as drinking water, groundwater, industrial wastewaters, surface waters, soils, solids, and sediments.

METHOD SUMMARY This is a GC method used to determine the concentration of 28 organophosphorus pesticides.

The use of Gel Permeation Cleanup (EPA Method 3640) for sample cleanup has been demonstrated to yield recoveries of less than 85% for many method analytes and is therefore not recommended for use with this method.

This method provides GC conditions for the detection of ppb concentrations of organophosphorus compounds. Prior to the use of this method, appropriate sample preparation techniques must be used. Water samples are extracted at a neutral pH with methylene chloride as a solvent by using a separatory funnel (EPA Method 3510) or a continuous liquid-liquid extractor (EPA Method 3520). Soxhlet extraction (EPA Method 3540) or ultrasonic extraction (EPA Method 3550) using methylene chloride/acetone (1:1) are used for solid samples. Both neat and diluted organic liquids (EPA Method 3580) may be analyzed by direct injection. Spiked samples are used to verify the applicability of the chosen extraction technique to each new sample type. A GC with a flame photometric (FPD) or nitrogen-phosphorus detector (NPD) is used for this multiresidue procedure.

INTERFERENCES The use of Florisil cleanup materials (EPA Method 3620) for some of the compounds in this method has been demonstrated to yield recoveries less than 85% and is therefore not recommended for all compounds. Use of phosphorus or halogen specific detectors, however, often obviates the necessity for cleanup for relatively clean sample matrices. If particular circumstances demand the use of an alternative cleanup procedure, the analyst must determine the elution profile and demonstrate that the recovery of each analyte is no less than 85%.

Use of a flame photometric detector (FPD) in the phosphorus mode will minimize interferences from materials that do not contain phosphorus. Elemental sulfur, however, may interfere with the determination of certain organophosphorus compounds by flame photometric gas chromatography. Sulfur cleanup using EPA Method 3660 may alleviate this interference. A nitrogen phosphorus detector (NPD) is also recommended.

A few analytes coelute on certain columns. Therefore, select a second column for confirmation where coelution of the analytes of interest does not occur.

Method interferences may be caused by contaminants in solvents, reagents, glassware, and other sample processing hardware that lead to discrete artifacts or elevated baselines in gas chromatograms. All these materials must be routinely demonstrated to be free from interferences under the conditions of the analysis by analyzing reagent blanks.

INSTRUMENTATION A GC with a NPD or a FPD will be needed. A data system or integrator is recommended for measuring peak areas and/or peak heights. A Kuderna-Danish (K-D) apparatus will be needed for extract concentration.

Column 1: 15 m × 0.53 mm megabore capillary column, 1.0 μm film thickness, DB-210.
Column 2: 15 m × 0.53 mm megabore capillary column, 1.5 μm film thickness, SPB-608.
Column 3: 15 m × 0.53 mm megabore capillary column, 1.0 μm film thickness, DB-5.

Three megabore capillary columns are included for analysis of organophosphates by this method. Column 1 (DB-210 or equivalent) and Column 2 (SPB-608 or equivalent) are recommended if a large number of organophosphorus analytes are to be determined. If the superior resolution offered by Column 1 and Column 2 is not required, Column 3 (DB-5 or equivalent) may be used. For megabore capillary columns, automatic injections of 1 μL are recommended.

PRECISION & ACCURACY The MDL actually achieved in a given analysis will vary, as it is dependent on instrument sensitivity and matrix effects. Single operator accuracy and precision studies have been conducted with spiked water and soil samples.

MULTIPLICATION FACTORS FOR OTHER MATRICES (a)

Matrix	Factor (b)
Groundwater (EPA Method 3510 or EPA Method 3520)	10
Low-concentration soil by Soxhlet and no cleanup	10 (c)
Low-concentration soil by ultrasonic extraction with GPC cleanup	6.7 (c)
High-concentration soil and sludges by ultrasonic extraction	500 (c)
Non-water miscible waste (EPA Method 3580)	1000 (c)

(a) Sample EQLs are highly matrix-dependent. The EQLs listed here are provided for guidance and may not always be achievable.
(b) EQL = [Method detection limit] × [Factor]. For non-aqueous samples the factor is on a wet-weight basis.
(c) Multiply this factory times the soil MDL.

The MDL (in μg/L) when reagent water was extracted using a separatory funnel was 0.20.
The MDL (in μg/kg) when soil was extracted using Soxhlet extraction (EPA Method 3540) was 10.0.
Accuracy (as % recovery) with separatory funnel extraction ranged from 82 (with low spikes) to 80 (with high spikes).
Accuracy (as % recovery) with continuous liquid-liquid extraction ranged from 39 (with low spikes) to 83 (with high spikes).
Accuracy (as % recovery) with Soxhlet extraction of soils ranged from 65 (with low spikes to 75 (with high spikes).
Accuracy (as % recovery) with ultrasonic extraction of soils ranged from 19 (with low spikes) to 35 (with high spikes).

SAMPLE COLLECTION, PRESERVATION & HANDLING
Containers used to collect samples for the determination of semivolatile organic compounds should be soap and water washed followed by methanol (or isopropanol) rinsing. The sample containers should be of glass or Teflon® and have screw-top covers with Teflon® liners.

No preservation is used with concentrated waste samples. With liquid samples containing no residual chlorine and with soil, sediment, and sludge samples, immediately cooling to 4°C is the only preservation used. When residual chlorine is present then 3 mL of 10% aqueous sodium sulfate is added for each gallon of sample collected, followed by cooling to 4°C.

Liquid samples must be extracted within 7 days and their extracts analyzed within 40 days. Concentrated waste, soil, sediment, and sludge samples must be extracted within 14 days and their extracts analyzed within 40 days.

SAMPLE PREPARATION In general, water samples are extracted at a neutral pH with methylene chloride, using either EPA Method 3510 or EPA Method 3520. Solid samples are

extracted using either EPA Method 3540 or EPA Method 3550 with methylene chloride/acetone (1:1) as the extraction solvent.

Prior to GC analysis, the extraction solvent may be exchanged to hexane. Single lab data indicates that samples should not be transferred with 100% hexane during sample workup as the more water soluble organophosphorus compounds may be lost.

If cleanup is performed on the samples, the analyst should analyze the samples by GC. This will confirm elution patterns and the absence of interferences from the reagents. If peak detection and identification is prevented by the presence of interferences, further cleanup is required.

QUALITY CONTROL The analyst should monitor the performance of the extraction, cleanup (when used), and analytical system and the effectiveness of the method in dealing with each sample matrix by spiking each sample, standard, and blank with one or two surrogates (e.g., organophosphorus compounds not expected to be present in the sample). Deuterated analogs of analytes should not be used as surrogates for gas chromatographic analysis due to coelution problems.

A minimum of five concentrations for each analyte of interest should be prepared through dilution of the stock standards with isooctane. One of the concentrations should be at a concentration near, but above, the MDL.

Include a mid-level check standard after each group of 10 samples in the analysis sequence. GC/MS techniques should be judiciously employed to support qualitative identifications made with this method. Follow the GC/MS operating requirements specified in EPA Method 8270.

When available, chemical ionization mass spectra may be employed to aid in the qualitative identification process. To confirm an identification of a compound, the background-corrected mass spectrum of the compound must be obtained from the sample extract and must be compared with a mass spectrum from a stock or calibration standard analyzed under the same chromatographic conditions. The molecular ion and all other ions present above 20% relative abundance in the mass spectrum of the standard must be present in the mass spectrum of the sample with agreement to ± 20%. The retention time of the compound in the sample must be within six seconds of the retention time for the same compound in the standard solution.

Should the MS procedure fail to provide satisfactory results, additional steps may be taken before reanalysis. These steps may include the use of alternate packed or capillary GC columns or additional sample cleanup.

REFERENCE Test Methods for Evaluating Solid Waste, Physical/Chemical Methods, SW-846, 3rd Edition, U.S. EPA, Office of Solid Waste, Washington, DC, EPA Method 8141 July 1992.

Ethoprop   **EPA Method 8140**
CAS #13194-48-4

TITLE Organophosphorus Pesticides

MATRIX Groundwater, soils, sludges, water miscible liquid wastes, and non-water miscible wastes.

APPLICATION This method is used for the analysis of 21 organophosphorus pesticides. Samples are extracted, concentrated, and analyzed using direct injection of both neat and diluted organic liquid into a gas chromatograph (GC).

INTERFERENCES Solvents, reagents, and glassware may introduce artifacts. Other interferences may come from coextracted compounds from samples. The use of Florisil cleanup materials may produce low recoveries. Elemental sulfur may interfere with some compounds when using a flame photometric detector. Sulfur cleanup (Method 3660) may alleviate sulfur interference.

INSTRUMENTATION GC capable of on-column injections and a flame photometric detector (FPD) or a thermionic detector. Column 1: 1.8 m by 2 mm with 5% SP-2401 on Supelcoport. Column 2: 1.8 m by 2 mm with 3% SP-2401 on Supelcoport. Column 3: 50 cm by ⅛ IN EFLON® with 15% SE-54 on Gas Chrom Q. The preferred column is Column Number 2.

RANGE 1.0–51.5 µg/L.

MDL 0.25 µg/L (in reagent water).

PQL FACTORS FOR MULTIPLYING × FID MDL VALUE

Matrix	Multiplication Factor
Groundwater	10
Low-level soil by sonication with GPC cleanup	670
High-level soil and sludge by sonication	10,000
Non-water miscible waste	100,000

PRECISION 4.1% (single operator standard deviation)

ACCURACY 100.5% (single operator average recovery)

SAMPLING METHOD Use 8-oz. widemouth glass bottles with Teflon®-lined caps for concentrated waste samples, soils, sediments, and sludges. Use 1 or 2½ gallon amber glass bottles with Teflon®-lined caps for liquid (water) samples.

STABILITY Cool soil, sediment, sludge, and liquid samples to 4°C. If residual chlorine is present in liquid samples add 3 mL of 10% sodium thiosulfate per gallon of sample and cool to 4°C.

MHT 14 days for concentrated waste, soil, sediment, or sludge; 7 days for liquid samples; all extracts must be analyzed within 40 days.

QUALITY CONTROL A quality control check sample concentrate containing this compound in acetone at a concentration 1,000 times more concentrated than the selected spike concentration is required. The QC check sample concentrate may be prepared from pure standard materials or purchased as certified solutions. Use appropriate trip, matrix, control site, method, reagent, and solvent blanks. Internal, surrogate, and five concentration level calibration standards are used.

REFERENCE Method 8140, SW-846, 3rd ed., Sept. 1986.

Ethyl carbamate **EPA Method 8270**
CAS #51-79-6

TITLE Semivolatile Organic Compounds by GC/MS

MATRIX This method is used to determine the concentration of semivolatile organic compounds in extracts prepared from all types of solid waste matrices, soils, and groundwater. Although surface waters are not specifically mentioned, this method should be applicable to water samples from rivers, lakes, etc.

METHOD SUMMARY This method covers 259 semivolatile organic compounds. In very limited applications direct injection of the sample into the GC/MS system may be appropriate, but this results in very high detection limits (approximately 10,000 µg/L). Typically, a 1-L liquid sample, containing surrogate, and matrix spiking standards, is extracted in a continuous extractor first under acid conditions and then under basic conditions. Typically 30 g of a solid sample, containing surrogate, and matrix spiking standards, is extracted ultrasonically. After concentrating the extract to 1 mL it is spiked with 10 µL of an internal standard solution just prior to analysis by GC/MS. The volume injected should contain about 100 ng of base/neutral and 200 ng of acid surrogates (for a 1-µL injection). Analysis is performed by GC/MS using a capillary GC column.

INTERFERENCES Raw GC/MS data from all blanks, samples, and spikes must be evaluated for interferences. Contamination by carryover can occur whenever high-concentration and low-concentration samples are sequentially analyzed. To reduce carryover, the sample syringe must be rinsed out between samples with solvent. Whenever an unusually concentrated sample is encountered, it should be followed by the analysis of blank solvent to check for cross-contamination.

INSTRUMENTATION A GC/MS and a data system are required. The GC column used is a 30 m × 0.25 mm I.D. (or 0.32 mm I.D.) 1um film thickness silicone-coated fused silica capillary column. A continuous liquid-liquid extractor equipped with Teflon® or glass connection joints and stopcocks requiring no lubrication, a K-D concentrating apparatus, water bath, and an ultrasonic disrupter with a minimum power of 300 W and with pulsing capability are also required.

PRECISION & ACCURACY The estimated quantitation limit (EQL) of Method 8270B for determining an individual compound is approximately 1 mg/kg (wet weight) for soil or sediment samples, 1–200 mg/kg for wastes (dependent on matrix and method of preparation), and 10 µg/L for groundwater samples. EQLs will be proportionately higher for sample extracts that require dilution to avoid saturation of the detector.

The EQL(b) for groundwater in µg/L is 50.
The EQL (a, b) for low concentrations in soil and sediment in µg/kg is not determined.
Accuracy as µg/L is not listed.
Overall precision in µg/L is not listed.

(a) *EQLs listed for soil/sediment are based on wet weight. Normally data is reported in a dry-weight basis; therefore, EQLs will be higher based on the % dry weight of each sample. This calculation is based on a 30-g sample and gel permeation chromatography cleanup.*

(b) *Sample EQLs are highly matrix-dependent. The EQLs are provided for guidance and may not always be achievable.*

C = *True value for concentration, in µg/L.*
X = *Average recovery found for measurements of samples containing a concentration of C, in µg/L.*

ESTIMATED QUANTITATION LIMIT

Other Matrices	Factor (a)
High-concentration soil and sludges by sonicator	7.5
Non-water miscible waste	75

(a) *EQL for other matrices = [EQL for low soil/sediment] × [Factor]. This estimated EQL is similar to an EPA "Practical Quantitation Limit."*

SAMPLING METHOD

Liquid samples — Use a 1 or 2½ gallon amber glass bottle with a screw-top Teflon®-lined cover that has been prewashed with detergent and rinsed with distilled water and methanol (or isopropanol).

Soils, sediments, or sludges — Use an 8-oz. widemouth glass with a screw-top Teflon®-lined cover that has been prewashed with detergent and rinsed with distilled water and methanol (or isopropanol).

SAMPLE PRESERVATION

Liquid samples — If residual chlorine is present, add 3 mL of 10% sodium thiosulfate per gallon, cool to 4°C and store in a solvent-free refrigerator until analysis; if chlorine is not present, then eliminate the sodium thiosulfate addition.

Soils, sediments, or sludges — Cool samples to 4°C and store in a solvent-free refrigerator.

MHT Liquid samples must be extracted within 7 days and the extracts analyzed within 40 days. Soils, sediments, or sludges may be stored for a maximum of 14 days and the extracts analyzed within 40 days.

SAMPLE PREPARATION

Liquid samples — Transfer 1 L quantitatively to a continuous extractor. If high concentrations are anticipated, a smaller volume may be used and then diluted with organic-free reagent water to 1 L. Adjust pH, if necessary, to pH <2 using 1:1 (V/V) sulfuric acid. Pipette 1.0 mL of a surrogate standard spiking solution into each sample. For the sample in each analytical batch selected for spiking, add 1.0 mL of a matrix spiking standard. For base/neutral acid analysis, the amount of the surrogates and matrix spiking compounds added to the sample should result in a final concentration of 100 ng/µL of each analyte in the extract to be analyzed (assuming a 1-µL injection). Extract with methylene chloride for 18–24 h. Next, adjust the pH of the aqueous phase to pH >11 using 10 N sodium hydroxide and extract it with methylene chloride again for 18–24 h. Dry the extract through a column containing anhydrous sodium sulfate and concentrate it to 1 mL using a K-D concentrator.

Soils, sediments, or sludges — Use 30 g of sample. Nonporous or wet samples (gummy or clay type) that do not have a free-flowing sandy texture must be mixed with anhydrous sodium sulfate until the sample is free flowing. Add 1 mL of surrogate standards to all samples, spikes, standards, and blanks. For the sample in each analytical batch selected for spiking, add 1.0 mL of a matrix spiking standard. For base/neutral acid analysis, the amount added of the surrogates and matrix spiking compounds should result in a final concentration of 100 ng/µL of each base/neutral analyte and 200 ng/µL of each acid analyte in the extract to be analyzed (assuming a 1-µL injection). Immediately add a 100-mL mixture of 1:1 methylene chloride:acetone and extract the sample ultrasonically for 3 min and then decant or filter the extracts. Repeat the extraction two or more times. Dry the extract using a column with anhydrous sodium sulfate and concentrate it to 1 mL in a K-D concentrator.

QUALITY CONTROL A methylene chloride solution containing 50 ng/µL of decafluorotriphenylphosphine (DFTPP) is used for tuning the GC/MS system each 12-h shift. A system performance check also must be made during every 12-h shift. A standard containing 50 ng/µL each of 4,4'-DDT, pentachlorophenol, and benzidine is required to verify injection port inertness and GC column performance. A calibration standard at mid-concentration, containing each compound of interest, including all required surrogates, must be performed every 12 h during analysis. After the system performance check is met, calibration check compounds (CCCs) are used to check the validity of the initial calibration.

The internal standard responses and retention times in the calibration check standard must be evaluated immediately after or during data acquisition. If the retention time for any internal standard changes by more than 30 seconds from the last check calibration (12 h), the chromatographic system must be inspected for malfunctions and corrections must be made, as required. If the electron ionization current plot (EICP) area for any of the internal standards changes by a factor of two from the last daily calibration standard check, the mass spectrometer must be inspected for malfunctions and corrections must be made, as appropriate.

Demonstrate, through the analysis of a reagent water blank, that interferences from the analytical system, glassware, and reagents are under control. The blank samples should be carried through all stages of the sample preparation and measurement steps. For each analytical batch (up to 20 samples), a reagent blank, matrix spike, and matrix spike duplicate/duplicate must be analyzed (the frequency of the spikes may be different for different monitoring programs). The blank and spiked samples must be carried through all stages of the sample preparation and measurement steps. A QC reference sample concentrate containing each analyte at a concentration of 100 mg/L in methanol is required.

REFERENCE Test Methods for Evaluating Solid Waste (SW-846). U.S. EPA 1983, Method 8270B, Rev. 2, Nov. 1990. Office of Solid Waste, Washington, DC.

Ethyl cyanide
CAS #107-12-0

EPA Method 1624

TITLE Volatile Organic Compounds by Isotope Dilution GC/MS

MATRIX Compounds may be determined in waters, soils, and municipal sludges by this method.

METHOD SUMMARY This method is used to determine 58 volatile toxic organic pollutants associated with the CWA (as amended 1987); the RCRA (as amended 1986); the CERCLA (as amended 1986); and other compounds amenable to purge-and-trap gas chromatography-mass spectrometry (GC/MS).

If the solids content is less than 1%, stable isotopically-labeled analogs of the compounds of interest are added to a 5-mL sample and the sample is purged with an inert gas at 20–25°C in a chamber designed for soil or water samples. If the solids content is greater than 1%, 5 mL of reagent water and the labeled compounds are added to a 5-g aliquot of sample and the mixture is purged at 40°C. Compounds that will not purge at 20–25°C or at 40°C are purged at 78–85°C. In the purging process, the volatile compounds are transferred from the aqueous phase into the gaseous phase where they are passed into a sorbent column, and trapped. After purging is completed, the trap is backflushed and heated rapidly to desorb the compounds into a GC. The compounds are separated by the GC and detected by a MS. The labeled compounds serve to correct the variability of the analytical technique.

INTERFERENCES Impurities in the purge gas, organic compounds outgassing from the plumbing upstream of the trap, and solvent vapors in the lab account for most problems. Samples can be contaminated by diffusion of volatile organic compounds (particularly methylene chloride) through the bottle seal during shipment and storage. Contamination by carryover can occur when high-level and low-level samples are analyzed sequentially. When an unusually concentrated sample is encountered, follow it by analysis of a reagent water blank to check for carryover.

INSTRUMENTATION Major equipment includes a GC with linear temperature programming and a glass jet separator as the MS interface, a MS with 70 eV electron impact ionization, and a data system to collect and record response factors.

Column: 2.8 m × 2 mm I.D. glass, packed with 1% SP-1000 on Carbopak B, 60/80 mesh, or equivalent.

PRECISION & ACCURACY The detection limits of the method are usually dependent on the level of interferences rather than instrumental limitations. The method detection limits were determined in digested sludge (low solids) and in filter cake or compost (high solids).

The MDL (in µg/kg) for low solids is not listed and for high solids is not listed.

Labeled and native compound precision (in µg/L) as standard deviation was not listed.

Labeled and native compound accuracy (in µg/L) as average recovery was not listed.

Acceptance criteria are at 20 µg/L for this compound.

SAMPLE COLLECTION, PRESERVATION & HANDLING
Grab samples are collected in glass containers having a total volume greater than 20 mL. Fill and seal each bottle so that no air bubbles are entrapped. Samples are maintained at 0 to 4°C from the time of collection until analysis. If an aqueous sample contains residual chlorine, add sodium thiosulfate preservative (10 mg/40 mL) to the empty sample bottles just prior to shipment to the sample site. All samples must be analyzed within 14 days of collection.

SAMPLE PREPARATION Samples containing less than 1% solids are analyzed directly as aqueous samples. Samples containing 1% solids or greater are analyzed as solid samples utilizing one of two methods, depending on the levels of pollutants, in the sample. Samples containing 1% solids or greater, and low to moderate levels of pollutants are analyzed by purging a known weight of sample added to 5 mL of reagent water. Samples containing 1% solids or greater, and high levels of pollutants, are extracted with methanol, and an aliquot of the methanol extract is added to reagent water and purged.

QUALITY CONTROL A field blank prepared from reagent water and carried through the sampling and handling protocol may serve as a check on contamination from shipment and storage.

The analyst is permitted to modify this method to improve separations or lower the costs of measurements, provided all performance specifications are met. Analyses of blanks are required. When results of spikes indicate atypical method performance for samples, the samples are diluted to bring method performance within acceptable limits. Analyze two sets of four 5-mL aliquots (8 aliquots total) of the aqueous performance standard. Spike all samples with labeled compounds to assess method performance on the sample matrix. Compute the percent recovery of the labeled compounds using the internal standard method. Compare the percent recovery for each compound with the corresponding labeled compound recovery. Reagent water blanks are analyzed to demonstrate freedom from carryover contamination. Field replicates may be collected to determine the precision of the sampling technique, and spiked samples may be required to determine the accuracy of the analysis when the internal method is used.

REFERENCE Volatile Organic Compounds by Isotope Dilution GC/MS. Office of Water Regulation and Standards, U.S. EPA Industrial Technology Division, Washington, DC, EPA Method 1624, Rev. C, June 1989 (contact W.A. Telliard, U.S. EPA, Office of Water Regulations and Standards, 401 M St., SW, Washington, DC, 20460. Phone: 202-382-7131).

Ethyl methacrylate   **EPA Method 1624**
CAS #97-63-2

TITLE Volatile Organic Compounds by Isotope Dilution GC/MS

MATRIX Compounds may be determined in waters, soils, and municipal sludges by this method.

METHOD SUMMARY This method is used to determine 58 volatile toxic organic pollutants associated with the CWA (as amended 1987); the RCRA (as amended 1986); the CERCLA (as amended 1986); and other compounds amenable to purge-and-trap gas chromatography-mass spectrometry (GC/MS).

If the solids content is less than 1%, stable isotopically-labeled analogs of the compounds of interest are added to a 5-mL sample and the sample is purged with an inert gas at 20–25°C in a chamber designed for soil or water samples. If the solids content is greater than 1%, 5 mL of reagent water and the labeled compounds are added to a 5-g aliquot of sample and the mixture is purged at 40°C. Compounds that will not purge at 20–25°C or at 40°C are purged at 78–85°C. In the purging process, the volatile compounds are transferred from the aqueous phase into the gaseous phase where they are passed into a sorbent column, and trapped. After purging is completed, the trap is backflushed and heated rapidly to desorb the compounds into a GC. The compounds are separated by the GC and detected by a MS. The labeled compounds serve to correct the variability of the analytical technique.

INTERFERENCES Impurities in the purge gas, organic compounds outgassing from the plumbing upstream of the trap, and solvent vapors in the lab account for most problems. Samples can be contaminated by diffusion of volatile organic compounds (particularly methylene chloride) through the bottle seal during shipment and storage. Contamination by carryover can occur when high-level and low-level samples are analyzed sequentially. When an unusually concentrated sample is encountered, follow it by analysis of a reagent water blank to check for carryover.

INSTRUMENTATION Major equipment includes a GC with linear temperature programming and a glass jet separator as the MS interface, a MS with 70 eV electron impact ionization, and a data system to collect and record response factors.

Column: 2.8 m × 2 mm I.D. glass, packed with 1% SP-1000 on Carbopak B, 60/80 mesh, or equivalent.

PRECISION & ACCURACY The detection limits of the method are usually dependent on the level of interferences rather than instrumental limitations. The method detection limits were determined in digested sludge (low solids) and in filter cake or compost (high solids).

The MDL (in µg/kg) for low solids is not listed and for high solids is not listed.
Labeled and native compound precision (in µg/L) as standard deviation was not listed.
Labeled and native compound accuracy (in µg/L) as average recovery was not listed.
Acceptance criteria are at 20 µg/L for this compound.

SAMPLE COLLECTION, PRESERVATION & HANDLING
Grab samples are collected in glass containers having a total volume greater than 20 mL. Fill and seal each bottle so that no air bubbles are entrapped. Samples are maintained at 0 to 4°C from the time of collection until analysis. If an aqueous sample contains residual chlorine, add sodium thiosulfate preservative (10 mg/40 mL) to the empty sample bottles just prior to shipment

to the sample site. All samples must be analyzed within 14 days of collection.

SAMPLE PREPARATION Samples containing less than 1% solids are analyzed directly as aqueous samples. Samples containing 1% solids or greater are analyzed as solid samples utilizing one of two methods, depending on the levels of pollutants, in the sample. Samples containing 1% solids or greater, and low to moderate levels of pollutants are analyzed by purging a known weight of sample added to 5 mL of reagent water. Samples containing 1% solids or greater, and high levels of pollutants, are extracted with methanol, and an aliquot of the methanol extract is added to reagent water and purged.

QUALITY CONTROL A field blank prepared from reagent water and carried through the sampling and handling protocol may serve as a check on contamination from shipment and storage.

The analyst is permitted to modify this method to improve separations or lower the costs of measurements, provided all performance specifications are met. Analyses of blanks are required. When results of spikes indicate atypical method performance for samples, the samples are diluted to bring method performance within acceptable limits. Analyze two sets of four 5-mL aliquots (8 aliquots total) of the aqueous performance standard. Spike all samples with labeled compounds to assess method performance on the sample matrix. Compute the percent recovery of the labeled compounds using the internal standard method. Compare the percent recovery for each compound with the corresponding labeled compound recovery. Reagent water blanks are analyzed to demonstrate freedom from carryover contamination. Field replicates may be collected to determine the precision of the sampling technique, and spiked samples may be required to determine the accuracy of the analysis when the internal method is used.

REFERENCE Volatile Organic Compounds by Isotope Dilution GC/MS. Office of Water Regulation and Standards, U.S. EPA Industrial Technology Division, Washington, DC, EPA Method 1624, Rev. C, June 1989 (contact W.A. Telliard, U.S. EPA, Office of Water Regulations and Standards, 401 M St., SW, Washington, DC, 20460. Phone: 202-382-7131).

Ethyl methacrylate   **EPA Method 8240**
CAS #97-63-2

TITLE Volatile Organics By GC/MS: Packed Column Technique

MATRIX Nearly all types of sample matarices, regardless of water content, can be analyzed using this method. This includes groundwater, aqueous sludges, caustic liquors, acid liquors, waste solvents, oily wastes, mousses, tars, fibrous wastes, polymetric emulsions, filter cakes, spent carbons, spent catalysts, soils, and sediments.

METHOD SUMMARY Method 8240B covers 80 volatile organic compounds that are introduced into a gas chromatograph by the purge-and-trap method or by direct injection (in limited applications). For the purge-and-trap method an inert gas (zero grade nitrogen or helium) is bubbled through a 5-mL solution at ambient temperature. Purged sample components are trapped in a tube of sorbent materials. When purging is complete, the sorbent tube is heated and backflushed with inert gas to desorb the trapped components onto a GC column.

INTERFERENCES Impurities in the purge gas and from organic compounds outgassing from the plumbing ahead of the trap account for many contamination problems. Interferences purged or coextracted from the samples will vary considerably from source to source. Cross-contamination can occur whenever high-level and low-level samples are analyzed sequentially. Whenever an unusually concentrated sample is analyzed, it should be followed by the analysis of organic-free reagent water to check for cross-contamination. Samples also can be contaminated by diffusion of volatile organics (particularly methylene chloride and fluorocarbons) through the septum seal into the sample during shipment and storage. A trip blank can serve as a check on such contamination. The lab where volatile analysis is performed and also the refrigerated storage area should be completely free of solvents.

INSTRUMENTATION A gas chromatograph/mass spectrometry/data system (GC/MS) equipped with a 6 ft × 0.1 in I.D. glass column packed with 1% SP-1000 on Carbopack-B (60/80 mesh) is required. Also needed is a 5-mL purging device, a sorbent trap, and a thermal desorption apparatus.

PRECISION & ACCURACY This method is reported to have been tested by 15 laboratories using organic-free reagent water, drinking water, surface water, and industrial wastewaters (not specified) fortified at six concentrations over the range 5–600 µg/L.

Sample estimated quantitation limits (EQLs) are highly matrix-dependent. The EQLs listed may not always be achievable. EQLs listed for soils or sediments are based on wet weight. Normally, data is reported on a dry-weight basis; therefore, EQLs will be higher, based on the percent dry weight of each sample. Note that EQLs are even more variable than MDLs and that they are highly variable depending on the matrix being analyzed.

EQL in groundwater in µg/L was 5.
EQL in low soil or sediment in µg/kg was 5.
Accuracy (a) in µg/L was not listed.
Precision (b) in µg/L was not listed.

(a) *Average recovery found for measurements of samples containing a concentration of C, in µg/L.*
(b) *Overall precision found for measurements of samples with average recovery X for samples containing a concentration of C in µg/L.*
X = *Average recovery found for measurement of samples containing a concentration of C in µg/L.*

MULTIPLICATION FACTORS FOR OTHER MATRICES

Other Matrices	Factor (a)
Waste miscible liquid waste	50
High-concentration soil and sludge	125
Non-water miscible waste	500

(a) *EQL = [EQL for low soil/sediment] × [Factor]. For non-aqueous samples, the factor is on a wet-weight basis.*

SAMPLING METHOD

Liquid samples — Use a 40-mL glass screw-cap VOA vial with a Teflon®-faced silicone septum that has been prewashed, rinsed with distilled deionized water, and oven dried. However, if residual chlorine is present, collect sample in a 40-oz. soil VOA container which has been pre-preserved with 4 drops of 10% sodium thiosulfate, mix gently, and then transfer the sample to a 40-mL VOA vial. Collect bubble-free samples in duplicate and seal them in separate plastic bags.

Soils or sediments, and sludges — Use an 8-oz. widemouth glass bottle with a Teflon®-faced silicone septum that has been prewashed with detergent, rinsed with distilled deionized water, and oven dried. Tap slightly to eliminate free air space. Collect samples in duplicate and seal them in separate plastic bags.

SAMPLE PRESERVATION

Liquid samples — Add 4 drops of concentrated HCL and immediately cool samples to 4°C and store in a solvent-free refrigerator.

Soils or sediments, and sludges — Cool samples to 4°C and store in a solvent-free refrigerator.

MHT Maximum holding time is 14 days from the date of sample collection.

SAMPLE PREPARATION

Liquid samples — Remove the plunger from a 5-mL syringe and carefully pour the sample into the syringe barrel to just short of overflowing. Replace the syringe plunger and compress the sample. Open the syringe valve and vent any residual air while adjusting the sample volume to 5.0 mL. If there is only one volatile organic analysis (VOA) vial, a second syringe should be filled at this time to protect against possible loss of sample integrity. Add 10 µL of surrogate spiking solution and 10 µL of internal standard spiking solution through the valve bore of the 5-mL syringe, then close the valve. The surrogate and internal standards may be mixed and added as a single spiking solution.

Sediments, soils, and waste samples — All samples of this type should be screened by GC analysis using a headspace method (EPA Method 3810) or the hexadecane extraction and screening method (EPA Method 3820). Use the screening data to determine whether to use the low-concentration method (0.005–1 mg/kg) or the high-concentration method (>1 mg/kg).

Low-concentration method — The low-concentration method is based on purging a heated sediment or soil sample mixed with organic-free reagent water containing the surrogate and internal standards. Analyze all reagent blanks and standards under the same conditions as the samples.

Use a 5-g sample if the expected concentration is <0.1 mg/kg or a 1-g sample for expected concentrations between 0.1 and 1 mg/kg. Mix the contents of the sample container with a narrow metal spatula. Weigh the amount of the sample into a tared purge device. Add the spiked water to the purge device, which contains the weighed amount of sample, and connect the device to the purge-and-trap system.

High-concentration method — This method is based on extracting the sediment or soil with methanol. A waste sample is either extracted or diluted, depending on its solubility in methanol. Wastes that are insoluble in methanol are diluted with reagent tetraglyme or possibly polyethylene glycol (PEG). An aliquot of the extract is added to organic-free reagent water containing surrogate and internal standards. This is purged at ambient temperature. All samples with an expected concentration of >1.0 mg/kg should be analyzed by this method.

Mix the contents of the sample container with a narrow metal spatula. For sediments or soils and solid wastes that are insoluble in methanol, weigh 4 g (wet weight) of sample into a tared 20-mL vial. For waste that is soluble in methanol, tetraglyme, or PEG, weigh 1 g (wet weight) into a tared scintillation vial or culture tube or a 10-mL volumetric flask. Quickly add 9.0 mL of appropriate solvent then add 1.0 mL of a surrogate spiking solution to the vial, cap it, and shake it for 2 min.

METHANOL EXTRACT REQUIRED FOR ANALYSIS OF HIGH-CONCENTRATION SOILS OR SEDIMENTS

Approximate Concentration Range	Volume of Methanol Extract (a)
500–10,000 µg/kg	100 µL
1,000–20,000 µg/kg	50 µL
5,000–100,000 µg/kg	10 µL
25,000–500,000 µg/kg	100 µL of 1/50 dilution (b)

Calculate appropriate dilution factor for concentrations exceeding this table.

(a) The volume of methanol added to 5 mL of water being purged should be kept constant. Therefore, add to the 5-mL syringe whatever volume of methanol is necessary to maintain a volume of 100 µL added to the syringe.
(b) Dilute an aliquot of the methanol extract and then take 100 µL for analysis.

QUALITY CONTROL Demonstrate, through the analysis of a reagent water blank, that interferences from the analytical system, glassware, and reagents are under control. Blank samples should be carried through all stages of the sample preparation and measurement steps. For each analytical batch (up to 20 samples), a reagent blank, matrix spike, and matrix spike duplicate must be analyzed (the frequency of the spikes may be different for different monitoring programs). The blank and spiked samples must be carried through all stages of the sample preparation and measurement steps. QC samples mentioned in the section on Interferences will also be needed as appropriate to those situations.

REFERENCE Test Methods for Evaluating Solid Waste (SW-846). U.S. EPA. 1983. Method 8240B, Rev. 2, Nov. 1990. Office of Solid Wastes, Washington, DC.

Ethyl methanesulfonate   **EPA Method 1625**
CAS #62-50-0

TITLE Semivolatile Organic Compounds by Isotope Dilution GC/MS

MATRIX The compounds may be determined in waters, soils, and municipal sludges by this method.

METHOD SUMMARY This method is used to determine 176 semivolatile toxic organic pollutants associated with the CWA (as amended 1987); the RCRA (as amended 1986); the CERCLA (as amended 1986); and other compounds amenable to extraction and analysis by capillary column gas chromatography-mass spectrometry (GC/MS).

Stable isotopically-labeled analogs of the compounds of interest are added to the sample. If the solids content is less than 1%, a 1-L sample is extracted at pH 12–13, then at pH <2 with methylene chloride using continuous extraction techniques.

If the solids content is 30% or less, the sample is diluted to 1% solids with reagent water, homogenized ultrasonically, and extracted at pH 12–13, then at pH <2 with methylene chloride using continuous extraction techniques. If the solids content is greater than 30%, the sample is extracted using ultrasonic techniques.

Each extract is dried over sodium sulfate, concentrated to a volume of 5 mL, cleaned up using GPC, if necessary, and concentrated. Extracts are concentrated to 1 mL if GPC is not performed, and to 0.5 mL if GPC is performed.

An internal standard is added to the extract, and a 1-mL aliquot of the extract is injected into the GC. The compounds are separated by GC and detected by a MS. The labeled compounds serve to correct the variability of the analytical technique.

INTERFERENCES Solvents, reagents, glassware, and other sample processing hardware may yield artifacts and/or elevated baselines causing misinterpretation of chromatograms and spectra. Materials used in the analysis must be demonstrated to be free from interferences under the conditions of analysis by running method blanks initially and with each sample lot (sample started through the extraction process on a given 8-h shift, to a maximum of 20). Specific selection of reagents and purification of solvents by distillation in all glass systems may be required. Glassware and, where possible, reagents are cleaned by solvent rinse and baking at 450°C for 1-h minimum. Interferences coextracted from samples will vary considerably from source to source, depending on the diversity of the site being sampled.

INSTRUMENTATION Major instrumentation includes a GC with a splitless or on-column injection port for capillary column, a MS with 70 eV electron impact ionization, and a data system to collect and record MS data, and process it. A K-D apparatus is used to concentrate extracts.

GC Column: 30 m × 0.25 mm I.D. 5% phenyl, 94% methyl, 1% vinyl silicone bonded phased fused silica capillary column.

PRECISION & ACCURACY The detection limits of the method are usually dependent on the level of interferences rather than instrumental limitations. The limits typify the minimum quantities that can be detected with no interferences present.

The minimum level (in µg/mL) was not listed. This is defined as a minimum level at which the analytical system shall give recognizable mass spectra (background corrected) and acceptable calibration points.

The MDL (in µg/kg) in low solids was not listed and in high solids was not listed; these were determined in digested sludge (low solids) and in filter cake or compost (high solids).

The labeled and native compound initial precision as standard deviation (in µg/L) was not listed.
The labeled and native compound initial accuracy as average recovery (in µg/L) was not listed.

SAMPLE COLLECTION, PRESERVATION & HANDLING Collect samples in glass containers. Aqueous samples which flow freely are collected in refrigerated bottles using automatic sampling equipment. Solid samples are collected as grab samples using widemouth jars. Maintain samples at 0 to 4°C from the time of collection until extraction. If residual chlorine is present in aqueous samples, add 80 mg sodium thiosulfate/L of water. Begin sample extraction within 7 days of collection, and analyze all extracts within 40 days of extraction.

SAMPLE PREPARATION Samples containing 1% solids or less are extracted directly using continuous liquid-liquid extraction techniques. Samples containing 1 to 30% solids are diluted to the 1% level with reagent water and extracted using continuous liquid-liquid extraction techniques. Samples containing greater than 30% solids are extracted using ultrasonic techniques.

- Base/neutral extraction — Adjust the pH of the waters in the extractors to 12–13 with 6 N NaOH. Extract with methylene chloride for 24–48 h.
- Acid extraction — Adjust the pH of the waters in the extractors to 2 or less using 6 N sulfuric acid. Extract with methylene chloride for 24–48 h.
- Ultrasonic extraction of high solids samples — Add anhydrous sodium sulfate to the sample and QC aliquot(s). Add acetone:methylene chloride (1:1) to the sample and mix thoroughly

Concentrate extracts using a K-D apparatus.

QUALITY CONTROL The analyst is permitted to modify this method to improve separations or lower the costs of measurements, provided all performance specifications are met. Analyses of blanks are required to demonstrate freedom from contamination. When results of spikes indicate atypical method performance for samples, the samples are diluted to bring method performance within acceptable limits.

For low solids (aqueous samples), extract, concentrate, and analyze two sets of four 1-L aliquots (8 aliquots total) of the precision and recovery standard. For high solids samples, two sets of four 30-g aliquots of the high solids reference matrix are used.

Spike all samples with labeled compounds to assess method performance. Compute percent recovery of the labeled compounds using the internal standard method. Compare the labeled compound recovery for each compound with the corresponding labeled compound recovery.

Reagent water and high solids reference matrix blanks are analyzed to demonstrate freedom from contamination. Extract and concentrate a 1-L reagent water blank or a high solids reference matrix blank with each sample's lot (samples started through the extraction process on the same 8-h shift, to a maximum of 20 samples).

Field replicates may be collected to determine the precision of the sampling technique, and spiked samples may be required to determine the accuracy of the analysis when the internal standard method is used.

REFERENCE Semivolatile Organic Compounds by Isotope Dilution GC/MS. Office of Water Regulation and Standards, U.S. EPA Industrial Technology Division, Washington, DC, EPA Method 1625, Rev. C, June 1989 (contact W.A. Telliard, U.S. EPA, Office of Water Regulations and Standards, 401 M St., SW, Washington, DC, 20460. Phone: 202-382-7131).

Ethyl methanesulfonate   **EPA Method 8270**
CAS #62-50-0

TITLE Semivolatile Organic Compounds by GC/MS

MATRIX This method is used to determine the concentration of semivolatile organic compounds in extracts prepared from all types of solid waste matrices, soils, and groundwater. Although surface waters are not specifically mentioned, this method should be applicable to water samples from rivers, lakes, etc.

METHOD SUMMARY This method covers 259 semivolatile organic compounds. In very limited applications direct injection of the sample into the GC/MS system may be appropriate, but this results in very high detection limits (approximately 10,000 µg/L). Typically, a 1-L liquid sample, containing surrogate, and matrix spiking standards, is extracted in a continuous extractor first under acid conditions and then under basic conditions. Typically 30 g of a solid sample, containing surrogate, and matrix spiking standards, is extracted ultrasonically. After concentrating the extract to 1 mL it is spiked with 10 µL of an internal standard solution just prior to analysis by GC/MS. The volume injected should contain about 100 ng of base/neutral and 200 ng of acid surrogates (for a 1-µL injection). Analysis is performed by GC/MS using a capillary GC column.

INTERFERENCES Raw GC/MS data from all blanks, samples, and spikes must be evaluated for interferences. Contamination by carryover can occur whenever high-concentration and low-concentration samples are sequentially analyzed. To reduce carryover, the sample syringe must be rinsed out between samples with solvent. Whenever an unusually concentrated sample is encountered, it should be followed by the analysis of blank solvent to check for cross-contamination.

INSTRUMENTATION A GC/MS and a data system are required. The GC column used is a 30 m × 0.25 mm I.D. (or 0.32 mm I.D.) 1um film thickness silicone-coated fused silica capillary column. A continuous liquid-liquid extractor equipped with Teflon® or glass connection joints and stopcocks requiring no lubrication, a K-D concentrating apparatus, water bath, and an ultrasonic disrupter with a minimum power of 300 W and with pulsing capability are also required.

PRECISION & ACCURACY The estimated quantitation limit (EQL) of Method 8270B for determining an individual compound is approximately 1 mg/kg (wet weight) for soil or sediment samples, 1–200 mg/kg for wastes (dependent on matrix and method of preparation), and 10 µg/L for groundwater samples. EQLs will be proportionately higher for sample extracts that require dilution to avoid saturation of the detector.

The EQL(b) for groundwater in µg/L is 20.
The EQL (a, b) for low concentrations in soil and sediment in µg/kg is not determined.
Accuracy as µg/L is not listed.
Overall precision in µg/L is not listed.

(a) EQLs listed for soil/sediment are based on wet weight. Normally data is reported in a dry-weight basis; therefore, EQLs will be higher based on the % dry weight of each sample. This calculation is based on a 30-g sample and gel permeation chromatography cleanup.
(b) Sample EQLs are highly matrix-dependent. The EQLs are provided for guidance and may not always be achievable.
C = True value for concentration, in µg/L.
X = Average recovery found for measurements of samples containing a concentration of C, in µg/L.

ESTIMATED QUANTITATION LIMIT

Other Matrices	Factor (a)
High-concentration soil and sludges by sonicator	7.5
Non-water miscible waste	75

(a) EQL for other matrices = [EQL for low soil/sediment] × [Factor]. This estimated EQL is similar to an EPA "Practical Quantitation Limit."

SAMPLING METHOD
Liquid samples — Use a 1 or 2½ gallon amber glass bottle with a screw-top Teflon®-lined cover that has been prewashed with detergent and rinsed with distilled water and methanol (or isopropanol).

Soils, sediments, or sludges — Use an 8-oz. widemouth glass with a screw-top Teflon®-lined cover that has been prewashed with detergent and rinsed with distilled water and methanol (or isopropanol).

SAMPLE PRESERVATION
Liquid samples — If residual chlorine is present, add 3 mL of 10% sodium thiosulfate per gallon, cool to 4°C and store in a solvent-free refrigerator until analysis; if chlorine is not present, then eliminate the sodium thiosulfate addition.

Soils, sediments, or sludges — Cool samples to 4°C and store in a solvent-free refrigerator.

MHT Liquid samples must be extracted within 7 days and the extracts analyzed within 40 days. Soils, sediments, or sludges may be stored for a maximum of 14 days and the extracts analyzed within 40 days.

SAMPLE PREPARATION

Liquid samples — Transfer 1 L quantitatively to a continuous extractor. If high concentrations are anticipated, a smaller volume may be used and then diluted with organic-free reagent water to 1 L. Adjust pH, if necessary, to pH <2 using 1:1 (V/V) sulfuric acid. Pipette 1.0 mL of a surrogate standard spiking solution into each sample. For the sample in each analytical batch selected for spiking, add 1.0 mL of a matrix spiking standard. For base/neutral acid analysis, the amount of the surrogates and matrix spiking compounds added to the sample should result in a final concentration of 100 ng/μL of each analyte in the extract to be analyzed (assuming a 1-μL injection). Extract with methylene chloride for 18–24 h. Next, adjust the pH of the aqueous phase to pH >11 using 10 N sodium hydroxide and extract it with methylene chloride again for 18–24 h. Dry the extract through a column containing anhydrous sodium sulfate and concentrate it to 1 mL using a K-D concentrator.

Soils, sediments, or sludges — Use 30 g of sample. Nonporous or wet samples (gummy or clay type) that do not have a free-flowing sandy texture must be mixed with anhydrous sodium sulfate until the sample is free flowing. Add 1 mL of surrogate standards to all samples, spikes, standards, and blanks. For the sample in each analytical batch selected for spiking, add 1.0 mL of a matrix spiking standard. For base/neutral acid analysis, the amount added of the surrogates and matrix spiking compounds should result in a final concentration of 100 ng/μL of each base/neutral analyte and 200 ng/μL of each acid analyte in the extract to be analyzed (assuming a 1-μL injection). Immediately add a 100-mL mixture of 1:1 methylene chloride:acetone and extract the sample ultrasonically for 3 min and then decant or filter the extracts. Repeat the extraction two or more times. Dry the extract using a column with anhydrous sodium sulfate and concentrate it to 1 mL in a K-D concentrator.

QUALITY CONTROL

A methylene chloride solution containing 50 ng/μL of decafluorotriphenylphosphine (DFTPP) is used for tuning the GC/MS system each 12-h shift. A system performance check also must be made during every 12-h shift. A standard containing 50 ng/μL each of 4,4′-DDT, pentachlorophenol, and benzidine is required to verify injection port inertness and GC column performance. A calibration standard at mid-concentration, containing each compound of interest, including all required surrogates, must be performed every 12 h during analysis. After the system performance check is met, calibration check compounds (CCCs) are used to check the validity of the initial calibration.

The internal standard responses and retention times in the calibration check standard must be evaluated immediately after or during data acquisition. If the retention time for any internal standard changes by more than 30 seconds from the last check calibration (12 h), the chromatographic system must be inspected for malfunctions and corrections must be made, as required. If the electron ionization current plot (EICP) area for any of the internal standards changes by a factor of two from the last daily calibration standard check, the mass spectrometer must be inspected for malfunctions and corrections must be made, as appropriate.

Demonstrate, through the analysis of a reagent water blank, that interferences from the analytical system, glassware, and reagents are under control. The blank samples should be carried through all stages of the sample preparation and measurement steps. For each analytical batch (up to 20 samples), a reagent blank, matrix spike, and matrix spike duplicate/duplicate must be analyzed (the frequency of the spikes may be different for different monitoring programs). The blank and spiked samples must be carried through all stages of the sample preparation and measurement steps. A QC reference sample concentrate containing each analyte at a concentration of 100 mg/L in methanol is required.

REFERENCE Test Methods for Evaluating Solid Waste (SW-846). U.S. EPA 1983, Method 8270B, Rev. 2, Nov. 1990. Office of Solid Waste, Washington, DC.

Ethyl parathion **EPA Method 8141**
CAS #56-38-2

TITLE Organophosphorus Compounds by Gas Chromatography: Capillary Column Technique

MATRIX This method covers aqueous and solid matrices. This includes a wide variety such as drinking water, groundwater, industrial wastewaters, surface waters, soils, solids, and sediments.

METHOD SUMMARY This is a GC method used to determine the concentration of 28 organophosphorus pesticides.

The use of Gel Permeation Cleanup (EPA Method 3640) for sample cleanup has been demonstrated to yield recoveries of less than 85% for many method analytes and is therefore not recommended for use with this method.

This method provides GC conditions for the detection of ppb concentrations of organophosphorus compounds. Prior to the use of this method, appropriate sample preparation techniques must be used. Water samples are extracted at a neutral pH with methylene chloride as a solvent by using a separatory funnel (EPA Method 3510) or a continuous liquid-liquid extractor (EPA Method 3520). Soxhlet extraction (EPA Method 3540) or ultrasonic extraction (EPA Method 3550) using methylene chloride/acetone (1:1) are used for solid samples. Both neat and diluted organic liquids (EPA Method 3580) may be analyzed by direct injection. Spiked samples are used to verify the applicability of the chosen extraction technique to each new sample type. A GC with a flame photometric (FPD) or nitrogen-phosphorus detector (NPD) is used for this multiresidue procedure.

INTERFERENCES The use of Florisil cleanup materials (EPA Method 3620) for some of the compounds in this method has been demonstrated to yield recoveries less than 85% and is therefore not recommended for all compounds. Use of phosphorus or halogen specific detectors, however, often obviates the necessity for cleanup for relatively clean sample matrices. If particular circumstances demand the use of an alternative cleanup procedure, the analyst must determine the elution profile and demonstrate that the recovery of each analyte is no less than 85%.

Use of a flame photometric detector (FPD) in the phosphorus mode will minimize interferences from materials that do not contain phosphorus. Elemental sulfur, however, may interfere with the determination of certain organophosphorus compounds by flame photometric gas chromatography. Sulfur cleanup using EPA Method 3660 may alleviate this interference. A nitrogen phosphorus detector (NPD) is also recommended.

A few analytes coelute on certain columns. Therefore, select a second column for confirmation where coelution of the analytes of interest does not occur.

Method interferences may be caused by contaminants in solvents, reagents, glassware, and other sample processing hardware that lead to discrete artifacts or elevated baselines in gas chromatograms. All these materials must be routinely demonstrated to be free from interferences under the conditions of the analysis by analyzing reagent blanks.

INSTRUMENTATION A GC with a NPD or a FPD will be needed. A data system or integrator is recommended for measuring peak areas and/or peak heights. A Kuderna-Danish (K-D) apparatus will be needed for extract concentration.

Column 1: 15 m × 0.53 mm megabore capillary column, 1.0 μm film thickness, DB-210.

Column 2: 15 m × 0.53 mm megabore capillary column, 1.5 μm film thickness, SPB-608.

Column 3: 15 m × 0.53 mm megabore capillary column, 1.0 μm film thickness, DB-5.

Three megabore capillary columns are included for analysis of organophosphates by this method. Column 1 (DB-210 or equivalent) and Column 2 (SPB-608 or equivalent) are recommended if a large number of organophosphorus analytes are to be determined. If the superior resolution offered by Column 1 and Column 2 is not required, Column 3 (DB-5 or equivalent) may be used. For megabore capillary columns, automatic injections of 1 μL are recommended.

PRECISION & ACCURACY The MDL actually achieved in a given analysis will vary, as it is dependent on instrument sensitivity and matrix effects. Single operator accuracy and precision studies have been conducted with spiked water and soil samples.

MULTIPLICATION FACTORS FOR OTHER MATRICES (a)

Matrix	Factor (b)
Groundwater (EPA Method 3510 or EPA Method 3520)	10
Low-concentration soil by Soxhlet and no cleanup	10 (c)
Low-concentration soil by ultrasonic extraction with GPC cleanup	6.7 (c)
High-concentration soil and sludges by ultrasonic extraction	500 (c)
Non-water miscible waste (EPA Method 3580)	1000 (c)

(a) Sample EQLs are highly matrix-dependent. The EQLs listed here are provided for guidance and may not always be achievable.
(b) EQL = [Method detection limit] × [Factor]. For non-aqueous samples the factor is on a wet-weight basis.
(c) Multiply this factory times the soil MDL.

The MDL (in μg/L) when reagent water was extracted using a separatory funnel was 0.06.

The MDL (in μg/kg) when soil was extracted using Soxhlet extraction (EPA Method 3540) was 3.0.

Accuracy (as % recovery) with separatory funnel extraction ranged from 101 (with low spikes) to 86 (with high spikes).

Accuracy (as % recovery) with continuous liquid-liquid extraction ranged from 106 (with low spikes) to 87 (with high spikes).

Accuracy (as % recovery) with Soxhlet extraction of soils ranged from 75 (with low spikes to 80 (with high spikes).

Accuracy (as % recovery) with ultrasonic extraction of soils ranged from not recovered (with low spikes) to 75 (with high spikes).

SAMPLE COLLECTION, PRESERVATION & HANDLING
Containers used to collect samples for the determination of semivolatile organic compounds should be soap and water washed followed by methanol (or isopropanol) rinsing. The sample containers should be of glass or Teflon® and have screw-top covers with Teflon® liners.

No preservation is used with concentrated waste samples. With liquid samples containing no residual chlorine and with soil, sediment, and sludge samples, immediately cooling to 4°C is the only preservation used. When residual chlorine is present then 3 mL of 10% aqueous sodium sulfate is added for each gallon of sample collected, followed by cooling to 4°C.

Liquid samples must be extracted within 7 days and their extracts analyzed within 40 days. Concentrated waste, soil, sediment, and sludge samples must be extracted within 14 days and their extracts analyzed within 40 days.

SAMPLE PREPARATION In general, water samples are extracted at a neutral pH with methylene chloride, using either EPA Method 3510 or EPA Method 3520. Solid samples are extracted using either EPA Method 3540 or EPA Method 3550 with methylene chloride/acetone (1:1) as the extraction solvent.

Prior to GC analysis, the extraction solvent may be exchanged to hexane. Single lab data indicates that samples should not be transferred with 100% hexane during sample workup as the more water soluble organophosphorus compounds may be lost.

If cleanup is performed on the samples, the analyst should analyze the samples by GC. This will confirm elution patterns and the absence of interferences from the reagents. If peak detection and identification is prevented by the presence of interferences, further cleanup is required.

QUALITY CONTROL The analyst should monitor the performance of the extraction, cleanup (when used), and analytical system and the effectiveness of the method in dealing with each sample matrix by spiking each sample, standard, and blank with one or two surrogates (e.g., organophosphorus compounds not expected to be present in the sample). Deuterated analogs of analytes should not be used as surrogates for gas chromatographic analysis due to coelution problems.

A minimum of five concentrations for each analyte of interest should be prepared through dilution of the stock standards with isooctane. One of the concentrations should be at a concentration near, but above, the MDL.

Include a mid-level check standard after each group of 10 samples in the analysis sequence. GC/MS techniques should be judiciously employed to support qualitative identifications made with this method. Follow the GC/MS operating requirements specified in EPA Method 8270.

When available, chemical ionization mass spectra may be employed to aid in the qualitative identification process. To confirm an identification of a compound, the background-corrected mass spectrum of the compound must be obtained from the sample extract and must be compared with a mass spectrum from a stock or calibration standard analyzed under the same chromatographic conditions. The molecular ion and all other ions present above 20% relative abundance in the mass spectrum of the standard must be present in the mass spectrum of the sample with agreement to ± 20%. The retention time of the compound in the sample must be within six seconds of the retention time for the same compound in the standard solution.

Should the MS procedure fail to provide satisfactory results, additional steps may be taken before reanalysis. These steps may include the use of alternate packed or capillary GC columns or additional sample cleanup.

REFERENCE Test Methods for Evaluating Solid Waste, Physical/Chemical Methods, SW-846, 3rd Edition, U.S. EPA, Office of Solid Waste, Washington, DC, EPA Method 8141 July 1992.

Ethyl parathion   **EPA Method 8270**
CAS #56-38-2

TITLE Semivolatile Organic Compounds by GC/MS

MATRIX This method is used to determine the concentration of semivolatile organic compounds in extracts prepared from all types of solid waste matrices, soils, and groundwater. Although surface waters are not specifically mentioned, this method should be applicable to water samples from rivers, lakes, etc.

METHOD SUMMARY This method covers 259 semivolatile organic compounds. In very limited applications direct injection of the sample into the GC/MS system may be appropriate, but this results in very high detection limits (approximately 10,000 µg/L). Typically, a 1-L liquid sample, containing surrogate, and matrix spiking standards, is extracted in a continuous extractor first under acid conditions and then under basic conditions. Typically 30 g of a solid sample, containing surrogate, and matrix spiking standards, is extracted ultrasonically. After concentrating the extract to 1 mL it is spiked with 10 µL of an internal standard solution just prior to analysis by GC/MS. The volume injected should contain about 100 ng of base/neutral and 200 ng of acid surrogates (for a 1-µL injection). Analysis is performed by GC/MS using a capillary GC column.

INTERFERENCES Raw GC/MS data from all blanks, samples, and spikes must be evaluated for interferences. Contamination by carryover can occur whenever high-concentration and low-concentration samples are sequentially analyzed. To reduce carryover, the sample syringe must be rinsed out between samples with solvent. Whenever an unusually concentrated sample is encountered, it should be followed by the analysis of blank solvent to check for cross-contamination.

INSTRUMENTATION A GC/MS and a data system are required. The GC column used is a 30 m × 0.25 mm I.D. (or 0.32 mm I.D.) 1um film thickness silicone-coated fused silica capillary column. A continuous liquid-liquid extractor equipped with Teflon® or glass connection joints and stopcocks requiring no lubrication, a K-D concentrating apparatus, water bath, and an ultrasonic disrupter with a minimum power of 300 W and with pulsing capability are also required.

PRECISION & ACCURACY The estimated quantitation limit (EQL) of Method 8270B for determining an individual compound is approximately 1 mg/kg (wet weight) for soil or sediment samples, 1–200 mg/kg for wastes (dependent on matrix and method of preparation), and 10 µg/L for groundwater samples. EQLs will be proportionately higher for sample extracts that require dilution to avoid saturation of the detector.

The EQL(b) for groundwater in µg/L is not listed.
The EQL (a, b) for low concentrations in soil and sediment in µg/kg is not listed.
Accuracy as µg/L is not listed.
Overall precision in µg/L is not listed.

(a) *EQLs listed for soil/sediment are based on wet weight. Normally data is reported in a dry-weight basis; therefore, EQLs will be higher based on the % dry weight of each sample. This calculation is based on a 30-g sample and gel permeation chromatography cleanup.*

(b) *Sample EQLs are highly matrix-dependent. The EQLs are provided for guidance and may not always be achievable.*

$C =$ *True value for concentration, in µg/L.*
$X =$ *Average recovery found for measurements of samples containing a concentration of C, in µg/L.*

ESTIMATED QUANTITATION LIMIT

Other Matrices	Factor (a)
High-concentration soil and sludges by sonicator	7.5
Non-water miscible waste	75

(a) *EQL for other matrices = [EQL for low soil/sediment] × [Factor]. This estimated EQL is similar to an EPA "Practical Quantitation Limit."*

SAMPLING METHOD

Liquid samples — Use a 1 or 2½ gallon amber glass bottle with a screw-top Teflon®-lined cover that has been prewashed with detergent and rinsed with distilled water and methanol (or isopropanol).

Soils, sediments, or sludges — Use an 8-oz. widemouth glass with a screw-top Teflon®-lined cover that has been prewashed with detergent and rinsed with distilled water and methanol (or isopropanol).

SAMPLE PRESERVATION

Liquid samples — If residual chlorine is present, add 3 mL of 10% sodium thiosulfate per gallon, cool to 4°C and store in a

solvent-free refrigerator until analysis; if chlorine is not present, then eliminate the sodium thiosulfate addition.

Soils, sediments, or sludges — Cool samples to 4°C and store in a solvent-free refrigerator.

MHT Liquid samples must be extracted within 7 days and the extracts analyzed within 40 days. Soils, sediments, or sludges may be stored for a maximum of 14 days and the extracts analyzed within 40 days.

SAMPLE PREPARATION

Liquid samples — Transfer 1 L quantitatively to a continuous extractor. If high concentrations are anticipated, a smaller volume may be used and then diluted with organic-free reagent water to 1 L. Adjust pH, if necessary, to pH <2 using 1:1 (V/V) sulfuric acid. Pipette 1.0 mL of a surrogate standard spiking solution into each sample. For the sample in each analytical batch selected for spiking, add 1.0 mL of a matrix spiking standard. For base/neutral acid analysis, the amount of the surrogates and matrix spiking compounds added to the sample should result in a final concentration of 100 ng/µL of each analyte in the extract to be analyzed (assuming a 1-µL injection). Extract with methylene chloride for 18–24 h. Next, adjust the pH of the aqueous phase to pH >11 using 10 N sodium hydroxide and extract it with methylene chloride again for 18–24 h. Dry the extract through a column containing anhydrous sodium sulfate and concentrate it to 1 mL using a K-D concentrator.

Soils, sediments, or sludges — Use 30 g of sample. Nonporous or wet samples (gummy or clay type) that do not have a free-flowing sandy texture must be mixed with anhydrous sodium sulfate until the sample is free flowing. Add 1 mL of surrogate standards to all samples, spikes, standards, and blanks. For the sample in each analytical batch selected for spiking, add 1.0 mL of a matrix spiking standard. For base/neutral acid analysis, the amount added of the surrogates and matrix spiking compounds should result in a final concentration of 100 ng/µL of each base/neutral analyte and 200 ng/µL of each acid analyte in the extract to be analyzed (assuming a 1-µL injection). Immediately add a 100-mL mixture of 1:1 methylene chloride:acetone and extract the sample ultrasonically for 3 min and then decant or filter the extracts. Repeat the extraction two or more times. Dry the extract using a column with anhydrous sodium sulfate and concentrate it to 1 mL in a K-D concentrator.

QUALITY CONTROL A methylene chloride solution containing 50 ng/µL of decafluorotriphenylphosphine (DFTPP) is used for tuning the GC/MS system each 12-h shift. A system performance check also must be made during every 12-h shift. A standard containing 50 ng/µL each of 4,4'-DDT, pentachlorophenol, and benzidine is required to verify injection port inertness and GC column performance. A calibration standard at mid-concentration, containing each compound of interest, including all required surrogates, must be performed every 12 h during analysis. After the system performance check is met, calibration check compounds (CCCs) are used to check the validity of the initial calibration.

The internal standard responses and retention times in the calibration check standard must be evaluated immediately after or during data acquisition. If the retention time for any internal standard changes by more than 30 seconds from the last check calibration (12 h), the chromatographic system must be inspected for malfunctions and corrections must be made, as required. If the electron ionization current plot (EICP) area for any of the internal standards changes by a factor of two from the last daily calibration standard check, the mass spectrometer must be inspected for malfunctions and corrections must be made, as appropriate.

Demonstrate, through the analysis of a reagent water blank, that interferences from the analytical system, glassware, and reagents are under control. The blank samples should be carried through all stages of the sample preparation and measurement steps. For each analytical batch (up to 20 samples), a reagent blank, matrix spike, and matrix spike duplicate/duplicate must be analyzed (the frequency of the spikes may be different for different monitoring programs). The blank and spiked samples must be carried through all stages of the sample preparation and measurement steps. A QC reference sample concentrate containing each analyte at a concentration of 100 mg/L in methanol is required.

REFERENCE Test Methods for Evaluating Solid Waste (SW-846). U.S. EPA 1983, Method 8270B, Rev. 2, Nov. 1990. Office of Solid Waste, Washington, DC.

Ethylbenzene   **EPA Method 1624**
CAS #100-41-4

TITLE Volatile Organic Compounds by Isotope Dilution GC/MS

MATRIX Compounds may be determined in waters, soils, and municipal sludges by this method.

METHOD SUMMARY This method is used to determine 58 volatile toxic organic pollutants associated with the CWA (as amended 1987); the RCRA (as amended 1986); the CERCLA (as amended 1986); and other compounds amenable to purge-and-trap gas chromatography-mass spectrometry (GC/MS).

If the solids content is less than 1%, stable isotopically-labeled analogs of the compounds of interest are added to a 5-mL sample and the sample is purged with an inert gas at 20–25°C in a chamber designed for soil or water samples. If the solids content is greater than 1%, 5 mL of reagent water and the labeled compounds are added to a 5-g aliquot of sample and the mixture is purged at 40°C. Compounds that will not purge at 20–25°C or at 40°C are purged at 78–85°C. In the purging process, the volatile compounds are transferred from the aqueous phase into the gaseous phase where they are passed into a sorbent column, and trapped. After purging is completed, the trap is backflushed and heated rapidly to desorb the compounds into a GC. The compounds are separated by the GC and detected by a MS. The labeled compounds serve to correct the variability of the analytical technique.

INTERFERENCES Impurities in the purge gas, organic compounds outgassing from the plumbing upstream of the trap, and solvent vapors in the lab account for most problems. Samples can be contaminated by diffusion of volatile organic compounds (particularly methylene chloride) through the bottle seal during shipment and storage. Contamination by carryover can occur when high-level and low-level samples are analyzed sequentially. When an unusually concentrated sample is encountered, follow it by analysis of a reagent water blank to check for carryover.

INSTRUMENTATION Major equipment includes a GC with linear temperature programming and a glass jet separator as the MS interface, a MS with 70 eV electron impact ionization, and a data system to collect and record response factors.

Column: 2.8 m × 2 mm I.D. glass, packed with 1% SP-1000 on Carbopak B, 60/80 mesh, or equivalent.

PRECISION & ACCURACY The detection limits of the method are usually dependent on the level of interferences rather than instrumental limitations. The method detection limits were determined in digested sludge (low solids) and in filter cake or compost (high solids).

The MDL (in µg/kg) for low solids is 28 and for high solids is 4.
Labeled and native compound precision (in µg/L) as standard deviation was 9.6.
Labeled and native compound accuracy (in µg/L) as average recovery was 16–29.
Acceptance criteria are at 20 µg/L for this compound.

SAMPLE COLLECTION, PRESERVATION & HANDLING Grab samples are collected in glass containers having a total volume greater than 20 mL. Fill and seal each bottle so that no air bubbles are entrapped. Samples are maintained at 0 to 4°C from the time of collection until analysis. If an aqueous sample contains residual chlorine, add sodium thiosulfate preservative (10 mg/40 mL) to the empty sample bottles just prior to shipment to the sample site. All samples must be analyzed within 14 days of collection.

SAMPLE PREPARATION Samples containing less than 1% solids are analyzed directly as aqueous samples. Samples containing 1% solids or greater are analyzed as solid samples utilizing one of two methods, depending on the levels of pollutants, in the sample. Samples containing 1% solids or greater, and low to moderate levels of pollutants are analyzed by purging a known weight of sample added to 5 mL of reagent water. Samples containing 1% solids or greater, and high levels of pollutants, are extracted with methanol, and an aliquot of the methanol extract is added to reagent water and purged.

QUALITY CONTROL A field blank prepared from reagent water and carried through the sampling and handling protocol may serve as a check on contamination from shipment and storage.

The analyst is permitted to modify this method to improve separations or lower the costs of measurements, provided all performance specifications are met. Analyses of blanks are required. When results of spikes indicate atypical method performance for samples, the samples are diluted to bring method performance within acceptable limits. Analyze two sets of four 5-mL aliquots (8 aliquots total) of the aqueous performance standard. Spike all samples with labeled compounds to assess method performance on the sample matrix. Compute the percent recovery of the labeled compounds using the internal standard method. Compare the percent recovery for each compound with the corresponding labeled compound recovery. Reagent water blanks are analyzed to demonstrate freedom from carryover contamination. Field replicates may be collected to determine the precision of the sampling technique, and spiked samples may be required to determine the accuracy of the analysis when the internal method is used.

REFERENCE Volatile Organic Compounds by Isotope Dilution GC/MS. Office of Water Regulation and Standards, U.S. EPA Industrial Technology Division, Washington, DC, EPA Method 1624, Rev. C, June 1989 (contact W.A. Telliard, U.S. EPA, Office of Water Regulations and Standards, 401 M St., SW, Washington, DC, 20460. Phone: 202-382-7131).

Ethylbenzene EPA Method 502
CAS #100-41-4

TITLE Volatile Organic Compounds in Water By Purge and Trap Capillary Column Gas Chromatography with Photoionization and Electrolytic Conductivity Detectors in Series. U.S. EPA Method 502.2, Rev. 2.0, 1989.

MATRIX Drinking water and raw source water. The latter should include most surface water and groundwater sources.

METHOD SUMMARY This method covers 60 volatile organic compounds that contain halogen atoms and/or that are aromatic. An inert gas (zero grade nitrogen or helium) is bubbled through a 25-mL or a 5-mL water sample (depending on the expected concentration of the analytes). Purged sample components are trapped in a tube of sorbent materials. When purging is complete, the sorbent tube is heated and backflushed with helium to desorb the trapped sample onto a capillary GC column. The column is temperature programmed to separate the method analytes which are then detected with a photoionization detector (PID) and a Hall electrolytic conductivity (HECD) placed in series. The PID is selective for aromatic compounds and the HECD is selective for halogenated compounds.

INTERFERENCES Impurities in the purge gas and from organic compounds outgassing from the plumbing ahead of the trap account for many contamination problems. Interferences purged or coextracted from the samples will vary considerably from source to source, depending upon the particular sample or extract being tested. Cross-contamination can occur whenever high-level and low-level samples are analyzed sequentially. Samples also can be contaminated by diffusion of volatile organics (particularly methylene chloride and fluorocarbons) through the septum seal into the sample during shipment and storage. The lab where volatile analysis is performed and also the refrigerated storage area should be completely free of solvents.

INSTRUMENTATION A GC containing a series configuration of a high temperature photoionization detector (PID) equipped with 10.0 eV (nominal) lamp and Hall electrolytic conductivity detector (HECD) is required. Also required is an all-glass 5-mL purging device, a sorbent trap, and a thermal desorption apparatus which is connected to the GC system.

Column 1: VOCOL glass wide-bore capillary column.
Column 2: RTX–502.2 mega-bore capillary column.
Column 3: DB-62 mega-bore capillary column.

PRECISION & ACCURACY Method detection limits are dependent upon the characteristics of the gas chromatographic system used. Analytes that are not separated chromatographically cannot be individually identified and used in the same calibration mixture or water samples unless an alternative technique for identification and quantification, such as mass spectrometry, is used.

Electrolytic conductivity detetor (c) range in µg/L (a) was 0.02–200.
Electrolytic conductivity detetor (c) MDL in µg/L (b) was not listed.
Electrolytic conductivity detetor (c) accuracy as % recovery was not listed.
Electrolytic conductivity detetor (c) precision as % RSD was not listed.
Photoionization detector (d) range in µg/L (a) was 0.02–200.
Photoionization detector (d) MDL in µg/L (b) was 0.01.
Photoionization detector (d) accuracy as % recovery was 101.
Photoionization detector (d) precision as % RSD was 1.4.

(a) *The applicable concentration range of this method is compound, instrument, and matrix-dependent. It is listed as being approximately 0.02 to 200 µg/L but no specific information is provided so caution should be observed.*
(b) *The method detection limits reports with this method are compound, instrument, and matrix-dependent. The values reported were calculated using reagent water fortified with the corresponding compounds at 10 µg/L and a GC-equipped with a 60 m × 0.75 mm VOLCOL wide bore capillary column with 1.5 µm film thickness and using helium carrier gas.*
(c) *Recoveries and relative standard deviations were determined from seven samples of reagent water fortified with 10 µg/L of each compound. 2-Bromo-1-chloropropane was used as the internal standard for calculating average recoveries.*
(d) *Recoveries and relative standard deviations were determined from seven samples of reagent water fortified with 10 µg/L of each compound. Fluorobenzene was used as the internal standard for calculating average recoveries.*

SAMPLING METHOD Collect samples using a 40- to 120-mL screw-cap vial (prewashed with detergent, rinsed with distilled water and oven dried at 105°C) with a Teflon®-faced silicone septum. Collect bubble-free samples and place the septum with the Teflon® side down on the water.

SAMPLE PRESERVATION If residual chlorine is present in the water add about 25 mg of ascorbic acid to each vial before samples are collected to remove the chlorine. Add hydrochloric acid to reduce pH to <2, immediately cool samples to 4°C, and store them in a solvent-free refrigerator at 4°C until analysis.

MHT The maximum holding time for samples is 14 days from the time they were collected.

SAMPLE PREPARATION Remove the plungers from two 5-mL syringes and attach a closed syringe valve to each. Warm the sample to room temperature, open the sample bottle, and carefully pour the sample into one of the syringe barrels to just short of overflowing. Replace the syringe plunger, invert the syringe, and compress the sample. Open the syringe valve and vent any residual air while adjusting the sample volume to 5.0 mL. Add 10 µL of the internal calibration standard to the sample through the syringe valve. Close the valve. Fill the second syringe in an identical manner from the same sample bottle. Reserve this second syringe for a reanalysis if necessary.

QUALITY CONTROL As an initial demonstration of lab accuracy and precision, analyze 4 to 7 replicates of a lab fortified blank containing analyte at 0.1–5 µg/L. Collect all samples in duplicate. Surrogate analytes (similar to those of the analytes of interest), whose concentration is known in every sample, are measured using the same internal standard calibration procedure. Duplicate field reagent water blanks (trip blanks) must be analyzed with each set of samples, lab reagent blanks (method blanks) must be analyzed with each batch of samples processed as a group within a work shift. Also, a single lab-fortified blank that contains each of the analytes of interest should be analyzed with each batch of samples processed as a group within a work shift. A 3- to 5-point calibration curve is needed depending on the calibration range factor required.

EPA CONTACT & HOTLINE For technical questions contact Dr. Baldev Bathija, U.S. EPA, Office of Ground Water and Drinking Water (WH-550D), 401 M St. SW, Washington, DC 20460. Tel. (202) 260-3040. For further information the EPA Safe Drinking Water Hotline may be called at: (800) 426-4791.

REFERENCE Methods for the Determination of Organic Compounds in Drinking Water, EPA/600/4-88/039 (revised July 1991; Final Rule for determination of compliance with the MCL for Total Trihalomethanes under 141.30, in 40 CFR Part 141, Vol. 58, No. 147, Fed. Reg., Tuesday Aug. 3, 1993). U.S. EPA Environmental Monitoring Systems Laboratory, Cincinnati, OH, 45268, U.S.A. Available from the National Technical Information Service (NTIS), 5285 Port Royal Road, Springfield, VA 22161; Tel. 800-553-6847. NTIS Order Number is PB91-231480.

Ethylbenzene   **EPA Method 524**
CAS #100-41-4

TITLE Measurement of Purgeable Organic Compounds in Water by Capillary Column GC/MS.

MATRIX Drinking water and raw source water; the latter should include most surface water and groundwater sources.

METHOD SUMMARY Method 524.2 covers 60 volatile organic compounds. An inert gas (zero grade nitrogen or helium) is bubbled through a 25-mL or a 5-mL water sample (depending on the expected concentration of the analytes). Purged sample components are trapped in a tube of sorbent materials. When purging is complete, the sorbent tube is heated and backflushed with helium to desorb the trapped sample onto a capillary GC column.

INTERFERENCES Impurities in the purge gas and from organic compounds outgassing from the plumbing ahead of the trap account for many contamination problems. Interferences purged or coextracted from the samples will vary considerably from source to source, depending upon the particular sample or extract being tested. Cross-contamination can occur whenever high-level and low-level samples are analyzed sequentially. Samples also can be contaminated by diffusion of volatile organics (particularly methylene chloride and fluorocarbons) through the septum seal into the sample during shipment and storage.

INSTRUMENTATION A GC/MS with a data system equipped with one of the following capillary GC columns:

Column 1: VOCOL glass wide bore capillary column.
Column 2: DB-624 fused silica capillary column.
Column 3: DB-5 fused silica capillary column.

Also required is an all-glass 25 mL or 5-mL purging device, a sorbent trap, and a thermal desorption apparatus which is connected to the GC/MS system.

PRECISION & ACCURACY Method detection limits are compound- and instrument-dependent, and may vary from approximately 0.02–0.35 µg/L. Note in the table below that the "true" concentration range used for accuracy and precision measurements was quite narrow. However, the applicable concentration range of this method is primarily column dependent and is approximately 0.02 to 200 µg/L for the wide-bore thick-film columns. Narrow-bore thin-film columns may have a capacity which limits the range to about 0.02 to 20 µg/L. Analytes that are inefficiently purged from water will not be detected when present at low concentrations, but they can be measured with acceptable accuracy and precision when present in sufficient amounts.

Analytes that are not separated chromatographically, but which have different mass spectra and non-interfering quantification ions, can be identified and measured in the same calibration mixture or water sample. Analytes which have very similar mass spectra cannot be individually identified and measured in the same calibration mixture or water samples unless they have different retention times. Co-eluting compounds with very similar mass spectra, typically many structural isomers, must be reported as an isomeric group or pair.

The range (in µg/L) was 0.1–10.
The Method Detection Limig (in µg/L) was 0.06.
The accuracy (as % recovery) was 99.
The precision (in %) was 8.6.

Note: Data were obtained from 16–31 determinations using a wide-bore capillary column and a jet separator interfaced to a quadrupole mass spectrometer. All analytes were in a reagent water matrix.

SAMPLING METHOD Collect samples using a 40- to 120-mL screw-cap vial (prewashed with detergent, rinsed with distilled water and oven dried at 105°C) with a Teflon®-faced silicone septum. Collect bubble-free samples and place the septum with the Teflon® side down on the water.

SAMPLE PRESERVATION If residual chlorine is present in the water add about 25 mg of ascorbic acid to each vial before samples are collected to remove the chlorine. Add hydrochloric acid to reduce pH to <2, and immediately cool samples to 4°C, and store them in a solvent-free refrigerator at 4°C until analysis.

MHT The maximum holding time for samples is 14 days from the time they were collected.

SAMPLE PREPARATION Remove the plungers from two 25-mL (or 5-mL depending on sample size) syringes and attach a closed syringe valve to each. Warm the sample to room temperature, open the sample bottle, and carefully pour the sample into one of the syringe barrels to just short of overflowing. Replace the syringe plunger, invert the syringe, and compress the sample. Open the syringe valve and vent any residual air while adjusting the sample volume to 25.0 mL (or 5 mL). For samples and blanks, add 5 µL of the fortification solution containing the internal standard and the surrogates to the sample through the syringe valve. For calibration standards and lab fortified blanks, add 5 µL of the fortification solution containing the internal standard only. Close the valve. Fill the second syringe in an identical manner from the same sample bottle. Reserve this second syringe for a reanalysis if necessary.

QUALITY CONTROL As an initial demonstration of lab accuracy and precision, analyze 4 to 7 replicates of a lab fortified blank containing analyte at 0.2–5 µg/L. Collect all samples in duplicate. Surrogate analytes (similar to those of the analytes of interest), whose concentration is known in every sample, are measured using the same internal standard calibration procedure. Duplicate field reagent water blanks (trip blanks) must be analyzed with each set of samples, lab reagent blanks (method blanks) must be analyzed with each batch of samples processed as a group within a work shift. Also, a single lab-fortified blank that contains each of the analytes of interest should be analyzed with each batch of samples processed as a group within a work shift. A 3- to 5-point calibration curve is needed depending on the calibration range factor required.

EPA CONTACT & HOTLINE For technical questions contact Dr. Baldev Bathija, U.S. EPA, Office of Ground Water and Drinking Water (WH-550D), 401 M St. SW, Washington, DC 20460. Tel. (202) 260-3040. For further information the EPA Safe Drinking Water Hotline may be called at: (800) 426-4791.

REFERENCE Methods for the Determination of Organic Compounds in Drinking Water, EPA/600/4-88/039 (revised July 1991; Final Rule for determination of compliance with the MCL for Total Trihalomethanes under 141.30, in 40 CFR Part 141, Vol. 58, No. 147, Fed. Reg., Tuesday Aug. 3, 1993). U.S. EPA Environmental Monitoring Systems Laboratory, Cincinnati, OH, 45268, U.S.A. Available from the National Technical Information Service (NTIS), 5285 Port Royal Road, Springfield, VA 22161; Tel. 800-553-6847. NTIS Order Number is PB91-231480.

Ethylbenzene   **EPA Method 8021**
CAS #100-41-4

TITLE Halogenated Volatile by Gas Chromatography Using Photoionization and Electrolytic Conductivity Detectors in Series: Capillary Column Technique

MATRIX This method is applicable to nearly all types of samples, regardless of water content, including groundwater, aqueous sludges, caustic liquors, acid liquors, waste solvents, oily wastes, mousses, tars, fibrous wastes, polymeric emulsions, filter cakes, spent carbons, spent catalysts, soils, and sediments.

METHOD SUMMARY This method is used to determine 60 volatile organic compounds in a variety of solid waste matrices. It provides GC conditions for the detection of halogenated and aromatic volatile organic compounds. Samples can be analyzed using direct injection or purge-and-trap (EPA Method 5030). Groundwater samples must be analyzed using EPA Method 5030 (where applicable). A temperature program is used with the GC. Detection is achieved by a photoionization detector (PID) and a Hall electrolytic conductivity detector (HECD) in series.

INTERFERENCES Samples can be contaminated by diffusion of volatile organics (particularly chlorofluorocarbons and methylene chloride) through the sample container septum during shipment and storage.

INSTRUMENTATION A GC-equipped with variable-constant differential flow controllers, subambient oven controller, PID and HECD detectors connected with a short piece of uncoated capillary tubing and a data system.

Column: 60 m × 0.75 mm I.D. VOCOL wide-bore capillary column with 1.5 µm film thickness.

PRECISION & ACCURACY MDLs are compound-dependent and vary with purging efficiency and concentration. The applicable concentration range of this method is compound- and instrument-dependent but is approximately 0.1 to 200 µg/L. Analytes that are inefficiently purged from water will not be detected when present at low concentrations, but they can be measured with acceptable accuracy and precision when present in sufficient amounts. The estimated quantitation limit (EQL) for an individual compound is approximately 1 µg/kg (wet weight) for soil/sediment samples, 100 µg/kg (wet weight) for wastes, and 1 µg/L for groundwater. EQLs will be proportionately higher for sample extracts and samples that require dilution or reduced sample size to avoid saturation of the detector.

MULTIPLICATION FACTORS FOR OTHER MATRICES (a)

Matrix	Factor (b)
Groundwater	10
Low-concentration soil	10
Water miscible liquid waste	500
High-concentration soil and sludge	1250
Non-water miscible waste	1250

(a) Sample EQLs are highly matrix-dependent. The EQLs listed herein are provided for guidance and may not always be achievable. (b) EQL = [Method detection limit] × [Factor]. For non-aqueous samples, the factor is on a wet-weight basis.

SINGLE LABORATORY ACCURACY & PRECISION DATA FOR VOCs IN WATER

This method was tested in a single lab using water spiked at 10 µg/L and the following data was reported:

Recoveries and standard deviations were determined from seven samples and spiked at 10 µg/L of each analyte. Recoveries were determined by the internal standard method. Internal standards were: Fluorobenzene for PID and 2-Bromo-1-chloropropane for HECD.

The average recovery (in percent) for the PID was 101.
The standard deviation of the recovery for the PID was 1.4.
The MDL (in µg/mL) for the PID was 0.005.
The average recovery (in percent) for the HECD was none (no response for this detector).
The standard deviation of the recovery for the HECD was none (no response for this detector)-.
The MDL (in µg/mL) for the HECD was none (no response for this detector).

SAMPLE COLLECTION, PRESERVATION & HANDLING
Volatile Organics — Standard 40-mL glass screw-cap VOA vials with Teflon®-faced silicone septum may be used for both liquid and solid matrices. When collecting samples, liquids and solids should be introduced into the vials gently to reduce agitation which might drive off volatile compounds. If there are any air bubbles present the sample must be retaken. Tap slightly as they are filled to try and eliminate as much free air space as possible. The two vials from each sampling locations should be sealed in separate plastic bags to prevent cross-contamination between samples particularly if the sampled waste is suspected of containing high levels of volatile organics.

Semivolatile organics — Containers used to collect samples for the determination of semivolatile organic compounds should be soap and water washed followed by methanol (or isopropanol) rinsing. The sample containers should be of glass or Teflon® and have screw-top covers with Teflon® liners.

Preservation for volatile organics — No preservation is used with concentrated waste samples. With liquid samples containing no residual chlorine, 4 drops of concentrated hydrochloric acid are added and the samples are immediately cooled to 4°C. When liquid samples contain residual chlorine, they are treated as above and, in addition, 4 drops of 4% aqueous sodium thiosulfate are added. Soil, sediment, and sludge samples are only cooled to 4°C.

Preservation for semivolatile organics — No preservation is used with concentrated waste samples. With liquid samples containing no residual chlorine and with soil, sediment, and sludge samples, immediately cooling to 4°C is the only preservation used. When residual chlorine is present then 3 mL of 10% aqueous sodium sulfate is added for each gallon of sample collected, followed by cooling to 4°C.

MHT The holding time for all volatile organics samples is 14 days. Liquid samples must be extracted within 7 days and their extracts analyzed within 40 days. Concentrated waste, soil, sediment, and sludge samples must be extracted within 14 days and their extracts analyzed within 40 days.

SAMPLE PREPARATION Volatile compounds are introduced into the gas chromatograph either by direct injector or purge-and-trap (EPA Method 5030). EPA Method 5030 may be used directly on groundwater samples or low-concentration contaminated soils and sediments. For medium-concentration

soils or sediments, methanolic extraction, as described in EPA Method 5030, may be necessary prior to purge-and-trap analysis.

QUALITY CONTROL Calculate surrogate standard recovery on all samples, blanks, and spikes. A trip blank is recommended to check on sampling, storage, and handling contamination. Calibration standards, at a minimum of five concentration levels, are prepared in organic-free reagent water. One of the concentration levels should be at a concentration near, but above, the method detection limit.

A combination of bromochloromethane, 2-bromo-1-chloropropane, 1,4-dichlorobutane, and bromochlorobenzene are recommended as surrogate standards to encompass the range of the temperature program used in this method.

REFERENCE Test Methods for Evaluating Solid Waste, Physical/Chemical Methods, SW-846, 3rd Edition, U.S. EPA, Office of Solid Waste, Washington, DC, EPA Method 8021A, Rev. 1, Nov. 1992.

Ethylbenzene   **EPA Method 8240**
CAS #100-41-4

TITLE Volatile Organics By GC/MS: Packed Column Technique

MATRIX Nearly all types of sample matarices, regardless of water content, can be analyzed using this method. This includes groundwater, aqueous sludges, caustic liquors, acid liquors, waste solvents, oily wastes, mousses, tars, fibrous wastes, polymetric emulsions, filter cakes, spent carbons, spent catalysts, soils, and sediments.

METHOD SUMMARY Method 8240B covers 80 volatile organic compounds that are introduced into a gas chromatograph by the purge-and-trap method or by direct injection (in limited applications). For the purge-and-trap method an inert gas (zero grade nitrogen or helium) is bubbled through a 5-mL solution at ambient temperature. Purged sample components are trapped in a tube of sorbent materials. When purging is complete, the sorbent tube is heated and backflushed with inert gas to desorb the trapped components onto a GC column.

INTERFERENCES Impurities in the purge gas and from organic compounds outgassing from the plumbing ahead of the trap account for many contamination problems. Interferences purged or coextracted from the samples will vary considerably from source to source. Cross-contamination can occur whenever high-level and low-level samples are analyzed sequentially. Whenever an unusually concentrated sample is analyzed, it should be followed by the analysis of organic-free reagent water to check for cross-contamination. Samples also can be contaminated by diffusion of volatile organics (particularly methylene chloride and fluorocarbons) through the septum seal into the sample during shipment and storage. A trip blank can serve as a check on such contamination. The lab where volatile analysis is performed and also the refrigerated storage area should be completely free of solvents.

INSTRUMENTATION A gas chromatograph/mass spectrometry/data system (GC/MS) equipped with a 6 ft × 0.1 in I.D. glass column packed with 1% SP-1000 on Carbopack-B (60/80 mesh) is required. Also needed is a 5-mL purging device, a sorbent trap, and a thermal desorption apparatus.

PRECISION & ACCURACY This method is reported to have been tested by 15 laboratories using organic-free reagent water, drinking water, surface water, and industrial wastewaters (not specified) fortified at six concentrations over the range 5–600 µg/L.

Sample estimated quantitation limits (EQLs) are highly matrix-dependent. The EQLs listed may not always be achievable. EQLs listed for soils or sediments are based on wet weight. Normally, data is reported on a dry-weight basis; therefore, EQLs will be higher, based on the percent dry weight of each sample. Note that EQLs are even more variable than MDLs and that they are highly variable depending on the matrix being analyzed.

EQL in groundwater in µg/L was 5.
EQL in low soil or sediment in µg/kg was 5.
Accuracy (a) in µg/L was $0.98C + 2.48$.
Precision (b) in µg/L was $0.26x - 1.72$.

(a) Average recovery found for measurements of samples containing a concentration of C, in µg/L.
(b) Overall precision found for measurements of samples with average recovery X for samples containing a concentration of C in µg/L.
$X = $ *Average recovery found for measurement of samples containing a concentration of C in µg/L.*

MULTIPLICATION FACTORS FOR OTHER MATRICES

Other Matrices	Factor (a)
Waste miscible liquid waste	50
High-concentration soil and sludge	125
Non-water miscible waste	500

(a) EQL = [EQL for low soil/sediment] × [Factor]. For non-aqueous samples, the factor is on a wet-weight basis.

SAMPLING METHOD

Liquid samples — Use a 40-mL glass screw-cap VOA vial with a Teflon®-faced silicone septum that has been prewashed, rinsed with distilled deionized water, and oven dried. However, if residual chlorine is present, collect sample in a 40-oz. soil VOA container which has been pre-preserved with 4 drops of 10% sodium thiosulfate, mix gently, and then transfer the sample to a 40-mL VOA vial. Collect bubble-free samples in duplicate and seal them in separate plastic bags.

Soils or sediments, and sludges — Use an 8-oz. widemouth glass bottle with a Teflon®-faced silicone septum that has been prewashed with detergent, rinsed with distilled deionized water, and oven dried. Tap slightly to eliminate free air space. Collect samples in duplicate and seal them in separate plastic bags.

SAMPLE PRESERVATION

Liquid samples — Add 4 drops of concentrated HCL and immediately cool samples to 4°C and store in a solvent-free refrigerator.

Soils or sediments, and sludges — Cool samples to 4°C and store in a solvent-free refrigerator.

MHT Maximum holding time is 14 days from the date of sample collection.

SAMPLE PREPARATION

Liquid samples — Remove the plunger from a 5-mL syringe and carefully pour the sample into the syringe barrel to just short of overflowing. Replace the syringe plunger and compress the sample. Open the syringe valve and vent any residual air while adjusting the sample volume to 5.0 mL. If there is only one volatile organic analysis (VOA) vial, a second syringe should be filled at this time to protect against possible loss of sample integrity. Add 10 µL of surrogate spiking solution and 10 µL of internal standard spiking solution through the valve bore of the 5-mL syringe, then close the valve. The surrogate and internal standards may be mixed and added as a single spiking solution.

Sediments, soils, and waste samples — All samples of this type should be screened by GC analysis using a headspace method (EPA Method 3810) or the hexadecane extraction and screening method (EPA Method 3820). Use the screening data to determine whether to use the low-concentration method (0.005–1 mg/kg) or the high-concentration method (>1 mg/kg).

Low-concentration method — The low-concentration method is based on purging a heated sediment or soil sample mixed with organic-free reagent water containing the surrogate and internal standards. Analyze all reagent blanks and standards under the same conditions as the samples.

Use a 5-g sample if the expected concentration is <0.1 mg/kg or a 1-g sample for expected concentrations between 0.1 and 1 mg/kg. Mix the contents of the sample container with a narrow metal spatula. Weigh the amount of the sample into a tared purge device. Add the spiked water to the purge device, which contains the weighed amount of sample, and connect the device to the purge-and-trap system.

High-concentration method — This method is based on extracting the sediment or soil with methanol. A waste sample is either extracted or diluted, depending on its solubility in methanol. Wastes that are insoluble in methanol are diluted with reagent tetraglyme or possibly polyethylene glycol (PEG). An aliquot of the extract is added to organic-free reagent water containing surrogate and internal standards. This is purged at ambient temperature. All samples with an expected concentration of >1.0 mg/kg should be analyzed by this method.

Mix the contents of the sample container with a narrow metal spatula. For sediments or soils and solid wastes that are insoluble in methanol, weigh 4 g (wet weight) of sample into a tared 20-mL vial. For waste that is soluble in methanol, tetraglyme, or PEG, weigh 1 g (wet weight) into a tared scintillation vial or culture tube or a 10-mL volumetric flask. Quickly add 9.0 mL of appropriate solvent then add 1.0 mL of a surrogate spiking solution to the vial, cap it, and shake it for 2 min.

METHANOL EXTRACT REQUIRED FOR ANALYSIS OF HIGH-CONCENTRATION SOILS OR SEDIMENTS

Approximate Concentration Range	Volume of Methanol Extract (a)
500–10,000 µg/kg	100 µL
1,000–20,000 µg/kg	50 µL
5,000–100,000 µg/kg	10 µL
25,000–500,000 µg/kg	100 µL of 1/50 dilution (b)

Calculate appropriate dilution factor for concentrations exceeding this table.

(a) The volume of methanol added to 5 mL of water being purged should be kept constant. Therefore, add to the 5-mL syringe whatever volume of methanol is necessary to maintain a volume of 100 µL added to the syringe.
(b) Dilute an aliquot of the methanol extract and then take 100 µL for analysis.

QUALITY CONTROL Demonstrate, through the analysis of a reagent water blank, that interferences from the analytical system, glassware, and reagents are under control. Blank samples should be carried through all stages of the sample preparation and measurement steps. For each analytical batch (up to 20 samples), a reagent blank, matrix spike, and matrix spike duplicate must be analyzed (the frequency of the spikes may be different for different monitoring programs). The blank and spiked samples must be carried through all stages of the sample preparation and measurement steps. QC samples mentioned in the section on Interferences will also be needed as appropriate to those situations.

REFERENCE Test Methods for Evaluating Solid Waste (SW-846). U.S. EPA. 1983. Method 8240B, Rev. 2, Nov. 1990. Office of Solid Wastes, Washington, DC.

Ethylbenzene   **EPA Method 8260**
CAS #100-41-4

TITLE Volatile Organic Compounds by GC/MS: Capillary Column Technique

MATRIX This method is applicable to nearly all types of samples, regardless of water content, including groundwater, soils, and sediments.

METHOD SUMMARY Method 8260A covers 58 volatile organic compounds that are introduced into a gas chromatograph by the purge-and-trap method or by direct injection (in limited applications). Zero-grade helium is bubbled through a 5-mL solution at ambient temperature. Purged sample components are trapped in a tube containing suitable sorbent materials. When purging is complete, the sorbent tube is heated and backflushed with helium to desorb trapped sample components. The analytes are desorbed directly to a large bore capillary or cryofocussed on a capillary precolumn before being flash evaporated to a narrow bore capillary for analysis.

INTERFERENCES Major contaminant sources are volatile materials in the lab and impurities in the inert purging gas and in the sorbent trap. Interfering contamination may occur when a sample containing low concentrations of volatile organic compounds is analyzed immediately after a sample containing high concentrations of volatile organic compounds. After analysis of a sample containing high concentrations of volatile organic compounds, one or more calibration blanks should be analyzed to check for cross-contamination. Screening of the samples prior to purge-and-trap GC/MS analysis is highly recommended to prevent contamination of the system. This is especially true for soil and waste samples.

Special precautions must be taken to analyze for methylene chloride. The analytical and sample storage area should be isolated from all atmospheric sources of methylene chloride. All gas chromatography carrier gas lines and purge gas plumbing should be constructed from stainless steel or copper tubing. Laboratory clothing previously exposed to methylene chloride fumes during liquid-liquid extraction procedures can contribute to sample contamination.

Samples can also be contaminated by diffusion of volatile organics (particularly methylene chloride and fluorocarbons) through the septum seal during shipment and storage. A trip blank can serve as a check on such contamination.

INSTRUMENTATION GC/MS with a temperature-programmable chromatograph suitable for splitless injection equipped with variable constant differential flow controllers, a subambient oven controller, a purging device, sorbent trap, a thermal desorption apparatus and a capillary precolumn interface when using cryogenic cooling will be needed. The following GC columns may be used:

Column 1: 60 m × 0.75 mm I.D. capillary column coated with VOCOL, 1.5 μm film thickness.
Column 2: 30 m × 0.53 mm capillary column coated with DB-624 or VOCOL, 3 μm film thickness.
Column 3: 30 m × 0.32 mm I.D. capillary column coated with DB-5 or SE-54, 1-μm film thickness.

PRECISION & ACCURACY This method has been tested in a single lab using spiked water. Using a wide-bore capillary column, water was spiked at concentrations between 0.5 and 10 μg/L. Single lab accuracy and precision data are presented. The MDL actually achieved in a given analysis will vary depending on instrument sensitivity and matrix effects.

The MDL (a) in μg/L was 0.06.
The concentration range in μg/L was 0.1–10.
The mean accuracy (% of true value) was 99.
The precision as relative standard deviation was 8.6.

Note: The MDL is based on a 25-mL sample volume instead of a 5-mL sample volume.

SAMPLING METHOD
Liquid samples — Use a 40-mL glass screw-cap VOA vial with a Teflon®-faced silicone septum that has been prewashed, rinsed with distilled deionized water, and oven dried. If residual chlorine is present, collect the sample in a 4-oz soil VOA container which has been pre-preserved with 4 drops of 10% sodium thiosulfate. Mix gently and transfer the sample to a 40-mL VOA vial. Collect bubble-free samples in duplicate and seal each sample in a separate plastic bag.

Soils, sediments and sludges — Use an 8-oz widemouth glass bottle with Teflon®-faced silicone septum that has been prewashed, rinsed with distilled deionized water, and oven dried. **Do not** heat the septum for more than 1 h. Tap slightly to eliminate any free air space. Collect samples in duplicate and seal each one in a separate plastic bag.

SAMPLE PRESERVATION
Liquid samples — Add 4 drops of concentrated HCL, cool to 4°C and store in a solvent-free refrigerator.

Soils, sediments and sludges — Cool samples to 4°C and store in a solvent-free refrigerator.

MHT The maximum holding time of any sample (liquids, soils, sediments, and sludges) is 14 days.

SAMPLE PREPARATION
Liquid samples — Remove the plunger from a 5-mL syringe and carefully pour the sample into the syringe barrel to just short of overflowing. Replace the syringe plunger and compress the sample. Open the syringe valve and vent any residual air while adjusting the sample volume to 5.0 mL. If there is only one volatile organic analysis (VOA) vial, a second syringe should be filled at this time to protect against possible loss of sample integrity. Add 10 μL of surrogate spiking solution and 10 μL of internal standard spiking solution through the valve bore of the 5-mL syringe, then close the valve. The surrogate and internal standards may be mixed and added as a single spiking solution.

Sediments, soils, and waste samples — All samples of this type should be screened by GC analysis using a headspace method (EPA Method 3810) or the hexadecane extraction and screening method (EPA Method 3820). Use the screening data to determine whether to use the low-concentration method (0.005–1 mg/kg) or the high-concentration method (>1 mg/kg).

Low-concentration method — The low-concentration method is based on purging a heated sediment or soil sample mixed with organic-free reagent water containing the surrogate and internal standards. Analyze all reagent blanks and standards under the same conditions as the samples.

Use a 5-g sample if the expected concentration is <0.1 mg/kg or a 1-g sample for expected concentrations between 0.1 and 1 mg/kg. Mix the contents of the sample container with a narrow metal spatula. Weigh the amount of the sample into a tared purge device. Add the spiked water to the purge device, which contains the weighed amount of sample, and connect the device to the purge-and-trap system.

High-concentration method — This method is based on extracting the sediment or soil with methanol. A waste sample is either extracted or diluted, depending on its solubility in methanol. Wastes that are insoluble in methanol are diluted with reagent tetraglyme or possibly polyethylene glycol (PEG). An aliquot of the extract is added to organic-free reagent water containing surrogate and internal standards. This is purged at ambient temperature. All samples with an expected concentration of >1.0 mg/kg should be analyzed by this method.

Mix the contents of the sample container with a narrow metal spatula. For sediments or soils and solid wastes that are insoluble in methanol, weigh 4 g (wet weight) of sample into a tared

20-mL vial. For waste that is soluble in methanol, tetraglyme, or PEG, weigh 1 g (wet weight) into a tared scintillation vial or culture tube or a 10-mL volumetric flask. Quickly add 9.0 mL of appropriate solvent then add 1.0 mL of a surrogate spiking solution to the vial, cap it, and shake it for 2 min.

METHANOL EXTRACT REQUIRED FOR ANALYSIS OF HIGH-CONCENTRATION SOILS OR SEDIMENTS

Approximate Concentration Range	Volume of Methanol Extract (a)
500–10,000 µg/kg	100 µL
1,000–20,000 µg/kg	50 µL
5,000–100,000 µg/kg	10 µL
25,000–500,000 µg/kg	100 µL of 1/50 dilution (b)

Calculate appropriate dilution factor for concentrations exceeding this table.

(a) The volume of methanol added to 5 mL of water being purged should be kept constant. Therefore, add to the 5-mL syringe whatever volume of methanol is necessary to maintain a volume of 100 µL added to the syringe.

(b) Dilute an aliquot of the methanol extract and then take 100 µL for analysis.

QUALITY CONTROL Demonstrate, through the analysis of a reagent water blank, that interferences from the analytical system, glassware, and reagents are under control. Blank samples should be carried through all stages of the sample preparation and measurement steps. For each analytical batch (up to 20 samples), a reagent blank, matrix spike, and matrix spike duplicate must be analyzed (the frequency of the spikes may be different for different monitoring programs). The blank and spiked samples must be carried through all stages of the sample preparation and measurement steps. QC samples mentioned in the section on Interferences will also be needed as appropriate to those situations.

Matrix spiking standards should be prepared from volatile organic compounds which will be representative of the compounds being investigated. The recommended internal standards are chlorobenzene-d5, 1,4-difluorobenzene, 1,4-dichlorobenzene-d4, and pentafluorobenzene. Using stock standard solutions, prepare secondary dilution standards containing the compounds of interest, either singly or mixed together in methanol. Store them in a vial with no headspace for no more than one week. Surrogates recommended are toluene-d8, 4-bromofluorobenzene, and dibromofluoromethane. Each sample undergoing GC/MS analysis must be spiked with 10 µL of the surrogate spiking solution prior to analysis.

REFERENCE Test Methods for Evaluating Solid Waste (SW-846). U.S. EPA 1983, Method 8260A, Rev. 1, Nov. 1990. Office of Solid Waste, Washington, DC.

Ethylbenzene   **EPA Method 503.1**
CAS #100-41-4

TITLE Aromatic & Unsaturated VOCs

MATRIX Drinking water (finished or in Water any treatment stage) and raw source water.

APPLICATION Method covers 28 aromatic and unsaturated VOCs. An inert gas is bubbled through a 5-mL water sample. Purged sample components are trapped in tube of sorbent materials. When purging is complete, sorbent tube is heated and backflushed with inert gas to desorb trapped sample onto a packed GC column.

INTERFERENCES During analysis, major contaminant sources are volatile materials in the lab and impurities in purging gas and sorbent trap. With high and low level samples, there can be carryover contamination. Excess water causes a negative baseline deflection.

INSTRUMENTATION Purge and Trap GC w/photoionization detector. (Two GC columns are recommended); Column 1: 5% SP-1200 and 1.75% Bentone 34 on Supelcoport; Column 2: 1,2,3-tris(2-cyanoethoxy)propane on Chromosorb W.

RANGE 2.2–600 µg/L. (Drinking water).

MDL 0.002 µg/L in water

PRECISION RSD = 8.5% at 0.40 µg/L conc.; 7 samples

ACCURACY Average recovery = 93% at 0.40 µg/L conc.; 7 samples

SAMPLING METHOD Use a 40–120-mL screw-cap vial (prewashed with detergent, rinsed with distilled water and oven dried at 105°C) with a PTFE-faced silicone septum. If residual chlorine is in the water add about 25 mg of ascorbic acid to each vial before sample collection. Collect bubble-free samples.

STABILITY Cool to 4°C; HCl to pH <2.

MHT 14 days.

QUALITY CONTROL As initial demonstration of lab accuracy and precision, analyze 4 to 7 replicates of a lab fortified blank containing the analyte at 0.1–5 µg/L. Collect all samples in duplicate.

REFERENCE Method 503.1, Volatile Aromatic & Unsaturated Organic Compounds in H2O by Purge and Trap GC, EPA 600/4-88/039.

Ethylbenzene   **EPA Method 602**
CAS #100-41-4

TITLE Purgeable Aromatics

MATRIX Wastewater

APPLICATION Method covers 7 purgeable aromatics. (Method 624 provides GC/MS conditions appropriate for the qualitative and quantitative confirmation of results). EPA Method describes conditions for a 2nd GC column to confirm measurements made with primary column.

INTERFERENCES Impurities in the purge gas and organic compounds outgassing from the plumbing ahead of the trap. With high- and low-level samples, there can be carryover contamination. Diffusion of volatile organics through the septum seal into the sample.

INSTRUMENTATION GC-equipped with photoionization detector. (With purge-and-trap unit)

RANGE 2.1–550 µg/L.

MDL 0.2 µg/L.

PRECISION 0.26X + 0.23 µg/L (overall precision).

ACCURACY 0.94C + 0.31 µg/L (as recovery).

SAMPLING METHOD 25-mL glass vial. Teflon®-lined septum.

STABILITY Cool, 4°C, 0.008% Na2S2O3. HCl to pH 2.

MHT 14 days.

QUALITY CONTROL The lab must on an ongoing basis, spike at least 10% of the samples from each sample site being monitored to assess accuracy.

REFERENCE Method 602, *Federal Register* Part VIII 40 CFR Part 136, Oct 26, 1984.

Ethylbenzene — EPA Method 624
CAS #100-41-4

TITLE Purgeables

MATRIX Wastewater

APPLICATION Method covers 31 purgeable organics. An inert gas is bubbled through a 5-mL water sample in a specially designed purging chamber. Here, purgeables are transferred from aqueous to gaseous phase, passed onto a sorbent column, and trapped. Trap is heated and backflushed with inert gas to desorb purgeables onto a GC column, where purgeables are separated.

INTERFERENCES Impurities in the purge gas, organic compounds outgassing from the plumbing ahead of the trap, and solvent vapors in the lab. With high- and low-level samples, there can be carryover contamination.

INSTRUMENTATION GC/MS with purge-and-trap unit.

RANGE 5–600 µg/L

MDL 7.2 µg/L

PRECISION 0.26X–1.72 µg/L (overall precision).

ACCURACY 0.98C + 2.48 µg/L (as recovery).

SAMPLING METHOD 25-mL glass vial. Teflon®-lined septum.

STABILITY Cool, 4°C, 0.008% Na2S2O3. HCl to pH 2.

MHT 14 days.

QUALITY CONTROL The lab must on an ongoing basis, spike at least 5% of the samples from each sample site being monitored to assess accuracy.

REFERENCE Method 624, *Federal Register* Part VIII 40 CFR Part 136, Oct 26, 1984.

Ethylbenzene — EPA Method 8020
CAS #100-41-4

TITLE Aromatic Volatile Organics

MATRIX Groundwater, soils, sludges, water miscible liquid wastes, and non-water miscible wastes.

APPLICATION This method is used to analyze for 8 aromatic VOCs. Samples are analyzed using direct injection or purge-and-trap methods. Groundwater must be analyzed by the purge-and-trap method. The method provides an optional GC column that is used for analyte confirmation and may also help resolve analytes from interferences.

INTERFERENCES There can be carryover contamination with high- and low-level samples. Impurities may come from the purge-and-trap apparatus, organic compounds outgassing from the plumbing ahead of trap, diffusion of VOCs through the sample bottle septum during shipping or storage, or from solvent vapors in the lab.

INSTRUMENTATION GC capable of on-column injections or purge-and-trap sample introduction and a photoionization detector (PID). Column 1: 6 ft by 0.082 in with 5% SP-1200 and 1.75% Bentone-34 on Supelcoport. Column 2: 8 ft by 0.1 in with 5% 1,2,3-tris(2-cyanoethoxy)propane on Chromosorb W-AW.

RANGE 2.1–500 µg/L

MDL 0.2 µg/L (reagent water).

PQL FACTORS FOR MULTIPLYING × MDL VALUE

Matrix	Multiplication Factor
Groundwater	10
Low-level soil	10
Water miscible liquid waste	500
High-level soil and sludge	1250
Non-water miscible waste	1250

PRECISION 0.26X + 0.23 µg/L (overall precision).

ACCURACY 0.94C + 0.31 µg/L (as recovery).

SAMPLING METHOD For water and liquid samples use glass 40-mL vials with Teflon®-lined septum caps and collect two vials per sample location with no headspace. For solids and concentrated waste samples use widemouth glass bottles with Teflon® liners. Cool all samples to 4°C

STABILITY For concentrated wastes, soils, sediments, or sludges cool to 4°C. For liquids, add 4 drops of concentrated hydrochloric acid and cool to 4°C.

MHT 14 days.

QUALITY CONTROL Analyze a reagent blank, matrix spike, and matrix spike duplicate/duplicate for each analytical batch (up to 20 samples). Demonstrate the purity of glassware and reagents by analyzing a reagent water method blank. Internal, surrogate, and five concentration level calibration standards are used. The QC check sample concentrate should contain this compound at 10 µg/mL in methanol.

REFERENCE Test Methods for Evaluating Solid Waste (SW-846), U.S. EPA Office of Solid Waste, Washington, DC, Method 8020A, Rev. 1, Nov. 1992.

Ethylene oxide EPA Method 8240
CAS #75-21-8

TITLE Volatile Organics By GC/MS: Packed Column Technique

MATRIX Nearly all types of sample matarices, regardless of water content, can be analyzed using this method. This includes groundwater, aqueous sludges, caustic liquors, acid liquors, waste solvents, oily wastes, mousses, tars, fibrous wastes, polymetric emulsions, filter cakes, spent carbons, spent catalysts, soils, and sediments.

METHOD SUMMARY Method 8240B covers 80 volatile organic compounds that are introduced into a gas chromatograph by the purge-and-trap method or by direct injection (in limited applications). For the purge-and-trap method an inert gas (zero grade nitrogen or helium) is bubbled through a 5-mL solution at ambient temperature. Purged sample components are trapped in a tube of sorbent materials. When purging is complete, the sorbent tube is heated and backflushed with inert gas to desorb the trapped components onto a GC column.

INTERFERENCES Impurities in the purge gas and from organic compounds outgassing from the plumbing ahead of the trap account for many contamination problems. Interferences purged or coextracted from the samples will vary considerably from source to source. Cross-contamination can occur whenever high-level and low-level samples are analyzed sequentially. Whenever an unusually concentrated sample is analyzed, it should be followed by the analysis of organic-free reagent water to check for cross-contamination. Samples also can be contaminated by diffusion of volatile organics (particularly methylene chloride and fluorocarbons) through the septum seal into the sample during shipment and storage. A trip blank can serve as a check on such contamination. The lab where volatile analysis is performed and also the refrigerated storage area should be completely free of solvents.

INSTRUMENTATION A gas chromatograph/mass spectrometry/data system (GC/MS) equipped with a 6 ft × 0.1 in I.D. glass column packed with 1% SP-1000 on Carbopack-B (60/80 mesh) is required. Also needed is a 5-mL purging device, a sorbent trap, and a thermal desorption apparatus.

PRECISION & ACCURACY This method is reported to have been tested by 15 laboratories using organic-free reagent water, drinking water, surface water, and industrial wastewaters (not specified) fortified at six concentrations over the range 5–600 µg/L.

Sample estimated quantitation limits (EQLs) are highly matrix-dependent. The EQLs listed may not always be achievable. EQLs listed for soils or sediments are based on wet weight. Normally, data is reported on a dry-weight basis; therefore, EQLs will be higher, based on the percent dry weight of each sample. Note that EQLs are even more variable than MDLs and that they are highly variable depending on the matrix being analyzed.

EQL in groundwater in µg/L was not listed.
EQL in low soil or sediment in µg/kg was not listed.
Accuracy (a) in µg/L was not listed.
Precision (b) in µg/L was not listed.

(a) *Average recovery found for measurements of samples containing a concentration of C, in µg/L.*
(b) *Overall precision found for measurements of samples with average recovery X for samples containing a concentration of C in µg/L.*
X = *Average recovery found for measurement of samples containing a concentration of C in µg/L.*

MULTIPLICATION FACTORS FOR OTHER MATRICES

Other Matrices	Factor (a)
Waste miscible liquid waste	50
High-concentration soil and sludge	125
Non-water miscible waste	500

(a) *EQL = [EQL for low soil/sediment] × [Factor]. For non-aqueous samples, the factor is on a wet-weight basis.*

SAMPLING METHOD
Liquid samples — Use a 40-mL glass screw-cap VOA vial with a Teflon®-faced silicone septum that has been prewashed, rinsed with distilled deionized water, and oven dried. However, if residual chlorine is present, collect sample in a 40-oz. soil VOA container which has been pre-preserved with 4 drops of 10% sodium thiosulfate, mix gently, and then transfer the sample to a 40-mL VOA vial. Collect bubble-free samples in duplicate and seal them in separate plastic bags.

Soils or sediments, and sludges — Use an 8-oz. widemouth glass bottle with a Teflon®-faced silicone septum that has been prewashed with detergent, rinsed with distilled deionized water, and oven dried. Tap slightly to eliminate free air space. Collect samples in duplicate and seal them in separate plastic bags.

SAMPLE PRESERVATION
Liquid samples — Add 4 drops of concentrated HCL and immediately cool samples to 4°C and store in a solvent-free refrigerator.

Soils or sediments, and sludges — Cool samples to 4°C and store in a solvent-free refrigerator.

MHT Maximum holding time is 14 days from the date of sample collection.

SAMPLE PREPARATION
Liquid samples — Remove the plunger from a 5-mL syringe and carefully pour the sample into the syringe barrel to just short of overflowing. Replace the syringe plunger and compress the sample. Open the syringe valve and vent any residual air

while adjusting the sample volume to 5.0 mL. If there is only one volatile organic analysis (VOA) vial, a second syringe should be filled at this time to protect against possible loss of sample integrity. Add 10 µL of surrogate spiking solution and 10 µL of internal standard spiking solution through the valve bore of the 5-mL syringe, then close the valve. The surrogate and internal standards may be mixed and added as a single spiking solution.

Sediments, soils, and waste samples — All samples of this type should be screened by GC analysis using a headspace method (EPA Method 3810) or the hexadecane extraction and screening method (EPA Method 3820). Use the screening data to determine whether to use the low-concentration method (0.005–1 mg/kg) or the high-concentration method (>1 mg/kg).

Low-concentration method — The low-concentration method is based on purging a heated sediment or soil sample mixed with organic-free reagent water containing the surrogate and internal standards. Analyze all reagent blanks and standards under the same conditions as the samples.

Use a 5-g sample if the expected concentration is <0.1 mg/kg or a 1-g sample for expected concentrations between 0.1 and 1 mg/kg. Mix the contents of the sample container with a narrow metal spatula. Weigh the amount of the sample into a tared purge device. Add the spiked water to the purge device, which contains the weighed amount of sample, and connect the device to the purge-and-trap system.

High-concentration method — This method is based on extracting the sediment or soil with methanol. A waste sample is either extracted or diluted, depending on its solubility in methanol. Wastes that are insoluble in methanol are diluted with reagent tetraglyme or possibly polyethylene glycol (PEG). An aliquot of the extract is added to organic-free reagent water containing surrogate and internal standards. This is purged at ambient temperature. All samples with an expected concentration of >1.0 mg/kg should be analyzed by this method.

Mix the contents of the sample container with a narrow metal spatula. For sediments or soils and solid wastes that are insoluble in methanol, weigh 4 g (wet weight) of sample into a tared 20-mL vial. For waste that is soluble in methanol, tetraglyme, or PEG, weigh 1 g (wet weight) into a tared scintillation vial or culture tube or a 10-mL volumetric flask. Quickly add 9.0 mL of appropriate solvent then add 1.0 mL of a surrogate spiking solution to the vial, cap it, and shake it for 2 min.

METHANOL EXTRACT REQUIRED FOR ANALYSIS OF HIGH-CONCENTRATION SOILS OR SEDIMENTS

Approximate Concentration Range	Volume of Methanol Extract (a)
500–10,000 µg/kg	100 µL
1,000–20,000 µg/kg	50 µL
5,000–100,000 µg/kg	10 µL
25,000–500,000 µg/kg	100 µL of 1/50 dilution (b)

Calculate appropriate dilution factor for concentrations exceeding this table.

(a) The volume of methanol added to 5 mL of water being purged should be kept constant. Therefore, add to the 5-mL syringe whatever volume of methanol is necessary to maintain a volume of 100 µL added to the syringe.
(b) Dilute an aliquot of the methanol extract and then take 100 µL for analysis.

QUALITY CONTROL Demonstrate, through the analysis of a reagent water blank, that interferences from the analytical system, glassware, and reagents are under control. Blank samples should be carried through all stages of the sample preparation and measurement steps. For each analytical batch (up to 20 samples), a reagent blank, matrix spike, and matrix spike duplicate must be analyzed (the frequency of the spikes may be different for different monitoring programs). The blank and spiked samples must be carried through all stages of the sample preparation and measurement steps. QC samples mentioned in the section on Interferences will also be needed as appropriate to those situations.

REFERENCE Test Methods for Evaluating Solid Waste (SW-846). U.S. EPA. 1983. Method 8240B, Rev. 2, Nov. 1990. Office of Solid Wastes, Washington, DC.

Ethylenethiourea EPA Method 1625
CAS #96-45-7

TITLE Semivolatile Organic Compounds by Isotope Dilution GC/MS

MATRIX The compounds may be determined in waters, soils, and municipal sludges by this method.

METHOD SUMMARY This method is used to determine 176 semivolatile toxic organic pollutants associated with the CWA (as amended 1987); the RCRA (as amended 1986); the CERCLA (as amended 1986); and other compounds amenable to extraction and analysis by capillary column gas chromatography-mass spectrometry (GC/MS).

Stable isotopically-labeled analogs of the compounds of interest are added to the sample. If the solids content is less than 1%, a 1-L sample is extracted at pH 12–13, then at pH <2 with methylene chloride using continuous extraction techniques.

If the solids content is 30% or less, the sample is diluted to 1% solids with reagent water, homogenized ultrasonically, and extracted at pH 12–13, then at pH <2 with methylene chloride using continuous extraction techniques. If the solids content is greater than 30%, the sample is extracted using ultrasonic techniques.

Each extract is dried over sodium sulfate, concentrated to a volume of 5 mL, cleaned up using GPC, if necessary, and concentrated. Extracts are concentrated to 1 mL if GPC is not performed, and to 0.5 mL if GPC is performed.

An internal standard is added to the extract, and a 1-mL aliquot of the extract is injected into the GC. The compounds are separated by GC and detected by a MS. The labeled compounds serve to correct the variability of the analytical technique.

INTERFERENCES Solvents, reagents, glassware, and other sample processing hardware may yield artifacts and/or elevated

baselines causing misinterpretation of chromatograms and spectra. Materials used in the analysis must be demonstrated to be free from interferences under the conditions of analysis by running method blanks initially and with each sample lot (sample started through the extraction process on a given 8-h shift, to a maximum of 20). Specific selection of reagents and purification of solvents by distillation in all glass systems may be required. Glassware and, where possible, reagents are cleaned by solvent rinse and baking at 450°C for 1-h minimum. Interferences coextracted from samples will vary considerably from source to source, depending on the diversity of the site being sampled.

INSTRUMENTATION Major instrumentation includes a GC with a splitless or on-column injection port for capillary column, a MS with 70 eV electron impact ionization, and a data system to collect and record MS data, and process it. A K-D apparatus is used to concentrate extracts.

GC Column: 30 m × 0.25 mm I.D. 5% phenyl, 94% methyl, 1% vinyl silicone bonded phased fused silica capillary column.

PRECISION & ACCURACY The detection limits of the method are usually dependent on the level of interferences rather than instrumental limitations. The limits typify the minimum quantities that can be detected with no interferences present.

The minimum level (in µg/mL) was not listed. This is defined as a minimum level at which the analytical system shall give recognizable mass spectra (background corrected) and acceptable calibration points.

The MDL (in µg/kg) in low solids was not listed and in high solids was not listed; these were determined in digested sludge (low solids) and in filter cake or compost (high solids).

The labeled and native compound initial precision as standard deviation (in µg/L) was not listed.
The labeled and native compound initial accuracy as average recovery (in µg/L) was not listed.

SAMPLE COLLECTION, PRESERVATION & HANDLING
Collect samples in glass containers. Aqueous samples which flow freely are collected in refrigerated bottles using automatic sampling equipment. Solid samples are collected as grab samples using widemouth jars. Maintain samples at 0 to 4°C from the time of collection until extraction. If residual chlorine is present in aqueous samples, add 80 mg sodium thiosulfate/L of water. Begin sample extraction within 7 days of collection, and analyze all extracts within 40 days of extraction.

SAMPLE PREPARATION Samples containing 1% solids or less are extracted directly using continuous liquid-liquid extraction techniques. Samples containing 1 to 30% solids are diluted to the 1% level with reagent water and extracted using continuous liquid-liquid extraction techniques. Samples containing greater than 30% solids are extracted using ultrasonic techniques.

Base/neutral extraction — Adjust the pH of the waters in the extractors to 12–13 with 6 N NaOH. Extract with methylene chloride for 24–48 h.

Acid extraction — Adjust the pH of the waters in the extractors to 2 or less using 6 N sulfuric acid. Extract with methylene chloride for 24–48 h.

Ultrasonic extraction of high solids samples — Add anhydrous sodium sulfate to the sample and QC aliquot(s). Add acetone:methylene chloride (1:1) to the sample and mix thoroughly

Concentrate extracts using a K-D apparatus.

QUALITY CONTROL The analyst is permitted to modify this method to improve separations or lower the costs of measurements, provided all performance specifications are met. Analyses of blanks are required to demonstrate freedom from contamination. When results of spikes indicate atypical method performance for samples, the samples are diluted to bring method performance within acceptable limits.

For low solids (aqueous samples), extract, concentrate, and analyze two sets of four 1-L aliquots (8 aliquots total) of the precision and recovery standard. For high solids samples, two sets of four 30-g aliquots of the high solids reference matrix are used.

Spike all samples with labeled compounds to assess method performance. Compute percent recovery of the labeled compounds using the internal standard method. Compare the labeled compound recovery for each compound with the corresponding labeled compound recovery.

Reagent water and high solids reference matrix blanks are analyzed to demonstrate freedom from contamination. Extract and concentrate a 1-L reagent water blank or a high solids reference matrix blank with each sample's lot (samples started through the extraction process on the same 8-h shift, to a maximum of 20 samples).

Field replicates may be collected to determine the precision of the sampling technique, and spiked samples may be required to determine the accuracy of the analysis when the internal standard method is used.

REFERENCE Semivolatile Organic Compounds by Isotope Dilution GC/MS. Office of Water Regulation and Standards, U.S. EPA Industrial Technology Division, Washington, DC, EPA Method 1625, Rev. C, June 1989 (contact W.A. Telliard, U.S. EPA, Office of Water Regulations and Standards, 401 M St., SW, Washington, DC, 20460. Phone: 202-382-7131).

Ethynylestradiol-3 methyl ether   **EPA Method 1625**
CAS #72-33-3

TITLE Semivolatile Organic Compounds by Isotope Dilution GC/MS

MATRIX The compounds may be determined in waters, soils, and municipal sludges by this method.

METHOD SUMMARY This method is used to determine 176 semivolatile toxic organic pollutants associated with the CWA (as amended 1987); the RCRA (as amended 1986); the

CERCLA (as amended 1986); and other compounds amenable to extraction and analysis by capillary column gas chromatography-mass spectrometry (GC/MS).

Stable isotopically-labeled analogs of the compounds of interest are added to the sample. If the solids content is less than 1%, a 1-L sample is extracted at pH 12–13, then at pH <2 with methylene chloride using continuous extraction techniques.

If the solids content is 30% or less, the sample is diluted to 1% solids with reagent water, homogenized ultrasonically, and extracted at pH 12–13, then at pH <2 with methylene chloride using continuous extraction techniques. If the solids content is greater than 30%, the sample is extracted using ultrasonic techniques.

Each extract is dried over sodium sulfate, concentrated to a volume of 5 mL, cleaned up using GPC, if necessary, and concentrated. Extracts are concentrated to 1 mL if GPC is not performed, and to 0.5 mL if GPC is performed.

An internal standard is added to the extract, and a 1-mL aliquot of the extract is injected into the GC. The compounds are separated by GC and detected by a MS. The labeled compounds serve to correct the variability of the analytical technique.

INTERFERENCES Solvents, reagents, glassware, and other sample processing hardware may yield artifacts and/or elevated baselines causing misinterpretation of chromatograms and spectra. Materials used in the analysis must be demonstrated to be free from interferences under the conditions of analysis by running method blanks initially and with each sample lot (sample started through the extraction process on a given 8-h shift, to a maximum of 20). Specific selection of reagents and purification of solvents by distillation in all glass systems may be required. Glassware and, where possible, reagents are cleaned by solvent rinse and baking at 450°C for 1-h minimum. Interferences coextracted from samples will vary considerably from source to source, depending on the diversity of the site being sampled.

INSTRUMENTATION Major instrumentation includes a GC with a splitless or on-column injection port for capillary column, a MS with 70 eV electron impact ionization, and a data system to collect and record MS data, and process it. A K-D apparatus is used to concentrate extracts.

GC Column: 30 m × 0.25 mm I.D. 5% phenyl, 94% methyl, 1% vinyl silicone bonded phased fused silica capillary column.

PRECISION & ACCURACY The detection limits of the method are usually dependent on the level of interferences rather than instrumental limitations. The limits typify the minimum quantities that can be detected with no interferences present.

The minimum level (in µg/mL) was not listed. This is defined as a minimum level at which the analytical system shall give recognizable mass spectra (background corrected) and acceptable calibration points.

The MDL (in µg/kg) in low solids was not listed and in high solids was not listed; these were determined in digested sludge (low solids) and in filter cake or compost (high solids).

The labeled and native compound initial precision as standard deviation (in µg/L) was not listed.
The labeled and native compound initial accuracy as average recovery (in µg/L) was not listed.

SAMPLE COLLECTION, PRESERVATION & HANDLING
Collect samples in glass containers. Aqueous samples which flow freely are collected in refrigerated bottles using automatic sampling equipment. Solid samples are collected as grab samples using widemouth jars. Maintain samples at 0 to 4°C from the time of collection until extraction. If residual chlorine is present in aqueous samples, add 80 mg sodium thiosulfate/L of water. Begin sample extraction within 7 days of collection, and analyze all extracts within 40 days of extraction.

SAMPLE PREPARATION Samples containing 1% solids or less are extracted directly using continuous liquid-liquid extraction techniques. Samples containing 1 to 30% solids are diluted to the 1% level with reagent water and extracted using continuous liquid-liquid extraction techniques. Samples containing greater than 30% solids are extracted using ultrasonic techniques.

Base/neutral extraction — Adjust the pH of the waters in the extractors to 12–13 with 6 N NaOH. Extract with methylene chloride for 24–48 h.
Acid extraction — Adjust the pH of the waters in the extractors to 2 or less using 6 N sulfuric acid. Extract with methylene chloride for 24–48 h.
Ultrasonic extraction of high solids samples — Add anhydrous sodium sulfate to the sample and QC aliquot(s). Add acetone:methylene chloride (1:1) to the sample and mix thoroughly

Concentrate extracts using a K-D apparatus.

QUALITY CONTROL The analyst is permitted to modify this method to improve separations or lower the costs of measurements, provided all performance specifications are met. Analyses of blanks are required to demonstrate freedom from contamination. When results of spikes indicate atypical method performance for samples, the samples are diluted to bring method performance within acceptable limits.

For low solids (aqueous samples), extract, concentrate, and analyze two sets of four 1-L aliquots (8 aliquots total) of the precision and recovery standard. For high solids samples, two sets of four 30-g aliquots of the high solids reference matrix are used.

Spike all samples with labeled compounds to assess method performance. Compute percent recovery of the labeled compounds using the internal standard method. Compare the labeled compound recovery for each compound with the corresponding labeled compound recovery.

Reagent water and high solids reference matrix blanks are analyzed to demonstrate freedom from contamination. Extract and concentrate a 1-L reagent water blank or a high solids reference matrix blank with each sample's lot (samples started through the extraction process on the same 8-h shift, to a maximum of 20 samples).

Field replicates may be collected to determine the precision of the sampling technique, and spiked samples may be required to determine the accuracy of the analysis when the internal standard method is used.

REFERENCE Semivolatile Organic Compounds by Isotope Dilution GC/MS. Office of Water Regulation and Standards, U.S. EPA Industrial Technology Division, Washington, DC, EPA Method 1625, Rev. C, June 1989 (contact W.A. Telliard, U.S. EPA, Office of Water Regulations and Standards, 401 M St., SW, Washington, DC, 20460. Phone: 202-382-7131).

F

Famphur
CAS #52-85-7

EPA Method 8270

TITLE Semivolatile Organic Compounds by GC/MS

MATRIX This method is used to determine the concentration of semivolatile organic compounds in extracts prepared from all types of solid waste matrices, soils, and groundwater. Although surface waters are not specifically mentioned, this method should be applicable to water samples from rivers, lakes, etc.

METHOD SUMMARY This method covers 259 semivolatile organic compounds. In very limited applications direct injection of the sample into the GC/MS system may be appropriate, but this results in very high detection limits (approximately 10,000 μg/L). Typically, a 1-L liquid sample, containing surrogate, and matrix spiking standards, is extracted in a continuous extractor first under acid conditions and then under basic conditions. Typically 30 g of a solid sample, containing surrogate, and matrix spiking standards, is extracted ultrasonically. After concentrating the extract to 1 mL it is spiked with 10 μL of an internal standard solution just prior to analysis by GC/MS. The volume injected should contain about 100 ng of base/neutral and 200 ng of acid surrogates (for a 1-μL injection). Analysis is performed by GC/MS using a capillary GC column.

INTERFERENCES Raw GC/MS data from all blanks, samples, and spikes must be evaluated for interferences. Contamination by carryover can occur whenever high-concentration and low-concentration samples are sequentially analyzed. To reduce carryover, the sample syringe must be rinsed out between samples with solvent. Whenever an unusually concentrated sample is encountered, it should be followed by the analysis of blank solvent to check for cross-contamination.

INSTRUMENTATION A GC/MS and a data system are required. The GC column used is a 30 m × 0.25 mm I.D. (or 0.32 mm I.D.) 1um film thickness silicone-coated fused silica capillary column. A continuous liquid-liquid extractor equipped with Teflon® or glass connection joints and stopcocks requiring no lubrication, a K-D concentrating apparatus, water bath, and an ultrasonic disrupter with a minimum power of 300 W and with pulsing capability are also required.

PRECISION & ACCURACY The estimated quantitation limit (EQL) of Method 8270B for determining an individual compound is approximately 1 mg/kg (wet weight) for soil or sediment samples, 1–200 mg/kg for wastes (dependent on matrix and method of preparation), and 10 μg/L for groundwater samples. EQLs will be proportionately higher for sample extracts that require dilution to avoid saturation of the detector.

The EQL(b) for groundwater in μg/L is 20.
The EQL (a, b) for low concentrations in soil and sediment in μg/kg is not determined.
Accuracy as μg/L is not listed.
Overall precision in μg/L is not listed.

(a) EQLs listed for soil/sediment are based on wet weight. Normally data is reported in a dry-weight basis; therefore, EQLs will be higher based on the % dry weight of each sample. This calculation is based on a 30-g sample and gel permeation chromatography cleanup.

(b) Sample EQLs are highly matrix-dependent. The EQLs are provided for guidance and may not always be achievable.

C = True value for concentration, in μg/L.
X = Average recovery found for measurements of samples containing a concentration of C, in μg/L.

ESTIMATED QUANTITATION LIMIT

Other Matrices	Factor (a)
High-concentration soil and sludges by sonicator	7.5
Non-water miscible waste	75

(a) EQL for other matrices = [EQL for low soil/sediment] × [Factor]. This estimated EQL is similar to an EPA "Practical Quantitation Limit."

SAMPLING METHOD
Liquid samples — Use a 1 or 2½ gallon amber glass bottle with a screw-top Teflon®-lined cover that has been prewashed with detergent and rinsed with distilled water and methanol (or isopropanol).

Soils, sediments, or sludges — Use an 8-oz. widemouth glass with a screw-top Teflon®-lined cover that has been prewashed with detergent and rinsed with distilled water and methanol (or isopropanol).

SAMPLE PRESERVATION
Liquid samples — If residual chlorine is present, add 3 mL of 10% sodium thiosulfate per gallon, cool to 4°C and store in a solvent-free refrigerator until analysis; if chlorine is not present, then eliminate the sodium thiosulfate addition.

Soils, sediments, or sludges — Cool samples to 4°C and store in a solvent-free refrigerator.

MHT Liquid samples must be extracted within 7 days and the extracts analyzed within 40 days. Soils, sediments, or sludges may be stored for a maximum of 14 days and the extracts analyzed within 40 days.

SAMPLE PREPARATION
Liquid samples — Transfer 1 L quantitatively to a continuous extractor. If high concentrations are anticipated, a smaller volume may be used and then diluted with organic-free reagent water to 1 L. Adjust pH, if necessary, to pH <2 using 1:1 (V/V) sulfuric acid. Pipette 1.0 mL of a surrogate standard spiking solution into each sample. For the sample in each analytical batch selected for spiking, add 1.0 mL of a matrix spiking standard. For base/neutral acid analysis, the amount of the surrogates and matrix spiking compounds added to the sample should result in a final concentration of 100 ng/μL of each analyte in the extract to be analyzed (assuming a 1-μL injection). Extract with methylene chloride for 18–24 h. Next, adjust the pH of the aqueous phase to pH >11 using 10 N sodium hydroxide and extract it with methylene chloride again for

18–24 h. Dry the extract through a column containing anhydrous sodium sulfate and concentrate it to 1 mL using a K-D concentrator.

Soils, sediments, or sludges — Use 30 g of sample. Nonporous or wet samples (gummy or clay type) that do not have a free-flowing sandy texture must be mixed with anhydrous sodium sulfate until the sample is free flowing. Add 1 mL of surrogate standards to all samples, spikes, standards, and blanks. For the sample in each analytical batch selected for spiking, add 1.0 mL of a matrix spiking standard. For base/neutral acid analysis, the amount added of the surrogates and matrix spiking compounds should result in a final concentration of 100 ng/μL of each base/neutral analyte and 200 ng/μL of each acid analyte in the extract to be analyzed (assuming a 1-μL injection). Immediately add a 100-mL mixture of 1:1 methylene chloride:acetone and extract the sample ultrasonically for 3 min and then decant or filter the extracts. Repeat the extraction two or more times. Dry the extract using a column with anhydrous sodium sulfate and concentrate it to 1 mL in a K-D concentrator.

QUALITY CONTROL A methylene chloride solution containing 50 ng/μL of decafluorotriphenylphosphine (DFTPP) is used for tuning the GC/MS system each 12-h shift. A system performance check also must be made during every 12-h shift. A standard containing 50 ng/μL each of 4,4′-DDT, pentachlorophenol, and benzidine is required to verify injection port inertness and GC column performance. A calibration standard at mid-concentration, containing each compound of interest, including all required surrogates, must be performed every 12 h during analysis. After the system performance check is met, calibration check compounds (CCCs) are used to check the validity of the initial calibration.

The internal standard responses and retention times in the calibration check standard must be evaluated immediately after or during data acquisition. If the retention time for any internal standard changes by more than 30 seconds from the last check calibration (12 h), the chromatographic system must be inspected for malfunctions and corrections must be made, as required. If the electron ionization current plot (EICP) area for any of the internal standards changes by a factor of two from the last daily calibration standard check, the mass spectrometer must be inspected for malfunctions and corrections must be made, as appropriate.

Demonstrate, through the analysis of a reagent water blank, that interferences from the analytical system, glassware, and reagents are under control. The blank samples should be carried through all stages of the sample preparation and measurement steps. For each analytical batch (up to 20 samples), a reagent blank, matrix spike, and matrix spike duplicate/duplicate must be analyzed (the frequency of the spikes may be different for different monitoring programs). The blank and spiked samples must be carried through all stages of the sample preparation and measurement steps. A QC reference sample concentrate containing each analyte at a concentration of 100 mg/L in methanol is required.

REFERENCE Test Methods for Evaluating Solid Waste (SW-846). U.S. EPA 1983, Method 8270B, Rev. 2, Nov. 1990. Office of Solid Waste, Washington, DC.

Fenamiphos   **EPA Method 507**
CAS #22224-92-6

TITLE Determination of Nitrogen and Phosphorus-Containing Pesticides in Water by GC/NPD

MATRIX This method is applicable to the determination of certain nitrogen and phosphorus-containing pesticides in finished drinking water and groundwater.

METHOD SUMMARY Method 507 covers 46 nitrogen- and phosphorus-containing pesticides. A 1-L sample is fortified with a surrogate standard, salted, buffered, extracted with methylene chloride, and concentrated; then the solvent is exchanged with methyl tert-butyl ether (MTBE) and concentrated again, and a 2-μL aliquot of a sample extract is injected into a GC system equipped with a selective nitrogen-phosphorus detector and a capillary column for analysis.

INTERFERENCES Method interferences may be caused by contaminants in solvents, reagents, glassware, and other sample processing apparatus. Interfering contamination may occur when a sample containing low concentrations of analytes is analyzed immediately following a sample containing relatively high concentrations. One or more injections of MTBE should be made following the analysis of a sample with high concentrations of analytes to check for analyte carryover. Matrix interferences may be caused by contaminants that are coextracted from the sample. The extent of matrix interferences will vary considerably from source to source, depending upon the water sampled.

INSTRUMENTATION A gas chromatograph system (GC) equipped with a nitrogen-phosphorus detector (NPD) is needed.

Column 1: 30 m × 0.25 mm I.D. DB-5 bonded fused silica column, 0.25 μm film thickness, or equivalent.
Column 2: 30 m × 0.25 mm I.D. DB-1701 bonded fused silica column, 0.25 μm film thickness, or equivalent.

PRECISION & ACCURACY This method has been validated in a single lab and estimated detection limits (EDLs) have been determined for each analyte. Observed detection limits may vary among waters, depending upon the nature of the interferences in the sample matrix and the specific instrumentation used. Analytes that are not separated chromatographically cannot be individually identified and measured unless an alternative technique for identification and quantification exist.

The estimated detection limit (in μg/L) was 1. The EDL is defined as either method detection limit or a level of compound in a sample yielding a peak in the final extract with signal-to-noise ratio of approximately 5, whichever value is higher.

The concentration used for these measurements (in μg/L) was 10.
The accuracy (as % recovery) was 90.
The precision (% RSD) was 8.

SAMPLING METHOD Grab samples are collected in 1-L glass sample bottles (prewashed with detergent and hot tap water, rinsed with reagent water, and dried in an oven at 400°C for 1 h) with screw caps lined with PTFE-fluorocarbon.

SAMPLE PRESERVATION Add mercuric chloride to the sample bottle in amounts to produce a concentration of 10 mg/L. If residual chlorine is present, add 80 mg of sodium thiosulfate/L of sample to the sample bottle prior to collection. After collection, seal bottle and shake vigorously for 1 min, then cool the sample to 4°C immediately and store it at 4°C in the dark until extraction.

MHT Maximum holding time of the samples, and in some cases the extracts, is 14 days.

SAMPLE PREPARATION Fortify the sample with 50 µL of the surrogate standard solution, adjust to pH 7 with phosphate buffer, add 100 g NaCl to the sample, and seal and shake to dissolve the salt; then extract with methylene chloride in a separatory funnel or in a mechanical tumbler bottle. Dry the extract by pouring it through a solvent-rinsed drying column containing about 10 cm of anhydrous sodium sulfate. Collect the extract in a Kuderna-Danish (K-D) concentrator and rinse the column with 20–30 mL methylene chloride. Concentrate the extract to about 2 mL and rinse the flask and its lower joint into the concentrator tube with 1 to 2 mL of methyl t-butyl ether (MTBE). Add 5–10 mL of MTBE and concentrate the extract twice (adding more MTBE) to a final volume of 5.0 mL and store it at 4°C until analysis.

Note: If methylene chloride is not completely removed from the final extract, it may cause detector problems.

QUALITY CONTROL Minimum quality control requirements are initial demonstration of lab capability, determination of surrogate compound recoveries in each sample and blank, monitoring internal standard peak area or height in each sample and blank, analysis of lab reagent blanks, lab fortified samples, lab fortified blanks, and other QC samples. A lab reagent blank is analyzed to demonstrate that all glassware and reagent interferences are under control.

Initial demonstration of capability is fulfilled by analyzing four fortified reagent water samples with the recovery value for each analyte falling within the acceptable range (±30% average recovery). Surrogate recoveries from samples or method blanks must be 70–130%. The internal standard response for any sample chromatogram should not deviate from the daily calibration check standard's internal standard response by more than 30% or lab fortified blanks and sample matrices are used to assess lab performance and analyte recovery, respectively.

If the response for the target analyte peak exceeds the working range of the system, dilute the extract and reanalyze. Alternative techniques such as an alternate detector or second chromatography column should be used to confirm peak identification when sample components are not resolved adequately.

EPA CONTACT & HOTLINE For technical questions contact Dr. Baldev Bathija, U.S. EPA, Office of Ground Water and Drinking Water (WH-550D), 401 M St. SW, Washington, DC 20460. Tel. (202) 260-3040. For further information the EPA Safe Drinking Water Hotline may be called at: (800) 426-4791.

REFERENCE Methods for the Determination of Organic Compounds in Drinking Water, EPA/600/4-88/039 (revised July 1991). U.S. EPA Environmental Monitoring Systems Laboratory, Cincinnati, OH, 45268, U.S.A. Available from the National Technical Information Service (NTIS), 5285 Port Royal Road, Springfield, VA 22161; Tel. 800-553-6847. NTIS Order Number is PB91-231480.

Fenarimol   **EPA Method 507**
CAS #60168-88-9

TITLE Determination of Nitrogen and Phosphorus-Containing Pesticides in Water by GC/NPD

MATRIX This method is applicable to the determination of certain nitrogen and phosphorus-containing pesticides in finished drinking water and groundwater.

METHOD SUMMARY Method 507 covers 46 nitrogen- and phosphorus-containing pesticides. A 1-L sample is fortified with a surrogate standard, salted, buffered, extracted with methylene chloride, and concentrated; then the solvent is exchanged with methyl tert-butyl ether (MTBE) and concentrated again, and a 2-µL aliquot of a sample extract is injected into a GC system equipped with a selective nitrogen-phosphorus detector and a capillary column for analysis.

INTERFERENCES Method interferences may be caused by contaminants in solvents, reagents, glassware, and other sample processing apparatus. Interfering contamination may occur when a sample containing low concentrations of analytes is analyzed immediately following a sample containing relatively high concentrations. One or more injections of MTBE should be made following the analysis of a sample with high concentrations of analytes to check for analyte carryover. Matrix interferences may be caused by contaminants that are coextracted from the sample. The extent of matrix interferences will vary considerably from source to source, depending upon the water sampled.

INSTRUMENTATION A gas chromatograph system (GC) equipped with a nitrogen-phosphorus detector (NPD) is needed.

Column 1: 30 m × 0.25 mm I.D. DB-5 bonded fused silica column, 0.25 µm film thickness, or equivalent.
Column 2: 30 m × 0.25 mm I.D. DB-1701 bonded fused silica column, 0.25 µm film thickness, or equivalent.

PRECISION & ACCURACY This method has been validated in a single lab and estimated detection limits (EDLs) have been determined for each analyte. Observed detection limits may vary among waters, depending upon the nature of the interferences in the sample matrix and the specific instrumentation used. Analytes that are not separated chromatographically cannot be individually identified and measured unless an alternative technique for identification and quantification exist.

The estimated detection limit (in µg/L) was 0.38. The EDL is defined as either method detection limit or a level of compound in a sample yielding a peak in the final extract with signal-to-noise ratio of approximately 5, whichever value is higher.

The concentration used for these measurements (in µg/L) was 3.8.

The accuracy (as % recovery) was 99.
The precision (% RSD) was 5.

SAMPLING METHOD Grab samples are collected in 1-L glass sample bottles (prewashed with detergent and hot tap water, rinsed with reagent water, and dried in an oven at 400°C for 1 h) with screw caps lined with PTFE-fluorocarbon.

SAMPLE PRESERVATION Add mercuric chloride to the sample bottle in amounts to produce a concentration of 10 mg/L. If residual chlorine is present, add 80 mg of sodium thiosulfate/L of sample to the sample bottle prior to collection. After collection, seal bottle and shake vigorously for 1 min, then cool the sample to 4°C immediately and store it at 4°C in the dark until extraction.

MHT Maximum holding time of the samples, and in some cases the extracts, is 14 days.

SAMPLE PREPARATION Fortify the sample with 50 µL of the surrogate standard solution, adjust to pH 7 with phosphate buffer, add 100 g NaCl to the sample, and seal and shake to dissolve the salt; then extract with methylene chloride in a separatory funnel or in a mechanical tumbler bottle. Dry the extract by pouring it through a solvent-rinsed drying column containing about 10 cm of anhydrous sodium sulfate. Collect the extract in a Kuderna-Danish (K-D) concentrator and rinse the column with 20–30 mL methylene chloride. Concentrate the extract to about 2 mL and rinse the flask and its lower joint into the concentrator tube with 1 to 2 mL of methyl t-butyl ether (MTBE). Add 5–10 mL of MTBE and concentrate the extract twice (adding more MTBE) to a final volume of 5.0 mL and store it at 4°C until analysis.

Note: If methylene chloride is not completely removed from the final extract, it may cause detector problems.

QUALITY CONTROL Minimum quality control requirements are initial demonstration of lab capability, determination of surrogate compound recoveries in each sample and blank, monitoring internal standard peak area or height in each sample and blank, analysis of lab reagent blanks, lab fortified samples, lab fortified blanks, and other QC samples. A lab reagent blank is analyzed to demonstrate that all glassware and reagent interferences are under control.

Initial demonstration of capability is fulfilled by analyzing four fortified reagent water samples with the recovery value for each analyte falling within the acceptable range (±30% average recovery). Surrogate recoveries from samples or method blanks must be 70–130%. The internal standard response for any sample chromatogram should not deviate from the daily calibration check standard's internal standard response by more than 30% or lab fortified blanks and sample matrices are used to assess lab performance and analyte recovery, respectively.

If the response for the target analyte peak exceeds the working range of the system, dilute the extract and reanalyze. Alternative techniques such as an alternate detector or second chromatography column should be used to confirm peak identification when sample components are not resolved adequately.

EPA CONTACT & HOTLINE For technical questions contact Dr. Baldev Bathija, U.S. EPA, Office of Ground Water and Drinking Water (WH-550D), 401 M St. SW, Washington, DC 20460. Tel. (202) 260-3040. For further information the EPA Safe Drinking Water Hotline may be called at: (800) 426-4791.

REFERENCE Methods for the Determination of Organic Compounds in Drinking Water, EPA/600/4-88/039 (revised July 1991). U.S. EPA Environmental Monitoring Systems Laboratory, Cincinnati, OH, 45268, U.S.A. Available from the National Technical Information Service (NTIS), 5285 Port Royal Road, Springfield, VA 22161; Tel. 800-553-6847. NTIS Order Number is PB91-231480.

Fensulfothion   **EPA Method 8141**
CAS #115-90-2

TITLE Organophosphorus Compounds by Gas Chromatography: Capillary Column Technique

MATRIX This method covers aqueous and solid matrices. This includes a wide variety such as drinking water, groundwater, industrial wastewaters, surface waters, soils, solids, and sediments.

METHOD SUMMARY This is a GC method used to determine the concentration of 28 organophosphorus pesticides.

The use of Gel Permeation Cleanup (EPA Method 3640) for sample cleanup has been demonstrated to yield recoveries of less than 85% for many method analytes and is therefore not recommended for use with this method.

This method provides GC conditions for the detection of ppb concentrations of organophosphorus compounds. Prior to the use of this method, appropriate sample preparation techniques must be used. Water samples are extracted at a neutral pH with methylene chloride as a solvent by using a separatory funnel (EPA Method 3510) or a continuous liquid-liquid extractor (EPA Method 3520). Soxhlet extraction (EPA Method 3540) or ultrasonic extraction (EPA Method 3550) using methylene chloride/acetone (1:1) are used for solid samples. Both neat and diluted organic liquids (EPA Method 3580) may be analyzed by direct injection. Spiked samples are used to verify the applicability of the chosen extraction technique to each new sample type. A GC with a flame photometric (FPD) or nitrogen-phosphorus detector (NPD) is used for this multiresidue procedure.

INTERFERENCES The use of Florisil cleanup materials (EPA Method 3620) for some of the compounds in this method has been demonstrated to yield recoveries less than 85% and is therefore not recommended for all compounds. Use of phosphorus or halogen specific detectors, however, often obviates the necessity for cleanup for relatively clean sample matrices. If particular circumstances demand the use of an alternative cleanup procedure, the analyst must determine the elution profile and demonstrate that the recovery of each analyte is no less than 85%.

Use of a flame photometric detector (FPD) in the phosphorus mode will minimize interferences from materials that do not contain phosphorus. Elemental sulfur, however, may interfere with the determination of certain organophosphorus compounds by flame photometric gas chromatography. Sulfur cleanup using EPA Method 3660 may alleviate this interference. A nitrogen phosphorus detector (NPD) is also recommended.

A few analytes coelute on certain columns. Therefore, select a second column for confirmation where coelution of the analytes of interest does not occur.

Method interferences may be caused by contaminants in solvents, reagents, glassware, and other sample processing hardware that lead to discrete artifacts or elevated baselines in gas chromatograms. All these materials must be routinely demonstrated to be free from interferences under the conditions of the analysis by analyzing reagent blanks.

INSTRUMENTATION A GC with a NPD or a FPD will be needed. A data system or integrator is recommended for measuring peak areas and/or peak heights. A Kuderna-Danish (K-D) apparatus will be needed for extract concentration.

Column 1: 15 m × 0.53 mm megabore capillary column, 1.0 μm film thickness, DB-210.
Column 2: 15 m × 0.53 mm megabore capillary column, 1.5 μm film thickness, SPB-608.
Column 3: 15 m × 0.53 mm megabore capillary column, 1.0 μm film thickness, DB-5.

Three megabore capillary columns are included for analysis of organophosphates by this method. Column 1 (DB-210 or equivalent) and Column 2 (SPB-608 or equivalent) are recommended if a large number of organophosphorus analytes are to be determined. If the superior resolution offered by Column 1 and Column 2 is not required, Column 3 (DB-5 or equivalent) may be used. For megabore capillary columns, automatic injections of 1 μL are recommended.

PRECISION & ACCURACY The MDL actually achieved in a given analysis will vary, as it is dependent on instrument sensitivity and matrix effects. Single operator accuracy and precision studies have been conducted with spiked water and soil samples.

MULTIPLICATION FACTORS FOR OTHER MATRICES (a)

Matrix	Factor (b)
Groundwater (EPA Method 3510 or EPA Method 3520)	10
Low-concentration soil by Soxhlet and no cleanup	10 (c)
Low-concentration soil by ultrasonic extraction with GPC cleanup	6.7 (c)
High-concentration soil and sludges by ultrasonic extraction	500 (c)
Non-water miscible waste (EPA Method 3580)	1000 (c)

(a) Sample EQLs are highly matrix-dependent. The EQLs listed here are provided for guidance and may not always be achievable.
(b) EQL = [Method detection limit] × [Factor]. For non-aqueous samples the factor is on a wet-weight basis.
(c) Multiply this factory times the soil MDL.

The MDL (in μg/L) when reagent water was extracted using a separatory funnel was 0.08.

The MDL (in μg/kg) when soil was extracted using Soxhlet extraction (EPA Method 3540) was 4.0.

Accuracy (as % recovery) with separatory funnel extraction ranged from 84 (with low spikes) to 96 (with high spikes).

Accuracy (as % recovery) with continuous liquid-liquid extraction ranged from 90 (with low spikes) to 90 (with high spikes).

Accuracy (as % recovery) with Soxhlet extraction of soils ranged from 72 (with low spikes to 111 (with high spikes).

Accuracy (as % recovery) with ultrasonic extraction of soils ranged from not recovered (with low spikes) to 2 (with high spikes).

SAMPLE COLLECTION, PRESERVATION & HANDLING
Containers used to collect samples for the determination of semivolatile organic compounds should be soap and water washed followed by methanol (or isopropanol) rinsing. The sample containers should be of glass or Teflon® and have screw-top covers with Teflon® liners.

No preservation is used with concentrated waste samples. With liquid samples containing no residual chlorine and with soil, sediment, and sludge samples, immediately cooling to 4°C is the only preservation used. When residual chlorine is present then 3 mL of 10% aqueous sodium sulfate is added for each gallon of sample collected, followed by cooling to 4°C.

Liquid samples must be extracted within 7 days and their extracts analyzed within 40 days. Concentrated waste, soil, sediment, and sludge samples must be extracted within 14 days and their extracts analyzed within 40 days.

SAMPLE PREPARATION In general, water samples are extracted at a neutral pH with methylene chloride, using either EPA Method 3510 or EPA Method 3520. Solid samples are extracted using either EPA Method 3540 or EPA Method 3550 with methylene chloride/acetone (1:1) as the extraction solvent.

Prior to GC analysis, the extraction solvent may be exchanged to hexane. Single lab data indicates that samples should not be transferred with 100% hexane during sample workup as the more water soluble organophosphorus compounds may be lost.

If cleanup is performed on the samples, the analyst should analyze the samples by GC. This will confirm elution patterns and the absence of interferences from the reagents. If peak detection and identification is prevented by the presence of interferences, further cleanup is required.

QUALITY CONTROL The analyst should monitor the performance of the extraction, cleanup (when used), and analytical system and the effectiveness of the method in dealing with each sample matrix by spiking each sample, standard, and blank with one or two surrogates (e.g., organophosphorus compounds not expected to be present in the sample). Deuterated analogs of analytes should not be used as surrogates for gas chromatographic analysis due to coelution problems.

A minimum of five concentrations for each analyte of interest should be prepared through dilution of the stock standards with isooctane. One of the concentrations should be at a concentration near, but above, the MDL.

Include a mid-level check standard after each group of 10 samples in the analysis sequence. GC/MS techniques should be judiciously employed to support qualitative identifications made with this method. Follow the GC/MS operating requirements specified in EPA Method 8270.

When available, chemical ionization mass spectra may be employed to aid in the qualitative identification process. To confirm an identification of a compound, the background-corrected mass spectrum of the compound must be obtained from the sample extract and must be compared with a mass spectrum from a stock or calibration standard analyzed under the same chromatographic conditions. The molecular ion and all other ions present above 20% relative abundance in the mass spectrum of the standard must be present in the mass spectrum of the sample with agreement to ± 20%. The retention time of the compound in the sample must be within six seconds of the retention time for the same compound in the standard solution.

Should the MS procedure fail to provide satisfactory results, additional steps may be taken before reanalysis. These steps may include the use of alternate packed or capillary GC columns or additional sample cleanup.

REFERENCE Test Methods for Evaluating Solid Waste, Physical/Chemical Methods, SW-846, 3rd Edition, U.S. EPA, Office of Solid Waste, Washington, DC, EPA Method 8141 July 1992.

Fensulfothion   **EPA Method 8270**
CAS #115-90-2

TITLE Semivolatile Organic Compounds by GC/MS

MATRIX This method is used to determine the concentration of semivolatile organic compounds in extracts prepared from all types of solid waste matrices, soils, and groundwater. Although surface waters are not specifically mentioned, this method should be applicable to water samples from rivers, lakes, etc.

METHOD SUMMARY This method covers 259 semivolatile organic compounds. In very limited applications direct injection of the sample into the GC/MS system may be appropriate, but this results in very high detection limits (approximately 10,000 µg/L). Typically, a 1-L liquid sample, containing surrogate, and matrix spiking standards, is extracted in a continuous extractor first under acid conditions and then under basic conditions. Typically 30 g of a solid sample, containing surrogate, and matrix spiking standards, is extracted ultrasonically. After concentrating the extract to 1 mL it is spiked with 10 µL of an internal standard solution just prior to analysis by GC/MS. The volume injected should contain about 100 ng of base/neutral and 200 ng of acid surrogates (for a 1-µL injection). Analysis is performed by GC/MS using a capillary GC column.

INTERFERENCES Raw GC/MS data from all blanks, samples, and spikes must be evaluated for interferences. Contamination by carryover can occur whenever high-concentration and low-concentration samples are sequentially analyzed. To reduce carryover, the sample syringe must be rinsed out between samples with solvent. Whenever an unusually concentrated sample is encountered, it should be followed by the analysis of blank solvent to check for cross-contamination.

INSTRUMENTATION A GC/MS and a data system are required. The GC column used is a 30 m × 0.25 mm I.D. (or 0.32 mm I.D.) 1um film thickness silicone-coated fused silica capillary column. A continuous liquid-liquid extractor equipped with Teflon® or glass connection joints and stopcocks requiring no lubrication, a K-D concentrating apparatus, water bath, and an ultrasonic disrupter with a minimum power of 300 W and with pulsing capability are also required.

PRECISION & ACCURACY The estimated quantitation limit (EQL) of Method 8270B for determining an individual compound is approximately 1 mg/kg (wet weight) for soil or sediment samples, 1–200 mg/kg for wastes (dependent on matrix and method of preparation), and 10 µg/L for groundwater samples. EQLs will be proportionately higher for sample extracts that require dilution to avoid saturation of the detector.

The EQL(b) for groundwater in µg/L is 40.
The EQL (a, b) for low concentrations in soil and sediment in µg/kg is not determined.
Accuracy as µg/L is not listed.
Overall precision in µg/L is not listed.

(a) *EQLs listed for soil/sediment are based on wet weight. Normally data is reported in a dry-weight basis; therefore, EQLs will be higher based on the % dry weight of each sample. This calculation is based on a 30-g sample and gel permeation chromatography cleanup.*
(b) *Sample EQLs are highly matrix-dependent. The EQLs are provided for guidance and may not always be achievable.*
$C =$ *True value for concentration, in µg/L.*
$X =$ *Average recovery found for measurements of samples containing a concentration of C, in µg/L.*

ESTIMATED QUANTITATION LIMIT

Other Matrices	Factor (a)
High-concentration soil and sludges by sonicator	7.5
Non-water miscible waste	75

(a) *EQL for other matrices = [EQL for low soil/sediment] × [Factor]. This estimated EQL is similar to an EPA "Practical Quantitation Limit."*

SAMPLING METHOD

Liquid samples — Use a 1 or 2½ gallon amber glass bottle with a screw-top Teflon®-lined cover that has been prewashed with detergent and rinsed with distilled water and methanol (or isopropanol).

Soils, sediments, or sludges — Use an 8-oz. widemouth glass with a screw-top Teflon®-lined cover that has been prewashed with detergent and rinsed with distilled water and methanol (or isopropanol).

SAMPLE PRESERVATION

Liquid samples — If residual chlorine is present, add 3 mL of 10% sodium thiosulfate per gallon, cool to 4°C and store in a

solvent-free refrigerator until analysis; if chlorine is not present, then eliminate the sodium thiosulfate addition.

Soils, sediments, or sludges — Cool samples to 4°C and store in a solvent-free refrigerator.

MHT Liquid samples must be extracted within 7 days and the extracts analyzed within 40 days. Soils, sediments, or sludges may be stored for a maximum of 14 days and the extracts analyzed within 40 days.

SAMPLE PREPARATION

Liquid samples — Transfer 1 L quantitatively to a continuous extractor. If high concentrations are anticipated, a smaller volume may be used and then diluted with organic-free reagent water to 1 L. Adjust pH, if necessary, to pH <2 using 1:1 (V/V) sulfuric acid. Pipette 1.0 mL of a surrogate standard spiking solution into each sample. For the sample in each analytical batch selected for spiking, add 1.0 mL of a matrix spiking standard. For base/neutral acid analysis, the amount of the surrogates and matrix spiking compounds added to the sample should result in a final concentration of 100 ng/µL of each analyte in the extract to be analyzed (assuming a 1-µL injection). Extract with methylene chloride for 18–24 h. Next, adjust the pH of the aqueous phase to pH >11 using 10 N sodium hydroxide and extract it with methylene chloride again for 18–24 h. Dry the extract through a column containing anhydrous sodium sulfate and concentrate it to 1 mL using a K-D concentrator.

Soils, sediments, or sludges — Use 30 g of sample. Nonporous or wet samples (gummy or clay type) that do not have a free-flowing sandy texture must be mixed with anhydrous sodium sulfate until the sample is free flowing. Add 1 mL of surrogate standards to all samples, spikes, standards, and blanks. For the sample in each analytical batch selected for spiking, add 1.0 mL of a matrix spiking standard. For base/neutral acid analysis, the amount added of the surrogates and matrix spiking compounds should result in a final concentration of 100 ng/µL of each base/neutral analyte and 200 ng/µL of each acid analyte in the extract to be analyzed (assuming a 1-µL injection). Immediately add a 100-mL mixture of 1:1 methylene chloride:acetone and extract the sample ultrasonically for 3 min and then decant or filter the extracts. Repeat the extraction two or more times. Dry the extract using a column with anhydrous sodium sulfate and concentrate it to 1 mL in a K-D concentrator.

QUALITY CONTROL A methylene chloride solution containing 50 ng/µL of decafluorotriphenylphosphine (DFTPP) is used for tuning the GC/MS system each 12-h shift. A system performance check also must be made during every 12-h shift. A standard containing 50 ng/µL each of 4,4'-DDT, pentachlorophenol, and benzidine is required to verify injection port inertness and GC column performance. A calibration standard at mid-concentration, containing each compound of interest, including all required surrogates, must be performed every 12 h during analysis. After the system performance check is met, calibration check compounds (CCCs) are used to check the validity of the initial calibration.

The internal standard responses and retention times in the calibration check standard must be evaluated immediately after or during data acquisition. If the retention time for any internal standard changes by more than 30 seconds from the last check calibration (12 h), the chromatographic system must be inspected for malfunctions and corrections must be made, as required. If the electron ionization current plot (EICP) area for any of the internal standards changes by a factor of two from the last daily calibration standard check, the mass spectrometer must be inspected for malfunctions and corrections must be made, as appropriate.

Demonstrate, through the analysis of a reagent water blank, that interferences from the analytical system, glassware, and reagents are under control. The blank samples should be carried through all stages of the sample preparation and measurement steps. For each analytical batch (up to 20 samples), a reagent blank, matrix spike, and matrix spike duplicate/duplicate must be analyzed (the frequency of the spikes may be different for different monitoring programs). The blank and spiked samples must be carried through all stages of the sample preparation and measurement steps. A QC reference sample concentrate containing each analyte at a concentration of 100 mg/L in methanol is required.

REFERENCE Test Methods for Evaluating Solid Waste (SW-846). U.S. EPA 1983, Method 8270B, Rev. 2, Nov. 1990. Office of Solid Waste, Washington, DC.

Fensulfothion **EPA Method 8140**
CAS #115-90-2

TITLE Organophosphorus Pesticides

MATRIX Groundwater, soils, sludges, water miscible liquid wastes, and non-water miscible wastes.

APPLICATION This method is used for the analysis of 21 organophosphorus pesticides. Samples are extracted, concentrated, and analyzed using direct injection of both neat and diluted organic liquid into a gas chromatograph (GC).

INTERFERENCES Solvents, reagents, and glassware may introduce artifacts. Other interferences may come from coextracted compounds from samples. The use of Florisil cleanup materials may produce low recoveries. Elemental sulfur may interfere with some compounds when using a flame photometric detector. Sulfur cleanup (Method 3660) may alleviate sulfur interference.

INSTRUMENTATION GC capable of on-column injections and a flame photometric detector (FPD) or a thermionic detector. Column 1: 1.8 m by 2 mm with 5% SP-2401 on Supelcoport. Column 2: 1.8 m by 2 mm with 3% SP-2401 on Supelcoport. Column 3: 50 cm by ⅛ IN EFLON® with 15% SE-54 on Gas Chrom Q. The preferred column is Column Number 1.

RANGE 23.9–110 µg/L

MDL 1.5 µg/L (in reagent water).

PQL FACTORS FOR MULTIPLYING × FID MDL VALUE

Matrix	Multiplication Factor
Groundwater	10
Low-level soil by sonication with GPC cleanup	670
High-level soil and sludge by sonication	10,000
Non-water miscible waste	100,000

PRECISION 17.1% (single operator standard deviation)

ACCURACY 94.1% (single operator average recovery)

SAMPLING METHOD Use 8-oz. widemouth glass bottles with Teflon®-lined caps for concentrated waste samples, soils, sediments, and sludges. Use 1 or 2½ gallon amber glass bottles with Teflon®-lined caps for liquid (water) samples.

STABILITY Cool soil, sediment, sludge, and liquid samples to 4°C. If residual chlorine is present in liquid samples add 3 mL of 10% sodium thiosulfate per gallon of sample and cool to 4°C.

MHT 14 days for concentrated waste, soil, sediment, or sludge; 7 days for liquid samples; all extracts must be analyzed within 40 days.

QUALITY CONTROL A quality control check sample concentrate containing this compound in acetone at a concentration 1,000 times more concentrated than the selected spike concentration is required. The QC check sample concentrate may be prepared from pure standard materials or purchased as certified solutions. Use appropriate trip, matrix, control site, method, reagent, and solvent blanks. Internal, surrogate, and five concentration level calibration standards are used.

REFERENCE Method 8140, SW-846, 3rd ed., Sept. 1986.

Fenthion   **EPA Method 8141**
CAS #55-38-9

TITLE Organophosphorus Compounds by Gas Chromatography: Capillary Column Technique

MATRIX This method covers aqueous and solid matrices. This includes a wide variety such as drinking water, groundwater, industrial wastewaters, surface waters, soils, solids, and sediments.

METHOD SUMMARY This is a GC method used to determine the concentration of 28 organophosphorus pesticides.

The use of Gel Permeation Cleanup (EPA Method 3640) for sample cleanup has been demonstrated to yield recoveries of less than 85% for many method analytes and is therefore not recommended for use with this method.

This method provides GC conditions for the detection of ppb concentrations of organophosphorus compounds. Prior to the use of this method, appropriate sample preparation techniques must be used. Water samples are extracted at a neutral pH with methylene chloride as a solvent by using a separatory funnel (EPA Method 3510) or a continuous liquid-liquid extractor (EPA Method 3520). Soxhlet extraction (EPA Method 3540) or ultrasonic extraction (EPA Method 3550) using methylene chloride/acetone (1:1) are used for solid samples. Both neat and diluted organic liquids (EPA Method 3580) may be analyzed by direct injection. Spiked samples are used to verify the applicability of the chosen extraction technique to each new sample type. A GC with a flame photometric (FPD) or nitrogen-phosphorus detector (NPD) is used for this multiresidue procedure.

INTERFERENCES The use of Florisil cleanup materials (EPA Method 3620) for some of the compounds in this method has been demonstrated to yield recoveries less than 85% and is therefore not recommended for all compounds. Use of phosphorus or halogen specific detectors, however, often obviates the necessity for cleanup for relatively clean sample matrices. If particular circumstances demand the use of an alternative cleanup procedure, the analyst must determine the elution profile and demonstrate that the recovery of each analyte is no less than 85%.

Use of a flame photometric detector (FPD) in the phosphorus mode will minimize interferences from materials that do not contain phosphorus. Elemental sulfur, however, may interfere with the determination of certain organophosphorus compounds by flame photometric gas chromatography. Sulfur cleanup using EPA Method 3660 may alleviate this interference. A nitrogen phosphorus detector (NPD) is also recommended.

A few analytes coelute on certain columns. Therefore, select a second column for confirmation where coelution of the analytes of interest does not occur.

Method interferences may be caused by contaminants in solvents, reagents, glassware, and other sample processing hardware that lead to discrete artifacts or elevated baselines in gas chromatograms. All these materials must be routinely demonstrated to be free from interferences under the conditions of the analysis by analyzing reagent blanks.

INSTRUMENTATION A GC with a NPD or a FPD will be needed. A data system or integrator is recommended for measuring peak areas and/or peak heights. A Kuderna-Danish (K-D) apparatus will be needed for extract concentration.

Column 1: 15 m × 0.53 mm megabore capillary column, 1.0 μm film thickness, DB-210.
Column 2: 15 m × 0.53 mm megabore capillary column, 1.5 μm film thickness, SPB-608.
Column 3: 15 m × 0.53 mm megabore capillary column, 1.0 μm film thickness, DB-5.

Three megabore capillary columns are included for analysis of organophosphates by this method. Column 1 (DB-210 or equivalent) and Column 2 (SPB-608 or equivalent) are recommended if a large number of organophosphorus analytes are to be determined. If the superior resolution offered by Column 1 and Column 2 is not required, Column 3 (DB-5 or equivalent) may be used. For megabore capillary columns, automatic injections of 1 μL are recommended.

PRECISION & ACCURACY The MDL actually achieved in a given analysis will vary, as it is dependent on instrument sensitivity and matrix effects. Single operator accuracy and precision studies have been conducted with spiked water and soil samples.

MULTIPLICATION FACTORS FOR OTHER MATRICES (a)

Matrix	Factor (b)
Groundwater (EPA Method 3510 or EPA Method 3520)	10
Low-concentration soil by Soxhlet and no cleanup	10 (c)
Low-concentration soil by ultrasonic extraction with GPC cleanup	6.7 (c)
High-concentration soil and sludges by ultrasonic extraction	500 (c)
Non-water miscible waste (EPA Method 3580)	1000 (c)

(a) Sample EQLs are highly matrix-dependent. The EQLs listed here are provided for guidance and may not always be achievable.
(b) EQL = [Method detection limit] × [Factor]. For non-aqueous samples the factor is on a wet-weight basis.
(c) Multiply this factory times the soil MDL.

The MDL (in µg/L) when reagent water was extracted using a separatory funnel was 0.08.

The MDL (in µg/kg) when soil was extracted using Soxhlet extraction (EPA Method 3540) was 5.0.

Accuracy (as % recovery) with separatory funnel extraction ranged from not recovered (with low spikes) to 89 (with high spikes).

Accuracy (as % recovery) with continuous liquid-liquid extraction ranged from 8 (with low spikes) to 86 (with high spikes).

Accuracy (as % recovery) with Soxhlet extraction of soils ranged from not recovered (with low spikes to 89 (with high spikes).

Accuracy (as % recovery) with ultrasonic extraction of soils ranged from not recovered (with low spikes) to 84 (with high spikes).

SAMPLE COLLECTION, PRESERVATION & HANDLING

Containers used to collect samples for the determination of semivolatile organic compounds should be soap and water washed followed by methanol (or isopropanol) rinsing. The sample containers should be of glass or Teflon® and have screw-top covers with Teflon® liners.

No preservation is used with concentrated waste samples. With liquid samples containing no residual chlorine and with soil, sediment, and sludge samples, immediately cooling to 4°C is the only preservation used. When residual chlorine is present then 3 mL of 10% aqueous sodium sulfate is added for each gallon of sample collected, followed by cooling to 4°C.

Liquid samples must be extracted within 7 days and their extracts analyzed within 40 days. Concentrated waste, soil, sediment, and sludge samples must be extracted within 14 days and their extracts analyzed within 40 days.

SAMPLE PREPARATION
In general, water samples are extracted at a neutral pH with methylene chloride, using either EPA Method 3510 or EPA Method 3520. Solid samples are extracted using either EPA Method 3540 or EPA Method 3550 with methylene chloride/acetone (1:1) as the extraction solvent.

Prior to GC analysis, the extraction solvent may be exchanged to hexane. Single lab data indicates that samples should not be transferred with 100% hexane during sample workup as the more water soluble organophosphorus compounds may be lost.

If cleanup is performed on the samples, the analyst should analyze the samples by GC. This will confirm elution patterns and the absence of interferences from the reagents. If peak detection and identification is prevented by the presence of interferences, further cleanup is required.

QUALITY CONTROL
The analyst should monitor the performance of the extraction, cleanup (when used), and analytical system and the effectiveness of the method in dealing with each sample matrix by spiking each sample, standard, and blank with one or two surrogates (e.g., organophosphorus compounds not expected to be present in the sample). Deuterated analogs of analytes should not be used as surrogates for gas chromatographic analysis due to coelution problems.

A minimum of five concentrations for each analyte of interest should be prepared through dilution of the stock standards with isooctane. One of the concentrations should be at a concentration near, but above, the MDL.

Include a mid-level check standard after each group of 10 samples in the analysis sequence. GC/MS techniques should be judiciously employed to support qualitative identifications made with this method. Follow the GC/MS operating requirements specified in EPA Method 8270.

When available, chemical ionization mass spectra may be employed to aid in the qualitative identification process. To confirm an identification of a compound, the background-corrected mass spectrum of the compound must be obtained from the sample extract and must be compared with a mass spectrum from a stock or calibration standard analyzed under the same chromatographic conditions. The molecular ion and all other ions present above 20% relative abundance in the mass spectrum of the standard must be present in the mass spectrum of the sample with agreement to ± 20%. The retention time of the compound in the sample must be within six seconds of the retention time for the same compound in the standard solution.

Should the MS procedure fail to provide satisfactory results, additional steps may be taken before reanalysis. These steps may include the use of alternate packed or capillary GC columns or additional sample cleanup.

REFERENCE
Test Methods for Evaluating Solid Waste, Physical/Chemical Methods, SW-846, 3rd Edition, U.S. EPA, Office of Solid Waste, Washington, DC, EPA Method 8141 July 1992.

Fenthion   **EPA Method 8270**
CAS #55-38-9

TITLE
Semivolatile Organic Compounds by GC/MS

MATRIX
This method is used to determine the concentration of semivolatile organic compounds in extracts prepared from all types of solid waste matrices, soils, and groundwater. Although surface waters are not specifically mentioned, this

method should be applicable to water samples from rivers, lakes, etc.

METHOD SUMMARY This method covers 259 semivolatile organic compounds. In very limited applications direct injection of the sample into the GC/MS system may be appropriate, but this results in very high detection limits (approximately 10,000 µg/L). Typically, a 1-L liquid sample, containing surrogate, and matrix spiking standards, is extracted in a continuous extractor first under acid conditions and then under basic conditions. Typically 30 g of a solid sample, containing surrogate, and matrix spiking standards, is extracted ultrasonically. After concentrating the extract to 1 mL it is spiked with 10 µL of an internal standard solution just prior to analysis by GC/MS. The volume injected should contain about 100 ng of base/neutral and 200 ng of acid surrogates (for a 1-µL injection). Analysis is performed by GC/MS using a capillary GC column.

INTERFERENCES Raw GC/MS data from all blanks, samples, and spikes must be evaluated for interferences. Contamination by carryover can occur whenever high-concentration and low-concentration samples are sequentially analyzed. To reduce carryover, the sample syringe must be rinsed out between samples with solvent. Whenever an unusually concentrated sample is encountered, it should be followed by the analysis of blank solvent to check for cross-contamination.

INSTRUMENTATION A GC/MS and a data system are required. The GC column used is a 30 m × 0.25 mm I.D. (or 0.32 mm I.D.) 1um film thickness silicone-coated fused silica capillary column. A continuous liquid-liquid extractor equipped with Teflon® or glass connection joints and stopcocks requiring no lubrication, a K-D concentrating apparatus, water bath, and an ultrasonic disrupter with a minimum power of 300 W and with pulsing capability are also required.

PRECISION & ACCURACY The estimated quantitation limit (EQL) of Method 8270B for determining an individual compound is approximately 1 mg/kg (wet weight) for soil or sediment samples, 1–200 mg/kg for wastes (dependent on matrix and method of preparation), and 10 µg/L for groundwater samples. EQLs will be proportionately higher for sample extracts that require dilution to avoid saturation of the detector.

The EQL(b) for groundwater in µg/L is 10.
The EQL (a, b) for low concentrations in soil and sediment in µg/kg is not determined.
Accuracy as µg/L is not listed.
Overall precision in µg/L is not listed.

(a) EQLs listed for soil/sediment are based on wet weight. Normally data is reported in a dry-weight basis; therefore, EQLs will be higher based on the % dry weight of each sample. This calculation is based on a 30-g sample and gel permeation chromatography cleanup.
(b) Sample EQLs are highly matrix-dependent. The EQLs are provided for guidance and may not always be achievable.
C = True value for concentration, in µg/L.
X = Average recovery found for measurements of samples containing a concentration of C, in µg/L.

ESTIMATED QUANTITATION LIMIT

Other Matrices	Factor (a)
High-concentration soil and sludges by sonicator	7.5
Non-water miscible waste	75

(a) EQL for other matrices = [EQL for low soil/sediment] × [Factor]. This estimated EQL is similar to an EPA "Practical Quantitation Limit."

SAMPLING METHOD
Liquid samples — Use a 1 or 2½ gallon amber glass bottle with a screw-top Teflon®-lined cover that has been prewashed with detergent and rinsed with distilled water and methanol (or isopropanol).

Soils, sediments, or sludges — Use an 8-oz. widemouth glass with a screw-top Teflon®-lined cover that has been prewashed with detergent and rinsed with distilled water and methanol (or isopropanol).

SAMPLE PRESERVATION
Liquid samples — If residual chlorine is present, add 3 mL of 10% sodium thiosulfate per gallon, cool to 4°C and store in a solvent-free refrigerator until analysis; if chlorine is not present, then eliminate the sodium thiosulfate addition.

Soils, sediments, or sludges — Cool samples to 4°C and store in a solvent-free refrigerator.

MHT Liquid samples must be extracted within 7 days and the extracts analyzed within 40 days. Soils, sediments, or sludges may be stored for a maximum of 14 days and the extracts analyzed within 40 days.

SAMPLE PREPARATION
Liquid samples — Transfer 1 L quantitatively to a continuous extractor. If high concentrations are anticipated, a smaller volume may be used and then diluted with organic-free reagent water to 1 L. Adjust pH, if necessary, to pH <2 using 1:1 (V/V) sulfuric acid. Pipette 1.0 mL of a surrogate standard spiking solution into each sample. For the sample in each analytical batch selected for spiking, add 1.0 mL of a matrix spiking standard. For base/neutral acid analysis, the amount of the surrogates and matrix spiking compounds added to the sample should result in a final concentration of 100 ng/µL of each analyte in the extract to be analyzed (assuming a 1-µL injection). Extract with methylene chloride for 18–24 h. Next, adjust the pH of the aqueous phase to pH >11 using 10 N sodium hydroxide and extract it with methylene chloride again for 18–24 h. Dry the extract through a column containing anhydrous sodium sulfate and concentrate it to 1 mL using a K-D concentrator.

Soils, sediments, or sludges — Use 30 g of sample. Nonporous or wet samples (gummy or clay type) that do not have a free-flowing sandy texture must be mixed with anhydrous sodium sulfate until the sample is free flowing. Add 1 mL of surrogate standards to all samples, spikes, standards, and blanks. For the sample in each analytical batch selected for spiking, add 1.0 mL of a matrix spiking standard. For base/neutral acid analysis, the amount added of the surrogates and matrix spiking compounds should result in a final concentration of 100 ng/µL of each base/neutral analyte and 200 ng/µL of each acid analyte

in the extract to be analyzed (assuming a 1-µL injection). Immediately add a 100-mL mixture of 1:1 methylene chloride:acetone and extract the sample ultrasonically for 3 min and then decant or filter the extracts. Repeat the extraction two or more times. Dry the extract using a column with anhydrous sodium sulfate and concentrate it to 1 mL in a K-D concentrator.

QUALITY CONTROL A methylene chloride solution containing 50 ng/µL of decafluorotriphenylphosphine (DFTPP) is used for tuning the GC/MS system each 12-h shift. A system performance check also must be made during every 12-h shift. A standard containing 50 ng/µL each of 4,4'-DDT, pentachlorophenol, and benzidine is required to verify injection port inertness and GC column performance. A calibration standard at mid-concentration, containing each compound of interest, including all required surrogates, must be performed every 12 h during analysis. After the system performance check is met, calibration check compounds (CCCs) are used to check the validity of the initial calibration.

The internal standard responses and retention times in the calibration check standard must be evaluated immediately after or during data acquisition. If the retention time for any internal standard changes by more than 30 seconds from the last check calibration (12 h), the chromatographic system must be inspected for malfunctions and corrections must be made, as required. If the electron ionization current plot (EICP) area for any of the internal standards changes by a factor of two from the last daily calibration standard check, the mass spectrometer must be inspected for malfunctions and corrections must be made, as appropriate.

Demonstrate, through the analysis of a reagent water blank, that interferences from the analytical system, glassware, and reagents are under control. The blank samples should be carried through all stages of the sample preparation and measurement steps. For each analytical batch (up to 20 samples), a reagent blank, matrix spike, and matrix spike duplicate/duplicate must be analyzed (the frequency of the spikes may be different for different monitoring programs). The blank and spiked samples must be carried through all stages of the sample preparation and measurement steps. A QC reference sample concentrate containing each analyte at a concentration of 100 mg/L in methanol is required.

REFERENCE Test Methods for Evaluating Solid Waste (SW-846). U.S. EPA 1983, Method 8270B, Rev. 2, Nov. 1990. Office of Solid Waste, Washington, DC.

Fenthion   **EPA Method 8140**
CAS #55-38-9

TITLE Organophosphorus Pesticides

MATRIX Groundwater, soils, sludges, water miscible liquid wastes, and non-water miscible wastes.

APPLICATION This method is used for the analysis of 21 organophosphorus pesticides. Samples are extracted, concentrated, and analyzed using direct injection of both neat and diluted organic liquid into a gas chromatograph (GC).

INTERFERENCES Solvents, reagents, and glassware may introduce artifacts. Other interferences may come from coextracted compounds from samples. The use of Florisil cleanup materials may produce low recoveries. Elemental sulfur may interfere with some compounds when using a flame photometric detector. Sulfur cleanup (Method 3660) may alleviate sulfur interference.

INSTRUMENTATION GC capable of on-column injections and a flame photometric detector (FPD) or a thermionic detector. Column 1: 1.8 m by 2 mm with 5% SP-2401 on Supelcoport. Column 2: 1.8 m by 2 mm with 3% SP-2401 on Supelcoport. Column 3: 50 cm by ⅛ IN EFLON® with 15% SE-54 on Gas Chrom Q. The preferred column is Column Number 1.

RANGE 5.3–64 µg/L

MDL 0.10 µg/L (in reagent water).

PQL FACTORS FOR MULTIPLYING × FID MDL VALUE

Matrix	Multiplication Factor
Groundwater	10
Low-level soil by sonication with GPC cleanup	670
High-level soil and sludge by sonication	10,000
Non-water miscible waste	100,000

PRECISION 19.9% (single operator standard deviation)

ACCURACY 68.7% (single operator average recovery)

SAMPLING METHOD Use 8-oz. widemouth glass bottles with Teflon®-lined caps for concentrated waste samples, soils, sediments, and sludges. Use 1 or 2½ gallon amber glass bottles with Teflon®-lined caps for liquid (water) samples.

STABILITY Cool soil, sediment, sludge, and liquid samples to 4°C. If residual chlorine is present in liquid samples add 3 mL of 10% sodium thiosulfate per gallon of sample and cool to 4°C.

MHT 14 days for concentrated waste, soil, sediment, or sludge; 7 days for liquid samples; all extracts must be analyzed within 40 days.

QUALITY CONTROL A quality control check sample concentrate containing this compound in acetone at a concentration 1,000 times more concentrated than the selected spike concentration is required. The QC check sample concentrate may be prepared from pure standard materials or purchased as certified solutions. Use appropriate trip, matrix, control site, method, reagent, and solvent blanks. Internal, surrogate, and five concentration level calibration standards are used.

REFERENCE Method 8140, SW-846, 3rd ed., Sept. 1986.

Fluchloralin   **EPA Method 8270**
CAS #33245-39-5

TITLE Semivolatile Organic Compounds by GC/MS

MATRIX This method is used to determine the concentration of semivolatile organic compounds in extracts prepared from all types of solid waste matrices, soils, and groundwater.

Although surface waters are not specifically mentioned, this method should be applicable to water samples from rivers, lakes, etc.

METHOD SUMMARY This method covers 259 semivolatile organic compounds. In very limited applications direct injection of the sample into the GC/MS system may be appropriate, but this results in very high detection limits (approximately 10,000 µg/L). Typically, a 1-L liquid sample, containing surrogate, and matrix spiking standards, is extracted in a continuous extractor first under acid conditions and then under basic conditions. Typically 30 g of a solid sample, containing surrogate, and matrix spiking standards, is extracted ultrasonically. After concentrating the extract to 1 mL it is spiked with 10 µL of an internal standard solution just prior to analysis by GC/MS. The volume injected should contain about 100 ng of base/neutral and 200 ng of acid surrogates (for a 1-µL injection). Analysis is performed by GC/MS using a capillary GC column.

INTERFERENCES Raw GC/MS data from all blanks, samples, and spikes must be evaluated for interferences. Contamination by carryover can occur whenever high-concentration and low-concentration samples are sequentially analyzed. To reduce carryover, the sample syringe must be rinsed out between samples with solvent. Whenever an unusually concentrated sample is encountered, it should be followed by the analysis of blank solvent to check for cross-contamination.

INSTRUMENTATION A GC/MS and a data system are required. The GC column used is a 30 m × 0.25 mm I.D. (or 0.32 mm I.D.) 1um film thickness silicone-coated fused silica capillary column. A continuous liquid-liquid extractor equipped with Teflon® or glass connection joints and stopcocks requiring no lubrication, a K-D concentrating apparatus, water bath, and an ultrasonic disrupter with a minimum power of 300 W and with pulsing capability are also required.

PRECISION & ACCURACY The estimated quantitation limit (EQL) of Method 8270B for determining an individual compound is approximately 1 mg/kg (wet weight) for soil or sediment samples, 1–200 mg/kg for wastes (dependent on matrix and method of preparation), and 10 µg/L for groundwater samples. EQLs will be proportionately higher for sample extracts that require dilution to avoid saturation of the detector.

The EQL(b) for groundwater in µg/L is 20.
The EQL (a, b) for low concentrations in soil and sediment in µg/kg is not determined.
Accuracy as µg/L is not listed.
Overall precision in µg/L is not listed.

(a) *EQLs listed for soil/sediment are based on wet weight. Normally data is reported in a dry-weight basis; therefore, EQLs will be higher based on the % dry weight of each sample. This calculation is based on a 30-g sample and gel permeation chromatography cleanup.*
(b) *Sample EQLs are highly matrix-dependent. The EQLs are provided for guidance and may not always be achievable.*
C = *True value for concentration, in µg/L.*
X = *Average recovery found for measurements of samples containing a concentration of C, in µg/L.*

ESTIMATED QUANTITATION LIMIT

Other Matrices	Factor (a)
High-concentration soil and sludges by sonicator	7.5
Non-water miscible waste	75

(a) *EQL for other matrices = [EQL for low soil/sediment] × [Factor]. This estimated EQL is similar to an EPA "Practical Quantitation Limit."*

SAMPLING METHOD
Liquid samples — Use a 1 or 2½ gallon amber glass bottle with a screw-top Teflon®-lined cover that has been prewashed with detergent and rinsed with distilled water and methanol (or isopropanol).

Soils, sediments, or sludges — Use an 8-oz. widemouth glass with a screw-top Teflon®-lined cover that has been prewashed with detergent and rinsed with distilled water and methanol (or isopropanol).

SAMPLE PRESERVATION
Liquid samples — If residual chlorine is present, add 3 mL of 10% sodium thiosulfate per gallon, cool to 4°C and store in a solvent-free refrigerator until analysis; if chlorine is not present, then eliminate the sodium thiosulfate addition.

Soils, sediments, or sludges — Cool samples to 4°C and store in a solvent-free refrigerator.

MHT Liquid samples must be extracted within 7 days and the extracts analyzed within 40 days. Soils, sediments, or sludges may be stored for a maximum of 14 days and the extracts analyzed within 40 days.

SAMPLE PREPARATION
Liquid samples — Transfer 1 L quantitatively to a continuous extractor. If high concentrations are anticipated, a smaller volume may be used and then diluted with organic-free reagent water to 1 L. Adjust pH, if necessary, to pH <2 using 1:1 (V/V) sulfuric acid. Pipette 1.0 mL of a surrogate standard spiking solution into each sample. For the sample in each analytical batch selected for spiking, add 1.0 mL of a matrix spiking standard. For base/neutral acid analysis, the amount of the surrogates and matrix spiking compounds added to the sample should result in a final concentration of 100 ng/µL of each analyte in the extract to be analyzed (assuming a 1-µL injection). Extract with methylene chloride for 18–24 h. Next, adjust the pH of the aqueous phase to pH >11 using 10 N sodium hydroxide and extract it with methylene chloride again for 18–24 h. Dry the extract through a column containing anhydrous sodium sulfate and concentrate it to 1 mL using a K-D concentrator.

Soils, sediments, or sludges — Use 30 g of sample. Nonporous or wet samples (gummy or clay type) that do not have a free-flowing sandy texture must be mixed with anhydrous sodium sulfate until the sample is free flowing. Add 1 mL of surrogate standards to all samples, spikes, standards, and blanks. For the sample in each analytical batch selected for spiking, add 1.0 mL of a matrix spiking standard. For base/neutral acid analysis, the amount added of the surrogates and matrix spiking compounds should result in a final concentration of 100 ng/µL of each base/neutral analyte and 200 ng/µL of each acid analyte

in the extract to be analyzed (assuming a 1-µL injection). Immediately add a 100-mL mixture of 1:1 methylene chloride:acetone and extract the sample ultrasonically for 3 min and then decant or filter the extracts. Repeat the extraction two or more times. Dry the extract using a column with anhydrous sodium sulfate and concentrate it to 1 mL in a K-D concentrator.

QUALITY CONTROL A methylene chloride solution containing 50 ng/µL of decafluorotriphenylphosphine (DFTPP) is used for tuning the GC/MS system each 12-h shift. A system performance check also must be made during every 12-h shift. A standard containing 50 ng/µL each of 4,4'-DDT, pentachlorophenol, and benzidine is required to verify injection port inertness and GC column performance. A calibration standard at mid-concentration, containing each compound of interest, including all required surrogates, must be performed every 12 h during analysis. After the system performance check is met, calibration check compounds (CCCs) are used to check the validity of the initial calibration.

The internal standard responses and retention times in the calibration check standard must be evaluated immediately after or during data acquisition. If the retention time for any internal standard changes by more than 30 seconds from the last check calibration (12 h), the chromatographic system must be inspected for malfunctions and corrections must be made, as required. If the electron ionization current plot (EICP) area for any of the internal standards changes by a factor of two from the last daily calibration standard check, the mass spectrometer must be inspected for malfunctions and corrections must be made, as appropriate.

Demonstrate, through the analysis of a reagent water blank, that interferences from the analytical system, glassware, and reagents are under control. The blank samples should be carried through all stages of the sample preparation and measurement steps. For each analytical batch (up to 20 samples), a reagent blank, matrix spike, and matrix spike duplicate/duplicate must be analyzed (the frequency of the spikes may be different for different monitoring programs). The blank and spiked samples must be carried through all stages of the sample preparation and measurement steps. A QC reference sample concentrate containing each analyte at a concentration of 100 mg/L in methanol is required.

REFERENCE Test Methods for Evaluating Solid Waste (SW-846). U.S. EPA 1983, Method 8270B, Rev. 2, Nov. 1990. Office of Solid Waste, Washington, DC.

Fluoranthene   **EPA Method 1625**
CAS #206-44-0

TITLE Semivolatile Organic Compounds by Isotope Dilution GC/MS

MATRIX The compounds may be determined in waters, soils, and municipal sludges by this method.

METHOD SUMMARY This method is used to determine 176 semivolatile toxic organic pollutants associated with the CWA (as amended 1987); the RCRA (as amended 1986); the CERCLA (as amended 1986); and other compounds amenable to extraction and analysis by capillary column gas chromatography-mass spectrometry (GC/MS).

Stable isotopically-labeled analogs of the compounds of interest are added to the sample. If the solids content is less than 1%, a 1-L sample is extracted at pH 12–13, then at pH <2 with methylene chloride using continuous extraction techniques.

If the solids content is 30% or less, the sample is diluted to 1% solids with reagent water, homogenized ultrasonically, and extracted at pH 12–13, then at pH <2 with methylene chloride using continuous extraction techniques. If the solids content is greater than 30%, the sample is extracted using ultrasonic techniques.

Each extract is dried over sodium sulfate, concentrated to a volume of 5 mL, cleaned up using GPC, if necessary, and concentrated. Extracts are concentrated to 1 mL if GPC is not performed, and to 0.5 mL if GPC is performed.

An internal standard is added to the extract, and a 1-mL aliquot of the extract is injected into the GC. The compounds are separated by GC and detected by a MS. The labeled compounds serve to correct the variability of the analytical technique.

INTERFERENCES Solvents, reagents, glassware, and other sample processing hardware may yield artifacts and/or elevated baselines causing misinterpretation of chromatograms and spectra. Materials used in the analysis must be demonstrated to be free from interferences under the conditions of analysis by running method blanks initially and with each sample lot (sample started through the extraction process on a given 8-h shift, to a maximum of 20). Specific selection of reagents and purification of solvents by distillation in all glass systems may be required. Glassware and, where possible, reagents are cleaned by solvent rinse and baking at 450°C for 1-h minimum. Interferences coextracted from samples will vary considerably from source to source, depending on the diversity of the site being sampled.

INSTRUMENTATION Major instrumentation includes a GC with a splitless or on-column injection port for capillary column, a MS with 70 eV electron impact ionization, and a data system to collect and record MS data, and process it. A K-D apparatus is used to concentrate extracts.

GC Column: 30 m × 0.25 mm I.D. 5% phenyl, 94% methyl, 1% vinyl silicone bonded phased fused silica capillary column.

PRECISION & ACCURACY The detection limits of the method are usually dependent on the level of interferences rather than instrumental limitations. The limits typify the minimum quantities that can be detected with no interferences present.

The minimum level (in µg/mL) was 10. This is defined as a minimum level at which the analytical system shall give recognizable mass spectra (background corrected) and acceptable calibration points.

The MDL (in µg/kg) in low solids was 54 and in high solids was 22; these were determined in digested sludge (low solids) and in filter cake or compost (high solids).

The labeled and native compound initial precision as standard deviation (in µg/L) was 33.

The labeled and native compound initial accuracy as average recovery (in µg/L) was 71–177.

SAMPLE COLLECTION, PRESERVATION & HANDLING
Collect samples in glass containers. Aqueous samples which flow freely are collected in refrigerated bottles using automatic sampling equipment. Solid samples are collected as grab samples using widemouth jars. Maintain samples at 0 to 4°C from the time of collection until extraction. If residual chlorine is present in aqueous samples, add 80 mg sodium thiosulfate/L of water. Begin sample extraction within 7 days of collection, and analyze all extracts within 40 days of extraction.

SAMPLE PREPARATION Samples containing 1% solids or less are extracted directly using continuous liquid-liquid extraction techniques. Samples containing 1 to 30% solids are diluted to the 1% level with reagent water and extracted using continuous liquid-liquid extraction techniques. Samples containing greater than 30% solids are extracted using ultrasonic techniques.

Base/neutral extraction — Adjust the pH of the waters in the extractors to 12–13 with 6 N NaOH. Extract with methylene chloride for 24–48 h.

Acid extraction — Adjust the pH of the waters in the extractors to 2 or less using 6 N sulfuric acid. Extract with methylene chloride for 24–48 h.

Ultrasonic extraction of high solids samples — Add anhydrous sodium sulfate to the sample and QC aliquot(s). Add acetone:methylene chloride (1:1) to the sample and mix thoroughly

Concentrate extracts using a K-D apparatus.

QUALITY CONTROL The analyst is permitted to modify this method to improve separations or lower the costs of measurements, provided all performance specifications are met. Analyses of blanks are required to demonstrate freedom from contamination. When results of spikes indicate atypical method performance for samples, the samples are diluted to bring method performance within acceptable limits.

For low solids (aqueous samples), extract, concentrate, and analyze two sets of four 1-L aliquots (8 aliquots total) of the precision and recovery standard. For high solids samples, two sets of four 30-g aliquots of the high solids reference matrix are used.

Spike all samples with labeled compounds to assess method performance. Compute percent recovery of the labeled compounds using the internal standard method. Compare the labeled compound recovery for each compound with the corresponding labeled compound recovery.

Reagent water and high solids reference matrix blanks are analyzed to demonstrate freedom from contamination. Extract and concentrate a 1-L reagent water blank or a high solids reference matrix blank with each sample's lot (samples started through the extraction process on the same 8-h shift, to a maximum of 20 samples).

Field replicates may be collected to determine the precision of the sampling technique, and spiked samples may be required to determine the accuracy of the analysis when the internal standard method is used.

REFERENCE Semivolatile Organic Compounds by Isotope Dilution GC/MS. Office of Water Regulation and Standards, U.S. EPA Industrial Technology Division, Washington, DC, EPA Method 1625, Rev. C, June 1989 (contact W.A. Telliard, U.S. EPA, Office of Water Regulations and Standards, 401 M St., SW, Washington, DC, 20460. Phone: 202-382-7131).

Fluoranthene   **EPA Method 625**
CAS #206-44-0

TITLE Base/Neutrals and Acids, U.S. EPA Method 625

MATRIX This methods covers municipal and industrial wastewaters.

METHOD SUMMARY Approximately 1 L of sample is serially extracted with methylene chloride at a pH greater than 11 and again at a pH less than 2 using a separatory funnel or a continuous extractor. The methylene chloride extract is dried, concentrated to a volume of 1 mL, and analyzed by GC/MS. Qualitative identification of the parameters in the extract is performed using the retention time and the relative abundance of three characteristic masses (m/z). Qualitative analysis is performed using either external or internal standard techniques with a single characteristic m/z.

INTERFERENCES Method interferences may be caused by contaminants in solvents, reagents, glassware, and other sample processing hardware. Glassware must be scrupulously cleaned. Glassware should be heated in a muffle furnace at 400°C for 5 to 30 min. Some thermally stable materials, such as PCBs, may not be eliminated by this treatment. Solvent rinses with acetone and pesticide quality hexane may be substituted for the muffle furnace heating. Matrix interferences may be caused by contaminants that are coextracted from the sample. The base-neutral extraction may cause significantly reduced recovery of phenols. The packed gas chromatographic columns recommended for the basic fraction may not exhibit sufficient resolution for some analytes.

INSTRUMENTATION A GC/MS system with an injection port designed for on-column injection when using packed columns and for splitless injection when using capillary columns.

Column for base/neutrals: 1.8 m long × 2 mm I.D. glass, packed with 3% SP-2550 on Supelcoport (100/120 mesh) or equivalent.

Column for acids: 1.8 m long × 2 mm I.D. glass, packed with 1% SP-1240DA on Supelcoport (100/120 mesh) or equivalent.

PRECISION & ACCURACY The MDL concentrations were obtained using reagent water. The MDL actually achieved in a given analysis will vary depending on instrument sensitivity and matrix effects. This method was tested by 15 laboratories using reagent water, drinking water, surface water, and industrial wastewaters spiked at six concentrations over the range 5

to 100 µg/L. Single operator precision, overall precision, and method accuracy were found to be directly related to the concentration of the parameter matrix.

The MDL (in µg/L) in reagent water was 2.2.
The standard deviation (in µg/L based on 4 recovery measurements) was 32.8.
The range (in µg/L) for average recovery for 4 measurements was 42.9–121.3.
The range (in %) for percent recovery was 26–137.
Accuracy (in µg/L) as expected recovery for one or more measurements of a sample containing a true concentration of C was 0.81C + 1.10.
Precision (in µg/L) as expected single analyst standard deviation of measurements at an average concentration found at X was 0.22X–0.73.
Overall precision (in µg/L) as expected interlaboratory standard deviation of measurements in an average concentration found at X was 0.28X–0.60.

C = *True value of the concentration in µg/L.*
X = *Average recovery found for measurements of samples containing a concentration at C in µg/L.*

SAMPLE PREPARATION   Adjust the pH to >11 with sodium hydroxide and serially extract in a separatory funnel with methylene chloride or else in a continuous extractor. Next, adjust the pH to <2 with sulfuric acid and serially extract in a separatory funnel with methylene chloride or else in a continuous extractor. Dry the extracts separately through a column of anhydrous sodium sulfate and then concentrate each of the extracts to 1.0 mL using a K-D apparatus.

SAMPLE COLLECTION, PRESERVATION & HANDLING
Grab samples must be collected in glass containers. All samples must be refrigerated at 4°C from the time of collection until extraction. If residual chlorine is present, add 80 mg of sodium thiosulfate/L of sample and mix well. All samples must be extracted within 7 days of collection and completely analyzed within 40 days of extraction.

QUALITY CONTROL   Make an initial, one-time, demonstration of the ability to generate acceptable accuracy and precision with this method. Before processing any samples, the analyst must analyze a reagent water blank to demonstrate that interferences from the analytical system and glassware are under control. Each time a set of samples is extracted or reagents are changed, a reagent water blank must be processed. Spike and analyze a minimum of 5% of all samples to monitor and evaluate lab data quality. A QC check sample concentrate that contains each parameter of interest at a concentration of 100 µg/mL in acetone is required. PCBs and multicomponent pesticides may be omitted from this test.

After analysis of five spiked wastewater samples, calculate the average percent recovery and the standard deviation of the percent recovery. Spike all samples with the surrogate standard spiking solution and calculate the percent recovery of each surrogate compound.

REFERENCE   *Federal Register*, Vol. 49, No. 209. Friday, Oct. 26, 1984.

Fluoranthene **EPA Method 8270**
CAS #206-44-0

TITLE   Semivolatile Organic Compounds by GC/MS

MATRIX   This method is used to determine the concentration of semivolatile organic compounds in extracts prepared from all types of solid waste matrices, soils, and groundwater. Although surface waters are not specifically mentioned, this method should be applicable to water samples from rivers, lakes, etc.

METHOD SUMMARY   This method covers 259 semivolatile organic compounds. In very limited applications direct injection of the sample into the GC/MS system may be appropriate, but this results in very high detection limits (approximately 10,000 µg/L). Typically, a 1-L liquid sample, containing surrogate, and matrix spiking standards, is extracted in a continuous extractor first under acid conditions and then under basic conditions. Typically 30 g of a solid sample, containing surrogate, and matrix spiking standards, is extracted ultrasonically. After concentrating the extract to 1 mL it is spiked with 10 µL of an internal standard solution just prior to analysis by GC/MS. The volume injected should contain about 100 ng of base/neutral and 200 ng of acid surrogates (for a 1-µL injection). Analysis is performed by GC/MS using a capillary GC column.

INTERFERENCES   Raw GC/MS data from all blanks, samples, and spikes must be evaluated for interferences. Contamination by carryover can occur whenever high-concentration and low-concentration samples are sequentially analyzed. To reduce carryover, the sample syringe must be rinsed out between samples with solvent. Whenever an unusually concentrated sample is encountered, it should be followed by the analysis of blank solvent to check for cross-contamination.

INSTRUMENTATION   A GC/MS and a data system are required. The GC column used is a 30 m × 0.25 mm I.D. (or 0.32 mm I.D.) 1um film thickness silicone-coated fused silica capillary column. A continuous liquid-liquid extractor equipped with Teflon® or glass connection joints and stopcocks requiring no lubrication, a K-D concentrating apparatus, water bath, and an ultrasonic disrupter with a minimum power of 300 W and with pulsing capability are also required.

PRECISION & ACCURACY   The estimated quantitation limit (EQL) of Method 8270B for determining an individual compound is approximately 1 mg/kg (wet weight) for soil or sediment samples, 1–200 mg/kg for wastes (dependent on matrix and method of preparation), and 10 µg/L for groundwater samples. EQLs will be proportionately higher for sample extracts that require dilution to avoid saturation of the detector.

The EQL(b) for groundwater in µg/L is 10.
The EQL (a, b) for low concentrations in soil and sediment in µg/kg is 660.
Accuracy as µg/L is 0.81C + 1.10.
Overall precision in µg/L is 0.28X–0.60.

(a)   *EQLs listed for soil/sediment are based on wet weight. Normally data is reported in a dry-weight basis; therefore, EQLs will be higher based on the % dry weight of each sample.*

This calculation is based on a 30-g sample and gel permeation chromatography cleanup.

(b) *Sample EQLs are highly matrix-dependent. The EQLs are provided for guidance and may not always be achievable.*

C = *True value for concentration, in µg/L.*

X = *Average recovery found for measurements of samples containing a concentration of C, in µg/L.*

ESTIMATED QUANTITATION LIMIT

Other Matrices	Factor (a)
High-concentration soil and sludges by sonicator	7.5
Non-water miscible waste	75

(a) *EQL for other matrices = [EQL for low soil/sediment] × [Factor]. This estimated EQL is similar to an EPA "Practical Quantitation Limit."*

SAMPLING METHOD

Liquid samples — Use a 1 or 2½ gallon amber glass bottle with a screw-top Teflon®-lined cover that has been prewashed with detergent and rinsed with distilled water and methanol (or isopropanol).

Soils, sediments, or sludges — Use an 8-oz. widemouth glass with a screw-top Teflon®-lined cover that has been prewashed with detergent and rinsed with distilled water and methanol (or isopropanol).

SAMPLE PRESERVATION

Liquid samples — If residual chlorine is present, add 3 mL of 10% sodium thiosulfate per gallon, cool to 4°C and store in a solvent-free refrigerator until analysis; if chlorine is not present, then eliminate the sodium thiosulfate addition.

Soils, sediments, or sludges — Cool samples to 4°C and store in a solvent-free refrigerator.

MHT Liquid samples must be extracted within 7 days and the extracts analyzed within 40 days. Soils, sediments, or sludges may be stored for a maximum of 14 days and the extracts analyzed within 40 days.

SAMPLE PREPARATION

Liquid samples — Transfer 1 L quantitatively to a continuous extractor. If high concentrations are anticipated, a smaller volume may be used and then diluted with organic-free reagent water to 1 L. Adjust pH, if necessary, to pH <2 using 1:1 (V/V) sulfuric acid. Pipette 1.0 mL of a surrogate standard spiking solution into each sample. For the sample in each analytical batch selected for spiking, add 1.0 mL of a matrix spiking standard. For base/neutral acid analysis, the amount of the surrogates and matrix spiking compounds added to the sample should result in a final concentration of 100 ng/µL of each analyte in the extract to be analyzed (assuming a 1-µL injection). Extract with methylene chloride for 18–24 h. Next, adjust the pH of the aqueous phase to pH >11 using 10 N sodium hydroxide and extract it with methylene chloride again for 18–24 h. Dry the extract through a column containing anhydrous sodium sulfate and concentrate it to 1 mL using a K-D concentrator.

Soils, sediments, or sludges — Use 30 g of sample. Nonporous or wet samples (gummy or clay type) that do not have a free-flowing sandy texture must be mixed with anhydrous sodium sulfate until the sample is free flowing. Add 1 mL of surrogate standards to all samples, spikes, standards, and blanks. For the sample in each analytical batch selected for spiking, add 1.0 mL of a matrix spiking standard. For base/neutral acid analysis, the amount added of the surrogates and matrix spiking compounds should result in a final concentration of 100 ng/µL of each base/neutral analyte and 200 ng/µL of each acid analyte in the extract to be analyzed (assuming a 1-µL injection). Immediately add a 100-mL mixture of 1:1 methylene chloride:acetone and extract the sample ultrasonically for 3 min and then decant or filter the extracts. Repeat the extraction two or more times. Dry the extract using a column with anhydrous sodium sulfate and concentrate it to 1 mL in a K-D concentrator.

QUALITY CONTROL A methylene chloride solution containing 50 ng/µL of decafluorotriphenylphosphine (DFTPP) is used for tuning the GC/MS system each 12-h shift. A system performance check also must be made during every 12-h shift. A standard containing 50 ng/µL each of 4,4'-DDT, pentachlorophenol, and benzidine is required to verify injection port inertness and GC column performance. A calibration standard at mid-concentration, containing each compound of interest, including all required surrogates, must be performed every 12 h during analysis. After the system performance check is met, calibration check compounds (CCCs) are used to check the validity of the initial calibration.

The internal standard responses and retention times in the calibration check standard must be evaluated immediately after or during data acquisition. If the retention time for any internal standard changes by more than 30 seconds from the last check calibration (12 h), the chromatographic system must be inspected for malfunctions and corrections must be made, as required. If the electron ionization current plot (EICP) area for any of the internal standards changes by a factor of two from the last daily calibration standard check, the mass spectrometer must be inspected for malfunctions and corrections must be made, as appropriate.

Demonstrate, through the analysis of a reagent water blank, that interferences from the analytical system, glassware, and reagents are under control. The blank samples should be carried through all stages of the sample preparation and measurement steps. For each analytical batch (up to 20 samples), a reagent blank, matrix spike, and matrix spike duplicate/duplicate must be analyzed (the frequency of the spikes may be different for different monitoring programs). The blank and spiked samples must be carried through all stages of the sample preparation and measurement steps. A QC reference sample concentrate containing each analyte at a concentration of 100 mg/L in methanol is required.

REFERENCE Test Methods for Evaluating Solid Waste (SW-846). U.S. EPA 1983, Method 8270B, Rev. 2, Nov. 1990. Office of Solid Waste, Washington, DC.

Fluoranthene   **EPA Method 8100**
CAS #206-44-0

TITLE Polynuclear Aromatic Hydrocarbons

MATRIX Groundwater, soils, sludges, water miscible liquid wastes, and non-water miscible wastes.

APPLICATION This method is used for the analysis of various PAHs. Samples are extracted, concentrated, and analyzed using direct injection of both neat and diluted organic liquids. The method provides two optional GC columns that are better than Column 1 and that may help resolve analytes from interferences.

INTERFERENCES Solvents, reagents, and glassware may introduce artifacts. Other interferences may come from coextracted compounds from samples.

INSTRUMENTATION GC capable of on-column injections and a flame with detector (FID). Column 1: a 1.8 m by 2 mm 3% OV-17 on Chromosorb W-AW-DCMS column. Column 2: a 30 m by 0.25 mm SE-54 fused silica capillary colunm. Column 3: a 30 m by 0.32 mm SE-54 fused silica capillary column.

RANGE 0.1–425 µg/L

MDL Not reported.

PQL FACTORS FOR MULTIPLYING × FID MDL VALUE Not available.

PRECISION 0.32X + 0.03 µg/L (overall precision).

ACCURACY 0.68C + 0.07 µg/L (as recovery).

SAMPLING METHOD Use 8-oz. widemouth glass bottles with Teflon®-lined caps for concentrated waste samples, soils, sediments, and sludges. Use 1 or 2½ gallon amber glass bottles with Teflon®-lined caps for liquid (water) samples.

STABILITY Cool soil, sediment, sludge, and liquid samples to 4°C. If residual chlorine is present in liquid samples add 3 mL of 10% sodium thiosulfate per gallon of sample and cool to 4°C.

MHT 14 days for concentrated waste, soil, sediment, or sludge; 7 days for liquid samples; all extracts must be analyzed within 40 days.

QUALITY CONTROL A quality control check sample concentrate containing each analyte of interest is required. The QC check sample concentrate may be prepared from pure standard materials or purchased as certified solutions Use appropriate trip, matrix, control site, method, reagent, and solvent blanks. Internal, surrogate, and five concentration level calibration standards are used. The quality control check sample concentrate should contain fluoranthene at 10 µg/mL in acetonitrile.

REFERENCE Test Methods for Evaluating Solid Waste (SW-846), U.S. EPA Office of Solid Waste, Washington, DC, Method 8100, Nov. 1986.

Fluoranthene   **EPA Method 8310**
CAS #206-44-0

TITLE Polynuclear Aromatic Hydrocarbons

MATRIX Groundwater, soils, sludges, water miscible liquid wastes, and non-water miscible wastes.

APPLICATION This method is used for the analysis of 16 polynuclear aromatic hydrocarbons (PAHs). Samples are extracted, concentrated, and analyzed using HPLC with detection by UV and fluorescence detectors.

INTERFERENCES Solvents, reagents, and glassware may introduce artifacts. Other interferences may come from coextracted compounds from samples.

INSTRUMENTATION HPLC with a gradient pumping system and a 250 mm by 2.6 mm reverse phase HC-ODS Sil-X 5-micron particle-size column. The fluorescence detector uses an excitation wavelength of 280 nm and emission greater than 389 nm cutoff with dispersive optics.

RANGE 0.1–425 µg/L

MDL 0.21 µg/L (fluorescence; reagent water).

PQL FACTORS FOR MULTIPLYING × FID MDL VALUE

Matrix	Multiplication Factor
Groundwater	10
Low-level soil by sonication with GPC cleanup	670
High-level soil and sludge by sonication	10,000
Non-water miscible waste	100,000

PRECISION 0.32X + 0.03 µg/L (overall precision).

ACCURACY 0.68C + 0.07 µg/L (as recovery).

SAMPLING METHOD Use 8-oz. widemouth glass bottles with Teflon®-lined caps for concentrated waste samples, soils, sediments, and sludges. Use 1 or 2½ gallon amber glass bottles with Teflon®-lined caps for liquid (water) samples.

STABILITY Cool soil, sediment, sludge, and liquid samples to 4°C. If residual chlorine is present in liquid samples add 3 mL of 10% sodium thiosulfate per gallon of sample and cool to 4°C.

MHT 14 days for concentrated waste, soil, sediment, or sludge; 7 days for liquid samples; all extracts must be analyzed within 40 days.

QUALITY CONTROL Internal, surrogate, and five concentration level calibration standards are used. The calibration standards must be used with the analytical method blank. A quality control check sample concentrate containing fluoranthene at 10 µg/mL is required. The QC check sample concentrate may be prepared from pure standard materials or purchased as certified solutions. Use appropriate trip, matrix, control site, method, reagent, and solvent blanks.

REFERENCE Test Methods for Evaluating Solid Waste (SW-846), U.S. EPA Office of Solid Waste, Washington, DC, Method 8310, Rev. 0, Nov. 1986.

Fluorene   **EPA Method 1625**
CAS #86-73-7

TITLE Semivolatile Organic Compounds by Isotope Dilution GC/MS

MATRIX The compounds may be determined in waters, soils, and municipal sludges by this method.

METHOD SUMMARY This method is used to determine 176 semivolatile toxic organic pollutants associated with the CWA (as amended 1987); the RCRA (as amended 1986); the CERCLA (as amended 1986); and other compounds amenable to extraction and analysis by capillary column gas chromatography-mass spectrometry (GC/MS).

Stable isotopically-labeled analogs of the compounds of interest are added to the sample. If the solids content is less than 1%, a 1-L sample is extracted at pH 12–13, then at pH <2 with methylene chloride using continuous extraction techniques.

If the solids content is 30% or less, the sample is diluted to 1% solids with reagent water, homogenized ultrasonically, and extracted at pH 12–13, then at pH <2 with methylene chloride using continuous extraction techniques. If the solids content is greater than 30%, the sample is extracted using ultrasonic techniques.

Each extract is dried over sodium sulfate, concentrated to a volume of 5 mL, cleaned up using GPC, if necessary, and concentrated. Extracts are concentrated to 1 mL if GPC is not performed, and to 0.5 mL if GPC is performed.

An internal standard is added to the extract, and a 1-mL aliquot of the extract is injected into the GC. The compounds are separated by GC and detected by a MS. The labeled compounds serve to correct the variability of the analytical technique.

INTERFERENCES Solvents, reagents, glassware, and other sample processing hardware may yield artifacts and/or elevated baselines causing misinterpretation of chromatograms and spectra. Materials used in the analysis must be demonstrated to be free from interferences under the conditions of analysis by running method blanks initially and with each sample lot (sample started through the extraction process on a given 8-h shift, to a maximum of 20). Specific selection of reagents and purification of solvents by distillation in all glass systems may be required. Glassware and, where possible, reagents are cleaned by solvent rinse and baking at 450°C for 1-h minimum. Interferences coextracted from samples will vary considerably from source to source, depending on the diversity of the site being sampled.

INSTRUMENTATION Major instrumentation includes a GC with a splitless or on-column injection port for capillary column, a MS with 70 eV electron impact ionization, and a data system to collect and record MS data, and process it. A K-D apparatus is used to concentrate extracts.

GC Column: 30 m × 0.25 mm I.D. 5% phenyl, 94% methyl, 1% vinyl silicone bonded phased fused silica capillary column.

PRECISION & ACCURACY The detection limits of the method are usually dependent on the level of interferences rather than instrumental limitations. The limits typify the minimum quantities that can be detected with no interferences present.

The minimum level (in µg/mL) was 10. This is defined as a minimum level at which the analytical system shall give recognizable mass spectra (background corrected) and acceptable calibration points.

The MDL (in µg/kg) in low solids was 69 and in high solids was 61; these were determined in digested sludge (low solids) and in filter cake or compost (high solids).

The labeled and native compound initial precision as standard deviation (in µg/L) was 29.
The labeled and native compound initial accuracy as average recovery (in µg/L) was 81–132.

SAMPLE COLLECTION, PRESERVATION & HANDLING Collect samples in glass containers. Aqueous samples which flow freely are collected in refrigerated bottles using automatic sampling equipment. Solid samples are collected as grab samples using widemouth jars. Maintain samples at 0 to 4°C from the time of collection until extraction. If residual chlorine is present in aqueous samples, add 80 mg sodium thiosulfate/L of water. Begin sample extraction within 7 days of collection, and analyze all extracts within 40 days of extraction.

SAMPLE PREPARATION Samples containing 1% solids or less are extracted directly using continuous liquid-liquid extraction techniques. Samples containing 1 to 30% solids are diluted to the 1% level with reagent water and extracted using continuous liquid-liquid extraction techniques. Samples containing greater than 30% solids are extracted using ultrasonic techniques.

Base/neutral extraction — Adjust the pH of the waters in the extractors to 12–13 with 6 N NaOH. Extract with methylene chloride for 24–48 h.
Acid extraction — Adjust the pH of the waters in the extractors to 2 or less using 6 N sulfuric acid. Extract with methylene chloride for 24–48 h.
Ultrasonic extraction of high solids samples — Add anhydrous sodium sulfate to the sample and QC aliquot(s). Add acetone:methylene chloride (1:1) to the sample and mix thoroughly

Concentrate extracts using a K-D apparatus.

QUALITY CONTROL The analyst is permitted to modify this method to improve separations or lower the costs of measurements, provided all performance specifications are met. Analyses of blanks are required to demonstrate freedom from contamination. When results of spikes indicate atypical method performance for samples, the samples are diluted to bring method performance within acceptable limits.

For low solids (aqueous samples), extract, concentrate, and analyze two sets of four 1-L aliquots (8 aliquots total) of the precision and recovery standard. For high solids samples, two sets of four 30-g aliquots of the high solids reference matrix are used.

Spike all samples with labeled compounds to assess method performance. Compute percent recovery of the labeled compounds using the internal standard method. Compare the labeled compound recovery for each compound with the corresponding labeled compound recovery.

Reagent water and high solids reference matrix blanks are analyzed to demonstrate freedom from contamination. Extract and concentrate a 1-L reagent water blank or a high solids reference matrix blank with each sample's lot (samples started through the extraction process on the same 8-h shift, to a maximum of 20 samples).

Field replicates may be collected to determine the precision of the sampling technique, and spiked samples may be required to determine the accuracy of the analysis when the internal standard method is used.

REFERENCE Semivolatile Organic Compounds by Isotope Dilution GC/MS. Office of Water Regulation and Standards, U.S. EPA Industrial Technology Division, Washington, DC, EPA Method 1625, Rev. C, June 1989 (contact W.A. Telliard, U.S. EPA, Office of Water Regulations and Standards, 401 M St., SW, Washington, DC, 20460. Phone: 202-382-7131).

Fluorene   **EPA Method 625**
CAS #86-73-7

TITLE Base/Neutrals and Acids, U.S. EPA Method 625

MATRIX This methods covers municipal and industrial wastewaters.

METHOD SUMMARY Approximately 1 L of sample is serially extracted with methylene chloride at a pH greater than 11 and again at a pH less than 2 using a separatory funnel or a continuous extractor. The methylene chloride extract is dried, concentrated to a volume of 1 mL, and analyzed by GC/MS. Qualitative identification of the parameters in the extract is performed using the retention time and the relative abundance of three characteristic masses (m/z). Qualitative analysis is performed using either external or internal standard techniques with a single characteristic m/z.

INTERFERENCES Method interferences may be caused by contaminants in solvents, reagents, glassware, and other sample processing hardware. Glassware must be scrupulously cleaned. Glassware should be heated in a muffle furnace at 400°C for 5 to 30 min. Some thermally stable materials, such as PCBs, may not be eliminated by this treatment. Solvent rinses with acetone and pesticide quality hexane may be substituted for the muffle furnace heating. Matrix interferences may be caused by contaminants that are coextracted from the sample. The base-neutral extraction may cause significantly reduced recovery of phenols. The packed gas chromatographic columns recommended for the basic fraction may not exhibit sufficient resolution for some analytes.

INSTRUMENTATION A GC/MS system with an injection port designed for on-column injection when using packed columns and for splitless injection when using capillary columns.

Column for base/neutrals: 1.8 m long × 2 mm I.D. glass, packed with 3% SP-2550 on Supelcoport (100/120 mesh) or equivalent.
Column for acids: 1.8 m long × 2 mm I.D. glass, packed with 1% SP-1240DA on Supelcoport (100/120 mesh) or equivalent.

PRECISION & ACCURACY The MDL concentrations were obtained using reagent water. The MDL actually achieved in a given analysis will vary depending on instrument sensitivity and matrix effects. This method was tested by 15 laboratories using reagent water, drinking water, surface water, and industrial wastewaters spiked at six concentrations over the range 5 to 100 µg/L. Single operator precision, overall precision, and method accuracy were found to be directly related to the concentration of the parameter matrix.

The MDL (in µg/L) in reagent water was 1.9.
The standard deviation (in µg/L based on 4 recovery measurements) was 20.7.
The range (in µg/L) for average recovery for 4 measurements was 71.6–108.4.
The range (in %) for percent recovery was 59–121.
Accuracy (in µg/L) as expected recovery for one or more measurements of a sample containing a true concentration of C was 0.90C-0.00.
Precision (in µg/L) as expected single analyst standard deviation of measurements at an average concentration found at X was 0.12X + 0.26.
Overall precision (in µg/L) as expected interlaboratory standard deviation of measurements in an average concentration found at X was 0.13X + 0.61.

$C = $ *True value of the concentration in µg/L.*
$X = $ *Average recovery found for measurements of samples containing a concentration at C in µg/L.*

SAMPLE PREPARATION Adjust the pH to >11 with sodium hydroxide and serially extract in a separatory funnel with methylene chloride or else in a continuous extractor. Next, adjust the pH to <2 with sulfuric acid and serially extract in a separatory funnel with methylene chloride or else in a continuous extractor. Dry the extracts separately through a column of anhydrous sodium sulfate and then concentrate each of the extracts to 1.0 mL using a K-D apparatus.

SAMPLE COLLECTION, PRESERVATION & HANDLING Grab samples must be collected in glass containers. All samples must be refrigerated at 4°C from the time of collection until extraction. If residual chlorine is present, add 80 mg of sodium thiosulfate/L of sample and mix well. All samples must be extracted within 7 days of collection and completely analyzed within 40 days of extraction.

QUALITY CONTROL Make an initial, one-time, demonstration of the ability to generate acceptable accuracy and precision with this method. Before processing any samples, the analyst must analyze a reagent water blank to demonstrate that interferences from the analytical system and glassware are under control. Each time a set of samples is extracted or reagents are changed, a reagent water blank must be processed. Spike and analyze a minimum of 5% of all samples to monitor and evaluate lab data quality. A QC check sample concentrate that contains each parameter of interest at a concentration of 100 µg/mL in acetone is required. PCBs and multicomponent pesticides may be omitted from this test.

After analysis of five spiked wastewater samples, calculate the average percent recovery and the standard deviation of the percent recovery. Spike all samples with the surrogate standard spiking solution and calculate the percent recovery of each surrogate compound.

REFERENCE *Federal Register*, Vol. 49, No. 209. Friday, Oct. 26, 1984.

Fluorene	EPA Method 8270
CAS #86-73-7	

TITLE Semivolatile Organic Compounds by GC/MS

MATRIX This method is used to determine the concentration of semivolatile organic compounds in extracts prepared from all types of solid waste matrices, soils, and groundwater. Although surface waters are not specifically mentioned, this method should be applicable to water samples from rivers, lakes, etc.

METHOD SUMMARY This method covers 259 semivolatile organic compounds. In very limited applications direct injection of the sample into the GC/MS system may be appropriate, but this results in very high detection limits (approximately 10,000 µg/L). Typically, a 1-L liquid sample, containing surrogate, and matrix spiking standards, is extracted in a continuous extractor first under acid conditions and then under basic conditions. Typically 30 g of a solid sample, containing surrogate, and matrix spiking standards, is extracted ultrasonically. After concentrating the extract to 1 mL it is spiked with 10 µL of an internal standard solution just prior to analysis by GC/MS. The volume injected should contain about 100 ng of base/neutral and 200 ng of acid surrogates (for a 1-µL injection). Analysis is performed by GC/MS using a capillary GC column.

INTERFERENCES Raw GC/MS data from all blanks, samples, and spikes must be evaluated for interferences. Contamination by carryover can occur whenever high-concentration and low-concentration samples are sequentially analyzed. To reduce carryover, the sample syringe must be rinsed out between samples with solvent. Whenever an unusually concentrated sample is encountered, it should be followed by the analysis of blank solvent to check for cross-contamination.

INSTRUMENTATION A GC/MS and a data system are required. The GC column used is a 30 m × 0.25 mm I.D. (or 0.32 mm I.D.) 1um film thickness silicone-coated fused silica capillary column. A continuous liquid-liquid extractor equipped with Teflon® or glass connection joints and stopcocks requiring no lubrication, a K-D concentrating apparatus, water bath, and an ultrasonic disrupter with a minimum power of 300 W and with pulsing capability are also required.

PRECISION & ACCURACY The estimated quantitation limit (EQL) of Method 8270B for determining an individual compound is approximately 1 mg/kg (wet weight) for soil or sediment samples, 1–200 mg/kg for wastes (dependent on matrix and method of preparation), and 10 µg/L for groundwater samples. EQLs will be proportionately higher for sample extracts that require dilution to avoid saturation of the detector.

The EQL(b) for groundwater in µg/L is 10.

The EQL (a, b) for low concentrations in soil and sediment in µg/kg is 660.
Accuracy as µg/L is 0.90C–0.00.
Overall precision in µg/L is 0.13X + 0.61.

(a) EQLs listed for soil/sediment are based on wet weight. Normally data is reported in a dry-weight basis; therefore, EQLs will be higher based on the % dry weight of each sample. This calculation is based on a 30-g sample and gel permeation chromatography cleanup.

(b) Sample EQLs are highly matrix-dependent. The EQLs are provided for guidance and may not always be achievable.

C = True value for concentration, in µg/L.
X = Average recovery found for measurements of samples containing a concentration of C, in µg/L.

ESTIMATED QUANTITATION LIMIT

Other Matrices	Factor (a)
High-concentration soil and sludges by sonicator	7.5
Non-water miscible waste	75

(a) EQL for other matrices = [EQL for low soil/sediment] × [Factor]. This estimated EQL is similar to an EPA "Practical Quantitation Limit."

SAMPLING METHOD

Liquid samples — Use a 1 or 2½ gallon amber glass bottle with a screw-top Teflon®-lined cover that has been prewashed with detergent and rinsed with distilled water and methanol (or isopropanol).

Soils, sediments, or sludges — Use an 8-oz. widemouth glass with a screw-top Teflon®-lined cover that has been prewashed with detergent and rinsed with distilled water and methanol (or isopropanol).

SAMPLE PRESERVATION

Liquid samples — If residual chlorine is present, add 3 mL of 10% sodium thiosulfate per gallon, cool to 4°C and store in a solvent-free refrigerator until analysis; if chlorine is not present, then eliminate the sodium thiosulfate addition.

Soils, sediments, or sludges — Cool samples to 4°C and store in a solvent-free refrigerator.

MHT Liquid samples must be extracted within 7 days and the extracts analyzed within 40 days. Soils, sediments, or sludges may be stored for a maximum of 14 days and the extracts analyzed within 40 days.

SAMPLE PREPARATION

Liquid samples — Transfer 1 L quantitatively to a continuous extractor. If high concentrations are anticipated, a smaller volume may be used and then diluted with organic-free reagent water to 1 L. Adjust pH, if necessary, to pH <2 using 1:1 (V/V) sulfuric acid. Pipette 1.0 mL of a surrogate standard spiking solution into each sample. For the sample in each analytical batch selected for spiking, add 1.0 mL of a matrix spiking standard. For base/neutral acid analysis, the amount of the surrogates and matrix spiking compounds added to the sample should result in a final concentration of 100 ng/µL of each analyte in the extract to be analyzed (assuming a 1-µL injection). Extract with methylene chloride for 18–24 h. Next, adjust the pH of the aqueous phase to pH >11 using 10 N sodium hydroxide and extract it with methylene chloride again for 18–24 h. Dry the extract through a column containing anhydrous sodium sulfate and concentrate it to 1 mL using a K-D concentrator.

Soils, sediments, or sludges — Use 30 g of sample. Nonporous or wet samples (gummy or clay type) that do not have a free-flowing sandy texture must be mixed with anhydrous sodium sulfate until the sample is free flowing. Add 1 mL of surrogate standards to all samples, spikes, standards, and blanks. For the sample in each analytical batch selected for spiking, add 1.0 mL of a matrix spiking standard. For base/neutral acid analysis, the amount added of the surrogates and matrix spiking compounds should result in a final concentration of 100 ng/µL of each base/neutral analyte and 200 ng/µL of each acid analyte in the extract to be analyzed (assuming a 1-µL injection). Immediately add a 100-mL mixture of 1:1 methylene chloride:acetone and extract the sample ultrasonically for 3 min and then decant or filter the extracts. Repeat the extraction two or more times. Dry the extract using a column with anhydrous sodium sulfate and concentrate it to 1 mL in a K-D concentrator.

QUALITY CONTROL A methylene chloride solution containing 50 ng/µL of decafluorotriphenylphosphine (DFTPP) is used for tuning the GC/MS system each 12-h shift. A system performance check also must be made during every 12-h shift. A standard containing 50 ng/µL each of 4,4′-DDT, pentachlorophenol, and benzidine is required to verify injection port inertness and GC column performance. A calibration standard at mid-concentration, containing each compound of interest, including all required surrogates, must be performed every 12 h during analysis. After the system performance check is met, calibration check compounds (CCCs) are used to check the validity of the initial calibration.

The internal standard responses and retention times in the calibration check standard must be evaluated immediately after or during data acquisition. If the retention time for any internal standard changes by more than 30 seconds from the last check calibration (12 h), the chromatographic system must be inspected for malfunctions and corrections must be made, as required. If the electron ionization current plot (EICP) area for any of the internal standards changes by a factor of two from the last daily calibration standard check, the mass spectrometer must be inspected for malfunctions and corrections must be made, as appropriate.

Demonstrate, through the analysis of a reagent water blank, that interferences from the analytical system, glassware, and reagents are under control. The blank samples should be carried through all stages of the sample preparation and measurement steps. For each analytical batch (up to 20 samples), a reagent blank, matrix spike, and matrix spike duplicate/duplicate must be analyzed (the frequency of the spikes may be different for different monitoring programs). The blank and spiked samples must be carried through all stages of the sample preparation and measurement steps. A QC reference sample concentrate containing each analyte at a concentration of 100 mg/L in methanol is required.

REFERENCE Test Methods for Evaluating Solid Waste (SW-846). U.S. EPA 1983, Method 8270B, Rev. 2, Nov. 1990. Office of Solid Waste, Washington, DC.

Fluorene — EPA Method 8100
CAS #86-73-7

TITLE Polynuclear Aromatic Hydrocarbons

MATRIX Groundwater, soils, sludges, water miscible liquid wastes, and non-water miscible wastes.

APPLICATION This method is used for the analysis of various PAHs. Samples are extracted, concentrated, and analyzed using direct injection of both neat and diluted organic liquids. The method provides two optional GC columns that are better than Column 1 and that may help resolve analytes from interferences.

INTERFERENCES Solvents, reagents, and glassware may introduce artifacts. Other interferences may come from coextracted compounds from samples.

INSTRUMENTATION GC capable of on-column injections and a flame with detector (FID). Column 1: a 1.8 m by 2 mm 3% OV-17 on Chromosorb W-AW-DCMS column. Column 2: a 30 m by 0.25 mm SE-54 fused silica capillary colunm. Column 3: a 30 m by 0.32 mm SE-54 fused silica capillary column.

RANGE 0.1–425 µg/L

MDL Not reported.

PQL FACTORS FOR MULTIPLYING × FID MDL VALUE Not available.

PRECISION 0.63X–0.65 µg/L (overall precision).

ACCURACY 0.56C–0.52 µg/L (as recovery).

SAMPLING METHOD Use 8-oz. widemouth glass bottles with Teflon®-lined caps for concentrated waste samples, soils, sediments, and sludges. Use 1 or 2½ gallon amber glass bottles with Teflon®-lined caps for liquid (water) samples.

STABILITY Cool soil, sediment, sludge, and liquid samples to 4°C. If residual chlorine is present in liquid samples add 3 mL of 10% sodium thiosulfate per gallon of sample and cool to 4°C.

MHT 14 days for concentrated waste, soil, sediment, or sludge; 7 days for liquid samples; all extracts must be analyzed within 40 days.

QUALITY CONTROL A quality control check sample concentrate containing each analyte of interest is required. The QC check sample concentrate may be prepared from pure standard materials or purchased as certified solutions Use appropriate trip, matrix, control site, method, reagent, and solvent blanks. Internal, surrogate, and five concentration level calibration standards are used. The quality control check sample concentrate should contain fluorene at 100 µg/mL in acetonitrile.

REFERENCE Test Methods for Evaluating Solid Waste (SW-846), U.S. EPA Office of Solid Waste, Washington, DC, Method 8100, Nov. 1986.

Fluorene   **EPA Method 8310**
CAS #86-73-7

TITLE Polynuclear Aromatic Hydrocarbons

MATRIX Groundwater, soils, sludges, water miscible liquid wastes, and non-water miscible wastes.

APPLICATION This method is used for the analysis of 16 polynuclear aromatic hydrocarbons (PAHs). Samples are extracted, concentrated, and analyzed using HPLC with detection by UV and fluorescence detectors.

INTERFERENCES Solvents, reagents, and glassware may introduce artifacts. Other interferences may come from coextracted compounds from samples.

INSTRUMENTATION HPLC with a gradient pumping system and a 250 mm by 2.6 mm reverse phase HC-ODS Sil-X 5-micron particle-size column. The UV detector uses an excitation wavelength of 254 nm coupled to the fluorescence detector. The fluorescence detector uses an excitation wavelength of 280 nm and emission greater than 389 nm cutoff with dispersive optics.

RANGE 0.1–425 µg/L

MDL 0.21 µg/L (UV detector; reagent water).

PQL FACTORS FOR MULTIPLYING × FID MDL VALUE

Matrix	Multiplication Factor
Groundwater	10
Low-level soil by sonication with GPC cleanup	670
High-level soil and sludge by sonication	10,000
Non-water miscible waste	100,000

PRECISION 0.63X–0.65 µg/L (overall precision).

ACCURACY 0.56C–0.52 µg/L (as recovery).

SAMPLING METHOD Use 8-oz. widemouth glass bottles with Teflon®-lined caps for concentrated waste samples, soils, sediments, and sludges. Use 1 or 2½ gallon amber glass bottles with Teflon®-lined caps for liquid (water) samples.

STABILITY Cool soil, sediment, sludge, and liquid samples to 4°C. If residual chlorine is present in liquid samples add 3 mL of 10% sodium thiosulfate per gallon of sample and cool to 4°C.

MHT 14 days for concentrated waste, soil, sediment, or sludge; 7 days for liquid samples; all extracts must be analyzed within 40 days.

QUALITY CONTROL Internal, surrogate, and five concentration level calibration standards are used. The calibration standards must be used with the analytical method blank. A quality control check sample concentrate containing fluorene at 100 µg/mL is required. The QC check sample concentrate may be prepared from pure standard materials or purchased as certified solutions. Use appropriate trip, matrix, control site, method, reagent, and solvent blanks.

REFERENCE Test Methods for Evaluating Solid Waste (SW-846), U.S. EPA Office of Solid Waste, Washington, DC, Method 8310, Rev. 0, Nov. 1986.

Fluoride   **EPA Method 340**

TITLE Fluoride (Potentiometric, Ion Selective Electrode)

MATRIX This method is applicable to the measurement of total and dissolved fluoride in drinking, surface and saline waters, domestic and industrial wastes.

METHOD SUMMARY Fluoride is determined potentiometrically using a fluoride electrode in conjunction with a standard single junction sleeve-type reference electrode and a pH meter with an expanded millivolt scale or a selective ion m with a direct concentration scale for fluoride. Check the pH first and if it is highly basic (>9), add 1N hydrochloric acid to adjust the pH to 8.3. Immerse the electrodes in a buffered solution and observe the m reading while mixing. The electrodes must remain in the solution for at least three min or until the reading has stabilized. At concentrations under 0.5 mg/L, it may require as long as five min to reach a stable m reading; high concentrations stabilize more quickly. If a pH meter is used, record the potential measurements for each unknown sample and convert the potential reading to the fluoride ion concentration of the unknown using a standard's concentration curve. If a selective ion m is used, read the fluoride level in the unknown sample directly in mg/L on the fluoride scale.

INTERFERENCES Extremes of pH interfere so the sample pH should be between 5 and 9. Polyvalent cations of silicon (Si^{+4}), iron (Fe^{+3}), and aluminum (Al^{+3}) interfere by forming complexes with fluoride. The degree of interference depends upon the concentration of the complexing cations, the concentration of fluoride and the pH of the sample. The addition of a pH 5.0 buffer containing a strong chelating agent preferentially complexes aluminum (the most common interference), silicon, and iron and eliminates the pH problem.

INSTRUMENTATION Electrometer (pH meter) with an expanded mv scale or a selective ion meter; fluoride ion activity electrode; and a single junction, sleeve-type reference electrode.

RANGE 0.1–1000 mg/L

MDL Not listed.

PRECISION When a synthetic sample containing 0.85 mg/L fluoride and no interferences was analyzed by 111 analysts, a mean of 0.84 mg/L with a standard deviation of ± 0.03 was obtained. On the same study, a synthetic sample containing 0.75 mg/L fluoride, 2.5 mg/L polyphosphate and 300 mg/L alkalinity, was analyzed by the same 111 analysts and a mean of 0.75 mg/L fluoride with a standard deviation of ±0.036 was obtained.

ACCURACY 99% of true value at a concentration of 0.75 mg/L or higher.

SAMPLING METHOD No special requirements; 50 to 300 mL of sample is needed; collect samples in a high density polyethylene (HDPE) bottle.

STABILITY No special requirements; no preservatives are needed.

MHT 28 days

QUALITY CONTROL A calibration standard curve as well as instrument calibration are required.

REFERENCE Methods for Chemical Analysis of Water and Wastes, EPA-600/4-79-020, Method 340.2, Fluoride (Potentiometric, Ion Selective Electrode), March 1983; EPA Environmental Monitoring Systems Laboratory, Cincinnati, OH, 45268.

Fluoride — EPA Method 340.2

TITLE Inorganics, Non-Metallics

MATRIX Drinking, surface, and saline waters. Wastewater.

APPLICATION Date issued 1971. Editorial Rev. 1974. (Potentiometric, ion selective electrode). Fluoride(F) is determined potentiometrically using a fluoride electrode in conjunction with a standard single junction sleeve type reference electrode and a pH meter.

INTERFERENCES pH extremes interfere; sample pH should be between 5 and 9. Polyvalent cations of silicon, iron and aluminum interfere (form complexes with fluoride). Degree of interference depends on complexing cations, concentration of fluoride and pH of sample.

INSTRUMENTATION Selective ion m with direct concentration scale for (F) or pH meter with expanded mv scale

RANGE 0.1–1000 mg/L F.

MDL Not listed.

PRECISION SD = ±0.03 at 0.85 mg/L F.

ACCURACY Mean = 0.84 mg/L at 0.85 mg/L F.

SAMPLING METHOD Plastic.

STABILITY No preservation required.

MHT 28 days.

QUALITY CONTROL For industrial waste samples, regular amount buffer may not be adequate. Analyst should check pH first. If highly basic (pH >9), add 1N HCl and adjust pH to 8.3. [Electrodes must remain in the solution at least 3 min or until reading has stabilized. (Up to 5 min)]

REFERENCE Methods for the Chemical Analysis of Water and Wastes, EPA-600/4-79-020, U.S. EPA, EMSL, 1979.

Fluoride — EPA Method 340.3

TITLE Inorganics, Non-Metallics

MATRIX Drinking, surface, and saline waters. Wastewater.

APPLICATION Date issued 1971. (Colorimetric, automated complexone).

Fluoride (F) ion reacts with the red cerous chelate of alizarin complexone. There is a positive color developed in contrast to a bleaching action in other fluoride Methods. For total or total dissolved fluoride, the bellack distillation must be performed prior to complexone analysis.

INTERFERENCES Aluminum forms an extremely stable fluoro compound, which is overcome by treatment with 8-hydroxyquinoline (complexes the aluminum) and extraction with chloroform. At aluminum levels <0.2 mg/L, extraction procedure is not required.

INSTRUMENTATION Technicon auto analyzer, 650 nm filters, 15 mm tubular flow cell.

RANGE 0.05–1.5 mg F/L.

MDL Not listed.

PRECISION SD = ±0.018 at Concentrations of 0.06, 0.15 and 1.08 mg F/L.

ACCURACY Recoveries = 89 and 102% at concentrations of 0.14 and 1.25 mg F/L.

SAMPLING METHOD Plastic.

STABILITY No preservation required.

MHT 28 Days.

QUALITY CONTROL Arrange fluoride standards in sampler in order of decreasing concentration.

REFERENCE Methods for the Chemical Analysis of Water and Wastes, EPA-600/4-79-020, U.S. EPA, EMSL, 1979.

Fluoride — EPA Method 340

TITLE Fluoride (Potentiometric, Ion Selective Electrode)

MATRIX This method is applicable to the measurement of total and dissolved fluoride in drinking, surface and saline waters, domestic and industrial wastes.

METHOD SUMMARY Fluoride is determined potentiometrically using a fluoride electrode in conjunction with a standard single junction sleeve-type reference electrode and a pH meter with an expanded millivolt scale or a selective ion m with a direct concentration scale for fluoride. Check the pH first and if it is highly basic (pH >9), add 1N hydrochloric acid to adjust the pH to 8.3. Immerse the electrodes in a buffered solution and observe the m reading while mixing. The electrodes must remain in the solution for at least three min or until the reading has stabilized. At concentrations under 0.5 mg/L, it may require as long as five min to reach a stable m reading; high concentrations stabilize more quickly. If a pH meter is used, record the potential measurements for each unknown sample and convert the potential reading to the fluoride ion concentration of the unknown using a standard's concentration curve. If a selective ion m is used, read the fluoride level in the unknown sample directly in mg/L on the fluoride scale.

INTERFERENCES Extremes of pH interfere so the sample pH should be between 5 and 9. Polyvalent cations of silicon (Si^{+4}), iron (Fe^{+3}), and aluminum (Al^{+3}) interfere by forming complexes with fluoride. The degree of interference depends upon the concentration of the complexing cations, the concentration of fluoride and the pH of the sample. The addition of a pH 5.0 buffer containing a strong chelating agent preferentially complexes aluminum (the most common interference), silicon, and iron and eliminates the pH problem.

INSTRUMENTATION Electrometer (pH meter) with an expanded mv scale or a selective ion meter; fluoride ion activity electrode; and a single junction, sleeve-type reference electrode.

PRECISION & ACCURACY

RANGE 0.1–1000 mg/L

MDL Not listed.

PRECISION When a synthetic sample containing 0.85 mg/L fluoride and no interferences was analyzed by 111 analysts, a mean of 0.84 mg/L with a standard deviation of ± 0.03 was obtained. On the same study, a synthetic sample containing 0.75 mg/L fluoride, 2.5 mg/L polyphosphate and 300 mg/L alkalinity, was analyzed by the same 111 analysts and a mean of 0.75 mg/L fluoride with a standard deviation of ± 0.036 was obtained.

ACCURACY 99% of true value at a concentration of 0.75 mg/L or higher.

SAMPLE COLLECTION, PRESERVATION & HANDLING No information was provided.

SAMPLING METHOD No special requirements; 50 to 300 mL of sample is needed; collect samples in a high density polyethylene (HDPE) bottle.

STABILITY No special requirements; no preservatives are needed.

MHT 28 days

QUALITY CONTROL A calibration standard curve as well as instrument calibration are required.

REFERENCE Methods for Chemical Analysis of Water and Wastes, EPA-600/4-79-020, Method 340.2, Fluoride (Potentiometric, Ion Selective Electrode), March 1983; EPA Environmental Monitoring Systems Laboratory, Cincinnati, OH, 45268.

Fluoride, Total   **EPA Method 340.1**

TITLE Inorganics, Non-Metallics

MATRIX Drinking, surface, and saline waters. Wastewater.

APPLICATION Date issued 1971. Editorial Rev. 1978. (Colorimetric, spadns with bellack distillation). After distillation to remove interferences, sample is treated with spadns reagent. Loss of color resulting from fluoride reaction with z-s dye is function of fluoride concentration.

INTERFERENCES A small error in reagent addition is most prominent source of error in this test. Care must be taken to avoid overheating flask above level of solution. (Maintain an even flame entirely under boiling flask). Extend range using fluoride ion selective method.

INSTRUMENTATION Distillation equipment. Spectrophotometer at 570 nm or filterphotometer at 550–580 nm.

RANGE 0.1–2.5 mg/L F. (Can be extended).

MDL Not listed.

PRECISION SD = ±0.089 mg/L F at 0.83 mg/L F.

ACCURACY Mean = 0.81 mg/L F at 0.83 m/L F

SAMPLING METHOD Plastic.

STABILITY No preservation required.

MHT 28 Days.

QUALITY CONTROL Plot absorbance vs. concentration. Prepare a new standard curve whenever fresh reagent is made. (Fluoride concentration is measured as the diffeence of absorbance in the blank and the sample).

REFERENCE EPA Methods for the Chemical Analysis of Water and Wastes, EPA-600/4-79-020, U.S. EPA, EMSL, 1979.

Fluridone   **EPA Method 507**
CAS #59756-60-4

TITLE Determination of Nitrogen and Phosphorus-Containing Pesticides in Water by GC/NPD

MATRIX This method is applicable to the determination of certain nitrogen and phosphorus-containing pesticides in finished drinking water and groundwater.

METHOD SUMMARY Method 507 covers 46 nitrogen- and phosphorus-containing pesticides. A 1-L sample is fortified with a surrogate standard, salted, buffered, extracted with methylene chloride, and concentrated; then the solvent is exchanged with methyl tert-butyl ether (MTBE) and concentrated again, and a 2-μL aliquot of a sample extract is injected into a GC system equipped with a selective nitrogen-phosphorus detector and a capillary column for analysis.

INTERFERENCES Method interferences may be caused by contaminants in solvents, reagents, glassware, and other sample processing apparatus. Interfering contamination may occur when a sample containing low concentrations of analytes is analyzed immediately following a sample containing relatively high concentrations. One or more injections of MTBE should be made following the analysis of a sample with high concentrations of analytes to check for analyte carryover. Matrix interferences may be caused by contaminants that are coextracted from the sample. The extent of matrix interferences will vary considerably from source to source, depending upon the water sampled.

INSTRUMENTATION A gas chromatograph system (GC) equipped with a nitrogen-phosphorus detector (NPD) is needed.

Column 1: 30 m × 0.25 mm I.D. DB-5 bonded fused silica column, 0.25 μm film thickness, or equivalent.
Column 2: 30 m × 0.25 mm I.D. DB-1701 bonded fused silica column, 0.25 μm film thickness, or equivalent.

PRECISION & ACCURACY This method has been validated in a single lab and estimated detection limits (EDLs) have been determined for each analyte. Observed detection limits may vary among waters, depending upon the nature of the interferences in the sample matrix and the specific instrumentation used. Analytes that are not separated chromatographically cannot be individually identified and measured unless an alternative technique for identification and quantification exist.

The estimated detection limit (in μg/L) was 3.8. The EDL is defined as either method detection limit or a level of compound in a sample yielding a peak in the final extract with signal-to-noise ratio of approximately 5, whichever value is higher.

The concentration used for these measurements (in μg/L) was 38.
The accuracy (as % recovery) was 87.
The precision (% RSD) was 9.

SAMPLING METHOD Grab samples are collected in 1-L glass sample bottles (prewashed with detergent and hot tap water, rinsed with reagent water, and dried in an oven at 400°C for 1 h) with screw caps lined with PTFE-fluorocarbon.

SAMPLE PRESERVATION Add mercuric chloride to the sample bottle in amounts to produce a concentration of 10 mg/L. If residual chlorine is present, add 80 mg of sodium thiosulfate/L of sample to the sample bottle prior to collection. After collection, seal bottle and shake vigorously for 1 min, then cool the sample to 4°C immediately and store it at 4°C in the dark until extraction.

MHT Maximum holding time of the samples, and in some cases the extracts, is 14 days.

Note: Samples with this compound exhibited recoveries of less than 60% after 14 days.

SAMPLE PREPARATION Fortify the sample with 50 μL of the surrogate standard solution, adjust to pH 7 with phosphate buffer, add 100 g NaCl to the sample, and seal and shake to dissolve the salt; then extract with methylene chloride in a separatory funnel or in a mechanical tumbler bottle. Dry the extract by pouring it through a solvent-rinsed drying column containing about 10 cm of anhydrous sodium sulfate. Collect the extract in a Kuderna-Danish (K-D) concentrator and rinse the column with 20–30 mL methylene chloride. Concentrate the extract to about 2 mL and rinse the flask and its lower joint into the concentrator tube with 1 to 2 mL of methyl t-butyl ether (MTBE). Add 5–10 mL of MTBE and concentrate the extract twice (adding more MTBE) to a final volume of 5.0 mL and store it at 4°C until analysis.

Note: If methylene chloride is not completely removed from the final extract, it may cause detector problems.

QUALITY CONTROL Minimum quality control requirements are initial demonstration of lab capability, determination of surrogate compound recoveries in each sample and blank, monitoring internal standard peak area or height in each sample and blank, analysis of lab reagent blanks, lab fortified samples, lab fortified blanks, and other QC samples. A lab reagent blank is analyzed to demonstrate that all glassware and reagent interferences are under control.

Initial demonstration of capability is fulfilled by analyzing four fortified reagent water samples with the recovery value for each analyte falling within the acceptable range (±30% average recovery). Surrogate recoveries from samples or method blanks must be 70–130%. The internal standard response for any sample chromatogram should not deviate from the daily calibration check standard's internal standard response by more than 30% or lab fortified blanks and sample matrices are used to assess lab performance and analyte recovery, respectively.

If the response for the target analyte peak exceeds the working range of the system, dilute the extract and reanalyze. Alternative techniques such as an alternate detector or second chromatography column should be used to confirm peak identification when sample components are not resolved adequately.

EPA CONTACT & HOTLINE For technical questions contact Dr. Baldev Bathija, U.S. EPA, Office of Ground Water and Drinking Water (WH-550D), 401 M St. SW, Washington, DC 20460. Tel. (202) 260-3040. For further information the EPA Safe Drinking Water Hotline may be called at: (800) 426-4791.

REFERENCE Methods for the Determination of Organic Compounds in Drinking Water, EPA/600/4-88/039 (revised July 1991). U.S. EPA Environmental Monitoring Systems Laboratory, Cincinnati, OH, 45268, U.S.A. Available from the National Technical Information Service (NTIS), 5285 Port Royal Road, Springfield, VA 22161; Tel. 800-553-6847. NTIS Order Number is PB91-231480.

H

Hardness, Total (mg/L as CaCO₃)   EPA Method 130.1

TITLE Physical Properties

MATRIX Drinking, surface, and saline waters.

APPLICATION Date issued 1971. (Colorimetric, automated EDTA). The magnesium edta exchanges magnesium on an equiv basis for any calcium and/or other cations to form a more stable EDTA chelate than magnesium. The free Mg reacts with calmagite at pH 10.

INTERFERENCES No significant interferences. (The free magnesium (Mg) plus calgamite reaction gives a red-violet complex, and by measuring only magnesium concentration in the final reaction stream, total hardness can be measured accurately).

INSTRUMENTATION Technicon auto analyzer, 520 nm filters, 15 mm tubular flow cell.

RANGE 10–400 mg/L as $CaCO_3$.

MDL Not listed.

PRECISION SD = ±1.5 at 19 and 120 mg/L as $CaCO_3$.

ACCURACY At concentrations, 39 and 296 mg/L as $CaCO_3$, recoveries, 89 and 93%.

SAMPLING METHOD Plastic or glass (100 mL).

STABILITY HNO_3 to pH <2. H_2SO_4 to pH <2.

MHT 6 months.

QUALITY CONTROL For most wastewaters and highly polluted waters, the sample must be digested as in atomic absorption method. Arrange working standards in sampler in order of decreasing concentrations.

REFERENCE EPA Methods for the Chemical Analysis of Water and Wastes, EPA-600/4-79-020, U.S. EPA, EMSL, 1979.

Hardness, Total (mg/L as CaCO₃)   EPA Method 130.2

TITLE Physical Properties

MATRIX Drinking, surface, and saline waters. Wastewater.

APPLICATION Date issued 1971. Editorial Rev. 1978. (Titrimetric, EDTA). Calcium (Ca) and magnesium (Mg) ions are sequestered upon addition of disodium-EDTA. Reaction end point, using indicator, has red color in presence of Ca and Mg, a blue color when they're sequestered.

INTERFERENCES Excessive amounts of heavy metals can interfere. This is usually overcome by complexing the metals with cyanide. Routine addition of sodium cyanide solution (caution:deadly poisin) to prevent potential metallic interference is recommended.

INSTRUMENTATION Standard lab titrimetric equipment.

RANGE All concentration ranges of hardness.

MDL Not listed.

PRECISION SD = 2.98 mg/L, $CaCO_3$ at 194 mg/L, $CaCO_3$.

ACCURACY As bias, –2.0 mg/L, $CaCO_3$ at 194 mg/L, $CaCO_3$.

SAMPLING METHOD Plastic or glass (100 mL).

STABILITY HNO3 to pH <2. H2SO4 to pH <2.

MHT 6 months.

QUALITY CONTROL Use a sample aliquot containing not more than 25 mg calcium carbonate($CaCO_3$). Use inhibitors to reduce metal ion interference during titration. Automated titration may be used.

REFERENCE Methods for the Chemical Analysis of Water and Wastes, EPA-600/4-79-020, U.S. EPA, EMSL, 1979.

Heptachlor   EPA Method 505
CAS #76-44-8

TITLE Analysis of Organohalide Pesticides and Commercial Polychlorinated Biphenyl (PCB) Products in Water by Microextraction and Gas Chromatography. U.S. EPA Method 505, Rev. 2.0, 1989.

MATRIX This method is applicable to drinking water and raw source water. The latter should include most surface water and groundwater sources.

METHOD SUMMARY Method 505 covers 25 pesticides and commercial PCB products. This is a very sensitive method that is more useful for monitoring than for exploratory analyses. 5-mL of water are saturated with sodium chloride and then extracted by shaking with 2 mL of hexane. The sample extracts are transferred to an autosampler setup to inject 1–2 µL portions into a gas chromatograph (GC) for analysis. Alternatively, 1–2 µL portions of samples, blanks, and standards may be manually injected. Each extract is analyzed by capillary GC/ECD with confirmation using either a second capillary column or GC/MS. The electron capture detector is easy to use, but it is a nonselective detector. The microextraction technique also eliminates the expensive sample preparation costs of other methods, but it has the disadvantage of being less sensitive than most because the extracts are not concentrated.

INTERFERENCES Method interferences may be caused by contaminants in solvents, reagents, glassware, and other sample processing apparatus that lead to discrete artifacts or elevated baselines. Interfering contamination may occur when a sample containing low concentrations of analytes is analyzed immediately following a sample containing relatively high concentrations of the analytes. Matrix interferences also may be caused by contaminants that are coextracted from the sample; cleanup of sample extracts may be necessary in these cases. Some pesticides and commercial PCB products from aqueous solutions adhere to glass surfaces, so sample transfers and contact with glass surfaces should be minimized. Some pesticides are rapidly oxidized by chlorine so dechlorination with sodium thiosulfate at the time of sample collection is important. Also, splitless injectors may cause degradation of some pesticides.

INSTRUMENTATION A gas chromatograph/electron capture detector/data system, with temperature programming and split/splitless injector suitable for use with capillary columns is needed.

Column 1: 0.32 mm I.D. × 30 m fused silica capillary with chemically bond methyl polysiloxane phase (DB-1, 1.0 µm film, or equivalent).

Column 2: 0.32 mm I.D. × 30 m fused silica capillary with 1:1 mixed phase of dimethyl silicone and polyethylene glycol (Durawax-DX3, 0.25 µm film, or equivalent).

Column 3: 0.32 mm I.D. × 25 m fused silica capillary with chemically bonded 50:50 methyl-phenyl silicone (OV-17, 1.5 µm film, or equivalent).

Column 1 should be used as the primary analytical column. Columns 2 and 3 are recommended for use as confirmatory columns when GC/MS confirmation is not available.

PRECISION & ACCURACY Method detection limits are dependent upon the characteristics of the gas chromatographic system used. Analytes that are not separated chromatographically cannot be individually identified and used in the same calibration mixture or water samples unless an alternative technique for identification and quantification, such as mass spectrometry, is used.

The concentration(s) (in µg/L) used for these QC measurements was 0.032 and 1.2.
The MDL (in µg/L) was 0.003.
The accuracy (% recovery) for reagent water at the above concentration(s) was 77 and 80 and the precision (%) was 10.2 and 7.4.
The accuracy (% recovery) for groundwater at the above concentration(s) was 37 and 71 and the precision (%) was 6.8 and 9.8.
The accuracy (% recovery) for tap water at the above concentration(s) was 200 and 106 and the precision (5) was 22.6 and 16.8.

Note: No range of concentrations is provided with this method.

SAMPLING METHOD Collect samples using a 40-mL screw-cap vial (prewashed with detergent, rinsed with distilled water and oven dried at 400°C for one h) with a Teflon®-faced silicone septum. Collect bubble-free samples and place the septum with the Teflon® side down on the water.

SAMPLE PRESERVATION If residual chlorine is present in the water add about 3 mg of sodium thiosulfate to each vial before samples are collected to remove the chlorine. Alternatively, add 75 µL of 0.04 g/mL solution of sodium thiosulfate to each vial just prior to sampling. Immediately cool samples to 4°C, and store them in a solvent-free refrigerator at 4°C until analysis.

MHT The maximum holding time is 14 days from the time the sample was collected until it must be analyzed.

SAMPLE PREPARATION Remove the sample from storage and allow it to come to room temperature. Remove a 5-mL volume from each container and weigh the container to the nearest 0.1 g. Add 6 g of sodium chloride and 2.0 mL of hexane to each sample bottle. Recap the sample and shake it vigorously for one min. Allow the water and hexane phases to separate, remove the cap, and transfer 0.5 mL of hexane into an autosampler vial using a disposable glass pipette. Transfer the remaining hexane phase into a second autosampler vial and store at 4°C for reanalysis, if necessary. Discard the remaining sample/hexane mixture and reweigh the empty container to determine net weight of sample.

QUALITY CONTROL Minimum quality control requirements are initial demonstration of lab capability, analysis of lab reagent blanks, fortified blanks, fortified sample matrix, and quality control samples. The lab must analyze at least one fortified blank per sample set, or at least one for every 20 samples. The fortifying concentration of each analyte should be 10 times the method detection limit or the maximum calibration limit (MCL), whichever is less. Calculate accuracy as percent recovery and develop control limits from the mean percent recovery and standard deviation.

The lab must add a known concentration of the analytes to a minimum of 10% of the routine samples, or one lab fortified sample matrix per sample set. Calculate the percent recovery for each analyte and compare to the control limits established from the analyses of the fortified blanks.

EPA CONTACT & HOTLINE For technical questions contact Dr. Baldev Bathija, U.S. EPA, Office of Ground Water and Drinking Water (WH-550D), 401 M St. SW, Washington, DC 20460. Tel. (202) 260-3040. For further information the EPA Safe Drinking Water Hotline may be called at: (800) 426-4791.

REFERENCE Methods for the Determination of Organic Compounds in Drinking Water, EPA/600/4-88/039 (revised July 1991). U.S. EPA Environmental Monitoring Systems Laboratory, Cincinnati, OH, 45268, U.S.A. Available from the National Technical Information Service (NTIS), 5285 Port Royal Road, Springfield, VA 22161; Tel. 800-553-6847. NTIS Order Number is PB91-231480.

Heptachlor   EPA Method 625
CAS #76-44-8

TITLE Base/Neutrals and Acids, U.S. EPA Method 625

MATRIX This methods covers municipal and industrial wastewaters.

METHOD SUMMARY Approximately 1 L of sample is serially extracted with methylene chloride at a pH greater than 11 and again at a pH less than 2 using a separatory funnel or a continuous extractor. The methylene chloride extract is dried, concentrated to a volume of 1 mL, and analyzed by GC/MS. Qualitative identification of the parameters in the extract is performed using the retention time and the relative abundance of three characteristic masses (m/z). Qualitative analysis is performed using either external or internal standard techniques with a single characteristic m/z.

INTERFERENCES Method interferences may be caused by contaminants in solvents, reagents, glassware, and other sample processing hardware. Glassware must be scrupulously cleaned. Glassware should be heated in a muffle furnace at 400°C for 5 to 30 min. Some thermally stable materials, such as PCBs, may not be eliminated by this treatment. Solvent rinses with acetone and pesticide quality hexane may be substituted for the muffle furnace heating. Matrix interferences may be caused by contaminants that are coextracted from the sample. The base-neutral extraction may cause significantly reduced recovery of phenols. The packed gas chromatographic columns recommended for the basic fraction may not exhibit sufficient resolution for some analytes.

INSTRUMENTATION A GC/MS system with an injection port designed for on-column injection when using packed columns and for splitless injection when using capillary columns.

Column for base/neutrals: 1.8 m long × 2 mm I.D. glass, packed with 3% SP-2550 on Supelcoport (100/120 mesh) or equivalent.
Column for acids: 1.8 m long × 2 mm I.D. glass, packed with 1% SP-1240DA on Supelcoport (100/120 mesh) or equivalent.

PRECISION & ACCURACY The MDL concentrations were obtained using reagent water. The MDL actually achieved in a given analysis will vary depending on instrument sensitivity and matrix effects. This method was tested by 15 laboratories using reagent water, drinking water, surface water, and industrial wastewaters spiked at six concentrations over the range 5 to 100 µg/L. Single operator precision, overall precision, and method accuracy were found to be directly related to the concentration of the parameter matrix.

The MDL (in µg/L) in reagent water was 1.9.
The standard deviation (in µg/L based on 4 recovery measurements) was 37.2.
The range (in µg/L) for average recovery for 4 measurements was D–172.2.
The range (in %) for percent recovery was D–192.
Accuracy (in µg/L) as expected recovery for one or more measurements of a sample containing a true concentration of C was 0.87C–2.97.
Precision (in µg/L) as expected single analyst standard deviation of measurements at an average concentration found at X was 0.24X–0.56.
Overall precision (in µg/L) as expected interlaboratory standard deviation of measurements in an average concentration found at X was 0.50X–0.23.

C = *True value of the concentration in µg/L.*
X = *Average recovery found for measurements of samples containing a concentration at C in µg/L.*

SAMPLE PREPARATION Adjust the pH to >11 with sodium hydroxide and serially extract in a separatory funnel with methylene chloride or else in a continuous extractor. Next, adjust the pH to <2 with sulfuric acid and serially extract in a separatory funnel with methylene chloride or else in a continuous extractor. Dry the extracts separately through a column of anhydrous sodium sulfate and then concentrate each of the extracts to 1.0 mL using a K-D apparatus.

SAMPLE COLLECTION, PRESERVATION & HANDLING Grab samples must be collected in glass containers. All samples must be refrigerated at 4°C from the time of collection until extraction. If residual chlorine is present, add 80 mg of sodium thiosulfate/L of sample and mix well. All samples must be extracted within 7 days of collection and completely analyzed within 40 days of extraction.

QUALITY CONTROL Make an initial, one-time, demonstration of the ability to generate acceptable accuracy and precision with this method. Before processing any samples, the analyst must analyze a reagent water blank to demonstrate that interferences from the analytical system and glassware are under control. Each time a set of samples is extracted or reagents are changed, a reagent water blank must be processed. Spike and analyze a minimum of 5% of all samples to monitor and evaluate lab data quality. A QC check sample concentrate that contains each parameter of interest at a concentration of 100 µg/mL in acetone is required. PCBs and multicomponent pesticides may be omitted from this test.

After analysis of five spiked wastewater samples, calculate the average percent recovery and the standard deviation of the percent recovery. Spike all samples with the surrogate standard spiking solution and calculate the percent recovery of each surrogate compound.

REFERENCE *Federal Register*, Vol. 49, No. 209. Friday, Oct. 26, 1984.

Heptachlor   **EPA Method 8080**
CAS #76-44-8

TITLE Organochlorine Pesticides and Polychlorinated Biphenyls By Gas Chromatography

MATRIX This method is used to determine the concentration of various organochlorine pesticides and polychlorinated biphenyls in extracts prepared from water, groundwater, soils, and sediments.

METHOD SUMMARY This method covers 26 pesticides and Aroclor (PCB) mixtures and it is suitable for monitoring-type analyses. After extraction, concentration and solvent exchange to hexane, a 2- to 5-µL sample aliquot is injected into a GC using the solvent flush technique, and the analytes are detected by an electron capture detector (ECD) or an electrolytic conductivity detector in the halogen mode (HECD). Both neat and diluted organic liquids may be analyzed by direct injection.

INTERFERENCES Interferences coextracted from the samples will vary considerably from source to source. Interferences by phthalate esters can pose a major problem in pesticide determinations when using the ECD. Cross-contamination of clean glassware routinely occurs when plastics are handled during extraction steps, especially when solvent-wetted surfaces are handled. The contamination from phthalate esters can be completely eliminated with a microcoulometric or electrolytic conductivity detector. Solvents, reagent, glassware, and other sample processing hardware may yield artifacts and/or interferences to sample analysis.

INSTRUMENTATION A gas chromatograph capable of on-column injections is needed. It must be equipped with an ECD or a HECD and one of the following GC columns:

Column 1: Supelcoport (100/120 mesh) coated with 1.5% SP-2250/1.95% SP-2401 packed in a 1.8 m × 4 mm I.D. glass column.

Column 2: Supelcoport (100/120 mesh) coated with 3% OV-1 in a 1.8 m × 4 mm I.D. glass column.

PRECISION & ACCURACY The method was tested by 20 laboratories using organic-free reagent water, drinking water, surface water, and three industrial wastewaters spiked at six concentrations. Concentrations used in the study ranged from 0.5 to 30 µg/L for single-component pesticides and from 8.5 to 400 µg/L for multicomponent parameters. Overall precision and method accuracy were found to be directly related to the concentration of the analyte and essentially independent of the sample matrix. The sensitivity of this method usually depends on the concentration of interferences rather than on instrumental limitations.

MDL in µg/L was 0.003.
Concentration range in µg/L was 0.5–30.
Accuracy as recovery (x^*) in µg/L was 0.69C + 0.04 .
Overall precision (S^*) in µg/L was 0.16x + 0.08.

x^* = Expected recovery for one or more measurements of a sample containing concentration C, in µg/L.
S^* = Expected interlaboratory standard deviation of measurements at an average concentration found of the analyte in µg/L.
C = True value for the concentration, in µg/L.
X = Average recovery found for measurements of samples containing a concentration of C, in µg/L.

SAMPLING METHOD
Liquid samples — Use a 1 or 2½ gallon amber glass bottle with a screw-top Teflon®-lined cover. Pre-wash the bottle with detergent, rinse with distilled water and methanol (or isopropanol).

Soil, sediments, and sludges — Use an 8-oz. widemouth glass with a screw-top Teflon®-lined cover. Pre-wash the bottle with detergent, rinse with distilled water and methanol (or isopropanol).

SAMPLE PRESERVATION Cool water, soil, sediment, or sludge samples immediately to 4°C.

Water samples — If residual chlorine is present, add 3 mL of 10% sodium thiosulfate per gallon and cool to 4°C. All extracts and samples should be stored under refrigeration.

MHT Liquid samples must be extracted within 7 days and the extracts must be analyzed within 40 days. Soils, sediments, and sludges may be stored for a maximum of 14 days prior to extraction.

SAMPLE PREPARATION
Liquid samples — Extract 1 L samples in a continuous extractor at pH 5–9 with methylene chloride after adding 1.0 mL of surrogate spiking solution to each sample. Pass the extract through a column of anhydrous sodium sulfate to dry and concentrate it in a K-D apparatus to 1 mL volume.

Soils, sediments and sludges — Rapidly weigh approximately 30 g of sample into a 400-mL beaker to avoid loss of the more volatile extractables. Nonporous or wet samples (gummy or clay type) that do not have a free-flowing sandy texture must be mixed with anhydrous sodium sulfate until the sample is free flowing. Add 1 mL of surrogate standards to all samples, spikes, standards, and blanks. Add 100 mL of 1:1 methylene chloride:acetone and extract ultrasonically. Decant and filter extracts, dry the extract by passing it through a drying column containing anhydrous sodium sulfate and concentrate to 1 mL in a K-D apparatus.

Hexane solvent exchange — Add 50 mL of hexane, a new boiling chip, and concentrate until the apparent volume of liquid reaches 1 mL. Adjust the extract volume to 10.0 mL. Stopper the concentration tube and store refrigerated at 4°C if further processing will not be performed immediately. If the extract will be stored longer than two days, transfer it to a vial with Teflon®-lined screw-cap or crimp top.

QUALITY CONTROL Demonstrate through the analysis of a reagent water blank, that all glassware and reagents are interference free. Each time a set of samples is processed, a method blank should be processed as a safeguard against chronic lab contamination. A reagent blank, a matrix spike, and a duplicate or matrix spike duplicate must be performed for each analytical batch (up to a maximum of 20 samples) analyzed.

Analytical system performance must be verified by analyzing QC check samples. The QC check sample concentration should contain each single-component analyte at the following concentrations in acetone: 4,4'-DDD, 10 µg/mL; 4,4'-DDT, 10 µg/mL; endosulfan II, 10 µg/mL; endosulfan sulfate, 10 µg/mL; and any other single-component pesticide at 2 µg/mL. If the method is only to be used to analyze PCBs, Chlordane, or Toxaphene, the QC check sample concentrate should contain the most representative multicomponent parameter at a concentration of 50 µg/mL in acetone.

REFERENCE Test Methods for Evaluating Solid Waste (SW-846). U.S. EPA. 1983. Method 8080B, Rev. 2, Nov. 1990. Office of Solid Wastes, Washington, DC.

Heptachlor   **EPA Method 8270**
CAS #76-44-8

TITLE Semivolatile Organic Compounds by GC/MS

MATRIX This method is used to determine the concentration of semivolatile organic compounds in extracts prepared from all types of solid waste matrices, soils, and groundwater. Although surface waters are not specifically mentioned, this method should be applicable to water samples from rivers, lakes, etc.

METHOD SUMMARY This method covers 259 semivolatile organic compounds. In very limited applications direct injection of the sample into the GC/MS system may be appropriate, but this results in very high detection limits (approximately

10,000 µg/L). Typically, a 1-L liquid sample, containing surrogate, and matrix spiking standards, is extracted in a continuous extractor first under acid conditions and then under basic conditions. Typically 30 g of a solid sample, containing surrogate, and matrix spiking standards, is extracted ultrasonically. After concentrating the extract to 1 mL it is spiked with 10 µL of an internal standard solution just prior to analysis by GC/MS. The volume injected should contain about 100 ng of base/neutral and 200 ng of acid surrogates (for a 1-µL injection). Analysis is performed by GC/MS using a capillary GC column.

INTERFERENCES Raw GC/MS data from all blanks, samples, and spikes must be evaluated for interferences. Contamination by carryover can occur whenever high-concentration and low-concentration samples are sequentially analyzed. To reduce carryover, the sample syringe must be rinsed out between samples with solvent. Whenever an unusually concentrated sample is encountered, it should be followed by the analysis of blank solvent to check for cross-contamination.

INSTRUMENTATION A GC/MS and a data system are required. The GC column used is a 30 m × 0.25 mm I.D. (or 0.32 mm I.D.) 1um film thickness silicone-coated fused silica capillary column. A continuous liquid-liquid extractor equipped with Teflon® or glass connection joints and stopcocks requiring no lubrication, a K-D concentrating apparatus, water bath, and an ultrasonic disrupter with a minimum power of 300 W and with pulsing capability are also required.

PRECISION & ACCURACY The estimated quantitation limit (EQL) of Method 8270B for determining an individual compound is approximately 1 mg/kg (wet weight) for soil or sediment samples, 1–200 mg/kg for wastes (dependent on matrix and method of preparation), and 10 µg/L for groundwater samples. EQLs will be proportionately higher for sample extracts that require dilution to avoid saturation of the detector.

The EQL(b) for groundwater in µg/L is not listed.
The EQL (a, b) for low concentrations in soil and sediment in µg/kg is not listed.
Accuracy as µg/L is $0.87C–2.97$.
Overall precision in µg/L is $0.50X–0.23$.

(a) EQLs listed for soil/sediment are based on wet weight. Normally data is reported in a dry-weight basis; therefore, EQLs will be higher based on the % dry weight of each sample. This calculation is based on a 30-g sample and gel permeation chromatography cleanup.
(b) Sample EQLs are highly matrix-dependent. The EQLs are provided for guidance and may not always be achievable.
$C =$ *True value for concentration, in µg/L.*
$X =$ *Average recovery found for measurements of samples containing a concentration of C, in µg/L.*

ESTIMATED QUANTITATION LIMIT

Other Matrices	Factor (a)
High-concentration soil and sludges by sonicator	7.5
Non-water miscible waste	75

(a) EQL for other matrices = [EQL for low soil/sediment] ×[Factor]. This estimated EQL is similar to an EPA "Practical Quantitation Limit."

SAMPLING METHOD
Liquid samples — Use a 1 or 2½ gallon amber glass bottle with a screw-top Teflon®-lined cover that has been prewashed with detergent and rinsed with distilled water and methanol (or isopropanol).

Soils, sediments, or sludges — Use an 8-oz. widemouth glass with a screw-top Teflon®-lined cover that has been prewashed with detergent and rinsed with distilled water and methanol (or isopropanol).

SAMPLE PRESERVATION
Liquid samples — If residual chlorine is present, add 3 mL of 10% sodium thiosulfate per gallon, cool to 4°C and store in a solvent-free refrigerator until analysis; if chlorine is not present, then eliminate the sodium thiosulfate addition.

Soils, sediments, or sludges — Cool samples to 4°C and store in a solvent-free refrigerator.

MHT Liquid samples must be extracted within 7 days and the extracts analyzed within 40 days. Soils, sediments, or sludges may be stored for a maximum of 14 days and the extracts analyzed within 40 days.

SAMPLE PREPARATION
Liquid samples — Transfer 1 L quantitatively to a continuous extractor. If high concentrations are anticipated, a smaller volume may be used and then diluted with organic-free reagent water to 1 L. Adjust pH, if necessary, to pH <2 using 1:1 (V/V) sulfuric acid. Pipette 1.0 mL of a surrogate standard spiking solution into each sample. For the sample in each analytical batch selected for spiking, add 1.0 mL of a matrix spiking standard. For base/neutral acid analysis, the amount of the surrogates and matrix spiking compounds added to the sample should result in a final concentration of 100 ng/µL of each analyte in the extract to be analyzed (assuming a 1-µL injection). Extract with methylene chloride for 18–24 h. Next, adjust the pH of the aqueous phase to pH >11 using 10 N sodium hydroxide and extract it with methylene chloride again for 18–24 h. Dry the extract through a column containing anhydrous sodium sulfate and concentrate it to 1 mL using a K-D concentrator.

Soils, sediments, or sludges — Use 30 g of sample. Nonporous or wet samples (gummy or clay type) that do not have a free-flowing sandy texture must be mixed with anhydrous sodium sulfate until the sample is free flowing. Add 1 mL of surrogate standards to all samples, spikes, standards, and blanks. For the sample in each analytical batch selected for spiking, add 1.0 mL of a matrix spiking standard. For base/neutral acid analysis, the amount added of the surrogates and matrix spiking compounds should result in a final concentration of 100 ng/µL of each base/neutral analyte and 200 ng/µL of each acid analyte in the extract to be analyzed (assuming a 1-µL injection). Immediately add a 100-mL mixture of 1:1 methylene chloride:acetone and extract the sample ultrasonically for 3 min and then decant or filter the extracts. Repeat the extraction two or more times. Dry the extract using a column with anhydrous sodium sulfate and concentrate it to 1 mL in a K-D concentrator.

QUALITY CONTROL A methylene chloride solution containing 50 ng/µL of decafluorotriphenylphosphine (DFTPP) is

used for tuning the GC/MS system each 12-h shift. A system performance check also must be made during every 12-h shift. A standard containing 50 ng/μL each of 4,4'-DDT, pentachlorophenol, and benzidine is required to verify injection port inertness and GC column performance. A calibration standard at mid-concentration, containing each compound of interest, including all required surrogates, must be performed every 12 h during analysis. After the system performance check is met, calibration check compounds (CCCs) are used to check the validity of the initial calibration.

The internal standard responses and retention times in the calibration check standard must be evaluated immediately after or during data acquisition. If the retention time for any internal standard changes by more than 30 seconds from the last check calibration (12 h), the chromatographic system must be inspected for malfunctions and corrections must be made, as required. If the electron ionization current plot (EICP) area for any of the internal standards changes by a factor of two from the last daily calibration standard check, the mass spectrometer must be inspected for malfunctions and corrections must be made, as appropriate.

Demonstrate, through the analysis of a reagent water blank, that interferences from the analytical system, glassware, and reagents are under control. The blank samples should be carried through all stages of the sample preparation and measurement steps. For each analytical batch (up to 20 samples), a reagent blank, matrix spike, and matrix spike duplicate/duplicate must be analyzed (the frequency of the spikes may be different for different monitoring programs). The blank and spiked samples must be carried through all stages of the sample preparation and measurement steps. A QC reference sample concentrate containing each analyte at a concentration of 100 mg/L in methanol is required.

REFERENCE Test Methods for Evaluating Solid Waste (SW-846). U.S. EPA 1983, Method 8270B, Rev. 2, Nov. 1990. Office of Solid Waste, Washington, DC.

Heptachlor epoxide   **EPA Method 505**
CAS #1024-57-3

TITLE Analysis of Organohalide Pesticides and Commercial Polychlorinated Biphenyl (PCB) Products in Water by Microextraction and Gas Chromatography. U.S. EPA Method 505, Rev. 2.0, 1989.

MATRIX This method is applicable to drinking water and raw source water. The latter should include most surface water and groundwater sources.

METHOD SUMMARY Method 505 covers 25 pesticides and commercial PCB products. This is a very sensitive method that is more useful for monitoring than for exploratory analyses. 5-mL of water are saturated with sodium chloride and then extracted by shaking with 2 mL of hexane. The sample extracts are transferred to an autosampler setup to inject 1–2 μL portions into a gas chromatograph (GC) for analysis. Alternatively, 1–2 μL portions of samples, blanks, and standards may be manually injected. Each extract is analyzed by capillary GC/ECD with confirmation using either a second capillary column or GC/MS. The electron capture detector is easy to use, but it is a nonselective detector. The microextraction technique also eliminates the expensive sample preparation costs of other methods, but it has the disadvantage of being less sensitive than most because the extracts are not concentrated.

INTERFERENCES Method interferences may be caused by contaminants in solvents, reagents, glassware, and other sample processing apparatus that lead to discrete artifacts or elevated baselines. Interfering contamination may occur when a sample containing low concentrations of analytes is analyzed immediately following a sample containing relatively high concentrations of the analytes. Matrix interferences also may be caused by contaminants that are coextracted from the sample; cleanup of sample extracts may be necessary in these cases. Some pesticides and commercial PCB products from aqueous solutions adhere to glass surfaces, so sample transfers and contact with glass surfaces should be minimized. Some pesticides are rapidly oxidized by chlorine so dechlorination with sodium thiosulfate at the time of sample collection is important. Also, splitless injectors may cause degradation of some pesticides.

INSTRUMENTATION A gas chromatograph/electron capture detector/data system, with temperature programming and split/splitless injector suitable for use with capillary columns is needed.

Column 1: 0.32 mm I.D. × 30 m fused silica capillary with chemically bond methyl polysiloxane phase (DB-1, 1.0 μm film, or equivalent).
Column 2: 0.32 mm I.D. × 30 m fused silica capillary with 1:1 mixed phase of dimethyl silicone and polyethylene glycol (Durawax-DX3, 0.25 μm film, or equivalent).
Column 3: 0.32 mm I.D. × 25 m fused silica capillary with chemically bonded 50:50 methyl-phenyl silicone (OV-17, 1.5 μm film, or equivalent).

Column 1 should be used as the primary analytical column. Columns 2 and 3 are recommended for use as confirmatory columns when GC/MS confirmation is not available.

PRECISION & ACCURACY Method detection limits are dependent upon the characteristics of the gas chromatographic system used. Analytes that are not separated chromatographically cannot be individually identified and used in the same calibration mixture or water samples unless an alternative technique for identification and quantification, such as mass spectrometry, is used.

The concentration(s) (in μg/L) used for these QC measurements was 0.04 and 1.4.
The MDL (in μg/L) was 0.004.
The accuracy (% recovery) for reagent water at the above concentration(s) was 100 and 115 and the precision (%) was 15.6 and 6.6.
The accuracy (% recovery) for groundwater at the above concentration(s) was 90 and 103 and the precision (%) was 14.2 and 6.9.
The accuracy (% recovery) for tap water at the above concentration(s) was 112 and 81 and the precision (5) was 7.5 and 5.9.

Note: No range of concentrations is provided with this method.

SAMPLING METHOD Collect samples using a 40-mL screw-cap vial (prewashed with detergent, rinsed with distilled water and oven dried at 400°C for one h) with a Teflon®-faced silicone septum. Collect bubble-free samples and place the septum with the Teflon® side down on the water.

SAMPLE PRESERVATION If residual chlorine is present in the water add about 3 mg of sodium thiosulfate to each vial before samples are collected to remove the chlorine. Alternatively, add 75 µL of 0.04 g/mL solution of sodium thiosulfate to each vial just prior to sampling. Immediately cool samples to 4°C, and store them in a solvent-free refrigerator at 4°C until analysis.

MHT The maximum holding time is 14 days from the time the sample was collected until it must be analyzed.

SAMPLE PREPARATION Remove the sample from storage and allow it to come to room temperature. Remove a 5-mL volume from each container and weigh the container to the nearest 0.1 g. Add 6 g of sodium chloride and 2.0 mL of hexane to each sample bottle. Recap the sample and shake it vigorously for one min. Allow the water and hexane phases to separate, remove the cap, and transfer 0.5 mL of hexane into an autosampler vial using a disposable glass pipette. Transfer the remaining hexane phase into a second autosampler vial and store at 4°C for reanalysis, if necessary. Discard the remaining sample/hexane mixture and reweigh the empty container to determine net weight of sample.

QUALITY CONTROL Minimum quality control requirements are initial demonstration of lab capability, analysis of lab reagent blanks, fortified blanks, fortified sample matrix, and quality control samples. The lab must analyze at least one fortified blank per sample set, or at least one for every 20 samples. The fortifying concentration of each analyte should be 10 times the method detection limit or the maximum calibration limit (MCL), whichever is less. Calculate accuracy as percent recovery and develop control limits from the mean percent recovery and standard deviation.

The lab must add a known concentration of the analytes to a minimum of 10% of the routine samples, or one lab fortified sample matrix per sample set. Calculate the percent recovery for each analyte and compare to the control limits established from the analyses of the fortified blanks.

EPA CONTACT & HOTLINE For technical questions contact Dr. Baldev Bathija, U.S. EPA, Office of Ground Water and Drinking Water (WH-550D), 401 M St. SW, Washington, DC 20460. Tel. (202) 260-3040. For further information the EPA Safe Drinking Water Hotline may be called at: (800) 426-4791.

REFERENCE Methods for the Determination of Organic Compounds in Drinking Water, EPA/600/4-88/039 (revised July 1991). U.S. EPA Environmental Monitoring Systems Laboratory, Cincinnati, OH, 45268, U.S.A. Available from the National Technical Information Service (NTIS), 5285 Port Royal Road, Springfield, VA 22161; Tel. 800-553-6847. NTIS Order Number is PB91-231480.

Heptachlor epoxide — EPA Method 625
CAS #1024-57-3

TITLE Base/Neutrals and Acids, U.S. EPA Method 625

MATRIX This methods covers municipal and industrial wastewaters.

METHOD SUMMARY Approximately 1 L of sample is serially extracted with methylene chloride at a pH greater than 11 and again at a pH less than 2 using a separatory funnel or a continuous extractor. The methylene chloride extract is dried, concentrated to a volume of 1 mL, and analyzed by GC/MS. Qualitative identification of the parameters in the extract is performed using the retention time and the relative abundance of three characteristic masses (m/z). Qualitative analysis is performed using either external or internal standard techniques with a single characteristic m/z.

INTERFERENCES Method interferences may be caused by contaminants in solvents, reagents, glassware, and other sample processing hardware. Glassware must be scrupulously cleaned. Glassware should be heated in a muffle furnace at 400°C for 5 to 30 min. Some thermally stable materials, such as PCBs, may not be eliminated by this treatment. Solvent rinses with acetone and pesticide quality hexane may be substituted for the muffle furnace heating. Matrix interferences may be caused by contaminants that are coextracted from the sample. The base-neutral extraction may cause significantly reduced recovery of phenols. The packed gas chromatographic columns recommended for the basic fraction may not exhibit sufficient resolution for some analytes.

INSTRUMENTATION A GC/MS system with an injection port designed for on-column injection when using packed columns and for splitless injection when using capillary columns.

Column for base/neutrals: 1.8 m long × 2 mm I.D. glass, packed with 3% SP-2550 on Supelcoport (100/120 mesh) or equivalent.

Column for acids: 1.8 m long × 2 mm I.D. glass, packed with 1% SP-1240DA on Supelcoport (100/120 mesh) or equivalent.

PRECISION & ACCURACY The MDL concentrations were obtained using reagent water. The MDL actually achieved in a given analysis will vary depending on instrument sensitivity and matrix effects. This method was tested by 15 laboratories using reagent water, drinking water, surface water, and industrial wastewaters spiked at six concentrations over the range 5 to 100 µg/L. Single operator precision, overall precision, and method accuracy were found to be directly related to the concentration of the parameter matrix.

The MDL (in µg/L) in reagent water was 2.2.

The standard deviation (in µg/L based on 4 recovery measurements) was 54.7.

The range (in µg/L) for average recovery for 4 measurements was 70.9–109.4.

The range (in %) for percent recovery was 26–155.

Accuracy (in µg/L) as expected recovery for one or more measurements of a sample containing a true concentration of C was 0.92C-1.87.

Precision (in µg/L) as expected single analyst standard deviation of measurements at an average concentration found at X was 0.33X–0.46.

Overall precision (in µg/L) as expected interlaboratory standard deviation of measurements in an average concentration found at X was 0.28X + 0.64.

C = *True value of the concentration in µg/L.*

X = *Average recovery found for measurements of samples containing a concentration at C in µg/L.*

SAMPLE PREPARATION Adjust the pH to >11 with sodium hydroxide and serially extract in a separatory funnel with methylene chloride or else in a continuous extractor. Next, adjust the pH to <2 with sulfuric acid and serially extract in a separatory funnel with methylene chloride or else in a continuous extractor. Dry the extracts separately through a column of anhydrous sodium sulfate and then concentrate each of the extracts to 1.0 mL using a K-D apparatus.

SAMPLE COLLECTION, PRESERVATION & HANDLING Grab samples must be collected in glass containers. All samples must be refrigerated at 4°C from the time of collection until extraction. If residual chlorine is present, add 80 mg of sodium thiosulfate/L of sample and mix well. All samples must be extracted within 7 days of collection and completely analyzed within 40 days of extraction.

QUALITY CONTROL Make an initial, one-time, demonstration of the ability to generate acceptable accuracy and precision with this method. Before processing any samples, the analyst must analyze a reagent water blank to demonstrate that interferences from the analytical system and glassware are under control. Each time a set of samples is extracted or reagents are changed, a reagent water blank must be processed. Spike and analyze a minimum of 5% of all samples to monitor and evaluate lab data quality. A QC check sample concentrate that contains each parameter of interest at a concentration of 100 µg/mL in acetone is required. PCBs and multicomponent pesticides may be omitted from this test.

After analysis of five spiked wastewater samples, calculate the average percent recovery and the standard deviation of the percent recovery. Spike all samples with the surrogate standard spiking solution and calculate the percent recovery of each surrogate compound.

REFERENCE *Federal Register*, Vol. 49, No. 209. Friday, Oct. 26, 1984.

Heptachlor epoxide   **EPA Method 8080**
CAS #1024-57-3

TITLE Organochlorine Pesticides and Polychlorinated Biphenyls By Gas Chromatography

MATRIX This method is used to determine the concentration of various organochlorine pesticides and polychlorinated biphenyls in extracts prepared from water, groundwater, soils, and sediments.

METHOD SUMMARY This method covers 26 pesticides and Aroclor (PCB) mixtures and it is suitable for monitoring-type analyses. After extraction, concentration and solvent exchange to hexane, a 2- to 5-µL sample aliquot is injected into a GC using the solvent flush technique, and the analytes are detected by an electron capture detector (ECD) or an electrolytic conductivity detector in the halogen mode (HECD). Both neat and diluted organic liquids may be analyzed by direct injection.

INTERFERENCES Interferences coextracted from the samples will vary considerably from source to source. Interferences by phthalate esters can pose a major problem in pesticide determinations when using the ECD. Cross-contamination of clean glassware routinely occurs when plastics are handled during extraction steps, especially when solvent-wetted surfaces are handled. The contamination from phthalate esters can be completely eliminated with a microcoulometric or electrolytic conductivity detector. Solvents, reagent, glassware, and other sample processing hardware may yield artifacts and/or interferences to sample analysis.

INSTRUMENTATION A gas chromatograph capable of on-column injections is needed. It must be equipped with an ECD or a HECD and one of the following GC columns:

Column 1: Supelcoport (100/120 mesh) coated with 1.5% SP-2250/1.95% SP-2401 packed in a 1.8 m × 4 mm I.D. glass column.

Column 2: Supelcoport (100/120 mesh) coated with 3% OV-1 in a 1.8 m × 4 mm I.D. glass column.

PRECISION & ACCURACY The method was tested by 20 laboratories using organic-free reagent water, drinking water, surface water, and three industrial wastewaters spiked at six concentrations. Concentrations used in the study ranged from 0.5 to 30 µg/L for single-component pesticides and from 8.5 to 400 µg/L for multicomponent parameters. Overall precision and method accuracy were found to be directly related to the concentration of the analyte and essentially independent of the sample matrix. The sensitivity of this method usually depends on the concentration of interferences rather than on instrumental limitations.

MDL in µg/L was 0.083.
Concentration range in µg/L was 0.5–30.
Accuracy as recovery (x*) in µg/L was 0.89C + 0.10 .
Overall precision (S*) in µg/L was 0.25x–0.08.

x* = *Expected recovery for one or more measurements of a sample containing concentration C, in µg/L.*

S* = *Expected interlaboratory standard deviation of measurements at an average concentration found of the analyte in µg/L.*

C = *True value for the concentration, in µg/L.*

X = *Average recovery found for measurements of samples containing a concentration of C, in µg/L.*

SAMPLING METHOD
Liquid samples — Use a 1 or 2½ gallon amber glass bottle with a screw-top Teflon®-lined cover. Pre-wash the bottle with detergent, rinse with distilled water and methanol (or isopropanol).

Soil, sediments, and sludges — Use an 8-oz. widemouth glass with a screw-top Teflon®-lined cover. Pre-wash the bottle with detergent, rinse with distilled water and methanol (or isopropanol).

SAMPLE PRESERVATION Cool water, soil, sediment, or sludge samples immediately to 4°C.

Water samples — If residual chlorine is present, add 3 mL of 10% sodium thiosulfate per gallon and cool to 4°C. All extracts and samples should be stored under refrigeration.

MHT Liquid samples must be extracted within 7 days and the extracts must be analyzed within 40 days. Soils, sediments, and sludges may be stored for a maximum of 14 days prior to extraction.

SAMPLE PREPARATION
Liquid samples — Extract 1 L samples in a continuous extractor at pH 5–9 with methylene chloride after adding 1.0 mL of surrogate spiking solution to each sample. Pass the extract through a column of anhydrous sodium sulfate to dry and concentrate it in a K-D apparatus to 1 mL volume.

Soils, sediments and sludges — Rapidly weigh approximately 30 g of sample into a 400-mL beaker to avoid loss of the more volatile extractables. Nonporous or wet samples (gummy or clay type) that do not have a free-flowing sandy texture must be mixed with anhydrous sodium sulfate until the sample is free flowing. Add 1 mL of surrogate standards to all samples, spikes, standards, and blanks. Add 100 mL of 1:1 methylene chloride:acetone and extract ultrasonically. Decant and filter extracts, dry the extract by passing it through a drying column containing anhydrous sodium sulfate and concentrate to 1 mL in a K-D apparatus.

Hexane solvent exchange — Add 50 mL of hexane, a new boiling chip, and concentrate until the apparent volume of liquid reaches 1 mL. Adjust the extract volume to 10.0 mL. Stopper the concentration tube and store refrigerated at 4°C if further processing will not be performed immediately. If the extract will be stored longer than two days, transfer it to a vial with Teflon®-lined screw-cap or crimp top.

QUALITY CONTROL Demonstrate through the analysis of a reagent water blank, that all glassware and reagents are interference free. Each time a set of samples is processed, a method blank should be processed as a safeguard against chronic lab contamination. A reagent blank, a matrix spike, and a duplicate or matrix spike duplicate must be performed for each analytical batch (up to a maximum of 20 samples) analyzed.

Analytical system performance must be verified by analyzing QC check samples. The QC check sample concentration should contain each single-component analyte at the following concentrations in acetone: 4,4'-DDD, 10 µg/mL; 4,4'-DDT, 10 µg/mL; endosulfan II, 10 µg/mL; endosulfan sulfate, 10 µg/mL; and any other single-component pesticide at 2 µg/mL. If the method is only to be used to analyze PCBs, Chlordane, or Toxaphene, the QC check sample concentrate should contain the most representative multicomponent parameter at a concentration of 50 µg/mL in acetone.

REFERENCE Test Methods for Evaluating Solid Waste (SW-846). U.S. EPA. 1983. Method 8080B, Rev. 2, Nov. 1990. Office of Solid Wastes, Washington, DC.

Heptachlor epoxide   **EPA Method 8270**
CAS #1024-57-3

TITLE Semivolatile Organic Compounds by GC/MS

MATRIX This method is used to determine the concentration of semivolatile organic compounds in extracts prepared from all types of solid waste matrices, soils, and groundwater. Although surface waters are not specifically mentioned, this method should be applicable to water samples from rivers, lakes, etc.

METHOD SUMMARY This method covers 259 semivolatile organic compounds. In very limited applications direct injection of the sample into the GC/MS system may be appropriate, but this results in very high detection limits (approximately 10,000 µg/L). Typically, a 1-L liquid sample, containing surrogate, and matrix spiking standards, is extracted in a continuous extractor first under acid conditions and then under basic conditions. Typically 30 g of a solid sample, containing surrogate, and matrix spiking standards, is extracted ultrasonically. After concentrating the extract to 1 mL it is spiked with 10 µL of an internal standard solution just prior to analysis by GC/MS. The volume injected should contain about 100 ng of base/neutral and 200 ng of acid surrogates (for a 1-µL injection). Analysis is performed by GC/MS using a capillary GC column.

INTERFERENCES Raw GC/MS data from all blanks, samples, and spikes must be evaluated for interferences. Contamination by carryover can occur whenever high-concentration and low-concentration samples are sequentially analyzed. To reduce carryover, the sample syringe must be rinsed out between samples with solvent. Whenever an unusually concentrated sample is encountered, it should be followed by the analysis of blank solvent to check for cross-contamination.

INSTRUMENTATION A GC/MS and a data system are required. The GC column used is a 30 m × 0.25 mm I.D. (or 0.32 mm I.D.) 1um film thickness silicone-coated fused silica capillary column. A continuous liquid-liquid extractor equipped with Teflon® or glass connection joints and stopcocks requiring no lubrication, a K-D concentrating apparatus, water bath, and an ultrasonic disrupter with a minimum power of 300 W and with pulsing capability are also required.

PRECISION & ACCURACY The estimated quantitation limit (EQL) of Method 8270B for determining an individual compound is approximately 1 mg/kg (wet weight) for soil or sediment samples, 1–200 mg/kg for wastes (dependent on matrix and method of preparation), and 10 µg/L for groundwater samples. EQLs will be proportionately higher for sample extracts that require dilution to avoid saturation of the detector.

The EQL(b) for groundwater in µg/L is not listed.
The EQL (a, b) for low concentrations in soil and sediment in µg/kg is not listed.
Accuracy as µg/L is 0.92C–1.87.

Overall precision in µg/L is 0.28X + 0.64.

(a) *EQLs listed for soil/sediment are based on wet weight. Normally data is reported in a dry-weight basis; therefore, EQLs will be higher based on the % dry weight of each sample. This calculation is based on a 30-g sample and gel permeation chromatography cleanup.*
(b) *Sample EQLs are highly matrix-dependent. The EQLs are provided for guidance and may not always be achievable.*
C = *True value for concentration, in µg/L.*
X = *Average recovery found for measurements of samples containing a concentration of C, in µg/L.*

ESTIMATED QUANTITATION LIMIT

Other Matrices	Factor (a)
High-concentration soil and sludges by sonicator	7.5
Non-water miscible waste	75

(a) *EQL for other matrices = [EQL for low soil/sediment] × [Factor]. This estimated EQL is similar to an EPA "Practical Quantitation Limit."*

SAMPLING METHOD
Liquid samples — Use a 1 or 2½ gallon amber glass bottle with a screw-top Teflon®-lined cover that has been prewashed with detergent and rinsed with distilled water and methanol (or isopropanol).

Soils, sediments, or sludges — Use an 8-oz. widemouth glass with a screw-top Teflon®-lined cover that has been prewashed with detergent and rinsed with distilled water and methanol (or isopropanol).

SAMPLE PRESERVATION
Liquid samples — If residual chlorine is present, add 3 mL of 10% sodium thiosulfate per gallon, cool to 4°C and store in a solvent-free refrigerator until analysis; if chlorine is not present, then eliminate the sodium thiosulfate addition.

Soils, sediments, or sludges — Cool samples to 4°C and store in a solvent-free refrigerator.

MHT Liquid samples must be extracted within 7 days and the extracts analyzed within 40 days. Soils, sediments, or sludges may be stored for a maximum of 14 days and the extracts analyzed within 40 days.

SAMPLE PREPARATION
Liquid samples — Transfer 1 L quantitatively to a continuous extractor. If high concentrations are anticipated, a smaller volume may be used and then diluted with organic-free reagent water to 1 L. Adjust pH, if necessary, to pH <2 using 1:1 (V/V) sulfuric acid. Pipette 1.0 mL of a surrogate standard spiking solution into each sample. For the sample in each analytical batch selected for spiking, add 1.0 mL of a matrix spiking standard. For base/neutral acid analysis, the amount of the surrogates and matrix spiking compounds added to the sample should result in a final concentration of 100 ng/µL of each analyte in the extract to be analyzed (assuming a 1-µL injection). Extract with methylene chloride for 18–24 h. Next, adjust the pH of the aqueous phase to pH >11 using 10 N sodium hydroxide and extract it with methylene chloride again for 18–24 h. Dry the extract through a column containing anhydrous sodium sulfate and concentrate it to 1 mL using a K-D concentrator.

Soils, sediments, or sludges — Use 30 g of sample. Nonporous or wet samples (gummy or clay type) that do not have a free-flowing sandy texture must be mixed with anhydrous sodium sulfate until the sample is free flowing. Add 1 mL of surrogate standards to all samples, spikes, standards, and blanks. For the sample in each analytical batch selected for spiking, add 1.0 mL of a matrix spiking standard. For base/neutral acid analysis, the amount added of the surrogates and matrix spiking compounds should result in a final concentration of 100 ng/µL of each base/neutral analyte and 200 ng/µL of each acid analyte in the extract to be analyzed (assuming a 1-µL injection). Immediately add a 100-mL mixture of 1:1 methylene chloride:acetone and extract the sample ultrasonically for 3 min and then decant or filter the extracts. Repeat the extraction two or more times. Dry the extract using a column with anhydrous sodium sulfate and concentrate it to 1 mL in a K-D concentrator.

QUALITY CONTROL A methylene chloride solution containing 50 ng/µL of decafluorotriphenylphosphine (DFTPP) is used for tuning the GC/MS system each 12-h shift. A system performance check also must be made during every 12-h shift. A standard containing 50 ng/µL each of 4,4'-DDT, pentachlorophenol, and benzidine is required to verify injection port inertness and GC column performance. A calibration standard at mid-concentration, containing each compound of interest, including all required surrogates, must be performed every 12 h during analysis. After the system performance check is met, calibration check compounds (CCCs) are used to check the validity of the initial calibration.

The internal standard responses and retention times in the calibration check standard must be evaluated immediately after or during data acquisition. If the retention time for any internal standard changes by more than 30 seconds from the last check calibration (12 h), the chromatographic system must be inspected for malfunctions and corrections must be made, as required. If the electron ionization current plot (EICP) area for any of the internal standards changes by a factor of two from the last daily calibration standard check, the mass spectrometer must be inspected for malfunctions and corrections must be made, as appropriate.

Demonstrate, through the analysis of a reagent water blank, that interferences from the analytical system, glassware, and reagents are under control. The blank samples should be carried through all stages of the sample preparation and measurement steps. For each analytical batch (up to 20 samples), a reagent blank, matrix spike, and matrix spike duplicate/duplicate must be analyzed (the frequency of the spikes may be different for different monitoring programs). The blank and spiked samples must be carried through all stages of the sample preparation and measurement steps. A QC reference sample concentrate containing each analyte at a concentration of 100 mg/L in methanol is required.

REFERENCE Test Methods for Evaluating Solid Waste (SW-846). U.S. EPA 1983, Method 8270B, Rev. 2, Nov. 1990. Office of Solid Waste, Washington, DC.

Hexachlorobenzene
CAS #118-74-1

EPA Method 1625

TITLE Semivolatile Organic Compounds by Isotope Dilution GC/MS

MATRIX The compounds may be determined in waters, soils, and municipal sludges by this method.

METHOD SUMMARY This method is used to determine 176 semivolatile toxic organic pollutants associated with the CWA (as amended 1987); the RCRA (as amended 1986); the CERCLA (as amended 1986); and other compounds amenable to extraction and analysis by capillary column gas chromatography-mass spectrometry (GC/MS).

Stable isotopically-labeled analogs of the compounds of interest are added to the sample. If the solids content is less than 1%, a 1-L sample is extracted at pH 12–13, then at pH <2 with methylene chloride using continuous extraction techniques.

If the solids content is 30% or less, the sample is diluted to 1% solids with reagent water, homogenized ultrasonically, and extracted at pH 12–13, then at pH <2 with methylene chloride using continuous extraction techniques. If the solids content is greater than 30%, the sample is extracted using ultrasonic techniques.

Each extract is dried over sodium sulfate, concentrated to a volume of 5 mL, cleaned up using GPC, if necessary, and concentrated. Extracts are concentrated to 1 mL if GPC is not performed, and to 0.5 mL if GPC is performed.

An internal standard is added to the extract, and a 1-mL aliquot of the extract is injected into the GC. The compounds are separated by GC and detected by a MS. The labeled compounds serve to correct the variability of the analytical technique.

INTERFERENCES Solvents, reagents, glassware, and other sample processing hardware may yield artifacts and/or elevated baselines causing misinterpretation of chromatograms and spectra. Materials used in the analysis must be demonstrated to be free from interferences under the conditions of analysis by running method blanks initially and with each sample lot (sample started through the extraction process on a given 8-h shift, to a maximum of 20). Specific selection of reagents and purification of solvents by distillation in all glass systems may be required. Glassware and, where possible, reagents are cleaned by solvent rinse and baking at 450°C for 1-h minimum. Interferences coextracted from samples will vary considerably from source to source, depending on the diversity of the site being sampled.

INSTRUMENTATION Major instrumentation includes a GC with a splitless or on-column injection port for capillary column, a MS with 70 eV electron impact ionization, and a data system to collect and record MS data, and process it. A K-D apparatus is used to concentrate extracts.

GC Column: 30 m × 0.25 mm I.D. 5% phenyl, 94% methyl, 1% vinyl silicone bonded phased fused silica capillary column.

PRECISION & ACCURACY The detection limits of the method are usually dependent on the level of interferences rather than instrumental limitations. The limits typify the minimum quantities that can be detected with no interferences present.

The minimum level (in µg/mL) was 10. This is defined as a minimum level at which the analytical system shall give recognizable mass spectra (background corrected) and acceptable calibration points.

The MDL (in µg/kg) in low solids was 51 and in high solids was 48; these were determined in digested sludge (low solids) and in filter cake or compost (high solids).

The labeled and native compound initial precision as standard deviation (in µg/L) was 16.

The labeled and native compound initial accuracy as average recovery (in µg/L) was 90–124.

SAMPLE COLLECTION, PRESERVATION & HANDLING
Collect samples in glass containers. Aqueous samples which flow freely are collected in refrigerated bottles using automatic sampling equipment. Solid samples are collected as grab samples using widemouth jars. Maintain samples at 0 to 4°C from the time of collection until extraction. If residual chlorine is present in aqueous samples, add 80 mg sodium thiosulfate/L of water. Begin sample extraction within 7 days of collection, and analyze all extracts within 40 days of extraction.

SAMPLE PREPARATION Samples containing 1% solids or less are extracted directly using continuous liquid-liquid extraction techniques. Samples containing 1 to 30% solids are diluted to the 1% level with reagent water and extracted using continuous liquid-liquid extraction techniques. Samples containing greater than 30% solids are extracted using ultrasonic techniques.

Base/neutral extraction — Adjust the pH of the waters in the extractors to 12–13 with 6 N NaOH. Extract with methylene chloride for 24–48 h.

Acid extraction — Adjust the pH of the waters in the extractors to 2 or less using 6 N sulfuric acid. Extract with methylene chloride for 24–48 h.

Ultrasonic extraction of high solids samples — Add anhydrous sodium sulfate to the sample and QC aliquot(s). Add acetone:methylene chloride (1:1) to the sample and mix thoroughly

Concentrate extracts using a K-D apparatus.

QUALITY CONTROL The analyst is permitted to modify this method to improve separations or lower the costs of measurements, provided all performance specifications are met. Analyses of blanks are required to demonstrate freedom from contamination. When results of spikes indicate atypical method performance for samples, the samples are diluted to bring method performance within acceptable limits.

For low solids (aqueous samples), extract, concentrate, and analyze two sets of four 1-L aliquots (8 aliquots total) of the precision and recovery standard. For high solids samples, two sets of four 30-g aliquots of the high solids reference matrix are used.

Spike all samples with labeled compounds to assess method performance. Compute percent recovery of the labeled compounds using the internal standard method. Compare the labeled compound recovery for each compound with the corresponding labeled compound recovery.

Reagent water and high solids reference matrix blanks are analyzed to demonstrate freedom from contamination. Extract and concentrate a 1-L reagent water blank or a high solids reference matrix blank with each sample's lot (samples started through the extraction process on the same 8-h shift, to a maximum of 20 samples).

Field replicates may be collected to determine the precision of the sampling technique, and spiked samples may be required to determine the accuracy of the analysis when the internal standard method is used.

REFERENCE Semivolatile Organic Compounds by Isotope Dilution GC/MS. Office of Water Regulation and Standards, U.S. EPA Industrial Technology Division, Washington, DC, EPA Method 1625, Rev. C, June 1989 (contact W.A. Telliard, U.S. EPA, Office of Water Regulations and Standards, 401 M St., SW, Washington, DC, 20460. Phone: 202-382-7131).

Hexachlorobenzene   **EPA Method 505**
CAS #118-74-1

TITLE Analysis of Organohalide Pesticides and Commercial Polychlorinated Biphenyl (PCB) Products in Water by Microextraction and Gas Chromatography. U.S. EPA Method 505, Rev. 2.0, 1989.

MATRIX This method is applicable to drinking water and raw source water. The latter should include most surface water and groundwater sources.

METHOD SUMMARY Method 505 covers 25 pesticides and commercial PCB products. This is a very sensitive method that is more useful for monitoring than for exploratory analyses. 5-mL of water are saturated with sodium chloride and then extracted by shaking with 2 mL of hexane. The sample extracts are transferred to an autosampler setup to inject 1–2 µL portions into a gas chromatograph (GC) for analysis. Alternatively, 1–2 µL portions of samples, blanks, and standards may be manually injected. Each extract is analyzed by capillary GC/ECD with confirmation using either a second capillary column or GC/MS. The electron capture detector is easy to use, but it is a nonselective detector. The microextraction technique also eliminates the expensive sample preparation costs of other methods, but it has the disadvantage of being less sensitive than most because the extracts are not concentrated.

INTERFERENCES Method interferences may be caused by contaminants in solvents, reagents, glassware, and other sample processing apparatus that lead to discrete artifacts or elevated baselines. Interfering contamination may occur when a sample containing low concentrations of analytes is analyzed immediately following a sample containing relatively high concentrations of the analytes. Matrix interferences also may be caused by contaminants that are coextracted from the sample; cleanup of sample extracts may be necessary in these cases. Some pesticides and commercial PCB products from aqueous solutions adhere to glass surfaces, so sample transfers and contact with glass surfaces should be minimized. Some pesticides are rapidly oxidized by chlorine so dechlorination with sodium thiosulfate at the time of sample collection is important. Also, splitless injectors may cause degradation of some pesticides.

INSTRUMENTATION A gas chromatograph/electron capture detector/data system, with temperature programming and split/splitless injector suitable for use with capillary columns is needed.

Column 1: 0.32 mm I.D. × 30 m fused silica capillary with chemically bond methyl polysiloxane phase (DB-1, 1.0 µm film, or equivalent).

Column 2: 0.32 mm I.D. × 30 m fused silica capillary with 1:1 mixed phase of dimethyl silicone and polyethylene glycol (Durawax-DX3, 0.25 µm film, or equivalent).

Column 3: 0.32 mm I.D. × 25 m fused silica capillary with chemically bonded 50:50 methyl-phenyl silicone (OV-17, 1.5 µm film, or equivalent).

Column 1 should be used as the primary analytical column. Columns 2 and 3 are recommended for use as confirmatory columns when GC/MS confirmation is not available.

PRECISION & ACCURACY Method detection limits are dependent upon the characteristics of the gas chromatographic system used. Analytes that are not separated chromatographically cannot be individually identified and used in the same calibration mixture or water samples unless an alternative technique for identification and quantification, such as mass spectrometry, is used.

The concentration(s) (in µg/L) used for these QC measurements was 0.003 and 0.09.

The MDL (in µg/L) was 0.002.

The accuracy (% recovery) for reagent water at the above concentration(s) was 104 and 103 and the precision (%) was 13.5 and 6.6.

The accuracy (% recovery) for groundwater at the above concentration(s) was 91 and 1.1 and the precision (%) was 10.9 and 4.4.

The accuracy (% recovery) for tap water at the above concentration(s) was 100 and 88 and the precision (5) was 15.6 and 13.4.

Note: No range of concentrations is provided with this method.

SAMPLING METHOD Collect samples using a 40-mL screw-cap vial (prewashed with detergent, rinsed with distilled water and oven dried at 400°C for one h) with a Teflon®-faced silicone septum. Collect bubble-free samples and place the septum with the Teflon® side down on the water.

SAMPLE PRESERVATION If residual chlorine is present in the water add about 3 mg of sodium thiosulfate to each vial before samples are collected to remove the chlorine. Alternatively, add 75 µL of 0.04 g/mL solution of sodium thiosulfate to each vial just prior to sampling. Immediately cool samples to 4°C, and store them in a solvent-free refrigerator at 4°C until analysis.

MHT The maximum holding time is 14 days from the time the sample was collected until it must be analyzed.

SAMPLE PREPARATION Remove the sample from storage and allow it to come to room temperature. Remove a 5-mL volume from each container and weigh the container to the nearest 0.1 g. Add 6 g of sodium chloride and 2.0 mL of hexane to each sample bottle. Recap the sample and shake it vigorously for one min. Allow the water and hexane phases to separate, remove the cap, and transfer 0.5 mL of hexane into an autosampler vial using a disposable glass pipette. Transfer the remaining hexane phase into a second autosampler vial and store at 4°C for reanalysis, if necessary. Discard the remaining sample/hexane mixture and reweigh the empty container to determine net weight of sample.

QUALITY CONTROL Minimum quality control requirements are initial demonstration of lab capability, analysis of lab reagent blanks, fortified blanks, fortified sample matrix, and quality control samples. The lab must analyze at least one fortified blank per sample set, or at least one for every 20 samples. The fortifying concentration of each analyte should be 10 times the method detection limit or the maximum calibration limit (MCL), whichever is less. Calculate accuracy as percent recovery and develop control limits from the mean percent recovery and standard deviation.

The lab must add a known concentration of the analytes to a minimum of 10% of the routine samples, or one lab fortified sample matrix per sample set. Calculate the percent recovery for each analyte and compare to the control limits established from the analyses of the fortified blanks.

EPA CONTACT & HOTLINE For technical questions contact Dr. Baldev Bathija, U.S. EPA, Office of Ground Water and Drinking Water (WH-550D), 401 M St. SW, Washington, DC 20460. Tel. (202) 260-3040. For further information the EPA Safe Drinking Water Hotline may be called at: (800) 426-4791.

REFERENCE Methods for the Determination of Organic Compounds in Drinking Water, EPA/600/4-88/039 (revised July 1991). U.S. EPA Environmental Monitoring Systems Laboratory, Cincinnati, OH, 45268, U.S.A. Available from the National Technical Information Service (NTIS), 5285 Port Royal Road, Springfield, VA 22161; Tel. 800-553-6847. NTIS Order Number is PB91-231480.

Hexachlorobenzene   **EPA Method 625**
CAS #118-74-1

TITLE Base/Neutrals and Acids, U.S. EPA Method 625

MATRIX This methods covers municipal and industrial wastewaters.

METHOD SUMMARY Approximately 1 L of sample is serially extracted with methylene chloride at a pH greater than 11 and again at a pH less than 2 using a separatory funnel or a continuous extractor. The methylene chloride extract is dried, concentrated to a volume of 1 mL, and analyzed by GC/MS. Qualitative identification of the parameters in the extract is performed using the retention time and the relative abundance of three characteristic masses (m/z). Qualitative analysis is performed using either external or internal standard techniques with a single characteristic m/z.

INTERFERENCES Method interferences may be caused by contaminants in solvents, reagents, glassware, and other sample processing hardware. Glassware must be scrupulously cleaned. Glassware should be heated in a muffle furnace at 400°C for 5 to 30 min. Some thermally stable materials, such as PCBs, may not be eliminated by this treatment. Solvent rinses with acetone and pesticide quality hexane may be substituted for the muffle furnace heating. Matrix interferences may be caused by contaminants that are coextracted from the sample. The base-neutral extraction may cause significantly reduced recovery of phenols. The packed gas chromatographic columns recommended for the basic fraction may not exhibit sufficient resolution for some analytes.

INSTRUMENTATION A GC/MS system with an injection port designed for on-column injection when using packed columns and for splitless injection when using capillary columns.

Column for base/neutrals: 1.8 m long × 2 mm I.D. glass, packed with 3% SP-2550 on Supelcoport (100/120 mesh) or equivalent.

Column for acids: 1.8 m long × 2 mm I.D. glass, packed with 1% SP-1240DA on Supelcoport (100/120 mesh) or equivalent.

PRECISION & ACCURACY The MDL concentrations were obtained using reagent water. The MDL actually achieved in a given analysis will vary depending on instrument sensitivity and matrix effects. This method was tested by 15 laboratories using reagent water, drinking water, surface water, and industrial wastewaters spiked at six concentrations over the range 5 to 100 µg/L. Single operator precision, overall precision, and method accuracy were found to be directly related to the concentration of the parameter matrix.

The MDL (in µg/L) in reagent water was 1.9.
The standard deviation (in µg/L based on 4 recovery measurements) was 24.9.
The range (in µg/L) for average recovery for 4 measurements was 7.8–141.5.
The range (in %) for percent recovery was D-152.
Accuracy (in µg/L) as expected recovery for one or more measurements of a sample containing a true concentration of C was 0.74C + 0.66.
Precision (in µg/L) as expected single analyst standard deviation of measurements at an average concentration found at X was 0.18X–0.10.
Overall precision (in µg/L) as expected interlaboratory standard deviation of measurements in an average concentration found at X was 0.43X–0.52.

C = *True value of the concentration in µg/L.*
X = *Average recovery found for measurements of samples containing a concentration at C in µg/L.*

SAMPLE PREPARATION Adjust the pH to >11 with sodium hydroxide and serially extract in a separatory funnel with methylene chloride or else in a continuous extractor. Next, adjust the pH to <2 with sulfuric acid and serially extract in a separatory

funnel with methylene chloride or else in a continuous extractor. Dry the extracts separately through a column of anhydrous sodium sulfate and then concentrate each of the extracts to 1.0 mL using a K-D apparatus.

SAMPLE COLLECTION, PRESERVATION & HANDLING
Grab samples must be collected in glass containers. All samples must be refrigerated at 4°C from the time of collection until extraction. If residual chlorine is present, add 80 mg of sodium thiosulfate/L of sample and mix well. All samples must be extracted within 7 days of collection and completely analyzed within 40 days of extraction.

QUALITY CONTROL Make an initial, one-time, demonstration of the ability to generate acceptable accuracy and precision with this method. Before processing any samples, the analyst must analyze a reagent water blank to demonstrate that interferences from the analytical system and glassware are under control. Each time a set of samples is extracted or reagents are changed, a reagent water blank must be processed. Spike and analyze a minimum of 5% of all samples to monitor and evaluate lab data quality. A QC check sample concentrate that contains each parameter of interest at a concentration of 100 µg/mL in acetone is required. PCBs and multicomponent pesticides may be omitted from this test.

After analysis of five spiked wastewater samples, calculate the average percent recovery and the standard deviation of the percent recovery. Spike all samples with the surrogate standard spiking solution and calculate the percent recovery of each surrogate compound.

REFERENCE *Federal Register*, Vol. 49, No. 209. Friday, Oct. 26, 1984.

| Hexachlorobenzene | EPA Method 8120 |
| CAS #118-74-1 | |

TITLE Chlorinated Hydrocarbons by Gas Chromatography

MATRIX This method covers aqueous and solid matrices. This includes a wide variety such as drinking water, groundwater, industrial wastewaters, surface waters, soils, solids, and sediments.

METHOD SUMMARY This method is used to determine the concentration of 14 chlorinated hydrocarbons. It provides gas chromatographic conditions for the detection of ppb concentrations of certain chlorinated hydrocarbons. Prior to use of this method, appropriate sample extraction techniques must be used. Both neat and diluted organic liquids (EPA Method 3580, Waste Dilution) may be analyzed by direct injection. A 2 to 5 µg/mL aliquot of the extract is injected into a gas chromatograph (GC) using the solvent flush technique, and compounds in the GC effluent are detected by an electron capture detector (ECD).

INTERFERENCES Solvents, reagents, glassware, and other sample processing hardware may yield discrete artifacts and/or elevated baselines causing misinterpretation of gas chromatograms. Interferences coextracted from samples will vary considerably from source to source, depending upon the waste being sampled.

INSTRUMENTATION An analytical system complete with GC suitable for on-column injections and accessories, including detectors, column supplies, recorder, gases and syringes is required. A data system for measuring peak areas and/or peak heights is recommended. The GC is equipped with an electron capture detector (ECD). A K-D apparatus is needed for sample preparation.

Column 1: 1.8 m × 2 mm I.D. glass column packed with 1% SP-1000 on Supelcoport (100/120 mesh) or equivalent.

Column 2: 1.8 m × 2 mm I.D. glass column packed with 1.5% OV-1/2.4% OV-225 on Supelcoport (80/100 mesh) or equivalent.

PRECISION & ACCURACY The method was tested by 20 laboratories using organic-free reagent water, drinking water, surface water, and three industrial wastewaters spiked at six concentrations over the range 1.0 to 356 µg/L. Single operator precision, overall precision, and method accuracy were found to be directly related to the concentration of the parameter and essentially independent of the sample matrix.

MULTIPLICATION FACTORS FOR OTHER MATRICES (a)

Matrix	Factor (b)
Groundwater	10
Low-concentration soil by ultrasonic cleanup extraction with GPC	670
High-concentration soil and sludges by ultrasonic extraction	10,000
Non-water miscible waste	100,000

(a) Sample EQLs are highly matrix-dependent. The EQLs listed are provided for guidance and may not always be achievable.
(b) EQL = [Method detection limit] × [Factor]. For non-aqueous samples, the factor is on a wet-weight basis.

PRECISION & ACCURACY The estimates below are based upon the performance in a single lab.

The accuracy (in µg/L) as expected recovery for one or more measurements of a sample containing a concentration of C was 0.87C-0.02.

The precision (in µg/L) as expected single analyst standard deviation of measurements at an average concentration of x" was 0.14 x"-0.07.

The precision (in µg/L) as expected interlaboratory standard deviation measurements at an average concentration found of x" was 0.36x" -0.19.

$C =$ *True value for the concentration, in µg/L.*
$x" =$ *Average recovery found for measurements of samples containing a concentration of C, in µg/L.*

SAMPLE COLLECTION, PRESERVATION & HANDLING
Extracts must be stored under refrigeration at 4°C and analyzed within 40 days of extraction.

SAMPLE PREPARATION In general, water samples are extracted at a neutral, or as is, pH with methylene chloride using either EPA Method 3510 or EPA Method 3520. Solid samples are extracted using either EPA Method 3540 or EPA Method 3550. Prior to gas chromatographic analysis, the extraction solvent must be exchanged to hexane.

QUALITY CONTROL The quality control check concentrate (EPA Method 8000) should contain each parameter of interest in acetone at the following concentrations: hexachloro-substituted hydrocarbon, 10 µg/mL; and any other chlorinated hydrocarbon, 100 µg/mL. Calculate surrogate standard recovery on all samples, blanks, and spikes.

Prepare stock standard solutions in isooctane or hexane. Calibration standards at a minimum of five concentrations should be prepared through dilution of the stock standards with isooctane or hexane. Internal standards and surrogate standards are also needed.

REFERENCE Test Methods for Evaluating Solid Waste, Physical/Chemical Methods, SW-846, 3rd Edition, U.S. EPA, Office of Solid Waste, Washington, DC, 1990. EPA Method 8120 A Rev. 1, Nov. 1990.

Hexachlorobenzene   **EPA Method 8121**
CAS #118-74-1

TITLE Chlorinated Hydrocarbons by GC: Capillary Column Technique

MATRIX This method covers aqueous and solid matrices. This includes a wide variety such as drinking water, groundwater, industrial wastewaters, surface waters, soils, solids, and sediments.

METHOD SUMMARY This method provides procedures for the determination of 22 chlorinated hydrocarbons in water, soil/sediment, and waste matrices. A measured volume or weight of sample is extracted by using one of the appropriate sample extraction techniques specified in EPA Method 3510, EPA Method 3520, EPA Method 3540, or EPA Method 3550, or diluted using EPA Method 3580. Aqueous samples are extracted at neutral pH with methylene chloride by using either a separatory funnel (EPA Method 3510) or a continuous liquid-liquid extractor (EPA Method 3520). Solid samples are extracted with hexane/acetone (1:1) by using a Soxhlet extractor (EPA Method 3540) or with methylene chloride/acetone (1:1) by using an ultrasonic extractor (EPA Method 3550). After cleanup, the extract or diluted sample is analyzed by gas chromatography with electron capture detection (GC/ECD).

The sensitivity level of this method usually depends on the level of interferences rather than on instrumental limitations. This method may be used in conjunction with EPA Method 3620, Florisil Column Cleanup, EPA Method 3660, Sulfur Cleanup, and EPA Method 3640, Gel Permeation Chromatography, to aid in the elimination of interferences.

INTERFERENCES Solvents, reagents, glassware, and other hardware used in sample processing may introduce artifacts which may result in elevated baselines, causing misinterpretation of gas chromatograms. Interferants coextracted from the samples will vary considerably from waste to waste. Glassware must be scrupulously clean. Phthalate esters, if present in a sample, will interfere only with the BHC isomers. The presence of elemental sulfur will result in large peaks, and can often mask the region of compounds eluting after 1,2,4,5-tetrachlorobenzene. The tetrabutylammonium (TBA)-sulfite procedure (EPA Method 3660) works well for the removal of elemental sulfur. Waxes and lipids can be removed by gel permeation chromatography (EPA Method 3640).

INSTRUMENTATION A GC suitable for on-column injections and all required accessories, including and electron capture detector (ECD), analytical columns, recorder, gases, and syringes are needed. A data system for measuring peak heights and/or peak areas is recommended. A Kuderna-Danish (K-D) apparatus will also be needed to concentrate extracts.

Column 1: 30 m × 0.53 mm I.D. fused-silica capillary column chemically bonded with trifluoropropyl methyl silicone (DB-210 or equivalent).
Column 2: 30 m × 0.53 mm I.D. fused-silica capillary column chemically bonded with polyethylene glycol (DB-WAX or equivalent).

PRECISION & ACCURACY This method has been tested in a single lab by using organic-free reagent water, sandy loam samples, and extracts which were spiked with the test compounds at one concentration. Single-operator precision and method accuracy were found to be related to the concentration of compound and the type of matrix. The accuracy and precision technique will be determined by the sample matrix, sample preparation technique, optional cleanup techniques, and calibration procedures used.

MULTIPLICATION FACTORS FOR OTHER MATRICES (a)

Matrix	Factor (b)
Groundwater	10
Low-concentration soil by ultrasonic cleanup extraction with GPC	670
High-concentration soil and sludges by ultrasonic extraction	10,000
Non-water miscible waste	100,000

(a) Sample EQLs are highly matrix-dependent. The EQLs listed are provided for guidance and may not always be achievable.
(b) EQL = [Method detection limit] × [Factor]. For non-aqueous samples, the factor is on a wet-weight basis.

PRECISION & ACCURACY MDL is the method detection limit for organic-free reagent water. MDL was determined from the analysis of eight replicate aliquots processed through the entire analytical method (extraction, Florisil cartridge cleanup, and GC/ECD analysis).

The MDL (in ng/L) was 5.6.

The accuracy (as average % recovery using 5 determinations and no Florisil cleanup) from a spike concentration of 1.0 µg/L and separatory funnel extraction was 92% with a final volume of 10 mL.

The precision (as RSD% using 5 determinations and no Florisil cleanup) from a spike concentration of 1.0 µg/L and separatory funnel extraction was 7.1% with a final volume of 10 mL.

The accuracy (as average % recovery using 5 determinations and no Florisil cleanup), from a spike concentration of 330 μg/L and ultrasonic extraction of solid samples using 1:1 methylene chloride and acetone, was 81% with a final volume of 10 mL.

The precision (as RSD% using 5 determinations and no Florisil cleanup), from a spike concentration of 330 μg/L and ultrasonic extraction of solid samples using 1:1 methylene chloride and acetone, was 3.2% with a final volume of 10 mL.

SAMPLE COLLECTION, PRESERVATION & HANDLING
Volatile Organics — Standard 40-mL glass screw-cap VOA vials with Teflon®-faced silicone septum may be used for both liquid and solid matrices. When collecting samples, liquids and solids should be introduced into the vials gently to reduce agitation which might drive off volatile compounds. If there are any air bubbles present the sample must be retaken. The vials with solids should be tapped slightly as they are filled to try and eliminate as much free air space as possible. Two vials from each sampling location should be sealed in separate plastic bags to prevent cross-contamination between samples.

Semivolatile organics — Containers used to collect samples for the determination of semivolatile organic compounds should be soap and water washed followed by methanol (or isopropanol) rinsing. The sample containers should be of glass or Teflon® and have screw-top covers with Teflon® liners.

Preservation for volatile organics — No preservation is used with concentrated waste samples. With liquid samples containing no residual chlorine, 4 drops of concentrated hydrochloric acid are added and the samples are immediately cooled to 4°C. When liquid samples contain residual chlorine, they are treated as above and, in addition, 4 drops of 4% aqueous sodium thiosulfate are added to remove the residual chlorine. Soil, sediment, and sludge samples are only cooled to 4°C.

Preservation for semivolatile organics — No preservation is used with concentrated waste samples. With liquid samples containing no residual chlorine and with soil, sediment, and sludge samples, immediately cooling to 4°C is the only preservation used. When residual chlorine is present then 3 mL of 10% aqueous sodium sulfate is added for each gallon of sample collected, followed by cooling to 4°C.

Holding times — The holding time for all volatile organics samples is 14 days. Liquid samples must be extracted within 7 days and their extracts analyzed within 40 days. Concentrated waste, soil, sediment, and sludge samples must be extracted within 14 days and their extracts analyzed within 40 days.

SAMPLE PREPARATION Prepare stock standard solutions in hexane. Calibration standards at a minimum of five concentrations should be prepared through dilution of the stock standards with hexane. The suggested internal standards are: 2,5-dibromotoluene, 1,3,5-tribromobenzene, and α, α–dibromo-m-xylene. The analyst can use any of the three compounds provided that they are resolved from matrix interferences. Recommended surrogate compounds are α-2,6-trichlorotoluene, 1,4-dichloronaphthalene, and 2,3,4,5,6-pentachlorotoluene.

In general, water samples are extracted at a neutral pH with methylene chloride using a separatory funnel (EPA Method 3510) or a continuous liquid-liquid extractor (EPA Method 3520). Solid samples are extracted with hexane/acetone (1:1 v:v) using a Soxhlet extractor (EPA Method 3540) or with methylene chloride/acetone (1:1 v:v) using an ultrasonic extractor (EPA Method 3550). Non-aqueous waste samples may be diluted using EPA Method 3580. Prior to Florisil cleanup or gas chromatographic analysis, the extraction solvent must be exchanged to hexane. Sample extracts that will be subjected to gel permeation chromatography do not need solvent exchange.

Cleanup procedures may not be necessary for a relatively clean matrix. If removal of interferences such as chlorinated phenols, phthalate esters, etc., is required, proceed with the procedure outlined in EPA Method 3620.

QUALITY CONTROL Analyze a quality control check standard to demonstrate that the operation of the GC is in control. The frequency of the check standard analysis is equivalent to 10% of the samples analyzed. If the recovery of any compound found in the check standard is less than 80% of the certified value, the problem must be corrected and a new set of calibration standards must be prepared and analyzed. Calculate surrogate standard recoveries for all samples, blanks, and spikes. An internal standard peak area check must be performed on all samples. The internal standard must be evaluated for acceptance by determining whether the measured area for the internal standard deviates by more than 30% from the average area for the internal standard in the calibration standards. When the internal standard peak area is outside that limit, all samples that fall outside the QC criteria must be reanalyzed. Any compound confirmed by two columns may also be confirmed by GC/MS (EPA Method 8270). The GC/MS would normally require a minimum concentration of 1 ng/μL in the final extract for each compound. Include a mid-concentration calibration standard after each group of 20 samples in the analysis sequence. The response factors for the mid-concentration calibration must be within 15% of the average values for the multiconcentration calibration.

REFERENCE Test Methods for Evaluating Solid Waste, Physical/Chemical Methods, SW-846, 3rd Edition, U.S. EPA, Office of Solid Waste, Washington, DC, 1990. EPA Method 8121, Rev. 0, Nov. 1990.

Hexachlorobenzene   **EPA Method 8270**
CAS #118-74-1

TITLE Semivolatile Organic Compounds by GC/MS

MATRIX This method is used to determine the concentration of semivolatile organic compounds in extracts prepared from all types of solid waste matrices, soils, and groundwater. Although surface waters are not specifically mentioned, this method should be applicable to water samples from rivers, lakes, etc.

METHOD SUMMARY This method covers 259 semivolatile organic compounds. In very limited applications direct injection of the sample into the GC/MS system may be appropriate, but this results in very high detection limits (approximately 10,000 μg/L). Typically, a 1-L liquid sample, containing surrogate, and matrix spiking standards, is extracted in a continuous extractor first under acid conditions and then under basic conditions. Typically 30 g of a solid sample, containing surrogate, and matrix spiking standards, is extracted ultrasonically. After concentrating the extract to 1 mL it is spiked with 10 µL of an internal standard solution just prior to analysis by GC/MS. The volume injected should contain about 100 ng of base/neutral and 200 ng of acid surrogates (for a 1-µL injection). Analysis is performed by GC/MS using a capillary GC column.

INTERFERENCES Raw GC/MS data from all blanks, samples, and spikes must be evaluated for interferences. Contamination by carryover can occur whenever high-concentration and low-concentration samples are sequentially analyzed. To reduce carryover, the sample syringe must be rinsed out between samples with solvent. Whenever an unusually concentrated sample is encountered, it should be followed by the analysis of blank solvent to check for cross-contamination.

INSTRUMENTATION A GC/MS and a data system are required. The GC column used is a 30 m × 0.25 mm I.D. (or 0.32 mm I.D.) 1um film thickness silicone-coated fused silica capillary column. A continuous liquid-liquid extractor equipped with Teflon® or glass connection joints and stopcocks requiring no lubrication, a K-D concentrating apparatus, water bath, and an ultrasonic disrupter with a minimum power of 300 W and with pulsing capability are also required.

PRECISION & ACCURACY The estimated quantitation limit (EQL) of Method 8270B for determining an individual compound is approximately 1 mg/kg (wet weight) for soil or sediment samples, 1–200 mg/kg for wastes (dependent on matrix and method of preparation), and 10 µg/L for groundwater samples. EQLs will be proportionately higher for sample extracts that require dilution to avoid saturation of the detector.

The EQL(b) for groundwater in µg/L is 10.
The EQL (a, b) for low concentrations in soil and sediment in µg/kg is 660.
Accuracy as µg/L is 0.74C + 0.66.
Overall precision in µg/L is 0.43X–0.52.

(a) *EQLs listed for soil/sediment are based on wet weight. Normally data is reported in a dry-weight basis; therefore, EQLs will be higher based on the % dry weight of each sample. This calculation is based on a 30-g sample and gel permeation chromatography cleanup.*
(b) *Sample EQLs are highly matrix-dependent. The EQLs are provided for guidance and may not always be achievable.*
C = *True value for concentration, in µg/L.*
X = *Average recovery found for measurements of samples containing a concentration of C, in µg/L.*

ESTIMATED QUANTITATION LIMIT

Other Matrices	Factor (a)
High-concentration soil and sludges by sonicator	7.5
Non-water miscible waste	75

(a) *EQL for other matrices = [EQL for low soil/sediment] × [Factor]. This estimated EQL is similar to an EPA "Practical Quantitation Limit."*

SAMPLING METHOD
Liquid samples — Use a 1 or 2½ gallon amber glass bottle with a screw-top Teflon®-lined cover that has been prewashed with detergent and rinsed with distilled water and methanol (or isopropanol).

Soils, sediments, or sludges — Use an 8-oz. widemouth glass with a screw-top Teflon®-lined cover that has been prewashed with detergent and rinsed with distilled water and methanol (or isopropanol).

SAMPLE PRESERVATION
Liquid samples — If residual chlorine is present, add 3 mL of 10% sodium thiosulfate per gallon, cool to 4°C and store in a solvent-free refrigerator until analysis; if chlorine is not present, then eliminate the sodium thiosulfate addition.

Soils, sediments, or sludges — Cool samples to 4°C and store in a solvent-free refrigerator.

MHT Liquid samples must be extracted within 7 days and the extracts analyzed within 40 days. Soils, sediments, or sludges may be stored for a maximum of 14 days and the extracts analyzed within 40 days.

SAMPLE PREPARATION
Liquid samples — Transfer 1 L quantitatively to a continuous extractor. If high concentrations are anticipated, a smaller volume may be used and then diluted with organic-free reagent water to 1 L. Adjust pH, if necessary, to pH <2 using 1:1 (V/V) sulfuric acid. Pipette 1.0 mL of a surrogate standard spiking solution into each sample. For the sample in each analytical batch selected for spiking, add 1.0 mL of a matrix spiking standard. For base/neutral acid analysis, the amount of the surrogates and matrix spiking compounds added to the sample should result in a final concentration of 100 ng/µL of each analyte in the extract to be analyzed (assuming a 1-µL injection). Extract with methylene chloride for 18–24 h. Next, adjust the pH of the aqueous phase to pH >11 using 10 N sodium hydroxide and extract it with methylene chloride again for 18–24 h. Dry the extract through a column containing anhydrous sodium sulfate and concentrate it to 1 mL using a K-D concentrator.

Soils, sediments, or sludges — Use 30 g of sample. Nonporous or wet samples (gummy or clay type) that do not have a free-flowing sandy texture must be mixed with anhydrous sodium sulfate until the sample is free flowing. Add 1 mL of surrogate standards to all samples, spikes, standards, and blanks. For the sample in each analytical batch selected for spiking, add 1.0 mL of a matrix spiking standard. For base/neutral acid analysis, the amount added of the surrogates and matrix spiking compounds should result in a final concentration of 100 ng/µL of each base/neutral analyte and 200 ng/µL of each acid analyte in the extract to be analyzed (assuming a 1-µL injection). Immediately add a 100-mL mixture of 1:1 methylene chloride:acetone and extract the sample ultrasonically for 3 min and then decant or filter the extracts. Repeat the extraction two or more times. Dry the extract using a column with anhydrous sodium sulfate and concentrate it to 1 mL in a K-D concentrator.

QUALITY CONTROL A methylene chloride solution containing 50 ng/µL of decafluorotriphenylphosphine (DFTPP) is

used for tuning the GC/MS system each 12-h shift. A system performance check also must be made during every 12-h shift. A standard containing 50 ng/μL each of 4,4'-DDT, pentachlorophenol, and benzidine is required to verify injection port inertness and GC column performance. A calibration standard at mid-concentration, containing each compound of interest, including all required surrogates, must be performed every 12 h during analysis. After the system performance check is met, calibration check compounds (CCCs) are used to check the validity of the initial calibration.

The internal standard responses and retention times in the calibration check standard must be evaluated immediately after or during data acquisition. If the retention time for any internal standard changes by more than 30 seconds from the last check calibration (12 h), the chromatographic system must be inspected for malfunctions and corrections must be made, as required. If the electron ionization current plot (EICP) area for any of the internal standards changes by a factor of two from the last daily calibration standard check, the mass spectrometer must be inspected for malfunctions and corrections must be made, as appropriate.

Demonstrate, through the analysis of a reagent water blank, that interferences from the analytical system, glassware, and reagents are under control. The blank samples should be carried through all stages of the sample preparation and measurement steps. For each analytical batch (up to 20 samples), a reagent blank, matrix spike, and matrix spike duplicate/duplicate must be analyzed (the frequency of the spikes may be different for different monitoring programs). The blank and spiked samples must be carried through all stages of the sample preparation and measurement steps. A QC reference sample concentrate containing each analyte at a concentration of 100 mg/L in methanol is required.

REFERENCE Test Methods for Evaluating Solid Waste (SW-846). U.S. EPA 1983, Method 8270B, Rev. 2, Nov. 1990. Office of Solid Waste, Washington, DC.

Hexachlorobutadiene   **EPA Method 1625**
CAS #87-68-3

TITLE Semivolatile Organic Compounds by Isotope Dilution GC/MS

MATRIX The compounds may be determined in waters, soils, and municipal sludges by this method.

METHOD SUMMARY This method is used to determine 176 semivolatile toxic organic pollutants associated with the CWA (as amended 1987); the RCRA (as amended 1986); the CERCLA (as amended 1986); and other compounds amenable to extraction and analysis by capillary column gas chromatography-mass spectrometry (GC/MS).

Stable isotopically-labeled analogs of the compounds of interest are added to the sample. If the solids content is less than 1%, a 1-L sample is extracted at pH 12–13, then at pH <2 with methylene chloride using continuous extraction techniques.

If the solids content is 30% or less, the sample is diluted to 1% solids with reagent water, homogenized ultrasonically, and extracted at pH 12–13, then at pH <2 with methylene chloride using continuous extraction techniques. If the solids content is greater than 30%, the sample is extracted using ultrasonic techniques.

Each extract is dried over sodium sulfate, concentrated to a volume of 5 mL, cleaned up using GPC, if necessary, and concentrated. Extracts are concentrated to 1 mL if GPC is not performed, and to 0.5 mL if GPC is performed.

An internal standard is added to the extract, and a 1-mL aliquot of the extract is injected into the GC. The compounds are separated by GC and detected by a MS. The labeled compounds serve to correct the variability of the analytical technique.

INTERFERENCES Solvents, reagents, glassware, and other sample processing hardware may yield artifacts and/or elevated baselines causing misinterpretation of chromatograms and spectra. Materials used in the analysis must be demonstrated to be free from interferences under the conditions of analysis by running method blanks initially and with each sample lot (sample started through the extraction process on a given 8-h shift, to a maximum of 20). Specific selection of reagents and purification of solvents by distillation in all glass systems may be required. Glassware and, where possible, reagents are cleaned by solvent rinse and baking at 450°C for 1-h minimum. Interferences coextracted from samples will vary considerably from source to source, depending on the diversity of the site being sampled.

INSTRUMENTATION Major instrumentation includes a GC with a splitless or on-column injection port for capillary column, a MS with 70 eV electron impact ionization, and a data system to collect and record MS data, and process it. A K-D apparatus is used to concentrate extracts.

GC Column: 30 m × 0.25 mm I.D. 5% phenyl, 94% methyl, 1% vinyl silicone bonded phased fused silica capillary column.

PRECISION & ACCURACY The detection limits of the method are usually dependent on the level of interferences rather than instrumental limitations. The limits typify the minimum quantities that can be detected with no interferences present.

The minimum level (in μg/mL) was 10. This is defined as a minimum level at which the analytical system shall give recognizable mass spectra (background corrected) and acceptable calibration points.

The MDL (in μg/kg) in low solids was 46 and in high solids was 22; these were determined in digested sludge (low solids) and in filter cake or compost (high solids).

The labeled and native compound initial precision as standard deviation (in μg/L) was 56.
The labeled and native compound initial accuracy as average recovery (in μg/L) was 51–251.

SAMPLE COLLECTION, PRESERVATION & HANDLING Collect samples in glass containers. Aqueous samples which flow freely are collected in refrigerated bottles using automatic

sampling equipment. Solid samples are collected as grab samples using widemouth jars. Maintain samples at 0 to 4°C from the time of collection until extraction. If residual chlorine is present in aqueous samples, add 80 mg sodium thiosulfate/L of water. Begin sample extraction within 7 days of collection, and analyze all extracts within 40 days of extraction.

SAMPLE PREPARATION Samples containing 1% solids or less are extracted directly using continuous liquid-liquid extraction techniques. Samples containing 1 to 30% solids are diluted to the 1% level with reagent water and extracted using continuous liquid-liquid extraction techniques. Samples containing greater than 30% solids are extracted using ultrasonic techniques.

Base/neutral extraction — Adjust the pH of the waters in the extractors to 12–13 with 6 N NaOH. Extract with methylene chloride for 24–48 h.

Acid extraction — Adjust the pH of the waters in the extractors to 2 or less using 6 N sulfuric acid. Extract with methylene chloride for 24–48 h.

Ultrasonic extraction of high solids samples — Add anhydrous sodium sulfate to the sample and QC aliquot(s). Add acetone:methylene chloride (1:1) to the sample and mix thoroughly

Concentrate extracts using a K-D apparatus.

QUALITY CONTROL The analyst is permitted to modify this method to improve separations or lower the costs of measurements, provided all performance specifications are met. Analyses of blanks are required to demonstrate freedom from contamination. When results of spikes indicate atypical method performance for samples, the samples are diluted to bring method performance within acceptable limits.

For low solids (aqueous samples), extract, concentrate, and analyze two sets of four 1-L aliquots (8 aliquots total) of the precision and recovery standard. For high solids samples, two sets of four 30-g aliquots of the high solids reference matrix are used.

Spike all samples with labeled compounds to assess method performance. Compute percent recovery of the labeled compounds using the internal standard method. Compare the labeled compound recovery for each compound with the corresponding labeled compound recovery.

Reagent water and high solids reference matrix blanks are analyzed to demonstrate freedom from contamination. Extract and concentrate a 1-L reagent water blank or a high solids reference matrix blank with each sample's lot (samples started through the extraction process on the same 8-h shift, to a maximum of 20 samples).

Field replicates may be collected to determine the precision of the sampling technique, and spiked samples may be required to determine the accuracy of the analysis when the internal standard method is used.

REFERENCE Semivolatile Organic Compounds by Isotope Dilution GC/MS. Office of Water Regulation and Standards, U.S. EPA Industrial Technology Division, Washington, DC, EPA Method 1625, Rev. C, June 1989 (contact W.A. Telliard, U.S. EPA, Office of Water Regulations and Standards, 401 M St., SW, Washington, DC, 20460. Phone: 202-382-7131).

Hexachlorobutadiene   EPA Method 502
CAS #87-68-3

TITLE Volatile Organic Compounds in Water By Purge and Trap Capillary Column Gas Chromatography with Photoionization and Electrolytic Conductivity Detectors in Series. U.S. EPA Method 502.2, Rev. 2.0, 1989.

MATRIX Drinking water and raw source water. The latter should include most surface water and groundwater sources.

METHOD SUMMARY This method covers 60 volatile organic compounds that contain halogen atoms and/or that are aromatic. An inert gas (zero grade nitrogen or helium) is bubbled through a 25-mL or a 5-mL water sample (depending on the expected concentration of the analytes). Purged sample components are trapped in a tube of sorbent materials. When purging is complete, the sorbent tube is heated and backflushed with helium to desorb the trapped sample onto a capillary GC column. The column is temperature programmed to separate the method analytes which are then detected with a photoionization detector (PID) and a Hall electrolytic conductivity (HECD) placed in series. The PID is selective for aromatic compounds and the HECD is selective for halogenated compounds.

INTERFERENCES Impurities in the purge gas and from organic compounds outgassing from the plumbing ahead of the trap account for many contamination problems. Interferences purged or coextracted from the samples will vary considerably from source to source, depending upon the particular sample or extract being tested. Cross-contamination can occur whenever high-level and low-level samples are analyzed sequentially. Samples also can be contaminated by diffusion of volatile organics (particularly methylene chloride and fluorocarbons) through the septum seal into the sample during shipment and storage. The lab where volatile analysis is performed and also the refrigerated storage area should be completely free of solvents.

INSTRUMENTATION A GC containing a series configuration of a high temperature photoionization detector (PID) equipped with 10.0 eV (nominal) lamp and Hall electrolytic conductivity detector (HECD) is required. Also required is an all-glass 5-mL purging device, a sorbent trap, and a thermal desorption apparatus which is connected to the GC system.

Column 1: VOCOL glass wide-bore capillary column.
Column 2: RTX–502.2 mega-bore capillary column.
Column 3: DB-62 mega-bore capillary column.

PRECISION & ACCURACY Method detection limits are dependent upon the characteristics of the gas chromatographic system used. Analytes that are not separated chromatographically cannot be individually identified and used in the same calibration mixture or water samples unless an alternative technique for identification and quantification, such as mass spectrometry, is used.

Electrolytic conductivity detetor (c) range in μg/L (a) was 0.02–200.
Electrolytic conductivity detetor (c) MDL in μg/L (b) was 0.02.
Electrolytic conductivity detetor (c) accuracy as % recovery was 98.
Electrolytic conductivity detetor (c) precision as % RSD was 8.3.
Photoionization detector (d) range in μg/L (a) was 0.02–200.
Photoionization detector (d) MDL in μg/L (b) was 0.06.
Photoionization detector (d) accuracy as % recovery was 99.
Photoionization detector (d) precision as % RSD was 9.5.

(a) *The applicable concentration range of this method is compound, instrument, and matrix-dependent. It is listed as being approximately 0.02 to 200 μg/L but no specific information is provided so caution should be observed.*

(b) *The method detection limits reports with this method are compound, instrument, and matrix-dependent. The values reported were calculated using reagent water fortified with the corresponding compounds at 10 μg/L and a GC-equipped with a 60 m × 0.75 mm VOLCOL wide bore capillary column with 1.5 μm film thickness and using helium carrier gas.*

(c) *Recoveries and relative standard deviations were determined from seven samples of reagent water fortified with 10 μg/L of each compound. 2-Bromo-1-chloropropane was used as the internal standard for calculating average recoveries.*

(d) *Recoveries and relative standard deviations were determined from seven samples of reagent water fortified with 10 μg/L of each compound. Fluorobenzene was used as the internal standard for calculating average recoveries.*

SAMPLING METHOD Collect samples using a 40- to 120-mL screw-cap vial (prewashed with detergent, rinsed with distilled water and oven dried at 105°C) with a Teflon®-faced silicone septum. Collect bubble-free samples and place the septum with the Teflon® side down on the water.

SAMPLE PRESERVATION If residual chlorine is present in the water add about 25 mg of ascorbic acid to each vial before samples are collected to remove the chlorine. Add hydrochloric acid to reduce pH to <2, immediately cool samples to 4°C, and store them in a solvent-free refrigerator at 4°C until analysis.

MHT The maximum holding time for samples is 14 days from the time they were collected.

SAMPLE PREPARATION Remove the plungers from two 5-mL syringes and attach a closed syringe valve to each. Warm the sample to room temperature, open the sample bottle, and carefully pour the sample into one of the syringe barrels to just short of overflowing. Replace the syringe plunger, invert the syringe, and compress the sample. Open the syringe valve and vent any residual air while adjusting the sample volume to 5.0 mL. Add 10 μL of the internal calibration standard to the sample through the syringe valve. Close the valve. Fill the second syringe in an identical manner from the same sample bottle. Reserve this second syringe for a reanalysis if necessary.

QUALITY CONTROL As an initial demonstration of lab accuracy and precision, analyze 4 to 7 replicates of a lab fortified blank containing analyte at 0.1–5 μg/L. Collect all samples in duplicate. Surrogate analytes (similar to those of the analytes of interest), whose concentration is known in every sample, are measured using the same internal standard calibration procedure. Duplicate field reagent water blanks (trip blanks) must be analyzed with each set of samples, lab reagent blanks (method blanks) must be analyzed with each batch of samples processed as a group within a work shift. Also, a single lab-fortified blank that contains each of the analytes of interest should be analyzed with each batch of samples processed as a group within a work shift. A 3- to 5-point calibration curve is needed depending on the calibration range factor required.

EPA CONTACT & HOTLINE For technical questions contact Dr. Baldev Bathija, U.S. EPA, Office of Ground Water and Drinking Water (WH-550D), 401 M St. SW, Washington, DC 20460. Tel. (202) 260-3040. For further information the EPA Safe Drinking Water Hotline may be called at: (800) 426-4791.

REFERENCE Methods for the Determination of Organic Compounds in Drinking Water, EPA/600/4-88/039 (revised July 1991; Final Rule for determination of compliance with the MCL for Total Trihalomethanes under 141.30, in 40 CFR Part 141, Vol. 58, No. 147, Fed. Reg., Tuesday Aug. 3, 1993). U.S. EPA Environmental Monitoring Systems Laboratory, Cincinnati, OH, 45268, U.S.A. Available from the National Technical Information Service (NTIS), 5285 Port Royal Road, Springfield, VA 22161; Tel. 800-553-6847. NTIS Order Number is PB91-231480.

Hexachlorobutadiene EPA Method 524
CAS #87-68-3

TITLE Measurement of Purgeable Organic Compounds in Water by Capillary Column GC/MS.

MATRIX Drinking water and raw source water; the latter should include most surface water and groundwater sources.

METHOD SUMMARY Method 524.2 covers 60 volatile organic compounds. An inert gas (zero grade nitrogen or helium) is bubbled through a 25-mL or a 5-mL water sample (depending on the expected concentration of the analytes). Purged sample components are trapped in a tube of sorbent materials. When purging is complete, the sorbent tube is heated and backflushed with helium to desorb the trapped sample onto a capillary GC column.

INTERFERENCES Impurities in the purge gas and from organic compounds outgassing from the plumbing ahead of the trap account for many contamination problems. Interferences purged or coextracted from the samples will vary considerably from source to source, depending upon the particular sample or extract being tested. Cross-contamination can occur whenever high-level and low-level samples are analyzed sequentially. Samples also can be contaminated by diffusion of volatile organics (particularly methylene chloride and fluorocarbons) through the septum seal into the sample during shipment and storage.

INSTRUMENTATION A GC/MS with a data system equipped with one of the following capillary GC columns:

Column 1: VOCOL glass wide bore capillary column.
Column 2: DB-624 fused silica capillary column.
Column 3: DB-5 fused silica capillary column.

Also required is an all-glass 25 mL or 5-mL purging device, a sorbent trap, and a thermal desorption apparatus which is connected to the GC/MS system.

PRECISION & ACCURACY Method detection limits are compound- and instrument-dependent, and may vary from approximately 0.02–0.35 µg/L. Note in the table below that the "true" concentration range used for accuracy and precision measurements was quite narrow. However, the applicable concentration range of this method is primarily column dependent and is approximately 0.02 to 200 µg/L for the wide-bore thick-film columns. Narrow-bore thin-film columns may have a capacity which limits the range to about 0.02 to 20 µg/L. Analytes that are inefficiently purged from water will not be detected when present at low concentrations, but they can be measured with acceptable accuracy and precision when present in sufficient amounts.

Analytes that are not separated chromatographically, but which have different mass spectra and non-interfering quantification ions, can be identified and measured in the same calibration mixture or water sample. Analytes which have very similar mass spectra cannot be individually identified and measured in the same calibration mixture or water samples unless they have different retention times. Co-eluting compounds with very similar mass spectra, typically many structural isomers, must be reported as an isomeric group or pair.

The range (in µg/L) was 0.5–10.
The Method Detection Limig (in µg/L) was 0.11.
The accuracy (as % recovery) was 100.
The precision (in %) was 6.8.

Note: Data were obtained from 16–31 determinations using a wide-bore capillary column and a jet separator interfaced to a quadrupole mass spectrometer. All analytes were in a reagent water matrix.

SAMPLING METHOD Collect samples using a 40- to 120-mL screw-cap vial (prewashed with detergent, rinsed with distilled water and oven dried at 105°C) with a Teflon®-faced silicone septum. Collect bubble-free samples and place the septum with the Teflon® side down on the water.

SAMPLE PRESERVATION If residual chlorine is present in the water add about 25 mg of ascorbic acid to each vial before samples are collected to remove the chlorine. Add hydrochloric acid to reduce pH to <2, and immediately cool samples to 4°C, and store them in a solvent-free refrigerator at 4°C until analysis.

MHT The maximum holding time for samples is 14 days from the time they were collected.

SAMPLE PREPARATION Remove the plungers from two 25-mL (or 5-mL depending on sample size) syringes and attach a closed syringe valve to each. Warm the sample to room temperature, open the sample bottle, and carefully pour the sample into one of the syringe barrels to just short of overflowing. Replace the syringe plunger, invert the syringe, and compress the sample. Open the syringe valve and vent any residual air while adjusting the sample volume to 25.0 mL (or 5 mL). For samples and blanks, add 5 µL of the fortification solution containing the internal standard and the surrogates to the sample through the syringe valve. For calibration standards and lab fortified blanks, add 5 µL of the fortification solution containing the internal standard only. Close the valve. Fill the second syringe in an identical manner from the same sample bottle. Reserve this second syringe for a reanalysis if necessary.

QUALITY CONTROL As an initial demonstration of lab accuracy and precision, analyze 4 to 7 replicates of a lab fortified blank containing analyte at 0.2–5 µg/L. Collect all samples in duplicate. Surrogate analytes (similar to those of the analytes of interest), whose concentration is known in every sample, are measured using the same internal standard calibration procedure. Duplicate field reagent water blanks (trip blanks) must be analyzed with each set of samples, lab reagent blanks (method blanks) must be analyzed with each batch of samples processed as a group within a work shift. Also, a single lab-fortified blank that contains each of the analytes of interest should be analyzed with each batch of samples processed as a group within a work shift. A 3- to 5-point calibration curve is needed depending on the calibration range factor required.

EPA CONTACT & HOTLINE For technical questions contact Dr. Baldev Bathija, U.S. EPA, Office of Ground Water and Drinking Water (WH-550D), 401 M St. SW, Washington, DC 20460. Tel. (202) 260-3040. For further information the EPA Safe Drinking Water Hotline may be called at: (800) 426-4791.

REFERENCE Methods for the Determination of Organic Compounds in Drinking Water, EPA/600/4-88/039 (revised July 1991; Final Rule for determination of compliance with the MCL for Total Trihalomethanes under 141.30, in 40 CFR Part 141, Vol. 58, No. 147, Fed. Reg., Tuesday Aug. 3, 1993). U.S. EPA Environmental Monitoring Systems Laboratory, Cincinnati, OH, 45268, U.S.A. Available from the National Technical Information Service (NTIS), 5285 Port Royal Road, Springfield, VA 22161; Tel. 800-553-6847. NTIS Order Number is PB91-231480.

Hexachlorobutadiene   **EPA Method 625**
CAS #87-68-3

TITLE Base/Neutrals and Acids, U.S. EPA Method 625

MATRIX This methods covers municipal and industrial wastewaters.

METHOD SUMMARY Approximately 1 L of sample is serially extracted with methylene chloride at a pH greater than 11 and again at a pH less than 2 using a separatory funnel or a continuous extractor. The methylene chloride extract is dried, concentrated to a volume of 1 mL, and analyzed by GC/MS. Qualitative identification of the parameters in the extract is performed using the retention time and the relative abundance of three characteristic masses (m/z). Qualitative analysis is performed using either external or internal standard techniques with a single characteristic m/z.

INTERFERENCES Method interferences may be caused by contaminants in solvents, reagents, glassware, and other sample processing hardware. Glassware must be scrupulously cleaned. Glassware should be heated in a muffle furnace at 400°C for 5 to 30 min. Some thermally stable materials, such as PCBs, may not be eliminated by this treatment. Solvent rinses with acetone and pesticide quality hexane may be substituted for the muffle furnace heating. Matrix interferences may be caused by contaminants that are coextracted from the sample. The base-neutral extraction may cause significantly reduced recovery of phenols. The packed gas chromatographic columns recommended for the basic fraction may not exhibit sufficient resolution for some analytes.

INSTRUMENTATION A GC/MS system with an injection port designed for on-column injection when using packed columns and for splitless injection when using capillary columns.

Column for base/neutrals: 1.8 m long × 2 mm I.D. glass, packed with 3% SP-2550 on Supelcoport (100/120 mesh) or equivalent.

Column for acids: 1.8 m long × 2 mm I.D. glass, packed with 1% SP-1240DA on Supelcoport (100/120 mesh) or equivalent.

PRECISION & ACCURACY The MDL concentrations were obtained using reagent water. The MDL actually achieved in a given analysis will vary depending on instrument sensitivity and matrix effects. This method was tested by 15 laboratories using reagent water, drinking water, surface water, and industrial wastewaters spiked at six concentrations over the range 5 to 100 µg/L. Single operator precision, overall precision, and method accuracy were found to be directly related to the concentration of the parameter matrix.

The MDL (in µg/L) in reagent water was not reported.

The standard deviation (in µg/L based on 4 recovery measurements) was 26.3.

The range (in µg/L) for average recovery for 4 measurements was 37.8–102.2.

The range (in %) for percent recovery was 24–116.

Accuracy (in µg/L) as expected recovery for one or more measurements of a sample containing a true concentration of C was 0.71C-1.01.

Precision (in µg/L) as expected single analyst standard deviation of measurements at an average concentration found at X was 0.19X + 0.92.

Overall precision (in µg/L) as expected interlaboratory standard deviation of measurements in an average concentration found at X was 0.26X + 0.49.

C = *True value of the concentration in µg/L.*

X = *Average recovery found for measurements of samples containing a concentration at C in µg/L.*

SAMPLE PREPARATION Adjust the pH to >11 with sodium hydroxide and serially extract in a separatory funnel with methylene chloride or else in a continuous extractor. Next, adjust the pH to <2 with sulfuric acid and serially extract in a separatory funnel with methylene chloride or else in a continuous extractor. Dry the extracts separately through a column of anhydrous sodium sulfate and then concentrate each of the extracts to 1.0 mL using a K-D apparatus.

SAMPLE COLLECTION, PRESERVATION & HANDLING Grab samples must be collected in glass containers. All samples must be refrigerated at 4°C from the time of collection until extraction. If residual chlorine is present, add 80 mg of sodium thiosulfate/L of sample and mix well. All samples must be extracted within 7 days of collection and completely analyzed within 40 days of extraction.

QUALITY CONTROL Make an initial, one-time, demonstration of the ability to generate acceptable accuracy and precision with this method. Before processing any samples, the analyst must analyze a reagent water blank to demonstrate that interferences from the analytical system and glassware are under control. Each time a set of samples is extracted or reagents are changed, a reagent water blank must be processed. Spike and analyze a minimum of 5% of all samples to monitor and evaluate lab data quality. A QC check sample concentrate that contains each parameter of interest at a concentration of 100 µg/mL in acetone is required. PCBs and multicomponent pesticides may be omitted from this test.

After analysis of five spiked wastewater samples, calculate the average percent recovery and the standard deviation of the percent recovery. Spike all samples with the surrogate standard spiking solution and calculate the percent recovery of each surrogate compound.

REFERENCE *Federal Register,* Vol. 49, No. 209. Friday, Oct. 26, 1984.

Hexachlorobutadiene   **EPA Method 8021**
CAS #87-68-3

TITLE Halogenated Volatile by Gas Chromatography Using Photoionization and Electrolytic Conductivity Detectors in Series: Capillary Column Technique

MATRIX This method is applicable to nearly all types of samples, regardless of water content, including groundwater, aqueous sludges, caustic liquors, acid liquors, waste solvents, oily wastes, mousses, tars, fibrous wastes, polymeric emulsions, filter cakes, spent carbons, spent catalysts, soils, and sediments.

METHOD SUMMARY This method is used to determine 60 volatile organic compounds in a variety of solid waste matrices. It provides GC conditions for the detection of halogenated and aromatic volatile organic compounds. Samples can be analyzed using direct injection or purge-and-trap (EPA Method 5030). Groundwater samples must be analyzed using EPA Method 5030 (where applicable). A temperature program is used with the GC. Detection is achieved by a photoionization detector (PID) and a Hall electrolytic conductivity detector (HECD) in series.

INTERFERENCES Samples can be contaminated by diffusion of volatile organics (particularly chlorofluorocarbons and methylene chloride) through the sample container septum during shipment and storage.

INSTRUMENTATION A GC-equipped with variable-constant differential flow controllers, subambient oven controller,

PID and HECD detectors connected with a short piece of uncoated capillary tubing and a data system.

Column: 60 m × 0.75 mm I.D. VOCOL wide-bore capillary column with 1.5 µm film thickness.

PRECISION & ACCURACY MDLs are compound-dependent and vary with purging efficiency and concentration. The applicable concentration range of this method is compound- and instrument-dependent but is approximately 0.1 to 200 µg/L. Analytes that are inefficiently purged from water will not be detected when present at low concentrations, but they can be measured with acceptable accuracy and precision when present in sufficient amounts. The estimated quantitation limit (EQL) for an individual compound is approximately 1 µg/kg (wet weight) for soil/sediment samples, 100 µg/kg (wet weight) for wastes, and 1 µg/L for groundwater. EQLs will be proportionately higher for sample extracts and samples that require dilution or reduced sample size to avoid saturation of the detector.

MULTIPLICATION FACTORS FOR OTHER MATRICES (a)

Matrix	Factor (b)
Groundwater	10
Low-concentration soil	10
Water miscible liquid waste	500
High-concentration soil and sludge	1250
Non-water miscible waste	1250

(a) Sample EQLs are highly matrix-dependent. The EQLs listed herein are provided for guidance and may not always be achievable.
(b) EQL = [Method detection limit] × [Factor]. For non-aqueous samples, the factor is on a wet-weight basis.

SINGLE LABORATORY ACCURACY & PRECISION DATA FOR VOCs IN WATER

This method was tested in a single lab using water spiked at 10 µg/L and the following data was reported:

Recoveries and standard deviations were determined from seven samples and spiked at 10 µg/L of each analyte. Recoveries were determined by the internal standard method. Internal standards were: Fluorobenzene for PID and 2-Bromo-1-chloropropane for HECD.

The average recovery (in percent) for the PID was 99.
The standard deviation of the recovery for the PID was 9.5.
The MDL (in µg/mL) for the PID was 0.06.
The average recovery (in percent) for the HECD was 98.
The standard deviation of the recovery for the HECD was 8.3.
The MDL (in µg/mL) for the HECD was 0.02.

SAMPLE COLLECTION, PRESERVATION & HANDLING
Volatile Organics — Standard 40-mL glass screw-cap VOA vials with Teflon®-faced silicone septum may be used for both liquid and solid matrices. When collecting samples, liquids and solids should be introduced into the vials gently to reduce agitation which might drive off volatile compounds. If there are any air bubbles present the sample must be retaken. Tap slightly as they are filled to try and eliminate as much free air space as possible. The two vials from each sampling locations should be sealed in separate plastic bags to prevent cross-contamination between samples particularly if the sampled waste is suspected of containing high levels of volatile organics.

Semivolatile organics — Containers used to collect samples for the determination of semivolatile organic compounds should be soap and water washed followed by methanol (or isopropanol) rinsing. The sample containers should be of glass or Teflon® and have screw-top covers with Teflon® liners.

Preservation for volatile organics — No preservation is used with concentrated waste samples. With liquid samples containing no residual chlorine, 4 drops of concentrated hydrochloric acid are added and the samples are immediately cooled to 4°C. When liquid samples contain residual chlorine, they are treated as above and, in addition, 4 drops of 4% aqueous sodium thiosulfate are added. Soil, sediment, and sludge samples are only cooled to 4°C.

Preservation for semivolatile organics — No preservation is used with concentrated waste samples. With liquid samples containing no residual chlorine and with soil, sediment, and sludge samples, immediately cooling to 4°C is the only preservation used. When residual chlorine is present then 3 mL of 10% aqueous sodium sulfate is added for each gallon of sample collected, followed by cooling to 4°C.

MHT The holding time for all volatile organics samples is 14 days. Liquid samples must be extracted within 7 days and their extracts analyzed within 40 days. Concentrated waste, soil, sediment, and sludge samples must be extracted within 14 days and their extracts analyzed within 40 days.

SAMPLE PREPARATION Volatile compounds are introduced into the gas chromatograph either by direct injector or purge-and-trap (EPA Method 5030). EPA Method 5030 may be used directly on groundwater samples or low-concentration contaminated soils and sediments. For medium-concentration soils or sediments, methanolic extraction, as described in EPA Method 5030, may be necessary prior to purge-and-trap analysis.

QUALITY CONTROL Calculate surrogate standard recovery on all samples, blanks, and spikes. A trip blank is recommended to check on sampling, storage, and handling contamination. Calibration standards, at a minimum of five concentration levels, are prepared in organic-free reagent water. One of the concentration levels should be at a concentration near, but above, the method detection limit.

A combination of bromochloromethane, 2-bromo-1-chloropropane, 1,4-dichlorobutane, and bromochlorobenzene are recommended as surrogate standards to encompass the range of the temperature program used in this method.

REFERENCE Test Methods for Evaluating Solid Waste, Physical/Chemical Methods, SW-846, 3rd Edition, U.S. EPA, Office of Solid Waste, Washington, DC, EPA Method 8021A, Rev. 1, Nov. 1992.

Hexachlorobutadiene   **EPA Method 8120**
CAS #87-68-3

TITLE Chlorinated Hydrocarbons by Gas Chromatography

MATRIX This method covers aqueous and solid matrices. This includes a wide variety such as drinking water, groundwater, industrial wastewaters, surface waters, soils, solids, and sediments.

METHOD SUMMARY This method is used to determine the concentration of 14 chlorinated hydrocarbons. It provides gas chromatographic conditions for the detection of ppb concentrations of certain chlorinated hydrocarbons. Prior to use of this method, appropriate sample extraction techniques must be used. Both neat and diluted organic liquids (EPA Method 3580, Waste Dilution) may be analyzed by direct injection. A 2 to 5 µg/mL aliquot of the extract is injected into a gas chromatograph (GC) using the solvent flush technique, and compounds in the GC effluent are detected by an electron capture detector (ECD).

INTERFERENCES Solvents, reagents, glassware, and other sample processing hardware may yield discrete artifacts and/or elevated baselines causing misinterpretation of gas chromatograms. Interferences coextracted from samples will vary considerably from source to source, depending upon the waste being sampled.

INSTRUMENTATION An analytical system complete with GC suitable for on-column injections and accessories, including detectors, column supplies, recorder, gases and syringes is required. A data system for measuring peak areas and/or peak heights is recommended. The GC is equipped with an electron capture detector (ECD). A K-D apparatus is needed for sample preparation.

Column 1: 1.8 m × 2 mm I.D. glass column packed with 1% SP-1000 on Supelcoport (100/120 mesh) or equivalent.

Column 2: 1.8 m × 2 mm I.D. glass column packed with 1.5% OV-1/2.4% OV-225 on Supelcoport (80/100 mesh) or equivalent.

PRECISION & ACCURACY The method was tested by 20 laboratories using organic-free reagent water, drinking water, surface water, and three industrial wastewaters spiked at six concentrations over the range 1.0 to 356 µg/L. Single operator precision, overall precision, and method accuracy were found to be directly related to the concentration of the parameter and essentially independent of the sample matrix.

MULTIPLICATION FACTORS FOR OTHER MATRICES (a)

Matrix	Factor (b)
Groundwater	10
Low-concentration soil by ultrasonic cleanup extraction with GPC	670
High-concentration soil and sludges by ultrasonic extraction	10,000
Non-water miscible waste	100,000

(a) *Sample EQLs are highly matrix-dependent. The EQLs listed are provided for guidance and may not always be achievable.*
(b) *EQL = [Method detection limit] × [Factor]. For non-aqueous samples, the factor is on a wet-weight basis.*

PRECISION & ACCURACY The estimates below are based upon the performance in a single lab.

The accuracy (in µg/L) as expected recovery for one or more measurements of a sample containing a concentration of C was $0.61C + 0.03$.

The precision (in µg/L) as expected single analyst standard deviation of measurements at an average concentration of x" was $0.18x" - 0.08$.

The precision (in µg/L) as expected interlaboratory standard deviation measurements at an average concentration found of x" was $0.53x" - 0.12$.

C = *True value for the concentration, in µg/L.*
$x"$ = *Average recovery found for measurements of samples containing a concentration of C, in µg/L.*

SAMPLE COLLECTION, PRESERVATION & HANDLING Extracts must be stored under refrigeration at 4°C and analyzed within 40 days of extraction.

SAMPLE PREPARATION In general, water samples are extracted at a neutral, or as is, pH with methylene chloride using either EPA Method 3510 or EPA Method 3520. Solid samples are extracted using either EPA Method 3540 or EPA Method 3550. Prior to gas chromatographic analysis, the extraction solvent must be exchanged to hexane.

QUALITY CONTROL The quality control check concentrate (EPA Method 8000) should contain each parameter of interest in acetone at the following concentrations: hexachloro-substituted hydrocarbon, 10 µg/mL; and any other chlorinated hydrocarbon, 100 µg/mL. Calculate surrogate standard recovery on all samples, blanks, and spikes.

Prepare stock standard solutions in isooctane or hexane. Calibration standards at a minimum of five concentrations should be prepared through dilution of the stock standards with isooctane or hexane. Internal standards and surrogate standards are also needed.

REFERENCE Test Methods for Evaluating Solid Waste, Physical/Chemical Methods, SW-846, 3rd Edition, U.S. EPA, Office of Solid Waste, Washington, DC, 1990. EPA Method 8120 A Rev. 1, Nov. 1990.

Hexachlorobutadiene   **EPA Method 8121**
CAS #87-68-3

TITLE Chlorinated Hydrocarbons by GC: Capillary Column Technique

MATRIX This method covers aqueous and solid matrices. This includes a wide variety such as drinking water, groundwater, industrial wastewaters, surface waters, soils, solids, and sediments.

METHOD SUMMARY This method provides procedures for the determination of 22 chlorinated hydrocarbons in water, soil/sediment, and waste matrices. A measured volume or weight of sample is extracted by using one of the appropriate sample extraction techniques specified in EPA Method 3510, EPA Method 3520, EPA Method 3540, or EPA Method 3550, or diluted using EPA Method 3580. Aqueous samples are extracted at neutral pH with methylene chloride by using either a separatory funnel (EPA Method 3510) or a continuous liquid-liquid extractor (EPA Method 3520). Solid samples are extracted with hexane/acetone (1:1) by using a Soxhlet extractor (EPA Method 3540) or with methylene chloride/acetone

(1:1) by using an ultrasonic extractor (EPA Method 3550). After cleanup, the extract or diluted sample is analyzed by gas chromatography with electron capture detection (GC/ECD).

The sensitivity level of this method usually depends on the level of interferences rather than on instrumental limitations. This method may be used in conjunction with EPA Method 3620, Florisil Column Cleanup, EPA Method 3660, Sulfur Cleanup, and EPA Method 3640, Gel Permeation Chromatography, to aid in the elimination of interferences.

INTERFERENCES Solvents, reagents, glassware, and other hardware used in sample processing may introduce artifacts which may result in elevated baselines, causing misinterpretation of gas chromatograms. Interferants coextracted from the samples will vary considerably from waste to waste. Glassware must be scrupulously clean. Phthalate esters, if present in a sample, will interfere only with the BHC isomers. The presence of elemental sulfur will result in large peaks, and can often mask the region of compounds eluting after 1,2,4,5-tetrachlorobenzene. The tetrabutylammonium (TBA)-sulfite procedure (EPA Method 3660) works well for the removal of elemental sulfur. Waxes and lipids can be removed by gel permeation chromatography (EPA Method 3640).

INSTRUMENTATION A GC suitable for on-column injections and all required accessories, including and electron capture detector (ECD), analytical columns, recorder, gases, and syringes are needed. A data system for measuring peak heights and/or peak areas is recommended. A Kuderna-Danish (K-D) apparatus will also be needed to concentrate extracts.

Column 1: 30 m × 0.53 mm I.D. fused-silica capillary column chemically bonded with trifluoropropyl methyl silicone (DB-210 or equivalent).

Column 2: 30 m × 0.53 mm I.D. fused-silica capillary column chemically bonded with polyethylene glycol (DB-WAX or equivalent).

PRECISION & ACCURACY This method has been tested in a single lab by using organic-free reagent water, sandy loam samples, and extracts which were spiked with the test compounds at one concentration. Single-operator precision and method accuracy were found to be related to the concentration of compound and the type of matrix. The accuracy and precision technique will be determined by the sample matrix, sample preparation technique, optional cleanup techniques, and calibration procedures used.

MULTIPLICATION FACTORS FOR OTHER MATRICES (a)

Matrix	Factor (b)
Groundwater	10
Low-concentration soil by ultrasonic cleanup extraction with GPC	670
High-concentration soil and sludges by ultrasonic extraction	10,000
Non-water miscible waste	100,000

(a) Sample EQLs are highly matrix-dependent. The EQLs listed are provided for guidance and may not always be achievable.
(b) EQL = [Method detection limit] × [Factor]. For non-aqueous samples, the factor is on a wet-weight basis.

PRECISION & ACCURACY MDL is the method detection limit for organic-free reagent water. MDL was determined from the analysis of eight replicate aliquots processed through the entire analytical method (extraction, Florisil cartridge cleanup, and GC/ECD analysis).

The MDL (in ng/L) was 1.4.

The accuracy (as average % recovery using 5 determinations and no Florisil cleanup) from a spike concentration of 1.0 µg/L and separatory funnel extraction was 95% with a final volume of 10 mL.

The precision (as RSD% using 5 determinations and no Florisil cleanup) from a spike concentration of 1.0 µg/L and separatory funnel extraction was 3.6% with a final volume of 10 mL.

The accuracy (as average % recovery using 5 determinations and no Florisil cleanup), from a spike concentration of 330 µg/L and ultrasonic extraction of solid samples using 1:1 methylene chloride and acetone, was 83% with a final volume of 10 mL.

The precision (as RSD% using 5 determinations and no Florisil cleanup), from a spike concentration of 330 µg/L and ultrasonic extraction of solid samples using 1:1 methylene chloride and acetone, was 4.7% with a final volume of 10 mL.

SAMPLE COLLECTION, PRESERVATION & HANDLING
Volatile Organics — Standard 40-mL glass screw-cap VOA vials with Teflon®-faced silicone septum may be used for both liquid and solid matrices. When collecting samples, liquids and solids should be introduced into the vials gently to reduce agitation which might drive off volatile compounds. If there are any air bubbles present the sample must be retaken. The vials with solids should be tapped slightly as they are filled to try and eliminate as much free air space as possible. Two vials from each sampling location should be sealed in separate plastic bags to prevent cross-contamination between samples.

Semivolatile organics — Containers used to collect samples for the determination of semivolatile organic compounds should be soap and water washed followed by methanol (or isopropanol) rinsing. The sample containers should be of glass or Teflon® and have screw-top covers with Teflon® liners.

Preservation for volatile organics — No preservation is used with concentrated waste samples. With liquid samples containing no residual chlorine, 4 drops of concentrated hydrochloric acid are added and the samples are immediately cooled to 4°C. When liquid samples contain residual chlorine, they are treated as above and, in addition, 4 drops of 4% aqueous sodium thiosulfate are added to remove the residual chlorine. Soil, sediment, and sludge samples are only cooled to 4°C.

Preservation for semivolatile organics — No preservation is used with concentrated waste samples. With liquid samples containing no residual chlorine and with soil, sediment, and sludge samples, immediately cooling to 4°C is the only preservation used. When residual chlorine is present then 3 mL of 10% aqueous sodium sulfate is added for each gallon of sample collected, followed by cooling to 4°C.

Holding times — The holding time for all volatile organics samples is 14 days. Liquid samples must be extracted within 7 days and their extracts analyzed within 40 days. Concentrated waste, soil, sediment, and sludge samples must be extracted within 14 days and their extracts analyzed within 40 days.

SAMPLE PREPARATION Prepare stock standard solutions in hexane. Calibration standards at a minimum of five concentrations should be prepared through dilution of the stock standards with hexane. The suggested internal standards are: 2,5-dibromotoluene, 1,3,5-tribromobenzene, and α, α–dibromo-m-xylene. The analyst can use any of the three compounds provided that they are resolved from matrix interferences. Recommended surrogate compounds are α-2,6-trichlorotoluene, 1,4-dichloronaphthalene, and 2,3,4,5,6-pentachlorotoluene.

In general, water samples are extracted at a neutral pH with methylene chloride using a separatory funnel (EPA Method 3510) or a continuous liquid-liquid extractor (EPA Method 3520). Solid samples are extracted with hexane/acetone (1:1 v:v) using a Soxhlet extractor (EPA Method 3540) or with methylene chloride/acetone (1:1 v:v) using an ultrasonic extractor (EPA Method 3550). Non-aqueous waste samples may be diluted using EPA Method 3580. Prior to Florisil cleanup or gas chromatographic analysis, the extraction solvent must be exchanged to hexane. Sample extracts that will be subjected to gel permeation chromatography do not need solvent exchange.

Cleanup procedures may not be necessary for a relatively clean matrix. If removal of interferences such as chlorinated phenols, phthalate esters, etc., is required, proceed with the procedure outlined in EPA Method 3620.

QUALITY CONTROL Analyze a quality control check standard to demonstrate that the operation of the GC is in control. The frequency of the check standard analysis is equivalent to 10% of the samples analyzed. If the recovery of any compound found in the check standard is less than 80% of the certified value, the problem must be corrected and a new set of calibration standards must be prepared and analyzed. Calculate surrogate standard recoveries for all samples, blanks, and spikes. An internal standard peak area check must be performed on all samples. The internal standard must be evaluated for acceptance by determining whether the measured area for the internal standard deviates by more than 30% from the average area for the internal standard in the calibration standards. When the internal standard peak area is outside that limit, all samples that fall outside the QC criteria must be reanalyzed. Any compound confirmed by two columns may also be confirmed by GC/MS (EPA Method 8270). The GC/MS would normally require a minimum concentration of 1 ng/μL in the final extract for each compound. Include a mid-concentration calibration standard after each group of 20 samples in the analysis sequence. The response factors for the mid-concentration calibration must be within 15% of the average values for the multiconcentration calibration.

REFERENCE Test Methods for Evaluating Solid Waste, Physical/Chemical Methods, SW-846, 3rd Edition, U.S. EPA, Office of Solid Waste, Washington, DC, 1990. EPA Method 8121, Rev. 0, Nov. 1990.

Hexachlorobutadiene   EPA Method 8260
CAS #87-68-3

TITLE Volatile Organic Compounds by GC/MS: Capillary Column Technique

MATRIX This method is applicable to nearly all types of samples, regardless of water content, including groundwater, soils, and sediments.

METHOD SUMMARY Method 8260A covers 58 volatile organic compounds that are introduced into a gas chromatograph by the purge-and-trap method or by direct injection (in limited applications). Zero-grade helium is bubbled through a 5-mL solution at ambient temperature. Purged sample components are trapped in a tube containing suitable sorbent materials. When purging is complete, the sorbent tube is heated and backflushed with helium to desorb trapped sample components. The analytes are desorbed directly to a large bore capillary or cryofocussed on a capillary precolumn before being flash evaporated to a narrow bore capillary for analysis.

INTERFERENCES Major contaminant sources are volatile materials in the lab and impurities in the inert purging gas and in the sorbent trap. Interfering contamination may occur when a sample containing low concentrations of volatile organic compounds is analyzed immediately after a sample containing high concentrations of volatile organic compounds. After analysis of a sample containing high concentrations of volatile organic compounds, one or more calibration blanks should be analyzed to check for cross-contamination. Screening of the samples prior to purge-and-trap GC/MS analysis is highly recommended to prevent contamination of the system. This is especially true for soil and waste samples.

Special precautions must be taken to analyze for methylene chloride. The analytical and sample storage area should be isolated from all atmospheric sources of methylene chloride. All gas chromatography carrier gas lines and purge gas plumbing should be constructed from stainless steel or copper tubing. Laboratory clothing previously exposed to methylene chloride fumes during liquid-liquid extraction procedures can contribute to sample contamination.

Samples can also be contaminated by diffusion of volatile organics (particularly methylene chloride and fluorocarbons) through the septum seal during shipment and storage. A trip blank can serve as a check on such contamination.

INSTRUMENTATION GC/MS with a temperature-programmable chromatograph suitable for splitless injection equipped with variable constant differential flow controllers, a subambient oven controller, a purging device, sorbent trap, a thermal desorption apparatus and a capillary precolumn interface when using cryogenic cooling will be needed. The following GC columns may be used:

Column 1: 60 m × 0.75 mm I.D. capillary column coated with VOCOL, 1.5 μm film thickness.
Column 2: 30 m × 0.53 mm capillary column coated with DB-624 or VOCOL, 3 μm film thickness.

Column 3: 30 m × 0.32 mm I.D. capillary column coated with DB-5 or SE-54, 1-μm film thickness.

PRECISION & ACCURACY This method has been tested in a single lab using spiked water. Using a wide-bore capillary column, water was spiked at concentrations between 0.5 and 10 μg/L. Single lab accuracy and precision data are presented. The MDL actually achieved in a given analysis will vary depending on instrument sensitivity and matrix effects.

The MDL (a) in μg/L was 0.11.
The concentration range in μg/L was 0.5–10.
The mean accuracy (% of true value) was 100.
The precision as relative standard deviation was 6.8.

Note: The MDL is based on a 25-mL sample volume instead of a 5-mL sample volume.

SAMPLING METHOD
Liquid samples — Use a 40-mL glass screw-cap VOA vial with a Teflon®-faced silicone septum that has been prewashed, rinsed with distilled deionized water, and oven dried. If residual chlorine is present, collect the sample in a 4-oz soil VOA container which has been pre-preserved with 4 drops of 10% sodium thiosulfate. Mix gently and transfer the sample to a 40-mL VOA vial. Collect bubble-free samples in duplicate and seal each sample in a separate plastic bag.

Soils, sediments and sludges — Use an 8-oz widemouth glass bottle with Teflon®-faced silicone septum that has been prewashed, rinsed with distilled deionized water, and oven dried. **Do not** heat the septum for more than 1 h. Tap slightly to eliminate any free air space. Collect samples in duplicate and seal each one in a separate plastic bag.

SAMPLE PRESERVATION
Liquid samples — Add 4 drops of concentrated HCL, cool to 4°C and store in a solvent-free refrigerator.

Soils, sediments and sludges — Cool samples to 4°C and store in a solvent-free refrigerator.

MHT The maximum holding time of any sample (liquids, soils, sediments, and sludges) is 14 days.

SAMPLE PREPARATION
Liquid samples — Remove the plunger from a 5-mL syringe and carefully pour the sample into the syringe barrel to just short of overflowing. Replace the syringe plunger and compress the sample. Open the syringe valve and vent any residual air while adjusting the sample volume to 5.0 mL. If there is only one volatile organic analysis (VOA) vial, a second syringe should be filled at this time to protect against possible loss of sample integrity. Add 10 μL of surrogate spiking solution and 10 μL of internal standard spiking solution through the valve bore of the 5-mL syringe, then close the valve. The surrogate and internal standards may be mixed and added as a single spiking solution.

Sediments, soils, and waste samples — All samples of this type should be screened by GC analysis using a headspace method (EPA Method 3810) or the hexadecane extraction and screening method (EPA Method 3820). Use the screening data to determine whether to use the low-concentration method (0.005–1 mg/kg) or the high-concentration method (>1 mg/kg).

Low-concentration method — The low-concentration method is based on purging a heated sediment or soil sample mixed with organic-free reagent water containing the surrogate and internal standards. Analyze all reagent blanks and standards under the same conditions as the samples.

Use a 5-g sample if the expected concentration is <0.1 mg/kg or a 1-g sample for expected concentrations between 0.1 and 1 mg/kg. Mix the contents of the sample container with a narrow metal spatula. Weigh the amount of the sample into a tared purge device. Add the spiked water to the purge device, which contains the weighed amount of sample, and connect the device to the purge-and-trap system.

High-concentration method — This method is based on extracting the sediment or soil with methanol. A waste sample is either extracted or diluted, depending on its solubility in methanol. Wastes that are insoluble in methanol are diluted with reagent tetraglyme or possibly polyethylene glycol (PEG). An aliquot of the extract is added to organic-free reagent water containing surrogate and internal standards. This is purged at ambient temperature. All samples with an expected concentration of >1.0 mg/kg should be analyzed by this method.

Mix the contents of the sample container with a narrow metal spatula. For sediments or soils and solid wastes that are insoluble in methanol, weigh 4 g (wet weight) of sample into a tared 20-mL vial. For waste that is soluble in methanol, tetraglyme, or PEG, weigh 1 g (wet weight) into a tared scintillation vial or culture tube or a 10-mL volumetric flask. Quickly add 9.0 mL of appropriate solvent then add 1.0 mL of a surrogate spiking solution to the vial, cap it, and shake it for 2 min.

METHANOL EXTRACT REQUIRED FOR ANALYSIS OF HIGH-CONCENTRATION SOILS OR SEDIMENTS

Approximate Concentration Range	Volume of Methanol Extract (a)
500–10,000 μg/kg	100 μL
1,000–20,000 μg/kg	50 μL
5,000–100,000 μg/kg	10 μL
25,000–500,000 μg/kg	100 μL of 1/50 dilution (b)

Calculate appropriate dilution factor for concentrations exceeding this table.

(a) The volume of methanol added to 5 mL of water being purged should be kept constant. Therefore, add to the 5-mL syringe whatever volume of methanol is necessary to maintain a volume of 100 μL added to the syringe.

(b) Dilute an aliquot of the methanol extract and then take 100 μL for analysis.

QUALITY CONTROL Demonstrate, through the analysis of a reagent water blank, that interferences from the analytical system, glassware, and reagents are under control. Blank samples should be carried through all stages of the sample preparation and measurement steps. For each analytical batch (up to 20 samples), a reagent blank, matrix spike, and matrix spike duplicate must be analyzed (the frequency of the spikes may be different for different monitoring programs). The blank and

spiked samples must be carried through all stages of the sample preparation and measurement steps. QC samples mentioned in the section on Interferences will also be needed as appropriate to those situations.

Matrix spiking standards should be prepared from volatile organic compounds which will be representative of the compounds being investigated. The recommended internal standards are chlorobenzene-d5, 1,4-difluorobenzene, 1,4-dichlorobenzene-d4, and pentafluorobenzene. Using stock standard solutions, prepare secondary dilution standards containing the compounds of interest, either singly or mixed together in methanol. Store them in a vial with no headspace for no more than one week. Surrogates recommended are toluene-d8, 4-bromofluorobenzene, and dibromofluoromethane. Each sample undergoing GC/MS analysis must be spiked with 10 μL of the surrogate spiking solution prior to analysis.

REFERENCE Test Methods for Evaluating Solid Waste (SW-846). U.S. EPA 1983, Method 8260A, Rev. 1, Nov. 1990. Office of Solid Waste, Washington, DC.

Hexachlorobutadiene   **EPA Method 8270**
CAS #87-68-3

TITLE Semivolatile Organic Compounds by GC/MS

MATRIX This method is used to determine the concentration of semivolatile organic compounds in extracts prepared from all types of solid waste matrices, soils, and groundwater. Although surface waters are not specifically mentioned, this method should be applicable to water samples from rivers, lakes, etc.

METHOD SUMMARY This method covers 259 semivolatile organic compounds. In very limited applications direct injection of the sample into the GC/MS system may be appropriate, but this results in very high detection limits (approximately 10,000 μg/L). Typically, a 1-L liquid sample, containing surrogate, and matrix spiking standards, is extracted in a continuous extractor first under acid conditions and then under basic conditions. Typically 30 g of a solid sample, containing surrogate, and matrix spiking standards, is extracted ultrasonically. After concentrating the extract to 1 mL it is spiked with 10 μL of an internal standard solution just prior to analysis by GC/MS. The volume injected should contain about 100 ng of base/neutral and 200 ng of acid surrogates (for a 1-μL injection). Analysis is performed by GC/MS using a capillary GC column.

INTERFERENCES Raw GC/MS data from all blanks, samples, and spikes must be evaluated for interferences. Contamination by carryover can occur whenever high-concentration and low-concentration samples are sequentially analyzed. To reduce carryover, the sample syringe must be rinsed out between samples with solvent. Whenever an unusually concentrated sample is encountered, it should be followed by the analysis of blank solvent to check for cross-contamination.

INSTRUMENTATION A GC/MS and a data system are required. The GC column used is a 30 m × 0.25 mm I.D. (or 0.32 mm I.D.) 1um film thickness silicone-coated fused silica capillary column. A continuous liquid-liquid extractor equipped with Teflon® or glass connection joints and stopcocks requiring no lubrication, a K-D concentrating apparatus, water bath, and an ultrasonic disrupter with a minimum power of 300 W and with pulsing capability are also required.

PRECISION & ACCURACY The estimated quantitation limit (EQL) of Method 8270B for determining an individual compound is approximately 1 mg/kg (wet weight) for soil or sediment samples, 1–200 mg/kg for wastes (dependent on matrix and method of preparation), and 10 μg/L for groundwater samples. EQLs will be proportionately higher for sample extracts that require dilution to avoid saturation of the detector.

The EQL(b) for groundwater in μg/L is 10.
The EQL (a, b) for low concentrations in soil and sediment in μg/kg is 660.
Accuracy as μg/L is $0.71C–1.01$.
Overall precision in μg/L is $0.26X + 0.49$.

(a) EQLs listed for soil/sediment are based on wet weight. Normally data is reported in a dry-weight basis; therefore, EQLs will be higher based on the % dry weight of each sample. This calculation is based on a 30-g sample and gel permeation chromatography cleanup.

(b) Sample EQLs are highly matrix-dependent. The EQLs are provided for guidance and may not always be achievable.

$C =$ True value for concentration, in μg/L.
$X =$ Average recovery found for measurements of samples containing a concentration of C, in μg/L.

ESTIMATED QUANTITATION LIMIT

Other Matrices	Factor (a)
High-concentration soil and sludges by sonicator	7.5
Non-water miscible waste	75

(a) EQL for other matrices = [EQL for low soil/sediment] × [Factor]. This estimated EQL is similar to an EPA "Practical Quantitation Limit."

SAMPLING METHOD

Liquid samples — Use a 1 or 2½ gallon amber glass bottle with a screw-top Teflon®-lined cover that has been prewashed with detergent and rinsed with distilled water and methanol (or isopropanol).

Soils, sediments, or sludges — Use an 8-oz. widemouth glass with a screw-top Teflon®-lined cover that has been prewashed with detergent and rinsed with distilled water and methanol (or isopropanol).

SAMPLE PRESERVATION

Liquid samples — If residual chlorine is present, add 3 mL of 10% sodium thiosulfate per gallon, cool to 4°C and store in a solvent-free refrigerator until analysis; if chlorine is not present, then eliminate the sodium thiosulfate addition.

Soils, sediments, or sludges — Cool samples to 4°C and store in a solvent-free refrigerator.

MHT Liquid samples must be extracted within 7 days and the extracts analyzed within 40 days. Soils, sediments, or sludges may

be stored for a maximum of 14 days and the extracts analyzed within 40 days.

SAMPLE PREPARATION

Liquid samples — Transfer 1 L quantitatively to a continuous extractor. If high concentrations are anticipated, a smaller volume may be used and then diluted with organic-free reagent water to 1 L. Adjust pH, if necessary, to pH <2 using 1:1 (V/V) sulfuric acid. Pipette 1.0 mL of a surrogate standard spiking solution into each sample. For the sample in each analytical batch selected for spiking, add 1.0 mL of a matrix spiking standard. For base/neutral acid analysis, the amount of the surrogates and matrix spiking compounds added to the sample should result in a final concentration of 100 ng/µL of each analyte in the extract to be analyzed (assuming a 1-µL injection). Extract with methylene chloride for 18–24 h. Next, adjust the pH of the aqueous phase to pH >11 using 10 N sodium hydroxide and extract it with methylene chloride again for 18–24 h. Dry the extract through a column containing anhydrous sodium sulfate and concentrate it to 1 mL using a K-D concentrator.

Soils, sediments, or sludges — Use 30 g of sample. Nonporous or wet samples (gummy or clay type) that do not have a free-flowing sandy texture must be mixed with anhydrous sodium sulfate until the sample is free flowing. Add 1 mL of surrogate standards to all samples, spikes, standards, and blanks. For the sample in each analytical batch selected for spiking, add 1.0 mL of a matrix spiking standard. For base/neutral acid analysis, the amount added of the surrogates and matrix spiking compounds should result in a final concentration of 100 ng/µL of each base/neutral analyte and 200 ng/µL of each acid analyte in the extract to be analyzed (assuming a 1-µL injection). Immediately add a 100-mL mixture of 1:1 methylene chloride:acetone and extract the sample ultrasonically for 3 min and then decant or filter the extracts. Repeat the extraction two or more times. Dry the extract using a column with anhydrous sodium sulfate and concentrate it to 1 mL in a K-D concentrator.

QUALITY CONTROL A methylene chloride solution containing 50 ng/µL of decafluorotriphenylphosphine (DFTPP) is used for tuning the GC/MS system each 12-h shift. A system performance check also must be made during every 12-h shift. A standard containing 50 ng/µL each of 4,4'-DDT, pentachlorophenol, and benzidine is required to verify injection port inertness and GC column performance. A calibration standard at mid-concentration, containing each compound of interest, including all required surrogates, must be performed every 12 h during analysis. After the system performance check is met, calibration check compounds (CCCs) are used to check the validity of the initial calibration.

The internal standard responses and retention times in the calibration check standard must be evaluated immediately after or during data acquisition. If the retention time for any internal standard changes by more than 30 seconds from the last check calibration (12 h), the chromatographic system must be inspected for malfunctions and corrections must be made, as required. If the electron ionization current plot (EICP) area for any of the internal standards changes by a factor of two from the last daily calibration standard check, the mass spectrometer must be inspected for malfunctions and corrections must be made, as appropriate.

Demonstrate, through the analysis of a reagent water blank, that interferences from the analytical system, glassware, and reagents are under control. The blank samples should be carried through all stages of the sample preparation and measurement steps. For each analytical batch (up to 20 samples), a reagent blank, matrix spike, and matrix spike duplicate/duplicate must be analyzed (the frequency of the spikes may be different for different monitoring programs). The blank and spiked samples must be carried through all stages of the sample preparation and measurement steps. A QC reference sample concentrate containing each analyte at a concentration of 100 mg/L in methanol is required.

REFERENCE Test Methods for Evaluating Solid Waste (SW-846). U.S. EPA 1983, Method 8270B, Rev. 2, Nov. 1990. Office of Solid Waste, Washington, DC.

Hexachlorobutadiene   **EPA Method 503.1**
CAS #87-68-3

TITLE Aromatic & Unsaturated VOCs

MATRIX Drinking water (finished or in Water any treatment stage) and raw source water.

APPLICATION Method covers 28 aromatic and unsaturated VOCs. An inert gas is bubbled through a 5-mL water sample. Purged sample components are trapped in tube of sorbent materials. When purging is complete, sorbent tube is heated and backflushed with inert gas to desorb trapped sample onto a packed GC column.

INTERFERENCES During analysis, major contaminant sources are volatile materials in the lab and impurities in purging gas and sorbent trap. With high and low level samples, there can be carryover contamination. Excess water causes a negative baseline deflection.

INSTRUMENTATION Purge and Trap GC w/photoionization detector. (Two GC columns are recommended); Column 1: 5% SP-1200 and 1.75% Bentone 34 on Supelcoport; 5% 1,2,3-tris(2-cyanoethoxy)propane on Chromosorb W.

RANGE 2.2–600 µg/L (Drinking water)

MDL 0.02 µg/L in water

PRECISION RSD = 16.8% at 0.50 µg/L; 10 samples

ACCURACY Average recovery = 74% at 0.50 µg/L; 10 samples

SAMPLING METHOD Use a 40–120-mL screw-cap vial (prewashed with detergent, rinsed with distilled water and oven dried at 105°C) with a PTFE-faced silicone septum. If residual chlorine is in the water add about 25 mg of ascorbic acid to each vial before sample collection. Collect bubble-free samples.

STABILITY Cool to 4°C; HCl to pH <2.

MHT 14 days.

QUALITY CONTROL As an initial demonstration of lab accuracy and precision, analyze 4 to 7 replicates of a lab fortified blank containing analyte at 0.1–5 µg/L. Collect all samples in duplicate.

REFERENCE Method 503.1, Volatile Aromatic & Unsaturated Organic Compounds in H2O by Purge and Trap GC, EPA 600/4-88/039.

Hexachlorocyclohexane EPA Method 8120
CAS #608-73-1

TITLE Chlorinated Hydrocarbons by Gas Chromatography

MATRIX This method covers aqueous and solid matrices. This includes a wide variety such as drinking water, groundwater, industrial wastewaters, surface waters, soils, solids, and sediments.

METHOD SUMMARY This method is used to determine the concentration of 14 chlorinated hydrocarbons. It provides gas chromatographic conditions for the detection of ppb concentrations of certain chlorinated hydrocarbons. Prior to use of this method, appropriate sample extraction techniques must be used. Both neat and diluted organic liquids (EPA Method 3580, Waste Dilution) may be analyzed by direct injection. A 2 to 5 µg/mL aliquot of the extract is injected into a gas chromatograph (GC) using the solvent flush technique, and compounds in the GC effluent are detected by an electron capture detector (ECD).

INTERFERENCES Solvents, reagents, glassware, and other sample processing hardware may yield discrete artifacts and/or elevated baselines causing misinterpretation of gas chromatograms. Interferences coextracted from samples will vary considerably from source to source, depending upon the waste being sampled.

INSTRUMENTATION An analytical system complete with GC suitable for on-column injections and accessories, including detectors, column supplies, recorder, gases and syringes is required. A data system for measuring peak areas and/or peak heights is recommended. The GC is equipped with an electron capture detector (ECD). A K-D apparatus is needed for sample preparation.

Column 1: 1.8 m × 2 mm I.D. glass column packed with 1% SP-1000 on Supelcoport (100/120 mesh) or equivalent.

Column 2: 1.8 m × 2 mm I.D. glass column packed with 1.5% OV-1/2.4% OV-225 on Supelcoport (80/100 mesh) or equivalent.

PRECISION & ACCURACY The method was tested by 20 laboratories using organic-free reagent water, drinking water, surface water, and three industrial wastewaters spiked at six concentrations over the range 1.0 to 356 µg/L. Single operator precision, overall precision, and method accuracy were found to be directly related to the concentration of the parameter and essentially independent of the sample matrix.

MULTIPLICATION FACTORS FOR OTHER MATRICES (a)

Matrix	Factor (b)
Groundwater	10
Low-concentration soil by ultrasonic cleanup extraction with GPC	670
High-concentration soil and sludges by ultrasonic extraction	10,000
Non-water miscible waste	100,000

(a) Sample EQLs are highly matrix-dependent. The EQLs listed are provided for guidance and may not always be achievable.
(b) EQL = [Method detection limit] × [Factor]. For non-aqueous samples, the factor is on a wet-weight basis.

PRECISION & ACCURACY The estimates below are based upon the performance in a single lab.

The accuracy (in µg/L) as expected recovery for one or more measurements of a sample containing a concentration of C was No Data.

The precision (in µg/L) as expected single analyst standard deviation of measurements at an average concentration of x" was No Data.

The precision (in µg/L) as expected interlaboratory standard deviation measurements at an average concentration found of x" was No Data.

C = True value for the concentration, in µg/L.
$x"$ = Average recovery found for measurements of samples containing a concentration of C, in µg/L.

SAMPLE COLLECTION, PRESERVATION & HANDLING Extracts must be stored under refrigeration at 4°C and analyzed within 40 days of extraction.

SAMPLE PREPARATION In general, water samples are extracted at a neutral, or as is, pH with methylene chloride using either EPA Method 3510 or EPA Method 3520. Solid samples are extracted using either EPA Method 3540 or EPA Method 3550. Prior to gas chromatographic analysis, the extraction solvent must be exchanged to hexane.

QUALITY CONTROL The quality control check concentrate (EPA Method 8000) should contain each parameter of interest in acetone at the following concentrations: hexachloro-substituted hydrocarbon, 10 µg/mL; and any other chlorinated hydrocarbon, 100 µg/mL. Calculate surrogate standard recovery on all samples, blanks, and spikes.

Prepare stock standard solutions in isooctane or hexane. Calibration standards at a minimum of five concentrations should be prepared through dilution of the stock standards with isooctane or hexane. Internal standards and surrogate standards are also needed.

REFERENCE Test Methods for Evaluating Solid Waste, Physical/Chemical Methods, SW-846, 3rd Edition, U.S. EPA, Office of Solid Waste, Washington, DC, 1990. EPA Method 8120 A Rev. 1, Nov. 1990.

Hexachlorocyclopentadiene
CAS #77-47-4
EPA Method 1625

TITLE Semivolatile Organic Compounds by Isotope Dilution GC/MS

MATRIX The compounds may be determined in waters, soils, and municipal sludges by this method.

METHOD SUMMARY This method is used to determine 176 semivolatile toxic organic pollutants associated with the CWA (as amended 1987); the RCRA (as amended 1986); the CERCLA (as amended 1986); and other compounds amenable to extraction and analysis by capillary column gas chromatography-mass spectrometry (GC/MS).

Stable isotopically-labeled analogs of the compounds of interest are added to the sample. If the solids content is less than 1%, a 1-L sample is extracted at pH 12–13, then at pH <2 with methylene chloride using continuous extraction techniques.

If the solids content is 30% or less, the sample is diluted to 1% solids with reagent water, homogenized ultrasonically, and extracted at pH 12–13, then at pH <2 with methylene chloride using continuous extraction techniques. If the solids content is greater than 30%, the sample is extracted using ultrasonic techniques.

Each extract is dried over sodium sulfate, concentrated to a volume of 5 mL, cleaned up using GPC, if necessary, and concentrated. Extracts are concentrated to 1 mL if GPC is not performed, and to 0.5 mL if GPC is performed.

An internal standard is added to the extract, and a 1-mL aliquot of the extract is injected into the GC. The compounds are separated by GC and detected by a MS. The labeled compounds serve to correct the variability of the analytical technique.

INTERFERENCES Solvents, reagents, glassware, and other sample processing hardware may yield artifacts and/or elevated baselines causing misinterpretation of chromatograms and spectra. Materials used in the analysis must be demonstrated to be free from interferences under the conditions of analysis by running method blanks initially and with each sample lot (sample started through the extraction process on a given 8-h shift, to a maximum of 20). Specific selection of reagents and purification of solvents by distillation in all glass systems may be required. Glassware and, where possible, reagents are cleaned by solvent rinse and baking at 450°C for 1-h minimum. Interferences coextracted from samples will vary considerably from source to source, depending on the diversity of the site being sampled.

INSTRUMENTATION Major instrumentation includes a GC with a splitless or on-column injection port for capillary column, a MS with 70 eV electron impact ionization, and a data system to collect and record MS data, and process it. A K-D apparatus is used to concentrate extracts.

GC Column: 30 m × 0.25 mm I.D. 5% phenyl, 94% methyl, 1% vinyl silicone bonded phased fused silica capillary column.

PRECISION & ACCURACY The detection limits of the method are usually dependent on the level of interferences rather than instrumental limitations. The limits typify the minimum quantities that can be detected with no interferences present.

The minimum level (in µg/mL) was 10. This is defined as a minimum level at which the analytical system shall give recognizable mass spectra (background corrected) and acceptable calibration points.

The MDL (in µg/kg) in low solids was not detected and in high solids was not detected; these were determined in digested sludge (low solids) and in filter cake or compost (high solids).

The labeled and native compound initial precision as standard deviation (in µg/L) was 15.

The labeled and native compound initial accuracy as average recovery (in µg/L) was 69–144.

SAMPLE COLLECTION, PRESERVATION & HANDLING Collect samples in glass containers. Aqueous samples which flow freely are collected in refrigerated bottles using automatic sampling equipment. Solid samples are collected as grab samples using widemouth jars. Maintain samples at 0 to 4°C from the time of collection until extraction. If residual chlorine is present in aqueous samples, add 80 mg sodium thiosulfate/L of water. Begin sample extraction within 7 days of collection, and analyze all extracts within 40 days of extraction.

SAMPLE PREPARATION Samples containing 1% solids or less are extracted directly using continuous liquid-liquid extraction techniques. Samples containing 1 to 30% solids are diluted to the 1% level with reagent water and extracted using continuous liquid-liquid extraction techniques. Samples containing greater than 30% solids are extracted using ultrasonic techniques.

Base/neutral extraction — Adjust the pH of the waters in the extractors to 12–13 with 6 N NaOH. Extract with methylene chloride for 24–48 h.

Acid extraction — Adjust the pH of the waters in the extractors to 2 or less using 6 N sulfuric acid. Extract with methylene chloride for 24–48 h.

Ultrasonic extraction of high solids samples — Add anhydrous sodium sulfate to the sample and QC aliquot(s). Add acetone:methylene chloride (1:1) to the sample and mix thoroughly

Concentrate extracts using a K-D apparatus.

QUALITY CONTROL The analyst is permitted to modify this method to improve separations or lower the costs of measurements, provided all performance specifications are met. Analyses of blanks are required to demonstrate freedom from contamination. When results of spikes indicate atypical method performance for samples, the samples are diluted to bring method performance within acceptable limits.

For low solids (aqueous samples), extract, concentrate, and analyze two sets of four 1-L aliquots (8 aliquots total) of the precision and recovery standard. For high solids samples, two sets of four 30-g aliquots of the high solids reference matrix are used.

Spike all samples with labeled compounds to assess method performance. Compute percent recovery of the labeled compounds using the internal standard method. Compare the labeled compound recovery for each compound with the corresponding labeled compound recovery.

Reagent water and high solids reference matrix blanks are analyzed to demonstrate freedom from contamination. Extract and concentrate a 1-L reagent water blank or a high solids reference matrix blank with each sample's lot (samples started through the extraction process on the same 8-h shift, to a maximum of 20 samples).

Field replicates may be collected to determine the precision of the sampling technique, and spiked samples may be required to determine the accuracy of the analysis when the internal standard method is used.

REFERENCE Semivolatile Organic Compounds by Isotope Dilution GC/MS. Office of Water Regulation and Standards, U.S. EPA Industrial Technology Division, Washington, DC, EPA Method 1625, Rev. C, June 1989 (contact W.A. Telliard, U.S. EPA, Office of Water Regulations and Standards, 401 M St., SW, Washington, DC, 20460. Phone: 202-382-7131).

Hexachlorocyclopentadiene   **EPA Method 505**
CAS #77-47-4

TITLE Analysis of Organohalide Pesticides and Commercial Polychlorinated Biphenyl (PCB) Products in Water by Microextraction and Gas Chromatography. U.S. EPA Method 505, Rev. 2.0, 1989.

MATRIX This method is applicable to drinking water and raw source water. The latter should include most surface water and groundwater sources.

METHOD SUMMARY Method 505 covers 25 pesticides and commercial PCB products. This is a very sensitive method that is more useful for monitoring than for exploratory analyses. 5-mL of water are saturated with sodium chloride and then extracted by shaking with 2 mL of hexane. The sample extracts are transferred to an autosampler setup to inject 1–2 µL portions into a gas chromatograph (GC) for analysis. Alternatively, 1–2 µL portions of samples, blanks, and standards may be manually injected. Each extract is analyzed by capillary GC/ECD with confirmation using either a second capillary column or GC/MS. The electron capture detector is easy to use, but it is a nonselective detector. The microextraction technique also eliminates the expensive sample preparation costs of other methods, but it has the disadvantage of being less sensitive than most because the extracts are not concentrated.

INTERFERENCES Method interferences may be caused by contaminants in solvents, reagents, glassware, and other sample processing apparatus that lead to discrete artifacts or elevated baselines. Interfering contamination may occur when a sample containing low concentrations of analytes is analyzed immediately following a sample containing relatively high concentrations of the analytes. Matrix interferences also may be caused by contaminants that are coextracted from the sample; cleanup of sample extracts may be necessary in these cases. Some pesticides and commercial PCB products from aqueous solutions adhere to glass surfaces, so sample transfers and contact with glass surfaces should be minimized. Some pesticides are rapidly oxidized by chlorine so dechlorination with sodium thiosulfate at the time of sample collection is important. Also, splitless injectors may cause degradation of some pesticides.

INSTRUMENTATION A gas chromatograph/electron capture detector/data system, with temperature programming and split/splitless injector suitable for use with capillary columns is needed.

Column 1: 0.32 mm I.D. × 30 m fused silica capillary with chemically bond methyl polysiloxane phase (DB-1, 1.0 µm film, or equivalent).
Column 2: 0.32 mm I.D. × 30 m fused silica capillary with 1:1 mixed phase of dimethyl silicone and polyethylene glycol (Durawax-DX3, 0.25 µm film, or equivalent).
Column 3: 0.32 mm I.D. × 25 m fused silica capillary with chemically bonded 50:50 methyl-phenyl silicone (OV-17, 1.5 µm film, or equivalent).

Column 1 should be used as the primary analytical column. Columns 2 and 3 are recommended for use as confirmatory columns when GC/MS confirmation is not available.

PRECISION & ACCURACY Method detection limits are dependent upon the characteristics of the gas chromatographic system used. Analytes that are not separated chromatographically cannot be individually identified and used in the same calibration mixture or water samples unless an alternative technique for identification and quantification, such as mass spectrometry, is used.

The concentration(s) (in µg/L) used for these QC measurements was 0.15 and 0.35.
The MDL (in µg/L) was 0.13.
The accuracy (% recovery) for reagent water at the above concentration(s) was 73 and 73 and the precision (%) was 5.1 and 11.7.
The accuracy (% recovery) for groundwater at the above concentration(s) was 87 and 69 and the precision (%) was 5.1 and 4.8.
The accuracy (% recovery) for tap water at the above concentration(s) was 191 and 109 and the precision (5) was 18.5 and 14.3.

Note: No range of concentrations is provided with this method.

SAMPLING METHOD Collect samples using a 40-mL screw-cap vial (prewashed with detergent, rinsed with distilled water and oven dried at 400°C for one h) with a Teflon®-faced silicone septum. Collect bubble-free samples and place the septum with the Teflon® side down on the water.

SAMPLE PRESERVATION If residual chlorine is present in the water add about 3 mg of sodium thiosulfate to each vial before samples are collected to remove the chlorine. Alternatively, add 75 µL of 0.04 g/mL solution of sodium thiosulfate to each vial just prior to sampling. Immediately cool samples to 4°C, and store them in a solvent-free refrigerator at 4°C until analysis.

MHT The maximum holding time is 14 days from the time the sample was collected until it must be analyzed.

SAMPLE PREPARATION Remove the sample from storage and allow it to come to room temperature. Remove a 5-mL volume from each container and weigh the container to the nearest 0.1 g. Add 6 g of sodium chloride and 2.0 mL of hexane to each sample bottle. Recap the sample and shake it vigorously for one min. Allow the water and hexane phases to separate, remove the cap, and transfer 0.5 mL of hexane into an autosampler vial using a disposable glass pipette. Transfer the remaining hexane phase into a second autosampler vial and store at 4°C for reanalysis, if necessary. Discard the remaining sample/hexane mixture and reweigh the empty container to determine net weight of sample.

QUALITY CONTROL Minimum quality control requirements are initial demonstration of lab capability, analysis of lab reagent blanks, fortified blanks, fortified sample matrix, and quality control samples. The lab must analyze at least one fortified blank per sample set, or at least one for every 20 samples. The fortifying concentration of each analyte should be 10 times the method detection limit or the maximum calibration limit (MCL), whichever is less. Calculate accuracy as percent recovery and develop control limits from the mean percent recovery and standard deviation.

The lab must add a known concentration of the analytes to a minimum of 10% of the routine samples, or one lab fortified sample matrix per sample set. Calculate the percent recovery for each analyte and compare to the control limits established from the analyses of the fortified blanks.

EPA CONTACT & HOTLINE For technical questions contact Dr. Baldev Bathija, U.S. EPA, Office of Ground Water and Drinking Water (WH-550D), 401 M St. SW, Washington, DC 20460. Tel. (202) 260-3040. For further information the EPA Safe Drinking Water Hotline may be called at: (800) 426-4791.

REFERENCE Methods for the Determination of Organic Compounds in Drinking Water, EPA/600/4-88/039 (revised July 1991). U.S. EPA Environmental Monitoring Systems Laboratory, Cincinnati, OH, 45268, U.S.A. Available from the National Technical Information Service (NTIS), 5285 Port Royal Road, Springfield, VA 22161; Tel. 800-553-6847. NTIS Order Number is PB91-231480.

Hexachlorocyclopentadiene   **EPA Method 8120**
CAS #77-47-4

TITLE Chlorinated Hydrocarbons by Gas Chromatography

MATRIX This method covers aqueous and solid matrices. This includes a wide variety such as drinking water, groundwater, industrial wastewaters, surface waters, soils, solids, and sediments.

METHOD SUMMARY This method is used to determine the concentration of 14 chlorinated hydrocarbons. It provides gas chromatographic conditions for the detection of ppb concentrations of certain chlorinated hydrocarbons. Prior to use of this method, appropriate sample extraction techniques must be used. Both neat and diluted organic liquids (EPA Method 3580, Waste Dilution) may be analyzed by direct injection. A 2 to 5 µg/mL aliquot of the extract is injected into a gas chromatograph (GC) using the solvent flush technique, and compounds in the GC effluent are detected by an electron capture detector (ECD).

INTERFERENCES Solvents, reagents, glassware, and other sample processing hardware may yield discrete artifacts and/or elevated baselines causing misinterpretation of gas chromatograms. Interferences coextracted from samples will vary considerably from source to source, depending upon the waste being sampled.

INSTRUMENTATION An analytical system complete with GC suitable for on-column injections and accessories, including detectors, column supplies, recorder, gases and syringes is required. A data system for measuring peak areas and/or peak heights is recommended. The GC is equipped with an electron capture detector (ECD). A K-D apparatus is needed for sample preparation.

Column 1: 1.8 m × 2 mm I.D. glass column packed with 1% SP-1000 on Supelcoport (100/120 mesh) or equivalent.

Column 2: 1.8 m × 2 mm I.D. glass column packed with 1.5% OV-1/2.4% OV-225 on Supelcoport (80/100 mesh) or equivalent.

PRECISION & ACCURACY The method was tested by 20 laboratories using organic-free reagent water, drinking water, surface water, and three industrial wastewaters spiked at six concentrations over the range 1.0 to 356 µg/L. Single operator precision, overall precision, and method accuracy were found to be directly related to the concentration of the parameter and essentially independent of the sample matrix.

MULTIPLICATION FACTORS FOR OTHER MATRICES (a)

Matrix	Factor (b)
Groundwater	10
Low-concentration soil by ultrasonic cleanup extraction with GPC	670
High-concentration soil and sludges by ultrasonic extraction	10,000
Non-water miscible waste	100,000

(a) Sample EQLs are highly matrix-dependent. The EQLs listed are provided for guidance and may not always be achievable.
(b) EQL = [Method detection limit] × [Factor]. For non-aqueous samples, the factor is on a wet-weight basis.

PRECISION & ACCURACY The estimates below are based upon the performance in a single lab.

The accuracy (in µg/L) as expected recovery for one or more measurements of a sample containing a concentration of C was 0.47C.

The precision (in µg/L) as expected single analyst standard deviation of measurements at an average concentration of x" was 0.24x".

The precision (in µg/L) as expected interlaboratory standard deviation measurements at an average concentration found of x" was 0.50x".

C = True value for the concentration, in µg/L.
x'' = Average recovery found for measurements of samples containing a concentration of C, in µg/L.

SAMPLE COLLECTION, PRESERVATION & HANDLING
Extracts must be stored under refrigeration at 4°C and analyzed within 40 days of extraction.

SAMPLE PREPARATION In general, water samples are extracted at a neutral, or as is, pH with methylene chloride using either EPA Method 3510 or EPA Method 3520. Solid samples are extracted using either EPA Method 3540 or EPA Method 3550. Prior to gas chromatographic analysis, the extraction solvent must be exchanged to hexane.

QUALITY CONTROL The quality control check concentrate (EPA Method 8000) should contain each parameter of interest in acetone at the following concentrations: hexachloro-substituted hydrocarbon, 10 µg/mL; and any other chlorinated hydrocarbon, 100 µg/mL. Calculate surrogate standard recovery on all samples, blanks, and spikes.

Prepare stock standard solutions in isooctane or hexane. Calibration standards at a minimum of five concentrations should be prepared through dilution of the stock standards with isooctane or hexane. Internal standards and surrogate standards are also needed.

REFERENCE Test Methods for Evaluating Solid Waste, Physical/Chemical Methods, SW-846, 3rd Edition, U.S. EPA, Office of Solid Waste, Washington, DC, 1990. EPA Method 8120 A Rev. 1, Nov. 1990.

Hexachlorocyclopentadiene   **EPA Method 8121**
CAS #77-47-4

TITLE Chlorinated Hydrocarbons by GC: Capillary Column Technique

MATRIX This method covers aqueous and solid matrices. This includes a wide variety such as drinking water, groundwater, industrial wastewaters, surface waters, soils, solids, and sediments.

METHOD SUMMARY This method provides procedures for the determination of 22 chlorinated hydrocarbons in water, soil/sediment, and waste matrices. A measured volume or weight of sample is extracted by using one of the appropriate sample extraction techniques specified in EPA Method 3510, EPA Method 3520, EPA Method 3540, or EPA Method 3550, or diluted using EPA Method 3580. Aqueous samples are extracted at neutral pH with methylene chloride by using either a separatory funnel (EPA Method 3510) or a continuous liquid-liquid extractor (EPA Method 3520). Solid samples are extracted with hexane/acetone (1:1) by using a Soxhlet extractor (EPA Method 3540) or with methylene chloride/acetone (1:1) by using an ultrasonic extractor (EPA Method 3550). After cleanup, the extract or diluted sample is analyzed by gas chromatography with electron capture detection (GC/ECD).

The sensitivity level of this method usually depends on the level of interferences rather than on instrumental limitations. This method may be used in conjunction with EPA Method 3620, Florisil Column Cleanup, EPA Method 3660, Sulfur Cleanup, and EPA Method 3640, Gel Permeation Chromatography, to aid in the elimination of interferences.

INTERFERENCES Solvents, reagents, glassware, and other hardware used in sample processing may introduce artifacts which may result in elevated baselines, causing misinterpretation of gas chromatograms. Interferants coextracted from the samples will vary considerably from waste to waste. Glassware must be scrupulously clean. Phthalate esters, if present in a sample, will interfere only with the BHC isomers. The presence of elemental sulfur will result in large peaks, and can often mask the region of compounds eluting after 1,2,4,5-tetrachlorobenzene. The tetrabutylammonium (TBA)-sulfite procedure (EPA Method 3660) works well for the removal of elemental sulfur. Waxes and lipids can be removed by gel permeation chromatography (EPA Method 3640).

INSTRUMENTATION A GC suitable for on-column injections and all required accessories, including and electron capture detector (ECD), analytical columns, recorder, gases, and syringes are needed. A data system for measuring peak heights and/or peak areas is recommended. A Kuderna-Danish (K-D) apparatus will also be needed to concentrate extracts.

Column 1: 30 m × 0.53 mm I.D. fused-silica capillary column chemically bonded with trifluoropropyl methyl silicone (DB-210 or equivalent).

Column 2: 30 m × 0.53 mm I.D. fused-silica capillary column chemically bonded with polyethylene glycol (DB-WAX or equivalent).

PRECISION & ACCURACY This method has been tested in a single lab by using organic-free reagent water, sandy loam samples, and extracts which were spiked with the test compounds at one concentration. Single-operator precision and method accuracy were found to be related to the concentration of compound and the type of matrix. The accuracy and precision technique will be determined by the sample matrix, sample preparation technique, optional cleanup techniques, and calibration procedures used.

MULTIPLICATION FACTORS FOR OTHER MATRICES (a)

Matrix	Factor (b)
Groundwater	10
Low-concentration soil by ultrasonic cleanup extraction with GPC	670
High-concentration soil and sludges by ultrasonic extraction	10,000
Non-water miscible waste	100,000

(a) Sample EQLs are highly matrix-dependent. The EQLs listed are provided for guidance and may not always be achievable.
(b) EQL = [Method detection limit] × [Factor]. For non-aqueous samples, the factor is on a wet-weight basis.

PRECISION & ACCURACY MDL is the method detection limit for organic-free reagent water. MDL was determined from the analysis of eight replicate aliquots processed through the entire analytical method (extraction, Florisil cartridge cleanup, and GC/ECD analysis).

The MDL (in ng/L) was 240.

The accuracy (as average % recovery using 5 determinations and no Florisil cleanup) from a spike concentration of 10 µg/L and separatory funnel extraction was 97% with a final volume of 10 mL.

The precision (as RSD% using 5 determinations and no Florisil cleanup) from a spike concentration of 10 µg/L and separatory funnel extraction was 5.1% with a final volume of 10 mL.

The accuracy (as average % recovery using 5 determinations and no Florisil cleanup), from a spike concentration of 330 µg/L and ultrasonic extraction of solid samples using 1:1 methylene chloride and acetone, was 44% with a final volume of 10 mL.

The precision (as RSD% using 5 determinations and no Florisil cleanup), from a spike concentration of 330 µg/L and ultrasonic extraction of solid samples using 1:1 methylene chloride and acetone, was 25.9% with a final volume of 10 mL.

SAMPLE COLLECTION, PRESERVATION & HANDLING

Volatile Organics — Standard 40-mL glass screw-cap VOA vials with Teflon®-faced silicone septum may be used for both liquid and solid matrices. When collecting samples, liquids and solids should be introduced into the vials gently to reduce agitation which might drive off volatile compounds. If there are any air bubbles present the sample must be retaken. The vials with solids should be tapped slightly as they are filled to try and eliminate as much free air space as possible. Two vials from each sampling location should be sealed in separate plastic bags to prevent cross-contamination between samples.

Semivolatile organics — Containers used to collect samples for the determination of semivolatile organic compounds should be soap and water washed followed by methanol (or isopropanol) rinsing. The sample containers should be of glass or Teflon® and have screw-top covers with Teflon® liners.

Preservation for volatile organics — No preservation is used with concentrated waste samples. With liquid samples containing no residual chlorine, 4 drops of concentrated hydrochloric acid are added and the samples are immediately cooled to 4°C. When liquid samples contain residual chlorine, they are treated as above and, in addition, 4 drops of 4% aqueous sodium thiosulfate are added to remove the residual chlorine. Soil, sediment, and sludge samples are only cooled to 4°C.

Preservation for semivolatile organics — No preservation is used with concentrated waste samples. With liquid samples containing no residual chlorine and with soil, sediment, and sludge samples, immediately cooling to 4°C is the only preservation used. When residual chlorine is present then 3 mL of 10% aqueous sodium sulfate is added for each gallon of sample collected, followed by cooling to 4°C.

Holding times — The holding time for all volatile organics samples is 14 days. Liquid samples must be extracted within 7 days and their extracts analyzed within 40 days. Concentrated waste, soil, sediment, and sludge samples must be extracted within 14 days and their extracts analyzed within 40 days.

SAMPLE PREPARATION
Prepare stock standard solutions in hexane. Calibration standards at a minimum of five concentrations should be prepared through dilution of the stock standards with hexane. The suggested internal standards are: 2,5-dibromotoluene, 1,3,5-tribromobenzene, and α,α–dibromo-m-xylene. The analyst can use any of the three compounds provided that they are resolved from matrix interferences. Recommended surrogate compounds are α-2,6-trichlorotoluene, 1,4-dichloronaphthalene, and 2,3,4,5,6-pentachlorotoluene.

In general, water samples are extracted at a neutral pH with methylene chloride using a separatory funnel (EPA Method 3510) or a continuous liquid-liquid extractor (EPA Method 3520). Solid samples are extracted with hexane/acetone (1:1 v:v) using a Soxhlet extractor (EPA Method 3540) or with methylene chloride/acetone (1:1 v:v) using an ultrasonic extractor (EPA Method 3550). Non-aqueous waste samples may be diluted using EPA Method 3580. Prior to Florisil cleanup or gas chromatographic analysis, the extraction solvent must be exchanged to hexane. Sample extracts that will be subjected to gel permeation chromatography do not need solvent exchange.

Cleanup procedures may not be necessary for a relatively clean matrix. If removal of interferences such as chlorinated phenols, phthalate esters, etc., is required, proceed with the procedure outlined in EPA Method 3620.

QUALITY CONTROL
Analyze a quality control check standard to demonstrate that the operation of the GC is in control. The frequency of the check standard analysis is equivalent to 10% of the samples analyzed. If the recovery of any compound found in the check standard is less than 80% of the certified value, the problem must be corrected and a new set of calibration standards must be prepared and analyzed. Calculate surrogate standard recoveries for all samples, blanks, and spikes. An internal standard peak area check must be performed on all samples. The internal standard must be evaluated for acceptance by determining whether the measured area for the internal standard deviates by more than 30% from the average area for the internal standard in the calibration standards. When the internal standard peak area is outside that limit, all samples that fall outside the QC criteria must be reanalyzed. Any compound confirmed by two columns may also be confirmed by GC/MS (EPA Method 8270). The GC/MS would normally require a minimum concentration of 1 ng/µL in the final extract for each compound. Include a mid-concentration calibration standard after each group of 20 samples in the analysis sequence. The response factors for the mid-concentration calibration must be within 15% of the average values for the multiconcentration calibration.

REFERENCE
Test Methods for Evaluating Solid Waste, Physical/Chemical Methods, SW-846, 3rd Edition, U.S. EPA, Office of Solid Waste, Washington, DC, 1990. EPA Method 8121, Rev. 0, Nov. 1990.

Hexachlorocyclopentadiene   **EPA Method 8270**
CAS #77-47-4

TITLE
Semivolatile Organic Compounds by GC/MS

MATRIX This method is used to determine the concentration of semivolatile organic compounds in extracts prepared from all types of solid waste matrices, soils, and groundwater. Although surface waters are not specifically mentioned, this method should be applicable to water samples from rivers, lakes, etc.

METHOD SUMMARY This method covers 259 semivolatile organic compounds. In very limited applications direct injection of the sample into the GC/MS system may be appropriate, but this results in very high detection limits (approximately 10,000 µg/L). Typically, a 1-L liquid sample, containing surrogate, and matrix spiking standards, is extracted in a continuous extractor first under acid conditions and then under basic conditions. Typically 30 g of a solid sample, containing surrogate, and matrix spiking standards, is extracted ultrasonically. After concentrating the extract to 1 mL it is spiked with 10 µL of an internal standard solution just prior to analysis by GC/MS. The volume injected should contain about 100 ng of base/neutral and 200 ng of acid surrogates (for a 1-µL injection). Analysis is performed by GC/MS using a capillary GC column.

INTERFERENCES Raw GC/MS data from all blanks, samples, and spikes must be evaluated for interferences. Contamination by carryover can occur whenever high-concentration and low-concentration samples are sequentially analyzed. To reduce carryover, the sample syringe must be rinsed out between samples with solvent. Whenever an unusually concentrated sample is encountered, it should be followed by the analysis of blank solvent to check for cross-contamination.

INSTRUMENTATION A GC/MS and a data system are required. The GC column used is a 30 m × 0.25 mm I.D. (or 0.32 mm I.D.) 1um film thickness silicone-coated fused silica capillary column. A continuous liquid-liquid extractor equipped with Teflon® or glass connection joints and stopcocks requiring no lubrication, a K-D concentrating apparatus, water bath, and an ultrasonic disrupter with a minimum power of 300 W and with pulsing capability are also required.

PRECISION & ACCURACY The estimated quantitation limit (EQL) of Method 8270B for determining an individual compound is approximately 1 mg/kg (wet weight) for soil or sediment samples, 1–200 mg/kg for wastes (dependent on matrix and method of preparation), and 10 µg/L for groundwater samples. EQLs will be proportionately higher for sample extracts that require dilution to avoid saturation of the detector.

The EQL(b) for groundwater in µg/L is 10.
The EQL (a, b) for low concentrations in soil and sediment in µg/kg is 660.
Accuracy as µg/L is not listed.
Overall precision in µg/L is not listed.

(a) EQLs listed for soil/sediment are based on wet weight. Normally data is reported in a dry-weight basis; therefore, EQLs will be higher based on the % dry weight of each sample. This calculation is based on a 30-g sample and gel permeation chromatography cleanup.
(b) Sample EQLs are highly matrix-dependent. The EQLs are provided for guidance and may not always be achievable.

$C =$ True value for concentration, in µg/L.

$X =$ Average recovery found for measurements of samples containing a concentration of C, in µg/L.

ESTIMATED QUANTITATION LIMIT

Other Matrices	Factor (a)
High-concentration soil and sludges by sonicator	7.5
Non-water miscible waste	75

(a) EQL for other matrices = [EQL for low soil/sediment] × [Factor]. This estimated EQL is similar to an EPA "Practical Quantitation Limit."

SAMPLING METHOD
Liquid samples — Use a 1 or 2½ gallon amber glass bottle with a screw-top Teflon®-lined cover that has been prewashed with detergent and rinsed with distilled water and methanol (or isopropanol).

Soils, sediments, or sludges — Use an 8-oz. widemouth glass with a screw-top Teflon®-lined cover that has been prewashed with detergent and rinsed with distilled water and methanol (or isopropanol).

SAMPLE PRESERVATION
Liquid samples — If residual chlorine is present, add 3 mL of 10% sodium thiosulfate per gallon, cool to 4°C and store in a solvent-free refrigerator until analysis; if chlorine is not present, then eliminate the sodium thiosulfate addition.

Soils, sediments, or sludges — Cool samples to 4°C and store in a solvent-free refrigerator.

MHT Liquid samples must be extracted within 7 days and the extracts analyzed within 40 days. Soils, sediments, or sludges may be stored for a maximum of 14 days and the extracts analyzed within 40 days.

SAMPLE PREPARATION
Liquid samples — Transfer 1 L quantitatively to a continuous extractor. If high concentrations are anticipated, a smaller volume may be used and then diluted with organic-free reagent water to 1 L. Adjust pH, if necessary, to pH <2 using 1:1 (V/V) sulfuric acid. Pipette 1.0 mL of a surrogate standard spiking solution into each sample. For the sample in each analytical batch selected for spiking, add 1.0 mL of a matrix spiking standard. For base/neutral acid analysis, the amount of the surrogates and matrix spiking compounds added to the sample should result in a final concentration of 100 ng/µL of each analyte in the extract to be analyzed (assuming a 1-µL injection). Extract with methylene chloride for 18–24 h. Next, adjust the pH of the aqueous phase to pH >11 using 10 N sodium hydroxide and extract it with methylene chloride again for 18–24 h. Dry the extract through a column containing anhydrous sodium sulfate and concentrate it to 1 mL using a K-D concentrator.

Soils, sediments, or sludges — Use 30 g of sample. Nonporous or wet samples (gummy or clay type) that do not have a free-flowing sandy texture must be mixed with anhydrous sodium sulfate until the sample is free flowing. Add 1 mL of surrogate standards to all samples, spikes, standards, and blanks. For the sample in each analytical batch selected for spiking, add 1.0 mL of a matrix spiking standard. For base/neutral acid analysis, the amount added of the surrogates and matrix spiking compounds should result in a final concentration of 100 ng/μL of each base/neutral analyte and 200 ng/μL of each acid analyte in the extract to be analyzed (assuming a 1-μL injection). Immediately add a 100-mL mixture of 1:1 methylene chloride:acetone and extract the sample ultrasonically for 3 min and then decant or filter the extracts. Repeat the extraction two or more times. Dry the extract using a column with anhydrous sodium sulfate and concentrate it to 1 mL in a K-D concentrator.

Note: Hexachlorocyclopentadiene is subject to thermal and photochemical decomposition so samples must be proteccted from light and heat.

QUALITY CONTROL A methylene chloride solution containing 50 ng/μL of decafluorotriphenylphosphine (DFTPP) is used for tuning the GC/MS system each 12-h shift. A system performance check also must be made during every 12-h shift. A standard containing 50 ng/μL each of 4,4′-DDT, pentachlorophenol, and benzidine is required to verify injection port inertness and GC column performance. A calibration standard at mid-concentration, containing each compound of interest, including all required surrogates, must be performed every 12 h during analysis. After the system performance check is met, calibration check compounds (CCCs) are used to check the validity of the initial calibration.

The internal standard responses and retention times in the calibration check standard must be evaluated immediately after or during data acquisition. If the retention time for any internal standard changes by more than 30 seconds from the last check calibration (12 h), the chromatographic system must be inspected for malfunctions and corrections must be made, as required. If the electron ionization current plot (EICP) area for any of the internal standards changes by a factor of two from the last daily calibration standard check, the mass spectrometer must be inspected for malfunctions and corrections must be made, as appropriate.

Demonstrate, through the analysis of a reagent water blank, that interferences from the analytical system, glassware, and reagents are under control. The blank samples should be carried through all stages of the sample preparation and measurement steps. For each analytical batch (up to 20 samples), a reagent blank, matrix spike, and matrix spike duplicate/duplicate must be analyzed (the frequency of the spikes may be different for different monitoring programs). The blank and spiked samples must be carried through all stages of the sample preparation and measurement steps. A QC reference sample concentrate containing each analyte at a concentration of 100 mg/L in methanol is required.

REFERENCE Test Methods for Evaluating Solid Waste (SW-846). U.S. EPA 1983, Method 8270B, Rev. 2, Nov. 1990. Office of Solid Waste, Washington, DC.

Hexachloroethane   **EPA Method 1625**
CAS #67-72-1

TITLE Semivolatile Organic Compounds by Isotope Dilution GC/MS

MATRIX The compounds may be determined in waters, soils, and municipal sludges by this method.

METHOD SUMMARY This method is used to determine 176 semivolatile toxic organic pollutants associated with the CWA (as amended 1987); the RCRA (as amended 1986); the CERCLA (as amended 1986); and other compounds amenable to extraction and analysis by capillary column gas chromatography-mass spectrometry (GC/MS).

Stable isotopically-labeled analogs of the compounds of interest are added to the sample. If the solids content is less than 1%, a 1-L sample is extracted at pH 12–13, then at pH <2 with methylene chloride using continuous extraction techniques.

If the solids content is 30% or less, the sample is diluted to 1% solids with reagent water, homogenized ultrasonically, and extracted at pH 12–13, then at pH <2 with methylene chloride using continuous extraction techniques. If the solids content is greater than 30%, the sample is extracted using ultrasonic techniques.

Each extract is dried over sodium sulfate, concentrated to a volume of 5 mL, cleaned up using GPC, if necessary, and concentrated. Extracts are concentrated to 1 mL if GPC is not performed, and to 0.5 mL if GPC is performed.

An internal standard is added to the extract, and a 1-mL aliquot of the extract is injected into the GC. The compounds are separated by GC and detected by a MS. The labeled compounds serve to correct the variability of the analytical technique.

INTERFERENCES Solvents, reagents, glassware, and other sample processing hardware may yield artifacts and/or elevated baselines causing misinterpretation of chromatograms and spectra. Materials used in the analysis must be demonstrated to be free from interferences under the conditions of analysis by running method blanks initially and with each sample lot (sample started through the extraction process on a given 8-h shift, to a maximum of 20). Specific selection of reagents and purification of solvents by distillation in all glass systems may be required. Glassware and, where possible, reagents are cleaned by solvent rinse and baking at 450°C for 1-h minimum. Interferences coextracted from samples will vary considerably from source to source, depending on the diversity of the site being sampled.

INSTRUMENTATION Major instrumentation includes a GC with a splitless or on-column injection port for capillary column, a MS with 70 eV electron impact ionization, and a data system to collect and record MS data, and process it. A K-D apparatus is used to concentrate extracts.

GC Column: 30 m × 0.25 mm I.D. 5% phenyl, 94% methyl, 1% vinyl silicone bonded phased fused silica capillary column.

PRECISION & ACCURACY The detection limits of the method are usually dependent on the level of interferences rather than instrumental limitations. The limits typify the minimum quantities that can be detected with no interferences present.

The minimum level (in μg/mL) was 10. This is defined as a minimum level at which the analytical system shall give recognizable mass spectra (background corrected) and acceptable calibration points.

The MDL (in µg/kg) in low solids was 58 and in high solids was 55; these were determined in digested sludge (low solids) and in filter cake or compost (high solids).

The labeled and native compound initial precision as standard deviation (in µg/L) was 227.

The labeled and native compound initial accuracy as average recovery (in µg/L) was 21-ns.

ns = *no specification; the limit was outside the range that could be reliably measured.*

SAMPLE COLLECTION, PRESERVATION & HANDLING
Collect samples in glass containers. Aqueous samples which flow freely are collected in refrigerated bottles using automatic sampling equipment. Solid samples are collected as grab samples using widemouth jars. Maintain samples at 0 to 4°C from the time of collection until extraction. If residual chlorine is present in aqueous samples, add 80 mg sodium thiosulfate/L of water. Begin sample extraction within 7 days of collection, and analyze all extracts within 40 days of extraction.

SAMPLE PREPARATION Samples containing 1% solids or less are extracted directly using continuous liquid-liquid extraction techniques. Samples containing 1 to 30% solids are diluted to the 1% level with reagent water and extracted using continuous liquid-liquid extraction techniques. Samples containing greater than 30% solids are extracted using ultrasonic techniques.

- Base/neutral extraction — Adjust the pH of the waters in the extractors to 12–13 with 6 *N* NaOH. Extract with methylene chloride for 24–48 h.
- Acid extraction — Adjust the pH of the waters in the extractors to 2 or less using 6 *N* sulfuric acid. Extract with methylene chloride for 24–48 h.
- Ultrasonic extraction of high solids samples — Add anhydrous sodium sulfate to the sample and QC aliquot(s). Add acetone:methylene chloride (1:1) to the sample and mix thoroughly

Concentrate extracts using a K-D apparatus.

QUALITY CONTROL The analyst is permitted to modify this method to improve separations or lower the costs of measurements, provided all performance specifications are met. Analyses of blanks are required to demonstrate freedom from contamination. When results of spikes indicate atypical method performance for samples, the samples are diluted to bring method performance within acceptable limits.

For low solids (aqueous samples), extract, concentrate, and analyze two sets of four 1-L aliquots (8 aliquots total) of the precision and recovery standard. For high solids samples, two sets of four 30-g aliquots of the high solids reference matrix are used.

Spike all samples with labeled compounds to assess method performance. Compute percent recovery of the labeled compounds using the internal standard method. Compare the labeled compound recovery for each compound with the corresponding labeled compound recovery.

Reagent water and high solids reference matrix blanks are analyzed to demonstrate freedom from contamination. Extract and concentrate a 1-L reagent water blank or a high solids reference matrix blank with each sample's lot (samples started through the extraction process on the same 8-h shift, to a maximum of 20 samples).

Field replicates may be collected to determine the precision of the sampling technique, and spiked samples may be required to determine the accuracy of the analysis when the internal standard method is used.

REFERENCE Semivolatile Organic Compounds by Isotope Dilution GC/MS. Office of Water Regulation and Standards, U.S. EPA Industrial Technology Division, Washington, DC, EPA Method 1625, Rev. C, June 1989 (contact W.A. Telliard, U.S. EPA, Office of Water Regulations and Standards, 401 M St., SW, Washington, DC, 20460. Phone: 202-382-7131).

Hexachloroethane   EPA Method 625
CAS #67-72-1

TITLE Base/Neutrals and Acids, U.S. EPA Method 625

MATRIX This methods covers municipal and industrial wastewaters.

METHOD SUMMARY Approximately 1 L of sample is serially extracted with methylene chloride at a pH greater than 11 and again at a pH less than 2 using a separatory funnel or a continuous extractor. The methylene chloride extract is dried, concentrated to a volume of 1 mL, and analyzed by GC/MS. Qualitative identification of the parameters in the extract is performed using the retention time and the relative abundance of three characteristic masses (m/z). Qualitative analysis is performed using either external or internal standard techniques with a single characteristic m/z.

INTERFERENCES Method interferences may be caused by contaminants in solvents, reagents, glassware, and other sample processing hardware. Glassware must be scrupulously cleaned. Glassware should be heated in a muffle furnace at 400°C for 5 to 30 min. Some thermally stable materials, such as PCBs, may not be eliminated by this treatment. Solvent rinses with acetone and pesticide quality hexane may be substituted for the muffle furnace heating. Matrix interferences may be caused by contaminants that are coextracted from the sample. The base-neutral extraction may cause significantly reduced recovery of phenols. The packed gas chromatographic columns recommended for the basic fraction may not exhibit sufficient resolution for some analytes.

INSTRUMENTATION A GC/MS system with an injection port designed for on-column injection when using packed columns and for splitless injection when using capillary columns.

Column for base/neutrals: 1.8 m long × 2 mm I.D. glass, packed with 3% SP-2550 on Supelcoport (100/120 mesh) or equivalent.

Column for acids: 1.8 m long × 2 mm I.D. glass, packed with 1% SP-1240DA on Supelcoport (100/120 mesh) or equivalent.

PRECISION & ACCURACY The MDL concentrations were obtained using reagent water. The MDL actually achieved in a given analysis will vary depending on instrument sensitivity and matrix effects. This method was tested by 15 laboratories using reagent water, drinking water, surface water, and industrial wastewaters spiked at six concentrations over the range 5 to 100 µg/L. Single operator precision, overall precision, and method accuracy were found to be directly related to the concentration of the parameter matrix.

The MDL (in µg/L) in reagent water was not reported.

The standard deviation (in µg/L based on 4 recovery measurements) was 24.5.

The range (in µg/L) for average recovery for 4 measurements was 55.2–100.0.

The range (in %) for percent recovery was 40–113.

Accuracy (in µg/L) as expected recovery for one or more measurements of a sample containing a true concentration of C was 0.73C-0.83.

Precision (in µg/L) as expected single analyst standard deviation of measurements at an average concentration found at X was 0.17X + 0.67.

Overall precision (in µg/L) as expected interlaboratory standard deviation of measurements in an average concentration found at X was 0.17X + 0.80.

C = *True value of the concentration in µg/L.*
X = *Average recovery found for measurements of samples containing a concentration at C in µg/L.*

SAMPLE PREPARATION Adjust the pH to >11 with sodium hydroxide and serially extract in a separatory funnel with methylene chloride or else in a continuous extractor. Next, adjust the pH to <2 with sulfuric acid and serially extract in a separatory funnel with methylene chloride or else in a continuous extractor. Dry the extracts separately through a column of anhydrous sodium sulfate and then concentrate each of the extracts to 1.0 mL using a K-D apparatus.

SAMPLE COLLECTION, PRESERVATION & HANDLING Grab samples must be collected in glass containers. All samples must be refrigerated at 4°C from the time of collection until extraction. If residual chlorine is present, add 80 mg of sodium thiosulfate/L of sample and mix well. All samples must be extracted within 7 days of collection and completely analyzed within 40 days of extraction.

QUALITY CONTROL Make an initial, one-time, demonstration of the ability to generate acceptable accuracy and precision with this method. Before processing any samples, the analyst must analyze a reagent water blank to demonstrate that interferences from the analytical system and glassware are under control. Each time a set of samples is extracted or reagents are changed, a reagent water blank must be processed. Spike and analyze a minimum of 5% of all samples to monitor and evaluate lab data quality. A QC check sample concentrate that contains each parameter of interest at a concentration of 100 µg/mL in acetone is required. PCBs and multicomponent pesticides may be omitted from this test.

After analysis of five spiked wastewater samples, calculate the average percent recovery and the standard deviation of the percent recovery. Spike all samples with the surrogate standard spiking solution and calculate the percent recovery of each surrogate compound.

REFERENCE *Federal Register*, Vol. 49, No. 209. Friday, Oct. 26, 1984.

Hexachloroethane — EPA Method 8120
CAS #67-72-1

TITLE Chlorinated Hydrocarbons by Gas Chromatography

MATRIX This method covers aqueous and solid matrices. This includes a wide variety such as drinking water, groundwater, industrial wastewaters, surface waters, soils, solids, and sediments.

METHOD SUMMARY This method is used to determine the concentration of 14 chlorinated hydrocarbons. It provides gas chromatographic conditions for the detection of ppb concentrations of certain chlorinated hydrocarbons. Prior to use of this method, appropriate sample extraction techniques must be used. Both neat and diluted organic liquids (EPA Method 3580, Waste Dilution) may be analyzed by direct injection. A 2 to 5 µg/mL aliquot of the extract is injected into a gas chromatograph (GC) using the solvent flush technique, and compounds in the GC effluent are detected by an electron capture detector (ECD).

INTERFERENCES Solvents, reagents, glassware, and other sample processing hardware may yield discrete artifacts and/or elevated baselines causing misinterpretation of gas chromatograms. Interferences coextracted from samples will vary considerably from source to source, depending upon the waste being sampled.

INSTRUMENTATION An analytical system complete with GC suitable for on-column injections and accessories, including detectors, column supplies, recorder, gases and syringes is required. A data system for measuring peak areas and/or peak heights is recommended. The GC is equipped with an electron capture detector (ECD). A K-D apparatus is needed for sample preparation.

Column 1: 1.8 m × 2 mm I.D. glass column packed with 1% SP-1000 on Supelcoport (100/120 mesh) or equivalent.

Column 2: 1.8 m × 2 mm I.D. glass column packed with 1.5% OV-1/2.4% OV-225 on Supelcoport (80/100 mesh) or equivalent.

PRECISION & ACCURACY The method was tested by 20 laboratories using organic-free reagent water, drinking water, surface water, and three industrial wastewaters spiked at six concentrations over the range 1.0 to 356 µg/L. Single operator precision, overall precision, and method accuracy were found to be directly related to the concentration of the parameter and essentially independent of the sample matrix.

MULTIPLICATION FACTORS FOR OTHER MATRICES (a)

Matrix	Factor (b)
Groundwater	10
Low-concentration soil by ultrasonic cleanup extraction with GPC	670
High-concentration soil and sludges by ultrasonic extraction	10,000
Non-water miscible waste	100,000

(a) Sample EQLs are highly matrix-dependent. The EQLs listed are provided for guidance and may not always be achievable.
(b) EQL = [Method detection limit] × [Factor]. For non-aqueous samples, the factor is on a wet-weight basis.

PRECISION & ACCURACY The estimates below are based upon the performance in a single lab.

The accuracy (in μg/L) as expected recovery for one or more measurements of a sample containing a concentration of C was 0.74C-0.02.

The precision (in μg/L) as expected single analyst standard deviation of measurements at an average concentration of x" was 0.23 x"-0.07.

The precision (in μg/L) as expected interlaboratory standard deviation measurements at an average concentration found of x" was 0.36 x"-0.00.

C = True value for the concentration, in μg/L.
$x"$ = Average recovery found for measurements of samples containing a concentration of C, in μg/L.

SAMPLE COLLECTION, PRESERVATION & HANDLING
Extracts must be stored under refrigeration at 4°C and analyzed within 40 days of extraction.

SAMPLE PREPARATION In general, water samples are extracted at a neutral, or as is, pH with methylene chloride using either EPA Method 3510 or EPA Method 3520. Solid samples are extracted using either EPA Method 3540 or EPA Method 3550. Prior to gas chromatographic analysis, the extraction solvent must be exchanged to hexane.

QUALITY CONTROL The quality control check concentrate (EPA Method 8000) should contain each parameter of interest in acetone at the following concentrations: hexachloro-substituted hydrocarbon, 10 μg/mL; and any other chlorinated hydrocarbon, 100 μg/mL. Calculate surrogate standard recovery on all samples, blanks, and spikes.

Prepare stock standard solutions in isooctane or hexane. Calibration standards at a minimum of five concentrations should be prepared through dilution of the stock standards with isooctane or hexane. Internal standards and surrogate standards are also needed.

REFERENCE Test Methods for Evaluating Solid Waste, Physical/Chemical Methods, SW-846, 3rd Edition, U.S. EPA, Office of Solid Waste, Washington, DC, 1990. EPA Method 8120 A Rev. 1, Nov. 1990.

Hexachloroethane — EPA Method 8121
CAS #67-72-1

TITLE Chlorinated Hydrocarbons by GC: Capillary Column Technique

MATRIX This method covers aqueous and solid matrices. This includes a wide variety such as drinking water, groundwater, industrial wastewaters, surface waters, soils, solids, and sediments.

METHOD SUMMARY This method provides procedures for the determination of 22 chlorinated hydrocarbons in water, soil/sediment, and waste matrices. A measured volume or weight of sample is extracted by using one of the appropriate sample extraction techniques specified in EPA Method 3510, EPA Method 3520, EPA Method 3540, or EPA Method 3550, or diluted using EPA Method 3580. Aqueous samples are extracted at neutral pH with methylene chloride by using either a separatory funnel (EPA Method 3510) or a continuous liquid-liquid extractor (EPA Method 3520). Solid samples are extracted with hexane/acetone (1:1) by using a Soxhlet extractor (EPA Method 3540) or with methylene chloride/acetone (1:1) by using an ultrasonic extractor (EPA Method 3550). After cleanup, the extract or diluted sample is analyzed by gas chromatography with electron capture detection (GC/ECD).

The sensitivity level of this method usually depends on the level of interferences rather than on instrumental limitations. This method may be used in conjunction with EPA Method 3620, Florisil Column Cleanup, EPA Method 3660, Sulfur Cleanup, and EPA Method 3640, Gel Permeation Chromatography, to aid in the elimination of interferences.

INTERFERENCES Solvents, reagents, glassware, and other hardware used in sample processing may introduce artifacts which may result in elevated baselines, causing misinterpretation of gas chromatograms. Interferants coextracted from the samples will vary considerably from waste to waste. Glassware must be scrupulously clean. Phthalate esters, if present in a sample, will interfere only with the BHC isomers. The presence of elemental sulfur will result in large peaks, and can often mask the region of compounds eluting after 1,2,4,5-tetrachlorobenzene. The tetrabutylammonium (TBA)-sulfite procedure (EPA Method 3660) works well for the removal of elemental sulfur. Waxes and lipids can be removed by gel permeation chromatography (EPA Method 3640).

INSTRUMENTATION A GC suitable for on-column injections and all required accessories, including and electron capture detector (ECD), analytical columns, recorder, gases, and syringes are needed. A data system for measuring peak heights and/or peak areas is recommended. A Kuderna-Danish (K-D) apparatus will also be needed to concentrate extracts.

Column 1: 30 m × 0.53 mm I.D. fused-silica capillary column chemically bonded with trifluoropropyl methyl silicone (DB-210 or equivalent).
Column 2: 30 m × 0.53 mm I.D. fused-silica capillary column chemically bonded with polyethylene glycol (DB-WAX or equivalent).

PRECISION & ACCURACY This method has been tested in a single lab by using organic-free reagent water, sandy loam samples, and extracts which were spiked with the test compounds at one concentration. Single-operator precision and method accuracy were found to be related to the concentration of compound and the type of matrix. The accuracy and precision technique will be determined by the sample matrix, sample preparation technique, optional cleanup techniques, and calibration procedures used.

MULTIPLICATION FACTORS FOR OTHER MATRICES (a)

Matrix	Factor (b)
Groundwater	10
Low-concentration soil by ultrasonic cleanup extraction with GPC	670
High-concentration soil and sludges by ultrasonic extraction	10,000
Non-water miscible waste	100,000

(a) Sample EQLs are highly matrix-dependent. The EQLs listed are provided for guidance and may not always be achievable.
(b) EQL = [Method detection limit] × [Factor]. For non-aqueous samples, the factor is on a wet-weight basis.

PRECISION & ACCURACY MDL is the method detection limit for organic-free reagent water. MDL was determined from the analysis of eight replicate aliquots processed through the entire analytical method (extraction, Florisil cartridge cleanup, and GC/ECD analysis).

The MDL (in ng/L) was 1.6.

The accuracy (as average % recovery using 5 determinations and no Florisil cleanup) from a spike concentration of 1.0 µg/L and separatory funnel extraction was 96% with a final volume of 10 mL.

The precision (as RSD% using 5 determinations and no Florisil cleanup) from a spike concentration of 1.0 µg/L and separatory funnel extraction was 4.0% with a final volume of 10 mL.

The accuracy (as average % recovery using 5 determinations and no Florisil cleanup), from a spike concentration of 330 µg/L and ultrasonic extraction of solid samples using 1:1 methylene chloride and acetone, was 83% with a final volume of 10 mL.

The precision (as RSD% using 5 determinations and no Florisil cleanup), from a spike concentration of 330 µg/L and ultrasonic extraction of solid samples using 1:1 methylene chloride and acetone, was 4.6% with a final volume of 10 mL.

SAMPLE COLLECTION, PRESERVATION & HANDLING
Volatile Organics — Standard 40-mL glass screw-cap VOA vials with Teflon®-faced silicone septum may be used for both liquid and solid matrices. When collecting samples, liquids and solids should be introduced into the vials gently to reduce agitation which might drive off volatile compounds. If there are any air bubbles present the sample must be retaken. The vials with solids should be tapped slightly as they are filled to try and eliminate as much free air space as possible. Two vials from each sampling location should be sealed in separate plastic bags to prevent cross-contamination between samples.

Semivolatile organics — Containers used to collect samples for the determination of semivolatile organic compounds should be soap and water washed followed by methanol (or isopropanol) rinsing. The sample containers should be of glass or Teflon® and have screw-top covers with Teflon® liners.

Preservation for volatile organics — No preservation is used with concentrated waste samples. With liquid samples containing no residual chlorine, 4 drops of concentrated hydrochloric acid are added and the samples are immediately cooled to 4°C. When liquid samples contain residual chlorine, they are treated as above and, in addition, 4 drops of 4% aqueous sodium thiosulfate are added to remove the residual chlorine. Soil, sediment, and sludge samples are only cooled to 4°C.

Preservation for semivolatile organics — No preservation is used with concentrated waste samples. With liquid samples containing no residual chlorine and with soil, sediment, and sludge samples, immediately cooling to 4°C is the only preservation used. When residual chlorine is present then 3 mL of 10% aqueous sodium sulfate is added for each gallon of sample collected, followed by cooling to 4°C.

Holding times — The holding time for all volatile organics samples is 14 days. Liquid samples must be extracted within 7 days and their extracts analyzed within 40 days. Concentrated waste, soil, sediment, and sludge samples must be extracted within 14 days and their extracts analyzed within 40 days.

SAMPLE PREPARATION Prepare stock standard solutions in hexane. Calibration standards at a minimum of five concentrations should be prepared through dilution of the stock standards with hexane. The suggested internal standards are: 2,5-dibromotoluene, 1,3,5-tribromobenzene, and α,α-dibromo-m-xylene. The analyst can use any of the three compounds provided that they are resolved from matrix interferences. Recommended surrogate compounds are α-2,6-trichlorotoluene, 1,4-dichloronaphthalene, and 2,3,4,5,6-pentachlorotoluene.

In general, water samples are extracted at a neutral pH with methylene chloride using a separatory funnel (EPA Method 3510) or a continuous liquid-liquid extractor (EPA Method 3520). Solid samples are extracted with hexane/acetone (1:1 v:v) using a Soxhlet extractor (EPA Method 3540) or with methylene chloride/acetone (1:1 v:v) using an ultrasonic extractor (EPA Method 3550). Non-aqueous waste samples may be diluted using EPA Method 3580. Prior to Florisil cleanup or gas chromatographic analysis, the extraction solvent must be exchanged to hexane. Sample extracts that will be subjected to gel permeation chromatography do not need solvent exchange.

Cleanup procedures may not be necessary for a relatively clean matrix. If removal of interferences such as chlorinated phenols, phthalate esters, etc., is required, proceed with the procedure outlined in EPA Method 3620.

QUALITY CONTROL Analyze a quality control check standard to demonstrate that the operation of the GC is in control. The frequency of the check standard analysis is equivalent to 10% of the samples analyzed. If the recovery of any compound found in the check standard is less than 80% of the certified value, the problem must be corrected and a new set of calibration standards must be prepared and analyzed. Calculate surrogate standard recoveries for all samples, blanks, and spikes. An internal standard peak area check must be performed on all samples. The internal standard must be evaluated for acceptance by determining whether the measured area for the internal standard deviates by more than 30% from the average area for the internal standard in the calibration standards. When the internal standard peak area is outside that limit, all samples that fall outside the QC criteria must be reanalyzed. Any compound confirmed by two columns may also be confirmed by GC/MS (EPA Method 8270). The GC/MS would normally require a minimum concentration of 1 ng/µL in the final extract for each compound. Include a mid-concentration calibration standard after each group of 20 samples in the analysis sequence. The response factors for the mid-concentration calibration must be within 15% of the average values for the multiconcentration calibration.

REFERENCE Test Methods for Evaluating Solid Waste, Physical/Chemical Methods, SW-846, 3rd Edition, U.S. EPA, Office of Solid Waste, Washington, DC, 1990. EPA Method 8121, Rev. 0, Nov. 1990.

Hexachloroethane   **EPA Method 8270**
CAS #67-72-1

TITLE Semivolatile Organic Compounds by GC/MS

MATRIX This method is used to determine the concentration of semivolatile organic compounds in extracts prepared from all types of solid waste matrices, soils, and groundwater. Although surface waters are not specifically mentioned, this method should be applicable to water samples from rivers, lakes, etc.

METHOD SUMMARY This method covers 259 semivolatile organic compounds. In very limited applications direct injection of the sample into the GC/MS system may be appropriate, but this results in very high detection limits (approximately 10,000 µg/L). Typically, a 1-L liquid sample, containing surrogate, and matrix spiking standards, is extracted in a continuous extractor first under acid conditions and then under basic conditions. Typically 30 g of a solid sample, containing surrogate, and matrix spiking standards, is extracted ultrasonically. After concentrating the extract to 1 mL it is spiked with 10 µL of an internal standard solution just prior to analysis by GC/MS. The volume injected should contain about 100 ng of base/neutral and 200 ng of acid surrogates (for a 1-µL injection). Analysis is performed by GC/MS using a capillary GC column.

INTERFERENCES Raw GC/MS data from all blanks, samples, and spikes must be evaluated for interferences. Contamination by carryover can occur whenever high-concentration and low-concentration samples are sequentially analyzed. To reduce carryover, the sample syringe must be rinsed out between samples with solvent. Whenever an unusually concentrated sample is encountered, it should be followed by the analysis of blank solvent to check for cross-contamination.

INSTRUMENTATION A GC/MS and a data system are required. The GC column used is a 30 m × 0.25 mm I.D. (or 0.32 mm I.D.) 1um film thickness silicone-coated fused silica capillary column. A continuous liquid-liquid extractor equipped with Teflon® or glass connection joints and stopcocks requiring no lubrication, a K-D concentrating apparatus, water bath, and an ultrasonic disrupter with a minimum power of 300 W and with pulsing capability are also required.

PRECISION & ACCURACY The estimated quantitation limit (EQL) of Method 8270B for determining an individual compound is approximately 1 mg/kg (wet weight) for soil or sediment samples, 1–200 mg/kg for wastes (dependent on matrix and method of preparation), and 10 µg/L for groundwater samples. EQLs will be proportionately higher for sample extracts that require dilution to avoid saturation of the detector.

The EQL(b) for groundwater in µg/L is 10.
The EQL (a, b) for low concentrations in soil and sediment in µg/kg is 660.
Accuracy as µg/L is 0.73C–0.83.
Overall precision in µg/L is 0.17X + 0.80.

(a) EQLs listed for soil/sediment are based on wet weight. Normally data is reported in a dry-weight basis; therefore, EQLs will be higher based on the % dry weight of each sample. This calculation is based on a 30-g sample and gel permeation chromatography cleanup.

(b) Sample EQLs are highly matrix-dependent. The EQLs are provided for guidance and may not always be achievable.

C = True value for concentration, in µg/L.
X = Average recovery found for measurements of samples containing a concentration of C, in µg/L.

ESTIMATED QUANTITATION LIMIT

Other Matrices	Factor (a)
High-concentration soil and sludges by sonicator	7.5
Non-water miscible waste	75

(a) EQL for other matrices = [EQL for low soil/sediment] × [Factor]. This estimated EQL is similar to an EPA "Practical Quantitation Limit."

SAMPLING METHOD

Liquid samples — Use a 1 or 2½ gallon amber glass bottle with a screw-top Teflon®-lined cover that has been prewashed with detergent and rinsed with distilled water and methanol (or isopropanol).

Soils, sediments, or sludges — Use an 8-oz. widemouth glass with a screw-top Teflon®-lined cover that has been prewashed with detergent and rinsed with distilled water and methanol (or isopropanol).

SAMPLE PRESERVATION

Liquid samples — If residual chlorine is present, add 3 mL of 10% sodium thiosulfate per gallon, cool to 4°C and store in a solvent-free refrigerator until analysis; if chlorine is not present, then eliminate the sodium thiosulfate addition.

Soils, sediments, or sludges — Cool samples to 4°C and store in a solvent-free refrigerator.

MHT Liquid samples must be extracted within 7 days and the extracts analyzed within 40 days. Soils, sediments, or sludges may be stored for a maximum of 14 days and the extracts analyzed within 40 days.

SAMPLE PREPARATION

Liquid samples — Transfer 1 L quantitatively to a continuous extractor. If high concentrations are anticipated, a smaller volume may be used and then diluted with organic-free reagent water to 1 L. Adjust pH, if necessary, to pH <2 using 1:1 (V/V) sulfuric acid. Pipette 1.0 mL of a surrogate standard spiking solution into each sample. For the sample in each analytical batch selected for spiking, add 1.0 mL of a matrix spiking standard. For base/neutral acid analysis, the amount of the surrogates and matrix spiking compounds added to the sample should result in a final concentration of 100 ng/μL of each analyte in the extract to be analyzed (assuming a 1-μL injection). Extract with methylene chloride for 18–24 h. Next, adjust the pH of the aqueous phase to pH >11 using 10 N sodium hydroxide and extract it with methylene chloride again for 18–24 h. Dry the extract through a column containing anhydrous sodium sulfate and concentrate it to 1 mL using a K-D concentrator.

Soils, sediments, or sludges — Use 30 g of sample. Nonporous or wet samples (gummy or clay type) that do not have a free-flowing sandy texture must be mixed with anhydrous sodium sulfate until the sample is free flowing. Add 1 mL of surrogate standards to all samples, spikes, standards, and blanks. For the sample in each analytical batch selected for spiking, add 1.0 mL of a matrix spiking standard. For base/neutral acid analysis, the amount added of the surrogates and matrix spiking compounds should result in a final concentration of 100 ng/μL of each base/neutral analyte and 200 ng/μL of each acid analyte in the extract to be analyzed (assuming a 1-μL injection). Immediately add a 100-mL mixture of 1:1 methylene chloride:acetone and extract the sample ultrasonically for 3 min and then decant or filter the extracts. Repeat the extraction two or more times. Dry the extract using a column with anhydrous sodium sulfate and concentrate it to 1 mL in a K-D concentrator.

QUALITY CONTROL A methylene chloride solution containing 50 ng/μL of decafluorotriphenylphosphine (DFTPP) is used for tuning the GC/MS system each 12-h shift. A system performance check also must be made during every 12-h shift. A standard containing 50 ng/μL each of 4,4′-DDT, pentachlorophenol, and benzidine is required to verify injection port inertness and GC column performance. A calibration standard at mid-concentration, containing each compound of interest, including all required surrogates, must be performed every 12 h during analysis. After the system performance check is met, calibration check compounds (CCCs) are used to check the validity of the initial calibration.

The internal standard responses and retention times in the calibration check standard must be evaluated immediately after or during data acquisition. If the retention time for any internal standard changes by more than 30 seconds from the last check calibration (12 h), the chromatographic system must be inspected for malfunctions and corrections must be made, as required. If the electron ionization current plot (EICP) area for any of the internal standards changes by a factor of two from the last daily calibration standard check, the mass spectrometer must be inspected for malfunctions and corrections must be made, as appropriate.

Demonstrate, through the analysis of a reagent water blank, that interferences from the analytical system, glassware, and reagents are under control. The blank samples should be carried through all stages of the sample preparation and measurement steps. For each analytical batch (up to 20 samples), a reagent blank, matrix spike, and matrix spike duplicate/duplicate must be analyzed (the frequency of the spikes may be different for different monitoring programs). The blank and spiked samples must be carried through all stages of the sample preparation and measurement steps. A QC reference sample concentrate containing each analyte at a concentration of 100 mg/L in methanol is required.

REFERENCE Test Methods for Evaluating Solid Waste (SW-846). U.S. EPA 1983, Method 8270B, Rev. 2, Nov. 1990. Office of Solid Waste, Washington, DC.

Hexachlorophene   EPA Method 8270
CAS #70-30-4

TITLE Semivolatile Organic Compounds by GC/MS

MATRIX This method is used to determine the concentration of semivolatile organic compounds in extracts prepared from all types of solid waste matrices, soils, and groundwater. Although surface waters are not specifically mentioned, this method should be applicable to water samples from rivers, lakes, etc.

METHOD SUMMARY This method covers 259 semivolatile organic compounds. In very limited applications direct injection of the sample into the GC/MS system may be appropriate, but this results in very high detection limits (approximately 10,000 μg/L). Typically, a 1-L liquid sample, containing surrogate, and matrix spiking standards, is extracted in a continuous extractor first under acid conditions and then under basic conditions. Typically 30 g of a solid sample, containing surrogate, and matrix spiking standards, is extracted ultrasonically. After concentrating the extract to 1 mL it is spiked with 10 μL of an internal standard solution just prior to analysis by GC/MS. The volume injected should contain about 100 ng of base/neutral and 200 ng of acid surrogates (for a 1-μL injection). Analysis is performed by GC/MS using a capillary GC column.

INTERFERENCES Raw GC/MS data from all blanks, samples, and spikes must be evaluated for interferences. Contamination by carryover can occur whenever high-concentration and low-concentration samples are sequentially analyzed. To reduce carryover, the sample syringe must be rinsed out between samples with solvent. Whenever an unusually concentrated sample is encountered, it should be followed by the analysis of blank solvent to check for cross-contamination.

INSTRUMENTATION A GC/MS and a data system are required. The GC column used is a 30 m × 0.25 mm I.D. (or 0.32 mm I.D.) 1um film thickness silicone-coated fused silica

capillary column. A continuous liquid-liquid extractor equipped with Teflon® or glass connection joints and stopcocks requiring no lubrication, a K-D concentrating apparatus, water bath, and an ultrasonic disrupter with a minimum power of 300 W and with pulsing capability are also required.

PRECISION & ACCURACY The estimated quantitation limit (EQL) of Method 8270B for determining an individual compound is approximately 1 mg/kg (wet weight) for soil or sediment samples, 1–200 mg/kg for wastes (dependent on matrix and method of preparation), and 10 µg/L for groundwater samples. EQLs will be proportionately higher for sample extracts that require dilution to avoid saturation of the detector.

The EQL(b) for groundwater in µg/L is 50.
The EQL (a, b) for low concentrations in soil and sediment in µg/kg is not determined.
Accuracy as µg/L is not listed.
Overall precision in µg/L is not listed.

(a) EQLs listed for soil/sediment are based on wet weight. Normally data is reported in a dry-weight basis; therefore, EQLs will be higher based on the % dry weight of each sample. This calculation is based on a 30-g sample and gel permeation chromatography cleanup.
(b) Sample EQLs are highly matrix-dependent. The EQLs are provided for guidance and may not always be achievable.
C = True value for concentration, in µg/L.
X = Average recovery found for measurements of samples containing a concentration of C, in µg/L.

ESTIMATED QUANTITATION LIMIT

Other Matrices	Factor (a)
High-concentration soil and sludges by sonicator	7.5
Non-water miscible waste	75

(a) EQL for other matrices = [EQL for low soil/sediment] × [Factor]. This estimated EQL is similar to an EPA "Practical Quantitation Limit."

SAMPLING METHOD

Liquid samples — Use a 1 or 2½ gallon amber glass bottle with a screw-top Teflon®-lined cover that has been prewashed with detergent and rinsed with distilled water and methanol (or isopropanol).

Soils, sediments, or sludges — Use an 8-oz. widemouth glass with a screw-top Teflon®-lined cover that has been prewashed with detergent and rinsed with distilled water and methanol (or isopropanol).

SAMPLE PRESERVATION

Liquid samples — If residual chlorine is present, add 3 mL of 10% sodium thiosulfate per gallon, cool to 4°C and store in a solvent-free refrigerator until analysis; if chlorine is not present, then eliminate the sodium thiosulfate addition.

Soils, sediments, or sludges — Cool samples to 4°C and store in a solvent-free refrigerator.

MHT Liquid samples must be extracted within 7 days and the extracts analyzed within 40 days. Soils, sediments, or sludges may be stored for a maximum of 14 days and the extracts analyzed within 40 days.

SAMPLE PREPARATION

Liquid samples — Transfer 1 L quantitatively to a continuous extractor. If high concentrations are anticipated, a smaller volume may be used and then diluted with organic-free reagent water to 1 L. Adjust pH, if necessary, to pH <2 using 1:1 (V/V) sulfuric acid. Pipette 1.0 mL of a surrogate standard spiking solution into each sample. For the sample in each analytical batch selected for spiking, add 1.0 mL of a matrix spiking standard. For base/neutral acid analysis, the amount of the surrogates and matrix spiking compounds added to the sample should result in a final concentration of 100 ng/µL of each analyte in the extract to be analyzed (assuming a 1-µL injection). Extract with methylene chloride for 18–24 h. Next, adjust the pH of the aqueous phase to pH >11 using 10 N sodium hydroxide and extract it with methylene chloride again for 18–24 h. Dry the extract through a column containing anhydrous sodium sulfate and concentrate it to 1 mL using a K-D concentrator.

Soils, sediments, or sludges — Use 30 g of sample. Nonporous or wet samples (gummy or clay type) that do not have a free-flowing sandy texture must be mixed with anhydrous sodium sulfate until the sample is free flowing. Add 1 mL of surrogate standards to all samples, spikes, standards, and blanks. For the sample in each analytical batch selected for spiking, add 1.0 mL of a matrix spiking standard. For base/neutral acid analysis, the amount added of the surrogates and matrix spiking compounds should result in a final concentration of 100 ng/µL of each base/neutral analyte and 200 ng/µL of each acid analyte in the extract to be analyzed (assuming a 1-µL injection). Immediately add a 100-mL mixture of 1:1 methylene chloride:acetone and extract the sample ultrasonically for 3 min and then decant or filter the extracts. Repeat the extraction two or more times. Dry the extract using a column with anhydrous sodium sulfate and concentrate it to 1 mL in a K-D concentrator.

QUALITY CONTROL A methylene chloride solution containing 50 ng/µL of decafluorotriphenylphosphine (DFTPP) is used for tuning the GC/MS system each 12-h shift. A system performance check also must be made during every 12-h shift. A standard containing 50 ng/µL each of 4,4′-DDT, pentachlorophenol, and benzidine is required to verify injection port inertness and GC column performance. A calibration standard at mid-concentration, containing each compound of interest, including all required surrogates, must be performed every 12 h during analysis. After the system performance check is met, calibration check compounds (CCCs) are used to check the validity of the initial calibration.

The internal standard responses and retention times in the calibration check standard must be evaluated immediately after or during data acquisition. If the retention time for any internal standard changes by more than 30 seconds from the last check calibration (12 h), the chromatographic system must be inspected for malfunctions and corrections must be made, as required. If the electron ionization current plot (EICP) area for any of the internal standards changes by a factor of two from the last daily calibration standard check, the mass spectrometer

must be inspected for malfunctions and corrections must be made, as appropriate.

Demonstrate, through the analysis of a reagent water blank, that interferences from the analytical system, glassware, and reagents are under control. The blank samples should be carried through all stages of the sample preparation and measurement steps. For each analytical batch (up to 20 samples), a reagent blank, matrix spike, and matrix spike duplicate/duplicate must be analyzed (the frequency of the spikes may be different for different monitoring programs). The blank and spiked samples must be carried through all stages of the sample preparation and measurement steps. A QC reference sample concentrate containing each analyte at a concentration of 100 mg/L in methanol is required.

REFERENCE Test Methods for Evaluating Solid Waste (SW-846). U.S. EPA 1983, Method 8270B, Rev. 2, Nov. 1990. Office of Solid Waste, Washington, DC.

Hexachloropropene   **EPA Method 1625**
CAS #1888-71-7

TITLE Semivolatile Organic Compounds by Isotope Dilution GC/MS

MATRIX The compounds may be determined in waters, soils, and municipal sludges by this method.

METHOD SUMMARY This method is used to determine 176 semivolatile toxic organic pollutants associated with the CWA (as amended 1987); the RCRA (as amended 1986); the CERCLA (as amended 1986); and other compounds amenable to extraction and analysis by capillary column gas chromatography-mass spectrometry (GC/MS).

Stable isotopically-labeled analogs of the compounds of interest are added to the sample. If the solids content is less than 1%, a 1-L sample is extracted at pH 12–13, then at pH <2 with methylene chloride using continuous extraction techniques.

If the solids content is 30% or less, the sample is diluted to 1% solids with reagent water, homogenized ultrasonically, and extracted at pH 12–13, then at pH <2 with methylene chloride using continuous extraction techniques. If the solids content is greater than 30%, the sample is extracted using ultrasonic techniques.

Each extract is dried over sodium sulfate, concentrated to a volume of 5 mL, cleaned up using GPC, if necessary, and concentrated. Extracts are concentrated to 1 mL if GPC is not performed, and to 0.5 mL if GPC is performed.

An internal standard is added to the extract, and a 1-mL aliquot of the extract is injected into the GC. The compounds are separated by GC and detected by a MS. The labeled compounds serve to correct the variability of the analytical technique.

INTERFERENCES Solvents, reagents, glassware, and other sample processing hardware may yield artifacts and/or elevated baselines causing misinterpretation of chromatograms and spectra. Materials used in the analysis must be demonstrated to be free from interferences under the conditions of analysis by running method blanks initially and with each sample lot (sample started through the extraction process on a given 8-h shift, to a maximum of 20). Specific selection of reagents and purification of solvents by distillation in all glass systems may be required. Glassware and, where possible, reagents are cleaned by solvent rinse and baking at 450°C for 1-h minimum. Interferences coextracted from samples will vary considerably from source to source, depending on the diversity of the site being sampled.

INSTRUMENTATION Major instrumentation includes a GC with a splitless or on-column injection port for capillary column, a MS with 70 eV electron impact ionization, and a data system to collect and record MS data, and process it. A K-D apparatus is used to concentrate extracts.

GC Column: 30 m × 0.25 mm I.D. 5% phenyl, 94% methyl, 1% vinyl silicone bonded phased fused silica capillary column.

PRECISION & ACCURACY The detection limits of the method are usually dependent on the level of interferences rather than instrumental limitations. The limits typify the minimum quantities that can be detected with no interferences present.

The minimum level (in μg/mL) was not listed. This is defined as a minimum level at which the analytical system shall give recognizable mass spectra (background corrected) and acceptable calibration points.

The MDL (in μg/kg) in low solids was not listed and in high solids was not listed; these were determined in digested sludge (low solids) and in filter cake or compost (high solids).

The labeled and native compound initial precision as standard deviation (in μg/L) was not listed.
The labeled and native compound initial accuracy as average recovery (in μg/L) was not listed.

SAMPLE COLLECTION, PRESERVATION & HANDLING
Collect samples in glass containers. Aqueous samples which flow freely are collected in refrigerated bottles using automatic sampling equipment. Solid samples are collected as grab samples using widemouth jars. Maintain samples at 0 to 4°C from the time of collection until extraction. If residual chlorine is present in aqueous samples, add 80 mg sodium thiosulfate/L of water. Begin sample extraction within 7 days of collection, and analyze all extracts within 40 days of extraction.

SAMPLE PREPARATION Samples containing 1% solids or less are extracted directly using continuous liquid-liquid extraction techniques. Samples containing 1 to 30% solids are diluted to the 1% level with reagent water and extracted using continuous liquid-liquid extraction techniques. Samples containing greater than 30% solids are extracted using ultrasonic techniques.

Base/neutral extraction — Adjust the pH of the waters in the extractors to 12–13 with 6 *N* NaOH. Extract with methylene chloride for 24–48 h.

Acid extraction — Adjust the pH of the waters in the extractors to 2 or less using 6 *N* sulfuric acid. Extract with methylene chloride for 24–48 h.

Ultrasonic extraction of high solids samples — Add anhydrous sodium sulfate to the sample and QC aliquot(s). Add acetone:methylene chloride (1:1) to the sample and mix thoroughly

Concentrate extracts using a K-D apparatus.

QUALITY CONTROL The analyst is permitted to modify this method to improve separations or lower the costs of measurements, provided all performance specifications are met. Analyses of blanks are required to demonstrate freedom from contamination. When results of spikes indicate atypical method performance for samples, the samples are diluted to bring method performance within acceptable limits.

For low solids (aqueous samples), extract, concentrate, and analyze two sets of four 1-L aliquots (8 aliquots total) of the precision and recovery standard. For high solids samples, two sets of four 30-g aliquots of the high solids reference matrix are used.

Spike all samples with labeled compounds to assess method performance. Compute percent recovery of the labeled compounds using the internal standard method. Compare the labeled compound recovery for each compound with the corresponding labeled compound recovery.

Reagent water and high solids reference matrix blanks are analyzed to demonstrate freedom from contamination. Extract and concentrate a 1-L reagent water blank or a high solids reference matrix blank with each sample's lot (samples started through the extraction process on the same 8-h shift, to a maximum of 20 samples).

Field replicates may be collected to determine the precision of the sampling technique, and spiked samples may be required to determine the accuracy of the analysis when the internal standard method is used.

REFERENCE Semivolatile Organic Compounds by Isotope Dilution GC/MS. Office of Water Regulation and Standards, U.S. EPA Industrial Technology Division, Washington, DC, EPA Method 1625, Rev. C, June 1989 (contact W.A. Telliard, U.S. EPA, Office of Water Regulations and Standards, 401 M St., SW, Washington, DC, 20460. Phone: 202-382-7131).

Hexachloropropene   **EPA Method 8270**
CAS #1888-71-7

TITLE Semivolatile Organic Compounds by GC/MS

MATRIX This method is used to determine the concentration of semivolatile organic compounds in extracts prepared from all types of solid waste matrices, soils, and groundwater. Although surface waters are not specifically mentioned, this method should be applicable to water samples from rivers, lakes, etc.

METHOD SUMMARY This method covers 259 semivolatile organic compounds. In very limited applications direct injection of the sample into the GC/MS system may be appropriate, but this results in very high detection limits (approximately 10,000 µg/L). Typically, a 1-L liquid sample, containing surrogate, and matrix spiking standards, is extracted in a continuous extractor first under acid conditions and then under basic conditions. Typically 30 g of a solid sample, containing surrogate, and matrix spiking standards, is extracted ultrasonically. After concentrating the extract to 1 mL it is spiked with 10 µL of an internal standard solution just prior to analysis by GC/MS. The volume injected should contain about 100 ng of base/neutral and 200 ng of acid surrogates (for a 1-µL injection). Analysis is performed by GC/MS using a capillary GC column.

INTERFERENCES Raw GC/MS data from all blanks, samples, and spikes must be evaluated for interferences. Contamination by carryover can occur whenever high-concentration and low-concentration samples are sequentially analyzed. To reduce carryover, the sample syringe must be rinsed out between samples with solvent. Whenever an unusually concentrated sample is encountered, it should be followed by the analysis of blank solvent to check for cross-contamination.

INSTRUMENTATION A GC/MS and a data system are required. The GC column used is a 30 m × 0.25 mm I.D. (or 0.32 mm I.D.) 1um film thickness silicone-coated fused silica capillary column. A continuous liquid-liquid extractor equipped with Teflon® or glass connection joints and stopcocks requiring no lubrication, a K-D concentrating apparatus, water bath, and an ultrasonic disrupter with a minimum power of 300 W and with pulsing capability are also required.

PRECISION & ACCURACY The estimated quantitation limit (EQL) of Method 8270B for determining an individual compound is approximately 1 mg/kg (wet weight) for soil or sediment samples, 1–200 mg/kg for wastes (dependent on matrix and method of preparation), and 10 µg/L for groundwater samples. EQLs will be proportionately higher for sample extracts that require dilution to avoid saturation of the detector.

The EQL(b) for groundwater in µg/L is 10.
The EQL (a, b) for low concentrations in soil and sediment in µg/kg is not determined.
Accuracy as µg/L is not listed.
Overall precision in µg/L is not listed.

(a) EQLs listed for soil/sediment are based on wet weight. Normally data is reported in a dry-weight basis; therefore, EQLs will be higher based on the % dry weight of each sample. This calculation is based on a 30-g sample and gel permeation chromatography cleanup.
(b) Sample EQLs are highly matrix-dependent. The EQLs are provided for guidance and may not always be achievable.
C = True value for concentration, in µg/L.
X = Average recovery found for measurements of samples containing a concentration of C, in µg/L.

ESTIMATED QUANTITATION LIMIT

Other Matrices	Factor (a)
High-concentration soil and sludges by sonicator	7.5
Non-water miscible waste	75

(a) EQL for other matrices = [EQL for low soil/sediment] × [Factor]. This estimated EQL is similar to an EPA "Practical Quantitation Limit."

SAMPLING METHOD

Liquid samples — Use a 1 or 2½ gallon amber glass bottle with a screw-top Teflon®-lined cover that has been prewashed with detergent and rinsed with distilled water and methanol (or isopropanol).

Soils, sediments, or sludges — Use an 8-oz. widemouth glass with a screw-top Teflon®-lined cover that has been prewashed with detergent and rinsed with distilled water and methanol (or isopropanol).

SAMPLE PRESERVATION

Liquid samples — If residual chlorine is present, add 3 mL of 10% sodium thiosulfate per gallon, cool to 4°C and store in a solvent-free refrigerator until analysis; if chlorine is not present, then eliminate the sodium thiosulfate addition.

Soils, sediments, or sludges — Cool samples to 4°C and store in a solvent-free refrigerator.

MHT Liquid samples must be extracted within 7 days and the extracts analyzed within 40 days. Soils, sediments, or sludges may be stored for a maximum of 14 days and the extracts analyzed within 40 days.

SAMPLE PREPARATION

Liquid samples — Transfer 1 L quantitatively to a continuous extractor. If high concentrations are anticipated, a smaller volume may be used and then diluted with organic-free reagent water to 1 L. Adjust pH, if necessary, to pH <2 using 1:1 (V/V) sulfuric acid. Pipette 1.0 mL of a surrogate standard spiking solution into each sample. For the sample in each analytical batch selected for spiking, add 1.0 mL of a matrix spiking standard. For base/neutral acid analysis, the amount of the surrogates and matrix spiking compounds added to the sample should result in a final concentration of 100 ng/µL of each analyte in the extract to be analyzed (assuming a 1-µL injection). Extract with methylene chloride for 18–24 h. Next, adjust the pH of the aqueous phase to pH >11 using 10 N sodium hydroxide and extract it with methylene chloride again for 18–24 h. Dry the extract through a column containing anhydrous sodium sulfate and concentrate it to 1 mL using a K-D concentrator.

Soils, sediments, or sludges — Use 30 g of sample. Nonporous or wet samples (gummy or clay type) that do not have a free-flowing sandy texture must be mixed with anhydrous sodium sulfate until the sample is free flowing. Add 1 mL of surrogate standards to all samples, spikes, standards, and blanks. For the sample in each analytical batch selected for spiking, add 1.0 mL of a matrix spiking standard. For base/neutral acid analysis, the amount added of the surrogates and matrix spiking compounds should result in a final concentration of 100 ng/µL of each base/neutral analyte and 200 ng/µL of each acid analyte in the extract to be analyzed (assuming a 1-µL injection). Immediately add a 100-mL mixture of 1:1 methylene chloride:acetone and extract the sample ultrasonically for 3 min and then decant or filter the extracts. Repeat the extraction two or more times. Dry the extract using a column with anhydrous sodium sulfate and concentrate it to 1 mL in a K-D concentrator.

QUALITY CONTROL A methylene chloride solution containing 50 ng/µL of decafluorotriphenylphosphine (DFTPP) is used for tuning the GC/MS system each 12-h shift. A system performance check also must be made during every 12-h shift. A standard containing 50 ng/µL each of 4,4'-DDT, pentachlorophenol, and benzidine is required to verify injection port inertness and GC column performance. A calibration standard at mid-concentration, containing each compound of interest, including all required surrogates, must be performed every 12 h during analysis. After the system performance check is met, calibration check compounds (CCCs) are used to check the validity of the initial calibration.

The internal standard responses and retention times in the calibration check standard must be evaluated immediately after or during data acquisition. If the retention time for any internal standard changes by more than 30 seconds from the last check calibration (12 h), the chromatographic system must be inspected for malfunctions and corrections must be made, as required. If the electron ionization current plot (EICP) area for any of the internal standards changes by a factor of two from the last daily calibration standard check, the mass spectrometer must be inspected for malfunctions and corrections must be made, as appropriate.

Demonstrate, through the analysis of a reagent water blank, that interferences from the analytical system, glassware, and reagents are under control. The blank samples should be carried through all stages of the sample preparation and measurement steps. For each analytical batch (up to 20 samples), a reagent blank, matrix spike, and matrix spike duplicate/duplicate must be analyzed (the frequency of the spikes may be different for different monitoring programs). The blank and spiked samples must be carried through all stages of the sample preparation and measurement steps. A QC reference sample concentrate containing each analyte at a concentration of 100 mg/L in methanol is required.

REFERENCE Test Methods for Evaluating Solid Waste (SW-846). U.S. EPA 1983, Method 8270B, Rev. 2, Nov. 1990. Office of Solid Waste, Washington, DC.

---

n-Hexacosane   EPA Method 1625
CAS #630-01-3

TITLE Semivolatile Organic Compounds by Isotope Dilution GC/MS

MATRIX The compounds may be determined in waters, soils, and municipal sludges by this method.

METHOD SUMMARY This method is used to determine 176 semivolatile toxic organic pollutants associated with the CWA (as amended 1987); the RCRA (as amended 1986); the CERCLA (as amended 1986); and other compounds amenable to extraction and analysis by capillary column gas chromatography-mass spectrometry (GC/MS).

Stable isotopically-labeled analogs of the compounds of interest are added to the sample. If the solids content is less than 1%, a 1-L sample is extracted at pH 12–13, then at pH <2 with methylene chloride using continuous extraction techniques.

If the solids content is 30% or less, the sample is diluted to 1% solids with reagent water, homogenized ultrasonically, and extracted at pH 12–13, then at pH <2 with methylene chloride using continuous extraction techniques. If the solids content is greater than 30%, the sample is extracted using ultrasonic techniques.

Each extract is dried over sodium sulfate, concentrated to a volume of 5 mL, cleaned up using GPC, if necessary, and concentrated. Extracts are concentrated to 1 mL if GPC is not performed, and to 0.5 mL if GPC is performed.

An internal standard is added to the extract, and a 1-mL aliquot of the extract is injected into the GC. The compounds are separated by GC and detected by a MS. The labeled compounds serve to correct the variability of the analytical technique.

INTERFERENCES Solvents, reagents, glassware, and other sample processing hardware may yield artifacts and/or elevated baselines causing misinterpretation of chromatograms and spectra. Materials used in the analysis must be demonstrated to be free from interferences under the conditions of analysis by running method blanks initially and with each sample lot (sample started through the extraction process on a given 8-h shift, to a maximum of 20). Specific selection of reagents and purification of solvents by distillation in all glass systems may be required. Glassware and, where possible, reagents are cleaned by solvent rinse and baking at 450°C for 1-h minimum. Interferences coextracted from samples will vary considerably from source to source, depending on the diversity of the site being sampled.

INSTRUMENTATION Major instrumentation includes a GC with a splitless or on-column injection port for capillary column, a MS with 70 eV electron impact ionization, and a data system to collect and record MS data, and process it. A K-D apparatus is used to concentrate extracts.

GC Column: 30 m × 0.25 mm I.D. 5% phenyl, 94% methyl, 1% vinyl silicone bonded phased fused silica capillary column.

PRECISION & ACCURACY The detection limits of the method are usually dependent on the level of interferences rather than instrumental limitations. The limits typify the minimum quantities that can be detected with no interferences present.

The minimum level (in μg/mL) was 10. This is defined as a minimum level at which the analytical system shall give recognizable mass spectra (background corrected) and acceptable calibration points.

The MDL (in μg/kg) in low solids was 609 and in high solids was 886; these were determined in digested sludge (low solids) and in filter cake or compost (high solids).

Note: Background levels of this compound were present in the sludge tested, resulting in higher than expected MDLs. The MDL for this compound is expected to be approximately 50 μg/kg with no interferences present.

The labeled and native compound initial precision as standard deviation (in μg/L) was 35.

The labeled and native compound initial accuracy as average recovery (in μg/L) was 35–193.

SAMPLE COLLECTION, PRESERVATION & HANDLING
Collect samples in glass containers. Aqueous samples which flow freely are collected in refrigerated bottles using automatic sampling equipment. Solid samples are collected as grab samples using widemouth jars. Maintain samples at 0 to 4°C from the time of collection until extraction. If residual chlorine is present in aqueous samples, add 80 mg sodium thiosulfate/L of water. Begin sample extraction within 7 days of collection, and analyze all extracts within 40 days of extraction.

SAMPLE PREPARATION Samples containing 1% solids or less are extracted directly using continuous liquid-liquid extraction techniques. Samples containing 1 to 30% solids are diluted to the 1% level with reagent water and extracted using continuous liquid-liquid extraction techniques. Samples containing greater than 30% solids are extracted using ultrasonic techniques.

Base/neutral extraction — Adjust the pH of the waters in the extractors to 12–13 with 6 N NaOH. Extract with methylene chloride for 24–48 h.

Acid extraction — Adjust the pH of the waters in the extractors to 2 or less using 6 N sulfuric acid. Extract with methylene chloride for 24–48 h.

Ultrasonic extraction of high solids samples — Add anhydrous sodium sulfate to the sample and QC aliquot(s). Add acetone:methylene chloride (1:1) to the sample and mix thoroughly

Concentrate extracts using a K-D apparatus.

QUALITY CONTROL The analyst is permitted to modify this method to improve separations or lower the costs of measurements, provided all performance specifications are met. Analyses of blanks are required to demonstrate freedom from contamination. When results of spikes indicate atypical method performance for samples, the samples are diluted to bring method performance within acceptable limits.

For low solids (aqueous samples), extract, concentrate, and analyze two sets of four 1-L aliquots (8 aliquots total) of the precision and recovery standard. For high solids samples, two sets of four 30-g aliquots of the high solids reference matrix are used.

Spike all samples with labeled compounds to assess method performance. Compute percent recovery of the labeled compounds using the internal standard method. Compare the labeled compound recovery for each compound with the corresponding labeled compound recovery.

Reagent water and high solids reference matrix blanks are analyzed to demonstrate freedom from contamination. Extract and concentrate a 1-L reagent water blank or a high solids reference matrix blank with each sample's lot (samples started through the extraction process on the same 8-h shift, to a maximum of 20 samples).

Field replicates may be collected to determine the precision of the sampling technique, and spiked samples may be required

to determine the accuracy of the analysis when the internal standard method is used.

REFERENCE Semivolatile Organic Compounds by Isotope Dilution GC/MS. Office of Water Regulation and Standards, U.S. EPA Industrial Technology Division, Washington, DC, EPA Method 1625, Rev. C, June 1989 (contact W.A. Telliard, U.S. EPA, Office of Water Regulations and Standards, 401 M St., SW, Washington, DC, 20460. Phone: 202-382-7131).

n-Hexadecane   EPA Method 1625
CAS #544-76-3

TITLE Semivolatile Organic Compounds by Isotope Dilution GC/MS

MATRIX The compounds may be determined in waters, soils, and municipal sludges by this method.

METHOD SUMMARY This method is used to determine 176 semivolatile toxic organic pollutants associated with the CWA (as amended 1987); the RCRA (as amended 1986); the CERCLA (as amended 1986); and other compounds amenable to extraction and analysis by capillary column gas chromatography-mass spectrometry (GC/MS).

Stable isotopically-labeled analogs of the compounds of interest are added to the sample. If the solids content is less than 1%, a 1-L sample is extracted at pH 12–13, then at pH <2 with methylene chloride using continuous extraction techniques.

If the solids content is 30% or less, the sample is diluted to 1% solids with reagent water, homogenized ultrasonically, and extracted at pH 12–13, then at pH <2 with methylene chloride using continuous extraction techniques. If the solids content is greater than 30%, the sample is extracted using ultrasonic techniques.

Each extract is dried over sodium sulfate, concentrated to a volume of 5 mL, cleaned up using GPC, if necessary, and concentrated. Extracts are concentrated to 1 mL if GPC is not performed, and to 0.5 mL if GPC is performed.

An internal standard is added to the extract, and a 1-mL aliquot of the extract is injected into the GC. The compounds are separated by GC and detected by a MS. The labeled compounds serve to correct the variability of the analytical technique.

INTERFERENCES Solvents, reagents, glassware, and other sample processing hardware may yield artifacts and/or elevated baselines causing misinterpretation of chromatograms and spectra. Materials used in the analysis must be demonstrated to be free from interferences under the conditions of analysis by running method blanks initially and with each sample lot (sample started through the extraction process on a given 8-h shift, to a maximum of 20). Specific selection of reagents and purification of solvents by distillation in all glass systems may be required. Glassware and, where possible, reagents are cleaned by solvent rinse and baking at 450°C for 1-h minimum. Interferences coextracted from samples will vary considerably from source to source, depending on the diversity of the site being sampled.

INSTRUMENTATION Major instrumentation includes a GC with a splitless or on-column injection port for capillary column, a MS with 70 eV electron impact ionization, and a data system to collect and record MS data, and process it. A K-D apparatus is used to concentrate extracts.

GC Column: 30 m × 0.25 mm I.D. 5% phenyl, 94% methyl, 1% vinyl silicone bonded phased fused silica capillary column.

PRECISION & ACCURACY The detection limits of the method are usually dependent on the level of interferences rather than instrumental limitations. The limits typify the minimum quantities that can be detected with no interferences present.

The minimum level (in µg/mL) was 10. This is defined as a minimum level at which the analytical system shall give recognizable mass spectra (background corrected) and acceptable calibration points.

The MDL (in µg/kg) in low solids was 116 and in high solids was 644; these were determined in digested sludge (low solids) and in filter cake or compost (high solids).

Note: Background levels of this compound were present in the sludge tested, resulting in higher than expected MDLs. The MDL for this compound is expected to be approximately 50 µg/kg with no interferences present.

The labeled and native compound initial precision as standard deviation (in µg/L) was 33.
The labeled and native compound initial accuracy as average recovery (in µg/L) was 80–162.

SAMPLE COLLECTION, PRESERVATION & HANDLING
Collect samples in glass containers. Aqueous samples which flow freely are collected in refrigerated bottles using automatic sampling equipment. Solid samples are collected as grab samples using widemouth jars. Maintain samples at 0 to 4°C from the time of collection until extraction. If residual chlorine is present in aqueous samples, add 80 mg sodium thiosulfate/L of water. Begin sample extraction within 7 days of collection, and analyze all extracts within 40 days of extraction.

SAMPLE PREPARATION Samples containing 1% solids or less are extracted directly using continuous liquid-liquid extraction techniques. Samples containing 1 to 30% solids are diluted to the 1% level with reagent water and extracted using continuous liquid-liquid extraction techniques. Samples containing greater than 30% solids are extracted using ultrasonic techniques.

Base/neutral extraction — Adjust the pH of the waters in the extractors to 12–13 with 6 N NaOH. Extract with methylene chloride for 24–48 h.
Acid extraction — Adjust the pH of the waters in the extractors to 2 or less using 6 N sulfuric acid. Extract with methylene chloride for 24–48 h.
Ultrasonic extraction of high solids samples — Add anhydrous sodium sulfate to the sample and QC aliquot(s). Add acetone:methylene chloride (1:1) to the sample and mix thoroughly

Concentrate extracts using a K-D apparatus.

QUALITY CONTROL The analyst is permitted to modify this method to improve separations or lower the costs of measurements, provided all performance specifications are met. Analyses of blanks are required to demonstrate freedom from contamination. When results of spikes indicate atypical method performance for samples, the samples are diluted to bring method performance within acceptable limits.

For low solids (aqueous samples), extract, concentrate, and analyze two sets of four 1-L aliquots (8 aliquots total) of the precision and recovery standard. For high solids samples, two sets of four 30-g aliquots of the high solids reference matrix are used.

Spike all samples with labeled compounds to assess method performance. Compute percent recovery of the labeled compounds using the internal standard method. Compare the labeled compound recovery for each compound with the corresponding labeled compound recovery.

Reagent water and high solids reference matrix blanks are analyzed to demonstrate freedom from contamination. Extract and concentrate a 1-L reagent water blank or a high solids reference matrix blank with each sample's lot (samples started through the extraction process on the same 8-h shift, to a maximum of 20 samples).

Field replicates may be collected to determine the precision of the sampling technique, and spiked samples may be required to determine the accuracy of the analysis when the internal standard method is used.

REFERENCE Semivolatile Organic Compounds by Isotope Dilution GC/MS. Office of Water Regulation and Standards, U.S. EPA Industrial Technology Division, Washington, DC, EPA Method 1625, Rev. C, June 1989 (contact W.A. Telliard, U.S. EPA, Office of Water Regulations and Standards, 401 M St., SW, Washington, DC, 20460. Phone: 202-382-7131).

Hexamethyl phosphoramide   **EPA Method 8270**
CAS #680-31-9

TITLE Semivolatile Organic Compounds by GC/MS

MATRIX This method is used to determine the concentration of semivolatile organic compounds in extracts prepared from all types of solid waste matrices, soils, and groundwater. Although surface waters are not specifically mentioned, this method should be applicable to water samples from rivers, lakes, etc.

METHOD SUMMARY This method covers 259 semivolatile organic compounds. In very limited applications direct injection of the sample into the GC/MS system may be appropriate, but this results in very high detection limits (approximately 10,000 µg/L). Typically, a 1-L liquid sample, containing surrogate, and matrix spiking standards, is extracted in a continuous extractor first under acid conditions and then under basic conditions. Typically 30 g of a solid sample, containing surrogate, and matrix spiking standards, is extracted ultrasonically. After concentrating the extract to 1 mL it is spiked with 10 µL of an internal standard solution just prior to analysis by GC/MS. The volume injected should contain about 100 ng of base/neutral and 200 ng of acid surrogates (for a 1-µL injection). Analysis is performed by GC/MS using a capillary GC column.

INTERFERENCES Raw GC/MS data from all blanks, samples, and spikes must be evaluated for interferences. Contamination by carryover can occur whenever high-concentration and low-concentration samples are sequentially analyzed. To reduce carryover, the sample syringe must be rinsed out between samples with solvent. Whenever an unusually concentrated sample is encountered, it should be followed by the analysis of blank solvent to check for cross-contamination.

INSTRUMENTATION A GC/MS and a data system are required. The GC column used is a 30 m × 0.25 mm I.D. (or 0.32 mm I.D.) 1um film thickness silicone-coated fused silica capillary column. A continuous liquid-liquid extractor equipped with Teflon® or glass connection joints and stopcocks requiring no lubrication, a K-D concentrating apparatus, water bath, and an ultrasonic disrupter with a minimum power of 300 W and with pulsing capability are also required.

PRECISION & ACCURACY The estimated quantitation limit (EQL) of Method 8270B for determining an individual compound is approximately 1 mg/kg (wet weight) for soil or sediment samples, 1–200 mg/kg for wastes (dependent on matrix and method of preparation), and 10 µg/L for groundwater samples. EQLs will be proportionately higher for sample extracts that require dilution to avoid saturation of the detector.

The EQL(b) for groundwater in µg/L is 20.
The EQL (a, b) for low concentrations in soil and sediment
   in µg/kg is not determined.
Accuracy as µg/L is not listed.
Overall precision in µg/L is not listed.

(a) *EQLs listed for soil/sediment are based on wet weight. Normally data is reported in a dry-weight basis; therefore, EQLs will be higher based on the % dry weight of each sample. This calculation is based on a 30-g sample and gel permeation chromatography cleanup.*
(b) *Sample EQLs are highly matrix-dependent. The EQLs are provided for guidance and may not always be achievable.*
C = *True value for concentration, in µg/L.*
X = *Average recovery found for measurements of samples containing a concentration of C, in µg/L.*

ESTIMATED QUANTITATION LIMIT

Other Matrices	Factor (a)
High-concentration soil and sludges by sonicator	7.5
Non-water miscible waste	75

(a) *EQL for other matrices = [EQL for low soil/sediment] × [Factor]. This estimated EQL is similar to an EPA "Practical Quantitation Limit."*

SAMPLING METHOD

Liquid samples — Use a 1 or 2½ gallon amber glass bottle with a screw-top Teflon®-lined cover that has been prewashed with detergent and rinsed with distilled water and methanol (or isopropanol).

Soils, sediments, or sludges — Use an 8-oz. widemouth glass with a screw-top Teflon®-lined cover that has been prewashed with detergent and rinsed with distilled water and methanol (or isopropanol).

SAMPLE PRESERVATION

Liquid samples — If residual chlorine is present, add 3 mL of 10% sodium thiosulfate per gallon, cool to 4°C and store in a solvent-free refrigerator until analysis; if chlorine is not present, then eliminate the sodium thiosulfate addition.

Soils, sediments, or sludges — Cool samples to 4°C and store in a solvent-free refrigerator.

MHT Liquid samples must be extracted within 7 days and the extracts analyzed within 40 days. Soils, sediments, or sludges may be stored for a maximum of 14 days and the extracts analyzed within 40 days.

SAMPLE PREPARATION

Liquid samples — Transfer 1 L quantitatively to a continuous extractor. If high concentrations are anticipated, a smaller volume may be used and then diluted with organic-free reagent water to 1 L. Adjust pH, if necessary, to pH <2 using 1:1 (V/V) sulfuric acid. Pipette 1.0 mL of a surrogate standard spiking solution into each sample. For the sample in each analytical batch selected for spiking, add 1.0 mL of a matrix spiking standard. For base/neutral acid analysis, the amount of the surrogates and matrix spiking compounds added to the sample should result in a final concentration of 100 ng/µL of each analyte in the extract to be analyzed (assuming a 1-µL injection). Extract with methylene chloride for 18–24 h. Next, adjust the pH of the aqueous phase to pH >11 using 10 N sodium hydroxide and extract it with methylene chloride again for 18–24 h. Dry the extract through a column containing anhydrous sodium sulfate and concentrate it to 1 mL using a K-D concentrator.

Soils, sediments, or sludges — Use 30 g of sample. Nonporous or wet samples (gummy or clay type) that do not have a free-flowing sandy texture must be mixed with anhydrous sodium sulfate until the sample is free flowing. Add 1 mL of surrogate standards to all samples, spikes, standards, and blanks. For the sample in each analytical batch selected for spiking, add 1.0 mL of a matrix spiking standard. For base/neutral acid analysis, the amount added of the surrogates and matrix spiking compounds should result in a final concentration of 100 ng/µL of each base/neutral analyte and 200 ng/µL of each acid analyte in the extract to be analyzed (assuming a 1-µL injection). Immediately add a 100-mL mixture of 1:1 methylene chloride:acetone and extract the sample ultrasonically for 3 min and then decant or filter the extracts. Repeat the extraction two or more times. Dry the extract using a column with anhydrous sodium sulfate and concentrate it to 1 mL in a K-D concentrator.

QUALITY CONTROL A methylene chloride solution containing 50 ng/µL of decafluorotriphenylphosphine (DFTPP) is used for tuning the GC/MS system each 12-h shift. A system performance check also must be made during every 12-h shift. A standard containing 50 ng/µL each of 4,4′-DDT, pentachlorophenol, and benzidine is required to verify injection port inertness and GC column performance. A calibration standard at mid-concentration, containing each compound of interest, including all required surrogates, must be performed every 12 h during analysis. After the system performance check is met, calibration check compounds (CCCs) are used to check the validity of the initial calibration.

The internal standard responses and retention times in the calibration check standard must be evaluated immediately after or during data acquisition. If the retention time for any internal standard changes by more than 30 seconds from the last check calibration (12 h), the chromatographic system must be inspected for malfunctions and corrections must be made, as required. If the electron ionization current plot (EICP) area for any of the internal standards changes by a factor of two from the last daily calibration standard check, the mass spectrometer must be inspected for malfunctions and corrections must be made, as appropriate.

Demonstrate, through the analysis of a reagent water blank, that interferences from the analytical system, glassware, and reagents are under control. The blank samples should be carried through all stages of the sample preparation and measurement steps. For each analytical batch (up to 20 samples), a reagent blank, matrix spike, and matrix spike duplicate/duplicate must be analyzed (the frequency of the spikes may be different for different monitoring programs). The blank and spiked samples must be carried through all stages of the sample preparation and measurement steps. A QC reference sample concentrate containing each analyte at a concentration of 100 mg/L in methanol is required.

REFERENCE Test Methods for Evaluating Solid Waste (SW-846). U.S. EPA 1983, Method 8270B, Rev. 2, Nov. 1990. Office of Solid Waste, Washington, DC.

Hexanoic acid   **EPA Method 1625**
CAS #142-62-1

TITLE Semivolatile Organic Compounds by Isotope Dilution GC/MS

MATRIX The compounds may be determined in waters, soils, and municipal sludges by this method.

METHOD SUMMARY This method is used to determine 176 semivolatile toxic organic pollutants associated with the CWA (as amended 1987); the RCRA (as amended 1986); the CERCLA (as amended 1986); and other compounds amenable to extraction and analysis by capillary column gas chromatography-mass spectrometry (GC/MS).

Stable isotopically-labeled analogs of the compounds of interest are added to the sample. If the solids content is less than 1%, a 1-L sample is extracted at pH 12–13, then at pH <2 with methylene chloride using continuous extraction techniques.

If the solids content is 30% or less, the sample is diluted to 1% solids with reagent water, homogenized ultrasonically, and extracted at pH 12–13, then at pH <2 with methylene chloride using continuous extraction techniques. If the solids content is

greater than 30%, the sample is extracted using ultrasonic techniques.

Each extract is dried over sodium sulfate, concentrated to a volume of 5 mL, cleaned up using GPC, if necessary, and concentrated. Extracts are concentrated to 1 mL if GPC is not performed, and to 0.5 mL if GPC is performed.

An internal standard is added to the extract, and a 1-mL aliquot of the extract is injected into the GC. The compounds are separated by GC and detected by a MS. The labeled compounds serve to correct the variability of the analytical technique.

INTERFERENCES Solvents, reagents, glassware, and other sample processing hardware may yield artifacts and/or elevated baselines causing misinterpretation of chromatograms and spectra. Materials used in the analysis must be demonstrated to be free from interferences under the conditions of analysis by running method blanks initially and with each sample lot (sample started through the extraction process on a given 8-h shift, to a maximum of 20). Specific selection of reagents and purification of solvents by distillation in all glass systems may be required. Glassware and, where possible, reagents are cleaned by solvent rinse and baking at 450°C for 1-h minimum. Interferences coextracted from samples will vary considerably from source to source, depending on the diversity of the site being sampled.

INSTRUMENTATION Major instrumentation includes a GC with a splitless or on-column injection port for capillary column, a MS with 70 eV electron impact ionization, and a data system to collect and record MS data, and process it. A K-D apparatus is used to concentrate extracts.

GC Column: 30 m × 0.25 mm I.D. 5% phenyl, 94% methyl, 1% vinyl silicone bonded phased fused silica capillary column.

PRECISION & ACCURACY The detection limits of the method are usually dependent on the level of interferences rather than instrumental limitations. The limits typify the minimum quantities that can be detected with no interferences present.

The minimum level (in µg/mL) was not listed. This is defined as a minimum level at which the analytical system shall give recognizable mass spectra (background corrected) and acceptable calibration points.

The MDL (in µg/kg) in low solids was not listed and in high solids was not listed; these were determined in digested sludge (low solids) and in filter cake or compost (high solids).

The labeled and native compound initial precision as standard deviation (in µg/L) was not listed.
The labeled and native compound initial accuracy as average recovery (in µg/L) was not listed.

SAMPLE COLLECTION, PRESERVATION & HANDLING Collect samples in glass containers. Aqueous samples which flow freely are collected in refrigerated bottles using automatic sampling equipment. Solid samples are collected as grab samples using widemouth jars. Maintain samples at 0 to 4°C from the time of collection until extraction. If residual chlorine is present in aqueous samples, add 80 mg sodium thiosulfate/L of water. Begin sample extraction within 7 days of collection, and analyze all extracts within 40 days of extraction.

SAMPLE PREPARATION Samples containing 1% solids or less are extracted directly using continuous liquid-liquid extraction techniques. Samples containing 1 to 30% solids are diluted to the 1% level with reagent water and extracted using continuous liquid-liquid extraction techniques. Samples containing greater than 30% solids are extracted using ultrasonic techniques.

Base/neutral extraction — Adjust the pH of the waters in the extractors to 12–13 with 6 N NaOH. Extract with methylene chloride for 24–48 h.
Acid extraction — Adjust the pH of the waters in the extractors to 2 or less using 6 N sulfuric acid. Extract with methylene chloride for 24–48 h.
Ultrasonic extraction of high solids samples — Add anhydrous sodium sulfate to the sample and QC aliquot(s). Add acetone:methylene chloride (1:1) to the sample and mix thoroughly

Concentrate extracts using a K-D apparatus.

QUALITY CONTROL The analyst is permitted to modify this method to improve separations or lower the costs of measurements, provided all performance specifications are met. Analyses of blanks are required to demonstrate freedom from contamination. When results of spikes indicate atypical method performance for samples, the samples are diluted to bring method performance within acceptable limits.

For low solids (aqueous samples), extract, concentrate, and analyze two sets of four 1-L aliquots (8 aliquots total) of the precision and recovery standard. For high solids samples, two sets of four 30-g aliquots of the high solids reference matrix are used.

Spike all samples with labeled compounds to assess method performance. Compute percent recovery of the labeled compounds using the internal standard method. Compare the labeled compound recovery for each compound with the corresponding labeled compound recovery.

Reagent water and high solids reference matrix blanks are analyzed to demonstrate freedom from contamination. Extract and concentrate a 1-L reagent water blank or a high solids reference matrix blank with each sample's lot (samples started through the extraction process on the same 8-h shift, to a maximum of 20 samples).

Field replicates may be collected to determine the precision of the sampling technique, and spiked samples may be required to determine the accuracy of the analysis when the internal standard method is used.

REFERENCE Semivolatile Organic Compounds by Isotope Dilution GC/MS. Office of Water Regulation and Standards, U.S. EPA Industrial Technology Division, Washington, DC, EPA Method 1625, Rev. C, June 1989 (contact W.A. Telliard, U.S. EPA, Office of Water Regulations and Standards, 401 M St., SW, Washington, DC, 20460. Phone: 202-382-7131).

2-Hexanone
CAS #591-78-6

EPA Method 1624

TITLE Volatile Organic Compounds by Isotope Dilution GC/MS

MATRIX Compounds may be determined in waters, soils, and municipal sludges by this method.

METHOD SUMMARY This method is used to determine 58 volatile toxic organic pollutants associated with the CWA (as amended 1987); the RCRA (as amended 1986); the CERCLA (as amended 1986); and other compounds amenable to purge-and-trap gas chromatography-mass spectrometry (GC/MS).

If the solids content is less than 1%, stable isotopically-labeled analogs of the compounds of interest are added to a 5-mL sample and the sample is purged with an inert gas at 20–25°C in a chamber designed for soil or water samples. If the solids content is greater than 1%, 5 mL of reagent water and the labeled compounds are added to a 5-g aliquot of sample and the mixture is purged at 40°C. Compounds that will not purge at 20–25°C or at 40°C are purged at 78–85°C. In the purging process, the volatile compounds are transferred from the aqueous phase into the gaseous phase where they are passed into a sorbent column, and trapped. After purging is completed, the trap is backflushed and heated rapidly to desorb the compounds into a GC. The compounds are separated by the GC and detected by a MS. The labeled compounds serve to correct the variability of the analytical technique.

INTERFERENCES Impurities in the purge gas, organic compounds outgassing from the plumbing upstream of the trap, and solvent vapors in the lab account for most problems. Samples can be contaminated by diffusion of volatile organic compounds (particularly methylene chloride) through the bottle seal during shipment and storage. Contamination by carryover can occur when high-level and low-level samples are analyzed sequentially. When an unusually concentrated sample is encountered, follow it by analysis of a reagent water blank to check for carryover.

INSTRUMENTATION Major equipment includes a GC with linear temperature programming and a glass jet separator as the MS interface, a MS with 70 eV electron impact ionization, and a data system to collect and record response factors.

Column: 2.8 m × 2 mm I.D. glass, packed with 1% SP-1000 on Carbopak B, 60/80 mesh, or equivalent.

PRECISION & ACCURACY The detection limits of the method are usually dependent on the level of interferences rather than instrumental limitations. The method detection limits were determined in digested sludge (low solids) and in filter cake or compost (high solids).

The MDL (in µg/kg) for low solids is not listed and for high solids is not listed.
Labeled and native compound precision (in µg/L) as standard deviation was not listed.
Labeled and native compound accuracy (in µg/L) as average recovery was not listed.
Acceptance criteria are at 20 µg/L for this compound.

SAMPLE COLLECTION, PRESERVATION & HANDLING Grab samples are collected in glass containers having a total volume greater than 20 mL. Fill and seal each bottle so that no air bubbles are entrapped. Samples are maintained at 0 to 4°C from the time of collection until analysis. If an aqueous sample contains residual chlorine, add sodium thiosulfate preservative (10 mg/40 mL) to the empty sample bottles just prior to shipment to the sample site. All samples must be analyzed within 14 days of collection.

SAMPLE PREPARATION Samples containing less than 1% solids are analyzed directly as aqueous samples. Samples containing 1% solids or greater are analyzed as solid samples utilizing one of two methods, depending on the levels of pollutants, in the sample. Samples containing 1% solids or greater, and low to moderate levels of pollutants are analyzed by purging a known weight of sample added to 5 mL of reagent water. Samples containing 1% solids or greater, and high levels of pollutants, are extracted with methanol, and an aliquot of the methanol extract is added to reagent water and purged.

QUALITY CONTROL A field blank prepared from reagent water and carried through the sampling and handling protocol may serve as a check on contamination from shipment and storage.

The analyst is permitted to modify this method to improve separations or lower the costs of measurements, provided all performance specifications are met. Analyses of blanks are required. When results of spikes indicate atypical method performance for samples, the samples are diluted to bring method performance within acceptable limits. Analyze two sets of four 5-mL aliquots (8 aliquots total) of the aqueous performance standard. Spike all samples with labeled compounds to assess method performance on the sample matrix. Compute the percent recovery of the labeled compounds using the internal standard method. Compare the percent recovery for each compound with the corresponding labeled compound recovery. Reagent water blanks are analyzed to demonstrate freedom from carryover contamination. Field replicates may be collected to determine the precision of the sampling technique, and spiked samples may be required to determine the accuracy of the analysis when the internal method is used.

REFERENCE Volatile Organic Compounds by Isotope Dilution GC/MS. Office of Water Regulation and Standards, U.S. EPA Industrial Technology Division, Washington, DC, EPA Method 1624, Rev. C, June 1989 (contact W.A. Telliard, U.S. EPA, Office of Water Regulations and Standards, 401 M St., SW, Washington, DC, 20460. Phone: 202-382-7131).

2-Hexanone
CAS #591-78-6

EPA Method 8240

TITLE Volatile Organics By GC/MS: Packed Column Technique

MATRIX Nearly all types of sample matarices, regardless of water content, can be analyzed using this method. This includes groundwater, aqueous sludges, caustic liquors, acid liquors, waste solvents, oily wastes, mousses, tars, fibrous wastes, polymetric

emulsions, filter cakes, spent carbons, spent catalysts, soils, and sediments.

METHOD SUMMARY Method 8240B covers 80 volatile organic compounds that are introduced into a gas chromatograph by the purge-and-trap method or by direct injection (in limited applications). For the purge-and-trap method an inert gas (zero grade nitrogen or helium) is bubbled through a 5-mL solution at ambient temperature. Purged sample components are trapped in a tube of sorbent materials. When purging is complete, the sorbent tube is heated and backflushed with inert gas to desorb the trapped components onto a GC column.

INTERFERENCES Impurities in the purge gas and from organic compounds outgassing from the plumbing ahead of the trap account for many contamination problems. Interferences purged or coextracted from the samples will vary considerably from source to source. Cross-contamination can occur whenever high-level and low-level samples are analyzed sequentially. Whenever an unusually concentrated sample is analyzed, it should be followed by the analysis of organic-free reagent water to check for cross-contamination. Samples also can be contaminated by diffusion of volatile organics (particularly methylene chloride and fluorocarbons) through the septum seal into the sample during shipment and storage. A trip blank can serve as a check on such contamination. The lab where volatile analysis is performed and also the refrigerated storage area should be completely free of solvents.

INSTRUMENTATION A gas chromatograph/mass spectrometry/data system (GC/MS) equipped with a 6 ft × 0.1 in I.D. glass column packed with 1% SP-1000 on Carbopack-B (60/80 mesh) is required. Also needed is a 5-mL purging device, a sorbent trap, and a thermal desorption apparatus.

PRECISION & ACCURACY This method is reported to have been tested by 15 laboratories using organic-free reagent water, drinking water, surface water, and industrial wastewaters (not specified) fortified at six concentrations over the range 5–600 µg/L.

Sample estimated quantitation limits (EQLs) are highly matrix-dependent. The EQLs listed may not always be achievable. EQLs listed for soils or sediments are based on wet weight. Normally, data is reported on a dry-weight basis; therefore, EQLs will be higher, based on the percent dry weight of each sample. Note that EQLs are even more variable than MDLs and that they are highly variable depending on the matrix being analyzed.

EQL in groundwater in µg/L was 50.
EQL in low soil or sediment in µg/kg was 50.
Accuracy (a) in µg/L was not listed.
Precision (b) in µg/L was not listed.

(a) *Average recovery found for measurements of samples containing a concentration of C, in µg/L.*
(b) *Overall precision found for measurements of samples with average recovery X for samples containing a concentration of C in µg/L.*
X = *Average recovery found for measurement of samples containing a concentration of C in µg/L.*

MULTIPLICATION FACTORS FOR OTHER MATRICES

Other Matrices	Factor (a)
Waste miscible liquid waste	50
High-concentration soil and sludge	125
Non-water miscible waste	500

(a) EQL = [EQL for low soil/sediment] × [Factor]. For non-aqueous samples, the factor is on a wet-weight basis.

SAMPLING METHOD
Liquid samples — Use a 40-mL glass screw-cap VOA vial with a Teflon®-faced silicone septum that has been prewashed, rinsed with distilled deionized water, and oven dried. However, if residual chlorine is present, collect sample in a 40-oz. soil VOA container which has been pre-preserved with 4 drops of 10% sodium thiosulfate, mix gently, and then transfer the sample to a 40-mL VOA vial. Collect bubble-free samples in duplicate and seal them in separate plastic bags.

Soils or sediments, and sludges — Use an 8-oz. widemouth glass bottle with a Teflon®-faced silicone septum that has been prewashed with detergent, rinsed with distilled deionized water, and oven dried. Tap slightly to eliminate free air space. Collect samples in duplicate and seal them in separate plastic bags.

SAMPLE PRESERVATION
Liquid samples — Add 4 drops of concentrated HCL and immediately cool samples to 4°C and store in a solvent-free refrigerator.

Soils or sediments, and sludges — Cool samples to 4°C and store in a solvent-free refrigerator.

MHT Maximum holding time is 14 days from the date of sample collection.

SAMPLE PREPARATION
Liquid samples — Remove the plunger from a 5-mL syringe and carefully pour the sample into the syringe barrel to just short of overflowing. Replace the syringe plunger and compress the sample. Open the syringe valve and vent any residual air while adjusting the sample volume to 5.0 mL. If there is only one volatile organic analysis (VOA) vial, a second syringe should be filled at this time to protect against possible loss of sample integrity. Add 10 µL of surrogate spiking solution and 10 µL of internal standard spiking solution through the valve bore of the 5-mL syringe, then close the valve. The surrogate and internal standards may be mixed and added as a single spiking solution.

Sediments, soils, and waste samples — All samples of this type should be screened by GC analysis using a headspace method (EPA Method 3810) or the hexadecane extraction and screening method (EPA Method 3820). Use the screening data to determine whether to use the low-concentration method (0.005–1 mg/kg) or the high-concentration method (>1 mg/kg).

Low-concentration method — The low-concentration method is based on purging a heated sediment or soil sample mixed with organic-free reagent water containing the surrogate and internal standards. Analyze all reagent blanks and standards under the same conditions as the samples.

Use a 5-g sample if the expected concentration is <0.1 mg/kg or a 1-g sample for expected concentrations between 0.1 and 1 mg/kg. Mix the contents of the sample container with a narrow metal spatula. Weigh the amount of the sample into a tared purge device. Add the spiked water to the purge device, which contains the weighed amount of sample, and connect the device to the purge-and-trap system.

High-concentration method — This method is based on extracting the sediment or soil with methanol. A waste sample is either extracted or diluted, depending on its solubility in methanol. Wastes that are insoluble in methanol are diluted with reagent tetraglyme or possibly polyethylene glycol (PEG). An aliquot of the extract is added to organic-free reagent water containing surrogate and internal standards. This is purged at ambient temperature. All samples with an expected concentration of >1.0 mg/kg should be analyzed by this method.

Mix the contents of the sample container with a narrow metal spatula. For sediments or soils and solid wastes that are insoluble in methanol, weigh 4 g (wet weight) of sample into a tared 20-mL vial. For waste that is soluble in methanol, tetraglyme, or PEG, weigh 1 g (wet weight) into a tared scintillation vial or culture tube or a 10-mL volumetric flask. Quickly add 9.0 mL of appropriate solvent then add 1.0 mL of a surrogate spiking solution to the vial, cap it, and shake it for 2 min.

METHANOL EXTRACT REQUIRED FOR ANALYSIS OF HIGH-CONCENTRATION SOILS OR SEDIMENTS

Approximate Concentration Range	Volume of Methanol Extract (a)
500–10,000 µg/kg	100 µL
1,000–20,000 µg/kg	50 µL
5,000–100,000 µg/kg	10 µL
25,000–500,000 µg/kg	100 µL of 1/50 dilution (b)

Calculate appropriate dilution factor for concentrations exceeding this table.

(a) The volume of methanol added to 5 mL of water being purged should be kept constant. Therefore, add to the 5-mL syringe whatever volume of methanol is necessary to maintain a volume of 100 µL added to the syringe.
(b) Dilute an aliquot of the methanol extract and then take 100 µL for analysis.

QUALITY CONTROL Demonstrate, through the analysis of a reagent water blank, that interferences from the analytical system, glassware, and reagents are under control. Blank samples should be carried through all stages of the sample preparation and measurement steps. For each analytical batch (up to 20 samples), a reagent blank, matrix spike, and matrix spike duplicate must be analyzed (the frequency of the spikes may be different for different monitoring programs). The blank and spiked samples must be carried through all stages of the sample preparation and measurement steps. QC samples mentioned in the section on Interferences will also be needed as appropriate to those situations.

REFERENCE Test Methods for Evaluating Solid Waste (SW-846). U.S. EPA. 1983. Method 8240B, Rev. 2, Nov. 1990. Office of Solid Wastes, Washington, DC.

Hexazinone   EPA Method 507
CAS #51235-04-2

TITLE Determination of Nitrogen and Phosphorus-Containing Pesticides in Water by GC/NPD

MATRIX This method is applicable to the determination of certain nitrogen and phosphorus-containing pesticides in finished drinking water and groundwater.

METHOD SUMMARY Method 507 covers 46 nitrogen- and phosphorus-containing pesticides. A 1-L sample is fortified with a surrogate standard, salted, buffered, extracted with methylene chloride, and concentrated; then the solvent is exchanged with methyl tert-butyl ether (MTBE) and concentrated again, and a 2-µL aliquot of a sample extract is injected into a GC system equipped with a selective nitrogen-phosphorus detector and a capillary column for analysis.

INTERFERENCES Method interferences may be caused by contaminants in solvents, reagents, glassware, and other sample processing apparatus. Interfering contamination may occur when a sample containing low concentrations of analytes is analyzed immediately following a sample containing relatively high concentrations. One or more injections of MTBE should be made following the analysis of a sample with high concentrations of analytes to check for analyte carryover. Matrix interferences may be caused by contaminants that are coextracted from the sample. The extent of matrix interferences will vary considerably from source to source, depending upon the water sampled.

INSTRUMENTATION A gas chromatograph system (GC) equipped with a nitrogen-phosphorus detector (NPD) is needed.

Column 1: 30 m × 0.25 mm I.D. DB-5 bonded fused silica column, 0.25 µm film thickness, or equivalent.
Column 2: 30 m × 0.25 mm I.D. DB-1701 bonded fused silica column, 0.25 µm film thickness, or equivalent.

PRECISION & ACCURACY This method has been validated in a single lab and estimated detection limits (EDLs) have been determined for each analyte. Observed detection limits may vary among waters, depending upon the nature of the interferences in the sample matrix and the specific instrumentation used. Analytes that are not separated chromatographically cannot be individually identified and measured unless an alternative technique for identification and quantification exist.

The estimated detection limit (in µg/L) was 0.76. The EDL is defined as either method detection limit or a level of compound in a sample yielding a peak in the final extract with signal-to-noise ratio of approximately 5, whichever value is higher.

The concentration used for these measurements (in µg/L) was 7.6.
The accuracy (as % recovery) was 90.
The precision (% RSD) was 7.

SAMPLING METHOD Grab samples are collected in 1-L glass sample bottles (prewashed with detergent and hot tap

water, rinsed with reagent water, and dried in an oven at 400°C for 1 h) with screw caps lined with PTFE-fluorocarbon.

SAMPLE PRESERVATION Add mercuric chloride to the sample bottle in amounts to produce a concentration of 10 mg/L. If residual chlorine is present, add 80 mg of sodium thiosulfate/L of sample to the sample bottle prior to collection. After collection, seal bottle and shake vigorously for 1 min, then cool the sample to 4°C immediately and store it at 4°C in the dark until extraction.

MHT Maximum holding time of the samples, and in some cases the extracts, is 14 days.

SAMPLE PREPARATION Fortify the sample with 50 µL of the surrogate standard solution, adjust to pH 7 with phosphate buffer, add 100 g NaCl to the sample, and seal and shake to dissolve the salt; then extract with methylene chloride in a separatory funnel or in a mechanical tumbler bottle. Dry the extract by pouring it through a solvent-rinsed drying column containing about 10 cm of anhydrous sodium sulfate. Collect the extract in a Kuderna-Danish (K-D) concentrator and rinse the column with 20–30 mL methylene chloride. Concentrate the extract to about 2 mL and rinse the flask and its lower joint into the concentrator tube with 1 to 2 mL of methyl t-butyl ether (MTBE). Add 5–10 mL of MTBE and concentrate the extract twice (adding more MTBE) to a final volume of 5.0 mL and store it at 4°C until analysis.

Note: If methylene chloride is not completely removed from the final extract, it may cause detector problems.

QUALITY CONTROL Minimum quality control requirements are initial demonstration of lab capability, determination of surrogate compound recoveries in each sample and blank, monitoring internal standard peak area or height in each sample and blank, analysis of lab reagent blanks, lab fortified samples, lab fortified blanks, and other QC samples. A lab reagent blank is analyzed to demonstrate that all glassware and reagent interferences are under control.

Initial demonstration of capability is fulfilled by analyzing four fortified reagent water samples with the recovery value for each analyte falling within the acceptable range (±30% average recovery). Surrogate recoveries from samples or method blanks must be 70–130%. The internal standard response for any sample chromatogram should not deviate from the daily calibration check standard's internal standard response by more than 30% or lab fortified blanks and sample matrices are used to assess lab performance and analyte recovery, respectively.

If the response for the target analyte peak exceeds the working range of the system, dilute the extract and reanalyze. Alternative techniques such as an alternate detector or second chromatography column should be used to confirm peak identification when sample components are not resolved adequately.

EPA CONTACT & HOTLINE For technical questions contact Dr. Baldev Bathija, U.S. EPA, Office of Ground Water and Drinking Water (WH-550D), 401 M St. SW, Washington, DC 20460. Tel. (202) 260-3040. For further information the EPA Safe Drinking Water Hotline may be called at: (800) 426-4791.

REFERENCE Methods for the Determination of Organic Compounds in Drinking Water, EPA/600/4-88/039 (revised July 1991). U.S. EPA Environmental Monitoring Systems Laboratory, Cincinnati, OH, 45268, U.S.A. Available from the National Technical Information Service (NTIS), 5285 Port Royal Road, Springfield, VA 22161; Tel. 800-553-6847. NTIS Order Number is PB91-231480.

Hexyl 2-ethylhexyl phthalate   **EPA Method 8061**
CAS #75673-16-4

TITLE Phthalate Esters by Capillary Gas Chromatography With Electron Capture Detection (GC/ECD)

MATRIX This method covers aqueous and solid matrices. This includes a wide variety such as drinking water, groundwater, industrial wastewaters, surface waters, soils, solids, and sediments.

METHOD SUMMARY This method is used to determine the identities and concentrations of phthalate esters in liquid, solid and sludge matrices. When used to analyze for any or all of the target analytes, compound identification should be supported by at least one additional qualitative technique. This method describes conditions for parallel column, dual electron capture detector analysis, which fulfills the above requirement. Alternatively, GC/MS could be used for compound confirmation.

A measured volume or weight of sample (approximately 1 L for liquids, 10 to 30 g for solids and sludges) is extracted by using the appropriate sample extraction technique specified in EPA Method 3510, EPA Method 3540, and EPA Method 3550. After cleanup, the extract is analyzed by GC/ECD.

INTERFERENCES The sensitivity of this method usually depends on the level of interferences rather than on instrumental limitations. If interferences prevent detection of the analytes, cleanup of the sample extracts is necessary. Either EPA Method 3610 or EPA Method 3620 alone or followed by EPA Method 3660, Sulfur Cleanup, may be used to eliminate interferences in the analysis. EPA Method 3640, Gel Permeation Cleanup, is applicable for samples that contain high amounts of lipids and waxes.

Interferences coextracted from the samples will vary considerably from waste to waste. Glassware must be scrupulously clean. All glassware require treatment in a muffle furnace at 400°C for 2 to 4 h, or thorough rinsing with pesticide-grade solvent, prior to use. Volumetric glassware should not be heated in a muffle furnace. Storage of glassware in the lab introduces contamination, even if the glassware is wrapped in aluminum foil. Sodium sulfate, Florisil, and alumina may be contaminated with phthalate esters and, therefore, use of these materials in sample cleanup should be employed cautiously. If these materials are used, they must be obtained packaged in glass. Heating at 400°C for sodium sulfate, 320°C for Florisil, and 210°C for alumina is recommended. Glass wool used in any step of sample preparation should be a specially treated pyrex wool, pesticide grade, and must be baked at 400°C for 4 h immediately prior to use.

Paper thimbles and filter paper must be exhaustively washed with the solvent that will be used in the sample extraction. Soxhlet extraction of paper thimbles and filter paper for 12 h with fresh solvent should be repeated for a minimum of three times. Method blanks should be obtained before any of the precleaned thimbles or filter papers are used.

INSTRUMENTATION Gas chromatograph suitable for on-column and split/splitless injections.

Column 1: 30 m × 0.53 mm ID, 5% phenyl/95% methyl silicone fused-silica open tubular column, DB-5, 1.5 µg film thickness.

Column 2: 30 m × 0.53 mm ID, 14% cyanopropyl phenyl silicone fused-silica open tubular column, DB-1701, 1.0 µg film thickness.

A dual electron capture detector (ECD) is used. A Kuderna-Danish (K-D) apparatus is required along with a vacuum manifold consisting of individually adjustable, easily accessible flow-control valves for up to 24 cartridges, sample rack, chemically resistant cover and seals, heavy-duty glass basin, removable stainless steel solvent guides, built-in vacuum gauge and valve. Also, 6-mL, 1-g solid-phase extraction cartridges, LC-Florisil or equivalent, prepackaged, ready to use will be needed.

PRECISION & ACCURACY The MDL actually achieved in a given analysis will vary, as it is dependent on instrument sensitivity and matrix effects. This method has been tested in a single lab. Single-operator precision, overall precision, and method accuracy were found to be related to the concentration of the compounds and the type of matrix.

MULTIPLICATION FACTORS FOR OTHER MATRICES (a)

Matrix	Factor (b)
Groundwater	10
Low-concentration soil by ultrasonic cleanup extraction with GPC	670
High-concentration soil and sludges by ultrasonic extraction	10,000
Non-water miscible waste	100,000

(a) Sample EQLs are highly matrix-dependent. The EQLs listed are provided for guidance and may not always be achievable.
(b) EQL = [Method detection limit] × [Factor]. For non-aqueous samples, the factor is on a wet-weight basis.

The MDL using 7 replicate determinations and a spike concentration of 100 µg/L was 130 ng/L.

The average recovery from HPLC-grade water using 4 determinatons and a spike concentration of 100 µg/L was 93.9%.

The precision (as RSD) from HPLC-grade water using 4 determinatons and a spike concentration of 100 µg/L was 22.4%.

The average recovery from groundwater using 4 determinatons and a spike concentration of 100 µg/L was 83.4%.

The precision (as RSD) from groundwater using 4 determinatons and a spike concentration of 100 µg/L was 8.8%.

The average recovery (in %) with %RSD (in parenthesis) from 3 determinations and a spike concentration of 20 µg/L in water was 84.7 (5.3) using 3M Empore Disks and EPA Method 8061.

The average recovery (in %) with %RSD (in parenthesis) from 3 determinations and a spike concentration of 20 µg/L in leachate was 91.1 (27.5) using 3M Empore Disks and EPA Method 8061.

The average recovery (in %) with %RSD (in parenthesis) from 3 determinations and a spike concentration of 20 µg/L in estuarine groundwater was 81.4 (17.6) using 3M Empore Disks and EPA Method 8061.

The average recovery (in %) with %RSD (in parenthesis) from 3 determinations and a spike concentration of 1 mg/kg in estuarine sediment was not determined (matrix interferant) after sulfur cleanup with EPA Method 3660.

The average recovery (in %) with %RSD (in parenthesis) from 3 determinations and a spike concentration of 1 mg/kg in municipal sludge was 114 (10.5).

The average recovery (in %) with %RSD (in parenthesis) from 3 determinations and a spike concentration of 1 mg/kg in sandy loam soil was 57.7 (2.8).

SAMPLE COLLECTION, PRESERVATION & HANDLING
Containers used to collect samples for the determination of semivolatile organic compounds should be soap and water washed followed by methanol (or isopropanol) rinsing. The sample containers should be of glass or Teflon® and have screw-top covers with Teflon® liners. Sample containers should be filled with care to prevent any portion of the collected sample coming in contact with the sampler's gloves.

No preservation is used with concentrated waste samples. With liquid samples containing no residual chlorine and with soil, sediment, and sludge samples, immediately cooling to 4°C is the only preservation used. When residual chlorine is present then 3 mL of 10% aqueous sodium sulfate is added for each gallon of sample collected, followed by cooling to 4°C.

MHT Liquid samples must be extracted within 7 days and their extracts analyzed within 40 days. Concentrated waste, soil, sediment, and sludge samples must be extracted within 14 days and their extracts analyzed within 40 days.

SAMPLE PREPARATION In general, water samples are extracted at a pH of 5 to 7 with methylene chloride in a separatory funnel (EPA Method 3510). EPA Method 3520 is not recommended for the extraction of aqueous samples because the longer chain esters tend to adsorb to the glassware and consequently, their extraction recoveries may be poor. Solid samples are extracted with hexane/acetone (1:) or methylene chloride/acetone (1:1) in a Soxhlet extractor (EPA Method 3540) or with an ultrasonic extractor (EPA Method 3550). Immediately prior to extraction, spike 500 µL of the surrogate standard spiking solution into 1-L aqueous sample or 30-g solid sample. Extraction of particulate-free aqueous samples using C-18 extraction disks is an optional method that can be used.

Prior to Florisil cleanup or GC analysis, the methylene chloride and methylene chloride/acetone extracts must be exchanged to hexane. Exchange is not required for the acetonitrile extracts. Cleanup may not be necessary for extracts from a relatively clean sample matrix. Florisil Cartridge Cleanup may be used for extract cleanup.

If PCBs and organochlorine pesticides are known to be present in the sample, and if Florisil Cartridge Cleanup is considered, then two fractions are collected: Fraction 1 is eluted with 5 mL of 20% methylene chloride in hexane and Fraction 2 is eluted with 5 mL of 10% acetone in hexane. Fraction 1 contains the organochlorine pesticides and PCBs, and can be discarded. Fraction 2 contains the phthalate esters and is analyzed by GC/ECD.

QUALITY CONTROL Identify compounds in the sample by comparing the retention times of the peaks in the sample chromatogram with those of the peaks in standard chromatograms. The retention time window used to make identification is based upon measurements of actual retention time variations over the course of 10 consecutive injections.

Calibration standards are prepared at a minimum of five concentrations for each parameter of interest through dilution of the stock standard solutions with hexane. One of the concentrations should be at a concentration near, but above, the method detection limit. Prepare stock standard solutions in hexane. Stock standards should be checked frequently for signs of degradation or evaporation, especially just prior to preparing calibration standards from them. Stock standard solutions must be replaced after one year, or sooner if comparison with check standards indicates a problem. The suggested internal standards are: 2,5-dibromotoluene, 1,3,5-tribromobenzene, and α, α'-dibromo-m-xylene. The analyst can use any of the three compounds provided that they are resolved from matrix interferences. Recommended surrogate compounds are α-2,6-trichlorotoluene, 1,4-dichloronaphthalene, and 2,3,4,5,6-pentachlorotoluene.

Spike each sample, standard, and blank with surrogate compounds. Three surrogates are suggested for this method: diphenyl phthalate, diphenyl isophthalate, and dibenzyl phthalate.

The quality control check sample concentrate should contain the test compounds at 5 to 10 ng/µL An internal standard peak area check must be performed on all samples. The internal standard must be evaluated for acceptance by determining whether the measured area for the internal standard deviates by more than 30% from the average are for the internal standard in the calibration standards. When the internal standard peak area is outside that limit, all samples that fall outside the QC criteria must be reanalyzed. Benzyl benzoate has been tested and found appropriate as an internal standard for this method.

Any compounds confirmed by two columns may also be confirmed by GC/MS. The sample extract and associated blank should be analyzed by GC/MS. A reference standard of the compound must also be analyzed by GC/MS. Include a mid-concentration calibration standard after each group of 20 samples. The response factors for the mid-concentration calibration must be within ± 15% of the average values for the multiconcentration calibration. Demonstrate through the analyses of standards that the Florisil fractionation scheme is reproducible.

REFERENCE Test Methods for Evaluating Solid Waste, Physical/Chemical Methods, SW-846, 3rd Edition, U.S. EPA, Office of Solid Waste, Washington, DC, EPA Method 8061, Nov. 1990.

1,2,3,4,6,7,8-HpCDD   EPA Method 8280
CAS #35822-46-9

TITLE The Analysis of Polychlorinated Dibenzo-P-Dioxins and Polychlorinated Dibenzofurans.

MATRIX This method is appropriate for the determination of tetra-, penta-, hexa-, hepta-, and octachlorinated dibenzo-p-dioxins (PCDDs) and dibenzofurans (PCDFs) in chemical wastes including still bottoms, fuel oils, sludges, fly ash, reactor residues, soil and water.

METHOD SUMMARY This method covers 22 PCDD and PCDF compounds and it uses a high resolution capillary GC column with low resolution mass spectrometry. Samples are extracted and concentrated by several methods that vary depending on the matrix involved. The organic extracts are cleaned-up by washing with aqueous basic and acid solutions and then separated into fractions using a column of neutral alumina. The fraction containing the PCDDs and PCDFs is then further cleaned-up using a column of activated carbon. The final extract is concentrated and Carbon-13 labeled internal standards are added prior to analysis by GC/MS using a capillary GC column and selected ion monitoring using five sets of ions that are detailed in the method. Certain 2,3,7,8-substituted congeners are used to provide calibration and method recovery information. Proper column selection and access to reference isomer standards, may in certain cases, provide isomer specific data.

INTERFERENCES Solvents, reagents, glassware, and other sample processing hardware may yield discrete artifacts and/or elevated baselines which may cause misinterpretation of chromatographic data. Use high purity reagents and solvents to minimize interference problems. Interferents coextracted from the sample will vary considerably from source to source. PCDDs and PCDFs are often associated with other interfering chlorinated compounds such as PCBs and polychlorinated diphenyl ethers which may be found at concentrations several orders of magnitude higher than that of the analytes of interest.

INSTRUMENTATION A low resolution GC/MS utilizing 70 ev must be capable of selected ion monitoring (SIM) for at least 11 ions simultaneously, with a cycle time of 1 second or less. Minimum integration time for SIM is 50 ms per m/z. Also required is a GC-to-MS interface constructed of all glass or glass-lined materials. One of the following GC columns is required:

Column 1: 50 m CP-Sil-88 fused silica capillary column.
Column 2: DB-5 (30 m × 0.25 mm I.D., 0.25-um film thickness) fused silica capillary column.
Column 3: 30 m SP-2250 fused silica capillary column.

When toluene is employed as the final solvent, use of a bonded phase column is recommended. Solvent exchange into tridecane is required for other liquid phases or nonbonded columns such as CP-Sil-88. Chromatographic conditions must be adjusted to account for solvent boiling points.

PRECISION & ACCURACY Accuracy, precision, MDLs and concentration ranges for the compounds covered by this method have not been determined or published by EPA yet. The sensitivity of this method is dependent upon the level of interferents within a given matrix. Proposed quantification levels for target analytes were 2 ppb in soil samples, up to 10 ppb in other solid wastes and 10 ppt in water. Actual values have been shown to vary by homologous series and, to a lesser degree, by individual isomer.

SAMPLING METHOD Grab and composite samples must be collected in 1 L or 1-quart amber glass bottles with Teflon®-lined screw-caps that have been acid-washed and solvent rinsed before use. If compositing equipment is used, the system must incorporate glass sample containers for the collection of a minimum of 250 mL. samples.

SAMPLE PRESERVATION Samples must be cooled and stored at 4°C.

MHT Samples must be extracted within 30 days and analyzed within 45 days of collection.

SAMPLE PREPARATION
Soil samples — Extract 20 g of a 1:1 soil and anhydrous sodium sulfate mixture with 1:4 methanol-petroleum ether for 2 h in a wrist-action shaker. Concentrate the extract in a K-D and then exchange the solvent with hexane in the K-D. The final volume is 15 mL of hexane.

Aqueous samples — Extract a 1 L sample with methylene chloride, dry it with anhydrous sodium sulfate and concentrate it in a K-D followed by exchange with hexane in the K-D. A continuous liquid-liquid extractor may be used in place of a separatory funnel to avoid emulsions. The final volume is 15 mL of hexane.

Alumina column clean-up — The clean-up procedure described below consists of two phases. The first phase involves a sequential basic and acid washing of the extract that contains the analytes.

Wash the 15 mL hexane extract with 20% potassium hydroxide in a separatory funnel. Repeat the washing until no color is visible in the bottom layer but no more than four times because strong base is known to degrade certain PCDDs/PCDFs, so contact time must be minimized. Next, partition the 15 mL hexane against 40 mL of 5% sodium chloride. Next, partition the 15 mL hexane against 40 mL of concentrated sulfuric acid. Repeat the acid washings until no color is visible in the acid layer (but no more than four times). Finally, partition the 15 mL hexane against 40 mL of 5% sodium chloride. Dry the organic layer with anhydrous sodium sulfate and concentrate it to near dryness with a rotary evaporator. Dissolve the hexane residue from the first phase of the clean-up in 2 mL of hexane and apply it carefully to the top of a pre-eluted Woelm super 1 neutral alumina column. Elute the column with 10 mL of 8 percent (v/v) methylene chloride in hexane. Check by GC/MS analysis that no PCDDs of PCDFs are elute in this fraction before discarding it. Elute the PCDDs and PCDFs from the column with 15 mL of 60 percent (v/v) methylene chloride in hexane and collect this second fraction in a conical shaped concentrator tube.

Carbon column clean-up — Using a carefully regulated stream of nitrogen, concentrate the first 8 percent fraction (methylene chloride in hexane) from the alumina column to about 1 mL. Save this 8 percent concentrate for GC/MS analysis to check for breakthrough of PCDDs and PCDFs. Concentrate the second 60 percent fraction (methylene chloride in hexane) to about 2 to 3 mL. Prepare a carbon column and rinse the carbon with 5 mL cyclohexane/methylene chloride (50:50 v/v) in the forward direction of flow and then in the reverse direction of flow. While still in the reverse direction of flow, transfer the sample concentrate to the column and elute with 10 mL of methylene chloride/methanol/benzene (75:20:5, v/v). Save all above eluates and combine them (this fraction may be used as a check on column efficiency). Next, turn the column over and, in the direction of forward flow, elute the PCDD/PCDF fraction with 20 mL toluene. Evaporate the toluene fraction to about 1 mL on a rotary evaporator and transfer this concentrate to a 2.0-mL Reacti-vial. Concentrate the sample using a stream of nitrogen gas. The final volume will depend on the relative concentration of target analytes but it is typically 100 µL for soil samples and 500 µL for sludge, still bottom, and fly ash samples. Extracts which are determined to be outside the calibration range for individual analytes must be diluted or a smaller portion of the sample must be re-extracted.

An alternate carbon column clean-up also may be used with a 1 mL HPLC injector loop. The injector loop is connected to the optional HPLC column.

QUALITY CONTROL Demonstrate, using a method blank, that all glassware and reagents are interferent-free at the MDL of the matrix of interest. A "method blank" must be run with each 20 or fewer samples. The method blank is also dosed with the internal standards. For water samples, 1 L of deionized and/or distilled water should be used as the method blank. Mineral oil may be used as the method blank for other matrices.

Calculate response factors for standards relative to the internal standards. Add a recovery standard to the samples prior to injection. The concentration of the recovery standard in the sample extract must be the same as that in the calibration standards used to measure the response factors.

Field duplicates (individual samples taken from the same location at the same time) should be analyzed periodically to determine the total precision (field and lab). Where appropriate, field blanks should be provided to monitor for possible cross-contamination of samples in the field. GC column performance must be demonstrated initially and verified prior to analyzing any sample in a 12-hr period. The GC column performance check solution must be analyzed under the same chromatographic and mass spectrometric conditions used for other samples and standards.

Retention times of target analytes must be verified using reference standards. These values must correspond to the retention time windows established. While certain cleanup techniques are provided as part of this method, unique samples may require additional cleanup techniques to achieve the method detection limit.

REFERENCE Test Methods for Evaluating Solid Waste (SW-846). U.S.E.P.A., 1986. Method 8280, Rev. 0, Sept. 1986. Office of Solid Wastes, Washington, DC.

1,2,3,4,6,7,8-HpCDD   EPA Method 8290
CAS #35822-46-9

TITLE Polychlorinated Dibenzodioxins (PCDDs) and Polychlorinated Dibenzofurans (PCDFs) by High-Resolution Gas Chromatography/High-Resolution Mass Spectrometry (HRGC/HRMS).

MATRIX This method is applicable with a variety of environmental matrices including: water, soil, sediment, paper pulp, fly ash, fish tissue, human adipose tissue, sludges, fuel oil, chemical reactor residue, and still bottoms.

METHOD SUMMARY This method provides procedures for the detection and quantitative measurement of polychlorinated dibenzo-p-dioxins (tetra- through octachlorinated homologues; PCDDs), and polychlorinated dibenzofurans (tetra- through octachlorinated homologues; PCDFs) in a variety of environmental matrices and at part-per-trillion (ppt) to part-per-quadrillion (ppq) concentrations. High-resolution gas chromatography and high-resolution mass spectrometry (HRGC/HRMS) on purified sample extracts provides highly specific identification of each analyte. Quantification is provided using calibration standards.

INTERFERENCES Solvents, reagents, glassware, and other sample processing hardware may yield discrete artifacts that may cause misinterpretation of the chromatographic data. Analysts should avoid using PVC gloves. Interferants coextracted from the sample will vary considerably from matrix to matrix. PCDDs and PCDFs are often associated with other interfering chlorinated substances such as polychlorinated biphenyls (PCBs), polychlorinated diphenyl ethers (PCDEs), polychlorinated naphthalenes, and polychlorinated alkyldibenzofurans that may be found at concentrations several orders of magnitude higher than the PCDDs or PCDFs.

A high-resolution capillary column is used in this method. However, no single column is known to resolve all isomers. The 60 m DB-5 GC column is capable of 2,3,7,8-TCDD isomer specificity. In order to determine the concentration of the 2,3,7,8-TCDD (if detected on the DB-5 column), the sample extract must be reanalyzed on a column capable of 2,3,7,8-TCDF isomer specificity (e.g., DB-225, SP-2330, SP-2331, or equivalent).

INSTRUMENTATION High-Resolution Gas Chromatograph/High-Resolution Mass Spectrometer/Data System (HRGC/HRMS/DS) equipped with a GC injection port designed so that the separation of 2,3,7,8-TCDD from the other TCDD isomers achieved in the gas chromatographic column is not appreciably degraded.

Column 1: 60 m DB-5 fused silica capillary column.

Column 2: 30 m DB-225 fused silica capillary column, or equivalent.

PRECISION & ACCURACY Precision, bias and concentration ranges for the compounds covered by this method have not been determined yet. The sensitivity of Method 8290 is dependent upon the level of interferences within a given matrix. Samples containing concentrations of specific congeneric analytes of PCDDs and PCDFs that are greater than ten times the upper method calibration limits must be analyzed by a protocol designed for such concentration levels, e.g., EPA Method 8280.

SAMPLE PREPARATION
Sludge/wet fuel oil — Extract aqueous sludge or wet fuel oil samples by refluxing a sample with toluene using a Dean-Stark water separator until all the water is removed. Filter the toluene extract through a glass fiber filter, or equivalent, and concentrate it to near dryness either on a rotary evaporator using an inert gas. Transfer the concentrate to a separatory funnel using hexane and wash it with 5% sodium chloride solution. Proceed to clean up.

Soil/sediment — If the sample is wet, add anhydrous powdered sodium sulfate to it until a free flowing mixture is obtained. Place the soil/sodium sulfate mixture in the Soxhlet apparatus, add toluene, and reflux for 16 h. The solvent must cycle completely through the system five times per h. Cool and filter the extract through a glass fiber filter and concentrate to near dryness on a rotary evaporator. Transfer the residue to a separatory funnel, using hexane. Proceed to clean up.

Aqueous samples — Use a 1-L sample; the method may require acetone to be added to it. When the sample is judged to contain 1% or more solids, it must be filtered through a glass fiber filter that has been rinsed with toluene. If the suspended solids content is too great to filter, centrifuge the sample, decant, and then filter the aqueous phase. Combine the solids from the centrifuge bottle(s) with the particulates on the filter and with the filter itself and proceed with Soxhlet extraction for soil/sediment. Extract the aqueous filtrate with methylene chloride in a separatory funnel, filter the extract through anhydrous sodium sulfate, and concentrate it using a K-D apparatus or a rotary evaporator. Exchange the solvent with hexane and proceed to clean up.

Clean up — The sample extract is cleaned up utilizing a number of different techniques. Partition cleanup is where the sample extract is partitioned with concentrated sulfuric acid, 5% aqueous sodium chloride, and 20% aqueous potassium hydroxide. Silica/alumina column cleanup involves packing gravity columns with silica gel and alumina and sequentially eluting the residue from the partition cleanup. Carbon column cleanup involves packing a column with a mixture of AX–21 and Celite 545 and sequentially eluting the sample concentrate from the silica/alumina cleanup with hexane, cyclohexane/methylene chloride (50:50), and methylene chloride/methanol/toluene (75:20:5). Then the column is turned upside down and the PCDD/PCDF fraction is eluted with toluene. The toluene fraction is concentrated and stored in the dark at room temperature until analysis.

QUALITY CONTROL Demonstrate, through the analysis of a reagent water blank, that interferences from the analytical system, glassware, and reagents are under control. For each analytical batch (up to 20 samples), a reagent blank, matrix spike, and matrix spike duplicate/duplicate must be analyzed (the frequency of the spikes may be different for different monitoring programs). The blank and spiked samples must be carried through all stages of the sample preparation and measurement steps.

A GC column performance check is required at the beginning of each 12-h period during which samples are analyzed. An HRGC/HRMS method blank run is required between a calibration run and the first sample run. The same method blank extract may thus be analyzed more than once if the number of samples within a batch requires more than 12 h of analyses.

At the beginning of each 12-h period during which samples are to be analyzed, an aliquot of the 1) GC column performance check solution and 2) a high-resolution concentration calibration must be analyzed to demonstrate adequate GC resolution and sensitivity, response factor reproducibility, and mass range calibration, and to establish the PCDD/PCDF retention time windows. A mass resolution check must also be performed to demonstrate adequate mass resolution using an appropriate reference compound (perfluorokerosene (PFK) is recommended). If the required criteria are not met, remedial action must be taken before any samples are analyzed.

Routine or continuing calibration (using a high resolution calibration solution) and the mass resolution check must also be performed at the end of each 12 h period. Furthermore, a HRGC/HRMS method blank analysis must be recorded following a calibration analysis and the first sample analysis.

To evaluate the performance of the analytical method, the QC check samples must be handled in exactly the same manner as actual samples. Therefore, 1.0 mL of the QC check sample concentrate is spiked into each of four 1 L aliquots of reagent water (which becomes the QC check sample), extracted, and then analyzed by GC. The variety of semivolatile analytes which may be analyzed by GC is such that the concentration of the QC check sample concentrate is different for the different analytical techniques presented in the full method.

The analyst must demonstrate also that the compounds of interest are being quantitatively recovered by the cleanup technique before the cleanup is applied to actual samples. For sample extracts that are cleaned up, the associated quality control samples (e.g., spikes, blanks, and duplicates) must also be processed through the same cleanup procedure. The analysis using each determinative method (GC, GC/MS, HPLC) specifies instrument calibration procedures using stock standards. It is recommended that cleanup also be performed on a series of the same type of standards to validate chromatographic elution patterns for the compounds of interest and to verify the absence of interferences from reagents.

REFERENCE Test Methods for Evaluating Solid Waste (SW-846). U.S. EPA. 1983. Method 8290, Rev. 0, Nov. 1990. Office of Solid Wastes, Washington, DC.

1,2,3,4,6,7,8-HpCDF   **EPA Method 8280**
CAS #67562-39-4

TITLE The Analysis of Polychlorinated Dibenzo-P-Dioxins and Polychlorinated Dibenzofurans.

MATRIX This method is appropriate for the determination of tetra-, penta-, hexa-, hepta-, and octachlorinated dibenzo-p-dioxins (PCDDs) and dibenzofurans (PCDFs) in chemical wastes including still bottoms, fuel oils, sludges, fly ash, reactor residues, soil and water.

METHOD SUMMARY This method covers 22 PCDD and PCDF compounds and it uses a high resolution capillary GC column with low resolution mass spectrometry. Samples are extracted and concentrated by several methods that vary depending on the matrix involved. The organic extracts are cleaned-up by washing with aqueous basic and acid solutions and then separated into fractions using a column of neutral alumina. The fraction containing the PCDDs and PCDFs is then further cleaned-up using a column of activated carbon. The final extract is concentrated and Carbon-13 labeled internal standards are added prior to analysis by GC/MS using a capillary GC column and selected ion monitoring using five sets of ions that are detailed in the method. Certain 2,3,7,8-substituted congeners are used to provide calibration and method recovery information. Proper column selection and access to reference isomer standards, may in certain cases, provide isomer specific data.

INTERFERENCES Solvents, reagents, glassware, and other sample processing hardware may yield discrete artifacts and/or elevated baselines which may cause misinterpretation of chromatographic data. Use high purity reagents and solvents to minimize interference problems. Interferents coextracted from the sample will vary considerably from source to source. PCDDs and PCDFs are often associated with other interfering chlorinated compounds such as PCBs and polychlorinated diphenyl ethers which may be found at concentrations several orders of magnitude higher than that of the analytes of interest.

INSTRUMENTATION A low resolution GC/MS utilizing 70 ev must be capable of selected ion monitoring (SIM) for at least 11 ions simultaneously, with a cycle time of 1 second or less. Minimum integration time for SIM is 50 ms per m/z. Also required is a GC-to-MS interface constructed of all glass or glass-lined materials. One of the following GC columns is required:

Column 1: 50 m CP-Sil-88 fused silica capillary column.
Column 2: DB-5 (30 m × 0.25 mm I.D., 0.25-um film thickness) fused silica capillary column.
Column 3: 30 m SP-2250 fused silica capillary column.

When toluene is employed as the final solvent, use of a bonded phase column is recommended. Solvent exchange into tridecane is required for other liquid phases or nonbonded columns such as CP-Sil-88. Chromatographic conditions must be adjusted to account for solvent boiling points.

PRECISION & ACCURACY Accuracy, precision, MDLs and concentration ranges for the compounds covered by this

method have not been determined or published by EPA yet. The sensitivity of this method is dependent upon the level of interferents within a given matrix. Proposed quantification levels for target analytes were 2 ppb in soil samples, up to 10 ppb in other solid wastes and 10 ppt in water. Actual values have been shown to vary by homologous series and, to a lesser degree, by individual isomer.

SAMPLING METHOD Grab and composite samples must be collected in 1 L or 1-quart amber glass bottles with Teflon®-lined screw-caps that have been acid-washed and solvent rinsed before use. If compositing equipment is used, the system must incorporate glass sample containers for the collection of a minimum of 250 mL. samples.

SAMPLE PRESERVATION Samples must be cooled and stored at 4°C.

MHT Samples must be extracted within 30 days and analyzed within 45 days of collection.

SAMPLE PREPARATION
Soil samples — Extract 20 g of a 1:1 soil and anhydrous sodium sulfate mixture with 1:4 methanol-petroleum ether for 2 h in a wrist-action shaker. Concentrate the extract in a K-D and then exchange the solvent with hexane in the K-D. The final volume is 15 mL of hexane.

Aqueous samples — Extract a 1 L sample with methylene chloride, dry it with anhydrous sodium sulfate and concentrate it in a K-D followed by exchange with hexane in the K-D. A continuous liquid-liquid extractor may be used in place of a separatory funnel to avoid emulsions. The final volume is 15 mL of hexane.

Alumina column clean-up — The clean-up procedure described below consists of two phases. The first phase involves a sequential basic and acid washing of the extract that contains the analytes.

Wash the 15 mL hexane extract with 20% potassium hydroxide in a separatory funnel. Repeat the washing until no color is visible in the bottom layer but no more than four times because strong base is known to degrade certain PCDDs/PCDFs, so contact time must be minimized. Next, partition the 15 mL hexane against 40 mL of 5% sodium chloride. Next, partition the 15 mL hexane against 40 mL of concentrated sulfuric acid. Repeat the acid washings until no color is visible in the acid layer (but no more than four times). Finally, partition the 15 mL hexane against 40 mL of 5% sodium chloride. Dry the organic layer with anhydrous sodium sulfate and concentrate it to near dryness with a rotary evaporator. Dissolve the hexane residue from the first phase of the clean-up in 2 mL of hexane and apply it carefully to the top of a pre-eluted Woelm super 1 neutral alumina column. Elute the column with 10 mL of 8 percent (v/v) methylene chloride in hexane. Check by GC/MS analysis that no PCDDs of PCDFs are elute in this fraction before discarding it. Elute the PCDDs and PCDFs from the column with 15 mL of 60 percent (v/v) methylene chloride in hexane and collect this second fraction in a conical shaped concentrator tube.

Carbon column clean-up — Using a carefully regulated stream of nitrogen, concentrate the first 8 percent fraction (methylene chloride in hexane) from the alumina column to about 1 mL. Save this 8 percent concentrate for GC/MS analysis to check for breakthrough of PCDDs and PCDFs. Concentrate the second 60 percent fraction (methylene chloride in hexane) to about 2 to 3 mL. Prepare a carbon column and rinse the carbon with 5 mL cyclohexane/methylene chloride (50:50 v/v) in the forward direction of flow and then in the reverse direction of flow. While still in the reverse direction of flow, transfer the sample concentrate to the column and elute with 10 mL of methylene chloride/methanol/benzene (75:20:5, v/v). Save all above eluates and combine them (this fraction may be used as a check on column efficiency). Next, turn the column over and, in the direction of forward flow, elute the PCDD/PCDF fraction with 20 mL toluene. Evaporate the toluene fraction to about 1 mL on a rotary evaporator and transfer this concentrate to a 2.0-mL Reacti-vial. Concentrate the sample using a stream of nitrogen gas. The final volume will depend on the relative concentration of target analytes but it is typically 100 µL for soil samples and 500 µL for sludge, still bottom, and fly ash samples. Extracts which are determined to be outside the calibration range for individual analytes must be diluted or a smaller portion of the sample must be re-extracted.

An alternate carbon column clean-up also may be used with a 1 mL HPLC injector loop. The injector loop is connected to the optional HPLC column.

QUALITY CONTROL Demonstrate, using a method blank, that all glassware and reagents are interferent-free at the MDL of the matrix of interest. A "method blank" must be run with each 20 or fewer samples. The method blank is also dosed with the internal standards. For water samples, 1 L of deionized and/or distilled water should be used as the method blank. Mineral oil may be used as the method blank for other matrices.

Calculate response factors for standards relative to the internal standards. Add a recovery standard to the samples prior to injection. The concentration of the recovery standard in the sample extract must be the same as that in the calibration standards used to measure the response factors.

Field duplicates (individual samples taken from the same location at the same time) should be analyzed periodically to determine the total precision (field and lab). Where appropriate, field blanks should be provided to monitor for possible cross-contamination of samples in the field. GC column performance must be demonstrated initially and verified prior to analyzing any sample in a 12-hr period. The GC column performance check solution must be analyzed under the same chromatographic and mass spectrometric conditions used for other samples and standards.

Retention times of target analytes must be verified using reference standards. These values must correspond to the retention time windows established. While certain cleanup techniques are provided as part of this method, unique samples may require additional cleanup techniques to achieve the method detection limit.

REFERENCE Test Methods for Evaluating Solid Waste (SW-846). U.S.E.P.A., 1986. Method 8280, Rev. 0, Sept. 1986. Office of Solid Wastes, Washington, DC.

1,2,3,4,6,7,8-HpCDF   EPA Method 8290
CAS #67562-39-4

TITLE Polychlorinated Dibenzodioxins (PCDDs) and Polychlorinated Dibenzofurans (PCDFs) by High-Resolution Gas Chromatography/High-Resolution Mass Spectrometry (HRGC/HRMS).

MATRIX This method is applicable with a variety of environmental matrices including: water, soil, sediment, paper pulp, fly ash, fish tissue, human adipose tissue, sludges, fuel oil, chemical reactor residue, and still bottoms.

METHOD SUMMARY This method provides procedures for the detection and quantitative measurement of polychlorinated dibenzo-p-dioxins (tetra- through octachlorinated homologues; PCDDs), and polychlorinated dibenzofurans (tetra-through octachlorinated homologues; PCDFs) in a variety of environmental matrices and at part-per-trillion (ppt) to part-per-quadrillion (ppq) concentrations. High-resolution gas chromatography and high-resolution mass spectrometry (HRGC/HRMS) on purified sample extracts provides highly specific identification of each analyte. Quantification is provided using calibration standards.

INTERFERENCES Solvents, reagents, glassware, and other sample processing hardware may yield discrete artifacts that may cause misinterpretation of the chromatographic data. Analysts should avoid using PVC gloves. Interferants coextracted from the sample will vary considerably from matrix to matrix. PCDDs and PCDFs are often associated with other interfering chlorinated substances such as polychlorinated biphenyls (PCBs), polychlorinated diphenyl ethers (PCDEs), polychlorinated naphthalenes, and polychlorinated alkyldibenzofurans that may be found at concentrations several orders of magnitude higher than the PCDDs or PCDFs.

A high-resolution capillary column is used in this method. However, no single column is known to resolve all isomers. The 60 m DB-5 GC column is capable of 2,3,7,8-TCDD isomer specificity. In order to determine the concentration of the 2,3,7,8-TCDD (if detected on the DB-5 column), the sample extract must be reanalyzed on a column capable of 2,3,7,8-TCDF isomer specificity (e.g., DB-225, SP-2330, SP-2331, or equivalent).

INSTRUMENTATION High-Resolution Gas Chromatograph/High-Resolution Mass Spectrometer/Data System (HRGC/HRMS/DS) equipped with a GC injection port designed so that the separation of 2,3,7,8-TCDD from the other TCDD isomers achieved in the gas chromatographic column is not appreciably degraded.

Column 1: 60 m DB-5 fused silica capillary column.
Column 2: 30 m DB-225 fused silica capillary column, or equivalent.

PRECISION & ACCURACY Precision, bias and concentration ranges for the compounds covered by this method have not been determined yet. The sensitivity of Method 8290 is dependent upon the level of interferences within a given matrix. Samples containing concentrations of specific congeneric analytes of PCDDs and PCDFs that are greater than ten times the upper method calibration limits must be analyzed by a protocol designed for such concentration levels, e.g., EPA Method 8280.

SAMPLE PREPARATION

Sludge/wet fuel oil — Extract aqueous sludge or wet fuel oil samples by refluxing a sample with toluene using a Dean-Stark water separator until all the water is removed. Filter the toluene extract through a glass fiber filter, or equivalent, and concentrate it to near dryness either on a rotary evaporator using an inert gas. Transfer the concentrate to a separatory funnel using hexane and wash it with 5% sodium chloride solution. Proceed to clean up.

Soil/sediment — If the sample is wet, add anhydrous powdered sodium sulfate to it until a free flowing mixture is obtained. Place the soil/sodium sulfate mixture in the Soxhlet apparatus, add toluene, and reflux for 16 h. The solvent must cycle completely through the system five times per h. Cool and filter the extract through a glass fiber filter and concentrate to near dryness on a rotary evaporator. Transfer the residue to a separatory funnel, using hexane. Proceed to clean up.

Aqueous samples — Use a 1-L sample; the method may require acetone to be added to it. When the sample is judged to contain 1% or more solids, it must be filtered through a glass fiber filter that has been rinsed with toluene. If the suspended solids content is too great to filter, centrifuge the sample, decant, and then filter the aqueous phase. Combine the solids from the centrifuge bottle(s) with the particulates on the filter and with the filter itself and proceed with Soxhlet extraction for soil/sediment. Extract the aqueous filtrate with methylene chloride in a separatory funnel, filter the extract through anhydrous sodium sulfate, and concentrate it using a K-D apparatus or a rotary evaporator. Exchange the solvent with hexane and proceed to clean up.

Clean up — The sample extract is cleaned up utilizing a number of different techniques. Partition cleanup is where the sample extract is partitioned with concentrated sulfuric acid, 5% aqueous sodium chloride, and 20% aqueous potassium hydroxide. Silica/alumina column cleanup involves packing gravity columns with silica gel and alumina and sequentially eluting the residue from the partition cleanup. Carbon column cleanup involves packing a column with a mixture of AX–21 and Celite 545 and sequentially eluting the sample concentrate from the silica/alumina cleanup with hexane, cyclohexane/methylene chloride (50:50), and methylene chloride/methanol/toluene (75:20:5). Then the column is turned upside down and the PCDD/PCDF fraction is eluted with toluene. The toluene fraction is concentrated and stored in the dark at room temperature until analysis.

QUALITY CONTROL Demonstrate, through the analysis of a reagent water blank, that interferences from the analytical system, glassware, and reagents are under control. For each

analytical batch (up to 20 samples), a reagent blank, matrix spike, and matrix spike duplicate/duplicate must be analyzed (the frequency of the spikes may be different for different monitoring programs). The blank and spiked samples must be carried through all stages of the sample preparation and measurement steps.

A GC column performance check is required at the beginning of each 12-h period during which samples are analyzed. An HRGC/HRMS method blank run is required between a calibration run and the first sample run. The same method blank extract may thus be analyzed more than once if the number of samples within a batch requires more than 12 h of analyses.

At the beginning of each 12-h period during which samples are to be analyzed, an aliquot of the 1) GC column performance check solution and 2) a high-resolution concentration calibration must be analyzed to demonstrate adequate GC resolution and sensitivity, response factor reproducibility, and mass range calibration, and to establish the PCDD/PCDF retention time windows. A mass resolution check must also be performed to demonstrate adequate mass resolution using an appropriate reference compound (perfluorokerosene (PFK) is recommended). If the required criteria are not met, remedial action must be taken before any samples are analyzed.

Routine or continuing calibration (using a high resolution calibration solution) and the mass resolution check must also be performed at the end of each 12 h period. Furthermore, a HRGC/HRMS method blank analysis must be recorded following a calibration analysis and the first sample analysis.

To evaluate the performance of the analytical method, the QC check samples must be handled in exactly the same manner as actual samples. Therefore, 1.0 mL of the QC check sample concentrate is spiked into each of four 1 L aliquots of reagent water (which becomes the QC check sample), extracted, and then analyzed by GC. The variety of semivolatile analytes which may be analyzed by GC is such that the concentration of the QC check sample concentrate is different for the different analytical techniques presented in the full method.

The analyst must demonstrate also that the compounds of interest are being quantitatively recovered by the cleanup technique before the cleanup is applied to actual samples. For sample extracts that are cleaned up, the associated quality control samples (e.g., spikes, blanks, and duplicates) must also be processed through the same cleanup procedure. The analysis using each determinative method (GC, GC/MS, HPLC) specifies instrument calibration procedures using stock standards. It is recommended that cleanup also be performed on a series of the same type of standards to validate chromatographic elution patterns for the compounds of interest and to verify the absence of interferences from reagents.

REFERENCE Test Methods for Evaluating Solid Waste (SW-846). U.S. EPA. 1983. Method 8290, Rev. 0, Nov. 1990. Office of Solid Wastes, Washington, DC.

1,2,3,4,7,8,9-HpCDF   EPA Method 8290
CAS #55673-89-7

TITLE Polychlorinated Dibenzodioxins (PCDDs) and Polychlorinated Dibenzofurans (PCDFs) by High-Resolution Gas Chromatography/High-Resolution Mass Spectrometry (HRGC/HRMS).

MATRIX This method is applicable with a variety of environmental matrices including: water, soil, sediment, paper pulp, fly ash, fish tissue, human adipose tissue, sludges, fuel oil, chemical reactor residue, and still bottoms.

METHOD SUMMARY This method provides procedures for the detection and quantitative measurement of polychlorinated dibenzo-p-dioxins (tetra- through octachlorinated homologues; PCDDs), and polychlorinated dibenzofurans (tetra- through octachlorinated homologues; PCDFs) in a variety of environmental matrices and at part-per-trillion (ppt) to part-per-quadrillion (ppq) concentrations. High-resolution gas chromatography and high-resolution mass spectrometry (HRGC/HRMS) on purified sample extracts provides highly specific identification of each analyte. Quantification is provided using calibration standards.

INTERFERENCES Solvents, reagents, glassware, and other sample processing hardware may yield discrete artifacts that may cause misinterpretation of the chromatographic data. Analysts should avoid using PVC gloves. Interferants coextracted from the sample will vary considerably from matrix to matrix. PCDDs and PCDFs are often associated with other interfering chlorinated substances such as polychlorinated biphenyls (PCBs), polychlorinated diphenyl ethers (PCDEs), polychlorinated naphthalenes, and polychlorinated alkyldibenzofurans that may be found at concentrations several orders of magnitude higher than the PCDDs or PCDFs.

A high-resolution capillary column is used in this method. However, no single column is known to resolve all isomers. The 60 m DB-5 GC column is capable of 2,3,7,8-TCDD isomer specificity. In order to determine the concentration of the 2,3,7,8-TCDD (if detected on the DB-5 column), the sample extract must be reanalyzed on a column capable of 2,3,7,8-TCDF isomer specificity (e.g., DB-225, SP-2330, SP-2331, or equivalent).

INSTRUMENTATION High-Resolution Gas Chromatograph/High-Resolution Mass Spectrometer/Data System (HRGC/HRMS/DS) equipped with a GC injection port designed so that the separation of 2,3,7,8-TCDD from the other TCDD isomers achieved in the gas chromatographic column is not appreciably degraded.

Column 1: 60 m DB-5 fused silica capillary column.
Column 2: 30 m DB-225 fused silica capillary column, or equivalent.

PRECISION & ACCURACY Precision, bias and concentration ranges for the compounds covered by this method have not been determined yet. The sensitivity of Method 8290 is dependent upon the level of interferences within a given matrix. Samples containing concentrations of specific congeneric

analytes of PCDDs and PCDFs that are greater than ten times the upper method calibration limits must be analyzed by a protocol designed for such concentration levels, e.g., EPA Method 8280.

SAMPLE PREPARATION

Sludge/wet fuel oil — Extract aqueous sludge or wet fuel oil samples by refluxing a sample with toluene using a Dean-Stark water separator until all the water is removed. Filter the toluene extract through a glass fiber filter, or equivalent, and concentrate it to near dryness either on a rotary evaporator using an inert gas. Transfer the concentrate to a separatory funnel using hexane and wash it with 5% sodium chloride solution. Proceed to clean up.

Soil/sediment — If the sample is wet, add anhydrous powdered sodium sulfate to it until a free flowing mixture is obtained. Place the soil/sodium sulfate mixture in the Soxhlet apparatus, add toluene, and reflux for 16 h. The solvent must cycle completely through the system five times per h. Cool and filter the extract through a glass fiber filter and concentrate to near dryness on a rotary evaporator. Transfer the residue to a separatory funnel, using hexane. Proceed to clean up.

Aqueous samples — Use a 1-L sample; the method may require acetone to be added to it. When the sample is judged to contain 1% or more solids, it must be filtered through a glass fiber filter that has been rinsed with toluene. If the suspended solids content is too great to filter, centrifuge the sample, decant, and then filter the aqueous phase. Combine the solids from the centrifuge bottle(s) with the particulates on the filter and with the filter itself and proceed with Soxhlet extraction for soil/sediment. Extract the aqueous filtrate with methylene chloride in a separatory funnel, filter the extract through anhydrous sodium sulfate, and concentrate it using a K-D apparatus or a rotary evaporator. Exchange the solvent with hexane and proceed to clean up.

Clean up — The sample extract is cleaned up utilizing a number of different techniques. Partition cleanup is where the sample extract is partitioned with concentrated sulfuric acid, 5% aqueous sodium chloride, and 20% aqueous potassium hydroxide. Silica/alumina column cleanup involves packing gravity columns with silica gel and alumina and sequentially eluting the residue from the partition cleanup. Carbon column cleanup involves packing a column with a mixture of AX–21 and Celite 545 and sequentially eluting the sample concentrate from the silica/alumina cleanup with hexane, cyclohexane/methylene chloride (50:50), and methylene chloride/methanol/toluene (75:20:5). Then the column is turned upside down and the PCDD/PCDF fraction is eluted with toluene. The toluene fraction is concentrated and stored in the dark at room temperature until analysis.

QUALITY CONTROL
Demonstrate, through the analysis of a reagent water blank, that interferences from the analytical system, glassware, and reagents are under control. For each analytical batch (up to 20 samples), a reagent blank, matrix spike, and matrix spike duplicate/duplicate must be analyzed (the frequency of the spikes may be different for different monitoring programs). The blank and spiked samples must be carried through all stages of the sample preparation and measurement steps.

A GC column performance check is required at the beginning of each 12-h period during which samples are analyzed. An HRGC/HRMS method blank run is required between a calibration run and the first sample run. The same method blank extract may thus be analyzed more than once if the number of samples within a batch requires more than 12 h of analyses.

At the beginning of each 12-h period during which samples are to be analyzed, an aliquot of the 1) GC column performance check solution and 2) a high-resolution concentration calibration must be analyzed to demonstrate adequate GC resolution and sensitivity, response factor reproducibility, and mass range calibration, and to establish the PCDD/PCDF retention time windows. A mass resolution check must also be performed to demonstrate adequate mass resolution using an appropriate reference compound (perfluorokerosene (PFK) is recommended). If the required criteria are not met, remedial action must be taken before any samples are analyzed.

Routine or continuing calibration (using a high resolution calibration solution) and the mass resolution check must also be performed at the end of each 12 h period. Furthermore, a HRGC/HRMS method blank analysis must be recorded following a calibration analysis and the first sample analysis.

To evaluate the performance of the analytical method, the QC check samples must be handled in exactly the same manner as actual samples. Therefore, 1.0 mL of the QC check sample concentrate is spiked into each of four 1 L aliquots of reagent water (which becomes the QC check sample), extracted, and then analyzed by GC. The variety of semivolatile analytes which may be analyzed by GC is such that the concentration of the QC check sample concentrate is different for the different analytical techniques presented in the full method.

The analyst must demonstrate also that the compounds of interest are being quantitatively recovered by the cleanup technique before the cleanup is applied to actual samples. For sample extracts that are cleaned up, the associated quality control samples (e.g., spikes, blanks, and duplicates) must also be processed through the same cleanup procedure. The analysis using each determinative method (GC, GC/MS, HPLC) specifies instrument calibration procedures using stock standards. It is recommended that cleanup also be performed on a series of the same type of standards to validate chromatographic elution patterns for the compounds of interest and to verify the absence of interferences from reagents.

REFERENCE Test Methods for Evaluating Solid Waste (SW-846). U.S. EPA. 1983. Method 8290, Rev. 0, Nov. 1990. Office of Solid Wastes, Washington, DC.

1,2,3,4,7,8-HxCDD   **EPA Method 8280**
CAS #57653-85-7

TITLE The Analysis of Polychlorinated Dibenzo-P-Dioxins and Polychlorinated Dibenzofurans.

MATRIX This method is appropriate for the determination of tetra-, penta-, hexa-, hepta-, and octachlorinated dibenzo-p-dioxins (PCDDs) and dibenzofurans (PCDFs) in chemical wastes including still bottoms, fuel oils, sludges, fly ash, reactor residues, soil and water.

METHOD SUMMARY This method covers 22 PCDD and PCDF compounds and it uses a high resolution capillary GC column with low resolution mass spectrometry. Samples are extracted and concentrated by several methods that vary depending on the matrix involved. The organic extracts are cleaned-up by washing with aqueous basic and acid solutions and then separated into fractions using a column of neutral alumina. The fraction containing the PCDDs and PCDFs is then further cleaned-up using a column of activated carbon. The final extract is concentrated and Carbon-13 labeled internal standards are added prior to analysis by GC/MS using a capillary GC column and selected ion monitoring using five sets of ions that are detailed in the method. Certain 2,3,7,8-substituted congeners are used to provide calibration and method recovery information. Proper column selection and access to reference isomer standards, may in certain cases, provide isomer specific data.

INTERFERENCES Solvents, reagents, glassware, and other sample processing hardware may yield discrete artifacts and/or elevated baselines which may cause misinterpretation of chromatographic data. Use high purity reagents and solvents to minimize interference problems. Interferents coextracted from the sample will vary considerably from source to source. PCDDs and PCDFs are often associated with other interfering chlorinated compounds such as PCBs and polychlorinated diphenyl ethers which may be found at concentrations several orders of magnitude higher than that of the analytes of interest.

INSTRUMENTATION A low resolution GC/MS utilizing 70 ev must be capable of selected ion monitoring (SIM) for at least 11 ions simultaneously, with a cycle time of 1 second or less. Minimum integration time for SIM is 50 ms per m/z. Also required is a GC-to-MS interface constructed of all glass or glass-lined materials. One of the following GC columns is required:

Column 1: 50 m CP-Sil-88 fused silica capillary column.
Column 2: DB-5 (30 m × 0.25 mm I.D., 0.25-um film thickness) fused silica capillary column.
Column 3: 30 m SP-2250 fused silica capillary column.

When toluene is employed as the final solvent, use of a bonded phase column is recommended. Solvent exchange into tridecane is required for other liquid phases or nonbonded columns such as CP-Sil-88. Chromatographic conditions must be adjusted to account for solvent boiling points.

PRECISION & ACCURACY Accuracy, precision, MDLs and concentration ranges for the compounds covered by this method have not been determined or published by EPA yet. The sensitivity of this method is dependent upon the level of interferents within a given matrix. Proposed quantification levels for target analytes were 2 ppb in soil samples, up to 10 ppb in other solid wastes and 10 ppt in water. Actual values have been shown to vary by homologous series and, to a lesser degree, by individual isomer.

SAMPLING METHOD Grab and composite samples must be collected in 1 L or 1-quart amber glass bottles with Teflon®-lined screw-caps that have been acid-washed and solvent rinsed before use. If compositing equipment is used, the system must incorporate glass sample containers for the collection of a minimum of 250 mL. samples.

SAMPLE PRESERVATION Samples must be cooled and stored at 4°C.

MHT Samples must be extracted within 30 days and analyzed within 45 days of collection.

SAMPLE PREPARATION
Soil samples — Extract 20 g of a 1:1 soil and anhydrous sodium sulfate mixture with 1:4 methanol-petroleum ether for 2 h in a wrist-action shaker. Concentrate the extract in a K-D and then exchange the solvent with hexane in the K-D. The final volume is 15 mL of hexane.

Aqueous samples — Extract a 1 L sample with methylene chloride, dry it with anhydrous sodium sulfate and concentrate it in a K-D followed by exchange with hexane in the K-D. A continuous liquid-liquid extractor may be used in place of a separatory funnel to avoid emulsions. The final volume is 15 mL of hexane.

Alumina column clean-up — The clean-up procedure described below consists of two phases. The first phase involves a sequential basic and acid washing of the extract that contains the analytes.

Wash the 15 mL hexane extract with 20% potassium hydroxide in a separatory funnel. Repeat the washing until no color is visible in the bottom layer but no more than four times because strong base is known to degrade certain PCDDs/PCDFs, so contact time must be minimized. Next, partition the 15 mL hexane against 40 mL of 5% sodium chloride. Next, partition the 15 mL hexane against 40 mL of concentrated sulfuric acid. Repeat the acid washings until no color is visible in the acid layer (but no more than four times). Finally, partition the 15 mL hexane against 40 mL of 5% sodium chloride. Dry the organic layer with anhydrous sodium sulfate and concentrate it to near dryness with a rotary evaporator. Dissolve the hexane residue from the first phase of the clean-up in 2 mL of hexane and apply it carefully to the top of a pre-eluted Woelm super 1 neutral alumina column. Elute the column with 10 mL of 8 percent (v/v) methylene chloride in hexane. Check by GC/MS analysis that no PCDDs of PCDFs are elute in this fraction before discarding it. Elute the PCDDs and PCDFs from the column with 15 mL of 60 percent (v/v) methylene chloride in hexane and collect this second fraction in a conical shaped concentrator tube.

Carbon column clean-up — Using a carefully regulated stream of nitrogen, concentrate the first 8 percent fraction (methylene chloride in hexane) from the alumina column to about 1 mL. Save this 8 percent concentrate for GC/MS analysis to check for breakthrough of PCDDs and PCDFs. Concentrate the second 60 percent fraction (methylene chloride in hexane) to about 2 to 3 mL. Prepare a carbon column and rinse the carbon with 5 mL cyclohexane/methylene chloride (50:50 v/v) in the forward direction of flow and then in the reverse direction of flow. While still in the reverse direction of flow, transfer the sample concentrate to the column and elute with 10 mL of methylene chloride/methanol/benzene (75:20:5, v/v). Save all above eluates and combine them (this fraction may be used as a check on column efficiency). Next, turn the column over and, in the direction of forward flow, elute the PCDD/PCDF fraction with 20 mL toluene. Evaporate the toluene fraction to about 1 mL on a rotary evaporator and transfer this concentrate to a 2.0-mL Reacti-vial. Concentrate the sample using a stream of nitrogen gas. The final volume will depend on the relative concentration of target analytes but it is typically 100 µL for soil samples and 500 µL for sludge, still bottom, and fly ash samples. Extracts which are determined to be outside the calibration range for individual analytes must be diluted or a smaller portion of the sample must be re-extracted.

An alternate carbon column clean-up also may be used with a 1 mL HPLC injector loop. The injector loop is connected to the optional HPLC column.

QUALITY CONTROL Demonstrate, using a method blank, that all glassware and reagents are interferent-free at the MDL of the matrix of interest. A "method blank" must be run with each 20 or fewer samples. The method blank is also dosed with the internal standards. For water samples, 1 L of deionized and/or distilled water should be used as the method blank. Mineral oil may be used as the method blank for other matrices.

Calculate response factors for standards relative to the internal standards. Add a recovery standard to the samples prior to injection. The concentration of the recovery standard in the sample extract must be the same as that in the calibration standards used to measure the response factors.

Field duplicates (individual samples taken from the same location at the same time) should be analyzed periodically to determine the total precision (field and lab). Where appropriate, field blanks should be provided to monitor for possible cross-contamination of samples in the field. GC column performance must be demonstrated initially and verified prior to analyzing any sample in a 12-hr period. The GC column performance check solution must be analyzed under the same chromatographic and mass spectrometric conditions used for other samples and standards.

Retention times of target analytes must be verified using reference standards. These values must correspond to the retention time windows established. While certain cleanup techniques are provided as part of this method, unique samples may require additional cleanup techniques to achieve the method detection limit.

REFERENCE Test Methods for Evaluating Solid Waste (SW-846). U.S.E.P.A., 1986. Method 8280, Rev. 0, Sept. 1986. Office of Solid Wastes, Washington, DC.

1,2,3,6,7,8-HxCDD   EPA Method 8280
CAS #34465-46-8

TITLE The Analysis of Polychlorinated Dibenzo-P-Dioxins and Polychlorinated Dibenzofurans.

MATRIX This method is appropriate for the determination of tetra-, penta-, hexa-, hepta-, and octachlorinated dibenzo-p-dioxins (PCDDs) and dibenzofurans (PCDFs) in chemical wastes including still bottoms, fuel oils, sludges, fly ash, reactor residues, soil and water.

METHOD SUMMARY This method covers 22 PCDD and PCDF compounds and it uses a high resolution capillary GC column with low resolution mass spectrometry. Samples are extracted and concentrated by several methods that vary depending on the matrix involved. The organic extracts are cleaned-up by washing with aqueous basic and acid solutions and then separated into fractions using a column of neutral alumina. The fraction containing the PCDDs and PCDFs is then further cleaned-up using a column of activated carbon. The final extract is concentrated and Carbon-13 labeled internal standards are added prior to analysis by GC/MS using a capillary GC column and selected ion monitoring using five sets of ions that are detailed in the method. Certain 2,3,7,8-substituted congeners are used to provide calibration and method recovery information. Proper column selection and access to reference isomer standards, may in certain cases, provide isomer specific data.

INTERFERENCES Solvents, reagents, glassware, and other sample processing hardware may yield discrete artifacts and/or elevated baselines which may cause misinterpretation of chromatographic data. Use high purity reagents and solvents to minimize interference problems. Interferents coextracted from the sample will vary considerably from source to source. PCDDs and PCDFs are often associated with other interfering chlorinated compounds such as PCBs and polychlorinated diphenyl ethers which may be found at concentrations several orders of magnitude higher than that of the analytes of interest.

INSTRUMENTATION A low resolution GC/MS utilizing 70 ev must be capable of selected ion monitoring (SIM) for at least 11 ions simultaneously, with a cycle time of 1 second or less. Minimum integration time for SIM is 50 ms per m/z. Also required is a GC-to-MS interface constructed of all glass or glass-lined materials. One of the following GC columns is required:

Column 1: 50 m CP-Sil-88 fused silica capillary column.
Column 2: DB-5 (30 m × 0.25 mm I.D., 0.25-um film thickness) fused silica capillary column.
Column 3: 30 m SP-2250 fused silica capillary column.

When toluene is employed as the final solvent, use of a bonded phase column is recommended. Solvent exchange into tridecane is required for other liquid phases or nonbonded columns such as CP-Sil-88. Chromatographic conditions must be adjusted to account for solvent boiling points.

PRECISION & ACCURACY Accuracy, precision, MDLs and concentration ranges for the compounds covered by this

method have not been determined or published by EPA yet. The sensitivity of this method is dependent upon the level of interferents within a given matrix. Proposed quantification levels for target analytes were 2 ppb in soil samples, up to 10 ppb in other solid wastes and 10 ppt in water. Actual values have been shown to vary by homologous series and, to a lesser degree, by individual isomer.

SAMPLING METHOD Grab and composite samples must be collected in 1 L or 1-quart amber glass bottles with Teflon®-lined screw-caps that have been acid-washed and solvent rinsed before use. If compositing equipment is used, the system must incorporate glass sample containers for the collection of a minimum of 250 mL. samples.

SAMPLE PRESERVATION Samples must be cooled and stored at 4°C.

MHT Samples must be extracted within 30 days and analyzed within 45 days of collection.

SAMPLE PREPARATION
Soil samples — Extract 20 g of a 1:1 soil and anhydrous sodium sulfate mixture with 1:4 methanol-petroleum ether for 2 h in a wrist-action shaker. Concentrate the extract in a K-D and then exchange the solvent with hexane in the K-D. The final volume is 15 mL of hexane.

Aqueous samples — Extract a 1 L sample with methylene chloride, dry it with anhydrous sodium sulfate and concentrate it in a K-D followed by exchange with hexane in the K-D. A continuous liquid-liquid extractor may be used in place of a separatory funnel to avoid emulsions. The final volume is 15 mL of hexane.

Alumina column clean-up — The clean-up procedure described below consists of two phases. The first phase involves a sequential basic and acid washing of the extract that contains the analytes.

Wash the 15 mL hexane extract with 20% potassium hydroxide in a separatory funnel. Repeat the washing until no color is visible in the bottom layer but no more than four times because strong base is known to degrade certain PCDDs/PCDFs, so contact time must be minimized. Next, partition the 15 mL hexane against 40 mL of 5% sodium chloride. Next, partition the 15 mL hexane against 40 mL of concentrated sulfuric acid. Repeat the acid washings until no color is visible in the acid layer (but no more than four times). Finally, partition the 15 mL hexane against 40 mL of 5% sodium chloride. Dry the organic layer with anhydrous sodium sulfate and concentrate it to near dryness with a rotary evaporator. Dissolve the hexane residue from the first phase of the clean-up in 2 mL of hexane and apply it carefully to the top of a pre-eluted Woelm super 1 neutral alumina column. Elute the column with 10 mL of 8 percent (v/v) methylene chloride in hexane. Check by GC/MS analysis that no PCDDs of PCDFs are elute in this fraction before discarding it. Elute the PCDDs and PCDFs from the column with 15 mL of 60 percent (v/v) methylene chloride in hexane and collect this second fraction in a conical shaped concentrator tube.

Carbon column clean-up — Using a carefully regulated stream of nitrogen, concentrate the first 8 percent fraction (methylene chloride in hexane) from the alumina column to about 1 mL. Save this 8 percent concentrate for GC/MS analysis to check for breakthrough of PCDDs and PCDFs. Concentrate the second 60 percent fraction (methylene chloride in hexane) to about 2 to 3 mL. Prepare a carbon column and rinse the carbon with 5 mL cyclohexane/methylene chloride (50:50 v/v) in the forward direction of flow and then in the reverse direction of flow. While still in the reverse direction of flow, transfer the sample concentrate to the column and elute with 10 mL of methylene chloride/methanol/benzene (75:20:5, v/v). Save all above eluates and combine them (this fraction may be used as a check on column efficiency). Next, turn the column over and, in the direction of forward flow, elute the PCDD/PCDF fraction with 20 mL toluene. Evaporate the toluene fraction to about 1 mL on a rotary evaporator and transfer this concentrate to a 2.0-mL Reacti-vial. Concentrate the sample using a stream of nitrogen gas. The final volume will depend on the relative concentration of target analytes but it is typically 100 µL for soil samples and 500 µL for sludge, still bottom, and fly ash samples. Extracts which are determined to be outside the calibration range for individual analytes must be diluted or a smaller portion of the sample must be re-extracted.

An alternate carbon column clean-up also may be used with a 1 mL HPLC injector loop. The injector loop is connected to the optional HPLC column.

QUALITY CONTROL Demonstrate, using a method blank, that all glassware and reagents are interferent-free at the MDL of the matrix of interest. A "method blank" must be run with each 20 or fewer samples. The method blank is also dosed with the internal standards. For water samples, 1 L of deionized and/or distilled water should be used as the method blank. Mineral oil may be used as the method blank for other matrices.

Calculate response factors for standards relative to the internal standards. Add a recovery standard to the samples prior to injection. The concentration of the recovery standard in the sample extract must be the same as that in the calibration standards used to measure the response factors.

Field duplicates (individual samples taken from the same location at the same time) should be analyzed periodically to determine the total precision (field and lab). Where appropriate, field blanks should be provided to monitor for possible cross-contamination of samples in the field. GC column performance must be demonstrated initially and verified prior to analyzing any sample in a 12-hr period. The GC column performance check solution must be analyzed under the same chromatographic and mass spectrometric conditions used for other samples and standards.

Retention times of target analytes must be verified using reference standards. These values must correspond to the retention time windows established. While certain cleanup techniques are provided as part of this method, unique samples may require additional cleanup techniques to achieve the method detection limit.

REFERENCE Test Methods for Evaluating Solid Waste (SW-846). U.S.E.P.A., 1986. Method 8280, Rev. 0, Sept. 1986. Office of Solid Wastes, Washington, DC.

1,2,3,4,7,8-HxCDD   **EPA Method 8290**
CAS #57653-85-7

TITLE Polychlorinated Dibenzodioxins (PCDDs) and Polychlorinated Dibenzofurans (PCDFs) by High-Resolution Gas Chromatography/High-Resolution Mass Spectrometry (HRGC/HRMS).

MATRIX This method is applicable with a variety of environmental matrices including: water, soil, sediment, paper pulp, fly ash, fish tissue, human adipose tissue, sludges, fuel oil, chemical reactor residue, and still bottoms.

METHOD SUMMARY This method provides procedures for the detection and quantitative measurement of polychlorinated dibenzo-p-dioxins (tetra- through octachlorinated homologues; PCDDs), and polychlorinated dibenzofurans (tetra- through octachlorinated homologues; PCDFs) in a variety of environmental matrices and at part-per-trillion (ppt) to part-per-quadrillion (ppq) concentrations. High-resolution gas chromatography and high-resolution mass spectrometry (HRGC/HRMS) on purified sample extracts provides highly specific identification of each analyte. Quantification is provided using calibration standards.

INTERFERENCES Solvents, reagents, glassware, and other sample processing hardware may yield discrete artifacts that may cause misinterpretation of the chromatographic data. Analysts should avoid using PVC gloves. Interferants coextracted from the sample will vary considerably from matrix to matrix. PCDDs and PCDFs are often associated with other interfering chlorinated substances such as polychlorinated biphenyls (PCBs), polychlorinated diphenyl ethers (PCDEs), polychlorinated naphthalenes, and polychlorinated alkyldibenzofurans that may be found at concentrations several orders of magnitude higher than the PCDDs or PCDFs.

A high-resolution capillary column is used in this method. However, no single column is known to resolve all isomers. The 60 m DB-5 GC column is capable of 2,3,7,8-TCDD isomer specificity. In order to determine the concentration of the 2,3,7,8-TCDD (if detected on the DB-5 column), the sample extract must be reanalyzed on a column capable of 2,3,7,8-TCDF isomer specificity (e.g., DB-225, SP-2330, SP-2331, or equivalent).

INSTRUMENTATION High-Resolution Gas Chromatograph/High-Resolution Mass Spectrometer/Data System (HRGC/HRMS/DS) equipped with a GC injection port designed so that the separation of 2,3,7,8-TCDD from the other TCDD isomers achieved in the gas chromatographic column is not appreciably degraded.

Column 1: 60 m DB-5 fused silica capillary column.
Column 2: 30 m DB-225 fused silica capillary column, or equivalent.

PRECISION & ACCURACY Precision, bias and concentration ranges for the compounds covered by this method have not been determined yet. The sensitivity of Method 8290 is dependent upon the level of interferences within a given matrix. Samples containing concentrations of specific congeneric analytes of PCDDs and PCDFs that are greater than ten times the upper method calibration limits must be analyzed by a protocol designed for such concentration levels, e.g., EPA Method 8280.

SAMPLE PREPARATION
Sludge/wet fuel oil — Extract aqueous sludge or wet fuel oil samples by refluxing a sample with toluene using a Dean-Stark water separator until all the water is removed. Filter the toluene extract through a glass fiber filter, or equivalent, and concentrate it to near dryness either on a rotary evaporator using an inert gas. Transfer the concentrate to a separatory funnel using hexane and wash it with 5% sodium chloride solution. Proceed to clean up.

Soil/sediment — If the sample is wet, add anhydrous powdered sodium sulfate to it until a free flowing mixture is obtained. Place the soil/sodium sulfate mixture in the Soxhlet apparatus, add toluene, and reflux for 16 h. The solvent must cycle completely through the system five times per h. Cool and filter the extract through a glass fiber filter and concentrate to near dryness on a rotary evaporator. Transfer the residue to a separatory funnel, using hexane. Proceed to clean up.

Aqueous samples — Use a 1-L sample; the method may require acetone to be added to it. When the sample is judged to contain 1% or more solids, it must be filtered through a glass fiber filter that has been rinsed with toluene. If the suspended solids content is too great to filter, centrifuge the sample, decant, and then filter the aqueous phase. Combine the solids from the centrifuge bottle(s) with the particulates on the filter and with the filter itself and proceed with Soxhlet extraction for soil/sediment. Extract the aqueous filtrate with methylene chloride in a separatory funnel, filter the extract through anhydrous sodium sulfate, and concentrate it using a K-D apparatus or a rotary evaporator. Exchange the solvent with hexane and proceed to clean up.

Clean up — The sample extract is cleaned up utilizing a number of different techniques. Partition cleanup is where the sample extract is partitioned with concentrated sulfuric acid, 5% aqueous sodium chloride, and 20% aqueous potassium hydroxide. Silica/alumina column cleanup involves packing gravity columns with silica gel and alumina and sequentially eluting the residue from the partition cleanup. Carbon column cleanup involves packing a column with a mixture of AX–21 and Celite 545 and sequentially eluting the sample concentrate from the silica/alumina cleanup with hexane, cyclohexane/methylene chloride (50:50), and methylene chloride/methanol/toluene (75:20:5). Then the column is turned upside down and the PCDD/PCDF fraction is eluted with toluene. The toluene fraction is concentrated and stored in the dark at room temperature until analysis.

QUALITY CONTROL Demonstrate, through the analysis of a reagent water blank, that interferences from the analytical system, glassware, and reagents are under control. For each

analytical batch (up to 20 samples), a reagent blank, matrix spike, and matrix spike duplicate/duplicate must be analyzed (the frequency of the spikes may be different for different monitoring programs). The blank and spiked samples must be carried through all stages of the sample preparation and measurement steps.

A GC column performance check is required at the beginning of each 12-h period during which samples are analyzed. An HRGC/HRMS method blank run is required between a calibration run and the first sample run. The same method blank extract may thus be analyzed more than once if the number of samples within a batch requires more than 12 h of analyses.

At the beginning of each 12-h period during which samples are to be analyzed, an aliquot of the 1) GC column performance check solution and 2) a high-resolution concentration calibration must be analyzed to demonstrate adequate GC resolution and sensitivity, response factor reproducibility, and mass range calibration, and to establish the PCDD/PCDF retention time windows. A mass resolution check must also be performed to demonstrate adequate mass resolution using an appropriate reference compound (perfluorokerosene (PFK) is recommended). If the required criteria are not met, remedial action must be taken before any samples are analyzed.

Routine or continuing calibration (using a high resolution calibration solution) and the mass resolution check must also be performed at the end of each 12 h period. Furthermore, a HRGC/HRMS method blank analysis must be recorded following a calibration analysis and the first sample analysis.

To evaluate the performance of the analytical method, the QC check samples must be handled in exactly the same manner as actual samples. Therefore, 1.0 mL of the QC check sample concentrate is spiked into each of four 1 L aliquots of reagent water (which becomes the QC check sample), extracted, and then analyzed by GC. The variety of semivolatile analytes which may be analyzed by GC is such that the concentration of the QC check sample concentrate is different for the different analytical techniques presented in the full method.

The analyst must demonstrate also that the compounds of interest are being quantitatively recovered by the cleanup technique before the cleanup is applied to actual samples. For sample extracts that are cleaned up, the associated quality control samples (e.g., spikes, blanks, and duplicates) must also be processed through the same cleanup procedure. The analysis using each determinative method (GC, GC/MS, HPLC) specifies instrument calibration procedures using stock standards. It is recommended that cleanup also be performed on a series of the same type of standards to validate chromatographic elution patterns for the compounds of interest and to verify the absence of interferences from reagents.

REFERENCE Test Methods for Evaluating Solid Waste (SW-846). U.S. EPA. 1983. Method 8290, Rev. 0, Nov. 1990. Office of Solid Wastes, Washington, DC.

1,2,3,6,7,8-HxCDD   **EPA Method 8290**
CAS #34465-46-8

TITLE Polychlorinated Dibenzodioxins (PCDDs) and Polychlorinated Dibenzofurans (PCDFs) by High-Resolution Gas Chromatography/High-Resolution Mass Spectrometry (HRGC/HRMS).

MATRIX This method is applicable with a variety of environmental matrices including: water, soil, sediment, paper pulp, fly ash, fish tissue, human adipose tissue, sludges, fuel oil, chemical reactor residue, and still bottoms.

METHOD SUMMARY This method provides procedures for the detection and quantitative measurement of polychlorinated dibenzo-p-dioxins (tetra- through octachlorinated homologues; PCDDs), and polychlorinated dibenzofurans (tetra- through octachlorinated homologues; PCDFs) in a variety of environmental matrices and at part-per-trillion (ppt) to part-per-quadrillion (ppq) concentrations. High-resolution gas chromatography and high-resolution mass spectrometry (HRGC/HRMS) on purified sample extracts provides highly specific identification of each analyte. Quantification is provided using calibration standards.

INTERFERENCES Solvents, reagents, glassware, and other sample processing hardware may yield discrete artifacts that may cause misinterpretation of the chromatographic data. Analysts should avoid using PVC gloves. Interferants coextracted from the sample will vary considerably from matrix to matrix. PCDDs and PCDFs are often associated with other interfering chlorinated substances such as polychlorinated biphenyls (PCBs), polychlorinated diphenyl ethers (PCDEs), polychlorinated naphthalenes, and polychlorinated alkyldibenzofurans that may be found at concentrations several orders of magnitude higher than the PCDDs or PCDFs.

A high-resolution capillary column is used in this method. However, no single column is known to resolve all isomers. The 60 m DB-5 GC column is capable of 2,3,7,8-TCDD isomer specificity. In order to determine the concentration of the 2,3,7,8-TCDD (if detected on the DB-5 column), the sample extract must be reanalyzed on a column capable of 2,3,7,8-TCDF isomer specificity (e.g., DB-225, SP-2330, SP-2331, or equivalent).

INSTRUMENTATION High-Resolution Gas Chromatograph/High-Resolution Mass Spectrometer/Data System (HRGC/HRMS/DS) equipped with a GC injection port designed so that the separation of 2,3,7,8-TCDD from the other TCDD isomers achieved in the gas chromatographic column is not appreciably degraded.

Column 1: 60 m DB-5 fused silica capillary column.
Column 2: 30 m DB-225 fused silica capillary column, or equivalent.

PRECISION & ACCURACY Precision, bias and concentration ranges for the compounds covered by this method have not been determined yet. The sensitivity of Method 8290 is dependent upon the level of interferences within a given matrix. Samples containing concentrations of specific congeneric

analytes of PCDDs and PCDFs that are greater than ten times the upper method calibration limits must be analyzed by a protocol designed for such concentration levels, e.g., EPA Method 8280.

SAMPLE PREPARATION
Sludge/wet fuel oil — Extract aqueous sludge or wet fuel oil samples by refluxing a sample with toluene using a Dean-Stark water separator until all the water is removed. Filter the toluene extract through a glass fiber filter, or equivalent, and concentrate it to near dryness either on a rotary evaporator using an inert gas. Transfer the concentrate to a separatory funnel using hexane and wash it with 5% sodium chloride solution. Proceed to clean up.

Soil/sediment — If the sample is wet, add anhydrous powdered sodium sulfate to it until a free flowing mixture is obtained. Place the soil/sodium sulfate mixture in the Soxhlet apparatus, add toluene, and reflux for 16 h. The solvent must cycle completely through the system five times per h. Cool and filter the extract through a glass fiber filter and concentrate to near dryness on a rotary evaporator. Transfer the residue to a separatory funnel, using hexane. Proceed to clean up.

Aqueous samples — Use a 1-L sample; the method may require acetone to be added to it. When the sample is judged to contain 1% or more solids, it must be filtered through a glass fiber filter that has been rinsed with toluene. If the suspended solids content is too great to filter, centrifuge the sample, decant, and then filter the aqueous phase. Combine the solids from the centrifuge bottle(s) with the particulates on the filter and with the filter itself and proceed with Soxhlet extraction for soil/sediment. Extract the aqueous filtrate with methylene chloride in a separatory funnel, filter the extract through anhydrous sodium sulfate, and concentrate it using a K-D apparatus or a rotary evaporator. Exchange the solvent with hexane and proceed to clean up.

Clean up — The sample extract is cleaned up utilizing a number of different techniques. Partition cleanup is where the sample extract is partitioned with concentrated sulfuric acid, 5% aqueous sodium chloride, and 20% aqueous potassium hydroxide. Silica/alumina column cleanup involves packing gravity columns with silica gel and alumina and sequentially eluting the residue from the partition cleanup. Carbon column cleanup involves packing a column with a mixture of AX–21 and Celite 545 and sequentially eluting the sample concentrate from the silica/alumina cleanup with hexane, cyclohexane/methylene chloride (50:50), and methylene chloride/methanol/toluene (75:20:5). Then the column is turned upside down and the PCDD/PCDF fraction is eluted with toluene. The toluene fraction is concentrated and stored in the dark at room temperature until analysis.

QUALITY CONTROL Demonstrate, through the analysis of a reagent water blank, that interferences from the analytical system, glassware, and reagents are under control. For each analytical batch (up to 20 samples), a reagent blank, matrix spike, and matrix spike duplicate/duplicate must be analyzed (the frequency of the spikes may be different for different monitoring programs). The blank and spiked samples must be carried through all stages of the sample preparation and measurement steps.

A GC column performance check is required at the beginning of each 12-h period during which samples are analyzed. An HRGC/HRMS method blank run is required between a calibration run and the first sample run. The same method blank extract may thus be analyzed more than once if the number of samples within a batch requires more than 12 h of analyses.

At the beginning of each 12-h period during which samples are to be analyzed, an aliquot of the 1) GC column performance check solution and 2) a high-resolution concentration calibration must be analyzed to demonstrate adequate GC resolution and sensitivity, response factor reproducibility, and mass range calibration, and to establish the PCDD/PCDF retention time windows. A mass resolution check must also be performed to demonstrate adequate mass resolution using an appropriate reference compound (perfluorokerosene (PFK) is recommended). If the required criteria are not met, remedial action must be taken before any samples are analyzed.

Routine or continuing calibration (using a high resolution calibration solution) and the mass resolution check must also be performed at the end of each 12 h period. Furthermore, a HRGC/HRMS method blank analysis must be recorded following a calibration analysis and the first sample analysis.

To evaluate the performance of the analytical method, the QC check samples must be handled in exactly the same manner as actual samples. Therefore, 1.0 mL of the QC check sample concentrate is spiked into each of four 1 L aliquots of reagent water (which becomes the QC check sample), extracted, and then analyzed by GC. The variety of semivolatile analytes which may be analyzed by GC is such that the concentration of the QC check sample concentrate is different for the different analytical techniques presented in the full method.

The analyst must demonstrate also that the compounds of interest are being quantitatively recovered by the cleanup technique before the cleanup is applied to actual samples. For sample extracts that are cleaned up, the associated quality control samples (e.g., spikes, blanks, and duplicates) must also be processed through the same cleanup procedure. The analysis using each determinative method (GC, GC/MS, HPLC) specifies instrument calibration procedures using stock standards. It is recommended that cleanup also be performed on a series of the same type of standards to validate chromatographic elution patterns for the compounds of interest and to verify the absence of interferences from reagents.

REFERENCE Test Methods for Evaluating Solid Waste (SW-846). U.S. EPA. 1983. Method 8290, Rev. 0, Nov. 1990. Office of Solid Wastes, Washington, DC.

1,2,3,7,8,9-HxCDD   **EPA Method 8290**
CAS #19408-74-3

TITLE Polychlorinated Dibenzodioxins (PCDDs) and Polychlorinated Dibenzofurans (PCDFs) by High-Resolution Gas

Chromatography/High-Resolution Mass Spectrometry (HRGC/HRMS).

MATRIX This method is applicable with a variety of environmental matrices including: water, soil, sediment, paper pulp, fly ash, fish tissue, human adipose tissue, sludges, fuel oil, chemical reactor residue, and still bottoms.

METHOD SUMMARY This method provides procedures for the detection and quantitative measurement of polychlorinated dibenzo-p-dioxins (tetra- through octachlorinated homologues; PCDDs), and polychlorinated dibenzofurans (tetra- through octachlorinated homologues; PCDFs) in a variety of environmental matrices and at part-per-trillion (ppt) to part-per-quadrillion (ppq) concentrations. High-resolution gas chromatography and high-resolution mass spectrometry (HRGC/HRMS) on purified sample extracts provides highly specific identification of each analyte. Quantification is provided using calibration standards.

INTERFERENCES Solvents, reagents, glassware, and other sample processing hardware may yield discrete artifacts that may cause misinterpretation of the chromatographic data. Analysts should avoid using PVC gloves. Interferants coextracted from the sample will vary considerably from matrix to matrix. PCDDs and PCDFs are often associated with other interfering chlorinated substances such as polychlorinated biphenyls (PCBs), polychlorinated diphenyl ethers (PCDEs), polychlorinated naphthalenes, and polychlorinated alkyldibenzofurans that may be found at concentrations several orders of magnitude higher than the PCDDs or PCDFs.

A high-resolution capillary column is used in this method. However, no single column is known to resolve all isomers. The 60 m DB-5 GC column is capable of 2,3,7,8-TCDD isomer specificity. In order to determine the concentration of the 2,3,7,8-TCDD (if detected on the DB-5 column), the sample extract must be reanalyzed on a column capable of 2,3,7,8-TCDF isomer specificity (e.g., DB-225, SP-2330, SP-2331, or equivalent).

INSTRUMENTATION High-Resolution Gas Chromatograph/High-Resolution Mass Spectrometer/Data System (HRGC/HRMS/DS) equipped with a GC injection port designed so that the separation of 2,3,7,8-TCDD from the other TCDD isomers achieved in the gas chromatographic column is not appreciably degraded.

Column 1: 60 m DB-5 fused silica capillary column.
Column 2: 30 m DB-225 fused silica capillary column, or equivalent.

PRECISION & ACCURACY Precision, bias and concentration ranges for the compounds covered by this method have not been determined yet. The sensitivity of Method 8290 is dependent upon the level of interferences within a given matrix. Samples containing concentrations of specific congeneric analytes of PCDDs and PCDFs that are greater than ten times the upper method calibration limits must be analyzed by a protocol designed for such concentration levels, e.g., EPA Method 8280.

SAMPLE PREPARATION
Sludge/wet fuel oil — Extract aqueous sludge or wet fuel oil samples by refluxing a sample with toluene using a Dean-Stark water separator until all the water is removed. Filter the toluene extract through a glass fiber filter, or equivalent, and concentrate it to near dryness either on a rotary evaporator using an inert gas. Transfer the concentrate to a separatory funnel using hexane and wash it with 5% sodium chloride solution. Proceed to clean up.

Soil/sediment — If the sample is wet, add anhydrous powdered sodium sulfate to it until a free flowing mixture is obtained. Place the soil/sodium sulfate mixture in the Soxhlet apparatus, add toluene, and reflux for 16 h. The solvent must cycle completely through the system five times per h. Cool and filter the extract through a glass fiber filter and concentrate to near dryness on a rotary evaporator. Transfer the residue to a separatory funnel, using hexane. Proceed to clean up.

Aqueous samples — Use a 1-L sample; the method may require acetone to be added to it. When the sample is judged to contain 1% or more solids, it must be filtered through a glass fiber filter that has been rinsed with toluene. If the suspended solids content is too great to filter, centrifuge the sample, decant, and then filter the aqueous phase. Combine the solids from the centrifuge bottle(s) with the particulates on the filter and with the filter itself and proceed with Soxhlet extraction for soil/sediment. Extract the aqueous filtrate with methylene chloride in a separatory funnel, filter the extract through anhydrous sodium sulfate, and concentrate it using a K-D apparatus or a rotary evaporator. Exchange the solvent with hexane and proceed to clean up.

Clean up — The sample extract is cleaned up utilizing a number of different techniques. Partition cleanup is where the sample extract is partitioned with concentrated sulfuric acid, 5% aqueous sodium chloride, and 20% aqueous potassium hydroxide. Silica/alumina column cleanup involves packing gravity columns with silica gel and alumina and sequentially eluting the residue from the partition cleanup. Carbon column cleanup involves packing a column with a mixture of AX–21 and Celite 545 and sequentially eluting the sample concentrate from the silica/alumina cleanup with hexane, cyclohexane/methylene chloride (50:50), and methylene chloride/methanol/toluene (75:20:5). Then the column is turned upside down and the PCDD/PCDF fraction is eluted with toluene. The toluene fraction is concentrated and stored in the dark at room temperature until analysis.

QUALITY CONTROL Demonstrate, through the analysis of a reagent water blank, that interferences from the analytical system, glassware, and reagents are under control. For each analytical batch (up to 20 samples), a reagent blank, matrix spike, and matrix spike duplicate/duplicate must be analyzed (the frequency of the spikes may be different for different monitoring programs). The blank and spiked samples must be carried through all stages of the sample preparation and measurement steps.

A GC column performance check is required at the beginning of each 12-h period during which samples are analyzed. An

HRGC/HRMS method blank run is required between a calibration run and the first sample run. The same method blank extract may thus be analyzed more than once if the number of samples within a batch requires more than 12 h of analyses.

At the beginning of each 12-h period during which samples are to be analyzed, an aliquot of the 1) GC column performance check solution and 2) a high-resolution concentration calibration must be analyzed to demonstrate adequate GC resolution and sensitivity, response factor reproducibility, and mass range calibration, and to establish the PCDD/PCDF retention time windows. A mass resolution check must also be performed to demonstrate adequate mass resolution using an appropriate reference compound (perfluorokerosene (PFK) is recommended). If the required criteria are not met, remedial action must be taken before any samples are analyzed.

Routine or continuing calibration (using a high resolution calibration solution) and the mass resolution check must also be performed at the end of each 12 h period. Furthermore, a HRGC/HRMS method blank analysis must be recorded following a calibration analysis and the first sample analysis.

To evaluate the performance of the analytical method, the QC check samples must be handled in exactly the same manner as actual samples. Therefore, 1.0 mL of the QC check sample concentrate is spiked into each of four 1 L aliquots of reagent water (which becomes the QC check sample), extracted, and then analyzed by GC. The variety of semivolatile analytes which may be analyzed by GC is such that the concentration of the QC check sample concentrate is different for the different analytical techniques presented in the full method.

The analyst must demonstrate also that the compounds of interest are being quantitatively recovered by the cleanup technique before the cleanup is applied to actual samples. For sample extracts that are cleaned up, the associated quality control samples (e.g., spikes, blanks, and duplicates) must also be processed through the same cleanup procedure. The analysis using each determinative method (GC, GC/MS, HPLC) specifies instrument calibration procedures using stock standards. It is recommended that cleanup also be performed on a series of the same type of standards to validate chromatographic elution patterns for the compounds of interest and to verify the absence of interferences from reagents.

REFERENCE Test Methods for Evaluating Solid Waste (SW-846). U.S. EPA. 1983. Method 8290, Rev. 0, Nov. 1990. Office of Solid Wastes, Washington, DC.

1,2,3,4,7,8-HxCDD   **EPA Method 8280**
CAS #39227-28-6

TITLE Analysis of PCDDs and PCDFs

MATRIX chemical wastes, fuel oils, still bottoms, sludges, water, soil, fly ash, reactor residues.

APPLICATION This method is used for the analysis of tetra-, penta-, hexa-, hepta-, and octachlorinated dibenzo-p-dioxins (PCDDs) and dibenzofurans (PCDFs). The sensitivity of the method is dependent on the level of interferents within the matrix. Only experienced analysts should be used. Special safety precautions must be observed and an EPA-approved sample disposal plan must be used.

INTERFERENCES Solvents, reagents, and glassware may introduce artifacts. Other interferences may come from coextracted compounds from samples; PCBs and polychlorinated diphenyl ethers are common interferents.

INSTRUMENTATION GC/MS with a fused silica capillary column. Also, solvent extraction and concentration glassware and either a gravity flow activated carbon AX–21/silica gel Type 60 EM reagent column or a HPLC with a 10 mm by 7 cm silanized glass column with active carbon AX–21 and Spherisorb S10W silica for sample cleanup. One of three fused silica capillary GC columns may be used: column 1: 50 m CP-Sil-88; Column 2: 30 m by 0.25 mm DB-5; colunm 3: 30 m SP-2250.

RANGE 50–6,000 picograms.

MDL Not determined.

PRECISION as (RSD) 38% with 5 ng/g in clay; 8.8% with 25 ng/g soil; 3.4% with 125 ng/g in sludge.

ACCURACY (as Mean % Recovery): 46.8% with 5 ng/g in clay; 65.0% with 25 ng/g in soil; 81.9% with 125 ng/g in sludge; 125.4% with 46 ng/g in fly ash; 89.1% with 2500 ng/g in still bottom.

SAMPLING METHOD Use 1 L (or quart) amber glass bottles with Teflon®-lined or solvent washed foil screw caps. Tape caps to bottle after sampling.

Compositing equipment must use glass containers and contain no Tygon or rubber tubing. Sample bottles must not be prewashed with the sample before its collection. Aqueous samples cannot be aliquoted from sample containers — the entire sample must be used and the container is washed out with the extracting solvent.

STABILITY Cool to 4°C and store at this temperature. When toluene is employed as the final solvent use a bonded phase GC column for separation. Otherwise, solvent exchange into tridecane is required for other liquid phases or the CP-Sil-88 GC column.

MHT 30 days; samples must be completely analyzed within 45 days of collection.

QUALITY CONTROL A method blank must be analyzed each time a set of samples is extracted or there is a change in reagents. A lab method blank must be run with each analytical batch of 20 or fewer samples. Field duplicates and field blanks should be analyzed periodically. GC column performance must be demonstrated initially and verified prior to analyzing any sample in a 12 h period. A series of calibration standards must be processed through the procedure to validate elution patterns and absence of interferents from reagents. Both the alumina column and carbon column performance must be routinely checked for presence of the analyte.

Performance evaluation samples and split samples with other laboratories are also expected to be periodically analyzed.

REFERENCE Method 8280, SW-846, 3rd ed., Nov.1986.

1,2,3,6,7,8-HxCDD — EPA Method 8280
CAS #57653-85-7

TITLE Analysis of PCDDs and PCDFs

MATRIX chemical wastes, fuel oils, still bottoms, sludges, water, soil, fly ash, reactor residues.

APPLICATION This method is used for the analysis of tetra-, penta-, hexa-, hepta-, and octachlorinated dibenzo-p-dioxins (PCDDs) and dibenzofurans (PCDFs). The sensitivity of the method is dependent on the level of interferents within the matrix. Only experienced analysts should be used. Special safety precautions must be observed and an EPA-approved sample disposal plan must be used.

INTERFERENCES Solvents, reagents, and glassware may introduce artifacts. Other interferences may come from coextracted compounds from samples.

PCBs and polychlorinated diphenyl ethers are common interferents.

INSTRUMENTATION GC/MS with a fused silica capillary column. Also, solvent extraction and concentration glassware and either a gravity flow activated carbon AX–21/silica gel Type 60 EM reagent column or a HPLC with a 10 mm by 7 cm silanized glass column with active carbon AX–21 and Spherisorb S10W silica for sample cleanup. One of three fused silica capillary GC columns may be used: column 1: 50 m CP-Sil-88; Column 2: 30 m by 0.25 mm DB-5; colunm 3: 30 m SP-2250.

RANGE 50–6,000 picograms.

MDL 2.21 ng/L (in reagent water); 1.25 µg/kg (in Missouri soil); 0.55 µg/kg (in fly ash); 2.30 µg/kg (in industrial sludge); 6.21 µg/kg (in still bottom); 5.02 µg/kg (in fuel oil) MDLs are for carbon-13 labeled analyte.

PRECISION (as RSD) Not determined.

ACCURACY (as Mean % Recovery) Not determined.

SAMPLING METHOD Use 1 L (or quart) amber glass bottles with Teflon®-lined or solvent washed foil screw caps. Tape caps to bottle after sampling.

Compositing equipment must use glass containers and contain no Tygon or rubber tubing. Sample bottles must not be prewashed with the sample before its collection. Aqueous samples cannot be aliquoted from sample containers — the entire sample must be used and the container is washed out with the extracting solvent.

STABILITY Cool to 4°C and store at this temperature. When toluene is employed as the final solvent use a bonded phase GC column for separation. Otherwise, solvent exchange into tridecane is required for other liquid phases or the CP-Sil-88 GC column.

MHT 30 days; samples must be completely analyzed within 45 days of collection.

QUALITY CONTROL A method blank must be analyzed each time a set of samples is extracted or there is a change in reagents. A lab method blank must be run with each analytical batch of 20 or fewer samples. Field duplicates and field blanks should be analyzed periodically. GC column performance must be demonstrated initially and verified prior to analyzing any sample in a 12 h period. A series of calibration standards must be processed through the procedure to validate elution patterns and absence of interferents from reagents. Both the alumina column and carbon column performance must be routinely checked for presence of the analyte.

Performance evaluation samples and split samples with other laboratories are also expected to be periodically analyzed.

REFERENCE Method 8280, SW-846, 3rd ed., Nov.1986.

1,2,3,4,7,8-HxCDF — EPA Method 8280
CAS #70648-26-9

TITLE The Analysis of Polychlorinated Dibenzo-P-Dioxins and Polychlorinated Dibenzofurans.

MATRIX This method is appropriate for the determination of tetra-, penta-, hexa-, hepta-, and octachlorinated dibenzo-p-dioxins (PCDDs) and dibenzofurans (PCDFs) in chemical wastes including still bottoms, fuel oils, sludges, fly ash, reactor residues, soil and water.

METHOD SUMMARY This method covers 22 PCDD and PCDF compounds and it uses a high resolution capillary GC column with low resolution mass spectrometry. Samples are extracted and concentrated by several methods that vary depending on the matrix involved. The organic extracts are cleaned-up by washing with aqueous basic and acid solutions and then separated into fractions using a column of neutral alumina. The fraction containing the PCDDs and PCDFs is then further cleaned-up using a column of activated carbon. The final extract is concentrated and Carbon-13 labeled internal standards are added prior to analysis by GC/MS using a capillary GC column and selected ion monitoring using five sets of ions that are detailed in the method. Certain 2,3,7,8-substituted congeners are used to provide calibration and method recovery information. Proper column selection and access to reference isomer standards, may in certain cases, provide isomer specific data.

INTERFERENCES Solvents, reagents, glassware, and other sample processing hardware may yield discrete artifacts and/or elevated baselines which may cause misinterpretation of chromatographic data. Use high purity reagents and solvents to minimize interference problems. Interferents coextracted from the sample will vary considerably from source to source. PCDDs and PCDFs are often associated with other interfering chlorinated compounds such as PCBs and polychlorinated diphenyl ethers which may be found at concentrations several orders of magnitude higher than that of the analytes of interest.

INSTRUMENTATION A low resolution GC/MS utilizing 70 ev must be capable of selected ion monitoring (SIM) for at least 11 ions simultaneously, with a cycle time of 1 second or less. Minimum integration time for SIM is 50 ms per m/z. Also required is a GC-to-MS interface constructed of all glass or glass-lined materials. One of the following GC columns is required:

Column 1: 50 m CP-Sil-88 fused silica capillary column.
Column 2: DB-5 (30 m × 0.25 mm I.D., 0.25-um film thickness) fused silica capillary column.
Column 3: 30 m SP-2250 fused silica capillary column.

When toluene is employed as the final solvent, use of a bonded phase column is recommended. Solvent exchange into tridecane is required for other liquid phases or nonbonded columns such as CP-Sil-88. Chromatographic conditions must be adjusted to account for solvent boiling points.

PRECISION & ACCURACY Accuracy, precision, MDLs and concentration ranges for the compounds covered by this method have not been determined or published by EPA yet. The sensitivity of this method is dependent upon the level of interferents within a given matrix. Proposed quantification levels for target analytes were 2 ppb in soil samples, up to 10 ppb in other solid wastes and 10 ppt in water. Actual values have been shown to vary by homologous series and, to a lesser degree, by individual isomer.

SAMPLING METHOD Grab and composite samples must be collected in 1 L or 1-quart amber glass bottles with Teflon®-lined screw-caps that have been acid-washed and solvent rinsed before use. If compositing equipment is used, the system must incorporate glass sample containers for the collection of a minimum of 250 mL. samples.

SAMPLE PRESERVATION Samples must be cooled and stored at 4°C.

MHT Samples must be extracted within 30 days and analyzed within 45 days of collection.

SAMPLE PREPARATION
Soil samples — Extract 20 g of a 1:1 soil and anhydrous sodium sulfate mixture with 1:4 methanol-petroleum ether for 2 h in a wrist-action shaker. Concentrate the extract in a K-D and then exchange the solvent with hexane in the K-D. The final volume is 15 mL of hexane.

Aqueous samples — Extract a 1 L sample with methylene chloride, dry it with anhydrous sodium sulfate and concentrate it in a K-D followed by exchange with hexane in the K-D. A continuous liquid-liquid extractor may be used in place of a separatory funnel to avoid emulsions. The final volume is 15 mL of hexane.

Alumina column clean-up — The clean-up procedure described below consists of two phases. The first phase involves a sequential basic and acid washing of the extract that contains the analytes.

Wash the 15 mL hexane extract with 20% potassium hydroxide in a separatory funnel. Repeat the washing until no color is visible in the bottom layer but no more than four times because strong base is known to degrade certain PCDDs/PCDFs, so contact time must be minimized. Next, partition the 15 mL hexane against 40 mL of 5% sodium chloride. Next, partition the 15 mL hexane against 40 mL of concentrated sulfuric acid. Repeat the acid washings until no color is visible in the acid layer (but no more than four times). Finally, partition the 15 mL hexane against 40 mL of 5% sodium chloride. Dry the organic layer with anhydrous sodium sulfate and concentrate it to near dryness with a rotary evaporator. Dissolve the hexane residue from the first phase of the clean-up in 2 mL of hexane and apply it carefully to the top of a pre-eluted Woelm super 1 neutral alumina column. Elute the column with 10 mL of 8 percent (v/v) methylene chloride in hexane. Check by GC/MS analysis that no PCDDs of PCDFs are elute in this fraction before discarding it. Elute the PCDDs and PCDFs from the column with 15 mL of 60 percent (v/v) methylene chloride in hexane and collect this second fraction in a conical shaped concentrator tube.

Carbon column clean-up — Using a carefully regulated stream of nitrogen, concentrate the first 8 percent fraction (methylene chloride in hexane) from the alumina column to about 1 mL. Save this 8 percent concentrate for GC/MS analysis to check for breakthrough of PCDDs and PCDFs. Concentrate the second 60 percent fraction (methylene chloride in hexane) to about 2 to 3 mL. Prepare a carbon column and rinse the carbon with 5 mL cyclohexane/methylene chloride (50:50 v/v) in the forward direction of flow and then in the reverse direction of flow. While still in the reverse direction of flow, transfer the sample concentrate to the column and elute with 10 mL of methylene chloride/methanol/benzene (75:20:5, v/v). Save all above eluates and combine them (this fraction may be used as a check on column efficiency). Next, turn the column over and, in the direction of forward flow, elute the PCDD/PCDF fraction with 20 mL toluene. Evaporate the toluene fraction to about 1 mL on a rotary evaporator and transfer this concentrate to a 2.0-mL Reacti-vial. Concentrate the sample using a stream of nitrogen gas. The final volume will depend on the relative concentration of target analytes but it is typically 100 µL for soil samples and 500 µL for sludge, still bottom, and fly ash samples. Extracts which are determined to be outside the calibration range for individual analytes must be diluted or a smaller portion of the sample must be re-extracted.

An alternate carbon column clean-up also may be used with a 1 mL HPLC injector loop. The injector loop is connected to the optional HPLC column.

QUALITY CONTROL Demonstrate, using a method blank, that all glassware and reagents are interferent-free at the MDL of the matrix of interest. A "method blank" must be run with each 20 or fewer samples. The method blank is also dosed with the internal standards. For water samples, 1 L of deionized and/or distilled water should be used as the method blank. Mineral oil may be used as the method blank for other matrices.

Calculate response factors for standards relative to the internal standards. Add a recovery standard to the samples prior to injection. The concentration of the recovery standard in the sample extract must be the same as that in the calibration standards used to measure the response factors.

Field duplicates (individual samples taken from the same location at the same time) should be analyzed periodically to determine the total precision (field and lab). Where appropriate, field blanks should be provided to monitor for possible cross-contamination of samples in the field. GC column performance must be demonstrated initially and verified prior to analyzing any sample in a 12-hr period. The GC column performance check solution must be analyzed under the same chromatographic and mass spectrometric conditions used for other samples and standards.

Retention times of target analytes must be verified using reference standards. These values must correspond to the retention time windows established. While certain cleanup techniques are provided as part of this method, unique samples may require additional cleanup techniques to achieve the method detection limit.

REFERENCE Test Methods for Evaluating Solid Waste (SW-846). U.S.E.P.A., 1986. Method 8280, Rev. 0, Sept. 1986. Office of Solid Wastes, Washington, DC.

1,2,3,4,7,8-HxCDF   **EPA Method 8290**
CAS #70648-26-9

TITLE Polychlorinated Dibenzodioxins (PCDDs) and Polychlorinated Dibenzofurans (PCDFs) by High-Resolution Gas Chromatography/High-Resolution Mass Spectrometry (HRGC/HRMS).

MATRIX This method is applicable with a variety of environmental matrices including: water, soil, sediment, paper pulp, fly ash, fish tissue, human adipose tissue, sludges, fuel oil, chemical reactor residue, and still bottoms.

METHOD SUMMARY This method provides procedures for the detection and quantitative measurement of polychlorinated dibenzo-p-dioxins (tetra- through octachlorinated homologues; PCDDs), and polychlorinated dibenzofurans (tetra- through octachlorinated homologues; PCDFs) in a variety of environmental matrices and at part-per-trillion (ppt) to part-per-quadrillion (ppq) concentrations. High-resolution gas chromatography and high-resolution mass spectrometry (HRGC/HRMS) on purified sample extracts provides highly specific identification of each analyte. Quantification is provided using calibration standards.

INTERFERENCES Solvents, reagents, glassware, and other sample processing hardware may yield discrete artifacts that may cause misinterpretation of the chromatographic data. Analysts should avoid using PVC gloves. Interferants coextracted from the sample will vary considerably from matrix to matrix. PCDDs and PCDFs are often associated with other interfering chlorinated substances such as polychlorinated biphenyls (PCBs), polychlorinated diphenyl ethers (PCDEs), polychlorinated naphthalenes, and polychlorinated alkyldibenzofurans that may be found at concentrations several orders of magnitude higher than the PCDDs or PCDFs.

A high-resolution capillary column is used in this method. However, no single column is known to resolve all isomers. The 60 m DB-5 GC column is capable of 2,3,7,8-TCDD isomer specificity. In order to determine the concentration of the 2,3,7,8-TCDD (if detected on the DB-5 column), the sample extract must be reanalyzed on a column capable of 2,3,7,8-TCDF isomer specificity (e.g., DB-225, SP-2330, SP-2331, or equivalent).

INSTRUMENTATION High-Resolution Gas Chromatograph/High-Resolution Mass Spectrometer/Data System (HRGC/HRMS/DS) equipped with a GC injection port designed so that the separation of 2,3,7,8-TCDD from the other TCDD isomers achieved in the gas chromatographic column is not appreciably degraded.

Column 1: 60 m DB-5 fused silica capillary column.
Column 2: 30 m DB-225 fused silica capillary column, or equivalent.

PRECISION & ACCURACY Precision, bias and concentration ranges for the compounds covered by this method have not been determined yet. The sensitivity of Method 8290 is dependent upon the level of interferences within a given matrix. Samples containing concentrations of specific congeneric analytes of PCDDs and PCDFs that are greater than ten times the upper method calibration limits must be analyzed by a protocol designed for such concentration levels, e.g., EPA Method 8280.

SAMPLE PREPARATION
Sludge/wet fuel oil — Extract aqueous sludge or wet fuel oil samples by refluxing a sample with toluene using a Dean-Stark water separator until all the water is removed. Filter the toluene extract through a glass fiber filter, or equivalent, and concentrate it to near dryness either on a rotary evaporator using an inert gas. Transfer the concentrate to a separatory funnel using hexane and wash it with 5% sodium chloride solution. Proceed to clean up.

Soil/sediment — If the sample is wet, add anhydrous powdered sodium sulfate to it until a free flowing mixture is obtained. Place the soil/sodium sulfate mixture in the Soxhlet apparatus, add toluene, and reflux for 16 h. The solvent must cycle completely through the system five times per h. Cool and filter the extract through a glass fiber filter and concentrate to near dryness on a rotary evaporator. Transfer the residue to a separatory funnel, using hexane. Proceed to clean up.

Aqueous samples — Use a 1-L sample; the method may require acetone to be added to it. When the sample is judged to contain 1% or more solids, it must be filtered through a glass fiber filter that has been rinsed with toluene. If the suspended solids content is too great to filter, centrifuge the sample, decant, and then filter the aqueous phase. Combine the solids from the centrifuge bottle(s) with the particulates on the filter and with the filter itself and proceed with Soxhlet extraction for soil/sediment. Extract the aqueous filtrate with methylene chloride in a separatory funnel, filter the extract through anhydrous sodium sulfate, and concentrate it using a K-D apparatus or a rotary evaporator. Exchange the solvent with hexane and proceed to clean up.

Clean up — The sample extract is cleaned up utilizing a number of different techniques. Partition cleanup is where the sample extract is partitioned with concentrated sulfuric acid, 5% aqueous sodium chloride, and 20% aqueous potassium hydroxide. Silica/alumina column cleanup involves packing gravity columns with silica gel and alumina and sequentially eluting the residue from the partition cleanup. Carbon column cleanup involves packing a column with a mixture of AX–21 and Celite 545 and sequentially eluting the sample concentrate from the silica/alumina cleanup with hexane, cyclohexane/methylene chloride (50:50), and methylene chloride/methanol/toluene (75:20:5). Then the column is turned upside down and the PCDD/PCDF fraction is eluted with toluene. The toluene fraction is concentrated and stored in the dark at room temperature until analysis.

QUALITY CONTROL Demonstrate, through the analysis of a reagent water blank, that interferences from the analytical system, glassware, and reagents are under control. For each analytical batch (up to 20 samples), a reagent blank, matrix spike, and matrix spike duplicate/duplicate must be analyzed (the frequency of the spikes may be different for different monitoring programs). The blank and spiked samples must be carried through all stages of the sample preparation and measurement steps.

A GC column performance check is required at the beginning of each 12-h period during which samples are analyzed. An HRGC/HRMS method blank run is required between a calibration run and the first sample run. The same method blank extract may thus be analyzed more than once if the number of samples within a batch requires more than 12 h of analyses.

At the beginning of each 12-h period during which samples are to be analyzed, an aliquot of the 1) GC column performance check solution and 2) a high-resolution concentration calibration must be analyzed to demonstrate adequate GC resolution and sensitivity, response factor reproducibility, and mass range calibration, and to establish the PCDD/PCDF retention time windows. A mass resolution check must also be performed to demonstrate adequate mass resolution using an appropriate reference compound (perfluorokerosene (PFK) is recommended). If the required criteria are not met, remedial action must be taken before any samples are analyzed.

Routine or continuing calibration (using a high resolution calibration solution) and the mass resolution check must also be performed at the end of each 12 h period. Furthermore, a HRGC/HRMS method blank analysis must be recorded following a calibration analysis and the first sample analysis.

To evaluate the performance of the analytical method, the QC check samples must be handled in exactly the same manner as actual samples. Therefore, 1.0 mL of the QC check sample concentrate is spiked into each of four 1 L aliquots of reagent water (which becomes the QC check sample), extracted, and then analyzed by GC. The variety of semivolatile analytes which may be analyzed by GC is such that the concentration of the QC check sample concentrate is different for the different analytical techniques presented in the full method.

The analyst must demonstrate also that the compounds of interest are being quantitatively recovered by the cleanup technique before the cleanup is applied to actual samples. For sample extracts that are cleaned up, the associated quality control samples (e.g., spikes, blanks, and duplicates) must also be processed through the same cleanup procedure. The analysis using each determinative method (GC, GC/MS, HPLC) specifies instrument calibration procedures using stock standards. It is recommended that cleanup also be performed on a series of the same type of standards to validate chromatographic elution patterns for the compounds of interest and to verify the absence of interferences from reagents.

REFERENCE Test Methods for Evaluating Solid Waste (SW-846). U.S. EPA. 1983. Method 8290, Rev. 0, Nov. 1990. Office of Solid Wastes, Washington, DC.

1,2,3,6,7,8-HxCDF   **EPA Method 8290**
CAS #57117-44-9

TITLE Polychlorinated Dibenzodioxins (PCDDs) and Polychlorinated Dibenzofurans (PCDFs) by High-Resolution Gas Chromatography/High-Resolution Mass Spectrometry (HRGC/HRMS).

MATRIX This method is applicable with a variety of environmental matrices including: water, soil, sediment, paper pulp, fly ash, fish tissue, human adipose tissue, sludges, fuel oil, chemical reactor residue, and still bottoms.

METHOD SUMMARY This method provides procedures for the detection and quantitative measurement of polychlorinated dibenzo-p-dioxins (tetra- through octachlorinated homologues; PCDDs), and polychlorinated dibenzofurans (tetra- through octachlorinated homologues; PCDFs) in a variety of environmental matrices and at part-per-trillion (ppt) to part-per-quadrillion (ppq) concentrations. High-resolution gas chromatography and high-resolution mass spectrometry (HRGC/HRMS) on purified sample extracts provides highly specific identification of each analyte. Quantification is provided using calibration standards.

INTERFERENCES Solvents, reagents, glassware, and other sample processing hardware may yield discrete artifacts that may cause misinterpretation of the chromatographic data. Analysts should avoid using PVC gloves. Interferants coextracted from the sample will vary considerably from matrix to matrix. PCDDs and PCDFs are often associated with other interfering chlorinated substances such as polychlorinated biphenyls (PCBs), polychlorinated diphenyl ethers (PCDEs), polychlorinated naphthalenes, and polychlorinated alkyldibenzofurans that may be found at concentrations several orders of magnitude higher than the PCDDs or PCDFs.

A high-resolution capillary column is used in this method. However, no single column is known to resolve all isomers. The 60 m DB-5 GC column is capable of 2,3,7,8-TCDD isomer specificity. In order to determine the concentration of the 2,3,7,8-TCDD (if detected on the DB-5 column), the sample extract must be reanalyzed on a column capable of 2,3,7,8-TCDF isomer specificity (e.g., DB-225, SP-2330, SP-2331, or equivalent).

INSTRUMENTATION High-Resolution Gas Chromatograph/High-Resolution Mass Spectrometer/Data System (HRGC/HRMS/DS) equipped with a GC injection port designed so that the separation of 2,3,7,8-TCDD from the other

TCDD isomers achieved in the gas chromatographic column is not appreciably degraded.

Column 1: 60 m DB-5 fused silica capillary column.
Column 2: 30 m DB-225 fused silica capillary column, or equivalent.

PRECISION & ACCURACY Precision, bias and concentration ranges for the compounds covered by this method have not been determined yet. The sensitivity of Method 8290 is dependent upon the level of interferences within a given matrix. Samples containing concentrations of specific congeneric analytes of PCDDs and PCDFs that are greater than ten times the upper method calibration limits must be analyzed by a protocol designed for such concentration levels, e.g., EPA Method 8280.

SAMPLE PREPARATION
Sludge/wet fuel oil — Extract aqueous sludge or wet fuel oil samples by refluxing a sample with toluene using a Dean-Stark water separator until all the water is removed. Filter the toluene extract through a glass fiber filter, or equivalent, and concentrate it to near dryness either on a rotary evaporator using an inert gas. Transfer the concentrate to a separatory funnel using hexane and wash it with 5% sodium chloride solution. Proceed to clean up.

Soil/sediment — If the sample is wet, add anhydrous powdered sodium sulfate to it until a free flowing mixture is obtained. Place the soil/sodium sulfate mixture in the Soxhlet apparatus, add toluene, and reflux for 16 h. The solvent must cycle completely through the system five times per h. Cool and filter the extract through a glass fiber filter and concentrate to near dryness on a rotary evaporator. Transfer the residue to a separatory funnel, using hexane. Proceed to clean up.

Aqueous samples — Use a 1-L sample; the method may require acetone to be added to it. When the sample is judged to contain 1% or more solids, it must be filtered through a glass fiber filter that has been rinsed with toluene. If the suspended solids content is too great to filter, centrifuge the sample, decant, and then filter the aqueous phase. Combine the solids from the centrifuge bottle(s) with the particulates on the filter and with the filter itself and proceed with Soxhlet extraction for soil/sediment. Extract the aqueous filtrate with methylene chloride in a separatory funnel, filter the extract through anhydrous sodium sulfate, and concentrate it using a K-D apparatus or a rotary evaporator. Exchange the solvent with hexane and proceed to clean up.

Clean up — The sample extract is cleaned up utilizing a number of different techniques. Partition cleanup is where the sample extract is partitioned with concentrated sulfuric acid, 5% aqueous sodium chloride, and 20% aqueous potassium hydroxide. Silica/alumina column cleanup involves packing gravity columns with silica gel and alumina and sequentially eluting the residue from the partition cleanup. Carbon column cleanup involves packing a column with a mixture of AX–21 and Celite 545 and sequentially eluting the sample concentrate from the silica/alumina cleanup with hexane, cyclohexane/methylene chloride (50:50), and methylene chloride/methanol/toluene (75:20:5). Then the column is turned upside down and the PCDD/PCDF fraction is eluted with toluene. The toluene fraction is concentrated and stored in the dark at room temperature until analysis.

QUALITY CONTROL Demonstrate, through the analysis of a reagent water blank, that interferences from the analytical system, glassware, and reagents are under control. For each analytical batch (up to 20 samples), a reagent blank, matrix spike, and matrix spike duplicate/duplicate must be analyzed (the frequency of the spikes may be different for different monitoring programs). The blank and spiked samples must be carried through all stages of the sample preparation and measurement steps.

A GC column performance check is required at the beginning of each 12-h period during which samples are analyzed. An HRGC/HRMS method blank run is required between a calibration run and the first sample run. The same method blank extract may thus be analyzed more than once if the number of samples within a batch requires more than 12 h of analyses.

At the beginning of each 12-h period during which samples are to be analyzed, an aliquot of the 1) GC column performance check solution and 2) a high-resolution concentration calibration must be analyzed to demonstrate adequate GC resolution and sensitivity, response factor reproducibility, and mass range calibration, and to establish the PCDD/PCDF retention time windows. A mass resolution check must also be performed to demonstrate adequate mass resolution using an appropriate reference compound (perfluorokerosene (PFK) is recommended). If the required criteria are not met, remedial action must be taken before any samples are analyzed.

Routine or continuing calibration (using a high resolution calibration solution) and the mass resolution check must also be performed at the end of each 12 h period. Furthermore, a HRGC/HRMS method blank analysis must be recorded following a calibration analysis and the first sample analysis.

To evaluate the performance of the analytical method, the QC check samples must be handled in exactly the same manner as actual samples. Therefore, 1.0 mL of the QC check sample concentrate is spiked into each of four 1 L aliquots of reagent water (which becomes the QC check sample), extracted, and then analyzed by GC. The variety of semivolatile analytes which may be analyzed by GC is such that the concentration of the QC check sample concentrate is different for the different analytical techniques presented in the full method.

The analyst must demonstrate also that the compounds of interest are being quantitatively recovered by the cleanup technique before the cleanup is applied to actual samples. For sample extracts that are cleaned up, the associated quality control samples (e.g., spikes, blanks, and duplicates) must also be processed through the same cleanup procedure. The analysis using each determinative method (GC, GC/MS, HPLC) specifies instrument calibration procedures using stock standards. It is recommended that cleanup also be performed on a series of the same type of standards to validate chromatographic elution patterns for the compounds of interest and to verify the absence of interferences from reagents.

REFERENCE Test Methods for Evaluating Solid Waste (SW-846). U.S. EPA. 1983. Method 8290, Rev. 0, Nov. 1990. Office of Solid Wastes, Washington, DC.

1,2,3,7,8,9-HxCDF   **EPA Method 8290**
CAS #72918-21-9

TITLE Polychlorinated Dibenzodioxins (PCDDs) and Polychlorinated Dibenzofurans (PCDFs) by High-Resolution Gas Chromatography/High-Resolution Mass Spectrometry (HRGC/HRMS).

MATRIX This method is applicable with a variety of environmental matrices including: water, soil, sediment, paper pulp, fly ash, fish tissue, human adipose tissue, sludges, fuel oil, chemical reactor residue, and still bottoms.

METHOD SUMMARY This method provides procedures for the detection and quantitative measurement of polychlorinated dibenzo-p-dioxins (tetra- through octachlorinated homologues; PCDDs), and polychlorinated dibenzofurans (tetra- through octachlorinated homologues; PCDFs) in a variety of environmental matrices and at part-per-trillion (ppt) to part-per-quadrillion (ppq) concentrations. High-resolution gas chromatography and high-resolution mass spectrometry (HRGC/HRMS) on purified sample extracts provides highly specific identification of each analyte. Quantification is provided using calibration standards.

INTERFERENCES Solvents, reagents, glassware, and other sample processing hardware may yield discrete artifacts that may cause misinterpretation of the chromatographic data. Analysts should avoid using PVC gloves. Interferants coextracted from the sample will vary considerably from matrix to matrix. PCDDs and PCDFs are often associated with other interfering chlorinated substances such as polychlorinated biphenyls (PCBs), polychlorinated diphenyl ethers (PCDEs), polychlorinated naphthalenes, and polychlorinated alkyldibenzofurans that may be found at concentrations several orders of magnitude higher than the PCDDs or PCDFs.

A high-resolution capillary column is used in this method. However, no single column is known to resolve all isomers. The 60 m DB-5 GC column is capable of 2,3,7,8-TCDD isomer specificity. In order to determine the concentration of the 2,3,7,8-TCDD (if detected on the DB-5 column), the sample extract must be reanalyzed on a column capable of 2,3,7,8-TCDF isomer specificity (e.g., DB-225, SP-2330, SP-2331, or equivalent).

INSTRUMENTATION High-Resolution Gas Chromatograph/High-Resolution Mass Spectrometer/Data System (HRGC/HRMS/DS) equipped with a GC injection port designed so that the separation of 2,3,7,8-TCDD from the other TCDD isomers achieved in the gas chromatographic column is not appreciably degraded.

Column 1: 60 m DB-5 fused silica capillary column.
Column 2: 30 m DB-225 fused silica capillary column, or equivalent.

PRECISION & ACCURACY Precision, bias and concentration ranges for the compounds covered by this method have not been determined yet. The sensitivity of Method 8290 is dependent upon the level of interferences within a given matrix. Samples containing concentrations of specific congeneric analytes of PCDDs and PCDFs that are greater than ten times the upper method calibration limits must be analyzed by a protocol designed for such concentration levels, e.g., EPA Method 8280.

SAMPLE PREPARATION
Sludge/wet fuel oil — Extract aqueous sludge or wet fuel oil samples by refluxing a sample with toluene using a Dean-Stark water separator until all the water is removed. Filter the toluene extract through a glass fiber filter, or equivalent, and concentrate it to near dryness either on a rotary evaporator using an inert gas. Transfer the concentrate to a separatory funnel using hexane and wash it with 5% sodium chloride solution. Proceed to clean up.

Soil/sediment — If the sample is wet, add anhydrous powdered sodium sulfate to it until a free flowing mixture is obtained. Place the soil/sodium sulfate mixture in the Soxhlet apparatus, add toluene, and reflux for 16 h. The solvent must cycle completely through the system five times per h. Cool and filter the extract through a glass fiber filter and concentrate to near dryness on a rotary evaporator. Transfer the residue to a separatory funnel, using hexane. Proceed to clean up.

Aqueous samples — Use a 1-L sample; the method may require acetone to be added to it. When the sample is judged to contain 1% or more solids, it must be filtered through a glass fiber filter that has been rinsed with toluene. If the suspended solids content is too great to filter, centrifuge the sample, decant, and then filter the aqueous phase. Combine the solids from the centrifuge bottle(s) with the particulates on the filter and with the filter itself and proceed with Soxhlet extraction for soil/sediment. Extract the aqueous filtrate with methylene chloride in a separatory funnel, filter the extract through anhydrous sodium sulfate, and concentrate it using a K-D apparatus or a rotary evaporator. Exchange the solvent with hexane and proceed to clean up.

Clean up — The sample extract is cleaned up utilizing a number of different techniques. Partition cleanup is where the sample extract is partitioned with concentrated sulfuric acid, 5% aqueous sodium chloride, and 20% aqueous potassium hydroxide. Silica/alumina column cleanup involves packing gravity columns with silica gel and alumina and sequentially eluting the residue from the partition cleanup. Carbon column cleanup involves packing a column with a mixture of AX–21 and Celite 545 and sequentially eluting the sample concentrate from the silica/alumina cleanup with hexane, cyclohexane/methylene chloride (50:50), and methylene chloride/methanol/toluene (75:20:5). Then the column is turned upside down and the PCDD/PCDF fraction is eluted with toluene. The toluene fraction is concentrated and stored in the dark at room temperature until analysis.

QUALITY CONTROL Demonstrate, through the analysis of a reagent water blank, that interferences from the analytical system, glassware, and reagents are under control. For each

analytical batch (up to 20 samples), a reagent blank, matrix spike, and matrix spike duplicate/duplicate must be analyzed (the frequency of the spikes may be different for different monitoring programs). The blank and spiked samples must be carried through all stages of the sample preparation and measurement steps.

A GC column performance check is required at the beginning of each 12-h period during which samples are analyzed. An HRGC/HRMS method blank run is required between a calibration run and the first sample run. The same method blank extract may thus be analyzed more than once if the number of samples within a batch requires more than 12 h of analyses.

At the beginning of each 12-h period during which samples are to be analyzed, an aliquot of the 1) GC column performance check solution and 2) a high-resolution concentration calibration must be analyzed to demonstrate adequate GC resolution and sensitivity, response factor reproducibility, and mass range calibration, and to establish the PCDD/PCDF retention time windows. A mass resolution check must also be performed to demonstrate adequate mass resolution using an appropriate reference compound (perfluorokerosene (PFK) is recommended). If the required criteria are not met, remedial action must be taken before any samples are analyzed.

Routine or continuing calibration (using a high resolution calibration solution) and the mass resolution check must also be performed at the end of each 12 h period. Furthermore, a HRGC/HRMS method blank analysis must be recorded following a calibration analysis and the first sample analysis.

To evaluate the performance of the analytical method, the QC check samples must be handled in exactly the same manner as actual samples. Therefore, 1.0 mL of the QC check sample concentrate is spiked into each of four 1 L aliquots of reagent water (which becomes the QC check sample), extracted, and then analyzed by GC. The variety of semivolatile analytes which may be analyzed by GC is such that the concentration of the QC check sample concentrate is different for the different analytical techniques presented in the full method.

The analyst must demonstrate also that the compounds of interest are being quantitatively recovered by the cleanup technique before the cleanup is applied to actual samples. For sample extracts that are cleaned up, the associated quality control samples (e.g., spikes, blanks, and duplicates) must also be processed through the same cleanup procedure. The analysis using each determinative method (GC, GC/MS, HPLC) specifies instrument calibration procedures using stock standards. It is recommended that cleanup also be performed on a series of the same type of standards to validate chromatographic elution patterns for the compounds of interest and to verify the absence of interferences from reagents.

REFERENCE Test Methods for Evaluating Solid Waste (SW-846). U.S. EPA. 1983. Method 8290, Rev. 0, Nov. 1990. Office of Solid Wastes, Washington, DC.

2,3,4,6,7,8-HxCDF   EPA Method 8290
CAS #60851-34-5

TITLE Polychlorinated Dibenzodioxins (PCDDs) and Polychlorinated Dibenzofurans (PCDFs) by High-Resolution Gas Chromatography/High-Resolution Mass Spectrometry (HRGC/HRMS).

MATRIX This method is applicable with a variety of environmental matrices including: water, soil, sediment, paper pulp, fly ash, fish tissue, human adipose tissue, sludges, fuel oil, chemical reactor residue, and still bottoms.

METHOD SUMMARY This method provides procedures for the detection and quantitative measurement of polychlorinated dibenzo-p-dioxins (tetra- through octachlorinated homologues; PCDDs), and polychlorinated dibenzofurans (tetra- through octachlorinated homologues; PCDFs) in a variety of environmental matrices and at part-per-trillion (ppt) to part-per-quadrillion (ppq) concentrations. High-resolution gas chromatography and high-resolution mass spectrometry (HRGC/HRMS) on purified sample extracts provides highly specific identification of each analyte. Quantification is provided using calibration standards.

INTERFERENCES Solvents, reagents, glassware, and other sample processing hardware may yield discrete artifacts that may cause misinterpretation of the chromatographic data. Analysts should avoid using PVC gloves. Interferants coextracted from the sample will vary considerably from matrix to matrix. PCDDs and PCDFs are often associated with other interfering chlorinated substances such as polychlorinated biphenyls (PCBs), polychlorinated diphenyl ethers (PCDEs), polychlorinated naphthalenes, and polychlorinated alkyldibenzofurans that may be found at concentrations several orders of magnitude higher than the PCDDs or PCDFs.

A high-resolution capillary column is used in this method. However, no single column is known to resolve all isomers. The 60 m DB-5 GC column is capable of 2,3,7,8-TCDD isomer specificity. In order to determine the concentration of the 2,3,7,8-TCDD (if detected on the DB-5 column), the sample extract must be reanalyzed on a column capable of 2,3,7,8-TCDF isomer specificity (e.g., DB-225, SP-2330, SP-2331, or equivalent).

INSTRUMENTATION High-Resolution Gas Chromatograph/High-Resolution Mass Spectrometer/Data System (HRGC/HRMS/DS) equipped with a GC injection port designed so that the separation of 2,3,7,8-TCDD from the other TCDD isomers achieved in the gas chromatographic column is not appreciably degraded.

Column 1: 60 m DB-5 fused silica capillary column.
Column 2: 30 m DB-225 fused silica capillary column, or equivalent.

PRECISION & ACCURACY Precision, bias and concentration ranges for the compounds covered by this method have not been determined yet. The sensitivity of Method 8290 is dependent upon the level of interferences within a given matrix. Samples containing concentrations of specific congeneric

analytes of PCDDs and PCDFs that are greater than ten times the upper method calibration limits must be analyzed by a protocol designed for such concentration levels, e.g., EPA Method 8280.

SAMPLE PREPARATION

Sludge/wet fuel oil — Extract aqueous sludge or wet fuel oil samples by refluxing a sample with toluene using a Dean-Stark water separator until all the water is removed. Filter the toluene extract through a glass fiber filter, or equivalent, and concentrate it to near dryness either on a rotary evaporator using an inert gas. Transfer the concentrate to a separatory funnel using hexane and wash it with 5% sodium chloride solution. Proceed to clean up.

Soil/sediment — If the sample is wet, add anhydrous powdered sodium sulfate to it until a free flowing mixture is obtained. Place the soil/sodium sulfate mixture in the Soxhlet apparatus, add toluene, and reflux for 16 h. The solvent must cycle completely through the system five times per h. Cool and filter the extract through a glass fiber filter and concentrate to near dryness on a rotary evaporator. Transfer the residue to a separatory funnel, using hexane. Proceed to clean up.

Aqueous samples — Use a 1-L sample; the method may require acetone to be added to it. When the sample is judged to contain 1% or more solids, it must be filtered through a glass fiber filter that has been rinsed with toluene. If the suspended solids content is too great to filter, centrifuge the sample, decant, and then filter the aqueous phase. Combine the solids from the centrifuge bottle(s) with the particulates on the filter and with the filter itself and proceed with Soxhlet extraction for soil/sediment. Extract the aqueous filtrate with methylene chloride in a separatory funnel, filter the extract through anhydrous sodium sulfate, and concentrate it using a K-D apparatus or a rotary evaporator. Exchange the solvent with hexane and proceed to clean up.

Clean up — The sample extract is cleaned up utilizing a number of different techniques. Partition cleanup is where the sample extract is partitioned with concentrated sulfuric acid, 5% aqueous sodium chloride, and 20% aqueous potassium hydroxide. Silica/alumina column cleanup involves packing gravity columns with silica gel and alumina and sequentially eluting the residue from the partition cleanup. Carbon column cleanup involves packing a column with a mixture of AX–21 and Celite 545 and sequentially eluting the sample concentrate from the silica/alumina cleanup with hexane, cyclohexane/methylene chloride (50:50), and methylene chloride/methanol/toluene (75:20:5). Then the column is turned upside down and the PCDD/PCDF fraction is eluted with toluene. The toluene fraction is concentrated and stored in the dark at room temperature until analysis.

QUALITY CONTROL
Demonstrate, through the analysis of a reagent water blank, that interferences from the analytical system, glassware, and reagents are under control. For each analytical batch (up to 20 samples), a reagent blank, matrix spike, and matrix spike duplicate/duplicate must be analyzed (the frequency of the spikes may be different for different monitoring programs). The blank and spiked samples must be carried through all stages of the sample preparation and measurement steps.

A GC column performance check is required at the beginning of each 12-h period during which samples are analyzed. An HRGC/HRMS method blank run is required between a calibration run and the first sample run. The same method blank extract may thus be analyzed more than once if the number of samples within a batch requires more than 12 h of analyses.

At the beginning of each 12-h period during which samples are to be analyzed, an aliquot of the 1) GC column performance check solution and 2) a high-resolution concentration calibration must be analyzed to demonstrate adequate GC resolution and sensitivity, response factor reproducibility, and mass range calibration, and to establish the PCDD/PCDF retention time windows. A mass resolution check must also be performed to demonstrate adequate mass resolution using an appropriate reference compound (perfluorokerosene (PFK) is recommended). If the required criteria are not met, remedial action must be taken before any samples are analyzed.

Routine or continuing calibration (using a high resolution calibration solution) and the mass resolution check must also be performed at the end of each 12 h period. Furthermore, a HRGC/HRMS method blank analysis must be recorded following a calibration analysis and the first sample analysis.

To evaluate the performance of the analytical method, the QC check samples must be handled in exactly the same manner as actual samples. Therefore, 1.0 mL of the QC check sample concentrate is spiked into each of four 1 L aliquots of reagent water (which becomes the QC check sample), extracted, and then analyzed by GC. The variety of semivolatile analytes which may be analyzed by GC is such that the concentration of the QC check sample concentrate is different for the different analytical techniques presented in the full method.

The analyst must demonstrate also that the compounds of interest are being quantitatively recovered by the cleanup technique before the cleanup is applied to actual samples. For sample extracts that are cleaned up, the associated quality control samples (e.g., spikes, blanks, and duplicates) must also be processed through the same cleanup procedure. The analysis using each determinative method (GC, GC/MS, HPLC) specifies instrument calibration procedures using stock standards. It is recommended that cleanup also be performed on a series of the same type of standards to validate chromatographic elution patterns for the compounds of interest and to verify the absence of interferences from reagents.

REFERENCE Test Methods for Evaluating Solid Waste (SW-846). U.S. EPA. 1983. Method 8290, Rev. 0, Nov. 1990. Office of Solid Wastes, Washington, DC.

1,2,3,4,7,8-HxCDF EPA Method 8280
CAS #70648-26-9

TITLE Analysis of PCDDs and PCDFs

MATRIX chemical wastes, fuel oils, still bottoms, sludges, water, soil, fly ash, reactor residues.

APPLICATION This method is used for the analysis of tetra-, penta-, hexa-, hepta-, and octachlorinated dibenzo-p-dioxins (PCDDs) and dibenzofurans (PCDFs). The sensitivity of the method is dependent on the level of interferents within the matrix. Only experienced analysts should be used. Special safety precautions must be observed and an EPA-approved sample disposal plan must be used.

INTERFERENCES Solvents, reagents, and glassware may introduce artifacts. Other interferences may come from coextracted compounds from samples; PCBs and polychlorinated diphenyl ethers are common interferents.

INSTRUMENTATION GC/MS with a fused silica capillary column. Also, solvent extraction and concentration glassware and either a gravity flow activated carbon AX–21/silica gel Type 60 EM reagent column or a HPLC with a 10 mm by 7 cm silanized glass column with active carbon AX–21 and Spherisorb S10W silica for sample cleanup. One of three fused silica capillary GC columns may be used: column 1: 50 m CP-Sil-88; Column 2: 30 m by 0.25 mm DB-5; colunm 3: 30 m SP-2250.

RANGE 50–6,000 picograms.

MDL 2.53 ng/L (in reagent water); 0.83 µg/kg (in Missouri soil); 0.30 µg/kg (in fly ash); 2.17 µg/kg (in industrial sludge); 2.27 µg/kg (in still bottom); 2.09 µg/kg (in fuel oil); MDLs are for carbon-13 labeled analyte.

PRECISION as (RSD) 26% with 5 ng/g in clay; 6.8% with 25 ng/g soil. 5.6% with 139 ng/g in sludge; 13.5% with 24.2 ng/g in fly ash.

ACCURACY (as Mean % Recovery) 54.2% with 5 ng/g in clay; 68.5% with 25 ng/g in soil; 82.2% with 125 ng/g in sludge; 91.0% with 46 ng/g in fly ash; 92.9% with 2500 ng/g in still bottom.

SAMPLING METHOD Use 1 L (or quart) amber glass bottles with Teflon®-lined or solvent washed foil screw caps. Tape caps to bottle after sampling.

Compositing equipment must use glass containers and contain no Tygon or rubber tubing. Sample bottles must not be prewashed with the sample before its collection. Aqueous samples cannot be aliquoted from sample containers — the entire sample must be used and the container is washed out with the extracting solvent.

STABILITY Cool to 4°C and store at this temperature. When toluene is employed as the final solvent use a bonded phase GC column for separation. Otherwise, solvent exchange into tridecane is required for other liquid phases or the CP-Sil-88 GC column.

MHT 30 days; samples must be completely analyzed within 45 days of collection.

QUALITY CONTROL A method blank must be analyzed each time a set of samples is extracted or there is a change in reagents. A lab method blank must be run with each analytical batch of 20 or fewer samples. Field duplicates and field blanks should be analyzed periodically. GC column performance must be demonstrated initially and verified prior to analyzing any sample in a 12 h period. A series of calibration standards must be processed through the procedure to validate elution patterns and absence of interferents from reagents. Both the alumina column and carbon column performance must be routinely checked for presence of the analyte.

Performance evaluation samples and split samples with other laboratories are also expected to be periodically analyzed.

REFERENCE Method 8280, SW-846, 3rd ed., Nov.1986.

Hydroquinone   **EPA Method 8270**
CAS #123-31-9

TITLE Semivolatile Organic Compounds by GC/MS

MATRIX This method is used to determine the concentration of semivolatile organic compounds in extracts prepared from all types of solid waste matrices, soils, and groundwater. Although surface waters are not specifically mentioned, this method should be applicable to water samples from rivers, lakes, etc.

METHOD SUMMARY This method covers 259 semivolatile organic compounds. In very limited applications direct injection of the sample into the GC/MS system may be appropriate, but this results in very high detection limits (approximately 10,000 µg/L). Typically, a 1-L liquid sample, containing surrogate, and matrix spiking standards, is extracted in a continuous extractor first under acid conditions and then under basic conditions. Typically 30 g of a solid sample, containing surrogate, and matrix spiking standards, is extracted ultrasonically. After concentrating the extract to 1 mL it is spiked with 10 µL of an internal standard solution just prior to analysis by GC/MS. The volume injected should contain about 100 ng of base/neutral and 200 ng of acid surrogates (for a 1-µL injection). Analysis is performed by GC/MS using a capillary GC column.

INTERFERENCES Raw GC/MS data from all blanks, samples, and spikes must be evaluated for interferences. Contamination by carryover can occur whenever high-concentration and low-concentration samples are sequentially analyzed. To reduce carryover, the sample syringe must be rinsed out between samples with solvent. Whenever an unusually concentrated sample is encountered, it should be followed by the analysis of blank solvent to check for cross-contamination.

INSTRUMENTATION A GC/MS and a data system are required. The GC column used is a 30 m × 0.25 mm I.D. (or 0.32 mm I.D.) 1um film thickness silicone-coated fused silica capillary column. A continuous liquid-liquid extractor equipped with Teflon® or glass connection joints and stopcocks requiring no lubrication, a K-D concentrating apparatus, water bath, and an ultrasonic disrupter with a minimum power of 300 W and with pulsing capability are also required.

PRECISION & ACCURACY The estimated quantitation limit (EQL) of Method 8270B for determining an individual compound is approximately 1 mg/kg (wet weight) for soil or sediment samples, 1–200 mg/kg for wastes (dependent on matrix and method of preparation), and 10 µg/L for groundwater samples. EQLs will be proportionately higher for sample extracts that require dilution to avoid saturation of the detector.

The EQL(b) for groundwater in µg/L is not determined.
The EQL (a, b) for low concentrations in soil and sediment in µg/kg is not determined.
Accuracy as µg/L is not listed.
Overall precision in µg/L is not listed.

(a) *EQLs listed for soil/sediment are based on wet weight. Normally data is reported in a dry-weight basis; therefore, EQLs will be higher based on the % dry weight of each sample. This calculation is based on a 30-g sample and gel permeation chromatography cleanup.*
(b) *Sample EQLs are highly matrix-dependent. The EQLs are provided for guidance and may not always be achievable.*
C = *True value for concentration, in µg/L.*
X = *Average recovery found for measurements of samples containing a concentration of C, in µg/L.*

ESTIMATED QUANTITATION LIMIT

Other Matrices	Factor (a)
High-concentration soil and sludges by sonicator	7.5
Non-water miscible waste	75

(a) *EQL for other matrices = [EQL for low soil/sediment] × [Factor]. This estimated EQL is similar to an EPA "Practical Quantitation Limit."*

SAMPLING METHOD

Liquid samples — Use a 1 or 2½ gallon amber glass bottle with a screw-top Teflon®-lined cover that has been prewashed with detergent and rinsed with distilled water and methanol (or isopropanol).

Soils, sediments, or sludges — Use an 8-oz. widemouth glass with a screw-top Teflon®-lined cover that has been prewashed with detergent and rinsed with distilled water and methanol (or isopropanol).

SAMPLE PRESERVATION

Liquid samples — If residual chlorine is present, add 3 mL of 10% sodium thiosulfate per gallon, cool to 4°C and store in a solvent-free refrigerator until analysis; if chlorine is not present, then eliminate the sodium thiosulfate addition.

Soils, sediments, or sludges — Cool samples to 4°C and store in a solvent-free refrigerator.

MHT Liquid samples must be extracted within 7 days and the extracts analyzed within 40 days. Soils, sediments, or sludges may be stored for a maximum of 14 days and the extracts analyzed within 40 days.

SAMPLE PREPARATION

Liquid samples — Transfer 1 L quantitatively to a continuous extractor. If high concentrations are anticipated, a smaller volume may be used and then diluted with organic-free reagent water to 1 L. Adjust pH, if necessary, to pH <2 using 1:1 (V/V) sulfuric acid. Pipette 1.0 mL of a surrogate standard spiking solution into each sample. For the sample in each analytical batch selected for spiking, add 1.0 mL of a matrix spiking standard. For base/neutral acid analysis, the amount of the surrogates and matrix spiking compounds added to the sample should result in a final concentration of 100 ng/µL of each analyte in the extract to be analyzed (assuming a 1-µL injection). Extract with methylene chloride for 18–24 h. Next, adjust the pH of the aqueous phase to pH >11 using 10 N sodium hydroxide and extract it with methylene chloride again for 18–24 h. Dry the extract through a column containing anhydrous sodium sulfate and concentrate it to 1 mL using a K-D concentrator.

Soils, sediments, or sludges — Use 30 g of sample. Nonporous or wet samples (gummy or clay type) that do not have a free-flowing sandy texture must be mixed with anhydrous sodium sulfate until the sample is free flowing. Add 1 mL of surrogate standards to all samples, spikes, standards, and blanks. For the sample in each analytical batch selected for spiking, add 1.0 mL of a matrix spiking standard. For base/neutral acid analysis, the amount added of the surrogates and matrix spiking compounds should result in a final concentration of 100 ng/µL of each base/neutral analyte and 200 ng/µL of each acid analyte in the extract to be analyzed (assuming a 1-µL injection). Immediately add a 100-mL mixture of 1:1 methylene chloride:acetone and extract the sample ultrasonically for 3 min and then decant or filter the extracts. Repeat the extraction two or more times. Dry the extract using a column with anhydrous sodium sulfate and concentrate it to 1 mL in a K-D concentrator.

QUALITY CONTROL A methylene chloride solution containing 50 ng/µL of decafluorotriphenylphosphine (DFTPP) is used for tuning the GC/MS system each 12-h shift. A system performance check also must be made during every 12-h shift. A standard containing 50 ng/µL each of 4,4'-DDT, pentachlorophenol, and benzidine is required to verify injection port inertness and GC column performance. A calibration standard at mid-concentration, containing each compound of interest, including all required surrogates, must be performed every 12 h during analysis. After the system performance check is met, calibration check compounds (CCCs) are used to check the validity of the initial calibration.

The internal standard responses and retention times in the calibration check standard must be evaluated immediately after or during data acquisition. If the retention time for any internal standard changes by more than 30 seconds from the last check calibration (12 h), the chromatographic system must be inspected for malfunctions and corrections must be made, as required. If the electron ionization current plot (EICP) area for any of the internal standards changes by a factor of two from the last daily calibration standard check, the mass spectrometer must be inspected for malfunctions and corrections must be made, as appropriate.

Demonstrate, through the analysis of a reagent water blank, that interferences from the analytical system, glassware, and reagents are under control. The blank samples should be carried through all stages of the sample preparation and measurement steps. For each analytical batch (up to 20 samples), a reagent blank, matrix spike, and matrix spike duplicate/duplicate must be analyzed (the frequency of the spikes may be different for different monitoring programs). The blank and spiked samples must be carried through all stages of the sample preparation and measurement steps. A QC reference sample concentrate containing each analyte at a concentration of 100 mg/L in methanol is required.

REFERENCE Test Methods for Evaluating Solid Waste (SW-846). U.S. EPA 1983, Method 8270B, Rev. 2, Nov. 1990. Office of Solid Waste, Washington, DC.

5-Hydroxydicamba   **EPA Method 8151**
CAS #7600-50-2

TITLE Chlorinated Herbicides by GC Using Methylation or Pentafluorobenzylation Derivatization: Capillary Column Technique.

MATRIX This method covers aqueous and solid matrices. This includes a wide variety such as drinking water, groundwater, industrial wastewaters, surface waters, soils, solids, and sediments.

METHOD SUMMARY This is a GC method for determining 19 chlorinated acid herbicides in aqueous, soil, and waste matrices. Because these compounds are produced and used in various forms (i.e., acid, salt, ester, etc.) a hydrolysis step is included to convert the herbicide to the acid form prior to analysis. This method provides hydrolysis, extraction, derivatization and GC conditions for the analysis of chlorinated acid herbicides in water, soil, and waste samples. Water samples are hydrolyzed *in situ*, extracted with diethyl ether, and then esterified with either diazomethane or pentafluorobenzyl bromide. The derivatives are determined by gas chromatography with an electron capture detector (GC/ECD). The results are reported as acid equivalents. The sensitivity of this method depends on the level of interferences in addition to instrumental limitations.

INTERFERENCES Method interferences may be caused by contaminants in solvents, reagents, glassware, and other sample processing hardware. Immediately prior to use, glassware should be rinsed with the next solvent to be used. Matrix interferences may be caused by contaminants that are coextracted from the sample. Organic acids, especially chlorinated acids, cause the most direct interference with the determination by methylation. Phenols, including chlorophenols, may also interfere with this procedure. The determination using pentafluorobenzylation is more sensitive, and more prone to interferences from the presence of organic acids of phenols than by methylation. Alkaline hydrolysis and subsequent extraction of the basic solution removes many chlorinated hydrocarbons and phthalate esters that might otherwise interfere with the ECD analysis. The herbicides, being strong organic acids, react readily with alkaline substances and may be lost during analysis. Therefore, glassware must be acid-rinsed and then rinsed to constant pH with organic-free reagent water.

INSTRUMENTATION A GC suitable for Grob-type injection using capillary columns. A data system for measuring peak heights and/or peak areas is recommended. An electron capture detector (ECD) is used. Also a K-D apparatus, a diazomethane generator, a centrifuge and an ultrasonic disrupter will be required.

Narrow Bore Columns:
Primary Column 1: 30 m × 0.25 mm, 5% phenyl/95% methyl silicone (DB-5), 0.25 µm film thickness.
Primary Column 1a (GC/MS): 30 m × 0.32 mm, 5% phenyl/95% methyl silicone (DB-5), 1-µm film thickness.
Column 2: 30 m × 0.25 mm DB-608 with a 25 µm film thickness.
Confirmation Column: 30 m × 0.25 mm, 14% cyanopropyl phenyl silicone (DB-1701), 0.25 µm film thickness.

Megabore Columns:
Primary Column: 30 m × 0.53 mm DB-608 with 0.83 µm film thickness.
Confirmation Column: 30 m × 0.53 mm, 14% cyanopropyl phenyl silicone (DB-1701), 1.0 µm film thickness.

PRECISION & ACCURACY Method detection limits (MDLs) are compound-dependent and vary with derivitization efficiency, derivative recovery, the matrix sampled, and herbicide concentration.

The estimated MDL (in µg/L) was 0.04 for aqueous samples using GC/ECD.

The estimated MDL (in µg/kg) was No Data for soil samples using GC/ECD when corrected back to 50 g samples extracted and concentrated to 10 mL with 5-µL injections.

The estimated GC/MS identification limit (in ng) was No Data for soil samples using GC/MS.

Mean percent recovery, calculated from 7–8 determinations of spiked reagent water, after diazomethane derivatization, from a spike concentration (in µg/L) of 0.2 was 103 with a standard deviation of the percent recovery of 16.5.

Mean percent recovery, calculated from 10 determinations of spiked clay and clay/still bottom samples over the linear concentration range (in ng/g) of no data was none reported with a percent relative standard deviation of none. The RSD % was calculated on 10 samples high in the linear concentration range and 10 low in the range. The linear concentration range was determined using standard solutions and corrected to 50 g soil samples.

SAMPLE COLLECTION, PRESERVATION & HANDLING Containers used to collect samples for the determination of semivolatile organic compounds should be soap and water washed followed by methanol (or isopropanol) rinsing. The sample containers should be of glass or Teflon® and have screw-top covers with Teflon® liners.

No preservation is used with concentrated waste samples. With liquid samples containing no residual chlorine and with soil, sediment, and sludge samples, immediately cooling to 4°C is the only preservation used. When residual chlorine is present then 3 mL of 10% aqueous sodium sulfate is added for each gallon of sample collected, followed by cooling to 4°C.

The holding time for all volatile organics samples is 14 days. Liquid samples must be extracted within 7 days and their extracts analyzed within 40 days. Concentrated waste, soil, sediment, and sludge samples must be extracted within 14 days and their extracts analyzed within 40 days.

SAMPLE PREPARATION
Preparation of soil, sediment, and other solid samples — Acidify 30 g (dry weight) solids with 0.1 M phosphate buffer (pH = 2.5) and thoroughly mix the contents. Spike the sample with surrogate compound(s). The ultrasonic extraction of solids must be optimized for each type of sample. In order for the ultrasonic extractor to efficiently extract solid samples, the sample must be free flowing when the solvent is added. Acidified anhydrous sodium sulfate should be added to clay-type soils, or any other solid that is not a free-flowing sandy texture, until a free flowing mixture is obtained. Add methylene chloride and perform ultrasonic extraction. Combine organic extracts from the repetitive extractings of the sample and centrifuge. Add aqueous potassium hydroxide, water, and methanol to the extract and reflux the mixture on a water bath. Extract the solution three times with methylene chloride and discard the methylene chloride phase. The basic solution contains the herbicide salts. Adjust the pH of the solution to <2 with cold sulfuric acid and extract three times with methylene chloride. Combine the extracts and pour them through a pre-rinsed drying column containing acidified anhydrous sodium sulfate. Collect the dried extracts in a K-D flask and concentrate them.

Preparation of aqueous samples — Measure 1 L of sample into a 2 L separatory funnel and spike it with surrogate compound(s). Add NaCl to the sample, then add 6 N NaOH to the sample to a pH of 12 or more and let the sample sit at room temperature for 1 h to hydrolyze esters. Extract the sample three times with methylene chloride and discard the extracts. Then add cold 12 N sulfuric acid to a pH less than or equal to 2, and extract the sample three times with ethyl ether. Collect the ether phase in a flask containing acidified anhydrous sodium sulfate and allow it to remain in contact with the sodium sulfate for a minimum of 2 h. The drying step is very critical to ensuring complete esterification; any moisture remaining in the ether will result in low herbicide recoveries.

Extract concentration and derivatization — The combined ether extract is concentrated to about 1 mL using a K-D apparatus followed by using a micro Snyder column or nitrogen gas blowdown. If methyl esters are to be produced, then dilute the concentrated ether extract with 1 mL of isooctane and 0.5 mL of methanol, dilute to a final volume of 4 mL, and esterify with diazomethane. If pentafluorobenzene esters are to be produced, then dilute concentrated ether extract with acetone to a final volume of 4 mL and esterify with pentafluorobenzyl bromide.

QUALITY CONTROL Select a representative spike concentration for each compound (acid or ester) to be measured. Using stock standard, prepare a quality control check sample concentrate, in acetone, that is 1000 times more concentrated than the selected concentrations. Use this quality control check sample concentrate to prepare quality control check samples. Calculate surrogate standard recovery on all standards, samples, blanks, and spikes. GC/MS techniques should be judiciously employed to support qualitative identifications made with this method. When available, chemical ionization mass spectra may be employed to aid the qualitative identification process.

REFERENCE Test Methods for Evaluating Solid Waste, Physical/Chemical Methods, SW-846, 3rd Edition, U.S. EPA, Office of Solid Waste, Washington, DC, EPA Method 8151, Nov. 1990.

2-Hydroxypropionitrile   **EPA Method 8240**
CAS #78-97-7

TITLE Volatile Organics By GC/MS: Packed Column Technique

MATRIX Nearly all types of sample matarices, regardless of water content, can be analyzed using this method. This includes groundwater, aqueous sludges, caustic liquors, acid liquors, waste solvents, oily wastes, mousses, tars, fibrous wastes, polymetric emulsions, filter cakes, spent carbons, spent catalysts, soils, and sediments.

METHOD SUMMARY Method 8240B covers 80 volatile organic compounds that are introduced into a gas chromatograph by the purge-and-trap method or by direct injection (in limited applications). For the purge-and-trap method an inert gas (zero grade nitrogen or helium) is bubbled through a 5-mL solution at ambient temperature. Purged sample components are trapped in a tube of sorbent materials. When purging is complete, the sorbent tube is heated and backflushed with inert gas to desorb the trapped components onto a GC column.

INTERFERENCES Impurities in the purge gas and from organic compounds outgassing from the plumbing ahead of the trap account for many contamination problems. Interferences purged or coextracted from the samples will vary considerably from source to source. Cross-contamination can occur whenever high-level and low-level samples are analyzed sequentially. Whenever an unusually concentrated sample is analyzed, it should be followed by the analysis of organic-free reagent water to check for cross-contamination. Samples also can be contaminated by diffusion of volatile organics (particularly methylene chloride and fluorocarbons) through the septum seal into the sample during shipment and storage. A trip blank can serve as a check on such contamination. The lab where volatile analysis is performed and also the refrigerated storage area should be completely free of solvents.

INSTRUMENTATION A gas chromatograph/mass spectrometry/data system (GC/MS) equipped with a 6 ft × 0.1 in I.D. glass column packed with 1% SP-1000 on Carbopack-B (60/80 mesh) is required. Also needed is a 5-mL purging device, a sorbent trap, and a thermal desorption apparatus.

PRECISION & ACCURACY This method is reported to have been tested by 15 laboratories using organic-free reagent water, drinking water, surface water, and industrial wastewaters (not specified) fortified at six concentrations over the range 5–600 µg/L.

Sample estimated quantitation limits (EQLs) are highly matrix-dependent. The EQLs listed may not always be achievable. EQLs listed for soils or sediments are based on wet weight. Normally, data is reported on a dry-weight basis; therefore,

EQLs will be higher, based on the percent dry weight of each sample. Note that EQLs are even more variable than MDLs and that they are highly variable depending on the matrix being analyzed.

EQL in groundwater in µg/L was not listed.
EQL in low soil or sediment in µg/kg was not listed.
Accuracy (a) in µg/L was not listed.
Precision (b) in µg/L was not listed.

(a) *Average recovery found for measurements of samples containing a concentration of C, in µg/L.*
(b) *Overall precision found for measurements of samples with average recovery X for samples containing a concentration of C in µg/L.*
X = *Average recovery found for measurement of samples containing a concentration of C in µg/L.*

MULTIPLICATION FACTORS FOR OTHER MATRICES

Other Matrices	Factor (a)
Waste miscible liquid waste	50
High-concentration soil and sludge	125
Non-water miscible waste	500

(a) EQL = [EQL for low soil/sediment] × [Factor]. For non-aqueous samples, the factor is on a wet-weight basis.

SAMPLING METHOD

Liquid samples — Use a 40-mL glass screw-cap VOA vial with a Teflon®-faced silicone septum that has been prewashed, rinsed with distilled deionized water, and oven dried. However, if residual chlorine is present, collect sample in a 40-oz. soil VOA container which has been pre-preserved with 4 drops of 10% sodium thiosulfate, mix gently, and then transfer the sample to a 40-mL VOA vial. Collect bubble-free samples in duplicate and seal them in separate plastic bags.

Soils or sediments, and sludges — Use an 8-oz. widemouth glass bottle with a Teflon®-faced silicone septum that has been prewashed with detergent, rinsed with distilled deionized water, and oven dried. Tap slightly to eliminate free air space. Collect samples in duplicate and seal them in separate plastic bags.

SAMPLE PRESERVATION

Liquid samples — Add 4 drops of concentrated HCL and immediately cool samples to 4°C and store in a solvent-free refrigerator.

Soils or sediments, and sludges — Cool samples to 4°C and store in a solvent-free refrigerator.

MHT Maximum holding time is 14 days from the date of sample collection.

SAMPLE PREPARATION

Liquid samples — Remove the plunger from a 5-mL syringe and carefully pour the sample into the syringe barrel to just short of overflowing. Replace the syringe plunger and compress the sample. Open the syringe valve and vent any residual air while adjusting the sample volume to 5.0 mL. If there is only one volatile organic analysis (VOA) vial, a second syringe should be filled at this time to protect against possible loss of sample integrity. Add 10 µL of surrogate spiking solution and 10 µL of internal standard spiking solution through the valve bore of the 5-mL syringe, then close the valve. The surrogate and internal standards may be mixed and added as a single spiking solution.

Sediments, soils, and waste samples — All samples of this type should be screened by GC analysis using a headspace method (EPA Method 3810) or the hexadecane extraction and screening method (EPA Method 3820). Use the screening data to determine whether to use the low-concentration method (0.005–1 mg/kg) or the high-concentration method (>1 mg/kg).

Low-concentration method — The low-concentration method is based on purging a heated sediment or soil sample mixed with organic-free reagent water containing the surrogate and internal standards. Analyze all reagent blanks and standards under the same conditions as the samples.

Use a 5-g sample if the expected concentration is <0.1 mg/kg or a 1-g sample for expected concentrations between 0.1 and 1 mg/kg. Mix the contents of the sample container with a narrow metal spatula. Weigh the amount of the sample into a tared purge device. Add the spiked water to the purge device, which contains the weighed amount of sample, and connect the device to the purge-and-trap system.

High-concentration method — This method is based on extracting the sediment or soil with methanol. A waste sample is either extracted or diluted, depending on its solubility in methanol. Wastes that are insoluble in methanol are diluted with reagent tetraglyme or possibly polyethylene glycol (PEG). An aliquot of the extract is added to organic-free reagent water containing surrogate and internal standards. This is purged at ambient temperature. All samples with an expected concentration of >1.0 mg/kg should be analyzed by this method.

Mix the contents of the sample container with a narrow metal spatula. For sediments or soils and solid wastes that are insoluble in methanol, weigh 4 g (wet weight) of sample into a tared 20-mL vial. For waste that is soluble in methanol, tetraglyme, or PEG, weigh 1 g (wet weight) into a tared scintillation vial or culture tube or a 10-mL volumetric flask. Quickly add 9.0 mL of appropriate solvent then add 1.0 mL of a surrogate spiking solution to the vial, cap it, and shake it for 2 min.

METHANOL EXTRACT REQUIRED FOR ANALYSIS OF HIGH-CONCENTRATION SOILS OR SEDIMENTS

Approximate Concentration Range	Volume of Methanol Extract (a)
500–10,000 µg/kg	100 µL
1,000–20,000 µg/kg	50 µL
5,000–100,000 µg/kg	10 µL
25,000–500,000 µg/kg	100 µL of 1/50 dilution (b)

Calculate appropriate dilution factor for concentrations exceeding this table.

(a) The volume of methanol added to 5 mL of water being purged should be kept constant. Therefore, add to the 5-mL syringe whatever volume of methanol is necessary to maintain a volume of 100 µL added to the syringe.
(b) Dilute an aliquot of the methanol extract and then take 100 µL for analysis.

QUALITY CONTROL Demonstrate, through the analysis of a reagent water blank, that interferences from the analytical system, glassware, and reagents are under control. Blank samples should be carried through all stages of the sample preparation and measurement steps. For each analytical batch (up to 20 samples), a reagent blank, matrix spike, and matrix spike duplicate must be analyzed (the frequency of the spikes may be different for different monitoring programs). The blank and spiked samples must be carried through all stages of the sample preparation and measurement steps. QC samples mentioned in the section on Interferences will also be needed as appropriate to those situations.

REFERENCE Test Methods for Evaluating Solid Waste (SW-846). U.S. EPA. 1983. Method 8240B, Rev. 2, Nov. 1990. Office of Solid Wastes, Washington, DC.

I

Indeno(1,2,3-cd)pyrene   **EPA Method 625**
CAS #193-39-5

TITLE Base/Neutrals and Acids, U.S. EPA Method 625

MATRIX This methods covers municipal and industrial wastewaters.

METHOD SUMMARY Approximately 1 L of sample is serially extracted with methylene chloride at a pH greater than 11 and again at a pH less than 2 using a separatory funnel or a continuous extractor. The methylene chloride extract is dried, concentrated to a volume of 1 mL, and analyzed by GC/MS. Qualitative identification of the parameters in the extract is performed using the retention time and the relative abundance of three characteristic masses (m/z). Qualitative analysis is performed using either external or internal standard techniques with a single characteristic m/z.

INTERFERENCES Method interferences may be caused by contaminants in solvents, reagents, glassware, and other sample processing hardware. Glassware must be scrupulously cleaned. Glassware should be heated in a muffle furnace at 400°C for 5 to 30 min. Some thermally stable materials, such as PCBs, may not be eliminated by this treatment. Solvent rinses with acetone and pesticide quality hexane may be substituted for the muffle furnace heating. Matrix interferences may be caused by contaminants that are coextracted from the sample. The base-neutral extraction may cause significantly reduced recovery of phenols. The packed gas chromatographic columns recommended for the basic fraction may not exhibit sufficient resolution for some analytes.

INSTRUMENTATION A GC/MS system with an injection port designed for on-column injection when using packed columns and for splitless injection when using capillary columns.

Column for base/neutrals: 1.8 m long × 2 mm I.D. glass, packed with 3% SP-2550 on Supelcoport (100/120 mesh) or equivalent.

Column for acids: 1.8 m long × 2 mm I.D. glass, packed with 1% SP-1240DA on Supelcoport (100/120 mesh) or equivalent.

PRECISION & ACCURACY The MDL concentrations were obtained using reagent water. The MDL actually achieved in a given analysis will vary depending on instrument sensitivity and matrix effects. This method was tested by 15 laboratories using reagent water, drinking water, surface water, and industrial wastewaters spiked at six concentrations over the range 5 to 100 µg/L. Single operator precision, overall precision, and method accuracy were found to be directly related to the concentration of the parameter matrix.

The MDL (in µg/L) in reagent water was 3.7.
The standard deviation (in µg/L based on 4 recovery measurements) was 44.6.
The range (in µg/L) for average recovery for 4 measurements was D-150.9.
The range (in %) for percent recovery was D-171.

Accuracy (in µg/L) as expected recovery for one or more measurements of a sample containing a true concentration of C was 0.78C-3.10.

Precision (in µg/L) as expected single analyst standard deviation of measurements at an average concentration found at X was 0.29X + 1.46.

Overall precision (in µg/L) as expected interlaboratory standard deviation of measurements in an average concentration found at X was 0.50X + 0.44.

C = *True value of the concentration in µg/L.*
X = *Average recovery found for measurements of samples containing a concentration at C in µg/L.*

SAMPLE PREPARATION Adjust the pH to >11 with sodium hydroxide and serially extract in a separatory funnel with methylene chloride or else in a continuous extractor. Next, adjust the pH to <2 with sulfuric acid and serially extract in a separatory funnel with methylene chloride or else in a continuous extractor. Dry the extracts separately through a column of anhydrous sodium sulfate and then concentrate each of the extracts to 1.0 mL using a K-D apparatus.

SAMPLE COLLECTION, PRESERVATION & HANDLING Grab samples must be collected in glass containers. All samples must be refrigerated at 4°C from the time of collection until extraction. If residual chlorine is present, add 80 mg of sodium thiosulfate/L of sample and mix well. All samples must be extracted within 7 days of collection and completely analyzed within 40 days of extraction.

QUALITY CONTROL Make an initial, one-time, demonstration of the ability to generate acceptable accuracy and precision with this method. Before processing any samples, the analyst must analyze a reagent water blank to demonstrate that interferences from the analytical system and glassware are under control. Each time a set of samples is extracted or reagents are changed, a reagent water blank must be processed. Spike and analyze a minimum of 5% of all samples to monitor and evaluate lab data quality. A QC check sample concentrate that contains each parameter of interest at a concentration of 100 µg/mL in acetone is required. PCBs and multicomponent pesticides may be omitted from this test.

After analysis of five spiked wastewater samples, calculate the average percent recovery and the standard deviation of the percent recovery. Spike all samples with the surrogate standard spiking solution and calculate the percent recovery of each surrogate compound.

REFERENCE *Federal Register*, Vol. 49, No. 209. Friday, Oct. 26, 1984.

Indeno(1,2,3-cd)pyrene   **EPA Method 1625**
CAS #193-39-5

TITLE Semivolatile Organic Compounds by Isotope Dilution GC/MS

MATRIX The compounds may be determined in waters, soils, and municipal sludges by this method.

METHOD SUMMARY This method is used to determine 176 semivolatile toxic organic pollutants associated with the CWA (as amended 1987); the RCRA (as amended 1986); the CERCLA (as amended 1986); and other compounds amenable to extraction and analysis by capillary column gas chromatography-mass spectrometry (GC/MS).

Stable isotopically-labeled analogs of the compounds of interest are added to the sample. If the solids content is less than 1%, a 1-L sample is extracted at pH 12–13, then at pH <2 with methylene chloride using continuous extraction techniques.

If the solids content is 30% or less, the sample is diluted to 1% solids with reagent water, homogenized ultrasonically, and extracted at pH 12–13, then at pH <2 with methylene chloride using continuous extraction techniques. If the solids content is greater than 30%, the sample is extracted using ultrasonic techniques.

Each extract is dried over sodium sulfate, concentrated to a volume of 5 mL, cleaned up using GPC, if necessary, and concentrated. Extracts are concentrated to 1 mL if GPC is not performed, and to 0.5 mL if GPC is performed.

An internal standard is added to the extract, and a 1-mL aliquot of the extract is injected into the GC. The compounds are separated by GC and detected by a MS. The labeled compounds serve to correct the variability of the analytical technique.

INTERFERENCES Solvents, reagents, glassware, and other sample processing hardware may yield artifacts and/or elevated baselines causing misinterpretation of chromatograms and spectra. Materials used in the analysis must be demonstrated to be free from interferences under the conditions of analysis by running method blanks initially and with each sample lot (sample started through the extraction process on a given 8-h shift, to a maximum of 20). Specific selection of reagents and purification of solvents by distillation in all glass systems may be required. Glassware and, where possible, reagents are cleaned by solvent rinse and baking at 450°C for 1-h minimum. Interferences coextracted from samples will vary considerably from source to source, depending on the diversity of the site being sampled.

INSTRUMENTATION Major instrumentation includes a GC with a splitless or on-column injection port for capillary column, a MS with 70 eV electron impact ionization, and a data system to collect and record MS data, and process it. A K-D apparatus is used to concentrate extracts.

GC Column: 30 m × 0.25 mm I.D. 5% phenyl, 94% methyl, 1% vinyl silicone bonded phased fused silica capillary column.

PRECISION & ACCURACY The detection limits of the method are usually dependent on the level of interferences rather than instrumental limitations. The limits typify the minimum quantities that can be detected with no interferences present.

The minimum level (in µg/mL) was 20. This is defined as a minimum level at which the analytical system shall give recognizable mass spectra (background corrected) and acceptable calibration points.

The MDL (in µg/kg) in low solids was 67 and in high solids was 263; these were determined in digested sludge (low solids) and in filter cake or compost (high solids).

Note: Background levels of this compound were present in the sludge tested, resulting in higher than expected MDLs. The MDL for this compound is expected to be approximately 50 µg/kg with no interferences present.

The labeled and native compound initial precision as standard deviation (in µg/L) was 55.

The labeled and native compound initial accuracy as average recovery (in µg/L) was 23–299.

SAMPLE COLLECTION, PRESERVATION & HANDLING
Collect samples in glass containers. Aqueous samples which flow freely are collected in refrigerated bottles using automatic sampling equipment. Solid samples are collected as grab samples using widemouth jars. Maintain samples at 0 to 4°C from the time of collection until extraction. If residual chlorine is present in aqueous samples, add 80 mg sodium thiosulfate/L of water. Begin sample extraction within 7 days of collection, and analyze all extracts within 40 days of extraction.

SAMPLE PREPARATION Samples containing 1% solids or less are extracted directly using continuous liquid-liquid extraction techniques. Samples containing 1 to 30% solids are diluted to the 1% level with reagent water and extracted using continuous liquid-liquid extraction techniques. Samples containing greater than 30% solids are extracted using ultrasonic techniques.

Base/neutral extraction — Adjust the pH of the waters in the extractors to 12–13 with 6 N NaOH. Extract with methylene chloride for 24–48 h.

Acid extraction — Adjust the pH of the waters in the extractors to 2 or less using 6 N sulfuric acid. Extract with methylene chloride for 24–48 h.

Ultrasonic extraction of high solids samples — Add anhydrous sodium sulfate to the sample and QC aliquot(s). Add acetone:methylene chloride (1:1) to the sample and mix thoroughly

Concentrate extracts using a K-D apparatus.

QUALITY CONTROL The analyst is permitted to modify this method to improve separations or lower the costs of measurements, provided all performance specifications are met. Analyses of blanks are required to demonstrate freedom from contamination. When results of spikes indicate atypical method performance for samples, the samples are diluted to bring method performance within acceptable limits.

For low solids (aqueous samples), extract, concentrate, and analyze two sets of four 1-L aliquots (8 aliquots total) of the precision and recovery standard. For high solids samples, two sets of four 30-g aliquots of the high solids reference matrix are used.

Spike all samples with labeled compounds to assess method performance. Compute percent recovery of the labeled compounds

using the internal standard method. Compare the labeled compound recovery for each compound with the corresponding labeled compound recovery.

Reagent water and high solids reference matrix blanks are analyzed to demonstrate freedom from contamination. Extract and concentrate a 1-L reagent water blank or a high solids reference matrix blank with each sample's lot (samples started through the extraction process on the same 8-h shift, to a maximum of 20 samples).

Field replicates may be collected to determine the precision of the sampling technique, and spiked samples may be required to determine the accuracy of the analysis when the internal standard method is used.

REFERENCE Semivolatile Organic Compounds by Isotope Dilution GC/MS. Office of Water Regulation and Standards, U.S. EPA Industrial Technology Division, Washington, DC, EPA Method 1625, Rev. C, June 1989 (contact W.A. Telliard, U.S. EPA, Office of Water Regulations and Standards, 401 M St., SW, Washington, DC, 20460. Phone: 202-382-7131).

Indeno(1,2,3-cd)pyrene   **EPA Method 8270**
CAS #193-39-5

TITLE Semivolatile Organic Compounds by GC/MS

MATRIX This method is used to determine the concentration of semivolatile organic compounds in extracts prepared from all types of solid waste matrices, soils, and groundwater. Although surface waters are not specifically mentioned, this method should be applicable to water samples from rivers, lakes, etc.

METHOD SUMMARY This method covers 259 semivolatile organic compounds. In very limited applications direct injection of the sample into the GC/MS system may be appropriate, but this results in very high detection limits (approximately 10,000 µg/L). Typically, a 1-L liquid sample, containing surrogate, and matrix spiking standards, is extracted in a continuous extractor first under acid conditions and then under basic conditions. Typically 30 g of a solid sample, containing surrogate, and matrix spiking standards, is extracted ultrasonically. After concentrating the extract to 1 mL it is spiked with 10 µL of an internal standard solution just prior to analysis by GC/MS. The volume injected should contain about 100 ng of base/neutral and 200 ng of acid surrogates (for a 1-µL injection). Analysis is performed by GC/MS using a capillary GC column.

INTERFERENCES Raw GC/MS data from all blanks, samples, and spikes must be evaluated for interferences. Contamination by carryover can occur whenever high-concentration and low-concentration samples are sequentially analyzed. To reduce carryover, the sample syringe must be rinsed out between samples with solvent. Whenever an unusually concentrated sample is encountered, it should be followed by the analysis of blank solvent to check for cross-contamination.

INSTRUMENTATION A GC/MS and a data system are required. The GC column used is a 30 m × 0.25 mm I.D. (or 0.32 mm I.D.) 1um film thickness silicone-coated fused silica capillary column. A continuous liquid-liquid extractor equipped with Teflon® or glass connection joints and stopcocks requiring no lubrication, a K-D concentrating apparatus, water bath, and an ultrasonic disrupter with a minimum power of 300 W and with pulsing capability are also required.

PRECISION & ACCURACY The estimated quantitation limit (EQL) of Method 8270B for determining an individual compound is approximately 1 mg/kg (wet weight) for soil or sediment samples, 1–200 mg/kg for wastes (dependent on matrix and method of preparation), and 10 µg/L for groundwater samples. EQLs will be proportionately higher for sample extracts that require dilution to avoid saturation of the detector.

The EQL(b) for groundwater in µg/L is 10.
The EQL (a, b) for low concentrations in soil and sediment in µg/kg is 660.
Accuracy as µg/L is 0.78C–3.10.
Overall precision in µg/L is 0.50X–0.44.

(a)   *EQLs listed for soil/sediment are based on wet weight. Normally data is reported in a dry-weight basis; therefore, EQLs will be higher based on the % dry weight of each sample. This calculation is based on a 30-g sample and gel permeation chromatography cleanup.*
(b)   *Sample EQLs are highly matrix-dependent. The EQLs are provided for guidance and may not always be achievable.*
C =   *True value for concentration, in µg/L.*
X =   *Average recovery found for measurements of samples containing a concentration of C, in µg/L.*

ESTIMATED QUANTITATION LIMIT

Other Matrices	Factor (a)
High-concentration soil and sludges by sonicator	7.5
Non-water miscible waste	75

(a) *EQL for other matrices = [EQL for low soil/sediment] × [Factor]. This estimated EQL is similar to an EPA "Practical Quantitation Limit."*

SAMPLING METHOD
Liquid samples — Use a 1 or 2½ gallon amber glass bottle with a screw-top Teflon®-lined cover that has been prewashed with detergent and rinsed with distilled water and methanol (or isopropanol).

Soils, sediments, or sludges — Use an 8-oz. widemouth glass with a screw-top Teflon®-lined cover that has been prewashed with detergent and rinsed with distilled water and methanol (or isopropanol).

SAMPLE PRESERVATION
Liquid samples — If residual chlorine is present, add 3 mL of 10% sodium thiosulfate per gallon, cool to 4°C and store in a solvent-free refrigerator until analysis; if chlorine is not present, then eliminate the sodium thiosulfate addition.

Soils, sediments, or sludges — Cool samples to 4°C and store in a solvent-free refrigerator.

MHT Liquid samples must be extracted within 7 days and the extracts analyzed within 40 days. Soils, sediments, or sludges may

be stored for a maximum of 14 days and the extracts analyzed within 40 days.

SAMPLE PREPARATION

Liquid samples — Transfer 1 L quantitatively to a continuous extractor. If high concentrations are anticipated, a smaller volume may be used and then diluted with organic-free reagent water to 1 L. Adjust pH, if necessary, to pH <2 using 1:1 (V/V) sulfuric acid. Pipette 1.0 mL of a surrogate standard spiking solution into each sample. For the sample in each analytical batch selected for spiking, add 1.0 mL of a matrix spiking standard. For base/neutral acid analysis, the amount of the surrogates and matrix spiking compounds added to the sample should result in a final concentration of 100 ng/µL of each analyte in the extract to be analyzed (assuming a 1-µL injection). Extract with methylene chloride for 18–24 h. Next, adjust the pH of the aqueous phase to pH >11 using 10 N sodium hydroxide and extract it with methylene chloride again for 18–24 h. Dry the extract through a column containing anhydrous sodium sulfate and concentrate it to 1 mL using a K-D concentrator.

Soils, sediments, or sludges — Use 30 g of sample. Nonporous or wet samples (gummy or clay type) that do not have a free-flowing sandy texture must be mixed with anhydrous sodium sulfate until the sample is free flowing. Add 1 mL of surrogate standards to all samples, spikes, standards, and blanks. For the sample in each analytical batch selected for spiking, add 1.0 mL of a matrix spiking standard. For base/neutral acid analysis, the amount added of the surrogates and matrix spiking compounds should result in a final concentration of 100 ng/µL of each base/neutral analyte and 200 ng/µL of each acid analyte in the extract to be analyzed (assuming a 1-µL injection). Immediately add a 100-mL mixture of 1:1 methylene chloride:acetone and extract the sample ultrasonically for 3 min and then decant or filter the extracts. Repeat the extraction two or more times. Dry the extract using a column with anhydrous sodium sulfate and concentrate it to 1 mL in a K-D concentrator.

QUALITY CONTROL A methylene chloride solution containing 50 ng/µL of decafluorotriphenylphosphine (DFTPP) is used for tuning the GC/MS system each 12-h shift. A system performance check also must be made during every 12-h shift. A standard containing 50 ng/µL each of 4,4′-DDT, pentachlorophenol, and benzidine is required to verify injection port inertness and GC column performance. A calibration standard at mid-concentration, containing each compound of interest, including all required surrogates, must be performed every 12 h during analysis. After the system performance check is met, calibration check compounds (CCCs) are used to check the validity of the initial calibration.

The internal standard responses and retention times in the calibration check standard must be evaluated immediately after or during data acquisition. If the retention time for any internal standard changes by more than 30 seconds from the last check calibration (12 h), the chromatographic system must be inspected for malfunctions and corrections must be made, as required. If the electron ionization current plot (EICP) area for any of the internal standards changes by a factor of two from the last daily calibration standard check, the mass spectrometer must be inspected for malfunctions and corrections must be made, as appropriate.

Demonstrate, through the analysis of a reagent water blank, that interferences from the analytical system, glassware, and reagents are under control. The blank samples should be carried through all stages of the sample preparation and measurement steps. For each analytical batch (up to 20 samples), a reagent blank, matrix spike, and matrix spike duplicate/duplicate must be analyzed (the frequency of the spikes may be different for different monitoring programs). The blank and spiked samples must be carried through all stages of the sample preparation and measurement steps. A QC reference sample concentrate containing each analyte at a concentration of 100 mg/L in methanol is required.

REFERENCE Test Methods for Evaluating Solid Waste (SW-846). U.S. EPA 1983, Method 8270B, Rev. 2, Nov. 1990. Office of Solid Waste, Washington, DC.

Indeno(1,2,3-cd)pyrene   **EPA Method 8100**
CAS #193-39-5

TITLE Polynuclear Aromatic Hydrocarbons

MATRIX Groundwater, soils, sludges, water miscible liquid wastes, and non-water miscible wastes.

APPLICATION This method is used for the analysis of various PAHs. Samples are extracted, concentrated, and analyzed using direct injection of both neat and diluted organic liquids. The method provides two optional GC columns that are better than Column 1 and that may help resolve analytes from interferences.

INTERFERENCES Solvents, reagents, and glassware may introduce artifacts. Other interferences may come from coextracted compounds from samples.

INSTRUMENTATION GC capable of on-column injections and a flame with detector (FID). Column 1: a 1.8 m by 2 mm 3% OV-17 on Chromosorb W-AW-DCMS column. Column 2: a 30 m by 0.25 mm SE-54 fused silica capillary colunm. Column 3: a 30 m by 0.32 mm SE-54 fused silica capillary column.

RANGE 0.1–425 µg/L

MDL Not reported.

PQL FACTORS FOR MULTIPLYING × FID MDL VALUE Not available.

PRECISION 0.42X + 0.01 µg/L (overall precision).

ACCURACY 0.54C + 0.06 µg/L (as recovery).

SAMPLING METHOD Use 8-oz. widemouth glass bottles with Teflon®-lined caps for concentrated waste samples, soils, sediments, and sludges. Use 1 or 2½ gallon amber glass bottles with Teflon®-lined caps for liquid (water) samples.

STABILITY Cool soil, sediment, sludge, and liquid samples to 4°C. If residual chlorine is present in liquid samples add 3 mL of 10% sodium thiosulfate per gallon of sample and cool to 4°C.

MHT 14 days for concentrated waste, soil, sediment, or sludge; 7 days for liquid samples; all extracts must be analyzed within 40 days.

QUALITY CONTROL A quality control check sample concentrate containing each analyte of interest is required. The QC check sample concentrate may be prepared from pure standard materials or purchased as certified solutions Use appropriate trip, matrix, control site, method, reagent, and solvent blanks. Internal, surrogate, and five concentration level calibration standards are used. The quality control check sample concentrate should contain indeno(1,2,3-cd)pyrene at 10 µg/mL in acetonitrile.

REFERENCE Test Methods for Evaluating Solid Waste (SW-846), U.S. EPA Office of Solid Waste, Washington, DC, Method 8100, Nov. 1986.

Indeno(1,2,3-cd)pyrene   **EPA Method 8310**
CAS #193-39-5

TITLE Polynuclear Aromatic Hydrocarbons

MATRIX Groundwater, soils, sludges, water miscible liquid wastes, and non-water miscible wastes.

APPLICATION This method is used for the analysis of 16 polynuclear aromatic hydrocarbons (PAHs). Samples are extracted, concentrated, and analyzed using HPLC with detection by UV and fluorescence detectors.

INTERFERENCES Solvents, reagents, and glassware may introduce artifacts. Other interferences may come from coextracted compounds from samples.

INSTRUMENTATION HPLC with a gradient pumping system and a 250 mm by 2.6 mm reverse phase HC-ODS Sil-X 5-micron particle-size column. The fluorescence detector uses an excitation wavelength of 280 nm and emission greater than 389 nm cutoff with dispersive optics.

RANGE 0.1–425 µg/L.

MDL 0.043 µg/L (fluorescence; reagent water).

PQL FACTORS FOR MULTIPLYING × FID MDL VALUE

Matrix	Multiplication Factor
Groundwater	10
Low-level soil by sonication with GPC cleanup	670
High-level soil and sludge by sonication	10,000
Non-water miscible waste	100,000

PRECISION 0.42X + 0.01 µg/L (overall precision).

ACCURACY 0.54C + 0.06 µg/L (as recovery).

SAMPLING METHOD Use 8-oz. widemouth glass bottles with Teflon®-lined caps for concentrated waste samples, soils, sediments, and sludges. Use 1 or 2½ gallon amber glass bottles with Teflon®-lined caps for liquid (water) samples.

STABILITY Cool soil, sediment, sludge, and liquid samples to 4°C. If residual chlorine is present in liquid samples add 3 mL of 10% sodium thiosulfate per gallon of sample and cool to 4°C.

MHT 14 days for concentrated waste, soil, sediment, or sludge; 7 days for liquid samples; all extracts must be analyzed within 40 days.

QUALITY CONTROL Internal, surrogate, and five concentration level calibration standards are used. The calibration standards must be used with the analytical method blank. A quality control check sample concentrate containing indeno(1,2,3-cd)pyrene at 10 µg/mL is required. The QC check sample concentrate may be prepared from pure standard materials or purchased as certified solutions. Use appropriate trip, matrix, control site, method, reagent, and solvent blanks.

REFERENCE Test Methods for Evaluating Solid Waste (SW-846), U.S. EPA Office of Solid Waste, Washington, DC, Method 8310, Rev. 0, Nov. 1986.

Iodide (Titrimetric)   **EPA Method 345.1**

TITLE Inorganics, Non-Metallics

MATRIX Drinking, surface, and saline waters. Wastewater.

APPLICATION Date issued 1974. After pretreatment to remove interferences, the sample is analyzed for iodide by converting the iodide to iodate with bromine water and titrating with phenylarsine oxide (PAO) or sodium thiosulfate.

INTERFERENCES Iron, manganese and organic matter can interfere. (Calcium oxide pretreatment nullifies this interference). Color interferes with observation of indicator and bromine-water color changes. (Overcome by using a pH meter and standardized amounts of reagents).

INSTRUMENTATION Laboratory iodometric titration equipment and glassware.

RANGE 2–20 mg/L of iodide.

MDL Not listed.

PRECISION SD = ±0.06 mg/L at 11.6 mg/L of iodide.

ACCURACY Recovery = 97% at 11.6 mg/L of iodide.

SAMPLING METHOD Plastic or glass (100 mL).

STABILITY No *Federal Register* rules apply.

QUALITY CONTROL A distilled water blank must be run with each set of samples because of iodide in reagents.

REFERENCE Methods for the Chemical Analysis of Water and Wastes, EPA-600/4-79-020, U.S. EPA, EMSL, 1979.

Iodomethane
CAS #74-88-4

EPA Method 1624

TITLE Volatile Organic Compounds by Isotope Dilution GC/MS

MATRIX Compounds may be determined in waters, soils, and municipal sludges by this method.

METHOD SUMMARY This method is used to determine 58 volatile toxic organic pollutants associated with the CWA (as amended 1987); the RCRA (as amended 1986); the CERCLA (as amended 1986); and other compounds amenable to purge-and-trap gas chromatography-mass spectrometry (GC/MS).

If the solids content is less than 1%, stable isotopically-labeled analogs of the compounds of interest are added to a 5-mL sample and the sample is purged with an inert gas at 20–25°C in a chamber designed for soil or water samples. If the solids content is greater than 1%, 5 mL of reagent water and the labeled compounds are added to a 5-g aliquot of sample and the mixture is purged at 40°C. Compounds that will not purge at 20–25°C or at 40°C are purged at 78–85°C. In the purging process, the volatile compounds are transferred from the aqueous phase into the gaseous phase where they are passed into a sorbent column, and trapped. After purging is completed, the trap is backflushed and heated rapidly to desorb the compounds into a GC. The compounds are separated by the GC and detected by a MS. The labeled compounds serve to correct the variability of the analytical technique.

INTERFERENCES Impurities in the purge gas, organic compounds outgassing from the plumbing upstream of the trap, and solvent vapors in the lab account for most problems. Samples can be contaminated by diffusion of volatile organic compounds (particularly methylene chloride) through the bottle seal during shipment and storage. Contamination by carryover can occur when high-level and low-level samples are analyzed sequentially. When an unusually concentrated sample is encountered, follow it by analysis of a reagent water blank to check for carryover.

INSTRUMENTATION Major equipment includes a GC with linear temperature programming and a glass jet separator as the MS interface, a MS with 70 eV electron impact ionization, and a data system to collect and record response factors.

Column: 2.8 m × 2 mm I.D. glass, packed with 1% SP-1000 on Carbopak B, 60/80 mesh, or equivalent.

PRECISION & ACCURACY The detection limits of the method are usually dependent on the level of interferences rather than instrumental limitations. The method detection limits were determined in digested sludge (low solids) and in filter cake or compost (high solids).

The MDL (in μg/kg) for low solids is not listed and for high solids is not listed.
Labeled and native compound precision (in μg/L) as standard deviation was not listed.
Labeled and native compound accuracy (in μg/L) as average recovery was not listed.
Acceptance criteria are at 20 μg/L for this compound.

SAMPLE COLLECTION, PRESERVATION & HANDLING Grab samples are collected in glass containers having a total volume greater than 20 mL. Fill and seal each bottle so that no air bubbles are entrapped. Samples are maintained at 0 to 4°C from the time of collection until analysis. If an aqueous sample contains residual chlorine, add sodium thiosulfate preservative (10 mg/40 mL) to the empty sample bottles just prior to shipment to the sample site. All samples must be analyzed within 14 days of collection.

SAMPLE PREPARATION Samples containing less than 1% solids are analyzed directly as aqueous samples. Samples containing 1% solids or greater are analyzed as solid samples utilizing one of two methods, depending on the levels of pollutants, in the sample. Samples containing 1% solids or greater, and low to moderate levels of pollutants are analyzed by purging a known weight of sample added to 5 mL of reagent water. Samples containing 1% solids or greater, and high levels of pollutants, are extracted with methanol, and an aliquot of the methanol extract is added to reagent water and purged.

QUALITY CONTROL A field blank prepared from reagent water and carried through the sampling and handling protocol may serve as a check on contamination from shipment and storage.

The analyst is permitted to modify this method to improve separations or lower the costs of measurements, provided all performance specifications are met. Analyses of blanks are required. When results of spikes indicate atypical method performance for samples, the samples are diluted to bring method performance within acceptable limits. Analyze two sets of four 5-mL aliquots (8 aliquots total) of the aqueous performance standard. Spike all samples with labeled compounds to assess method performance on the sample matrix. Compute the percent recovery of the labeled compounds using the internal standard method. Compare the percent recovery for each compound with the corresponding labeled compound recovery. Reagent water blanks are analyzed to demonstrate freedom from carryover contamination. Field replicates may be collected to determine the precision of the sampling technique, and spiked samples may be required to determine the accuracy of the analysis when the internal method is used.

REFERENCE Volatile Organic Compounds by Isotope Dilution GC/MS. Office of Water Regulation and Standards, U.S. EPA Industrial Technology Division, Washington, DC, EPA Method 1624, Rev. C, June 1989 (contact W.A. Telliard, U.S. EPA, Office of Water Regulations and Standards, 401 M St., SW, Washington, DC, 20460. Phone: 202-382-7131).

Iodomethane
CAS #74-88-4

EPA Method 8240

TITLE Volatile Organics By GC/MS: Packed Column Technique

MATRIX Nearly all types of sample matarices, regardless of water content, can be analyzed using this method. This includes groundwater, aqueous sludges, caustic liquors, acid liquors, waste solvents, oily wastes, mousses, tars, fibrous wastes, polymetric

emulsions, filter cakes, spent carbons, spent catalysts, soils, and sediments.

METHOD SUMMARY Method 8240B covers 80 volatile organic compounds that are introduced into a gas chromatograph by the purge-and-trap method or by direct injection (in limited applications). For the purge-and-trap method an inert gas (zero grade nitrogen or helium) is bubbled through a 5-mL solution at ambient temperature. Purged sample components are trapped in a tube of sorbent materials. When purging is complete, the sorbent tube is heated and backflushed with inert gas to desorb the trapped components onto a GC column.

INTERFERENCES Impurities in the purge gas and from organic compounds outgassing from the plumbing ahead of the trap account for many contamination problems. Interferences purged or coextracted from the samples will vary considerably from source to source. Cross-contamination can occur whenever high-level and low-level samples are analyzed sequentially. Whenever an unusually concentrated sample is analyzed, it should be followed by the analysis of organic-free reagent water to check for cross-contamination. Samples also can be contaminated by diffusion of volatile organics (particularly methylene chloride and fluorocarbons) through the septum seal into the sample during shipment and storage. A trip blank can serve as a check on such contamination. The lab where volatile analysis is performed and also the refrigerated storage area should be completely free of solvents.

INSTRUMENTATION A gas chromatograph/mass spectrometry/data system (GC/MS) equipped with a 6 ft × 0.1 in I.D. glass column packed with 1% SP-1000 on Carbopack-B (60/80 mesh) is required. Also needed is a 5-mL purging device, a sorbent trap, and a thermal desorption apparatus.

PRECISION & ACCURACY This method is reported to have been tested by 15 laboratories using organic-free reagent water, drinking water, surface water, and industrial wastewaters (not specified) fortified at six concentrations over the range 5–600 µg/L.

Sample estimated quantitation limits (EQLs) are highly matrix-dependent. The EQLs listed may not always be achievable. EQLs listed for soils or sediments are based on wet weight. Normally, data is reported on a dry-weight basis; therefore, EQLs will be higher, based on the percent dry weight of each sample. Note that EQLs are even more variable than MDLs and that they are highly variable depending on the matrix being analyzed.

EQL in groundwater in µg/L was not listed.
EQL in low soil or sediment in µg/kg was not listed.
Accuracy (a) in µg/L was not listed.
Precision (b) in µg/L was not listed.

(a) *Average recovery found for measurements of samples containing a concentration of C, in µg/L.*
(b) *Overall precision found for measurements of samples with average recovery X for samples containing a concentration of C in µg/L.*
X = *Average recovery found for measurement of samples containing a concentration of C in µg/L.*

MULTIPLICATION FACTORS FOR OTHER MATRICES

Other Matrices	Factor (a)
Waste miscible liquid waste	50
High-concentration soil and sludge	125
Non-water miscible waste	500

(a) *EQL = [EQL for low soil/sediment] × [Factor]. For non-aqueous samples, the factor is on a wet-weight basis.*

SAMPLING METHOD
Liquid samples — Use a 40-mL glass screw-cap VOA vial with a Teflon®-faced silicone septum that has been prewashed, rinsed with distilled deionized water, and oven dried. However, if residual chlorine is present, collect sample in a 40-oz. soil VOA container which has been pre-preserved with 4 drops of 10% sodium thiosulfate, mix gently, and then transfer the sample to a 40-mL VOA vial. Collect bubble-free samples in duplicate and seal them in separate plastic bags.

Soils or sediments, and sludges — Use an 8-oz. widemouth glass bottle with a Teflon®-faced silicone septum that has been prewashed with detergent, rinsed with distilled deionized water, and oven dried. Tap slightly to eliminate free air space. Collect samples in duplicate and seal them in separate plastic bags.

SAMPLE PRESERVATION
Liquid samples — Add 4 drops of concentrated HCL and immediately cool samples to 4°C and store in a solvent-free refrigerator.

Soils or sediments, and sludges — Cool samples to 4°C and store in a solvent-free refrigerator.

MHT Maximum holding time is 14 days from the date of sample collection.

SAMPLE PREPARATION
Liquid samples — Remove the plunger from a 5-mL syringe and carefully pour the sample into the syringe barrel to just short of overflowing. Replace the syringe plunger and compress the sample. Open the syringe valve and vent any residual air while adjusting the sample volume to 5.0 mL. If there is only one volatile organic analysis (VOA) vial, a second syringe should be filled at this time to protect against possible loss of sample integrity. Add 10 µL of surrogate spiking solution and 10 µL of internal standard spiking solution through the valve bore of the 5-mL syringe, then close the valve. The surrogate and internal standards may be mixed and added as a single spiking solution.

Sediments, soils, and waste samples — All samples of this type should be screened by GC analysis using a headspace method (EPA Method 3810) or the hexadecane extraction and screening method (EPA Method 3820). Use the screening data to determine whether to use the low-concentration method (0.005–1 mg/kg) or the high-concentration method (>1 mg/kg).

Low-concentration method — The low-concentration method is based on purging a heated sediment or soil sample mixed with organic-free reagent water containing the surrogate and internal standards. Analyze all reagent blanks and standards under the same conditions as the samples.

Use a 5-g sample if the expected concentration is <0.1 mg/kg or a 1-g sample for expected concentrations between 0.1 and 1 mg/kg. Mix the contents of the sample container with a narrow metal spatula. Weigh the amount of the sample into a tared purge device. Add the spiked water to the purge device, which contains the weighed amount of sample, and connect the device to the purge-and-trap system.

High-concentration method — This method is based on extracting the sediment or soil with methanol. A waste sample is either extracted or diluted, depending on its solubility in methanol. Wastes that are insoluble in methanol are diluted with reagent tetraglyme or possibly polyethylene glycol (PEG). An aliquot of the extract is added to organic-free reagent water containing surrogate and internal standards. This is purged at ambient temperature. All samples with an expected concentration of >1.0 mg/kg should be analyzed by this method.

Mix the contents of the sample container with a narrow metal spatula. For sediments or soils and solid wastes that are insoluble in methanol, weigh 4 g (wet weight) of sample into a tared 20-mL vial. For waste that is soluble in methanol, tetraglyme, or PEG, weigh 1 g (wet weight) into a tared scintillation vial or culture tube or a 10-mL volumetric flask. Quickly add 9.0 mL of appropriate solvent then add 1.0 mL of a surrogate spiking solution to the vial, cap it, and shake it for 2 min.

METHANOL EXTRACT REQUIRED FOR ANALYSIS OF HIGH-CONCENTRATION SOILS OR SEDIMENTS

Approximate Concentration Range	Volume of Methanol Extract (a)
500–10,000 µg/kg	100 µL
1,000–20,000 µg/kg	50 µL
5,000–100,000 µg/kg	10 µL
25,000–500,000 µg/kg	100 µL of 1/50 dilution (b)

Calculate appropriate dilution factor for concentrations exceeding this table.

(a) The volume of methanol added to 5 mL of water being purged should be kept constant. Therefore, add to the 5-mL syringe whatever volume of methanol is necessary to maintain a volume of 100 µL added to the syringe.
(b) Dilute an aliquot of the methanol extract and then take 100 µL for analysis.

QUALITY CONTROL Demonstrate, through the analysis of a reagent water blank, that interferences from the analytical system, glassware, and reagents are under control. Blank samples should be carried through all stages of the sample preparation and measurement steps. For each analytical batch (up to 20 samples), a reagent blank, matrix spike, and matrix spike duplicate must be analyzed (the frequency of the spikes may be different for different monitoring programs). The blank and spiked samples must be carried through all stages of the sample preparation and measurement steps. QC samples mentioned in the section on Interferences will also be needed as appropriate to those situations.

REFERENCE Test Methods for Evaluating Solid Waste (SW-846). U.S. EPA. 1983. Method 8240B, Rev. 2, Nov. 1990. Office of Solid Wastes, Washington, DC.

Iron
CAS #7439-89-6
EPA Method 6010

TITLE Inductively Coupled Plasma-Atomic Emission Spectroscopy

MATRIX This method is applicable to the determination of trace elements, including metals, in groundwater, soils, sludges, sediments, and other solid wastes. All matrices require digestion prior to analysis. The method of standard addition must be used for the analysis of all sample digests unless either serial dilution or matrix spike addition demonstrates it is not required.

METHOD SUMMARY Method 6010 covers 25 elements using ICP analysis. It measures element-emitted light by optical spectrometry. Samples, following an appropriate acid digestion, are nebulized and the resulting aerosol is transported to the plasma torch. Element-specific atomic line emission spectra are produced by a radio-frequency inductively coupled plasma.

INTERFERENCES Interferences may be categorized as spectral or non-spectral. Spectral interferences are caused by overlap of a spectral line from another element, unresolved overlap of molecular band spectra, background contribution from continuous or recombination phenomenon, and stray light from the line emission of high concentration elements. Non-spectral interferences include physical and chemical interferences. Physical interferences are effects associated with the sample nebulization and transport processes. Changes in viscosity and surface tension can cause significant inaccuracies. Chemical interferences include molecular compound formation, ionization effects, and solute vaporization effects. Normally these effects are not significant and can be minimized by careful selection of operating conditions. Chemical interferences are highly dependent on matrix type and the specific analyte element.

INSTRUMENTATION An inductively coupled argon plasma emission spectrometer (ICP) capable of background correction is required.

PRECISION & ACCURACY Detection limits, sensitivity, and optimum ranges of the metals will vary with the matrices and model of the spectrometer. In a single lab evaluation, seven wastes were analyzed for 22 elements. The mean percent relative standard deviation from triplicate analyses for all elements and wastes was 9 ± 2%. The mean percent recovery of spiked elements for all wastes was 93 ± 6%. Spike levels ranged from 100 µg/L to 100 mg/L. The wastes included sludges and industrial wastewaters.

Estimated instrument detection limit in µg/L is 7.
Spiked concentration in µg/L is 20.
Mean reported value in µg/L is 19.
Precision as RSD % is 15.

SAMPLING METHOD Samples should be collected in borosilicate glass, linear polyethylene, polypropylene, or Teflon® bottles that have been prewashed with detergent and tap water, and rinsed with 1:1 nitric acid and tap water or 1:1 hydrochloric acid and tap water. Collect at least 2 g of solids and 200 mL of aqueous samples.

SAMPLE PRESERVATION Add nitric acid to make the samples pH <2.

MHT The maximum holding time for properly preserved samples is 6 months.

SAMPLE PREPARATION Preliminary treatment of most matrices is necessary because of the complexity and variability of sample matrices. Water samples that have been prefiltered and acidified will not need acid digestion. Methods for acid digestion of waters for total recoverable or dissolved metals, acid digestions of aqueous samples and extracts for total metals, and acid digestion of sediments, sludges, and soils are summarized below.

Total recoverable or dissolved metals in water — To prepare surface and groundwater samples for determination of total recoverable and dissolved metals, a 100-mL aliquot of well-mixed sample is acidified with concentrated nitric acid and concentrated hydrochloric acid, then heated until the volume is reduced to 15–20 mL. Adjust the final volume to 100 mL with reagent water.

Total metals in aqueous samples, soil and sediment extracts — To prepare aqueous samples, soil and sediment extracts, and wastes that contain suspended solids, a 100-mL aliquot is made acidic with concentrated nitric acid and the solution is evaporated to about 5 mL on a hot plate. Continue heating and adding additional acid until sample digestion is complete, which is usually indicated when the digestate is light in color or does not change in appearance. Evaporate the solution to about 3 mL and cool it and add a small quantity of 1:1 hydrochloric acid (10 mL/100 mL of final solution). Cover the beaker and reflux for 15 min. Wash down the beaker walls and filters or centrifuge the sample to remove silicates and other insoluble material. Filter the sample and adjust the final volume to 100 mL with reagent water and the final acid concentration to 10%.

Sediments, sludges, and soils — To prepare sediments, sludges and soil samples, transfer 1–2 g to a conical beaker and add 10 mL of 1:1 nitric acid, mix the slurry, and cover it with a watch glass. Heat the sample and reflux for 10–15 min without boiling. Allow it to cool, then add 5 mL of concentrated nitric acid and reflux for 30 min. Repeat last step and then allow the solution to evaporate to 5 mL without boiling. Cool and add 2 mL of water and 3 mL of 30% hydrogen peroxide. Cover and place the beaker on the hot plate. Heat and add 30% hydrogen peroxide in 1-mL aliquots with warming until the effervescence is minimal but do not add more than a total of 10 mL of 30% hydrogen peroxide. If the sample is being prepared for the analysis of Ag, Al, As, Ba, Be, Ca, Cd, Co, Cr, Cu, Fe, K, Mg, Mn, Mo, Na, Ni, Os, Pb, Se, Tl, V, and Zn, then add 5 mL of concentrated hydrochloric acid and 10 mL of water and return the covered beaker to a hot plate for 15 min of additional refluxing without boiling. Dilute the sample to a 100 mL volume with water after cooling and filter or centrifuge to remove particulates.

QUALITY CONTROL Laboratory control samples must be analyzed for each analytical method. A method blank should be analyzed with each batch of samples. The effect of the matrix on method performance must be demonstrated: when appropriate, there should be at least one matrix spike and either one matrix duplicate or one matrix spike duplicate per analytical batch. The bias and precision of the method, as well as the method detection limit for each specific matrix type, must be measured.

Dilute and reanalyze samples that are more concentrated than the linear calibration limit. Employ a minimum of one reagent blank per sample batch to determine if contamination or any memory effects are occurring. Whenever a new or unusual sample matrix is encountered, perform either a serial dilution test or a matrix spike addition test to ensure that neither positive or negative interferences are operating on any of the analyte elements. Check the instrument standardization by verifying calibration every 10 samples using a calibration blank and a check standard.

REFERENCE Test Methods for Evaluating Solid Waste (SW-846). U.S. EPA. 1983. Method 6010, Rev. 0, Sept. 1986. Office of Solid Wastes, Washington, DC.

Iron   **EPA Method 200.7**
CAS #7439-89-6

TITLE Inductively Coupled Plasma (ICP)

MATRIX Dissolved, suspended, or total element in drinking and surface waters and in domestic and industrial wastewaters.

APPLICATION The method covers the determination of 25 metals. Dissolved elements are determined in filtered and acidified samples after appropriate digestion (which increases dissolved solids). Its primary advantage is that ICP instruments allow simultaneous or rapid sequential determination of many elements in a short time. Samples are first nebulized and the aerosol is transported to a plasma torch in which element specific atomic line emission spectra are produced by a radio frequency inductively coupled plasma. Background correction is required for trace element detection except in the case of line broadning.

INTERFERENCES There are spectral, physical, and chemical interferences. The primary disadvantage of ICP instruments is background radiation from other elements and the plasma gases (spectral interferences). Changes in sample viscosity and surface tension with samples containing high dissolved solids (especially those exceeding 1500 mg/L) or high acid concentrations can cause physical interferences. Ionization effects, solute vaporization and molecular compound formation can cause chemical interferences. Manganese can cause interference at the 100 mg/L level.

INSTRUMENTATION Inductively coupled argon plasma emission spectroscopy. 259.940 nm wavelength

RANGE Not listed.

MDL 7 µg/L.

PRECISION SD = 3.0% Mean at true value 600 µg/L.

ACCURACY Mean recovery = 93% ± 6% of spiked elements for all wastes.

SAMPLING METHOD Wash sample container with detergent and tap water, rinse with 1 + 1 nitric acid and tap water, then rinse with 1 + 1 hydrochloric acid and tap water, then rinse with deionized, distilled water in that order. Perform any filtration or acid preservation steps when the sample is collected or as soon as possible thereafter.

STABILITY Cool samples to 4°C.

MHT 24 h.

QUALITY CONTROL Mixed calibration standards, an instrument check standard, and an interference check solution are used in addition to a quality control sample. The quality control sample should be prepared in the same acid matrix as the calibration standards at 10 times the instrumental detection limits and in accordance with the instructions provided by the supplier. Furthermore, two types of blanks are required: a calibration blank and a reagent blank.

REFERENCE Method 200.7, U.S. EPA, EMSL-Cincinnati, OH, Nov. 1980

Iron **EPA Method 236.2**
CAS #7439-89-6

TITLE Metals (Total, Dissolved, Suspended) AA Furnace Technique

MATRIX Drinking, surface and saline, waters. Wastewater

APPLICATION Date issued 1978. A representative sample aliquot is placed in a graphite tube in furnace, evaporated to dryness, charred and atomized. Radiation from excited element is passed through vapor and radiation intensity decreases proportional to amount of Fe in vapor.

INTERFERENCES Furnace technique subject to chemical and matrix interferences. Furnace gases may have molecular absorption bands enclosing analytical wavelength. Smoke-producing sample matrix can interfere. If Fe isn't volitalized and removed from furnace, memory effects occur.

INSTRUMENTATION AAS. Iron (Fe) hollow cathode lamp or EDL. Graphite furnace. Pipets.

RANGE 5–100 µg/L.

MDL 1 µg/L.

PRECISION Not listed.

ACCURACY Not listed.

SAMPLING METHOD Prewashed plastic or glass containers.

STABILITY HNO3 to pH <2.

MHT 6 months.

QUALITY CONTROL A check standard should be run approximately after every 10 sample injections. Standards are run in part to monitor the life and performance of the graphite tube. Lack of reproducibility or significant change in the signal for the standard indicates tube should be replaced.

REFERENCE EPA Methods for the Chemical Analysis of Water and Wastes, EPA-600/4-79-020, U.S. EPA, EMSL, 1979.

Iron **EPA Method 315 B**
CAS #7439-89-6

TITLE Phenanthroline Method

MATRIX Natural or treated waters total dissolved ferrous

APPLICATION Iron is brought into solution, reduced to ferrous state by boiling with acid and hydroxylamine and treated with 1,10-phenanthroline at pH 3.2 to 3.3. Three molecules of phenanthroline chelate each atom of ferrous iron to form an orange-red complex.

INTERFERENCES Strong oxidizing agents, cyanide, nitrite, and phosphates (particularly polyphosphates), chromium, zinc, cobalt, copper, and nickel interfere. Bismuth, cadmium, mercury, molybdate and silver precipitate phenanthroline.

INSTRUMENTATION Spectro (or filter) photometer with green filter. 510 nm. Light path > 1 cm.

RANGE Not listed.

MDL 10 µg/L.

PRECISION SD = 25.5% at 300 µg Fe/L (aqueous mixture of 8 metals)

ACCURACY Relative error = 13.3% at 300 µg Fe/L (aqueous mix of 8 metals)

SAMPLING METHOD Plastic or glass. Clean with acid. Rinse with distilled water.

STABILITY HNO3 to pH <2.

MHT 6 months.

QUALITY CONTROL Use reagents low in iron. Use iron-free distilled water in preparing standards and reagent solutions. Store reagents in glass stoppered bottles. Calculate ferric iron by subtracting ferrous iron from total iron. Don't expose phenanthroline solutions to sunlight.

REFERENCE Standard Methods for the Examination of Water and Waste Water, 16th ed., Page 215, 1985.

Iron **EPA Method 7380**
CAS #7439-89-6

TITLE Atomic Absorption (AA) Direct Aspiration

MATRIX Drinking, surface, and saline waters, wastewater

APPLICATION Sample is aspirated and atomized in a flame. A light beam from an Fe hollow cathode lamp is directed through the flame into a monochromator and onto a detector.

Since wavelength of light beam is specific for Fe, light energy absorbed by detector is measure of iron.

INTERFERENCES The most troublesomee type is chemical, caused by lack of absorption of atoms bound in molecular combination in the flame. High dissolved solids in sample may result in nonatomic absorbance interference. Iron is a universal contaminant; avoid contamination.

INSTRUMENTATION Atomic absorption spectrometer. Iron hollow cathode lamp. [248.3 nm wavelength(primary)]

RANGE 0.3–5 mg/L.

MDL 0.03 mg/L

PRECISION Standard deviation = 173 µg/L at 840 µg/L (true value) 82 labs

ACCURACY As bias = +1.8% at 840 µg/L (true value) 82 labs

SAMPLING METHOD Use glass or plastic containers. Collect 200 g of solids and 600 mL of liquid samples.

STABILITY Cool solid samples to 4°C and analyze as soon as possible. Add nitric acid to liquid samples to pH <2.

MHT 6 months.

QUALITY CONTROL At least one duplicate and one spike sample should be run every 20 samples or with each matrix type to verify precision of the method. For 20 or more samples per day, verify working standard curve. Run an additional standard at or near mid-range every 10 samples.

REFERENCE Method 7380, SW-846, 3rd ed., Nov. 1986.

Iron   **EPA Method 7381**
CAS #7439-89-6

TITLE Atomic Absorption (AA) Furnace Technique

MATRIX Wastes, mobility procedure extracts, soils and groundwater

APPLICATION Aqueous samples, EP extracts, industrial wastes, soils, sludges, sediments, and solid wastes require digestion before analysis. An aliquot of sample is placed in the graphite tube in the furnace and slowly evaporated, charred and atomized. Absorption of lamp radiation during atomization is proportional to iron concentration.

INTERFERENCES The furnace technique is subject to chemical interferences. Composition of sample matrix can have major effect on analysis. Modify matrix to remove interferences. Iron is a universal contaminant. Use great care to avoid contamination.

INSTRUMENTATION Atomic absorption spectrometer. Iron hollow cathode lamp or electrodeless discharge lamp. Graphite furnace. Strip-chart recorder

RANGE 5–100 µg/L.

MDL 1 µg/L (248.3 nm wavelength)

PRECISION Not listed.

ACCURACY Not listed.

SAMPLING METHOD Use glass or plastic containers. Collect 200 g of solids and 600 mL of liquid samples.

STABILITY Cool solid samples to 4°C and analyze as soon as possible. Add nitric acid to liquid samples to pH <2.

MHT 6 months.

QUALITY CONTROL At least one duplicate and one spike sample should be run every 20 samples, or with each matrix type to verify method precision. If 20 or more samples are run a day, run a standard (at or near mid-range) every 10 samples.

REFERENCE Method 7381, SW-846, 3rd ed., (Included as Rev. 0, Dec. 1987)

Isobutyl alcohol   **EPA Method 1624**
CAS #78-83-1

TITLE Volatile Organic Compounds by Isotope Dilution GC/MS

MATRIX Compounds may be determined in waters, soils, and municipal sludges by this method.

METHOD SUMMARY This method is used to determine 58 volatile toxic organic pollutants associated with the CWA (as amended 1987); the RCRA (as amended 1986); the CERCLA (as amended 1986); and other compounds amenable to purge-and-trap gas chromatography-mass spectrometry (GC/MS).

If the solids content is less than 1%, stable isotopically-labeled analogs of the compounds of interest are added to a 5-mL sample and the sample is purged with an inert gas at 20–25°C in a chamber designed for soil or water samples. If the solids content is greater than 1%, 5 mL of reagent water and the labeled compounds are added to a 5-g aliquot of sample and the mixture is purged at 40°C. Compounds that will not purge at 20–25°C or at 40°C are purged at 78–85°C. In the purging process, the volatile compounds are transferred from the aqueous phase into the gaseous phase where they are passed into a sorbent column, and trapped. After purging is completed, the trap is backflushed and heated rapidly to desorb the compounds into a GC. The compounds are separated by the GC and detected by a MS. The labeled compounds serve to correct the variability of the analytical technique.

INTERFERENCES Impurities in the purge gas, organic compounds outgassing from the plumbing upstream of the trap, and solvent vapors in the lab account for most problems. Samples can be contaminated by diffusion of volatile organic compounds (particularly methylene chloride) through the bottle seal during shipment and storage. Contamination by carryover can occur when high-level and low-level samples are analyzed sequentially. When an unusually concentrated sample is encountered, follow it by analysis of a reagent water blank to check for carryover.

INSTRUMENTATION Major equipment includes a GC with linear temperature programming and a glass jet separator as the MS interface, a MS with 70 eV electron impact ionization, and a data system to collect and record response factors.

Column: 2.8 m × 2 mm I.D. glass, packed with 1% SP-1000 on Carbopak B, 60/80 mesh, or equivalent.

PRECISION & ACCURACY The detection limits of the method are usually dependent on the level of interferences rather than instrumental limitations. The method detection limits were determined in digested sludge (low solids) and in filter cake or compost (high solids).

The MDL (in µg/kg) for low solids is not listed and for high solids is not listed.
Labeled and native compound precision (in µg/L) as standard deviation was not listed.
Labeled and native compound accuracy (in µg/L) as average recovery was not listed.
Acceptance criteria are at 20 µg/L for this compound.

SAMPLE COLLECTION, PRESERVATION & HANDLING Grab samples are collected in glass containers having a total volume greater than 20 mL. Fill and seal each bottle so that no air bubbles are entrapped. Samples are maintained at 0 to 4°C from the time of collection until analysis. If an aqueous sample contains residual chlorine, add sodium thiosulfate preservative (10 mg/40 mL) to the empty sample bottles just prior to shipment to the sample site. All samples must be analyzed within 14 days of collection.

SAMPLE PREPARATION Samples containing less than 1% solids are analyzed directly as aqueous samples. Samples containing 1% solids or greater are analyzed as solid samples utilizing one of two methods, depending on the levels of pollutants, in the sample. Samples containing 1% solids or greater, and low to moderate levels of pollutants are analyzed by purging a known weight of sample added to 5 mL of reagent water. Samples containing 1% solids or greater, and high levels of pollutants, are extracted with methanol, and an aliquot of the methanol extract is added to reagent water and purged.

QUALITY CONTROL A field blank prepared from reagent water and carried through the sampling and handling protocol may serve as a check on contamination from shipment and storage.

The analyst is permitted to modify this method to improve separations or lower the costs of measurements, provided all performance specifications are met. Analyses of blanks are required. When results of spikes indicate atypical method performance for samples, the samples are diluted to bring method performance within acceptable limits. Analyze two sets of four 5-mL aliquots (8 aliquots total) of the aqueous performance standard. Spike all samples with labeled compounds to assess method performance on the sample matrix. Compute the percent recovery of the labeled compounds using the internal standard method. Compare the percent recovery for each compound with the corresponding labeled compound recovery. Reagent water blanks are analyzed to demonstrate freedom from carryover contamination. Field replicates may be collected to determine the precision of the sampling technique, and spiked samples may be required to determine the accuracy of the analysis when the internal method is used.

REFERENCE Volatile Organic Compounds by Isotope Dilution GC/MS. Office of Water Regulation and Standards, U.S. EPA Industrial Technology Division, Washington, DC, EPA Method 1624, Rev. C, June 1989 (contact W.A. Telliard, U.S. EPA, Office of Water Regulations and Standards, 401 M St., SW, Washington, DC, 20460. Phone: 202-382-7131).

Isobutyl alcohol   **EPA Method 8240**
CAS #78-83-1

TITLE Volatile Organics By GC/MS: Packed Column Technique

MATRIX Nearly all types of sample matarices, regardless of water content, can be analyzed using this method. This includes groundwater, aqueous sludges, caustic liquors, acid liquors, waste solvents, oily wastes, mousses, tars, fibrous wastes, polymetric emulsions, filter cakes, spent carbons, spent catalysts, soils, and sediments.

METHOD SUMMARY Method 8240B covers 80 volatile organic compounds that are introduced into a gas chromatograph by the purge-and-trap method or by direct injection (in limited applications). For the purge-and-trap method an inert gas (zero grade nitrogen or helium) is bubbled through a 5-mL solution at ambient temperature. Purged sample components are trapped in a tube of sorbent materials. When purging is complete, the sorbent tube is heated and backflushed with inert gas to desorb the trapped components onto a GC column.

INTERFERENCES Impurities in the purge gas and from organic compounds outgassing from the plumbing ahead of the trap account for many contamination problems. Interferences purged or coextracted from the samples will vary considerably from source to source. Cross-contamination can occur whenever high-level and low-level samples are analyzed sequentially. Whenever an unusually concentrated sample is analyzed, it should be followed by the analysis of organic-free reagent water to check for cross-contamination. Samples also can be contaminated by diffusion of volatile organics (particularly methylene chloride and fluorocarbons) through the septum seal into the sample during shipment and storage. A trip blank can serve as a check on such contamination. The lab where volatile analysis is performed and also the refrigerated storage area should be completely free of solvents.

INSTRUMENTATION A gas chromatograph/mass spectrometry/data system (GC/MS) equipped with a 6 ft × 0.1 in I.D. glass column packed with 1% SP-1000 on Carbopack-B (60/80 mesh) is required. Also needed is a 5-mL purging device, a sorbent trap, and a thermal desorption apparatus.

PRECISION & ACCURACY This method is reported to have been tested by 15 laboratories using organic-free reagent water, drinking water, surface water, and industrial wastewaters (not specified) fortified at six concentrations over the range 5–600 µg/L.

Sample estimated quantitation limits (EQLs) are highly matrix-dependent. The EQLs listed may not always be achievable. EQLs listed for soils or sediments are based on wet weight. Normally, data is reported on a dry-weight basis; therefore, EQLs will be higher, based on the percent dry weight of each sample. Note that EQLs are even more variable than MDLs and that they are highly variable depending on the matrix being analyzed.

EQL in groundwater in µg/L was 100.
EQL in low soil or sediment in µg/kg was 100.
Accuracy (a) in µg/L was not listed.
Precision (b) in µg/L was not listed.

(a) *Average recovery found for measurements of samples containing a concentration of C, in µg/L.*
(b) *Overall precision found for measurements of samples with average recovery X for samples containing a concentration of C in µg/L.*
X = *Average recovery found for measurement of samples containing a concentration of C in µg/L.*

MULTIPLICATION FACTORS FOR OTHER MATRICES

Other Matrices	Factor (a)
Waste miscible liquid waste	50
High-concentration soil and sludge	125
Non-water miscible waste	500

(a) *EQL = [EQL for low soil/sediment] × [Factor]. For non-aqueous samples, the factor is on a wet-weight basis.*

SAMPLING METHOD
Liquid samples — Use a 40-mL glass screw-cap VOA vial with a Teflon®-faced silicone septum that has been prewashed, rinsed with distilled deionized water, and oven dried. However, if residual chlorine is present, collect sample in a 40-oz. soil VOA container which has been pre-preserved with 4 drops of 10% sodium thiosulfate, mix gently, and then transfer the sample to a 40-mL VOA vial. Collect bubble-free samples in duplicate and seal them in separate plastic bags.

Soils or sediments, and sludges — Use an 8-oz. widemouth glass bottle with a Teflon®-faced silicone septum that has been prewashed with detergent, rinsed with distilled deionized water, and oven dried. Tap slightly to eliminate free air space. Collect samples in duplicate and seal them in separate plastic bags.

SAMPLE PRESERVATION
Liquid samples — Add 4 drops of concentrated HCL and immediately cool samples to 4°C and store in a solvent-free refrigerator.

Soils or sediments, and sludges — Cool samples to 4°C and store in a solvent-free refrigerator.

MHT Maximum holding time is 14 days from the date of sample collection.

SAMPLE PREPARATION
Liquid samples — Remove the plunger from a 5-mL syringe and carefully pour the sample into the syringe barrel to just short of overflowing. Replace the syringe plunger and compress the sample. Open the syringe valve and vent any residual air while adjusting the sample volume to 5.0 mL. If there is only one volatile organic analysis (VOA) vial, a second syringe should be filled at this time to protect against possible loss of sample integrity. Add 10 µL of surrogate spiking solution and 10 µL of internal standard spiking solution through the valve bore of the 5-mL syringe, then close the valve. The surrogate and internal standards may be mixed and added as a single spiking solution.

Sediments, soils, and waste samples — All samples of this type should be screened by GC analysis using a headspace method (EPA Method 3810) or the hexadecane extraction and screening method (EPA Method 3820). Use the screening data to determine whether to use the low-concentration method (0.005–1 mg/kg) or the high-concentration method (>1 mg/kg).

Low-concentration method — The low-concentration method is based on purging a heated sediment or soil sample mixed with organic-free reagent water containing the surrogate and internal standards. Analyze all reagent blanks and standards under the same conditions as the samples.

Use a 5-g sample if the expected concentration is <0.1 mg/kg or a 1-g sample for expected concentrations between 0.1 and 1 mg/kg. Mix the contents of the sample container with a narrow metal spatula. Weigh the amount of the sample into a tared purge device. Add the spiked water to the purge device, which contains the weighed amount of sample, and connect the device to the purge-and-trap system.

High-concentration method — This method is based on extracting the sediment or soil with methanol. A waste sample is either extracted or diluted, depending on its solubility in methanol. Wastes that are insoluble in methanol are diluted with reagent tetraglyme or possibly polyethylene glycol (PEG). An aliquot of the extract is added to organic-free reagent water containing surrogate and internal standards. This is purged at ambient temperature. All samples with an expected concentration of >1.0 mg/kg should be analyzed by this method.

Mix the contents of the sample container with a narrow metal spatula. For sediments or soils and solid wastes that are insoluble in methanol, weigh 4 g (wet weight) of sample into a tared 20-mL vial. For waste that is soluble in methanol, tetraglyme, or PEG, weigh 1 g (wet weight) into a tared scintillation vial or culture tube or a 10-mL volumetric flask. Quickly add 9.0 mL of appropriate solvent then add 1.0 mL of a surrogate spiking solution to the vial, cap it, and shake it for 2 min.

METHANOL EXTRACT REQUIRED FOR ANALYSIS OF HIGH-CONCENTRATION SOILS OR SEDIMENTS

Approximate Concentration Range	Volume of Methanol Extract (a)
500–10,000 µg/kg	100 µL
1,000–20,000 µg/kg	50 µL
5,000–100,000 µg/kg	10 µL
25,000–500,000 µg/kg	100 µL of 1/50 dilution (b)

Calculate appropriate dilution factor for concentrations exceeding this table.

(a) *The volume of methanol added to 5 mL of water being purged should be kept constant. Therefore, add to the 5-mL syringe whatever volume of methanol is necessary to maintain a volume of 100 µL added to the syringe.*

(b) Dilute an aliquot of the methanol extract and then take 100 µL for analysis.

QUALITY CONTROL Demonstrate, through the analysis of a reagent water blank, that interferences from the analytical system, glassware, and reagents are under control. Blank samples should be carried through all stages of the sample preparation and measurement steps. For each analytical batch (up to 20 samples), a reagent blank, matrix spike, and matrix spike duplicate must be analyzed (the frequency of the spikes may be different for different monitoring programs). The blank and spiked samples must be carried through all stages of the sample preparation and measurement steps. QC samples mentioned in the section on Interferences will also be needed as appropriate to those situations.

REFERENCE Test Methods for Evaluating Solid Waste (SW-846). U.S. EPA. 1983. Method 8240B, Rev. 2, Nov. 1990. Office of Solid Wastes, Washington, DC.

Isodrin   **EPA Method 8270**
CAS #465-73-6

TITLE Semivolatile Organic Compounds by GC/MS

MATRIX This method is used to determine the concentration of semivolatile organic compounds in extracts prepared from all types of solid waste matrices, soils, and groundwater. Although surface waters are not specifically mentioned, this method should be applicable to water samples from rivers, lakes, etc.

METHOD SUMMARY This method covers 259 semivolatile organic compounds. In very limited applications direct injection of the sample into the GC/MS system may be appropriate, but this results in very high detection limits (approximately 10,000 µg/L). Typically, a 1-L liquid sample, containing surrogate, and matrix spiking standards, is extracted in a continuous extractor first under acid conditions and then under basic conditions. Typically 30 g of a solid sample, containing surrogate, and matrix spiking standards, is extracted ultrasonically. After concentrating the extract to 1 mL it is spiked with 10 µL of an internal standard solution just prior to analysis by GC/MS. The volume injected should contain about 100 ng of base/neutral and 200 ng of acid surrogates (for a 1-µL injection). Analysis is performed by GC/MS using a capillary GC column.

INTERFERENCES Raw GC/MS data from all blanks, samples, and spikes must be evaluated for interferences. Contamination by carryover can occur whenever high-concentration and low-concentration samples are sequentially analyzed. To reduce carryover, the sample syringe must be rinsed out between samples with solvent. Whenever an unusually concentrated sample is encountered, it should be followed by the analysis of blank solvent to check for cross-contamination.

INSTRUMENTATION A GC/MS and a data system are required. The GC column used is a 30 m × 0.25 mm I.D. (or 0.32 mm I.D.) 1um film thickness silicone-coated fused silica capillary column. A continuous liquid-liquid extractor equipped with Teflon® or glass connection joints and stopcocks requiring no lubrication, a K-D concentrating apparatus, water bath, and an ultrasonic disrupter with a minimum power of 300 W and with pulsing capability are also required.

PRECISION & ACCURACY The estimated quantitation limit (EQL) of Method 8270B for determining an individual compound is approximately 1 mg/kg (wet weight) for soil or sediment samples, 1–200 mg/kg for wastes (dependent on matrix and method of preparation), and 10 µg/L for groundwater samples. EQLs will be proportionately higher for sample extracts that require dilution to avoid saturation of the detector.

The EQL(b) for groundwater in µg/L is 20.
The EQL (a, b) for low concentrations in soil and sediment in µg/kg is not determined.
Accuracy as µg/L is not listed.
Overall precision in µg/L is not listed.

(a) *EQLs listed for soil/sediment are based on wet weight. Normally data is reported in a dry-weight basis; therefore, EQLs will be higher based on the % dry weight of each sample. This calculation is based on a 30-g sample and gel permeation chromatography cleanup.*

(b) *Sample EQLs are highly matrix-dependent. The EQLs are provided for guidance and may not always be achievable.*

C = *True value for concentration, in µg/L.*
X = *Average recovery found for measurements of samples containing a concentration of C, in µg/L.*

ESTIMATED QUANTITATION LIMIT

Other Matrices	Factor (a)
High-concentration soil and sludges by sonicator	7.5
Non-water miscible waste	75

(a) EQL for other matrices = [EQL for low soil/sediment] × [Factor]. This estimated EQL is similar to an EPA "Practical Quantitation Limit."

SAMPLING METHOD

Liquid samples — Use a 1 or 2½ gallon amber glass bottle with a screw-top Teflon®-lined cover that has been prewashed with detergent and rinsed with distilled water and methanol (or isopropanol).

Soils, sediments, or sludges — Use an 8-oz. widemouth glass with a screw-top Teflon®-lined cover that has been prewashed with detergent and rinsed with distilled water and methanol (or isopropanol).

SAMPLE PRESERVATION

Liquid samples — If residual chlorine is present, add 3 mL of 10% sodium thiosulfate per gallon, cool to 4°C and store in a solvent-free refrigerator until analysis; if chlorine is not present, then eliminate the sodium thiosulfate addition.

Soils, sediments, or sludges — Cool samples to 4°C and store in a solvent-free refrigerator.

MHT Liquid samples must be extracted within 7 days and the extracts analyzed within 40 days. Soils, sediments, or sludges may

be stored for a maximum of 14 days and the extracts analyzed within 40 days.

SAMPLE PREPARATION
Liquid samples — Transfer 1 L quantitatively to a continuous extractor. If high concentrations are anticipated, a smaller volume may be used and then diluted with organic-free reagent water to 1 L. Adjust pH, if necessary, to pH <2 using 1:1 (V/V) sulfuric acid. Pipette 1.0 mL of a surrogate standard spiking solution into each sample. For the sample in each analytical batch selected for spiking, add 1.0 mL of a matrix spiking standard. For base/neutral acid analysis, the amount of the surrogates and matrix spiking compounds added to the sample should result in a final concentration of 100 ng/µL of each analyte in the extract to be analyzed (assuming a 1-µL injection). Extract with methylene chloride for 18–24 h. Next, adjust the pH of the aqueous phase to pH >11 using 10 N sodium hydroxide and extract it with methylene chloride again for 18–24 h. Dry the extract through a column containing anhydrous sodium sulfate and concentrate it to 1 mL using a K-D concentrator.

Soils, sediments, or sludges — Use 30 g of sample. Nonporous or wet samples (gummy or clay type) that do not have a free-flowing sandy texture must be mixed with anhydrous sodium sulfate until the sample is free flowing. Add 1 mL of surrogate standards to all samples, spikes, standards, and blanks. For the sample in each analytical batch selected for spiking, add 1.0 mL of a matrix spiking standard. For base/neutral acid analysis, the amount added of the surrogates and matrix spiking compounds should result in a final concentration of 100 ng/µL of each base/neutral analyte and 200 ng/µL of each acid analyte in the extract to be analyzed (assuming a 1-µL injection). Immediately add a 100-mL mixture of 1:1 methylene chloride:acetone and extract the sample ultrasonically for 3 min and then decant or filter the extracts. Repeat the extraction two or more times. Dry the extract using a column with anhydrous sodium sulfate and concentrate it to 1 mL in a K-D concentrator.

QUALITY CONTROL A methylene chloride solution containing 50 ng/µL of decafluorotriphenylphosphine (DFTPP) is used for tuning the GC/MS system each 12-h shift. A system performance check also must be made during every 12-h shift. A standard containing 50 ng/µL each of 4,4′-DDT, pentachlorophenol, and benzidine is required to verify injection port inertness and GC column performance. A calibration standard at mid-concentration, containing each compound of interest, including all required surrogates, must be performed every 12 h during analysis. After the system performance check is met, calibration check compounds (CCCs) are used to check the validity of the initial calibration.

The internal standard responses and retention times in the calibration check standard must be evaluated immediately after or during data acquisition. If the retention time for any internal standard changes by more than 30 seconds from the last check calibration (12 h), the chromatographic system must be inspected for malfunctions and corrections must be made, as required. If the electron ionization current plot (EICP) area for any of the internal standards changes by a factor of two from the last daily calibration standard check, the mass spectrometer must be inspected for malfunctions and corrections must be made, as appropriate.

Demonstrate, through the analysis of a reagent water blank, that interferences from the analytical system, glassware, and reagents are under control. The blank samples should be carried through all stages of the sample preparation and measurement steps. For each analytical batch (up to 20 samples), a reagent blank, matrix spike, and matrix spike duplicate/duplicate must be analyzed (the frequency of the spikes may be different for different monitoring programs). The blank and spiked samples must be carried through all stages of the sample preparation and measurement steps. A QC reference sample concentrate containing each analyte at a concentration of 100 mg/L in methanol is required.

REFERENCE Test Methods for Evaluating Solid Waste (SW-846). U.S. EPA 1983, Method 8270B, Rev. 2, Nov. 1990. Office of Solid Waste, Washington, DC.

Isophorone   **EPA Method 1625**
CAS #78-59-1

TITLE Semivolatile Organic Compounds by Isotope Dilution GC/MS

MATRIX The compounds may be determined in waters, soils, and municipal sludges by this method.

METHOD SUMMARY This method is used to determine 176 semivolatile toxic organic pollutants associated with the CWA (as amended 1987); the RCRA (as amended 1986); the CERCLA (as amended 1986); and other compounds amenable to extraction and analysis by capillary column gas chromatography-mass spectrometry (GC/MS).

Stable isotopically-labeled analogs of the compounds of interest are added to the sample. If the solids content is less than 1%, a 1-L sample is extracted at pH 12–13, then at pH <2 with methylene chloride using continuous extraction techniques.

If the solids content is 30% or less, the sample is diluted to 1% solids with reagent water, homogenized ultrasonically, and extracted at pH 12–13, then at pH <2 with methylene chloride using continuous extraction techniques. If the solids content is greater than 30%, the sample is extracted using ultrasonic techniques.

Each extract is dried over sodium sulfate, concentrated to a volume of 5 mL, cleaned up using GPC, if necessary, and concentrated. Extracts are concentrated to 1 mL if GPC is not performed, and to 0.5 mL if GPC is performed.

An internal standard is added to the extract, and a 1-mL aliquot of the extract is injected into the GC. The compounds are separated by GC and detected by a MS. The labeled compounds serve to correct the variability of the analytical technique.

INTERFERENCES Solvents, reagents, glassware, and other sample processing hardware may yield artifacts and/or elevated baselines causing misinterpretation of chromatograms and spectra. Materials used in the analysis must be demonstrated

to be free from interferences under the conditions of analysis by running method blanks initially and with each sample lot (sample started through the extraction process on a given 8-h shift, to a maximum of 20). Specific selection of reagents and purification of solvents by distillation in all glass systems may be required. Glassware and, where possible, reagents are cleaned by solvent rinse and baking at 450°C for 1-h minimum. Interferences coextracted from samples will vary considerably from source to source, depending on the diversity of the site being sampled.

INSTRUMENTATION Major instrumentation includes a GC with a splitless or on-column injection port for capillary column, a MS with 70 eV electron impact ionization, and a data system to collect and record MS data, and process it. A K-D apparatus is used to concentrate extracts.

GC Column: 30 m × 0.25 mm I.D. 5% phenyl, 94% methyl, 1% vinyl silicone bonded phased fused silica capillary column.

PRECISION & ACCURACY The detection limits of the method are usually dependent on the level of interferences rather than instrumental limitations. The limits typify the minimum quantities that can be detected with no interferences present.

The minimum level (in µg/mL) was 10. This is defined as a minimum level at which the analytical system shall give recognizable mass spectra (background corrected) and acceptable calibration points.

The MDL (in µg/kg) in low solids was 8 and in high solids was 5; these were determined in digested sludge (low solids) and in filter cake or compost (high solids).

The labeled and native compound initial precision as standard deviation (in µg/L) was 25.
The labeled and native compound initial accuracy as average recovery (in µg/L) was 76–156.

SAMPLE COLLECTION, PRESERVATION & HANDLING Collect samples in glass containers. Aqueous samples which flow freely are collected in refrigerated bottles using automatic sampling equipment. Solid samples are collected as grab samples using widemouth jars. Maintain samples at 0 to 4°C from the time of collection until extraction. If residual chlorine is present in aqueous samples, add 80 mg sodium thiosulfate/L of water. Begin sample extraction within 7 days of collection, and analyze all extracts within 40 days of extraction.

SAMPLE PREPARATION Samples containing 1% solids or less are extracted directly using continuous liquid-liquid extraction techniques. Samples containing 1 to 30% solids are diluted to the 1% level with reagent water and extracted using continuous liquid-liquid extraction techniques. Samples containing greater than 30% solids are extracted using ultrasonic techniques.

Base/neutral extraction — Adjust the pH of the waters in the extractors to 12–13 with 6 *N* NaOH. Extract with methylene chloride for 24–48 h.
Acid extraction — Adjust the pH of the waters in the extractors to 2 or less using 6 *N* sulfuric acid. Extract with methylene chloride for 24–48 h.
Ultrasonic extraction of high solids samples — Add anhydrous sodium sulfate to the sample and QC aliquot(s). Add acetone:methylene chloride (1:1) to the sample and mix thoroughly

Concentrate extracts using a K-D apparatus.

QUALITY CONTROL The analyst is permitted to modify this method to improve separations or lower the costs of measurements, provided all performance specifications are met. Analyses of blanks are required to demonstrate freedom from contamination. When results of spikes indicate atypical method performance for samples, the samples are diluted to bring method performance within acceptable limits.

For low solids (aqueous samples), extract, concentrate, and analyze two sets of four 1-L aliquots (8 aliquots total) of the precision and recovery standard. For high solids samples, two sets of four 30-g aliquots of the high solids reference matrix are used.

Spike all samples with labeled compounds to assess method performance. Compute percent recovery of the labeled compounds using the internal standard method. Compare the labeled compound recovery for each compound with the corresponding labeled compound recovery.

Reagent water and high solids reference matrix blanks are analyzed to demonstrate freedom from contamination. Extract and concentrate a 1-L reagent water blank or a high solids reference matrix blank with each sample's lot (samples started through the extraction process on the same 8-h shift, to a maximum of 20 samples).

Field replicates may be collected to determine the precision of the sampling technique, and spiked samples may be required to determine the accuracy of the analysis when the internal standard method is used.

REFERENCE Semivolatile Organic Compounds by Isotope Dilution GC/MS. Office of Water Regulation and Standards, U.S. EPA Industrial Technology Division, Washington, DC, EPA Method 1625, Rev. C, June 1989 (contact W.A. Telliard, U.S. EPA, Office of Water Regulations and Standards, 401 M St., SW, Washington, DC, 20460. Phone: 202-382-7131).

Isophorone   **EPA Method 625**
CAS #78-59-1

TITLE Base/Neutrals and Acids, U.S. EPA Method 625

MATRIX This methods covers municipal and industrial wastewaters.

METHOD SUMMARY Approximately 1 L of sample is serially extracted with methylene chloride at a pH greater than 11 and again at a pH less than 2 using a separatory funnel or a continuous extractor. The methylene chloride extract is dried, concentrated to a volume of 1 mL, and analyzed by GC/MS. Qualitative identification of the parameters in the extract is performed using the retention time and the relative abundance of three characteristic masses (m/z). Qualitative analysis is performed using either external or internal standard techniques with a single characteristic m/z.

INTERFERENCES Method interferences may be caused by contaminants in solvents, reagents, glassware, and other sample processing hardware. Glassware must be scrupulously cleaned. Glassware should be heated in a muffle furnace at 400°C for 5 to 30 min. Some thermally stable materials, such as PCBs, may not be eliminated by this treatment. Solvent rinses with acetone and pesticide quality hexane may be substituted for the muffle furnace heating. Matrix interferences may be caused by contaminants that are coextracted from the sample. The base-neutral extraction may cause significantly reduced recovery of phenols. The packed gas chromatographic columns recommended for the basic fraction may not exhibit sufficient resolution for some analytes.

INSTRUMENTATION A GC/MS system with an injection port designed for on-column injection when using packed columns and for splitless injection when using capillary columns.

Column for base/neutrals: 1.8 m long × 2 mm I.D. glass, packed with 3% SP-2550 on Supelcoport (100/120 mesh) or equivalent.

Column for acids: 1.8 m long × 2 mm I.D. glass, packed with 1% SP-1240DA on Supelcoport (100/120 mesh) or equivalent.

PRECISION & ACCURACY The MDL concentrations were obtained using reagent water. The MDL actually achieved in a given analysis will vary depending on instrument sensitivity and matrix effects. This method was tested by 15 laboratories using reagent water, drinking water, surface water, and industrial wastewaters spiked at six concentrations over the range 5 to 100 µg/L. Single operator precision, overall precision, and method accuracy were found to be directly related to the concentration of the parameter matrix.

The MDL (in µg/L) in reagent water was not reported.

The standard deviation (in µg/L based on 4 recovery measurements) was 63.3.

The range (in µg/L) for average recovery for 4 measurements was 46.6–180.2.

The range (in %) for percent recovery was 21–196.

Accuracy (in µg/L) as expected recovery for one or more measurements of a sample containing a true concentration of C was 1.12C + 1.41.

Precision (in µg/L) as expected single analyst standard deviation of measurements at an average concentration found at X was 0.27X + 0.77.

Overall precision (in µg/L) as expected interlaboratory standard deviation of measurements in an average concentration found at X was 0.33X + 0.26.

$C =$ *True value of the concentration in µg/L.*
$X =$ *Average recovery found for measurements of samples containing a concentration at C in µg/L.*

SAMPLE PREPARATION Adjust the pH to >11 with sodium hydroxide and serially extract in a separatory funnel with methylene chloride or else in a continuous extractor. Next, adjust the pH to <2 with sulfuric acid and serially extract in a separatory funnel with methylene chloride or else in a continuous extractor. Dry the extracts separately through a column of anhydrous sodium sulfate and then concentrate each of the extracts to 1.0 mL using a K-D apparatus.

SAMPLE COLLECTION, PRESERVATION & HANDLING Grab samples must be collected in glass containers. All samples must be refrigerated at 4°C from the time of collection until extraction. If residual chlorine is present, add 80 mg of sodium thiosulfate/L of sample and mix well. All samples must be extracted within 7 days of collection and completely analyzed within 40 days of extraction.

QUALITY CONTROL Make an initial, one-time, demonstration of the ability to generate acceptable accuracy and precision with this method. Before processing any samples, the analyst must analyze a reagent water blank to demonstrate that interferences from the analytical system and glassware are under control. Each time a set of samples is extracted or reagents are changed, a reagent water blank must be processed. Spike and analyze a minimum of 5% of all samples to monitor and evaluate lab data quality. A QC check sample concentrate that contains each parameter of interest at a concentration of 100 µg/mL in acetone is required. PCBs and multicomponent pesticides may be omitted from this test.

After analysis of five spiked wastewater samples, calculate the average percent recovery and the standard deviation of the percent recovery. Spike all samples with the surrogate standard spiking solution and calculate the percent recovery of each surrogate compound.

REFERENCE *Federal Register*, Vol. 49, No. 209. Friday, Oct. 26, 1984.

Isophorone   EPA Method 8270
CAS #78-59-1

TITLE Semivolatile Organic Compounds by GC/MS

MATRIX This method is used to determine the concentration of semivolatile organic compounds in extracts prepared from all types of solid waste matrices, soils, and groundwater. Although surface waters are not specifically mentioned, this method should be applicable to water samples from rivers, lakes, etc.

METHOD SUMMARY This method covers 259 semivolatile organic compounds. In very limited applications direct injection of the sample into the GC/MS system may be appropriate, but this results in very high detection limits (approximately 10,000 µg/L). Typically, a 1-L liquid sample, containing surrogate, and matrix spiking standards, is extracted in a continuous extractor first under acid conditions and then under basic conditions. Typically 30 g of a solid sample, containing surrogate, and matrix spiking standards, is extracted ultrasonically. After concentrating the extract to 1 mL it is spiked with 10 µL of an internal standard solution just prior to analysis by GC/MS. The volume injected should contain about 100 ng of base/neutral and 200 ng of acid surrogates (for a 1-µL injection). Analysis is performed by GC/MS using a capillary GC column.

INTERFERENCES Raw GC/MS data from all blanks, samples, and spikes must be evaluated for interferences. Contamination by carryover can occur whenever high-concentration and low-concentration samples are sequentially analyzed. To reduce carryover, the sample syringe must be rinsed out between samples with solvent. Whenever an unusually concentrated sample is encountered, it should be followed by the analysis of blank solvent to check for cross-contamination.

INSTRUMENTATION A GC/MS and a data system are required. The GC column used is a 30 m × 0.25 mm I.D. (or 0.32 mm I.D.) 1um film thickness silicone-coated fused silica capillary column. A continuous liquid-liquid extractor equipped with Teflon® or glass connection joints and stopcocks requiring no lubrication, a K-D concentrating apparatus, water bath, and an ultrasonic disrupter with a minimum power of 300 W and with pulsing capability are also required.

PRECISION & ACCURACY The estimated quantitation limit (EQL) of Method 8270B for determining an individual compound is approximately 1 mg/kg (wet weight) for soil or sediment samples, 1–200 mg/kg for wastes (dependent on matrix and method of preparation), and 10 µg/L for groundwater samples. EQLs will be proportionately higher for sample extracts that require dilution to avoid saturation of the detector.

The EQL(b) for groundwater in µg/L is 10.
The EQL (a, b) for low concentrations in soil and sediment in µg/kg is 660.
Accuracy as µg/L is 1.12C + 1.14.
Overall precision in µg/L is 0.33X + 0.26.

(a) EQLs listed for soil/sediment are based on wet weight. Normally data is reported in a dry-weight basis; therefore, EQLs will be higher based on the % dry weight of each sample. This calculation is based on a 30-g sample and gel permeation chromatography cleanup.
(b) Sample EQLs are highly matrix-dependent. The EQLs are provided for guidance and may not always be achievable.
C = True value for concentration, in µg/L.
X = Average recovery found for measurements of samples containing a concentration of C, in µg/L.

ESTIMATED QUANTITATION LIMIT

Other Matrices	Factor (a)
High-concentration soil and sludges by sonicator	7.5
Non-water miscible waste	75

(a) EQL for other matrices = [EQL for low soil/sediment] × [Factor]. This estimated EQL is similar to an EPA "Practical Quantitation Limit."

SAMPLING METHOD
Liquid samples — Use a 1 or 2½ gallon amber glass bottle with a screw-top Teflon®-lined cover that has been prewashed with detergent and rinsed with distilled water and methanol (or isopropanol).

Soils, sediments, or sludges — Use an 8-oz. widemouth glass with a screw-top Teflon®-lined cover that has been prewashed with detergent and rinsed with distilled water and methanol (or isopropanol).

SAMPLE PRESERVATION
Liquid samples — If residual chlorine is present, add 3 mL of 10% sodium thiosulfate per gallon, cool to 4°C and store in a solvent-free refrigerator until analysis; if chlorine is not present, then eliminate the sodium thiosulfate addition.

Soils, sediments, or sludges — Cool samples to 4°C and store in a solvent-free refrigerator.

MHT Liquid samples must be extracted within 7 days and the extracts analyzed within 40 days. Soils, sediments, or sludges may be stored for a maximum of 14 days and the extracts analyzed within 40 days.

SAMPLE PREPARATION
Liquid samples — Transfer 1 L quantitatively to a continuous extractor. If high concentrations are anticipated, a smaller volume may be used and then diluted with organic-free reagent water to 1 L. Adjust pH, if necessary, to pH <2 using 1:1 (V/V) sulfuric acid. Pipette 1.0 mL of a surrogate standard spiking solution into each sample. For the sample in each analytical batch selected for spiking, add 1.0 mL of a matrix spiking standard. For base/neutral acid analysis, the amount of the surrogates and matrix spiking compounds added to the sample should result in a final concentration of 100 ng/µL of each analyte in the extract to be analyzed (assuming a 1-µL injection). Extract with methylene chloride for 18–24 h. Next, adjust the pH of the aqueous phase to pH >11 using 10 N sodium hydroxide and extract it with methylene chloride again for 18–24 h. Dry the extract through a column containing anhydrous sodium sulfate and concentrate it to 1 mL using a K-D concentrator.

Soils, sediments, or sludges — Use 30 g of sample. Nonporous or wet samples (gummy or clay type) that do not have a free-flowing sandy texture must be mixed with anhydrous sodium sulfate until the sample is free flowing. Add 1 mL of surrogate standards to all samples, spikes, standards, and blanks. For the sample in each analytical batch selected for spiking, add 1.0 mL of a matrix spiking standard. For base/neutral acid analysis, the amount added of the surrogates and matrix spiking compounds should result in a final concentration of 100 ng/µL of each base/neutral analyte and 200 ng/µL of each acid analyte in the extract to be analyzed (assuming a 1-µL injection). Immediately add a 100-mL mixture of 1:1 methylene chloride:acetone and extract the sample ultrasonically for 3 min and then decant or filter the extracts. Repeat the extraction two or more times. Dry the extract using a column with anhydrous sodium sulfate and concentrate it to 1 mL in a K-D concentrator.

QUALITY CONTROL A methylene chloride solution containing 50 ng/µL of decafluorotriphenylphosphine (DFTPP) is used for tuning the GC/MS system each 12-h shift. A system performance check also must be made during every 12-h shift. A standard containing 50 ng/µL each of 4,4′-DDT, pentachlorophenol, and benzidine is required to verify injection port inertness and GC column performance. A calibration standard at mid-concentration, containing each compound of interest, including all required surrogates, must be performed every 12 h during analysis. After the system performance check is met, calibration check compounds (CCCs) are used to check the validity of the initial calibration.

The internal standard responses and retention times in the calibration check standard must be evaluated immediately after or during data acquisition. If the retention time for any internal standard changes by more than 30 seconds from the last check calibration (12 h), the chromatographic system must be inspected for malfunctions and corrections must be made, as required. If the electron ionization current plot (EICP) area for any of the internal standards changes by a factor of two from the last daily calibration standard check, the mass spectrometer must be inspected for malfunctions and corrections must be made, as appropriate.

Demonstrate, through the analysis of a reagent water blank, that interferences from the analytical system, glassware, and reagents are under control. The blank samples should be carried through all stages of the sample preparation and measurement steps. For each analytical batch (up to 20 samples), a reagent blank, matrix spike, and matrix spike duplicate/duplicate must be analyzed (the frequency of the spikes may be different for different monitoring programs). The blank and spiked samples must be carried through all stages of the sample preparation and measurement steps. A QC reference sample concentrate containing each analyte at a concentration of 100 mg/L in methanol is required.

REFERENCE Test Methods for Evaluating Solid Waste (SW-846). U.S. EPA 1983, Method 8270B, Rev. 2, Nov. 1990. Office of Solid Waste, Washington, DC.

Isophorone   **EPA Method 8090**
CAS #78-59-1

TITLE Nitroaromatics & Cyclic Ketones

MATRIX Groundwater, soils, sludges, water miscible liquid wastes, and non-water miscible wastes.

APPLICATION This method is used for the analysis of various nitroaromatic and cyclic ketone compounds. Samples are extracted, concentrated, and analyzed using direct injection of both neat and diluted organic liquids.

Dinitrotoluenes are determined using ECD and the other compounds amenable to this method are determined using FID. The method provides an optional GC column which is used for analyte confirmation and that may help resolve analytes from interferences.

INTERFERENCES Solvents, reagents, and glassware may introduce artifacts. Other interferences may come from coextracted compounds from samples.

INSTRUMENTATION GC capable of on-column injections and a flame with detector (FID) or electron capture detector (ECD). Column 1: a 1.2 m by 2 mm or 4 mm with 1.95% QF-1/1.5% OV-17 on Gas-Chrom Q. Column 2: a 3 meter by 2 mm or 4 mm with 3% OV-101 on Gas-Chrom Q.

RANGE 1–515 µg/L.

MDL 5.7 µg/L (FID) and 15.7 µg/L (ECD)

PQL FACTORS FOR MULTIPLYING × FID MDL VALUE

Matrix	Multiplication Factor
Groundwater	10
Low-level soil by sonication with GPC cleanup	670
High-level soil and sludge by sonication	10,000
Non-water miscible waste	100,000

PRECISION 0.46X + 0.31 µg/L (overall precision).

ACCURACY 0.49C + 2.93 µg/L (as recovery).

SAMPLING METHOD Use 8-oz. widemouth glass bottles with Teflon®-lined caps for concentrated waste samples, soils, sediments, and sludges. Use 1 or 2½ gallon amber glass bottles with Teflon®-lined caps for liquid (water) samples.

STABILITY Cool soil, sediment, sludge, and liquid samples to 4°C. If residual chlorine is present in liquid samples add 3 mL of 10% sodium thiosulfate per gallon of sample and cool to 4°C.

MHT 14 days for concentrated waste, soil, sediment, or sludge; 7 days for liquid samples; all extracts must be analyzed within 40 days.

QUALITY CONTROL A quality control check sample concentrate containing each analyte of interest is required. The QC check sample concentrate may be prepared from pure standard materials or purchased as certified solutions Use appropriate trip, matrix, control site, method, reagent, and solvent blanks. Internal, surrogate, and five concentration level calibration standards are used. The QC check sample concentrate should contain this compound at 100 µg/mL in acetone.

REFERENCE Method 8090, SW-846, 3rd ed., Nov.1986.

Isopropylbenzene   **EPA Method 502**
CAS #98-82-8

TITLE Volatile Organic Compounds in Water By Purge and Trap Capillary Column Gas Chromatography with Photoionization and Electrolytic Conductivity Detectors in Series. U.S. EPA Method 502.2, Rev. 2.0, 1989.

MATRIX Drinking water and raw source water. The latter should include most surface water and groundwater sources.

METHOD SUMMARY This method covers 60 volatile organic compounds that contain halogen atoms and/or that are aromatic. An inert gas (zero grade nitrogen or helium) is bubbled through a 25-mL or a 5-mL water sample (depending on the expected concentration of the analytes). Purged sample components are trapped in a tube of sorbent materials. When purging is complete, the sorbent tube is heated and backflushed with helium to desorb the trapped sample onto a capillary GC column. The column is temperature programmed to separate the method analytes which are then detected with a photoionization detector (PID) and a Hall electrolytic conductivity (HECD) placed in series. The PID is selective for aromatic compounds and the HECD is selective for halogenated compounds.

INTERFERENCES Impurities in the purge gas and from organic compounds outgassing from the plumbing ahead of the trap account for many contamination problems. Interferences purged or coextracted from the samples will vary considerably from source to source, depending upon the particular sample or extract being tested. Cross-contamination can occur whenever high-level and low-level samples are analyzed sequentially. Samples also can be contaminated by diffusion of volatile organics (particularly methylene chloride and fluorocarbons) through the septum seal into the sample during shipment and storage. The lab where volatile analysis is performed and also the refrigerated storage area should be completely free of solvents.

INSTRUMENTATION A GC containing a series configuration of a high temperature photoionization detector (PID) equipped with 10.0 eV (nominal) lamp and Hall electrolytic conductivity detector (HECD) is required. Also required is an all-glass 5-mL purging device, a sorbent trap, and a thermal desorption apparatus which is connected to the GC system.

Column 1: VOCOL glass wide-bore capillary column.
Column 2: RTX–502.2 mega-bore capillary column.
Column 3: DB-62 mega-bore capillary column.

PRECISION & ACCURACY Method detection limits are dependent upon the characteristics of the gas chromatographic system used. Analytes that are not separated chromatographically cannot be individually identified and used in the same calibration mixture or water samples unless an alternative technique for identification and quantification, such as mass spectrometry, is used.

Electrolytic conductivity detetor (c) range in µg/L (a) was 0.02–200.
Electrolytic conductivity detetor (c) MDL in µg/L (b) was not listed.
Electrolytic conductivity detetor (c) accuracy as % recovery was not listed.
Electrolytic conductivity detetor (c) precision as % RSD was not listed.
Photoionization detector (d) range in µg/L (a) was 0.02–200.
Photoionization detector (d) MDL in µg/L (b) was 0.05.
Photoionization detector (d) accuracy as % recovery was 98.
Photoionization detector (d) precision as % RSD was 0.9.

(a) The applicable concentration range of this method is compound, instrument, and matrix-dependent. It is listed as being approximately 0.02 to 200 µg/L but no specific information is provided so caution should be observed.
(b) The method detection limits reports with this method are compound, instrument, and matrix-dependent. The values reported were calculated using reagent water fortified with the corresponding compounds at 10 µg/L and a GC-equipped with a 60 m × 0.75 mm VOLCOL wide bore capillary column with 1.5 µm film thickness and using helium carrier gas.
(c) Recoveries and relative standard deviations were determined from seven samples of reagent water fortified with 10 µg/L of each compound. 2-Bromo-1-chloropropane was used as the internal standard for calculating average recoveries.
(d) Recoveries and relative standard deviations were determined from seven samples of reagent water fortified with 10 µg/L of each compound. Fluorobenzene was used as the internal standard for calculating average recoveries.

SAMPLING METHOD Collect samples using a 40- to 120-mL screw-cap vial (prewashed with detergent, rinsed with distilled water and oven dried at 105°C) with a Teflon®-faced silicone septum. Collect bubble-free samples and place the septum with the Teflon® side down on the water.

SAMPLE PRESERVATION If residual chlorine is present in the water add about 25 mg of ascorbic acid to each vial before samples are collected to remove the chlorine. Add hydrochloric acid to reduce pH to <2, immediately cool samples to 4°C, and store them in a solvent-free refrigerator at 4°C until analysis.

MHT The maximum holding time for samples is 14 days from the time they were collected.

SAMPLE PREPARATION Remove the plungers from two 5-mL syringes and attach a closed syringe valve to each. Warm the sample to room temperature, open the sample bottle, and carefully pour the sample into one of the syringe barrels to just short of overflowing. Replace the syringe plunger, invert the syringe, and compress the sample. Open the syringe valve and vent any residual air while adjusting the sample volume to 5.0 mL. Add 10 µL of the internal calibration standard to the sample through the syringe valve. Close the valve. Fill the second syringe in an identical manner from the same sample bottle. Reserve this second syringe for a reanalysis if necessary.

QUALITY CONTROL As an initial demonstration of lab accuracy and precision, analyze 4 to 7 replicates of a lab fortified blank containing analyte at 0.1–5 µg/L. Collect all samples in duplicate. Surrogate analytes (similar to those of the analytes of interest), whose concentration is known in every sample, are measured using the same internal standard calibration procedure. Duplicate field reagent water blanks (trip blanks) must be analyzed with each set of samples, lab reagent blanks (method blanks) must be analyzed with each batch of samples processed as a group within a work shift. Also, a single lab-fortified blank that contains each of the analytes of interest should be analyzed with each batch of samples processed as a group within a work shift. A 3- to 5-point calibration curve is needed depending on the calibration range factor required.

EPA CONTACT & HOTLINE For technical questions contact Dr. Baldev Bathija, U.S. EPA, Office of Ground Water and Drinking Water (WH-550D), 401 M St. SW, Washington, DC 20460. Tel. (202) 260-3040. For further information the EPA Safe Drinking Water Hotline may be called at: (800) 426-4791.

REFERENCE Methods for the Determination of Organic Compounds in Drinking Water, EPA/600/4-88/039 (revised July 1991; Final Rule for determination of compliance with the MCL for Total Trihalomethanes under 141.30, in 40 CFR Part 141, Vol. 58, No. 147, Fed. Reg., Tuesday Aug. 3, 1993). U.S. EPA Environmental Monitoring Systems Laboratory, Cincinnati, OH, 45268, U.S.A. Available from the National Technical Information Service (NTIS), 5285 Port Royal Road, Springfield, VA 22161; Tel. 800-553-6847. NTIS Order Number is PB91-231480.

Isopropylbenzene
CAS #98-82-8

EPA Method 524

TITLE Measurement of Purgeable Organic Compounds in Water by Capillary Column GC/MS.

MATRIX Drinking water and raw source water; the latter should include most surface water and groundwater sources.

METHOD SUMMARY Method 524.2 covers 60 volatile organic compounds. An inert gas (zero grade nitrogen or helium) is bubbled through a 25-mL or a 5-mL water sample (depending on the expected concentration of the analytes). Purged sample components are trapped in a tube of sorbent materials. When purging is complete, the sorbent tube is heated and backflushed with helium to desorb the trapped sample onto a capillary GC column.

INTERFERENCES Impurities in the purge gas and from organic compounds outgassing from the plumbing ahead of the trap account for many contamination problems. Interferences purged or coextracted from the samples will vary considerably from source to source, depending upon the particular sample or extract being tested. Cross-contamination can occur whenever high-level and low-level samples are analyzed sequentially. Samples also can be contaminated by diffusion of volatile organics (particularly methylene chloride and fluorocarbons) through the septum seal into the sample during shipment and storage.

INSTRUMENTATION A GC/MS with a data system equipped with one of the following capillary GC columns:

Column 1: VOCOL glass wide bore capillary column.
Column 2: DB-624 fused silica capillary column.
Column 3: DB-5 fused silica capillary column.

Also required is an all-glass 25 mL or 5-mL purging device, a sorbent trap, and a thermal desorption apparatus which is connected to the GC/MS system.

PRECISION & ACCURACY Method detection limits are compound- and instrument-dependent, and may vary from approximately 0.02–0.35 µg/L. Note in the table below that the "true" concentration range used for accuracy and precision measurements was quite narrow. However, the applicable concentration range of this method is primarily column dependent and is approximately 0.02 to 200 µg/L for the wide-bore thick-film columns. Narrow-bore thin-film columns may have a capacity which limits the range to about 0.02 to 20 µg/L. Analytes that are inefficiently purged from water will not be detected when present at low concentrations, but they can be measured with acceptable accuracy and precision when present in sufficient amounts.

Analytes that are not separated chromatographically, but which have different mass spectra and non-interfering quantification ions, can be identified and measured in the same calibration mixture or water sample. Analytes which have very similar mass spectra cannot be individually identified and measured in the same calibration mixture or water samples unless they have different retention times. Co-eluting compounds with very similar mass spectra, typically many structural isomers, must be reported as an isomeric group or pair.

The range (in µg/L) was 0.5–10.
The Method Detection Limig (in µg/L) was 0.15.
The accuracy (as % recovery) was 101.
The precision (in %) was 7.6.

Note: Data were obtained from 16–31 determinations using a wide-bore capillary column and a jet separator interfaced to a quadrupole mass spectrometer. All analytes were in a reagent water matrix.

SAMPLING METHOD Collect samples using a 40- to 120-mL screw-cap vial (prewashed with detergent, rinsed with distilled water and oven dried at 105°C) with a Teflon®-faced silicone septum. Collect bubble-free samples and place the septum with the Teflon® side down on the water.

SAMPLE PRESERVATION If residual chlorine is present in the water add about 25 mg of ascorbic acid to each vial before samples are collected to remove the chlorine. Add hydrochloric acid to reduce pH to <2, and immediately cool samples to 4°C, and store them in a solvent-free refrigerator at 4°C until analysis.

MHT The maximum holding time for samples is 14 days from the time they were collected.

SAMPLE PREPARATION Remove the plungers from two 25-mL (or 5-mL depending on sample size) syringes and attach a closed syringe valve to each. Warm the sample to room temperature, open the sample bottle, and carefully pour the sample into one of the syringe barrels to just short of overflowing. Replace the syringe plunger, invert the syringe, and compress the sample. Open the syringe valve and vent any residual air while adjusting the sample volume to 25.0 mL (or 5 mL). For samples and blanks, add 5 µL of the fortification solution containing the internal standard and the surrogates to the sample through the syringe valve. For calibration standards and lab fortified blanks, add 5 µL of the fortification solution containing the internal standard only. Close the valve. Fill the second syringe in an identical manner from the same sample bottle. Reserve this second syringe for a reanalysis if necessary.

QUALITY CONTROL As an initial demonstration of lab accuracy and precision, analyze 4 to 7 replicates of a lab fortified blank containing analyte at 0.2–5 µg/L. Collect all samples in duplicate. Surrogate analytes (similar to those of the analytes of interest), whose concentration is known in every sample, are measured using the same internal standard calibration procedure. Duplicate field reagent water blanks (trip blanks) must be analyzed with each set of samples, lab reagent blanks (method blanks) must be analyzed with each batch of samples processed as a group within a work shift. Also, a single lab-fortified blank that contains each of the analytes of interest should be analyzed with each batch of samples processed as a group within a work shift. A 3- to 5-point calibration curve is needed depending on the calibration range factor required.

EPA CONTACT & HOTLINE For technical questions contact Dr. Baldev Bathija, U.S. EPA, Office of Ground Water and Drinking Water (WH-550D), 401 M St. SW, Washington, DC

20460. Tel. (202) 260-3040. For further information the EPA Safe Drinking Water Hotline may be called at: (800) 426-4791.

REFERENCE Methods for the Determination of Organic Compounds in Drinking Water, EPA/600/4-88/039 (revised July 1991; Final Rule for determination of compliance with the MCL for Total Trihalomethanes under 141.30, in 40 CFR Part 141, Vol. 58, No. 147, Fed. Reg., Tuesday Aug. 3, 1993). U.S. EPA Environmental Monitoring Systems Laboratory, Cincinnati, OH, 45268, U.S.A. Available from the National Technical Information Service (NTIS), 5285 Port Royal Road, Springfield, VA 22161; Tel. 800-553-6847. NTIS Order Number is PB91-231480.

Isopropylbenzene   **EPA Method 8021**
CAS #98-82-8

TITLE Halogenated Volatile by Gas Chromatography Using Photoionization and Electrolytic Conductivity Detectors in Series: Capillary Column Technique

MATRIX This method is applicable to nearly all types of samples, regardless of water content, including groundwater, aqueous sludges, caustic liquors, acid liquors, waste solvents, oily wastes, mousses, tars, fibrous wastes, polymeric emulsions, filter cakes, spent carbons, spent catalysts, soils, and sediments.

METHOD SUMMARY This method is used to determine 60 volatile organic compounds in a variety of solid waste matrices. It provides GC conditions for the detection of halogenated and aromatic volatile organic compounds. Samples can be analyzed using direct injection or purge-and-trap (EPA Method 5030). Groundwater samples must be analyzed using EPA Method 5030 (where applicable). A temperature program is used with the GC. Detection is achieved by a photoionization detector (PID) and a Hall electrolytic conductivity detector (HECD) in series.

INTERFERENCES Samples can be contaminated by diffusion of volatile organics (particularly chlorofluorocarbons and methylene chloride) through the sample container septum during shipment and storage.

INSTRUMENTATION A GC-equipped with variable-constant differential flow controllers, subambient oven controller, PID and HECD detectors connected with a short piece of uncoated capillary tubing and a data system.

Column: 60 m × 0.75 mm I.D. VOCOL wide-bore capillary column with 1.5 µm film thickness.

PRECISION & ACCURACY MDLs are compound-dependent and vary with purging efficiency and concentration. The applicable concentration range of this method is compound- and instrument-dependent but is approximately 0.1 to 200 µg/L. Analytes that are inefficiently purged from water will not be detected when present at low concentrations, but they can be measured with acceptable accuracy and precision when present in sufficient amounts. The estimated quantitation limit (EQL) for an individual compound is approximately 1 µg/kg (wet weight) for soil/sediment samples, 100 µg/kg (wet weight) for wastes, and 1 µg/L for groundwater. EQLs will be proportionately higher for sample extracts and samples that require dilution or reduced sample size to avoid saturation of the detector.

MULTIPLICATION FACTORS FOR OTHER MATRICES (a)

Matrix	Factor (b)
Groundwater	10
Low-concentration soil	10
Water miscible liquid waste	500
High-concentration soil and sludge	1250
Non-water miscible waste	1250

(a) Sample EQLs are highly matrix-dependent. The EQLs listed herein are provided for guidance and may not always be achievable. (b) EQL = [Method detection limit] × [Factor]. For non-aqueous samples, the factor is on a wet-weight basis.

SINGLE LABORATORY ACCURACY & PRECISION DATA FOR VOCs IN WATER

This method was tested in a single lab using water spiked at 10 µg/L and the following data was reported:

Recoveries and standard deviations were determined from seven samples and spiked at 10 µg/L of each analyte. Recoveries were determined by the internal standard method. Internal standards were: Fluorobenzene for PID and 2-Bromo-1-chloropropane for HECD.

The average recovery (in percent) for the PID was 98.
The standard deviation of the recovery for the PID was 0.9.
The MDL (in µg/mL) for the PID was 0.05.
The average recovery (in percent) for the HECD was none (no response for this detector).
The standard deviation of the recovery for the HECD was none (no response for this detector)-.
The MDL (in µg/mL) for the HECD was none (no response for this detector).

SAMPLE COLLECTION, PRESERVATION & HANDLING
Volatile Organics — Standard 40-mL glass screw-cap VOA vials with Teflon®-faced silicone septum may be used for both liquid and solid matrices. When collecting samples, liquids and solids should be introduced into the vials gently to reduce agitation which might drive off volatile compounds. If there are any air bubbles present the sample must be retaken. Tap slightly as they are filled to try and eliminate as much free air space as possible. The two vials from each sampling locations should be sealed in separate plastic bags to prevent cross-contamination between samples particularly if the sampled waste is suspected of containing high levels of volatile organics.

Semivolatile organics — Containers used to collect samples for the determination of semivolatile organic compounds should be soap and water washed followed by methanol (or isopropanol) rinsing. The sample containers should be of glass or Teflon® and have screw-top covers with Teflon® liners.

Preservation for volatile organics — No preservation is used with concentrated waste samples. With liquid samples containing no residual chlorine, 4 drops of concentrated hydrochloric acid are added and the samples are immediately cooled to 4°C. When liquid samples contain residual chlorine, they are treated as above and, in addition, 4 drops of 4% aqueous sodium thiosulfate are added. Soil, sediment, and sludge samples are only cooled to 4°C.

Preservation for semivolatile organics — No preservation is used with concentrated waste samples. With liquid samples containing no residual chlorine and with soil, sediment, and sludge samples, immediately cooling to 4°C is the only preservation used. When residual chlorine is present then 3 mL of 10% aqueous sodium sulfate is added for each gallon of sample collected, followed by cooling to 4°C.

MHT The holding time for all volatile organics samples is 14 days. Liquid samples must be extracted within 7 days and their extracts analyzed within 40 days. Concentrated waste, soil, sediment, and sludge samples must be extracted within 14 days and their extracts analyzed within 40 days.

SAMPLE PREPARATION Volatile compounds are introduced into the gas chromatograph either by direct injector or purge-and-trap (EPA Method 5030). EPA Method 5030 may be used directly on groundwater samples or low-concentration contaminated soils and sediments. For medium-concentration soils or sediments, methanolic extraction, as described in EPA Method 5030, may be necessary prior to purge-and-trap analysis.

QUALITY CONTROL Calculate surrogate standard recovery on all samples, blanks, and spikes. A trip blank is recommended to check on sampling, storage, and handling contamination. Calibration standards, at a minimum of five concentration levels, are prepared in organic-free reagent water. One of the concentration levels should be at a concentration near, but above, the method detection limit.

A combination of bromochloromethane, 2-bromo-1-chloropropane, 1,4-dichlorobutane, and bromochlorobenzene are recommended as surrogate standards to encompass the range of the temperature program used in this method.

REFERENCE Test Methods for Evaluating Solid Waste, Physical/Chemical Methods, SW-846, 3rd Edition, U.S. EPA, Office of Solid Waste, Washington, DC, EPA Method 8021A, Rev. 1, Nov. 1992.

Isopropylbenzene **EPA Method 8260**
CAS #98-82-8

TITLE Volatile Organic Compounds by GC/MS: Capillary Column Technique

MATRIX This method is applicable to nearly all types of samples, regardless of water content, including groundwater, soils, and sediments.

METHOD SUMMARY Method 8260A covers 58 volatile organic compounds that are introduced into a gas chromatograph by the purge-and-trap method or by direct injection (in limited applications). Zero-grade helium is bubbled through a 5-mL solution at ambient temperature. Purged sample components are trapped in a tube containing suitable sorbent materials. When purging is complete, the sorbent tube is heated and backflushed with helium to desorb trapped sample components. The analytes are desorbed directly to a large bore capillary or cryofocussed on a capillary precolumn before being flash evaporated to a narrow bore capillary for analysis.

INTERFERENCES Major contaminant sources are volatile materials in the lab and impurities in the inert purging gas and in the sorbent trap. Interfering contamination may occur when a sample containing low concentrations of volatile organic compounds is analyzed immediately after a sample containing high concentrations of volatile organic compounds. After analysis of a sample containing high concentrations of volatile organic compounds, one or more calibration blanks should be analyzed to check for cross-contamination. Screening of the samples prior to purge-and-trap GC/MS analysis is highly recommended to prevent contamination of the system. This is especially true for soil and waste samples.

Special precautions must be taken to analyze for methylene chloride. The analytical and sample storage area should be isolated from all atmospheric sources of methylene chloride. All gas chromatography carrier gas lines and purge gas plumbing should be constructed from stainless steel or copper tubing. Laboratory clothing previously exposed to methylene chloride fumes during liquid-liquid extraction procedures can contribute to sample contamination.

Samples can also be contaminated by diffusion of volatile organics (particularly methylene chloride and fluorocarbons) through the septum seal during shipment and storage. A trip blank can serve as a check on such contamination.

INSTRUMENTATION GC/MS with a temperature-programmable chromatograph suitable for splitless injection equipped with variable constant differential flow controllers, a subambient oven controller, a purging device, sorbent trap, a thermal desorption apparatus and a capillary precolumn interface when using cryogenic cooling will be needed. The following GC columns may be used:

Column 1: 60 m × 0.75 mm I.D. capillary column coated with VOCOL, 1.5 μm film thickness.
Column 2: 30 m × 0.53 mm capillary column coated with DB-624 or VOCOL, 3 μm film thickness.
Column 3: 30 m × 0.32 mm I.D. capillary column coated with DB-5 or SE-54, 1-μm film thickness.

PRECISION & ACCURACY This method has been tested in a single lab using spiked water. Using a wide-bore capillary column, water was spiked at concentrations between 0.5 and 10 μg/L. Single lab accuracy and precision data are presented. The MDL actually achieved in a given analysis will vary depending on instrument sensitivity and matrix effects.

The MDL (a) in μg/L was 0.15.
The concentration range in μg/L was 0.5–10.
The mean accuracy (% of true value) was 101.
The precision as relative standard deviation was 7.6.

Note: The MDL is based on a 25-mL sample volume instead of a 5-mL sample volume.

SAMPLING METHOD

Liquid samples — Use a 40-mL glass screw-cap VOA vial with a Teflon®-faced silicone septum that has been prewashed, rinsed with distilled deionized water, and oven dried. If residual chlorine is present, collect the sample in a 4-oz soil VOA container which has been pre-preserved with 4 drops of 10% sodium thiosulfate. Mix gently and transfer the sample to a 40-mL VOA vial. Collect bubble-free samples in duplicate and seal each sample in a separate plastic bag.

Soils, sediments and sludges — Use an 8-oz widemouth glass bottle with Teflon®-faced silicone septum that has been pre-washed, rinsed with distilled deionized water, and oven dried. **Do not** heat the septum for more than 1 h. Tap slightly to eliminate any free air space. Collect samples in duplicate and seal each one in a separate plastic bag.

SAMPLE PRESERVATION
Liquid samples — Add 4 drops of concentrated HCL, cool to 4°C and store in a solvent-free refrigerator.

Soils, sediments and sludges — Cool samples to 4°C and store in a solvent-free refrigerator.

MHT The maximum holding time of any sample (liquids, soils, sediments, and sludges) is 14 days.

SAMPLE PREPARATION
Liquid samples — Remove the plunger from a 5-mL syringe and carefully pour the sample into the syringe barrel to just short of overflowing. Replace the syringe plunger and compress the sample. Open the syringe valve and vent any residual air while adjusting the sample volume to 5.0 mL. If there is only one volatile organic analysis (VOA) vial, a second syringe should be filled at this time to protect against possible loss of sample integrity. Add 10 µL of surrogate spiking solution and 10 µL of internal standard spiking solution through the valve bore of the 5-mL syringe, then close the valve. The surrogate and internal standards may be mixed and added as a single spiking solution.

Sediments, soils, and waste samples — All samples of this type should be screened by GC analysis using a headspace method (EPA Method 3810) or the hexadecane extraction and screening method (EPA Method 3820). Use the screening data to determine whether to use the low-concentration method (0.005–1 mg/kg) or the high-concentration method (>1 mg/kg).

Low-concentration method — The low-concentration method is based on purging a heated sediment or soil sample mixed with organic-free reagent water containing the surrogate and internal standards. Analyze all reagent blanks and standards under the same conditions as the samples.

Use a 5-g sample if the expected concentration is <0.1 mg/kg or a 1-g sample for expected concentrations between 0.1 and 1 mg/kg. Mix the contents of the sample container with a narrow metal spatula. Weigh the amount of the sample into a tared purge device. Add the spiked water to the purge device, which contains the weighed amount of sample, and connect the device to the purge-and-trap system.

High-concentration method — This method is based on extracting the sediment or soil with methanol. A waste sample is either extracted or diluted, depending on its solubility in methanol. Wastes that are insoluble in methanol are diluted with reagent tetraglyme or possibly polyethylene glycol (PEG).

An aliquot of the extract is added to organic-free reagent water containing surrogate and internal standards. This is purged at ambient temperature. All samples with an expected concentration of >1.0 mg/kg should be analyzed by this method.

Mix the contents of the sample container with a narrow metal spatula. For sediments or soils and solid wastes that are insoluble in methanol, weigh 4 g (wet weight) of sample into a tared 20-mL vial. For waste that is soluble in methanol, tetraglyme, or PEG, weigh 1 g (wet weight) into a tared scintillation vial or culture tube or a 10-mL volumetric flask. Quickly add 9.0 mL of appropriate solvent then add 1.0 mL of a surrogate spiking solution to the vial, cap it, and shake it for 2 min.

METHANOL EXTRACT REQUIRED FOR ANALYSIS OF HIGH-CONCENTRATION SOILS OR SEDIMENTS

Approximate Concentration Range	Volume of Methanol Extract (a)
500–10,000 µg/kg	100 µL
1,000–20,000 µg/kg	50 µL
5,000–100,000 µg/kg	10 µL
25,000–500,000 µg/kg	100 µL of 1/50 dilution (b)

Calculate appropriate dilution factor for concentrations exceeding this table.

(a) The volume of methanol added to 5 mL of water being purged should be kept constant. Therefore, add to the 5-mL syringe whatever volume of methanol is necessary to maintain a volume of 100 µL added to the syringe.
(b) Dilute an aliquot of the methanol extract and then take 100 µL for analysis.

QUALITY CONTROL Demonstrate, through the analysis of a reagent water blank, that interferences from the analytical system, glassware, and reagents are under control. Blank samples should be carried through all stages of the sample preparation and measurement steps. For each analytical batch (up to 20 samples), a reagent blank, matrix spike, and matrix spike duplicate must be analyzed (the frequency of the spikes may be different for different monitoring programs). The blank and spiked samples must be carried through all stages of the sample preparation and measurement steps. QC samples mentioned in the section on Interferences will also be needed as appropriate to those situations.

Matrix spiking standards should be prepared from volatile organic compounds which will be representative of the compounds being investigated. The recommended internal standards are chlorobenzene-d5, 1,4-difluorobenzene, 1,4-dichlorobenzene-d4, and pentafluorobenzene. Using stock standard solutions, prepare secondary dilution standards containing the compounds of interest, either singly or mixed together in methanol. Store them in a vial with no headspace for no more than one week. Surrogates recommended are toluene-d8, 4-bromofluorobenzene, and dibromofluoromethane. Each sample undergoing GC/MS analysis must be spiked with 10 µL of the surrogate spiking solution prior to analysis.

REFERENCE Test Methods for Evaluating Solid Waste (SW-846). U.S. EPA 1983, Method 8260A, Rev. 1, Nov. 1990. Office of Solid Waste, Washington, DC.

Isopropylbenzene **EPA Method 503.1**
CAS #98-82-8

TITLE Aromatic & Unsaturated VOCs in Water

MATRIX Drinking water (finished or any treatment stage) and raw source water.

APPLICATION Method covers 28 aromatic and unsaturated VOCs. An inert gas is bubbled through a 5-mL water sample. Purged sample components are trapped in tube of sorbent materials. When purging is complete, sorbent tube is heated and backflushed with inert gas to desorb trapped sample onto a packed GC column.

INTERFERENCES During analysis, major contaminant sources are volatile materials in the lab and impurities in purging gas and sorbent trap. With high and low level samples, there can be carryover contamination. Excess water causes a negative baseline deflection.

INSTRUMENTATION Purge and Trap GC w/photoionization detector. (Two GC columns are recommended); Column 1: 5% SP-1200 and 1.75% Bentone 34 on Supelcoport; Column 2: 1,2,3-tris(2-cyanoethoxy)propane on Chromosorb W.

RANGE 2.2–600 µg/L. (Drinking water)

MDL 0.005 µg/L in water

PRECISION RSD = 8.7% at 0.40 µg/L conc.; 7 samples

ACCURACY Average recovery = 88% at 0.40 µg/L conc.; 7 samples

SAMPLING METHOD Use a 40–120-mL screw-cap vial (prewashed with detergent, rinsed with distilled water and oven dried at 105°C) with a PTFE-faced silicone septum. If residual chlorine is in the water add about 25 mg of ascorbic acid to each vial before sample collection. Collect bubble-free samples.

STABILITY Cool to 4°C; HCl to pH <2.

MHT 14 days.

QUALITY CONTROL As initial demonstration of lab accuracy and precision, analyze 4 to 7 replicates of a lab fortified blank containing the analyte at 0.1–5 µg/L. Collect all samples in duplicate.

REFERENCE Method 503.1, Volatile Aromatic & Unsaturated Organic Compounds in H2O by Purge and Trap GC, EPA 600/4-88/039.

2-Isopropylnaphthalene **EPA Method 1625**
CAS #2027-17-0

TITLE Semivolatile Organic Compounds by Isotope Dilution GC/MS

MATRIX The compounds may be determined in waters, soils, and municipal sludges by this method.

METHOD SUMMARY This method is used to determine 176 semivolatile toxic organic pollutants associated with the CWA (as amended 1987); the RCRA (as amended 1986); the CERCLA (as amended 1986); and other compounds amenable to extraction and analysis by capillary column gas chromatography-mass spectrometry (GC/MS).

Stable isotopically-labeled analogs of the compounds of interest are added to the sample. If the solids content is less than 1%, a 1-L sample is extracted at pH 12–13, then at pH <2 with methylene chloride using continuous extraction techniques.

If the solids content is 30% or less, the sample is diluted to 1% solids with reagent water, homogenized ultrasonically, and extracted at pH 12–13, then at pH <2 with methylene chloride using continuous extraction techniques. If the solids content is greater than 30%, the sample is extracted using ultrasonic techniques.

Each extract is dried over sodium sulfate, concentrated to a volume of 5 mL, cleaned up using GPC, if necessary, and concentrated. Extracts are concentrated to 1 mL if GPC is not performed, and to 0.5 mL if GPC is performed.

An internal standard is added to the extract, and a 1-mL aliquot of the extract is injected into the GC. The compounds are separated by GC and detected by a MS. The labeled compounds serve to correct the variability of the analytical technique.

INTERFERENCES Solvents, reagents, glassware, and other sample processing hardware may yield artifacts and/or elevated baselines causing misinterpretation of chromatograms and spectra. Materials used in the analysis must be demonstrated to be free from interferences under the conditions of analysis by running method blanks initially and with each sample lot (sample started through the extraction process on a given 8-h shift, to a maximum of 20). Specific selection of reagents and purification of solvents by distillation in all glass systems may be required. Glassware and, where possible, reagents are cleaned by solvent rinse and baking at 450°C for 1-h minimum. Interferences coextracted from samples will vary considerably from source to source, depending on the diversity of the site being sampled.

INSTRUMENTATION Major instrumentation includes a GC with a splitless or on-column injection port for capillary column, a MS with 70 eV electron impact ionization, and a data system to collect and record MS data, and process it. A K-D apparatus is used to concentrate extracts.

GC Column: 30 m × 0.25 mm I.D. 5% phenyl, 94% methyl, 1% vinyl silicone bonded phased fused silica capillary column.

PRECISION & ACCURACY The detection limits of the method are usually dependent on the level of interferences rather than instrumental limitations. The limits typify the minimum quantities that can be detected with no interferences present.

The minimum level (in µg/mL) was not listed. This is defined as a minimum level at which the analytical system shall give recognizable mass spectra (background corrected) and acceptable calibration points.

The MDL (in μg/kg) in low solids was not listed and in high solids was not listed; these were determined in digested sludge (low solids) and in filter cake or compost (high solids).

The labeled and native compound initial precision as standard deviation (in μg/L) was not listed.

The labeled and native compound initial accuracy as average recovery (in μg/L) was not listed.

SAMPLE COLLECTION, PRESERVATION & HANDLING
Collect samples in glass containers. Aqueous samples which flow freely are collected in refrigerated bottles using automatic sampling equipment. Solid samples are collected as grab samples using widemouth jars. Maintain samples at 0 to 4°C from the time of collection until extraction. If residual chlorine is present in aqueous samples, add 80 mg sodium thiosulfate/L of water. Begin sample extraction within 7 days of collection, and analyze all extracts within 40 days of extraction.

SAMPLE PREPARATION Samples containing 1% solids or less are extracted directly using continuous liquid-liquid extraction techniques. Samples containing 1 to 30% solids are diluted to the 1% level with reagent water and extracted using continuous liquid-liquid extraction techniques. Samples containing greater than 30% solids are extracted using ultrasonic techniques.

- Base/neutral extraction — Adjust the pH of the waters in the extractors to 12–13 with 6 N NaOH. Extract with methylene chloride for 24–48 h.
- Acid extraction — Adjust the pH of the waters in the extractors to 2 or less using 6 N sulfuric acid. Extract with methylene chloride for 24–48 h.
- Ultrasonic extraction of high solids samples — Add anhydrous sodium sulfate to the sample and QC aliquot(s). Add acetone:methylene chloride (1:1) to the sample and mix thoroughly

Concentrate extracts using a K-D apparatus.

QUALITY CONTROL The analyst is permitted to modify this method to improve separations or lower the costs of measurements, provided all performance specifications are met. Analyses of blanks are required to demonstrate freedom from contamination. When results of spikes indicate atypical method performance for samples, the samples are diluted to bring method performance within acceptable limits.

For low solids (aqueous samples), extract, concentrate, and analyze two sets of four 1-L aliquots (8 aliquots total) of the precision and recovery standard. For high solids samples, two sets of four 30-g aliquots of the high solids reference matrix are used.

Spike all samples with labeled compounds to assess method performance. Compute percent recovery of the labeled compounds using the internal standard method. Compare the labeled compound recovery for each compound with the corresponding labeled compound recovery.

Reagent water and high solids reference matrix blanks are analyzed to demonstrate freedom from contamination. Extract and concentrate a 1-L reagent water blank or a high solids reference matrix blank with each sample's lot (samples started through the extraction process on the same 8-h shift, to a maximum of 20 samples).

Field replicates may be collected to determine the precision of the sampling technique, and spiked samples may be required to determine the accuracy of the analysis when the internal standard method is used.

REFERENCE Semivolatile Organic Compounds by Isotope Dilution GC/MS. Office of Water Regulation and Standards, U.S. EPA Industrial Technology Division, Washington, DC, EPA Method 1625, Rev. C, June 1989 (contact W.A. Telliard, U.S. EPA, Office of Water Regulations and Standards, 401 M St., SW, Washington, DC, 20460. Phone: 202-382-7131).

4-Isopropyltoluene   **EPA Method 502**
CAS #99-87-6

TITLE Volatile Organic Compounds in Water By Purge and Trap Capillary Column Gas Chromatography with Photoionization and Electrolytic Conductivity Detectors in Series. U.S. EPA Method 502.2, Rev. 2.0, 1989.

MATRIX Drinking water and raw source water. The latter should include most surface water and groundwater sources.

METHOD SUMMARY This method covers 60 volatile organic compounds that contain halogen atoms and/or that are aromatic. An inert gas (zero grade nitrogen or helium) is bubbled through a 25-mL or a 5-mL water sample (depending on the expected concentration of the analytes). Purged sample components are trapped in a tube of sorbent materials. When purging is complete, the sorbent tube is heated and backflushed with helium to desorb the trapped sample onto a capillary GC column. The column is temperature programmed to separate the method analytes which are then detected with a photoionization detector (PID) and a Hall electrolytic conductivity (HECD) placed in series. The PID is selective for aromatic compounds and the HECD is selective for halogenated compounds.

INTERFERENCES Impurities in the purge gas and from organic compounds outgassing from the plumbing ahead of the trap account for many contamination problems. Interferences purged or coextracted from the samples will vary considerably from source to source, depending upon the particular sample or extract being tested. Cross-contamination can occur whenever high-level and low-level samples are analyzed sequentially. Samples also can be contaminated by diffusion of volatile organics (particularly methylene chloride and fluorocarbons) through the septum seal into the sample during shipment and storage. The lab where volatile analysis is performed and also the refrigerated storage area should be completely free of solvents.

INSTRUMENTATION A GC containing a series configuration of a high temperature photoionization detector (PID) equipped with 10.0 eV (nominal) lamp and Hall electrolytic conductivity detector (HECD) is required. Also required is an all-glass 5-mL purging device, a sorbent trap, and a thermal desorption apparatus which is connected to the GC system.

Column 1: VOCOL glass wide-bore capillary column.
Column 2: RTX–502.2 mega-bore capillary column.
Column 3: DB-62 mega-bore capillary column.

PRECISION & ACCURACY Method detection limits are dependent upon the characteristics of the gas chromatographic system used. Analytes that are not separated chromatographically cannot be individually identified and used in the same calibration mixture or water samples unless an alternative technique for identification and quantification, such as mass spectrometry, is used.

Electrolytic conductivity detetor (c) range in µg/L (a) was 0.02–200.

Electrolytic conductivity detetor (c) MDL in µg/L (b) was not listed.

Electrolytic conductivity detetor (c) accuracy as % recovery was not listed.

Electrolytic conductivity detetor (c) precision as % RSD was not listed.

Photoionization detector (d) range in µg/L (a) was 0.02–200.
Photoionization detector (d) MDL in µg/L (b) was 0.01.
Photoionization detector (d) accuracy as % recovery was 98.
Photoionization detector (d) precision as % RSD was 2.4.

(a) *The applicable concentration range of this method is compound, instrument, and matrix-dependent. It is listed as being approximately 0.02 to 200 µg/L but no specific information is provided so caution should be observed.*
(b) *The method detection limits reports with this method are compound, instrument, and matrix-dependent. The values reported were calculated using reagent water fortified with the corresponding compounds at 10 µg/L and a GC-equipped with a 60 m × 0.75 mm VOLCOL wide bore capillary column with 1.5 µm film thickness and using helium carrier gas.*
(c) *Recoveries and relative standard deviations were determined from seven samples of reagent water fortified with 10 µg/L of each compound. 2-Bromo-1-chloropropane was used as the internal standard for calculating average recoveries.*
(d) *Recoveries and relative standard deviations were determined from seven samples of reagent water fortified with 10 µg/L of each compound. Fluorobenzene was used as the internal standard for calculating average recoveries.*

SAMPLING METHOD Collect samples using a 40- to 120-mL screw-cap vial (prewashed with detergent, rinsed with distilled water and oven dried at 105°C) with a Teflon®-faced silicone septum. Collect bubble-free samples and place the septum with the Teflon® side down on the water.

SAMPLE PRESERVATION If residual chlorine is present in the water add about 25 mg of ascorbic acid to each vial before samples are collected to remove the chlorine. Add hydrochloric acid to reduce pH to <2, immediately cool samples to 4°C, and store them in a solvent-free refrigerator at 4°C until analysis.

MHT The maximum holding time for samples is 14 days from the time they were collected.

SAMPLE PREPARATION Remove the plungers from two 5-mL syringes and attach a closed syringe valve to each. Warm the sample to room temperature, open the sample bottle, and carefully pour the sample into one of the syringe barrels to just short of overflowing. Replace the syringe plunger, invert the syringe, and compress the sample. Open the syringe valve and vent any residual air while adjusting the sample volume to 5.0 mL. Add 10 µL of the internal calibration standard to the sample through the syringe valve. Close the valve. Fill the second syringe in an identical manner from the same sample bottle. Reserve this second syringe for a reanalysis if necessary.

QUALITY CONTROL As an initial demonstration of lab accuracy and precision, analyze 4 to 7 replicates of a lab fortified blank containing analyte at 0.1–5 µg/L. Collect all samples in duplicate. Surrogate analytes (similar to those of the analytes of interest), whose concentration is known in every sample, are measured using the same internal standard calibration procedure. Duplicate field reagent water blanks (trip blanks) must be analyzed with each set of samples, lab reagent blanks (method blanks) must be analyzed with each batch of samples processed as a group within a work shift. Also, a single lab-fortified blank that contains each of the analytes of interest should be analyzed with each batch of samples processed as a group within a work shift. A 3- to 5-point calibration curve is needed depending on the calibration range factor required.

EPA CONTACT & HOTLINE For technical questions contact Dr. Baldev Bathija, U.S. EPA, Office of Ground Water and Drinking Water (WH-550D), 401 M St. SW, Washington, DC 20460. Tel. (202) 260-3040. For further information the EPA Safe Drinking Water Hotline may be called at: (800) 426-4791.

REFERENCE Methods for the Determination of Organic Compounds in Drinking Water, EPA/600/4-88/039 (revised July 1991; Final Rule for determination of compliance with the MCL for Total Trihalomethanes under 141.30, in 40 CFR Part 141, Vol. 58, No. 147, Fed. Reg., Tuesday Aug. 3, 1993). U.S. EPA Environmental Monitoring Systems Laboratory, Cincinnati, OH, 45268, U.S.A. Available from the National Technical Information Service (NTIS), 5285 Port Royal Road, Springfield, VA 22161; Tel. 800-553-6847. NTIS Order Number is PB91-231480.

4-Isopropyltoluene   **EPA Method 524**
CAS #99-87-6

TITLE Measurement of Purgeable Organic Compounds in Water by Capillary Column GC/MS.

MATRIX Drinking water and raw source water; the latter should include most surface water and groundwater sources.

METHOD SUMMARY Method 524.2 covers 60 volatile organic compounds. An inert gas (zero grade nitrogen or helium) is bubbled through a 25-mL or a 5-mL water sample (depending on the expected concentration of the analytes). Purged sample components are trapped in a tube of sorbent materials. When purging is complete, the sorbent tube is heated and backflushed with helium to desorb the trapped sample onto a capillary GC column.

INTERFERENCES Impurities in the purge gas and from organic compounds outgassing from the plumbing ahead of

the trap account for many contamination problems. Interferences purged or coextracted from the samples will vary considerably from source to source, depending upon the particular sample or extract being tested. Cross-contamination can occur whenever high-level and low-level samples are analyzed sequentially. Samples also can be contaminated by diffusion of volatile organics (particularly methylene chloride and fluorocarbons) through the septum seal into the sample during shipment and storage.

INSTRUMENTATION A GC/MS with a data system equipped with one of the following capillary GC columns:

Column 1: VOCOL glass wide bore capillary column.
Column 2: DB-624 fused silica capillary column.
Column 3: DB-5 fused silica capillary column.

Also required is an all-glass 25 mL or 5-mL purging device, a sorbent trap, and a thermal desorption apparatus which is connected to the GC/MS system.

PRECISION & ACCURACY Method detection limits are compound- and instrument-dependent, and may vary from approximately 0.02–0.35 µg/L. Note in the table below that the "true" concentration range used for accuracy and precision measurements was quite narrow. However, the applicable concentration range of this method is primarily column dependent and is approximately 0.02 to 200 µg/L for the wide-bore thick-film columns. Narrow-bore thin-film columns may have a capacity which limits the range to about 0.02 to 20 µg/L. Analytes that are inefficiently purged from water will not be detected when present at low concentrations, but they can be measured with acceptable accuracy and precision when present in sufficient amounts.

Analytes that are not separated chromatographically, but which have different mass spectra and non-interfering quantification ions, can be identified and measured in the same calibration mixture or water sample. Analytes which have very similar mass spectra cannot be individually identified and measured in the same calibration mixture or water samples unless they have different retention times. Co-eluting compounds with very similar mass spectra, typically many structural isomers, must be reported as an isomeric group or pair.

The range (in µg/L) was 0.1–10.
The Method Detection Limig (in µg/L) was 0.12.
The accuracy (as % recovery) was 99.
The precision (in %) was 6.7.

Note: Data were obtained from 16–31 determinations using a wide-bore capillary column and a jet separator interfaced to a quadrupole mass spectrometer. All analytes were in a reagent water matrix.

SAMPLING METHOD Collect samples using a 40- to 120-mL screw-cap vial (prewashed with detergent, rinsed with distilled water and oven dried at 105°C) with a Teflon®-faced silicone septum. Collect bubble-free samples and place the septum with the Teflon® side down on the water.

SAMPLE PRESERVATION If residual chlorine is present in the water add about 25 mg of ascorbic acid to each vial before samples are collected to remove the chlorine. Add hydrochloric acid to reduce pH to <2, and immediately cool samples to 4°C, and store them in a solvent-free refrigerator at 4°C until analysis.

MHT The maximum holding time for samples is 14 days from the time they were collected.

SAMPLE PREPARATION Remove the plungers from two 25-mL (or 5-mL depending on sample size) syringes and attach a closed syringe valve to each. Warm the sample to room temperature, open the sample bottle, and carefully pour the sample into one of the syringe barrels to just short of overflowing. Replace the syringe plunger, invert the syringe, and compress the sample. Open the syringe valve and vent any residual air while adjusting the sample volume to 25.0 mL (or 5 mL). For samples and blanks, add 5 µL of the fortification solution containing the internal standard and the surrogates to the sample through the syringe valve. For calibration standards and lab fortified blanks, add 5 µL of the fortification solution containing the internal standard only. Close the valve. Fill the second syringe in an identical manner from the same sample bottle. Reserve this second syringe for a reanalysis if necessary.

QUALITY CONTROL As an initial demonstration of lab accuracy and precision, analyze 4 to 7 replicates of a lab fortified blank containing analyte at 0.2–5 µg/L. Collect all samples in duplicate. Surrogate analytes (similar to those of the analytes of interest), whose concentration is known in every sample, are measured using the same internal standard calibration procedure. Duplicate field reagent water blanks (trip blanks) must be analyzed with each set of samples, lab reagent blanks (method blanks) must be analyzed with each batch of samples processed as a group within a work shift. Also, a single lab-fortified blank that contains each of the analytes of interest should be analyzed with each batch of samples processed as a group within a work shift. A 3- to 5-point calibration curve is needed depending on the calibration range factor required.

EPA CONTACT & HOTLINE For technical questions contact Dr. Baldev Bathija, U.S. EPA, Office of Ground Water and Drinking Water (WH-550D), 401 M St. SW, Washington, DC 20460. Tel. (202) 260-3040. For further information the EPA Safe Drinking Water Hotline may be called at: (800) 426-4791.

REFERENCE Methods for the Determination of Organic Compounds in Drinking Water, EPA/600/4-88/039 (revised July 1991; Final Rule for determination of compliance with the MCL for Total Trihalomethanes under 141.30, in 40 CFR Part 141, Vol. 58, No. 147, Fed. Reg., Tuesday Aug. 3, 1993). U.S. EPA Environmental Monitoring Systems Laboratory, Cincinnati, OH, 45268, U.S.A. Available from the National Technical Information Service (NTIS), 5285 Port Royal Road, Springfield, VA 22161; Tel. 800-553-6847. NTIS Order Number is PB91-231480.

4-Isopropyltoluene   EPA Method 8021
CAS #99-87-6

TITLE Halogenated Volatile by Gas Chromatography Using Photoionization and Electrolytic Conductivity Detectors in Series: Capillary Column Technique

MATRIX This method is applicable to nearly all types of samples, regardless of water content, including groundwater, aqueous sludges, caustic liquors, acid liquors, waste solvents, oily wastes, mousses, tars, fibrous wastes, polymeric emulsions, filter cakes, spent carbons, spent catalysts, soils, and sediments.

METHOD SUMMARY This method is used to determine 60 volatile organic compounds in a variety of solid waste matrices. It provides GC conditions for the detection of halogenated and aromatic volatile organic compounds. Samples can be analyzed using direct injection or purge-and-trap (EPA Method 5030). Groundwater samples must be analyzed using EPA Method 5030 (where applicable). A temperature program is used with the GC. Detection is achieved by a photoionization detector (PID) and a Hall electrolytic conductivity detector (HECD) in series.

INTERFERENCES Samples can be contaminated by diffusion of volatile organics (particularly chlorofluorocarbons and methylene chloride) through the sample container septum during shipment and storage.

INSTRUMENTATION A GC-equipped with variable-constant differential flow controllers, subambient oven controller, PID and HECD detectors connected with a short piece of uncoated capillary tubing and a data system.

Column: 60 m × 0.75 mm I.D. VOCOL wide-bore capillary column with 1.5 µm film thickness.

PRECISION & ACCURACY MDLs are compound-dependent and vary with purging efficiency and concentration. The applicable concentration range of this method is compound- and instrument-dependent but is approximately 0.1 to 200 µg/L. Analytes that are inefficiently purged from water will not be detected when present at low concentrations, but they can be measured with acceptable accuracy and precision when present in sufficient amounts. The estimated quantitation limit (EQL) for an individual compound is approximately 1 µg/kg (wet weight) for soil/sediment samples, 100 µg/kg (wet weight) for wastes, and 1 µg/L for groundwater. EQLs will be proportionately higher for sample extracts and samples that require dilution or reduced sample size to avoid saturation of the detector.

MULTIPLICATION FACTORS FOR OTHER MATRICES (a)

Matrix	Factor (b)
Groundwater	10
Low-concentration soil	10
Water miscible liquid waste	500
High-concentration soil and sludge	1250
Non-water miscible waste	1250

(a) Sample EQLs are highly matrix-dependent. The EQLs listed herein are provided for guidance and may not always be achievable. (b) EQL = [Method detection limit] × [Factor]. For non-aqueous samples, the factor is on a wet-weight basis.

SINGLE LABORATORY ACCURACY & PRECISION DATA FOR VOCs IN WATER

This method was tested in a single lab using water spiked at 10 µg/L and the following data was reported:

Recoveries and standard deviations were determined from seven samples and spiked at 10 µg/L of each analyte. Recoveries were determined by the internal standard method. Internal standards were: Fluorobenzene for PID and 2-Bromo-1-chloropropane for HECD.

The average recovery (in percent) for the PID was 98.
The standard deviation of the recovery for the PID was 2.4.
The MDL (in µg/mL) for the PID was 0.01.
The average recovery (in percent) for the HECD was none (no response for this detector).
The standard deviation of the recovery for the HECD was none (no response for this detector).
The MDL (in µg/mL) for the HECD was none (no response for this detector).

SAMPLE COLLECTION, PRESERVATION & HANDLING
Volatile Organics — Standard 40-mL glass screw-cap VOA vials with Teflon®-faced silicone septum may be used for both liquid and solid matrices. When collecting samples, liquids and solids should be introduced into the vials gently to reduce agitation which might drive off volatile compounds. If there are any air bubbles present the sample must be retaken. Tap slightly as they are filled to try and eliminate as much free air space as possible. The two vials from each sampling locations should be sealed in separate plastic bags to prevent cross-contamination between samples particularly if the sampled waste is suspected of containing high levels of volatile organics.

Semivolatile organics — Containers used to collect samples for the determination of semivolatile organic compounds should be soap and water washed followed by methanol (or isopropanol) rinsing. The sample containers should be of glass or Teflon® and have screw-top covers with Teflon® liners.

Preservation for volatile organics — No preservation is used with concentrated waste samples. With liquid samples containing no residual chlorine, 4 drops of concentrated hydrochloric acid are added and the samples are immediately cooled to 4°C. When liquid samples contain residual chlorine, they are treated as above and, in addition, 4 drops of 4% aqueous sodium thiosulfate are added. Soil, sediment, and sludge samples are only cooled to 4°C.

Preservation for semivolatile organics — No preservation is used with concentrated waste samples. With liquid samples containing no residual chlorine and with soil, sediment, and sludge samples, immediately cooling to 4°C is the only preservation used. When residual chlorine is present then 3 mL of 10% aqueous sodium sulfate is added for each gallon of sample collected, followed by cooling to 4°C.

MHT The holding time for all volatile organics samples is 14 days. Liquid samples must be extracted within 7 days and their extracts analyzed within 40 days. Concentrated waste, soil, sediment, and sludge samples must be extracted within 14 days and their extracts analyzed within 40 days.

SAMPLE PREPARATION Volatile compounds are introduced into the gas chromatograph either by direct injector or purge-and-trap (EPA Method 5030). EPA Method 5030 may be used directly on groundwater samples or low-concentration contaminated soils and sediments. For medium-concentration

soils or sediments, methanolic extraction, as described in EPA Method 5030, may be necessary prior to purge-and-trap analysis.

QUALITY CONTROL Calculate surrogate standard recovery on all samples, blanks, and spikes. A trip blank is recommended to check on sampling, storage, and handling contamination. Calibration standards, at a minimum of five concentration levels, are prepared in organic-free reagent water. One of the concentration levels should be at a concentration near, but above, the method detection limit.

A combination of bromochloromethane, 2-bromo-1-chloropropane, 1,4-dichlorobutane, and bromochlorobenzene are recommended as surrogate standards to encompass the range of the temperature program used in this method.

REFERENCE Test Methods for Evaluating Solid Waste, Physical/Chemical Methods, SW-846, 3rd Edition, U.S. EPA, Office of Solid Waste, Washington, DC, EPA Method 8021A, Rev. 1, Nov. 1992.

---

p-Isopropyltoluene **EPA Method 8260**
CAS #99-87-6

TITLE Volatile Organic Compounds by GC/MS: Capillary Column Technique

MATRIX This method is applicable to nearly all types of samples, regardless of water content, including groundwater, soils, and sediments.

METHOD SUMMARY Method 8260A covers 58 volatile organic compounds that are introduced into a gas chromatograph by the purge-and-trap method or by direct injection (in limited applications). Zero-grade helium is bubbled through a 5-mL solution at ambient temperature. Purged sample components are trapped in a tube containing suitable sorbent materials. When purging is complete, the sorbent tube is heated and backflushed with helium to desorb trapped sample components. The analytes are desorbed directly to a large bore capillary or cryofocussed on a capillary precolumn before being flash evaporated to a narrow bore capillary for analysis.

INTERFERENCES Major contaminant sources are volatile materials in the lab and impurities in the inert purging gas and in the sorbent trap. Interfering contamination may occur when a sample containing low concentrations of volatile organic compounds is analyzed immediately after a sample containing high concentrations of volatile organic compounds. After analysis of a sample containing high concentrations of volatile organic compounds, one or more calibration blanks should be analyzed to check for cross-contamination. Screening of the samples prior to purge-and-trap GC/MS analysis is highly recommended to prevent contamination of the system. This is especially true for soil and waste samples.

Special precautions must be taken to analyze for methylene chloride. The analytical and sample storage area should be isolated from all atmospheric sources of methylene chloride. All gas chromatography carrier gas lines and purge gas plumbing should be constructed from stainless steel or copper tubing. Laboratory clothing previously exposed to methylene chloride fumes during liquid-liquid extraction procedures can contribute to sample contamination.

Samples can also be contaminated by diffusion of volatile organics (particularly methylene chloride and fluorocarbons) through the septum seal during shipment and storage. A trip blank can serve as a check on such contamination.

INSTRUMENTATION GC/MS with a temperature-programmable chromatograph suitable for splitless injection equipped with variable constant differential flow controllers, a subambient oven controller, a purging device, sorbent trap, a thermal desorption apparatus and a capillary precolumn interface when using cryogenic cooling will be needed. The following GC columns may be used:

Column 1: 60 m × 0.75 mm I.D. capillary column coated with VOCOL, 1.5 μm film thickness.
Column 2: 30 m × 0.53 mm capillary column coated with DB-624 or VOCOL, 3 μm film thickness.
Column 3: 30 m × 0.32 mm I.D. capillary column coated with DB-5 or SE-54, 1-μm film thickness.

PRECISION & ACCURACY This method has been tested in a single lab using spiked water. Using a wide-bore capillary column, water was spiked at concentrations between 0.5 and 10 μg/L. Single lab accuracy and precision data are presented. The MDL actually achieved in a given analysis will vary depending on instrument sensitivity and matrix effects.

The MDL (a) in μg/L was 0.12.
The concentration range in μg/L was 0.1–10.
The mean accuracy (% of true value) was 99.
The precision as relative standard deviation was 6.7.

Note: The MDL is based on a 25-mL sample volume instead of a 5-mL sample volume.

SAMPLING METHOD
Liquid samples — Use a 40-mL glass screw-cap VOA vial with a Teflon®-faced silicone septum that has been prewashed, rinsed with distilled deionized water, and oven dried. If residual chlorine is present, collect the sample in a 4-oz soil VOA container which has been pre-preserved with 4 drops of 10% sodium thiosulfate. Mix gently and transfer the sample to a 40-mL VOA vial. Collect bubble-free samples in duplicate and seal each sample in a separate plastic bag.

Soils, sediments and sludges — Use an 8-oz widemouth glass bottle with Teflon®-faced silicone septum that has been prewashed, rinsed with distilled deionized water, and oven dried. **Do not** heat the septum for more than 1 h. Tap slightly to eliminate any free air space. Collect samples in duplicate and seal each one in a separate plastic bag.

SAMPLE PRESERVATION
Liquid samples — Add 4 drops of concentrated HCL, cool to 4°C and store in a solvent-free refrigerator.

Soils, sediments and sludges — Cool samples to 4°C and store in a solvent-free refrigerator.

MHT The maximum holding time of any sample (liquids, soils, sediments, and sludges) is 14 days.

SAMPLE PREPARATION

Liquid samples — Remove the plunger from a 5-mL syringe and carefully pour the sample into the syringe barrel to just short of overflowing. Replace the syringe plunger and compress the sample. Open the syringe valve and vent any residual air while adjusting the sample volume to 5.0 mL. If there is only one volatile organic analysis (VOA) vial, a second syringe should be filled at this time to protect against possible loss of sample integrity. Add 10 µL of surrogate spiking solution and 10 µL of internal standard spiking solution through the valve bore of the 5-mL syringe, then close the valve. The surrogate and internal standards may be mixed and added as a single spiking solution.

Sediments, soils, and waste samples — All samples of this type should be screened by GC analysis using a headspace method (EPA Method 3810) or the hexadecane extraction and screening method (EPA Method 3820). Use the screening data to determine whether to use the low-concentration method (0.005–1 mg/kg) or the high-concentration method (>1 mg/kg).

Low-concentration method — The low-concentration method is based on purging a heated sediment or soil sample mixed with organic-free reagent water containing the surrogate and internal standards. Analyze all reagent blanks and standards under the same conditions as the samples.

Use a 5-g sample if the expected concentration is <0.1 mg/kg or a 1-g sample for expected concentrations between 0.1 and 1 mg/kg. Mix the contents of the sample container with a narrow metal spatula. Weigh the amount of the sample into a tared purge device. Add the spiked water to the purge device, which contains the weighed amount of sample, and connect the device to the purge-and-trap system.

High-concentration method — This method is based on extracting the sediment or soil with methanol. A waste sample is either extracted or diluted, depending on its solubility in methanol. Wastes that are insoluble in methanol are diluted with reagent tetraglyme or possibly polyethylene glycol (PEG). An aliquot of the extract is added to organic-free reagent water containing surrogate and internal standards. This is purged at ambient temperature. All samples with an expected concentration of >1.0 mg/kg should be analyzed by this method.

Mix the contents of the sample container with a narrow metal spatula. For sediments or soils and solid wastes that are insoluble in methanol, weigh 4 g (wet weight) of sample into a tared 20-mL vial. For waste that is soluble in methanol, tetraglyme, or PEG, weigh 1 g (wet weight) into a tared scintillation vial or culture tube or a 10-mL volumetric flask. Quickly add 9.0 mL of appropriate solvent then add 1.0 mL of a surrogate spiking solution to the vial, cap it, and shake it for 2 min.

METHANOL EXTRACT REQUIRED FOR ANALYSIS OF HIGH-CONCENTRATION SOILS OR SEDIMENTS

Approximate Concentration Range	Volume of Methanol Extract (a)
500–10,000 µg/kg	100 µL
1,000–20,000 µg/kg	50 µL
5,000–100,000 µg/kg	10 µL
25,000–500,000 µg/kg	100 µL of 1/50 dilution (b)

Calculate appropriate dilution factor for concentrations exceeding this table.

(a) The volume of methanol added to 5 mL of water being purged should be kept constant. Therefore, add to the 5-mL syringe whatever volume of methanol is necessary to maintain a volume of 100 µL added to the syringe.

(b) Dilute an aliquot of the methanol extract and then take 100 µL for analysis.

QUALITY CONTROL Demonstrate, through the analysis of a reagent water blank, that interferences from the analytical system, glassware, and reagents are under control. Blank samples should be carried through all stages of the sample preparation and measurement steps. For each analytical batch (up to 20 samples), a reagent blank, matrix spike, and matrix spike duplicate must be analyzed (the frequency of the spikes may be different for different monitoring programs). The blank and spiked samples must be carried through all stages of the sample preparation and measurement steps. QC samples mentioned in the section on Interferences will also be needed as appropriate to those situations.

Matrix spiking standards should be prepared from volatile organic compounds which will be representative of the compounds being investigated. The recommended internal standards are chlorobenzene-d5, 1,4-difluorobenzene, 1,4-dichlorobenzene-d4, and pentafluorobenzene. Using stock standard solutions, prepare secondary dilution standards containing the compounds of interest, either singly or mixed together in methanol. Store them in a vial with no headspace for no more than one week. Surrogates recommended are toluene-d8, 4-bromofluorobenzene, and dibromofluoromethane. Each sample undergoing GC/MS analysis must be spiked with 10 µL of the surrogate spiking solution prior to analysis.

REFERENCE Test Methods for Evaluating Solid Waste (SW-846). U.S. EPA 1983, Method 8260A, Rev. 1, Nov. 1990. Office of Solid Waste, Washington, DC.

4-Isopropyltoluene   **EPA Method 503.1**
CAS #99-87-6

TITLE Aromatic & Unsaturated VOCs in Water

MATRIX Drinking water (finished or any treatment stage) and raw source water.

APPLICATION Method covers 28 aromatic and unsaturated VOCs. An inert gas is bubbled through a 5-mL water sample. Purged sample components are trapped in tube of sorbent materials. When purging is complete, sorbent tube is heated and backflushed with inert gas to desorb trapped sample onto a packed GC column.

INTERFERENCES During analysis, major contaminant sources are volatile materials in the lab and impurities in purging gas and sorbent trap. With high and low level samples, there can be carryover contamination. Excess water causes a negative baseline deflection.

INSTRUMENTATION Purge and Trap GC w/photoionization detector. (Two GC columns are recommended); Column 1: 5% SP-1200 and 1.75% Bentone 34 on Supelcoport; Column 2: 1,2,3-tris(2-cyanoethoxy)propane on Chromosorb W.

RANGE 2.2–600 µg/L. (Drinking water).

MDL 0.009 µg/L in water

PRECISION Not listed.

ACCURACY Not listed.

SAMPLING METHOD Use a 40–120-mL screw-cap vial (prewashed with detergent, rinsed with distilled water and oven dried at 105°C) with a PTFE-faced silicone septum. If residual chlorine is in the water add about 25 mg of ascorbic acid to each vial before sample collection. Collect bubble-free samples.

STABILITY Cool to 4°C; HCl to pH <2.

MHT 14 days.

QUALITY CONTROL As initial demonstration of lab accuracy and precision, analyze 4 to 7 replicates of a lab fortified blank containing the analyte at 0.1–5 µg/L. Collect all samples in duplicate.

REFERENCE Method 503.1, Volatile Aromatic & Unsaturated Organic Compounds in H2O by Purge and Trap GC, EPA 600/4-88/039.

Isosafrole   **EPA Method 1625**
CAS #120-58-1

TITLE Semivolatile Organic Compounds by Isotope Dilution GC/MS

MATRIX The compounds may be determined in waters, soils, and municipal sludges by this method.

METHOD SUMMARY This method is used to determine 176 semivolatile toxic organic pollutants associated with the CWA (as amended 1987); the RCRA (as amended 1986); the CERCLA (as amended 1986); and other compounds amenable to extraction and analysis by capillary column gas chromatography-mass spectrometry (GC/MS).

Stable isotopically-labeled analogs of the compounds of interest are added to the sample. If the solids content is less than 1%, a 1-L sample is extracted at pH 12–13, then at pH <2 with methylene chloride using continuous extraction techniques.

If the solids content is 30% or less, the sample is diluted to 1% solids with reagent water, homogenized ultrasonically, and extracted at pH 12–13, then at pH <2 with methylene chloride using continuous extraction techniques. If the solids content is greater than 30%, the sample is extracted using ultrasonic techniques.

Each extract is dried over sodium sulfate, concentrated to a volume of 5 mL, cleaned up using GPC, if necessary, and concentrated. Extracts are concentrated to 1 mL if GPC is not performed, and to 0.5 mL if GPC is performed.

An internal standard is added to the extract, and a 1-mL aliquot of the extract is injected into the GC. The compounds are separated by GC and detected by a MS. The labeled compounds serve to correct the variability of the analytical technique.

INTERFERENCES Solvents, reagents, glassware, and other sample processing hardware may yield artifacts and/or elevated baselines causing misinterpretation of chromatograms and spectra. Materials used in the analysis must be demonstrated to be free from interferences under the conditions of analysis by running method blanks initially and with each sample lot (sample started through the extraction process on a given 8-h shift, to a maximum of 20). Specific selection of reagents and purification of solvents by distillation in all glass systems may be required. Glassware and, where possible, reagents are cleaned by solvent rinse and baking at 450°C for 1-h minimum. Interferences coextracted from samples will vary considerably from source to source, depending on the diversity of the site being sampled.

INSTRUMENTATION Major instrumentation includes a GC with a splitless or on-column injection port for capillary column, a MS with 70 eV electron impact ionization, and a data system to collect and record MS data, and process it. A K-D apparatus is used to concentrate extracts.

GC Column: 30 m × 0.25 mm I.D. 5% phenyl, 94% methyl, 1% vinyl silicone bonded phased fused silica capillary column.

PRECISION & ACCURACY The detection limits of the method are usually dependent on the level of interferences rather than instrumental limitations. The limits typify the minimum quantities that can be detected with no interferences present.

The minimum level (in µg/mL) was not listed. This is defined as a minimum level at which the analytical system shall give recognizable mass spectra (background corrected) and acceptable calibration points.

The MDL (in µg/kg) in low solids was not listed and in high solids was not listed; these were determined in digested sludge (low solids) and in filter cake or compost (high solids).

The labeled and native compound initial precision as standard deviation (in µg/L) was not listed.
The labeled and native compound initial accuracy as average recovery (in µg/L) was not listed.

SAMPLE COLLECTION, PRESERVATION & HANDLING Collect samples in glass containers. Aqueous samples which flow freely are collected in refrigerated bottles using automatic sampling equipment. Solid samples are collected as grab samples using widemouth jars. Maintain samples at 0 to 4°C from the time of collection until extraction. If residual chlorine is present in aqueous samples, add 80 mg sodium thiosulfate/L of water. Begin sample extraction within 7 days of collection, and analyze all extracts within 40 days of extraction.

SAMPLE PREPARATION Samples containing 1% solids or less are extracted directly using continuous liquid-liquid extraction techniques. Samples containing 1 to 30% solids are diluted to the 1% level with reagent water and extracted using continuous

liquid-liquid extraction techniques. Samples containing greater than 30% solids are extracted using ultrasonic techniques.

Base/neutral extraction — Adjust the pH of the waters in the extractors to 12–13 with 6 N NaOH. Extract with methylene chloride for 24–48 h.

Acid extraction — Adjust the pH of the waters in the extractors to 2 or less using 6 N sulfuric acid. Extract with methylene chloride for 24–48 h.

Ultrasonic extraction of high solids samples — Add anhydrous sodium sulfate to the sample and QC aliquot(s). Add acetone:methylene chloride (1:1) to the sample and mix thoroughly

Concentrate extracts using a K-D apparatus.

QUALITY CONTROL The analyst is permitted to modify this method to improve separations or lower the costs of measurements, provided all performance specifications are met. Analyses of blanks are required to demonstrate freedom from contamination. When results of spikes indicate atypical method performance for samples, the samples are diluted to bring method performance within acceptable limits.

For low solids (aqueous samples), extract, concentrate, and analyze two sets of four 1-L aliquots (8 aliquots total) of the precision and recovery standard. For high solids samples, two sets of four 30-g aliquots of the high solids reference matrix are used.

Spike all samples with labeled compounds to assess method performance. Compute percent recovery of the labeled compounds using the internal standard method. Compare the labeled compound recovery for each compound with the corresponding labeled compound recovery.

Reagent water and high solids reference matrix blanks are analyzed to demonstrate freedom from contamination. Extract and concentrate a 1-L reagent water blank or a high solids reference matrix blank with each sample's lot (samples started through the extraction process on the same 8-h shift, to a maximum of 20 samples).

Field replicates may be collected to determine the precision of the sampling technique, and spiked samples may be required to determine the accuracy of the analysis when the internal standard method is used.

REFERENCE Semivolatile Organic Compounds by Isotope Dilution GC/MS. Office of Water Regulation and Standards, U.S. EPA Industrial Technology Division, Washington, DC, EPA Method 1625, Rev. C, June 1989 (contact W.A. Telliard, U.S. EPA, Office of Water Regulations and Standards, 401 M St., SW, Washington, DC, 20460. Phone: 202-382-7131).

Isosafrole **EPA Method 8270**
CAS #120-58-1

TITLE Semivolatile Organic Compounds by GC/MS

MATRIX This method is used to determine the concentration of semivolatile organic compounds in extracts prepared from all types of solid waste matrices, soils, and groundwater. Although surface waters are not specifically mentioned, this method should be applicable to water samples from rivers, lakes, etc.

METHOD SUMMARY This method covers 259 semivolatile organic compounds. In very limited applications direct injection of the sample into the GC/MS system may be appropriate, but this results in very high detection limits (approximately 10,000 µg/L). Typically, a 1-L liquid sample, containing surrogate, and matrix spiking standards, is extracted in a continuous extractor first under acid conditions and then under basic conditions. Typically 30 g of a solid sample, containing surrogate, and matrix spiking standards, is extracted ultrasonically. After concentrating the extract to 1 mL it is spiked with 10 µL of an internal standard solution just prior to analysis by GC/MS. The volume injected should contain about 100 ng of base/neutral and 200 ng of acid surrogates (for a 1-µL injection). Analysis is performed by GC/MS using a capillary GC column.

INTERFERENCES Raw GC/MS data from all blanks, samples, and spikes must be evaluated for interferences. Contamination by carryover can occur whenever high-concentration and low-concentration samples are sequentially analyzed. To reduce carryover, the sample syringe must be rinsed out between samples with solvent. Whenever an unusually concentrated sample is encountered, it should be followed by the analysis of blank solvent to check for cross-contamination.

INSTRUMENTATION A GC/MS and a data system are required. The GC column used is a 30 m × 0.25 mm I.D. (or 0.32 mm I.D.) 1um film thickness silicone-coated fused silica capillary column. A continuous liquid-liquid extractor equipped with Teflon® or glass connection joints and stopcocks requiring no lubrication, a K-D concentrating apparatus, water bath, and an ultrasonic disrupter with a minimum power of 300 W and with pulsing capability are also required.

PRECISION & ACCURACY The estimated quantitation limit (EQL) of Method 8270B for determining an individual compound is approximately 1 mg/kg (wet weight) for soil or sediment samples, 1–200 mg/kg for wastes (dependent on matrix and method of preparation), and 10 µg/L for groundwater samples. EQLs will be proportionately higher for sample extracts that require dilution to avoid saturation of the detector.

The EQL(b) for groundwater in µg/L is 10.
The EQL (a, b) for low concentrations in soil and sediment in µg/kg is not determined.
Accuracy as µg/L is not listed.
Overall precision in µg/L is not listed.

(a) *EQLs listed for soil/sediment are based on wet weight. Normally data is reported in a dry-weight basis; therefore, EQLs will be higher based on the % dry weight of each sample. This calculation is based on a 30-g sample and gel permeation chromatography cleanup.*
(b) *Sample EQLs are highly matrix-dependent. The EQLs are provided for guidance and may not always be achievable.*
C = *True value for concentration, in µg/L.*
X = *Average recovery found for measurements of samples containing a concentration of C, in µg/L.*

ESTIMATED QUANTITATION LIMIT

Other Matrices	Factor (a)
High-concentration soil and sludges by sonicator	7.5
Non-water miscible waste	75

(a) EQL for other matrices = [EQL for low soil/sediment] × [Factor]. This estimated EQL is similar to an EPA "Practical Quantitation Limit."

SAMPLING METHOD

Liquid samples — Use a 1 or 2½ gallon amber glass bottle with a screw-top Teflon®-lined cover that has been prewashed with detergent and rinsed with distilled water and methanol (or isopropanol).

Soils, sediments, or sludges — Use an 8-oz. widemouth glass with a screw-top Teflon®-lined cover that has been prewashed with detergent and rinsed with distilled water and methanol (or isopropanol).

SAMPLE PRESERVATION

Liquid samples — If residual chlorine is present, add 3 mL of 10% sodium thiosulfate per gallon, cool to 4°C and store in a solvent-free refrigerator until analysis; if chlorine is not present, then eliminate the sodium thiosulfate addition.

Soils, sediments, or sludges — Cool samples to 4°C and store in a solvent-free refrigerator.

MHT Liquid samples must be extracted within 7 days and the extracts analyzed within 40 days. Soils, sediments, or sludges may be stored for a maximum of 14 days and the extracts analyzed within 40 days.

SAMPLE PREPARATION

Liquid samples — Transfer 1 L quantitatively to a continuous extractor. If high concentrations are anticipated, a smaller volume may be used and then diluted with organic-free reagent water to 1 L. Adjust pH, if necessary, to pH <2 using 1:1 (V/V) sulfuric acid. Pipette 1.0 mL of a surrogate standard spiking solution into each sample. For the sample in each analytical batch selected for spiking, add 1.0 mL of a matrix spiking standard. For base/neutral acid analysis, the amount of the surrogates and matrix spiking compounds added to the sample should result in a final concentration of 100 ng/µL of each analyte in the extract to be analyzed (assuming a 1-µL injection). Extract with methylene chloride for 18–24 h. Next, adjust the pH of the aqueous phase to pH >11 using 10 N sodium hydroxide and extract it with methylene chloride again for 18–24 h. Dry the extract through a column containing anhydrous sodium sulfate and concentrate it to 1 mL using a K-D concentrator.

Soils, sediments, or sludges — Use 30 g of sample. Nonporous or wet samples (gummy or clay type) that do not have a free-flowing sandy texture must be mixed with anhydrous sodium sulfate until the sample is free flowing. Add 1 mL of surrogate standards to all samples, spikes, standards, and blanks. For the sample in each analytical batch selected for spiking, add 1.0 mL of a matrix spiking standard. For base/neutral acid analysis, the amount added of the surrogates and matrix spiking compounds should result in a final concentration of 100 ng/µL of each base/neutral analyte and 200 ng/µL of each acid analyte in the extract to be analyzed (assuming a 1-µL injection). Immediately add a 100-mL mixture of 1:1 methylene chloride:acetone and extract the sample ultrasonically for 3 min and then decant or filter the extracts. Repeat the extraction two or more times. Dry the extract using a column with anhydrous sodium sulfate and concentrate it to 1 mL in a K-D concentrator.

QUALITY CONTROL A methylene chloride solution containing 50 ng/µL of decafluorotriphenylphosphine (DFTPP) is used for tuning the GC/MS system each 12-h shift. A system performance check also must be made during every 12-h shift. A standard containing 50 ng/µL each of 4,4'-DDT, pentachlorophenol, and benzidine is required to verify injection port inertness and GC column performance. A calibration standard at mid-concentration, containing each compound of interest, including all required surrogates, must be performed every 12 h during analysis. After the system performance check is met, calibration check compounds (CCCs) are used to check the validity of the initial calibration.

The internal standard responses and retention times in the calibration check standard must be evaluated immediately after or during data acquisition. If the retention time for any internal standard changes by more than 30 seconds from the last check calibration (12 h), the chromatographic system must be inspected for malfunctions and corrections must be made, as required. If the electron ionization current plot (EICP) area for any of the internal standards changes by a factor of two from the last daily calibration standard check, the mass spectrometer must be inspected for malfunctions and corrections must be made, as appropriate.

Demonstrate, through the analysis of a reagent water blank, that interferences from the analytical system, glassware, and reagents are under control. The blank samples should be carried through all stages of the sample preparation and measurement steps. For each analytical batch (up to 20 samples), a reagent blank, matrix spike, and matrix spike duplicate/duplicate must be analyzed (the frequency of the spikes may be different for different monitoring programs). The blank and spiked samples must be carried through all stages of the sample preparation and measurement steps. A QC reference sample concentrate containing each analyte at a concentration of 100 mg/L in methanol is required.

REFERENCE Test Methods for Evaluating Solid Waste (SW-846). U.S. EPA 1983, Method 8270B, Rev. 2, Nov. 1990. Office of Solid Waste, Washington, DC.

K

Kepone
CAS #143-50-0

EPA Method 8270

TITLE Semivolatile Organic Compounds by GC/MS

MATRIX This method is used to determine the concentration of semivolatile organic compounds in extracts prepared from all types of solid waste matrices, soils, and groundwater. Although surface waters are not specifically mentioned, this method should be applicable to water samples from rivers, lakes, etc.

METHOD SUMMARY This method covers 259 semivolatile organic compounds. In very limited applications direct injection of the sample into the GC/MS system may be appropriate, but this results in very high detection limits (approximately 10,000 µg/L). Typically, a 1-L liquid sample, containing surrogate, and matrix spiking standards, is extracted in a continuous extractor first under acid conditions and then under basic conditions. Typically 30 g of a solid sample, containing surrogate, and matrix spiking standards, is extracted ultrasonically. After concentrating the extract to 1 mL it is spiked with 10 µL of an internal standard solution just prior to analysis by GC/MS. The volume injected should contain about 100 ng of base/neutral and 200 ng of acid surrogates (for a 1-µL injection). Analysis is performed by GC/MS using a capillary GC column.

INTERFERENCES Raw GC/MS data from all blanks, samples, and spikes must be evaluated for interferences. Contamination by carryover can occur whenever high-concentration and low-concentration samples are sequentially analyzed. To reduce carryover, the sample syringe must be rinsed out between samples with solvent. Whenever an unusually concentrated sample is encountered, it should be followed by the analysis of blank solvent to check for cross-contamination.

INSTRUMENTATION A GC/MS and a data system are required. The GC column used is a 30 m × 0.25 mm I.D. (or 0.32 mm I.D.) 1um film thickness silicone-coated fused silica capillary column. A continuous liquid-liquid extractor equipped with Teflon® or glass connection joints and stopcocks requiring no lubrication, a K-D concentrating apparatus, water bath, and an ultrasonic disrupter with a minimum power of 300 W and with pulsing capability are also required.

PRECISION & ACCURACY The estimated quantitation limit (EQL) of Method 8270B for determining an individual compound is approximately 1 mg/kg (wet weight) for soil or sediment samples, 1–200 mg/kg for wastes (dependent on matrix and method of preparation), and 10 µg/L for groundwater samples. EQLs will be proportionately higher for sample extracts that require dilution to avoid saturation of the detector.

The EQL(b) for groundwater in µg/L is 20.
The EQL (a, b) for low concentrations in soil and sediment in µg/kg is not determined.
Accuracy as µg/L is not listed.
Overall precision in µg/L is not listed.

(a) EQLs listed for soil/sediment are based on wet weight. Normally data is reported in a dry-weight basis; therefore, EQLs will be higher based on the % dry weight of each sample. This calculation is based on a 30-g sample and gel permeation chromatography cleanup.

(b) Sample EQLs are highly matrix-dependent. The EQLs are provided for guidance and may not always be achievable.

C = True value for concentration, in µg/L.

X = Average recovery found for measurements of samples containing a concentration of C, in µg/L.

ESTIMATED QUANTITATION LIMIT

Other Matrices	Factor (a)
High-concentration soil and sludges by sonicator	7.5
Non-water miscible waste	75

(a) EQL for other matrices = [EQL for low soil/sediment] × [Factor]. This estimated EQL is similar to an EPA "Practical Quantitation Limit."

SAMPLING METHOD

Liquid samples — Use a 1 or 2½ gallon amber glass bottle with a screw-top Teflon®-lined cover that has been prewashed with detergent and rinsed with distilled water and methanol (or isopropanol).

Soils, sediments, or sludges — Use an 8-oz. widemouth glass with a screw-top Teflon®-lined cover that has been prewashed with detergent and rinsed with distilled water and methanol (or isopropanol).

SAMPLE PRESERVATION

Liquid samples — If residual chlorine is present, add 3 mL of 10% sodium thiosulfate per gallon, cool to 4°C and store in a solvent-free refrigerator until analysis; if chlorine is not present, then eliminate the sodium thiosulfate addition.

Soils, sediments, or sludges — Cool samples to 4°C and store in a solvent-free refrigerator.

MHT Liquid samples must be extracted within 7 days and the extracts analyzed within 40 days. Soils, sediments, or sludges may be stored for a maximum of 14 days and the extracts analyzed within 40 days.

SAMPLE PREPARATION

Liquid samples — Transfer 1 L quantitatively to a continuous extractor. If high concentrations are anticipated, a smaller volume may be used and then diluted with organic-free reagent water to 1 L. Adjust pH, if necessary, to pH <2 using 1:1 (V/V) sulfuric acid. Pipette 1.0 mL of a surrogate standard spiking solution into each sample. For the sample in each analytical batch selected for spiking, add 1.0 mL of a matrix spiking standard. For base/neutral acid analysis, the amount of the surrogates and matrix spiking compounds added to the sample should result in a final concentration of 100 ng/µL of each analyte in the extract to be analyzed (assuming a 1-µL injection). Extract with methylene chloride for 18–24 h. Next, adjust the pH of the aqueous phase to pH >11 using 10 N sodium hydroxide and extract it with methylene chloride again for

18–24 h. Dry the extract through a column containing anhydrous sodium sulfate and concentrate it to 1 mL using a K-D concentrator.

Soils, sediments, or sludges — Use 30 g of sample. Nonporous or wet samples (gummy or clay type) that do not have a free-flowing sandy texture must be mixed with anhydrous sodium sulfate until the sample is free flowing. Add 1 mL of surrogate standards to all samples, spikes, standards, and blanks. For the sample in each analytical batch selected for spiking, add 1.0 mL of a matrix spiking standard. For base/neutral acid analysis, the amount added of the surrogates and matrix spiking compounds should result in a final concentration of 100 ng/µL of each base/neutral analyte and 200 ng/µL of each acid analyte in the extract to be analyzed (assuming a 1-µL injection). Immediately add a 100-mL mixture of 1:1 methylene chloride:acetone and extract the sample ultrasonically for 3 min and then decant or filter the extracts. Repeat the extraction two or more times. Dry the extract using a column with anhydrous sodium sulfate and concentrate it to 1 mL in a K-D concentrator.

QUALITY CONTROL A methylene chloride solution containing 50 ng/µL of decafluorotriphenylphosphine (DFTPP) is used for tuning the GC/MS system each 12-h shift. A system performance check also must be made during every 12-h shift. A standard containing 50 ng/µL each of 4,4'-DDT, pentachlorophenol, and benzidine is required to verify injection port inertness and GC column performance. A calibration standard at mid-concentration, containing each compound of interest, including all required surrogates, must be performed every 12 h during analysis. After the system performance check is met, calibration check compounds (CCCs) are used to check the validity of the initial calibration.

The internal standard responses and retention times in the calibration check standard must be evaluated immediately after or during data acquisition. If the retention time for any internal standard changes by more than 30 seconds from the last check calibration (12 h), the chromatographic system must be inspected for malfunctions and corrections must be made, as required. If the electron ionization current plot (EICP) area for any of the internal standards changes by a factor of two from the last daily calibration standard check, the mass spectrometer must be inspected for malfunctions and corrections must be made, as appropriate.

Demonstrate, through the analysis of a reagent water blank, that interferences from the analytical system, glassware, and reagents are under control. The blank samples should be carried through all stages of the sample preparation and measurement steps. For each analytical batch (up to 20 samples), a reagent blank, matrix spike, and matrix spike duplicate/duplicate must be analyzed (the frequency of the spikes may be different for different monitoring programs). The blank and spiked samples must be carried through all stages of the sample preparation and measurement steps. A QC reference sample concentrate containing each analyte at a concentration of 100 mg/L in methanol is required.

REFERENCE Test Methods for Evaluating Solid Waste (SW-846). U.S. EPA 1983, Method 8270B, Rev. 2, Nov. 1990. Office of Solid Waste, Washington, DC.

L

Lead
CAS #7439-92-1

EPA Method 6010

TITLE Inductively Coupled Plasma-Atomic Emission Spectroscopy

MATRIX This method is applicable to the determination of trace elements, including metals, in groundwater, soils, sludges, sediments, and other solid wastes. All matrices require digestion prior to analysis. The method of standard addition must be used for the analysis of all sample digests unless either serial dilution or matrix spike addition demonstrates it is not required.

METHOD SUMMARY Method 6010 covers 25 elements using ICP analysis. It measures element-emitted light by optical spectrometry. Samples, following an appropriate acid digestion, are nebulized and the resulting aerosol is transported to the plasma torch. Element-specific atomic line emission spectra are produced by a radio-frequency inductively coupled plasma.

INTERFERENCES Interferences may be categorized as spectral or non-spectral. Spectral interferences are caused by overlap of a spectral line from another element, unresolved overlap of molecular band spectra, background contribution from continuous or recombination phenomenon, and stray light from the line emission of high concentration elements. Non-spectral interferences include physical and chemical interferences. Physical interferences are effects associated with the sample nebulization and transport processes. Changes in viscosity and surface tension can cause significant inaccuracies. Chemical interferences include molecular compound formation, ionization effects, and solute vaporization effects. Normally these effects are not significant and can be minimized by careful selection of operating conditions. Chemical interferences are highly dependent on matrix type and the specific analyte element.

INSTRUMENTATION An inductively coupled argon plasma emission spectrometer (ICP) capable of background correction is required.

PRECISION & ACCURACY Detection limits, sensitivity, and optimum ranges of the metals will vary with the matrices and model of the spectrometer. In a single lab evaluation, seven wastes were analyzed for 22 elements. The mean percent relative standard deviation from triplicate analyses for all elements and wastes was 9 ± 2%. The mean percent recovery of spiked elements for all wastes was 93 ± 6%. Spike levels ranged from 100 µg/L to 100 mg/L. The wastes included sludges and industrial wastewaters.

Estimated instrument detection limit in µg/L is 42.
Spiked concentration in µg/L is 24.
Mean reported value in µg/L is 30.
Precision as RSD % is 32.

SAMPLING METHOD Samples should be collected in borosilicate glass, linear polyethylene, polypropylene, or Teflon® bottles that have been prewashed with detergent and tap water, and rinsed with 1:1 nitric acid and tap water or 1:1 hydrochloric acid and tap water. Collect at least 2 g of solids and 200 mL of aqueous samples.

SAMPLE PRESERVATION Add nitric acid to make the samples pH <2.

MHT The maximum holding time for properly preserved samples is 6 months.

SAMPLE PREPARATION Preliminary treatment of most matrices is necessary because of the complexity and variability of sample matrices. Water samples that have been prefiltered and acidified will not need acid digestion. Methods for acid digestion of waters for total recoverable or dissolved metals, acid digestions of aqueous samples and extracts for total metals, and acid digestion of sediments, sludges, and soils are summarized below.

Total recoverable or dissolved metals in water — To prepare surface and groundwater samples for determination of total recoverable and dissolved metals, a 100-mL aliquot of well-mixed sample is acidified with concentrated nitric acid and concentrated hydrochloric acid, then heated until the volume is reduced to 15–20 mL. Adjust the final volume to 100 mL with reagent water.

Total metals in aqueous samples, soil and sediment extracts — To prepare aqueous samples, soil and sediment extracts, and wastes that contain suspended solids, a 100-mL aliquot is made acidic with concentrated nitric acid and the solution is evaporated to about 5 mL on a hot plate. Continue heating and adding additional acid until sample digestion is complete, which is usually indicated when the digestate is light in color or does not change in appearance. Evaporate the solution to about 3 mL and cool it and add a small quantity of 1:1 hydrochloric acid (10 mL/100 mL of final solution). Cover the beaker and reflux for 15 min. Wash down the beaker walls and filters or centrifuge the sample to remove silicates and other insoluble material. Filter the sample and adjust the final volume to 100 mL with reagent water and the final acid concentration to 10%.

Sediments, sludges, and soils — To prepare sediments, sludges and soil samples, transfer 1–2 g to a conical beaker and add 10 mL of 1:1 nitric acid, mix the slurry, and cover it with a watch glass. Heat the sample and reflux for 10–15 min without boiling. Allow it to cool, then add 5 mL of concentrated nitric acid and reflux for 30 min. Repeat last step and then allow the solution to evaporate to 5 mL without boiling. Cool and add 2 mL of water and 3 mL of 30% hydrogen peroxide. Cover and place the beaker on the hot plate. Heat and add 30% hydrogen peroxide in 1-mL aliquots with warming until the effervescence is minimal but do not add more than a total of 10 mL of 30% hydrogen peroxide. If the sample is being prepared for the analysis of Ag, Al, As, Ba, Be, Ca, Cd, Co, Cr, Cu, Fe, K, Mg, Mn, Mo, Na, Ni, Os, Pb, Se, Tl, V, and Zn, then add 5 mL of concentrated hydrochloric acid and 10 mL of water and return the covered beaker to a hot plate for 15 min of additional refluxing without boiling. Dilute the sample to a 100 mL volume with

water after cooling and filter or centrifuge to remove particulates.

QUALITY CONTROL Laboratory control samples must be analyzed for each analytical method. A method blank should be analyzed with each batch of samples. The effect of the matrix on method performance must be demonstrated: when appropriate, there should be at least one matrix spike and either one matrix duplicate or one matrix spike duplicate per analytical batch. The bias and precision of the method, as well as the method detection limit for each specific matrix type, must be measured.

Dilute and reanalyze samples that are more concentrated than the linear calibration limit. Employ a minimum of one reagent blank per sample batch to determine if contamination or any memory effects are occurring. Whenever a new or unusual sample matrix is encountered, perform either a serial dilution test or a matrix spike addition test to ensure that neither positive or negative interferences are operating on any of the analyte elements. Check the instrument standardization by verifying calibration every 10 samples using a calibration blank and a check standard.

REFERENCE Test Methods for Evaluating Solid Waste (SW-846). U.S. EPA. 1983. Method 6010, Rev. 0, Sept. 1986. Office of Solid Wastes, Washington, DC.

Lead   **EPA Method 200.7**
CAS #7439-92-1

TITLE Inductively Coupled Plasma

MATRIX Dissolved, suspended or (ICP) total element in drinking and surface waters and in domestic and industrial wastewaters.

APPLICATION The method covers the determination of 25 metals. Dissolved elements are determined in filtered and acidified samples after appropriate digestion (which increases dissolved solids). Its primary advantage is that ICP instruments allow simultaneous or rapid sequential determination of many elements in a short time. Samples are first nebulized and the aerosol is transported to a plasma torch in which element specific atomic line emission spectra are produced by a radio frequency inductively coupled plasma. Background correction is required for trace element detection except in the case of line broadning.

INTERFERENCES There are spectral, physical, and chemical interferences. The primary disadvantage of ICP instruments is background radiation from other elements and the plasma gases (spectral interferences). Changes in sample viscosity and surface tension with samples containing high dissolved solids (especially those exceeding 1500 mg/L) or high acid concentrations can cause physical interferences. Ionization effects, solute vaporization and molecular compound formation can cause chemical interferences. Aluminum can cause interference at the 100 mg/L level.

INSTRUMENTATION Inductively coupled argon plasma emission spectroscopy. 220.353 nm wavelength

RANGE Not listed.

MDL 42 µg/L.

PRECISION SD = 16% Mean at true value 250 µg/L.

ACCURACY Mean recovery = 93% ± 6% of spiked elements for all wastes.

SAMPLING METHOD Wash sample container with detergent and tap water, rinse with 1 + 1 nitric acid and tap water, then rinse with 1 + 1 hydrochloric acid and tap water, then rinse with deionized, distilled water in that order. Perform any filtration or acid preservation steps when the sample is collected or as soon as possible thereafter.

STABILITY Cool samples to 4°C.

MHT 24 h.

QUALITY CONTROL Mixed calibration standards, an instrument check standard, and an interference check solution are used in addition to a quality control sample. The quality control sample should be prepared in the same acid matrix as the calibration standards at 10 times the instrumental detection limits and in accordance with the instructions provided by the supplier. Furthermore, two types of blanks are required: a calibration blank and a reagent blank.

REFERENCE Method 200.7, U.S. EPA, EMSL-Cincinnati, OH, Nov. 1980

Lead   **EPA Method 7420**
CAS #7439-92-1

TITLE Atomic Absorption (AA) Direct Aspiration

MATRIX Drinking, surface, and saline waters. Wastewater

APPLICATION Sample is aspirated and atomized in a flame. A light beam from a Pb hollow cathode lamp is directed through the flame into a monochromator and onto a detector. Since wavelength of light beam is specific for Pb, light energy absorbed by detector is measure of lead.

INTERFERENCES The most troublesomee type is chemical, caused by lack of absorption of atoms bound in molecular combination in the flame. High dissolved solids in sample may result in nonatomic absorbance interference. Background correction is required.

INSTRUMENTATION Atomic absorption spectrometer. Lead hollow cathode lamp. [283.3 nm wavelength(primary)]

RANGE 1–20 mg/L.

MDL 0.1 mg/L

PRECISION Standard deviation = 128 µg/L at 367 µg/L (true value) 74 labs

ACCURACY As bias = +2.9% at 367 µg/L (true value) 74 labs

SAMPLING METHOD Use glass or plastic containers. Collect 200 g of solids and 600 mL of liquid samples.

STABILITY Cool solid samples to 4°C and analyze as soon as possible. Add nitric acid to liquid samples to pH <2.

MHT 6 months.

QUALITY CONTROL At least one duplicate and one spike sample should be run every 20 samples or with each matrix type to verify precision of the method. For 20 or more samples per day, verify working standard curve. Run an additional standard at or near mid-range every 10 samples.

REFERENCE Method 7420, SW-846, 3rd ed., Nov.1986.

Lead   EPA Method 7421
CAS #7439-92-1

TITLE Atomic Absorption (AA)

MATRIX Drinking, surface, and furnace technique saline waters. Wastewater.

APPLICATION Pb in solution is readily determined by atomic absorption spectrometer, but detection limits, sensitivity and optimum range vary with the matrices and models of AA spectrophotometers. While drinking water may be analyzed directly, ground water, other aqueous samples, EP extracts, industrial wastes, soils, sludges, and sediments require digestion.

INTERFERENCES "Chemical" interference is caused by lack of absorption of atoms bound in molecular combination in the flame. High dissolved solids in sample may cause interference from non atomic absorbance. Ionization and spectral interferences can occur.

INSTRUMENTATION Atomic absorption spectrometer. Lead (Pb) hollow cathode lamp. Graphite furnace. 283 nm wavelength.

RANGE 5–100 µg/L.

MDL 1 µg/L

PRECISION Standard deviation = ±3.7 at 100 µg Pb/L.

ACCURACY Recovery = 95% at 100 µg Pb/L.

SAMPLING METHOD Use glass or plastic containers. Collect 200 g of solids and 600 mL of liquid samples.

STABILITY Cool solid samples to 4°C and analyze as soon as possible. Add nitric acid to liquid samples to pH <2.

MHT 6 months.

QUALITY CONTROL At least one duplicate and one spike sample should be run every 20 samples, or with each matrix type to verify method precision. If 20 or more samples are run, run a standard (at or near mid-range) every 10 samples.

REFERENCE Method 7421, SW-846, 3rd ed., Nov.1986.

Leptophos   EPA Method 8270
CAS #21609-90-5

TITLE Semivolatile Organic Compounds by GC/MS

MATRIX This method is used to determine the concentration of semivolatile organic compounds in extracts prepared from all types of solid waste matrices, soils, and groundwater. Although surface waters are not specifically mentioned, this method should be applicable to water samples from rivers, lakes, etc.

METHOD SUMMARY This method covers 259 semivolatile organic compounds. In very limited applications direct injection of the sample into the GC/MS system may be appropriate, but this results in very high detection limits (approximately 10,000 µg/L). Typically, a 1-L liquid sample, containing surrogate, and matrix spiking standards, is extracted in a continuous extractor first under acid conditions and then under basic conditions. Typically 30 g of a solid sample, containing surrogate, and matrix spiking standards, is extracted ultrasonically. After concentrating the extract to 1 mL it is spiked with 10 µL of an internal standard solution just prior to analysis by GC/MS. The volume injected should contain about 100 ng of base/neutral and 200 ng of acid surrogates (for a 1-µL injection). Analysis is performed by GC/MS using a capillary GC column.

INTERFERENCES Raw GC/MS data from all blanks, samples, and spikes must be evaluated for interferences. Contamination by carryover can occur whenever high-concentration and low-concentration samples are sequentially analyzed. To reduce carryover, the sample syringe must be rinsed out between samples with solvent. Whenever an unusually concentrated sample is encountered, it should be followed by the analysis of blank solvent to check for cross-contamination.

INSTRUMENTATION A GC/MS and a data system are required. The GC column used is a 30 m × 0.25 mm I.D. (or 0.32 mm I.D.) 1um film thickness silicone-coated fused silica capillary column. A continuous liquid-liquid extractor equipped with Teflon® or glass connection joints and stopcocks requiring no lubrication, a K-D concentrating apparatus, water bath, and an ultrasonic disrupter with a minimum power of 300 W and with pulsing capability are also required.

PRECISION & ACCURACY The estimated quantitation limit (EQL) of Method 8270B for determining an individual compound is approximately 1 mg/kg (wet weight) for soil or sediment samples, 1–200 mg/kg for wastes (dependent on matrix and method of preparation), and 10 µg/L for groundwater samples. EQLs will be proportionately higher for sample extracts that require dilution to avoid saturation of the detector.

The EQL(b) for groundwater in µg/L is 10.
The EQL (a, b) for low concentrations in soil and sediment in µg/kg is not determined.
Accuracy as µg/L is not listed.
Overall precision in µg/L is not listed.

(a) *EQLs listed for soil/sediment are based on wet weight. Normally data is reported in a dry-weight basis; therefore, EQLs* will be higher based on the % dry weight of each sample. This calculation is based on a 30-g sample and gel permeation chromatography cleanup.

(b) Sample EQLs are highly matrix-dependent. The EQLs are provided for guidance and may not always be achievable.

C = True value for concentration, in µg/L.

X = Average recovery found for measurements of samples containing a concentration of C, in µg/L.

ESTIMATED QUANTITATION LIMIT

Other Matrices	Factor (a)
High-concentration soil and sludges by sonicator	7.5
Non-water miscible waste	75

(a) EQL for other matrices = [EQL for low soil/sediment] × [Factor]. This estimated EQL is similar to an EPA "Practical Quantitation Limit."

SAMPLING METHOD

Liquid samples — Use a 1 or 2½ gallon amber glass bottle with a screw-top Teflon®-lined cover that has been prewashed with detergent and rinsed with distilled water and methanol (or isopropanol).

Soils, sediments, or sludges — Use an 8-oz. widemouth glass with a screw-top Teflon®-lined cover that has been prewashed with detergent and rinsed with distilled water and methanol (or isopropanol).

SAMPLE PRESERVATION

Liquid samples — If residual chlorine is present, add 3 mL of 10% sodium thiosulfate per gallon, cool to 4°C and store in a solvent-free refrigerator until analysis; if chlorine is not present, then eliminate the sodium thiosulfate addition.

Soils, sediments, or sludges — Cool samples to 4°C and store in a solvent-free refrigerator.

MHT Liquid samples must be extracted within 7 days and the extracts analyzed within 40 days. Soils, sediments, or sludges may be stored for a maximum of 14 days and the extracts analyzed within 40 days.

SAMPLE PREPARATION

Liquid samples — Transfer 1 L quantitatively to a continuous extractor. If high concentrations are anticipated, a smaller volume may be used and then diluted with organic-free reagent water to 1 L. Adjust pH, if necessary, to pH <2 using 1:1 (V/V) sulfuric acid. Pipette 1.0 mL of a surrogate standard spiking solution into each sample. For the sample in each analytical batch selected for spiking, add 1.0 mL of a matrix spiking standard. For base/neutral acid analysis, the amount of the surrogates and matrix spiking compounds added to the sample should result in a final concentration of 100 ng/µL of each analyte in the extract to be analyzed (assuming a 1-µL injection). Extract with methylene chloride for 18–24 h. Next, adjust the pH of the aqueous phase to pH >11 using 10 N sodium hydroxide and extract it with methylene chloride again for 18–24 h. Dry the extract through a column containing anhydrous sodium sulfate and concentrate it to 1 mL using a K-D concentrator.

Soils, sediments, or sludges — Use 30 g of sample. Nonporous or wet samples (gummy or clay type) that do not have a free-flowing sandy texture must be mixed with anhydrous sodium sulfate until the sample is free flowing. Add 1 mL of surrogate standards to all samples, spikes, standards, and blanks. For the sample in each analytical batch selected for spiking, add 1.0 mL of a matrix spiking standard. For base/neutral acid analysis, the amount added of the surrogates and matrix spiking compounds should result in a final concentration of 100 ng/µL of each base/neutral analyte and 200 ng/µL of each acid analyte in the extract to be analyzed (assuming a 1-µL injection). Immediately add a 100-mL mixture of 1:1 methylene chloride:acetone and extract the sample ultrasonically for 3 min and then decant or filter the extracts. Repeat the extraction two or more times. Dry the extract using a column with anhydrous sodium sulfate and concentrate it to 1 mL in a K-D concentrator.

QUALITY CONTROL A methylene chloride solution containing 50 ng/µL of decafluorotriphenylphosphine (DFTPP) is used for tuning the GC/MS system each 12-h shift. A system performance check also must be made during every 12-h shift. A standard containing 50 ng/µL each of 4,4'-DDT, pentachlorophenol, and benzidine is required to verify injection port inertness and GC column performance. A calibration standard at mid-concentration, containing each compound of interest, including all required surrogates, must be performed every 12 h during analysis. After the system performance check is met, calibration check compounds (CCCs) are used to check the validity of the initial calibration.

The internal standard responses and retention times in the calibration check standard must be evaluated immediately after or during data acquisition. If the retention time for any internal standard changes by more than 30 seconds from the last check calibration (12 h), the chromatographic system must be inspected for malfunctions and corrections must be made, as required. If the electron ionization current plot (EICP) area for any of the internal standards changes by a factor of two from the last daily calibration standard check, the mass spectrometer must be inspected for malfunctions and corrections must be made, as appropriate.

Demonstrate, through the analysis of a reagent water blank, that interferences from the analytical system, glassware, and reagents are under control. The blank samples should be carried through all stages of the sample preparation and measurement steps. For each analytical batch (up to 20 samples), a reagent blank, matrix spike, and matrix spike duplicate/duplicate must be analyzed (the frequency of the spikes may be different for different monitoring programs). The blank and spiked samples must be carried through all stages of the sample preparation and measurement steps. A QC reference sample concentrate containing each analyte at a concentration of 100 mg/L in methanol is required.

REFERENCE Test Methods for Evaluating Solid Waste (SW-846). U.S. EPA 1983, Method 8270B, Rev. 2, Nov. 1990. Office of Solid Waste, Washington, DC.

Longifolene
CAS #475-20-7
EPA Method 1625

TITLE Semivolatile Organic Compounds by Isotope Dilution GC/MS

MATRIX The compounds may be determined in waters, soils, and municipal sludges by this method.

METHOD SUMMARY This method is used to determine 176 semivolatile toxic organic pollutants associated with the CWA (as amended 1987); the RCRA (as amended 1986); the CERCLA (as amended 1986); and other compounds amenable to extraction and analysis by capillary column gas chromatography-mass spectrometry (GC/MS).

Stable isotopically-labeled analogs of the compounds of interest are added to the sample. If the solids content is less than 1%, a 1-L sample is extracted at pH 12–13, then at pH <2 with methylene chloride using continuous extraction techniques.

If the solids content is 30% or less, the sample is diluted to 1% solids with reagent water, homogenized ultrasonically, and extracted at pH 12–13, then at pH <2 with methylene chloride using continuous extraction techniques. If the solids content is greater than 30%, the sample is extracted using ultrasonic techniques.

Each extract is dried over sodium sulfate, concentrated to a volume of 5 mL, cleaned up using GPC, if necessary, and concentrated. Extracts are concentrated to 1 mL if GPC is not performed, and to 0.5 mL if GPC is performed.

An internal standard is added to the extract, and a 1-mL aliquot of the extract is injected into the GC. The compounds are separated by GC and detected by a MS. The labeled compounds serve to correct the variability of the analytical technique.

INTERFERENCES Solvents, reagents, glassware, and other sample processing hardware may yield artifacts and/or elevated baselines causing misinterpretation of chromatograms and spectra. Materials used in the analysis must be demonstrated to be free from interferences under the conditions of analysis by running method blanks initially and with each sample lot (sample started through the extraction process on a given 8-h shift, to a maximum of 20). Specific selection of reagents and purification of solvents by distillation in all glass systems may be required. Glassware and, where possible, reagents are cleaned by solvent rinse and baking at 450°C for 1-h minimum. Interferences coextracted from samples will vary considerably from source to source, depending on the diversity of the site being sampled.

INSTRUMENTATION Major instrumentation includes a GC with a splitless or on-column injection port for capillary column, a MS with 70 eV electron impact ionization, and a data system to collect and record MS data, and process it. A K-D apparatus is used to concentrate extracts.

GC Column: 30 m × 0.25 mm I.D. 5% phenyl, 94% methyl, 1% vinyl silicone bonded phased fused silica capillary column.

PRECISION & ACCURACY The detection limits of the method are usually dependent on the level of interferences rather than instrumental limitations. The limits typify the minimum quantities that can be detected with no interferences present.

The minimum level (in µg/mL) was not listed. This is defined as a minimum level at which the analytical system shall give recognizable mass spectra (background corrected) and acceptable calibration points.

The MDL (in µg/kg) in low solids was not listed and in high solids was not listed; these were determined in digested sludge (low solids) and in filter cake or compost (high solids).

The labeled and native compound initial precision as standard deviation (in µg/L) was not listed.
The labeled and native compound initial accuracy as average recovery (in µg/L) was not listed.

SAMPLE COLLECTION, PRESERVATION & HANDLING
Collect samples in glass containers. Aqueous samples which flow freely are collected in refrigerated bottles using automatic sampling equipment. Solid samples are collected as grab samples using widemouth jars. Maintain samples at 0 to 4°C from the time of collection until extraction. If residual chlorine is present in aqueous samples, add 80 mg sodium thiosulfate/L of water. Begin sample extraction within 7 days of collection, and analyze all extracts within 40 days of extraction.

SAMPLE PREPARATION Samples containing 1% solids or less are extracted directly using continuous liquid-liquid extraction techniques. Samples containing 1 to 30% solids are diluted to the 1% level with reagent water and extracted using continuous liquid-liquid extraction techniques. Samples containing greater than 30% solids are extracted using ultrasonic techniques.

Base/neutral extraction — Adjust the pH of the waters in the extractors to 12–13 with 6 N NaOH. Extract with methylene chloride for 24–48 h.
Acid extraction — Adjust the pH of the waters in the extractors to 2 or less using 6 N sulfuric acid. Extract with methylene chloride for 24–48 h.
Ultrasonic extraction of high solids samples — Add anhydrous sodium sulfate to the sample and QC aliquot(s). Add acetone:methylene chloride (1:1) to the sample and mix thoroughly

Concentrate extracts using a K-D apparatus.

QUALITY CONTROL The analyst is permitted to modify this method to improve separations or lower the costs of measurements, provided all performance specifications are met. Analyses of blanks are required to demonstrate freedom from contamination. When results of spikes indicate atypical method performance for samples, the samples are diluted to bring method performance within acceptable limits.

For low solids (aqueous samples), extract, concentrate, and analyze two sets of four 1-L aliquots (8 aliquots total) of the precision and recovery standard. For high solids samples, two sets of four 30-g aliquots of the high solids reference matrix are used.

Spike all samples with labeled compounds to assess method performance. Compute percent recovery of the labeled compounds using the internal standard method. Compare the labeled compound recovery for each compound with the corresponding labeled compound recovery.

Reagent water and high solids reference matrix blanks are analyzed to demonstrate freedom from contamination. Extract and concentrate a 1-L reagent water blank or a high solids reference matrix blank with each sample's lot (samples started through the extraction process on the same 8-h shift, to a maximum of 20 samples).

Field replicates may be collected to determine the precision of the sampling technique, and spiked samples may be required to determine the accuracy of the analysis when the internal standard method is used.

REFERENCE Semivolatile Organic Compounds by Isotope Dilution GC/MS. Office of Water Regulation and Standards, U.S. EPA Industrial Technology Division, Washington, DC, EPA Method 1625, Rev. C, June 1989 (contact W.A. Telliard, U.S. EPA, Office of Water Regulations and Standards, 401 M St., SW, Washington, DC, 20460. Phone: 202-382-7131).

M

Magnesium
CAS #7439-95-4

EPA Method 6010

TITLE Inductively Coupled Plasma-Atomic Emission Spectroscopy

MATRIX This method is applicable to the determination of trace elements, including metals, in groundwater, soils, sludges, sediments, and other solid wastes. All matrices require digestion prior to analysis. The method of standard addition must be used for the analysis of all sample digests unless either serial dilution or matrix spike addition demonstrates it is not required.

METHOD SUMMARY Method 6010 covers 25 elements using ICP analysis. It measures element-emitted light by optical spectrometry. Samples, following an appropriate acid digestion, are nebulized and the resulting aerosol is transported to the plasma torch. Element-specific atomic line emission spectra are produced by a radio-frequency inductively coupled plasma.

INTERFERENCES Interferences may be categorized as spectral or non-spectral. Spectral interferences are caused by overlap of a spectral line from another element, unresolved overlap of molecular band spectra, background contribution from continuous or recombination phenomenon, and stray light from the line emission of high concentration elements. Non-spectral interferences include physical and chemical interferences. Physical interferences are effects associated with the sample nebulization and transport processes. Changes in viscosity and surface tension can cause significant inaccuracies. Chemical interferences include molecular compound formation, ionization effects, and solute vaporization effects. Normally these effects are not significant and can be minimized by careful selection of operating conditions. Chemical interferences are highly dependent on matrix type and the specific analyte element.

INSTRUMENTATION An inductively coupled argon plasma emission spectrometer (ICP) capable of background correction is required.

PRECISION & ACCURACY Detection limits, sensitivity, and optimum ranges of the metals will vary with the matrices and model of the spectrometer. In a single lab evaluation, seven wastes were analyzed for 22 elements. The mean percent relative standard deviation from triplicate analyses for all elements and wastes was 9 ± 2%. The mean percent recovery of spiked elements for all wastes was 93 ± 6%. Spike levels ranged from 100 μg/L to 100 mg/L. The wastes included sludges and industrial wastewaters.

Estimated instrument detection limit in μg/L is 30.
Spiked concentration in μg/L is not listed.
Mean reported value in μg/L is not listed.
Precision as RSD % is not listed.

SAMPLING METHOD Samples should be collected in borosilicate glass, linear polyethylene, polypropylene, or Teflon® bottles that have been prewashed with detergent and tap water, and rinsed with 1:1 nitric acid and tap water or 1:1 hydrochloric acid and tap water. Collect at least 2 g of solids and 200 mL of aqueous samples.

SAMPLE PRESERVATION Add nitric acid to make the samples pH <2.

MHT The maximum holding time for properly preserved samples is 6 months.

SAMPLE PREPARATION Preliminary treatment of most matrices is necessary because of the complexity and variability of sample matrices. Water samples that have been prefiltered and acidified will not need acid digestion. Methods for acid digestion of waters for total recoverable or dissolved metals, acid digestions of aqueous samples and extracts for total metals, and acid digestion of sediments, sludges, and soils are summarized below.

Total recoverable or dissolved metals in water — To prepare surface and groundwater samples for determination of total recoverable and dissolved metals, a 100-mL aliquot of well-mixed sample is acidified with concentrated nitric acid and concentrated hydrochloric acid, then heated until the volume is reduced to 15–20 mL. Adjust the final volume to 100 mL with reagent water.

Total metals in aqueous samples, soil, and sediment extracts — To prepare aqueous samples, soil and sediment extracts, and wastes that contain suspended solids, a 100-mL aliquot is made acidic with concentrated nitric acid and the solution is evaporated to about 5 mL on a hot plate. Continue heating and adding additional acid until sample digestion is complete, which is usually indicated when the digestate is light in color or does not change in appearance. Evaporate the solution to about 3 mL and cool it and add a small quantity of 1:1 hydrochloric acid (10 mL/100 mL of final solution). Cover the beaker and reflux for 15 min. Wash down the beaker walls and filters or centrifuge the sample to remove silicates and other insoluble material. Filter the sample and adjust the final volume to 100 mL with reagent water and the final acid concentration to 10%.

Sediments, sludges, and soils — To prepare sediments, sludges and soil samples, transfer 1–2 g to a conical beaker and add 10 mL of 1:1 nitric acid, mix the slurry, and cover it with a watch glass. Heat the sample and reflux for 10–15 min without boiling. Allow it to cool, then add 5 mL of concentrated nitric acid and reflux for 30 min. Repeat last step and then allow the solution to evaporate to 5 mL without boiling. Cool and add 2 mL of water and 3 mL of 30% hydrogen peroxide. Cover and place the beaker on the hot plate. Heat and add 30% hydrogen peroxide in 1-mL aliquots with warming until the effervescence is minimal but do not add more than a total of 10 mL of 30% hydrogen peroxide. If the sample is being prepared for the analysis of Ag, Al, As, Ba, Be, Ca, Cd, Co, Cr, Cu, Fe, K, Mg, Mn, Mo, Na, Ni, Os, Pb, Se, Tl, V, and Zn, then add 5 mL of concentrated hydrochloric acid and 10 mL of water and return the covered beaker to a hot plate for 15 min of additional refluxing without boiling. Dilute the sample to a 100 mL volume with water after cooling and filter or centrifuge to remove particulates.

QUALITY CONTROL Laboratory control samples must be analyzed for each analytical method. A method blank should be analyzed with each batch of samples. The effect of the matrix on method performance must be demonstrated: when appropriate, there should be at least one matrix spike and either one matrix duplicate or one matrix spike duplicate per analytical batch. The bias and precision of the method, as well as the method detection limit for each specific matrix type, must be measured.

Dilute and reanalyze samples that are more concentrated than the linear calibration limit. Employ a minimum of one reagent blank per sample batch to determine if contamination or any memory effects are occurring. Whenever a new or unusual sample matrix is encountered, perform either a serial dilution test or a matrix spike addition test to ensure that neither positive or negative interferences are operating on any of the analyte elements. Check the instrument standardization by verifying calibration every 10 samples using a calibration blank and a check standard.

REFERENCE Test Methods for Evaluating Solid Waste (SW-846). U.S. EPA. 1983. Method 6010, Rev. 0, Sept. 1986. Office of Solid Wastes, Washington, DC.

Magnesium EPA Method 200.7

TITLE Inductively Coupled Plasma

MATRIX Dissolved, suspended or (ICP) total element in drinking and surface waters and in domestic and industrial wastewaters.

APPLICATION The method covers the determination of 25 metals. Dissolved elements are determined in filtered and acidified samples after appropriate digestion (which increases dissolved solids). Its primary advantage is that ICP instruments allow simultaneous or rapid sequential determination of many elements in a short time. Samples are first nebulized and the aerosol is transported to a plasma torch in which element specific atomic line emission spectra are produced by a radio frequency inductively coupled plasma. Background correction is required for trace element detection except in the case of line broadning.

INTERFERENCES There are spectral, physical, and chemical interferences. The primary disadvantage of ICP instruments is background radiation from other elements and the plasma gases (spectral interferences). Changes in sample viscosity and surface tension with samples containing high dissolved solids (especially those exceeding 1500 mg/L) or high acid concentrations can cause physical interferences. Ionization effects, solute vaporization and molecular compound formation can cause chemical interferences. Calcium, chromium, iron, manganese, thallium and vanadium can cause interference at the 100 mg/L level.

INSTRUMENTATION Inductively coupled argon plasma emission spectroscopy. 279.079 nm wavelength.

RANGE Not listed.

MDL 30 µg/L.

PRECISION Not listed.

ACCURACY Mean recovery = 93% ± 6% of spiked elements for all wastes.

SAMPLING METHOD Wash sample container with detergent and tap water, rinse with 1 + 1 nitric acid and tap water, then rinse with 1 + 1 hydrochloric acid and tap water, then rinse with deionized, distilled water in that order. Perform any filtration or acid preservation steps when the sample is collected or as soon as possible thereafter.

STABILITY Cool samples to 4°C.

MHT 24 h.

QUALITY CONTROL Mixed calibration standards, an instrument check standard, and an interference check solution are used in addition to a quality control sample. The quality control sample should be prepared in the same acid matrix as the calibration standards at 10 times the instrumental detection limits and in accordance with the instructions provided by the supplier. Furthermore, two types of blanks are required: a calibration blank and a reagent blank.

REFERENCE Method 200.7, U.S. EPA, EMSL-Cincinnati, OH, Nov. 1980

Magnesium EPA Method 242.1

TITLE Metals (Total, Dissolved, Atomic Absorption (AA))

MATRIX Drinking, surface, and saline suspended) direct aspiration waters. Wastewater.

APPLICATION Date issued 1971. Editorial Rev. 1974 and 1978. Sample is aspirated and atomized in a flame. Light beam from hollow cathode (made of Mg) lamp is directed through the flame into monochromator, then to detector which measures amount absorbed light.

INTERFERENCES Aluminum interference at conc's > 2 mg/L masked by lanthanum addition. Phosphate interference is overcome by lanthanum addition. There can be interference from presence of high dissolved solids. Chemical and ionization interferences can occur.

INSTRUMENTATION AAS. Magnesium (Mg) hollow cathode lamp. Burner. Pipets. Strip chart recorder.

RANGE 0.02–0.5 mg/L at 282.5 nm wavelength.

MDL 0.001 mg/L.

PRECISION SD = ±(0.1 and 0.2) at 2.1 and 8.2 mg Mg/L.

ACCURACY Recoveries = 100% at 2.1 and 8.2 mg Mg/L.

SAMPLING METHOD Prewashed plastic or glass containers.

STABILITY HNO3 to pH <2.

MHT 6 months.

QUALITY CONTROL After calibration curve composed of a minimum of a reagent blank and 3 standards has been prepared, subsequent calibration curves must be verified by use of at least a reagent blank and one standard near MCL. Must check within 10% of original curve. (For drinking water analysis).

REFERENCE Methods for the Chemical Analysis of Water and Wastes, EPA-600/4-79-020, U.S. EPA, EMSL, 1979.

Magnesium   **EPA Method 7450**
CAS #7439-95-4

TITLE Atomic Absorption (AA)

MATRIX Drinking, surface, and direct aspiration saline waters. Wastewater.

APPLICATION Sample is aspirated and atomized in a flame. A light beam from a Mg hollow cathode lamp is directed through the flame into a monochromator and onto a detector. Since wavelength of light beam is specific for Mg, light energy absorbed by detector is measure of magnesium.

INTERFERENCES The most troublesomee type is chemical, caused by lack of absorption of atoms bound in molecular combination in the flame. High dissolved solids in sample may result in nonatomic absorbance interference. Add lanthanum to prevent complexing problems.

INSTRUMENTATION Atomic absorption spectrometer. Magnesium hollow cathode lamp. (285.2 nm wavelength).

RANGE 0.02–0.05 mg/L.

MDL 0.001 mg/L

PRECISION Standard deviation = ± (0.1 and 0.2) at (2.1 and 8.2) mg Mg/L

ACCURACY Recoveries = (100 and 100)% at (2.1 and 8.2) mg Mg/L

SAMPLING METHOD Use glass or plastic containers. Collect 200 g of solids and 600 mL of liquid samples.

STABILITY Cool solid samples to 4°C and analyze as soon as possible. Add nitric acid to liquid samples to pH <2.

MHT 6 months.

QUALITY CONTROL At least one duplicate and one spike sample should be run every 20 samples or with each matrix type to verify precision of the method. For 20 or more samples per day, verify working standard curve. Run an additional standard at or near mid-range every 10 samples.

REFERENCE Method 7450, SW-846, 3rd ed., Nov.1986.

Malachite green   **EPA Method 1625**
CAS #569-64-2

TITLE Semivolatile Organic Compounds by Isotope Dilution GC/MS

MATRIX The compounds may be determined in waters, soils, and municipal sludges by this method.

METHOD SUMMARY This method is used to determine 176 semivolatile toxic organic pollutants associated with the CWA (as amended 1987); the RCRA (as amended 1986); the CERCLA (as amended 1986); and other compounds amenable to extraction and analysis by capillary column gas chromatography-mass spectrometry (GC/MS).

Stable isotopically-labeled analogs of the compounds of interest are added to the sample. If the solids content is less than 1%, a 1-L sample is extracted at pH 12–13, then at pH <2 with methylene chloride using continuous extraction techniques.

If the solids content is 30% or less, the sample is diluted to 1% solids with reagent water, homogenized ultrasonically, and extracted at pH 12–13, then at pH <2 with methylene chloride using continuous extraction techniques. If the solids content is greater than 30%, the sample is extracted using ultrasonic techniques.

Each extract is dried over sodium sulfate, concentrated to a volume of 5 mL, cleaned up using GPC, if necessary, and concentrated. Extracts are concentrated to 1 mL if GPC is not performed, and to 0.5 mL if GPC is performed.

An internal standard is added to the extract, and a 1-mL aliquot of the extract is injected into the GC. The compounds are separated by GC and detected by a MS. The labeled compounds serve to correct the variability of the analytical technique.

INTERFERENCES Solvents, reagents, glassware, and other sample processing hardware may yield artifacts and/or elevated baselines causing misinterpretation of chromatograms and spectra. Materials used in the analysis must be demonstrated to be free from interferences under the conditions of analysis by running method blanks initially and with each sample lot (sample started through the extraction process on a given 8-h shift, to a maximum of 20). Specific selection of reagents and purification of solvents by distillation in all glass systems may be required. Glassware and, where possible, reagents are cleaned by solvent rinse and baking at 450°C for 1-h minimum. Interferences coextracted from samples will vary considerably from source to source, depending on the diversity of the site being sampled.

INSTRUMENTATION Major instrumentation includes a GC with a splitless or on-column injection port for capillary column, a MS with 70 eV electron impact ionization, and a data system to collect and record MS data, and process it. A K-D apparatus is used to concentrate extracts.

GC Column: 30 m × 0.25 mm I.D. 5% phenyl, 94% methyl, 1% vinyl silicone bonded phased fused silica capillary column.

PRECISION & ACCURACY The detection limits of the method are usually dependent on the level of interferences rather than instrumental limitations. The limits typify the minimum quantities that can be detected with no interferences present.

The minimum level (in µg/mL) was not listed. This is defined as a minimum level at which the analytical system shall give

recognizable mass spectra (background corrected) and acceptable calibration points.

The MDL (in μg/kg) in low solids was not listed and in high solids was not listed; these were determined in digested sludge (low solids) and in filter cake or compost (high solids).

The labeled and native compound initial precision as standard deviation (in μg/L) was not listed.

The labeled and native compound initial accuracy as average recovery (in μg/L) was not listed.

SAMPLE COLLECTION, PRESERVATION & HANDLING
Collect samples in glass containers. Aqueous samples which flow freely are collected in refrigerated bottles using automatic sampling equipment. Solid samples are collected as grab samples using widemouth jars. Maintain samples at 0 to 4°C from the time of collection until extraction. If residual chlorine is present in aqueous samples, add 80 mg sodium thiosulfate/L of water. Begin sample extraction within 7 days of collection, and analyze all extracts within 40 days of extraction.

SAMPLE PREPARATION Samples containing 1% solids or less are extracted directly using continuous liquid-liquid extraction techniques. Samples containing 1 to 30% solids are diluted to the 1% level with reagent water and extracted using continuous liquid-liquid extraction techniques. Samples containing greater than 30% solids are extracted using ultrasonic techniques.

- Base/neutral extraction — Adjust the pH of the waters in the extractors to 12–13 with 6 *N* NaOH. Extract with methylene chloride for 24–48 h.
- Acid extraction — Adjust the pH of the waters in the extractors to 2 or less using 6 *N* sulfuric acid. Extract with methylene chloride for 24–48 h.
- Ultrasonic extraction of high solids samples — Add anhydrous sodium sulfate to the sample and QC aliquot(s). Add acetone:methylene chloride (1:1) to the sample and mix thoroughly

Concentrate extracts using a K-D apparatus.

QUALITY CONTROL The analyst is permitted to modify this method to improve separations or lower the costs of measurements, provided all performance specifications are met. Analyses of blanks are required to demonstrate freedom from contamination. When results of spikes indicate atypical method performance for samples, the samples are diluted to bring method performance within acceptable limits.

For low solids (aqueous samples), extract, concentrate, and analyze two sets of four 1-L aliquots (8 aliquots total) of the precision and recovery standard. For high solids samples, two sets of four 30-g aliquots of the high solids reference matrix are used.

Spike all samples with labeled compounds to assess method performance. Compute percent recovery of the labeled compounds using the internal standard method. Compare the labeled compound recovery for each compound with the corresponding labeled compound recovery.

Reagent water and high solids reference matrix blanks are analyzed to demonstrate freedom from contamination. Extract and concentrate a 1-L reagent water blank or a high solids reference matrix blank with each sample's lot (samples started through the extraction process on the same 8-h shift, to a maximum of 20 samples).

Field replicates may be collected to determine the precision of the sampling technique, and spiked samples may be required to determine the accuracy of the analysis when the internal standard method is used.

REFERENCE Semivolatile Organic Compounds by Isotope Dilution GC/MS. Office of Water Regulation and Standards, U.S. EPA Industrial Technology Division, Washington, DC, EPA Method 1625, Rev. C, June 1989 (contact W.A. Telliard, U.S. EPA, Office of Water Regulations and Standards, 401 M St., SW, Washington, DC, 20460. Phone: 202-382-7131).

Malathion   **EPA Method 8141**
CAS #121-75-5

TITLE Organophosphorus Compounds by Gas Chromatography: Capillary Column Technique

MATRIX This method covers aqueous and solid matrices. This includes a wide variety such as drinking water, groundwater, industrial wastewaters, surface waters, soils, solids, and sediments.

METHOD SUMMARY This is a GC method used to determine the concentration of 28 organophosphorus pesticides.

The use of Gel Permeation Cleanup (EPA Method 3640) for sample cleanup has been demonstrated to yield recoveries of less than 85% for many method analytes and is therefore not recommended for use with this method.

This method provides GC conditions for the detection of ppb concentrations of organophosphorus compounds. Prior to the use of this method, appropriate sample preparation techniques must be used. Water samples are extracted at a neutral pH with methylene chloride as a solvent by using a separatory funnel (EPA Method 3510) or a continuous liquid-liquid extractor (EPA Method 3520). Soxhlet extraction (EPA Method 3540) or ultrasonic extraction (EPA Method 3550) using methylene chloride/acetone (1:1) are used for solid samples. Both neat and diluted organic liquids (EPA Method 3580) may be analyzed by direct injection. Spiked samples are used to verify the applicability of the chosen extraction technique to each new sample type. A GC with a flame photometric (FPD) or nitrogen-phosphorus detector (NPD) is used for this multiresidue procedure.

INTERFERENCES The use of Florisil cleanup materials (EPA Method 3620) for some of the compounds in this method has been demonstrated to yield recoveries less than 85% and is therefore not recommended for all compounds. Use of phosphorus or halogen specific detectors, however, often obviates the necessity for cleanup for relatively clean sample matrices. If particular circumstances demand the use of an alternative cleanup procedure, the analyst must determine the elution profile and demonstrate that the recovery of each analyte is no less than 85%.

Use of a flame photometric detector (FPD) in the phosphorus mode will minimize interferences from materials that do not contain phosphorus. Elemental sulfur, however, may interfere with the determination of certain organophosphorus compounds by flame photometric gas chromatography. Sulfur cleanup using EPA Method 3660 may alleviate this interference. A nitrogen phosphorus detector (NPD) is also recommended.

A few analytes coelute on certain columns. Therefore, select a second column for confirmation where coelution of the analytes of interest does not occur.

Method interferences may be caused by contaminants in solvents, reagents, glassware, and other sample processing hardware that lead to discrete artifacts or elevated baselines in gas chromatograms. All these materials must be routinely demonstrated to be free from interferences under the conditions of the analysis by analyzing reagent blanks.

INSTRUMENTATION A GC with a NPD or a FPD will be needed. A data system or integrator is recommended for measuring peak areas and/or peak heights. A Kuderna-Danish (K-D) apparatus will be needed for extract concentration.

Column 1: 15 m × 0.53 mm megabore capillary column, 1.0 μm film thickness, DB-210.
Column 2: 15 m × 0.53 mm megabore capillary column, 1.5 μm film thickness, SPB-608.
Column 3: 15 m × 0.53 mm megabore capillary column, 1.0 μm film thickness, DB-5.

Three megabore capillary columns are included for analysis of organophosphates by this method. Column 1 (DB-210 or equivalent) and Column 2 (SPB-608 or equivalent) are recommended if a large number of organophosphorus analytes are to be determined. If the superior resolution offered by Column 1 and Column 2 is not required, Column 3 (DB-5 or equivalent) may be used. For megabore capillary columns, automatic injections of 1 μL are recommended.

PRECISION & ACCURACY The MDL actually achieved in a given analysis will vary, as it is dependent on instrument sensitivity and matrix effects. Single operator accuracy and precision studies have been conducted with spiked water and soil samples.

MULTIPLICATION FACTORS FOR OTHER MATRICES (a)

Matrix	Factor (b)
Groundwater (EPA Method 3510 or EPA Method 3520)	10
Low-concentration soil by Soxhlet and no cleanup	10 (c)
Low-concentration soil by ultrasonic extraction with GPC cleanup	6.7 (c)
High-concentration soil and sludges by ultrasonic extraction	500 (c)
Non-water miscible waste (EPA Method 3580)	1000 (c)

(a) Sample EQLs are highly matrix-dependent. The EQLs listed here are provided for guidance and may not always be achievable.
(b) EQL = [Method detection limit] × [Factor]. For non-aqueous samples the factor is on a wet-weight basis.
(c) Multiply this factory times the soil MDL.

The MDL (in μg/L) when reagent water was extracted using a separatory funnel was 0.11.
The MDL (in μg/kg) when soil was extracted using Soxhlet extraction (EPA Method 3540) was 5.5.
Accuracy (as % recovery) with separatory funnel extraction ranged from 127 (with low spikes) to 86 (with high spikes).
Accuracy (as % recovery) with continuous liquid-liquid extraction ranged from 105 (with low spikes) to 86 (with high spikes).
Accuracy (as % recovery) with Soxhlet extraction of soils ranged from 100 (with low spikes to 81 (with high spikes).
Accuracy (as % recovery) with ultrasonic extraction of soils ranged from 55 (with low spikes) to 31 (with high spikes).

SAMPLE COLLECTION, PRESERVATION & HANDLING
Containers used to collect samples for the determination of semivolatile organic compounds should be soap and water washed followed by methanol (or isopropanol) rinsing. The sample containers should be of glass or Teflon® and have screw-top covers with Teflon® liners.

No preservation is used with concentrated waste samples. With liquid samples containing no residual chlorine and with soil, sediment, and sludge samples, immediately cooling to 4°C is the only preservation used. When residual chlorine is present then 3 mL of 10% aqueous sodium sulfate is added for each gallon of sample collected, followed by cooling to 4°C.

Liquid samples must be extracted within 7 days and their extracts analyzed within 40 days. Concentrated waste, soil, sediment, and sludge samples must be extracted within 14 days and their extracts analyzed within 40 days.

SAMPLE PREPARATION In general, water samples are extracted at a neutral pH with methylene chloride, using either EPA Method 3510 or EPA Method 3520. Solid samples are extracted using either EPA Method 3540 or EPA Method 3550 with methylene chloride/acetone (1:1) as the extraction solvent.

Prior to GC analysis, the extraction solvent may be exchanged to hexane. Single lab data indicates that samples should not be transferred with 100% hexane during sample workup as the more water soluble organophosphorus compounds may be lost.

If cleanup is performed on the samples, the analyst should analyze the samples by GC. This will confirm elution patterns and the absence of interferences from the reagents. If peak detection and identification is prevented by the presence of interferences, further cleanup is required.

QUALITY CONTROL The analyst should monitor the performance of the extraction, cleanup (when used), and analytical system and the effectiveness of the method in dealing with each sample matrix by spiking each sample, standard, and blank with one or two surrogates (e.g., organophosphorus compounds not expected to be present in the sample). Deuterated analogs of analytes should not be used as surrogates for gas chromatographic analysis due to coelution problems.

A minimum of five concentrations for each analyte of interest should be prepared through dilution of the stock standards with isooctane. One of the concentrations should be at a concentration near, but above, the MDL.

Include a mid-level check standard after each group of 10 samples in the analysis sequence. GC/MS techniques should be judiciously employed to support qualitative identifications made with this method. Follow the GC/MS operating requirements specified in EPA Method 8270.

When available, chemical ionization mass spectra may be employed to aid in the qualitative identification process. To confirm an identification of a compound, the background-corrected mass spectrum of the compound must be obtained from the sample extract and must be compared with a mass spectrum from a stock or calibration standard analyzed under the same chromatographic conditions. The molecular ion and all other ions present above 20% relative abundance in the mass spectrum of the standard must be present in the mass spectrum of the sample with agreement to ±20%. The retention time of the compound in the sample must be within six seconds of the retention time for the same compound in the standard solution.

Should the MS procedure fail to provide satisfactory results, additional steps may be taken before reanalysis. These steps may include the use of alternate packed or capillary GC columns or additional sample cleanup.

REFERENCE Test Methods for Evaluating Solid Waste, Physical/Chemical Methods, SW-846, 3rd Edition, U.S. EPA, Office of Solid Waste, Washington, DC, EPA Method 8141 July 1992.

Malathion   **EPA Method 8270**
CAS #121-75-5

TITLE Semivolatile Organic Compounds by GC/MS

MATRIX This method is used to determine the concentration of semivolatile organic compounds in extracts prepared from all types of solid waste matrices, soils, and groundwater. Although surface waters are not specifically mentioned, this method should be applicable to water samples from rivers, lakes, etc.

METHOD SUMMARY This method covers 259 semivolatile organic compounds. In very limited applications direct injection of the sample into the GC/MS system may be appropriate, but this results in very high detection limits (approximately 10,000 µg/L). Typically, a 1-L liquid sample, containing surrogate, and matrix spiking standards, is extracted in a continuous extractor first under acid conditions and then under basic conditions. Typically 30 g of a solid sample, containing surrogate, and matrix spiking standards, is extracted ultrasonically. After concentrating the extract to 1 mL it is spiked with 10 µL of an internal standard solution just prior to analysis by GC/MS. The volume injected should contain about 100 ng of base/neutral and 200 ng of acid surrogates (for a 1-µL injection). Analysis is performed by GC/MS using a capillary GC column.

INTERFERENCES Raw GC/MS data from all blanks, samples, and spikes must be evaluated for interferences. Contamination by carryover can occur whenever high-concentration and low-concentration samples are sequentially analyzed. To reduce carryover, the sample syringe must be rinsed out between samples with solvent. Whenever an unusually concentrated sample is encountered, it should be followed by the analysis of blank solvent to check for cross-contamination.

INSTRUMENTATION A GC/MS and a data system are required. The GC column used is a 30 m × 0.25 mm I.D. (or 0.32 mm I.D.) 1um film thickness silicone-coated fused silica capillary column. A continuous liquid-liquid extractor equipped with Teflon® or glass connection joints and stopcocks requiring no lubrication, a K-D concentrating apparatus, water bath, and an ultrasonic disrupter with a minimum power of 300 W and with pulsing capability are also required.

PRECISION & ACCURACY The estimated quantitation limit (EQL) of Method 8270B for determining an individual compound is approximately 1 mg/kg (wet weight) for soil or sediment samples, 1–200 mg/kg for wastes (dependent on matrix and method of preparation), and 10 µg/L for groundwater samples. EQLs will be proportionately higher for sample extracts that require dilution to avoid saturation of the detector.

The EQL(b) for groundwater in µg/L is 50.
The EQL (a, b) for low concentrations in soil and sediment in µg/kg is not determined.
Accuracy as µg/L is not listed.
Overall precision in µg/L is not listed.

(a) *EQLs listed for soil/sediment are based on wet weight. Normally data is reported in a dry-weight basis; therefore, EQLs will be higher based on the % dry weight of each sample. This calculation is based on a 30 g sample and gel permeation chromatography cleanup.*

(b) *Sample EQLs are highly matrix-dependent. The EQLs are provided for guidance and may not always be achievable.*

$C =$ *True value for concentration, in µg/L.*
$X =$ *Average recovery found for measurements of samples containing a concentration of C, in µg/L.*

ESTIMATED QUANTITATION LIMIT

Other Matrices	Factor (a)
High-concentration soil and sludges by sonicator	7.5
Non-water miscible waste	75

(a) EQL for other matrices = [EQL for low soil/sediment] × [Factor]. This estimated EQL is similar to an EPA "Practical Quantitation Limit."

SAMPLING METHOD

Liquid samples — Use a 1 or 2½ gallon amber glass bottle with a screw-top Teflon®-lined cover that has been prewashed with detergent and rinsed with distilled water and methanol (or isopropanol).

Soils, sediments, or sludges — Use an 8-oz. widemouth glass with a screw-top Teflon®-lined cover that has been prewashed with detergent and rinsed with distilled water and methanol (or isopropanol).

SAMPLE PRESERVATION

Liquid samples — If residual chlorine is present, add 3 mL of 10% sodium thiosulfate per gallon, cool to 4°C and store in a solvent-free refrigerator until analysis; if chlorine is not present, then eliminate the sodium thiosulfate addition.

Soils, sediments, or sludges — Cool samples to 4°C and store in a solvent-free refrigerator.

MHT Liquid samples must be extracted within 7 days and the extracts analyzed within 40 days. Soils, sediments, or sludges may be stored for a maximum of 14 days and the extracts analyzed within 40 days.

SAMPLE PREPARATION
Liquid samples — Transfer 1 L quantitatively to a continuous extractor. If high concentrations are anticipated, a smaller volume may be used and then diluted with organic-free reagent water to 1 L. Adjust pH, if necessary, to pH <2 using 1:1 (V/V) sulfuric acid. Pipette 1.0 mL of a surrogate standard spiking solution into each sample. For the sample in each analytical batch selected for spiking, add 1.0 mL of a matrix spiking standard. For base/neutral acid analysis, the amount of the surrogates and matrix spiking compounds added to the sample should result in a final concentration of 100 ng/µL of each analyte in the extract to be analyzed (assuming a 1-µL injection). Extract with methylene chloride for 18–24 h. Next, adjust the pH of the aqueous phase to pH >11 using 10 N sodium hydroxide and extract it with methylene chloride again for 18–24 h. Dry the extract through a column containing anhydrous sodium sulfate and concentrate it to 1 mL using a K-D concentrator.

Soils, sediments, or sludges — Use 30 g of sample. Nonporous or wet samples (gummy or clay type) that do not have a free-flowing sandy texture must be mixed with anhydrous sodium sulfate until the sample is free flowing. Add 1 mL of surrogate standards to all samples, spikes, standards, and blanks. For the sample in each analytical batch selected for spiking, add 1.0 mL of a matrix spiking standard. For base/neutral acid analysis, the amount added of the surrogates and matrix spiking compounds should result in a final concentration of 100 ng/µL of each base/neutral analyte and 200 ng/µL of each acid analyte in the extract to be analyzed (assuming a 1-µL injection). Immediately add a 100-mL mixture of 1:1 methylene chloride:acetone and extract the sample ultrasonically for 3 min and then decant or filter the extracts. Repeat the extraction two or more times. Dry the extract using a column with anhydrous sodium sulfate and concentrate it to 1 mL in a K-D concentrator.

QUALITY CONTROL A methylene chloride solution containing 50 ng/µL of decafluorotriphenylphosphine (DFTPP) is used for tuning the GC/MS system each 12-h shift. A system performance check also must be made during every 12-h shift. A standard containing 50 ng/µL each of 4,4′-DDT, pentachlorophenol, and benzidine is required to verify injection port inertness and GC column performance. A calibration standard at mid-concentration, containing each compound of interest, including all required surrogates, must be performed every 12 h during analysis. After the system performance check is met, calibration check compounds (CCCs) are used to check the validity of the initial calibration.

The internal standard responses and retention times in the calibration check standard must be evaluated immediately after or during data acquisition. If the retention time for any internal standard changes by more than 30 seconds from the last check calibration (12 h), the chromatographic system must be inspected for malfunctions and corrections must be made, as required. If the electron ionization current plot (EICP) area for any of the internal standards changes by a factor of two from the last daily calibration standard check, the mass spectrometer must be inspected for malfunctions and corrections must be made, as appropriate.

Demonstrate, through the analysis of a reagent water blank, that interferences from the analytical system, glassware, and reagents are under control. The blank samples should be carried through all stages of the sample preparation and measurement steps. For each analytical batch (up to 20 samples), a reagent blank, matrix spike, and matrix spike duplicate/duplicate must be analyzed (the frequency of the spikes may be different for different monitoring programs). The blank and spiked samples must be carried through all stages of the sample preparation and measurement steps. A QC reference sample concentrate containing each analyte at a concentration of 100 mg/L in methanol is required.

REFERENCE Test Methods for Evaluating Solid Waste (SW-846). U.S. EPA 1983, Method 8270B, Rev. 2, Nov. 1990. Office of Solid Waste, Washington, DC.

Maleic anhydride   EPA Method 8270
CAS #108-31-6

TITLE Semivolatile Organic Compounds by GC/MS

MATRIX This method is used to determine the concentration of semivolatile organic compounds in extracts prepared from all types of solid waste matrices, soils, and groundwater. Although surface waters are not specifically mentioned, this method should be applicable to water samples from rivers, lakes, etc.

METHOD SUMMARY This method covers 259 semivolatile organic compounds. In very limited applications direct injection of the sample into the GC/MS system may be appropriate, but this results in very high detection limits (approximately 10,000 µg/L). Typically, a 1-L liquid sample, containing surrogate, and matrix spiking standards, is extracted in a continuous extractor first under acid conditions and then under basic conditions. Typically 30 g of a solid sample, containing surrogate, and matrix spiking standards, is extracted ultrasonically. After concentrating the extract to 1 mL it is spiked with 10 µL of an internal standard solution just prior to analysis by GC/MS. The volume injected should contain about 100 ng of base/neutral and 200 ng of acid surrogates (for a 1-µL injection). Analysis is performed by GC/MS using a capillary GC column.

INTERFERENCES Raw GC/MS data from all blanks, samples, and spikes must be evaluated for interferences. Contamination by carryover can occur whenever high-concentration and low-concentration samples are sequentially analyzed. To reduce carryover, the sample syringe must be rinsed out between samples with solvent. Whenever an unusually concentrated sample is encountered, it should be followed by the analysis of blank solvent to check for cross-contamination.

INSTRUMENTATION A GC/MS and a data system are required. The GC column used is a 30 m × 0.25 mm I.D. (or 0.32 mm I.D.) 1um film thickness silicone-coated fused silica capillary column. A continuous liquid-liquid extractor equipped with Teflon® or glass connection joints and stopcocks requiring no lubrication, a K-D concentrating apparatus, water bath, and an ultrasonic disrupter with a minimum power of 300 W and with pulsing capability are also required.

PRECISION & ACCURACY The estimated quantitation limit (EQL) of Method 8270B for determining an individual compound is approximately 1 mg/kg (wet weight) for soil or sediment samples, 1–200 mg/kg for wastes (dependent on matrix and method of preparation), and 10 µg/L for groundwater samples. EQLs will be proportionately higher for sample extracts that require dilution to avoid saturation of the detector.

The EQL(b) for groundwater in µg/L is Not Applicable.
The EQL (a, b) for low concentrations in soil and sediment in µg/kg is not determined.
Accuracy as µg/L is not listed.
Overall precision in µg/L is not listed.

(a) *EQLs listed for soil/sediment are based on wet weight. Normally data is reported in a dry-weight basis; therefore, EQLs will be higher based on the % dry weight of each sample. This calculation is based on a 30 g sample and gel permeation chromatography cleanup.*
(b) *Sample EQLs are highly matrix-dependent. The EQLs are provided for guidance and may not always be achievable.*
C = *True value for concentration, in µg/L.*
X = *Average recovery found for measurements of samples containing a concentration of C, in µg/L.*

ESTIMATED QUANTITATION LIMIT

Other Matrices	Factor (a)
High-concentration soil and sludges by sonicator	7.5
Non-water miscible waste	75

(a) *EQL for other matrices = [EQL for low soil/sediment] × [Factor]. This estimated EQL is similar to an EPA "Practical Quantitation Limit."*

SAMPLING METHOD

Liquid samples — Use a 1 or 2½ gallon amber glass bottle with a screw-top Teflon®-lined cover that has been prewashed with detergent and rinsed with distilled water and methanol (or isopropanol).

Soils, sediments, or sludges — Use an 8-oz. widemouth glass with a screw-top Teflon®-lined cover that has been prewashed with detergent and rinsed with distilled water and methanol (or isopropanol).

SAMPLE PRESERVATION

Liquid samples — If residual chlorine is present, add 3 mL of 10% sodium thiosulfate per gallon, cool to 4°C and store in a solvent-free refrigerator until analysis; if chlorine is not present, then eliminate the sodium thiosulfate addition.

Soils, sediments, or sludges — Cool samples to 4°C and store in a solvent-free refrigerator.

MHT Liquid samples must be extracted within 7 days and the extracts analyzed within 40 days. Soils, sediments, or sludges may be stored for a maximum of 14 days and the extracts analyzed within 40 days.

SAMPLE PREPARATION

Liquid samples — Transfer 1 L quantitatively to a continuous extractor. If high concentrations are anticipated, a smaller volume may be used and then diluted with organic-free reagent water to 1 L. Adjust pH, if necessary, to pH <2 using 1:1 (V/V) sulfuric acid. Pipette 1.0 mL of a surrogate standard spiking solution into each sample. For the sample in each analytical batch selected for spiking, add 1.0 mL of a matrix spiking standard. For base/neutral acid analysis, the amount of the surrogates and matrix spiking compounds added to the sample should result in a final concentration of 100 ng/µL of each analyte in the extract to be analyzed (assuming a 1-µL injection). Extract with methylene chloride for 18–24 h. Next, adjust the pH of the aqueous phase to pH >11 using 10 N sodium hydroxide and extract it with methylene chloride again for 18–24 h. Dry the extract through a column containing anhydrous sodium sulfate and concentrate it to 1 mL using a K-D concentrator.

Soils, sediments, or sludges — Use 30 g of sample. Nonporous or wet samples (gummy or clay type) that do not have a free-flowing sandy texture must be mixed with anhydrous sodium sulfate until the sample is free flowing. Add 1 mL of surrogate standards to all samples, spikes, standards, and blanks. For the sample in each analytical batch selected for spiking, add 1.0 mL of a matrix spiking standard. For base/neutral acid analysis, the amount added of the surrogates and matrix spiking compounds should result in a final concentration of 100 ng/µL of each base/neutral analyte and 200 ng/µL of each acid analyte in the extract to be analyzed (assuming a 1-µL injection). Immediately add a 100-mL mixture of 1:1 methylene chloride:acetone and extract the sample ultrasonically for 3 min and then decant or filter the extracts. Repeat the extraction two or more times. Dry the extract using a column with anhydrous sodium sulfate and concentrate it to 1 mL in a K-D concentrator.

QUALITY CONTROL A methylene chloride solution containing 50 ng/µL of decafluorotriphenylphosphine (DFTPP) is used for tuning the GC/MS system each 12-h shift. A system performance check also must be made during every 12-h shift. A standard containing 50 ng/µL each of 4,4′-DDT, pentachlorophenol, and benzidine is required to verify injection port inertness and GC column performance. A calibration standard at mid-concentration, containing each compound of interest, including all required surrogates, must be performed every 12 h during analysis. After the system performance check is met, calibration check compounds (CCCs) are used to check the validity of the initial calibration.

The internal standard responses and retention times in the calibration check standard must be evaluated immediately after or during data acquisition. If the retention time for any internal standard changes by more than 30 seconds from the last check calibration (12 h), the chromatographic system must be inspected for malfunctions and corrections must be made, as required. If the electron ionization current plot (EICP) area for

any of the internal standards changes by a factor of two from the last daily calibration standard check, the mass spectrometer must be inspected for malfunctions and corrections must be made, as appropriate.

Demonstrate, through the analysis of a reagent water blank, that interferences from the analytical system, glassware, and reagents are under control. The blank samples should be carried through all stages of the sample preparation and measurement steps. For each analytical batch (up to 20 samples), a reagent blank, matrix spike, and matrix spike duplicate/duplicate must be analyzed (the frequency of the spikes may be different for different monitoring programs). The blank and spiked samples must be carried through all stages of the sample preparation and measurement steps. A QC reference sample concentrate containing each analyte at a concentration of 100 mg/L in methanol is required.

REFERENCE Test Methods for Evaluating Solid Waste (SW-846). U.S. EPA 1983, Method 8270B, Rev. 2, Nov. 1990. Office of Solid Waste, Washington, DC.

Malononitrile EPA Method 8240
CAS #109-77-3

TITLE Volatile Organics By GC/MS: Packed Column Technique

MATRIX Nearly all types of sample matarices, regardless of water content, can be analyzed using this method. This includes groundwater, aqueous sludges, caustic liquors, acid liquors, waste solvents, oily wastes, mousses, tars, fibrous wastes, polymetric emulsions, filter cakes, spent carbons, spent catalysts, soils, and sediments.

METHOD SUMMARY Method 8240B covers 80 volatile organic compounds that are introduced into a gas chromatograph by the purge-and-trap method or by direct injection (in limited applications). For the purge-and-trap method an inert gas (zero grade nitrogen or helium) is bubbled through a 5-mL solution at ambient temperature. Purged sample components are trapped in a tube of sorbent materials. When purging is complete, the sorbent tube is heated and backflushed with inert gas to desorb the trapped components onto a GC column.

INTERFERENCES Impurities in the purge gas and from organic compounds outgassing from the plumbing ahead of the trap account for many contamination problems. Interferences purged or coextracted from the samples will vary considerably from source to source. Cross-contamination can occur whenever high-level and low-level samples are analyzed sequentially. Whenever an unusually concentrated sample is analyzed, it should be followed by the analysis of organic-free reagent water to check for cross-contamination. Samples also can be contaminated by diffusion of volatile organics (particularly methylene chloride and fluorocarbons) through the septum seal into the sample during shipment and storage. A trip blank can serve as a check on such contamination. The lab where volatile analysis is performed and also the refrigerated storage area should be completely free of solvents.

INSTRUMENTATION A gas chromatograph/mass spectrometry/data system (GC/MS) equipped with a 6 ft × 0.1 in I.D. glass column packed with 1% SP-1000 on Carbopack-B (60/80 mesh) is required. Also needed is a 5-mL purging device, a sorbent trap, and a thermal desorption apparatus.

PRECISION & ACCURACY This method is reported to have been tested by 15 laboratories using organic-free reagent water, drinking water, surface water, and industrial wastewaters (not specified) fortified at six concentrations over the range 5–600 µg/L.

Sample estimated quantitation limits (EQLs) are highly matrix-dependent. The EQLs listed may not always be achievable. EQLs listed for soils or sediments are based on wet weight. Normally, data is reported on a dry-weight basis; therefore, EQLs will be higher, based on the percent dry weight of each sample. Note that EQLs are even more variable than MDLs and that they are highly variable depending on the matrix being analyzed.

EQL in groundwater in µg/L was not listed.
EQL in low soil or sediment in µg/kg was not listed.
Accuracy (a) in µg/L was not listed.
Precision (b) in µg/L was not listed.

(a) *Average recovery found for measurements of samples containing a concentration of C, in µg/L.*
(b) *Overall precision found for measurements of samples with average recovery X for samples containing a concentration of C in µg/L.*
X = *Average recovery found for measurement of samples containing a concentration of C in µg/L.*

MULTIPLICATION FACTORS FOR OTHER MATRICES

Other Matrices	Factor (a)
Waste miscible liquid waste	50
High-concentration soil and sludge	125
Non-water miscible waste	500

(a) EQL = [EQL for low soil/sediment] × [Factor]. For non-aqueous samples, the factor is on a wet-weight basis.

SAMPLING METHOD
Liquid samples — Use a 40-mL glass screw-cap VOA vial with a Teflon®-faced silicone septum that has been prewashed, rinsed with distilled deionized water, and oven dried. However, if residual chlorine is present, collect sample in a 40-oz. soil VOA container which has been pre-preserved with 4 drops of 10% sodium thiosulfate, mix gently, and then transfer the sample to a 40-mL VOA vial. Collect bubble-free samples in duplicate and seal them in separate plastic bags.

Soils or sediments, and sludges — Use an 8-oz. widemouth glass bottle with a Teflon®-faced silicone septum that has been prewashed, rinsed with distilled deionized water, and oven dried. Tap slightly to eliminate free air space. Collect samples in duplicate and seal them in separate plastic bags.

SAMPLE PRESERVATION
Liquid samples — Add 4 drops of concentrated HCL and immediately cool samples to 4°C and store in a solvent-free refrigerator.

Soils or sediments, and sludges — Cool samples to 4°C and store in a solvent-free refrigerator.

MHT Maximum holding time is 14 days from the date of sample collection.

SAMPLE PREPARATION
Liquid samples — Remove the plunger from a 5-mL syringe and carefully pour the sample into the syringe barrel to just short of overflowing. Replace the syringe plunger and compress the sample. Open the syringe valve and vent any residual air while adjusting the sample volume to 5.0 mL. If there is only one volatile organic analysis (VOA) vial, a second syringe should be filled at this time to protect against possible loss of sample integrity. Add 10 µL of surrogate spiking solution and 10 µL of internal standard spiking solution through the valve bore of the 5-mL syringe, then close the valve. The surrogate and internal standards may be mixed and added as a single spiking solution.

Sediments, soils, and waste samples — All samples of this type should be screened by GC analysis using a headspace method (EPA Method 3810) or the hexadecane extraction and screening method (EPA Method 3820). Use the screening data to determine whether to use the low-concentration method (0.005–1 mg/kg) or the high-concentration method (>1 mg/kg).

Low-concentration method — The low-concentration method is based on purging a heated sediment or soil sample mixed with organic-free reagent water containing the surrogate and internal standards. Analyze all reagent blanks and standards under the same conditions as the samples.

Use a 5-g sample if the expected concentration is <0.1 mg/kg or a 1-g sample for expected concentrations between 0.1 and 1 mg/kg. Mix the contents of the sample container with a narrow metal spatula. Weigh the amount of the sample into a tared purge device. Add the spiked water to the purge device, which contains the weighed amount of sample, and connect the device to the purge-and-trap system.

High-concentration method — This method is based on extracting the sediment or soil with methanol. A waste sample is either extracted or diluted, depending on its solubility in methanol. Wastes that are insoluble in methanol are diluted with reagent tetraglyme or possibly polyethylene glycol (PEG). An aliquot of the extract is added to organic-free reagent water containing surrogate and internal standards. This is purged at ambient temperature. All samples with an expected concentration of >1.0 mg/kg should be analyzed by this method.

Mix the contents of the sample container with a narrow metal spatula. For sediments or soils and solid wastes that are insoluble in methanol, weigh 4 g (wet weight) of sample into a tared 20-mL vial. For waste that is soluble in methanol, tetraglyme, or PEG, weigh 1 g (wet weight) into a tared scintillation vial or culture tube or a 10-mL volumetric flask. Quickly add 9.0 mL of appropriate solvent then add 1.0 mL of a surrogate spiking solution to the vial, cap it, and shake it for 2 min.

METHANOL EXTRACT REQUIRED FOR ANALYSIS OF HIGH-CONCENTRATION SOILS OR SEDIMENTS

Approximate Concentration Range	Volume of Methanol Extract (a)
500–10,000 µg/kg	100 µL
1,000–20,000 µg/kg	50 µL
5,000–100,000 µg/kg	10 µL
25,000–500,000 µg/kg	100 µL of 1/50 dilution (b)

Calculate appropriate dilution factor for concentrations exceeding this table.

(a) The volume of methanol added to 5 mL of water being purged should be kept constant. Therefore, add to the 5-mL syringe whatever volume of methanol is necessary to maintain a volume of 100 µL added to the syringe.
(b) Dilute an aliquot of the methanol extract and then take 100 µL for analysis.

QUALITY CONTROL Demonstrate, through the analysis of a reagent water blank, that interferences from the analytical system, glassware, and reagents are under control. Blank samples should be carried through all stages of the sample preparation and measurement steps. For each analytical batch (up to 20 samples), a reagent blank, matrix spike, and matrix spike duplicate must be analyzed (the frequency of the spikes may be different for different monitoring programs). The blank and spiked samples must be carried through all stages of the sample preparation and measurement steps. QC samples mentioned in the section on Interferences will also be needed as appropriate to those situations.

REFERENCE Test Methods for Evaluating Solid Waste (SW-846). U.S. EPA. 1983. Method 8240B, Rev. 2, Nov. 1990. Office of Solid Wastes, Washington, DC.

Manganese
CAS #7439-96-5
EPA Method 6010

TITLE Inductively Coupled Plasma-Atomic Emission Spectroscopy

MATRIX This method is applicable to the determination of trace elements, including metals, in groundwater, soils, sludges, sediments, and other solid wastes. All matrices require digestion prior to analysis. The method of standard addition must be used for the analysis of all sample digests unless either serial dilution or matrix spike addition demonstrates it is not required.

METHOD SUMMARY Method 6010 covers 25 elements using ICP analysis. It measures element-emitted light by optical spectrometry. Samples, following an appropriate acid digestion, are nebulized and the resulting aerosol is transported to the plasma torch. Element-specific atomic line emission spectra are produced by a radio-frequency inductively coupled plasma.

INTERFERENCES Interferences may be categorized as spectral or non-spectral. Spectral interferences are caused by overlap of a spectral line from another element, unresolved overlap of molecular band spectra, background contribution from continuous or recombination phenomenon, and stray light from the line emission of high concentration elements. Non-spectral interferences include physical and chemical interferences. Physical interferences are effects associated with the sample nebulization and transport processes. Changes in viscosity and surface tension can cause significant inaccuracies. Chemical interferences include molecular compound formation, ionization effects, and solute vaporization effects. Normally these effects are not significant and can be minimized by careful selection of operating conditions. Chemical interferences are highly dependent on matrix type and the specific analyte element.

INSTRUMENTATION An inductively coupled argon plasma emission spectrometer (ICP) capable of background correction is required.

PRECISION & ACCURACY Detection limits, sensitivity, and optimum ranges of the metals will vary with the matrices and model of the spectrometer. In a single lab evaluation, seven wastes were analyzed for 22 elements. The mean percent relative standard deviation from triplicate analyses for all elements and wastes was 9 ± 2%. The mean percent recovery of spiked elements for all wastes was 93 ± 6%. Spike levels ranged from 100 µg/L to 100 mg/L. The wastes included sludges and industrial wastewaters.

Estimated instrument detection limit in µg/L is 2.
Spiked concentration in µg/L is 15.
Mean reported value in µg/L is 15.
Precision as RSD % is 6.7.

SAMPLING METHOD Samples should be collected in borosilicate glass, linear polyethylene, polypropylene, or Teflon® bottles that have been prewashed with detergent and tap water, and rinsed with 1:1 nitric acid and tap water or 1:1 hydrochloric acid and tap water. Collect at least 2 g of solids and 200 mL of aqueous samples.

SAMPLE PRESERVATION Add nitric acid to make the samples pH <2.

MHT The maximum holding time for properly preserved samples is 6 months.

SAMPLE PREPARATION Preliminary treatment of most matrices is necessary because of the complexity and variability of sample matrices. Water samples that have been prefiltered and acidified will not need acid digestion. Methods for acid digestion of waters for total recoverable or dissolved metals, acid digestions of aqueous samples and extracts for total metals, and acid digestion of sediments, sludges, and soils are summarized below.

Total recoverable or dissolved metals in water — To prepare surface and groundwater samples for determination of total recoverable and dissolved metals, a 100-mL aliquot of well-mixed sample is acidified with concentrated nitric acid and concentrated hydrochloric acid, then heated until the volume is reduced to 15–20 mL. Adjust the final volume to 100 mL with reagent water.

Total metals in aqueous samples, soil, and sediment extracts — To prepare aqueous samples, soil and sediment extracts, and wastes that contain suspended solids, a 100-mL aliquot is made acidic with concentrated nitric acid and the solution is evaporated to about 5 mL on a hot plate. Continue heating and adding additional acid until sample digestion is complete, which is usually indicated when the digestate is light in color or does not change in appearance. Evaporate the solution to about 3 mL and cool it and add a small quantity of 1:1 hydrochloric acid (10 mL/100 mL of final solution). Cover the beaker and reflux for 15 min. Wash down the beaker walls and filters or centrifuge the sample to remove silicates and other insoluble material. Filter the sample and adjust the final volume to 100 mL with reagent water and the final acid concentration to 10%.

Sediments, sludges, and soils — To prepare sediments, sludges and soil samples, transfer 1–2 g to a conical beaker and add 10 mL of 1:1 nitric acid, mix the slurry, and cover it with a watch glass. Heat the sample and reflux for 10–15 min without boiling. Allow it to cool, then add 5 mL of concentrated nitric acid and reflux for 30 min. Repeat last step and then allow the solution to evaporate to 5 mL without boiling. Cool and add 2 mL of water and 3 mL of 30% hydrogen peroxide. Cover and place the beaker on the hot plate. Heat and add 30% hydrogen peroxide in 1-mL aliquots with warming until the effervescence is minimal but do not add more than a total of 10 mL of 30% hydrogen peroxide. If the sample is being prepared for the analysis of Ag, Al, As, Ba, Be, Ca, Cd, Co, Cr, Cu, Fe, K, Mg, Mn, Mo, Na, Ni, Os, Pb, Se, Tl, V, and Zn, then add 5 mL of concentrated hydrochloric acid and 10 mL of water and return the covered beaker to a hot plate for 15 min of additional refluxing without boiling. Dilute the sample to a 100 mL volume with water after cooling and filter or centrifuge to remove particulates.

QUALITY CONTROL Laboratory control samples must be analyzed for each analytical method. A method blank should be analyzed with each batch of samples. The effect of the matrix on method performance must be demonstrated: when appropriate, there should be at least one matrix spike and either one matrix duplicate or one matrix spike duplicate per analytical batch. The bias and precision of the method, as well as the method detection limit for each specific matrix type, must be measured.

Dilute and reanalyze samples that are more concentrated than the linear calibration limit. Employ a minimum of one reagent blank per sample batch to determine if contamination or any memory effects are occurring. Whenever a new or unusual sample matrix is encountered, perform either a serial dilution test or a matrix spike addition test to ensure that neither positive or negative interferences are operating on any of the analyte elements. Check the instrument standardization by verifying calibration every 10 samples using a calibration blank and a check standard.

REFERENCE Test Methods for Evaluating Solid Waste (SW-846). U.S. EPA. 1983. Method 6010, Rev. 0, Sept. 1986. Office of Solid Wastes, Washington, DC.

Manganese
CAS #7439-96-5
EPA Method 200.7

TITLE Inductively Coupled Plasma

MATRIX Dissolved, suspended or (ICP) total element in drinking and surface waters and in domestic and industrial wastewaters.

APPLICATION The method covers the determination of 25 metals. Dissolved elements are determined in filtered and acidified samples after appropriate digestion (which increases dissolved solids). Its primary advantage is that ICP instruments allow simultaneous or rapid sequential determination of many elements in a short time. Samples are first nebulized and the aerosol is transported to a plasma torch in which element specific atomic line emission spectra are produced by a radio frequency inductively coupled plasma. Background correction is required for trace element detection except in the case of line broadning.

INTERFERENCES There are spectral, physical, and chemical interferences. The primary disadvantage of ICP instruments is background radiation from other elements and the plasma gases (spectral interferences). Changes in sample viscosity and surface tension with samples containing high dissolved solids (especially those exceeding 1500 mg/L) or high acid concentrations can cause physical interferences. Ionization effects, solute vaporization and molecular compound formation can cause chemical interferences. Aluminum, chromium, iron, and magnesium can cause interference at the 100 mg/L level.

INSTRUMENTATION Inductively coupled argon plasma emission spectroscopy. 257.610 nm wavelength.

RANGE Not listed.

MDL 2 µg/L.

PRECISION SD = 2.7% Mean at true value 350 µg/L.

ACCURACY Mean recovery = 93% ± 6% of spiked elements for all wastes.

SAMPLING METHOD Wash sample container with detergent and tap water, rinse with 1 + 1 nitric acid and tap water, then rinse with 1 + 1 hydrochloric acid and tap water, then rinse with deionized, distilled water in that order. Perform any filtration or acid preservation steps when the sample is collected or as soon as possible thereafter.

STABILITY Cool samples to 4°C.24 h.

QUALITY CONTROL Mixed calibration standards, an instrument check standard, and an interference check solution are used in addition to a quality control sample. The quality control sample should be prepared in the same acid matrix as the calibration standards at 10 times the instrumental detection limits and in accordance with the instructions provided by the supplier. Furthermore, two types of blanks are required: a calibration blank and a reagent blank.

REFERENCE Method 200.7, U.S. EPA, EMSL-Cincinnati, OH, Nov. 1980

Manganese
CAS #7439-96-5
EPA Method 243.1

TITLE Metals (Total, Dissolved, Suspended Direct) Atomic Absorption (AA)

MATRIX Drinking, surface, and saline aspiration waters. Wastewater.

APPLICATION Date issued 1971. Editorial Rev. 1974 and 1978. Sample is aspirated and atomized in a flame. Light beam from hollow cathode (made of Mn) lamp is directed through the flame into monochromator, then to detector which measures amount absorbed light.

INTERFERENCES Silica interference in determination of manganese is eliminated by adding calcium. There can be interference from presence of high dissolved solids. Chemical and ionization interferences can occur. (Use special procedure when Mn <25 µg/L).

INSTRUMENTATION AAS. Manganese (Mn) hollow cathode lamp. Burner. Pipets, strip chart recorder.

RANGE 0.1–3 mg/L at 279.5 nm wavelength.

MDL 0.01 mg/L.

PRECISION SD = 31 µg/L at 106 µg Mn/L. (70 Labs).

ACCURACY As bias, −2.1 at 106 µg Mn/L. (70 Labs).

SAMPLING METHOD Prewashed plastic or glass containers.

STABILITY HNO_3 to pH <2.6 months.

QUALITY CONTROL After calibration curve composed of a minimum of a reagent blank and 3 standards has been prepared, subsequent calibration curves must be verified by use of at least a reagent blank and one standard near MCL. Must check within 10% of original curve. (For drinking water analysis).

REFERENCE EPA Methods for the Chemical Analysis of Water and Wastes, EPA-600/4-79-020, U.S. EPA, EMSL, 1979.

Manganese
CAS #7439-96-5
EPA Method 7460

TITLE Atomic Absorption (AA)

MATRIX Drinking, surface, and direct aspiration saline waters. Wastewater.

APPLICATION Sample is aspirated and atomized in a flame. A light beam from a Mn hollow cathode lamp is directed through the flame into a monochromator and onto a detector. Since wavelength of light beam is specific for Mn, light energy absorbed by detector is measure of manganese.

INTERFERENCES The most troublesomee type is chemical, caused by lack of absorption of atoms bound in molecular combination in the flame. High dissolved solids in sample may result in nonatomic absorbance interference. Background correction is required.

INSTRUMENTATION Atomic absorption spectrometer. Manganese hollow cathode lamp. [279.5 nm (primary)]

RANGE 0.1–3 mg/L.

MDL 0.01 mg/L

PRECISION Standard deviation = 70 µg/L at 426 µg/L (true value) 77 labs

ACCURACY As bias = +1.5% at 426 µg/L (true value) 77 labs

SAMPLING METHOD Use glass or plastic containers. Collect 200 g of solids and 600 mL of liquid samples.

STABILITY Cool solid samples to 4°C and analyze as soon as possible. Add nitric acid to liquid samples to pH <2.

MHT 6 months.

QUALITY CONTROL At least one duplicate and one spike sample should be run every 20 samples or with each matrix type to verify precision of the method. For 20 or more samples per day, verify working standard curve. Run an additional standard at or near mid-range every 10 samples.

REFERENCE Method 7460, SW-846, 3rd ed., Nov.1986.

Manganese — EPA Method 7461
CAS #7439-96-5

TITLE Atomic Absorption (AA) Furnace Technique

MATRIX Wastes, mobility procedure, extracts, soils, and groundwater

APPLICATION Aqueous samples, EP extracts, industrial wastes, soils, sludges, sediments, and solid wastes require digestion before analysis. An aliquot of sample is placed in the graphite tube in the furnace and slowly evaporated, charred and atomized. Absorption of lamp radiation during atomization is proportional to (Mn) concentration.

INTERFERENCES The furnace technique is subject to chemical interferences. Composition of sample matrix can have major effect on analysis. Modify matrix to remove interferences. Background correction must be used. Cross-contamination may be major error source.

INSTRUMENTATION Atomic absorption spectrometer. Manganese hollow cathode lamp or electrodeless discharge lamp. Graphite furnace. Strip-chart recorder.

RANGE 1–30 µg/L.

MDL 0.2 µg/L (279.5 nm wavelength)

PRECISION Not listed.

ACCURACY Not listed.

SAMPLING METHOD Use glass or plastic containers. Collect 200 g of solids and 600 mL of liquid samples.

STABILITY Cool solid samples to 4°C and analyze as soon as possible. Add nitric acid to liquid samples to pH <2.

MHT 6 months.

QUALITY CONTROL At least one duplicate and one spike sample should be run every 20 samples, or with each matrix type to verify method precision. If 20 or more samples are run a day, run a standard (at or near mid-range) every 10 samples.

REFERENCE Method 7461, SW-846, 3rd ed., (Included as Rev. 0, Dec. 1987)

MCPA — EPA Method 8151
CAS #94-74-6

TITLE Chlorinated Herbicides by GC Using Methylation or Pentafluorobenzylation Derivatization: Capillary Column Technique.

MATRIX This method covers aqueous and solid matrices. This includes a wide variety such as drinking water, groundwater, industrial wastewaters, surface waters, soils, solids, and sediments.

METHOD SUMMARY This is a GC method for determining 19 chlorinated acid herbicides in aqueous, soil, and waste matrices. Because these compounds are produced and used in various forms (i.e., acid, salt, ester, etc.) a hydrolysis step is included to convert the herbicide to the acid form prior to analysis. This method provides hydrolysis, extraction, derivatization and GC conditions for the analysis of chlorinated acid herbicides in water, soil, and waste samples. Water samples are hydrolyzed *in situ*, extracted with diethyl ether, and then esterified with either diazomethane or pentafluorobenzyl bromide. The derivatives are determined by gas chromatography with an electron capture detector (GC/ECD). The results are reported as acid equivalents. The sensitivity of this method depends on the level of interferences in addition to instrumental limitations.

INTERFERENCES Method interferences may be caused by contaminants in solvents, reagents, glassware, and other sample processing hardware. Immediately prior to use, glassware should be rinsed with the next solvent to be used. Matrix interferences may be caused by contaminants that are coextracted from the sample. Organic acids, especially chlorinated acids, cause the most direct interference with the determination by methylation. Phenols, including chlorophenols, may also interfere with this procedure. The determination using pentafluorobenzylation is more sensitive, and more prone to interferences from the presence of organic acids of phenols than by methylation. Alkaline hydrolysis and subsequent extraction of the basic solution removes many chlorinated hydrocarbons and phthalate esters that might otherwise interfere with the ECD analysis. The herbicides, being strong organic acids, react readily with alkaline substances and may be lost during analysis. Therefore, glassware must be acid-rinsed and then rinsed to constant pH with organic-free reagent water.

INSTRUMENTATION A GC suitable for Grob-type injection using capillary columns. A data system for measuring peak heights and/or peak areas is recommended. An electron capture detector (ECD) is used. Also a K-D apparatus, a diazomethane generator, a centrifuge and an ultrasonic disrupter will be required.

Narrow Bore Columns:

Primary Column 1: 30 m × 0.25 mm, 5% phenyl/95% methyl silicone (DB-5), 0.25 µm film thickness.

Primary Column 1a (GC/MS): 30 m × 0.32 mm, 5% phenyl/95% methyl silicone (DB-5), 1-µm film thickness.

Column 2: 30 m × 0.25 mm DB-608 with a 25 µm film thickness.

Confirmation Column: 30 m × 0.25 mm, 14% cyanopropyl phenyl silicone (DB-1701), 0.25 µm film thickness.

Megabore Columns:

Primary Column: 30 m × 0.53 mm DB-608 with 0.83 µm film thickness.

Confirmation Column: 30 m × 0.53 mm, 14% cyanopropyl phenyl silicone (DB-1701), 1.0 µm film thickness.

PRECISION & ACCURACY Method detection limits (MDLs) are compound-dependent and vary with derivitization efficiency, derivative recovery, the matrix sampled, and herbicide concentration.

The estimated MDL (in µg/L) was 0.056 for aqueous samples using GC/ECD.

The estimated MDL (in µg/kg) was 43 for soil samples using GC/ECD when corrected back to 50 g samples extracted and concentrated to 10 mL with 5-µL injections.

The estimated GC/MS identification limit (in ng) was 0.3 for soil samples using GC/MS.

Mean percent recovery, calculated from 7–8 determinations of spiked reagent water, after diazomethane derivatization, from a spike concentration (in µg/L) of none reported was no data with a standard deviation of the percent recovery of none reported.

Mean percent recovery, calculated from 10 determinations of spiked clay and clay/still bottom samples over the linear concentration range (in ng/g) of 620–61,200 was 96.9 with a percent relative standard deviation of 5.3. The RSD % was calculated on 10 samples high in the linear concentration range and 10 low in the range. The linear concentration range was determined using standard solutions and corrected to 50 g soil samples.

SAMPLE COLLECTION, PRESERVATION & HANDLING
Containers used to collect samples for the determination of semivolatile organic compounds should be soap and water washed followed by methanol (or isopropanol) rinsing. The sample containers should be of glass or Teflon® and have screw-top covers with Teflon® liners.

No preservation is used with concentrated waste samples. With liquid samples containing no residual chlorine and with soil, sediment, and sludge samples, immediately cooling to 4°C is the only preservation used. When residual chlorine is present then 3 mL of 10% aqueous sodium sulfate is added for each gallon of sample collected, followed by cooling to 4°C.

The holding time for all volatile organics samples is 14 days. Liquid samples must be extracted within 7 days and their extracts analyzed within 40 days. Concentrated waste, soil, sediment, and sludge samples must be extracted within 14 days and their extracts analyzed within 40 days.

SAMPLE PREPARATION
Preparation of soil, sediment, and other solid samples — Acidify 30 g (dry weight) solids with 0.1 M phosphate buffer (pH = 2.5) and thoroughly mix the contents. Spike the sample with surrogate compound(s). The ultrasonic extraction of solids must be optimized for each type of sample. In order for the ultrasonic extractor to efficiently extract solid samples, the sample must be free flowing when the solvent is added. Acidified anhydrous sodium sulfate should be added to clay-type soils, or any other solid that is not a free-flowing sandy texture, until a free flowing mixture is obtained. Add methylene chloride and perform ultrasonic extraction. Combine organic extracts from the repetitive extractings of the sample and centrifuge. Add aqueous potassium hydroxide, water, and methanol to the extract and reflux the mixture on a water bath. Extract the solution three times with methylene chloride and discard the methylene chloride phase. The basic solution contains the herbicide salts. Adjust the pH of the solution to <2 with cold sulfuric acid and extract three times with methylene chloride. Combine the extracts and pour them through a pre-rinsed drying column containing acidified anhydrous sodium sulfate. Collect the dried extracts in a K-D flask and concentrate them.

Preparation of aqueous samples — Measure 1 L of sample into a 2 L separatory funnel and spike it with surrogate compound(s). Add NaCl to the sample, then add 6 N NaOH to the sample to a pH of 12 or more and let the sample sit at room temperature for 1 h to hydrolyze esters. Extract the sample three times with methylene chloride and discard the extracts. Then add cold 12 N sulfuric acid to a pH less than or equal to 2, and extract the sample three times with ethyl ether. Collect the ether phase in a flask containing acidified anhydrous sodium sulfate and allow it to remain in contact with the sodium sulfate for a minimum of 2 h. The drying step is very critical to ensuring complete esterification; any moisture remaining in the ether will result in low herbicide recoveries.

Extract concentration and derivatization — The combined ether extract is concentrated to about 1 mL using a K-D apparatus followed by using a micro Snyder column or nitrogen gas blowdown. If methyl esters are to be produced, then dilute the concentrated ether extract with 1 mL of isooctane and 0.5 mL of methanol, dilute to a final volume of 4 mL, and esterify with diazomethane. If pentafluorobenzene esters are to be produced, then dilute concentrated ether extract with acetone to a final volume of 4 mL and esterify with pentafluorobenzyl bromide.

QUALITY CONTROL Select a representative spike concentration for each compound (acid or ester) to be measured. Using stock standard, prepare a quality control check sample concentrate, in acetone, that is 1000 times more concentrated than the selected concentrations. Use this quality control check sample concentrate to prepare quality control check samples. Calculate surrogate standard recovery on all standards, samples, blanks, and spikes. GC/MS techniques should be judiciously employed to support qualitative identifications made with this method. When available, chemical ionization mass spectra may be employed to aid the qualitative identification process.

REFERENCE Test Methods for Evaluating Solid Waste, Physical/Chemical Methods, SW-846, 3rd Edition, U.S. EPA, Office of Solid Waste, Washington, DC, EPA Method 8151, Nov. 1990.

MCPA   EPA Method 8150
CAS #94-74-6

TITLE Chlorinated Herbicides

MATRIX Groundwater, soils, sludges, water miscible liquid wastes, and non-water miscible wastes.

APPLICATION This method is used for the analysis of 10 chlorinated herbicides. Samples are extracted, hydrolyzed with potassium hydroxide, and extraneous organics are removed by a solvent wash. After acidification, the acids are extracted, concentrated and converted to their methyl esters using diazomethane. They are then analyzed using direct injection into a gas chromatograph (GC). Be very careful because diazomethane can explode under certain conditions and it is also a carcinogen.

INTERFERENCES Organic acids and phenols (especially chlorinated acids and phenols) may cause interferences. Phthalate esters are not as significant an interference as with other GC-ECD methods if an electron capture detector is used. The herbicides may react readily with alkaline substances and be lost during analysis so all glassware and glass wool must be acid rinsed and sodium sulfate must be acidified with sulfuric acid prior to use. Sensitivity usually depends on the level of interferences rather than on instrumentation.

INSTRUMENTATION GC capable of on-column injections and an electron capture detector (ECD)or a halogen specific detector. Column 1: 1.8 m by 4 mm with 1.5% SP-2250/1.95% SP-2401 on Supelcoport. Column 2: 1.8 m by 4 mm with 5% OV-210 on Gas Chrom Q. Column 3: 1.98 m by 2 mm with 0.1%. SP-1000 on Carbopack C. The preferred column is Column Number 1.

RANGE Not listed.

MDL 249 µg/L (in reagent water; ECD)

PQL FACTORS FOR MULTIPLYING × FID MDL VALUE

Matrix	Multiplication Factor
Groundwater	10
Low-level soil by sonication with GPC cleanup	670
High-level soil and sludge by sonication	10,000
Non-water miscible waste	100,000

PRECISION (as standard deviation) 4% with 2020 µg/L spike in drinking water; 3% with 2020 µg/L in municipal water.

ACCURACY (as mean recovery) 98% with 2020 µg/L spike in drinking water; 73% with 2020 µg/L in municipal water.

SAMPLING METHOD Use 8-oz. widemouth glass bottles with Teflon®-lined caps for concentrated waste samples, soils, sediments, and sludges. Use 1 or 2½ gallon amber glass bottles with Teflon®-lined caps for liquid (water) samples.

STABILITY Cool soil, sediment, sludge, and liquid samples to 4°C. If residual chlorine is present in liquid samples add 3 mL of 10% sodium thiosulfate per gallon of sample and cool to 4°C.

MHT 14 days for concentrated waste, soil, sediment, or sludge; 7 days for liquid samples; all extracts must be analyzed within 40 days.

QUALITY CONTROL A quality control check sample concentrate containing this compound in acetone at a concentration 1,000 times more concentrated than the selected spike concentration is required. The QC check sample concentrate may be prepared from pure standard materials or purchased as certified solutions. Use appropriate trip, matrix, control site, method, reagent, and solvent blanks. Internal, surrogate, and five concentration level calibration standards are used.

REFERENCE Method 8150, SW-846, 3rd ed., Sept. 1986.

MCPP   EPA Method 8151
CAS #93-65-2

TITLE Chlorinated Herbicides by GC Using Methylation or Pentafluorobenzylation Derivatization: Capillary Column Technique.

MATRIX This method covers aqueous and solid matrices. This includes a wide variety such as drinking water, groundwater, industrial wastewaters, surface waters, soils, solids, and sediments.

METHOD SUMMARY This is a GC method for determining 19 chlorinated acid herbicides in aqueous, soil, and waste matrices. Because these compounds are produced and used in various forms (i.e., acid, salt, ester, etc.) a hydrolysis step is included to convert the herbicide to the acid form prior to analysis. This method provides hydrolysis, extraction, derivatization and GC conditions for the analysis of chlorinated acid herbicides in water, soil, and waste samples. Water samples are hydrolyzed *in situ*, extracted with diethyl ether, and then esterified with either diazomethane or pentafluorobenzyl bromide. The derivatives are determined by gas chromatography with an electron capture detector (GC/ECD). The results are reported as acid equivalents. The sensitivity of this method depends on the level of interferences in addition to instrumental limitations.

INTERFERENCES Method interferences may be caused by contaminants in solvents, reagents, glassware, and other sample processing hardware. Immediately prior to use, glassware should be rinsed with the next solvent to be used. Matrix interferences may be caused by contaminants that are coextracted from the sample. Organic acids, especially chlorinated acids, cause the most direct interference with the determination by methylation. Phenols, including chlorophenols, may also interfere with this procedure. The determination using pentafluorobenzylation is more sensitive, and more prone to interferences from the presence of organic acids of phenols than by methylation. Alkaline hydrolysis and subsequent extraction of the basic solution removes many chlorinated hydrocarbons and phthalate esters that might otherwise interfere with the ECD analysis. The herbicides, being strong organic acids, react readily with alkaline substances and may be lost during analysis. Therefore, glassware must be acid-rinsed and then rinsed to constant pH with organic-free reagent water.

INSTRUMENTATION A GC suitable for Grob-type injection using capillary columns. A data system for measuring peak heights and/or peak areas is recommended. An electron capture detector (ECD) is used. Also a K-D apparatus, a diazomethane generator, a centrifuge and an ultrasonic disrupter will be required.

Narrow Bore Columns:
Primary Column 1: 30 m × 0.25 mm, 5% phenyl/95% methyl silicone (DB-5), 0.25 μm film thickness.
Primary Column 1a (GC/MS): 30 m × 0.32 mm, 5% phenyl/95% methyl silicone (DB-5), 1-μm film thickness.
Column 2: 30 m × 0.25 mm DB-608 with a 25 μm film thickness.
Confirmation Column: 30 m × 0.25 mm, 14% cyanopropyl phenyl silicone (DB-1701), 0.25 μm film thickness.

Megabore Columns:
Primary Column: 30 m × 0.53 mm DB-608 with 0.83 μm film thickness.
Confirmation Column: 30 m × 0.53 mm, 14% cyanopropyl phenyl silicone (DB-1701), 1.0 μm film thickness.

PRECISION & ACCURACY Method detection limits (MDLs) are compound-dependent and vary with derivitization efficiency, derivative recovery, the matrix sampled, and herbicide concentration.

The estimated MDL (in μg/L) was 0.09 for aqueous samples using GC/ECD.

The estimated MDL (in μg/kg) was 66 for soil samples using GC/ECD when corrected back to 50 g samples extracted and concentrated to 10 mL with 5-μL injections.

The estimated GC/MS identification limit (in ng) was 0.43 for soil samples using GC/MS.

Mean percent recovery, calculated from 7–8 determinations of spiked reagent water, after diazomethane derivatization, from a spike concentration (in μg/L) of none reported was no data with a standard deviation of the percent recovery of none reported.

Mean percent recovery, calculated from 10 determinations of spiked clay and clay/still bottom samples over the linear concentration range (in ng/g) of 620–61,800 was 98.3 with a percent relative standard deviation of 3.4. The RSD % was calculated on 10 samples high in the linear concentration range and 10 low in the range. The linear concentration range was determined using standard solutions and corrected to 50 g soil samples.

SAMPLE COLLECTION, PRESERVATION & HANDLING
Containers used to collect samples for the determination of semivolatile organic compounds should be soap and water washed followed by methanol (or isopropanol) rinsing. The sample containers should be of glass or Teflon® and have screw-top covers with Teflon® liners.

No preservation is used with concentrated waste samples. With liquid samples containing no residual chlorine and with soil, sediment, and sludge samples, immediately cooling to 4°C is the only preservation used. When residual chlorine is present then 3 mL of 10% aqueous sodium sulfate is added for each gallon of sample collected, followed by cooling to 4°C.

The holding time for all volatile organics samples is 14 days. Liquid samples must be extracted within 7 days and their extracts analyzed within 40 days. Concentrated waste, soil, sediment, and sludge samples must be extracted within 14 days and their extracts analyzed within 40 days.

SAMPLE PREPARATION
Preparation of soil, sediment, and other solid samples — Acidify 30 g (dry weight) solids with 0.1 M phosphate buffer (pH = 2.5) and thoroughly mix the contents. Spike the sample with surrogate compound(s). The ultrasonic extraction of solids must be optimized for each type of sample. In order for the ultrasonic extractor to efficiently extract solid samples, the sample must be free flowing when the solvent is added. Acidified anhydrous sodium sulfate should be added to clay-type soils, or any other solid that is not a free-flowing sandy texture, until a free flowing mixture is obtained. Add methylene chloride and perform ultrasonic extraction. Combine organic extracts from the repetitive extractings of the sample and centrifuge. Add aqueous potassium hydroxide, water, and methanol to the extract and reflux the mixture on a water bath. Extract the solution three times with methylene chloride and discard the methylene chloride phase. The basic solution contains the herbicide salts. Adjust the pH of the solution to <2 with cold sulfuric acid and extract three times with methylene chloride. Combine the extracts and pour them through a pre-rinsed drying column containing acidified anhydrous sodium sulfate. Collect the dried extracts in a K-D flask and concentrate them.

Preparation of aqueous samples — Measure 1 L of sample into a 2 L separatory funnel and spike it with surrogate compound(s). Add NaCl to the sample, then add 6 N NaOH to the sample to a pH of 12 or more and let the sample sit at room temperature for 1 h to hydrolyze esters. Extract the sample three times with methylene chloride and discard the extracts. Then add cold 12 N sulfuric acid to a pH less than or equal to 2, and extract the sample three times with ethyl ether. Collect the ether phase in a flask containing acidified anhydrous sodium sulfate and allow it to remain in contact with the sodium sulfate for a minimum of 2 h. The drying step is very critical to ensuring complete esterification; any moisture remaining in the ether will result in low herbicide recoveries.

Extract concentration and derivatization — The combined ether extract is concentrated to about 1 mL using a K-D apparatus followed by using a micro Snyder column or nitrogen gas blowdown. If methyl esters are to be produced, then dilute the concentrated ether extract with 1 mL of isooctane and 0.5 mL of methanol, dilute to a final volume of 4 mL, and esterify with diazomethane. If pentafluorobenzene esters are to be produced, then dilute concentrated ether extract with acetone to a final volume of 4 mL and esterify with pentafluorobenzyl bromide.

QUALITY CONTROL Select a representative spike concentration for each compound (acid or ester) to be measured. Using stock standard, prepare a quality control check sample concentrate, in acetone, that is 1000 times more concentrated than the selected concentrations. Use this quality control check sample concentrate to prepare quality control check samples. Calculate surrogate standard recovery on all standards, samples, blanks, and spikes. GC/MS techniques should be judiciously employed to support qualitative identifications made with this method. When available, chemical ionization mass spectra may be employed to aid the qualitative identification process.

REFERENCE Test Methods for Evaluating Solid Waste, Physical/Chemical Methods, SW-846, 3rd Edition, U.S. EPA, Office of Solid Waste, Washington, DC, EPA Method 8151, Nov. 1990.

MCPP   **EPA Method 8150**
CAS #93-65-2

TITLE Chlorinated Herbicides

MATRIX Groundwater, soils, sludges, water miscible liquid wastes, and non-water miscible wastes.

APPLICATION This method is used for the analysis of 10 chlorinated herbicides. Samples are extracted, hydrolyzed with potassium hydroxide, and extraneous organics are removed by a solvent wash. After acidification, the acids are extracted, concentrated and converted to their methyl esters using diazomethane. They are then analyzed using direct injection into a gas chromatograph (GC). Be very careful because diazomethane can explode under certain conditions and it is also a carcinogen.

INTERFERENCES Organic acids and phenols (especially chlorinated acids and phenols) may cause interferences. Phthalate esters are not as significant an interference as with other GC-ECD methods if an electron capture detector is used. The herbicides may react readily with alkaline substances and be lost during analysis so all glassware and glass wool must be acid rinsed and sodium sulfate must be acidified with sulfuric acid prior to use. Sensitivity usually depends on the level of interferences rather than on instrumentation.

INSTRUMENTATION GC capable of on-column injections and an electron capture detector (ECD) or a halogen specific detector. Column 1: 1.8 m by 4 mm with 1.5% SP-2250/1.95% SP-2401 on Supelcoport. Column 2: 1.8 m by 4 mm with 5% OV-210 on Gas Chrom Q. Column 3: 1.98 m by 2 mm with 0.1%. SP-1000 on Carbopack C. The preferred column is Column Number 1.

RANGE Not listed.

MDL 192 µg/L (in reagent water; ECD)

PQL FACTORS FOR MULTIPLYING × FID MDL VALUE

Matrix	Multiplication Factor
Groundwater	10
Low-level soil by sonication with GPC cleanup	670
High-level soil and sludge by sonication	10,000
Non-water miscible waste	100,000

PRECISION (as standard deviation) 4% with 2080 µg/L spike in drinking water; 3% with 2100 µg/L in municipal water.

ACCURACY (as mean recovery) 94% with 2080 µg/L spike in drinking water; 97% with 2100 µg/L in municipal water.

SAMPLING METHOD Use 8-oz. widemouth glass bottles with Teflon®-lined caps for concentrated waste samples, soils, sediments, and sludges. Use 1 or 2½ gallon amber glass bottles with Teflon®-lined caps for liquid (water) samples.

STABILITY Cool soil, sediment, sludge, and liquid samples to 4°C. If residual chlorine is present in liquid samples add 3 mL of 10% sodium thiosulfate per gallon of sample and cool to 4°C.

MHT 14 days for concentrated waste, soil, sediment, or sludge; 7 days for liquid samples; all extracts must be analyzed within 40 days.

QUALITY CONTROL A quality control check sample concentrate containing this compound in acetone at a concentration 1,000 times more concentrated than the selected spike concentration is required. The QC check sample concentrate may be prepared from pure standard materials or purchased as certified solutions. Use appropriate trip, matrix, control site, method, reagent, and solvent blanks. Internal, surrogate, and five concentration level calibration standards are used.

REFERENCE Method 8150, SW-846, 3rd ed., Sept. 1986.

Mercury   **EPA Method 7470**
CAS #7439-97-6

TITLE Mercury in Liquid Waste (Manual Cold Vapor Technique)

MATRIX This method is applicable to the determination of mercury in mobility- procedure extracts, aqueous wastes, and groundwater. It can also be used for analyzing certain solid and sludge-type wastes. This method may be applicable for analyses of some sediment samples but it is primarily used for aqueous samples.

METHOD SUMMARY This method is based on the absorption of radiation at 253.7 nm by mercury vapor. The mercury is reduced to the elemental state and aerated from solution in a closed system. A sample is digested in strong sulfuric and nitric acid with potassium permanganate added to eliminate interferences from sulfide. After removal of excess potassium permanganage stannous sulfate is added and elemental mercury is volatilized in an aeration apparatus. The sample is allowed to stand without manual agitation while a circulating pump is allowed to run continuously. As mercury is volatilized, it is swept into an absorption cell and the absorbance increases as mercury is analyzed with an atomic absorption spectrophotometer. The increase is followed using a recorder; as the recorder levels off, a bypass valve is opened and aeration is continued until the absorbance returns to its minimum value on the recorder. Then the bypass valve is closed, the stopper and frit from the BOD bottle is removed, and aeration is continued to measure mercury in the sample by comparing response to a calibration curve prepared from known standard concentrations.

INTERFERENCES Potassium permanganate is added to eliminate possible interference from sulfide. Concentrations as high as 20 mg/kg of sulfide as sodium sulfide do not interfere with the recovery of added inorganic mercury from reagent water. Copper has also been reported to interfere, but concentrations as high as 10 mg/kg had no effect in recovery studies. Seawaters, brines, and industrial effluents high in chlorides require additional permanganate because, during the oxidation step, chlorides are converted to free chlorine which also absorbs at 253.7 nm. Remove all free chlorine by using an excess (25 mL) of hydroxylamine sulfate reagent. Some volatile organic materials that absorb at this wavelength may also cause interference.

INSTRUMENTATION An atomic absorption (AA) spectrophotometer equipped with a mercury hollow cathode lamp or electrodeless discharge lamp is used for analysis. Also required is a cold vapor generator for mercury vapor.

PRECISION & ACCURACY The typical detection limit for this method is 0.2 µg/L.

CONCENTRATION VS. BIAS

Spiked Conc. % Bias	Accuracy as Deviation as µg/L	Standard as µg/L
0.21	66	0.276
0.27	53	0.279
0.51	32	0.541
0.60	18	0.390
3.4	0.34	1.49
4.1	−7.1	1.12
8.8	−0.4	3.69
9.6	−5.2	3.57

SAMPLING METHOD All sample containers must be prewashed with detergents, acids, and reagent water. Plastic and glass containers are both suitable although plastic is generally used.

SAMPLE PRESERVATION Aqueous samples must be acidified to a pH <2 with nitric acid. Non-aqueous samples should be stored at 4°C.

MHT The maximum hold time for mercury samples is 28 days.

SAMPLE PREPARATION Transfer 100 mL of a sample containing <1.0 g of mercury to a 300 mL biological oxygen demand (BOD) bottle or equivalent. Add 5 mL of sulfuric acid and 2.5 mL of concentrated nitric acid, mixing after each addition. Then add 15 mL of potassium permanganate. Shake and add additional portions until the purple color persists for at least 15 min. Add 8 mL of potassium persulfate to each bottle and heat for 2 h in a water bath at 95°C. Cool and add 6 mL of sodium chloride-hydroxylamine sulfate to reduce the excess permanganate. After at least 30 seconds, add 5 mL of stannous sulfate and immediately attach to an aeration apparatus.

QUALITY CONTROL A calibration curve must be prepared each day with a minimum of a calibration blank and three standards at different concentrations of mercury. If more than 10 samples per day are analyzed, the working standard curve must be verified by measuring a mid-range standard after every 10 samples. The value of this mid-range standard must be within 20% of the true value. At least one matrix spike and one matrix spike duplicate sample must be included in each analytical batch, as well as a lab control sample. For each analytical batch select one typical sample for serial dilution to determine whether interferences are present. The concentration of the analyte should be at least 25 times the estimated detection limit. If all of the samples in the batch are below 10 times the detection limits, perform a spike recovery analysis.

REFERENCE Test Methods for Evaluating Solid Waste (SW-846). U.S. EPA. 1983. Method 7470A, Rev. 1, Nov. 1990. Office of Solid Wastes, Washington, DC.

Mercury EPA Method 7471
CAS #7439-97-6

TITLE Mercury in Solid or Semisolid Waste (Manual Cold Vapor Technique)

MATRIX This method is applicable to the determination of total mercury (organic and inorganic) in soils, sediments, bottom deposits, and sludge-type materials. All samples must be subjected to an appropriate dissolution step prior to analysis. If this dissolution procedure is not sufficient to dissolve a specific matrix type or sample, then this method is not applicable for that matrix.

METHOD SUMMARY This method is based on the absorption of radiation at 253.7 nm by mercury vapor and is limited to mercury in soils and sediments. The mercury is reduced to the elemental state and aerated from solution in a closed system. A sample is digested in strong sulfuric and nitric acid with potassium permanganate added to eliminate interferences from sulfide. After removal of excess potassium permanganage stannous sulfate is added and elemental mercury is volatilized in an aeration apparatus. The sample is allowed to stand without manual agitation while a circulating pump is allowed to run continuously. As mercury is volatilized, it is swept into an absorption cell and the absorbance increases as mercury is analyzed with an atomic absorption spectrophotometer. The increase is followed using a recorder; as the recorder levels off, a bypass valve is opened and aeration is continued until the absorbance returns to its minimum value on the recorder. Then the bypass valve is closed, the stopper and frit from the BOD bottle is removed, and aeration is continued to measure mercury in the sample by comparing response to a calibration curve prepared from known standard concentrations.

INTERFERENCES Potassium permanganate is added to eliminate possible interference from sulfide. Concentrations as high as 20 mg/kg of sulfide as sodium sulfide do not interfere with the recovery of added inorganic mercury from reagent water. Copper has also been reported to interfere, but concentrations as high as 10 mg/kg had no effect in recovery studies. Seawaters, brines, and industrial effluents high in chlorides require additional permanganate because, during the oxidation step, chlorides are converted to free chlorine which also absorbs at 253.7 nm. Remove all free chlorine by using an excess (25 mL) of hydroxylamine sulfate reagent. Some volatile organic materials that absorb at this wavelength may also cause interference.

INSTRUMENTATION An atomic absorption spectrometer (AA) equipped with a mercury hollow cathode lamp or electrodeless discharge lamp is used for the analysis; also required is a cold vapor generator.

PRECISION & ACCURACY The typical detection limit for this method is 0.2 µg/L.

CONCENTRATION VS. BIAS

Spiked Conc. (µg/L)	Recovery (% of True Value)	Standard Deviation (µg/g)
0.30	97	0.02
0.87	94	0.03

SAMPLING METHOD All sample containers must be prewashed with detergents, acid, and reagent water. Plastic and glass containers are both suitable, but plastic containers are usually used.

SAMPLE PRESERVATION Aqueous samples must be acidified to a pH <2 with nitric acid. Nonaqueous samples must be stored at 4°C.

MHT The maximum holding time for mercury samples is 28 days.

SAMPLE PREPARATION Weigh triplicate 0.2 g portions of untreated sample and place each of them in the bottom of separate biological oxygen demand (BOD) bottles. Add 5 mL of reagent water and 5 mL of aqua regia to each bottle and heat for 2 min in a water bath at 95°C. Cool and add 50 mL reagent water and 15 mL potassium permanganate solution to each sample bottle. Mix thoroughly and place the samples in a water bath for 30 min at 95°C. Cool and add 6 mL of sodium chloride-hydroxylamine sulfate to reduce the excess permanganate. Then add 55 mL of reagent water and 5 mL of stannous sulfate to each bottle and immediately attach each bottle to the aeration apparatus.

QUALITY CONTROL A calibration curve must be prepared each day with a minimum of a calibration blank and three standards. If more than 10 samples per day are analyzed, the working standard curve must be verified by measuring a midrange standard after every 10 samples. This value must be within 20% of the true value. At least one matrix spike and one matrix spike duplicate sample must be included in each analytical batch, as well as a lab control sample. For each analytical batch select one typical sample for serial dilution to determine whether interferences are present. The concentration of the analyte should be at least 25 times the estimated detection limit. If all of the samples in the batch are below 10 times the detection limits, perform the spike recovery analysis.

REFERENCE Test Methods for Evaluating Solid Waste (SW-846). U.S. EPA. 1983. Method 7471A, Rev. 1, Nov. 1990. Office of Solid Wastes, Washington, DC.

Merphos   **EPA Method 507**
CAS #150-50-5

TITLE Determination of Nitrogen and Phosphorus-Containing Pesticides in Water by GC/NPD

MATRIX This method is applicable to the determination of certain nitrogen and phosphorus-containing pesticides in finished drinking water and groundwater.

METHOD SUMMARY Method 507 covers 46 nitrogen- and phosphorus-containing pesticides. A 1-L sample is fortified with a surrogate standard, salted, buffered, extracted with methylene chloride, and concentrated; then the solvent is exchanged with methyl tert-butyl ether (MTBE) and concentrated again, and a 2-µL aliquot of a sample extract is injected into a GC system equipped with a selective nitrogen-phosphorus detector and a capillary column for analysis.

INTERFERENCES Method interferences may be caused by contaminants in solvents, reagents, glassware, and other sample processing apparatus. Interfering contamination may occur when a sample containing low concentrations of analytes is analyzed immediately following a sample containing relatively high concentrations. One or more injections of MTBE should be made following the analysis of a sample with high concentrations of analytes to check for analyte carryover. Matrix interferences may be caused by contaminants that are coextracted from the sample. The extent of matrix interferences will vary considerably from source to source, depending upon the water sampled.

INSTRUMENTATION A gas chromatograph system (GC) equipped with a nitrogen-phosphorus detector (NPD) is needed.

Column 1: 30 m × 0.25 mm I.D. DB-5 bonded fused silica column, 0.25 µm film thickness, or equivalent.
Column 2: 30 m × 0.25 mm I.D. DB-1701 bonded fused silica column, 0.25 µm film thickness, or equivalent.

PRECISION & ACCURACY This method has been validated in a single lab and estimated detection limits (EDLs) have been determined for each analyte. Observed detection limits may vary among waters, depending upon the nature of the interferences in the sample matrix and the specific instrumentation used. Analytes that are not separated chromatographically cannot be individually identified and measured unless an alternative technique for identification and quantification exist.

The estimated detection limit (in µg/L) was 0.25. The EDL is defined as either method detection limit or a level of compound in a sample yielding a peak in the final extract with signal-to-noise ratio of approximately 5, whichever value is higher.

The concentration used for these measurements (in μg/L) was 2.5.

The accuracy (as % recovery) was 96.

The precision (% RSD) was 8.

SAMPLING METHOD Grab samples are collected in 1-L glass sample bottles (prewashed with detergent and hot tap water, rinsed with reagent water, and dried in an oven at 400°C for 1 h) with screw caps lined with PTFE-fluorocarbon.

SAMPLE PRESERVATION Add mercuric chloride to the sample bottle in amounts to produce a concentration of 10 mg/L. If residual chlorine is present, add 80 mg of sodium thiosulfate/L of sample to the sample bottle prior to collection. After collection, seal bottle and shake vigorously for 1 min, then cool the sample to 4°C immediately and store it at 4°C in the dark until extraction.

MHT Maximum holding time of the samples, and in some cases the extracts, is 14 days.

SAMPLE PREPARATION Fortify the sample with 50 μL of the surrogate standard solution, adjust to pH 7 with phosphate buffer, add 100 g NaCl to the sample, and seal and shake to dissolve the salt; then extract with methylene chloride in a separatory funnel or in a mechanical tumbler bottle. Dry the extract by pouring it through a solvent-rinsed drying column containing about 10 cm of anhydrous sodium sulfate. Collect the extract in a Kuderna-Danish (K-D) concentrator and rinse the column with 20–30 mL methylene chloride. Concentrate the extract to about 2 mL and rinse the flask and its lower joint into the concentrator tube with 1 to 2 mL of methyl t-butyl ether (MTBE). Add 5–10 mL of MTBE and concentrate the extract twice (adding more MTBE) to a final volume of 5.0 mL and store it at 4°C until analysis.

Note: If methylene chloride is not completely removed from the final extract, it may cause detector problems.

QUALITY CONTROL Minimum quality control requirements are initial demonstration of lab capability, determination of surrogate compound recoveries in each sample and blank, monitoring internal standard peak area or height in each sample and blank, analysis of lab reagent blanks, lab fortified samples, lab fortified blanks, and other QC samples. A lab reagent blank is analyzed to demonstrate that all glassware and reagent interferences are under control.

Initial demonstration of capability is fulfilled by analyzing four fortified reagent water samples with the recovery value for each analyte falling within the acceptable range (±30% average recovery). Surrogate recoveries from samples or method blanks must be 70–130%. The internal standard response for any sample chromatogram should not deviate from the daily calibration check standard's internal standard response by more than 30% or lab fortified blanks and sample matrices are used to assess lab performance and analyte recovery, respectively.

If the response for the target analyte peak exceeds the working range of the system, dilute the extract and reanalyze. Alternative techniques such as an alternate detector or second chromatography column should be used to confirm peak identification when sample components are not resolved adequately.

EPA CONTACT & HOTLINE For technical questions contact Dr. Baldev Bathija, U.S. EPA, Office of Ground Water and Drinking Water (WH-550D), 401 M St. SW, Washington, DC 20460. Tel. (202) 260-3040. For further information the EPA Safe Drinking Water Hotline may be called at: (800) 426-4791.

REFERENCE Methods for the Determination of Organic Compounds in Drinking Water, EPA/600/4-88/039 (revised July 1991). U.S. EPA Environmental Monitoring Systems Laboratory, Cincinnati, OH, 45268, U.S.A. Available from the National Technical Information Service (NTIS), 5285 Port Royal Road, Springfield, VA 22161; Tel. 800-553-6847. NTIS Order Number is PB91-231480.

Merphos   **EPA Method 8141**
CAS #150-50-5

TITLE Organophosphorus Compounds by Gas Chromatography: Capillary Column Technique

MATRIX This method covers aqueous and solid matrices. This includes a wide variety such as drinking water, groundwater, industrial wastewaters, surface waters, soils, solids, and sediments.

METHOD SUMMARY This is a GC method used to determine the concentration of 28 organophosphorus pesticides.

The use of Gel Permeation Cleanup (EPA Method 3640) for sample cleanup has been demonstrated to yield recoveries of less than 85% for many method analytes and is therefore not recommended for use with this method.

This method provides GC conditions for the detection of ppb concentrations of organophosphorus compounds. Prior to the use of this method, appropriate sample preparation techniques must be used. Water samples are extracted at a neutral pH with methylene chloride as a solvent by using a separatory funnel (EPA Method 3510) or a continuous liquid-liquid extractor (EPA Method 3520). Soxhlet extraction (EPA Method 3540) or ultrasonic extraction (EPA Method 3550) using methylene chloride/acetone (1:1) are used for solid samples. Both neat and diluted organic liquids (EPA Method 3580) may be analyzed by direct injection. Spiked samples are used to verify the applicability of the chosen extraction technique to each new sample type. A GC with a flame photometric (FPD) or nitrogen-phosphorus detector (NPD) is used for this multiresidue procedure.

INTERFERENCES The use of Florisil cleanup materials (EPA Method 3620) for some of the compounds in this method has been demonstrated to yield recoveries less than 85% and is therefore not recommended for all compounds. Use of phosphorus or halogen specific detectors, however, often obviates the necessity for cleanup for relatively clean sample matrices. If particular circumstances demand the use of an alternative cleanup procedure, the analyst must determine the elution profile and demonstrate that the recovery of each analyte is no less than 85%.

Use of a flame photometric detector (FPD) in the phosphorus mode will minimize interferences from materials that do not contain phosphorus. Elemental sulfur, however, may interfere with the determination of certain organophosphorus compounds by flame photometric gas chromatography. Sulfur cleanup using EPA Method 3660 may alleviate this interference. A nitrogen phosphorus detector (NPD) is also recommended.

A few analytes coelute on certain columns. Therefore, select a second column for confirmation where coelution of the analytes of interest does not occur.

Method interferences may be caused by contaminants in solvents, reagents, glassware, and other sample processing hardware that lead to discrete artifacts or elevated baselines in gas chromatograms. All these materials must be routinely demonstrated to be free from interferences under the conditions of the analysis by analyzing reagent blanks.

INSTRUMENTATION A GC with a NPD or a FPD will be needed. A data system or integrator is recommended for measuring peak areas and/or peak heights. A Kuderna-Danish (K-D) apparatus will be needed for extract concentration.

Column 1: 15 m × 0.53 mm megabore capillary column, 1.0 μm film thickness, DB-210.
Column 2: 15 m × 0.53 mm megabore capillary column, 1.5 μm film thickness, SPB-608.
Column 3: 15 m × 0.53 mm megabore capillary column, 1.0 μm film thickness, DB-5.

Three megabore capillary columns are included for analysis of organophosphates by this method. Column 1 (DB-210 or equivalent) and Column 2 (SPB-608 or equivalent) are recommended if a large number of organophosphorus analytes are to be determined. If the superior resolution offered by Column 1 and Column 2 is not required, Column 3 (DB-5 or equivalent) may be used. For megabore capillary columns, automatic injections of 1 μL are recommended.

PRECISION & ACCURACY The MDL actually achieved in a given analysis will vary, as it is dependent on instrument sensitivity and matrix effects. Single operator accuracy and precision studies have been conducted with spiked water and soil samples.

MULTIPLICATION FACTORS FOR OTHER MATRICES (a)

Matrix	Factor (b)
Groundwater (EPA Method 3510 or EPA Method 3520)	10
Low-concentration soil by Soxhlet and no cleanup	10 (c)
Low-concentration soil by ultrasonic extraction with GPC cleanup	6.7 (c)
High-concentration soil and sludges by ultrasonic extraction	500 (c)
Non-water miscible waste (EPA Method 3580)	1000 (c)

(a) Sample EQLs are highly matrix-dependent. The EQLs listed here are provided for guidance and may not always be achievable.
(b) EQL = [Method detection limit] × [Factor]. For non-aqueous samples the factor is on a wet-weight basis.
(c) Multiply this factory times the soil MDL.

The MDL (in μg/L) when reagent water was extracted using a separatory funnel was 0.20.
The MDL (in μg/kg) when soil was extracted using Soxhlet extraction (EPA Method 3540) was 10.0.
Accuracy (as % recovery) with separatory funnel extraction ranged from not recovered (with low spikes) to 81 (with high spikes).
Accuracy (as % recovery) with continuous liquid-liquid extraction ranged from not recovered (with low spikes) to 79 (with high spikes).
Accuracy (as % recovery) with Soxhlet extraction of soils ranged from 62 (with low spikes to 60 (with high spikes).
Accuracy (as % recovery) with ultrasonic extraction of soils ranged from not recovered (with low spikes) to 155 (with high spikes).

SAMPLE COLLECTION, PRESERVATION & HANDLING
Containers used to collect samples for the determination of semivolatile organic compounds should be soap and water washed followed by methanol (or isopropanol) rinsing. The sample containers should be of glass or Teflon® and have screw-top covers with Teflon® liners.

No preservation is used with concentrated waste samples. With liquid samples containing no residual chlorine and with soil, sediment, and sludge samples, immediately cooling to 4°C is the only preservation used. When residual chlorine is present then 3 mL of 10% aqueous sodium sulfate is added for each gallon of sample collected, followed by cooling to 4°C.

Liquid samples must be extracted within 7 days and their extracts analyzed within 40 days. Concentrated waste, soil, sediment, and sludge samples must be extracted within 14 days and their extracts analyzed within 40 days.

SAMPLE PREPARATION In general, water samples are extracted at a neutral pH with methylene chloride, using either EPA Method 3510 or EPA Method 3520. Solid samples are extracted using either EPA Method 3540 or EPA Method 3550 with methylene chloride/acetone (1:1) as the extraction solvent.

Prior to GC analysis, the extraction solvent may be exchanged to hexane. Single lab data indicates that samples should not be transferred with 100% hexane during sample workup as the more water soluble organophosphorus compounds may be lost.

If cleanup is performed on the samples, the analyst should analyze the samples by GC. This will confirm elution patterns and the absence of interferences from the reagents. If peak detection and identification is prevented by the presence of interferences, further cleanup is required.

QUALITY CONTROL The analyst should monitor the performance of the extraction, cleanup (when used), and analytical system and the effectiveness of the method in dealing with each sample matrix by spiking each sample, standard, and blank with one or two surrogates (e.g., organophosphorus compounds not expected to be present in the sample). Deuterated analogs of analytes should not be used as surrogates for gas chromatographic analysis due to coelution problems.

A minimum of five concentrations for each analyte of interest should be prepared through dilution of the stock standards with isooctane. One of the concentrations should be at a concentration near, but above, the MDL.

Include a mid-level check standard after each group of 10 samples in the analysis sequence. GC/MS techniques should be judiciously employed to support qualitative identifications made with this method. Follow the GC/MS operating requirements specified in EPA Method 8270.

When available, chemical ionization mass spectra may be employed to aid in the qualitative identification process. To confirm an identification of a compound, the background-corrected mass spectrum of the compound must be obtained from the sample extract and must be compared with a mass spectrum from a stock or calibration standard analyzed under the same chromatographic conditions. The molecular ion and all other ions present above 20% relative abundance in the mass spectrum of the standard must be present in the mass spectrum of the sample with agreement to ±20%. The retention time of the compound in the sample must be within six seconds of the retention time for the same compound in the standard solution.

Should the MS procedure fail to provide satisfactory results, additional steps may be taken before reanalysis. These steps may include the use of alternate packed or capillary GC columns or additional sample cleanup.

REFERENCE Test Methods for Evaluating Solid Waste, Physical/Chemical Methods, SW-846, 3rd Edition, U.S. EPA, Office of Solid Waste, Washington, DC, EPA Method 8141 July 1992.

Merphos   **EPA Method 8140**
CAS #150-50-5

TITLE Organophosphorus Pesticides

MATRIX Groundwater, soils, sludges, water miscible liquid wastes, and non-water miscible wastes.

APPLICATION This method is used for the analysis of 21 organophosphorus pesticides. Samples are extracted, concentrated, and analyzed using direct injection of both neat and diluted organic liquid into a gas chromatograph (GC).

INTERFERENCES Solvents, reagents, and glassware may introduce artifacts. Other interferences may come from coextracted compounds from samples. The use of Florisil cleanup materials may produce low recoveries. Elemental sulfur may interfere with some compounds when using a flame photometric detector. Sulfur cleanup (Method 3660) may alleviate sulfur interference.

INSTRUMENTATION GC capable of on-column injections and a flame photometric detector (FPD) or a thermionic detector. Column 1: 1.8 m by 2 mm with 5% SP-2401 on Supelcoport. Column 2: 1.8 m by 2 mm with 3% SP-2401 on Supelcoport. Column 3: 50 cm by ⅛ in Teflon® with 15% SE-54 on Gas Chrom Q. The preferred column is Column Number 2.

RANGE 1.0–50 µg/L.

MDL 0.25 µg/L (in reagent water).

PQL FACTORS FOR MULTIPLYING × FID MDL VALUE

Matrix	Multiplication Factor
Groundwater	10
Low-level soil by sonication with GPC cleanup	670
High-level soil and sludge by sonication	10,000
Non-water miscible waste	100,000

PRECISION 7.9% (single operator standard deviation)

ACCURACY 120.7% (single operator average recovery)

SAMPLING METHOD Use 8-oz. widemouth glass bottles with Teflon®-lined caps for concentrated waste samples, soils, sediments, and sludges. Use 1 or 2½ gallon amber glass bottles with Teflon®-lined caps for liquid (water) samples.

STABILITY Cool soil, sediment, sludge, and liquid samples to 4°C. If residual chlorine is present in liquid samples add 3 mL of 10% sodium thiosulfate per gallon of sample and cool to 4°C.

MHT 14 days for concentrated waste, soil, sediment, or sludge; 7 days for liquid samples; all extracts must be analyzed within 40 days.

QUALITY CONTROL A quality control check sample concentrate containing this compound in acetone at a concentration 1,000 times more concentrated than the selected spike concentration is required. The QC check sample concentrate may be prepared from pure standard materials or purchased as certified solutions. Use appropriate trip, matrix, control site, method, reagent, and solvent blanks. Internal, surrogate, and five concentration level calibration standards are used.

REFERENCE Method 8140, SW-846, 3rd ed., Sept. 1986.

Mestranol   **EPA Method 8270**
CAS #72-33-3

TITLE Semivolatile Organic Compounds by GC/MS

MATRIX This method is used to determine the concentration of semivolatile organic compounds in extracts prepared from all types of solid waste matrices, soils, and groundwater. Although surface waters are not specifically mentioned, this method should be applicable to water samples from rivers, lakes, etc.

METHOD SUMMARY This method covers 259 semivolatile organic compounds. In very limited applications direct injection of the sample into the GC/MS system may be appropriate, but this results in very high detection limits (approximately 10,000 µg/L). Typically, a 1-L liquid sample, containing surrogate, and matrix spiking standards, is extracted in a continuous extractor first under acid conditions and then under basic conditions. Typically 30 g of a solid sample, containing surrogate, and matrix spiking standards, is extracted ultrasonically. After concentrating the extract to 1 mL it is spiked with 10 µL of an internal standard solution just prior to analysis by GC/MS. The volume injected should contain about 100 ng of base/neutral

and 200 ng of acid surrogates (for a 1-μL injection). Analysis is performed by GC/MS using a capillary GC column.

INTERFERENCES Raw GC/MS data from all blanks, samples, and spikes must be evaluated for interferences. Contamination by carryover can occur whenever high-concentration and low-concentration samples are sequentially analyzed. To reduce carryover, the sample syringe must be rinsed out between samples with solvent. Whenever an unusually concentrated sample is encountered, it should be followed by the analysis of blank solvent to check for cross-contamination.

INSTRUMENTATION A GC/MS and a data system are required. The GC column used is a 30 m × 0.25 mm I.D. (or 0.32 mm I.D.) 1um film thickness silicone-coated fused silica capillary column. A continuous liquid-liquid extractor equipped with Teflon® or glass connection joints and stopcocks requiring no lubrication, a K-D concentrating apparatus, water bath, and an ultrasonic disrupter with a minimum power of 300 W and with pulsing capability are also required.

PRECISION & ACCURACY The estimated quantitation limit (EQL) of Method 8270B for determining an individual compound is approximately 1 mg/kg (wet weight) for soil or sediment samples, 1–200 mg/kg for wastes (dependent on matrix and method of preparation), and 10 μg/L for groundwater samples. EQLs will be proportionately higher for sample extracts that require dilution to avoid saturation of the detector.

The EQL(b) for groundwater in μg/L is 20.
The EQL (a, b) for low concentrations in soil and sediment in μg/kg is not determined.
Accuracy as μg/L is not listed.
Overall precision in μg/L is not listed.

(a) *EQLs listed for soil/sediment are based on wet weight. Normally data is reported in a dry-weight basis; therefore, EQLs will be higher based on the % dry weight of each sample. This calculation is based on a 30 g sample and gel permeation chromatography cleanup.*
(b) *Sample EQLs are highly matrix-dependent. The EQLs are provided for guidance and may not always be achievable.*
C = *True value for concentration, in μg/L.*
X = *Average recovery found for measurements of samples containing a concentration of C, in μg/L.*

ESTIMATED QUANTITATION LIMIT

Other Matrices	Factor (a)
High-concentration soil and sludges by sonicator	7.5
Non-water miscible waste	75

(a) *EQL for other matrices = [EQL for low soil/sediment] × [Factor]. This estimated EQL is similar to an EPA "Practical Quantitation Limit."*

SAMPLING METHOD
Liquid samples — Use a 1 or 2½ gallon amber glass bottle with a screw-top Teflon®-lined cover that has been prewashed with detergent and rinsed with distilled water and methanol (or isopropanol).

Soils, sediments, or sludges — Use an 8-oz. widemouth glass with a screw-top Teflon®-lined cover that has been prewashed with detergent and rinsed with distilled water and methanol (or isopropanol).

SAMPLE PRESERVATION
Liquid samples — If residual chlorine is present, add 3 mL of 10% sodium thiosulfate per gallon, cool to 4°C and store in a solvent-free refrigerator until analysis; if chlorine is not present, then eliminate the sodium thiosulfate addition.

Soils, sediments, or sludges — Cool samples to 4°C and store in a solvent-free refrigerator.

MHT Liquid samples must be extracted within 7 days and the extracts analyzed within 40 days. Soils, sediments, or sludges may be stored for a maximum of 14 days and the extracts analyzed within 40 days.

SAMPLE PREPARATION
Liquid samples — Transfer 1 L quantitatively to a continuous extractor. If high concentrations are anticipated, a smaller volume may be used and then diluted with organic-free reagent water to 1 L. Adjust pH, if necessary, to pH <2 using 1:1 (V/V) sulfuric acid. Pipette 1.0 mL of a surrogate standard spiking solution into each sample. For the sample in each analytical batch selected for spiking, add 1.0 mL of a matrix spiking standard. For base/neutral acid analysis, the amount of the surrogates and matrix spiking compounds added to the sample should result in a final concentration of 100 ng/μL of each analyte in the extract to be analyzed (assuming a 1-μL injection). Extract with methylene chloride for 18–24 h. Next, adjust the pH of the aqueous phase to pH >11 using 10 N sodium hydroxide and extract it with methylene chloride again for 18–24 h. Dry the extract through a column containing anhydrous sodium sulfate and concentrate it to 1 mL using a K-D concentrator.

Soils, sediments, or sludges — Use 30 g of sample. Nonporous or wet samples (gummy or clay type) that do not have a free-flowing sandy texture must be mixed with anhydrous sodium sulfate until the sample is free flowing. Add 1 mL of surrogate standards to all samples, spikes, standards, and blanks. For the sample in each analytical batch selected for spiking, add 1.0 mL of a matrix spiking standard. For base/neutral acid analysis, the amount added of the surrogates and matrix spiking compounds should result in a final concentration of 100 ng/μL of each base/neutral analyte and 200 ng/μL of each acid analyte in the extract to be analyzed (assuming a 1-μL injection). Immediately add a 100-mL mixture of 1:1 methylene chloride:acetone and extract the sample ultrasonically for 3 min and then decant or filter the extracts. Repeat the extraction two or more times. Dry the extract using a column with anhydrous sodium sulfate and concentrate it to 1 mL in a K-D concentrator.

QUALITY CONTROL A methylene chloride solution containing 50 ng/μL of decafluorotriphenylphosphine (DFTPP) is used for tuning the GC/MS system each 12-h shift. A system performance check also must be made during every 12-h shift. A standard containing 50 ng/μL each of 4,4′-DDT, pentachlorophenol, and benzidine is required to verify injection port inertness and GC column performance. A calibration standard at mid-concentration, containing each compound of interest, including all required surrogates, must be performed every 12 h

during analysis. After the system performance check is met, calibration check compounds (CCCs) are used to check the validity of the initial calibration.

The internal standard responses and retention times in the calibration check standard must be evaluated immediately after or during data acquisition. If the retention time for any internal standard changes by more than 30 seconds from the last check calibration (12 h), the chromatographic system must be inspected for malfunctions and corrections must be made, as required. If the electron ionization current plot (EICP) area for any of the internal standards changes by a factor of two from the last daily calibration standard check, the mass spectrometer must be inspected for malfunctions and corrections must be made, as appropriate.

Demonstrate, through the analysis of a reagent water blank, that interferences from the analytical system, glassware, and reagents are under control. The blank samples should be carried through all stages of the sample preparation and measurement steps. For each analytical batch (up to 20 samples), a reagent blank, matrix spike, and matrix spike duplicate/duplicate must be analyzed (the frequency of the spikes may be different for different monitoring programs). The blank and spiked samples must be carried through all stages of the sample preparation and measurement steps. A QC reference sample concentrate containing each analyte at a concentration of 100 mg/L in methanol is required.

REFERENCE   Test Methods for Evaluating Solid Waste (SW-846). U.S. EPA 1983, Method 8270B, Rev. 2, Nov. 1990. Office of Solid Waste, Washington, DC.

Methacrylonitrile EPA Method 1624
CAS #126-98-7

TITLE   Volatile Organic Compounds by Isotope Dilution GC/MS

MATRIX   Compounds may be determined in waters, soils, and municipal sludges by this method.

METHOD SUMMARY   This method is used to determine 58 volatile toxic organic pollutants associated with the CWA (as amended 1987); the RCRA (as amended 1986); the CERCLA (as amended 1986); and other compounds amenable to purge-and-trap gas chromatography-mass spectrometry (GC/MS).

If the solids content is less than 1%, stable isotopically-labeled analogs of the compounds of interest are added to a 5-mL sample and the sample is purged with an inert gas at 20–25°C in a chamber designed for soil or water samples. If the solids content is greater than 1%, 5 mL of reagent water and the labeled compounds are added to a 5-g aliquot of sample and the mixture is purged at 40°C. Compounds that will not purge at 20–25°C or at 40°C are purged at 78–85°C. In the purging process, the volatile compounds are transferred from the aqueous phase into the gaseous phase where they are passed into a sorbent column, and trapped. After purging is completed, the trap is backflushed and heated rapidly to desorb the compounds into a GC. The compounds are separated by the GC and detected by a MS. The labeled compounds serve to correct the variability of the analytical technique.

INTERFERENCES   Impurities in the purge gas, organic compounds outgassing from the plumbing upstream of the trap, and solvent vapors in the lab account for most problems. Samples can be contaminated by diffusion of volatile organic compounds (particularly methylene chloride) through the bottle seal during shipment and storage. Contamination by carryover can occur when high-level and low-level samples are analyzed sequentially. When an unusually concentrated sample is encountered, follow it by analysis of a reagent water blank to check for carryover.

INSTRUMENTATION   Major equipment includes a GC with linear temperature programming and a glass jet separator as the MS interface, a MS with 70 eV electron impact ionization, and a data system to collect and record response factors.

Column: 2.8 m × 2 mm I.D. glass, packed with 1% SP-1000 on Carbopak B, 60/80 mesh, or equivalent.

PRECISION & ACCURACY   The detection limits of the method are usually dependent on the level of interferences rather than instrumental limitations. The method detection limits were determined in digested sludge (low solids) and in filter cake or compost (high solids).

The MDL (in µg/kg) for low solids is not listed and for high solids is not listed.
Labeled and native compound precision (in µg/L) as standard deviation was not listed.
Labeled and native compound accuracy (in µg/L) as average recovery was not listed.
Acceptance criteria are at 20 µg/L for this compound.

SAMPLE COLLECTION, PRESERVATION & HANDLING   Grab samples are collected in glass containers having a total volume greater than 20 mL. Fill and seal each bottle so that no air bubbles are entrapped. Samples are maintained at 0 to 4°C from the time of collection until analysis. If an aqueous sample contains residual chlorine, add sodium thiosulfate preservative (10 mg/40 mL) to the empty sample bottles just prior to shipment to the sample site. All samples must be analyzed within 14 days of collection.

SAMPLE PREPARATION   Samples containing less than 1% solids are analyzed directly as aqueous samples. Samples containing 1% solids or greater are analyzed as solid samples utilizing one of two methods, depending on the levels of pollutants, in the sample. Samples containing 1% solids or greater, and low to moderate levels of pollutants are analyzed by purging a known weight of sample added to 5 mL of reagent water. Samples containing 1% solids or greater, and high levels of pollutants, are extracted with methanol, and an aliquot of the methanol extract is added to reagent water and purged.

QUALITY CONTROL   A field blank prepared from reagent water and carried through the sampling and handling protocol may serve as a check on contamination from shipment and storage.

The analyst is permitted to modify this method to improve separations or lower the costs of measurements, provided all performance specifications are met. Analyses of blanks are required. When results of spikes indicate atypical method performance for samples, the samples are diluted to bring method performance within acceptable limits. Analyze two sets of four 5-mL aliquots (8 aliquots total) of the aqueous performance standard. Spike all samples with labeled compounds to assess method performance on the sample matrix. Compute the percent recovery of the labeled compounds using the internal standard method. Compare the percent recovery for each compound with the corresponding labeled compound recovery. Reagent water blanks are analyzed to demonstrate freedom from carryover contamination. Field replicates may be collected to determine the precision of the sampling technique, and spiked samples may be required to determine the accuracy of the analysis when the internal method is used.

REFERENCE Volatile Organic Compounds by Isotope Dilution GC/MS. Office of Water Regulation and Standards, U.S. EPA Industrial Technology Division, Washington, DC, EPA Method 1624, Rev. C, June 1989 (contact W.A. Telliard, U.S. EPA, Office of Water Regulations and Standards, 401 M St., SW, Washington, DC, 20460. Phone: 202-382-7131).

Methacrylonitrile **EPA Method 8240**
CAS #126-98-7

TITLE Volatile Organics By GC/MS: Packed Column Technique

MATRIX Nearly all types of sample matarices, regardless of water content, can be analyzed using this method. This includes groundwater, aqueous sludges, caustic liquors, acid liquors, waste solvents, oily wastes, mousses, tars, fibrous wastes, polymetric emulsions, filter cakes, spent carbons, spent catalysts, soils, and sediments.

METHOD SUMMARY Method 8240B covers 80 volatile organic compounds that are introduced into a gas chromatograph by the purge-and-trap method or by direct injection (in limited applications). For the purge-and-trap method an inert gas (zero grade nitrogen or helium) is bubbled through a 5-mL solution at ambient temperature. Purged sample components are trapped in a tube of sorbent materials. When purging is complete, the sorbent tube is heated and backflushed with inert gas to desorb the trapped components onto a GC column.

INTERFERENCES Impurities in the purge gas and from organic compounds outgassing from the plumbing ahead of the trap account for many contamination problems. Interferences purged or coextracted from the samples will vary considerably from source to source. Cross-contamination can occur whenever high-level and low-level samples are analyzed sequentially. Whenever an unusually concentrated sample is analyzed, it should be followed by the analysis of organic-free reagent water to check for cross-contamination. Samples also can be contaminated by diffusion of volatile organics (particularly methylene chloride and fluorocarbons) through the septum seal into the sample during shipment and storage. A trip blank can serve as a check on such contamination. The lab where volatile analysis is performed and also the refrigerated storage area should be completely free of solvents.

INSTRUMENTATION A gas chromatograph/mass spectrometry/data system (GC/MS) equipped with a 6 ft × 0.1 in I.D. glass column packed with 1% SP-1000 on Carbopack-B (60/80 mesh) is required. Also needed is a 5-mL purging device, a sorbent trap, and a thermal desorption apparatus.

PRECISION & ACCURACY This method is reported to have been tested by 15 laboratories using organic-free reagent water, drinking water, surface water, and industrial wastewaters (not specified) fortified at six concentrations over the range 5–600 µg/L.

Sample estimated quantitation limits (EQLs) are highly matrix-dependent. The EQLs listed may not always be achievable. EQLs listed for soils or sediments are based on wet weight. Normally, data is reported on a dry-weight basis; therefore, EQLs will be higher, based on the percent dry weight of each sample. Note that EQLs are even more variable than MDLs and that they are highly variable depending on the matrix being analyzed.

EQL in groundwater in µg/L was 100.
EQL in low soil or sediment in µg/kg was 100.
Accuracy (a) in µg/L was not listed.
Precision (b) in µg/L was not listed.

(a) *Average recovery found for measurements of samples containing a concentration of C, in µg/L.*
(b) *Overall precision found for measurements of samples with average recovery X for samples containing a concentration of C in µg/L.*
X = *Average recovery found for measurement of samples containing a concentration of C in µg/L.*

MULTIPLICATION FACTORS FOR OTHER MATRICES

Other Matrices	Factor (a)
Waste miscible liquid waste	50
High-concentration soil and sludge	125
Non-water miscible waste	500

(a) *EQL = [EQL for low soil/sediment] × [Factor]. For non-aqueous samples, the factor is on a wet-weight basis.*

SAMPLING METHOD
Liquid samples — Use a 40-mL glass screw-cap VOA vial with a Teflon®-faced silicone septum that has been prewashed, rinsed with distilled deionized water, and oven dried. However, if residual chlorine is present, collect sample in a 40-oz. soil VOA container which has been pre-preserved with 4 drops of 10% sodium thiosulfate, mix gently, and then transfer the sample to a 40-mL VOA vial. Collect bubble-free samples in duplicate and seal them in separate plastic bags.

Soils or sediments, and sludges — Use an 8-oz. widemouth glass bottle with a Teflon®-faced silicone septum that has been prewashed with detergent, rinsed with distilled deionized water, and oven dried. Tap slightly to eliminate free air space. Collect samples in duplicate and seal them in separate plastic bags.

SAMPLE PRESERVATION
Liquid samples — Add 4 drops of concentrated HCL and immediately cool samples to 4°C and store in a solvent-free refrigerator.

Soils or sediments, and sludges — Cool samples to 4°C and store in a solvent-free refrigerator.

MHT Maximum holding time is 14 days from the date of sample collection.

SAMPLE PREPARATION

Liquid samples — Remove the plunger from a 5-mL syringe and carefully pour the sample into the syringe barrel to just short of overflowing. Replace the syringe plunger and compress the sample. Open the syringe valve and vent any residual air while adjusting the sample volume to 5.0 mL. If there is only one volatile organic analysis (VOA) vial, a second syringe should be filled at this time to protect against possible loss of sample integrity. Add 10 µL of surrogate spiking solution and 10 µL of internal standard spiking solution through the valve bore of the 5-mL syringe, then close the valve. The surrogate and internal standards may be mixed and added as a single spiking solution.

Sediments, soils, and waste samples — All samples of this type should be screened by GC analysis using a headspace method (EPA Method 3810) or the hexadecane extraction and screening method (EPA Method 3820). Use the screening data to determine whether to use the low-concentration method (0.005–1 mg/kg) or the high-concentration method (>1 mg/kg).

Low-concentration method — The low-concentration method is based on purging a heated sediment or soil sample mixed with organic-free reagent water containing the surrogate and internal standards. Analyze all reagent blanks and standards under the same conditions as the samples.

Use a 5-g sample if the expected concentration is <0.1 mg/kg or a 1-g sample for expected concentrations between 0.1 and 1 mg/kg. Mix the contents of the sample container with a narrow metal spatula. Weigh the amount of the sample into a tared purge device. Add the spiked water to the purge device, which contains the weighed amount of sample, and connect the device to the purge-and-trap system.

High-concentration method — This method is based on extracting the sediment or soil with methanol. A waste sample is either extracted or diluted, depending on its solubility in methanol. Wastes that are insoluble in methanol are diluted with reagent tetraglyme or possibly polyethylene glycol (PEG). An aliquot of the extract is added to organic-free reagent water containing surrogate and internal standards. This is purged at ambient temperature. All samples with an expected concentration of >1.0 mg/kg should be analyzed by this method.

Mix the contents of the sample container with a narrow metal spatula. For sediments or soils and solid wastes that are insoluble in methanol, weigh 4 g (wet weight) of sample into a tared 20-mL vial. For waste that is soluble in methanol, tetraglyme, or PEG, weigh 1 g (wet weight) into a tared scintillation vial or culture tube or a 10-mL volumetric flask. Quickly add 9.0 mL of appropriate solvent then add 1.0 mL of a surrogate spiking solution to the vial, cap it, and shake it for 2 min.

METHANOL EXTRACT REQUIRED FOR ANALYSIS OF HIGH-CONCENTRATION SOILS OR SEDIMENTS

Approximate Concentration Range	Volume of Methanol Extract (a)
500–10,000 µg/kg	100 µL
1,000–20,000 µg/kg	50 µL
5,000–100,000 µg/kg	10 µL
25,000–500,000 µg/kg	100 µL of 1/50 dilution (b)

Calculate appropriate dilution factor for concentrations exceeding this table.

(a) The volume of methanol added to 5 mL of water being purged should be kept constant. Therefore, add to the 5-mL syringe whatever volume of methanol is necessary to maintain a volume of 100 µL added to the syringe.
(b) Dilute an aliquot of the methanol extract and then take 100 µL for analysis.

QUALITY CONTROL Demonstrate, through the analysis of a reagent water blank, that interferences from the analytical system, glassware, and reagents are under control. Blank samples should be carried through all stages of the sample preparation and measurement steps. For each analytical batch (up to 20 samples), a reagent blank, matrix spike, and matrix spike duplicate must be analyzed (the frequency of the spikes may be different for different monitoring programs). The blank and spiked samples must be carried through all stages of the sample preparation and measurement steps. QC samples mentioned in the section on Interferences will also be needed as appropriate to those situations.

REFERENCE Test Methods for Evaluating Solid Waste (SW-846). U.S. EPA. 1983. Method 8240B, Rev. 2, Nov. 1990. Office of Solid Wastes, Washington, DC.

Methapyrilene   **EPA Method 1625**
CAS #91-80-5

TITLE Semivolatile Organic Compounds by Isotope Dilution GC/MS

MATRIX The compounds may be determined in waters, soils, and municipal sludges by this method.

METHOD SUMMARY This method is used to determine 176 semivolatile toxic organic pollutants associated with the CWA (as amended 1987); the RCRA (as amended 1986); the CERCLA (as amended 1986); and other compounds amenable to extraction and analysis by capillary column gas chromatography-mass spectrometry (GC/MS).

Stable isotopically-labeled analogs of the compounds of interest are added to the sample. If the solids content is less than 1%, a 1-L sample is extracted at pH 12–13, then at pH <2 with methylene chloride using continuous extraction techniques.

If the solids content is 30% or less, the sample is diluted to 1% solids with reagent water, homogenized ultrasonically, and extracted at pH 12–13, then at pH <2 with methylene chloride using continuous extraction techniques. If the solids content is

greater than 30%, the sample is extracted using ultrasonic techniques.

Each extract is dried over sodium sulfate, concentrated to a volume of 5 mL, cleaned up using GPC, if necessary, and concentrated. Extracts are concentrated to 1 mL if GPC is not performed, and to 0.5 mL if GPC is performed.

An internal standard is added to the extract, and a 1-mL aliquot of the extract is injected into the GC. The compounds are separated by GC and detected by a MS. The labeled compounds serve to correct the variability of the analytical technique.

INTERFERENCES Solvents, reagents, glassware, and other sample processing hardware may yield artifacts and/or elevated baselines causing misinterpretation of chromatograms and spectra. Materials used in the analysis must be demonstrated to be free from interferences under the conditions of analysis by running method blanks initially and with each sample lot (sample started through the extraction process on a given 8-h shift, to a maximum of 20). Specific selection of reagents and purification of solvents by distillation in all glass systems may be required. Glassware and, where possible, reagents are cleaned by solvent rinse and baking at 450°C for 1-h minimum. Interferences coextracted from samples will vary considerably from source to source, depending on the diversity of the site being sampled.

INSTRUMENTATION Major instrumentation includes a GC with a splitless or on-column injection port for capillary column, a MS with 70 eV electron impact ionization, and a data system to collect and record MS data, and process it. A K-D apparatus is used to concentrate extracts.

GC Column: 30 m × 0.25 mm I.D. 5% phenyl, 94% methyl, 1% vinyl silicone bonded phased fused silica capillary column.

PRECISION & ACCURACY The detection limits of the method are usually dependent on the level of interferences rather than instrumental limitations. The limits typify the minimum quantities that can be detected with no interferences present.

The minimum level (in µg/mL) was not listed. This is defined as a minimum level at which the analytical system shall give recognizable mass spectra (background corrected) and acceptable calibration points.

The MDL (in µg/kg) in low solids was not listed and in high solids was not listed; these were determined in digested sludge (low solids) and in filter cake or compost (high solids).

The labeled and native compound initial precision as standard deviation (in µg/L) was not listed.
The labeled and native compound initial accuracy as average recovery (in µg/L) was not listed.

SAMPLE COLLECTION, PRESERVATION & HANDLING
Collect samples in glass containers. Aqueous samples which flow freely are collected in refrigerated bottles using automatic sampling equipment. Solid samples are collected as grab samples using widemouth jars. Maintain samples at 0 to 4°C from the time of collection until extraction. If residual chlorine is present in aqueous samples, add 80 mg sodium thiosulfate/L of water. Begin sample extraction within 7 days of collection, and analyze all extracts within 40 days of extraction.

SAMPLE PREPARATION Samples containing 1% solids or less are extracted directly using continuous liquid-liquid extraction techniques. Samples containing 1 to 30% solids are diluted to the 1% level with reagent water and extracted using continuous liquid-liquid extraction techniques. Samples containing greater than 30% solids are extracted using ultrasonic techniques.

Base/neutral extraction — Adjust the pH of the waters in the extractors to 12–13 with 6 N NaOH. Extract with methylene chloride for 24–48 h.
Acid extraction — Adjust the pH of the waters in the extractors to 2 or less using 6 N sulfuric acid. Extract with methylene chloride for 24–48 h.
Ultrasonic extraction of high solids samples — Add anhydrous sodium sulfate to the sample and QC aliquot(s). Add acetone:methylene chloride (1:1) to the sample and mix thoroughly

Concentrate extracts using a K-D apparatus.

QUALITY CONTROL The analyst is permitted to modify this method to improve separations or lower the costs of measurements, provided all performance specifications are met. Analyses of blanks are required to demonstrate freedom from contamination. When results of spikes indicate atypical method performance for samples, the samples are diluted to bring method performance within acceptable limits.

For low solids (aqueous samples), extract, concentrate, and analyze two sets of four 1-L aliquots (8 aliquots total) of the precision and recovery standard. For high solids samples, two sets of four 30-g aliquots of the high solids reference matrix are used.

Spike all samples with labeled compounds to assess method performance. Compute percent recovery of the labeled compounds using the internal standard method. Compare the labeled compound recovery for each compound with the corresponding labeled compound recovery.

Reagent water and high solids reference matrix blanks are analyzed to demonstrate freedom from contamination. Extract and concentrate a 1-L reagent water blank or a high solids reference matrix blank with each sample's lot (samples started through the extraction process on the same 8-h shift, to a maximum of 20 samples).

Field replicates may be collected to determine the precision of the sampling technique, and spiked samples may be required to determine the accuracy of the analysis when the internal standard method is used.

REFERENCE Semivolatile Organic Compounds by Isotope Dilution GC/MS. Office of Water Regulation and Standards, U.S. EPA Industrial Technology Division, Washington, DC, EPA Method 1625, Rev. C, June 1989 (contact W.A. Telliard, U.S. EPA, Office of Water Regulations and Standards, 401 M St., SW, Washington, DC, 20460. Phone: 202-382-7131).

Methapyrilene
CAS #91-80-5
EPA Method 8270

TITLE Semivolatile Organic Compounds by GC/MS

MATRIX This method is used to determine the concentration of semivolatile organic compounds in extracts prepared from all types of solid waste matrices, soils, and groundwater. Although surface waters are not specifically mentioned, this method should be applicable to water samples from rivers, lakes, etc.

METHOD SUMMARY This method covers 259 semivolatile organic compounds. In very limited applications direct injection of the sample into the GC/MS system may be appropriate, but this results in very high detection limits (approximately 10,000 µg/L). Typically, a 1-L liquid sample, containing surrogate, and matrix spiking standards, is extracted in a continuous extractor first under acid conditions and then under basic conditions. Typically 30 g of a solid sample, containing surrogate, and matrix spiking standards, is extracted ultrasonically. After concentrating the extract to 1 mL it is spiked with 10 µL of an internal standard solution just prior to analysis by GC/MS. The volume injected should contain about 100 ng of base/neutral and 200 ng of acid surrogates (for a 1-µL injection). Analysis is performed by GC/MS using a capillary GC column.

INTERFERENCES Raw GC/MS data from all blanks, samples, and spikes must be evaluated for interferences. Contamination by carryover can occur whenever high-concentration and low-concentration samples are sequentially analyzed. To reduce carryover, the sample syringe must be rinsed out between samples with solvent. Whenever an unusually concentrated sample is encountered, it should be followed by the analysis of blank solvent to check for cross-contamination.

INSTRUMENTATION A GC/MS and a data system are required. The GC column used is a 30 m × 0.25 mm I.D. (or 0.32 mm I.D.) 1um film thickness silicone-coated fused silica capillary column. A continuous liquid-liquid extractor equipped with Teflon® or glass connection joints and stopcocks requiring no lubrication, a K-D concentrating apparatus, water bath, and an ultrasonic disrupter with a minimum power of 300 W and with pulsing capability are also required.

PRECISION & ACCURACY The estimated quantitation limit (EQL) of Method 8270B for determining an individual compound is approximately 1 mg/kg (wet weight) for soil or sediment samples, 1–200 mg/kg for wastes (dependent on matrix and method of preparation), and 10 µg/L for groundwater samples. EQLs will be proportionately higher for sample extracts that require dilution to avoid saturation of the detector.

The EQL(b) for groundwater in µg/L is 100.
The EQL (a, b) for low concentrations in soil and sediment in µg/kg is not determined.
Accuracy as µg/L is not listed.
Overall precision in µg/L is not listed.

(a) EQLs listed for soil/sediment are based on wet weight. Normally data is reported in a dry-weight basis; therefore, EQLs will be higher based on the % dry weight of each sample. This calculation is based on a 30 g sample and gel permeation chromatography cleanup.

(b) Sample EQLs are highly matrix-dependent. The EQLs are provided for guidance and may not always be achievable.

C = True value for concentration, in µg/L.
X = Average recovery found for measurements of samples containing a concentration of C, in µg/L.

ESTIMATED QUANTITATION LIMIT

Other Matrices	Factor (a)
High-concentration soil and sludges by sonicator	7.5
Non-water miscible waste	75

(a) EQL for other matrices = [EQL for low soil/sediment] × [Factor]. This estimated EQL is similar to an EPA "Practical Quantitation Limit."

SAMPLING METHOD
Liquid samples — Use a 1 or 2½ gallon amber glass bottle with a screw-top Teflon®-lined cover that has been prewashed with detergent and rinsed with distilled water and methanol (or isopropanol).

Soils, sediments, or sludges — Use an 8-oz. widemouth glass with a screw-top Teflon®-lined cover that has been prewashed with detergent and rinsed with distilled water and methanol (or isopropanol).

SAMPLE PRESERVATION
Liquid samples — If residual chlorine is present, add 3 mL of 10% sodium thiosulfate per gallon, cool to 4°C and store in a solvent-free refrigerator until analysis; if chlorine is not present, then eliminate the sodium thiosulfate addition.

Soils, sediments, or sludges — Cool samples to 4°C and store in a solvent-free refrigerator.

MHT Liquid samples must be extracted within 7 days and the extracts analyzed within 40 days. Soils, sediments, or sludges may be stored for a maximum of 14 days and the extracts analyzed within 40 days.

SAMPLE PREPARATION
Liquid samples — Transfer 1 L quantitatively to a continuous extractor. If high concentrations are anticipated, a smaller volume may be used and then diluted with organic-free reagent water to 1 L. Adjust pH, if necessary, to pH <2 using 1:1 (V/V) sulfuric acid. Pipette 1.0 mL of a surrogate standard spiking solution into each sample. For the sample in each analytical batch selected for spiking, add 1.0 mL of a matrix spiking standard. For base/neutral acid analysis, the amount of the surrogates and matrix spiking compounds added to the sample should result in a final concentration of 100 ng/µL of each analyte in the extract to be analyzed (assuming a 1-µL injection). Extract with methylene chloride for 18–24 h. Next, adjust the pH of the aqueous phase to pH >11 using 10 N sodium hydroxide and extract it with methylene chloride again for 18–24 h. Dry the extract through a column containing anhydrous sodium sulfate and concentrate it to 1 mL using a K-D concentrator.

Soils, sediments, or sludges — Use 30 g of sample. Nonporous or wet samples (gummy or clay type) that do not have a free-flowing

sandy texture must be mixed with anhydrous sodium sulfate until the sample is free flowing. Add 1 mL of surrogate standards to all samples, spikes, standards, and blanks. For the sample in each analytical batch selected for spiking, add 1.0 mL of a matrix spiking standard. For base/neutral acid analysis, the amount added of the surrogates and matrix spiking compounds should result in a final concentration of 100 ng/µL of each base/neutral analyte and 200 ng/µL of each acid analyte in the extract to be analyzed (assuming a 1-µL injection). Immediately add a 100-mL mixture of 1:1 methylene chloride:acetone and extract the sample ultrasonically for 3 min and then decant or filter the extracts. Repeat the extraction two or more times. Dry the extract using a column with anhydrous sodium sulfate and concentrate it to 1 mL in a K-D concentrator.

QUALITY CONTROL A methylene chloride solution containing 50 ng/µL of decafluorotriphenylphosphine (DFTPP) is used for tuning the GC/MS system each 12-h shift. A system performance check also must be made during every 12-h shift. A standard containing 50 ng/µL each of 4,4'-DDT, pentachlorophenol, and benzidine is required to verify injection port inertness and GC column performance. A calibration standard at mid-concentration, containing each compound of interest, including all required surrogates, must be performed every 12 h during analysis. After the system performance check is met, calibration check compounds (CCCs) are used to check the validity of the initial calibration.

The internal standard responses and retention times in the calibration check standard must be evaluated immediately after or during data acquisition. If the retention time for any internal standard changes by more than 30 seconds from the last check calibration (12 h), the chromatographic system must be inspected for malfunctions and corrections must be made, as required. If the electron ionization current plot (EICP) area for any of the internal standards changes by a factor of two from the last daily calibration standard check, the mass spectrometer must be inspected for malfunctions and corrections must be made, as appropriate.

Demonstrate, through the analysis of a reagent water blank, that interferences from the analytical system, glassware, and reagents are under control. The blank samples should be carried through all stages of the sample preparation and measurement steps. For each analytical batch (up to 20 samples), a reagent blank, matrix spike, and matrix spike duplicate/duplicate must be analyzed (the frequency of the spikes may be different for different monitoring programs). The blank and spiked samples must be carried through all stages of the sample preparation and measurement steps. A QC reference sample concentrate containing each analyte at a concentration of 100 mg/L in methanol is required.

REFERENCE Test Methods for Evaluating Solid Waste (SW-846). U.S. EPA 1983, Method 8270B, Rev. 2, Nov. 1990. Office of Solid Waste, Washington, DC.

Metholachlor — EPA Method 507
CAS #51218-45-2

TITLE Determination of Nitrogen and Phosphorus-Containing Pesticides in Water by GC/NPD

MATRIX This method is applicable to the determination of certain nitrogen and phosphorus-containing pesticides in finished drinking water and groundwater.

METHOD SUMMARY Method 507 covers 46 nitrogen- and phosphorus-containing pesticides. A 1-L sample is fortified with a surrogate standard, salted, buffered, extracted with methylene chloride, and concentrated; then the solvent is exchanged with methyl tert-butyl ether (MTBE) and concentrated again, and a 2-µL aliquot of a sample extract is injected into a GC system equipped with a selective nitrogen-phosphorus detector and a capillary column for analysis.

INTERFERENCES Method interferences may be caused by contaminants in solvents, reagents, glassware, and other sample processing apparatus. Interfering contamination may occur when a sample containing low concentrations of analytes is analyzed immediately following a sample containing relatively high concentrations. One or more injections of MTBE should be made following the analysis of a sample with high concentrations of analytes to check for analyte carryover. Matrix interferences may be caused by contaminants that are coextracted from the sample. The extent of matrix interferences will vary considerably from source to source, depending upon the water sampled.

INSTRUMENTATION A gas chromatograph system (GC) equipped with a nitrogen-phosphorus detector (NPD) is needed.

Column 1: 30 m × 0.25 mm I.D. DB-5 bonded fused silica column, 0.25 µm film thickness, or equivalent.

Column 2: 30 m × 0.25 mm I.D. DB-1701 bonded fused silica column, 0.25 µm film thickness, or equivalent.

PRECISION & ACCURACY This method has been validated in a single lab and estimated detection limits (EDLs) have been determined for each analyte. Observed detection limits may vary among waters, depending upon the nature of the interferences in the sample matrix and the specific instrumentation used. Analytes that are not separated chromatographically cannot be individually identified and measured unless an alternative technique for identification and quantification exist.

The estimated detection limit (in µg/L) was 0.75. The EDL is defined as either method detection limit or a level of compound in a sample yielding a peak in the final extract with signal-to-noise ratio of approximately 5, whichever value is higher.

The concentration used for these measurements (in µg/L) was 7.5.
The accuracy (as % recovery) was 93.
The precision (% RSD) was 4.

SAMPLING METHOD Grab samples are collected in 1-L glass sample bottles (prewashed with detergent and hot tap

water, rinsed with reagent water, and dried in an oven at 400°C for 1 h) with screw caps lined with PTFE-fluorocarbon.

SAMPLE PRESERVATION Add mercuric chloride to the sample bottle in amounts to produce a concentration of 10 mg/L. If residual chlorine is present, add 80 mg of sodium thiosulfate/L of sample to the sample bottle prior to collection. After collection, seal bottle and shake vigorously for 1 min, then cool the sample to 4°C immediately and store it at 4°C in the dark until extraction.

MHT Maximum holding time of the samples, and in some cases the extracts, is 14 days.

Note: Samples with this compound exhibited recoveries of less than 60% after 14 days.

SAMPLE PREPARATION Fortify the sample with 50 µL of the surrogate standard solution, adjust to pH 7 with phosphate buffer, add 100 g NaCl to the sample, and seal and shake to dissolve the salt; then extract with methylene chloride in a separatory funnel or in a mechanical tumbler bottle. Dry the extract by pouring it through a solvent-rinsed drying column containing about 10 cm of anhydrous sodium sulfate. Collect the extract in a Kuderna-Danish (K-D) concentrator and rinse the column with 20–30 mL methylene chloride. Concentrate the extract to about 2 mL and rinse the flask and its lower joint into the concentrator tube with 1 to 2 mL of methyl t-butyl ether (MTBE). Add 5–10 mL of MTBE and concentrate the extract twice (adding more MTBE) to a final volume of 5.0 mL and store it at 4°C until analysis.

Note: If methylene chloride is not completely removed from the final extract, it may cause detector problems.

QUALITY CONTROL Minimum quality control requirements are initial demonstration of lab capability, determination of surrogate compound recoveries in each sample and blank, monitoring internal standard peak area or height in each sample and blank, analysis of lab reagent blanks, lab fortified samples, lab fortified blanks, and other QC samples. A lab reagent blank is analyzed to demonstrate that all glassware and reagent interferences are under control.

Initial demonstration of capability is fulfilled by analyzing four fortified reagent water samples with the recovery value for each analyte falling within the acceptable range (±30% average recovery). Surrogate recoveries from samples or method blanks must be 70–130%. The internal standard response for any sample chromatogram should not deviate from the daily calibration check standard's internal standard response by more than 30% or lab fortified blanks and sample matrices are used to assess lab performance and analyte recovery, respectively.

If the response for the target analyte peak exceeds the working range of the system, dilute the extract and reanalyze. Alternative techniques such as an alternate detector or second chromatography column should be used to confirm peak identification when sample components are not resolved adequately.

EPA CONTACT & HOTLINE For technical questions contact Dr. Baldev Bathija, U.S. EPA, Office of Ground Water and Drinking Water (WH-550D), 401 M St. SW, Washington, DC 20460. Tel. (202) 260-3040. For further information the EPA Safe Drinking Water Hotline may be called at: (800) 426-4791.

REFERENCE Methods for the Determination of Organic Compounds in Drinking Water, EPA/600/4-88/039 (revised July 1991). U.S. EPA Environmental Monitoring Systems Laboratory, Cincinnati, OH, 45268, U.S.A. Available from the National Technical Information Service (NTIS), 5285 Port Royal Road, Springfield, VA 22161; Tel. 800-553-6847. NTIS Order Number is PB91-231480.

Methoxychlor EPA Method 505
CAS #72-43-5

TITLE Analysis of Organohalide Pesticides and Commercial Polychlorinated Biphenyl (PCB) Products in Water by Microextraction and Gas Chromatography. U.S. EPA Method 505, Rev. 2.0, 1989.

MATRIX This method is applicable to drinking water and raw source water. The latter should include most surface water and groundwater sources.

METHOD SUMMARY Method 505 covers 25 pesticides and commercial PCB products. This is a very sensitive method that is more useful for monitoring than for exploratory analyses. 5-mL of water are saturated with sodium chloride and then extracted by shaking with 2 mL of hexane. The sample extracts are transferred to an autosampler setup to inject 1–2 µL portions into a gas chromatograph (GC) for analysis. Alternatively, 1–2 µL portions of samples, blanks, and standards may be manually injected. Each extract is analyzed by capillary GC/ECD with confirmation using either a second capillary column or GC/MS. The electron capture detector is easy to use, but it is a nonselective detector. The microextraction technique also eliminates the expensive sample preparation costs of other methods, but it has the disadvantage of being less sensitive than most because the extracts are not concentrated.

INTERFERENCES Method interferences may be caused by contaminants in solvents, reagents, glassware, and other sample processing apparatus that lead to discrete artifacts or elevated baselines. Interfering contamination may occur when a sample containing low concentrations of analytes is analyzed immediately following a sample containing relatively high concentrations of the analytes. Matrix interferences also may be caused by contaminants that are coextracted from the sample; cleanup of sample extracts may be necessary in these cases. Some pesticides and commercial PCB products from aqueous solutions adhere to glass surfaces, so sample transfers and contact with glass surfaces should be minimized. Some pesticides are rapidly oxidized by chlorine so dechlorination with sodium thiosulfate at the time of sample collection is important. Also, splitless injectors may cause degradation of some pesticides.

INSTRUMENTATION A gas chromatograph/electron capture detector/data system, with temperature programming and split/splitless injector suitable for use with capillary columns is needed.

Column 1: 0.32 mm I.D. × 30 m fused silica capillary with chemically bond methyl polysiloxane phase (DB-1, 1.0 µm film, or equivalent).

Column 2: 0.32 mm I.D. × 30 m fused silica capillary with 1:1 mixed phase of dimethyl silicone and polyethylene glycol (Durawax-DX3, 0.25 µm film, or equivalent).

Column 3: 0.32 mm I.D. × 25 m fused silica capillary with chemically bonded 50:50 methyl-phenyl silicone (OV-17, 1.5 µm film, or equivalent).

Column 1 should be used as the primary analytical column. Columns 2 and 3 are recommended for use as confirmatory columns when GC/MS confirmation is not available.

PRECISION & ACCURACY Method detection limits are dependent upon the characteristics of the gas chromatographic system used. Analytes that are not separated chromatographically cannot be individually identified and used in the same calibration mixture or water samples unless an alternative technique for identification and quantification, such as mass spectrometry, is used.

The concentration(s) (in µg/L) used for these QC measurements was 2.10 and 7.03.

The MDL (in µg/L) was 0.96.

The accuracy (% recovery) for reagent water at the above concentration(s) was 100 and 98 and the precision (%) was 21.0 and 10.9.

The accuracy (% recovery) for groundwater at the above concentration(s) was not listed and not listed and the precision (%) was not listed and not listed.

The accuracy (% recovery) for tap water at the above concentration(s) was not listed and not listed and the precision (5) was not listed and not listed.

Note: No range of concentrations is provided with this method.

SAMPLING METHOD Collect samples using a 40-mL screw-cap vial (prewashed with detergent, rinsed with distilled water and oven dried at 400°C for one h) with a Teflon®-faced silicone septum. Collect bubble-free samples and place the septum with the Teflon® side down on the water.

SAMPLE PRESERVATION If residual chlorine is present in the water add about 3 mg of sodium thiosulfate to each vial before samples are collected to remove the chlorine. Alternatively, add 75 µL of 0.04 g/mL solution of sodium thiosulfate to each vial just prior to sampling. Immediately cool samples to 4°C, and store them in a solvent-free refrigerator at 4°C until analysis.

MHT The maximum holding time is 14 days from the time the sample was collected until it must be analyzed.

SAMPLE PREPARATION Remove the sample from storage and allow it to come to room temperature. Remove a 5-mL volume from each container and weigh the container to the nearest 0.1 g. Add 6 g of sodium chloride and 2.0 mL of hexane to each sample bottle. Recap the sample and shake it vigorously for one min. Allow the water and hexane phases to separate, remove the cap, and transfer 0.5 mL of hexane into an autosampler vial using a disposable glass pipette. Transfer the remaining hexane phase into a second autosampler vial and store at 4°C for reanalysis, if necessary. Discard the remaining sample/hexane mixture and reweigh the empty container to determine net weight of sample.

QUALITY CONTROL Minimum quality control requirements are initial demonstration of lab capability, analysis of lab reagent blanks, fortified blanks, fortified sample matrix, and quality control samples. The lab must analyze at least one fortified blank per sample set, or at least one for every 20 samples. The fortifying concentration of each analyte should be 10 times the method detection limit or the maximum calibration limit (MCL), whichever is less. Calculate accuracy as percent recovery and develop control limits from the mean percent recovery and standard deviation.

The lab must add a known concentration of the analytes to a minimum of 10% of the routine samples, or one lab fortified sample matrix per sample set. Calculate the percent recovery for each analyte and compare to the control limits established from the analyses of the fortified blanks.

EPA CONTACT & HOTLINE For technical questions contact Dr. Baldev Bathija, U.S. EPA, Office of Ground Water and Drinking Water (WH-550D), 401 M St. SW, Washington, DC 20460. Tel. (202) 260-3040. For further information the EPA Safe Drinking Water Hotline may be called at: (800) 426-4791.

REFERENCE Methods for the Determination of Organic Compounds in Drinking Water, EPA/600/4-88/039 (revised July 1991). U.S. EPA Environmental Monitoring Systems Laboratory, Cincinnati, OH, 45268, U.S.A. Available from the National Technical Information Service (NTIS), 5285 Port Royal Road, Springfield, VA 22161; Tel. 800-553-6847. NTIS Order Number is PB91-231480.

4,4'-Methoxychlor   EPA Method 8080
CAS #72-43-5

TITLE Organochlorine Pesticides and Polychlorinated Biphenyls By Gas Chromatography

MATRIX This method is used to determine the concentration of various organochlorine pesticides and polychlorinated biphenyls in extracts prepared from water, groundwater, soils, and sediments.

METHOD SUMMARY This method covers 26 pesticides and Aroclor (PCB) mixtures and it is suitable for monitoring-type analyses. After extraction, concentration and solvent exchange to hexane, a 2- to 5-µL sample aliquot is injected into a GC using the solvent flush technique, and the analytes are detected by an electron capture detector (ECD) or an electrolytic conductivity detector in the halogen mode (HECD). Both neat and diluted organic liquids may be analyzed by direct injection.

INTERFERENCES Interferences coextracted from the samples will vary considerably from source to source. Interferences by phthalate esters can pose a major problem in pesticide determinations when using the ECD. Cross-contamination of clean glassware routinely occurs when plastics are handled during extraction steps, especially when solvent-wetted surfaces are handled. The contamination from phthalate esters can be completely eliminated with a microcoulometric or electrolytic conductivity detector. Solvents, reagent, glassware, and other sample processing hardware may yield artifacts and/or interferences to sample analysis.

INSTRUMENTATION A gas chromatograph capable of on-column injections is needed. It must be equipped with an ECD or a HECD and one of the following GC columns:

Column 1: Supelcoport (100/120 mesh) coated with 1.5% SP-2250/1.95% SP-2401 packed in a 1.8 m × 4 mm I.D. glass column.

Column 2: Supelcoport (100/120 mesh) coated with 3% OV-1 in a 1.8 m × 4 mm I.D. glass column.

PRECISION & ACCURACY The method was tested by 20 laboratories using organic-free reagent water, drinking water, surface water, and three industrial wastewaters spiked at six concentrations. Concentrations used in the study ranged from 0.5 to 30 µg/L for single-component pesticides and from 8.5 to 400 µg/L for multicomponent parameters. Overall precision and method accuracy were found to be directly related to the concentration of the analyte and essentially independent of the sample matrix. The sensitivity of this method usually depends on the concentration of interferences rather than on instrumental limitations.

MDL in µg/L was 0.176.
Concentration range in µg/L was 0.5–30.
Accuracy as recovery (x*) in µg/L was not listed.
Overall precision (S*) in µg/L was not listed.

x^* *Expected recovery for one or more measurements of a sample containing concentration C, in µg/L.*
S^* = *Expected interlaboratory standard deviation of measurements at an average concentration found of the analyte in µg/L.*
C = *True value for the concentration, in µg/L.*
X = *Average recovery found for measurements of samples containing a concentration of C, in µg/L.*

SAMPLING METHOD
Liquid samples — Use a 1 or 2½ gallon amber glass bottle with a screw-top Teflon®-lined cover. Pre-wash the bottle with detergent, rinse with distilled water and methanol (or isopropanol).

Soil, sediments, and sludges — Use an 8-oz. widemouth glass with a screw-top Teflon®-lined cover. Pre-wash the bottle with detergent, rinse with distilled water and methanol (or isopropanol).

SAMPLE PRESERVATION Cool water, soil, sediment, or sludge samples immediately to 4°C.

Water samples — If residual chlorine is present, add 3 mL of 10% sodium thiosulfate per gallon and cool to 4°C. All extracts and samples should be stored under refrigeration.

MHT Liquid samples must be extracted within 7 days and the extracts must be analyzed within 40 days. Soils, sediments, and sludges may be stored for a maximum of 14 days prior to extraction.

SAMPLE PREPARATION
Liquid samples — Extract 1 L samples in a continuous extractor at pH 5–9 with methylene chloride after adding 1.0 mL of surrogate spiking solution to each sample. Pass the extract through a column of anhydrous sodium sulfate to dry and concentrate it in a K-D apparatus to 1 mL volume.

Soils, sediments and sludges — Rapidly weigh approximately 30 g of sample into a 400-mL beaker to avoid loss of the more volatile extractables. Nonporous or wet samples (gummy or clay type) that do not have a free-flowing sandy texture must be mixed with anhydrous sodium sulfate until the sample is free flowing. Add 1 mL of surrogate standards to all samples, spikes, standards, and blanks. Add 100 mL of 1:1 methylene chloride:acetone and extract ultrasonically. Decant and filter extracts, dry the extract by passing it through a drying column containing anhydrous sodium sulfate and concentrate to 1 mL in a K-D apparatus.

Hexane solvent exchange — Add 50 mL of hexane, a new boiling chip, and concentrate until the apparent volume of liquid reaches 1 mL. Adjust the extract volume to 10.0 mL. Stopper the concentration tube and store refrigerated at 4°C if further processing will not be performed immediately. If the extract will be stored longer than two days, transfer it to a vial with Teflon®-lined screw-cap or crimp top.

QUALITY CONTROL Demonstrate through the analysis of a reagent water blank, that all glassware and reagents are interference free. Each time a set of samples is processed, a method blank should be processed as a safeguard against chronic lab contamination. A reagent blank, a matrix spike, and a duplicate or matrix spike duplicate must be performed for each analytical batch (up to a maximum of 20 samples) analyzed.

Analytical system performance must be verified by analyzing QC check samples. The QC check sample concentration should contain each single-component analyte at the following concentrations in acetone: 4,4'-DDD, 10 µg/mL; 4,4'-DDT, 10 µg/mL; endosulfan II, 10 µg/mL; endosulfan sulfate, 10 µg/mL; and any other single-component pesticide at 2 µg/mL. If the method is only to be used to analyze PCBs, Chlordane, or Toxaphene, the QC check sample concentrate should contain the most representative multicomponent parameter at a concentration of 50 µg/mL in acetone.

REFERENCE Test Methods for Evaluating Solid Waste (SW-846). U.S. EPA. 1983. Method 8080B, Rev. 2, Nov. 1990. Office of Solid Wastes, Washington, DC.

Methoxychlor   **EPA Method 8270**
CAS #72-43-5

TITLE Semivolatile Organic Compounds by GC/MS

MATRIX This method is used to determine the concentration of semivolatile organic compounds in extracts prepared from all types of solid waste matrices, soils, and groundwater. Although surface waters are not specifically mentioned, this method should be applicable to water samples from rivers, lakes, etc.

METHOD SUMMARY This method covers 259 semivolatile organic compounds. In very limited applications direct injection of the sample into the GC/MS system may be appropriate, but this results in very high detection limits (approximately 10,000 μg/L). Typically, a 1-L liquid sample, containing surrogate, and matrix spiking standards, is extracted in a continuous extractor first under acid conditions and then under basic conditions. Typically 30 g of a solid sample, containing surrogate, and matrix spiking standards, is extracted ultrasonically. After concentrating the extract to 1 mL it is spiked with 10 μL of an internal standard solution just prior to analysis by GC/MS. The volume injected should contain about 100 ng of base/neutral and 200 ng of acid surrogates (for a 1-μL injection). Analysis is performed by GC/MS using a capillary GC column.

INTERFERENCES Raw GC/MS data from all blanks, samples, and spikes must be evaluated for interferences. Contamination by carryover can occur whenever high-concentration and low-concentration samples are sequentially analyzed. To reduce carryover, the sample syringe must be rinsed out between samples with solvent. Whenever an unusually concentrated sample is encountered, it should be followed by the analysis of blank solvent to check for cross-contamination.

INSTRUMENTATION A GC/MS and a data system are required. The GC column used is a 30 m × 0.25 mm I.D. (or 0.32 mm I.D.) 1um film thickness silicone-coated fused silica capillary column. A continuous liquid-liquid extractor equipped with Teflon® or glass connection joints and stopcocks requiring no lubrication, a K-D concentrating apparatus, water bath, and an ultrasonic disrupter with a minimum power of 300 W and with pulsing capability are also required.

PRECISION & ACCURACY The estimated quantitation limit (EQL) of Method 8270B for determining an individual compound is approximately 1 mg/kg (wet weight) for soil or sediment samples, 1–200 mg/kg for wastes (dependent on matrix and method of preparation), and 10 μg/L for groundwater samples. EQLs will be proportionately higher for sample extracts that require dilution to avoid saturation of the detector.

The EQL(b) for groundwater in μg/L is 10.
The EQL (a, b) for low concentrations in soil and sediment in μg/kg is not determined.
Accuracy as μg/L is not listed.
Overall precision in μg/L is not listed.

(a) *EQLs listed for soil/sediment are based on wet weight. Normally data is reported in a dry-weight basis; therefore, EQLs will be higher based on the % dry weight of each sample. This calculation is based on a 30 g sample and gel permeation chromatography cleanup.*
(b) *Sample EQLs are highly matrix-dependent. The EQLs are provided for guidance and may not always be achievable.*
C = *True value for concentration, in μg/L.*
X = *Average recovery found for measurements of samples containing a concentration of C, in μg/L.*

ESTIMATED QUANTITATION LIMIT

Other Matrices	Factor (a)
High-concentration soil and sludges by sonicator	7.5
Non-water miscible waste	75

(a) *EQL for other matrices = [EQL for low soil/sediment] × [Factor]. This estimated EQL is similar to an EPA "Practical Quantitation Limit."*

SAMPLING METHOD
Liquid samples — Use a 1 or 2½ gallon amber glass bottle with a screw-top Teflon®-lined cover that has been prewashed with detergent and rinsed with distilled water and methanol (or isopropanol).

Soils, sediments, or sludges — Use an 8-oz. widemouth glass with a screw-top Teflon®-lined cover that has been prewashed with detergent and rinsed with distilled water and methanol (or isopropanol).

SAMPLE PRESERVATION
Liquid samples — If residual chlorine is present, add 3 mL of 10% sodium thiosulfate per gallon, cool to 4°C and store in a solvent-free refrigerator until analysis; if chlorine is not present, then eliminate the sodium thiosulfate addition.

Soils, sediments, or sludges — Cool samples to 4°C and store in a solvent-free refrigerator.

MHT Liquid samples must be extracted within 7 days and the extracts analyzed within 40 days. Soils, sediments, or sludges may be stored for a maximum of 14 days and the extracts analyzed within 40 days.

SAMPLE PREPARATION
Liquid samples — Transfer 1 L quantitatively to a continuous extractor. If high concentrations are anticipated, a smaller volume may be used and then diluted with organic-free reagent water to 1 L. Adjust pH, if necessary, to pH <2 using 1:1 (V/V) sulfuric acid. Pipette 1.0 mL of a surrogate standard spiking solution into each sample. For the sample in each analytical batch selected for spiking, add 1.0 mL of a matrix spiking standard. For base/neutral acid analysis, the amount of the surrogates and matrix spiking compounds added to the sample should result in a final concentration of 100 ng/μL of each analyte in the extract to be analyzed (assuming a 1-μL injection). Extract with methylene chloride for 18–24 h. Next, adjust the pH of the aqueous phase to pH >11 using 10 N sodium hydroxide and extract it with methylene chloride again for 18–24 h. Dry the extract through a column containing anhydrous sodium sulfate and concentrate it to 1 mL using a K-D concentrator.

Soils, sediments, or sludges — Use 30 g of sample. Nonporous or wet samples (gummy or clay type) that do not have a free-flowing sandy texture must be mixed with anhydrous sodium sulfate until the sample is free flowing. Add 1 mL of surrogate standards to all samples, spikes, standards, and blanks. For the sample in each analytical batch selected for spiking, add 1.0 mL of a matrix spiking standard. For base/neutral acid analysis, the amount added of the surrogates and matrix spiking compounds

should result in a final concentration of 100 ng/µL of each base/neutral analyte and 200 ng/µL of each acid analyte in the extract to be analyzed (assuming a 1-µL injection). Immediately add a 100-mL mixture of 1:1 methylene chloride:acetone and extract the sample ultrasonically for 3 min and then decant or filter the extracts. Repeat the extraction two or more times. Dry the extract using a column with anhydrous sodium sulfate and concentrate it to 1 mL in a K-D concentrator.

QUALITY CONTROL A methylene chloride solution containing 50 ng/µL of decafluorotriphenylphosphine (DFTPP) is used for tuning the GC/MS system each 12-h shift. A system performance check also must be made during every 12-h shift. A standard containing 50 ng/µL each of 4,4'-DDT, pentachlorophenol, and benzidine is required to verify injection port inertness and GC column performance. A calibration standard at mid-concentration, containing each compound of interest, including all required surrogates, must be performed every 12 h during analysis. After the system performance check is met, calibration check compounds (CCCs) are used to check the validity of the initial calibration.

The internal standard responses and retention times in the calibration check standard must be evaluated immediately after or during data acquisition. If the retention time for any internal standard changes by more than 30 seconds from the last check calibration (12 h), the chromatographic system must be inspected for malfunctions and corrections must be made, as required. If the electron ionization current plot (EICP) area for any of the internal standards changes by a factor of two from the last daily calibration standard check, the mass spectrometer must be inspected for malfunctions and corrections must be made, as appropriate.

Demonstrate, through the analysis of a reagent water blank, that interferences from the analytical system, glassware, and reagents are under control. The blank samples should be carried through all stages of the sample preparation and measurement steps. For each analytical batch (up to 20 samples), a reagent blank, matrix spike, and matrix spike duplicate/duplicate must be analyzed (the frequency of the spikes may be different for different monitoring programs). The blank and spiked samples must be carried through all stages of the sample preparation and measurement steps. A QC reference sample concentrate containing each analyte at a concentration of 100 mg/L in methanol is required.

REFERENCE Test Methods for Evaluating Solid Waste (SW-846). U.S. EPA 1983, Method 8270B, Rev. 2, Nov. 1990. Office of Solid Waste, Washington, DC.

Methyl ethyl ketone (MEK)   **EPA Method 1624**
CAS #78-93-3

TITLE Volatile Organic Compounds by Isotope Dilution GC/MS

MATRIX Compounds may be determined in waters, soils, and municipal sludges by this method.

METHOD SUMMARY This method is used to determine 58 volatile toxic organic pollutants associated with the CWA (as amended 1987); the RCRA (as amended 1986); the CERCLA (as amended 1986); and other compounds amenable to purge-and-trap gas chromatography-mass spectrometry (GC/MS).

If the solids content is less than 1%, stable isotopically-labeled analogs of the compounds of interest are added to a 5-mL sample and the sample is purged with an inert gas at 20–25°C in a chamber designed for soil or water samples. If the solids content is greater than 1%, 5 mL of reagent water and the labeled compounds are added to a 5-g aliquot of sample and the mixture is purged at 40°C. Compounds that will not purge at 20–25°C or at 40°C are purged at 78–85°C. In the purging process, the volatile compounds are transferred from the aqueous phase into the gaseous phase where they are passed into a sorbent column, and trapped. After purging is completed, the trap is backflushed and heated rapidly to desorb the compounds into a GC. The compounds are separated by the GC and detected by a MS. The labeled compounds serve to correct the variability of the analytical technique.

INTERFERENCES Impurities in the purge gas, organic compounds outgassing from the plumbing upstream of the trap, and solvent vapors in the lab account for most problems. Samples can be contaminated by diffusion of volatile organic compounds (particularly methylene chloride) through the bottle seal during shipment and storage. Contamination by carryover can occur when high-level and low-level samples are analyzed sequentially. When an unusually concentrated sample is encountered, follow it by analysis of a reagent water blank to check for carryover.

INSTRUMENTATION Major equipment includes a GC with linear temperature programming and a glass jet separator as the MS interface, a MS with 70 eV electron impact ionization, and a data system to collect and record response factors.

Column: 2.8 m × 2 mm I.D. glass, packed with 1% SP-1000 on Carbopak B, 60/80 mesh, or equivalent.

PRECISION & ACCURACY The detection limits of the method are usually dependent on the level of interferences rather than instrumental limitations. The method detection limits were determined in digested sludge (low solids) and in filter cake or compost (high solids).

The MDL (in µg/kg) for low solids is 241 and for high solids is 80.

Background levels of this compound were present in the sludge with low solids, resulting in a higher than expected MDL. Background levels of this compound were present in the sludge with high solids, resulting in a higher than expected MDL.

Labeled and native compound precision (in µg/L) as standard deviation was 57.0.
Labeled and native compound accuracy (in µg/L) as average recovery was 66–159.

Acceptance criteria are at 100 µg/L for this compound.

SAMPLE COLLECTION, PRESERVATION & HANDLING
Grab samples are collected in glass containers having a total

volume greater than 20 mL. Fill and seal each bottle so that no air bubbles are entrapped. Samples are maintained at 0 to 4°C from the time of collection until analysis. If an aqueous sample contains residual chlorine, add sodium thiosulfate preservative (10 mg/40 mL) to the empty sample bottles just prior to shipment to the sample site. All samples must be analyzed within 14 days of collection.

SAMPLE PREPARATION Samples containing less than 1% solids are analyzed directly as aqueous samples. Samples containing 1% solids or greater are analyzed as solid samples utilizing one of two methods, depending on the levels of pollutants, in the sample. Samples containing 1% solids or greater, and low to moderate levels of pollutants are analyzed by purging a known weight of sample added to 5 mL of reagent water. Samples containing 1% solids or greater, and high levels of pollutants, are extracted with methanol, and an aliquot of the methanol extract is added to reagent water and purged.

QUALITY CONTROL A field blank prepared from reagent water and carried through the sampling and handling protocol may serve as a check on contamination from shipment and storage.

The analyst is permitted to modify this method to improve separations or lower the costs of measurements, provided all performance specifications are met. Analyses of blanks are required. When results of spikes indicate atypical method performance for samples, the samples are diluted to bring method performance within acceptable limits. Analyze two sets of four 5-mL aliquots (8 aliquots total) of the aqueous performance standard. Spike all samples with labeled compounds to assess method performance on the sample matrix. Compute the percent recovery of the labeled compounds using the internal standard method. Compare the percent recovery for each compound with the corresponding labeled compound recovery. Reagent water blanks are analyzed to demonstrate freedom from carryover contamination. Field replicates may be collected to determine the precision of the sampling technique, and spiked samples may be required to determine the accuracy of the analysis when the internal method is used.

REFERENCE Volatile Organic Compounds by Isotope Dilution GC/MS. Office of Water Regulation and Standards, U.S. EPA Industrial Technology Division, Washington, DC, EPA Method 1624, Rev. C, June 1989 (contact W.A. Telliard, U.S. EPA, Office of Water Regulations and Standards, 401 M St., SW, Washington, DC, 20460. Phone: 202-382-7131).

Methyl ethyl ketone (MEK)   **EPA Method 8015**
CAS #78-93-3

TITLE Nonhalogenated Volatile Organics

MATRIX Groundwater, soils, sludges, water miscible liquid wastes, and non-water miscible wastes.

APPLICATION This method is used for the analysis of 6 nonhalogenated VOCs. Samples are analyzed using direct injection or purge-and-trap methods. Groundwater must be analyzed by the purge-and-trap method. The method provides an optional GC column that may help resolve analytes from interferences and which is also used for analyte confirmation.

INTERFERENCES There can be carryover contamination with high- and low-level samples. Impurities may come from the purge-and-trap apparatus, organic compounds outgassing from the plumbing ahead of trap, diffusion of VOCs through the sample bottle septum during shipping or storage, or from solvent vapors in the lab.

INSTRUMENTATION GC capable of on-column injections or purge-and-trap sample introduction and a flame ionization detector (FID). Column 1: an 8 ft by 0.1 in 1% SP-1000 on Carbopack-B column. Column 2: a 6 ft by 0.1 in bonded n-octane on Porasil-C.

RANGE Not available.

MDL Not available.

PRECISION Not available.

ACCURACY Not available.

SAMPLING METHOD For water and liquid samples; use glass 40-mL vials with Teflon®-lined septum caps and collect two vials per sample location with no headspace. For solids and concentrated waste samples; use widemouth glass bottles with Teflon® liners. Cool all samples to 4°C

STABILITY For concentrated wastes, soils, sediments, or sludges: cool to 4°C. For liquids: add 4 drops of concentrated hydrochloric acid, cool to 4°C.

MHT 14 days.

QUALITY CONTROL Analyze a reagent blank, matrix spike, and matrix spike duplicate/duplicate for each analytical batch (up to 20 samples). Demonstrate the purity of glassware and reagents by analyzing a reagent water method blank. Internal, surrogate, and five concentration level calibration standards are used.

REFERENCE Method 8015, SW-846, 3rd ed., Nov. 1986.

Methyl iodide   **EPA Method 8240**
CAS #74-88-4

TITLE Volatile Organics By GC/MS: Packed Column Technique

MATRIX Nearly all types of sample matarices, regardless of water content, can be analyzed using this method. This includes groundwater, aqueous sludges, caustic liquors, acid liquors, waste solvents, oily wastes, mousses, tars, fibrous wastes, polymetric emulsions, filter cakes, spent carbons, spent catalysts, soils, and sediments.

METHOD SUMMARY Method 8240B covers 80 volatile organic compounds that are introduced into a gas chromatograph by the purge-and-trap method or by direct injection (in limited applications). For the purge-and-trap method an inert gas (zero grade nitrogen or helium) is bubbled through a 5-mL solution at ambient temperature. Purged sample components are trapped in a tube of sorbent materials. When purging is complete, the sorbent tube is heated and backflushed with inert gas to desorb the trapped components onto a GC column.

INTERFERENCES Impurities in the purge gas and from organic compounds outgassing from the plumbing ahead of the trap account for many contamination problems. Interferences purged or coextracted from the samples will vary considerably from source to source. Cross-contamination can occur whenever high-level and low-level samples are analyzed sequentially. Whenever an unusually concentrated sample is analyzed, it should be followed by the analysis of organic-free reagent water to check for cross-contamination. Samples also can be contaminated by diffusion of volatile organics (particularly methylene chloride and fluorocarbons) through the septum seal into the sample during shipment and storage. A trip blank can serve as a check on such contamination. The lab where volatile analysis is performed and also the refrigerated storage area should be completely free of solvents.

INSTRUMENTATION A gas chromatograph/mass spectrometry/data system (GC/MS) equipped with a 6 ft × 0.1 in I.D. glass column packed with 1% SP-1000 on Carbopack-B (60/80 mesh) is required. Also needed is a 5-mL purging device, a sorbent trap, and a thermal desorption apparatus.

PRECISION & ACCURACY This method is reported to have been tested by 15 laboratories using organic-free reagent water, drinking water, surface water, and industrial wastewaters (not specified) fortified at six concentrations over the range 5–600 µg/L.

Sample estimated quantitation limits (EQLs) are highly matrix-dependent. The EQLs listed may not always be achievable. EQLs listed for soils or sediments are based on wet weight. Normally, data is reported on a dry-weight basis; therefore, EQLs will be higher, based on the percent dry weight of each sample. Note that EQLs are even more variable than MDLs and that they are highly variable depending on the matrix being analyzed.

EQL in groundwater in µg/L was 5.
EQL in low soil or sediment in µg/kg was 5.
Accuracy (a) in µg/L was not listed.
Precision (b) in µg/L was not listed.

(a) Average recovery found for measurements of samples containing a concentration of C, in µg/L.
(b) Overall precision found for measurements of samples with average recovery X for samples containing a concentration of C in µg/L.
X = Average recovery found for measurement of samples containing a concentration of C in µg/L.

MULTIPLICATION FACTORS FOR OTHER MATRICES

Other Matrices	Factor (a)
Waste miscible liquid waste	50
High-concentration soil and sludge	125
Non-water miscible waste	500

(a) EQL = [EQL for low soil/sediment] × [Factor]. For non-aqueous samples, the factor is on a wet-weight basis.

SAMPLING METHOD
Liquid samples — Use a 40-mL glass screw-cap VOA vial with a Teflon®-faced silicone septum that has been prewashed, rinsed with distilled deionized water, and oven dried. However, if residual chlorine is present, collect sample in a 40-oz. soil VOA container which has been pre-preserved with 4 drops of 10% sodium thiosulfate, mix gently, and then transfer the sample to a 40-mL VOA vial. Collect bubble-free samples in duplicate and seal them in separate plastic bags.

Soils or sediments, and sludges — Use an 8-oz. widemouth glass bottle with a Teflon®-faced silicone septum that has been prewashed with detergent, rinsed with distilled deionized water, and oven dried. Tap slightly to eliminate free air space. Collect samples in duplicate and seal them in separate plastic bags.

SAMPLE PRESERVATION
Liquid samples — Add 4 drops of concentrated HCL and immediately cool samples to 4°C and store in a solvent-free refrigerator.

Soils or sediments, and sludges — Cool samples to 4°C and store in a solvent-free refrigerator.

MHT Maximum holding time is 14 days from the date of sample collection.

SAMPLE PREPARATION
Liquid samples — Remove the plunger from a 5-mL syringe and carefully pour the sample into the syringe barrel to just short of overflowing. Replace the syringe plunger and compress the sample. Open the syringe valve and vent any residual air while adjusting the sample volume to 5.0 mL. If there is only one volatile organic analysis (VOA) vial, a second syringe should be filled at this time to protect against possible loss of sample integrity. Add 10 µL of surrogate spiking solution and 10 µL of internal standard spiking solution through the valve bore of the 5-mL syringe, then close the valve. The surrogate and internal standards may be mixed and added as a single spiking solution.

Sediments, soils, and waste samples — All samples of this type should be screened by GC analysis using a headspace method (EPA Method 3810) or the hexadecane extraction and screening method (EPA Method 3820). Use the screening data to determine whether to use the low-concentration method (0.005–1 mg/kg) or the high-concentration method (>1 mg/kg).

Low-concentration method — The low-concentration method is based on purging a heated sediment or soil sample mixed with organic-free reagent water containing the surrogate and internal standards. Analyze all reagent blanks and standards under the same conditions as the samples.

Use a 5-g sample if the expected concentration is <0.1 mg/kg or a 1-g sample for expected concentrations between 0.1 and 1 mg/kg. Mix the contents of the sample container with a narrow metal spatula. Weigh the amount of the sample into a tared purge device. Add the spiked water to the purge device, which contains the weighed amount of sample, and connect the device to the purge-and-trap system.

High-concentration method — This method is based on extracting the sediment or soil with methanol. A waste sample

is either extracted or diluted, depending on its solubility in methanol. Wastes that are insoluble in methanol are diluted with reagent tetraglyme or possibly polyethylene glycol (PEG). An aliquot of the extract is added to organic-free reagent water containing surrogate and internal standards. This is purged at ambient temperature. All samples with an expected concentration of >1.0 mg/kg should be analyzed by this method.

Mix the contents of the sample container with a narrow metal spatula. For sediments or soils and solid wastes that are insoluble in methanol, weigh 4 g (wet weight) of sample into a tared 20-mL vial. For waste that is soluble in methanol, tetraglyme, or PEG, weigh 1 g (wet weight) into a tared scintillation vial or culture tube or a 10-mL volumetric flask. Quickly add 9.0 mL of appropriate solvent then add 1.0 mL of a surrogate spiking solution to the vial, cap it, and shake it for 2 min.

METHANOL EXTRACT REQUIRED FOR ANALYSIS OF HIGH-CONCENTRATION SOILS OR SEDIMENTS

Approximate Concentration Range	Volume of Methanol Extract (a)
500–10,000 µg/kg	100 µL
1,000–20,000 µg/kg	50 µL
5,000–100,000 µg/kg	10 µL
25,000–500,000 µg/kg	100 µL of 1/50 dilution (b)

Calculate appropriate dilution factor for concentrations exceeding this table.

(a) The volume of methanol added to 5 mL of water being purged should be kept constant. Therefore, add to the 5-mL syringe whatever volume of methanol is necessary to maintain a volume of 100 µL added to the syringe.
(b) Dilute an aliquot of the methanol extract and then take 100 µL for analysis.

QUALITY CONTROL Demonstrate, through the analysis of a reagent water blank, that interferences from the analytical system, glassware, and reagents are under control. Blank samples should be carried through all stages of the sample preparation and measurement steps. For each analytical batch (up to 20 samples), a reagent blank, matrix spike, and matrix spike duplicate must be analyzed (the frequency of the spikes may be different for different monitoring programs). The blank and spiked samples must be carried through all stages of the sample preparation and measurement steps. QC samples mentioned in the section on Interferences will also be needed as appropriate to those situations.

REFERENCE Test Methods for Evaluating Solid Waste (SW-846). U.S. EPA. 1983. Method 8240B, Rev. 2, Nov. 1990. Office of Solid Wastes, Washington, DC.

Methyl isobutyl ketone (MIBK)   **EPA Method 8015**
CAS #108-10-1

TITLE Nonhalogenated Volatile Organics

MATRIX Groundwater, soils, sludges, water miscible liquid wastes, and non-water miscible wastes.

APPLICATION This method is used for the analysis of 6 nonhalogenated VOCs. Samples are analyzed using direct injection or purge-and-trap methods. Groundwater must be analyzed by the purge-and-trap method. The method provides an optional GC column that may help resolve analytes from interferences and which is also used for analyte confirmation.

INTERFERENCES There can be carryover contamination with high- and low-level samples. Impurities may come from the purge-and-trap apparatus, organic compounds outgassing from the plumbing ahead of trap, diffusion of VOCs through the sample bottle septum during shipping or storage, or from solvent vapors in the lab.

INSTRUMENTATION GC capable of on-column injections or purge-and-trap sample introduction and a flame ionization detector (FID). Column 1: an 8 ft by 0.1 in 1% SP-1000 on Carbopack-B column. Column 2: a 6 ft by 0.1 in bonded n-octane on Porasil-C.

RANGE Not available.

MDL Not available.

PRECISION Not available.

ACCURACY Not available.

SAMPLING METHOD For water and liquid samples; use glass 40-mL vials with Teflon®-lined septum caps and collect two vials per sample location with no headspace. For solids and concentrated waste samples; use widemouth glass bottles with Teflon® liners. Cool all samples to 4°C

STABILITY For concentrated wastes, soils, sediments, or sludges: cool to 4°C. For liquids: add 4 drops of concentrated hydrochloric acid, cool to 4°C.

MHT 14 days.

QUALITY CONTROL Analyze a reagent blank, matrix spike, and matrix spike duplicate/duplicate for each analytical batch (up to 20 samples). Demonstrate the purity of glassware and reagents by analyzing a reagent water method blank. Internal, surrogate, and five concentration level calibration standards are used.

REFERENCE Method 8015, SW-846, 3rd ed., Nov.1986.

Methyl methacrylate   **EPA Method 1624**
CAS #80-62-6

TITLE Volatile Organic Compounds by Isotope Dilution GC/MS

MATRIX Compounds may be determined in waters, soils, and municipal sludges by this method.

METHOD SUMMARY This method is used to determine 58 volatile toxic organic pollutants associated with the CWA (as amended 1987); the RCRA (as amended 1986); the CERCLA (as amended 1986); and other compounds amenable to purge-and-trap gas chromatography-mass spectrometry (GC/MS).

If the solids content is less than 1%, stable isotopically-labeled analogs of the compounds of interest are added to a 5-mL sample and the sample is purged with an inert gas at 20–25°C in a chamber designed for soil or water samples. If the solids content is greater than 1%, 5 mL of reagent water and the labeled compounds are added to a 5-g aliquot of sample and the mixture is purged at 40°C. Compounds that will not purge at 20–25°C or at 40°C are purged at 78–85°C. In the purging process, the volatile compounds are transferred from the aqueous phase into the gaseous phase where they are passed into a sorbent column, and trapped. After purging is completed, the trap is backflushed and heated rapidly to desorb the compounds into a GC. The compounds are separated by the GC and detected by a MS. The labeled compounds serve to correct the variability of the analytical technique.

INTERFERENCES Impurities in the purge gas, organic compounds outgassing from the plumbing upstream of the trap, and solvent vapors in the lab account for most problems. Samples can be contaminated by diffusion of volatile organic compounds (particularly methylene chloride) through the bottle seal during shipment and storage. Contamination by carryover can occur when high-level and low-level samples are analyzed sequentially. When an unusually concentrated sample is encountered, follow it by analysis of a reagent water blank to check for carryover.

INSTRUMENTATION Major equipment includes a GC with linear temperature programming and a glass jet separator as the MS interface, a MS with 70 eV electron impact ionization, and a data system to collect and record response factors.

Column: 2.8 m × 2 mm I.D. glass, packed with 1% SP-1000 on Carbopak B, 60/80 mesh, or equivalent.

PRECISION & ACCURACY The detection limits of the method are usually dependent on the level of interferences rather than instrumental limitations. The method detection limits were determined in digested sludge (low solids) and in filter cake or compost (high solids).

The MDL (in µg/kg) for low solids is not listed and for high solids is not listed.
Labeled and native compound precision (in µg/L) as standard deviation was not listed.
Labeled and native compound accuracy (in µg/L) as average recovery was not listed.

Acceptance criteria are at 20 µg/L for this compound.

SAMPLE COLLECTION, PRESERVATION & HANDLING Grab samples are collected in glass containers having a total volume greater than 20 mL. Fill and seal each bottle so that no air bubbles are entrapped. Samples are maintained at 0 to 4°C from the time of collection until analysis. If an aqueous sample contains residual chlorine, add sodium thiosulfate preservative (10 mg/40 mL) to the empty sample bottles just prior to shipment to the sample site. All samples must be analyzed within 14 days of collection.

SAMPLE PREPARATION Samples containing less than 1% solids are analyzed directly as aqueous samples. Samples containing 1% solids or greater are analyzed as solid samples utilizing one of two methods, depending on the levels of pollutants, in the sample. Samples containing 1% solids or greater, and low to moderate levels of pollutants are analyzed by purging a known weight of sample added to 5 mL of reagent water. Samples containing 1% solids or greater, and high levels of pollutants, are extracted with methanol, and an aliquot of the methanol extract is added to reagent water and purged.

QUALITY CONTROL A field blank prepared from reagent water and carried through the sampling and handling protocol may serve as a check on contamination from shipment and storage.

The analyst is permitted to modify this method to improve separations or lower the costs of measurements, provided all performance specifications are met. Analyses of blanks are required. When results of spikes indicate atypical method performance for samples, the samples are diluted to bring method performance within acceptable limits. Analyze two sets of four 5-mL aliquots (8 aliquots total) of the aqueous performance standard. Spike all samples with labeled compounds to assess method performance on the sample matrix. Compute the percent recovery of the labeled compounds using the internal standard method. Compare the percent recovery for each compound with the corresponding labeled compound recovery. Reagent water blanks are analyzed to demonstrate freedom from carryover contamination. Field replicates may be collected to determine the precision of the sampling technique, and spiked samples may be required to determine the accuracy of the analysis when the internal method is used.

REFERENCE Volatile Organic Compounds by Isotope Dilution GC/MS. Office of Water Regulation and Standards, U.S. EPA Industrial Technology Division, Washington, DC, EPA Method 1624, Rev. C, June 1989 (contact W.A. Telliard, U.S. EPA, Office of Water Regulations and Standards, 401 M St., SW, Washington, DC, 20460. Phone: 202-382-7131).

Methyl methacrylate   **EPA Method 8240**
CAS #80-62-6

TITLE Volatile Organics By GC/MS: Packed Column Technique

MATRIX Nearly all types of sample matarices, regardless of water content, can be analyzed using this method. This includes groundwater, aqueous sludges, caustic liquors, acid liquors, waste solvents, oily wastes, mousses, tars, fibrous wastes, polymetric emulsions, filter cakes, spent carbons, spent catalysts, soils, and sediments.

METHOD SUMMARY Method 8240B covers 80 volatile organic compounds that are introduced into a gas chromatograph by the purge-and-trap method or by direct injection (in limited applications). For the purge-and-trap method an inert gas (zero grade nitrogen or helium) is bubbled through a 5-mL solution at ambient temperature. Purged sample components are trapped in a tube of sorbent materials. When purging is complete, the sorbent tube is heated and backflushed with inert gas to desorb the trapped components onto a GC column.

INTERFERENCES Impurities in the purge gas and from organic compounds outgassing from the plumbing ahead of the trap account for many contamination problems. Interferences purged or coextracted from the samples will vary considerably from source to source. Cross-contamination can occur whenever high-level and low-level samples are analyzed sequentially. Whenever an unusually concentrated sample is analyzed, it should be followed by the analysis of organic-free reagent water to check for cross-contamination. Samples also can be contaminated by diffusion of volatile organics (particularly methylene chloride and fluorocarbons) through the septum seal into the sample during shipment and storage. A trip blank can serve as a check on such contamination. The lab where volatile analysis is performed and also the refrigerated storage area should be completely free of solvents.

INSTRUMENTATION A gas chromatograph/mass spectrometry/data system (GC/MS) equipped with a 6 ft × 0.1 in I.D. glass column packed with 1% SP-1000 on Carbopack-B (60/80 mesh) is required. Also needed is a 5-mL purging device, a sorbent trap, and a thermal desorption apparatus.

PRECISION & ACCURACY This method is reported to have been tested by 15 laboratories using organic-free reagent water, drinking water, surface water, and industrial wastewaters (not specified) fortified at six concentrations over the range 5–600 µg/L.

Sample estimated quantitation limits (EQLs) are highly matrix-dependent. The EQLs listed may not always be achievable. EQLs listed for soils or sediments are based on wet weight. Normally, data is reported on a dry-weight basis; therefore, EQLs will be higher, based on the percent dry weight of each sample. Note that EQLs are even more variable than MDLs and that they are highly variable depending on the matrix being analyzed.

EQL in groundwater in µg/L was 5.
EQL in low soil or sediment in µg/kg was 50.
Accuracy (a) in µg/L was not listed.
Precision (b) in µg/L was not listed.

(a) Average recovery found for measurements of samples containing a concentration of C, in µg/L.
(b) Overall precision found for measurements of samples with average recovery X for samples containing a concentration of C in µg/L.
X = Average recovery found for measurement of samples containing a concentration of C in µg/L.

MULTIPLICATION FACTORS FOR OTHER MATRICES

Other Matrices	Factor (a)
Waste miscible liquid waste	50
High-concentration soil and sludge	125
Non-water miscible waste	500

(a) EQL = [EQL for low soil/sediment] × [Factor]. For non-aqueous samples, the factor is on a wet-weight basis.

SAMPLING METHOD
Liquid samples — Use a 40-mL glass screw-cap VOA vial with a Teflon®-faced silicone septum that has been prewashed, rinsed with distilled deionized water, and oven dried. However, if residual chlorine is present, collect sample in a 40-oz. soil VOA container which has been pre-preserved with 4 drops of 10% sodium thiosulfate, mix gently, and then transfer the sample to a 40-mL VOA vial. Collect bubble-free samples in duplicate and seal them in separate plastic bags.

Soils or sediments, and sludges — Use an 8-oz. widemouth glass bottle with a Teflon®-faced silicone septum that has been prewashed with detergent, rinsed with distilled deionized water, and oven dried. Tap slightly to eliminate free air space. Collect samples in duplicate and seal them in separate plastic bags.

SAMPLE PRESERVATION
Liquid samples — Add 4 drops of concentrated HCL and immediately cool samples to 4°C and store in a solvent-free refrigerator.

Soils or sediments, and sludges — Cool samples to 4°C and store in a solvent-free refrigerator.

MHT Maximum holding time is 14 days from the date of sample collection.

SAMPLE PREPARATION
Liquid samples — Remove the plunger from a 5-mL syringe and carefully pour the sample into the syringe barrel to just short of overflowing. Replace the syringe plunger and compress the sample. Open the syringe valve and vent any residual air while adjusting the sample volume to 5.0 mL. If there is only one volatile organic analysis (VOA) vial, a second syringe should be filled at this time to protect against possible loss of sample integrity. Add 10 µL of surrogate spiking solution and 10 µL of internal standard spiking solution through the valve bore of the 5-mL syringe, then close the valve. The surrogate and internal standards may be mixed and added as a single spiking solution.

Sediments, soils, and waste samples — All samples of this type should be screened by GC analysis using a headspace method (EPA Method 3810) or the hexadecane extraction and screening method (EPA Method 3820). Use the screening data to determine whether to use the low-concentration method (0.005–1 mg/kg) or the high-concentration method (>1 mg/kg).

Low-concentration method — The low-concentration method is based on purging a heated sediment or soil sample mixed with organic-free reagent water containing the surrogate and internal standards. Analyze all reagent blanks and standards under the same conditions as the samples.

Use a 5-g sample if the expected concentration is <0.1 mg/kg or a 1-g sample for expected concentrations between 0.1 and 1 mg/kg. Mix the contents of the sample container with a narrow metal spatula. Weigh the amount of the sample into a tared purge device. Add the spiked water to the purge device, which contains the weighed amount of sample, and connect the device to the purge-and-trap system.

High-concentration method — This method is based on extracting the sediment or soil with methanol. A waste sample is either extracted or diluted, depending on its solubility in methanol. Wastes that are insoluble in methanol are diluted with reagent tetraglyme or possibly polyethylene glycol (PEG). An aliquot of the extract is added to organic-free reagent water

containing surrogate and internal standards. This is purged at ambient temperature. All samples with an expected concentration of >1.0 mg/kg should be analyzed by this method.

Mix the contents of the sample container with a narrow metal spatula. For sediments or soils and solid wastes that are insoluble in methanol, weigh 4 g (wet weight) of sample into a tared 20-mL vial. For waste that is soluble in methanol, tetraglyme, or PEG, weigh 1 g (wet weight) into a tared scintillation vial or culture tube or a 10-mL volumetric flask. Quickly add 9.0 mL of appropriate solvent then add 1.0 mL of a surrogate spiking solution to the vial, cap it, and shake it for 2 min.

METHANOL EXTRACT REQUIRED FOR ANALYSIS OF HIGH-CONCENTRATION SOILS OR SEDIMENTS

Approximate Concentration Range	Volume of Methanol Extract (a)
500–10,000 µg/kg	100 µL
1,000–20,000 µg/kg	50 µL
5,000–100,000 µg/kg	10 µL
25,000–500,000 µg/kg	100 µL of 1/50 dilution (b)

Calculate appropriate dilution factor for concentrations exceeding this table.

(a) The volume of methanol added to 5 mL of water being purged should be kept constant. Therefore, add to the 5-mL syringe whatever volume of methanol is necessary to maintain a volume of 100 µL added to the syringe.
(b) Dilute an aliquot of the methanol extract and then take 100 µL for analysis.

QUALITY CONTROL Demonstrate, through the analysis of a reagent water blank, that interferences from the analytical system, glassware, and reagents are under control. Blank samples should be carried through all stages of the sample preparation and measurement steps. For each analytical batch (up to 20 samples), a reagent blank, matrix spike, and matrix spike duplicate must be analyzed (the frequency of the spikes may be different for different monitoring programs). The blank and spiked samples must be carried through all stages of the sample preparation and measurement steps. QC samples mentioned in the section on Interferences will also be needed as appropriate to those situations.

REFERENCE Test Methods for Evaluating Solid Waste (SW-846). U.S. EPA. 1983. Method 8240B, Rev. 2, Nov. 1990. Office of Solid Wastes, Washington, DC.

Methyl methanesulfonate   **EPA Method 1625**
CAS #66-27-3

TITLE Semivolatile Organic Compounds by Isotope Dilution GC/MS

MATRIX The compounds may be determined in waters, soils, and municipal sludges by this method.

METHOD SUMMARY This method is used to determine 176 semivolatile toxic organic pollutants associated with the CWA (as amended 1987); the RCRA (as amended 1986); the CERCLA (as amended 1986); and other compounds amenable to extraction and analysis by capillary column gas chromatography-mass spectrometry (GC/MS).

Stable isotopically-labeled analogs of the compounds of interest are added to the sample. If the solids content is less than 1%, a 1-L sample is extracted at pH 12–13, then at pH <2 with methylene chloride using continuous extraction techniques.

If the solids content is 30% or less, the sample is diluted to 1% solids with reagent water, homogenized ultrasonically, and extracted at pH 12–13, then at pH <2 with methylene chloride using continuous extraction techniques. If the solids content is greater than 30%, the sample is extracted using ultrasonic techniques.

Each extract is dried over sodium sulfate, concentrated to a volume of 5 mL, cleaned up using GPC, if necessary, and concentrated. Extracts are concentrated to 1 mL if GPC is not performed, and to 0.5 mL if GPC is performed.

An internal standard is added to the extract, and a 1-mL aliquot of the extract is injected into the GC. The compounds are separated by GC and detected by a MS. The labeled compounds serve to correct the variability of the analytical technique.

INTERFERENCES Solvents, reagents, glassware, and other sample processing hardware may yield artifacts and/or elevated baselines causing misinterpretation of chromatograms and spectra. Materials used in the analysis must be demonstrated to be free from interferences under the conditions of analysis by running method blanks initially and with each sample lot (sample started through the extraction process on a given 8-h shift, to a maximum of 20). Specific selection of reagents and purification of solvents by distillation in all glass systems may be required. Glassware and, where possible, reagents are cleaned by solvent rinse and baking at 450°C for 1-h minimum. Interferences coextracted from samples will vary considerably from source to source, depending on the diversity of the site being sampled.

INSTRUMENTATION Major instrumentation includes a GC with a splitless or on-column injection port for capillary column, a MS with 70 eV electron impact ionization, and a data system to collect and record MS data, and process it. A K-D apparatus is used to concentrate extracts.

GC Column: 30 m × 0.25 mm I.D. 5% phenyl, 94% methyl, 1% vinyl silicone bonded phased fused silica capillary column.

PRECISION & ACCURACY The detection limits of the method are usually dependent on the level of interferences rather than instrumental limitations. The limits typify the minimum quantities that can be detected with no interferences present.

The minimum level (in µg/mL) was not listed. This is defined as a minimum level at which the analytical system shall give recognizable mass spectra (background corrected) and acceptable calibration points.

The MDL (in µg/kg) in low solids was not listed and in high solids was not listed; these were determined in digested sludge (low solids) and in filter cake or compost (high solids).

The labeled and native compound initial precision as standard deviation (in μg/L) was not listed.

The labeled and native compound initial accuracy as average recovery (in μg/L) was not listed.

SAMPLE COLLECTION, PRESERVATION & HANDLING
Collect samples in glass containers. Aqueous samples which flow freely are collected in refrigerated bottles using automatic sampling equipment. Solid samples are collected as grab samples using widemouth jars. Maintain samples at 0 to 4°C from the time of collection until extraction. If residual chlorine is present in aqueous samples, add 80 mg sodium thiosulfate/L of water. Begin sample extraction within 7 days of collection, and analyze all extracts within 40 days of extraction.

SAMPLE PREPARATION Samples containing 1% solids or less are extracted directly using continuous liquid-liquid extraction techniques. Samples containing 1 to 30% solids are diluted to the 1% level with reagent water and extracted using continuous liquid-liquid extraction techniques. Samples containing greater than 30% solids are extracted using ultrasonic techniques.

Base/neutral extraction — Adjust the pH of the waters in the extractors to 12–13 with 6 N NaOH. Extract with methylene chloride for 24–48 h.

Acid extraction — Adjust the pH of the waters in the extractors to 2 or less using 6 N sulfuric acid. Extract with methylene chloride for 24–48 h.

Ultrasonic extraction of high solids samples — Add anhydrous sodium sulfate to the sample and QC aliquot(s). Add acetone:methylene chloride (1:1) to the sample and mix thoroughly

Concentrate extracts using a K-D apparatus.

QUALITY CONTROL The analyst is permitted to modify this method to improve separations or lower the costs of measurements, provided all performance specifications are met. Analyses of blanks are required to demonstrate freedom from contamination. When results of spikes indicate atypical method performance for samples, the samples are diluted to bring method performance within acceptable limits.

For low solids (aqueous samples), extract, concentrate, and analyze two sets of four 1-L aliquots (8 aliquots total) of the precision and recovery standard. For high solids samples, two sets of four 30-g aliquots of the high solids reference matrix are used.

Spike all samples with labeled compounds to assess method performance. Compute percent recovery of the labeled compounds using the internal standard method. Compare the labeled compound recovery for each compound with the corresponding labeled compound recovery.

Reagent water and high solids reference matrix blanks are analyzed to demonstrate freedom from contamination. Extract and concentrate a 1-L reagent water blank or a high solids reference matrix blank with each sample's lot (samples started through the extraction process on the same 8-h shift, to a maximum of 20 samples).

Field replicates may be collected to determine the precision of the sampling technique, and spiked samples may be required to determine the accuracy of the analysis when the internal standard method is used.

REFERENCE Semivolatile Organic Compounds by Isotope Dilution GC/MS. Office of Water Regulation and Standards, U.S. EPA Industrial Technology Division, Washington, DC, EPA Method 1625, Rev. C, June 1989 (contact W.A. Telliard, U.S. EPA, Office of Water Regulations and Standards, 401 M St., SW, Washington, DC, 20460. Phone: 202-382-7131).

Methyl methanesulfonate   EPA Method 8270
CAS #66-27-3

TITLE Semivolatile Organic Compounds by GC/MS

MATRIX This method is used to determine the concentration of semivolatile organic compounds in extracts prepared from all types of solid waste matrices, soils, and groundwater. Although surface waters are not specifically mentioned, this method should be applicable to water samples from rivers, lakes, etc.

METHOD SUMMARY This method covers 259 semivolatile organic compounds. In very limited applications direct injection of the sample into the GC/MS system may be appropriate, but this results in very high detection limits (approximately 10,000 μg/L). Typically, a 1-L liquid sample, containing surrogate, and matrix spiking standards, is extracted in a continuous extractor first under acid conditions and then under basic conditions. Typically 30 g of a solid sample, containing surrogate, and matrix spiking standards, is extracted ultrasonically. After concentrating the extract to 1 mL it is spiked with 10 μL of an internal standard solution just prior to analysis by GC/MS. The volume injected should contain about 100 ng of base/neutral and 200 ng of acid surrogates (for a 1-μL injection). Analysis is performed by GC/MS using a capillary GC column.

INTERFERENCES Raw GC/MS data from all blanks, samples, and spikes must be evaluated for interferences. Contamination by carryover can occur whenever high-concentration and low-concentration samples are sequentially analyzed. To reduce carryover, the sample syringe must be rinsed out between samples with solvent. Whenever an unusually concentrated sample is encountered, it should be followed by the analysis of blank solvent to check for cross-contamination.

INSTRUMENTATION A GC/MS and a data system are required. The GC column used is a 30 m × 0.25 mm I.D. (or 0.32 mm I.D.) 1um film thickness silicone-coated fused silica capillary column. A continuous liquid-liquid extractor equipped with Teflon® or glass connection joints and stopcocks requiring no lubrication, a K-D concentrating apparatus, water bath, and an ultrasonic disrupter with a minimum power of 300 W and with pulsing capability are also required.

PRECISION & ACCURACY The estimated quantitation limit (EQL) of Method 8270B for determining an individual compound is approximately 1 mg/kg (wet weight) for soil or sediment samples, 1–200 mg/kg for wastes (dependent on

matrix and method of preparation), and 10 µg/L for groundwater samples. EQLs will be proportionately higher for sample extracts that require dilution to avoid saturation of the detector.

The EQL(b) for groundwater in µg/L is 10.
The EQL (a, b) for low concentrations in soil and sediment in µg/kg is not determined.
Accuracy as µg/L is not listed.
Overall precision in µg/L is not listed.

(a) *EQLs listed for soil/sediment are based on wet weight. Normally data is reported in a dry-weight basis; therefore, EQLs will be higher based on the % dry weight of each sample. This calculation is based on a 30 g sample and gel permeation chromatography cleanup.*
(b) *Sample EQLs are highly matrix-dependent. The EQLs are provided for guidance and may not always be achievable.*
C = *True value for concentration, in µg/L.*
X = *Average recovery found for measurements of samples containing a concentration of C, in µg/L.*

ESTIMATED QUANTITATION LIMIT

Other Matrices	Factor (a)
High-concentration soil and sludges by sonicator	7.5
Non-water miscible waste	75

(a) *EQL for other matrices = [EQL for low soil/sediment] × [Factor]. This estimated EQL is similar to an EPA "Practical Quantitation Limit."*

SAMPLING METHOD

Liquid samples — Use a 1 or 2½ gallon amber glass bottle with a screw-top Teflon®-lined cover that has been prewashed with detergent and rinsed with distilled water and methanol (or isopropanol).

Soils, sediments, or sludges — Use an 8-oz. widemouth glass with a screw-top Teflon®-lined cover that has been prewashed with detergent and rinsed with distilled water and methanol (or isopropanol).

SAMPLE PRESERVATION

Liquid samples — If residual chlorine is present, add 3 mL of 10% sodium thiosulfate per gallon, cool to 4°C and store in a solvent-free refrigerator until analysis; if chlorine is not present, then eliminate the sodium thiosulfate addition.

Soils, sediments, or sludges — Cool samples to 4°C and store in a solvent-free refrigerator.

MHT Liquid samples must be extracted within 7 days and the extracts analyzed within 40 days. Soils, sediments, or sludges may be stored for a maximum of 14 days and the extracts analyzed within 40 days.

SAMPLE PREPARATION

Liquid samples — Transfer 1 L quantitatively to a continuous extractor. If high concentrations are anticipated, a smaller volume may be used and then diluted with organic-free reagent water to 1 L. Adjust pH, if necessary, to pH <2 using 1:1 (V/V) sulfuric acid. Pipette 1.0 mL of a surrogate standard spiking solution into each sample. For the sample in each analytical batch selected for spiking, add 1.0 mL of a matrix spiking standard. For base/neutral acid analysis, the amount of the surrogates and matrix spiking compounds added to the sample should result in a final concentration of 100 ng/µL of each analyte in the extract to be analyzed (assuming a 1-µL injection). Extract with methylene chloride for 18–24 h. Next, adjust the pH of the aqueous phase to pH >11 using 10 *N* sodium hydroxide and extract it with methylene chloride again for 18–24 h. Dry the extract through a column containing anhydrous sodium sulfate and concentrate it to 1 mL using a K-D concentrator.

Soils, sediments, or sludges — Use 30 g of sample. Nonporous or wet samples (gummy or clay type) that do not have a free-flowing sandy texture must be mixed with anhydrous sodium sulfate until the sample is free flowing. Add 1 mL of surrogate standards to all samples, spikes, standards, and blanks. For the sample in each analytical batch selected for spiking, add 1.0 mL of a matrix spiking standard. For base/neutral acid analysis, the amount added of the surrogates and matrix spiking compounds should result in a final concentration of 100 ng/µL of each base/neutral analyte and 200 ng/µL of each acid analyte in the extract to be analyzed (assuming a 1-µL injection). Immediately add a 100-mL mixture of 1:1 methylene chloride:acetone and extract the sample ultrasonically for 3 min and then decant or filter the extracts. Repeat the extraction two or more times. Dry the extract using a column with anhydrous sodium sulfate and concentrate it to 1 mL in a K-D concentrator.

QUALITY CONTROL A methylene chloride solution containing 50 ng/µL of decafluorotriphenylphosphine (DFTPP) is used for tuning the GC/MS system each 12-h shift. A system performance check also must be made during every 12-h shift. A standard containing 50 ng/µL each of 4,4′-DDT, pentachlorophenol, and benzidine is required to verify injection port inertness and GC column performance. A calibration standard at mid-concentration, containing each compound of interest, including all required surrogates, must be performed every 12 h during analysis. After the system performance check is met, calibration check compounds (CCCs) are used to check the validity of the initial calibration.

The internal standard responses and retention times in the calibration check standard must be evaluated immediately after or during data acquisition. If the retention time for any internal standard changes by more than 30 seconds from the last check calibration (12 h), the chromatographic system must be inspected for malfunctions and corrections must be made, as required. If the electron ionization current plot (EICP) area for any of the internal standards changes by a factor of two from the last daily calibration standard check, the mass spectrometer must be inspected for malfunctions and corrections must be made, as appropriate.

Demonstrate, through the analysis of a reagent water blank, that interferences from the analytical system, glassware, and reagents are under control. The blank samples should be carried through all stages of the sample preparation and measurement steps. For each analytical batch (up to 20 samples), a reagent blank, matrix spike, and matrix spike duplicate/duplicate must be analyzed (the frequency of the spikes may be different for different monitoring programs). The blank and

spiked samples must be carried through all stages of the sample preparation and measurement steps. A QC reference sample concentrate containing each analyte at a concentration of 100 mg/L in methanol is required.

REFERENCE Test Methods for Evaluating Solid Waste (SW-846). U.S. EPA 1983, Method 8270B, Rev. 2, Nov. 1990. Office of Solid Waste, Washington, DC.

Methyl paraoxon   **EPA Method 507**
CAS #950-35-6

TITLE Determination of Nitrogen and Phosphorus-Containing Pesticides in Water by GC/NPD

MATRIX This method is applicable to the determination of certain nitrogen and phosphorus-containing pesticides in finished drinking water and groundwater.

METHOD SUMMARY Method 507 covers 46 nitrogen- and phosphorus-containing pesticides. A 1-L sample is fortified with a surrogate standard, salted, buffered, extracted with methylene chloride, and concentrated; then the solvent is exchanged with methyl tert-butyl ether (MTBE) and concentrated again, and a 2-μL aliquot of a sample extract is injected into a GC system equipped with a selective nitrogen-phosphorus detector and a capillary column for analysis.

INTERFERENCES Method interferences may be caused by contaminants in solvents, reagents, glassware, and other sample processing apparatus. Interfering contamination may occur when a sample containing low concentrations of analytes is analyzed immediately following a sample containing relatively high concentrations. One or more injections of MTBE should be made following the analysis of a sample with high concentrations of analytes to check for analyte carryover. Matrix interferences may be caused by contaminants that are coextracted from the sample. The extent of matrix interferences will vary considerably from source to source, depending upon the water sampled.

INSTRUMENTATION A gas chromatograph system (GC) equipped with a nitrogen-phosphorus detector (NPD) is needed.

Column 1: 30 m × 0.25 mm I.D. DB-5 bonded fused silica column, 0.25 μm film thickness, or equivalent.
Column 2: 30 m × 0.25 mm I.D. DB-1701 bonded fused silica column, 0.25 μm film thickness, or equivalent.

PRECISION & ACCURACY This method has been validated in a single lab and estimated detection limits (EDLs) have been determined for each analyte. Observed detection limits may vary among waters, depending upon the nature of the interferences in the sample matrix and the specific instrumentation used. Analytes that are not separated chromatographically cannot be individually identified and measured unless an alternative technique for identification and quantification exist.

The estimated detection limit (in μg/L) was 2.5. The EDL is defined as either method detection limit or a level of compound in a sample yielding a peak in the final extract with signal-to-noise ratio of approximately 5, whichever value is higher.

The concentration used for these measurements (in μg/L) was 25.
The accuracy (as % recovery) was 98.
The precision (% RSD) was 10.

SAMPLING METHOD Grab samples are collected in 1-L glass sample bottles (prewashed with detergent and hot tap water, rinsed with reagent water, and dried in an oven at 400°C for 1 h) with screw caps lined with PTFE-fluorocarbon.

SAMPLE PRESERVATION Add mercuric chloride to the sample bottle in amounts to produce a concentration of 10 mg/L. If residual chlorine is present, add 80 mg of sodium thiosulfate/L of sample to the sample bottle prior to collection. After collection, seal bottle and shake vigorously for 1 min, then cool the sample to 4°C immediately and store it at 4°C in the dark until extraction.

MHT Maximum holding time of the samples, and in some cases the extracts, is 14 days.

SAMPLE PREPARATION Fortify the sample with 50 μL of the surrogate standard solution, adjust to pH 7 with phosphate buffer, add 100 g NaCl to the sample, and seal and shake to dissolve the salt; then extract with methylene chloride in a separatory funnel or in a mechanical tumbler bottle. Dry the extract by pouring it through a solvent-rinsed drying column containing about 10 cm of anhydrous sodium sulfate. Collect the extract in a Kuderna-Danish (K-D) concentrator and rinse the column with 20–30 mL methylene chloride. Concentrate the extract to about 2 mL and rinse the flask and its lower joint into the concentrator tube with 1 to 2 mL of methyl t-butyl ether (MTBE). Add 5–10 mL of MTBE and concentrate the extract twice (adding more MTBE) to a final volume of 5.0 mL and store it at 4°C until analysis.

Note: If methylene chloride is not completely removed from the final extract, it may cause detector problems.

QUALITY CONTROL Minimum quality control requirements are initial demonstration of lab capability, determination of surrogate compound recoveries in each sample and blank, monitoring internal standard peak area or height in each sample and blank, analysis of lab reagent blanks, lab fortified samples, lab fortified blanks, and other QC samples. A lab reagent blank is analyzed to demonstrate that all glassware and reagent interferences are under control.

Initial demonstration of capability is fulfilled by analyzing four fortified reagent water samples with the recovery value for each analyte falling within the acceptable range (±30% average recovery). Surrogate recoveries from samples or method blanks must be 70–130%. The internal standard response for any sample chromatogram should not deviate from the daily calibration check standard's internal standard response by more than 30% or lab fortified blanks and sample matrices are used to assess lab performance and analyte recovery, respectively.

If the response for the target analyte peak exceeds the working range of the system, dilute the extract and reanalyze. Alternative techniques such as an alternate detector or second chromatography column should be used to confirm peak identification when sample components are not resolved adequately.

EPA CONTACT & HOTLINE For technical questions contact Dr. Baldev Bathija, U.S. EPA, Office of Ground Water and Drinking Water (WH-550D), 401 M St. SW, Washington, DC 20460. Tel. (202) 260-3040. For further information the EPA Safe Drinking Water Hotline may be called at: (800) 426-4791.

REFERENCE Methods for the Determination of Organic Compounds in Drinking Water, EPA/600/4-88/039 (revised July 1991). U.S. EPA Environmental Monitoring Systems Laboratory, Cincinnati, OH, 45268, U.S.A. Available from the National Technical Information Service (NTIS), 5285 Port Royal Road, Springfield, VA 22161; Tel. 800-553-6847. NTIS Order Number is PB91-231480.

Methyl parathion   **EPA Method 8141**
CAS #298-00-0

TITLE Organophosphorus Compounds by Gas Chromatography: Capillary Column Technique

MATRIX This method covers aqueous and solid matrices. This includes a wide variety such as drinking water, groundwater, industrial wastewaters, surface waters, soils, solids, and sediments.

METHOD SUMMARY This is a GC method used to determine the concentration of 28 organophosphorus pesticides.

The use of Gel Permeation Cleanup (EPA Method 3640) for sample cleanup has been demonstrated to yield recoveries of less than 85% for many method analytes and is therefore not recommended for use with this method.

This method provides GC conditions for the detection of ppb concentrations of organophosphorus compounds. Prior to the use of this method, appropriate sample preparation techniques must be used. Water samples are extracted at a neutral pH with methylene chloride as a solvent by using a separatory funnel (EPA Method 3510) or a continuous liquid-liquid extractor (EPA Method 3520). Soxhlet extraction (EPA Method 3540) or ultrasonic extraction (EPA Method 3550) using methylene chloride/acetone (1:1) are used for solid samples. Both neat and diluted organic liquids (EPA Method 3580) may be analyzed by direct injection. Spiked samples are used to verify the applicability of the chosen extraction technique to each new sample type. A GC with a flame photometric (FPD) or nitrogen-phosphorus detector (NPD) is used for this multiresidue procedure.

INTERFERENCES The use of Florisil cleanup materials (EPA Method 3620) for some of the compounds in this method has been demonstrated to yield recoveries less than 85% and is therefore not recommended for all compounds. Use of phosphorus or halogen specific detectors, however, often obviates the necessity for cleanup for relatively clean sample matrices. If particular circumstances demand the use of an alternative cleanup procedure, the analyst must determine the elution profile and demonstrate that the recovery of each analyte is no less than 85%.

Use of a flame photometric detector (FPD) in the phosphorus mode will minimize interferences from materials that do not contain phosphorus. Elemental sulfur, however, may interfere with the determination of certain organophosphorus compounds by flame photometric gas chromatography. Sulfur cleanup using EPA Method 3660 may alleviate this interference. A nitrogen phosphorus detector (NPD) is also recommended.

A few analytes coelute on certain columns. Therefore, select a second column for confirmation where coelution of the analytes of interest does not occur.

Method interferences may be caused by contaminants in solvents, reagents, glassware, and other sample processing hardware that lead to discrete artifacts or elevated baselines in gas chromatograms. All these materials must be routinely demonstrated to be free from interferences under the conditions of the analysis by analyzing reagent blanks.

INSTRUMENTATION A GC with a NPD or a FPD will be needed. A data system or integrator is recommended for measuring peak areas and/or peak heights. A Kuderna-Danish (K-D) apparatus will be needed for extract concentration.

Column 1: 15 m × 0.53 mm megabore capillary column, 1.0 µm film thickness, DB-210.
Column 2: 15 m × 0.53 mm megabore capillary column, 1.5 µm film thickness, SPB-608.
Column 3: 15 m × 0.53 mm megabore capillary column, 1.0 µm film thickness, DB-5.

Three megabore capillary columns are included for analysis of organophosphates by this method. Column 1 (DB-210 or equivalent) and Column 2 (SPB-608 or equivalent) are recommended if a large number of organophosphorus analytes are to be determined. If the superior resolution offered by Column 1 and Column 2 is not required, Column 3 (DB-5 or equivalent) may be used. For megabore capillary columns, automatic injections of 1 µL are recommended.

PRECISION & ACCURACY The MDL actually achieved in a given analysis will vary, as it is dependent on instrument sensitivity and matrix effects. Single operator accuracy and precision studies have been conducted with spiked water and soil samples.

MULTIPLICATION FACTORS FOR OTHER MATRICES (a)

Matrix	Factor (b)
Groundwater	10
(EPA Method 3510 or EPA Method 3520)	
Low-concentration soil by Soxhlet and no cleanup	10 (c)
Low-concentration soil by ultrasonic extraction with GPC cleanup	6.7 (c)
High-concentration soil and sludges by ultrasonic extraction	500 (c)
Non-water miscible waste (EPA Method 3580)	1000 (c)

(a) Sample EQLs are highly matrix-dependent. The EQLs listed here are provided for guidance and may not always be achievable.
(b) EQL = [Method detection limit] × [Factor]. For non-aqueous samples the factor is on a wet-weight basis.
(c) Multiply this factory times the soil MDL.

The MDL (in µg/L) when reagent water was extracted using a separatory funnel was 0.12.

The MDL (in μg/kg) when soil was extracted using Soxhlet extraction (EPA Method 3540) was 6.0.

Accuracy (as % recovery) with separatory funnel extraction ranged from not recovered (with low spikes) to 44 (with high spikes).

Accuracy (as % recovery) with continuous liquid-liquid extraction ranged from not recovered (with low spikes) to 43 (with high spikes).

Accuracy (as % recovery) with Soxhlet extraction of soils ranged from not recovered (with low spikes to 28 (with high spikes).

Accuracy (as % recovery) with ultrasonic extraction of soils ranged from 63 (with low spikes) to 17 (with high spikes).

SAMPLE COLLECTION, PRESERVATION & HANDLING
Containers used to collect samples for the determination of semivolatile organic compounds should be soap and water washed followed by methanol (or isopropanol) rinsing. The sample containers should be of glass or Teflon® and have screw-top covers with Teflon® liners.

No preservation is used with concentrated waste samples. With liquid samples containing no residual chlorine and with soil, sediment, and sludge samples, immediately cooling to 4°C is the only preservation used. When residual chlorine is present then 3 mL of 10% aqueous sodium sulfate is added for each gallon of sample collected, followed by cooling to 4°C.

Liquid samples must be extracted within 7 days and their extracts analyzed within 40 days. Concentrated waste, soil, sediment, and sludge samples must be extracted within 14 days and their extracts analyzed within 40 days.

SAMPLE PREPARATION In general, water samples are extracted at a neutral pH with methylene chloride, using either EPA Method 3510 or EPA Method 3520. Solid samples are extracted using either EPA Method 3540 or EPA Method 3550 with methylene chloride/acetone (1:1) as the extraction solvent.

Prior to GC analysis, the extraction solvent may be exchanged to hexane. Single lab data indicates that samples should not be transferred with 100% hexane during sample workup as the more water soluble organophosphorus compounds may be lost.

If cleanup is performed on the samples, the analyst should analyze the samples by GC. This will confirm elution patterns and the absence of interferences from the reagents. If peak detection and identification is prevented by the presence of interferences, further cleanup is required.

QUALITY CONTROL The analyst should monitor the performance of the extraction, cleanup (when used), and analytical system and the effectiveness of the method in dealing with each sample matrix by spiking each sample, standard, and blank with one or two surrogates (e.g., organophosphorus compounds not expected to be present in the sample). Deuterated analogs of analytes should not be used as surrogates for gas chromatographic analysis due to coelution problems.

A minimum of five concentrations for each analyte of interest should be prepared through dilution of the stock standards with isooctane. One of the concentrations should be at a concentration near, but above, the MDL.

Include a mid-level check standard after each group of 10 samples in the analysis sequence. GC/MS techniques should be judiciously employed to support qualitative identifications made with this method. Follow the GC/MS operating requirements specified in EPA Method 8270.

When available, chemical ionization mass spectra may be employed to aid in the qualitative identification process. To confirm an identification of a compound, the background-corrected mass spectrum of the compound must be obtained from the sample extract and must be compared with a mass spectrum from a stock or calibration standard analyzed under the same chromatographic conditions. The molecular ion and all other ions present above 20% relative abundance in the mass spectrum of the standard must be present in the mass spectrum of the sample with agreement to ±20%. The retention time of the compound in the sample must be within six seconds of the retention time for the same compound in the standard solution.

Should the MS procedure fail to provide satisfactory results, additional steps may be taken before reanalysis. These steps may include the use of alternate packed or capillary GC columns or additional sample cleanup.

REFERENCE Test Methods for Evaluating Solid Waste, Physical/Chemical Methods, SW-846, 3rd Edition, U.S. EPA, Office of Solid Waste, Washington, DC, EPA Method 8141 July 1992.

Methyl parathion   **EPA Method 8270**
CAS #298-00-0

TITLE Semivolatile Organic Compounds by GC/MS

MATRIX This method is used to determine the concentration of semivolatile organic compounds in extracts prepared from all types of solid waste matrices, soils, and groundwater. Although surface waters are not specifically mentioned, this method should be applicable to water samples from rivers, lakes, etc.

METHOD SUMMARY This method covers 259 semivolatile organic compounds. In very limited applications direct injection of the sample into the GC/MS system may be appropriate, but this results in very high detection limits (approximately 10,000 μg/L). Typically, a 1-L liquid sample, containing surrogate, and matrix spiking standards, is extracted in a continuous extractor first under acid conditions and then under basic conditions. Typically 30 g of a solid sample, containing surrogate, and matrix spiking standards, is extracted ultrasonically. After concentrating the extract to 1 mL it is spiked with 10 μL of an internal standard solution just prior to analysis by GC/MS. The volume injected should contain about 100 ng of base/neutral and 200 ng of acid surrogates (for a 1-μL injection). Analysis is performed by GC/MS using a capillary GC column.

INTERFERENCES Raw GC/MS data from all blanks, samples, and spikes must be evaluated for interferences. Contamination by carryover can occur whenever high-concentration and low-concentration samples are sequentially analyzed. To reduce carryover, the sample syringe must be rinsed out between samples with solvent. Whenever an unusually concentrated

sample is encountered, it should be followed by the analysis of blank solvent to check for cross-contamination.

INSTRUMENTATION A GC/MS and a data system are required. The GC column used is a 30 m × 0.25 mm I.D. (or 0.32 mm I.D.) 1um film thickness silicone-coated fused silica capillary column. A continuous liquid-liquid extractor equipped with Teflon® or glass connection joints and stopcocks requiring no lubrication, a K-D concentrating apparatus, water bath, and an ultrasonic disrupter with a minimum power of 300 W and with pulsing capability are also required.

PRECISION & ACCURACY The estimated quantitation limit (EQL) of Method 8270B for determining an individual compound is approximately 1 mg/kg (wet weight) for soil or sediment samples, 1–200 mg/kg for wastes (dependent on matrix and method of preparation), and 10 μg/L for groundwater samples. EQLs will be proportionately higher for sample extracts that require dilution to avoid saturation of the detector.

The EQL(b) for groundwater in μg/L is 10.
The EQL (a, b) for low concentrations in soil and sediment in μg/kg is not determined.
Accuracy as μg/L is not listed.
Overall precision in μg/L is not listed.

(a) EQLs listed for soil/sediment are based on wet weight. Normally data is reported in a dry-weight basis; therefore, EQLs will be higher based on the % dry weight of each sample. This calculation is based on a 30 g sample and gel permeation chromatography cleanup.

(b) Sample EQLs are highly matrix-dependent. The EQLs are provided for guidance and may not always be achievable.

C = True value for concentration, in μg/L.
X = Average recovery found for measurements of samples containing a concentration of C, in μg/L.

ESTIMATED QUANTITATION LIMIT

Other Matrices	Factor (a)
High-concentration soil and sludges by sonicator	7.5
Non-water miscible waste	75

(a) EQL for other matrices = [EQL for low soil/sediment] × [Factor]. This estimated EQL is similar to an EPA "Practical Quantitation Limit."

SAMPLING METHOD
Liquid samples — Use a 1 or 2½ gallon amber glass bottle with a screw-top Teflon®-lined cover that has been prewashed with detergent and rinsed with distilled water and methanol (or isopropanol).

Soils, sediments, or sludges — Use an 8-oz. widemouth glass with a screw-top Teflon®-lined cover that has been prewashed with detergent and rinsed with distilled water and methanol (or isopropanol).

SAMPLE PRESERVATION
Liquid samples — If residual chlorine is present, add 3 mL of 10% sodium thiosulfate per gallon, cool to 4°C and store in a solvent-free refrigerator until analysis; if chlorine is not present, then eliminate the sodium thiosulfate addition.

Soils, sediments, or sludges — Cool samples to 4°C and store in a solvent-free refrigerator.

MHT Liquid samples must be extracted within 7 days and the extracts analyzed within 40 days. Soils, sediments, or sludges may be stored for a maximum of 14 days and the extracts analyzed within 40 days.

SAMPLE PREPARATION
Liquid samples — Transfer 1 L quantitatively to a continuous extractor. If high concentrations are anticipated, a smaller volume may be used and then diluted with organic-free reagent water to 1 L. Adjust pH, if necessary, to pH <2 using 1:1 (V/V) sulfuric acid. Pipette 1.0 mL of a surrogate standard spiking solution into each sample. For the sample in each analytical batch selected for spiking, add 1.0 mL of a matrix spiking standard. For base/neutral acid analysis, the amount of the surrogates and matrix spiking compounds added to the sample should result in a final concentration of 100 ng/μL of each analyte in the extract to be analyzed (assuming a 1-μL injection). Extract with methylene chloride for 18–24 h. Next, adjust the pH of the aqueous phase to pH >11 using 10 N sodium hydroxide and extract it with methylene chloride again for 18–24 h. Dry the extract through a column containing anhydrous sodium sulfate and concentrate it to 1 mL using a K-D concentrator.

Soils, sediments, or sludges — Use 30 g of sample. Nonporous or wet samples (gummy or clay type) that do not have a free-flowing sandy texture must be mixed with anhydrous sodium sulfate until the sample is free flowing. Add 1 mL of surrogate standards to all samples, spikes, standards, and blanks. For the sample in each analytical batch selected for spiking, add 1.0 mL of a matrix spiking standard. For base/neutral acid analysis, the amount added of the surrogates and matrix spiking compounds should result in a final concentration of 100 ng/μL of each base/neutral analyte and 200 ng/μL of each acid analyte in the extract to be analyzed (assuming a 1-μL injection). Immediately add a 100-mL mixture of 1:1 methylene chloride:acetone and extract the sample ultrasonically for 3 min and then decant or filter the extracts. Repeat the extraction two or more times. Dry the extract using a column with anhydrous sodium sulfate and concentrate it to 1 mL in a K-D concentrator.

QUALITY CONTROL A methylene chloride solution containing 50 ng/μL of decafluorotriphenylphosphine (DFTPP) is used for tuning the GC/MS system each 12-h shift. A system performance check also must be made during every 12-h shift. A standard containing 50 ng/μL each of 4,4'-DDT, pentachlorophenol, and benzidine is required to verify injection port inertness and GC column performance. A calibration standard at mid-concentration, containing each compound of interest, including all required surrogates, must be performed every 12 h during analysis. After the system performance check is met, calibration check compounds (CCCs) are used to check the validity of the initial calibration.

The internal standard responses and retention times in the calibration check standard must be evaluated immediately after or during data acquisition. If the retention time for any internal standard changes by more than 30 seconds from the last check calibration (12 h), the chromatographic system must be

inspected for malfunctions and corrections must be made, as required. If the electron ionization current plot (EICP) area for any of the internal standards changes by a factor of two from the last daily calibration standard check, the mass spectrometer must be inspected for malfunctions and corrections must be made, as appropriate.

Demonstrate, through the analysis of a reagent water blank, that interferences from the analytical system, glassware, and reagents are under control. The blank samples should be carried through all stages of the sample preparation and measurement steps. For each analytical batch (up to 20 samples), a reagent blank, matrix spike, and matrix spike duplicate/duplicate must be analyzed (the frequency of the spikes may be different for different monitoring programs). The blank and spiked samples must be carried through all stages of the sample preparation and measurement steps. A QC reference sample concentrate containing each analyte at a concentration of 100 mg/L in methanol is required.

REFERENCE Test Methods for Evaluating Solid Waste (SW-846). U.S. EPA 1983, Method 8270B, Rev. 2, Nov. 1990. Office of Solid Waste, Washington, DC.

Methyl parathion   **EPA Method 8140**
CAS #298-00-0

TITLE Organophosphorus Pesticides

MATRIX Groundwater, soils, sludges, water miscible liquid wastes, and non-water miscible wastes.

APPLICATION This method is used for the analysis of 21 organophosphorus pesticides. Samples are extracted, concentrated, and analyzed using direct injection of both neat and diluted organic liquid into a gas chromatograph (GC).

INTERFERENCES Solvents, reagents, and glassware may introduce artifacts. Other interferences may come from coextracted compounds from samples. The use of Florisil cleanup materials may produce low recoveries. Elemental sulfur may interfere with some compounds when using a flame photometric detector. Sulfur cleanup (Method 3660) may alleviate sulfur interference.

INSTRUMENTATION GC capable of on-column injections and a flame photometric detector (FPD) or a thermionic detector. Column 1: 1.8 m by 2 mm with 5% SP-2401 on Supelcoport. Column 2: 1.8 m by 2 mm with 3% SP-2401 on Supelcoport. Column 3: 50 cm by ⅛ in Teflon® with 15% SE-54 on Gas Chrom Q. The preferred column is Column Number 2.

RANGE 0.5–500 µg/L.

MDL 0.03 µg/L (in reagent water).

PQL FACTORS FOR MULTIPLYING × FID MDL VALUE

Matrix	Multiplication Factor
Groundwater	10
Low-level soil by sonication with GPC cleanup	670
High-level soil and sludge by sonication	10,000
Non-water miscible waste	100,000

PRECISION 5.3% (single operator standard deviation)

ACCURACY 96.0% (single operator average recovery)

SAMPLING METHOD Use 8-oz. widemouth glass bottles with Teflon®-lined caps for concentrated waste samples, soils, sediments, and sludges. Use 1 or 2½ gallon amber glass bottles with Teflon®-lined caps for liquid (water) samples.

STABILITY Cool soil, sediment, sludge, and liquid samples to 4°C. If residual chlorine is present in liquid samples add 3 mL of 10% sodium thiosulfate per gallon of sample and cool to 4°C.

MHT 14 days for concentrated waste, soil, sediment, or sludge; 7 days for liquid samples; all extracts must be analyzed within 40 days.

QUALITY CONTROL A quality control check sample concentrate containing this compound in acetone at a concentration 1,000 times more concentrated than the selected spike concentration is required. The QC check sample concentrate may be prepared from pure standard materials or purchased as certified solutions. Use appropriate trip, matrix, control site, method, reagent, and solvent blanks. Internal, surrogate, and five concentration level calibration standards are used.

REFERENCE Method 8140, SW-846, 3rd ed., Sept. 1986.

4-Methyl-2-pentanone   **EPA Method 1624**
CAS #108-10-1

TITLE Volatile Organic Compounds by Isotope Dilution GC/MS

MATRIX Compounds may be determined in waters, soils, and municipal sludges by this method.

METHOD SUMMARY This method is used to determine 58 volatile toxic organic pollutants associated with the CWA (as amended 1987); the RCRA (as amended 1986); the CERCLA (as amended 1986); and other compounds amenable to purge-and-trap gas chromatography-mass spectrometry (GC/MS).

If the solids content is less than 1%, stable isotopically-labeled analogs of the compounds of interest are added to a 5-mL sample and the sample is purged with an inert gas at 20–25°C in a chamber designed for soil or water samples. If the solids content is greater than 1%, 5 mL of reagent water and the labeled compounds are added to a 5-g aliquot of sample and the mixture is purged at 40°C. Compounds that will not purge at 20–25°C or at 40°C are purged at 78–85°C. In the purging process, the volatile compounds are transferred from the aqueous phase into the gaseous phase where they are passed into a sorbent column, and trapped. After purging is completed, the trap is backflushed and heated rapidly to desorb the compounds into a GC. The compounds are separated by the GC and detected by a MS. The labeled compounds serve to correct the variability of the analytical technique.

INTERFERENCES Impurities in the purge gas, organic compounds outgassing from the plumbing upstream of the trap, and solvent vapors in the lab account for most problems. Samples can be contaminated by diffusion of volatile organic compounds (particularly methylene chloride) through the bottle seal during shipment and storage. Contamination by carryover can occur when high-level and low-level samples are analyzed sequentially. When an unusually concentrated sample is encountered, follow it by analysis of a reagent water blank to check for carryover.

INSTRUMENTATION Major equipment includes a GC with linear temperature programming and a glass jet separator as the MS interface, a MS with 70 eV electron impact ionization, and a data system to collect and record response factors.

Column: 2.8 m × 2 mm I.D. glass, packed with 1% SP-1000 on Carbopak B, 60/80 mesh, or equivalent.

PRECISION & ACCURACY The detection limits of the method are usually dependent on the level of interferences rather than instrumental limitations. The method detection limits were determined in digested sludge (low solids) and in filter cake or compost (high solids).

The MDL (in µg/kg) for low solids is not listed and for high solids is not listed.
Labeled and native compound precision (in µg/L) as standard deviation was not listed.
Labeled and native compound accuracy (in µg/L) as average recovery was not listed.
Acceptance criteria are at 20 µg/L for this compound.

SAMPLE COLLECTION, PRESERVATION & HANDLING Grab samples are collected in glass containers having a total volume greater than 20 mL. Fill and seal each bottle so that no air bubbles are entrapped. Samples are maintained at 0 to 4°C from the time of collection until analysis. If an aqueous sample contains residual chlorine, add sodium thiosulfate preservative (10 mg/40 mL) to the empty sample bottles just prior to shipment to the sample site. All samples must be analyzed within 14 days of collection.

SAMPLE PREPARATION Samples containing less than 1% solids are analyzed directly as aqueous samples. Samples containing 1% solids or greater are analyzed as solid samples utilizing one of two methods, depending on the levels of pollutants, in the sample. Samples containing 1% solids or greater, and low to moderate levels of pollutants are analyzed by purging a known weight of sample added to 5 mL of reagent water. Samples containing 1% solids or greater, and high levels of pollutants, are extracted with methanol, and an aliquot of the methanol extract is added to reagent water and purged.

QUALITY CONTROL A field blank prepared from reagent water and carried through the sampling and handling protocol may serve as a check on contamination from shipment and storage.

The analyst is permitted to modify this method to improve separations or lower the costs of measurements, provided all performance specifications are met. Analyses of blanks are required. When results of spikes indicate atypical method performance for samples, the samples are diluted to bring method performance within acceptable limits. Analyze two sets of four 5-mL aliquots (8 aliquots total) of the aqueous performance standard. Spike all samples with labeled compounds to assess method performance on the sample matrix. Compute the percent recovery of the labeled compounds using the internal standard method. Compare the percent recovery for each compound with the corresponding labeled compound recovery. Reagent water blanks are analyzed to demonstrate freedom from carryover contamination. Field replicates may be collected to determine the precision of the sampling technique, and spiked samples may be required to determine the accuracy of the analysis when the internal method is used.

REFERENCE Volatile Organic Compounds by Isotope Dilution GC/MS. Office of Water Regulation and Standards, U.S. EPA Industrial Technology Division, Washington, DC, EPA Method 1624, Rev. C, June 1989 (contact W.A. Telliard, U.S. EPA, Office of Water Regulations and Standards, 401 M St., SW, Washington, DC, 20460. Phone: 202-382-7131).

4-Methyl-2-pentanone   **EPA Method 8240**
CAS #108-10-1

TITLE Volatile Organics By GC/MS: Packed Column Technique

MATRIX Nearly all types of sample matarices, regardless of water content, can be analyzed using this method. This includes groundwater, aqueous sludges, caustic liquors, acid liquors, waste solvents, oily wastes, mousses, tars, fibrous wastes, polymetric emulsions, filter cakes, spent carbons, spent catalysts, soils, and sediments.

METHOD SUMMARY Method 8240B covers 80 volatile organic compounds that are introduced into a gas chromatograph by the purge-and-trap method or by direct injection (in limited applications). For the purge-and-trap method an inert gas (zero grade nitrogen or helium) is bubbled through a 5-mL solution at ambient temperature. Purged sample components are trapped in a tube of sorbent materials. When purging is complete, the sorbent tube is heated and backflushed with inert gas to desorb the trapped components onto a GC column.

INTERFERENCES Impurities in the purge gas and from organic compounds outgassing from the plumbing ahead of the trap account for many contamination problems. Interferences purged or coextracted from the samples will vary considerably from source to source. Cross-contamination can occur whenever high-level and low-level samples are analyzed sequentially. Whenever an unusually concentrated sample is analyzed, it should be followed by the analysis of organic-free reagent water to check for cross-contamination. Samples also can be contaminated by diffusion of volatile organics (particularly methylene chloride and fluorocarbons) through the septum seal into the sample during shipment and storage. A trip blank can serve as a check on such contamination. The lab where volatile analysis is performed and also the refrigerated storage area should be completely free of solvents.

INSTRUMENTATION A gas chromatograph/mass spectrometry/data system (GC/MS) equipped with a 6 ft × 0.1 in I.D. glass column packed with 1% SP-1000 on Carbopack-B (60/80 mesh) is required. Also needed is a 5-mL purging device, a sorbent trap, and a thermal desorption apparatus.

PRECISION & ACCURACY This method is reported to have been tested by 15 laboratories using organic-free reagent water, drinking water, surface water, and industrial wastewaters (not specified) fortified at six concentrations over the range 5–600 µg/L.

Sample estimated quantitation limits (EQLs) are highly matrix-dependent. The EQLs listed may not always be achievable. EQLs listed for soils or sediments are based on wet weight. Normally, data is reported on a dry-weight basis; therefore, EQLs will be higher, based on the percent dry weight of each sample. Note that EQLs are even more variable than MDLs and that they are highly variable depending on the matrix being analyzed.

EQL in groundwater in µg/L was 50.
EQL in low soil or sediment in µg/kg was 50.
Accuracy (a) in µg/L was not listed.
Precision (b) in µg/L was not listed.

(a) *Average recovery found for measurements of samples containing a concentration of C, in µg/L.*
(b) *Overall precision found for measurements of samples with average recovery X for samples containing a concentration of C in µg/L.*
X = *Average recovery found for measurement of samples containing a concentration of C in µg/L.*

MULTIPLICATION FACTORS FOR OTHER MATRICES

Other Matrices	Factor (a)
Waste miscible liquid waste	50
High-concentration soil and sludge	125
Non-water miscible waste	500

(a) *EQL = [EQL for low soil/sediment] × [Factor]. For non-aqueous samples, the factor is on a wet-weight basis.*

SAMPLING METHOD

Liquid samples — Use a 40-mL glass screw-cap VOA vial with a Teflon®-faced silicone septum that has been prewashed, rinsed with distilled deionized water, and oven dried. However, if residual chlorine is present, collect sample in a 40-oz. soil VOA container which has been pre-preserved with 4 drops of 10% sodium thiosulfate, mix gently, and then transfer the sample to a 40-mL VOA vial. Collect bubble-free samples in duplicate and seal them in separate plastic bags.

Soils or sediments, and sludges — Use an 8-oz. widemouth glass bottle with a Teflon®-faced silicone septum that has been prewashed with detergent, rinsed with distilled deionized water, and oven dried. Tap slightly to eliminate free air space. Collect samples in duplicate and seal them in separate plastic bags.

SAMPLE PRESERVATION

Liquid samples — Add 4 drops of concentrated HCL and immediately cool samples to 4°C and store in a solvent-free refrigerator.

Soils or sediments, and sludges — Cool samples to 4°C and store in a solvent-free refrigerator.

MHT Maximum holding time is 14 days from the date of sample collection.

SAMPLE PREPARATION

Liquid samples — Remove the plunger from a 5-mL syringe and carefully pour the sample into the syringe barrel to just short of overflowing. Replace the syringe plunger and compress the sample. Open the syringe valve and vent any residual air while adjusting the sample volume to 5.0 mL. If there is only one volatile organic analysis (VOA) vial, a second syringe should be filled at this time to protect against possible loss of sample integrity. Add 10 µL of surrogate spiking solution and 10 µL of internal standard spiking solution through the valve bore of the 5-mL syringe, then close the valve. The surrogate and internal standards may be mixed and added as a single spiking solution.

Sediments, soils, and waste samples — All samples of this type should be screened by GC analysis using a headspace method (EPA Method 3810) or the hexadecane extraction and screening method (EPA Method 3820). Use the screening data to determine whether to use the low-concentration method (0.005–1 mg/kg) or the high-concentration method (>1 mg/kg).

Low-concentration method — The low-concentration method is based on purging a heated sediment or soil sample mixed with organic-free reagent water containing the surrogate and internal standards. Analyze all reagent blanks and standards under the same conditions as the samples.

Use a 5-g sample if the expected concentration is <0.1 mg/kg or a 1-g sample for expected concentrations between 0.1 and 1 mg/kg. Mix the contents of the sample container with a narrow metal spatula. Weigh the amount of the sample into a tared purge device. Add the spiked water to the purge device, which contains the weighed amount of sample, and connect the device to the purge-and-trap system.

High-concentration method — This method is based on extracting the sediment or soil with methanol. A waste sample is either extracted or diluted, depending on its solubility in methanol. Wastes that are insoluble in methanol are diluted with reagent tetraglyme or possibly polyethylene glycol (PEG). An aliquot of the extract is added to organic-free reagent water containing surrogate and internal standards. This is purged at ambient temperature. All samples with an expected concentration of >1.0 mg/kg should be analyzed by this method.

Mix the contents of the sample container with a narrow metal spatula. For sediments or soils and solid wastes that are insoluble in methanol, weigh 4 g (wet weight) of sample into a tared 20-mL vial. For waste that is soluble in methanol, tetraglyme, or PEG, weigh 1 g (wet weight) into a tared scintillation vial or culture tube or a 10-mL volumetric flask. Quickly add 9.0 mL of appropriate solvent then add 1.0 mL of a surrogate spiking solution to the vial, cap it, and shake it for 2 min.

METHANOL EXTRACT REQUIRED FOR ANALYSIS OF HIGH-CONCENTRATION SOILS OR SEDIMENTS

Approximate Concentration Range	Volume of Methanol Extract (a)
500–10,000 µg/kg	100 µL
1,000–20,000 µg/kg	50 µL
5,000–100,000 µg/kg	10 µL
25,000–500,000 µg/kg	100 µL of 1/50 dilution (b)

Calculate appropriate dilution factor for concentrations exceeding this table.

(a) The volume of methanol added to 5 mL of water being purged should be kept constant. Therefore, add to the 5-mL syringe whatever volume of methanol is necessary to maintain a volume of 100 µL added to the syringe.

(b) Dilute an aliquot of the methanol extract and then take 100 µL for analysis.

QUALITY CONTROL Demonstrate, through the analysis of a reagent water blank, that interferences from the analytical system, glassware, and reagents are under control. Blank samples should be carried through all stages of the sample preparation and measurement steps. For each analytical batch (up to 20 samples), a reagent blank, matrix spike, and matrix spike duplicate must be analyzed (the frequency of the spikes may be different for different monitoring programs). The blank and spiked samples must be carried through all stages of the sample preparation and measurement steps. QC samples mentioned in the section on Interferences will also be needed as appropriate to those situations.

REFERENCE Test Methods for Evaluating Solid Waste (SW-846). U.S. EPA. 1983. Method 8240B, Rev. 2, Nov. 1990. Office of Solid Wastes, Washington, DC.

2-Methyl-4,6-dinitrophenol **EPA Method 1625**
CAS #534-52-1

TITLE Semivolatile Organic Compounds by Isotope Dilution GC/MS

MATRIX The compounds may be determined in waters, soils, and municipal sludges by this method.

METHOD SUMMARY This method is used to determine 176 semivolatile toxic organic pollutants associated with the CWA (as amended 1987); the RCRA (as amended 1986); the CERCLA (as amended 1986); and other compounds amenable to extraction and analysis by capillary column gas chromatography-mass spectrometry (GC/MS).

Stable isotopically-labeled analogs of the compounds of interest are added to the sample. If the solids content is less than 1%, a 1-L sample is extracted at pH 12–13, then at pH <2 with methylene chloride using continuous extraction techniques.

If the solids content is 30% or less, the sample is diluted to 1% solids with reagent water, homogenized ultrasonically, and extracted at pH 12–13, then at pH <2 with methylene chloride using continuous extraction techniques. If the solids content is greater than 30%, the sample is extracted using ultrasonic techniques.

Each extract is dried over sodium sulfate, concentrated to a volume of 5 mL, cleaned up using GPC, if necessary, and concentrated. Extracts are concentrated to 1 mL if GPC is not performed, and to 0.5 mL if GPC is performed.

An internal standard is added to the extract, and a 1-mL aliquot of the extract is injected into the GC. The compounds are separated by GC and detected by a MS. The labeled compounds serve to correct the variability of the analytical technique.

INTERFERENCES Solvents, reagents, glassware, and other sample processing hardware may yield artifacts and/or elevated baselines causing misinterpretation of chromatograms and spectra. Materials used in the analysis must be demonstrated to be free from interferences under the conditions of analysis by running method blanks initially and with each sample lot (sample started through the extraction process on a given 8-h shift, to a maximum of 20). Specific selection of reagents and purification of solvents by distillation in all glass systems may be required. Glassware and, where possible, reagents are cleaned by solvent rinse and baking at 450°C for 1-h minimum. Interferences coextracted from samples will vary considerably from source to source, depending on the diversity of the site being sampled.

INSTRUMENTATION Major instrumentation includes a GC with a splitless or on-column injection port for capillary column, a MS with 70 eV electron impact ionization, and a data system to collect and record MS data, and process it. A K-D apparatus is used to concentrate extracts.

GC Column: 30 m × 0.25 mm I.D. 5% phenyl, 94% methyl, 1% vinyl silicone bonded phased fused silica capillary column.

PRECISION & ACCURACY The detection limits of the method are usually dependent on the level of interferences rather than instrumental limitations. The limits typify the minimum quantities that can be detected with no interferences present.

The minimum level (in µg/mL) was 20. This is defined as a minimum level at which the analytical system shall give recognizable mass spectra (background corrected) and acceptable calibration points.

The MDL (in µg/kg) in low solids was 385 and in high solids was 83; these were determined in digested sludge (low solids) and in filter cake or compost (high solids).

The labeled and native compound initial precision as standard deviation (in µg/L) was 19.

The labeled and native compound initial accuracy as average recovery (in µg/L) was 77–133.

SAMPLE COLLECTION, PRESERVATION & HANDLING Collect samples in glass containers. Aqueous samples which flow freely are collected in refrigerated bottles using automatic sampling equipment. Solid samples are collected as grab samples using widemouth jars. Maintain samples at 0 to 4°C from the time of collection until extraction. If residual chlorine is present in aqueous samples, add 80 mg sodium thiosulfate/L of water. Begin sample extraction within 7 days of collection, and analyze all extracts within 40 days of extraction.

SAMPLE PREPARATION Samples containing 1% solids or less are extracted directly using continuous liquid-liquid extraction techniques. Samples containing 1 to 30% solids are diluted to the 1% level with reagent water and extracted using continuous liquid-liquid extraction techniques. Samples containing greater than 30% solids are extracted using ultrasonic techniques.

Base/neutral extraction — Adjust the pH of the waters in the extractors to 12–13 with 6 N NaOH. Extract with methylene chloride for 24–48 h.

Acid extraction — Adjust the pH of the waters in the extractors to 2 or less using 6 N sulfuric acid. Extract with methylene chloride for 24–48 h.

Ultrasonic extraction of high solids samples — Add anhydrous sodium sulfate to the sample and QC aliquot(s). Add acetone:methylene chloride (1:1) to the sample and mix thoroughly

Concentrate extracts using a K-D apparatus.

QUALITY CONTROL The analyst is permitted to modify this method to improve separations or lower the costs of measurements, provided all performance specifications are met. Analyses of blanks are required to demonstrate freedom from contamination. When results of spikes indicate atypical method performance for samples, the samples are diluted to bring method performance within acceptable limits.

For low solids (aqueous samples), extract, concentrate, and analyze two sets of four 1-L aliquots (8 aliquots total) of the precision and recovery standard. For high solids samples, two sets of four 30-g aliquots of the high solids reference matrix are used.

Spike all samples with labeled compounds to assess method performance. Compute percent recovery of the labeled compounds using the internal standard method. Compare the labeled compound recovery for each compound with the corresponding labeled compound recovery.

Reagent water and high solids reference matrix blanks are analyzed to demonstrate freedom from contamination. Extract and concentrate a 1-L reagent water blank or a high solids reference matrix blank with each sample's lot (samples started through the extraction process on the same 8-h shift, to a maximum of 20 samples).

Field replicates may be collected to determine the precision of the sampling technique, and spiked samples may be required to determine the accuracy of the analysis when the internal standard method is used.

REFERENCE Semivolatile Organic Compounds by Isotope Dilution GC/MS. Office of Water Regulation and Standards, U.S. EPA Industrial Technology Division, Washington, DC, EPA Method 1625, Rev. C, June 1989 (contact W.A. Telliard, U.S. EPA, Office of Water Regulations and Standards, 401 M St., SW, Washington, DC, 20460. Phone: 202-382-7131).

2-Methyl-4,6-dinitrophenol **EPA Method 625**
CAS #534-52-1

TITLE Base/Neutrals and Acids, U.S. EPA Method 625

MATRIX This methods covers municipal and industrial wastewaters.

METHOD SUMMARY Approximately 1 L of sample is serially extracted with methylene chloride at a pH greater than 11 and again at a pH less than 2 using a separatory funnel or a continuous extractor. The methylene chloride extract is dried, concentrated to a volume of 1 mL, and analyzed by GC/MS. Qualitative identification of the parameters in the extract is performed using the retention time and the relative abundance of three characteristic masses (m/z). Qualitative analysis is performed using either external or internal standard techniques with a single characteristic m/z.

INTERFERENCES Method interferences may be caused by contaminants in solvents, reagents, glassware, and other sample processing hardware. Glassware must be scrupulously cleaned. Glassware should be heated in a muffle furnace at 400°C for 5 to 30 min. Some thermally stable materials, such as PCBs, may not be eliminated by this treatment. Solvent rinses with acetone and pesticide quality hexane may be substituted for the muffle furnace heating. Matrix interferences may be caused by contaminants that are coextracted from the sample. The base-neutral extraction may cause significantly reduced recovery of phenols. The packed gas chromatographic columns recommended for the basic fraction may not exhibit sufficient resolution for some analytes.

INSTRUMENTATION A GC/MS system with an injection port designed for on-column injection when using packed columns and for splitless injection when using capillary columns.

Column for base/neutrals: 1.8 m long × 2 mm I.D. glass, packed with 3% SP-2550 on Supelcoport (100/120 mesh) or equivalent.

Column for acids: 1.8 m long × 2 mm I.D. glass, packed with 1% SP-1240DA on Supelcoport (100/120 mesh) or equivalent.

PRECISION & ACCURACY The MDL concentrations were obtained using reagent water. The MDL actually achieved in a given analysis will vary depending on instrument sensitivity and matrix effects. This method was tested by 15 laboratories using reagent water, drinking water, surface water, and industrial wastewaters spiked at six concentrations over the range 5 to 100 µg/L. Single operator precision, overall precision, and method accuracy were found to be directly related to the concentration of the parameter matrix.

The MDL (in µg/L) in reagent water was 24.

The standard deviation (in µg/L based on 4 recovery measurements) was 93.2.

The range (in µg/L) for average recovery for 4 measurements was 53.0–100.0.

The range (in %) for percent recovery was D-181.

Accuracy (in µg/L) as expected recovery for one or more measurements of a sample containing a true concentration of C was $1.04C - 28.04$.

Precision (in µg/L) as expected single analyst standard deviation of measurements at an average concentration found at X was $0.10X + 42.29$.

Overall precision (in µg/L) as expected interlaboratory standard deviation of measurements in an average concentration found at X was $0.26X + 23.10$.

$C =$ *True value of the concentration in µg/L.*
$X =$ *Average recovery found for measurements of samples containing a concentration at C in µg/L.*

SAMPLE PREPARATION Adjust the pH to >11 with sodium hydroxide and serially extract in a separatory funnel with methylene chloride or else in a continuous extractor. Next, adjust the pH to <2 with sulfuric acid and serially extract in a separatory funnel with methylene chloride or else in a continuous extractor. Dry the extracts separately through a column of anhydrous sodium sulfate and then concentrate each of the extracts to 1.0 mL using a K-D apparatus.

SAMPLE COLLECTION, PRESERVATION & HANDLING Grab samples must be collected in glass containers. All samples must be refrigerated at 4°C from the time of collection until extraction. If residual chlorine is present, add 80 mg of sodium thiosulfate/L of sample and mix well. All samples must be extracted within 7 days of collection and completely analyzed within 40 days of extraction.

QUALITY CONTROL Make an initial, one-time, demonstration of the ability to generate acceptable accuracy and precision with this method. Before processing any samples, the analyst must analyze a reagent water blank to demonstrate that interferences from the analytical system and glassware are under control. Each time a set of samples is extracted or reagents are changed, a reagent water blank must be processed. Spike and analyze a minimum of 5% of all samples to monitor and evaluate lab data quality. A QC check sample concentrate that contains each parameter of interest at a concentration of 100 µg/mL in acetone is required. PCBs and multicomponent pesticides may be omitted from this test.

After analysis of five spiked wastewater samples, calculate the average percent recovery and the standard deviation of the percent recovery. Spike all samples with the surrogate standard spiking solution and calculate the percent recovery of each surrogate compound.

REFERENCE Federal Register, Vol. 49, No. 209. Friday, Oct. 26, 1984.

2-Methyl-4,6-dinitrophenol   **EPA Method 8040**
CAS #534-52-1

TITLE Phenols

MATRIX Groundwater, soils, sludges, water miscible liquid wastes, and non-water miscible wastes.

APPLICATION This method is used for the analysis of 17 phenols. Samples are extracted, concentrated, and analyzed using direct injection of both neat and diluted organic liquids. Pentafluorobenzylbromide (PFB) derivatives also may be made to increase sensitivity of the method.

INTERFERENCES There can be carryover contamination with high- and low-level samples. Solvents, reagents, and glassware may introduce artifacts. Other interferences may come from coextracted compounds from samples.

INSTRUMENTATION GC capable of on-column injections and a flame with detector (FID) or electron capture detector (ECD). Column for underivatized phenol: 1.8 m by 2.0 mm with 1% SP-1240DA on Supelcoport. Column for derivatized phenols: 1.8 m by 2.0 mm with 5% OV-17 on Chromosorb W-AW-DMCS.

RANGE 12–450 µg/L

MDL 16.0 µg/L (FID)

PQL FACTORS FOR MULTIPLYING × FID MDL VALUE

Matrix	Multiplication Factor
Groundwater	10
Low-level soil by sonication with GPC cleanup	670
High-level soil and sludge by sonication	10,000
Non-water miscible waste	100,000

PRECISION 0.19X + 5.85 µg/L (overall precision using FID)

ACCURACY 0.84C–1.01 µg/L (as recovery using FID)

SAMPLING METHOD Use 8-oz. widemouth glass bottles with Teflon®-lined caps for concentrated waste samples, soils, sediments, and sludges. Use 1 or 2½ gallon amber glass bottles with Teflon®-lined caps for liquid (water) samples.

STABILITY Cool soil, sediment, sludge, and liquid samples to 4°C. If residual chlorine is present in liquid samples add 3 mL of 10% sodium thiosulfate per gallon of sample and cool to 4°C.

MHT 14 days for concentrated waste, soil, sediment, or sludge; 7 days for liquid samples; all extracts must be analyzed within 40 days.

QUALITY CONTROL A quality control check sample concentrate containing each analyte of interest is required. The QC check sample concentrate may be prepared from pure standard materials or purchased as certified solutions Use appropriate trip, matrix, control site, method, reagent, and solvent blanks. Internal, surrogate, and five concentration level calibration standards are used. The QC check sample concentrate should contain this compound at 100 µg/mL in 2-propanol.

REFERENCE Test Methods for Evaluating Solid Waste (SW-846), U.S. EPA Office of Solid Waste, Washington, DC, Method 8040A, Rev. 1, Nov. 1990.

2-Methyl-5-nitroaniline   **EPA Method 8270**
CAS #99-55-8

TITLE Semivolatile Organic Compounds by GC/MS

MATRIX This method is used to determine the concentration of semivolatile organic compounds in extracts prepared from all types of solid waste matrices, soils, and groundwater. Although surface waters are not specifically mentioned, this method should be applicable to water samples from rivers, lakes, etc.

METHOD SUMMARY This method covers 259 semivolatile organic compounds. In very limited applications direct injection of the sample into the GC/MS system may be appropriate, but this results in very high detection limits (approximately 10,000 µg/L). Typically, a 1-L liquid sample, containing surrogate, and matrix spiking standards, is extracted in a continuous extractor first under acid conditions and then under basic conditions. Typically 30 g of a solid sample, containing surrogate, and matrix spiking standards, is extracted ultrasonically. After concentrating the extract to 1 mL it is spiked with 10 µL of an internal standard solution just prior to analysis by GC/MS. The volume injected should contain about 100 ng of base/neutral and 200 ng of acid surrogates (for a 1-µL injection). Analysis is performed by GC/MS using a capillary GC column.

INTERFERENCES Raw GC/MS data from all blanks, samples, and spikes must be evaluated for interferences. Contamination by carryover can occur whenever high-concentration and low-concentration samples are sequentially analyzed. To reduce carryover, the sample syringe must be rinsed out between samples with solvent. Whenever an unusually concentrated sample is encountered, it should be followed by the analysis of blank solvent to check for cross-contamination.

INSTRUMENTATION A GC/MS and a data system are required. The GC column used is a 30 m × 0.25 mm I.D. (or 0.32 mm I.D.) 1um film thickness silicone-coated fused silica capillary column. A continuous liquid-liquid extractor equipped with Teflon® or glass connection joints and stopcocks requiring no lubrication, a K-D concentrating apparatus, water bath, and an ultrasonic disrupter with a minimum power of 300 W and with pulsing capability are also required.

PRECISION & ACCURACY The estimated quantitation limit (EQL) of Method 8270B for determining an individual compound is approximately 1 mg/kg (wet weight) for soil or sediment samples, 1–200 mg/kg for wastes (dependent on matrix and method of preparation), and 10 µg/L for groundwater samples. EQLs will be proportionately higher for sample extracts that require dilution to avoid saturation of the detector.

The EQL(b) for groundwater in µg/L is not listed.
The EQL (a, b) for low concentrations in soil and sediment in µg/kg is not listed.
Accuracy as µg/L is not listed.
Overall precision in µg/L is not listed.

(a) EQLs listed for soil/sediment are based on wet weight. Normally data is reported in a dry-weight basis; therefore, EQLs will be higher based on the % dry weight of each sample. This calculation is based on a 30 g sample and gel permeation chromatography cleanup.
(b) Sample EQLs are highly matrix-dependent. The EQLs are provided for guidance and may not always be achievable.
C = True value for concentration, in µg/L.
X = Average recovery found for measurements of samples containing a concentration of C, in µg/L.

ESTIMATED QUANTITATION LIMIT

Other Matrices	Factor (a)
High-concentration soil and sludges by sonicator	7.5
Non-water miscible waste	75

(a) EQL for other matrices = [EQL for low soil/sediment] × [Factor]. This estimated EQL is similar to an EPA "Practical Quantitation Limit."

SAMPLING METHOD
Liquid samples — Use a 1 or 2½ gallon amber glass bottle with a screw-top Teflon®-lined cover that has been prewashed with detergent and rinsed with distilled water and methanol (or isopropanol).

Soils, sediments, or sludges — Use an 8-oz. widemouth glass with a screw-top Teflon®-lined cover that has been prewashed with detergent and rinsed with distilled water and methanol (or isopropanol).

SAMPLE PRESERVATION
Liquid samples — If residual chlorine is present, add 3 mL of 10% sodium thiosulfate per gallon, cool to 4°C and store in a solvent-free refrigerator until analysis; if chlorine is not present, then eliminate the sodium thiosulfate addition.

Soils, sediments, or sludges — Cool samples to 4°C and store in a solvent-free refrigerator.

MHT Liquid samples must be extracted within 7 days and the extracts analyzed within 40 days. Soils, sediments, or sludges may be stored for a maximum of 14 days and the extracts analyzed within 40 days.

SAMPLE PREPARATION
Liquid samples — Transfer 1 L quantitatively to a continuous extractor. If high concentrations are anticipated, a smaller volume may be used and then diluted with organic-free reagent water to 1 L. Adjust pH, if necessary, to pH <2 using 1:1 (V/V) sulfuric acid. Pipette 1.0 mL of a surrogate standard spiking solution into each sample. For the sample in each analytical batch selected for spiking, add 1.0 mL of a matrix spiking standard. For base/neutral acid analysis, the amount of the surrogates and matrix spiking compounds added to the sample should result in a final concentration of 100 ng/µL of each analyte in the extract to be analyzed (assuming a 1-µL injection). Extract with methylene chloride for 18–24 h. Next, adjust the pH of the aqueous phase to pH >11 using 10 N sodium hydroxide and extract it with methylene chloride again for 18–24 h. Dry the extract through a column containing anhydrous sodium sulfate and concentrate it to 1 mL using a K-D concentrator.

Soils, sediments, or sludges — Use 30 g of sample. Nonporous or wet samples (gummy or clay type) that do not have a free-flowing sandy texture must be mixed with anhydrous sodium sulfate until the sample is free flowing. Add 1 mL of surrogate standards to all samples, spikes, standards, and blanks. For the sample in each analytical batch selected for spiking, add 1.0 mL of a matrix spiking standard. For base/neutral acid analysis, the amount added of the surrogates and matrix spiking compounds should result in a final concentration of 100 ng/µL of each base/neutral analyte and 200 ng/µL of each acid analyte in the extract to be analyzed (assuming a 1-µL injection). Immediately add a 100-mL mixture of 1:1 methylene chloride:acetone and extract the sample ultrasonically for 3 min and then decant or filter the extracts. Repeat the extraction two or more times. Dry the extract using a column with anhydrous sodium sulfate and concentrate it to 1 mL in a K-D concentrator.

QUALITY CONTROL A methylene chloride solution containing 50 ng/µL of decafluorotriphenylphosphine (DFTPP) is

used for tuning the GC/MS system each 12-h shift. A system performance check also must be made during every 12-h shift. A standard containing 50 ng/μL each of 4,4′-DDT, pentachlorophenol, and benzidine is required to verify injection port inertness and GC column performance. A calibration standard at mid-concentration, containing each compound of interest, including all required surrogates, must be performed every 12 h during analysis. After the system performance check is met, calibration check compounds (CCCs) are used to check the validity of the initial calibration.

The internal standard responses and retention times in the calibration check standard must be evaluated immediately after or during data acquisition. If the retention time for any internal standard changes by more than 30 seconds from the last check calibration (12 h), the chromatographic system must be inspected for malfunctions and corrections must be made, as required. If the electron ionization current plot (EICP) area for any of the internal standards changes by a factor of two from the last daily calibration standard check, the mass spectrometer must be inspected for malfunctions and corrections must be made, as appropriate.

Demonstrate, through the analysis of a reagent water blank, that interferences from the analytical system, glassware, and reagents are under control. The blank samples should be carried through all stages of the sample preparation and measurement steps. For each analytical batch (up to 20 samples), a reagent blank, matrix spike, and matrix spike duplicate/duplicate must be analyzed (the frequency of the spikes may be different for different monitoring programs). The blank and spiked samples must be carried through all stages of the sample preparation and measurement steps. A QC reference sample concentrate containing each analyte at a concentration of 100 mg/L in methanol is required.

REFERENCE Test Methods for Evaluating Solid Waste (SW-846). U.S. EPA 1983, Method 8270B, Rev. 2, Nov. 1990. Office of Solid Waste, Washington, DC.

2-Methylbenzothioazole　　　　　　　**EPA Method 1625**
CAS #120-75-2

TITLE Semivolatile Organic Compounds by Isotope Dilution GC/MS

MATRIX The compounds may be determined in waters, soils, and municipal sludges by this method.

METHOD SUMMARY This method is used to determine 176 semivolatile toxic organic pollutants associated with the CWA (as amended 1987); the RCRA (as amended 1986); the CERCLA (as amended 1986); and other compounds amenable to extraction and analysis by capillary column gas chromatography-mass spectrometry (GC/MS).

Stable isotopically-labeled analogs of the compounds of interest are added to the sample. If the solids content is less than 1%, a 1-L sample is extracted at pH 12–13, then at pH <2 with methylene chloride using continuous extraction techniques.

If the solids content is 30% or less, the sample is diluted to 1% solids with reagent water, homogenized ultrasonically, and extracted at pH 12–13, then at pH <2 with methylene chloride using continuous extraction techniques. If the solids content is greater than 30%, the sample is extracted using ultrasonic techniques.

Each extract is dried over sodium sulfate, concentrated to a volume of 5 mL, cleaned up using GPC, if necessary, and concentrated. Extracts are concentrated to 1 mL if GPC is not performed, and to 0.5 mL if GPC is performed.

An internal standard is added to the extract, and a 1-mL aliquot of the extract is injected into the GC. The compounds are separated by GC and detected by a MS. The labeled compounds serve to correct the variability of the analytical technique.

INTERFERENCES Solvents, reagents, glassware, and other sample processing hardware may yield artifacts and/or elevated baselines causing misinterpretation of chromatograms and spectra. Materials used in the analysis must be demonstrated to be free from interferences under the conditions of analysis by running method blanks initially and with each sample lot (sample started through the extraction process on a given 8-h shift, to a maximum of 20). Specific selection of reagents and purification of solvents by distillation in all glass systems may be required. Glassware and, where possible, reagents are cleaned by solvent rinse and baking at 450°C for 1-h minimum. Interferences coextracted from samples will vary considerably from source to source, depending on the diversity of the site being sampled.

INSTRUMENTATION Major instrumentation includes a GC with a splitless or on-column injection port for capillary column, a MS with 70 eV electron impact ionization, and a data system to collect and record MS data, and process it. A K-D apparatus is used to concentrate extracts.

GC Column: 30 m × 0.25 mm I.D. 5% phenyl, 94% methyl, 1% vinyl silicone bonded phased fused silica capillary column.

PRECISION & ACCURACY The detection limits of the method are usually dependent on the level of interferences rather than instrumental limitations. The limits typify the minimum quantities that can be detected with no interferences present.

The minimum level (in μg/mL) was not listed. This is defined as a minimum level at which the analytical system shall give recognizable mass spectra (background corrected) and acceptable calibration points.

The MDL (in μg/kg) in low solids was not listed and in high solids was not listed; these were determined in digested sludge (low solids) and in filter cake or compost (high solids).

The labeled and native compound initial precision as standard deviation (in μg/L) was not listed.
The labeled and native compound initial accuracy as average recovery (in μg/L) was not listed.

SAMPLE COLLECTION, PRESERVATION & HANDLING
Collect samples in glass containers. Aqueous samples which flow freely are collected in refrigerated bottles using automatic

sampling equipment. Solid samples are collected as grab samples using widemouth jars. Maintain samples at 0 to 4°C from the time of collection until extraction. If residual chlorine is present in aqueous samples, add 80 mg sodium thiosulfate/L of water. Begin sample extraction within 7 days of collection, and analyze all extracts within 40 days of extraction.

SAMPLE PREPARATION Samples containing 1% solids or less are extracted directly using continuous liquid-liquid extraction techniques. Samples containing 1 to 30% solids are diluted to the 1% level with reagent water and extracted using continuous liquid-liquid extraction techniques. Samples containing greater than 30% solids are extracted using ultrasonic techniques.

Base/neutral extraction — Adjust the pH of the waters in the extractors to 12–13 with 6 N NaOH. Extract with methylene chloride for 24–48 h.

Acid extraction — Adjust the pH of the waters in the extractors to 2 or less using 6 N sulfuric acid. Extract with methylene chloride for 24–48 h.

Ultrasonic extraction of high solids samples — Add anhydrous sodium sulfate to the sample and QC aliquot(s). Add acetone:methylene chloride (1:1) to the sample and mix thoroughly

Concentrate extracts using a K-D apparatus.

QUALITY CONTROL The analyst is permitted to modify this method to improve separations or lower the costs of measurements, provided all performance specifications are met. Analyses of blanks are required to demonstrate freedom from contamination. When results of spikes indicate atypical method performance for samples, the samples are diluted to bring method performance within acceptable limits.

For low solids (aqueous samples), extract, concentrate, and analyze two sets of four 1-L aliquots (8 aliquots total) of the precision and recovery standard. For high solids samples, two sets of four 30-g aliquots of the high solids reference matrix are used.

Spike all samples with labeled compounds to assess method performance. Compute percent recovery of the labeled compounds using the internal standard method. Compare the labeled compound recovery for each compound with the corresponding labeled compound recovery.

Reagent water and high solids reference matrix blanks are analyzed to demonstrate freedom from contamination. Extract and concentrate a 1-L reagent water blank or a high solids reference matrix blank with each sample's lot (samples started through the extraction process on the same 8-h shift, to a maximum of 20 samples).

Field replicates may be collected to determine the precision of the sampling technique, and spiked samples may be required to determine the accuracy of the analysis when the internal standard method is used.

REFERENCE Semivolatile Organic Compounds by Isotope Dilution GC/MS. Office of Water Regulation and Standards, U.S. EPA Industrial Technology Division, Washington, DC, EPA Method 1625, Rev. C, June 1989 (contact W.A. Telliard, U.S. EPA, Office of Water Regulations and Standards, 401 M St., SW, Washington, DC, 20460. Phone: 202-382-7131).

3-Methylcholanthrene   EPA Method 1625
CAS #56-49-5

TITLE Semivolatile Organic Compounds by Isotope Dilution GC/MS

MATRIX The compounds may be determined in waters, soils, and municipal sludges by this method.

METHOD SUMMARY This method is used to determine 176 semivolatile toxic organic pollutants associated with the CWA (as amended 1987); the RCRA (as amended 1986); the CERCLA (as amended 1986); and other compounds amenable to extraction and analysis by capillary column gas chromatography-mass spectrometry (GC/MS).

Stable isotopically-labeled analogs of the compounds of interest are added to the sample. If the solids content is less than 1%, a 1-L sample is extracted at pH 12–13, then at pH <2 with methylene chloride using continuous extraction techniques.

If the solids content is 30% or less, the sample is diluted to 1% solids with reagent water, homogenized ultrasonically, and extracted at pH 12–13, then at pH <2 with methylene chloride using continuous extraction techniques. If the solids content is greater than 30%, the sample is extracted using ultrasonic techniques.

Each extract is dried over sodium sulfate, concentrated to a volume of 5 mL, cleaned up using GPC, if necessary, and concentrated. Extracts are concentrated to 1 mL if GPC is not performed, and to 0.5 mL if GPC is performed.

An internal standard is added to the extract, and a 1-mL aliquot of the extract is injected into the GC. The compounds are separated by GC and detected by a MS. The labeled compounds serve to correct the variability of the analytical technique.

INTERFERENCES Solvents, reagents, glassware, and other sample processing hardware may yield artifacts and/or elevated baselines causing misinterpretation of chromatograms and spectra. Materials used in the analysis must be demonstrated to be free from interferences under the conditions of analysis by running method blanks initially and with each sample lot (sample started through the extraction process on a given 8-h shift, to a maximum of 20). Specific selection of reagents and purification of solvents by distillation in all glass systems may be required. Glassware and, where possible, reagents are cleaned by solvent rinse and baking at 450°C for 1-h minimum. Interferences coextracted from samples will vary considerably from source to source, depending on the diversity of the site being sampled.

INSTRUMENTATION Major instrumentation includes a GC with a splitless or on-column injection port for capillary column, a MS with 70 eV electron impact ionization, and a data system to collect and record MS data, and process it. A K-D apparatus is used to concentrate extracts.

GC Column: 30 m × 0.25 mm I.D. 5% phenyl, 94% methyl, 1% vinyl silicone bonded phased fused silica capillary column.

PRECISION & ACCURACY The detection limits of the method are usually dependent on the level of interferences rather than instrumental limitations. The limits typify the minimum quantities that can be detected with no interferences present.

The minimum level (in µg/mL) was not listed. This is defined as a minimum level at which the analytical system shall give recognizable mass spectra (background corrected) and acceptable calibration points.

The MDL (in µg/kg) in low solids was not listed and in high solids was not listed; these were determined in digested sludge (low solids) and in filter cake or compost (high solids).

The labeled and native compound initial precision as standard deviation (in µg/L) was not listed.
The labeled and native compound initial accuracy as average recovery (in µg/L) was not listed.

SAMPLE COLLECTION, PRESERVATION & HANDLING Collect samples in glass containers. Aqueous samples which flow freely are collected in refrigerated bottles using automatic sampling equipment. Solid samples are collected as grab samples using widemouth jars. Maintain samples at 0 to 4°C from the time of collection until extraction. If residual chlorine is present in aqueous samples, add 80 mg sodium thiosulfate/L of water. Begin sample extraction within 7 days of collection, and analyze all extracts within 40 days of extraction.

SAMPLE PREPARATION Samples containing 1% solids or less are extracted directly using continuous liquid-liquid extraction techniques. Samples containing 1 to 30% solids are diluted to the 1% level with reagent water and extracted using continuous liquid-liquid extraction techniques. Samples containing greater than 30% solids are extracted using ultrasonic techniques.

- Base/neutral extraction — Adjust the pH of the waters in the extractors to 12–13 with 6 *N* NaOH. Extract with methylene chloride for 24–48 h.
- Acid extraction — Adjust the pH of the waters in the extractors to 2 or less using 6 *N* sulfuric acid. Extract with methylene chloride for 24–48 h.
- Ultrasonic extraction of high solids samples — Add anhydrous sodium sulfate to the sample and QC aliquot(s). Add acetone:methylene chloride (1:1) to the sample and mix thoroughly

Concentrate extracts using a K-D apparatus.

QUALITY CONTROL The analyst is permitted to modify this method to improve separations or lower the costs of measurements, provided all performance specifications are met. Analyses of blanks are required to demonstrate freedom from contamination. When results of spikes indicate atypical method performance for samples, the samples are diluted to bring method performance within acceptable limits.

For low solids (aqueous samples), extract, concentrate, and analyze two sets of four 1-L aliquots (8 aliquots total) of the precision and recovery standard. For high solids samples, two sets of four 30-g aliquots of the high solids reference matrix are used.

Spike all samples with labeled compounds to assess method performance. Compute percent recovery of the labeled compounds using the internal standard method. Compare the labeled compound recovery for each compound with the corresponding labeled compound recovery.

Reagent water and high solids reference matrix blanks are analyzed to demonstrate freedom from contamination. Extract and concentrate a 1-L reagent water blank or a high solids reference matrix blank with each sample's lot (samples started through the extraction process on the same 8-h shift, to a maximum of 20 samples).

Field replicates may be collected to determine the precision of the sampling technique, and spiked samples may be required to determine the accuracy of the analysis when the internal standard method is used.

REFERENCE Semivolatile Organic Compounds by Isotope Dilution GC/MS. Office of Water Regulation and Standards, U.S. EPA Industrial Technology Division, Washington, DC, EPA Method 1625, Rev. C, June 1989 (contact W.A. Telliard, U.S. EPA, Office of Water Regulations and Standards, 401 M St., SW, Washington, DC, 20460. Phone: 202-382-7131).

3-Methylcholanthrene   **EPA Method 8270**
CAS #56-49-5

TITLE Semivolatile Organic Compounds by GC/MS

MATRIX This method is used to determine the concentration of semivolatile organic compounds in extracts prepared from all types of solid waste matrices, soils, and groundwater. Although surface waters are not specifically mentioned, this method should be applicable to water samples from rivers, lakes, etc.

METHOD SUMMARY This method covers 259 semivolatile organic compounds. In very limited applications direct injection of the sample into the GC/MS system may be appropriate, but this results in very high detection limits (approximately 10,000 µg/L). Typically, a 1-L liquid sample, containing surrogate, and matrix spiking standards, is extracted in a continuous extractor first under acid conditions and then under basic conditions. Typically 30 g of a solid sample, containing surrogate, and matrix spiking standards, is extracted ultrasonically. After concentrating the extract to 1 mL it is spiked with 10 µL of an internal standard solution just prior to analysis by GC/MS. The volume injected should contain about 100 ng of base/neutral and 200 ng of acid surrogates (for a 1-µL injection). Analysis is performed by GC/MS using a capillary GC column.

INTERFERENCES Raw GC/MS data from all blanks, samples, and spikes must be evaluated for interferences. Contamination by carryover can occur whenever high-concentration and low-concentration samples are sequentially analyzed. To reduce carryover, the sample syringe must be rinsed out between samples with solvent. Whenever an unusually concentrated sample is encountered, it should be followed by the analysis of blank solvent to check for cross-contamination.

INSTRUMENTATION A GC/MS and a data system are required. The GC column used is a 30 m × 0.25 mm I.D. (or 0.32 mm I.D.) 1um film thickness silicone-coated fused silica capillary column. A continuous liquid-liquid extractor equipped with Teflon® or glass connection joints and stopcocks requiring no lubrication, a K-D concentrating apparatus, water bath, and an ultrasonic disrupter with a minimum power of 300 W and with pulsing capability are also required.

PRECISION & ACCURACY The estimated quantitation limit (EQL) of Method 8270B for determining an individual compound is approximately 1 mg/kg (wet weight) for soil or sediment samples, 1–200 mg/kg for wastes (dependent on matrix and method of preparation), and 10 µg/L for groundwater samples. EQLs will be proportionately higher for sample extracts that require dilution to avoid saturation of the detector.

The EQL(b) for groundwater in µg/L is 10.
The EQL (a, b) for low concentrations in soil and sediment in µg/kg is not determined.
Accuracy as µg/L is not listed.
Overall precision in µg/L is not listed.

(a) *EQLs listed for soil/sediment are based on wet weight. Normally data is reported in a dry-weight basis; therefore, EQLs will be higher based on the % dry weight of each sample. This calculation is based on a 30 g sample and gel permeation chromatography cleanup.*
(b) *Sample EQLs are highly matrix-dependent. The EQLs are provided for guidance and may not always be achievable.*
C = *True value for concentration, in µg/L.*
X = *Average recovery found for measurements of samples containing a concentration of C, in µg/L.*

ESTIMATED QUANTITATION LIMIT

Other Matrices	Factor (a)
High-concentration soil and sludges by sonicator	7.5
Non-water miscible waste	75

(a) EQL for other matrices = [EQL for low soil/sediment] × [Factor]. This estimated EQL is similar to an EPA "Practical Quantitation Limit."

SAMPLING METHOD
Liquid samples — Use a 1 or 2½ gallon amber glass bottle with a screw-top Teflon®-lined cover that has been prewashed with detergent and rinsed with distilled water and methanol (or isopropanol).

Soils, sediments, or sludges — Use an 8-oz. widemouth glass with a screw-top Teflon®-lined cover that has been prewashed with detergent and rinsed with distilled water and methanol (or isopropanol).

SAMPLE PRESERVATION
Liquid samples — If residual chlorine is present, add 3 mL of 10% sodium thiosulfate per gallon, cool to 4°C and store in a solvent-free refrigerator until analysis; if chlorine is not present, then eliminate the sodium thiosulfate addition.

Soils, sediments, or sludges — Cool samples to 4°C and store in a solvent-free refrigerator.

MHT Liquid samples must be extracted within 7 days and the extracts analyzed within 40 days. Soils, sediments, or sludges may be stored for a maximum of 14 days and the extracts analyzed within 40 days.

SAMPLE PREPARATION
Liquid samples — Transfer 1 L quantitatively to a continuous extractor. If high concentrations are anticipated, a smaller volume may be used and then diluted with organic-free reagent water to 1 L. Adjust pH, if necessary, to pH <2 using 1:1 (V/V) sulfuric acid. Pipette 1.0 mL of a surrogate standard spiking solution into each sample. For the sample in each analytical batch selected for spiking, add 1.0 mL of a matrix spiking standard. For base/neutral acid analysis, the amount of the surrogates and matrix spiking compounds added to the sample should result in a final concentration of 100 ng/µL of each analyte in the extract to be analyzed (assuming a 1-µL injection). Extract with methylene chloride for 18–24 h. Next, adjust the pH of the aqueous phase to pH >11 using 10 N sodium hydroxide and extract it with methylene chloride again for 18–24 h. Dry the extract through a column containing anhydrous sodium sulfate and concentrate it to 1 mL using a K-D concentrator.

Soils, sediments, or sludges — Use 30 g of sample. Nonporous or wet samples (gummy or clay type) that do not have a free-flowing sandy texture must be mixed with anhydrous sodium sulfate until the sample is free flowing. Add 1 mL of surrogate standards to all samples, spikes, standards, and blanks. For the sample in each analytical batch selected for spiking, add 1.0 mL of a matrix spiking standard. For base/neutral acid analysis, the amount added of the surrogates and matrix spiking compounds should result in a final concentration of 100 ng/µL of each base/neutral analyte and 200 ng/µL of each acid analyte in the extract to be analyzed (assuming a 1-µL injection). Immediately add a 100-mL mixture of 1:1 methylene chloride:acetone and extract the sample ultrasonically for 3 min and then decant or filter the extracts. Repeat the extraction two or more times. Dry the extract using a column with anhydrous sodium sulfate and concentrate it to 1 mL in a K-D concentrator.

QUALITY CONTROL A methylene chloride solution containing 50 ng/µL of decafluorotriphenylphosphine (DFTPP) is used for tuning the GC/MS system each 12-h shift. A system performance check also must be made during every 12-h shift. A standard containing 50 ng/µL each of 4,4'-DDT, pentachlorophenol, and benzidine is required to verify injection port inertness and GC column performance. A calibration standard at mid-concentration, containing each compound of interest, including all required surrogates, must be performed every 12 h during analysis. After the system performance check is met, calibration check compounds (CCCs) are used to check the validity of the initial calibration.

The internal standard responses and retention times in the calibration check standard must be evaluated immediately after or during data acquisition. If the retention time for any internal standard changes by more than 30 seconds from the last check calibration (12 h), the chromatographic system must be

inspected for malfunctions and corrections must be made, as required. If the electron ionization current plot (EICP) area for any of the internal standards changes by a factor of two from the last daily calibration standard check, the mass spectrometer must be inspected for malfunctions and corrections must be made, as appropriate.

Demonstrate, through the analysis of a reagent water blank, that interferences from the analytical system, glassware, and reagents are under control. The blank samples should be carried through all stages of the sample preparation and measurement steps. For each analytical batch (up to 20 samples), a reagent blank, matrix spike, and matrix spike duplicate/duplicate must be analyzed (the frequency of the spikes may be different for different monitoring programs). The blank and spiked samples must be carried through all stages of the sample preparation and measurement steps. A QC reference sample concentrate containing each analyte at a concentration of 100 mg/L in methanol is required.

REFERENCE Test Methods for Evaluating Solid Waste (SW-846). U.S. EPA 1983, Method 8270B, Rev. 2, Nov. 1990. Office of Solid Waste, Washington, DC.

Methylene chloride   **EPA Method 1624**
CAS #75-09-2

TITLE Volatile Organic Compounds by Isotope Dilution GC/MS

MATRIX Compounds may be determined in waters, soils, and municipal sludges by this method.

METHOD SUMMARY This method is used to determine 58 volatile toxic organic pollutants associated with the CWA (as amended 1987); the RCRA (as amended 1986); the CERCLA (as amended 1986); and other compounds amenable to purge-and-trap gas chromatography-mass spectrometry (GC/MS).

If the solids content is less than 1%, stable isotopically-labeled analogs of the compounds of interest are added to a 5-mL sample and the sample is purged with an inert gas at 20–25°C in a chamber designed for soil or water samples. If the solids content is greater than 1%, 5 mL of reagent water and the labeled compounds are added to a 5-g aliquot of sample and the mixture is purged at 40°C. Compounds that will not purge at 20–25°C or at 40°C are purged at 78–85°C. In the purging process, the volatile compounds are transferred from the aqueous phase into the gaseous phase where they are passed into a sorbent column, and trapped. After purging is completed, the trap is backflushed and heated rapidly to desorb the compounds into a GC. The compounds are separated by the GC and detected by a MS. The labeled compounds serve to correct the variability of the analytical technique.

INTERFERENCES Impurities in the purge gas, organic compounds outgassing from the plumbing upstream of the trap, and solvent vapors in the lab account for most problems. Samples can be contaminated by diffusion of volatile organic compounds (particularly methylene chloride) through the bottle seal during shipment and storage. Contamination by carryover can occur when high-level and low-level samples are analyzed sequentially. When an unusually concentrated sample is encountered, follow it by analysis of a reagent water blank to check for carryover.

INSTRUMENTATION Major equipment includes a GC with linear temperature programming and a glass jet separator as the MS interface, a MS with 70 eV electron impact ionization, and a data system to collect and record response factors.

Column: 2.8 m × 2 mm I.D. glass, packed with 1% SP-1000 on Carbopak B, 60/80 mesh, or equivalent.

PRECISION & ACCURACY The detection limits of the method are usually dependent on the level of interferences rather than instrumental limitations. The method detection limits were determined in digested sludge (low solids) and in filter cake or compost (high solids).

The MDL (in µg/kg) for low solids is 566 and for high solids is 280.

Background levels of this compound were present in the sludge with low solids, resulting in a higher than expected MDL. Background levels of this compound were present in the sludge with high solids, resulting in a higher than expected MDL.

Labeled and native compound precision (in µg/L) as standard deviation was 9.7.
Labeled and native compound accuracy (in µg/L) as average recovery was detected to 50.

Acceptance criteria are at 20 µg/L for this compound.

SAMPLE COLLECTION, PRESERVATION & HANDLING Grab samples are collected in glass containers having a total volume greater than 20 mL. Fill and seal each bottle so that no air bubbles are entrapped. Samples are maintained at 0 to 4°C from the time of collection until analysis. If an aqueous sample contains residual chlorine, add sodium thiosulfate preservative (10 mg/40 mL) to the empty sample bottles just prior to shipment to the sample site. All samples must be analyzed within 14 days of collection.

SAMPLE PREPARATION Samples containing less than 1% solids are analyzed directly as aqueous samples. Samples containing 1% solids or greater are analyzed as solid samples utilizing one of two methods, depending on the levels of pollutants, in the sample. Samples containing 1% solids or greater, and low to moderate levels of pollutants are analyzed by purging a known weight of sample added to 5 mL of reagent water. Samples containing 1% solids or greater, and high levels of pollutants, are extracted with methanol, and an aliquot of the methanol extract is added to reagent water and purged.

QUALITY CONTROL A field blank prepared from reagent water and carried through the sampling and handling protocol may serve as a check on contamination from shipment and storage.

The analyst is permitted to modify this method to improve separations or lower the costs of measurements, provided all performance specifications are met. Analyses of blanks are required. When results of spikes indicate atypical method performance for samples, the samples are diluted to bring method performance within acceptable limits. Analyze two sets of four 5-mL aliquots (8 aliquots total) of the aqueous performance standard. Spike all samples with labeled compounds to assess method performance on the sample matrix. Compute the percent recovery of the labeled compounds using the internal standard method. Compare the percent recovery for each compound with the corresponding labeled compound recovery. Reagent water blanks are analyzed to demonstrate freedom from carryover contamination. Field replicates may be collected to determine the precision of the sampling technique, and spiked samples may be required to determine the accuracy of the analysis when the internal method is used.

REFERENCE Volatile Organic Compounds by Isotope Dilution GC/MS. Office of Water Regulation and Standards, U.S. EPA Industrial Technology Division, Washington, DC, EPA Method 1624, Rev. C, June 1989 (contact W.A. Telliard, U.S. EPA, Office of Water Regulations and Standards, 401 M St., SW, Washington, DC, 20460. Phone: 202-382-7131).

Methylene chloride	EPA Method 502
CAS #75-09-2	

TITLE Volatile Organic Compounds in Water By Purge and Trap Capillary Column Gas Chromatography with Photoionization and Electrolytic Conductivity Detectors in Series. U.S. EPA Method 502.2, Rev. 2.0, 1989.

MATRIX Drinking water and raw source water. The latter should include most surface water and groundwater sources.

METHOD SUMMARY This method covers 60 volatile organic compounds that contain halogen atoms and/or that are aromatic. An inert gas (zero grade nitrogen or helium) is bubbled through a 25-mL or a 5-mL water sample (depending on the expected concentration of the analytes). Purged sample components are trapped in a tube of sorbent materials. When purging is complete, the sorbent tube is heated and backflushed with helium to desorb the trapped sample onto a capillary GC column. The column is temperature programmed to separate the method analytes which are then detected with a photoionization detector (PID) and a Hall electrolytic conductivity (HECD) placed in series. The PID is selective for aromatic compounds and the HECD is selective for halogenated compounds.

INTERFERENCES Impurities in the purge gas and from organic compounds outgassing from the plumbing ahead of the trap account for many contamination problems. Interferences purged or coextracted from the samples will vary considerably from source to source, depending upon the particular sample or extract being tested. Cross-contamination can occur whenever high-level and low-level samples are analyzed sequentially. Samples also can be contaminated by diffusion of volatile organics (particularly methylene chloride and fluorocarbons) through the septum seal into the sample during shipment and storage. The lab where volatile analysis is performed and also the refrigerated storage area should be completely free of solvents.

INSTRUMENTATION A GC containing a series configuration of a high temperature photoionization detector (PID) equipped with 10.0 eV (nominal) lamp and Hall electrolytic conductivity detector (HECD) is required. Also required is an all-glass 5-mL purging device, a sorbent trap, and a thermal desorption apparatus which is connected to the GC system.

Column 1: VOCOL glass wide-bore capillary column.
Column 2: RTX–502.2 mega-bore capillary column.
Column 3: DB-62 mega-bore capillary column.

PRECISION & ACCURACY Method detection limits are dependent upon the characteristics of the gas chromatographic system used. Analytes that are not separated chromatographically cannot be individually identified and used in the same calibration mixture or water samples unless an alternative technique for identification and quantification, such as mass spectrometry, is used.

Electrolytic conductivity detetor (c) range in µg/L (a) was 0.02–200.
Electrolytic conductivity detetor (c) MDL in µg/L (b) was 0.02.
Electrolytic conductivity detetor (c) accuracy as % recovery was 97.
Electrolytic conductivity detetor (c) precision as % RSD was 2.9.
Photoionization detector (d) range in µg/L (a) was 0.02–200.
Photoionization detector (d) MDL in µg/L (b) was not listed.
Photoionization detector (d) accuracy as % recovery was not listed.
Photoionization detector (d) precision as % RSD was not listed.

(a) The applicable concentration range of this method is compound, instrument, and matrix-dependent. It is listed as being approximately 0.02 to 200 µg/L but no specific information is provided so caution should be observed.

(b) The method detection limits reports with this method are compound, instrument, and matrix-dependent. The values reported were calculated using reagent water fortified with the corresponding compounds at 10 µg/L and a GC-equipped with a 60 m × 0.75 mm VOLCOL wide bore capillary column with 1.5 µm film thickness and using helium carrier gas.

(c) Recoveries and relative standard deviations were determined from seven samples of reagent water fortified with 10 µg/L of each compound. 2-Bromo-1-chloropropane was used as the internal standard for calculating average recoveries.

(d) Recoveries and relative standard deviations were determined from seven samples of reagent water fortified with 10 µg/L of each compound. Fluorobenzene was used as the internal standard for calculating average recoveries.

SAMPLING METHOD Collect samples using a 40- to 120-mL screw-cap vial (prewashed with detergent, rinsed with distilled water and oven dried at 105°C) with a Teflon®-faced silicone septum. Collect bubble-free samples and place the septum with the Teflon® side down on the water.

SAMPLE PRESERVATION If residual chlorine is present in the water add about 25 mg of ascorbic acid to each vial before samples are collected to remove the chlorine. Add hydrochloric

acid to reduce pH to <2, immediately cool samples to 4°C, and store them in a solvent-free refrigerator at 4°C until analysis.

MHT The maximum holding time for samples is 14 days from the time they were collected.

SAMPLE PREPARATION Remove the plungers from two 5-mL syringes and attach a closed syringe valve to each. Warm the sample to room temperature, open the sample bottle, and carefully pour the sample into one of the syringe barrels to just short of overflowing. Replace the syringe plunger, invert the syringe, and compress the sample. Open the syringe valve and vent any residual air while adjusting the sample volume to 5.0 mL. Add 10 µL of the internal calibration standard to the sample through the syringe valve. Close the valve. Fill the second syringe in an identical manner from the same sample bottle. Reserve this second syringe for a reanalysis if necessary.

QUALITY CONTROL As an initial demonstration of lab accuracy and precision, analyze 4 to 7 replicates of a lab fortified blank containing analyte at 0.1–5 µg/L. Collect all samples in duplicate. Surrogate analytes (similar to those of the analytes of interest), whose concentration is known in every sample, are measured using the same internal standard calibration procedure. Duplicate field reagent water blanks (trip blanks) must be analyzed with each set of samples, lab reagent blanks (method blanks) must be analyzed with each batch of samples processed as a group within a work shift. Also, a single lab-fortified blank that contains each of the analytes of interest should be analyzed with each batch of samples processed as a group within a work shift. A 3- to 5-point calibration curve is needed depending on the calibration range factor required.

EPA CONTACT & HOTLINE For technical questions contact Dr. Baldev Bathija, U.S. EPA, Office of Ground Water and Drinking Water (WH-550D), 401 M St. SW, Washington, DC 20460. Tel. (202) 260-3040. For further information the EPA Safe Drinking Water Hotline may be called at: (800) 426-4791.

REFERENCE Methods for the Determination of Organic Compounds in Drinking Water, EPA/600/4-88/039 (revised July 1991; Final Rule for determination of compliance with the MCL for Total Trihalomethanes under 141.30, in 40 CFR Part 141, Vol. 58, No. 147, Fed. Reg., Tuesday Aug. 3, 1993). U.S. EPA Environmental Monitoring Systems Laboratory, Cincinnati, OH, 45268, U.S.A. Available from the National Technical Information Service (NTIS), 5285 Port Royal Road, Springfield, VA 22161; Tel. 800-553-6847. NTIS Order Number is PB91-231480.

Methylene chloride   **EPA Method 524**
CAS #75-09-2

TITLE Measurement of Purgeable Organic Compounds in Water by Capillary Column GC/MS.

MATRIX Drinking water and raw source water; the latter should include most surface water and groundwater sources.

METHOD SUMMARY Method 524.2 covers 60 volatile organic compounds. An inert gas (zero grade nitrogen or helium) is bubbled through a 25-mL or a 5-mL water sample (depending on the expected concentration of the analytes). Purged sample components are trapped in a tube of sorbent materials. When purging is complete, the sorbent tube is heated and backflushed with helium to desorb the trapped sample onto a capillary GC column.

INTERFERENCES Impurities in the purge gas and from organic compounds outgassing from the plumbing ahead of the trap account for many contamination problems. Interferences purged or coextracted from the samples will vary considerably from source to source, depending upon the particular sample or extract being tested. Cross-contamination can occur whenever high-level and low-level samples are analyzed sequentially. Samples also can be contaminated by diffusion of volatile organics (particularly methylene chloride and fluorocarbons) through the septum seal into the sample during shipment and storage.

INSTRUMENTATION A GC/MS with a data system equipped with one of the following capillary GC columns:

Column 1: VOCOL glass wide bore capillary column.
Column 2: DB-624 fused silica capillary column.
Column 3: DB-5 fused silica capillary column.

Also required is an all-glass 25 mL or 5-mL purging device, a sorbent trap, and a thermal desorption apparatus which is connected to the GC/MS system.

PRECISION & ACCURACY Method detection limits are compound- and instrument-dependent, and may vary from approximately 0.02–0.35 µg/L. Note in the table below that the "true" concentration range used for accuracy and precision measurements was quite narrow. However, the applicable concentration range of this method is primarily column dependent and is approximately 0.02 to 200 µg/L for the wide-bore thick-film columns. Narrow-bore thin-film columns may have a capacity which limits the range to about 0.02 to 20 µg/L. Analytes that are inefficiently purged from water will not be detected when present at low concentrations, but they can be measured with acceptable accuracy and precision when present in sufficient amounts.

Analytes that are not separated chromatographically, but which have different mass spectra and non-interfering quantification ions, can be identified and measured in the same calibration mixture or water sample. Analytes which have very similar mass spectra cannot be individually identified and measured in the same calibration mixture or water samples unless they have different retention times. Co-eluting compounds with very similar mass spectra, typically many structural isomers, must be reported as an isomeric group or pair.

The range (in µg/L) was 0.1–10.
The Method Detection Limig (in µg/L) was 0.03.
The accuracy (as % recovery) was 95.
The precision (in %) was 5.3.

Note: Data were obtained from 16–31 determinations using a wide-bore capillary column and a jet separator interfaced to a quadrupole mass spectrometer. All analytes were in a reagent water matrix.

SAMPLING METHOD Collect samples using a 40- to 120-mL screw-cap vial (prewashed with detergent, rinsed with distilled water and oven dried at 105°C) with a Teflon®-faced silicone septum. Collect bubble-free samples and place the septum with the Teflon® side down on the water.

SAMPLE PRESERVATION If residual chlorine is present in the water add about 25 mg of ascorbic acid to each vial before samples are collected to remove the chlorine. Add hydrochloric acid to reduce pH to <2, and immediately cool samples to 4°C, and store them in a solvent-free refrigerator at 4°C until analysis.

MHT The maximum holding time for samples is 14 days from the time they were collected.

SAMPLE PREPARATION Remove the plungers from two 25-mL (or 5-mL depending on sample size) syringes and attach a closed syringe valve to each. Warm the sample to room temperature, open the sample bottle, and carefully pour the sample into one of the syringe barrels to just short of overflowing. Replace the syringe plunger, invert the syringe, and compress the sample. Open the syringe valve and vent any residual air while adjusting the sample volume to 25.0 mL (or 5 mL). For samples and blanks, add 5 µL of the fortification solution containing the internal standard and the surrogates to the sample through the syringe valve. For calibration standards and lab fortified blanks, add 5 µL of the fortification solution containing the internal standard only. Close the valve. Fill the second syringe in an identical manner from the same sample bottle. Reserve this second syringe for a reanalysis if necessary.

QUALITY CONTROL As an initial demonstration of lab accuracy and precision, analyze 4 to 7 replicates of a lab fortified blank containing analyte at 0.2–5 µg/L. Collect all samples in duplicate. Surrogate analytes (similar to those of the analytes of interest), whose concentration is known in every sample, are measured using the same internal standard calibration procedure. Duplicate field reagent water blanks (trip blanks) must be analyzed with each set of samples, lab reagent blanks (method blanks) must be analyzed with each batch of samples processed as a group within a work shift. Also, a single lab-fortified blank that contains each of the analytes of interest should be analyzed with each batch of samples processed as a group within a work shift. A 3- to 5-point calibration curve is needed depending on the calibration range factor required.

EPA CONTACT & HOTLINE For technical questions contact Dr. Baldev Bathija, U.S. EPA, Office of Ground Water and Drinking Water (WH-550D), 401 M St. SW, Washington, DC 20460. Tel. (202) 260-3040. For further information the EPA Safe Drinking Water Hotline may be called at: (800) 426-4791.

REFERENCE Methods for the Determination of Organic Compounds in Drinking Water, EPA/600/4-88/039 (revised July 1991; Final Rule for determination of compliance with the MCL for Total Trihalomethanes under 141.30, in 40 CFR Part 141, Vol. 58, No. 147, Fed. Reg., Tuesday Aug. 3, 1993). U.S. EPA Environmental Monitoring Systems Laboratory, Cincinnati, OH, 45268, U.S.A. Available from the National Technical Information Service (NTIS), 5285 Port Royal Road, Springfield, VA 22161; Tel. 800-553-6847. NTIS Order Number is PB91-231480.

Methylene chloride   **EPA Method 8021**
CAS #75-09-2

TITLE Halogenated Volatile by Gas Chromatography Using Photoionization and Electrolytic Conductivity Detectors in Series: Capillary Column Technique

MATRIX This method is applicable to nearly all types of samples, regardless of water content, including groundwater, aqueous sludges, caustic liquors, acid liquors, waste solvents, oily wastes, mousses, tars, fibrous wastes, polymeric emulsions, filter cakes, spent carbons, spent catalysts, soils, and sediments.

METHOD SUMMARY This method is used to determine 60 volatile organic compounds in a variety of solid waste matrices. It provides GC conditions for the detection of halogenated and aromatic volatile organic compounds. Samples can be analyzed using direct injection or purge-and-trap (EPA Method 5030). Groundwater samples must be analyzed using EPA Method 5030 (where applicable). A temperature program is used with the GC. Detection is achieved by a photoionization detector (PID) and a Hall electrolytic conductivity detector (HECD) in series.

INTERFERENCES Samples can be contaminated by diffusion of volatile organics (particularly chlorofluorocarbons and methylene chloride) through the sample container septum during shipment and storage.

INSTRUMENTATION A GC-equipped with variable-constant differential flow controllers, subambient oven controller, PID and HECD detectors connected with a short piece of uncoated capillary tubing and a data system.

Column: 60 m × 0.75 mm I.D. VOCOL wide-bore capillary column with 1.5 µm film thickness.

PRECISION & ACCURACY MDLs are compound-dependent and vary with purging efficiency and concentration. The applicable concentration range of this method is compound- and instrument-dependent but is approximately 0.1 to 200 µg/L. Analytes that are inefficiently purged from water will not be detected when present at low concentrations, but they can be measured with acceptable accuracy and precision when present in sufficient amounts. The estimated quantitation limit (EQL) for an individual compound is approximately 1 µg/kg (wet weight) for soil/sediment samples, 100 µg/kg (wet weight) for wastes, and 1 µg/L for groundwater. EQLs will be proportionately higher for sample extracts and samples that require dilution or reduced sample size to avoid saturation of the detector.

MULTIPLICATION FACTORS FOR OTHER MATRICES (a)

Matrix	Factor (b)
Groundwater	10
Low-concentration soil	10
Water miscible liquid waste	500
High-concentration soil and sludge	1250
Non-water miscible waste	1250

(a) Sample EQLs are highly matrix-dependent. The EQLs listed herein are provided for guidance and may not always be achievable.
(b) EQL = [Method detection limit] × [Factor]. For non-aqueous samples, the factor is on a wet-weight basis.

SINGLE LABORATORY ACCURACY & PRECISION DATA FOR VOCs IN WATER

This method was tested in a single lab using water spiked at 10 µg/L and the following data was reported:

Recoveries and standard deviations were determined from seven samples and spiked at 10 µg/L of each analyte. Recoveries were determined by the internal standard method. Internal standards were: Fluorobenzene for PID and 2-Bromo-1-chloropropane for HECD.

The average recovery (in percent) for the PID was none (no response for this detector).
The standard deviation of the recovery for the PID was none (no response for this detector).
The MDL (in µg/mL) for the PID was none (no response for this detector).
The average recovery (in percent) for the HECD was 97.
The standard deviation of the recovery for the HECD was 2.8.
The MDL (in µg/mL) for the HECD was 0.02.

SAMPLE COLLECTION, PRESERVATION & HANDLING

Volatile Organics — Standard 40-mL glass screw-cap VOA vials with Teflon®-faced silicone septum may be used for both liquid and solid matrices. When collecting samples, liquids and solids should be introduced into the vials gently to reduce agitation which might drive off volatile compounds. If there are any air bubbles present the sample must be retaken. Tap slightly as they are filled to try and eliminate as much free air space as possible. The two vials from each sampling locations should be sealed in separate plastic bags to prevent cross-contamination between samples particularly if the sampled waste is suspected of containing high levels of volatile organics.

Semivolatile organics — Containers used to collect samples for the determination of semivolatile organic compounds should be soap and water washed followed by methanol (or isopropanol) rinsing. The sample containers should be of glass or Teflon® and have screw-top covers with Teflon® liners.

Preservation for volatile organics — No preservation is used with concentrated waste samples. With liquid samples containing no residual chlorine, 4 drops of concentrated hydrochloric acid are added and the samples are immediately cooled to 4°C. When liquid samples contain residual chlorine, they are treated as above and, in addition, 4 drops of 4% aqueous sodium thiosulfate are added. Soil, sediment, and sludge samples are only cooled to 4°C.

Preservation for semivolatile organics — No preservation is used with concentrated waste samples. With liquid samples containing no residual chlorine and with soil, sediment, and sludge samples, immediately cooling to 4°C is the only preservation used. When residual chlorine is present then 3 mL of 10% aqueous sodium sulfate is added for each gallon of sample collected, followed by cooling to 4°C.

MHT The holding time for all volatile organics samples is 14 days. Liquid samples must be extracted within 7 days and their extracts analyzed within 40 days. Concentrated waste, soil, sediment, and sludge samples must be extracted within 14 days and their extracts analyzed within 40 days.

SAMPLE PREPARATION Volatile compounds are introduced into the gas chromatograph either by direct injector or purge-and-trap (EPA Method 5030). EPA Method 5030 may be used directly on groundwater samples or low-concentration contaminated soils and sediments. For medium-concentration soils or sediments, methanolic extraction, as described in EPA Method 5030, may be necessary prior to purge-and-trap analysis.

QUALITY CONTROL Calculate surrogate standard recovery on all samples, blanks, and spikes. A trip blank is recommended to check on sampling, storage, and handling contamination. Calibration standards, at a minimum of five concentration levels, are prepared in organic-free reagent water. One of the concentration levels should be at a concentration near, but above, the method detection limit.

A combination of bromochloromethane, 2-bromo-1-chloropropane, 1,4-dichlorobutane, and bromochlorobenzene are recommended as surrogate standards to encompass the range of the temperature program used in this method.

REFERENCE Test Methods for Evaluating Solid Waste, Physical/Chemical Methods, SW-846, 3rd Edition, U.S. EPA, Office of Solid Waste, Washington, DC, EPA Method 8021A, Rev. 1, Nov. 1992.

Methylene chloride   **EPA Method 8240**
CAS #75-09-2

TITLE Volatile Organics By GC/MS: Packed Column Technique

MATRIX Nearly all types of sample matarices, regardless of water content, can be analyzed using this method. This includes groundwater, aqueous sludges, caustic liquors, acid liquors, waste solvents, oily wastes, mousses, tars, fibrous wastes, polymetric emulsions, filter cakes, spent carbons, spent catalysts, soils, and sediments.

METHOD SUMMARY Method 8240B covers 80 volatile organic compounds that are introduced into a gas chromatograph by the purge-and-trap method or by direct injection (in limited applications). For the purge-and-trap method an inert gas (zero grade nitrogen or helium) is bubbled through a 5-mL solution at ambient temperature. Purged sample components are trapped in a tube of sorbent materials. When purging is complete, the sorbent tube is heated and backflushed with inert gas to desorb the trapped components onto a GC column.

INTERFERENCES Impurities in the purge gas and from organic compounds outgassing from the plumbing ahead of the trap account for many contamination problems. Interferences purged or coextracted from the samples will vary considerably from source to source. Cross-contamination can occur whenever high-level and low-level samples are analyzed sequentially. Whenever an unusually concentrated sample is analyzed, it should be followed by the analysis of organic-free reagent water to check for cross-contamination. Samples also can be contaminated by diffusion of volatile organics (particularly methylene chloride and fluorocarbons) through the septum seal into the sample during shipment and storage. A trip blank can serve as a check on such contamination. The lab

where volatile analysis is performed and also the refrigerated storage area should be completely free of solvents.

INSTRUMENTATION A gas chromatograph/mass spectrometry/data system (GC/MS) equipped with a 6 ft × 0.1 in I.D. glass column packed with 1% SP-1000 on Carbopack-B (60/80 mesh) is required. Also needed is a 5-mL purging device, a sorbent trap, and a thermal desorption apparatus.

PRECISION & ACCURACY This method is reported to have been tested by 15 laboratories using organic-free reagent water, drinking water, surface water, and industrial wastewaters (not specified) fortified at six concentrations over the range 5–600 µg/L.

Sample estimated quantitation limits (EQLs) are highly matrix-dependent. The EQLs listed may not always be achievable. EQLs listed for soils or sediments are based on wet weight. Normally, data is reported on a dry-weight basis; therefore, EQLs will be higher, based on the percent dry weight of each sample. Note that EQLs are even more variable than MDLs and that they are highly variable depending on the matrix being analyzed.

EQL in groundwater in µg/L was 5.
EQL in low soil or sediment in µg/kg was 5.
Accuracy (a) in µg/L was $0.87C + 1.88$.
Precision (b) in µg/L was $0.32x + 4.00$.

(a) *Average recovery found for measurements of samples containing a concentration of C, in µg/L.*
(b) *Overall precision found for measurements of samples with average recovery X for samples containing a concentration of C in µg/L.*
X = *Average recovery found for measurement of samples containing a concentration of C in µg/L.*

MULTIPLICATION FACTORS FOR OTHER MATRICES

Other Matrices	Factor (a)
Waste miscible liquid waste	50
High-concentration soil and sludge	125
Non-water miscible waste	500

(a) *EQL = [EQL for low soil/sediment] × [Factor]. For non-aqueous samples, the factor is on a wet-weight basis.*

SAMPLING METHOD
Liquid samples — Use a 40-mL glass screw-cap VOA vial with a Teflon®-faced silicone septum that has been prewashed, rinsed with distilled deionized water, and oven dried. However, if residual chlorine is present, collect sample in a 40-oz. soil VOA container which has been pre-preserved with 4 drops of 10% sodium thiosulfate, mix gently, and then transfer the sample to a 40-mL VOA vial. Collect bubble-free samples in duplicate and seal them in separate plastic bags.

Soils or sediments, and sludges — Use an 8-oz. widemouth glass bottle with a Teflon®-faced silicone septum that has been prewashed with detergent, rinsed with distilled deionized water, and oven dried. Tap slightly to eliminate free air space. Collect samples in duplicate and seal them in separate plastic bags.

SAMPLE PRESERVATION
Liquid samples — Add 4 drops of concentrated HCL and immediately cool samples to 4°C and store in a solvent-free refrigerator.

Soils or sediments, and sludges — Cool samples to 4°C and store in a solvent-free refrigerator.

MHT Maximum holding time is 14 days from the date of sample collection.

SAMPLE PREPARATION
Liquid samples — Remove the plunger from a 5-mL syringe and carefully pour the sample into the syringe barrel to just short of overflowing. Replace the syringe plunger and compress the sample. Open the syringe valve and vent any residual air while adjusting the sample volume to 5.0 mL. If there is only one volatile organic analysis (VOA) vial, a second syringe should be filled at this time to protect against possible loss of sample integrity. Add 10 µL of surrogate spiking solution and 10 µL of internal standard spiking solution through the valve bore of the 5-mL syringe, then close the valve. The surrogate and internal standards may be mixed and added as a single spiking solution.

Sediments, soils, and waste samples — All samples of this type should be screened by GC analysis using a headspace method (EPA Method 3810) or the hexadecane extraction and screening method (EPA Method 3820). Use the screening data to determine whether to use the low-concentration method (0.005–1 mg/kg) or the high-concentration method (>1 mg/kg).

Low-concentration method — The low-concentration method is based on purging a heated sediment or soil sample mixed with organic-free reagent water containing the surrogate and internal standards. Analyze all reagent blanks and standards under the same conditions as the samples.

Use a 5-g sample if the expected concentration is <0.1 mg/kg or a 1-g sample for expected concentrations between 0.1 and 1 mg/kg. Mix the contents of the sample container with a narrow metal spatula. Weigh the amount of the sample into a tared purge device. Add the spiked water to the purge device, which contains the weighed amount of sample, and connect the device to the purge-and-trap system.

High-concentration method — This method is based on extracting the sediment or soil with methanol. A waste sample is either extracted or diluted, depending on its solubility in methanol. Wastes that are insoluble in methanol are diluted with reagent tetraglyme or possibly polyethylene glycol (PEG). An aliquot of the extract is added to organic-free reagent water containing surrogate and internal standards. This is purged at ambient temperature. All samples with an expected concentration of >1.0 mg/kg should be analyzed by this method.

Mix the contents of the sample container with a narrow metal spatula. For sediments or soils and solid wastes that are insoluble in methanol, weigh 4 g (wet weight) of sample into a tared 20-mL vial. For waste that is soluble in methanol, tetraglyme, or PEG, weigh 1 g (wet weight) into a tared scintillation vial or culture tube or a 10-mL volumetric flask. Quickly add

9.0 mL of appropriate solvent then add 1.0 mL of a surrogate spiking solution to the vial, cap it, and shake it for 2 min.

METHANOL EXTRACT REQUIRED FOR ANALYSIS OF HIGH-CONCENTRATION SOILS OR SEDIMENTS

Approximate Concentration Range	Volume of Methanol Extract (a)
500–10,000 μg/kg	100 μL
1,000–20,000 μg/kg	50 μL
5,000–100,000 μg/kg	10 μL
25,000–500,000 μg/kg	100 μL of 1/50 dilution (b)

Calculate appropriate dilution factor for concentrations exceeding this table.

(a) The volume of methanol added to 5 mL of water being purged should be kept constant. Therefore, add to the 5-mL syringe whatever volume of methanol is necessary to maintain a volume of 100 μL added to the syringe.
(b) Dilute an aliquot of the methanol extract and then take 100 μL for analysis.

QUALITY CONTROL Demonstrate, through the analysis of a reagent water blank, that interferences from the analytical system, glassware, and reagents are under control. Blank samples should be carried through all stages of the sample preparation and measurement steps. For each analytical batch (up to 20 samples), a reagent blank, matrix spike, and matrix spike duplicate must be analyzed (the frequency of the spikes may be different for different monitoring programs). The blank and spiked samples must be carried through all stages of the sample preparation and measurement steps. QC samples mentioned in the section on Interferences will also be needed as appropriate to those situations.

REFERENCE Test Methods for Evaluating Solid Waste (SW-846). U.S. EPA. 1983. Method 8240B, Rev. 2, Nov. 1990. Office of Solid Wastes, Washington, DC.

Methylene chloride   **EPA Method 8260**
CAS #75-09-2

TITLE Volatile Organic Compounds by GC/MS: Capillary Column Technique

MATRIX This method is applicable to nearly all types of samples, regardless of water content, including groundwater, soils, and sediments.

METHOD SUMMARY Method 8260A covers 58 volatile organic compounds that are introduced into a gas chromatograph by the purge-and-trap method or by direct injection (in limited applications). Zero-grade helium is bubbled through a 5-mL solution at ambient temperature. Purged sample components are trapped in a tube containing suitable sorbent materials. When purging is complete, the sorbent tube is heated and backflushed with helium to desorb trapped sample components. The analytes are desorbed directly to a large bore capillary or cryofocussed on a capillary precolumn before being flash evaporated to a narrow bore capillary for analysis.

INTERFERENCES Major contaminant sources are volatile materials in the lab and impurities in the inert purging gas and in the sorbent trap. Interfering contamination may occur when a sample containing low concentrations of volatile organic compounds is analyzed immediately after a sample containing high concentrations of volatile organic compounds. After analysis of a sample containing high concentrations of volatile organic compounds, one or more calibration blanks should be analyzed to check for cross-contamination. Screening of the samples prior to purge-and-trap GC/MS analysis is highly recommended to prevent contamination of the system. This is especially true for soil and waste samples.

Special precautions must be taken to analyze for methylene chloride. The analytical and sample storage area should be isolated from all atmospheric sources of methylene chloride. All gas chromatography carrier gas lines and purge gas plumbing should be constructed from stainless steel or copper tubing. Laboratory clothing previously exposed to methylene chloride fumes during liquid-liquid extraction procedures can contribute to sample contamination.

Samples can also be contaminated by diffusion of volatile organics (particularly methylene chloride and fluorocarbons) through the septum seal during shipment and storage. A trip blank can serve as a check on such contamination.

INSTRUMENTATION GC/MS with a temperature-programmable chromatograph suitable for splitless injection equipped with variable constant differential flow controllers, a subambient oven controller, a purging device, sorbent trap, a thermal desorption apparatus and a capillary precolumn interface when using cryogenic cooling will be needed. The following GC columns may be used:

Column 1: 60 m × 0.75 mm I.D. capillary column coated with VOCOL, 1.5 μm film thickness.
Column 2: 30 m × 0.53 mm capillary column coated with DB-624 or VOCOL, 3 μm film thickness.
Column 3: 30 m × 0.32 mm I.D. capillary column coated with DB-5 or SE-54, 1-μm film thickness.

PRECISION & ACCURACY This method has been tested in a single lab using spiked water. Using a wide-bore capillary column, water was spiked at concentrations between 0.5 and 10 μg/L. Single lab accuracy and precision data are presented. The MDL actually achieved in a given analysis will vary depending on instrument sensitivity and matrix effects.

The MDL (a) in μg/L was 0.03.
The concentration range in μg/L was 0.1–10.
The mean accuracy (% of true value) was 95.
The precision as relative standard deviation was 5.3.

Note: The MDL is based on a 25-mL sample volume instead of a 5-mL sample volume.

SAMPLING METHOD

Liquid samples — Use a 40-mL glass screw-cap VOA vial with a Teflon®-faced silicone septum that has been prewashed, rinsed with distilled deionized water, and oven dried. If residual chlorine is present, collect the sample in a 4-oz soil VOA container which has been pre-preserved with 4 drops of 10%

sodium thiosulfate. Mix gently and transfer the sample to a 40-mL VOA vial. Collect bubble-free samples in duplicate and seal each sample in a separate plastic bag.

Soils, sediments and sludges — Use an 8-oz widemouth glass bottle with Teflon®-faced silicone septum that has been prewashed, rinsed with distilled deionized water, and oven dried. **Do not** heat the septum for more than 1 h. Tap slightly to eliminate any free air space. Collect samples in duplicate and seal each one in a separate plastic bag.

SAMPLE PRESERVATION
Liquid samples — Add 4 drops of concentrated HCL, cool to 4°C and store in a solvent-free refrigerator.

Soils, sediments and sludges — Cool samples to 4°C and store in a solvent-free refrigerator.

MHT The maximum holding time of any sample (liquids, soils, sediments, and sludges) is 14 days.

SAMPLE PREPARATION
Liquid samples — Remove the plunger from a 5-mL syringe and carefully pour the sample into the syringe barrel to just short of overflowing. Replace the syringe plunger and compress the sample. Open the syringe valve and vent any residual air while adjusting the sample volume to 5.0 mL. If there is only one volatile organic analysis (VOA) vial, a second syringe should be filled at this time to protect against possible loss of sample integrity. Add 10 μL of surrogate spiking solution and 10 μL of internal standard spiking solution through the valve bore of the 5-mL syringe, then close the valve. The surrogate and internal standards may be mixed and added as a single spiking solution.

Sediments, soils, and waste samples — All samples of this type should be screened by GC analysis using a headspace method (EPA Method 3810) or the hexadecane extraction and screening method (EPA Method 3820). Use the screening data to determine whether to use the low-concentration method (0.005–1 mg/kg) or the high-concentration method (>1 mg/kg).

Low-concentration method — The low-concentration method is based on purging a heated sediment or soil sample mixed with organic-free reagent water containing the surrogate and internal standards. Analyze all reagent blanks and standards under the same conditions as the samples.

Use a 5-g sample if the expected concentration is <0.1 mg/kg or a 1-g sample for expected concentrations between 0.1 and 1 mg/kg. Mix the contents of the sample container with a narrow metal spatula. Weigh the amount of the sample into a tared purge device. Add the spiked water to the purge device, which contains the weighed amount of sample, and connect the device to the purge-and-trap system.

High-concentration method — This method is based on extracting the sediment or soil with methanol. A waste sample is either extracted or diluted, depending on its solubility in methanol. Wastes that are insoluble in methanol are diluted with reagent tetraglyme or possibly polyethylene glycol (PEG). An aliquot of the extract is added to organic-free reagent water containing surrogate and internal standards. This is purged at ambient temperature. All samples with an expected concentration of >1.0 mg/kg should be analyzed by this method.

Mix the contents of the sample container with a narrow metal spatula. For sediments or soils and solid wastes that are insoluble in methanol, weigh 4 g (wet weight) of sample into a tared 20-mL vial. For waste that is soluble in methanol, tetraglyme, or PEG, weigh 1 g (wet weight) into a tared scintillation vial or culture tube or a 10-mL volumetric flask. Quickly add 9.0 mL of appropriate solvent then add 1.0 mL of a surrogate spiking solution to the vial, cap it, and shake it for 2 min.

METHANOL EXTRACT REQUIRED FOR ANALYSIS OF HIGH-CONCENTRATION SOILS OR SEDIMENTS

Approximate Concentration Range	Volume of Methanol Extract (a)
500–10,000 μg/kg	100 μL
1,000–20,000 μg/kg	50 μL
5,000–100,000 μg/kg	10 μL
25,000–500,000 μg/kg	100 μL of 1/50 dilution (b)

Calculate appropriate dilution factor for concentrations exceeding this table.

(a) The volume of methanol added to 5 mL of water being purged should be kept constant. Therefore, add to the 5-mL syringe whatever volume of methanol is necessary to maintain a volume of 100 μL added to the syringe.
(b) Dilute an aliquot of the methanol extract and then take 100 μL for analysis.

QUALITY CONTROL Demonstrate, through the analysis of a reagent water blank, that interferences from the analytical system, glassware, and reagents are under control. Blank samples should be carried through all stages of the sample preparation and measurement steps. For each analytical batch (up to 20 samples), a reagent blank, matrix spike, and matrix spike duplicate must be analyzed (the frequency of the spikes may be different for different monitoring programs). The blank and spiked samples must be carried through all stages of the sample preparation and measurement steps. QC samples mentioned in the section on Interferences will also be needed as appropriate to those situations.

Matrix spiking standards should be prepared from volatile organic compounds which will be representative of the compounds being investigated. The recommended internal standards are chlorobenzene-d5, 1,4-difluorobenzene, 1,4-dichlorobenzene-d4, and pentafluorobenzene. Using stock standard solutions, prepare secondary dilution standards containing the compounds of interest, either singly or mixed together in methanol. Store them in a vial with no headspace for no more than one week. Surrogates recommended are toluene-d8, 4-bromofluorobenzene, and dibromofluoromethane. Each sample undergoing GC/MS analysis must be spiked with 10 μL of the surrogate spiking solution prior to analysis.

REFERENCE Test Methods for Evaluating Solid Waste (SW-846). U.S. EPA 1983, Method 8260A, Rev. 1, Nov. 1990. Office of Solid Waste, Washington, DC.

Methylene chloride	EPA Method 601

CAS #75-09-2

TITLE Purgeable Halocarbons

MATRIX Wastewater.

APPLICATION Method covers 29 purgeable halocarbons. (Method 624 provides GC/MS conditions appropriate for the qualitative and quantitative confirmation of results). Method describes conditions for a 2nd GC column to confirm measurements made with primary column.

INTERFERENCES Impurities in the purge gas and organic compounds outgassing from the plumbing ahead of the trap. With high- and low-level samples, there can be carryover contamination. Diffusion of volatile organics through the septum seal into the sample.

INSTRUMENTATION GC-equipped with halide-specific detector. (With purge-and-trap unit).

RANGE 8.0–500 µg/L.

MDL 0.25 µg/L.

PRECISION 0.21X + 1.43 µg/L (overall precision).

ACCURACY 0.91C–0.93 µg/L (as recovery).

SAMPLING METHOD 25-mL glass vial. Teflon®-lined septum.

STABILITY Cool, 4°C, 0.008% sodium thiosulfate.

MHT 14 days.

QUALITY CONTROL The lab must on an ongoing basis, spike at least 10% of the samples from each sample site being monitored to assess accuracy.

REFERENCE Method 601, *Federal Register* Part VIII 40 CFR Part 136, Oct 26, 1984.

Methylene chloride	EPA Method 624

CAS #75-09-2

TITLE Purgeables

MATRIX Wastewater.

APPLICATION Method covers 31 purgeable organics. An inert gas is bubbled through a 5-mL water sample in a specially designed purging chamber. Here, purgeables are transferred from aqueous to gaseous phase, passed onto a sorbent column, and trapped. Trap is heated and backflushed with inert gas to desorb purgeables onto a GC column, where purgeables are separated.

INTERFERENCES Impurities in the purge gas, organic compounds outgassing from the plumbing ahead of the trap, and solvent vapors in the lab. With high- and low-level samples, there can be carryover contamination.

INSTRUMENTATION GC/MS with purge-and-trap unit.

RANGE 5–600 µg/L.

MDL 2.8 µg/L.

PRECISION 0.32X + 4.00 µg/L (overall precision).

ACCURACY 0.87C + 1.88 µg/L (as recovery).

SAMPLING METHOD 25-mL glass vial. Teflon®-lined septum.

STABILITY Cool, 4°C, 0.008% Sodium thiosulfate.

MHT 14 days.

QUALITY CONTROL The lab must on an ongoing basis, spike at least 5% of the samples from each sample site being monitored to assess accuracy.

REFERENCE Method 624, *Federal Register* Part VIII 40 CFR Part 136, Oct 26, 1984.

Methylene chloride	EPA Method 8010

CAS #75-09-2

TITLE Halogenated Volatile Organics

MATRIX Groundwater, soils, sludges, water miscible liquid wastes, and non-water miscible wastes.

APPLICATION This method is used for the analysis of 39 halogenated VOCs. Samples are analyzed using direct injection or purge-and-trap methods. Groundwater must be analyzed by the purge-and-trap method. The method provides an optional GC column which is used for analyte confirmation and that may help resolve analytes from interferences.

INTERFERENCES There can be carryover contamination with high- and low-level samples. Impurities may come from the purge-and-trap apparatus, organic compounds outgassing from the plumbing ahead of trap, diffusion of VOCs through the sample bottle septum during shipping or storage, or from solvent vapors in the lab.

INSTRUMENTATION GC capable of on-column injections or purge-and-trap sample introduction and a halogen specific detector. Column 1: 8 ft by 0.1 in 1%. SP-1000 on Carbopack-B. Column 2: 6 ft by 0.1 in bonded n-octane on Porasil-C.

RANGE 8–500 µg/L (reagent water).

MDL Not determined.

PQL FACTORS FOR MULTIPLYING × MDL VALUE

Matrix	Multiplication Factor
Groundwater	10
Low-level soil	10
Water miscible liquid waste	500
High-level soil and sludge	1250
Non-water miscible waste	1250

PRECISION 0.21X + 1.43 µg/L (overall precision).

ACCURACY 0.91C–0.93 µg/L (as recovery).

SAMPLING METHOD For water and liquid samples; use glass 40-mL vials with Teflon®-lined septum caps and collect two vials per sample location with no headspace. For solids and concentrated waste samples; use widemouth glass bottles with Teflon® liners.

STABILITY For concentrated wastes, soils, sediments, or sludges: cool to 4°C. For liquids: add 4 drops of concentrated hydrochloric acid and cool to 4°C.

MHT 14 days.

QUALITY CONTROL Analyze a reagent blank, matrix spike, and matrix spike duplicate/duplicate for each analytical batch (up to 20 samples). Demonstrate the purity of glassware and reagents by analyzing a reagent water method blank. Internal, surrogate, and five concentration level calibration standards are used.

REFERENCE Test Methods for Evaluating Solid Waste (SW-846), U.S. EPA Office of Solid Waste, Washington, DC, Method 8010B, Rev. 2, Nov. 1992.

4,4′-Methylenebis(2-chloraniline) EPA Method 8270
CAS #101-14-4

TITLE Semivolatile Organic Compounds by GC/MS

MATRIX This method is used to determine the concentration of semivolatile organic compounds in extracts prepared from all types of solid waste matrices, soils, and groundwater. Although surface waters are not specifically mentioned, this method should be applicable to water samples from rivers, lakes, etc.

METHOD SUMMARY This method covers 259 semivolatile organic compounds. In very limited applications direct injection of the sample into the GC/MS system may be appropriate, but this results in very high detection limits (approximately 10,000 µg/L). Typically, a 1-L liquid sample, containing surrogate, and matrix spiking standards, is extracted in a continuous extractor first under acid conditions and then under basic conditions. Typically 30 g of a solid sample, containing surrogate, and matrix spiking standards, is extracted ultrasonically. After concentrating the extract to 1 mL it is spiked with 10 µL of an internal standard solution just prior to analysis by GC/MS. The volume injected should contain about 100 ng of base/neutral and 200 ng of acid surrogates (for a 1-µL injection). Analysis is performed by GC/MS using a capillary GC column.

INTERFERENCES Raw GC/MS data from all blanks, samples, and spikes must be evaluated for interferences. Contamination by carryover can occur whenever high-concentration and low-concentration samples are sequentially analyzed. To reduce carryover, the sample syringe must be rinsed out between samples with solvent. Whenever an unusually concentrated sample is encountered, it should be followed by the analysis of blank solvent to check for cross-contamination.

INSTRUMENTATION A GC/MS and a data system are required. The GC column used is a 30 m × 0.25 mm I.D. (or 0.32 mm I.D.) 1um film thickness silicone-coated fused silica capillary column. A continuous liquid-liquid extractor equipped with Teflon® or glass connection joints and stopcocks requiring no lubrication, a K-D concentrating apparatus, water bath, and an ultrasonic disrupter with a minimum power of 300 W and with pulsing capability are also required.

PRECISION & ACCURACY The estimated quantitation limit (EQL) of Method 8270B for determining an individual compound is approximately 1 mg/kg (wet weight) for soil or sediment samples, 1–200 mg/kg for wastes (dependent on matrix and method of preparation), and 10 µg/L for groundwater samples. EQLs will be proportionately higher for sample extracts that require dilution to avoid saturation of the detector.

The EQL(b) for groundwater in µg/L is Not Applicable.
The EQL (a, b) for low concentrations in soil and sediment in µg/kg is not determined.
Accuracy as µg/L is not listed.
Overall precision in µg/L is not listed.

(a) EQLs listed for soil/sediment are based on wet weight. Normally data is reported in a dry-weight basis; therefore, EQLs will be higher based on the % dry weight of each sample. This calculation is based on a 30 g sample and gel permeation chromatography cleanup.

(b) Sample EQLs are highly matrix-dependent. The EQLs are provided for guidance and may not always be achievable.

C = True value for concentration, in µg/L.
X = Average recovery found for measurements of samples containing a concentration of C, in µg/L.

ESTIMATED QUANTITATION LIMIT

Other Matrices	Factor (a)
High-concentration soil and sludges by sonicator	7.5
Non-water miscible waste	75

(a) EQL for other matrices = [EQL for low soil/sediment] × [Factor]. This estimated EQL is similar to an EPA "Practical Quantitation Limit."

SAMPLING METHOD
Liquid samples — Use a 1 or 2½ gallon amber glass bottle with a screw-top Teflon®-lined cover that has been prewashed with detergent and rinsed with distilled water and methanol (or isopropanol).

Soils, sediments, or sludges — Use an 8-oz. widemouth glass with a screw-top Teflon®-lined cover that has been prewashed with detergent and rinsed with distilled water and methanol (or isopropanol).

SAMPLE PRESERVATION
Liquid samples — If residual chlorine is present, add 3 mL of 10% sodium thiosulfate per gallon, cool to 4°C and store in a solvent-free refrigerator until analysis; if chlorine is not present, then eliminate the sodium thiosulfate addition.

Soils, sediments, or sludges — Cool samples to 4°C and store in a solvent-free refrigerator.

MHT Liquid samples must be extracted within 7 days and the extracts analyzed within 40 days. Soils, sediments, or sludges may be stored for a maximum of 14 days and the extracts analyzed within 40 days.

SAMPLE PREPARATION

Liquid samples — Transfer 1 L quantitatively to a continuous extractor. If high concentrations are anticipated, a smaller volume may be used and then diluted with organic-free reagent water to 1 L. Adjust pH, if necessary, to pH <2 using 1:1 (V/V) sulfuric acid. Pipette 1.0 mL of a surrogate standard spiking solution into each sample. For the sample in each analytical batch selected for spiking, add 1.0 mL of a matrix spiking standard. For base/neutral acid analysis, the amount of the surrogates and matrix spiking compounds added to the sample should result in a final concentration of 100 ng/μL of each analyte in the extract to be analyzed (assuming a 1-μL injection). Extract with methylene chloride for 18–24 h. Next, adjust the pH of the aqueous phase to pH >11 using 10 N sodium hydroxide and extract it with methylene chloride again for 18–24 h. Dry the extract through a column containing anhydrous sodium sulfate and concentrate it to 1 mL using a K-D concentrator.

Soils, sediments, or sludges — Use 30 g of sample. Nonporous or wet samples (gummy or clay type) that do not have a free-flowing sandy texture must be mixed with anhydrous sodium sulfate until the sample is free flowing. Add 1 mL of surrogate standards to all samples, spikes, standards, and blanks. For the sample in each analytical batch selected for spiking, add 1.0 mL of a matrix spiking standard. For base/neutral acid analysis, the amount added of the surrogates and matrix spiking compounds should result in a final concentration of 100 ng/μL of each base/neutral analyte and 200 ng/μL of each acid analyte in the extract to be analyzed (assuming a 1-μL injection). Immediately add a 100-mL mixture of 1:1 methylene chloride:acetone and extract the sample ultrasonically for 3 min and then decant or filter the extracts. Repeat the extraction two or more times. Dry the extract using a column with anhydrous sodium sulfate and concentrate it to 1 mL in a K-D concentrator.

QUALITY CONTROL A methylene chloride solution containing 50 ng/μL of decafluorotriphenylphosphine (DFTPP) is used for tuning the GC/MS system each 12-h shift. A system performance check also must be made during every 12-h shift. A standard containing 50 ng/μL each of 4,4′-DDT, pentachlorophenol, and benzidine is required to verify injection port inertness and GC column performance. A calibration standard at mid-concentration, containing each compound of interest, including all required surrogates, must be performed every 12 h during analysis. After the system performance check is met, calibration check compounds (CCCs) are used to check the validity of the initial calibration.

The internal standard responses and retention times in the calibration check standard must be evaluated immediately after or during data acquisition. If the retention time for any internal standard changes by more than 30 seconds from the last check calibration (12 h), the chromatographic system must be inspected for malfunctions and corrections must be made, as required. If the electron ionization current plot (EICP) area for any of the internal standards changes by a factor of two from the last daily calibration standard check, the mass spectrometer must be inspected for malfunctions and corrections must be made, as appropriate.

Demonstrate, through the analysis of a reagent water blank, that interferences from the analytical system, glassware, and reagents are under control. The blank samples should be carried through all stages of the sample preparation and measurement steps. For each analytical batch (up to 20 samples), a reagent blank, matrix spike, and matrix spike duplicate/duplicate must be analyzed (the frequency of the spikes may be different for different monitoring programs). The blank and spiked samples must be carried through all stages of the sample preparation and measurement steps. A QC reference sample concentrate containing each analyte at a concentration of 100 mg/L in methanol is required.

REFERENCE Test Methods for Evaluating Solid Waste (SW-846). U.S. EPA 1983, Method 8270B, Rev. 2, Nov. 1990. Office of Solid Waste, Washington, DC.

4,4′-Methylenebis(2-chloroaniline)   EPA Method 1625
CAS #101-14-4

TITLE Semivolatile Organic Compounds by Isotope Dilution GC/MS

MATRIX The compounds may be determined in waters, soils, and municipal sludges by this method.

METHOD SUMMARY This method is used to determine 176 semivolatile toxic organic pollutants associated with the CWA (as amended 1987); the RCRA (as amended 1986); the CERCLA (as amended 1986); and other compounds amenable to extraction and analysis by capillary column gas chromatography-mass spectrometry (GC/MS).

Stable isotopically-labeled analogs of the compounds of interest are added to the sample. If the solids content is less than 1%, a 1-L sample is extracted at pH 12–13, then at pH <2 with methylene chloride using continuous extraction techniques.

If the solids content is 30% or less, the sample is diluted to 1% solids with reagent water, homogenized ultrasonically, and extracted at pH 12–13, then at pH <2 with methylene chloride using continuous extraction techniques. If the solids content is greater than 30%, the sample is extracted using ultrasonic techniques.

Each extract is dried over sodium sulfate, concentrated to a volume of 5 mL, cleaned up using GPC, if necessary, and concentrated. Extracts are concentrated to 1 mL if GPC is not performed, and to 0.5 mL if GPC is performed.

An internal standard is added to the extract, and a 1-mL aliquot of the extract is injected into the GC. The compounds are separated by GC and detected by a MS. The labeled compounds serve to correct the variability of the analytical technique.

INTERFERENCES Solvents, reagents, glassware, and other sample processing hardware may yield artifacts and/or elevated baselines causing misinterpretation of chromatograms and spectra. Materials used in the analysis must be demonstrated to be free from interferences under the conditions of analysis by running method blanks initially and with each sample lot (sample started through the extraction process on a given 8-h shift, to a maximum of 20). Specific selection of reagents and purification of solvents by distillation in all glass systems may be required. Glassware and, where possible, reagents are cleaned by solvent rinse and baking at 450°C for 1-h minimum. Interferences coextracted from samples will vary considerably from source to source, depending on the diversity of the site being sampled.

INSTRUMENTATION Major instrumentation includes a GC with a splitless or on-column injection port for capillary column, a MS with 70 eV electron impact ionization, and a data system to collect and record MS data, and process it. A K-D apparatus is used to concentrate extracts.

GC Column: 30 m × 0.25 mm I.D. 5% phenyl, 94% methyl, 1% vinyl silicone bonded phased fused silica capillary column.

PRECISION & ACCURACY The detection limits of the method are usually dependent on the level of interferences rather than instrumental limitations. The limits typify the minimum quantities that can be detected with no interferences present.

The minimum level (in µg/mL) was not listed. This is defined as a minimum level at which the analytical system shall give recognizable mass spectra (background corrected) and acceptable calibration points.

The MDL (in µg/kg) in low solids was not listed and in high solids was not listed; these were determined in digested sludge (low solids) and in filter cake or compost (high solids).

The labeled and native compound initial precision as standard deviation (in µg/L) was not listed.
The labeled and native compound initial accuracy as average recovery (in µg/L) was not listed.

SAMPLE COLLECTION, PRESERVATION & HANDLING Collect samples in glass containers. Aqueous samples which flow freely are collected in refrigerated bottles using automatic sampling equipment. Solid samples are collected as grab samples using widemouth jars. Maintain samples at 0 to 4°C from the time of collection until extraction. If residual chlorine is present in aqueous samples, add 80 mg sodium thiosulfate/L of water. Begin sample extraction within 7 days of collection, and analyze all extracts within 40 days of extraction.

SAMPLE PREPARATION Samples containing 1% solids or less are extracted directly using continuous liquid-liquid extraction techniques. Samples containing 1 to 30% solids are diluted to the 1% level with reagent water and extracted using continuous liquid-liquid extraction techniques. Samples containing greater than 30% solids are extracted using ultrasonic techniques.

Base/neutral extraction — Adjust the pH of the waters in the extractors to 12–13 with 6 N NaOH. Extract with methylene chloride for 24–48 h.
Acid extraction — Adjust the pH of the waters in the extractors to 2 or less using 6 N sulfuric acid. Extract with methylene chloride for 24–48 h.
Ultrasonic extraction of high solids samples — Add anhydrous sodium sulfate to the sample and QC aliquot(s).

Add acetone:methylene chloride (1:1) to the sample and mix thoroughly

Concentrate extracts using a K-D apparatus.

QUALITY CONTROL The analyst is permitted to modify this method to improve separations or lower the costs of measurements, provided all performance specifications are met. Analyses of blanks are required to demonstrate freedom from contamination. When results of spikes indicate atypical method performance for samples, the samples are diluted to bring method performance within acceptable limits.

For low solids (aqueous samples), extract, concentrate, and analyze two sets of four 1-L aliquots (8 aliquots total) of the precision and recovery standard. For high solids samples, two sets of four 30-g aliquots of the high solids reference matrix are used.

Spike all samples with labeled compounds to assess method performance. Compute percent recovery of the labeled compounds using the internal standard method. Compare the labeled compound recovery for each compound with the corresponding labeled compound recovery.

Reagent water and high solids reference matrix blanks are analyzed to demonstrate freedom from contamination. Extract and concentrate a 1-L reagent water blank or a high solids reference matrix blank with each sample's lot (samples started through the extraction process on the same 8-h shift, to a maximum of 20 samples).

Field replicates may be collected to determine the precision of the sampling technique, and spiked samples may be required to determine the accuracy of the analysis when the internal standard method is used.

REFERENCE Semivolatile Organic Compounds by Isotope Dilution GC/MS. Office of Water Regulation and Standards, U.S. EPA Industrial Technology Division, Washington, DC, EPA Method 1625, Rev. C, June 1989 (contact W.A. Telliard, U.S. EPA, Office of Water Regulations and Standards, 401 M St., SW, Washington, DC, 20460. Phone: 202-382-7131).

4,4'-Methylenebis(N,N-dimethylaniline) EPA Method 8270 CAS #101-61-1

TITLE Semivolatile Organic Compounds by GC/MS

MATRIX This method is used to determine the concentration of semivolatile organic compounds in extracts prepared from all types of solid waste matrices, soils, and groundwater. Although surface waters are not specifically mentioned, this method should be applicable to water samples from rivers, lakes, etc.

METHOD SUMMARY This method covers 259 semivolatile organic compounds. In very limited applications direct injection of the sample into the GC/MS system may be appropriate, but this results in very high detection limits (approximately 10,000 µg/L). Typically, a 1-L liquid sample, containing surrogate, and matrix spiking standards, is extracted in a continuous extractor first under acid conditions and then under basic conditions. Typically 30 g of a solid sample, containing surrogate, and matrix spiking standards, is extracted ultrasonically. After concentrating the extract to 1 mL it is spiked with 10 μL of an internal standard solution just prior to analysis by GC/MS. The volume injected should contain about 100 ng of base/neutral and 200 ng of acid surrogates (for a 1-μL injection). Analysis is performed by GC/MS using a capillary GC column.

INTERFERENCES Raw GC/MS data from all blanks, samples, and spikes must be evaluated for interferences. Contamination by carryover can occur whenever high-concentration and low-concentration samples are sequentially analyzed. To reduce carryover, the sample syringe must be rinsed out between samples with solvent. Whenever an unusually concentrated sample is encountered, it should be followed by the analysis of blank solvent to check for cross-contamination.

INSTRUMENTATION A GC/MS and a data system are required. The GC column used is a 30 m × 0.25 mm I.D. (or 0.32 mm I.D.) 1um film thickness silicone-coated fused silica capillary column. A continuous liquid-liquid extractor equipped with Teflon® or glass connection joints and stopcocks requiring no lubrication, a K-D concentrating apparatus, water bath, and an ultrasonic disrupter with a minimum power of 300 W and with pulsing capability are also required.

PRECISION & ACCURACY The estimated quantitation limit (EQL) of Method 8270B for determining an individual compound is approximately 1 mg/kg (wet weight) for soil or sediment samples, 1–200 mg/kg for wastes (dependent on matrix and method of preparation), and 10 μg/L for groundwater samples. EQLs will be proportionately higher for sample extracts that require dilution to avoid saturation of the detector.

The EQL(b) for groundwater in μg/L is not listed.
The EQL (a, b) for low concentrations in soil and sediment in μg/kg is not listed.
Accuracy as μg/L is not listed.
Overall precision in μg/L is not listed.

(a) *EQLs listed for soil/sediment are based on wet weight. Normally data is reported in a dry-weight basis; therefore, EQLs will be higher based on the % dry weight of each sample. This calculation is based on a 30 g sample and gel permeation chromatography cleanup.*
(b) *Sample EQLs are highly matrix-dependent. The EQLs are provided for guidance and may not always be achievable.*
C = *True value for concentration, in μg/L.*
X = *Average recovery found for measurements of samples containing a concentration of C, in μg/L.*

ESTIMATED QUANTITATION LIMIT

Other Matrices	Factor (a)
High-concentration soil and sludges by sonicator	7.5
Non-water miscible waste	75

(a) *EQL for other matrices = [EQL for low soil/sediment] × [Factor]. This estimated EQL is similar to an EPA "Practical Quantitation Limit."*

SAMPLING METHOD
Liquid samples — Use a 1 or 2½ gallon amber glass bottle with a screw-top Teflon®-lined cover that has been prewashed with detergent and rinsed with distilled water and methanol (or isopropanol).

Soils, sediments, or sludges — Use an 8-oz. widemouth glass with a screw-top Teflon®-lined cover that has been prewashed with detergent and rinsed with distilled water and methanol (or isopropanol).

SAMPLE PRESERVATION
Liquid samples — If residual chlorine is present, add 3 mL of 10% sodium thiosulfate per gallon, cool to 4°C and store in a solvent-free refrigerator until analysis; if chlorine is not present, then eliminate the sodium thiosulfate addition.

Soils, sediments, or sludges — Cool samples to 4°C and store in a solvent-free refrigerator.

MHT Liquid samples must be extracted within 7 days and the extracts analyzed within 40 days. Soils, sediments, or sludges may be stored for a maximum of 14 days and the extracts analyzed within 40 days.

SAMPLE PREPARATION
Liquid samples — Transfer 1 L quantitatively to a continuous extractor. If high concentrations are anticipated, a smaller volume may be used and then diluted with organic-free reagent water to 1 L. Adjust pH, if necessary, to pH <2 using 1:1 (V/V) sulfuric acid. Pipette 1.0 mL of a surrogate standard spiking solution into each sample. For the sample in each analytical batch selected for spiking, add 1.0 mL of a matrix spiking standard. For base/neutral acid analysis, the amount of the surrogates and matrix spiking compounds added to the sample should result in a final concentration of 100 ng/μL of each analyte in the extract to be analyzed (assuming a 1-μL injection). Extract with methylene chloride for 18–24 h. Next, adjust the pH of the aqueous phase to pH >11 using 10 N sodium hydroxide and extract it with methylene chloride again for 18–24 h. Dry the extract through a column containing anhydrous sodium sulfate and concentrate it to 1 mL using a K-D concentrator.

Soils, sediments, or sludges — Use 30 g of sample. Nonporous or wet samples (gummy or clay type) that do not have a free-flowing sandy texture must be mixed with anhydrous sodium sulfate until the sample is free flowing. Add 1 mL of surrogate standards to all samples, spikes, standards, and blanks. For the sample in each analytical batch selected for spiking, add 1.0 mL of a matrix spiking standard. For base/neutral acid analysis, the amount added of the surrogates and matrix spiking compounds should result in a final concentration of 100 ng/μL of each base/neutral analyte and 200 ng/μL of each acid analyte in the extract to be analyzed (assuming a 1-μL injection). Immediately add a 100-mL mixture of 1:1 methylene chloride:acetone and extract the sample ultrasonically for 3 min and then decant or filter the extracts. Repeat the extraction two or more times. Dry the extract using a column with anhydrous sodium sulfate and concentrate it to 1 mL in a K-D concentrator.

QUALITY CONTROL A methylene chloride solution containing 50 ng/μL of decafluorotriphenylphosphine (DFTPP) is

used for tuning the GC/MS system each 12-h shift. A system performance check also must be made during every 12-h shift. A standard containing 50 ng/μL each of 4,4′-DDT, pentachlorophenol, and benzidine is required to verify injection port inertness and GC column performance. A calibration standard at mid-concentration, containing each compound of interest, including all required surrogates, must be performed every 12 h during analysis. After the system performance check is met, calibration check compounds (CCCs) are used to check the validity of the initial calibration.

The internal standard responses and retention times in the calibration check standard must be evaluated immediately after or during data acquisition. If the retention time for any internal standard changes by more than 30 seconds from the last check calibration (12 h), the chromatographic system must be inspected for malfunctions and corrections must be made, as required. If the electron ionization current plot (EICP) area for any of the internal standards changes by a factor of two from the last daily calibration standard check, the mass spectrometer must be inspected for malfunctions and corrections must be made, as appropriate.

Demonstrate, through the analysis of a reagent water blank, that interferences from the analytical system, glassware, and reagents are under control. The blank samples should be carried through all stages of the sample preparation and measurement steps. For each analytical batch (up to 20 samples), a reagent blank, matrix spike, and matrix spike duplicate/duplicate must be analyzed (the frequency of the spikes may be different for different monitoring programs). The blank and spiked samples must be carried through all stages of the sample preparation and measurement steps. A QC reference sample concentrate containing each analyte at a concentration of 100 mg/L in methanol is required.

REFERENCE   Test Methods for Evaluating Solid Waste (SW-846). U.S. EPA 1983, Method 8270B, Rev. 2, Nov. 1990. Office of Solid Waste, Washington, DC.

4,5-Methylenephenanthrene　　　　　**EPA Method 1625**
CAS #203-64-5

TITLE   Semivolatile Organic Compounds by Isotope Dilution GC/MS

MATRIX   The compounds may be determined in waters, soils, and municipal sludges by this method.

METHOD SUMMARY   This method is used to determine 176 semivolatile toxic organic pollutants associated with the CWA (as amended 1987); the RCRA (as amended 1986); the CERCLA (as amended 1986); and other compounds amenable to extraction and analysis by capillary column gas chromatography-mass spectrometry (GC/MS).

Stable isotopically-labeled analogs of the compounds of interest are added to the sample. If the solids content is less than 1%, a 1-L sample is extracted at pH 12–13, then at pH <2 with methylene chloride using continuous extraction techniques.

If the solids content is 30% or less, the sample is diluted to 1% solids with reagent water, homogenized ultrasonically, and extracted at pH 12–13, then at pH <2 with methylene chloride using continuous extraction techniques. If the solids content is greater than 30%, the sample is extracted using ultrasonic techniques.

Each extract is dried over sodium sulfate, concentrated to a volume of 5 mL, cleaned up using GPC, if necessary, and concentrated. Extracts are concentrated to 1 mL if GPC is not performed, and to 0.5 mL if GPC is performed.

An internal standard is added to the extract, and a 1-mL aliquot of the extract is injected into the GC. The compounds are separated by GC and detected by a MS. The labeled compounds serve to correct the variability of the analytical technique.

INTERFERENCES   Solvents, reagents, glassware, and other sample processing hardware may yield artifacts and/or elevated baselines causing misinterpretation of chromatograms and spectra. Materials used in the analysis must be demonstrated to be free from interferences under the conditions of analysis by running method blanks initially and with each sample lot (sample started through the extraction process on a given 8-h shift, to a maximum of 20). Specific selection of reagents and purification of solvents by distillation in all glass systems may be required. Glassware and, where possible, reagents are cleaned by solvent rinse and baking at 450°C for 1-h minimum. Interferences coextracted from samples will vary considerably from source to source, depending on the diversity of the site being sampled.

INSTRUMENTATION   Major instrumentation includes a GC with a splitless or on-column injection port for capillary column, a MS with 70 eV electron impact ionization, and a data system to collect and record MS data, and process it. A K-D apparatus is used to concentrate extracts.

GC Column: 30 m × 0.25 mm I.D. 5% phenyl, 94% methyl, 1% vinyl silicone bonded phased fused silica capillary column.

PRECISION & ACCURACY   The detection limits of the method are usually dependent on the level of interferences rather than instrumental limitations. The limits typify the minimum quantities that can be detected with no interferences present.

The minimum level (in μg/mL) was not listed. This is defined as a minimum level at which the analytical system shall give recognizable mass spectra (background corrected) and acceptable calibration points.

The MDL (in μg/kg) in low solids was not listed and in high solids was not listed; these were determined in digested sludge (low solids) and in filter cake or compost (high solids).

The labeled and native compound initial precision as standard deviation (in μg/L) was not listed.
The labeled and native compound initial accuracy as average recovery (in μg/L) was not listed.

SAMPLE COLLECTION, PRESERVATION & HANDLING
Collect samples in glass containers. Aqueous samples which flow freely are collected in refrigerated bottles using automatic

sampling equipment. Solid samples are collected as grab samples using widemouth jars. Maintain samples at 0 to 4°C from the time of collection until extraction. If residual chlorine is present in aqueous samples, add 80 mg sodium thiosulfate/L of water. Begin sample extraction within 7 days of collection, and analyze all extracts within 40 days of extraction.

SAMPLE PREPARATION Samples containing 1% solids or less are extracted directly using continuous liquid-liquid extraction techniques. Samples containing 1 to 30% solids are diluted to the 1% level with reagent water and extracted using continuous liquid-liquid extraction techniques. Samples containing greater than 30% solids are extracted using ultrasonic techniques.

- Base/neutral extraction — Adjust the pH of the waters in the extractors to 12–13 with 6 N NaOH. Extract with methylene chloride for 24–48 h.
- Acid extraction — Adjust the pH of the waters in the extractors to 2 or less using 6 N sulfuric acid. Extract with methylene chloride for 24–48 h.
- Ultrasonic extraction of high solids samples — Add anhydrous sodium sulfate to the sample and QC aliquot(s). Add acetone:methylene chloride (1:1) to the sample and mix thoroughly

Concentrate extracts using a K-D apparatus.

QUALITY CONTROL The analyst is permitted to modify this method to improve separations or lower the costs of measurements, provided all performance specifications are met. Analyses of blanks are required to demonstrate freedom from contamination. When results of spikes indicate atypical method performance for samples, the samples are diluted to bring method performance within acceptable limits.

For low solids (aqueous samples), extract, concentrate, and analyze two sets of four 1-L aliquots (8 aliquots total) of the precision and recovery standard. For high solids samples, two sets of four 30-g aliquots of the high solids reference matrix are used.

Spike all samples with labeled compounds to assess method performance. Compute percent recovery of the labeled compounds using the internal standard method. Compare the labeled compound recovery for each compound with the corresponding labeled compound recovery.

Reagent water and high solids reference matrix blanks are analyzed to demonstrate freedom from contamination. Extract and concentrate a 1-L reagent water blank or a high solids reference matrix blank with each sample's lot (samples started through the extraction process on the same 8-h shift, to a maximum of 20 samples).

Field replicates may be collected to determine the precision of the sampling technique, and spiked samples may be required to determine the accuracy of the analysis when the internal standard method is used.

REFERENCE Semivolatile Organic Compounds by Isotope Dilution GC/MS. Office of Water Regulation and Standards, U.S. EPA Industrial Technology Division, Washington, DC, EPA Method 1625, Rev. C, June 1989 (contact W.A. Telliard, U.S. EPA, Office of Water Regulations and Standards, 401 M St., SW, Washington, DC, 20460. Phone: 202-382-7131).

1-Methylfluorene   EPA Method 1625
CAS #1730-37-6

TITLE Semivolatile Organic Compounds by Isotope Dilution GC/MS

MATRIX The compounds may be determined in waters, soils, and municipal sludges by this method.

METHOD SUMMARY This method is used to determine 176 semivolatile toxic organic pollutants associated with the CWA (as amended 1987); the RCRA (as amended 1986); the CERCLA (as amended 1986); and other compounds amenable to extraction and analysis by capillary column gas chromatography-mass spectrometry (GC/MS).

Stable isotopically-labeled analogs of the compounds of interest are added to the sample. If the solids content is less than 1%, a 1-L sample is extracted at pH 12–13, then at pH <2 with methylene chloride using continuous extraction techniques.

If the solids content is 30% or less, the sample is diluted to 1% solids with reagent water, homogenized ultrasonically, and extracted at pH 12–13, then at pH <2 with methylene chloride using continuous extraction techniques. If the solids content is greater than 30%, the sample is extracted using ultrasonic techniques.

Each extract is dried over sodium sulfate, concentrated to a volume of 5 mL, cleaned up using GPC, if necessary, and concentrated. Extracts are concentrated to 1 mL if GPC is not performed, and to 0.5 mL if GPC is performed.

An internal standard is added to the extract, and a 1-mL aliquot of the extract is injected into the GC. The compounds are separated by GC and detected by a MS. The labeled compounds serve to correct the variability of the analytical technique.

INTERFERENCES Solvents, reagents, glassware, and other sample processing hardware may yield artifacts and/or elevated baselines causing misinterpretation of chromatograms and spectra. Materials used in the analysis must be demonstrated to be free from interferences under the conditions of analysis by running method blanks initially and with each sample lot (sample started through the extraction process on a given 8-h shift, to a maximum of 20). Specific selection of reagents and purification of solvents by distillation in all glass systems may be required. Glassware and, where possible, reagents are cleaned by solvent rinse and baking at 450°C for 1-h minimum. Interferences coextracted from samples will vary considerably from source to source, depending on the diversity of the site being sampled.

INSTRUMENTATION Major instrumentation includes a GC with a splitless or on-column injection port for capillary column, a MS with 70 eV electron impact ionization, and a data system to collect and record MS data, and process it. A K-D apparatus is used to concentrate extracts.

GC Column: 30 m × 0.25 mm I.D. 5% phenyl, 94% methyl, 1% vinyl silicone bonded phased fused silica capillary column.

PRECISION & ACCURACY The detection limits of the method are usually dependent on the level of interferences rather than instrumental limitations. The limits typify the minimum quantities that can be detected with no interferences present.

The minimum level (in µg/mL) was not listed. This is defined as a minimum level at which the analytical system shall give recognizable mass spectra (background corrected) and acceptable calibration points.

The MDL (in µg/kg) in low solids was not listed and in high solids was not listed; these were determined in digested sludge (low solids) and in filter cake or compost (high solids).

The labeled and native compound initial precision as standard deviation (in µg/L) was not listed.
The labeled and native compound initial accuracy as average recovery (in µg/L) was not listed.

SAMPLE COLLECTION, PRESERVATION & HANDLING
Collect samples in glass containers. Aqueous samples which flow freely are collected in refrigerated bottles using automatic sampling equipment. Solid samples are collected as grab samples using widemouth jars. Maintain samples at 0 to 4°C from the time of collection until extraction. If residual chlorine is present in aqueous samples, add 80 mg sodium thiosulfate/L of water. Begin sample extraction within 7 days of collection, and analyze all extracts within 40 days of extraction.

SAMPLE PREPARATION Samples containing 1% solids or less are extracted directly using continuous liquid-liquid extraction techniques. Samples containing 1 to 30% solids are diluted to the 1% level with reagent water and extracted using continuous liquid-liquid extraction techniques. Samples containing greater than 30% solids are extracted using ultrasonic techniques.

Base/neutral extraction — Adjust the pH of the waters in the extractors to 12–13 with 6 N NaOH. Extract with methylene chloride for 24–48 h.
Acid extraction — Adjust the pH of the waters in the extractors to 2 or less using 6 N sulfuric acid. Extract with methylene chloride for 24–48 h.
Ultrasonic extraction of high solids samples — Add anhydrous sodium sulfate to the sample and QC aliquot(s). Add acetone:methylene chloride (1:1) to the sample and mix thoroughly

Concentrate extracts using a K-D apparatus.

QUALITY CONTROL The analyst is permitted to modify this method to improve separations or lower the costs of measurements, provided all performance specifications are met. Analyses of blanks are required to demonstrate freedom from contamination. When results of spikes indicate atypical method performance for samples, the samples are diluted to bring method performance within acceptable limits.

For low solids (aqueous samples), extract, concentrate, and analyze two sets of four 1-L aliquots (8 aliquots total) of the precision and recovery standard. For high solids samples, two sets of four 30-g aliquots of the high solids reference matrix are used.

Spike all samples with labeled compounds to assess method performance. Compute percent recovery of the labeled compounds using the internal standard method. Compare the labeled compound recovery for each compound with the corresponding labeled compound recovery.

Reagent water and high solids reference matrix blanks are analyzed to demonstrate freedom from contamination. Extract and concentrate a 1-L reagent water blank or a high solids reference matrix blank with each sample's lot (samples started through the extraction process on the same 8-h shift, to a maximum of 20 samples).

Field replicates may be collected to determine the precision of the sampling technique, and spiked samples may be required to determine the accuracy of the analysis when the internal standard method is used.

REFERENCE Semivolatile Organic Compounds by Isotope Dilution GC/MS. Office of Water Regulation and Standards, U.S. EPA Industrial Technology Division, Washington, DC, EPA Method 1625, Rev. C, June 1989 (contact W.A. Telliard, U.S. EPA, Office of Water Regulations and Standards, 401 M St., SW, Washington, DC, 20460. Phone: 202-382-7131).

2-Methylnaphthalene   EPA Method 1625
CAS #91-57-6

TITLE Semivolatile Organic Compounds by Isotope Dilution GC/MS

MATRIX The compounds may be determined in waters, soils, and municipal sludges by this method.

METHOD SUMMARY This method is used to determine 176 semivolatile toxic organic pollutants associated with the CWA (as amended 1987); the RCRA (as amended 1986); the CERCLA (as amended 1986); and other compounds amenable to extraction and analysis by capillary column gas chromatography-mass spectrometry (GC/MS).

Stable isotopically-labeled analogs of the compounds of interest are added to the sample. If the solids content is less than 1%, a 1-L sample is extracted at pH 12–13, then at pH <2 with methylene chloride using continuous extraction techniques.

If the solids content is 30% or less, the sample is diluted to 1% solids with reagent water, homogenized ultrasonically, and extracted at pH 12–13, then at pH <2 with methylene chloride using continuous extraction techniques. If the solids content is greater than 30%, the sample is extracted using ultrasonic techniques.

Each extract is dried over sodium sulfate, concentrated to a volume of 5 mL, cleaned up using GPC, if necessary, and concentrated. Extracts are concentrated to 1 mL if GPC is not performed, and to 0.5 mL if GPC is performed.

An internal standard is added to the extract, and a 1-mL aliquot of the extract is injected into the GC. The compounds are separated by GC and detected by a MS. The labeled compounds serve to correct the variability of the analytical technique.

INTERFERENCES Solvents, reagents, glassware, and other sample processing hardware may yield artifacts and/or elevated baselines causing misinterpretation of chromatograms and spectra. Materials used in the analysis must be demonstrated to be free from interferences under the conditions of analysis by running method blanks initially and with each sample lot (sample started through the extraction process on a given 8-h shift, to a maximum of 20). Specific selection of reagents and purification of solvents by distillation in all glass systems may be required. Glassware and, where possible, reagents are cleaned by solvent rinse and baking at 450°C for 1-h minimum. Interferences coextracted from samples will vary considerably from source to source, depending on the diversity of the site being sampled.

INSTRUMENTATION Major instrumentation includes a GC with a splitless or on-column injection port for capillary column, a MS with 70 eV electron impact ionization, and a data system to collect and record MS data, and process it. A K-D apparatus is used to concentrate extracts.

GC Column: 30 m × 0.25 mm I.D. 5% phenyl, 94% methyl, 1% vinyl silicone bonded phased fused silica capillary column.

PRECISION & ACCURACY The detection limits of the method are usually dependent on the level of interferences rather than instrumental limitations. The limits typify the minimum quantities that can be detected with no interferences present.

The minimum level (in µg/mL) was not listed. This is defined as a minimum level at which the analytical system shall give recognizable mass spectra (background corrected) and acceptable calibration points.

The MDL (in µg/kg) in low solids was not listed and in high solids was not listed; these were determined in digested sludge (low solids) and in filter cake or compost (high solids).

The labeled and native compound initial precision as standard deviation (in µg/L) was not listed.

The labeled and native compound initial accuracy as average recovery (in µg/L) was not listed.

SAMPLE COLLECTION, PRESERVATION & HANDLING Collect samples in glass containers. Aqueous samples which flow freely are collected in refrigerated bottles using automatic sampling equipment. Solid samples are collected as grab samples using widemouth jars. Maintain samples at 0 to 4°C from the time of collection until extraction. If residual chlorine is present in aqueous samples, add 80 mg sodium thiosulfate/L of water. Begin sample extraction within 7 days of collection, and analyze all extracts within 40 days of extraction.

SAMPLE PREPARATION Samples containing 1% solids or less are extracted directly using continuous liquid-liquid extraction techniques. Samples containing 1 to 30% solids are diluted to the 1% level with reagent water and extracted using continuous liquid-liquid extraction techniques. Samples containing greater than 30% solids are extracted using ultrasonic techniques.

Base/neutral extraction — Adjust the pH of the waters in the extractors to 12–13 with 6 N NaOH. Extract with methylene chloride for 24–48 h.

Acid extraction — Adjust the pH of the waters in the extractors to 2 or less using 6 N sulfuric acid. Extract with methylene chloride for 24–48 h.

Ultrasonic extraction of high solids samples — Add anhydrous sodium sulfate to the sample and QC aliquot(s). Add acetone:methylene chloride (1:1) to the sample and mix thoroughly

Concentrate extracts using a K-D apparatus.

QUALITY CONTROL The analyst is permitted to modify this method to improve separations or lower the costs of measurements, provided all performance specifications are met. Analyses of blanks are required to demonstrate freedom from contamination. When results of spikes indicate atypical method performance for samples, the samples are diluted to bring method performance within acceptable limits.

For low solids (aqueous samples), extract, concentrate, and analyze two sets of four 1-L aliquots (8 aliquots total) of the precision and recovery standard. For high solids samples, two sets of four 30-g aliquots of the high solids reference matrix are used.

Spike all samples with labeled compounds to assess method performance. Compute percent recovery of the labeled compounds using the internal standard method. Compare the labeled compound recovery for each compound with the corresponding labeled compound recovery.

Reagent water and high solids reference matrix blanks are analyzed to demonstrate freedom from contamination. Extract and concentrate a 1-L reagent water blank or a high solids reference matrix blank with each sample's lot (samples started through the extraction process on the same 8-h shift, to a maximum of 20 samples).

Field replicates may be collected to determine the precision of the sampling technique, and spiked samples may be required to determine the accuracy of the analysis when the internal standard method is used.

REFERENCE Semivolatile Organic Compounds by Isotope Dilution GC/MS. Office of Water Regulation and Standards, U.S. EPA Industrial Technology Division, Washington, DC, EPA Method 1625, Rev. C, June 1989 (contact W.A. Telliard, U.S. EPA, Office of Water Regulations and Standards, 401 M St., SW, Washington, DC, 20460. Phone: 202-382-7131).

2-Methylnaphthalene   **EPA Method 8270**
CAS #91-57-6

TITLE Semivolatile Organic Compounds by GC/MS

MATRIX This method is used to determine the concentration of semivolatile organic compounds in extracts prepared

from all types of solid waste matrices, soils, and groundwater. Although surface waters are not specifically mentioned, this method should be applicable to water samples from rivers, lakes, etc.

METHOD SUMMARY This method covers 259 semivolatile organic compounds. In very limited applications direct injection of the sample into the GC/MS system may be appropriate, but this results in very high detection limits (approximately 10,000 µg/L). Typically, a 1-L liquid sample, containing surrogate, and matrix spiking standards, is extracted in a continuous extractor first under acid conditions and then under basic conditions. Typically 30 g of a solid sample, containing surrogate, and matrix spiking standards, is extracted ultrasonically. After concentrating the extract to 1 mL it is spiked with 10 µL of an internal standard solution just prior to analysis by GC/MS. The volume injected should contain about 100 ng of base/neutral and 200 ng of acid surrogates (for a 1-µL injection). Analysis is performed by GC/MS using a capillary GC column.

INTERFERENCES Raw GC/MS data from all blanks, samples, and spikes must be evaluated for interferences. Contamination by carryover can occur whenever high-concentration and low-concentration samples are sequentially analyzed. To reduce carryover, the sample syringe must be rinsed out between samples with solvent. Whenever an unusually concentrated sample is encountered, it should be followed by the analysis of blank solvent to check for cross-contamination.

INSTRUMENTATION A GC/MS and a data system are required. The GC column used is a 30 m × 0.25 mm I.D. (or 0.32 mm I.D.) 1um film thickness silicone-coated fused silica capillary column. A continuous liquid-liquid extractor equipped with Teflon® or glass connection joints and stopcocks requiring no lubrication, a K-D concentrating apparatus, water bath, and an ultrasonic disrupter with a minimum power of 300 W and with pulsing capability are also required.

PRECISION & ACCURACY The estimated quantitation limit (EQL) of Method 8270B for determining an individual compound is approximately 1 mg/kg (wet weight) for soil or sediment samples, 1–200 mg/kg for wastes (dependent on matrix and method of preparation), and 10 µg/L for groundwater samples. EQLs will be proportionately higher for sample extracts that require dilution to avoid saturation of the detector.

The EQL(b) for groundwater in µg/L is 10.
The EQL (a, b) for low concentrations in soil and sediment in µg/kg is 660.
Accuracy as µg/L is not listed.
Overall precision in µg/L is not listed.

(a) EQLs listed for soil/sediment are based on wet weight. Normally data is reported in a dry-weight basis; therefore, EQLs will be higher based on the % dry weight of each sample. This calculation is based on a 30 g sample and gel permeation chromatography cleanup.
(b) Sample EQLs are highly matrix-dependent. The EQLs are provided for guidance and may not always be achievable.
$C =$ *True value for concentration, in µg/L.*
$X =$ *Average recovery found for measurements of samples containing a concentration of C, in µg/L.*

ESTIMATED QUANTITATION LIMIT

Other Matrices	Factor (a)
High-concentration soil and sludges by sonicator	7.5
Non-water miscible waste	75

(a) EQL for other matrices = [EQL for low soil/sediment] × [Factor]. This estimated EQL is similar to an EPA "Practical Quantitation Limit."

SAMPLING METHOD
Liquid samples — Use a 1 or 2½ gallon amber glass bottle with a screw-top Teflon®-lined cover that has been prewashed with detergent and rinsed with distilled water and methanol (or isopropanol).

Soils, sediments, or sludges — Use an 8-oz. widemouth glass with a screw-top Teflon®-lined cover that has been prewashed with detergent and rinsed with distilled water and methanol (or isopropanol).

SAMPLE PRESERVATION
Liquid samples — If residual chlorine is present, add 3 mL of 10% sodium thiosulfate per gallon, cool to 4°C and store in a solvent-free refrigerator until analysis; if chlorine is not present, then eliminate the sodium thiosulfate addition.

Soils, sediments, or sludges — Cool samples to 4°C and store in a solvent-free refrigerator.

MHT Liquid samples must be extracted within 7 days and the extracts analyzed within 40 days. Soils, sediments, or sludges may be stored for a maximum of 14 days and the extracts analyzed within 40 days.

SAMPLE PREPARATION
Liquid samples — Transfer 1 L quantitatively to a continuous extractor. If high concentrations are anticipated, a smaller volume may be used and then diluted with organic-free reagent water to 1 L. Adjust pH, if necessary, to pH <2 using 1:1 (V/V) sulfuric acid. Pipette 1.0 mL of a surrogate standard spiking solution into each sample. For the sample in each analytical batch selected for spiking, add 1.0 mL of a matrix spiking standard. For base/neutral acid analysis, the amount of the surrogates and matrix spiking compounds added to the sample should result in a final concentration of 100 ng/µL of each analyte in the extract to be analyzed (assuming a 1-µL injection). Extract with methylene chloride for 18–24 h. Next, adjust the pH of the aqueous phase to pH >11 using 10 N sodium hydroxide and extract it with methylene chloride again for 18–24 h. Dry the extract through a column containing anhydrous sodium sulfate and concentrate it to 1 mL using a K-D concentrator.

Soils, sediments, or sludges — Use 30 g of sample. Nonporous or wet samples (gummy or clay type) that do not have a free-flowing sandy texture must be mixed with anhydrous sodium sulfate until the sample is free flowing. Add 1 mL of surrogate standards to all samples, spikes, standards, and blanks. For the sample in each analytical batch selected for spiking, add 1.0 mL of a matrix spiking standard. For base/neutral acid analysis, the amount added of the surrogates and matrix spiking compounds should result in a final concentration of 100 ng/µL of each base/neutral analyte and 200 ng/µL of each acid analyte

in the extract to be analyzed (assuming a 1-µL injection). Immediately add a 100-mL mixture of 1:1 methylene chloride:acetone and extract the sample ultrasonically for 3 min and then decant or filter the extracts. Repeat the extraction two or more times. Dry the extract using a column with anhydrous sodium sulfate and concentrate it to 1 mL in a K-D concentrator.

QUALITY CONTROL A methylene chloride solution containing 50 ng/µL of decafluorotriphenylphosphine (DFTPP) is used for tuning the GC/MS system each 12-h shift. A system performance check also must be made during every 12-h shift. A standard containing 50 ng/µL each of 4,4′-DDT, pentachlorophenol, and benzidine is required to verify injection port inertness and GC column performance. A calibration standard at mid-concentration, containing each compound of interest, including all required surrogates, must be performed every 12 h during analysis. After the system performance check is met, calibration check compounds (CCCs) are used to check the validity of the initial calibration.

The internal standard responses and retention times in the calibration check standard must be evaluated immediately after or during data acquisition. If the retention time for any internal standard changes by more than 30 seconds from the last check calibration (12 h), the chromatographic system must be inspected for malfunctions and corrections must be made, as required. If the electron ionization current plot (EICP) area for any of the internal standards changes by a factor of two from the last daily calibration standard check, the mass spectrometer must be inspected for malfunctions and corrections must be made, as appropriate.

Demonstrate, through the analysis of a reagent water blank, that interferences from the analytical system, glassware, and reagents are under control. The blank samples should be carried through all stages of the sample preparation and measurement steps. For each analytical batch (up to 20 samples), a reagent blank, matrix spike, and matrix spike duplicate/duplicate must be analyzed (the frequency of the spikes may be different for different monitoring programs). The blank and spiked samples must be carried through all stages of the sample preparation and measurement steps. A QC reference sample concentrate containing each analyte at a concentration of 100 mg/L in methanol is required.

REFERENCE Test Methods for Evaluating Solid Waste (SW-846). U.S. EPA 1983, Method 8270B, Rev. 2, Nov. 1990. Office of Solid Waste, Washington, DC.

1-Methylphenanthrene **EPA Method 1625**
CAS #832-69-9

TITLE Semivolatile Organic Compounds by Isotope Dilution GC/MS

MATRIX The compounds may be determined in waters, soils, and municipal sludges by this method.

METHOD SUMMARY This method is used to determine 176 semivolatile toxic organic pollutants associated with the CWA (as amended 1987); the RCRA (as amended 1986); the CERCLA (as amended 1986); and other compounds amenable to extraction and analysis by capillary column gas chromatography-mass spectrometry (GC/MS).

Stable isotopically-labeled analogs of the compounds of interest are added to the sample. If the solids content is less than 1%, a 1-L sample is extracted at pH 12–13, then at pH <2 with methylene chloride using continuous extraction techniques.

If the solids content is 30% or less, the sample is diluted to 1% solids with reagent water, homogenized ultrasonically, and extracted at pH 12–13, then at pH <2 with methylene chloride using continuous extraction techniques. If the solids content is greater than 30%, the sample is extracted using ultrasonic techniques.

Each extract is dried over sodium sulfate, concentrated to a volume of 5 mL, cleaned up using GPC, if necessary, and concentrated. Extracts are concentrated to 1 mL if GPC is not performed, and to 0.5 mL if GPC is performed.

An internal standard is added to the extract, and a 1-mL aliquot of the extract is injected into the GC. The compounds are separated by GC and detected by a MS. The labeled compounds serve to correct the variability of the analytical technique.

INTERFERENCES Solvents, reagents, glassware, and other sample processing hardware may yield artifacts and/or elevated baselines causing misinterpretation of chromatograms and spectra. Materials used in the analysis must be demonstrated to be free from interferences under the conditions of analysis by running method blanks initially and with each sample lot (sample started through the extraction process on a given 8-h shift, to a maximum of 20). Specific selection of reagents and purification of solvents by distillation in all glass systems may be required. Glassware and, where possible, reagents are cleaned by solvent rinse and baking at 450°C for 1-h minimum. Interferences coextracted from samples will vary considerably from source to source, depending on the diversity of the site being sampled.

INSTRUMENTATION Major instrumentation includes a GC with a splitless or on-column injection port for capillary column, a MS with 70 eV electron impact ionization, and a data system to collect and record MS data, and process it. A K-D apparatus is used to concentrate extracts.

GC Column: 30 m × 0.25 mm I.D. 5% phenyl, 94% methyl, 1% vinyl silicone bonded phased fused silica capillary column.

PRECISION & ACCURACY The detection limits of the method are usually dependent on the level of interferences rather than instrumental limitations. The limits typify the minimum quantities that can be detected with no interferences present.

The minimum level (in µg/mL) was not listed. This is defined as a minimum level at which the analytical system shall give recognizable mass spectra (background corrected) and acceptable calibration points.

The MDL (in µg/kg) in low solids was not listed and in high solids was not listed; these were determined in digested sludge (low solids) and in filter cake or compost (high solids).

The labeled and native compound initial precision as standard deviation (in μg/L) was not listed.

The labeled and native compound initial accuracy as average recovery (in μg/L) was not listed.

SAMPLE COLLECTION, PRESERVATION & HANDLING Collect samples in glass containers. Aqueous samples which flow freely are collected in refrigerated bottles using automatic sampling equipment. Solid samples are collected as grab samples using widemouth jars. Maintain samples at 0 to 4°C from the time of collection until extraction. If residual chlorine is present in aqueous samples, add 80 mg sodium thiosulfate/L of water. Begin sample extraction within 7 days of collection, and analyze all extracts within 40 days of extraction.

SAMPLE PREPARATION Samples containing 1% solids or less are extracted directly using continuous liquid-liquid extraction techniques. Samples containing 1 to 30% solids are diluted to the 1% level with reagent water and extracted using continuous liquid-liquid extraction techniques. Samples containing greater than 30% solids are extracted using ultrasonic techniques.

Base/neutral extraction — Adjust the pH of the waters in the extractors to 12–13 with 6 N NaOH. Extract with methylene chloride for 24–48 h.

Acid extraction — Adjust the pH of the waters in the extractors to 2 or less using 6 N sulfuric acid. Extract with methylene chloride for 24–48 h.

Ultrasonic extraction of high solids samples — Add anhydrous sodium sulfate to the sample and QC aliquot(s). Add acetone:methylene chloride (1:1) to the sample and mix thoroughly

Concentrate extracts using a K-D apparatus.

QUALITY CONTROL The analyst is permitted to modify this method to improve separations or lower the costs of measurements, provided all performance specifications are met. Analyses of blanks are required to demonstrate freedom from contamination. When results of spikes indicate atypical method performance for samples, the samples are diluted to bring method performance within acceptable limits.

For low solids (aqueous samples), extract, concentrate, and analyze two sets of four 1-L aliquots (8 aliquots total) of the precision and recovery standard. For high solids samples, two sets of four 30-g aliquots of the high solids reference matrix are used.

Spike all samples with labeled compounds to assess method performance. Compute percent recovery of the labeled compounds using the internal standard method. Compare the labeled compound recovery for each compound with the corresponding labeled compound recovery.

Reagent water and high solids reference matrix blanks are analyzed to demonstrate freedom from contamination. Extract and concentrate a 1-L reagent water blank or a high solids reference matrix blank with each sample's lot (samples started through the extraction process on the same 8-h shift, to a maximum of 20 samples).

Field replicates may be collected to determine the precision of the sampling technique, and spiked samples may be required to determine the accuracy of the analysis when the internal standard method is used.

REFERENCE Semivolatile Organic Compounds by Isotope Dilution GC/MS. Office of Water Regulation and Standards, U.S. EPA Industrial Technology Division, Washington, DC, EPA Method 1625, Rev. C, June 1989 (contact W.A. Telliard, U.S. EPA, Office of Water Regulations and Standards, 401 M St., SW, Washington, DC, 20460. Phone: 202-382-7131).

2-Methylphenol   EPA Method 8270
CAS #95-48-7

TITLE Semivolatile Organic Compounds by GC/MS

MATRIX This method is used to determine the concentration of semivolatile organic compounds in extracts prepared from all types of solid waste matrices, soils, and groundwater. Although surface waters are not specifically mentioned, this method should be applicable to water samples from rivers, lakes, etc.

METHOD SUMMARY This method covers 259 semivolatile organic compounds. In very limited applications direct injection of the sample into the GC/MS system may be appropriate, but this results in very high detection limits (approximately 10,000 μg/L). Typically, a 1-L liquid sample, containing surrogate, and matrix spiking standards, is extracted in a continuous extractor first under acid conditions and then under basic conditions. Typically 30 g of a solid sample, containing surrogate, and matrix spiking standards, is extracted ultrasonically. After concentrating the extract to 1 mL it is spiked with 10 μL of an internal standard solution just prior to analysis by GC/MS. The volume injected should contain about 100 ng of base/neutral and 200 ng of acid surrogates (for a 1-μL injection). Analysis is performed by GC/MS using a capillary GC column.

INTERFERENCES Raw GC/MS data from all blanks, samples, and spikes must be evaluated for interferences. Contamination by carryover can occur whenever high-concentration and low-concentration samples are sequentially analyzed. To reduce carryover, the sample syringe must be rinsed out between samples with solvent. Whenever an unusually concentrated sample is encountered, it should be followed by the analysis of blank solvent to check for cross-contamination.

INSTRUMENTATION A GC/MS and a data system are required. The GC column used is a 30 m × 0.25 mm I.D. (or 0.32 mm I.D.) 1um film thickness silicone-coated fused silica capillary column. A continuous liquid-liquid extractor equipped with Teflon® or glass connection joints and stopcocks requiring no lubrication, a K-D concentrating apparatus, water bath, and an ultrasonic disrupter with a minimum power of 300 W and with pulsing capability are also required.

PRECISION & ACCURACY The estimated quantitation limit (EQL) of Method 8270B for determining an individual compound is approximately 1 mg/kg (wet weight) for soil or sediment samples, 1–200 mg/kg for wastes (dependent on

matrix and method of preparation), and 10 μg/L for groundwater samples. EQLs will be proportionately higher for sample extracts that require dilution to avoid saturation of the detector.

The EQL(b) for groundwater in μg/L is 10.
The EQL (a, b) for low concentrations in soil and sediment in μg/kg is 660.
Accuracy as μg/L is not listed.
Overall precision in μg/L is not listed.

(a) *EQLs listed for soil/sediment are based on wet weight. Normally data is reported in a dry-weight basis; therefore, EQLs will be higher based on the % dry weight of each sample. This calculation is based on a 30 g sample and gel permeation chromatography cleanup.*
(b) *Sample EQLs are highly matrix-dependent. The EQLs are provided for guidance and may not always be achievable.*
C = *True value for concentration, in μg/L.*
X = *Average recovery found for measurements of samples containing a concentration of C, in μg/L.*

ESTIMATED QUANTITATION LIMIT

Other Matrices	Factor (a)
High-concentration soil and sludges by sonicator	7.5
Non-water miscible waste	75

(a) *EQL for other matrices = [EQL for low soil/sediment] × [Factor]. This estimated EQL is similar to an EPA "Practical Quantitation Limit."*

SAMPLING METHOD

Liquid samples — Use a 1 or 2½ gallon amber glass bottle with a screw-top Teflon®-lined cover that has been prewashed with detergent and rinsed with distilled water and methanol (or isopropanol).

Soils, sediments, or sludges — Use an 8-oz. widemouth glass with a screw-top Teflon®-lined cover that has been prewashed with detergent and rinsed with distilled water and methanol (or isopropanol).

SAMPLE PRESERVATION

Liquid samples — If residual chlorine is present, add 3 mL of 10% sodium thiosulfate per gallon, cool to 4°C and store in a solvent-free refrigerator until analysis; if chlorine is not present, then eliminate the sodium thiosulfate addition.

Soils, sediments, or sludges — Cool samples to 4°C and store in a solvent-free refrigerator.

MHT Liquid samples must be extracted within 7 days and the extracts analyzed within 40 days. Soils, sediments, or sludges may be stored for a maximum of 14 days and the extracts analyzed within 40 days.

SAMPLE PREPARATION

Liquid samples — Transfer 1 L quantitatively to a continuous extractor. If high concentrations are anticipated, a smaller volume may be used and then diluted with organic-free reagent water to 1 L. Adjust pH, if necessary, to pH <2 using 1:1 (V/V) sulfuric acid. Pipette 1.0 mL of a surrogate standard spiking solution into each sample. For the sample in each analytical batch selected for spiking, add 1.0 mL of a matrix spiking standard. For base/neutral acid analysis, the amount of the surrogates and matrix spiking compounds added to the sample should result in a final concentration of 100 ng/μL of each analyte in the extract to be analyzed (assuming a 1-μL injection). Extract with methylene chloride for 18–24 h. Next, adjust the pH of the aqueous phase to pH >11 using 10 N sodium hydroxide and extract it with methylene chloride again for 18–24 h. Dry the extract through a column containing anhydrous sodium sulfate and concentrate it to 1 mL using a K-D concentrator.

Soils, sediments, or sludges — Use 30 g of sample. Nonporous or wet samples (gummy or clay type) that do not have a free-flowing sandy texture must be mixed with anhydrous sodium sulfate until the sample is free flowing. Add 1 mL of surrogate standards to all samples, spikes, standards, and blanks. For the sample in each analytical batch selected for spiking, add 1.0 mL of a matrix spiking standard. For base/neutral acid analysis, the amount added of the surrogates and matrix spiking compounds should result in a final concentration of 100 ng/μL of each base/neutral analyte and 200 ng/μL of each acid analyte in the extract to be analyzed (assuming a 1-μL injection). Immediately add a 100-mL mixture of 1:1 methylene chloride:acetone and extract the sample ultrasonically for 3 min and then decant or filter the extracts. Repeat the extraction two or more times. Dry the extract using a column with anhydrous sodium sulfate and concentrate it to 1 mL in a K-D concentrator.

QUALITY CONTROL A methylene chloride solution containing 50 ng/μL of decafluorotriphenylphosphine (DFTPP) is used for tuning the GC/MS system each 12-h shift. A system performance check also must be made during every 12-h shift. A standard containing 50 ng/μL each of 4,4'-DDT, pentachlorophenol, and benzidine is required to verify injection port inertness and GC column performance. A calibration standard at mid-concentration, containing each compound of interest, including all required surrogates, must be performed every 12 h during analysis. After the system performance check is met, calibration check compounds (CCCs) are used to check the validity of the initial calibration.

The internal standard responses and retention times in the calibration check standard must be evaluated immediately after or during data acquisition. If the retention time for any internal standard changes by more than 30 seconds from the last check calibration (12 h), the chromatographic system must be inspected for malfunctions and corrections must be made, as required. If the electron ionization current plot (EICP) area for any of the internal standards changes by a factor of two from the last daily calibration standard check, the mass spectrometer must be inspected for malfunctions and corrections must be made, as appropriate.

Demonstrate, through the analysis of a reagent water blank, that interferences from the analytical system, glassware, and reagents are under control. The blank samples should be carried through all stages of the sample preparation and measurement steps. For each analytical batch (up to 20 samples), a reagent blank, matrix spike, and matrix spike duplicate/duplicate must be analyzed (the frequency of the spikes may be different for different monitoring programs). The blank and

spiked samples must be carried through all stages of the sample preparation and measurement steps. A QC reference sample concentrate containing each analyte at a concentration of 100 mg/L in methanol is required.

REFERENCE Test Methods for Evaluating Solid Waste (SW-846). U.S. EPA 1983, Method 8270B, Rev. 2, Nov. 1990. Office of Solid Waste, Washington, DC.

3-Methylphenol   **EPA Method 8270**
CAS #108-39-4

TITLE Semivolatile Organic Compounds by GC/MS

MATRIX This method is used to determine the concentration of semivolatile organic compounds in extracts prepared from all types of solid waste matrices, soils, and groundwater. Although surface waters are not specifically mentioned, this method should be applicable to water samples from rivers, lakes, etc.

METHOD SUMMARY This method covers 259 semivolatile organic compounds. In very limited applications direct injection of the sample into the GC/MS system may be appropriate, but this results in very high detection limits (approximately 10,000 µg/L). Typically, a 1-L liquid sample, containing surrogate, and matrix spiking standards, is extracted in a continuous extractor first under acid conditions and then under basic conditions. Typically 30 g of a solid sample, containing surrogate, and matrix spiking standards, is extracted ultrasonically. After concentrating the extract to 1 mL it is spiked with 10 µL of an internal standard solution just prior to analysis by GC/MS. The volume injected should contain about 100 ng of base/neutral and 200 ng of acid surrogates (for a 1-µL injection). Analysis is performed by GC/MS using a capillary GC column.

INTERFERENCES Raw GC/MS data from all blanks, samples, and spikes must be evaluated for interferences. Contamination by carryover can occur whenever high-concentration and low-concentration samples are sequentially analyzed. To reduce carryover, the sample syringe must be rinsed out between samples with solvent. Whenever an unusually concentrated sample is encountered, it should be followed by the analysis of blank solvent to check for cross-contamination.

INSTRUMENTATION A GC/MS and a data system are required. The GC column used is a 30 m × 0.25 mm I.D. (or 0.32 mm I.D.) 1um film thickness silicone-coated fused silica capillary column. A continuous liquid-liquid extractor equipped with Teflon® or glass connection joints and stopcocks requiring no lubrication, a K-D concentrating apparatus, water bath, and an ultrasonic disrupter with a minimum power of 300 W and with pulsing capability are also required.

PRECISION & ACCURACY The estimated quantitation limit (EQL) of Method 8270B for determining an individual compound is approximately 1 mg/kg (wet weight) for soil or sediment samples, 1–200 mg/kg for wastes (dependent on matrix and method of preparation), and 10 µg/L for groundwater samples. EQLs will be proportionately higher for sample extracts that require dilution to avoid saturation of the detector.

The EQL(b) for groundwater in µg/L is 10.
The EQL (a, b) for low concentrations in soil and sediment in µg/kg is not determined.
Accuracy as µg/L is not listed.
Overall precision in µg/L is not listed.

(a) *EQLs listed for soil/sediment are based on wet weight. Normally data is reported in a dry-weight basis; therefore, EQLs will be higher based on the % dry weight of each sample. This calculation is based on a 30 g sample and gel permeation chromatography cleanup.*
(b) *Sample EQLs are highly matrix-dependent. The EQLs are provided for guidance and may not always be achievable.*
$C =$ *True value for concentration, in µg/L.*
$X =$ *Average recovery found for measurements of samples containing a concentration of C, in µg/L.*

ESTIMATED QUANTITATION LIMIT

Other Matrices	Factor (a)
High-concentration soil and sludges by sonicator	7.5
Non-water miscible waste	75

(a) *EQL for other matrices = [EQL for low soil/sediment] × [Factor]. This estimated EQL is similar to an EPA "Practical Quantitation Limit."*

SAMPLING METHOD
Liquid samples — Use a 1 or 2½ gallon amber glass bottle with a screw-top Teflon®-lined cover that has been prewashed with detergent and rinsed with distilled water and methanol (or isopropanol).

Soils, sediments, or sludges — Use an 8-oz. widemouth glass with a screw-top Teflon®-lined cover that has been prewashed with detergent and rinsed with distilled water and methanol (or isopropanol).

SAMPLE PRESERVATION
Liquid samples — If residual chlorine is present, add 3 mL of 10% sodium thiosulfate per gallon, cool to 4°C and store in a solvent-free refrigerator until analysis; if chlorine is not present, then eliminate the sodium thiosulfate addition.

Soils, sediments, or sludges — Cool samples to 4°C and store in a solvent-free refrigerator.

MHT Liquid samples must be extracted within 7 days and the extracts analyzed within 40 days. Soils, sediments, or sludges may be stored for a maximum of 14 days and the extracts analyzed within 40 days.

SAMPLE PREPARATION
Liquid samples — Transfer 1 L quantitatively to a continuous extractor. If high concentrations are anticipated, a smaller volume may be used and then diluted with organic-free reagent water to 1 L. Adjust pH, if necessary, to pH <2 using 1:1 (V/V) sulfuric acid. Pipette 1.0 mL of a surrogate standard spiking solution into each sample. For the sample in each analytical batch selected for spiking, add 1.0 mL of a matrix spiking standard. For base/neutral acid analysis, the amount of the surrogates and matrix spiking compounds added to the sample should result in a final concentration of 100 ng/µL of each analyte in the extract to be analyzed (assuming a 1-µL injection). Extract

with methylene chloride for 18–24 h. Next, adjust the pH of the aqueous phase to pH >11 using 10 N sodium hydroxide and extract it with methylene chloride again for 18–24 h. Dry the extract through a column containing anhydrous sodium sulfate and concentrate it to 1 mL using a K-D concentrator.

Soils, sediments, or sludges — Use 30 g of sample. Nonporous or wet samples (gummy or clay type) that do not have a free-flowing sandy texture must be mixed with anhydrous sodium sulfate until the sample is free flowing. Add 1 mL of surrogate standards to all samples, spikes, standards, and blanks. For the sample in each analytical batch selected for spiking, add 1.0 mL of a matrix spiking standard. For base/neutral acid analysis, the amount added of the surrogates and matrix spiking compounds should result in a final concentration of 100 ng/µL of each base/neutral analyte and 200 ng/µL of each acid analyte in the extract to be analyzed (assuming a 1-µL injection). Immediately add a 100-mL mixture of 1:1 methylene chloride:acetone and extract the sample ultrasonically for 3 min and then decant or filter the extracts. Repeat the extraction two or more times. Dry the extract using a column with anhydrous sodium sulfate and concentrate it to 1 mL in a K-D concentrator.

QUALITY CONTROL A methylene chloride solution containing 50 ng/µL of decafluorotriphenylphosphine (DFTPP) is used for tuning the GC/MS system each 12-h shift. A system performance check also must be made during every 12-h shift. A standard containing 50 ng/µL each of 4,4′-DDT, pentachlorophenol, and benzidine is required to verify injection port inertness and GC column performance. A calibration standard at mid-concentration, containing each compound of interest, including all required surrogates, must be performed every 12 h during analysis. After the system performance check is met, calibration check compounds (CCCs) are used to check the validity of the initial calibration.

The internal standard responses and retention times in the calibration check standard must be evaluated immediately after or during data acquisition. If the retention time for any internal standard changes by more than 30 seconds from the last check calibration (12 h), the chromatographic system must be inspected for malfunctions and corrections must be made, as required. If the electron ionization current plot (EICP) area for any of the internal standards changes by a factor of two from the last daily calibration standard check, the mass spectrometer must be inspected for malfunctions and corrections must be made, as appropriate.

Demonstrate, through the analysis of a reagent water blank, that interferences from the analytical system, glassware, and reagents are under control. The blank samples should be carried through all stages of the sample preparation and measurement steps. For each analytical batch (up to 20 samples), a reagent blank, matrix spike, and matrix spike duplicate/duplicate must be analyzed (the frequency of the spikes may be different for different monitoring programs). The blank and spiked samples must be carried through all stages of the sample preparation and measurement steps. A QC reference sample concentrate containing each analyte at a concentration of 100 mg/L in methanol is required.

REFERENCE Test Methods for Evaluating Solid Waste (SW-846). U.S. EPA 1983, Method 8270B, Rev. 2, Nov. 1990. Office of Solid Waste, Washington, DC.

4-Methylphenol — EPA Method 8270
CAS #106-44-5

TITLE Semivolatile Organic Compounds by GC/MS

MATRIX This method is used to determine the concentration of semivolatile organic compounds in extracts prepared from all types of solid waste matrices, soils, and groundwater. Although surface waters are not specifically mentioned, this method should be applicable to water samples from rivers, lakes, etc.

METHOD SUMMARY This method covers 259 semivolatile organic compounds. In very limited applications direct injection of the sample into the GC/MS system may be appropriate, but this results in very high detection limits (approximately 10,000 µg/L). Typically, a 1-L liquid sample, containing surrogate, and matrix spiking standards, is extracted in a continuous extractor first under acid conditions and then under basic conditions. Typically 30 g of a solid sample, containing surrogate, and matrix spiking standards, is extracted ultrasonically. After concentrating the extract to 1 mL it is spiked with 10 µL of an internal standard solution just prior to analysis by GC/MS. The volume injected should contain about 100 ng of base/neutral and 200 ng of acid surrogates (for a 1-µL injection). Analysis is performed by GC/MS using a capillary GC column.

INTERFERENCES Raw GC/MS data from all blanks, samples, and spikes must be evaluated for interferences. Contamination by carryover can occur whenever high-concentration and low-concentration samples are sequentially analyzed. To reduce carryover, the sample syringe must be rinsed out between samples with solvent. Whenever an unusually concentrated sample is encountered, it should be followed by the analysis of blank solvent to check for cross-contamination.

INSTRUMENTATION A GC/MS and a data system are required. The GC column used is a 30 m × 0.25 mm I.D. (or 0.32 mm I.D.) 1um film thickness silicone-coated fused silica capillary column. A continuous liquid-liquid extractor equipped with Teflon® or glass connection joints and stopcocks requiring no lubrication, a K-D concentrating apparatus, water bath, and an ultrasonic disrupter with a minimum power of 300 W and with pulsing capability are also required.

PRECISION & ACCURACY The estimated quantitation limit (EQL) of Method 8270B for determining an individual compound is approximately 1 mg/kg (wet weight) for soil or sediment samples, 1–200 mg/kg for wastes (dependent on matrix and method of preparation), and 10 µg/L for groundwater samples. EQLs will be proportionately higher for sample extracts that require dilution to avoid saturation of the detector.

The EQL(b) for groundwater in µg/L is 10.
The EQL (a, b) for low concentrations in soil and sediment in µg/kg is 660.
Accuracy as µg/L is not listed.

Overall precision in µg/L is not listed.

(a) *EQLs listed for soil/sediment are based on wet weight. Normally data is reported in a dry-weight basis; therefore, EQLs will be higher based on the % dry weight of each sample. This calculation is based on a 30 g sample and gel permeation chromatography cleanup.*
(b) *Sample EQLs are highly matrix-dependent. The EQLs are provided for guidance and may not always be achievable.*
C = *True value for concentration, in µg/L.*
X = *Average recovery found for measurements of samples containing a concentration of C, in µg/L.*

ESTIMATED QUANTITATION LIMIT

Other Matrices	Factor (a)
High-concentration soil and sludges by sonicator	7.5
Non-water miscible waste	75

(a) *EQL for other matrices = [EQL for low soil/sediment] × [Factor]. This estimated EQL is similar to an EPA "Practical Quantitation Limit."*

SAMPLING METHOD

Liquid samples — Use a 1 or 2½ gallon amber glass bottle with a screw-top Teflon®-lined cover that has been prewashed with detergent and rinsed with distilled water and methanol (or isopropanol).

Soils, sediments, or sludges — Use an 8-oz. widemouth glass with a screw-top Teflon®-lined cover that has been prewashed with detergent and rinsed with distilled water and methanol (or isopropanol).

SAMPLE PRESERVATION

Liquid samples — If residual chlorine is present, add 3 mL of 10% sodium thiosulfate per gallon, cool to 4°C and store in a solvent-free refrigerator until analysis; if chlorine is not present, then eliminate the sodium thiosulfate addition.

Soils, sediments, or sludges — Cool samples to 4°C and store in a solvent-free refrigerator.

MHT Liquid samples must be extracted within 7 days and the extracts analyzed within 40 days. Soils, sediments, or sludges may be stored for a maximum of 14 days and the extracts analyzed within 40 days.

SAMPLE PREPARATION

Liquid samples — Transfer 1 L quantitatively to a continuous extractor. If high concentrations are anticipated, a smaller volume may be used and then diluted with organic-free reagent water to 1 L. Adjust pH, if necessary, to pH <2 using 1:1 (V/V) sulfuric acid. Pipette 1.0 mL of a surrogate standard spiking solution into each sample. For the sample in each analytical batch selected for spiking, add 1.0 mL of a matrix spiking standard. For base/neutral acid analysis, the amount of the surrogates and matrix spiking compounds added to the sample should result in a final concentration of 100 ng/µL of each analyte in the extract to be analyzed (assuming a 1-µL injection). Extract with methylene chloride for 18–24 h. Next, adjust the pH of the aqueous phase to pH >11 using 10 N sodium hydroxide and extract it with methylene chloride again for 18–24 h. Dry the extract through a column containing anhydrous sodium sulfate and concentrate it to 1 mL using a K-D concentrator.

Soils, sediments, or sludges — Use 30 g of sample. Nonporous or wet samples (gummy or clay type) that do not have a free-flowing sandy texture must be mixed with anhydrous sodium sulfate until the sample is free flowing. Add 1 mL of surrogate standards to all samples, spikes, standards, and blanks. For the sample in each analytical batch selected for spiking, add 1.0 mL of a matrix spiking standard. For base/neutral acid analysis, the amount added of the surrogates and matrix spiking compounds should result in a final concentration of 100 ng/µL of each base/neutral analyte and 200 ng/µL of each acid analyte in the extract to be analyzed (assuming a 1-µL injection). Immediately add a 100-mL mixture of 1:1 methylene chloride:acetone and extract the sample ultrasonically for 3 min and then decant or filter the extracts. Repeat the extraction two or more times. Dry the extract using a column with anhydrous sodium sulfate and concentrate it to 1 mL in a K-D concentrator.

QUALITY CONTROL A methylene chloride solution containing 50 ng/µL of decafluorotriphenylphosphine (DFTPP) is used for tuning the GC/MS system each 12-h shift. A system performance check also must be made during every 12-h shift. A standard containing 50 ng/µL each of 4,4′-DDT, pentachlorophenol, and benzidine is required to verify injection port inertness and GC column performance. A calibration standard at mid-concentration, containing each compound of interest, including all required surrogates, must be performed every 12 h during analysis. After the system performance check is met, calibration check compounds (CCCs) are used to check the validity of the initial calibration.

The internal standard responses and retention times in the calibration check standard must be evaluated immediately after or during data acquisition. If the retention time for any internal standard changes by more than 30 seconds from the last check calibration (12 h), the chromatographic system must be inspected for malfunctions and corrections must be made, as required. If the electron ionization current plot (EICP) area for any of the internal standards changes by a factor of two from the last daily calibration standard check, the mass spectrometer must be inspected for malfunctions and corrections must be made, as appropriate.

Demonstrate, through the analysis of a reagent water blank, that interferences from the analytical system, glassware, and reagents are under control. The blank samples should be carried through all stages of the sample preparation and measurement steps. For each analytical batch (up to 20 samples), a reagent blank, matrix spike, and matrix spike duplicate/duplicate must be analyzed (the frequency of the spikes may be different for different monitoring programs). The blank and spiked samples must be carried through all stages of the sample preparation and measurement steps. A QC reference sample concentrate containing each analyte at a concentration of 100 mg/L in methanol is required.

REFERENCE Test Methods for Evaluating Solid Waste (SW-846). U.S. EPA 1983, Method 8270B, Rev. 2, Nov. 1990. Office of Solid Waste, Washington, DC.

2-Methylpyridine
CAS #109-06-8
EPA Method 8270

TITLE Semivolatile Organic Compounds by GC/MS

MATRIX This method is used to determine the concentration of semivolatile organic compounds in extracts prepared from all types of solid waste matrices, soils, and groundwater. Although surface waters are not specifically mentioned, this method should be applicable to water samples from rivers, lakes, etc.

METHOD SUMMARY This method covers 259 semivolatile organic compounds. In very limited applications direct injection of the sample into the GC/MS system may be appropriate, but this results in very high detection limits (approximately 10,000 µg/L). Typically, a 1-L liquid sample, containing surrogate, and matrix spiking standards, is extracted in a continuous extractor first under acid conditions and then under basic conditions. Typically 30 g of a solid sample, containing surrogate, and matrix spiking standards, is extracted ultrasonically. After concentrating the extract to 1 mL it is spiked with 10 µL of an internal standard solution just prior to analysis by GC/MS. The volume injected should contain about 100 ng of base/neutral and 200 ng of acid surrogates (for a 1-µL injection). Analysis is performed by GC/MS using a capillary GC column.

INTERFERENCES Raw GC/MS data from all blanks, samples, and spikes must be evaluated for interferences. Contamination by carryover can occur whenever high-concentration and low-concentration samples are sequentially analyzed. To reduce carryover, the sample syringe must be rinsed out between samples with solvent. Whenever an unusually concentrated sample is encountered, it should be followed by the analysis of blank solvent to check for cross-contamination.

INSTRUMENTATION A GC/MS and a data system are required. The GC column used is a 30 m × 0.25 mm I.D. (or 0.32 mm I.D.) 1um film thickness silicone-coated fused silica capillary column. A continuous liquid-liquid extractor equipped with Teflon® or glass connection joints and stopcocks requiring no lubrication, a K-D concentrating apparatus, water bath, and an ultrasonic disrupter with a minimum power of 300 W and with pulsing capability are also required.

PRECISION & ACCURACY The estimated quantitation limit (EQL) of Method 8270B for determining an individual compound is approximately 1 mg/kg (wet weight) for soil or sediment samples, 1–200 mg/kg for wastes (dependent on matrix and method of preparation), and 10 µg/L for groundwater samples. EQLs will be proportionately higher for sample extracts that require dilution to avoid saturation of the detector.

The EQL(b) for groundwater in µg/L is not listed.
The EQL (a, b) for low concentrations in soil and sediment in µg/kg is not listed.
Accuracy as µg/L is not listed.
Overall precision in µg/L is not listed.

(a) EQLs listed for soil/sediment are based on wet weight. Normally data is reported in a dry-weight basis; therefore, EQLs will be higher based on the % dry weight of each sample. This calculation is based on a 30 g sample and gel permeation chromatography cleanup.

(b) Sample EQLs are highly matrix-dependent. The EQLs are provided for guidance and may not always be achievable.

C = True value for concentration, in µg/L.
X = Average recovery found for measurements of samples containing a concentration of C, in µg/L.

ESTIMATED QUANTITATION LIMIT

Other Matrices	Factor (a)
High-concentration soil and sludges by sonicator	7.5
Non-water miscible waste	75

(a) EQL for other matrices = [EQL for low soil/sediment] × [Factor]. This estimated EQL is similar to an EPA "Practical Quantitation Limit."

SAMPLING METHOD
Liquid samples — Use a 1 or 2½ gallon amber glass bottle with a screw-top Teflon®-lined cover that has been prewashed with detergent and rinsed with distilled water and methanol (or isopropanol).

Soils, sediments, or sludges — Use an 8-oz. widemouth glass with a screw-top Teflon®-lined cover that has been prewashed with detergent and rinsed with distilled water and methanol (or isopropanol).

SAMPLE PRESERVATION
Liquid samples — If residual chlorine is present, add 3 mL of 10% sodium thiosulfate per gallon, cool to 4°C and store in a solvent-free refrigerator until analysis; if chlorine is not present, then eliminate the sodium thiosulfate addition.

Soils, sediments, or sludges — Cool samples to 4°C and store in a solvent-free refrigerator.

MHT Liquid samples must be extracted within 7 days and the extracts analyzed within 40 days. Soils, sediments, or sludges may be stored for a maximum of 14 days and the extracts analyzed within 40 days.

SAMPLE PREPARATION
Liquid samples — Transfer 1 L quantitatively to a continuous extractor. If high concentrations are anticipated, a smaller volume may be used and then diluted with organic-free reagent water to 1 L. Adjust pH, if necessary, to pH <2 using 1:1 (V/V) sulfuric acid. Pipette 1.0 mL of a surrogate standard spiking solution into each sample. For the sample in each analytical batch selected for spiking, add 1.0 mL of a matrix spiking standard. For base/neutral acid analysis, the amount of the surrogates and matrix spiking compounds added to the sample should result in a final concentration of 100 ng/µL of each analyte in the extract to be analyzed (assuming a 1-µL injection). Extract with methylene chloride for 18–24 h. Next, adjust the pH of the aqueous phase to pH >11 using 10 N sodium hydroxide and extract it with methylene chloride again for 18–24 h. Dry the extract through a column containing anhydrous sodium sulfate and concentrate it to 1 mL using a K-D concentrator.

Soils, sediments, or sludges — Use 30 g of sample. Nonporous or wet samples (gummy or clay type) that do not have a free-flowing

sandy texture must be mixed with anhydrous sodium sulfate until the sample is free flowing. Add 1 mL of surrogate standards to all samples, spikes, standards, and blanks. For the sample in each analytical batch selected for spiking, add 1.0 mL of a matrix spiking standard. For base/neutral acid analysis, the amount added of the surrogates and matrix spiking compounds should result in a final concentration of 100 ng/µL of each base/neutral analyte and 200 ng/µL of each acid analyte in the extract to be analyzed (assuming a 1-µL injection). Immediately add a 100-mL mixture of 1:1 methylene chloride:acetone and extract the sample ultrasonically for 3 min and then decant or filter the extracts. Repeat the extraction two or more times. Dry the extract using a column with anhydrous sodium sulfate and concentrate it to 1 mL in a K-D concentrator.

QUALITY CONTROL A methylene chloride solution containing 50 ng/µL of decafluorotriphenylphosphine (DFTPP) is used for tuning the GC/MS system each 12-h shift. A system performance check also must be made during every 12-h shift. A standard containing 50 ng/µL each of 4,4′-DDT, pentachlorophenol, and benzidine is required to verify injection port inertness and GC column performance. A calibration standard at mid-concentration, containing each compound of interest, including all required surrogates, must be performed every 12 h during analysis. After the system performance check is met, calibration check compounds (CCCs) are used to check the validity of the initial calibration.

The internal standard responses and retention times in the calibration check standard must be evaluated immediately after or during data acquisition. If the retention time for any internal standard changes by more than 30 seconds from the last check calibration (12 h), the chromatographic system must be inspected for malfunctions and corrections must be made, as required. If the electron ionization current plot (EICP) area for any of the internal standards changes by a factor of two from the last daily calibration standard check, the mass spectrometer must be inspected for malfunctions and corrections must be made, as appropriate.

Demonstrate, through the analysis of a reagent water blank, that interferences from the analytical system, glassware, and reagents are under control. The blank samples should be carried through all stages of the sample preparation and measurement steps. For each analytical batch (up to 20 samples), a reagent blank, matrix spike, and matrix spike duplicate/duplicate must be analyzed (the frequency of the spikes may be different for different monitoring programs). The blank and spiked samples must be carried through all stages of the sample preparation and measurement steps. A QC reference sample concentrate containing each analyte at a concentration of 100 mg/L in methanol is required.

REFERENCE Test Methods for Evaluating Solid Waste (SW-846). U.S. EPA 1983, Method 8270B, Rev. 2, Nov. 1990. Office of Solid Waste, Washington, DC.

2-(Methylthio)-benzothiazole   EPA Method 1625
CAS #615-22-5

TITLE Semivolatile Organic Compounds by Isotope Dilution GC/MS

MATRIX The compounds may be determined in waters, soils, and municipal sludges by this method.

METHOD SUMMARY This method is used to determine 176 semivolatile toxic organic pollutants associated with the CWA (as amended 1987); the RCRA (as amended 1986); the CERCLA (as amended 1986); and other compounds amenable to extraction and analysis by capillary column gas chromatography-mass spectrometry (GC/MS).

Stable isotopically-labeled analogs of the compounds of interest are added to the sample. If the solids content is less than 1%, a 1-L sample is extracted at pH 12–13, then at pH <2 with methylene chloride using continuous extraction techniques.

If the solids content is 30% or less, the sample is diluted to 1% solids with reagent water, homogenized ultrasonically, and extracted at pH 12–13, then at pH <2 with methylene chloride using continuous extraction techniques. If the solids content is greater than 30%, the sample is extracted using ultrasonic techniques.

Each extract is dried over sodium sulfate, concentrated to a volume of 5 mL, cleaned up using GPC, if necessary, and concentrated. Extracts are concentrated to 1 mL if GPC is not performed, and to 0.5 mL if GPC is performed.

An internal standard is added to the extract, and a 1-mL aliquot of the extract is injected into the GC. The compounds are separated by GC and detected by a MS. The labeled compounds serve to correct the variability of the analytical technique.

INTERFERENCES Solvents, reagents, glassware, and other sample processing hardware may yield artifacts and/or elevated baselines causing misinterpretation of chromatograms and spectra. Materials used in the analysis must be demonstrated to be free from interferences under the conditions of analysis by running method blanks initially and with each sample lot (sample started through the extraction process on a given 8-h shift, to a maximum of 20). Specific selection of reagents and purification of solvents by distillation in all glass systems may be required. Glassware and, where possible, reagents are cleaned by solvent rinse and baking at 450°C for 1-h minimum. Interferences coextracted from samples will vary considerably from source to source, depending on the diversity of the site being sampled.

INSTRUMENTATION Major instrumentation includes a GC with a splitless or on-column injection port for capillary column, a MS with 70 eV electron impact ionization, and a data system to collect and record MS data, and process it. A K-D apparatus is used to concentrate extracts.

GC Column: 30 m × 0.25 mm I.D. 5% phenyl, 94% methyl, 1% vinyl silicone bonded phased fused silica capillary column.

PRECISION & ACCURACY The detection limits of the method are usually dependent on the level of interferences rather than instrumental limitations. The limits typify the minimum quantities that can be detected with no interferences present.

The minimum level (in µg/mL) was not listed. This is defined as a minimum level at which the analytical system shall give recognizable mass spectra (background corrected) and acceptable calibration points.

The MDL (in µg/kg) in low solids was not listed and in high solids was not listed; these were determined in digested sludge (low solids) and in filter cake or compost (high solids).

The labeled and native compound initial precision as standard deviation (in µg/L) was not listed.

The labeled and native compound initial accuracy as average recovery (in µg/L) was not listed.

SAMPLE COLLECTION, PRESERVATION & HANDLING Collect samples in glass containers. Aqueous samples which flow freely are collected in refrigerated bottles using automatic sampling equipment. Solid samples are collected as grab samples using widemouth jars. Maintain samples at 0 to 4°C from the time of collection until extraction. If residual chlorine is present in aqueous samples, add 80 mg sodium thiosulfate/L of water. Begin sample extraction within 7 days of collection, and analyze all extracts within 40 days of extraction.

SAMPLE PREPARATION Samples containing 1% solids or less are extracted directly using continuous liquid-liquid extraction techniques. Samples containing 1 to 30% solids are diluted to the 1% level with reagent water and extracted using continuous liquid-liquid extraction techniques. Samples containing greater than 30% solids are extracted using ultrasonic techniques.

- Base/neutral extraction — Adjust the pH of the waters in the extractors to 12–13 with 6 N NaOH. Extract with methylene chloride for 24–48 h.
- Acid extraction — Adjust the pH of the waters in the extractors to 2 or less using 6 N sulfuric acid. Extract with methylene chloride for 24–48 h.
- Ultrasonic extraction of high solids samples — Add anhydrous sodium sulfate to the sample and QC aliquot(s). Add acetone:methylene chloride (1:1) to the sample and mix thoroughly

Concentrate extracts using a K-D apparatus.

QUALITY CONTROL The analyst is permitted to modify this method to improve separations or lower the costs of measurements, provided all performance specifications are met. Analyses of blanks are required to demonstrate freedom from contamination. When results of spikes indicate atypical method performance for samples, the samples are diluted to bring method performance within acceptable limits.

For low solids (aqueous samples), extract, concentrate, and analyze two sets of four 1-L aliquots (8 aliquots total) of the precision and recovery standard. For high solids samples, two sets of four 30-g aliquots of the high solids reference matrix are used.

Spike all samples with labeled compounds to assess method performance. Compute percent recovery of the labeled compounds using the internal standard method. Compare the labeled compound recovery for each compound with the corresponding labeled compound recovery.

Reagent water and high solids reference matrix blanks are analyzed to demonstrate freedom from contamination. Extract and concentrate a 1-L reagent water blank or a high solids reference matrix blank with each sample's lot (samples started through the extraction process on the same 8-h shift, to a maximum of 20 samples).

Field replicates may be collected to determine the precision of the sampling technique, and spiked samples may be required to determine the accuracy of the analysis when the internal standard method is used.

REFERENCE Semivolatile Organic Compounds by Isotope Dilution GC/MS. Office of Water Regulation and Standards, U.S. EPA Industrial Technology Division, Washington, DC, EPA Method 1625, Rev. C, June 1989 (contact W.A. Telliard, U.S. EPA, Office of Water Regulations and Standards, 401 M St., SW, Washington, DC, 20460. Phone: 202-382-7131).

Metribuzin   **EPA Method 507**
CAS #21087-64-9

TITLE Determination of Nitrogen and Phosphorus-Containing Pesticides in Water by GC/NPD

MATRIX This method is applicable to the determination of certain nitrogen and phosphorus-containing pesticides in finished drinking water and groundwater.

METHOD SUMMARY Method 507 covers 46 nitrogen- and phosphorus-containing pesticides. A 1-L sample is fortified with a surrogate standard, salted, buffered, extracted with methylene chloride, and concentrated; then the solvent is exchanged with methyl tert-butyl ether (MTBE) and concentrated again, and a 2-µL aliquot of a sample extract is injected into a GC system equipped with a selective nitrogen-phosphorus detector and a capillary column for analysis.

INTERFERENCES Method interferences may be caused by contaminants in solvents, reagents, glassware, and other sample processing apparatus. Interfering contamination may occur when a sample containing low concentrations of analytes is analyzed immediately following a sample containing relatively high concentrations. One or more injections of MTBE should be made following the analysis of a sample with high concentrations of analytes to check for analyte carryover. Matrix interferences may be caused by contaminants that are coextracted from the sample. The extent of matrix interferences will vary considerably from source to source, depending upon the water sampled.

INSTRUMENTATION A gas chromatograph system (GC) equipped with a nitrogen-phosphorus detector (NPD) is needed.

Column 1: 30 m × 0.25 mm I.D. DB-5 bonded fused silica column, 0.25 μm film thickness, or equivalent.

Column 2: 30 m × 0.25 mm I.D. DB-1701 bonded fused silica column, 0.25 μm film thickness, or equivalent.

PRECISION & ACCURACY This method has been validated in a single lab and estimated detection limits (EDLs) have been determined for each analyte. Observed detection limits may vary among waters, depending upon the nature of the interferences in the sample matrix and the specific instrumentation used. Analytes that are not separated chromatographically cannot be individually identified and measured unless an alternative technique for identification and quantification exist.

The estimated detection limit (in μg/L) was 0.15. The EDL is defined as either method detection limit or a level of compound in a sample yielding a peak in the final extract with signal-to-noise ratio of approximately 5, whichever value is higher.

The concentration used for these measurements (in μg/L) was 1.5.
The accuracy (as % recovery) was 101.
The precision (% RSD) was 5.

SAMPLING METHOD Grab samples are collected in 1-L glass sample bottles (prewashed with detergent and hot tap water, rinsed with reagent water, and dried in an oven at 400°C for 1 h) with screw caps lined with PTFE-fluorocarbon.

SAMPLE PRESERVATION Add mercuric chloride to the sample bottle in amounts to produce a concentration of 10 mg/L. If residual chlorine is present, add 80 mg of sodium thiosulfate/L of sample to the sample bottle prior to collection. After collection, seal bottle and shake vigorously for 1 min, then cool the sample to 4°C immediately and store it at 4°C in the dark until extraction.

MHT Maximum holding time of the samples, and in some cases the extracts, is 14 days.

SAMPLE PREPARATION Fortify the sample with 50 μL of the surrogate standard solution, adjust to pH 7 with phosphate buffer, add 100 g NaCl to the sample, and seal and shake to dissolve the salt; then extract with methylene chloride in a separatory funnel or in a mechanical tumbler bottle. Dry the extract by pouring it through a solvent-rinsed drying column containing about 10 cm of anhydrous sodium sulfate. Collect the extract in a Kuderna-Danish (K-D) concentrator and rinse the column with 20–30 mL methylene chloride. Concentrate the extract to about 2 mL and rinse the flask and its lower joint into the concentrator tube with 1 to 2 mL of methyl t-butyl ether (MTBE). Add 5–10 mL of MTBE and concentrate the extract twice (adding more MTBE) to a final volume of 5.0 mL and store it at 4°C until analysis.

Note: If methylene chloride is not completely removed from the final extract, it may cause detector problems.

QUALITY CONTROL Minimum quality control requirements are initial demonstration of lab capability, determination of surrogate compound recoveries in each sample and blank, monitoring internal standard peak area or height in each sample and blank, analysis of lab reagent blanks, lab fortified samples, lab fortified blanks, and other QC samples. A lab reagent blank is analyzed to demonstrate that all glassware and reagent interferences are under control.

Initial demonstration of capability is fulfilled by analyzing four fortified reagent water samples with the recovery value for each analyte falling within the acceptable range (±30% average recovery). Surrogate recoveries from samples or method blanks must be 70–130%. The internal standard response for any sample chromatogram should not deviate from the daily calibration check standard's internal standard response by more than 30% or lab fortified blanks and sample matrices are used to assess lab performance and analyte recovery, respectively.

If the response for the target analyte peak exceeds the working range of the system, dilute the extract and reanalyze. Alternative techniques such as an alternate detector or second chromatography column should be used to confirm peak identification when sample components are not resolved adequately.

EPA CONTACT & HOTLINE For technical questions contact Dr. Baldev Bathija, U.S. EPA, Office of Ground Water and Drinking Water (WH-550D), 401 M St. SW, Washington, DC 20460. Tel. (202) 260-3040. For further information the EPA Safe Drinking Water Hotline may be called at: (800) 426-4791.

REFERENCE Methods for the Determination of Organic Compounds in Drinking Water, EPA/600/4-88/039 (revised July 1991). U.S. EPA Environmental Monitoring Systems Laboratory, Cincinnati, OH, 45268, U.S.A. Available from the National Technical Information Service (NTIS), 5285 Port Royal Road, Springfield, VA 22161; Tel. 800-553-6847. NTIS Order Number is PB91-231480.

Mevinphos   **EPA Method 507**
CAS #7786-34-7

TITLE Determination of Nitrogen and Phosphorus-Containing Pesticides in Water by GC/NPD

MATRIX This method is applicable to the determination of certain nitrogen and phosphorus-containing pesticides in finished drinking water and groundwater.

METHOD SUMMARY Method 507 covers 46 nitrogen- and phosphorus-containing pesticides. A 1-L sample is fortified with a surrogate standard, salted, buffered, extracted with methylene chloride, and concentrated; then the solvent is exchanged with methyl tert-butyl ether (MTBE) and concentrated again, and a 2-μL aliquot of a sample extract is injected into a GC system equipped with a selective nitrogen-phosphorus detector and a capillary column for analysis.

INTERFERENCES Method interferences may be caused by contaminants in solvents, reagents, glassware, and other sample processing apparatus. Interfering contamination may occur when a sample containing low concentrations of analytes is analyzed immediately following a sample containing relatively high concentrations. One or more injections of MTBE should be made following the analysis of a sample with high concentrations of analytes to check for analyte carryover. Matrix interferences may be caused by contaminants that are coextracted from the sample. The extent of matrix interferences will vary considerably from source to source, depending upon the water sampled.

INSTRUMENTATION A gas chromatograph system (GC) equipped with a nitrogen-phosphorus detector (NPD) is needed.

Column 1: 30 m × 0.25 mm I.D. DB-5 bonded fused silica column, 0.25 μm film thickness, or equivalent.
Column 2: 30 m × 0.25 mm I.D. DB-1701 bonded fused silica column, 0.25 μm film thickness, or equivalent.

PRECISION & ACCURACY This method has been validated in a single lab and estimated detection limits (EDLs) have been determined for each analyte. Observed detection limits may vary among waters, depending upon the nature of the interferences in the sample matrix and the specific instrumentation used. Analytes that are not separated chromatographically cannot be individually identified and measured unless an alternative technique for identification and quantification exist.

The estimated detection limit (in μg/L) was 5. The EDL is defined as either method detection limit or a level of compound in a sample yielding a peak in the final extract with signal-to-noise ratio of approximately 5, whichever value is higher.

The concentration used for these measurements (in μg/L) was 50.
The accuracy (as % recovery) was 95.
The precision (% RSD) was 11.

SAMPLING METHOD Grab samples are collected in 1-L glass sample bottles (prewashed with detergent and hot tap water, rinsed with reagent water, and dried in an oven at 400°C for 1 h) with screw caps lined with PTFE-fluorocarbon.

SAMPLE PRESERVATION Add mercuric chloride to the sample bottle in amounts to produce a concentration of 10 mg/L. If residual chlorine is present, add 80 mg of sodium thiosulfate/L of sample to the sample bottle prior to collection. After collection, seal bottle and shake vigorously for 1 min, then cool the sample to 4°C immediately and store it at 4°C in the dark until extraction.

MHT Maximum holding time of the samples, and in some cases the extracts, is 14 days.

SAMPLE PREPARATION Fortify the sample with 50 μL of the surrogate standard solution, adjust to pH 7 with phosphate buffer, add 100 g NaCl to the sample, and seal and shake to dissolve the salt; then extract with methylene chloride in a separatory funnel or in a mechanical tumbler bottle. Dry the extract by pouring it through a solvent-rinsed drying column containing about 10 cm of anhydrous sodium sulfate. Collect the extract in a Kuderna-Danish (K-D) concentrator and rinse the column with 20–30 mL methylene chloride. Concentrate the extract to about 2 mL and rinse the flask and its lower joint into the concentrator tube with 1 to 2 mL of methyl t-butyl ether (MTBE). Add 5–10 mL of MTBE and concentrate the extract twice (adding more MTBE) to a final volume of 5.0 mL and store it at 4°C until analysis.

Note: If methylene chloride is not completely removed from the final extract, it may cause detector problems.

QUALITY CONTROL Minimum quality control requirements are initial demonstration of lab capability, determination of surrogate compound recoveries in each sample and blank, monitoring internal standard peak area or height in each sample and blank, analysis of lab reagent blanks, lab fortified samples, lab fortified blanks, and other QC samples. A lab reagent blank is analyzed to demonstrate that all glassware and reagent interferences are under control.

Initial demonstration of capability is fulfilled by analyzing four fortified reagent water samples with the recovery value for each analyte falling within the acceptable range (±30% average recovery). Surrogate recoveries from samples or method blanks must be 70–130%. The internal standard response for any sample chromatogram should not deviate from the daily calibration check standard's internal standard response by more than 30% or lab fortified blanks and sample matrices are used to assess lab performance and analyte recovery, respectively.

If the response for the target analyte peak exceeds the working range of the system, dilute the extract and reanalyze. Alternative techniques such as an alternate detector or second chromatography column should be used to confirm peak identification when sample components are not resolved adequately.

EPA CONTACT & HOTLINE For technical questions contact Dr. Baldev Bathija, U.S. EPA, Office of Ground Water and Drinking Water (WH-550D), 401 M St. SW, Washington, DC 20460. Tel. (202) 260-3040. For further information the EPA Safe Drinking Water Hotline may be called at: (800) 426-4791.

REFERENCE Methods for the Determination of Organic Compounds in Drinking Water, EPA/600/4-88/039 (revised July 1991). U.S. EPA Environmental Monitoring Systems Laboratory, Cincinnati, OH, 45268, U.S.A. Available from the National Technical Information Service (NTIS), 5285 Port Royal Road, Springfield, VA 22161; Tel. 800-553-6847. NTIS Order Number is PB91-231480.

Mevinphos   **EPA Method 8141**
CAS #7786-34-7

TITLE Organophosphorus Compounds by Gas Chromatography: Capillary Column Technique

MATRIX This method covers aqueous and solid matrices. This includes a wide variety such as drinking water, groundwater, industrial wastewaters, surface waters, soils, solids, and sediments.

METHOD SUMMARY This is a GC method used to determine the concentration of 28 organophosphorus pesticides.

The use of Gel Permeation Cleanup (EPA Method 3640) for sample cleanup has been demonstrated to yield recoveries of less than 85% for many method analytes and is therefore not recommended for use with this method.

This method provides GC conditions for the detection of ppb concentrations of organophosphorus compounds. Prior to the use of this method, appropriate sample preparation techniques must be used. Water samples are extracted at a neutral pH with methylene chloride as a solvent by using a separatory funnel (EPA Method 3510) or a continuous liquid-liquid extractor (EPA Method 3520). Soxhlet extraction (EPA Method 3540) or ultrasonic extraction (EPA Method 3550) using methylene chloride/acetone (1:1) are used for solid samples. Both neat and diluted organic liquids (EPA Method 3580) may be analyzed by direct injection. Spiked samples are used to verify the applicability of the chosen extraction technique to each new sample type. A GC with a flame photometric (FPD) or nitrogen-phosphorus detector (NPD) is used for this multiresidue procedure.

INTERFERENCES The use of Florisil cleanup materials (EPA Method 3620) for some of the compounds in this method has been demonstrated to yield recoveries less than 85% and is therefore not recommended for all compounds. Use of phosphorus or halogen specific detectors, however, often obviates the necessity for cleanup for relatively clean sample matrices. If particular circumstances demand the use of an alternative cleanup procedure, the analyst must determine the elution profile and demonstrate that the recovery of each analyte is no less than 85%.

Use of a flame photometric detector (FPD) in the phosphorus mode will minimize interferences from materials that do not contain phosphorus. Elemental sulfur, however, may interfere with the determination of certain organophosphorus compounds by flame photometric gas chromatography. Sulfur cleanup using EPA Method 3660 may alleviate this interference. A nitrogen phosphorus detector (NPD) is also recommended.

A few analytes coelute on certain columns. Therefore, select a second column for confirmation where coelution of the analytes of interest does not occur.

Method interferences may be caused by contaminants in solvents, reagents, glassware, and other sample processing hardware that lead to discrete artifacts or elevated baselines in gas chromatograms. All these materials must be routinely demonstrated to be free from interferences under the conditions of the analysis by analyzing reagent blanks.

INSTRUMENTATION A GC with a NPD or a FPD will be needed. A data system or integrator is recommended for measuring peak areas and/or peak heights. A Kuderna-Danish (K-D) apparatus will be needed for extract concentration.

Column 1: 15 m × 0.53 mm megabore capillary column, 1.0 µm film thickness, DB-210.
Column 2: 15 m × 0.53 mm megabore capillary column, 1.5 µm film thickness, SPB-608.
Column 3: 15 m × 0.53 mm megabore capillary column, 1.0 µm film thickness, DB-5.

Three megabore capillary columns are included for analysis of organophosphates by this method. Column 1 (DB-210 or equivalent) and Column 2 (SPB-608 or equivalent) are recommended if a large number of organophosphorus analytes are to be determined. If the superior resolution offered by Column 1 and Column 2 is not required, Column 3 (DB-5 or equivalent) may be used. For megabore capillary columns, automatic injections of 1 µL are recommended.

PRECISION & ACCURACY The MDL actually achieved in a given analysis will vary, as it is dependent on instrument sensitivity and matrix effects. Single operator accuracy and precision studies have been conducted with spiked water and soil samples.

MULTIPLICATION FACTORS FOR OTHER MATRICES (a)

Matrix	Factor (b)
Groundwater	10
(EPA Method 3510 or EPA Method 3520)	
Low-concentration soil by Soxhlet and no cleanup	10 (c)
Low-concentration soil by ultrasonic extraction with GPC cleanup	6.7 (c)
High-concentration soil and sludges by ultrasonic extraction	500 (c)
Non-water miscible waste (EPA Method 3580)	1000 (c)

(a) Sample EQLs are highly matrix-dependent. The EQLs listed here are provided for guidance and may not always be achievable.
(b) EQL = [Method detection limit] × [Factor]. For non-aqueous samples the factor is on a wet-weight basis.
(c) Multiply this factory times the soil MDL.

The MDL (in µg/L) when reagent water was extracted using a separatory funnel was 0.50.
The MDL (in µg/kg) when soil was extracted using Soxhlet extraction (EPA Method 3540) was 25.0.
Accuracy (as % recovery) with separatory funnel extraction ranged from not recovered (with low spikes) to 55 (with high spikes).
Accuracy (as % recovery) with continuous liquid-liquid extraction ranged from not recovered (with low spikes) to 49 (with high spikes).
Accuracy (as % recovery) with Soxhlet extraction of soils ranged from not recovered (with low spikes to 63 (with high spikes).
Accuracy (as % recovery) with ultrasonic extraction of soils ranged from not recovered (with low spikes) to 23 (with high spikes).

SAMPLE COLLECTION, PRESERVATION & HANDLING
Containers used to collect samples for the determination of semivolatile organic compounds should be soap and water washed followed by methanol (or isopropanol) rinsing. The sample containers should be of glass or Teflon® and have screw-top covers with Teflon® liners.

No preservation is used with concentrated waste samples. With liquid samples containing no residual chlorine and with soil, sediment, and sludge samples, immediately cooling to 4°C is the only preservation used. When residual chlorine is present then 3 mL of 10% aqueous sodium sulfate is added for each gallon of sample collected, followed by cooling to 4°C.

Liquid samples must be extracted within 7 days and their extracts analyzed within 40 days. Concentrated waste, soil, sediment, and sludge samples must be extracted within 14 days and their extracts analyzed within 40 days.

SAMPLE PREPARATION In general, water samples are extracted at a neutral pH with methylene chloride, using either EPA Method 3510 or EPA Method 3520. Solid samples are extracted using either EPA Method 3540 or EPA Method 3550 with methylene chloride/acetone (1:1) as the extraction solvent.

Prior to GC analysis, the extraction solvent may be exchanged to hexane. Single lab data indicates that samples should not be transferred with 100% hexane during sample workup as the more water soluble organophosphorus compounds may be lost.

If cleanup is performed on the samples, the analyst should analyze the samples by GC. This will confirm elution patterns and the absence of interferences from the reagents. If peak detection and identification is prevented by the presence of interferences, further cleanup is required.

QUALITY CONTROL The analyst should monitor the performance of the extraction, cleanup (when used), and analytical system and the effectiveness of the method in dealing with each sample matrix by spiking each sample, standard, and blank with one or two surrogates (e.g., organophosphorus compounds not expected to be present in the sample). Deuterated analogs of analytes should not be used as surrogates for gas chromatographic analysis due to coelution problems.

A minimum of five concentrations for each analyte of interest should be prepared through dilution of the stock standards with isooctane. One of the concentrations should be at a concentration near, but above, the MDL.

Include a mid-level check standard after each group of 10 samples in the analysis sequence. GC/MS techniques should be judiciously employed to support qualitative identifications made with this method. Follow the GC/MS operating requirements specified in EPA Method 8270.

When available, chemical ionization mass spectra may be employed to aid in the qualitative identification process. To confirm an identification of a compound, the background-corrected mass spectrum of the compound must be obtained from the sample extract and must be compared with a mass spectrum from a stock or calibration standard analyzed under the same chromatographic conditions. The molecular ion and all other ions present above 20% relative abundance in the mass spectrum of the standard must be present in the mass spectrum of the sample with agreement to ±20%. The retention time of the compound in the sample must be within six seconds of the retention time for the same compound in the standard solution.

Should the MS procedure fail to provide satisfactory results, additional steps may be taken before reanalysis. These steps may include the use of alternate packed or capillary GC columns or additional sample cleanup.

REFERENCE Test Methods for Evaluating Solid Waste, Physical/Chemical Methods, SW-846, 3rd Edition, U.S. EPA, Office of Solid Waste, Washington, DC, EPA Method 8141 July 1992.

Mevinphos   **EPA Method 8270**
CAS #7786-34-7

TITLE Semivolatile Organic Compounds by GC/MS

MATRIX This method is used to determine the concentration of semivolatile organic compounds in extracts prepared from all types of solid waste matrices, soils, and groundwater. Although surface waters are not specifically mentioned, this method should be applicable to water samples from rivers, lakes, etc.

METHOD SUMMARY This method covers 259 semivolatile organic compounds. In very limited applications direct injection of the sample into the GC/MS system may be appropriate, but this results in very high detection limits (approximately 10,000 µg/L). Typically, a 1-L liquid sample, containing surrogate, and matrix spiking standards, is extracted in a continuous extractor first under acid conditions and then under basic conditions. Typically 30 g of a solid sample, containing surrogate, and matrix spiking standards, is extracted ultrasonically. After concentrating the extract to 1 mL it is spiked with 10 µL of an internal standard solution just prior to analysis by GC/MS. The volume injected should contain about 100 ng of base/neutral and 200 ng of acid surrogates (for a 1-µL injection). Analysis is performed by GC/MS using a capillary GC column.

INTERFERENCES Raw GC/MS data from all blanks, samples, and spikes must be evaluated for interferences. Contamination by carryover can occur whenever high-concentration and low-concentration samples are sequentially analyzed. To reduce carryover, the sample syringe must be rinsed out between samples with solvent. Whenever an unusually concentrated sample is encountered, it should be followed by the analysis of blank solvent to check for cross-contamination.

INSTRUMENTATION A GC/MS and a data system are required. The GC column used is a 30 m × 0.25 mm I.D. (or 0.32 mm I.D.) 1um film thickness silicone-coated fused silica capillary column. A continuous liquid-liquid extractor equipped with Teflon® or glass connection joints and stopcocks requiring no lubrication, a K-D concentrating apparatus, water bath, and an ultrasonic disrupter with a minimum power of 300 W and with pulsing capability are also required.

PRECISION & ACCURACY The estimated quantitation limit (EQL) of Method 8270B for determining an individual compound is approximately 1 mg/kg (wet weight) for soil or sediment samples, 1–200 mg/kg for wastes (dependent on matrix and method of preparation), and 10 µg/L for groundwater samples. EQLs will be proportionately higher for sample extracts that require dilution to avoid saturation of the detector.

The EQL(b) for groundwater in µg/L is 10.
The EQL (a, b) for low concentrations in soil and sediment in µg/kg is not determined.
Accuracy as µg/L is not listed.
Overall precision in µg/L is not listed.

(a) *EQLs listed for soil/sediment are based on wet weight. Normally data is reported in a dry-weight basis; therefore, EQLs will be higher based on the % dry weight of each sample. This calculation is based on a 30 g sample and gel permeation chromatography cleanup.*

(b) *Sample EQLs are highly matrix-dependent. The EQLs are provided for guidance and may not always be achievable.*

$C =$ *True value for concentration, in µg/L.*

$X =$ Average recovery found for measurements of samples containing a concentration of C, in µg/L.

ESTIMATED QUANTITATION LIMIT

Other Matrices	Factor (a)
High-concentration soil and sludges by sonicator	7.5
Non-water miscible waste	75

(a) EQL for other matrices = [EQL for low soil/sediment] × [Factor]. This estimated EQL is similar to an EPA "Practical Quantitation Limit."

SAMPLING METHOD

Liquid samples — Use a 1 or 2½ gallon amber glass bottle with a screw-top Teflon®-lined cover that has been prewashed with detergent and rinsed with distilled water and methanol (or isopropanol).

Soils, sediments, or sludges — Use an 8-oz. widemouth glass with a screw-top Teflon®-lined cover that has been prewashed with detergent and rinsed with distilled water and methanol (or isopropanol).

SAMPLE PRESERVATION

Liquid samples — If residual chlorine is present, add 3 mL of 10% sodium thiosulfate per gallon, cool to 4°C and store in a solvent-free refrigerator until analysis; if chlorine is not present, then eliminate the sodium thiosulfate addition.

Soils, sediments, or sludges — Cool samples to 4°C and store in a solvent-free refrigerator.

MHT Liquid samples must be extracted within 7 days and the extracts analyzed within 40 days. Soils, sediments, or sludges may be stored for a maximum of 14 days and the extracts analyzed within 40 days.

SAMPLE PREPARATION

Liquid samples — Transfer 1 L quantitatively to a continuous extractor. If high concentrations are anticipated, a smaller volume may be used and then diluted with organic-free reagent water to 1 L. Adjust pH, if necessary, to pH <2 using 1:1 (V/V) sulfuric acid. Pipette 1.0 mL of a surrogate standard spiking solution into each sample. For the sample in each analytical batch selected for spiking, add 1.0 mL of a matrix spiking standard. For base/neutral acid analysis, the amount of the surrogates and matrix spiking compounds added to the sample should result in a final concentration of 100 ng/µL of each analyte in the extract to be analyzed (assuming a 1-µL injection). Extract with methylene chloride for 18–24 h. Next, adjust the pH of the aqueous phase to pH >11 using 10 N sodium hydroxide and extract it with methylene chloride again for 18–24 h. Dry the extract through a column containing anhydrous sodium sulfate and concentrate it to 1 mL using a K-D concentrator.

Soils, sediments, or sludges — Use 30 g of sample. Nonporous or wet samples (gummy or clay type) that do not have a free-flowing sandy texture must be mixed with anhydrous sodium sulfate until the sample is free flowing. Add 1 mL of surrogate standards to all samples, spikes, standards, and blanks. For the sample in each analytical batch selected for spiking, add 1.0 mL of a matrix spiking standard. For base/neutral acid analysis, the amount added of the surrogates and matrix spiking compounds should result in a final concentration of 100 ng/µL of each base/neutral analyte and 200 ng/µL of each acid analyte in the extract to be analyzed (assuming a 1-µL injection). Immediately add a 100-mL mixture of 1:1 methylene chloride:acetone and extract the sample ultrasonically for 3 min and then decant or filter the extracts. Repeat the extraction two or more times. Dry the extract using a column with anhydrous sodium sulfate and concentrate it to 1 mL in a K-D concentrator.

QUALITY CONTROL A methylene chloride solution containing 50 ng/µL of decafluorotriphenylphosphine (DFTPP) is used for tuning the GC/MS system each 12-h shift. A system performance check also must be made during every 12-h shift. A standard containing 50 ng/µL each of 4,4'-DDT, pentachlorophenol, and benzidine is required to verify injection port inertness and GC column performance. A calibration standard at mid-concentration, containing each compound of interest, including all required surrogates, must be performed every 12 h during analysis. After the system performance check is met, calibration check compounds (CCCs) are used to check the validity of the initial calibration.

The internal standard responses and retention times in the calibration check standard must be evaluated immediately after or during data acquisition. If the retention time for any internal standard changes by more than 30 seconds from the last check calibration (12 h), the chromatographic system must be inspected for malfunctions and corrections must be made, as required. If the electron ionization current plot (EICP) area for any of the internal standards changes by a factor of two from the last daily calibration standard check, the mass spectrometer must be inspected for malfunctions and corrections must be made, as appropriate.

Demonstrate, through the analysis of a reagent water blank, that interferences from the analytical system, glassware, and reagents are under control. The blank samples should be carried through all stages of the sample preparation and measurement steps. For each analytical batch (up to 20 samples), a reagent blank, matrix spike, and matrix spike duplicate/duplicate must be analyzed (the frequency of the spikes may be different for different monitoring programs). The blank and spiked samples must be carried through all stages of the sample preparation and measurement steps. A QC reference sample concentrate containing each analyte at a concentration of 100 mg/L in methanol is required.

REFERENCE Test Methods for Evaluating Solid Waste (SW-846). U.S. EPA 1983, Method 8270B, Rev. 2, Nov. 1990. Office of Solid Waste, Washington, DC.

Mevinphos   **EPA Method 8140**
CAS #7786-34-7

TITLE Organophosphorus Pesticides

MATRIX Groundwater, soils, sludges, water miscible liquid wastes, and non-water miscible wastes.

APPLICATION This method is used for the analysis of 21 organophosphorus pesticides. Samples are extracted, concentrated, and analyzed using direct injection of both neat and diluted organic liquid into a gas chromatograph (GC).

INTERFERENCES Solvents, reagents, and glassware may introduce artifacts. Other interferences may come from coextracted compounds from samples. The use of Florisil cleanup materials may produce low recoveries. Elemental sulfur may interfere with some compounds when using a flame photometric detector. Sulfur cleanup (Method 3660) may alleviate sulfur interference.

INSTRUMENTATION GC capable of on-column injections and a flame photometric detector (FPD) or a thermionic detector. Column 1: 1.8 m by 2 mm with 5% SP-2401 on Supelcoport. Column 2: 1.8 m by 2 mm with 3% SP-2401 on Supelcoport. Column 3: 50 cm by ⅛ in Teflon® with 15% SE-54 on Gas Chrom Q. The preferred column is Column Number 1.

RANGE 15.5–520 µg/L.

MDL 0.3 µg/L (in reagent water).

PQL FACTORS FOR MULTIPLYING × FID MDL VALUE

Matrix	Multiplication Factor
Groundwater	10
Low-level soil by sonication with GPC cleanup	670
High-level soil and sludge by sonication	10,000
Non-water miscible waste	100,000

PRECISION 7.8% (single operator standard deviation)

ACCURACY 56.5% (single operator average recovery)

SAMPLING METHOD Use 8-oz. widemouth glass bottles with Teflon®-lined caps for concentrated waste samples, soils, sediments, and sludges. Use 1 or 2½ gallon amber glass bottles with Teflon®-lined caps for liquid (water) samples.

STABILITY Cool soil, sediment, sludge, and liquid samples to 4°C. If residual chlorine is present in liquid samples add 3 mL of 10% sodium thiosulfate per gallon of sample and cool to 4°C.

MHT 14 days for concentrated waste, soil, sediment, or sludge; 7 days for liquid samples; all extracts must be analyzed within 40 days.

QUALITY CONTROL A quality control check sample concentrate containing this compound in acetone at a concentration 1,000 times more concentrated than the selected spike concentration is required. The QC check sample concentrate may be prepared from pure standard materials or purchased as certified solutions. Use appropriate trip, matrix, control site, method, reagent, and solvent blanks. Internal, surrogate, and five concentration level calibration standards are used.

REFERENCE Method 8140, SW-846, 3rd ed., Sept. 1986.

Mexacarbate EPA Method 8270
CAS #315-18-4

TITLE Semivolatile Organic Compounds by GC/MS

MATRIX This method is used to determine the concentration of semivolatile organic compounds in extracts prepared from all types of solid waste matrices, soils, and groundwater. Although surface waters are not specifically mentioned, this method should be applicable to water samples from rivers, lakes, etc.

METHOD SUMMARY This method covers 259 semivolatile organic compounds. In very limited applications direct injection of the sample into the GC/MS system may be appropriate, but this results in very high detection limits (approximately 10,000 µg/L). Typically, a 1-L liquid sample, containing surrogate, and matrix spiking standards, is extracted in a continuous extractor first under acid conditions and then under basic conditions. Typically 30 g of a solid sample, containing surrogate, and matrix spiking standards, is extracted ultrasonically. After concentrating the extract to 1 mL it is spiked with 10 µL of an internal standard solution just prior to analysis by GC/MS. The volume injected should contain about 100 ng of base/neutral and 200 ng of acid surrogates (for a 1-µL injection). Analysis is performed by GC/MS using a capillary GC column.

INTERFERENCES Raw GC/MS data from all blanks, samples, and spikes must be evaluated for interferences. Contamination by carryover can occur whenever high-concentration and low-concentration samples are sequentially analyzed. To reduce carryover, the sample syringe must be rinsed out between samples with solvent. Whenever an unusually concentrated sample is encountered, it should be followed by the analysis of blank solvent to check for cross-contamination.

INSTRUMENTATION A GC/MS and a data system are required. The GC column used is a 30 m × 0.25 mm I.D. (or 0.32 mm I.D.) 1um film thickness silicone-coated fused silica capillary column. A continuous liquid-liquid extractor equipped with Teflon® or glass connection joints and stopcocks requiring no lubrication, a K-D concentrating apparatus, water bath, and an ultrasonic disrupter with a minimum power of 300 W and with pulsing capability are also required.

PRECISION & ACCURACY The estimated quantitation limit (EQL) of Method 8270B for determining an individual compound is approximately 1 mg/kg (wet weight) for soil or sediment samples, 1–200 mg/kg for wastes (dependent on matrix and method of preparation), and 10 µg/L for groundwater samples. EQLs will be proportionately higher for sample extracts that require dilution to avoid saturation of the detector.

The EQL(b) for groundwater in µg/L is 20.
The EQL (a, b) for low concentrations in soil and sediment in µg/kg is not determined.
Accuracy as µg/L is not listed.
Overall precision in µg/L is not listed.

(a) EQLs listed for soil/sediment are based on wet weight. Normally data is reported in a dry-weight basis; therefore, EQLs will be higher based on the % dry weight of each sample. This calculation is based on a 30 g sample and gel permeation chromatography cleanup.

(b) Sample EQLs are highly matrix-dependent. The EQLs are provided for guidance and may not always be achievable.

C = True value for concentration, in µg/L.

X = Average recovery found for measurements of samples containing a concentration of C, in µg/L.

ESTIMATED QUANTITATION LIMIT

Other Matrices	Factor (a)
High-concentration soil and sludges by sonicator	7.5
Non-water miscible waste	75

(a) EQL for other matrices = [EQL for low soil/sediment] × [Factor]. This estimated EQL is similar to an EPA "Practical Quantitation Limit."

SAMPLING METHOD

Liquid samples — Use a 1 or 2½ gallon amber glass bottle with a screw-top Teflon®-lined cover that has been prewashed with detergent and rinsed with distilled water and methanol (or isopropanol).

Soils, sediments, or sludges — Use an 8-oz. widemouth glass with a screw-top Teflon®-lined cover that has been prewashed with detergent and rinsed with distilled water and methanol (or isopropanol).

SAMPLE PRESERVATION

Liquid samples — If residual chlorine is present, add 3 mL of 10% sodium thiosulfate per gallon, cool to 4°C and store in a solvent-free refrigerator until analysis; if chlorine is not present, then eliminate the sodium thiosulfate addition.

Soils, sediments, or sludges — Cool samples to 4°C and store in a solvent-free refrigerator.

MHT Liquid samples must be extracted within 7 days and the extracts analyzed within 40 days. Soils, sediments, or sludges may be stored for a maximum of 14 days and the extracts analyzed within 40 days.

SAMPLE PREPARATION

Liquid samples — Transfer 1 L quantitatively to a continuous extractor. If high concentrations are anticipated, a smaller volume may be used and then diluted with organic-free reagent water to 1 L. Adjust pH, if necessary, to pH <2 using 1:1 (V/V) sulfuric acid. Pipette 1.0 mL of a surrogate standard spiking solution into each sample. For the sample in each analytical batch selected for spiking, add 1.0 mL of a matrix spiking standard. For base/neutral acid analysis, the amount of the surrogates and matrix spiking compounds added to the sample should result in a final concentration of 100 ng/µL of each analyte in the extract to be analyzed (assuming a 1-µL injection). Extract with methylene chloride for 18–24 h. Next, adjust the pH of the aqueous phase to pH >11 using 10 N sodium hydroxide and extract it with methylene chloride again for 18–24 h. Dry the extract through a column containing anhydrous sodium sulfate and concentrate it to 1 mL using a K-D concentrator.

Soils, sediments, or sludges — Use 30 g of sample. Nonporous or wet samples (gummy or clay type) that do not have a free-flowing sandy texture must be mixed with anhydrous sodium sulfate until the sample is free flowing. Add 1 mL of surrogate standards to all samples, spikes, standards, and blanks. For the sample in each analytical batch selected for spiking, add 1.0 mL of a matrix spiking standard. For base/neutral acid analysis, the amount added of the surrogates and matrix spiking compounds should result in a final concentration of 100 ng/µL of each base/neutral analyte and 200 ng/µL of each acid analyte in the extract to be analyzed (assuming a 1-µL injection). Immediately add a 100-mL mixture of 1:1 methylene chloride:acetone and extract the sample ultrasonically for 3 min and then decant or filter the extracts. Repeat the extraction two or more times. Dry the extract using a column with anhydrous sodium sulfate and concentrate it to 1 mL in a K-D concentrator.

QUALITY CONTROL A methylene chloride solution containing 50 ng/µL of decafluorotriphenylphosphine (DFTPP) is used for tuning the GC/MS system each 12-h shift. A system performance check also must be made during every 12-h shift. A standard containing 50 ng/µL each of 4,4'-DDT, pentachlorophenol, and benzidine is required to verify injection port inertness and GC column performance. A calibration standard at mid-concentration, containing each compound of interest, including all required surrogates, must be performed every 12 h during analysis. After the system performance check is met, calibration check compounds (CCCs) are used to check the validity of the initial calibration.

The internal standard responses and retention times in the calibration check standard must be evaluated immediately after or during data acquisition. If the retention time for any internal standard changes by more than 30 seconds from the last check calibration (12 h), the chromatographic system must be inspected for malfunctions and corrections must be made, as required. If the electron ionization current plot (EICP) area for any of the internal standards changes by a factor of two from the last daily calibration standard check, the mass spectrometer must be inspected for malfunctions and corrections must be made, as appropriate.

Demonstrate, through the analysis of a reagent water blank, that interferences from the analytical system, glassware, and reagents are under control. The blank samples should be carried through all stages of the sample preparation and measurement steps. For each analytical batch (up to 20 samples), a reagent blank, matrix spike, and matrix spike duplicate/duplicate must be analyzed (the frequency of the spikes may be different for different monitoring programs). The blank and spiked samples must be carried through all stages of the sample preparation and measurement steps. A QC reference sample concentrate containing each analyte at a concentration of 100 mg/L in methanol is required.

REFERENCE Test Methods for Evaluating Solid Waste (SW-846). U.S. EPA 1983, Method 8270B, Rev. 2, Nov. 1990. Office of Solid Waste, Washington, DC.

MGK 264   **EPA Method 507**
CAS #113-48-4

TITLE Determination of Nitrogen and Phosphorus-Containing Pesticides in Water by GC/NPD

MATRIX This method is applicable to the determination of certain nitrogen and phosphorus-containing pesticides in finished drinking water and groundwater.

METHOD SUMMARY Method 507 covers 46 nitrogen- and phosphorus-containing pesticides. A 1-L sample is fortified with a surrogate standard, salted, buffered, extracted with methylene chloride, and concentrated; then the solvent is exchanged with methyl tert-butyl ether (MTBE) and concentrated again, and a 2-µL aliquot of a sample extract is injected into a GC system equipped with a selective nitrogen-phosphorus detector and a capillary column for analysis.

INTERFERENCES Method interferences may be caused by contaminants in solvents, reagents, glassware, and other sample processing apparatus. Interfering contamination may occur when a sample containing low concentrations of analytes is analyzed immediately following a sample containing relatively high concentrations. One or more injections of MTBE should be made following the analysis of a sample with high concentrations of analytes to check for analyte carryover. Matrix interferences may be caused by contaminants that are coextracted from the sample. The extent of matrix interferences will vary considerably from source to source, depending upon the water sampled.

INSTRUMENTATION A gas chromatograph system (GC) equipped with a nitrogen-phosphorus detector (NPD) is needed.

Column 1: 30 m × 0.25 mm I.D. DB-5 bonded fused silica column, 0.25 µm film thickness, or equivalent.
Column 2: 30 m × 0.25 mm I.D. DB-1701 bonded fused silica column, 0.25 µm film thickness, or equivalent.

PRECISION & ACCURACY This method has been validated in a single lab and estimated detection limits (EDLs) have been determined for each analyte. Observed detection limits may vary among waters, depending upon the nature of the interferences in the sample matrix and the specific instrumentation used. Analytes that are not separated chromatographically cannot be individually identified and measured unless an alternative technique for identification and quantification exist.

The estimated detection limit (in µg/L) was 0.5. The EDL is defined as either method detection limit or a level of compound in a sample yielding a peak in the final extract with signal-to-noise ratio of approximately 5, whichever value is higher.

The concentration used for these measurements (in µg/L) was 5.
The accuracy (as % recovery) was 100.
The precision (% RSD) was 4.

SAMPLING METHOD Grab samples are collected in 1-L glass sample bottles (prewashed with detergent and hot tap water, rinsed with reagent water, and dried in an oven at 400°C for 1 h) with screw caps lined with PTFE-fluorocarbon.

SAMPLE PRESERVATION Add mercuric chloride to the sample bottle in amounts to produce a concentration of 10 mg/L. If residual chlorine is present, add 80 mg of sodium thiosulfate/L of sample to the sample bottle prior to collection. After collection, seal bottle and shake vigorously for 1 min, then cool the sample to 4°C immediately and store it at 4°C in the dark until extraction.

MHT Maximum holding time of the samples, and in some cases the extracts, is 14 days.

SAMPLE PREPARATION Fortify the sample with 50 µL of the surrogate standard solution, adjust to pH 7 with phosphate buffer, add 100 g NaCl to the sample, and seal and shake to dissolve the salt; then extract with methylene chloride in a separatory funnel or in a mechanical tumbler bottle. Dry the extract by pouring it through a solvent-rinsed drying column containing about 10 cm of anhydrous sodium sulfate. Collect the extract in a Kuderna-Danish (K-D) concentrator and rinse the column with 20–30 mL methylene chloride. Concentrate the extract to about 2 mL and rinse the flask and its lower joint into the concentrator tube with 1 to 2 mL of methyl t-butyl ether (MTBE). Add 5–10 mL of MTBE and concentrate the extract twice (adding more MTBE) to a final volume of 5.0 mL and store it at 4°C until analysis.

Note: If methylene chloride is not completely removed from the final extract, it may cause detector problems.

QUALITY CONTROL Minimum quality control requirements are initial demonstration of lab capability, determination of surrogate compound recoveries in each sample and blank, monitoring internal standard peak area or height in each sample and blank, analysis of lab reagent blanks, lab fortified samples, lab fortified blanks, and other QC samples. A lab reagent blank is analyzed to demonstrate that all glassware and reagent interferences are under control.

Initial demonstration of capability is fulfilled by analyzing four fortified reagent water samples with the recovery value for each analyte falling within the acceptable range (±30% average recovery). Surrogate recoveries from samples or method blanks must be 70–130%. The internal standard response for any sample chromatogram should not deviate from the daily calibration check standard's internal standard response by more than 30% or lab fortified blanks and sample matrices are used to assess lab performance and analyte recovery, respectively.

If the response for the target analyte peak exceeds the working range of the system, dilute the extract and reanalyze. Alternative techniques such as an alternate detector or second chromatography column should be used to confirm peak identification when sample components are not resolved adequately.

EPA CONTACT & HOTLINE For technical questions contact Dr. Baldev Bathija, U.S. EPA, Office of Ground Water and Drinking Water (WH-550D), 401 M St. SW, Washington, DC 20460. Tel. (202) 260-3040. For further information the EPA Safe Drinking Water Hotline may be called at: (800) 426-4791.

REFERENCE Methods for the Determination of Organic Compounds in Drinking Water, EPA/600/4-88/039 (revised July 1991). U.S. EPA Environmental Monitoring Systems Laboratory, Cincinnati, OH, 45268, U.S.A. Available from the National Technical Information Service (NTIS), 5285 Port Royal Road, Springfield, VA 22161; Tel. 800-553-6847. NTIS Order Number is PB91-231480.

Mirex
CAS #2385-85-5

EPA Method 8270

TITLE Semivolatile Organic Compounds by GC/MS

MATRIX This method is used to determine the concentration of semivolatile organic compounds in extracts prepared from all types of solid waste matrices, soils, and groundwater. Although surface waters are not specifically mentioned, this method should be applicable to water samples from rivers, lakes, etc.

METHOD SUMMARY This method covers 259 semivolatile organic compounds. In very limited applications direct injection of the sample into the GC/MS system may be appropriate, but this results in very high detection limits (approximately 10,000 µg/L). Typically, a 1-L liquid sample, containing surrogate, and matrix spiking standards, is extracted in a continuous extractor first under acid conditions and then under basic conditions. Typically 30 g of a solid sample, containing surrogate, and matrix spiking standards, is extracted ultrasonically. After concentrating the extract to 1 mL it is spiked with 10 µL of an internal standard solution just prior to analysis by GC/MS. The volume injected should contain about 100 ng of base/neutral and 200 ng of acid surrogates (for a 1-µL injection). Analysis is performed by GC/MS using a capillary GC column.

INTERFERENCES Raw GC/MS data from all blanks, samples, and spikes must be evaluated for interferences. Contamination by carryover can occur whenever high-concentration and low-concentration samples are sequentially analyzed. To reduce carryover, the sample syringe must be rinsed out between samples with solvent. Whenever an unusually concentrated sample is encountered, it should be followed by the analysis of blank solvent to check for cross-contamination.

INSTRUMENTATION A GC/MS and a data system are required. The GC column used is a 30 m × 0.25 mm I.D. (or 0.32 mm I.D.) 1um film thickness silicone-coated fused silica capillary column. A continuous liquid-liquid extractor equipped with Teflon® or glass connection joints and stopcocks requiring no lubrication, a K-D concentrating apparatus, water bath, and an ultrasonic disrupter with a minimum power of 300 W and with pulsing capability are also required.

PRECISION & ACCURACY The estimated quantitation limit (EQL) of Method 8270B for determining an individual compound is approximately 1 mg/kg (wet weight) for soil or sediment samples, 1–200 mg/kg for wastes (dependent on matrix and method of preparation), and 10 µg/L for groundwater samples. EQLs will be proportionately higher for sample extracts that require dilution to avoid saturation of the detector.

The EQL(b) for groundwater in µg/L is 10.
The EQL (a, b) for low concentrations in soil and sediment in µg/kg is not determined.
Accuracy as µg/L is not listed.
Overall precision in µg/L is not listed.

(a) *EQLs listed for soil/sediment are based on wet weight. Normally data is reported in a dry-weight basis; therefore, EQLs will be higher based on the % dry weight of each sample.* *This calculation is based on a 30 g sample and gel permeation chromatography cleanup.*
(b) *Sample EQLs are highly matrix-dependent. The EQLs are provided for guidance and may not always be achievable.*

C = *True value for concentration, in µg/L.*
X = *Average recovery found for measurements of samples containing a concentration of C, in µg/L.*

ESTIMATED QUANTITATION LIMIT

Other Matrices	Factor (a)
High-concentration soil and sludges by sonicator	7.5
Non-water miscible waste	75

(a) *EQL for other matrices = [EQL for low soil/sediment] × [Factor]. This estimated EQL is similar to an EPA "Practical Quantitation Limit."*

SAMPLING METHOD
Liquid samples — Use a 1 or 2½ gallon amber glass bottle with a screw-top Teflon®-lined cover that has been prewashed with detergent and rinsed with distilled water and methanol (or isopropanol).

Soils, sediments, or sludges — Use an 8-oz. widemouth glass with a screw-top Teflon®-lined cover that has been prewashed with detergent and rinsed with distilled water and methanol (or isopropanol).

SAMPLE PRESERVATION
Liquid samples — If residual chlorine is present, add 3 mL of 10% sodium thiosulfate per gallon, cool to 4°C and store in a solvent-free refrigerator until analysis; if chlorine is not present, then eliminate the sodium thiosulfate addition.

Soils, sediments, or sludges — Cool samples to 4°C and store in a solvent-free refrigerator.

MHT Liquid samples must be extracted within 7 days and the extracts analyzed within 40 days. Soils, sediments, or sludges may be stored for a maximum of 14 days and the extracts analyzed within 40 days.

SAMPLE PREPARATION
Liquid samples — Transfer 1 L quantitatively to a continuous extractor. If high concentrations are anticipated, a smaller volume may be used and then diluted with organic-free reagent water to 1 L. Adjust pH, if necessary, to pH <2 using 1:1 (V/V) sulfuric acid. Pipette 1.0 mL of a surrogate standard spiking solution into each sample. For the sample in each analytical batch selected for spiking, add 1.0 mL of a matrix spiking standard. For base/neutral acid analysis, the amount of the surrogates and matrix spiking compounds added to the sample should result in a final concentration of 100 ng/µL of each analyte in the extract to be analyzed (assuming a 1-µL injection). Extract with methylene chloride for 18–24 h. Next, adjust the pH of the aqueous phase to pH >11 using 10 N sodium hydroxide and extract it with methylene chloride again for 18–24 h. Dry the extract through a column containing anhydrous sodium sulfate and concentrate it to 1 mL using a K-D concentrator.

Soils, sediments, or sludges — Use 30 g of sample. Nonporous or wet samples (gummy or clay type) that do not have a free-flowing

sandy texture must be mixed with anhydrous sodium sulfate until the sample is free flowing. Add 1 mL of surrogate standards to all samples, spikes, standards, and blanks. For the sample in each analytical batch selected for spiking, add 1.0 mL of a matrix spiking standard. For base/neutral acid analysis, the amount added of the surrogates and matrix spiking compounds should result in a final concentration of 100 ng/µL of each base/neutral analyte and 200 ng/µL of each acid analyte in the extract to be analyzed (assuming a 1-µL injection). Immediately add a 100-mL mixture of 1:1 methylene chloride:acetone and extract the sample ultrasonically for 3 min and then decant or filter the extracts. Repeat the extraction two or more times. Dry the extract using a column with anhydrous sodium sulfate and concentrate it to 1 mL in a K-D concentrator.

QUALITY CONTROL A methylene chloride solution containing 50 ng/µL of decafluorotriphenylphosphine (DFTPP) is used for tuning the GC/MS system each 12-h shift. A system performance check also must be made during every 12-h shift. A standard containing 50 ng/µL each of 4,4'-DDT, pentachlorophenol, and benzidine is required to verify injection port inertness and GC column performance. A calibration standard at mid-concentration, containing each compound of interest, including all required surrogates, must be performed every 12 h during analysis. After the system performance check is met, calibration check compounds (CCCs) are used to check the validity of the initial calibration.

The internal standard responses and retention times in the calibration check standard must be evaluated immediately after or during data acquisition. If the retention time for any internal standard changes by more than 30 seconds from the last check calibration (12 h), the chromatographic system must be inspected for malfunctions and corrections must be made, as required. If the electron ionization current plot (EICP) area for any of the internal standards changes by a factor of two from the last daily calibration standard check, the mass spectrometer must be inspected for malfunctions and corrections must be made, as appropriate.

Demonstrate, through the analysis of a reagent water blank, that interferences from the analytical system, glassware, and reagents are under control. The blank samples should be carried through all stages of the sample preparation and measurement steps. For each analytical batch (up to 20 samples), a reagent blank, matrix spike, and matrix spike duplicate/duplicate must be analyzed (the frequency of the spikes may be different for different monitoring programs). The blank and spiked samples must be carried through all stages of the sample preparation and measurement steps. A QC reference sample concentrate containing each analyte at a concentration of 100 mg/L in methanol is required.

REFERENCE Test Methods for Evaluating Solid Waste (SW-846). U.S. EPA 1983, Method 8270B, Rev. 2, Nov. 1990. Office of Solid Waste, Washington, DC.

Molinate
CAS #2212-67-1

EPA Method 507

TITLE Determination of Nitrogen and Phosphorus-Containing Pesticides in Water by GC/NPD

MATRIX This method is applicable to the determination of certain nitrogen and phosphorus-containing pesticides in finished drinking water and groundwater.

METHOD SUMMARY Method 507 covers 46 nitrogen- and phosphorus-containing pesticides. A 1-L sample is fortified with a surrogate standard, salted, buffered, extracted with methylene chloride, and concentrated; then the solvent is exchanged with methyl tert-butyl ether (MTBE) and concentrated again, and a 2-µL aliquot of a sample extract is injected into a GC system equipped with a selective nitrogen-phosphorus detector and a capillary column for analysis.

INTERFERENCES Method interferences may be caused by contaminants in solvents, reagents, glassware, and other sample processing apparatus. Interfering contamination may occur when a sample containing low concentrations of analytes is analyzed immediately following a sample containing relatively high concentrations. One or more injections of MTBE should be made following the analysis of a sample with high concentrations of analytes to check for analyte carryover. Matrix interferences may be caused by contaminants that are coextracted from the sample. The extent of matrix interferences will vary considerably from source to source, depending upon the water sampled.

INSTRUMENTATION A gas chromatograph system (GC) equipped with a nitrogen-phosphorus detector (NPD) is needed.

Column 1: 30 m × 0.25 mm I.D. DB-5 bonded fused silica column, 0.25 µm film thickness, or equivalent.
Column 2: 30 m × 0.25 mm I.D. DB-1701 bonded fused silica column, 0.25 µm film thickness, or equivalent.

PRECISION & ACCURACY This method has been validated in a single lab and estimated detection limits (EDLs) have been determined for each analyte. Observed detection limits may vary among waters, depending upon the nature of the interferences in the sample matrix and the specific instrumentation used. Analytes that are not separated chromatographically cannot be individually identified and measured unless an alternative technique for identification and quantification exist.

The estimated detection limit (in µg/L) was 0.15. The EDL is defined as either method detection limit or a level of compound in a sample yielding a peak in the final extract with signal-to-noise ratio of approximately 5, whichever value is higher.

The concentration used for these measurements (in µg/L) was 1.5.
The accuracy (as % recovery) was 98.
The precision (% RSD) was 18.

SAMPLING METHOD Grab samples are collected in 1-L glass sample bottles (prewashed with detergent and hot tap

water, rinsed with reagent water, and dried in an oven at 400°C for 1 h) with screw caps lined with PTFE-fluorocarbon.

SAMPLE PRESERVATION Add mercuric chloride to the sample bottle in amounts to produce a concentration of 10 mg/L. If residual chlorine is present, add 80 mg of sodium thiosulfate/L of sample to the sample bottle prior to collection. After collection, seal bottle and shake vigorously for 1 min, then cool the sample to 4°C immediately and store it at 4°C in the dark until extraction.

MHT Maximum holding time of the samples, and in some cases the extracts, is 14 days.

SAMPLE PREPARATION Fortify the sample with 50 µL of the surrogate standard solution, adjust to pH 7 with phosphate buffer, add 100 g NaCl to the sample, and seal and shake to dissolve the salt; then extract with methylene chloride in a separatory funnel or in a mechanical tumbler bottle. Dry the extract by pouring it through a solvent-rinsed drying column containing about 10 cm of anhydrous sodium sulfate. Collect the extract in a Kuderna-Danish (K-D) concentrator and rinse the column with 20–30 mL methylene chloride. Concentrate the extract to about 2 mL and rinse the flask and its lower joint into the concentrator tube with 1 to 2 mL of methyl t-butyl ether (MTBE). Add 5–10 mL of MTBE and concentrate the extract twice (adding more MTBE) to a final volume of 5.0 mL and store it at 4°C until analysis.

Note: If methylene chloride is not completely removed from the final extract, it may cause detector problems.

QUALITY CONTROL Minimum quality control requirements are initial demonstration of lab capability, determination of surrogate compound recoveries in each sample and blank, monitoring internal standard peak area or height in each sample and blank, analysis of lab reagent blanks, lab fortified samples, lab fortified blanks, and other QC samples. A lab reagent blank is analyzed to demonstrate that all glassware and reagent interferences are under control.

Initial demonstration of capability is fulfilled by analyzing four fortified reagent water samples with the recovery value for each analyte falling within the acceptable range (±30% average recovery). Surrogate recoveries from samples or method blanks must be 70–130%. The internal standard response for any sample chromatogram should not deviate from the daily calibration check standard's internal standard response by more than 30% or lab fortified blanks and sample matrices are used to assess lab performance and analyte recovery, respectively.

If the response for the target analyte peak exceeds the working range of the system, dilute the extract and reanalyze. Alternative techniques such as an alternate detector or second chromatography column should be used to confirm peak identification when sample components are not resolved adequately.

EPA CONTACT & HOTLINE For technical questions contact Dr. Baldev Bathija, U.S. EPA, Office of Ground Water and Drinking Water (WH-550D), 401 M St. SW, Washington, DC 20460. Tel. (202) 260-3040. For further information the EPA Safe Drinking Water Hotline may be called at: (800) 426-4791.

REFERENCE Methods for the Determination of Organic Compounds in Drinking Water, EPA/600/4-88/039 (revised July 1991). U.S. EPA Environmental Monitoring Systems Laboratory, Cincinnati, OH, 45268, U.S.A. Available from the National Technical Information Service (NTIS), 5285 Port Royal Road, Springfield, VA 22161; Tel. 800-553-6847. NTIS Order Number is PB91-231480.

Molybdenum   EPA Method 6010
CAS #7439-98-7

TITLE Inductively Coupled Plasma-Atomic Emission Spectroscopy

MATRIX This method is applicable to the determination of trace elements, including metals, in groundwater, soils, sludges, sediments, and other solid wastes. All matrices require digestion prior to analysis. The method of standard addition must be used for the analysis of all sample digests unless either serial dilution or matrix spike addition demonstrates it is not required.

METHOD SUMMARY Method 6010 covers 25 elements using ICP analysis. It measures element-emitted light by optical spectrometry. Samples, following an appropriate acid digestion, are nebulized and the resulting aerosol is transported to the plasma torch. Element-specific atomic line emission spectra are produced by a radio-frequency inductively coupled plasma.

INTERFERENCES Interferences may be categorized as spectral or non-spectral. Spectral interferences are caused by overlap of a spectral line from another element, unresolved overlap of molecular band spectra, background contribution from continuous or recombination phenomenon, and stray light from the line emission of high concentration elements. Non-spectral interferences include physical and chemical interferences. Physical interferences are effects associated with the sample nebulization and transport processes. Changes in viscosity and surface tension can cause significant inaccuracies. Chemical interferences include molecular compound formation, ionization effects, and solute vaporization effects. Normally these effects are not significant and can be minimized by careful selection of operating conditions. Chemical interferences are highly dependent on matrix type and the specific analyte element.

INSTRUMENTATION An inductively coupled argon plasma emission spectrometer (ICP) capable of background correction is required.

PRECISION & ACCURACY Detection limits, sensitivity, and optimum ranges of the metals will vary with the matrices and model of the spectrometer. In a single lab evaluation, seven wastes were analyzed for 22 elements. The mean percent relative standard deviation from triplicate analyses for all elements and wastes was 9 ± 2%. The mean percent recovery of spiked elements for all wastes was 93 ± 6%. Spike levels ranged from 100 µg/L to 100 mg/L. The wastes included sludges and industrial wastewaters.

Estimated instrument detection limit in µg/L is 8.
Spiked concentration in µg/L is not listed.

Mean reported value in µg/L is not listed.
Precision as RSD % is not listed.

SAMPLING METHOD Samples should be collected in borosilicate glass, linear polyethylene, polypropylene, or Teflon® bottles that have been prewashed with detergent and tap water, and rinsed with 1:1 nitric acid and tap water or 1:1 hydrochloric acid and tap water. Collect at least 2 g of solids and 200 mL of aqueous samples.

SAMPLE PRESERVATION Add nitric acid to make the samples pH <2.

MHT The maximum holding time for properly preserved samples is 6 months.

SAMPLE PREPARATION Preliminary treatment of most matrices is necessary because of the complexity and variability of sample matrices. Water samples that have been prefiltered and acidified will not need acid digestion. Methods for acid digestion of waters for total recoverable or dissolved metals, acid digestions of aqueous samples and extracts for total metals, and acid digestion of sediments, sludges, and soils are summarized below.

Total recoverable or dissolved metals in water — To prepare surface and groundwater samples for determination of total recoverable and dissolved metals, a 100-mL aliquot of well-mixed sample is acidified with concentrated nitric acid and concentrated hydrochloric acid, then heated until the volume is reduced to 15–20 mL. Adjust the final volume to 100 mL with reagent water.

Total metals in aqueous samples, soil, and sediment extracts — To prepare aqueous samples, soil and sediment extracts, and wastes that contain suspended solids, a 100-mL aliquot is made acidic with concentrated nitric acid and the solution is evaporated to about 5 mL on a hot plate. Continue heating and adding additional acid until sample digestion is complete, which is usually indicated when the digestate is light in color or does not change in appearance. Evaporate the solution to about 3 mL and cool it and add a small quantity of 1:1 hydrochloric acid (10 mL/100 mL of final solution). Cover the beaker and reflux for 15 min. Wash down the beaker walls and filters or centrifuge the sample to remove silicates and other insoluble material. Filter the sample and adjust the final volume to 100 mL with reagent water and the final acid concentration to 10%.

Sediments, sludges, and soils — To prepare sediments, sludges and soil samples, transfer 1–2 g to a conical beaker and add 10 mL of 1:1 nitric acid, mix the slurry, and cover it with a watch glass. Heat the sample and reflux for 10–15 min without boiling. Allow it to cool, then add 5 mL of concentrated nitric acid and reflux for 30 min. Repeat last step and then allow the solution to evaporate to 5 mL without boiling. Cool and add 2 mL of water and 3 mL of 30% hydrogen peroxide. Cover and place the beaker on the hot plate. Heat and add 30% hydrogen peroxide in 1-mL aliquots with warming until the effervescence is minimal but do not add more than a total of 10 mL of 30% hydrogen peroxide. If the sample is being prepared for the analysis of Ag, Al, As, Ba, Be, Ca, Cd, Co, Cr, Cu, Fe, K, Mg, Mn, Mo, Na, Ni, Os, Pb, Se, Tl, V, and Zn, then add 5 mL of concentrated hydrochloric acid and 10 mL of water and return the covered beaker to a hot plate for 15 min of additional refluxing without boiling. Dilute the sample to a 100 mL volume with water after cooling and filter or centrifuge to remove particulates.

QUALITY CONTROL Laboratory control samples must be analyzed for each analytical method. A method blank should be analyzed with each batch of samples. The effect of the matrix on method performance must be demonstrated: when appropriate, there should be at least one matrix spike and either one matrix duplicate or one matrix spike duplicate per analytical batch. The bias and precision of the method, as well as the method detection limit for each specific matrix type, must be measured.

Dilute and reanalyze samples that are more concentrated than the linear calibration limit. Employ a minimum of one reagent blank per sample batch to determine if contamination or any memory effects are occurring. Whenever a new or unusual sample matrix is encountered, perform either a serial dilution test or a matrix spike addition test to ensure that neither positive or negative interferences are operating on any of the analyte elements. Check the instrument standardization by verifying calibration every 10 samples using a calibration blank and a check standard.

REFERENCE Test Methods for Evaluating Solid Waste (SW-846). U.S. EPA. 1983. Method 6010, Rev. 0, Sept. 1986. Office of Solid Wastes, Washington, DC.

Molybdenum   EPA Method 7481
CAS #7439-98-7

TITLE Atomic Absorption (AA) Furnace Technique

MATRIX Wastes, mobility procedure, extracts, soils, and groundwater

APPLICATION Aqueous samples, EP extracts, industrial wastes, soils, sludges, sediments, and solid wastes require digestion before analysis. An aliquot of sample is placed in the graphite tube in the furnace and slowly evaporated, charred and atomized. Absorption of lamp radiation during atomization is proportional to molybdenum concentration.

INTERFERENCES The furnace technique is subject to chemical interferences.

Composition of sample matrix can effect analysis. Molybdenum is prone to carbide formation. Use pyrolitically coated graphite tube. Memory effects are possible; clean furnace after concentration sample analysis.

INSTRUMENTATION Atomic absorption spectrometer. Molybdenum hollow cathode lamp or electrodeless discharge lamp. Graphite furnace. Strip-chart recorder

RANGE 3–60 µg/L.

MDL 1 µg/L (313.3 nm wavelength)

PRECISION Not listed.

ACCURACY Not listed.

SAMPLING METHOD Use glass or plastic containers. Collect 200 g of solids and 600 mL of liquid samples.

STABILITY Cool solid samples to 4°C and analyze as soon as possible. Add nitric acid to liquid samples to pH <2.

MHT 6 months.

QUALITY CONTROL At least one duplicate and one spike sample should be run every 20 samples, or with each matrix type to verify method precision. If 20 or more samples are run a day, run a standard (at or near mid-range) every 10 samples.

REFERENCE Method 7481, SW-846, 3rd ed., Nov.1986.

Molybdenum   EPA Method 200.7
CAS #7439-98-7

TITLE Inductively Coupled Plasma

MATRIX Dissolved, suspended or (ICP) total element in drinking and surface waters and in domestic and industrial wastewaters.

APPLICATION The method covers the determination of 25 metals. Dissolved elements are determined in filtered and acidified samples after appropriate digestion (which increases dissolved solids). Its primary advantage is that ICP instruments allow simultaneous or rapid sequential determination of many elements in a short time. Samples are first nebulized and the aerosol is transported to a plasma torch in which element specific atomic line emission spectra are produced by a radio frequency inductively coupled plasma. Background correction is required for trace element detection except in the case of line broadning.

INTERFERENCES There are spectral, physical, and chemical interferences. The primary disadvantage of ICP instruments is background radiation from other elements and the plasma gases (spectral interferences). Changes in sample viscosity and surface tension with samples containing high dissolved solids (especially those exceeding 1500 mg/L) or high acid concentrations can cause physical interferences. Ionization effects, solute vaporization and molecular compound formation can cause chemical interferences. Aluminum and iron can cause interference at the 100 mg/L level.

INSTRUMENTATION Inductively coupled argon plasma emission spectroscopy. 202.030 nm wavelength

RANGE Not listed.

MDL 8 µg/L.

PRECISION Not listed.

ACCURACY Mean recovery = 93% ± 6% of spiked elements for all wastes.

SAMPLING METHOD Wash sample container with detergent and tap water, rinse with 1 + 1 nitric acid and tap water, then rinse with 1 + 1 hydrochloric acid and tap water, then rinse with deionized, distilled water in that order. Perform any filtration or acid preservation steps when the sample is collected or as soon as possible thereafter.

STABILITY Cool samples to 4°C.

MHT 24 h.

QUALITY CONTROL Mixed calibration standards, an instrument check standard, and an interference check solution are used in addition to a quality control sample. The quality control sample should be prepared in the same acid matrix as the calibration standards at 10 times the instrumental detection limits and in accordance with the instructions provided by the supplier. Furthermore, two types of blanks are required: a calibration blank and a reagent blank.

REFERENCE Method 200.7, U.S. EPA, EMSL-Cincinnati, OH, Nov. 1980

Molybdenum   EPA Method 7480
CAS #7439-98-7

TITLE Atomic Absorption (AA)

MATRIX Drinking, surface, and direct aspiration saline waters. Wastewater.

APPLICATION Sample is aspirated and atomized in a flame. A light beam from a Mo hollow cathode lamp is directed through the flame into a monochromator and onto a detector. Since wavelength of light beam is specific for Mo, light energy absorbed by detector is measure of molybdenum.

INTERFERENCES The most troublesomee type is chemical, caused by lack of absorption of atoms bound in molecular combination in the flame. High dissolved solids in sample may result in nonatomic absorbance interference. Addition of aluminum greatly reduces flame interference.

INSTRUMENTATION Atomic absorption spectrometer. Molybdenum hollow cathode lamp. (313.3 nm wavelength)

RANGE 1–40 mg/L.

MDL 0.1 mg/L

PRECISION Standard deviation = ±0.007, 0.02, 0.07 at 0.30, 1.5, 7.5 mg Mo/L

ACCURACY Recoveries = 100, 96, 95% at 0.30, 1.5, 7.5 mg Mo/L

SAMPLING METHOD Use glass or plastic containers. Collect 200 g of solids and 600 mL of liquid samples.

STABILITY Cool solid samples to 4°C and analyze as soon as possible. Add nitric acid to liquid samples to pH <2.

MHT 6 months.

QUALITY CONTROL At least one duplicate and one spike sample should be run every 20 samples or with each matrix type to verify precision of the method. For 20 or more samples per day, verify working standard curve. Run an additional standard at or near mid-range every 10 samples.

REFERENCE Method 7480, SW-846, 3rd ed., Nov.1986.

Monocrotophos EPA Method 8141
CAS #6923-22-4

TITLE Organophosphorus Compounds by Gas Chromatography: Capillary Column Technique

MATRIX This method covers aqueous and solid matrices. This includes a wide variety such as drinking water, groundwater, industrial wastewaters, surface waters, soils, solids, and sediments.

METHOD SUMMARY This is a GC method used to determine the concentration of 28 organophosphorus pesticides.

The use of Gel Permeation Cleanup (EPA Method 3640) for sample cleanup has been demonstrated to yield recoveries of less than 85% for many method analytes and is therefore not recommended for use with this method.

This method provides GC conditions for the detection of ppb concentrations of organophosphorus compounds. Prior to the use of this method, appropriate sample preparation techniques must be used. Water samples are extracted at a neutral pH with methylene chloride as a solvent by using a separatory funnel (EPA Method 3510) or a continuous liquid-liquid extractor (EPA Method 3520). Soxhlet extraction (EPA Method 3540) or ultrasonic extraction (EPA Method 3550) using methylene chloride/acetone (1:1) are used for solid samples. Both neat and diluted organic liquids (EPA Method 3580) may be analyzed by direct injection. Spiked samples are used to verify the applicability of the chosen extraction technique to each new sample type. A GC with a flame photometric (FPD) or nitrogen-phosphorus detector (NPD) is used for this multiresidue procedure.

INTERFERENCES The use of Florisil cleanup materials (EPA Method 3620) for some of the compounds in this method has been demonstrated to yield recoveries less than 85% and is therefore not recommended for all compounds. Use of phosphorus or halogen specific detectors, however, often obviates the necessity for cleanup for relatively clean sample matrices. If particular circumstances demand the use of an alternative cleanup procedure, the analyst must determine the elution profile and demonstrate that the recovery of each analyte is no less than 85%.

Use of a flame photometric detector (FPD) in the phosphorus mode will minimize interferences from materials that do not contain phosphorus. Elemental sulfur, however, may interfere with the determination of certain organophosphorus compounds by flame photometric gas chromatography. Sulfur cleanup using EPA Method 3660 may alleviate this interference. A nitrogen phosphorus detector (NPD) is also recommended.

A few analytes coelute on certain columns. Therefore, select a second column for confirmation where coelution of the analytes of interest does not occur.

Method interferences may be caused by contaminants in solvents, reagents, glassware, and other sample processing hardware that lead to discrete artifacts or elevated baselines in gas chromatograms. All these materials must be routinely demonstrated to be free from interferences under the conditions of the analysis by analyzing reagent blanks.

INSTRUMENTATION A GC with a NPD or a FPD will be needed. A data system or integrator is recommended for measuring peak areas and/or peak heights. A Kuderna-Danish (K-D) apparatus will be needed for extract concentration.

Column 1: 15 m × 0.53 mm megabore capillary column, 1.0 μm film thickness, DB-210.
Column 2: 15 m × 0.53 mm megabore capillary column, 1.5 μm film thickness, SPB-608.
Column 3: 15 m × 0.53 mm megabore capillary column, 1.0 μm film thickness, DB-5.

Three megabore capillary columns are included for analysis of organophosphates by this method. Column 1 (DB-210 or equivalent) and Column 2 (SPB-608 or equivalent) are recommended if a large number of organophosphorus analytes are to be determined. If the superior resolution offered by Column 1 and Column 2 is not required, Column 3 (DB-5 or equivalent) may be used. For megabore capillary columns, automatic injections of 1 μL are recommended.

PRECISION & ACCURACY The MDL actually achieved in a given analysis will vary, as it is dependent on instrument sensitivity and matrix effects. Single operator accuracy and precision studies have been conducted with spiked water and soil samples.

MULTIPLICATION FACTORS FOR OTHER MATRICES (a)

Matrix	Factor (b)
Groundwater (EPA Method 3510 or EPA Method 3520)	10
Low-concentration soil by Soxhlet and no cleanup	10 (c)
Low-concentration soil by ultrasonic extraction with GPC cleanup	6.7 (c)
High-concentration soil and sludges by ultrasonic extraction	500 (c)
Non-water miscible waste (EPA Method 3580)	1000 (c)

(a) Sample EQLs are highly matrix-dependent. The EQLs listed here are provided for guidance and may not always be achievable.
(b) EQL = [Method detection limit] × [Factor]. For non-aqueous samples the factor is on a wet-weight basis.
(c) Multiply this factory times the soil MDL.

The MDL (in μg/L) when reagent water was extracted using a separatory funnel was not reported.
The MDL (in μg/kg) when soil was extracted using Soxhlet extraction (EPA Method 3540) was not reported.
Accuracy (as % recovery) with separatory funnel extraction ranged from not recovered (with low spikes) to not recovered (with high spikes).
Accuracy (as % recovery) with continuous liquid-liquid extraction ranged from not recovered (with low spikes) to 1 (with high spikes).

Accuracy (as % recovery) with Soxhlet extraction of soils ranged from not recovered (with low spikes to not recovered (with high spikes).

Accuracy (as % recovery) with ultrasonic extraction of soils ranged from not recovered (with low spikes) to not recovered (with high spikes).

SAMPLE COLLECTION, PRESERVATION & HANDLING
Containers used to collect samples for the determination of semivolatile organic compounds should be soap and water washed followed by methanol (or isopropanol) rinsing. The sample containers should be of glass or Teflon® and have screw-top covers with Teflon® liners.

No preservation is used with concentrated waste samples. With liquid samples containing no residual chlorine and with soil, sediment, and sludge samples, immediately cooling to 4°C is the only preservation used. When residual chlorine is present then 3 mL of 10% aqueous sodium sulfate is added for each gallon of sample collected, followed by cooling to 4°C.

Liquid samples must be extracted within 7 days and their extracts analyzed within 40 days. Concentrated waste, soil, sediment, and sludge samples must be extracted within 14 days and their extracts analyzed within 40 days.

SAMPLE PREPARATION In general, water samples are extracted at a neutral pH with methylene chloride, using either EPA Method 3510 or EPA Method 3520. Solid samples are extracted using either EPA Method 3540 or EPA Method 3550 with methylene chloride/acetone (1:1) as the extraction solvent.

Prior to GC analysis, the extraction solvent may be exchanged to hexane. Single lab data indicates that samples should not be transferred with 100% hexane during sample workup as the more water soluble organophosphorus compounds may be lost.

If cleanup is performed on the samples, the analyst should analyze the samples by GC. This will confirm elution patterns and the absence of interferences from the reagents. If peak detection and identification is prevented by the presence of interferences, further cleanup is required.

QUALITY CONTROL The analyst should monitor the performance of the extraction, cleanup (when used), and analytical system and the effectiveness of the method in dealing with each sample matrix by spiking each sample, standard, and blank with one or two surrogates (e.g., organophosphorus compounds not expected to be present in the sample). Deuterated analogs of analytes should not be used as surrogates for gas chromatographic analysis due to coelution problems.

A minimum of five concentrations for each analyte of interest should be prepared through dilution of the stock standards with isooctane. One of the concentrations should be at a concentration near, but above, the MDL.

Include a mid-level check standard after each group of 10 samples in the analysis sequence. GC/MS techniques should be judiciously employed to support qualitative identifications made with this method. Follow the GC/MS operating requirements specified in EPA Method 8270.

When available, chemical ionization mass spectra may be employed to aid in the qualitative identification process. To confirm an identification of a compound, the background-corrected mass spectrum of the compound must be obtained from the sample extract and must be compared with a mass spectrum from a stock or calibration standard analyzed under the same chromatographic conditions. The molecular ion and all other ions present above 20% relative abundance in the mass spectrum of the standard must be present in the mass spectrum of the sample with agreement to ±20%. The retention time of the compound in the sample must be within six seconds of the retention time for the same compound in the standard solution.

Should the MS procedure fail to provide satisfactory results, additional steps may be taken before reanalysis. These steps may include the use of alternate packed or capillary GC columns or additional sample cleanup.

REFERENCE Test Methods for Evaluating Solid Waste, Physical/Chemical Methods, SW-846, 3rd Edition, U.S. EPA, Office of Solid Waste, Washington, DC, EPA Method 8141 July 1992.

Monocrotophos **EPA Method 8270**
CAS #6923-22-4

TITLE Semivolatile Organic Compounds by GC/MS

MATRIX This method is used to determine the concentration of semivolatile organic compounds in extracts prepared from all types of solid waste matrices, soils, and groundwater. Although surface waters are not specifically mentioned, this method should be applicable to water samples from rivers, lakes, etc.

METHOD SUMMARY This method covers 259 semivolatile organic compounds. In very limited applications direct injection of the sample into the GC/MS system may be appropriate, but this results in very high detection limits (approximately 10,000 µg/L). Typically, a 1-L liquid sample, containing surrogate, and matrix spiking standards, is extracted in a continuous extractor first under acid conditions and then under basic conditions. Typically 30 g of a solid sample, containing surrogate, and matrix spiking standards, is extracted ultrasonically. After concentrating the extract to 1 mL it is spiked with 10 µL of an internal standard solution just prior to analysis by GC/MS. The volume injected should contain about 100 ng of base/neutral and 200 ng of acid surrogates (for a 1-µL injection). Analysis is performed by GC/MS using a capillary GC column.

INTERFERENCES Raw GC/MS data from all blanks, samples, and spikes must be evaluated for interferences. Contamination by carryover can occur whenever high-concentration and low-concentration samples are sequentially analyzed. To reduce carryover, the sample syringe must be rinsed out between samples with solvent. Whenever an unusually concentrated sample is encountered, it should be followed by the analysis of blank solvent to check for cross-contamination.

INSTRUMENTATION A GC/MS and a data system are required. The GC column used is a 30 m × 0.25 mm I.D. (or 0.32 mm I.D.) 1um film thickness silicone-coated fused silica

capillary column. A continuous liquid-liquid extractor equipped with Teflon® or glass connection joints and stopcocks requiring no lubrication, a K-D concentrating apparatus, water bath, and an ultrasonic disrupter with a minimum power of 300 W and with pulsing capability are also required.

PRECISION & ACCURACY The estimated quantitation limit (EQL) of Method 8270B for determining an individual compound is approximately 1 mg/kg (wet weight) for soil or sediment samples, 1–200 mg/kg for wastes (dependent on matrix and method of preparation), and 10 µg/L for groundwater samples. EQLs will be proportionately higher for sample extracts that require dilution to avoid saturation of the detector.

The EQL(b) for groundwater in µg/L is 40.
The EQL (a, b) for low concentrations in soil and sediment in µg/kg is not determined.
Accuracy as µg/L is not listed.
Overall precision in µg/L is not listed.

(a) *EQLs listed for soil/sediment are based on wet weight. Normally data is reported in a dry-weight basis; therefore, EQLs will be higher based on the % dry weight of each sample. This calculation is based on a 30 g sample and gel permeation chromatography cleanup.*
(b) *Sample EQLs are highly matrix-dependent. The EQLs are provided for guidance and may not always be achievable.*
C = *True value for concentration, in µg/L.*
X = *Average recovery found for measurements of samples containing a concentration of C, in µg/L.*

ESTIMATED QUANTITATION LIMIT

Other Matrices	Factor (a)
High-concentration soil and sludges by sonicator	7.5
Non-water miscible waste	75

(a) *EQL for other matrices = [EQL for low soil/sediment] × [Factor]. This estimated EQL is similar to an EPA "Practical Quantitation Limit."*

SAMPLING METHOD
Liquid samples — Use a 1 or 2½ gallon amber glass bottle with a screw-top Teflon®-lined cover that has been prewashed with detergent and rinsed with distilled water and methanol (or isopropanol).

Soils, sediments, or sludges — Use an 8-oz. widemouth glass with a screw-top Teflon®-lined cover that has been prewashed with detergent and rinsed with distilled water and methanol (or isopropanol).

SAMPLE PRESERVATION
Liquid samples — If residual chlorine is present, add 3 mL of 10% sodium thiosulfate per gallon, cool to 4°C and store in a solvent-free refrigerator until analysis; if chlorine is not present, then eliminate the sodium thiosulfate addition.

Soils, sediments, or sludges — Cool samples to 4°C and store in a solvent-free refrigerator.

MHT Liquid samples must be extracted within 7 days and the extracts analyzed within 40 days. Soils, sediments, or sludges may be stored for a maximum of 14 days and the extracts analyzed within 40 days.

SAMPLE PREPARATION
Liquid samples — Transfer 1 L quantitatively to a continuous extractor. If high concentrations are anticipated, a smaller volume may be used and then diluted with organic-free reagent water to 1 L. Adjust pH, if necessary, to pH <2 using 1:1 (V/V) sulfuric acid. Pipette 1.0 mL of a surrogate standard spiking solution into each sample. For the sample in each analytical batch selected for spiking, add 1.0 mL of a matrix spiking standard. For base/neutral acid analysis, the amount of the surrogates and matrix spiking compounds added to the sample should result in a final concentration of 100 ng/µL of each analyte in the extract to be analyzed (assuming a 1-µL injection). Extract with methylene chloride for 18–24 h. Next, adjust the pH of the aqueous phase to pH >11 using 10 N sodium hydroxide and extract it with methylene chloride again for 18–24 h. Dry the extract through a column containing anhydrous sodium sulfate and concentrate it to 1 mL using a K-D concentrator.

Soils, sediments, or sludges — Use 30 g of sample. Nonporous or wet samples (gummy or clay type) that do not have a free-flowing sandy texture must be mixed with anhydrous sodium sulfate until the sample is free flowing. Add 1 mL of surrogate standards to all samples, spikes, standards, and blanks. For the sample in each analytical batch selected for spiking, add 1.0 mL of a matrix spiking standard. For base/neutral acid analysis, the amount added of the surrogates and matrix spiking compounds should result in a final concentration of 100 ng/µL of each base/neutral analyte and 200 ng/µL of each acid analyte in the extract to be analyzed (assuming a 1-µL injection). Immediately add a 100-mL mixture of 1:1 methylene chloride:acetone and extract the sample ultrasonically for 3 min and then decant or filter the extracts. Repeat the extraction two or more times. Dry the extract using a column with anhydrous sodium sulfate and concentrate it to 1 mL in a K-D concentrator.

QUALITY CONTROL A methylene chloride solution containing 50 ng/µL of decafluorotriphenylphosphine (DFTPP) is used for tuning the GC/MS system each 12-h shift. A system performance check also must be made during every 12-h shift. A standard containing 50 ng/µL each of 4,4′-DDT, pentachlorophenol, and benzidine is required to verify injection port inertness and GC column performance. A calibration standard at mid-concentration, containing each compound of interest, including all required surrogates, must be performed every 12 h during analysis. After the system performance check is met, calibration check compounds (CCCs) are used to check the validity of the initial calibration.

The internal standard responses and retention times in the calibration check standard must be evaluated immediately after or during data acquisition. If the retention time for any internal standard changes by more than 30 seconds from the last check calibration (12 h), the chromatographic system must be inspected for malfunctions and corrections must be made, as required. If the electron ionization current plot (EICP) area for any of the internal standards changes by a factor of two from the last daily calibration standard check, the mass spectrometer must be inspected for malfunctions and corrections must be made, as appropriate.

Demonstrate, through the analysis of a reagent water blank, that interferences from the analytical system, glassware, and reagents are under control. The blank samples should be carried through all stages of the sample preparation and measurement steps. For each analytical batch (up to 20 samples), a reagent blank, matrix spike, and matrix spike duplicate/duplicate must be analyzed (the frequency of the spikes may be different for different monitoring programs). The blank and spiked samples must be carried through all stages of the sample preparation and measurement steps. A QC reference sample concentrate containing each analyte at a concentration of 100 mg/L in methanol is required.

REFERENCE Test Methods for Evaluating Solid Waste (SW-846). U.S. EPA 1983, Method 8270B, Rev. 2, Nov. 1990. Office of Solid Waste, Washington, DC.

N

Naled
CAS #300-76-5

EPA Method 8141

TITLE Organophosphorus Compounds by Gas Chromatography: Capillary Column Technique

MATRIX This method covers aqueous and solid matrices. This includes a wide variety such as drinking water, groundwater, industrial wastewaters, surface waters, soils, solids, and sediments.

METHOD SUMMARY This is a GC method used to determine the concentration of 28 organophosphorus pesticides.

The use of Gel Permeation Cleanup (EPA Method 3640) for sample cleanup has been demonstrated to yield recoveries of less than 85% for many method analytes and is therefore not recommended for use with this method.

This method provides GC conditions for the detection of ppb concentrations of organophosphorus compounds. Prior to the use of this method, appropriate sample preparation techniques must be used. Water samples are extracted at a neutral pH with methylene chloride as a solvent by using a separatory funnel (EPA Method 3510) or a continuous liquid-liquid extractor (EPA Method 3520). Soxhlet extraction (EPA Method 3540) or ultrasonic extraction (EPA Method 3550) using methylene chloride/acetone (1:1) are used for solid samples. Both neat and diluted organic liquids (EPA Method 3580) may be analyzed by direct injection. Spiked samples are used to verify the applicability of the chosen extraction technique to each new sample type. A GC with a flame photometric (FPD) or nitrogen-phosphorus detector (NPD) is used for this multiresidue procedure.

INTERFERENCES The use of Florisil cleanup materials (EPA Method 3620) for some of the compounds in this method has been demonstrated to yield recoveries less than 85% and is therefore not recommended for all compounds. Use of phosphorus or halogen specific detectors, however, often obviates the necessity for cleanup for relatively clean sample matrices. If particular circumstances demand the use of an alternative cleanup procedure, the analyst must determine the elution profile and demonstrate that the recovery of each analyte is no less than 85%.

Use of a flame photometric detector (FPD) in the phosphorus mode will minimize interferences from materials that do not contain phosphorus. Elemental sulfur, however, may interfere with the determination of certain organophosphorus compounds by flame photometric gas chromatography. Sulfur cleanup using EPA Method 3660 may alleviate this interference. A nitrogen phosphorus detector (NPD) is also recommended.

A few analytes coelute on certain columns. Therefore, select a second column for confirmation where coelution of the analytes of interest does not occur.

Method interferences may be caused by contaminants in solvents, reagents, glassware, and other sample processing hardware that lead to discrete artifacts or elevated baselines in gas chromatograms. All these materials must be routinely demonstrated to be free from interferences under the conditions of the analysis by analyzing reagent blanks.

INSTRUMENTATION A GC with a NPD or a FPD will be needed. A data system or integrator is recommended for measuring peak areas and/or peak heights. A Kuderna-Danish (K-D) apparatus will be needed for extract concentration.

Column 1: 15 m × 0.53 mm megabore capillary column, 1.0 μm film thickness, DB-210.

Column 2: 15 m × 0.53 mm megabore capillary column, 1.5 μm film thickness, SPB-608.

Column 3: 15 m × 0.53 mm megabore capillary column, 1.0 μm film thickness, DB-5.

Three megabore capillary columns are included for analysis of organophosphates by this method. Column 1 (DB-210 or equivalent) and Column 2 (SPB-608 or equivalent) are recommended if a large number of organophosphorus analytes are to be determined. If the superior resolution offered by Column 1 and Column 2 is not required, Column 3 (DB-5 or equivalent) may be used. For megabore capillary columns, automatic injections of 1 μL are recommended.

PRECISION & ACCURACY The MDL actually achieved in a given analysis will vary, as it is dependent on instrument sensitivity and matrix effects. Single operator accuracy and precision studies have been conducted with spiked water and soil samples.

MULTIPLICATION FACTORS FOR OTHER MATRICES (a)

Matrix	Factor (b)
Groundwater (EPA Method 3510 or EPA Method 3520)	10
Low-concentration soil by Soxhlet and no cleanup	10 (c)
Low-concentration soil by ultrasonic extraction with GPC cleanup	6.7 (c)
High-concentration soil and sludges by ultrasonic extraction	500 (c)
Non-water miscible waste (EPA Method 3580)	1000 (c)

(a) Sample EQLs are highly matrix-dependent. The EQLs listed here are provided for guidance and may not always be achievable.
(b) EQL = [Method detection limit] × [Factor]. For non-aqueous samples the factor is on a wet-weight basis.
(c) Multiply this factory times the soil MDL.

The MDL (in μg/L) when reagent water was extracted using a separatory funnel was 0.50.

The MDL (in μg/kg) when soil was extracted using Soxhlet extraction (EPA Method 3540) was 25.0.

Accuracy (as % recovery) with separatory funnel extraction ranged from not recovered (with low spikes) to not recovered (with high spikes).

Accuracy (as % recovery) with continuous liquid-liquid extraction ranged from not recovered (with low spikes) to 74 (with high spikes).

Accuracy (as % recovery) with Soxhlet extraction of soils ranged from not recovered (with low spikes to not recovered (with high spikes).

Accuracy (as % recovery) with ultrasonic extraction of soils ranged from 82 (with low spikes) to 33 (with high spikes).

SAMPLE COLLECTION, PRESERVATION & HANDLING
Containers used to collect samples for the determination of semivolatile organic compounds should be soap and water washed followed by methanol (or isopropanol) rinsing. The sample containers should be of glass or Teflon® and have screw-top covers with Teflon® liners.

No preservation is used with concentrated waste samples. With liquid samples containing no residual chlorine and with soil, sediment, and sludge samples, immediately cooling to 4°C is the only preservation used. When residual chlorine is present then 3 mL of 10% aqueous sodium sulfate is added for each gallon of sample collected, followed by cooling to 4°C.

Liquid samples must be extracted within 7 days and their extracts analyzed within 40 days. Concentrated waste, soil, sediment, and sludge samples must be extracted within 14 days and their extracts analyzed within 40 days.

SAMPLE PREPARATION In general, water samples are extracted at a neutral pH with methylene chloride, using either EPA Method 3510 or EPA Method 3520. Solid samples are extracted using either EPA Method 3540 or EPA Method 3550 with methylene chloride/acetone (1:1) as the extraction solvent.

Prior to GC analysis, the extraction solvent may be exchanged to hexane. Single lab data indicates that samples should not be transferred with 100% hexane during sample workup as the more water soluble organophosphorus compounds may be lost.

If cleanup is performed on the samples, the analyst should analyze the samples by GC. This will confirm elution patterns and the absence of interferences from the reagents. If peak detection and identification is prevented by the presence of interferences, further cleanup is required.

QUALITY CONTROL The analyst should monitor the performance of the extraction, cleanup (when used), and analytical system and the effectiveness of the method in dealing with each sample matrix by spiking each sample, standard, and blank with one or two surrogates (e.g., organophosphorus compounds not expected to be present in the sample). Deuterated analogs of analytes should not be used as surrogates for gas chromatographic analysis due to coelution problems.

A minimum of five concentrations for each analyte of interest should be prepared through dilution of the stock standards with isooctane. One of the concentrations should be at a concentration near, but above, the MDL.

Include a mid-level check standard after each group of 10 samples in the analysis sequence. GC/MS techniques should be judiciously employed to support qualitative identifications made with this method. Follow the GC/MS operating requirements specified in EPA Method 8270.

When available, chemical ionization mass spectra may be employed to aid in the qualitative identification process. To confirm an identification of a compound, the background-corrected mass spectrum of the compound must be obtained from the sample extract and must be compared with a mass spectrum from a stock or calibration standard analyzed under the same chromatographic conditions. The molecular ion and all other ions present above 20% relative abundance in the mass spectrum of the standard must be present in the mass spectrum of the sample with agreement to ±20%. The retention time of the compound in the sample must be within six seconds of the retention time for the same compound in the standard solution.

Should the MS procedure fail to provide satisfactory results, additional steps may be taken before reanalysis. These steps may include the use of alternate packed or capillary GC columns or additional sample cleanup.

REFERENCE Test Methods for Evaluating Solid Waste, Physical/Chemical Methods, SW-846, 3rd Edition, U.S. EPA, Office of Solid Waste, Washington, DC, EPA Method 8141 July 1992.

Naled   **EPA Method 8270**
CAS #300-76-5

TITLE Semivolatile Organic Compounds by GC/MS

MATRIX This method is used to determine the concentration of semivolatile organic compounds in extracts prepared from all types of solid waste matrices, soils, and groundwater. Although surface waters are not specifically mentioned, this method should be applicable to water samples from rivers, lakes, etc.

METHOD SUMMARY This method covers 259 semivolatile organic compounds. In very limited applications direct injection of the sample into the GC/MS system may be appropriate, but this results in very high detection limits (approximately 10,000 µg/L). Typically, a 1-L liquid sample, containing surrogate, and matrix spiking standards, is extracted in a continuous extractor first under acid conditions and then under basic conditions. Typically 30 g of a solid sample, containing surrogate, and matrix spiking standards, is extracted ultrasonically. After concentrating the extract to 1 mL it is spiked with 10 µL of an internal standard solution just prior to analysis by GC/MS. The volume injected should contain about 100 ng of base/neutral and 200 ng of acid surrogates (for a 1-µL injection). Analysis is performed by GC/MS using a capillary GC column.

INTERFERENCES Raw GC/MS data from all blanks, samples, and spikes must be evaluated for interferences. Contamination by carryover can occur whenever high-concentration and low-concentration samples are sequentially analyzed. To reduce carryover, the sample syringe must be rinsed out between samples with solvent. Whenever an unusually concentrated sample is encountered, it should be followed by the analysis of blank solvent to check for cross-contamination.

INSTRUMENTATION A GC/MS and a data system are required. The GC column used is a 30 m × 0.25 mm I.D. (or 0.32 mm I.D.) 1um film thickness silicone-coated fused silica capillary column. A continuous liquid-liquid extractor equipped with Teflon® or glass connection joints and stopcocks requiring no lubrication, a K-D concentrating apparatus, water bath, and an ultrasonic disrupter with a minimum power of 300 W and with pulsing capability are also required.

PRECISION & ACCURACY The estimated quantitation limit (EQL) of Method 8270B for determining an individual compound is approximately 1 mg/kg (wet weight) for soil or sediment samples, 1–200 mg/kg for wastes (dependent on matrix and method of preparation), and 10 µg/L for groundwater samples. EQLs will be proportionately higher for sample extracts that require dilution to avoid saturation of the detector.

The EQL(b) for groundwater in µg/L is 20.
The EQL (a, b) for low concentrations in soil and sediment in µg/kg is not determined.
Accuracy as µg/L is not listed.
Overall precision in µg/L is not listed.

(a) EQLs listed for soil/sediment are based on wet weight. Normally data is reported in a dry-weight basis; therefore, EQLs will be higher based on the % dry weight of each sample. This calculation is based on a 30 g sample and gel permeation chromatography cleanup.
(b) Sample EQLs are highly matrix-dependent. The EQLs are provided for guidance and may not always be achievable.
C = *True value for concentration, in µg/L.*
X = *Average recovery found for measurements of samples containing a concentration of C, in µg/L.*

ESTIMATED QUANTITATION LIMIT

Other Matrices	Factor (a)
High-concentration soil and sludges by sonicator	7.5
Non-water miscible waste	75

(a) EQL for other matrices = [EQL for low soil/sediment] × [Factor]. This estimated EQL is similar to an EPA "Practical Quantitation Limit."

SAMPLING METHOD

Liquid samples — Use a 1 or 2½ gallon amber glass bottle with a screw-top Teflon®-lined cover that has been prewashed with detergent and rinsed with distilled water and methanol (or isopropanol).

Soils, sediments, or sludges — Use an 8-oz. widemouth glass with a screw-top Teflon®-lined cover that has been prewashed with detergent and rinsed with distilled water and methanol (or isopropanol).

SAMPLE PRESERVATION

Liquid samples — If residual chlorine is present, add 3 mL of 10% sodium thiosulfate per gallon, cool to 4°C and store in a solvent-free refrigerator until analysis; if chlorine is not present, then eliminate the sodium thiosulfate addition.

Soils, sediments, or sludges — Cool samples to 4°C and store in a solvent-free refrigerator.

MHT Liquid samples must be extracted within 7 days and the extracts analyzed within 40 days. Soils, sediments, or sludges may be stored for a maximum of 14 days and the extracts analyzed within 40 days.

SAMPLE PREPARATION

Liquid samples — Transfer 1 L quantitatively to a continuous extractor. If high concentrations are anticipated, a smaller volume may be used and then diluted with organic-free reagent water to 1 L. Adjust pH, if necessary, to pH <2 using 1:1 (V/V) sulfuric acid. Pipette 1.0 mL of a surrogate standard spiking solution into each sample. For the sample in each analytical batch selected for spiking, add 1.0 mL of a matrix spiking standard. For base/neutral acid analysis, the amount of the surrogates and matrix spiking compounds added to the sample should result in a final concentration of 100 ng/µL of each analyte in the extract to be analyzed (assuming a 1-µL injection). Extract with methylene chloride for 18–24 h. Next, adjust the pH of the aqueous phase to pH >11 using 10 N sodium hydroxide and extract it with methylene chloride again for 18–24 h. Dry the extract through a column containing anhydrous sodium sulfate and concentrate it to 1 mL using a K-D concentrator.

Soils, sediments, or sludges — Use 30 g of sample. Nonporous or wet samples (gummy or clay type) that do not have a free-flowing sandy texture must be mixed with anhydrous sodium sulfate until the sample is free flowing. Add 1 mL of surrogate standards to all samples, spikes, standards, and blanks. For the sample in each analytical batch selected for spiking, add 1.0 mL of a matrix spiking standard. For base/neutral acid analysis, the amount added of the surrogates and matrix spiking compounds should result in a final concentration of 100 ng/µL of each base/neutral analyte and 200 ng/µL of each acid analyte in the extract to be analyzed (assuming a 1-µL injection). Immediately add a 100-mL mixture of 1:1 methylene chloride:acetone and extract the sample ultrasonically for 3 min and then decant or filter the extracts. Repeat the extraction two or more times. Dry the extract using a column with anhydrous sodium sulfate and concentrate it to 1 mL in a K-D concentrator.

QUALITY CONTROL A methylene chloride solution containing 50 ng/µL of decafluorotriphenylphosphine (DFTPP) is used for tuning the GC/MS system each 12-h shift. A system performance check also must be made during every 12-h shift. A standard containing 50 ng/µL each of 4,4′-DDT, pentachlorophenol, and benzidine is required to verify injection port inertness and GC column performance. A calibration standard at mid-concentration, containing each compound of interest, including all required surrogates, must be performed every 12 h during analysis. After the system performance check is met, calibration check compounds (CCCs) are used to check the validity of the initial calibration.

The internal standard responses and retention times in the calibration check standard must be evaluated immediately after or during data acquisition. If the retention time for any internal standard changes by more than 30 seconds from the last check calibration (12 h), the chromatographic system must be inspected for malfunctions and corrections must be made, as required. If the electron ionization current plot (EICP) area for any of the internal standards changes by a factor of two from the last daily calibration standard check, the mass spectrometer must be inspected for malfunctions and corrections must be made, as appropriate.

Demonstrate, through the analysis of a reagent water blank, that interferences from the analytical system, glassware, and reagents are under control. The blank samples should be carried through all stages of the sample preparation and measurement steps. For each analytical batch (up to 20 samples), a reagent blank, matrix spike, and matrix spike duplicate/duplicate must be analyzed (the frequency of the spikes may be different for different monitoring programs). The blank and spiked samples must be carried through all stages of the sample preparation and measurement steps. A QC reference sample concentrate containing each analyte at a concentration of 100 mg/L in methanol is required.

REFERENCE Test Methods for Evaluating Solid Waste (SW-846). U.S. EPA 1983, Method 8270B, Rev. 2, Nov. 1990. Office of Solid Waste, Washington, DC.

Naled   EPA Method 8140
CAS #300-76-5

TITLE Organophosphorus Pesticides

MATRIX Groundwater, soils, sludges, water miscible liquid wastes, and non-water miscible wastes.

APPLICATION This method is used for the analysis of 21 organophosphorus pesticides. Samples are extracted, concentrated, and analyzed using direct injection of both neat and diluted organic liquid into a gas chromatograph (GC).

INTERFERENCES Solvents, reagents, and glassware may introduce artifacts. Other interferences may come from coextracted compounds from samples. The use of Florisil cleanup materials may produce low recoveries. Elemental sulfur may interfere with some compounds when using a flame photometric detector. Sulfur cleanup (Method 3660) may alleviate sulfur interference.

INSTRUMENTATION GC capable of on-column injections and a flame photometric detector (FPD) or a thermionic detector. A halogen specific detector may also be used and may have the advantage of fewer interferences. Column 1: 1.8 meter by 2 mm with 5% SP-2401 on Supelcoport. Column 2: 1.8 m by 2 mm with 3% SP-2401 on Supelcoport. Column 3: 50 cm by ⅛ in Teflon® with 15% SE-54 on Gas Chrom Q. The preferred column is Column Number 3.

RANGE 25.8–294 µg/L.

MDL 0.1 µg/L (in reagent water).

PQL FACTORS FOR MULTIPLYING × FID MDL VALUE

Matrix	Multiplication Factor
Groundwater	10
Low-level soil by sonication with GPC cleanup	670
High-level soil and sludge by sonication	10,000
Non-water miscible waste	100,000

PRECISION 8.1% (single operator standard deviation)

ACCURACY 78.0% (single operator average recovery)

SAMPLING METHOD Use 8-oz. widemouth glass bottles with Teflon®-lined caps for concentrated waste samples, soils, sediments, and sludges. Use 1 or 2½ gallon amber glass bottles with Teflon®-lined caps for liquid (water) samples.

STABILITY Cool soil, sediment, sludge, and liquid samples to 4°C. If residual chlorine is present in liquid samples add 3 mL of 10% sodium thiosulfate per gallon of sample and cool to 4°C.

MHT 14 days for concentrated waste, soil, sediment, or sludge; 7 days for liquid samples; all extracts must be analyzed within 40 days.

QUALITY CONTROL A quality control check sample concentrate containing this compound in acetone at a concentration 1,000 times more concentrated than the selected spike concentration is required. The QC check sample concentrate may be prepared from pure standard materials or purchased as certified solutions. Use appropriate trip, matrix, control site, method, reagent, and solvent blanks. Internal, surrogate, and five concentration level calibration standards are used.

REFERENCE Method 8140, SW-846, 3rd ed., Sept. 1986.

Naphthalene   EPA Method 1625
CAS #91-20-3

TITLE Semivolatile Organic Compounds by Isotope Dilution GC/MS

MATRIX The compounds may be determined in waters, soils, and municipal sludges by this method.

METHOD SUMMARY This method is used to determine 176 semivolatile toxic organic pollutants associated with the CWA (as amended 1987); the RCRA (as amended 1986); the CERCLA (as amended 1986); and other compounds amenable to extraction and analysis by capillary column gas chromatography-mass spectrometry (GC/MS).

Stable isotopically-labeled analogs of the compounds of interest are added to the sample. If the solids content is less than 1%, a 1-L sample is extracted at pH 12–13, then at pH <2 with methylene chloride using continuous extraction techniques.

If the solids content is 30% or less, the sample is diluted to 1% solids with reagent water, homogenized ultrasonically, and extracted at pH 12–13, then at pH <2 with methylene chloride using continuous extraction techniques. If the solids content is greater than 30%, the sample is extracted using ultrasonic techniques.

Each extract is dried over sodium sulfate, concentrated to a volume of 5 mL, cleaned up using GPC, if necessary, and concentrated. Extracts are concentrated to 1 mL if GPC is not performed, and to 0.5 mL if GPC is performed.

An internal standard is added to the extract, and a 1-mL aliquot of the extract is injected into the GC. The compounds are separated by GC and detected by a MS. The labeled compounds serve to correct the variability of the analytical technique.

INTERFERENCES Solvents, reagents, glassware, and other sample processing hardware may yield artifacts and/or elevated baselines causing misinterpretation of chromatograms and spectra. Materials used in the analysis must be demonstrated to be free from interferences under the conditions of analysis by running method blanks initially and with each sample lot (sample started through the extraction process on a given 8-h shift, to a maximum of 20). Specific selection of reagents and purification of solvents by distillation in all glass systems may be required. Glassware and, where possible, reagents are cleaned by solvent rinse and baking at 450°C for 1-h minimum. Interferences coextracted from samples will vary considerably from source to source, depending on the diversity of the site being sampled.

INSTRUMENTATION Major instrumentation includes a GC with a splitless or on-column injection port for capillary column, a MS with 70 eV electron impact ionization, and a data system to collect and record MS data, and process it. A K-D apparatus is used to concentrate extracts.

GC Column: 30 m × 0.25 mm I.D. 5% phenyl, 94% methyl, 1% vinyl silicone bonded phased fused silica capillary column.

PRECISION & ACCURACY The detection limits of the method are usually dependent on the level of interferences rather than instrumental limitations. The limits typify the minimum quantities that can be detected with no interferences present.

The minimum level (in µg/mL) was 10. This is defined as a minimum level at which the analytical system shall give recognizable mass spectra (background corrected) and acceptable calibration points.

The MDL (in µg/kg) in low solids was 62 and in high solids was 42; these were determined in digested sludge (low solids) and in filter cake or compost (high solids).

The labeled and native compound initial precision as standard deviation (in µg/L) was 20.
The labeled and native compound initial accuracy as average recovery (in µg/L) was 80–139.

SAMPLE COLLECTION, PRESERVATION & HANDLING Collect samples in glass containers. Aqueous samples which flow freely are collected in refrigerated bottles using automatic sampling equipment. Solid samples are collected as grab samples using widemouth jars. Maintain samples at 0 to 4°C from the time of collection until extraction. If residual chlorine is present in aqueous samples, add 80 mg sodium thiosulfate/L of water. Begin sample extraction within 7 days of collection, and analyze all extracts within 40 days of extraction.

SAMPLE PREPARATION Samples containing 1% solids or less are extracted directly using continuous liquid-liquid extraction techniques. Samples containing 1 to 30% solids are diluted to the 1% level with reagent water and extracted using continuous liquid-liquid extraction techniques. Samples containing greater than 30% solids are extracted using ultrasonic techniques.

Base/neutral extraction — Adjust the pH of the waters in the extractors to 12–13 with 6 N NaOH. Extract with methylene chloride for 24–48 h.
Acid extraction — Adjust the pH of the waters in the extractors to 2 or less using 6 N sulfuric acid. Extract with methylene chloride for 24–48 h.
Ultrasonic extraction of high solids samples — Add anhydrous sodium sulfate to the sample and QC aliquot(s). Add acetone:methylene chloride (1:1) to the sample and mix thoroughly

Concentrate extracts using a K-D apparatus.

QUALITY CONTROL The analyst is permitted to modify this method to improve separations or lower the costs of measurements, provided all performance specifications are met. Analyses of blanks are required to demonstrate freedom from contamination. When results of spikes indicate atypical method performance for samples, the samples are diluted to bring method performance within acceptable limits.

For low solids (aqueous samples), extract, concentrate, and analyze two sets of four 1-L aliquots (8 aliquots total) of the precision and recovery standard. For high solids samples, two sets of four 30-g aliquots of the high solids reference matrix are used.

Spike all samples with labeled compounds to assess method performance. Compute percent recovery of the labeled compounds using the internal standard method. Compare the labeled compound recovery for each compound with the corresponding labeled compound recovery.

Reagent water and high solids reference matrix blanks are analyzed to demonstrate freedom from contamination. Extract and concentrate a 1-L reagent water blank or a high solids reference matrix blank with each sample's lot (samples started through the extraction process on the same 8-h shift, to a maximum of 20 samples).

Field replicates may be collected to determine the precision of the sampling technique, and spiked samples may be required to determine the accuracy of the analysis when the internal standard method is used.

REFERENCE Semivolatile Organic Compounds by Isotope Dilution GC/MS. Office of Water Regulation and Standards, U.S. EPA Industrial Technology Division, Washington, DC, EPA Method 1625, Rev. C, June 1989 (contact W.A. Telliard, U.S. EPA, Office of Water Regulations and Standards, 401 M St., SW, Washington, DC, 20460. Phone: 202-382-7131).

Naphthalene   **EPA Method 502**
CAS #91-20-3

TITLE Volatile Organic Compounds in Water By Purge and Trap Capillary Column Gas Chromatography with Photoionization and Electrolytic Conductivity Detectors in Series. U.S. EPA Method 502.2, Rev. 2.0, 1989.

MATRIX Drinking water and raw source water. The latter should include most surface water and groundwater sources.

METHOD SUMMARY This method covers 60 volatile organic compounds that contain halogen atoms and/or that are aromatic. An inert gas (zero grade nitrogen or helium) is bubbled through a 25-mL or a 5-mL water sample (depending on the expected concentration of the analytes). Purged sample components are trapped in a tube of sorbent materials. When purging is complete, the sorbent tube is heated and backflushed with helium to desorb the trapped sample onto a capillary GC column. The column is temperature programmed to separate the method analytes which are then detected with a photoionization detector (PID) and a Hall electrolytic conductivity (HECD) placed in series. The PID is selective for aromatic compounds and the HECD is selective for halogenated compounds.

INTERFERENCES Impurities in the purge gas and from organic compounds outgassing from the plumbing ahead of the trap account for many contamination problems. Interferences purged or coextracted from the samples will vary considerably from source to source, depending upon the particular sample or extract being tested. Cross-contamination can occur whenever high-level and low-level samples are analyzed sequentially. Samples also can be contaminated by diffusion of volatile organics (particularly methylene chloride and fluorocarbons) through the septum seal into the sample during shipment and storage. The lab where volatile analysis is performed and also the refrigerated storage area should be completely free of solvents.

INSTRUMENTATION A GC containing a series configuration of a high temperature photoionization detector (PID) equipped with 10.0 eV (nominal) lamp and Hall electrolytic conductivity detector (HECD) is required. Also required is an all-glass 5-mL purging device, a sorbent trap, and a thermal desorption apparatus which is connected to the GC system.

Column 1: VOCOL glass wide-bore capillary column.
Column 2: RTX–502.2 mega-bore capillary column.
Column 3: DB-62 mega-bore capillary column.

PRECISION & ACCURACY Method detection limits are dependent upon the characteristics of the gas chromatographic system used. Analytes that are not separated chromatographically cannot be individually identified and used in the same calibration mixture or water samples unless an alternative technique for identification and quantification, such as mass spectrometry, is used.

Electrolytic conductivity detetor (c) range in µg/L (a) was 0.02–2000.
Electrolytic conductivity detetor (c) MDL in µg/L (b) was not listed.
Electrolytic conductivity detetor (c) accuracy as % recovery was not listed.
Electrolytic conductivity detetor (c) precision as % RSD was not listed.
Photoionization detector (d) range in µg/L (a) was 0.02–2000.
Photoionization detector (d) MDL in µg/L (b) was 0.06.
Photoionization detector (d) accuracy as % recovery was 102.
Photoionization detector (d) precision as % RSD was 6.2.

(a) The applicable concentration range of this method is compound, instrument, and matrix-dependent. It is listed as being approximately 0.02 to 200 µg/L but no specific information is provided so caution should be observed.
(b) The method detection limits reports with this method are compound, instrument, and matrix-dependent. The values reported were calculated using reagent water fortified with the corresponding compounds at 10 µg/L and a GC-equipped with a 60 m × 0.75 mm VOLCOL wide bore capillary column with 1.5 µm film thickness and using helium carrier gas.
(c) Recoveries and relative standard deviations were determined from seven samples of reagent water fortified with 10 µg/L of each compound. 2-Bromo-1-chloropropane was used as the internal standard for calculating average recoveries.
(d) Recoveries and relative standard deviations were determined from seven samples of reagent water fortified with 10 µg/L of each compound. Fluorobenzene was used as the internal standard for calculating average recoveries.

SAMPLING METHOD Collect samples using a 40- to 120-mL screw-cap vial (prewashed with detergent, rinsed with distilled water and oven dried at 105°C) with a Teflon®-faced silicone septum. Collect bubble-free samples and place the septum with the Teflon® side down on the water.

SAMPLE PRESERVATION If residual chlorine is present in the water add about 25 mg of ascorbic acid to each vial before samples are collected to remove the chlorine. Add hydrochloric acid to reduce pH to <2, immediately cool samples to 4°C, and store them in a solvent-free refrigerator at 4°C until analysis.

MHT The maximum holding time for samples is 14 days from the time they were collected.

SAMPLE PREPARATION Remove the plungers from two 5-mL syringes and attach a closed syringe valve to each. Warm the sample to room temperature, open the sample bottle, and carefully pour the sample into one of the syringe barrels to just short of overflowing. Replace the syringe plunger, invert the syringe, and compress the sample. Open the syringe valve and vent any residual air while adjusting the sample volume to 5.0 mL. Add 10 µL of the internal calibration standard to the sample through the syringe valve. Close the valve. Fill the second syringe in an identical manner from the same sample bottle. Reserve this second syringe for a reanalysis if necessary.

QUALITY CONTROL As an initial demonstration of lab accuracy and precision, analyze 4 to 7 replicates of a lab fortified blank containing analyte at 0.1–5 µg/L. Collect all samples in duplicate. Surrogate analytes (similar to those of the analytes of interest), whose concentration is known in every sample, are measured using the same internal standard calibration procedure. Duplicate field reagent water blanks (trip blanks) must be analyzed with each set of samples, lab reagent blanks (method blanks) must be analyzed with each batch of samples processed as a group within a work shift. Also, a single lab-fortified blank that contains each of the analytes of interest should be analyzed with each batch of samples processed as a group within a work shift. A 3- to 5-point calibration curve is needed depending on the calibration range factor required.

EPA CONTACT & HOTLINE For technical questions contact Dr. Baldev Bathija, U.S. EPA, Office of Ground Water and Drinking Water (WH-550D), 401 M St. SW, Washington, DC 20460. Tel. (202) 260-3040. For further information the EPA Safe Drinking Water Hotline may be called at: (800) 426-4791.

REFERENCE Methods for the Determination of Organic Compounds in Drinking Water, EPA/600/4-88/039 (revised July 1991; Final Rule for determination of compliance with the MCL for Total Trihalomethanes under 141.30, in 40 CFR Part 141, Vol. 58, No. 147, Fed. Reg., Tuesday Aug. 3, 1993). U.S. EPA Environmental Monitoring Systems Laboratory, Cincinnati, OH, 45268, U.S.A. Available from the National Technical Information Service (NTIS), 5285 Port Royal Road, Springfield, VA 22161; Tel. 800-553-6847. NTIS Order Number is PB91-231480.

Naphthalene	EPA Method 524
CAS #91-20-3	

TITLE Measurement of Purgeable Organic Compounds in Water by Capillary Column GC/MS.

MATRIX Drinking water and raw source water; the latter should include most surface water and groundwater sources.

METHOD SUMMARY Method 524.2 covers 60 volatile organic compounds. An inert gas (zero grade nitrogen or helium) is bubbled through a 25-mL or a 5-mL water sample (depending on the expected concentration of the analytes). Purged sample components are trapped in a tube of sorbent materials. When purging is complete, the sorbent tube is heated and backflushed with helium to desorb the trapped sample onto a capillary GC column.

INTERFERENCES Impurities in the purge gas and from organic compounds outgassing from the plumbing ahead of the trap account for many contamination problems. Interferences purged or coextracted from the samples will vary considerably from source to source, depending upon the particular sample or extract being tested. Cross-contamination can occur whenever high-level and low-level samples are analyzed sequentially. Samples also can be contaminated by diffusion of volatile organics (particularly methylene chloride and fluorocarbons) through the septum seal into the sample during shipment and storage.

INSTRUMENTATION A GC/MS with a data system equipped with one of the following capillary GC columns:

Column 1: VOCOL glass wide bore capillary column.
Column 2: DB-624 fused silica capillary column.
Column 3: DB-5 fused silica capillary column.

Also required is an all-glass 25 mL or 5-mL purging device, a sorbent trap, and a thermal desorption apparatus which is connected to the GC/MS system.

PRECISION & ACCURACY Method detection limits are compound- and instrument-dependent, and may vary from approximately 0.02–0.35 µg/L. Note in the table below that the "true" concentration range used for accuracy and precision measurements was quite narrow. However, the applicable concentration range of this method is primarily column dependent and is approximately 0.02 to 200 µg/L for the wide-bore thick-film columns. Narrow-bore thin-film columns may have a capacity which limits the range to about 0.02 to 20 µg/L. Analytes that are inefficiently purged from water will not be detected when present at low concentrations, but they can be measured with acceptable accuracy and precision when present in sufficient amounts.

Analytes that are not separated chromatographically, but which have different mass spectra and non-interfering quantification ions, can be identified and measured in the same calibration mixture or water sample. Analytes which have very similar mass spectra cannot be individually identified and measured in the same calibration mixture or water samples unless they have different retention times. Co-eluting compounds with very similar mass spectra, typically many structural isomers, must be reported as an isomeric group or pair.

The range (in µg/L) was 0.1–100.
The Method Detection Limig (in µg/L) was 0.04.
The accuracy (as % recovery) was 104.
The precision (in %) was 8.2.

Note: Data were obtained from 16–31 determinations using a wide-bore capillary column and a jet separator interfaced to a quadrupole mass spectrometer. All analytes were in a reagent water matrix.

SAMPLING METHOD Collect samples using a 40- to 120-mL screw-cap vial (prewashed with detergent, rinsed with distilled water and oven dried at 105°C) with a Teflon®-faced silicone septum. Collect bubble-free samples and place the septum with the Teflon® side down on the water.

SAMPLE PRESERVATION If residual chlorine is present in the water add about 25 mg of ascorbic acid to each vial before samples are collected to remove the chlorine. Add hydrochloric acid to reduce pH to <2, and immediately cool samples to 4°C, and store them in a solvent-free refrigerator at 4°C until analysis.

MHT The maximum holding time for samples is 14 days from the time they were collected.

SAMPLE PREPARATION Remove the plungers from two 25-mL (or 5-mL depending on sample size) syringes and attach a closed syringe valve to each. Warm the sample to room temperature, open the sample bottle, and carefully pour the sample into one of the syringe barrels to just short of overflowing. Replace the syringe plunger, invert the syringe, and compress the sample. Open the syringe valve and vent any residual air while adjusting the sample volume to 25.0 mL (or 5 mL). For samples and blanks, add 5 µL of the fortification solution containing the internal standard and the surrogates to the sample through the syringe valve. For calibration standards and lab fortified blanks, add 5 µL of the fortification solution containing the internal standard only. Close the valve. Fill the second syringe in an identical manner from the same sample bottle. Reserve this second syringe for a reanalysis if necessary.

QUALITY CONTROL As an initial demonstration of lab accuracy and precision, analyze 4 to 7 replicates of a lab fortified

blank containing analyte at 0.2–5 µg/L. Collect all samples in duplicate. Surrogate analytes (similar to those of the analytes of interest), whose concentration is known in every sample, are measured using the same internal standard calibration procedure. Duplicate field reagent water blanks (trip blanks) must be analyzed with each set of samples, lab reagent blanks (method blanks) must be analyzed with each batch of samples processed as a group within a work shift. Also, a single lab-fortified blank that contains each of the analytes of interest should be analyzed with each batch of samples processed as a group within a work shift. A 3- to 5-point calibration curve is needed depending on the calibration range factor required.

EPA CONTACT & HOTLINE For technical questions contact Dr. Baldev Bathija, U.S. EPA, Office of Ground Water and Drinking Water (WH-550D), 401 M St. SW, Washington, DC 20460. Tel. (202) 260-3040. For further information the EPA Safe Drinking Water Hotline may be called at: (800) 426-4791.

REFERENCE Methods for the Determination of Organic Compounds in Drinking Water, EPA/600/4-88/039 (revised July 1991; Final Rule for determination of compliance with the MCL for Total Trihalomethanes under 141.30, in 40 CFR Part 141, Vol. 58, No. 147, Fed. Reg., Tuesday Aug. 3, 1993). U.S. EPA Environmental Monitoring Systems Laboratory, Cincinnati, OH, 45268, U.S.A. Available from the National Technical Information Service (NTIS), 5285 Port Royal Road, Springfield, VA 22161; Tel. 800-553-6847. NTIS Order Number is PB91-231480.

Naphthalene **EPA Method 625**
CAS #91-20-3

TITLE Base/Neutrals and Acids, U.S. EPA Method 625

MATRIX This methods covers municipal and industrial wastewaters.

METHOD SUMMARY Approximately 1 L of sample is serially extracted with methylene chloride at a pH greater than 11 and again at a pH less than 2 using a separatory funnel or a continuous extractor. The methylene chloride extract is dried, concentrated to a volume of 1 mL, and analyzed by GC/MS. Qualitative identification of the parameters in the extract is performed using the retention time and the relative abundance of three characteristic masses (m/z). Qualitative analysis is performed using either external or internal standard techniques with a single characteristic m/z.

INTERFERENCES Method interferences may be caused by contaminants in solvents, reagents, glassware, and other sample processing hardware. Glassware must be scrupulously cleaned. Glassware should be heated in a muffle furnace at 400°C for 5 to 30 min. Some thermally stable materials, such as PCBs, may not be eliminated by this treatment. Solvent rinses with acetone and pesticide quality hexane may be substituted for the muffle furnace heating. Matrix interferences may be caused by contaminants that are coextracted from the sample. The base-neutral extraction may cause significantly reduced recovery of phenols. The packed gas chromatographic columns recommended for the basic fraction may not exhibit sufficient resolution for some analytes.

INSTRUMENTATION A GC/MS system with an injection port designed for on-column injection when using packed columns and for splitless injection when using capillary columns.

Column for base/neutrals: 1.8 m long × 2 mm I.D. glass, packed with 3% SP-2550 on Supelcoport (100/120 mesh) or equivalent.

Column for acids: 1.8 m long × 2 mm I.D. glass, packed with 1% SP-1240DA on Supelcoport (100/120 mesh) or equivalent.

PRECISION & ACCURACY The MDL concentrations were obtained using reagent water. The MDL actually achieved in a given analysis will vary depending on instrument sensitivity and matrix effects. This method was tested by 15 laboratories using reagent water, drinking water, surface water, and industrial wastewaters spiked at six concentrations over the range 5 to 100 µg/L. Single operator precision, overall precision, and method accuracy were found to be directly related to the concentration of the parameter matrix.

The MDL (in µg/L) in reagent water was not reported.

The standard deviation (in µg/L based on 4 recovery measurements) was 30.1.

The range (in µg/L) for average recovery for 4 measurements was 35.6–119.6.

The range (in %) for percent recovery was 21–133.

Accuracy (in µg/L) as expected recovery for one or more measurements of a sample containing a true concentration of C was 0.76C + 1.58.

Precision (in µg/L) as expected single analyst standard deviation of measurements at an average concentration found at X was 0.21X–0.41.

Overall precision (in µg/L) as expected interlaboratory standard deviation of measurements in an average concentration found at X was 0.30X–0.68.

$C = $ *True value of the concentration in µg/L.*
$X = $ *Average recovery found for measurements of samples containing a concentration at C in µg/L.*

SAMPLE PREPARATION Adjust the pH to >11 with sodium hydroxide and serially extract in a separatory funnel with methylene chloride or else in a continuous extractor. Next, adjust the pH to <2 with sulfuric acid and serially extract in a separatory funnel with methylene chloride or else in a continuous extractor. Dry the extracts separately through a column of anhydrous sodium sulfate and then concentrate each of the extracts to 1.0 mL using a K-D apparatus.

SAMPLE COLLECTION, PRESERVATION & HANDLING Grab samples must be collected in glass containers. All samples must be refrigerated at 4°C from the time of collection until extraction. If residual chlorine is present, add 80 mg of sodium thiosulfate/L of sample and mix well. All samples must be extracted within 7 days of collection and completely analyzed within 40 days of extraction.

QUALITY CONTROL Make an initial, one-time, demonstration of the ability to generate acceptable accuracy and precision with this method. Before processing any samples, the analyst must analyze a reagent water blank to demonstrate that interferences from the analytical system and glassware are under control. Each time a set of samples is extracted or reagents are changed, a reagent water blank must be processed. Spike and analyze a minimum of 5% of all samples to monitor and evaluate lab data quality. A QC check sample concentrate that contains each parameter of interest at a concentration of 100 µg/mL in acetone is required. PCBs and multicomponent pesticides may be omitted from this test.

After analysis of five spiked wastewater samples, calculate the average percent recovery and the standard deviation of the percent recovery. Spike all samples with the surrogate standard spiking solution and calculate the percent recovery of each surrogate compound.

REFERENCE Federal Register, Vol. 49, No. 209. Friday, Oct. 26, 1984.

Naphthalene   **EPA Method 8021**
CAS #91-20-3

TITLE Halogenated Volatile by Gas Chromatography Using Photoionization and Electrolytic Conductivity Detectors in Series: Capillary Column Technique

MATRIX This method is applicable to nearly all types of samples, regardless of water content, including groundwater, aqueous sludges, caustic liquors, acid liquors, waste solvents, oily wastes, mousses, tars, fibrous wastes, polymeric emulsions, filter cakes, spent carbons, spent catalysts, soils, and sediments.

METHOD SUMMARY This method is used to determine 60 volatile organic compounds in a variety of solid waste matrices. It provides GC conditions for the detection of halogenated and aromatic volatile organic compounds. Samples can be analyzed using direct injection or purge-and-trap (EPA Method 5030). Groundwater samples must be analyzed using EPA Method 5030 (where applicable). A temperature program is used with the GC. Detection is achieved by a photoionization detector (PID) and a Hall electrolytic conductivity detector (HECD) in series.

INTERFERENCES Samples can be contaminated by diffusion of volatile organics (particularly chlorofluorocarbons and methylene chloride) through the sample container septum during shipment and storage.

INSTRUMENTATION A GC-equipped with variable-constant differential flow controllers, subambient oven controller, PID and HECD detectors connected with a short piece of uncoated capillary tubing and a data system.

Column: 60 m × 0.75 mm I.D. VOCOL wide-bore capillary column with 1.5 µm film thickness.

PRECISION & ACCURACY MDLs are compound-dependent and vary with purging efficiency and concentration. The applicable concentration range of this method is compound- and instrument-dependent but is approximately 0.1 to 200 µg/L. Analytes that are inefficiently purged from water will not be detected when present at low concentrations, but they can be measured with acceptable accuracy and precision when present in sufficient amounts. The estimated quantitation limit (EQL) for an individual compound is approximately 1 µg/kg (wet weight) for soil/sediment samples, 100 µg/kg (wet weight) for wastes, and 1 µg/L for groundwater. EQLs will be proportionately higher for sample extracts and samples that require dilution or reduced sample size to avoid saturation of the detector.

MULTIPLICATION FACTORS FOR OTHER MATRICES (a)

Matrix	Factor (b)
Groundwater	10
Low-concentration soil	10
Water miscible liquid waste	500
High-concentration soil and sludge	1250
Non-water miscible waste	1250

(a) Sample EQLs are highly matrix-dependent. The EQLs listed herein are provided for guidance and may not always be achievable.
(b) EQL = [Method detection limit] × [Factor]. For non-aqueous samples, the factor is on a wet-weight basis.

SINGLE LABORATORY ACCURACY & PRECISION DATA FOR VOCs IN WATER
This method was tested in a single lab using water spiked at 10 µg/L and the following data was reported:

Recoveries and standard deviations were determined from seven samples and spiked at 10 µg/L of each analyte. Recoveries were determined by the internal standard method. Internal standards were: Fluorobenzene for PID and 2-Bromo-1-chloropropane for HECD.

The average recovery (in percent) for the PID was 102.
The standard deviation of the recovery for the PID was 6.3.
The MDL (in µg/mL) for the PID was 0.06.
The average recovery (in percent) for the HECD was none (no response for this detector).
The standard deviation of the recovery for the HECD was none (no response for this detector)-.
The MDL (in µg/mL) for the HECD was none (no response for this detector).

SAMPLE COLLECTION, PRESERVATION & HANDLING
Volatile Organics — Standard 40-mL glass screw-cap VOA vials with Teflon®-faced silicone septum may be used for both liquid and solid matrices. When collecting samples, liquids and solids should be introduced into the vials gently to reduce agitation which might drive off volatile compounds. If there are any air bubbles present the sample must be retaken. Tap slightly as they are filled to try and eliminate as much free air space as possible. The two vials from each sampling locations should be sealed in separate plastic bags to prevent cross-contamination between samples particularly if the sampled waste is suspected of containing high levels of volatile organics.

Semivolatile organics — Containers used to collect samples for the determination of semivolatile organic compounds should be soap and water washed followed by methanol (or isopropanol) rinsing. The sample containers should be of glass or Teflon® and have screw-top covers with Teflon® liners.

Preservation for volatile organics — No preservation is used with concentrated waste samples. With liquid samples containing no

residual chlorine, 4 drops of concentrated hydrochloric acid are added and the samples are immediately cooled to 4°C. When liquid samples contain residual chlorine, they are treated as above and, in addition, 4 drops of 4% aqueous sodium thiosulfate are added. Soil, sediment, and sludge samples are only cooled to 4°C.

Preservation for semivolatile organics — No preservation is used with concentrated waste samples. With liquid samples containing no residual chlorine and with soil, sediment, and sludge samples, immediately cooling to 4°C is the only preservation used. When residual chlorine is present then 3 mL of 10% aqueous sodium sulfate is added for each gallon of sample collected, followed by cooling to 4°C.

MHT The holding time for all volatile organics samples is 14 days. Liquid samples must be extracted within 7 days and their extracts analyzed within 40 days. Concentrated waste, soil, sediment, and sludge samples must be extracted within 14 days and their extracts analyzed within 40 days.

SAMPLE PREPARATION Volatile compounds are introduced into the gas chromatograph either by direct injector or purge-and-trap (EPA Method 5030). EPA Method 5030 may be used directly on groundwater samples or low-concentration contaminated soils and sediments. For medium-concentration soils or sediments, methanolic extraction, as described in EPA Method 5030, may be necessary prior to purge-and-trap analysis.

QUALITY CONTROL Calculate surrogate standard recovery on all samples, blanks, and spikes. A trip blank is recommended to check on sampling, storage, and handling contamination. Calibration standards, at a minimum of five concentration levels, are prepared in organic-free reagent water. One of the concentration levels should be at a concentration near, but above, the method detection limit.

A combination of bromochloromethane, 2-bromo-1-chloropropane, 1,4-dichlorobutane, and bromochlorobenzene are recommended as surrogate standards to encompass the range of the temperature program used in this method.

REFERENCE Test Methods for Evaluating Solid Waste, Physical/Chemical Methods, SW-846, 3rd Edition, U.S. EPA, Office of Solid Waste, Washington, DC, EPA Method 8021A, Rev. 1, Nov. 1992.

Naphthalene   **EPA Method 8260**
CAS #91-20-3

TITLE Volatile Organic Compounds by GC/MS: Capillary Column Technique

MATRIX This method is applicable to nearly all types of samples, regardless of water content, including groundwater, soils, and sediments.

METHOD SUMMARY Method 8260A covers 58 volatile organic compounds that are introduced into a gas chromatograph by the purge-and-trap method or by direct injection (in limited applications). Zero-grade helium is bubbled through a 5-mL solution at ambient temperature. Purged sample components are trapped in a tube containing suitable sorbent materials. When purging is complete, the sorbent tube is heated and backflushed with helium to desorb trapped sample components. The analytes are desorbed directly to a large bore capillary or cryofocussed on a capillary precolumn before being flash evaporated to a narrow bore capillary for analysis.

INTERFERENCES Major contaminant sources are volatile materials in the lab and impurities in the inert purging gas and in the sorbent trap. Interfering contamination may occur when a sample containing low concentrations of volatile organic compounds is analyzed immediately after a sample containing high concentrations of volatile organic compounds. After analysis of a sample containing high concentrations of volatile organic compounds, one or more calibration blanks should be analyzed to check for cross-contamination. Screening of the samples prior to purge-and-trap GC/MS analysis is highly recommended to prevent contamination of the system. This is especially true for soil and waste samples.

Special precautions must be taken to analyze for methylene chloride. The analytical and sample storage area should be isolated from all atmospheric sources of methylene chloride. All gas chromatography carrier gas lines and purge gas plumbing should be constructed from stainless steel or copper tubing. Laboratory clothing previously exposed to methylene chloride fumes during liquid-liquid extraction procedures can contribute to sample contamination.

Samples can also be contaminated by diffusion of volatile organics (particularly methylene chloride and fluorocarbons) through the septum seal during shipment and storage. A trip blank can serve as a check on such contamination.

INSTRUMENTATION GC/MS with a temperature-programmable chromatograph suitable for splitless injection equipped with variable constant differential flow controllers, a subambient oven controller, a purging device, sorbent trap, a thermal desorption apparatus and a capillary precolumn interface when using cryogenic cooling will be needed. The following GC columns may be used:

Column 1: 60 m × 0.75 mm I.D. capillary column coated with VOCOL, 1.5 µm film thickness.
Column 2: 30 m × 0.53 mm capillary column coated with DB-624 or VOCOL, 3 µm film thickness.
Column 3: 30 m × 0.32 mm I.D. capillary column coated with DB-5 or SE-54, 1-µm film thickness.

PRECISION & ACCURACY This method has been tested in a single lab using spiked water. Using a wide-bore capillary column, water was spiked at concentrations between 0.5 and 10 µg/L. Single lab accuracy and precision data are presented. The MDL actually achieved in a given analysis will vary depending on instrument sensitivity and matrix effects.

The MDL (a) in µg/L was 0.04.
The concentration range in µg/L was 0.1–100.
The mean accuracy (% of true value) was 104.
The precision as relative standard deviation was 8.2.

Note: The MDL is based on a 25-mL sample volume instead of a 5-mL sample volume.

SAMPLING METHOD

Liquid samples — Use a 40-mL glass screw-cap VOA vial with a Teflon®-faced silicone septum that has been prewashed, rinsed with distilled deionized water, and oven dried. If residual chlorine is present, collect the sample in a 4-oz soil VOA container which has been pre-preserved with 4 drops of 10% sodium thiosulfate. Mix gently and transfer the sample to a 40-mL VOA vial. Collect bubble-free samples in duplicate and seal each sample in a separate plastic bag.

Soils, sediments and sludges — Use an 8-oz widemouth glass bottle with Teflon®-faced silicone septum that has been prewashed, rinsed with distilled deionized water, and oven dried. **Do not** heat the septum for more than 1 h. Tap slightly to eliminate any free air space. Collect samples in duplicate and seal each one in a separate plastic bag.

SAMPLE PRESERVATION

Liquid samples — Add 4 drops of concentrated HCL, cool to 4°C and store in a solvent-free refrigerator.

Soils, sediments and sludges — Cool samples to 4°C and store in a solvent-free refrigerator.

MHT The maximum holding time of any sample (liquids, soils, sediments, and sludges) is 14 days.

SAMPLE PREPARATION

Liquid samples — Remove the plunger from a 5-mL syringe and carefully pour the sample into the syringe barrel to just short of overflowing. Replace the syringe plunger and compress the sample. Open the syringe valve and vent any residual air while adjusting the sample volume to 5.0 mL. If there is only one volatile organic analysis (VOA) vial, a second syringe should be filled at this time to protect against possible loss of sample integrity. Add 10 µL of surrogate spiking solution and 10 µL of internal standard spiking solution through the valve bore of the 5-mL syringe, then close the valve. The surrogate and internal standards may be mixed and added as a single spiking solution.

Sediments, soils, and waste samples — All samples of this type should be screened by GC analysis using a headspace method (EPA Method 3810) or the hexadecane extraction and screening method (EPA Method 3820). Use the screening data to determine whether to use the low-concentration method (0.005–1 mg/kg) or the high-concentration method (>1 mg/kg).

Low-concentration method — The low-concentration method is based on purging a heated sediment or soil sample mixed with organic-free reagent water containing the surrogate and internal standards. Analyze all reagent blanks and standards under the same conditions as the samples.

Use a 5-g sample if the expected concentration is <0.1 mg/kg or a 1-g sample for expected concentrations between 0.1 and 1 mg/kg. Mix the contents of the sample container with a narrow metal spatula. Weigh the amount of the sample into a tared purge device. Add the spiked water to the purge device, which contains the weighed amount of sample, and connect the device to the purge-and-trap system.

High-concentration method — This method is based on extracting the sediment or soil with methanol. A waste sample is either extracted or diluted, depending on its solubility in methanol. Wastes that are insoluble in methanol are diluted with reagent tetraglyme or possibly polyethylene glycol (PEG). An aliquot of the extract is added to organic-free reagent water containing surrogate and internal standards. This is purged at ambient temperature. All samples with an expected concentration of >1.0 mg/kg should be analyzed by this method.

Mix the contents of the sample container with a narrow metal spatula. For sediments or soils and solid wastes that are insoluble in methanol, weigh 4 g (wet weight) of sample into a tared 20-mL vial. For waste that is soluble in methanol, tetraglyme, or PEG, weigh 1 g (wet weight) into a tared scintillation vial or culture tube or a 10-mL volumetric flask. Quickly add 9.0 mL of appropriate solvent then add 1.0 mL of a surrogate spiking solution to the vial, cap it, and shake it for 2 min.

METHANOL EXTRACT REQUIRED FOR ANALYSIS OF HIGH-CONCENTRATION SOILS OR SEDIMENTS

Approximate Concentration Range	Volume of Methanol Extract (a)
500–10,000 µg/kg	100 µL
1,000–20,000 µg/kg	50 µL
5,000–100,000 µg/kg	10 µL
25,000–500,000 µg/kg	100 µL of 1/50 dilution (b)

Calculate appropriate dilution factor for concentrations exceeding this table.

(a) The volume of methanol added to 5 mL of water being purged should be kept constant. Therefore, add to the 5-mL syringe whatever volume of methanol is necessary to maintain a volume of 100 µL added to the syringe.
(b) Dilute an aliquot of the methanol extract and then take 100 µL for analysis.

QUALITY CONTROL Demonstrate, through the analysis of a reagent water blank, that interferences from the analytical system, glassware, and reagents are under control. Blank samples should be carried through all stages of the sample preparation and measurement steps. For each analytical batch (up to 20 samples), a reagent blank, matrix spike, and matrix spike duplicate must be analyzed (the frequency of the spikes may be different for different monitoring programs). The blank and spiked samples must be carried through all stages of the sample preparation and measurement steps. QC samples mentioned in the section on Interferences will also be needed as appropriate to those situations.

Matrix spiking standards should be prepared from volatile organic compounds which will be representative of the compounds being investigated. The recommended internal standards are chlorobenzene-d5, 1,4-difluorobenzene, 1,4-dichlorobenzene-d4, and pentafluorobenzene. Using stock standard solutions, prepare secondary dilution standards containing the compounds of interest, either singly or mixed together in methanol. Store them in a vial with no headspace for no more than one week. Surrogates recommended are toluene-d8, 4-bromofluorobenzene, and dibromofluoromethane. Each sample undergoing GC/MS analysis must be spiked with 10 µL of the surrogate spiking solution prior to analysis.

REFERENCE Test Methods for Evaluating Solid Waste (SW-846). U.S. EPA 1983, Method 8260A, Rev. 1, Nov. 1990. Office of Solid Waste, Washington, DC.

Naphthalene   **EPA Method 8270**
CAS #91-20-3

TITLE Semivolatile Organic Compounds by GC/MS

MATRIX This method is used to determine the concentration of semivolatile organic compounds in extracts prepared from all types of solid waste matrices, soils, and groundwater. Although surface waters are not specifically mentioned, this method should be applicable to water samples from rivers, lakes, etc.

METHOD SUMMARY This method covers 259 semivolatile organic compounds. In very limited applications direct injection of the sample into the GC/MS system may be appropriate, but this results in very high detection limits (approximately 10,000 µg/L). Typically, a 1-L liquid sample, containing surrogate, and matrix spiking standards, is extracted in a continuous extractor first under acid conditions and then under basic conditions. Typically 30 g of a solid sample, containing surrogate, and matrix spiking standards, is extracted ultrasonically. After concentrating the extract to 1 mL it is spiked with 10 µL of an internal standard solution just prior to analysis by GC/MS. The volume injected should contain about 100 ng of base/neutral and 200 ng of acid surrogates (for a 1-µL injection). Analysis is performed by GC/MS using a capillary GC column.

INTERFERENCES Raw GC/MS data from all blanks, samples, and spikes must be evaluated for interferences. Contamination by carryover can occur whenever high-concentration and low-concentration samples are sequentially analyzed. To reduce carryover, the sample syringe must be rinsed out between samples with solvent. Whenever an unusually concentrated sample is encountered, it should be followed by the analysis of blank solvent to check for cross-contamination.

INSTRUMENTATION A GC/MS and a data system are required. The GC column used is a 30 m × 0.25 mm I.D. (or 0.32 mm I.D.) 1um film thickness silicone-coated fused silica capillary column. A continuous liquid-liquid extractor equipped with Teflon® or glass connection joints and stopcocks requiring no lubrication, a K-D concentrating apparatus, water bath, and an ultrasonic disrupter with a minimum power of 300 W and with pulsing capability are also required.

PRECISION & ACCURACY The estimated quantitation limit (EQL) of Method 8270B for determining an individual compound is approximately 1 mg/kg (wet weight) for soil or sediment samples, 1–200 mg/kg for wastes (dependent on matrix and method of preparation), and 10 µg/L for groundwater samples. EQLs will be proportionately higher for sample extracts that require dilution to avoid saturation of the detector.

The EQL(b) for groundwater in µg/L is 10.
The EQL (a, b) for low concentrations in soil and sediment in µg/kg is 660.
Accuracy as µg/L is 0.76C +1.58.

Overall precision in µg/L is 0.30X–0.68.

(a) *EQLs listed for soil/sediment are based on wet weight. Normally data is reported in a dry-weight basis; therefore, EQLs will be higher based on the % dry weight of each sample. This calculation is based on a 30 g sample and gel permeation chromatography cleanup.*

(b) *Sample EQLs are highly matrix-dependent. The EQLs are provided for guidance and may not always be achievable.*

$C =$ *True value for concentration, in µg/L.*
$X =$ *Average recovery found for measurements of samples containing a concentration of C, in µg/L.*

ESTIMATED QUANTITATION LIMIT

Other Matrices	Factor (a)
High-concentration soil and sludges by sonicator	7.5
Non-water miscible waste	75

(a) *EQL for other matrices = [EQL for low soil/sediment] × [Factor]. This estimated EQL is similar to an EPA "Practical Quantitation Limit."*

SAMPLING METHOD
Liquid samples — Use a 1 or 2½ gallon amber glass bottle with a screw-top Teflon®-lined cover that has been prewashed with detergent and rinsed with distilled water and methanol (or isopropanol).

Soils, sediments, or sludges — Use an 8-oz. widemouth glass with a screw-top Teflon®-lined cover that has been prewashed with detergent and rinsed with distilled water and methanol (or isopropanol).

SAMPLE PRESERVATION
Liquid samples — If residual chlorine is present, add 3 mL of 10% sodium thiosulfate per gallon, cool to 4°C and store in a solvent-free refrigerator until analysis; if chlorine is not present, then eliminate the sodium thiosulfate addition.

Soils, sediments, or sludges — Cool samples to 4°C and store in a solvent-free refrigerator.

MHT Liquid samples must be extracted within 7 days and the extracts analyzed within 40 days. Soils, sediments, or sludges may be stored for a maximum of 14 days and the extracts analyzed within 40 days.

SAMPLE PREPARATION
Liquid samples — Transfer 1 L quantitatively to a continuous extractor. If high concentrations are anticipated, a smaller volume may be used and then diluted with organic-free reagent water to 1 L. Adjust pH, if necessary, to pH <2 using 1:1 (V/V) sulfuric acid. Pipette 1.0 mL of a surrogate standard spiking solution into each sample. For the sample in each analytical batch selected for spiking, add 1.0 mL of a matrix spiking standard. For base/neutral acid analysis, the amount of the surrogates and matrix spiking compounds added to the sample should result in a final concentration of 100 ng/µL of each analyte in the extract to be analyzed (assuming a 1-µL injection). Extract with methylene chloride for 18–24 h. Next, adjust the pH of the aqueous phase to pH >11 using 10 N sodium hydroxide and extract it with methylene chloride again for 18–24 h. Dry the extract through a column containing anhydrous

sodium sulfate and concentrate it to 1 mL using a K-D concentrator.

Soils, sediments, or sludges — Use 30 g of sample. Nonporous or wet samples (gummy or clay type) that do not have a free-flowing sandy texture must be mixed with anhydrous sodium sulfate until the sample is free flowing. Add 1 mL of surrogate standards to all samples, spikes, standards, and blanks. For the sample in each analytical batch selected for spiking, add 1.0 mL of a matrix spiking standard. For base/neutral acid analysis, the amount added of the surrogates and matrix spiking compounds should result in a final concentration of 100 ng/µL of each base/neutral analyte and 200 ng/µL of each acid analyte in the extract to be analyzed (assuming a 1-µL injection). Immediately add a 100-mL mixture of 1:1 methylene chloride:acetone and extract the sample ultrasonically for 3 min and then decant or filter the extracts. Repeat the extraction two or more times. Dry the extract using a column with anhydrous sodium sulfate and concentrate it to 1 mL in a K-D concentrator.

QUALITY CONTROL A methylene chloride solution containing 50 ng/µL of decafluorotriphenylphosphine (DFTPP) is used for tuning the GC/MS system each 12-h shift. A system performance check also must be made during every 12-h shift. A standard containing 50 ng/µL each of 4,4'-DDT, pentachlorophenol, and benzidine is required to verify injection port inertness and GC column performance. A calibration standard at mid-concentration, containing each compound of interest, including all required surrogates, must be performed every 12 h during analysis. After the system performance check is met, calibration check compounds (CCCs) are used to check the validity of the initial calibration.

The internal standard responses and retention times in the calibration check standard must be evaluated immediately after or during data acquisition. If the retention time for any internal standard changes by more than 30 seconds from the last check calibration (12 h), the chromatographic system must be inspected for malfunctions and corrections must be made, as required. If the electron ionization current plot (EICP) area for any of the internal standards changes by a factor of two from the last daily calibration standard check, the mass spectrometer must be inspected for malfunctions and corrections must be made, as appropriate.

Demonstrate, through the analysis of a reagent water blank, that interferences from the analytical system, glassware, and reagents are under control. The blank samples should be carried through all stages of the sample preparation and measurement steps. For each analytical batch (up to 20 samples), a reagent blank, matrix spike, and matrix spike duplicate/duplicate must be analyzed (the frequency of the spikes may be different for different monitoring programs). The blank and spiked samples must be carried through all stages of the sample preparation and measurement steps. A QC reference sample concentrate containing each analyte at a concentration of 100 mg/L in methanol is required.

REFERENCE Test Methods for Evaluating Solid Waste (SW-846). U.S. EPA 1983, Method 8270B, Rev. 2, Nov. 1990. Office of Solid Waste, Washington, DC.

Naphthalene   **EPA Method 503.1**
CAS #91-20-3

TITLE Aromatic & Unsaturated VOCs Water

MATRIX Drinking water (finished or in any treatment stage) and raw source water.

APPLICATION Method covers 28 aromatic and unsaturated VOCs. An inert gas is bubbled through a 5-mL water sample. Purged sample components are trapped in tube of sorbent materials. When purging is complete, sorbent tube is heated and backflushed with inert gas to desorb trapped sample onto a packed GC column.

INTERFERENCES During analysis, major contaminant sources are volatile materials in the lab and impurities in purging gas and sorbent trap. With high and low level samples, there can be carryover contamination. Excess water causes a negative baseline deflection.

INSTRUMENTATION Purge and Trap GC w/photoionization detector. (Two GC columns are recommended); Column 1: 5% SP-1200 and 1.75% Bentone 34 on Supelcoport; Column 2: 1,2,3-tris(2-cyanoethoxy)propane on Chromosorb W.

RANGE 2.2–600 µg/L. (Drinking water)

MDL 0.04 µg/L in water

PRECISION RSD = 14.8% at 0.50 µg/L conc.; 16 samples

ACCURACY Average recovery = 92% at 0.50 µg/L conc.; 16 samples

SAMPLING METHOD Use a 40–120-mL screw-cap vial (prewashed with detergent, rinsed with distilled water and oven dried at 105°C) with a PTFE-faced silicone septum. If residual chlorine is in the water add about 25 mg of ascorbic acid to each vial before sample collection. Collect bubble-free samples.

STABILITY Cool to 4°C; HCl to pH <2.

MHT 14 days.

QUALITY CONTROL As initial demonstration of lab accuracy and precision, analyze 4 to 7 replicates of a lab fortified blank containing the analyte at 0.1–5 µg/L. Collect all samples in duplicate.

REFERENCE Method 503.1, Volatile Aromatic & Unsaturated Organic Compounds in H2O by Purge and Trap GC, EPA 600/4-88/039.

Naphthalene   **EPA Method 8100**
CAS #91-20-3

TITLE Polynuclear Aromatic Hydrocarbons

MATRIX Groundwater, soils, sludges, water miscible liquid wastes, and non-water miscible wastes.

APPLICATION This method is used for the analysis of various PAHs. Samples are extracted, concentrated, and analyzed using direct injection of both neat and diluted organic liquids. The method provides two optional GC columns that are better than Column 1 and that may help resolve analytes from interferences.

INTERFERENCES Solvents, reagents, and glassware may introduce artifacts. Other interferences may come from coextracted compounds from samples.

INSTRUMENTATION GC capable of on-column injections and a flame with detector (FID). Column 1: a 1.8 m by 2 mm 3% OV-17 on Chromosorb W-AW-DCMS column. Column 2: a 30 m by 0.25 mm SE-54 fused silica capillary colunm. Column 3: a 30 m by 0.32 mm SE-54 fused silica capillary column.

RANGE 0.1–425 µg/L.

MDL Not reported.

PQL FACTORS FOR MULTIPLYING × FID MDL VALUE Not available.

PRECISION 0.41X + 0.74 µg/L (overall precision).

ACCURACY 0.57C–0.70 µg/L (as recovery).

SAMPLING METHOD Use 8-oz. widemouth glass bottles with Teflon®-lined caps for concentrated waste samples, soils, sediments, and sludges. Use 1 or 2½ gallon amber glass bottles with Teflon®-lined caps for liquid (water) samples.

STABILITY Cool soil, sediment, sludge, and liquid samples to 4°C. If residual chlorine is present in liquid samples add 3 mL of 10% sodium thiosulfate per gallon of sample and cool to 4°C.

MHT 14 days for concentrated waste, soil, sediment, or sludge; 7 days for liquid samples; all extracts must be analyzed within 40 days.

QUALITY CONTROL A quality control check sample concentrate containing each analyte of interest is required. The QC check sample concentrate may be prepared from pure standard materials or purchased as certified solutions Use appropriate trip, matrix, control site, method, reagent, and solvent blanks. Internal, surrogate, and five concentration level calibration standards are used. The quality control check sample concentrate should contain naphthalene at 100 µg/mL in acetonitrile.

REFERENCE Test Methods for Evaluating Solid Waste (SW-846), U.S. EPA Office of Solid Waste, Washington, DC, Method 8100, Nov. 1986.

Naphthalene   **EPA Method 8310**
CAS #91-20-3

TITLE Polynuclear Aromatic Hydrocarbons

MATRIX Groundwater, soils, sludges, water miscible liquid wastes, and non-water miscible wastes.

APPLICATION This method is used for the analysis of 16 polynuclear aromatic hydrocarbons (PAHs). Samples are extracted, concentrated, and analyzed using HPLC with detection by UV and fluorescence detectors.

INTERFERENCES Solvents, reagents, and glassware may introduce artifacts. Other interferences may come from coextracted compounds from samples.

INSTRUMENTATION HPLC with a gradient pumping system and a 250 mm by 2.6 mm reverse phase HC-ODS Sil-X 5-micron particle-size column. The UV detector uses an excitation wavelength of 254 nm coupled to the fluorescence detector. The fluorescence detector uses an excitation wavelength of 280 nm and emission greater than 389 nm cutoff with dispersive optics.

RANGE 0.1–425 µg/L.

MDL 1.8 µg/L (UV detector; reagent water).

PQL FACTORS FOR MULTIPLYING × FID MDL VALUE

Matrix	Multiplication Factor
Groundwater	10
Low-level soil by sonication with GPC cleanup	670
High-level soil and sludge by sonication	10,000
Non-water miscible waste	100,000

PRECISION 0.41X + 0.74 µg/L (overall precision).

ACCURACY 0.57C–0.70 µg/L (as recovery).

SAMPLING METHOD Use 8-oz. widemouth glass bottles with Teflon®-lined caps for concentrated waste samples, soils, sediments, and sludges. Use 1 or 2½ gallon amber glass bottles with Teflon®-lined caps for liquid (water) samples.

STABILITY Cool soil, sediment, sludge, and liquid samples to 4°C. If residual chlorine is present in liquid samples add 3 mL of 10% sodium thiosulfate per gallon of sample and cool to 4°C.

MHT 14 days for concentrated waste, soil, sediment, or sludge; 7 days for liquid samples; all extracts must be analyzed within 40 days.

QUALITY CONTROL Internal, surrogate, and five concentration level calibration standards are used. The calibration standards must be used with the analytical method blank. A quality control check sample concentrate containing naphthalene at 100 µg/mL is required. The QC check sample concentrate may be prepared from pure standard materials or purchased as certified solutions. Use appropriate trip, matrix, control site, method, reagent, and solvent blanks.

REFERENCE Test Methods for Evaluating Solid Waste (SW-846), U.S. EPA Office of Solid Waste, Washington, DC, Method 8310, Rev. 0, Nov. 1986.

1,5-Naphthalenediamine   **EPA Method 1625**
CAS #2243-62-1

TITLE Semivolatile Organic Compounds by Isotope Dilution GC/MS

MATRIX The compounds may be determined in waters, soils, and municipal sludges by this method.

METHOD SUMMARY This method is used to determine 176 semivolatile toxic organic pollutants associated with the CWA (as amended 1987); the RCRA (as amended 1986); the CERCLA (as amended 1986); and other compounds amenable to extraction and analysis by capillary column gas chromatography-mass spectrometry (GC/MS).

Stable isotopically-labeled analogs of the compounds of interest are added to the sample. If the solids content is less than 1%, a 1-L sample is extracted at pH 12–13, then at pH <2 with methylene chloride using continuous extraction techniques.

If the solids content is 30% or less, the sample is diluted to 1% solids with reagent water, homogenized ultrasonically, and extracted at pH 12–13, then at pH <2 with methylene chloride using continuous extraction techniques. If the solids content is greater than 30%, the sample is extracted using ultrasonic techniques.

Each extract is dried over sodium sulfate, concentrated to a volume of 5 mL, cleaned up using GPC, if necessary, and concentrated. Extracts are concentrated to 1 mL if GPC is not performed, and to 0.5 mL if GPC is performed.

An internal standard is added to the extract, and a 1-mL aliquot of the extract is injected into the GC. The compounds are separated by GC and detected by a MS. The labeled compounds serve to correct the variability of the analytical technique.

INTERFERENCES Solvents, reagents, glassware, and other sample processing hardware may yield artifacts and/or elevated baselines causing misinterpretation of chromatograms and spectra. Materials used in the analysis must be demonstrated to be free from interferences under the conditions of analysis by running method blanks initially and with each sample lot (sample started through the extraction process on a given 8-h shift, to a maximum of 20). Specific selection of reagents and purification of solvents by distillation in all glass systems may be required. Glassware and, where possible, reagents are cleaned by solvent rinse and baking at 450°C for 1-h minimum. Interferences coextracted from samples will vary considerably from source to source, depending on the diversity of the site being sampled.

INSTRUMENTATION Major instrumentation includes a GC with a splitless or on-column injection port for capillary column, a MS with 70 eV electron impact ionization, and a data system to collect and record MS data, and process it. A K-D apparatus is used to concentrate extracts.

GC Column: 30 m × 0.25 mm I.D. 5% phenyl, 94% methyl, 1% vinyl silicone bonded phased fused silica capillary column.

PRECISION & ACCURACY The detection limits of the method are usually dependent on the level of interferences rather than instrumental limitations. The limits typify the minimum quantities that can be detected with no interferences present.

The minimum level (in µg/mL) was not listed. This is defined as a minimum level at which the analytical system shall give recognizable mass spectra (background corrected) and acceptable calibration points.

The MDL (in µg/kg) in low solids was not listed and in high solids was not listed; these were determined in digested sludge (low solids) and in filter cake or compost (high solids).

The labeled and native compound initial precision as standard deviation (in µg/L) was not listed.
The labeled and native compound initial accuracy as average recovery (in µg/L) was not listed.

SAMPLE COLLECTION, PRESERVATION & HANDLING
Collect samples in glass containers. Aqueous samples which flow freely are collected in refrigerated bottles using automatic sampling equipment. Solid samples are collected as grab samples using widemouth jars. Maintain samples at 0 to 4°C from the time of collection until extraction. If residual chlorine is present in aqueous samples, add 80 mg sodium thiosulfate/L of water. Begin sample extraction within 7 days of collection, and analyze all extracts within 40 days of extraction.

SAMPLE PREPARATION Samples containing 1% solids or less are extracted directly using continuous liquid-liquid extraction techniques. Samples containing 1 to 30% solids are diluted to the 1% level with reagent water and extracted using continuous liquid-liquid extraction techniques. Samples containing greater than 30% solids are extracted using ultrasonic techniques.

Base/neutral extraction — Adjust the pH of the waters in the extractors to 12–13 with 6 N NaOH. Extract with methylene chloride for 24–48 h.
Acid extraction — Adjust the pH of the waters in the extractors to 2 or less using 6 N sulfuric acid. Extract with methylene chloride for 24–48 h.
Ultrasonic extraction of high solids samples — Add anhydrous sodium sulfate to the sample and QC aliquot(s). Add acetone:methylene chloride (1:1) to the sample and mix thoroughly

Concentrate extracts using a K-D apparatus.

QUALITY CONTROL The analyst is permitted to modify this method to improve separations or lower the costs of measurements, provided all performance specifications are met. Analyses of blanks are required to demonstrate freedom from contamination. When results of spikes indicate atypical method performance for samples, the samples are diluted to bring method performance within acceptable limits.

For low solids (aqueous samples), extract, concentrate, and analyze two sets of four 1-L aliquots (8 aliquots total) of the precision and recovery standard. For high solids samples, two sets of four 30-g aliquots of the high solids reference matrix are used.

Spike all samples with labeled compounds to assess method performance. Compute percent recovery of the labeled compounds using the internal standard method. Compare the labeled compound recovery for each compound with the corresponding labeled compound recovery.

Reagent water and high solids reference matrix blanks are analyzed to demonstrate freedom from contamination. Extract and concentrate a 1-L reagent water blank or a high solids reference matrix blank with each sample's lot (samples started through the extraction process on the same 8-h shift, to a maximum of 20 samples).

Field replicates may be collected to determine the precision of the sampling technique, and spiked samples may be required to determine the accuracy of the analysis when the internal standard method is used.

REFERENCE Semivolatile Organic Compounds by Isotope Dilution GC/MS. Office of Water Regulation and Standards, U.S. EPA Industrial Technology Division, Washington, DC, EPA Method 1625, Rev. C, June 1989 (contact W.A. Telliard, U.S. EPA, Office of Water Regulations and Standards, 401 M St., SW, Washington, DC, 20460. Phone: 202-382-7131).

1,4-Naphthoquinone   **EPA Method 1625**
CAS #130-15-4

TITLE Semivolatile Organic Compounds by Isotope Dilution GC/MS

MATRIX The compounds may be determined in waters, soils, and municipal sludges by this method.

METHOD SUMMARY This method is used to determine 176 semivolatile toxic organic pollutants associated with the CWA (as amended 1987); the RCRA (as amended 1986); the CERCLA (as amended 1986); and other compounds amenable to extraction and analysis by capillary column gas chromatography-mass spectrometry (GC/MS).

Stable isotopically-labeled analogs of the compounds of interest are added to the sample. If the solids content is less than 1%, a 1-L sample is extracted at pH 12–13, then at pH <2 with methylene chloride using continuous extraction techniques.

If the solids content is 30% or less, the sample is diluted to 1% solids with reagent water, homogenized ultrasonically, and extracted at pH 12–13, then at pH <2 with methylene chloride using continuous extraction techniques. If the solids content is greater than 30%, the sample is extracted using ultrasonic techniques.

Each extract is dried over sodium sulfate, concentrated to a volume of 5 mL, cleaned up using GPC, if necessary, and concentrated. Extracts are concentrated to 1 mL if GPC is not performed, and to 0.5 mL if GPC is performed.

An internal standard is added to the extract, and a 1-mL aliquot of the extract is injected into the GC. The compounds are separated by GC and detected by a MS. The labeled compounds serve to correct the variability of the analytical technique.

INTERFERENCES Solvents, reagents, glassware, and other sample processing hardware may yield artifacts and/or elevated baselines causing misinterpretation of chromatograms and spectra. Materials used in the analysis must be demonstrated to be free from interferences under the conditions of analysis by running method blanks initially and with each sample lot (sample started through the extraction process on a given 8-h shift, to a maximum of 20). Specific selection of reagents and purification of solvents by distillation in all glass systems may be required. Glassware and, where possible, reagents are cleaned by solvent rinse and baking at 450°C for 1-h minimum. Interferences coextracted from samples will vary considerably from source to source, depending on the diversity of the site being sampled.

INSTRUMENTATION Major instrumentation includes a GC with a splitless or on-column injection port for capillary column, a MS with 70 eV electron impact ionization, and a data system to collect and record MS data, and process it. A K-D apparatus is used to concentrate extracts.

GC Column: 30 m × 0.25 mm I.D. 5% phenyl, 94% methyl, 1% vinyl silicone bonded phased fused silica capillary column.

PRECISION & ACCURACY The detection limits of the method are usually dependent on the level of interferences rather than instrumental limitations. The limits typify the minimum quantities that can be detected with no interferences present.

The minimum level (in µg/mL) was not listed. This is defined as a minimum level at which the analytical system shall give recognizable mass spectra (background corrected) and acceptable calibration points.

The MDL (in µg/kg) in low solids was not listed and in high solids was not listed; these were determined in digested sludge (low solids) and in filter cake or compost (high solids).

The labeled and native compound initial precision as standard deviation (in µg/L) was not listed.
The labeled and native compound initial accuracy as average recovery (in µg/L) was not listed.

SAMPLE COLLECTION, PRESERVATION & HANDLING Collect samples in glass containers. Aqueous samples which flow freely are collected in refrigerated bottles using automatic sampling equipment. Solid samples are collected as grab samples using widemouth jars. Maintain samples at 0 to 4°C from the time of collection until extraction. If residual chlorine is present in aqueous samples, add 80 mg sodium thiosulfate/L of water. Begin sample extraction within 7 days of collection, and analyze all extracts within 40 days of extraction.

SAMPLE PREPARATION Samples containing 1% solids or less are extracted directly using continuous liquid-liquid extraction techniques. Samples containing 1 to 30% solids are diluted to the 1% level with reagent water and extracted using continuous liquid-liquid extraction techniques. Samples containing greater than 30% solids are extracted using ultrasonic techniques.

Base/neutral extraction — Adjust the pH of the waters in the extractors to 12–13 with 6 N NaOH. Extract with methylene chloride for 24–48 h.

Acid extraction — Adjust the pH of the waters in the extractors to 2 or less using 6 N sulfuric acid. Extract with methylene chloride for 24–48 h.

Ultrasonic extraction of high solids samples — Add anhydrous sodium sulfate to the sample and QC aliquot(s). Add acetone:methylene chloride (1:1) to the sample and mix thoroughly

Concentrate extracts using a K-D apparatus.

QUALITY CONTROL The analyst is permitted to modify this method to improve separations or lower the costs of measurements, provided all performance specifications are met. Analyses of blanks are required to demonstrate freedom from contamination. When results of spikes indicate atypical method performance for samples, the samples are diluted to bring method performance within acceptable limits.

For low solids (aqueous samples), extract, concentrate, and analyze two sets of four 1-L aliquots (8 aliquots total) of the precision and recovery standard. For high solids samples, two sets of four 30-g aliquots of the high solids reference matrix are used.

Spike all samples with labeled compounds to assess method performance. Compute percent recovery of the labeled compounds using the internal standard method. Compare the labeled compound recovery for each compound with the corresponding labeled compound recovery.

Reagent water and high solids reference matrix blanks are analyzed to demonstrate freedom from contamination. Extract and concentrate a 1-L reagent water blank or a high solids reference matrix blank with each sample's lot (samples started through the extraction process on the same 8-h shift, to a maximum of 20 samples).

Field replicates may be collected to determine the precision of the sampling technique, and spiked samples may be required to determine the accuracy of the analysis when the internal standard method is used.

REFERENCE Semivolatile Organic Compounds by Isotope Dilution GC/MS. Office of Water Regulation and Standards, U.S. EPA Industrial Technology Division, Washington, DC, EPA Method 1625, Rev. C, June 1989 (contact W.A. Telliard, U.S. EPA, Office of Water Regulations and Standards, 401 M St., SW, Washington, DC, 20460. Phone: 202-382-7131).

1,4-Naphthoquinone   **EPA Method 8270**
CAS #130-15-4

TITLE Semivolatile Organic Compounds by GC/MS

MATRIX This method is used to determine the concentration of semivolatile organic compounds in extracts prepared from all types of solid waste matrices, soils, and groundwater. Although surface waters are not specifically mentioned, this method should be applicable to water samples from rivers, lakes, etc.

METHOD SUMMARY This method covers 259 semivolatile organic compounds. In very limited applications direct injection of the sample into the GC/MS system may be appropriate, but this results in very high detection limits (approximately 10,000 µg/L). Typically, a 1-L liquid sample, containing surrogate, and matrix spiking standards, is extracted in a continuous extractor first under acid conditions and then under basic conditions. Typically 30 g of a solid sample, containing surrogate, and matrix spiking standards, is extracted ultrasonically. After concentrating the extract to 1 mL it is spiked with 10 µL of an internal standard solution just prior to analysis by GC/MS. The volume injected should contain about 100 ng of base/neutral and 200 ng of acid surrogates (for a 1-µL injection). Analysis is performed by GC/MS using a capillary GC column.

INTERFERENCES Raw GC/MS data from all blanks, samples, and spikes must be evaluated for interferences. Contamination by carryover can occur whenever high-concentration and low-concentration samples are sequentially analyzed. To reduce carryover, the sample syringe must be rinsed out between samples with solvent. Whenever an unusually concentrated sample is encountered, it should be followed by the analysis of blank solvent to check for cross-contamination.

INSTRUMENTATION A GC/MS and a data system are required. The GC column used is a 30 m × 0.25 mm I.D. (or 0.32 mm I.D.) 1um film thickness silicone-coated fused silica capillary column. A continuous liquid-liquid extractor equipped with Teflon® or glass connection joints and stopcocks requiring no lubrication, a K-D concentrating apparatus, water bath, and an ultrasonic disrupter with a minimum power of 300 W and with pulsing capability are also required.

PRECISION & ACCURACY The estimated quantitation limit (EQL) of Method 8270B for determining an individual compound is approximately 1 mg/kg (wet weight) for soil or sediment samples, 1–200 mg/kg for wastes (dependent on matrix and method of preparation), and 10 µg/L for groundwater samples. EQLs will be proportionately higher for sample extracts that require dilution to avoid saturation of the detector.

The EQL(b) for groundwater in µg/L is 10.
The EQL (a, b) for low concentrations in soil and sediment in µg/kg is not determined.
Accuracy as µg/L is not listed.
Overall precision in µg/L is not listed.

(a) *EQLs listed for soil/sediment are based on wet weight. Normally data is reported in a dry-weight basis; therefore, EQLs will be higher based on the % dry weight of each sample. This calculation is based on a 30 g sample and gel permeation chromatography cleanup.*
(b) *Sample EQLs are highly matrix-dependent. The EQLs are provided for guidance and may not always be achievable.*
C = *True value for concentration, in µg/L.*
X = *Average recovery found for measurements of samples containing a concentration of C, in µg/L.*

ESTIMATED QUANTITATION LIMIT

Other Matrices	Factor (a)
High-concentration soil and sludges by sonicator	7.5
Non-water miscible waste	75

(a) *EQL for other matrices = [EQL for low soil/sediment] × [Factor]. This estimated EQL is similar to an EPA "Practical Quantitation Limit."*

SAMPLING METHOD

Liquid samples — Use a 1 or 2½ gallon amber glass bottle with a screw-top Teflon®-lined cover that has been prewashed with detergent and rinsed with distilled water and methanol (or isopropanol).

Soils, sediments, or sludges — Use an 8-oz. widemouth glass with a screw-top Teflon®-lined cover that has been prewashed with detergent and rinsed with distilled water and methanol (or isopropanol).

SAMPLE PRESERVATION

Liquid samples — If residual chlorine is present, add 3 mL of 10% sodium thiosulfate per gallon, cool to 4°C and store in a solvent-free refrigerator until analysis; if chlorine is not present, then eliminate the sodium thiosulfate addition.

Soils, sediments, or sludges — Cool samples to 4°C and store in a solvent-free refrigerator.

MHT Liquid samples must be extracted within 7 days and the extracts analyzed within 40 days. Soils, sediments, or sludges may be stored for a maximum of 14 days and the extracts analyzed within 40 days.

SAMPLE PREPARATION

Liquid samples — Transfer 1 L quantitatively to a continuous extractor. If high concentrations are anticipated, a smaller volume may be used and then diluted with organic-free reagent water to 1 L. Adjust pH, if necessary, to pH <2 using 1:1 (V/V) sulfuric acid. Pipette 1.0 mL of a surrogate standard spiking solution into each sample. For the sample in each analytical batch selected for spiking, add 1.0 mL of a matrix spiking standard. For base/neutral acid analysis, the amount of the surrogates and matrix spiking compounds added to the sample should result in a final concentration of 100 ng/µL of each analyte in the extract to be analyzed (assuming a 1-µL injection). Extract with methylene chloride for 18–24 h. Next, adjust the pH of the aqueous phase to pH >11 using 10 N sodium hydroxide and extract it with methylene chloride again for 18–24 h. Dry the extract through a column containing anhydrous sodium sulfate and concentrate it to 1 mL using a K-D concentrator.

Soils, sediments, or sludges — Use 30 g of sample. Nonporous or wet samples (gummy or clay type) that do not have a free-flowing sandy texture must be mixed with anhydrous sodium sulfate until the sample is free flowing. Add 1 mL of surrogate standards to all samples, spikes, standards, and blanks. For the sample in each analytical batch selected for spiking, add 1.0 mL of a matrix spiking standard. For base/neutral acid analysis, the amount added of the surrogates and matrix spiking compounds should result in a final concentration of 100 ng/µL of each base/neutral analyte and 200 ng/µL of each acid analyte in the extract to be analyzed (assuming a 1-µL injection). Immediately add a 100-mL mixture of 1:1 methylene chloride:acetone and extract the sample ultrasonically for 3 min and then decant or filter the extracts. Repeat the extraction two or more times. Dry the extract using a column with anhydrous sodium sulfate and concentrate it to 1 mL in a K-D concentrator.

QUALITY CONTROL A methylene chloride solution containing 50 ng/µL of decafluorotriphenylphosphine (DFTPP) is used for tuning the GC/MS system each 12-h shift. A system performance check also must be made during every 12-h shift. A standard containing 50 ng/µL each of 4,4′-DDT, pentachlorophenol, and benzidine is required to verify injection port inertness and GC column performance. A calibration standard at mid-concentration, containing each compound of interest, including all required surrogates, must be performed every 12 h during analysis. After the system performance check is met, calibration check compounds (CCCs) are used to check the validity of the initial calibration.

The internal standard responses and retention times in the calibration check standard must be evaluated immediately after or during data acquisition. If the retention time for any internal standard changes by more than 30 seconds from the last check calibration (12 h), the chromatographic system must be inspected for malfunctions and corrections must be made, as required. If the electron ionization current plot (EICP) area for any of the internal standards changes by a factor of two from the last daily calibration standard check, the mass spectrometer must be inspected for malfunctions and corrections must be made, as appropriate.

Demonstrate, through the analysis of a reagent water blank, that interferences from the analytical system, glassware, and reagents are under control. The blank samples should be carried through all stages of the sample preparation and measurement steps. For each analytical batch (up to 20 samples), a reagent blank, matrix spike, and matrix spike duplicate/duplicate must be analyzed (the frequency of the spikes may be different for different monitoring programs). The blank and spiked samples must be carried through all stages of the sample preparation and measurement steps. A QC reference sample concentrate containing each analyte at a concentration of 100 mg/L in methanol is required.

REFERENCE Test Methods for Evaluating Solid Waste (SW-846). U.S. EPA 1983, Method 8270B, Rev. 2, Nov. 1990. Office of Solid Waste, Washington, DC.

1-Naphthylamine (α-Naphthylamine)   EPA Method 1625
CAS #134-32-7

TITLE Semivolatile Organic Compounds by Isotope Dilution GC/MS

MATRIX The compounds may be determined in waters, soils, and municipal sludges by this method.

METHOD SUMMARY This method is used to determine 176 semivolatile toxic organic pollutants associated with the CWA (as amended 1987); the RCRA (as amended 1986); the CERCLA (as amended 1986); and other compounds amenable to extraction and analysis by capillary column gas chromatography-mass spectrometry (GC/MS).

Stable isotopically-labeled analogs of the compounds of interest are added to the sample. If the solids content is less than 1%, a 1-L sample is extracted at pH 12–13, then at pH <2 with methylene chloride using continuous extraction techniques.

If the solids content is 30% or less, the sample is diluted to 1% solids with reagent water, homogenized ultrasonically, and extracted at pH 12–13, then at pH <2 with methylene chloride using continuous extraction techniques. If the solids content is greater than 30%, the sample is extracted using ultrasonic techniques.

Each extract is dried over sodium sulfate, concentrated to a volume of 5 mL, cleaned up using GPC, if necessary, and concentrated. Extracts are concentrated to 1 mL if GPC is not performed, and to 0.5 mL if GPC is performed.

An internal standard is added to the extract, and a 1-mL aliquot of the extract is injected into the GC. The compounds are separated by GC and detected by a MS. The labeled compounds serve to correct the variability of the analytical technique.

INTERFERENCES Solvents, reagents, glassware, and other sample processing hardware may yield artifacts and/or elevated baselines causing misinterpretation of chromatograms and spectra. Materials used in the analysis must be demonstrated to be free from interferences under the conditions of analysis by running method blanks initially and with each sample lot (sample started through the extraction process on a given 8-h shift, to a maximum of 20). Specific selection of reagents and purification of solvents by distillation in all glass systems may be required. Glassware and, where possible, reagents are cleaned by solvent rinse and baking at 450°C for 1-h minimum. Interferences coextracted from samples will vary considerably from source to source, depending on the diversity of the site being sampled.

INSTRUMENTATION Major instrumentation includes a GC with a splitless or on-column injection port for capillary column, a MS with 70 eV electron impact ionization, and a data system to collect and record MS data, and process it. A K-D apparatus is used to concentrate extracts.

GC Column: 30 m × 0.25 mm I.D. 5% phenyl, 94% methyl, 1% vinyl silicone bonded phased fused silica capillary column.

PRECISION & ACCURACY The detection limits of the method are usually dependent on the level of interferences rather than instrumental limitations. The limits typify the minimum quantities that can be detected with no interferences present.

The minimum level (in µg/mL) was not listed. This is defined as a minimum level at which the analytical system shall give recognizable mass spectra (background corrected) and acceptable calibration points.

The MDL (in µg/kg) in low solids was not listed and in high solids was not listed; these were determined in digested sludge (low solids) and in filter cake or compost (high solids).

The labeled and native compound initial precision as standard deviation (in µg/L) was not listed.
The labeled and native compound initial accuracy as average recovery (in µg/L) was not listed.

SAMPLE COLLECTION, PRESERVATION & HANDLING Collect samples in glass containers. Aqueous samples which flow freely are collected in refrigerated bottles using automatic sampling equipment. Solid samples are collected as grab samples using widemouth jars. Maintain samples at 0 to 4°C from the time of collection until extraction. If residual chlorine is present in aqueous samples, add 80 mg sodium thiosulfate/L of water. Begin sample extraction within 7 days of collection, and analyze all extracts within 40 days of extraction.

SAMPLE PREPARATION Samples containing 1% solids or less are extracted directly using continuous liquid-liquid extraction techniques. Samples containing 1 to 30% solids are diluted to the 1% level with reagent water and extracted using continuous liquid-liquid extraction techniques. Samples containing greater than 30% solids are extracted using ultrasonic techniques.

Base/neutral extraction — Adjust the pH of the waters in the extractors to 12–13 with 6 N NaOH. Extract with methylene chloride for 24–48 h.
Acid extraction — Adjust the pH of the waters in the extractors to 2 or less using 6 N sulfuric acid. Extract with methylene chloride for 24–48 h.
Ultrasonic extraction of high solids samples — Add anhydrous sodium sulfate to the sample and QC aliquot(s). Add acetone:methylene chloride (1:1) to the sample and mix thoroughly

Concentrate extracts using a K-D apparatus.

QUALITY CONTROL The analyst is permitted to modify this method to improve separations or lower the costs of measurements, provided all performance specifications are met. Analyses of blanks are required to demonstrate freedom from contamination. When results of spikes indicate atypical method performance for samples, the samples are diluted to bring method performance within acceptable limits.

For low solids (aqueous samples), extract, concentrate, and analyze two sets of four 1-L aliquots (8 aliquots total) of the precision and recovery standard. For high solids samples, two sets of four 30-g aliquots of the high solids reference matrix are used.

Spike all samples with labeled compounds to assess method performance. Compute percent recovery of the labeled compounds using the internal standard method. Compare the labeled compound recovery for each compound with the corresponding labeled compound recovery.

Reagent water and high solids reference matrix blanks are analyzed to demonstrate freedom from contamination. Extract and concentrate a 1-L reagent water blank or a high solids reference matrix blank with each sample's lot (samples started through the extraction process on the same 8-h shift, to a maximum of 20 samples).

Field replicates may be collected to determine the precision of the sampling technique, and spiked samples may be required to determine the accuracy of the analysis when the internal standard method is used.

REFERENCE Semivolatile Organic Compounds by Isotope Dilution GC/MS. Office of Water Regulation and Standards, U.S. EPA Industrial Technology Division, Washington, DC, EPA Method 1625, Rev. C, June 1989 (contact W.A. Telliard,

U.S. EPA, Office of Water Regulations and Standards, 401 M St., SW, Washington, DC, 20460. Phone: 202-382-7131).

2-Naphthylamine (β-Naphthylamine) EPA Method 1625
CAS #91-59-8

TITLE Semivolatile Organic Compounds by Isotope Dilution GC/MS

MATRIX The compounds may be determined in waters, soils, and municipal sludges by this method.

METHOD SUMMARY This method is used to determine 176 semivolatile toxic organic pollutants associated with the CWA (as amended 1987); the RCRA (as amended 1986); the CERCLA (as amended 1986); and other compounds amenable to extraction and analysis by capillary column gas chromatography-mass spectrometry (GC/MS).

Stable isotopically-labeled analogs of the compounds of interest are added to the sample. If the solids content is less than 1%, a 1-L sample is extracted at pH 12–13, then at pH <2 with methylene chloride using continuous extraction techniques.

If the solids content is 30% or less, the sample is diluted to 1% solids with reagent water, homogenized ultrasonically, and extracted at pH 12–13, then at pH <2 with methylene chloride using continuous extraction techniques. If the solids content is greater than 30%, the sample is extracted using ultrasonic techniques.

Each extract is dried over sodium sulfate, concentrated to a volume of 5 mL, cleaned up using GPC, if necessary, and concentrated. Extracts are concentrated to 1 mL if GPC is not performed, and to 0.5 mL if GPC is performed.

An internal standard is added to the extract, and a 1-mL aliquot of the extract is injected into the GC. The compounds are separated by GC and detected by a MS. The labeled compounds serve to correct the variability of the analytical technique.

INTERFERENCES Solvents, reagents, glassware, and other sample processing hardware may yield artifacts and/or elevated baselines causing misinterpretation of chromatograms and spectra. Materials used in the analysis must be demonstrated to be free from interferences under the conditions of analysis by running method blanks initially and with each sample lot (sample started through the extraction process on a given 8-h shift, to a maximum of 20). Specific selection of reagents and purification of solvents by distillation in all glass systems may be required. Glassware and, where possible, reagents are cleaned by solvent rinse and baking at 450°C for 1-h minimum. Interferences coextracted from samples will vary considerably from source to source, depending on the diversity of the site being sampled.

INSTRUMENTATION Major instrumentation includes a GC with a splitless or on-column injection port for capillary column, a MS with 70 eV electron impact ionization, and a data system to collect and record MS data, and process it. A K-D apparatus is used to concentrate extracts.

GC Column: 30 m × 0.25 mm I.D. 5% phenyl, 94% methyl, 1% vinyl silicone bonded phased fused silica capillary column.

PRECISION & ACCURACY The detection limits of the method are usually dependent on the level of interferences rather than instrumental limitations. The limits typify the minimum quantities that can be detected with no interferences present.

The minimum level (in μg/mL) was 50. This is defined as a minimum level at which the analytical system shall give recognizable mass spectra (background corrected) and acceptable calibration points.

The MDL (in μg/kg) in low solids was 49 and in high solids was 37; these were determined in digested sludge (low solids) and in filter cake or compost (high solids).

The labeled and native compound initial precision as standard deviation (in μg/L) was 49.
The labeled and native compound initial accuracy as average recovery (in μg/L) was 10-ns.

ns = no specification; the limit was outside the range that could be reliably measured.

SAMPLE COLLECTION, PRESERVATION & HANDLING
Collect samples in glass containers. Aqueous samples which flow freely are collected in refrigerated bottles using automatic sampling equipment. Solid samples are collected as grab samples using widemouth jars. Maintain samples at 0 to 4°C from the time of collection until extraction. If residual chlorine is present in aqueous samples, add 80 mg sodium thiosulfate/L of water. Begin sample extraction within 7 days of collection, and analyze all extracts within 40 days of extraction.

SAMPLE PREPARATION Samples containing 1% solids or less are extracted directly using continuous liquid-liquid extraction techniques. Samples containing 1 to 30% solids are diluted to the 1% level with reagent water and extracted using continuous liquid-liquid extraction techniques. Samples containing greater than 30% solids are extracted using ultrasonic techniques.

Base/neutral extraction — Adjust the pH of the waters in the extractors to 12–13 with 6 *N* NaOH. Extract with methylene chloride for 24–48 h.
Acid extraction — Adjust the pH of the waters in the extractors to 2 or less using 6 *N* sulfuric acid. Extract with methylene chloride for 24–48 h.
Ultrasonic extraction of high solids samples — Add anhydrous sodium sulfate to the sample and QC aliquot(s). Add acetone:methylene chloride (1:1) to the sample and mix thoroughly

Concentrate extracts using a K-D apparatus.

QUALITY CONTROL The analyst is permitted to modify this method to improve separations or lower the costs of measurements, provided all performance specifications are met. Analyses of blanks are required to demonstrate freedom from contamination. When results of spikes indicate atypical method performance for samples, the samples are diluted to bring method performance within acceptable limits.

For low solids (aqueous samples), extract, concentrate, and analyze two sets of four 1-L aliquots (8 aliquots total) of the precision and recovery standard. For high solids samples, two sets of four 30-g aliquots of the high solids reference matrix are used.

Spike all samples with labeled compounds to assess method performance. Compute percent recovery of the labeled compounds using the internal standard method. Compare the labeled compound recovery for each compound with the corresponding labeled compound recovery.

Reagent water and high solids reference matrix blanks are analyzed to demonstrate freedom from contamination. Extract and concentrate a 1-L reagent water blank or a high solids reference matrix blank with each sample's lot (samples started through the extraction process on the same 8-h shift, to a maximum of 20 samples).

Field replicates may be collected to determine the precision of the sampling technique, and spiked samples may be required to determine the accuracy of the analysis when the internal standard method is used.

REFERENCE Semivolatile Organic Compounds by Isotope Dilution GC/MS. Office of Water Regulation and Standards, U.S. EPA Industrial Technology Division, Washington, DC, EPA Method 1625, Rev. C, June 1989 (contact W.A. Telliard, U.S. EPA, Office of Water Regulations and Standards, 401 M St., SW, Washington, DC, 20460. Phone: 202-382-7131).

1-Naphthylaminel (α-Naphthylamine)   EPA Method 8270
CAS #134-32-7

TITLE Semivolatile Organic Compounds by GC/MS

MATRIX This method is used to determine the concentration of semivolatile organic compounds in extracts prepared from all types of solid waste matrices, soils, and groundwater. Although surface waters are not specifically mentioned, this method should be applicable to water samples from rivers, lakes, etc.

METHOD SUMMARY This method covers 259 semivolatile organic compounds. In very limited applications direct injection of the sample into the GC/MS system may be appropriate, but this results in very high detection limits (approximately 10,000 µg/L). Typically, a 1-L liquid sample, containing surrogate, and matrix spiking standards, is extracted in a continuous extractor first under acid conditions and then under basic conditions. Typically 30 g of a solid sample, containing surrogate, and matrix spiking standards, is extracted ultrasonically. After concentrating the extract to 1 mL it is spiked with 10 µL of an internal standard solution just prior to analysis by GC/MS. The volume injected should contain about 100 ng of base/neutral and 200 ng of acid surrogates (for a 1-µL injection). Analysis is performed by GC/MS using a capillary GC column.

INTERFERENCES Raw GC/MS data from all blanks, samples, and spikes must be evaluated for interferences. Contamination by carryover can occur whenever high-concentration and low-concentration samples are sequentially analyzed. To reduce carryover, the sample syringe must be rinsed out between samples with solvent. Whenever an unusually concentrated sample is encountered, it should be followed by the analysis of blank solvent to check for cross-contamination.

INSTRUMENTATION A GC/MS and a data system are required. The GC column used is a 30 m × 0.25 mm I.D. (or 0.32 mm I.D.) 1um film thickness silicone-coated fused silica capillary column. A continuous liquid-liquid extractor equipped with Teflon® or glass connection joints and stopcocks requiring no lubrication, a K-D concentrating apparatus, water bath, and an ultrasonic disrupter with a minimum power of 300 W and with pulsing capability are also required.

PRECISION & ACCURACY The estimated quantitation limit (EQL) of Method 8270B for determining an individual compound is approximately 1 mg/kg (wet weight) for soil or sediment samples, 1–200 mg/kg for wastes (dependent on matrix and method of preparation), and 10 µg/L for groundwater samples. EQLs will be proportionately higher for sample extracts that require dilution to avoid saturation of the detector.

The EQL(b) for groundwater in µg/L is 10.
The EQL (a, b) for low concentrations in soil and sediment in µg/kg is not determined.
Accuracy as µg/L is not listed.
Overall precision in µg/L is not listed.

(a) *EQLs listed for soil/sediment are based on wet weight. Normally data is reported in a dry-weight basis; therefore, EQLs will be higher based on the % dry weight of each sample. This calculation is based on a 30 g sample and gel permeation chromatography cleanup.*
(b) *Sample EQLs are highly matrix-dependent. The EQLs are provided for guidance and may not always be achievable.*
C = *True value for concentration, in µg/L.*
X = *Average recovery found for measurements of samples containing a concentration of C, in µg/L.*

ESTIMATED QUANTITATION LIMIT

Other Matrices	Factor (a)
High-concentration soil and sludges by sonicator	7.5
Non-water miscible waste	75

(a) *EQL for other matrices = [EQL for low soil/sediment] × [Factor]. This estimated EQL is similar to an EPA "Practical Quantitation Limit."*

SAMPLING METHOD
Liquid samples — Use a 1 or 2½ gallon amber glass bottle with a screw-top Teflon®-lined cover that has been prewashed with detergent and rinsed with distilled water and methanol (or isopropanol).

Soils, sediments, or sludges — Use an 8-oz. widemouth glass with a screw-top Teflon®-lined cover that has been prewashed with detergent and rinsed with distilled water and methanol (or isopropanol).

SAMPLE PRESERVATION
Liquid samples — If residual chlorine is present, add 3 mL of 10% sodium thiosulfate per gallon, cool to 4°C and store in a

solvent-free refrigerator until analysis; if chlorine is not present, then eliminate the sodium thiosulfate addition.

Soils, sediments, or sludges — Cool samples to 4°C and store in a solvent-free refrigerator.

MHT Liquid samples must be extracted within 7 days and the extracts analyzed within 40 days. Soils, sediments, or sludges may be stored for a maximum of 14 days and the extracts analyzed within 40 days.

SAMPLE PREPARATION
Liquid samples — Transfer 1 L quantitatively to a continuous extractor. If high concentrations are anticipated, a smaller volume may be used and then diluted with organic-free reagent water to 1 L. Adjust pH, if necessary, to pH <2 using 1:1 (V/V) sulfuric acid. Pipette 1.0 mL of a surrogate standard spiking solution into each sample. For the sample in each analytical batch selected for spiking, add 1.0 mL of a matrix spiking standard. For base/neutral acid analysis, the amount of the surrogates and matrix spiking compounds added to the sample should result in a final concentration of 100 ng/μL of each analyte in the extract to be analyzed (assuming a 1-μL injection). Extract with methylene chloride for 18–24 h. Next, adjust the pH of the aqueous phase to pH >11 using 10 N sodium hydroxide and extract it with methylene chloride again for 18–24 h. Dry the extract through a column containing anhydrous sodium sulfate and concentrate it to 1 mL using a K-D concentrator.

Soils, sediments, or sludges — Use 30 g of sample. Nonporous or wet samples (gummy or clay type) that do not have a free-flowing sandy texture must be mixed with anhydrous sodium sulfate until the sample is free flowing. Add 1 mL of surrogate standards to all samples, spikes, standards, and blanks. For the sample in each analytical batch selected for spiking, add 1.0 mL of a matrix spiking standard. For base/neutral acid analysis, the amount added of the surrogates and matrix spiking compounds should result in a final concentration of 100 ng/μL of each base/neutral analyte and 200 ng/μL of each acid analyte in the extract to be analyzed (assuming a 1-μL injection). Immediately add a 100-mL mixture of 1:1 methylene chloride:acetone and extract the sample ultrasonically for 3 min and then decant or filter the extracts. Repeat the extraction two or more times. Dry the extract using a column with anhydrous sodium sulfate and concentrate it to 1 mL in a K-D concentrator.

QUALITY CONTROL A methylene chloride solution containing 50 ng/μL of decafluorotriphenylphosphine (DFTPP) is used for tuning the GC/MS system each 12-h shift. A system performance check also must be made during every 12-h shift. A standard containing 50 ng/μL each of 4,4'-DDT, pentachlorophenol, and benzidine is required to verify injection port inertness and GC column performance. A calibration standard at mid-concentration, containing each compound of interest, including all required surrogates, must be performed every 12 h during analysis. After the system performance check is met, calibration check compounds (CCCs) are used to check the validity of the initial calibration.

The internal standard responses and retention times in the calibration check standard must be evaluated immediately after or during data acquisition. If the retention time for any internal standard changes by more than 30 seconds from the last check calibration (12 h), the chromatographic system must be inspected for malfunctions and corrections must be made, as required. If the electron ionization current plot (EICP) area for any of the internal standards changes by a factor of two from the last daily calibration standard check, the mass spectrometer must be inspected for malfunctions and corrections must be made, as appropriate.

Demonstrate, through the analysis of a reagent water blank, that interferences from the analytical system, glassware, and reagents are under control. The blank samples should be carried through all stages of the sample preparation and measurement steps. For each analytical batch (up to 20 samples), a reagent blank, matrix spike, and matrix spike duplicate/duplicate must be analyzed (the frequency of the spikes may be different for different monitoring programs). The blank and spiked samples must be carried through all stages of the sample preparation and measurement steps. A QC reference sample concentrate containing each analyte at a concentration of 100 mg/L in methanol is required.

REFERENCE Test Methods for Evaluating Solid Waste (SW-846). U.S. EPA 1983, Method 8270B, Rev. 2, Nov. 1990. Office of Solid Waste, Washington, DC.

2-Naphthylamine (β-Naphthylamine)   EPA Method 8270
CAS #91-59-8

TITLE Semivolatile Organic Compounds by GC/MS

MATRIX This method is used to determine the concentration of semivolatile organic compounds in extracts prepared from all types of solid waste matrices, soils, and groundwater. Although surface waters are not specifically mentioned, this method should be applicable to water samples from rivers, lakes, etc.

METHOD SUMMARY This method covers 259 semivolatile organic compounds. In very limited applications direct injection of the sample into the GC/MS system may be appropriate, but this results in very high detection limits (approximately 10,000 μg/L). Typically, a 1-L liquid sample, containing surrogate, and matrix spiking standards, is extracted in a continuous extractor first under acid conditions and then under basic conditions. Typically 30 g of a solid sample, containing surrogate, and matrix spiking standards, is extracted ultrasonically. After concentrating the extract to 1 mL it is spiked with 10 μL of an internal standard solution just prior to analysis by GC/MS. The volume injected should contain about 100 ng of base/neutral and 200 ng of acid surrogates (for a 1-μL injection). Analysis is performed by GC/MS using a capillary GC column.

INTERFERENCES Raw GC/MS data from all blanks, samples, and spikes must be evaluated for interferences. Contamination by carryover can occur whenever high-concentration and low-concentration samples are sequentially analyzed. To reduce carryover, the sample syringe must be rinsed out between samples with solvent. Whenever an unusually concentrated

sample is encountered, it should be followed by the analysis of blank solvent to check for cross-contamination.

INSTRUMENTATION A GC/MS and a data system are required. The GC column used is a 30 m × 0.25 mm I.D. (or 0.32 mm I.D.) 1um film thickness silicone-coated fused silica capillary column. A continuous liquid-liquid extractor equipped with Teflon® or glass connection joints and stopcocks requiring no lubrication, a K-D concentrating apparatus, water bath, and an ultrasonic disrupter with a minimum power of 300 W and with pulsing capability are also required.

PRECISION & ACCURACY The estimated quantitation limit (EQL) of Method 8270B for determining an individual compound is approximately 1 mg/kg (wet weight) for soil or sediment samples, 1–200 mg/kg for wastes (dependent on matrix and method of preparation), and 10 µg/L for groundwater samples. EQLs will be proportionately higher for sample extracts that require dilution to avoid saturation of the detector.

The EQL(b) for groundwater in µg/L is 10.
The EQL (a, b) for low concentrations in soil and sediment in µg/kg is not determined.
Accuracy as µg/L is not listed.
Overall precision in µg/L is not listed.

(a) EQLs listed for soil/sediment are based on wet weight. Normally data is reported in a dry-weight basis; therefore, EQLs will be higher based on the % dry weight of each sample. This calculation is based on a 30 g sample and gel permeation chromatography cleanup.
(b) Sample EQLs are highly matrix-dependent. The EQLs are provided for guidance and may not always be achievable.
C = True value for concentration, in µg/L.
X = Average recovery found for measurements of samples containing a concentration of C, in µg/L.

ESTIMATED QUANTITATION LIMIT

Other Matrices	Factor (a)
High-concentration soil and sludges by sonicator	7.5
Non-water miscible waste	75

(a) EQL for other matrices = [EQL for low soil/sediment] × [Factor]. This estimated EQL is similar to an EPA "Practical Quantitation Limit."

SAMPLING METHOD
Liquid samples — Use a 1 or 2½ gallon amber glass bottle with a screw-top Teflon®-lined cover that has been prewashed with detergent and rinsed with distilled water and methanol (or isopropanol).

Soils, sediments, or sludges — Use an 8-oz. widemouth glass with a screw-top Teflon®-lined cover that has been prewashed with detergent and rinsed with distilled water and methanol (or isopropanol).

SAMPLE PRESERVATION
Liquid samples — If residual chlorine is present, add 3 mL of 10% sodium thiosulfate per gallon, cool to 4°C and store in a solvent-free refrigerator until analysis; if chlorine is not present, then eliminate the sodium thiosulfate addition.

Soils, sediments, or sludges — Cool samples to 4°C and store in a solvent-free refrigerator.

MHT Liquid samples must be extracted within 7 days and the extracts analyzed within 40 days. Soils, sediments, or sludges may be stored for a maximum of 14 days and the extracts analyzed within 40 days.

SAMPLE PREPARATION
Liquid samples — Transfer 1 L quantitatively to a continuous extractor. If high concentrations are anticipated, a smaller volume may be used and then diluted with organic-free reagent water to 1 L. Adjust pH, if necessary, to pH <2 using 1:1 (V/V) sulfuric acid. Pipette 1.0 mL of a surrogate standard spiking solution into each sample. For the sample in each analytical batch selected for spiking, add 1.0 mL of a matrix spiking standard. For base/neutral acid analysis, the amount of the surrogates and matrix spiking compounds added to the sample should result in a final concentration of 100 ng/µL of each analyte in the extract to be analyzed (assuming a 1-µL injection). Extract with methylene chloride for 18–24 h. Next, adjust the pH of the aqueous phase to pH >11 using 10 N sodium hydroxide and extract it with methylene chloride again for 18–24 h. Dry the extract through a column containing anhydrous sodium sulfate and concentrate it to 1 mL using a K-D concentrator.

Soils, sediments, or sludges — Use 30 g of sample. Nonporous or wet samples (gummy or clay type) that do not have a free-flowing sandy texture must be mixed with anhydrous sodium sulfate until the sample is free flowing. Add 1 mL of surrogate standards to all samples, spikes, standards, and blanks. For the sample in each analytical batch selected for spiking, add 1.0 mL of a matrix spiking standard. For base/neutral acid analysis, the amount added of the surrogates and matrix spiking compounds should result in a final concentration of 100 ng/µL of each base/neutral analyte and 200 ng/µL of each acid analyte in the extract to be analyzed (assuming a 1-µL injection). Immediately add a 100-mL mixture of 1:1 methylene chloride:acetone and extract the sample ultrasonically for 3 min and then decant or filter the extracts. Repeat the extraction two or more times. Dry the extract using a column with anhydrous sodium sulfate and concentrate it to 1 mL in a K-D concentrator.

QUALITY CONTROL A methylene chloride solution containing 50 ng/µL of decafluorotriphenylphosphine (DFTPP) is used for tuning the GC/MS system each 12-h shift. A system performance check also must be made during every 12-h shift. A standard containing 50 ng/µL each of 4,4'-DDT, pentachlorophenol, and benzidine is required to verify injection port inertness and GC column performance. A calibration standard at mid-concentration, containing each compound of interest, including all required surrogates, must be performed every 12 h during analysis. After the system performance check is met, calibration check compounds (CCCs) are used to check the validity of the initial calibration.

The internal standard responses and retention times in the calibration check standard must be evaluated immediately after or during data acquisition. If the retention time for any internal standard changes by more than 30 seconds from the last check calibration (12 h), the chromatographic system must be

inspected for malfunctions and corrections must be made, as required. If the electron ionization current plot (EICP) area for any of the internal standards changes by a factor of two from the last daily calibration standard check, the mass spectrometer must be inspected for malfunctions and corrections must be made, as appropriate.

Demonstrate, through the analysis of a reagent water blank, that interferences from the analytical system, glassware, and reagents are under control. The blank samples should be carried through all stages of the sample preparation and measurement steps. For each analytical batch (up to 20 samples), a reagent blank, matrix spike, and matrix spike duplicate/duplicate must be analyzed (the frequency of the spikes may be different for different monitoring programs). The blank and spiked samples must be carried through all stages of the sample preparation and measurement steps. A QC reference sample concentrate containing each analyte at a concentration of 100 mg/L in methanol is required.

REFERENCE Test Methods for Evaluating Solid Waste (SW-846). U.S. EPA 1983, Method 8270B, Rev. 2, Nov. 1990. Office of Solid Waste, Washington, DC.

Napropamide   **EPA Method 507**
CAS #15299-99-7

TITLE Determination of Nitrogen and Phosphorus-Containing Pesticides in Water by GC/NPD

MATRIX This method is applicable to the determination of certain nitrogen and phosphorus-containing pesticides in finished drinking water and groundwater.

METHOD SUMMARY Method 507 covers 46 nitrogen- and phosphorus-containing pesticides. A 1-L sample is fortified with a surrogate standard, salted, buffered, extracted with methylene chloride, and concentrated; then the solvent is exchanged with methyl tert-butyl ether (MTBE) and concentrated again, and a 2-μL aliquot of a sample extract is injected into a GC system equipped with a selective nitrogen-phosphorus detector and a capillary column for analysis.

INTERFERENCES Method interferences may be caused by contaminants in solvents, reagents, glassware, and other sample processing apparatus. Interfering contamination may occur when a sample containing low concentrations of analytes is analyzed immediately following a sample containing relatively high concentrations. One or more injections of MTBE should be made following the analysis of a sample with high concentrations of analytes to check for analyte carryover. Matrix interferences may be caused by contaminants that are coextracted from the sample. The extent of matrix interferences will vary considerably from source to source, depending upon the water sampled.

INSTRUMENTATION A gas chromatograph system (GC) equipped with a nitrogen-phosphorus detector (NPD) is needed.

Column 1: 30 m × 0.25 mm I.D. DB-5 bonded fused silica column, 0.25 μm film thickness, or equivalent.
Column 2: 30 m × 0.25 mm I.D. DB-1701 bonded fused silica column, 0.25 μm film thickness, or equivalent.

PRECISION & ACCURACY This method has been validated in a single lab and estimated detection limits (EDLs) have been determined for each analyte. Observed detection limits may vary among waters, depending upon the nature of the interferences in the sample matrix and the specific instrumentation used. Analytes that are not separated chromatographically cannot be individually identified and measured unless an alternative technique for identification and quantification exist.

The estimated detection limit (in μg/L) was 0.25. The EDL is defined as either method detection limit or a level of compound in a sample yielding a peak in the final extract with signal-to-noise ratio of approximately 5, whichever value is higher.

The concentration used for these measurements (in μg/L) was 2.5.
The accuracy (as % recovery) was 101.
The precision (% RSD) was 6.

SAMPLING METHOD Grab samples are collected in 1-L glass sample bottles (prewashed with detergent and hot tap water, rinsed with reagent water, and dried in an oven at 400°C for 1 h) with screw caps lined with PTFE-fluorocarbon.

SAMPLE PRESERVATION Add mercuric chloride to the sample bottle in amounts to produce a concentration of 10 mg/L. If residual chlorine is present, add 80 mg of sodium thiosulfate/L of sample to the sample bottle prior to collection. After collection, seal bottle and shake vigorously for 1 min, then cool the sample to 4°C immediately and store it at 4°C in the dark until extraction.

MHT Maximum holding time of the samples, and in some cases the extracts, is 14 days.

SAMPLE PREPARATION Fortify the sample with 50 μL of the surrogate standard solution, adjust to pH 7 with phosphate buffer, add 100 g NaCl to the sample, and seal and shake to dissolve the salt; then extract with methylene chloride in a separatory funnel or in a mechanical tumbler bottle. Dry the extract by pouring it through a solvent-rinsed drying column containing about 10 cm of anhydrous sodium sulfate. Collect the extract in a Kuderna-Danish (K-D) concentrator and rinse the column with 20–30 mL methylene chloride. Concentrate the extract to about 2 mL and rinse the flask and its lower joint into the concentrator tube with 1 to 2 mL of methyl t-butyl ether (MTBE). Add 5–10 mL of MTBE and concentrate the extract twice (adding more MTBE) to a final volume of 5.0 mL and store it at 4°C until analysis.

Note: If methylene chloride is not completely removed from the final extract, it may cause detector problems.

QUALITY CONTROL Minimum quality control requirements are initial demonstration of lab capability, determination of surrogate compound recoveries in each sample and blank, monitoring internal standard peak area or height in each sample and blank, analysis of lab reagent blanks, lab fortified samples, lab fortified blanks, and other QC samples. A lab reagent blank is analyzed to demonstrate that all glassware and reagent interferences are under control.

Initial demonstration of capability is fulfilled by analyzing four fortified reagent water samples with the recovery value for each analyte falling within the acceptable range (±30% average recovery). Surrogate recoveries from samples or method blanks must be 70–130%. The internal standard response for any sample chromatogram should not deviate from the daily calibration check standard's internal standard response by more than 30% or lab fortified blanks and sample matrices are used to assess lab performance and analyte recovery, respectively.

If the response for the target analyte peak exceeds the working range of the system, dilute the extract and reanalyze. Alternative techniques such as an alternate detector or second chromatography column should be used to confirm peak identification when sample components are not resolved adequately.

EPA CONTACT & HOTLINE For technical questions contact Dr. Baldev Bathija, U.S. EPA, Office of Ground Water and Drinking Water (WH-550D), 401 M St. SW, Washington, DC 20460. Tel. (202) 260-3040. For further information the EPA Safe Drinking Water Hotline may be called at: (800) 426-4791.

REFERENCE Methods for the Determination of Organic Compounds in Drinking Water, EPA/600/4-88/039 (revised July 1991). U.S. EPA Environmental Monitoring Systems Laboratory, Cincinnati, OH, 45268, U.S.A. Available from the National Technical Information Service (NTIS), 5285 Port Royal Road, Springfield, VA 22161; Tel. 800-553-6847. NTIS Order Number is PB91-231480.

Nickel **EPA Method 6010**
CAS #7440-02-0

TITLE Inductively Coupled Plasma-Atomic Emission Spectroscopy

MATRIX This method is applicable to the determination of trace elements, including metals, in groundwater, soils, sludges, sediments, and other solid wastes. All matrices require digestion prior to analysis. The method of standard addition must be used for the analysis of all sample digests unless either serial dilution or matrix spike addition demonstrates it is not required.

METHOD SUMMARY Method 6010 covers 25 elements using ICP analysis. It measures element-emitted light by optical spectrometry. Samples, following an appropriate acid digestion, are nebulized and the resulting aerosol is transported to the plasma torch. Element-specific atomic line emission spectra are produced by a radio-frequency inductively coupled plasma.

INTERFERENCES Interferences may be categorized as spectral or non-spectral. Spectral interferences are caused by overlap of a spectral line from another element, unresolved overlap of molecular band spectra, background contribution from continuous or recombination phenomenon, and stray light from the line emission of high concentration elements. Non-spectral interferences include physical and chemical interferences. Physical interferences are effects associated with the sample nebulization and transport processes. Changes in viscosity and surface tension can cause significant inaccuracies. Chemical interferences include molecular compound formation, ionization effects, and solute vaporization effects. Normally these effects are not significant and can be minimized by careful selection of operating conditions. Chemical interferences are highly dependent on matrix type and the specific analyte element.

INSTRUMENTATION An inductively coupled argon plasma emission spectrometer (ICP) capable of background correction is required.

PRECISION & ACCURACY Detection limits, sensitivity, and optimum ranges of the metals will vary with the matrices and model of the spectrometer. In a single lab evaluation, seven wastes were analyzed for 22 elements. The mean percent relative standard deviation from triplicate analyses for all elements and wastes was 9 ± 2%. The mean percent recovery of spiked elements for all wastes was 93 ± 6%. Spike levels ranged from 100 µg/L to 100 mg/L. The wastes included sludges and industrial wastewaters.

Estimated instrument detection limit in µg/L is 15.
Spiked concentration in µg/L is 30.
Mean reported value in µg/L is 28.
Precision as RSD % is 11.

SAMPLING METHOD Samples should be collected in borosilicate glass, linear polyethylene, polypropylene, or Teflon® bottles that have been prewashed with detergent and tap water, and rinsed with 1:1 nitric acid and tap water or 1:1 hydrochloric acid and tap water. Collect at least 2 g of solids and 200 mL of aqueous samples.

SAMPLE PRESERVATION Add nitric acid to make the samples pH <2.

MHT The maximum holding time for properly preserved samples is 6 months.

SAMPLE PREPARATION Preliminary treatment of most matrices is necessary because of the complexity and variability of sample matrices. Water samples that have been prefiltered and acidified will not need acid digestion. Methods for acid digestion of waters for total recoverable or dissolved metals, acid digestions of aqueous samples and extracts for total metals, and acid digestion of sediments, sludges, and soils are summarized below.

Total recoverable or dissolved metals in water — To prepare surface and groundwater samples for determination of total recoverable and dissolved metals, a 100-mL aliquot of well-mixed sample is acidified with concentrated nitric acid and concentrated hydrochloric acid, then heated until the volume is reduced to 15–20 mL. Adjust the final volume to 100 mL with reagent water.

Total metals in aqueous samples, soil, and sediment extracts — To prepare aqueous samples, soil and sediment extracts, and wastes that contain suspended solids, a 100-mL aliquot is made acidic with concentrated nitric acid and the solution is evaporated to about 5 mL on a hot plate. Continue heating and adding additional acid until sample digestion is complete, which is usually indicated when the digestate is light in color or does not change in appearance. Evaporate the solution to about 3 mL and cool it and add a small quantity of 1:1 hydrochloric acid (10 mL/100 mL of final solution). Cover the beaker and reflux for 15 min. Wash down the beaker walls and filters or centrifuge the sample to remove silicates and other insoluble material. Filter the sample and adjust the final volume to 100 mL with reagent water and the final acid concentration to 10%.

Sediments, sludges, and soils — To prepare sediments, sludges and soil samples, transfer 1–2 g to a conical beaker and add 10 mL of 1:1 nitric acid, mix the slurry, and cover it with a watch glass. Heat the sample and reflux for 10–15 min without boiling. Allow it to cool, then add 5 mL of concentrated nitric acid and reflux for 30 min. Repeat last step and then allow the solution to evaporate to 5 mL without boiling. Cool and add 2 mL of water and 3 mL of 30% hydrogen peroxide. Cover and place the beaker on the hot plate. Heat and add 30% hydrogen peroxide in 1-mL aliquots with warming until the effervescence is minimal but do not add more than a total of 10 mL of 30% hydrogen peroxide. If the sample is being prepared for the analysis of Ag, Al, As, Ba, Be, Ca, Cd, Co, Cr, Cu, Fe, K, Mg, Mn, Mo, Na, Ni, Os, Pb, Se, Tl, V, and Zn, then add 5 mL of concentrated hydrochloric acid and 10 mL of water and return the covered beaker to a hot plate for 15 min of additional refluxing without boiling. Dilute the sample to a 100 mL volume with water after cooling and filter or centrifuge to remove particulates.

QUALITY CONTROL Laboratory control samples must be analyzed for each analytical method. A method blank should be analyzed with each batch of samples. The effect of the matrix on method performance must be demonstrated: when appropriate, there should be at least one matrix spike and either one matrix duplicate or one matrix spike duplicate per analytical batch. The bias and precision of the method, as well as the method detection limit for each specific matrix type, must be measured.

Dilute and reanalyze samples that are more concentrated than the linear calibration limit. Employ a minimum of one reagent blank per sample batch to determine if contamination or any memory effects are occurring. Whenever a new or unusual sample matrix is encountered, perform either a serial dilution test or a matrix spike addition test to ensure that neither positive or negative interferences are operating on any of the analyte elements. Check the instrument standardization by verifying calibration every 10 samples using a calibration blank and a check standard.

REFERENCE Test Methods for Evaluating Solid Waste (SW-846). U.S. EPA. 1983. Method 6010, Rev. 0, Sept. 1986. Office of Solid Wastes, Washington, DC.

Nickel   EPA Method 200.7
CAS #7440-02-0

TITLE Inductively Coupled Plasma

MATRIX Dissolved, suspended or (ICP) total element in drinking and surface waters and in domestic and industrial wastewaters.

APPLICATION The method covers the determination of 25 metals. Dissolved elements are determined in filtered and acidified samples after appropriate digestion (which increases dissolved solids). Its primary advantage is that ICP instruments allow simultaneous or rapid sequential determination of many elements in a short time. Samples are first nebulized and the aerosol is transported to a plasma torch in which element specific atomic line emission spectra are produced by a radio frequency inductively coupled plasma. Background correction is required for trace element detection except in the case of line broadning.

INTERFERENCES There are spectral, physical, and chemical interferences. The primary disadvantage of ICP instruments is background radiation from other elements and the plasma gases (spectral interferences). Changes in sample viscosity and surface tension with samples containing high dissolved solids (especially those exceeding 1500 mg/L) or high acid concentrations can cause physical interferences. Ionization effects, solute vaporization and molecular compound formation can cause chemical interferences. No other elements cause interference at the 100 mg/L level.

INSTRUMENTATION Inductively coupled argon plasma emission spectroscopy. 231.604 nm wavelength.

RANGE Not listed.

MDL 15 µg/L.

PRECISION SD = 5.8% Mean at true value 250 µg/L.

ACCURACY Mean recovery = 93% ± 6% of spiked elements for all wastes.

SAMPLING METHOD Wash sample container with detergent and tap water, rinse with 1 + 1 nitric acid and tap water, then rinse with 1 + 1 hydrochloric acid and tap water, then rinse with deionized, distilled water in that order. Perform any filtration or acid preservation steps when the sample is collected or as soon as possible thereafter.

STABILITY Cool samples to 4°C.

MHT 24 h.

QUALITY CONTROL Mixed calibration standards, an instrument check standard, and an interference check solution are used in addition to a quality control sample. The quality control sample should be prepared in the same acid matrix as the calibration standards at 10 times the instrumental detection limits and in accordance with the instructions provided by the supplier. Furthermore, two types of blanks are required: a calibration blank and a reagent blank.

REFERENCE Method 200.7, U.S. EPA, EMSL-Cincinnati, OH, Nov. 1980

Nickel
CAS #7440-02-0
EPA Method 7520

TITLE Atomic Absorption (AA)

MATRIX Drinking, surface, and direct aspiration saline waters. Wastewater.

APPLICATION Sample is aspirated and atomized in a flame. A light beam from a Ni hollow cathode lamp is directed through the flame into a monochromator and onto a detector. Since wavelength of light beam is specific for Ni, light energy absorbed by detector is measure of nickel.

INTERFERENCES The most troublesomee type is chemical, caused by lack of absorption of atoms bound in molecular combination in the flame. High dissolved solids in sample may result in nonatomic absorbance interference. High concentrations of iron, cobalt and chromium can interfere.

INSTRUMENTATION Atomic absorption spectrometer. Nickel hollow cathode lamp. [232.0 nm (primary)].

RANGE 0.3–5 mg/L.

MDL 0.04 mg/L

PRECISION Standard deviation = ±0.011, 0.02, 0.04 at 0.20, 1.5, 7.5 mg Mo/L

ACCURACY Recoveries = 100, 97, 93% at 0.20, 1.5, 7.5 mg Mo/L

SAMPLING METHOD Use glass or plastic containers. Collect 200 g of solids and 600 mL of liquid samples.

STABILITY Cool solid samples to 4°C and analyze as soon as possible. Add nitric acid to liquid samples to pH <2.

MHT 6 months.

QUALITY CONTROL At least one duplicate and one spike sample should be run every 20 samples or with each matrix type to verify precision of the method. For 20 or more samples per day, verify working standard curve. Run an additional standard at or near mid-range every 10 samples.

REFERENCE Method 7520, SW-846, 3rd ed., Nov.1986.

Nicotine
CAS #54-11-5
EPA Method 8270

TITLE Semivolatile Organic Compounds by GC/MS

MATRIX This method is used to determine the concentration of semivolatile organic compounds in extracts prepared from all types of solid waste matrices, soils, and groundwater. Although surface waters are not specifically mentioned, this method should be applicable to water samples from rivers, lakes, etc.

METHOD SUMMARY This method covers 259 semivolatile organic compounds. In very limited applications direct injection of the sample into the GC/MS system may be appropriate, but this results in very high detection limits (approximately 10,000 μg/L). Typically, a 1-L liquid sample, containing surrogate, and matrix spiking standards, is extracted in a continuous extractor first under acid conditions and then under basic conditions. Typically 30 g of a solid sample, containing surrogate, and matrix spiking standards, is extracted ultrasonically. After concentrating the extract to 1 mL it is spiked with 10 μL of an internal standard solution just prior to analysis by GC/MS. The volume injected should contain about 100 ng of base/neutral and 200 ng of acid surrogates (for a 1-μL injection). Analysis is performed by GC/MS using a capillary GC column.

INTERFERENCES Raw GC/MS data from all blanks, samples, and spikes must be evaluated for interferences. Contamination by carryover can occur whenever high-concentration and low-concentration samples are sequentially analyzed. To reduce carryover, the sample syringe must be rinsed out between samples with solvent. Whenever an unusually concentrated sample is encountered, it should be followed by the analysis of blank solvent to check for cross-contamination.

INSTRUMENTATION A GC/MS and a data system are required. The GC column used is a 30 m × 0.25 mm I.D. (or 0.32 mm I.D.) 1um film thickness silicone-coated fused silica capillary column. A continuous liquid-liquid extractor equipped with Teflon® or glass connection joints and stopcocks requiring no lubrication, a K-D concentrating apparatus, water bath, and an ultrasonic disrupter with a minimum power of 300 W and with pulsing capability are also required.

PRECISION & ACCURACY The estimated quantitation limit (EQL) of Method 8270B for determining an individual compound is approximately 1 mg/kg (wet weight) for soil or sediment samples, 1–200 mg/kg for wastes (dependent on matrix and method of preparation), and 10 μg/L for groundwater samples. EQLs will be proportionately higher for sample extracts that require dilution to avoid saturation of the detector.

The EQL(b) for groundwater in μg/L is 20.
The EQL (a, b) for low concentrations in soil and sediment in μg/kg is not determined.
Accuracy as μg/L is not listed.
Overall precision in μg/L is not listed.

(a) *EQLs listed for soil/sediment are based on wet weight. Normally data is reported in a dry-weight basis; therefore, EQLs will be higher based on the % dry weight of each sample. This calculation is based on a 30 g sample and gel permeation chromatography cleanup.*

(b) *Sample EQLs are highly matrix-dependent. The EQLs are provided for guidance and may not always be achievable.*

C = True value for concentration, in μg/L.
X = Average recovery found for measurements of samples containing a concentration of C, in μg/L.

ESTIMATED QUANTITATION LIMIT

Other Matrices	Factor (a)
High-concentration soil and sludges by sonicator	7.5
Non-water miscible waste	75

(a) *EQL for other matrices = [EQL for low soil/sediment] × [Factor]. This estimated EQL is similar to an EPA "Practical Quantitation Limit."*

SAMPLING METHOD

Liquid samples — Use a 1 or 2½ gallon amber glass bottle with a screw-top Teflon®-lined cover that has been prewashed with detergent and rinsed with distilled water and methanol (or isopropanol).

Soils, sediments, or sludges — Use an 8-oz. widemouth glass with a screw-top Teflon®-lined cover that has been prewashed with detergent and rinsed with distilled water and methanol (or isopropanol).

SAMPLE PRESERVATION

Liquid samples — If residual chlorine is present, add 3 mL of 10% sodium thiosulfate per gallon, cool to 4°C and store in a solvent-free refrigerator until analysis; if chlorine is not present, then eliminate the sodium thiosulfate addition.

Soils, sediments, or sludges — Cool samples to 4°C and store in a solvent-free refrigerator.

MHT Liquid samples must be extracted within 7 days and the extracts analyzed within 40 days. Soils, sediments, or sludges may be stored for a maximum of 14 days and the extracts analyzed within 40 days.

SAMPLE PREPARATION

Liquid samples — Transfer 1 L quantitatively to a continuous extractor. If high concentrations are anticipated, a smaller volume may be used and then diluted with organic-free reagent water to 1 L. Adjust pH, if necessary, to pH <2 using 1:1 (V/V) sulfuric acid. Pipette 1.0 mL of a surrogate standard spiking solution into each sample. For the sample in each analytical batch selected for spiking, add 1.0 mL of a matrix spiking standard. For base/neutral acid analysis, the amount of the surrogates and matrix spiking compounds added to the sample should result in a final concentration of 100 ng/µL of each analyte in the extract to be analyzed (assuming a 1-µL injection). Extract with methylene chloride for 18–24 h. Next, adjust the pH of the aqueous phase to pH >11 using 10 N sodium hydroxide and extract it with methylene chloride again for 18–24 h. Dry the extract through a column containing anhydrous sodium sulfate and concentrate it to 1 mL using a K-D concentrator.

Soils, sediments, or sludges — Use 30 g of sample. Nonporous or wet samples (gummy or clay type) that do not have a free-flowing sandy texture must be mixed with anhydrous sodium sulfate until the sample is free flowing. Add 1 mL of surrogate standards to all samples, spikes, standards, and blanks. For the sample in each analytical batch selected for spiking, add 1.0 mL of a matrix spiking standard. For base/neutral acid analysis, the amount added of the surrogates and matrix spiking compounds should result in a final concentration of 100 ng/µL of each base/neutral analyte and 200 ng/µL of each acid analyte in the extract to be analyzed (assuming a 1-µL injection). Immediately add a 100-mL mixture of 1:1 methylene chloride:acetone and extract the sample ultrasonically for 3 min and then decant or filter the extracts. Repeat the extraction two or more times. Dry the extract using a column with anhydrous sodium sulfate and concentrate it to 1 mL in a K-D concentrator.

QUALITY CONTROL A methylene chloride solution containing 50 ng/µL of decafluorotriphenylphosphine (DFTPP) is used for tuning the GC/MS system each 12-h shift. A system performance check also must be made during every 12-h shift. A standard containing 50 ng/µL each of 4,4′-DDT, pentachlorophenol, and benzidine is required to verify injection port inertness and GC column performance. A calibration standard at mid-concentration, containing each compound of interest, including all required surrogates, must be performed every 12 h during analysis. After the system performance check is met, calibration check compounds (CCCs) are used to check the validity of the initial calibration.

The internal standard responses and retention times in the calibration check standard must be evaluated immediately after or during data acquisition. If the retention time for any internal standard changes by more than 30 seconds from the last check calibration (12 h), the chromatographic system must be inspected for malfunctions and corrections must be made, as required. If the electron ionization current plot (EICP) area for any of the internal standards changes by a factor of two from the last daily calibration standard check, the mass spectrometer must be inspected for malfunctions and corrections must be made, as appropriate.

Demonstrate, through the analysis of a reagent water blank, that interferences from the analytical system, glassware, and reagents are under control. The blank samples should be carried through all stages of the sample preparation and measurement steps. For each analytical batch (up to 20 samples), a reagent blank, matrix spike, and matrix spike duplicate/duplicate must be analyzed (the frequency of the spikes may be different for different monitoring programs). The blank and spiked samples must be carried through all stages of the sample preparation and measurement steps. A QC reference sample concentrate containing each analyte at a concentration of 100 mg/L in methanol is required.

REFERENCE Test Methods for Evaluating Solid Waste (SW-846). U.S. EPA 1983, Method 8270B, Rev. 2, Nov. 1990. Office of Solid Waste, Washington, DC.

Nitrate-N (Total)   EPA Method 300.0

TITLE Inorganic Anions in Water

MATRIX Drinking, surface and mixed wastewater.

APPLICATION A small volume of sample, typically 2 to 3 mL, is introduced into an ion chromatograph. The anions of interest are separated and measured using a system comprised of a guard column, separator column, suppressor column and conductivity detector.

INTERFERENCES Interferences can be caused by substances with retention times similar to and overlapping those of ion of interest. Large amounts of an anion can interfere with peak resolution of adjacent anion. EPA Method interference can be caused by reagent or equipment contamination.

INSTRUMENTATION Ion chromatograph. Analytical balance. Guard, separator and suppressor columns.

RANGE Not listed.

MDL 0.013 mg/L.

PRECISION SD = 0.356 mg/L at 31.0 mg/L Nitrate-A (Drinking water).

ACCURACY Mean recovery = 100.7% at 31.0 mg/L Nitrate-N (Drinking water).

SAMPLING METHOD Plastic or glass.

STABILITY Cool, 4°C.

MHT 48 h.

QUALITY CONTROL The lab should spike and analyze a minimum of 10% of all samples to monitor continuing lab performance. Field and lab duplicates should be analyzed. Measure retention times of stds. (Nitratesexhibit great changes in retention times).

REFERENCE Test Method — The Determination of Inorganic Anions in Water by Ion Chromatography, (EPA-600/4-84-017).

5-Nitro-o-anisidine EPA Method 8270
CAS #99-59-2

TITLE Semivolatile Organic Compounds by GC/MS

MATRIX This method is used to determine the concentration of semivolatile organic compounds in extracts prepared from all types of solid waste matrices, soils, and groundwater. Although surface waters are not specifically mentioned, this method should be applicable to water samples from rivers, lakes, etc.

METHOD SUMMARY This method covers 259 semivolatile organic compounds. In very limited applications direct injection of the sample into the GC/MS system may be appropriate, but this results in very high detection limits (approximately 10,000 µg/L). Typically, a 1-L liquid sample, containing surrogate, and matrix spiking standards, is extracted in a continuous extractor first under acid conditions and then under basic conditions. Typically 30 g of a solid sample, containing surrogate, and matrix spiking standards, is extracted ultrasonically. After concentrating the extract to 1 mL it is spiked with 10 µL of an internal standard solution just prior to analysis by GC/MS. The volume injected should contain about 100 ng of base/neutral and 200 ng of acid surrogates (for a 1-µL injection). Analysis is performed by GC/MS using a capillary GC column.

INTERFERENCES Raw GC/MS data from all blanks, samples, and spikes must be evaluated for interferences. Contamination by carryover can occur whenever high-concentration and low-concentration samples are sequentially analyzed. To reduce carryover, the sample syringe must be rinsed out between samples with solvent. Whenever an unusually concentrated sample is encountered, it should be followed by the analysis of blank solvent to check for cross-contamination.

INSTRUMENTATION A GC/MS and a data system are required. The GC column used is a 30 m × 0.25 mm I.D. (or 0.32 mm I.D.) 1um film thickness silicone-coated fused silica capillary column. A continuous liquid-liquid extractor equipped with Teflon® or glass connection joints and stopcocks requiring no lubrication, a K-D concentrating apparatus, water bath, and an ultrasonic disrupter with a minimum power of 300 W and with pulsing capability are also required.

PRECISION & ACCURACY The estimated quantitation limit (EQL) of Method 8270B for determining an individual compound is approximately 1 mg/kg (wet weight) for soil or sediment samples, 1–200 mg/kg for wastes (dependent on matrix and method of preparation), and 10 µg/L for groundwater samples. EQLs will be proportionately higher for sample extracts that require dilution to avoid saturation of the detector.

The EQL(b) for groundwater in µg/L is 10.
The EQL (a, b) for low concentrations in soil and sediment in µg/kg is not determined.
Accuracy as µg/L is not listed.
Overall precision in µg/L is not listed.

(a) EQLs listed for soil/sediment are based on wet weight. Normally data is reported in a dry-weight basis; therefore, EQLs will be higher based on the % dry weight of each sample. This calculation is based on a 30 g sample and gel permeation chromatography cleanup.
(b) Sample EQLs are highly matrix-dependent. The EQLs are provided for guidance and may not always be achievable.
C = True value for concentration, in µg/L.
X = Average recovery found for measurements of samples containing a concentration of C, in µg/L.

ESTIMATED QUANTITATION LIMIT

Other Matrices	Factor (a)
High-concentration soil and sludges by sonicator	7.5
Non-water miscible waste	75

(a) EQL for other matrices = [EQL for low soil/sediment] ×[Factor]. This estimated EQL is similar to an EPA "Practical Quantitation Limit."

SAMPLING METHOD
Liquid samples — Use a 1 or 2½ gallon amber glass bottle with a screw-top Teflon®-lined cover that has been prewashed with detergent and rinsed with distilled water and methanol (or isopropanol).

Soils, sediments, or sludges — Use an 8-oz. widemouth glass with a screw-top Teflon®-lined cover that has been prewashed with detergent and rinsed with distilled water and methanol (or isopropanol).

SAMPLE PRESERVATION
Liquid samples — If residual chlorine is present, add 3 mL of 10% sodium thiosulfate per gallon, cool to 4°C and store in a solvent-free refrigerator until analysis; if chlorine is not present, then eliminate the sodium thiosulfate addition.

Soils, sediments, or sludges — Cool samples to 4°C and store in a solvent-free refrigerator.

MHT Liquid samples must be extracted within 7 days and the extracts analyzed within 40 days. Soils, sediments, or sludges may

be stored for a maximum of 14 days and the extracts analyzed within 40 days.

SAMPLE PREPARATION

Liquid samples — Transfer 1 L quantitatively to a continuous extractor. If high concentrations are anticipated, a smaller volume may be used and then diluted with organic-free reagent water to 1 L. Adjust pH, if necessary, to pH <2 using 1:1 (V/V) sulfuric acid. Pipette 1.0 mL of a surrogate standard spiking solution into each sample. For the sample in each analytical batch selected for spiking, add 1.0 mL of a matrix spiking standard. For base/neutral acid analysis, the amount of the surrogates and matrix spiking compounds added to the sample should result in a final concentration of 100 ng/µL of each analyte in the extract to be analyzed (assuming a 1-µL injection). Extract with methylene chloride for 18–24 h. Next, adjust the pH of the aqueous phase to pH >11 using 10 N sodium hydroxide and extract it with methylene chloride again for 18–24 h. Dry the extract through a column containing anhydrous sodium sulfate and concentrate it to 1 mL using a K-D concentrator.

Soils, sediments, or sludges — Use 30 g of sample. Nonporous or wet samples (gummy or clay type) that do not have a free-flowing sandy texture must be mixed with anhydrous sodium sulfate until the sample is free flowing. Add 1 mL of surrogate standards to all samples, spikes, standards, and blanks. For the sample in each analytical batch selected for spiking, add 1.0 mL of a matrix spiking standard. For base/neutral acid analysis, the amount added of the surrogates and matrix spiking compounds should result in a final concentration of 100 ng/µL of each base/neutral analyte and 200 ng/µL of each acid analyte in the extract to be analyzed (assuming a 1-µL injection). Immediately add a 100-mL mixture of 1:1 methylene chloride:acetone and extract the sample ultrasonically for 3 min and then decant or filter the extracts. Repeat the extraction two or more times. Dry the extract using a column with anhydrous sodium sulfate and concentrate it to 1 mL in a K-D concentrator.

QUALITY CONTROL A methylene chloride solution containing 50 ng/µL of decafluorotriphenylphosphine (DFTPP) is used for tuning the GC/MS system each 12-h shift. A system performance check also must be made during every 12-h shift. A standard containing 50 ng/µL each of 4,4′-DDT, pentachlorophenol, and benzidine is required to verify injection port inertness and GC column performance. A calibration standard at mid-concentration, containing each compound of interest, including all required surrogates, must be performed every 12 h during analysis. After the system performance check is met, calibration check compounds (CCCs) are used to check the validity of the initial calibration.

The internal standard responses and retention times in the calibration check standard must be evaluated immediately after or during data acquisition. If the retention time for any internal standard changes by more than 30 seconds from the last check calibration (12 h), the chromatographic system must be inspected for malfunctions and corrections must be made, as required. If the electron ionization current plot (EICP) area for any of the internal standards changes by a factor of two from the last daily calibration standard check, the mass spectrometer must be inspected for malfunctions and corrections must be made, as appropriate.

Demonstrate, through the analysis of a reagent water blank, that interferences from the analytical system, glassware, and reagents are under control. The blank samples should be carried through all stages of the sample preparation and measurement steps. For each analytical batch (up to 20 samples), a reagent blank, matrix spike, and matrix spike duplicate/duplicate must be analyzed (the frequency of the spikes may be different for different monitoring programs). The blank and spiked samples must be carried through all stages of the sample preparation and measurement steps. A QC reference sample concentrate containing each analyte at a concentration of 100 mg/L in methanol is required.

REFERENCE Test Methods for Evaluating Solid Waste (SW-846). U.S. EPA 1983, Method 8270B, Rev. 2, Nov. 1990. Office of Solid Waste, Washington, DC.

5-Nitro-o-toluidine   **EPA Method 1625**
CAS #99-55-8

TITLE Semivolatile Organic Compounds by Isotope Dilution GC/MS

MATRIX The compounds may be determined in waters, soils, and municipal sludges by this method.

METHOD SUMMARY This method is used to determine 176 semivolatile toxic organic pollutants associated with the CWA (as amended 1987); the RCRA (as amended 1986); the CERCLA (as amended 1986); and other compounds amenable to extraction and analysis by capillary column gas chromatography-mass spectrometry (GC/MS).

Stable isotopically-labeled analogs of the compounds of interest are added to the sample. If the solids content is less than 1%, a 1-L sample is extracted at pH 12–13, then at pH <2 with methylene chloride using continuous extraction techniques.

If the solids content is 30% or less, the sample is diluted to 1% solids with reagent water, homogenized ultrasonically, and extracted at pH 12–13, then at pH <2 with methylene chloride using continuous extraction techniques. If the solids content is greater than 30%, the sample is extracted using ultrasonic techniques.

Each extract is dried over sodium sulfate, concentrated to a volume of 5 mL, cleaned up using GPC, if necessary, and concentrated. Extracts are concentrated to 1 mL if GPC is not performed, and to 0.5 mL if GPC is performed.

An internal standard is added to the extract, and a 1-mL aliquot of the extract is injected into the GC. The compounds are separated by GC and detected by a MS. The labeled compounds serve to correct the variability of the analytical technique.

INTERFERENCES Solvents, reagents, glassware, and other sample processing hardware may yield artifacts and/or elevated baselines causing misinterpretation of chromatograms and spectra. Materials used in the analysis must be demonstrated

to be free from interferences under the conditions of analysis by running method blanks initially and with each sample lot (sample started through the extraction process on a given 8-h shift, to a maximum of 20). Specific selection of reagents and purification of solvents by distillation in all glass systems may be required. Glassware and, where possible, reagents are cleaned by solvent rinse and baking at 450°C for 1-h minimum. Interferences coextracted from samples will vary considerably from source to source, depending on the diversity of the site being sampled.

INSTRUMENTATION Major instrumentation includes a GC with a splitless or on-column injection port for capillary column, a MS with 70 eV electron impact ionization, and a data system to collect and record MS data, and process it. A K-D apparatus is used to concentrate extracts.

GC Column: 30 m × 0.25 mm I.D. 5% phenyl, 94% methyl, 1% vinyl silicone bonded phased fused silica capillary column.

PRECISION & ACCURACY The detection limits of the method are usually dependent on the level of interferences rather than instrumental limitations. The limits typify the minimum quantities that can be detected with no interferences present.

The minimum level (in µg/mL) was not listed. This is defined as a minimum level at which the analytical system shall give recognizable mass spectra (background corrected) and acceptable calibration points.

The MDL (in µg/kg) in low solids was not listed and in high solids was not listed; these were determined in digested sludge (low solids) and in filter cake or compost (high solids).

The labeled and native compound initial precision as standard deviation (in µg/L) was not listed.
The labeled and native compound initial accuracy as average recovery (in µg/L) was not listed.

SAMPLE COLLECTION, PRESERVATION & HANDLING Collect samples in glass containers. Aqueous samples which flow freely are collected in refrigerated bottles using automatic sampling equipment. Solid samples are collected as grab samples using widemouth jars. Maintain samples at 0 to 4°C from the time of collection until extraction. If residual chlorine is present in aqueous samples, add 80 mg sodium thiosulfate/L of water. Begin sample extraction within 7 days of collection, and analyze all extracts within 40 days of extraction.

SAMPLE PREPARATION Samples containing 1% solids or less are extracted directly using continuous liquid-liquid extraction techniques. Samples containing 1 to 30% solids are diluted to the 1% level with reagent water and extracted using continuous liquid-liquid extraction techniques. Samples containing greater than 30% solids are extracted using ultrasonic techniques.

Base/neutral extraction — Adjust the pH of the waters in the extractors to 12–13 with 6 N NaOH. Extract with methylene chloride for 24–48 h.
Acid extraction — Adjust the pH of the waters in the extractors to 2 or less using 6 N sulfuric acid. Extract with methylene chloride for 24–48 h.
Ultrasonic extraction of high solids samples — Add anhydrous sodium sulfate to the sample and QC aliquot(s). Add acetone:methylene chloride (1:1) to the sample and mix thoroughly

Concentrate extracts using a K-D apparatus.

QUALITY CONTROL The analyst is permitted to modify this method to improve separations or lower the costs of measurements, provided all performance specifications are met. Analyses of blanks are required to demonstrate freedom from contamination. When results of spikes indicate atypical method performance for samples, the samples are diluted to bring method performance within acceptable limits.

For low solids (aqueous samples), extract, concentrate, and analyze two sets of four 1-L aliquots (8 aliquots total) of the precision and recovery standard. For high solids samples, two sets of four 30-g aliquots of the high solids reference matrix are used.

Spike all samples with labeled compounds to assess method performance. Compute percent recovery of the labeled compounds using the internal standard method. Compare the labeled compound recovery for each compound with the corresponding labeled compound recovery.

Reagent water and high solids reference matrix blanks are analyzed to demonstrate freedom from contamination. Extract and concentrate a 1-L reagent water blank or a high solids reference matrix blank with each sample's lot (samples started through the extraction process on the same 8-h shift, to a maximum of 20 samples).

Field replicates may be collected to determine the precision of the sampling technique, and spiked samples may be required to determine the accuracy of the analysis when the internal standard method is used.

REFERENCE Semivolatile Organic Compounds by Isotope Dilution GC/MS. Office of Water Regulation and Standards, U.S. EPA Industrial Technology Division, Washington, DC, EPA Method 1625, Rev. C, June 1989 (contact W.A. Telliard, U.S. EPA, Office of Water Regulations and Standards, 401 M St., SW, Washington, DC, 20460. Phone: 202-382-7131).

5-Nitro-o-toluidine   **EPA Method 8270**
CAS #99-55-8

TITLE Semivolatile Organic Compounds by GC/MS

MATRIX This method is used to determine the concentration of semivolatile organic compounds in extracts prepared from all types of solid waste matrices, soils, and groundwater. Although surface waters are not specifically mentioned, this method should be applicable to water samples from rivers, lakes, etc.

METHOD SUMMARY This method covers 259 semivolatile organic compounds. In very limited applications direct injection of the sample into the GC/MS system may be appropriate, but this results in very high detection limits (approximately

10,000 μg/L). Typically, a 1-L liquid sample, containing surrogate, and matrix spiking standards, is extracted in a continuous extractor first under acid conditions and then under basic conditions. Typically 30 g of a solid sample, containing surrogate, and matrix spiking standards, is extracted ultrasonically. After concentrating the extract to 1 mL it is spiked with 10 μL of an internal standard solution just prior to analysis by GC/MS. The volume injected should contain about 100 ng of base/neutral and 200 ng of acid surrogates (for a 1-μL injection). Analysis is performed by GC/MS using a capillary GC column.

INTERFERENCES Raw GC/MS data from all blanks, samples, and spikes must be evaluated for interferences. Contamination by carryover can occur whenever high-concentration and low-concentration samples are sequentially analyzed. To reduce carryover, the sample syringe must be rinsed out between samples with solvent. Whenever an unusually concentrated sample is encountered, it should be followed by the analysis of blank solvent to check for cross-contamination.

INSTRUMENTATION A GC/MS and a data system are required. The GC column used is a 30 m × 0.25 mm I.D. (or 0.32 mm I.D.) 1um film thickness silicone-coated fused silica capillary column. A continuous liquid-liquid extractor equipped with Teflon® or glass connection joints and stopcocks requiring no lubrication, a K-D concentrating apparatus, water bath, and an ultrasonic disrupter with a minimum power of 300 W and with pulsing capability are also required.

PRECISION & ACCURACY The estimated quantitation limit (EQL) of Method 8270B for determining an individual compound is approximately 1 mg/kg (wet weight) for soil or sediment samples, 1–200 mg/kg for wastes (dependent on matrix and method of preparation), and 10 μg/L for groundwater samples. EQLs will be proportionally higher for sample extracts that require dilution to avoid saturation of the detector.

The EQL(b) for groundwater in μg/L is 10.
The EQL (a, b) for low concentrations in soil and sediment in μg/kg is not determined.
Accuracy as μg/L is not listed.
Overall precision in μg/L is not listed.

(a) *EQLs listed for soil/sediment are based on wet weight. Normally data is reported in a dry-weight basis; therefore, EQLs will be higher based on the % dry weight of each sample. This calculation is based on a 30 g sample and gel permeation chromatography cleanup.*
(b) *Sample EQLs are highly matrix-dependent. The EQLs are provided for guidance and may not always be achievable.*
C = *True value for concentration, in μg/L.*
X = *Average recovery found for measurements of samples containing a concentration of C, in μg/L.*

ESTIMATED QUANTITATION LIMIT

Other Matrices	Factor (a)
High-concentration soil and sludges by sonicator	7.5
Non-water miscible waste	75

(a) *EQL for other matrices = [EQL for low soil/sediment] × [Factor]. This estimated EQL is similar to an EPA "Practical Quantitation Limit."*

SAMPLING METHOD
Liquid samples — Use a 1 or 2½ gallon amber glass bottle with a screw-top Teflon®-lined cover that has been prewashed with detergent and rinsed with distilled water and methanol (or isopropanol).

Soils, sediments, or sludges — Use an 8-oz. widemouth glass with a screw-top Teflon®-lined cover that has been prewashed with detergent and rinsed with distilled water and methanol (or isopropanol).

SAMPLE PRESERVATION
Liquid samples — If residual chlorine is present, add 3 mL of 10% sodium thiosulfate per gallon, cool to 4°C and store in a solvent-free refrigerator until analysis; if chlorine is not present, then eliminate the sodium thiosulfate addition.

Soils, sediments, or sludges — Cool samples to 4°C and store in a solvent-free refrigerator.

MHT Liquid samples must be extracted within 7 days and the extracts analyzed within 40 days. Soils, sediments, or sludges may be stored for a maximum of 14 days and the extracts analyzed within 40 days.

SAMPLE PREPARATION
Liquid samples — Transfer 1 L quantitatively to a continuous extractor. If high concentrations are anticipated, a smaller volume may be used and then diluted with organic-free reagent water to 1 L. Adjust pH, if necessary, to pH <2 using 1:1 (V/V) sulfuric acid. Pipette 1.0 mL of a surrogate standard spiking solution into each sample. For the sample in each analytical batch selected for spiking, add 1.0 mL of a matrix spiking standard. For base/neutral acid analysis, the amount of the surrogates and matrix spiking compounds added to the sample should result in a final concentration of 100 ng/μL of each analyte in the extract to be analyzed (assuming a 1-μL injection). Extract with methylene chloride for 18–24 h. Next, adjust the pH of the aqueous phase to pH >11 using 10 N sodium hydroxide and extract it with methylene chloride again for 18–24 h. Dry the extract through a column containing anhydrous sodium sulfate and concentrate it to 1 mL using a K-D concentrator.

Soils, sediments, or sludges — Use 30 g of sample. Nonporous or wet samples (gummy or clay type) that do not have a free-flowing sandy texture must be mixed with anhydrous sodium sulfate until the sample is free flowing. Add 1 mL of surrogate standards to all samples, spikes, standards, and blanks. For the sample in each analytical batch selected for spiking, add 1.0 mL of a matrix spiking standard. For base/neutral acid analysis, the amount added of the surrogates and matrix spiking compounds should result in a final concentration of 100 ng/μL of each base/neutral analyte and 200 ng/μL of each acid analyte in the extract to be analyzed (assuming a 1-μL injection). Immediately add a 100-mL mixture of 1:1 methylene chloride:acetone and extract the sample ultrasonically for 3 min and then decant or filter the extracts. Repeat the extraction two or more times. Dry the extract using a column with anhydrous sodium sulfate and concentrate it to 1 mL in a K-D concentrator.

QUALITY CONTROL A methylene chloride solution containing 50 ng/μL of decafluorotriphenylphosphine (DFTPP) is

used for tuning the GC/MS system each 12-h shift. A system performance check also must be made during every 12-h shift. A standard containing 50 ng/μL each of 4,4′-DDT, pentachlorophenol, and benzidine is required to verify injection port inertness and GC column performance. A calibration standard at mid-concentration, containing each compound of interest, including all required surrogates, must be performed every 12 h during analysis. After the system performance check is met, calibration check compounds (CCCs) are used to check the validity of the initial calibration.

The internal standard responses and retention times in the calibration check standard must be evaluated immediately after or during data acquisition. If the retention time for any internal standard changes by more than 30 seconds from the last check calibration (12 h), the chromatographic system must be inspected for malfunctions and corrections must be made, as required. If the electron ionization current plot (EICP) area for any of the internal standards changes by a factor of two from the last daily calibration standard check, the mass spectrometer must be inspected for malfunctions and corrections must be made, as appropriate.

Demonstrate, through the analysis of a reagent water blank, that interferences from the analytical system, glassware, and reagents are under control. The blank samples should be carried through all stages of the sample preparation and measurement steps. For each analytical batch (up to 20 samples), a reagent blank, matrix spike, and matrix spike duplicate/duplicate must be analyzed (the frequency of the spikes may be different for different monitoring programs). The blank and spiked samples must be carried through all stages of the sample preparation and measurement steps. A QC reference sample concentrate containing each analyte at a concentration of 100 mg/L in methanol is required.

REFERENCE Test Methods for Evaluating Solid Waste (SW-846). U.S. EPA 1983, Method 8270B, Rev. 2, Nov. 1990. Office of Solid Waste, Washington, DC.

5-Nitroacenaphthene **EPA Method 8270**
CAS #602-87-9

TITLE Semivolatile Organic Compounds by GC/MS

MATRIX This method is used to determine the concentration of semivolatile organic compounds in extracts prepared from all types of solid waste matrices, soils, and groundwater. Although surface waters are not specifically mentioned, this method should be applicable to water samples from rivers, lakes, etc.

METHOD SUMMARY This method covers 259 semivolatile organic compounds. In very limited applications direct injection of the sample into the GC/MS system may be appropriate, but this results in very high detection limits (approximately 10,000 μg/L). Typically, a 1-L liquid sample, containing surrogate, and matrix spiking standards, is extracted in a continuous extractor first under acid conditions and then under basic conditions. Typically 30 g of a solid sample, containing surrogate, and matrix spiking standards, is extracted ultrasonically. After concentrating the extract to 1 mL it is spiked with 10 μL of an internal standard solution just prior to analysis by GC/MS. The volume injected should contain about 100 ng of base/neutral and 200 ng of acid surrogates (for a 1-μL injection). Analysis is performed by GC/MS using a capillary GC column.

INTERFERENCES Raw GC/MS data from all blanks, samples, and spikes must be evaluated for interferences. Contamination by carryover can occur whenever high-concentration and low-concentration samples are sequentially analyzed. To reduce carryover, the sample syringe must be rinsed out between samples with solvent. Whenever an unusually concentrated sample is encountered, it should be followed by the analysis of blank solvent to check for cross-contamination.

INSTRUMENTATION A GC/MS and a data system are required. The GC column used is a 30 m × 0.25 mm I.D. (or 0.32 mm I.D.) 1um film thickness silicone-coated fused silica capillary column. A continuous liquid-liquid extractor equipped with Teflon® or glass connection joints and stopcocks requiring no lubrication, a K-D concentrating apparatus, water bath, and an ultrasonic disrupter with a minimum power of 300 W and with pulsing capability are also required.

PRECISION & ACCURACY The estimated quantitation limit (EQL) of Method 8270B for determining an individual compound is approximately 1 mg/kg (wet weight) for soil or sediment samples, 1–200 mg/kg for wastes (dependent on matrix and method of preparation), and 10 μg/L for groundwater samples. EQLs will be proportionately higher for sample extracts that require dilution to avoid saturation of the detector.

The EQL(b) for groundwater in μg/L is 10.
The EQL (a, b) for low concentrations in soil and sediment in μg/kg is not determined.
Accuracy as μg/L is not listed.
Overall precision in μg/L is not listed.

(a) *EQLs listed for soil/sediment are based on wet weight. Normally data is reported in a dry-weight basis; therefore, EQLs will be higher based on the % dry weight of each sample. This calculation is based on a 30 g sample and gel permeation chromatography cleanup.*
(b) *Sample EQLs are highly matrix-dependent. The EQLs are provided for guidance and may not always be achievable.*
C = *True value for concentration, in μg/L.*
X = *Average recovery found for measurements of samples containing a concentration of C, in μg/L.*

ESTIMATED QUANTITATION LIMIT

Other Matrices	Factor (a)
High-concentration soil and sludges by sonicator	7.5
Non-water miscible waste	75

(a) *EQL for other matrices = [EQL for low soil/sediment] × [Factor]. This estimated EQL is similar to an EPA "Practical Quantitation Limit."*

SAMPLING METHOD

Liquid samples — Use a 1 or 2½ gallon amber glass bottle with a screw-top Teflon®-lined cover that has been prewashed with

detergent and rinsed with distilled water and methanol (or isopropanol).

Soils, sediments, or sludges — Use an 8-oz. widemouth glass with a screw-top Teflon®-lined cover that has been prewashed with detergent and rinsed with distilled water and methanol (or isopropanol).

SAMPLE PRESERVATION
Liquid samples — If residual chlorine is present, add 3 mL of 10% sodium thiosulfate per gallon, cool to 4°C and store in a solvent-free refrigerator until analysis; if chlorine is not present, then eliminate the sodium thiosulfate addition.

Soils, sediments, or sludges — Cool samples to 4°C and store in a solvent-free refrigerator.

MHT Liquid samples must be extracted within 7 days and the extracts analyzed within 40 days. Soils, sediments, or sludges may be stored for a maximum of 14 days and the extracts analyzed within 40 days.

SAMPLE PREPARATION
Liquid samples — Transfer 1 L quantitatively to a continuous extractor. If high concentrations are anticipated, a smaller volume may be used and then diluted with organic-free reagent water to 1 L. Adjust pH, if necessary, to pH <2 using 1:1 (V/V) sulfuric acid. Pipette 1.0 mL of a surrogate standard spiking solution into each sample. For the sample in each analytical batch selected for spiking, add 1.0 mL of a matrix spiking standard. For base/neutral acid analysis, the amount of the surrogates and matrix spiking compounds added to the sample should result in a final concentration of 100 ng/µL of each analyte in the extract to be analyzed (assuming a 1-µL injection). Extract with methylene chloride for 18–24 h. Next, adjust the pH of the aqueous phase to pH >11 using 10 N sodium hydroxide and extract it with methylene chloride again for 18–24 h. Dry the extract through a column containing anhydrous sodium sulfate and concentrate it to 1 mL using a K-D concentrator.

Soils, sediments, or sludges — Use 30 g of sample. Nonporous or wet samples (gummy or clay type) that do not have a free-flowing sandy texture must be mixed with anhydrous sodium sulfate until the sample is free flowing. Add 1 mL of surrogate standards to all samples, spikes, standards, and blanks. For the sample in each analytical batch selected for spiking, add 1.0 mL of a matrix spiking standard. For base/neutral acid analysis, the amount added of the surrogates and matrix spiking compounds should result in a final concentration of 100 ng/µL of each base/neutral analyte and 200 ng/µL of each acid analyte in the extract to be analyzed (assuming a 1-µL injection). Immediately add a 100-mL mixture of 1:1 methylene chloride:acetone and extract the sample ultrasonically for 3 min and then decant or filter the extracts. Repeat the extraction two or more times. Dry the extract using a column with anhydrous sodium sulfate and concentrate it to 1 mL in a K-D concentrator.

QUALITY CONTROL A methylene chloride solution containing 50 ng/µL of decafluorotriphenylphosphine (DFTPP) is used for tuning the GC/MS system each 12-h shift. A system performance check also must be made during every 12-h shift. A standard containing 50 ng/µL each of 4,4′-DDT, pentachlorophenol, and benzidine is required to verify injection port inertness and GC column performance. A calibration standard at mid-concentration, containing each compound of interest, including all required surrogates, must be performed every 12 h during analysis. After the system performance check is met, calibration check compounds (CCCs) are used to check the validity of the initial calibration.

The internal standard responses and retention times in the calibration check standard must be evaluated immediately after or during data acquisition. If the retention time for any internal standard changes by more than 30 seconds from the last check calibration (12 h), the chromatographic system must be inspected for malfunctions and corrections must be made, as required. If the electron ionization current plot (EICP) area for any of the internal standards changes by a factor of two from the last daily calibration standard check, the mass spectrometer must be inspected for malfunctions and corrections must be made, as appropriate.

Demonstrate, through the analysis of a reagent water blank, that interferences from the analytical system, glassware, and reagents are under control. The blank samples should be carried through all stages of the sample preparation and measurement steps. For each analytical batch (up to 20 samples), a reagent blank, matrix spike, and matrix spike duplicate/duplicate must be analyzed (the frequency of the spikes may be different for different monitoring programs). The blank and spiked samples must be carried through all stages of the sample preparation and measurement steps. A QC reference sample concentrate containing each analyte at a concentration of 100 mg/L in methanol is required.

REFERENCE Test Methods for Evaluating Solid Waste (SW-846). U.S. EPA 1983, Method 8270B, Rev. 2, Nov. 1990. Office of Solid Waste, Washington, DC.

2-Nitroaniline   **EPA Method 1625**
CAS #88-74-4

TITLE Semivolatile Organic Compounds by Isotope Dilution GC/MS

MATRIX The compounds may be determined in waters, soils, and municipal sludges by this method.

METHOD SUMMARY This method is used to determine 176 semivolatile toxic organic pollutants associated with the CWA (as amended 1987); the RCRA (as amended 1986); the CERCLA (as amended 1986); and other compounds amenable to extraction and analysis by capillary column gas chromatography-mass spectrometry (GC/MS).

Stable isotopically-labeled analogs of the compounds of interest are added to the sample. If the solids content is less than 1%, a 1-L sample is extracted at pH 12–13, then at pH <2 with methylene chloride using continuous extraction techniques.

If the solids content is 30% or less, the sample is diluted to 1% solids with reagent water, homogenized ultrasonically, and extracted at pH 12–13, then at pH <2 with methylene chloride using continuous extraction techniques. If the solids content is greater than 30%, the sample is extracted using ultrasonic techniques.

Each extract is dried over sodium sulfate, concentrated to a volume of 5 mL, cleaned up using GPC, if necessary, and concentrated. Extracts are concentrated to 1 mL if GPC is not performed, and to 0.5 mL if GPC is performed.

An internal standard is added to the extract, and a 1-mL aliquot of the extract is injected into the GC. The compounds are separated by GC and detected by a MS. The labeled compounds serve to correct the variability of the analytical technique.

INTERFERENCES Solvents, reagents, glassware, and other sample processing hardware may yield artifacts and/or elevated baselines causing misinterpretation of chromatograms and spectra. Materials used in the analysis must be demonstrated to be free from interferences under the conditions of analysis by running method blanks initially and with each sample lot (sample started through the extraction process on a given 8-h shift, to a maximum of 20). Specific selection of reagents and purification of solvents by distillation in all glass systems may be required. Glassware and, where possible, reagents are cleaned by solvent rinse and baking at 450°C for 1-h minimum. Interferences coextracted from samples will vary considerably from source to source, depending on the diversity of the site being sampled.

INSTRUMENTATION Major instrumentation includes a GC with a splitless or on-column injection port for capillary column, a MS with 70 eV electron impact ionization, and a data system to collect and record MS data, and process it. A K-D apparatus is used to concentrate extracts.

GC Column: 30 m × 0.25 mm I.D. 5% phenyl, 94% methyl, 1% vinyl silicone bonded phased fused silica capillary column.

PRECISION & ACCURACY The detection limits of the method are usually dependent on the level of interferences rather than instrumental limitations. The limits typify the minimum quantities that can be detected with no interferences present.

The minimum level (in µg/mL) was not listed. This is defined as a minimum level at which the analytical system shall give recognizable mass spectra (background corrected) and acceptable calibration points.

The MDL (in µg/kg) in low solids was not listed and in high solids was not listed; these were determined in digested sludge (low solids) and in filter cake or compost (high solids).

The labeled and native compound initial precision as standard deviation (in µg/L) was not listed.
The labeled and native compound initial accuracy as average recovery (in µg/L) was not listed.

SAMPLE COLLECTION, PRESERVATION & HANDLING
Collect samples in glass containers. Aqueous samples which flow freely are collected in refrigerated bottles using automatic sampling equipment. Solid samples are collected as grab samples using widemouth jars. Maintain samples at 0 to 4°C from the time of collection until extraction. If residual chlorine is present in aqueous samples, add 80 mg sodium thiosulfate/L of water. Begin sample extraction within 7 days of collection, and analyze all extracts within 40 days of extraction.

SAMPLE PREPARATION Samples containing 1% solids or less are extracted directly using continuous liquid-liquid extraction techniques. Samples containing 1 to 30% solids are diluted to the 1% level with reagent water and extracted using continuous liquid-liquid extraction techniques. Samples containing greater than 30% solids are extracted using ultrasonic techniques.

Base/neutral extraction — Adjust the pH of the waters in the extractors to 12–13 with 6 N NaOH. Extract with methylene chloride for 24–48 h.
Acid extraction — Adjust the pH of the waters in the extractors to 2 or less using 6 N sulfuric acid. Extract with methylene chloride for 24–48 h.
Ultrasonic extraction of high solids samples — Add anhydrous sodium sulfate to the sample and QC aliquot(s). Add acetone:methylene chloride (1:1) to the sample and mix thoroughly

Concentrate extracts using a K-D apparatus.

QUALITY CONTROL The analyst is permitted to modify this method to improve separations or lower the costs of measurements, provided all performance specifications are met. Analyses of blanks are required to demonstrate freedom from contamination. When results of spikes indicate atypical method performance for samples, the samples are diluted to bring method performance within acceptable limits.

For low solids (aqueous samples), extract, concentrate, and analyze two sets of four 1-L aliquots (8 aliquots total) of the precision and recovery standard. For high solids samples, two sets of four 30-g aliquots of the high solids reference matrix are used.

Spike all samples with labeled compounds to assess method performance. Compute percent recovery of the labeled compounds using the internal standard method. Compare the labeled compound recovery for each compound with the corresponding labeled compound recovery.

Reagent water and high solids reference matrix blanks are analyzed to demonstrate freedom from contamination. Extract and concentrate a 1-L reagent water blank or a high solids reference matrix blank with each sample's lot (samples started through the extraction process on the same 8-h shift, to a maximum of 20 samples).

Field replicates may be collected to determine the precision of the sampling technique, and spiked samples may be required to determine the accuracy of the analysis when the internal standard method is used.

REFERENCE Semivolatile Organic Compounds by Isotope Dilution GC/MS. Office of Water Regulation and Standards, U.S. EPA Industrial Technology Division, Washington, DC, EPA Method 1625, Rev. C, June 1989 (contact W.A. Telliard, U.S. EPA, Office of Water Regulations and Standards, 401 M St., SW, Washington, DC, 20460. Phone: 202-382-7131).

3-Nitroaniline
CAS #99-09-2

EPA Method 1625

TITLE Semivolatile Organic Compounds by Isotope Dilution GC/MS

MATRIX The compounds may be determined in waters, soils, and municipal sludges by this method.

METHOD SUMMARY This method is used to determine 176 semivolatile toxic organic pollutants associated with the CWA (as amended 1987); the RCRA (as amended 1986); the CERCLA (as amended 1986); and other compounds amenable to extraction and analysis by capillary column gas chromatography-mass spectrometry (GC/MS).

Stable isotopically-labeled analogs of the compounds of interest are added to the sample. If the solids content is less than 1%, a 1-L sample is extracted at pH 12–13, then at pH <2 with methylene chloride using continuous extraction techniques.

If the solids content is 30% or less, the sample is diluted to 1% solids with reagent water, homogenized ultrasonically, and extracted at pH 12–13, then at pH <2 with methylene chloride using continuous extraction techniques. If the solids content is greater than 30%, the sample is extracted using ultrasonic techniques.

Each extract is dried over sodium sulfate, concentrated to a volume of 5 mL, cleaned up using GPC, if necessary, and concentrated. Extracts are concentrated to 1 mL if GPC is not performed, and to 0.5 mL if GPC is performed.

An internal standard is added to the extract, and a 1-mL aliquot of the extract is injected into the GC. The compounds are separated by GC and detected by a MS. The labeled compounds serve to correct the variability of the analytical technique.

INTERFERENCES Solvents, reagents, glassware, and other sample processing hardware may yield artifacts and/or elevated baselines causing misinterpretation of chromatograms and spectra. Materials used in the analysis must be demonstrated to be free from interferences under the conditions of analysis by running method blanks initially and with each sample lot (sample started through the extraction process on a given 8-h shift, to a maximum of 20). Specific selection of reagents and purification of solvents by distillation in all glass systems may be required. Glassware and, where possible, reagents are cleaned by solvent rinse and baking at 450°C for 1-h minimum. Interferences coextracted from samples will vary considerably from source to source, depending on the diversity of the site being sampled.

INSTRUMENTATION Major instrumentation includes a GC with a splitless or on-column injection port for capillary column, a MS with 70 eV electron impact ionization, and a data system to collect and record MS data, and process it. A K-D apparatus is used to concentrate extracts.

GC Column: 30 m × 0.25 mm I.D. 5% phenyl, 94% methyl, 1% vinyl silicone bonded phased fused silica capillary column.

PRECISION & ACCURACY The detection limits of the method are usually dependent on the level of interferences rather than instrumental limitations. The limits typify the minimum quantities that can be detected with no interferences present.

The minimum level (in µg/mL) was not listed. This is defined as a minimum level at which the analytical system shall give recognizable mass spectra (background corrected) and acceptable calibration points.

The MDL (in µg/kg) in low solids was not listed and in high solids was not listed; these were determined in digested sludge (low solids) and in filter cake or compost (high solids).

The labeled and native compound initial precision as standard deviation (in µg/L) was not listed.
The labeled and native compound initial accuracy as average recovery (in µg/L) was not listed.

SAMPLE COLLECTION, PRESERVATION & HANDLING
Collect samples in glass containers. Aqueous samples which flow freely are collected in refrigerated bottles using automatic sampling equipment. Solid samples are collected as grab samples using widemouth jars. Maintain samples at 0 to 4°C from the time of collection until extraction. If residual chlorine is present in aqueous samples, add 80 mg sodium thiosulfate/L of water. Begin sample extraction within 7 days of collection, and analyze all extracts within 40 days of extraction.

SAMPLE PREPARATION Samples containing 1% solids or less are extracted directly using continuous liquid-liquid extraction techniques. Samples containing 1 to 30% solids are diluted to the 1% level with reagent water and extracted using continuous liquid-liquid extraction techniques. Samples containing greater than 30% solids are extracted using ultrasonic techniques.

Base/neutral extraction — Adjust the pH of the waters in the extractors to 12–13 with 6 N NaOH. Extract with methylene chloride for 24–48 h.
Acid extraction — Adjust the pH of the waters in the extractors to 2 or less using 6 N sulfuric acid. Extract with methylene chloride for 24–48 h.
Ultrasonic extraction of high solids samples — Add anhydrous sodium sulfate to the sample and QC aliquot(s). Add acetone:methylene chloride (1:1) to the sample and mix thoroughly

Concentrate extracts using a K-D apparatus.

QUALITY CONTROL The analyst is permitted to modify this method to improve separations or lower the costs of measurements, provided all performance specifications are met. Analyses of blanks are required to demonstrate freedom from contamination. When results of spikes indicate atypical method performance for samples, the samples are diluted to bring method performance within acceptable limits.

For low solids (aqueous samples), extract, concentrate, and analyze two sets of four 1-L aliquots (8 aliquots total) of the precision and recovery standard. For high solids samples, two sets of four 30-g aliquots of the high solids reference matrix are used.

Spike all samples with labeled compounds to assess method performance. Compute percent recovery of the labeled compounds using the internal standard method. Compare the labeled compound recovery for each compound with the corresponding labeled compound recovery.

Reagent water and high solids reference matrix blanks are analyzed to demonstrate freedom from contamination. Extract and concentrate a 1-L reagent water blank or a high solids reference matrix blank with each sample's lot (samples started through the extraction process on the same 8-h shift, to a maximum of 20 samples).

Field replicates may be collected to determine the precision of the sampling technique, and spiked samples may be required to determine the accuracy of the analysis when the internal standard method is used.

REFERENCE Semivolatile Organic Compounds by Isotope Dilution GC/MS. Office of Water Regulation and Standards, U.S. EPA Industrial Technology Division, Washington, DC, EPA Method 1625, Rev. C, June 1989 (contact W.A. Telliard, U.S. EPA, Office of Water Regulations and Standards, 401 M St., SW, Washington, DC, 20460. Phone: 202-382-7131).

4-Nitroaniline EPA Method 1625
CAS #100-01-6

TITLE Semivolatile Organic Compounds by Isotope Dilution GC/MS

MATRIX The compounds may be determined in waters, soils, and municipal sludges by this method.

METHOD SUMMARY This method is used to determine 176 semivolatile toxic organic pollutants associated with the CWA (as amended 1987); the RCRA (as amended 1986); the CERCLA (as amended 1986); and other compounds amenable to extraction and analysis by capillary column gas chromatography-mass spectrometry (GC/MS).

Stable isotopically-labeled analogs of the compounds of interest are added to the sample. If the solids content is less than 1%, a 1-L sample is extracted at pH 12–13, then at pH <2 with methylene chloride using continuous extraction techniques.

If the solids content is 30% or less, the sample is diluted to 1% solids with reagent water, homogenized ultrasonically, and extracted at pH 12–13, then at pH <2 with methylene chloride using continuous extraction techniques. If the solids content is greater than 30%, the sample is extracted using ultrasonic techniques.

Each extract is dried over sodium sulfate, concentrated to a volume of 5 mL, cleaned up using GPC, if necessary, and concentrated. Extracts are concentrated to 1 mL if GPC is not performed, and to 0.5 mL if GPC is performed.

An internal standard is added to the extract, and a 1-mL aliquot of the extract is injected into the GC. The compounds are separated by GC and detected by a MS. The labeled compounds serve to correct the variability of the analytical technique.

INTERFERENCES Solvents, reagents, glassware, and other sample processing hardware may yield artifacts and/or elevated baselines causing misinterpretation of chromatograms and spectra. Materials used in the analysis must be demonstrated to be free from interferences under the conditions of analysis by running method blanks initially and with each sample lot (sample started through the extraction process on a given 8-h shift, to a maximum of 20). Specific selection of reagents and purification of solvents by distillation in all glass systems may be required. Glassware and, where possible, reagents are cleaned by solvent rinse and baking at 450°C for 1-h minimum. Interferences coextracted from samples will vary considerably from source to source, depending on the diversity of the site being sampled.

INSTRUMENTATION Major instrumentation includes a GC with a splitless or on-column injection port for capillary column, a MS with 70 eV electron impact ionization, and a data system to collect and record MS data, and process it. A K-D apparatus is used to concentrate extracts.

GC Column: 30 m × 0.25 mm I.D. 5% phenyl, 94% methyl, 1% vinyl silicone bonded phased fused silica capillary column.

PRECISION & ACCURACY The detection limits of the method are usually dependent on the level of interferences rather than instrumental limitations. The limits typify the minimum quantities that can be detected with no interferences present.

The minimum level (in µg/mL) was not listed. This is defined as a minimum level at which the analytical system shall give recognizable mass spectra (background corrected) and acceptable calibration points.

The MDL (in µg/kg) in low solids was not listed and in high solids was not listed; these were determined in digested sludge (low solids) and in filter cake or compost (high solids).

The labeled and native compound initial precision as standard deviation (in µg/L) was not listed.
The labeled and native compound initial accuracy as average recovery (in µg/L) was not listed.

SAMPLE COLLECTION, PRESERVATION & HANDLING Collect samples in glass containers. Aqueous samples which flow freely are collected in refrigerated bottles using automatic sampling equipment. Solid samples are collected as grab samples using widemouth jars. Maintain samples at 0 to 4°C from the time of collection until extraction. If residual chlorine is present in aqueous samples, add 80 mg sodium thiosulfate/L of water. Begin sample extraction within 7 days of collection, and analyze all extracts within 40 days of extraction.

SAMPLE PREPARATION Samples containing 1% solids or less are extracted directly using continuous liquid-liquid extraction techniques. Samples containing 1 to 30% solids are diluted to the 1% level with reagent water and extracted using continuous liquid-liquid extraction techniques. Samples containing greater than 30% solids are extracted using ultrasonic techniques.

Base/neutral extraction — Adjust the pH of the waters in the extractors to 12–13 with 6 N NaOH. Extract with methylene chloride for 24–48 h.

Acid extraction — Adjust the pH of the waters in the extractors to 2 or less using 6 N sulfuric acid. Extract with methylene chloride for 24–48 h.

Ultrasonic extraction of high solids samples — Add anhydrous sodium sulfate to the sample and QC aliquot(s). Add acetone:methylene chloride (1:1) to the sample and mix thoroughly

Concentrate extracts using a K-D apparatus.

QUALITY CONTROL The analyst is permitted to modify this method to improve separations or lower the costs of measurements, provided all performance specifications are met. Analyses of blanks are required to demonstrate freedom from contamination. When results of spikes indicate atypical method performance for samples, the samples are diluted to bring method performance within acceptable limits.

For low solids (aqueous samples), extract, concentrate, and analyze two sets of four 1-L aliquots (8 aliquots total) of the precision and recovery standard. For high solids samples, two sets of four 30-g aliquots of the high solids reference matrix are used.

Spike all samples with labeled compounds to assess method performance. Compute percent recovery of the labeled compounds using the internal standard method. Compare the labeled compound recovery for each compound with the corresponding labeled compound recovery.

Reagent water and high solids reference matrix blanks are analyzed to demonstrate freedom from contamination. Extract and concentrate a 1-L reagent water blank or a high solids reference matrix blank with each sample's lot (samples started through the extraction process on the same 8-h shift, to a maximum of 20 samples).

Field replicates may be collected to determine the precision of the sampling technique, and spiked samples may be required to determine the accuracy of the analysis when the internal standard method is used.

REFERENCE Semivolatile Organic Compounds by Isotope Dilution GC/MS. Office of Water Regulation and Standards, U.S. EPA Industrial Technology Division, Washington, DC, EPA Method 1625, Rev. C, June 1989 (contact W.A. Telliard, U.S. EPA, Office of Water Regulations and Standards, 401 M St., SW, Washington, DC, 20460. Phone: 202-382-7131).

2-Nitroaniline — EPA Method 8270
CAS #88-74-4

TITLE Semivolatile Organic Compounds by GC/MS

MATRIX This method is used to determine the concentration of semivolatile organic compounds in extracts prepared from all types of solid waste matrices, soils, and groundwater. Although surface waters are not specifically mentioned, this method should be applicable to water samples from rivers, lakes, etc.

METHOD SUMMARY This method covers 259 semivolatile organic compounds. In very limited applications direct injection of the sample into the GC/MS system may be appropriate, but this results in very high detection limits (approximately 10,000 µg/L). Typically, a 1-L liquid sample, containing surrogate, and matrix spiking standards, is extracted in a continuous extractor first under acid conditions and then under basic conditions. Typically 30 g of a solid sample, containing surrogate, and matrix spiking standards, is extracted ultrasonically. After concentrating the extract to 1 mL it is spiked with 10 µL of an internal standard solution just prior to analysis by GC/MS. The volume injected should contain about 100 ng of base/neutral and 200 ng of acid surrogates (for a 1-µL injection). Analysis is performed by GC/MS using a capillary GC column.

INTERFERENCES Raw GC/MS data from all blanks, samples, and spikes must be evaluated for interferences. Contamination by carryover can occur whenever high-concentration and low-concentration samples are sequentially analyzed. To reduce carryover, the sample syringe must be rinsed out between samples with solvent. Whenever an unusually concentrated sample is encountered, it should be followed by the analysis of blank solvent to check for cross-contamination.

INSTRUMENTATION A GC/MS and a data system are required. The GC column used is a 30 m × 0.25 mm I.D. (or 0.32 mm I.D.) 1um film thickness silicone-coated fused silica capillary column. A continuous liquid-liquid extractor equipped with Teflon® or glass connection joints and stopcocks requiring no lubrication, a K-D concentrating apparatus, water bath, and an ultrasonic disrupter with a minimum power of 300 W and with pulsing capability are also required.

PRECISION & ACCURACY The estimated quantitation limit (EQL) of Method 8270B for determining an individual compound is approximately 1 mg/kg (wet weight) for soil or sediment samples, 1–200 mg/kg for wastes (dependent on matrix and method of preparation), and 10 µg/L for groundwater samples. EQLs will be proportionately higher for sample extracts that require dilution to avoid saturation of the detector.

The EQL(b) for groundwater in µg/L is 50.
The EQL (a, b) for low concentrations in soil and sediment in µg/kg is 3300.
Accuracy as µg/L is not listed.
Overall precision in µg/L is not listed.

(a) *EQLs listed for soil/sediment are based on wet weight. Normally data is reported in a dry-weight basis; therefore, EQLs will be higher based on the % dry weight of each sample. This calculation is based on a 30 g sample and gel permeation chromatography cleanup.*
(b) *Sample EQLs are highly matrix-dependent. The EQLs are provided for guidance and may not always be achievable.*
C = *True value for concentration, in µg/L.*
X = *Average recovery found for measurements of samples containing a concentration of C, in µg/L.*

ESTIMATED QUANTITATION LIMIT

Other Matrices	Factor (a)
High-concentration soil and sludges by sonicator	7.5
Non-water miscible waste	75

(a) EQL for other matrices = [EQL for low soil/sediment] × [Factor]. This estimated EQL is similar to an EPA "Practical Quantitation Limit."

SAMPLING METHOD

Liquid samples — Use a 1 or 2½ gallon amber glass bottle with a screw-top Teflon®-lined cover that has been prewashed with detergent and rinsed with distilled water and methanol (or isopropanol).

Soils, sediments, or sludges — Use an 8-oz. widemouth glass with a screw-top Teflon®-lined cover that has been prewashed with detergent and rinsed with distilled water and methanol (or isopropanol).

SAMPLE PRESERVATION

Liquid samples — If residual chlorine is present, add 3 mL of 10% sodium thiosulfate per gallon, cool to 4°C and store in a solvent-free refrigerator until analysis; if chlorine is not present, then eliminate the sodium thiosulfate addition.

Soils, sediments, or sludges — Cool samples to 4°C and store in a solvent-free refrigerator.

MHT Liquid samples must be extracted within 7 days and the extracts analyzed within 40 days. Soils, sediments, or sludges may be stored for a maximum of 14 days and the extracts analyzed within 40 days.

SAMPLE PREPARATION

Liquid samples — Transfer 1 L quantitatively to a continuous extractor. If high concentrations are anticipated, a smaller volume may be used and then diluted with organic-free reagent water to 1 L. Adjust pH, if necessary, to pH <2 using 1:1 (V/V) sulfuric acid. Pipette 1.0 mL of a surrogate standard spiking solution into each sample. For the sample in each analytical batch selected for spiking, add 1.0 mL of a matrix spiking standard. For base/neutral acid analysis, the amount of the surrogates and matrix spiking compounds added to the sample should result in a final concentration of 100 ng/µL of each analyte in the extract to be analyzed (assuming a 1-µL injection). Extract with methylene chloride for 18–24 h. Next, adjust the pH of the aqueous phase to pH >11 using 10 N sodium hydroxide and extract it with methylene chloride again for 18–24 h. Dry the extract through a column containing anhydrous sodium sulfate and concentrate it to 1 mL using a K-D concentrator.

Soils, sediments, or sludges — Use 30 g of sample. Nonporous or wet samples (gummy or clay type) that do not have a free-flowing sandy texture must be mixed with anhydrous sodium sulfate until the sample is free flowing. Add 1 mL of surrogate standards to all samples, spikes, standards, and blanks. For the sample in each analytical batch selected for spiking, add 1.0 mL of a matrix spiking standard. For base/neutral acid analysis, the amount added of the surrogates and matrix spiking compounds should result in a final concentration of 100 ng/µL of each base/neutral analyte and 200 ng/µL of each acid analyte in the extract to be analyzed (assuming a 1-µL injection). Immediately add a 100-mL mixture of 1:1 methylene chloride:acetone and extract the sample ultrasonically for 3 min and then decant or filter the extracts. Repeat the extraction two or more times. Dry the extract using a column with anhydrous sodium sulfate and concentrate it to 1 mL in a K-D concentrator.

Note: This compound may be exhibit erratic chromatographic behavior, especially if the GC system is contaminated with high boiling material.

QUALITY CONTROL A methylene chloride solution containing 50 ng/µL of decafluorotriphenylphosphine (DFTPP) is used for tuning the GC/MS system each 12-h shift. A system performance check also must be made during every 12-h shift. A standard containing 50 ng/µL each of 4,4′-DDT, pentachlorophenol, and benzidine is required to verify injection port inertness and GC column performance. A calibration standard at mid-concentration, containing each compound of interest, including all required surrogates, must be performed every 12 h during analysis. After the system performance check is met, calibration check compounds (CCCs) are used to check the validity of the initial calibration.

The internal standard responses and retention times in the calibration check standard must be evaluated immediately after or during data acquisition. If the retention time for any internal standard changes by more than 30 seconds from the last check calibration (12 h), the chromatographic system must be inspected for malfunctions and corrections must be made, as required. If the electron ionization current plot (EICP) area for any of the internal standards changes by a factor of two from the last daily calibration standard check, the mass spectrometer must be inspected for malfunctions and corrections must be made, as appropriate.

Demonstrate, through the analysis of a reagent water blank, that interferences from the analytical system, glassware, and reagents are under control. The blank samples should be carried through all stages of the sample preparation and measurement steps. For each analytical batch (up to 20 samples), a reagent blank, matrix spike, and matrix spike duplicate/duplicate must be analyzed (the frequency of the spikes may be different for different monitoring programs). The blank and spiked samples must be carried through all stages of the sample preparation and measurement steps. A QC reference sample concentrate containing each analyte at a concentration of 100 mg/L in methanol is required.

REFERENCE Test Methods for Evaluating Solid Waste (SW-846). U.S. EPA 1983, Method 8270B, Rev. 2, Nov. 1990. Office of Solid Waste, Washington, DC.

3-Nitroaniline   **EPA Method 8270**
CAS #99-09-2

TITLE Semivolatile Organic Compounds by GC/MS

MATRIX This method is used to determine the concentration of semivolatile organic compounds in extracts prepared from all types of solid waste matrices, soils, and groundwater.

Although surface waters are not specifically mentioned, this method should be applicable to water samples from rivers, lakes, etc.

METHOD SUMMARY This method covers 259 semivolatile organic compounds. In very limited applications direct injection of the sample into the GC/MS system may be appropriate, but this results in very high detection limits (approximately 10,000 µg/L). Typically, a 1-L liquid sample, containing surrogate, and matrix spiking standards, is extracted in a continuous extractor first under acid conditions and then under basic conditions. Typically 30 g of a solid sample, containing surrogate, and matrix spiking standards, is extracted ultrasonically. After concentrating the extract to 1 mL it is spiked with 10 µL of an internal standard solution just prior to analysis by GC/MS. The volume injected should contain about 100 ng of base/neutral and 200 ng of acid surrogates (for a 1-µL injection). Analysis is performed by GC/MS using a capillary GC column.

INTERFERENCES Raw GC/MS data from all blanks, samples, and spikes must be evaluated for interferences. Contamination by carryover can occur whenever high-concentration and low-concentration samples are sequentially analyzed. To reduce carryover, the sample syringe must be rinsed out between samples with solvent. Whenever an unusually concentrated sample is encountered, it should be followed by the analysis of blank solvent to check for cross-contamination.

INSTRUMENTATION A GC/MS and a data system are required. The GC column used is a 30 m × 0.25 mm I.D. (or 0.32 mm I.D.) 1um film thickness silicone-coated fused silica capillary column. A continuous liquid-liquid extractor equipped with Teflon® or glass connection joints and stopcocks requiring no lubrication, a K-D concentrating apparatus, water bath, and an ultrasonic disrupter with a minimum power of 300 W and with pulsing capability are also required.

PRECISION & ACCURACY The estimated quantitation limit (EQL) of Method 8270B for determining an individual compound is approximately 1 mg/kg (wet weight) for soil or sediment samples, 1–200 mg/kg for wastes (dependent on matrix and method of preparation), and 10 µg/L for groundwater samples. EQLs will be proportionately higher for sample extracts that require dilution to avoid saturation of the detector.

The EQL(b) for groundwater in µg/L is 50.
The EQL (a, b) for low concentrations in soil and sediment in µg/kg is 3300.
Accuracy as µg/L is not listed.
Overall precision in µg/L is not listed.

(a) EQLs listed for soil/sediment are based on wet weight. Normally data is reported in a dry-weight basis; therefore, EQLs will be higher based on the % dry weight of each sample. This calculation is based on a 30 g sample and gel permeation chromatography cleanup.
(b) Sample EQLs are highly matrix-dependent. The EQLs are provided for guidance and may not always be achievable.
C = True value for concentration, in µg/L.
X = Average recovery found for measurements of samples containing a concentration of C, in µg/L.

ESTIMATED QUANTITATION LIMIT

Other Matrices	Factor (a)
High-concentration soil and sludges by sonicator	7.5
Non-water miscible waste	75

(a) EQL for other matrices = [EQL for low soil/sediment] × [Factor]. This estimated EQL is similar to an EPA "Practical Quantitation Limit."

SAMPLING METHOD
Liquid samples — Use a 1 or 2½ gallon amber glass bottle with a screw-top Teflon®-lined cover that has been prewashed with detergent and rinsed with distilled water and methanol (or isopropanol).

Soils, sediments, or sludges — Use an 8-oz. widemouth glass with a screw-top Teflon®-lined cover that has been prewashed with detergent and rinsed with distilled water and methanol (or isopropanol).

SAMPLE PRESERVATION
Liquid samples — If residual chlorine is present, add 3 mL of 10% sodium thiosulfate per gallon, cool to 4°C and store in a solvent-free refrigerator until analysis; if chlorine is not present, then eliminate the sodium thiosulfate addition.

Soils, sediments, or sludges — Cool samples to 4°C and store in a solvent-free refrigerator.

MHT Liquid samples must be extracted within 7 days and the extracts analyzed within 40 days. Soils, sediments, or sludges may be stored for a maximum of 14 days and the extracts analyzed within 40 days.

SAMPLE PREPARATION
Liquid samples — Transfer 1 L quantitatively to a continuous extractor. If high concentrations are anticipated, a smaller volume may be used and then diluted with organic-free reagent water to 1 L. Adjust pH, if necessary, to pH <2 using 1:1 (V/V) sulfuric acid. Pipette 1.0 mL of a surrogate standard spiking solution into each sample. For the sample in each analytical batch selected for spiking, add 1.0 mL of a matrix spiking standard. For base/neutral acid analysis, the amount of the surrogates and matrix spiking compounds added to the sample should result in a final concentration of 100 ng/µL of each analyte in the extract to be analyzed (assuming a 1-µL injection). Extract with methylene chloride for 18–24 h. Next, adjust the pH of the aqueous phase to pH >11 using 10 N sodium hydroxide and extract it with methylene chloride again for 18–24 h. Dry the extract through a column containing anhydrous sodium sulfate and concentrate it to 1 mL using a K-D concentrator.

Soils, sediments, or sludges — Use 30 g of sample. Nonporous or wet samples (gummy or clay type) that do not have a free-flowing sandy texture must be mixed with anhydrous sodium sulfate until the sample is free flowing. Add 1 mL of surrogate standards to all samples, spikes, standards, and blanks. For the sample in each analytical batch selected for spiking, add 1.0 mL of a matrix spiking standard. For base/neutral acid analysis, the amount added of the surrogates and matrix spiking compounds should result in a final concentration of 100 ng/µL of each base/neutral analyte and 200 ng/µL of each acid analyte

in the extract to be analyzed (assuming a 1-μL injection). Immediately add a 100-mL mixture of 1:1 methylene chloride:acetone and extract the sample ultrasonically for 3 min and then decant or filter the extracts. Repeat the extraction two or more times. Dry the extract using a column with anhydrous sodium sulfate and concentrate it to 1 mL in a K-D concentrator.

Note: This compound may be exhibit erratic chromatographic behavior, especially if the GC system is contaminated with high boiling material.

QUALITY CONTROL A methylene chloride solution containing 50 ng/μL of decafluorotriphenylphosphine (DFTPP) is used for tuning the GC/MS system each 12-h shift. A system performance check also must be made during every 12-h shift. A standard containing 50 ng/μL each of 4,4'-DDT, pentachlorophenol, and benzidine is required to verify injection port inertness and GC column performance. A calibration standard at mid-concentration, containing each compound of interest, including all required surrogates, must be performed every 12 h during analysis. After the system performance check is met, calibration check compounds (CCCs) are used to check the validity of the initial calibration.

The internal standard responses and retention times in the calibration check standard must be evaluated immediately after or during data acquisition. If the retention time for any internal standard changes by more than 30 seconds from the last check calibration (12 h), the chromatographic system must be inspected for malfunctions and corrections must be made, as required. If the electron ionization current plot (EICP) area for any of the internal standards changes by a factor of two from the last daily calibration standard check, the mass spectrometer must be inspected for malfunctions and corrections must be made, as appropriate.

Demonstrate, through the analysis of a reagent water blank, that interferences from the analytical system, glassware, and reagents are under control. The blank samples should be carried through all stages of the sample preparation and measurement steps. For each analytical batch (up to 20 samples), a reagent blank, matrix spike, and matrix spike duplicate/duplicate must be analyzed (the frequency of the spikes may be different for different monitoring programs). The blank and spiked samples must be carried through all stages of the sample preparation and measurement steps. A QC reference sample concentrate containing each analyte at a concentration of 100 mg/L in methanol is required.

REFERENCE Test Methods for Evaluating Solid Waste (SW-846). U.S. EPA 1983, Method 8270B, Rev. 2, Nov. 1990. Office of Solid Waste, Washington, DC.

4-Nitroaniline
CAS #100-01-6
EPA Method 8270

TITLE Semivolatile Organic Compounds by GC/MS

MATRIX This method is used to determine the concentration of semivolatile organic compounds in extracts prepared from all types of solid waste matrices, soils, and groundwater. Although surface waters are not specifically mentioned, this method should be applicable to water samples from rivers, lakes, etc.

METHOD SUMMARY This method covers 259 semivolatile organic compounds. In very limited applications direct injection of the sample into the GC/MS system may be appropriate, but this results in very high detection limits (approximately 10,000 μg/L). Typically, a 1-L liquid sample, containing surrogate, and matrix spiking standards, is extracted in a continuous extractor first under acid conditions and then under basic conditions. Typically 30 g of a solid sample, containing surrogate, and matrix spiking standards, is extracted ultrasonically. After concentrating the extract to 1 mL it is spiked with 10 μL of an internal standard solution just prior to analysis by GC/MS. The volume injected should contain about 100 ng of base/neutral and 200 ng of acid surrogates (for a 1-μL injection). Analysis is performed by GC/MS using a capillary GC column.

INTERFERENCES Raw GC/MS data from all blanks, samples, and spikes must be evaluated for interferences. Contamination by carryover can occur whenever high-concentration and low-concentration samples are sequentially analyzed. To reduce carryover, the sample syringe must be rinsed out between samples with solvent. Whenever an unusually concentrated sample is encountered, it should be followed by the analysis of blank solvent to check for cross-contamination.

INSTRUMENTATION A GC/MS and a data system are required. The GC column used is a 30 m × 0.25 mm I.D. (or 0.32 mm I.D.) 1um film thickness silicone-coated fused silica capillary column. A continuous liquid-liquid extractor equipped with Teflon® or glass connection joints and stopcocks requiring no lubrication, a K-D concentrating apparatus, water bath, and an ultrasonic disrupter with a minimum power of 300 W and with pulsing capability are also required.

PRECISION & ACCURACY The estimated quantitation limit (EQL) of Method 8270B for determining an individual compound is approximately 1 mg/kg (wet weight) for soil or sediment samples, 1–200 mg/kg for wastes (dependent on matrix and method of preparation), and 10 μg/L for groundwater samples. EQLs will be proportionately higher for sample extracts that require dilution to avoid saturation of the detector.

The EQL(b) for groundwater in μg/L is 20.
The EQL (a, b) for low concentrations in soil and sediment
  in μg/kg is not determined.
Accuracy as μg/L is not listed.
Overall precision in μg/L is not listed.

(a) *EQLs listed for soil/sediment are based on wet weight. Normally data is reported in a dry-weight basis; therefore, EQLs will be higher based on the % dry weight of each sample. This calculation is based on a 30 g sample and gel permeation chromatography cleanup.*

(b) *Sample EQLs are highly matrix-dependent. The EQLs are provided for guidance and may not always be achievable.*

C = *True value for concentration, in μg/L.*
X = *Average recovery found for measurements of samples containing a concentration of C, in μg/L.*

ESTIMATED QUANTITATION LIMIT

Other Matrices	Factor (a)
High-concentration soil and sludges by sonicator	7.5
Non-water miscible waste	75

(a) EQL for other matrices = [EQL for low soil/sediment] × [Factor]. This estimated EQL is similar to an EPA "Practical Quantitation Limit."

SAMPLING METHOD

Liquid samples — Use a 1 or 2½ gallon amber glass bottle with a screw-top Teflon®-lined cover that has been prewashed with detergent and rinsed with distilled water and methanol (or isopropanol).

Soils, sediments, or sludges — Use an 8-oz. widemouth glass with a screw-top Teflon®-lined cover that has been prewashed with detergent and rinsed with distilled water and methanol (or isopropanol).

SAMPLE PRESERVATION

Liquid samples — If residual chlorine is present, add 3 mL of 10% sodium thiosulfate per gallon, cool to 4°C and store in a solvent-free refrigerator until analysis; if chlorine is not present, then eliminate the sodium thiosulfate addition.

Soils, sediments, or sludges — Cool samples to 4°C and store in a solvent-free refrigerator.

MHT Liquid samples must be extracted within 7 days and the extracts analyzed within 40 days. Soils, sediments, or sludges may be stored for a maximum of 14 days and the extracts analyzed within 40 days.

SAMPLE PREPARATION

Liquid samples — Transfer 1 L quantitatively to a continuous extractor. If high concentrations are anticipated, a smaller volume may be used and then diluted with organic-free reagent water to 1 L. Adjust pH, if necessary, to pH <2 using 1:1 (V/V) sulfuric acid. Pipette 1.0 mL of a surrogate standard spiking solution into each sample. For the sample in each analytical batch selected for spiking, add 1.0 mL of a matrix spiking standard. For base/neutral acid analysis, the amount of the surrogates and matrix spiking compounds added to the sample should result in a final concentration of 100 ng/µL of each analyte in the extract to be analyzed (assuming a 1-µL injection). Extract with methylene chloride for 18–24 h. Next, adjust the pH of the aqueous phase to pH >11 using 10 N sodium hydroxide and extract it with methylene chloride again for 18–24 h. Dry the extract through a column containing anhydrous sodium sulfate and concentrate it to 1 mL using a K-D concentrator.

Soils, sediments, or sludges — Use 30 g of sample. Nonporous or wet samples (gummy or clay type) that do not have a free-flowing sandy texture must be mixed with anhydrous sodium sulfate until the sample is free flowing. Add 1 mL of surrogate standards to all samples, spikes, standards, and blanks. For the sample in each analytical batch selected for spiking, add 1.0 mL of a matrix spiking standard. For base/neutral acid analysis, the amount added of the surrogates and matrix spiking compounds should result in a final concentration of 100 ng/µL of each base/neutral analyte and 200 ng/µL of each acid analyte in the extract to be analyzed (assuming a 1-µL injection). Immediately add a 100-mL mixture of 1:1 methylene chloride:acetone and extract the sample ultrasonically for 3 min and then decant or filter the extracts. Repeat the extraction two or more times. Dry the extract using a column with anhydrous sodium sulfate and concentrate it to 1 mL in a K-D concentrator.

QUALITY CONTROL A methylene chloride solution containing 50 ng/µL of decafluorotriphenylphosphine (DFTPP) is used for tuning the GC/MS system each 12-h shift. A system performance check also must be made during every 12-h shift. A standard containing 50 ng/µL each of 4,4'-DDT, pentachlorophenol, and benzidine is required to verify injection port inertness and GC column performance. A calibration standard at mid-concentration, containing each compound of interest, including all required surrogates, must be performed every 12 h during analysis. After the system performance check is met, calibration check compounds (CCCs) are used to check the validity of the initial calibration.

The internal standard responses and retention times in the calibration check standard must be evaluated immediately after or during data acquisition. If the retention time for any internal standard changes by more than 30 seconds from the last check calibration (12 h), the chromatographic system must be inspected for malfunctions and corrections must be made, as required. If the electron ionization current plot (EICP) area for any of the internal standards changes by a factor of two from the last daily calibration standard check, the mass spectrometer must be inspected for malfunctions and corrections must be made, as appropriate.

Demonstrate, through the analysis of a reagent water blank, that interferences from the analytical system, glassware, and reagents are under control. The blank samples should be carried through all stages of the sample preparation and measurement steps. For each analytical batch (up to 20 samples), a reagent blank, matrix spike, and matrix spike duplicate/duplicate must be analyzed (the frequency of the spikes may be different for different monitoring programs). The blank and spiked samples must be carried through all stages of the sample preparation and measurement steps. A QC reference sample concentrate containing each analyte at a concentration of 100 mg/L in methanol is required.

REFERENCE Test Methods for Evaluating Solid Waste (SW-846). U.S. EPA 1983, Method 8270B, Rev. 2, Nov. 1990. Office of Solid Waste, Washington, DC.

Nitrobenzene   **EPA Method 1625**
CAS #98-95-3

TITLE Semivolatile Organic Compounds by Isotope Dilution GC/MS

MATRIX The compounds may be determined in waters, soils, and municipal sludges by this method.

METHOD SUMMARY This method is used to determine 176 semivolatile toxic organic pollutants associated with the CWA (as amended 1987); the RCRA (as amended 1986); the

CERCLA (as amended 1986); and other compounds amenable to extraction and analysis by capillary column gas chromatography-mass spectrometry (GC/MS).

Stable isotopically-labeled analogs of the compounds of interest are added to the sample. If the solids content is less than 1%, a 1-L sample is extracted at pH 12–13, then at pH <2 with methylene chloride using continuous extraction techniques.

If the solids content is 30% or less, the sample is diluted to 1% solids with reagent water, homogenized ultrasonically, and extracted at pH 12–13, then at pH <2 with methylene chloride using continuous extraction techniques. If the solids content is greater than 30%, the sample is extracted using ultrasonic techniques.

Each extract is dried over sodium sulfate, concentrated to a volume of 5 mL, cleaned up using GPC, if necessary, and concentrated. Extracts are concentrated to 1 mL if GPC is not performed, and to 0.5 mL if GPC is performed.

An internal standard is added to the extract, and a 1-mL aliquot of the extract is injected into the GC. The compounds are separated by GC and detected by a MS. The labeled compounds serve to correct the variability of the analytical technique.

INTERFERENCES Solvents, reagents, glassware, and other sample processing hardware may yield artifacts and/or elevated baselines causing misinterpretation of chromatograms and spectra. Materials used in the analysis must be demonstrated to be free from interferences under the conditions of analysis by running method blanks initially and with each sample lot (sample started through the extraction process on a given 8-h shift, to a maximum of 20). Specific selection of reagents and purification of solvents by distillation in all glass systems may be required. Glassware and, where possible, reagents are cleaned by solvent rinse and baking at 450°C for 1-h minimum. Interferences coextracted from samples will vary considerably from source to source, depending on the diversity of the site being sampled.

INSTRUMENTATION Major instrumentation includes a GC with a splitless or on-column injection port for capillary column, a MS with 70 eV electron impact ionization, and a data system to collect and record MS data, and process it. A K-D apparatus is used to concentrate extracts.

GC Column: 30 m × 0.25 mm I.D. 5% phenyl, 94% methyl, 1% vinyl silicone bonded phased fused silica capillary column.

PRECISION & ACCURACY The detection limits of the method are usually dependent on the level of interferences rather than instrumental limitations. The limits typify the minimum quantities that can be detected with no interferences present.

The minimum level (in µg/mL) was 10. This is defined as a minimum level at which the analytical system shall give recognizable mass spectra (background corrected) and acceptable calibration points.

The MDL (in µg/kg) in low solids was 39 and in high solids was 28; these were determined in digested sludge (low solids) and in filter cake or compost (high solids).

The labeled and native compound initial precision as standard deviation (in µg/L) was 25.

The labeled and native compound initial accuracy as average recovery (in µg/L) was 69–161.

SAMPLE COLLECTION, PRESERVATION & HANDLING
Collect samples in glass containers. Aqueous samples which flow freely are collected in refrigerated bottles using automatic sampling equipment. Solid samples are collected as grab samples using widemouth jars. Maintain samples at 0 to 4°C from the time of collection until extraction. If residual chlorine is present in aqueous samples, add 80 mg sodium thiosulfate/L of water. Begin sample extraction within 7 days of collection, and analyze all extracts within 40 days of extraction.

SAMPLE PREPARATION Samples containing 1% solids or less are extracted directly using continuous liquid-liquid extraction techniques. Samples containing 1 to 30% solids are diluted to the 1% level with reagent water and extracted using continuous liquid-liquid extraction techniques. Samples containing greater than 30% solids are extracted using ultrasonic techniques.

Base/neutral extraction — Adjust the pH of the waters in the extractors to 12–13 with 6 N NaOH. Extract with methylene chloride for 24–48 h.

Acid extraction — Adjust the pH of the waters in the extractors to 2 or less using 6 N sulfuric acid. Extract with methylene chloride for 24–48 h.

Ultrasonic extraction of high solids samples — Add anhydrous sodium sulfate to the sample and QC aliquot(s). Add acetone:methylene chloride (1:1) to the sample and mix thoroughly

Concentrate extracts using a K-D apparatus.

QUALITY CONTROL The analyst is permitted to modify this method to improve separations or lower the costs of measurements, provided all performance specifications are met. Analyses of blanks are required to demonstrate freedom from contamination. When results of spikes indicate atypical method performance for samples, the samples are diluted to bring method performance within acceptable limits.

For low solids (aqueous samples), extract, concentrate, and analyze two sets of four 1-L aliquots (8 aliquots total) of the precision and recovery standard. For high solids samples, two sets of four 30-g aliquots of the high solids reference matrix are used.

Spike all samples with labeled compounds to assess method performance. Compute percent recovery of the labeled compounds using the internal standard method. Compare the labeled compound recovery for each compound with the corresponding labeled compound recovery.

Reagent water and high solids reference matrix blanks are analyzed to demonstrate freedom from contamination. Extract and concentrate a 1-L reagent water blank or a high solids reference matrix blank with each sample's lot (samples started through the extraction process on the same 8-h shift, to a maximum of 20 samples).

Field replicates may be collected to determine the precision of the sampling technique, and spiked samples may be required to determine the accuracy of the analysis when the internal standard method is used.

REFERENCE Semivolatile Organic Compounds by Isotope Dilution GC/MS. Office of Water Regulation and Standards, U.S. EPA Industrial Technology Division, Washington, DC, EPA Method 1625, Rev. C, June 1989 (contact W.A. Telliard, U.S. EPA, Office of Water Regulations and Standards, 401 M St., SW, Washington, DC, 20460. Phone: 202-382-7131).

Nitrobenzene   **EPA Method 625**
CAS #98-95-3

TITLE Base/Neutrals and Acids, U.S. EPA Method 625

MATRIX This methods covers municipal and industrial wastewaters.

METHOD SUMMARY Approximately 1 L of sample is serially extracted with methylene chloride at a pH greater than 11 and again at a pH less than 2 using a separatory funnel or a continuous extractor. The methylene chloride extract is dried, concentrated to a volume of 1 mL, and analyzed by GC/MS. Qualitative identification of the parameters in the extract is performed using the retention time and the relative abundance of three characteristic masses (m/z). Qualitative analysis is performed using either external or internal standard techniques with a single characteristic m/z.

INTERFERENCES Method interferences may be caused by contaminants in solvents, reagents, glassware, and other sample processing hardware. Glassware must be scrupulously cleaned. Glassware should be heated in a muffle furnace at 400°C for 5 to 30 min. Some thermally stable materials, such as PCBs, may not be eliminated by this treatment. Solvent rinses with acetone and pesticide quality hexane may be substituted for the muffle furnace heating. Matrix interferences may be caused by contaminants that are coextracted from the sample. The base-neutral extraction may cause significantly reduced recovery of phenols. The packed gas chromatographic columns recommended for the basic fraction may not exhibit sufficient resolution for some analytes.

INSTRUMENTATION A GC/MS system with an injection port designed for on-column injection when using packed columns and for splitless injection when using capillary columns.

Column for base/neutrals: 1.8 m long × 2 mm I.D. glass, packed with 3% SP-2550 on Supelcoport (100/120 mesh) or equivalent.
Column for acids: 1.8 m long × 2 mm I.D. glass, packed with 1% SP-1240DA on Supelcoport (100/120 mesh) or equivalent.

PRECISION & ACCURACY The MDL concentrations were obtained using reagent water. The MDL actually achieved in a given analysis will vary depending on instrument sensitivity and matrix effects. This method was tested by 15 laboratories using reagent water, drinking water, surface water, and industrial wastewaters spiked at six concentrations over the range 5 to 100 µg/L. Single operator precision, overall precision, and method accuracy were found to be directly related to the concentration of the parameter matrix.

The MDL (in µg/L) in reagent water was not reported.
The standard deviation (in µg/L based on 4 recovery measurements) was 39.3.
The range (in µg/L) for average recovery for 4 measurements was 54.3–157.6.
The range (in %) for percent recovery was 35–180.
Accuracy (in µg/L) as expected recovery for one or more measurements of a sample containing a true concentration of C was 1.09C-3.05.
Precision (in µg/L) as expected single analyst standard deviation of measurements at an average concentration found at X was 0.19X + 0.92.
Overall precision (in µg/L) as expected interlaboratory standard deviation of measurements in an average concentration found at X was 0.27X + 0.21.

C = *True value of the concentration in µg/L.*
X = *Average recovery found for measurements of samples containing a concentration at C in µg/L.*

SAMPLE PREPARATION Adjust the pH to >11 with sodium hydroxide and serially extract in a separatory funnel with methylene chloride or else in a continuous extractor. Next, adjust the pH to <2 with sulfuric acid and serially extract in a separatory funnel with methylene chloride or else in a continuous extractor. Dry the extracts separately through a column of anhydrous sodium sulfate and then concentrate each of the extracts to 1.0 mL using a K-D apparatus.

SAMPLE COLLECTION, PRESERVATION & HANDLING Grab samples must be collected in glass containers. All samples must be refrigerated at 4°C from the time of collection until extraction. If residual chlorine is present, add 80 mg of sodium thiosulfate/L of sample and mix well. All samples must be extracted within 7 days of collection and completely analyzed within 40 days of extraction.

QUALITY CONTROL Make an initial, one-time, demonstration of the ability to generate acceptable accuracy and precision with this method. Before processing any samples, the analyst must analyze a reagent water blank to demonstrate that interferences from the analytical system and glassware are under control. Each time a set of samples is extracted or reagents are changed, a reagent water blank must be processed. Spike and analyze a minimum of 5% of all samples to monitor and evaluate lab data quality. A QC check sample concentrate that contains each parameter of interest at a concentration of 100 µg/mL in acetone is required. PCBs and multicomponent pesticides may be omitted from this test.

After analysis of five spiked wastewater samples, calculate the average percent recovery and the standard deviation of the percent recovery. Spike all samples with the surrogate standard spiking solution and calculate the percent recovery of each surrogate compound.

REFERENCE *Federal Register*, Vol. 49, No. 209. Friday, Oct. 26, 1984.

Nitrobenzene
CAS #98-95-3

EPA Method 8270

TITLE Semivolatile Organic Compounds by GC/MS

MATRIX This method is used to determine the concentration of semivolatile organic compounds in extracts prepared from all types of solid waste matrices, soils, and groundwater. Although surface waters are not specifically mentioned, this method should be applicable to water samples from rivers, lakes, etc.

METHOD SUMMARY This method covers 259 semivolatile organic compounds. In very limited applications direct injection of the sample into the GC/MS system may be appropriate, but this results in very high detection limits (approximately 10,000 µg/L). Typically, a 1-L liquid sample, containing surrogate, and matrix spiking standards, is extracted in a continuous extractor first under acid conditions and then under basic conditions. Typically 30 g of a solid sample, containing surrogate, and matrix spiking standards, is extracted ultrasonically. After concentrating the extract to 1 mL it is spiked with 10 µL of an internal standard solution just prior to analysis by GC/MS. The volume injected should contain about 100 ng of base/neutral and 200 ng of acid surrogates (for a 1-µL injection). Analysis is performed by GC/MS using a capillary GC column.

INTERFERENCES Raw GC/MS data from all blanks, samples, and spikes must be evaluated for interferences. Contamination by carryover can occur whenever high-concentration and low-concentration samples are sequentially analyzed. To reduce carryover, the sample syringe must be rinsed out between samples with solvent. Whenever an unusually concentrated sample is encountered, it should be followed by the analysis of blank solvent to check for cross-contamination.

INSTRUMENTATION A GC/MS and a data system are required. The GC column used is a 30 m × 0.25 mm I.D. (or 0.32 mm I.D.) 1um film thickness silicone-coated fused silica capillary column. A continuous liquid-liquid extractor equipped with Teflon® or glass connection joints and stopcocks requiring no lubrication, a K-D concentrating apparatus, water bath, and an ultrasonic disrupter with a minimum power of 300 W and with pulsing capability are also required.

PRECISION & ACCURACY The estimated quantitation limit (EQL) of Method 8270B for determining an individual compound is approximately 1 mg/kg (wet weight) for soil or sediment samples, 1–200 mg/kg for wastes (dependent on matrix and method of preparation), and 10 µg/L for groundwater samples. EQLs will be proportionately higher for sample extracts that require dilution to avoid saturation of the detector.

The EQL(b) for groundwater in µg/L is 10.
The EQL (a, b) for low concentrations in soil and sediment in µg/kg is 660.
Accuracy as µg/L is $1.09C - 3.05$.
Overall precision in µg/L is $0.27X + 0.21$.

(a) EQLs listed for soil/sediment are based on wet weight. Normally data is reported in a dry-weight basis; therefore, EQLs will be higher based on the % dry weight of each sample. This calculation is based on a 30 g sample and gel permeation chromatography cleanup.

(b) Sample EQLs are highly matrix-dependent. The EQLs are provided for guidance and may not always be achievable.

C = True value for concentration, in µg/L.
X = Average recovery found for measurements of samples containing a concentration of C, in µg/L.

ESTIMATED QUANTITATION LIMIT

Other Matrices	Factor (a)
High-concentration soil and sludges by sonicator	7.5
Non-water miscible waste	75

(a) EQL for other matrices = [EQL for low soil/sediment] × [Factor]. This estimated EQL is similar to an EPA "Practical Quantitation Limit."

SAMPLING METHOD

Liquid samples — Use a 1 or 2½ gallon amber glass bottle with a screw-top Teflon®-lined cover that has been prewashed with detergent and rinsed with distilled water and methanol (or isopropanol).

Soils, sediments, or sludges — Use an 8-oz. widemouth glass with a screw-top Teflon®-lined cover that has been prewashed with detergent and rinsed with distilled water and methanol (or isopropanol).

SAMPLE PRESERVATION

Liquid samples — If residual chlorine is present, add 3 mL of 10% sodium thiosulfate per gallon, cool to 4°C and store in a solvent-free refrigerator until analysis; if chlorine is not present, then eliminate the sodium thiosulfate addition.

Soils, sediments, or sludges — Cool samples to 4°C and store in a solvent-free refrigerator.

MHT Liquid samples must be extracted within 7 days and the extracts analyzed within 40 days. Soils, sediments, or sludges may be stored for a maximum of 14 days and the extracts analyzed within 40 days.

SAMPLE PREPARATION

Liquid samples — Transfer 1 L quantitatively to a continuous extractor. If high concentrations are anticipated, a smaller volume may be used and then diluted with organic-free reagent water to 1 L. Adjust pH, if necessary, to pH <2 using 1:1 (V/V) sulfuric acid. Pipette 1.0 mL of a surrogate standard spiking solution into each sample. For the sample in each analytical batch selected for spiking, add 1.0 mL of a matrix spiking standard. For base/neutral acid analysis, the amount of the surrogates and matrix spiking compounds added to the sample should result in a final concentration of 100 ng/µL of each analyte in the extract to be analyzed (assuming a 1-µL injection). Extract with methylene chloride for 18–24 h. Next, adjust the pH of the aqueous phase to pH >11 using 10 N sodium hydroxide and extract it with methylene chloride again for 18–24 h. Dry the extract through a column containing anhydrous sodium sulfate and concentrate it to 1 mL using a K-D concentrator.

Soils, sediments, or sludges — Use 30 g of sample. Nonporous or wet samples (gummy or clay type) that do not have a free-flowing

sandy texture must be mixed with anhydrous sodium sulfate until the sample is free flowing. Add 1 mL of surrogate standards to all samples, spikes, standards, and blanks. For the sample in each analytical batch selected for spiking, add 1.0 mL of a matrix spiking standard. For base/neutral acid analysis, the amount added of the surrogates and matrix spiking compounds should result in a final concentration of 100 ng/µL of each base/neutral analyte and 200 ng/µL of each acid analyte in the extract to be analyzed (assuming a 1-µL injection). Immediately add a 100-mL mixture of 1:1 methylene chloride:acetone and extract the sample ultrasonically for 3 min and then decant or filter the extracts. Repeat the extraction two or more times. Dry the extract using a column with anhydrous sodium sulfate and concentrate it to 1 mL in a K-D concentrator.

QUALITY CONTROL A methylene chloride solution containing 50 ng/µL of decafluorotriphenylphosphine (DFTPP) is used for tuning the GC/MS system each 12-h shift. A system performance check also must be made during every 12-h shift. A standard containing 50 ng/µL each of 4,4′-DDT, pentachlorophenol, and benzidine is required to verify injection port inertness and GC column performance. A calibration standard at mid-concentration, containing each compound of interest, including all required surrogates, must be performed every 12 h during analysis. After the system performance check is met, calibration check compounds (CCCs) are used to check the validity of the initial calibration.

The internal standard responses and retention times in the calibration check standard must be evaluated immediately after or during data acquisition. If the retention time for any internal standard changes by more than 30 seconds from the last check calibration (12 h), the chromatographic system must be inspected for malfunctions and corrections must be made, as required. If the electron ionization current plot (EICP) area for any of the internal standards changes by a factor of two from the last daily calibration standard check, the mass spectrometer must be inspected for malfunctions and corrections must be made, as appropriate.

Demonstrate, through the analysis of a reagent water blank, that interferences from the analytical system, glassware, and reagents are under control. The blank samples should be carried through all stages of the sample preparation and measurement steps. For each analytical batch (up to 20 samples), a reagent blank, matrix spike, and matrix spike duplicate/duplicate must be analyzed (the frequency of the spikes may be different for different monitoring programs). The blank and spiked samples must be carried through all stages of the sample preparation and measurement steps. A QC reference sample concentrate containing each analyte at a concentration of 100 mg/L in methanol is required.

REFERENCE Test Methods for Evaluating Solid Waste (SW-846). U.S. EPA 1983, Method 8270B, Rev. 2, Nov. 1990. Office of Solid Waste, Washington, DC.

Nitrobenzene — EPA Method 8090
CAS #98-95-3

TITLE Nitroaromatics & Cyclic Ketones

MATRIX Groundwater, soils, sludges, water miscible liquid wastes, and non-water miscible wastes.

APPLICATION This method is used for the analysis of various nitroaromatic and cyclic ketone compounds. Samples are extracted, concentrated, and analyzed using direct injection of both neat and diluted organic liquids. Dinitrotoluenes are determined using ECD and the other compounds amenable to this method are determined using FID. The method provides an optional GC column which is used for analyte confirmation and that may help resolve analytes from interferences.

INTERFERENCES Solvents, reagents, and glassware may introduce artifacts. Other interferences may come from coextracted compounds from samples.

INSTRUMENTATION GC capable of on-column injections and a flame with detector (FID) or electron capture detector (ECD). Column 1: a 1.2 m by 2 mm or 4 mm with 1.95% QF-1/1.5% OV-17 on Gas-Chrom Q. Column 2: a 3 meter by 2 mm or 4 mm with 3% OV-101 on Gas-Chrom Q.

RANGE 1–515 µg/L.

MDL 3.6 µg/L (FID) and 13.7 µg/L (ECD)

PQL FACTORS FOR MULTIPLYING × FID MDL VALUE

Matrix	Multiplication Factor
Groundwater	10
Low-level soil by sonication with GPC cleanup	670
High-level soil and sludge by sonication	10,000
Non-water miscible waste	100,000

PRECISION 0.37X–0.78 µg/L (overall precision).

ACCURACY 0.60C + 2.00 µg/L (as recovery).

SAMPLING METHOD Use 8-oz. widemouth glass bottles with Teflon®-lined caps for concentrated waste samples, soils, sediments, and sludges. Use 1 or 2½ gallon amber glass bottles with Teflon®-lined caps for liquid (water) samples.

STABILITY Cool soil, sediment, sludge, and liquid samples to 4°C. If residual chlorine is present in liquid samples add 3 mL of 10% sodium thiosulfate per gallon of sample and cool to 4°C.

MHT 14 days for concentrated waste, soil, sediment, or sludge; 7 days for liquid samples; all extracts must be analyzed within 40 days.

QUALITY CONTROL A quality control check sample concentrate containing each analyte of interest is required. The QC check sample concentrate may be prepared from pure standard materials or purchased as certified solutions Use appropriate trip, matrix, control site, method, reagent, and solvent blanks. Internal, surrogate, and five concentration level calibration standards are used. The QC check sample concentrate should contain this compound at 100 µg/mL in acetone.

REFERENCE Method 8090, SW-846, 3rd ed., Nov.1986.

4-Nitrobiphenyl   **EPA Method 8270**
CAS #92-93-3

TITLE Semivolatile Organic Compounds by GC/MS

MATRIX This method is used to determine the concentration of semivolatile organic compounds in extracts prepared from all types of solid waste matrices, soils, and groundwater. Although surface waters are not specifically mentioned, this method should be applicable to water samples from rivers, lakes, etc.

METHOD SUMMARY This method covers 259 semivolatile organic compounds. In very limited applications direct injection of the sample into the GC/MS system may be appropriate, but this results in very high detection limits (approximately 10,000 µg/L). Typically, a 1-L liquid sample, containing surrogate, and matrix spiking standards, is extracted in a continuous extractor first under acid conditions and then under basic conditions. Typically 30 g of a solid sample, containing surrogate, and matrix spiking standards, is extracted ultrasonically. After concentrating the extract to 1 mL it is spiked with 10 µL of an internal standard solution just prior to analysis by GC/MS. The volume injected should contain about 100 ng of base/neutral and 200 ng of acid surrogates (for a 1-µL injection). Analysis is performed by GC/MS using a capillary GC column.

INTERFERENCES Raw GC/MS data from all blanks, samples, and spikes must be evaluated for interferences. Contamination by carryover can occur whenever high-concentration and low-concentration samples are sequentially analyzed. To reduce carryover, the sample syringe must be rinsed out between samples with solvent. Whenever an unusually concentrated sample is encountered, it should be followed by the analysis of blank solvent to check for cross-contamination.

INSTRUMENTATION A GC/MS and a data system are required. The GC column used is a 30 m × 0.25 mm I.D. (or 0.32 mm I.D.) 1um film thickness silicone-coated fused silica capillary column. A continuous liquid-liquid extractor equipped with Teflon® or glass connection joints and stopcocks requiring no lubrication, a K-D concentrating apparatus, water bath, and an ultrasonic disrupter with a minimum power of 300 W and with pulsing capability are also required.

PRECISION & ACCURACY The estimated quantitation limit (EQL) of Method 8270B for determining an individual compound is approximately 1 mg/kg (wet weight) for soil or sediment samples, 1–200 mg/kg for wastes (dependent on matrix and method of preparation), and 10 µg/L for groundwater samples. EQLs will be proportionately higher for sample extracts that require dilution to avoid saturation of the detector.

The EQL(b) for groundwater in µg/L is 10.
The EQL (a, b) for low concentrations in soil and sediment in µg/kg is not determined.
Accuracy as µg/L is not listed.
Overall precision in µg/L is not listed.

(a) EQLs listed for soil/sediment are based on wet weight. Normally data is reported in a dry-weight basis; therefore, EQLs will be higher based on the % dry weight of each sample. This calculation is based on a 30 g sample and gel permeation chromatography cleanup.

(b) Sample EQLs are highly matrix-dependent. The EQLs are provided for guidance and may not always be achievable.

C = True value for concentration, in µg/L.
X = Average recovery found for measurements of samples containing a concentration of C, in µg/L.

ESTIMATED QUANTITATION LIMIT

Other Matrices	Factor (a)
High-concentration soil and sludges by sonicator	7.5
Non-water miscible waste	75

(a) EQL for other matrices = [EQL for low soil/sediment] × [Factor]. This estimated EQL is similar to an EPA "Practical Quantitation Limit."

SAMPLING METHOD

Liquid samples — Use a 1 or 2½ gallon amber glass bottle with a screw-top Teflon®-lined cover that has been prewashed with detergent and rinsed with distilled water and methanol (or isopropanol).

Soils, sediments, or sludges — Use an 8-oz. widemouth glass with a screw-top Teflon®-lined cover that has been prewashed with detergent and rinsed with distilled water and methanol (or isopropanol).

SAMPLE PRESERVATION

Liquid samples — If residual chlorine is present, add 3 mL of 10% sodium thiosulfate per gallon, cool to 4°C and store in a solvent-free refrigerator until analysis; if chlorine is not present, then eliminate the sodium thiosulfate addition.

Soils, sediments, or sludges — Cool samples to 4°C and store in a solvent-free refrigerator.

MHT Liquid samples must be extracted within 7 days and the extracts analyzed within 40 days. Soils, sediments, or sludges may be stored for a maximum of 14 days and the extracts analyzed within 40 days.

SAMPLE PREPARATION

Liquid samples — Transfer 1 L quantitatively to a continuous extractor. If high concentrations are anticipated, a smaller volume may be used and then diluted with organic-free reagent water to 1 L. Adjust pH, if necessary, to pH <2 using 1:1 (V/V) sulfuric acid. Pipette 1.0 mL of a surrogate standard spiking solution into each sample. For the sample in each analytical batch selected for spiking, add 1.0 mL of a matrix spiking standard. For base/neutral acid analysis, the amount of the surrogates and matrix spiking compounds added to the sample should result in a final concentration of 100 ng/µL of each analyte in the extract to be analyzed (assuming a 1-µL injection). Extract with methylene chloride for 18–24 h. Next, adjust the pH of the aqueous phase to pH >11 using 10 N sodium hydroxide and extract it with methylene chloride again for

18–24 h. Dry the extract through a column containing anhydrous sodium sulfate and concentrate it to 1 mL using a K-D concentrator.

Soils, sediments, or sludges — Use 30 g of sample. Nonporous or wet samples (gummy or clay type) that do not have a free-flowing sandy texture must be mixed with anhydrous sodium sulfate until the sample is free flowing. Add 1 mL of surrogate standards to all samples, spikes, standards, and blanks. For the sample in each analytical batch selected for spiking, add 1.0 mL of a matrix spiking standard. For base/neutral acid analysis, the amount added of the surrogates and matrix spiking compounds should result in a final concentration of 100 ng/µL of each base/neutral analyte and 200 ng/µL of each acid analyte in the extract to be analyzed (assuming a 1-µL injection). Immediately add a 100-mL mixture of 1:1 methylene chloride:acetone and extract the sample ultrasonically for 3 min and then decant or filter the extracts. Repeat the extraction two or more times. Dry the extract using a column with anhydrous sodium sulfate and concentrate it to 1 mL in a K-D concentrator.

QUALITY CONTROL A methylene chloride solution containing 50 ng/µL of decafluorotriphenylphosphine (DFTPP) is used for tuning the GC/MS system each 12-h shift. A system performance check also must be made during every 12-h shift. A standard containing 50 ng/µL each of 4,4′-DDT, pentachlorophenol, and benzidine is required to verify injection port inertness and GC column performance. A calibration standard at mid-concentration, containing each compound of interest, including all required surrogates, must be performed every 12 h during analysis. After the system performance check is met, calibration check compounds (CCCs) are used to check the validity of the initial calibration.

The internal standard responses and retention times in the calibration check standard must be evaluated immediately after or during data acquisition. If the retention time for any internal standard changes by more than 30 seconds from the last check calibration (12 h), the chromatographic system must be inspected for malfunctions and corrections must be made, as required. If the electron ionization current plot (EICP) area for any of the internal standards changes by a factor of two from the last daily calibration standard check, the mass spectrometer must be inspected for malfunctions and corrections must be made, as appropriate.

Demonstrate, through the analysis of a reagent water blank, that interferences from the analytical system, glassware, and reagents are under control. The blank samples should be carried through all stages of the sample preparation and measurement steps. For each analytical batch (up to 20 samples), a reagent blank, matrix spike, and matrix spike duplicate/duplicate must be analyzed (the frequency of the spikes may be different for different monitoring programs). The blank and spiked samples must be carried through all stages of the sample preparation and measurement steps. A QC reference sample concentrate containing each analyte at a concentration of 100 mg/L in methanol is required.

REFERENCE Test Methods for Evaluating Solid Waste (SW-846). U.S. EPA 1983, Method 8270B, Rev. 2, Nov. 1990. Office of Solid Waste, Washington, DC.

Nitrofen   **EPA Method 8270**
CAS #1836-75-5

TITLE Semivolatile Organic Compounds by GC/MS

MATRIX This method is used to determine the concentration of semivolatile organic compounds in extracts prepared from all types of solid waste matrices, soils, and groundwater. Although surface waters are not specifically mentioned, this method should be applicable to water samples from rivers, lakes, etc.

METHOD SUMMARY This method covers 259 semivolatile organic compounds. In very limited applications direct injection of the sample into the GC/MS system may be appropriate, but this results in very high detection limits (approximately 10,000 µg/L). Typically, a 1-L liquid sample, containing surrogate, and matrix spiking standards, is extracted in a continuous extractor first under acid conditions and then under basic conditions. Typically 30 g of a solid sample, containing surrogate, and matrix spiking standards, is extracted ultrasonically. After concentrating the extract to 1 mL it is spiked with 10 µL of an internal standard solution just prior to analysis by GC/MS. The volume injected should contain about 100 ng of base/neutral and 200 ng of acid surrogates (for a 1-µL injection). Analysis is performed by GC/MS using a capillary GC column.

INTERFERENCES Raw GC/MS data from all blanks, samples, and spikes must be evaluated for interferences. Contamination by carryover can occur whenever high-concentration and low-concentration samples are sequentially analyzed. To reduce carryover, the sample syringe must be rinsed out between samples with solvent. Whenever an unusually concentrated sample is encountered, it should be followed by the analysis of blank solvent to check for cross-contamination.

INSTRUMENTATION A GC/MS and a data system are required. The GC column used is a 30 m × 0.25 mm I.D. (or 0.32 mm I.D.) 1um film thickness silicone-coated fused silica capillary column. A continuous liquid-liquid extractor equipped with Teflon® or glass connection joints and stopcocks requiring no lubrication, a K-D concentrating apparatus, water bath, and an ultrasonic disrupter with a minimum power of 300 W and with pulsing capability are also required.

PRECISION & ACCURACY The estimated quantitation limit (EQL) of Method 8270B for determining an individual compound is approximately 1 mg/kg (wet weight) for soil or sediment samples, 1–200 mg/kg for wastes (dependent on matrix and method of preparation), and 10 µg/L for groundwater samples. EQLs will be proportionately higher for sample extracts that require dilution to avoid saturation of the detector.

The EQL(b) for groundwater in µg/L is 20.
The EQL (a, b) for low concentrations in soil and sediment in µg/kg is not determined.
Accuracy as µg/L is not listed.
Overall precision in µg/L is not listed.

(a) *EQLs listed for soil/sediment are based on wet weight. Normally data is reported in a dry-weight basis; therefore, EQLs will be higher based on the % dry weight of each sample.*

This calculation is based on a 30 g sample and gel permeation chromatography cleanup.

(b) *Sample EQLs are highly matrix-dependent. The EQLs are provided for guidance and may not always be achievable.*

C = *True value for concentration, in µg/L.*

X = *Average recovery found for measurements of samples containing a concentration of C, in µg/L.*

ESTIMATED QUANTITATION LIMIT

Other Matrices	Factor (a)
High-concentration soil and sludges by sonicator	7.5
Non-water miscible waste	75

(a) *EQL for other matrices = [EQL for low soil/sediment] × [Factor]. This estimated EQL is similar to an EPA "Practical Quantitation Limit."*

SAMPLING METHOD

Liquid samples — Use a 1 or 2½ gallon amber glass bottle with a screw-top Teflon®-lined cover that has been prewashed with detergent and rinsed with distilled water and methanol (or isopropanol).

Soils, sediments, or sludges — Use an 8-oz. widemouth glass with a screw-top Teflon®-lined cover that has been prewashed with detergent and rinsed with distilled water and methanol (or isopropanol).

SAMPLE PRESERVATION

Liquid samples — If residual chlorine is present, add 3 mL of 10% sodium thiosulfate per gallon, cool to 4°C and store in a solvent-free refrigerator until analysis; if chlorine is not present, then eliminate the sodium thiosulfate addition.

Soils, sediments, or sludges — Cool samples to 4°C and store in a solvent-free refrigerator.

MHT Liquid samples must be extracted within 7 days and the extracts analyzed within 40 days. Soils, sediments, or sludges may be stored for a maximum of 14 days and the extracts analyzed within 40 days.

SAMPLE PREPARATION

Liquid samples — Transfer 1 L quantitatively to a continuous extractor. If high concentrations are anticipated, a smaller volume may be used and then diluted with organic-free reagent water to 1 L. Adjust pH, if necessary, to pH <2 using 1:1 (V/V) sulfuric acid. Pipette 1.0 mL of a surrogate standard spiking solution into each sample. For the sample in each analytical batch selected for spiking, add 1.0 mL of a matrix spiking standard. For base/neutral acid analysis, the amount of the surrogates and matrix spiking compounds added to the sample should result in a final concentration of 100 ng/µL of each analyte in the extract to be analyzed (assuming a 1-µL injection). Extract with methylene chloride for 18–24 h. Next, adjust the pH of the aqueous phase to pH >11 using 10 N sodium hydroxide and extract it with methylene chloride again for 18–24 h. Dry the extract through a column containing anhydrous sodium sulfate and concentrate it to 1 mL using a K-D concentrator.

Soils, sediments, or sludges — Use 30 g of sample. Nonporous or wet samples (gummy or clay type) that do not have a free-flowing sandy texture must be mixed with anhydrous sodium sulfate until the sample is free flowing. Add 1 mL of surrogate standards to all samples, spikes, standards, and blanks. For the sample in each analytical batch selected for spiking, add 1.0 mL of a matrix spiking standard. For base/neutral acid analysis, the amount added of the surrogates and matrix spiking compounds should result in a final concentration of 100 ng/µL of each base/neutral analyte and 200 ng/µL of each acid analyte in the extract to be analyzed (assuming a 1-µL injection). Immediately add a 100-mL mixture of 1:1 methylene chloride:acetone and extract the sample ultrasonically for 3 min and then decant or filter the extracts. Repeat the extraction two or more times. Dry the extract using a column with anhydrous sodium sulfate and concentrate it to 1 mL in a K-D concentrator.

QUALITY CONTROL A methylene chloride solution containing 50 ng/µL of decafluorotriphenylphosphine (DFTPP) is used for tuning the GC/MS system each 12-h shift. A system performance check also must be made during every 12-h shift. A standard containing 50 ng/µL each of 4,4′-DDT, pentachlorophenol, and benzidine is required to verify injection port inertness and GC column performance. A calibration standard at mid-concentration, containing each compound of interest, including all required surrogates, must be performed every 12 h during analysis. After the system performance check is met, calibration check compounds (CCCs) are used to check the validity of the initial calibration.

The internal standard responses and retention times in the calibration check standard must be evaluated immediately after or during data acquisition. If the retention time for any internal standard changes by more than 30 seconds from the last check calibration (12 h), the chromatographic system must be inspected for malfunctions and corrections must be made, as required. If the electron ionization current plot (EICP) area for any of the internal standards changes by a factor of two from the last daily calibration standard check, the mass spectrometer must be inspected for malfunctions and corrections must be made, as appropriate.

Demonstrate, through the analysis of a reagent water blank, that interferences from the analytical system, glassware, and reagents are under control. The blank samples should be carried through all stages of the sample preparation and measurement steps. For each analytical batch (up to 20 samples), a reagent blank, matrix spike, and matrix spike duplicate/duplicate must be analyzed (the frequency of the spikes may be different for different monitoring programs). The blank and spiked samples must be carried through all stages of the sample preparation and measurement steps. A QC reference sample concentrate containing each analyte at a concentration of 100 mg/L in methanol is required.

REFERENCE Test Methods for Evaluating Solid Waste (SW-846). U.S. EPA 1983, Method 8270B, Rev. 2, Nov. 1990. Office of Solid Waste, Washington, DC.

Nitrogen, Ammonia   EPA Method 350.1

TITLE   Inorganics, Non-Metallics

MATRIX Drinking, surface, and saline waters. Wastewater.

APPLICATION Date issued 1974. Editorial Rev. 1978. (Colorimetric, automated phenate). Alkaline phenol and hypochlorite react with ammonia to form indophenol blue that is proportional to the ammonia concentration. (Blue color formed is intensified with sodium nitroprusside).

INTERFERENCES Calcium and magnesium ions may be present in concentration sufficient to cause precipitation problems during analysis. Sample turbidity and color may interfere with this method.

INSTRUMENTATION Technicon auto analyzer, 630–660 nm, 15 mm or 50 mm tubular flow cell.

RANGE 0.01–2.0 mg/L NH_3 as N.

MDL Not listed.

PRECISION SD = ±0.005 At 4 conc's (0.43–1.41 mg NH3-N/L).

ACCURACY At concentrations 0.16 and 1.44, Recoveries were 107% and 99%.

SAMPLING METHOD Plastic or glass. (400 mL).

STABILITY Cool, 4°C. H2SO4 to pH <2.

MHT 28 days.

QUALITY CONTROL Approximately 20–60 samples per h can be analyzed. Arrange ammonia standards in sampler in order of decreasing concentration. Of nitrogen. All solutions must be made using ammonia-free water. When saline waters are analyzed, substitute ocean water is used to prepare standards.

REFERENCE Methods for the Chemical Analysis of Water and Wastes, EPA-600/4-79-020, U.S. EPA, EMSL, 1979.

Nitrogen, Ammonia   **EPA Method 350.2**

TITLE Inorganics, Non-Metallics

MATRIX Drinking, surface, and saline waters. Wastewater.

APPLICATION Date issued 1971. Editorial Rev. 1974. (Colorimetric; titimetric; potentiometric distillation procedure). Method covers the range of 0.05 to 1.0 mg NH3-N/L for colorimetric; 1.0 to 25 mg/L for titrimetric; and 0.05 to 1400 mg/L for electrode method.

INTERFERENCES Cyanate hydrolyzes slightly, even at pH 9.5. Volatile alkaline compounds may cause an off-color upon nesslerization. (Some compounds may be eliminated by boiling off at low pH, 2–3). Residual chlorine must be removed by sample pretreatment with Na2S2O3.

INSTRUMENTATION Spectrophotometer or filter photometer. 425 nm. Light path of 1 cm or more.

RANGE See Applicability.

MDL Not listed.

PRECISION SD = 0.244 mg NH3-N/L at 1.71 mg NH3-N/L increment.

ACCURACY As bias, +0.01 mg NH3-N/L at 1.71 mg NH3-N/L increment.

SAMPLING METHOD Plastic or glass (400 mL).

STABILITY Cool, 4°C. H2SO4 to pH <2.

MHT 28 days.

QUALITY CONTROL 1) colorimetric determination is prepared in matched nessler tubes and absorbance read at 425 nm. 2) For titrimetric determination, the distillate in receiving flask is titrated with standard H2SO4. 3) for potential determination, consult Method 350.3:Selective ion electrode Method.

REFERENCE EPA Methods for the Chemical Analysis of Water and Wastes, EPA-600/4-79-020, U.S. EPA, EMSL, 1979.

Nitrogen, Ammonia   **EPA Method 350.3**

TITLE Inorganics, Non-Metallics

MATRIX Drinking, surface, and saline waters. Wastewater.

APPLICATION Date issued 1974. (Potentiometric,ion selective electrode). The ammonia is determined potentiometrically using an ion selective ammonia electrode. The NH3 electrode uses a hydrophobic gas-permeable membrane to separate the sample from NH4°Cl internal soln.

INTERFERENCES Volatile amines act as a positive interference. Mercury interferes by forming a complex with ammonia. Thus the sample can not be preserved with mercuric chloride.

INSTRUMENTATION pH meter with expanded mv scale or specific ion meter.

RANGE 0.03–1400 mg NH3-N/L.

MDL Not listed.

PRECISION SD = ±0.038 at 1.00 mg NH3-N/L.

ACCURACY Recoveries = 96 and 91% at 0.19 and 0.13 mg NH3-N/L

SAMPLING METHOD Plastic or glass (400 mL).

STABILITY Cool, 4°C. H2SO4 to pH <2.

MHT 28 days.

QUALITY CONTROL Distilled water must be ammonia free. When analyzing saline waters, standards must be made up in synthetic ocean water. See Method 350.1 for preparation directions.

REFERENCE EPA Methods for the Chemical Analysis of Water and Wastes, EPA-600/4-79-020, U.S. EPA, EMSL, 1979.

Nitrogen, Kjeldahl, Total (TKN) — EPA Method 351.1

TITLE Inorganics, Non-Metallics

MATRIX Surface and saline waters.

APPLICATION Date issued 1971. Editorial Rev. 1974 and 1978. (Colorimetric, automated phenate). Total Kjeldahl nitrogen is defined as the sum of free-ammonia and of organic nitrogen compounds which are converted to ammonium sulfate under the conditions of digestion. Sample is automatically digested with a sulfuric acid solution containing catalyst. Organic nitrogen is converted to ammonium sulfate.

INTERFERENCES Iron and chromium ions tend to catalyze while copper ions tend to inhibit the indophenol color reaction.

INSTRUMENTATION Technicon auto analyzer 630 nm filters 50 mm tubular flow cell.

RANGE 0.05–2.0 mg N/L.

MDL Not listed.

PRECISION SD = 0.61 K-N mg N/L at 2.18 K-N mg N/L.

ACCURACY As bias, –0.62 mg N/L at 2.18 K-N mg N/L.

SAMPLING METHOD Plastic or glass (500 mL).

STABILITY Cool, 4°C. H_2SO_4 to pH <2.

MHT 28 days.

QUALITY CONTROL All solutions must be made using ammonia free water. Arrange standards in sampler cups in order of increasing concentration.

REFERENCE Methods for the Chemical Analysis of Water and Wastes, EPA-600/4-79-020, U.S. EPA, EMSL, 1979.

Nitrogen, Kjeldahl, Total (TKN) — EPA Method 351.2

TITLE Inorganics, Non-Metallics

MATRIX Drinking, surface, and wastewaters.

APPLICATION Date issued 1978. (Colorimetric, semi-automated block digester, AAII). The sample is heated in the presence of sulfuric acid, potassium sulfate and mercuric sulfate for 2.5 Hrs. Residue is cooled, diluted to 25 mL and placed on autoanalyzer for NH_3 detmn. (The digested sample may also be used for phosphorus determination).

INTERFERENCES The procedure converts nitrogen components of biological origin such as amino acids, proteins and peptides to ammonia, but may not convert the nitrogenous compounds of some industrial wastes such as amines, nitro compounds, hydrazones, semicarbazones and some amines.

INSTRUMENTATION Block digester-40. Technicon manifold for ammonia (NH_3).

RANGE 0.1–20 mg/L TKN.

MDL Not listed.

PRECISION Not listed.

ACCURACY Not listed.

SAMPLING METHOD Plastic or glass (500 mL).

STABILITY Cool, 4°C. H_2SO_4 to pH <2.

MHT 28 days.

QUALITY CONTROL All solutions must be made using ammonia free water. Use Teflon® boiling stones. The range may be extended with sample dilution.

REFERENCE EPA Methods for the Chemical Analysis of Water and Wastes, EPA-600/4-79-020, U.S. EPA, EMSL, 1979.

Nitrogen, Kjeldahl, Total (TKN) — EPA Method 351.3

TITLE Inorganics, Non-Metallics

MATRIX Drinking, surface, and saline waters. Wastewater.

APPLICATION Date issued 1971. Editorial Rev. 1974 and 1978. (Colorimetric; titrimetric; potentiometric). Three alternatives are listed for the determination of NH_3 after distillation. Titrimetric Method-concentrations above 1 mg N/L; colorimetric- concentrations <1 mg N/L; potentiometric-0.05–1400 mg N/L.

INTERFERENCES High nitrate concentrations (10× or more than TKN level) result in low TKN values. The reaction between nitrate and ammonia (NH_3) can be prevented by the use of an ion exchange resin (chloride form) to remove the nitrate prior to the TKN analysis.

INSTRUMENTATION Spectrophotometer. 425 nm. Light path 1 cm or longer.

RANGE See Applicability.

MDL Not listed.

PRECISION SD = 1.056 mg N/L at 4.10 K-N mg N/L.

ACCURACY As Bias, +0.4 mg N/L at 4.10 K-N mg N/L.

SAMPLING METHOD Plastic or glass (500 mL).

STABILITY Cool, 4°C. H_2SO_4 to pH <2.

MHT 28 days.

QUALITY CONTROL Colorimetric determination is prepared in Nessler tubes and absorbance read at 425 nm. For titrimetric determination, distillate in receiving flask is titrated with standard H_2SO_4. For potentiometric determination, consult Method 350.3. All solution must be made with ammonia free water.

REFERENCE EPA Methods for the Chemical Analysis of Water and Wastes, EPA-600/4-79-020, U.S. EPA, EMSL, 1979.

Nitrogen, Kjeldahl, Total (TKN)   EPA Method 351.4

TITLE   Inorganics, Non-Metallics

MATRIX   Drinking, surface, and wastewaters.

APPLICATION   Date issued 1978. (Potentiometric, ion selective electrode). Following digestion and cooling, distilled water is added to digestion flask and pH adjusted between 3 and 4.5 Using 10 N sodium hydroxide. Sample is cooled and transferred to a 100 mL beaker. After inserting electrode into sample, sodium hydroxide-sodium iodide-EDTA solution is added and ammonia measured.

INTERFERENCES   Interference from metals are eliminated with addition of sodium iodide. High nitrate concentrations (10× or more than TKN level) result in low TKN values. The nitrate and ammonia (NH3) reaction is prevented by using an anion exchange resin to remove nitrate.

INSTRUMENTATION   pH meter (with expanded mv scale). NH3 selective electrode. Digestion apparatus.

RANGE   0.03–25 mg TKN/L.

MDL   Not listed.

PRECISION   Not listed.

ACCURACY   Not listed.

SAMPLING METHOD   Plastic or glass (500 mL).

STABILITY   Cool, 4°C. H2SO4 to pH <2.

MHT   28 days.

QUALITY CONTROL   Either macro or micro Kjeldahl system or block digestor can be used for digestion. All solutions must be made using ammonia-free water.

REFERENCE   Methods for the Chemical Analysis of Water and Wastes, EPA-600/4-79-020, U.S. EPA, EMSL, 1979.

Nitrogen, Nitrate   EPA Method 352.1

TITLE   Inorganics, Non-Metallics

MATRIX   Drinking, surface, and saline waters. Wastewater.

APPLICATION   Date issued 1971. (Colorimetric, brucine). Method is based upon reaction of nitrate ion with brucine sulfate in a 13 N H2SO4 solution @ 100°C. The color of resulting complex is measured at 410 nm. Temperature control of reaction is extremely critical.

INTERFERENCES   Dissolved organic matter causes an off color in 13 N H2SO4. Effects of salinity and residual chlorine interference can be eliminated. All strong oxidizing or reducing agents interfere. Ferrous and ferric iron and quadrivalent manganese interfere slightly.

INSTRUMENTATION   Spectro (or filter) photometer at 410 nm. Water bath at 100c with stirring mechanism

RANGE   0.1–2.0 mg NO3-N/L.

MDL   Not listed.

PRECISION   SD = 0.214 mg N/L at 1.24 mg N/L (as nitrogen, nitrate)

ACCURACY   As bias, +0.04 mg N/L at 1.24 mg N/L (as NO3-N)

SAMPLING METHOD   Plastic or glass (100 mL).

STABILITY   Cool, 4°C.

MHT   48 h.

QUALITY CONTROL   Uneven heating of samples and standards during reaction time results in erratic values. Absolute control of temperature during critical color development period cannot be too strongly emphasized. Use distilled water free of nitrite and nitrate to prepare reagents and standards.

REFERENCE   EPA Methods for the Chemical Analysis of Water and Wastes, EPA-600/4-79-020, U.S. EPA, EMSL, 1979.

Nitrogen, Nitrate-Nitrite   EPA Method 353.1

TITLE   Inorganics, Non-Metallics

MATRIX   Drinking, surface, and wastewaters.

APPLICATION   Date issued 1971. Reissued with Rev. 1978. (Colorimetric, automated, hydrazine reduction). Nitrate is reduced to nitrite with hydrazine sulfate and the nitrite determined by diazotizing to form a highly colored dye which is measured colorimetrically.

INTERFERENCES   Sample color that absorbs in the photometric range used for analysis will interfere. The apparent nitrate (NO3) and nitrite (NO2) concentrations varied ±10% with concentrations of sulfide ion up to 10 mg/L.

INSTRUMENTATION   Colorimeter equipped with an 8, 15, or 50 mm flow cell. 529 nm Filters.

RANGE   0.01–10 mg/L NO3-NO2 nitrogen.

MDL   Not listed.

PRECISION   SD = ±0.03 at 4.75 µg NO3-N/L.

ACCURACY   Recoveries were 99 and 101% at 0.75 and 2.97 µg NO3-N/L.

SAMPLING METHOD   Plastic or glass (100 mL).

STABILITY   Cool, 4°C. H2SO4 to pH <2.

MHT   28 days.

QUALITY CONTROL   Use continuous filter to remove precipitate. Place appropriate nitrate standards in sampler in order of decreasing concentration of nitrogen.

REFERENCE   EPA Methods for the Chemical Analysis of Water and Wastes, EPA-600/4-79-020, U.S. EPA, EMSL, 1979.

Nitrogen, Nitrate-Nitrite — EPA Method 353.2

TITLE Inorganics, Non-Metallics

MATRIX Surface, saline, and wastewater.

APPLICATION Date issued 1971. Editorial Rev. 1974 and 1978. (Colorimetric, automated, cadmium reduction). EPA Method pertains to determination of nitrite singly or nitrite and nitrate combined. A filtered sample is passed through a column containing granulated copper-cadmium.

INTERFERENCES Build-up of suspended matter in reduction column restricts flow, low results may be found on samples with high concentrations of iron, copper or other metals, and samples with large concentrations of oil and grease will coat the surface of the cadmium.

INSTRUMENTATION Technicon auto analyzer, 540 nm filters, 15 or 50 mm tubular flow cell.

RANGE 0.05–10.0 mg/L NO3-NO2 N.

MDL Not listed.

PRECISION SD = 0.176 mg N/L at 2.48 NO3-N mg N/L.

ACCURACY As bias, −0.067 mg N/L at 2.48 NO3-N mg N/L.

SAMPLING METHOD Plastic or glass (100 mL).

STABILITY Cool, 4°C. H2SO4 to pH <2.

MHT 28 days.

QUALITY CONTROL Caution: samples for reduction column must not be preserved with mercuric chloride. When samples to be analyzed are saline waters, substitute ocean water should be used.(See Method 350.1). The range may be extended with sample dilution.

REFERENCE Methods for the Chemical Analysis of Water and Wastes, EPA-600/4-79-020, U.S. EPA, EMSL,

Nitrogen, Nitrate-Nitrite — EPA Method 353.3

TITLE Inorganics, Non-Metallics

MATRIX Drinking, surface, and saline waters. Wastewater.

APPLICATION Date issued 1974. (Spectrophotometric, cadmium reduction). A filtered sample is passed through a column containing granulated copper-cadmium to reduce nitrate to nitrite. The nitrite is determined by diazotizing to form a highly colored azo dye.

INTERFERENCES Build-up of suspended matter in the reduction column restricts sample flow. Low results may be obtained on samples with high concentrations of iron, copper or other metals. Samples with large amounts of oil and grease coat the surface of the cadmium.

INSTRUMENTATION Spectrophotometer. 540 nm. Light path of 1 cm or longer.

RANGE 0.01–1.0 mg/L NO3-NO2 N.

MDL Not listed.

PRECISION SD = ±0.004 and 0.005 at 0.24 and 0.55 mg NO3 + NO2-N/L.

ACCURACY Recoveries were, 100 and 102% at 0.24 and 0.55 mg NO3 + NO2-N/L

SAMPLING METHOD Plastic or glass (100 mL).

STABILITY Cool, 4°C. H2SO4 to pH <2.

MHT 28 days.

QUALITY CONTROL Caution: samples for reduction must not be preserved with mercuric chloride. Carry out procedures for turbidity removal, oil and grease removal and add EDTA to eliminate high concentrations of metals interference. The range may be extended with sample dilution.

REFERENCE Methods for the Chemical Analysis of Water and Wastes, EPA-600/4-79-020, U.S. EPA,EMSL, 1979.

Nitrogen, Nitrite — EPA Method 354.1

TITLE Inorganics, Non-Metallics

MATRIX Drinking, surface, and saline waters. Wastewater.

APPLICATION Date issued 1971. (Spectrophotometric). The diazonium compound formed by diazotation of sulfanilamide by nitrite in water under acid conditions is coupled with n-(1-naphthyl)-ethylene diamine dihydrochloride to produce a reddish-purple color.

INTERFERENCES There are very few known interferences at concentrations <1000 times that of nitrite. Strong oxidants or reductants readily affect nitrite concentrations. High alkalinity (>600 mg/L) give low results due to a pH shift.

INSTRUMENTATION Spectrophotometer at 540 nm. 1cm or larger cells.

RANGE 0.01–1.0 mg NO2-N/L.

MDL Not listed.

PRECISION Not listed.

ACCURACY Not listed.

SAMPLING METHOD Plastic or glass (50 mL).

STABILITY Cool, 4°C.

MHT 48 h.

QUALITY CONTROL Use distilled water free of nitrite and nitrate to prepare all reagents and standards. If sample pH is > 10 or total alkalinity is >600 mg/L, adjust pH to 6 with 1:3 HCl. If necessary, filter sample through 0.45 μm filter using first portion of filtrate to rinse filter flask.

REFERENCE Methods for the Chemical Analysis of Water and Wastes, EPA-600/4-79-020, U.S. EPA, EMSL, 1979.

2-Nitrophenol
CAS #88-75-5

EPA Method 1625

TITLE Semivolatile Organic Compounds by Isotope Dilution GC/MS

MATRIX The compounds may be determined in waters, soils, and municipal sludges by this method.

METHOD SUMMARY This method is used to determine 176 semivolatile toxic organic pollutants associated with the CWA (as amended 1987); the RCRA (as amended 1986); the CERCLA (as amended 1986); and other compounds amenable to extraction and analysis by capillary column gas chromatography-mass spectrometry (GC/MS).

Stable isotopically-labeled analogs of the compounds of interest are added to the sample. If the solids content is less than 1%, a 1-L sample is extracted at pH 12–13, then at pH <2 with methylene chloride using continuous extraction techniques.

If the solids content is 30% or less, the sample is diluted to 1% solids with reagent water, homogenized ultrasonically, and extracted at pH 12–13, then at pH <2 with methylene chloride using continuous extraction techniques. If the solids content is greater than 30%, the sample is extracted using ultrasonic techniques.

Each extract is dried over sodium sulfate, concentrated to a volume of 5 mL, cleaned up using GPC, if necessary, and concentrated. Extracts are concentrated to 1 mL if GPC is not performed, and to 0.5 mL if GPC is performed.

An internal standard is added to the extract, and a 1-mL aliquot of the extract is injected into the GC. The compounds are separated by GC and detected by a MS. The labeled compounds serve to correct the variability of the analytical technique.

INTERFERENCES Solvents, reagents, glassware, and other sample processing hardware may yield artifacts and/or elevated baselines causing misinterpretation of chromatograms and spectra. Materials used in the analysis must be demonstrated to be free from interferences under the conditions of analysis by running method blanks initially and with each sample lot (sample started through the extraction process on a given 8-h shift, to a maximum of 20). Specific selection of reagents and purification of solvents by distillation in all glass systems may be required. Glassware and, where possible, reagents are cleaned by solvent rinse and baking at 450°C for 1-h minimum. Interferences coextracted from samples will vary considerably from source to source, depending on the diversity of the site being sampled.

INSTRUMENTATION Major instrumentation includes a GC with a splitless or on-column injection port for capillary column, a MS with 70 eV electron impact ionization, and a data system to collect and record MS data, and process it. A K-D apparatus is used to concentrate extracts.

GC Column: 30 m × 0.25 mm I.D. 5% phenyl, 94% methyl, 1% vinyl silicone bonded phased fused silica capillary column.

PRECISION & ACCURACY The detection limits of the method are usually dependent on the level of interferences rather than instrumental limitations. The limits typify the minimum quantities that can be detected with no interferences present.

The minimum level (in µg/mL) was 20. This is defined as a minimum level at which the analytical system shall give recognizable mass spectra (background corrected) and acceptable calibration points.

The MDL (in µg/kg) in low solids was 49 and in high solids was 44; these were determined in digested sludge (low solids) and in filter cake or compost (high solids).

The labeled and native compound initial precision as standard deviation (in µg/L) was 15.

The labeled and native compound initial accuracy as average recovery (in µg/L) was 78–140.

SAMPLE COLLECTION, PRESERVATION & HANDLING
Collect samples in glass containers. Aqueous samples which flow freely are collected in refrigerated bottles using automatic sampling equipment. Solid samples are collected as grab samples using widemouth jars. Maintain samples at 0 to 4°C from the time of collection until extraction. If residual chlorine is present in aqueous samples, add 80 mg sodium thiosulfate/L of water. Begin sample extraction within 7 days of collection, and analyze all extracts within 40 days of extraction.

SAMPLE PREPARATION Samples containing 1% solids or less are extracted directly using continuous liquid-liquid extraction techniques. Samples containing 1 to 30% solids are diluted to the 1% level with reagent water and extracted using continuous liquid-liquid extraction techniques. Samples containing greater than 30% solids are extracted using ultrasonic techniques.

- Base/neutral extraction — Adjust the pH of the waters in the extractors to 12–13 with 6 N NaOH. Extract with methylene chloride for 24–48 h.
- Acid extraction — Adjust the pH of the waters in the extractors to 2 or less using 6 N sulfuric acid. Extract with methylene chloride for 24–48 h.
- Ultrasonic extraction of high solids samples — Add anhydrous sodium sulfate to the sample and QC aliquot(s). Add acetone:methylene chloride (1:1) to the sample and mix thoroughly

Concentrate extracts using a K-D apparatus.

QUALITY CONTROL The analyst is permitted to modify this method to improve separations or lower the costs of measurements, provided all performance specifications are met. Analyses of blanks are required to demonstrate freedom from contamination. When results of spikes indicate atypical method performance for samples, the samples are diluted to bring method performance within acceptable limits.

For low solids (aqueous samples), extract, concentrate, and analyze two sets of four 1-L aliquots (8 aliquots total) of the precision and recovery standard. For high solids samples, two sets of four 30-g aliquots of the high solids reference matrix are used.

Spike all samples with labeled compounds to assess method performance. Compute percent recovery of the labeled compounds using the internal standard method. Compare the labeled compound recovery for each compound with the corresponding labeled compound recovery.

Reagent water and high solids reference matrix blanks are analyzed to demonstrate freedom from contamination. Extract and concentrate a 1-L reagent water blank or a high solids reference matrix blank with each sample's lot (samples started through the extraction process on the same 8-h shift, to a maximum of 20 samples).

Field replicates may be collected to determine the precision of the sampling technique, and spiked samples may be required to determine the accuracy of the analysis when the internal standard method is used.

REFERENCE Semivolatile Organic Compounds by Isotope Dilution GC/MS. Office of Water Regulation and Standards, U.S. EPA Industrial Technology Division, Washington, DC, EPA Method 1625, Rev. C, June 1989 (contact W.A. Telliard, U.S. EPA, Office of Water Regulations and Standards, 401 M St., SW, Washington, DC, 20460. Phone: 202-382-7131).

4-Nitrophenol **EPA Method 1625**
CAS #100-02-7

TITLE Semivolatile Organic Compounds by Isotope Dilution GC/MS

MATRIX The compounds may be determined in waters, soils, and municipal sludges by this method.

METHOD SUMMARY This method is used to determine 176 semivolatile toxic organic pollutants associated with the CWA (as amended 1987); the RCRA (as amended 1986); the CERCLA (as amended 1986); and other compounds amenable to extraction and analysis by capillary column gas chromatography-mass spectrometry (GC/MS).

Stable isotopically-labeled analogs of the compounds of interest are added to the sample. If the solids content is less than 1%, a 1-L sample is extracted at pH 12–13, then at pH <2 with methylene chloride using continuous extraction techniques.

If the solids content is 30% or less, the sample is diluted to 1% solids with reagent water, homogenized ultrasonically, and extracted at pH 12–13, then at pH <2 with methylene chloride using continuous extraction techniques. If the solids content is greater than 30%, the sample is extracted using ultrasonic techniques.

Each extract is dried over sodium sulfate, concentrated to a volume of 5 mL, cleaned up using GPC, if necessary, and concentrated. Extracts are concentrated to 1 mL if GPC is not performed, and to 0.5 mL if GPC is performed.

An internal standard is added to the extract, and a 1-mL aliquot of the extract is injected into the GC. The compounds are separated by GC and detected by a MS. The labeled compounds serve to correct the variability of the analytical technique.

INTERFERENCES Solvents, reagents, glassware, and other sample processing hardware may yield artifacts and/or elevated baselines causing misinterpretation of chromatograms and spectra. Materials used in the analysis must be demonstrated to be free from interferences under the conditions of analysis by running method blanks initially and with each sample lot (sample started through the extraction process on a given 8-h shift, to a maximum of 20). Specific selection of reagents and purification of solvents by distillation in all glass systems may be required. Glassware and, where possible, reagents are cleaned by solvent rinse and baking at 450°C for 1-h minimum. Interferences coextracted from samples will vary considerably from source to source, depending on the diversity of the site being sampled.

INSTRUMENTATION Major instrumentation includes a GC with a splitless or on-column injection port for capillary column, a MS with 70 eV electron impact ionization, and a data system to collect and record MS data, and process it. A K-D apparatus is used to concentrate extracts.

GC Column: 30 m × 0.25 mm I.D. 5% phenyl, 94% methyl, 1% vinyl silicone bonded phased fused silica capillary column.

PRECISION & ACCURACY The detection limits of the method are usually dependent on the level of interferences rather than instrumental limitations. The limits typify the minimum quantities that can be detected with no interferences present.

The minimum level (in µg/mL) was 50. This is defined as a minimum level at which the analytical system shall give recognizable mass spectra (background corrected) and acceptable calibration points.

The MDL (in µg/kg) in low solids was 287 and in high solids was 11; these were determined in digested sludge (low solids) and in filter cake or compost (high solids).

The labeled and native compound initial precision as standard deviation (in µg/L) was 42.
The labeled and native compound initial accuracy as average recovery (in µg/L) was 62–146.

SAMPLE COLLECTION, PRESERVATION & HANDLING Collect samples in glass containers. Aqueous samples which flow freely are collected in refrigerated bottles using automatic sampling equipment. Solid samples are collected as grab samples using widemouth jars. Maintain samples at 0 to 4°C from the time of collection until extraction. If residual chlorine is present in aqueous samples, add 80 mg sodium thiosulfate/L of water. Begin sample extraction within 7 days of collection, and analyze all extracts within 40 days of extraction.

SAMPLE PREPARATION Samples containing 1% solids or less are extracted directly using continuous liquid-liquid extraction techniques. Samples containing 1 to 30% solids are diluted to the 1% level with reagent water and extracted using continuous liquid-liquid extraction techniques. Samples containing greater than 30% solids are extracted using ultrasonic techniques.

Base/neutral extraction — Adjust the pH of the waters in the extractors to 12–13 with 6 N NaOH. Extract with methylene chloride for 24–48 h.

Acid extraction — Adjust the pH of the waters in the extractors to 2 or less using 6 N sulfuric acid. Extract with methylene chloride for 24–48 h.

Ultrasonic extraction of high solids samples — Add anhydrous sodium sulfate to the sample and QC aliquot(s). Add acetone:methylene chloride (1:1) to the sample and mix thoroughly

Concentrate extracts using a K-D apparatus.

QUALITY CONTROL The analyst is permitted to modify this method to improve separations or lower the costs of measurements, provided all performance specifications are met. Analyses of blanks are required to demonstrate freedom from contamination. When results of spikes indicate atypical method performance for samples, the samples are diluted to bring method performance within acceptable limits.

For low solids (aqueous samples), extract, concentrate, and analyze two sets of four 1-L aliquots (8 aliquots total) of the precision and recovery standard. For high solids samples, two sets of four 30-g aliquots of the high solids reference matrix are used.

Spike all samples with labeled compounds to assess method performance. Compute percent recovery of the labeled compounds using the internal standard method. Compare the labeled compound recovery for each compound with the corresponding labeled compound recovery.

Reagent water and high solids reference matrix blanks are analyzed to demonstrate freedom from contamination. Extract and concentrate a 1-L reagent water blank or a high solids reference matrix blank with each sample's lot (samples started through the extraction process on the same 8-h shift, to a maximum of 20 samples).

Field replicates may be collected to determine the precision of the sampling technique, and spiked samples may be required to determine the accuracy of the analysis when the internal standard method is used.

REFERENCE Semivolatile Organic Compounds by Isotope Dilution GC/MS. Office of Water Regulation and Standards, U.S. EPA Industrial Technology Division, Washington, DC, EPA Method 1625, Rev. C, June 1989 (contact W.A. Telliard, U.S. EPA, Office of Water Regulations and Standards, 401 M St., SW, Washington, DC, 20460. Phone: 202-382-7131).

2-Nitrophenol EPA Method 625
CAS #88-75-5

TITLE Base/Neutrals and Acids, U.S. EPA Method 625

MATRIX This methods covers municipal and industrial wastewaters.

METHOD SUMMARY Approximately 1 L of sample is serially extracted with methylene chloride at a pH greater than 11 and again at a pH less than 2 using a separatory funnel or a continuous extractor. The methylene chloride extract is dried, concentrated to a volume of 1 mL, and analyzed by GC/MS. Qualitative identification of the parameters in the extract is performed using the retention time and the relative abundance of three characteristic masses (m/z). Qualitative analysis is performed using either external or internal standard techniques with a single characteristic m/z.

INTERFERENCES Method interferences may be caused by contaminants in solvents, reagents, glassware, and other sample processing hardware. Glassware must be scrupulously cleaned. Glassware should be heated in a muffle furnace at 400°C for 5 to 30 min. Some thermally stable materials, such as PCBs, may not be eliminated by this treatment. Solvent rinses with acetone and pesticide quality hexane may be substituted for the muffle furnace heating. Matrix interferences may be caused by contaminants that are coextracted from the sample. The base-neutral extraction may cause significantly reduced recovery of phenols. The packed gas chromatographic columns recommended for the basic fraction may not exhibit sufficient resolution for some analytes.

INSTRUMENTATION A GC/MS system with an injection port designed for on-column injection when using packed columns and for splitless injection when using capillary columns.

Column for base/neutrals: 1.8 m long × 2 mm I.D. glass, packed with 3% SP-2550 on Supelcoport (100/120 mesh) or equivalent.

Column for acids: 1.8 m long × 2 mm I.D. glass, packed with 1% SP-1240DA on Supelcoport (100/120 mesh) or equivalent.

PRECISION & ACCURACY The MDL concentrations were obtained using reagent water. The MDL actually achieved in a given analysis will vary depending on instrument sensitivity and matrix effects. This method was tested by 15 laboratories using reagent water, drinking water, surface water, and industrial wastewaters spiked at six concentrations over the range 5 to 100 µg/L. Single operator precision, overall precision, and method accuracy were found to be directly related to the concentration of the parameter matrix.

The MDL (in µg/L) in reagent water was 3.6.

The standard deviation (in µg/L based on 4 recovery measurements) was 35.2.

The range (in µg/L) for average recovery for 4 measurements was 45.0–166.7.

The range (in %) for percent recovery was 29–182.

Accuracy (in µg/L) as expected recovery for one or more measurements of a sample containing a true concentration of C was 1.07C-1.15.

Precision (in µg/L) as expected single analyst standard deviation of measurements at an average concentration found at X was 0.16X + 1.94.

Overall precision (in µg/L) as expected interlaboratory standard deviation of measurements in an average concentration found at X was 0.27X + 2.60.

$C = $ *True value of the concentration in µg/L.*

$X = $ *Average recovery found for measurements of samples containing a concentration at C in µg/L.*

SAMPLE PREPARATION Adjust the pH to >11 with sodium hydroxide and serially extract in a separatory funnel with methylene chloride or else in a continuous extractor. Next, adjust the pH to <2 with sulfuric acid and serially extract in a separatory funnel with methylene chloride or else in a continuous extractor. Dry the extracts separately through a column of anhydrous sodium sulfate and then concentrate each of the extracts to 1.0 mL using a K-D apparatus.

SAMPLE COLLECTION, PRESERVATION & HANDLING Grab samples must be collected in glass containers. All samples must be refrigerated at 4°C from the time of collection until extraction. If residual chlorine is present, add 80 mg of sodium thiosulfate/L of sample and mix well. All samples must be extracted within 7 days of collection and completely analyzed within 40 days of extraction.

QUALITY CONTROL Make an initial, one-time, demonstration of the ability to generate acceptable accuracy and precision with this method. Before processing any samples, the analyst must analyze a reagent water blank to demonstrate that interferences from the analytical system and glassware are under control. Each time a set of samples is extracted or reagents are changed, a reagent water blank must be processed. Spike and analyze a minimum of 5% of all samples to monitor and evaluate lab data quality. A QC check sample concentrate that contains each parameter of interest at a concentration of 100 µg/mL in acetone is required. PCBs and multicomponent pesticides may be omitted from this test.

After analysis of five spiked wastewater samples, calculate the average percent recovery and the standard deviation of the percent recovery. Spike all samples with the surrogate standard spiking solution and calculate the percent recovery of each surrogate compound.

REFERENCE Federal Register, Vol. 49, No. 209. Friday, Oct. 26, 1984.

4-Nitrophenol EPA Method 625
CAS #100-02-7

TITLE Base/Neutrals and Acids, U.S. EPA Method 625

MATRIX This methods covers municipal and industrial wastewaters.

METHOD SUMMARY Approximately 1 L of sample is serially extracted with methylene chloride at a pH greater than 11 and again at a pH less than 2 using a separatory funnel or a continuous extractor. The methylene chloride extract is dried, concentrated to a volume of 1 mL, and analyzed by GC/MS. Qualitative identification of the parameters in the extract is performed using the retention time and the relative abundance of three characteristic masses (m/z). Qualitative analysis is performed using either external or internal standard techniques with a single characteristic m/z.

INTERFERENCES Method interferences may be caused by contaminants in solvents, reagents, glassware, and other sample processing hardware. Glassware must be scrupulously cleaned. Glassware should be heated in a muffle furnace at 400°C for 5 to 30 min. Some thermally stable materials, such as PCBs, may not be eliminated by this treatment. Solvent rinses with acetone and pesticide quality hexane may be substituted for the muffle furnace heating. Matrix interferences may be caused by contaminants that are coextracted from the sample. The base-neutral extraction may cause significantly reduced recovery of phenols. The packed gas chromatographic columns recommended for the basic fraction may not exhibit sufficient resolution for some analytes.

INSTRUMENTATION A GC/MS system with an injection port designed for on-column injection when using packed columns and for splitless injection when using capillary columns.

Column for base/neutrals: 1.8 m long × 2 mm I.D. glass, packed with 3% SP-2550 on Supelcoport (100/120 mesh) or equivalent.

Column for acids: 1.8 m long × 2 mm I.D. glass, packed with 1% SP-1240DA on Supelcoport (100/120 mesh) or equivalent.

PRECISION & ACCURACY The MDL concentrations were obtained using reagent water. The MDL actually achieved in a given analysis will vary depending on instrument sensitivity and matrix effects. This method was tested by 15 laboratories using reagent water, drinking water, surface water, and industrial wastewaters spiked at six concentrations over the range 5 to 100 µg/L. Single operator precision, overall precision, and method accuracy were found to be directly related to the concentration of the parameter matrix.

The MDL (in µg/L) in reagent water was 2.4.

The standard deviation (in µg/L based on 4 recovery measurements) was 47.2.

The range (in µg/L) for average recovery for 4 measurements was 13.0–106.5.

The range (in %) for percent recovery was D-132.

Accuracy (in µg/L) as expected recovery for one or more measurements of a sample containing a true concentration of C was $0.61C - 1.22$.

Precision (in µg/L) as expected single analyst standard deviation of measurements at an average concentration found at X was $0.38X + 2.57$.

Overall precision (in µg/L) as expected interlaboratory standard deviation of measurements in an average concentration found at X was $0.44X + 3.24$.

C = *True value of the concentration in µg/L.*
X = *Average recovery found for measurements of samples containing a concentration at C in µg/L.*

SAMPLE PREPARATION Adjust the pH to >11 with sodium hydroxide and serially extract in a separatory funnel with methylene chloride or else in a continuous extractor. Next, adjust the pH to <2 with sulfuric acid and serially extract in a separatory funnel with methylene chloride or else in a continuous extractor. Dry the extracts separately through a column of anhydrous sodium sulfate and then concentrate each of the extracts to 1.0 mL using a K-D apparatus.

SAMPLE COLLECTION, PRESERVATION & HANDLING Grab samples must be collected in glass containers. All samples must be refrigerated at 4°C from the time of collection until extraction. If residual chlorine is present, add 80 mg of sodium thiosulfate/L of sample and mix well. All samples must be extracted within 7 days of collection and completely analyzed within 40 days of extraction.

QUALITY CONTROL Make an initial, one-time, demonstration of the ability to generate acceptable accuracy and precision with this method. Before processing any samples, the analyst must analyze a reagent water blank to demonstrate that interferences from the analytical system and glassware are under control. Each time a set of samples is extracted or reagents are changed, a reagent water blank must be processed. Spike and analyze a minimum of 5% of all samples to monitor and evaluate lab data quality. A QC check sample concentrate that contains each parameter of interest at a concentration of 100 µg/mL in acetone is required. PCBs and multicomponent pesticides may be omitted from this test.

After analysis of five spiked wastewater samples, calculate the average percent recovery and the standard deviation of the percent recovery. Spike all samples with the surrogate standard spiking solution and calculate the percent recovery of each surrogate compound.

REFERENCE Federal Register, Vol. 49, No. 209. Friday, Oct. 26, 1984.

4-Nitrophenol **EPA Method 8151**
CAS #100-02-7

TITLE Chlorinated Herbicides by GC Using Methylation or Pentafluorobenzylation Derivatization: Capillary Column Technique.

MATRIX This method covers aqueous and solid matrices. This includes a wide variety such as drinking water, groundwater, industrial wastewaters, surface waters, soils, solids, and sediments.

METHOD SUMMARY This is a GC method for determining 19 chlorinated acid herbicides in aqueous, soil, and waste matrices. Because these compounds are produced and used in various forms (i.e., acid, salt, ester, etc.) a hydrolysis step is included to convert the herbicide to the acid form prior to analysis. This method provides hydrolysis, extraction, derivatization and GC conditions for the analysis of chlorinated acid herbicides in water, soil, and waste samples. Water samples are hydrolyzed *in situ*, extracted with diethyl ether, and then esterified with either diazomethane or pentafluorobenzyl bromide. The derivatives are determined by gas chromatography with an electron capture detector (GC/ECD). The results are reported as acid equivalents. The sensitivity of this method depends on the level of interferences in addition to instrumental limitations.

INTERFERENCES Method interferences may be caused by contaminants in solvents, reagents, glassware, and other sample processing hardware. Immediately prior to use, glassware should be rinsed with the next solvent to be used. Matrix interferences may be caused by contaminants that are coextracted from the sample. Organic acids, especially chlorinated acids, cause the most direct interference with the determination by methylation. Phenols, including chlorophenols, may also interfere with this procedure. The determination using pentafluorobenzylation is more sensitive, and more prone to interferences from the presence of organic acids of phenols than by methylation. Alkaline hydrolysis and subsequent extraction of the basic solution removes many chlorinated hydrocarbons and phthalate esters that might otherwise interfere with the ECD analysis. The herbicides, being strong organic acids, react readily with alkaline substances and may be lost during analysis. Therefore, glassware must be acid-rinsed and then rinsed to constant pH with organic-free reagent water.

INSTRUMENTATION A GC suitable for Grob-type injection using capillary columns. A data system for measuring peak heights and/or peak areas is recommended. An electron capture detector (ECD) is used. Also a K-D apparatus, a diazomethane generator, a centrifuge and an ultrasonic disrupter will be required.

Narrow Bore Columns:
Primary Column 1: 30 m × 0.25 mm, 5% phenyl/95% methyl silicone (DB-5), 0.25 µm film thickness.
Primary Column 1a (GC/MS): 30 m × 0.32 mm, 5% phenyl/95% methyl silicone (DB-5), 1-µm film thickness.
Column 2: 30 m × 0.25 mm DB-608 with a 25 µm film thickness.
Confirmation Column: 30 m × 0.25 mm, 14% cyanopropyl phenyl silicone (DB-1701), 0.25 µm film thickness.

Megabore Columns:
Primary Column: 30 m × 0.53 mm DB-608 with 0.83 µm film thickness.
Confirmation Column: 30 m × 0.53 mm, 14% cyanopropyl phenyl silicone (DB-1701), 1.0 µm film thickness.

PRECISION & ACCURACY Method detection limits (MDLs) are compound-dependent and vary with derivitization efficiency, derivative recovery, the matrix sampled, and herbicide concentration.

The estimated MDL (in µg/L) was 0.13 for aqueous samples using GC/ECD.

The estimated MDL (in µg/kg) was 0.34 for soil samples using GC/ECD when corrected back to 50 g samples extracted and concentrated to 10 mL with 5-µL injections.

The estimated GC/MS identification limit (in ng) was 0.44 for soil samples using GC/MS.

Mean percent recovery, calculated from 7–8 determinations of spiked reagent water, after diazomethane derivatization, from a spike concentration (in µg/L) of 1 was 131 with a standard deviation of the percent recovery of 23.6.

Mean percent recovery, calculated from 10 determinations of spiked clay and clay/still bottom samples over the linear concentration range (in ng/g) of no data was none reported with a percent relative standard deviation of none. The RSD % was calculated on 10 samples high in the linear concentration range and 10 low in the range. The linear concentration range was determined using standard solutions and corrected to 50 g soil samples.

SAMPLE COLLECTION, PRESERVATION & HANDLING

Containers used to collect samples for the determination of semivolatile organic compounds should be soap and water washed followed by methanol (or isopropanol) rinsing. The sample containers should be of glass or Teflon® and have screw-top covers with Teflon® liners.

No preservation is used with concentrated waste samples. With liquid samples containing no residual chlorine and with soil, sediment, and sludge samples, immediately cooling to 4°C is the only preservation used. When residual chlorine is present then 3 mL of 10% aqueous sodium sulfate is added for each gallon of sample collected, followed by cooling to 4°C.

The holding time for all volatile organics samples is 14 days. Liquid samples must be extracted within 7 days and their extracts analyzed within 40 days. Concentrated waste, soil, sediment, and sludge samples must be extracted within 14 days and their extracts analyzed within 40 days.

SAMPLE PREPARATION

Preparation of soil, sediment, and other solid samples — Acidify 30 g (dry weight) solids with 0.1 M phosphate buffer (pH = 2.5) and thoroughly mix the contents. Spike the sample with surrogate compound(s). The ultrasonic extraction of solids must be optimized for each type of sample. In order for the ultrasonic extractor to efficiently extract solid samples, the sample must be free flowing when the solvent is added. Acidified anhydrous sodium sulfate should be added to clay-type soils, or any other solid that is not a free-flowing sandy texture, until a free flowing mixture is obtained. Add methylene chloride and perform ultrasonic extraction. Combine organic extracts from the repetitive extractings of the sample and centrifuge. Add aqueous potassium hydroxide, water, and methanol to the extract and reflux the mixture on a water bath. Extract the solution three times with methylene chloride and discard the methylene chloride phase. The basic solution contains the herbicide salts. Adjust the pH of the solution to <2 with cold sulfuric acid and extract three times with methylene chloride. Combine the extracts and pour them through a pre-rinsed drying column containing acidified anhydrous sodium sulfate. Collect the dried extracts in a K-D flask and concentrate them.

Preparation of aqueous samples — Measure 1 L of sample into a 2 L separatory funnel and spike it with surrogate compound(s). Add NaCl to the sample, then add 6 N NaOH to the sample to a pH of 12 or more and let the sample sit at room temperature for 1 h to hydrolyze esters. Extract the sample three times with methylene chloride and discard the extracts. Then add cold 12 N sulfuric acid to a pH less than or equal to 2, and extract the sample three times with ethyl ether. Collect the ether phase in a flask containing acidified anhydrous sodium sulfate and allow it to remain in contact with the sodium sulfate for a minimum of 2 h. The drying step is very critical to ensuring complete esterification; any moisture remaining in the ether will result in low herbicide recoveries.

Extract concentration and derivatization — The combined ether extract is concentrated to about 1 mL using a K-D apparatus followed by a micro Snyder column or nitrogen gas blowdown. If methyl esters are to be produced, then dilute the concentrated ether extract with 1 mL of isooctane and 0.5 mL of methanol, dilute to a final volume of 4 mL, and esterify with diazomethane. If pentafluorobenzene esters are to be produced, then dilute concentrated ether extract with acetone to a final volume of 4 mL and esterify with pentafluorobenzyl bromide.

QUALITY CONTROL
Select a representative spike concentration for each compound (acid or ester) to be measured. Using stock standard, prepare a quality control check sample concentrate, in acetone, that is 1000 times more concentrated than the selected concentrations. Use this quality control check sample concentrate to prepare quality control check samples. Calculate surrogate standard recovery on all standards, samples, blanks, and spikes. GC/MS techniques should be judiciously employed to support qualitative identifications made with this method. When available, chemical ionization mass spectra may be employed to aid the qualitative identification process.

REFERENCE
Test Methods for Evaluating Solid Waste, Physical/Chemical Methods, SW-846, 3rd Edition, U.S. EPA, Office of Solid Waste, Washington, DC, EPA Method 8151, Nov. 1990.

2-Nitrophenol
CAS #88-75-5
EPA Method 8270

TITLE Semivolatile Organic Compounds by GC/MS

MATRIX This method is used to determine the concentration of semivolatile organic compounds in extracts prepared from all types of solid waste matrices, soils, and groundwater. Although surface waters are not specifically mentioned, this method should be applicable to water samples from rivers, lakes, etc.

METHOD SUMMARY This method covers 259 semivolatile organic compounds. In very limited applications direct injection of the sample into the GC/MS system may be appropriate, but this results in very high detection limits (approximately 10,000 µg/L). Typically, a 1-L liquid sample, containing surrogate, and matrix spiking standards, is extracted in a continuous extractor first under acid conditions and then under basic conditions. Typically 30 g of a solid sample, containing surrogate, and matrix spiking standards, is extracted ultrasonically. After concentrating the extract to 1 mL it is spiked with 10 µL of an internal standard solution just prior to analysis by GC/MS. The volume injected should contain about 100 ng of base/neutral and 200 ng of acid surrogates (for a 1-µL injection). Analysis is performed by GC/MS using a capillary GC column.

INTERFERENCES Raw GC/MS data from all blanks, samples, and spikes must be evaluated for interferences. Contamination by carryover can occur whenever high-concentration and low-concentration samples are sequentially analyzed. To reduce carryover, the sample syringe must be rinsed out between samples with solvent. Whenever an unusually concentrated sample is encountered, it should be followed by the analysis of blank solvent to check for cross-contamination.

INSTRUMENTATION A GC/MS and a data system are required. The GC column used is a 30 m × 0.25 mm I.D. (or

0.32 mm I.D.) 1um film thickness silicone-coated fused silica capillary column. A continuous liquid-liquid extractor equipped with Teflon® or glass connection joints and stopcocks requiring no lubrication, a K-D concentrating apparatus, water bath, and an ultrasonic disrupter with a minimum power of 300 W and with pulsing capability are also required.

PRECISION & ACCURACY The estimated quantitation limit (EQL) of Method 8270B for determining an individual compound is approximately 1 mg/kg (wet weight) for soil or sediment samples, 1–200 mg/kg for wastes (dependent on matrix and method of preparation), and 10 µg/L for groundwater samples. EQLs will be proportionately higher for sample extracts that require dilution to avoid saturation of the detector.

The EQL(b) for groundwater in µg/L is 10.
The EQL (a, b) for low concentrations in soil and sediment in µg/kg is 660.
Accuracy as µg/L is $0.07C - 1.15$.
Overall precision in µg/L is $0.27X + 2.60$.

(a) *EQLs listed for soil/sediment are based on wet weight. Normally data is reported in a dry-weight basis; therefore, EQLs will be higher based on the % dry weight of each sample. This calculation is based on a 30 g sample and gel permeation chromatography cleanup.*
(b) *Sample EQLs are highly matrix-dependent. The EQLs are provided for guidance and may not always be achievable.*
$C =$ *True value for concentration, in µg/L.*
$X =$ *Average recovery found for measurements of samples containing a concentration of C, in µg/L.*

ESTIMATED QUANTITATION LIMIT

Other Matrices	Factor (a)
High-concentration soil and sludges by sonicator	7.5
Non-water miscible waste	75

(a) EQL for other matrices = [EQL for low soil/sediment] × [Factor]. This estimated EQL is similar to an EPA "Practical Quantitation Limit."

SAMPLING METHOD
Liquid samples — Use a 1 or 2½ gallon amber glass bottle with a screw-top Teflon®-lined cover that has been prewashed with detergent and rinsed with distilled water and methanol (or isopropanol).

Soils, sediments, or sludges — Use an 8-oz. widemouth glass with a screw-top Teflon®-lined cover that has been prewashed with detergent and rinsed with distilled water and methanol (or isopropanol).

SAMPLE PRESERVATION
Liquid samples — If residual chlorine is present, add 3 mL of 10% sodium thiosulfate per gallon, cool to 4°C and store in a solvent-free refrigerator until analysis; if chlorine is not present, then eliminate the sodium thiosulfate addition.

Soils, sediments, or sludges — Cool samples to 4°C and store in a solvent-free refrigerator.

MHT Liquid samples must be extracted within 7 days and the extracts analyzed within 40 days. Soils, sediments, or sludges may be stored for a maximum of 14 days and the extracts analyzed within 40 days.

SAMPLE PREPARATION
Liquid samples — Transfer 1 L quantitatively to a continuous extractor. If high concentrations are anticipated, a smaller volume may be used and then diluted with organic-free reagent water to 1 L. Adjust pH, if necessary, to pH <2 using 1:1 (V/V) sulfuric acid. Pipette 1.0 mL of a surrogate standard spiking solution into each sample. For the sample in each analytical batch selected for spiking, add 1.0 mL of a matrix spiking standard. For base/neutral acid analysis, the amount of the surrogates and matrix spiking compounds added to the sample should result in a final concentration of 100 ng/µL of each analyte in the extract to be analyzed (assuming a 1-µL injection). Extract with methylene chloride for 18–24 h. Next, adjust the pH of the aqueous phase to pH >11 using 10 *N* sodium hydroxide and extract it with methylene chloride again for 18–24 h. Dry the extract through a column containing anhydrous sodium sulfate and concentrate it to 1 mL using a K-D concentrator.

Soils, sediments, or sludges — Use 30 g of sample. Nonporous or wet samples (gummy or clay type) that do not have a free-flowing sandy texture must be mixed with anhydrous sodium sulfate until the sample is free flowing. Add 1 mL of surrogate standards to all samples, spikes, standards, and blanks. For the sample in each analytical batch selected for spiking, add 1.0 mL of a matrix spiking standard. For base/neutral acid analysis, the amount added of the surrogates and matrix spiking compounds should result in a final concentration of 100 ng/µL of each base/neutral analyte and 200 ng/µL of each acid analyte in the extract to be analyzed (assuming a 1-µL injection). Immediately add a 100-mL mixture of 1:1 methylene chloride:acetone and extract the sample ultrasonically for 3 min and then decant or filter the extracts. Repeat the extraction two or more times. Dry the extract using a column with anhydrous sodium sulfate and concentrate it to 1 mL in a K-D concentrator.

QUALITY CONTROL A methylene chloride solution containing 50 ng/µL of decafluorotriphenylphosphine (DFTPP) is used for tuning the GC/MS system each 12-h shift. A system performance check also must be made during every 12-h shift. A standard containing 50 ng/µL each of 4,4′-DDT, pentachlorophenol, and benzidine is required to verify injection port inertness and GC column performance. A calibration standard at mid-concentration, containing each compound of interest, including all required surrogates, must be performed every 12 h during analysis. After the system performance check is met, calibration check compounds (CCCs) are used to check the validity of the initial calibration.

The internal standard responses and retention times in the calibration check standard must be evaluated immediately after or during data acquisition. If the retention time for any internal standard changes by more than 30 seconds from the last check calibration (12 h), the chromatographic system must be inspected for malfunctions and corrections must be made, as required. If the electron ionization current plot (EICP) area for any of the internal standards changes by a factor of two from the last daily calibration standard check, the mass spectrometer

must be inspected for malfunctions and corrections must be made, as appropriate.

Demonstrate, through the analysis of a reagent water blank, that interferences from the analytical system, glassware, and reagents are under control. The blank samples should be carried through all stages of the sample preparation and measurement steps. For each analytical batch (up to 20 samples), a reagent blank, matrix spike, and matrix spike duplicate/duplicate must be analyzed (the frequency of the spikes may be different for different monitoring programs). The blank and spiked samples must be carried through all stages of the sample preparation and measurement steps. A QC reference sample concentrate containing each analyte at a concentration of 100 mg/L in methanol is required.

REFERENCE Test Methods for Evaluating Solid Waste (SW-846). U.S. EPA 1983, Method 8270B, Rev. 2, Nov. 1990. Office of Solid Waste, Washington, DC.

4-Nitrophenol   **EPA Method 8270**
CAS #100-02-7

TITLE Semivolatile Organic Compounds by GC/MS

MATRIX This method is used to determine the concentration of semivolatile organic compounds in extracts prepared from all types of solid waste matrices, soils, and groundwater. Although surface waters are not specifically mentioned, this method should be applicable to water samples from rivers, lakes, etc.

METHOD SUMMARY This method covers 259 semivolatile organic compounds. In very limited applications direct injection of the sample into the GC/MS system may be appropriate, but this results in very high detection limits (approximately 10,000 µg/L). Typically, a 1-L liquid sample, containing surrogate, and matrix spiking standards, is extracted in a continuous extractor first under acid conditions and then under basic conditions. Typically 30 g of a solid sample, containing surrogate, and matrix spiking standards, is extracted ultrasonically. After concentrating the extract to 1 mL it is spiked with 10 µL of an internal standard solution just prior to analysis by GC/MS. The volume injected should contain about 100 ng of base/neutral and 200 ng of acid surrogates (for a 1-µL injection). Analysis is performed by GC/MS using a capillary GC column.

INTERFERENCES Raw GC/MS data from all blanks, samples, and spikes must be evaluated for interferences. Contamination by carryover can occur whenever high-concentration and low-concentration samples are sequentially analyzed. To reduce carryover, the sample syringe must be rinsed out between samples with solvent. Whenever an unusually concentrated sample is encountered, it should be followed by the analysis of blank solvent to check for cross-contamination.

INSTRUMENTATION A GC/MS and a data system are required. The GC column used is a 30 m × 0.25 mm I.D. (or 0.32 mm I.D.) 1um film thickness silicone-coated fused silica capillary column. A continuous liquid-liquid extractor equipped with Teflon® or glass connection joints and stopcocks requiring no lubrication, a K-D concentrating apparatus, water bath, and an ultrasonic disrupter with a minimum power of 300 W and with pulsing capability are also required.

PRECISION & ACCURACY The estimated quantitation limit (EQL) of Method 8270B for determining an individual compound is approximately 1 mg/kg (wet weight) for soil or sediment samples, 1–200 mg/kg for wastes (dependent on matrix and method of preparation), and 10 µg/L for groundwater samples. EQLs will be proportionately higher for sample extracts that require dilution to avoid saturation of the detector.

The EQL(b) for groundwater in µg/L is 50.
The EQL (a, b) for low concentrations in soil and sediment in µg/kg is 3300.
Accuracy as µg/L is $0.61C - 1.22$.
Overall precision in µg/L is $0.44X + 3.24$.

(a) *EQLs listed for soil/sediment are based on wet weight. Normally data is reported in a dry-weight basis; therefore, EQLs will be higher based on the % dry weight of each sample. This calculation is based on a 30 g sample and gel permeation chromatography cleanup.*
(b) *Sample EQLs are highly matrix-dependent. The EQLs are provided for guidance and may not always be achievable.*
$C =$ *True value for concentration, in µg/L.*
$X =$ *Average recovery found for measurements of samples containing a concentration of C, in µg/L.*

ESTIMATED QUANTITATION LIMIT

Other Matrices	Factor (a)
High-concentration soil and sludges by sonicator	7.5
Non-water miscible waste	75

(a) *EQL for other matrices = [EQL for low soil/sediment] × [Factor]. This estimated EQL is similar to an EPA "Practical Quantitation Limit."*

SAMPLING METHOD
Liquid samples — Use a 1 or 2½ gallon amber glass bottle with a screw-top Teflon®-lined cover that has been prewashed with detergent and rinsed with distilled water and methanol (or isopropanol).

Soils, sediments, or sludges — Use an 8-oz. widemouth glass with a screw-top Teflon®-lined cover that has been prewashed with detergent and rinsed with distilled water and methanol (or isopropanol).

SAMPLE PRESERVATION
Liquid samples — If residual chlorine is present, add 3 mL of 10% sodium thiosulfate per gallon, cool to 4°C and store in a solvent-free refrigerator until analysis; if chlorine is not present, then eliminate the sodium thiosulfate addition.

Soils, sediments, or sludges — Cool samples to 4°C and store in a solvent-free refrigerator.

MHT Liquid samples must be extracted within 7 days and the extracts analyzed within 40 days. Soils, sediments, or sludges may be stored for a maximum of 14 days and the extracts analyzed within 40 days.

SAMPLE PREPARATION

Liquid samples — Transfer 1 L quantitatively to a continuous extractor. If high concentrations are anticipated, a smaller volume may be used and then diluted with organic-free reagent water to 1 L. Adjust pH, if necessary, to pH <2 using 1:1 (V/V) sulfuric acid. Pipette 1.0 mL of a surrogate standard spiking solution into each sample. For the sample in each analytical batch selected for spiking, add 1.0 mL of a matrix spiking standard. For base/neutral acid analysis, the amount of the surrogates and matrix spiking compounds added to the sample should result in a final concentration of 100 ng/µL of each analyte in the extract to be analyzed (assuming a 1-µL injection). Extract with methylene chloride for 18–24 h. Next, adjust the pH of the aqueous phase to pH >11 using 10 N sodium hydroxide and extract it with methylene chloride again for 18–24 h. Dry the extract through a column containing anhydrous sodium sulfate and concentrate it to 1 mL using a K-D concentrator.

Soils, sediments, or sludges — Use 30 g of sample. Nonporous or wet samples (gummy or clay type) that do not have a free-flowing sandy texture must be mixed with anhydrous sodium sulfate until the sample is free flowing. Add 1 mL of surrogate standards to all samples, spikes, standards, and blanks. For the sample in each analytical batch selected for spiking, add 1.0 mL of a matrix spiking standard. For base/neutral acid analysis, the amount added of the surrogates and matrix spiking compounds should result in a final concentration of 100 ng/µL of each base/neutral analyte and 200 ng/µL of each acid analyte in the extract to be analyzed (assuming a 1-µL injection). Immediately add a 100-mL mixture of 1:1 methylene chloride:acetone and extract the sample ultrasonically for 3 min and then decant or filter the extracts. Repeat the extraction two or more times. Dry the extract using a column with anhydrous sodium sulfate and concentrate it to 1 mL in a K-D concentrator.

Note: This compound may be exhibit erratic chromatographic behavior, especially if the GC system is contaminated with high boiling material.

QUALITY CONTROL

A methylene chloride solution containing 50 ng/µL of decafluorotriphenylphosphine (DFTPP) is used for tuning the GC/MS system each 12-h shift. A system performance check also must be made during every 12-h shift. A standard containing 50 ng/µL each of 4,4'-DDT, pentachlorophenol, and benzidine is required to verify injection port inertness and GC column performance. A calibration standard at mid-concentration, containing each compound of interest, including all required surrogates, must be performed every 12 h during analysis. After the system performance check is met, calibration check compounds (CCCs) are used to check the validity of the initial calibration.

The internal standard responses and retention times in the calibration check standard must be evaluated immediately after or during data acquisition. If the retention time for any internal standard changes by more than 30 seconds from the last check calibration (12 h), the chromatographic system must be inspected for malfunctions and corrections must be made, as required. If the electron ionization current plot (EICP) area for any of the internal standards changes by a factor of two from the last daily calibration standard check, the mass spectrometer must be inspected for malfunctions and corrections must be made, as appropriate.

Demonstrate, through the analysis of a reagent water blank, that interferences from the analytical system, glassware, and reagents are under control. The blank samples should be carried through all stages of the sample preparation and measurement steps. For each analytical batch (up to 20 samples), a reagent blank, matrix spike, and matrix spike duplicate/duplicate must be analyzed (the frequency of the spikes may be different for different monitoring programs). The blank and spiked samples must be carried through all stages of the sample preparation and measurement steps. A QC reference sample concentrate containing each analyte at a concentration of 100 mg/L in methanol is required.

REFERENCE Test Methods for Evaluating Solid Waste (SW-846). U.S. EPA 1983, Method 8270B, Rev. 2, Nov. 1990. Office of Solid Waste, Washington, DC.

2-Nitrophenol   EPA Method 8040
CAS #88-75-5

TITLE Phenols

MATRIX Groundwater, soils, sludges, water miscible liquid wastes, and non-water miscible wastes.

APPLICATION This method is used for the analysis of 17 phenols. Samples are extracted, concentrated, and analyzed using direct injection of both neat and diluted organic liquids. Pentafluorobenzylbromide (PFB) derivatives also may be made to increase sensitivity of the method.

INTERFERENCES There can be carryover contamination with high- and low-level samples. Solvents, reagents, and glassware may introduce artifacts. Other interferences may come from coextracted compounds from samples.

INSTRUMENTATION GC capable of on-column injections and a flame with detector (FID) or electron capture detector (ECD). Column for underivatized phenol: 1.8 m by 2.0 mm with 1% SP-1240DA on Supelcoport. Column for derivatized phenols: 1.8 m by 2.0 mm with 5% OV-17 on Chromosorb W-AW-DMCS.

RANGE 12–450 µg/L.

MDL 0.45 µg/L (FID) and 0.77 µg/L (ECD)

PQL FACTORS FOR MULTIPLYING × FID MDL VALUE

Matrix	Multiplication Factor
Groundwater	10
Low-level soil by sonication with GPC cleanup	670
High-level soil and sludge by sonication	10,000
Non-water miscible waste	100,000

PRECISION 0.14X + 3.84 µg/L (overall precision using FID)

ACCURACY 0.81C–0.76 µg/L (as recovery using FID)

SAMPLING METHOD Use 8-oz. widemouth glass bottles with Teflon®-lined caps for concentrated waste samples, soils, sediments, and sludges. Use 1 or 2½ gallon amber glass bottles with Teflon®-lined caps for liquid (water) samples.

STABILITY Cool soil, sediment, sludge, and liquid samples to 4°C. If residual chlorine is present in liquid samples add 3 mL of 10% sodium thiosulfate per gallon of sample and cool to 4°C.

MHT 14 days for concentrated waste, soil, sediment, or sludge; 7 days for liquid samples; all extracts must be analyzed within 40 days.

QUALITY CONTROL A quality control check sample concentrate containing each analyte of interest is required. The QC check sample concentrate may be prepared from pure standard materials or purchased as certified solutions Use appropriate trip, matrix, control site, method, reagent, and solvent blanks. Internal, surrogate, and five concentration level calibration standards are used. The QC check sample concentrate should contain this compound at 100 µg/mL in 2-propanol.

REFERENCE Test Methods for Evaluating Solid Waste (SW-846), U.S. EPA Office of Solid Waste, Washington, DC, Method 8040A, Rev. 1, Nov. 1990.

4-Nitrophenol   EPA Method 8040
CAS #100-02-7

TITLE Phenols

MATRIX Groundwater, soils, sludges, water miscible liquid wastes, and non-water miscible wastes.

APPLICATION This method is used for the analysis of 17 phenols. Samples are extracted, concentrated, and analyzed using direct injection of both neat and diluted organic liquids. Pentafluorobenzylbromide (PFB) derivatives also may be made to increase sensitivity of the method.

INTERFERENCES There can be carryover contamination with high- and low-level samples. Solvents, reagents, and glassware may introduce artifacts. Other interferences may come from coextracted compounds from samples.

INSTRUMENTATION GC capable of on-column injections and a flame with detector (FID) or electron capture detector (ECD). Column for underivatized phenol: 1.8 m by 2.0 mm with 1% SP-1240DA on Supelcoport. Column for derivatized phenols: 1.8 m by 2.0 mm with 5% OV-17 on Chromosorb W-AW-DMCS.

RANGE 12–450 µg/L.

MDL 2.8 µg/L (FID) and 0.70 µg/L (ECD)

PQL FACTORS FOR MULTIPLYING × FID MDL VALUE

Matrix	Multiplication Factor
Groundwater	10
Low-level soil by sonication with GPC cleanup	670
High-level soil and sludge by sonication	10,000
Non-water miscible waste	100,000

PRECISION 0.19X + 4.79 µg/L (overall precision using FID)

ACCURACY 0.46C + 0.18 µg/L (as recovery using FID)

SAMPLING METHOD Use 8-oz. widemouth glass bottles with Teflon®-lined caps for concentrated waste samples, soils, sediments, and sludges. Use 1 or 2½ gallon amber glass bottles with Teflon®-lined caps for liquid (water) samples.

STABILITY Cool soil, sediment, sludge, and liquid samples to 4°C. If residual chlorine is present in liquid samples add 3 mL of 10% sodium thiosulfate per gallon of sample and cool to 4°C.

MHT 14 days for concentrated waste, soil, sediment, or sludge; 7 days for liquid samples; all extracts must be analyzed within 40 days.

QUALITY CONTROL A quality control check sample concentrate containing each analyte of interest is required. The QC check sample concentrate may be prepared from pure standard materials or purchased as certified solutions Use appropriate trip, matrix, control site, method, reagent, and solvent blanks. Internal, surrogate, and five concentration level calibration standards are used. The QC check sample concentrate should contain this compound at 100 µg/mL in 2-propanol.

REFERENCE Test Methods for Evaluating Solid Waste (SW-846), U.S. EPA Office of Solid Waste, Washington, DC, Method 8040A, Rev. 1, Nov. 1990.

Nitroquinoline-1-oxide   EPA Method 8270
CAS #56-57-5

TITLE Semivolatile Organic Compounds by GC/MS

MATRIX This method is used to determine the concentration of semivolatile organic compounds in extracts prepared from all types of solid waste matrices, soils, and groundwater. Although surface waters are not specifically mentioned, this method should be applicable to water samples from rivers, lakes, etc.

METHOD SUMMARY This method covers 259 semivolatile organic compounds. In very limited applications direct injection of the sample into the GC/MS system may be appropriate, but this results in very high detection limits (approximately 10,000 µg/L). Typically, a 1-L liquid sample, containing surrogate, and matrix spiking standards, is extracted in a continuous extractor first under acid conditions and then under basic conditions. Typically 30 g of a solid sample, containing surrogate, and matrix spiking standards, is extracted ultrasonically. After concentrating the extract to 1 mL it is spiked with 10 µL of an internal standard solution just prior to analysis by GC/MS. The volume injected should contain about 100 ng of base/neutral and 200 ng of acid surrogates (for a 1-µL injection). Analysis is performed by GC/MS using a capillary GC column.

INTERFERENCES Raw GC/MS data from all blanks, samples, and spikes must be evaluated for interferences. Contamination by carryover can occur whenever high-concentration and low-concentration samples are sequentially analyzed. To reduce carryover, the sample syringe must be rinsed out between samples with solvent. Whenever an unusually concentrated sample is encountered, it should be followed by the analysis of blank solvent to check for cross-contamination.

INSTRUMENTATION A GC/MS and a data system are required. The GC column used is a 30 m × 0.25 mm I.D. (or 0.32 mm I.D.) 1um film thickness silicone-coated fused silica capillary column. A continuous liquid-liquid extractor equipped with Teflon® or glass connection joints and stopcocks requiring no lubrication, a K-D concentrating apparatus, water bath, and an ultrasonic disrupter with a minimum power of 300 W and with pulsing capability are also required.

PRECISION & ACCURACY The estimated quantitation limit (EQL) of Method 8270B for determining an individual compound is approximately 1 mg/kg (wet weight) for soil or sediment samples, 1–200 mg/kg for wastes (dependent on matrix and method of preparation), and 10 µg/L for groundwater samples. EQLs will be proportionately higher for sample extracts that require dilution to avoid saturation of the detector.

The EQL(b) for groundwater in µg/L is 40.
The EQL (a, b) for low concentrations in soil and sediment in µg/kg is not determined.
Accuracy as µg/L is not listed.
Overall precision in µg/L is not listed.

(a) EQLs listed for soil/sediment are based on wet weight. Normally data is reported in a dry-weight basis; therefore, EQLs will be higher based on the % dry weight of each sample. This calculation is based on a 30 g sample and gel permeation chromatography cleanup.
(b) Sample EQLs are highly matrix-dependent. The EQLs are provided for guidance and may not always be achievable.
C = True value for concentration, in µg/L.
X = Average recovery found for measurements of samples containing a concentration of C, in µg/L.

ESTIMATED QUANTITATION LIMIT

Other Matrices	Factor (a)
High-concentration soil and sludges by sonicator	7.5
Non-water miscible waste	75

(a) EQL for other matrices = [EQL for low soil/sediment] × [Factor]. This estimated EQL is similar to an EPA "Practical Quantitation Limit."

SAMPLING METHOD
Liquid samples — Use a 1 or 2½ gallon amber glass bottle with a screw-top Teflon®-lined cover that has been prewashed with detergent and rinsed with distilled water and methanol (or isopropanol).

Soils, sediments, or sludges — Use an 8-oz. widemouth glass with a screw-top Teflon®-lined cover that has been prewashed with detergent and rinsed with distilled water and methanol (or isopropanol).

SAMPLE PRESERVATION
Liquid samples — If residual chlorine is present, add 3 mL of 10% sodium thiosulfate per gallon, cool to 4°C and store in a solvent-free refrigerator until analysis; if chlorine is not present, then eliminate the sodium thiosulfate addition.

Soils, sediments, or sludges — Cool samples to 4°C and store in a solvent-free refrigerator.

MHT Liquid samples must be extracted within 7 days and the extracts analyzed within 40 days. Soils, sediments, or sludges may be stored for a maximum of 14 days and the extracts analyzed within 40 days.

SAMPLE PREPARATION
Liquid samples — Transfer 1 L quantitatively to a continuous extractor. If high concentrations are anticipated, a smaller volume may be used and then diluted with organic-free reagent water to 1 L. Adjust pH, if necessary, to pH <2 using 1:1 (V/V) sulfuric acid. Pipette 1.0 mL of a surrogate standard spiking solution into each sample. For the sample in each analytical batch selected for spiking, add 1.0 mL of a matrix spiking standard. For base/neutral acid analysis, the amount of the surrogates and matrix spiking compounds added to the sample should result in a final concentration of 100 ng/µL of each analyte in the extract to be analyzed (assuming a 1-µL injection). Extract with methylene chloride for 18–24 h. Next, adjust the pH of the aqueous phase to pH >11 using 10 N sodium hydroxide and extract it with methylene chloride again for 18–24 h. Dry the extract through a column containing anhydrous sodium sulfate and concentrate it to 1 mL using a K-D concentrator.

Soils, sediments, or sludges — Use 30 g of sample. Nonporous or wet samples (gummy or clay type) that do not have a free-flowing sandy texture must be mixed with anhydrous sodium sulfate until the sample is free flowing. Add 1 mL of surrogate standards to all samples, spikes, standards, and blanks. For the sample in each analytical batch selected for spiking, add 1.0 mL of a matrix spiking standard. For base/neutral acid analysis, the amount added of the surrogates and matrix spiking compounds should result in a final concentration of 100 ng/µL of each base/neutral analyte and 200 ng/µL of each acid analyte in the extract to be analyzed (assuming a 1-µL injection). Immediately add a 100-mL mixture of 1:1 methylene chloride:acetone and extract the sample ultrasonically for 3 min and then decant or filter the extracts. Repeat the extraction two or more times. Dry the extract using a column with anhydrous sodium sulfate and concentrate it to 1 mL in a K-D concentrator.

QUALITY CONTROL A methylene chloride solution containing 50 ng/µL of decafluorotriphenylphosphine (DFTPP) is used for tuning the GC/MS system each 12-h shift. A system performance check also must be made during every 12-h shift. A standard containing 50 ng/µL each of 4,4'-DDT, pentachlorophenol, and benzidine is required to verify injection port inertness and GC column performance. A calibration standard at mid-concentration, containing each compound of interest, including all required surrogates, must be performed every 12 h during analysis. After the system performance check is met, calibration check compounds (CCCs) are used to check the validity of the initial calibration.

The internal standard responses and retention times in the calibration check standard must be evaluated immediately after or during data acquisition. If the retention time for any internal standard changes by more than 30 seconds from the last check calibration (12 h), the chromatographic system must be

inspected for malfunctions and corrections must be made, as required. If the electron ionization current plot (EICP) area for any of the internal standards changes by a factor of two from the last daily calibration standard check, the mass spectrometer must be inspected for malfunctions and corrections must be made, as appropriate.

Demonstrate, through the analysis of a reagent water blank, that interferences from the analytical system, glassware, and reagents are under control. The blank samples should be carried through all stages of the sample preparation and measurement steps. For each analytical batch (up to 20 samples), a reagent blank, matrix spike, and matrix spike duplicate/duplicate must be analyzed (the frequency of the spikes may be different for different monitoring programs). The blank and spiked samples must be carried through all stages of the sample preparation and measurement steps. A QC reference sample concentrate containing each analyte at a concentration of 100 mg/L in methanol is required.

REFERENCE Test Methods for Evaluating Solid Waste (SW-846). U.S. EPA 1983, Method 8270B, Rev. 2, Nov. 1990. Office of Solid Waste, Washington, DC.

N-Nitrosodi-n-butylamine   **EPA Method 1625**
CAS #924-16-3

TITLE Semivolatile Organic Compounds by Isotope Dilution GC/MS

MATRIX The compounds may be determined in waters, soils, and municipal sludges by this method.

METHOD SUMMARY This method is used to determine 176 semivolatile toxic organic pollutants associated with the CWA (as amended 1987); the RCRA (as amended 1986); the CERCLA (as amended 1986); and other compounds amenable to extraction and analysis by capillary column gas chromatography-mass spectrometry (GC/MS).

Stable isotopically-labeled analogs of the compounds of interest are added to the sample. If the solids content is less than 1%, a 1-L sample is extracted at pH 12–13, then at pH <2 with methylene chloride using continuous extraction techniques.

If the solids content is 30% or less, the sample is diluted to 1% solids with reagent water, homogenized ultrasonically, and extracted at pH 12–13, then at pH <2 with methylene chloride using continuous extraction techniques. If the solids content is greater than 30%, the sample is extracted using ultrasonic techniques.

Each extract is dried over sodium sulfate, concentrated to a volume of 5 mL, cleaned up using GPC, if necessary, and concentrated. Extracts are concentrated to 1 mL if GPC is not performed, and to 0.5 mL if GPC is performed.

An internal standard is added to the extract, and a 1-mL aliquot of the extract is injected into the GC. The compounds are separated by GC and detected by a MS. The labeled compounds serve to correct the variability of the analytical technique.

INTERFERENCES Solvents, reagents, glassware, and other sample processing hardware may yield artifacts and/or elevated baselines causing misinterpretation of chromatograms and spectra. Materials used in the analysis must be demonstrated to be free from interferences under the conditions of analysis by running method blanks initially and with each sample lot (sample started through the extraction process on a given 8-h shift, to a maximum of 20). Specific selection of reagents and purification of solvents by distillation in all glass systems may be required. Glassware and, where possible, reagents are cleaned by solvent rinse and baking at 450°C for 1-h minimum. Interferences coextracted from samples will vary considerably from source to source, depending on the diversity of the site being sampled.

INSTRUMENTATION Major instrumentation includes a GC with a splitless or on-column injection port for capillary column, a MS with 70 eV electron impact ionization, and a data system to collect and record MS data, and process it. A K-D apparatus is used to concentrate extracts.

GC Column: 30 m × 0.25 mm I.D. 5% phenyl, 94% methyl, 1% vinyl silicone bonded phased fused silica capillary column.

PRECISION & ACCURACY The detection limits of the method are usually dependent on the level of interferences rather than instrumental limitations. The limits typify the minimum quantities that can be detected with no interferences present.

The minimum level (in µg/mL) was not listed. This is defined as a minimum level at which the analytical system shall give recognizable mass spectra (background corrected) and acceptable calibration points.

The MDL (in µg/kg) in low solids was not listed and in high solids was not listed; these were determined in digested sludge (low solids) and in filter cake or compost (high solids).

The labeled and native compound initial precision as standard deviation (in µg/L) was not listed.
The labeled and native compound initial accuracy as average recovery (in µg/L) was not listed.

SAMPLE COLLECTION, PRESERVATION & HANDLING Collect samples in glass containers. Aqueous samples which flow freely are collected in refrigerated bottles using automatic sampling equipment. Solid samples are collected as grab samples using widemouth jars. Maintain samples at 0 to 4°C from the time of collection until extraction. If residual chlorine is present in aqueous samples, add 80 mg sodium thiosulfate/L of water. Begin sample extraction within 7 days of collection, and analyze all extracts within 40 days of extraction.

SAMPLE PREPARATION Samples containing 1% solids or less are extracted directly using continuous liquid-liquid extraction techniques. Samples containing 1 to 30% solids are diluted to the 1% level with reagent water and extracted using continuous liquid-liquid extraction techniques. Samples containing greater than 30% solids are extracted using ultrasonic techniques.

Base/neutral extraction — Adjust the pH of the waters in the extractors to 12–13 with 6 N NaOH. Extract with methylene chloride for 24–48 h.

Acid extraction — Adjust the pH of the waters in the extractors to 2 or less using 6 N sulfuric acid. Extract with methylene chloride for 24–48 h.

Ultrasonic extraction of high solids samples — Add anhydrous sodium sulfate to the sample and QC aliquot(s). Add acetone:methylene chloride (1:1) to the sample and mix thoroughly

Concentrate extracts using a K-D apparatus.

QUALITY CONTROL The analyst is permitted to modify this method to improve separations or lower the costs of measurements, provided all performance specifications are met. Analyses of blanks are required to demonstrate freedom from contamination. When results of spikes indicate atypical method performance for samples, the samples are diluted to bring method performance within acceptable limits.

For low solids (aqueous samples), extract, concentrate, and analyze two sets of four 1-L aliquots (8 aliquots total) of the precision and recovery standard. For high solids samples, two sets of four 30-g aliquots of the high solids reference matrix are used.

Spike all samples with labeled compounds to assess method performance. Compute percent recovery of the labeled compounds using the internal standard method. Compare the labeled compound recovery for each compound with the corresponding labeled compound recovery.

Reagent water and high solids reference matrix blanks are analyzed to demonstrate freedom from contamination. Extract and concentrate a 1-L reagent water blank or a high solids reference matrix blank with each sample's lot (samples started through the extraction process on the same 8-h shift, to a maximum of 20 samples).

Field replicates may be collected to determine the precision of the sampling technique, and spiked samples may be required to determine the accuracy of the analysis when the internal standard method is used.

REFERENCE Semivolatile Organic Compounds by Isotope Dilution GC/MS. Office of Water Regulation and Standards, U.S. EPA Industrial Technology Division, Washington, DC, EPA Method 1625, Rev. C, June 1989 (contact W.A. Telliard, U.S. EPA, Office of Water Regulations and Standards, 401 M St., SW, Washington, DC, 20460. Phone: 202-382-7131).

N-Nitrosodi-n-propylamine   **EPA Method 1625**
CAS #621-64-7

TITLE Semivolatile Organic Compounds by Isotope Dilution GC/MS

MATRIX The compounds may be determined in waters, soils, and municipal sludges by this method.

METHOD SUMMARY This method is used to determine 176 semivolatile toxic organic pollutants associated with the CWA (as amended 1987); the RCRA (as amended 1986); the CERCLA (as amended 1986); and other compounds amenable to extraction and analysis by capillary column gas chromatography-mass spectrometry (GC/MS).

Stable isotopically-labeled analogs of the compounds of interest are added to the sample. If the solids content is less than 1%, a 1-L sample is extracted at pH 12–13, then at pH <2 with methylene chloride using continuous extraction techniques.

If the solids content is 30% or less, the sample is diluted to 1% solids with reagent water, homogenized ultrasonically, and extracted at pH 12–13, then at pH <2 with methylene chloride using continuous extraction techniques. If the solids content is greater than 30%, the sample is extracted using ultrasonic techniques.

Each extract is dried over sodium sulfate, concentrated to a volume of 5 mL, cleaned up using GPC, if necessary, and concentrated. Extracts are concentrated to 1 mL if GPC is not performed, and to 0.5 mL if GPC is performed.

An internal standard is added to the extract, and a 1-mL aliquot of the extract is injected into the GC. The compounds are separated by GC and detected by a MS. The labeled compounds serve to correct the variability of the analytical technique.

INTERFERENCES Solvents, reagents, glassware, and other sample processing hardware may yield artifacts and/or elevated baselines causing misinterpretation of chromatograms and spectra. Materials used in the analysis must be demonstrated to be free from interferences under the conditions of analysis by running method blanks initially and with each sample lot (sample started through the extraction process on a given 8-h shift, to a maximum of 20). Specific selection of reagents and purification of solvents by distillation in all glass systems may be required. Glassware and, where possible, reagents are cleaned by solvent rinse and baking at 450°C for 1-h minimum. Interferences coextracted from samples will vary considerably from source to source, depending on the diversity of the site being sampled.

INSTRUMENTATION Major instrumentation includes a GC with a splitless or on-column injection port for capillary column, a MS with 70 eV electron impact ionization, and a data system to collect and record MS data, and process it. A K-D apparatus is used to concentrate extracts.

GC Column: 30 m × 0.25 mm I.D. 5% phenyl, 94% methyl, 1% vinyl silicone bonded phased fused silica capillary column.

PRECISION & ACCURACY The detection limits of the method are usually dependent on the level of interferences rather than instrumental limitations. The limits typify the minimum quantities that can be detected with no interferences present.

The minimum level (in µg/mL) was 20. This is defined as a minimum level at which the analytical system shall give recognizable mass spectra (background corrected) and acceptable calibration points.

The MDL (in μg/kg) in low solids was 46 and in high solids was 47; these were determined in digested sludge (low solids) and in filter cake or compost (high solids).

The labeled and native compound initial precision as standard deviation (in μg/L) was 45.

The labeled and native compound initial accuracy as average recovery (in μg/L) was 65–142.

SAMPLE COLLECTION, PRESERVATION & HANDLING
Collect samples in glass containers. Aqueous samples which flow freely are collected in refrigerated bottles using automatic sampling equipment. Solid samples are collected as grab samples using widemouth jars. Maintain samples at 0 to 4°C from the time of collection until extraction. If residual chlorine is present in aqueous samples, add 80 mg sodium thiosulfate/L of water. Begin sample extraction within 7 days of collection, and analyze all extracts within 40 days of extraction.

SAMPLE PREPARATION
Samples containing 1% solids or less are extracted directly using continuous liquid-liquid extraction techniques. Samples containing 1 to 30% solids are diluted to the 1% level with reagent water and extracted using continuous liquid-liquid extraction techniques. Samples containing greater than 30% solids are extracted using ultrasonic techniques.

Base/neutral extraction — Adjust the pH of the waters in the extractors to 12–13 with 6 N NaOH. Extract with methylene chloride for 24–48 h.

Acid extraction — Adjust the pH of the waters in the extractors to 2 or less using 6 N sulfuric acid. Extract with methylene chloride for 24–48 h.

Ultrasonic extraction of high solids samples — Add anhydrous sodium sulfate to the sample and QC aliquot(s). Add acetone:methylene chloride (1:1) to the sample and mix thoroughly

Concentrate extracts using a K-D apparatus.

QUALITY CONTROL
The analyst is permitted to modify this method to improve separations or lower the costs of measurements, provided all performance specifications are met. Analyses of blanks are required to demonstrate freedom from contamination. When results of spikes indicate atypical method performance for samples, the samples are diluted to bring method performance within acceptable limits.

For low solids (aqueous samples), extract, concentrate, and analyze two sets of four 1-L aliquots (8 aliquots total) of the precision and recovery standard. For high solids samples, two sets of four 30-g aliquots of the high solids reference matrix are used.

Spike all samples with labeled compounds to assess method performance. Compute percent recovery of the labeled compounds using the internal standard method. Compare the labeled compound recovery for each compound with the corresponding labeled compound recovery.

Reagent water and high solids reference matrix blanks are analyzed to demonstrate freedom from contamination. Extract and concentrate a 1-L reagent water blank or a high solids reference matrix blank with each sample's lot (samples started through the extraction process on the same 8-h shift, to a maximum of 20 samples).

Field replicates may be collected to determine the precision of the sampling technique, and spiked samples may be required to determine the accuracy of the analysis when the internal standard method is used.

REFERENCE Semivolatile Organic Compounds by Isotope Dilution GC/MS. Office of Water Regulation and Standards, U.S. EPA Industrial Technology Division, Washington, DC, EPA Method 1625, Rev. C, June 1989 (contact W.A. Telliard, U.S. EPA, Office of Water Regulations and Standards, 401 M St., SW, Washington, DC, 20460. Phone: 202-382-7131).

N-Nitrosodi-n-propylamine   EPA Method 625
CAS #621-64-7

TITLE Base/Neutrals and Acids, U.S. EPA Method 625

MATRIX This methods covers municipal and industrial wastewaters.

METHOD SUMMARY Approximately 1 L of sample is serially extracted with methylene chloride at a pH greater than 11 and again at a pH less than 2 using a separatory funnel or a continuous extractor. The methylene chloride extract is dried, concentrated to a volume of 1 mL, and analyzed by GC/MS. Qualitative identification of the parameters in the extract is performed using the retention time and the relative abundance of three characteristic masses (m/z). Qualitative analysis is performed using either external or internal standard techniques with a single characteristic m/z.

INTERFERENCES Method interferences may be caused by contaminants in solvents, reagents, glassware, and other sample processing hardware. Glassware must be scrupulously cleaned. Glassware should be heated in a muffle furnace at 400°C for 5 to 30 min. Some thermally stable materials, such as PCBs, may not be eliminated by this treatment. Solvent rinses with acetone and pesticide quality hexane may be substituted for the muffle furnace heating. Matrix interferences may be caused by contaminants that are coextracted from the sample. The base-neutral extraction may cause significantly reduced recovery of phenols. The packed gas chromatographic columns recommended for the basic fraction may not exhibit sufficient resolution for some analytes.

INSTRUMENTATION A GC/MS system with an injection port designed for on-column injection when using packed columns and for splitless injection when using capillary columns.

Column for base/neutrals: 1.8 m long × 2 mm I.D. glass, packed with 3% SP-2550 on Supelcoport (100/120 mesh) or equivalent.

Column for acids: 1.8 m long × 2 mm I.D. glass, packed with 1% SP-1240DA on Supelcoport (100/120 mesh) or equivalent.

PRECISION & ACCURACY The MDL concentrations were obtained using reagent water. The MDL actually achieved in a given analysis will vary depending on instrument sensitivity and matrix effects. This method was tested by 15 laboratories

using reagent water, drinking water, surface water, and industrial wastewaters spiked at six concentrations over the range 5 to 100 µg/L. Single operator precision, overall precision, and method accuracy were found to be directly related to the concentration of the parameter matrix.

The MDL (in µg/L) in reagent water was not reported.
The standard deviation (in µg/L based on 4 recovery measurements) was 55.4.
The range (in µg/L) for average recovery for 4 measurements was 13.6–197.9.
The range (in %) for percent recovery was D–230.
Accuracy (in µg/L) as expected recovery for one or more measurements of a sample containing a true concentration of C was $1.12C-6.22$.
Precision (in µg/L) as expected single analyst standard deviation of measurements at an average concentration found at X was $0.27X + 0.68$.
Overall precision (in µg/L) as expected interlaboratory standard deviation of measurements in an average concentration found at X was $0.44X + 0.47$.

C = True value of the concentration in µg/L.
X = Average recovery found for measurements of samples containing a concentration at C in µg/L.

SAMPLE PREPARATION Adjust the pH to >11 with sodium hydroxide and serially extract in a separatory funnel with methylene chloride or else in a continuous extractor. Next, adjust the pH to <2 with sulfuric acid and serially extract in a separatory funnel with methylene chloride or else in a continuous extractor. Dry the extracts separately through a column of anhydrous sodium sulfate and then concentrate each of the extracts to 1.0 mL using a K-D apparatus.

SAMPLE COLLECTION, PRESERVATION & HANDLING Grab samples must be collected in glass containers. All samples must be refrigerated at 4°C from the time of collection until extraction. If residual chlorine is present, add 80 mg of sodium thiosulfate/L of sample and mix well. All samples must be extracted within 7 days of collection and completely analyzed within 40 days of extraction.

QUALITY CONTROL Make an initial, one-time, demonstration of the ability to generate acceptable accuracy and precision with this method. Before processing any samples, the analyst must analyze a reagent water blank to demonstrate that interferences from the analytical system and glassware are under control. Each time a set of samples is extracted or reagents are changed, a reagent water blank must be processed. Spike and analyze a minimum of 5% of all samples to monitor and evaluate lab data quality. A QC check sample concentrate that contains each parameter of interest at a concentration of 100 µg/mL in acetone is required. PCBs and multicomponent pesticides may be omitted from this test.

After analysis of five spiked wastewater samples, calculate the average percent recovery and the standard deviation of the percent recovery. Spike all samples with the surrogate standard spiking solution and calculate the percent recovery of each surrogate compound.

REFERENCE *Federal Register*, Vol. 49, No. 209. Friday, Oct. 26, 1984.

N-Nitrosodi-n-propylamine   **EPA Method 8270**
CAS #621-64-7

TITLE Semivolatile Organic Compounds by GC/MS

MATRIX This method is used to determine the concentration of semivolatile organic compounds in extracts prepared from all types of solid waste matrices, soils, and groundwater. Although surface waters are not specifically mentioned, this method should be applicable to water samples from rivers, lakes, etc.

METHOD SUMMARY This method covers 259 semivolatile organic compounds. In very limited applications direct injection of the sample into the GC/MS system may be appropriate, but this results in very high detection limits (approximately 10,000 µg/L). Typically, a 1-L liquid sample, containing surrogate, and matrix spiking standards, is extracted in a continuous extractor first under acid conditions and then under basic conditions. Typically 30 g of a solid sample, containing surrogate, and matrix spiking standards, is extracted ultrasonically. After concentrating the extract to 1 mL it is spiked with 10 µL of an internal standard solution just prior to analysis by GC/MS. The volume injected should contain about 100 ng of base/neutral and 200 ng of acid surrogates (for a 1-µL injection). Analysis is performed by GC/MS using a capillary GC column.

INTERFERENCES Raw GC/MS data from all blanks, samples, and spikes must be evaluated for interferences. Contamination by carryover can occur whenever high-concentration and low-concentration samples are sequentially analyzed. To reduce carryover, the sample syringe must be rinsed out between samples with solvent. Whenever an unusually concentrated sample is encountered, it should be followed by the analysis of blank solvent to check for cross-contamination.

INSTRUMENTATION A GC/MS and a data system are required. The GC column used is a 30 m × 0.25 mm I.D. (or 0.32 mm I.D.) 1um film thickness silicone-coated fused silica capillary column. A continuous liquid-liquid extractor equipped with Teflon® or glass connection joints and stopcocks requiring no lubrication, a K-D concentrating apparatus, water bath, and an ultrasonic disrupter with a minimum power of 300 W and with pulsing capability are also required.

PRECISION & ACCURACY The estimated quantitation limit (EQL) of Method 8270B for determining an individual compound is approximately 1 mg/kg (wet weight) for soil or sediment samples, 1–200 mg/kg for wastes (dependent on matrix and method of preparation), and 10 µg/L for groundwater samples. EQLs will be proportionately higher for sample extracts that require dilution to avoid saturation of the detector.

The EQL(b) for groundwater in µg/L is 10.
The EQL (a, b) for low concentrations in soil and sediment in µg/kg is 660.
Accuracy as µg/L is $1.12C-6.22$.
Overall precision in µg/L is $0.44X + 0.47$.

(a) *EQLs listed for soil/sediment are based on wet weight. Normally data is reported in a dry-weight basis; therefore, EQLs will be higher based on the % dry weight of each sample. This calculation is based on a 30 g sample and gel permeation chromatography cleanup.*
(b) *Sample EQLs are highly matrix-dependent. The EQLs are provided for guidance and may not always be achievable.*
C = *True value for concentration, in µg/L.*
X = *Average recovery found for measurements of samples containing a concentration of C, in µg/L.*

ESTIMATED QUANTITATION LIMIT

Other Matrices	Factor (a)
High-concentration soil and sludges by sonicator	7.5
Non-water miscible waste	75

(a) EQL for other matrices = [EQL for low soil/sediment] × [Factor]. This estimated EQL is similar to an EPA "Practical Quantitation Limit."

SAMPLING METHOD

Liquid samples — Use a 1 or 2½ gallon amber glass bottle with a screw-top Teflon®-lined cover that has been prewashed with detergent and rinsed with distilled water and methanol (or isopropanol).

Soils, sediments, or sludges — Use an 8-oz. widemouth glass with a screw-top Teflon®-lined cover that has been prewashed with detergent and rinsed with distilled water and methanol (or isopropanol).

SAMPLE PRESERVATION

Liquid samples — If residual chlorine is present, add 3 mL of 10% sodium thiosulfate per gallon, cool to 4°C and store in a solvent-free refrigerator until analysis; if chlorine is not present, then eliminate the sodium thiosulfate addition.

Soils, sediments, or sludges — Cool samples to 4°C and store in a solvent-free refrigerator.

MHT Liquid samples must be extracted within 7 days and the extracts analyzed within 40 days. Soils, sediments, or sludges may be stored for a maximum of 14 days and the extracts analyzed within 40 days.

SAMPLE PREPARATION

Liquid samples — Transfer 1 L quantitatively to a continuous extractor. If high concentrations are anticipated, a smaller volume may be used and then diluted with organic-free reagent water to 1 L. Adjust pH, if necessary, to pH <2 using 1:1 (V/V) sulfuric acid. Pipette 1.0 mL of a surrogate standard spiking solution into each sample. For the sample in each analytical batch selected for spiking, add 1.0 mL of a matrix spiking standard. For base/neutral acid analysis, the amount of the surrogates and matrix spiking compounds added to the sample should result in a final concentration of 100 ng/µL of each analyte in the extract to be analyzed (assuming a 1-µL injection). Extract with methylene chloride for 18–24 h. Next, adjust the pH of the aqueous phase to pH >11 using 10 N sodium hydroxide and extract it with methylene chloride again for 18–24 h. Dry the extract through a column containing anhydrous sodium sulfate and concentrate it to 1 mL using a K-D concentrator.

Soils, sediments, or sludges — Use 30 g of sample. Nonporous or wet samples (gummy or clay type) that do not have a free-flowing sandy texture must be mixed with anhydrous sodium sulfate until the sample is free flowing. Add 1 mL of surrogate standards to all samples, spikes, standards, and blanks. For the sample in each analytical batch selected for spiking, add 1.0 mL of a matrix spiking standard. For base/neutral acid analysis, the amount added of the surrogates and matrix spiking compounds should result in a final concentration of 100 ng/µL of each base/neutral analyte and 200 ng/µL of each acid analyte in the extract to be analyzed (assuming a 1-µL injection). Immediately add a 100-mL mixture of 1:1 methylene chloride:acetone and extract the sample ultrasonically for 3 min and then decant or filter the extracts. Repeat the extraction two or more times. Dry the extract using a column with anhydrous sodium sulfate and concentrate it to 1 mL in a K-D concentrator.

QUALITY CONTROL A methylene chloride solution containing 50 ng/µL of decafluorotriphenylphosphine (DFTPP) is used for tuning the GC/MS system each 12-h shift. A system performance check also must be made during every 12-h shift. A standard containing 50 ng/µL each of 4,4'-DDT, pentachlorophenol, and benzidine is required to verify injection port inertness and GC column performance. A calibration standard at mid-concentration, containing each compound of interest, including all required surrogates, must be performed every 12 h during analysis. After the system performance check is met, calibration check compounds (CCCs) are used to check the validity of the initial calibration.

The internal standard responses and retention times in the calibration check standard must be evaluated immediately after or during data acquisition. If the retention time for any internal standard changes by more than 30 seconds from the last check calibration (12 h), the chromatographic system must be inspected for malfunctions and corrections must be made, as required. If the electron ionization current plot (EICP) area for any of the internal standards changes by a factor of two from the last daily calibration standard check, the mass spectrometer must be inspected for malfunctions and corrections must be made, as appropriate.

Demonstrate, through the analysis of a reagent water blank, that interferences from the analytical system, glassware, and reagents are under control. The blank samples should be carried through all stages of the sample preparation and measurement steps. For each analytical batch (up to 20 samples), a reagent blank, matrix spike, and matrix spike duplicate/duplicate must be analyzed (the frequency of the spikes may be different for different monitoring programs). The blank and spiked samples must be carried through all stages of the sample preparation and measurement steps. A QC reference sample concentrate containing each analyte at a concentration of 100 mg/L in methanol is required.

REFERENCE Test Methods for Evaluating Solid Waste (SW-846). U.S. EPA 1983, Method 8270B, Rev. 2, Nov. 1990. Office of Solid Waste, Washington, DC.

N-Nitrosodibutylamine
CAS #924-16-3

EPA Method 8270

TITLE Semivolatile Organic Compounds by GC/MS

MATRIX This method is used to determine the concentration of semivolatile organic compounds in extracts prepared from all types of solid waste matrices, soils, and groundwater. Although surface waters are not specifically mentioned, this method should be applicable to water samples from rivers, lakes, etc.

METHOD SUMMARY This method covers 259 semivolatile organic compounds. In very limited applications direct injection of the sample into the GC/MS system may be appropriate, but this results in very high detection limits (approximately 10,000 μg/L). Typically, a 1-L liquid sample, containing surrogate, and matrix spiking standards, is extracted in a continuous extractor first under acid conditions and then under basic conditions. Typically 30 g of a solid sample, containing surrogate, and matrix spiking standards, is extracted ultrasonically. After concentrating the extract to 1 mL it is spiked with 10 μL of an internal standard solution just prior to analysis by GC/MS. The volume injected should contain about 100 ng of base/neutral and 200 ng of acid surrogates (for a 1-μL injection). Analysis is performed by GC/MS using a capillary GC column.

INTERFERENCES Raw GC/MS data from all blanks, samples, and spikes must be evaluated for interferences. Contamination by carryover can occur whenever high-concentration and low-concentration samples are sequentially analyzed. To reduce carryover, the sample syringe must be rinsed out between samples with solvent. Whenever an unusually concentrated sample is encountered, it should be followed by the analysis of blank solvent to check for cross-contamination.

INSTRUMENTATION A GC/MS and a data system are required. The GC column used is a 30 m × 0.25 mm I.D. (or 0.32 mm I.D.) 1um film thickness silicone-coated fused silica capillary column. A continuous liquid-liquid extractor equipped with Teflon® or glass connection joints and stopcocks requiring no lubrication, a K-D concentrating apparatus, water bath, and an ultrasonic disrupter with a minimum power of 300 W and with pulsing capability are also required.

PRECISION & ACCURACY The estimated quantitation limit (EQL) of Method 8270B for determining an individual compound is approximately 1 mg/kg (wet weight) for soil or sediment samples, 1–200 mg/kg for wastes (dependent on matrix and method of preparation), and 10 μg/L for groundwater samples. EQLs will be proportionately higher for sample extracts that require dilution to avoid saturation of the detector.

The EQL(b) for groundwater in μg/L is 10.
The EQL (a, b) for low concentrations in soil and sediment in μg/kg is not determined.
Accuracy as μg/L is not listed.
Overall precision in μg/L is not listed.

(a) *EQLs listed for soil/sediment are based on wet weight. Normally data is reported in a dry-weight basis; therefore, EQLs will be higher based on the % dry weight of each sample.*

This calculation is based on a 30 g sample and gel permeation chromatography cleanup.

(b) *Sample EQLs are highly matrix-dependent. The EQLs are provided for guidance and may not always be achievable.*

C = *True value for concentration, in μg/L.*
X = *Average recovery found for measurements of samples containing a concentration of C, in μg/L.*

ESTIMATED QUANTITATION LIMIT

Other Matrices	Factor (a)
High-concentration soil and sludges by sonicator	7.5
Non-water miscible waste	75

(a) *EQL for other matrices = [EQL for low soil/sediment] × [Factor]. This estimated EQL is similar to an EPA "Practical Quantitation Limit."*

SAMPLING METHOD
Liquid samples — Use a 1 or 2½ gallon amber glass bottle with a screw-top Teflon®-lined cover that has been prewashed with detergent and rinsed with distilled water and methanol (or isopropanol).

Soils, sediments, or sludges — Use an 8-oz. widemouth glass with a screw-top Teflon®-lined cover that has been prewashed with detergent and rinsed with distilled water and methanol (or isopropanol).

SAMPLE PRESERVATION
Liquid samples — If residual chlorine is present, add 3 mL of 10% sodium thiosulfate per gallon, cool to 4°C and store in a solvent-free refrigerator until analysis; if chlorine is not present, then eliminate the sodium thiosulfate addition.

Soils, sediments, or sludges — Cool samples to 4°C and store in a solvent-free refrigerator.

MHT Liquid samples must be extracted within 7 days and the extracts analyzed within 40 days. Soils, sediments, or sludges may be stored for a maximum of 14 days and the extracts analyzed within 40 days.

SAMPLE PREPARATION
Liquid samples — Transfer 1 L quantitatively to a continuous extractor. If high concentrations are anticipated, a smaller volume may be used and then diluted with organic-free reagent water to 1 L. Adjust pH, if necessary, to pH <2 using 1:1 (V/V) sulfuric acid. Pipette 1.0 mL of a surrogate standard spiking solution into each sample. For the sample in each analytical batch selected for spiking, add 1.0 mL of a matrix spiking standard. For base/neutral acid analysis, the amount of the surrogates and matrix spiking compounds added to the sample should result in a final concentration of 100 ng/μL of each analyte in the extract to be analyzed (assuming a 1-μL injection). Extract with methylene chloride for 18–24 h. Next, adjust the pH of the aqueous phase to pH >11 using 10 N sodium hydroxide and extract it with methylene chloride again for 18–24 h. Dry the extract through a column containing anhydrous sodium sulfate and concentrate it to 1 mL using a K-D concentrator.

Soils, sediments, or sludges — Use 30 g of sample. Nonporous or wet samples (gummy or clay type) that do not have a free-flowing

sandy texture must be mixed with anhydrous sodium sulfate until the sample is free flowing. Add 1 mL of surrogate standards to all samples, spikes, standards, and blanks. For the sample in each analytical batch selected for spiking, add 1.0 mL of a matrix spiking standard. For base/neutral acid analysis, the amount added of the surrogates and matrix spiking compounds should result in a final concentration of 100 ng/μL of each base/neutral analyte and 200 ng/μL of each acid analyte in the extract to be analyzed (assuming a 1-μL injection). Immediately add a 100-mL mixture of 1:1 methylene chloride:acetone and extract the sample ultrasonically for 3 min and then decant or filter the extracts. Repeat the extraction two or more times. Dry the extract using a column with anhydrous sodium sulfate and concentrate it to 1 mL in a K-D concentrator.

QUALITY CONTROL A methylene chloride solution containing 50 ng/μL of decafluorotriphenylphosphine (DFTPP) is used for tuning the GC/MS system each 12-h shift. A system performance check also must be made during every 12-h shift. A standard containing 50 ng/μL each of 4,4'-DDT, pentachlorophenol, and benzidine is required to verify injection port inertness and GC column performance. A calibration standard at mid-concentration, containing each compound of interest, including all required surrogates, must be performed every 12 h during analysis. After the system performance check is met, calibration check compounds (CCCs) are used to check the validity of the initial calibration.

The internal standard responses and retention times in the calibration check standard must be evaluated immediately after or during data acquisition. If the retention time for any internal standard changes by more than 30 seconds from the last check calibration (12 h), the chromatographic system must be inspected for malfunctions and corrections must be made, as required. If the electron ionization current plot (EICP) area for any of the internal standards changes by a factor of two from the last daily calibration standard check, the mass spectrometer must be inspected for malfunctions and corrections must be made, as appropriate.

Demonstrate, through the analysis of a reagent water blank, that interferences from the analytical system, glassware, and reagents are under control. The blank samples should be carried through all stages of the sample preparation and measurement steps. For each analytical batch (up to 20 samples), a reagent blank, matrix spike, and matrix spike duplicate/duplicate must be analyzed (the frequency of the spikes may be different for different monitoring programs). The blank and spiked samples must be carried through all stages of the sample preparation and measurement steps. A QC reference sample concentrate containing each analyte at a concentration of 100 mg/L in methanol is required.

REFERENCE Test Methods for Evaluating Solid Waste (SW-846). U.S. EPA 1983, Method 8270B, Rev. 2, Nov. 1990. Office of Solid Waste, Washington, DC.

N-Nitrosodiethylamine EPA Method 1625
CAS #55-18-5

TITLE Semivolatile Organic Compounds by Isotope Dilution GC/MS

MATRIX The compounds may be determined in waters, soils, and municipal sludges by this method.

METHOD SUMMARY This method is used to determine 176 semivolatile toxic organic pollutants associated with the CWA (as amended 1987); the RCRA (as amended 1986); the CERCLA (as amended 1986); and other compounds amenable to extraction and analysis by capillary column gas chromatography-mass spectrometry (GC/MS).

Stable isotopically-labeled analogs of the compounds of interest are added to the sample. If the solids content is less than 1%, a 1-L sample is extracted at pH 12–13, then at pH <2 with methylene chloride using continuous extraction techniques.

If the solids content is 30% or less, the sample is diluted to 1% solids with reagent water, homogenized ultrasonically, and extracted at pH 12–13, then at pH <2 with methylene chloride using continuous extraction techniques. If the solids content is greater than 30%, the sample is extracted using ultrasonic techniques.

Each extract is dried over sodium sulfate, concentrated to a volume of 5 mL, cleaned up using GPC, if necessary, and concentrated. Extracts are concentrated to 1 mL if GPC is not performed, and to 0.5 mL if GPC is performed.

An internal standard is added to the extract, and a 1-mL aliquot of the extract is injected into the GC. The compounds are separated by GC and detected by a MS. The labeled compounds serve to correct the variability of the analytical technique.

INTERFERENCES Solvents, reagents, glassware, and other sample processing hardware may yield artifacts and/or elevated baselines causing misinterpretation of chromatograms and spectra. Materials used in the analysis must be demonstrated to be free from interferences under the conditions of analysis by running method blanks initially and with each sample lot (sample started through the extraction process on a given 8-h shift, to a maximum of 20). Specific selection of reagents and purification of solvents by distillation in all glass systems may be required. Glassware and, where possible, reagents are cleaned by solvent rinse and baking at 450°C for 1-h minimum. Interferences coextracted from samples will vary considerably from source to source, depending on the diversity of the site being sampled.

INSTRUMENTATION Major instrumentation includes a GC with a splitless or on-column injection port for capillary column, a MS with 70 eV electron impact ionization, and a data system to collect and record MS data, and process it. A K-D apparatus is used to concentrate extracts.

GC Column: 30 m × 0.25 mm I.D. 5% phenyl, 94% methyl, 1% vinyl silicone bonded phased fused silica capillary column.

PRECISION & ACCURACY The detection limits of the method are usually dependent on the level of interferences rather than instrumental limitations. The limits typify the minimum quantities that can be detected with no interferences present.

The minimum level (in µg/mL) was 50. This is defined as a minimum level at which the analytical system shall give recognizable mass spectra (background corrected) and acceptable calibration points.

The MDL (in µg/kg) in low solids was 16 and in high solids was 27; these were determined in digested sludge (low solids) and in filter cake or compost (high solids).

The labeled and native compound initial precision as standard deviation (in µg/L) was not listed.

The labeled and native compound initial accuracy as average recovery (in µg/L) was not listed.

SAMPLE COLLECTION, PRESERVATION & HANDLING Collect samples in glass containers. Aqueous samples which flow freely are collected in refrigerated bottles using automatic sampling equipment. Solid samples are collected as grab samples using widemouth jars. Maintain samples at 0 to 4°C from the time of collection until extraction. If residual chlorine is present in aqueous samples, add 80 mg sodium thiosulfate/L of water. Begin sample extraction within 7 days of collection, and analyze all extracts within 40 days of extraction.

SAMPLE PREPARATION Samples containing 1% solids or less are extracted directly using continuous liquid-liquid extraction techniques. Samples containing 1 to 30% solids are diluted to the 1% level with reagent water and extracted using continuous liquid-liquid extraction techniques. Samples containing greater than 30% solids are extracted using ultrasonic techniques.

- Base/neutral extraction — Adjust the pH of the waters in the extractors to 12–13 with 6 N NaOH. Extract with methylene chloride for 24–48 h.
- Acid extraction — Adjust the pH of the waters in the extractors to 2 or less using 6 N sulfuric acid. Extract with methylene chloride for 24–48 h.
- Ultrasonic extraction of high solids samples — Add anhydrous sodium sulfate to the sample and QC aliquot(s). Add acetone:methylene chloride (1:1) to the sample and mix thoroughly

Concentrate extracts using a K-D apparatus.

QUALITY CONTROL The analyst is permitted to modify this method to improve separations or lower the costs of measurements, provided all performance specifications are met. Analyses of blanks are required to demonstrate freedom from contamination. When results of spikes indicate atypical method performance for samples, the samples are diluted to bring method performance within acceptable limits.

For low solids (aqueous samples), extract, concentrate, and analyze two sets of four 1-L aliquots (8 aliquots total) of the precision and recovery standard. For high solids samples, two sets of four 30-g aliquots of the high solids reference matrix are used.

Spike all samples with labeled compounds to assess method performance. Compute percent recovery of the labeled compounds using the internal standard method. Compare the labeled compound recovery for each compound with the corresponding labeled compound recovery.

Reagent water and high solids reference matrix blanks are analyzed to demonstrate freedom from contamination. Extract and concentrate a 1-L reagent water blank or a high solids reference matrix blank with each sample's lot (samples started through the extraction process on the same 8-h shift, to a maximum of 20 samples).

Field replicates may be collected to determine the precision of the sampling technique, and spiked samples may be required to determine the accuracy of the analysis when the internal standard method is used.

REFERENCE Semivolatile Organic Compounds by Isotope Dilution GC/MS. Office of Water Regulation and Standards, U.S. EPA Industrial Technology Division, Washington, DC, EPA Method 1625, Rev. C, June 1989 (contact W.A. Telliard, U.S. EPA, Office of Water Regulations and Standards, 401 M St., SW, Washington, DC, 20460. Phone: 202-382-7131).

N-Nitrosodiethylamine   **EPA Method 8270**
CAS #55-18-5

TITLE Semivolatile Organic Compounds by GC/MS

MATRIX This method is used to determine the concentration of semivolatile organic compounds in extracts prepared from all types of solid waste matrices, soils, and groundwater. Although surface waters are not specifically mentioned, this method should be applicable to water samples from rivers, lakes, etc.

METHOD SUMMARY This method covers 259 semivolatile organic compounds. In very limited applications direct injection of the sample into the GC/MS system may be appropriate, but this results in very high detection limits (approximately 10,000 µg/L). Typically, a 1-L liquid sample, containing surrogate, and matrix spiking standards, is extracted in a continuous extractor first under acid conditions and then under basic conditions. Typically 30 g of a solid sample, containing surrogate, and matrix spiking standards, is extracted ultrasonically. After concentrating the extract to 1 mL it is spiked with 10 µL of an internal standard solution just prior to analysis by GC/MS. The volume injected should contain about 100 ng of base/neutral and 200 ng of acid surrogates (for a 1-µL injection). Analysis is performed by GC/MS using a capillary GC column.

INTERFERENCES Raw GC/MS data from all blanks, samples, and spikes must be evaluated for interferences. Contamination by carryover can occur whenever high-concentration and low-concentration samples are sequentially analyzed. To reduce carryover, the sample syringe must be rinsed out between samples with solvent. Whenever an unusually concentrated sample is encountered, it should be followed by the analysis of blank solvent to check for cross-contamination.

INSTRUMENTATION A GC/MS and a data system are required. The GC column used is a 30 m × 0.25 mm I.D. (or 0.32 mm I.D.) 1um film thickness silicone-coated fused silica capillary column. A continuous liquid-liquid extractor equipped with Teflon® or glass connection joints and stopcocks requiring no lubrication, a K-D concentrating apparatus, water bath, and an ultrasonic disrupter with a minimum power of 300 W and with pulsing capability are also required.

PRECISION & ACCURACY The estimated quantitation limit (EQL) of Method 8270B for determining an individual compound is approximately 1 mg/kg (wet weight) for soil or sediment samples, 1–200 mg/kg for wastes (dependent on matrix and method of preparation), and 10 µg/L for groundwater samples. EQLs will be proportionately higher for sample extracts that require dilution to avoid saturation of the detector.

The EQL(b) for groundwater in µg/L is 20.
The EQL (a, b) for low concentrations in soil and sediment in µg/kg is not determined.
Accuracy as µg/L is not listed.
Overall precision in µg/L is not listed.

(a) *EQLs listed for soil/sediment are based on wet weight. Normally data is reported in a dry-weight basis; therefore, EQLs will be higher based on the % dry weight of each sample. This calculation is based on a 30 g sample and gel permeation chromatography cleanup.*
(b) *Sample EQLs are highly matrix-dependent. The EQLs are provided for guidance and may not always be achievable.*
C = *True value for concentration, in µg/L.*
X = *Average recovery found for measurements of samples containing a concentration of C, in µg/L.*

ESTIMATED QUANTITATION LIMIT

Other Matrices	Factor (a)
High-concentration soil and sludges by sonicator	7.5
Non-water miscible waste	75

(a) *EQL for other matrices = [EQL for low soil/sediment] × [Factor]. This estimated EQL is similar to an EPA "Practical Quantitation Limit."*

SAMPLING METHOD
Liquid samples — Use a 1 or 2½ gallon amber glass bottle with a screw-top Teflon®-lined cover that has been prewashed with detergent and rinsed with distilled water and methanol (or isopropanol).

Soils, sediments, or sludges — Use an 8-oz. widemouth glass with a screw-top Teflon®-lined cover that has been prewashed with detergent and rinsed with distilled water and methanol (or isopropanol).

SAMPLE PRESERVATION
Liquid samples — If residual chlorine is present, add 3 mL of 10% sodium thiosulfate per gallon, cool to 4°C and store in a solvent-free refrigerator until analysis; if chlorine is not present, then eliminate the sodium thiosulfate addition.

Soils, sediments, or sludges — Cool samples to 4°C and store in a solvent-free refrigerator.

MHT Liquid samples must be extracted within 7 days and the extracts analyzed within 40 days. Soils, sediments, or sludges may be stored for a maximum of 14 days and the extracts analyzed within 40 days.

SAMPLE PREPARATION
Liquid samples — Transfer 1 L quantitatively to a continuous extractor. If high concentrations are anticipated, a smaller volume may be used and then diluted with organic-free reagent water to 1 L. Adjust pH, if necessary, to pH <2 using 1:1 (V/V) sulfuric acid. Pipette 1.0 mL of a surrogate standard spiking solution into each sample. For the sample in each analytical batch selected for spiking, add 1.0 mL of a matrix spiking standard. For base/neutral acid analysis, the amount of the surrogates and matrix spiking compounds added to the sample should result in a final concentration of 100 ng/µL of each analyte in the extract to be analyzed (assuming a 1-µL injection). Extract with methylene chloride for 18–24 h. Next, adjust the pH of the aqueous phase to pH >11 using 10 N sodium hydroxide and extract it with methylene chloride again for 18–24 h. Dry the extract through a column containing anhydrous sodium sulfate and concentrate it to 1 mL using a K-D concentrator.

Soils, sediments, or sludges — Use 30 g of sample. Nonporous or wet samples (gummy or clay type) that do not have a free-flowing sandy texture must be mixed with anhydrous sodium sulfate until the sample is free flowing. Add 1 mL of surrogate standards to all samples, spikes, standards, and blanks. For the sample in each analytical batch selected for spiking, add 1.0 mL of a matrix spiking standard. For base/neutral acid analysis, the amount added of the surrogates and matrix spiking compounds should result in a final concentration of 100 ng/µL of each base/neutral analyte and 200 ng/µL of each acid analyte in the extract to be analyzed (assuming a 1-µL injection). Immediately add a 100-mL mixture of 1:1 methylene chloride:acetone and extract the sample ultrasonically for 3 min and then decant or filter the extracts. Repeat the extraction two or more times. Dry the extract using a column with anhydrous sodium sulfate and concentrate it to 1 mL in a K-D concentrator.

QUALITY CONTROL A methylene chloride solution containing 50 ng/µL of decafluorotriphenylphosphine (DFTPP) is used for tuning the GC/MS system each 12-h shift. A system performance check also must be made during every 12-h shift. A standard containing 50 ng/µL each of 4,4′-DDT, pentachlorophenol, and benzidine is required to verify injection port inertness and GC column performance. A calibration standard at mid-concentration, containing each compound of interest, including all required surrogates, must be performed every 12 h during analysis. After the system performance check is met, calibration check compounds (CCCs) are used to check the validity of the initial calibration.

The internal standard responses and retention times in the calibration check standard must be evaluated immediately after or during data acquisition. If the retention time for any internal standard changes by more than 30 seconds from the last check calibration (12 h), the chromatographic system must be inspected for malfunctions and corrections must be made, as required. If the electron ionization current plot (EICP) area for

any of the internal standards changes by a factor of two from the last daily calibration standard check, the mass spectrometer must be inspected for malfunctions and corrections must be made, as appropriate.

Demonstrate, through the analysis of a reagent water blank, that interferences from the analytical system, glassware, and reagents are under control. The blank samples should be carried through all stages of the sample preparation and measurement steps. For each analytical batch (up to 20 samples), a reagent blank, matrix spike, and matrix spike duplicate/duplicate must be analyzed (the frequency of the spikes may be different for different monitoring programs). The blank and spiked samples must be carried through all stages of the sample preparation and measurement steps. A QC reference sample concentrate containing each analyte at a concentration of 100 mg/L in methanol is required.

REFERENCE Test Methods for Evaluating Solid Waste (SW-846). U.S. EPA 1983, Method 8270B, Rev. 2, Nov. 1990. Office of Solid Waste, Washington, DC.

N-Nitrosodimethylamine EPA Method 1625
CAS #62-75-9

TITLE Semivolatile Organic Compounds by Isotope Dilution GC/MS

MATRIX The compounds may be determined in waters, soils, and municipal sludges by this method.

METHOD SUMMARY This method is used to determine 176 semivolatile toxic organic pollutants associated with the CWA (as amended 1987); the RCRA (as amended 1986); the CERCLA (as amended 1986); and other compounds amenable to extraction and analysis by capillary column gas chromatography-mass spectrometry (GC/MS).

Stable isotopically-labeled analogs of the compounds of interest are added to the sample. If the solids content is less than 1%, a 1-L sample is extracted at pH 12–13, then at pH <2 with methylene chloride using continuous extraction techniques.

If the solids content is 30% or less, the sample is diluted to 1% solids with reagent water, homogenized ultrasonically, and extracted at pH 12–13, then at pH <2 with methylene chloride using continuous extraction techniques. If the solids content is greater than 30%, the sample is extracted using ultrasonic techniques.

Each extract is dried over sodium sulfate, concentrated to a volume of 5 mL, cleaned up using GPC, if necessary, and concentrated. Extracts are concentrated to 1 mL if GPC is not performed, and to 0.5 mL if GPC is performed.

An internal standard is added to the extract, and a 1-mL aliquot of the extract is injected into the GC. The compounds are separated by GC and detected by a MS. The labeled compounds serve to correct the variability of the analytical technique.

INTERFERENCES Solvents, reagents, glassware, and other sample processing hardware may yield artifacts and/or elevated baselines causing misinterpretation of chromatograms and spectra. Materials used in the analysis must be demonstrated to be free from interferences under the conditions of analysis by running method blanks initially and with each sample lot (sample started through the extraction process on a given 8-h shift, to a maximum of 20). Specific selection of reagents and purification of solvents by distillation in all glass systems may be required. Glassware and, where possible, reagents are cleaned by solvent rinse and baking at 450°C for 1-h minimum. Interferences coextracted from samples will vary considerably from source to source, depending on the diversity of the site being sampled.

INSTRUMENTATION Major instrumentation includes a GC with a splitless or on-column injection port for capillary column, a MS with 70 eV electron impact ionization, and a data system to collect and record MS data, and process it. A K-D apparatus is used to concentrate extracts.

GC Column: 30 m × 0.25 mm I.D. 5% phenyl, 94% methyl, 1% vinyl silicone bonded phased fused silica capillary column.

PRECISION & ACCURACY The detection limits of the method are usually dependent on the level of interferences rather than instrumental limitations. The limits typify the minimum quantities that can be detected with no interferences present.

The minimum level (in µg/mL) was not listed. This is defined as a minimum level at which the analytical system shall give recognizable mass spectra (background corrected) and acceptable calibration points.

The MDL (in µg/kg) in low solids was not listed and in high solids was not listed; these were determined in digested sludge (low solids) and in filter cake or compost (high solids).

The labeled and native compound initial precision as standard deviation (in µg/L) was 49.
The labeled and native compound initial accuracy as average recovery (in µg/L) was 10-ns.

ns = no specification; the limit was outside the range that could be reliably measured.

SAMPLE COLLECTION, PRESERVATION & HANDLING Collect samples in glass containers. Aqueous samples which flow freely are collected in refrigerated bottles using automatic sampling equipment. Solid samples are collected as grab samples using widemouth jars. Maintain samples at 0 to 4°C from the time of collection until extraction. If residual chlorine is present in aqueous samples, add 80 mg sodium thiosulfate/L of water. Begin sample extraction within 7 days of collection, and analyze all extracts within 40 days of extraction.

SAMPLE PREPARATION Samples containing 1% solids or less are extracted directly using continuous liquid-liquid extraction techniques. Samples containing 1 to 30% solids are diluted to the 1% level with reagent water and extracted using continuous liquid-liquid extraction techniques. Samples containing greater than 30% solids are extracted using ultrasonic techniques.

Base/neutral extraction — Adjust the pH of the waters in the extractors to 12–13 with 6 N NaOH. Extract with methylene chloride for 24–48 h.

Acid extraction — Adjust the pH of the waters in the extractors to 2 or less using 6 N sulfuric acid. Extract with methylene chloride for 24–48 h.

Ultrasonic extraction of high solids samples — Add anhydrous sodium sulfate to the sample and QC aliquot(s). Add acetone:methylene chloride (1:1) to the sample and mix thoroughly

Concentrate extracts using a K-D apparatus.

QUALITY CONTROL The analyst is permitted to modify this method to improve separations or lower the costs of measurements, provided all performance specifications are met. Analyses of blanks are required to demonstrate freedom from contamination. When results of spikes indicate atypical method performance for samples, the samples are diluted to bring method performance within acceptable limits.

For low solids (aqueous samples), extract, concentrate, and analyze two sets of four 1-L aliquots (8 aliquots total) of the precision and recovery standard. For high solids samples, two sets of four 30-g aliquots of the high solids reference matrix are used.

Spike all samples with labeled compounds to assess method performance. Compute percent recovery of the labeled compounds using the internal standard method. Compare the labeled compound recovery for each compound with the corresponding labeled compound recovery.

Reagent water and high solids reference matrix blanks are analyzed to demonstrate freedom from contamination. Extract and concentrate a 1-L reagent water blank or a high solids reference matrix blank with each sample's lot (samples started through the extraction process on the same 8-h shift, to a maximum of 20 samples).

Field replicates may be collected to determine the precision of the sampling technique, and spiked samples may be required to determine the accuracy of the analysis when the internal standard method is used.

REFERENCE Semivolatile Organic Compounds by Isotope Dilution GC/MS. Office of Water Regulation and Standards, U.S. EPA Industrial Technology Division, Washington, DC, EPA Method 1625, Rev. C, June 1989 (contact W.A. Telliard, U.S. EPA, Office of Water Regulations and Standards, 401 M St., SW, Washington, DC, 20460. Phone: 202-382-7131).

N-Nitrosodimethylamine   EPA Method 625
CAS #62-75-9

TITLE Base/Neutrals and Acids, U.S. EPA Method 625

MATRIX This methods covers municipal and industrial wastewaters.

METHOD SUMMARY Approximately 1 L of sample is serially extracted with methylene chloride at a pH greater than 11 and again at a pH less than 2 using a separatory funnel or a continuous extractor. The methylene chloride extract is dried, concentrated to a volume of 1 mL, and analyzed by GC/MS. Qualitative identification of the parameters in the extract is performed using the retention time and the relative abundance of three characteristic masses (m/z). Qualitative analysis is performed using either external or internal standard techniques with a single characteristic m/z.

INTERFERENCES Method interferences may be caused by contaminants in solvents, reagents, glassware, and other sample processing hardware. Glassware must be scrupulously cleaned. Glassware should be heated in a muffle furnace at 400°C for 5 to 30 min. Some thermally stable materials, such as PCBs, may not be eliminated by this treatment. Solvent rinses with acetone and pesticide quality hexane may be substituted for the muffle furnace heating. Matrix interferences may be caused by contaminants that are coextracted from the sample. The base-neutral extraction may cause significantly reduced recovery of phenols. The packed gas chromatographic columns recommended for the basic fraction may not exhibit sufficient resolution for some analytes.

INSTRUMENTATION A GC/MS system with an injection port designed for on-column injection when using packed columns and for splitless injection when using capillary columns.

Column for base/neutrals: 1.8 m long × 2 mm I.D. glass, packed with 3% SP-2550 on Supelcoport (100/120 mesh) or equivalent.

Column for acids: 1.8 m long × 2 mm I.D. glass, packed with 1% SP-1240DA on Supelcoport (100/120 mesh) or equivalent.

PRECISION & ACCURACY The MDL concentrations were obtained using reagent water. The MDL actually achieved in a given analysis will vary depending on instrument sensitivity and matrix effects. This method was tested by 15 laboratories using reagent water, drinking water, surface water, and industrial wastewaters spiked at six concentrations over the range 5 to 100 µg/L. Single operator precision, overall precision, and method accuracy were found to be directly related to the concentration of the parameter matrix.

The MDL (in µg/L) in reagent water was not detected.
The standard deviation (in µg/L based on 4 recovery measurements) was not reported.
The range (in µg/L) for average recovery for 4 measurements was not reported.
The range (in %) for percent recovery was not reported.
Accuracy (in µg/L) as expected recovery for one or more measurements of a sample containing a true concentration of C was not reported.
Precision (in µg/L) as expected single analyst standard deviation of measurements at an average concentration found at X was not reported.
Overall precision (in µg/L) as expected interlaboratory standard deviation of measurements in an average concentration found at X was not reported.

C = *True value of the concentration in µg/L.*
X = *Average recovery found for measurements of samples containing a concentration at C in µg/L.*

SAMPLE PREPARATION Adjust the pH to >11 with sodium hydroxide and serially extract in a separatory funnel with methylene chloride or else in a continuous extractor. Next, adjust the pH to <2 with sulfuric acid and serially extract in a separatory funnel with methylene chloride or else in a continuous extractor. Dry the extracts separately through a column of anhydrous sodium sulfate and then concentrate each of the extracts to 1.0 mL using a K-D apparatus.

SAMPLE COLLECTION, PRESERVATION & HANDLING Grab samples must be collected in glass containers. All samples must be refrigerated at 4°C from the time of collection until extraction. If residual chlorine is present, add 80 mg of sodium thiosulfate/L of sample and mix well. All samples must be extracted within 7 days of collection and completely analyzed within 40 days of extraction.

QUALITY CONTROL Make an initial, one-time, demonstration of the ability to generate acceptable accuracy and precision with this method. Before processing any samples, the analyst must analyze a reagent water blank to demonstrate that interferences from the analytical system and glassware are under control. Each time a set of samples is extracted or reagents are changed, a reagent water blank must be processed. Spike and analyze a minimum of 5% of all samples to monitor and evaluate lab data quality. A QC check sample concentrate that contains each parameter of interest at a concentration of 100 µg/mL in acetone is required. PCBs and multicomponent pesticides may be omitted from this test.

After analysis of five spiked wastewater samples, calculate the average percent recovery and the standard deviation of the percent recovery. Spike all samples with the surrogate standard spiking solution and calculate the percent recovery of each surrogate compound.

REFERENCE Federal Register, Vol. 49, No. 209. Friday, Oct. 26, 1984.

N-Nitrosodimethylamine **EPA Method 8270**
CAS #62-75-9

TITLE Semivolatile Organic Compounds by GC/MS

MATRIX This method is used to determine the concentration of semivolatile organic compounds in extracts prepared from all types of solid waste matrices, soils, and groundwater. Although surface waters are not specifically mentioned, this method should be applicable to water samples from rivers, lakes, etc.

METHOD SUMMARY This method covers 259 semivolatile organic compounds. In very limited applications direct injection of the sample into the GC/MS system may be appropriate, but this results in very high detection limits (approximately 10,000 µg/L). Typically, a 1-L liquid sample, containing surrogate, and matrix spiking standards, is extracted in a continuous extractor first under acid conditions and then under basic conditions. Typically 30 g of a solid sample, containing surrogate, and matrix spiking standards, is extracted ultrasonically. After concentrating the extract to 1 mL it is spiked with 10 µL of an internal standard solution just prior to analysis by GC/MS. The volume injected should contain about 100 ng of base/neutral and 200 ng of acid surrogates (for a 1-µL injection). Analysis is performed by GC/MS using a capillary GC column.

INTERFERENCES Raw GC/MS data from all blanks, samples, and spikes must be evaluated for interferences. Contamination by carryover can occur whenever high-concentration and low-concentration samples are sequentially analyzed. To reduce carryover, the sample syringe must be rinsed out between samples with solvent. Whenever an unusually concentrated sample is encountered, it should be followed by the analysis of blank solvent to check for cross-contamination.

INSTRUMENTATION A GC/MS and a data system are required. The GC column used is a 30 m × 0.25 mm I.D. (or 0.32 mm I.D.) 1um film thickness silicone-coated fused silica capillary column. A continuous liquid-liquid extractor equipped with Teflon® or glass connection joints and stopcocks requiring no lubrication, a K-D concentrating apparatus, water bath, and an ultrasonic disrupter with a minimum power of 300 W and with pulsing capability are also required.

PRECISION & ACCURACY The estimated quantitation limit (EQL) of Method 8270B for determining an individual compound is approximately 1 mg/kg (wet weight) for soil or sediment samples, 1–200 mg/kg for wastes (dependent on matrix and method of preparation), and 10 µg/L for groundwater samples. EQLs will be proportionately higher for sample extracts that require dilution to avoid saturation of the detector.

The EQL(b) for groundwater in µg/L is not listed.
The EQL (a, b) for low concentrations in soil and sediment in µg/kg is not listed.
Accuracy as µg/L is not listed.
Overall precision in µg/L is not listed.

(a) *EQLs listed for soil/sediment are based on wet weight. Normally data is reported in a dry-weight basis; therefore, EQLs will be higher based on the % dry weight of each sample. This calculation is based on a 30 g sample and gel permeation chromatography cleanup.*
(b) *Sample EQLs are highly matrix-dependent. The EQLs are provided for guidance and may not always be achievable.*
C = *True value for concentration, in µg/L.*
X = *Average recovery found for measurements of samples containing a concentration of C, in µg/L.*

ESTIMATED QUANTITATION LIMIT

Other Matrices	Factor (a)
High-concentration soil and sludges by sonicator	7.5
Non-water miscible waste	75

(a) *EQL for other matrices = [EQL for low soil/sediment] × [Factor]. This estimated EQL is similar to an EPA "Practical Quantitation Limit."*

SAMPLING METHOD

Liquid samples — Use a 1 or 2½ gallon amber glass bottle with a screw-top Teflon®-lined cover that has been prewashed with detergent and rinsed with distilled water and methanol (or isopropanol).

Soils, sediments, or sludges — Use an 8-oz. widemouth glass with a screw-top Teflon®-lined cover that has been prewashed with detergent and rinsed with distilled water and methanol (or isopropanol).

SAMPLE PRESERVATION

Liquid samples — If residual chlorine is present, add 3 mL of 10% sodium thiosulfate per gallon, cool to 4°C and store in a solvent-free refrigerator until analysis; if chlorine is not present, then eliminate the sodium thiosulfate addition.

Soils, sediments, or sludges — Cool samples to 4°C and store in a solvent-free refrigerator.

MHT Liquid samples must be extracted within 7 days and the extracts analyzed within 40 days. Soils, sediments, or sludges may be stored for a maximum of 14 days and the extracts analyzed within 40 days.

SAMPLE PREPARATION

Liquid samples — Transfer 1 L quantitatively to a continuous extractor. If high concentrations are anticipated, a smaller volume may be used and then diluted with organic-free reagent water to 1 L. Adjust pH, if necessary, to pH <2 using 1:1 (V/V) sulfuric acid. Pipette 1.0 mL of a surrogate standard spiking solution into each sample. For the sample in each analytical batch selected for spiking, add 1.0 mL of a matrix spiking standard. For base/neutral acid analysis, the amount of the surrogates and matrix spiking compounds added to the sample should result in a final concentration of 100 ng/µL of each analyte in the extract to be analyzed (assuming a 1-µL injection). Extract with methylene chloride for 18–24 h. Next, adjust the pH of the aqueous phase to pH >11 using 10 N sodium hydroxide and extract it with methylene chloride again for 18–24 h. Dry the extract through a column containing anhydrous sodium sulfate and concentrate it to 1 mL using a K-D concentrator.

Soils, sediments, or sludges — Use 30 g of sample. Nonporous or wet samples (gummy or clay type) that do not have a free-flowing sandy texture must be mixed with anhydrous sodium sulfate until the sample is free flowing. Add 1 mL of surrogate standards to all samples, spikes, standards, and blanks. For the sample in each analytical batch selected for spiking, add 1.0 mL of a matrix spiking standard. For base/neutral acid analysis, the amount added of the surrogates and matrix spiking compounds should result in a final concentration of 100 ng/µL of each base/neutral analyte and 200 ng/µL of each acid analyte in the extract to be analyzed (assuming a 1-µL injection). Immediately add a 100-mL mixture of 1:1 methylene chloride:acetone and extract the sample ultrasonically for 3 min and then decant or filter the extracts. Repeat the extraction two or more times. Dry the extract using a column with anhydrous sodium sulfate and concentrate it to 1 mL in a K-D concentrator.

Note: N-nitrosodimethylamine is difficult to separate from the solvent under the chromatographic conditions described above so they may need to be modified. It also decomposes in the gas chromatographic inlet and cannot be separated from diphenylamine.

QUALITY CONTROL
A methylene chloride solution containing 50 ng/µL of decafluorotriphenylphosphine (DFTPP) is used for tuning the GC/MS system each 12-h shift. A system performance check also must be made during every 12-h shift. A standard containing 50 ng/µL each of 4,4'-DDT, pentachlorophenol, and benzidine is required to verify injection port inertness and GC column performance. A calibration standard at mid-concentration, containing each compound of interest, including all required surrogates, must be performed every 12 h during analysis. After the system performance check is met, calibration check compounds (CCCs) are used to check the validity of the initial calibration.

The internal standard responses and retention times in the calibration check standard must be evaluated immediately after or during data acquisition. If the retention time for any internal standard changes by more than 30 seconds from the last check calibration (12 h), the chromatographic system must be inspected for malfunctions and corrections must be made, as required. If the electron ionization current plot (EICP) area for any of the internal standards changes by a factor of two from the last daily calibration standard check, the mass spectrometer must be inspected for malfunctions and corrections must be made, as appropriate.

Demonstrate, through the analysis of a reagent water blank, that interferences from the analytical system, glassware, and reagents are under control. The blank samples should be carried through all stages of the sample preparation and measurement steps. For each analytical batch (up to 20 samples), a reagent blank, matrix spike, and matrix spike duplicate/duplicate must be analyzed (the frequency of the spikes may be different for different monitoring programs). The blank and spiked samples must be carried through all stages of the sample preparation and measurement steps. A QC reference sample concentrate containing each analyte at a concentration of 100 mg/L in methanol is required.

REFERENCE Test Methods for Evaluating Solid Waste (SW-846). U.S. EPA 1983, Method 8270B, Rev. 2, Nov. 1990. Office of Solid Waste, Washington, DC.

N-Nitrosodiphenylamine   **EPA Method 1625**
CAS #86-30-6

TITLE Semivolatile Organic Compounds by Isotope Dilution GC/MS

MATRIX The compounds may be determined in waters, soils, and municipal sludges by this method.

METHOD SUMMARY This method is used to determine 176 semivolatile toxic organic pollutants associated with the CWA (as amended 1987); the RCRA (as amended 1986); CERCLA (as amended 1986); and other compounds amenable to extraction and analysis by capillary column gas chromatography-mass spectrometry (GC/MS).

Stable isotopically-labeled analogs of the compounds of interest are added to the sample. If the solids content is less than 1%, a 1-L sample is extracted at pH 12–13, then at pH <2 with methylene chloride using continuous extraction techniques.

If the solids content is 30% or less, the sample is diluted to 1% solids with reagent water, homogenized ultrasonically, and extracted at pH 12–13, then at pH <2 with methylene chloride using continuous extraction techniques. If the solids content is greater than 30%, the sample is extracted using ultrasonic techniques.

Each extract is dried over sodium sulfate, concentrated to a volume of 5 mL, cleaned up using GPC, if necessary, and concentrated. Extracts are concentrated to 1 mL if GPC is not performed, and to 0.5 mL if GPC is performed.

An internal standard is added to the extract, and a 1-mL aliquot of the extract is injected into the GC. The compounds are separated by GC and detected by a MS. The labeled compounds serve to correct the variability of the analytical technique.

INTERFERENCES Solvents, reagents, glassware, and other sample processing hardware may yield artifacts and/or elevated baselines causing misinterpretation of chromatograms and spectra. Materials used in the analysis must be demonstrated to be free from interferences under the conditions of analysis by running method blanks initially and with each sample lot (sample started through the extraction process on a given 8-h shift, to a maximum of 20). Specific selection of reagents and purification of solvents by distillation in all glass systems may be required. Glassware and, where possible, reagents are cleaned by solvent rinse and baking at 450°C for 1-h minimum. Interferences coextracted from samples will vary considerably from source to source, depending on the diversity of the site being sampled.

INSTRUMENTATION Major instrumentation includes a GC with a splitless or on-column injection port for capillary column, a MS with 70 eV electron impact ionization, and a data system to collect and record MS data, and process it. A K-D apparatus is used to concentrate extracts.

GC Column: 30 m × 0.25 mm I.D. 5% phenyl, 94% methyl, 1% vinyl silicone bonded phased fused silica capillary column.

PRECISION & ACCURACY The detection limits of the method are usually dependent on the level of interferences rather than instrumental limitations. The limits typify the minimum quantities that can be detected with no interferences present.

The minimum level (in µg/mL) was 20. This is defined as a minimum level at which the analytical system shall give recognizable mass spectra (background corrected) and acceptable calibration points.

The MDL (in µg/kg) in low solids was 55 and in high solids was 36; these were determined in digested sludge (low solids) and in filter cake or compost (high solids).

The labeled and native compound initial precision as standard deviation (in µg/L) was 45.
The labeled and native compound initial accuracy as average recovery (in µg/L) was 65–142.

SAMPLE COLLECTION, PRESERVATION & HANDLING Collect samples in glass containers. Aqueous samples which flow freely are collected in refrigerated bottles using automatic sampling equipment. Solid samples are collected as grab samples using widemouth jars. Maintain samples at 0 to 4°C from the time of collection until extraction. If residual chlorine is present in aqueous samples, add 80 mg sodium thiosulfate/L of water. Begin sample extraction within 7 days of collection, and analyze all extracts within 40 days of extraction.

SAMPLE PREPARATION Samples containing 1% solids or less are extracted directly using continuous liquid-liquid extraction techniques. Samples containing 1 to 30% solids are diluted to the 1% level with reagent water and extracted using continuous liquid-liquid extraction techniques. Samples containing greater than 30% solids are extracted using ultrasonic techniques.

Base/neutral extraction — Adjust the pH of the waters in the extractors to 12–13 with 6 N NaOH. Extract with methylene chloride for 24–48 h.
Acid extraction — Adjust the pH of the waters in the extractors to 2 or less using 6 N sulfuric acid. Extract with methylene chloride for 24–48 h.
Ultrasonic extraction of high solids samples — Add anhydrous sodium sulfate to the sample and QC aliquot(s). Add acetone:methylene chloride (1:1) to the sample and mix thoroughly

Concentrate extracts using a K-D apparatus.

QUALITY CONTROL The analyst is permitted to modify this method to improve separations or lower the costs of measurements, provided all performance specifications are met. Analyses of blanks are required to demonstrate freedom from contamination. When results of spikes indicate atypical method performance for samples, the samples are diluted to bring method performance within acceptable limits.

For low solids (aqueous samples), extract, concentrate, and analyze two sets of four 1-L aliquots (8 aliquots total) of the precision and recovery standard. For high solids samples, two sets of four 30-g aliquots of the high solids reference matrix are used.

Spike all samples with labeled compounds to assess method performance. Compute percent recovery of the labeled compounds using the internal standard method. Compare the labeled compound recovery for each compound with the corresponding labeled compound recovery.

Reagent water and high solids reference matrix blanks are analyzed to demonstrate freedom from contamination. Extract and concentrate a 1-L reagent water blank or a high solids reference matrix blank with each sample's lot (samples started through the extraction process on the same 8-h shift, to a maximum of 20 samples).

Field replicates may be collected to determine the precision of the sampling technique, and spiked samples may be required to determine the accuracy of the analysis when the internal standard method is used.

REFERENCE Semivolatile Organic Compounds by Isotope Dilution GC/MS. Office of Water Regulation and Standards, U.S. EPA Industrial Technology Division, Washington, DC, EPA Method 1625, Rev. C, June 1989 (contact W.A. Telliard,

U.S. EPA, Office of Water Regulations and Standards, 401 M St., SW, Washington, DC, 20460. Phone: 202-382-7131).

N-Nitrosodiphenylamine   **EPA Method 625**
CAS #86-30-6

TITLE Base/Neutrals and Acids, U.S. EPA Method 625

MATRIX This methods covers municipal and industrial wastewaters.

METHOD SUMMARY Approximately 1 L of sample is serially extracted with methylene chloride at a pH greater than 11 and again at a pH less than 2 using a separatory funnel or a continuous extractor. The methylene chloride extract is dried, concentrated to a volume of 1 mL, and analyzed by GC/MS. Qualitative identification of the parameters in the extract is performed using the retention time and the relative abundance of three characteristic masses (m/z). Qualitative analysis is performed using either external or internal standard techniques with a single characteristic m/z.

INTERFERENCES Method interferences may be caused by contaminants in solvents, reagents, glassware, and other sample processing hardware. Glassware must be scrupulously cleaned. Glassware should be heated in a muffle furnace at 400°C for 5 to 30 min. Some thermally stable materials, such as PCBs, may not be eliminated by this treatment. Solvent rinses with acetone and pesticide quality hexane may be substituted for the muffle furnace heating. Matrix interferences may be caused by contaminants that are coextracted from the sample. The base-neutral extraction may cause significantly reduced recovery of phenols. The packed gas chromatographic columns recommended for the basic fraction may not exhibit sufficient resolution for some analytes.

INSTRUMENTATION A GC/MS system with an injection port designed for on-column injection when using packed columns and for splitless injection when using capillary columns.

Column for base/neutrals: 1.8 m long × 2 mm I.D. glass, packed with 3% SP-2550 on Supelcoport (100/120 mesh) or equivalent.

Column for acids: 1.8 m long × 2 mm I.D. glass, packed with 1% SP-1240DA on Supelcoport (100/120 mesh) or equivalent.

PRECISION & ACCURACY The MDL concentrations were obtained using reagent water. The MDL actually achieved in a given analysis will vary depending on instrument sensitivity and matrix effects. This method was tested by 15 laboratories using reagent water, drinking water, surface water, and industrial wastewaters spiked at six concentrations over the range 5 to 100 µg/L. Single operator precision, overall precision, and method accuracy were found to be directly related to the concentration of the parameter matrix.

The MDL (in µg/L) in reagent water was 1.9.

The standard deviation (in µg/L based on 4 recovery measurements) was not reported.

The range (in µg/L) for average recovery for 4 measurements was not reported.

The range (in %) for percent recovery was not reported.

Accuracy (in µg/L) as expected recovery for one or more measurements of a sample containing a true concentration of C was not reported.

Precision (in µg/L) as expected single analyst standard deviation of measurements at an average concentration found at X was not reported.

Overall precision (in µg/L) as expected interlaboratory standard deviation of measurements in an average concentration found at X was not reported.

$C =$ *True value of the concentration in µg/L.*
$X =$ *Average recovery found for measurements of samples containing a concentration at C in µg/L.*

SAMPLE PREPARATION Adjust the pH to >11 with sodium hydroxide and serially extract in a separatory funnel with methylene chloride or else in a continuous extractor. Next, adjust the pH to <2 with sulfuric acid and serially extract in a separatory funnel with methylene chloride or else in a continuous extractor. Dry the extracts separately through a column of anhydrous sodium sulfate and then concentrate each of the extracts to 1.0 mL using a K-D apparatus.

SAMPLE COLLECTION, PRESERVATION & HANDLING Grab samples must be collected in glass containers. All samples must be refrigerated at 4°C from the time of collection until extraction. If residual chlorine is present, add 80 mg of sodium thiosulfate/L of sample and mix well. All samples must be extracted within 7 days of collection and completely analyzed within 40 days of extraction.

QUALITY CONTROL Make an initial, one-time, demonstration of the ability to generate acceptable accuracy and precision with this method. Before processing any samples, the analyst must analyze a reagent water blank to demonstrate that interferences from the analytical system and glassware are under control. Each time a set of samples is extracted or reagents are changed, a reagent water blank must be processed. Spike and analyze a minimum of 5% of all samples to monitor and evaluate lab data quality. A QC check sample concentrate that contains each parameter of interest at a concentration of 100 µg/mL in acetone is required. PCBs and multicomponent pesticides may be omitted from this test.

After analysis of five spiked wastewater samples, calculate the average percent recovery and the standard deviation of the percent recovery. Spike all samples with the surrogate standard spiking solution and calculate the percent recovery of each surrogate compound.

REFERENCE *Federal Register*, Vol. 49, No. 209. Friday, Oct. 26, 1984.

N-Nitrosodiphenylamine   **EPA Method 8270**
CAS #86-30-6

TITLE Semivolatile Organic Compounds by GC/MS

MATRIX This method is used to determine the concentration of semivolatile organic compounds in extracts prepared

from all types of solid waste matrices, soils, and groundwater. Although surface waters are not specifically mentioned, this method should be applicable to water samples from rivers, lakes, etc.

METHOD SUMMARY This method covers 259 semivolatile organic compounds. In very limited applications direct injection of the sample into the GC/MS system may be appropriate, but this results in very high detection limits (approximately 10,000 µg/L). Typically, a 1-L liquid sample, containing surrogate, and matrix spiking standards, is extracted in a continuous extractor first under acid conditions and then under basic conditions. Typically 30 g of a solid sample, containing surrogate, and matrix spiking standards, is extracted ultrasonically. After concentrating the extract to 1 mL it is spiked with 10 µL of an internal standard solution just prior to analysis by GC/MS. The volume injected should contain about 100 ng of base/neutral and 200 ng of acid surrogates (for a 1-µL injection). Analysis is performed by GC/MS using a capillary GC column.

INTERFERENCES Raw GC/MS data from all blanks, samples, and spikes must be evaluated for interferences. Contamination by carryover can occur whenever high-concentration and low-concentration samples are sequentially analyzed. To reduce carryover, the sample syringe must be rinsed out between samples with solvent. Whenever an unusually concentrated sample is encountered, it should be followed by the analysis of blank solvent to check for cross-contamination.

INSTRUMENTATION A GC/MS and a data system are required. The GC column used is a 30 m × 0.25 mm I.D. (or 0.32 mm I.D.) 1um film thickness silicone-coated fused silica capillary column. A continuous liquid-liquid extractor equipped with Teflon® or glass connection joints and stopcocks requiring no lubrication, a K-D concentrating apparatus, water bath, and an ultrasonic disrupter with a minimum power of 300 W and with pulsing capability are also required.

PRECISION & ACCURACY The estimated quantitation limit (EQL) of Method 8270B for determining an individual compound is approximately 1 mg/kg (wet weight) for soil or sediment samples, 1–200 mg/kg for wastes (dependent on matrix and method of preparation), and 10 µg/L for groundwater samples. EQLs will be proportionately higher for sample extracts that require dilution to avoid saturation of the detector.

The EQL(b) for groundwater in µg/L is 10.
The EQL (a, b) for low concentrations in soil and sediment in µg/kg is 660.
Accuracy as µg/L is not listed.
Overall precision in µg/L is not listed.

(a) *EQLs listed for soil/sediment are based on wet weight. Normally data is reported in a dry-weight basis; therefore, EQLs will be higher based on the % dry weight of each sample. This calculation is based on a 30 g sample and gel permeation chromatography cleanup.*
(b) *Sample EQLs are highly matrix-dependent. The EQLs are provided for guidance and may not always be achievable.*
C = *True value for concentration, in µg/L.*
X = *Average recovery found for measurements of samples containing a concentration of C, in µg/L.*

ESTIMATED QUANTITATION LIMIT

Other Matrices	Factor (a)
High-concentration soil and sludges by sonicator	7.5
Non-water miscible waste	75

(a) *EQL for other matrices = [EQL for low soil/sediment] × [Factor]. This estimated EQL is similar to an EPA "Practical Quantitation Limit."*

SAMPLING METHOD

Liquid samples — Use a 1 or 2½ gallon amber glass bottle with a screw-top Teflon®-lined cover that has been prewashed with detergent and rinsed with distilled water and methanol (or isopropanol).

Soils, sediments, or sludges — Use an 8-oz. widemouth glass with a screw-top Teflon®-lined cover that has been prewashed with detergent and rinsed with distilled water and methanol (or isopropanol).

SAMPLE PRESERVATION

Liquid samples — If residual chlorine is present, add 3 mL of 10% sodium thiosulfate per gallon, cool to 4°C and store in a solvent-free refrigerator until analysis; if chlorine is not present, then eliminate the sodium thiosulfate addition.

Soils, sediments, or sludges — Cool samples to 4°C and store in a solvent-free refrigerator.

MHT Liquid samples must be extracted within 7 days and the extracts analyzed within 40 days. Soils, sediments, or sludges may be stored for a maximum of 14 days and the extracts analyzed within 40 days.

SAMPLE PREPARATION

Liquid samples — Transfer 1 L quantitatively to a continuous extractor. If high concentrations are anticipated, a smaller volume may be used and then diluted with organic-free reagent water to 1 L. Adjust pH, if necessary, to pH <2 using 1:1 (V/V) sulfuric acid. Pipette 1.0 mL of a surrogate standard spiking solution into each sample. For the sample in each analytical batch selected for spiking, add 1.0 mL of a matrix spiking standard. For base/neutral acid analysis, the amount of the surrogates and matrix spiking compounds added to the sample should result in a final concentration of 100 ng/µL of each analyte in the extract to be analyzed (assuming a 1-µL injection). Extract with methylene chloride for 18–24 h. Next, adjust the pH of the aqueous phase to pH >11 using 10 N sodium hydroxide and extract it with methylene chloride again for 18–24 h. Dry the extract through a column containing anhydrous sodium sulfate and concentrate it to 1 mL using a K-D concentrator.

Soils, sediments, or sludges — Use 30 g of sample. Nonporous or wet samples (gummy or clay type) that do not have a free-flowing sandy texture must be mixed with anhydrous sodium sulfate until the sample is free flowing. Add 1 mL of surrogate standards to all samples, spikes, standards, and blanks. For the sample in each analytical batch selected for spiking, add 1.0 mL of a matrix spiking standard. For base/neutral acid analysis, the amount added of the surrogates and matrix spiking compounds should result in a final concentration of 100 ng/µL of each base/neutral analyte and 200 ng/µL of each acid analyte

in the extract to be analyzed (assuming a 1-μL injection). Immediately add a 100-mL mixture of 1:1 methylene chloride:acetone and extract the sample ultrasonically for 3 min and then decant or filter the extracts. Repeat the extraction two or more times. Dry the extract using a column with anhydrous sodium sulfate and concentrate it to 1 mL in a K-D concentrator.

Note: N-nitrosodiphenylamine decomposes in the gas chromatographic inlet and cannot be separated from diphenylamine.

QUALITY CONTROL A methylene chloride solution containing 50 ng/μL of decafluorotriphenylphosphine (DFTPP) is used for tuning the GC/MS system each 12-h shift. A system performance check also must be made during every 12-h shift. A standard containing 50 ng/μL each of 4,4′-DDT, pentachlorophenol, and benzidine is required to verify injection port inertness and GC column performance. A calibration standard at mid-concentration, containing each compound of interest, including all required surrogates, must be performed every 12 h during analysis. After the system performance check is met, calibration check compounds (CCCs) are used to check the validity of the initial calibration.

The internal standard responses and retention times in the calibration check standard must be evaluated immediately after or during data acquisition. If the retention time for any internal standard changes by more than 30 seconds from the last check calibration (12 h), the chromatographic system must be inspected for malfunctions and corrections must be made, as required. If the electron ionization current plot (EICP) area for any of the internal standards changes by a factor of two from the last daily calibration standard check, the mass spectrometer must be inspected for malfunctions and corrections must be made, as appropriate.

Demonstrate, through the analysis of a reagent water blank, that interferences from the analytical system, glassware, and reagents are under control. The blank samples should be carried through all stages of the sample preparation and measurement steps. For each analytical batch (up to 20 samples), a reagent blank, matrix spike, and matrix spike duplicate/duplicate must be analyzed (the frequency of the spikes may be different for different monitoring programs). The blank and spiked samples must be carried through all stages of the sample preparation and measurement steps. A QC reference sample concentrate containing each analyte at a concentration of 100 mg/L in methanol is required.

REFERENCE Test Methods for Evaluating Solid Waste (SW-846). U.S. EPA 1983, Method 8270B, Rev. 2, Nov. 1990. Office of Solid Waste, Washington, DC.

N-Nitrosomethylethylamine **EPA Method 1625**
CAS #10595-95-6

TITLE Semivolatile Organic Compounds by Isotope Dilution GC/MS

MATRIX The compounds may be determined in waters, soils, and municipal sludges by this method.

METHOD SUMMARY This method is used to determine 176 semivolatile toxic organic pollutants associated with the CWA (as amended 1987); the RCRA (as amended 1986); the CERCLA (as amended 1986); and other compounds amenable to extraction and analysis by capillary column gas chromatography-mass spectrometry (GC/MS).

Stable isotopically-labeled analogs of the compounds of interest are added to the sample. If the solids content is less than 1%, a 1-L sample is extracted at pH 12–13, then at pH <2 with methylene chloride using continuous extraction techniques.

If the solids content is 30% or less, the sample is diluted to 1% solids with reagent water, homogenized ultrasonically, and extracted at pH 12–13, then at pH <2 with methylene chloride using continuous extraction techniques. If the solids content is greater than 30%, the sample is extracted using ultrasonic techniques.

Each extract is dried over sodium sulfate, concentrated to a volume of 5 mL, cleaned up using GPC, if necessary, and concentrated. Extracts are concentrated to 1 mL if GPC is not performed, and to 0.5 mL if GPC is performed.

An internal standard is added to the extract, and a 1-mL aliquot of the extract is injected into the GC. The compounds are separated by GC and detected by a MS. The labeled compounds serve to correct the variability of the analytical technique.

INTERFERENCES Solvents, reagents, glassware, and other sample processing hardware may yield artifacts and/or elevated baselines causing misinterpretation of chromatograms and spectra. Materials used in the analysis must be demonstrated to be free from interferences under the conditions of analysis by running method blanks initially and with each sample lot (sample started through the extraction process on a given 8-h shift, to a maximum of 20). Specific selection of reagents and purification of solvents by distillation in all glass systems may be required. Glassware and, where possible, reagents are cleaned by solvent rinse and baking at 450°C for 1-h minimum. Interferences coextracted from samples will vary considerably from source to source, depending on the diversity of the site being sampled.

INSTRUMENTATION Major instrumentation includes a GC with a splitless or on-column injection port for capillary column, a MS with 70 eV electron impact ionization, and a data system to collect and record MS data, and process it. A K-D apparatus is used to concentrate extracts.

GC Column: 30 m × 0.25 mm I.D. 5% phenyl, 94% methyl, 1% vinyl silicone bonded phased fused silica capillary column.

PRECISION & ACCURACY The detection limits of the method are usually dependent on the level of interferences rather than instrumental limitations. The limits typify the minimum quantities that can be detected with no interferences present.

The minimum level (in μg/mL) was not listed. This is defined as a minimum level at which the analytical system shall give recognizable mass spectra (background corrected) and acceptable calibration points.

The MDL (in μg/kg) in low solids was not listed and in high solids was not listed; these were determined in digested sludge (low solids) and in filter cake or compost (high solids).

The labeled and native compound initial precision as standard deviation (in μg/L) was not listed.

The labeled and native compound initial accuracy as average recovery (in μg/L) was not listed.

SAMPLE COLLECTION, PRESERVATION & HANDLING

Collect samples in glass containers. Aqueous samples which flow freely are collected in refrigerated bottles using automatic sampling equipment. Solid samples are collected as grab samples using widemouth jars. Maintain samples at 0 to 4°C from the time of collection until extraction. If residual chlorine is present in aqueous samples, add 80 mg sodium thiosulfate/L of water. Begin sample extraction within 7 days of collection, and analyze all extracts within 40 days of extraction.

SAMPLE PREPARATION

Samples containing 1% solids or less are extracted directly using continuous liquid-liquid extraction techniques. Samples containing 1 to 30% solids are diluted to the 1% level with reagent water and extracted using continuous liquid-liquid extraction techniques. Samples containing greater than 30% solids are extracted using ultrasonic techniques.

- Base/neutral extraction — Adjust the pH of the waters in the extractors to 12–13 with 6 N NaOH. Extract with methylene chloride for 24–48 h.
- Acid extraction — Adjust the pH of the waters in the extractors to 2 or less using 6 N sulfuric acid. Extract with methylene chloride for 24–48 h.
- Ultrasonic extraction of high solids samples — Add anhydrous sodium sulfate to the sample and QC aliquot(s). Add acetone:methylene chloride (1:1) to the sample and mix thoroughly

Concentrate extracts using a K-D apparatus.

QUALITY CONTROL

The analyst is permitted to modify this method to improve separations or lower the costs of measurements, provided all performance specifications are met. Analyses of blanks are required to demonstrate freedom from contamination. When results of spikes indicate atypical method performance for samples, the samples are diluted to bring method performance within acceptable limits.

For low solids (aqueous samples), extract, concentrate, and analyze two sets of four 1-L aliquots (8 aliquots total) of the precision and recovery standard. For high solids samples, two sets of four 30-g aliquots of the high solids reference matrix are used.

Spike all samples with labeled compounds to assess method performance. Compute percent recovery of the labeled compounds using the internal standard method. Compare the labeled compound recovery for each compound with the corresponding labeled compound recovery.

Reagent water and high solids reference matrix blanks are analyzed to demonstrate freedom from contamination. Extract and concentrate a 1-L reagent water blank or a high solids reference matrix blank with each sample's lot (samples started through the extraction process on the same 8-h shift, to a maximum of 20 samples).

Field replicates may be collected to determine the precision of the sampling technique, and spiked samples may be required to determine the accuracy of the analysis when the internal standard method is used.

REFERENCE Semivolatile Organic Compounds by Isotope Dilution GC/MS. Office of Water Regulation and Standards, U.S. EPA Industrial Technology Division, Washington, DC, EPA Method 1625, Rev. C, June 1989 (contact W.A. Telliard, U.S. EPA, Office of Water Regulations and Standards, 401 M St., SW, Washington, DC, 20460. Phone: 202-382-7131).

N-Nitrosomethylethylamine   **EPA Method 8270**
CAS #10595-95-6

TITLE Semivolatile Organic Compounds by GC/MS

MATRIX This method is used to determine the concentration of semivolatile organic compounds in extracts prepared from all types of solid waste matrices, soils, and groundwater. Although surface waters are not specifically mentioned, this method should be applicable to water samples from rivers, lakes, etc.

METHOD SUMMARY This method covers 259 semivolatile organic compounds. In very limited applications direct injection of the sample into the GC/MS system may be appropriate, but this results in very high detection limits (approximately 10,000 μg/L). Typically, a 1-L liquid sample, containing surrogate, and matrix spiking standards, is extracted in a continuous extractor first under acid conditions and then under basic conditions. Typically 30 g of a solid sample, containing surrogate, and matrix spiking standards, is extracted ultrasonically. After concentrating the extract to 1 mL it is spiked with 10 μL of an internal standard solution just prior to analysis by GC/MS. The volume injected should contain about 100 ng of base/neutral and 200 ng of acid surrogates (for a 1-μL injection). Analysis is performed by GC/MS using a capillary GC column.

INTERFERENCES Raw GC/MS data from all blanks, samples, and spikes must be evaluated for interferences. Contamination by carryover can occur whenever high-concentration and low-concentration samples are sequentially analyzed. To reduce carryover, the sample syringe must be rinsed out between samples with solvent. Whenever an unusually concentrated sample is encountered, it should be followed by the analysis of blank solvent to check for cross-contamination.

INSTRUMENTATION A GC/MS and a data system are required. The GC column used is a 30 m × 0.25 mm I.D. (or 0.32 mm I.D.) 1um film thickness silicone-coated fused silica capillary column. A continuous liquid-liquid extractor equipped with Teflon® or glass connection joints and stopcocks requiring no lubrication, a K-D concentrating apparatus, water bath, and an ultrasonic disrupter with a minimum power of 300 W and with pulsing capability are also required.

PRECISION & ACCURACY The estimated quantitation limit (EQL) of Method 8270B for determining an individual compound is approximately 1 mg/kg (wet weight) for soil or sediment samples, 1–200 mg/kg for wastes (dependent on matrix and method of preparation), and 10 µg/L for groundwater samples. EQLs will be proportionately higher for sample extracts that require dilution to avoid saturation of the detector.

The EQL(b) for groundwater in µg/L is not listed.
The EQL (a, b) for low concentrations in soil and sediment in µg/kg is not listed.
Accuracy as µg/L is not listed.
Overall precision in µg/L is not listed.

(a) *EQLs listed for soil/sediment are based on wet weight. Normally data is reported in a dry-weight basis; therefore, EQLs will be higher based on the % dry weight of each sample. This calculation is based on a 30 g sample and gel permeation chromatography cleanup.*
(b) *Sample EQLs are highly matrix-dependent. The EQLs are provided for guidance and may not always be achievable.*
C = *True value for concentration, in µg/L.*
X = *Average recovery found for measurements of samples containing a concentration of C, in µg/L.*

ESTIMATED QUANTITATION LIMIT

Other Matrices	Factor (a)
High-concentration soil and sludges by sonicator	7.5
Non-water miscible waste	75

(a) *EQL for other matrices = [EQL for low soil/sediment] × [Factor]. This estimated EQL is similar to an EPA "Practical Quantitation Limit."*

SAMPLING METHOD
Liquid samples — Use a 1 or 2½ gallon amber glass bottle with a screw-top Teflon®-lined cover that has been prewashed with detergent and rinsed with distilled water and methanol (or isopropanol).

Soils, sediments, or sludges — Use an 8-oz. widemouth glass with a screw-top Teflon®-lined cover that has been prewashed with detergent and rinsed with distilled water and methanol (or isopropanol).

SAMPLE PRESERVATION
Liquid samples — If residual chlorine is present, add 3 mL of 10% sodium thiosulfate per gallon, cool to 4°C and store in a solvent-free refrigerator until analysis; if chlorine is not present, then eliminate the sodium thiosulfate addition.

Soils, sediments, or sludges — Cool samples to 4°C and store in a solvent-free refrigerator.

MHT Liquid samples must be extracted within 7 days and the extracts analyzed within 40 days. Soils, sediments, or sludges may be stored for a maximum of 14 days and the extracts analyzed within 40 days.

SAMPLE PREPARATION
Liquid samples — Transfer 1 L quantitatively to a continuous extractor. If high concentrations are anticipated, a smaller volume may be used and then diluted with organic-free reagent water to 1 L. Adjust pH, if necessary, to pH <2 using 1:1 (V/V) sulfuric acid. Pipette 1.0 mL of a surrogate standard spiking solution into each sample. For the sample in each analytical batch selected for spiking, add 1.0 mL of a matrix spiking standard. For base/neutral acid analysis, the amount of the surrogates and matrix spiking compounds added to the sample should result in a final concentration of 100 ng/µL of each analyte in the extract to be analyzed (assuming a 1-µL injection). Extract with methylene chloride for 18–24 h. Next, adjust the pH of the aqueous phase to pH >11 using 10 N sodium hydroxide and extract it with methylene chloride again for 18–24 h. Dry the extract through a column containing anhydrous sodium sulfate and concentrate it to 1 mL using a K-D concentrator.

Soils, sediments, or sludges — Use 30 g of sample. Nonporous or wet samples (gummy or clay type) that do not have a free-flowing sandy texture must be mixed with anhydrous sodium sulfate until the sample is free flowing. Add 1 mL of surrogate standards to all samples, spikes, standards, and blanks. For the sample in each analytical batch selected for spiking, add 1.0 mL of a matrix spiking standard. For base/neutral acid analysis, the amount added of the surrogates and matrix spiking compounds should result in a final concentration of 100 ng/µL of each base/neutral analyte and 200 ng/µL of each acid analyte in the extract to be analyzed (assuming a 1-µL injection). Immediately add a 100-mL mixture of 1:1 methylene chloride:acetone and extract the sample ultrasonically for 3 min and then decant or filter the extracts. Repeat the extraction two or more times. Dry the extract using a column with anhydrous sodium sulfate and concentrate it to 1 mL in a K-D concentrator.

QUALITY CONTROL A methylene chloride solution containing 50 ng/µL of decafluorotriphenylphosphine (DFTPP) is used for tuning the GC/MS system each 12-h shift. A system performance check also must be made during every 12-h shift. A standard containing 50 ng/µL each of 4,4'-DDT, pentachlorophenol, and benzidine is required to verify injection port inertness and GC column performance. A calibration standard at mid-concentration, containing each compound of interest, including all required surrogates, must be performed every 12 h during analysis. After the system performance check is met, calibration check compounds (CCCs) are used to check the validity of the initial calibration.

The internal standard responses and retention times in the calibration check standard must be evaluated immediately after or during data acquisition. If the retention time for any internal standard changes by more than 30 seconds from the last check calibration (12 h), the chromatographic system must be inspected for malfunctions and corrections must be made, as required. If the electron ionization current plot (EICP) area for any of the internal standards changes by a factor of two from the last daily calibration standard check, the mass spectrometer must be inspected for malfunctions and corrections must be made, as appropriate.

Demonstrate, through the analysis of a reagent water blank, that interferences from the analytical system, glassware, and reagents are under control. The blank samples should be carried through all stages of the sample preparation and measurement

steps. For each analytical batch (up to 20 samples), a reagent blank, matrix spike, and matrix spike duplicate/duplicate must be analyzed (the frequency of the spikes may be different for different monitoring programs). The blank and spiked samples must be carried through all stages of the sample preparation and measurement steps. A QC reference sample concentrate containing each analyte at a concentration of 100 mg/L in methanol is required.

REFERENCE Test Methods for Evaluating Solid Waste (SW-846). U.S. EPA 1983, Method 8270B, Rev. 2, Nov. 1990. Office of Solid Waste, Washington, DC.

N-Nitrosomethylphenylamine EPA Method 1625
CAS #614-00-6

TITLE Semivolatile Organic Compounds by Isotope Dilution GC/MS

MATRIX The compounds may be determined in waters, soils, and municipal sludges by this method.

METHOD SUMMARY This method is used to determine 176 semivolatile toxic organic pollutants associated with the CWA (as amended 1987); the RCRA (as amended 1986); the CERCLA (as amended 1986); and other compounds amenable to extraction and analysis by capillary column gas chromatography-mass spectrometry (GC/MS).

Stable isotopically-labeled analogs of the compounds of interest are added to the sample. If the solids content is less than 1%, a 1-L sample is extracted at pH 12–13, then at pH <2 with methylene chloride using continuous extraction techniques.

If the solids content is 30% or less, the sample is diluted to 1% solids with reagent water, homogenized ultrasonically, and extracted at pH 12–13, then at pH <2 with methylene chloride using continuous extraction techniques. If the solids content is greater than 30%, the sample is extracted using ultrasonic techniques.

Each extract is dried over sodium sulfate, concentrated to a volume of 5 mL, cleaned up using GPC, if necessary, and concentrated. Extracts are concentrated to 1 mL if GPC is not performed, and to 0.5 mL if GPC is performed.

An internal standard is added to the extract, and a 1-mL aliquot of the extract is injected into the GC. The compounds are separated by GC and detected by a MS. The labeled compounds serve to correct the variability of the analytical technique.

INTERFERENCES Solvents, reagents, glassware, and other sample processing hardware may yield artifacts and/or elevated baselines causing misinterpretation of chromatograms and spectra. Materials used in the analysis must be demonstrated to be free from interferences under the conditions of analysis by running method blanks initially and with each sample lot (sample started through the extraction process on a given 8-h shift, to a maximum of 20). Specific selection of reagents and purification of solvents by distillation in all glass systems may be required. Glassware and, where possible, reagents are cleaned by solvent rinse and baking at 450°C for 1-h minimum.

Interferences coextracted from samples will vary considerably from source to source, depending on the diversity of the site being sampled.

INSTRUMENTATION Major instrumentation includes a GC with a splitless or on-column injection port for capillary column, a MS with 70 eV electron impact ionization, and a data system to collect and record MS data, and process it. A K-D apparatus is used to concentrate extracts.

GC Column: 30 m × 0.25 mm I.D. 5% phenyl, 94% methyl, 1% vinyl silicone bonded phased fused silica capillary column.

PRECISION & ACCURACY The detection limits of the method are usually dependent on the level of interferences rather than instrumental limitations. The limits typify the minimum quantities that can be detected with no interferences present.

The minimum level (in µg/mL) was not listed. This is defined as a minimum level at which the analytical system shall give recognizable mass spectra (background corrected) and acceptable calibration points.

The MDL (in µg/kg) in low solids was not listed and in high solids was not listed; these were determined in digested sludge (low solids) and in filter cake or compost (high solids).

The labeled and native compound initial precision as standard deviation (in µg/L) was not listed.
The labeled and native compound initial accuracy as average recovery (in µg/L) was not listed.

SAMPLE COLLECTION, PRESERVATION & HANDLING
Collect samples in glass containers. Aqueous samples which flow freely are collected in refrigerated bottles using automatic sampling equipment. Solid samples are collected as grab samples using widemouth jars. Maintain samples at 0 to 4°C from the time of collection until extraction. If residual chlorine is present in aqueous samples, add 80 mg sodium thiosulfate/L of water. Begin sample extraction within 7 days of collection, and analyze all extracts within 40 days of extraction.

SAMPLE PREPARATION Samples containing 1% solids or less are extracted directly using continuous liquid-liquid extraction techniques. Samples containing 1 to 30% solids are diluted to the 1% level with reagent water and extracted using continuous liquid-liquid extraction techniques. Samples containing greater than 30% solids are extracted using ultrasonic techniques.

Base/neutral extraction — Adjust the pH of the waters in the extractors to 12–13 with 6 N NaOH. Extract with methylene chloride for 24–48 h.
Acid extraction — Adjust the pH of the waters in the extractors to 2 or less using 6 N sulfuric acid. Extract with methylene chloride for 24–48 h.
Ultrasonic extraction of high solids samples — Add anhydrous sodium sulfate to the sample and QC aliquot(s). Add acetone:methylene chloride (1:1) to the sample and mix thoroughly

Concentrate extracts using a K-D apparatus.

QUALITY CONTROL The analyst is permitted to modify this method to improve separations or lower the costs of measurements, provided all performance specifications are met. Analyses of blanks are required to demonstrate freedom from contamination. When results of spikes indicate atypical method performance for samples, the samples are diluted to bring method performance within acceptable limits.

For low solids (aqueous samples), extract, concentrate, and analyze two sets of four 1-L aliquots (8 aliquots total) of the precision and recovery standard. For high solids samples, two sets of four 30-g aliquots of the high solids reference matrix are used.

Spike all samples with labeled compounds to assess method performance. Compute percent recovery of the labeled compounds using the internal standard method. Compare the labeled compound recovery for each compound with the corresponding labeled compound recovery.

Reagent water and high solids reference matrix blanks are analyzed to demonstrate freedom from contamination. Extract and concentrate a 1-L reagent water blank or a high solids reference matrix blank with each sample's lot (samples started through the extraction process on the same 8-h shift, to a maximum of 20 samples).

Field replicates may be collected to determine the precision of the sampling technique, and spiked samples may be required to determine the accuracy of the analysis when the internal standard method is used.

REFERENCE Semivolatile Organic Compounds by Isotope Dilution GC/MS. Office of Water Regulation and Standards, U.S. EPA Industrial Technology Division, Washington, DC, EPA Method 1625, Rev. C, June 1989 (contact W.A. Telliard, U.S. EPA, Office of Water Regulations and Standards, 401 M St., SW, Washington, DC, 20460. Phone: 202-382-7131).

N-Nitrosomorpholine   **EPA Method 1625**
CAS #59-89-2

TITLE Semivolatile Organic Compounds by Isotope Dilution GC/MS

MATRIX The compounds may be determined in waters, soils, and municipal sludges by this method.

METHOD SUMMARY This method is used to determine 176 semivolatile toxic organic pollutants associated with the CWA (as amended 1987); the RCRA (as amended 1986); the CERCLA (as amended 1986); and other compounds amenable to extraction and analysis by capillary column gas chromatography-mass spectrometry (GC/MS).

Stable isotopically-labeled analogs of the compounds of interest are added to the sample. If the solids content is less than 1%, a 1-L sample is extracted at pH 12–13, then at pH <2 with methylene chloride using continuous extraction techniques.

If the solids content is 30% or less, the sample is diluted to 1% solids with reagent water, homogenized ultrasonically, and extracted at pH 12–13, then at pH <2 with methylene chloride using continuous extraction techniques. If the solids content is greater than 30%, the sample is extracted using ultrasonic techniques.

Each extract is dried over sodium sulfate, concentrated to a volume of 5 mL, cleaned up using GPC, if necessary, and concentrated. Extracts are concentrated to 1 mL if GPC is not performed, and to 0.5 mL if GPC is performed.

An internal standard is added to the extract, and a 1-mL aliquot of the extract is injected into the GC. The compounds are separated by GC and detected by a MS. The labeled compounds serve to correct the variability of the analytical technique.

INTERFERENCES Solvents, reagents, glassware, and other sample processing hardware may yield artifacts and/or elevated baselines causing misinterpretation of chromatograms and spectra. Materials used in the analysis must be demonstrated to be free from interferences under the conditions of analysis by running method blanks initially and with each sample lot (sample started through the extraction process on a given 8-h shift, to a maximum of 20). Specific selection of reagents and purification of solvents by distillation in all glass systems may be required. Glassware and, where possible, reagents are cleaned by solvent rinse and baking at 450°C for 1-h minimum. Interferences coextracted from samples will vary considerably from source to source, depending on the diversity of the site being sampled.

INSTRUMENTATION Major instrumentation includes a GC with a splitless or on-column injection port for capillary column, a MS with 70 eV electron impact ionization, and a data system to collect and record MS data, and process it. A K-D apparatus is used to concentrate extracts.

GC Column: 30 m × 0.25 mm I.D. 5% phenyl, 94% methyl, 1% vinyl silicone bonded phased fused silica capillary column.

PRECISION & ACCURACY The detection limits of the method are usually dependent on the level of interferences rather than instrumental limitations. The limits typify the minimum quantities that can be detected with no interferences present.

The minimum level (in µg/mL) was not listed. This is defined as a minimum level at which the analytical system shall give recognizable mass spectra (background corrected) and acceptable calibration points.

The MDL (in µg/kg) in low solids was not listed and in high solids was not listed; these were determined in digested sludge (low solids) and in filter cake or compost (high solids).

The labeled and native compound initial precision as standard deviation (in µg/L) was not listed.
The labeled and native compound initial accuracy as average recovery (in µg/L) was not listed.

SAMPLE COLLECTION, PRESERVATION & HANDLING
Collect samples in glass containers. Aqueous samples which flow freely are collected in refrigerated bottles using automatic sampling equipment. Solid samples are collected as grab samples using widemouth jars. Maintain samples at 0 to 4°C from

the time of collection until extraction. If residual chlorine is present in aqueous samples, add 80 mg sodium thiosulfate/L of water. Begin sample extraction within 7 days of collection, and analyze all extracts within 40 days of extraction.

SAMPLE PREPARATION Samples containing 1% solids or less are extracted directly using continuous liquid-liquid extraction techniques. Samples containing 1 to 30% solids are diluted to the 1% level with reagent water and extracted using continuous liquid-liquid extraction techniques. Samples containing greater than 30% solids are extracted using ultrasonic techniques.

Base/neutral extraction — Adjust the pH of the waters in the extractors to 12–13 with 6 *N* NaOH. Extract with methylene chloride for 24–48 h.

Acid extraction — Adjust the pH of the waters in the extractors to 2 or less using 6 *N* sulfuric acid. Extract with methylene chloride for 24–48 h.

Ultrasonic extraction of high solids samples — Add anhydrous sodium sulfate to the sample and QC aliquot(s). Add acetone:methylene chloride (1:1) to the sample and mix thoroughly

Concentrate extracts using a K-D apparatus.

QUALITY CONTROL The analyst is permitted to modify this method to improve separations or lower the costs of measurements, provided all performance specifications are met. Analyses of blanks are required to demonstrate freedom from contamination. When results of spikes indicate atypical method performance for samples, the samples are diluted to bring method performance within acceptable limits.

For low solids (aqueous samples), extract, concentrate, and analyze two sets of four 1-L aliquots (8 aliquots total) of the precision and recovery standard. For high solids samples, two sets of four 30-g aliquots of the high solids reference matrix are used.

Spike all samples with labeled compounds to assess method performance. Compute percent recovery of the labeled compounds using the internal standard method. Compare the labeled compound recovery for each compound with the corresponding labeled compound recovery.

Reagent water and high solids reference matrix blanks are analyzed to demonstrate freedom from contamination. Extract and concentrate a 1-L reagent water blank or a high solids reference matrix blank with each sample's lot (samples started through the extraction process on the same 8-h shift, to a maximum of 20 samples).

Field replicates may be collected to determine the precision of the sampling technique, and spiked samples may be required to determine the accuracy of the analysis when the internal standard method is used.

REFERENCE Semivolatile Organic Compounds by Isotope Dilution GC/MS. Office of Water Regulation and Standards, U.S. EPA Industrial Technology Division, Washington, DC, EPA Method 1625, Rev. C, June 1989 (contact W.A. Telliard, U.S. EPA, Office of Water Regulations and Standards, 401 M St., SW, Washington, DC, 20460. Phone: 202-382-7131).

N-Nitrosomorpholine   EPA Method 8270
CAS #59-89-2

TITLE Semivolatile Organic Compounds by GC/MS

MATRIX This method is used to determine the concentration of semivolatile organic compounds in extracts prepared from all types of solid waste matrices, soils, and groundwater. Although surface waters are not specifically mentioned, this method should be applicable to water samples from rivers, lakes, etc.

METHOD SUMMARY This method covers 259 semivolatile organic compounds. In very limited applications direct injection of the sample into the GC/MS system may be appropriate, but this results in very high detection limits (approximately 10,000 μg/L). Typically, a 1-L liquid sample, containing surrogate, and matrix spiking standards, is extracted in a continuous extractor first under acid conditions and then under basic conditions. Typically 30 g of a solid sample, containing surrogate, and matrix spiking standards, is extracted ultrasonically. After concentrating the extract to 1 mL it is spiked with 10 μL of an internal standard solution just prior to analysis by GC/MS. The volume injected should contain about 100 ng of base/neutral and 200 ng of acid surrogates (for a 1-μL injection). Analysis is performed by GC/MS using a capillary GC column.

INTERFERENCES Raw GC/MS data from all blanks, samples, and spikes must be evaluated for interferences. Contamination by carryover can occur whenever high-concentration and low-concentration samples are sequentially analyzed. To reduce carryover, the sample syringe must be rinsed out between samples with solvent. Whenever an unusually concentrated sample is encountered, it should be followed by the analysis of blank solvent to check for cross-contamination.

INSTRUMENTATION A GC/MS and a data system are required. The GC column used is a 30 m × 0.25 mm I.D. (or 0.32 mm I.D.) 1um film thickness silicone-coated fused silica capillary column. A continuous liquid-liquid extractor equipped with Teflon® or glass connection joints and stopcocks requiring no lubrication, a K-D concentrating apparatus, water bath, and an ultrasonic disrupter with a minimum power of 300 W and with pulsing capability are also required.

PRECISION & ACCURACY The estimated quantitation limit (EQL) of Method 8270B for determining an individual compound is approximately 1 mg/kg (wet weight) for soil or sediment samples, 1–200 mg/kg for wastes (dependent on matrix and method of preparation), and 10 μg/L for groundwater samples. EQLs will be proportionately higher for sample extracts that require dilution to avoid saturation of the detector.

The EQL(b) for groundwater in μg/L is not listed.
The EQL (a, b) for low concentrations in soil and sediment in μg/kg is not listed.
Accuracy as μg/L is not listed.
Overall precision in μg/L is not listed.

(a) *EQLs listed for soil/sediment are based on wet weight. Normally data is reported in a dry-weight basis; therefore, EQLs will be higher based on the % dry weight of each sample.*

This calculation is based on a 30 g sample and gel permeation chromatography cleanup.

(b) *Sample EQLs are highly matrix-dependent. The EQLs are provided for guidance and may not always be achievable.*

C = *True value for concentration, in µg/L.*

X = *Average recovery found for measurements of samples containing a concentration of C, in µg/L.*

ESTIMATED QUANTITATION LIMIT

Other Matrices	Factor (a)
High-concentration soil and sludges by sonicator	7.5
Non-water miscible waste	75

(a) *EQL for other matrices = [EQL for low soil/sediment] × [Factor]. This estimated EQL is similar to an EPA "Practical Quantitation Limit."*

SAMPLING METHOD

Liquid samples — Use a 1 or 2½ gallon amber glass bottle with a screw-top Teflon®-lined cover that has been prewashed with detergent and rinsed with distilled water and methanol (or isopropanol).

Soils, sediments, or sludges — Use an 8-oz. widemouth glass with a screw-top Teflon®-lined cover that has been prewashed with detergent and rinsed with distilled water and methanol (or isopropanol).

SAMPLE PRESERVATION

Liquid samples — If residual chlorine is present, add 3 mL of 10% sodium thiosulfate per gallon, cool to 4°C and store in a solvent-free refrigerator until analysis; if chlorine is not present, then eliminate the sodium thiosulfate addition.

Soils, sediments, or sludges — Cool samples to 4°C and store in a solvent-free refrigerator.

MHT Liquid samples must be extracted within 7 days and the extracts analyzed within 40 days. Soils, sediments, or sludges may be stored for a maximum of 14 days and the extracts analyzed within 40 days.

SAMPLE PREPARATION

Liquid samples — Transfer 1 L quantitatively to a continuous extractor. If high concentrations are anticipated, a smaller volume may be used and then diluted with organic-free reagent water to 1 L. Adjust pH, if necessary, to pH <2 using 1:1 (V/V) sulfuric acid. Pipette 1.0 mL of a surrogate standard spiking solution into each sample. For the sample in each analytical batch selected for spiking, add 1.0 mL of a matrix spiking standard. For base/neutral acid analysis, the amount of the surrogates and matrix spiking compounds added to the sample should result in a final concentration of 100 ng/µL of each analyte in the extract to be analyzed (assuming a 1-µL injection). Extract with methylene chloride for 18–24 h. Next, adjust the pH of the aqueous phase to pH >11 using 10 N sodium hydroxide and extract it with methylene chloride again for 18–24 h. Dry the extract through a column containing anhydrous sodium sulfate and concentrate it to 1 mL using a K-D concentrator.

Soils, sediments, or sludges — Use 30 g of sample. Nonporous or wet samples (gummy or clay type) that do not have a free-flowing sandy texture must be mixed with anhydrous sodium sulfate until the sample is free flowing. Add 1 mL of surrogate standards to all samples, spikes, standards, and blanks. For the sample in each analytical batch selected for spiking, add 1.0 mL of a matrix spiking standard. For base/neutral acid analysis, the amount added of the surrogates and matrix spiking compounds should result in a final concentration of 100 ng/µL of each base/neutral analyte and 200 ng/µL of each acid analyte in the extract to be analyzed (assuming a 1-µL injection). Immediately add a 100-mL mixture of 1:1 methylene chloride:acetone and extract the sample ultrasonically for 3 min and then decant or filter the extracts. Repeat the extraction two or more times. Dry the extract using a column with anhydrous sodium sulfate and concentrate it to 1 mL in a K-D concentrator.

QUALITY CONTROL A methylene chloride solution containing 50 ng/µL of decafluorotriphenylphosphine (DFTPP) is used for tuning the GC/MS system each 12-h shift. A system performance check also must be made during every 12-h shift. A standard containing 50 ng/µL each of 4,4'-DDT, pentachlorophenol, and benzidine is required to verify injection port inertness and GC column performance. A calibration standard at mid-concentration, containing each compound of interest, including all required surrogates, must be performed every 12 h during analysis. After the system performance check is met, calibration check compounds (CCCs) are used to check the validity of the initial calibration.

The internal standard responses and retention times in the calibration check standard must be evaluated immediately after or during data acquisition. If the retention time for any internal standard changes by more than 30 seconds from the last check calibration (12 h), the chromatographic system must be inspected for malfunctions and corrections must be made, as required. If the electron ionization current plot (EICP) area for any of the internal standards changes by a factor of two from the last daily calibration standard check, the mass spectrometer must be inspected for malfunctions and corrections must be made, as appropriate.

Demonstrate, through the analysis of a reagent water blank, that interferences from the analytical system, glassware, and reagents are under control. The blank samples should be carried through all stages of the sample preparation and measurement steps. For each analytical batch (up to 20 samples), a reagent blank, matrix spike, and matrix spike duplicate/duplicate must be analyzed (the frequency of the spikes may be different for different monitoring programs). The blank and spiked samples must be carried through all stages of the sample preparation and measurement steps. A QC reference sample concentrate containing each analyte at a concentration of 100 mg/L in methanol is required.

REFERENCE Test Methods for Evaluating Solid Waste (SW-846). U.S. EPA 1983, Method 8270B, Rev. 2, Nov. 1990. Office of Solid Waste, Washington, DC.

N-Nitrosopiperidine   **EPA Method 1625**
CAS #100-75-4

TITLE Semivolatile Organic Compounds by Isotope Dilution GC/MS

MATRIX The compounds may be determined in waters, soils, and municipal sludges by this method.

METHOD SUMMARY This method is used to determine 176 semivolatile toxic organic pollutants associated with the CWA (as amended 1987); the RCRA (as amended 1986); the CERCLA (as amended 1986); and other compounds amenable to extraction and analysis by capillary column gas chromatography-mass spectrometry (GC/MS).

Stable isotopically-labeled analogs of the compounds of interest are added to the sample. If the solids content is less than 1%, a 1-L sample is extracted at pH 12–13, then at pH <2 with methylene chloride using continuous extraction techniques.

If the solids content is 30% or less, the sample is diluted to 1% solids with reagent water, homogenized ultrasonically, and extracted at pH 12–13, then at pH <2 with methylene chloride using continuous extraction techniques. If the solids content is greater than 30%, the sample is extracted using ultrasonic techniques.

Each extract is dried over sodium sulfate, concentrated to a volume of 5 mL, cleaned up using GPC, if necessary, and concentrated. Extracts are concentrated to 1 mL if GPC is not performed, and to 0.5 mL if GPC is performed.

An internal standard is added to the extract, and a 1-mL aliquot of the extract is injected into the GC. The compounds are separated by GC and detected by a MS. The labeled compounds serve to correct the variability of the analytical technique.

INTERFERENCES Solvents, reagents, glassware, and other sample processing hardware may yield artifacts and/or elevated baselines causing misinterpretation of chromatograms and spectra. Materials used in the analysis must be demonstrated to be free from interferences under the conditions of analysis by running method blanks initially and with each sample lot (sample started through the extraction process on a given 8-h shift, to a maximum of 20). Specific selection of reagents and purification of solvents by distillation in all glass systems may be required. Glassware and, where possible, reagents are cleaned by solvent rinse and baking at 450°C for 1-h minimum. Interferences coextracted from samples will vary considerably from source to source, depending on the diversity of the site being sampled.

INSTRUMENTATION Major instrumentation includes a GC with a splitless or on-column injection port for capillary column, a MS with 70 eV electron impact ionization, and a data system to collect and record MS data, and process it. A K-D apparatus is used to concentrate extracts.

GC Column: 30 m × 0.25 mm I.D. 5% phenyl, 94% methyl, 1% vinyl silicone bonded phased fused silica capillary column.

PRECISION & ACCURACY The detection limits of the method are usually dependent on the level of interferences rather than instrumental limitations. The limits typify the minimum quantities that can be detected with no interferences present.

The minimum level (in µg/mL) was not listed. This is defined as a minimum level at which the analytical system shall give recognizable mass spectra (background corrected) and acceptable calibration points.

The MDL (in µg/kg) in low solids was not listed and in high solids was not listed; these were determined in digested sludge (low solids) and in filter cake or compost (high solids).

The labeled and native compound initial precision as standard deviation (in µg/L) was not listed.
The labeled and native compound initial accuracy as average recovery (in µg/L) was not listed.

SAMPLE COLLECTION, PRESERVATION & HANDLING Collect samples in glass containers. Aqueous samples which flow freely are collected in refrigerated bottles using automatic sampling equipment. Solid samples are collected as grab samples using widemouth jars. Maintain samples at 0 to 4°C from the time of collection until extraction. If residual chlorine is present in aqueous samples, add 80 mg sodium thiosulfate/L of water. Begin sample extraction within 7 days of collection, and analyze all extracts within 40 days of extraction.

SAMPLE PREPARATION Samples containing 1% solids or less are extracted directly using continuous liquid-liquid extraction techniques. Samples containing 1 to 30% solids are diluted to the 1% level with reagent water and extracted using continuous liquid-liquid extraction techniques. Samples containing greater than 30% solids are extracted using ultrasonic techniques.

- Base/neutral extraction — Adjust the pH of the waters in the extractors to 12–13 with 6 N NaOH. Extract with methylene chloride for 24–48 h.
- Acid extraction — Adjust the pH of the waters in the extractors to 2 or less using 6 N sulfuric acid. Extract with methylene chloride for 24–48 h.
- Ultrasonic extraction of high solids samples — Add anhydrous sodium sulfate to the sample and QC aliquot(s). Add acetone:methylene chloride (1:1) to the sample and mix thoroughly

Concentrate extracts using a K-D apparatus.

QUALITY CONTROL The analyst is permitted to modify this method to improve separations or lower the costs of measurements, provided all performance specifications are met. Analyses of blanks are required to demonstrate freedom from contamination. When results of spikes indicate atypical method performance for samples, the samples are diluted to bring method performance within acceptable limits.

For low solids (aqueous samples), extract, concentrate, and analyze two sets of four 1-L aliquots (8 aliquots total) of the precision and recovery standard. For high solids samples, two sets of four 30-g aliquots of the high solids reference matrix are used.

Spike all samples with labeled compounds to assess method performance. Compute percent recovery of the labeled compounds using the internal standard method. Compare the labeled compound recovery for each compound with the corresponding labeled compound recovery.

Reagent water and high solids reference matrix blanks are analyzed to demonstrate freedom from contamination. Extract and concentrate a 1-L reagent water blank or a high solids reference matrix blank with each sample's lot (samples started through the extraction process on the same 8-h shift, to a maximum of 20 samples).

Field replicates may be collected to determine the precision of the sampling technique, and spiked samples may be required to determine the accuracy of the analysis when the internal standard method is used.

REFERENCE Semivolatile Organic Compounds by Isotope Dilution GC/MS. Office of Water Regulation and Standards, U.S. EPA Industrial Technology Division, Washington, DC, EPA Method 1625, Rev. C, June 1989 (contact W.A. Telliard, U.S. EPA, Office of Water Regulations and Standards, 401 M St., SW, Washington, DC, 20460. Phone: 202-382-7131).

N-Nitrosopiperidine   EPA Method 8270
CAS #100-75-4

TITLE Semivolatile Organic Compounds by GC/MS

MATRIX This method is used to determine the concentration of semivolatile organic compounds in extracts prepared from all types of solid waste matrices, soils, and groundwater. Although surface waters are not specifically mentioned, this method should be applicable to water samples from rivers, lakes, etc.

METHOD SUMMARY This method covers 259 semivolatile organic compounds. In very limited applications direct injection of the sample into the GC/MS system may be appropriate, but this results in very high detection limits (approximately 10,000 µg/L). Typically, a 1-L liquid sample, containing surrogate, and matrix spiking standards, is extracted in a continuous extractor first under acid conditions and then under basic conditions. Typically 30 g of a solid sample, containing surrogate, and matrix spiking standards, is extracted ultrasonically. After concentrating the extract to 1 mL it is spiked with 10 µL of an internal standard solution just prior to analysis by GC/MS. The volume injected should contain about 100 ng of base/neutral and 200 ng of acid surrogates (for a 1-µL injection). Analysis is performed by GC/MS using a capillary GC column.

INTERFERENCES Raw GC/MS data from all blanks, samples, and spikes must be evaluated for interferences. Contamination by carryover can occur whenever high-concentration and low-concentration samples are sequentially analyzed. To reduce carryover, the sample syringe must be rinsed out between samples with solvent. Whenever an unusually concentrated sample is encountered, it should be followed by the analysis of blank solvent to check for cross-contamination.

INSTRUMENTATION A GC/MS and a data system are required. The GC column used is a 30 m × 0.25 mm I.D. (or 0.32 mm I.D.) 1um film thickness silicone-coated fused silica capillary column. A continuous liquid-liquid extractor equipped with Teflon® or glass connection joints and stopcocks requiring no lubrication, a K-D concentrating apparatus, water bath, and an ultrasonic disrupter with a minimum power of 300 W and with pulsing capability are also required.

PRECISION & ACCURACY The estimated quantitation limit (EQL) of Method 8270B for determining an individual compound is approximately 1 mg/kg (wet weight) for soil or sediment samples, 1–200 mg/kg for wastes (dependent on matrix and method of preparation), and 10 µg/L for groundwater samples. EQLs will be proportionately higher for sample extracts that require dilution to avoid saturation of the detector.

The EQL(b) for groundwater in µg/L is 20.
The EQL (a, b) for low concentrations in soil and sediment in µg/kg is not determined.
Accuracy as µg/L is not listed.
Overall precision in µg/L is not listed.

(a) *EQLs listed for soil/sediment are based on wet weight. Normally data is reported in a dry-weight basis; therefore, EQLs will be higher based on the % dry weight of each sample. This calculation is based on a 30 g sample and gel permeation chromatography cleanup.*

(b) *Sample EQLs are highly matrix-dependent. The EQLs are provided for guidance and may not always be achievable.*

C = *True value for concentration, in µg/L.*
X = *Average recovery found for measurements of samples containing a concentration of C, in µg/L.*

ESTIMATED QUANTITATION LIMIT

Other Matrices	Factor (a)
High-concentration soil and sludges by sonicator	7.5
Non-water miscible waste	75

(a) *EQL for other matrices = [EQL for low soil/sediment] × [Factor]. This estimated EQL is similar to an EPA "Practical Quantitation Limit."*

SAMPLING METHOD
Liquid samples — Use a 1 or 2½ gallon amber glass bottle with a screw-top Teflon®-lined cover that has been prewashed with detergent and rinsed with distilled water and methanol (or isopropanol).

Soils, sediments, or sludges — Use an 8-oz. widemouth glass with a screw-top Teflon®-lined cover that has been prewashed with detergent and rinsed with distilled water and methanol (or isopropanol).

SAMPLE PRESERVATION
Liquid samples — If residual chlorine is present, add 3 mL of 10% sodium thiosulfate per gallon, cool to 4°C and store in a solvent-free refrigerator until analysis; if chlorine is not present, then eliminate the sodium thiosulfate addition.

Soils, sediments, or sludges — Cool samples to 4°C and store in a solvent-free refrigerator.

MHT Liquid samples must be extracted within 7 days and the extracts analyzed within 40 days. Soils, sediments, or sludges may be stored for a maximum of 14 days and the extracts analyzed within 40 days.

SAMPLE PREPARATION

Liquid samples — Transfer 1 L quantitatively to a continuous extractor. If high concentrations are anticipated, a smaller volume may be used and then diluted with organic-free reagent water to 1 L. Adjust pH, if necessary, to pH <2 using 1:1 (V/V) sulfuric acid. Pipette 1.0 mL of a surrogate standard spiking solution into each sample. For the sample in each analytical batch selected for spiking, add 1.0 mL of a matrix spiking standard. For base/neutral acid analysis, the amount of the surrogates and matrix spiking compounds added to the sample should result in a final concentration of 100 ng/µL of each analyte in the extract to be analyzed (assuming a 1-µL injection). Extract with methylene chloride for 18–24 h. Next, adjust the pH of the aqueous phase to pH >11 using 10 N sodium hydroxide and extract it with methylene chloride again for 18–24 h. Dry the extract through a column containing anhydrous sodium sulfate and concentrate it to 1 mL using a K-D concentrator.

Soils, sediments, or sludges — Use 30 g of sample. Nonporous or wet samples (gummy or clay type) that do not have a free-flowing sandy texture must be mixed with anhydrous sodium sulfate until the sample is free flowing. Add 1 mL of surrogate standards to all samples, spikes, standards, and blanks. For the sample in each analytical batch selected for spiking, add 1.0 mL of a matrix spiking standard. For base/neutral acid analysis, the amount added of the surrogates and matrix spiking compounds should result in a final concentration of 100 ng/µL of each base/neutral analyte and 200 ng/µL of each acid analyte in the extract to be analyzed (assuming a 1-µL injection). Immediately add a 100-mL mixture of 1:1 methylene chloride:acetone and extract the sample ultrasonically for 3 min and then decant or filter the extracts. Repeat the extraction two or more times. Dry the extract using a column with anhydrous sodium sulfate and concentrate it to 1 mL in a K-D concentrator.

QUALITY CONTROL A methylene chloride solution containing 50 ng/µL of decafluorotriphenylphosphine (DFTPP) is used for tuning the GC/MS system each 12-h shift. A system performance check also must be made during every 12-h shift. A standard containing 50 ng/µL each of 4,4'-DDT, pentachlorophenol, and benzidine is required to verify injection port inertness and GC column performance. A calibration standard at mid-concentration, containing each compound of interest, including all required surrogates, must be performed every 12 h during analysis. After the system performance check is met, calibration check compounds (CCCs) are used to check the validity of the initial calibration.

The internal standard responses and retention times in the calibration check standard must be evaluated immediately after or during data acquisition. If the retention time for any internal standard changes by more than 30 seconds from the last check calibration (12 h), the chromatographic system must be inspected for malfunctions and corrections must be made, as required. If the electron ionization current plot (EICP) area for any of the internal standards changes by a factor of two from the last daily calibration standard check, the mass spectrometer must be inspected for malfunctions and corrections must be made, as appropriate.

Demonstrate, through the analysis of a reagent water blank, that interferences from the analytical system, glassware, and reagents are under control. The blank samples should be carried through all stages of the sample preparation and measurement steps. For each analytical batch (up to 20 samples), a reagent blank, matrix spike, and matrix spike duplicate/duplicate must be analyzed (the frequency of the spikes may be different for different monitoring programs). The blank and spiked samples must be carried through all stages of the sample preparation and measurement steps. A QC reference sample concentrate containing each analyte at a concentration of 100 mg/L in methanol is required.

REFERENCE Test Methods for Evaluating Solid Waste (SW-846). U.S. EPA 1983, Method 8270B, Rev. 2, Nov. 1990. Office of Solid Waste, Washington, DC.

N-Nitrosopyrrolidine   **EPA Method 8270**
CAS #930-55-2

TITLE Semivolatile Organic Compounds by GC/MS

MATRIX This method is used to determine the concentration of semivolatile organic compounds in extracts prepared from all types of solid waste matrices, soils, and groundwater. Although surface waters are not specifically mentioned, this method should be applicable to water samples from rivers, lakes, etc.

METHOD SUMMARY This method covers 259 semivolatile organic compounds. In very limited applications direct injection of the sample into the GC/MS system may be appropriate, but this results in very high detection limits (approximately 10,000 µg/L). Typically, a 1-L liquid sample, containing surrogate, and matrix spiking standards, is extracted in a continuous extractor first under acid conditions and then under basic conditions. Typically 30 g of a solid sample, containing surrogate, and matrix spiking standards, is extracted ultrasonically. After concentrating the extract to 1 mL it is spiked with 10 µL of an internal standard solution just prior to analysis by GC/MS. The volume injected should contain about 100 ng of base/neutral and 200 ng of acid surrogates (for a 1-µL injection). Analysis is performed by GC/MS using a capillary GC column.

INTERFERENCES Raw GC/MS data from all blanks, samples, and spikes must be evaluated for interferences. Contamination by carryover can occur whenever high-concentration and low-concentration samples are sequentially analyzed. To reduce carryover, the sample syringe must be rinsed out between samples with solvent. Whenever an unusually concentrated sample is encountered, it should be followed by the analysis of blank solvent to check for cross-contamination.

INSTRUMENTATION A GC/MS and a data system are required. The GC column used is a 30 m × 0.25 mm I.D. (or 0.32 mm I.D.) 1um film thickness silicone-coated fused silica capillary column. A continuous liquid-liquid extractor equipped with Teflon® or glass connection joints and stopcocks requiring no lubrication, a K-D concentrating apparatus, water

bath, and an ultrasonic disrupter with a minimum power of 300 W and with pulsing capability are also required.

PRECISION & ACCURACY The estimated quantitation limit (EQL) of Method 8270B for determining an individual compound is approximately 1 mg/kg (wet weight) for soil or sediment samples, 1–200 mg/kg for wastes (dependent on matrix and method of preparation), and 10 µg/L for groundwater samples. EQLs will be proportionately higher for sample extracts that require dilution to avoid saturation of the detector.

The EQL(b) for groundwater in µg/L is 40.
The EQL (a, b) for low concentrations in soil and sediment in µg/kg is not determined.
Accuracy as µg/L is not listed.
Overall precision in µg/L is not listed.

(a) *EQLs listed for soil/sediment are based on wet weight. Normally data is reported in a dry-weight basis; therefore, EQLs will be higher based on the % dry weight of each sample. This calculation is based on a 30 g sample and gel permeation chromatography cleanup.*
(b) *Sample EQLs are highly matrix-dependent. The EQLs are provided for guidance and may not always be achievable.*
C = *True value for concentration, in µg/L.*
X = *Average recovery found for measurements of samples containing a concentration of C, in µg/L.*

ESTIMATED QUANTITATION LIMIT

Other Matrices	Factor (a)
High-concentration soil and sludges by sonicator	7.5
Non-water miscible waste	75

(a) *EQL for other matrices = [EQL for low soil/sediment] × [Factor]. This estimated EQL is similar to an EPA "Practical Quantitation Limit."*

SAMPLING METHOD
Liquid samples — Use a 1 or 2½ gallon amber glass bottle with a screw-top Teflon®-lined cover that has been prewashed with detergent and rinsed with distilled water and methanol (or isopropanol).

Soils, sediments, or sludges — Use an 8-oz. widemouth glass with a screw-top Teflon®-lined cover that has been prewashed with detergent and rinsed with distilled water and methanol (or isopropanol).

SAMPLE PRESERVATION
Liquid samples — If residual chlorine is present, add 3 mL of 10% sodium thiosulfate per gallon, cool to 4°C and store in a solvent-free refrigerator until analysis; if chlorine is not present, then eliminate the sodium thiosulfate addition.

Soils, sediments, or sludges — Cool samples to 4°C and store in a solvent-free refrigerator.

MHT Liquid samples must be extracted within 7 days and the extracts analyzed within 40 days. Soils, sediments, or sludges may be stored for a maximum of 14 days and the extracts analyzed within 40 days.

SAMPLE PREPARATION
Liquid samples — Transfer 1 L quantitatively to a continuous extractor. If high concentrations are anticipated, a smaller volume may be used and then diluted with organic-free reagent water to 1 L. Adjust pH, if necessary, to pH <2 using 1:1 (V/V) sulfuric acid. Pipette 1.0 mL of a surrogate standard spiking solution into each sample. For the sample in each analytical batch selected for spiking, add 1.0 mL of a matrix spiking standard. For base/neutral acid analysis, the amount of the surrogates and matrix spiking compounds added to the sample should result in a final concentration of 100 ng/µL of each analyte in the extract to be analyzed (assuming a 1-µL injection). Extract with methylene chloride for 18–24 h. Next, adjust the pH of the aqueous phase to pH >11 using 10 N sodium hydroxide and extract it with methylene chloride again for 18–24 h. Dry the extract through a column containing anhydrous sodium sulfate and concentrate it to 1 mL using a K-D concentrator.

Soils, sediments, or sludges — Use 30 g of sample. Nonporous or wet samples (gummy or clay type) that do not have a free-flowing sandy texture must be mixed with anhydrous sodium sulfate until the sample is free flowing. Add 1 mL of surrogate standards to all samples, spikes, standards, and blanks. For the sample in each analytical batch selected for spiking, add 1.0 mL of a matrix spiking standard. For base/neutral acid analysis, the amount added of the surrogates and matrix spiking compounds should result in a final concentration of 100 ng/µL of each base/neutral analyte and 200 ng/µL of each acid analyte in the extract to be analyzed (assuming a 1-µL injection). Immediately add a 100-mL mixture of 1:1 methylene chloride:acetone and extract the sample ultrasonically for 3 min and then decant or filter the extracts. Repeat the extraction two or more times. Dry the extract using a column with anhydrous sodium sulfate and concentrate it to 1 mL in a K-D concentrator.

QUALITY CONTROL A methylene chloride solution containing 50 ng/µL of decafluorotriphenylphosphine (DFTPP) is used for tuning the GC/MS system each 12-h shift. A system performance check also must be made during every 12-h shift. A standard containing 50 ng/µL each of 4,4'-DDT, pentachlorophenol, and benzidine is required to verify injection port inertness and GC column performance. A calibration standard at mid-concentration, containing each compound of interest, including all required surrogates, must be performed every 12 h during analysis. After the system performance check is met, calibration check compounds (CCCs) are used to check the validity of the initial calibration.

The internal standard responses and retention times in the calibration check standard must be evaluated immediately after or during data acquisition. If the retention time for any internal standard changes by more than 30 seconds from the last check calibration (12 h), the chromatographic system must be inspected for malfunctions and corrections must be made, as required. If the electron ionization current plot (EICP) area for any of the internal standards changes by a factor of two from the last daily calibration standard check, the mass spectrometer must be inspected for malfunctions and corrections must be made, as appropriate.

Demonstrate, through the analysis of a reagent water blank, that interferences from the analytical system, glassware, and reagents are under control. The blank samples should be carried through all stages of the sample preparation and measurement steps. For each analytical batch (up to 20 samples), a reagent blank, matrix spike, and matrix spike duplicate/duplicate must be analyzed (the frequency of the spikes may be different for different monitoring programs). The blank and spiked samples must be carried through all stages of the sample preparation and measurement steps. A QC reference sample concentrate containing each analyte at a concentration of 100 mg/L in methanol is required.

REFERENCE Test Methods for Evaluating Solid Waste (SW-846). U.S. EPA 1983, Method 8270B, Rev. 2, Nov. 1990. Office of Solid Waste, Washington, DC.

cis-Nonachlor EPA Method 505
CAS #39765-80-5

TITLE Analysis of Organohalide Pesticides and Commercial Polychlorinated Biphenyl (PCB) Products in Water by Microextraction and Gas Chromatography. U.S. EPA Method 505, Rev. 2.0, 1989.

MATRIX This method is applicable to drinking water and raw source water. The latter should include most surface water and groundwater sources.

METHOD SUMMARY Method 505 covers 25 pesticides and commercial PCB products. This is a very sensitive method that is more useful for monitoring than for exploratory analyses. 5-mL of water are saturated with sodium chloride and then extracted by shaking with 2 mL of hexane. The sample extracts are transferred to an autosampler setup to inject 1–2 µL portions into a gas chromatograph (GC) for analysis. Alternatively, 1–2 µL portions of samples, blanks, and standards may be manually injected. Each extract is analyzed by capillary GC/ECD with confirmation using either a second capillary column or GC/MS. The electron capture detector is easy to use, but it is a nonselective detector. The microextraction technique also eliminates the expensive sample preparation costs of other methods, but it has the disadvantage of being less sensitive than most because the extracts are not concentrated.

INTERFERENCES Method interferences may be caused by contaminants in solvents, reagents, glassware, and other sample processing apparatus that lead to discrete artifacts or elevated baselines. Interfering contamination may occur when a sample containing low concentrations of analytes is analyzed immediately following a sample containing relatively high concentrations of the analytes. Matrix interferences also may be caused by contaminants that are coextracted from the sample; cleanup of sample extracts may be necessary in these cases. Some pesticides and commercial PCB products from aqueous solutions adhere to glass surfaces, so sample transfers and contact with glass surfaces should be minimized. Some pesticides are rapidly oxidized by chlorine so dechlorination with sodium thiosulfate at the time of sample collection is important. Also, splitless injectors may cause degradation of some pesticides.

INSTRUMENTATION A gas chromatograph/electron capture detector/data system, with temperature programming and split/splitless injector suitable for use with capillary columns is needed.

Column 1: 0.32 mm I.D. × 30 m fused silica capillary with chemically bond methyl polysiloxane phase (DB-1, 1.0 µm film, or equivalent);

Column 2: 0.32 mm I.D. × 30 m fused silica capillary with 1:1 mixed phase of dimethyl silicone and polyethylene glycol (Durawax-DX3, 0.25 µm film, or equivalent);

Column 3: 0.32 mm I.D. × 25 m fused silica capillary with chemically bonded 50:50 methyl-phenyl silicone (OV-17, 1.5 µm film, or equivalent).

Column 1 should be used as the primary analytical column. Columns 2 and 3 are recommended for use as confirmatory columns when GC/MS confirmation is not available.

PRECISION & ACCURACY Method detection limits are dependent upon the characteristics of the gas chromatographic system used. Analytes that are not separated chromatographically cannot be individually identified and used in the same calibration mixture or water samples unless an alternative technique for identification and quantification, such as mass spectrometry, is used.

The concentration(s) (in µg/L) used for these QC measurements was 0.06 and 0.45.

The MDL (in µg/L) was 0.027.

The accuracy (% recovery) for reagent water at the above concentration(s) was 110 and 81 and the precision (%) was 15.2 and 21.3.

The accuracy (% recovery) for groundwater at the above concentration(s) was 101 and 93 and the precision (%) was 7.2 and 18.3.

The accuracy (% recovery) for tap water at the above concentration(s) was 93 and 87 and the precision (5) was 14.3 and 5.4.

Note: No range of concentrations is provided with this method.

SAMPLING METHOD Collect samples using a 40-mL screw-cap vial (prewashed with detergent, rinsed with distilled water and oven dried at 400°C for one h) with a Teflon®-faced silicone septum. Collect bubble-free samples and place the septum with the Teflon® side down on the water.

SAMPLE PRESERVATION If residual chlorine is present in the water add about 3 mg of sodium thiosulfate to each vial before samples are collected to remove the chlorine. Alternatively, add 75 µL of 0.04 g/mL solution of sodium thiosulfate to each vial just prior to sampling. Immediately cool samples to 4°C, and store them in a solvent-free refrigerator at 4°C until analysis.

MHT The maximum holding time is 14 days from the time the sample was collected until it must be analyzed.

SAMPLE PREPARATION Remove the sample from storage and allow it to come to room temperature. Remove a 5-mL volume from each container and weigh the container to the nearest 0.1 g. Add 6 g of sodium chloride and 2.0 mL of hexane to each sample bottle. Recap the sample and shake it vigorously for one min. Allow the water and hexane phases to separate, remove the cap, and transfer 0.5 mL of hexane into an autosampler vial using a disposable glass pipette. Transfer the remaining hexane phase into a second autosampler vial and store at 4°C for reanalysis, if necessary. Discard the remaining sample/hexane mixture and reweigh the empty container to determine net weight of sample.

QUALITY CONTROL Minimum quality control requirements are initial demonstration of lab capability, analysis of lab reagent blanks, fortified blanks, fortified sample matrix, and quality control samples. The lab must analyze at least one fortified blank per sample set, or at least one for every 20 samples. The fortifying concentration of each analyte should be 10 times the method detection limit or the maximum calibration limit (MCL), whichever is less. Calculate accuracy as percent recovery and develop control limits from the mean percent recovery and standard deviation.

The lab must add a known concentration of the analytes to a minimum of 10% of the routine samples, or one lab fortified sample matrix per sample set. Calculate the percent recovery for each analyte and compare to the control limits established from the analyses of the fortified blanks.

EPA CONTACT & HOTLINE For technical questions contact Dr. Baldev Bathija, U.S. EPA, Office of Ground Water and Drinking Water (WH-550D), 401 M St. SW, Washington, DC 20460. Tel. (202) 260-3040. For further information the EPA Safe Drinking Water Hotline may be called at: (800) 426-4791.

REFERENCE Methods for the Determination of Organic Compounds in Drinking Water, EPA/600/4-88/039 (revised July 1991). U.S. EPA Environmental Monitoring Systems Laboratory, Cincinnati, OH, 45268, U.S.A. Available from the National Technical Information Service (NTIS), 5285 Port Royal Road, Springfield, VA 22161; Tel. 800-553-6847. NTIS Order Number is PB91-231480.

---

trans-Nonachlor   **EPA Method 505**
CAS #39765-80-5

TITLE Analysis of Organohalide Pesticides and Commercial Polychlorinated Biphenyl (PCB) Products in Water by Microextraction and Gas Chromatography. U.S. EPA Method 505, Rev. 2.0, 1989.

MATRIX This method is applicable to drinking water and raw source water. The latter should include most surface water and groundwater sources.

METHOD SUMMARY Method 505 covers 25 pesticides and commercial PCB products. This is a very sensitive method that is more useful for monitoring than for exploratory analyses. 5-mL of water are saturated with sodium chloride and then extracted by shaking with 2 mL of hexane. The sample extracts are transferred to an autosampler setup to inject 1–2 μL portions into a gas chromatograph (GC) for analysis. Alternatively, 1–2 μL portions of samples, blanks, and standards may be manually injected. Each extract is analyzed by capillary GC/ECD with confirmation using either a second capillary column or GC/MS. The electron capture detector is easy to use, but it is a nonselective detector. The microextraction technique also eliminates the expensive sample preparation costs of other methods, but it has the disadvantage of being less sensitive than most because the extracts are not concentrated.

INTERFERENCES Method interferences may be caused by contaminants in solvents, reagents, glassware, and other sample processing apparatus that lead to discrete artifacts or elevated baselines. Interfering contamination may occur when a sample containing low concentrations of analytes is analyzed immediately following a sample containing relatively high concentrations of the analytes. Matrix interferences also may be caused by contaminants that are coextracted from the sample; cleanup of sample extracts may be necessary in these cases. Some pesticides and commercial PCB products from aqueous solutions adhere to glass surfaces, so sample transfers and contact with glass surfaces should be minimized. Some pesticides are rapidly oxidized by chlorine so dechlorination with sodium thiosulfate at the time of sample collection is important. Also, splitless injectors may cause degradation of some pesticides.

INSTRUMENTATION A gas chromatograph/electron capture detector/data system, with temperature programming and split/splitless injector suitable for use with capillary columns is needed.

Column 1: 0.32 mm I.D. × 30 m fused silica capillary with chemically bond methyl polysiloxane phase (DB-1, 1.0 μm film, or equivalent);

Column 2: 0.32 mm I.D. × 30 m fused silica capillary with 1:1 mixed phase of dimethyl silicone and polyethylene glycol (Durawax-DX3, 0.25 μm film, or equivalent);

Column 3: 0.32 mm I.D. × 25 m fused silica capillary with chemically bonded 50:50 methyl-phenyl silicone (OV-17, 1.5 μm film, or equivalent).

Column 1 should be used as the primary analytical column. Columns 2 and 3 are recommended for use as confirmatory columns when GC/MS confirmation is not available.

PRECISION & ACCURACY Method detection limits are dependent upon the characteristics of the gas chromatographic system used. Analytes that are not separated chromatographically cannot be individually identified and used in the same calibration mixture or water samples unless an alternative technique for identification and quantification, such as mass spectrometry, is used.

The concentration(s) (in μg/L) used for these QC measurements was 0.06 and 0.35.

The MDL (in μg/L) was 0.011.

The accuracy (% recovery) for reagent water at the above concentration(s) was 95 and 86 and the precision (%) was 9.6 and 21.8.

The accuracy (% recovery) for groundwater at the above concentration(s) was 83 and 94 and the precision (%) was 7.1 and 17.2.

The accuracy (% recovery) for tap water at the above concentration(s) was 73 and 86 and the precision (5) was 4.1 and 5.1.

Note: No range of concentrations is provided with this method.

SAMPLING METHOD Collect samples using a 40-mL screw-cap vial (prewashed with detergent, rinsed with distilled water and oven dried at 400°C for one h) with a Teflon®-faced silicone septum. Collect bubble-free samples and place the septum with the Teflon® side down on the water.

SAMPLE PRESERVATION If residual chlorine is present in the water add about 3 mg of sodium thiosulfate to each vial before samples are collected to remove the chlorine. Alternatively, add 75 µL of 0.04 g/mL solution of sodium thiosulfate to each vial just prior to sampling. Immediately cool samples to 4°C, and store them in a solvent-free refrigerator at 4°C until analysis.

MHT The maximum holding time is 14 days from the time the sample was collected until it must be analyzed.

SAMPLE PREPARATION Remove the sample from storage and allow it to come to room temperature. Remove a 5-mL volume from each container and weigh the container to the nearest 0.1 g. Add 6 g of sodium chloride and 2.0 mL of hexane to each sample bottle. Recap the sample and shake it vigorously for one min. Allow the water and hexane phases to separate, remove the cap, and transfer 0.5 mL of hexane into an autosampler vial using a disposable glass pipette. Transfer the remaining hexane phase into a second autosampler vial and store at 4°C for reanalysis, if necessary. Discard the remaining sample/hexane mixture and reweigh the empty container to determine net weight of sample.

QUALITY CONTROL Minimum quality control requirements are initial demonstration of lab capability, analysis of lab reagent blanks, fortified blanks, fortified sample matrix, and quality control samples. The lab must analyze at least one fortified blank per sample set, or at least one for every 20 samples. The fortifying concentration of each analyte should be 10 times the method detection limit or the maximum calibration limit (MCL), whichever is less. Calculate accuracy as percent recovery and develop control limits from the mean percent recovery and standard deviation.

The lab must add a known concentration of the analytes to a minimum of 10% of the routine samples, or one lab fortified sample matrix per sample set. Calculate the percent recovery for each analyte and compare to the control limits established from the analyses of the fortified blanks.

EPA CONTACT & HOTLINE For technical questions contact Dr. Baldev Bathija, U.S. EPA, Office of Ground Water and Drinking Water (WH-550D), 401 M St. SW, Washington, DC 20460. Tel. (202) 260-3040. For further information the EPA Safe Drinking Water Hotline may be called at: (800) 426-4791.

REFERENCE Methods for the Determination of Organic Compounds in Drinking Water, EPA/600/4-88/039 (revised July 1991). U.S. EPA Environmental Monitoring Systems Laboratory, Cincinnati, OH, 45268, U.S.A. Available from the National Technical Information Service (NTIS), 5285 Port Royal Road, Springfield, VA 22161; Tel. 800-553-6847. NTIS Order Number is PB91-231480.

Norflurazon   **EPA Method 507**
CAS #27314-13-2

TITLE Determination of Nitrogen and Phosphorus-Containing Pesticides in Water by GC/NPD

MATRIX This method is applicable to the determination of certain nitrogen and phosphorus-containing pesticides in finished drinking water and groundwater.

METHOD SUMMARY Method 507 covers 46 nitrogen- and phosphorus-containing pesticides. A 1-L sample is fortified with a surrogate standard, salted, buffered, extracted with methylene chloride, and concentrated; then the solvent is exchanged with methyl tert-butyl ether (MTBE) and concentrated again, and a 2-µL aliquot of a sample extract is injected into a GC system equipped with a selective nitrogen-phosphorus detector and a capillary column for analysis.

INTERFERENCES Method interferences may be caused by contaminants in solvents, reagents, glassware, and other sample processing apparatus. Interfering contamination may occur when a sample containing low concentrations of analytes is analyzed immediately following a sample containing relatively high concentrations. One or more injections of MTBE should be made following the analysis of a sample with high concentrations of analytes to check for analyte carryover. Matrix interferences may be caused by contaminants that are coextracted from the sample. The extent of matrix interferences will vary considerably from source to source, depending upon the water sampled.

INSTRUMENTATION A gas chromatograph system (GC) equipped with a nitrogen-phosphorus detector (NPD) is needed.

Column 1: 30 m × 0.25 mm I.D. DB-5 bonded fused silica column, 0.25 µm film thickness, or equivalent;
Column 2: 30 m × 0.25 mm I.D. DB-1701 bonded fused silica column, 0.25 µm film thickness, or equivalent.

PRECISION & ACCURACY This method has been validated in a single lab and estimated detection limits (EDLs) have been determined for each analyte. Observed detection limits may vary among waters, depending upon the nature of the interferences in the sample matrix and the specific instrumentation used. Analytes that are not separated chromatographically cannot be individually identified and measured unless an alternative technique for identification and quantification exist.

The estimated detection limit (in µg/L) was 0.5. The EDL is defined as either method detection limit or a level of compound in a sample yielding a peak in the final extract with signal-to-noise ratio of approximately 5, whichever value is higher.

The concentration used for these measurements (in µg/L) was 5.
The accuracy (as % recovery) was 94.
The precision (% RSD) was 5.

SAMPLING METHOD Grab samples are collected in 1-L glass sample bottles (prewashed with detergent and hot tap water, rinsed with reagent water, and dried in an oven at 400°C for 1 h) with screw caps lined with PTFE-fluorocarbon.

SAMPLE PRESERVATION Add mercuric chloride to the sample bottle in amounts to produce a concentration of 10 mg/L. If residual chlorine is present, add 80 mg of sodium thiosulfate/L of sample to the sample bottle prior to collection. After collection, seal bottle and shake vigorously for 1 min, then cool the sample to 4°C immediately and store it at 4°C in the dark until extraction.

MHT Maximum holding time of the samples, and in some cases the extracts, is 14 days.

SAMPLE PREPARATION Fortify the sample with 50 µL of the surrogate standard solution, adjust to pH 7 with phosphate buffer, add 100 g NaCl to the sample, and seal and shake to dissolve the salt; then extract with methylene chloride in a separatory funnel or in a mechanical tumbler bottle. Dry the extract by pouring it through a solvent-rinsed drying column containing about 10 cm of anhydrous sodium sulfate. Collect the extract in a Kuderna-Danish (K-D) concentrator and rinse the column with 20–30 mL methylene chloride. Concentrate the extract to about 2 mL and rinse the flask and its lower joint into the concentrator tube with 1 to 2 mL of methyl t-butyl ether (MTBE). Add 5–10 mL of MTBE and concentrate the extract twice (adding more MTBE) to a final volume of 5.0 mL and store it at 4°C until analysis.

Note: If methylene chloride is not completely removed from the final extract, it may cause detector problems.

QUALITY CONTROL Minimum quality control requirements are initial demonstration of lab capability, determination of surrogate compound recoveries in each sample and blank, monitoring internal standard peak area or height in each sample and blank, analysis of lab reagent blanks, lab fortified samples, lab fortified blanks, and other QC samples. A lab reagent blank is analyzed to demonstrate that all glassware and reagent interferences are under control.

Initial demonstration of capability is fulfilled by analyzing four fortified reagent water samples with the recovery value for each analyte falling within the acceptable range (±30% average recovery). Surrogate recoveries from samples or method blanks must be 70–130%. The internal standard response for any sample chromatogram should not deviate from the daily calibration check standard's internal standard response by more than 30% or lab fortified blanks or sample matrices are used to assess lab performance and analyte recovery, respectively.

If the response for the target analyte peak exceeds the working range of the system, dilute the extract and reanalyze. Alternative techniques such as an alternate detector or second chromatography column should be used to confirm peak identification when sample components are not resolved adequately.

EPA CONTACT & HOTLINE For technical questions contact Dr. Baldev Bathija, U.S. EPA, Office of Ground Water and Drinking Water (WH-550D), 401 M St. SW, Washington, DC 20460. Tel. (202) 260-3040. For further information the EPA Safe Drinking Water Hotline may be called at: (800) 426-4791.

REFERENCE Methods for the Determination of Organic Compounds in Drinking Water, EPA/600/4-88/039 (revised July 1991). U.S. EPA Environmental Monitoring Systems Laboratory, Cincinnati, OH, 45268, U.S.A. Available from the National Technical Information Service (NTIS), 5285 Port Royal Road, Springfield, VA 22161; Tel. 800-553-6847. NTIS Order Number is PB91-231480.

O

OCDD
CAS #3268-87-9

EPA Method 8280

TITLE The Analysis of Polychlorinated Dibenzo-P-Dioxins and Polychlorinated Dibenzofurans

MATRIX This method is appropriate for the determination of tetra-, penta-, hexa-, hepta-, and octachlorinated dibenzo-p-dioxins (PCDDs) and dibenzofurans (PCDFs) in chemical wastes including still bottoms, fuel oils, sludges, fly ash, reactor residues, soil and water.

METHOD SUMMARY This method covers 22 PCDD and PCDF compounds and it uses a high resolution capillary GC column with low resolution mass spectrometry. Samples are extracted and concentrated by several methods that vary depending on the matrix involved. The organic extracts are cleaned-up by washing with aqueous basic and acid solutions and then separated into fractions using a column of neutral alumina. The fraction containing the PCDDs and PCDFs is then further cleaned-up using a column of activated carbon. The final extract is concentrated and Carbon-13 labeled internal standards are added prior to analysis by GC/MS using a capillary GC column and selected ion monitoring using five sets of ions that are detailed in the method. Certain 2,3,7,8-substituted congeners are used to provide calibration and method recovery information. Proper column selection and access to reference isomer standards, may in certain cases, provide isomer specific data.

INTERFERENCES Solvents, reagents, glassware, and other sample processing hardware may yield discrete artifacts and/or elevated baselines which may cause misinterpretation of chromatographic data. Use high purity reagents and solvents to minimize interference problems. Interferents coextracted from the sample will vary considerably from source to source. PCDDs and PCDFs are often associated with other interfering chlorinated compounds such as PCBs and polychlorinated diphenyl ethers which may be found at concentrations several orders of magnitude higher than that of the analytes of interest.

INSTRUMENTATION A low resolution GC/MS utilizing 70 ev must be capable of selected ion monitoring (SIM) for at least 11 ions simultaneously, with a cycle time of 1 second or less. Minimum integration time for SIM is 50 ms per m/z. Also required is a GC-to-MS interface constructed of all glass or glass-lined materials. One of the following GC columns is required:

Column 1: 50 m CP-Sil-88 fused silica capillary column.
Column 2: DB-5 (30 m × 0.25 mm I.D., 0.25-um film thickness) fused silica capillary column.
Column 3: 30 m SP-2250 fused silica capillary column.

When toluene is employed as the final solvent, use of a bonded phase column is recommended. Solvent exchange into tridecane is required for other liquid phases or nonbonded columns such as CP-Sil-88. Chromatographic conditions must be adjusted to account for solvent boiling points.

PRECISION & ACCURACY Accuracy, precision, MDLs and concentration ranges for the compounds covered by this method have not been determined or published by EPA yet. The sensitivity of this method is dependent upon the level of interferents within a given matrix. Proposed quantification levels for target analytes were 2 ppb in soil samples, up to 10 ppb in other solid wastes and 10 ppt in water. Actual values have been shown to vary by homologous series and, to a lesser degree, by individual isomer.

SAMPLING METHOD Grab and composite samples must be collected in 1 L or 1-quart amber glass bottles with Teflon®-lined screw-caps that have been acid-washed and solvent rinsed before use. If compositing equipment is used, the system must incorporate glass sample containers for the collection of a minimum of 250 mL. samples.

SAMPLE PRESERVATION Samples must be cooled and stored at 4°C.

MHT Samples must be extracted within 30 days and analyzed within 45 days of collection.

SAMPLE PREPARATION

Soil samples — Extract 20 g of a 1:1 soil and anhydrous sodium sulfate mixture with 1:4 methanol-petroleum ether for 2 h in a wrist-action shaker. Concentrate the extract in a K-D and then exchange the solvent with hexane in the K-D. The final volume is 15 mL of hexane.

Aqueous samples — Extract a 1 L sample with methylene chloride, dry it with anhydrous sodium sulfate and concentrate it in a K-D followed by exchange with hexane in the K-D. A continuous liquid-liquid extractor may be used in place of a separatory funnel to avoid emulsions. The final volume is 15 mL of hexane.

Alumina column clean-up — The clean-up procedure described below consists of two phases. The first phase involves a sequential basic and acid washing of the extract that contains the analytes.

Wash the 15 mL hexane extract with 20% potassium hydroxide in a separatory funnel. Repeat the washing until no color is visible in the bottom layer but no more than four times because strong base is known to degrade certain PCDDs/PCDFs, so contact time must be minimized. Next, partition the 15 mL hexane against 40 mL of 5% sodium chloride. Next, partition the 15 mL hexane against 40 mL of concentrated sulfuric acid. Repeat the acid washings until no color is visible in the acid layer (but no more than four times). Finally, partition the 15 mL hexane against 40 mL of 5% sodium chloride. Dry the organic layer with anhydrous sodium sulfate and concentrate it to near dryness with a rotary evaporator. Dissolve the hexane residue from the first phase of the clean-up in 2 mL of hexane and apply it carefully to the top of a pre-eluted Woelm super 1 neutral alumina column. Elute the column with 10 mL of 8 percent (v/v) methylene chloride in hexane. Check by GC/MS analysis that no PCDDs of PCDFs are elute in this fraction before discarding it. Elute the PCDDs and PCDFs from the

column with 15 mL of 60 percent (v/v) methylene chloride in hexane and collect this second fraction in a conical shaped concentrator tube.

Carbon Column Clean-up — Using a carefully regulated stream of nitrogen, concentrate the first 8 percent fraction (methylene chloride in hexane) from the alumina column to about 1 mL. Save this 8 percent concentrate for GC/MS analysis to check for breakthrough of PCDDs and PCDFs. Concentrate the second 60 percent fraction (methylene chloride in hexane) to about 2 to 3 mL. Prepare a carbon column and rinse the carbon with 5 mL cyclohexane/methylene chloride (50:50 v/v) in the forward direction of flow and then in the reverse direction of flow. While still in the reverse direction of flow, transfer the sample concentrate to the column and elute with 10 mL of methylene chloride/methanol/benzene (75:20:5, v/v). Save all above eluates and combine them (this fraction may be used as a check on column efficiency). Next, turn the column over and, in the direction of forward flow, elute the PCDD/PCDF fraction with 20 mL toluene. Evaporate the toluene fraction to about 1 mL on a rotary evaporator and transfer this concentrate to a 2.0-mL Reacti-vial. Concentrate the sample using a stream of nitrogen gas. The final volume will depend on the relative concentration of target analytes but it is typically 100 µL for soil samples and 500 µL for sludge, still bottom, and fly ash samples. Extracts which are determined to be outside the calibration range for individual analytes must be diluted or a smaller portion of the sample must be re-extracted.

An alternate carbon column clean-up also may be used with a 1 mL HPLC injector loop. The injector loop is connected to the optional HPLC column.

QUALITY CONTROL Demonstrate, using a method blank, that all glassware and reagents are interferent-free at the MDL of the matrix of interest. A "method blank" must be run with each 20 or fewer samples. The method blank is also dosed with the internal standards. For water samples, 1 L of deionized and/or distilled water should be used as the method blank. Mineral oil may be used as the method blank for other matrices.

Calculate response factors for standards relative to the internal standards. Add a recovery standard to the samples prior to injection. The concentration of the recovery standard in the sample extract must be the same as that in the calibration standards used to measure the response factors.

Field duplicates (individual samples taken from the same location at the same time) should be analyzed periodically to determine the total precision (field and lab). Where appropriate, field blanks should be provided to monitor for possible cross-contamination of samples in the field. GC column performance must be demonstrated initially and verified prior to analyzing any sample in a 12-hr period. The GC column performance check solution must be analyzed under the same chromatographic and mass spectrometric conditions used for other samples and standards.

Retention times of target analytes must be verified using reference standards. These values must correspond to the retention time windows established. While certain cleanup techniques are provided as part of this method, unique samples may require additional cleanup techniques to achieve the method detection limit.

REFERENCE Test Methods for Evaluating Solid Waste (SW-846). U.S.E.P.A., 1986. Method 8280, Rev. 0, Sept. 1986. Office of Solid Wastes, Washington, DC.

OCDD   **EPA Method 8290**
CAS #3268-87-9

TITLE Polychlorinated Dibenzodioxins (PCDDs) and Polychlorinated Dibenzofurans (PCDFs) by High-Resolution Gas Chromatography/High-Resolution Mass Spectrometry (HRGC/HRMS).

MATRIX: This method is applicable with a variety of environmental matrices including: water, soil, sediment, paper pulp, fly ash, fish tissue, human adipose tissue, sludges, fuel oil, chemical reactor residue, and still bottoms.

METHOD SUMMARY This method provides procedures for the detection and quantitative measurement of polychlorinated dibenzo-p-dioxins (tetra- through octachlorinated homologues; PCDDs), and polychlorinated dibenzofurans (tetra- through octachlorinated homologues; PCDFs) in a variety of environmental matrices and at part-per-trillion (ppt) to part-per-quadrillion (ppq) concentrations. High-resolution gas chromatography and high-resolution mass spectrometry (HRGC/HRMS) on purified sample extracts provides highly specific identification of each analyte. Quantification is provided using calibration standards.

INTERFERENCES Solvents, reagents, glassware, and other sample processing hardware may yield discrete artifacts that may cause misinterpretation of the chromatographic data. Analysts should avoid using PVC gloves. Interferants coextracted from the sample will vary considerably from matrix to matrix. PCDDs and PCDFs are often associated with other interfering chlorinated substances such as polychlorinated biphenyls (PCBs), polychlorinated diphenyl ethers (PCDEs), polychlorinated naphthalenes, and polychlorinated alkyldibenzofurans that may be found at concentrations several orders of magnitude higher than the PCDDs or PCDFs.

A high-resolution capillary column is used in this method. However, no single column is known to resolve all isomers. The 60 m DB-5 GC column is capable of 2,3,7,8-TCDD isomer specificity. In order to determine the concentration of the 2,3,7,8-TCDD (if detected on the DB-5 column), the sample extract must be reanalyzed on a column capable of 2,3,7,8-TCDF isomer specificity (e.g., DB-225, SP-2330, SP-2331, or equivalent).

INSTRUMENTATION High-Resolution Gas Chromatograph/High-Resolution Mass Spectrometer/Data System (HRGC/HRMS/DS) equipped with a GC injection port designed so that the separation of 2,3,7,8-TCDD from the other TCDD isomers achieved in the gas chromatographic column is not appreciably degraded.

Column 1: 60 m DB-5 fused silica capillary column.

Column 2: 30 m DB-225 fused silica capillary column, or equivalent.

PRECISION & ACCURACY Precision, bias and concentration ranges for the compounds covered by this method have not been determined yet. The sensitivity of Method 8290 is dependent upon the level of interferences within a given matrix. Samples containing concentrations of specific congeneric analytes of PCDDs and PCDFs that are greater than ten times the upper method calibration limits must be analyzed by a protocol designed for such concentration levels, e.g., EPA Method 8280.

SAMPLE PREPARATION
Sludge/wet fuel oil — Extract aqueous sludge or wet fuel oil samples by refluxing a sample with toluene using a Dean-Stark water separator until all the water is removed. Filter the toluene extract through a glass fiber filter, or equivalent, and concentrate it to near dryness either on a rotary evaporator using an inert gas. Transfer the concentrate to a separatory funnel using hexane and wash it with 5% sodium chloride solution. Proceed to clean up.

Soil/sediment — If the sample is wet, add anhydrous powdered sodium sulfate to it until a free flowing mixture is obtained. Place the soil/sodium sulfate mixture in the Soxhlet apparatus, add toluene, and reflux for 16 h. The solvent must cycle completely through the system five times per h. Cool and filter the extract through a glass fiber filter and concentrate to near dryness on a rotary evaporator. Transfer the residue to a separatory funnel, using hexane. Proceed to clean up.

Aqueous samples — Use a 1-L sample; the method may require acetone to be added to it. When the sample is judged to contain 1% or more solids, it must be filtered through a glass fiber filter that has been rinsed with toluene. If the suspended solids content is too great to filter, centrifuge the sample, decant, and then filter the aqueous phase. Combine the solids from the centrifuge bottle(s) with the particulates on the filter and with the filter itself and proceed with Soxhlet extraction for soil/sediment. Extract the aqueous filtrate with methylene chloride in a separatory funnel, filter the extract through anhydrous sodium sulfate, and concentrate it using a K-D apparatus or a rotary evaporator. Exchange the solvent with hexane and proceed to clean up.

Clean up — The sample extract is cleaned up utilizing a number of different techniques. Partition cleanup is where the sample extract is partitioned with concentrated sulfuric acid, 5% aqueous sodium chloride, and 20% aqueous potassium hydroxide. Silica/alumina column cleanup involves packing gravity columns with silica gel and alumina and sequentially eluting the residue from the partition cleanup. Carbon column cleanup involves packing a column with a mixture of AX–21 and Celite 545 and sequentially eluting the sample concentrate from the silica/alumina cleanup with hexane, cyclohexane/methylene chloride (50:50), and methylene chloride/methanol/toluene (75:20:5). Then the column is turned upside down and the PCDD/PCDF fraction is eluted with toluene. The toluene fraction is concentrated and stored in the dark at room temperature until analysis.

QUALITY CONTROL Demonstrate, through the analysis of a reagent water blank, that interferences from the analytical system, glassware, and reagents are under control. For each analytical batch (up to 20 samples), a reagent blank, matrix spike, and matrix spike duplicate/duplicate must be analyzed (the frequency of the spikes may be different for different monitoring programs). The blank and spiked samples must be carried through all stages of the sample preparation and measurement steps.

A GC column performance check is required at the beginning of each 12-h period during which samples are analyzed. An HRGC/HRMS method blank run is required between a calibration run and the first sample run. The same method blank extract may thus be analyzed more than once if the number of samples within a batch requires more than 12 h of analyses.

At the beginning of each 12-h period during which samples are to be analyzed, an aliquot of the 1) GC column performance check solution and 2) a high-resolution concentration calibration must be analyzed to demonstrate adequate GC resolution and sensitivity, response factor reproducibility, and mass range calibration, and to establish the PCDD/PCDF retention time windows. A mass resolution check must also be performed to demonstrate adequate mass resolution using an appropriate reference compound (perfluorokerosene (PFK) is recommended). If the required criteria are not met, remedial action must be taken before any samples are analyzed.

Routine or continuing calibration (using a high resolution calibration solution) and the mass resolution check must also be performed at the end of each 12 h period. Furthermore, a HRGC/HRMS method blank analysis must be recorded following a calibration analysis and the first sample analysis.

To evaluate the performance of the analytical method, the QC check samples must be handled in exactly the same manner as actual samples. Therefore, 1.0 mL of the QC check sample concentrate is spiked into each of four 1 L aliquots of reagent water (which becomes the QC check sample), extracted, and then analyzed by GC. The variety of semivolatile analytes which may be analyzed by GC is such that the concentration of the QC check sample concentrate is different for the different analytical techniques presented in the full method.

The analyst must demonstrate also that the compounds of interest are being quantitatively recovered by the cleanup technique before the cleanup is applied to actual samples. For sample extracts that are cleaned up, the associated quality control samples (e.g., spikes, blanks, and duplicates) must also be processed through the same cleanup procedure. The analysis using each determinative method (GC, GC/MS, HPLC) specifies instrument calibration procedures using stock standards. It is recommended that cleanup also be performed on a series of the same type of standards to validate chromatographic elution patterns for the compounds of interest and to verify the absence of interferences from reagents.

REFERENCE Test Methods for Evaluating Solid Waste (SW-846). U.S. EPA. 1983. Method 8290, Rev. 0, Nov. 1990. Office of Solid Wastes, Washington, DC.

OCDF   **EPA Method 8280**
CAS #39001-02-0

TITLE The Analysis of Polychlorinated Dibenzo-P-Dioxins and Polychlorinated Dibenzofurans.

MATRIX This method is appropriate for the determination of tetra-, penta-, hexa-, hepta-, and octachlorinated dibenzo-p-dioxins (PCDDs) and dibenzofurans (PCDFs) in chemical wastes including still bottoms, fuel oils, sludges, fly ash, reactor residues, soil and water.

METHOD SUMMARY This method covers 22 PCDD and PCDF compounds and it uses a high resolution capillary GC column with low resolution mass spectrometry. Samples are extracted and concentrated by several methods that vary depending on the matrix involved. The organic extracts are cleaned-up by washing with aqueous basic and acid solutions and then separated into fractions using a column of neutral alumina. The fraction containing the PCDDs and PCDFs is then further cleaned-up using a column of activated carbon. The final extract is concentrated and Carbon-13 labeled internal standards are added prior to analysis by GC/MS using a capillary GC column and selected ion monitoring using five sets of ions that are detailed in the method. Certain 2,3,7,8-substituted congeners are used to provide calibration and method recovery information. Proper column selection and access to reference isomer standards, may in certain cases, provide isomer specific data.

INTERFERENCES Solvents, reagents, glassware, and other sample processing hardware may yield discrete artifacts and/or elevated baselines which may cause misinterpretation of chromatographic data. Use high purity reagents and solvents to minimize interference problems. Interferents coextracted from the sample will vary considerably from source to source. PCDDs and PCDFs are often associated with other interfering chlorinated compounds such as PCBs and polychlorinated diphenyl ethers which may be found at concentrations several orders of magnitude higher than that of the analytes of interest.

INSTRUMENTATION A low resolution GC/MS utilizing 70 ev must be capable of selected ion monitoring (SIM) for at least 11 ions simultaneously, with a cycle time of 1 second or less. Minimum integration time for SIM is 50 ms per m/z. Also required is a GC-to-MS interface constructed of all glass or glass-lined materials. One of the following GC columns is required:

Column 1: 50 m CP-Sil-88 fused silica capillary column.
Column 2: DB-5 (30 m × 0.25 mm I.D., 0.25-um film thickness) fused silica capillary column.
Column 3: 30 m SP-2250 fused silica capillary column.

When toluene is employed as the final solvent, use of a bonded phase column is recommended. Solvent exchange into tridecane is required for other liquid phases or nonbonded columns such as CP-Sil-88. Chromatographic conditions must be adjusted to account for solvent boiling points.

PRECISION & ACCURACY Accuracy, precision, MDLs and concentration ranges for the compounds covered by this method have not been determined or published by EPA yet. The sensitivity of this method is dependent upon the level of interferents within a given matrix. Proposed quantification levels for target analytes were 2 ppb in soil samples, up to 10 ppb in other solid wastes and 10 ppt in water. Actual values have been shown to vary by homologous series and, to a lesser degree, by individual isomer.

SAMPLING METHOD Grab and composite samples must be collected in 1 L or 1-quart amber glass bottles with Teflon®-lined screw-caps that have been acid-washed and solvent rinsed before use. If compositing equipment is used, the system must incorporate glass sample containers for the collection of a minimum of 250 mL. samples.

SAMPLE PRESERVATION Samples must be cooled and stored at 4°C.

MHT Samples must be extracted within 30 days and analyzed within 45 days of collection.

SAMPLE PREPARATION
Soil samples — Extract 20 g of a 1:1 soil and anhydrous sodium sulfate mixture with 1:4 methanol-petroleum ether for 2 h in a wrist-action shaker. Concentrate the extract in a K-D and then exchange the solvent with hexane in the K-D. The final volume is 15 mL of hexane.

Aqueous samples — Extract a 1 L sample with methylene chloride, dry it with anhydrous sodium sulfate and concentrate it in a K-D followed by exchange with hexane in the K-D. A continuous liquid-liquid extractor may be used in place of a separatory funnel to avoid emulsions. The final volume is 15 mL of hexane.

Alumina column clean-up — The clean-up procedure described below consists of two phases. The first phase involves a sequential basic and acid washing of the extract that contains the analytes.

Wash the 15 mL hexane extract with 20% potassium hydroxide in a separatory funnel. Repeat the washing until no color is visible in the bottom layer but no more than four times because strong base is known to degrade certain PCDDs/PCDFs, so contact time must be minimized. Next, partition the 15 mL hexane against 40 mL of 5% sodium chloride. Next, partition the 15 mL hexane against 40 mL of concentrated sulfuric acid. Repeat the acid washings until no color is visible in the acid layer (but no more than four times). Finally, partition the 15 mL hexane against 40 mL of 5% sodium chloride. Dry the organic layer with anhydrous sodium sulfate and concentrate it to near dryness with a rotary evaporator. Dissolve the hexane residue from the first phase of the clean-up in 2 mL of hexane and apply it carefully to the top of a pre-eluted Woelm super 1 neutral alumina column. Elute the column with 10 mL of 8 percent (v/v) methylene chloride in hexane. Check by GC/MS analysis that no PCDDs of PCDFs are elute in this fraction before discarding it. Elute the PCDDs and PCDFs from the column with 15 mL of 60 percent (v/v) methylene chloride in hexane and collect this second fraction in a conical shaped concentrator tube.

Carbon Column Clean-up — Using a carefully regulated stream of nitrogen, concentrate the first 8 percent fraction (methylene chloride in hexane) from the alumina column to about 1 mL. Save this 8 percent concentrate for GC/MS analysis to check for breakthrough of PCDDs and PCDFs. Concentrate the second 60 percent fraction (methylene chloride in hexane) to about 2 to 3 mL. Prepare a carbon column and rinse the carbon with 5 mL cyclohexane/methylene chloride (50:50 v/v) in the forward direction of flow and then in the reverse direction of flow. While still in the reverse direction of flow, transfer the sample concentrate to the column and elute with 10 mL of methylene chloride/methanol/benzene (75:20:5, v/v/v). Save all above eluates and combine them (this fraction may be used as a check on column efficiency). Next, turn the column over and, in the direction of forward flow, elute the PCDD/PCDF fraction with 20 mL toluene. Evaporate the toluene fraction to about 1 mL on a rotary evaporator and transfer this concentrate to a 2.0-mL Reacti-vial. Concentrate the sample using a stream of nitrogen gas. The final volume will depend on the relative concentration of target analytes but it is typically 100 µL for soil samples and 500 µL for sludge, still bottom, and fly ash samples. Extracts which are determined to be outside the calibration range for individual analytes must be diluted or a smaller portion of the sample must be re-extracted.

An alternate carbon column clean-up also may be used with a 1 mL HPLC injector loop. The injector loop is connected to the optional HPLC column.

QUALITY CONTROL Demonstrate, using a method blank, that all glassware and reagents are interferent-free at the MDL of the matrix of interest. A "method blank" must be run with each 20 or fewer samples. The method blank is also dosed with the internal standards. For water samples, 1 L of deionized and/or distilled water should be used as the method blank. Mineral oil may be used as the method blank for other matrices.

Calculate response factors for standards relative to the internal standards. Add a recovery standard to the samples prior to injection. The concentration of the recovery standard in the sample extract must be the same as that in the calibration standards used to measure the response factors.

Field duplicates (individual samples taken from the same location at the same time) should be analyzed periodically to determine the total precision (field and lab). Where appropriate, field blanks should be provided to monitor for possible cross-contamination of samples in the field. GC column performance must be demonstrated initially and verified prior to analyzing any sample in a 12-hr period. The GC column performance check solution must be analyzed under the same chromatographic and mass spectrometric conditions used for other samples and standards.

Retention times of target analytes must be verified using reference standards. These values must correspond to the retention time windows established. While certain cleanup techniques are provided as part of this method, unique samples may require additional cleanup techniques to achieve the method detection limit.

REFERENCE Test Methods for Evaluating Solid Waste (SW-846). U.S.E.P.A., 1986. Method 8280, Rev. 0, Sept. 1986. Office of Solid Wastes, Washington, DC.

OCDF   EPA Method 8290
CAS #39001-02-0

TITLE Polychlorinated Dibenzodioxins (PCDDs) and Polychlorinated Dibenzofurans (PCDFs) by High-Resolution Gas Chromatography/High-Resolution Mass Spectrometry (HRGC/HRMS).

MATRIX: This method is applicable with a variety of environmental matrices including: water, soil, sediment, paper pulp, fly ash, fish tissue, human adipose tissue, sludges, fuel oil, chemical reactor residue, and still bottoms.

METHOD SUMMARY This method provides procedures for the detection and quantitative measurement of polychlorinated dibenzo-p-dioxins (tetra- through octachlorinated homologues; PCDDs), and polychlorinated dibenzofurans (tetra- through octachlorinated homologues; PCDFs) in a variety of environmental matrices and at part-per-trillion (ppt) to part-per-quadrillion (ppq) concentrations. High-resolution gas chromatography and high-resolution mass spectrometry (HRGC/HRMS) on purified sample extracts provides highly specific identification of each analyte. Quantification is provided using calibration standards.

INTERFERENCES Solvents, reagents, glassware, and other sample processing hardware may yield discrete artifacts that may cause misinterpretation of the chromatographic data. Analysts should avoid using PVC gloves. Interferants coextracted from the sample will vary considerably from matrix to matrix. PCDDs and PCDFs are often associated with other interfering chlorinated substances such as polychlorinated biphenyls (PCBs), polychlorinated diphenyl ethers (PCDEs), polychlorinated naphthalenes, and polychlorinated alkyldibenzofurans that may be found at concentrations several orders of magnitude higher than the PCDDs or PCDFs.

A high-resolution capillary column is used in this method. However, no single column is known to resolve all isomers. The 60 m DB-5 GC column is capable of 2,3,7,8-TCDD isomer specificity. In order to determine the concentration of the 2,3,7,8-TCDD (if detected on the DB-5 column), the sample extract must be reanalyzed on a column capable of 2,3,7,8-TCDF isomer specificity (e.g., DB-225, SP-2330, SP-2331, or equivalent).

INSTRUMENTATION High-Resolution Gas Chromatograph/High-Resolution Mass Spectrometer/Data System (HRGC/HRMS/DS) equipped with a GC injection port designed so that the separation of 2,3,7,8-TCDD from the other TCDD isomers achieved in the gas chromatographic column is not appreciably degraded.

Column 1: 60 m DB-5 fused silica capillary column.
Column 2: 30 m DB-225 fused silica capillary column, or equivalent.

PRECISION & ACCURACY Precision, bias and concentration ranges for the compounds covered by this method have not been determined yet. The sensitivity of Method 8290 is dependent upon the level of interferences within a given matrix. Samples containing concentrations of specific congeneric analytes of PCDDs and PCDFs that are greater than ten times the upper method calibration limits must be analyzed by a protocol designed for such concentration levels, e.g., EPA Method 8280.

SAMPLE PREPARATION
Sludge/wet fuel oil — Extract aqueous sludge or wet fuel oil samples by refluxing a sample with toluene using a Dean-Stark water separator until all the water is removed. Filter the toluene extract through a glass fiber filter, or equivalent, and concentrate it to near dryness either on a rotary evaporator using an inert gas. Transfer the concentrate to a separatory funnel using hexane and wash it with 5% sodium chloride solution. Proceed to clean up.

Soil/sediment — If the sample is wet, add anhydrous powdered sodium sulfate to it until a free flowing mixture is obtained. Place the soil/sodium sulfate mixture in the Soxhlet apparatus, add toluene, and reflux for 16 h. The solvent must cycle completely through the system five times per h. Cool and filter the extract through a glass fiber filter and concentrate to near dryness on a rotary evaporator. Transfer the residue to a separatory funnel, using hexane. Proceed to clean up.

Aqueous samples — Use a 1-L sample; the method may require acetone to be added to it. When the sample is judged to contain 1% or more solids, it must be filtered through a glass fiber filter that has been rinsed with toluene. If the suspended solids content is too great to filter, centrifuge the sample, decant, and then filter the aqueous phase. Combine the solids from the centrifuge bottle(s) with the particulates on the filter and with the filter itself and proceed with Soxhlet extraction for soil/sediment. Extract the aqueous filtrate with methylene chloride in a separatory funnel, filter the extract through anhydrous sodium sulfate, and concentrate it using a K-D apparatus or a rotary evaporator. Exchange the solvent with hexane and proceed to clean up.

Clean up — The sample extract is cleaned up utilizing a number of different techniques. Partition cleanup is where the sample extract is partitioned with concentrated sulfuric acid, 5% aqueous sodium chloride, and 20% aqueous potassium hydroxide. Silica/alumina column cleanup involves packing gravity columns with silica gel and alumina and sequentially eluting the residue from the partition cleanup. Carbon column cleanup involves packing a column with a mixture of AX–21 and Celite 545 and sequentially eluting the sample concentrate from the silica/alumina cleanup with hexane, cyclohexane/methylene chloride (50:50), and methylene chloride/methanol/toluene (75:20:5). Then the column is turned upside down and the PCDD/PCDF fraction is eluted with toluene. The toluene fraction is concentrated and stored in the dark at room temperature until analysis.

QUALITY CONTROL Demonstrate, through the analysis of a reagent water blank, that interferences from the analytical system, glassware, and reagents are under control. For each analytical batch (up to 20 samples), a reagent blank, matrix spike, and matrix spike duplicate/duplicate must be analyzed (the frequency of the spikes may be different for different monitoring programs). The blank and spiked samples must be carried through all stages of the sample preparation and measurement steps.

A GC column performance check is required at the beginning of each 12-h period during which samples are analyzed. An HRGC/HRMS method blank run is required between a calibration run and the first sample run. The same method blank extract may thus be analyzed more than once if the number of samples within a batch requires more than 12 h of analyses.

At the beginning of each 12-h period during which samples are to be analyzed, an aliquot of the 1) GC column performance check solution and 2) a high-resolution concentration calibration must be analyzed to demonstrate adequate GC resolution and sensitivity, response factor reproducibility, and mass range calibration, and to establish the PCDD/PCDF retention time windows. A mass resolution check must also be performed to demonstrate adequate mass resolution using an appropriate reference compound (perfluorokerosene (PFK) is recommended). If the required criteria are not met, remedial action must be taken before any samples are analyzed.

Routine or continuing calibration (using a high resolution calibration solution) and the mass resolution check must also be performed at the end of each 12 h period. Furthermore, a HRGC/HRMS method blank analysis must be recorded following a calibration analysis and the first sample analysis.

To evaluate the performance of the analytical method, the QC check samples must be handled in exactly the same manner as actual samples. Therefore, 1.0 mL of the QC check sample concentrate is spiked into each of four 1 L aliquots of reagent water (which becomes the QC check sample), extracted, and then analyzed by GC. The variety of semivolatile analytes which may be analyzed by GC is such that the concentration of the QC check sample concentrate is different for the different analytical techniques presented in the full method.

The analyst must demonstrate also that the compounds of interest are being quantitatively recovered by the cleanup technique before the cleanup is applied to actual samples. For sample extracts that are cleaned up, the associated quality control samples (e.g., spikes, blanks, and duplicates) must also be processed through the same cleanup procedure. The analysis using each determinative method (GC, GC/MS, HPLC) specifies instrument calibration procedures using stock standards. It is recommended that cleanup also be performed on a series of the same type of standards to validate chromatographic elution patterns for the compounds of interest and to verify the absence of interferences from reagents.

REFERENCE Test Methods for Evaluating Solid Waste (SW-846). U.S. EPA. 1983. Method 8290, Rev. 0, Nov. 1990. Office of Solid Wastes, Washington, DC.

n-Octacosane
CAS #630-02-4

EPA Method 1625

TITLE Semivolatile Organic Compounds by Isotope Dilution GC/MS

MATRIX The compounds may be determined in waters, soils, and municipal sludges by this method.

METHOD SUMMARY This method is used to determine 176 semivolatile toxic organic pollutants associated with the CWA (as amended 1987); the RCRA (as amended 1986); the CERCLA (as amended 1986); and other compounds amenable to extraction and analysis by capillary column gas chromatography-mass spectrometry (GC/MS).

Stable isotopically-labeled analogs of the compounds of interest are added to the sample. If the solids content is less than 1%, a 1-L sample is extracted at pH 12–13, then at pH <2 with methylene chloride using continuous extraction techniques.

If the solids content is 30% or less, the sample is diluted to 1% solids with reagent water, homogenized ultrasonically, and extracted at pH 12–13, then at pH <2 with methylene chloride using continuous extraction techniques. If the solids content is greater than 30%, the sample is extracted using ultrasonic techniques.

Each extract is dried over sodium sulfate, concentrated to a volume of 5 mL, cleaned up using GPC, if necessary, and concentrated. Extracts are concentrated to 1 mL if GPC is not performed, and to 0.5 mL if GPC is performed.

An internal standard is added to the extract, and a 1-mL aliquot of the extract is injected into the GC. The compounds are separated by GC and detected by a MS. The labeled compounds serve to correct the variability of the analytical technique.

INTERFERENCES Solvents, reagents, glassware, and other sample processing hardware may yield artifacts and/or elevated baselines causing misinterpretation of chromatograms and spectra. Materials used in the analysis must be demonstrated to be free from interferences under the conditions of analysis by running method blanks initially and with each sample lot (sample started through the extraction process on a given 8-h shift, to a maximum of 20). Specific selection of reagents and purification of solvents by distillation in all glass systems may be required. Glassware and, where possible, reagents are cleaned by solvent rinse and baking at 450°C for 1-h minimum. Interferences coextracted from samples will vary considerably from source to source, depending on the diversity of the site being sampled.

INSTRUMENTATION Major instrumentation includes a GC with a splitless or on-column injection port for capillary column, a MS with 70 eV electron impact ionization, and a data system to collect and record MS data, and process it. A K-D apparatus is used to concentrate extracts.

GC Column: 30 m × 0.25 mm I.D. 5% phenyl, 94% methyl, 1% vinyl silicone bonded phased fused silica capillary column.

PRECISION & ACCURACY The detection limits of the method are usually dependent on the level of interferences rather than instrumental limitations. The limits typify the minimum quantities that can be detected with no interferences present.

The minimum level (in µg/mL) was 10. This is defined as a minimum level at which the analytical system shall give recognizable mass spectra (background corrected) and acceptable calibration points.

The MDL (in µg/kg) in low solids was 492 and in high solids was 1810; these were determined in digested sludge (low solids) and in filter cake or compost (high solids).

Note: Background levels of this compound were present in the sludge tested, resulting in higher than expected MDLs. The MDL for this compound is expected to be approximately 50 µg/kg with no interferences present.

The labeled and native compound initial precision as standard deviation (in µg/L) was 35.
The labeled and native compound initial accuracy as average recovery (in µg/L) was 35–193.

SAMPLE COLLECTION, PRESERVATION & HANDLING Collect samples in glass containers. Aqueous samples which flow freely are collected in refrigerated bottles using automatic sampling equipment. Solid samples are collected as grab samples using widemouth jars. Maintain samples at 0 to 4°C from the time of collection until extraction. If residual chlorine is present in aqueous samples, add 80 mg sodium thiosulfate/L of water. Begin sample extraction within 7 days of collection, and analyze all extracts within 40 days of extraction.

SAMPLE PREPARATION Samples containing 1% solids or less are extracted directly using continuous liquid-liquid extraction techniques. Samples containing 1 to 30% solids are diluted to the 1% level with reagent water and extracted using continuous liquid-liquid extraction techniques. Samples containing greater than 30% solids are extracted using ultrasonic techniques.

Base/neutral extraction — Adjust the pH of the waters in the extractors to 12–13 with 6 *N* NaOH. Extract with methylene chloride for 24–48 h.
Acid extraction — Adjust the pH of the waters in the extractors to 2 or less using 6 *N* sulfuric acid. Extract with methylene chloride for 24–48 h.
Ultrasonic extraction of high solids samples — Add anhydrous sodium sulfate to the sample and QC aliquot(s). Add acetone:methylene chloride (1:1) to the sample and mix thoroughly

Concentrate extracts using a K-D apparatus.

QUALITY CONTROL The analyst is permitted to modify this method to improve separations or lower the costs of measurements, provided all performance specifications are met. Analyses of blanks are required to demonstrate freedom from contamination. When results of spikes indicate atypical method performance for samples, the samples are diluted to bring method performance within acceptable limits.

For low solids (aqueous samples), extract, concentrate, and analyze two sets of four 1-L aliquots (8 aliquots total) of the

precision and recovery standard. For high solids samples, two sets of four 30-g aliquots of the high solids reference matrix are used.

Spike all samples with labeled compounds to assess method performance. Compute percent recovery of the labeled compounds using the internal standard method. Compare the labeled compound recovery for each compound with the corresponding labeled compound recovery.

Reagent water and high solids reference matrix blanks are analyzed to demonstrate freedom from contamination. Extract and concentrate a 1-L reagent water blank or a high solids reference matrix blank with each sample's lot (samples started through the extraction process on the same 8-h shift, to a maximum of 20 samples).

Field replicates may be collected to determine the precision of the sampling technique, and spiked samples may be required to determine the accuracy of the analysis when the internal standard method is used.

REFERENCE Semivolatile Organic Compounds by Isotope Dilution GC/MS. Office of Water Regulation and Standards, U.S. EPA Industrial Technology Division, Washington, DC, EPA Method 1625, Rev. C, June 1989 (contact W.A. Telliard, U.S. EPA, Office of Water Regulations and Standards, 401 M St., SW, Washington, DC, 20460. Phone: 202-382-7131).

---

n-Octadecane   EPA Method 1625
CAS #593-45-3

TITLE Semivolatile Organic Compounds by Isotope Dilution GC/MS

MATRIX The compounds may be determined in waters, soils, and municipal sludges by this method.

METHOD SUMMARY This method is used to determine 176 semivolatile toxic organic pollutants associated with the CWA (as amended 1987); the RCRA (as amended 1986); the CERCLA (as amended 1986); and other compounds amenable to extraction and analysis by capillary column gas chromatography-mass spectrometry (GC/MS).

Stable isotopically-labeled analogs of the compounds of interest are added to the sample. If the solids content is less than 1%, a 1-L sample is extracted at pH 12–13, then at pH <2 with methylene chloride using continuous extraction techniques.

If the solids content is 30% or less, the sample is diluted to 1% solids with reagent water, homogenized ultrasonically, and extracted at pH 12–13, then at pH <2 with methylene chloride using continuous extraction techniques. If the solids content is greater than 30%, the sample is extracted using ultrasonic techniques.

Each extract is dried over sodium sulfate, concentrated to a volume of 5 mL, cleaned up using GPC, if necessary, and concentrated. Extracts are concentrated to 1 mL if GPC is not performed, and to 0.5 mL if GPC is performed.

An internal standard is added to the extract, and a 1-mL aliquot of the extract is injected into the GC. The compounds are separated by GC and detected by a MS. The labeled compounds serve to correct the variability of the analytical technique.

INTERFERENCES Solvents, reagents, glassware, and other sample processing hardware may yield artifacts and/or elevated baselines causing misinterpretation of chromatograms and spectra. Materials used in the analysis must be demonstrated to be free from interferences under the conditions of analysis by running method blanks initially and with each sample lot (sample started through the extraction process on a given 8-h shift, to a maximum of 20). Specific selection of reagents and purification of solvents by distillation in all glass systems may be required. Glassware and, where possible, reagents are cleaned by solvent rinse and baking at 450°C for 1-h minimum. Interferences coextracted from samples will vary considerably from source to source, depending on the diversity of the site being sampled.

INSTRUMENTATION Major instrumentation includes a GC with a splitless or on-column injection port for capillary column, a MS with 70 eV electron impact ionization, and a data system to collect and record MS data, and process it. A K-D apparatus is used to concentrate extracts.

GC Column: 30 m × 0.25 mm I.D. 5% phenyl, 94% methyl, 1% vinyl silicone bonded phased fused silica capillary column.

PRECISION & ACCURACY The detection limits of the method are usually dependent on the level of interferences rather than instrumental limitations. The limits typify the minimum quantities that can be detected with no interferences present.

The minimum level (in µg/mL) was 10. This is defined as a minimum level at which the analytical system shall give recognizable mass spectra (background corrected) and acceptable calibration points.

The MDL (in µg/kg) in low solids was 134 and in high solids was 844; these were determined in digested sludge (low solids) and in filter cake or compost (high solids).

Note: Background levels of this compound were present in the sludge tested, resulting in higher than expected MDLs. The MDL for this compound is expected to be approximately 50 µg/kg with no interferences present.

The labeled and native compound initial precision as standard deviation (in µg/L) was 39.
The labeled and native compound initial accuracy as average recovery (in µg/L) was 42–131.

SAMPLE COLLECTION, PRESERVATION & HANDLING Collect samples in glass containers. Aqueous samples which flow freely are collected in refrigerated bottles using automatic sampling equipment. Solid samples are collected as grab samples using widemouth jars. Maintain samples at 0 to 4°C from the time of collection until extraction. If residual chlorine is present in aqueous samples, add 80 mg sodium thiosulfate/L of water. Begin sample extraction within 7 days of collection, and analyze all extracts within 40 days of extraction.

SAMPLE PREPARATION Samples containing 1% solids or less are extracted directly using continuous liquid-liquid extraction techniques. Samples containing 1 to 30% solids are diluted to the 1% level with reagent water and extracted using continuous liquid-liquid extraction techniques. Samples containing greater than 30% solids are extracted using ultrasonic techniques.

- Base/neutral extraction — Adjust the pH of the waters in the extractors to 12–13 with 6 N NaOH. Extract with methylene chloride for 24–48 h.
- Acid extraction — Adjust the pH of the waters in the extractors to 2 or less using 6 N sulfuric acid. Extract with methylene chloride for 24–48 h.
- Ultrasonic extraction of high solids samples — Add anhydrous sodium sulfate to the sample and QC aliquot(s). Add acetone:methylene chloride (1:1) to the sample and mix thoroughly

Concentrate extracts using a K-D apparatus.

QUALITY CONTROL The analyst is permitted to modify this method to improve separations or lower the costs of measurements, provided all performance specifications are met. Analyses of blanks are required to demonstrate freedom from contamination. When results of spikes indicate atypical method performance for samples, the samples are diluted to bring method performance within acceptable limits.

For low solids (aqueous samples), extract, concentrate, and analyze two sets of four 1-L aliquots (8 aliquots total) of the precision and recovery standard. For high solids samples, two sets of four 30-g aliquots of the high solids reference matrix are used.

Spike all samples with labeled compounds to assess method performance. Compute percent recovery of the labeled compounds using the internal standard method. Compare the labeled compound recovery for each compound with the corresponding labeled compound recovery.

Reagent water and high solids reference matrix blanks are analyzed to demonstrate freedom from contamination. Extract and concentrate a 1-L reagent water blank or a high solids reference matrix blank with each sample's lot (samples started through the extraction process on the same 8-h shift, to a maximum of 20 samples).

Field replicates may be collected to determine the precision of the sampling technique, and spiked samples may be required to determine the accuracy of the analysis when the internal standard method is used.

REFERENCE Semivolatile Organic Compounds by Isotope Dilution GC/MS. Office of Water Regulation and Standards, U.S. EPA Industrial Technology Division, Washington, DC, EPA Method 1625, Rev. C, June 1989 (contact W.A. Telliard, U.S. EPA, Office of Water Regulations and Standards, 401 M St., SW, Washington, DC, 20460. Phone: 202-382-7131).

Organic Carbon, Total EPA Method 415.1

TITLE Organics — Combustion or

MATRIX Drinking, surface, and saline oxidation products. Waters. Wastewater.

APPLICATION Date issued 1971. Editorial Rev. 1974. Organic carbon is converted to carbon dioxide (CO_2) by catalytic combustion or wet chemical oxidation. CO_2 formed can be measured by infrared detector or converted to methane (CH_4) and measured by flame ionization.

INTERFERENCES Carbonate and bicarbonate can interfere and must be removed or accounted for in calculation. This proceedure is applicable only to homogeneous samples which can be injected into apparatus reproducibly by syringe or pipet. Applies to TOC level above 1 mg/L.

INSTRUMENTATION Apparatus for total and dissolved organic carbon. Blender apparatus.

RANGE Not listed.

MDL Not listed.

PRECISION SD = 8.32 TOC mg/L at 107 TOC mg/L.

ACCURACY As bias, +1.08 mg/L at 107 TOC mg/L.

SAMPLING METHOD Plastic or glass (25 mL).

STABILITY Cool, 4°C. HCl or H_2SO_4 to pH<2.

MHT 28 days.

QUALITY CONTROL Protect samples from sunlight and atmospheric oxygen. for instrument calibration, series of standards should encompass the expected concentration range of the samples. Instrument manufacturer's instructions should be followed.

REFERENCE EPA Methods for the Chemical Analysis of Water and Wastes, EPA-600/4-79-020, U.S. EPA, EMSL, 1979.

Osmium EPA Method 7550
CAS #7440-04-2

TITLE Atomic Absorption (AA) Direct Aspiration

MATRIX Wastes, mobility procedure extracts, soils, and groundwater.

APPLICATION Prior to analysis, samples must be prepared for direct aspiration. sample preparation method varies with the matrix. Following appropriate dissolution of the sample, an osmium lamp light beam is directed through an aspirated aliquot in a flame, into a monochromator and detector. Due to extreme toxicity of osmium and its compounds extreme care must be taken to handle it safely.

INTERFERENCES Background correction is required. Monitor samples and standards for viscosity differences since viscosity may alter aspiration rate.

INSTRUMENTATION Atomic absorption spectrometer. Osmium hollow cathode lamp. (290.0 nm wavelength). Strip-chart recorder.

RANGE 2–100 mg/L.

MDL 0.3 mg/L

PRECISION Not listed.

ACCURACY Not listed.

SAMPLING METHOD Use plastic or glass containers (prewashed). Sample as per chapter 9.

STABILITY Non aqueous samples: cool to 4°C and analyze as soon as possible. Aqueous samples: add nitric acid to pH <2.

MHT 6 months.

QUALITY CONTROL Run one spike duplicate sample for every 10 samples. Verify calibration with an independently prepared check standard every 15 samples. Due to the very volatile nature of some osmium compounds, the applicable method must be verified by spiked samples or standard reference materials or both. Make osmium standards on daily basis, because of osmium volatility.

REFERENCE Method 7550, SW-846, 3rd ed., Nov.1986.

Oxygen EPA Method 360.1

TITLE Inorganics, Non-Metallics

MATRIX Wastewaters and streams. Dissolved oxygen (DO)

APPLICATION Date issued 1971. (Membrane electrode). This probe method is recommended for those samples containing materials which interfere with modified Winkler procedure. It is recommended for monitoring streams, lakes, outfalls, etc., with continuous record of DO.

INTERFERENCES Probes with membranes respond to partial pressure of oxygen which in turn is a function of dissolved organic salts. Conversion factors for sea and brackish waters may be calculated. (Conversion factors for specific inorganic salts may be determined experimentally)

INSTRUMENTATION Dissolved oxygen probe. Weston and Stack, YSI or Beckman.

RANGE Not listed.

MDL Not listed.

PRECISION Manufacturer's Specifications Claim 0.1 mg/L repeatability.

ACCURACY Manufacturer's Specifications Claim 0.1 mg/L repeatability.

SAMPLING METHOD Glass container only (both bottle and top). sample collection from shallow depths (less than 5 ft), use an apha type sampler. A Kemmerer type sampler is recommended for samples collected at depths > 5 ft. Fill 300 mL bottle to overflowing.

STABILITY No preservation required. Analyze immediately.

QUALITY CONTROL Record temperature at time of sampling.

REFERENCE EPA Methods for the Chemical Analysis of Water and Wastes, EPA-600/4-79-020, U.S. EPA, EMSL, 1979.

Oxygen EPA Method 360.2

TITLE Inorganics, Non-Metallics

MATRIX Wastewater and streams. Dissolved oxygen (DO)

APPLICATION Date issued 1971. (Modified Winkler, full bottle technique). This Method is applicable for use with most wastewaters and streams that contain nitrate nitrogen and not more than 1 mg/L ferrous iron. The DO probe technique gives comparable results.

INTERFERENCES There are a number of interferences to the dissolved oxygen test, including oxidizing and reducing agents, nitrate ion, ferrous iron and organic matter. Most common interferences in the Winkler procedure may be overcome by use of the DO probe.

INSTRUMENTATION A titration with 0.0375N sodium thiosulfate to a starch-iodine end point.

RANGE Not listed.

MDL Not listed.

PRECISION Exact data unavailable.

ACCURACY Exact data unavailable.

SAMPLING METHOD Glass container only (both bottle and top). sample collection from shallow depths (less than 5 ft) and use an apha type sampler.

A Kemmerer type sampler is recommended for samples collected at depths > 5ft. Fill a 300 mL bottle to overflowing.

STABILITY Fix on site and store in dark.

MHT 8 h.

QUALITY CONTROL Record temperature at time of sampling.

REFERENCE Methods for the Chemical Analysis of Water and Wastes, EPA-600/4-79-020, U.S. EPA, EMSL, 1979.

Oxygen, Dissolved (DO) EPA Method 360.1

TITLE Inorganics, Non-Metallics

MATRIX Wastewaters and streams.

APPLICATION Date issued 1971. (Membrane electrode). This probe method is recommended for those samples containing materials which interfere with modified Winkler procedure. It is recommended for monitoring streams, lakes, outfalls, etc., with continuous record of DO.

INTERFERENCES Probes with membranes respond to partial pressure of oxygen which in turn is a function of dissolved organic salts. Conversion factors for sea and brackish waters may be calculated (conversion factors for specific inorganic salts may be determined experimentally).

INSTRUMENTATION Dissolved oxygen probe. Weston and Stack, YSI or Beckman.

RANGE Not listed.

MDL Not listed.

PRECISION Manufacturer's Specifications Claim 0.1 mg/L repeatability.

ACCURACY Manufacturer's Specifications Claim 0.1 mg/L repeatability.

SAMPLING METHOD Glass container only (both bottle and top). sample collection from shallow depths (less than 5 ft), use an apha type sampler. A kemmerer type sampler is recommended for samples collected at depths > 5 ft. Fill 300 mL bottle to overflowing.

STABILITY No preservation required. Analyze immediately.

QUALITY CONTROL Record temperature at time of sampling.

REFERENCE EPA Methods for the Chemical Analysis of Water and Wastes, EPA-600/4-79-020, U.S. EPA, EMSL, 1979.

Oxygen, Dissolved (DO) — EPA Method 360.2

TITLE Inorganics, Non-Metallics

MATRIX Wastewater and streams.

APPLICATION Date issued 1971. (Modified Winkler, full bottle technique). This Method is applicable for use with most wastewaters and streams that contain nitrate nitrogen and not more than 1 mg/L ferrous iron. The DO probe technique gives comparable results.

INTERFERENCES There are a number of interferences to the dissolved oxygen test, including oxidizing and reducing agents, nitrate ion, ferrous iron and organic matter. Most common interferences in the Winkler procedure may be overcome by use of the DO probe.

INSTRUMENTATION A titration with 0.0375N sodium thiosulfate to a starch-iodine end point.

RANGE Not listed.

MDL Not listed.

PRECISION Exact data unavailable.

ACCURACY Exact data unavailable.

SAMPLING METHOD Glass container only (both bottle and top). sample collection from shallow depths (less than 5 ft), use an apha type sampler. A Kemmerer type sampler is recommended for samples collected at depths > 5 ft. Fill a 300 mL bottle to overflowing.

STABILITY Fix on site and store in dark.

MHT 8 h.

QUALITY CONTROL Record temperature at time of sampling.

REFERENCE Methods for the Chemical Analysis of Water and Wastes, EPA-600/4-79-020, U.S. EPA, EMSL, 1979.

P

Paraldehyde
CAS #123-3-7

EPA Method 8015

TITLE Nonhalogenated Volatile Organics

MATRIX Groundwater, soils, sludges, water miscible liquid wastes, and non-water miscible wastes.

APPLICATION This method is used for the analysis of 6 nonhalogenated VOCs. Samples are analyzed using direct injection or purge-and-trap methods. Groundwater must be analyzed by the purge-and-trap method. The method provides an optional GC column that may help resolve analytes from interferences and which is also used for analyte confirmation.

INTERFERENCES There can be carryover contamination with high- and low-level samples. Impurities may come from the purge-and-trap apparatus, organic compounds outgassing from the plumbing ahead of trap, diffusion of VOCs through the sample bottle septum during shipping or storage, or from solvent vapors in the lab.

INSTRUMENTATION GC capable of on-column injections or purge-and-trap sample introduction and a flame ionization detector (FID). Column 1: an 8 ft by 0.1 in 1% SP-1000 on Carbopack-B column. Column 2: a 6 ft by 0.1 in bonded n-octane on Porasil-C.

RANGE Not available.

MDL Not available.

PRECISION Not available.

ACCURACY Not available.

SAMPLING METHOD For water and liquid samples; use glass 40-mL vials with Teflon®-lined septum caps and collect two vials per sample location with no headspace. For solids and concentrated waste samples; use widemouth glass bottles with Teflon® liners. Cool all samples to 4°C

STABILITY For concentrated wastes, soils, sediments, or sludges: cool to 4°C. For liquids: add 4 drops of concentrated hydrochloric acid, cool to 4°C.

MHT 14 days.

QUALITY CONTROL Analyze a reagent blank, matrix spike, and matrix spike duplicate/duplicate for each analytical batch (up to 20 samples). Demonstrate the purity of glassware and reagents by analyzing a reagent water method blank. Internal, surrogate, and five concentration level calibration standards are used.

REFERENCE Method 8015, SW-846, 3rd ed., Nov.1986.

Pebulate
CAS #1114-71-2

EPA Method 507

TITLE Determination of Nitrogen and Phosphorus-Containing Pesticides in Water by GC/NPD

MATRIX This method is applicable to the determination of certain nitrogen and phosphorus-containing pesticides in finished drinking water and groundwater.

METHOD SUMMARY Method 507 covers 46 nitrogen- and phosphorus-containing pesticides. A 1-L sample is fortified with a surrogate standard, salted, buffered, extracted with methylene chloride, and concentrated; then the solvent is exchanged with methyl tert-butyl ether (MTBE) and concentrated again, and a 2-µL aliquot of a sample extract is injected into a GC system equipped with a selective nitrogen-phosphorus detector and a capillary column for analysis.

INTERFERENCES Method interferences may be caused by contaminants in solvents, reagents, glassware, and other sample processing apparatus. Interfering contamination may occur when a sample containing low concentrations of analytes is analyzed immediately following a sample containing relatively high concentrations. One or more injections of MTBE should be made following the analysis of a sample with high concentrations of analytes to check for analyte carryover. Matrix interferences may be caused by contaminants that are coextracted from the sample. The extent of matrix interferences will vary considerably from source to source, depending upon the water sampled.

INSTRUMENTATION A gas chromatograph system (GC) equipped with a nitrogen-phosphorus detector (NPD) is needed.

Column 1: 30 m × 0.25 mm I.D. DB-5 bonded fused silica column, 0.25 µm film thickness, or equivalent.
Column 2: 30 m × 0.25 mm I.D. DB-1701 bonded fused silica column, 0.25 µm film thickness, or equivalent.

PRECISION & ACCURACY This method has been validated in a single lab and estimated detection limits (EDLs) have been determined for each analyte. Observed detection limits may vary among waters, depending upon the nature of the interferences in the sample matrix and the specific instrumentation used. Analytes that are not separated chromatographically cannot be individually identified and measured unless an alternative technique for identification and quantification exist.

The estimated detection limit (in µg/L) was 0.13. The EDL is defined as either method detection limit or a level of compound in a sample yielding a peak in the final extract with signal-to-noise ratio of approximately 5, whichever value is higher.

The concentration used for these measurements (in µg/L) was 1.3.
The accuracy (as % recovery) was 94.
The precision (% RSD) was 9.

SAMPLING METHOD Grab samples are collected in 1-L glass sample bottles (prewashed with detergent and hot tap water, rinsed with reagent water, and dried in an oven at 400°C for 1 h) with screw caps lined with PTFE-fluorocarbon.

SAMPLE PRESERVATION Add mercuric chloride to the sample bottle in amounts to produce a concentration of 10 mg/L. If residual chlorine is present, add 80 mg of sodium thiosulfate/L of sample to the sample bottle prior to collection. After collection, seal bottle and shake vigorously for 1 min, then cool the sample to 4°C immediately and store it at 4°C in the dark until extraction.

MHT Maximum holding time of the samples, and in some cases the extracts, is 14 days.

SAMPLE PREPARATION Fortify the sample with 50 μL of the surrogate standard solution, adjust to pH 7 with phosphate buffer, add 100 g NaCl to the sample, and seal and shake to dissolve the salt; then extract with methylene chloride in a separatory funnel or in a mechanical tumbler bottle. Dry the extract by pouring it through a solvent-rinsed drying column containing about 10 cm of anhydrous sodium sulfate. Collect the extract in a Kuderna-Danish (K-D) concentrator and rinse the column with 20–30 mL methylene chloride. Concentrate the extract to about 2 mL and rinse the flask and its lower joint into the concentrator tube with 1 to 2 mL of methyl t-butyl ether (MTBE). Add 5–10 mL of MTBE and concentrate the extract twice (adding more MTBE) to a final volume of 5.0 mL and store it at 4°C until analysis.

Note: If methylene chloride is not completely removed from the final extract, it may cause detector problems.

QUALITY CONTROL Minimum quality control requirements are initial demonstration of lab capability, determination of surrogate compound recoveries in each sample and blank, monitoring internal standard peak area or height in each sample and blank, analysis of lab reagent blanks, lab fortified samples, lab fortified blanks, and other QC samples. A lab reagent blank is analyzed to demonstrate that all glassware and reagent interferences are under control.

Initial demonstration of capability is fulfilled by analyzing four fortified reagent water samples with the recovery value for each analyte falling within the acceptable range (±30% average recovery). Surrogate recoveries from samples or method blanks must be 70–130%. The internal standard response for any sample chromatogram should not deviate from the daily calibration check standard's internal standard response by more than 30% or lab fortified blanks and sample matrices are used to assess lab performance and analyte recovery, respectively.

If the response for the target analyte peak exceeds the working range of the system, dilute the extract and reanalyze. Alternative techniques such as an alternate detector or second chromatography column should be used to confirm peak identification when sample components are not resolved adequately.

EPA CONTACT & HOTLINE For technical questions contact Dr. Baldev Bathija, U.S. EPA, Office of Ground Water and Drinking Water (WH-550D), 401 M St. SW, Washington, DC 20460. Tel. (202) 260-3040. For further information the EPA Safe Drinking Water Hotline may be called at: (800) 426-4791.

REFERENCE Methods for the Determination of Organic Compounds in Drinking Water, EPA/600/4-88/039 (revised July 1991). U.S. EPA Environmental Monitoring Systems Laboratory, Cincinnati, OH, 45268, U.S.A. Available from the National Technical Information Service (NTIS), 5285 Port Royal Road, Springfield, VA 22161; Tel. 800-553-6847. NTIS Order Number is PB91-231480.

1,2,3,4,7-PeCDD   **EPA Method 8280**
CAS #39227-61-7

TITLE The Analysis of Polychlorinated Dibenzo-P-Dioxins and Polychlorinated Dibenzofurans.

MATRIX This method is appropriate for the determination of tetra-, penta-, hexa-, hepta-, and octachlorinated dibenzo-p-dioxins (PCDDs) and dibenzofurans (PCDFs) in chemical wastes including still bottoms, fuel oils, sludges, fly ash, reactor residues, soil and water.

METHOD SUMMARY This method covers 22 PCDD and PCDF compounds and it uses a high resolution capillary GC column with low resolution mass spectrometry. Samples are extracted and concentrated by several methods that vary depending on the matrix involved. The organic extracts are cleaned-up by washing with aqueous basic and acid solutions and then separated into fractions using a column of neutral alumina. The fraction containing the PCDDs and PCDFs is then further cleaned-up using a column of activated carbon. The final extract is concentrated and Carbon-13 labeled internal standards are added prior to analysis by GC/MS using a capillary GC column and selected ion monitoring using five sets of ions that are detailed in the method. Certain 2,3,7,8-substituted congeners are used to provide calibration and method recovery information. Proper column selection and access to reference isomer standards, may in certain cases, provide isomer specific data.

INTERFERENCES Solvents, reagents, glassware, and other sample processing hardware may yield discrete artifacts and/or elevated baselines which may cause misinterpretation of chromatographic data. Use high purity reagents and solvents to minimize interference problems. Interferents coextracted from the sample will vary considerably from source to source. PCDDs and PCDFs are often associated with other interfering chlorinated compounds such as PCBs and polychlorinated diphenyl ethers which may be found at concentrations several orders of magnitude higher than that of the analytes of interest.

INSTRUMENTATION A low resolution GC/MS utilizing 70 ev must be capable of selected ion monitoring (SIM) for at least 11 ions simultaneously, with a cycle time of 1 second or less. Minimum integration time for SIM is 50 ms per m/z. Also required is a GC-to-MS interface constructed of all glass or glass-lined materials. One of the following GC columns is required:

Column 1: 50 m CP-Sil-88 fused silica capillary column.
Column 2: DB-5 (30 m × 0.25 mm I.D., 0.25-um film thickness) fused silica capillary column.
Column 3: 30 m SP-2250 fused silica capillary column.

When toluene is employed as the final solvent, use of a bonded phase column is recommended. Solvent exchange into tridecane is required for other liquid phases or nonbonded columns such as CP-Sil-88. Chromatographic conditions must be adjusted to account for solvent boiling points.

PRECISION & ACCURACY Accuracy, precision, MDLs and concentration ranges for the compounds covered by this method have not been determined or published by EPA yet. The sensitivity of this method is dependent upon the level of interferents within a given matrix. Proposed quantification levels for target analytes were 2 ppb in soil samples, up to 10 ppb in other solid wastes and 10 ppt in water. Actual values have been shown to vary by homologous series and, to a lesser degree, by individual isomer.

SAMPLING METHOD Grab and composite samples must be collected in 1 L or 1-quart amber glass bottles with Teflon®-lined screw-caps that have been acid-washed and solvent rinsed before use. If compositing equipment is used, the system must incorporate glass sample containers for the collection of a minimum of 250 mL. samples.

SAMPLE PRESERVATION Samples must be cooled and stored at 4°C.

MHT Samples must be extracted within 30 days and analyzed within 45 days of collection.

SAMPLE PREPARATION

Soil Samples — Extract 20 g of a 1:1 soil and anhydrous sodium sulfate mixture with 1:4 methanol-petroleum ether for 2 h in a wrist-action shaker. Concentrate the extract in a K-D and then exchange the solvent with hexane in the K-D. The final volume is 15 mL of hexane.

Aqueous samples — Extract a 1 L sample with methylene chloride, dry it with anhydrous sodium sulfate and concentrate it in a K-D followed by exchange with hexane in the K-D. A continuous liquid-liquid extractor may be used in place of a separatory funnel to avoid emulsions. The final volume is 15 mL of hexane.

Alumina column clean-up — The clean-up procedure described below consists of two phases. The first phase involves a sequential basic and acid washing of the extract that contains the analytes.

Wash the 15 mL hexane extract with 20% potassium hydroxide in a separatory funnel. Repeat the washing until no color is visible in the bottom layer but no more than four times because strong base is known to degrade certain PCDDs/PCDFs, so contact time must be minimized. Next, partition the 15 mL hexane against 40 mL of 5% sodium chloride. Next, partition the 15 mL hexane against 40 mL of concentrated sulfuric acid. Repeat the acid washings until no color is visible in the acid layer (but no more than four times). Finally, partition the 15 mL hexane against 40 mL of 5% sodium chloride. Dry the organic layer with anhydrous sodium sulfate and concentrate it to near dryness with a rotary evaporator. Dissolve the hexane residue from the first phase of the clean-up in 2 mL of hexane and apply it carefully to the top of a pre-eluted Woelm super 1 neutral alumina column. Elute the column with 10 mL of 8 percent (v/v) methylene chloride in hexane. Check by GC/MS analysis that no PCDDs of PCDFs are elute in this fraction before discarding it. Elute the PCDDs and PCDFs from the column with 15 mL of 60 percent (v/v) methylene chloride in hexane and collect this second fraction in a conical shaped concentrator tube.

Carbon column clean-up — Using a carefully regulated stream of nitrogen, concentrate the first 8 percent fraction (methylene chloride in hexane) from the alumina column to about 1 mL. Save this 8 percent concentrate for GC/MS analysis to check for breakthrough of PCDDs and PCDFs. Concentrate the second 60 percent fraction (methylene chloride in hexane) to about 2 to 3 mL. Prepare a carbon column and rinse the carbon with 5 mL cyclohexane/methylene chloride (50:50 v/v) in the forward direction of flow and then in the reverse direction of flow. While still in the reverse direction of flow, transfer the sample concentrate to the column and elute with 10 mL of methylene chloride/methanol/benzene (75:20:5, v/v). Save all above eluates and combine them (this fraction may be used as a check on column efficiency). Next, turn the column over and, in the direction of forward flow, elute the PCDD/PCDF fraction with 20 mL toluene. Evaporate the toluene fraction to about 1 mL on a rotary evaporator and transfer this concentrate to a 2.0-mL Reacti-vial. Concentrate the sample using a stream of nitrogen gas. The final volume will depend on the relative concentration of target analytes but it is typically 100 µL for soil samples and 500 µL for sludge, still bottom, and fly ash samples. Extracts which are determined to be outside the calibration range for individual analytes must be diluted or a smaller portion of the sample must be re-extracted.

An alternate carbon column clean-up also may be used with a 1 mL HPLC injector loop. The injector loop is connected to the optional HPLC column.

QUALITY CONTROL Demonstrate, using a method blank, that all glassware and reagents are interferent-free at the MDL of the matrix of interest. A "method blank" must be run with each 20 or fewer samples. The method blank is also dosed with the internal standards. For water samples, 1 L of deionized and/or distilled water should be used as the method blank. Mineral oil may be used as the method blank for other matrices.

Calculate response factors for standards relative to the internal standards. Add a recovery standard to the samples prior to injection. The concentration of the recovery standard in the sample extract must be the same as that in the calibration standards used to measure the response factors.

Field duplicates (individual samples taken from the same location at the same time) should be analyzed periodically to determine the total precision (field and lab). Where appropriate, field blanks should be provided to monitor for possible cross-contamination of samples in the field. GC column performance must be demonstrated initially and verified prior to analyzing any sample in a 12-hr period. The GC column performance check solution must be analyzed under the same chromatographic and mass spectrometric conditions used for other samples and standards.

Retention times of target analytes must be verified using reference standards. These values must correspond to the retention time windows established. While certain cleanup techniques are provided as part of this method, unique samples may require additional cleanup techniques to achieve the method detection limit.

REFERENCE Test Methods for Evaluating Solid Waste (SW-846). U.S.E.P.A., 1986. Method 8280, Rev. 0, Sept. 1986. Office of Solid Wastes, Washington, DC.

1,2,3,7,8-PeCDD   EPA Method 8280
CAS #40321-76-4

TITLE The Analysis of Polychlorinated Dibenzo-P-Dioxins and Polychlorinated Dibenzofurans.

MATRIX This method is appropriate for the determination of tetra-, penta-, hexa-, hepta-, and octachlorinated dibenzo-p-dioxins (PCDDs) and dibenzofurans (PCDFs) in chemical wastes including still bottoms, fuel oils, sludges, fly ash, reactor residues, soil and water.

METHOD SUMMARY This method covers 22 PCDD and PCDF compounds and it uses a high resolution capillary GC column with low resolution mass spectrometry. Samples are extracted and concentrated by several methods that vary depending on the matrix involved. The organic extracts are cleaned-up by washing with aqueous basic and acid solutions and then separated into fractions using a column of neutral alumina. The fraction containing the PCDDs and PCDFs is then further cleaned-up using a column of activated carbon. The final extract is concentrated and Carbon-13 labeled internal standards are added prior to analysis by GC/MS using a capillary GC column and selected ion monitoring using five sets of ions that are detailed in the method. Certain 2,3,7,8-substituted congeners are used to provide calibration and method recovery information. Proper column selection and access to reference isomer standards, may in certain cases, provide isomer specific data.

INTERFERENCES Solvents, reagents, glassware, and other sample processing hardware may yield discrete artifacts and/or elevated baselines which may cause misinterpretation of chromatographic data. Use high purity reagents and solvents to minimize interference problems. Interferents coextracted from the sample will vary considerably from source to source. PCDDs and PCDFs are often associated with other interfering chlorinated compounds such as PCBs and polychlorinated diphenyl ethers which may be found at concentrations several orders of magnitude higher than that of the analytes of interest.

INSTRUMENTATION A low resolution GC/MS utilizing 70 ev must be capable of selected ion monitoring (SIM) for at least 11 ions simultaneously, with a cycle time of 1 second or less. Minimum integration time for SIM is 50 ms per m/z. Also required is a GC-to-MS interface constructed of all glass or glass-lined materials. One of the following GC columns is required:

Column 1: 50 m CP-Sil-88 fused silica capillary column.
Column 2: DB-5 (30 m × 0.25 mm I.D., 0.25-um film thickness) fused silica capillary column.
Column 3: 30 m SP-2250 fused silica capillary column.

When toluene is employed as the final solvent, use of a bonded phase column is recommended. Solvent exchange into tridecane is required for other liquid phases or nonbonded columns such as CP-Sil-88. Chromatographic conditions must be adjusted to account for solvent boiling points.

PRECISION & ACCURACY Accuracy, precision, MDLs and concentration ranges for the compounds covered by this method have not been determined or published by EPA yet. The sensitivity of this method is dependent upon the level of interferents within a given matrix. Proposed quantification levels for target analytes were 2 ppb in soil samples, up to 10 ppb in other solid wastes and 10 ppt in water. Actual values have been shown to vary by homologous series and, to a lesser degree, by individual isomer.

SAMPLING METHOD Grab and composite samples must be collected in 1 L or 1-quart amber glass bottles with Teflon®-lined screw-caps that have been acid-washed and solvent rinsed before use. If compositing equipment is used, the system must incorporate glass sample containers for the collection of a minimum of 250 mL. samples.

SAMPLE PRESERVATION Samples must be cooled and stored at 4°C.

MHT Samples must be extracted within 30 days and analyzed within 45 days of collection.

SAMPLE PREPARATION
Soil Samples — Extract 20 g of a 1:1 soil and anhydrous sodium sulfate mixture with 1:4 methanol-petroleum ether for 2 h in a wrist-action shaker. Concentrate the extract in a K-D and then exchange the solvent with hexane in the K-D. The final volume is 15 mL of hexane.

Aqueous samples — Extract a 1 L sample with methylene chloride, dry it with anhydrous sodium sulfate and concentrate it in a K-D followed by exchange with hexane in the K-D. A continuous liquid-liquid extractor may be used in place of a separatory funnel to avoid emulsions. The final volume is 15 mL of hexane.

Alumina column clean-up — The clean-up procedure described below consists of two phases. The first phase involves a sequential basic and acid washing of the extract that contains the analytes.

Wash the 15 mL hexane extract with 20% potassium hydroxide in a separatory funnel. Repeat the washing until no color is visible in the bottom layer but no more than four times because strong base is known to degrade certain PCDDs/PCDFs, so contact time must be minimized. Next, partition the 15 mL hexane against 40 mL of 5% sodium chloride. Next, partition the 15 mL hexane against 40 mL of concentrated sulfuric acid. Repeat the acid washings until no color is visible in the acid layer (but no more than four times). Finally, partition the 15 mL hexane against 40 mL of 5% sodium chloride. Dry the organic layer with anhydrous sodium sulfate and concentrate it to near dryness with a rotary evaporator. Dissolve the hexane residue from the first phase of the clean-up in 2 mL of hexane and apply it carefully to the top of a pre-eluted Woelm super 1 neutral alumina column. Elute the column with 10 mL of 8 percent (v/v) methylene chloride in hexane. Check by GC/MS analysis that no PCDDs of PCDFs are elute in this fraction before discarding it. Elute the PCDDs and PCDFs from the column with 15 mL of 60 percent (v/v) methylene chloride in hexane and collect this second fraction in a conical shaped concentrator tube.

Carbon column clean-up — Using a carefully regulated stream of nitrogen, concentrate the first 8 percent fraction (methylene chloride in hexane) from the alumina column to about 1 mL. Save this 8 percent concentrate for GC/MS analysis to check for breakthrough of PCDDs and PCDFs. Concentrate the second 60 percent fraction (methylene chloride in hexane) to about 2 to 3 mL. Prepare a carbon column and rinse the carbon with 5 mL cyclohexane/methylene chloride (50:50 v/v) in the forward direction of flow and then in the reverse direction of flow. While still in the reverse direction of flow, transfer the sample concentrate to the column and elute with 10 mL of methylene chloride/methanol/benzene (75:20:5, v/v). Save all above eluates and combine them (this fraction may be used as a check on column efficiency). Next, turn the column over and, in the direction of forward flow, elute the PCDD/PCDF fraction with 20 mL toluene. Evaporate the toluene fraction to about 1 mL on a rotary evaporator and transfer this concentrate to a 2.0-mL Reacti-vial. Concentrate the sample using a stream of nitrogen gas. The final volume will depend on the relative concentration of target analytes but it is typically 100 µL for soil samples and 500 µL for sludge, still bottom, and fly ash samples. Extracts which are determined to be outside the calibration range for individual analytes must be diluted or a smaller portion of the sample must be re-extracted.

An alternate carbon column clean-up also may be used with a 1 mL HPLC injector loop. The injector loop is connected to the optional HPLC column.

QUALITY CONTROL Demonstrate, using a method blank, that all glassware and reagents are interferent-free at the MDL of the matrix of interest. A "method blank" must be run with each 20 or fewer samples. The method blank is also dosed with the internal standards. For water samples, 1 L of deionized and/or distilled water should be used as the method blank. Mineral oil may be used as the method blank for other matrices.

Calculate response factors for standards relative to the internal standards. Add a recovery standard to the samples prior to injection. The concentration of the recovery standard in the sample extract must be the same as that in the calibration standards used to measure the response factors.

Field duplicates (individual samples taken from the same location at the same time) should be analyzed periodically to determine the total precision (field and lab). Where appropriate, field blanks should be provided to monitor for possible cross-contamination of samples in the field. GC column performance must be demonstrated initially and verified prior to analyzing any sample in a 12-hr period. The GC column performance check solution must be analyzed under the same chromatographic and mass spectrometric conditions used for other samples and standards.

Retention times of target analytes must be verified using reference standards. These values must correspond to the retention time windows established. While certain cleanup techniques are provided as part of this method, unique samples may require additional cleanup techniques to achieve the method detection limit.

REFERENCE Test Methods for Evaluating Solid Waste (SW-846). U.S.E.P.A., 1986. Method 8280, Rev. 0, Sept. 1986. Office of Solid Wastes, Washington, DC.

1,2,3,7,8-PeCDD   EPA Method 8290
CAS #40321-76-4

TITLE Polychlorinated Dibenzodioxins (PCDDs) and Polychlorinated Dibenzofurans (PCDFs) by High-Resolution Gas Chromatography/High-Resolution Mass Spectrometry (HRGC/HRMS).

MATRIX This method is applicable with a variety of environmental matrices including: water, soil, sediment, paper pulp, fly ash, fish tissue, human adipose tissue, sludges, fuel oil, chemical reactor residue, and still bottoms.

METHOD SUMMARY This method provides procedures for the detection and quantitative measurement of polychlorinated dibenzo-p-dioxins (tetra- through octachlorinated homologues; PCDDs), and polychlorinated dibenzofurans (tetra- through octachlorinated homologues; PCDFs) in a variety of environmental matrices and at part-per-trillion (ppt) to part-per-quadrillion (ppq) concentrations. High-resolution gas chromatography and high-resolution mass spectrometry (HRGC/HRMS) on purified sample extracts provides highly specific identification of each analyte. Quantification is provided using calibration standards.

INTERFERENCES Solvents, reagents, glassware, and other sample processing hardware may yield discrete artifacts that may cause misinterpretation of the chromatographic data. Analysts should avoid using PVC gloves. Interferants coextracted from the sample will vary considerably from matrix to matrix. PCDDs and PCDFs are often associated with other interfering chlorinated substances such as polychlorinated biphenyls (PCBs), polychlorinated diphenyl ethers (PCDEs), polychlorinated naphthalenes, and polychlorinated alkyldibenzofurans that may be found at concentrations several orders of magnitude higher than the PCDDs or PCDFs.

A high-resolution capillary column is used in this method. However, no single column is known to resolve all isomers. The 60 m DB-5 GC column is capable of 2,3,7,8-TCDD isomer specificity. In order to determine the concentration of the 2,3,7,8-TCDD (if detected on the DB-5 column), the sample extract must be reanalyzed on a column capable of 2,3,7,8-TCDF isomer specificity (e.g., DB-225, SP-2330, SP-2331, or equivalent).

INSTRUMENTATION High-Resolution Gas Chromatograph/High-Resolution Mass Spectrometer/Data System (HRGC/HRMS/DS) equipped with a GC injection port designed so that the separation of 2,3,7,8-TCDD from the other

TCDD isomers achieved in the gas chromatographic column is not appreciably degraded.

Column 1: 60 m DB-5 fused silica capillary column.
Column 2: 30 m DB-225 fused silica capillary column, or equivalent.

PRECISION & ACCURACY Precision, bias and concentration ranges for the compounds covered by this method have not been determined yet. The sensitivity of Method 8290 is dependent upon the level of interferences within a given matrix. Samples containing concentrations of specific congeneric analytes of PCDDs and PCDFs that are greater than ten times the upper method calibration limits must be analyzed by a protocol designed for such concentration levels, e.g., EPA Method 8280.

SAMPLE PREPARATION

Sludge/wet fuel oil — Extract aqueous sludge or wet fuel oil samples by refluxing a sample with toluene using a Dean-Stark water separator until all the water is removed. Filter the toluene extract through a glass fiber filter, or equivalent, and concentrate it to near dryness either on a rotary evaporator using an inert gas. Transfer the concentrate to a separatory funnel using hexane and wash it with 5% sodium chloride solution. Proceed to clean up.

Soil/sediment — If the sample is wet, add anhydrous powdered sodium sulfate to it until a free flowing mixture is obtained. Place the soil/sodium sulfate mixture in the Soxhlet apparatus, add toluene, and reflux for 16 h. The solvent must cycle completely through the system five times per h. Cool and filter the extract through a glass fiber filter and concentrate to near dryness on a rotary evaporator. Transfer the residue to a separatory funnel, using hexane. Proceed to clean up.

Aqueous samples — Use a 1-L sample; the method may require acetone to be added to it. When the sample is judged to contain 1% or more solids, it must be filtered through a glass fiber filter that has been rinsed with toluene. If the suspended solids content is too great to filter, centrifuge the sample, decant, and then filter the aqueous phase. Combine the solids from the centrifuge bottle(s) with the particulates on the filter and with the filter itself and proceed with Soxhlet extraction for soil/sediment. Extract the aqueous filtrate with methylene chloride in a separatory funnel, filter the extract through anhydrous sodium sulfate, and concentrate it using a K-D apparatus or a rotary evaporator. Exchange the solvent with hexane and proceed to clean up.

Clean up — The sample extract is cleaned up utilizing a number of different techniques. Partition cleanup is where the sample extract is partitioned with concentrated sulfuric acid, 5% aqueous sodium chloride, and 20% aqueous potassium hydroxide. Silica/alumina column cleanup involves packing gravity columns with silica gel and alumina and sequentially eluting the residue from the partition cleanup. Carbon column cleanup involves packing a column with a mixture of AX–21 and Celite 545 and sequentially eluting the sample concentrate from the silica/alumina cleanup with hexane, cyclohexane/methylene chloride (50:50), and methylene chloride/methanol/toluene (75:20:5). Then the column is turned upside down and the PCDD/PCDF fraction is eluted with toluene. The toluene fraction is concentrated and stored in the dark at room temperature until analysis.

QUALITY CONTROL Demonstrate, through the analysis of a reagent water blank, that interferences from the analytical system, glassware, and reagents are under control. For each analytical batch (up to 20 samples), a reagent blank, matrix spike, and matrix spike duplicate/duplicate must be analyzed (the frequency of the spikes may be different for different monitoring programs). The blank and spiked samples must be carried through all stages of the sample preparation and measurement steps.

A GC column performance check is required at the beginning of each 12-h period during which samples are analyzed. An HRGC/HRMS method blank run is required between a calibration run and the first sample run. The same method blank extract may thus be analyzed more than once if the number of samples within a batch requires more than 12 h of analyses.

At the beginning of each 12-h period during which samples are to be analyzed, an aliquot of the 1) GC column performance check solution and 2) a high-resolution concentration calibration must be analyzed to demonstrate adequate GC resolution and sensitivity, response factor reproducibility, and mass range calibration, and to establish the PCDD/PCDF retention time windows. A mass resolution check must also be performed to demonstrate adequate mass resolution using an appropriate reference compound (perfluorokerosene (PFK) is recommended). If the required criteria are not met, remedial action must be taken before any samples are analyzed.

Routine or continuing calibration (using a high resolution calibration solution) and the mass resolution check must also be performed at the end of each 12 h period. Furthermore, a HRGC/HRMS method blank analysis must be recorded following a calibration analysis and the first sample analysis.

To evaluate the performance of the analytical method, the QC check samples must be handled in exactly the same manner as actual samples. Therefore, 1.0 mL of the QC check sample concentrate is spiked into each of four 1 L aliquots of reagent water (which becomes the QC check sample), extracted, and then analyzed by GC. The variety of semivolatile analytes which may be analyzed by GC is such that the concentration of the QC check sample concentrate is different for the different analytical techniques presented in the full method.

The analyst must demonstrate also that the compounds of interest are being quantitatively recovered by the cleanup technique before the cleanup is applied to actual samples. For sample extracts that are cleaned up, the associated quality control samples (e.g., spikes, blanks, and duplicates) must also be processed through the same cleanup procedure. The analysis using each determinative method (GC, GC/MS, HPLC) specifies instrument calibration procedures using stock standards. It is recommended that cleanup also be performed on a series of the same type of standards to validate chromatographic elution patterns for the compounds of interest and to verify the absence of interferences from reagents.

REFERENCE Test Methods for Evaluating Solid Waste (SW-846). U.S. EPA. 1983. Method 8290, Rev. 0, Nov. 1990. Office of Solid Wastes, Washington, DC.

1,2,3,7,8-PeCDF   **EPA Method 8280**
CAS #57117-41-6

TITLE The Analysis of Polychlorinated Dibenzo-P-Dioxins and Polychlorinated Dibenzofurans.

MATRIX This method is appropriate for the determination of tetra-, penta-, hexa-, hepta-, and octachlorinated dibenzo-p-dioxins (PCDDs) and dibenzofurans (PCDFs) in chemical wastes including still bottoms, fuel oils, sludges, fly ash, reactor residues, soil and water.

METHOD SUMMARY This method covers 22 PCDD and PCDF compounds and it uses a high resolution capillary GC column with low resolution mass spectrometry. Samples are extracted and concentrated by several methods that vary depending on the matrix involved. The organic extracts are cleaned-up by washing with aqueous basic and acid solutions and then separated into fractions using a column of neutral alumina. The fraction containing the PCDDs and PCDFs is then further cleaned-up using a column of activated carbon. The final extract is concentrated and Carbon-13 labeled internal standards are added prior to analysis by GC/MS using a capillary GC column and selected ion monitoring using five sets of ions that are detailed in the method. Certain 2,3,7,8-substituted congeners are used to provide calibration and method recovery information. Proper column selection and access to reference isomer standards, may in certain cases, provide isomer specific data.

INTERFERENCES Solvents, reagents, glassware, and other sample processing hardware may yield discrete artifacts and/or elevated baselines which may cause misinterpretation of chromatographic data. Use high purity reagents and solvents to minimize interference problems. Interferents coextracted from the sample will vary considerably from source to source. PCDDs and PCDFs are often associated with other interfering chlorinated compounds such as PCBs and polychlorinated diphenyl ethers which may be found at concentrations several orders of magnitude higher than that of the analytes of interest.

INSTRUMENTATION A low resolution GC/MS utilizing 70 ev must be capable of selected ion monitoring (SIM) for at least 11 ions simultaneously, with a cycle time of 1 second or less. Minimum integration time for SIM is 50 ms per m/z. Also required is a GC-to-MS interface constructed of all glass or glass-lined materials. One of the following GC columns is required:

Column 1: 50 m CP-Sil-88 fused silica capillary column.
Column 2: DB-5 (30 m × 0.25 mm I.D., 0.25-um film thickness) fused silica capillary column.
Column 3: 30 m SP-2250 fused silica capillary column.

When toluene is employed as the final solvent, use of a bonded phase column is recommended. Solvent exchange into tridecane is required for other liquid phases or nonbonded columns such as CP-Sil-88. Chromatographic conditions must be adjusted to account for solvent boiling points.

PRECISION & ACCURACY Accuracy, precision, MDLs and concentration ranges for the compounds covered by this method have not been determined or published by EPA yet. The sensitivity of this method is dependent upon the level of interferents within a given matrix. Proposed quantification levels for target analytes were 2 ppb in soil samples, up to 10 ppb in other solid wastes and 10 ppt in water. Actual values have been shown to vary by homologous series and, to a lesser degree, by individual isomer.

SAMPLING METHOD Grab and composite samples must be collected in 1 L or 1-quart amber glass bottles with Teflon®-lined screw-caps that have been acid-washed and solvent rinsed before use. If compositing equipment is used, the system must incorporate glass sample containers for the collection of a minimum of 250 mL. samples.

SAMPLE PRESERVATION Samples must be cooled and stored at 4°C.

MHT Samples must be extracted within 30 days and analyzed within 45 days of collection.

SAMPLE PREPARATION
Soil Samples — Extract 20 g of a 1:1 soil and anhydrous sodium sulfate mixture with 1:4 methanol-petroleum ether for 2 h in a wrist-action shaker. Concentrate the extract in a K-D and then exchange the solvent with hexane in the K-D. The final volume is 15 mL of hexane.

Aqueous samples — Extract a 1 L sample with methylene chloride, dry it with anhydrous sodium sulfate and concentrate it in a K-D followed by exchange with hexane in the K-D. A continuous liquid-liquid extractor may be used in place of a separatory funnel to avoid emulsions. The final volume is 15 mL of hexane.

Alumina column clean-up — The clean-up procedure described below consists of two phases. The first phase involves a sequential basic and acid washing of the extract that contains the analytes.

Wash the 15 mL hexane extract with 20% potassium hydroxide in a separatory funnel. Repeat the washing until no color is visible in the bottom layer but no more than four times because strong base is known to degrade certain PCDDs/PCDFs, so contact time must be minimized. Next, partition the 15 mL hexane against 40 mL of 5% sodium chloride. Next, partition the 15 mL hexane against 40 mL of concentrated sulfuric acid. Repeat the acid washings until no color is visible in the acid layer (but no more than four times). Finally, partition the 15 mL hexane against 40 mL of 5% sodium chloride. Dry the organic layer with anhydrous sodium sulfate and concentrate it to near dryness with a rotary evaporator. Dissolve the hexane residue from the first phase of the clean-up in 2 mL of hexane and apply it carefully to the top of a pre-eluted Woelm super 1 neutral alumina column. Elute the column with 10 mL of 8 percent (v/v) methylene chloride in hexane. Check by GC/MS analysis that no PCDDs of PCDFs are elute in this fraction before discarding it. Elute the PCDDs and PCDFs from the

column with 15 mL of 60 percent (v/v) methylene chloride in hexane and collect this second fraction in a conical shaped concentrator tube.

Carbon column clean-up — Using a carefully regulated stream of nitrogen, concentrate the first 8 percent fraction (methylene chloride in hexane) from the alumina column to about 1 mL. Save this 8 percent concentrate for GC/MS analysis to check for breakthrough of PCDDs and PCDFs. Concentrate the second 60 percent fraction (methylene chloride in hexane) to about 2 to 3 mL. Prepare a carbon column and rinse the carbon with 5 mL cyclohexane/methylene chloride (50:50 v/v) in the forward direction of flow and then in the reverse direction of flow. While still in the reverse direction of flow, transfer the sample concentrate to the column and elute with 10 mL of methylene chloride/methanol/benzene (75:20:5, v/v). Save all above eluates and combine them (this fraction may be used as a check on column efficiency). Next, turn the column over and, in the direction of forward flow, elute the PCDD/PCDF fraction with 20 mL toluene. Evaporate the toluene fraction to about 1 mL on a rotary evaporator and transfer this concentrate to a 2.0-mL Reacti-vial. Concentrate the sample using a stream of nitrogen gas. The final volume will depend on the relative concentration of target analytes but it is typically 100 µL for soil samples and 500 µL for sludge, still bottom, and fly ash samples. Extracts which are determined to be outside the calibration range for individual analytes must be diluted or a smaller portion of the sample must be re-extracted.

An alternate carbon column clean-up also may be used with a 1 mL HPLC injector loop. The injector loop is connected to the optional HPLC column.

QUALITY CONTROL Demonstrate, using a method blank, that all glassware and reagents are interferent-free at the MDL of the matrix of interest. A "method blank" must be run with each 20 or fewer samples. The method blank is also dosed with the internal standards. For water samples, 1 L of deionized and/or distilled water should be used as the method blank. Mineral oil may be used as the method blank for other matrices.

Calculate response factors for standards relative to the internal standards. Add a recovery standard to the samples prior to injection. The concentration of the recovery standard in the sample extract must be the same as that in the calibration standards used to measure the response factors.

Field duplicates (individual samples taken from the same location at the same time) should be analyzed periodically to determine the total precision (field and lab). Where appropriate, field blanks should be provided to monitor for possible cross-contamination of samples in the field. GC column performance must be demonstrated initially and verified prior to analyzing any sample in a 12-hr period. The GC column performance check solution must be analyzed under the same chromatographic and mass spectrometric conditions used for other samples and standards.

Retention times of target analytes must be verified using reference standards. These values must correspond to the retention time windows established. While certain cleanup techniques are provided as part of this method, unique samples may require additional cleanup techniques to achieve the method detection limit.

REFERENCE Test Methods for Evaluating Solid Waste (SW-846). U.S.E.P.A., 1986. Method 8280, Rev. 0, Sept. 1986. Office of Solid Wastes, Washington, DC.

1,2,3,7,8-PeCDF   EPA Method 8290
CAS #57117-41-6

TITLE Polychlorinated Dibenzodioxins (PCDDs) and Polychlorinated Dibenzofurans (PCDFs) by High-Resolution Gas Chromatography/High-Resolution Mass Spectrometry (HRGC/HRMS).

MATRIX This method is applicable with a variety of environmental matrices including: water, soil, sediment, paper pulp, fly ash, fish tissue, human adipose tissue, sludges, fuel oil, chemical reactor residue, and still bottoms.

METHOD SUMMARY This method provides procedures for the detection and quantitative measurement of polychlorinated dibenzo-p-dioxins (tetra- through octachlorinated homologues; PCDDs), and polychlorinated dibenzofurans (tetra- through octachlorinated homologues; PCDFs) in a variety of environmental matrices and at part-per-trillion (ppt) to part-per-quadrillion (ppq) concentrations. High-resolution gas chromatography and high-resolution mass spectrometry (HRGC/HRMS) on purified sample extracts provides highly specific identification of each analyte. Quantification is provided using calibration standards.

INTERFERENCES Solvents, reagents, glassware, and other sample processing hardware may yield discrete artifacts that may cause misinterpretation of the chromatographic data. Analysts should avoid using PVC gloves. Interferants coextracted from the sample will vary considerably from matrix to matrix. PCDDs and PCDFs are often associated with other interfering chlorinated substances such as polychlorinated biphenyls (PCBs), polychlorinated diphenyl ethers (PCDEs), polychlorinated naphthalenes, and polychlorinated alkyldibenzofurans that may be found at concentrations several orders of magnitude higher than the PCDDs or PCDFs.

A high-resolution capillary column is used in this method. However, no single column is known to resolve all isomers. The 60 m DB-5 GC column is capable of 2,3,7,8-TCDD isomer specificity. In order to determine the concentration of the 2,3,7,8-TCDD (if detected on the DB-5 column), the sample extract must be reanalyzed on a column capable of 2,3,7,8-TCDF isomer specificity (e.g., DB-225, SP-2330, SP-2331, or equivalent).

INSTRUMENTATION High-Resolution Gas Chromatograph/High-Resolution Mass Spectrometer/Data System (HRGC/HRMS/DS) equipped with a GC injection port designed so that the separation of 2,3,7,8-TCDD from the other TCDD isomers achieved in the gas chromatographic column is not appreciably degraded.

Column 1: 60 m DB-5 fused silica capillary column.

Column 2: 30 m DB-225 fused silica capillary column, or equivalent.

PRECISION & ACCURACY Precision, bias and concentration ranges for the compounds covered by this method have not been determined yet. The sensitivity of Method 8290 is dependent upon the level of interferences within a given matrix. Samples containing concentrations of specific congeneric analytes of PCDDs and PCDFs that are greater than ten times the upper method calibration limits must be analyzed by a protocol designed for such concentration levels, e.g., EPA Method 8280.

SAMPLE PREPARATION
Sludge/wet fuel oil — Extract aqueous sludge or wet fuel oil samples by refluxing a sample with toluene using a Dean-Stark water separator until all the water is removed. Filter the toluene extract through a glass fiber filter, or equivalent, and concentrate it to near dryness either on a rotary evaporator using an inert gas. Transfer the concentrate to a separatory funnel using hexane and wash it with 5% sodium chloride solution. Proceed to clean up.

Soil/sediment — If the sample is wet, add anhydrous powdered sodium sulfate to it until a free flowing mixture is obtained. Place the soil/sodium sulfate mixture in the Soxhlet apparatus, add toluene, and reflux for 16 h. The solvent must cycle completely through the system five times per h. Cool and filter the extract through a glass fiber filter and concentrate to near dryness on a rotary evaporator. Transfer the residue to a separatory funnel, using hexane. Proceed to clean up.

Aqueous samples — Use a 1-L sample; the method may require acetone to be added to it. When the sample is judged to contain 1% or more solids, it must be filtered through a glass fiber filter that has been rinsed with toluene. If the suspended solids content is too great to filter, centrifuge the sample, decant, and then filter the aqueous phase. Combine the solids from the centrifuge bottle(s) with the particulates on the filter and with the filter itself and proceed with Soxhlet extraction for soil/sediment. Extract the aqueous filtrate with methylene chloride in a separatory funnel, filter the extract through anhydrous sodium sulfate, and concentrate it using a K-D apparatus or a rotary evaporator. Exchange the solvent with hexane and proceed to clean up.

Clean up — The sample extract is cleaned up utilizing a number of different techniques. Partition cleanup is where the sample extract is partitioned with concentrated sulfuric acid, 5% aqueous sodium chloride, and 20% aqueous potassium hydroxide. Silica/alumina column cleanup involves packing gravity columns with silica gel and alumina and sequentially eluting the residue from the partition cleanup. Carbon column cleanup involves packing a column with a mixture of AX–21 and Celite 545 and sequentially eluting the sample concentrate from the silica/alumina cleanup with hexane, cyclohexane/methylene chloride (50:50), and methylene chloride/methanol/toluene (75:20:5). Then the column is turned upside down and the PCDD/PCDF fraction is eluted with toluene. The toluene fraction is concentrated and stored in the dark at room temperature until analysis.

QUALITY CONTROL Demonstrate, through the analysis of a reagent water blank, that interferences from the analytical system, glassware, and reagents are under control. For each analytical batch (up to 20 samples), a reagent blank, matrix spike, and matrix spike duplicate/duplicate must be analyzed (the frequency of the spikes may be different for different monitoring programs). The blank and spiked samples must be carried through all stages of the sample preparation and measurement steps.

A GC column performance check is required at the beginning of each 12-h period during which samples are analyzed. An HRGC/HRMS method blank run is required between a calibration run and the first sample run. The same method blank extract may thus be analyzed more than once if the number of samples within a batch requires more than 12 h of analyses.

At the beginning of each 12-h period during which samples are to be analyzed, an aliquot of the 1) GC column performance check solution and 2) a high-resolution concentration calibration must be analyzed to demonstrate adequate GC resolution and sensitivity, response factor reproducibility, and mass range calibration, and to establish the PCDD/PCDF retention time windows. A mass resolution check must also be performed to demonstrate adequate mass resolution using an appropriate reference compound (perfluorokerosene (PFK) is recommended). If the required criteria are not met, remedial action must be taken before any samples are analyzed.

Routine or continuing calibration (using a high resolution calibration solution) and the mass resolution check must also be performed at the end of each 12 h period. Furthermore, a HRGC/HRMS method blank analysis must be recorded following a calibration analysis and the first sample analysis.

To evaluate the performance of the analytical method, the QC check samples must be handled in exactly the same manner as actual samples. Therefore, 1.0 mL of the QC check sample concentrate is spiked into each of four 1 L aliquots of reagent water (which becomes the QC check sample), extracted, and then analyzed by GC. The variety of semivolatile analytes which may be analyzed by GC is such that the concentration of the QC check sample concentrate is different for the different analytical techniques presented in the full method.

The analyst must demonstrate also that the compounds of interest are being quantitatively recovered by the cleanup technique before the cleanup is applied to actual samples. For sample extracts that are cleaned up, the associated quality control samples (e.g., spikes, blanks, and duplicates) must also be processed through the same cleanup procedure. The analysis using each determinative method (GC, GC/MS, HPLC) specifies instrument calibration procedures using stock standards. It is recommended that cleanup also be performed on a series of the same type of standards to validate chromatographic elution patterns for the compounds of interest and to verify the absence of interferences from reagents.

REFERENCE Test Methods for Evaluating Solid Waste (SW-846). U.S. EPA. 1983. Method 8290, Rev. 0, Nov. 1990. Office of Solid Wastes, Washington, DC.

2,3,4,7,8-PeCDF
CAS #57117-31-4

EPA Method 8290

TITLE Polychlorinated Dibenzodioxins (PCDDs) and Polychlorinated Dibenzofurans (PCDFs) by High-Resolution Gas Chromatography/High-Resolution Mass Spectrometry (HRGC/HRMS).

MATRIX This method is applicable with a variety of environmental matrices including: water, soil, sediment, paper pulp, fly ash, fish tissue, human adipose tissue, sludges, fuel oil, chemical reactor residue, and still bottoms.

METHOD SUMMARY This method provides procedures for the detection and quantitative measurement of polychlorinated dibenzo-p-dioxins (tetra- through octachlorinated homologues; PCDDs), and polychlorinated dibenzofurans (tetra- through octachlorinated homologues; PCDFs) in a variety of environmental matrices and at part-per-trillion (ppt) to part-per-quadrillion (ppq) concentrations. High-resolution gas chromatography and high-resolution mass spectrometry (HRGC/HRMS) on purified sample extracts provides highly specific identification of each analyte. Quantification is provided using calibration standards.

INTERFERENCES Solvents, reagents, glassware, and other sample processing hardware may yield discrete artifacts that may cause misinterpretation of the chromatographic data. Analysts should avoid using PVC gloves. Interferants coextracted from the sample will vary considerably from matrix to matrix. PCDDs and PCDFs are often associated with other interfering chlorinated substances such as polychlorinated biphenyls (PCBs), polychlorinated diphenyl ethers (PCDEs), polychlorinated naphthalenes, and polychlorinated alkyldibenzofurans that may be found at concentrations several orders of magnitude higher than the PCDDs or PCDFs.

A high-resolution capillary column is used in this method. However, no single column is known to resolve all isomers. The 60 m DB-5 GC column is capable of 2,3,7,8-TCDD isomer specificity. In order to determine the concentration of the 2,3,7,8-TCDD (if detected on the DB-5 column), the sample extract must be reanalyzed on a column capable of 2,3,7,8-TCDF isomer specificity (e.g., DB-225, SP-2330, SP-2331, or equivalent).

INSTRUMENTATION High-Resolution Gas Chromatograph/High-Resolution Mass Spectrometer/Data System (HRGC/HRMS/DS) equipped with a GC injection port designed so that the separation of 2,3,7,8-TCDD from the other TCDD isomers achieved in the gas chromatographic column is not appreciably degraded.

Column 1: 60 m DB-5 fused silica capillary column.
Column 2: 30 m DB-225 fused silica capillary column, or equivalent.

PRECISION & ACCURACY Precision, bias and concentration ranges for the compounds covered by this method have not been determined yet. The sensitivity of Method 8290 is dependent upon the level of interferences within a given matrix. Samples containing concentrations of specific congeneric analytes of PCDDs and PCDFs that are greater than ten times the upper method calibration limits must be analyzed by a protocol designed for such concentration levels, e.g., EPA Method 8280.

SAMPLE PREPARATION
Sludge/wet fuel oil — Extract aqueous sludge or wet fuel oil samples by refluxing a sample with toluene using a Dean-Stark water separator until all the water is removed. Filter the toluene extract through a glass fiber filter, or equivalent, and concentrate it to near dryness either on a rotary evaporator using an inert gas. Transfer the concentrate to a separatory funnel using hexane and wash it with 5% sodium chloride solution. Proceed to clean up.

Soil/sediment — If the sample is wet, add anhydrous powdered sodium sulfate to it until a free flowing mixture is obtained. Place the soil/sodium sulfate mixture in the Soxhlet apparatus, add toluene, and reflux for 16 h. The solvent must cycle completely through the system five times per h. Cool and filter the extract through a glass fiber filter and concentrate to near dryness on a rotary evaporator. Transfer the residue to a separatory funnel, using hexane. Proceed to clean up.

Aqueous samples — Use a 1-L sample; the method may require acetone to be added to it. When the sample is judged to contain 1% or more solids, it must be filtered through a glass fiber filter that has been rinsed with toluene. If the suspended solids content is too great to filter, centrifuge the sample, decant, and then filter the aqueous phase. Combine the solids from the centrifuge bottle(s) with the particulates on the filter and with the filter itself and proceed with Soxhlet extraction for soil/sediment. Extract the aqueous filtrate with methylene chloride in a separatory funnel, filter the extract through anhydrous sodium sulfate, and concentrate it using a K-D apparatus or a rotary evaporator. Exchange the solvent with hexane and proceed to clean up.

Clean up — The sample extract is cleaned up utilizing a number of different techniques. Partition cleanup is where the sample extract is partitioned with concentrated sulfuric acid, 5% aqueous sodium chloride, and 20% aqueous potassium hydroxide. Silica/alumina column cleanup involves packing gravity columns with silica gel and alumina and sequentially eluting the residue from the partition cleanup. Carbon column cleanup involves packing a column with a mixture of AX-21 and Celite 545 and sequentially eluting the sample concentrate from the silica/alumina cleanup with hexane, cyclohexane/methylene chloride (50:50), and methylene chloride/methanol/toluene (75:20:5). Then the column is turned upside down and the PCDD/PCDF fraction is eluted with toluene. The toluene fraction is concentrated and stored in the dark at room temperature until analysis.

QUALITY CONTROL Demonstrate, through the analysis of a reagent water blank, that interferences from the analytical system, glassware, and reagents are under control. For each analytical batch (up to 20 samples), a reagent blank, matrix spike, and matrix spike duplicate/duplicate must be analyzed (the frequency of the spikes may be different for different monitoring programs). The blank and spiked samples must be carried through all stages of the sample preparation and measurement steps.

A GC column performance check is required at the beginning of each 12-h period during which samples are analyzed. An HRGC/HRMS method blank run is required between a calibration run and the first sample run. The same method blank extract may thus be analyzed more than once if the number of samples within a batch requires more than 12 h of analyses.

At the beginning of each 12-h period during which samples are to be analyzed, an aliquot of the 1) GC column performance check solution and 2) a high-resolution concentration calibration must be analyzed to demonstrate adequate GC resolution and sensitivity, response factor reproducibility, and mass range calibration, and to establish the PCDD/PCDF retention time windows. A mass resolution check must also be performed to demonstrate adequate mass resolution using an appropriate reference compound (perfluorokerosene (PFK) is recommended). If the required criteria are not met, remedial action must be taken before any samples are analyzed.

Routine or continuing calibration (using a high resolution calibration solution) and the mass resolution check must also be performed at the end of each 12 h period. Furthermore, a HRGC/HRMS method blank analysis must be recorded following a calibration analysis and the first sample analysis.

To evaluate the performance of the analytical method, the QC check samples must be handled in exactly the same manner as actual samples. Therefore, 1.0 mL of the QC check sample concentrate is spiked into each of four 1 L aliquots of reagent water (which becomes the QC check sample), extracted, and then analyzed by GC. The variety of semivolatile analytes which may be analyzed by GC is such that the concentration of the QC check sample concentrate is different for the different analytical techniques presented in the full method.

The analyst must demonstrate also that the compounds of interest are being quantitatively recovered by the cleanup technique before the cleanup is applied to actual samples. For sample extracts that are cleaned up, the associated quality control samples (e.g., spikes, blanks, and duplicates) must also be processed through the same cleanup procedure. The analysis using each determinative method (GC, GC/MS, HPLC) specifies instrument calibration procedures using stock standards. It is recommended that cleanup also be performed on a series of the same type of standards to validate chromatographic elution patterns for the compounds of interest and to verify the absence of interferences from reagents.

REFERENCE Test Methods for Evaluating Solid Waste (SW-846). U.S. EPA. 1983. Method 8290, Rev. 0, Nov. 1990. Office of Solid Wastes, Washington, DC.

1,2,3,7,8-PeCDF   **EPA Method 8280**
CAS #57117-41-6

TITLE Analysis of PCDDs and PCDFs

MATRIX Chemical wastes, fuel oils, still bottoms, sludges, water, soil, fly ash, reactor residues.

APPLICATION This method is used for the analysis of tetra-, penta-, hexa-, hepta-, and octachlorinated dibenzo-p-dioxins (PCDDs) and dibenzofurans (PCDFs). The sensitivity of the method is dependent on the level of interferents within the matrix. Only experienced analysts should be used.

Special safety precautions must be observed and an EPA-approved sample disposal plan must be used.

INTERFERENCES Solvents, reagents, and glassware may introduce artifacts. Other interferences may come from coextracted compounds from samples; PCBs and polychlorinated diphenyl ethers are common interferents.

INSTRUMENTATION GC/MS with a fused silica capillary column. Also, solvent extraction and concentration glassware and either a gravity flow activated carbon AX–21/silica gel Type 60 EM reagent column or a HPLC with a 10 mm by 7 cm silanized glass column with active carbon AX–21 and Spherisorb S10W silica for sample cleanup. One of three fused silica capillary GC columns may be used: column 1: 50 m CP-Sil-88; Column 2: 30 m by 0.25 mm DB-5; colunm 3: 30 m SP-2250.

RANGE 50–6,000 picograms

MDL 1.64 ng/L (in reagent water); 0.33 µg/kg (in Missouri soil); 0.16 µg/kg (in fly ash); 0.92 µg/kg (in industrial sludge); 1.61 µg/kg (in still bottom); 0.80 µg/kg (in fuel oil); MDLs are for carbon-13 labeled analyte.

PRECISION (as RSD): 6.1% with 5 ng/g in clay; 5.0% with 25 ng/g soil; 4.8% with 125 ng/g in sludge.

ACCURACY (as Mean % Recovery): 57.4% with 5 ng/g in clay; 64.4% with 25 ng/g in soil; 84.8% with 125 ng/g in sludge; 105.8% with 46 ng/g in fly ash.

SAMPLING METHOD Use 1 L (or quart) amber glass bottles with Teflon®-lined or solvent washed foil screw caps. Tape caps to bottle after sampling.

Compositing equipment must use glass containers and contain no Tygon or rubber tubing. Sample bottles must not be prewashed with the sample before its collection. Aqueous samples cannot be aliquoted from sample containers — the entire sample must be used and the container is washed out with the extracting solvent.

STABILITY Cool to 4°C and store at this temperature. When toluene is employed as the final solvent use a bonded phase GC column for separation. Otherwise, solvent exchange into tridecane is required for other liquid phases or the CP-Sil-88 GC column.

MHT 30 days; samples must be completely analyzed within 45 days of collection.

QUALITY CONTROL A method blank must be analyzed each time a set of samples is extracted or there is a change in reagents. A lab method blank must be run with each analytical batch of 20 or fewer samples. Field duplicates and field blanks should be analyzed periodically. GC column performance must be demonstrated initially and verified prior to analyzing any sample in a 12 h period. A series of calibration standards must be processed through the procedure to validate elution patterns and absence of interferents from reagents. Both the alumina column and carbon column performance must be routinely checked for presence of the analyte.

Performance evaluation samples and split samples with other laboratories are also expected to be periodically analyzed.

REFERENCE Method 8280, SW-846, 3rd ed., Nov.1986.

Pentachlorobenzene EPA Method 1625
CAS #608-93-5

TITLE Semivolatile Organic Compounds by Isotope Dilution GC/MS

MATRIX The compounds may be determined in waters, soils, and municipal sludges by this method.

METHOD SUMMARY This method is used to determine 176 semivolatile toxic organic pollutants associated with the CWA (as amended 1987); the RCRA (as amended 1986); the CERCLA (as amended 1986); and other compounds amenable to extraction and analysis by capillary column gas chromatography-mass spectrometry (GC/MS).

Stable isotopically-labeled analogs of the compounds of interest are added to the sample. If the solids content is less than 1%, a 1-L sample is extracted at pH 12–13, then at pH <2 with methylene chloride using continuous extraction techniques.

If the solids content is 30% or less, the sample is diluted to 1% solids with reagent water, homogenized ultrasonically, and extracted at pH 12–13, then at pH <2 with methylene chloride using continuous extraction techniques. If the solids content is greater than 30%, the sample is extracted using ultrasonic techniques.

Each extract is dried over sodium sulfate, concentrated to a volume of 5 mL, cleaned up using GPC, if necessary, and concentrated. Extracts are concentrated to 1 mL if GPC is not performed, and to 0.5 mL if GPC is performed.

An internal standard is added to the extract, and a 1-mL aliquot of the extract is injected into the GC. The compounds are separated by GC and detected by a MS. The labeled compounds serve to correct the variability of the analytical technique.

INTERFERENCES Solvents, reagents, glassware, and other sample processing hardware may yield artifacts and/or elevated baselines causing misinterpretation of chromatograms and spectra. Materials used in the analysis must be demonstrated to be free from interferences under the conditions of analysis by running method blanks initially and with each sample lot (sample started through the extraction process on a given 8-h shift, to a maximum of 20). Specific selection of reagents and purification of solvents by distillation in all glass systems may be required. Glassware and, where possible, reagents are cleaned by solvent rinse and baking at 450°C for 1-h minimum. Interferences coextracted from samples will vary considerably from source to source, depending on the diversity of the site being sampled.

INSTRUMENTATION Major instrumentation includes a GC with a splitless or on-column injection port for capillary column, a MS with 70 eV electron impact ionization, and a data system to collect and record MS data, and process it. A K-D apparatus is used to concentrate extracts.

GC Column: 30 m × 0.25 mm I.D. 5% phenyl, 94% methyl, 1% vinyl silicone bonded phased fused silica capillary column.

PRECISION & ACCURACY The detection limits of the method are usually dependent on the level of interferences rather than instrumental limitations. The limits typify the minimum quantities that can be detected with no interferences present.

The minimum level (in µg/mL) was not listed. This is defined as a minimum level at which the analytical system shall give recognizable mass spectra (background corrected) and acceptable calibration points.

The MDL (in µg/kg) in low solids was not listed and in high solids was not listed; these were determined in digested sludge (low solids) and in filter cake or compost (high solids).

The labeled and native compound initial precision as standard deviation (in µg/L) was not listed.
The labeled and native compound initial accuracy as average recovery (in µg/L) was not listed.

SAMPLE COLLECTION, PRESERVATION & HANDLING Collect samples in glass containers. Aqueous samples which flow freely are collected in refrigerated bottles using automatic sampling equipment. Solid samples are collected as grab samples using widemouth jars. Maintain samples at 0 to 4°C from the time of collection until extraction. If residual chlorine is present in aqueous samples, add 80 mg sodium thiosulfate/L of water. Begin sample extraction within 7 days of collection, and analyze all extracts within 40 days of extraction.

SAMPLE PREPARATION Samples containing 1% solids or less are extracted directly using continuous liquid-liquid extraction techniques. Samples containing 1 to 30% solids are diluted to the 1% level with reagent water and extracted using continuous liquid-liquid extraction techniques. Samples containing greater than 30% solids are extracted using ultrasonic techniques.

Base/neutral extraction — Adjust the pH of the waters in the extractors to 12–13 with 6 N NaOH. Extract with methylene chloride for 24–48 h.
Acid extraction — Adjust the pH of the waters in the extractors to 2 or less using 6 N sulfuric acid. Extract with methylene chloride for 24–48 h.
Ultrasonic extraction of high solids samples — Add anhydrous sodium sulfate to the sample and QC aliquot(s). Add acetone:methylene chloride (1:1) to the sample and mix thoroughly

Concentrate extracts using a K-D apparatus.

QUALITY CONTROL The analyst is permitted to modify this method to improve separations or lower the costs of measurements, provided all performance specifications are met. Analyses of blanks are required to demonstrate freedom from contamination. When results of spikes indicate atypical method performance for samples, the samples are diluted to bring method performance within acceptable limits.

For low solids (aqueous samples), extract, concentrate, and analyze two sets of four 1-L aliquots (8 aliquots total) of the precision and recovery standard. For high solids samples, two sets of four 30-g aliquots of the high solids reference matrix are used.

Spike all samples with labeled compounds to assess method performance. Compute percent recovery of the labeled compounds using the internal standard method. Compare the labeled compound recovery for each compound with the corresponding labeled compound recovery.

Reagent water and high solids reference matrix blanks are analyzed to demonstrate freedom from contamination. Extract and concentrate a 1-L reagent water blank or a high solids reference matrix blank with each sample's lot (samples started through the extraction process on the same 8-h shift, to a maximum of 20 samples).

Field replicates may be collected to determine the precision of the sampling technique, and spiked samples may be required to determine the accuracy of the analysis when the internal standard method is used.

REFERENCE Semivolatile Organic Compounds by Isotope Dilution GC/MS. Office of Water Regulation and Standards, U.S. EPA Industrial Technology Division, Washington, DC, EPA Method 1625, Rev. C, June 1989 (contact W.A. Telliard, U.S. EPA, Office of Water Regulations and Standards, 401 M St., SW, Washington, DC, 20460. Phone: 202-382-7131).

Pentachlorobenzene   **EPA Method 8121**
CAS #608-93-5

TITLE Chlorinated Hydrocarbons by GC: Capillary Column Technique

MATRIX This method covers aqueous and solid matrices. This includes a wide variety such as drinking water, groundwater, industrial wastewaters, surface waters, soils, solids, and sediments.

METHOD SUMMARY This method provides procedures for the determination of 22 chlorinated hydrocarbons in water, soil/sediment, and waste matrices. A measured volume or weight of sample is extracted by using one of the appropriate sample extraction techniques specified in EPA Method 3510, EPA Method 3520, EPA Method 3540, or EPA Method 3550, or diluted using EPA Method 3580. Aqueous samples are extracted at neutral pH with methylene chloride by using either a separatory funnel (EPA Method 3510) or a continuous liquid-liquid extractor (EPA Method 3520). Solid samples are extracted with hexane/acetone (1:1) by using a Soxhlet extractor (EPA Method 3540) or with methylene chloride/acetone (1:1) by using an ultrasonic extractor (EPA Method 3550). After cleanup, the extract or diluted sample is analyzed by gas chromatography with electron capture detection (GC/ECD).

The sensitivity level of this method usually depends on the level of interferences rather than on instrumental limitations. This method may be used in conjunction with EPA Method 3620, Florisil Column Cleanup, EPA Method 3660, Sulfur Cleanup, and EPA Method 3640, Gel Permeation Chromatography, to aid in the elimination of interferences.

INTERFERENCES Solvents, reagents, glassware, and other hardware used in sample processing may introduce artifacts which may result in elevated baselines, causing misinterpretation of gas chromatograms. Interferants coextracted from the samples will vary considerably from waste to waste. Glassware must be scrupulously clean. Phthalate esters, if present in a sample, will interfere only with the BHC isomers. The presence of elemental sulfur will result in large peaks, and can often mask the region of compounds eluting after 1,2,4,5-tetrachlorobenzene. The tetrabutylammonium (TBA)-sulfite procedure (EPA Method 3660) works well for the removal of elemental sulfur. Waxes and lipids can be removed by gel permeation chromatography (EPA Method 3640).

INSTRUMENTATION A GC suitable for on-column injections and all required accessories, including and electron capture detector (ECD), analytical columns, recorder, gases, and syringes are needed. A data system for measuring peak heights and/or peak areas is recommended. A Kuderna-Danish (K-D) apparatus will also be needed to concentrate extracts.

Column 1: 30 m × 0.53 mm I.D. fused-silica capillary column chemically bonded with trifluoropropyl methyl silicone (DB-210 or equivalent).
Column 2: 30 m × 0.53 mm I.D. fused-silica capillary column chemically bonded with polyethylene glycol (DB-WAX or equivalent).

PRECISION & ACCURACY This method has been tested in a single lab by using organic-free reagent water, sandy loam samples, and extracts which were spiked with the test compounds at one concentration. Single-operator precision and method accuracy were found to be related to the concentration of compound and the type of matrix. The accuracy and precision technique will be determined by the sample matrix, sample preparation technique, optional cleanup techniques, and calibration procedures used.

MULTIPLICATION FACTORS FOR OTHER MATRICES (a)

Matrix	Factor (b)
Groundwater	10
Low-concentration soil by ultrasonic extraction with GPC cleanup	670
High-concentration soil and sludges by ultrasonic extraction	10,000
Waste not miscible with water	100,000

(a) Sample EQLs are highly matrix-dependent. The EQLs listed herein are provided for guidance and may not always be achievable. (b) EQL = [Method detection limit] × [Factor]. For nonaqueous samples, the factor is on a wet-weight basis.

PRECISION & ACCURACY MDL is the method detection limit for organic-free reagent water. MDL was determined from the analysis of eight replicate aliquots processed through the entire analytical method (extraction, Florisil cartridge cleanup, and GC/ECD analysis).

The MDL (in ng/L) was 38.

The accuracy (as average % recovery using 5 determinations and no Florisil cleanup) from a spike concentration of 1.0 μg/L and separatory funnel extraction was 89% with a final volume of 10 mL.

The precision (as RSD% using 5 determinations and no Florisil cleanup) from a spike concentration of 1.0 μg/L and separatory funnel extraction was 6.5% with a final volume of 10 mL.

The accuracy (as average % recovery using 5 determinations and no Florisil cleanup), from a spike concentration of 330μg/L and ultrasonic extraction of solid samples using 1:1 methylene chloride and acetone, was 81% with a final volume of 10 mL.

The precision (as RSD% using 5 determinations and no Florisil cleanup), from a spike concentration of 330μg/L and ultrasonic extraction of solid samples using 1:1 methylene chloride and acetone, was 3.5% with a final volume of 10 mL.

SAMPLE COLLECTION, PRESERVATION & HANDLING
Volatile organics — Standard 40-mL glass screw-cap VOA vials with Teflon®-faced silicone septum may be used for both liquid and solid matrices. When collecting samples, liquids and solids should be introduced into the vials gently to reduce agitation which might drive off volatile compounds. If there are any air bubbles present the sample must be retaken. The vials with solids should be tapped slightly as they are filled to try and eliminate as much free air space as possible. Two vials from each sampling location should be sealed in separate plastic bags to prevent cross-contamination between samples.

Semivolatile organics — Containers used to collect samples for the determination of semivolatile organic compounds should be soap and water washed followed by methanol (or isopropanol) rinsing. The sample containers should be of glass or Teflon® and have screw-top covers with Teflon® liners.

Preservation for volatile organics — No preservation is used with concentrated waste samples. With liquid samples containing no residual chlorine, 4 drops of concentrated hydrochloric acid are added and the samples are immediately cooled to 4°C. When liquid samples contain residual chlorine, they are treated as above and, in addition, 4 drops of 4% aqueous sodium thiosulfate are added to remove the residual chlorine. Soil, sediment, and sludge samples are only cooled to 4°C.

Preservation for semivolatile organics — No preservation is used with concentrated waste samples. With liquid samples containing no residual chlorine and with soil, sediment, and sludge samples, immediately cooling to 4°C is the only preservation used. When residual chlorine is present then 3 mL of 10% aqueous sodium sulfate is added for each gallon of sample collected, followed by cooling to 4°C.

Holding times — The holding time for all volatile organics samples is 14 days. Liquid samples must be extracted within 7 days and their extracts analyzed within 40 days. Concentrated waste, soil, sediment, and sludge samples must be extracted within 14 days and their extracts analyzed within 40 days.

SAMPLE PREPARATION Prepare stock standard solutions in hexane. Calibration standards at a minimum of five concentrations should be prepared through dilution of the stock standards with hexane. The suggested internal standards are: 2,5-dibromotoluene, 1,3,5-tribromobenzene, and α, α-dibromo-m-xylene. The analyst can use any of the three compounds provided that they are resolved from matrix interferences. Recommended surrogate compounds are α-2,6-trichlorotoluene, 1,4-dichloronaphthalene, and 2,3,4,5,6-pentachlorotoluene.

In general, water samples are extracted at a neutral pH with methylene chloride using a separatory funnel (EPA Method 3510) or a continuous liquid-liquid extractor (EPA Method 3520). Solid samples are extracted with hexane/acetone (1:1 v:v) using a Soxhlet extractor (EPA Method 3540) or with methylene chloride/acetone (1:1 v:v) using an ultrasonic extractor (EPA Method 3550). Non-aqueous waste samples may be diluted using EPA Method 3580. Prior to Florisil cleanup or gas chromatographic analysis, the extraction solvent must be exchanged to hexane. Sample extracts that will be subjected to gel permeation chromatography do not need solvent exchange.

Cleanup procedures may not be necessary for a relatively clean matrix. If removal of interferences such as chlorinated phenols, phthalate esters, etc., is required, proceed with the procedure outlined in EPA Method 3620.

QUALITY CONTROL Analyze a quality control check standard to demonstrate that the operation of the GC is in control. The frequency of the check standard analysis is equivalent to 10% of the samples analyzed. If the recovery of any compound found in the check standard is less than 80% of the certified value, the problem must be corrected and a new set of calibration standards must be prepared and analyzed. Calculate surrogate standard recoveries for all samples, blanks, and spikes. An internal standard peak area check must be performed on all samples. The internal standard must be evaluated for acceptance by determining whether the measured area for the internal standard deviates by more than 30% from the average area for the internal standard in the calibration standards. When the internal standard peak area is outside that limit, all samples that fall outside the QC criteria must be reanalyzed. Any compound confirmed by two columns may also be confirmed by GC/MS (EPA Method 8270). The GC/MS would normally require a minimum concentration of 1 ng/μL in the final extract for each compound. Include a mid-concentration calibration standard after each group of 20 samples in the analysis sequence. The response factors for the mid-concentration calibration must be within 15% of the average values for the multiconcentration calibration.

REFERENCE Test Methods for Evaluating Solid Waste, Physical/Chemical Methods, SW-846, 3rd Edition, U.S. EPA, Office of Solid Waste, Washington, DC, 1990. EPA Method 8121, Rev. 0, Nov. 1990.

Pentachlorobenzene
CAS #608-93-5

EPA Method 8270

TITLE Semivolatile Organic Compounds by GC/MS

MATRIX This method is used to determine the concentration of semivolatile organic compounds in extracts prepared from all types of solid waste matrices, soils, and groundwater. Although surface waters are not specifically mentioned, this method should be applicable to water samples from rivers, lakes, etc.

METHOD SUMMARY This method covers 259 semivolatile organic compounds. In very limited applications direct injection of the sample into the GC/MS system may be appropriate, but this results in very high detection limits (approximately 10,000 µg/L). Typically, a 1-L liquid sample, containing surrogate, and matrix spiking standards, is extracted in a continuous extractor first under acid conditions and then under basic conditions. Typically 30 g of a solid sample, containing surrogate, and matrix spiking standards, is extracted ultrasonically. After concentrating the extract to 1 mL it is spiked with 10 µL of an internal standard solution just prior to analysis by GC/MS. The volume injected should contain about 100 ng of base/neutral and 200 ng of acid surrogates (for a 1-µL injection). Analysis is performed by GC/MS using a capillary GC column.

INTERFERENCES Raw GC/MS data from all blanks, samples, and spikes must be evaluated for interferences. Contamination by carryover can occur whenever high-concentration and low-concentration samples are sequentially analyzed. To reduce carryover, the sample syringe must be rinsed out between samples with solvent. Whenever an unusually concentrated sample is encountered, it should be followed by the analysis of blank solvent to check for cross-contamination.

INSTRUMENTATION A GC/MS and a data system are required. The GC column used is a 30 m × 0.25 mm I.D. (or 0.32 mm I.D.) 1um film thickness silicone-coated fused silica capillary column. A continuous liquid-liquid extractor equipped with Teflon® or glass connection joints and stopcocks requiring no lubrication, a K-D concentrating apparatus, water bath, and an ultrasonic disrupter with a minimum power of 300 W and with pulsing capability are also required.

PRECISION & ACCURACY The estimated quantitation limit (EQL) of Method 8270B for determining an individual compound is approximately 1 mg/kg (wet weight) for soil or sediment samples, 1–200 mg/kg for wastes (dependent on matrix and method of preparation), and 10 µg/L for groundwater samples. EQLs will be proportionately higher for sample extracts that require dilution to avoid saturation of the detector.

The EQL(b) for groundwater in µg/L is 10.
The EQL (a, b) for low concentrations in soil and sediment in µg/kg is not determined.
Accuracy as µg/L is not listed.
Overall precision in µg/L is not listed.

(a) *EQLs listed for soil/sediment are based on wet weight. Normally data is reported in a dry-weight basis; therefore, EQLs will be higher based on the % dry weight of each sample.*

This calculation is based on a 30 g sample and gel permeation chromatography cleanup.

(b) *Sample EQLs are highly matrix-dependent. The EQLs are provided for guidance and may not always be achievable.*

C = *True value for concentration, in µg/L.*
X = *Average recovery found for measurements of samples containing a concentration of C, in µg/L.*

ESTIMATED QUANTITATION LIMIT

Other Matrices	Factor (a)
High-concentration soil and sludges by sonicator	7.5
Non-water miscible waste	75

(a) *EQL for other matrices = [EQL for low soil/sediment] × [Factor]. This estimated EQL is similar to an EPA "Practical Quantitation Limit."*

SAMPLING METHOD

Liquid samples — Use a 1 or 2½ gallon amber glass bottle with a screw-top Teflon®-lined cover that has been prewashed with detergent and rinsed with distilled water and methanol (or isopropanol).

Soils, sediments, or sludges — Use an 8-oz. widemouth glass with a screw-top Teflon®-lined cover that has been prewashed with detergent and rinsed with distilled water and methanol (or isopropanol).

SAMPLE PRESERVATION

Liquid samples — If residual chlorine is present, add 3 mL of 10% sodium thiosulfate per gallon, cool to 4°C and store in a solvent-free refrigerator until analysis; if chlorine is not present, then eliminate the sodium thiosulfate addition.

Soils, sediments, or sludges — Cool samples to 4°C and store in a solvent-free refrigerator.

MHT Liquid samples must be extracted within 7 days and the extracts analyzed within 40 days. Soils, sediments, or sludges may be stored for a maximum of 14 days and the extracts analyzed within 40 days.

SAMPLE PREPARATION

Liquid samples — Transfer 1 L quantitatively to a continuous extractor. If high concentrations are anticipated, a smaller volume may be used and then diluted with organic-free reagent water to 1 L. Adjust pH, if necessary, to pH <2 using 1:1 (V/V) sulfuric acid. Pipette 1.0 mL of a surrogate standard spiking solution into each sample. For the sample in each analytical batch selected for spiking, add 1.0 mL of a matrix spiking standard. For base/neutral acid analysis, the amount of the surrogates and matrix spiking compounds added to the sample should result in a final concentration of 100 ng/µL of each analyte in the extract to be analyzed (assuming a 1-µL injection). Extract with methylene chloride for 18–24 h. Next, adjust the pH of the aqueous phase to pH >11 using 10 N sodium hydroxide and extract it with methylene chloride again for 18–24 h. Dry the extract through a column containing anhydrous sodium sulfate and concentrate it to 1 mL using a K-D concentrator.

Soils, sediments, or sludges — Use 30 g of sample. Nonporous or wet samples (gummy or clay type) that do not have a free-flowing

sandy texture must be mixed with anhydrous sodium sulfate until the sample is free flowing. Add 1 mL of surrogate standards to all samples, spikes, standards, and blanks. For the sample in each analytical batch selected for spiking, add 1.0 mL of a matrix spiking standard. For base/neutral acid analysis, the amount added of the surrogates and matrix spiking compounds should result in a final concentration of 100 ng/μL of each base/neutral analyte and 200 ng/μL of each acid analyte in the extract to be analyzed (assuming a 1-μL injection). Immediately add a 100-mL mixture of 1:1 methylene chloride:acetone and extract the sample ultrasonically for 3 min and then decant or filter the extracts. Repeat the extraction two or more times. Dry the extract using a column with anhydrous sodium sulfate and concentrate it to 1 mL in a K-D concentrator.

QUALITY CONTROL A methylene chloride solution containing 50 ng/μL of decafluorotriphenylphosphine (DFTPP) is used for tuning the GC/MS system each 12-h shift. A system performance check also must be made during every 12-h shift. A standard containing 50 ng/μL each of 4,4′-DDT, pentachlorophenol, and benzidine is required to verify injection port inertness and GC column performance. A calibration standard at mid-concentration, containing each compound of interest, including all required surrogates, must be performed every 12 h during analysis. After the system performance check is met, calibration check compounds (CCCs) are used to check the validity of the initial calibration.

The internal standard responses and retention times in the calibration check standard must be evaluated immediately after or during data acquisition. If the retention time for any internal standard changes by more than 30 seconds from the last check calibration (12 h), the chromatographic system must be inspected for malfunctions and corrections must be made, as required. If the electron ionization current plot (EICP) area for any of the internal standards changes by a factor of two from the last daily calibration standard check, the mass spectrometer must be inspected for malfunctions and corrections must be made, as appropriate.

Demonstrate, through the analysis of a reagent water blank, that interferences from the analytical system, glassware, and reagents are under control. The blank samples should be carried through all stages of the sample preparation and measurement steps. For each analytical batch (up to 20 samples), a reagent blank, matrix spike, and matrix spike duplicate/duplicate must be analyzed (the frequency of the spikes may be different for different monitoring programs). The blank and spiked samples must be carried through all stages of the sample preparation and measurement steps. A QC reference sample concentrate containing each analyte at a concentration of 100 mg/L in methanol is required.

REFERENCE Test Methods for Evaluating Solid Waste (SW-846). U.S. EPA 1983, Method 8270B, Rev. 2, Nov. 1990. Office of Solid Waste, Washington, DC.

Pentachloroethane — EPA Method 1625
CAS #76-01-7

TITLE Semivolatile Organic Compounds by Isotope Dilution GC/MS

MATRIX The compounds may be determined in waters, soils, and municipal sludges by this method.

METHOD SUMMARY This method is used to determine 176 semivolatile toxic organic pollutants associated with the CWA (as amended 1987); the RCRA (as amended 1986); the CERCLA (as amended 1986); and other compounds amenable to extraction and analysis by capillary column gas chromatography-mass spectrometry (GC/MS).

Stable isotopically-labeled analogs of the compounds of interest are added to the sample. If the solids content is less than 1%, a 1-L sample is extracted at pH 12–13, then at pH <2 with methylene chloride using continuous extraction techniques.

If the solids content is 30% or less, the sample is diluted to 1% solids with reagent water, homogenized ultrasonically, and extracted at pH 12–13, then at pH <2 with methylene chloride using continuous extraction techniques. If the solids content is greater than 30%, the sample is extracted using ultrasonic techniques.

Each extract is dried over sodium sulfate, concentrated to a volume of 5 mL, cleaned up using GPC, if necessary, and concentrated. Extracts are concentrated to 1 mL if GPC is not performed, and to 0.5 mL if GPC is performed.

An internal standard is added to the extract, and a 1-mL aliquot of the extract is injected into the GC. The compounds are separated by GC and detected by a MS. The labeled compounds serve to correct the variability of the analytical technique.

INTERFERENCES Solvents, reagents, glassware, and other sample processing hardware may yield artifacts and/or elevated baselines causing misinterpretation of chromatograms and spectra. Materials used in the analysis must be demonstrated to be free from interferences under the conditions of analysis by running method blanks initially and with each sample lot (sample started through the extraction process on a given 8-h shift, to a maximum of 20). Specific selection of reagents and purification of solvents by distillation in all glass systems may be required. Glassware and, where possible, reagents are cleaned by solvent rinse and baking at 450°C for 1-h minimum. Interferences coextracted from samples will vary considerably from source to source, depending on the diversity of the site being sampled.

INSTRUMENTATION Major instrumentation includes a GC with a splitless or on-column injection port for capillary column, a MS with 70 eV electron impact ionization, and a data system to collect and record MS data, and process it. A K-D apparatus is used to concentrate extracts.

GC Column: 30 m × 0.25 mm I.D. 5% phenyl, 94% methyl, 1% vinyl silicone bonded phased fused silica capillary column.

PRECISION & ACCURACY The detection limits of the method are usually dependent on the level of interferences rather than instrumental limitations. The limits typify the minimum quantities that can be detected with no interferences present.

The minimum level (in µg/mL) was not listed. This is defined as a minimum level at which the analytical system shall give recognizable mass spectra (background corrected) and acceptable calibration points.

The MDL (in µg/kg) in low solids was not listed and in high solids was not listed; these were determined in digested sludge (low solids) and in filter cake or compost (high solids).

The labeled and native compound initial precision as standard deviation (in µg/L) was not listed.
The labeled and native compound initial accuracy as average recovery (in µg/L) was not listed.

SAMPLE COLLECTION, PRESERVATION & HANDLING Collect samples in glass containers. Aqueous samples which flow freely are collected in refrigerated bottles using automatic sampling equipment. Solid samples are collected as grab samples using widemouth jars. Maintain samples at 0 to 4°C from the time of collection until extraction. If residual chlorine is present in aqueous samples, add 80 mg sodium thiosulfate/L of water. Begin sample extraction within 7 days of collection, and analyze all extracts within 40 days of extraction.

SAMPLE PREPARATION Samples containing 1% solids or less are extracted directly using continuous liquid-liquid extraction techniques. Samples containing 1 to 30% solids are diluted to the 1% level with reagent water and extracted using continuous liquid-liquid extraction techniques. Samples containing greater than 30% solids are extracted using ultrasonic techniques.

- Base/neutral extraction — Adjust the pH of the waters in the extractors to 12–13 with 6 *N* NaOH. Extract with methylene chloride for 24–48 h.
- Acid extraction — Adjust the pH of the waters in the extractors to 2 or less using 6 *N* sulfuric acid. Extract with methylene chloride for 24–48 h.
- Ultrasonic extraction of high solids samples — Add anhydrous sodium sulfate to the sample and QC aliquot(s). Add acetone:methylene chloride (1:1) to the sample and mix thoroughly

Concentrate extracts using a K-D apparatus.

QUALITY CONTROL The analyst is permitted to modify this method to improve separations or lower the costs of measurements, provided all performance specifications are met. Analyses of blanks are required to demonstrate freedom from contamination. When results of spikes indicate atypical method performance for samples, the samples are diluted to bring method performance within acceptable limits.

For low solids (aqueous samples), extract, concentrate, and analyze two sets of four 1-L aliquots (8 aliquots total) of the precision and recovery standard. For high solids samples, two sets of four 30-g aliquots of the high solids reference matrix are used.

Spike all samples with labeled compounds to assess method performance. Compute percent recovery of the labeled compounds using the internal standard method. Compare the labeled compound recovery for each compound with the corresponding labeled compound recovery.

Reagent water and high solids reference matrix blanks are analyzed to demonstrate freedom from contamination. Extract and concentrate a 1-L reagent water blank or a high solids reference matrix blank with each sample's lot (samples started through the extraction process on the same 8-h shift, to a maximum of 20 samples).

Field replicates may be collected to determine the precision of the sampling technique, and spiked samples may be required to determine the accuracy of the analysis when the internal standard method is used.

REFERENCE Semivolatile Organic Compounds by Isotope Dilution GC/MS. Office of Water Regulation and Standards, U.S. EPA Industrial Technology Division, Washington, DC, EPA Method 1625, Rev. C, June 1989 (contact W.A. Telliard, U.S. EPA, Office of Water Regulations and Standards, 401 M St., SW, Washington, DC, 20460. Phone: 202-382-7131).

Pentachloroethane   **EPA Method 8240**
CAS #76-01-7

TITLE Volatile Organics By GC/MS: Packed Column Technique

MATRIX Nearly all types of sample matarices, regardless of water content, can be analyzed using this method. This includes groundwater, aqueous sludges, caustic liquors, acid liquors, waste solvents, oily wastes, mousses, tars, fibrous wastes, polymetric emulsions, filter cakes, spent carbons, spent catalysts, soils, and sediments.

METHOD SUMMARY Method 8240B covers 80 volatile organic compounds that are introduced into a gas chromatograph by the purge-and-trap method or by direct injection (in limited applications). For the purge-and-trap method an inert gas (zero grade nitrogen or helium) is bubbled through a 5-mL solution at ambient temperature. Purged sample components are trapped in a tube of sorbent materials. When purging is complete, the sorbent tube is heated and backflushed with inert gas to desorb the trapped components onto a GC column.

INTERFERENCES Impurities in the purge gas and from organic compounds outgassing from the plumbing ahead of the trap account for many contamination problems. Interferences purged or coextracted from the samples will vary considerably from source to source. Cross-contamination can occur whenever high-level and low-level samples are analyzed sequentially. Whenever an unusually concentrated sample is analyzed, it should be followed by the analysis of organic-free reagent water to check for cross-contamination. Samples also can be contaminated by diffusion of volatile organics (particularly methylene chloride and fluorocarbons) through the septum seal into the sample during shipment and storage. A trip blank can serve as a check on such contamination. The lab

where volatile analysis is performed and also the refrigerated storage area should be completely free of solvents.

INSTRUMENTATION A gas chromatograph/mass spectrometry/data system (GC/MS) equipped with a 6 ft × 0.1 in I.D. glass column packed with 1% SP-1000 on Carbopack-B (60/80 mesh) is required. Also needed is a 5-mL purging device, a sorbent trap, and a thermal desorption apparatus.

PRECISION & ACCURACY This method is reported to have been tested by 15 laboratories using organic-free reagent water, drinking water, surface water, and industrial wastewaters (not specified) fortified at six concentrations over the range 5–600 µg/L.

Sample estimated quantitation limits (EQLs) are highly matrix-dependent. The EQLs listed may not always be achievable. EQLs listed for soils or sediments are based on wet weight. Normally, data is reported on a dry-weight basis; therefore, EQLs will be higher, based on the percent dry weight of each sample. Note that EQLs are even more variable than MDLs and that they are highly variable depending on the matrix being analyzed.

EQL in groundwater in µg/L was 10.
EQL in low soil or sediment in µg/kg was 10.
Accuracy (a) in µg/L was not listed.
Precision (b) in µg/L was not listed.

(a) Average recovery found for measurements of samples containing a concentration of C, in µg/L.
(b) Overall precision found for measurements of samples with average recovery X for samples containing a concentration of C in µg/L.
X = Average recovery found for measurement of samples containing a concentration of C in µg/L.

MULTIPLICATION FACTORS FOR OTHER MATRICES

Other Matrices	Factor (a)
Waste miscible liquid waste	50
High-concentration soil and sludge	125
Non-water miscible waste	500

(a) EQL = [EQL for low soil/sediment] × [Factor]. For non-aqueous samples, the factor is on a wet-weight basis.

SAMPLING METHOD

Liquid samples — Use a 40-mL glass screw-cap VOA vial with a Teflon®-faced silicone septum that has been prewashed, rinsed with distilled deionized water, and oven dried. However, if residual chlorine is present, collect sample in a 40-oz. soil VOA container which has been pre-preserved with 4 drops of 10% sodium thiosulfate, mix gently, and then transfer the sample to a 40-mL VOA vial. Collect bubble-free samples in duplicate and seal them in separate plastic bags.

Soils or sediments, and sludges — Use an 8-oz. widemouth glass bottle with a Teflon®-faced silicone septum that has been prewashed with detergent, rinsed with distilled deionized water, and oven dried. Tap slightly to eliminate free air space. Collect samples in duplicate and seal them in separate plastic bags.

SAMPLE PRESERVATION

Liquid samples — Add 4 drops of concentrated HCL and immediately cool samples to 4°C and store in a solvent-free refrigerator.

Soils or sediments, and sludges — Cool samples to 4°C and store in a solvent-free refrigerator.

MHT Maximum holding time is 14 days from the date of sample collection.

SAMPLE PREPARATION

Liquid samples — Remove the plunger from a 5-mL syringe and carefully pour the sample into the syringe barrel to just short of overflowing. Replace the syringe plunger and compress the sample. Open the syringe valve and vent any residual air while adjusting the sample volume to 5.0 mL. If there is only one volatile organic analysis (VOA) vial, a second syringe should be filled at this time to protect against possible loss of sample integrity. Add 10 µL of surrogate spiking solution and 10 µL of internal standard spiking solution through the valve bore of the 5-mL syringe, then close the valve. The surrogate and internal standards may be mixed and added as a single spiking solution.

Sediments, soils, and waste samples — All samples of this type should be screened by GC analysis using a headspace method (EPA Method 3810) or the hexadecane extraction and screening method (EPA Method 3820). Use the screening data to determine whether to use the low-concentration method (0.005–1 mg/kg) or the high-concentration method (>1 mg/kg).

Low-concentration method — The low-concentration method is based on purging a heated sediment or soil sample mixed with organic-free reagent water containing the surrogate and internal standards. Analyze all reagent blanks and standards under the same conditions as the samples.

Use a 5-g sample if the expected concentration is <0.1 mg/kg or a 1-g sample for expected concentrations between 0.1 and 1 mg/kg. Mix the contents of the sample container with a narrow metal spatula. Weigh the amount of the sample into a tared purge device. Add the spiked water to the purge device, which contains the weighed amount of sample, and connect the device to the purge-and-trap system.

High-concentration method — This method is based on extracting the sediment or soil with methanol. A waste sample is either extracted or diluted, depending on its solubility in methanol. Wastes that are insoluble in methanol are diluted with reagent tetraglyme or possibly polyethylene glycol (PEG). An aliquot of the extract is added to organic-free reagent water containing surrogate and internal standards. This is purged at ambient temperature. All samples with an expected concentration of >1.0 mg/kg should be analyzed by this method.

Mix the contents of the sample container with a narrow metal spatula. For sediments or soils and solid wastes that are insoluble in methanol, weigh 4 g (wet weight) of sample into a tared 20-mL vial. For waste that is soluble in methanol, tetraglyme, or PEG, weigh 1 g (wet weight) into a tared scintillation vial or culture tube or a 10-mL volumetric flask. Quickly add

9.0 mL of appropriate solvent then add 1.0 mL of a surrogate spiking solution to the vial, cap it, and shake it for 2 min.

METHANOL EXTRACT REQUIRED FOR ANALYSIS OF HIGH-CONCENTRATION SOILS OR SEDIMENTS

Approximate Concentration Range	Volume of Methanol Extract (a)
500–10,000 µg/kg	100 µL
1,000–20,000 µg/kg	50 µL
5,000–100,000 µg/kg	10 µL
25,000–500,000 µg/kg	100 µL of 1/50 dilution (b)

Calculate appropriate dilution factor for concentrations exceeding this table.

(a) The volume of methanol added to 5 mL of water being purged should be kept constant. Therefore, add to the 5-mL syringe whatever volume of methanol is necessary to maintain a volume of 100 µL added to the syringe.
(b) Dilute an aliquot of the methanol extract and then take 100 µL for analysis.

QUALITY CONTROL Demonstrate, through the analysis of a reagent water blank, that interferences from the analytical system, glassware, and reagents are under control. Blank samples should be carried through all stages of the sample preparation and measurement steps. For each analytical batch (up to 20 samples), a reagent blank, matrix spike, and matrix spike duplicate must be analyzed (the frequency of the spikes may be different for different monitoring programs). The blank and spiked samples must be carried through all stages of the sample preparation and measurement steps. QC samples mentioned in the section on Interferences will also be needed as appropriate to those situations.

REFERENCE Test Methods for Evaluating Solid Waste (SW-846). U.S. EPA. 1983. Method 8240B, Rev. 2, Nov. 1990. Office of Solid Wastes, Washington, DC.

Pentachlorohexane   **EPA Method 8120**
CAS #96989-91-2

TITLE Chlorinated Hydrocarbons by Gas Chromatography

MATRIX This method covers aqueous and solid matrices. This includes a wide variety such as drinking water, groundwater, industrial wastewaters, surface waters, soils, solids, and sediments.

METHOD SUMMARY This method is used to determine the concentration of 14 chlorinated hydrocarbons. It provides gas chromatographic conditions for the detection of ppb concentrations of certain chlorinated hydrocarbons. Prior to use of this method, appropriate sample extraction techniques must be used. Both neat and diluted organic liquids (EPA Method 3580, Waste Dilution) may be analyzed by direct injection. A 2 to 5 µg/mL aliquot of the extract is injected into a gas chromatograph (GC) using the solvent flush technique, and compounds in the GC effluent are detected by an electron capture detector (ECD).

INTERFERENCES Solvents, reagents, glassware, and other sample processing hardware may yield discrete artifacts and/or elevated baselines causing misinterpretation of gas chromatograms. Interferences coextracted from samples will vary considerably from source to source, depending upon the waste being sampled.

INSTRUMENTATION An analytical system complete with GC suitable for on-column injections and accessories, including detectors, column supplies, recorder, gases and syringes is required. A data system for measuring peak areas and/or peak heights is recommended. The GC is equipped with an electron capture detector (ECD). A K-D apparatus is needed for sample preparation.

Column 1: 1.8 m × 2 mm I.D. glass column packed with 1% SP-1000 on Supelcoport (100/120 mesh) or equivalent.
Column 2: 1.8 m × 2 mm I.D. glass column packed with 1.5% OV-1/2.4% OV-225 on Supelcoport (80/100 mesh) or equivalent.

PRECISION & ACCURACY The method was tested by 20 laboratories using organic-free reagent water, drinking water, surface water, and three industrial wastewaters spiked at six concentrations over the range 1.0 to 356 µg/L. Single operator precision, overall precision, and method accuracy were found to be directly related to the concentration of the parameter and essentially independent of the sample matrix.

MULTIPLICATION FACTORS FOR OTHER MATRICES (a)

Matrix	Factor (b)
Groundwater	10
Low-concentration soil by ultrasonic extraction with GPC cleanup	670
High-concentration soil and sludges by ultrasonic extraction	10,000
Waste not miscible with water	100,000

(a) Sample EQLs are highly matrix-dependent. The EQLs listed herein are provided for guidance and may not always be achievable.
(b) EQL = [Method detection limit] × [Factor]. For nonaqueous samples, the factor is on a wet-weight basis.

PRECISION & ACCURACY The estimates below are based upon the performance in a single lab.

The accuracy (in µg/L) as expected recovery for one or more measurements of a sample containing a concentration of C was No Data.
The precision (in µg/L) as expected single analyst standard deviation of measurements at an average concentration of x" was No Data.
The precision (in µg/L) as expected interlaboratory standard deviation measurements at an average concentration found of x" was No Data.

C = *True value for the concentration, in µg/L.*
$x"$ = *Average recovery found for measurements of samples containing a concentration of C, in µg/L.*

SAMPLE COLLECTION, PRESERVATION & HANDLING
Extracts must be stored under refrigeration at 4°C and analyzed within 40 days of extraction.

SAMPLE PREPARATION In general, water samples are extracted at a neutral, or as is, pH with methylene chloride using either EPA Method 3510 or EPA Method 3520. Solid samples are extracted using either EPA Method 3540 or EPA Method 3550. Prior to gas chromatographic analysis, the extraction solvent must be exchanged to hexane.

QUALITY CONTROL The quality control check concentrate (EPA Method 8000) should contain each parameter of interest in acetone at the following concentrations: hexachloro-substituted hydrocarbon, 10 µg/mL; and any other chlorinated hydrocarbon, 100 µg/mL. Calculate surrogate standard recovery on all samples, blanks, and spikes.

Prepare stock standard solutions in isooctane or hexane. Calibration standards at a minimum of five concentrations should be prepared through dilution of the stock standards with isooctane or hexane. Internal standards and surrogate standards are also needed.

REFERENCE Test Methods for Evaluating Solid Waste, Physical/Chemical Methods, SW-846, 3rd Edition, U.S. EPA, Office of Solid Waste, Washington, DC, 1990. EPA Method 8120 A Rev. 1, Nov. 1990.

Pentachloronitrobenzene   **EPA Method 8270**
CAS #82-68-8

TITLE Semivolatile Organic Compounds by GC/MS

MATRIX This method is used to determine the concentration of semivolatile organic compounds in extracts prepared from all types of solid waste matrices, soils, and groundwater. Although surface waters are not specifically mentioned, this method should be applicable to water samples from rivers, lakes, etc.

METHOD SUMMARY This method covers 259 semivolatile organic compounds. In very limited applications direct injection of the sample into the GC/MS system may be appropriate, but this results in very high detection limits (approximately 10,000 µg/L). Typically, a 1-L liquid sample, containing surrogate, and matrix spiking standards, is extracted in a continuous extractor first under acid conditions and then under basic conditions. Typically 30 g of a solid sample, containing surrogate, and matrix spiking standards, is extracted ultrasonically. After concentrating the extract to 1 mL it is spiked with 10 µL of an internal standard solution just prior to analysis by GC/MS. The volume injected should contain about 100 ng of base/neutral and 200 ng of acid surrogates (for a 1-µL injection). Analysis is performed by GC/MS using a capillary GC column.

INTERFERENCES Raw GC/MS data from all blanks, samples, and spikes must be evaluated for interferences. Contamination by carryover can occur whenever high-concentration and low-concentration samples are sequentially analyzed. To reduce carryover, the sample syringe must be rinsed out between samples with solvent. Whenever an unusually concentrated sample is encountered, it should be followed by the analysis of blank solvent to check for cross-contamination.

INSTRUMENTATION A GC/MS and a data system are required. The GC column used is a 30 m × 0.25 mm I.D. (or 0.32 mm I.D.) 1um film thickness silicone-coated fused silica capillary column. A continuous liquid-liquid extractor equipped with Teflon® or glass connection joints and stopcocks requiring no lubrication, a K-D concentrating apparatus, water bath, and an ultrasonic disrupter with a minimum power of 300 W and with pulsing capability are also required.

PRECISION & ACCURACY The estimated quantitation limit (EQL) of Method 8270B for determining an individual compound is approximately 1 mg/kg (wet weight) for soil or sediment samples, 1–200 mg/kg for wastes (dependent on matrix and method of preparation), and 10 µg/L for groundwater samples. EQLs will be proportionately higher for sample extracts that require dilution to avoid saturation of the detector.

The EQL(b) for groundwater in µg/L is 20.
The EQL (a, b) for low concentrations in soil and sediment in µg/kg is not determined.
Accuracy as µg/L is not listed.
Overall precision in µg/L is not listed.

(a) *EQLs listed for soil/sediment are based on wet weight. Normally data is reported in a dry-weight basis; therefore, EQLs will be higher based on the % dry weight of each sample. This calculation is based on a 30 g sample and gel permeation chromatography cleanup.*
(b) *Sample EQLs are highly matrix-dependent. The EQLs are provided for guidance and may not always be achievable.*
$C =$ *True value for concentration, in µg/L.*
$X =$ *Average recovery found for measurements of samples containing a concentration of C, in µg/L.*

ESTIMATED QUANTITATION LIMIT

Other Matrices	Factor (a)
High-concentration soil and sludges by sonicator	7.5
Non-water miscible waste	75

(a) *EQL for other matrices = [EQL for low soil/sediment] × [Factor]. This estimated EQL is similar to an EPA "Practical Quantitation Limit."*

SAMPLING METHOD
Liquid samples — Use a 1 or 2½ gallon amber glass bottle with a screw-top Teflon®-lined cover that has been prewashed with detergent and rinsed with distilled water and methanol (or isopropanol).

Soils, sediments, or sludges — Use an 8-oz. widemouth glass with a screw-top Teflon®-lined cover that has been prewashed with detergent and rinsed with distilled water and methanol (or isopropanol).

SAMPLE PRESERVATION
Liquid samples — If residual chlorine is present, add 3 mL of 10% sodium thiosulfate per gallon, cool to 4°C and store in a solvent-free refrigerator until analysis; if chlorine is not present, then eliminate the sodium thiosulfate addition.

Soils, sediments, or sludges — Cool samples to 4°C and store in a solvent-free refrigerator.

MHT Liquid samples must be extracted within 7 days and the extracts analyzed within 40 days. Soils, sediments, or sludges may be stored for a maximum of 14 days and the extracts analyzed within 40 days.

SAMPLE PREPARATION

Liquid samples — Transfer 1 L quantitatively to a continuous extractor. If high concentrations are anticipated, a smaller volume may be used and then diluted with organic-free reagent water to 1 L. Adjust pH, if necessary, to pH <2 using 1:1 (V/V) sulfuric acid. Pipette 1.0 mL of a surrogate standard spiking solution into each sample. For the sample in each analytical batch selected for spiking, add 1.0 mL of a matrix spiking standard. For base/neutral acid analysis, the amount of the surrogates and matrix spiking compounds added to the sample should result in a final concentration of 100 ng/µL of each analyte in the extract to be analyzed (assuming a 1-µL injection). Extract with methylene chloride for 18–24 h. Next, adjust the pH of the aqueous phase to pH >11 using 10 N sodium hydroxide and extract it with methylene chloride again for 18–24 h. Dry the extract through a column containing anhydrous sodium sulfate and concentrate it to 1 mL using a K-D concentrator.

Soils, sediments, or sludges — Use 30 g of sample. Nonporous or wet samples (gummy or clay type) that do not have a free-flowing sandy texture must be mixed with anhydrous sodium sulfate until the sample is free flowing. Add 1 mL of surrogate standards to all samples, spikes, standards, and blanks. For the sample in each analytical batch selected for spiking, add 1.0 mL of a matrix spiking standard. For base/neutral acid analysis, the amount added of the surrogates and matrix spiking compounds should result in a final concentration of 100 ng/µL of each base/neutral analyte and 200 ng/µL of each acid analyte in the extract to be analyzed (assuming a 1-µL injection). Immediately add a 100-mL mixture of 1:1 methylene chloride:acetone and extract the sample ultrasonically for 3 min and then decant or filter the extracts. Repeat the extraction two or more times. Dry the extract using a column with anhydrous sodium sulfate and concentrate it to 1 mL in a K-D concentrator.

QUALITY CONTROL
A methylene chloride solution containing 50 ng/µL of decafluorotriphenylphosphine (DFTPP) is used for tuning the GC/MS system each 12-h shift. A system performance check also must be made during every 12-h shift. A standard containing 50 ng/µL each of 4,4′-DDT, pentachlorophenol, and benzidine is required to verify injection port inertness and GC column performance. A calibration standard at mid-concentration, containing each compound of interest, including all required surrogates, must be performed every 12 h during analysis. After the system performance check is met, calibration check compounds (CCCs) are used to check the validity of the initial calibration.

The internal standard responses and retention times in the calibration check standard must be evaluated immediately after or during data acquisition. If the retention time for any internal standard changes by more than 30 seconds from the last check calibration (12 h), the chromatographic system must be inspected for malfunctions and corrections must be made, as required. If the electron ionization current plot (EICP) area for any of the internal standards changes by a factor of two from the last daily calibration standard check, the mass spectrometer must be inspected for malfunctions and corrections must be made, as appropriate.

Demonstrate, through the analysis of a reagent water blank, that interferences from the analytical system, glassware, and reagents are under control. The blank samples should be carried through all stages of the sample preparation and measurement steps. For each analytical batch (up to 20 samples), a reagent blank, matrix spike, and matrix spike duplicate/duplicate must be analyzed (the frequency of the spikes may be different for different monitoring programs). The blank and spiked samples must be carried through all stages of the sample preparation and measurement steps. A QC reference sample concentrate containing each analyte at a concentration of 100 mg/L in methanol is required.

REFERENCE Test Methods for Evaluating Solid Waste (SW-846). U.S. EPA 1983, Method 8270B, Rev. 2, Nov. 1990. Office of Solid Waste, Washington, DC.

Pentachlorophenol EPA Method 1625
CAS #87-86-5

TITLE Semivolatile Organic Compounds by Isotope Dilution GC/MS

MATRIX The compounds may be determined in waters, soils, and municipal sludges by this method.

METHOD SUMMARY This method is used to determine 176 semivolatile toxic organic pollutants associated with the CWA (as amended 1987); the RCRA (as amended 1986); the CERCLA (as amended 1986); and other compounds amenable to extraction and analysis by capillary column gas chromatography-mass spectrometry (GC/MS).

Stable isotopically-labeled analogs of the compounds of interest are added to the sample. If the solids content is less than 1%, a 1-L sample is extracted at pH 12–13, then at pH <2 with methylene chloride using continuous extraction techniques.

If the solids content is 30% or less, the sample is diluted to 1% solids with reagent water, homogenized ultrasonically, and extracted at pH 12–13, then at pH <2 with methylene chloride using continuous extraction techniques. If the solids content is greater than 30%, the sample is extracted using ultrasonic techniques.

Each extract is dried over sodium sulfate, concentrated to a volume of 5 mL, cleaned up using GPC, if necessary, and concentrated. Extracts are concentrated to 1 mL if GPC is not performed, and to 0.5 mL if GPC is performed.

An internal standard is added to the extract, and a 1-mL aliquot of the extract is injected into the GC. The compounds are separated by GC and detected by a MS. The labeled compounds serve to correct the variability of the analytical technique.

INTERFERENCES Solvents, reagents, glassware, and other sample processing hardware may yield artifacts and/or elevated

baselines causing misinterpretation of chromatograms and spectra. Materials used in the analysis must be demonstrated to be free from interferences under the conditions of analysis by running method blanks initially and with each sample lot (sample started through the extraction process on a given 8-h shift, to a maximum of 20). Specific selection of reagents and purification of solvents by distillation in all glass systems may be required. Glassware and, where possible, reagents are cleaned by solvent rinse and baking at 450°C for 1-h minimum. Interferences coextracted from samples will vary considerably from source to source, depending on the diversity of the site being sampled.

INSTRUMENTATION Major instrumentation includes a GC with a splitless or on-column injection port for capillary column, a MS with 70 eV electron impact ionization, and a data system to collect and record MS data, and process it. A K-D apparatus is used to concentrate extracts.

GC Column: 30 m × 0.25 mm I.D. 5% phenyl, 94% methyl, 1% vinyl silicone bonded phased fused silica capillary column.

PRECISION & ACCURACY The detection limits of the method are usually dependent on the level of interferences rather than instrumental limitations. The limits typify the minimum quantities that can be detected with no interferences present.

The minimum level (in µg/mL) was 50. This is defined as a minimum level at which the analytical system shall give recognizable mass spectra (background corrected) and acceptable calibration points.

The MDL (in µg/kg) in low solids was 51 and in high solids was 207; these were determined in digested sludge (low solids) and in filter cake or compost (high solids).

The labeled and native compound initial precision as standard deviation (in µg/L) was 21.

The labeled and native compound initial accuracy as average recovery (in µg/L) was 76–140.

SAMPLE COLLECTION, PRESERVATION & HANDLING Collect samples in glass containers. Aqueous samples which flow freely are collected in refrigerated bottles using automatic sampling equipment. Solid samples are collected as grab samples using widemouth jars. Maintain samples at 0 to 4°C from the time of collection until extraction. If residual chlorine is present in aqueous samples, add 80 mg sodium thiosulfate/L of water. Begin sample extraction within 7 days of collection, and analyze all extracts within 40 days of extraction.

SAMPLE PREPARATION Samples containing 1% solids or less are extracted directly using continuous liquid-liquid extraction techniques. Samples containing 1 to 30% solids are diluted to the 1% level with reagent water and extracted using continuous liquid-liquid extraction techniques. Samples containing greater than 30% solids are extracted using ultrasonic techniques.

Base/neutral extraction — Adjust the pH of the waters in the extractors to 12–13 with 6 N NaOH. Extract with methylene chloride for 24–48 h.

Acid extraction — Adjust the pH of the waters in the extractors to 2 or less using 6 N sulfuric acid. Extract with methylene chloride for 24–48 h.

Ultrasonic extraction of high solids samples — Add anhydrous sodium sulfate to the sample and QC aliquot(s). Add acetone:methylene chloride (1:1) to the sample and mix thoroughly

Concentrate extracts using a K-D apparatus.

QUALITY CONTROL The analyst is permitted to modify this method to improve separations or lower the costs of measurements, provided all performance specifications are met. Analyses of blanks are required to demonstrate freedom from contamination. When results of spikes indicate atypical method performance for samples, the samples are diluted to bring method performance within acceptable limits.

For low solids (aqueous samples), extract, concentrate, and analyze two sets of four 1-L aliquots (8 aliquots total) of the precision and recovery standard. For high solids samples, two sets of four 30-g aliquots of the high solids reference matrix are used.

Spike all samples with labeled compounds to assess method performance. Compute percent recovery of the labeled compounds using the internal standard method. Compare the labeled compound recovery for each compound with the corresponding labeled compound recovery.

Reagent water and high solids reference matrix blanks are analyzed to demonstrate freedom from contamination. Extract and concentrate a 1-L reagent water blank or a high solids reference matrix blank with each sample's lot (samples started through the extraction process on the same 8-h shift, to a maximum of 20 samples).

Field replicates may be collected to determine the precision of the sampling technique, and spiked samples may be required to determine the accuracy of the analysis when the internal standard method is used.

REFERENCE Semivolatile Organic Compounds by Isotope Dilution GC/MS. Office of Water Regulation and Standards, U.S. EPA Industrial Technology Division, Washington, DC, EPA Method 1625, Rev. C, June 1989 (contact W.A. Telliard, U.S. EPA, Office of Water Regulations and Standards, 401 M St., SW, Washington, DC, 20460. Phone: 202-382-7131).

Pentachlorophenol   **EPA Method 625**
CAS #87-86-5

TITLE Base/Neutrals and Acids, U.S. EPA Method 625

MATRIX This methods covers municipal and industrial wastewaters.

METHOD SUMMARY Approximately 1 L of sample is serially extracted with methylene chloride at a pH greater than 11 and again at a pH less than 2 using a separatory funnel or a continuous extractor. The methylene chloride extract is dried, concentrated to a volume of 1 mL, and analyzed by GC/MS.

Qualitative identification of the parameters in the extract is performed using the retention time and the relative abundance of three characteristic masses (m/z). Qualitative analysis is performed using either external or internal standard techniques with a single characteristic m/z.

INTERFERENCES Method interferences may be caused by contaminants in solvents, reagents, glassware, and other sample processing hardware. Glassware must be scrupulously cleaned. Glassware should be heated in a muffle furnace at 400°C for 5 to 30 min. Some thermally stable materials, such as PCBs, may not be eliminated by this treatment. Solvent rinses with acetone and pesticide quality hexane may be substituted for the muffle furnace heating. Matrix interferences may be caused by contaminants that are coextracted from the sample. The base-neutral extraction may cause significantly reduced recovery of phenols. The packed gas chromatographic columns recommended for the basic fraction may not exhibit sufficient resolution for some analytes.

INSTRUMENTATION A GC/MS system with an injection port designed for on-column injection when using packed columns and for splitless injection when using capillary columns.

Column for base/neutrals: 1.8 m long × 2 mm I.D. glass, packed with 3% SP-2550 on Supelcoport (100/120 mesh) or equivalent.

Column for acids: 1.8 m long × 2 mm I.D. glass, packed with 1% SP-1240DA on Supelcoport (100/120 mesh) or equivalent.

PRECISION & ACCURACY The MDL concentrations were obtained using reagent water. The MDL actually achieved in a given analysis will vary depending on instrument sensitivity and matrix effects. This method was tested by 15 laboratories using reagent water, drinking water, surface water, and industrial wastewaters spiked at six concentrations over the range 5 to 100 µg/L. Single operator precision, overall precision, and method accuracy were found to be directly related to the concentration of the parameter matrix.

The MDL (in µg/L) in reagent water was 3.6.

The standard deviation (in µg/L based on 4 recovery measurements) was 48.9.

The range (in µg/L) for average recovery for 4 measurements was 38.1–151.8.

The range (in %) for percent recovery was 14–176.

Accuracy (in µg/L) as expected recovery for one or more measurements of a sample containing a true concentration of C was 0.93C + 1.99.

Precision (in µg/L) as expected single analyst standard deviation of measurements at an average concentration found at X was 0.24X + 3.03.

Overall precision (in µg/L) as expected interlaboratory standard deviation of measurements in an average concentration found at X was 0.30X + 4.33.

C = *True value of the concentration in µg/L.*
X = *Average recovery found for measurements of samples containing a concentration at C in µg/L.*

SAMPLE PREPARATION Adjust the pH to >11 with sodium hydroxide and serially extract in a separatory funnel with methylene chloride or else in a continuous extractor. Next, adjust the pH to <2 with sulfuric acid and serially extract in a separatory funnel with methylene chloride or else in a continuous extractor. Dry the extracts separately through a column of anhydrous sodium sulfate and then concentrate each of the extracts to 1.0 mL using a K-D apparatus.

SAMPLE COLLECTION, PRESERVATION & HANDLING Grab samples must be collected in glass containers. All samples must be refrigerated at 4°C from the time of collection until extraction. If residual chlorine is present, add 80 mg of sodium thiosulfate/L of sample and mix well. All samples must be extracted within 7 days of collection and completely analyzed within 40 days of extraction.

QUALITY CONTROL Make an initial, one-time, demonstration of the ability to generate acceptable accuracy and precision with this method. Before processing any samples, the analyst must analyze a reagent water blank to demonstrate that interferences from the analytical system and glassware are under control. Each time a set of samples is extracted or reagents are changed, a reagent water blank must be processed. Spike and analyze a minimum of 5% of all samples to monitor and evaluate lab data quality. A QC check sample concentrate that contains each parameter of interest at a concentration of 100 µg/mL in acetone is required. PCBs and multicomponent pesticides may be omitted from this test.

After analysis of five spiked wastewater samples, calculate the average percent recovery and the standard deviation of the percent recovery. Spike all samples with the surrogate standard spiking solution and calculate the percent recovery of each surrogate compound.

REFERENCE Federal Register, Vol. 49, No. 209. Friday, Oct. 26, 1984.

Pentachlorophenol   **EPA Method 8151**
CAS #87-86-5

TITLE Chlorinated Herbicides by GC Using Methylation or Pentafluorobenzylation Derivatization: Capillary Column Technique.

MATRIX This method covers aqueous and solid matrices. This includes a wide variety such as drinking water, groundwater, industrial wastewaters, surface waters, soils, solids, and sediments.

METHOD SUMMARY This is a GC method for determining 19 chlorinated acid herbicides in aqueous, soil, and waste matrices. Because these compounds are produced and used in various forms (i.e., acid, salt, ester, etc.) a hydrolysis step is included to convert the herbicide to the acid form prior to analysis. This method provides hydrolysis, extraction, derivatization and GC conditions for the analysis of chlorinated acid herbicides in water, soil, and waste samples. Water samples are hydrolyzed *in situ*, extracted with diethyl ether, and then esterified with either diazomethane or pentafluorobenzyl bromide. The derivatives are determined by gas chromatography with an electron capture detector (GC/ECD). The results are reported as acid equivalents. The sensitivity of this method depends on the level of interferences in addition to instrumental limitations.

INTERFERENCES Method interferences may be caused by contaminants in solvents, reagents, glassware, and other sample processing hardware. Immediately prior to use, glassware should be rinsed with the next solvent to be used. Matrix interferences may be caused by contaminants that are coextracted from the sample. Organic acids, especially chlorinated acids, cause the most direct interference with the determination by methylation. Phenols, including chlorophenols, may also interfere with this procedure. The determination using pentafluorobenzylation is more sensitive, and more prone to interferences from the presence of organic acids of phenols than by methylation. Alkaline hydrolysis and subsequent extraction of the basic solution removes many chlorinated hydrocarbons and phthalate esters that might otherwise interfere with the ECD analysis. The herbicides, being strong organic acids, react readily with alkaline substances and may be lost during analysis. Therefore, glassware must be acid-rinsed and then rinsed to constant pH with organic-free reagent water.

INSTRUMENTATION A GC suitable for Grob-type injection using capillary columns. A data system for measuring peak heights and/or peak areas is recommended. An electron capture detector (ECD) is used. Also a K-D apparatus, a diazomethane generator, a centrifuge and an ultrasonic disrupter will be required.

Narrow Bore Columns:
Primary Column 1: 30 m × 0.25 mm, 5% phenyl/95% methyl silicone (DB-5), 0.25 μm film thickness.
Primary Column 1a (GC/MS): 30 m × 0.32 mm, 5% phenyl/95% methyl silicone (DB-5), 1-μm film thickness.
Column 2: 30 m × 0.25 mm DB-608 with a 25 μm film thickness.
Confirmation Column: 30 m × 0.25 mm, 14% cyanopropyl phenyl silicone (DB-1701), 0.25 μm film thickness.

Megabore Columns:
Primary Column: 30 m × 0.53 mm DB-608 with 0.83 μm film thickness.
Confirmation Column: 30 m × 0.53 mm, 14% cyanopropyl phenyl silicone (DB-1701), 1.0 μm film thickness.

PRECISION & ACCURACY Method detection limits (MDLs) are compound-dependent and vary with derivitization efficiency, derivative recovery, the matrix sampled, and herbicide concentration.

The estimated MDL (in μg/L) was 0.076 for aqueous samples using GC/ECD.

The estimated MDL (in μg/kg) was 0.16 for soil samples using GC/ECD when corrected back to 50 g samples extracted and concentrated to 10 mL with 5-μL injections.

The estimated GC/MS identification limit (in ng) was 1.3 for soil samples using GC/MS.

Mean percent recovery, calculated from 7–8 determinations of spiked reagent water, after diazomethane derivatization, from a spike concentration (in μg/L) of 0.04 was 130 with a standard deviation of the percent recovery of 31.2.

Mean percent recovery, calculated from 10 determinations of spiked clay and clay/still bottom samples over the linear concentration range (in ng/g) of no data was none reported with a percent relative standard deviation of none. The RSD % was calculated on 10 samples high in the linear concentration range and 10 low in the range. The linear concentration range was determined using standard solutions and corrected to 50 g soil samples.

SAMPLE COLLECTION, PRESERVATION & HANDLING
Containers used to collect samples for the determination of semivolatile organic compounds should be soap and water washed followed by methanol (or isopropanol) rinsing. The sample containers should be of glass or Teflon® and have screw-top covers with Teflon® liners.

No preservation is used with concentrated waste samples. With liquid samples containing no residual chlorine and with soil, sediment, and sludge samples, immediately cooling to 4°C is the only preservation used. When residual chlorine is present then 3 mL of 10% aqueous sodium sulfate is added for each gallon of sample collected, followed by cooling to 4°C.

The holding time for all volatile organics samples is 14 days. Liquid samples must be extracted within 7 days and their extracts analyzed within 40 days. Concentrated waste, soil, sediment, and sludge samples must be extracted within 14 days and their extracts analyzed within 40 days.

SAMPLE PREPARATION
Preparation of soil, sediment, and other solid samples — Acidify 30 g (dry weight) solids with 0.1 M phosphate buffer (pH = 2.5) and thoroughly mix the contents. Spike the sample with surrogate compound(s). The ultrasonic extraction of solids must be optimized for each type of sample. In order for the ultrasonic extractor to efficiently extract solid samples, the sample must be free flowing when the solvent is added. Acidified anhydrous sodium sulfate should be added to clay-type soils, or any other solid that is not a free-flowing sandy texture, until a free flowing mixture is obtained. Add methylene chloride and perform ultrasonic extraction. Combine organic extracts from the repetitive extractings of the sample and centrifuge. Add aqueous potassium hydroxide, water, and methanol to the extract and reflux the mixture on a water bath. Extract the solution three times with methylene chloride and discard the methylene chloride phase. The basic solution contains the herbicide salts. Adjust the pH of the solution to <2 with cold sulfuric acid and extract three times with methylene chloride. Combine the extracts and pour them through a pre-rinsed drying column containing acidified anhydrous sodium sulfate. Collect the dried extracts in a K-D flask and concentrate them.

Preparation of aqueous samples — Measure 1 L of sample into a 2 L separatory funnel and spike it with surrogate compound(s). Add NaCl to the sample, then add 6 N NaOH to the sample to a pH of 12 or more and let the sample sit at room temperature for 1 h to hydrolyze esters. Extract the sample three times with methylene chloride and discard the extracts. Then add cold 12 N sulfuric acid to a pH less than or equal to 2, and extract the sample three times with ethyl ether. Collect the ether phase in a flask containing acidified anhydrous sodium sulfate and allow it to remain in contact with the sodium sulfate for a minimum of 2 h. The drying step is very critical to ensuring complete esterification; any moisture remaining in the ether will result in low herbicide recoveries.

Extract concentration and derivatization — The combined ether extract is concentrated to about 1 mL using a K-D apparatus followed by using a micro Snyder column or nitrogen gas blowdown. If methyl esters are to be produced, then dilute the concentrated ether extract with 1 mL of isooctane and 0.5 mL of methanol, dilute to a final volume of 4 mL, and esterify with diazomethane. If pentafluorobenzene esters are to be produced, then dilute concentrated ether extract with acetone to a final volume of 4 mL and esterify with pentafluorobenzyl bromide.

QUALITY CONTROL Select a representative spike concentration for each compound (acid or ester) to be measured. Using stock standard, prepare a quality control check sample concentrate, in acetone, that is 1000 times more concentrated than the selected concentrations. Use this quality control check sample concentrate to prepare quality control check samples. Calculate surrogate standard recovery on all standards, samples, blanks, and spikes. GC/MS techniques should be judiciously employed to support qualitative identifications made with this method. When available, chemical ionization mass spectra may be employed to aid the qualitative identification process.

REFERENCE Test Methods for Evaluating Solid Waste, Physical/Chemical Methods, SW-846, 3rd Edition, U.S. EPA, Office of Solid Waste, Washington, DC, EPA Method 8151, Nov. 1990.

Pentachlorophenol **EPA Method 8270**
CAS #87-86-5

TITLE Semivolatile Organic Compounds by GC/MS

MATRIX This method is used to determine the concentration of semivolatile organic compounds in extracts prepared from all types of solid waste matrices, soils, and groundwater. Although surface waters are not specifically mentioned, this method should be applicable to water samples from rivers, lakes, etc.

METHOD SUMMARY This method covers 259 semivolatile organic compounds. In very limited applications direct injection of the sample into the GC/MS system may be appropriate, but this results in very high detection limits (approximately 10,000 µg/L). Typically, a 1-L liquid sample, containing surrogate, and matrix spiking standards, is extracted in a continuous extractor first under acid conditions and then under basic conditions. Typically 30 g of a solid sample, containing surrogate, and matrix spiking standards, is extracted ultrasonically. After concentrating the extract to 1 mL it is spiked with 10 µL of an internal standard solution just prior to analysis by GC/MS. The volume injected should contain about 100 ng of base/neutral and 200 ng of acid surrogates (for a 1-µL injection). Analysis is performed by GC/MS using a capillary GC column.

INTERFERENCES Raw GC/MS data from all blanks, samples, and spikes must be evaluated for interferences. Contamination by carryover can occur whenever high-concentration and low-concentration samples are sequentially analyzed. To reduce carryover, the sample syringe must be rinsed out between samples with solvent. Whenever an unusually concentrated sample is encountered, it should be followed by the analysis of blank solvent to check for cross-contamination.

INSTRUMENTATION A GC/MS and a data system are required. The GC column used is a 30 m × 0.25 mm I.D. (or 0.32 mm I.D.) 1um film thickness silicone-coated fused silica capillary column. A continuous liquid-liquid extractor equipped with Teflon® or glass connection joints and stopcocks requiring no lubrication, a K-D concentrating apparatus, water bath, and an ultrasonic disrupter with a minimum power of 300 W and with pulsing capability are also required.

PRECISION & ACCURACY The estimated quantitation limit (EQL) of Method 8270B for determining an individual compound is approximately 1 mg/kg (wet weight) for soil or sediment samples, 1–200 mg/kg for wastes (dependent on matrix and method of preparation), and 10 µg/L for groundwater samples. EQLs will be proportionately higher for sample extracts that require dilution to avoid saturation of the detector.

The EQL(b) for groundwater in µg/L is 50.
The EQL (a, b) for low concentrations in soil and sediment in µg/kg is 3300.
Accuracy as µg/L is $0.93C + 1.99$.
Overall precision in µg/L is $0.30X + 4.33$.

(a) *EQLs listed for soil/sediment are based on wet weight. Normally data is reported in a dry-weight basis; therefore, EQLs will be higher based on the % dry weight of each sample. This calculation is based on a 30 g sample and gel permeation chromatography cleanup.*
(b) *Sample EQLs are highly matrix-dependent. The EQLs are provided for guidance and may not always be achievable.*
C = *True value for concentration, in µg/L.*
X = *Average recovery found for measurements of samples containing a concentration of C, in µg/L.*

ESTIMATED QUANTITATION LIMIT

Other Matrices	Factor (a)
High-concentration soil and sludges by sonicator	7.5
Non-water miscible waste	75

(a) *EQL for other matrices = [EQL for low soil/sediment] × [Factor]. This estimated EQL is similar to an EPA "Practical Quantitation Limit."*

SAMPLING METHOD

Liquid samples — Use a 1 or 2½ gallon amber glass bottle with a screw-top Teflon®-lined cover that has been prewashed with detergent and rinsed with distilled water and methanol (or isopropanol).

Soils, sediments, or sludges — Use an 8-oz. widemouth glass with a screw-top Teflon®-lined cover that has been prewashed with detergent and rinsed with distilled water and methanol (or isopropanol).

SAMPLE PRESERVATION

Liquid samples — If residual chlorine is present, add 3 mL of 10% sodium thiosulfate per gallon, cool to 4°C and store in a solvent-free refrigerator until analysis; if chlorine is not present, then eliminate the sodium thiosulfate addition.

Soils, sediments, or sludges — Cool samples to 4°C and store in a solvent-free refrigerator.

MHT Liquid samples must be extracted within 7 days and the extracts analyzed within 40 days. Soils, sediments, or sludges may be stored for a maximum of 14 days and the extracts analyzed within 40 days.

SAMPLE PREPARATION

Liquid samples — Transfer 1 L quantitatively to a continuous extractor. If high concentrations are anticipated, a smaller volume may be used and then diluted with organic-free reagent water to 1 L. Adjust pH, if necessary, to pH <2 using 1:1 (V/V) sulfuric acid. Pipette 1.0 mL of a surrogate standard spiking solution into each sample. For the sample in each analytical batch selected for spiking, add 1.0 mL of a matrix spiking standard. For base/neutral acid analysis, the amount of the surrogates and matrix spiking compounds added to the sample should result in a final concentration of 100 ng/µL of each analyte in the extract to be analyzed (assuming a 1-µL injection). Extract with methylene chloride for 18–24 h. Next, adjust the pH of the aqueous phase to pH >11 using 10 N sodium hydroxide and extract it with methylene chloride again for 18–24 h. Dry the extract through a column containing anhydrous sodium sulfate and concentrate it to 1 mL using a K-D concentrator.

Soils, sediments, or sludges — Use 30 g of sample. Nonporous or wet samples (gummy or clay type) that do not have a free-flowing sandy texture must be mixed with anhydrous sodium sulfate until the sample is free flowing. Add 1 mL of surrogate standards to all samples, spikes, standards, and blanks. For the sample in each analytical batch selected for spiking, add 1.0 mL of a matrix spiking standard. For base/neutral acid analysis, the amount added of the surrogates and matrix spiking compounds should result in a final concentration of 100 ng/µL of each base/neutral analyte and 200 ng/µL of each acid analyte in the extract to be analyzed (assuming a 1-µL injection). Immediately add a 100-mL mixture of 1:1 methylene chloride:acetone and extract the sample ultrasonically for 3 min and then decant or filter the extracts. Repeat the extraction two or more times. Dry the extract using a column with anhydrous sodium sulfate and concentrate it to 1 mL in a K-D concentrator.

Note: N-nitrosodiphenylamine decomposes in the gas chromatographic inlet and cannot be separated from diphenylamine.

QUALITY CONTROL

A methylene chloride solution containing 50 ng/µL of decafluorotriphenylphosphine (DFTPP) is used for tuning the GC/MS system each 12-h shift. A system performance check also must be made during every 12-h shift. A standard containing 50 ng/µL each of 4,4'-DDT, pentachlorophenol, and benzidine is required to verify injection port inertness and GC column performance. A calibration standard at mid-concentration, containing each compound of interest, including all required surrogates, must be performed every 12 h during analysis. After the system performance check is met, calibration check compounds (CCCs) are used to check the validity of the initial calibration.

The internal standard responses and retention times in the calibration check standard must be evaluated immediately after or during data acquisition. If the retention time for any internal standard changes by more than 30 seconds from the last check calibration (12 h), the chromatographic system must be inspected for malfunctions and corrections must be made, as required. If the electron ionization current plot (EICP) area for any of the internal standards changes by a factor of two from the last daily calibration standard check, the mass spectrometer must be inspected for malfunctions and corrections must be made, as appropriate.

Demonstrate, through the analysis of a reagent water blank, that interferences from the analytical system, glassware, and reagents are under control. The blank samples should be carried through all stages of the sample preparation and measurement steps. For each analytical batch (up to 20 samples), a reagent blank, matrix spike, and matrix spike duplicate/duplicate must be analyzed (the frequency of the spikes may be different for different monitoring programs). The blank and spiked samples must be carried through all stages of the sample preparation and measurement steps. A QC reference sample concentrate containing each analyte at a concentration of 100 mg/L in methanol is required.

REFERENCE Test Methods for Evaluating Solid Waste (SW-846). U.S. EPA 1983, Method 8270B, Rev. 2, Nov. 1990. Office of Solid Waste, Washington, DC.

Pentachlorophenol   **EPA Method 8040**
CAS #87-86-5

TITLE Phenols

MATRIX Groundwater, soils, sludges, water miscible liquid wastes, and non-water miscible wastes.

APPLICATION This method is used for the analysis of 17 phenols. Samples are extracted, concentrated, and analyzed using direct injection of both neat and diluted organic liquids. Pentafluorobenzylbromide (PFB) derivatives also may be made to increase sensitivity of the method.

INTERFERENCES There can be carryover contamination with high- and low-level samples. Solvents, reagents, and glassware may introduce artifacts. Other interferences may come from coextracted compounds from samples.

INSTRUMENTATION GC capable of on-column injections and a flame with detector (FID) or electron capture detector (ECD). Column for underivatized phenol: 1.8 m by 2.0 mm with 1% SP-1240DA on Supelcoport. Column for derivatized phenols: 1.8 m by 2.0 mm with 5% OV-17 on Chromosorb W-AW-DMCS.

RANGE 12–450 µg/L

MDL   7.4 µg/L (FID) and 0.59 µg/L (ECD)

PQL FACTORS FOR MULTIPLYING × FID MDL VALUE

Matrix	Multiplication Factor
Groundwater	10
Low-level soil by sonication with GPC cleanup	670
High-level soil and sludge by sonication	10,000
Non-water miscible waste	100,000

PRECISION   0.23X + 0.57 µg/L (overall precision using FID)

ACCURACY   0.83C + 2.07 µg/L (as recovery using FID)

SAMPLING METHOD   Use 8-oz. widemouth glass bottles with Teflon®-lined caps for concentrated waste samples, soils, sediments, and sludges. Use 1 or 2½ gallon amber glass bottles with Teflon®-lined caps for liquid (water) samples.

STABILITY   Cool soil, sediment, sludge, and liquid samples to 4°C. If residual chlorine is present in liquid samples add 3 mL of 10% sodium thiosulfate per gallon of sample and cool to 4°C.

MHT   14 days for concentrated waste, soil, sediment, or sludge; 7 days for liquid samples; all extracts must be analyzed within 40 days.

QUALITY CONTROL   A quality control check sample concentrate containing each analyte of interest is required. The QC check sample concentrate may be prepared from pure standard materials or purchased as certified solutions Use appropriate trip, matrix, control site, method, reagent, and solvent blanks. Internal, surrogate, and five concentration level calibration standards are used. The QC check sample concentrate should contain this compound at 100 µg/mL in 2-propanol.

REFERENCE   Test Methods for Evaluating Solid Waste (SW-846), U.S. EPA Office of Solid Waste, Washington, DC, Method 8040A, Rev. 1, Nov. 1990.

Pentamethylbenzene **EPA Method 1625**
CAS #700-12-9

TITLE   Semivolatile Organic Compounds by Isotope Dilution GC/MS

MATRIX   The compounds may be determined in waters, soils, and municipal sludges by this method.

METHOD SUMMARY   This method is used to determine 176 semivolatile toxic organic pollutants associated with the CWA (as amended 1987); the RCRA (as amended 1986); the CERCLA (as amended 1986); and other compounds amenable to extraction and analysis by capillary column gas chromatography-mass spectrometry (GC/MS).

Stable isotopically-labeled analogs of the compounds of interest are added to the sample. If the solids content is less than 1%, a 1-L sample is extracted at pH 12–13, then at pH <2 with methylene chloride using continuous extraction techniques.

If the solids content is 30% or less, the sample is diluted to 1% solids with reagent water, homogenized ultrasonically, and extracted at pH 12–13, then at pH <2 with methylene chloride using continuous extraction techniques. If the solids content is greater than 30%, the sample is extracted using ultrasonic techniques.

Each extract is dried over sodium sulfate, concentrated to a volume of 5 mL, cleaned up using GPC, if necessary, and concentrated. Extracts are concentrated to 1 mL if GPC is not performed, and to 0.5 mL if GPC is performed.

An internal standard is added to the extract, and a 1-mL aliquot of the extract is injected into the GC. The compounds are separated by GC and detected by a MS. The labeled compounds serve to correct the variability of the analytical technique.

INTERFERENCES   Solvents, reagents, glassware, and other sample processing hardware may yield artifacts and/or elevated baselines causing misinterpretation of chromatograms and spectra. Materials used in the analysis must be demonstrated to be free from interferences under the conditions of analysis by running method blanks initially and with each sample lot (sample started through the extraction process on a given 8-h shift, to a maximum of 20). Specific selection of reagents and purification of solvents by distillation in all glass systems may be required. Glassware and, where possible, reagents are cleaned by solvent rinse and baking at 450°C for 1-h minimum. Interferences coextracted from samples will vary considerably from source to source, depending on the diversity of the site being sampled.

INSTRUMENTATION   Major instrumentation includes a GC with a splitless or on-column injection port for capillary column, a MS with 70 eV electron impact ionization, and a data system to collect and record MS data, and process it. A K-D apparatus is used to concentrate extracts.

GC Column: 30 m × 0.25 mm I.D. 5% phenyl, 94% methyl, 1% vinyl silicone bonded phased fused silica capillary column.

PRECISION & ACCURACY   The detection limits of the method are usually dependent on the level of interferences rather than instrumental limitations. The limits typify the minimum quantities that can be detected with no interferences present.

The minimum level (in µg/mL) was not listed. This is defined as a minimum level at which the analytical system shall give recognizable mass spectra (background corrected) and acceptable calibration points.

The MDL (in µg/kg) in low solids was not listed and in high solids was not listed; these were determined in digested sludge (low solids) and in filter cake or compost (high solids).

The labeled and native compound initial precision as standard deviation (in µg/L) was not listed.
The labeled and native compound initial accuracy as average recovery (in µg/L) was not listed.

SAMPLE COLLECTION, PRESERVATION & HANDLING
Collect samples in glass containers. Aqueous samples which flow freely are collected in refrigerated bottles using automatic sampling equipment. Solid samples are collected as grab samples using widemouth jars. Maintain samples at 0 to 4°C from

the time of collection until extraction. If residual chlorine is present in aqueous samples, add 80 mg sodium thiosulfate/L of water. Begin sample extraction within 7 days of collection, and analyze all extracts within 40 days of extraction.

SAMPLE PREPARATION Samples containing 1% solids or less are extracted directly using continuous liquid-liquid extraction techniques. Samples containing 1 to 30% solids are diluted to the 1% level with reagent water and extracted using continuous liquid-liquid extraction techniques. Samples containing greater than 30% solids are extracted using ultrasonic techniques.

Base/neutral extraction — Adjust the pH of the waters in the extractors to 12–13 with 6 N NaOH. Extract with methylene chloride for 24–48 h.

Acid extraction — Adjust the pH of the waters in the extractors to 2 or less using 6 N sulfuric acid. Extract with methylene chloride for 24–48 h.

Ultrasonic extraction of high solids samples — Add anhydrous sodium sulfate to the sample and QC aliquot(s). Add acetone:methylene chloride (1:1) to the sample and mix thoroughly

Concentrate extracts using a K-D apparatus.

QUALITY CONTROL The analyst is permitted to modify this method to improve separations or lower the costs of measurements, provided all performance specifications are met. Analyses of blanks are required to demonstrate freedom from contamination. When results of spikes indicate atypical method performance for samples, the samples are diluted to bring method performance within acceptable limits.

For low solids (aqueous samples), extract, concentrate, and analyze two sets of four 1-L aliquots (8 aliquots total) of the precision and recovery standard. For high solids samples, two sets of four 30-g aliquots of the high solids reference matrix are used.

Spike all samples with labeled compounds to assess method performance. Compute percent recovery of the labeled compounds using the internal standard method. Compare the labeled compound recovery for each compound with the corresponding labeled compound recovery.

Reagent water and high solids reference matrix blanks are analyzed to demonstrate freedom from contamination. Extract and concentrate a 1-L reagent water blank or a high solids reference matrix blank with each sample's lot (samples started through the extraction process on the same 8-h shift, to a maximum of 20 samples).

Field replicates may be collected to determine the precision of the sampling technique, and spiked samples may be required to determine the accuracy of the analysis when the internal standard method is used.

REFERENCE Semivolatile Organic Compounds by Isotope Dilution GC/MS. Office of Water Regulation and Standards, U.S. EPA Industrial Technology Division, Washington, DC, EPA Method 1625, Rev. C, June 1989 (contact W.A. Telliard, U.S. EPA, Office of Water Regulations and Standards, 401 M St., SW, Washington, DC, 20460. Phone: 202-382-7131).

Perylene   **EPA Method 1625**
CAS #198-55-0

TITLE Semivolatile Organic Compounds by Isotope Dilution GC/MS

MATRIX The compounds may be determined in waters, soils, and municipal sludges by this method.

METHOD SUMMARY This method is used to determine 176 semivolatile toxic organic pollutants associated with the CWA (as amended 1987); the RCRA (as amended 1986); the CERCLA (as amended 1986); and other compounds amenable to extraction and analysis by capillary column gas chromatography-mass spectrometry (GC/MS).

Stable isotopically-labeled analogs of the compounds of interest are added to the sample. If the solids content is less than 1%, a 1-L sample is extracted at pH 12–13, then at pH <2 with methylene chloride using continuous extraction techniques.

If the solids content is 30% or less, the sample is diluted to 1% solids with reagent water, homogenized ultrasonically, and extracted at pH 12–13, then at pH <2 with methylene chloride using continuous extraction techniques. If the solids content is greater than 30%, the sample is extracted using ultrasonic techniques.

Each extract is dried over sodium sulfate, concentrated to a volume of 5 mL, cleaned up using GPC, if necessary, and concentrated. Extracts are concentrated to 1 mL if GPC is not performed, and to 0.5 mL if GPC is performed.

An internal standard is added to the extract, and a 1-mL aliquot of the extract is injected into the GC. The compounds are separated by GC and detected by a MS. The labeled compounds serve to correct the variability of the analytical technique.

INTERFERENCES Solvents, reagents, glassware, and other sample processing hardware may yield artifacts and/or elevated baselines causing misinterpretation of chromatograms and spectra. Materials used in the analysis must be demonstrated to be free from interferences under the conditions of analysis by running method blanks initially and with each sample lot (sample started through the extraction process on a given 8-h shift, to a maximum of 20). Specific selection of reagents and purification of solvents by distillation in all glass systems may be required. Glassware and, where possible, reagents are cleaned by solvent rinse and baking at 450°C for 1-h minimum. Interferences coextracted from samples will vary considerably from source to source, depending on the diversity of the site being sampled.

INSTRUMENTATION Major instrumentation includes a GC with a splitless or on-column injection port for capillary column, a MS with 70 eV electron impact ionization, and a data system to collect and record MS data, and process it. A K-D apparatus is used to concentrate extracts.

GC Column: 30 m × 0.25 mm I.D. 5% phenyl, 94% methyl, 1% vinyl silicone bonded phased fused silica capillary column.

PRECISION & ACCURACY The detection limits of the method are usually dependent on the level of interferences

rather than instrumental limitations. The limits typify the minimum quantities that can be detected with no interferences present.

The minimum level (in µg/mL) was not listed. This is defined as a minimum level at which the analytical system shall give recognizable mass spectra (background corrected) and acceptable calibration points.

The MDL (in µg/kg) in low solids was not listed and in high solids was not listed; these were determined in digested sludge (low solids) and in filter cake or compost (high solids).

The labeled and native compound initial precision as standard deviation (in µg/L) was not listed.
The labeled and native compound initial accuracy as average recovery (in µg/L) was not listed.

SAMPLE COLLECTION, PRESERVATION & HANDLING
Collect samples in glass containers. Aqueous samples which flow freely are collected in refrigerated bottles using automatic sampling equipment. Solid samples are collected as grab samples using widemouth jars. Maintain samples at 0 to 4°C from the time of collection until extraction. If residual chlorine is present in aqueous samples, add 80 mg sodium thiosulfate/L of water. Begin sample extraction within 7 days of collection, and analyze all extracts within 40 days of extraction.

SAMPLE PREPARATION Samples containing 1% solids or less are extracted directly using continuous liquid-liquid extraction techniques. Samples containing 1 to 30% solids are diluted to the 1% level with reagent water and extracted using continuous liquid-liquid extraction techniques. Samples containing greater than 30% solids are extracted using ultrasonic techniques.

Base/neutral extraction — Adjust the pH of the waters in the extractors to 12–13 with 6 N NaOH. Extract with methylene chloride for 24–48 h.
Acid extraction — Adjust the pH of the waters in the extractors to 2 or less using 6 N sulfuric acid. Extract with methylene chloride for 24–48 h.
Ultrasonic extraction of high solids samples — Add anhydrous sodium sulfate to the sample and QC aliquot(s). Add acetone:methylene chloride (1:1) to the sample and mix thoroughly

Concentrate extracts using a K-D apparatus.

QUALITY CONTROL The analyst is permitted to modify this method to improve separations or lower the costs of measurements, provided all performance specifications are met. Analyses of blanks are required to demonstrate freedom from contamination. When results of spikes indicate atypical method performance for samples, the samples are diluted to bring method performance within acceptable limits.

For low solids (aqueous samples), extract, concentrate, and analyze two sets of four 1-L aliquots (8 aliquots total) of the precision and recovery standard. For high solids samples, two sets of four 30-g aliquots of the high solids reference matrix are used.

Spike all samples with labeled compounds to assess method performance. Compute percent recovery of the labeled compounds using the internal standard method. Compare the labeled compound recovery for each compound with the corresponding labeled compound recovery.

Reagent water and high solids reference matrix blanks are analyzed to demonstrate freedom from contamination. Extract and concentrate a 1-L reagent water blank or a high solids reference matrix blank with each sample's lot (samples started through the extraction process on the same 8-h shift, to a maximum of 20 samples).

Field replicates may be collected to determine the precision of the sampling technique, and spiked samples may be required to determine the accuracy of the analysis when the internal standard method is used.

REFERENCE Semivolatile Organic Compounds by Isotope Dilution GC/MS. Office of Water Regulation and Standards, U.S. EPA Industrial Technology Division, Washington, DC, EPA Method 1625, Rev. C, June 1989 (contact W.A. Telliard, U.S. EPA, Office of Water Regulations and Standards, 401 M St., SW, Washington, DC, 20460. Phone: 202-382-7131).

pH — EPA Method 9040

TITLE pH Electrometric Measurement

MATRIX This method is applicable to the measurement of pH of aqueous wastes and those multi-phase wastes where the aqueous phase constitutes at least 20% of the total volume of the waste. The pH measurement requires minimum water content to cover the electrodes (about 25 mL).

METHOD SUMMARY The pH of the sample is determined electrometrically using either a glass electrode in combination with a reference electrode or with a combination electrode. The measuring device is calibrated using a series of standard solutions of known pH. Samples are placed in a clean glass beaker using a sufficient volume to cover the sensing elements of the electrodes and to give adequate clearance for the magnetic stirring bar. The corrosivity of concentrated acids and bases cannot be measured.

INTERFERENCES The glass electrode is not generally subject to solution interferences from color, turbidity, collidal matter, oxidants, reductants, or high salinity. Sodium error at pH levels > 10 can be reduced or eliminated by using a low sodium error electrode. Coatings of oily material or particulate matter can impair electrode response. Temperature effects on the electrometric determination of pH arise from two sources. The first is caused by the change in electrode output at various temperatures. This interference can be controlled with instruments having temperature compensation. The second source of temperature effects is the change of pH due to changes in the sample as the temperature changes; this is sample-dependent and cannot be controlled.

INSTRUMENTATION A pH meter with glass electrode and a reference electrode is required.

PRECISION & ACCURACY

Variation of accuracy as a function of pH units is listed below.

PRECISION & ACCURACY VERSUS pH

Std. Dev. (pH Units)	Accuracy as Bias (%)	Accuracy as pH Units Bias (pH Units)
3.5	0.10	−0.29	−0.01
3.5	0.11	−0.00	Not listed.
7.1	0.20	+1.01	0.07
7.2	0.18	−0.03	−0.002
8.0	0.13	−0.12	−0.01
8.0	0.12	+0.16	+0.01

SAMPLING METHOD Collect samples in 60 mL or larger, plastic or glass bottles. All bottles must be thoroughly cleaned and rinsed to remove soluble materials.

SAMPLE PRESERVATION No preservation is required.

MHT Analyze samples immediately.

SAMPLE PREPARATION No specific sample preparation is required when determining the pH of an aqueous sample.

QUALITY CONTROL No specific quality control procedures were listed for this method. Electrodes must be rinsed thoroughly between samples.

REFERENCE Test Methods for Evaluating Solid Waste (SW-846). U.S. EPA. 1983. Method 9040A, Rev. 1, Nov. 1990. Office of Solid Wastes, Washington, DC.

---

pH (electrometric)   EPA Method 150.1

TITLE Physical Properties

MATRIX Drinking, surface, and saline waters. Wastewater.

APPLICATION Date issued 1971. Editorial Rev. 1978. At a given temperature the intensity of the acidic or basic nature of a solution is indicated by pH (hydrogen ion activity). Alkalinity and acidity are acid-and-base neutralizing abilities of water usually expressed as mg $CaCO_3$/L. pH is determined electrometrically using a glass electrode with a reference potential or a combination electrode.

INTERFERENCES Coatings of oily or particulate matter, temperature effects, and sodium errors at pH levels >10 are interferences.

INSTRUMENTATION pH meter, lab or field model. Magnetic stirrer and Teflon® stirring bar.

RANGE pH meter range (0–14).

MDL Report pH to nearest 0.1 unit.

PRECISION Not listed.

ACCURACY Limit of accuracy, ±0.1 pH unit.

SAMPLING METHOD Plastic or glass. (25 mL).

STABILITY No preservation required. Analyze immediately.

QUALITY CONTROL Calibrate at minimum of two points that bracket expected pH of the samples and are approximately 3 pH units or more apart. Sample should be within 2°C of buffers, if automatic temperature compensation is not provided.

REFERENCE EPA Methods for the Chemical Analysis of Water and Wastes, EPA-600/4-79-020, U.S. EPA, EMSL, 1979.

Phenacetin   EPA Method 1625
CAS #62-44-2

TITLE Semivolatile Organic Compounds by Isotope Dilution GC/MS

MATRIX The compounds may be determined in waters, soils, and municipal sludges by this method.

METHOD SUMMARY This method is used to determine 176 semivolatile toxic organic pollutants associated with the CWA (as amended 1987); the RCRA (as amended 1986); the CERCLA (as amended 1986); and other compounds amenable to extraction and analysis by capillary column gas chromatography-mass spectrometry (GC/MS).

Stable isotopically-labeled analogs of the compounds of interest are added to the sample. If the solids content is less than 1%, a 1-L sample is extracted at pH 12–13, then at pH <2 with methylene chloride using continuous extraction techniques.

If the solids content is 30% or less, the sample is diluted to 1% solids with reagent water, homogenized ultrasonically, and extracted at pH 12–13, then at pH <2 with methylene chloride using continuous extraction techniques. If the solids content is greater than 30%, the sample is extracted using ultrasonic techniques.

Each extract is dried over sodium sulfate, concentrated to a volume of 5 mL, cleaned up using GPC, if necessary, and concentrated. Extracts are concentrated to 1 mL if GPC is not performed, and to 0.5 mL if GPC is performed.

An internal standard is added to the extract, and a 1-mL aliquot of the extract is injected into the GC. The compounds are separated by GC and detected by a MS. The labeled compounds serve to correct the variability of the analytical technique.

INTERFERENCES Solvents, reagents, glassware, and other sample processing hardware may yield artifacts and/or elevated baselines causing misinterpretation of chromatograms and spectra. Materials used in the analysis must be demonstrated to be free from interferences under the conditions of analysis by running method blanks initially and with each sample lot (sample started through the extraction process on a given 8-h shift, to a maximum of 20). Specific selection of reagents and purification of solvents by distillation in all glass systems may be required. Glassware and, where possible, reagents are cleaned by solvent rinse and baking at 450°C for 1-h minimum. Interferences coextracted from samples will vary considerably from source to source, depending on the diversity of the site being sampled.

INSTRUMENTATION Major instrumentation includes a GC with a splitless or on-column injection port for capillary column, a MS with 70 eV electron impact ionization, and a data system to collect and record MS data, and process it. A K-D apparatus is used to concentrate extracts.

GC Column: 30 m × 0.25 mm I.D. 5% phenyl, 94% methyl, 1% vinyl silicone bonded phased fused silica capillary column.

PRECISION & ACCURACY The detection limits of the method are usually dependent on the level of interferences rather than instrumental limitations. The limits typify the minimum quantities that can be detected with no interferences present.

The minimum level (in µg/mL) was not listed. This is defined as a minimum level at which the analytical system shall give recognizable mass spectra (background corrected) and acceptable calibration points.

The MDL (in µg/kg) in low solids was not listed and in high solids was not listed; these were determined in digested sludge (low solids) and in filter cake or compost (high solids).

The labeled and native compound initial precision as standard deviation (in µg/L) was not listed.
The labeled and native compound initial accuracy as average recovery (in µg/L) was not listed.

SAMPLE COLLECTION, PRESERVATION & HANDLING Collect samples in glass containers. Aqueous samples which flow freely are collected in refrigerated bottles using automatic sampling equipment. Solid samples are collected as grab samples using widemouth jars. Maintain samples at 0 to 4°C from the time of collection until extraction. If residual chlorine is present in aqueous samples, add 80 mg sodium thiosulfate/L of water. Begin sample extraction within 7 days of collection, and analyze all extracts within 40 days of extraction.

SAMPLE PREPARATION Samples containing 1% solids or less are extracted directly using continuous liquid-liquid extraction techniques. Samples containing 1 to 30% solids are diluted to the 1% level with reagent water and extracted using continuous liquid-liquid extraction techniques. Samples containing greater than 30% solids are extracted using ultrasonic techniques.

- Base/neutral extraction — Adjust the pH of the waters in the extractors to 12–13 with 6 N NaOH. Extract with methylene chloride for 24–48 h.
- Acid extraction — Adjust the pH of the waters in the extractors to 2 or less using 6 N sulfuric acid. Extract with methylene chloride for 24–48 h.
- Ultrasonic extraction of high solids samples — Add anhydrous sodium sulfate to the sample and QC aliquot(s). Add acetone:methylene chloride (1:1) to the sample and mix thoroughly

Concentrate extracts using a K-D apparatus.

QUALITY CONTROL The analyst is permitted to modify this method to improve separations or lower the costs of measurements, provided all performance specifications are met. Analyses of blanks are required to demonstrate freedom from contamination. When results of spikes indicate atypical method performance for samples, the samples are diluted to bring method performance within acceptable limits.

For low solids (aqueous samples), extract, concentrate, and analyze two sets of four 1-L aliquots (8 aliquots total) of the precision and recovery standard. For high solids samples, two sets of four 30-g aliquots of the high solids reference matrix are used.

Spike all samples with labeled compounds to assess method performance. Compute percent recovery of the labeled compounds using the internal standard method. Compare the labeled compound recovery for each compound with the corresponding labeled compound recovery.

Reagent water and high solids reference matrix blanks are analyzed to demonstrate freedom from contamination. Extract and concentrate a 1-L reagent water blank or a high solids reference matrix blank with each sample's lot (samples started through the extraction process on the same 8-h shift, to a maximum of 20 samples).

Field replicates may be collected to determine the precision of the sampling technique, and spiked samples may be required to determine the accuracy of the analysis when the internal standard method is used.

REFERENCE Semivolatile Organic Compounds by Isotope Dilution GC/MS. Office of Water Regulation and Standards, U.S. EPA Industrial Technology Division, Washington, DC, EPA Method 1625, Rev. C, June 1989 (contact W.A. Telliard, U.S. EPA, Office of Water Regulations and Standards, 401 M St., SW, Washington, DC, 20460. Phone: 202-382-7131).

Phenacetin   **EPA Method 8270**
CAS #62-44-2

TITLE Semivolatile Organic Compounds by GC/MS

MATRIX This method is used to determine the concentration of semivolatile organic compounds in extracts prepared from all types of solid waste matrices, soils, and groundwater. Although surface waters are not specifically mentioned, this method should be applicable to water samples from rivers, lakes, etc.

METHOD SUMMARY This method covers 259 semivolatile organic compounds. In very limited applications direct injection of the sample into the GC/MS system may be appropriate, but this results in very high detection limits (approximately 10,000 µg/L). Typically, a 1-L liquid sample, containing surrogate, and matrix spiking standards, is extracted in a continuous extractor first under acid conditions and then under basic conditions. Typically 30 g of a solid sample, containing surrogate, and matrix spiking standards, is extracted ultrasonically. After concentrating the extract to 1 mL it is spiked with 10 µL of an internal standard solution just prior to analysis by GC/MS. The volume injected should contain about 100 ng of base/neutral and 200 ng of acid surrogates (for a 1-µL injection). Analysis is performed by GC/MS using a capillary GC column.

INTERFERENCES Raw GC/MS data from all blanks, samples, and spikes must be evaluated for interferences. Contamination by carryover can occur whenever high-concentration and low-concentration samples are sequentially analyzed. To reduce carryover, the sample syringe must be rinsed out between samples with solvent. Whenever an unusually concentrated sample is encountered, it should be followed by the analysis of blank solvent to check for cross-contamination.

INSTRUMENTATION A GC/MS and a data system are required. The GC column used is a 30 m × 0.25 mm I.D. (or 0.32 mm I.D.) 1um film thickness silicone-coated fused silica capillary column. A continuous liquid-liquid extractor equipped with Teflon® or glass connection joints and stopcocks requiring no lubrication, a K-D concentrating apparatus, water bath, and an ultrasonic disrupter with a minimum power of 300 W and with pulsing capability are also required.

PRECISION & ACCURACY The estimated quantitation limit (EQL) of Method 8270B for determining an individual compound is approximately 1 mg/kg (wet weight) for soil or sediment samples, 1–200 mg/kg for wastes (dependent on matrix and method of preparation), and 10 µg/L for groundwater samples. EQLs will be proportionately higher for sample extracts that require dilution to avoid saturation of the detector.

The EQL(b) for groundwater in µg/L is 20.
The EQL (a, b) for low concentrations in soil and sediment in µg/kg is not determined.
Accuracy as µg/L is not listed.
Overall precision in µg/L is not listed.

(a) *EQLs listed for soil/sediment are based on wet weight. Normally data is reported in a dry-weight basis; therefore, EQLs will be higher based on the % dry weight of each sample. This calculation is based on a 30 g sample and gel permeation chromatography cleanup.*

(b) *Sample EQLs are highly matrix-dependent. The EQLs are provided for guidance and may not always be achievable.*

C = *True value for concentration, in µg/L.*
X = *Average recovery found for measurements of samples containing a concentration of C, in µg/L.*

ESTIMATED QUANTITATION LIMIT

Other Matrices	Factor (a)
High-concentration soil and sludges by sonicator	7.5
Non-water miscible waste	75

(a) *EQL for other matrices = [EQL for low soil/sediment] × [Factor]. This estimated EQL is similar to an EPA "Practical Quantitation Limit."*

SAMPLING METHOD

Liquid samples — Use a 1 or 2½ gallon amber glass bottle with a screw-top Teflon®-lined cover that has been prewashed with detergent and rinsed with distilled water and methanol (or isopropanol).

Soils, sediments, or sludges — Use an 8-oz. widemouth glass with a screw-top Teflon®-lined cover that has been prewashed with detergent and rinsed with distilled water and methanol (or isopropanol).

SAMPLE PRESERVATION

Liquid samples — If residual chlorine is present, add 3 mL of 10% sodium thiosulfate per gallon, cool to 4°C and store in a solvent-free refrigerator until analysis; if chlorine is not present, then eliminate the sodium thiosulfate addition.

Soils, sediments, or sludges — Cool samples to 4°C and store in a solvent-free refrigerator.

MHT Liquid samples must be extracted within 7 days and the extracts analyzed within 40 days. Soils, sediments, or sludges may be stored for a maximum of 14 days and the extracts analyzed within 40 days.

SAMPLE PREPARATION

Liquid samples — Transfer 1 L quantitatively to a continuous extractor. If high concentrations are anticipated, a smaller volume may be used and then diluted with organic-free reagent water to 1 L. Adjust pH, if necessary, to pH <2 using 1:1 (V/V) sulfuric acid. Pipette 1.0 mL of a surrogate standard spiking solution into each sample. For the sample in each analytical batch selected for spiking, add 1.0 mL of a matrix spiking standard. For base/neutral acid analysis, the amount of the surrogates and matrix spiking compounds added to the sample should result in a final concentration of 100 ng/µL of each analyte in the extract to be analyzed (assuming a 1-µL injection). Extract with methylene chloride for 18–24 h. Next, adjust the pH of the aqueous phase to pH >11 using 10 N sodium hydroxide and extract it with methylene chloride again for 18–24 h. Dry the extract through a column containing anhydrous sodium sulfate and concentrate it to 1 mL using a K-D concentrator.

Soils, sediments, or sludges — Use 30 g of sample. Nonporous or wet samples (gummy or clay type) that do not have a free-flowing sandy texture must be mixed with anhydrous sodium sulfate until the sample is free flowing. Add 1 mL of surrogate standards to all samples, spikes, standards, and blanks. For the sample in each analytical batch selected for spiking, add 1.0 mL of a matrix spiking standard. For base/neutral acid analysis, the amount added of the surrogates and matrix spiking compounds should result in a final concentration of 100 ng/µL of each base/neutral analyte and 200 ng/µL of each acid analyte in the extract to be analyzed (assuming a 1-µL injection). Immediately add a 100-mL mixture of 1:1 methylene chloride:acetone and extract the sample ultrasonically for 3 min and then decant or filter the extracts. Repeat the extraction two or more times. Dry the extract using a column with anhydrous sodium sulfate and concentrate it to 1 mL in a K-D concentrator.

QUALITY CONTROL A methylene chloride solution containing 50 ng/µL of decafluorotriphenylphosphine (DFTPP) is used for tuning the GC/MS system each 12-h shift. A system performance check also must be made during every 12-h shift. A standard containing 50 ng/µL each of 4,4'-DDT, pentachlorophenol, and benzidine is required to verify injection port inertness and GC column performance. A calibration standard at mid-concentration, containing each compound of interest, including all required surrogates, must be performed every 12 h during analysis. After the system performance check is met, calibration check compounds (CCCs) are used to check the validity of the initial calibration.

The internal standard responses and retention times in the calibration check standard must be evaluated immediately after or during data acquisition. If the retention time for any internal standard changes by more than 30 seconds from the last check calibration (12 h), the chromatographic system must be inspected for malfunctions and corrections must be made, as required. If the electron ionization current plot (EICP) area for any of the internal standards changes by a factor of two from the last daily calibration standard check, the mass spectrometer must be inspected for malfunctions and corrections must be made, as appropriate.

Demonstrate, through the analysis of a reagent water blank, that interferences from the analytical system, glassware, and reagents are under control. The blank samples should be carried through all stages of the sample preparation and measurement steps. For each analytical batch (up to 20 samples), a reagent blank, matrix spike, and matrix spike duplicate/duplicate must be analyzed (the frequency of the spikes may be different for different monitoring programs). The blank and spiked samples must be carried through all stages of the sample preparation and measurement steps. A QC reference sample concentrate containing each analyte at a concentration of 100 mg/L in methanol is required.

REFERENCE Test Methods for Evaluating Solid Waste (SW-846). U.S. EPA 1983, Method 8270B, Rev. 2, Nov. 1990. Office of Solid Waste, Washington, DC.

Phenanthrene   **EPA Method 1625**
CAS #85-01-8

TITLE Semivolatile Organic Compounds by Isotope Dilution GC/MS

MATRIX The compounds may be determined in waters, soils, and municipal sludges by this method.

METHOD SUMMARY This method is used to determine 176 semivolatile toxic organic pollutants associated with the CWA (as amended 1987); the RCRA (as amended 1986); the CERCLA (as amended 1986); and other compounds amenable to extraction and analysis by capillary column gas chromatography-mass spectrometry (GC/MS).

Stable isotopically-labeled analogs of the compounds of interest are added to the sample. If the solids content is less than 1%, a 1-L sample is extracted at pH 12–13, then at pH <2 with methylene chloride using continuous extraction techniques.

If the solids content is 30% or less, the sample is diluted to 1% solids with reagent water, homogenized ultrasonically, and extracted at pH 12–13, then at pH <2 with methylene chloride using continuous extraction techniques. If the solids content is greater than 30%, the sample is extracted using ultrasonic techniques.

Each extract is dried over sodium sulfate, concentrated to a volume of 5 mL, cleaned up using GPC, if necessary, and concentrated. Extracts are concentrated to 1 mL if GPC is not performed, and to 0.5 mL if GPC is performed.

An internal standard is added to the extract, and a 1-mL aliquot of the extract is injected into the GC. The compounds are separated by GC and detected by a MS. The labeled compounds serve to correct the variability of the analytical technique.

INTERFERENCES Solvents, reagents, glassware, and other sample processing hardware may yield artifacts and/or elevated baselines causing misinterpretation of chromatograms and spectra. Materials used in the analysis must be demonstrated to be free from interferences under the conditions of analysis by running method blanks initially and with each sample lot (sample started through the extraction process on a given 8-h shift, to a maximum of 20). Specific selection of reagents and purification of solvents by distillation in all glass systems may be required. Glassware and, where possible, reagents are cleaned by solvent rinse and baking at 450°C for 1-h minimum. Interferences coextracted from samples will vary considerably from source to source, depending on the diversity of the site being sampled.

INSTRUMENTATION Major instrumentation includes a GC with a splitless or on-column injection port for capillary column, a MS with 70 eV electron impact ionization, and a data system to collect and record MS data, and process it. A K-D apparatus is used to concentrate extracts.

GC Column: 30 m × 0.25 mm I.D. 5% phenyl, 94% methyl, 1% vinyl silicone bonded phased fused silica capillary column.

PRECISION & ACCURACY The detection limits of the method are usually dependent on the level of interferences rather than instrumental limitations. The limits typify the minimum quantities that can be detected with no interferences present.

The minimum level (in µg/mL) was 10. This is defined as a minimum level at which the analytical system shall give recognizable mass spectra (background corrected) and acceptable calibration points.

The MDL (in µg/kg) in low solids was 42 and in high solids was 22; these were determined in digested sludge (low solids) and in filter cake or compost (high solids).

The labeled and native compound initial precision as standard deviation (in µg/L) was 13.
The labeled and native compound initial accuracy as average recovery (in µg/L) was 93–119.

SAMPLE COLLECTION, PRESERVATION & HANDLING
Collect samples in glass containers. Aqueous samples which flow freely are collected in refrigerated bottles using automatic sampling equipment. Solid samples are collected as grab samples using widemouth jars. Maintain samples at 0 to 4°C from the time of collection until extraction. If residual chlorine is present in aqueous samples, add 80 mg sodium thiosulfate/L of water. Begin sample extraction within 7 days of collection, and analyze all extracts within 40 days of extraction.

SAMPLE PREPARATION Samples containing 1% solids or less are extracted directly using continuous liquid-liquid extraction techniques. Samples containing 1 to 30% solids are diluted to the 1% level with reagent water and extracted using continuous

liquid-liquid extraction techniques. Samples containing greater than 30% solids are extracted using ultrasonic techniques.

Base/neutral extraction — Adjust the pH of the waters in the extractors to 12–13 with 6 N NaOH. Extract with methylene chloride for 24–48 h.

Acid extraction — Adjust the pH of the waters in the extractors to 2 or less using 6 N sulfuric acid. Extract with methylene chloride for 24–48 h.

Ultrasonic extraction of high solids samples — Add anhydrous sodium sulfate to the sample and QC aliquot(s). Add acetone:methylene chloride (1:1) to the sample and mix thoroughly

Concentrate extracts using a K-D apparatus.

QUALITY CONTROL The analyst is permitted to modify this method to improve separations or lower the costs of measurements, provided all performance specifications are met. Analyses of blanks are required to demonstrate freedom from contamination. When results of spikes indicate atypical method performance for samples, the samples are diluted to bring method performance within acceptable limits.

For low solids (aqueous samples), extract, concentrate, and analyze two sets of four 1-L aliquots (8 aliquots total) of the precision and recovery standard. For high solids samples, two sets of four 30-g aliquots of the high solids reference matrix are used.

Spike all samples with labeled compounds to assess method performance. Compute percent recovery of the labeled compounds using the internal standard method. Compare the labeled compound recovery for each compound with the corresponding labeled compound recovery.

Reagent water and high solids reference matrix blanks are analyzed to demonstrate freedom from contamination. Extract and concentrate a 1-L reagent water blank or a high solids reference matrix blank with each sample's lot (samples started through the extraction process on the same 8-h shift, to a maximum of 20 samples).

Field replicates may be collected to determine the precision of the sampling technique, and spiked samples may be required to determine the accuracy of the analysis when the internal standard method is used.

REFERENCE Semivolatile Organic Compounds by Isotope Dilution GC/MS. Office of Water Regulation and Standards, U.S. EPA Industrial Technology Division, Washington, DC, EPA Method 1625, Rev. C, June 1989 (contact W.A. Telliard, U.S. EPA, Office of Water Regulations and Standards, 401 M St., SW, Washington, DC, 20460. Phone: 202-382-7131).

Phenanthrene   **EPA Method 625**
CAS #85-01-8

TITLE Base/Neutrals and Acids, U.S. EPA Method 625

MATRIX This methods covers municipal and industrial wastewaters.

METHOD SUMMARY Approximately 1 L of sample is serially extracted with methylene chloride at a pH greater than 11 and again at a pH less than 2 using a separatory funnel or a continuous extractor. The methylene chloride extract is dried, concentrated to a volume of 1 mL, and analyzed by GC/MS. Qualitative identification of the parameters in the extract is performed using the retention time and the relative abundance of three characteristic masses (m/z). Qualitative analysis is performed using either external or internal standard techniques with a single characteristic m/z.

INTERFERENCES Method interferences may be caused by contaminants in solvents, reagents, glassware, and other sample processing hardware. Glassware must be scrupulously cleaned. Glassware should be heated in a muffle furnace at 400°C for 5 to 30 min. Some thermally stable materials, such as PCBs, may not be eliminated by this treatment. Solvent rinses with acetone and pesticide quality hexane may be substituted for the muffle furnace heating. Matrix interferences may be caused by contaminants that are coextracted from the sample. The baseneutral extraction may cause significantly reduced recovery of phenols. The packed gas chromatographic columns recommended for the basic fraction may not exhibit sufficient resolution for some analytes.

INSTRUMENTATION A GC/MS system with an injection port designed for on-column injection when using packed columns and for splitless injection when using capillary columns.

Column for base/neutrals: 1.8 m long × 2 mm I.D. glass, packed with 3% SP-2550 on Supelcoport (100/120 mesh) or equivalent.

Column for acids: 1.8 m long × 2 mm I.D. glass, packed with 1% SP-1240DA on Supelcoport (100/120 mesh) or equivalent.

PRECISION & ACCURACY The MDL concentrations were obtained using reagent water. The MDL actually achieved in a given analysis will vary depending on instrument sensitivity and matrix effects. This method was tested by 15 laboratories using reagent water, drinking water, surface water, and industrial wastewaters spiked at six concentrations over the range 5 to 100 µg/L. Single operator precision, overall precision, and method accuracy were found to be directly related to the concentration of the parameter matrix.

The MDL (in µg/L) in reagent water was 5.4.

The standard deviation (in µg/L based on 4 recovery measurements) was 20.6.

The range (in µg/L) for average recovery for 4 measurements was 65.2–08.7.

The range (in %) for percent recovery was 54–120.

Accuracy (in µg/L) as expected recovery for one or more measurements of a sample containing a true concentration of C was 0.87C−0.06.

Precision (in µg/L) as expected single analyst standard deviation of measurements at an average concentration found at X was 0.12X + 0.57.

Overall precision (in µg/L) as expected interlaboratory standard deviation of measurements in an average concentration found at X was 0.15X + .025.

C = *True value of the concentration in µg/L.*

X = Average recovery found for measurements of samples containing a concentration at C in µg/L.

SAMPLE PREPARATION Adjust the pH to >11 with sodium hydroxide and serially extract in a separatory funnel with methylene chloride or else in a continuous extractor. Next, adjust the pH to <2 with sulfuric acid and serially extract in a separatory funnel with methylene chloride or else in a continuous extractor. Dry the extracts separately through a column of anhydrous sodium sulfate and then concentrate each of the extracts to 1.0 mL using a K-D apparatus.

SAMPLE COLLECTION, PRESERVATION & HANDLING Grab samples must be collected in glass containers. All samples must be refrigerated at 4°C from the time of collection until extraction. If residual chlorine is present, add 80 mg of sodium thiosulfate/L of sample and mix well. All samples must be extracted within 7 days of collection and completely analyzed within 40 days of extraction.

QUALITY CONTROL Make an initial, one-time, demonstration of the ability to generate acceptable accuracy and precision with this method. Before processing any samples, the analyst must analyze a reagent water blank to demonstrate that interferences from the analytical system and glassware are under control. Each time a set of samples is extracted or reagents are changed, a reagent water blank must be processed. Spike and analyze a minimum of 5% of all samples to monitor and evaluate lab data quality. A QC check sample concentrate that contains each parameter of interest at a concentration of 100 µg/mL in acetone is required. PCBs and multicomponent pesticides may be omitted from this test.

After analysis of five spiked wastewater samples, calculate the average percent recovery and the standard deviation of the percent recovery. Spike all samples with the surrogate standard spiking solution and calculate the percent recovery of each surrogate compound.

REFERENCE Federal Register, Vol. 49, No. 209. Friday, Oct. 26, 1984.

Phenanthrene   **EPA Method 8270**
CAS #85-01-8

TITLE Semivolatile Organic Compounds by GC/MS

MATRIX This method is used to determine the concentration of semivolatile organic compounds in extracts prepared from all types of solid waste matrices, soils, and groundwater. Although surface waters are not specifically mentioned, this method should be applicable to water samples from rivers, lakes, etc.

METHOD SUMMARY This method covers 259 semivolatile organic compounds. In very limited applications direct injection of the sample into the GC/MS system may be appropriate, but this results in very high detection limits (approximately 10,000 µg/L). Typically, a 1-L liquid sample, containing surrogate, and matrix spiking standards, is extracted in a continuous extractor first under acid conditions and then under basic conditions. Typically 30 g of a solid sample, containing surrogate, and matrix spiking standards, is extracted ultrasonically. After concentrating the extract to 1 mL it is spiked with 10 µL of an internal standard solution just prior to analysis by GC/MS. The volume injected should contain about 100 ng of base/neutral and 200 ng of acid surrogates (for a 1-µL injection). Analysis is performed by GC/MS using a capillary GC column.

INTERFERENCES Raw GC/MS data from all blanks, samples, and spikes must be evaluated for interferences. Contamination by carryover can occur whenever high-concentration and low-concentration samples are sequentially analyzed. To reduce carryover, the sample syringe must be rinsed out between samples with solvent. Whenever an unusually concentrated sample is encountered, it should be followed by the analysis of blank solvent to check for cross-contamination.

INSTRUMENTATION A GC/MS and a data system are required. The GC column used is a 30 m × 0.25 mm I.D. (or 0.32 mm I.D.) 1um film thickness silicone-coated fused silica capillary column. A continuous liquid-liquid extractor equipped with Teflon® or glass connection joints and stopcocks requiring no lubrication, a K-D concentrating apparatus, water bath, and an ultrasonic disrupter with a minimum power of 300 W and with pulsing capability are also required.

PRECISION & ACCURACY The estimated quantitation limit (EQL) of Method 8270B for determining an individual compound is approximately 1 mg/kg (wet weight) for soil or sediment samples, 1–200 mg/kg for wastes (dependent on matrix and method of preparation), and 10 µg/L for groundwater samples. EQLs will be proportionately higher for sample extracts that require dilution to avoid saturation of the detector.

The EQL(b) for groundwater in µg/L is 10.
The EQL (a, b) for low concentrations in soil and sediment in µg/kg is 660.
Accuracy as µg/L is $0.87C + 0.06$.
Overall precision in µg/L is $0.15X + 0.25$.

(a) EQLs listed for soil/sediment are based on wet weight. Normally data is reported in a dry-weight basis; therefore, EQLs will be higher based on the % dry weight of each sample. This calculation is based on a 30 g sample and gel permeation chromatography cleanup.
(b) Sample EQLs are highly matrix-dependent. The EQLs are provided for guidance and may not always be achievable.
C = True value for concentration, in µg/L.
X = Average recovery found for measurements of samples containing a concentration of C, in µg/L.

ESTIMATED QUANTITATION LIMIT

Other Matrices	Factor (a)
High-concentration soil and sludges by sonicator	7.5
Non-water miscible waste	75

(a) EQL for other matrices = [EQL for low soil/sediment] × [Factor]. This estimated EQL is similar to an EPA "Practical Quantitation Limit."

SAMPLING METHOD

Liquid samples — Use a 1 or 2½ gallon amber glass bottle with a screw-top Teflon®-lined cover that has been prewashed with

detergent and rinsed with distilled water and methanol (or isopropanol).

Soils, sediments, or sludges — Use an 8-oz. widemouth glass with a screw-top Teflon®-lined cover that has been prewashed with detergent and rinsed with distilled water and methanol (or isopropanol).

SAMPLE PRESERVATION
Liquid samples — If residual chlorine is present, add 3 mL of 10% sodium thiosulfate per gallon, cool to 4°C and store in a solvent-free refrigerator until analysis; if chlorine is not present, then eliminate the sodium thiosulfate addition.

Soils, sediments, or sludges — Cool samples to 4°C and store in a solvent-free refrigerator.

MHT Liquid samples must be extracted within 7 days and the extracts analyzed within 40 days. Soils, sediments, or sludges may be stored for a maximum of 14 days and the extracts analyzed within 40 days.

SAMPLE PREPARATION
Liquid samples — Transfer 1 L quantitatively to a continuous extractor. If high concentrations are anticipated, a smaller volume may be used and then diluted with organic-free reagent water to 1 L. Adjust pH, if necessary, to pH <2 using 1:1 (V/V) sulfuric acid. Pipette 1.0 mL of a surrogate standard spiking solution into each sample. For the sample in each analytical batch selected for spiking, add 1.0 mL of a matrix spiking standard. For base/neutral acid analysis, the amount of the surrogates and matrix spiking compounds added to the sample should result in a final concentration of 100 ng/μL of each analyte in the extract to be analyzed (assuming a 1-μL injection). Extract with methylene chloride for 18–24 h. Next, adjust the pH of the aqueous phase to pH >11 using 10 N sodium hydroxide and extract it with methylene chloride again for 18–24 h. Dry the extract through a column containing anhydrous sodium sulfate and concentrate it to 1 mL using a K-D concentrator.

Soils, sediments, or sludges — Use 30 g of sample. Nonporous or wet samples (gummy or clay type) that do not have a free-flowing sandy texture must be mixed with anhydrous sodium sulfate until the sample is free flowing. Add 1 mL of surrogate standards to all samples, spikes, standards, and blanks. For the sample in each analytical batch selected for spiking, add 1.0 mL of a matrix spiking standard. For base/neutral acid analysis, the amount added of the surrogates and matrix spiking compounds should result in a final concentration of 100 ng/μL of each base/neutral analyte and 200 ng/μL of each acid analyte in the extract to be analyzed (assuming a 1-μL injection). Immediately add a 100-mL mixture of 1:1 methylene chloride:acetone and extract the sample ultrasonically for 3 min and then decant or filter the extracts. Repeat the extraction two or more times. Dry the extract using a column with anhydrous sodium sulfate and concentrate it to 1 mL in a K-D concentrator.

QUALITY CONTROL A methylene chloride solution containing 50 ng/μL of decafluorotriphenylphosphine (DFTPP) is used for tuning the GC/MS system each 12-h shift. A system performance check also must be made during every 12-h shift. A standard containing 50 ng/μL each of 4,4'-DDT, pentachlorophenol, and benzidine is required to verify injection port inertness and GC column performance. A calibration standard at mid-concentration, containing each compound of interest, including all required surrogates, must be performed every 12 h during analysis. After the system performance check is met, calibration check compounds (CCCs) are used to check the validity of the initial calibration.

The internal standard responses and retention times in the calibration check standard must be evaluated immediately after or during data acquisition. If the retention time for any internal standard changes by more than 30 seconds from the last check calibration (12 h), the chromatographic system must be inspected for malfunctions and corrections must be made, as required. If the electron ionization current plot (EICP) area for any of the internal standards changes by a factor of two from the last daily calibration standard check, the mass spectrometer must be inspected for malfunctions and corrections must be made, as appropriate.

Demonstrate, through the analysis of a reagent water blank, that interferences from the analytical system, glassware, and reagents are under control. The blank samples should be carried through all stages of the sample preparation and measurement steps. For each analytical batch (up to 20 samples), a reagent blank, matrix spike, and matrix spike duplicate/duplicate must be analyzed (the frequency of the spikes may be different for different monitoring programs). The blank and spiked samples must be carried through all stages of the sample preparation and measurement steps. A QC reference sample concentrate containing each analyte at a concentration of 100 mg/L in methanol is required.

Phenanthrene   **EPA Method 8100**
CAS #85-01-8

TITLE Polynuclear Aromatic Hydrocarbons

MATRIX Groundwater, soils, sludges, water miscible liquid wastes, and non-water miscible wastes.

APPLICATION This method is used for the analysis of various PAHs. Samples are extracted, concentrated, and analyzed using direct injection of both neat and diluted organic liquids. The method provides two optional GC columns that are better than Column 1 and that may help resolve analytes from interferences.

INTERFERENCES Solvents, reagents, and glassware may introduce artifacts. Other interferences may come from coextracted compounds from samples.

INSTRUMENTATION GC capable of on-column injections and a flame with detector (FID). Column 1: a 1.8 m by 2 mm 3% OV-17 on Chromosorb W-AW-DCMS column. Column 2: a 30 m by 0.25 mm SE-54 fused silica capillary colunm. Column 3: a 30 m by 0.32 mm SE-54 fused silica capillary column.

RANGE 0.1–425 μg/L

MDL Not reported.

PQL FACTORS FOR MULTIPLYING × FID MDL VALUE
Not available.

PRECISION 0.47X–0.25 µg/L (overall precision).

ACCURACY 0.72C–0.95 µg/L (as recovery).

SAMPLING METHOD Use 8-oz. widemouth glass bottles with Teflon®-lined caps for concentrated waste samples, soils, sediments, and sludges. Use 1 or 2½ gallon amber glass bottles with Teflon®-lined caps for liquid (water) samples.

STABILITY Cool soil, sediment, sludge, and liquid samples to 4°C. If residual chlorine is present in liquid samples add 3 mL of 10% sodium thiosulfate per gallon of sample and cool to 4°C.

MHT 14 days for concentrated waste, soil, sediment, or sludge; 7 days for liquid samples; all extracts must be analyzed within 40 days.

QUALITY CONTROL A quality control check sample concentrate containing each analyte of interest is required. The QC check sample concentrate may be prepared from pure standard materials or purchased as certified solutions Use appropriate trip, matrix, control site, method, reagent, and solvent blanks. Internal, surrogate, and five concentration level calibration standards are used. The quality control check sample concentrate should contain phenanthrene at 100 µg/mL in acetonitrile.

REFERENCE Test Methods for Evaluating Solid Waste (SW-846), U.S. EPA Office of Solid Waste, Washington, DC, Method 8100, Nov. 1986.

Phenanthrene
CAS #85-01-8
EPA Method 8310

TITLE Polynuclear Aromatic Hydrocarbons

MATRIX Groundwater, soils, sludges, water miscible liquid wastes, and non-water miscible wastes.

APPLICATION This method is used for the analysis of 16 polynuclear aromatic hydrocarbons (PAHs). Samples are extracted, concentrated, and analyzed using HPLC with detection by UV and fluorescence detectors.

INTERFERENCES Solvents, reagents, and glassware may introduce artifacts. Other interferences may come from coextracted compounds from samples.

INSTRUMENTATION HPLC with a gradient pumping system and a 250 mm by 2.6 mm reverse phase HC-ODS Sil-X 5-micron particle-size column. The fluorescence detector uses an excitation wavelength of 280 nm and emission greater than 389 nm cutoff with dispersive optics.

RANGE 0.1–425 µg/L

MDL 0.64 µg/L (fluorescence; reagent water).

PQL FACTORS FOR MULTIPLYING × FID MDL VALUE

Matrix	Multiplication Factor
Groundwater	10
Low-level soil by sonication with GPC cleanup	670
High-level soil and sludge by sonication	10,000
Non-water miscible waste	100,000

PRECISION 0.47X–0.25 µg/L (overall precision).

ACCURACY 0.72C–0.95 µg/L (as recovery).

SAMPLING METHOD Use 8-oz. widemouth glass bottles with Teflon®-lined caps for concentrated waste samples, soils, sediments, and sludges. Use 1 or 2½ gallon amber glass bottles with Teflon®-lined caps for liquid (water) samples.

STABILITY Cool soil, sediment, sludge, and liquid samples to 4°C. If residual chlorine is present in liquid samples add 3 mL of 10% sodium thiosulfate per gallon of sample and cool to 4°C.

MHT 14 days for concentrated waste, soil, sediment, or sludge; 7 days for liquid samples; all extracts must be analyzed within 40 days.

QUALITY CONTROL Internal, surrogate, and five concentration level calibration standards are used. The calibration standards must be used with the analytical method blank. A quality control check sample concentrate containing phenanthrene at 100 µg/mL is required. The QC check sample concentrate may be prepared from pure standard materials or purchased as certified solutions. Use appropriate trip, matrix, control site, method, reagent, and solvent blanks.

REFERENCE Test Methods for Evaluating Solid Waste (SW-846), U.S. EPA Office of Solid Waste, Washington, DC, Method 8310, Rev. 0, Nov. 1986.

Phenobarbital
CAS #50-06-6
EPA Method 8270

TITLE Semivolatile Organic Compounds by GC/MS

MATRIX This method is used to determine the concentration of semivolatile organic compounds in extracts prepared from all types of solid waste matrices, soils, and groundwater. Although surface waters are not specifically mentioned, this method should be applicable to water samples from rivers, lakes, etc.

METHOD SUMMARY This method covers 259 semivolatile organic compounds. In very limited applications direct injection of the sample into the GC/MS system may be appropriate, but this results in very high detection limits (approximately 10,000 µg/L). Typically, a 1-L liquid sample, containing surrogate, and matrix spiking standards, is extracted in a continuous extractor first under acid conditions and then under basic conditions. Typically 30 g of a solid sample, containing surrogate, and matrix spiking standards, is extracted ultrasonically. After concentrating the extract to 1 mL it is spiked with 10 µL of an

internal standard solution just prior to analysis by GC/MS. The volume injected should contain about 100 ng of base/neutral and 200 ng of acid surrogates (for a 1-μL injection). Analysis is performed by GC/MS using a capillary GC column.

INTERFERENCES Raw GC/MS data from all blanks, samples, and spikes must be evaluated for interferences. Contamination by carryover can occur whenever high-concentration and low-concentration samples are sequentially analyzed. To reduce carryover, the sample syringe must be rinsed out between samples with solvent. Whenever an unusually concentrated sample is encountered, it should be followed by the analysis of blank solvent to check for cross-contamination.

INSTRUMENTATION A GC/MS and a data system are required. The GC column used is a 30 m × 0.25 mm I.D. (or 0.32 mm I.D.) 1um film thickness silicone-coated fused silica capillary column. A continuous liquid-liquid extractor equipped with Teflon® or glass connection joints and stopcocks requiring no lubrication, a K-D concentrating apparatus, water bath, and an ultrasonic disrupter with a minimum power of 300 W and with pulsing capability are also required.

PRECISION & ACCURACY The estimated quantitation limit (EQL) of Method 8270B for determining an individual compound is approximately 1 mg/kg (wet weight) for soil or sediment samples, 1–200 mg/kg for wastes (dependent on matrix and method of preparation), and 10 μg/L for groundwater samples. EQLs will be proportionately higher for sample extracts that require dilution to avoid saturation of the detector.

The EQL(b) for groundwater in μg/L is 10.
The EQL (a, b) for low concentrations in soil and sediment in μg/kg is not determined.
Accuracy as μg/L is not listed.
Overall precision in μg/L is not listed.

(a) *EQLs listed for soil/sediment are based on wet weight. Normally data is reported in a dry-weight basis; therefore, EQLs will be higher based on the % dry weight of each sample. This calculation is based on a 30 g sample and gel permeation chromatography cleanup.*
(b) *Sample EQLs are highly matrix-dependent. The EQLs are provided for guidance and may not always be achievable.*
C = *True value for concentration, in μg/L.*
X = *Average recovery found for measurements of samples containing a concentration of C, in μg/L.*

ESTIMATED QUANTITATION LIMIT

Other Matrices	Factor (a)
High-concentration soil and sludges by sonicator	7.5
Non-water miscible waste	75

(a) *EQL for other matrices = [EQL for low soil/sediment] × [Factor]. This estimated EQL is similar to an EPA "Practical Quantitation Limit."*

SAMPLING METHOD
Liquid samples — Use a 1 or 2½ gallon amber glass bottle with a screw-top Teflon®-lined cover that has been prewashed with detergent and rinsed with distilled water and methanol (or isopropanol).

Soils, sediments, or sludges — Use an 8-oz. widemouth glass with a screw-top Teflon®-lined cover that has been prewashed with detergent and rinsed with distilled water and methanol (or isopropanol).

SAMPLE PRESERVATION
Liquid samples — If residual chlorine is present, add 3 mL of 10% sodium thiosulfate per gallon, cool to 4°C and store in a solvent-free refrigerator until analysis; if chlorine is not present, then eliminate the sodium thiosulfate addition.

Soils, sediments, or sludges — Cool samples to 4°C and store in a solvent-free refrigerator.

MHT Liquid samples must be extracted within 7 days and the extracts analyzed within 40 days. Soils, sediments, or sludges may be stored for a maximum of 14 days and the extracts analyzed within 40 days.

SAMPLE PREPARATION
Liquid samples — Transfer 1 L quantitatively to a continuous extractor. If high concentrations are anticipated, a smaller volume may be used and then diluted with organic-free reagent water to 1 L. Adjust pH, if necessary, to pH <2 using 1:1 (V/V) sulfuric acid. Pipette 1.0 mL of a surrogate standard spiking solution into each sample. For the sample in each analytical batch selected for spiking, add 1.0 mL of a matrix spiking standard. For base/neutral acid analysis, the amount of the surrogates and matrix spiking compounds added to the sample should result in a final concentration of 100 ng/μL of each analyte in the extract to be analyzed (assuming a 1-μL injection). Extract with methylene chloride for 18–24 h. Next, adjust the pH of the aqueous phase to pH >11 using 10 N sodium hydroxide and extract it with methylene chloride again for 18–24 h. Dry the extract through a column containing anhydrous sodium sulfate and concentrate it to 1 mL using a K-D concentrator.

Soils, sediments, or sludges — Use 30 g of sample. Nonporous or wet samples (gummy or clay type) that do not have a free-flowing sandy texture must be mixed with anhydrous sodium sulfate until the sample is free flowing. Add 1 mL of surrogate standards to all samples, spikes, standards, and blanks. For the sample in each analytical batch selected for spiking, add 1.0 mL of a matrix spiking standard. For base/neutral acid analysis, the amount added of the surrogates and matrix spiking compounds should result in a final concentration of 100 ng/μL of each base/neutral analyte and 200 ng/μL of each acid analyte in the extract to be analyzed (assuming a 1-μL injection). Immediately add a 100-mL mixture of 1:1 methylene chloride:acetone and extract the sample ultrasonically for 3 min and then decant or filter the extracts. Repeat the extraction two or more times. Dry the extract using a column with anhydrous sodium sulfate and concentrate it to 1 mL in a K-D concentrator.

QUALITY CONTROL A methylene chloride solution containing 50 ng/μL of decafluorotriphenylphosphine (DFTPP) is used for tuning the GC/MS system each 12-h shift. A system performance check also must be made during every 12-h shift. A standard containing 50 ng/μL each of 4,4'-DDT, pentachlorophenol, and benzidine is required to verify injection port inertness and GC column performance. A calibration standard

at mid-concentration, containing each compound of interest, including all required surrogates, must be performed every 12 h during analysis. After the system performance check is met, calibration check compounds (CCCs) are used to check the validity of the initial calibration.

The internal standard responses and retention times in the calibration check standard must be evaluated immediately after or during data acquisition. If the retention time for any internal standard changes by more than 30 seconds from the last check calibration (12 h), the chromatographic system must be inspected for malfunctions and corrections must be made, as required. If the electron ionization current plot (EICP) area for any of the internal standards changes by a factor of two from the last daily calibration standard check, the mass spectrometer must be inspected for malfunctions and corrections must be made, as appropriate.

Demonstrate, through the analysis of a reagent water blank, that interferences from the analytical system, glassware, and reagents are under control. The blank samples should be carried through all stages of the sample preparation and measurement steps. For each analytical batch (up to 20 samples), a reagent blank, matrix spike, and matrix spike duplicate/duplicate must be analyzed (the frequency of the spikes may be different for different monitoring programs). The blank and spiked samples must be carried through all stages of the sample preparation and measurement steps. A QC reference sample concentrate containing each analyte at a concentration of 100 mg/L in methanol is required.

REFERENCE Test Methods for Evaluating Solid Waste (SW-846). U.S. EPA 1983, Method 8270B, Rev. 2, Nov. 1990. Office of Solid Waste, Washington, DC.

Phenol
CAS #108-95-2
EPA Method 1625

TITLE Semivolatile Organic Compounds by Isotope Dilution GC/MS

MATRIX The compounds may be determined in waters, soils, and municipal sludges by this method.

METHOD SUMMARY This method is used to determine 176 semivolatile toxic organic pollutants associated with the CWA (as amended 1987); the RCRA (as amended 1986); the CERCLA (as amended 1986); and other compounds amenable to extraction and analysis by capillary column gas chromatography-mass spectrometry (GC/MS).

Stable isotopically-labeled analogs of the compounds of interest are added to the sample. If the solids content is less than 1%, a 1-L sample is extracted at pH 12–13, then at pH <2 with methylene chloride using continuous extraction techniques.

If the solids content is 30% or less, the sample is diluted to 1% solids with reagent water, homogenized ultrasonically, and extracted at pH 12–13, then at pH <2 with methylene chloride using continuous extraction techniques. If the solids content is greater than 30%, the sample is extracted using ultrasonic techniques.

Each extract is dried over sodium sulfate, concentrated to a volume of 5 mL, cleaned up using GPC, if necessary, and concentrated. Extracts are concentrated to 1 mL if GPC is not performed, and to 0.5 mL if GPC is performed.

An internal standard is added to the extract, and a 1-mL aliquot of the extract is injected into the GC. The compounds are separated by GC and detected by a MS. The labeled compounds serve to correct the variability of the analytical technique.

INTERFERENCES Solvents, reagents, glassware, and other sample processing hardware may yield artifacts and/or elevated baselines causing misinterpretation of chromatograms and spectra. Materials used in the analysis must be demonstrated to be free from interferences under the conditions of analysis by running method blanks initially and with each sample lot (sample started through the extraction process on a given 8-h shift, to a maximum of 20). Specific selection of reagents and purification of solvents by distillation in all glass systems may be required. Glassware and, where possible, reagents are cleaned by solvent rinse and baking at 450°C for 1-h minimum. Interferences coextracted from samples will vary considerably from source to source, depending on the diversity of the site being sampled.

INSTRUMENTATION Major instrumentation includes a GC with a splitless or on-column injection port for capillary column, a MS with 70 eV electron impact ionization, and a data system to collect and record MS data, and process it. A K-D apparatus is used to concentrate extracts.

GC Column: 30 m × 0.25 mm I.D. 5% phenyl, 94% methyl, 1% vinyl silicone bonded phased fused silica capillary column.

PRECISION & ACCURACY The detection limits of the method are usually dependent on the level of interferences rather than instrumental limitations. The limits typify the minimum quantities that can be detected with no interferences present.

The minimum level (in µg/mL) was 10. This is defined as a minimum level at which the analytical system shall give recognizable mass spectra (background corrected) and acceptable calibration points.

The MDL (in µg/kg) in low solids was 2501 and in high solids was 757; these were determined in digested sludge (low solids) and in filter cake or compost (high solids).

Note: Background levels of this compound were present in the sludge tested, resulting in higher than expected MDLs. The MDL for this compound is expected to be approximately 50 µg/kg with no interferences present.

The labeled and native compound initial precision as standard deviation (in µg/L) was 36.

The labeled and native compound initial accuracy as average recovery (in µg/L) was 77–127.

SAMPLE COLLECTION, PRESERVATION & HANDLING Collect samples in glass containers. Aqueous samples which flow freely are collected in refrigerated bottles using automatic

sampling equipment. Solid samples are collected as grab samples using widemouth jars. Maintain samples at 0 to 4°C from the time of collection until extraction. If residual chlorine is present in aqueous samples, add 80 mg sodium thiosulfate/L of water. Begin sample extraction within 7 days of collection, and analyze all extracts within 40 days of extraction.

SAMPLE PREPARATION Samples containing 1% solids or less are extracted directly using continuous liquid-liquid extraction techniques. Samples containing 1 to 30% solids are diluted to the 1% level with reagent water and extracted using continuous liquid-liquid extraction techniques. Samples containing greater than 30% solids are extracted using ultrasonic techniques.

Base/neutral extraction — Adjust the pH of the waters in the extractors to 12–13 with 6 N NaOH. Extract with methylene chloride for 24–48 h.

Acid extraction — Adjust the pH of the waters in the extractors to 2 or less using 6 N sulfuric acid. Extract with methylene chloride for 24–48 h.

Ultrasonic extraction of high solids samples — Add anhydrous sodium sulfate to the sample and QC aliquot(s). Add acetone:methylene chloride (1:1) to the sample and mix thoroughly

Concentrate extracts using a K-D apparatus.

QUALITY CONTROL The analyst is permitted to modify this method to improve separations or lower the costs of measurements, provided all performance specifications are met. Analyses of blanks are required to demonstrate freedom from contamination. When results of spikes indicate atypical method performance for samples, the samples are diluted to bring method performance within acceptable limits.

For low solids (aqueous samples), extract, concentrate, and analyze two sets of four 1-L aliquots (8 aliquots total) of the precision and recovery standard. For high solids samples, two sets of four 30-g aliquots of the high solids reference matrix are used.

Spike all samples with labeled compounds to assess method performance. Compute percent recovery of the labeled compounds using the internal standard method. Compare the labeled compound recovery for each compound with the corresponding labeled compound recovery.

Reagent water and high solids reference matrix blanks are analyzed to demonstrate freedom from contamination. Extract and concentrate a 1-L reagent water blank or a high solids reference matrix blank with each sample's lot (samples started through the extraction process on the same 8-h shift, to a maximum of 20 samples).

Field replicates may be collected to determine the precision of the sampling technique, and spiked samples may be required to determine the accuracy of the analysis when the internal standard method is used.

REFERENCE Semivolatile Organic Compounds by Isotope Dilution GC/MS. Office of Water Regulation and Standards, U.S. EPA Industrial Technology Division, Washington, DC, EPA Method 1625, Rev. C, June 1989 (contact W.A. Telliard, U.S. EPA, Office of Water Regulations and Standards, 401 M St., SW, Washington, DC, 20460. Phone: 202-382-7131).

Phenol   EPA Method 625
CAS #108-95-2

TITLE Base/Neutrals and Acids, U.S. EPA Method 625

MATRIX This methods covers municipal and industrial wastewaters.

METHOD SUMMARY Approximately 1 L of sample is serially extracted with methylene chloride at a pH greater than 11 and again at a pH less than 2 using a separatory funnel or a continuous extractor. The methylene chloride extract is dried, concentrated to a volume of 1 mL, and analyzed by GC/MS. Qualitative identification of the parameters in the extract is performed using the retention time and the relative abundance of three characteristic masses (m/z). Qualitative analysis is performed using either external or internal standard techniques with a single characteristic m/z.

INTERFERENCES Method interferences may be caused by contaminants in solvents, reagents, glassware, and other sample processing hardware. Glassware must be scrupulously cleaned. Glassware should be heated in a muffle furnace at 400°C for 5 to 30 min. Some thermally stable materials, such as PCBs, may not be eliminated by this treatment. Solvent rinses with acetone and pesticide quality hexane may be substituted for the muffle furnace heating. Matrix interferences may be caused by contaminants that are coextracted from the sample. The base-neutral extraction may cause significantly reduced recovery of phenols. The packed gas chromatographic columns recommended for the basic fraction may not exhibit sufficient resolution for some analytes.

INSTRUMENTATION A GC/MS system with an injection port designed for on-column injection when using packed columns and for splitless injection when using capillary columns.

Column for base/neutrals: 1.8 m long × 2 mm I.D. glass, packed with 3% SP-2550 on Supelcoport (100/120 mesh) or equivalent.

Column for acids: 1.8 m long × 2 mm I.D. glass, packed with 1% SP-1240DA on Supelcoport (100/120 mesh) or equivalent.

PRECISION & ACCURACY The MDL concentrations were obtained using reagent water. The MDL actually achieved in a given analysis will vary depending on instrument sensitivity and matrix effects. This method was tested by 15 laboratories using reagent water, drinking water, surface water, and industrial wastewaters spiked at six concentrations over the range 5 to 100 µg/L. Single operator precision, overall precision, and method accuracy were found to be directly related to the concentration of the parameter matrix.

The MDL (in µg/L) in reagent water was 1.5.

The standard deviation (in µg/L based on 4 recovery measurements) was 22.6.

The range (in µg/L) for average recovery for 4 measurements was 16.6–100.0.

The range (in %) for percent recovery was 5–112.

Accuracy (in µg/L) as expected recovery for one or more measurements of a sample containing a true concentration of C was 0.43C + 1.26.

Precision (in µg/L) as expected single analyst standard deviation of measurements at an average concentration found at X was 0.28X + 0.73.

Overall precision (in µg/L) as expected interlaboratory standard deviation of measurements in an average concentration found at X was 0.35X + 0.58.

C = *True value of the concentration in µg/L.*
X = *Average recovery found for measurements of samples containing a concentration at C in µg/L.*

SAMPLE PREPARATION Adjust the pH to >11 with sodium hydroxide and serially extract in a separatory funnel with methylene chloride or else in a continuous extractor. Next, adjust the pH to <2 with sulfuric acid and serially extract in a separatory funnel with methylene chloride or else in a continuous extractor. Dry the extracts separately through a column of anhydrous sodium sulfate and then concentrate each of the extracts to 1.0 mL using a K-D apparatus.

SAMPLE COLLECTION, PRESERVATION & HANDLING
Grab samples must be collected in glass containers. All samples must be refrigerated at 4°C from the time of collection until extraction. If residual chlorine is present, add 80 mg of sodium thiosulfate/L of sample and mix well. All samples must be extracted within 7 days of collection and completely analyzed within 40 days of extraction.

QUALITY CONTROL Make an initial, one-time, demonstration of the ability to generate acceptable accuracy and precision with this method. Before processing any samples, the analyst must analyze a reagent water blank to demonstrate that interferences from the analytical system and glassware are under control. Each time a set of samples is extracted or reagents are changed, a reagent water blank must be processed. Spike and analyze a minimum of 5% of all samples to monitor and evaluate lab data quality. A QC check sample concentrate that contains each parameter of interest at a concentration of 100 µg/mL in acetone is required. PCBs and multicomponent pesticides may be omitted from this test.

After analysis of five spiked wastewater samples, calculate the average percent recovery and the standard deviation of the percent recovery. Spike all samples with the surrogate standard spiking solution and calculate the percent recovery of each surrogate compound.

REFERENCE *Federal Register*, Vol. 49, No. 209. Friday, Oct. 26, 1984.

Phenol   **EPA Method 8270**
CAS #108-95-2

TITLE Semivolatile Organic Compounds by GC/MS

MATRIX This method is used to determine the concentration of semivolatile organic compounds in extracts prepared from all types of solid waste matrices, soils, and groundwater. Although surface waters are not specifically mentioned, this method should be applicable to water samples from rivers, lakes, etc.

METHOD SUMMARY This method covers 259 semivolatile organic compounds. In very limited applications direct injection of the sample into the GC/MS system may be appropriate, but this results in very high detection limits (approximately 10,000 µg/L). Typically, a 1-L liquid sample, containing surrogate, and matrix spiking standards, is extracted in a continuous extractor first under acid conditions and then under basic conditions. Typically 30 g of a solid sample, containing surrogate, and matrix spiking standards, is extracted ultrasonically. After concentrating the extract to 1 mL it is spiked with 10 µL of an internal standard solution just prior to analysis by GC/MS. The volume injected should contain about 100 ng of base/neutral and 200 ng of acid surrogates (for a 1-µL injection). Analysis is performed by GC/MS using a capillary GC column.

INTERFERENCES Raw GC/MS data from all blanks, samples, and spikes must be evaluated for interferences. Contamination by carryover can occur whenever high-concentration and low-concentration samples are sequentially analyzed. To reduce carryover, the sample syringe must be rinsed out between samples with solvent. Whenever an unusually concentrated sample is encountered, it should be followed by the analysis of blank solvent to check for cross-contamination.

INSTRUMENTATION A GC/MS and a data system are required. The GC column used is a 30 m × 0.25 mm I.D. (or 0.32 mm I.D.) 1um film thickness silicone-coated fused silica capillary column. A continuous liquid-liquid extractor equipped with Teflon® or glass connection joints and stopcocks requiring no lubrication, a K-D concentrating apparatus, water bath, and an ultrasonic disrupter with a minimum power of 300 W and with pulsing capability are also required.

PRECISION & ACCURACY The estimated quantitation limit (EQL) of Method 8270B for determining an individual compound is approximately 1 mg/kg (wet weight) for soil or sediment samples, 1–200 mg/kg for wastes (dependent on matrix and method of preparation), and 10 µg/L for groundwater samples. EQLs will be proportionately higher for sample extracts that require dilution to avoid saturation of the detector.

The EQL(b) for groundwater in µg/L is 10.
The EQL (a, b) for low concentrations in soil and sediment in µg/kg is 660.
Accuracy as µg/L is 0.43C + 1.26.
Overall precision in µg/L is 0.35X + 0.58.

(a) *EQLs listed for soil/sediment are based on wet weight. Normally data is reported in a dry-weight basis; therefore, EQLs will be higher based on the % dry weight of each sample. This calculation is based on a 30 g sample and gel permeation chromatography cleanup.*
(b) *Sample EQLs are highly matrix-dependent. The EQLs are provided for guidance and may not always be achievable.*
C = *True value for concentration, in µg/L.*
X = *Average recovery found for measurements of samples containing a concentration of C, in µg/L.*

ESTIMATED QUANTITATION LIMIT

Other Matrices	Factor (a)
High-concentration soil and sludges by sonicator	7.5
Non-water miscible waste	75

(a) EQL for other matrices = [EQL for low soil/sediment] × [Factor]. This estimated EQL is similar to an EPA "Practical Quantitation Limit."

SAMPLING METHOD
Liquid samples — Use a 1 or 2½ gallon amber glass bottle with a screw-top Teflon®-lined cover that has been prewashed with detergent and rinsed with distilled water and methanol (or isopropanol).

Soils, sediments, or sludges — Use an 8-oz. widemouth glass with a screw-top Teflon®-lined cover that has been prewashed with detergent and rinsed with distilled water and methanol (or isopropanol).

SAMPLE PRESERVATION
Liquid samples — If residual chlorine is present, add 3 mL of 10% sodium thiosulfate per gallon, cool to 4°C and store in a solvent-free refrigerator until analysis; if chlorine is not present, then eliminate the sodium thiosulfate addition.

Soils, sediments, or sludges — Cool samples to 4°C and store in a solvent-free refrigerator.

MHT Liquid samples must be extracted within 7 days and the extracts analyzed within 40 days. Soils, sediments, or sludges may be stored for a maximum of 14 days and the extracts analyzed within 40 days.

SAMPLE PREPARATION
Liquid samples — Transfer 1 L quantitatively to a continuous extractor. If high concentrations are anticipated, a smaller volume may be used and then diluted with organic-free reagent water to 1 L. Adjust pH, if necessary, to pH <2 using 1:1 (V/V) sulfuric acid. Pipette 1.0 mL of a surrogate standard spiking solution into each sample. For the sample in each analytical batch selected for spiking, add 1.0 mL of a matrix spiking standard. For base/neutral acid analysis, the amount of the surrogates and matrix spiking compounds added to the sample should result in a final concentration of 100 ng/µL of each analyte in the extract to be analyzed (assuming a 1-µL injection). Extract with methylene chloride for 18–24 h. Next, adjust the pH of the aqueous phase to pH >11 using 10 N sodium hydroxide and extract it with methylene chloride again for 18–24 h. Dry the extract through a column containing anhydrous sodium sulfate and concentrate it to 1 mL using a K-D concentrator.

Soils, sediments, or sludges — Use 30 g of sample. Nonporous or wet samples (gummy or clay type) that do not have a free-flowing sandy texture must be mixed with anhydrous sodium sulfate until the sample is free flowing. Add 1 mL of surrogate standards to all samples, spikes, standards, and blanks. For the sample in each analytical batch selected for spiking, add 1.0 mL of a matrix spiking standard. For base/neutral acid analysis, the amount added of the surrogates and matrix spiking compounds should result in a final concentration of 100 ng/µL of each base/neutral analyte and 200 ng/µL of each acid analyte in the extract to be analyzed (assuming a 1-µL injection). Immediately add a 100-mL mixture of 1:1 methylene chloride:acetone and extract the sample ultrasonically for 3 min and then decant or filter the extracts. Repeat the extraction two or more times. Dry the extract using a column with anhydrous sodium sulfate and concentrate it to 1 mL in a K-D concentrator.

QUALITY CONTROL A methylene chloride solution containing 50 ng/µL of decafluorotriphenylphosphine (DFTPP) is used for tuning the GC/MS system each 12-h shift. A system performance check also must be made during every 12-h shift. A standard containing 50 ng/µL each of 4,4'-DDT, pentachlorophenol, and benzidine is required to verify injection port inertness and GC column performance. A calibration standard at mid-concentration, containing each compound of interest, including all required surrogates, must be performed every 12 h during analysis. After the system performance check is met, calibration check compounds (CCCs) are used to check the validity of the initial calibration.

The internal standard responses and retention times in the calibration check standard must be evaluated immediately after or during data acquisition. If the retention time for any internal standard changes by more than 30 seconds from the last check calibration (12 h), the chromatographic system must be inspected for malfunctions and corrections must be made, as required. If the electron ionization current plot (EICP) area for any of the internal standards changes by a factor of two from the last daily calibration standard check, the mass spectrometer must be inspected for malfunctions and corrections must be made, as appropriate.

Demonstrate, through the analysis of a reagent water blank, that interferences from the analytical system, glassware, and reagents are under control. The blank samples should be carried through all stages of the sample preparation and measurement steps. For each analytical batch (up to 20 samples), a reagent blank, matrix spike, and matrix spike duplicate/duplicate must be analyzed (the frequency of the spikes may be different for different monitoring programs). The blank and spiked samples must be carried through all stages of the sample preparation and measurement steps. A QC reference sample concentrate containing each analyte at a concentration of 100 mg/L in methanol is required.

REFERENCE Test Methods for Evaluating Solid Waste (SW-846). U.S. EPA 1983, Method 8270B, Rev. 2, Nov. 1990. Office of Solid Waste, Washington, DC.

Phenol   **EPA Method 8040**
CAS #108-95-2

TITLE Phenols

MATRIX Groundwater, soils, sludges, water miscible liquid wastes, and non-water miscible wastes.

APPLICATION This method is used for the analysis of 17 phenols. Samples are extracted, concentrated, and analyzed using direct injection of both neat and diluted organic liquids.

Pentafluorobenzylbromide (PFB) derivatives also may be made to increase sensitivity of the method.

INTERFERENCES There can be carryover contamination with high- and low-level samples. Solvents, reagents, and glassware may introduce artifacts. Other interferences may come from coextracted compounds from samples.

INSTRUMENTATION GC capable of on-column injections and a flame with detector (FID) or electron capture detector (ECD). Column for underivatized phenol: 1.8 m by 2.0 mm with 1% SP-1240DA on Supelcoport. Column for derivatized phenols: 1.8 m by 2.0 mm with 5% OV-17 on Chromosorb W-AW-DMCS.

RANGE 12–450 µg/L

MDL 0.14 µg/L (FID) and 2.2 µg/L (ECD)

PQL FACTORS FOR MULTIPLYING × FID MDL VALUE

Matrix	Multiplication Factor
Groundwater	10
Low-level soil by sonication with GPC cleanup	670
High-level soil and sludge by sonication	10,000
Non-water miscible waste	100,000

PRECISION $0.17X + 0.77$ µg/L (overall precision using FID)

ACCURACY $0.43C + 0.11$ µg/L (as recovery using FID)

SAMPLING METHOD Use 8-oz. widemouth glass bottles with Teflon®-lined caps for concentrated waste samples, soils, sediments, and sludges. Use 1 or 2½ gallon amber glass bottles with Teflon®-lined caps for liquid (water) samples.

STABILITY Cool soil, sediment, sludge, and liquid samples to 4°C. If residual chlorine is present in liquid samples add 3 mL of 10% sodium thiosulfate per gallon of sample and cool to 4°C.

MHT 14 days for concentrated waste, soil, sediment, or sludge; 7 days for liquid samples; all extracts must be analyzed within 40 days.

QUALITY CONTROL A quality control check sample concentrate containing each analyte of interest is required. The QC check sample concentrate may be prepared from pure standard materials or purchased as certified solutions Use appropriate trip, matrix, control site, method, reagent, and solvent blanks. Internal, surrogate, and five concentration level calibration standards are used. The QC check sample concentrate should contain this compound at 100 µg/mL in 2-propanol.

REFERENCE Test Methods for Evaluating Solid Waste (SW-846), U.S. EPA Office of Solid Waste, Washington, DC, Method 8040A, Rev. 1, Nov. 1990.

Phenothiazine EPA Method 1625
CAS #92-84-2

TITLE Semivolatile Organic Compounds by Isotope Dilution GC/MS

MATRIX The compounds may be determined in waters, soils, and municipal sludges by this method.

METHOD SUMMARY This method is used to determine 176 semivolatile toxic organic pollutants associated with the CWA (as amended 1987); the RCRA (as amended 1986); the CERCLA (as amended 1986); and other compounds amenable to extraction and analysis by capillary column gas chromatography-mass spectrometry (GC/MS).

Stable isotopically-labeled analogs of the compounds of interest are added to the sample. If the solids content is less than 1%, a 1-L sample is extracted at pH 12–13, then at pH <2 with methylene chloride using continuous extraction techniques.

If the solids content is 30% or less, the sample is diluted to 1% solids with reagent water, homogenized ultrasonically, and extracted at pH 12–13, then at pH <2 with methylene chloride using continuous extraction techniques. If the solids content is greater than 30%, the sample is extracted using ultrasonic techniques.

Each extract is dried over sodium sulfate, concentrated to a volume of 5 mL, cleaned up using GPC, if necessary, and concentrated. Extracts are concentrated to 1 mL if GPC is not performed, and to 0.5 mL if GPC is performed.

An internal standard is added to the extract, and a 1-mL aliquot of the extract is injected into the GC. The compounds are separated by GC and detected by a MS. The labeled compounds serve to correct the variability of the analytical technique.

INTERFERENCES Solvents, reagents, glassware, and other sample processing hardware may yield artifacts and/or elevated baselines causing misinterpretation of chromatograms and spectra. Materials used in the analysis must be demonstrated to be free from interferences under the conditions of analysis by running method blanks initially and with each sample lot (sample started through the extraction process on a given 8-h shift, to a maximum of 20). Specific selection of reagents and purification of solvents by distillation in all glass systems may be required. Glassware and, where possible, reagents are cleaned by solvent rinse and baking at 450°C for 1-h minimum. Interferences coextracted from samples will vary considerably from source to source, depending on the diversity of the site being sampled.

INSTRUMENTATION Major instrumentation includes a GC with a splitless or on-column injection port for capillary column, a MS with 70 eV electron impact ionization, and a data system to collect and record MS data, and process it. A K-D apparatus is used to concentrate extracts.

GC Column: 30 m × 0.25 mm I.D. 5% phenyl, 94% methyl, 1% vinyl silicone bonded phased fused silica capillary column.

PRECISION & ACCURACY The detection limits of the method are usually dependent on the level of interferences rather than instrumental limitations. The limits typify the minimum quantities that can be detected with no interferences present.

The minimum level (in µg/mL) was not listed. This is defined as a minimum level at which the analytical system shall give

recognizable mass spectra (background corrected) and acceptable calibration points.

The MDL (in μg/kg) in low solids was not listed and in high solids was not listed; these were determined in digested sludge (low solids) and in filter cake or compost (high solids).

The labeled and native compound initial precision as standard deviation (in μg/L) was not listed.
The labeled and native compound initial accuracy as average recovery (in μg/L) was not listed.

SAMPLE COLLECTION, PRESERVATION & HANDLING
Collect samples in glass containers. Aqueous samples which flow freely are collected in refrigerated bottles using automatic sampling equipment. Solid samples are collected as grab samples using widemouth jars. Maintain samples at 0 to 4°C from the time of collection until extraction. If residual chlorine is present in aqueous samples, add 80 mg sodium thiosulfate/L of water. Begin sample extraction within 7 days of collection, and analyze all extracts within 40 days of extraction.

SAMPLE PREPARATION Samples containing 1% solids or less are extracted directly using continuous liquid-liquid extraction techniques. Samples containing 1 to 30% solids are diluted to the 1% level with reagent water and extracted using continuous liquid-liquid extraction techniques. Samples containing greater than 30% solids are extracted using ultrasonic techniques.

Base/neutral extraction — Adjust the pH of the waters in the extractors to 12–13 with 6 N NaOH. Extract with methylene chloride for 24–48 h.
Acid extraction — Adjust the pH of the waters in the extractors to 2 or less using 6 N sulfuric acid. Extract with methylene chloride for 24–48 h.
Ultrasonic extraction of high solids samples — Add anhydrous sodium sulfate to the sample and QC aliquot(s). Add acetone:methylene chloride (1:1) to the sample and mix thoroughly

Concentrate extracts using a K-D apparatus.

QUALITY CONTROL The analyst is permitted to modify this method to improve separations or lower the costs of measurements, provided all performance specifications are met. Analyses of blanks are required to demonstrate freedom from contamination. When results of spikes indicate atypical method performance for samples, the samples are diluted to bring method performance within acceptable limits.

For low solids (aqueous samples), extract, concentrate, and analyze two sets of four 1-L aliquots (8 aliquots total) of the precision and recovery standard. For high solids samples, two sets of four 30-g aliquots of the high solids reference matrix are used.

Spike all samples with labeled compounds to assess method performance. Compute percent recovery of the labeled compounds using the internal standard method. Compare the labeled compound recovery for each compound with the corresponding labeled compound recovery.

Reagent water and high solids reference matrix blanks are analyzed to demonstrate freedom from contamination. Extract and concentrate a 1-L reagent water blank or a high solids reference matrix blank with each sample's lot (samples started through the extraction process on the same 8-h shift, to a maximum of 20 samples).

Field replicates may be collected to determine the precision of the sampling technique, and spiked samples may be required to determine the accuracy of the analysis when the internal standard method is used.

REFERENCE Semivolatile Organic Compounds by Isotope Dilution GC/MS. Office of Water Regulation and Standards, U.S. EPA Industrial Technology Division, Washington, DC, EPA Method 1625, Rev. C, June 1989 (contact W.A. Telliard, U.S. EPA, Office of Water Regulations and Standards, 401 M St., SW, Washington, DC, 20460. Phone: 202-382-7131).

1,4-Phenylenediamine   **EPA Method 8270**
CAS #106-50-3

TITLE Semivolatile Organic Compounds by GC/MS

MATRIX This method is used to determine the concentration of semivolatile organic compounds in extracts prepared from all types of solid waste matrices, soils, and groundwater. Although surface waters are not specifically mentioned, this method should be applicable to water samples from rivers, lakes, etc.

METHOD SUMMARY This method covers 259 semivolatile organic compounds. In very limited applications direct injection of the sample into the GC/MS system may be appropriate, but this results in very high detection limits (approximately 10,000 μg/L). Typically, a 1-L liquid sample, containing surrogate, and matrix spiking standards, is extracted in a continuous extractor first under acid conditions and then under basic conditions. Typically 30 g of a solid sample, containing surrogate, and matrix spiking standards, is extracted ultrasonically. After concentrating the extract to 1 mL it is spiked with 10 μL of an internal standard solution just prior to analysis by GC/MS. The volume injected should contain about 100 ng of base/neutral and 200 ng of acid surrogates (for a 1-μL injection). Analysis is performed by GC/MS using a capillary GC column.

INTERFERENCES Raw GC/MS data from all blanks, samples, and spikes must be evaluated for interferences. Contamination by carryover can occur whenever high-concentration and low-concentration samples are sequentially analyzed. To reduce carryover, the sample syringe must be rinsed out between samples with solvent. Whenever an unusually concentrated sample is encountered, it should be followed by the analysis of blank solvent to check for cross-contamination.

INSTRUMENTATION A GC/MS and a data system are required. The GC column used is a 30 m × 0.25 mm I.D. (or 0.32 mm I.D.) 1um film thickness silicone-coated fused silica capillary column. A continuous liquid-liquid extractor equipped with Teflon® or glass connection joints and stopcocks requiring no lubrication, a K-D concentrating apparatus, water

bath, and an ultrasonic disrupter with a minimum power of 300 W and with pulsing capability are also required.

PRECISION & ACCURACY The estimated quantitation limit (EQL) of Method 8270B for determining an individual compound is approximately 1 mg/kg (wet weight) for soil or sediment samples, 1–200 mg/kg for wastes (dependent on matrix and method of preparation), and 10 µg/L for groundwater samples. EQLs will be proportionately higher for sample extracts that require dilution to avoid saturation of the detector.

The EQL(b) for groundwater in µg/L is 10.
The EQL (a, b) for low concentrations in soil and sediment in µg/kg is not determined.
Accuracy as µg/L is not listed.
Overall precision in µg/L is not listed.

(a) *EQLs listed for soil/sediment are based on wet weight. Normally data is reported in a dry-weight basis; therefore, EQLs will be higher based on the % dry weight of each sample. This calculation is based on a 30 g sample and gel permeation chromatography cleanup.*
(b) *Sample EQLs are highly matrix-dependent. The EQLs are provided for guidance and may not always be achievable.*
C = *True value for concentration, in µg/L.*
X = *Average recovery found for measurements of samples containing a concentration of C, in µg/L.*

ESTIMATED QUANTITATION LIMIT

Other Matrices	Factor (a)
High-concentration soil and sludges by sonicator	7.5
Non-water miscible waste	75

(a) *EQL for other matrices = [EQL for low soil/sediment] × [Factor]. This estimated EQL is similar to an EPA "Practical Quantitation Limit."*

SAMPLING METHOD
Liquid samples — Use a 1 or 2½ gallon amber glass bottle with a screw-top Teflon®-lined cover that has been prewashed with detergent and rinsed with distilled water and methanol (or isopropanol).

Soils, sediments, or sludges — Use an 8-oz. widemouth glass with a screw-top Teflon®-lined cover that has been prewashed with detergent and rinsed with distilled water and methanol (or isopropanol).

SAMPLE PRESERVATION
Liquid samples — If residual chlorine is present, add 3 mL of 10% sodium thiosulfate per gallon, cool to 4°C and store in a solvent-free refrigerator until analysis; if chlorine is not present, then eliminate the sodium thiosulfate addition.

Soils, sediments, or sludges — Cool samples to 4°C and store in a solvent-free refrigerator.

MHT Liquid samples must be extracted within 7 days and the extracts analyzed within 40 days. Soils, sediments, or sludges may be stored for a maximum of 14 days and the extracts analyzed within 40 days.

SAMPLE PREPARATION
Liquid samples — Transfer 1 L quantitatively to a continuous extractor. If high concentrations are anticipated, a smaller volume may be used and then diluted with organic-free reagent water to 1 L. Adjust pH, if necessary, to pH <2 using 1:1 (V/V) sulfuric acid. Pipette 1.0 mL of a surrogate standard spiking solution into each sample. For the sample in each analytical batch selected for spiking, add 1.0 mL of a matrix spiking standard. For base/neutral acid analysis, the amount of the surrogates and matrix spiking compounds added to the sample should result in a final concentration of 100 ng/µL of each analyte in the extract to be analyzed (assuming a 1-µL injection). Extract with methylene chloride for 18–24 h. Next, adjust the pH of the aqueous phase to pH >11 using 10 N sodium hydroxide and extract it with methylene chloride again for 18–24 h. Dry the extract through a column containing anhydrous sodium sulfate and concentrate it to 1 mL using a K-D concentrator.

Soils, sediments, or sludges — Use 30 g of sample. Nonporous or wet samples (gummy or clay type) that do not have a free-flowing sandy texture must be mixed with anhydrous sodium sulfate until the sample is free flowing. Add 1 mL of surrogate standards to all samples, spikes, standards, and blanks. For the sample in each analytical batch selected for spiking, add 1.0 mL of a matrix spiking standard. For base/neutral acid analysis, the amount added of the surrogates and matrix spiking compounds should result in a final concentration of 100 ng/µL of each base/neutral analyte and 200 ng/µL of each acid analyte in the extract to be analyzed (assuming a 1-µL injection). Immediately add a 100-mL mixture of 1:1 methylene chloride:acetone and extract the sample ultrasonically for 3 min and then decant or filter the extracts. Repeat the extraction two or more times. Dry the extract using a column with anhydrous sodium sulfate and concentrate it to 1 mL in a K-D concentrator.

QUALITY CONTROL A methylene chloride solution containing 50 ng/µL of decafluorotriphenylphosphine (DFTPP) is used for tuning the GC/MS system each 12-h shift. A system performance check also must be made during every 12-h shift. A standard containing 50 ng/µL each of 4,4′-DDT, pentachlorophenol, and benzidine is required to verify injection port inertness and GC column performance. A calibration standard at mid-concentration, containing each compound of interest, including all required surrogates, must be performed every 12 h during analysis. After the system performance check is met, calibration check compounds (CCCs) are used to check the validity of the initial calibration.

The internal standard responses and retention times in the calibration check standard must be evaluated immediately after or during data acquisition. If the retention time for any internal standard changes by more than 30 seconds from the last check calibration (12 h), the chromatographic system must be inspected for malfunctions and corrections must be made, as required. If the electron ionization current plot (EICP) area for any of the internal standards changes by a factor of two from the last daily calibration standard check, the mass spectrometer must be inspected for malfunctions and corrections must be made, as appropriate.

Demonstrate, through the analysis of a reagent water blank, that interferences from the analytical system, glassware, and reagents are under control. The blank samples should be carried through all stages of the sample preparation and measurement steps. For each analytical batch (up to 20 samples), a reagent blank, matrix spike, and matrix spike duplicate/duplicate must be analyzed (the frequency of the spikes may be different for different monitoring programs). The blank and spiked samples must be carried through all stages of the sample preparation and measurement steps. A QC reference sample concentrate containing each analyte at a concentration of 100 mg/L in methanol is required.

REFERENCE Test Methods for Evaluating Solid Waste (SW-846). U.S. EPA 1983, Method 8270B, Rev. 2, Nov. 1990. Office of Solid Waste, Washington, DC.

1-Phenylnaphthalene   **EPA Method 1625**
CAS #605-02-7

TITLE Semivolatile Organic Compounds by Isotope Dilution GC/MS

MATRIX The compounds may be determined in waters, soils, and municipal sludges by this method.

METHOD SUMMARY This method is used to determine 176 semivolatile toxic organic pollutants associated with the CWA (as amended 1987); the RCRA (as amended 1986); the CERCLA (as amended 1986); and other compounds amenable to extraction and analysis by capillary column gas chromatography-mass spectrometry (GC/MS).

Stable isotopically-labeled analogs of the compounds of interest are added to the sample. If the solids content is less than 1%, a 1-L sample is extracted at pH 12–13, then at pH <2 with methylene chloride using continuous extraction techniques.

If the solids content is 30% or less, the sample is diluted to 1% solids with reagent water, homogenized ultrasonically, and extracted at pH 12–13, then at pH <2 with methylene chloride using continuous extraction techniques. If the solids content is greater than 30%, the sample is extracted using ultrasonic techniques.

Each extract is dried over sodium sulfate, concentrated to a volume of 5 mL, cleaned up using GPC, if necessary, and concentrated. Extracts are concentrated to 1 mL if GPC is not performed, and to 0.5 mL if GPC is performed.

An internal standard is added to the extract, and a 1-mL aliquot of the extract is injected into the GC. The compounds are separated by GC and detected by a MS. The labeled compounds serve to correct the variability of the analytical technique.

INTERFERENCES Solvents, reagents, glassware, and other sample processing hardware may yield artifacts and/or elevated baselines causing misinterpretation of chromatograms and spectra. Materials used in the analysis must be demonstrated to be free from interferences under the conditions of analysis by running method blanks initially and with each sample lot (sample started through the extraction process on a given 8-h shift, to a maximum of 20). Specific selection of reagents and purification of solvents by distillation in all glass systems may be required. Glassware and, where possible, reagents are cleaned by solvent rinse and baking at 450°C for 1-h minimum. Interferences coextracted from samples will vary considerably from source to source, depending on the diversity of the site being sampled.

INSTRUMENTATION Major instrumentation includes a GC with a splitless or on-column injection port for capillary column, a MS with 70 eV electron impact ionization, and a data system to collect and record MS data, and process it. A K-D apparatus is used to concentrate extracts.

GC Column: 30 m × 0.25 mm I.D. 5% phenyl, 94% methyl, 1% vinyl silicone bonded phased fused silica capillary column.

PRECISION & ACCURACY The detection limits of the method are usually dependent on the level of interferences rather than instrumental limitations. The limits typify the minimum quantities that can be detected with no interferences present.

The minimum level (in µg/mL) was not listed. This is defined as a minimum level at which the analytical system shall give recognizable mass spectra (background corrected) and acceptable calibration points.

The MDL (in µg/kg) in low solids was not listed and in high solids was not listed; these were determined in digested sludge (low solids) and in filter cake or compost (high solids).

The labeled and native compound initial precision as standard deviation (in µg/L) was not listed.
The labeled and native compound initial accuracy as average recovery (in µg/L) was not listed.

SAMPLE COLLECTION, PRESERVATION & HANDLING
Collect samples in glass containers. Aqueous samples which flow freely are collected in refrigerated bottles using automatic sampling equipment. Solid samples are collected as grab samples using widemouth jars. Maintain samples at 0 to 4°C from the time of collection until extraction. If residual chlorine is present in aqueous samples, add 80 mg sodium thiosulfate/L of water. Begin sample extraction within 7 days of collection, and analyze all extracts within 40 days of extraction.

SAMPLE PREPARATION Samples containing 1% solids or less are extracted directly using continuous liquid-liquid extraction techniques. Samples containing 1 to 30% solids are diluted to the 1% level with reagent water and extracted using continuous liquid-liquid extraction techniques. Samples containing greater than 30% solids are extracted using ultrasonic techniques.

Base/neutral extraction — Adjust the pH of the waters in the extractors to 12–13 with 6 N NaOH. Extract with methylene chloride for 24–48 h.
Acid extraction — Adjust the pH of the waters in the extractors to 2 or less using 6 N sulfuric acid. Extract with methylene chloride for 24–48 h.
Ultrasonic extraction of high solids samples — Add anhydrous sodium sulfate to the sample and QC aliquot(s).

Add acetone:methylene chloride (1:1) to the sample and mix thoroughly

Concentrate extracts using a K-D apparatus.

QUALITY CONTROL The analyst is permitted to modify this method to improve separations or lower the costs of measurements, provided all performance specifications are met. Analyses of blanks are required to demonstrate freedom from contamination. When results of spikes indicate atypical method performance for samples, the samples are diluted to bring method performance within acceptable limits.

For low solids (aqueous samples), extract, concentrate, and analyze two sets of four 1-L aliquots (8 aliquots total) of the precision and recovery standard. For high solids samples, two sets of four 30-g aliquots of the high solids reference matrix are used.

Spike all samples with labeled compounds to assess method performance. Compute percent recovery of the labeled compounds using the internal standard method. Compare the labeled compound recovery for each compound with the corresponding labeled compound recovery.

Reagent water and high solids reference matrix blanks are analyzed to demonstrate freedom from contamination. Extract and concentrate a 1-L reagent water blank or a high solids reference matrix blank with each sample's lot (samples started through the extraction process on the same 8-h shift, to a maximum of 20 samples).

Field replicates may be collected to determine the precision of the sampling technique, and spiked samples may be required to determine the accuracy of the analysis when the internal standard method is used.

REFERENCE Semivolatile Organic Compounds by Isotope Dilution GC/MS. Office of Water Regulation and Standards, U.S. EPA Industrial Technology Division, Washington, DC, EPA Method 1625, Rev. C, June 1989 (contact W.A. Telliard, U.S. EPA, Office of Water Regulations and Standards, 401 M St., SW, Washington, DC, 20460. Phone: 202-382-7131).

2-Phenylnaphthalene **EPA Method 1625**
CAS #612-94-2

TITLE Semivolatile Organic Compounds by Isotope Dilution GC/MS

MATRIX The compounds may be determined in waters, soils, and municipal sludges by this method.

METHOD SUMMARY This method is used to determine 176 semivolatile toxic organic pollutants associated with the CWA (as amended 1987); the RCRA (as amended 1986); the CERCLA (as amended 1986); and other compounds amenable to extraction and analysis by capillary column gas chromatography-mass spectrometry (GC/MS).

Stable isotopically-labeled analogs of the compounds of interest are added to the sample. If the solids content is less than 1%, a 1-L sample is extracted at pH 12–13, then at pH <2 with methylene chloride using continuous extraction techniques.

If the solids content is 30% or less, the sample is diluted to 1% solids with reagent water, homogenized ultrasonically, and extracted at pH 12–13, then at pH <2 with methylene chloride using continuous extraction techniques. If the solids content is greater than 30%, the sample is extracted using ultrasonic techniques.

Each extract is dried over sodium sulfate, concentrated to a volume of 5 mL, cleaned up using GPC, if necessary, and concentrated. Extracts are concentrated to 1 mL if GPC is not performed, and to 0.5 mL if GPC is performed.

An internal standard is added to the extract, and a 1-mL aliquot of the extract is injected into the GC. The compounds are separated by GC and detected by a MS. The labeled compounds serve to correct the variability of the analytical technique.

INTERFERENCES Solvents, reagents, glassware, and other sample processing hardware may yield artifacts and/or elevated baselines causing misinterpretation of chromatograms and spectra. Materials used in the analysis must be demonstrated to be free from interferences under the conditions of analysis by running method blanks initially and with each sample lot (sample started through the extraction process on a given 8-h shift, to a maximum of 20). Specific selection of reagents and purification of solvents by distillation in all glass systems may be required. Glassware and, where possible, reagents are cleaned by solvent rinse and baking at 450°C for 1-h minimum. Interferences coextracted from samples will vary considerably from source to source, depending on the diversity of the site being sampled.

INSTRUMENTATION Major instrumentation includes a GC with a splitless or on-column injection port for capillary column, a MS with 70 eV electron impact ionization, and a data system to collect and record MS data, and process it. A K-D apparatus is used to concentrate extracts.

GC Column: 30 m × 0.25 mm I.D. 5% phenyl, 94% methyl, 1% vinyl silicone bonded phased fused silica capillary column.

PRECISION & ACCURACY The detection limits of the method are usually dependent on the level of interferences rather than instrumental limitations. The limits typify the minimum quantities that can be detected with no interferences present.

The minimum level (in µg/mL) was not listed. This is defined as a minimum level at which the analytical system shall give recognizable mass spectra (background corrected) and acceptable calibration points.

The MDL (in µg/kg) in low solids was not listed and in high solids was not listed; these were determined in digested sludge (low solids) and in filter cake or compost (high solids).

The labeled and native compound initial precision as standard deviation (in µg/L) was not listed.

The labeled and native compound initial accuracy as average recovery (in µg/L) was not listed.

SAMPLE COLLECTION, PRESERVATION & HANDLING
Collect samples in glass containers. Aqueous samples which flow freely are collected in refrigerated bottles using automatic sampling equipment. Solid samples are collected as grab samples using widemouth jars. Maintain samples at 0 to 4°C from the time of collection until extraction. If residual chlorine is present in aqueous samples, add 80 mg sodium thiosulfate/L of water. Begin sample extraction within 7 days of collection, and analyze all extracts within 40 days of extraction.

SAMPLE PREPARATION Samples containing 1% solids or less are extracted directly using continuous liquid-liquid extraction techniques. Samples containing 1 to 30% solids are diluted to the 1% level with reagent water and extracted using continuous liquid-liquid extraction techniques. Samples containing greater than 30% solids are extracted using ultrasonic techniques.

Base/neutral extraction — Adjust the pH of the waters in the extractors to 12–13 with 6 N NaOH. Extract with methylene chloride for 24–48 h.

Acid extraction — Adjust the pH of the waters in the extractors to 2 or less using 6 N sulfuric acid. Extract with methylene chloride for 24–48 h.

Ultrasonic extraction of high solids samples — Add anhydrous sodium sulfate to the sample and QC aliquot(s). Add acetone:methylene chloride (1:1) to the sample and mix thoroughly

Concentrate extracts using a K-D apparatus.

QUALITY CONTROL The analyst is permitted to modify this method to improve separations or lower the costs of measurements, provided all performance specifications are met. Analyses of blanks are required to demonstrate freedom from contamination. When results of spikes indicate atypical method performance for samples, the samples are diluted to bring method performance within acceptable limits.

For low solids (aqueous samples), extract, concentrate, and analyze two sets of four 1-L aliquots (8 aliquots total) of the precision and recovery standard. For high solids samples, two sets of four 30-g aliquots of the high solids reference matrix are used.

Spike all samples with labeled compounds to assess method performance. Compute percent recovery of the labeled compounds using the internal standard method. Compare the labeled compound recovery for each compound with the corresponding labeled compound recovery.

Reagent water and high solids reference matrix blanks are analyzed to demonstrate freedom from contamination. Extract and concentrate a 1-L reagent water blank or a high solids reference matrix blank with each sample's lot (samples started through the extraction process on the same 8-h shift, to a maximum of 20 samples).

Field replicates may be collected to determine the precision of the sampling technique, and spiked samples may be required to determine the accuracy of the analysis when the internal standard method is used.

REFERENCE Semivolatile Organic Compounds by Isotope Dilution GC/MS. Office of Water Regulation and Standards, U.S. EPA Industrial Technology Division, Washington, DC, EPA Method 1625, Rev. C, June 1989 (contact W.A. Telliard, U.S. EPA, Office of Water Regulations and Standards, 401 M St., SW, Washington, DC, 20460. Phone: 202-382-7131).

Phorate   EPA Method 8141
CAS #298-02-2

TITLE Organophosphorus Compounds by Gas Chromatography: Capillary Column Technique

MATRIX This method covers aqueous and solid matrices. This includes a wide variety such as drinking water, groundwater, industrial wastewaters, surface waters, soils, solids, and sediments.

METHOD SUMMARY This is a GC method used to determine the concentration of 28 organophosphorus pesticides.

The use of Gel Permeation Cleanup (EPA Method 3640) for sample cleanup has been demonstrated to yield recoveries of less than 85% for many method analytes and is therefore not recommended for use with this method.

This method provides GC conditions for the detection of ppb concentrations of organophosphorus compounds. Prior to the use of this method, appropriate sample preparation techniques must be used. Water samples are extracted at a neutral pH with methylene chloride as a solvent by using a separatory funnel (EPA Method 3510) or a continuous liquid-liquid extractor (EPA Method 3520). Soxhlet extraction (EPA Method 3540) or ultrasonic extraction (EPA Method 3550) using methylene chloride/acetone (1:1) are used for solid samples. Both neat and diluted organic liquids (EPA Method 3580) may be analyzed by direct injection. Spiked samples are used to verify the applicability of the chosen extraction technique to each new sample type. A GC with a flame photometric (FPD) or nitrogen-phosphorus detector (NPD) is used for this multiresidue procedure.

INTERFERENCES The use of Florisil cleanup materials (EPA Method 3620) for some of the compounds in this method has been demonstrated to yield recoveries less than 85% and is therefore not recommended for all compounds. Use of phosphorus or halogen specific detectors, however, often obviates the necessity for cleanup for relatively clean sample matrices. If particular circumstances demand the use of an alternative cleanup procedure, the analyst must determine the elution profile and demonstrate that the recovery of each analyte is no less than 85%.

Use of a flame photometric detector (FPD) in the phosphorus mode will minimize interferences from materials that do not contain phosphorus. Elemental sulfur, however, may interfere with the determination of certain organophosphorus compounds by flame photometric gas chromatography. Sulfur cleanup using EPA Method 3660 may alleviate this interference. A nitrogen phosphorus detector (NPD) is also recommended.

A few analytes coelute on certain columns. Therefore, select a second column for confirmation where coelution of the analytes of interest does not occur.

Method interferences may be caused by contaminants in solvents, reagents, glassware, and other sample processing hardware that lead to discrete artifacts or elevated baselines in gas chromatograms. All these materials must be routinely demonstrated to be free from interferences under the conditions of the analysis by analyzing reagent blanks.

INSTRUMENTATION A GC with a NPD or a FPD will be needed. A data system or integrator is recommended for measuring peak areas and/or peak heights. A Kuderna-Danish (K-D) apparatus will be needed for extract concentration.

Column 1: 15 m × 0.53 mm megabore capillary column, 1.0 μm film thickness, DB-210.
Column 2: 15 m × 0.53 mm megabore capillary column, 1.5 μm film thickness, SPB-608.
Column 3: 15 m × 0.53 mm megabore capillary column, 1.0 μm film thickness, DB-5.

Three megabore capillary columns are included for analysis of organophosphates by this method. Column 1 (DB-210 or equivalent) and Column 2 (SPB-608 or equivalent) are recommended if a large number of organophosphorus analytes are to be determined. If the superior resolution offered by Column 1 and Column 2 is not required, Column 3 (DB-5 or equivalent) may be used. For megabore capillary columns, automatic injections of 1 μL are recommended.

PRECISION & ACCURACY The MDL actually achieved in a given analysis will vary, as it is dependent on instrument sensitivity and matrix effects. Single operator accuracy and precision studies have been conducted with spiked water and soil samples.

MULTIPLICATION FACTORS FOR OTHER MATRICES (a)

Matrix	Factor (b)
Groundwater (EPA Method 3510 or EPA Method 3520)	10
Low-concentration soil by Soxhlet and no cleanup	10 (c)
Low-concentration soil by ultrasonic extraction with GPC cleanup	6.7 (c)
High-concentration soil and sludges by ultrasonic extraction	500 (c)
Non-water miscible waste (EPA Method 3580)	1000 (c)

(a) Sample EQLs are highly matrix-dependent. The EQLs listed here are provided for guidance and may not always be achievable.
(b) EQL = [Method detection limit] × [Factor]. For non-aqueous samples the factor is on a wet-weight basis.
(c) Multiply this factory times the soil MDL.

The MDL (in μg/L) when reagent water was extracted using a separatory funnel was 0.04.
The MDL (in μg/kg) when soil was extracted using Soxhlet extraction (EPA Method 3540) was 2.0.
Accuracy (as % recovery) with separatory funnel extraction ranged from 94 (with low spikes) to 73 (with high spikes).
Accuracy (as % recovery) with continuous liquid-liquid extraction ranged from 84 (with low spikes) to 74 (with high spikes).
Accuracy (as % recovery) with Soxhlet extraction of soils ranged from 75 (with low spikes to 78 (with high spikes).
Accuracy (as % recovery) with ultrasonic extraction of soils ranged from not recovered (with low spikes) to 64 (with high spikes).

SAMPLE COLLECTION, PRESERVATION & HANDLING
Containers used to collect samples for the determination of semivolatile organic compounds should be soap and water washed followed by methanol (or isopropanol) rinsing. The sample containers should be of glass or Teflon® and have screw-top covers with Teflon® liners.

No preservation is used with concentrated waste samples. With liquid samples containing no residual chlorine and with soil, sediment, and sludge samples, immediately cooling to 4°C is the only preservation used. When residual chlorine is present then 3 mL of 10% aqueous sodium sulfate is added for each gallon of sample collected, followed by cooling to 4°C.

Liquid samples must be extracted within 7 days and their extracts analyzed within 40 days. Concentrated waste, soil, sediment, and sludge samples must be extracted within 14 days and their extracts analyzed within 40 days.

SAMPLE PREPARATION In general, water samples are extracted at a neutral pH with methylene chloride, using either EPA Method 3510 or EPA Method 3520. Solid samples are extracted using either EPA Method 3540 or EPA Method 3550 with methylene chloride/acetone (1:1) as the extraction solvent.

Prior to GC analysis, the extraction solvent may be exchanged to hexane. Single lab data indicates that samples should not be transferred with 100% hexane during sample workup as the more water soluble organophosphorus compounds may be lost.

If cleanup is performed on the samples, the analyst should analyze the samples by GC. This will confirm elution patterns and the absence of interferences from the reagents. If peak detection and identification is prevented by the presence of interferences, further cleanup is required.

QUALITY CONTROL The analyst should monitor the performance of the extraction, cleanup (when used), and analytical system and the effectiveness of the method in dealing with each sample matrix by spiking each sample, standard, and blank with one or two surrogates (e.g., organophosphorus compounds not expected to be present in the sample). Deuterated analogs of analytes should not be used as surrogates for gas chromatographic analysis due to coelution problems.

A minimum of five concentrations for each analyte of interest should be prepared through dilution of the stock standards with isooctane. One of the concentrations should be at a concentration near, but above, the MDL.

Include a mid-level check standard after each group of 10 samples in the analysis sequence. GC/MS techniques should be judiciously employed to support qualitative identifications made with this method. Follow the GC/MS operating requirements specified in EPA Method 8270.

When available, chemical ionization mass spectra may be employed to aid in the qualitative identification process. To confirm an identification of a compound, the background-corrected mass spectrum of the compound must be obtained from the sample extract and must be compared with a mass spectrum from a stock or calibration standard analyzed under the same chromatographic conditions. The molecular ion and all other ions present above 20% relative abundance in the mass spectrum of the standard must be present in the mass spectrum of the sample with agreement to ±20%. The retention time of the compound in the sample must be within six seconds of the retention time for the same compound in the standard solution.

Should the MS procedure fail to provide satisfactory results, additional steps may be taken before reanalysis. These steps may include the use of alternate packed or capillary GC columns or additional sample cleanup.

REFERENCE Test Methods for Evaluating Solid Waste, Physical/Chemical Methods, SW-846, 3rd Edition, U.S. EPA, Office of Solid Waste, Washington, DC, EPA Method 8141 July 1992.

Phorate   **EPA Method 8270**
CAS #298-02-2

TITLE Semivolatile Organic Compounds by GC/MS

MATRIX This method is used to determine the concentration of semivolatile organic compounds in extracts prepared from all types of solid waste matrices, soils, and groundwater. Although surface waters are not specifically mentioned, this method should be applicable to water samples from rivers, lakes, etc.

METHOD SUMMARY This method covers 259 semivolatile organic compounds. In very limited applications direct injection of the sample into the GC/MS system may be appropriate, but this results in very high detection limits (approximately 10,000 µg/L). Typically, a 1-L liquid sample, containing surrogate, and matrix spiking standards, is extracted in a continuous extractor first under acid conditions and then under basic conditions. Typically 30 g of a solid sample, containing surrogate, and matrix spiking standards, is extracted ultrasonically. After concentrating the extract to 1 mL it is spiked with 10 µL of an internal standard solution just prior to analysis by GC/MS. The volume injected should contain about 100 ng of base/neutral and 200 ng of acid surrogates (for a 1-µL injection). Analysis is performed by GC/MS using a capillary GC column.

INTERFERENCES Raw GC/MS data from all blanks, samples, and spikes must be evaluated for interferences. Contamination by carryover can occur whenever high-concentration and low-concentration samples are sequentially analyzed. To reduce carryover, the sample syringe must be rinsed out between samples with solvent. Whenever an unusually concentrated sample is encountered, it should be followed by the analysis of blank solvent to check for cross-contamination.

INSTRUMENTATION A GC/MS and a data system are required. The GC column used is a 30 m × 0.25 mm I.D. (or 0.32 mm I.D.) 1um film thickness silicone-coated fused silica capillary column. A continuous liquid-liquid extractor equipped with Teflon® or glass connection joints and stopcocks requiring no lubrication, a K-D concentrating apparatus, water bath, and an ultrasonic disrupter with a minimum power of 300 W and with pulsing capability are also required.

PRECISION & ACCURACY The estimated quantitation limit (EQL) of Method 8270B for determining an individual compound is approximately 1 mg/kg (wet weight) for soil or sediment samples, 1–200 mg/kg for wastes (dependent on matrix and method of preparation), and 10 µg/L for groundwater samples. EQLs will be proportionately higher for sample extracts that require dilution to avoid saturation of the detector.

The EQL(b) for groundwater in µg/L is 10.
The EQL (a, b) for low concentrations in soil and sediment in µg/kg is not determined.
Accuracy as µg/L is not listed.
Overall precision in µg/L is not listed.

(a) EQLs listed for soil/sediment are based on wet weight. Normally data is reported in a dry-weight basis; therefore, EQLs will be higher based on the % dry weight of each sample. This calculation is based on a 30 g sample and gel permeation chromatography cleanup.
(b) Sample EQLs are highly matrix-dependent. The EQLs are provided for guidance and may not always be achievable.
$C =$ True value for concentration, in µg/L.
$X =$ Average recovery found for measurements of samples containing a concentration of C, in µg/L.

ESTIMATED QUANTITATION LIMIT

Other Matrices	Factor (a)
High-concentration soil and sludges by sonicator	7.5
Non-water miscible waste	75

(a) EQL for other matrices = [EQL for low soil/sediment] × [Factor]. This estimated EQL is similar to an EPA "Practical Quantitation Limit."

SAMPLING METHOD
Liquid samples — Use a 1 or 2½ gallon amber glass bottle with a screw-top Teflon®-lined cover that has been prewashed with detergent and rinsed with distilled water and methanol (or isopropanol).

Soils, sediments, or sludges — Use an 8-oz. widemouth glass with a screw-top Teflon®-lined cover that has been prewashed with detergent and rinsed with distilled water and methanol (or isopropanol).

SAMPLE PRESERVATION
Liquid samples — If residual chlorine is present, add 3 mL of 10% sodium thiosulfate per gallon, cool to 4°C and store in a solvent-free refrigerator until analysis; if chlorine is not present, then eliminate the sodium thiosulfate addition.

Soils, sediments, or sludges — Cool samples to 4°C and store in a solvent-free refrigerator.

MHT Liquid samples must be extracted within 7 days and the extracts analyzed within 40 days. Soils, sediments, or sludges may

be stored for a maximum of 14 days and the extracts analyzed within 40 days.

SAMPLE PREPARATION

Liquid samples — Transfer 1 L quantitatively to a continuous extractor. If high concentrations are anticipated, a smaller volume may be used and then diluted with organic-free reagent water to 1 L. Adjust pH, if necessary, to pH <2 using 1:1 (V/V) sulfuric acid. Pipette 1.0 mL of a surrogate standard spiking solution into each sample. For the sample in each analytical batch selected for spiking, add 1.0 mL of a matrix spiking standard. For base/neutral acid analysis, the amount of the surrogates and matrix spiking compounds added to the sample should result in a final concentration of 100 ng/µL of each analyte in the extract to be analyzed (assuming a 1-µL injection). Extract with methylene chloride for 18–24 h. Next, adjust the pH of the aqueous phase to pH >11 using 10 N sodium hydroxide and extract it with methylene chloride again for 18–24 h. Dry the extract through a column containing anhydrous sodium sulfate and concentrate it to 1 mL using a K-D concentrator.

Soils, sediments, or sludges — Use 30 g of sample. Nonporous or wet samples (gummy or clay type) that do not have a free-flowing sandy texture must be mixed with anhydrous sodium sulfate until the sample is free flowing. Add 1 mL of surrogate standards to all samples, spikes, standards, and blanks. For the sample in each analytical batch selected for spiking, add 1.0 mL of a matrix spiking standard. For base/neutral acid analysis, the amount added of the surrogates and matrix spiking compounds should result in a final concentration of 100 ng/µL of each base/neutral analyte and 200 ng/µL of each acid analyte in the extract to be analyzed (assuming a 1-µL injection). Immediately add a 100-mL mixture of 1:1 methylene chloride:acetone and extract the sample ultrasonically for 3 min and then decant or filter the extracts. Repeat the extraction two or more times. Dry the extract using a column with anhydrous sodium sulfate and concentrate it to 1 mL in a K-D concentrator.

QUALITY CONTROL A methylene chloride solution containing 50 ng/µL of decafluorotriphenylphosphine (DFTPP) is used for tuning the GC/MS system each 12-h shift. A system performance check also must be made during every 12-h shift. A standard containing 50 ng/µL each of 4,4'-DDT, pentachlorophenol, and benzidine is required to verify injection port inertness and GC column performance. A calibration standard at mid-concentration, containing each compound of interest, including all required surrogates, must be performed every 12 h during analysis. After the system performance check is met, calibration check compounds (CCCs) are used to check the validity of the initial calibration.

The internal standard responses and retention times in the calibration check standard must be evaluated immediately after or during data acquisition. If the retention time for any internal standard changes by more than 30 seconds from the last check calibration (12 h), the chromatographic system must be inspected for malfunctions and corrections must be made, as required. If the electron ionization current plot (EICP) area for any of the internal standards changes by a factor of two from the last daily calibration standard check, the mass spectrometer must be inspected for malfunctions and corrections must be made, as appropriate.

Demonstrate, through the analysis of a reagent water blank, that interferences from the analytical system, glassware, and reagents are under control. The blank samples should be carried through all stages of the sample preparation and measurement steps. For each analytical batch (up to 20 samples), a reagent blank, matrix spike, and matrix spike duplicate/duplicate must be analyzed (the frequency of the spikes may be different for different monitoring programs). The blank and spiked samples must be carried through all stages of the sample preparation and measurement steps. A QC reference sample concentrate containing each analyte at a concentration of 100 mg/L in methanol is required.

REFERENCE Test Methods for Evaluating Solid Waste (SW-846). U.S. EPA 1983, Method 8270B, Rev. 2, Nov. 1990. Office of Solid Waste, Washington, DC.

Phorate   EPA Method 8140
CAS #298-02-2

TITLE Organophosphorus Pesticides

MATRIX Groundwater, soils, sludges, water miscible liquid wastes, and non-water miscible wastes.

APPLICATION This method is used for the analysis of 21 organophosphorus pesticides. Samples are extracted, concentrated, and analyzed using direct injection of both neat and diluted organic liquid into a gas chromatograph (GC).

INTERFERENCES Solvents, reagents, and glassware may introduce artifacts. Other interferences may come from coextracted compounds from samples. The use of Florisil cleanup materials may produce low recoveries. Elemental sulfur may interfere with some compounds when using a flame photometric detector. Sulfur cleanup (Method 3660) may alleviate sulfur interference.

INSTRUMENTATION GC capable of on-column injections and a flame photometric detector (FPD) or a thermionic detector. Column 1: 1.8 m by 2 mm with 5% SP-2401 on Supelcoport. Column 2: 1.8 m by 2 mm with 3% SP-2401 on Supelcoport. Column 3: 50 cm by ⅛ in Teflon® with 15% SE-54 on Gas Chrom Q. The preferred column is Column Number 1.

RANGE 4.9–47 µg/L

MDL 0.15 µg/L (in reagent water).

PQL FACTORS FOR MULTIPLYING × FID MDL VALUE

Matrix	Multiplication Factor
Groundwater	10
Low-level soil by sonication with GPC cleanup	670
High-level soil and sludge by sonication	10,000
Non-water miscible waste	100,000

PRECISION 8.9% (single operator standard deviation)

ACCURACY 62.7% (single operator average recovery)

SAMPLING METHOD Use 8-oz. widemouth glass bottles with Teflon®-lined caps for concentrated waste samples, soils, sediments, and sludges. Use 1 or 2½ gallon amber glass bottles with Teflon®-lined caps for liquid (water) samples.

STABILITY Cool soil, sediment, sludge, and liquid samples to 4°C. If residual chlorine is present in liquid samples add 3 mL of 10% sodium thiosulfate per gallon of sample and cool to 4°C.

MHT 14 days for concentrated waste, soil, sediment, or sludge; 7 days for liquid samples; all extracts must be analyzed within 40 days.

QUALITY CONTROL A quality control check sample concentrate containing this compound in acetone at a concentration 1,000 times more concentrated than the selected spike concentration is required. The QC check sample concentrate may be prepared from pure standard materials or purchased as certified solutions. Use appropriate trip, matrix, control site, method, reagent, and solvent blanks. Internal, surrogate, and five concentration level calibration standards are used.

REFERENCE Method 8140, SW-846, 3rd ed., Sept. 1986.

Phosalone   **EPA Method 8270**
CAS #2310-17-0

TITLE Semivolatile Organic Compounds by GC/MS

MATRIX This method is used to determine the concentration of semivolatile organic compounds in extracts prepared from all types of solid waste matrices, soils, and groundwater. Although surface waters are not specifically mentioned, this method should be applicable to water samples from rivers, lakes, etc.

METHOD SUMMARY This method covers 259 semivolatile organic compounds. In very limited applications direct injection of the sample into the GC/MS system may be appropriate, but this results in very high detection limits (approximately 10,000 µg/L). Typically, a 1-L liquid sample, containing surrogate, and matrix spiking standards, is extracted in a continuous extractor first under acid conditions and then under basic conditions. Typically 30 g of a solid sample, containing surrogate, and matrix spiking standards, is extracted ultrasonically. After concentrating the extract to 1 mL it is spiked with 10 µL of an internal standard solution just prior to analysis by GC/MS. The volume injected should contain about 100 ng of base/neutral and 200 ng of acid surrogates (for a 1-µL injection). Analysis is performed by GC/MS using a capillary GC column.

INTERFERENCES Raw GC/MS data from all blanks, samples, and spikes must be evaluated for interferences. Contamination by carryover can occur whenever high-concentration and low-concentration samples are sequentially analyzed. To reduce carryover, the sample syringe must be rinsed out between samples with solvent. Whenever an unusually concentrated sample is encountered, it should be followed by the analysis of blank solvent to check for cross-contamination.

INSTRUMENTATION A GC/MS and a data system are required. The GC column used is a 30 m × 0.25 mm I.D. (or 0.32 mm I.D.) 1um film thickness silicone-coated fused silica capillary column. A continuous liquid-liquid extractor equipped with Teflon® or glass connection joints and stopcocks requiring no lubrication, a K-D concentrating apparatus, water bath, and an ultrasonic disrupter with a minimum power of 300 W and with pulsing capability are also required.

PRECISION & ACCURACY The estimated quantitation limit (EQL) of Method 8270B for determining an individual compound is approximately 1 mg/kg (wet weight) for soil or sediment samples, 1–200 mg/kg for wastes (dependent on matrix and method of preparation), and 10 µg/L for groundwater samples. EQLs will be proportionately higher for sample extracts that require dilution to avoid saturation of the detector.

The EQL(b) for groundwater in µg/L is 100.
The EQL (a, b) for low concentrations in soil and sediment in µg/kg is not determined.
Accuracy as µg/L is not listed.
Overall precision in µg/L is not listed.

(a) EQLs listed for soil/sediment are based on wet weight. Normally data is reported in a dry-weight basis; therefore, EQLs will be higher based on the % dry weight of each sample. This calculation is based on a 30 g sample and gel permeation chromatography cleanup.
(b) Sample EQLs are highly matrix-dependent. The EQLs are provided for guidance and may not always be achievable.
C = True value for concentration, in µg/L.
X = Average recovery found for measurements of samples containing a concentration of C, in µg/L.

ESTIMATED QUANTITATION LIMIT

Other Matrices	Factor (a)
High-concentration soil and sludges by sonicator	7.5
Non-water miscible waste	75

(a) EQL for other matrices = [EQL for low soil/sediment] × [Factor]. This estimated EQL is similar to an EPA "Practical Quantitation Limit."

SAMPLING METHOD

Liquid samples — Use a 1 or 2½ gallon amber glass bottle with a screw-top Teflon®-lined cover that has been prewashed with detergent and rinsed with distilled water and methanol (or isopropanol).

Soils, sediments, or sludges — Use an 8-oz. widemouth glass with a screw-top Teflon®-lined cover that has been prewashed with detergent and rinsed with distilled water and methanol (or isopropanol).

SAMPLE PRESERVATION

Liquid samples — If residual chlorine is present, add 3 mL of 10% sodium thiosulfate per gallon, cool to 4°C and store in a solvent-free refrigerator until analysis; if chlorine is not present, then eliminate the sodium thiosulfate addition.

Soils, sediments, or sludges — Cool samples to 4°C and store in a solvent-free refrigerator.

MHT Liquid samples must be extracted within 7 days and the extracts analyzed within 40 days. Soils, sediments, or sludges may be stored for a maximum of 14 days and the extracts analyzed within 40 days.

SAMPLE PREPARATION

Liquid samples — Transfer 1 L quantitatively to a continuous extractor. If high concentrations are anticipated, a smaller volume may be used and then diluted with organic-free reagent water to 1 L. Adjust pH, if necessary, to pH <2 using 1:1 (V/V) sulfuric acid. Pipette 1.0 mL of a surrogate standard spiking solution into each sample. For the sample in each analytical batch selected for spiking, add 1.0 mL of a matrix spiking standard. For base/neutral acid analysis, the amount of the surrogates and matrix spiking compounds added to the sample should result in a final concentration of 100 ng/µL of each analyte in the extract to be analyzed (assuming a 1-µL injection). Extract with methylene chloride for 18–24 h. Next, adjust the pH of the aqueous phase to pH >11 using 10 N sodium hydroxide and extract it with methylene chloride again for 18–24 h. Dry the extract through a column containing anhydrous sodium sulfate and concentrate it to 1 mL using a K-D concentrator.

Soils, sediments, or sludges — Use 30 g of sample. Nonporous or wet samples (gummy or clay type) that do not have a free-flowing sandy texture must be mixed with anhydrous sodium sulfate until the sample is free flowing. Add 1 mL of surrogate standards to all samples, spikes, standards, and blanks. For the sample in each analytical batch selected for spiking, add 1.0 mL of a matrix spiking standard. For base/neutral acid analysis, the amount added of the surrogates and matrix spiking compounds should result in a final concentration of 100 ng/µL of each base/neutral analyte and 200 ng/µL of each acid analyte in the extract to be analyzed (assuming a 1-µL injection). Immediately add a 100-mL mixture of 1:1 methylene chloride:acetone and extract the sample ultrasonically for 3 min and then decant or filter the extracts. Repeat the extraction two or more times. Dry the extract using a column with anhydrous sodium sulfate and concentrate it to 1 mL in a K-D concentrator.

QUALITY CONTROL A methylene chloride solution containing 50 ng/µL of decafluorotriphenylphosphine (DFTPP) is used for tuning the GC/MS system each 12-h shift. A system performance check also must be made during every 12-h shift. A standard containing 50 ng/µL each of 4,4'-DDT, pentachlorophenol, and benzidine is required to verify injection port inertness and GC column performance. A calibration standard at mid-concentration, containing each compound of interest, including all required surrogates, must be performed every 12 h during analysis. After the system performance check is met, calibration check compounds (CCCs) are used to check the validity of the initial calibration.

The internal standard responses and retention times in the calibration check standard must be evaluated immediately after or during data acquisition. If the retention time for any internal standard changes by more than 30 seconds from the last check calibration (12 h), the chromatographic system must be inspected for malfunctions and corrections must be made, as required. If the electron ionization current plot (EICP) area for any of the internal standards changes by a factor of two from the last daily calibration standard check, the mass spectrometer must be inspected for malfunctions and corrections must be made, as appropriate.

Demonstrate, through the analysis of a reagent water blank, that interferences from the analytical system, glassware, and reagents are under control. The blank samples should be carried through all stages of the sample preparation and measurement steps. For each analytical batch (up to 20 samples), a reagent blank, matrix spike, and matrix spike duplicate/duplicate must be analyzed (the frequency of the spikes may be different for different monitoring programs). The blank and spiked samples must be carried through all stages of the sample preparation and measurement steps. A QC reference sample concentrate containing each analyte at a concentration of 100 mg/L in methanol is required.

REFERENCE Test Methods for Evaluating Solid Waste (SW-846). U.S. EPA 1983, Method 8270B, Rev. 2, Nov. 1990. Office of Solid Waste, Washington, DC.

Phosmet   **EPA Method 8270**
CAS #732-11-6

TITLE Semivolatile Organic Compounds by GC/MS

MATRIX This method is used to determine the concentration of semivolatile organic compounds in extracts prepared from all types of solid waste matrices, soils, and groundwater. Although surface waters are not specifically mentioned, this method should be applicable to water samples from rivers, lakes, etc.

METHOD SUMMARY This method covers 259 semivolatile organic compounds. In very limited applications direct injection of the sample into the GC/MS system may be appropriate, but this results in very high detection limits (approximately 10,000 µg/L). Typically, a 1-L liquid sample, containing surrogate, and matrix spiking standards, is extracted in a continuous extractor first under acid conditions and then under basic conditions. Typically 30 g of a solid sample, containing surrogate, and matrix spiking standards, is extracted ultrasonically. After concentrating the extract to 1 mL it is spiked with 10 µL of an internal standard solution just prior to analysis by GC/MS. The volume injected should contain about 100 ng of base/neutral and 200 ng of acid surrogates (for a 1-µL injection). Analysis is performed by GC/MS using a capillary GC column.

INTERFERENCES Raw GC/MS data from all blanks, samples, and spikes must be evaluated for interferences. Contamination by carryover can occur whenever high-concentration and low-concentration samples are sequentially analyzed. To reduce carryover, the sample syringe must be rinsed out between samples with solvent. Whenever an unusually concentrated sample is encountered, it should be followed by the analysis of blank solvent to check for cross-contamination.

INSTRUMENTATION A GC/MS and a data system are required. The GC column used is a 30 m × 0.25 mm I.D. (or 0.32 mm I.D.) 1um film thickness silicone-coated fused silica

capillary column. A continuous liquid-liquid extractor equipped with Teflon® or glass connection joints and stopcocks requiring no lubrication, a K-D concentrating apparatus, water bath, and an ultrasonic disrupter with a minimum power of 300 W and with pulsing capability are also required.

PRECISION & ACCURACY The estimated quantitation limit (EQL) of Method 8270B for determining an individual compound is approximately 1 mg/kg (wet weight) for soil or sediment samples, 1–200 mg/kg for wastes (dependent on matrix and method of preparation), and 10 µg/L for groundwater samples. EQLs will be proportionately higher for sample extracts that require dilution to avoid saturation of the detector.

The EQL(b) for groundwater in µg/L is 40.
The EQL (a, b) for low concentrations in soil and sediment in µg/kg is not determined.
Accuracy as µg/L is not listed.
Overall precision in µg/L is not listed.

(a) *EQLs listed for soil/sediment are based on wet weight. Normally data is reported in a dry-weight basis; therefore, EQLs will be higher based on the % dry weight of each sample. This calculation is based on a 30 g sample and gel permeation chromatography cleanup.*
(b) *Sample EQLs are highly matrix-dependent. The EQLs are provided for guidance and may not always be achievable.*
$C =$ *True value for concentration, in µg/L.*
$X =$ *Average recovery found for measurements of samples containing a concentration of C, in µg/L.*

ESTIMATED QUANTITATION LIMIT

Other Matrices	Factor (a)
High-concentration soil and sludges by sonicator	7.5
Non-water miscible waste	75

(a) *EQL for other matrices = [EQL for low soil/sediment] × [Factor]. This estimated EQL is similar to an EPA "Practical Quantitation Limit."*

SAMPLING METHOD
Liquid samples — Use a 1 or 2½ gallon amber glass bottle with a screw-top Teflon®-lined cover that has been prewashed with detergent and rinsed with distilled water and methanol (or isopropanol).

Soils, sediments, or sludges — Use an 8-oz. widemouth glass with a screw-top Teflon®-lined cover that has been prewashed with detergent and rinsed with distilled water and methanol (or isopropanol).

SAMPLE PRESERVATION
Liquid samples — If residual chlorine is present, add 3 mL of 10% sodium thiosulfate per gallon, cool to 4°C and store in a solvent-free refrigerator until analysis; if chlorine is not present, then eliminate the sodium thiosulfate addition.

Soils, sediments, or sludges — Cool samples to 4°C and store in a solvent-free refrigerator.

MHT Liquid samples must be extracted within 7 days and the extracts analyzed within 40 days. Soils, sediments, or sludges may be stored for a maximum of 14 days and the extracts analyzed within 40 days.

SAMPLE PREPARATION
Liquid samples — Transfer 1 L quantitatively to a continuous extractor. If high concentrations are anticipated, a smaller volume may be used and then diluted with organic-free reagent water to 1 L. Adjust pH, if necessary, to pH <2 using 1:1 (V/V) sulfuric acid. Pipette 1.0 mL of a surrogate standard spiking solution into each sample. For the sample in each analytical batch selected for spiking, add 1.0 mL of a matrix spiking standard. For base/neutral acid analysis, the amount of the surrogates and matrix spiking compounds added to the sample should result in a final concentration of 100 ng/µL of each analyte in the extract to be analyzed (assuming a 1-µL injection). Extract with methylene chloride for 18–24 h. Next, adjust the pH of the aqueous phase to pH >11 using 10 N sodium hydroxide and extract it with methylene chloride again for 18–24 h. Dry the extract through a column containing anhydrous sodium sulfate and concentrate it to 1 mL using a K-D concentrator.

Soils, sediments, or sludges — Use 30 g of sample. Nonporous or wet samples (gummy or clay type) that do not have a free-flowing sandy texture must be mixed with anhydrous sodium sulfate until the sample is free flowing. Add 1 mL of surrogate standards to all samples, spikes, standards, and blanks. For the sample in each analytical batch selected for spiking, add 1.0 mL of a matrix spiking standard. For base/neutral acid analysis, the amount added of the surrogates and matrix spiking compounds should result in a final concentration of 100 ng/µL of each base/neutral analyte and 200 ng/µL of each acid analyte in the extract to be analyzed (assuming a 1-µL injection). Immediately add a 100-mL mixture of 1:1 methylene chloride:acetone and extract the sample ultrasonically for 3 min and then decant or filter the extracts. Repeat the extraction two or more times. Dry the extract using a column with anhydrous sodium sulfate and concentrate it to 1 mL in a K-D concentrator.

QUALITY CONTROL A methylene chloride solution containing 50 ng/µL of decafluorotriphenylphosphine (DFTPP) is used for tuning the GC/MS system each 12-h shift. A system performance check also must be made during every 12-h shift. A standard containing 50 ng/µL each of 4,4'-DDT, pentachlorophenol, and benzidine is required to verify injection port inertness and GC column performance. A calibration standard at mid-concentration, containing each compound of interest, including all required surrogates, must be performed every 12 h during analysis. After the system performance check is met, calibration check compounds (CCCs) are used to check the validity of the initial calibration.

The internal standard responses and retention times in the calibration check standard must be evaluated immediately after or during data acquisition. If the retention time for any internal standard changes by more than 30 seconds from the last check calibration (12 h), the chromatographic system must be inspected for malfunctions and corrections must be made, as required. If the electron ionization current plot (EICP) area for any of the internal standards changes by a factor of two from the last daily calibration standard check, the mass spectrometer

must be inspected for malfunctions and corrections must be made, as appropriate.

Demonstrate, through the analysis of a reagent water blank, that interferences from the analytical system, glassware, and reagents are under control. The blank samples should be carried through all stages of the sample preparation and measurement steps. For each analytical batch (up to 20 samples), a reagent blank, matrix spike, and matrix spike duplicate/duplicate must be analyzed (the frequency of the spikes may be different for different monitoring programs). The blank and spiked samples must be carried through all stages of the sample preparation and measurement steps. A QC reference sample concentrate containing each analyte at a concentration of 100 mg/L in methanol is required.

REFERENCE Test Methods for Evaluating Solid Waste (SW-846). U.S. EPA 1983, Method 8270B, Rev. 2, Nov. 1990. Office of Solid Waste, Washington, DC.

Phosphamidon   **EPA Method 8270**
CAS #13171-21-6

TITLE Semivolatile Organic Compounds by GC/MS

MATRIX This method is used to determine the concentration of semivolatile organic compounds in extracts prepared from all types of solid waste matrices, soils, and groundwater. Although surface waters are not specifically mentioned, this method should be applicable to water samples from rivers, lakes, etc.

METHOD SUMMARY This method covers 259 semivolatile organic compounds. In very limited applications direct injection of the sample into the GC/MS system may be appropriate, but this results in very high detection limits (approximately 10,000 µg/L). Typically, a 1-L liquid sample, containing surrogate, and matrix spiking standards, is extracted in a continuous extractor first under acid conditions and then under basic conditions. Typically 30 g of a solid sample, containing surrogate, and matrix spiking standards, is extracted ultrasonically. After concentrating the extract to 1 mL it is spiked with 10 µL of an internal standard solution just prior to analysis by GC/MS. The volume injected should contain about 100 ng of base/neutral and 200 ng of acid surrogates (for a 1-µL injection). Analysis is performed by GC/MS using a capillary GC column.

INTERFERENCES Raw GC/MS data from all blanks, samples, and spikes must be evaluated for interferences. Contamination by carryover can occur whenever high-concentration and low-concentration samples are sequentially analyzed. To reduce carryover, the sample syringe must be rinsed out between samples with solvent. Whenever an unusually concentrated sample is encountered, it should be followed by the analysis of blank solvent to check for cross-contamination.

INSTRUMENTATION A GC/MS and a data system are required. The GC column used is a 30 m × 0.25 mm I.D. (or 0.32 mm I.D.) 1um film thickness silicone-coated fused silica capillary column. A continuous liquid-liquid extractor equipped with Teflon® or glass connection joints and stopcocks requiring no lubrication, a K-D concentrating apparatus, water bath, and an ultrasonic disrupter with a minimum power of 300 W and with pulsing capability are also required.

PRECISION & ACCURACY The estimated quantitation limit (EQL) of Method 8270B for determining an individual compound is approximately 1 mg/kg (wet weight) for soil or sediment samples, 1–200 mg/kg for wastes (dependent on matrix and method of preparation), and 10 µg/L for groundwater samples. EQLs will be proportionately higher for sample extracts that require dilution to avoid saturation of the detector.

The EQL(b) for groundwater in µg/L is 100.
The EQL (a, b) for low concentrations in soil and sediment in µg/kg is not determined.
Accuracy as µg/L is not listed.
Overall precision in µg/L is not listed.

(a) *EQLs listed for soil/sediment are based on wet weight. Normally data is reported in a dry-weight basis; therefore, EQLs will be higher based on the % dry weight of each sample. This calculation is based on a 30 g sample and gel permeation chromatography cleanup.*

(b) *Sample EQLs are highly matrix-dependent. The EQLs are provided for guidance and may not always be achievable.*

C = *True value for concentration, in µg/L.*

X = *Average recovery found for measurements of samples containing a concentration of C, in µg/L.*

ESTIMATED QUANTITATION LIMIT

Other Matrices	Factor (a)
High-concentration soil and sludges by sonicator	7.5
Non-water miscible waste	75

(a) *EQL for other matrices = [EQL for low soil/sediment] × [Factor]. This estimated EQL is similar to an EPA "Practical Quantitation Limit."*

SAMPLING METHOD
Liquid samples — Use a 1 or 2½ gallon amber glass bottle with a screw-top Teflon®-lined cover that has been prewashed with detergent and rinsed with distilled water and methanol (or isopropanol).

Soils, sediments, or sludges — Use an 8-oz. widemouth glass with a screw-top Teflon®-lined cover that has been prewashed with detergent and rinsed with distilled water and methanol (or isopropanol).

SAMPLE PRESERVATION
Liquid samples — If residual chlorine is present, add 3 mL of 10% sodium thiosulfate per gallon, cool to 4°C and store in a solvent-free refrigerator until analysis; if chlorine is not present, then eliminate the sodium thiosulfate addition.

Soils, sediments, or sludges — Cool samples to 4°C and store in a solvent-free refrigerator.

MHT Liquid samples must be extracted within 7 days and the extracts analyzed within 40 days. Soils, sediments, or sludges may be stored for a maximum of 14 days and the extracts analyzed within 40 days.

SAMPLE PREPARATION

Liquid samples — Transfer 1 L quantitatively to a continuous extractor. If high concentrations are anticipated, a smaller volume may be used and then diluted with organic-free reagent water to 1 L. Adjust pH, if necessary, to pH <2 using 1:1 (V/V) sulfuric acid. Pipette 1.0 mL of a surrogate standard spiking solution into each sample. For the sample in each analytical batch selected for spiking, add 1.0 mL of a matrix spiking standard. For base/neutral acid analysis, the amount of the surrogates and matrix spiking compounds added to the sample should result in a final concentration of 100 ng/µL of each analyte in the extract to be analyzed (assuming a 1-µL injection). Extract with methylene chloride for 18–24 h. Next, adjust the pH of the aqueous phase to pH >11 using 10 N sodium hydroxide and extract it with methylene chloride again for 18–24 h. Dry the extract through a column containing anhydrous sodium sulfate and concentrate it to 1 mL using a K-D concentrator.

Soils, sediments, or sludges — Use 30 g of sample. Nonporous or wet samples (gummy or clay type) that do not have a free-flowing sandy texture must be mixed with anhydrous sodium sulfate until the sample is free flowing. Add 1 mL of surrogate standards to all samples, spikes, standards, and blanks. For the sample in each analytical batch selected for spiking, add 1.0 mL of a matrix spiking standard. For base/neutral acid analysis, the amount added of the surrogates and matrix spiking compounds should result in a final concentration of 100 ng/µL of each base/neutral analyte and 200 ng/µL of each acid analyte in the extract to be analyzed (assuming a 1-µL injection). Immediately add a 100-mL mixture of 1:1 methylene chloride:acetone and extract the sample ultrasonically for 3 min and then decant or filter the extracts. Repeat the extraction two or more times. Dry the extract using a column with anhydrous sodium sulfate and concentrate it to 1 mL in a K-D concentrator.

QUALITY CONTROL A methylene chloride solution containing 50 ng/µL of decafluorotriphenylphosphine (DFTPP) is used for tuning the GC/MS system each 12-h shift. A system performance check also must be made during every 12-h shift. A standard containing 50 ng/µL each of 4,4'-DDT, pentachlorophenol, and benzidine is required to verify injection port inertness and GC column performance. A calibration standard at mid-concentration, containing each compound of interest, including all required surrogates, must be performed every 12 h during analysis. After the system performance check is met, calibration check compounds (CCCs) are used to check the validity of the initial calibration.

The internal standard responses and retention times in the calibration check standard must be evaluated immediately after or during data acquisition. If the retention time for any internal standard changes by more than 30 seconds from the last check calibration (12 h), the chromatographic system must be inspected for malfunctions and corrections must be made, as required. If the electron ionization current plot (EICP) area for any of the internal standards changes by a factor of two from the last daily calibration standard check, the mass spectrometer must be inspected for malfunctions and corrections must be made, as appropriate.

Demonstrate, through the analysis of a reagent water blank, that interferences from the analytical system, glassware, and reagents are under control. The blank samples should be carried through all stages of the sample preparation and measurement steps. For each analytical batch (up to 20 samples), a reagent blank, matrix spike, and matrix spike duplicate/duplicate must be analyzed (the frequency of the spikes may be different for different monitoring programs). The blank and spiked samples must be carried through all stages of the sample preparation and measurement steps. A QC reference sample concentrate containing each analyte at a concentration of 100 mg/L in methanol is required.

REFERENCE Test Methods for Evaluating Solid Waste (SW-846). U.S. EPA 1983, Method 8270B, Rev. 2, Nov. 1990. Office of Solid Waste, Washington, DC.

Phosphorus   EPA Method 6010
CAS #7723-14-0

TITLE Inductively Coupled Plasma-Atomic Emission Spectroscopy

MATRIX This method is applicable to the determination of trace elements, including metals, in groundwater, soils, sludges, sediments, and other solid wastes. All matrices require digestion prior to analysis. The method of standard addition must be used for the analysis of all sample digests unless either serial dilution or matrix spike addition demonstrates it is not required.

METHOD SUMMARY Method 6010 covers 25 elements using ICP analysis. It measures element-emitted light by optical spectrometry. Samples, following an appropriate acid digestion, are nebulized and the resulting aerosol is transported to the plasma torch. Element-specific atomic line emission spectra are produced by a radio-frequency inductively coupled plasma.

INTERFERENCES Interferences may be categorized as spectral or non-spectral. Spectral interferences are caused by overlap of a spectral line from another element, unresolved overlap of molecular band spectra, background contribution from continuous or recombination phenomenon, and stray light from the line emission of high concentration elements. Non-spectral interferences include physical and chemical interferences. Physical interferences are effects associated with the sample nebulization and transport processes. Changes in viscosity and surface tension can cause significant inaccuracies. Chemical interferences include molecular compound formation, ionization effects, and solute vaporization effects. Normally these effects are not significant and can be minimized by careful selection of operating conditions. Chemical interferences are highly dependent on matrix type and the specific analyte element.

INSTRUMENTATION An inductively coupled argon plasma emission spectrometer (ICP) capable of background correction is required.

PRECISION & ACCURACY Detection limits, sensitivity, and optimum ranges of the metals will vary with the matrices and model of the spectrometer. In a single lab evaluation, seven wastes were analyzed for 22 elements. The mean percent relative standard deviation from triplicate analyses for all elements and wastes was 9 ± 2%. The mean percent recovery of spiked elements for all wastes was 93 ± 6%. Spike levels ranged from 100 µg/L to 100 mg/L. The wastes included sludges and industrial wastewaters.

Estimated instrument detection limit in µg/L is 51.
Spiked concentration in µg/L is not listed.
Mean reported value in µg/L is not listed.
Precision as RSD % is not listed.

SAMPLING METHOD   Samples should be collected in borosilicate glass, linear polyethylene, polypropylene, or Teflon® bottles that have been prewashed with detergent and tap water, and rinsed with 1:1 nitric acid and tap water or 1:1 hydrochloric acid and tap water. Collect at least 2 g of solids and 200 mL of aqueous samples.

SAMPLE PRESERVATION   Add nitric acid to make the samples pH <2.

MHT   The maximum holding time for properly preserved samples is 6 months.

SAMPLE PREPARATION   Preliminary treatment of most matrices is necessary because of the complexity and variability of sample matrices. Water samples that have been prefiltered and acidified will not need acid digestion. Methods for acid digestion of waters for total recoverable or dissolved metals, acid digestions of aqueous samples and extracts for total metals, and acid digestion of sediments, sludges, and soils are summarized below.

Total recoverable or dissolved metals in water — To prepare surface and groundwater samples for determination of total recoverable and dissolved metals, a 100-mL aliquot of well-mixed sample is acidified with concentrated nitric acid and concentrated hydrochloric acid, then heated until the volume is reduced to 15–20 mL. Adjust the final volume to 100 mL with reagent water.

Total metals in aqueous samples, soil and sediment extracts — To prepare aqueous samples, soil and sediment extracts, and wastes that contain suspended solids, a 100-mL aliquot is made acidic with concentrated nitric acid and the solution is evaporated to about 5 mL on a hot plate. Continue heating and adding additional acid until sample digestion is complete, which is usually indicated when the digestate is light in color or does not change in appearance. Evaporate the solution to about 3 mL and cool it and add a small quantity of 1:1 hydrochloric acid (10 mL/100 mL of final solution). Cover the beaker and reflux for 15 min. Wash down the beaker walls and filters or centrifuge the sample to remove silicates and other insoluble material. Filter the sample and adjust the final volume to 100 mL with reagent water and the final acid concentration to 10%.

Sediments, sludges, and soils — To prepare sediments, sludges and soil samples, transfer 1–2 g to a conical beaker and add 10 mL of 1:1 nitric acid, mix the slurry, and cover it with a watch glass. Heat the sample and reflux for 10–15 min without boiling. Allow it to cool, then add 5 mL of concentrated nitric acid and reflux for 30 min. Repeat last step and then allow the solution to evaporate to 5 mL without boiling. Cool and add 2 mL of water and 3 mL of 30% hydrogen peroxide. Cover and place the beaker on the hot plate. Heat and add 30% hydrogen peroxide in 1-mL aliquots with warming until the effervescence is minimal but do not add more than a total of 10 mL of 30% hydrogen peroxide. If the sample is being prepared for the analysis of Ag, Al, As, Ba, Be, Ca, Cd, Co, Cr, Cu, Fe, K, Mg, Mn, Mo, Na, Ni, Os, Pb, Se, Tl, V, and Zn, then add 5 mL of concentrated hydrochloric acid and 10 mL of water and return the covered beaker to a hot plate for 15 min of additional refluxing without boiling. Dilute the sample to a 100 mL volume with water after cooling and filter or centrifuge to remove particulates.

QUALITY CONTROL   Laboratory control samples must be analyzed for each analytical method. A method blank should be analyzed with each batch of samples. The effect of the matrix on method performance must be demonstrated: when appropriate, there should be at least one matrix spike and either one matrix duplicate or one matrix spike duplicate per analytical batch. The bias and precision of the method, as well as the method detection limit for each specific matrix type, must be measured.

Dilute and reanalyze samples that are more concentrated than the linear calibration limit. Employ a minimum of one reagent blank per sample batch to determine if contamination or any memory effects are occurring. Whenever a new or unusual sample matrix is encountered, perform either a serial dilution test or a matrix spike addition test to ensure that neither positive or negative interferences are operating on any of the analyte elements. Check the instrument standardization by verifying calibration every 10 samples using a calibration blank and a check standard.

REFERENCE   Test Methods for Evaluating Solid Waste (SW-846). U.S. EPA. 1983. Method 6010, Rev. 0, Sept. 1986. Office of Solid Wastes, Washington, DC.

Phosphorus **EPA Method 365.1**
CAS #7723-14-0

TITLE   Inorganics, Non-Metallics

MATRIX   Drinking, surface and saline waters. Wastewater.

APPLICATION   Date issued 1971. Editorial Rev. 1974 and 1978. (Colorimetric, automated, ascorbic acid). Applies to specified forms of phosphorus (P), based on reactions for the orthophosphate ion. Most analyses are run for phosphorus and dissolved P, orthophosphate, and dissolved orthophosphate.

INTERFERENCES   High iron concentrations cause precipitation of and loss of phosphorus. Sample turbidity must be removed by filtration prior to analysis for orthophosphate. Salt error for samples 5 to 20% salt, <1%. Arsenic concentrations > phosphorus concentration, may interfere.

INSTRUMENTATION   Technicon auto analyzer. 650–660 or 880 nm filter.

RANGE 0.001–1.0 mg P/L range.

MDL Not listed.

PRECISION 0.066 mg P/L at 0.30 mg P/L (as orthophosphate).

ACCURACY As bias, –0.04 mg P/L at 0.30 mg P/L (as orthophosphate).

SAMPLING METHOD Plastic or glass. (50 mL).

STABILITY Cool, 4°C. H2SO4 to pH <2.

MHT 28 days.

QUALITY CONTROL This method is based on reactions specific for the orthophosphate ion in which an antimony-phospho-molybdate complex is reduced to an intensely blue colored complex by ascorbic acid. Color is proportional to phosphorus concentration. Measure color on auto analyzer

REFERENCE EPA Methods for the Chemical Analysis of Water and Wastes, EPA-600/4-79-020, U.S. EPA, EMSL, 1979.

Phosphorus — EPA Method 365.2
CAS #7723-14-0

TITLE Inorganics, Non-Metallics

MATRIX Drinking, surface and saline waters. Wastewater.

APPLICATION Date issued 1971. (Colorimetric, ascorbic acid, single reagent). Applies to specified forms of phosphorus (P), based on reactions for orthophosphate ion. Most commonly measured forms are; phosphorus and dissolved P, orthophosphate and dissolved orthophosphate.

INTERFERENCES High iron concentrations cause precipitation of and loss of phosphorus. Salt error for samples 5 to 20% salt, <1%. Arsenic concentration> phosphorus concentration, may interfere. (Only orthophosphate turns blue color in this test. Other (P) forms are converted to orthophosphate).

INSTRUMENTATION Spectro(or filter)photometer. 650 or 880 nm. Light path of 1 cm or longer.

RANGE 0.01–0.5 mg P/L.

MDL Not listed.

PRECISION SD = 0.018 mg P/L at 0.335 mg P/L (as orthophosphate)

ACCURACY As bias, –0.009 mg P/L L 0.335 mg P/L (as orthophosphate).

SAMPLING METHOD Plastic or glass (50 mL).

STABILITY Cool, 4°C. H2SO4 to pH <2.

MHT 28 days.

QUALITY CONTROL This method is based on reactions specific for the orthophosphate ion in which an antimony-phospho-molybdate complex is reduced to an intensely blue colored complex by ascorbic acid. Color is proportional to phosphorus concentration. Measure color on auto analyzer.

REFERENCE Methods for the Chemical Analysis of Water and Wastes, EPA-600/4-79-020, U.S. EPA, EMSL, 1979.

Phosphorus — EPA Method 365.3
CAS #7723-14-0

TITLE Inorganics, Non-Metallics

MATRIX Drinking, surface and saline waters. Wastewater.

APPLICATION Date issued 1978. (Colorimetric, ascorbic acid, two reagent).

Applies to specified forms of phosphorus (P), based on reactions for the orthophosphate ion. Most commonly measured forms are; phosphorus and dissolved P, orthophosphate and dissolved orthophosphate.

INTERFERENCES Arsenate interference is eliminated by reducing arsenic acid to arsenious acid using sodium bisulfite. High concentrations of iron cause low phosphorus recovery. The bisulfite treatment also eliminates this interference.

INSTRUMENTATION Spectro(or filter)photometer. 660 or 880 nm. Light path of 1 cm or longer.

RANGE 0.01–1.2 mg P/L.

MDL Not listed.

PRECISION Not listed.

ACCURACY Recoveries = 99 and 100% at 7.6 and 0.55 mg P/L (waste and sewage)

SAMPLING METHOD Plastic or glass (50 mL).

STABILITY Cool, 4°C. H2SO4 to pH <2.

MHT 28 days.

QUALITY CONTROL This Method is based on reactions specific for the orthophosphate ion in which an antimony-phospho-molybdate complex is reduced to an intensely blue colored complex by ascorbic acid. Color is proportional to phosphorus concentration. Measure color in auto analyzer.

REFERENCE EPA Methods for the Chemical Analysis of Water and Wastes, EPA-600/4-79-020, U.S. EPA, EMSL, 1979.

Phosphorus — EPA Method 365.4
CAS #7723-14-0

TITLE Inorganics, Non-Metallics

MATRIX Drinking, surface, and wastewaters.

APPLICATION Date issued 1974. (Colorimetric, automated, block digestor AAII). Sample is heated in presence of sulfuric acid, potassium sulfate and mercuric sulfate for 2 1/2 h. Residue is cooled, diluted to 25 mL and placed on auto analyzer for phosphorus determination. Temperature of block digester during 2 1/2 h digestion is 380 deg C.

INTERFERENCES Only add 4–8 Teflon® boiling chips during digestion. Too many boiling chips will cause sample to boil over.

INSTRUMENTATION block digestor BD-40. Technicon auto analyzer and method no. 327-74W.

RANGE 0.01–20 mg P/L.

MDL Not listed.

PRECISION SD = ±0.06 at 2.0 mg P/L (sewage sample)

ACCURACY Recoveries = 95 and 98% at 1.84 and 1.89 mg P/L (sewage samples)

SAMPLING METHOD Plastic or glass (50 mL).

STABILITY Cool, 4°C. H2SO4 to pH <2.

MHT 28 days.

QUALITY CONTROL This Method covers the determination of total phosphorus. Prepare standard curve plotting peak heights of standards against concentration values. Compare sample peak heights with standard curve to compute concentration. (If TKN is determined, the sample should be diluted with ammonia-free water).

REFERENCE Methods for the Chemical Analysis of Water and Wastes, EPA-600/4-79-020, U.S. EPA, EMSL, 1979.

Phthalic anhydride   **EPA Method 8270**
CAS #85-44-9

TITLE Semivolatile Organic Compounds by GC/MS

MATRIX This method is used to determine the concentration of semivolatile organic compounds in extracts prepared from all types of solid waste matrices, soils, and groundwater. Although surface waters are not specifically mentioned, this method should be applicable to water samples from rivers, lakes, etc.

METHOD SUMMARY This method covers 259 semivolatile organic compounds. In very limited applications direct injection of the sample into the GC/MS system may be appropriate, but this results in very high detection limits (approximately 10,000 µg/L). Typically, a 1-L liquid sample, containing surrogate, and matrix spiking standards, is extracted in a continuous extractor first under acid conditions and then under basic conditions. Typically 30 g of a solid sample, containing surrogate, and matrix spiking standards, is extracted ultrasonically. After concentrating the extract to 1 mL it is spiked with 10 µL of an internal standard solution just prior to analysis by GC/MS. The volume injected should contain about 100 ng of base/neutral and 200 ng of acid surrogates (for a 1-µL injection). Analysis is performed by GC/MS using a capillary GC column.

INTERFERENCES Raw GC/MS data from all blanks, samples, and spikes must be evaluated for interferences. Contamination by carryover can occur whenever high-concentration and low-concentration samples are sequentially analyzed. To reduce carryover, the sample syringe must be rinsed out between samples with solvent. Whenever an unusually concentrated sample is encountered, it should be followed by the analysis of blank solvent to check for cross-contamination.

INSTRUMENTATION A GC/MS and a data system are required. The GC column used is a 30 m × 0.25 mm I.D. (or 0.32 mm I.D.) 1um film thickness silicone-coated fused silica capillary column. A continuous liquid-liquid extractor equipped with Teflon® or glass connection joints and stopcocks requiring no lubrication, a K-D concentrating apparatus, water bath, and an ultrasonic disrupter with a minimum power of 300 W and with pulsing capability are also required.

PRECISION & ACCURACY The estimated quantitation limit (EQL) of Method 8270B for determining an individual compound is approximately 1 mg/kg (wet weight) for soil or sediment samples, 1–200 mg/kg for wastes (dependent on matrix and method of preparation), and 10 µg/L for groundwater samples. EQLs will be proportionately higher for sample extracts that require dilution to avoid saturation of the detector.

The EQL(b) for groundwater in µg/L is 100.
The EQL (a, b) for low concentrations in soil and sediment in µg/kg is not determined.
Accuracy as µg/L is not listed.
Overall precision in µg/L is not listed.

(a) *EQLs listed for soil/sediment are based on wet weight. Normally data is reported in a dry-weight basis; therefore, EQLs will be higher based on the % dry weight of each sample. This calculation is based on a 30 g sample and gel permeation chromatography cleanup.*
(b) *Sample EQLs are highly matrix-dependent. The EQLs are provided for guidance and may not always be achievable.*
C = *True value for concentration, in µg/L.*
X = *Average recovery found for measurements of samples containing a concentration of C, in µg/L.*

ESTIMATED QUANTITATION LIMIT

Other Matrices	Factor (a)
High-concentration soil and sludges by sonicator	7.5
Non-water miscible waste	75

(a) *EQL for other matrices = [EQL for low soil/sediment] × [Factor]. This estimated EQL is similar to an EPA "Practical Quantitation Limit."*

SAMPLING METHOD

Liquid samples — Use a 1 or 2½ gallon amber glass bottle with a screw-top Teflon®-lined cover that has been prewashed with detergent and rinsed with distilled water and methanol (or isopropanol).

Soils, sediments, or sludges — Use an 8-oz. widemouth glass with a screw-top Teflon®-lined cover that has been prewashed with detergent and rinsed with distilled water and methanol (or isopropanol).

SAMPLE PRESERVATION

Liquid samples — If residual chlorine is present, add 3 mL of 10% sodium thiosulfate per gallon, cool to 4°C and store in a solvent-free refrigerator until analysis; if chlorine is not present, then eliminate the sodium thiosulfate addition.

Soils, sediments, or sludges — Cool samples to 4°C and store in a solvent-free refrigerator.

MHT Liquid samples must be extracted within 7 days and the extracts analyzed within 40 days. Soils, sediments, or sludges may be stored for a maximum of 14 days and the extracts analyzed within 40 days.

SAMPLE PREPARATION

Liquid samples — Transfer 1 L quantitatively to a continuous extractor. If high concentrations are anticipated, a smaller volume may be used and then diluted with organic-free reagent water to 1 L. Adjust pH, if necessary, to pH <2 using 1:1 (V/V) sulfuric acid. Pipette 1.0 mL of a surrogate standard spiking solution into each sample. For the sample in each analytical batch selected for spiking, add 1.0 mL of a matrix spiking standard. For base/neutral acid analysis, the amount of the surrogates and matrix spiking compounds added to the sample should result in a final concentration of 100 ng/μL of each analyte in the extract to be analyzed (assuming a 1-μL injection). Extract with methylene chloride for 18–24 h. Next, adjust the pH of the aqueous phase to pH >11 using 10 N sodium hydroxide and extract it with methylene chloride again for 18–24 h. Dry the extract through a column containing anhydrous sodium sulfate and concentrate it to 1 mL using a K-D concentrator.

Soils, sediments, or sludges — Use 30 g of sample. Nonporous or wet samples (gummy or clay type) that do not have a free-flowing sandy texture must be mixed with anhydrous sodium sulfate until the sample is free flowing. Add 1 mL of surrogate standards to all samples, spikes, standards, and blanks. For the sample in each analytical batch selected for spiking, add 1.0 mL of a matrix spiking standard. For base/neutral acid analysis, the amount added of the surrogates and matrix spiking compounds should result in a final concentration of 100 ng/μL of each base/neutral analyte and 200 ng/μL of each acid analyte in the extract to be analyzed (assuming a 1-μL injection). Immediately add a 100-mL mixture of 1:1 methylene chloride:acetone and extract the sample ultrasonically for 3 min and then decant or filter the extracts. Repeat the extraction two or more times. Dry the extract using a column with anhydrous sodium sulfate and concentrate it to 1 mL in a K-D concentrator.

QUALITY CONTROL A methylene chloride solution containing 50 ng/μL of decafluorotriphenylphosphine (DFTPP) is used for tuning the GC/MS system each 12-h shift. A system performance check also must be made during every 12-h shift. A standard containing 50 ng/μL each of 4,4'-DDT, pentachlorophenol, and benzidine is required to verify injection port inertness and GC column performance. A calibration standard at mid-concentration, containing each compound of interest, including all required surrogates, must be performed every 12 h during analysis. After the system performance check is met, calibration check compounds (CCCs) are used to check the validity of the initial calibration.

The internal standard responses and retention times in the calibration check standard must be evaluated immediately after or during data acquisition. If the retention time for any internal standard changes by more than 30 seconds from the last check calibration (12 h), the chromatographic system must be inspected for malfunctions and corrections must be made, as required. If the electron ionization current plot (EICP) area for any of the internal standards changes by a factor of two from the last daily calibration standard check, the mass spectrometer must be inspected for malfunctions and corrections must be made, as appropriate.

Demonstrate, through the analysis of a reagent water blank, that interferences from the analytical system, glassware, and reagents are under control. The blank samples should be carried through all stages of the sample preparation and measurement steps. For each analytical batch (up to 20 samples), a reagent blank, matrix spike, and matrix spike duplicate/duplicate must be analyzed (the frequency of the spikes may be different for different monitoring programs). The blank and spiked samples must be carried through all stages of the sample preparation and measurement steps. A QC reference sample concentrate containing each analyte at a concentration of 100 mg/L in methanol is required.

REFERENCE Test Methods for Evaluating Solid Waste (SW-846). U.S. EPA 1983, Method 8270B, Rev. 2, Nov. 1990. Office of Solid Waste, Washington, DC.

Picloram   EPA Method 8151
CAS #1918-02-1

TITLE Chlorinated Herbicides by GC Using Methylation or Pentafluorobenzylation Derivatization: Capillary Column Technique.

MATRIX This method covers aqueous and solid matrices. This includes a wide variety such as drinking water, groundwater, industrial wastewaters, surface waters, soils, solids, and sediments.

METHOD SUMMARY This is a GC method for determining 19 chlorinated acid herbicides in aqueous, soil, and waste matrices. Because these compounds are produced and used in various forms (i.e., acid, salt, ester, etc.) a hydrolysis step is included to convert the herbicide to the acid form prior to analysis. This method provides hydrolysis, extraction, derivatization and GC conditions for the analysis of chlorinated acid herbicides in water, soil, and waste samples. Water samples are hydrolyzed *in situ*, extracted with diethyl ether, and then esterified with either diazomethane or pentafluorobenzyl bromide. The derivatives are determined by gas chromatography with an electron capture detector (GC/ECD). The results are reported as acid equivalents. The sensitivity of this method depends on the level of interferences in addition to instrumental limitations.

INTERFERENCES Method interferences may be caused by contaminants in solvents, reagents, glassware, and other sample processing hardware. Immediately prior to use, glassware should be rinsed with the next solvent to be used. Matrix interferences may be caused by contaminants that are coextracted from the sample. Organic acids, especially chlorinated acids, cause the most direct interference with the determination by methylation. Phenols, including chlorophenols, may also interfere with this procedure. The determination using pentafluorobenzylation is more sensitive, and more prone to interferences from the presence of organic acids of phenols than by methylation. Alkaline hydrolysis and subsequent extraction of the basic solution removes many chlorinated hydrocarbons and phthalate esters that might otherwise interfere with the ECD analysis. The herbicides, being strong organic acids, react readily with alkaline substances and may be lost during analysis. Therefore, glassware must be acid-rinsed and then rinsed to constant pH with organic-free reagent water.

INSTRUMENTATION A GC suitable for Grob-type injection using capillary columns. A data system for measuring peak heights and/or peak areas is recommended. An electron capture detector (ECD) is used. Also a K-D apparatus, a diazomethane generator, a centrifuge and an ultrasonic disrupter will be required.

Narrow Bore Columns:
Primary Column 1: 30 m × 0.25 mm, 5% phenyl/95% methyl silicone (DB-5), 0.25 µm film thickness.
Primary Column 1a (GC/MS): 30 m × 0.32 mm, 5% phenyl/95% methyl silicone (DB-5), 1-µm film thickness.
Column 2: 30 m × 0.25 mm DB-608 with a 25 µm film thickness.
Confirmation Column: 30 m × 0.25 mm, 14% cyanopropyl phenyl silicone (DB-1701), 0.25 µm film thickness.

Megabore Columns:
Primary Column: 30 m × 0.53 mm DB-608 with 0.83 µm film thickness.
Confirmation Column: 30 m × 0.53 mm, 14% cyanopropyl phenyl silicone (DB-1701), 1.0 µm film thickness.

PRECISION & ACCURACY Method detection limits (MDLs) are compound-dependent and vary with derivitization efficiency, derivative recovery, the matrix sampled, and herbicide concentration.

The estimated MDL (in µg/L) was 0.14 for aqueous samples using GC/ECD.

The estimated MDL (in µg/kg) was not reported for soil samples using GC/ECD when corrected back to 50 g samples extracted and concentrated to 10 mL with 5-µL injections.

The estimated GC/MS identification limit (in ng) was not reported for soil samples using GC/MS.

Mean percent recovery, calculated from 7–8 determinations of spiked reagent water, after diazomethane derivatization, from a spike concentration (in µg/L) of 0.6 was 91 with a standard deviation of the percent recovery of 15.5.

Mean percent recovery, calculated from 10 determinations of spiked clay and clay/still bottom samples over the linear concentration range (in ng/g) of no data was none reported with a percent relative standard deviation of none. The RSD % was calculated on 10 samples high in the linear concentration range and 10 low in the range. The linear concentration range was determined using standard solutions and corrected to 50 g soil samples.

SAMPLE COLLECTION, PRESERVATION & HANDLING
Containers used to collect samples for the determination of semivolatile organic compounds should be soap and water washed followed by methanol (or isopropanol) rinsing. The sample containers should be of glass or Teflon® and have screw-top covers with Teflon® liners.

No preservation is used with concentrated waste samples. With liquid samples containing no residual chlorine and with soil, sediment, and sludge samples, immediately cooling to 4°C is the only preservation used. When residual chlorine is present then 3 mL of 10% aqueous sodium sulfate is added for each gallon of sample collected, followed by cooling to 4°C.

The holding time for all volatile organics samples is 14 days. Liquid samples must be extracted within 7 days and their extracts analyzed within 40 days. Concentrated waste, soil, sediment, and sludge samples must be extracted within 14 days and their extracts analyzed within 40 days.

SAMPLE PREPARATION
Preparation of soil, sediment, and other solid samples — Acidify 30 g (dry weight) solids with 0.1 M phosphate buffer (pH = 2.5) and thoroughly mix the contents. Spike the sample with surrogate compound(s). The ultrasonic extraction of solids must be optimized for each type of sample. In order for the ultrasonic extractor to efficiently extract solid samples, the sample must be free flowing when the solvent is added. Acidified anhydrous sodium sulfate should be added to clay-type soils, or any other solid that is not a free-flowing sandy texture, until a free flowing mixture is obtained. Add methylene chloride and perform ultrasonic extraction. Combine organic extracts from the repetitive extractings of the sample and centrifuge. Add aqueous potassium hydroxide, water, and methanol to the extract and reflux the mixture on a water bath. Extract the solution three times with methylene chloride and discard the methylene chloride phase. The basic solution contains the herbicide salts. Adjust the pH of the solution to <2 with cold sulfuric acid and extract three times with methylene chloride. Combine the extracts and pour them through a pre-rinsed drying column containing acidified anhydrous sodium sulfate. Collect the dried extracts in a K-D flask and concentrate them.

Preparation of aqueous samples — Measure 1 L of sample into a 2 L separatory funnel and spike it with surrogate compound(s). Add NaCl to the sample, then add 6 N NaOH to the sample to a pH of 12 or more and let the sample sit at room temperature for 1 h to hydrolyze esters. Extract the sample three times with methylene chloride and discard the extracts. Then add cold 12 N sulfuric acid to a pH less than or equal to 2, and extract the sample three times with ethyl ether. Collect the ether phase in a flask containing acidified anhydrous sodium sulfate and allow it to remain in contact with the sodium sulfate for a minimum of 2 h. The drying step is very critical to ensuring complete esterification; any moisture remaining in the ether will result in low herbicide recoveries.

Extract concentration and derivatization — The combined ether extract is concentrated to about 1 mL using a K-D apparatus followed by using a micro Snyder column or nitrogen gas blowdown. If methyl esters are to be produced, then dilute the concentrated ether extract with 1 mL of isooctane and 0.5 mL of methanol, dilute to a final volume of 4 mL, and esterify with diazomethane. If pentafluorobenzene esters are to be produced, then dilute concentrated ether extract with acetone to a final volume of 4 mL and esterify with pentafluorobenzyl bromide.

QUALITY CONTROL Select a representative spike concentration for each compound (acid or ester) to be measured. Using stock standard, prepare a quality control check sample concentrate, in acetone, that is 1000 times more concentrated than the selected concentrations. Use this quality control check sample concentrate to prepare quality control check samples. Calculate surrogate standard recovery on all standards, samples, blanks, and spikes. GC/MS techniques should be judiciously employed to support qualitative identifications made with this method. When available, chemical ionization mass spectra may be employed to aid the qualitative identification process.

REFERENCE Test Methods for Evaluating Solid Waste, Physical/Chemical Methods, SW-846, 3rd Edition, U.S. EPA, Office of Solid Waste, Washington, DC, EPA Method 8151, Nov. 1990.

2-Picoline   **EPA Method 1625**
CAS #109-06-8

TITLE Semivolatile Organic Compounds by Isotope Dilution GC/MS

MATRIX The compounds may be determined in waters, soils, and municipal sludges by this method.

METHOD SUMMARY This method is used to determine 176 semivolatile toxic organic pollutants associated with the CWA (as amended 1987); the RCRA (as amended 1986); the CERCLA (as amended 1986); and other compounds amenable to extraction and analysis by capillary column gas chromatography-mass spectrometry (GC/MS).

Stable isotopically-labeled analogs of the compounds of interest are added to the sample. If the solids content is less than 1%, a 1-L sample is extracted at pH 12–13, then at pH <2 with methylene chloride using continuous extraction techniques.

If the solids content is 30% or less, the sample is diluted to 1% solids with reagent water, homogenized ultrasonically, and extracted at pH 12–13, then at pH <2 with methylene chloride using continuous extraction techniques. If the solids content is greater than 30%, the sample is extracted using ultrasonic techniques.

Each extract is dried over sodium sulfate, concentrated to a volume of 5 mL, cleaned up using GPC, if necessary, and concentrated. Extracts are concentrated to 1 mL if GPC is not performed, and to 0.5 mL if GPC is performed.

An internal standard is added to the extract, and a 1-mL aliquot of the extract is injected into the GC. The compounds are separated by GC and detected by a MS. The labeled compounds serve to correct the variability of the analytical technique.

INTERFERENCES Solvents, reagents, glassware, and other sample processing hardware may yield artifacts and/or elevated baselines causing misinterpretation of chromatograms and spectra. Materials used in the analysis must be demonstrated to be free from interferences under the conditions of analysis by running method blanks initially and with each sample lot (sample started through the extraction process on a given 8-h shift, to a maximum of 20). Specific selection of reagents and purification of solvents by distillation in all glass systems may be required. Glassware and, where possible, reagents are cleaned by solvent rinse and baking at 450°C for 1-h minimum. Interferences coextracted from samples will vary considerably from source to source, depending on the diversity of the site being sampled.

INSTRUMENTATION Major instrumentation includes a GC with a splitless or on-column injection port for capillary column, a MS with 70 eV electron impact ionization, and a data system to collect and record MS data, and process it. A K-D apparatus is used to concentrate extracts.

GC Column: 30 m × 0.25 mm I.D. 5% phenyl, 94% methyl, 1% vinyl silicone bonded phased fused silica capillary column.

PRECISION & ACCURACY The detection limits of the method are usually dependent on the level of interferences rather than instrumental limitations. The limits typify the minimum quantities that can be detected with no interferences present.

The minimum level (in µg/mL) was 50. This is defined as a minimum level at which the analytical system shall give recognizable mass spectra (background corrected) and acceptable calibration points.

The MDL (in µg/kg) in low solids was 25 and in high solids was 87; these were determined in digested sludge (low solids) and in filter cake or compost (high solids).

The labeled and native compound initial precision as standard deviation (in µg/L) was 38.

The labeled and native compound initial accuracy as average recovery (in µg/L) was 59–149.

SAMPLE COLLECTION, PRESERVATION & HANDLING Collect samples in glass containers. Aqueous samples which flow freely are collected in refrigerated bottles using automatic sampling equipment. Solid samples are collected as grab samples using widemouth jars. Maintain samples at 0 to 4°C from the time of collection until extraction. If residual chlorine is present in aqueous samples, add 80 mg sodium thiosulfate/L of water. Begin sample extraction within 7 days of collection, and analyze all extracts within 40 days of extraction.

SAMPLE PREPARATION Samples containing 1% solids or less are extracted directly using continuous liquid-liquid extraction techniques. Samples containing 1 to 30% solids are diluted to the 1% level with reagent water and extracted using continuous liquid-liquid extraction techniques. Samples containing greater than 30% solids are extracted using ultrasonic techniques.

Base/neutral extraction — Adjust the pH of the waters in the extractors to 12–13 with 6 N NaOH. Extract with methylene chloride for 24–48 h.

Acid extraction — Adjust the pH of the waters in the extractors to 2 or less using 6 N sulfuric acid. Extract with methylene chloride for 24–48 h.

Ultrasonic extraction of high solids samples — Add anhydrous sodium sulfate to the sample and QC aliquot(s). Add acetone:methylene chloride (1:1) to the sample and mix thoroughly

Concentrate extracts using a K-D apparatus.

QUALITY CONTROL The analyst is permitted to modify this method to improve separations or lower the costs of measurements, provided all performance specifications are met. Analyses of blanks are required to demonstrate freedom from contamination. When results of spikes indicate atypical method performance for samples, the samples are diluted to bring method performance within acceptable limits.

For low solids (aqueous samples), extract, concentrate, and analyze two sets of four 1-L aliquots (8 aliquots total) of the precision and recovery standard. For high solids samples, two sets of four 30-g aliquots of the high solids reference matrix are used.

Spike all samples with labeled compounds to assess method performance. Compute percent recovery of the labeled compounds using the internal standard method. Compare the labeled compound recovery for each compound with the corresponding labeled compound recovery.

Reagent water and high solids reference matrix blanks are analyzed to demonstrate freedom from contamination. Extract and concentrate a 1-L reagent water blank or a high solids reference matrix blank with each sample's lot (samples started through the extraction process on the same 8-h shift, to a maximum of 20 samples).

Field replicates may be collected to determine the precision of the sampling technique, and spiked samples may be required to determine the accuracy of the analysis when the internal standard method is used.

REFERENCE Semivolatile Organic Compounds by Isotope Dilution GC/MS. Office of Water Regulation and Standards, U.S. EPA Industrial Technology Division, Washington, DC, EPA Method 1625, Rev. C, June 1989 (contact W.A. Telliard, U.S. EPA, Office of Water Regulations and Standards, 401 M St., SW, Washington, DC, 20460. Phone: 202-382-7131).

2-Picoline   **EPA Method 1625**
CAS #109-06-8

TITLE Semivolatile Organic Compounds by Isotope Dilution GC/MS

MATRIX The compounds may be determined in waters, soils, and municipal sludges by this method.

METHOD SUMMARY This method is used to determine 176 semivolatile toxic organic pollutants associated with the CWA (as amended 1987); the RCRA (as amended 1986); the CERCLA (as amended 1986); and other compounds amenable to extraction and analysis by capillary column gas chromatography-mass spectrometry (GC/MS).

Stable isotopically-labeled analogs of the compounds of interest are added to the sample. If the solids content is less than 1%, a 1-L sample is extracted at pH 12–13, then at pH <2 with methylene chloride using continuous extraction techniques.

If the solids content is 30% or less, the sample is diluted to 1% solids with reagent water, homogenized ultrasonically, and extracted at pH 12–13, then at pH <2 with methylene chloride using continuous extraction techniques. If the solids content is greater than 30%, the sample is extracted using ultrasonic techniques.

Each extract is dried over sodium sulfate, concentrated to a volume of 5 mL, cleaned up using GPC, if necessary, and concentrated. Extracts are concentrated to 1 mL if GPC is not performed, and to 0.5 mL if GPC is performed.

An internal standard is added to the extract, and a 1-mL aliquot of the extract is injected into the GC. The compounds are separated by GC and detected by a MS. The labeled compounds serve to correct the variability of the analytical technique.

INTERFERENCES Solvents, reagents, glassware, and other sample processing hardware may yield artifacts and/or elevated baselines causing misinterpretation of chromatograms and spectra. Materials used in the analysis must be demonstrated to be free from interferences under the conditions of analysis by running method blanks initially and with each sample lot (sample started through the extraction process on a given 8-h shift, to a maximum of 20). Specific selection of reagents and purification of solvents by distillation in all glass systems may be required. Glassware and, where possible, reagents are cleaned by solvent rinse and baking at 450°C for 1-h minimum. Interferences coextracted from samples will vary considerably from source to source, depending on the diversity of the site being sampled.

INSTRUMENTATION Major instrumentation includes a GC with a splitless or on-column injection port for capillary column, a MS with 70 eV electron impact ionization, and a data system to collect and record MS data, and process it. A K-D apparatus is used to concentrate extracts.

GC Column: 30 m × 0.25 mm I.D. 5% phenyl, 94% methyl, 1% vinyl silicone bonded phased fused silica capillary column.

PRECISION & ACCURACY The detection limits of the method are usually dependent on the level of interferences rather than instrumental limitations. The limits typify the minimum quantities that can be detected with no interferences present.

The minimum level (in µg/mL) was 50. This is defined as a minimum level at which the analytical system shall give recognizable mass spectra (background corrected) and acceptable calibration points.

The MDL (in μg/kg) in low solids was 25 and in high solids was 87; these were determined in digested sludge (low solids) and in filter cake or compost (high solids).

The labeled and native compound initial precision as standard deviation (in μg/L) was 38.

The labeled and native compound initial accuracy as average recovery (in μg/L) was 59–149.

SAMPLE COLLECTION, PRESERVATION & HANDLING
Collect samples in glass containers. Aqueous samples which flow freely are collected in refrigerated bottles using automatic sampling equipment. Solid samples are collected as grab samples using widemouth jars. Maintain samples at 0 to 4°C from the time of collection until extraction. If residual chlorine is present in aqueous samples, add 80 mg sodium thiosulfate/L of water. Begin sample extraction within 7 days of collection, and analyze all extracts within 40 days of extraction.

SAMPLE PREPARATION Samples containing 1% solids or less are extracted directly using continuous liquid-liquid extraction techniques. Samples containing 1 to 30% solids are diluted to the 1% level with reagent water and extracted using continuous liquid-liquid extraction techniques. Samples containing greater than 30% solids are extracted using ultrasonic techniques.

- Base/neutral extraction — Adjust the pH of the waters in the extractors to 12–13 with 6 N NaOH. Extract with methylene chloride for 24–48 h.
- Acid extraction — Adjust the pH of the waters in the extractors to 2 or less using 6 N sulfuric acid. Extract with methylene chloride for 24–48 h.
- Ultrasonic extraction of high solids samples — Add anhydrous sodium sulfate to the sample and QC aliquot(s). Add acetone:methylene chloride (1:1) to the sample and mix thoroughly

Concentrate extracts using a K-D apparatus.

QUALITY CONTROL The analyst is permitted to modify this method to improve separations or lower the costs of measurements, provided all performance specifications are met. Analyses of blanks are required to demonstrate freedom from contamination. When results of spikes indicate atypical method performance for samples, the samples are diluted to bring method performance within acceptable limits.

For low solids (aqueous samples), extract, concentrate, and analyze two sets of four 1-L aliquots (8 aliquots total) of the precision and recovery standard. For high solids samples, two sets of four 30-g aliquots of the high solids reference matrix are used.

Spike all samples with labeled compounds to assess method performance. Compute percent recovery of the labeled compounds using the internal standard method. Compare the labeled compound recovery for each compound with the corresponding labeled compound recovery.

Reagent water and high solids reference matrix blanks are analyzed to demonstrate freedom from contamination. Extract and concentrate a 1-L reagent water blank or a high solids reference matrix blank with each sample's lot (samples started through the extraction process on the same 8-h shift, to a maximum of 20 samples).

Field replicates may be collected to determine the precision of the sampling technique, and spiked samples may be required to determine the accuracy of the analysis when the internal standard method is used.

REFERENCE Semivolatile Organic Compounds by Isotope Dilution GC/MS. Office of Water Regulation and Standards, U.S. EPA Industrial Technology Division, Washington, DC, EPA Method 1625, Rev. C, June 1989 (contact W.A. Telliard, U.S. EPA, Office of Water Regulations and Standards, 401 M St., SW, Washington, DC, 20460. Phone: 202-382-7131).

2-Picoline   **EPA Method 8240**
CAS #109-06-8

TITLE Volatile Organics By GC/MS: Packed Column Technique

MATRIX Nearly all types of sample matarices, regardless of water content, can be analyzed using this method. This includes groundwater, aqueous sludges, caustic liquors, acid liquors, waste solvents, oily wastes, mousses, tars, fibrous wastes, polymetric emulsions, filter cakes, spent carbons, spent catalysts, soils, and sediments.

METHOD SUMMARY Method 8240B covers 80 volatile organic compounds that are introduced into a gas chromatograph by the purge-and-trap method or by direct injection (in limited applications). For the purge-and-trap method an inert gas (zero grade nitrogen or helium) is bubbled through a 5-mL solution at ambient temperature. Purged sample components are trapped in a tube of sorbent materials. When purging is complete, the sorbent tube is heated and backflushed with inert gas to desorb the trapped components onto a GC column.

INTERFERENCES Impurities in the purge gas and from organic compounds outgassing from the plumbing ahead of the trap account for many contamination problems. Interferences purged or coextracted from the samples will vary considerably from source to source. Cross-contamination can occur whenever high-level and low-level samples are analyzed sequentially. Whenever an unusually concentrated sample is analyzed, it should be followed by the analysis of organic-free reagent water to check for cross-contamination. Samples also can be contaminated by diffusion of volatile organics (particularly methylene chloride and fluorocarbons) through the septum seal into the sample during shipment and storage. A trip blank can serve as a check on such contamination. The lab where volatile analysis is performed and also the refrigerated storage area should be completely free of solvents.

INSTRUMENTATION A gas chromatograph/mass spectrometry/data system (GC/MS) equipped with a 6 ft × 0.1 in I.D. glass column packed with 1% SP-1000 on Carbopack-B (60/80 mesh) is required. Also needed is a 5-mL purging device, a sorbent trap, and a thermal desorption apparatus.

PRECISION & ACCURACY This method is reported to have been tested by 15 laboratories using organic-free reagent water, drinking water, surface water, and industrial wastewaters (not specified) fortified at six concentrations over the range 5–600 µg/L.

Sample estimated quantitation limits (EQLs) are highly matrix-dependent. The EQLs listed may not always be achievable. EQLs listed for soils or sediments are based on wet weight. Normally, data is reported on a dry-weight basis; therefore, EQLs will be higher, based on the percent dry weight of each sample. Note that EQLs are even more variable than MDLs and that they are highly variable depending on the matrix being analyzed.

EQL in groundwater in µg/L was not listed.
EQL in low soil or sediment in µg/kg was not listed.
Accuracy (a) in µg/L was not listed.
Precision (b) in µg/L was not listed.

(a) Average recovery found for measurements of samples containing a concentration of C, in µg/L.
(b) Overall precision found for measurements of samples with average recovery X for samples containing a concentration of C in µg/L.
X = Average recovery found for measurement of samples containing a concentration of C in µg/L.

MULTIPLICATION FACTORS FOR OTHER MATRICES

Other Matrices	Factor (a)
Waste miscible liquid waste	50
High-concentration soil and sludge	125
Non-water miscible waste	500

(a) EQL = [EQL for low soil/sediment] × [Factor]. For non-aqueous samples, the factor is on a wet-weight basis.

SAMPLING METHOD

Liquid samples — Use a 40-mL glass screw-cap VOA vial with a Teflon®-faced silicone septum that has been prewashed, rinsed with distilled deionized water, and oven dried. However, if residual chlorine is present, collect sample in a 40-oz. soil VOA container which has been pre-preserved with 4 drops of 10% sodium thiosulfate, mix gently, and then transfer the sample to a 40-mL VOA vial. Collect bubble-free samples in duplicate and seal them in separate plastic bags.

Soils or sediments, and sludges — Use an 8-oz. widemouth glass bottle with a Teflon®-faced silicone septum that has been prewashed with detergent, rinsed with distilled deionized water, and oven dried. Tap slightly to eliminate free air space. Collect samples in duplicate and seal them in separate plastic bags.

SAMPLE PRESERVATION

Liquid samples — Add 4 drops of concentrated HCL and immediately cool samples to 4°C and store in a solvent-free refrigerator.

Soils or sediments, and sludges — Cool samples to 4°C and store in a solvent-free refrigerator.

MHT Maximum holding time is 14 days from the date of sample collection.

SAMPLE PREPARATION

Liquid samples — Remove the plunger from a 5-mL syringe and carefully pour the sample into the syringe barrel to just short of overflowing. Replace the syringe plunger and compress the sample. Open the syringe valve and vent any residual air while adjusting the sample volume to 5.0 mL. If there is only one volatile organic analysis (VOA) vial, a second syringe should be filled at this time to protect against possible loss of sample integrity. Add 10 µL of surrogate spiking solution and 10 µL of internal standard spiking solution through the valve bore of the 5-mL syringe, then close the valve. The surrogate and internal standards may be mixed and added as a single spiking solution.

Sediments, soils, and waste samples — All samples of this type should be screened by GC analysis using a headspace method (EPA Method 3810) or the hexadecane extraction and screening method (EPA Method 3820). Use the screening data to determine whether to use the low-concentration method (0.005–1 mg/kg) or the high-concentration method (>1 mg/kg).

Low-concentration method — The low-concentration method is based on purging a heated sediment or soil sample mixed with organic-free reagent water containing the surrogate and internal standards. Analyze all reagent blanks and standards under the same conditions as the samples.

Use a 5-g sample if the expected concentration is <0.1 mg/kg or a 1-g sample for expected concentrations between 0.1 and 1 mg/kg. Mix the contents of the sample container with a narrow metal spatula. Weigh the amount of the sample into a tared purge device. Add the spiked water to the purge device, which contains the weighed amount of sample, and connect the device to the purge-and-trap system.

High-concentration method — This method is based on extracting the sediment or soil with methanol. A waste sample is either extracted or diluted, depending on its solubility in methanol. Wastes that are insoluble in methanol are diluted with reagent tetraglyme or possibly polyethylene glycol (PEG). An aliquot of the extract is added to organic-free reagent water containing surrogate and internal standards. This is purged at ambient temperature. All samples with an expected concentration of >1.0 mg/kg should be analyzed by this method.

Mix the contents of the sample container with a narrow metal spatula. For sediments or soils and solid wastes that are insoluble in methanol, weigh 4 g (wet weight) of sample into a tared 20-mL vial. For waste that is soluble in methanol, tetraglyme, or PEG, weigh 1 g (wet weight) into a tared scintillation vial or culture tube or a 10-mL volumetric flask. Quickly add 9.0 mL of appropriate solvent then add 1.0 mL of a surrogate spiking solution to the vial, cap it, and shake it for 2 min.

METHANOL EXTRACT REQUIRED FOR ANALYSIS OF HIGH-CONCENTRATION SOILS OR SEDIMENTS

Approximate Concentration Range	Volume of Methanol Extract (a)
500–10,000 µg/kg	100 µL
1,000–20,000 µg/kg	50 µL
5,000–100,000 µg/kg	10 µL
25,000–500,000 µg/kg	100 µL of 1/50 dilution (b)

Calculate appropriate dilution factor for concentrations exceeding this table.

(a) The volume of methanol added to 5 mL of water being purged should be kept constant. Therefore, add to the 5-mL syringe whatever volume of methanol is necessary to maintain a volume of 100 µL added to the syringe.

(b) Dilute an aliquot of the methanol extract and then take 100 µL for analysis.

QUALITY CONTROL Demonstrate, through the analysis of a reagent water blank, that interferences from the analytical system, glassware, and reagents are under control. Blank samples should be carried through all stages of the sample preparation and measurement steps. For each analytical batch (up to 20 samples), a reagent blank, matrix spike, and matrix spike duplicate must be analyzed (the frequency of the spikes may be different for different monitoring programs). The blank and spiked samples must be carried through all stages of the sample preparation and measurement steps. QC samples mentioned in the section on Interferences will also be needed as appropriate to those situations.

REFERENCE Test Methods for Evaluating Solid Waste (SW-846). U.S. EPA. 1983. Method 8240B, Rev. 2, Nov. 1990. Office of Solid Wastes, Washington, DC.

2-Picoline **EPA Method 8270**
CAS #109-06-8

TITLE Semivolatile Organic Compounds by GC/MS

MATRIX This method is used to determine the concentration of semivolatile organic compounds in extracts prepared from all types of solid waste matrices, soils, and groundwater. Although surface waters are not specifically mentioned, this method should be applicable to water samples from rivers, lakes, etc.

METHOD SUMMARY This method covers 259 semivolatile organic compounds. In very limited applications direct injection of the sample into the GC/MS system may be appropriate, but this results in very high detection limits (approximately 10,000 µg/L). Typically, a 1-L liquid sample, containing surrogate, and matrix spiking standards, is extracted in a continuous extractor first under acid conditions and then under basic conditions. Typically 30 g of a solid sample, containing surrogate, and matrix spiking standards, is extracted ultrasonically. After concentrating the extract to 1 mL it is spiked with 10 µL of an internal standard solution just prior to analysis by GC/MS. The volume injected should contain about 100 ng of base/neutral and 200 ng of acid surrogates (for a 1-µL injection). Analysis is performed by GC/MS using a capillary GC column.

INTERFERENCES Raw GC/MS data from all blanks, samples, and spikes must be evaluated for interferences. Contamination by carryover can occur whenever high-concentration and low-concentration samples are sequentially analyzed. To reduce carryover, the sample syringe must be rinsed out between samples with solvent. Whenever an unusually concentrated sample is encountered, it should be followed by the analysis of blank solvent to check for cross-contamination.

INSTRUMENTATION A GC/MS and a data system are required. The GC column used is a 30 m × 0.25 mm I.D. (or 0.32 mm I.D.) 1um film thickness silicone-coated fused silica capillary column. A continuous liquid-liquid extractor equipped with Teflon® or glass connection joints and stopcocks requiring no lubrication, a K-D concentrating apparatus, water bath, and an ultrasonic disrupter with a minimum power of 300 W and with pulsing capability are also required.

PRECISION & ACCURACY The estimated quantitation limit (EQL) of Method 8270B for determining an individual compound is approximately 1 mg/kg (wet weight) for soil or sediment samples, 1–200 mg/kg for wastes (dependent on matrix and method of preparation), and 10 µg/L for groundwater samples. EQLs will be proportionately higher for sample extracts that require dilution to avoid saturation of the detector.

The EQL(b) for groundwater in µg/L is not determined.
The EQL (a, b) for low concentrations in soil and sediment in µg/kg is not determined.
Accuracy as µg/L is not listed.
Overall precision in µg/L is not listed.

(a) EQLs listed for soil/sediment are based on wet weight. Normally data is reported in a dry-weight basis; therefore, EQLs will be higher based on the % dry weight of each sample. This calculation is based on a 30 g sample and gel permeation chromatography cleanup.

(b) Sample EQLs are highly matrix-dependent. The EQLs are provided for guidance and may not always be achievable.

C = True value for concentration, in µg/L.
X = Average recovery found for measurements of samples containing a concentration of C, in µg/L.

ESTIMATED QUANTITATION LIMIT

Other Matrices	Factor (a)
High-concentration soil and sludges by sonicator	7.5
Non-water miscible waste	75

(a) EQL for other matrices = [EQL for low soil/sediment] × [Factor]. This estimated EQL is similar to an EPA "Practical Quantitation Limit."

SAMPLING METHOD

Liquid samples — Use a 1 or 2½ gallon amber glass bottle with a screw-top Teflon®-lined cover that has been prewashed with detergent and rinsed with distilled water and methanol (or isopropanol).

Soils, sediments, or sludges — Use an 8-oz. widemouth glass with a screw-top Teflon®-lined cover that has been prewashed with detergent and rinsed with distilled water and methanol (or isopropanol).

SAMPLE PRESERVATION

Liquid samples — If residual chlorine is present, add 3 mL of 10% sodium thiosulfate per gallon, cool to 4°C and store in a solvent-free refrigerator until analysis; if chlorine is not present, then eliminate the sodium thiosulfate addition.

Soils, sediments, or sludges — Cool samples to 4°C and store in a solvent-free refrigerator.

MHT Liquid samples must be extracted within 7 days and the extracts analyzed within 40 days. Soils, sediments, or sludges may be stored for a maximum of 14 days and the extracts analyzed within 40 days.

SAMPLE PREPARATION
Liquid samples — Transfer 1 L quantitatively to a continuous extractor. If high concentrations are anticipated, a smaller volume may be used and then diluted with organic-free reagent water to 1 L. Adjust pH, if necessary, to pH <2 using 1:1 (V/V) sulfuric acid. Pipette 1.0 mL of a surrogate standard spiking solution into each sample. For the sample in each analytical batch selected for spiking, add 1.0 mL of a matrix spiking standard. For base/neutral acid analysis, the amount of the surrogates and matrix spiking compounds added to the sample should result in a final concentration of 100 ng/µL of each analyte in the extract to be analyzed (assuming a 1-µL injection). Extract with methylene chloride for 18–24 h. Next, adjust the pH of the aqueous phase to pH >11 using 10 N sodium hydroxide and extract it with methylene chloride again for 18–24 h. Dry the extract through a column containing anhydrous sodium sulfate and concentrate it to 1 mL using a K-D concentrator.

Soils, sediments, or sludges — Use 30 g of sample. Nonporous or wet samples (gummy or clay type) that do not have a free-flowing sandy texture must be mixed with anhydrous sodium sulfate until the sample is free flowing. Add 1 mL of surrogate standards to all samples, spikes, standards, and blanks. For the sample in each analytical batch selected for spiking, add 1.0 mL of a matrix spiking standard. For base/neutral acid analysis, the amount added of the surrogates and matrix spiking compounds should result in a final concentration of 100 ng/µL of each base/neutral analyte and 200 ng/µL of each acid analyte in the extract to be analyzed (assuming a 1-µL injection). Immediately add a 100-mL mixture of 1:1 methylene chloride:acetone and extract the sample ultrasonically for 3 min and then decant or filter the extracts. Repeat the extraction two or more times. Dry the extract using a column with anhydrous sodium sulfate and concentrate it to 1 mL in a K-D concentrator.

QUALITY CONTROL A methylene chloride solution containing 50 ng/µL of decafluorotriphenylphosphine (DFTPP) is used for tuning the GC/MS system each 12-h shift. A system performance check also must be made during every 12-h shift. A standard containing 50 ng/µL each of 4,4′-DDT, pentachlorophenol, and benzidine is required to verify injection port inertness and GC column performance. A calibration standard at mid-concentration, containing each compound of interest, including all required surrogates, must be performed every 12 h during analysis. After the system performance check is met, calibration check compounds (CCCs) are used to check the validity of the initial calibration.

The internal standard responses and retention times in the calibration check standard must be evaluated immediately after or during data acquisition. If the retention time for any internal standard changes by more than 30 seconds from the last check calibration (12 h), the chromatographic system must be inspected for malfunctions and corrections must be made, as required. If the electron ionization current plot (EICP) area for any of the internal standards changes by a factor of two from the last daily calibration standard check, the mass spectrometer must be inspected for malfunctions and corrections must be made, as appropriate.

Demonstrate, through the analysis of a reagent water blank, that interferences from the analytical system, glassware, and reagents are under control. The blank samples should be carried through all stages of the sample preparation and measurement steps. For each analytical batch (up to 20 samples), a reagent blank, matrix spike, and matrix spike duplicate/duplicate must be analyzed (the frequency of the spikes may be different for different monitoring programs). The blank and spiked samples must be carried through all stages of the sample preparation and measurement steps. A QC reference sample concentrate containing each analyte at a concentration of 100 mg/L in methanol is required.

REFERENCE Test Methods for Evaluating Solid Waste (SW-846). U.S. EPA 1983, Method 8270B, Rev. 2, Nov. 1990. Office of Solid Waste, Washington, DC.

Piperonyl sulfoxide EPA Method 8270
CAS #120-62-7

TITLE Semivolatile Organic Compounds by GC/MS

MATRIX This method is used to determine the concentration of semivolatile organic compounds in extracts prepared from all types of solid waste matrices, soils, and groundwater. Although surface waters are not specifically mentioned, this method should be applicable to water samples from rivers, lakes, etc.

METHOD SUMMARY This method covers 259 semivolatile organic compounds. In very limited applications direct injection of the sample into the GC/MS system may be appropriate, but this results in very high detection limits (approximately 10,000 µg/L). Typically, a 1-L liquid sample, containing surrogate, and matrix spiking standards, is extracted in a continuous extractor first under acid conditions and then under basic conditions. Typically 30 g of a solid sample, containing surrogate, and matrix spiking standards, is extracted ultrasonically. After concentrating the extract to 1 mL it is spiked with 10 µL of an internal standard solution just prior to analysis by GC/MS. The volume injected should contain about 100 ng of base/neutral and 200 ng of acid surrogates (for a 1-µL injection). Analysis is performed by GC/MS using a capillary GC column.

INTERFERENCES Raw GC/MS data from all blanks, samples, and spikes must be evaluated for interferences. Contamination by carryover can occur whenever high-concentration and low-concentration samples are sequentially analyzed. To reduce carryover, the sample syringe must be rinsed out between samples with solvent. Whenever an unusually concentrated sample is encountered, it should be followed by the analysis of blank solvent to check for cross-contamination.

INSTRUMENTATION A GC/MS and a data system are required. The GC column used is a 30 m × 0.25 mm I.D. (or 0.32 mm I.D.) 1um film thickness silicone-coated fused silica capillary column. A continuous liquid-liquid extractor equipped with Teflon® or glass connection joints and stopcocks requiring no lubrication, a K-D concentrating apparatus, water bath, and an ultrasonic disrupter with a minimum power of 300 W and with pulsing capability are also required.

PRECISION & ACCURACY The estimated quantitation limit (EQL) of Method 8270B for determining an individual compound is approximately 1 mg/kg (wet weight) for soil or sediment samples, 1–200 mg/kg for wastes (dependent on matrix and method of preparation), and 10 µg/L for groundwater samples. EQLs will be proportionately higher for sample extracts that require dilution to avoid saturation of the detector.

The EQL(b) for groundwater in µg/L is 100.
The EQL (a, b) for low concentrations in soil and sediment in µg/kg is not determined.
Accuracy as µg/L is not listed.
Overall precision in µg/L is not listed.

(a) EQLs listed for soil/sediment are based on wet weight. Normally data is reported in a dry-weight basis; therefore, EQLs will be higher based on the % dry weight of each sample. This calculation is based on a 30 g sample and gel permeation chromatography cleanup.
(b) Sample EQLs are highly matrix-dependent. The EQLs are provided for guidance and may not always be achievable.
C = *True value for concentration, in µg/L.*
X = *Average recovery found for measurements of samples containing a concentration of C, in µg/L.*

ESTIMATED QUANTITATION LIMIT

Other Matrices	Factor (a)
High-concentration soil and sludges by sonicator	7.5
Non-water miscible waste	75

(a) EQL for other matrices = [EQL for low soil/sediment] × [Factor]. This estimated EQL is similar to an EPA "Practical Quantitation Limit."

SAMPLING METHOD
Liquid samples — Use a 1 or 2½ gallon amber glass bottle with a screw-top Teflon®-lined cover that has been prewashed with detergent and rinsed with distilled water and methanol (or isopropanol).

Soils, sediments, or sludges — Use an 8-oz. widemouth glass with a screw-top Teflon®-lined cover that has been prewashed with detergent and rinsed with distilled water and methanol (or isopropanol).

SAMPLE PRESERVATION
Liquid samples — If residual chlorine is present, add 3 mL of 10% sodium thiosulfate per gallon, cool to 4°C and store in a solvent-free refrigerator until analysis; if chlorine is not present, then eliminate the sodium thiosulfate addition.

Soils, sediments, or sludges — Cool samples to 4°C and store in a solvent-free refrigerator.

MHT Liquid samples must be extracted within 7 days and the extracts analyzed within 40 days. Soils, sediments, or sludges may be stored for a maximum of 14 days and the extracts analyzed within 40 days.

SAMPLE PREPARATION
Liquid samples — Transfer 1 L quantitatively to a continuous extractor. If high concentrations are anticipated, a smaller volume may be used and then diluted with organic-free reagent water to 1 L. Adjust pH, if necessary, to pH <2 using 1:1 (V/V) sulfuric acid. Pipette 1.0 mL of a surrogate standard spiking solution into each sample. For the sample in each analytical batch selected for spiking, add 1.0 mL of a matrix spiking standard. For base/neutral acid analysis, the amount of the surrogates and matrix spiking compounds added to the sample should result in a final concentration of 100 ng/µL of each analyte in the extract to be analyzed (assuming a 1-µL injection). Extract with methylene chloride for 18–24 h. Next, adjust the pH of the aqueous phase to pH >11 using 10 N sodium hydroxide and extract it with methylene chloride again for 18–24 h. Dry the extract through a column containing anhydrous sodium sulfate and concentrate it to 1 mL using a K-D concentrator.

Soils, sediments, or sludges — Use 30 g of sample. Nonporous or wet samples (gummy or clay type) that do not have a free-flowing sandy texture must be mixed with anhydrous sodium sulfate until the sample is free flowing. Add 1 mL of surrogate standards to all samples, spikes, standards, and blanks. For the sample in each analytical batch selected for spiking, add 1.0 mL of a matrix spiking standard. For base/neutral acid analysis, the amount added of the surrogates and matrix spiking compounds should result in a final concentration of 100 ng/µL of each base/neutral analyte and 200 ng/µL of each acid analyte in the extract to be analyzed (assuming a 1-µL injection). Immediately add a 100-mL mixture of 1:1 methylene chloride:acetone and extract the sample ultrasonically for 3 min and then decant or filter the extracts. Repeat the extraction two or more times. Dry the extract using a column with anhydrous sodium sulfate and concentrate it to 1 mL in a K-D concentrator.

QUALITY CONTROL A methylene chloride solution containing 50 ng/µL of decafluorotriphenylphosphine (DFTPP) is used for tuning the GC/MS system each 12-h shift. A system performance check also must be made during every 12-h shift. A standard containing 50 ng/µL each of 4,4'-DDT, pentachlorophenol, and benzidine is required to verify injection port inertness and GC column performance. A calibration standard at mid-concentration, containing each compound of interest, including all required surrogates, must be performed every 12 h during analysis. After the system performance check is met, calibration check compounds (CCCs) are used to check the validity of the initial calibration.

The internal standard responses and retention times in the calibration check standard must be evaluated immediately after or during data acquisition. If the retention time for any internal standard changes by more than 30 seconds from the last check calibration (12 h), the chromatographic system must be inspected for malfunctions and corrections must be made, as required. If the electron ionization current plot (EICP) area for

any of the internal standards changes by a factor of two from the last daily calibration standard check, the mass spectrometer must be inspected for malfunctions and corrections must be made, as appropriate.

Demonstrate, through the analysis of a reagent water blank, that interferences from the analytical system, glassware, and reagents are under control. The blank samples should be carried through all stages of the sample preparation and measurement steps. For each analytical batch (up to 20 samples), a reagent blank, matrix spike, and matrix spike duplicate/duplicate must be analyzed (the frequency of the spikes may be different for different monitoring programs). The blank and spiked samples must be carried through all stages of the sample preparation and measurement steps. A QC reference sample concentrate containing each analyte at a concentration of 100 mg/L in methanol is required.

REFERENCE Test Methods for Evaluating Solid Waste (SW-846). U.S. EPA 1983, Method 8270B, Rev. 2, Nov. 1990. Office of Solid Waste, Washington, DC.

Potassium   EPA Method 200.7
CAS #7440-09-7

TITLE Inductively Coupled Plasma

MATRIX Dissolved, suspended or (ICP) total element in drinking and surface waters and in domestic and industrial wastewaters.

APPLICATION The method covers the determination of 25 metals. Dissolved elements are determined in filtered and acidified samples after appropriate digestion (which increases dissolved solids). Its primary advantage is that ICP instruments allow simultaneous or rapid sequential determination of many elements in a short time. Samples are first nebulized and the aerosol is transported to a plasma torch in which element specific atomic line emission spectra are produced by a radio frequency inductively coupled plasma. Background correction is required for trace element detection except in the case of line broadning.

INTERFERENCES There are spectral, physical, and chemical interferences. The primary disadvantage of ICP instruments is background radiation from other elements and the plasma gases (spectral interferences). Changes in sample viscosity and surface tension with samples containing high dissolved solids (especially those exceeding 1500 mg/L) or high acid concentrations can cause physical interferences. Ionization effects, solute vaporization and molecular compound formation can cause chemical interferences.

INSTRUMENTATION Inductively coupled argon plasma emission spectroscopy. 766.491 nm wavelength

RANGE Not listed.

MDL Highly dependent on operating conditions and plasma position.

PRECISION Not listed.

ACCURACY Mean recovery = 93% ± 6% of spiked elements for all wastes.

SAMPLING METHOD Wash sample container with detergent and tap water, rinse with 1 + 1 nitric acid and tap water, then rinse with 1 + 1 hydrochloric acid and tap water, then rinse with deionized, distilled water in that order. Perform any filtration or acid preservation steps when the sample is collected or as soon as possible thereafter.

STABILITY Cool samples to 4°C.

MHT 24 h.

QUALITY CONTROL Mixed calibration standards, an instrument check standard, and an interference check solution are used in addition to a quality control sample. The quality control sample should be prepared in the same acid matrix as the calibration standards at 10 times the instrumental detection limits and in accordance with the instructions provided by the supplier. Furthermore, two types of blanks are required: a calibration blank and a reagent blank.

REFERENCE Method 200.7, U.S. EPA, EMSL-Cincinnati, OH, Nov. 1980

Potassium   EPA Method 6010
CAS #7440-09-7

TITLE Inductively Coupled Plasma

MATRIX Applies to all matrices; (ICP) groundwater, EP extracts, soils, sludges, sediments, aqueous samples, solid and industrial wastes.

APPLICATION The method covers the determination of 25 metals. Its primary advantage is that ICP instruments allow simultaneous or rapid sequential determination of many elements in a short time. Samples require digestion prior to analysis and are first nebulized and the aerosol is transported to a plasma torch in which element specific atomic line emission spectra are produced by a radio frequency inductively coupled plasma. Background correction is required for trace element detection except in the case of line broadning.

INTERFERENCES There are spectral, physical, and chemical interferences. The primary disadvantage of ICP instruments is background radiation from other elements and the plasma gases (spectral interferences). Changes in sample viscosity and surface tension with samples containing high dissolved solids or high acid concentrations can cause physical interferences. With effects, solute vaporization and molecular compound formation can cause chemical interferences.

INSTRUMENTATION Inductively coupled argon plasma emission spectroscopy. 766.491 nm wavelength

RANGE Not listed.

MDL Highly dependent on operating conditions and plasma position.

PRECISION Not listed.

ACCURACY Mean recovery = 93% ± 6% of spiked elements for all wastes.

SAMPLING METHOD Collect 400 mL of sample in plastic or glass containers.

STABILITY Cool samples to 4°C.

MHT 6 months

QUALITY CONTROL Mixed calibration standards, an instrument check standard, and an interference check solution are used in addition to a quality control sample. The quality control sample should be prepared in the same acid matrix as the calibration standards at 10 times the instrumental detection limits and in accordance with the instructions provided by the supplier. Furthermore, two types of blanks are required: a calibration blank and a reagent blank.

REFERENCE Method 6010, SW-846, 3rd ed., Nov.1986.

Potassium   EPA Method 7610
CAS #7440-09-7

TITLE Atomic Absorption (AA)

MATRIX Drinking, surface, and direct aspiration saline waters, wastewater.

APPLICATION Sample is aspirated and atomized in a flame. A light beam from a (K) hollow cathode lamp is directed through the flame into monochromator and onto detector. Since wavelength of light beam is specific for potassium, light energy absorbed by detector is measure of potassium.

INTERFERENCES The most troublesomee type is chemical, caused by lack of absorption of atoms bound in molecular combination in the flame. High dissolved solids in sample may result in nonatomic absorbance interference. Other alkali salts in the sample and sodium can interfere.

INSTRUMENTATION Atomic absorption spectrometer. Potassium hollow cathode lamp. (766.5 nm wavelength)

RANGE 0.1–2 mg/L

MDL 0.01 mg/L

PRECISION Standard deviation = ±0.20 and 0.50 at 1.60 and 6.30 mg K/L

ACCURACY Recoveries = 103 and 102% at 1.60 and 6.30 mg K/L

SAMPLING METHOD Use glass or plastic containers. Collect 200 g of solids and 600 mL of liquid samples.

STABILITY Cool solid samples to 4°C and analyze as soon as possible. Add nitric acid to liquid samples to pH <2.

MHT 6 months.

QUALITY CONTROL At least one duplicate and one spike sample should be run every 20 samples or with each matrix type to verify precision of the method. For 20 or more samples per day, verify working standard curve. Run an additional standard at or near mid-range every 10 samples.

REFERENCE Method 7610, SW-846, 3rd ed., Nov.1986.

Prometon   EPA Method 507
CAS #1610-18-0

TITLE Determination of Nitrogen and Phosphorus-Containing Pesticides in Water by GC/NPD

MATRIX This method is applicable to the determination of certain nitrogen and phosphorus-containing pesticides in finished drinking water and groundwater.

METHOD SUMMARY Method 507 covers 46 nitrogen- and phosphorus-containing pesticides. A 1-L sample is fortified with a surrogate standard, salted, buffered, extracted with methylene chloride, and concentrated; then the solvent is exchanged with methyl tert-butyl ether (MTBE) and concentrated again, and a 2-µL aliquot of a sample extract is injected into a GC system equipped with a selective nitrogen-phosphorus detector and a capillary column for analysis.

INTERFERENCES Method interferences may be caused by contaminants in solvents, reagents, glassware, and other sample processing apparatus. Interfering contamination may occur when a sample containing low concentrations of analytes is analyzed immediately following a sample containing relatively high concentrations. One or more injections of MTBE should be made following the analysis of a sample with high concentrations of analytes to check for analyte carryover. Matrix interferences may be caused by contaminants that are coextracted from the sample. The extent of matrix interferences will vary considerably from source to source, depending upon the water sampled.

INSTRUMENTATION A gas chromatograph system (GC) equipped with a nitrogen-phosphorus detector (NPD) is needed.

Column 1: 30 m × 0.25 mm I.D. DB-5 bonded fused silica column, 0.25 µm film thickness, or equivalent.

Column 2: 30 m × 0.25 mm I.D. DB-1701 bonded fused silica column, 0.25 µm film thickness, or equivalent.

PRECISION & ACCURACY This method has been validated in a single lab and estimated detection limits (EDLs) have been determined for each analyte. Observed detection limits may vary among waters, depending upon the nature of the interferences in the sample matrix and the specific instrumentation used. Analytes that are not separated chromatographically cannot be individually identified and measured unless an alternative technique for identification and quantification exist.

The estimated detection limit (in µg/L) was 0.3. The EDL is defined as either method detection limit or a level of compound in a sample yielding a peak in the final extract with signal-to-noise ratio of approximately 5, whichever value is higher.

The concentration used for these measurements (in µg/L) was 3. The accuracy (as % recovery) was 78.

The precision (% RSD) was 9.

SAMPLING METHOD Grab samples are collected in 1-L glass sample bottles (prewashed with detergent and hot tap water, rinsed with reagent water, and dried in an oven at 400°C for 1 h) with screw caps lined with PTFE-fluorocarbon.

SAMPLE PRESERVATION Add mercuric chloride to the sample bottle in amounts to produce a concentration of 10 mg/L. If residual chlorine is present, add 80 mg of sodium thiosulfate/L of sample to the sample bottle prior to collection. After collection, seal bottle and shake vigorously for 1 min, then cool the sample to 4°C immediately and store it at 4°C in the dark until extraction.

MHT Maximum holding time of the samples, and in some cases the extracts, is 14 days.

SAMPLE PREPARATION Fortify the sample with 50 µL of the surrogate standard solution, adjust to pH 7 with phosphate buffer, add 100 g NaCl to the sample, and seal and shake to dissolve the salt; then extract with methylene chloride in a separatory funnel or in a mechanical tumbler bottle. Dry the extract by pouring it through a solvent-rinsed drying column containing about 10 cm of anhydrous sodium sulfate. Collect the extract in a Kuderna-Danish (K-D) concentrator and rinse the column with 20–30 mL methylene chloride. Concentrate the extract to about 2 mL and rinse the flask and its lower joint into the concentrator tube with 1 to 2 mL of methyl t-butyl ether (MTBE). Add 5–10 mL of MTBE and concentrate the extract twice (adding more MTBE) to a final volume of 5.0 mL and store it at 4°C until analysis.

Note: If methylene chloride is not completely removed from the final extract, it may cause detector problems.

QUALITY CONTROL Minimum quality control requirements are initial demonstration of lab capability, determination of surrogate compound recoveries in each sample and blank, monitoring internal standard peak area or height in each sample and blank, analysis of lab reagent blanks, lab fortified samples, lab fortified blanks, and other QC samples. A lab reagent blank is analyzed to demonstrate that all glassware and reagent interferences are under control.

Initial demonstration of capability is fulfilled by analyzing four fortified reagent water samples with the recovery value for each analyte falling within the acceptable range (±30% average recovery). Surrogate recoveries from samples or method blanks must be 70–130%. The internal standard response for any sample chromatogram should not deviate from the daily calibration check standard's internal standard response by more than 30% or lab fortified blanks and sample matrices are used to assess lab performance and analyte recovery, respectively.

If the response for the target analyte peak exceeds the working range of the system, dilute the extract and reanalyze. Alternative techniques such as an alternate detector or second chromatography column should be used to confirm peak identification when sample components are not resolved adequately.

EPA CONTACT & HOTLINE For technical questions contact Dr. Baldev Bathija, U.S. EPA, Office of Ground Water and Drinking Water (WH-550D), 401 M St. SW, Washington, DC 20460. Tel. (202) 260-3040. For further information the EPA Safe Drinking Water Hotline may be called at: (800) 426-4791.

REFERENCE Methods for the Determination of Organic Compounds in Drinking Water, EPA/600/4-88/039 (revised July 1991). U.S. EPA Environmental Monitoring Systems Laboratory, Cincinnati, OH, 45268, U.S.A. Available from the National Technical Information Service (NTIS), 5285 Port Royal Road, Springfield, VA 22161; Tel. 800-553-6847. NTIS Order Number is PB91-231480.

Prometryn — EPA Method 507
CAS #7287-19-6

TITLE Determination of Nitrogen and Phosphorus-Containing Pesticides in Water by GC/NPD

MATRIX This method is applicable to the determination of certain nitrogen and phosphorus-containing pesticides in finished drinking water and groundwater.

METHOD SUMMARY Method 507 covers 46 nitrogen- and phosphorus-containing pesticides. A 1-L sample is fortified with a surrogate standard, salted, buffered, extracted with methylene chloride, and concentrated; then the solvent is exchanged with methyl tert-butyl ether (MTBE) and concentrated again, and a 2-µL aliquot of a sample extract is injected into a GC system equipped with a selective nitrogen-phosphorus detector and a capillary column for analysis.

INTERFERENCES Method interferences may be caused by contaminants in solvents, reagents, glassware, and other sample processing apparatus. Interfering contamination may occur when a sample containing low concentrations of analytes is analyzed immediately following a sample containing relatively high concentrations. One or more injections of MTBE should be made following the analysis of a sample with high concentrations of analytes to check for analyte carryover. Matrix interferences may be caused by contaminants that are coextracted from the sample. The extent of matrix interferences will vary considerably from source to source, depending upon the water sampled.

INSTRUMENTATION A gas chromatograph system (GC) equipped with a nitrogen-phosphorus detector (NPD) is needed.

Column 1: 30 m × 0.25 mm I.D. DB-5 bonded fused silica column, 0.25 µm film thickness, or equivalent.

Column 2: 30 m × 0.25 mm I.D. DB-1701 bonded fused silica column, 0.25 µm film thickness, or equivalent.

PRECISION & ACCURACY This method has been validated in a single lab and estimated detection limits (EDLs) have been determined for each analyte. Observed detection limits may vary among waters, depending upon the nature of the interferences in the sample matrix and the specific instrumentation used. Analytes that are not separated chromatographically cannot be individually identified and measured unless an alternative technique for identification and quantification exist.

The estimated detection limit (in µg/L) was 0.19. The EDL is defined as either method detection limit or a level of compound in a sample yielding a peak in the final extract with signal-to-noise ratio of approximately 5, whichever value is higher.

The concentration used for these measurements (in µg/L) was 1.9.
The accuracy (as % recovery) was 93.
The precision (% RSD) was 8.

SAMPLING METHOD Grab samples are collected in 1-L glass sample bottles (prewashed with detergent and hot tap water, rinsed with reagent water, and dried in an oven at 400°C for 1 h) with screw caps lined with PTFE-fluorocarbon.

SAMPLE PRESERVATION Add mercuric chloride to the sample bottle in amounts to produce a concentration of 10 mg/L. If residual chlorine is present, add 80 mg of sodium thiosulfate/L of sample to the sample bottle prior to collection. After collection, seal bottle and shake vigorously for 1 min, then cool the sample to 4°C immediately and store it at 4°C in the dark until extraction.

MHT Maximum holding time of the samples, and in some cases the extracts, is 14 days.

SAMPLE PREPARATION Fortify the sample with 50 µL of the surrogate standard solution, adjust to pH 7 with phosphate buffer, add 100 g NaCl to the sample, and seal and shake to dissolve the salt; then extract with methylene chloride in a separatory funnel or in a mechanical tumbler bottle. Dry the extract by pouring it through a solvent-rinsed drying column containing about 10 cm of anhydrous sodium sulfate. Collect the extract in a Kuderna-Danish (K-D) concentrator and rinse the column with 20–30 mL methylene chloride. Concentrate the extract to about 2 mL and rinse the flask and its lower joint into the concentrator tube with 1 to 2 mL of methyl t-butyl ether (MTBE). Add 5–10 mL of MTBE and concentrate the extract twice (adding more MTBE) to a final volume of 5.0 mL and store it at 4°C until analysis.

Note: If methylene chloride is not completely removed from the final extract, it may cause detector problems.

QUALITY CONTROL Minimum quality control requirements are initial demonstration of lab capability, determination of surrogate compound recoveries in each sample and blank, monitoring internal standard peak area or height in each sample and blank, analysis of lab reagent blanks, lab fortified samples, lab fortified blanks, and other QC samples. A lab reagent blank is analyzed to demonstrate that all glassware and reagent interferences are under control.

Initial demonstration of capability is fulfilled by analyzing four fortified reagent water samples with the recovery value for each analyte falling within the acceptable range (±30% average recovery). Surrogate recoveries from samples or method blanks must be 70–130%. The internal standard response for any sample chromatogram should not deviate from the daily calibration check standard's internal standard response by more than 30% or lab fortified blanks and sample matrices are used to assess lab performance and analyte recovery, respectively.

If the response for the target analyte peak exceeds the working range of the system, dilute the extract and reanalyze. Alternative techniques such as an alternate detector or second chromatography column should be used to confirm peak identification when sample components are not resolved adequately.

EPA CONTACT & HOTLINE For technical questions contact Dr. Baldev Bathija, U.S. EPA, Office of Ground Water and Drinking Water (WH-550D), 401 M St. SW, Washington, DC 20460. Tel. (202) 260-3040. For further information the EPA Safe Drinking Water Hotline may be called at: (800) 426-4791.

REFERENCE Methods for the Determination of Organic Compounds in Drinking Water, EPA/600/4-88/039 (revised July 1991). U.S. EPA Environmental Monitoring Systems Laboratory, Cincinnati, OH, 45268, U.S.A. Available from the National Technical Information Service (NTIS), 5285 Port Royal Road, Springfield, VA 22161; Tel. 800-553-6847. NTIS Order Number is PB91-231480.

Pronamide — EPA Method 1625
CAS #23950-58-5

TITLE Semivolatile Organic Compounds by Isotope Dilution GC/MS

MATRIX The compounds may be determined in waters, soils, and municipal sludges by this method.

METHOD SUMMARY This method is used to determine 176 semivolatile toxic organic pollutants associated with the CWA (as amended 1987); the RCRA (as amended 1986); the CERCLA (as amended 1986); and other compounds amenable to extraction and analysis by capillary column gas chromatography-mass spectrometry (GC/MS).

Stable isotopically-labeled analogs of the compounds of interest are added to the sample. If the solids content is less than 1%, a 1-L sample is extracted at pH 12–13, then at pH <2 with methylene chloride using continuous extraction techniques.

If the solids content is 30% or less, the sample is diluted to 1% solids with reagent water, homogenized ultrasonically, and extracted at pH 12–13, then at pH <2 with methylene chloride using continuous extraction techniques. If the solids content is greater than 30%, the sample is extracted using ultrasonic techniques.

Each extract is dried over sodium sulfate, concentrated to a volume of 5 mL, cleaned up using GPC, if necessary, and concentrated. Extracts are concentrated to 1 mL if GPC is not performed, and to 0.5 mL if GPC is performed.

An internal standard is added to the extract, and a 1-mL aliquot of the extract is injected into the GC. The compounds are separated by GC and detected by a MS. The labeled compounds serve to correct the variability of the analytical technique.

INTERFERENCES Solvents, reagents, glassware, and other sample processing hardware may yield artifacts and/or elevated baselines causing misinterpretation of chromatograms and spectra. Materials used in the analysis must be demonstrated

to be free from interferences under the conditions of analysis by running method blanks initially and with each sample lot (sample started through the extraction process on a given 8-h shift, to a maximum of 20). Specific selection of reagents and purification of solvents by distillation in all glass systems may be required. Glassware and, where possible, reagents are cleaned by solvent rinse and baking at 450°C for 1-h minimum. Interferences coextracted from samples will vary considerably from source to source, depending on the diversity of the site being sampled.

INSTRUMENTATION Major instrumentation includes a GC with a splitless or on-column injection port for capillary column, a MS with 70 eV electron impact ionization, and a data system to collect and record MS data, and process it. A K-D apparatus is used to concentrate extracts.

GC Column: 30 m × 0.25 mm I.D. 5% phenyl, 94% methyl, 1% vinyl silicone bonded phased fused silica capillary column.

PRECISION & ACCURACY The detection limits of the method are usually dependent on the level of interferences rather than instrumental limitations. The limits typify the minimum quantities that can be detected with no interferences present.

The minimum level (in µg/mL) was not listed. This is defined as a minimum level at which the analytical system shall give recognizable mass spectra (background corrected) and acceptable calibration points.

The MDL (in µg/kg) in low solids was not listed and in high solids was not listed; these were determined in digested sludge (low solids) and in filter cake or compost (high solids).

The labeled and native compound initial precision as standard deviation (in µg/L) was not listed.
The labeled and native compound initial accuracy as average recovery (in µg/L) was not listed.

SAMPLE COLLECTION, PRESERVATION & HANDLING Collect samples in glass containers. Aqueous samples which flow freely are collected in refrigerated bottles using automatic sampling equipment. Solid samples are collected as grab samples using widemouth jars. Maintain samples at 0 to 4°C from the time of collection until extraction. If residual chlorine is present in aqueous samples, add 80 mg sodium thiosulfate/L of water. Begin sample extraction within 7 days of collection, and analyze all extracts within 40 days of extraction.

SAMPLE PREPARATION Samples containing 1% solids or less are extracted directly using continuous liquid-liquid extraction techniques. Samples containing 1 to 30% solids are diluted to the 1% level with reagent water and extracted using continuous liquid-liquid extraction techniques. Samples containing greater than 30% solids are extracted using ultrasonic techniques.

Base/neutral extraction — Adjust the pH of the waters in the extractors to 12–13 with 6 N NaOH. Extract with methylene chloride for 24–48 h.
Acid extraction — Adjust the pH of the waters in the extractors to 2 or less using 6 N sulfuric acid. Extract with methylene chloride for 24–48 h.
Ultrasonic extraction of high solids samples — Add anhydrous sodium sulfate to the sample and QC aliquot(s). Add acetone:methylene chloride (1:1) to the sample and mix thoroughly

Concentrate extracts using a K-D apparatus.

QUALITY CONTROL The analyst is permitted to modify this method to improve separations or lower the costs of measurements, provided all performance specifications are met. Analyses of blanks are required to demonstrate freedom from contamination. When results of spikes indicate atypical method performance for samples, the samples are diluted to bring method performance within acceptable limits.

For low solids (aqueous samples), extract, concentrate, and analyze two sets of four 1-L aliquots (8 aliquots total) of the precision and recovery standard. For high solids samples, two sets of four 30-g aliquots of the high solids reference matrix are used.

Spike all samples with labeled compounds to assess method performance. Compute percent recovery of the labeled compounds using the internal standard method. Compare the labeled compound recovery for each compound with the corresponding labeled compound recovery.

Reagent water and high solids reference matrix blanks are analyzed to demonstrate freedom from contamination. Extract and concentrate a 1-L reagent water blank or a high solids reference matrix blank with each sample's lot (samples started through the extraction process on the same 8-h shift, to a maximum of 20 samples).

Field replicates may be collected to determine the precision of the sampling technique, and spiked samples may be required to determine the accuracy of the analysis when the internal standard method is used.

REFERENCE Semivolatile Organic Compounds by Isotope Dilution GC/MS. Office of Water Regulation and Standards, U.S. EPA Industrial Technology Division, Washington, DC, EPA Method 1625, Rev. C, June 1989 (contact W.A. Telliard, U.S. EPA, Office of Water Regulations and Standards, 401 M St., SW, Washington, DC, 20460. Phone: 202-382-7131).

Pronamide   EPA Method 1625
CAS #23950-58-5

TITLE Semivolatile Organic Compounds by Isotope Dilution GC/MS

MATRIX The compounds may be determined in waters, soils, and municipal sludges by this method.

METHOD SUMMARY This method is used to determine 176 semivolatile toxic organic pollutants associated with the CWA (as amended 1987); the RCRA (as amended 1986); the CERCLA (as amended 1986); and other compounds amenable to extraction and analysis by capillary column gas chromatography-mass spectrometry (GC/MS).

Stable isotopically-labeled analogs of the compounds of interest are added to the sample. If the solids content is less than 1%, a 1-L sample is extracted at pH 12–13, then at pH <2 with methylene chloride using continuous extraction techniques.

If the solids content is 30% or less, the sample is diluted to 1% solids with reagent water, homogenized ultrasonically, and extracted at pH 12–13, then at pH <2 with methylene chloride using continuous extraction techniques. If the solids content is greater than 30%, the sample is extracted using ultrasonic techniques.

Each extract is dried over sodium sulfate, concentrated to a volume of 5 mL, cleaned up using GPC, if necessary, and concentrated. Extracts are concentrated to 1 mL if GPC is not performed, and to 0.5 mL if GPC is performed.

An internal standard is added to the extract, and a 1-mL aliquot of the extract is injected into the GC. The compounds are separated by GC and detected by a MS. The labeled compounds serve to correct the variability of the analytical technique.

INTERFERENCES Solvents, reagents, glassware, and other sample processing hardware may yield artifacts and/or elevated baselines causing misinterpretation of chromatograms and spectra. Materials used in the analysis must be demonstrated to be free from interferences under the conditions of analysis by running method blanks initially and with each sample lot (sample started through the extraction process on a given 8-h shift, to a maximum of 20). Specific selection of reagents and purification of solvents by distillation in all glass systems may be required. Glassware and, where possible, reagents are cleaned by solvent rinse and baking at 450°C for 1-h minimum. Interferences coextracted from samples will vary considerably from source to source, depending on the diversity of the site being sampled.

INSTRUMENTATION Major instrumentation includes a GC with a splitless or on-column injection port for capillary column, a MS with 70 eV electron impact ionization, and a data system to collect and record MS data, and process it. A K-D apparatus is used to concentrate extracts.

GC Column: 30 m × 0.25 mm I.D. 5% phenyl, 94% methyl, 1% vinyl silicone bonded phased fused silica capillary column.

PRECISION & ACCURACY The detection limits of the method are usually dependent on the level of interferences rather than instrumental limitations. The limits typify the minimum quantities that can be detected with no interferences present.

The minimum level (in µg/mL) was not listed. This is defined as a minimum level at which the analytical system shall give recognizable mass spectra (background corrected) and acceptable calibration points.

The MDL (in µg/kg) in low solids was not listed and in high solids was not listed; these were determined in digested sludge (low solids) and in filter cake or compost (high solids).

The labeled and native compound initial precision as standard deviation (in µg/L) was not listed.

The labeled and native compound initial accuracy as average recovery (in µg/L) was not listed.

SAMPLE COLLECTION, PRESERVATION & HANDLING
Collect samples in glass containers. Aqueous samples which flow freely are collected in refrigerated bottles using automatic sampling equipment. Solid samples are collected as grab samples using widemouth jars. Maintain samples at 0 to 4°C from the time of collection until extraction. If residual chlorine is present in aqueous samples, add 80 mg sodium thiosulfate/L of water. Begin sample extraction within 7 days of collection, and analyze all extracts within 40 days of extraction.

SAMPLE PREPARATION Samples containing 1% solids or less are extracted directly using continuous liquid-liquid extraction techniques. Samples containing 1 to 30% solids are diluted to the 1% level with reagent water and extracted using continuous liquid-liquid extraction techniques. Samples containing greater than 30% solids are extracted using ultrasonic techniques.

Base/neutral extraction — Adjust the pH of the waters in the extractors to 12–13 with 6 N NaOH. Extract with methylene chloride for 24–48 h.

Acid extraction — Adjust the pH of the waters in the extractors to 2 or less using 6 N sulfuric acid. Extract with methylene chloride for 24–48 h.

Ultrasonic extraction of high solids samples — Add anhydrous sodium sulfate to the sample and QC aliquot(s). Add acetone:methylene chloride (1:1) to the sample and mix thoroughly

Concentrate extracts using a K-D apparatus.

QUALITY CONTROL The analyst is permitted to modify this method to improve separations or lower the costs of measurements, provided all performance specifications are met. Analyses of blanks are required to demonstrate freedom from contamination. When results of spikes indicate atypical method performance for samples, the samples are diluted to bring method performance within acceptable limits.

For low solids (aqueous samples), extract, concentrate, and analyze two sets of four 1-L aliquots (8 aliquots total) of the precision and recovery standard. For high solids samples, two sets of four 30-g aliquots of the high solids reference matrix are used.

Spike all samples with labeled compounds to assess method performance. Compute percent recovery of the labeled compounds using the internal standard method. Compare the labeled compound recovery for each compound with the corresponding labeled compound recovery.

Reagent water and high solids reference matrix blanks are analyzed to demonstrate freedom from contamination. Extract and concentrate a 1-L reagent water blank or a high solids reference matrix blank with each sample's lot (samples started through the extraction process on the same 8-h shift, to a maximum of 20 samples).

Field replicates may be collected to determine the precision of the sampling technique, and spiked samples may be required

to determine the accuracy of the analysis when the internal standard method is used.

REFERENCE Semivolatile Organic Compounds by Isotope Dilution GC/MS. Office of Water Regulation and Standards, U.S. EPA Industrial Technology Division, Washington, DC, EPA Method 1625, Rev. C, June 1989 (contact W.A. Telliard, U.S. EPA, Office of Water Regulations and Standards, 401 M St., SW, Washington, DC, 20460. Phone: 202-382-7131).

Pronamide   EPA Method 507
CAS #23950-58-5

TITLE Determination of Nitrogen and Phosphorus-Containing Pesticides in Water by GC/NPD

MATRIX This method is applicable to the determination of certain nitrogen and phosphorus-containing pesticides in finished drinking water and groundwater.

METHOD SUMMARY Method 507 covers 46 nitrogen- and phosphorus-containing pesticides. A 1-L sample is fortified with a surrogate standard, salted, buffered, extracted with methylene chloride, and concentrated; then the solvent is exchanged with methyl tert-butyl ether (MTBE) and concentrated again, and a 2-µL aliquot of a sample extract is injected into a GC system equipped with a selective nitrogen-phosphorus detector and a capillary column for analysis.

INTERFERENCES Method interferences may be caused by contaminants in solvents, reagents, glassware, and other sample processing apparatus. Interfering contamination may occur when a sample containing low concentrations of analytes is analyzed immediately following a sample containing relatively high concentrations. One or more injections of MTBE should be made following the analysis of a sample with high concentrations of analytes to check for analyte carryover. Matrix interferences may be caused by contaminants that are coextracted from the sample. The extent of matrix interferences will vary considerably from source to source, depending upon the water sampled.

INSTRUMENTATION A gas chromatograph system (GC) equipped with a nitrogen-phosphorus detector (NPD) is needed.

Column 1: 30 m × 0.25 mm I.D. DB-5 bonded fused silica column, 0.25 µm film thickness, or equivalent.

Column 2: 30 m × 0.25 mm I.D. DB-1701 bonded fused silica column, 0.25 µm film thickness, or equivalent.

PRECISION & ACCURACY This method has been validated in a single lab and estimated detection limits (EDLs) have been determined for each analyte. Observed detection limits may vary among waters, depending upon the nature of the interferences in the sample matrix and the specific instrumentation used. Analytes that are not separated chromatographically cannot be individually identified and measured unless an alternative technique for identification and quantification exist.

The estimated detection limit (in µg/L) was 0.76. The EDL is defined as either method detection limit or a level of compound in a sample yielding a peak in the final extract with signal-to-noise ratio of approximately 5, whichever value is higher.

The concentration used for these measurements (in µg/L) was 7.6.

The accuracy (as % recovery) was 91.

The precision (% RSD) was 10.

SAMPLING METHOD Grab samples are collected in 1-L glass sample bottles (prewashed with detergent and hot tap water, rinsed with reagent water, and dried in an oven at 400°C for 1 h) with screw caps lined with PTFE-fluorocarbon.

SAMPLE PRESERVATION Add mercuric chloride to the sample bottle in amounts to produce a concentration of 10 mg/L. If residual chlorine is present, add 80 mg of sodium thiosulfate/L of sample to the sample bottle prior to collection. After collection, seal bottle and shake vigorously for 1 min, then cool the sample to 4°C immediately and store it at 4°C in the dark until extraction.

MHT Maximum holding time of the samples, and in some cases the extracts, is 14 days.

Note: Samples with this compound must be extracted immediately.

SAMPLE PREPARATION Fortify the sample with 50 µL of the surrogate standard solution, adjust to pH 7 with phosphate buffer, add 100 g NaCl to the sample, and seal and shake to dissolve the salt; then extract with methylene chloride in a separatory funnel or in a mechanical tumbler bottle. Dry the extract by pouring it through a solvent-rinsed drying column containing about 10 cm of anhydrous sodium sulfate. Collect the extract in a Kuderna-Danish (K-D) concentrator and rinse the column with 20–30 mL methylene chloride. Concentrate the extract to about 2 mL and rinse the flask and its lower joint into the concentrator tube with 1 to 2 mL of methyl t-butyl ether (MTBE). Add 5–10 mL of MTBE and concentrate the extract twice (adding more MTBE) to a final volume of 5.0 mL and store it at 4°C until analysis.

Note: If methylene chloride is not completely removed from the final extract, it may cause detector problems.

QUALITY CONTROL Minimum quality control requirements are initial demonstration of lab capability, determination of surrogate compound recoveries in each sample and blank, monitoring internal standard peak area or height in each sample and blank, analysis of lab reagent blanks, lab fortified samples, lab fortified blanks, and other QC samples. A lab reagent blank is analyzed to demonstrate that all glassware and reagent interferences are under control.

Initial demonstration of capability is fulfilled by analyzing four fortified reagent water samples with the recovery value for each analyte falling within the acceptable range (±30% average recovery). Surrogate recoveries from samples or method blanks must be 70–130%. The internal standard response for any sample chromatogram should not deviate from the daily calibration check standard's internal standard response by more than 30% or lab fortified blanks and sample matrices are used to assess lab performance and analyte recovery, respectively.

If the response for the target analyte peak exceeds the working range of the system, dilute the extract and reanalyze. Alternative techniques such as an alternate detector or second chromatography column should be used to confirm peak identification when sample components are not resolved adequately.

EPA CONTACT & HOTLINE For technical questions contact Dr. Baldev Bathija, U.S. EPA, Office of Ground Water and Drinking Water (WH-550D), 401 M St. SW, Washington, DC 20460. Tel. (202) 260-3040. For further information the EPA Safe Drinking Water Hotline may be called at: (800) 426-4791.

REFERENCE Methods for the Determination of Organic Compounds in Drinking Water, EPA/600/4-88/039 (revised July 1991). U.S. EPA Environmental Monitoring Systems Laboratory, Cincinnati, OH, 45268, U.S.A. Available from the National Technical Information Service (NTIS), 5285 Port Royal Road, Springfield, VA 22161; Tel. 800-553-6847. NTIS Order Number is PB91-231480.

Pronamide   **EPA Method 8270**
CAS #23950-58-5

TITLE Semivolatile Organic Compounds by GC/MS

MATRIX This method is used to determine the concentration of semivolatile organic compounds in extracts prepared from all types of solid waste matrices, soils, and groundwater. Although surface waters are not specifically mentioned, this method should be applicable to water samples from rivers, lakes, etc.

METHOD SUMMARY This method covers 259 semivolatile organic compounds. In very limited applications direct injection of the sample into the GC/MS system may be appropriate, but this results in very high detection limits (approximately 10,000 µg/L). Typically, a 1-L liquid sample, containing surrogate, and matrix spiking standards, is extracted in a continuous extractor first under acid conditions and then under basic conditions. Typically 30 g of a solid sample, containing surrogate, and matrix spiking standards, is extracted ultrasonically. After concentrating the extract to 1 mL it is spiked with 10 µL of an internal standard solution just prior to analysis by GC/MS. The volume injected should contain about 100 ng of base/neutral and 200 ng of acid surrogates (for a 1-µL injection). Analysis is performed by GC/MS using a capillary GC column.

INTERFERENCES Raw GC/MS data from all blanks, samples, and spikes must be evaluated for interferences. Contamination by carryover can occur whenever high-concentration and low-concentration samples are sequentially analyzed. To reduce carryover, the sample syringe must be rinsed out between samples with solvent. Whenever an unusually concentrated sample is encountered, it should be followed by the analysis of blank solvent to check for cross-contamination.

INSTRUMENTATION A GC/MS and a data system are required. The GC column used is a 30 m × 0.25 mm I.D. (or 0.32 mm I.D.) 1um film thickness silicone-coated fused silica capillary column. A continuous liquid-liquid extractor equipped with Teflon® or glass connection joints and stopcocks requiring no lubrication, a K-D concentrating apparatus, water bath, and an ultrasonic disrupter with a minimum power of 300 W and with pulsing capability are also required.

PRECISION & ACCURACY The estimated quantitation limit (EQL) of Method 8270B for determining an individual compound is approximately 1 mg/kg (wet weight) for soil or sediment samples, 1–200 mg/kg for wastes (dependent on matrix and method of preparation), and 10 µg/L for groundwater samples. EQLs will be proportionately higher for sample extracts that require dilution to avoid saturation of the detector.

The EQL(b) for groundwater in µg/L is 10.
The EQL (a, b) for low concentrations in soil and sediment in µg/kg is not determined.
Accuracy as µg/L is not listed.
Overall precision in µg/L is not listed.

(a) EQLs listed for soil/sediment are based on wet weight. Normally data is reported in a dry-weight basis; therefore, EQLs will be higher based on the % dry weight of each sample. This calculation is based on a 30 g sample and gel permeation chromatography cleanup.
(b) Sample EQLs are highly matrix-dependent. The EQLs are provided for guidance and may not always be achievable.
$C =$ True value for concentration, in µg/L.
$X =$ Average recovery found for measurements of samples containing a concentration of C, in µg/L.

ESTIMATED QUANTITATION LIMIT

Other Matrices	Factor (a)
High-concentration soil and sludges by sonicator	7.5
Non-water miscible waste	75

(a) EQL for other matrices = [EQL for low soil/sediment] × [Factor]. This estimated EQL is similar to an EPA "Practical Quantitation Limit."

SAMPLING METHOD
Liquid samples — Use a 1 or 2½ gallon amber glass bottle with a screw-top Teflon®-lined cover that has been prewashed with detergent and rinsed with distilled water and methanol (or isopropanol).

Soils, sediments, or sludges — Use an 8-oz. widemouth glass with a screw-top Teflon®-lined cover that has been prewashed with detergent and rinsed with distilled water and methanol (or isopropanol).

SAMPLE PRESERVATION
Liquid samples — If residual chlorine is present, add 3 mL of 10% sodium thiosulfate per gallon, cool to 4°C and store in a solvent-free refrigerator until analysis; if chlorine is not present, then eliminate the sodium thiosulfate addition.

Soils, sediments, or sludges — Cool samples to 4°C and store in a solvent-free refrigerator.

MHT Liquid samples must be extracted within 7 days and the extracts analyzed within 40 days. Soils, sediments, or sludges may be stored for a maximum of 14 days and the extracts analyzed within 40 days.

SAMPLE PREPARATION

Liquid samples — Transfer 1 L quantitatively to a continuous extractor. If high concentrations are anticipated, a smaller volume may be used and then diluted with organic-free reagent water to 1 L. Adjust pH, if necessary, to pH <2 using 1:1 (V/V) sulfuric acid. Pipette 1.0 mL of a surrogate standard spiking solution into each sample. For the sample in each analytical batch selected for spiking, add 1.0 mL of a matrix spiking standard. For base/neutral acid analysis, the amount of the surrogates and matrix spiking compounds added to the sample should result in a final concentration of 100 ng/µL of each analyte in the extract to be analyzed (assuming a 1-µL injection). Extract with methylene chloride for 18–24 h. Next, adjust the pH of the aqueous phase to pH >11 using 10 N sodium hydroxide and extract it with methylene chloride again for 18–24 h. Dry the extract through a column containing anhydrous sodium sulfate and concentrate it to 1 mL using a K-D concentrator.

Soils, sediments, or sludges — Use 30 g of sample. Nonporous or wet samples (gummy or clay type) that do not have a free-flowing sandy texture must be mixed with anhydrous sodium sulfate until the sample is free flowing. Add 1 mL of surrogate standards to all samples, spikes, standards, and blanks. For the sample in each analytical batch selected for spiking, add 1.0 mL of a matrix spiking standard. For base/neutral acid analysis, the amount added of the surrogates and matrix spiking compounds should result in a final concentration of 100 ng/µL of each base/neutral analyte and 200 ng/µL of each acid analyte in the extract to be analyzed (assuming a 1-µL injection). Immediately add a 100-mL mixture of 1:1 methylene chloride:acetone and extract the sample ultrasonically for 3 min and then decant or filter the extracts. Repeat the extraction two or more times. Dry the extract using a column with anhydrous sodium sulfate and concentrate it to 1 mL in a K-D concentrator.

QUALITY CONTROL

A methylene chloride solution containing 50 ng/µL of decafluorotriphenylphosphine (DFTPP) is used for tuning the GC/MS system each 12-h shift. A system performance check also must be made during every 12-h shift. A standard containing 50 ng/µL each of 4,4′-DDT, pentachlorophenol, and benzidine is required to verify injection port inertness and GC column performance. A calibration standard at mid-concentration, containing each compound of interest, including all required surrogates, must be performed every 12 h during analysis. After the system performance check is met, calibration check compounds (CCCs) are used to check the validity of the initial calibration.

The internal standard responses and retention times in the calibration check standard must be evaluated immediately after or during data acquisition. If the retention time for any internal standard changes by more than 30 seconds from the last check calibration (12 h), the chromatographic system must be inspected for malfunctions and corrections must be made, as required. If the electron ionization current plot (EICP) area for any of the internal standards changes by a factor of two from the last daily calibration standard check, the mass spectrometer must be inspected for malfunctions and corrections must be made, as appropriate.

Demonstrate, through the analysis of a reagent water blank, that interferences from the analytical system, glassware, and reagents are under control. The blank samples should be carried through all stages of the sample preparation and measurement steps. For each analytical batch (up to 20 samples), a reagent blank, matrix spike, and matrix spike duplicate/duplicate must be analyzed (the frequency of the spikes may be different for different monitoring programs). The blank and spiked samples must be carried through all stages of the sample preparation and measurement steps. A QC reference sample concentrate containing each analyte at a concentration of 100 mg/L in methanol is required.

REFERENCE Test Methods for Evaluating Solid Waste (SW-846). U.S. EPA 1983, Method 8270B, Rev. 2, Nov. 1990. Office of Solid Waste, Washington, DC.

Propargyl alcohol   **EPA Method 8240**
CAS #107-19-7

TITLE   Volatile Organics By GC/MS: Packed Column Technique

MATRIX   Nearly all types of sample mataraces, regardless of water content, can be analyzed using this method. This includes groundwater, aqueous sludges, caustic liquors, acid liquors, waste solvents, oily wastes, mousses, tars, fibrous wastes, polymetric emulsions, filter cakes, spent carbons, spent catalysts, soils, and sediments.

METHOD SUMMARY   Method 8240B covers 80 volatile organic compounds that are introduced into a gas chromatograph by the purge-and-trap method or by direct injection (in limited applications). For the purge-and-trap method an inert gas (zero grade nitrogen or helium) is bubbled through a 5-mL solution at ambient temperature. Purged sample components are trapped in a tube of sorbent materials. When purging is complete, the sorbent tube is heated and backflushed with inert gas to desorb the trapped components onto a GC column.

INTERFERENCES   Impurities in the purge gas and from organic compounds outgassing from the plumbing ahead of the trap account for many contamination problems. Interferences purged or coextracted from the samples will vary considerably from source to source. Cross-contamination can occur whenever high-level and low-level samples are analyzed sequentially. Whenever an unusually concentrated sample is analyzed, it should be followed by the analysis of organic-free reagent water to check for cross-contamination. Samples also can be contaminated by diffusion of volatile organics (particularly methylene chloride and fluorocarbons) through the septum seal into the sample during shipment and storage. A trip blank can serve as a check on such contamination. The lab where volatile analysis is performed and also the refrigerated storage area should be completely free of solvents.

INSTRUMENTATION   A gas chromatograph/mass spectrometry/data system (GC/MS) equipped with a 6 ft × 0.1 in I.D. glass column packed with 1% SP-1000 on Carbopack-B (60/80 mesh) is required. Also needed is a 5-mL purging device, a sorbent trap, and a thermal desorption apparatus.

PRECISION & ACCURACY This method is reported to have been tested by 15 laboratories using organic-free reagent water, drinking water, surface water, and industrial wastewaters (not specified) fortified at six concentrations over the range 5–600 µg/L.

Sample estimated quantitation limits (EQLs) are highly matrix-dependent. The EQLs listed may not always be achievable. EQLs listed for soils or sediments are based on wet weight. Normally, data is reported on a dry-weight basis; therefore, EQLs will be higher, based on the percent dry weight of each sample. Note that EQLs are even more variable than MDLs and that they are highly variable depending on the matrix being analyzed.

EQL in groundwater in µg/L was not listed.
EQL in low soil or sediment in µg/kg was not listed.
Accuracy (a) in µg/L was not listed.
Precision (b) in µg/L was not listed.

(a) Average recovery found for measurements of samples containing a concentration of C, in µg/L.
(b) Overall precision found for measurements of samples with average recovery X for samples containing a concentration of C in µg/L.
X = Average recovery found for measurement of samples containing a concentration of C in µg/L.

MULTIPLICATION FACTORS FOR OTHER MATRICES

Other Matrices	Factor (a)
Waste miscible liquid waste	50
High-concentration soil and sludge	125
Non-water miscible waste	500

(a) EQL = [EQL for low soil/sediment] × [Factor]. For non-aqueous samples, the factor is on a wet-weight basis.

SAMPLING METHOD
Liquid samples — Use a 40-mL glass screw-cap VOA vial with a Teflon®-faced silicone septum that has been prewashed, rinsed with distilled deionized water, and oven dried. However, if residual chlorine is present, collect sample in a 40-oz. soil VOA container which has been pre-preserved with 4 drops of 10% sodium thiosulfate, mix gently, and then transfer the sample to a 40-mL VOA vial. Collect bubble-free samples in duplicate and seal them in separate plastic bags.

Soils or sediments, and sludges — Use an 8-oz. widemouth glass bottle with a Teflon®-faced silicone septum that has been prewashed with detergent, rinsed with distilled deionized water, and oven dried. Tap slightly to eliminate free air space. Collect samples in duplicate and seal them in separate plastic bags.

SAMPLE PRESERVATION
Liquid samples — Add 4 drops of concentrated HCL and immediately cool samples to 4°C and store in a solvent-free refrigerator.

Soils or sediments, and sludges — Cool samples to 4°C and store in a solvent-free refrigerator.

MHT Maximum holding time is 14 days from the date of sample collection.

SAMPLE PREPARATION
Liquid samples — Remove the plunger from a 5-mL syringe and carefully pour the sample into the syringe barrel to just short of overflowing. Replace the syringe plunger and compress the sample. Open the syringe valve and vent any residual air while adjusting the sample volume to 5.0 mL. If there is only one volatile organic analysis (VOA) vial, a second syringe should be filled at this time to protect against possible loss of sample integrity. Add 10 µL of surrogate spiking solution and 10 µL of internal standard spiking solution through the valve bore of the 5-mL syringe, then close the valve. The surrogate and internal standards may be mixed and added as a single spiking solution.

Sediments, soils, and waste samples — All samples of this type should be screened by GC analysis using a headspace method (EPA Method 3810) or the hexadecane extraction and screening method (EPA Method 3820). Use the screening data to determine whether to use the low-concentration method (0.005–1 mg/kg) or the high-concentration method (>1 mg/kg).

Low-concentration method — The low-concentration method is based on purging a heated sediment or soil sample mixed with organic-free reagent water containing the surrogate and internal standards. Analyze all reagent blanks and standards under the same conditions as the samples.

Use a 5-g sample if the expected concentration is <0.1 mg/kg or a 1-g sample for expected concentrations between 0.1 and 1 mg/kg. Mix the contents of the sample container with a narrow metal spatula. Weigh the amount of the sample into a tared purge device. Add the spiked water to the purge device, which contains the weighed amount of sample, and connect the device to the purge-and-trap system.

High-concentration method — This method is based on extracting the sediment or soil with methanol. A waste sample is either extracted or diluted, depending on its solubility in methanol. Wastes that are insoluble in methanol are diluted with reagent tetraglyme or possibly polyethylene glycol (PEG). An aliquot of the extract is added to organic-free reagent water containing surrogate and internal standards. This is purged at ambient temperature. All samples with an expected concentration of >1.0 mg/kg should be analyzed by this method.

Mix the contents of the sample container with a narrow metal spatula. For sediments or soils and solid wastes that are insoluble in methanol, weigh 4 g (wet weight) of sample into a tared 20-mL vial. For waste that is soluble in methanol, tetraglyme, or PEG, weigh 1 g (wet weight) into a tared scintillation vial or culture tube or a 10-mL volumetric flask. Quickly add 9.0 mL of appropriate solvent then add 1.0 mL of a surrogate spiking solution to the vial, cap it, and shake it for 2 min.

METHANOL EXTRACT REQUIRED FOR ANALYSIS OF HIGH-CONCENTRATION SOILS OR SEDIMENTS

Approximate Concentration Range	Volume of Methanol Extract (a)
500–10,000 µg/kg	100 µL
1,000–20,000 µg/kg	50 µL
5,000–100,000 µg/kg	10 µL
25,000–500,000 µg/kg	100 µL of 1/50 dilution (b)

Calculate appropriate dilution factor for concentrations exceeding this table.

(a) The volume of methanol added to 5 mL of water being purged should be kept constant. Therefore, add to the 5-mL syringe whatever volume of methanol is necessary to maintain a volume of 100 µL added to the syringe.

(b) Dilute an aliquot of the methanol extract and then take 100 µL for analysis.

QUALITY CONTROL Demonstrate, through the analysis of a reagent water blank, that interferences from the analytical system, glassware, and reagents are under control. Blank samples should be carried through all stages of the sample preparation and measurement steps. For each analytical batch (up to 20 samples), a reagent blank, matrix spike, and matrix spike duplicate must be analyzed (the frequency of the spikes may be different for different monitoring programs). The blank and spiked samples must be carried through all stages of the sample preparation and measurement steps. QC samples mentioned in the section on Interferences will also be needed as appropriate to those situations.

REFERENCE Test Methods for Evaluating Solid Waste (SW-846). U.S. EPA. 1983. Method 8240B, Rev. 2, Nov. 1990. Office of Solid Wastes, Washington, DC.

Propazine **EPA Method 507**
CAS #139-40-2

TITLE Determination of Nitrogen and Phosphorus-Containing Pesticides in Water by GC/NPD

MATRIX This method is applicable to the determination of certain nitrogen and phosphorus-containing pesticides in finished drinking water and groundwater.

METHOD SUMMARY Method 507 covers 46 nitrogen- and phosphorus-containing pesticides. A 1-L sample is fortified with a surrogate standard, salted, buffered, extracted with methylene chloride, and concentrated; then the solvent is exchanged with methyl tert-butyl ether (MTBE) and concentrated again, and a 2-µL aliquot of a sample extract is injected into a GC system equipped with a selective nitrogen-phosphorus detector and a capillary column for analysis.

INTERFERENCES Method interferences may be caused by contaminants in solvents, reagents, glassware, and other sample processing apparatus. Interfering contamination may occur when a sample containing low concentrations of analytes is analyzed immediately following a sample containing relatively high concentrations. One or more injections of MTBE should be made following the analysis of a sample with high concentrations of analytes to check for analyte carryover. Matrix interferences may be caused by contaminants that are coextracted from the sample. The extent of matrix interferences will vary considerably from source to source, depending upon the water sampled.

INSTRUMENTATION A gas chromatograph system (GC) equipped with a nitrogen-phosphorus detector (NPD) is needed.

Column 1: 30 m × 0.25 mm I.D. DB-5 bonded fused silica column, 0.25 µm film thickness, or equivalent.

Column 2: 30 m × 0.25 mm I.D. DB-1701 bonded fused silica column, 0.25 µm film thickness, or equivalent.

PRECISION & ACCURACY This method has been validated in a single lab and estimated detection limits (EDLs) have been determined for each analyte. Observed detection limits may vary among waters, depending upon the nature of the interferences in the sample matrix and the specific instrumentation used. Analytes that are not separated chromatographically cannot be individually identified and measured unless an alternative technique for identification and quantification exist.

The estimated detection limit (in µg/L) was 0.13. The EDL is defined as either method detection limit or a level of compound in a sample yielding a peak in the final extract with signal-to-noise ratio of approximately 5, whichever value is higher.

The concentration used for these measurements (in µg/L) was 1.3.
The accuracy (as % recovery) was 92.
The precision (% RSD) was 8.

SAMPLING METHOD Grab samples are collected in 1-L glass sample bottles (prewashed with detergent and hot tap water, rinsed with reagent water, and dried in an oven at 400°C for 1 h) with screw caps lined with PTFE-fluorocarbon.

SAMPLE PRESERVATION Add mercuric chloride to the sample bottle in amounts to produce a concentration of 10 mg/L. If residual chlorine is present, add 80 mg of sodium thiosulfate/L of sample to the sample bottle prior to collection. After collection, seal bottle and shake vigorously for 1 min, then cool the sample to 4°C immediately and store it at 4°C in the dark until extraction.

MHT Maximum holding time of the samples, and in some cases the extracts, is 14 days.

SAMPLE PREPARATION Fortify the sample with 50 µL of the surrogate standard solution, adjust to pH 7 with phosphate buffer, add 100 g NaCl to the sample, and seal and shake to dissolve the salt; then extract with methylene chloride in a separatory funnel or in a mechanical tumbler bottle. Dry the extract by pouring it through a solvent-rinsed drying column containing about 10 cm of anhydrous sodium sulfate. Collect the extract in a Kuderna-Danish (K-D) concentrator and rinse the column with 20–30 mL methylene chloride. Concentrate the extract to about 2 mL and rinse the flask and its lower joint into the concentrator tube with 1 to 2 mL of methyl t-butyl ether (MTBE). Add 5–10 mL of MTBE and concentrate the extract twice (adding more MTBE) to a final volume of 5.0 mL and store it at 4°C until analysis.

Note: If methylene chloride is not completely removed from the final extract, it may cause detector problems.

QUALITY CONTROL Minimum quality control requirements are initial demonstration of lab capability, determination of surrogate compound recoveries in each sample and blank, monitoring internal standard peak area or height in each sample and blank, analysis of lab reagent blanks, lab fortified samples, lab fortified blanks, and other QC samples. A lab reagent blank is analyzed to demonstrate that all glassware and reagent interferences are under control.

Initial demonstration of capability is fulfilled by analyzing four fortified reagent water samples with the recovery value for each analyte falling within the acceptable range (±30% average recovery). Surrogate recoveries from samples or method blanks must be 70–130%. The internal standard response for any sample chromatogram should not deviate from the daily calibration check standard's internal standard response by more than 30% or lab fortified blanks and sample matrices are used to assess lab performance and analyte recovery, respectively.

If the response for the target analyte peak exceeds the working range of the system, dilute the extract and reanalyze. Alternative techniques such as an alternate detector or second chromatography column should be used to confirm peak identification when sample components are not resolved adequately.

EPA CONTACT & HOTLINE For technical questions contact Dr. Baldev Bathija, U.S. EPA, Office of Ground Water and Drinking Water (WH-550D), 401 M St. SW, Washington, DC 20460. Tel. (202) 260-3040. For further information the EPA Safe Drinking Water Hotline may be called at: (800) 426-4791.

REFERENCE Methods for the Determination of Organic Compounds in Drinking Water, EPA/600/4-88/039 (revised July 1991). U.S. EPA Environmental Monitoring Systems Laboratory, Cincinnati, OH, 45268, U.S.A. Available from the National Technical Information Service (NTIS), 5285 Port Royal Road, Springfield, VA 22161; Tel. 800-553-6847. NTIS Order Number is PB91-231480.

β-Propiolactone	EPA Method 8240
CAS #57-57-8	

TITLE Volatile Organics By GC/MS: Packed Column Technique

MATRIX Nearly all types of sample matarices, regardless of water content, can be analyzed using this method. This includes groundwater, aqueous sludges, caustic liquors, acid liquors, waste solvents, oily wastes, mousses, tars, fibrous wastes, polymetric emulsions, filter cakes, spent carbons, spent catalysts, soils, and sediments.

METHOD SUMMARY Method 8240B covers 80 volatile organic compounds that are introduced into a gas chromatograph by the purge-and-trap method or by direct injection (in limited applications). For the purge-and-trap method an inert gas (zero grade nitrogen or helium) is bubbled through a 5-mL solution at ambient temperature. Purged sample components are trapped in a tube of sorbent materials. When purging is complete, the sorbent tube is heated and backflushed with inert gas to desorb the trapped components onto a GC column.

INTERFERENCES Impurities in the purge gas and from organic compounds outgassing from the plumbing ahead of the trap account for many contamination problems. Interferences purged or coextracted from the samples will vary considerably from source to source. Cross-contamination can occur whenever high-level and low-level samples are analyzed sequentially. Whenever an unusually concentrated sample is analyzed, it should be followed by the analysis of organic-free reagent water to check for cross-contamination. Samples also can be contaminated by diffusion of volatile organics (particularly methylene chloride and fluorocarbons) through the septum seal into the sample during shipment and storage. A trip blank can serve as a check on such contamination. The lab where volatile analysis is performed and also the refrigerated storage area should be completely free of solvents.

INSTRUMENTATION A gas chromatograph/mass spectrometry/data system (GC/MS) equipped with a 6 ft × 0.1 in I.D. glass column packed with 1% SP-1000 on Carbopack-B (60/80 mesh) is required. Also needed is a 5-mL purging device, a sorbent trap, and a thermal desorption apparatus.

PRECISION & ACCURACY This method is reported to have been tested by 15 laboratories using organic-free reagent water, drinking water, surface water, and industrial wastewaters (not specified) fortified at six concentrations over the range 5–600 µg/L.

Sample estimated quantitation limits (EQLs) are highly matrix-dependent. The EQLs listed may not always be achievable. EQLs listed for soils or sediments are based on wet weight. Normally, data is reported on a dry-weight basis; therefore, EQLs will be higher, based on the percent dry weight of each sample. Note that EQLs are even more variable than MDLs and that they are highly variable depending on the matrix being analyzed.

EQL in groundwater in µg/L was not listed.
EQL in low soil or sediment in µg/kg was not listed.
Accuracy (a) in µg/L was not listed.
Precision (b) in µg/L was not listed.

(a) Average recovery found for measurements of samples containing a concentration of C, in µg/L.
(b) Overall precision found for measurements of samples with average recovery X for samples containing a concentration of C in µg/L.
X = Average recovery found for measurement of samples containing a concentration of C in µg/L.

MULTIPLICATION FACTORS FOR OTHER MATRICES

Other Matrices	Factor (a)
Waste miscible liquid waste	50
High-concentration soil and sludge	125
Non-water miscible waste	500

(a) EQL = [EQL for low soil/sediment] × [Factor]. For non-aqueous samples, the factor is on a wet-weight basis.

SAMPLING METHOD

Liquid samples — Use a 40-mL glass screw-cap VOA vial with a Teflon®-faced silicone septum that has been prewashed, rinsed with distilled deionized water, and oven dried. However, if residual chlorine is present, collect sample in a 40-oz. soil VOA container which has been pre-preserved with 4 drops of

10% sodium thiosulfate, mix gently, and then transfer the sample to a 40-mL VOA vial. Collect bubble-free samples in duplicate and seal them in separate plastic bags.

Soils or sediments, and sludges — Use an 8-oz. widemouth glass bottle with a Teflon®-faced silicone septum that has been prewashed with detergent, rinsed with distilled deionized water, and oven dried. Tap slightly to eliminate free air space. Collect samples in duplicate and seal them in separate plastic bags.

SAMPLE PRESERVATION
Liquid samples — Add 4 drops of concentrated HCL and immediately cool samples to 4°C and store in a solvent-free refrigerator.

Soils or sediments, and sludges — Cool samples to 4°C and store in a solvent-free refrigerator.

MHT Maximum holding time is 14 days from the date of sample collection.

SAMPLE PREPARATION
Liquid samples — Remove the plunger from a 5-mL syringe and carefully pour the sample into the syringe barrel to just short of overflowing. Replace the syringe plunger and compress the sample. Open the syringe valve and vent any residual air while adjusting the sample volume to 5.0 mL. If there is only one volatile organic analysis (VOA) vial, a second syringe should be filled at this time to protect against possible loss of sample integrity. Add 10 µL of surrogate spiking solution and 10 µL of internal standard spiking solution through the valve bore of the 5-mL syringe, then close the valve. The surrogate and internal standards may be mixed and added as a single spiking solution.

Sediments, soils, and waste samples — All samples of this type should be screened by GC analysis using a headspace method (EPA Method 3810) or the hexadecane extraction and screening method (EPA Method 3820). Use the screening data to determine whether to use the low-concentration method (0.005–1 mg/kg) or the high-concentration method (>1 mg/kg).

Low-concentration method — The low-concentration method is based on purging a heated sediment or soil sample mixed with organic-free reagent water containing the surrogate and internal standards. Analyze all reagent blanks and standards under the same conditions as the samples.

Use a 5-g sample if the expected concentration is <0.1 mg/kg or a 1 g sample for expected concentrations between 0.1 and 1 mg/kg. Mix the contents of the sample container with a narrow metal spatula. Weigh the amount of the sample into a tared purge device. Add the spiked water to the purge device, which contains the weighed amount of sample, and connect the device to the purge-and-trap system.

High-concentration method — This method is based on extracting the sediment or soil with methanol. A waste sample is either extracted or diluted, depending on its solubility in methanol. Wastes that are insoluble in methanol are diluted with reagent tetraglyme or possibly polyethylene glycol (PEG). An aliquot of the extract is added to organic-free reagent water containing surrogate and internal standards. This is purged at ambient temperature. All samples with an expected concentration of >1.0 mg/kg should be analyzed by this method.

Mix the contents of the sample container with a narrow metal spatula. For sediments or soils and solid wastes that are insoluble in methanol, weigh 4 g (wet weight) of sample into a tared 20-mL vial. For waste that is soluble in methanol, tetraglyme, or PEG, weigh 1 g (wet weight) into a tared scintillation vial or culture tube or a 10-mL volumetric flask. Quickly add 9.0 mL of appropriate solvent then add 1.0 mL of a surrogate spiking solution to the vial, cap it, and shake it for 2 min.

METHANOL EXTRACT REQUIRED FOR ANALYSIS OF HIGH-CONCENTRATION SOILS OR SEDIMENTS

Approximate Concentration Range	Volume of Methanol Extract (a)
500–10,000 µg/kg	100 µL
1,000–20,000 µg/kg	50 µL
5,000–100,000 µg/kg	10 µL
25,000–500,000 µg/kg	100 µL of 1/50 dilution (b)

Calculate appropriate dilution factor for concentrations exceeding this table.

(a) The volume of methanol added to 5 mL of water being purged should be kept constant. Therefore, add to the 5-mL syringe whatever volume of methanol is necessary to maintain a volume of 100 µL added to the syringe.
(b) Dilute an aliquot of the methanol extract and then take 100 µL for analysis.

QUALITY CONTROL Demonstrate, through the analysis of a reagent water blank, that interferences from the analytical system, glassware, and reagents are under control. Blank samples should be carried through all stages of the sample preparation and measurement steps. For each analytical batch (up to 20 samples), a reagent blank, matrix spike, and matrix spike duplicate must be analyzed (the frequency of the spikes may be different for different monitoring programs). The blank and spiked samples must be carried through all stages of the sample preparation and measurement steps. QC samples mentioned in the section on Interferences will also be needed as appropriate to those situations.

REFERENCE Test Methods for Evaluating Solid Waste (SW-846). U.S. EPA. 1983. Method 8240B, Rev. 2, Nov. 1990. Office of Solid Wastes, Washington, DC.

Propionitrile EPA Method 8240
CAS #107-12-0

TITLE Volatile Organics By GC/MS: Packed Column Technique

MATRIX Nearly all types of sample matarices, regardless of water content, can be analyzed using this method. This includes groundwater, aqueous sludges, caustic liquors, acid liquors, waste solvents, oily wastes, mousses, tars, fibrous wastes, polymetric emulsions, filter cakes, spent carbons, spent catalysts, soils, and sediments.

METHOD SUMMARY Method 8240B covers 80 volatile organic compounds that are introduced into a gas chromatograph by the purge-and-trap method or by direct injection (in limited applications). For the purge-and-trap method an inert gas (zero grade nitrogen or helium) is bubbled through a 5-mL solution at ambient temperature. Purged sample components are trapped in a tube of sorbent materials. When purging is complete, the sorbent tube is heated and backflushed with inert gas to desorb the trapped components onto a GC column.

INTERFERENCES Impurities in the purge gas and from organic compounds outgassing from the plumbing ahead of the trap account for many contamination problems. Interferences purged or coextracted from the samples will vary considerably from source to source. Cross-contamination can occur whenever high-level and low-level samples are analyzed sequentially. Whenever an unusually concentrated sample is analyzed, it should be followed by the analysis of organic-free reagent water to check for cross-contamination. Samples also can be contaminated by diffusion of volatile organics (particularly methylene chloride and fluorocarbons) through the septum seal into the sample during shipment and storage. A trip blank can serve as a check on such contamination. The lab where volatile analysis is performed and also the refrigerated storage area should be completely free of solvents.

INSTRUMENTATION A gas chromatograph/mass spectrometry/data system (GC/MS) equipped with a 6 ft × 0.1 in I.D. glass column packed with 1% SP-1000 on Carbopack-B (60/80 mesh) is required. Also needed is a 5-mL purging device, a sorbent trap, and a thermal desorption apparatus.

PRECISION & ACCURACY This method is reported to have been tested by 15 laboratories using organic-free reagent water, drinking water, surface water, and industrial wastewaters (not specified) fortified at six concentrations over the range 5–600 µg/L.

Sample estimated quantitation limits (EQLs) are highly matrix-dependent. The EQLs listed may not always be achievable. EQLs listed for soils or sediments are based on wet weight. Normally, data is reported on a dry-weight basis; therefore, EQLs will be higher, based on the percent dry weight of each sample. Note that EQLs are even more variable than MDLs and that they are highly variable depending on the matrix being analyzed.

EQL in groundwater in µg/L was 100.
EQL in low soil or sediment in µg/kg was 100.
Accuracy (a) in µg/L was not listed.
Precision (b) in µg/L was not listed.

(a) Average recovery found for measurements of samples containing a concentration of C, in µg/L.
(b) Overall precision found for measurements of samples with average recovery X for samples containing a concentration of C in µg/L.
X = Average recovery found for measurement of samples containing a concentration of C in µg/L.

MULTIPLICATION FACTORS FOR OTHER MATRICES

Other Matrices	Factor (a)
Waste miscible liquid waste	50
High-concentration soil and sludge	125
Non-water miscible waste	500

(a) EQL = [EQL for low soil/sediment] × [Factor]. For non-aqueous samples, the factor is on a wet-weight basis.

SAMPLING METHOD
Liquid samples — Use a 40-mL glass screw-cap VOA vial with a Teflon®-faced silicone septum that has been prewashed, rinsed with distilled deionized water, and oven dried. However, if residual chlorine is present, collect sample in a 40-oz. soil VOA container which has been pre-preserved with 4 drops of 10% sodium thiosulfate, mix gently, and then transfer the sample to a 40-mL VOA vial. Collect bubble-free samples in duplicate and seal them in separate plastic bags.

Soils or sediments, and sludges — Use an 8-oz. widemouth glass bottle with a Teflon®-faced silicone septum that has been prewashed with detergent, rinsed with distilled deionized water, and oven dried. Tap slightly to eliminate free air space. Collect samples in duplicate and seal them in separate plastic bags.

SAMPLE PRESERVATION
Liquid samples — Add 4 drops of concentrated HCL and immediately cool samples to 4°C and store in a solvent-free refrigerator.

Soils or sediments, and sludges — Cool samples to 4°C and store in a solvent-free refrigerator.

MHT Maximum holding time is 14 days from the date of sample collection.

SAMPLE PREPARATION
Liquid samples — Remove the plunger from a 5-mL syringe and carefully pour the sample into the syringe barrel to just short of overflowing. Replace the syringe plunger and compress the sample. Open the syringe valve and vent any residual air while adjusting the sample volume to 5.0 mL. If there is only one volatile organic analysis (VOA) vial, a second syringe should be filled at this time to protect against possible loss of sample integrity. Add 10 µL of surrogate spiking solution and 10 µL of internal standard spiking solution through the valve bore of the 5-mL syringe, then close the valve. The surrogate and internal standards may be mixed and added as a single spiking solution.

Sediments, soils, and waste samples — All samples of this type should be screened by GC analysis using a headspace method (EPA Method 3810) or the hexadecane extraction and screening method (EPA Method 3820). Use the screening data to determine whether to use the low-concentration method (0.005–1 mg/kg) or the high-concentration method (>1 mg/kg).

Low-concentration method — The low-concentration method is based on purging a heated sediment or soil sample mixed with organic-free reagent water containing the surrogate and internal standards. Analyze all reagent blanks and standards under the same conditions as the samples.

Use a 5-g sample if the expected concentration is <0.1 mg/kg or a 1-g sample for expected concentrations between 0.1 and 1 mg/kg. Mix the contents of the sample container with a narrow metal spatula. Weigh the amount of the sample into a tared purge device. Add the spiked water to the purge device, which contains the weighed amount of sample, and connect the device to the purge-and-trap system.

High-concentration method — This method is based on extracting the sediment or soil with methanol. A waste sample is either extracted or diluted, depending on its solubility in methanol. Wastes that are insoluble in methanol are diluted with reagent tetraglyme or possibly polyethylene glycol (PEG). An aliquot of the extract is added to organic-free reagent water containing surrogate and internal standards. This is purged at ambient temperature. All samples with an expected concentration of >1.0 mg/kg should be analyzed by this method.

Mix the contents of the sample container with a narrow metal spatula. For sediments or soils and solid wastes that are insoluble in methanol, weigh 4 g (wet weight) of sample into a tared 20-mL vial. For waste that is soluble in methanol, tetraglyme, or PEG, weigh 1 g (wet weight) into a tared scintillation vial or culture tube or a 10-mL volumetric flask. Quickly add 9.0 mL of appropriate solvent then add 1.0 mL of a surrogate spiking solution to the vial, cap it, and shake it for 2 min.

METHANOL EXTRACT REQUIRED FOR ANALYSIS OF HIGH-CONCENTRATION SOILS OR SEDIMENTS

Approximate Concentration Range	Volume of Methanol Extract (a)
500–10,000 µg/kg	100 µL
1,000–20,000 µg/kg	50 µL
5,000–100,000 µg/kg	10 µL
25,000–500,000 µg/kg	100 µL of 1/50 dilution (b)

Calculate appropriate dilution factor for concentrations exceeding this table.

(a) The volume of methanol added to 5 mL of water being purged should be kept constant. Therefore, add to the 5-mL syringe whatever volume of methanol is necessary to maintain a volume of 100 µL added to the syringe.
(b) Dilute an aliquot of the methanol extract and then take 100 µL for analysis.

QUALITY CONTROL Demonstrate, through the analysis of a reagent water blank, that interferences from the analytical system, glassware, and reagents are under control. Blank samples should be carried through all stages of the sample preparation and measurement steps. For each analytical batch (up to 20 samples), a reagent blank, matrix spike, and matrix spike duplicate must be analyzed (the frequency of the spikes may be different for different monitoring programs). The blank and spiked samples must be carried through all stages of the sample preparation and measurement steps. QC samples mentioned in the section on Interferences will also be needed as appropriate to those situations.

REFERENCE Test Methods for Evaluating Solid Waste (SW-846). U.S. EPA. 1983. Method 8240B, Rev. 2, Nov. 1990. Office of Solid Wastes, Washington, DC.

---

n-Propylamine   **EPA Method 8240**
CAS #107-10-8

TITLE Volatile Organics By GC/MS: Packed Column Technique

MATRIX Nearly all types of sample matarices, regardless of water content, can be analyzed using this method. This includes groundwater, aqueous sludges, caustic liquors, acid liquors, waste solvents, oily wastes, mousses, tars, fibrous wastes, polymetric emulsions, filter cakes, spent carbons, spent catalysts, soils, and sediments.

METHOD SUMMARY Method 8240B covers 80 volatile organic compounds that are introduced into a gas chromatograph by the purge-and-trap method or by direct injection (in limited applications). For the purge-and-trap method an inert gas (zero grade nitrogen or helium) is bubbled through a 5-mL solution at ambient temperature. Purged sample components are trapped in a tube of sorbent materials. When purging is complete, the sorbent tube is heated and backflushed with inert gas to desorb the trapped components onto a GC column.

INTERFERENCES Impurities in the purge gas and from organic compounds outgassing from the plumbing ahead of the trap account for many contamination problems. Interferences purged or coextracted from the samples will vary considerably from source to source. Cross-contamination can occur whenever high-level and low-level samples are analyzed sequentially. Whenever an unusually concentrated sample is analyzed, it should be followed by the analysis of organic-free reagent water to check for cross-contamination. Samples also can be contaminated by diffusion of volatile organics (particularly methylene chloride and fluorocarbons) through the septum seal into the sample during shipment and storage. A trip blank can serve as a check on such contamination. The lab where volatile analysis is performed and also the refrigerated storage area should be completely free of solvents.

INSTRUMENTATION A gas chromatograph/mass spectrometry/data system (GC/MS) equipped with a 6 ft × 0.1 in I.D. glass column packed with 1% SP-1000 on Carbopack-B (60/80 mesh) is required. Also needed is a 5-mL purging device, a sorbent trap, and a thermal desorption apparatus.

PRECISION & ACCURACY This method is reported to have been tested by 15 laboratories using organic-free reagent water, drinking water, surface water, and industrial wastewaters (not specified) fortified at six concentrations over the range 5–600 µg/L.

Sample estimated quantitation limits (EQLs) are highly matrix-dependent. The EQLs listed may not always be achievable. EQLs listed for soils or sediments are based on wet weight. Normally, data is reported on a dry-weight basis; therefore, EQLs will be higher, based on the percent dry weight of each sample. Note that EQLs are even more variable than MDLs and that they are highly variable depending on the matrix being analyzed.

EQL in groundwater in µg/L was not listed.
EQL in low soil or sediment in µg/kg was not listed.
Accuracy (a) in µg/L was not listed.

Precision (b) in µg/L was not listed.

(a) *Average recovery found for measurements of samples containing a concentration of C, in µg/L.*
(b) *Overall precision found for measurements of samples with average recovery X for samples containing a concentration of C in µg/L.*
X = *Average recovery found for measurement of samples containing a concentration of C in µg/L.*

MULTIPLICATION FACTORS FOR OTHER MATRICES

Other Matrices	Factor (a)
Waste miscible liquid waste	50
High-concentration soil and sludge	125
Non-water miscible waste	500

(a) EQL = [EQL for low soil/sediment] × [Factor]. For non-aqueous samples, the factor is on a wet-weight basis.

SAMPLING METHOD

Liquid samples — Use a 40-mL glass screw-cap VOA vial with a Teflon®-faced silicone septum that has been prewashed, rinsed with distilled deionized water, and oven dried. However, if residual chlorine is present, collect sample in a 40-oz. soil VOA container which has been pre-preserved with 4 drops of 10% sodium thiosulfate, mix gently, and then transfer the sample to a 40-mL VOA vial. Collect bubble-free samples in duplicate and seal them in separate plastic bags.

Soils or sediments, and sludges — Use an 8-oz. widemouth glass bottle with a Teflon®-faced silicone septum that has been prewashed with detergent, rinsed with distilled deionized water, and oven dried. Tap slightly to eliminate free air space. Collect samples in duplicate and seal them in separate plastic bags.

SAMPLE PRESERVATION

Liquid samples — Add 4 drops of concentrated HCL and immediately cool samples to 4°C and store in a solvent-free refrigerator.

Soils or sediments, and sludges — Cool samples to 4°C and store in a solvent-free refrigerator.

MHT Maximum holding time is 14 days from the date of sample collection.

SAMPLE PREPARATION

Liquid samples — Remove the plunger from a 5-mL syringe and carefully pour the sample into the syringe barrel to just short of overflowing. Replace the syringe plunger and compress the sample. Open the syringe valve and vent any residual air while adjusting the sample volume to 5.0 mL. If there is only one volatile organic analysis (VOA) vial, a second syringe should be filled at this time to protect against possible loss of sample integrity. Add 10 µL of surrogate spiking solution and 10 µL of internal standard spiking solution through the valve bore of the 5-mL syringe, then close the valve. The surrogate and internal standards may be mixed and added as a single spiking solution.

Sediments, soils, and waste samples — All samples of this type should be screened by GC analysis using a headspace method (EPA Method 3810) or the hexadecane extraction and screening method (EPA Method 3820). Use the screening data to determine whether to use the low-concentration method (0.005–1 mg/kg) or the high-concentration method (>1 mg/kg).

Low-concentration method — The low-concentration method is based on purging a heated sediment or soil sample mixed with organic-free reagent water containing the surrogate and internal standards. Analyze all reagent blanks and standards under the same conditions as the samples.

Use a 5-g sample if the expected concentration is <0.1 mg/kg or a 1-g sample for expected concentrations between 0.1 and 1 mg/kg. Mix the contents of the sample container with a narrow metal spatula. Weigh the amount of the sample into a tared purge device. Add the spiked water to the purge device, which contains the weighed amount of sample, and connect the device to the purge-and-trap system.

High-concentration method — This method is based on extracting the sediment or soil with methanol. A waste sample is either extracted or diluted, depending on its solubility in methanol. Wastes that are insoluble in methanol are diluted with reagent tetraglyme or possibly polyethylene glycol (PEG). An aliquot of the extract is added to organic-free reagent water containing surrogate and internal standards. This is purged at ambient temperature. All samples with an expected concentration of >1.0 mg/kg should be analyzed by this method.

Mix the contents of the sample container with a narrow metal spatula. For sediments or soils and solid wastes that are insoluble in methanol, weigh 4 g (wet weight) of sample into a tared 20-mL vial. For waste that is soluble in methanol, tetraglyme, or PEG, weigh 1 g (wet weight) into a tared scintillation vial or culture tube or a 10-mL volumetric flask. Quickly add 9.0 mL of appropriate solvent then add 1.0 mL of a surrogate spiking solution to the vial, cap it, and shake it for 2 min.

METHANOL EXTRACT REQUIRED FOR ANALYSIS OF HIGH-CONCENTRATION SOILS OR SEDIMENTS

Approximate Concentration Range	Volume of Methanol Extract (a)
500–10,000 µg/kg	100 µL
1,000–20,000 µg/kg	50 µL
5,000–100,000 µg/kg	10 µL
25,000–500,000 µg/kg	100 µL of 1/50 dilution (b)

Calculate appropriate dilution factor for concentrations exceeding this table.

(a) The volume of methanol added to 5 mL of water being purged should be kept constant. Therefore, add to the 5-mL syringe whatever volume of methanol is necessary to maintain a volume of 100 µL added to the syringe.
(b) Dilute an aliquot of the methanol extract and then take 100 µL for analysis.

QUALITY CONTROL Demonstrate, through the analysis of a reagent water blank, that interferences from the analytical system, glassware, and reagents are under control. Blank samples should be carried through all stages of the sample preparation and measurement steps. For each analytical batch (up to 20 samples), a reagent blank, matrix spike, and matrix spike duplicate must be analyzed (the frequency of the spikes may be different for different monitoring programs). The blank and spiked samples must be carried through all stages of the sample preparation and measurement steps. QC samples mentioned in the section on Interferences will also be needed as appropriate to those situations.

REFERENCE Test Methods for Evaluating Solid Waste (SW-846). U.S. EPA. 1983. Method 8240B, Rev. 2, Nov. 1990. Office of Solid Wastes, Washington, DC.

---

n-Propylbenzene   EPA Method 502
CAS #103-65-1

TITLE Volatile Organic Compounds in Water By Purge and Trap Capillary Column Gas Chromatography with Photoionization and Electrolytic Conductivity Detectors in Series. U.S. EPA Method 502.2, Rev. 2.0, 1989.

MATRIX Drinking water and raw source water. The latter should include most surface water and groundwater sources.

METHOD SUMMARY This method covers 60 volatile organic compounds that contain halogen atoms and/or that are aromatic. An inert gas (zero grade nitrogen or helium) is bubbled through a 25-mL or a 5-mL water sample (depending on the expected concentration of the analytes). Purged sample components are trapped in a tube of sorbent materials. When purging is complete, the sorbent tube is heated and backflushed with helium to desorb the trapped sample onto a capillary GC column. The column is temperature programmed to separate the method analytes which are then detected with a photoionization detector (PID) and a Hall electrolytic conductivity (HECD) placed in series. The PID is selective for aromatic compounds and the HECD is selective for halogenated compounds.

INTERFERENCES Impurities in the purge gas and from organic compounds outgassing from the plumbing ahead of the trap account for many contamination problems. Interferences purged or coextracted from the samples will vary considerably from source to source, depending upon the particular sample or extract being tested. Cross-contamination can occur whenever high-level and low-level samples are analyzed sequentially. Samples also can be contaminated by diffusion of volatile organics (particularly methylene chloride and fluorocarbons) through the septum seal into the sample during shipment and storage. The lab where volatile analysis is performed and also the refrigerated storage area should be completely free of solvents.

INSTRUMENTATION A GC containing a series configuration of a high temperature photoionization detector (PID) equipped with 10.0 eV (nominal) lamp and Hall electrolytic conductivity detector (HECD) is required. Also required is an all-glass 5-mL purging device, a sorbent trap, and a thermal desorption apparatus which is connected to the GC system.

Column 1: VOCOL glass wide-bore capillary column.
Column 2: RTX–502.2 mega-bore capillary column.
Column 3: DB-62 mega-bore capillary column.

PRECISION & ACCURACY Method detection limits are dependent upon the characteristics of the gas chromatographic system used. Analytes that are not separated chromatographically cannot be individually identified and used in the same calibration mixture or water samples unless an alternative technique for identification and quantification, such as mass spectrometry, is used.

Electrolytic conductivity detetor (c) range in µg/L (a) was 0.02–200.
Electrolytic conductivity detetor (c) MDL in µg/L (b) was not listed.
Electrolytic conductivity detetor (c) accuracy as % recovery was not listed.
Electrolytic conductivity detetor (c) precision as % RSD was not listed.
Photoionization detector (d) range in µg/L (a) was 0.02–200.
Photoionization detector (d) MDL in µg/L (b) was 0.01.
Photoionization detector (d) accuracy as % recovery was 103.
Photoionization detector (d) precision as % RSD was 2.0.

(a) *The applicable concentration range of this method is compound, instrument, and matrix-dependent. It is listed as being approximately 0.02 to 200 µg/L but no specific information is provided so caution should be observed.*
(b) *The method detection limits reports with this method are compound, instrument, and matrix-dependent. The values reported were calculated using reagent water fortified with the corresponding compounds at 10 µg/L and a GC-equipped with a 60 m × 0.75 mm VOLCOL wide bore capillary column with 1.5 µm film thickness and using helium carrier gas.*
(c) *Recoveries and relative standard deviations were determined from seven samples of reagent water fortified with 10 µg/L of each compound. 2-Bromo-1-chloropropane was used as the internal standard for calculating average recoveries.*
(d) *Recoveries and relative standard deviations were determined from seven samples of reagent water fortified with 10 µg/L of each compound. Fluorobenzene was used as the internal standard for calculating average recoveries.*

SAMPLING METHOD Collect samples using a 40- to 120-mL screw-cap vial (prewashed with detergent, rinsed with distilled water and oven dried at 105°C) with a Teflon®-faced silicone septum. Collect bubble-free samples and place the septum with the Teflon® side down on the water.

SAMPLE PRESERVATION If residual chlorine is present in the water add about 25 mg of ascorbic acid to each vial before samples are collected to remove the chlorine. Add hydrochloric acid to reduce pH to <2, immediately cool samples to 4°C, and store them in a solvent-free refrigerator at 4°C until analysis.

MHT The maximum holding time for samples is 14 days from the time they were collected.

SAMPLE PREPARATION Remove the plungers from two 5-mL syringes and attach a closed syringe valve to each. Warm the sample to room temperature, open the sample bottle, and carefully pour the sample into one of the syringe barrels to just short of overflowing. Replace the syringe plunger, invert the syringe, and compress the sample. Open the syringe valve and vent any residual air while adjusting the sample volume to 5.0 mL. Add 10 µL of the internal calibration standard to the sample through the syringe valve. Close the valve. Fill the second syringe in an identical manner from the same sample bottle. Reserve this second syringe for a reanalysis if necessary.

QUALITY CONTROL As an initial demonstration of lab accuracy and precision, analyze 4 to 7 replicates of a lab fortified blank containing analyte at 0.1–5 µg/L. Collect all samples in duplicate. Surrogate analytes (similar to those of the analytes of interest), whose concentration is known in every sample, are measured using the same internal standard calibration procedure. Duplicate field reagent water blanks (trip blanks) must be analyzed with each set of samples, lab reagent blanks (method blanks) must be analyzed with each batch of samples processed as a group within a work shift. Also, a single lab-fortified blank that contains each of the analytes of interest should be analyzed with each batch of samples processed as a group within a work shift. A 3- to 5-point calibration curve is needed depending on the calibration range factor required.

EPA CONTACT & HOTLINE For technical questions contact Dr. Baldev Bathija, U.S. EPA, Office of Ground Water and Drinking Water (WH-550D), 401 M St. SW, Washington, DC 20460. Tel. (202) 260-3040. For further information the EPA Safe Drinking Water Hotline may be called at: (800) 426-4791.

REFERENCE Methods for the Determination of Organic Compounds in Drinking Water, EPA/600/4-88/039 (revised July 1991; Final Rule for determination of compliance with the MCL for Total Trihalomethanes under 141.30, in 40 CFR Part 141, Vol. 58, No. 147, Fed. Reg., Tuesday Aug. 3, 1993). U.S. EPA Environmental Monitoring Systems Laboratory, Cincinnati, OH, 45268, U.S.A. Available from the National Technical Information Service (NTIS), 5285 Port Royal Road, Springfield, VA 22161; Tel. 800-553-6847. NTIS Order Number is PB91-231480.

---

n-Propylbenzene   EPA Method 524
CAS #103-65-1

TITLE Measurement of Purgeable Organic Compounds in Water by Capillary Column GC/MS.

MATRIX Drinking water and raw source water; the latter should include most surface water and groundwater sources.

METHOD SUMMARY Method 524.2 covers 60 volatile organic compounds. An inert gas (zero grade nitrogen or helium) is bubbled through a 25-mL or a 5-mL water sample (depending on the expected concentration of the analytes). Purged sample components are trapped in a tube of sorbent materials. When purging is complete, the sorbent tube is heated and backflushed with helium to desorb the trapped sample onto a capillary GC column.

INTERFERENCES Impurities in the purge gas and from organic compounds outgassing from the plumbing ahead of the trap account for many contamination problems. Interferences purged or coextracted from the samples will vary considerably from source to source, depending upon the particular sample or extract being tested. Cross-contamination can occur whenever high-level and low-level samples are analyzed sequentially. Samples also can be contaminated by diffusion of volatile organics (particularly methylene chloride and fluorocarbons) through the septum seal into the sample during shipment and storage.

INSTRUMENTATION A GC/MS with a data system equipped with one of the following capillary GC columns:

Column 1: VOCOL glass wide bore capillary column.
Column 2: DB-624 fused silica capillary column.
Column 3: DB-5 fused silica capillary column.

Also required is an all-glass 25 mL or 5-mL purging device, a sorbent trap, and a thermal desorption apparatus which is connected to the GC/MS system.

PRECISION & ACCURACY Method detection limits are compound- and instrument-dependent, and may vary from approximately 0.02–0.35 µg/L. Note in the table below that the "true" concentration range used for accuracy and precision measurements was quite narrow. However, the applicable concentration range of this method is primarily column dependent and is approximately 0.02 to 200 µg/L for the wide-bore thick-film columns. Narrow-bore thin-film columns may have a capacity which limits the range to about 0.02 to 20 µg/L. Analytes that are inefficiently purged from water will not be detected when present at low concentrations, but they can be measured with acceptable accuracy and precision when present in sufficient amounts.

Analytes that are not separated chromatographically, but which have different mass spectra and non-interfering quantification ions, can be identified and measured in the same calibration mixture or water sample. Analytes which have very similar mass spectra cannot be individually identified and measured in the same calibration mixture or water samples unless they have different retention times. Co-eluting compounds with very similar mass spectra, typically many structural isomers, must be reported as an isomeric group or pair.

The range (in µg/L) was 0.1–10.
The Method Detection Limig (in µg/L) was 0.04.
The accuracy (as % recovery) was 100.
The precision (in %) was 5.8.

Note: Data were obtained from 16–31 determinations using a wide-bore capillary column and a jet separator interfaced to a quadrupole mass spectrometer. All analytes were in a reagent water matrix.

SAMPLING METHOD Collect samples using a 40- to 120-mL screw-cap vial (prewashed with detergent, rinsed with distilled water and oven dried at 105°C) with a Teflon®-faced silicone septum. Collect bubble-free samples and place the septum with the Teflon® side down on the water.

SAMPLE PRESERVATION If residual chlorine is present in the water add about 25 mg of ascorbic acid to each vial before samples are collected to remove the chlorine. Add hydrochloric acid to reduce pH to <2, and immediately cool samples to 4°C, and store them in a solvent-free refrigerator at 4°C until analysis.

MHT The maximum holding time for samples is 14 days from the time they were collected.

SAMPLE PREPARATION Remove the plungers from two 25-mL (or 5-mL depending on sample size) syringes and attach a closed syringe valve to each. Warm the sample to room temperature, open the sample bottle, and carefully pour the sample into one of the syringe barrels to just short of overflowing. Replace the syringe plunger, invert the syringe, and compress the sample. Open the syringe valve and vent any residual air while adjusting the sample volume to 25.0 mL (or 5 mL). For samples and blanks, add 5 µL of the fortification solution containing the internal standard and the surrogates to the sample through the syringe valve. For calibration standards and lab fortified blanks, add 5 µL of the fortification solution containing the internal standard only. Close the valve. Fill the second syringe in an identical manner from the same sample bottle. Reserve this second syringe for a reanalysis if necessary.

QUALITY CONTROL As an initial demonstration of lab accuracy and precision, analyze 4 to 7 replicates of a lab fortified blank containing analyte at 0.2–5 µg/L. Collect all samples in duplicate. Surrogate analytes (similar to those of the analytes of interest), whose concentration is known in every sample, are measured using the same internal standard calibration procedure. Duplicate field reagent water blanks (trip blanks) must be analyzed with each set of samples, lab reagent blanks (method blanks) must be analyzed with each batch of samples processed as a group within a work shift. Also, a single lab-fortified blank that contains each of the analytes of interest should be analyzed with each batch of samples processed as a group within a work shift. A 3- to 5-point calibration curve is needed depending on the calibration range factor required.

EPA CONTACT & HOTLINE For technical questions contact Dr. Baldev Bathija, U.S. EPA, Office of Ground Water and Drinking Water (WH-550D), 401 M St. SW, Washington, DC 20460. Tel. (202) 260-3040. For further information the EPA Safe Drinking Water Hotline may be called at: (800) 426-4791.

REFERENCE Methods for the Determination of Organic Compounds in Drinking Water, EPA/600/4-88/039 (revised July 1991; Final Rule for determination of compliance with the MCL for Total Trihalomethanes under 141.30, in 40 CFR Part 141, Vol. 58, No. 147, Fed. Reg., Tuesday Aug. 3, 1993). U.S. EPA Environmental Monitoring Systems Laboratory, Cincinnati, OH, 45268, U.S.A. Available from the National Technical Information Service (NTIS), 5285 Port Royal Road, Springfield, VA 22161; Tel. 800-553-6847. NTIS Order Number is PB91-231480.

---

n-Propylbenzene   EPA Method 8021
CAS #103-65-1

TITLE Halogenated Volatile by Gas Chromatography Using Photoionization and Electrolytic Conductivity Detectors in Series: Capillary Column Technique

MATRIX This method is applicable to nearly all types of samples, regardless of water content, including groundwater, aqueous sludges, caustic liquors, acid liquors, waste solvents, oily wastes, mousses, tars, fibrous wastes, polymeric emulsions, filter cakes, spent carbons, spent catalysts, soils, and sediments.

METHOD SUMMARY This method is used to determine 60 volatile organic compounds in a variety of solid waste matrices. It provides GC conditions for the detection of halogenated and aromatic volatile organic compounds. Samples can be analyzed using direct injection or purge-and-trap (EPA Method 5030). Groundwater samples must be analyzed using EPA Method 5030 (where applicable). A temperature program is used with the GC. Detection is achieved by a photoionization detector (PID) and a Hall electrolytic conductivity detector (HECD) in series.

INTERFERENCES Samples can be contaminated by diffusion of volatile organics (particularly chlorofluorocarbons and methylene chloride) through the sample container septum during shipment and storage.

INSTRUMENTATION A GC-equipped with variable-constant differential flow controllers, subambient oven controller, PID and HECD detectors connected with a short piece of uncoated capillary tubing and a data system.

Column: 60 m × 0.75 mm I.D. VOCOL wide-bore capillary column with 1.5 µm film thickness.

PRECISION & ACCURACY MDLs are compound-dependent and vary with purging efficiency and concentration. The applicable concentration range of this method is compound- and instrument-dependent but is approximately 0.1 to 200 µg/L. Analytes that are inefficiently purged from water will not be detected when present at low concentrations, but they can be measured with acceptable accuracy and precision when present in sufficient amounts. The estimated quantitation limit (EQL) for an individual compound is approximately 1 µg/kg (wet weight) for soil/sediment samples, 100 µg/kg (wet weight) for wastes, and 1 µg/L for groundwater. EQLs will be proportionately higher for sample extracts and samples that require dilution or reduced sample size to avoid saturation of the detector.

MULTIPLICATION FACTORS FOR OTHER MATRICES (a)

Matrix	Factor (b)
Groundwater	10
Low-concentration soil	10
Water miscible liquid waste	500
High-concentration soil and sludge	1250
Non-water miscible waste	1250

(a) Sample EQLs are highly matrix-dependent. The EQLs listed herein are provided for guidance and may not always be achievable. (b) EQL = [Method detection limit] × [Factor]. For non-aqueous samples, the factor is on a wet-weight basis.

SINGLE LABORATORY ACCURACY & PRECISION DATA FOR VOCs IN WATER

This method was tested in a single lab using water spiked at 10 µg/L and the following data was reported:

Recoveries and standard deviations were determined from seven samples and spiked at 10 µg/L of each analyte. Recoveries were determined by the internal standard method. Internal standards were: Fluorobenzene for PID and 2-Bromo-1-chloropropane for HECD.

The average recovery (in percent) for the PID was 103.
The standard deviation of the recovery for the PID was 2.0.
The MDL (in µg/mL) for the PID was 0.004.
The average recovery (in percent) for the HECD was none (no response for this detector).
The standard deviation of the recovery for the HECD was none (no response for this detector).
The MDL (in µg/mL) for the HECD was none (no response for this detector).

SAMPLE COLLECTION, PRESERVATION & HANDLING
Volatile Organics — Standard 40-mL glass screw-cap VOA vials with Teflon®-faced silicone septum may be used for both liquid and solid matrices. When collecting samples, liquids and solids should be introduced into the vials gently to reduce agitation which might drive off volatile compounds. If there are any air bubbles present the sample must be retaken. Tap slightly as they are filled to try and eliminate as much free air space as possible. The two vials from each sampling locations should be sealed in separate plastic bags to prevent cross-contamination between samples particularly if the sampled waste is suspected of containing high levels of volatile organics.

Semivolatile organics — Containers used to collect samples for the determination of semivolatile organic compounds should be soap and water washed followed by methanol (or isopropanol) rinsing. The sample containers should be of glass or Teflon® and have screw-top covers with Teflon® liners.

Preservation for volatile organics — No preservation is used with concentrated waste samples. With liquid samples containing no residual chlorine, 4 drops of concentrated hydrochloric acid are added and the samples are immediately cooled to 4°C. When liquid samples contain residual chlorine, they are treated as above and, in addition, 4 drops of 4% aqueous sodium thiosulfate are added. Soil, sediment, and sludge samples are only cooled to 4°C.

Preservation for semivolatile organics — No preservation is used with concentrated waste samples. With liquid samples containing no residual chlorine and with soil, sediment, and sludge samples, immediately cooling to 4°C is the only preservation used. When residual chlorine is present then 3 mL of 10% aqueous sodium sulfate is added for each gallon of sample collected, followed by cooling to 4°C.

MHT The holding time for all volatile organics samples is 14 days. Liquid samples must be extracted within 7 days and their extracts analyzed within 40 days. Concentrated waste, soil, sediment, and sludge samples must be extracted within 14 days and their extracts analyzed within 40 days.

SAMPLE PREPARATION Volatile compounds are introduced into the gas chromatograph either by direct injector or purge-and-trap (EPA Method 5030). EPA Method 5030 may be used directly on groundwater samples or low-concentration contaminated soils and sediments. For medium-concentration soils or sediments, methanolic extraction, as described in EPA Method 5030, may be necessary prior to purge-and-trap analysis.

QUALITY CONTROL Calculate surrogate standard recovery on all samples, blanks, and spikes. A trip blank is recommended to check on sampling, storage, and handling contamination. Calibration standards, at a minimum of five concentration levels, are prepared in organic-free reagent water. One of the concentration levels should be at a concentration near, but above, the method detection limit.

A combination of bromochloromethane, 2-bromo-1-chloropropane, 1,4-dichlorobutane, and bromochlorobenzene are recommended as surrogate standards to encompass the range of the temperature program used in this method.

REFERENCE Test Methods for Evaluating Solid Waste, Physical/Chemical Methods, SW-846, 3rd Edition, U.S. EPA, Office of Solid Waste, Washington, DC, EPA Method 8021A, Rev. 1, Nov. 1992.

---

n-Propylbenzene   **EPA Method 8260**
CAS #103-65-1

TITLE Volatile Organic Compounds by GC/MS: Capillary Column Technique

MATRIX This method is applicable to nearly all types of samples, regardless of water content, including groundwater, soils, and sediments.

METHOD SUMMARY Method 8260A covers 58 volatile organic compounds that are introduced into a gas chromatograph by the purge-and-trap method or by direct injection (in limited applications). Zero-grade helium is bubbled through a 5-mL solution at ambient temperature. Purged sample components are trapped in a tube containing suitable sorbent materials. When purging is complete, the sorbent tube is heated and backflushed with helium to desorb trapped sample components. The analytes are desorbed directly to a large bore capillary or cryofocussed on a capillary precolumn before being flash evaporated to a narrow bore capillary for analysis.

INTERFERENCES Major contaminant sources are volatile materials in the lab and impurities in the inert purging gas and in the sorbent trap. Interfering contamination may occur when a sample containing low concentrations of volatile organic compounds is analyzed immediately after a sample containing high concentrations of volatile organic compounds. After analysis of a sample containing high concentrations of volatile organic compounds, one or more calibration blanks should be analyzed to check for cross-contamination. Screening of the samples prior to purge-and-trap GC/MS analysis is highly recommended to prevent contamination of the system. This is especially true for soil and waste samples.

Special precautions must be taken to analyze for methylene chloride. The analytical and sample storage area should be isolated from all atmospheric sources of methylene chloride. All gas chromatography carrier gas lines and purge gas plumbing should be constructed from stainless steel or copper tubing. Laboratory clothing previously exposed to methylene chloride fumes during liquid-liquid extraction procedures can contribute to sample contamination.

Samples can also be contaminated by diffusion of volatile organics (particularly methylene chloride and fluorocarbons) through the septum seal during shipment and storage. A trip blank can serve as a check on such contamination.

INSTRUMENTATION GC/MS with a temperature-programmable chromatograph suitable for splitless injection equipped with variable constant differential flow controllers, a subambient oven controller, a purging device, sorbent trap, a thermal desorption apparatus and a capillary precolumn interface when using cryogenic cooling will be needed. The following GC columns may be used:

Column 1: 60 m × 0.75mm I.D. capillary column coated with VOCOL, 1.5 µm film thickness.
Column 2: 30 m × 0.53mm capillary column coated with DB-624 or VOCOL, 3 µm film thickness.
Column 3: 30 m × 0.32mm I.D. capillary column coated with DB-5 or SE-54, 1-µm film thickness.

PRECISION & ACCURACY This method has been tested in a single lab using spiked water. Using a wide-bore capillary column, water was spiked at concentrations between 0.5 and 10 µg/L. Single lab accuracy and precision data are presented. The MDL actually achieved in a given analysis will vary depending on instrument sensitivity and matrix effects.

The MDL (a) in µg/L was 0.04.
The concentration range in µg/L was 0.1–10.
The mean accuracy (% of true value) was 100.
The precision as relative standard deviation was 5.8.

Note: The MDL is based on a 25-mL sample volume instead of a 5-mL sample volume.

SAMPLING METHOD
Liquid samples — Use a 40-mL glass screw-cap VOA vial with a Teflon®-faced silicone septum that has been prewashed, rinsed with distilled deionized water, and oven dried. If residual chlorine is present, collect the sample in a 4-oz soil VOA container which has been pre-preserved with 4 drops of 10% sodium thiosulfate. Mix gently and transfer the sample to a 40-mL VOA vial. Collect bubble-free samples in duplicate and seal each sample in a separate plastic bag.

Soils, sediments and sludges — Use an 8-oz widemouth glass bottle with Teflon®-faced silicone septum that has been prewashed, rinsed with distilled deionized water, and oven dried. **Do not** heat the septum for more than 1 h. Tap slightly to eliminate any free air space. Collect samples in duplicate and seal each one in a separate plastic bag.

SAMPLE PRESERVATION
Liquid samples — Add 4 drops of concentrated HCL, cool to 4°C and store in a solvent-free refrigerator.

Soils, sediments and sludges — Cool samples to 4°C and store in a solvent-free refrigerator.

MHT The maximum holding time of any sample (liquids, soils, sediments, and sludges) is 14 days.

SAMPLE PREPARATION
Liquid samples — Remove the plunger from a 5-mL syringe and carefully pour the sample into the syringe barrel to just short of overflowing. Replace the syringe plunger and compress the sample. Open the syringe valve and vent any residual air while adjusting the sample volume to 5.0 mL. If there is only one volatile organic analysis (VOA) vial, a second syringe should be filled at this time to protect against possible loss of sample integrity. Add 10 µL of surrogate spiking solution and 10 µL of internal standard spiking solution through the valve bore of the 5-mL syringe, then close the valve. The surrogate and internal standards may be mixed and added as a single spiking solution.

Sediments, soils, and waste samples — All samples of this type should be screened by GC analysis using a headspace method (EPA Method 3810) or the hexadecane extraction and screening method (EPA Method 3820). Use the screening data to determine whether to use the low-concentration method (0.005–1 mg/kg) or the high-concentration method (>1 mg/kg).

Low-concentration method — The low-concentration method is based on purging a heated sediment or soil sample mixed with organic-free reagent water containing the surrogate and internal standards. Analyze all reagent blanks and standards under the same conditions as the samples.

Use a 5-g sample if the expected concentration is <0.1 mg/kg or a 1-g sample for expected concentrations between 0.1 and 1 mg/kg. Mix the contents of the sample container with a narrow metal spatula. Weigh the amount of the sample into a tared purge device. Add the spiked water to the purge device, which contains the weighed amount of sample, and connect the device to the purge-and-trap system.

High-concentration method — This method is based on extracting the sediment or soil with methanol. A waste sample is either extracted or diluted, depending on its solubility in methanol. Wastes that are insoluble in methanol are diluted with reagent tetraglyme or possibly polyethylene glycol (PEG). An aliquot of the extract is added to organic-free reagent water containing surrogate and internal standards. This is purged at ambient temperature. All samples with an expected concentration of >1.0 mg/kg should be analyzed by this method.

Mix the contents of the sample container with a narrow metal spatula. For sediments or soils and solid wastes that are insoluble in methanol, weigh 4 g (wet weight) of sample into a tared 20-mL vial. For waste that is soluble in methanol, tetraglyme, or PEG, weigh 1 g (wet weight) into a tared scintillation vial or culture tube or a 10-mL volumetric flask. Quickly add 9.0 mL of appropriate solvent then add 1.0 mL of a surrogate spiking solution to the vial, cap it, and shake it for 2 min.

METHANOL EXTRACT REQUIRED FOR ANALYSIS OF HIGH-CONCENTRATION SOILS OR SEDIMENTS

Approximate Concentration Range	Volume of Methanol Extract (a)
500–10,000 µg/kg	100 µL
1,000–20,000 µg/kg	50 µL
5,000–100,000 µg/kg	10 µL
25,000–500,000 µg/kg	100 µL of 1/50 dilution (b)

Calculate appropriate dilution factor for concentrations exceeding this table.

(a) The volume of methanol added to 5 mL of water being purged should be kept constant. Therefore, add to the 5-mL syringe whatever volume of methanol is necessary to maintain a volume of 100 µL added to the syringe.
(b) Dilute an aliquot of the methanol extract and then take 100 µL for analysis.

QUALITY CONTROL Demonstrate, through the analysis of a reagent water blank, that interferences from the analytical system, glassware, and reagents are under control. Blank samples should be carried through all stages of the sample preparation and measurement steps. For each analytical batch (up to 20 samples), a reagent blank, matrix spike, and matrix spike duplicate must be analyzed (the frequency of the spikes may be different for different monitoring programs). The blank and spiked samples must be carried through all stages of the sample preparation and measurement steps. QC samples mentioned in the section on Interferences will also be needed as appropriate to those situations.

Matrix spiking standards should be prepared from volatile organic compounds which will be representative of the compounds being investigated. The recommended internal standards are chlorobenzene-d5, 1,4-difluorobenzene, 1,4-dichlorobenzene-d4, and pentafluorobenzene. Using stock standard solutions, prepare secondary dilution standards containing the compounds of interest, either singly or mixed together in methanol. Store them in a vial with no headspace for no more than one week. Surrogates recommended are toluene-d8, 4-bromofluorobenzene, and dibromofluoromethane. Each sample undergoing GC/MS analysis must be spiked with 10 µL of the surrogate spiking solution prior to analysis.

REFERENCE Test Methods for Evaluating Solid Waste (SW-846). U.S. EPA 1983, Method 8260A, Rev. 1, Nov. 1990. Office of Solid Waste, Washington, DC.

---

n-Propylbenzene   **EPA Method 503.1**
CAS #103-65-1

TITLE Aromatic & Unsaturated VOCs Water

MATRIX Drinking water (finished or in any treatment stage) and raw source water.

APPLICATION Method covers 28 aromatic and unsaturated VOCs. An inert gas is bubbled through a 5-mL water sample. Purged sample components are trapped in tube of sorbent materials. When purging is complete, sorbent tube is heated and backflushed with inert gas to desorb trapped sample onto a packed GC column.

INTERFERENCES During analysis, major contaminant sources are volatile materials in the lab and impurities in purging gas and sorbent trap. With high and low level samples, there can be carryover contamination. Excess water causes a negative baseline deflection.

INSTRUMENTATION Purge and Trap GC w/photoionization detector. (Two GC columns are recommended); Column 1: 5% SP-1200 and 1.75% Bentone 34 on Supelcoport; Column 2: 1,2,3-tris(2-cyanoethoxy)propane on Chromosorb W.

RANGE 2.2–600 µg/L. (Drinking water)

MDL 0.009 µg/L in water

PRECISION RSD = 9.3% at 0.40 µg/L conc.; 7 samples

ACCURACY Average recovery = 83% at 0.40 µg/L conc.; 7 samples

SAMPLING METHOD Use a 40–120-mL screw-cap vial (prewashed with detergent, rinsed with distilled water and oven dried at 105°C) with a PTFE-faced silicone septum. If residual chlorine is in the water add about 25 mg of ascorbic acid to each vial before sample collection. Collect bubble-free samples.

STABILITY Cool to 4°C; HCl to pH <2.

MHT 14 days.

QUALITY CONTROL As initial demonstration of lab accuracy and precision, analyze 4 to 7 replicates of a lab fortified blank containing the analyte at 0.1–5 µg/L. Collect all samples in duplicate.

REFERENCE Method 503.1, Volatile Aromatic & Unsaturated Organic Compounds in H2O by Purge and Trap GC, EPA 600/4-88/039.

Propylthiouracil   **EPA Method 8270**
CAS #51-52-5

TITLE Semivolatile Organic Compounds by GC/MS

MATRIX This method is used to determine the concentration of semivolatile organic compounds in extracts prepared from all types of solid waste matrices, soils, and groundwater. Although surface waters are not specifically mentioned, this method should be applicable to water samples from rivers, lakes, etc.

METHOD SUMMARY This method covers 259 semivolatile organic compounds. In very limited applications direct injection of the sample into the GC/MS system may be appropriate, but this results in very high detection limits (approximately 10,000 µg/L). Typically, a 1-L liquid sample, containing surrogate, and matrix spiking standards, is extracted in a continuous extractor first under acid conditions and then under basic conditions. Typically 30 g of a solid sample, containing surrogate, and matrix spiking standards, is extracted ultrasonically. After concentrating the extract to 1 mL it is spiked with 10 µL of an internal standard solution just prior to analysis by GC/MS. The volume injected should contain about 100 ng of base/neutral and 200 ng of acid surrogates (for a 1-µL injection). Analysis is performed by GC/MS using a capillary GC column.

INTERFERENCES Raw GC/MS data from all blanks, samples, and spikes must be evaluated for interferences. Contamination by carryover can occur whenever high-concentration

and low-concentration samples are sequentially analyzed. To reduce carryover, the sample syringe must be rinsed out between samples with solvent. Whenever an unusually concentrated sample is encountered, it should be followed by the analysis of blank solvent to check for cross-contamination.

INSTRUMENTATION A GC/MS and a data system are required. The GC column used is a 30 m × 0.25 mm I.D. (or 0.32 mm I.D.) 1um film thickness silicone-coated fused silica capillary column. A continuous liquid-liquid extractor equipped with Teflon® or glass connection joints and stopcocks requiring no lubrication, a K-D concentrating apparatus, water bath, and an ultrasonic disrupter with a minimum power of 300 W and with pulsing capability are also required.

PRECISION & ACCURACY The estimated quantitation limit (EQL) of Method 8270B for determining an individual compound is approximately 1 mg/kg (wet weight) for soil or sediment samples, 1–200 mg/kg for wastes (dependent on matrix and method of preparation), and 10 µg/L for groundwater samples. EQLs will be proportionately higher for sample extracts that require dilution to avoid saturation of the detector.

The EQL(b) for groundwater in µg/L is 100.
The EQL (a, b) for low concentrations in soil and sediment in µg/kg is not determined.
Accuracy as µg/L is not listed.
Overall precision in µg/L is not listed.

(a) *EQLs listed for soil/sediment are based on wet weight. Normally data is reported in a dry-weight basis; therefore, EQLs will be higher based on the % dry weight of each sample. This calculation is based on a 30 g sample and gel permeation chromatography cleanup.*
(b) *Sample EQLs are highly matrix-dependent. The EQLs are provided for guidance and may not always be achievable.*
C = *True value for concentration, in µg/L.*
X = *Average recovery found for measurements of samples containing a concentration of C, in µg/L.*

ESTIMATED QUANTITATION LIMIT

Other Matrices	Factor (a)
High-concentration soil and sludges by sonicator	7.5
Non-water miscible waste	75

(a) EQL for other matrices = [EQL for low soil/sediment] × [Factor]. This estimated EQL is similar to an EPA "Practical Quantitation Limit."

SAMPLING METHOD
Liquid samples — Use a 1 or 2½ gallon amber glass bottle with a screw-top Teflon®-lined cover that has been prewashed with detergent and rinsed with distilled water and methanol (or isopropanol).

Soils, sediments, or sludges — Use an 8-oz. widemouth glass with a screw-top Teflon®-lined cover that has been prewashed with detergent and rinsed with distilled water and methanol (or isopropanol).

SAMPLE PRESERVATION
Liquid samples — If residual chlorine is present, add 3 mL of 10% sodium thiosulfate per gallon, cool to 4°C and store in a solvent-free refrigerator until analysis; if chlorine is not present, then eliminate the sodium thiosulfate addition.

Soils, sediments, or sludges — Cool samples to 4°C and store in a solvent-free refrigerator.

MHT Liquid samples must be extracted within 7 days and the extracts analyzed within 40 days. Soils, sediments, or sludges may be stored for a maximum of 14 days and the extracts analyzed within 40 days.

SAMPLE PREPARATION
Liquid samples — Transfer 1 L quantitatively to a continuous extractor. If high concentrations are anticipated, a smaller volume may be used and then diluted with organic-free reagent water to 1 L. Adjust pH, if necessary, to pH <2 using 1:1 (V/V) sulfuric acid. Pipette 1.0 mL of a surrogate standard spiking solution into each sample. For the sample in each analytical batch selected for spiking, add 1.0 mL of a matrix spiking standard. For base/neutral acid analysis, the amount of the surrogates and matrix spiking compounds added to the sample should result in a final concentration of 100 ng/µL of each analyte in the extract to be analyzed (assuming a 1-µL injection). Extract with methylene chloride for 18–24 h. Next, adjust the pH of the aqueous phase to pH >11 using 10 N sodium hydroxide and extract it with methylene chloride again for 18–24 h. Dry the extract through a column containing anhydrous sodium sulfate and concentrate it to 1 mL using a K-D concentrator.

Soils, sediments, or sludges — Use 30 g of sample. Nonporous or wet samples (gummy or clay type) that do not have a free-flowing sandy texture must be mixed with anhydrous sodium sulfate until the sample is free flowing. Add 1 mL of surrogate standards to all samples, spikes, standards, and blanks. For the sample in each analytical batch selected for spiking, add 1.0 mL of a matrix spiking standard. For base/neutral acid analysis, the amount added of the surrogates and matrix spiking compounds should result in a final concentration of 100 ng/µL of each base/neutral analyte and 200 ng/µL of each acid analyte in the extract to be analyzed (assuming a 1-µL injection). Immediately add a 100-mL mixture of 1:1 methylene chloride:acetone and extract the sample ultrasonically for 3 min and then decant or filter the extracts. Repeat the extraction two or more times. Dry the extract using a column with anhydrous sodium sulfate and concentrate it to 1 mL in a K-D concentrator.

QUALITY CONTROL A methylene chloride solution containing 50 ng/µL of decafluorotriphenylphosphine (DFTPP) is used for tuning the GC/MS system each 12-h shift. A system performance check also must be made during every 12-h shift. A standard containing 50 ng/µL each of 4,4'-DDT, pentachlorophenol, and benzidine is required to verify injection port inertness and GC column performance. A calibration standard at mid-concentration, containing each compound of interest, including all required surrogates, must be performed every 12 h during analysis. After the system performance check is met, calibration check compounds (CCCs) are used to check the validity of the initial calibration.

The internal standard responses and retention times in the calibration check standard must be evaluated immediately after

or during data acquisition. If the retention time for any internal standard changes by more than 30 seconds from the last check calibration (12 h), the chromatographic system must be inspected for malfunctions and corrections must be made, as required. If the electron ionization current plot (EICP) area for any of the internal standards changes by a factor of two from the last daily calibration standard check, the mass spectrometer must be inspected for malfunctions and corrections must be made, as appropriate.

Demonstrate, through the analysis of a reagent water blank, that interferences from the analytical system, glassware, and reagents are under control. The blank samples should be carried through all stages of the sample preparation and measurement steps. For each analytical batch (up to 20 samples), a reagent blank, matrix spike, and matrix spike duplicate/duplicate must be analyzed (the frequency of the spikes may be different for different monitoring programs). The blank and spiked samples must be carried through all stages of the sample preparation and measurement steps. A QC reference sample concentrate containing each analyte at a concentration of 100 mg/L in methanol is required.

REFERENCE Test Methods for Evaluating Solid Waste (SW-846). U.S. EPA 1983, Method 8270B, Rev. 2, Nov. 1990. Office of Solid Waste, Washington, DC.

Pyrene — EPA Method 1625
CAS #129-00-0

TITLE Semivolatile Organic Compounds by Isotope Dilution GC/MS

MATRIX The compounds may be determined in waters, soils, and municipal sludges by this method.

METHOD SUMMARY This method is used to determine 176 semivolatile toxic organic pollutants associated with the CWA (as amended 1987); the RCRA (as amended 1986); the CERCLA (as amended 1986); and other compounds amenable to extraction and analysis by capillary column gas chromatography-mass spectrometry (GC/MS).

Stable isotopically-labeled analogs of the compounds of interest are added to the sample. If the solids content is less than 1%, a 1-L sample is extracted at pH 12–13, then at pH <2 with methylene chloride using continuous extraction techniques.

If the solids content is 30% or less, the sample is diluted to 1% solids with reagent water, homogenized ultrasonically, and extracted at pH 12–13, then at pH <2 with methylene chloride using continuous extraction techniques. If the solids content is greater than 30%, the sample is extracted using ultrasonic techniques.

Each extract is dried over sodium sulfate, concentrated to a volume of 5 mL, cleaned up using GPC, if necessary, and concentrated. Extracts are concentrated to 1 mL if GPC is not performed, and to 0.5 mL if GPC is performed.

An internal standard is added to the extract, and a 1-mL aliquot of the extract is injected into the GC. The compounds are separated by GC and detected by a MS. The labeled compounds serve to correct the variability of the analytical technique.

INTERFERENCES Solvents, reagents, glassware, and other sample processing hardware may yield artifacts and/or elevated baselines causing misinterpretation of chromatograms and spectra. Materials used in the analysis must be demonstrated to be free from interferences under the conditions of analysis by running method blanks initially and with each sample lot (sample started through the extraction process on a given 8-h shift, to a maximum of 20). Specific selection of reagents and purification of solvents by distillation in all glass systems may be required. Glassware and, where possible, reagents are cleaned by solvent rinse and baking at 450°C for 1-h minimum. Interferences coextracted from samples will vary considerably from source to source, depending on the diversity of the site being sampled.

INSTRUMENTATION Major instrumentation includes a GC with a splitless or on-column injection port for capillary column, a MS with 70 eV electron impact ionization, and a data system to collect and record MS data, and process it. A K-D apparatus is used to concentrate extracts.

GC Column: 30 m × 0.25 mm I.D. 5% phenyl, 94% methyl, 1% vinyl silicone bonded phased fused silica capillary column.

PRECISION & ACCURACY The detection limits of the method are usually dependent on the level of interferences rather than instrumental limitations. The limits typify the minimum quantities that can be detected with no interferences present.

The minimum level (in µg/mL) was 10. This is defined as a minimum level at which the analytical system shall give recognizable mass spectra (background corrected) and acceptable calibration points.

The MDL (in µg/kg) in low solids was 40 and in high solids was 48; these were determined in digested sludge (low solids) and in filter cake or compost (high solids).

The labeled and native compound initial precision as standard deviation (in µg/L) was 19.
The labeled and native compound initial accuracy as average recovery (in µg/L) was 76–152.

SAMPLE COLLECTION, PRESERVATION & HANDLING Collect samples in glass containers. Aqueous samples which flow freely are collected in refrigerated bottles using automatic sampling equipment. Solid samples are collected as grab samples using widemouth jars. Maintain samples at 0 to 4°C from the time of collection until extraction. If residual chlorine is present in aqueous samples, add 80 mg sodium thiosulfate/L of water. Begin sample extraction within 7 days of collection, and analyze all extracts within 40 days of extraction.

SAMPLE PREPARATION Samples containing 1% solids or less are extracted directly using continuous liquid-liquid extraction techniques. Samples containing 1 to 30% solids are diluted to the 1% level with reagent water and extracted using continuous liquid-liquid extraction techniques. Samples containing greater than 30% solids are extracted using ultrasonic techniques.

Base/neutral extraction — Adjust the pH of the waters in the extractors to 12–13 with 6 *N* NaOH. Extract with methylene chloride for 24–48 h.

Acid extraction — Adjust the pH of the waters in the extractors to 2 or less using 6 *N* sulfuric acid. Extract with methylene chloride for 24–48 h.

Ultrasonic extraction of high solids samples — Add anhydrous sodium sulfate to the sample and QC aliquot(s). Add acetone:methylene chloride (1:1) to the sample and mix thoroughly

Concentrate extracts using a K-D apparatus.

QUALITY CONTROL The analyst is permitted to modify this method to improve separations or lower the costs of measurements, provided all performance specifications are met. Analyses of blanks are required to demonstrate freedom from contamination. When results of spikes indicate atypical method performance for samples, the samples are diluted to bring method performance within acceptable limits.

For low solids (aqueous samples), extract, concentrate, and analyze two sets of four 1-L aliquots (8 aliquots total) of the precision and recovery standard. For high solids samples, two sets of four 30-g aliquots of the high solids reference matrix are used.

Spike all samples with labeled compounds to assess method performance. Compute percent recovery of the labeled compounds using the internal standard method. Compare the labeled compound recovery for each compound with the corresponding labeled compound recovery.

Reagent water and high solids reference matrix blanks are analyzed to demonstrate freedom from contamination. Extract and concentrate a 1-L reagent water blank or a high solids reference matrix blank with each sample's lot (samples started through the extraction process on the same 8-h shift, to a maximum of 20 samples).

Field replicates may be collected to determine the precision of the sampling technique, and spiked samples may be required to determine the accuracy of the analysis when the internal standard method is used.

REFERENCE Semivolatile Organic Compounds by Isotope Dilution GC/MS. Office of Water Regulation and Standards, U.S. EPA Industrial Technology Division, Washington, DC, EPA Method 1625, Rev. C, June 1989 (contact W.A. Telliard, U.S. EPA, Office of Water Regulations and Standards, 401 M St., SW, Washington, DC, 20460. Phone: 202-382-7131).

Pyrene   **EPA Method 1625**
CAS #129-00-0

TITLE Semivolatile Organic Compounds by Isotope Dilution GC/MS

MATRIX The compounds may be determined in waters, soils, and municipal sludges by this method.

METHOD SUMMARY This method is used to determine 176 semivolatile toxic organic pollutants associated with the CWA (as amended 1987); the RCRA (as amended 1986); the CERCLA (as amended 1986); and other compounds amenable to extraction and analysis by capillary column gas chromatography-mass spectrometry (GC/MS).

Stable isotopically-labeled analogs of the compounds of interest are added to the sample. If the solids content is less than 1%, a 1-L sample is extracted at pH 12–13, then at pH <2 with methylene chloride using continuous extraction techniques.

If the solids content is 30% or less, the sample is diluted to 1% solids with reagent water, homogenized ultrasonically, and extracted at pH 12–13, then at pH <2 with methylene chloride using continuous extraction techniques. If the solids content is greater than 30%, the sample is extracted using ultrasonic techniques.

Each extract is dried over sodium sulfate, concentrated to a volume of 5 mL, cleaned up using GPC, if necessary, and concentrated. Extracts are concentrated to 1 mL if GPC is not performed, and to 0.5 mL if GPC is performed.

An internal standard is added to the extract, and a 1-mL aliquot of the extract is injected into the GC. The compounds are separated by GC and detected by a MS. The labeled compounds serve to correct the variability of the analytical technique.

INTERFERENCES Solvents, reagents, glassware, and other sample processing hardware may yield artifacts and/or elevated baselines causing misinterpretation of chromatograms and spectra. Materials used in the analysis must be demonstrated to be free from interferences under the conditions of analysis by running method blanks initially and with each sample lot (sample started through the extraction process on a given 8-h shift, to a maximum of 20). Specific selection of reagents and purification of solvents by distillation in all glass systems may be required. Glassware and, where possible, reagents are cleaned by solvent rinse and baking at 450°C for 1-h minimum. Interferences coextracted from samples will vary considerably from source to source, depending on the diversity of the site being sampled.

INSTRUMENTATION Major instrumentation includes a GC with a splitless or on-column injection port for capillary column, a MS with 70 eV electron impact ionization, and a data system to collect and record MS data, and process it. A K-D apparatus is used to concentrate extracts.

GC Column: 30 m × 0.25 mm I.D. 5% phenyl, 94% methyl, 1% vinyl silicone bonded phased fused silica capillary column.

PRECISION & ACCURACY The detection limits of the method are usually dependent on the level of interferences rather than instrumental limitations. The limits typify the minimum quantities that can be detected with no interferences present.

The minimum level (in μg/mL) was 10. This is defined as a minimum level at which the analytical system shall give recognizable mass spectra (background corrected) and acceptable calibration points.

The MDL (in μg/kg) in low solids was 40 and in high solids was 48; these were determined in digested sludge (low solids) and in filter cake or compost (high solids).

The labeled and native compound initial precision as standard deviation (in μg/L) was 19.

The labeled and native compound initial accuracy as average recovery (in μg/L) was 76–152.

SAMPLE COLLECTION, PRESERVATION & HANDLING Collect samples in glass containers. Aqueous samples which flow freely are collected in refrigerated bottles using automatic sampling equipment. Solid samples are collected as grab samples using widemouth jars. Maintain samples at 0 to 4°C from the time of collection until extraction. If residual chlorine is present in aqueous samples, add 80 mg sodium thiosulfate/L of water. Begin sample extraction within 7 days of collection, and analyze all extracts within 40 days of extraction.

SAMPLE PREPARATION Samples containing 1% solids or less are extracted directly using continuous liquid-liquid extraction techniques. Samples containing 1 to 30% solids are diluted to the 1% level with reagent water and extracted using continuous liquid-liquid extraction techniques. Samples containing greater than 30% solids are extracted using ultrasonic techniques.

Base/neutral extraction — Adjust the pH of the waters in the extractors to 12–13 with 6 N NaOH. Extract with methylene chloride for 24–48 h.

Acid extraction — Adjust the pH of the waters in the extractors to 2 or less using 6 N sulfuric acid. Extract with methylene chloride for 24–48 h.

Ultrasonic extraction of high solids samples — Add anhydrous sodium sulfate to the sample and QC aliquot(s). Add acetone:methylene chloride (1:1) to the sample and mix thoroughly

Concentrate extracts using a K-D apparatus.

QUALITY CONTROL The analyst is permitted to modify this method to improve separations or lower the costs of measurements, provided all performance specifications are met. Analyses of blanks are required to demonstrate freedom from contamination. When results of spikes indicate atypical method performance for samples, the samples are diluted to bring method performance within acceptable limits.

For low solids (aqueous samples), extract, concentrate, and analyze two sets of four 1-L aliquots (8 aliquots total) of the precision and recovery standard. For high solids samples, two sets of four 30-g aliquots of the high solids reference matrix are used.

Spike all samples with labeled compounds to assess method performance. Compute percent recovery of the labeled compounds using the internal standard method. Compare the labeled compound recovery for each compound with the corresponding labeled compound recovery.

Reagent water and high solids reference matrix blanks are analyzed to demonstrate freedom from contamination. Extract and concentrate a 1-L reagent water blank or a high solids reference matrix blank with each sample's lot (samples started through the extraction process on the same 8-h shift, to a maximum of 20 samples).

Field replicates may be collected to determine the precision of the sampling technique, and spiked samples may be required to determine the accuracy of the analysis when the internal standard method is used.

REFERENCE Semivolatile Organic Compounds by Isotope Dilution GC/MS. Office of Water Regulation and Standards, U.S. EPA Industrial Technology Division, Washington, DC, EPA Method 1625, Rev. C, June 1989 (contact W.A. Telliard, U.S. EPA, Office of Water Regulations and Standards, 401 M St., SW, Washington, DC, 20460. Phone: 202-382-7131).

Pyrene — EPA Method 625
CAS #129-00-0

TITLE Base/Neutrals and Acids, U.S. EPA Method 625

MATRIX This methods covers municipal and industrial wastewaters.

METHOD SUMMARY Approximately 1 L of sample is serially extracted with methylene chloride at a pH greater than 11 and again at a pH less than 2 using a separatory funnel or a continuous extractor. The methylene chloride extract is dried, concentrated to a volume of 1 mL, and analyzed by GC/MS. Qualitative identification of the parameters in the extract is performed using the retention time and the relative abundance of three characteristic masses (m/z). Qualitative analysis is performed using either external or internal standard techniques with a single characteristic m/z.

INTERFERENCES Method interferences may be caused by contaminants in solvents, reagents, glassware, and other sample processing hardware. Glassware must be scrupulously cleaned. Glassware should be heated in a muffle furnace at 400°C for 5 to 30 min. Some thermally stable materials, such as PCBs, may not be eliminated by this treatment. Solvent rinses with acetone and pesticide quality hexane may be substituted for the muffle furnace heating. Matrix interferences may be caused by contaminants that are coextracted from the sample. The base-neutral extraction may cause significantly reduced recovery of phenols. The packed gas chromatographic columns recommended for the basic fraction may not exhibit sufficient resolution for some analytes.

INSTRUMENTATION A GC/MS system with an injection port designed for on-column injection when using packed columns and for splitless injection when using capillary columns.

Column for base/neutrals: 1.8 m long × 2 mm I.D. glass, packed with 3% SP-2550 on Supelcoport (100/120 mesh) or equivalent.

Column for acids: 1.8 m long × 2 mm I.D. glass, packed with 1% SP-1240DA on Supelcoport (100/120 mesh) or equivalent.

PRECISION & ACCURACY The MDL concentrations were obtained using reagent water. The MDL actually achieved in a given analysis will vary depending on instrument sensitivity and matrix effects. This method was tested by 15 laboratories

using reagent water, drinking water, surface water, and industrial wastewaters spiked at six concentrations over the range 5 to 100 µg/L. Single operator precision, overall precision, and method accuracy were found to be directly related to the concentration of the parameter matrix.

The MDL (in µg/L) in reagent water was 1.9.
The standard deviation (in µg/L based on 4 recovery measurements) was 25.2.
The range (in µg/L) for average recovery for 4 measurements was 69.6–100.0.
The range (in %) for percent recovery was 52–115.
Accuracy (in µg/L) as expected recovery for one or more measurements of a sample containing a true concentration of C was 0.84C-0.16.
Precision (in µg/L) as expected single analyst standard deviation of measurements at an average concentration found at X was 0.16X + 0.06.
Overall precision (in µg/L) as expected interlaboratory standard deviation of measurements in an average concentration found at X was 0.15X + 0.31.

C = *True value of the concentration in µg/L.*
X = *Average recovery found for measurements of samples containing a concentration at C in µg/L.*

SAMPLE PREPARATION Adjust the pH to >11 with sodium hydroxide and serially extract in a separatory funnel with methylene chloride or else in a continuous extractor. Next, adjust the pH to <2 with sulfuric acid and serially extract in a separatory funnel with methylene chloride or else in a continuous extractor. Dry the extracts separately through a column of anhydrous sodium sulfate and then concentrate each of the extracts to 1.0 mL using a K-D apparatus.

SAMPLE COLLECTION, PRESERVATION & HANDLING Grab samples must be collected in glass containers. All samples must be refrigerated at 4°C from the time of collection until extraction. If residual chlorine is present, add 80 mg of sodium thiosulfate/L of sample and mix well. All samples must be extracted within 7 days of collection and completely analyzed within 40 days of extraction.

QUALITY CONTROL Make an initial, one-time, demonstration of the ability to generate acceptable accuracy and precision with this method. Before processing any samples, the analyst must analyze a reagent water blank to demonstrate that interferences from the analytical system and glassware are under control. Each time a set of samples is extracted or reagents are changed, a reagent water blank must be processed. Spike and analyze a minimum of 5% of all samples to monitor and evaluate lab data quality. A QC check sample concentrate that contains each parameter of interest at a concentration of 100 µg/mL in acetone is required. PCBs and multicomponent pesticides may be omitted from this test.

After analysis of five spiked wastewater samples, calculate the average percent recovery and the standard deviation of the percent recovery. Spike all samples with the surrogate standard spiking solution and calculate the percent recovery of each surrogate compound.

REFERENCE *Federal Register*, Vol. 49, No. 209. Friday, Oct. 26, 1984.

Pyrene   EPA Method 8270
CAS #129-00-0

TITLE Semivolatile Organic Compounds by GC/MS

MATRIX This method is used to determine the concentration of semivolatile organic compounds in extracts prepared from all types of solid waste matrices, soils, and groundwater. Although surface waters are not specifically mentioned, this method should be applicable to water samples from rivers, lakes, etc.

METHOD SUMMARY This method covers 259 semivolatile organic compounds. In very limited applications direct injection of the sample into the GC/MS system may be appropriate, but this results in very high detection limits (approximately 10,000 µg/L). Typically, a 1-L liquid sample, containing surrogate, and matrix spiking standards, is extracted in a continuous extractor first under acid conditions and then under basic conditions. Typically 30 g of a solid sample, containing surrogate, and matrix spiking standards, is extracted ultrasonically. After concentrating the extract to 1 mL it is spiked with 10 µL of an internal standard solution just prior to analysis by GC/MS. The volume injected should contain about 100 ng of base/neutral and 200 ng of acid surrogates (for a 1-µL injection). Analysis is performed by GC/MS using a capillary GC column.

INTERFERENCES Raw GC/MS data from all blanks, samples, and spikes must be evaluated for interferences. Contamination by carryover can occur whenever high-concentration and low-concentration samples are sequentially analyzed. To reduce carryover, the sample syringe must be rinsed out between samples with solvent. Whenever an unusually concentrated sample is encountered, it should be followed by the analysis of blank solvent to check for cross-contamination.

INSTRUMENTATION A GC/MS and a data system are required. The GC column used is a 30 m × 0.25 mm I.D. (or 0.32 mm I.D.) 1um film thickness silicone-coated fused silica capillary column. A continuous liquid-liquid extractor equipped with Teflon® or glass connection joints and stopcocks requiring no lubrication, a K-D concentrating apparatus, water bath, and an ultrasonic disrupter with a minimum power of 300 W and with pulsing capability are also required.

PRECISION & ACCURACY The estimated quantitation limit (EQL) of Method 8270B for determining an individual compound is approximately 1 mg/kg (wet weight) for soil or sediment samples, 1–200 mg/kg for wastes (dependent on matrix and method of preparation), and 10 µg/L for groundwater samples. EQLs will be proportionately higher for sample extracts that require dilution to avoid saturation of the detector.

The EQL(b) for groundwater in µg/L is 10.
The EQL (a, b) for low concentrations in soil and sediment in µg/kg is 660.
Accuracy as µg/L is 0.84C–0.16.
Overall precision in µg/L is 0.15X + 0.31.

(a) *EQLs listed for soil/sediment are based on wet weight. Normally data is reported in a dry-weight basis; therefore, EQLs will be higher based on the % dry weight of each sample. This calculation is based on a 30 g sample and gel permeation chromatography cleanup.*
(b) *Sample EQLs are highly matrix-dependent. The EQLs are provided for guidance and may not always be achievable.*

C = *True value for concentration, in μg/L.*
X = *Average recovery found for measurements of samples containing a concentration of C, in μg/L.*

ESTIMATED QUANTITATION LIMIT

Other Matrices	Factor (a)
High-concentration soil and sludges by sonicator	7.5
Non-water miscible waste	75

(a) EQL for other matrices = [EQL for low soil/sediment] × [Factor]. This estimated EQL is similar to an EPA "Practical Quantitation Limit."

SAMPLING METHOD

Liquid samples — Use a 1 or 2½ gallon amber glass bottle with a screw-top Teflon®-lined cover that has been prewashed with detergent and rinsed with distilled water and methanol (or isopropanol).

Soils, sediments, or sludges — Use an 8-oz. widemouth glass with a screw-top Teflon®-lined cover that has been prewashed with detergent and rinsed with distilled water and methanol (or isopropanol).

SAMPLE PRESERVATION

Liquid samples — If residual chlorine is present, add 3 mL of 10% sodium thiosulfate per gallon, cool to 4°C and store in a solvent-free refrigerator until analysis; if chlorine is not present, then eliminate the sodium thiosulfate addition.

Soils, sediments, or sludges — Cool samples to 4°C and store in a solvent-free refrigerator.

MHT Liquid samples must be extracted within 7 days and the extracts analyzed within 40 days. Soils, sediments, or sludges may be stored for a maximum of 14 days and the extracts analyzed within 40 days.

SAMPLE PREPARATION

Liquid samples — Transfer 1 L quantitatively to a continuous extractor. If high concentrations are anticipated, a smaller volume may be used and then diluted with organic-free reagent water to 1 L. Adjust pH, if necessary, to pH <2 using 1:1 (V/V) sulfuric acid. Pipette 1.0 mL of a surrogate standard spiking solution into each sample. For the sample in each analytical batch selected for spiking, add 1.0 mL of a matrix spiking standard. For base/neutral acid analysis, the amount of the surrogates and matrix spiking compounds added to the sample should result in a final concentration of 100 ng/μL of each analyte in the extract to be analyzed (assuming a 1-μL injection). Extract with methylene chloride for 18–24 h. Next, adjust the pH of the aqueous phase to pH >11 using 10 N sodium hydroxide and extract it with methylene chloride again for 18–24 h. Dry the extract through a column containing anhydrous sodium sulfate and concentrate it to 1 mL using a K-D concentrator.

Soils, sediments, or sludges — Use 30 g of sample. Nonporous or wet samples (gummy or clay type) that do not have a free-flowing sandy texture must be mixed with anhydrous sodium sulfate until the sample is free flowing. Add 1 mL of surrogate standards to all samples, spikes, standards, and blanks. For the sample in each analytical batch selected for spiking, add 1.0 mL of a matrix spiking standard. For base/neutral acid analysis, the amount added of the surrogates and matrix spiking compounds should result in a final concentration of 100 ng/μL of each base/neutral analyte and 200 ng/μL of each acid analyte in the extract to be analyzed (assuming a 1-μL injection). Immediately add a 100-mL mixture of 1:1 methylene chloride:acetone and extract the sample ultrasonically for 3 min and then decant or filter the extracts. Repeat the extraction two or more times. Dry the extract using a column with anhydrous sodium sulfate and concentrate it to 1 mL in a K-D concentrator.

QUALITY CONTROL A methylene chloride solution containing 50 ng/μL of decafluorotriphenylphosphine (DFTPP) is used for tuning the GC/MS system each 12-h shift. A system performance check also must be made during every 12-h shift. A standard containing 50 ng/μL each of 4,4′-DDT, pentachlorophenol, and benzidine is required to verify injection port inertness and GC column performance. A calibration standard at mid-concentration, containing each compound of interest, including all required surrogates, must be performed every 12 h during analysis. After the system performance check is met, calibration check compounds (CCCs) are used to check the validity of the initial calibration.

The internal standard responses and retention times in the calibration check standard must be evaluated immediately after or during data acquisition. If the retention time for any internal standard changes by more than 30 seconds from the last check calibration (12 h), the chromatographic system must be inspected for malfunctions and corrections must be made, as required. If the electron ionization current plot (EICP) area for any of the internal standards changes by a factor of two from the last daily calibration standard check, the mass spectrometer must be inspected for malfunctions and corrections must be made, as appropriate.

Demonstrate, through the analysis of a reagent water blank, that interferences from the analytical system, glassware, and reagents are under control. The blank samples should be carried through all stages of the sample preparation and measurement steps. For each analytical batch (up to 20 samples), a reagent blank, matrix spike, and matrix spike duplicate/duplicate must be analyzed (the frequency of the spikes may be different for different monitoring programs). The blank and spiked samples must be carried through all stages of the sample preparation and measurement steps. A QC reference sample concentrate containing each analyte at a concentration of 100 mg/L in methanol is required.

REFERENCE Test Methods for Evaluating Solid Waste (SW-846). U.S. EPA 1983, Method 8270B, Rev. 2, Nov. 1990. Office of Solid Waste, Washington, DC.

Pyrene
CAS #129-00-0

EPA Method 8100

TITLE Polynuclear Aromatic Hydrocarbons

MATRIX Groundwater, soils, sludges, water miscible liquid wastes, and non-water miscible wastes.

APPLICATION This method is used for the analysis of various PAHs. Samples are extracted, concentrated, and analyzed using direct injection of both neat and diluted organic liquids. The method provides two optional GC columns that are better than Column 1 and that may help resolve analytes from interferences.

INTERFERENCES Solvents, reagents, and glassware may introduce artifacts. Other interferences may come from coextracted compounds from samples.

INSTRUMENTATION GC capable of on-column injections and a flame with detector (FID). Column 1: a 1.8 m by 2 mm 3% OV-17 on Chromosorb W-AW-DCMS column. Column 2: a 30 m by 0.25 mm SE-54 fused silica capillary colunm. Column 3: a 30 m by 0.32 mm SE-54 fused silica capillary column.

RANGE 0.1–425 µg/L

MDL Not reported.

PQL FACTORS FOR MULTIPLYING × FID MDL VALUE Not available.

PRECISION 0.42X–0.00 µg/L (overall precision).

ACCURACY 0.69C–0.12 µg/L (as recovery).

SAMPLING METHOD Use 8-oz. widemouth glass bottles with Teflon®-lined caps for concentrated waste samples, soils, sediments, and sludges. Use 1 or 2½ gallon amber glass bottles with Teflon®-lined caps for liquid (water) samples.

STABILITY Cool soil, sediment, sludge, and liquid samples to 4°C. If residual chlorine is present in liquid samples add 3 mL of 10% sodium thiosulfate per gallon of sample and cool to 4°C.

MHT 14 days for concentrated waste, soil, sediment, or sludge; 7 days for liquid samples; all extracts must be analyzed within 40 days.

QUALITY CONTROL A quality control check sample concentrate containing each analyte of interest is required. The QC check sample concentrate may be prepared from pure standard materials or purchased as certified solutions Use appropriate trip, matrix, control site, method, reagent, and solvent blanks. Internal, surrogate, and five concentration level calibration standards are used. The quality control check sample concentrate should contain pyrene at 10 µg/mL in acetonitrile.

REFERENCE Test Methods for Evaluating Solid Waste (SW-846), U.S. EPA Office of Solid Waste, Washington, DC, Method 8100, Nov. 1986.

Pyrene
CAS #129-00-0

EPA Method 8310

TITLE Polynuclear Aromatic Hydrocarbons

MATRIX Groundwater, soils, sludges, water miscible liquid wastes, and non-water miscible wastes.

APPLICATION This method is used for the analysis of 16 polynuclear aromatic hydrocarbons (PAHs). Samples are extracted, concentrated, and analyzed using HPLC with detection by UV and fluorescence detectors.

INTERFERENCES Solvents, reagents, and glassware may introduce artifacts. Other interferences may come from coextracted compounds from samples.

INSTRUMENTATION HPLC with a gradient pumping system and a 250 mm by 2.6 mm reverse phase HC-ODS Sil-X 5-micron particle-size column. The fluorescence detector uses an excitation wavelength of 280 nm and emission greater than 389 nm cutoff with dispersive optics.

RANGE 0.1–425 µg/L

MDL 0.27 µg/L (fluorescence; reagent water).

PQL FACTORS FOR MULTIPLYING × FID MDL VALUE

Matrix	Multiplication Factor
Groundwater	10
Low-level soil by sonication with GPC cleanup	670
High-level soil and sludge by sonication	10,000
Non-water miscible waste	100,000

PRECISION 0.42X–0.00 µg/L (overall precision).

ACCURACY 0.69C–0.12 µg/L (as recovery).

SAMPLING METHOD Use 8-oz. widemouth glass bottles with Teflon®-lined caps for concentrated waste samples, soils, sediments, and sludges. Use 1 or 2½ gallon amber glass bottles with Teflon®-lined caps for liquid (water) samples.

STABILITY Cool soil, sediment, sludge, and liquid samples to 4°C. If residual chlorine is present in liquid samples add 3 mL of 10% sodium thiosulfate per gallon of sample and cool to 4°C.

MHT 14 days for concentrated waste, soil, sediment, or sludge; 7 days for liquid samples; all extracts must be analyzed within 40 days.

QUALITY CONTROL Internal, surrogate, and five concentration level calibration standards are used. The calibration standards must be used with the analytical method blank. A quality control check sample concentrate containing pyrene at 10 µg/mL is required. The QC check sample concentrate may be prepared from pure standard materials or purchased as certified solutions. Use appropriate trip, matrix, control site, method, reagent, and solvent blanks.

REFERENCE Test Methods for Evaluating Solid Waste (SW-846), U.S. EPA Office of Solid Waste, Washington, DC, Method 8310, Rev. 0, Nov. 1986.

Pyridine
CAS #110-86-1
EPA Method 1625

TITLE Semivolatile Organic Compounds by Isotope Dilution GC/MS

MATRIX The compounds may be determined in waters, soils, and municipal sludges by this method.

METHOD SUMMARY This method is used to determine 176 semivolatile toxic organic pollutants associated with the CWA (as amended 1987); the RCRA (as amended 1986); the CERCLA (as amended 1986); and other compounds amenable to extraction and analysis by capillary column gas chromatography-mass spectrometry (GC/MS).

Stable isotopically-labeled analogs of the compounds of interest are added to the sample. If the solids content is less than 1%, a 1-L sample is extracted at pH 12–13, then at pH <2 with methylene chloride using continuous extraction techniques.

If the solids content is 30% or less, the sample is diluted to 1% solids with reagent water, homogenized ultrasonically, and extracted at pH 12–13, then at pH <2 with methylene chloride using continuous extraction techniques. If the solids content is greater than 30%, the sample is extracted using ultrasonic techniques.

Each extract is dried over sodium sulfate, concentrated to a volume of 5 mL, cleaned up using GPC, if necessary, and concentrated. Extracts are concentrated to 1 mL if GPC is not performed, and to 0.5 mL if GPC is performed.

An internal standard is added to the extract, and a 1-mL aliquot of the extract is injected into the GC. The compounds are separated by GC and detected by a MS. The labeled compounds serve to correct the variability of the analytical technique.

INTERFERENCES Solvents, reagents, glassware, and other sample processing hardware may yield artifacts and/or elevated baselines causing misinterpretation of chromatograms and spectra. Materials used in the analysis must be demonstrated to be free from interferences under the conditions of analysis by running method blanks initially and with each sample lot (sample started through the extraction process on a given 8-h shift, to a maximum of 20). Specific selection of reagents and purification of solvents by distillation in all glass systems may be required. Glassware and, where possible, reagents are cleaned by solvent rinse and baking at 450°C for 1-h minimum. Interferences coextracted from samples will vary considerably from source to source, depending on the diversity of the site being sampled.

INSTRUMENTATION Major instrumentation includes a GC with a splitless or on-column injection port for capillary column, a MS with 70 eV electron impact ionization, and a data system to collect and record MS data, and process it. A K-D apparatus is used to concentrate extracts.

GC Column: 30 m × 0.25 mm I.D. 5% phenyl, 94% methyl, 1% vinyl silicone bonded phased fused silica capillary column.

PRECISION & ACCURACY The detection limits of the method are usually dependent on the level of interferences rather than instrumental limitations. The limits typify the minimum quantities that can be detected with no interferences present.

The minimum level (in µg/mL) was not listed. This is defined as a minimum level at which the analytical system shall give recognizable mass spectra (background corrected) and acceptable calibration points.

The MDL (in µg/kg) in low solids was not listed and in high solids was not listed; these were determined in digested sludge (low solids) and in filter cake or compost (high solids).

The labeled and native compound initial precision as standard deviation (in µg/L) was not listed.
The labeled and native compound initial accuracy as average recovery (in µg/L) was not listed.

SAMPLE COLLECTION, PRESERVATION & HANDLING Collect samples in glass containers. Aqueous samples which flow freely are collected in refrigerated bottles using automatic sampling equipment. Solid samples are collected as grab samples using widemouth jars. Maintain samples at 0 to 4°C from the time of collection until extraction. If residual chlorine is present in aqueous samples, add 80 mg sodium thiosulfate/L of water. Begin sample extraction within 7 days of collection, and analyze all extracts within 40 days of extraction.

SAMPLE PREPARATION Samples containing 1% solids or less are extracted directly using continuous liquid-liquid extraction techniques. Samples containing 1 to 30% solids are diluted to the 1% level with reagent water and extracted using continuous liquid-liquid extraction techniques. Samples containing greater than 30% solids are extracted using ultrasonic techniques.

- Base/neutral extraction — Adjust the pH of the waters in the extractors to 12–13 with 6 N NaOH. Extract with methylene chloride for 24–48 h.
- Acid extraction — Adjust the pH of the waters in the extractors to 2 or less using 6 N sulfuric acid. Extract with methylene chloride for 24–48 h.
- Ultrasonic extraction of high solids samples — Add anhydrous sodium sulfate to the sample and QC aliquot(s). Add acetone:methylene chloride (1:1) to the sample and mix thoroughly

Concentrate extracts using a K-D apparatus.

QUALITY CONTROL The analyst is permitted to modify this method to improve separations or lower the costs of measurements, provided all performance specifications are met. Analyses of blanks are required to demonstrate freedom from contamination. When results of spikes indicate atypical method performance for samples, the samples are diluted to bring method performance within acceptable limits.

For low solids (aqueous samples), extract, concentrate, and analyze two sets of four 1-L aliquots (8 aliquots total) of the precision and recovery standard. For high solids samples, two sets of four 30-g aliquots of the high solids reference matrix are used.

Spike all samples with labeled compounds to assess method performance. Compute percent recovery of the labeled compounds using the internal standard method. Compare the labeled compound recovery for each compound with the corresponding labeled compound recovery.

Reagent water and high solids reference matrix blanks are analyzed to demonstrate freedom from contamination. Extract and concentrate a 1-L reagent water blank or a high solids reference matrix blank with each sample's lot (samples started through the extraction process on the same 8-h shift, to a maximum of 20 samples).

Field replicates may be collected to determine the precision of the sampling technique, and spiked samples may be required to determine the accuracy of the analysis when the internal standard method is used.

REFERENCE Semivolatile Organic Compounds by Isotope Dilution GC/MS. Office of Water Regulation and Standards, U.S. EPA Industrial Technology Division, Washington, DC, EPA Method 1625, Rev. C, June 1989 (contact W.A. Telliard, U.S. EPA, Office of Water Regulations and Standards, 401 M St., SW, Washington, DC, 20460. Phone: 202-382-7131).

Pyridine **EPA Method 1625**
CAS #110-86-1

TITLE Semivolatile Organic Compounds by Isotope Dilution GC/MS

MATRIX The compounds may be determined in waters, soils, and municipal sludges by this method.

METHOD SUMMARY This method is used to determine 176 semivolatile toxic organic pollutants associated with the CWA (as amended 1987); the RCRA (as amended 1986); the CERCLA (as amended 1986); and other compounds amenable to extraction and analysis by capillary column gas chromatography-mass spectrometry (GC/MS).

Stable isotopically-labeled analogs of the compounds of interest are added to the sample. If the solids content is less than 1%, a 1-L sample is extracted at pH 12–13, then at pH <2 with methylene chloride using continuous extraction techniques.

If the solids content is 30% or less, the sample is diluted to 1% solids with reagent water, homogenized ultrasonically, and extracted at pH 12–13, then at pH <2 with methylene chloride using continuous extraction techniques. If the solids content is greater than 30%, the sample is extracted using ultrasonic techniques.

Each extract is dried over sodium sulfate, concentrated to a volume of 5 mL, cleaned up using GPC, if necessary, and concentrated. Extracts are concentrated to 1 mL if GPC is not performed, and to 0.5 mL if GPC is performed.

An internal standard is added to the extract, and a 1-mL aliquot of the extract is injected into the GC. The compounds are separated by GC and detected by a MS. The labeled compounds serve to correct the variability of the analytical technique.

INTERFERENCES Solvents, reagents, glassware, and other sample processing hardware may yield artifacts and/or elevated baselines causing misinterpretation of chromatograms and spectra. Materials used in the analysis must be demonstrated to be free from interferences under the conditions of analysis by running method blanks initially and with each sample lot (sample started through the extraction process on a given 8-h shift, to a maximum of 20). Specific selection of reagents and purification of solvents by distillation in all glass systems may be required. Glassware and, where possible, reagents are cleaned by solvent rinse and baking at 450°C for 1-h minimum. Interferences coextracted from samples will vary considerably from source to source, depending on the diversity of the site being sampled.

INSTRUMENTATION Major instrumentation includes a GC with a splitless or on-column injection port for capillary column, a MS with 70 eV electron impact ionization, and a data system to collect and record MS data, and process it. A K-D apparatus is used to concentrate extracts.

GC Column: 30 m × 0.25 mm I.D. 5% phenyl, 94% methyl, 1% vinyl silicone bonded phased fused silica capillary column.

PRECISION & ACCURACY The detection limits of the method are usually dependent on the level of interferences rather than instrumental limitations. The limits typify the minimum quantities that can be detected with no interferences present.

The minimum level (in µg/mL) was not listed. This is defined as a minimum level at which the analytical system shall give recognizable mass spectra (background corrected) and acceptable calibration points.

The MDL (in µg/kg) in low solids was not listed and in high solids was not listed; these were determined in digested sludge (low solids) and in filter cake or compost (high solids).

The labeled and native compound initial precision as standard deviation (in µg/L) was not listed.
The labeled and native compound initial accuracy as average recovery (in µg/L) was not listed.

SAMPLE COLLECTION, PRESERVATION & HANDLING Collect samples in glass containers. Aqueous samples which flow freely are collected in refrigerated bottles using automatic sampling equipment. Solid samples are collected as grab samples using widemouth jars. Maintain samples at 0 to 4°C from the time of collection until extraction. If residual chlorine is present in aqueous samples, add 80 mg sodium thiosulfate/L of water. Begin sample extraction within 7 days of collection, and analyze all extracts within 40 days of extraction.

SAMPLE PREPARATION Samples containing 1% solids or less are extracted directly using continuous liquid-liquid extraction techniques. Samples containing 1 to 30% solids are diluted to the 1% level with reagent water and extracted using continuous liquid-liquid extraction techniques. Samples containing greater than 30% solids are extracted using ultrasonic techniques.

Base/neutral extraction — Adjust the pH of the waters in the extractors to 12–13 with 6 N NaOH. Extract with methylene chloride for 24–48 h.

Acid extraction — Adjust the pH of the waters in the extractors to 2 or less using 6 N sulfuric acid. Extract with methylene chloride for 24–48 h.

Ultrasonic extraction of high solids samples — Add anhydrous sodium sulfate to the sample and QC aliquot(s). Add acetone:methylene chloride (1:1) to the sample and mix thoroughly

Concentrate extracts using a K-D apparatus.

QUALITY CONTROL The analyst is permitted to modify this method to improve separations or lower the costs of measurements, provided all performance specifications are met. Analyses of blanks are required to demonstrate freedom from contamination. When results of spikes indicate atypical method performance for samples, the samples are diluted to bring method performance within acceptable limits.

For low solids (aqueous samples), extract, concentrate, and analyze two sets of four 1-L aliquots (8 aliquots total) of the precision and recovery standard. For high solids samples, two sets of four 30-g aliquots of the high solids reference matrix are used.

Spike all samples with labeled compounds to assess method performance. Compute percent recovery of the labeled compounds using the internal standard method. Compare the labeled compound recovery for each compound with the corresponding labeled compound recovery.

Reagent water and high solids reference matrix blanks are analyzed to demonstrate freedom from contamination. Extract and concentrate a 1-L reagent water blank or a high solids reference matrix blank with each sample's lot (samples started through the extraction process on the same 8-h shift, to a maximum of 20 samples).

Field replicates may be collected to determine the precision of the sampling technique, and spiked samples may be required to determine the accuracy of the analysis when the internal standard method is used.

REFERENCE Semivolatile Organic Compounds by Isotope Dilution GC/MS. Office of Water Regulation and Standards, U.S. EPA Industrial Technology Division, Washington, DC, EPA Method 1625, Rev. C, June 1989 (contact W.A. Telliard, U.S. EPA, Office of Water Regulations and Standards, 401 M St., SW, Washington, DC, 20460. Phone: 202-382-7131).

Pyridine   EPA Method 8240
CAS #110-86-1

TITLE Volatile Organics By GC/MS: Packed Column Technique

MATRIX Nearly all types of sample matarices, regardless of water content, can be analyzed using this method. This includes groundwater, aqueous sludges, caustic liquors, acid liquors, waste solvents, oily wastes, mousses, tars, fibrous wastes, polymetric emulsions, filter cakes, spent carbons, spent catalysts, soils, and sediments.

METHOD SUMMARY Method 8240B covers 80 volatile organic compounds that are introduced into a gas chromatograph by the purge-and-trap method or by direct injection (in limited applications). For the purge-and-trap method an inert gas (zero grade nitrogen or helium) is bubbled through a 5-mL solution at ambient temperature. Purged sample components are trapped in a tube of sorbent materials. When purging is complete, the sorbent tube is heated and backflushed with inert gas to desorb the trapped components onto a GC column.

INTERFERENCES Impurities in the purge gas and from organic compounds outgassing from the plumbing ahead of the trap account for many contamination problems. Interferences purged or coextracted from the samples will vary considerably from source to source. Cross-contamination can occur whenever high-level and low-level samples are analyzed sequentially. Whenever an unusually concentrated sample is analyzed, it should be followed by the analysis of organic-free reagent water to check for cross-contamination. Samples also can be contaminated by diffusion of volatile organics (particularly methylene chloride and fluorocarbons) through the septum seal into the sample during shipment and storage. A trip blank can serve as a check on such contamination. The lab where volatile analysis is performed and also the refrigerated storage area should be completely free of solvents.

INSTRUMENTATION A gas chromatograph/mass spectrometry/data system (GC/MS) equipped with a 6 ft × 0.1 in I.D. glass column packed with 1% SP-1000 on Carbopack-B (60/80 mesh) is required. Also needed is a 5-mL purging device, a sorbent trap, and a thermal desorption apparatus.

PRECISION & ACCURACY This method is reported to have been tested by 15 laboratories using organic-free reagent water, drinking water, surface water, and industrial wastewaters (not specified) fortified at six concentrations over the range 5–600 µg/L.

Sample estimated quantitation limits (EQLs) are highly matrix-dependent. The EQLs listed may not always be achievable. EQLs listed for soils or sediments are based on wet weight. Normally, data is reported on a dry-weight basis; therefore, EQLs will be higher, based on the percent dry weight of each sample. Note that EQLs are even more variable than MDLs and that they are highly variable depending on the matrix being analyzed.

EQL in groundwater in µg/L was not listed.
EQL in low soil or sediment in µg/kg was not listed.
Accuracy (a) in µg/L was not listed.
Precision (b) in µg/L was not listed.

(a) *Average recovery found for measurements of samples containing a concentration of C, in µg/L.*
(b) *Overall precision found for measurements of samples with average recovery X for samples containing a concentration of C in µg/L.*
X = *Average recovery found for measurement of samples containing a concentration of C in µg/L.*

MULTIPLICATION FACTORS FOR OTHER MATRICES

Other Matrices	Factor (a)
Waste miscible liquid waste	50
High-concentration soil and sludge	125
Non-water miscible waste	500

(a) EQL = [EQL for low soil/sediment] × [Factor]. For non-aqueous samples, the factor is on a wet-weight basis.

SAMPLING METHOD

Liquid samples — Use a 40-mL glass screw-cap VOA vial with a Teflon®-faced silicone septum that has been prewashed, rinsed with distilled deionized water, and oven dried. However, if residual chlorine is present, collect sample in a 40-oz. soil VOA container which has been pre-preserved with 4 drops of 10% sodium thiosulfate, mix gently, and then transfer the sample to a 40-mL VOA vial. Collect bubble-free samples in duplicate and seal them in separate plastic bags.

Soils or sediments, and sludges — Use an 8-oz. widemouth glass bottle with a Teflon®-faced silicone septum that has been prewashed with detergent, rinsed with distilled deionized water, and oven dried. Tap slightly to eliminate free air space. Collect samples in duplicate and seal them in separate plastic bags.

SAMPLE PRESERVATION

Liquid samples — Add 4 drops of concentrated HCL and immediately cool samples to 4°C and store in a solvent-free refrigerator.

Soils or sediments, and sludges — Cool samples to 4°C and store in a solvent-free refrigerator.

MHT Maximum holding time is 14 days from the date of sample collection.

SAMPLE PREPARATION

Liquid samples — Remove the plunger from a 5-mL syringe and carefully pour the sample into the syringe barrel to just short of overflowing. Replace the syringe plunger and compress the sample. Open the syringe valve and vent any residual air while adjusting the sample volume to 5.0 mL. If there is only one volatile organic analysis (VOA) vial, a second syringe should be filled at this time to protect against possible loss of sample integrity. Add 10 µL of surrogate spiking solution and 10 µL of internal standard spiking solution through the valve bore of the 5-mL syringe, then close the valve. The surrogate and internal standards may be mixed and added as a single spiking solution.

Sediments, soils, and waste samples — All samples of this type should be screened by GC analysis using a headspace method (EPA Method 3810) or the hexadecane extraction and screening method (EPA Method 3820). Use the screening data to determine whether to use the low-concentration method (0.005–1 mg/kg) or the high-concentration method (>1 mg/kg).

Low-concentration method — The low-concentration method is based on purging a heated sediment or soil sample mixed with organic-free reagent water containing the surrogate and internal standards. Analyze all reagent blanks and standards under the same conditions as the samples.

Use a 5-g sample if the expected concentration is <0.1 mg/kg or a 1-g sample for expected concentrations between 0.1 and 1 mg/kg. Mix the contents of the sample container with a narrow metal spatula. Weigh the amount of the sample into a tared purge device. Add the spiked water to the purge device, which contains the weighed amount of sample, and connect the device to the purge-and-trap system.

High-concentration method — This method is based on extracting the sediment or soil with methanol. A waste sample is either extracted or diluted, depending on its solubility in methanol. Wastes that are insoluble in methanol are diluted with reagent tetraglyme or possibly polyethylene glycol (PEG). An aliquot of the extract is added to organic-free reagent water containing surrogate and internal standards. This is purged at ambient temperature. All samples with an expected concentration of >1.0 mg/kg should be analyzed by this method.

Mix the contents of the sample container with a narrow metal spatula. For sediments or soils and solid wastes that are insoluble in methanol, weigh 4 g (wet weight) of sample into a tared 20-mL vial. For waste that is soluble in methanol, tetraglyme, or PEG, weigh 1 g (wet weight) into a tared scintillation vial or culture tube or a 10-mL volumetric flask. Quickly add 9.0 mL of appropriate solvent then add 1.0 mL of a surrogate spiking solution to the vial, cap it, and shake it for 2 min.

METHANOL EXTRACT REQUIRED FOR ANALYSIS OF HIGH-CONCENTRATION SOILS OR SEDIMENTS

Approximate Concentration Range	Volume of Methanol Extract (a)
500–10,000 µg/kg	100 µL
1,000–20,000 µg/kg	50 µL
5,000–100,000 µg/kg	10 µL
25,000–500,000 µg/kg	100 µL of 1/50 dilution (b)

Calculate appropriate dilution factor for concentrations exceeding this table.

(a) The volume of methanol added to 5 mL of water being purged should be kept constant. Therefore, add to the 5-mL syringe whatever volume of methanol is necessary to maintain a volume of 100 µL added to the syringe.

(b) Dilute an aliquot of the methanol extract and then take 100 µL for analysis.

QUALITY CONTROL Demonstrate, through the analysis of a reagent water blank, that interferences from the analytical system, glassware, and reagents are under control. Blank samples should be carried through all stages of the sample preparation and measurement steps. For each analytical batch (up to 20 samples), a reagent blank, matrix spike, and matrix spike duplicate must be analyzed (the frequency of the spikes may be different for different monitoring programs). The blank and spiked samples must be carried through all stages of the sample preparation and measurement steps. QC samples mentioned in the section on Interferences will also be needed as appropriate to those situations.

REFERENCE Test Methods for Evaluating Solid Waste (SW-846). U.S. EPA. 1983. Method 8240B, Rev. 2, Nov. 1990. Office of Solid Wastes, Washington, DC.

Pyridine
CAS #110-86-1
EPA Method 8270

TITLE Semivolatile Organic Compounds by GC/MS

MATRIX This method is used to determine the concentration of semivolatile organic compounds in extracts prepared from all types of solid waste matrices, soils, and groundwater. Although surface waters are not specifically mentioned, this method should be applicable to water samples from rivers, lakes, etc.

METHOD SUMMARY This method covers 259 semivolatile organic compounds. In very limited applications direct injection of the sample into the GC/MS system may be appropriate, but this results in very high detection limits (approximately 10,000 µg/L). Typically, a 1-L liquid sample, containing surrogate, and matrix spiking standards, is extracted in a continuous extractor first under acid conditions and then under basic conditions. Typically 30 g of a solid sample, containing surrogate, and matrix spiking standards, is extracted ultrasonically. After concentrating the extract to 1 mL it is spiked with 10 µL of an internal standard solution just prior to analysis by GC/MS. The volume injected should contain about 100 ng of base/neutral and 200 ng of acid surrogates (for a 1-µL injection). Analysis is performed by GC/MS using a capillary GC column.

INTERFERENCES Raw GC/MS data from all blanks, samples, and spikes must be evaluated for interferences. Contamination by carryover can occur whenever high-concentration and low-concentration samples are sequentially analyzed. To reduce carryover, the sample syringe must be rinsed out between samples with solvent. Whenever an unusually concentrated sample is encountered, it should be followed by the analysis of blank solvent to check for cross-contamination.

INSTRUMENTATION A GC/MS and a data system are required. The GC column used is a 30 m × 0.25 mm I.D. (or 0.32 mm I.D.) 1um film thickness silicone-coated fused silica capillary column. A continuous liquid-liquid extractor equipped with Teflon® or glass connection joints and stopcocks requiring no lubrication, a K-D concentrating apparatus, water bath, and an ultrasonic disrupter with a minimum power of 300 W and with pulsing capability are also required.

PRECISION & ACCURACY The estimated quantitation limit (EQL) of Method 8270B for determining an individual compound is approximately 1 mg/kg (wet weight) for soil or sediment samples, 1–200 mg/kg for wastes (dependent on matrix and method of preparation), and 10 µg/L for groundwater samples. EQLs will be proportionately higher for sample extracts that require dilution to avoid saturation of the detector.

The EQL(b) for groundwater in µg/L is not determined.
The EQL (a, b) for low concentrations in soil and sediment in µg/kg is not determined.
Accuracy as µg/L is not listed.
Overall precision in µg/L is not listed.

(a) *EQLs listed for soil/sediment are based on wet weight. Normally data is reported in a dry-weight basis; therefore, EQLs will be higher based on the % dry weight of each sample. This calculation is based on a 30 g sample and gel permeation chromatography cleanup.*
(b) *Sample EQLs are highly matrix-dependent. The EQLs are provided for guidance and may not always be achievable.*
$C =$ *True value for concentration, in µg/L.*
$X =$ *Average recovery found for measurements of samples containing a concentration of C, in µg/L.*

ESTIMATED QUANTITATION LIMIT

Other Matrices	Factor (a)
High-concentration soil and sludges by sonicator	7.5
Non-water miscible waste	75

(a) *EQL for other matrices = [EQL for low soil/sediment] × [Factor]. This estimated EQL is similar to an EPA "Practical Quantitation Limit."*

SAMPLING METHOD
Liquid samples — Use a 1 or 2½ gallon amber glass bottle with a screw-top Teflon®-lined cover that has been prewashed with detergent and rinsed with distilled water and methanol (or isopropanol).

Soils, sediments, or sludges — Use an 8-oz. widemouth glass with a screw-top Teflon®-lined cover that has been prewashed with detergent and rinsed with distilled water and methanol (or isopropanol).

SAMPLE PRESERVATION
Liquid samples — If residual chlorine is present, add 3 mL of 10% sodium thiosulfate per gallon, cool to 4°C and store in a solvent-free refrigerator until analysis; if chlorine is not present, then eliminate the sodium thiosulfate addition.

Soils, sediments, or sludges — Cool samples to 4°C and store in a solvent-free refrigerator.

MHT Liquid samples must be extracted within 7 days and the extracts analyzed within 40 days. Soils, sediments, or sludges may be stored for a maximum of 14 days and the extracts analyzed within 40 days.

SAMPLE PREPARATION
Liquid samples — Transfer 1 L quantitatively to a continuous extractor. If high concentrations are anticipated, a smaller volume may be used and then diluted with organic-free reagent water to 1 L. Adjust pH, if necessary, to pH <2 using 1:1 (V/V) sulfuric acid. Pipette 1.0 mL of a surrogate standard spiking solution into each sample. For the sample in each analytical batch selected for spiking, add 1.0 mL of a matrix spiking standard. For base/neutral acid analysis, the amount of the surrogates and matrix spiking compounds added to the sample should result in a final concentration of 100 ng/µL of each analyte in the extract to be analyzed (assuming a 1-µL injection). Extract with methylene chloride for 18–24 h. Next, adjust the pH of the aqueous phase to pH >11 using 10 N sodium hydroxide and extract it with methylene chloride again for 18–24 h. Dry the extract through a column containing anhydrous sodium sulfate and concentrate it to 1 mL using a K-D concentrator.

Soils, sediments, or sludges — Use 30 g of sample. Nonporous or wet samples (gummy or clay type) that do not have a free-flowing

sandy texture must be mixed with anhydrous sodium sulfate until the sample is free flowing. Add 1 mL of surrogate standards to all samples, spikes, standards, and blanks. For the sample in each analytical batch selected for spiking, add 1.0 mL of a matrix spiking standard. For base/neutral acid analysis, the amount added of the surrogates and matrix spiking compounds should result in a final concentration of 100 ng/µL of each base/neutral analyte and 200 ng/µL of each acid analyte in the extract to be analyzed (assuming a 1-µL injection). Immediately add a 100-mL mixture of 1:1 methylene chloride:acetone and extract the sample ultrasonically for 3 min and then decant or filter the extracts. Repeat the extraction two or more times. Dry the extract using a column with anhydrous sodium sulfate and concentrate it to 1 mL in a K-D concentrator.

QUALITY CONTROL A methylene chloride solution containing 50 ng/µL of decafluorotriphenylphosphine (DFTPP) is used for tuning the GC/MS system each 12-h shift. A system performance check also must be made during every 12-h shift. A standard containing 50 ng/µL each of 4,4′-DDT, pentachlorophenol, and benzidine is required to verify injection port inertness and GC column performance. A calibration standard at mid-concentration, containing each compound of interest, including all required surrogates, must be performed every 12 h during analysis. After the system performance check is met, calibration check compounds (CCCs) are used to check the validity of the initial calibration.

The internal standard responses and retention times in the calibration check standard must be evaluated immediately after or during data acquisition. If the retention time for any internal standard changes by more than 30 seconds from the last check calibration (12 h), the chromatographic system must be inspected for malfunctions and corrections must be made, as required. If the electron ionization current plot (EICP) area for any of the internal standards changes by a factor of two from the last daily calibration standard check, the mass spectrometer must be inspected for malfunctions and corrections must be made, as appropriate.

Demonstrate, through the analysis of a reagent water blank, that interferences from the analytical system, glassware, and reagents are under control. The blank samples should be carried through all stages of the sample preparation and measurement steps. For each analytical batch (up to 20 samples), a reagent blank, matrix spike, and matrix spike duplicate/duplicate must be analyzed (the frequency of the spikes may be different for different monitoring programs). The blank and spiked samples must be carried through all stages of the sample preparation and measurement steps. A QC reference sample concentrate containing each analyte at a concentration of 100 mg/L in methanol is required.

REFERENCE Test Methods for Evaluating Solid Waste (SW-846). U.S. EPA 1983, Method 8270B, Rev. 2, Nov. 1990. Office of Solid Waste, Washington, DC.

R

Residue, Filterable (TDS) — EPA Method 160.1

TITLE Physical Properties

MATRIX Drinking, surface, and saline waters. Wastewater.

APPLICATION Date issued 1971. Also referred to as total dissolved solids. A well mixed sample is filtered through a standard glass fiber filter. The filtrate is evaporated and dried to constant weight at 180°C. The increase in dish weight represents the total dissolved solids. (The filtrate from residue, non-filterable may be used).

INTERFERENCES Highly mineralized waters with considerable calcium, magnesium, chloride, and/or sulfate content may be hygroscopic and will require prolonged drying, desiccation and rapid weighing. Samples with high concentrations of bicarbonates require prolonged drying.

INSTRUMENTATION Glass fiber filter discs (Reeves Angel 934-AH, or equiv). Drying oven @180c.

RANGE 10–20,000 mg/L.

MDL Not listed.

PRECISION Not listed.

ACCURACY Not listed.

SAMPLING METHOD Plastic or glass. (100 mL).

STABILITY Cool, 4°C.

MHT 48 h.

QUALITY CONTROL Too much residue in evaporating dish will crust over and entrap water that will not be driven off during drying. Limit total residue to 200 mg.

REFERENCE EPA Methods for the Chemical Analysis of Water and Wastes, EPA-600/4-79-020, U.S. EPA, EMSL, 1979.

Residue, Volatile (VSS) and (VS) — EPA Method 160.4

TITLE Physical Properties

MATRIX Sewage, sludge, waste, and sediments.

APPLICATION Date issued 1971. Also referred to as volatile suspended solids and soluble solids. Residue from determination of total (a), filterable (b), or non-filterable (c) residue is ignited in a muffle furnace. (The loss of wt on ignition is reported as volatile solids). VSS = ignited solids from (c). VS = ignited solids from (a).

INTERFERENCES Big source of error is failure to obtain representative sample. The test is subject to errors due to loss of water of crystall, loss of volatile organic matter prior to combustion, incomplete oxidation of certain complex organics and decomposition of mineral salts.

INSTRUMENTATION Muffle furnace at 550°C

RANGE Not listed.

MDL Not listed.

PRECISION SD = ±11 mg/L at 170 mg/L volatile residue concentration.

ACCURACY Not listed.

SAMPLING METHOD Plastic or glass. (100 mL).

STABILITY Cool, 4°C.

MHT 7 days.

QUALITY CONTROL Ignite residue from (a), (b), or (c) (see applicability) to constant weight in a muffle furnace at 550 c. Usually 15 to 20 min ignition is required. Cool and weigh. Repeat cycle of igniting, cooling, desiccating and weighing to constant weight.

REFERENCE Methods for the Chemical Analysis of Water and Wastes, EPA-600/4-79-020, U.S. EPA, EMSL, 1979.

Resorcinol — EPA Method 8270
CAS #108-46-3

TITLE Semivolatile Organic Compounds by GC/MS

MATRIX This method is used to determine the concentration of semivolatile organic compounds in extracts prepared from all types of solid waste matrices, soils, and groundwater. Although surface waters are not specifically mentioned, this method should be applicable to water samples from rivers, lakes, etc.

METHOD SUMMARY This method covers 259 semivolatile organic compounds. In very limited applications direct injection of the sample into the GC/MS system may be appropriate, but this results in very high detection limits (approximately 10,000 µg/L). Typically, a 1-L liquid sample, containing surrogate, and matrix spiking standards, is extracted in a continuous extractor first under acid conditions and then under basic conditions. Typically 30 g of a solid sample, containing surrogate, and matrix spiking standards, is extracted ultrasonically. After concentrating the extract to 1 mL it is spiked with 10 µL of an internal standard solution just prior to analysis by GC/MS. The volume injected should contain about 100 ng of base/neutral and 200 ng of acid surrogates (for a 1-µL injection). Analysis is performed by GC/MS using a capillary GC column.

INTERFERENCES Raw GC/MS data from all blanks, samples, and spikes must be evaluated for interferences. Contamination by carryover can occur whenever high-concentration and low-concentration samples are sequentially analyzed. To reduce carryover, the sample syringe must be rinsed out between samples with solvent. Whenever an unusually concentrated sample is encountered, it should be followed by the analysis of blank solvent to check for cross-contamination.

INSTRUMENTATION A GC/MS and a data system are required. The GC column used is a 30 m × 0.25 mm I.D. (or 0.32 mm I.D.) 1um film thickness silicone-coated fused silica

capillary column. A continuous liquid-liquid extractor equipped with Teflon® or glass connection joints and stopcocks requiring no lubrication, a K-D concentrating apparatus, water bath, and an ultrasonic disrupter with a minimum power of 300 W and with pulsing capability are also required.

PRECISION & ACCURACY The estimated quantitation limit (EQL) of Method 8270B for determining an individual compound is approximately 1 mg/kg (wet weight) for soil or sediment samples, 1–200 mg/kg for wastes (dependent on matrix and method of preparation), and 10 µg/L for groundwater samples. EQLs will be proportionately higher for sample extracts that require dilution to avoid saturation of the detector.

The EQL(b) for groundwater in µg/L is 100.
The EQL (a, b) for low concentrations in soil and sediment in µg/kg is not determined.
Accuracy as µg/L is not listed.
Overall precision in µg/L is not listed.

(a) *EQLs listed for soil/sediment are based on wet weight. Normally data is reported in a dry-weight basis; therefore, EQLs will be higher based on the % dry weight of each sample. This calculation is based on a 30 g sample and gel permeation chromatography cleanup.*
(b) *Sample EQLs are highly matrix-dependent. The EQLs are provided for guidance and may not always be achievable.*
C = *True value for concentration, in µg/L.*
X = *Average recovery found for measurements of samples containing a concentration of C, in µg/L.*

ESTIMATED QUANTITATION LIMIT

Other Matrices	Factor (a)
High-concentration soil and sludges by sonicator	7.5
Non-water miscible waste	75

(a) *EQL for other matrices = [EQL for low soil/sediment] × [Factor]. This estimated EQL is similar to an EPA "Practical Quantitation Limit."*

SAMPLING METHOD
Liquid samples — Use a 1 or 2½ gallon amber glass bottle with a screw-top Teflon®-lined cover that has been prewashed with detergent and rinsed with distilled water and methanol (or isopropanol).

Soils, sediments, or sludges — Use an 8-oz. widemouth glass with a screw-top Teflon®-lined cover that has been prewashed with detergent and rinsed with distilled water and methanol (or isopropanol).

SAMPLE PRESERVATION
Liquid samples — If residual chlorine is present, add 3 mL of 10% sodium thiosulfate per gallon, cool to 4°C and store in a solvent-free refrigerator until analysis; if chlorine is not present, then eliminate the sodium thiosulfate addition.

Soils, sediments, or sludges — Cool samples to 4°C and store in a solvent-free refrigerator.

MHT Liquid samples must be extracted within 7 days and the extracts analyzed within 40 days. Soils, sediments, or sludges may be stored for a maximum of 14 days and the extracts analyzed within 40 days.

SAMPLE PREPARATION
Liquid samples — Transfer 1 L quantitatively to a continuous extractor. If high concentrations are anticipated, a smaller volume may be used and then diluted with organic-free reagent water to 1 L. Adjust pH, if necessary, to pH <2 using 1:1 (V/V) sulfuric acid. Pipette 1.0 mL of a surrogate standard spiking solution into each sample. For the sample in each analytical batch selected for spiking, add 1.0 mL of a matrix spiking standard. For base/neutral acid analysis, the amount of the surrogates and matrix spiking compounds added to the sample should result in a final concentration of 100 ng/µL of each analyte in the extract to be analyzed (assuming a 1-µL injection). Extract with methylene chloride for 18–24 h. Next, adjust the pH of the aqueous phase to pH >11 using 10 N sodium hydroxide and extract it with methylene chloride again for 18–24 h. Dry the extract through a column containing anhydrous sodium sulfate and concentrate it to 1 mL using a K-D concentrator.

Soils, sediments, or sludges — Use 30 g of sample. Nonporous or wet samples (gummy or clay type) that do not have a free-flowing sandy texture must be mixed with anhydrous sodium sulfate until the sample is free flowing. Add 1 mL of surrogate standards to all samples, spikes, standards, and blanks. For the sample in each analytical batch selected for spiking, add 1.0 mL of a matrix spiking standard. For base/neutral acid analysis, the amount added of the surrogates and matrix spiking compounds should result in a final concentration of 100 ng/µL of each base/neutral analyte and 200 ng/µL of each acid analyte in the extract to be analyzed (assuming a 1-µL injection). Immediately add a 100-mL mixture of 1:1 methylene chloride:acetone and extract the sample ultrasonically for 3 min and then decant or filter the extracts. Repeat the extraction two or more times. Dry the extract using a column with anhydrous sodium sulfate and concentrate it to 1 mL in a K-D concentrator.

QUALITY CONTROL A methylene chloride solution containing 50 ng/µL of decafluorotriphenylphosphine (DFTPP) is used for tuning the GC/MS system each 12-h shift. A system performance check also must be made during every 12-h shift. A standard containing 50 ng/µL each of 4,4'-DDT, pentachlorophenol, and benzidine is required to verify injection port inertness and GC column performance. A calibration standard at mid-concentration, containing each compound of interest, including all required surrogates, must be performed every 12 h during analysis. After the system performance check is met, calibration check compounds (CCCs) are used to check the validity of the initial calibration.

The internal standard responses and retention times in the calibration check standard must be evaluated immediately after or during data acquisition. If the retention time for any internal standard changes by more than 30 seconds from the last check calibration (12 h), the chromatographic system must be inspected for malfunctions and corrections must be made, as required. If the electron ionization current plot (EICP) area for any of the internal standards changes by a factor of two from the last daily calibration standard check, the mass spectrometer must be inspected for malfunctions and corrections must be made, as appropriate.

Demonstrate, through the analysis of a reagent water blank, that interferences from the analytical system, glassware, and reagents are under control. The blank samples should be carried through all stages of the sample preparation and measurement steps. For each analytical batch (up to 20 samples), a reagent blank, matrix spike, and matrix spike duplicate/duplicate must be analyzed (the frequency of the spikes may be different for different monitoring programs). The blank and spiked samples must be carried through all stages of the sample preparation and measurement steps. A QC reference sample concentrate containing each analyte at a concentration of 100 mg/L in methanol is required.

REFERENCE Test Methods for Evaluating Solid Waste (SW-846). U.S. EPA 1983, Method 8270B, Rev. 2, Nov. 1990. Office of Solid Waste, Washington, DC.

Ronnel   EPA Method 8141
CAS #299-84-3

TITLE Organophosphorus Compounds by Gas Chromatography: Capillary Column Technique

MATRIX This method covers aqueous and solid matrices. This includes a wide variety such as drinking water, groundwater, industrial wastewaters, surface waters, soils, solids, and sediments.

METHOD SUMMARY This is a GC method used to determine the concentration of 28 organophosphorus pesticides.

The use of Gel Permeation Cleanup (EPA Method 3640) for sample cleanup has been demonstrated to yield recoveries of less than 85% for many method analytes and is therefore not recommended for use with this method.

This method provides GC conditions for the detection of ppb concentrations of organophosphorus compounds. Prior to the use of this method, appropriate sample preparation techniques must be used. Water samples are extracted at a neutral pH with methylene chloride as a solvent by using a separatory funnel (EPA Method 3510) or a continuous liquid-liquid extractor (EPA Method 3520). Soxhlet extraction (EPA Method 3540) or ultrasonic extraction (EPA Method 3550) using methylene chloride/acetone (1:1) are used for solid samples. Both neat and diluted organic liquids (EPA Method 3580) may be analyzed by direct injection. Spiked samples are used to verify the applicability of the chosen extraction technique to each new sample type. A GC with a flame photometric (FPD) or nitrogen-phosphorus detector (NPD) is used for this multiresidue procedure.

INTERFERENCES The use of Florisil cleanup materials (EPA Method 3620) for some of the compounds in this method has been demonstrated to yield recoveries less than 85% and is therefore not recommended for all compounds. Use of phosphorus or halogen specific detectors, however, often obviates the necessity for cleanup for relatively clean sample matrices. If particular circumstances demand the use of an alternative cleanup procedure, the analyst must determine the elution profile and demonstrate that the recovery of each analyte is no less than 85%.

Use of a flame photometric detector (FPD) in the phosphorus mode will minimize interferences from materials that do not contain phosphorus. Elemental sulfur, however, may interfere with the determination of certain organophosphorus compounds by flame photometric gas chromatography. Sulfur cleanup using EPA Method 3660 may alleviate this interference. A nitrogen phosphorus detector (NPD) is also recommended.

A few analytes coelute on certain columns. Therefore, select a second column for confirmation where coelution of the analytes of interest does not occur.

Method interferences may be caused by contaminants in solvents, reagents, glassware, and other sample processing hardware that lead to discrete artifacts or elevated baselines in gas chromatograms. All these materials must be routinely demonstrated to be free from interferences under the conditions of the analysis by analyzing reagent blanks.

INSTRUMENTATION A GC with a NPD or a FPD will be needed. A data system or integrator is recommended for measuring peak areas and/or peak heights. A Kuderna-Danish (K-D) apparatus will be needed for extract concentration.

Column 1: 15 m × 0.53 mm megabore capillary column, 1.0 µm film thickness, DB-210.
Column 2: 15 m × 0.53 mm megabore capillary column, 1.5 µm film thickness, SPB-608.
Column 3: 15 m × 0.53 mm megabore capillary column, 1.0 µm film thickness, DB-5.

Three megabore capillary columns are included for analysis of organophosphates by this method. Column 1 (DB-210 or equivalent) and Column 2 (SPB-608 or equivalent) are recommended if a large number of organophosphorus analytes are to be determined. If the superior resolution offered by Column 1 and Column 2 is not required, Column 3 (DB-5 or equivalent) may be used. For megabore capillary columns, automatic injections of 1 µL are recommended.

PRECISION & ACCURACY The MDL actually achieved in a given analysis will vary, as it is dependent on instrument sensitivity and matrix effects. Single operator accuracy and precision studies have been conducted with spiked water and soil samples.

MULTIPLICATION FACTORS FOR OTHER MATRICES (a)

Matrix	Factor (b)
Groundwater (EPA Method 3510 or EPA Method 3520)	10
Low-concentration soil by Soxhlet and no cleanup	10 (c)
Low-concentration soil by ultrasonic extraction with GPC cleanup	6.7 (c)
High-concentration soil and sludges by ultrasonic extraction	500 (c)
Non-water miscible waste (EPA Method 3580)	1000 (c)

(a) Sample EQLs are highly matrix-dependent. The EQLs listed here are provided for guidance and may not always be achievable.
(b) EQL = [Method detection limit] × [Factor]. For non-aqueous samples the factor is on a wet-weight basis.
(c) Multiply this factory times the soil MDL.

The MDL (in µg/L) when reagent water was extracted using a separatory funnel was 0.07.

The MDL (in µg/kg) when soil was extracted using Soxhlet extraction (EPA Method 3540) was 3.5.

Accuracy (as % recovery) with separatory funnel extraction ranged from 67 (with low spikes) to 87 (with high spikes).

Accuracy (as % recovery) with continuous liquid-liquid extraction ranged from 82 (with low spikes) to 89 (with high spikes).

Accuracy (as % recovery) with Soxhlet extraction of soils ranged from not recovered (with low spikes to 79 (with high spikes).

Accuracy (as % recovery) with ultrasonic extraction of soils ranged from 70 (with low spikes) to 81 (with high spikes).

SAMPLE COLLECTION, PRESERVATION & HANDLING Containers used to collect samples for the determination of semivolatile organic compounds should be soap and water washed followed by methanol (or isopropanol) rinsing. The sample containers should be of glass or Teflon® and have screw-top covers with Teflon® liners.

No preservation is used with concentrated waste samples. With liquid samples containing no residual chlorine and with soil, sediment, and sludge samples, immediately cooling to 4°C is the only preservation used. When residual chlorine is present then 3 mL of 10% aqueous sodium sulfate is added for each gallon of sample collected, followed by cooling to 4°C.

Liquid samples must be extracted within 7 days and their extracts analyzed within 40 days. Concentrated waste, soil, sediment, and sludge samples must be extracted within 14 days and their extracts analyzed within 40 days.

SAMPLE PREPARATION In general, water samples are extracted at a neutral pH with methylene chloride, using either EPA Method 3510 or EPA Method 3520. Solid samples are extracted using either EPA Method 3540 or EPA Method 3550 with methylene chloride/acetone (1:1) as the extraction solvent.

Prior to GC analysis, the extraction solvent may be exchanged to hexane. Single lab data indicates that samples should not be transferred with 100% hexane during sample workup as the more water soluble organophosphorus compounds may be lost.

If cleanup is performed on the samples, the analyst should analyze the samples by GC. This will confirm elution patterns and the absence of interferences from the reagents. If peak detection and identification is prevented by the presence of interferences, further cleanup is required.

QUALITY CONTROL The analyst should monitor the performance of the extraction, cleanup (when used), and analytical system and the effectiveness of the method in dealing with each sample matrix by spiking each sample, standard, and blank with one or two surrogates (e.g., organophosphorus compounds not expected to be present in the sample). Deuterated analogs of analytes should not be used as surrogates for gas chromatographic analysis due to coelution problems.

A minimum of five concentrations for each analyte of interest should be prepared through dilution of the stock standards with isooctane. One of the concentrations should be at a concentration near, but above, the MDL.

Include a mid-level check standard after each group of 10 samples in the analysis sequence. GC/MS techniques should be judiciously employed to support qualitative identifications made with this method. Follow the GC/MS operating requirements specified in EPA Method 8270.

When available, chemical ionization mass spectra may be employed to aid in the qualitative identification process. To confirm an identification of a compound, the background-corrected mass spectrum of the compound must be obtained from the sample extract and must be compared with a mass spectrum from a stock or calibration standard analyzed under the same chromatographic conditions. The molecular ion and all other ions present above 20% relative abundance in the mass spectrum of the standard must be present in the mass spectrum of the sample with agreement to ±20%. The retention time of the compound in the sample must be within six seconds of the retention time for the same compound in the standard solution.

Should the MS procedure fail to provide satisfactory results, additional steps may be taken before reanalysis. These steps may include the use of alternate packed or capillary GC columns or additional sample cleanup.

REFERENCE Test Methods for Evaluating Solid Waste, Physical/Chemical Methods, SW-846, 3rd Edition, U.S. EPA, Office of Solid Waste, Washington, DC, EPA Method 8141 July 1992.

Ronnel   **EPA Method 8140**
CAS #299-84-3

TITLE Organophosphorus Pesticides

MATRIX Groundwater, soils, sludges, water miscible liquid wastes, and non-water miscible wastes.

APPLICATION This method is used for the analysis of 21 organophosphorus pesticides. Samples are extracted, concentrated, and analyzed using direct injection of both neat and diluted organic liquid into a gas chromatograph (GC).

INTERFERENCES Solvents, reagents, and glassware may introduce artifacts. Other interferences may come from coextracted compounds from samples. The use of Florisil cleanup materials may produce low recoveries. Elemental sulfur may interfere with some compounds when using a flame photometric detector. Sulfur cleanup (Method 3660) may alleviate sulfur interference.

INSTRUMENTATION GC capable of on-column injections and a flame photometric detector (FPD) or a thermionic detector. Column 1: 1.8 m by 2 mm with 5% SP-2401 on Supelcoport. Column 2: 1.8 m by 2 mm with 3% SP-2401 on Supelcoport. Column 3: 50 cm by ⅛ in Teflon® with 15% SE-54 on Gas Chrom Q. The preferred column is Column Number 2.

RANGE 1.0–50 µg/L

MDL 0.3 µg/L (in reagent water).

PQL FACTORS FOR MULTIPLYING × FID MDL VALUE

Matrix	Multiplication Factor
Groundwater	10
Low-level soil by sonication with GPC cleanup	670
High-level soil and sludge by sonication	10,000
Non-water miscible waste	100,000

PRECISION 5.6% (single operator standard deviation)

ACCURACY 99.2% (single operator average recovery)

SAMPLING METHOD Use 8-oz. widemouth glass bottles with Teflon®-lined caps for concentrated waste samples, soils, sediments, and sludges. Use 1 or 2½ gallon amber glass bottles with Teflon®-lined caps for liquid (water) samples.

STABILITY Cool soil, sediment, sludge, and liquid samples to 4°C. If residual chlorine is present in liquid samples add 3 mL of 10% sodium thiosulfate per gallon of sample and cool to 4°C.

MHT 14 days for concentrated waste, soil, sediment, or sludge; 7 days for liquid samples; all extracts must be analyzed within 40 days.

QUALITY CONTROL A quality control check sample concentrate containing this compound in acetone at a concentration 1,000 times more concentrated than the selected spike concentration is required. The QC check sample concentrate may be prepared from pure standard materials or purchased as certified solutions. Use appropriate trip, matrix, control site, method, reagent, and solvent blanks. Internal, surrogate, and five concentration level calibration standards are used.

REFERENCE Method 8140, SW-846, 3rd ed., Sept. 1986.

S

Safrole
CAS #94-59-7

EPA Method 1625

TITLE Semivolatile Organic Compounds by Isotope Dilution GC/MS

MATRIX The compounds may be determined in waters, soils, and municipal sludges by this method.

METHOD SUMMARY This method is used to determine 176 semivolatile toxic organic pollutants associated with the CWA (as amended 1987); the RCRA (as amended 1986); the CERCLA (as amended 1986); and other compounds amenable to extraction and analysis by capillary column gas chromatography-mass spectrometry (GC/MS).

Stable isotopically-labeled analogs of the compounds of interest are added to the sample. If the solids content is less than 1%, a 1-L sample is extracted at pH 12–13, then at pH <2 with methylene chloride using continuous extraction techniques.

If the solids content is 30% or less, the sample is diluted to 1% solids with reagent water, homogenized ultrasonically, and extracted at pH 12–13, then at pH <2 with methylene chloride using continuous extraction techniques. If the solids content is greater than 30%, the sample is extracted using ultrasonic techniques.

Each extract is dried over sodium sulfate, concentrated to a volume of 5 mL, cleaned up using GPC, if necessary, and concentrated. Extracts are concentrated to 1 mL if GPC is not performed, and to 0.5 mL if GPC is performed.

An internal standard is added to the extract, and a 1-mL aliquot of the extract is injected into the GC. The compounds are separated by GC and detected by a MS. The labeled compounds serve to correct the variability of the analytical technique.

INTERFERENCES Solvents, reagents, glassware, and other sample processing hardware may yield artifacts and/or elevated baselines causing misinterpretation of chromatograms and spectra. Materials used in the analysis must be demonstrated to be free from interferences under the conditions of analysis by running method blanks initially and with each sample lot (sample started through the extraction process on a given 8-h shift, to a maximum of 20). Specific selection of reagents and purification of solvents by distillation in all glass systems may be required. Glassware and, where possible, reagents are cleaned by solvent rinse and baking at 450°C for 1-h minimum. Interferences coextracted from samples will vary considerably from source to source, depending on the diversity of the site being sampled.

INSTRUMENTATION Major instrumentation includes a GC with a splitless or on-column injection port for capillary column, a MS with 70 eV electron impact ionization, and a data system to collect and record MS data, and process it. A K-D apparatus is used to concentrate extracts.

GC Column: 30 m × 0.25 mm I.D. 5% phenyl, 94% methyl, 1% vinyl silicone bonded phased fused silica capillary column.

PRECISION & ACCURACY The detection limits of the method are usually dependent on the level of interferences rather than instrumental limitations. The limits typify the minimum quantities that can be detected with no interferences present.

The minimum level (in μg/mL) was not listed. This is defined as a minimum level at which the analytical system shall give recognizable mass spectra (background corrected) and acceptable calibration points.

The MDL (in μg/kg) in low solids was not listed and in high solids was not listed; these were determined in digested sludge (low solids) and in filter cake or compost (high solids).

The labeled and native compound initial precision as standard deviation (in μg/L) was not listed.
The labeled and native compound initial accuracy as average recovery (in μg/L) was not listed.

SAMPLE COLLECTION, PRESERVATION & HANDLING Collect samples in glass containers. Aqueous samples which flow freely are collected in refrigerated bottles using automatic sampling equipment. Solid samples are collected as grab samples using widemouth jars. Maintain samples at 0 to 4°C from the time of collection until extraction. If residual chlorine is present in aqueous samples, add 80 mg sodium thiosulfate/L of water. Begin sample extraction within 7 days of collection, and analyze all extracts within 40 days of extraction.

SAMPLE PREPARATION Samples containing 1% solids or less are extracted directly using continuous liquid-liquid extraction techniques. Samples containing 1 to 30% solids are diluted to the 1% level with reagent water and extracted using continuous liquid-liquid extraction techniques. Samples containing greater than 30% solids are extracted using ultrasonic techniques.

Base/neutral extraction — Adjust the pH of the waters in the extractors to 12–13 with 6 N NaOH. Extract with methylene chloride for 24–48 h.
Acid extraction — Adjust the pH of the waters in the extractors to 2 or less using 6 N sulfuric acid. Extract with methylene chloride for 24–48 h.
Ultrasonic extraction of high solids samples — Add anhydrous sodium sulfate to the sample and QC aliquot(s). Add acetone:methylene chloride (1:1) to the sample and mix thoroughly

Concentrate extracts using a K-D apparatus.

QUALITY CONTROL The analyst is permitted to modify this method to improve separations or lower the costs of measurements, provided all performance specifications are met. Analyses of blanks are required to demonstrate freedom from contamination. When results of spikes indicate atypical method performance for samples, the samples are diluted to bring method performance within acceptable limits.

For low solids (aqueous samples), extract, concentrate, and analyze two sets of four 1-L aliquots (8 aliquots total) of the

precision and recovery standard. For high solids samples, two sets of four 30-g aliquots of the high solids reference matrix are used.

Spike all samples with labeled compounds to assess method performance. Compute percent recovery of the labeled compounds using the internal standard method. Compare the labeled compound recovery for each compound with the corresponding labeled compound recovery.

Reagent water and high solids reference matrix blanks are analyzed to demonstrate freedom from contamination. Extract and concentrate a 1-L reagent water blank or a high solids reference matrix blank with each sample's lot (samples started through the extraction process on the same 8-h shift, to a maximum of 20 samples).

Field replicates may be collected to determine the precision of the sampling technique, and spiked samples may be required to determine the accuracy of the analysis when the internal standard method is used.

REFERENCE Semivolatile Organic Compounds by Isotope Dilution GC/MS. Office of Water Regulation and Standards, U.S. EPA Industrial Technology Division, Washington, DC, EPA Method 1625, Rev. C, June 1989 (contact W.A. Telliard, U.S. EPA, Office of Water Regulations and Standards, 401 M St., SW, Washington, DC, 20460. Phone: 202-382-7131).

Safrole   **EPA Method 1625**
CAS #94-59-7

TITLE Semivolatile Organic Compounds by Isotope Dilution GC/MS

MATRIX The compounds may be determined in waters, soils, and municipal sludges by this method.

METHOD SUMMARY This method is used to determine 176 semivolatile toxic organic pollutants associated with the CWA (as amended 1987); the RCRA (as amended 1986); the CERCLA (as amended 1986); and other compounds amenable to extraction and analysis by capillary column gas chromatography-mass spectrometry (GC/MS).

Stable isotopically-labeled analogs of the compounds of interest are added to the sample. If the solids content is less than 1%, a 1-L sample is extracted at pH 12–13, then at pH <2 with methylene chloride using continuous extraction techniques.

If the solids content is 30% or less, the sample is diluted to 1% solids with reagent water, homogenized ultrasonically, and extracted at pH 12–13, then at pH <2 with methylene chloride using continuous extraction techniques. If the solids content is greater than 30%, the sample is extracted using ultrasonic techniques.

Each extract is dried over sodium sulfate, concentrated to a volume of 5 mL, cleaned up using GPC, if necessary, and concentrated. Extracts are concentrated to 1 mL if GPC is not performed, and to 0.5 mL if GPC is performed.

An internal standard is added to the extract, and a 1-mL aliquot of the extract is injected into the GC. The compounds are separated by GC and detected by a MS. The labeled compounds serve to correct the variability of the analytical technique.

INTERFERENCES Solvents, reagents, glassware, and other sample processing hardware may yield artifacts and/or elevated baselines causing misinterpretation of chromatograms and spectra. Materials used in the analysis must be demonstrated to be free from interferences under the conditions of analysis by running method blanks initially and with each sample lot (sample started through the extraction process on a given 8-h shift, to a maximum of 20). Specific selection of reagents and purification of solvents by distillation in all glass systems may be required. Glassware and, where possible, reagents are cleaned by solvent rinse and baking at 450°C for 1-h minimum. Interferences coextracted from samples will vary considerably from source to source, depending on the diversity of the site being sampled.

INSTRUMENTATION Major instrumentation includes a GC with a splitless or on-column injection port for capillary column, a MS with 70 eV electron impact ionization, and a data system to collect and record MS data, and process it. A K-D apparatus is used to concentrate extracts.

GC Column: 30 m × 0.25 mm I.D. 5% phenyl, 94% methyl, 1% vinyl silicone bonded phased fused silica capillary column.

PRECISION & ACCURACY The detection limits of the method are usually dependent on the level of interferences rather than instrumental limitations. The limits typify the minimum quantities that can be detected with no interferences present.

The minimum level (in µg/mL) was not listed. This is defined as a minimum level at which the analytical system shall give recognizable mass spectra (background corrected) and acceptable calibration points.

The MDL (in µg/kg) in low solids was not listed and in high solids was not listed; these were determined in digested sludge (low solids) and in filter cake or compost (high solids).

The labeled and native compound initial precision as standard deviation (in µg/L) was not listed.
The labeled and native compound initial accuracy as average recovery (in µg/L) was not listed.

SAMPLE COLLECTION, PRESERVATION & HANDLING Collect samples in glass containers. Aqueous samples which flow freely are collected in refrigerated bottles using automatic sampling equipment. Solid samples are collected as grab samples using widemouth jars. Maintain samples at 0 to 4°C from the time of collection until extraction. If residual chlorine is present in aqueous samples, add 80 mg sodium thiosulfate/L of water. Begin sample extraction within 7 days of collection, and analyze all extracts within 40 days of extraction.

SAMPLE PREPARATION Samples containing 1% solids or less are extracted directly using continuous liquid-liquid extraction techniques. Samples containing 1 to 30% solids are diluted to the 1% level with reagent water and extracted using continuous

liquid-liquid extraction techniques. Samples containing greater than 30% solids are extracted using ultrasonic techniques.

Base/neutral extraction — Adjust the pH of the waters in the extractors to 12–13 with 6 N NaOH. Extract with methylene chloride for 24–48 h.

Acid extraction — Adjust the pH of the waters in the extractors to 2 or less using 6 N sulfuric acid. Extract with methylene chloride for 24–48 h.

Ultrasonic extraction of high solids samples — Add anhydrous sodium sulfate to the sample and QC aliquot(s). Add acetone:methylene chloride (1:1) to the sample and mix thoroughly

Concentrate extracts using a K-D apparatus.

QUALITY CONTROL The analyst is permitted to modify this method to improve separations or lower the costs of measurements, provided all performance specifications are met. Analyses of blanks are required to demonstrate freedom from contamination. When results of spikes indicate atypical method performance for samples, the samples are diluted to bring method performance within acceptable limits.

For low solids (aqueous samples), extract, concentrate, and analyze two sets of four 1-L aliquots (8 aliquots total) of the precision and recovery standard. For high solids samples, two sets of four 30-g aliquots of the high solids reference matrix are used.

Spike all samples with labeled compounds to assess method performance. Compute percent recovery of the labeled compounds using the internal standard method. Compare the labeled compound recovery for each compound with the corresponding labeled compound recovery.

Reagent water and high solids reference matrix blanks are analyzed to demonstrate freedom from contamination. Extract and concentrate a 1-L reagent water blank or a high solids reference matrix blank with each sample's lot (samples started through the extraction process on the same 8-h shift, to a maximum of 20 samples).

Field replicates may be collected to determine the precision of the sampling technique, and spiked samples may be required to determine the accuracy of the analysis when the internal standard method is used.

REFERENCE Semivolatile Organic Compounds by Isotope Dilution GC/MS. Office of Water Regulation and Standards, U.S. EPA Industrial Technology Division, Washington, DC, EPA Method 1625, Rev. C, June 1989 (contact W.A. Telliard, U.S. EPA, Office of Water Regulations and Standards, 401 M St., SW, Washington, DC, 20460. Phone: 202-382-7131).

Safrole   **EPA Method 8270**
CAS #94-59-7

TITLE Semivolatile Organic Compounds by GC/MS

MATRIX This method is used to determine the concentration of semivolatile organic compounds in extracts prepared from all types of solid waste matrices, soils, and groundwater. Although surface waters are not specifically mentioned, this method should be applicable to water samples from rivers, lakes, etc.

METHOD SUMMARY This method covers 259 semivolatile organic compounds. In very limited applications direct injection of the sample into the GC/MS system may be appropriate, but this results in very high detection limits (approximately 10,000 µg/L). Typically, a 1-L liquid sample, containing surrogate, and matrix spiking standards, is extracted in a continuous extractor first under acid conditions and then under basic conditions. Typically 30 g of a solid sample, containing surrogate, and matrix spiking standards, is extracted ultrasonically. After concentrating the extract to 1 mL it is spiked with 10 µL of an internal standard solution just prior to analysis by GC/MS. The volume injected should contain about 100 ng of base/neutral and 200 ng of acid surrogates (for a 1-µL injection). Analysis is performed by GC/MS using a capillary GC column.

INTERFERENCES Raw GC/MS data from all blanks, samples, and spikes must be evaluated for interferences. Contamination by carryover can occur whenever high-concentration and low-concentration samples are sequentially analyzed. To reduce carryover, the sample syringe must be rinsed out between samples with solvent. Whenever an unusually concentrated sample is encountered, it should be followed by the analysis of blank solvent to check for cross-contamination.

INSTRUMENTATION A GC/MS and a data system are required. The GC column used is a 30 m × 0.25 mm I.D. (or 0.32 mm I.D.) 1um film thickness silicone-coated fused silica capillary column. A continuous liquid-liquid extractor equipped with Teflon® or glass connection joints and stopcocks requiring no lubrication, a K-D concentrating apparatus, water bath, and an ultrasonic disrupter with a minimum power of 300 W and with pulsing capability are also required.

PRECISION & ACCURACY The estimated quantitation limit (EQL) of Method 8270B for determining an individual compound is approximately 1 mg/kg (wet weight) for soil or sediment samples, 1–200 mg/kg for wastes (dependent on matrix and method of preparation), and 10 µg/L for groundwater samples. EQLs will be proportionately higher for sample extracts that require dilution to avoid saturation of the detector.

The EQL(b) for groundwater in µg/L is 10.
The EQL (a, b) for low concentrations in soil and sediment in µg/kg is not determined.
Accuracy as µg/L is not listed.
Overall precision in µg/L is not listed.

(a) *EQLs listed for soil/sediment are based on wet weight. Normally data is reported in a dry-weight basis; therefore, EQLs will be higher based on the % dry weight of each sample. This calculation is based on a 30 g sample and gel permeation chromatography cleanup.*
(b) *Sample EQLs are highly matrix-dependent. The EQLs are provided for guidance and may not always be achievable.*
$C =$ *True value for concentration, in µg/L.*
$X =$ *Average recovery found for measurements of samples containing a concentration of C, in µg/L.*

ESTIMATED QUANTITATION LIMIT

Other Matrices	Factor (a)
High-concentration soil and sludges by sonicator	7.5
Non-water miscible waste	75

(a) EQL for other matrices = [EQL for low soil/sediment] ×[Factor]. This estimated EQL is similar to an EPA "Practical Quantitation Limit."

SAMPLING METHOD

Liquid samples — Use a 1 or 2½ gallon amber glass bottle with a screw-top Teflon®-lined cover that has been prewashed with detergent and rinsed with distilled water and methanol (or isopropanol).

Soils, sediments, or sludges — Use an 8-oz. widemouth glass with a screw-top Teflon®-lined cover that has been prewashed with detergent and rinsed with distilled water and methanol (or isopropanol).

SAMPLE PRESERVATION

Liquid samples — If residual chlorine is present, add 3 mL of 10% sodium thiosulfate per gallon, cool to 4°C and store in a solvent-free refrigerator until analysis; if chlorine is not present, then eliminate the sodium thiosulfate addition.

Soils, sediments, or sludges — Cool samples to 4°C and store in a solvent-free refrigerator.

MHT Liquid samples must be extracted within 7 days and the extracts analyzed within 40 days. Soils, sediments, or sludges may be stored for a maximum of 14 days and the extracts analyzed within 40 days.

SAMPLE PREPARATION

Liquid samples — Transfer 1 L quantitatively to a continuous extractor. If high concentrations are anticipated, a smaller volume may be used and then diluted with organic-free reagent water to 1 L. Adjust pH, if necessary, to pH <2 using 1:1 (V/V) sulfuric acid. Pipette 1.0 mL of a surrogate standard spiking solution into each sample. For the sample in each analytical batch selected for spiking, add 1.0 mL of a matrix spiking standard. For base/neutral acid analysis, the amount of the surrogates and matrix spiking compounds added to the sample should result in a final concentration of 100 ng/μL of each analyte in the extract to be analyzed (assuming a 1-μL injection). Extract with methylene chloride for 18–24 h. Next, adjust the pH of the aqueous phase to pH >11 using 10 N sodium hydroxide and extract it with methylene chloride again for 18–24 h. Dry the extract through a column containing anhydrous sodium sulfate and concentrate it to 1 mL using a K-D concentrator.

Soils, sediments, or sludges — Use 30 g of sample. Nonporous or wet samples (gummy or clay type) that do not have a free-flowing sandy texture must be mixed with anhydrous sodium sulfate until the sample is free flowing. Add 1 mL of surrogate standards to all samples, spikes, standards, and blanks. For the sample in each analytical batch selected for spiking, add 1.0 mL of a matrix spiking standard. For base/neutral acid analysis, the amount added of the surrogates and matrix spiking compounds should result in a final concentration of 100 ng/μL of each base/neutral analyte and 200 ng/μL of each acid analyte in the extract to be analyzed (assuming a 1-μL injection). Immediately add a 100-mL mixture of 1:1 methylene chloride:acetone and extract the sample ultrasonically for 3 min and then decant or filter the extracts. Repeat the extraction two or more times. Dry the extract using a column with anhydrous sodium sulfate and concentrate it to 1 mL in a K-D concentrator.

QUALITY CONTROL A methylene chloride solution containing 50 ng/μL of decafluorotriphenylphosphine (DFTPP) is used for tuning the GC/MS system each 12-h shift. A system performance check also must be made during every 12-h shift. A standard containing 50 ng/μL each of 4,4′-DDT, pentachlorophenol, and benzidine is required to verify injection port inertness and GC column performance. A calibration standard at mid-concentration, containing each compound of interest, including all required surrogates, must be performed every 12 h during analysis. After the system performance check is met, calibration check compounds (CCCs) are used to check the validity of the initial calibration.

The internal standard responses and retention times in the calibration check standard must be evaluated immediately after or during data acquisition. If the retention time for any internal standard changes by more than 30 seconds from the last check calibration (12 h), the chromatographic system must be inspected for malfunctions and corrections must be made, as required. If the electron ionization current plot (EICP) area for any of the internal standards changes by a factor of two from the last daily calibration standard check, the mass spectrometer must be inspected for malfunctions and corrections must be made, as appropriate.

Demonstrate, through the analysis of a reagent water blank, that interferences from the analytical system, glassware, and reagents are under control. The blank samples should be carried through all stages of the sample preparation and measurement steps. For each analytical batch (up to 20 samples), a reagent blank, matrix spike, and matrix spike duplicate/duplicate must be analyzed (the frequency of the spikes may be different for different monitoring programs). The blank and spiked samples must be carried through all stages of the sample preparation and measurement steps. A QC reference sample concentrate containing each analyte at a concentration of 100 mg/L in methanol is required.

REFERENCE Test Methods for Evaluating Solid Waste (SW-846). U.S. EPA 1983, Method 8270B, Rev. 2, Nov. 1990. Office of Solid Waste, Washington, DC.

Selenium   **EPA Method 6010**
CAS #7782-49-2

TITLE Inductively Coupled Plasma-Atomic Emission Spectroscopy

MATRIX This method is applicable to the determination of trace elements, including metals, in groundwater, soils, sludges, sediments, and other solid wastes. All matrices require digestion prior to analysis. The method of standard addition must be used for the analysis of all sample digests unless either serial dilution or matrix spike addition demonstrates it is not required.

METHOD SUMMARY Method 6010 covers 25 elements using ICP analysis. It measures element-emitted light by optical spectrometry. Samples, following an appropriate acid digestion, are nebulized and the resulting aerosol is transported to the plasma torch. Element-specific atomic line emission spectra are produced by a radio-frequency inductively coupled plasma.

INTERFERENCES Interferences may be categorized as spectral or non-spectral. Spectral interferences are caused by overlap of a spectral line from another element, unresolved overlap of molecular band spectra, background contribution from continuous or recombination phenomenon, and stray light from the line emission of high concentration elements. Non-spectral interferences include physical and chemical interferences. Physical interferences are effects associated with the sample nebulization and transport processes. Changes in viscosity and surface tension can cause significant inaccuracies. Chemical interferences include molecular compound formation, ionization effects, and solute vaporization effects. Normally these effects are not significant and can be minimized by careful selection of operating conditions. Chemical interferences are highly dependent on matrix type and the specific analyte element.

INSTRUMENTATION An inductively coupled argon plasma emission spectrometer (ICP) capable of background correction is required.

PRECISION & ACCURACY Detection limits, sensitivity, and optimum ranges of the metals will vary with the matrices and model of the spectrometer. In a single lab evaluation, seven wastes were analyzed for 22 elements. The mean percent relative standard deviation from triplicate analyses for all elements and wastes was 9 ± 2%. The mean percent recovery of spiked elements for all wastes was 93 ± 6%. Spike levels ranged from 100 µg/L to 100 mg/L. The wastes included sludges and industrial wastewaters.

Estimated instrument detection limit in µg/L is 75.
Spiked concentration in µg/L is 6.
Mean reported value in µg/L is 8.5.
Precision as RSD % is 42.

SAMPLING METHOD Samples should be collected in borosilicate glass, linear polyethylene, polypropylene, or Teflon® bottles that have been prewashed with detergent and tap water, and rinsed with 1:1 nitric acid and tap water or 1:1 hydrochloric acid and tap water. Collect at least 2 g of solids and 200 mL of aqueous samples.

SAMPLE PRESERVATION Add nitric acid to make the samples pH <2.

MHT The maximum holding time for properly preserved samples is 6 months.

SAMPLE PREPARATION Preliminary treatment of most matrices is necessary because of the complexity and variability of sample matrices. Water samples that have been prefiltered and acidified will not need acid digestion. Methods for acid digestion of waters for total recoverable or dissolved metals, acid digestions of aqueous samples and extracts for total metals, and acid digestion of sediments, sludges, and soils are summarized below.

Total recoverable or dissolved metals in water — To prepare surface and groundwater samples for determination of total recoverable and dissolved metals, a 100-mL aliquot of well-mixed sample is acidified with concentrated nitric acid and concentrated hydrochloric acid, then heated until the volume is reduced to 15–20 mL. Adjust the final volume to 100 mL with reagent water.

Total metals in aqueous samples, soil and sediment extracts — To prepare aqueous samples, soil and sediment extracts, and wastes that contain suspended solids, a 100-mL aliquot is made acidic with concentrated nitric acid and the solution is evaporated to about 5 mL on a hot plate. Continue heating and adding additional acid until sample digestion is complete, which is usually indicated when the digestate is light in color or does not change in appearance. Evaporate the solution to about 3 mL and cool it and add a small quantity of 1:1 hydrochloric acid (10 mL/100 mL of final solution). Cover the beaker and reflux for 15 min. Wash down the beaker walls and filters or centrifuge the sample to remove silicates and other insoluble material. Filter the sample and adjust the final volume to 100 mL with reagent water and the final acid concentration to 10%.

Sediments, sludges, and soils — To prepare sediments, sludges and soil samples, transfer 1–2 g to a conical beaker and add 10 mL of 1:1 nitric acid, mix the slurry, and cover it with a watch glass. Heat the sample and reflux for 10–15 min without boiling. Allow it to cool, then add 5 mL of concentrated nitric acid and reflux for 30 min. Repeat last step and then allow the solution to evaporate to 5 mL without boiling. Cool and add 2 mL of water and 3 mL of 30% hydrogen peroxide. Cover and place the beaker on the hot plate. Heat and add 30% hydrogen peroxide in 1-mL aliquots with warming until the effervescence is minimal but do not add more than a total of 10 mL of 30% hydrogen peroxide. If the sample is being prepared for the analysis of Ag, Al, As, Ba, Be, Ca, Cd, Co, Cr, Cu, Fe, K, Mg, Mn, Mo, Na, Ni, Os, Pb, Se, Tl, V, and Zn, then add 5 mL of concentrated hydrochloric acid and 10 mL of water and return the covered beaker to a hot plate for 15 min of additional refluxing without boiling. Dilute the sample to a 100 mL volume with water after cooling and filter or centrifuge to remove particulates.

QUALITY CONTROL Laboratory control samples must be analyzed for each analytical method. A method blank should be analyzed with each batch of samples. The effect of the matrix on method performance must be demonstrated: when appropriate, there should be at least one matrix spike and either one matrix duplicate or one matrix spike duplicate per analytical batch. The bias and precision of the method, as well as the method detection limit for each specific matrix type, must be measured.

Dilute and reanalyze samples that are more concentrated than the linear calibration limit. Employ a minimum of one reagent blank per sample batch to determine if contamination or any memory effects are occurring. Whenever a new or unusual sample matrix is encountered, perform either a serial dilution test or a matrix spike addition test to ensure that neither positive or negative interferences are operating on any of the analyte elements. Check the instrument standardization by verifying calibration every 10 samples using a calibration blank and a check standard.

REFERENCE Test Methods for Evaluating Solid Waste (SW-846). U.S. EPA. 1983. Method 6010, Rev. 0, Sept. 1986. Office of Solid Wastes, Washington, DC.

Selenium — EPA Method 200.7
CAS #7782-49-2

TITLE Inductively Coupled Plasma

MATRIX Dissolved, suspended or (ICP) total element in drinking and surface waters and in domestic and industrial wastewaters.

APPLICATION The method covers the determination of 25 metals. Dissolved elements are determined in filtered and acidified samples after appropriate digestion (which increases dissolved solids). Its primary advantage is that ICP instruments allow simultaneous or rapid sequential determination of many elements in a short time. Samples are first nebulized and the aerosol is transported to a plasma torch in which element specific atomic line emission spectra are produced by a radio frequency inductively coupled plasma. Background correction is required for trace element detection except in the case of line broadning.

INTERFERENCES There are spectral, physical, and chemical interferences. The primary disadvantage of ICP instruments is background radiation from other elements and the plasma gases (spectral interferences). Changes in sample viscosity and surface tension with samples containing high dissolved solids (especially those exceeding 1500 mg/L) or high acid concentrations can cause physical interferences. Ionization effects, solute vaporization and molecular compound formation can cause chemical interferences. Aluminum and iron can cause interference at the 100 mg/L level.

INSTRUMENTATION Inductively coupled argon plasma emission spectroscopy. 196.026 nm wavelength

RANGE Not listed.

MDL 75 µg/L.

PRECISION SD = 21.9% Mean at true value 40 µg/L (results from only 2 labs).

ACCURACY Mean recovery = 93% ± 6% of spiked elements for all wastes.

SAMPLING METHOD Wash sample container with detergent and tap water, rinse with 1 + 1 nitric acid and tap water, then rinse with 1 + 1 hydrochloric acid and tap water, then rinse with deionized, distilled water in that order. Perform any filtration or acid preservation steps when the sample is collected or as soon as possible thereafter.

STABILITY Cool samples to 4°C.

MHT 24 h.

QUALITY CONTROL Mixed calibration standards, an instrument check standard, and an interference check solution are used in addition to a quality control sample. The quality control sample should be prepared in the same acid matrix as the calibration standards at 10 times the instrumental detection limits and in accordance with the instructions provided by the supplier. Furthermore, two types of blanks are required: a calibration blank and a reagent blank.

REFERENCE Method 200.7, U.S. EPA, EMSL-Cincinnati, OH, Nov. 1980

Selenium — EPA Method 7740
CAS #7782-49-2

TITLE Atomic Absorption (AA) Furnace Technique

MATRIX Wastes, mobility procedure extracts, soils and groundwater.

APPLICATION Sample preparation converts organic Se to inorganic forms.

SAMPLE PREPARATION varies with the matrix. An aliquot of digestate is placed in a graphite tube in the furnace and slowly evaporated, charred and atomized. Absorption of lamp radiation during atomization is proportional to selenium concentration.

INTERFERENCES Elemental selenium and many of its compounds are volatile so there may be losses in selenium during sample preparation. There can be severe nonspecific absorption and light scattering caused by matrix components during atomization. Memory effects occur if selenium is not volatilized and removed from the furnace. Use of low wavelength (196.0 nm) makes selenium analysis susceptible to analytical problems.

INSTRUMENTATION Atomic absorption spectrometer. Selenium hollow cathode lamp or electrodeless discharge lamp. Graphite furnace. Strip-chart recorder

RANGE 5–100 µg/L

MDL 2 µg/L

PRECISION Standard deviation = ±0.60, 0.40, 0.50 at 5.0, 10, 20 µg Se/L

ACCURACY Recoveries = 92, 98, 100% at 5.0, 10, 20 µg Se/L

SAMPLING METHOD Use prewashed plastic or glass containers.

STABILITY Non aqueous samples: cool to 4°C. Aqueous samples: add nitric acid to pH <2.

MHT 6 months

QUALITY CONTROL Run one spike duplicate sample for every 10 samples. Verify calibration with an independently prepared check standard every 15 samples.

REFERENCE Method 7740, SW-846, 3rd ed., Nov.1986.

Selenium EPA Method 7741
CAS #7782-49-2

TITLE Atomic Absorption (AA) Gaseous Hydride

MATRIX Wastes, mobility procedure extracts, soils and groundwater.

APPLICATION Method approved only for sample matrices without high concentrations of Cr, Cu, Hg, Ni, Ag, Co and Mo. After sample preparation with HNO3/H2SO4 digestion, Se in digestate is reduced to Se(IV) with SnCl2. Se(IV) is then converted to a volatile hydride using hydrogen and is swept into an argon-hydrogen flame located in the optical path of an atomic absorption spectrometer. Absorption of the lamp radiation is proportional to the selenium concentration.

INTERFERENCES Traces of nitric acid left following sample work-up can result in analytical interferences. Elemental Se and many of its compounds are volatile; may be subject to losses during sample preparation. High concentrations of Cr, Cu, Hg, Ni, Ag, Co and Mo cause analytical interferences.

INSTRUMENTATION Atomic absorption spectrometer. Selenium (Se) hollow cathode lamp or electrodeless discharge lamp. Burner (for argon-hydrogen flame)

RANGE 2–20 µg/L

MDL 0.002 mg/L

PRECISION Standard deviation = ±1.1 at 10 µg/L on selenium oxide solution.

ACCURACY Recovery = 100% at 10 µg/L on selenium oxide solution.

SAMPLING METHOD Use prewashed plastic or glass containers. Collect 100 mL of sample.

STABILITY Add nitric acid to pH <2.

MHT 6 months.

QUALITY CONTROL Run one spike duplicate sample for every 10 samples. Verify calibration with an independently prepared check standard every 15 samples.

REFERENCE Method 7741, SW-846, 3rd ed., Nov.1986.

Silicon EPA Method 6010
CAS #7440-21-3

TITLE Inductively Coupled Plasma-Atomic Emission Spectroscopy

MATRIX This method is applicable to the determination of trace elements, including metals, in groundwater, soils, sludges, sediments, and other solid wastes. All matrices require digestion prior to analysis. The method of standard addition must be used for the analysis of all sample digests unless either serial dilution or matrix spike addition demonstrates it is not required.

METHOD SUMMARY Method 6010 covers 25 elements using ICP analysis. It measures element-emitted light by optical spectrometry. Samples, following an appropriate acid digestion, are nebulized and the resulting aerosol is transported to the plasma torch. Element-specific atomic line emission spectra are produced by a radio-frequency inductively coupled plasma.

INTERFERENCES Interferences may be categorized as spectral or non-spectral. Spectral interferences are caused by overlap of a spectral line from another element, unresolved overlap of molecular band spectra, background contribution from continuous or recombination phenomenon, and stray light from the line emission of high concentration elements. Non-spectral interferences include physical and chemical interferences. Physical interferences are effects associated with the sample nebulization and transport processes. Changes in viscosity and surface tension can cause significant inaccuracies. Chemical interferences include molecular compound formation, ionization effects, and solute vaporization effects. Normally these effects are not significant and can be minimized by careful selection of operating conditions. Chemical interferences are highly dependent on matrix type and the specific analyte element.

INSTRUMENTATION An inductively coupled argon plasma emission spectrometer (ICP) capable of background correction is required.

PRECISION & ACCURACY Detection limits, sensitivity, and optimum ranges of the metals will vary with the matrices and model of the spectrometer. In a single lab evaluation, seven wastes were analyzed for 22 elements. The mean percent relative standard deviation from triplicate analyses for all elements and wastes was 9 ± 2%. The mean percent recovery of spiked elements for all wastes was 93 ± 6%. Spike levels ranged from 100 µg/L to 100 mg/L. The wastes included sludges and industrial wastewaters.

Estimated instrument detection limit in µg/L is 58.
Spiked concentration in µg/L is not listed.
Mean reported value in µg/L is not listed.
Precision as RSD % is not listed.

SAMPLING METHOD Samples should be collected in borosilicate glass, linear polyethylene, polypropylene, or Teflon® bottles that have been prewashed with detergent and tap water, and rinsed with 1:1 nitric acid and tap water or 1:1 hydrochloric acid and tap water. Collect at least 2 g of solids and 200 mL of aqueous samples.

SAMPLE PRESERVATION Add nitric acid to make the samples pH <2.

MHT The maximum holding time for properly preserved samples is 6 months.

SAMPLE PREPARATION Preliminary treatment of most matrices is necessary because of the complexity and variability of sample matrices. Water samples that have been prefiltered and acidified will not need acid digestion. Methods for acid digestion of waters for total recoverable or dissolved metals, acid digestions of aqueous samples and extracts for total metals, and acid digestion of sediments, sludges, and soils are summarized below.

Total recoverable or dissolved metals in water — To prepare surface and groundwater samples for determination of total recoverable and dissolved metals, a 100-mL aliquot of well-mixed sample is acidified with concentrated nitric acid and concentrated hydrochloric acid, then heated until the volume is reduced to 15–20 mL. Adjust the final volume to 100 mL with reagent water.

Total metals in aqueous samples, soil and sediment extracts — To prepare aqueous samples, soil and sediment extracts, and wastes that contain suspended solids, a 100-mL aliquot is made acidic with concentrated nitric acid and the solution is evaporated to about 5 mL on a hot plate. Continue heating and adding additional acid until sample digestion is complete, which is usually indicated when the digestate is light in color or does not change in appearance. Evaporate the solution to about 3 mL and cool it and add a small quantity of 1:1 hydrochloric acid (10 mL/100 mL of final solution). Cover the beaker and reflux for 15 min. Wash down the beaker walls and filters or centrifuge the sample to remove silicates and other insoluble material. Filter the sample and adjust the final volume to 100 mL with reagent water and the final acid concentration to 10%.

Sediments, sludges, and soils — To prepare sediments, sludges and soil samples, transfer 1–2 g to a conical beaker and add 10 mL of 1:1 nitric acid, mix the slurry, and cover it with a watch glass. Heat the sample and reflux for 10–15 min without boiling. Allow it to cool, then add 5 mL of concentrated nitric acid and reflux for 30 min. Repeat last step and then allow the solution to evaporate to 5 mL without boiling. Cool and add 2 mL of water and 3 mL of 30% hydrogen peroxide. Cover and place the beaker on the hot plate. Heat and add 30% hydrogen peroxide in 1-mL aliquots with warming until the effervescence is minimal but do not add more than a total of 10 mL of 30% hydrogen peroxide. If the sample is being prepared for the analysis of Ag, Al, As, Ba, Be, Ca, Cd, Co, Cr, Cu, Fe, K, Mg, Mn, Mo, Na, Ni, Os, Pb, Se, Tl, V, and Zn, then add 5 mL of concentrated hydrochloric acid and 10 mL of water and return the covered beaker to a hot plate for 15 min of additional refluxing without boiling. Dilute the sample to a 100 mL volume with water after cooling and filter or centrifuge to remove particulates.

QUALITY CONTROL Laboratory control samples must be analyzed for each analytical method. A method blank should be analyzed with each batch of samples. The effect of the matrix on method performance must be demonstrated: when appropriate, there should be at least one matrix spike and either one matrix duplicate or one matrix spike duplicate per analytical batch. The bias and precision of the method, as well as the method detection limit for each specific matrix type, must be measured.

Dilute and reanalyze samples that are more concentrated than the linear calibration limit. Employ a minimum of one reagent blank per sample batch to determine if contamination or any memory effects are occurring. Whenever a new or unusual sample matrix is encountered, perform either a serial dilution test or a matrix spike addition test to ensure that neither positive or negative interferences are operating on any of the analyte elements. Check the instrument standardization by verifying calibration every 10 samples using a calibration blank and a check standard.

REFERENCE Test Methods for Evaluating Solid Waste (SW-846). U.S. EPA. 1983. Method 6010, Rev. 0, Sept. 1986. Office of Solid Wastes, Washington, DC.

Silicon EPA Method 200.7
CAS #7440-21-3

TITLE Inductively Coupled Plasma

MATRIX Dissolved, suspended or (ICP) total element in drinking and surface waters and in domestic and industrial wastewaters.

APPLICATION The method covers the determination of 25 metals. Dissolved elements are determined in filtered and acidified samples after appropriate digestion (which increases dissolved solids). Its primary advantage is that ICP instruments allow simultaneous or rapid sequential determination of many elements in a short time. Samples are first nebulized and the aerosol is transported to a plasma torch in which element specific atomic line emission spectra are produced by a radio frequency inductively coupled plasma. Background correction is required for trace element detection except in the case of line broadning.

INTERFERENCES There are spectral, physical, and chemical interferences. The primary disadvantage of ICP instruments is background radiation from other elements and the plasma gases (spectral interferences). Changes in sample viscosity and surface tension with samples containing high dissolved solids (especially those exceeding 1500 mg/L) or high acid concentrations can cause physical interferences. Ionization effects, solute vaporization and molecular compound formation can cause chemical interferences. Chromium and vanadium can cause interference at the 100 mg/L level.

INSTRUMENTATION Inductively coupled argon plasma emission spectroscopy. 288.158 nm wavelength

RANGE Not listed.

MDL 58 µg/L.

PRECISION Not listed.

ACCURACY Mean recovery = 93% ± 6% of spiked elements for all wastes.

SAMPLING METHOD Wash sample container with detergent and tap water, rinse with 1 + 1 nitric acid and tap water, then rinse with 1 + 1 hydrochloric acid and tap water, then rinse with deionized, distilled water in that order. Perform any filtration or acid preservation steps when the sample is collected or as soon as possible thereafter.

STABILITY Cool samples to 4°C.

MHT 24 h.

QUALITY CONTROL Mixed calibration standards, an instrument check standard, and an interference check solution are used in addition to a quality control sample. The quality control sample should be prepared in the same acid matrix as the calibration standards at 10 times the instrumental detection limits and in accordance with the instructions provided by the supplier. Furthermore, two types of blanks are required: a calibration blank and a reagent blank.

REFERENCE Method 200.7, U.S. EPA, EMSL-Cincinnati, OH, Nov. 1980

Silver   **EPA Method 6010**
CAS #7440-22-4

TITLE Inductively Coupled Plasma-Atomic Emission Spectroscopy

MATRIX This method is applicable to the determination of trace elements, including metals, in groundwater, soils, sludges, sediments, and other solid wastes. All matrices require digestion prior to analysis. The method of standard addition must be used for the analysis of all sample digests unless either serial dilution or matrix spike addition demonstrates it is not required.

METHOD SUMMARY Method 6010 covers 25 elements using ICP analysis. It measures element-emitted light by optical spectrometry. Samples, following an appropriate acid digestion, are nebulized and the resulting aerosol is transported to the plasma torch. Element-specific atomic line emission spectra are produced by a radio-frequency inductively coupled plasma.

INTERFERENCES Interferences may be categorized as spectral or non-spectral. Spectral interferences are caused by overlap of a spectral line from another element, unresolved overlap of molecular band spectra, background contribution from continuous or recombination phenomenon, and stray light from the line emission of high concentration elements. Non-spectral interferences include physical and chemical interferences. Physical interferences are effects associated with the sample nebulization and transport processes. Changes in viscosity and surface tension can cause significant inaccuracies. Chemical interferences include molecular compound formation, ionization effects, and solute vaporization effects. Normally these effects are not significant and can be minimized by careful selection of operating conditions. Chemical interferences are highly dependent on matrix type and the specific analyte element.

INSTRUMENTATION An inductively coupled argon plasma emission spectrometer (ICP) capable of background correction is required.

PRECISION & ACCURACY Detection limits, sensitivity, and optimum ranges of the metals will vary with the matrices and model of the spectrometer. In a single lab evaluation, seven wastes were analyzed for 22 elements. The mean percent relative standard deviation from triplicate analyses for all elements and wastes was 9 ± 2%. The mean percent recovery of spiked elements for all wastes was 93 ± 6%. Spike levels ranged from 100 µg/L to 100 mg/L. The wastes included sludges and industrial wastewaters.

Estimated instrument detection limit in µg/L is 7.
Spiked concentration in µg/L is not listed.
Mean reported value in µg/L is not listed.
Precision as RSD % is not listed.

SAMPLING METHOD Samples should be collected in borosilicate glass, linear polyethylene, polypropylene, or Teflon® bottles that have been prewashed with detergent and tap water, and rinsed with 1:1 nitric acid and tap water or 1:1 hydrochloric acid and tap water. Collect at least 2 g of solids and 200 mL of aqueous samples.

SAMPLE PRESERVATION Add nitric acid to make the samples pH <2.

MHT The maximum holding time for properly preserved samples is 6 months.

SAMPLE PREPARATION Preliminary treatment of most matrices is necessary because of the complexity and variability of sample matrices. Water samples that have been prefiltered and acidified will not need acid digestion. Methods for acid digestion of waters for total recoverable or dissolved metals, acid digestions of aqueous samples and extracts for total metals, and acid digestion of sediments, sludges, and soils are summarized below.

Total recoverable or dissolved metals in water — To prepare surface and groundwater samples for determination of total recoverable and dissolved metals, a 100-mL aliquot of well-mixed sample is acidified with concentrated nitric acid and concentrated hydrochloric acid, then heated until the volume is reduced to 15–20 mL. Adjust the final volume to 100 mL with reagent water.

Total metals in aqueous samples, soil and sediment extracts — To prepare aqueous samples, soil and sediment extracts, and wastes that contain suspended solids, a 100-mL aliquot is made acidic with concentrated nitric acid and the solution is evaporated to about 5 mL on a hot plate. Continue heating and adding additional acid until sample digestion is complete, which is usually indicated when the digestate is light in color or does not change in appearance. Evaporate the solution to about 3 mL and cool it and add a small quantity of 1:1 hydrochloric acid (10 mL/100 mL of final solution). Cover the beaker and reflux for 15 min. Wash down the beaker walls and filters or centrifuge the sample to remove silicates and other insoluble material. Filter the sample and adjust the final volume to 100 mL with reagent water and the final acid concentration to 10%.

Sediments, sludges, and soils — To prepare sediments, sludges and soil samples, transfer 1–2 g to a conical beaker and add 10 mL of 1:1 nitric acid, mix the slurry, and cover it with a watch glass. Heat the sample and reflux for 10–15 min without boiling. Allow it to cool, then add 5 mL of concentrated nitric acid and reflux for 30 min. Repeat last step and then allow the solution to evaporate to 5 mL without boiling. Cool and add 2 mL of water and 3 mL of 30% hydrogen peroxide. Cover and place the beaker on the hot plate. Heat and add 30% hydrogen peroxide in 1-mL aliquots with warming until the effervescence is minimal but do not add more than a total of 10 mL of 30% hydrogen peroxide. If the sample is being prepared for the analysis of Ag, Al, As, Ba, Be, Ca, Cd, Co, Cr, Cu, Fe, K, Mg, Mn, Mo, Na, Ni, Os, Pb, Se, Tl, V, and Zn, then add 5 mL of concentrated hydrochloric acid and 10 mL of water and return the covered beaker to a hot plate for 15 min of additional refluxing without boiling. Dilute the sample to a 100 mL volume with water after cooling and filter or centrifuge to remove particulates.

QUALITY CONTROL Laboratory control samples must be analyzed for each analytical method. A method blank should be analyzed with each batch of samples. The effect of the matrix on method performance must be demonstrated: when appropriate, there should be at least one matrix spike and either one matrix duplicate or one matrix spike duplicate per analytical batch. The bias and precision of the method, as well as the method detection limit for each specific matrix type, must be measured.

Dilute and reanalyze samples that are more concentrated than the linear calibration limit. Employ a minimum of one reagent blank per sample batch to determine if contamination or any memory effects are occurring. Whenever a new or unusual sample matrix is encountered, perform either a serial dilution test or a matrix spike addition test to ensure that neither positive or negative interferences are operating on any of the analyte elements. Check the instrument standardization by verifying calibration every 10 samples using a calibration blank and a check standard.

REFERENCE Test Methods for Evaluating Solid Waste (SW-846). U.S. EPA. 1983. Method 6010, Rev. 0, Sept. 1986. Office of Solid Wastes, Washington, DC.

Silver   **EPA Method 200.7**
CAS #7440-22-4

TITLE Inductively Coupled Plasma

MATRIX Dissolved, suspended or (ICP) total element in drinking and surface waters and in domestic and industrial wastewaters.

APPLICATION The method covers the determination of 25 metals. Dissolved elements are determined in filtered and acidified samples after appropriate digestion (which increases dissolved solids). Its primary advantage is that ICP instruments allow simultaneous or rapid sequential determination of many elements in a short time. Samples are first nebulized and the aerosol is transported to a plasma torch in which element specific atomic line emission spectra are produced by a radio frequency inductively coupled plasma. Background correction is required for trace element detection except in the case of line broadning.

INTERFERENCES There are spectral, physical, and chemical interferences. The primary disadvantage of ICP instruments is background radiation from other elements and the plasma gases (spectral interferences). Changes in sample viscosity and surface tension with samples containing high dissolved solids (especially those exceeding 1500 mg/L) or high acid concentrations can cause physical interferences. Ionization effects, solute vaporization and molecular compound formation can cause chemical interferences.

INSTRUMENTATION Inductively coupled argon plasma emission spectroscopy. 328.068 nm wavelength

RANGE Not listed.

MDL 7 µg/L.

PRECISION Not listed.

ACCURACY Mean recovery = 93% ± 6% of spiked elements for all wastes.

SAMPLING METHOD Wash sample container with detergent and tap water, rinse with 1 + 1 nitric acid and tap water, then rinse with 1 + 1 hydrochloric acid and tap water, then rinse with deionized, distilled water in that order. Perform any filtration or acid preservation steps when the sample is collected or as soon as possible thereafter.

STABILITY Cool samples to 4°C.

MHT 24 h.

QUALITY CONTROL Mixed calibration standards, an instrument check standard, and an interference check solution are used in addition to a quality control sample. The quality control sample should be prepared in the same acid matrix as the calibration standards at 10 times the instrumental detection limits and in accordance with the instructions provided by the supplier. Furthermore, two types of blanks are required: a calibration blank and a reagent blank.

REFERENCE Method 200.7, U.S. EPA, EMSL-Cincinnati, OH, Nov. 1980

Silver   **EPA Method 7760**
CAS #7440-22-4

TITLE Atomic Absorption (AA) Direct Aspiration

MATRIX Wastes, mobility procedure extracts, soils and groundwater.

APPLICATION Prior to analysis, samples must be prepared for direct aspiration. Sample preparation method varies with the matrix. Following appropriate dissolution of the sample, a silver lamp light beam is directed through aspirated aliquot in the flame, into a monochromator and detector. The resulting absorption of hollow cathode radiation is proportional to the silver concentration.

INTERFERENCES Background correction is required. Store ag standards and samples in brown bottles if possible. Silver chloride is insoluble, so avoid hydrochloric acid unless silver is in solution as a chloride complex. Monitor samples and standards for viscosity differences since this may alter the aspiration rate.

INSTRUMENTATION Atomic absorption spectrometer. Silver hollow cathode lamp. (328.1 nm wavelength). Strip-chart recorder.

RANGE 0.1–4 mg/L

MDL 0.01 mg/L

PRECISION Standard deviation = ±8.8 at 50 µg Ag/L (50 labs)

ACCURACY Relative error = 10.6% at 50 µg Ag/L (50 labs)

SAMPLING METHOD Use prewashed plastic or glass containers.

STABILITY Non aqueous samples: cool to 4°C and analyze as soon as possible. Aqueous samples: add nitric acid to pH <2.

MHT 6 months

QUALITY CONTROL Run one spike duplicate sample for every 10 samples. Verify calibration with an independently prepared check standard every 15 samples.

REFERENCE Method 7760, SW-846, 3rd ed., Nov.1986.

Silver   **EPA Method 7761**
CAS #7440-22-4

TITLE Atomic Absorption (AA) Furnace Technique

MATRIX Wastes, mobility procedure extracts, soils and groundwater.

APPLICATION Sample preparation (digestion) of matrices is always necessary and varies with the matrix. An aliquot of digestate is placed in a graphite tube in the furnace and slowly evaporated, charred and atomized. Absorption of lamp radiation during atomization is proportional to silver concentration.

INTERFERENCES The furnace technique is subject to chemical interferences.

Composition of the sample matrix can effect analysis. Modify matrix to remove interferences. Avoid nonspecific absorption and scattering. Memory effects occur if silver is not volatilized and removed from the furnace.

INSTRUMENTATION Atomic absorption spectrometer. Silver hollow cathode lamp or electrodeless discharge lamp. Graphite furnace. Strip-chart recorder. (328.1 nm wavelength).

RANGE 1–25 µg/L

MDL 0.2 µg/L

PRECISION Standard deviation = ±0.40, 0.70, 0.90 at 25, 50, 75 µg Ag/L

ACCURACY Recoveries = 94, 100, 104% at 25, 50, 75 µg Ag/L

SAMPLING METHOD Use prewashed plastic or glass containers.

STABILITY Non aqueous samples: cool to 4°C. Aqueous samples: add nitric acid to pH <2.

MHT 6 months

QUALITY CONTROL Run one spike replicate sample for every 10 samples or per analytical batch, whichever is more frequent. Verify calibration with an independently prepared check standard every 15 samples.

REFERENCE Method 7761, SW-846, 3rd ed., (Included as Rev. 0, Dec. 1987)

Simazine   **EPA Method 505**
CAS #122-34-9

TITLE Analysis of Organohalide Pesticides and Commercial Polychlorinated Biphenyl (PCB) Products in Water by Microextraction and Gas Chromatography. U.S. EPA Method 505, Rev. 2.0, 1989.

MATRIX This method is applicable to drinking water and raw source water. The latter should include most surface water and groundwater sources.

METHOD SUMMARY Method 505 covers 25 pesticides and commercial PCB products. This is a very sensitive method that is more useful for monitoring than for exploratory analyses. 5-mL of water are saturated with sodium chloride and then extracted by shaking with 2 mL of hexane. The sample extracts are transferred to an autosampler setup to inject 1–2 µL portions into a gas chromatograph (GC) for analysis. Alternatively, 1–2 µL portions of samples, blanks, and standards may be manually injected. Each extract is analyzed by capillary GC/ECD with confirmation using either a second capillary column or GC/MS. The electron capture detector is easy to use, but it is a nonselective detector. The microextraction technique also eliminates the expensive sample preparation costs of other methods, but it has the disadvantage of being less sensitive than most because the extracts are not concentrated.

INTERFERENCES Method interferences may be caused by contaminants in solvents, reagents, glassware, and other sample processing apparatus that lead to discrete artifacts or elevated baselines. Interfering contamination may occur when a sample containing low concentrations of analytes is analyzed immediately following a sample containing relatively high concentrations of the analytes. Matrix interferences also may be caused by contaminants that are coextracted from the sample; cleanup of sample extracts may be necessary in these cases. Some pesticides and commercial PCB products from aqueous solutions adhere to glass surfaces, so sample transfers and contact with glass surfaces should be minimized. Some pesticides are rapidly oxidized by chlorine so dechlorination with sodium thiosulfate at the time of sample collection is important. Also, splitless injectors may cause degradation of some pesticides.

INSTRUMENTATION A gas chromatograph/electron capture detector/data system, with temperature programming and split/splitless injector suitable for use with capillary columns is needed.

Column 1: 0.32 mm I.D. × 30 m fused silica capillary with chemically bond methyl polysiloxane phase (DB-1, 1.0 µm film, or equivalent).

Column 2: 0.32 mm I.D. × 30 m fused silica capillary with 1:1 mixed phase of dimethyl silicone and polyethylene glycol (Durawax-DX3, 0.25 µm film, or equivalent).

Column 3: 0.32 mm I.D. × 25 m fused silica capillary with chemically bonded 50:50 methyl-phenyl silicone (OV-17, 1.5 µm film, or equivalent).

Column 1 should be used as the primary analytical column. Columns 2 and 3 are recommended for use as confirmatory columns when GC/MS confirmation is not available.

PRECISION & ACCURACY Method detection limits are dependent upon the characteristics of the gas chromatographic system used. Analytes that are not separated chromatographically cannot be individually identified and used in the same calibration mixture or water samples unless an alternative technique for identification and quantification, such as mass spectrometry, is used.

The concentration(s) (in µg/L) used for these QC measurements was 25 and 60.

The MDL (in µg/L) was 6.8.

The accuracy (% recovery) for reagent water at the above concentration(s) was 99 and 65 and the precision (%) was 8.3 and 3.6.

The accuracy (% recovery) for groundwater at the above concentration(s) was 97 and 59 and the precision (%) was 9.2 and 18.0.

The accuracy (% recovery) for tap water at the above concentration(s) was 102 and 67 and the precision (5) was 13.4 and 6.2.

Note: No range of concentrations is provided with this method.

SAMPLING METHOD Collect samples using a 40-mL screw-cap vial (prewashed with detergent, rinsed with distilled water and oven dried at 400°C for one h) with a Teflon®-faced silicone septum. Collect bubble-free samples and place the septum with the Teflon® side down on the water.

SAMPLE PRESERVATION If residual chlorine is present in the water add about 3 mg of sodium thiosulfate to each vial before samples are collected to remove the chlorine. Alternatively, add 75 µL of 0.04 g/mL solution of sodium thiosulfate to each vial just prior to sampling. Immediately cool samples to 4°C, and store them in a solvent-free refrigerator at 4°C until analysis.

MHT The maximum holding time is 14 days from the time the sample was collected until it must be analyzed.

SAMPLE PREPARATION Remove the sample from storage and allow it to come to room temperature. Remove a 5-mL volume from each container and weigh the container to the nearest 0.1 g. Add 6 g of sodium chloride and 2.0 mL of hexane to each sample bottle. Recap the sample and shake it vigorously for one min. Allow the water and hexane phases to separate, remove the cap, and transfer 0.5 mL of hexane into an autosampler vial using a disposable glass pipette. Transfer the remaining hexane phase into a second autosampler vial and store at 4°C for reanalysis, if necessary. Discard the remaining sample/hexane mixture and reweigh the empty container to determine net weight of sample.

QUALITY CONTROL Minimum quality control requirements are initial demonstration of lab capability, analysis of lab reagent blanks, fortified blanks, fortified sample matrix, and quality control samples. The lab must analyze at least one fortified blank per sample set, or at least one for every 20 samples. The fortifying concentration of each analyte should be 10 times the method detection limit or the maximum calibration limit (MCL), whichever is less. Calculate accuracy as percent recovery and develop control limits from the mean percent recovery and standard deviation.

The lab must add a known concentration of the analytes to a minimum of 10% of the routine samples, or one lab fortified sample matrix per sample set. Calculate the percent recovery for each analyte and compare to the control limits established from the analyses of the fortified blanks.

EPA CONTACT & HOTLINE For technical questions contact Dr. Baldev Bathija, U.S. EPA, Office of Ground Water and Drinking Water (WH-550D), 401 M St. SW, Washington, DC 20460. Tel. (202) 260-3040. For further information the EPA Safe Drinking Water Hotline may be called at: (800) 426-4791.

REFERENCE Methods for the Determination of Organic Compounds in Drinking Water, EPA/600/4-88/039 (revised July 1991). U.S. EPA Environmental Monitoring Systems Laboratory, Cincinnati, OH, 45268, U.S.A. Available from the National Technical Information Service (NTIS), 5285 Port Royal Road, Springfield, VA 22161; Tel. 800-553-6847. NTIS Order Number is PB91-231480.

Simazine   **EPA Method 507**
CAS #122-34-9

TITLE Determination of Nitrogen and Phosphorus-Containing Pesticides in Water by GC/NPD

MATRIX This method is applicable to the determination of certain nitrogen and phosphorus-containing pesticides in finished drinking water and groundwater.

METHOD SUMMARY Method 507 covers 46 nitrogen- and phosphorus-containing pesticides. A 1-L sample is fortified with a surrogate standard, salted, buffered, extracted with methylene chloride, and concentrated; then the solvent is exchanged with methyl tert-butyl ether (MTBE) and concentrated again, and a 2-µL aliquot of a sample extract is injected into a GC system equipped with a selective nitrogen-phosphorus detector and a capillary column for analysis.

INTERFERENCES Method interferences may be caused by contaminants in solvents, reagents, glassware, and other sample processing apparatus. Interfering contamination may occur when a sample containing low concentrations of analytes is analyzed immediately following a sample containing relatively high concentrations. One or more injections of MTBE should be made following the analysis of a sample with high concentrations of analytes to check for analyte carryover. Matrix interferences may be caused by contaminants that are coextracted from the sample. The extent of matrix interferences will vary considerably from source to source, depending upon the water sampled.

INSTRUMENTATION A gas chromatograph system (GC) equipped with a nitrogen-phosphorus detector (NPD) is needed.

Column 1: 30 m × 0.25 mm I.D. DB-5 bonded fused silica column, 0.25 μm film thickness, or equivalent.

Column 2: 30 m × 0.25 mm I.D. DB-1701 bonded fused silica column, 0.25 μm film thickness, or equivalent.

PRECISION & ACCURACY This method has been validated in a single lab and estimated detection limits (EDLs) have been determined for each analyte. Observed detection limits may vary among waters, depending upon the nature of the interferences in the sample matrix and the specific instrumentation used. Analytes that are not separated chromatographically cannot be individually identified and measured unless an alternative technique for identification and quantification exist.

The estimated detection limit (in μg/L) was 0.075. The EDL is defined as either method detection limit or a level of compound in a sample yielding a peak in the final extract with signal-to-noise ratio of approximately 5, whichever value is higher.

The concentration used for these measurements (in μg/L) was 0.75.
The accuracy (as % recovery) was 100.
The precision (% RSD) was 7.

SAMPLING METHOD Grab samples are collected in 1-L glass sample bottles (prewashed with detergent and hot tap water, rinsed with reagent water, and dried in an oven at 400°C for 1 h) with screw caps lined with PTFE-fluorocarbon.

SAMPLE PRESERVATION Add mercuric chloride to the sample bottle in amounts to produce a concentration of 10 mg/L. If residual chlorine is present, add 80 mg of sodium thiosulfate/L of sample to the sample bottle prior to collection. After collection, seal bottle and shake vigorously for 1 min, then cool the sample to 4°C immediately and store it at 4°C in the dark until extraction.

MHT Maximum holding time of the samples, and in some cases the extracts, is 14 days.

SAMPLE PREPARATION Fortify the sample with 50 μL of the surrogate standard solution, adjust to pH 7 with phosphate buffer, add 100 g NaCl to the sample, and seal and shake to dissolve the salt; then extract with methylene chloride in a separatory funnel or in a mechanical tumbler bottle. Dry the extract by pouring it through a solvent-rinsed drying column containing about 10 cm of anhydrous sodium sulfate. Collect the extract in a Kuderna-Danish (K-D) concentrator and rinse the column with 20–30 mL methylene chloride. Concentrate the extract to about 2 mL and rinse the flask and its lower joint into the concentrator tube with 1 to 2 mL of methyl t-butyl ether (MTBE). Add 5–10 mL of MTBE and concentrate the extract twice (adding more MTBE) to a final volume of 5.0 mL and store it at 4°C until analysis.

Note: If methylene chloride is not completely removed from the final extract, it may cause detector problems.

QUALITY CONTROL Minimum quality control requirements are initial demonstration of lab capability, determination of surrogate compound recoveries in each sample and blank, monitoring internal standard peak area or height in each sample and blank, analysis of lab reagent blanks, lab fortified samples, lab fortified blanks, and other QC samples. A lab reagent blank is analyzed to demonstrate that all glassware and reagent interferences are under control.

Initial demonstration of capability is fulfilled by analyzing four fortified reagent water samples with the recovery value for each analyte falling within the acceptable range (±30% average recovery). Surrogate recoveries from samples or method blanks must be 70–130%. The internal standard response for any sample chromatogram should not deviate from the daily calibration check standard's internal standard response by more than 30% or lab fortified blanks and sample matrices are used to assess lab performance and analyte recovery, respectively.

If the response for the target analyte peak exceeds the working range of the system, dilute the extract and reanalyze. Alternative techniques such as an alternate detector or second chromatography column should be used to confirm peak identification when sample components are not resolved adequately.

EPA CONTACT & HOTLINE For technical questions contact Dr. Baldev Bathija, U.S. EPA, Office of Ground Water and Drinking Water (WH-550D), 401 M St. SW, Washington, DC 20460. Tel. (202) 260-3040. For further information the EPA Safe Drinking Water Hotline may be called at: (800) 426-4791.

REFERENCE Methods for the Determination of Organic Compounds in Drinking Water, EPA/600/4-88/039 (revised July 1991). U.S. EPA Environmental Monitoring Systems Laboratory, Cincinnati, OH, 45268, U.S.A. Available from the National Technical Information Service (NTIS), 5285 Port Royal Road, Springfield, VA 22161; Tel. 800-553-6847. NTIS Order Number is PB91-231480.

Simetryn   **EPA Method 507**
CAS #1014-70-6

TITLE Determination of Nitrogen and Phosphorus-Containing Pesticides in Water by GC/NPD

MATRIX This method is applicable to the determination of certain nitrogen and phosphorus-containing pesticides in finished drinking water and groundwater.

METHOD SUMMARY Method 507 covers 46 nitrogen- and phosphorus-containing pesticides. A 1-L sample is fortified with a surrogate standard, salted, buffered, extracted with methylene chloride, and concentrated; then the solvent is exchanged with methyl tert-butyl ether (MTBE) and concentrated again, and a 2-μL aliquot of a sample extract is injected into a GC system equipped with a selective nitrogen-phosphorus detector and a capillary column for analysis.

INTERFERENCES Method interferences may be caused by contaminants in solvents, reagents, glassware, and other sample processing apparatus. Interfering contamination may occur when a sample containing low concentrations of analytes is analyzed immediately following a sample containing relatively high concentrations. One or more injections of MTBE should be made following the analysis of a sample with high concentrations of analytes to check for analyte carryover. Matrix interferences may be caused by contaminants that are coextracted from the sample. The extent of matrix interferences will vary considerably from source to source, depending upon the water sampled.

INSTRUMENTATION A gas chromatograph system (GC) equipped with a nitrogen-phosphorus detector (NPD) is needed.

Column 1: 30 m × 0.25 mm I.D. DB-5 bonded fused silica column, 0.25 μm film thickness, or equivalent.

Column 2: 30 m × 0.25 mm I.D. DB-1701 bonded fused silica column, 0.25 μm film thickness, or equivalent.

PRECISION & ACCURACY This method has been validated in a single lab and estimated detection limits (EDLs) have been determined for each analyte. Observed detection limits may vary among waters, depending upon the nature of the interferences in the sample matrix and the specific instrumentation used. Analytes that are not separated chromatographically cannot be individually identified and measured unless an alternative technique for identification and quantification exist.

The estimated detection limit (in μg/L) was 0.25. The EDL is defined as either method detection limit or a level of compound in a sample yielding a peak in the final extract with signal-to-noise ratio of approximately 5, whichever value is higher.

The concentration used for these measurements (in μg/L) was 2.5.
The accuracy (as % recovery) was 99.
The precision (% RSD) was 5.

SAMPLING METHOD Grab samples are collected in 1-L glass sample bottles (prewashed with detergent and hot tap water, rinsed with reagent water, and dried in an oven at 400°C for 1 h) with screw caps lined with PTFE-fluorocarbon.

SAMPLE PRESERVATION Add mercuric chloride to the sample bottle in amounts to produce a concentration of 10 mg/L. If residual chlorine is present, add 80 mg of sodium thiosulfate/L of sample to the sample bottle prior to collection. After collection, seal bottle and shake vigorously for 1 min, then cool the sample to 4°C immediately and store it at 4°C in the dark until extraction.

MHT Maximum holding time of the samples, and in some cases the extracts, is 14 days.

SAMPLE PREPARATION Fortify the sample with 50 μL of the surrogate standard solution, adjust to pH 7 with phosphate buffer, add 100 g NaCl to the sample, and seal and shake to dissolve the salt; then extract with methylene chloride in a separatory funnel or in a mechanical tumbler bottle. Dry the extract by pouring it through a solvent-rinsed drying column containing about 10 cm of anhydrous sodium sulfate. Collect the extract in a Kuderna-Danish (K-D) concentrator and rinse the column with 20–30 mL methylene chloride. Concentrate the extract to about 2 mL and rinse the flask and its lower joint into the concentrator tube with 1 to 2 mL of methyl t-butyl ether (MTBE). Add 5–10 mL of MTBE and concentrate the extract twice (adding more MTBE) to a final volume of 5.0 mL and store it at 4°C until analysis.

Note: If methylene chloride is not completely removed from the final extract, it may cause detector problems.

QUALITY CONTROL Minimum quality control requirements are initial demonstration of lab capability, determination of surrogate compound recoveries in each sample and blank, monitoring internal standard peak area or height in each sample and blank, analysis of lab reagent blanks, lab fortified samples, lab fortified blanks, and other QC samples. A lab reagent blank is analyzed to demonstrate that all glassware and reagent interferences are under control.

Initial demonstration of capability is fulfilled by analyzing four fortified reagent water samples with the recovery value for each analyte falling within the acceptable range (±30% average recovery). Surrogate recoveries from samples or method blanks must be 70–130%. The internal standard response for any sample chromatogram should not deviate from the daily calibration check standard's internal standard response by more than 30% or lab fortified blanks and sample matrices are used to assess lab performance and analyte recovery, respectively.

If the response for the target analyte peak exceeds the working range of the system, dilute the extract and reanalyze. Alternative techniques such as an alternate detector or second chromatography column should be used to confirm peak identification when sample components are not resolved adequately.

EPA CONTACT & HOTLINE For technical questions contact Dr. Baldev Bathija, U.S. EPA, Office of Ground Water and Drinking Water (WH-550D), 401 M St. SW, Washington, DC 20460. Tel. (202) 260-3040. For further information the EPA Safe Drinking Water Hotline may be called at: (800) 426-4791.

REFERENCE Methods for the Determination of Organic Compounds in Drinking Water, EPA/600/4-88/039 (revised July 1991). U.S. EPA Environmental Monitoring Systems Laboratory, Cincinnati, OH, 45268, U.S.A. Available from the National Technical Information Service (NTIS), 5285 Port Royal Road, Springfield, VA 22161; Tel. 800-553-6847. NTIS Order Number is PB91-231480.

Sodium   **EPA Method 6010**
CAS #7440-23-5

TITLE Inductively Coupled Plasma-Atomic Emission Spectroscopy

MATRIX This method is applicable to the determination of trace elements, including metals, in groundwater, soils, sludges, sediments, and other solid wastes. All matrices require digestion prior to analysis. The method of standard addition must be used for the analysis of all sample digests unless either serial dilution or matrix spike addition demonstrates it is not required.

METHOD SUMMARY Method 6010 covers 25 elements using ICP analysis. It measures element-emitted light by optical spectrometry. Samples, following an appropriate acid digestion, are nebulized and the resulting aerosol is transported to the plasma torch. Element-specific atomic line emission spectra are produced by a radio-frequency inductively coupled plasma.

INTERFERENCES Interferences may be categorized as spectral or non-spectral. Spectral interferences are caused by overlap of a spectral line from another element, unresolved overlap of molecular band spectra, background contribution from continuous or recombination phenomenon, and stray light from the line emission of high concentration elements. Non-spectral interferences include physical and chemical interferences. Physical interferences are effects associated with the sample nebulization and transport processes. Changes in viscosity and surface tension can cause significant inaccuracies. Chemical interferences include molecular compound formation, ionization effects, and solute vaporization effects. Normally these effects are not significant and can be minimized by careful selection of operating conditions. Chemical interferences are highly dependent on matrix type and the specific analyte element.

INSTRUMENTATION An inductively coupled argon plasma emission spectrometer (ICP) capable of background correction is required.

PRECISION & ACCURACY Detection limits, sensitivity, and optimum ranges of the metals will vary with the matrices and model of the spectrometer. In a single lab evaluation, seven wastes were analyzed for 22 elements. The mean percent relative standard deviation from triplicate analyses for all elements and wastes was 9 ± 2%. The mean percent recovery of spiked elements for all wastes was 93 ± 6%. Spike levels ranged from 100 µg/L to 100 mg/L. The wastes included sludges and industrial wastewaters.

Estimated instrument detection limit in µg/L is 29.
Spiked concentration in µg/L is not listed.
Mean reported value in µg/L is not listed.
Precision as RSD % is not listed.

SAMPLING METHOD Samples should be collected in borosilicate glass, linear polyethylene, polypropylene, or Teflon® bottles that have been prewashed with detergent and tap water, and rinsed with 1:1 nitric acid and tap water or 1:1 hydrochloric acid and tap water. Collect at least 2 g of solids and 200 mL of aqueous samples.

SAMPLE PRESERVATION Add nitric acid to make the samples pH <2.

MHT The maximum holding time for properly preserved samples is 6 months.

SAMPLE PREPARATION Preliminary treatment of most matrices is necessary because of the complexity and variability of sample matrices. Water samples that have been prefiltered and acidified will not need acid digestion. Methods for acid digestion of waters for total recoverable or dissolved metals, acid digestions of aqueous samples and extracts for total metals, and acid digestion of sediments, sludges, and soils are summarized below.

Total recoverable or dissolved metals in water — To prepare surface and groundwater samples for determination of total recoverable and dissolved metals, a 100-mL aliquot of well-mixed sample is acidified with concentrated nitric acid and concentrated hydrochloric acid, then heated until the volume is reduced to 15–20 mL. Adjust the final volume to 100 mL with reagent water.

Total metals in aqueous samples, soil and sediment extracts — To prepare aqueous samples, soil and sediment extracts, and wastes that contain suspended solids, a 100-mL aliquot is made acidic with concentrated nitric acid and the solution is evaporated to about 5 mL on a hot plate. Continue heating and adding additional acid until sample digestion is complete, which is usually indicated when the digestate is light in color or does not change in appearance. Evaporate the solution to about 3 mL and cool it and add a small quantity of 1:1 hydrochloric acid (10 mL/100 mL of final solution). Cover the beaker and reflux for 15 min. Wash down the beaker walls and filters or centrifuge the sample to remove silicates and other insoluble material. Filter the sample and adjust the final volume to 100 mL with reagent water and the final acid concentration to 10%.

Sediments, sludges, and soils — To prepare sediments, sludges and soil samples, transfer 1–2 g to a conical beaker and add 10 mL of 1:1 nitric acid, mix the slurry, and cover it with a watch glass. Heat the sample and reflux for 10–15 min without boiling. Allow it to cool, then add 5 mL of concentrated nitric acid and reflux for 30 min. Repeat last step and then allow the solution to evaporate to 5 mL without boiling. Cool and add 2 mL of water and 3 mL of 30% hydrogen peroxide. Cover and place the beaker on the hot plate. Heat and add 30% hydrogen peroxide in 1-mL aliquots with warming until the effervescence is minimal but do not add more than a total of 10 mL of 30% hydrogen peroxide. If the sample is being prepared for the analysis of Ag, Al, As, Ba, Be, Ca, Cd, Co, Cr, Cu, Fe, K, Mg, Mn, Mo, Na, Ni, Os, Pb, Se, Tl, V, and Zn, then add 5 mL of concentrated hydrochloric acid and 10 mL of water and return the covered beaker to a hot plate for 15 min of additional refluxing without boiling. Dilute the sample to a 100 mL volume with water after cooling and filter or centrifuge to remove particulates.

QUALITY CONTROL Laboratory control samples must be analyzed for each analytical method. A method blank should be analyzed with each batch of samples. The effect of the matrix on method performance must be demonstrated: when appropriate, there should be at least one matrix spike and either one matrix duplicate or one matrix spike duplicate per analytical batch. The bias and precision of the method, as well as the method detection limit for each specific matrix type, must be measured.

Dilute and reanalyze samples that are more concentrated than the linear calibration limit. Employ a minimum of one reagent blank per sample batch to determine if contamination or any memory effects are occurring. Whenever a new or unusual sample matrix is encountered, perform either a serial dilution test or a matrix spike addition test to ensure that neither positive or negative interferences are operating on any of the analyte elements. Check the instrument standardization by verifying calibration every 10 samples using a calibration blank and a check standard.

REFERENCE Test Methods for Evaluating Solid Waste (SW-846). U.S. EPA. 1983. Method 6010, Rev. 0, Sept. 1986. Office of Solid Wastes, Washington, DC.

Sodium   EPA Method 200.7
CAS #7440-23-5

TITLE Inductively Coupled Plasma

MATRIX Dissolved, suspended or (ICP) total element in drinking and surface waters and in domestic and industrial wastewaters.

APPLICATION The method covers the determination of 25 metals. Dissolved elements are determined in filtered and acidified samples after appropriate digestion (which increases dissolved solids). Its primary advantage is that ICP instruments allow simultaneous or rapid sequential determination of many elements in a short time. Samples are first nebulized and the aerosol is transported to a plasma torch in which element specific atomic line emission spectra are produced by a radio frequency inductively coupled plasma. Background correction is required for trace element detection except in the case of line broadning.

INTERFERENCES There are spectral, physical, and chemical interferences. The primary disadvantage of ICP instruments is background radiation from other elements and the plasma gases (spectral interferences). Changes in sample viscosity and surface tension with samples containing high dissolved solids (especially those exceeding 1500 mg/L) or high acid concentrations can cause physical interferences. Ionization effects, solute vaporization and molecular compound formation can cause chemical interferences. Thallium can cause interference at the 100 mg/L level.

INSTRUMENTATION Inductively coupled argon plasma emission spectroscopy. 588.995 nm wavelength.

RANGE Not listed.

MDL 29 µg/L.

PRECISION Not listed.

ACCURACY Mean recovery = 93% ± 6% of spiked elements for all wastes.

SAMPLING METHOD Wash sample container with detergent and tap water, rinse with 1 + 1 nitric acid and tap water, then rinse with 1 + 1 hydrochloric acid and tap water, then rinse with deionized, distilled water in that order. Perform any filtration or acid preservation steps when the sample is collected or as soon as possible thereafter.

STABILITY Cool samples to 4°C.

MHT 24 h.

QUALITY CONTROL Mixed calibration standards, an instrument check standard, and an interference check solution are used in addition to a quality control sample. The quality control sample should be prepared in the same acid matrix as the calibration standards at 10 times the instrumental detection limits and in accordance with the instructions provided by the supplier. Furthermore, two types of blanks are required: a calibration blank and a reagent blank.

REFERENCE Method 200.7, U.S. EPA, EMSL-Cincinnati, OH, Nov. 1980

Sodium   EPA Method 273.1
CAS #7440-23-5

TITLE Metals (Total, Dissolved, Suspended) (AA)

MATRIX Drinking, surface, and saline direct aspiration waters. Wastewater.

APPLICATION Date issued 1971. Editorial Rev. 1974. Sample is aspirated and atomized in a flame. Light beam from hollow cathode (made of Na) lamp is directed through flame into monochromator, then to detector which measures amount absorbed light.

INTERFERENCES There can be interference from presence of high dissolved solids in sample. Chemical and ionization interferences can occur. Ionization may be controlled by adding potassium (1000 mg/L) to both standards and samples.

INSTRUMENTATION AAS. Sodium (Na) hollow cathode lamp. Burner. Pipets. Strip chart recorder.

RANGE 0.03–1 mg/L at 589.6 nm wavelength

MDL 0.002 mg/L.

PRECISION SD = ±(0.1 and 0.8) at 8.2 and 52 mg Na/L.

ACCURACY Recoveries = (102 and 100)% at 8.2 and 52 mg Na/L.

SAMPLING METHOD Use prewashed plastic or glass containers.

STABILITY HNO_3 to pH <2.

MHT 6 months.

QUALITY CONTROL After calibration curve composed of a minimum of a reagent blank and 3 standards has been prepared, subsequent calibration curves must be verified by use of at least a reagent blank and one standard near MCL. Must check within 10% of original curve. (For drinking water analysis)

REFERENCE EPA Methods for the Chemical Analysis of Water and Wastes.

EPA-600/4-79-020, U.S. EPA, EMSL, 1979.

Sodium   EPA Method 7770
CAS #7440-23-5

TITLE Atomic Absorption (AA)

MATRIX Drinking, surface, and direct aspiration saline waters. Wastewater.

APPLICATION Sample is aspirated and atomized in a flame. A light beam from a sodium hollow cathode lamp is directed through the flame into a monochromator and onto a detector. Since wavelength of light beam is specific for na, light energy absorbed by detector is measure of sodium.

INTERFERENCES The most troublesomee type is chemical, caused by lack of absorption of atoms bound in molecular combination in the flame. High dissolved solids in sample may result in nonatomic absorbance interference. Sodium is a universal contaminant; use the method with great care.

INSTRUMENTATION Atomic absorption spectrometer. Sodium hollow cathode lamp. (589.6 nm wavelength)

RANGE 0.03–1 mg/L

MDL 0.002 mg/L

PRECISION Standard deviation = ±0.10 and 0.80 at 8.20 and 52 mg Na/L

ACCURACY Recoveries = 102 and 100% at 8.20 and 52 mg Na/L

SAMPLING METHOD Use glass or plastic containers. Collect 200 g of solids and 600 mL of liquid samples.

STABILITY Cool solid samples to 4°C and analyze as soon as possible. Add nitric acid to liquid samples to pH <2.

MHT 6 months.

QUALITY CONTROL At least one duplicate and one spike sample should be run every 20 samples or with each matrix type to verify precision of the method.

For 20 or more samples per day, verify working standard curve. Run an additional standard at or near mid-range every 10 samples.

REFERENCE Method 7770, SW-846, 3rd ed., Nov.1986.

Specific Conductance   **EPA Method 9050**
CAS #None

TITLE Specific Conductance

MATRIX This method is applicable to the measurement of the specific conductance of drinking waters, groundwaters, surface waters, saline waters, and domestic and industrial aqueous wastes. It is not applicable to solid samples or to organic samples.

METHOD SUMMARY The specific conductance of a sample is measured using a self-contained conductivity meter. When possible, samples are analyzed at 25°C. If samples are analyzed at different temperatures, temperature corrections must be made; many instruments provide automatic temperature-corrected readings.

To determine the cell constant, the conductivity cell is rinsed with at least three portions of 0.01N potassium chloride solution; the temperature of a fourth portion is adjusted to 25.0°C. The resistance of this portion is measured and temperature is noted. The cell constant, C, is computed using the equation below.

$$C = (0.001413)(RKCl)(1 + 0.0191)(t-25)$$

where: RKCl = measured resistance in ohms; and t = observed temperature in degrees C.

INTERFERENCES Platinum electrodes can degrade and cause erratic results. When this happens the electrode should be replatinized. The specific conductance cell can become coated with oil and other materials. It is essential that the cell be thoroughly rinsed between samples.

INSTRUMENTATION A self-contained conductivity instrument consisting of a source of alternating current, a Wheatstone bridge, null indicator, and a platinum electrode or non-platinum electrode specific conductance cell are required.

PRECISION & ACCURACY

Variation of relative standard deviation and error as a function of conductivity units is listed below.

CONDUCTIVITY PRECISION & ACCURACY

Conductivity unhos/cm	No. of Results	Rel. Std. Dev. %	Relative Error %
147.0	117	8.6	9.4
303.0	120	7.8	1.9
228.0	120	8.4	3.0

Three synthetic samples were tested to produce the data in the above table.

SAMPLE PREPARATION No specific sample preparation procedure is required when measuring the conductance of water samples.

QUALITY CONTROL An independently prepared check standard should be analyzed with each batch of samples to verify calibration. Analyze one duplicate sample for every 10 samples.

REFERENCE Test Methods for Evaluating Solid Waste (SW-846). U.S. EPA. 1983. Method 9050A, Rev. 1, Nov. 1990. Office of Solid Wastes, Washington, DC.

Squalene   **EPA Method 1625**
CAS #7683-64-9

TITLE Semivolatile Organic Compounds by Isotope Dilution GC/MS

MATRIX The compounds may be determined in waters, soils, and municipal sludges by this method.

METHOD SUMMARY This method is used to determine 176 semivolatile toxic organic pollutants associated with the CWA (as amended 1987); the RCRA (as amended 1986); the CERCLA (as amended 1986); and other compounds amenable to extraction and analysis by capillary column gas chromatography-mass spectrometry (GC/MS).

Stable isotopically-labeled analogs of the compounds of interest are added to the sample. If the solids content is less than 1%, a 1-L sample is extracted at pH 12–13, then at pH <2 with methylene chloride using continuous extraction techniques.

If the solids content is 30% or less, the sample is diluted to 1% solids with reagent water, homogenized ultrasonically, and extracted at pH 12–13, then at pH <2 with methylene chloride using continuous extraction techniques. If the solids content is greater than 30%, the sample is extracted using ultrasonic techniques.

Each extract is dried over sodium sulfate, concentrated to a volume of 5 mL, cleaned up using GPC, if necessary, and concentrated. Extracts are concentrated to 1 mL if GPC is not performed, and to 0.5 mL if GPC is performed.

An internal standard is added to the extract, and a 1-mL aliquot of the extract is injected into the GC. The compounds are separated by GC and detected by a MS. The labeled compounds serve to correct the variability of the analytical technique.

INTERFERENCES Solvents, reagents, glassware, and other sample processing hardware may yield artifacts and/or elevated baselines causing misinterpretation of chromatograms and spectra. Materials used in the analysis must be demonstrated to be free from interferences under the conditions of analysis by running method blanks initially and with each sample lot (sample started through the extraction process on a given 8-h shift, to a maximum of 20). Specific selection of reagents and purification of solvents by distillation in all glass systems may be required. Glassware and, where possible, reagents are cleaned by solvent rinse and baking at 450°C for 1-h minimum. Interferences coextracted from samples will vary considerably from source to source, depending on the diversity of the site being sampled.

INSTRUMENTATION Major instrumentation includes a GC with a splitless or on-column injection port for capillary column, a MS with 70 eV electron impact ionization, and a data system to collect and record MS data, and process it. A K-D apparatus is used to concentrate extracts.

GC Column: 30 m × 0.25 mm I.D. 5% phenyl, 94% methyl, 1% vinyl silicone bonded phased fused silica capillary column.

PRECISION & ACCURACY The detection limits of the method are usually dependent on the level of interferences rather than instrumental limitations. The limits typify the minimum quantities that can be detected with no interferences present.

The minimum level (in µg/mL) was not listed. This is defined as a minimum level at which the analytical system shall give recognizable mass spectra (background corrected) and acceptable calibration points.

The MDL (in µg/kg) in low solids was not listed and in high solids was not listed; these were determined in digested sludge (low solids) and in filter cake or compost (high solids).

The labeled and native compound initial precision as standard deviation (in µg/L) was not listed.

The labeled and native compound initial accuracy as average recovery (in µg/L) was not listed.

SAMPLE COLLECTION, PRESERVATION & HANDLING
Collect samples in glass containers. Aqueous samples which flow freely are collected in refrigerated bottles using automatic sampling equipment. Solid samples are collected as grab samples using widemouth jars. Maintain samples at 0 to 4°C from the time of collection until extraction. If residual chlorine is present in aqueous samples, add 80 mg sodium thiosulfate/L of water. Begin sample extraction within 7 days of collection, and analyze all extracts within 40 days of extraction.

SAMPLE PREPARATION Samples containing 1% solids or less are extracted directly using continuous liquid-liquid extraction techniques. Samples containing 1 to 30% solids are diluted to the 1% level with reagent water and extracted using continuous liquid-liquid extraction techniques. Samples containing greater than 30% solids are extracted using ultrasonic techniques.

Base/neutral extraction — Adjust the pH of the waters in the extractors to 12–13 with 6 N NaOH. Extract with methylene chloride for 24–48 h.

Acid extraction — Adjust the pH of the waters in the extractors to 2 or less using 6 N sulfuric acid. Extract with methylene chloride for 24–48 h.

Ultrasonic extraction of high solids samples — Add anhydrous sodium sulfate to the sample and QC aliquot(s). Add acetone:methylene chloride (1:1) to the sample and mix thoroughly

Concentrate extracts using a K-D apparatus.

QUALITY CONTROL The analyst is permitted to modify this method to improve separations or lower the costs of measurements, provided all performance specifications are met. Analyses of blanks are required to demonstrate freedom from contamination. When results of spikes indicate atypical method performance for samples, the samples are diluted to bring method performance within acceptable limits.

For low solids (aqueous samples), extract, concentrate, and analyze two sets of four 1-L aliquots (8 aliquots total) of the precision and recovery standard. For high solids samples, two sets of four 30-g aliquots of the high solids reference matrix are used.

Spike all samples with labeled compounds to assess method performance. Compute percent recovery of the labeled compounds using the internal standard method. Compare the labeled compound recovery for each compound with the corresponding labeled compound recovery.

Reagent water and high solids reference matrix blanks are analyzed to demonstrate freedom from contamination. Extract and concentrate a 1-L reagent water blank or a high solids reference matrix blank with each sample's lot (samples started through the extraction process on the same 8-h shift, to a maximum of 20 samples).

Field replicates may be collected to determine the precision of the sampling technique, and spiked samples may be required

to determine the accuracy of the analysis when the internal standard method is used.

REFERENCE Semivolatile Organic Compounds by Isotope Dilution GC/MS. Office of Water Regulation and Standards, U.S. EPA Industrial Technology Division, Washington, DC, EPA Method 1625, Rev. C, June 1989 (contact W.A. Telliard, U.S. EPA, Office of Water Regulations and Standards, 401 M St., SW, Washington, DC, 20460. Phone: 202-382-7131).

Stirophos (Tetrachlorovinphos)   EPA Method 507
CAS #22248-79-9

TITLE Determination of Nitrogen and Phosphorus-Containing Pesticides in Water by GC/NPD

MATRIX This method is applicable to the determination of certain nitrogen and phosphorus-containing pesticides in finished drinking water and groundwater.

METHOD SUMMARY Method 507 covers 46 nitrogen- and phosphorus-containing pesticides. A 1-L sample is fortified with a surrogate standard, salted, buffered, extracted with methylene chloride, and concentrated; then the solvent is exchanged with methyl tert-butyl ether (MTBE) and concentrated again, and a 2-µL aliquot of a sample extract is injected into a GC system equipped with a selective nitrogen-phosphorus detector and a capillary column for analysis.

INTERFERENCES Method interferences may be caused by contaminants in solvents, reagents, glassware, and other sample processing apparatus. Interfering contamination may occur when a sample containing low concentrations of analytes is analyzed immediately following a sample containing relatively high concentrations. One or more injections of MTBE should be made following the analysis of a sample with high concentrations of analytes to check for analyte carryover. Matrix interferences may be caused by contaminants that are coextracted from the sample. The extent of matrix interferences will vary considerably from source to source, depending upon the water sampled.

INSTRUMENTATION A gas chromatograph system (GC) equipped with a nitrogen-phosphorus detector (NPD) is needed.

Column 1: 30 m × 0.25 mm I.D. DB-5 bonded fused silica column, 0.25 µm film thickness, or equivalent.

Column 2: 30 m × 0.25 mm I.D. DB-1701 bonded fused silica column, 0.25 µm film thickness, or equivalent.

PRECISION & ACCURACY This method has been validated in a single lab and estimated detection limits (EDLs) have been determined for each analyte. Observed detection limits may vary among waters, depending upon the nature of the interferences in the sample matrix and the specific instrumentation used. Analytes that are not separated chromatographically cannot be individually identified and measured unless an alternative technique for identification and quantification exist.

The estimated detection limit (in µg/L) was 0.76. The EDL is defined as either method detection limit or a level of compound in a sample yielding a peak in the final extract with signal-to-noise ratio of approximately 5, whichever value is higher.

The concentration used for these measurements (in µg/L) was 7.6.

The accuracy (as % recovery) was 98.

The precision (% RSD) was 6.

SAMPLING METHOD Grab samples are collected in 1-L glass sample bottles (prewashed with detergent and hot tap water, rinsed with reagent water, and dried in an oven at 400°C for 1 h) with screw caps lined with PTFE-fluorocarbon.

SAMPLE PRESERVATION Add mercuric chloride to the sample bottle in amounts to produce a concentration of 10 mg/L. If residual chlorine is present, add 80 mg of sodium thiosulfate/L of sample to the sample bottle prior to collection. After collection, seal bottle and shake vigorously for 1 min, then cool the sample to 4°C immediately and store it at 4°C in the dark until extraction.

MHT Maximum holding time of the samples, and in some cases the extracts, is 14 days.

SAMPLE PREPARATION Fortify the sample with 50 µL of the surrogate standard solution, adjust to pH 7 with phosphate buffer, add 100 g NaCl to the sample, and seal and shake to dissolve the salt; then extract with methylene chloride in a separatory funnel or in a mechanical tumbler bottle. Dry the extract by pouring it through a solvent-rinsed drying column containing about 10 cm of anhydrous sodium sulfate. Collect the extract in a Kuderna-Danish (K-D) concentrator and rinse the column with 20–30 mL methylene chloride. Concentrate the extract to about 2 mL and rinse the flask and its lower joint into the concentrator tube with 1 to 2 mL of methyl t-butyl ether (MTBE). Add 5–10 mL of MTBE and concentrate the extract twice (adding more MTBE) to a final volume of 5.0 mL and store it at 4°C until analysis.

Note: If methylene chloride is not completely removed from the final extract, it may cause detector problems.

QUALITY CONTROL Minimum quality control requirements are initial demonstration of lab capability, determination of surrogate compound recoveries in each sample and blank, monitoring internal standard peak area or height in each sample and blank, analysis of lab reagent blanks, lab fortified samples, lab fortified blanks, and other QC samples. A lab reagent blank is analyzed to demonstrate that all glassware and reagent interferences are under control.

Initial demonstration of capability is fulfilled by analyzing four fortified reagent water samples with the recovery value for each analyte falling within the acceptable range (±30% average recovery). Surrogate recoveries from samples or method blanks must be 70–130%. The internal standard response for any sample chromatogram should not deviate from the daily calibration check standard's internal standard response by more than 30% or lab fortified blanks and sample matrices are used to assess lab performance and analyte recovery, respectively.

If the response for the target analyte peak exceeds the working range of the system, dilute the extract and reanalyze. Alternative techniques such as an alternate detector or second chromatography column should be used to confirm peak identification when sample components are not resolved adequately.

EPA CONTACT & HOTLINE For technical questions contact Dr. Baldev Bathija, U.S. EPA, Office of Ground Water and Drinking Water (WH-550D), 401 M St. SW, Washington, DC 20460. Tel. (202) 260-3040. For further information the EPA Safe Drinking Water Hotline may be called at: (800) 426-4791.

REFERENCE Methods for the Determination of Organic Compounds in Drinking Water, EPA/600/4-88/039 (revised July 1991). U.S. EPA Environmental Monitoring Systems Laboratory, Cincinnati, OH, 45268, U.S.A. Available from the National Technical Information Service (NTIS), 5285 Port Royal Road, Springfield, VA 22161; Tel. 800-553-6847. NTIS Order Number is PB91-231480.

Stirophos (Tetrachlorovinphos)   **EPA Method 8141**
CAS #22248-79-9

TITLE Organophosphorus Compounds by Gas Chromatography: Capillary Column Technique

MATRIX This method covers aqueous and solid matrices. This includes a wide variety such as drinking water, groundwater, industrial wastewaters, surface waters, soils, solids, and sediments.

METHOD SUMMARY This is a GC method used to determine the concentration of 28 organophosphorus pesticides.

The use of Gel Permeation Cleanup (EPA Method 3640) for sample cleanup has been demonstrated to yield recoveries of less than 85% for many method analytes and is therefore not recommended for use with this method.

This method provides GC conditions for the detection of ppb concentrations of organophosphorus compounds. Prior to the use of this method, appropriate sample preparation techniques must be used. Water samples are extracted at a neutral pH with methylene chloride as a solvent by using a separatory funnel (EPA Method 3510) or a continuous liquid-liquid extractor (EPA Method 3520). Soxhlet extraction (EPA Method 3540) or ultrasonic extraction (EPA Method 3550) using methylene chloride/acetone (1:1) are used for solid samples. Both neat and diluted organic liquids (EPA Method 3580) may be analyzed by direct injection. Spiked samples are used to verify the applicability of the chosen extraction technique to each new sample type. A GC with a flame photometric (FPD) or nitrogen-phosphorus detector (NPD) is used for this multiresidue procedure.

INTERFERENCES The use of Florisil cleanup materials (EPA Method 3620) for some of the compounds in this method has been demonstrated to yield recoveries less than 85% and is therefore not recommended for all compounds. Use of phosphorus or halogen specific detectors, however, often obviates the necessity for cleanup for relatively clean sample matrices. If particular circumstances demand the use of an alternative cleanup procedure, the analyst must determine the elution profile and demonstrate that the recovery of each analyte is no less than 85%.

Use of a flame photometric detector (FPD) in the phosphorus mode will minimize interferences from materials that do not contain phosphorus. Elemental sulfur, however, may interfere with the determination of certain organophosphorus compounds by flame photometric gas chromatography. Sulfur cleanup using EPA Method 3660 may alleviate this interference. A nitrogen phosphorus detector (NPD) is also recommended.

A few analytes coelute on certain columns. Therefore, select a second column for confirmation where coelution of the analytes of interest does not occur.

Method interferences may be caused by contaminants in solvents, reagents, glassware, and other sample processing hardware that lead to discrete artifacts or elevated baselines in gas chromatograms. All these materials must be routinely demonstrated to be free from interferences under the conditions of the analysis by analyzing reagent blanks.

INSTRUMENTATION A GC with a NPD or a FPD will be needed. A data system or integrator is recommended for measuring peak areas and/or peak heights. A Kuderna-Danish (K-D) apparatus will be needed for extract concentration.

Column 1: 15 m × 0.53 mm megabore capillary column, 1.0 µm film thickness, DB-210.
Column 2: 15 m × 0.53 mm megabore capillary column, 1.5 µm film thickness, SPB-608.
Column 3: 15 m × 0.53 mm megabore capillary column, 1.0 µm film thickness, DB-5.

Three megabore capillary columns are included for analysis of organophosphates by this method. Column 1 (DB-210 or equivalent) and Column 2 (SPB-608 or equivalent) are recommended if a large number of organophosphorus analytes are to be determined. If the superior resolution offered by Column 1 and Column 2 is not required, Column 3 (DB-5 or equivalent) may be used. For megabore capillary columns, automatic injections of 1 µL are recommended.

PRECISION & ACCURACY The MDL actually achieved in a given analysis will vary, as it is dependent on instrument sensitivity and matrix effects. Single operator accuracy and precision studies have been conducted with spiked water and soil samples.

MULTIPLICATION FACTORS FOR OTHER MATRICES (a)

Matrix	Factor (b)
Groundwater (EPA Method 3510 or EPA Method 3520)	10
Low-concentration soil by Soxhlet and no cleanup	10 (c)
Low-concentration soil by ultrasonic extraction with GPC cleanup	6.7 (c)
High-concentration soil and sludges by ultrasonic extraction	500 (c)
Non-water miscible waste (EPA Method 3580)	1000 (c)

(a) Sample EQLs are highly matrix-dependent. The EQLs listed here are provided for guidance and may not always be achievable.
(b) EQL = [Method detection limit] × [Factor]. For non-aqueous samples the factor is on a wet-weight basis.
(c) Multiply this factory times the soil MDL.

The MDL (in µg/L) when reagent water was extracted using a separatory funnel was 0.80.

The MDL (in µg/kg) when soil was extracted using Soxhlet extraction (EPA Method 3540) was 40.0.

Accuracy (as % recovery) with separatory funnel extraction ranged from 79 (with low spikes) to 80 (with high spikes).

Accuracy (as % recovery) with continuous liquid-liquid extraction ranged from 56 (with low spikes) to 83 (with high spikes).

Accuracy (as % recovery) with Soxhlet extraction of soils ranged from 50 (with low spikes to 83 (with high spikes).

Accuracy (as % recovery) with ultrasonic extraction of soils ranged from not recovered (with low spikes) to 69 (with high spikes).

SAMPLE COLLECTION, PRESERVATION & HANDLING Containers used to collect samples for the determination of semivolatile organic compounds should be soap and water washed followed by methanol (or isopropanol) rinsing. The sample containers should be of glass or Teflon® and have screw-top covers with Teflon® liners.

No preservation is used with concentrated waste samples. With liquid samples containing no residual chlorine and with soil, sediment, and sludge samples, immediately cooling to 4°C is the only preservation used. When residual chlorine is present then 3 mL of 10% aqueous sodium sulfate is added for each gallon of sample collected, followed by cooling to 4°C.

Liquid samples must be extracted within 7 days and their extracts analyzed within 40 days. Concentrated waste, soil, sediment, and sludge samples must be extracted within 14 days and their extracts analyzed within 40 days.

SAMPLE PREPARATION In general, water samples are extracted at a neutral pH with methylene chloride, using either EPA Method 3510 or EPA Method 3520. Solid samples are extracted using either EPA Method 3540 or EPA Method 3550 with methylene chloride/acetone (1:1) as the extraction solvent.

Prior to GC analysis, the extraction solvent may be exchanged to hexane. Single lab data indicates that samples should not be transferred with 100% hexane during sample workup as the more water soluble organophosphorus compounds may be lost.

If cleanup is performed on the samples, the analyst should analyze the samples by GC. This will confirm elution patterns and the absence of interferences from the reagents. If peak detection and identification is prevented by the presence of interferences, further cleanup is required.

QUALITY CONTROL The analyst should monitor the performance of the extraction, cleanup (when used), and analytical system and the effectiveness of the method in dealing with each sample matrix by spiking each sample, standard, and blank with one or two surrogates (e.g., organophosphorus compounds not expected to be present in the sample). Deuterated analogs of analytes should not be used as surrogates for gas chromatographic analysis due to coelution problems.

A minimum of five concentrations for each analyte of interest should be prepared through dilution of the stock standards with isooctane. One of the concentrations should be at a concentration near, but above, the MDL.

Include a mid-level check standard after each group of 10 samples in the analysis sequence. GC/MS techniques should be judiciously employed to support qualitative identifications made with this method. Follow the GC/MS operating requirements specified in EPA Method 8270.

When available, chemical ionization mass spectra may be employed to aid in the qualitative identification process. To confirm an identification of a compound, the background-corrected mass spectrum of the compound must be obtained from the sample extract and must be compared with a mass spectrum from a stock or calibration standard analyzed under the same chromatographic conditions. The molecular ion and all other ions present above 20% relative abundance in the mass spectrum of the standard must be present in the mass spectrum of the sample with agreement to ±20%. The retention time of the compound in the sample must be within six seconds of the retention time for the same compound in the standard solution.

Should the MS procedure fail to provide satisfactory results, additional steps may be taken before reanalysis. These steps may include the use of alternate packed or capillary GC columns or additional sample cleanup.

REFERENCE Test Methods for Evaluating Solid Waste, Physical/Chemical Methods, SW-846, 3rd Edition, U.S. EPA, Office of Solid Waste, Washington, DC, EPA Method 8141 July 1992.

Stirophos (Tetrachlorovinphos)   **EPA Method 8140**
CAS #22248-79-9

TITLE Organophosphorus Pesticides

MATRIX Groundwater, soils, sludges, water miscible liquid wastes, and non-water miscible wastes.

APPLICATION This method is used for the analysis of 21 organophosphorus pesticides. Samples are extracted, concentrated, and analyzed using direct injection of both neat and diluted organic liquid into a gas chromatograph (GC).

INTERFERENCES Solvents, reagents, and glassware may introduce artifacts. Other interferences may come from coextracted compounds from samples. The use of Florisil cleanup materials may produce low recoveries. Elemental sulfur may interfere with some compounds when using a flame photometric detector. Sulfur cleanup (Method 3660) may alleviate sulfur interference.

INSTRUMENTATION GC capable of on-column injections and a flame photometric detector (FPD) or a thermionic detector. A halogen specific detector may also be used and may have the advantage of fewer interferences. Column 1: 1.8 meter by 2 mm with 5% SP-2401 on Supelcoport. Column 2: 1.8 m by 2 mm with 3% SP-2401 on Supelcoport. Column 3: 50 cm by ⅛ in Teflon® with 15% SE-54 on Gas Chrom Q. The preferred column is Column Number 1 or 3.

RANGE 30.3–505 µg/L

MDL 5.0 µg/L (in reagent water).

PQL FACTORS FOR MULTIPLYING × FID MDL VALUE

Matrix	Multiplication Factor
Groundwater	10
Low-level soil by sonication with GPC cleanup	670
High-level soil and sludge by sonication	10,000
Non-water miscible waste	100,000

PRECISION 5.9% (single operator standard deviation)

ACCURACY 66.1% (single operator average recovery)

SAMPLING METHOD Use 8-oz. widemouth glass bottles with Teflon®-lined caps for concentrated waste samples, soils, sediments, and sludges. Use 1 or 2½ gallon amber glass bottles with Teflon®-lined caps for liquid (water) samples.

STABILITY Cool soil, sediment, sludge, and liquid samples to 4°C. If residual chlorine is present in liquid samples add 3 mL of 10% sodium thiosulfate per gallon of sample and cool to 4°C.

MHT 14 days for concentrated waste, soil, sediment, or sludge; 7 days for liquid samples; all extracts must be analyzed within 40 days.

QUALITY CONTROL A quality control check sample concentrate containing this compound in acetone at a concentration 1,000 times more concentrated than the selected spike concentration is required. The QC check sample concentrate may be prepared from pure standard materials or purchased as certified solutions. Use appropriate trip, matrix, control site, method, reagent, and solvent blanks. Internal, surrogate, and five concentration level calibration standards are used.

REFERENCE Method 8140, SW-846, 3rd ed., Sept. 1986.

Strychnine   **EPA Method 8270**
CAS #60-41-3

TITLE Semivolatile Organic Compounds by GC/MS

MATRIX This method is used to determine the concentration of semivolatile organic compounds in extracts prepared from all types of solid waste matrices, soils, and groundwater. Although surface waters are not specifically mentioned, this method should be applicable to water samples from rivers, lakes, etc.

METHOD SUMMARY This method covers 259 semivolatile organic compounds. In very limited applications direct injection of the sample into the GC/MS system may be appropriate, but this results in very high detection limits (approximately 10,000 µg/L). Typically, a 1-L liquid sample, containing surrogate, and matrix spiking standards, is extracted in a continuous extractor first under acid conditions and then under basic conditions. Typically 30 g of a solid sample, containing surrogate, and matrix spiking standards, is extracted ultrasonically. After concentrating the extract to 1 mL it is spiked with 10 µL of an internal standard solution just prior to analysis by GC/MS. The volume injected should contain about 100 ng of base/neutral and 200 ng of acid surrogates (for a 1-µL injection). Analysis is performed by GC/MS using a capillary GC column.

INTERFERENCES Raw GC/MS data from all blanks, samples, and spikes must be evaluated for interferences. Contamination by carryover can occur whenever high-concentration and low-concentration samples are sequentially analyzed. To reduce carryover, the sample syringe must be rinsed out between samples with solvent. Whenever an unusually concentrated sample is encountered, it should be followed by the analysis of blank solvent to check for cross-contamination.

INSTRUMENTATION A GC/MS and a data system are required. The GC column used is a 30 m × 0.25 mm I.D. (or 0.32 mm I.D.) 1um film thickness silicone-coated fused silica capillary column. A continuous liquid-liquid extractor equipped with Teflon® or glass connection joints and stopcocks requiring no lubrication, a K-D concentrating apparatus, water bath, and an ultrasonic disrupter with a minimum power of 300 W and with pulsing capability are also required.

PRECISION & ACCURACY The estimated quantitation limit (EQL) of Method 8270B for determining an individual compound is approximately 1 mg/kg (wet weight) for soil or sediment samples, 1–200 mg/kg for wastes (dependent on matrix and method of preparation), and 10 µg/L for groundwater samples. EQLs will be proportionately higher for sample extracts that require dilution to avoid saturation of the detector.

The EQL(b) for groundwater in µg/L is 40.
The EQL (a, b) for low concentrations in soil and sediment in µg/kg is not determined.
Accuracy as µg/L is not listed.
Overall precision in µg/L is not listed.

(a) EQLs listed for soil/sediment are based on wet weight. Normally data is reported in a dry-weight basis; therefore, EQLs will be higher based on the % dry weight of each sample. This calculation is based on a 30 g sample and gel permeation chromatography cleanup.
(b) Sample EQLs are highly matrix-dependent. The EQLs are provided for guidance and may not always be achievable.
C = True value for concentration, in µg/L.
X = Average recovery found for measurements of samples containing a concentration of C, in µg/L.

ESTIMATED QUANTITATION LIMIT

Other Matrices	Factor (a)
High-concentration soil and sludges by sonicator	7.5
Non-water miscible waste	75

(a) EQL for other matrices = [EQL for low soil/sediment] × [Factor]. This estimated EQL is similar to an EPA "Practical Quantitation Limit."

SAMPLING METHOD

Liquid samples — Use a 1 or 2½ gallon amber glass bottle with a screw-top Teflon®-lined cover that has been prewashed with detergent and rinsed with distilled water and methanol (or isopropanol).

Soils, sediments, or sludges — Use an 8-oz. widemouth glass with a screw-top Teflon®-lined cover that has been prewashed

with detergent and rinsed with distilled water and methanol (or isopropanol).

SAMPLE PRESERVATION

Liquid samples — If residual chlorine is present, add 3 mL of 10% sodium thiosulfate per gallon, cool to 4°C and store in a solvent-free refrigerator until analysis; if chlorine is not present, then eliminate the sodium thiosulfate addition.

Soils, sediments, or sludges — Cool samples to 4°C and store in a solvent-free refrigerator.

MHT Liquid samples must be extracted within 7 days and the extracts analyzed within 40 days. Soils, sediments, or sludges may be stored for a maximum of 14 days and the extracts analyzed within 40 days.

SAMPLE PREPARATION

Liquid samples — Transfer 1 L quantitatively to a continuous extractor. If high concentrations are anticipated, a smaller volume may be used and then diluted with organic-free reagent water to 1 L. Adjust pH, if necessary, to pH <2 using 1:1 (V/V) sulfuric acid. Pipette 1.0 mL of a surrogate standard spiking solution into each sample. For the sample in each analytical batch selected for spiking, add 1.0 mL of a matrix spiking standard. For base/neutral acid analysis, the amount of the surrogates and matrix spiking compounds added to the sample should result in a final concentration of 100 ng/µL of each analyte in the extract to be analyzed (assuming a 1-µL injection). Extract with methylene chloride for 18–24 h. Next, adjust the pH of the aqueous phase to pH >11 using 10 N sodium hydroxide and extract it with methylene chloride again for 18–24 h. Dry the extract through a column containing anhydrous sodium sulfate and concentrate it to 1 mL using a K-D concentrator.

Soils, sediments, or sludges — Use 30 g of sample. Nonporous or wet samples (gummy or clay type) that do not have a free-flowing sandy texture must be mixed with anhydrous sodium sulfate until the sample is free flowing. Add 1 mL of surrogate standards to all samples, spikes, standards, and blanks. For the sample in each analytical batch selected for spiking, add 1.0 mL of a matrix spiking standard. For base/neutral acid analysis, the amount added of the surrogates and matrix spiking compounds should result in a final concentration of 100 ng/µL of each base/neutral analyte and 200 ng/µL of each acid analyte in the extract to be analyzed (assuming a 1-µL injection). Immediately add a 100-mL mixture of 1:1 methylene chloride:acetone and extract the sample ultrasonically for 3 min and then decant or filter the extracts. Repeat the extraction two or more times. Dry the extract using a column with anhydrous sodium sulfate and concentrate it to 1 mL in a K-D concentrator.

QUALITY CONTROL A methylene chloride solution containing 50 ng/µL of decafluorotriphenylphosphine (DFTPP) is used for tuning the GC/MS system each 12-h shift. A system performance check also must be made during every 12-h shift. A standard containing 50 ng/µL each of 4,4'-DDT, pentachlorophenol, and benzidine is required to verify injection port inertness and GC column performance. A calibration standard at mid-concentration, containing each compound of interest, including all required surrogates, must be performed every 12 h during analysis. After the system performance check is met, calibration check compounds (CCCs) are used to check the validity of the initial calibration.

The internal standard responses and retention times in the calibration check standard must be evaluated immediately after or during data acquisition. If the retention time for any internal standard changes by more than 30 seconds from the last check calibration (12 h), the chromatographic system must be inspected for malfunctions and corrections must be made, as required. If the electron ionization current plot (EICP) area for any of the internal standards changes by a factor of two from the last daily calibration standard check, the mass spectrometer must be inspected for malfunctions and corrections must be made, as appropriate.

Demonstrate, through the analysis of a reagent water blank, that interferences from the analytical system, glassware, and reagents are under control. The blank samples should be carried through all stages of the sample preparation and measurement steps. For each analytical batch (up to 20 samples), a reagent blank, matrix spike, and matrix spike duplicate/duplicate must be analyzed (the frequency of the spikes may be different for different monitoring programs). The blank and spiked samples must be carried through all stages of the sample preparation and measurement steps. A QC reference sample concentrate containing each analyte at a concentration of 100 mg/L in methanol is required.

REFERENCE Test Methods for Evaluating Solid Waste (SW-846). U.S. EPA 1983, Method 8270B, Rev. 2, Nov. 1990. Office of Solid Waste, Washington, DC.

Styrene
CAS #100-42-5

EPA Method 1625

TITLE Semivolatile Organic Compounds by Isotope Dilution GC/MS

MATRIX The compounds may be determined in waters, soils, and municipal sludges by this method.

METHOD SUMMARY This method is used to determine 176 semivolatile toxic organic pollutants associated with the CWA (as amended 1987); the RCRA (as amended 1986); the CERCLA (as amended 1986); and other compounds amenable to extraction and analysis by capillary column gas chromatography-mass spectrometry (GC/MS).

Stable isotopically-labeled analogs of the compounds of interest are added to the sample. If the solids content is less than 1%, a 1-L sample is extracted at pH 12–13, then at pH <2 with methylene chloride using continuous extraction techniques.

If the solids content is 30% or less, the sample is diluted to 1% solids with reagent water, homogenized ultrasonically, and extracted at pH 12–13, then at pH <2 with methylene chloride using continuous extraction techniques. If the solids content is greater than 30%, the sample is extracted using ultrasonic techniques.

Each extract is dried over sodium sulfate, concentrated to a volume of 5 mL, cleaned up using GPC, if necessary, and concentrated. Extracts are concentrated to 1 mL if GPC is not performed, and to 0.5 mL if GPC is performed.

An internal standard is added to the extract, and a 1-mL aliquot of the extract is injected into the GC. The compounds are separated by GC and detected by a MS. The labeled compounds serve to correct the variability of the analytical technique.

INTERFERENCES Solvents, reagents, glassware, and other sample processing hardware may yield artifacts and/or elevated baselines causing misinterpretation of chromatograms and spectra. Materials used in the analysis must be demonstrated to be free from interferences under the conditions of analysis by running method blanks initially and with each sample lot (sample started through the extraction process on a given 8-h shift, to a maximum of 20). Specific selection of reagents and purification of solvents by distillation in all glass systems may be required. Glassware and, where possible, reagents are cleaned by solvent rinse and baking at 450°C for 1-h minimum. Interferences coextracted from samples will vary considerably from source to source, depending on the diversity of the site being sampled.

INSTRUMENTATION Major instrumentation includes a GC with a splitless or on-column injection port for capillary column, a MS with 70 eV electron impact ionization, and a data system to collect and record MS data, and process it. A K-D apparatus is used to concentrate extracts.

GC Column: 30 m × 0.25 mm I.D. 5% phenyl, 94% methyl, 1% vinyl silicone bonded phased fused silica capillary column.

PRECISION & ACCURACY The detection limits of the method are usually dependent on the level of interferences rather than instrumental limitations. The limits typify the minimum quantities that can be detected with no interferences present.

The minimum level (in μg/mL) was 10. This is defined as a minimum level at which the analytical system shall give recognizable mass spectra (background corrected) and acceptable calibration points.

The MDL (in μg/kg) in low solids was 149 and in high solids was 17; these were determined in digested sludge (low solids) and in filter cake or compost (high solids).

Note: Background levels of this compound were present in the sludge tested, resulting in higher than expected MDLs. The MDL for this compound is expected to be approximately 50 μg/kg with no interferences present.

The labeled and native compound initial precision as standard deviation (in μg/L) was 42.
The labeled and native compound initial accuracy as average recovery (in μg/L) was 53–221.

SAMPLE COLLECTION, PRESERVATION & HANDLING Collect samples in glass containers. Aqueous samples which flow freely are collected in refrigerated bottles using automatic sampling equipment. Solid samples are collected as grab samples using widemouth jars. Maintain samples at 0 to 4°C from the time of collection until extraction. If residual chlorine is present in aqueous samples, add 80 mg sodium thiosulfate/L of water. Begin sample extraction within 7 days of collection, and analyze all extracts within 40 days of extraction.

SAMPLE PREPARATION Samples containing 1% solids or less are extracted directly using continuous liquid-liquid extraction techniques. Samples containing 1 to 30% solids are diluted to the 1% level with reagent water and extracted using continuous liquid-liquid extraction techniques. Samples containing greater than 30% solids are extracted using ultrasonic techniques.

Base/neutral extraction — Adjust the pH of the waters in the extractors to 12–13 with 6 N NaOH. Extract with methylene chloride for 24–48 h.
Acid extraction — Adjust the pH of the waters in the extractors to 2 or less using 6 N sulfuric acid. Extract with methylene chloride for 24–48 h.
Ultrasonic extraction of high solids samples — Add anhydrous sodium sulfate to the sample and QC aliquot(s). Add acetone:methylene chloride (1:1) to the sample and mix thoroughly

Concentrate extracts using a K-D apparatus.

QUALITY CONTROL The analyst is permitted to modify this method to improve separations or lower the costs of measurements, provided all performance specifications are met. Analyses of blanks are required to demonstrate freedom from contamination. When results of spikes indicate atypical method performance for samples, the samples are diluted to bring method performance within acceptable limits.

For low solids (aqueous samples), extract, concentrate, and analyze two sets of four 1-L aliquots (8 aliquots total) of the precision and recovery standard. For high solids samples, two sets of four 30-g aliquots of the high solids reference matrix are used.

Spike all samples with labeled compounds to assess method performance. Compute percent recovery of the labeled compounds using the internal standard method. Compare the labeled compound recovery for each compound with the corresponding labeled compound recovery.

Reagent water and high solids reference matrix blanks are analyzed to demonstrate freedom from contamination. Extract and concentrate a 1-L reagent water blank or a high solids reference matrix blank with each sample's lot (samples started through the extraction process on the same 8-h shift, to a maximum of 20 samples).

Field replicates may be collected to determine the precision of the sampling technique, and spiked samples may be required to determine the accuracy of the analysis when the internal standard method is used.

REFERENCE Semivolatile Organic Compounds by Isotope Dilution GC/MS. Office of Water Regulation and Standards, U.S. EPA Industrial Technology Division, Washington, DC, EPA Method 1625, Rev. C, June 1989 (contact W.A. Telliard, U.S. EPA, Office of Water Regulations and Standards, 401 M St., SW, Washington, DC, 20460. Phone: 202-382-7131).

Styrene
CAS #100-42-5

EPA Method 1625

TITLE Semivolatile Organic Compounds by Isotope Dilution GC/MS

MATRIX The compounds may be determined in waters, soils, and municipal sludges by this method.

METHOD SUMMARY This method is used to determine 176 semivolatile toxic organic pollutants associated with the CWA (as amended 1987); the RCRA (as amended 1986); the CERCLA (as amended 1986); and other compounds amenable to extraction and analysis by capillary column gas chromatography-mass spectrometry (GC/MS).

Stable isotopically-labeled analogs of the compounds of interest are added to the sample. If the solids content is less than 1%, a 1-L sample is extracted at pH 12–13, then at pH <2 with methylene chloride using continuous extraction techniques.

If the solids content is 30% or less, the sample is diluted to 1% solids with reagent water, homogenized ultrasonically, and extracted at pH 12–13, then at pH <2 with methylene chloride using continuous extraction techniques. If the solids content is greater than 30%, the sample is extracted using ultrasonic techniques.

Each extract is dried over sodium sulfate, concentrated to a volume of 5 mL, cleaned up using GPC, if necessary, and concentrated. Extracts are concentrated to 1 mL if GPC is not performed, and to 0.5 mL if GPC is performed.

An internal standard is added to the extract, and a 1-mL aliquot of the extract is injected into the GC. The compounds are separated by GC and detected by a MS. The labeled compounds serve to correct the variability of the analytical technique.

INTERFERENCES Solvents, reagents, glassware, and other sample processing hardware may yield artifacts and/or elevated baselines causing misinterpretation of chromatograms and spectra. Materials used in the analysis must be demonstrated to be free from interferences under the conditions of analysis by running method blanks initially and with each sample lot (sample started through the extraction process on a given 8-h shift, to a maximum of 20). Specific selection of reagents and purification of solvents by distillation in all glass systems may be required. Glassware and, where possible, reagents are cleaned by solvent rinse and baking at 450°C for 1-h minimum. Interferences coextracted from samples will vary considerably from source to source, depending on the diversity of the site being sampled.

INSTRUMENTATION Major instrumentation includes a GC with a splitless or on-column injection port for capillary column, a MS with 70 eV electron impact ionization, and a data system to collect and record MS data, and process it. A K-D apparatus is used to concentrate extracts.

GC Column: 30 m × 0.25 mm I.D. 5% phenyl, 94% methyl, 1% vinyl silicone bonded phased fused silica capillary column.

PRECISION & ACCURACY The detection limits of the method are usually dependent on the level of interferences rather than instrumental limitations. The limits typify the minimum quantities that can be detected with no interferences present.

The minimum level (in µg/mL) was 10. This is defined as a minimum level at which the analytical system shall give recognizable mass spectra (background corrected) and acceptable calibration points.

The MDL (in µg/kg) in low solids was 149 and in high solids was 17; these were determined in digested sludge (low solids) and in filter cake or compost (high solids).

Note: Background levels of this compound were present in the sludge tested, resulting in higher than expected MDLs. The MDL for this compound is expected to be approximately 50 µg/kg with no interferences present.

The labeled and native compound initial precision as standard deviation (in µg/L) was 42.
The labeled and native compound initial accuracy as average recovery (in µg/L) was 53–221.

SAMPLE COLLECTION, PRESERVATION & HANDLING
Collect samples in glass containers. Aqueous samples which flow freely are collected in refrigerated bottles using automatic sampling equipment. Solid samples are collected as grab samples using widemouth jars. Maintain samples at 0 to 4°C from the time of collection until extraction. If residual chlorine is present in aqueous samples, add 80 mg sodium thiosulfate/L of water. Begin sample extraction within 7 days of collection, and analyze all extracts within 40 days of extraction.

SAMPLE PREPARATION Samples containing 1% solids or less are extracted directly using continuous liquid-liquid extraction techniques. Samples containing 1 to 30% solids are diluted to the 1% level with reagent water and extracted using continuous liquid-liquid extraction techniques. Samples containing greater than 30% solids are extracted using ultrasonic techniques.

Base/neutral extraction — Adjust the pH of the waters in the extractors to 12–13 with 6 N NaOH. Extract with methylene chloride for 24–48 h.
Acid extraction — Adjust the pH of the waters in the extractors to 2 or less using 6 N sulfuric acid. Extract with methylene chloride for 24–48 h.
Ultrasonic extraction of high solids samples — Add anhydrous sodium sulfate to the sample and QC aliquot(s). Add acetone:methylene chloride (1:1) to the sample and mix thoroughly

Concentrate extracts using a K-D apparatus.

QUALITY CONTROL The analyst is permitted to modify this method to improve separations or lower the costs of measurements, provided all performance specifications are met. Analyses of blanks are required to demonstrate freedom from contamination. When results of spikes indicate atypical method performance for samples, the samples are diluted to bring method performance within acceptable limits.

For low solids (aqueous samples), extract, concentrate, and analyze two sets of four 1-L aliquots (8 aliquots total) of the

precision and recovery standard. For high solids samples, two sets of four 30-g aliquots of the high solids reference matrix are used.

Spike all samples with labeled compounds to assess method performance. Compute percent recovery of the labeled compounds using the internal standard method. Compare the labeled compound recovery for each compound with the corresponding labeled compound recovery.

Reagent water and high solids reference matrix blanks are analyzed to demonstrate freedom from contamination. Extract and concentrate a 1-L reagent water blank or a high solids reference matrix blank with each sample's lot (samples started through the extraction process on the same 8-h shift, to a maximum of 20 samples).

Field replicates may be collected to determine the precision of the sampling technique, and spiked samples may be required to determine the accuracy of the analysis when the internal standard method is used.

REFERENCE Semivolatile Organic Compounds by Isotope Dilution GC/MS. Office of Water Regulation and Standards, U.S. EPA Industrial Technology Division, Washington, DC, EPA Method 1625, Rev. C, June 1989 (contact W.A. Telliard, U.S. EPA, Office of Water Regulations and Standards, 401 M St., SW, Washington, DC, 20460. Phone: 202-382-7131).

Styrene	EPA Method 502
CAS #100-42-5	

TITLE Volatile Organic Compounds in Water By Purge and Trap Capillary Column Gas Chromatography with Photoionization and Electrolytic Conductivity Detectors in Series. U.S. EPA Method 502.2, Rev. 2.0, 1989.

MATRIX Drinking water and raw source water. The latter should include most surface water and groundwater sources.

METHOD SUMMARY This method covers 60 volatile organic compounds that contain halogen atoms and/or that are aromatic. An inert gas (zero grade nitrogen or helium) is bubbled through a 25-mL or a 5-mL water sample (depending on the expected concentration of the analytes). Purged sample components are trapped in a tube of sorbent materials. When purging is complete, the sorbent tube is heated and backflushed with helium to desorb the trapped sample onto a capillary GC column. The column is temperature programmed to separate the method analytes which are then detected with a photoionization detector (PID) and a Hall electrolytic conductivity (HECD) placed in series. The PID is selective for aromatic compounds and the HECD is selective for halogenated compounds.

INTERFERENCES Impurities in the purge gas and from organic compounds outgassing from the plumbing ahead of the trap account for many contamination problems. Interferences purged or coextracted from the samples will vary considerably from source to source, depending upon the particular sample or extract being tested. Cross-contamination can occur whenever high-level and low-level samples are analyzed sequentially. Samples also can be contaminated by diffusion of volatile organics (particularly methylene chloride and fluorocarbons) through the septum seal into the sample during shipment and storage. The lab where volatile analysis is performed and also the refrigerated storage area should be completely free of solvents.

INSTRUMENTATION A GC containing a series configuration of a high temperature photoionization detector (PID) equipped with 10.0 eV (nominal) lamp and Hall electrolytic conductivity detector (HECD) is required. Also required is an all-glass 5-mL purging device, a sorbent trap, and a thermal desorption apparatus which is connected to the GC system.

Column 1: VOCOL glass wide-bore capillary column.
Column 2: RTX–502.2 mega-bore capillary column.
Column 3: DB-62 mega-bore capillary column.

PRECISION & ACCURACY Method detection limits are dependent upon the characteristics of the gas chromatographic system used. Analytes that are not separated chromatographically cannot be individually identified and used in the same calibration mixture or water samples unless an alternative technique for identification and quantification, such as mass spectrometry, is used.

Electrolytic conductivity detetor (c) range in µg/L (a) was 0.02–2000.
Electrolytic conductivity detetor (c) MDL in µg/L (b) was not listed.
Electrolytic conductivity detetor (c) accuracy as % recovery was not listed.
Electrolytic conductivity detetor (c) precision as % RSD was not listed.
Photoionization detector (d) range in µg/L (a) was 0.02–2000.
Photoionization detector (d) MDL in µg/L (b) was 0.01.
Photoionization detector (d) accuracy as % recovery was 104.
Photoionization detector (d) precision as % RSD was 1.3.

(a) *The applicable concentration range of this method is compound, instrument, and matrix-dependent. It is listed as being approximately 0.02 to 200 µg/L but no specific information is provided so caution should be observed.*
(b) *The method detection limits reports with this method are compound, instrument, and matrix-dependent. The values reported were calculated using reagent water fortified with the corresponding compounds at 10 µg/L and a GC-equipped with a 60 m × 0.75 mm VOLCOL wide bore capillary column with 1.5 µm film thickness and using helium carrier gas.*
(c) *Recoveries and relative standard deviations were determined from seven samples of reagent water fortified with 10 µg/L of each compound. 2-Bromo-1-chloropropane was used as the internal standard for calculating average recoveries.*
(d) *Recoveries and relative standard deviations were determined from seven samples of reagent water fortified with 10 µg/L of each compound. Fluorobenzene was used as the internal standard for calculating average recoveries.*

SAMPLING METHOD Collect samples using a 40- to 120-mL screw-cap vial (prewashed with detergent, rinsed with distilled water and oven dried at 105°C) with a Teflon®-faced silicone septum. Collect bubble-free samples and place the septum with the Teflon® side down on the water.

SAMPLE PRESERVATION If residual chlorine is present in the water add about 25 mg of ascorbic acid to each vial before samples are collected to remove the chlorine. Add hydrochloric acid to reduce pH to <2, immediately cool samples to 4°C, and store them in a solvent-free refrigerator at 4°C until analysis.

MHT The maximum holding time for samples is 14 days from the time they were collected.

SAMPLE PREPARATION Remove the plungers from two 5-mL syringes and attach a closed syringe valve to each. Warm the sample to room temperature, open the sample bottle, and carefully pour the sample into one of the syringe barrels to just short of overflowing. Replace the syringe plunger, invert the syringe, and compress the sample. Open the syringe valve and vent any residual air while adjusting the sample volume to 5.0 mL. Add 10 µL of the internal calibration standard to the sample through the syringe valve. Close the valve. Fill the second syringe in an identical manner from the same sample bottle. Reserve this second syringe for a reanalysis if necessary.

QUALITY CONTROL As an initial demonstration of lab accuracy and precision, analyze 4 to 7 replicates of a lab fortified blank containing analyte at 0.1–5 µg/L. Collect all samples in duplicate. Surrogate analytes (similar to those of the analytes of interest), whose concentration is known in every sample, are measured using the same internal standard calibration procedure. Duplicate field reagent water blanks (trip blanks) must be analyzed with each set of samples, lab reagent blanks (method blanks) must be analyzed with each batch of samples processed as a group within a work shift. Also, a single lab-fortified blank that contains each of the analytes of interest should be analyzed with each batch of samples processed as a group within a work shift. A 3- to 5-point calibration curve is needed depending on the calibration range factor required.

EPA CONTACT & HOTLINE For technical questions contact Dr. Baldev Bathija, U.S. EPA, Office of Ground Water and Drinking Water (WH-550D), 401 M St. SW, Washington, DC 20460. Tel. (202) 260-3040. For further information the EPA Safe Drinking Water Hotline may be called at: (800) 426-4791.

REFERENCE Methods for the Determination of Organic Compounds in Drinking Water, EPA/600/4-88/039 (revised July 1991; Final Rule for determination of compliance with the MCL for Total Trihalomethanes under 141.30, in 40 CFR Part 141, Vol. 58, No. 147, Fed. Reg., Tuesday Aug. 3, 1993). U.S. EPA Environmental Monitoring Systems Laboratory, Cincinnati, OH, 45268, U.S.A. Available from the National Technical Information Service (NTIS), 5285 Port Royal Road, Springfield, VA 22161; Tel. 800-553-6847. NTIS Order Number is PB91-231480.

Styrene
CAS #100-42-5
EPA Method 524

TITLE Measurement of Purgeable Organic Compounds in Water by Capillary Column GC/MS.

MATRIX Drinking water and raw source water; the latter should include most surface water and groundwater sources.

METHOD SUMMARY Method 524.2 covers 60 volatile organic compounds. An inert gas (zero grade nitrogen or helium) is bubbled through a 25-mL or a 5-mL water sample (depending on the expected concentration of the analytes). Purged sample components are trapped in a tube of sorbent materials. When purging is complete, the sorbent tube is heated and backflushed with helium to desorb the trapped sample onto a capillary GC column.

INTERFERENCES Impurities in the purge gas and from organic compounds outgassing from the plumbing ahead of the trap account for many contamination problems. Interferences purged or coextracted from the samples will vary considerably from source to source, depending upon the particular sample or extract being tested. Cross-contamination can occur whenever high-level and low-level samples are analyzed sequentially. Samples also can be contaminated by diffusion of volatile organics (particularly methylene chloride and fluorocarbons) through the septum seal into the sample during shipment and storage.

INSTRUMENTATION A GC/MS with a data system equipped with one of the following capillary GC columns:

Column 1: VOCOL glass wide bore capillary column.
Column 2: DB-624 fused silica capillary column.
Column 3: DB-5 fused silica capillary column.

Also required is an all-glass 25 mL or 5-mL purging device, a sorbent trap, and a thermal desorption apparatus which is connected to the GC/MS system.

PRECISION & ACCURACY Method detection limits are compound- and instrument-dependent, and may vary from approximately 0.02–0.35 µg/L. Note in the table below that the "true" concentration range used for accuracy and precision measurements was quite narrow. However, the applicable concentration range of this method is primarily column dependent and is approximately 0.02 to 200 µg/L for the wide-bore thick-film columns. Narrow-bore thin-film columns may have a capacity which limits the range to about 0.02 to 20 µg/L. Analytes that are inefficiently purged from water will not be detected when present at low concentrations, but they can be measured with acceptable accuracy and precision when present in sufficient amounts.

Analytes that are not separated chromatographically, but which have different mass spectra and non-interfering quantification ions, can be identified and measured in the same calibration mixture or water sample. Analytes which have very similar mass spectra cannot be individually identified and measured in the same calibration mixture or water samples unless they have different retention times. Co-eluting compounds with very

similar mass spectra, typically many structural isomers, must be reported as an isomeric group or pair.

The range (in µg/L) was 0.1–100.
The Method Detection Limig (in µg/L) was 0.04.
The accuracy (as % recovery) was 102.
The precision (in %) was 7.2.

Note: Data were obtained from 16–31 determinations using a wide-bore capillary column and a jet separator interfaced to a quadrupole mass spectrometer. All analytes were in a reagent water matrix.

SAMPLING METHOD Collect samples using a 40- to 120-mL screw-cap vial (prewashed with detergent, rinsed with distilled water and oven dried at 105°C) with a Teflon®-faced silicone septum. Collect bubble-free samples and place the septum with the Teflon® side down on the water.

SAMPLE PRESERVATION If residual chlorine is present in the water add about 25 mg of ascorbic acid to each vial before samples are collected to remove the chlorine. Add hydrochloric acid to reduce pH to <2, and immediately cool samples to 4°C, and store them in a solvent-free refrigerator at 4°C until analysis.

MHT The maximum holding time for samples is 14 days from the time they were collected.

SAMPLE PREPARATION Remove the plungers from two 25-mL (or 5-mL depending on sample size) syringes and attach a closed syringe valve to each. Warm the sample to room temperature, open the sample bottle, and carefully pour the sample into one of the syringe barrels to just short of overflowing. Replace the syringe plunger, invert the syringe, and compress the sample. Open the syringe valve and vent any residual air while adjusting the sample volume to 25.0 mL (or 5 mL). For samples and blanks, add 5 µL of the fortification solution containing the internal standard and the surrogates to the sample through the syringe valve. For calibration standards and lab fortified blanks, add 5 µL of the fortification solution containing the internal standard only. Close the valve. Fill the second syringe in an identical manner from the same sample bottle. Reserve this second syringe for a reanalysis if necessary.

QUALITY CONTROL As an initial demonstration of lab accuracy and precision, analyze 4 to 7 replicates of a lab fortified blank containing analyte at 0.2–5 µg/L. Collect all samples in duplicate. Surrogate analytes (similar to those of the analytes of interest), whose concentration is known in every sample, are measured using the same internal standard calibration procedure. Duplicate field reagent water blanks (trip blanks) must be analyzed with each set of samples, lab reagent blanks (method blanks) must be analyzed with each batch of samples processed as a group within a work shift. Also, a single lab-fortified blank that contains each of the analytes of interest should be analyzed with each batch of samples processed as a group within a work shift. A 3- to 5-point calibration curve is needed depending on the calibration range factor required.

EPA CONTACT & HOTLINE For technical questions contact Dr. Baldev Bathija, U.S. EPA, Office of Ground Water and Drinking Water (WH-550D), 401 M St. SW, Washington, DC 20460. Tel. (202) 260-3040. For further information the EPA Safe Drinking Water Hotline may be called at: (800) 426-4791.

REFERENCE Methods for the Determination of Organic Compounds in Drinking Water, EPA/600/4-88/039 (revised July 1991; Final Rule for determination of compliance with the MCL for Total Trihalomethanes under 141.30, in 40 CFR Part 141, Vol. 58, No. 147, Fed. Reg., Tuesday Aug. 3, 1993). U.S. EPA Environmental Monitoring Systems Laboratory, Cincinnati, OH, 45268, U.S.A. Available from the National Technical Information Service (NTIS), 5285 Port Royal Road, Springfield, VA 22161; Tel. 800-553-6847. NTIS Order Number is PB91-231480.

Styrene   **EPA Method 8021**
CAS #100-42-5

TITLE Halogenated Volatile by Gas Chromatography Using Photoionization and Electrolytic Conductivity Detectors in Series: Capillary Column Technique

MATRIX This method is applicable to nearly all types of samples, regardless of water content, including groundwater, aqueous sludges, caustic liquors, acid liquors, waste solvents, oily wastes, mousses, tars, fibrous wastes, polymeric emulsions, filter cakes, spent carbons, spent catalysts, soils, and sediments.

METHOD SUMMARY This method is used to determine 60 volatile organic compounds in a variety of solid waste matrices. It provides GC conditions for the detection of halogenated and aromatic volatile organic compounds. Samples can be analyzed using direct injection or purge-and-trap (EPA Method 5030). Groundwater samples must be analyzed using EPA Method 5030 (where applicable). A temperature program is used with the GC. Detection is achieved by a photoionization detector (PID) and a Hall electrolytic conductivity detector (HECD) in series.

INTERFERENCES Samples can be contaminated by diffusion of volatile organics (particularly chlorofluorocarbons and methylene chloride) through the sample container septum during shipment and storage.

INSTRUMENTATION A GC-equipped with variable-constant differential flow controllers, subambient oven controller, PID and HECD detectors connected with a short piece of uncoated capillary tubing and a data system.

Column: 60 m × 0.75 mm I.D. VOCOL wide-bore capillary column with 1.5 µm film thickness.

PRECISION & ACCURACY MDLs are compound-dependent and vary with purging efficiency and concentration. The applicable concentration range of this method is compound- and instrument-dependent but is approximately 0.1 to 200 µg/L. Analytes that are inefficiently purged from water will not be detected when present at low concentrations, but they can be measured with acceptable accuracy and precision when present in sufficient amounts. The estimated quantitation limit (EQL) for an individual compound is approximately 1 µg/kg (wet weight) for soil/sediment samples, 100 µg/kg (wet weight) for wastes, and 1 µg/L for groundwater. EQLs will be proportionately higher for sample extracts and samples that require dilution or reduced sample size to avoid saturation of the detector.

MULTIPLICATION FACTORS FOR OTHER MATRICES (a)

Matrix	Factor (b)
Groundwater	10
Low-concentration soil	10
Water miscible liquid waste	500
High-concentration soil and sludge	1250
Non-water miscible waste	1250

(a) Sample EQLs are highly matrix-dependent. The EQLs listed herein are provided for guidance and may not always be achievable. (b) EQL = [Method detection limit] × [Factor]. For non-aqueous samples, the factor is on a wet-weight basis.

SINGLE LABORATORY ACCURACY & PRECISION DATA FOR VOCs IN WATER

This method was tested in a single lab using water spiked at 10 µg/L and the following data was reported:

Recoveries and standard deviations were determined from seven samples and spiked at 10 µg/L of each analyte. Recoveries were determined by the internal standard method. Internal standards were: Fluorobenzene for PID and 2-Bromo-1-chloropropane for HECD.

The average recovery (in percent) for the PID was 104.
The standard deviation of the recovery for the PID was 1.4.
The MDL (in µg/mL) for the PID was 0.01.
The average recovery (in percent) for the HECD was none (no response for this detector).
The standard deviation of the recovery for the HECD was none (no response for this detector)-.
The MDL (in µg/mL) for the HECD was none (no response for this detector).

SAMPLE COLLECTION, PRESERVATION & HANDLING

Volatile Organics — Standard 40-mL glass screw-cap VOA vials with Teflon®-faced silicone septum may be used for both liquid and solid matrices. When collecting samples, liquids and solids should be introduced into the vials gently to reduce agitation which might drive off volatile compounds. If there are any air bubbles present the sample must be retaken. Tap slightly as they are filled to try and eliminate as much free air space as possible. The two vials from each sampling locations should be sealed in separate plastic bags to prevent cross-contamination between samples particularly if the sampled waste is suspected of containing high levels of volatile organics.

Semivolatile organics — Containers used to collect samples for the determination of semivolatile organic compounds should be soap and water washed followed by methanol (or isopropanol) rinsing. The sample containers should be of glass or Teflon® and have screw-top covers with Teflon® liners.

Preservation for volatile organics — No preservation is used with concentrated waste samples. With liquid samples containing no residual chlorine, 4 drops of concentrated hydrochloric acid are added and the samples are immediately cooled to 4°C. When liquid samples contain residual chlorine, they are treated as above and, in addition, 4 drops of 4% aqueous sodium thiosulfate are added. Soil, sediment, and sludge samples are only cooled to 4°C.

Preservation for semivolatile organics — No preservation is used with concentrated waste samples. With liquid samples containing no residual chlorine and with soil, sediment, and sludge samples, immediately cooling to 4°C is the only preservation used. When residual chlorine is present then 3 mL of 10% aqueous sodium sulfate is added for each gallon of sample collected, followed by cooling to 4°C.

MHT The holding time for all volatile organics samples is 14 days. Liquid samples must be extracted within 7 days and their extracts analyzed within 40 days. Concentrated waste, soil, sediment, and sludge samples must be extracted within 14 days and their extracts analyzed within 40 days.

SAMPLE PREPARATION Volatile compounds are introduced into the gas chromatograph either by direct injector or purge-and-trap (EPA Method 5030). EPA Method 5030 may be used directly on groundwater samples or low-concentration contaminated soils and sediments. For medium-concentration soils or sediments, methanolic extraction, as described in EPA Method 5030, may be necessary prior to purge-and-trap analysis.

QUALITY CONTROL Calculate surrogate standard recovery on all samples, blanks, and spikes. A trip blank is recommended to check on sampling, storage, and handling contamination. Calibration standards, at a minimum of five concentration levels, are prepared in organic-free reagent water. One of the concentration levels should be at a concentration near, but above, the method detection limit.

A combination of bromochloromethane, 2-bromo-1-chloropropane, 1,4-dichlorobutane, and bromochlorobenzene are recommended as surrogate standards to encompass the range of the temperature program used in this method.

REFERENCE Test Methods for Evaluating Solid Waste, Physical/Chemical Methods, SW-846, 3rd Edition, U.S. EPA, Office of Solid Waste, Washington, DC, EPA Method 8021A, Rev. 1, Nov. 1992.

Styrene   **EPA Method 8240**
CAS #100-42-5

TITLE Volatile Organics By GC/MS: Packed Column Technique

MATRIX Nearly all types of sample matarices, regardless of water content, can be analyzed using this method. This includes groundwater, aqueous sludges, caustic liquors, acid liquors, waste solvents, oily wastes, mousses, tars, fibrous wastes, polymetric emulsions, filter cakes, spent carbons, spent catalysts, soils, and sediments.

METHOD SUMMARY Method 8240B covers 80 volatile organic compounds that are introduced into a gas chromatograph by the purge-and-trap method or by direct injection (in limited applications). For the purge-and-trap method an inert gas (zero grade nitrogen or helium) is bubbled through a 5-mL solution at ambient temperature. Purged sample components are trapped in a tube of sorbent materials. When purging is complete, the sorbent tube is heated and backflushed with inert gas to desorb the trapped components onto a GC column.

INTERFERENCES Impurities in the purge gas and from organic compounds outgassing from the plumbing ahead of the trap account for many contamination problems. Interferences purged or coextracted from the samples will vary considerably from source to source. Cross-contamination can occur whenever high-level and low-level samples are analyzed sequentially. Whenever an unusually concentrated sample is analyzed, it should be followed by the analysis of organic-free reagent water to check for cross-contamination. Samples also can be contaminated by diffusion of volatile organics (particularly methylene chloride and fluorocarbons) through the septum seal into the sample during shipment and storage. A trip blank can serve as a check on such contamination. The lab where volatile analysis is performed and also the refrigerated storage area should be completely free of solvents.

INSTRUMENTATION A gas chromatograph/mass spectrometry/data system (GC/MS) equipped with a 6 ft × 0.1 in I.D. glass column packed with 1% SP-1000 on Carbopack-B (60/80 mesh) is required. Also needed is a 5-mL purging device, a sorbent trap, and a thermal desorption apparatus.

PRECISION & ACCURACY This method is reported to have been tested by 15 laboratories using organic-free reagent water, drinking water, surface water, and industrial wastewaters (not specified) fortified at six concentrations over the range 5–600 µg/L.

Sample estimated quantitation limits (EQLs) are highly matrix-dependent. The EQLs listed may not always be achievable. EQLs listed for soils or sediments are based on wet weight. Normally, data is reported on a dry-weight basis; therefore, EQLs will be higher, based on the percent dry weight of each sample. Note that EQLs are even more variable than MDLs and that they are highly variable depending on the matrix being analyzed.

EQL in groundwater in µg/L was 5.
EQL in low soil or sediment in µg/kg was 5.
Accuracy (a) in µg/L was not listed.
Precision (b) in µg/L was not listed.

(a) *Average recovery found for measurements of samples containing a concentration of C, in µg/L.*
(b) *Overall precision found for measurements of samples with average recovery X for samples containing a concentration of C in µg/L.*
X = *Average recovery found for measurement of samples containing a concentration of C in µg/L.*

MULTIPLICATION FACTORS FOR OTHER MATRICES

Other Matrices	Factor (a)
Waste miscible liquid waste	50
High-concentration soil and sludge	125
Non-water miscible waste	500

(a) *EQL = [EQL for low soil/sediment] × [Factor]. For non-aqueous samples, the factor is on a wet-weight basis.*

SAMPLING METHOD
Liquid samples — Use a 40-mL glass screw-cap VOA vial with a Teflon®-faced silicone septum that has been prewashed, rinsed with distilled deionized water, and oven dried. However, if residual chlorine is present, collect sample in a 40-oz. soil VOA container which has been pre-preserved with 4 drops of 10% sodium thiosulfate, mix gently, and then transfer the sample to a 40-mL VOA vial. Collect bubble-free samples in duplicate and seal them in separate plastic bags.

Soils or sediments, and sludges — Use an 8-oz. widemouth glass bottle with a Teflon®-faced silicone septum that has been prewashed with detergent, rinsed with distilled deionized water, and oven dried. Tap slightly to eliminate free air space. Collect samples in duplicate and seal them in separate plastic bags.

SAMPLE PRESERVATION
Liquid samples — Add 4 drops of concentrated HCL and immediately cool samples to 4°C and store in a solvent-free refrigerator.

Soils or sediments, and sludges — Cool samples to 4°C and store in a solvent-free refrigerator.

MHT Maximum holding time is 14 days from the date of sample collection.

SAMPLE PREPARATION
Liquid samples — Remove the plunger from a 5-mL syringe and carefully pour the sample into the syringe barrel to just short of overflowing. Replace the syringe plunger and compress the sample. Open the syringe valve and vent any residual air while adjusting the sample volume to 5.0 mL. If there is only one volatile organic analysis (VOA) vial, a second syringe should be filled at this time to protect against possible loss of sample integrity. Add 10 µL of surrogate spiking solution and 10 µL of internal standard spiking solution through the valve bore of the 5-mL syringe, then close the valve. The surrogate and internal standards may be mixed and added as a single spiking solution.

Sediments, soils, and waste samples — All samples of this type should be screened by GC analysis using a headspace method (EPA Method 3810) or the hexadecane extraction and screening method (EPA Method 3820). Use the screening data to determine whether to use the low-concentration method (0.005–1 mg/kg) or the high-concentration method (>1 mg/kg).

Low-concentration method — The low-concentration method is based on purging a heated sediment or soil sample mixed with organic-free reagent water containing the surrogate and internal standards. Analyze all reagent blanks and standards under the same conditions as the samples.

Use a 5-g sample if the expected concentration is <0.1 mg/kg or a 1-g sample for expected concentrations between 0.1 and 1 mg/kg. Mix the contents of the sample container with a narrow metal spatula. Weigh the amount of the sample into a tared purge device. Add the spiked water to the purge device, which contains the weighed amount of sample, and connect the device to the purge-and-trap system.

High-concentration method — This method is based on extracting the sediment or soil with methanol. A waste sample

is either extracted or diluted, depending on its solubility in methanol. Wastes that are insoluble in methanol are diluted with reagent tetraglyme or possibly polyethylene glycol (PEG). An aliquot of the extract is added to organic-free reagent water containing surrogate and internal standards. This is purged at ambient temperature. All samples with an expected concentration of >1.0 mg/kg should be analyzed by this method.

Mix the contents of the sample container with a narrow metal spatula. For sediments or soils and solid wastes that are insoluble in methanol, weigh 4 g (wet weight) of sample into a tared 20-mL vial. For waste that is soluble in methanol, tetraglyme, or PEG, weigh 1 g (wet weight) into a tared scintillation vial or culture tube or a 10-mL volumetric flask. Quickly add 9.0 mL of appropriate solvent then add 1.0 mL of a surrogate spiking solution to the vial, cap it, and shake it for 2 min.

METHANOL EXTRACT REQUIRED FOR ANALYSIS OF HIGH-CONCENTRATION SOILS OR SEDIMENTS

Approximate Concentration Range	Volume of Methanol Extract (a)
500–10,000 µg/kg	100 µL
1,000–20,000 µg/kg	50 µL
5,000–100,000 µg/kg	10 µL
25,000–500,000 µg/kg	100 µL of 1/50 dilution (b)

Calculate appropriate dilution factor for concentrations exceeding this table.

(a) The volume of methanol added to 5 mL of water being purged should be kept constant. Therefore, add to the 5-mL syringe whatever volume of methanol is necessary to maintain a volume of 100 µL added to the syringe.
(b) Dilute an aliquot of the methanol extract and then take 100 µL for analysis.

QUALITY CONTROL Demonstrate, through the analysis of a reagent water blank, that interferences from the analytical system, glassware, and reagents are under control. Blank samples should be carried through all stages of the sample preparation and measurement steps. For each analytical batch (up to 20 samples), a reagent blank, matrix spike, and matrix spike duplicate must be analyzed (the frequency of the spikes may be different for different monitoring programs). The blank and spiked samples must be carried through all stages of the sample preparation and measurement steps. QC samples mentioned in the section on Interferences will also be needed as appropriate to those situations.

REFERENCE Test Methods for Evaluating Solid Waste (SW-846). U.S. EPA. 1983. Method 8240B, Rev. 2, Nov. 1990. Office of Solid Wastes, Washington, DC.

Styrene   EPA Method 8260
CAS #100-42-5

TITLE Volatile Organic Compounds by GC/MS: Capillary Column Technique

MATRIX This method is applicable to nearly all types of samples, regardless of water content, including groundwater, soils, and sediments.

METHOD SUMMARY Method 8260A covers 58 volatile organic compounds that are introduced into a gas chromatograph by the purge-and-trap method or by direct injection (in limited applications). Zero-grade helium is bubbled through a 5-mL solution at ambient temperature. Purged sample components are trapped in a tube containing suitable sorbent materials. When purging is complete, the sorbent tube is heated and backflushed with helium to desorb trapped sample components. The analytes are desorbed directly to a large bore capillary or cryofocussed on a capillary precolumn before being flash evaporated to a narrow bore capillary for analysis.

INTERFERENCES Major contaminant sources are volatile materials in the lab and impurities in the inert purging gas and in the sorbent trap. Interfering contamination may occur when a sample containing low concentrations of volatile organic compounds is analyzed immediately after a sample containing high concentrations of volatile organic compounds. After analysis of a sample containing high concentrations of volatile organic compounds, one or more calibration blanks should be analyzed to check for cross-contamination. Screening of the samples prior to purge-and-trap GC/MS analysis is highly recommended to prevent contamination of the system. This is especially true for soil and waste samples.

Special precautions must be taken to analyze for methylene chloride. The analytical and sample storage area should be isolated from all atmospheric sources of methylene chloride. All gas chromatography carrier gas lines and purge gas plumbing should be constructed from stainless steel or copper tubing. Laboratory clothing previously exposed to methylene chloride fumes during liquid-liquid extraction procedures can contribute to sample contamination.

Samples can also be contaminated by diffusion of volatile organics (particularly methylene chloride and fluorocarbons) through the septum seal during shipment and storage. A trip blank can serve as a check on such contamination.

INSTRUMENTATION GC/MS with a temperature-programmable chromatograph suitable for splitless injection equipped with variable constant differential flow controllers, a subambient oven controller, a purging device, sorbent trap, a thermal desorption apparatus and a capillary precolumn interface when using cryogenic cooling will be needed. The following GC columns may be used:

Column 1: 60 m × 0.75mm I.D. capillary column coated with VOCOL, 1.5 µm film thickness.
Column 2: 30 m × 0.53mm capillary column coated with DB-624 or VOCOL, 3 µm film thickness.
Column 3: 30 m × 0.32mm I.D. capillary column coated with DB-5 or SE-54, 1-µm film thickness.

PRECISION & ACCURACY This method has been tested in a single lab using spiked water. Using a wide-bore capillary column, water was spiked at concentrations between 0.5 and 10 µg/L. Single lab accuracy and precision data are presented.

The MDL actually achieved in a given analysis will vary depending on instrument sensitivity and matrix effects.

The MDL (a) in μg/L was 0.04.
The concentration range in μg/L was 0.1–100.
The mean accuracy (% of true value) was 102.
The precision as relative standard deviation was 7.2.

Note: The MDL is based on a 25-mL sample volume instead of a 5-mL sample volume.

SAMPLING METHOD

Liquid samples — Use a 40-mL glass screw-cap VOA vial with a Teflon®-faced silicone septum that has been prewashed, rinsed with distilled deionized water, and oven dried. If residual chlorine is present, collect the sample in a 4-oz soil VOA container which has been pre-preserved with 4 drops of 10% sodium thiosulfate. Mix gently and transfer the sample to a 40-mL VOA vial. Collect bubble-free samples in duplicate and seal each sample in a separate plastic bag.

Soils, sediments and sludges — Use an 8-oz widemouth glass bottle with Teflon®-faced silicone septum that has been prewashed, rinsed with distilled deionized water, and oven dried. **Do not** heat the septum for more than 1 h. Tap slightly to eliminate any free air space. Collect samples in duplicate and seal each one in a separate plastic bag.

SAMPLE PRESERVATION

Liquid samples — Add 4 drops of concentrated HCL, cool to 4°C and store in a solvent-free refrigerator.

Soils, sediments and sludges — Cool samples to 4°C and store in a solvent-free refrigerator.

MHT The maximum holding time of any sample (liquids, soils, sediments, and sludges) is 14 days.

SAMPLE PREPARATION

Liquid samples — Remove the plunger from a 5-mL syringe and carefully pour the sample into the syringe barrel to just short of overflowing. Replace the syringe plunger and compress the sample. Open the syringe valve and vent any residual air while adjusting the sample volume to 5.0 mL. If there is only one volatile organic analysis (VOA) vial, a second syringe should be filled at this time to protect against possible loss of sample integrity. Add 10 μL of surrogate spiking solution and 10 μL of internal standard spiking solution through the valve bore of the 5-mL syringe, then close the valve. The surrogate and internal standards may be mixed and added as a single spiking solution.

Sediments, soils, and waste samples — All samples of this type should be screened by GC analysis using a headspace method (EPA Method 3810) or the hexadecane extraction and screening method (EPA Method 3820). Use the screening data to determine whether to use the low-concentration method (0.005–1 mg/kg) or the high-concentration method (>1 mg/kg).

Low-concentration method — The low-concentration method is based on purging a heated sediment or soil sample mixed with organic-free reagent water containing the surrogate and internal standards. Analyze all reagent blanks and standards under the same conditions as the samples.

Use a 5-g sample if the expected concentration is <0.1 mg/kg or a 1-g sample for expected concentrations between 0.1 and 1 mg/kg. Mix the contents of the sample container with a narrow metal spatula. Weigh the amount of the sample into a tared purge device. Add the spiked water to the purge device, which contains the weighed amount of sample, and connect the device to the purge-and-trap system.

High-concentration method — This method is based on extracting the sediment or soil with methanol. A waste sample is either extracted or diluted, depending on its solubility in methanol. Wastes that are insoluble in methanol are diluted with reagent tetraglyme or possibly polyethylene glycol (PEG). An aliquot of the extract is added to organic-free reagent water containing surrogate and internal standards. This is purged at ambient temperature. All samples with an expected concentration of >1.0 mg/kg should be analyzed by this method.

Mix the contents of the sample container with a narrow metal spatula. For sediments or soils and solid wastes that are insoluble in methanol, weigh 4 g (wet weight) of sample into a tared 20-mL vial. For waste that is soluble in methanol, tetraglyme, or PEG, weigh 1 g (wet weight) into a tared scintillation vial or culture tube or a 10-mL volumetric flask. Quickly add 9.0 mL of appropriate solvent then add 1.0 mL of a surrogate spiking solution to the vial, cap it, and shake it for 2 min.

METHANOL EXTRACT REQUIRED FOR ANALYSIS OF HIGH-CONCENTRATION SOILS OR SEDIMENTS

Approximate Concentration Range	Volume of Methanol Extract (a)
500–10,000 μg/kg	100 μL
1,000–20,000 μg/kg	50 μL
5,000–100,000 μg/kg	10 μL
25,000–500,000 μg/kg	100 μL of 1/50 dilution (b)

Calculate appropriate dilution factor for concentrations exceeding this table.

(a) The volume of methanol added to 5 mL of water being purged should be kept constant. Therefore, add to the 5-mL syringe whatever volume of methanol is necessary to maintain a volume of 100 μL added to the syringe.
(b) Dilute an aliquot of the methanol extract and then take 100 μL for analysis.

QUALITY CONTROL Demonstrate, through the analysis of a reagent water blank, that interferences from the analytical system, glassware, and reagents are under control. Blank samples should be carried through all stages of the sample preparation and measurement steps. For each analytical batch (up to 20 samples), a reagent blank, matrix spike, and matrix spike duplicate must be analyzed (the frequency of the spikes may be different for different monitoring programs). The blank and spiked samples must be carried through all stages of the sample preparation and measurement steps. QC samples mentioned in the section on Interferences will also be needed as appropriate to those situations.

Matrix spiking standards should be prepared from volatile organic compounds which will be representative of the compounds being investigated. The recommended internal standards

are chlorobenzene-d5, 1,4-difluorobenzene, 1,4-dichlorobenzene-d4, and pentafluorobenzene. Using stock standard solutions, prepare secondary dilution standards containing the compounds of interest, either singly or mixed together in methanol. Store them in a vial with no headspace for no more than one week. Surrogates recommended are toluene-d8, 4-bromofluorobenzene, and dibromofluoromethane. Each sample undergoing GC/MS analysis must be spiked with 10 µL of the surrogate spiking solution prior to analysis.

REFERENCE Test Methods for Evaluating Solid Waste (SW-846). U.S. EPA 1983, Method 8260A, Rev. 1, Nov. 1990. Office of Solid Waste, Washington, DC.

Styrene EPA Method 503.1
CAS #100-42-5

TITLE Aromatic & Unsaturated VOCs Water

MATRIX Drinking water (finished or in any treatment stage) and raw source water.

APPLICATION Method covers 28 aromatic and unsaturated VOCs. An inert gas is bubbled through a 5-mL water sample. Purged sample components are trapped in tube of sorbent materials. When purging is complete, sorbent tube is heated and backflushed with inert gas to desorb trapped sample onto a packed GC column.

INTERFERENCES During analysis, major contaminant sources are volatile materials in the lab and impurities in purging gas and sorbent trap. With high and low level samples, there can be carryover contamination. Excess water causes a negative baseline deflection.

INSTRUMENTATION Purge and Trap GC w/photoionization detector. (Two GC columns are recommended); Column 1: 5% SP-1200 and 1.75% Bentone 34 on Supelcoport; Column 2: 1,2,3-tris(2-cyanoethoxy)propane on Chromosorb W.

RANGE 2.2–600 µg/L. (Drinking water)

MDL 0.008 µg/L in water

PRECISION Not listed.

ACCURACY Not listed.

SAMPLING METHOD Use a 40–120-mL screw-cap vial (prewashed with detergent, rinsed with distilled water and oven dried at 105°C) with a PTFE-faced silicone septum. If residual chlorine is in the water add about 25 mg of ascorbic acid to each vial before sample collection. Collect bubble-free samples.

STABILITY Cool to 4°C; HCl to pH <2.

MHT 14 days.

QUALITY CONTROL As initial demonstration of lab accuracy and precision, analyze 4 to 7 replicates of a lab fortified blank containing the analyte at 0.1–5 µg/L. Collect all samples in duplicate.

REFERENCE Method 503.1, Volatile Aromatic & Unsaturated Organic Compounds in H2O by Purge and Trap GC, EPA 600/4-88/039.

Sulfallate EPA Method 8270
CAS #95-06-7

TITLE Semivolatile Organic Compounds by GC/MS

MATRIX This method is used to determine the concentration of semivolatile organic compounds in extracts prepared from all types of solid waste matrices, soils, and groundwater. Although surface waters are not specifically mentioned, this method should be applicable to water samples from rivers, lakes, etc.

METHOD SUMMARY This method covers 259 semivolatile organic compounds. In very limited applications direct injection of the sample into the GC/MS system may be appropriate, but this results in very high detection limits (approximately 10,000 µg/L). Typically, a 1-L liquid sample, containing surrogate, and matrix spiking standards, is extracted in a continuous extractor first under acid conditions and then under basic conditions. Typically 30 g of a solid sample, containing surrogate, and matrix spiking standards, is extracted ultrasonically. After concentrating the extract to 1 mL it is spiked with 10 µL of an internal standard solution just prior to analysis by GC/MS. The volume injected should contain about 100 ng of base/neutral and 200 ng of acid surrogates (for a 1-µL injection). Analysis is performed by GC/MS using a capillary GC column.

INTERFERENCES Raw GC/MS data from all blanks, samples, and spikes must be evaluated for interferences. Contamination by carryover can occur whenever high-concentration and low-concentration samples are sequentially analyzed. To reduce carryover, the sample syringe must be rinsed out between samples with solvent. Whenever an unusually concentrated sample is encountered, it should be followed by the analysis of blank solvent to check for cross-contamination.

INSTRUMENTATION A GC/MS and a data system are required. The GC column used is a 30 m × 0.25 mm I.D. (or 0.32 mm I.D.) 1um film thickness silicone-coated fused silica capillary column. A continuous liquid-liquid extractor equipped with Teflon® or glass connection joints and stopcocks requiring no lubrication, a K-D concentrating apparatus, water bath, and an ultrasonic disrupter with a minimum power of 300 W and with pulsing capability are also required.

PRECISION & ACCURACY The estimated quantitation limit (EQL) of Method 8270B for determining an individual compound is approximately 1 mg/kg (wet weight) for soil or sediment samples, 1–200 mg/kg for wastes (dependent on matrix and method of preparation), and 10 µg/L for groundwater samples. EQLs will be proportionately higher for sample extracts that require dilution to avoid saturation of the detector.

The EQL(b) for groundwater in µg/L is 10.
The EQL (a, b) for low concentrations in soil and sediment in µg/kg is not determined.
Accuracy as µg/L is not listed.

Overall precision in µg/L is not listed.

(a) *EQLs listed for soil/sediment are based on wet weight. Normally data is reported in a dry-weight basis; therefore, EQLs will be higher based on the % dry weight of each sample. This calculation is based on a 30 g sample and gel permeation chromatography cleanup.*

(b) *Sample EQLs are highly matrix-dependent. The EQLs are provided for guidance and may not always be achievable.*

C = *True value for concentration, in µg/L.*

X = *Average recovery found for measurements of samples containing a concentration of C, in µg/L.*

ESTIMATED QUANTITATION LIMIT

Other Matrices	Factor (a)
High-concentration soil and sludges by sonicator	7.5
Non-water miscible waste	75

(a) *EQL for other matrices = [EQL for low soil/sediment] × [Factor]. This estimated EQL is similar to an EPA "Practical Quantitation Limit."*

SAMPLING METHOD

Liquid samples — Use a 1 or 2½ gallon amber glass bottle with a screw-top Teflon®-lined cover that has been prewashed with detergent and rinsed with distilled water and methanol (or isopropanol).

Soils, sediments, or sludges — Use an 8-oz. widemouth glass with a screw-top Teflon®-lined cover that has been prewashed with detergent and rinsed with distilled water and methanol (or isopropanol).

SAMPLE PRESERVATION

Liquid samples — If residual chlorine is present, add 3 mL of 10% sodium thiosulfate per gallon, cool to 4°C and store in a solvent-free refrigerator until analysis; if chlorine is not present, then eliminate the sodium thiosulfate addition.

Soils, sediments, or sludges — Cool samples to 4°C and store in a solvent-free refrigerator.

MHT Liquid samples must be extracted within 7 days and the extracts analyzed within 40 days. Soils, sediments, or sludges may be stored for a maximum of 14 days and the extracts analyzed within 40 days.

SAMPLE PREPARATION

Liquid samples — Transfer 1 L quantitatively to a continuous extractor. If high concentrations are anticipated, a smaller volume may be used and then diluted with organic-free reagent water to 1 L. Adjust pH, if necessary, to pH <2 using 1:1 (V/V) sulfuric acid. Pipette 1.0 mL of a surrogate standard spiking solution into each sample. For the sample in each analytical batch selected for spiking, add 1.0 mL of a matrix spiking standard. For base/neutral acid analysis, the amount of the surrogates and matrix spiking compounds added to the sample should result in a final concentration of 100 ng/µL of each analyte in the extract to be analyzed (assuming a 1-µL injection). Extract with methylene chloride for 18–24 h. Next, adjust the pH of the aqueous phase to pH >11 using 10 N sodium hydroxide and extract it with methylene chloride again for 18–24 h. Dry the extract through a column containing anhydrous sodium sulfate and concentrate it to 1 mL using a K-D concentrator.

Soils, sediments, or sludges — Use 30 g of sample. Nonporous or wet samples (gummy or clay type) that do not have a free-flowing sandy texture must be mixed with anhydrous sodium sulfate until the sample is free flowing. Add 1 mL of surrogate standards to all samples, spikes, standards, and blanks. For the sample in each analytical batch selected for spiking, add 1.0 mL of a matrix spiking standard. For base/neutral acid analysis, the amount added of the surrogates and matrix spiking compounds should result in a final concentration of 100 ng/µL of each base/neutral analyte and 200 ng/µL of each acid analyte in the extract to be analyzed (assuming a 1-µL injection). Immediately add a 100-mL mixture of 1:1 methylene chloride:acetone and extract the sample ultrasonically for 3 min and then decant or filter the extracts. Repeat the extraction two or more times. Dry the extract using a column with anhydrous sodium sulfate and concentrate it to 1 mL in a K-D concentrator.

QUALITY CONTROL A methylene chloride solution containing 50 ng/µL of decafluorotriphenylphosphine (DFTPP) is used for tuning the GC/MS system each 12-h shift. A system performance check also must be made during every 12-h shift. A standard containing 50 ng/µL each of 4,4'-DDT, pentachlorophenol, and benzidine is required to verify injection port inertness and GC column performance. A calibration standard at mid-concentration, containing each compound of interest, including all required surrogates, must be performed every 12 h during analysis. After the system performance check is met, calibration check compounds (CCCs) are used to check the validity of the initial calibration.

The internal standard responses and retention times in the calibration check standard must be evaluated immediately after or during data acquisition. If the retention time for any internal standard changes by more than 30 seconds from the last check calibration (12 h), the chromatographic system must be inspected for malfunctions and corrections must be made, as required. If the electron ionization current plot (EICP) area for any of the internal standards changes by a factor of two from the last daily calibration standard check, the mass spectrometer must be inspected for malfunctions and corrections must be made, as appropriate.

Demonstrate, through the analysis of a reagent water blank, that interferences from the analytical system, glassware, and reagents are under control. The blank samples should be carried through all stages of the sample preparation and measurement steps. For each analytical batch (up to 20 samples), a reagent blank, matrix spike, and matrix spike duplicate/duplicate must be analyzed (the frequency of the spikes may be different for different monitoring programs). The blank and spiked samples must be carried through all stages of the sample preparation and measurement steps. A QC reference sample concentrate containing each analyte at a concentration of 100 mg/L in methanol is required.

REFERENCE Test Methods for Evaluating Solid Waste (SW-846). U.S. EPA 1983, Method 8270B, Rev. 2, Nov. 1990. Office of Solid Waste, Washington, DC.

Sulfate — EPA Method 375.1

TITLE Inorganics, Non-Metallics

MATRIX Drinking, surface and wastewaters.

APPLICATION Date issued 1971. (Colorimetric, automated, chloranilate). When solid barium chloranilate is added to a solution containing sulfate, barium sulfate is precipitated, releasing the highly colored acid chloranilate ion. Color intensity = amount of sulfate present.

INTERFERENCES Cations, such as calcium, aluminum and iron interfere by precipitating the chloranilate. These ions are removed automatically by passage through an ion exchange column.

INSTRUMENTATION Technicon auto analyzer. 520 nm filters. 15 mm Tubular flow cells.

RANGE 10–400 mg SO4/L.

MDL Not listed.

PRECISION SD = ±0.8 at 294 mg SO4/L.

ACCURACY Recoveries = 99 and 102% at 82 and 295 mg SO4/L.

SAMPLING METHOD Plastic or glass (50 mL).

STABILITY Cool, 4°C.

MHT 28 days.

QUALITY CONTROL Place working standards in sampler in order of decreasing concentration. Approximately 15 samples per h can be analyzed.

REFERENCE EPA Methods for the Chemical Analysis of Water and Wastes, EPA-600/4-79-020, U.S. EPA, EMSL, 1979.

Sulfate — EPA Method 375.2

TITLE Inorganics, Non-Metallics

MATRIX Drinking, surface, and wastewaters.

APPLICATION Date issued 1978. (Colorimetric, automated, methylthymol blue, AAII). After being passed through cation-exchange column, sample is reacted with alcohol solution of barium chloride and MTB at pH 2.5–3.0 To form barium sulfate. This solution is raised to pH 12.5–13.0 so that excess barium reacts with MTB. Uncomplexed MTB = amount sulfate present]

INTERFERENCES Multivalent cation interferences are eliminated by the ion exchange column. Samples with pH below 2 should be neutralized since high acid concentrations elute cations from ion exchange resin. Filter or centrifuge turbid samples.

INSTRUMENTATION Technicon auto analyzer. 460 nm Interference filters. 15 nm Flow cell.

RANGE 3–300 mg SO4/L (or) 0.5–30 mg SO4/L

MDL Not listed.

PRECISION SD = ±1.6 at Mean conc of 110 mg/L (26 samples).

ACCURACY Mean recovery = 102% (on 24 surface and waste waters)

SAMPLING METHOD Plastic or glass (50 mL).

STABILITY Cool, 4°C.

MHT 28 days.

QUALITY CONTROL Analyze all working standards in duplicate at beginning of a run to develop standard curve. Approx 30 samples an h can be analyzed.

REFERENCE EPA Methods for the Chemical Analysis of Water and Wastes, EPA-600/4-79-020, U.S. EPA, EMSL, 1979.

Sulfate (Gravimetric) — EPA Method 375.3

TITLE Inorganics, Non-Metallics

MATRIX Drinking, surface and saline waters. Wastewater.

APPLICATION Date issued 1974. Editorial Rev. 1978. Sulfate is precipitated as barium sulfate (BaSO4) in HCl medium by the addition of barium chloride. After digestion period, precipitate is filtered, washed with hot water until chloride free, ignited and weighed as BaSO4.

INTERFERENCES High results may be obtained for samples containing suspended matter, nitrate, sulfite and silica. Alkali metal sulfates frequently yield low results. This is especially true of alkali hydrogen sulfates. Heavy metals such as chromium and iron can interfere.

INSTRUMENTATION Steam bath. Drying oven. Muffle furnace. Analytical balance. Filter paper (ashless)

RANGE Not listed.

MDL Not listed.

PRECISION SD = 4.7% at 259 mg/L sulfate (aqueous mix of 9 ions)

ACCURACY Relative error = 1.9% at 259 mg/L SO4 (aqueous mix, 9 ions)

SAMPLING METHOD Plastic or glass (50 mL).

STABILITY Cool, 4°C.

MHT 28 days.

QUALITY CONTROL This is most accurate method for sulfate concentrations above 10 mg/L. Use this method when greatest accuracy is required. Make sure precipitate is washed free of chloride. **Do not** let filter paper flame during ashing of precipitate.

REFERENCE Methods for the Chemical Analysis of Water and Wastes, EPA-600/4-79-020, U.S. EPA, EMSL, 1979.

Sulfate (Total) — EPA Method 300.0

TITLE Inorganic Anions in Water

MATRIX Drinking, surface and mixed wastewater.

APPLICATION A small volume of sample, typically 2 to 3 mL is introduced into an ion chromatograph. The anions of interest are separated and measured using a system comprised of a guard column, separator column, suppressor column and conductivity detector.

INTERFERENCES Interferences can be caused by substances with retention times similar to and overlaping those of ion of interest. Large amounts of an anion can interfere with peak resolution of adjacent anion. EPA Method interference can be caused by reagent or equipment contamination.

INSTRUMENTATION Ion chromatograph. Analytical balance. Guard, separator and suppressor columns.

RANGE (Report results in mg/L).

MDL 0.206 mg/L.

PRECISION SD = 1.475 mg/L at 98.5 mg/L sulfate (drinking water).

ACCURACY Recovery = 104.3% at 98.5 mg/L sulfate (drinking water)

SAMPLING METHOD Plastic or glass.

STABILITY Cool, 4°C.

MHT 28 days.

QUALITY CONTROL The lab should spike and analyze a minimum of 10% of all samples to monitor continuing lab performance. Field and lab duplicates should be analyzed. Measure retention times of standards. (Sulfates exhibit great changes in retention times).

REFERENCE Test Method-The Determination of Inorganic Anions in Water by Ion Chromatography, (EPA-600/4-84-017).

Sulfate (Turbidometric) — EPA Method 375.4

TITLE Inorganics, Non-Metallics

MATRIX Drinking, surface and wastewaters.

APPLICATION Date issued 1971. Editorial Rev. 1978. Sulfate ion is converted to a barium sulfate suspension under controlled conditions. The resulting turbidity is determined using a photometer and compared to a curve prepared from standard sulfate solutions

INTERFERENCES Suspended matter and color interfere. Silica in concentrations over 500 mg/L will interfere.

INSTRUMENTATION Nephelometer or [spectrophotometer at 420 nm. Light path of 4–5cm].

RANGE Not listed.

MDL 1 mg/L sulfate.

PRECISION SD = 7.86 mg/L at 110 mg SO4/L.

ACCURACY As bias, −3.3 mg/L at 110 mg SO4/L.

SAMPLING METHOD Plastic or glass (50 mL).

STABILITY Cool, 4°C.

MHT 28 days.

QUALITY CONTROL Correct for sample color and turbidity by running blanks from which barium chloride has been omitted. Suitable for all ranges of sulfate, but use sample aliquot with not more than 40 mg SO4/L. Above 50 mg/L the accuracy decreases and suspensions lose stability.

REFERENCE Methods for the Chemical Analysis of Water and Wastes, EPA-600/4-79-020, U.S. EPA, EMSL, 1979.

Sulfotep — EPA Method 8141
CAS #3689-24-5

TITLE Organophosphorus Compounds by Gas Chromatography: Capillary Column Technique

MATRIX This method covers aqueous and solid matrices. This includes a wide variety such as drinking water, groundwater, industrial wastewaters, surface waters, soils, solids, and sediments.

METHOD SUMMARY This is a GC method used to determine the concentration of 28 organophosphorus pesticides.

The use of Gel Permeation Cleanup (EPA Method 3640) for sample cleanup has been demonstrated to yield recoveries of less than 85% for many method analytes and is therefore not recommended for use with this method.

This method provides GC conditions for the detection of ppb concentrations of organophosphorus compounds. Prior to the use of this method, appropriate sample preparation techniques must be used. Water samples are extracted at a neutral pH with methylene chloride as a solvent by using a separatory funnel (EPA Method 3510) or a continuous liquid-liquid extractor (EPA Method 3520). Soxhlet extraction (EPA Method 3540) or ultrasonic extraction (EPA Method 3550) using methylene chloride/acetone (1:1) are used for solid samples. Both neat and diluted organic liquids (EPA Method 3580) may be analyzed by direct injection. Spiked samples are used to verify the applicability of the chosen extraction technique to each new sample type. A GC with a flame photometric (FPD) or nitrogen-phosphorus detector (NPD) is used for this multiresidue procedure.

INTERFERENCES The use of Florisil cleanup materials (EPA Method 3620) for some of the compounds in this method has been demonstrated to yield recoveries less than 85% and is therefore not recommended for all compounds. Use of phosphorus or halogen specific detectors, however, often obviates the necessity for cleanup for relatively clean sample matrices. If particular circumstances demand the use of an alternative cleanup procedure, the analyst must determine the elution profile and demonstrate that the recovery of each analyte is no less than 85%.

Use of a flame photometric detector (FPD) in the phosphorus mode will minimize interferences from materials that do not contain phosphorus. Elemental sulfur, however, may interfere with the determination of certain organophosphorus compounds by flame photometric gas chromatography. Sulfur cleanup using EPA Method 3660 may alleviate this interference. A nitrogen phosphorus detector (NPD) is also recommended.

A few analytes coelute on certain columns. Therefore, select a second column for confirmation where coelution of the analytes of interest does not occur.

Method interferences may be caused by contaminants in solvents, reagents, glassware, and other sample processing hardware that lead to discrete artifacts or elevated baselines in gas chromatograms. All these materials must be routinely demonstrated to be free from interferences under the conditions of the analysis by analyzing reagent blanks.

INSTRUMENTATION A GC with a NPD or a FPD will be needed. A data system or integrator is recommended for measuring peak areas and/or peak heights. A Kuderna-Danish (K-D) apparatus will be needed for extract concentration.

Column 1: 15 m × 0.53 mm megabore capillary column, 1.0 μm film thickness, DB-210.
Column 2: 15 m × 0.53 mm megabore capillary column, 1.5 μm film thickness, SPB-608.
Column 3: 15 m × 0.53 mm megabore capillary column, 1.0 μm film thickness, DB-5.

Three megabore capillary columns are included for analysis of organophosphates by this method. Column 1 (DB-210 or equivalent) and Column 2 (SPB-608 or equivalent) are recommended if a large number of organophosphorus analytes are to be determined. If the superior resolution offered by Column 1 and Column 2 is not required, Column 3 (DB-5 or equivalent) may be used. For megabore capillary columns, automatic injections of 1 μL are recommended.

PRECISION & ACCURACY The MDL actually achieved in a given analysis will vary, as it is dependent on instrument sensitivity and matrix effects. Single operator accuracy and precision studies have been conducted with spiked water and soil samples.

MULTIPLICATION FACTORS FOR OTHER MATRICES (a)

Matrix	Factor (b)
Groundwater (EPA Method 3510 or EPA Method 3520)	10
Low-concentration soil by Soxhlet and no cleanup	10 (c)
Low-concentration soil by ultrasonic extraction with GPC cleanup	6.7 (c)
High-concentration soil and sludges by ultrasonic extraction	500 (c)
Non-water miscible waste (EPA Method 3580)	1000 (c)

(a) Sample EQLs are highly matrix-dependent. The EQLs listed here are provided for guidance and may not always be achievable.
(b) EQL = [Method detection limit] × [Factor]. For non-aqueous samples the factor is on a wet-weight basis.
(c) Multiply this factory times the soil MDL.

The MDL (in μg/L) when reagent water was extracted using a separatory funnel was 0.07.
The MDL (in μg/kg) when soil was extracted using Soxhlet extraction (EPA Method 3540) was 3.5.
Accuracy (as % recovery) with separatory funnel extraction ranged from 87 (with low spikes) to 83 (with high spikes).
Accuracy (as % recovery) with continuous liquid-liquid extraction ranged from 40 (with low spikes) to 85 (with high spikes).
Accuracy (as % recovery) with Soxhlet extraction of soils ranged from 67 (with low spikes to 78 (with high spikes).
Accuracy (as % recovery) with ultrasonic extraction of soils ranged from not recovered (with low spikes) to 76 (with high spikes).

SAMPLE COLLECTION, PRESERVATION & HANDLING
Containers used to collect samples for the determination of semivolatile organic compounds should be soap and water washed followed by methanol (or isopropanol) rinsing. The sample containers should be of glass or Teflon® and have screw-top covers with Teflon® liners.

No preservation is used with concentrated waste samples. With liquid samples containing no residual chlorine and with soil, sediment, and sludge samples, immediately cooling to 4°C is the only preservation used. When residual chlorine is present then 3 mL of 10% aqueous sodium sulfate is added for each gallon of sample collected, followed by cooling to 4°C.

Liquid samples must be extracted within 7 days and their extracts analyzed within 40 days. Concentrated waste, soil, sediment, and sludge samples must be extracted within 14 days and their extracts analyzed within 40 days.

SAMPLE PREPARATION In general, water samples are extracted at a neutral pH with methylene chloride, using either EPA Method 3510 or EPA Method 3520. Solid samples are extracted using either EPA Method 3540 or EPA Method 3550 with methylene chloride/acetone (1:1) as the extraction solvent.

Prior to GC analysis, the extraction solvent may be exchanged to hexane. Single lab data indicates that samples should not be transferred with 100% hexane during sample workup as the more water soluble organophosphorus compounds may be lost.

If cleanup is performed on the samples, the analyst should analyze the samples by GC. This will confirm elution patterns and the absence of interferences from the reagents. If peak detection and identification is prevented by the presence of interferences, further cleanup is required.

QUALITY CONTROL The analyst should monitor the performance of the extraction, cleanup (when used), and analytical system and the effectiveness of the method in dealing with each sample matrix by spiking each sample, standard, and blank with one or two surrogates (e.g., organophosphorus compounds not expected to be present in the sample). Deuterated analogs of analytes should not be used as surrogates for gas chromatographic analysis due to coelution problems.

A minimum of five concentrations for each analyte of interest should be prepared through dilution of the stock standards with isooctane. One of the concentrations should be at a concentration near, but above, the MDL.

Include a mid-level check standard after each group of 10 samples in the analysis sequence. GC/MS techniques should be judiciously employed to support qualitative identifications made with this method. Follow the GC/MS operating requirements specified in EPA Method 8270.

When available, chemical ionization mass spectra may be employed to aid in the qualitative identification process. To confirm an identification of a compound, the background-corrected mass spectrum of the compound must be obtained from the sample extract and must be compared with a mass spectrum from a stock or calibration standard analyzed under the same chromatographic conditions. The molecular ion and all other ions present above 20% relative abundance in the mass spectrum of the standard must be present in the mass spectrum of the sample with agreement to ±20%. The retention time of the compound in the sample must be within six seconds of the retention time for the same compound in the standard solution.

Should the MS procedure fail to provide satisfactory results, additional steps may be taken before reanalysis. These steps may include the use of alternate packed or capillary GC columns or additional sample cleanup.

REFERENCE Test Methods for Evaluating Solid Waste, Physical/Chemical Methods, SW-846, 3rd Edition, U.S. EPA, Office of Solid Waste, Washington, DC, EPA Method 8141 July 1992.

T

2,4,5-T
CAS #93-76-5

EPA Method 8151

TITLE Chlorinated Herbicides by GC Using Methylation or Pentafluorobenzylation Derivatization: Capillary Column Technique.

MATRIX This method covers aqueous and solid matrices. This includes a wide variety such as drinking water, groundwater, industrial wastewaters, surface waters, soils, solids, and sediments.

METHOD SUMMARY This is a GC method for determining 19 chlorinated acid herbicides in aqueous, soil, and waste matrices. Because these compounds are produced and used in various forms (i.e., acid, salt, ester, etc.) a hydrolysis step is included to convert the herbicide to the acid form prior to analysis. This method provides hydrolysis, extraction, derivatization and GC conditions for the analysis of chlorinated acid herbicides in water, soil, and waste samples. Water samples are hydrolyzed *in situ*, extracted with diethyl ether, and then esterified with either diazomethane or pentafluorobenzyl bromide. The derivatives are determined by gas chromatography with an electron capture detector (GC/ECD). The results are reported as acid equivalents. The sensitivity of this method depends on the level of interferences in addition to instrumental limitations.

INTERFERENCES Method interferences may be caused by contaminants in solvents, reagents, glassware, and other sample processing hardware. Immediately prior to use, glassware should be rinsed with the next solvent to be used. Matrix interferences may be caused by contaminants that are coextracted from the sample. Organic acids, especially chlorinated acids, cause the most direct interference with the determination by methylation. Phenols, including chlorophenols, may also interfere with this procedure. The determination using pentafluorobenzylation is more sensitive, and more prone to interferences from the presence of organic acids of phenols than by methylation. Alkaline hydrolysis and subsequent extraction of the basic solution removes many chlorinated hydrocarbons and phthalate esters that might otherwise interfere with the ECD analysis. The herbicides, being strong organic acids, react readily with alkaline substances and may be lost during analysis. Therefore, glassware must be acid-rinsed and then rinsed to constant pH with organic-free reagent water.

INSTRUMENTATION A GC suitable for Grob-type injection using capillary columns. A data system for measuring peak heights and/or peak areas is recommended. An electron capture detector (ECD) is used. Also a K-D apparatus, a diazomethane generator, a centrifuge and an ultrasonic disrupter will be required.

Narrow Bore Columns:
Primary Column 1: 30 m × 0.25 mm, 5% phenyl/95% methyl silicone (DB-5), 0.25 μm film thickness.
Primary Column 1a (GC/MS): 30 m × 0.32 mm, 5% phenyl/95% methyl silicone (DB-5), 1-μm film thickness.
Column 2: 30 m × 0.25 mm DB-608 with a 25 μm film thickness.
Confirmation Column: 30 m × 0.25 mm, 14% cyanopropyl phenyl silicone (DB-1701), 0.25 μm film thickness.

Megabore Columns:
Primary Column: 30 m × 0.53 mm DB-608 with 0.83 μm film thickness.
Confirmation Column: 30 m × 0.53 mm, 14% cyanopropyl phenyl silicone (DB-1701), 1.0 μm film thickness.

PRECISION & ACCURACY Method detection limits (MDLs) are compound-dependent and vary with derivitization efficiency, derivative recovery, the matrix sampled, and herbicide concentration.

The estimated MDL (in μg/L) was 0.08 for aqueous samples using GC/ECD.

The estimated MDL (in μg/kg) was not reported for soil samples using GC/ECD when corrected back to 50 g samples extracted and concentrated to 10 mL with 5-μL injections.

The estimated GC/MS identification limit (in ng) was not reported for soil samples using GC/MS.

Mean percent recovery, calculated from 7–8 determinations of spiked reagent water, after diazomethane derivatization, from a spike concentration (in μg/L) of 0.2 was 134 with a standard deviation of the percent recovery of 30.8.

Mean percent recovery, calculated from 10 determinations of spiked clay and clay/still bottom samples over the linear concentration range (in ng/g) of 0.42–828 was 93.1 with a percent relative standard deviation of 7.3. The RSD % was calculated on 10 samples high in the linear concentration range and 10 low in the range. The linear concentration range was determined using standard solutions and corrected to 50 g soil samples.

SAMPLE COLLECTION, PRESERVATION & HANDLING
Containers used to collect samples for the determination of semivolatile organic compounds should be soap and water washed followed by methanol (or isopropanol) rinsing. The sample containers should be of glass or Teflon® and have screw-top covers with Teflon® liners.

No preservation is used with concentrated waste samples. With liquid samples containing no residual chlorine and with soil, sediment, and sludge samples, immediately cooling to 4°C is the only preservation used. When residual chlorine is present then 3 mL of 10% aqueous sodium sulfate is added for each gallon of sample collected, followed by cooling to 4°C.

The holding time for all volatile organics samples is 14 days. Liquid samples must be extracted within 7 days and their extracts analyzed within 40 days. Concentrated waste, soil, sediment, and sludge samples must be extracted within 14 days and their extracts analyzed within 40 days.

SAMPLE PREPARATION
Preparation of soil, sediment, and other solid samples — Acidify 30 g (dry weight) solids with 0.1 M phosphate buffer (pH = 2.5) and thoroughly mix the contents. Spike the sample with surrogate compound(s). The ultrasonic extraction of solids

must be optimized for each type of sample. In order for the ultrasonic extractor to efficiently extract solid samples, the sample must be free flowing when the solvent is added. Acidified anhydrous sodium sulfate should be added to clay-type soils, or any other solid that is not a free-flowing sandy texture, until a free flowing mixture is obtained. Add methylene chloride and perform ultrasonic extraction. Combine organic extracts from the repetitive extractings of the sample and centrifuge. Add aqueous potassium hydroxide, water, and methanol to the extract and reflux the mixture on a water bath. Extract the solution three times with methylene chloride and discard the methylene chloride phase. The basic solution contains the herbicide salts. Adjust the pH of the solution to <2 with cold sulfuric acid and extract three times with methylene chloride. Combine the extracts and pour them through a pre-rinsed drying column containing acidified anhydrous sodium sulfate. Collect the dried extracts in a K-D flask and concentrate them.

Preparation of aqueous samples — Measure 1 L of sample into a 2 L separatory funnel and spike it with surrogate compound(s). Add NaCl to the sample, then add 6 N NaOH to the sample to a pH of 12 or more and let the sample sit at room temperature for 1 h to hydrolyze esters. Extract the sample three times with methylene chloride and discard the extracts. Then add cold 12 N sulfuric acid to a pH less than or equal to 2, and extract the sample three times with ethyl ether. Collect the ether phase in a flask containing acidified anhydrous sodium sulfate and allow it to remain in contact with the sodium sulfate for a minimum of 2 h. The drying step is very critical to ensuring complete esterification; any moisture remaining in the ether will result in low herbicide recoveries.

Extract concentration and derivatization — The combined ether extract is concentrated to about 1 mL using a K-D apparatus followed by using a micro Snyder column or nitrogen gas blowdown. If methyl esters are to be produced, then dilute the concentrated ether extract with 1 mL of isooctane and 0.5 mL of methanol, dilute to a final volume of 4 mL, and esterify with diazomethane. If pentafluorobenzene esters are to be produced, then dilute concentrated ether extract with acetone to a final volume of 4 mL and esterify with pentafluorobenzyl bromide.

QUALITY CONTROL Select a representative spike concentration for each compound (acid or ester) to be measured. Using stock standard, prepare a quality control check sample concentrate, in acetone, that is 1000 times more concentrated than the selected concentrations. Use this quality control check sample concentrate to prepare quality control check samples. Calculate surrogate standard recovery on all standards, samples, blanks, and spikes. GC/MS techniques should be judiciously employed to support qualitative identifications made with this method. When available, chemical ionization mass spectra may be employed to aid the qualitative identification process.

REFERENCE Test Methods for Evaluating Solid Waste, Physical/Chemical Methods, SW-846, 3rd Edition, U.S. EPA, Office of Solid Waste, Washington, DC, EPA Method 8151, Nov. 1990.

1,2,3,4-TCDD **EPA Method 8280**
CAS #30746-58-8

TITLE The Analysis of Polychlorinated Dibenzo-P-Dioxins and Polychlorinated Dibenzofurans.

MATRIX This method is appropriate for the determination of tetra-, penta-, hexa-, hepta-, and octachlorinated dibenzo-p-dioxins (PCDDs) and dibenzofurans (PCDFs) in chemical wastes including still bottoms, fuel oils, sludges, fly ash, reactor residues, soil and water.

METHOD SUMMARY This method covers 22 PCDD and PCDF compounds and it uses a high resolution capillary GC column with low resolution mass spectrometry. Samples are extracted and concentrated by several methods that vary depending on the matrix involved. The organic extracts are cleaned-up by washing with aqueous basic and acid solutions and then separated into fractions using a column of neutral alumina. The fraction containing the PCDDs and PCDFs is then further cleaned-up using a column of activated carbon. The final extract is concentrated and Carbon-13 labeled internal standards are added prior to analysis by GC/MS using a capillary GC column and selected ion monitoring using five sets of ions that are detailed in the method. Certain 2,3,7,8-substituted congeners are used to provide calibration and method recovery information. Proper column selection and access to reference isomer standards, may in certain cases, provide isomer specific data.

INTERFERENCES Solvents, reagents, glassware, and other sample processing hardware may yield discrete artifacts and/or elevated baselines which may cause misinterpretation of chromatographic data. Use high purity reagents and solvents to minimize interference problems. Interferents coextracted from the sample will vary considerably from source to source. PCDDs and PCDFs are often associated with other interfering chlorinated compounds such as PCBs and polychlorinated diphenyl ethers which may be found at concentrations several orders of magnitude higher than that of the analytes of interest.

INSTRUMENTATION A low resolution GC/MS utilizing 70 ev must be capable of selected ion monitoring (SIM) for at least 11 ions simultaneously, with a cycle time of 1 second or less. Minimum integration time for SIM is 50 ms per m/z. Also required is a GC-to-MS interface constructed of all glass or glass-lined materials. One of the following GC columns is required:

Column 1: 50 m CP-Sil-88 fused silica capillary column.
Column 2: DB-5 (30 m × 0.25 mm I.D., 0.25-um film thickness) fused silica capillary column.
Column 3: 30 m SP-2250 fused silica capillary column.

When toluene is employed as the final solvent, use of a bonded phase column is recommended. Solvent exchange into tridecane is required for other liquid phases or nonbonded columns such as CP-Sil-88. Chromatographic conditions must be adjusted to account for solvent boiling points.

PRECISION & ACCURACY Accuracy, precision, MDLs and concentration ranges for the compounds covered by this

method have not been determined or published by EPA yet. The sensitivity of this method is dependent upon the level of interferents within a given matrix. Proposed quantification levels for target analytes were 2 ppb in soil samples, up to 10 ppb in other solid wastes and 10 ppt in water. Actual values have been shown to vary by homologous series and, to a lesser degree, by individual isomer.

SAMPLING METHOD Grab and composite samples must be collected in 1 L or 1-quart amber glass bottles with Teflon®-lined screw-caps that have been acid-washed and solvent rinsed before use. If compositing equipment is used, the system must incorporate glass sample containers for the collection of a minimum of 250 mL. samples.

SAMPLE PRESERVATION Samples must be cooled and stored at 4°C.

MHT Samples must be extracted within 30 days and analyzed within 45 days of collection.

SAMPLE PREPARATION
Soil samples — Extract 20 g of a 1:1 soil and anhydrous sodium sulfate mixture with 1:4 methanol-petroleum ether for 2 h in a wrist-action shaker. Concentrate the extract in a K-D and then exchange the solvent with hexane in the K-D. The final volume is 15 mL of hexane.

Aqueous samples — Extract a 1 L sample with methylene chloride, dry it with anhydrous sodium sulfate and concentrate it in a K-D followed by exchange with hexane in the K-D. A continuous liquid-liquid extractor may be used in place of a separatory funnel to avoid emulsions. The final volume is 15 mL of hexane.

Alumina column clean-up — The clean-up procedure described below consists of two phases. The first phase involves a sequential basic and acid washing of the extract that contains the analytes.

Wash the 15 mL hexane extract with 20% potassium hydroxide in a separatory funnel. Repeat the washing until no color is visible in the bottom layer but no more than four times because strong base is known to degrade certain PCDDs/PCDFs, so contact time must be minimized. Next, partition the 15 mL hexane against 40 mL of 5% sodium chloride. Next, partition the 15 mL hexane against 40 mL of concentrated sulfuric acid. Repeat the acid washings until no color is visible in the acid layer (but no more than four times). Finally, partition the 15 mL hexane against 40 mL of 5% sodium chloride. Dry the organic layer with anhydrous sodium sulfate and concentrate it to near dryness with a rotary evaporator. Dissolve the hexane residue from the first phase of the clean-up in 2 mL of hexane and apply it carefully to the top of a pre-eluted Woelm super 1 neutral alumina column. Elute the column with 10 mL of 8 percent (v/v) methylene chloride in hexane. Check by GC/MS analysis that no PCDDs of PCDFs are elute in this fraction before discarding it. Elute the PCDDs and PCDFs from the column with 15 mL of 60 percent (v/v) methylene chloride in hexane and collect this second fraction in a conical shaped concentrator tube.

Carbon column clean-up — Using a carefully regulated stream of nitrogen, concentrate the first 8 percent fraction (methylene chloride in hexane) from the alumina column to about 1 mL. Save this 8 percent concentrate for GC/MS analysis to check for breakthrough of PCDDs and PCDFs. Concentrate the second 60 percent fraction (methylene chloride in hexane) to about 2 to 3 mL. Prepare a carbon column and rinse the carbon with 5 mL cyclohexane/methylene chloride (50:50 v/v) in the forward direction of flow and then in the reverse direction of flow. While still in the reverse direction of flow, transfer the sample concentrate to the column and elute with 10 mL of methylene chloride/methanol/benzene (75:20:5, v/v). Save all above eluates and combine them (this fraction may be used as a check on column efficiency). Next, turn the column over and, in the direction of forward flow, elute the PCDD/PCDF fraction with 20 mL toluene. Evaporate the toluene fraction to about 1 mL on a rotary evaporator and transfer this concentrate to a 2.0-mL Reacti-vial. Concentrate the sample using a stream of nitrogen gas. The final volume will depend on the relative concentration of target analytes but it is typically 100 µL for soil samples and 500 µL for sludge, still bottom, and fly ash samples. Extracts which are determined to be outside the calibration range for individual analytes must be diluted or a smaller portion of the sample must be re-extracted.

An alternate carbon column clean-up also may be used with a 1 mL HPLC injector loop. The injector loop is connected to the optional HPLC column.

QUALITY CONTROL Demonstrate, using a method blank, that all glassware and reagents are interferent-free at the MDL of the matrix of interest. A "method blank" must be run with each 20 or fewer samples. The method blank is also dosed with the internal standards. For water samples, 1 L of deionized and/or distilled water should be used as the method blank. Mineral oil may be used as the method blank for other matrices.

Calculate response factors for standards relative to the internal standards. Add a recovery standard to the samples prior to injection. The concentration of the recovery standard in the sample extract must be the same as that in the calibration standards used to measure the response factors.

Field duplicates (individual samples taken from the same location at the same time) should be analyzed periodically to determine the total precision (field and lab). Where appropriate, field blanks should be provided to monitor for possible cross-contamination of samples in the field. GC column performance must be demonstrated initially and verified prior to analyzing any sample in a 12-hr period. The GC column performance check solution must be analyzed under the same chromatographic and mass spectrometric conditions used for other samples and standards.

Retention times of target analytes must be verified using reference standards. These values must correspond to the retention time windows established. While certain cleanup techniques are provided as part of this method, unique samples may require additional cleanup techniques to achieve the method detection limit.

REFERENCE Test Methods for Evaluating Solid Waste (SW-846). U.S.E.P.A., 1986. Method 8280, Rev. 0, Sept. 1986. Office of Solid Wastes, Washington, DC.

1,2,7,8-TCDD   **EPA Method 8280**
CAS #34816-53-0

TITLE The Analysis of Polychlorinated Dibenzo-P-Dioxins and Polychlorinated Dibenzofurans.

MATRIX This method is appropriate for the determination of tetra-, penta-, hexa-, hepta-, and octachlorinated dibenzo-p-dioxins (PCDDs) and dibenzofurans (PCDFs) in chemical wastes including still bottoms, fuel oils, sludges, fly ash, reactor residues, soil and water.

METHOD SUMMARY This method covers 22 PCDD and PCDF compounds and it uses a high resolution capillary GC column with low resolution mass spectrometry. Samples are extracted and concentrated by several methods that vary depending on the matrix involved. The organic extracts are cleaned-up by washing with aqueous basic and acid solutions and then separated into fractions using a column of neutral alumina. The fraction containing the PCDDs and PCDFs is then further cleaned-up using a column of activated carbon. The final extract is concentrated and Carbon-13 labeled internal standards are added prior to analysis by GC/MS using a capillary GC column and selected ion monitoring using five sets of ions that are detailed in the method. Certain 2,3,7,8-substituted congeners are used to provide calibration and method recovery information. Proper column selection and access to reference isomer standards, may in certain cases, provide isomer specific data.

INTERFERENCES Solvents, reagents, glassware, and other sample processing hardware may yield discrete artifacts and/or elevated baselines which may cause misinterpretation of chromatographic data. Use high purity reagents and solvents to minimize interference problems. Interferents coextracted from the sample will vary considerably from source to source. PCDDs and PCDFs are often associated with other interfering chlorinated compounds such as PCBs and polychlorinated diphenyl ethers which may be found at concentrations several orders of magnitude higher than that of the analytes of interest.

INSTRUMENTATION A low resolution GC/MS utilizing 70 ev must be capable of selected ion monitoring (SIM) for at least 11 ions simultaneously, with a cycle time of 1 second or less. Minimum integration time for SIM is 50 ms per m/z. Also required is a GC-to-MS interface constructed of all glass or glass-lined materials. One of the following GC columns is required:

Column 1: 50 m CP-Sil-88 fused silica capillary column.
Column 2: DB-5 (30 m × 0.25 mm I.D., 0.25-um film thickness) fused silica capillary column.
Column 3: 30 m SP-2250 fused silica capillary column.

When toluene is employed as the final solvent, use of a bonded phase column is recommended. Solvent exchange into tridecane is required for other liquid phases or nonbonded columns such as CP-Sil-88. Chromatographic conditions must be adjusted to account for solvent boiling points.

PRECISION & ACCURACY Accuracy, precision, MDLs and concentration ranges for the compounds covered by this method have not been determined or published by EPA yet. The sensitivity of this method is dependent upon the level of interferents within a given matrix. Proposed quantification levels for target analytes were 2 ppb in soil samples, up to 10 ppb in other solid wastes and 10 ppt in water. Actual values have been shown to vary by homologous series and, to a lesser degree, by individual isomer.

SAMPLING METHOD Grab and composite samples must be collected in 1 L or 1-quart amber glass bottles with Teflon®-lined screw-caps that have been acid-washed and solvent rinsed before use. If compositing equipment is used, the system must incorporate glass sample containers for the collection of a minimum of 250 mL. samples.

SAMPLE PRESERVATION Samples must be cooled and stored at 4°C.

MHT Samples must be extracted within 30 days and analyzed within 45 days of collection.

SAMPLE PREPARATION
Soil samples — Extract 20 g of a 1:1 soil and anhydrous sodium sulfate mixture with 1:4 methanol-petroleum ether for 2 h in a wrist-action shaker. Concentrate the extract in a K-D and then exchange the solvent with hexane in the K-D. The final volume is 15 mL of hexane.

Aqueous samples — Extract a 1 L sample with methylene chloride, dry it with anhydrous sodium sulfate and concentrate it in a K-D followed by exchange with hexane in the K-D. A continuous liquid-liquid extractor may be used in place of a separatory funnel to avoid emulsions. The final volume is 15 mL of hexane.

Alumina column clean-up — The clean-up procedure described below consists of two phases. The first phase involves a sequential basic and acid washing of the extract that contains the analytes.

Wash the 15 mL hexane extract with 20% potassium hydroxide in a separatory funnel. Repeat the washing until no color is visible in the bottom layer but no more than four times because strong base is known to degrade certain PCDDs/PCDFs, so contact time must be minimized. Next, partition the 15 mL hexane against 40 mL of 5% sodium chloride. Next, partition the 15 mL hexane against 40 mL of concentrated sulfuric acid. Repeat the acid washings until no color is visible in the acid layer (but no more than four times). Finally, partition the 15 mL hexane against 40 mL of 5% sodium chloride. Dry the organic layer with anhydrous sodium sulfate and concentrate it to near dryness with a rotary evaporator. Dissolve the hexane residue from the first phase of the clean-up in 2 mL of hexane and apply it carefully to the top of a pre-eluted Woelm super 1 neutral alumina column. Elute the column with 10 mL of 8 percent (v/v) methylene chloride in hexane. Check by GC/MS analysis that no PCDDs of PCDFs are elute in this fraction before discarding it. Elute the PCDDs and PCDFs from the

column with 15 mL of 60 percent (v/v) methylene chloride in hexane and collect this second fraction in a conical shaped concentrator tube.

Carbon column clean-up — Using a carefully regulated stream of nitrogen, concentrate the first 8 percent fraction (methylene chloride in hexane) from the alumina column to about 1 mL. Save this 8 percent concentrate for GC/MS analysis to check for breakthrough of PCDDs and PCDFs. Concentrate the second 60 percent fraction (methylene chloride in hexane) to about 2 to 3 mL. Prepare a carbon column and rinse the carbon with 5 mL cyclohexane/methylene chloride (50:50 v/v) in the forward direction of flow and then in the reverse direction of flow. While still in the reverse direction of flow, transfer the sample concentrate to the column and elute with 10 mL of methylene chloride/methanol/benzene (75:20:5, v/v). Save all above eluates and combine them (this fraction may be used as a check on column efficiency). Next, turn the column over and, in the direction of forward flow, elute the PCDD/PCDF fraction with 20 mL toluene. Evaporate the toluene fraction to about 1 mL on a rotary evaporator and transfer this concentrate to a 2.0-mL Reacti-vial. Concentrate the sample using a stream of nitrogen gas. The final volume will depend on the relative concentration of target analytes but it is typically 100 µL for soil samples and 500 µL for sludge, still bottom, and fly ash samples. Extracts which are determined to be outside the calibration range for individual analytes must be diluted or a smaller portion of the sample must be re-extracted.

An alternate carbon column clean-up also may be used with a 1 mL HPLC injector loop. The injector loop is connected to the optional HPLC column.

QUALITY CONTROL Demonstrate, using a method blank, that all glassware and reagents are interferent-free at the MDL of the matrix of interest. A "method blank" must be run with each 20 or fewer samples. The method blank is also dosed with the internal standards. For water samples, 1 L of deionized and/or distilled water should be used as the method blank. Mineral oil may be used as the method blank for other matrices.

Calculate response factors for standards relative to the internal standards. Add a recovery standard to the samples prior to injection. The concentration of the recovery standard in the sample extract must be the same as that in the calibration standards used to measure the response factors.

Field duplicates (individual samples taken from the same location at the same time) should be analyzed periodically to determine the total precision (field and lab). Where appropriate, field blanks should be provided to monitor for possible cross-contamination of samples in the field. GC column performance must be demonstrated initially and verified prior to analyzing any sample in a 12-hr period. The GC column performance check solution must be analyzed under the same chromatographic and mass spectrometric conditions used for other samples and standards.

Retention times of target analytes must be verified using reference standards. These values must correspond to the retention time windows established. While certain cleanup techniques are provided as part of this method, unique samples may require additional cleanup techniques to achieve the method detection limit.

REFERENCE Test Methods for Evaluating Solid Waste (SW-846). U.S.E.P.A., 1986. Method 8280, Rev. 0, Sept. 1986. Office of Solid Wastes, Washington, DC.

1,2,8,9-TCDD   EPA Method 8280
CAS #62470-54-6

TITLE The Analysis of Polychlorinated Dibenzo-P-Dioxins and Polychlorinated Dibenzofurans.

MATRIX This method is appropriate for the determination of tetra-, penta-, hexa-, hepta-, and octachlorinated dibenzo-p-dioxins (PCDDs) and dibenzofurans (PCDFs) in chemical wastes including still bottoms, fuel oils, sludges, fly ash, reactor residues, soil and water.

METHOD SUMMARY This method covers 22 PCDD and PCDF compounds and it uses a high resolution capillary GC column with low resolution mass spectrometry. Samples are extracted and concentrated by several methods that vary depending on the matrix involved. The organic extracts are cleaned-up by washing with aqueous basic and acid solutions and then separated into fractions using a column of neutral alumina. The fraction containing the PCDDs and PCDFs is then further cleaned-up using a column of activated carbon. The final extract is concentrated and Carbon-13 labeled internal standards are added prior to analysis by GC/MS using a capillary GC column and selected ion monitoring using five sets of ions that are detailed in the method. Certain 2,3,7,8-substituted congeners are used to provide calibration and method recovery information. Proper column selection and access to reference isomer standards, may in certain cases, provide isomer specific data.

INTERFERENCES Solvents, reagents, glassware, and other sample processing hardware may yield discrete artifacts and/or elevated baselines which may cause misinterpretation of chromatographic data. Use high purity reagents and solvents to minimize interference problems. Interferents coextracted from the sample will vary considerably from source to source. PCDDs and PCDFs are often associated with other interfering chlorinated compounds such as PCBs and polychlorinated diphenyl ethers which may be found at concentrations several orders of magnitude higher than that of the analytes of interest.

INSTRUMENTATION A low resolution GC/MS utilizing 70 ev must be capable of selected ion monitoring (SIM) for at least 11 ions simultaneously, with a cycle time of 1 second or less. Minimum integration time for SIM is 50 ms per m/z. Also required is a GC-to-MS interface constructed of all glass or glass-lined materials. One of the following GC columns is required:

Column 1: 50 m CP-Sil-88 fused silica capillary column.
Column 2: DB-5 (30 m × 0.25 mm I.D., 0.25-um film thickness) fused silica capillary column.
Column 3: 30 m SP-2250 fused silica capillary column.

When toluene is employed as the final solvent, use of a bonded phase column is recommended. Solvent exchange into tridecane is required for other liquid phases or nonbonded columns such as CP-Sil-88. Chromatographic conditions must be adjusted to account for solvent boiling points.

PRECISION & ACCURACY Accuracy, precision, MDLs and concentration ranges for the compounds covered by this method have not been determined or published by EPA yet. The sensitivity of this method is dependent upon the level of interferents within a given matrix. Proposed quantification levels for target analytes were 2 ppb in soil samples, up to 10 ppb in other solid wastes and 10 ppt in water. Actual values have been shown to vary by homologous series and, to a lesser degree, by individual isomer.

SAMPLING METHOD Grab and composite samples must be collected in 1 L or 1-quart amber glass bottles with Teflon®-lined screw-caps that have been acid-washed and solvent rinsed before use. If compositing equipment is used, the system must incorporate glass sample containers for the collection of a minimum of 250 mL. samples.

SAMPLE PRESERVATION Samples must be cooled and stored at 4°C.

MHT Samples must be extracted within 30 days and analyzed within 45 days of collection.

SAMPLE PREPARATION
Soil samples — Extract 20 g of a 1:1 soil and anhydrous sodium sulfate mixture with 1:4 methanol-petroleum ether for 2 h in a wrist-action shaker. Concentrate the extract in a K-D and then exchange the solvent with hexane in the K-D. The final volume is 15 mL of hexane.

Aqueous samples — Extract a 1 L sample with methylene chloride, dry it with anhydrous sodium sulfate and concentrate it in a K-D followed by exchange with hexane in the K-D. A continuous liquid-liquid extractor may be used in place of a separatory funnel to avoid emulsions. The final volume is 15 mL of hexane.

Alumina column clean-up — The clean-up procedure described below consists of two phases. The first phase involves a sequential basic and acid washing of the extract that contains the analytes.

Wash the 15 mL hexane extract with 20% potassium hydroxide in a separatory funnel. Repeat the washing until no color is visible in the bottom layer but no more than four times because strong base is known to degrade certain PCDDs/PCDFs, so contact time must be minimized. Next, partition the 15 mL hexane against 40 mL of 5% sodium chloride. Next, partition the 15 mL hexane against 40 mL of concentrated sulfuric acid. Repeat the acid washings until no color is visible in the acid layer (but no more than four times). Finally, partition the 15 mL hexane against 40 mL of 5% sodium chloride. Dry the organic layer with anhydrous sodium sulfate and concentrate it to near dryness with a rotary evaporator. Dissolve the hexane residue from the first phase of the clean-up in 2 mL of hexane and apply it carefully to the top of a pre-eluted Woelm super 1 neutral alumina column. Elute the column with 10 mL of 8 percent (v/v) methylene chloride in hexane. Check by GC/MS analysis that no PCDDs of PCDFs are elute in this fraction before discarding it. Elute the PCDDs and PCDFs from the column with 15 mL of 60 percent (v/v) methylene chloride in hexane and collect this second fraction in a conical shaped concentrator tube.

Carbon column clean-up — Using a carefully regulated stream of nitrogen, concentrate the first 8 percent fraction (methylene chloride in hexane) from the alumina column to about 1 mL. Save this 8 percent concentrate for GC/MS analysis to check for breakthrough of PCDDs and PCDFs. Concentrate the second 60 percent fraction (methylene chloride in hexane) to about 2 to 3 mL. Prepare a carbon column and rinse the carbon with 5 mL cyclohexane/methylene chloride (50:50 v/v) in the forward direction of flow and then in the reverse direction of flow. While still in the reverse direction of flow, transfer the sample concentrate to the column and elute with 10 mL of methylene chloride/methanol/benzene (75:20:5, v/v). Save all above eluates and combine them (this fraction may be used as a check on column efficiency). Next, turn the column over and, in the direction of forward flow, elute the PCDD/PCDF fraction with 20 mL toluene. Evaporate the toluene fraction to about 1 mL on a rotary evaporator and transfer this concentrate to a 2.0-mL Reacti-vial. Concentrate the sample using a stream of nitrogen gas. The final volume will depend on the relative concentration of target analytes but it is typically 100 µL for soil samples and 500 µL for sludge, still bottom, and fly ash samples. Extracts which are determined to be outside the calibration range for individual analytes must be diluted or a smaller portion of the sample must be re-extracted.

An alternate carbon column clean-up also may be used with a 1 mL HPLC injector loop. The injector loop is connected to the optional HPLC column.

QUALITY CONTROL Demonstrate, using a method blank, that all glassware and reagents are interferent-free at the MDL of the matrix of interest. A "method blank" must be run with each 20 or fewer samples. The method blank is also dosed with the internal standards. For water samples, 1 L of deionized and/or distilled water should be used as the method blank. Mineral oil may be used as the method blank for other matrices.

Calculate response factors for standards relative to the internal standards. Add a recovery standard to the samples prior to injection. The concentration of the recovery standard in the sample extract must be the same as that in the calibration standards used to measure the response factors.

Field duplicates (individual samples taken from the same location at the same time) should be analyzed periodically to determine the total precision (field and lab). Where appropriate, field blanks should be provided to monitor for possible cross-contamination of samples in the field. GC column performance must be demonstrated initially and verified prior to analyzing any sample in a 12-hr period. The GC column performance check solution must be analyzed under the same chromatographic and mass spectrometric conditions used for other samples and standards.

Retention times of target analytes must be verified using reference standards. These values must correspond to the retention time windows established. While certain cleanup techniques are provided as part of this method, unique samples may require additional cleanup techniques to achieve the method detection limit.

REFERENCE Test Methods for Evaluating Solid Waste (SW-846). U.S.E.P.A., 1986. Method 8280, Rev. 0, Sept. 1986. Office of Solid Wastes, Washington, DC.

1,3,6,8-TCDD — EPA Method 8280
CAS #33423-92-6

TITLE The Analysis of Polychlorinated Dibenzo-P-Dioxins and Polychlorinated Dibenzofurans.

MATRIX This method is appropriate for the determination of tetra-, penta-, hexa-, hepta-, and octachlorinated dibenzo-p-dioxins (PCDDs) and dibenzofurans (PCDFs) in chemical wastes including still bottoms, fuel oils, sludges, fly ash, reactor residues, soil and water.

METHOD SUMMARY This method covers 22 PCDD and PCDF compounds and it uses a high resolution capillary GC column with low resolution mass spectrometry. Samples are extracted and concentrated by several methods that vary depending on the matrix involved. The organic extracts are cleaned-up by washing with aqueous basic and acid solutions and then separated into fractions using a column of neutral alumina. The fraction containing the PCDDs and PCDFs is then further cleaned-up using a column of activated carbon. The final extract is concentrated and Carbon-13 labeled internal standards are added prior to analysis by GC/MS using a capillary GC column and selected ion monitoring using five sets of ions that are detailed in the method. Certain 2,3,7,8-substituted congeners are used to provide calibration and method recovery information. Proper column selection and access to reference isomer standards, may in certain cases, provide isomer specific data.

INTERFERENCES Solvents, reagents, glassware, and other sample processing hardware may yield discrete artifacts and/or elevated baselines which may cause misinterpretation of chromatographic data. Use high purity reagents and solvents to minimize interference problems. Interferents coextracted from the sample will vary considerably from source to source. PCDDs and PCDFs are often associated with other interfering chlorinated compounds such as PCBs and polychlorinated diphenyl ethers which may be found at concentrations several orders of magnitude higher than that of the analytes of interest.

INSTRUMENTATION A low resolution GC/MS utilizing 70 ev must be capable of selected ion monitoring (SIM) for at least 11 ions simultaneously, with a cycle time of 1 second or less. Minimum integration time for SIM is 50 ms per m/z. Also required is a GC-to-MS interface constructed of all glass or glass-lined materials. One of the following GC columns is required:

Column 1: 50 m CP-Sil-88 fused silica capillary column.

Column 2: DB-5 (30 m × 0.25 mm I.D., 0.25-um film thickness) fused silica capillary column.

Column 3: 30 m SP-2250 fused silica capillary column.

When toluene is employed as the final solvent, use of a bonded phase column is recommended. Solvent exchange into tridecane is required for other liquid phases or nonbonded columns such as CP-Sil-88. Chromatographic conditions must be adjusted to account for solvent boiling points.

PRECISION & ACCURACY Accuracy, precision, MDLs and concentration ranges for the compounds covered by this method have not been determined or published by EPA yet. The sensitivity of this method is dependent upon the level of interferents within a given matrix. Proposed quantification levels for target analytes were 2 ppb in soil samples, up to 10 ppb in other solid wastes and 10 ppt in water. Actual values have been shown to vary by homologous series and, to a lesser degree, by individual isomer.

SAMPLING METHOD Grab and composite samples must be collected in 1 L or 1-quart amber glass bottles with Teflon®-lined screw-caps that have been acid-washed and solvent rinsed before use. If compositing equipment is used, the system must incorporate glass sample containers for the collection of a minimum of 250 mL. samples.

SAMPLE PRESERVATION Samples must be cooled and stored at 4°C.

MHT Samples must be extracted within 30 days and analyzed within 45 days of collection.

SAMPLE PREPARATION

Soil samples — Extract 20 g of a 1:1 soil and anhydrous sodium sulfate mixture with 1:4 methanol-petroleum ether for 2 h in a wrist-action shaker. Concentrate the extract in a K-D and then exchange the solvent with hexane in the K-D. The final volume is 15 mL of hexane.

Aqueous samples — Extract a 1 L sample with methylene chloride, dry it with anhydrous sodium sulfate and concentrate it in a K-D followed by exchange with hexane in the K-D. A continuous liquid-liquid extractor may be used in place of a separatory funnel to avoid emulsions. The final volume is 15 mL of hexane.

Alumina column clean-up — The clean-up procedure described below consists of two phases. The first phase involves a sequential basic and acid washing of the extract that contains the analytes.

Wash the 15 mL hexane extract with 20% potassium hydroxide in a separatory funnel. Repeat the washing until no color is visible in the bottom layer but no more than four times because strong base is known to degrade certain PCDDs/PCDFs, so contact time must be minimized. Next, partition the 15 mL hexane against 40 mL of 5% sodium chloride. Next, partition the 15 mL hexane against 40 mL of concentrated sulfuric acid. Repeat the acid washings until no color is visible in the acid layer (but no more than four times). Finally, partition the 15 mL hexane against 40 mL of 5% sodium chloride. Dry the organic layer with anhydrous sodium sulfate and concentrate it to near dryness with a rotary evaporator. Dissolve the hexane residue from the first phase of the clean-up in 2 mL of hexane and apply it carefully to the top of a pre-eluted Woelm super 1 neutral alumina column. Elute the column with 10 mL of 8 percent (v/v) methylene chloride in hexane. Check by GC/MS analysis that no PCDDs of PCDFs are elute in this fraction before discarding it. Elute the PCDDs and PCDFs from the column with 15 mL of 60 percent (v/v) methylene chloride in hexane and collect this second fraction in a conical shaped concentrator tube.

Carbon column clean-up — Using a carefully regulated stream of nitrogen, concentrate the first 8 percent fraction (methylene chloride in hexane) from the alumina column to about 1 mL. Save this 8 percent concentrate for GC/MS analysis to check for breakthrough of PCDDs and PCDFs. Concentrate the second 60 percent fraction (methylene chloride in hexane) to about 2 to 3 mL. Prepare a carbon column and rinse the carbon with 5 mL cyclohexane/methylene chloride (50:50 v/v) in the forward direction of flow and then in the reverse direction of flow. While still in the reverse direction of flow, transfer the sample concentrate to the column and elute with 10 mL of methylene chloride/methanol/benzene (75:20:5, v/v). Save all above eluates and combine them (this fraction may be used as a check on column efficiency). Next, turn the column over and, in the direction of forward flow, elute the PCDD/PCDF fraction with 20 mL toluene. Evaporate the toluene fraction to about 1 mL on a rotary evaporator and transfer this concentrate to a 2.0-mL Reacti-vial. Concentrate the sample using a stream of nitrogen gas. The final volume will depend on the relative concentration of target analytes but it is typically 100 µL for soil samples and 500 µL for sludge, still bottom, and fly ash samples. Extracts which are determined to be outside the calibration range for individual analytes must be diluted or a smaller portion of the sample must be re-extracted.

An alternate carbon column clean-up also may be used with a 1 mL HPLC injector loop. The injector loop is connected to the optional HPLC column.

QUALITY CONTROL Demonstrate, using a method blank, that all glassware and reagents are interferent-free at the MDL of the matrix of interest. A "method blank" must be run with each 20 or fewer samples. The method blank is also dosed with the internal standards. For water samples, 1 L of deionized and/or distilled water should be used as the method blank. Mineral oil may be used as the method blank for other matrices.

Calculate response factors for standards relative to the internal standards. Add a recovery standard to the samples prior to injection. The concentration of the recovery standard in the sample extract must be the same as that in the calibration standards used to measure the response factors.

Field duplicates (individual samples taken from the same location at the same time) should be analyzed periodically to determine the total precision (field and lab). Where appropriate, field blanks should be provided to monitor for possible cross-contamination of samples in the field. GC column performance must be demonstrated initially and verified prior to analyzing any sample in a 12-hr period. The GC column performance check solution must be analyzed under the same chromatographic and mass spectrometric conditions used for other samples and standards.

Retention times of target analytes must be verified using reference standards. These values must correspond to the retention time windows established. While certain cleanup techniques are provided as part of this method, unique samples may require additional cleanup techniques to achieve the method detection limit.

REFERENCE Test Methods for Evaluating Solid Waste (SW-846). U.S.E.P.A., 1986. Method 8280, Rev. 0, Sept. 1986. Office of Solid Wastes, Washington, DC.

1,3,7,8-TCDD   **EPA Method 8280**
CAS #50585-46-1

TITLE The Analysis of Polychlorinated Dibenzo-P-Dioxins and Polychlorinated Dibenzofurans.

MATRIX This method is appropriate for the determination of tetra-, penta-, hexa-, hepta-, and octachlorinated dibenzo-p-dioxins (PCDDs) and dibenzofurans (PCDFs) in chemical wastes including still bottoms, fuel oils, sludges, fly ash, reactor residues, soil and water.

METHOD SUMMARY This method covers 22 PCDD and PCDF compounds and it uses a high resolution capillary GC column with low resolution mass spectrometry. Samples are extracted and concentrated by several methods that vary depending on the matrix involved. The organic extracts are cleaned-up by washing with aqueous basic and acid solutions and then separated into fractions using a column of neutral alumina. The fraction containing the PCDDs and PCDFs is then further cleaned-up using a column of activated carbon. The final extract is concentrated and Carbon-13 labeled internal standards are added prior to analysis by GC/MS using a capillary GC column and selected ion monitoring using five sets of ions that are detailed in the method. Certain 2,3,7,8-substituted congeners are used to provide calibration and method recovery information. Proper column selection and access to reference isomer standards, may in certain cases, provide isomer specific data.

INTERFERENCES Solvents, reagents, glassware, and other sample processing hardware may yield discrete artifacts and/or elevated baselines which may cause misinterpretation of chromatographic data. Use high purity reagents and solvents to minimize interference problems. Interferents coextracted from the sample will vary considerably from source to source. PCDDs and PCDFs are often associated with other interfering chlorinated compounds such as PCBs and polychlorinated diphenyl ethers which may be found at concentrations several orders of magnitude higher than that of the analytes of interest.

INSTRUMENTATION A low resolution GC/MS utilizing 70 ev must be capable of selected ion monitoring (SIM) for at least 11 ions simultaneously, with a cycle time of 1 second or less. Minimum integration time for SIM is 50 ms per m/z. Also required is a GC-to-MS interface constructed of all glass or glass-lined materials. One of the following GC columns is required:

Column 1: 50 m CP-Sil-88 fused silica capillary column.
Column 2: DB-5 (30 m × 0.25 mm I.D., 0.25-um film thickness) fused silica capillary column.
Column 3: 30 m SP-2250 fused silica capillary column.

When toluene is employed as the final solvent, use of a bonded phase column is recommended. Solvent exchange into tridecane is required for other liquid phases or nonbonded columns such as CP-Sil-88. Chromatographic conditions must be adjusted to account for solvent boiling points.

PRECISION & ACCURACY Accuracy, precision, MDLs and concentration ranges for the compounds covered by this method have not been determined or published by EPA yet. The sensitivity of this method is dependent upon the level of interferents within a given matrix. Proposed quantification levels for target analytes were 2 ppb in soil samples, up to 10 ppb in other solid wastes and 10 ppt in water. Actual values have been shown to vary by homologous series and, to a lesser degree, by individual isomer.

SAMPLING METHOD Grab and composite samples must be collected in 1 L or 1-quart amber glass bottles with Teflon®-lined screw-caps that have been acid-washed and solvent rinsed before use. If compositing equipment is used, the system must incorporate glass sample containers for the collection of a minimum of 250 mL. samples.

SAMPLE PRESERVATION Samples must be cooled and stored at 4°C.

MHT Samples must be extracted within 30 days and analyzed within 45 days of collection.

SAMPLE PREPARATION
Soil samples — Extract 20 g of a 1:1 soil and anhydrous sodium sulfate mixture with 1:4 methanol-petroleum ether for 2 h in a wrist-action shaker. Concentrate the extract in a K-D and then exchange the solvent with hexane in the K-D. The final volume is 15 mL of hexane.

Aqueous samples — Extract a 1 L sample with methylene chloride, dry it with anhydrous sodium sulfate and concentrate it in a K-D followed by exchange with hexane in the K-D. A continuous liquid-liquid extractor may be used in place of a separatory funnel to avoid emulsions. The final volume is 15 mL of hexane.

Alumina column clean-up — The clean-up procedure described below consists of two phases. The first phase involves a sequential basic and acid washing of the extract that contains the analytes.

Wash the 15 mL hexane extract with 20% potassium hydroxide in a separatory funnel. Repeat the washing until no color is visible in the bottom layer but no more than four times because strong base is known to degrade certain PCDDs/PCDFs, so contact time must be minimized. Next, partition the 15 mL hexane against 40 mL of 5% sodium chloride. Next, partition the 15 mL hexane against 40 mL of concentrated sulfuric acid. Repeat the acid washings until no color is visible in the acid layer (but no more than four times). Finally, partition the 15 mL hexane against 40 mL of 5% sodium chloride. Dry the organic layer with anhydrous sodium sulfate and concentrate it to near dryness with a rotary evaporator. Dissolve the hexane residue from the first phase of the clean-up in 2 mL of hexane and apply it carefully to the top of a pre-eluted Woelm super 1 neutral alumina column. Elute the column with 10 mL of 8 percent (v/v) methylene chloride in hexane. Check by GC/MS analysis that no PCDDs of PCDFs are elute in this fraction before discarding it. Elute the PCDDs and PCDFs from the column with 15 mL of 60 percent (v/v) methylene chloride in hexane and collect this second fraction in a conical shaped concentrator tube.

Carbon column clean-up — Using a carefully regulated stream of nitrogen, concentrate the first 8 percent fraction (methylene chloride in hexane) from the alumina column to about 1 mL. Save this 8 percent concentrate for GC/MS analysis to check for breakthrough of PCDDs and PCDFs. Concentrate the second 60 percent fraction (methylene chloride in hexane) to about 2 to 3 mL. Prepare a carbon column and rinse the carbon with 5 mL cyclohexane/methylene chloride (50:50 v/v) in the forward direction of flow and then in the reverse direction of flow. While still in the reverse direction of flow, transfer the sample concentrate to the column and elute with 10 mL of methylene chloride/methanol/benzene (75:20:5, v/v). Save all above eluates and combine them (this fraction may be used as a check on column efficiency). Next, turn the column over and, in the direction of forward flow, elute the PCDD/PCDF fraction with 20 mL toluene. Evaporate the toluene fraction to about 1 mL on a rotary evaporator and transfer this concentrate to a 2.0-mL Reacti-vial. Concentrate the sample using a stream of nitrogen gas. The final volume will depend on the relative concentration of target analytes but it is typically 100 µL for soil samples and 500 µL for sludge, still bottom, and fly ash samples. Extracts which are determined to be outside the calibration range for individual analytes must be diluted or a smaller portion of the sample must be re-extracted.

An alternate carbon column clean-up also may be used with a 1 mL HPLC injector loop. The injector loop is connected to the optional HPLC column.

QUALITY CONTROL Demonstrate, using a method blank, that all glassware and reagents are interferent-free at the MDL of the matrix of interest. A "method blank" must be run with each 20 or fewer samples. The method blank is also dosed with the internal standards. For water samples, 1 L of deionized and/or distilled water should be used as the method blank. Mineral oil may be used as the method blank for other matrices.

Calculate response factors for standards relative to the internal standards. Add a recovery standard to the samples prior to injection. The concentration of the recovery standard in the sample extract must be the same as that in the calibration standards used to measure the response factors.

Field duplicates (individual samples taken from the same location at the same time) should be analyzed periodically to determine the total precision (field and lab). Where appropriate, field blanks should be provided to monitor for possible cross-contamination of samples in the field. GC column performance must be demonstrated initially and verified prior to analyzing any sample in a 12-hr period. The GC column performance check solution must be analyzed under the same chromatographic and mass spectrometric conditions used for other samples and standards.

Retention times of target analytes must be verified using reference standards. These values must correspond to the retention time windows established. While certain cleanup techniques are provided as part of this method, unique samples may require additional cleanup techniques to achieve the method detection limit.

REFERENCE Test Methods for Evaluating Solid Waste (SW-846). U.S.E.P.A., 1986. Method 8280, Rev. 0, Sept. 1986. Office of Solid Wastes, Washington, DC.

1,3,7,9-TCDD   **EPA Method 8280**
CAS #62470-53-5

TITLE The Analysis of Polychlorinated Dibenzo-P-Dioxins and Polychlorinated Dibenzofurans.

MATRIX This method is appropriate for the determination of tetra-, penta-, hexa-, hepta-, and octachlorinated dibenzo-p-dioxins (PCDDs) and dibenzofurans (PCDFs) in chemical wastes including still bottoms, fuel oils, sludges, fly ash, reactor residues, soil and water.

METHOD SUMMARY This method covers 22 PCDD and PCDF compounds and it uses a high resolution capillary GC column with low resolution mass spectrometry. Samples are extracted and concentrated by several methods that vary depending on the matrix involved. The organic extracts are cleaned-up by washing with aqueous basic and acid solutions and then separated into fractions using a column of neutral alumina. The fraction containing the PCDDs and PCDFs is then further cleaned-up using a column of activated carbon. The final extract is concentrated and Carbon-13 labeled internal standards are added prior to analysis by GC/MS using a capillary GC column and selected ion monitoring using five sets of ions that are detailed in the method. Certain 2,3,7,8-substituted congeners are used to provide calibration and method recovery information. Proper column selection and access to reference isomer standards, may in certain cases, provide isomer specific data.

INTERFERENCES Solvents, reagents, glassware, and other sample processing hardware may yield discrete artifacts and/or elevated baselines which may cause misinterpretation of chromatographic data. Use high purity reagents and solvents to minimize interference problems. Interferents coextracted from the sample will vary considerably from source to source. PCDDs and PCDFs are often associated with other interfering chlorinated compounds such as PCBs and polychlorinated diphenyl ethers which may be found at concentrations several orders of magnitude higher than that of the analytes of interest.

INSTRUMENTATION A low resolution GC/MS utilizing 70 ev must be capable of selected ion monitoring (SIM) for at least 11 ions simultaneously, with a cycle time of 1 second or less. Minimum integration time for SIM is 50 ms per m/z. Also required is a GC-to-MS interface constructed of all glass or glass-lined materials. One of the following GC columns is required:

Column 1: 50 m CP-Sil-88 fused silica capillary column.
Column 2: DB-5 (30 m × 0.25 mm I.D., 0.25-um film thickness) fused silica capillary column.
Column 3: 30 m SP-2250 fused silica capillary column.

When toluene is employed as the final solvent, use of a bonded phase column is recommended. Solvent exchange into tridecane is required for other liquid phases or nonbonded columns such as CP-Sil-88. Chromatographic conditions must be adjusted to account for solvent boiling points.

PRECISION & ACCURACY Accuracy, precision, MDLs and concentration ranges for the compounds covered by this method have not been determined or published by EPA yet. The sensitivity of this method is dependent upon the level of interferents within a given matrix. Proposed quantification levels for target analytes were 2 ppb in soil samples, up to 10 ppb in other solid wastes and 10 ppt in water. Actual values have been shown to vary by homologous series and, to a lesser degree, by individual isomer.

SAMPLING METHOD Grab and composite samples must be collected in 1 L or 1-quart amber glass bottles with Teflon®-lined screw-caps that have been acid-washed and solvent rinsed before use. If compositing equipment is used, the system must incorporate glass sample containers for the collection of a minimum of 250 mL. samples.

SAMPLE PRESERVATION Samples must be cooled and stored at 4°C.

MHT Samples must be extracted within 30 days and analyzed within 45 days of collection.

SAMPLE PREPARATION
Soil samples — Extract 20 g of a 1:1 soil and anhydrous sodium sulfate mixture with 1:4 methanol-petroleum ether for 2 h in a wrist-action shaker. Concentrate the extract in a K-D and then exchange the solvent with hexane in the K-D. The final volume is 15 mL of hexane.

Aqueous samples — Extract a 1 L sample with methylene chloride, dry it with anhydrous sodium sulfate and concentrate it in a K-D followed by exchange with hexane in the K-D. A continuous liquid-liquid extractor may be used in place of a separatory funnel to avoid emulsions. The final volume is 15 mL of hexane.

Alumina column clean-up — The clean-up procedure described below consists of two phases. The first phase involves a sequential basic and acid washing of the extract that contains the analytes.

Wash the 15 mL hexane extract with 20% potassium hydroxide in a separatory funnel. Repeat the washing until no color is visible in the bottom layer but no more than four times because strong base is known to degrade certain PCDDs/PCDFs, so contact time must be minimized. Next, partition the 15 mL hexane against 40 mL of 5% sodium chloride. Next, partition the 15 mL hexane against 40 mL of concentrated sulfuric acid. Repeat the acid washings until no color is visible in the acid layer (but no more than four times). Finally, partition the 15 mL hexane against 40 mL of 5% sodium chloride. Dry the organic layer with anhydrous sodium sulfate and concentrate it to near dryness with a rotary evaporator. Dissolve the hexane residue from the first phase of the clean-up in 2 mL of hexane and apply it carefully to the top of a pre-eluted Woelm super 1 neutral alumina column. Elute the column with 10 mL of 8 percent (v/v) methylene chloride in hexane. Check by GC/MS analysis that no PCDDs of PCDFs are elute in this fraction before discarding it. Elute the PCDDs and PCDFs from the column with 15 mL of 60 percent (v/v) methylene chloride in hexane and collect this second fraction in a conical shaped concentrator tube.

Carbon column clean-up — Using a carefully regulated stream of nitrogen, concentrate the first 8 percent fraction (methylene chloride in hexane) from the alumina column to about 1 mL. Save this 8 percent concentrate for GC/MS analysis to check for breakthrough of PCDDs and PCDFs. Concentrate the second 60 percent fraction (methylene chloride in hexane) to about 2 to 3 mL. Prepare a carbon column and rinse the carbon with 5 mL cyclohexane/methylene chloride (50:50 v/v) in the forward direction of flow and then in the reverse direction of flow. While still in the reverse direction of flow, transfer the sample concentrate to the column and elute with 10 mL of methylene chloride/methanol/benzene (75:20:5, v/v). Save all above eluates and combine them (this fraction may be used as a check on column efficiency). Next, turn the column over and, in the direction of forward flow, elute the PCDD/PCDF fraction with 20 mL toluene. Evaporate the toluene fraction to about 1 mL on a rotary evaporator and transfer this concentrate to a 2.0-mL Reacti-vial. Concentrate the sample using a stream of nitrogen gas. The final volume will depend on the relative concentration of target analytes but it is typically 100 µL for soil samples and 500 µL for sludge, still bottom, and fly ash samples. Extracts which are determined to be outside the calibration range for individual analytes must be diluted or a smaller portion of the sample must be re-extracted.

An alternate carbon column clean-up also may be used with a 1 mL HPLC injector loop. The injector loop is connected to the optional HPLC column.

QUALITY CONTROL Demonstrate, using a method blank, that all glassware and reagents are interferent-free at the MDL of the matrix of interest. A "method blank" must be run with each 20 or fewer samples. The method blank is also dosed with the internal standards. For water samples, 1 L of deionized and/or distilled water should be used as the method blank. Mineral oil may be used as the method blank for other matrices.

Calculate response factors for standards relative to the internal standards. Add a recovery standard to the samples prior to injection. The concentration of the recovery standard in the sample extract must be the same as that in the calibration standards used to measure the response factors.

Field duplicates (individual samples taken from the same location at the same time) should be analyzed periodically to determine the total precision (field and lab). Where appropriate, field blanks should be provided to monitor for possible cross-contamination of samples in the field. GC column performance must be demonstrated initially and verified prior to analyzing any sample in a 12-hr period. The GC column performance check solution must be analyzed under the same chromatographic and mass spectrometric conditions used for other samples and standards.

Retention times of target analytes must be verified using reference standards. These values must correspond to the retention time windows established. While certain cleanup techniques are provided as part of this method, unique samples may require additional cleanup techniques to achieve the method detection limit.

REFERENCE Test Methods for Evaluating Solid Waste (SW-846). U.S.E.P.A., 1986. Method 8280, Rev. 0, Sept. 1986. Office of Solid Wastes, Washington, DC.

2,3,7,8-TCDD   **EPA Method 8280**
CAS #1746-01-6

TITLE The Analysis of Polychlorinated Dibenzo-P-Dioxins and Polychlorinated Dibenzofurans.

MATRIX This method is appropriate for the determination of tetra-, penta-, hexa-, hepta-, and octachlorinated dibenzo-p-dioxins (PCDDs) and dibenzofurans (PCDFs) in chemical wastes including still bottoms, fuel oils, sludges, fly ash, reactor residues, soil and water.

METHOD SUMMARY This method covers 22 PCDD and PCDF compounds and it uses a high resolution capillary GC column with low resolution mass spectrometry. Samples are extracted and concentrated by several methods that vary depending on the matrix involved. The organic extracts are cleaned-up by washing with aqueous basic and acid solutions and then separated into fractions using a column of neutral alumina. The fraction containing the PCDDs and PCDFs is then further cleaned-up using a column of activated carbon. The final extract is concentrated and Carbon-13 labeled internal standards are added prior to analysis by GC/MS using a capillary GC column and selected ion monitoring using five sets of ions that are detailed in the method. Certain 2,3,7,8-substituted congeners are used to provide calibration and method recovery information. Proper column selection and access to reference isomer standards, may in certain cases, provide isomer specific data.

INTERFERENCES Solvents, reagents, glassware, and other sample processing hardware may yield discrete artifacts and/or elevated baselines which may cause misinterpretation of chromatographic data. Use high purity reagents and solvents to minimize interference problems. Interferents coextracted from the sample will vary considerably from source to source. PCDDs and PCDFs are often associated with other interfering chlorinated compounds such as PCBs and polychlorinated diphenyl ethers which may be found at concentrations several orders of magnitude higher than that of the analytes of interest.

INSTRUMENTATION A low resolution GC/MS utilizing 70 ev must be capable of selected ion monitoring (SIM) for at least 11 ions simultaneously, with a cycle time of 1 second or less. Minimum integration time for SIM is 50 ms per m/z. Also required is a GC-to-MS interface constructed of all glass or glass-lined materials. One of the following GC columns is required:

Column 1: 50 m CP-Sil-88 fused silica capillary column.
Column 2: DB-5 (30 m × 0.25 mm I.D., 0.25-um film thickness) fused silica capillary column.
Column 3: 30 m SP-2250 fused silica capillary column.

When toluene is employed as the final solvent, use of a bonded phase column is recommended. Solvent exchange into tridecane is required for other liquid phases or nonbonded columns such as CP-Sil-88. Chromatographic conditions must be adjusted to account for solvent boiling points.

PRECISION & ACCURACY Accuracy, precision, MDLs and concentration ranges for the compounds covered by this method have not been determined or published by EPA yet. The sensitivity of this method is dependent upon the level of interferents within a given matrix. Proposed quantification levels for target analytes were 2 ppb in soil samples, up to 10 ppb in other solid wastes and 10 ppt in water. Actual values have been shown to vary by homologous series and, to a lesser degree, by individual isomer.

SAMPLING METHOD Grab and composite samples must be collected in 1 L or 1-quart amber glass bottles with Teflon®-lined screw-caps that have been acid-washed and solvent rinsed before use. If compositing equipment is used, the system must incorporate glass sample containers for the collection of a minimum of 250 mL. samples.

SAMPLE PRESERVATION Samples must be cooled and stored at 4°C.

MHT Samples must be extracted within 30 days and analyzed within 45 days of collection.

SAMPLE PREPARATION
Soil samples — Extract 20 g of a 1:1 soil and anhydrous sodium sulfate mixture with 1:4 methanol-petroleum ether for 2 h in a wrist-action shaker. Concentrate the extract in a K-D and then exchange the solvent with hexane in the K-D. The final volume is 15 mL of hexane.

Aqueous samples — Extract a 1 L sample with methylene chloride, dry it with anhydrous sodium sulfate and concentrate it in a K-D followed by exchange with hexane in the K-D. A continuous liquid-liquid extractor may be used in place of a separatory funnel to avoid emulsions. The final volume is 15 mL of hexane.

Alumina column clean-up — The clean-up procedure described below consists of two phases. The first phase involves a sequential basic and acid washing of the extract that contains the analytes.

Wash the 15 mL hexane extract with 20% potassium hydroxide in a separatory funnel. Repeat the washing until no color is visible in the bottom layer but no more than four times because strong base is known to degrade certain PCDDs/PCDFs, so contact time must be minimized. Next, partition the 15 mL hexane against 40 mL of 5% sodium chloride. Next, partition the 15 mL hexane against 40 mL of concentrated sulfuric acid. Repeat the acid washings until no color is visible in the acid layer (but no more than four times). Finally, partition the 15 mL hexane against 40 mL of 5% sodium chloride. Dry the organic layer with anhydrous sodium sulfate and concentrate it to near dryness with a rotary evaporator. Dissolve the hexane residue from the first phase of the clean-up in 2 mL of hexane and apply it carefully to the top of a pre-eluted Woelm super 1 neutral alumina column. Elute the column with 10 mL of 8 percent (v/v) methylene chloride in hexane. Check by GC/MS analysis that no PCDDs of PCDFs are elute in this fraction before discarding it. Elute the PCDDs and PCDFs from the column with 15 mL of 60 percent (v/v) methylene chloride in hexane and collect this second fraction in a conical shaped concentrator tube.

Carbon column clean-up — Using a carefully regulated stream of nitrogen, concentrate the first 8 percent fraction (methylene chloride in hexane) from the alumina column to about 1 mL. Save this 8 percent concentrate for GC/MS analysis to check for breakthrough of PCDDs and PCDFs. Concentrate the second 60 percent fraction (methylene chloride in hexane) to about 2 to 3 mL. Prepare a carbon column and rinse the carbon with 5 mL cyclohexane/methylene chloride (50:50 v/v) in the forward direction of flow and then in the reverse direction of flow. While still in the reverse direction of flow, transfer the sample concentrate to the column and elute with 10 mL of methylene chloride/methanol/benzene (75:20:5, v/v). Save all above eluates and combine them (this fraction may be used as a check on column efficiency). Next, turn the column over and, in the direction of forward flow, elute the PCDD/PCDF fraction with 20 mL toluene. Evaporate the toluene fraction to about 1 mL on a rotary evaporator and transfer this concentrate to a 2.0-mL Reacti-vial. Concentrate the sample using a stream of nitrogen gas. The final volume will depend on the relative concentration of target analytes but it is typically 100 µL for soil samples and 500 µL for sludge, still bottom, and fly ash samples. Extracts which are determined to be outside the calibration range for individual analytes must be diluted or a smaller portion of the sample must be re-extracted.

An alternate carbon column clean-up also may be used with a 1 mL HPLC injector loop. The injector loop is connected to the optional HPLC column.

QUALITY CONTROL Demonstrate, using a method blank, that all glassware and reagents are interferent-free at the MDL of the matrix of interest. A "method blank" must be run with each 20 or fewer samples. The method blank is also dosed with the internal standards. For water samples, 1 L of deionized and/or distilled water should be used as the method blank. Mineral oil may be used as the method blank for other matrices.

Calculate response factors for standards relative to the internal standards. Add a recovery standard to the samples prior to injection. The concentration of the recovery standard in the sample extract must be the same as that in the calibration standards used to measure the response factors.

Field duplicates (individual samples taken from the same location at the same time) should be analyzed periodically to determine the total precision (field and lab). Where appropriate, field blanks should be provided to monitor for possible cross-contamination of samples in the field. GC column performance must be demonstrated initially and verified prior to analyzing any sample in a 12-hr period. The GC column performance check solution must be analyzed under the same chromatographic and mass spectrometric conditions used for other samples and standards.

Retention times of target analytes must be verified using reference standards. These values must correspond to the retention time windows established. While certain cleanup techniques are provided as part of this method, unique samples may require additional cleanup techniques to achieve the method detection limit.

REFERENCE Test Methods for Evaluating Solid Waste (SW-846). U.S.E.P.A., 1986. Method 8280, Rev. 0, Sept. 1986. Office of Solid Wastes, Washington, DC.

2,3,7,8-TCDD   **EPA Method 8290**
CAS #1746-01-6

TITLE Polychlorinated Dibenzodioxins (PCDDs) and Polychlorinated Dibenzofurans (PCDFs) by High-Resolution Gas Chromatography/High-Resolution Mass Spectrometry (HRGC/HRMS).

MATRIX This method is applicable with a variety of environmental matrices including: water, soil, sediment, paper pulp, fly ash, fish tissue, human adipose tissue, sludges, fuel oil, chemical reactor residue, and still bottoms.

METHOD SUMMARY This method provides procedures for the detection and quantitative measurement of polychlorinated dibenzo-p-dioxins (tetra- through octachlorinated homologues; PCDDs), and polychlorinated dibenzofurans (tetra- through octachlorinated homologues; PCDFs) in a variety of environmental matrices and at part-per-trillion (ppt) to part-per-quadrillion (ppq) concentrations. High-resolution gas chromatography and high-resolution mass spectrometry (HRGC/HRMS) on purified sample extracts provides highly specific identification of each analyte. Quantification is provided using calibration standards.

INTERFERENCES Solvents, reagents, glassware, and other sample processing hardware may yield discrete artifacts that may cause misinterpretation of the chromatographic data. Analysts should avoid using PVC gloves. Interferants coextracted from the sample will vary considerably from matrix to matrix. PCDDs and PCDFs are often associated with other interfering chlorinated substances such as polychlorinated biphenyls (PCBs), polychlorinated diphenyl ethers (PCDEs), polychlorinated naphthalenes, and polychlorinated alkyldibenzofurans that may be found at concentrations several orders of magnitude higher than the PCDDs or PCDFs.

A high-resolution capillary column is used in this method. However, no single column is known to resolve all isomers. The 60 m DB-5 GC column is capable of 2,3,7,8-TCDD isomer specificity. In order to determine the concentration of the 2,3,7,8-TCDD (if detected on the DB-5 column), the sample extract must be reanalyzed on a column capable of 2,3,7,8-TCDF isomer specificity (e.g., DB-225, SP-2330, SP-2331, or equivalent).

INSTRUMENTATION High-Resolution Gas Chromatograph/High-Resolution Mass Spectrometer/Data System (HRGC/HRMS/DS) equipped with a GC injection port designed so that the separation of 2,3,7,8-TCDD from the other TCDD isomers achieved in the gas chromatographic column is not appreciably degraded.

Column 1: 60 m DB-5 fused silica capillary column.
Column 2: 30 m DB-225 fused silica capillary column, or equivalent.

PRECISION & ACCURACY Precision, bias and concentration ranges for the compounds covered by this method have not been determined yet. The sensitivity of Method 8290 is dependent upon the level of interferences within a given matrix. Samples containing concentrations of specific congeneric analytes of PCDDs and PCDFs that are greater than ten times the upper method calibration limits must be analyzed by a protocol designed for such concentration levels, e.g., EPA Method 8280.

SAMPLE PREPARATION

Sludge/Wet Fuel Oil: — Extract aqueous sludge or wet fuel oil samples by refluxing a sample with toluene using a Dean-Stark water separator until all the water is removed. Filter the toluene extract through a glass fiber filter, or equivalent, and concentrate it to near dryness either on a rotary evaporator using an inert gas. Transfer the concentrate to a separatory funnel using hexane and wash it with 5% sodium chloride solution. Proceed to clean up.

Soil/Sediment: — If the sample is wet, add anhydrous powdered sodium sulfate to it until a free flowing mixture is obtained. Place the soil/sodium sulfate mixture in the Soxhlet apparatus, add toluene, and reflux for 16 h. The solvent must cycle completely through the system five times per h. Cool and filter the extract through a glass fiber filter and concentrate to near dryness on a rotary evaporator. Transfer the residue to a separatory funnel, using hexane. Proceed to clean up.

Aqueous samples — Use a 1-L sample; the method may require acetone to be added to it. When the sample is judged to contain 1% or more solids, it must be filtered through a glass fiber filter that has been rinsed with toluene. If the suspended solids content is too great to filter, centrifuge the sample, decant, and then filter the aqueous phase. Combine the solids from the centrifuge bottle(s) with the particulates on the filter and with the filter itself and proceed with Soxhlet extraction for soil/sediment. Extract the aqueous filtrate with methylene chloride in a separatory funnel, filter the extract through anhydrous sodium sulfate, and concentrate it using a K-D apparatus or a rotary evaporator. Exchange the solvent with hexane and proceed to clean up.

Clean up — The sample extract is cleaned up utilizing a number of different techniques. Partition cleanup is where the sample extract is partitioned with concentrated sulfuric acid, 5% aqueous sodium chloride, and 20% aqueous potassium hydroxide. Silica/alumina column cleanup involves packing gravity columns with silica gel and alumina and sequentially eluting the residue from the partition cleanup. Carbon column cleanup involves packing a column with a mixture of AX–21 and Celite 545 and sequentially eluting the sample concentrate from the silica/alumina cleanup with hexane, cyclohexane/methylene chloride (50:50), and methylene chloride/methanol/toluene (75:20:5). Then the column is turned upside down and the PCDD/PCDF fraction is eluted with toluene. The toluene fraction is concentrated and stored in the dark at room temperature until analysis.

QUALITY CONTROL Demonstrate, through the analysis of a reagent water blank, that interferences from the analytical system, glassware, and reagents are under control. For each analytical batch (up to 20 samples), a reagent blank, matrix spike, and matrix spike duplicate/duplicate must be analyzed (the frequency of the spikes may be different for different monitoring programs). The blank and spiked samples must be carried through all stages of the sample preparation and measurement steps.

A GC column performance check is required at the beginning of each 12-h period during which samples are analyzed. An HRGC/HRMS method blank run is required between a calibration run and the first sample run. The same method blank extract may thus be analyzed more than once if the number of samples within a batch requires more than 12 h of analyses.

At the beginning of each 12-h period during which samples are to be analyzed, an aliquot of the 1) GC column performance check solution and 2) a high-resolution concentration calibration must be analyzed to demonstrate adequate GC resolution and sensitivity, response factor reproducibility, and mass range calibration, and to establish the PCDD/PCDF retention time windows. A mass resolution check must also be performed to demonstrate adequate mass resolution using an appropriate reference compound (perfluorokerosene (PFK) is recommended). If the required criteria are not met, remedial action must be taken before any samples are analyzed.

Routine or continuing calibration (using a high resolution calibration solution) and the mass resolution check must also be performed at the end of each 12 h period. Furthermore, a HRGC/HRMS method blank analysis must be recorded following a calibration analysis and the first sample analysis.

To evaluate the performance of the analytical method, the QC check samples must be handled in exactly the same manner as actual samples. Therefore, 1.0 mL of the QC check sample concentrate is spiked into each of four 1 L aliquots of reagent water (which becomes the QC check sample), extracted, and then analyzed by GC. The variety of semivolatile analytes which may be analyzed by GC is such that the concentration of the QC check sample concentrate is different for the different analytical techniques presented in the full method.

The analyst must demonstrate also that the compounds of interest are being quantitatively recovered by the cleanup technique before the cleanup is applied to actual samples. For sample extracts that are cleaned up, the associated quality control samples (e.g., spikes, blanks, and duplicates) must also be processed through the same cleanup procedure. The analysis using each determinative method (GC, GC/MS, HPLC) specifies instrument calibration procedures using stock standards. It is recommended that cleanup also be performed on a series of the same type of standards to validate chromatographic elution patterns for the compounds of interest and to verify the absence of interferences from reagents.

REFERENCE Test Methods for Evaluating Solid Waste (SW-846). U.S. EPA. 1983. Method 8290, Rev. 0, Nov. 1990. Office of Solid Wastes, Washington, DC.

1,2,7,8-TCDF   **EPA Method 8280**
CAS #58802-20-3

TITLE The Analysis of Polychlorinated Dibenzo-P-Dioxins and Polychlorinated Dibenzofurans.

MATRIX This method is appropriate for the determination of tetra-, penta-, hexa-, hepta-, and octachlorinated dibenzo-p-dioxins (PCDDs) and dibenzofurans (PCDFs) in chemical wastes including still bottoms, fuel oils, sludges, fly ash, reactor residues, soil and water.

METHOD SUMMARY This method covers 22 PCDD and PCDF compounds and it uses a high resolution capillary GC column with low resolution mass spectrometry. Samples are extracted and concentrated by several methods that vary depending on the matrix involved. The organic extracts are cleaned-up by washing with aqueous basic and acid solutions and then separated into fractions using a column of neutral alumina. The fraction containing the PCDDs and PCDFs is then further cleaned-up using a column of activated carbon. The final extract is concentrated and Carbon-13 labeled internal standards are added prior to analysis by GC/MS using a capillary GC column and selected ion monitoring using five sets of ions that are detailed in the method. Certain 2,3,7,8-substituted congeners are used to provide calibration and method recovery information. Proper column selection and access to reference isomer standards, may in certain cases, provide isomer specific data.

INTERFERENCES Solvents, reagents, glassware, and other sample processing hardware may yield discrete artifacts and/or elevated baselines which may cause misinterpretation of chromatographic data. Use high purity reagents and solvents to minimize interference problems. Interferents coextracted from the sample will vary considerably from source to source. PCDDs and PCDFs are often associated with other interfering chlorinated compounds such as PCBs and polychlorinated diphenyl ethers which may be found at concentrations several orders of magnitude higher than that of the analytes of interest.

INSTRUMENTATION A low resolution GC/MS utilizing 70 ev must be capable of selected ion monitoring (SIM) for at least 11 ions simultaneously, with a cycle time of 1 second or less. Minimum integration time for SIM is 50 ms per m/z. Also required is a GC-to-MS interface constructed of all glass or glass-lined materials. One of the following GC columns is required:

Column 1: 50 m CP-Sil-88 fused silica capillary column.
Column 2: DB-5 (30 m × 0.25 mm I.D., 0.25-um film thickness) fused silica capillary column.
Column 3: 30 m SP-2250 fused silica capillary column.

When toluene is employed as the final solvent, use of a bonded phase column is recommended. Solvent exchange into tridecane is required for other liquid phases or nonbonded columns such as CP-Sil-88. Chromatographic conditions must be adjusted to account for solvent boiling points.

PRECISION & ACCURACY Accuracy, precision, MDLs and concentration ranges for the compounds covered by this method have not been determined or published by EPA yet. The sensitivity of this method is dependent upon the level of interferents within a given matrix. Proposed quantification levels for target analytes were 2 ppb in soil samples, up to 10 ppb in other solid wastes and 10 ppt in water. Actual values have been shown to vary by homologous series and, to a lesser degree, by individual isomer.

SAMPLING METHOD Grab and composite samples must be collected in 1 L or 1-quart amber glass bottles with Teflon®-lined screw-caps that have been acid-washed and solvent rinsed before use. If compositing equipment is used, the system must incorporate glass sample containers for the collection of a minimum of 250 mL. samples.

SAMPLE PRESERVATION Samples must be cooled and stored at 4°C.

MHT Samples must be extracted within 30 days and analyzed within 45 days of collection.

SAMPLE PREPARATION
Soil samples — Extract 20 g of a 1:1 soil and anhydrous sodium sulfate mixture with 1:4 methanol-petroleum ether for 2 h in a wrist-action shaker. Concentrate the extract in a K-D and then exchange the solvent with hexane in the K-D. The final volume is 15 mL of hexane.

Aqueous samples — Extract a 1 L sample with methylene chloride, dry it with anhydrous sodium sulfate and concentrate it in a K-D followed by exchange with hexane in the K-D. A continuous liquid-liquid extractor may be used in place of a separatory funnel to avoid emulsions. The final volume is 15 mL of hexane.

Alumina column clean-up — The clean-up procedure described below consists of two phases. The first phase involves a sequential basic and acid washing of the extract that contains the analytes.

Wash the 15 mL hexane extract with 20% potassium hydroxide in a separatory funnel. Repeat the washing until no color is visible in the bottom layer but no more than four times because strong base is known to degrade certain PCDDs/PCDFs, so contact time must be minimized. Next, partition the 15 mL hexane against 40 mL of 5% sodium chloride. Next, partition the 15 mL hexane against 40 mL of concentrated sulfuric acid. Repeat the acid washings until no color is visible in the acid layer (but no more than four times). Finally, partition the 15 mL hexane against 40 mL of 5% sodium chloride. Dry the organic layer with anhydrous sodium sulfate and concentrate it to near dryness with a rotary evaporator. Dissolve the hexane residue from the first phase of the clean-up in 2 mL of hexane and apply it carefully to the top of a pre-eluted Woelm super 1 neutral alumina column. Elute the column with 10 mL of 8 percent (v/v) methylene chloride in hexane. Check by GC/MS analysis that no PCDDs of PCDFs are elute in this fraction before discarding it. Elute the PCDDs and PCDFs from the column with 15 mL of 60 percent (v/v) methylene chloride in hexane and collect this second fraction in a conical shaped concentrator tube.

Carbon column clean-up — Using a carefully regulated stream of nitrogen, concentrate the first 8 percent fraction (methylene chloride in hexane) from the alumina column to about 1 mL. Save this 8 percent concentrate for GC/MS analysis to check for breakthrough of PCDDs and PCDFs. Concentrate the second 60 percent fraction (methylene chloride in hexane) to about 2 to 3 mL. Prepare a carbon column and rinse the carbon with 5 mL cyclohexane/methylene chloride (50:50 v/v) in the forward direction of flow and then in the reverse direction of flow. While still in the reverse direction of flow, transfer the sample concentrate to the column and elute with 10 mL of methylene chloride/methanol/benzene (75:20:5, v/v). Save all above eluates and combine them (this fraction may be used as a check on column efficiency). Next, turn the column over and, in the direction of forward flow, elute the PCDD/PCDF fraction with 20 mL toluene. Evaporate the toluene fraction to about 1 mL on a rotary evaporator and transfer this concentrate to a 2.0-mL Reacti-vial. Concentrate the sample using a stream of nitrogen gas. The final volume will depend on the relative concentration of target analytes but it is typically 100 µL for soil samples and 500 µL for sludge, still bottom, and fly ash samples. Extracts which are determined to be outside the calibration range for individual analytes must be diluted or a smaller portion of the sample must be re-extracted.

An alternate carbon column clean-up also may be used with a 1 mL HPLC injector loop. The injector loop is connected to the optional HPLC column.

QUALITY CONTROL Demonstrate, using a method blank, that all glassware and reagents are interferent-free at the MDL of the matrix of interest. A "method blank" must be run with each 20 or fewer samples. The method blank is also dosed with the internal standards. For water samples, 1 L of deionized and/or distilled water should be used as the method blank. Mineral oil may be used as the method blank for other matrices.

Calculate response factors for standards relative to the internal standards. Add a recovery standard to the samples prior to injection. The concentration of the recovery standard in the sample extract must be the same as that in the calibration standards used to measure the response factors.

Field duplicates (individual samples taken from the same location at the same time) should be analyzed periodically to determine the total precision (field and lab). Where appropriate, field blanks should be provided to monitor for possible cross-contamination of samples in the field. GC column performance must be demonstrated initially and verified prior to analyzing any sample in a 12-hr period. The GC column performance check solution must be analyzed under the same chromatographic and mass spectrometric conditions used for other samples and standards.

Retention times of target analytes must be verified using reference standards. These values must correspond to the retention time windows established. While certain cleanup techniques are provided as part of this method, unique samples may require additional cleanup techniques to achieve the method detection limit.

REFERENCE Test Methods for Evaluating Solid Waste (SW-846). U.S.E.P.A., 1986. Method 8280, Rev. 0, Sept. 1986. Office of Solid Wastes, Washington, DC.

2,3,7,8-TCDF   **EPA Method 8280**
CAS #51207-31-9

TITLE The Analysis of Polychlorinated Dibenzo-P-Dioxins and Polychlorinated Dibenzofurans.

MATRIX This method is appropriate for the determination of tetra-, penta-, hexa-, hepta-, and octachlorinated dibenzo-p-dioxins (PCDDs) and dibenzofurans (PCDFs) in chemical wastes including still bottoms, fuel oils, sludges, fly ash, reactor residues, soil and water.

METHOD SUMMARY This method covers 22 PCDD and PCDF compounds and it uses a high resolution capillary GC column with low resolution mass spectrometry. Samples are extracted and concentrated by several methods that vary depending on the matrix involved. The organic extracts are cleaned-up by washing with aqueous basic and acid solutions and then separated into fractions using a column of neutral alumina. The fraction containing the PCDDs and PCDFs is then further cleaned-up using a column of activated carbon. The final extract is concentrated and Carbon-13 labeled internal standards are added prior to analysis by GC/MS using a capillary GC column and selected ion monitoring using five sets of ions that are detailed in the method. Certain 2,3,7,8-substituted congeners are used to provide calibration and method recovery information. Proper column selection and access to reference isomer standards, may in certain cases, provide isomer specific data.

INTERFERENCES Solvents, reagents, glassware, and other sample processing hardware may yield discrete artifacts and/or elevated baselines which may cause misinterpretation of chromatographic data. Use high purity reagents and solvents to minimize interference problems. Interferents coextracted from the sample will vary considerably from source to source. PCDDs and PCDFs are often associated with other interfering chlorinated compounds such as PCBs and polychlorinated diphenyl ethers which may be found at concentrations several orders of magnitude higher than that of the analytes of interest.

INSTRUMENTATION A low resolution GC/MS utilizing 70 ev must be capable of selected ion monitoring (SIM) for at least 11 ions simultaneously, with a cycle time of 1 second or less. Minimum integration time for SIM is 50 ms per m/z. Also required is a GC-to-MS interface constructed of all glass or glass-lined materials. One of the following GC columns is required:

Column 1: 50 m CP-Sil-88 fused silica capillary column.
Column 2: DB-5 (30 m × 0.25 mm I.D., 0.25-um film thickness) fused silica capillary column.
Column 3: 30 m SP-2250 fused silica capillary column.

When toluene is employed as the final solvent, use of a bonded phase column is recommended. Solvent exchange into tridecane is required for other liquid phases or nonbonded columns such as CP-Sil-88. Chromatographic conditions must be adjusted to account for solvent boiling points.

PRECISION & ACCURACY Accuracy, precision, MDLs and concentration ranges for the compounds covered by this method have not been determined or published by EPA yet. The sensitivity of this method is dependent upon the level of interferents within a given matrix. Proposed quantification levels for target analytes were 2 ppb in soil samples, up to 10 ppb in other solid wastes and 10 ppt in water. Actual values have been shown to vary by homologous series and, to a lesser degree, by individual isomer.

SAMPLING METHOD Grab and composite samples must be collected in 1 L or 1-quart amber glass bottles with Teflon®-lined screw-caps that have been acid-washed and solvent rinsed before use. If compositing equipment is used, the system must incorporate glass sample containers for the collection of a minimum of 250 mL. samples.

SAMPLE PRESERVATION Samples must be cooled and stored at 4°C.

MHT Samples must be extracted within 30 days and analyzed within 45 days of collection.

SAMPLE PREPARATION
Soil samples — Extract 20 g of a 1:1 soil and anhydrous sodium sulfate mixture with 1:4 methanol-petroleum ether for 2 h in a wrist-action shaker. Concentrate the extract in a K-D and then exchange the solvent with hexane in the K-D. The final volume is 15 mL of hexane.

Aqueous samples — Extract a 1 L sample with methylene chloride, dry it with anhydrous sodium sulfate and concentrate it in a K-D followed by exchange with hexane in the K-D. A continuous liquid-liquid extractor may be used in place of a separatory funnel to avoid emulsions. The final volume is 15 mL of hexane.

Alumina column clean-up — The clean-up procedure described below consists of two phases. The first phase involves a sequential basic and acid washing of the extract that contains the analytes.

Wash the 15 mL hexane extract with 20% potassium hydroxide in a separatory funnel. Repeat the washing until no color is visible in the bottom layer but no more than four times because strong base is known to degrade certain PCDDs/PCDFs, so contact time must be minimized. Next, partition the 15 mL hexane against 40 mL of 5% sodium chloride. Next, partition the 15 mL hexane against 40 mL of concentrated sulfuric acid. Repeat the acid washings until no color is visible in the acid layer (but no more than four times). Finally, partition the 15 mL hexane against 40 mL of 5% sodium chloride. Dry the organic layer with anhydrous sodium sulfate and concentrate it to near dryness with a rotary evaporator. Dissolve the hexane residue from the first phase of the clean-up in 2 mL of hexane and apply it carefully to the top of a pre-eluted Woelm super 1 neutral alumina column. Elute the column with 10 mL of 8 percent (v/v) methylene chloride in hexane. Check by GC/MS analysis that no PCDDs of PCDFs are elute in this fraction before discarding it. Elute the PCDDs and PCDFs from the column with 15 mL of 60 percent (v/v) methylene chloride in hexane and collect this second fraction in a conical shaped concentrator tube.

Carbon column clean-up — Using a carefully regulated stream of nitrogen, concentrate the first 8 percent fraction (methylene chloride in hexane) from the alumina column to about 1 mL. Save this 8 percent concentrate for GC/MS analysis to check for breakthrough of PCDDs and PCDFs. Concentrate the second 60 percent fraction (methylene chloride in hexane) to about 2 to 3 mL. Prepare a carbon column and rinse the carbon with 5 mL cyclohexane/methylene chloride (50:50 v/v) in the forward direction of flow and then in the reverse direction of flow. While still in the reverse direction of flow, transfer the sample concentrate to the column and elute with 10 mL of methylene chloride/methanol/benzene (75:20:5, v/v). Save all above eluates and combine them (this fraction may be used as a check on column efficiency). Next, turn the column over and, in the direction of forward flow, elute the PCDD/PCDF fraction with 20 mL toluene. Evaporate the toluene fraction to about 1 mL on a rotary evaporator and transfer this concentrate to a 2.0-mL Reacti-vial. Concentrate the sample using a stream of nitrogen gas. The final volume will depend on the relative concentration of target analytes but it is typically 100 µL for soil samples and 500 µL for sludge, still bottom, and fly ash samples. Extracts which are determined to be outside the calibration range for individual analytes must be diluted or a smaller portion of the sample must be re-extracted.

An alternate carbon column clean-up also may be used with a 1 mL HPLC injector loop. The injector loop is connected to the optional HPLC column.

QUALITY CONTROL Demonstrate, using a method blank, that all glassware and reagents are interferent-free at the MDL of the matrix of interest. A "method blank" must be run with each 20 or fewer samples. The method blank is also dosed with the internal standards. For water samples, 1 L of deionized and/or distilled water should be used as the method blank. Mineral oil may be used as the method blank for other matrices.

Calculate response factors for standards relative to the internal standards. Add a recovery standard to the samples prior to injection. The concentration of the recovery standard in the sample extract must be the same as that in the calibration standards used to measure the response factors.

Field duplicates (individual samples taken from the same location at the same time) should be analyzed periodically to determine the total precision (field and lab). Where appropriate, field blanks should be provided to monitor for possible cross-contamination of samples in the field. GC column performance must be demonstrated initially and verified prior to analyzing any sample in a 12-hr period. The GC column performance check solution must be analyzed under the same chromatographic and mass spectrometric conditions used for other samples and standards.

Retention times of target analytes must be verified using reference standards. These values must correspond to the retention time windows established. While certain cleanup techniques are provided as part of this method, unique samples may require additional cleanup techniques to achieve the method detection limit.

REFERENCE Test Methods for Evaluating Solid Waste (SW-846). U.S.E.P.A., 1986. Method 8280, Rev. 0, Sept. 1986. Office of Solid Wastes, Washington, DC.

2,3,7,8-TCDF
CAS #51207-31-9
EPA Method 8290

TITLE Polychlorinated Dibenzodioxins (PCDDs) and Polychlorinated Dibenzofurans (PCDFs) by High-Resolution Gas Chromatography/High-Resolution Mass Spectrometry (HRGC/HRMS).

MATRIX This method is applicable with a variety of environmental matrices including: water, soil, sediment, paper pulp, fly ash, fish tissue, human adipose tissue, sludges, fuel oil, chemical reactor residue, and still bottoms.

METHOD SUMMARY This method provides procedures for the detection and quantitative measurement of polychlorinated dibenzo-p-dioxins (tetra- through octachlorinated homologues; PCDDs), and polychlorinated dibenzofurans (tetra- through octachlorinated homologues; PCDFs) in a variety of environmental matrices and at part-per-trillion (ppt) to part-per-quadrillion (ppq) concentrations. High-resolution gas chromatography and high-resolution mass spectrometry (HRGC/HRMS) on purified sample extracts provides highly specific identification of each analyte. Quantification is provided using calibration standards.

INTERFERENCES Solvents, reagents, glassware, and other sample processing hardware may yield discrete artifacts that may cause misinterpretation of the chromatographic data. Analysts should avoid using PVC gloves. Interferants coextracted from the sample will vary considerably from matrix to matrix. PCDDs and PCDFs are often associated with other interfering chlorinated substances such as polychlorinated biphenyls (PCBs), polychlorinated diphenyl ethers (PCDEs), polychlorinated naphthalenes, and polychlorinated alkyldibenzofurans that may be found at concentrations several orders of magnitude higher than the PCDDs or PCDFs.

A high-resolution capillary column is used in this method. However, no single column is known to resolve all isomers. The 60 m DB-5 GC column is capable of 2,3,7,8-TCDD isomer specificity. In order to determine the concentration of the 2,3,7,8-TCDD (if detected on the DB-5 column), the sample extract must be reanalyzed on a column capable of 2,3,7,8-TCDF isomer specificity (e.g., DB-225, SP-2330, SP-2331, or equivalent).

INSTRUMENTATION High-Resolution Gas Chromatograph/High-Resolution Mass Spectrometer/Data System (HRGC/HRMS/DS) equipped with a GC injection port designed so that the separation of 2,3,7,8-TCDD from the other TCDD isomers achieved in the gas chromatographic column is not appreciably degraded.

Column 1: 60 m DB-5 fused silica capillary column.
Column 2: 30 m DB-225 fused silica capillary column, or equivalent.

PRECISION & ACCURACY Precision, bias and concentration ranges for the compounds covered by this method have not been determined yet. The sensitivity of Method 8290 is dependent upon the level of interferences within a given matrix. Samples containing concentrations of specific congeneric analytes of PCDDs and PCDFs that are greater than ten times the upper method calibration limits must be analyzed by a protocol designed for such concentration levels, e.g., EPA Method 8280.

SAMPLE PREPARATION
Sludge/Wet Fuel Oil: — Extract aqueous sludge or wet fuel oil samples by refluxing a sample with toluene using a Dean-Stark water separator until all the water is removed. Filter the toluene extract through a glass fiber filter, or equivalent, and concentrate it to near dryness either on a rotary evaporator using an inert gas. Transfer the concentrate to a separatory funnel using hexane and wash it with 5% sodium chloride solution. Proceed to clean up.

Soil/Sediment: — If the sample is wet, add anhydrous powdered sodium sulfate to it until a free flowing mixture is obtained. Place the soil/sodium sulfate mixture in the Soxhlet apparatus, add toluene, and reflux for 16 h. The solvent must cycle completely through the system five times per h. Cool and filter the extract through a glass fiber filter and concentrate to near dryness on a rotary evaporator. Transfer the residue to a separatory funnel, using hexane. Proceed to clean up.

Aqueous samples — Use a 1-L sample; the method may require acetone to be added to it. When the sample is judged to contain 1% or more solids, it must be filtered through a glass fiber filter that has been rinsed with toluene. If the suspended solids content is too great to filter, centrifuge the sample, decant, and then filter the aqueous phase. Combine the solids from the centrifuge bottle(s) with the particulates on the filter and with the filter itself and proceed with Soxhlet extraction for soil/sediment. Extract the aqueous filtrate with methylene chloride in a separatory funnel, filter the extract through anhydrous sodium sulfate, and concentrate it using a K-D apparatus or a rotary evaporator. Exchange the solvent with hexane and proceed to clean up.

Clean up — The sample extract is cleaned up utilizing a number of different techniques. Partition cleanup is where the sample extract is partitioned with concentrated sulfuric acid, 5% aqueous sodium chloride, and 20% aqueous potassium hydroxide. Silica/alumina column cleanup involves packing gravity columns with silica gel and alumina and sequentially eluting the residue from the partition cleanup. Carbon column cleanup involves packing a column with a mixture of AX-21 and Celite 545 and sequentially eluting the sample concentrate from the silica/alumina cleanup with hexane, cyclohexane/methylene chloride (50:50), and methylene chloride/methanol/toluene (75:20:5). Then the column is turned upside down and the PCDD/PCDF fraction is eluted with toluene. The toluene fraction is concentrated and stored in the dark at room temperature until analysis.

QUALITY CONTROL Demonstrate, through the analysis of a reagent water blank, that interferences from the analytical system, glassware, and reagents are under control. For each analytical batch (up to 20 samples), a reagent blank, matrix spike, and matrix spike duplicate/duplicate must be analyzed (the frequency of the spikes may be different for different monitoring programs). The blank and spiked samples must be carried through all stages of the sample preparation and measurement steps.

A GC column performance check is required at the beginning of each 12-h period during which samples are analyzed. An HRGC/HRMS method blank run is required between a calibration run and the first sample run. The same method blank extract may thus be analyzed more than once if the number of samples within a batch requires more than 12 h of analyses.

At the beginning of each 12-h period during which samples are to be analyzed, an aliquot of the 1) GC column performance check solution and 2) a high-resolution concentration calibration must be analyzed to demonstrate adequate GC resolution and sensitivity, response factor reproducibility, and mass range calibration, and to establish the PCDD/PCDF retention time windows. A mass resolution check must also be performed to demonstrate adequate mass resolution using an appropriate reference compound (perfluorokerosene (PFK) is recommended). If the required criteria are not met, remedial action must be taken before any samples are analyzed.

Routine or continuing calibration (using a high resolution calibration solution) and the mass resolution check must also be performed at the end of each 12 h period. Furthermore, a HRGC/HRMS method blank analysis must be recorded following a calibration analysis and the first sample analysis.

To evaluate the performance of the analytical method, the QC check samples must be handled in exactly the same manner as actual samples. Therefore, 1.0 mL of the QC check sample concentrate is spiked into each of four 1 L aliquots of reagent water (which becomes the QC check sample), extracted, and then analyzed by GC. The variety of semivolatile analytes which may be analyzed by GC is such that the concentration of the QC check sample concentrate is different for the different analytical techniques presented in the full method.

The analyst must demonstrate also that the compounds of interest are being quantitatively recovered by the cleanup technique before the cleanup is applied to actual samples. For sample extracts that are cleaned up, the associated quality control samples (e.g., spikes, blanks, and duplicates) must also be processed through the same cleanup procedure. The analysis using each determinative method (GC, GC/MS, HPLC) specifies instrument calibration procedures using stock standards. It is recommended that cleanup also be performed on a series of the same type of standards to validate chromatographic elution patterns for the compounds of interest and to verify the absence of interferences from reagents.

REFERENCE Test Methods for Evaluating Solid Waste (SW-846). U.S. EPA. 1983. Method 8290, Rev. 0, Nov. 1990. Office of Solid Wastes, Washington, DC.

Tebuthiuron   EPA Method 507
CAS #34014-18-1

TITLE Determination of Nitrogen and Phosphorus-Containing Pesticides in Water by GC/NPD

MATRIX This method is applicable to the determination of certain nitrogen and phosphorus-containing pesticides in finished drinking water and groundwater.

METHOD SUMMARY Method 507 covers 46 nitrogen- and phosphorus-containing pesticides. A 1-L sample is fortified with a surrogate standard, salted, buffered, extracted with methylene chloride, and concentrated; then the solvent is exchanged with methyl tert-butyl ether (MTBE) and concentrated again, and a 2-µL aliquot of a sample extract is injected into a GC system equipped with a selective nitrogen-phosphorus detector and a capillary column for analysis.

INTERFERENCES Method interferences may be caused by contaminants in solvents, reagents, glassware, and other sample processing apparatus. Interfering contamination may occur when a sample containing low concentrations of analytes is analyzed immediately following a sample containing relatively high concentrations. One or more injections of MTBE should be made following the analysis of a sample with high concentrations of analytes to check for analyte carryover. Matrix interferences may be caused by contaminants that are coextracted from the sample. The extent of matrix interferences will vary considerably from source to source, depending upon the water sampled.

INSTRUMENTATION A gas chromatograph system (GC) equipped with a nitrogen-phosphorus detector (NPD) is needed.

Column 1: 30 m × 0.25 mm I.D. DB-5 bonded fused silica column, 0.25 µm film thickness, or equivalent.

Column 2: 30 m × 0.25 mm I.D. DB-1701 bonded fused silica column, 0.25 µm film thickness, or equivalent.

PRECISION & ACCURACY This method has been validated in a single lab and estimated detection limits (EDLs) have been determined for each analyte. Observed detection limits may vary among waters, depending upon the nature of the interferences in the sample matrix and the specific instrumentation used. Analytes that are not separated chromatographically cannot be individually identified and measured unless an alternative technique for identification and quantification exist.

The estimated detection limit (in µg/L) was 1.3. The EDL is defined as either method detection limit or a level of compound in a sample yielding a peak in the final extract with signal-to-noise ratio of approximately 5, whichever value is higher.

The concentration used for these measurements (in µg/L) was 13.
The accuracy (as % recovery) was 84.
The precision (% RSD) was 9.

SAMPLING METHOD Grab samples are collected in 1-L glass sample bottles (prewashed with detergent and hot tap water, rinsed with reagent water, and dried in an oven at 400°C for 1 h) with screw caps lined with PTFE-fluorocarbon.

SAMPLE PRESERVATION Add mercuric chloride to the sample bottle in amounts to produce a concentration of 10 mg/L. If residual chlorine is present, add 80 mg of sodium thiosulfate/L of sample to the sample bottle prior to collection. After collection, seal bottle and shake vigorously for 1 min, then cool the sample to 4°C immediately and store it at 4°C in the dark until extraction.

MHT Maximum holding time of the samples, and in some cases the extracts, is 14 days.

SAMPLE PREPARATION Fortify the sample with 50 µL of the surrogate standard solution, adjust to pH 7 with phosphate buffer, add 100 g NaCl to the sample, and seal and shake to dissolve the salt; then extract with methylene chloride in a separatory funnel or in a mechanical tumbler bottle. Dry the extract by pouring it through a solvent-rinsed drying column containing about 10 cm of anhydrous sodium sulfate. Collect the extract in a Kuderna-Danish (K-D) concentrator and rinse the column with 20–30 mL methylene chloride. Concentrate the extract to about 2 mL and rinse the flask and its lower joint into the concentrator tube with 1 to 2 mL of methyl t-butyl ether (MTBE). Add 5–10 mL of MTBE and concentrate the extract twice (adding more MTBE) to a final volume of 5.0 mL and store it at 4°C until analysis.

Note: If methylene chloride is not completely removed from the final extract, it may cause detector problems.

QUALITY CONTROL Minimum quality control requirements are initial demonstration of lab capability, determination of surrogate compound recoveries in each sample and blank, monitoring internal standard peak area or height in each sample and blank, analysis of lab reagent blanks, lab fortified samples, lab fortified blanks, and other QC samples. A lab reagent blank is analyzed to demonstrate that all glassware and reagent interferences are under control.

Initial demonstration of capability is fulfilled by analyzing four fortified reagent water samples with the recovery value for each analyte falling within the acceptable range (±30% average recovery). Surrogate recoveries from samples or method blanks must be 70–130%. The internal standard response for any sample chromatogram should not deviate from the daily calibration check standard's internal standard response by more than 30% or lab fortified blanks and sample matrices are used to assess lab performance and analyte recovery, respectively.

If the response for the target analyte peak exceeds the working range of the system, dilute the extract and reanalyze. Alternative techniques such as an alternate detector or second chromatography column should be used to confirm peak identification when sample components are not resolved adequately.

EPA CONTACT & HOTLINE For technical questions contact Dr. Baldev Bathija, U.S. EPA, Office of Ground Water and Drinking Water (WH-550D), 401 M St. SW, Washington, DC 20460. Tel. (202) 260-3040. For further information the EPA Safe Drinking Water Hotline may be called at: (800) 426-4791.

REFERENCE Methods for the Determination of Organic Compounds in Drinking Water, EPA/600/4-88/039 (revised July 1991). U.S. EPA Environmental Monitoring Systems Laboratory, Cincinnati, OH, 45268, U.S.A. Available from the National Technical Information Service (NTIS), 5285 Port Royal Road, Springfield, VA 22161; Tel. 800-553-6847. NTIS Order Number is PB91-231480.

Temperature (Thermometric)   **EPA Method 170.1**

TITLE Physical Properties

MATRIX Drinking, surface, and saline waters. Wastewater.

APPLICATION Date issued 1971. Temperature readings are used in analyses calculations and general lab operations. In limnological studies, water temperatures as a function of depth often are required. Identification of deep wells is often possible by temperature measurements.

INTERFERENCES For field operations, to prevent breakage, use a thermometer with a metal case. Thermometer should have a minimal thermal capacity to permit rapid equilibration.

INSTRUMENTATION Thermometers (mercury-filled) or metallic (dial type) or a thermistor.

RANGE Not listed.

MDL Not listed.

PRECISION Not determined.

ACCURACY Not determined.

SAMPLING METHOD Plastic or glass (1000 mL).

STABILITY No preservation required. Analyze immediately.

QUALITY CONTROL Thermometers should have a scale marked for every 0.1C with markings etched on the capillary glass. Periodically check the thermometer against an NBS certified precision thermometer that is used with its certificate and collection chart.

REFERENCE Methods for the Chemical Analysis of Water and Wastes, EPA-600/4-79-020, U.S. EPA, EMSL, 1979.

TEPP   **EPA Method 8141**
CAS #21646-99-1

TITLE Organophosphorus Compounds by Gas Chromatography: Capillary Column Technique

MATRIX This method covers aqueous and solid matrices. This includes a wide variety such as drinking water, groundwater, industrial wastewaters, surface waters, soils, solids, and sediments.

METHOD SUMMARY This is a GC method used to determine the concentration of 28 organophosphorus pesticides.

The use of Gel Permeation Cleanup (EPA Method 3640) for sample cleanup has been demonstrated to yield recoveries of less than 85% for many method analytes and is therefore not recommended for use with this method.

This method provides GC conditions for the detection of ppb concentrations of organophosphorus compounds. Prior to the use of this method, appropriate sample preparation techniques must be used. Water samples are extracted at a neutral pH with methylene chloride as a solvent by using a separatory funnel (EPA Method 3510) or a continuous liquid-liquid extractor (EPA Method 3520). Soxhlet extraction (EPA Method 3540) or ultrasonic extraction (EPA Method 3550) using methylene chloride/acetone (1:1) are used for solid samples. Both neat and diluted organic liquids (EPA Method 3580) may be analyzed by direct injection. Spiked samples are used to verify the applicability of the chosen extraction technique to each new sample type. A GC with a flame photometric (FPD) or nitrogen-phosphorus detector (NPD) is used for this multiresidue procedure.

INTERFERENCES The use of Florisil cleanup materials (EPA Method 3620) for some of the compounds in this method has been demonstrated to yield recoveries less than 85% and is therefore not recommended for all compounds. Use of phosphorus or halogen specific detectors, however, often obviates the necessity for cleanup for relatively clean sample matrices. If particular circumstances demand the use of an alternative cleanup procedure, the analyst must determine the elution profile and demonstrate that the recovery of each analyte is no less than 85%.

Use of a flame photometric detector (FPD) in the phosphorus mode will minimize interferences from materials that do not contain phosphorus. Elemental sulfur, however, may interfere with the determination of certain organophosphorus compounds by flame photometric gas chromatography. Sulfur cleanup using EPA Method 3660 may alleviate this interference. A nitrogen phosphorus detector (NPD) is also recommended.

A few analytes coelute on certain columns. Therefore, select a second column for confirmation where coelution of the analytes of interest does not occur.

Method interferences may be caused by contaminants in solvents, reagents, glassware, and other sample processing hardware that lead to discrete artifacts or elevated baselines in gas chromatograms. All these materials must be routinely demonstrated to be free from interferences under the conditions of the analysis by analyzing reagent blanks.

INSTRUMENTATION A GC with a NPD or a FPD will be needed. A data system or integrator is recommended for measuring peak areas and/or peak heights. A Kuderna-Danish (K-D) apparatus will be needed for extract concentration.

Column 1: 15 m × 0.53 mm megabore capillary column, 1.0 μm film thickness, DB-210.
Column 2: 15 m × 0.53 mm megabore capillary column, 1.5 μm film thickness, SPB-608.
Column 3: 15 m × 0.53 mm megabore capillary column, 1.0 μm film thickness, DB-5.

Three megabore capillary columns are included for analysis of organophosphates by this method. Column 1 (DB-210 or equivalent) and Column 2 (SPB-608 or equivalent) are recommended if a large number of organophosphorus analytes are to be determined. If the superior resolution offered by Column 1 and Column 2 is not required, Column 3 (DB-5 or equivalent) may be used. For megabore capillary columns, automatic injections of 1 μL are recommended.

PRECISION & ACCURACY The MDL actually achieved in a given analysis will vary, as it is dependent on instrument sensitivity and matrix effects. Single operator accuracy and precision studies have been conducted with spiked water and soil samples.

MULTIPLICATION FACTORS FOR OTHER MATRICES (a)

Matrix	Factor (b)
Groundwater (EPA Method 3510 or EPA Method 3520)	10
Low-concentration soil by Soxhlet and no cleanup	10 (c)
Low-concentration soil by ultrasonic extraction with GPC cleanup	6.7 (c)
High-concentration soil and sludges by ultrasonic extraction	500 (c)
Non-Water miscible waste (EPA Method 3580)	1000 (c)

(a) Sample EQLs are highly matrix-dependent. The EQLs listed here are provided for guidance and may not always be achievable.
(b) EQL = [Method detection limit] × [Factor]. For non-aqueous samples the factor is on a wet-weight basis.
(c) Multiply this factory times the soil MDL.

The MDL (in μg/L) when reagent water was extracted using a separatory funnel was 0.80.
The MDL (in μg/kg) when soil was extracted using Soxhlet extraction (EPA Method 3540) was 40.0.
Accuracy (as % recovery) with separatory funnel extraction ranged from 96 (with low spikes) to 63 (with high spikes).
Accuracy (as % recovery) with continuous liquid-liquid extraction ranged from 39 (with low spikes) to 70 (with high spikes).
Accuracy (as % recovery) with Soxhlet extraction of soils ranged from 36 (with low spikes to 63 (with high spikes).
Accuracy (as % recovery) with ultrasonic extraction of soils ranged from 43 (with low spikes) to 3 (with high spikes).

SAMPLE COLLECTION, PRESERVATION & HANDLING
Containers used to collect samples for the determination of semivolatile organic compounds should be soap and water washed followed by methanol (or isopropanol) rinsing. The sample containers should be of glass or Teflon® and have screw-top covers with Teflon® liners.

No preservation is used with concentrated waste samples. With liquid samples containing no residual chlorine and with soil, sediment, and sludge samples, immediately cooling to 4°C is the only preservation used. When residual chlorine is present then 3 mL of 10% aqueous sodium sulfate is added for each gallon of sample collected, followed by cooling to 4°C.

Liquid samples must be extracted within 7 days and their extracts analyzed within 40 days. Concentrated waste, soil, sediment, and sludge samples must be extracted within 14 days and their extracts analyzed within 40 days.

SAMPLE PREPARATION In general, water samples are extracted at a neutral pH with methylene chloride, using either EPA Method 3510 or EPA Method 3520. Solid samples are extracted using either EPA Method 3540 or EPA Method 3550 with methylene chloride/acetone (1:1) as the extraction solvent.

Prior to GC analysis, the extraction solvent may be exchanged to hexane. Single lab data indicates that samples should not be transferred with 100% hexane during sample workup as the more water soluble organophosphorus compounds may be lost.

If cleanup is performed on the samples, the analyst should analyze the samples by GC. This will confirm elution patterns and the absence of interferences from the reagents. If peak detection and identification is prevented by the presence of interferences, further cleanup is required.

QUALITY CONTROL The analyst should monitor the performance of the extraction, cleanup (when used), and analytical system and the effectiveness of the method in dealing with each sample matrix by spiking each sample, standard, and blank with one or two surrogates (e.g., organophosphorus compounds not expected to be present in the sample). Deuterated analogs of analytes should not be used as surrogates for gas chromatographic analysis due to coelution problems.

A minimum of five concentrations for each analyte of interest should be prepared through dilution of the stock standards with isooctane. One of the concentrations should be at a concentration near, but above, the MDL.

Include a mid-level check standard after each group of 10 samples in the analysis sequence. GC/MS techniques should be judiciously employed to support qualitative identifications made with this method. Follow the GC/MS operating requirements specified in EPA Method 8270.

When available, chemical ionization mass spectra may be employed to aid in the qualitative identification process. To confirm an identification of a compound, the background-corrected mass spectrum of the compound must be obtained from the sample extract and must be compared with a mass spectrum from a stock or calibration standard analyzed under the same chromatographic conditions. The molecular ion and all other ions present above 20% relative abundance in the mass spectrum of the standard must be present in the mass spectrum of the sample with agreement to ±20%. The retention time of the compound in the sample must be within six seconds of the retention time for the same compound in the standard solution.

Should the MS procedure fail to provide satisfactory results, additional steps may be taken before reanalysis. These steps may include the use of alternate packed or capillary GC columns or additional sample cleanup.

REFERENCE Test Methods for Evaluating Solid Waste, Physical/Chemical Methods, SW-846, 3rd Edition, U.S. EPA, Office of Solid Waste, Washington, DC, EPA Method 8141 July 1992.

Terbacil   EPA Method 507
CAS #5902-51-2

TITLE Determination of Nitrogen and Phosphorus-Containing Pesticides in Water by GC/NPD

MATRIX This method is applicable to the determination of certain nitrogen and phosphorus-containing pesticides in finished drinking water and groundwater.

METHOD SUMMARY Method 507 covers 46 nitrogen- and phosphorus-containing pesticides. A 1-L sample is fortified with a surrogate standard, salted, buffered, extracted with methylene chloride, and concentrated; then the solvent is exchanged with methyl tert-butyl ether (MTBE) and concentrated again, and a 2-µL aliquot of a sample extract is injected into a GC system equipped with a selective nitrogen-phosphorus detector and a capillary column for analysis.

INTERFERENCES Method interferences may be caused by contaminants in solvents, reagents, glassware, and other sample processing apparatus. Interfering contamination may occur when a sample containing low concentrations of analytes is analyzed immediately following a sample containing relatively high concentrations. One or more injections of MTBE should be made following the analysis of a sample with high concentrations of analytes to check for analyte carryover. Matrix interferences may be caused by contaminants that are coextracted from the sample. The extent of matrix interferences will vary considerably from source to source, depending upon the water sampled.

INSTRUMENTATION A gas chromatograph system (GC) equipped with a nitrogen-phosphorus detector (NPD) is needed.

Column 1: 30 m × 0.25 mm I.D. DB-5 bonded fused silica column, 0.25 µm film thickness, or equivalent.

Column 2: 30 m × 0.25 mm I.D. DB-1701 bonded fused silica column, 0.25 µm film thickness, or equivalent.

PRECISION & ACCURACY This method has been validated in a single lab and estimated detection limits (EDLs) have been determined for each analyte. Observed detection limits may vary among waters, depending upon the nature of the interferences in the sample matrix and the specific instrumentation used. Analytes that are not separated chromatographically cannot be individually identified and measured unless an alternative technique for identification and quantification exist.

The estimated detection limit (in µg/L) was 4.5. The EDL is defined as either method detection limit or a level of compound in a sample yielding a peak in the final extract with signal-to-noise ratio of approximately 5, whichever value is higher.

The concentration used for these measurements (in µg/L) was 45.
The accuracy (as % recovery) was 97.
The precision (% RSD) was 6.

SAMPLING METHOD Grab samples are collected in 1-L glass sample bottles (prewashed with detergent and hot tap water, rinsed with reagent water, and dried in an oven at 400°C for 1 h) with screw caps lined with PTFE-fluorocarbon.

SAMPLE PRESERVATION Add mercuric chloride to the sample bottle in amounts to produce a concentration of 10 mg/L. If residual chlorine is present, add 80 mg of sodium thiosulfate/L of sample to the sample bottle prior to collection. After collection, seal bottle and shake vigorously for 1 min, then cool the sample to 4°C immediately and store it at 4°C in the dark until extraction.

MHT Maximum holding time of the samples, and in some cases the extracts, is 14 days.

Note: Samples with this compound exhibited recoveries of less than 60% after 14 days.

SAMPLE PREPARATION Fortify the sample with 50 µL of the surrogate standard solution, adjust to pH 7 with phosphate buffer, add 100 g NaCl to the sample, and seal and shake to dissolve the salt; then extract with methylene chloride in a separatory funnel or in a mechanical tumbler bottle. Dry the extract by pouring it through a solvent-rinsed drying column containing about 10 cm of anhydrous sodium sulfate. Collect the extract in a Kuderna-Danish (K-D) concentrator and rinse the column with 20–30 mL methylene chloride. Concentrate the extract to about 2 mL and rinse the flask and its lower joint into the concentrator tube with 1 to 2 mL of methyl t-butyl ether (MTBE). Add 5–10 mL of MTBE and concentrate the extract twice (adding more MTBE) to a final volume of 5.0 mL and store it at 4°C until analysis.

Note: If methylene chloride is not completely removed from the final extract, it may cause detector problems.

QUALITY CONTROL Minimum quality control requirements are initial demonstration of lab capability, determination of surrogate compound recoveries in each sample and blank, monitoring internal standard peak area or height in each sample and blank, analysis of lab reagent blanks, lab fortified samples, lab fortified blanks, and other QC samples. A lab reagent blank is analyzed to demonstrate that all glassware and reagent interferences are under control.

Initial demonstration of capability is fulfilled by analyzing four fortified reagent water samples with the recovery value for each analyte falling within the acceptable range (±30% average recovery). Surrogate recoveries from samples or method blanks must be 70–130%. The internal standard response for any sample chromatogram should not deviate from the daily calibration check standard's internal standard response by more than 30% or lab fortified blanks and sample matrices are used to assess lab performance and analyte recovery, respectively.

If the response for the target analyte peak exceeds the working range of the system, dilute the extract and reanalyze. Alternative techniques such as an alternate detector or second chromatography column should be used to confirm peak identification when sample components are not resolved adequately.

EPA CONTACT & HOTLINE For technical questions contact Dr. Baldev Bathija, U.S. EPA, Office of Ground Water and

Drinking Water (WH-550D), 401 M St. SW, Washington, DC 20460. Tel. (202) 260-3040. For further information the EPA Safe Drinking Water Hotline may be called at: (800) 426-4791.

REFERENCE Methods for the Determination of Organic Compounds in Drinking Water, EPA/600/4-88/039 (revised July 1991). U.S. EPA Environmental Monitoring Systems Laboratory, Cincinnati, OH, 45268, U.S.A. Available from the National Technical Information Service (NTIS), 5285 Port Royal Road, Springfield, VA 22161; Tel. 800-553-6847. NTIS Order Number is PB91-231480.

Terbufos
CAS #13071-79-9

EPA Method 507

TITLE Determination of Nitrogen and Phosphorus-Containing Pesticides in Water by GC/NPD

MATRIX This method is applicable to the determination of certain nitrogen and phosphorus-containing pesticides in finished drinking water and groundwater.

METHOD SUMMARY Method 507 covers 46 nitrogen- and phosphorus-containing pesticides. A 1-L sample is fortified with a surrogate standard, salted, buffered, extracted with methylene chloride, and concentrated; then the solvent is exchanged with methyl tert-butyl ether (MTBE) and concentrated again, and a 2-μL aliquot of a sample extract is injected into a GC system equipped with a selective nitrogen-phosphorus detector and a capillary column for analysis.

INTERFERENCES Method interferences may be caused by contaminants in solvents, reagents, glassware, and other sample processing apparatus. Interfering contamination may occur when a sample containing low concentrations of analytes is analyzed immediately following a sample containing relatively high concentrations. One or more injections of MTBE should be made following the analysis of a sample with high concentrations of analytes to check for analyte carryover. Matrix interferences may be caused by contaminants that are coextracted from the sample. The extent of matrix interferences will vary considerably from source to source, depending upon the water sampled.

INSTRUMENTATION A gas chromatograph system (GC) equipped with a nitrogen-phosphorus detector (NPD) is needed.

Column 1: 30 m × 0.25 mm I.D. DB-5 bonded fused silica column, 0.25 μm film thickness, or equivalent.
Column 2: 30 m × 0.25 mm I.D. DB-1701 bonded fused silica column, 0.25 μm film thickness, or equivalent.

PRECISION & ACCURACY This method has been validated in a single lab and estimated detection limits (EDLs) have been determined for each analyte. Observed detection limits may vary among waters, depending upon the nature of the interferences in the sample matrix and the specific instrumentation used. Analytes that are not separated chromatographically cannot be individually identified and measured unless an alternative technique for identification and quantification exist.

The estimated detection limit (in μg/L) was 0.5. The EDL is defined as either method detection limit or a level of compound in a sample yielding a peak in the final extract with signal-to-noise ratio of approximately 5, whichever value is higher.

The concentration used for these measurements (in μg/L) was 5.
The accuracy (as % recovery) was 97.
The precision (% RSD) was 4.

SAMPLING METHOD Grab samples are collected in 1-L glass sample bottles (prewashed with detergent and hot tap water, rinsed with reagent water, and dried in an oven at 400°C for 1 h) with screw caps lined with PTFE-fluorocarbon.

SAMPLE PRESERVATION Add mercuric chloride to the sample bottle in amounts to produce a concentration of 10 mg/L. If residual chlorine is present, add 80 mg of sodium thiosulfate/L of sample to the sample bottle prior to collection. After collection, seal bottle and shake vigorously for 1 min, then cool the sample to 4°C immediately and store it at 4°C in the dark until extraction.

MHT Maximum holding time of the samples, and in some cases the extracts, is 14 days.

Note: Samples with this compound must be extracted immediately.

SAMPLE PREPARATION Fortify the sample with 50 μL of the surrogate standard solution, adjust to pH 7 with phosphate buffer, add 100 g NaCl to the sample, and seal and shake to dissolve the salt; then extract with methylene chloride in a separatory funnel or in a mechanical tumbler bottle. Dry the extract by pouring it through a solvent-rinsed drying column containing about 10 cm of anhydrous sodium sulfate. Collect the extract in a Kuderna-Danish (K-D) concentrator and rinse the column with 20–30 mL methylene chloride. Concentrate the extract to about 2 mL and rinse the flask and its lower joint into the concentrator tube with 1 to 2 mL of methyl t-butyl ether (MTBE). Add 5–10 mL of MTBE and concentrate the extract twice (adding more MTBE) to a final volume of 5.0 mL and store it at 4°C until analysis.

Note: If methylene chloride is not completely removed from the final extract, it may cause detector problems.

QUALITY CONTROL Minimum quality control requirements are initial demonstration of lab capability, determination of surrogate compound recoveries in each sample and blank, monitoring internal standard peak area or height in each sample and blank, analysis of lab reagent blanks, lab fortified samples, lab fortified blanks, and other QC samples. A lab reagent blank is analyzed to demonstrate that all glassware and reagent interferences are under control.

Initial demonstration of capability is fulfilled by analyzing four fortified reagent water samples with the recovery value for each analyte falling within the acceptable range (±30% average recovery). Surrogate recoveries from samples or method blanks must be 70–130%. The internal standard response for any sample chromatogram should not deviate from the daily calibration check standard's internal standard response by more than 30% or lab fortified blanks and sample matrices are used to assess lab performance and analyte recovery, respectively.

If the response for the target analyte peak exceeds the working range of the system, dilute the extract and reanalyze. Alternative techniques such as an alternate detector or second chromatography column should be used to confirm peak identification when sample components are not resolved adequately.

EPA CONTACT & HOTLINE For technical questions contact Dr. Baldev Bathija, U.S. EPA, Office of Ground Water and Drinking Water (WH-550D), 401 M St. SW, Washington, DC 20460. Tel. (202) 260-3040. For further information the EPA Safe Drinking Water Hotline may be called at: (800) 426-4791.

REFERENCE Methods for the Determination of Organic Compounds in Drinking Water, EPA/600/4-88/039 (revised July 1991). U.S. EPA Environmental Monitoring Systems Laboratory, Cincinnati, OH, 45268, U.S.A. Available from the National Technical Information Service (NTIS), 5285 Port Royal Road, Springfield, VA 22161; Tel. 800-553-6847. NTIS Order Number is PB91-231480.

Terbufos   **EPA Method 8270**
CAS #13071-79-9

TITLE Semivolatile Organic Compounds by GC/MS

MATRIX This method is used to determine the concentration of semivolatile organic compounds in extracts prepared from all types of solid waste matrices, soils, and groundwater. Although surface waters are not specifically mentioned, this method should be applicable to water samples from rivers, lakes, etc.

METHOD SUMMARY This method covers 259 semivolatile organic compounds. In very limited applications direct injection of the sample into the GC/MS system may be appropriate, but this results in very high detection limits (approximately 10,000 μg/L). Typically, a 1-L liquid sample, containing surrogate, and matrix spiking standards, is extracted in a continuous extractor first under acid conditions and then under basic conditions. Typically 30 g of a solid sample, containing surrogate, and matrix spiking standards, is extracted ultrasonically. After concentrating the extract to 1 mL it is spiked with 10 μL of an internal standard solution just prior to analysis by GC/MS. The volume injected should contain about 100 ng of base/neutral and 200 ng of acid surrogates (for a 1-μL injection). Analysis is performed by GC/MS using a capillary GC column.

INTERFERENCES Raw GC/MS data from all blanks, samples, and spikes must be evaluated for interferences. Contamination by carryover can occur whenever high-concentration and low-concentration samples are sequentially analyzed. To reduce carryover, the sample syringe must be rinsed out between samples with solvent. Whenever an unusually concentrated sample is encountered, it should be followed by the analysis of blank solvent to check for cross-contamination.

INSTRUMENTATION A GC/MS and a data system are required. The GC column used is a 30 m × 0.25 mm I.D. (or 0.32 mm I.D.) 1um film thickness silicone-coated fused silica capillary column. A continuous liquid-liquid extractor equipped with Teflon® or glass connection joints and stopcocks requiring no lubrication, a K-D concentrating apparatus, water bath, and an ultrasonic disrupter with a minimum power of 300 W and with pulsing capability are also required.

PRECISION & ACCURACY The estimated quantitation limit (EQL) of Method 8270B for determining an individual compound is approximately 1 mg/kg (wet weight) for soil or sediment samples, 1–200 mg/kg for wastes (dependent on matrix and method of preparation), and 10 μg/L for groundwater samples. EQLs will be proportionately higher for sample extracts that require dilution to avoid saturation of the detector.

The EQL(b) for groundwater in μg/L is 20.
The EQL (a, b) for low concentrations in soil and sediment in μg/kg is not determined.
Accuracy as μg/L is not listed.
Overall precision in μg/L is not listed.

(a) EQLs listed for soil/sediment are based on wet weight. Normally data is reported in a dry-weight basis; therefore, EQLs will be higher based on the % dry weight of each sample. This calculation is based on a 30 g sample and gel permeation chromatography cleanup.
(b) Sample EQLs are highly matrix-dependent. The EQLs are provided for guidance and may not always be achievable.
C = True value for concentration, in μg/L.
X = Average recovery found for measurements of samples containing a concentration of C, in μg/L.

ESTIMATED QUANTITATION LIMIT

Other Matrices	Factor (a)
High-concentration soil and sludges by sonicator	7.5
Non-water miscible waste	75

(a) EQL for other matrices = [EQL for low soil/sediment × [Factor]. This estimated EQL is similar to an EPA "Practical Quantitation Limit."

SAMPLING METHOD
Liquid samples — Use a 1 or 2½ gallon amber glass bottle with a screw-top Teflon®-lined cover that has been prewashed with detergent and rinsed with distilled water and methanol (or isopropanol).

Soils, sediments, or sludges — Use an 8-oz. widemouth glass with a screw-top Teflon®-lined cover that has been prewashed with detergent and rinsed with distilled water and methanol (or isopropanol).

SAMPLE PRESERVATION
Liquid samples — If residual chlorine is present, add 3 mL of 10% sodium thiosulfate per gallon, cool to 4°C and store in a solvent-free refrigerator until analysis; if chlorine is not present, then eliminate the sodium thiosulfate addition.

Soils, sediments, or sludges — Cool samples to 4°C and store in a solvent-free refrigerator.

MHT Liquid samples must be extracted within 7 days and the extracts analyzed within 40 days. Soils, sediments, or sludges may be stored for a maximum of 14 days and the extracts analyzed within 40 days.

SAMPLE PREPARATION

Liquid samples — Transfer 1 L quantitatively to a continuous extractor. If high concentrations are anticipated, a smaller volume may be used and then diluted with organic-free reagent water to 1 L. Adjust pH, if necessary, to pH <2 using 1:1 (V/V) sulfuric acid. Pipette 1.0 mL of a surrogate standard spiking solution into each sample. For the sample in each analytical batch selected for spiking, add 1.0 mL of a matrix spiking standard. For base/neutral acid analysis, the amount of the surrogates and matrix spiking compounds added to the sample should result in a final concentration of 100 ng/µL of each analyte in the extract to be analyzed (assuming a 1-µL injection). Extract with methylene chloride for 18–24 h. Next, adjust the pH of the aqueous phase to pH >11 using 10 N sodium hydroxide and extract it with methylene chloride again for 18–24 h. Dry the extract through a column containing anhydrous sodium sulfate and concentrate it to 1 mL using a K-D concentrator.

Soils, sediments, or sludges — Use 30 g of sample. Nonporous or wet samples (gummy or clay type) that do not have a free-flowing sandy texture must be mixed with anhydrous sodium sulfate until the sample is free flowing. Add 1 mL of surrogate standards to all samples, spikes, standards, and blanks. For the sample in each analytical batch selected for spiking, add 1.0 mL of a matrix spiking standard. For base/neutral acid analysis, the amount added of the surrogates and matrix spiking compounds should result in a final concentration of 100 ng/µL of each base/neutral analyte and 200 ng/µL of each acid analyte in the extract to be analyzed (assuming a 1-µL injection). Immediately add a 100-mL mixture of 1:1 methylene chloride:acetone and extract the sample ultrasonically for 3 min and then decant or filter the extracts. Repeat the extraction two or more times. Dry the extract using a column with anhydrous sodium sulfate and concentrate it to 1 mL in a K-D concentrator.

QUALITY CONTROL

A methylene chloride solution containing 50 ng/µL of decafluorotriphenylphosphine (DFTPP) is used for tuning the GC/MS system each 12-h shift. A system performance check also must be made during every 12-h shift. A standard containing 50 ng/µL each of 4,4′-DDT, pentachlorophenol, and benzidine is required to verify injection port inertness and GC column performance. A calibration standard at mid-concentration, containing each compound of interest, including all required surrogates, must be performed every 12 h during analysis. After the system performance check is met, calibration check compounds (CCCs) are used to check the validity of the initial calibration.

The internal standard responses and retention times in the calibration check standard must be evaluated immediately after or during data acquisition. If the retention time for any internal standard changes by more than 30 seconds from the last check calibration (12 h), the chromatographic system must be inspected for malfunctions and corrections must be made, as required. If the electron ionization current plot (EICP) area for any of the internal standards changes by a factor of two from the last daily calibration standard check, the mass spectrometer must be inspected for malfunctions and corrections must be made, as appropriate.

Demonstrate, through the analysis of a reagent water blank, that interferences from the analytical system, glassware, and reagents are under control. The blank samples should be carried through all stages of the sample preparation and measurement steps. For each analytical batch (up to 20 samples), a reagent blank, matrix spike, and matrix spike duplicate/duplicate must be analyzed (the frequency of the spikes may be different for different monitoring programs). The blank and spiked samples must be carried through all stages of the sample preparation and measurement steps. A QC reference sample concentrate containing each analyte at a concentration of 100 mg/L in methanol is required.

REFERENCE Test Methods for Evaluating Solid Waste (SW-846). U.S. EPA 1983, Method 8270B, Rev. 2, Nov. 1990. Office of Solid Waste, Washington, DC.

Terbutryn　　　　　　　　　　　　　　　EPA Method 507
CAS #886-50-0

TITLE Determination of Nitrogen and Phosphorus-Containing Pesticides in Water by GC/NPD

MATRIX This method is applicable to the determination of certain nitrogen and phosphorus-containing pesticides in finished drinking water and groundwater.

METHOD SUMMARY Method 507 covers 46 nitrogen- and phosphorus-containing pesticides. A 1-L sample is fortified with a surrogate standard, salted, buffered, extracted with methylene chloride, and concentrated; then the solvent is exchanged with methyl tert-butyl ether (MTBE) and concentrated again, and a 2-µL aliquot of a sample extract is injected into a GC system equipped with a selective nitrogen-phosphorus detector and a capillary column for analysis.

INTERFERENCES Method interferences may be caused by contaminants in solvents, reagents, glassware, and other sample processing apparatus. Interfering contamination may occur when a sample containing low concentrations of analytes is analyzed immediately following a sample containing relatively high concentrations. One or more injections of MTBE should be made following the analysis of a sample with high concentrations of analytes to check for analyte carryover. Matrix interferences may be caused by contaminants that are coextracted from the sample. The extent of matrix interferences will vary considerably from source to source, depending upon the water sampled.

INSTRUMENTATION A gas chromatograph system (GC) equipped with a nitrogen-phosphorus detector (NPD) is needed.

Column 1: 30 m × 0.25 mm I.D. DB-5 bonded fused silica column, 0.25 µm film thickness, or equivalent.
Column 2: 30 m × 0.25 mm I.D. DB-1701 bonded fused silica column, 0.25 µm film thickness, or equivalent.

PRECISION & ACCURACY This method has been validated in a single lab and estimated detection limits (EDLs) have been determined for each analyte. Observed detection limits may vary among waters, depending upon the nature of the interferences in the sample matrix and the specific instrumentation used. Analytes that are not separated chromatographically cannot be individually identified and measured unless an alternative technique for identification and quantification exist.

The estimated detection limit (in µg/L) was 0.25. The EDL is defined as either method detection limit or a level of compound in a sample yielding a peak in the final extract with signal-to-noise ratio of approximately 5, whichever value is higher.

The concentration used for these measurements (in µg/L) was 2.5.

The accuracy (as % recovery) was 94.

The precision (% RSD) was 9.

SAMPLING METHOD Grab samples are collected in 1-L glass sample bottles (prewashed with detergent and hot tap water, rinsed with reagent water, and dried in an oven at 400°C for 1 h) with screw caps lined with PTFE-fluorocarbon.

SAMPLE PRESERVATION Add mercuric chloride to the sample bottle in amounts to produce a concentration of 10 mg/L. If residual chlorine is present, add 80 mg of sodium thiosulfate/L of sample to the sample bottle prior to collection. After collection, seal bottle and shake vigorously for 1 min, then cool the sample to 4°C immediately and store it at 4°C in the dark until extraction.

MHT Maximum holding time of the samples, and in some cases the extracts, is 14 days.

Note: Samples with this compound exhibited recoveries of less than 60% after 14 days.

SAMPLE PREPARATION Fortify the sample with 50 µL of the surrogate standard solution, adjust to pH 7 with phosphate buffer, add 100 g NaCl to the sample, and seal and shake to dissolve the salt; then extract with methylene chloride in a separatory funnel or in a mechanical tumbler bottle. Dry the extract by pouring it through a solvent-rinsed drying column containing about 10 cm of anhydrous sodium sulfate. Collect the extract in a Kuderna-Danish (K-D) concentrator and rinse the column with 20–30 mL methylene chloride. Concentrate the extract to about 2 mL and rinse the flask and its lower joint into the concentrator tube with 1 to 2 mL of methyl t-butyl ether (MTBE). Add 5–10 mL of MTBE and concentrate the extract twice (adding more MTBE) to a final volume of 5.0 mL and store it at 4°C until analysis.

Note: If methylene chloride is not completely removed from the final extract, it may cause detector problems.

QUALITY CONTROL Minimum quality control requirements are initial demonstration of lab capability, determination of surrogate compound recoveries in each sample and blank, monitoring internal standard peak area or height in each sample and blank, analysis of lab reagent blanks, lab fortified samples, lab fortified blanks, and other QC samples. A lab reagent blank is analyzed to demonstrate that all glassware and reagent interferences are under control.

Initial demonstration of capability is fulfilled by analyzing four fortified reagent water samples with the recovery value for each analyte falling within the acceptable range (±30% average recovery). Surrogate recoveries from samples or method blanks must be 70–130%. The internal standard response for any sample chromatogram should not deviate from the daily calibration check standard's internal standard response by more than 30% or lab fortified blanks and sample matrices are used to assess lab performance and analyte recovery, respectively.

If the response for the target analyte peak exceeds the working range of the system, dilute the extract and reanalyze. Alternative techniques such as an alternate detector or second chromatography column should be used to confirm peak identification when sample components are not resolved adequately.

EPA CONTACT & HOTLINE For technical questions contact Dr. Baldev Bathija, U.S. EPA, Office of Ground Water and Drinking Water (WH-550D), 401 M St. SW, Washington, DC 20460. Tel. (202) 260-3040. For further information the EPA Safe Drinking Water Hotline may be called at: (800) 426-4791.

REFERENCE Methods for the Determination of Organic Compounds in Drinking Water, EPA/600/4-88/039 (revised July 1991). U.S. EPA Environmental Monitoring Systems Laboratory, Cincinnati, OH, 45268, U.S.A. Available from the National Technical Information Service (NTIS), 5285 Port Royal Road, Springfield, VA 22161; Tel. 800-553-6847. NTIS Order Number is PB91-231480.

α-Terpineol
CAS #98-55-5

EPA Method 1625

TITLE Semivolatile Organic Compounds by Isotope Dilution GC/MS

MATRIX The compounds may be determined in waters, soils, and municipal sludges by this method.

METHOD SUMMARY This method is used to determine 176 semivolatile toxic organic pollutants associated with the CWA (as amended 1987); the RCRA (as amended 1986); the CERCLA (as amended 1986); and other compounds amenable to extraction and analysis by capillary column gas chromatography-mass spectrometry (GC/MS).

Stable isotopically-labeled analogs of the compounds of interest are added to the sample. If the solids content is less than 1%, a 1-L sample is extracted at pH 12–13, then at pH <2 with methylene chloride using continuous extraction techniques.

If the solids content is 30% or less, the sample is diluted to 1% solids with reagent water, homogenized ultrasonically, and extracted at pH 12–13, then at pH <2 with methylene chloride using continuous extraction techniques. If the solids content is greater than 30%, the sample is extracted using ultrasonic techniques.

Each extract is dried over sodium sulfate, concentrated to a volume of 5 mL, cleaned up using GPC, if necessary, and concentrated. Extracts are concentrated to 1 mL if GPC is not performed, and to 0.5 mL if GPC is performed.

An internal standard is added to the extract, and a 1-mL aliquot of the extract is injected into the GC. The compounds are separated by GC and detected by a MS. The labeled compounds serve to correct the variability of the analytical technique.

INTERFERENCES Solvents, reagents, glassware, and other sample processing hardware may yield artifacts and/or elevated baselines causing misinterpretation of chromatograms and spectra. Materials used in the analysis must be demonstrated to be free from interferences under the conditions of analysis by running method blanks initially and with each sample lot (sample started through the extraction process on a given 8-h shift, to a maximum of 20). Specific selection of reagents and purification of solvents by distillation in all glass systems may be required. Glassware and, where possible, reagents are cleaned by solvent rinse and baking at 450°C for 1-h minimum. Interferences coextracted from samples will vary considerably from source to source, depending on the diversity of the site being sampled.

INSTRUMENTATION Major instrumentation includes a GC with a splitless or on-column injection port for capillary column, a MS with 70 eV electron impact ionization, and a data system to collect and record MS data, and process it. A K-D apparatus is used to concentrate extracts.

GC Column: 30 m × 0.25 mm I.D. 5% phenyl, 94% methyl, 1% vinyl silicone bonded phased fused silica capillary column.

PRECISION & ACCURACY The detection limits of the method are usually dependent on the level of interferences rather than instrumental limitations. The limits typify the minimum quantities that can be detected with no interferences present.

The minimum level (in μg/mL) was 10. This is defined as a minimum level at which the analytical system shall give recognizable mass spectra (background corrected) and acceptable calibration points.

The MDL (in μg/kg) in low solids was not detected and in high solids was not detected; these were determined in digested sludge (low solids) and in filter cake or compost (high solids).

The labeled and native compound initial precision as standard deviation (in μg/L) was 44.

The labeled and native compound initial accuracy as average recovery (in μg/L) was 42–234.

SAMPLE COLLECTION, PRESERVATION & HANDLING Collect samples in glass containers. Aqueous samples which flow freely are collected in refrigerated bottles using automatic sampling equipment. Solid samples are collected as grab samples using widemouth jars. Maintain samples at 0 to 4°C from the time of collection until extraction. If residual chlorine is present in aqueous samples, add 80 mg sodium thiosulfate/L of water. Begin sample extraction within 7 days of collection, and analyze all extracts within 40 days of extraction.

SAMPLE PREPARATION Samples containing 1% solids or less are extracted directly using continuous liquid-liquid extraction techniques. Samples containing 1 to 30% solids are diluted to the 1% level with reagent water and extracted using continuous liquid-liquid extraction techniques. Samples containing greater than 30% solids are extracted using ultrasonic techniques.

Base/neutral extraction — Adjust the pH of the waters in the extractors to 12–13 with 6 N NaOH. Extract with methylene chloride for 24–48 h.

Acid extraction — Adjust the pH of the waters in the extractors to 2 or less using 6 N sulfuric acid. Extract with methylene chloride for 24–48 h.

Ultrasonic extraction of high solids samples — Add anhydrous sodium sulfate to the sample and QC aliquot(s). Add acetone:methylene chloride (1:1) to the sample and mix thoroughly

Concentrate extracts using a K-D apparatus.

QUALITY CONTROL The analyst is permitted to modify this method to improve separations or lower the costs of measurements, provided all performance specifications are met. Analyses of blanks are required to demonstrate freedom from contamination. When results of spikes indicate atypical method performance for samples, the samples are diluted to bring method performance within acceptable limits.

For low solids (aqueous samples), extract, concentrate, and analyze two sets of four 1-L aliquots (8 aliquots total) of the precision and recovery standard. For high solids samples, two sets of four 30-g aliquots of the high solids reference matrix are used.

Spike all samples with labeled compounds to assess method performance. Compute percent recovery of the labeled compounds using the internal standard method. Compare the labeled compound recovery for each compound with the corresponding labeled compound recovery.

Reagent water and high solids reference matrix blanks are analyzed to demonstrate freedom from contamination. Extract and concentrate a 1-L reagent water blank or a high solids reference matrix blank with each sample's lot (samples started through the extraction process on the same 8-h shift, to a maximum of 20 samples).

Field replicates may be collected to determine the precision of the sampling technique, and spiked samples may be required to determine the accuracy of the analysis when the internal standard method is used.

REFERENCE Semivolatile Organic Compounds by Isotope Dilution GC/MS. Office of Water Regulation and Standards, U.S. EPA Industrial Technology Division, Washington, DC, EPA Method 1625, Rev. C, June 1989 (contact W.A. Telliard, U.S. EPA, Office of Water Regulations and Standards, 401 M St., SW, Washington, DC, 20460. Phone: 202-382-7131).

1,2,4,5-Tetrachlorobenzene　　　　　　　　EPA Method 1625
CAS #95-94-3

TITLE Semivolatile Organic Compounds by Isotope Dilution GC/MS

MATRIX The compounds may be determined in waters, soils, and municipal sludges by this method.

METHOD SUMMARY This method is used to determine 176 semivolatile toxic organic pollutants associated with the CWA (as amended 1987); the RCRA (as amended 1986); the CERCLA (as amended 1986); and other compounds amenable to extraction and analysis by capillary column gas chromatography-mass spectrometry (GC/MS).

Stable isotopically-labeled analogs of the compounds of interest are added to the sample. If the solids content is less than 1%, a 1-L sample is extracted at pH 12–13, then at pH <2 with methylene chloride using continuous extraction techniques.

If the solids content is 30% or less, the sample is diluted to 1% solids with reagent water, homogenized ultrasonically, and extracted at pH 12–13, then at pH <2 with methylene chloride using continuous extraction techniques. If the solids content is greater than 30%, the sample is extracted using ultrasonic techniques.

Each extract is dried over sodium sulfate, concentrated to a volume of 5 mL, cleaned up using GPC, if necessary, and concentrated. Extracts are concentrated to 1 mL if GPC is not performed, and to 0.5 mL if GPC is performed.

An internal standard is added to the extract, and a 1-mL aliquot of the extract is injected into the GC. The compounds are separated by GC and detected by a MS. The labeled compounds serve to correct the variability of the analytical technique.

INTERFERENCES Solvents, reagents, glassware, and other sample processing hardware may yield artifacts and/or elevated baselines causing misinterpretation of chromatograms and spectra. Materials used in the analysis must be demonstrated to be free from interferences under the conditions of analysis by running method blanks initially and with each sample lot (sample started through the extraction process on a given 8-h shift, to a maximum of 20). Specific selection of reagents and purification of solvents by distillation in all glass systems may be required. Glassware and, where possible, reagents are cleaned by solvent rinse and baking at 450°C for 1-h minimum. Interferences coextracted from samples will vary considerably from source to source, depending on the diversity of the site being sampled.

INSTRUMENTATION Major instrumentation includes a GC with a splitless or on-column injection port for capillary column, a MS with 70 eV electron impact ionization, and a data system to collect and record MS data, and process it. A K-D apparatus is used to concentrate extracts.

GC Column: 30 m × 0.25 mm I.D. 5% phenyl, 94% methyl, 1% vinyl silicone bonded phased fused silica capillary column.

PRECISION & ACCURACY The detection limits of the method are usually dependent on the level of interferences rather than instrumental limitations. The limits typify the minimum quantities that can be detected with no interferences present.

The minimum level (in µg/mL) was not listed. This is defined as a minimum level at which the analytical system shall give recognizable mass spectra (background corrected) and acceptable calibration points.

The MDL (in µg/kg) in low solids was not listed and in high solids was not listed; these were determined in digested sludge (low solids) and in filter cake or compost (high solids).

The labeled and native compound initial precision as standard deviation (in µg/L) was not listed.
The labeled and native compound initial accuracy as average recovery (in µg/L) was not listed.

SAMPLE COLLECTION, PRESERVATION & HANDLING Collect samples in glass containers. Aqueous samples which flow freely are collected in refrigerated bottles using automatic sampling equipment. Solid samples are collected as grab samples using widemouth jars. Maintain samples at 0 to 4°C from the time of collection until extraction. If residual chlorine is present in aqueous samples, add 80 mg sodium thiosulfate/L of water. Begin sample extraction within 7 days of collection, and analyze all extracts within 40 days of extraction.

SAMPLE PREPARATION Samples containing 1% solids or less are extracted directly using continuous liquid-liquid extraction techniques. Samples containing 1 to 30% solids are diluted to the 1% level with reagent water and extracted using continuous liquid-liquid extraction techniques. Samples containing greater than 30% solids are extracted using ultrasonic techniques.

- Base/neutral extraction — Adjust the pH of the waters in the extractors to 12–13 with 6 N NaOH. Extract with methylene chloride for 24–48 h.
- Acid extraction — Adjust the pH of the waters in the extractors to 2 or less using 6 N sulfuric acid. Extract with methylene chloride for 24–48 h.
- Ultrasonic extraction of high solids samples — Add anhydrous sodium sulfate to the sample and QC aliquot(s). Add acetone:methylene chloride (1:1) to the sample and mix thoroughly

Concentrate extracts using a K-D apparatus.

QUALITY CONTROL The analyst is permitted to modify this method to improve separations or lower the costs of measurements, provided all performance specifications are met. Analyses of blanks are required to demonstrate freedom from contamination. When results of spikes indicate atypical method performance for samples, the samples are diluted to bring method performance within acceptable limits.

For low solids (aqueous samples), extract, concentrate, and analyze two sets of four 1-L aliquots (8 aliquots total) of the precision and recovery standard. For high solids samples, two sets of four 30-g aliquots of the high solids reference matrix are used.

Spike all samples with labeled compounds to assess method performance. Compute percent recovery of the labeled compounds using the internal standard method. Compare the labeled compound recovery for each compound with the corresponding labeled compound recovery.

Reagent water and high solids reference matrix blanks are analyzed to demonstrate freedom from contamination. Extract and concentrate a 1-L reagent water blank or a high solids reference matrix blank with each sample's lot (samples started through the extraction process on the same 8-h shift, to a maximum of 20 samples).

Field replicates may be collected to determine the precision of the sampling technique, and spiked samples may be required to determine the accuracy of the analysis when the internal standard method is used.

REFERENCE Semivolatile Organic Compounds by Isotope Dilution GC/MS. Office of Water Regulation and Standards, U.S. EPA Industrial Technology Division, Washington, DC, EPA Method 1625, Rev. C, June 1989 (contact W.A. Telliard, U.S. EPA, Office of Water Regulations and Standards, 401 M St., SW, Washington, DC, 20460. Phone: 202-382-7131).

| 1,2,3,4-Tetrachlorobenzene | EPA Method 8121 |
| CAS #634-66-2 | |

TITLE Chlorinated Hydrocarbons by GC: Capillary Column Technique

MATRIX This method covers aqueous and solid matrices. This includes a wide variety such as drinking water, groundwater, industrial wastewaters, surface waters, soils, solids, and sediments.

METHOD SUMMARY This method provides procedures for the determination of 22 chlorinated hydrocarbons in water, soil/sediment, and waste matrices. A measured volume or weight of sample is extracted by using one of the appropriate sample extraction techniques specified in EPA Method 3510, EPA Method 3520, EPA Method 3540, or EPA Method 3550, or diluted using EPA Method 3580. Aqueous samples are extracted at neutral pH with methylene chloride by using either a separatory funnel (EPA Method 3510) or a continuous liquid-liquid extractor (EPA Method 3520). Solid samples are extracted with hexane/acetone (1:1) by using a Soxhlet extractor (EPA Method 3540) or with methylene chloride/acetone (1:1) by using an ultrasonic extractor (EPA Method 3550). After cleanup, the extract or diluted sample is analyzed by gas chromatography with electron capture detection (GC/ECD).

The sensitivity level of this method usually depends on the level of interferences rather than on instrumental limitations. This method may be used in conjunction with EPA Method 3620, Florisil Column Cleanup, EPA Method 3660, Sulfur Cleanup, and EPA Method 3640, Gel Permeation Chromatography, to aid in the elimination of interferences.

INTERFERENCES Solvents, reagents, glassware, and other hardware used in sample processing may introduce artifacts which may result in elevated baselines, causing misinterpretation of gas chromatograms. Interferants coextracted from the samples will vary considerably from waste to waste. Glassware must be scrupulously clean. Phthalate esters, if present in a sample, will interfere only with the BHC isomers. The presence of elemental sulfur will result in large peaks, and can often mask the region of compounds eluting after 1,2,4,5-tetrachlorobenzene. The tetrabutylammonium (TBA)-sulfite procedure (EPA Method 3660) works well for the removal of elemental sulfur. Waxes and lipids can be removed by gel permeation chromatography (EPA Method 3640).

INSTRUMENTATION A GC suitable for on-column injections and all required accessories, including and electron capture detector (ECD), analytical columns, recorder, gases, and syringes are needed. A data system for measuring peak heights and/or peak areas is recommended. A Kuderna-Danish (K-D) apparatus will also be needed to concentrate extracts.

Column 1: 30 m × 0.53 mm I.D. fused-silica capillary column chemically bonded with trifluoropropyl methyl silicone (DB-210 or equivalent).

Column 2: 30 m × 0.53 mm I.D. fused-silica capillary column chemically bonded with polyethylene glycol (DB-WAX or equivalent).

PRECISION & ACCURACY This method has been tested in a single lab by using organic-free reagent water, sandy loam samples, and extracts which were spiked with the test compounds at one concentration. Single-operator precision and method accuracy were found to be related to the concentration of compound and the type of matrix. The accuracy and precision technique will be determined by the sample matrix, sample preparation technique, optional cleanup techniques, and calibration procedures used.

MULTIPLICATION FACTORS FOR OTHER MATRICES (a)

Matrix	Factor (b)
Groundwater	10
Low-concentration soil by ultrasonic extraction with GPC cleanup	670
High-concentration soil and sludges by ultrasonic extraction	10,000
Waste not miscible with water	100,000

(a) Sample EQLs are highly matrix-dependent. The EQLs listed herein are provided for guidance and may not always be achievable. (b) EQL = [Method detection limit] × [Factor]. For nonaqueous samples, the factor is on a wet-weight basis.

PRECISION & ACCURACY MDL is the method detection limit for organic-free reagent water. MDL was determined from the analysis of eight replicate aliquots processed through the entire analytical method (extraction, Florisil cartridge cleanup, and GC/ECD analysis).

The MDL (in ng/L) was 11.

The accuracy (as average % recovery using 5 determinations and no Florisil cleanup) from a spike concentration of 10 µg/L and separatory funnel extraction was 96% with a final volume of 10 mL.

The precision (as RSD% using 5 determinations and no Florisil cleanup) from a spike concentration of 10 µg/L and separatory funnel extraction was 3.4% with a final volume of 10 mL.

The accuracy (as average % recovery using 5 determinations and no Florisil cleanup), from a spike concentration of 3300µg/L and ultrasonic extraction of solid samples using 1:1 methylene chloride and acetone, was 88% with a final volume of 10 mL.

The precision (as RSD% using 5 determinations and no Florisil cleanup), from a spike concentration of 3300µg/L and ultrasonic extraction of solid samples using 1:1 methylene chloride and acetone, was 2.9% with a final volume of 10 mL.

SAMPLE COLLECTION, PRESERVATION & HANDLING
Volatile organics — Standard 40-mL glass screw-cap VOA vials with Teflon®-faced silicone septum may be used for both liquid and solid matrices. When collecting samples, liquids and solids should be introduced into the vials gently to reduce agitation which might drive off volatile compounds. If there are any air bubbles present the sample must be retaken. The vials with solids should be tapped slightly as they are filled to try and eliminate as much free air space as possible. Two vials from each sampling location should be sealed in separate plastic bags to prevent cross-contamination between samples.

Semivolatile organics — Containers used to collect samples for the determination of semivolatile organic compounds should be soap and water washed followed by methanol (or isopropanol) rinsing. The sample containers should be of glass or Teflon® and have screw-top covers with Teflon® liners.

Preservation for volatile organics — No preservation is used with concentrated waste samples. With liquid samples containing no residual chlorine, 4 drops of concentrated hydrochloric acid are added and the samples are immediately cooled to 4°C. When liquid samples contain residual chlorine, they are treated as above and, in addition, 4 drops of 4% aqueous sodium thiosulfate are added to remove the residual chlorine. Soil, sediment, and sludge samples are only cooled to 4°C.

Preservation for semivolatile organics — No preservation is used with concentrated waste samples. With liquid samples containing no residual chlorine and with soil, sediment, and sludge samples, immediately cooling to 4°C is the only preservation used. When residual chlorine is present then 3 mL of 10% aqueous sodium sulfate is added for each gallon of sample collected, followed by cooling to 4°C.

Holding times — The holding time for all volatile organics samples is 14 days. Liquid samples must be extracted within 7 days and their extracts analyzed within 40 days. Concentrated waste, soil, sediment, and sludge samples must be extracted within 14 days and their extracts analyzed within 40 days.

SAMPLE PREPARATION Prepare stock standard solutions in hexane. Calibration standards at a minimum of five concentrations should be prepared through dilution of the stock standards with hexane. The suggested internal standards are: 2,5-dibromotoluene, 1,3,5-tribromobenzene, and α, α-dibromo-m-xylene. The analyst can use any of the three compounds provided that they are resolved from matrix interferences. Recommended surrogate compounds are α-2,6-trichlorotoluene, 1,4-dichloronaphthalene, and 2,3,4,5,6-pentachlorotoluene.

In general, water samples are extracted at a neutral pH with methylene chloride using a separatory funnel (EPA Method 3510) or a continuous liquid-liquid extractor (EPA Method 3520). Solid samples are extracted with hexane/acetone (1:1 v:v) using a Soxhlet extractor (EPA Method 3540) or with methylene chloride/acetone (1:1 v:v) using an ultrasonic extractor (EPA Method 3550). Non-aqueous waste samples may be diluted using EPA Method 3580. Prior to Florisil cleanup or gas chromatographic analysis, the extraction solvent must be exchanged to hexane. Sample extracts that will be subjected to gel permeation chromatography do not need solvent exchange.

Cleanup procedures may not be necessary for a relatively clean matrix. If removal of interferences such as chlorinated phenols, phthalate esters, etc., is required, proceed with the procedure outlined in EPA Method 3620.

QUALITY CONTROL Analyze a quality control check standard to demonstrate that the operation of the GC is in control. The frequency of the check standard analysis is equivalent to 10% of the samples analyzed. If the recovery of any compound found in the check standard is less than 80% of the certified value, the problem must be corrected and a new set of calibration standards must be prepared and analyzed. Calculate surrogate standard recoveries for all samples, blanks, and spikes. An internal standard peak area check must be performed on all samples. The internal standard must be evaluated for acceptance by determining whether the measured area for the internal standard deviates by more than 30% from the average area for the internal standard in the calibration standards. When the internal standard peak area is outside that limit, all samples that fall outside the QC criteria must be reanalyzed. Any compound confirmed by two columns may also be confirmed by GC/MS (EPA Method 8270). The GC/MS would normally require a minimum concentration of 1 ng/µL in the final extract for each compound. Include a mid-concentration calibration standard after each group of 20 samples in the analysis sequence. The response factors for the mid-concentration calibration must be within 15% of the average values for the multiconcentration calibration.

REFERENCE Test Methods for Evaluating Solid Waste, Physical/Chemical Methods, SW-846, 3rd Edition, U.S. EPA, Office of Solid Waste, Washington, DC, 1990. EPA Method 8121, Rev. 0, Nov. 1990.

1,2,3,5-Tetrachlorobenzene   **EPA Method 8121**
CAS #634-90-2

TITLE Chlorinated Hydrocarbons by GC: Capillary Column Technique

MATRIX This method covers aqueous and solid matrices. This includes a wide variety such as drinking water, groundwater, industrial wastewaters, surface waters, soils, solids, and sediments.

METHOD SUMMARY This method provides procedures for the determination of 22 chlorinated hydrocarbons in water, soil/sediment, and waste matrices. A measured volume or weight of sample is extracted by using one of the appropriate sample extraction techniques specified in EPA Method 3510, EPA Method 3520, EPA Method 3540, or EPA Method 3550, or diluted using EPA Method 3580. Aqueous samples are extracted at neutral pH with methylene chloride by using either a separatory funnel (EPA Method 3510) or a continuous liquid-liquid extractor (EPA Method 3520). Solid samples are extracted with hexane/acetone (1:1) by using a Soxhlet extractor (EPA Method 3540) or with methylene chloride/acetone (1:1) by using an ultrasonic extractor (EPA Method 3550). After cleanup, the extract or diluted sample is analyzed by gas chromatography with electron capture detection (GC/ECD).

The sensitivity level of this method usually depends on the level of interferences rather than on instrumental limitations. This method may be used in conjunction with EPA Method 3620, Florisil Column Cleanup, EPA Method 3660, Sulfur Cleanup, and EPA Method 3640, Gel Permeation Chromatography, to aid in the elimination of interferences.

INTERFERENCES Solvents, reagents, glassware, and other hardware used in sample processing may introduce artifacts which may result in elevated baselines, causing misinterpretation of gas chromatograms. Interferants coextracted from the samples will vary considerably from waste to waste. Glassware must be scrupulously clean. Phthalate esters, if present in a sample, will interfere only with the BHC isomers. The presence of elemental sulfur will result in large peaks, and can often mask the region of compounds eluting after 1,2,4,5-tetrachlorobenzene. The tetrabutylammonium (TBA)-sulfite procedure (EPA Method 3660) works well for the removal of elemental sulfur. Waxes and lipids can be removed by gel permeation chromatography (EPA Method 3640).

INSTRUMENTATION A GC suitable for on-column injections and all required accessories, including and electron capture detector (ECD), analytical columns, recorder, gases, and syringes are needed. A data system for measuring peak heights and/or peak areas is recommended. A Kuderna-Danish (K-D) apparatus will also be needed to concentrate extracts.

Column 1: 30 m × 0.53 mm I.D. fused-silica capillary column chemically bonded with trifluoropropyl methyl silicone (DB-210 or equivalent).
Column 2: 30 m × 0.53 mm I.D. fused-silica capillary column chemically bonded with polyethylene glycol (DB-WAX or equivalent).

PRECISION & ACCURACY This method has been tested in a single lab by using organic-free reagent water, sandy loam samples, and extracts which were spiked with the test compounds at one concentration. Single-operator precision and method accuracy were found to be related to the concentration of compound and the type of matrix. The accuracy and precision technique will be determined by the sample matrix, sample preparation technique, optional cleanup techniques, and calibration procedures used.

MULTIPLICATION FACTORS FOR OTHER MATRICES (a)

Matrix	Factor (b)
Groundwater	10
Low-concentration soil by ultrasonic extraction with GPC cleanup	670
High-concentration soil and sludges by ultrasonic extraction	10,000
Waste not miscible with water	100,000

(a) Sample EQLs are highly matrix-dependent. The EQLs listed herein are provided for guidance and may not always be achievable.
(b) EQL = [Method detection limit] × [Factor]. For nonaqueous samples, the factor is on a wet-weight basis.

PRECISION & ACCURACY MDL is the method detection limit for organic-free reagent water. MDL was determined from the analysis of eight replicate aliquots processed through the entire analytical method (extraction, Florisil cartridge cleanup, and GC/ECD analysis).

The MDL (in ng/L) was 8.1.

The accuracy (as average % recovery using 5 determinations and no Florisil cleanup) from a spike concentration of 10 µg/L and separatory funnel extraction was 93% with a final volume of 10 mL.

The precision (as RSD% using 5 determinations and no Florisil cleanup) from a spike concentration of 10 µg/L and separatory funnel extraction was 4.6% with a final volume of 10 mL.

The accuracy (as average % recovery using 5 determinations and no Florisil cleanup), from a spike concentration of 3300µg/L and ultrasonic extraction of solid samples using 1:1 methylene chloride and acetone, was 80% with a final volume of 10 mL.

The precision (as RSD% using 5 determinations and no Florisil cleanup), from a spike concentration of 3300µg/L and ultrasonic extraction of solid samples using 1:1 methylene chloride and acetone, was 4.4% with a final volume of 10 mL.

SAMPLE COLLECTION, PRESERVATION & HANDLING
Volatile organics — Standard 40-mL glass screw-cap VOA vials with Teflon®-faced silicone septum may be used for both liquid and solid matrices. When collecting samples, liquids and solids should be introduced into the vials gently to reduce agitation which might drive off volatile compounds. If there are any air bubbles present the sample must be retaken. The vials with solids should be tapped slightly as they are filled to try and eliminate as much free air space as possible. Two vials from each sampling location should be sealed in separate plastic bags to prevent cross-contamination between samples.

Semivolatile organics — Containers used to collect samples for the determination of semivolatile organic compounds should be soap and water washed followed by methanol (or isopropanol) rinsing. The sample containers should be of glass or Teflon® and have screw-top covers with Teflon® liners.

Preservation for volatile organics — No preservation is used with concentrated waste samples. With liquid samples containing no

residual chlorine, 4 drops of concentrated hydrochloric acid are added and the samples are immediately cooled to 4°C. When liquid samples contain residual chlorine, they are treated as above and, in addition, 4 drops of 4% aqueous sodium thiosulfate are added to remove the residual chlorine. Soil, sediment, and sludge samples are only cooled to 4°C.

Preservation for semivolatile organics — No preservation is used with concentrated waste samples. With liquid samples containing no residual chlorine and with soil, sediment, and sludge samples, immediately cooling to 4°C is the only preservation used. When residual chlorine is present then 3 mL of 10% aqueous sodium sulfate is added for each gallon of sample collected, followed by cooling to 4°C.

Holding times — The holding time for all volatile organics samples is 14 days. Liquid samples must be extracted within 7 days and their extracts analyzed within 40 days. Concentrated waste, soil, sediment, and sludge samples must be extracted within 14 days and their extracts analyzed within 40 days.

SAMPLE PREPARATION Prepare stock standard solutions in hexane. Calibration standards at a minimum of five concentrations should be prepared through dilution of the stock standards with hexane. The suggested internal standards are: 2,5-dibromotoluene, 1,3,5-tribromobenzene, and α, α-dibromo-m-xylene. The analyst can use any of the three compounds provided that they are resolved from matrix interferences. Recommended surrogate compounds are α-2,6-trichlorotoluene, 1,4-dichloronaphthalene, and 2,3,4,5,6-pentachlorotoluene.

In general, water samples are extracted at a neutral pH with methylene chloride using a separatory funnel (EPA Method 3510) or a continuous liquid-liquid extractor (EPA Method 3520). Solid samples are extracted with hexane/acetone (1:1 v:v) using a Soxhlet extractor (EPA Method 3540) or with methylene chloride/acetone (1:1 v:v) using an ultrasonic extractor (EPA Method 3550). Non-aqueous waste samples may be diluted using EPA Method 3580. Prior to Florisil cleanup or gas chromatographic analysis, the extraction solvent must be exchanged to hexane. Sample extracts that will be subjected to gel permeation chromatography do not need solvent exchange.

Cleanup procedures may not be necessary for a relatively clean matrix. If removal of interferences such as chlorinated phenols, phthalate esters, etc., is required, proceed with the procedure outlined in EPA Method 3620.

QUALITY CONTROL Analyze a quality control check standard to demonstrate that the operation of the GC is in control. The frequency of the check standard analysis is equivalent to 10% of the samples analyzed. If the recovery of any compound found in the check standard is less than 80% of the certified value, the problem must be corrected and a new set of calibration standards must be prepared and analyzed. Calculate surrogate standard recoveries for all samples, blanks, and spikes. An internal standard peak area check must be performed on all samples. The internal standard must be evaluated for acceptance by determining whether the measured area for the internal standard deviates by more than 30% from the average area for the internal standard in the calibration standards. When the internal standard peak area is outside that limit, all samples that fall outside the QC criteria must be reanalyzed. Any compound confirmed by two columns may also be confirmed by GC/MS (EPA Method 8270). The GC/MS would normally require a minimum concentration of 1 ng/μL in the final extract for each compound. Include a mid-concentration calibration standard after each group of 20 samples in the analysis sequence. The response factors for the mid-concentration calibration must be within 15% of the average values for the multiconcentration calibration.

REFERENCE Test Methods for Evaluating Solid Waste, Physical/Chemical Methods, SW-846, 3rd Edition, U.S. EPA, Office of Solid Waste, Washington, DC, 1990. EPA Method 8121, Rev. 0, Nov. 1990.

1,2,4,5-Tetrachlorobenzene   EPA Method 8121
CAS #95-94-2

TITLE Chlorinated Hydrocarbons by GC: Capillary Column Technique

MATRIX This method covers aqueous and solid matrices. This includes a wide variety such as drinking water, groundwater, industrial wastewaters, surface waters, soils, solids, and sediments.

METHOD SUMMARY This method provides procedures for the determination of 22 chlorinated hydrocarbons in water, soil/sediment, and waste matrices. A measured volume or weight of sample is extracted by using one of the appropriate sample extraction techniques specified in EPA Method 3510, EPA Method 3520, EPA Method 3540, or EPA Method 3550, or diluted using EPA Method 3580. Aqueous samples are extracted at neutral pH with methylene chloride by using either a separatory funnel (EPA Method 3510) or a continuous liquid-liquid extractor (EPA Method 3520). Solid samples are extracted with hexane/acetone (1:1) by using a Soxhlet extractor (EPA Method 3540) or with methylene chloride/acetone (1:1) by using an ultrasonic extractor (EPA Method 3550). After cleanup, the extract or diluted sample is analyzed by gas chromatography with electron capture detection (GC/ECD).

The sensitivity level of this method usually depends on the level of interferences rather than on instrumental limitations. This method may be used in conjunction with EPA Method 3620, Florisil Column Cleanup, EPA Method 3660, Sulfur Cleanup, and EPA Method 3640, Gel Permeation Chromatography, to aid in the elimination of interferences.

INTERFERENCES Solvents, reagents, glassware, and other hardware used in sample processing may introduce artifacts which may result in elevated baselines, causing misinterpretation of gas chromatograms. Interferants coextracted from the samples will vary considerably from waste to waste. Glassware must be scrupulously clean. Phthalate esters, if present in a sample, will interfere only with the BHC isomers. The presence of elemental sulfur will result in large peaks, and can often mask the region of compounds eluting after 1,2,4,5-tetrachlorobenzene. The tetrabutylammonium (TBA)-sulfite procedure

(EPA Method 3660) works well for the removal of elemental sulfur. Waxes and lipids can be removed by gel permeation chromatography (EPA Method 3640).

INSTRUMENTATION   A GC suitable for on-column injections and all required accessories, including and electron capture detector (ECD), analytical columns, recorder, gases, and syringes are needed. A data system for measuring peak heights and/or peak areas is recommended. A Kuderna-Danish (K-D) apparatus will also be needed to concentrate extracts.

Column 1: 30 m × 0.53 mm I.D. fused-silica capillary column chemically bonded with trifluoropropyl methyl silicone (DB-210 or equivalent).

Column 2: 30 m × 0.53 mm I.D. fused-silica capillary column chemically bonded with polyethylene glycol (DB-WAX or equivalent).

PRECISION & ACCURACY   This method has been tested in a single lab by using organic-free reagent water, sandy loam samples, and extracts which were spiked with the test compounds at one concentration. Single-operator precision and method accuracy were found to be related to the concentration of compound and the type of matrix. The accuracy and precision technique will be determined by the sample matrix, sample preparation technique, optional cleanup techniques, and calibration procedures used.

MULTIPLICATION FACTORS FOR OTHER MATRICES (a)

Matrix	Factor (b)
Groundwater	10
Low-concentration soil by ultrasonic extraction with GPC cleanup	670
High-concentration soil and sludges by ultrasonic extraction	10,000
Waste not miscible with water	100,000

(a) Sample EQLs are highly matrix-dependent. The EQLs listed herein are provided for guidance and may not always be achievable. (b) EQL = [Method detection limit] × [Factor]. For nonaqueous samples, the factor is on a wet-weight basis.

PRECISION & ACCURACY   MDL is the method detection limit for organic-free reagent water. MDL was determined from the analysis of eight replicate aliquots processed through the entire analytical method (extraction, Florisil cartridge cleanup, and GC/ECD analysis).

The MDL (in ng/L) was 9.5.

The accuracy (as average % recovery using 5 determinations and no Florisil cleanup) from a spike concentration of 10 µg/L and separatory funnel extraction was 93% with a final volume of 10 mL.

The precision (as RSD% using 5 determinations and no Florisil cleanup) from a spike concentration of 10 µg/L and separatory funnel extraction was 4.6% with a final volume of 10 mL.

The accuracy (as average % recovery using 5 determinations and no Florisil cleanup), from a spike concentration of 3300µg/L and ultrasonic extraction of solid samples using 1:1 methylene chloride and acetone, was 80% with a final volume of 10 mL.

The precision (as RSD% using 5 determinations and no Florisil cleanup), from a spike concentration of 3300µg/L and ultrasonic extraction of solid samples using 1:1 methylene chloride and acetone, was 4.4% with a final volume of 10 mL.

SAMPLE COLLECTION, PRESERVATION & HANDLING
Volatile organics — Standard 40-mL glass screw-cap VOA vials with Teflon®-faced silicone septum may be used for both liquid and solid matrices. When collecting samples, liquids and solids should be introduced into the vials gently to reduce agitation which might drive off volatile compounds. If there are any air bubbles present the sample must be retaken. The vials with solids should be tapped slightly as they are filled to try and eliminate as much free air space as possible. Two vials from each sampling location should be sealed in separate plastic bags to prevent cross-contamination between samples.

Semivolatile organics — Containers used to collect samples for the determination of semivolatile organic compounds should be soap and water washed followed by methanol (or isopropanol) rinsing. The sample containers should be of glass or Teflon® and have screw-top covers with Teflon® liners.

Preservation for volatile organics — No preservation is used with concentrated waste samples. With liquid samples containing no residual chlorine, 4 drops of concentrated hydrochloric acid are added and the samples are immediately cooled to 4°C. When liquid samples contain residual chlorine, they are treated as above and, in addition, 4 drops of 4% aqueous sodium thiosulfate are added to remove the residual chlorine. Soil, sediment, and sludge samples are only cooled to 4°C.

Preservation for semivolatile organics — No preservation is used with concentrated waste samples. With liquid samples containing no residual chlorine and with soil, sediment, and sludge samples, immediately cooling to 4°C is the only preservation used. When residual chlorine is present then 3 mL of 10% aqueous sodium sulfate is added for each gallon of sample collected, followed by cooling to 4°C.

Holding times — The holding time for all volatile organics samples is 14 days. Liquid samples must be extracted within 7 days and their extracts analyzed within 40 days. Concentrated waste, soil, sediment, and sludge samples must be extracted within 14 days and their extracts analyzed within 40 days.

SAMPLE PREPARATION   Prepare stock standard solutions in hexane. Calibration standards at a minimum of five concentrations should be prepared through dilution of the stock standards with hexane. The suggested internal standards are: 2,5-dibromotoluene, 1,3,5-tribromobenzene, and α, α-dibromo-m-xylene. The analyst can use any of the three compounds provided that they are resolved from matrix interferences. Recommended surrogate compounds are α-2,6-trichlorotoluene, 1,4-dichloronaphthalene, and 2,3,4,5,6-pentachlorotoluene.

In general, water samples are extracted at a neutral pH with methylene chloride using a separatory funnel (EPA Method 3510) or a continuous liquid-liquid extractor (EPA Method 3520). Solid samples are extracted with hexane/acetone (1:1

v:v) using a Soxhlet extractor (EPA Method 3540) or with methylene chloride/acetone (1:1 v:v) using an ultrasonic extractor (EPA Method 3550). Non-aqueous waste samples may be diluted using EPA Method 3580. Prior to Florisil cleanup or gas chromatographic analysis, the extraction solvent must be exchanged to hexane. Sample extracts that will be subjected to gel permeation chromatography do not need solvent exchange.

Cleanup procedures may not be necessary for a relatively clean matrix. If removal of interferences such as chlorinated phenols, phthalate esters, etc., is required, proceed with the procedure outlined in EPA Method 3620.

QUALITY CONTROL Analyze a quality control check standard to demonstrate that the operation of the GC is in control. The frequency of the check standard analysis is equivalent to 10% of the samples analyzed. If the recovery of any compound found in the check standard is less than 80% of the certified value, the problem must be corrected and a new set of calibration standards must be prepared and analyzed. Calculate surrogate standard recoveries for all samples, blanks, and spikes. An internal standard peak area check must be performed on all samples. The internal standard must be evaluated for acceptance by determining whether the measured area for the internal standard deviates by more than 30% from the average area for the internal standard in the calibration standards. When the internal standard peak area is outside that limit, all samples that fall outside the QC criteria must be reanalyzed. Any compound confirmed by two columns may also be confirmed by GC/MS (EPA Method 8270). The GC/MS would normally require a minimum concentration of 1 ng/µL in the final extract for each compound. Include a mid-concentration calibration standard after each group of 20 samples in the analysis sequence. The response factors for the mid-concentration calibration must be within 15% of the average values for the multiconcentration calibration.

REFERENCE Test Methods for Evaluating Solid Waste, Physical/Chemical Methods, SW-846, 3rd Edition, U.S. EPA, Office of Solid Waste, Washington, DC, 1990. EPA Method 8121, Rev. 0, Nov. 1990.

| 1,2,4,5-Tetrachlorobenzene | EPA Method 8270 |
| CAS #95-94-3 | |

TITLE Semivolatile Organic Compounds by GC/MS

MATRIX This method is used to determine the concentration of semivolatile organic compounds in extracts prepared from all types of solid waste matrices, soils, and groundwater. Although surface waters are not specifically mentioned, this method should be applicable to water samples from rivers, lakes, etc.

METHOD SUMMARY This method covers 259 semivolatile organic compounds. In very limited applications direct injection of the sample into the GC/MS system may be appropriate, but this results in very high detection limits (approximately 10,000 µg/L). Typically, a 1-L liquid sample, containing surrogate, and matrix spiking standards, is extracted in a continuous extractor first under acid conditions and then under basic conditions. Typically 30 g of a solid sample, containing surrogate, and matrix spiking standards, is extracted ultrasonically. After concentrating the extract to 1 mL it is spiked with 10 µL of an internal standard solution just prior to analysis by GC/MS. The volume injected should contain about 100 ng of base/neutral and 200 ng of acid surrogates (for a 1-µL injection). Analysis is performed by GC/MS using a capillary GC column.

INTERFERENCES Raw GC/MS data from all blanks, samples, and spikes must be evaluated for interferences. Contamination by carryover can occur whenever high-concentration and low-concentration samples are sequentially analyzed. To reduce carryover, the sample syringe must be rinsed out between samples with solvent. Whenever an unusually concentrated sample is encountered, it should be followed by the analysis of blank solvent to check for cross-contamination.

INSTRUMENTATION A GC/MS and a data system are required. The GC column used is a 30 m × 0.25 mm I.D. (or 0.32 mm I.D.) 1um film thickness silicone-coated fused silica capillary column. A continuous liquid-liquid extractor equipped with Teflon® or glass connection joints and stopcocks requiring no lubrication, a K-D concentrating apparatus, water bath, and an ultrasonic disrupter with a minimum power of 300 W and with pulsing capability are also required.

PRECISION & ACCURACY The estimated quantitation limit (EQL) of Method 8270B for determining an individual compound is approximately 1 mg/kg (wet weight) for soil or sediment samples, 1–200 mg/kg for wastes (dependent on matrix and method of preparation), and 10 µg/L for groundwater samples. EQLs will be proportionately higher for sample extracts that require dilution to avoid saturation of the detector.

The EQL(b) for groundwater in µg/L is 10.
The EQL (a, b) for low concentrations in soil and sediment in µg/kg is not determined.
Accuracy as µg/L is not listed.
Overall precision in µg/L is not listed.

(a) *EQLs listed for soil/sediment are based on wet weight. Normally data is reported in a dry-weight basis; therefore, EQLs will be higher based on the % dry weight of each sample. This calculation is based on a 30 g sample and gel permeation chromatography cleanup.*
(b) *Sample EQLs are highly matrix-dependent. The EQLs are provided for guidance and may not always be achievable.*
C = *True value for concentration, in µg/L.*
X = *Average recovery found for measurements of samples containing a concentration of C, in µg/L.*

ESTIMATED QUANTITATION LIMIT

Other Matrices	Factor (a)
High-concentration soil and sludges by sonicator	7.5
Non-water miscible waste	75

(a) *EQL for other matrices = [EQL for low soil/sediment × [Factor]. This estimated EQL is similar to an EPA "Practical Quantitation Limit."*

SAMPLING METHOD

Liquid samples — Use a 1 or 2½ gallon amber glass bottle with a screw-top Teflon®-lined cover that has been prewashed with detergent and rinsed with distilled water and methanol (or isopropanol).

Soils, sediments, or sludges — Use an 8-oz. widemouth glass with a screw-top Teflon®-lined cover that has been prewashed with detergent and rinsed with distilled water and methanol (or isopropanol).

SAMPLE PRESERVATION

Liquid samples — If residual chlorine is present, add 3 mL of 10% sodium thiosulfate per gallon, cool to 4°C and store in a solvent-free refrigerator until analysis; if chlorine is not present, then eliminate the sodium thiosulfate addition.

Soils, sediments, or sludges — Cool samples to 4°C and store in a solvent-free refrigerator.

MHT Liquid samples must be extracted within 7 days and the extracts analyzed within 40 days. Soils, sediments, or sludges may be stored for a maximum of 14 days and the extracts analyzed within 40 days.

SAMPLE PREPARATION

Liquid samples — Transfer 1 L quantitatively to a continuous extractor. If high concentrations are anticipated, a smaller volume may be used and then diluted with organic-free reagent water to 1 L. Adjust pH, if necessary, to pH <2 using 1:1 (V/V) sulfuric acid. Pipette 1.0 mL of a surrogate standard spiking solution into each sample. For the sample in each analytical batch selected for spiking, add 1.0 mL of a matrix spiking standard. For base/neutral acid analysis, the amount of the surrogates and matrix spiking compounds added to the sample should result in a final concentration of 100 ng/µL of each analyte in the extract to be analyzed (assuming a 1-µL injection). Extract with methylene chloride for 18–24 h. Next, adjust the pH of the aqueous phase to pH >11 using 10 N sodium hydroxide and extract it with methylene chloride again for 18–24 h. Dry the extract through a column containing anhydrous sodium sulfate and concentrate it to 1 mL using a K-D concentrator.

Soils, sediments, or sludges — Use 30 g of sample. Nonporous or wet samples (gummy or clay type) that do not have a free-flowing sandy texture must be mixed with anhydrous sodium sulfate until the sample is free flowing. Add 1 mL of surrogate standards to all samples, spikes, standards, and blanks. For the sample in each analytical batch selected for spiking, add 1.0 mL of a matrix spiking standard. For base/neutral acid analysis, the amount added of the surrogates and matrix spiking compounds should result in a final concentration of 100 ng/µL of each base/neutral analyte and 200 ng/µL of each acid analyte in the extract to be analyzed (assuming a 1-µL injection). Immediately add a 100-mL mixture of 1:1 methylene chloride:acetone and extract the sample ultrasonically for 3 min and then decant or filter the extracts. Repeat the extraction two or more times. Dry the extract using a column with anhydrous sodium sulfate and concentrate it to 1 mL in a K-D concentrator.

QUALITY CONTROL A methylene chloride solution containing 50 ng/µL of decafluorotriphenylphosphine (DFTPP) is used for tuning the GC/MS system each 12-h shift. A system performance check also must be made during every 12-h shift. A standard containing 50 ng/µL each of 4,4′-DDT, pentachlorophenol, and benzidine is required to verify injection port inertness and GC column performance. A calibration standard at mid-concentration, containing each compound of interest, including all required surrogates, must be performed every 12 h during analysis. After the system performance check is met, calibration check compounds (CCCs) are used to check the validity of the initial calibration.

The internal standard responses and retention times in the calibration check standard must be evaluated immediately after or during data acquisition. If the retention time for any internal standard changes by more than 30 seconds from the last check calibration (12 h), the chromatographic system must be inspected for malfunctions and corrections must be made, as required. If the electron ionization current plot (EICP) area for any of the internal standards changes by a factor of two from the last daily calibration standard check, the mass spectrometer must be inspected for malfunctions and corrections must be made, as appropriate.

Demonstrate, through the analysis of a reagent water blank, that interferences from the analytical system, glassware, and reagents are under control. The blank samples should be carried through all stages of the sample preparation and measurement steps. For each analytical batch (up to 20 samples), a reagent blank, matrix spike, and matrix spike duplicate/duplicate must be analyzed (the frequency of the spikes may be different for different monitoring programs). The blank and spiked samples must be carried through all stages of the sample preparation and measurement steps. A QC reference sample concentrate containing each analyte at a concentration of 100 mg/L in methanol is required.

REFERENCE Test Methods for Evaluating Solid Waste (SW-846). U.S. EPA 1983, Method 8270B, Rev. 2, Nov. 1990. Office of Solid Waste, Washington, DC.

Tetrachlorobenzenes (3 isomers)   **EPA Method 8120**
CAS #Various

TITLE Chlorinated Hydrocarbons by Gas Chromatography

MATRIX This method covers aqueous and solid matrices. This includes a wide variety such as drinking water, groundwater, industrial wastewaters, surface waters, soils, solids, and sediments.

METHOD SUMMARY This method is used to determine the concentration of 14 chlorinated hydrocarbons. It provides gas chromatographic conditions for the detection of ppb concentrations of certain chlorinated hydrocarbons. Prior to use of this method, appropriate sample extraction techniques must be used. Both neat and diluted organic liquids (EPA Method 3580, Waste Dilution) may be analyzed by direct injection. A 2 to 5 µg/mL aliquot of the extract is injected into a gas chromatograph (GC) using the solvent flush technique, and compounds in the GC effluent are detected by an electron capture detector (ECD).

INTERFERENCES Solvents, reagents, glassware, and other sample processing hardware may yield discrete artifacts and/or elevated baselines causing misinterpretation of gas chromatograms. Interferences coextracted from samples will vary considerably from source to source, depending upon the waste being sampled.

INSTRUMENTATION An analytical system complete with GC suitable for on-column injections and accessories, including detectors, column supplies, recorder, gases and syringes is required. A data system for measuring peak areas and/or peak heights is recommended. The GC is equipped with an electron capture detector (ECD). A K-D apparatus is needed for sample preparation.

Column 1: 1.8 m × 2 mm I.D. glass column packed with 1% SP-1000 on Supelcoport (100/120 mesh) or equivalent.

Column 2: 1.8 m × 2 mm I.D. glass column packed with 1.5% OV-1/2.4% OV-225 on Supelcoport (80/100 mesh) or equivalent.

PRECISION & ACCURACY The method was tested by 20 laboratories using organic-free reagent water, drinking water, surface water, and three industrial wastewaters spiked at six concentrations over the range 1.0 to 356 µg/L. Single operator precision, overall precision, and method accuracy were found to be directly related to the concentration of the parameter and essentially independent of the sample matrix.

MULTIPLICATION FACTORS FOR OTHER MATRICES (a)

Matrix	Factor (b)
Groundwater	10
Low-concentration soil by ultrasonic extraction with GPC cleanup	670
High-concentration soil and sludges by ultrasonic extraction	10,000
Waste not miscible with water	100,000

(a) Sample EQLs are highly matrix-dependent. The EQLs listed herein are provided for guidance and may not always be achievable.
(b) EQL = [Method detection limit] × [Factor]. For nonaqueous samples, the factor is on a wet-weight basis.

PRECISION & ACCURACY The estimates below are based upon the performance in a single lab.

The accuracy (in µg/L) as expected recovery for one or more measurements of a sample containing a concentration of C was No Data.

The precision (in µg/L) as expected single analyst standard deviation of measurements at an average concentration of x" was No Data.

The precision (in µg/L) as expected interlaboratory standard deviation measurements at an average concentration found of x" was No Data.

C = True value for the concentration, in µg/L.
$x"$ = Average recovery found for measurements of samples containing a concentration of C, in µg/L.

SAMPLE COLLECTION, PRESERVATION & HANDLING Extracts must be stored under refrigeration at 4°C and analyzed within 40 days of extraction.

SAMPLE PREPARATION In general, water samples are extracted at a neutral, or as is, pH with methylene chloride using either EPA Method 3510 or EPA Method 3520. Solid samples are extracted using either EPA Method 3540 or EPA Method 3550. Prior to gas chromatographic analysis, the extraction solvent must be exchanged to hexane.

QUALITY CONTROL The quality control check concentrate (EPA Method 8000) should contain each parameter of interest in acetone at the following concentrations: hexachloro-substituted hydrocarbon, 10 µg/mL; and any other chlorinated hydrocarbon, 100 µg/mL. Calculate surrogate standard recovery on all samples, blanks, and spikes.

Prepare stock standard solutions in isooctane or hexane. Calibration standards at a minimum of five concentrations should be prepared through dilution of the stock standards with isooctane or hexane. Internal standards and surrogate standards are also needed.

REFERENCE Test Methods for Evaluating Solid Waste, Physical/Chemical Methods, SW-846, 3rd Edition, U.S. EPA, Office of Solid Waste, Washington, DC, 1990. EPA Method 8120 A Rev. 1, Nov. 1990.

1,1,1,2-Tetrachloroethane   **EPA Method 1624**
CAS #630-20-6

TITLE Volatile Organic Compounds by Isotope Dilution GC/MS

MATRIX Compounds may be determined in waters, soils, and municipal sludges by this method.

METHOD SUMMARY This method is used to determine 58 volatile toxic organic pollutants associated with the CWA (as amended 1987); the RCRA (as amended 1986); the CERCLA (as amended 1986); and other compounds amenable to purge-and-trap gas chromatography-mass spectrometry (GC/MS).

If the solids content is less than 1%, stable isotopically-labeled analogs of the compounds of interest are added to a 5-mL sample and the sample is purged with an inert gas at 20–25°C in a chamber designed for soil or water samples. If the solids content is greater than 1%, 5 mL of reagent water and the labeled compounds are added to a 5-g aliquot of sample and the mixture is purged at 40°C. Compounds that will not purge at 20–25°C or at 40°C are purged at 78–85°C. In the purging process, the volatile compounds are transferred from the aqueous phase into the gaseous phase where they are passed into a sorbent column, and trapped. After purging is completed, the trap is backflushed and heated rapidly to desorb the compounds into a GC. The compounds are separated by the GC and detected by a MS. The labeled compounds serve to correct the variability of the analytical technique.

INTERFERENCES Impurities in the purge gas, organic compounds outgassing from the plumbing upstream of the trap, and solvent vapors in the lab account for most problems. Samples can be contaminated by diffusion of volatile organic compounds (particularly methylene chloride) through the bottle seal during shipment and storage. Contamination by carryover can occur when high-level and low-level samples are analyzed sequentially. When an unusually concentrated sample is encountered, follow it by analysis of a reagent water blank to check for carryover.

INSTRUMENTATION Major equipment includes a GC with linear temperature programming and a glass jet separator as the MS interface, a MS with 70 eV electron impact ionization, and a data system to collect and record response factors.

Column: 2.8 m × 2 mm I.D. glass, packed with 1% SP-1000 on Carbopak B, 60/80 mesh, or equivalent.

PRECISION & ACCURACY The detection limits of the method are usually dependent on the level of interferences rather than instrumental limitations. The method detection limits were determined in digested sludge (low solids) and in filter cake or compost (high solids).

The MDL (in µg/kg) for low solids is not listed and for high solids is not listed.

Labeled and native compound precision (in µg/L) as standard deviation was not listed.

Labeled and native compound accuracy (in µg/L) as average recovery was not listed.

Acceptance criteria are at 20 µg/L for this compound.

SAMPLE COLLECTION, PRESERVATION & HANDLING Grab samples are collected in glass containers having a total volume greater than 20 mL. Fill and seal each bottle so that no air bubbles are entrapped. Samples are maintained at 0 to 4°C from the time of collection until analysis. If an aqueous sample contains residual chlorine, add sodium thiosulfate preservative (10 mg/40 mL) to the empty sample bottles just prior to shipment to the sample site. All samples must be analyzed within 14 days of collection.

SAMPLE PREPARATION Samples containing less than 1% solids are analyzed directly as aqueous samples. Samples containing 1% solids or greater are analyzed as solid samples utilizing one of two methods, depending on the levels of pollutants, in the sample. Samples containing 1% solids or greater, and low to moderate levels of pollutants are analyzed by purging a known weight of sample added to 5 mL of reagent water. Samples containing 1% solids or greater, and high levels of pollutants, are extracted with methanol, and an aliquot of the methanol extract is added to reagent water and purged.

QUALITY CONTROL A field blank prepared from reagent water and carried through the sampling and handling protocol may serve as a check on contamination from shipment and storage.

The analyst is permitted to modify this method to improve separations or lower the costs of measurements, provided all performance specifications are met. Analyses of blanks are required. When results of spikes indicate atypical method performance for samples, the samples are diluted to bring method performance within acceptable limits. Analyze two sets of four 5-mL aliquots (8 aliquots total) of the aqueous performance standard. Spike all samples with labeled compounds to assess method performance on the sample matrix. Compute the percent recovery of the labeled compounds using the internal standard method. Compare the percent recovery for each compound with the corresponding labeled compound recovery. Reagent water blanks are analyzed to demonstrate freedom from carryover contamination. Field replicates may be collected to determine the precision of the sampling technique, and spiked samples may be required to determine the accuracy of the analysis when the internal method is used.

REFERENCE Volatile Organic Compounds by Isotope Dilution GC/MS. Office of Water Regulation and Standards, U.S. EPA Industrial Technology Division, Washington, DC, EPA Method 1624, Rev. C, June 1989 (contact W.A. Telliard, U.S. EPA, Office of Water Regulations and Standards, 401 M St., SW, Washington, DC, 20460. Phone: 202-382-7131).

1,1,2,2-Tetrachloroethane   **EPA Method 1624**
CAS #79-34-5

TITLE Volatile Organic Compounds by Isotope Dilution GC/MS

MATRIX Compounds may be determined in waters, soils, and municipal sludges by this method.

METHOD SUMMARY This method is used to determine 58 volatile toxic organic pollutants associated with the CWA (as amended 1987); the RCRA (as amended 1986); the CERCLA (as amended 1986); and other compounds amenable to purge-and-trap gas chromatography-mass spectrometry (GC/MS).

If the solids content is less than 1%, stable isotopically-labeled analogs of the compounds of interest are added to a 5-mL sample and the sample is purged with an inert gas at 20–25°C in a chamber designed for soil or water samples. If the solids content is greater than 1%, 5 mL of reagent water and the labeled compounds are added to a 5-g aliquot of sample and the mixture is purged at 40°C. Compounds that will not purge at 20–25°C or at 40°C are purged at 78–85°C. In the purging process, the volatile compounds are transferred from the aqueous phase into the gaseous phase where they are passed into a sorbent column, and trapped. After purging is completed, the trap is backflushed and heated rapidly to desorb the compounds into a GC. The compounds are separated by the GC and detected by a MS. The labeled compounds serve to correct the variability of the analytical technique.

INTERFERENCES Impurities in the purge gas, organic compounds outgassing from the plumbing upstream of the trap, and solvent vapors in the lab account for most problems. Samples can be contaminated by diffusion of volatile organic compounds (particularly methylene chloride) through the bottle seal during shipment and storage. Contamination by carryover can occur when high-level and low-level samples are analyzed sequentially. When an unusually concentrated sample is encountered, follow it by analysis of a reagent water blank to check for carryover.

INSTRUMENTATION Major equipment includes a GC with linear temperature programming and a glass jet separator as the MS interface, a MS with 70 eV electron impact ionization, and a data system to collect and record response factors.

Column: 2.8 m × 2 mm I.D. glass, packed with 1% SP-1000 on Carbopak B, 60/80 mesh, or equivalent.

PRECISION & ACCURACY The detection limits of the method are usually dependent on the level of interferences rather than instrumental limitations. The method detection limits were determined in digested sludge (low solids) and in filter cake or compost (high solids).

The MDL (in µg/kg) for low solids is 20 and for high solids is 6.
Labeled and native compound precision (in µg/L) as standard deviation was 9.6.
Labeled and native compound accuracy (in µg/L) as average recovery was 11–30.

Acceptance criteria are at 20 µg/L for this compound.

SAMPLE COLLECTION, PRESERVATION & HANDLING Grab samples are collected in glass containers having a total volume greater than 20 mL. Fill and seal each bottle so that no air bubbles are entrapped. Samples are maintained at 0 to 4°C from the time of collection until analysis. If an aqueous sample contains residual chlorine, add sodium thiosulfate preservative (10 mg/40 mL) to the empty sample bottles just prior to shipment to the sample site. All samples must be analyzed within 14 days of collection.

SAMPLE PREPARATION Samples containing less than 1% solids are analyzed directly as aqueous samples. Samples containing 1% solids or greater are analyzed as solid samples utilizing one of two methods, depending on the levels of pollutants, in the sample. Samples containing 1% solids or greater, and low to moderate levels of pollutants are analyzed by purging a known weight of sample added to 5 mL of reagent water. Samples containing 1% solids or greater, and high levels of pollutants, are extracted with methanol, and an aliquot of the methanol extract is added to reagent water and purged.

QUALITY CONTROL A field blank prepared from reagent water and carried through the sampling and handling protocol may serve as a check on contamination from shipment and storage.

The analyst is permitted to modify this method to improve separations or lower the costs of measurements, provided all performance specifications are met. Analyses of blanks are required. When results of spikes indicate atypical method performance for samples, the samples are diluted to bring method performance within acceptable limits. Analyze two sets of four 5-mL aliquots (8 aliquots total) of the aqueous performance standard. Spike all samples with labeled compounds to assess method performance on the sample matrix. Compute the percent recovery of the labeled compounds using the internal standard method. Compare the percent recovery for each compound with the corresponding labeled compound recovery. Reagent water blanks are analyzed to demonstrate freedom from carryover contamination. Field replicates may be collected to determine the precision of the sampling technique, and spiked samples may be required to determine the accuracy of the analysis when the internal method is used.

REFERENCE Volatile Organic Compounds by Isotope Dilution GC/MS. Office of Water Regulation and Standards, U.S. EPA Industrial Technology Division, Washington, DC, EPA Method 1624, Rev. C, June 1989 (contact W.A. Telliard, U.S. EPA, Office of Water Regulations and Standards, 401 M St., SW, Washington, DC, 20460. Phone: 202-382-7131).

1,1,1,2-Tetrachloroethane   EPA Method 502
CAS #630-20-6

TITLE Volatile Organic Compounds in Water By Purge and Trap Capillary Column Gas Chromatography with Photoionization and Electrolytic Conductivity Detectors in Series. U.S. EPA Method 502.2, Rev. 2.0, 1989.

MATRIX Drinking water and raw source water. The latter should include most surface water and groundwater sources.

METHOD SUMMARY This method covers 60 volatile organic compounds that contain halogen atoms and/or that are aromatic. An inert gas (zero grade nitrogen or helium) is bubbled through a 25-mL or a 5-mL water sample (depending on the expected concentration of the analytes). Purged sample components are trapped in a tube of sorbent materials. When purging is complete, the sorbent tube is heated and backflushed with helium to desorb the trapped sample onto a capillary GC column. The column is temperature programmed to separate the method analytes which are then detected with a photoionization detector (PID) and a Hall electrolytic conductivity (HECD) placed in series. The PID is selective for aromatic compounds and the HECD is selective for halogenated compounds.

INTERFERENCES Impurities in the purge gas and from organic compounds outgassing from the plumbing ahead of the trap account for many contamination problems. Interferences purged or coextracted from the samples will vary considerably from source to source, depending upon the particular sample or extract being tested. Cross-contamination can occur whenever high-level and low-level samples are analyzed sequentially. Samples also can be contaminated by diffusion of volatile organics (particularly methylene chloride and fluorocarbons) through the septum seal into the sample during shipment and storage. The lab where volatile analysis is performed and also the refrigerated storage area should be completely free of solvents.

INSTRUMENTATION A GC containing a series configuration of a high temperature photoionization detector (PID) equipped with 10.0 eV (nominal) lamp and Hall electrolytic conductivity detector (HECD) is required. Also required is an all-glass 5-mL purging device, a sorbent trap, and a thermal desorption apparatus which is connected to the GC system.

Column 1: VOCOL glass wide-bore capillary column.
Column 2: RTX–502.2 mega-bore capillary column.
Column 3: DB-62 mega-bore capillary column.

PRECISION & ACCURACY Method detection limits are dependent upon the characteristics of the gas chromatographic system used. Analytes that are not separated chromatographically cannot be individually identified and used in the same calibration mixture or water samples unless an alternative technique for identification and quantification, such as mass spectrometry, is used.

Electrolytic conductivity detetor (c) range in µg/L (a) was 0.02–200.
Electrolytic conductivity detetor (c) MDL in µg/L (b) was 0.01.
Electrolytic conductivity detetor (c) accuracy as % recovery was 99.
Electrolytic conductivity detetor (c) precision as % RSD was 2.3.
Photoionization detector (d) range in µg/L (a) was 0.02–200.
Photoionization detector (d) MDL in µg/L (b) was not listed.
Photoionization detector (d) accuracy as % recovery was not listed.
Photoionization detector (d) precision as % RSD was not listed.

(a) The applicable concentration range of this method is compound, instrument, and matrix-dependent. It is listed as being approximately 0.02 to 200 µg/L but no specific information is provided so caution should be observed.
(b) The method detection limits reports with this method are compound, instrument, and matrix-dependent. The values reported were calculated using reagent water fortified with the corresponding compounds at 10 µg/L and a GC-equipped with a 60 m × 0.75 mm VOLCOL wide bore capillary column with 1.5 µm film thickness and using helium carrier gas.
(c) Recoveries and relative standard deviations were determined from seven samples of reagent water fortified with 10 µg/L of each compound. 2-Bromo-1-chloropropane was used as the internal standard for calculating average recoveries.
(d) Recoveries and relative standard deviations were determined from seven samples of reagent water fortified with 10 µg/L of each compound. Fluorobenzene was used as the internal standard for calculating average recoveries.

SAMPLING METHOD Collect samples using a 40- to 120-mL screw-cap vial (prewashed with detergent, rinsed with distilled water and oven dried at 105°C) with a Teflon®-faced silicone septum. Collect bubble-free samples and place the septum with the Teflon® side down on the water.

SAMPLE PRESERVATION If residual chlorine is present in the water add about 25 mg of ascorbic acid to each vial before samples are collected to remove the chlorine. Add hydrochloric acid to reduce pH to <2, immediately cool samples to 4°C, and store them in a solvent-free refrigerator at 4°C until analysis.

MHT The maximum holding time for samples is 14 days from the time they were collected.

SAMPLE PREPARATION Remove the plungers from two 5-mL syringes and attach a closed syringe valve to each. Warm the sample to room temperature, open the sample bottle, and carefully pour the sample into one of the syringe barrels to just short of overflowing. Replace the syringe plunger, invert the syringe, and compress the sample. Open the syringe valve and vent any residual air while adjusting the sample volume to 5.0 mL. Add 10 µL of the internal calibration standard to the sample through the syringe valve. Close the valve. Fill the second syringe in an identical manner from the same sample bottle. Reserve this second syringe for a reanalysis if necessary.

QUALITY CONTROL As an initial demonstration of lab accuracy and precision, analyze 4 to 7 replicates of a lab fortified blank containing analyte at 0.1–5 µg/L. Collect all samples in duplicate. Surrogate analytes (similar to those of the analytes of interest), whose concentration is known in every sample, are measured using the same internal standard calibration procedure. Duplicate field reagent water blanks (trip blanks) must be analyzed with each set of samples, lab reagent blanks (method blanks) must be analyzed with each batch of samples processed as a group within a work shift. Also, a single lab-fortified blank that contains each of the analytes of interest should be analyzed with each batch of samples processed as a group within a work shift. A 3- to 5-point calibration curve is needed depending on the calibration range factor required.

EPA CONTACT & HOTLINE For technical questions contact Dr. Baldev Bathija, U.S. EPA, Office of Ground Water and Drinking Water (WH-550D), 401 M St. SW, Washington, DC 20460. Tel. (202) 260-3040. For further information the EPA Safe Drinking Water Hotline may be called at: (800) 426-4791.

REFERENCE Methods for the Determination of Organic Compounds in Drinking Water, EPA/600/4-88/039 (revised July 1991; Final Rule for determination of compliance with the MCL for Total Trihalomethanes under 141.30, in 40 CFR Part 141, Vol. 58, No. 147, Fed. Reg., Tuesday Aug. 3, 1993). U.S. EPA Environmental Monitoring Systems Laboratory, Cincinnati, OH, 45268, U.S.A. Available from the National Technical Information Service (NTIS), 5285 Port Royal Road, Springfield, VA 22161; Tel. 800-553-6847. NTIS Order Number is PB91-231480.

1,1,2,2-Tetrachloroethane   **EPA Method 502**
CAS #79-34-5

TITLE Volatile Organic Compounds in Water By Purge and Trap Capillary Column Gas Chromatography with Photoionization and Electrolytic Conductivity Detectors in Series. U.S. EPA Method 502.2, Rev. 2.0, 1989.

MATRIX Drinking water and raw source water. The latter should include most surface water and groundwater sources.

METHOD SUMMARY This method covers 60 volatile organic compounds that contain halogen atoms and/or that are aromatic. An inert gas (zero grade nitrogen or helium) is bubbled through a 25-mL or a 5-mL water sample (depending on the expected concentration of the analytes). Purged sample components are trapped in a tube of sorbent materials. When purging is complete, the sorbent tube is heated and backflushed with helium to desorb the trapped sample onto a capillary GC column. The column is temperature programmed to separate the method analytes which are then detected with a photoionization detector (PID) and a Hall electrolytic conductivity

(HECD) placed in series. The PID is selective for aromatic compounds and the HECD is selective for halogenated compounds.

INTERFERENCES Impurities in the purge gas and from organic compounds outgassing from the plumbing ahead of the trap account for many contamination problems. Interferences purged or coextracted from the samples will vary considerably from source to source, depending upon the particular sample or extract being tested. Cross-contamination can occur whenever high-level and low-level samples are analyzed sequentially. Samples also can be contaminated by diffusion of volatile organics (particularly methylene chloride and fluorocarbons) through the septum seal into the sample during shipment and storage. The lab where volatile analysis is performed and also the refrigerated storage area should be completely free of solvents.

INSTRUMENTATION A GC containing a series configuration of a high temperature photoionization detector (PID) equipped with 10.0 eV (nominal) lamp and Hall electrolytic conductivity detector (HECD) is required. Also required is an all-glass 5-mL purging device, a sorbent trap, and a thermal desorption apparatus which is connected to the GC system.

Column 1: VOCOL glass wide-bore capillary column.
Column 2: RTX–502.2 mega-bore capillary column.
Column 3: DB-62 mega-bore capillary column.

PRECISION & ACCURACY Method detection limits are dependent upon the characteristics of the gas chromatographic system used. Analytes that are not separated chromatographically cannot be individually identified and used in the same calibration mixture or water samples unless an alternative technique for identification and quantification, such as mass spectrometry, is used.

Electrolytic conductivity detetor (c) range in µg/L (a) was 0.02–200.
Electrolytic conductivity detetor (c) MDL in µg/L (b) was 0.01.
Electrolytic conductivity detetor (c) accuracy as % recovery was 99.
Electrolytic conductivity detetor (c) precision as % RSD was 6.8.
Photoionization detector (d) range in µg/L (a) was 0.02–200.
Photoionization detector (d) MDL in µg/L (b) was not listed.
Photoionization detector (d) accuracy as % recovery was not listed.
Photoionization detector (d) precision as % RSD was not listed.

(a) The applicable concentration range of this method is compound, instrument, and matrix-dependent. It is listed as being approximately 0.02 to 200 µg/L but no specific information is provided so caution should be observed.
(b) The method detection limits reports with this method are compound, instrument, and matrix-dependent. The values reported were calculated using reagent water fortified with the corresponding compounds at 10 µg/L and a GC-equipped with a 60 m × 0.75 mm VOLCOL wide bore capillary column with 1.5 µm film thickness and using helium carrier gas.
(c) Recoveries and relative standard deviations were determined from seven samples of reagent water fortified with 10 µg/L of each compound. 2-Bromo-1-chloropropane was used as the internal standard for calculating average recoveries.
(d) Recoveries and relative standard deviations were determined from seven samples of reagent water fortified with 10 µg/L of each compound. Fluorobenzene was used as the internal standard for calculating average recoveries.

SAMPLING METHOD Collect samples using a 40- to 120-mL screw-cap vial (prewashed with detergent, rinsed with distilled water and oven dried at 105°C) with a Teflon®-faced silicone septum. Collect bubble-free samples and place the septum with the Teflon® side down on the water.

SAMPLE PRESERVATION If residual chlorine is present in the water add about 25 mg of ascorbic acid to each vial before samples are collected to remove the chlorine. Add hydrochloric acid to reduce pH to <2, immediately cool samples to 4°C, and store them in a solvent-free refrigerator at 4°C until analysis.

MHT The maximum holding time for samples is 14 days from the time they were collected.

SAMPLE PREPARATION Remove the plungers from two 5-mL syringes and attach a closed syringe valve to each. Warm the sample to room temperature, open the sample bottle, and carefully pour the sample into one of the syringe barrels to just short of overflowing. Replace the syringe plunger, invert the syringe, and compress the sample. Open the syringe valve and vent any residual air while adjusting the sample volume to 5.0 mL. Add 10 µL of the internal calibration standard to the sample through the syringe valve. Close the valve. Fill the second syringe in an identical manner from the same sample bottle. Reserve this second syringe for a reanalysis if necessary.

QUALITY CONTROL As an initial demonstration of lab accuracy and precision, analyze 4 to 7 replicates of a lab fortified blank containing analyte at 0.1–5 µg/L. Collect all samples in duplicate. Surrogate analytes (similar to those of the analytes of interest), whose concentration is known in every sample, are measured using the same internal standard calibration procedure. Duplicate field reagent water blanks (trip blanks) must be analyzed with each set of samples, lab reagent blanks (method blanks) must be analyzed with each batch of samples processed as a group within a work shift. Also, a single lab-fortified blank that contains each of the analytes of interest should be analyzed with each batch of samples processed as a group within a work shift. A 3- to 5-point calibration curve is needed depending on the calibration range factor required.

EPA CONTACT & HOTLINE For technical questions contact Dr. Baldev Bathija, U.S. EPA, Office of Ground Water and Drinking Water (WH-550D), 401 M St. SW, Washington, DC 20460. Tel. (202) 260-3040. For further information the EPA Safe Drinking Water Hotline may be called at: (800) 426-4791.

REFERENCE Methods for the Determination of Organic Compounds in Drinking Water, EPA/600/4-88/039 (revised July 1991; Final Rule for determination of compliance with the MCL for Total Trihalomethanes under 141.30, in 40 CFR Part 141, Vol. 58, No. 147, Fed. Reg., Tuesday Aug. 3, 1993). U.S. EPA Environmental Monitoring Systems Laboratory, Cincinnati, OH, 45268, U.S.A. Available from the National Technical Information Service (NTIS), 5285 Port Royal Road, Springfield,

1,1,1,2-Tetrachloroethane   EPA Method 524
CAS #630-20-6

TITLE Measurement of Purgeable Organic Compounds in Water by Capillary Column GC/MS.

MATRIX Drinking water and raw source water; the latter should include most surface water and groundwater sources.

METHOD SUMMARY Method 524.2 covers 60 volatile organic compounds. An inert gas (zero grade nitrogen or helium) is bubbled through a 25-mL or a 5-mL water sample (depending on the expected concentration of the analytes). Purged sample components are trapped in a tube of sorbent materials. When purging is complete, the sorbent tube is heated and backflushed with helium to desorb the trapped sample onto a capillary GC column.

INTERFERENCES Impurities in the purge gas and from organic compounds outgassing from the plumbing ahead of the trap account for many contamination problems. Interferences purged or coextracted from the samples will vary considerably from source to source, depending upon the particular sample or extract being tested. Cross-contamination can occur whenever high-level and low-level samples are analyzed sequentially. Samples also can be contaminated by diffusion of volatile organics (particularly methylene chloride and fluorocarbons) through the septum seal into the sample during shipment and storage.

INSTRUMENTATION A GC/MS with a data system equipped with one of the following capillary GC columns:

Column 1: VOCOL glass wide bore capillary column.
Column 2: DB-624 fused silica capillary column.
Column 3: DB-5 fused silica capillary column.

Also required is an all-glass 25 mL or 5-mL purging device, a sorbent trap, and a thermal desorption apparatus which is connected to the GC/MS system.

PRECISION & ACCURACY Method detection limits are compound- and instrument-dependent, and may vary from approximately 0.02–0.35 µg/L. Note in the table below that the "true" concentration range used for accuracy and precision measurements was quite narrow. However, the applicable concentration range of this method is primarily column dependent and is approximately 0.02 to 200 µg/L for the wide-bore thick-film columns. Narrow-bore thin-film columns may have a capacity which limits the range to about 0.02 to 20 µg/L. Analytes that are inefficiently purged from water will not be detected when present at low concentrations, but they can be measured with acceptable accuracy and precision when present in sufficient amounts.

Analytes that are not separated chromatographically, but which have different mass spectra and non-interfering quantification ions, can be identified and measured in the same calibration mixture or water sample. Analytes which have very similar mass spectra cannot be individually identified and measured in the same calibration mixture or water samples unless they have different retention times. Co-eluting compounds with very similar mass spectra, typically many structural isomers, must be reported as an isomeric group or pair.

The range (in µg/L) was 0.5–10.
The Method Detection Limig (in µg/L) was 0.05.
The accuracy (as % recovery) was 90.
The precision (in %) was 6.8.

Note: Data were obtained from 16–31 determinations using a wide-bore capillary column and a jet separator interfaced to a quadrupole mass spectrometer. All analytes were in a reagent water matrix.

SAMPLING METHOD Collect samples using a 40- to 120-mL screw-cap vial (prewashed with detergent, rinsed with distilled water and oven dried at 105°C) with a Teflon®-faced silicone septum. Collect bubble-free samples and place the septum with the Teflon® side down on the water.

SAMPLE PRESERVATION If residual chlorine is present in the water add about 25 mg of ascorbic acid to each vial before samples are collected to remove the chlorine. Add hydrochloric acid to reduce pH to <2, and immediately cool samples to 4°C, and store them in a solvent-free refrigerator at 4°C until analysis.

MHT The maximum holding time for samples is 14 days from the time they were collected.

SAMPLE PREPARATION Remove the plungers from two 25-mL (or 5-mL depending on sample size) syringes and attach a closed syringe valve to each. Warm the sample to room temperature, open the sample bottle, and carefully pour the sample into one of the syringe barrels to just short of overflowing. Replace the syringe plunger, invert the syringe, and compress the sample. Open the syringe valve and vent any residual air while adjusting the sample volume to 25.0 mL (or 5 mL). For samples and blanks, add 5 µL of the fortification solution containing the internal standard and the surrogates to the sample through the syringe valve. For calibration standards and lab fortified blanks, add 5 µL of the fortification solution containing the internal standard only. Close the valve. Fill the second syringe in an identical manner from the same sample bottle. Reserve this second syringe for a reanalysis if necessary.

QUALITY CONTROL As an initial demonstration of lab accuracy and precision, analyze 4 to 7 replicates of a lab fortified blank containing analyte at 0.2–5 µg/L. Collect all samples in duplicate. Surrogate analytes (similar to those of the analytes of interest), whose concentration is known in every sample, are measured using the same internal standard calibration procedure. Duplicate field reagent water blanks (trip blanks) must be analyzed with each set of samples, lab reagent blanks (method blanks) must be analyzed with each batch of samples processed as a group within a work shift. Also, a single lab-fortified blank that contains each of the analytes of interest should be analyzed with each batch of samples processed as a group within a work shift. A 3- to 5-point calibration curve is needed depending on the calibration range factor required.

EPA CONTACT & HOTLINE For technical questions contact Dr. Baldev Bathija, U.S. EPA, Office of Ground Water and Drinking Water (WH-550D), 401 M St. SW, Washington, DC 20460. Tel. (202) 260-3040. For further information the EPA Safe Drinking Water Hotline may be called at: (800) 426-4791.

REFERENCE Methods for the Determination of Organic Compounds in Drinking Water, EPA/600/4-88/039 (revised July 1991; Final Rule for determination of compliance with the MCL for Total Trihalomethanes under 141.30, in 40 CFR Part 141, Vol. 58, No. 147, Fed. Reg., Tuesday Aug. 3, 1993). U.S. EPA Environmental Monitoring Systems Laboratory, Cincinnati, OH, 45268, U.S.A. Available from the National Technical Information Service (NTIS), 5285 Port Royal Road, Springfield, VA 22161; Tel. 800-553-6847. NTIS Order Number is PB91-231480.

1,1,2,2-Tetrachloroethane   EPA Method 524
CAS #79-34-5

TITLE Measurement of Purgeable Organic Compounds in Water by Capillary Column GC/MS.

MATRIX Drinking water and raw source water; the latter should include most surface water and groundwater sources.

METHOD SUMMARY Method 524.2 covers 60 volatile organic compounds. An inert gas (zero grade nitrogen or helium) is bubbled through a 25-mL or a 5-mL water sample (depending on the expected concentration of the analytes). Purged sample components are trapped in a tube of sorbent materials. When purging is complete, the sorbent tube is heated and backflushed with helium to desorb the trapped sample onto a capillary GC column.

INTERFERENCES Impurities in the purge gas and from organic compounds outgassing from the plumbing ahead of the trap account for many contamination problems. Interferences purged or coextracted from the samples will vary considerably from source to source, depending upon the particular sample or extract being tested. Cross-contamination can occur whenever high-level and low-level samples are analyzed sequentially. Samples also can be contaminated by diffusion of volatile organics (particularly methylene chloride and fluorocarbons) through the septum seal into the sample during shipment and storage.

INSTRUMENTATION A GC/MS with a data system equipped with one of the following capillary GC columns:

Column 1: VOCOL glass wide bore capillary column.
Column 2: DB-624 fused silica capillary column.
Column 3: DB-5 fused silica capillary column.

Also required is an all-glass 25 mL or 5-mL purging device, a sorbent trap, and a thermal desorption apparatus which is connected to the GC/MS system.

PRECISION & ACCURACY Method detection limits are compound- and instrument-dependent, and may vary from approximately 0.02–0.35 µg/L. Note in the table below that the "true" concentration range used for accuracy and precision measurements was quite narrow. However, the applicable concentration range of this method is primarily column dependent and is approximately 0.02 to 200 µg/L for the wide-bore thick-film columns. Narrow-bore thin-film columns may have a capacity which limits the range to about 0.02 to 20 µg/L. Analytes that are inefficiently purged from water will not be detected when present at low concentrations, but they can be measured with acceptable accuracy and precision when present in sufficient amounts.

Analytes that are not separated chromatographically, but which have different mass spectra and non-interfering quantification ions, can be identified and measured in the same calibration mixture or water sample. Analytes which have very similar mass spectra cannot be individually identified and measured in the same calibration mixture or water samples unless they have different retention times. Co-eluting compounds with very similar mass spectra, typically many structural isomers, must be reported as an isomeric group or pair.

The range (in µg/L) was 0.1–10.
The Method Detection Limig (in µg/L) was 0.04.
The accuracy (as % recovery) was 91.
The precision (in %) was 6.3.

Note: Data were obtained from 16–31 determinations using a wide-bore capillary column and a jet separator interfaced to a quadrupole mass spectrometer. All analytes were in a reagent water matrix.

SAMPLING METHOD Collect samples using a 40- to 120-mL screw-cap vial (prewashed with detergent, rinsed with distilled water and oven dried at 105°C) with a Teflon®-faced silicone septum. Collect bubble-free samples and place the septum with the Teflon® side down on the water.

SAMPLE PRESERVATION If residual chlorine is present in the water add about 25 mg of ascorbic acid to each vial before samples are collected to remove the chlorine. Add hydrochloric acid to reduce pH to <2, and immediately cool samples to 4°C, and store them in a solvent-free refrigerator at 4°C until analysis.

MHT The maximum holding time for samples is 14 days from the time they were collected.

SAMPLE PREPARATION Remove the plungers from two 25-mL (or 5-mL depending on sample size) syringes and attach a closed syringe valve to each. Warm the sample to room temperature, open the sample bottle, and carefully pour the sample into one of the syringe barrels to just short of overflowing. Replace the syringe plunger, invert the syringe, and compress the sample. Open the syringe valve and vent any residual air while adjusting the sample volume to 25.0 mL (or 5 mL). For samples and blanks, add 5 µL of the fortification solution containing the internal standard and the surrogates to the sample through the syringe valve. For calibration standards and lab fortified blanks, add 5 µL of the fortification solution containing the internal standard only. Close the valve. Fill the second syringe in an identical manner from the same sample bottle. Reserve this second syringe for a reanalysis if necessary.

QUALITY CONTROL As an initial demonstration of lab accuracy and precision, analyze 4 to 7 replicates of a lab fortified

blank containing analyte at 0.2–5 µg/L. Collect all samples in duplicate. Surrogate analytes (similar to those of the analytes of interest), whose concentration is known in every sample, are measured using the same internal standard calibration procedure. Duplicate field reagent water blanks (trip blanks) must be analyzed with each set of samples, lab reagent blanks (method blanks) must be analyzed with each batch of samples processed as a group within a work shift. Also, a single lab-fortified blank that contains each of the analytes of interest should be analyzed with each batch of samples processed as a group within a work shift. A 3- to 5-point calibration curve is needed depending on the calibration range factor required.

EPA CONTACT & HOTLINE For technical questions contact Dr. Baldev Bathija, U.S. EPA, Office of Ground Water and Drinking Water (WH-550D), 401 M St. SW, Washington, DC 20460. Tel. (202) 260-3040. For further information the EPA Safe Drinking Water Hotline may be called at: (800) 426-4791.

REFERENCE Methods for the Determination of Organic Compounds in Drinking Water, EPA/600/4-88/039 (revised July 1991; Final Rule for determination of compliance with the MCL for Total Trihalomethanes under 141.30, in 40 CFR Part 141, Vol. 58, No. 147, Fed. Reg., Tuesday Aug. 3, 1993). U.S. EPA Environmental Monitoring Systems Laboratory, Cincinnati, OH, 45268, U.S.A. Available from the National Technical Information Service (NTIS), 5285 Port Royal Road, Springfield, VA 22161; Tel. 800-553-6847. NTIS Order Number is PB91-231480.

1,1,1,2-Tetrachloroethane   **EPA Method 8021**
CAS #630-20-6

TITLE Halogenated Volatile by Gas Chromatography Using Photoionization and Electrolytic Conductivity Detectors in Series: Capillary Column Technique

MATRIX This method is applicable to nearly all types of samples, regardless of water content, including groundwater, aqueous sludges, caustic liquors, acid liquors, waste solvents, oily wastes, mousses, tars, fibrous wastes, polymeric emulsions, filter cakes, spent carbons, spent catalysts, soils, and sediments.

METHOD SUMMARY This method is used to determine 60 volatile organic compounds in a variety of solid waste matrices. It provides GC conditions for the detection of halogenated and aromatic volatile organic compounds. Samples can be analyzed using direct injection or purge-and-trap (EPA Method 5030). Groundwater samples must be analyzed using EPA Method 5030 (where applicable). A temperature program is used with the GC. Detection is achieved by a photoionization detector (PID) and a Hall electrolytic conductivity detector (HECD) in series.

INTERFERENCES Samples can be contaminated by diffusion of volatile organics (particularly chlorofluorocarbons and methylene chloride) through the sample container septum during shipment and storage.

INSTRUMENTATION A GC-equipped with variable-constant differential flow controllers, subambient oven controller, PID and HECD detectors connected with a short piece of uncoated capillary tubing and a data system.

Column: 60 m × 0.75 mm I.D. VOCOL wide-bore capillary column with 1.5 µm film thickness.

PRECISION & ACCURACY MDLs are compound-dependent and vary with purging efficiency and concentration. The applicable concentration range of this method is compound- and instrument-dependent but is approximately 0.1 to 200 µg/L. Analytes that are inefficiently purged from water will not be detected when present at low concentrations, but they can be measured with acceptable accuracy and precision when present in sufficient amounts. The estimated quantitation limit (EQL) for an individual compound is approximately 1 µg/kg (wet weight) for soil/sediment samples, 100 µg/kg (wet weight) for wastes, and 1 µg/L for groundwater. EQLs will be proportionately higher for sample extracts and samples that require dilution or reduced sample size to avoid saturation of the detector.

MULTIPLICATION FACTORS FOR OTHER MATRICES (a)

Matrix	Factor (b)
Groundwater	10
Low-concentration soil	10
Water miscible liquid waste	500
High-concentration soil and sludge	1250
Non-water miscible waste	1250

(a) Sample EQLs are highly matrix-dependent. The EQLs listed herein are provided for guidance and may not always be achievable. (b) EQL = [Method detection limit] × [Factor]. For non-aqueous samples, the factor is on a wet-weight basis.

SINGLE LABORATORY ACCURACY & PRECISION DATA FOR VOCs IN WATER
This method was tested in a single lab using water spiked at 10 µg/L and the following data was reported:

Recoveries and standard deviations were determined from seven samples and spiked at 10 µg/L of each analyte. Recoveries were determined by the internal standard method. Internal standards were: Fluorobenzene for PID and 2-Bromo-1-chloropropane for HECD.

The average recovery (in percent) for the PID was none (no response for this detector).
The standard deviation of the recovery for the PID was none (no response for this detector)-.
The MDL (in µg/mL) for the PID was none (no response for this detector).
The average recovery (in percent) for the HECD was 99.
The standard deviation of the recovery for the HECD was 2.3.
The MDL (in µg/mL) for the HECD was 0.005.

SAMPLE COLLECTION, PRESERVATION & HANDLING
Volatile organics — Standard 40-mL glass screw-cap VOA vials with Teflon®-faced silicone septum may be used for both liquid and solid matrices. When collecting samples, liquids and solids should be introduced into the vials gently to reduce agitation

which might drive off volatile compounds. If there are any air bubbles present the sample must be retaken. Tap slightly as they are filled to try and eliminate as much free air space as possible. The two vials from each sampling locations should be sealed in separate plastic bags to prevent cross-contamination between samples particularly if the sampled waste is suspected of containing high levels of volatile organics.

Semivolatile organics — Containers used to collect samples for the determination of semivolatile organic compounds should be soap and water washed followed by methanol (or isopropanol) rinsing. The sample containers should be of glass or Teflon® and have screw-top covers with Teflon® liners.

Preservation for volatile organics — No preservation is used with concentrated waste samples. With liquid samples containing no residual chlorine, 4 drops of concentrated hydrochloric acid are added and the samples are immediately cooled to 4°C. When liquid samples contain residual chlorine, they are treated as above and, in addition, 4 drops of 4% aqueous sodium thiosulfate are added. Soil, sediment, and sludge samples are only cooled to 4°C.

Preservation for semivolatile organics — No preservation is used with concentrated waste samples. With liquid samples containing no residual chlorine and with soil, sediment, and sludge samples, immediately cooling to 4°C is the only preservation used. When residual chlorine is present then 3 mL of 10% aqueous sodium sulfate is added for each gallon of sample collected, followed by cooling to 4°C.

MHT The holding time for all volatile organics samples is 14 days. Liquid samples must be extracted within 7 days and their extracts analyzed within 40 days. Concentrated waste, soil, sediment, and sludge samples must be extracted within 14 days and their extracts analyzed within 40 days.

SAMPLE PREPARATION Volatile compounds are introduced into the gas chromatograph either by direct injector or purge-and-trap (EPA Method 5030). EPA Method 5030 may be used directly on groundwater samples or low-concentration contaminated soils and sediments. For medium-concentration soils or sediments, methanolic extraction, as described in EPA Method 5030, may be necessary prior to purge-and-trap analysis.

QUALITY CONTROL Calculate surrogate standard recovery on all samples, blanks, and spikes. A trip blank is recommended to check on sampling, storage, and handling contamination. Calibration standards, at a minimum of five concentration levels, are prepared in organic-free reagent water. One of the concentration levels should be at a concentration near, but above, the method detection limit.

A combination of bromochloromethane, 2-bromo-1-chloropropane, 1,4-dichlorobutane, and bromochlorobenzene are recommended as surrogate standards to encompass the range of the temperature program used in this method.

REFERENCE Test Methods for Evaluating Solid Waste, Physical/Chemical Methods, SW-846, 3rd Edition, U.S. EPA, Office of Solid Waste, Washington, DC, EPA Method 8021A, Rev. 1, Nov. 1992.

1,1,2,2-Tetrachloroethane   EPA Method 8021
CAS #79-34-5

TITLE Halogenated Volatile by Gas Chromatography Using Photoionization and Electrolytic Conductivity Detectors in Series: Capillary Column Technique

MATRIX This method is applicable to nearly all types of samples, regardless of water content, including groundwater, aqueous sludges, caustic liquors, acid liquors, waste solvents, oily wastes, mousses, tars, fibrous wastes, polymeric emulsions, filter cakes, spent carbons, spent catalysts, soils, and sediments.

METHOD SUMMARY This method is used to determine 60 volatile organic compounds in a variety of solid waste matrices. It provides GC conditions for the detection of halogenated and aromatic volatile organic compounds. Samples can be analyzed using direct injection or purge-and-trap (EPA Method 5030). Groundwater samples must be analyzed using EPA Method 5030 (where applicable). A temperature program is used with the GC. Detection is achieved by a photoionization detector (PID) and a Hall electrolytic conductivity detector (HECD) in series.

INTERFERENCES Samples can be contaminated by diffusion of volatile organics (particularly chlorofluorocarbons and methylene chloride) through the sample container septum during shipment and storage.

INSTRUMENTATION A GC-equipped with variable-constant differential flow controllers, subambient oven controller, PID and HECD detectors connected with a short piece of uncoated capillary tubing and a data system.

Column: 60 m × 0.75 mm I.D. VOCOL wide-bore capillary column with 1.5 μm film thickness.

PRECISION & ACCURACY MDLs are compound-dependent and vary with purging efficiency and concentration. The applicable concentration range of this method is compound- and instrument-dependent but is approximately 0.1 to 200 μg/L. Analytes that are inefficiently purged from water will not be detected when present at low concentrations, but they can be measured with acceptable accuracy and precision when present in sufficient amounts. The estimated quantitation limit (EQL) for an individual compound is approximately 1 μg/kg (wet weight) for soil/sediment samples, 100 μg/kg (wet weight) for wastes, and 1 μg/L for groundwater. EQLs will be proportionately higher for sample extracts and samples that require dilution or reduced sample size to avoid saturation of the detector.

MULTIPLICATION FACTORS FOR OTHER MATRICES (a)

Matrix	Factor (b)
Groundwater	10
Low-concentration soil	10
Water miscible liquid waste	500
High-concentration soil and sludge	1250
Non-water miscible waste	1250

(a) Sample EQLs are highly matrix-dependent. The EQLs listed herein are provided for guidance and may not always be achievable.
(b) EQL = [Method detection limit] × [Factor]. For non-aqueous samples, the factor is on a wet-weight basis.

SINGLE LABORATORY ACCURACY & PRECISION DATA FOR VOCs IN WATER

This method was tested in a single lab using water spiked at 10 µg/L and the following data was reported:

Recoveries and standard deviations were determined from seven samples and spiked at 10 µg/L of each analyte. Recoveries were determined by the internal standard method. Internal standards were: Fluorobenzene for PID and 2-Bromo-1-chloropropane for HECD.

The average recovery (in percent) for the PID was none (no response for this detector).
The standard deviation of the recovery for the PID was none (no response for this detector).
The MDL (in µg/mL) for the PID was none (no response for this detector).
The average recovery (in percent) for the HECD was 99.
The standard deviation of the recovery for the HECD was 6.8.
The MDL (in µg/mL) for the HECD was 0.01.

SAMPLE COLLECTION, PRESERVATION & HANDLING

Volatile organics — Standard 40-mL glass screw-cap VOA vials with Teflon®-faced silicone septum may be used for both liquid and solid matrices. When collecting samples, liquids and solids should be introduced into the vials gently to reduce agitation which might drive off volatile compounds. If there are any air bubbles present the sample must be retaken. Tap slightly as they are filled to try and eliminate as much free air space as possible. The two vials from each sampling locations should be sealed in separate plastic bags to prevent cross-contamination between samples particularly if the sampled waste is suspected of containing high levels of volatile organics.

Semivolatile organics — Containers used to collect samples for the determination of semivolatile organic compounds should be soap and water washed followed by methanol (or isopropanol) rinsing. The sample containers should be of glass or Teflon® and have screw-top covers with Teflon® liners.

Preservation for volatile organics — No preservation is used with concentrated waste samples. With liquid samples containing no residual chlorine, 4 drops of concentrated hydrochloric acid are added and the samples are immediately cooled to 4°C. When liquid samples contain residual chlorine, they are treated as above and, in addition, 4 drops of 4% aqueous sodium thiosulfate are added. Soil, sediment, and sludge samples are only cooled to 4°C.

Preservation for semivolatile organics — No preservation is used with concentrated waste samples. With liquid samples containing no residual chlorine and with soil, sediment, and sludge samples, immediately cooling to 4°C is the only preservation used. When residual chlorine is present then 3 mL of 10% aqueous sodium sulfate is added for each gallon of sample collected, followed by cooling to 4°C.

MHT The holding time for all volatile organics samples is 14 days. Liquid samples must be extracted within 7 days and their extracts analyzed within 40 days. Concentrated waste, soil, sediment, and sludge samples must be extracted within 14 days and their extracts analyzed within 40 days.

SAMPLE PREPARATION Volatile compounds are introduced into the gas chromatograph either by direct injector or purge-and-trap (EPA Method 5030). EPA Method 5030 may be used directly on groundwater samples or low-concentration contaminated soils and sediments. For medium-concentration soils or sediments, methanolic extraction, as described in EPA Method 5030, may be necessary prior to purge-and-trap analysis.

QUALITY CONTROL Calculate surrogate standard recovery on all samples, blanks, and spikes. A trip blank is recommended to check on sampling, storage, and handling contamination. Calibration standards, at a minimum of five concentration levels, are prepared in organic-free reagent water. One of the concentration levels should be at a concentration near, but above, the method detection limit.

A combination of bromochloromethane, 2-bromo-1-chloropropane, 1,4-dichlorobutane, and bromochlorobenzene are recommended as surrogate standards to encompass the range of the temperature program used in this method.

REFERENCE Test Methods for Evaluating Solid Waste, Physical/Chemical Methods, SW-846, 3rd Edition, U.S. EPA, Office of Solid Waste, Washington, DC, EPA Method 8021A, Rev. 1, Nov. 1992.

1,1,1,2-Tetrachloroethane   EPA Method 8240
CAS #630-20-6

TITLE Volatile Organics By GC/MS: Packed Column Technique

MATRIX Nearly all types of sample matarices, regardless of water content, can be analyzed using this method. This includes groundwater, aqueous sludges, caustic liquors, acid liquors, waste solvents, oily wastes, mousses, tars, fibrous wastes, polymetric emulsions, filter cakes, spent carbons, spent catalysts, soils, and sediments.

METHOD SUMMARY Method 8240B covers 80 volatile organic compounds that are introduced into a gas chromatograph by the purge-and-trap method or by direct injection (in limited applications). For the purge-and-trap method an inert gas (zero grade nitrogen or helium) is bubbled through a 5-mL solution at ambient temperature. Purged sample components are trapped in a tube of sorbent materials. When purging is complete, the sorbent tube is heated and backflushed with inert gas to desorb the trapped components onto a GC column.

INTERFERENCES Impurities in the purge gas and from organic compounds outgassing from the plumbing ahead of the trap account for many contamination problems. Interferences purged or coextracted from the samples will vary considerably from source to source. Cross-contamination can occur whenever high-level and low-level samples are analyzed sequentially. Whenever an unusually concentrated sample is analyzed, it should be followed by the analysis of organic-free reagent water to check for cross-contamination. Samples also can be contaminated by diffusion of volatile organics (particularly methylene chloride and fluorocarbons) through the septum seal into the sample during shipment and storage. A trip blank can serve as a check on such contamination. The lab

where volatile analysis is performed and also the refrigerated storage area should be completely free of solvents.

INSTRUMENTATION A gas chromatograph/mass spectrometry/data system (GC/MS) equipped with a 6 ft × 0.1 in I.D. glass column packed with 1% SP-1000 on Carbopack-B (60/80 mesh) is required. Also needed is a 5-mL purging device, a sorbent trap, and a thermal desorption apparatus.

PRECISION & ACCURACY This method is reported to have been tested by 15 laboratories using organic-free reagent water, drinking water, surface water, and industrial wastewaters (not specified) fortified at six concentrations over the range 5–600 µg/L.

Sample estimated quantitation limits (EQLs) are highly matrix-dependent. The EQLs listed may not always be achievable. EQLs listed for soils or sediments are based on wet weight. Normally, data is reported on a dry-weight basis; therefore, EQLs will be higher, based on the percent dry weight of each sample. Note that EQLs are even more variable than MDLs and that they are highly variable depending on the matrix being analyzed.

EQL in groundwater in µg/L was 5.
EQL in low soil or sediment in µg/kg was 5.
Accuracy (a) in µg/L was not listed.
Precision (b) in µg/L was not listed.

(a) *Average recovery found for measurements of samples containing a concentration of C, in µg/L.*
(b) *Overall precision found for measurements of samples with average recovery X for samples containing a concentration of C in µg/L.*
X = *Average recovery found for measurement of samples containing a concentration of C in µg/L.*

MULTIPLICATION FACTORS FOR OTHER MATRICES

Other Matrices	Factor (a)
Waste miscible liquid waste	50
High-concentration soil and sludge	125
Non-water miscible waste	500

(a) *EQL = [EQL for low soil sediment X [Factor]. For non-aqueous samples, the factor is on a wet-weight basis.*

SAMPLING METHOD

Liquid samples — Use a 40-mL glass screw-cap VOA vial with a Teflon®-faced silicone septum that has been prewashed, rinsed with distilled deionized water, and oven dried. However, if residual chlorine is present, collect sample in a 40-oz. soil VOA container which has been pre-preserved with 4 drops of 10% sodium thiosulfate, mix gently, and then transfer the sample to a 40-mL VOA vial. Collect bubble-free samples in duplicate and seal them in separate plastic bags.

Soils or sediments, and sludges — Use an 8-oz. widemouth glass bottle with a Teflon®-faced silicone septum that has been prewashed with detergent, rinsed with distilled deionized water, and oven dried. Tap slightly to eliminate free air space. Collect samples in duplicate and seal them in separate plastic bags.

SAMPLE PRESERVATION
Liquid samples — Add 4 drops of concentrated HCL and immediately cool samples to 4°C and store in a solvent-free refrigerator.

Soils or sediments, and sludges — Cool samples to 4°C and store in a solvent-free refrigerator.

MHT Maximum holding time is 14 days from the date of sample collection.

SAMPLE PREPARATION
Liquid samples — Remove the plunger from a 5-mL syringe and carefully pour the sample into the syringe barrel to just short of overflowing. Replace the syringe plunger and compress the sample. Open the syringe valve and vent any residual air while adjusting the sample volume to 5.0 mL. If there is only one volatile organic analysis (VOA) vial, a second syringe should be filled at this time to protect against possible loss of sample integrity. Add 10 µL of surrogate spiking solution and 10 µL of internal standard spiking solution through the valve bore of the 5-mL syringe, then close the valve. The surrogate and internal standards may be mixed and added as a single spiking solution.

Sediments, soils, and waste samples — All samples of this type should be screened by GC analysis using a headspace method (EPA Method 3810) or the hexadecane extraction and screening method (EPA Method 3820). Use the screening data to determine whether to use the low-concentration method (0.005–1 mg/kg) or the high-concentration method (>1 mg/kg).

Low-concentration method — The low-concentration method is based on purging a heated sediment or soil sample mixed with organic-free reagent water containing the surrogate and internal standards. Analyze all reagent blanks and standards under the same conditions as the samples.

Use a 5-g sample if the expected concentration is <0.1 mg/kg or a 1-g sample for expected concentrations between 0.1 and 1 mg/kg. Mix the contents of the sample container with a narrow metal spatula. Weigh the amount of the sample into a tared purge device. Add the spiked water to the purge device, which contains the weighed amount of sample, and connect the device to the purge-and-trap system.

High-concentration method — This method is based on extracting the sediment or soil with methanol. A waste sample is either extracted or diluted, depending on its solubility in methanol. Wastes that are insoluble in methanol are diluted with reagent tetraglyme or possibly polyethylene glycol (PEG). An aliquot of the extract is added to organic-free reagent water containing surrogate and internal standards. This is purged at ambient temperature. All samples with an expected concentration of >1.0 mg/kg should be analyzed by this method.

Mix the contents of the sample container with a narrow metal spatula. For sediments or soils and solid wastes that are insoluble in methanol, weigh 4 g (wet weight) of sample into a tared 20-mL vial. For waste that is soluble in methanol, tetraglyme, or PEG, weigh 1 g (wet weight) into a tared scintillation vial or culture tube or a 10-mL volumetric flask. Quickly add

9.0 mL of appropriate solvent then add 1.0 mL of a surrogate spiking solution to the vial, cap it, and shake it for 2 min.

METHANOL EXTRACT REQUIRED FOR ANALYSIS OF HIGH-CONCENTRATION SOILS OR SEDIMENTS

Approximate Concentration Range	Volume of Methanol Extract (a)
500–10,000 µg/kg	100 µL
1,000–20,000 µg/kg	50 µL
5,000–100,000 µg/kg	10 µL
25,000–500,000 µg/kg	100 µL of 1/50 dilution (b)

Calculate appropriate dilution factor for concentrations exceeding this table.

(a) The volume of methanol added to 5 mL of water being purged should be kept constant. Therefore, add to the 5-mL syringe whatever volume of methanol is necessary to maintain a volume of 100 µL added to the syringe.

(b) Dilute an aliquot of the methanol extract and then take 100 µL for analysis.

QUALITY CONTROL Demonstrate, through the analysis of a reagent water blank, that interferences from the analytical system, glassware, and reagents are under control. Blank samples should be carried through all stages of the sample preparation and measurement steps. For each analytical batch (up to 20 samples), a reagent blank, matrix spike, and matrix spike duplicate must be analyzed (the frequency of the spikes may be different for different monitoring programs). The blank and spiked samples must be carried through all stages of the sample preparation and measurement steps. QC samples mentioned in the section on Interferences will also be needed as appropriate to those situations.

REFERENCE Test Methods for Evaluating Solid Waste (SW-846). U.S. EPA. 1983. Method 8240B, Rev. 2, Nov. 1990. Office of Solid Wastes, Washington, DC.

1,1,2,2-Tetrachloroethane **EPA Method 8240**
CAS #79-34-5

TITLE Volatile Organics By GC/MS: Packed Column Technique

MATRIX Nearly all types of sample matarices, regardless of water content, can be analyzed using this method. This includes groundwater, aqueous sludges, caustic liquors, acid liquors, waste solvents, oily wastes, mousses, tars, fibrous wastes, polymetric emulsions, filter cakes, spent carbons, spent catalysts, soils, and sediments.

METHOD SUMMARY Method 8240B covers 80 volatile organic compounds that are introduced into a gas chromatograph by the purge-and-trap method or by direct injection (in limited applications). For the purge-and-trap method an inert gas (zero grade nitrogen or helium) is bubbled through a 5-mL solution at ambient temperature. Purged sample components are trapped in a tube of sorbent materials. When purging is complete, the sorbent tube is heated and backflushed with inert gas to desorb the trapped components onto a GC column.

INTERFERENCES Impurities in the purge gas and from organic compounds outgassing from the plumbing ahead of the trap account for many contamination problems. Interferences purged or coextracted from the samples will vary considerably from source to source. Cross-contamination can occur whenever high-level and low-level samples are analyzed sequentially. Whenever an unusually concentrated sample is analyzed, it should be followed by the analysis of organic-free reagent water to check for cross-contamination. Samples also can be contaminated by diffusion of volatile organics (particularly methylene chloride and fluorocarbons) through the septum seal into the sample during shipment and storage. A trip blank can serve as a check on such contamination. The lab where volatile analysis is performed and also the refrigerated storage area should be completely free of solvents.

INSTRUMENTATION A gas chromatograph/mass spectrometry/data system (GC/MS) equipped with a 6 ft × 0.1 in I.D. glass column packed with 1% SP-1000 on Carbopack-B (60/80 mesh) is required. Also needed is a 5-mL purging device, a sorbent trap, and a thermal desorption apparatus.

PRECISION & ACCURACY This method is reported to have been tested by 15 laboratories using organic-free reagent water, drinking water, surface water, and industrial wastewaters (not specified) fortified at six concentrations over the range 5–600 µg/L.

Sample estimated quantitation limits (EQLs) are highly matrix-dependent. The EQLs listed may not always be achievable. EQLs listed for soils or sediments are based on wet weight. Normally, data is reported on a dry-weight basis; therefore, EQLs will be higher, based on the percent dry weight of each sample. Note that EQLs are even more variable than MDLs and that they are highly variable depending on the matrix being analyzed.

EQL in groundwater in µg/L was 5.
EQL in low soil or sediment in µg/kg was 5.
Accuracy (a) in µg/L was $0.93C + 1.76$.
Precision (b) in µg/L was $0.20x + 0.41$.

(a) Average recovery found for measurements of samples containing a concentration of C, in µg/L.

(b) Overall precision found for measurements of samples with average recovery X for samples containing a concentration of C in µg/L.

$X =$ *Average recovery found for measurement of samples containing a concentration of C in µg/L.*

MULTIPLICATION FACTORS FOR OTHER MATRICES

Other Matrices	Factor (a)
Waste miscible liquid waste	50
High-concentration soil and sludge	125
Non-water miscible waste	500

(a) EQL = [EQL for low soil sediment X [Factor]. For non-aqueous samples, the factor is on a wet-weight basis.

SAMPLING METHOD

Liquid samples — Use a 40-mL glass screw-cap VOA vial with a Teflon®-faced silicone septum that has been prewashed, rinsed with distilled deionized water, and oven dried. However,

if residual chlorine is present, collect sample in a 40-oz. soil VOA container which has been pre-preserved with 4 drops of 10% sodium thiosulfate, mix gently, and then transfer the sample to a 40-mL VOA vial. Collect bubble-free samples in duplicate and seal them in separate plastic bags.

Soils or sediments, and sludges — Use an 8-oz. widemouth glass bottle with a Teflon®-faced silicone septum that has been prewashed with detergent, rinsed with distilled deionized water, and oven dried. Tap slightly to eliminate free air space. Collect samples in duplicate and seal them in separate plastic bags.

SAMPLE PRESERVATION
Liquid samples — Add 4 drops of concentrated HCL and immediately cool samples to 4°C and store in a solvent-free refrigerator.

Soils or sediments, and sludges — Cool samples to 4°C and store in a solvent-free refrigerator.

MHT Maximum holding time is 14 days from the date of sample collection.

SAMPLE PREPARATION
Liquid samples — Remove the plunger from a 5-mL syringe and carefully pour the sample into the syringe barrel to just short of overflowing. Replace the syringe plunger and compress the sample. Open the syringe valve and vent any residual air while adjusting the sample volume to 5.0 mL. If there is only one volatile organic analysis (VOA) vial, a second syringe should be filled at this time to protect against possible loss of sample integrity. Add 10 µL of surrogate spiking solution and 10 µL of internal standard spiking solution through the valve bore of the 5-mL syringe, then close the valve. The surrogate and internal standards may be mixed and added as a single spiking solution.

Sediments, soils, and waste samples — All samples of this type should be screened by GC analysis using a headspace method (EPA Method 3810) or the hexadecane extraction and screening method (EPA Method 3820). Use the screening data to determine whether to use the low-concentration method (0.005–1 mg/kg) or the high-concentration method (>1 mg/kg).

Low-concentration method — The low-concentration method is based on purging a heated sediment or soil sample mixed with organic-free reagent water containing the surrogate and internal standards. Analyze all reagent blanks and standards under the same conditions as the samples.

Use a 5-g sample if the expected concentration is <0.1 mg/kg or a 1-g sample for expected concentrations between 0.1 and 1 mg/kg. Mix the contents of the sample container with a narrow metal spatula. Weigh the amount of the sample into a tared purge device. Add the spiked water to the purge device, which contains the weighed amount of sample, and connect the device to the purge-and-trap system.

High-concentration method — This method is based on extracting the sediment or soil with methanol. A waste sample is either extracted or diluted, depending on its solubility in methanol. Wastes that are insoluble in methanol are diluted with reagent tetraglyme or possibly polyethylene glycol (PEG). An aliquot of the extract is added to organic-free reagent water containing surrogate and internal standards. This is purged at ambient temperature. All samples with an expected concentration of >1.0 mg/kg should be analyzed by this method.

Mix the contents of the sample container with a narrow metal spatula. For sediments or soils and solid wastes that are insoluble in methanol, weigh 4 g (wet weight) of sample into a tared 20-mL vial. For waste that is soluble in methanol, tetraglyme, or PEG, weigh 1 g (wet weight) into a tared scintillation vial or culture tube or a 10-mL volumetric flask. Quickly add 9.0 mL of appropriate solvent then add 1.0 mL of a surrogate spiking solution to the vial, cap it, and shake it for 2 min.

METHANOL EXTRACT REQUIRED FOR ANALYSIS OF HIGH-CONCENTRATION SOILS OR SEDIMENTS

Approximate Concentration Range	Volume of Methanol Extract (a)
500–10,000 µg/kg	100 µL
1,000–20,000 µg/kg	50 µL
5,000–100,000 µg/kg	10 µL
25,000–500,000 µg/kg	100 µL of 1/50 dilution (b)

Calculate appropriate dilution factor for concentrations exceeding this table.

(a) The volume of methanol added to 5 mL of water being purged should be kept constant. Therefore, add to the 5-mL syringe whatever volume of methanol is necessary to maintain a volume of 100 µL added to the syringe.

(b) Dilute an aliquot of the methanol extract and then take 100 µL for analysis.

QUALITY CONTROL Demonstrate, through the analysis of a reagent water blank, that interferences from the analytical system, glassware, and reagents are under control. Blank samples should be carried through all stages of the sample preparation and measurement steps. For each analytical batch (up to 20 samples), a reagent blank, matrix spike, and matrix spike duplicate must be analyzed (the frequency of the spikes may be different for different monitoring programs). The blank and spiked samples must be carried through all stages of the sample preparation and measurement steps. QC samples mentioned in the section on Interferences will also be needed as appropriate to those situations.

REFERENCE Test Methods for Evaluating Solid Waste (SW-846). U.S. EPA. 1983. Method 8240B, Rev. 2, Nov. 1990. Office of Solid Wastes, Washington, DC.

1,1,1,2-Tetrachloroethane   **EPA Method 8260**
CAS #630-20-6

TITLE Volatile Organic Compounds by GC/MS: Capillary Column Technique

MATRIX This method is applicable to nearly all types of samples, regardless of water content, including groundwater, soils, and sediments.

METHOD SUMMARY Method 8260A covers 58 volatile organic compounds that are introduced into a gas chromatograph by the purge-and-trap method or by direct injection (in limited applications). Zero-grade helium is bubbled through a

5-mL solution at ambient temperature. Purged sample components are trapped in a tube containing suitable sorbent materials. When purging is complete, the sorbent tube is heated and backflushed with helium to desorb trapped sample components. The analytes are desorbed directly to a large bore capillary or cryofocussed on a capillary precolumn before being flash evaporated to a narrow bore capillary for analysis.

INTERFERENCES Major contaminant sources are volatile materials in the lab and impurities in the inert purging gas and in the sorbent trap. Interfering contamination may occur when a sample containing low concentrations of volatile organic compounds is analyzed immediately after a sample containing high concentrations of volatile organic compounds. After analysis of a sample containing high concentrations of volatile organic compounds, one or more calibration blanks should be analyzed to check for cross-contamination. Screening of the samples prior to purge-and-trap GC/MS analysis is highly recommended to prevent contamination of the system. This is especially true for soil and waste samples.

Special precautions must be taken to analyze for methylene chloride. The analytical and sample storage area should be isolated from all atmospheric sources of methylene chloride. All gas chromatography carrier gas lines and purge gas plumbing should be constructed from stainless steel or copper tubing. Laboratory clothing previously exposed to methylene chloride fumes during liquid-liquid extraction procedures can contribute to sample contamination.

Samples can also be contaminated by diffusion of volatile organics (particularly methylene chloride and fluorocarbons) through the septum seal during shipment and storage. A trip blank can serve as a check on such contamination.

INSTRUMENTATION GC/MS with a temperature-programmable chromatograph suitable for splitless injection equipped with variable constant differential flow controllers, a subambient oven controller, a purging device, sorbent trap, a thermal desorption apparatus and a capillary precolumn interface when using cryogenic cooling will be needed. The following GC columns may be used:

Column 1: 60 m × 0.75mm I.D. capillary column coated with VOCOL, 1.5 µm film thickness.
Column 2: 30 m × 0.53mm capillary column coated with DB-624 or VOCOL, 3 µm film thickness.
Column 3: 30 m × 0.32mm I.D. capillary column coated with DB-5 or SE-54, 1-µm film thickness.

PRECISION & ACCURACY This method has been tested in a single lab using spiked water. Using a wide-bore capillary column, water was spiked at concentrations between 0.5 and 10 µg/L. Single lab accuracy and precision data are presented. The MDL actually achieved in a given analysis will vary depending on instrument sensitivity and matrix effects.

The MDL (a) in µg/L was 0.05.
The concentration range in µg/L was 0.5–10.
The mean accuracy (% of true value) was 90.
The precision as relative standard deviation was 6.8.

Note: The MDL is based on a 25-mL sample volume instead of a 5-mL sample volume.

SAMPLING METHOD
Liquid samples — Use a 40-mL glass screw-cap VOA vial with a Teflon®-faced silicone septum that has been prewashed, rinsed with distilled deionized water, and oven dried. If residual chlorine is present, collect the sample in a 4-oz soil VOA container which has been pre-preserved with 4 drops of 10% sodium thiosulfate. Mix gently and transfer the sample to a 40-mL VOA vial. Collect bubble-free samples in duplicate and seal each sample in a separate plastic bag.

Soils, sediments and sludges — Use an 8-oz widemouth glass bottle with Teflon®-faced silicone septum that has been prewashed, rinsed with distilled deionized water, and oven dried. **Do not** heat the septum for more than 1 h. Tap slightly to eliminate any free air space. Collect samples in duplicate and seal each one in a separate plastic bag.

SAMPLE PRESERVATION
Liquid samples — Add 4 drops of concentrated HCL, cool to 4°C and store in a solvent-free refrigerator.

Soils, sediments and sludges — Cool samples to 4°C and store in a solvent-free refrigerator.

MHT The maximum holding time of any sample (liquids, soils, sediments, and sludges) is 14 days.

SAMPLE PREPARATION
Liquid samples — Remove the plunger from a 5-mL syringe and carefully pour the sample into the syringe barrel to just short of overflowing. Replace the syringe plunger and compress the sample. Open the syringe valve and vent any residual air while adjusting the sample volume to 5.0 mL. If there is only one volatile organic analysis (VOA) vial, a second syringe should be filled at this time to protect against possible loss of sample integrity. Add 10 µL of surrogate spiking solution and 10 µL of internal standard spiking solution through the valve bore of the 5-mL syringe, then close the valve. The surrogate and internal standards may be mixed and added as a single spiking solution.

Sediments, soils, and waste samples — All samples of this type should be screened by GC analysis using a headspace method (EPA Method 3810) or the hexadecane extraction and screening method (EPA Method 3820). Use the screening data to determine whether to use the low-concentration method (0.005–1 mg/kg) or the high-concentration method (>1 mg/kg).

Low-concentration method — The low-concentration method is based on purging a heated sediment or soil sample mixed with organic-free reagent water containing the surrogate and internal standards. Analyze all reagent blanks and standards under the same conditions as the samples.

Use a 5-g sample if the expected concentration is <0.1 mg/kg or a 1-g sample for expected concentrations between 0.1 and 1 mg/kg. Mix the contents of the sample container with a narrow metal spatula. Weigh the amount of the sample into a tared purge device. Add the spiked water to the purge device, which contains the weighed amount of sample, and connect the device to the purge-and-trap system.

High-concentration method — This method is based on extracting the sediment or soil with methanol. A waste sample is either extracted or diluted, depending on its solubility in methanol. Wastes that are insoluble in methanol are diluted with reagent tetraglyme or possibly polyethylene glycol (PEG). An aliquot of the extract is added to organic-free reagent water containing surrogate and internal standards. This is purged at ambient temperature. All samples with an expected concentration of >1.0 mg/kg should be analyzed by this method.

Mix the contents of the sample container with a narrow metal spatula. For sediments or soils and solid wastes that are insoluble in methanol, weigh 4 g (wet weight) of sample into a tared 20-mL vial. For waste that is soluble in methanol, tetraglyme, or PEG, weigh 1 g (wet weight) into a tared scintillation vial or culture tube or a 10-mL volumetric flask. Quickly add 9.0 mL of appropriate solvent then add 1.0 mL of a surrogate spiking solution to the vial, cap it, and shake it for 2 min.

METHANOL EXTRACT REQUIRED FOR ANALYSIS OF HIGH-CONCENTRATION SOILS OR SEDIMENTS

Approximate Concentration Range	Volume of Methanol Extract (a)
500–10,000 µg/kg	100 µL
1,000–20,000 µg/kg	50 µL
5,000–100,000 µg/kg	10 µL
25,000–500,000 µg/kg	100 µL of 1/50 dilution (b)

Calculate appropriate dilution factor for concentrations exceeding this table.

(a) The volume of methanol added to 5 mL of water being purged should be kept constant. Therefore, add to the 5-mL syringe whatever volume of methanol is necessary to maintain a volume of 100 µL added to the syringe.
(b) Dilute an aliquot of the methanol extract and then take 100 µL for analysis.

QUALITY CONTROL Demonstrate, through the analysis of a reagent water blank, that interferences from the analytical system, glassware, and reagents are under control. Blank samples should be carried through all stages of the sample preparation and measurement steps. For each analytical batch (up to 20 samples), a reagent blank, matrix spike, and matrix spike duplicate must be analyzed (the frequency of the spikes may be different for different monitoring programs). The blank and spiked samples must be carried through all stages of the sample preparation and measurement steps. QC samples mentioned in the section on Interferences will also be needed as appropriate to those situations.

Matrix spiking standards should be prepared from volatile organic compounds which will be representative of the compounds being investigated. The recommended internal standards are chlorobenzene-d5, 1,4-difluorobenzene, 1,4-dichlorobenzene-d4, and pentafluorobenzene. Using stock standard solutions, prepare secondary dilution standards containing the compounds of interest, either singly or mixed together in methanol. Store them in a vial with no headspace for no more than one week. Surrogates recommended are toluene-d8, 4-bromofluorobenzene, and dibromofluoromethane. Each sample undergoing GC/MS analysis must be spiked with 10 µL of the surrogate spiking solution prior to analysis.

REFERENCE Test Methods for Evaluating Solid Waste (SW-846). U.S. EPA 1983, Method 8260A, Rev. 1, Nov. 1990. Office of Solid Waste, Washington, DC.

1,1,2,2-Tetrachloroethane **EPA Method 8260**
CAS #79-34-5

TITLE Volatile Organic Compounds by GC/MS: Capillary Column Technique

MATRIX This method is applicable to nearly all types of samples, regardless of water content, including groundwater, soils, and sediments.

METHOD SUMMARY Method 8260A covers 58 volatile organic compounds that are introduced into a gas chromatograph by the purge-and-trap method or by direct injection (in limited applications). Zero-grade helium is bubbled through a 5-mL solution at ambient temperature. Purged sample components are trapped in a tube containing suitable sorbent materials. When purging is complete, the sorbent tube is heated and backflushed with helium to desorb trapped sample components. The analytes are desorbed directly to a large bore capillary or cryofocussed on a capillary precolumn before being flash evaporated to a narrow bore capillary for analysis.

INTERFERENCES Major contaminant sources are volatile materials in the lab and impurities in the inert purging gas and in the sorbent trap. Interfering contamination may occur when a sample containing low concentrations of volatile organic compounds is analyzed immediately after a sample containing high concentrations of volatile organic compounds. After analysis of a sample containing high concentrations of volatile organic compounds, one or more calibration blanks should be analyzed to check for cross-contamination. Screening of the samples prior to purge-and-trap GC/MS analysis is highly recommended to prevent contamination of the system. This is especially true for soil and waste samples.

Special precautions must be taken to analyze for methylene chloride. The analytical and sample storage area should be isolated from all atmospheric sources of methylene chloride. All gas chromatography carrier gas lines and purge gas plumbing should be constructed from stainless steel or copper tubing. Laboratory clothing previously exposed to methylene chloride fumes during liquid-liquid extraction procedures can contribute to sample contamination.

Samples can also be contaminated by diffusion of volatile organics (particularly methylene chloride and fluorocarbons) through the septum seal during shipment and storage. A trip blank can serve as a check on such contamination.

INSTRUMENTATION GC/MS with a temperature-programmable chromatograph suitable for splitless injection equipped with variable constant differential flow controllers, a subambient oven controller, a purging device, sorbent trap, a thermal desorption apparatus and a capillary precolumn interface when using cryogenic cooling will be needed. The following GC columns may be used:

Column 1: 60 m × 0.75mm I.D. capillary column coated with VOCOL, 1.5 μm film thickness.
Column 2: 30 m × 0.53mm capillary column coated with DB-624 or VOCOL, 3 μm film thickness.
Column 3: 30 m × 0.32mm I.D. capillary column coated with DB-5 or SE-54, 1-μm film thickness.

PRECISION & ACCURACY This method has been tested in a single lab using spiked water. Using a wide-bore capillary column, water was spiked at concentrations between 0.5 and 10 μg/L. Single lab accuracy and precision data are presented. The MDL actually achieved in a given analysis will vary depending on instrument sensitivity and matrix effects.

The MDL (a) in μg/L was 0.04.
The concentration range in μg/L was 0.1–10.
The mean accuracy (% of true value) was 91.
The precision as relative standard deviation was 6.3.

Note: The MDL is based on a 25-mL sample volume instead of a 5-mL sample volume.

SAMPLING METHOD
Liquid samples — Use a 40-mL glass screw-cap VOA vial with a Teflon®-faced silicone septum that has been prewashed, rinsed with distilled deionized water, and oven dried. If residual chlorine is present, collect the sample in a 4-oz soil VOA container which has been pre-preserved with 4 drops of 10% sodium thiosulfate. Mix gently and transfer the sample to a 40-mL VOA vial. Collect bubble-free samples in duplicate and seal each sample in a separate plastic bag.

Soils, sediments and sludges — Use an 8-oz widemouth glass bottle with Teflon®-faced silicone septum that has been prewashed, rinsed with distilled deionized water, and oven dried. **Do not** heat the septum for more than 1 h. Tap slightly to eliminate any free air space. Collect samples in duplicate and seal each one in a separate plastic bag.

SAMPLE PRESERVATION
Liquid samples — Add 4 drops of concentrated HCL, cool to 4°C and store in a solvent-free refrigerator.

Soils, sediments and sludges — Cool samples to 4°C and store in a solvent-free refrigerator.

MHT The maximum holding time of any sample (liquids, soils, sediments, and sludges) is 14 days.

SAMPLE PREPARATION
Liquid samples — Remove the plunger from a 5-mL syringe and carefully pour the sample into the syringe barrel to just short of overflowing. Replace the syringe plunger and compress the sample. Open the syringe valve and vent any residual air while adjusting the sample volume to 5.0 mL. If there is only one volatile organic analysis (VOA) vial, a second syringe should be filled at this time to protect against possible loss of sample integrity. Add 10 μL of surrogate spiking solution and 10 μL of internal standard spiking solution through the valve bore of the 5-mL syringe, then close the valve. The surrogate and internal standards may be mixed and added as a single spiking solution.

Sediments, soils, and waste samples — All samples of this type should be screened by GC analysis using a headspace method (EPA Method 3810) or the hexadecane extraction and screening method (EPA Method 3820). Use the screening data to determine whether to use the low-concentration method (0.005–1 mg/kg) or the high-concentration method (>1 mg/kg).

Low-concentration method — The low-concentration method is based on purging a heated sediment or soil sample mixed with organic-free reagent water containing the surrogate and internal standards. Analyze all reagent blanks and standards under the same conditions as the samples.

Use a 5-g sample if the expected concentration is <0.1 mg/kg or a 1-g sample for expected concentrations between 0.1 and 1 mg/kg. Mix the contents of the sample container with a narrow metal spatula. Weigh the amount of the sample into a tared purge device. Add the spiked water to the purge device, which contains the weighed amount of sample, and connect the device to the purge-and-trap system.

High-concentration method — This method is based on extracting the sediment or soil with methanol. A waste sample is either extracted or diluted, depending on its solubility in methanol. Wastes that are insoluble in methanol are diluted with reagent tetraglyme or possibly polyethylene glycol (PEG). An aliquot of the extract is added to organic-free reagent water containing surrogate and internal standards. This is purged at ambient temperature. All samples with an expected concentration of >1.0 mg/kg should be analyzed by this method.

Mix the contents of the sample container with a narrow metal spatula. For sediments or soils and solid wastes that are insoluble in methanol, weigh 4 g (wet weight) of sample into a tared 20-mL vial. For waste that is soluble in methanol, tetraglyme, or PEG, weigh 1 g (wet weight) into a tared scintillation vial or culture tube or a 10-mL volumetric flask. Quickly add 9.0 mL of appropriate solvent then add 1.0 mL of a surrogate spiking solution to the vial, cap it, and shake it for 2 min.

METHANOL EXTRACT REQUIRED FOR ANALYSIS OF HIGH-CONCENTRATION SOILS OR SEDIMENTS

Approximate Concentration Range	Volume of Methanol Extract (a)
500–10,000 μg/kg	100 μL
1,000–20,000 μg/kg	50 μL
5,000–100,000 μg/kg	10 μL
25,000–500,000 μg/kg	100 μL of 1/50 dilution (b)

Calculate appropriate dilution factor for concentrations exceeding this table.

(a) The volume of methanol added to 5 mL of water being purged should be kept constant. Therefore, add to the 5-mL syringe whatever volume of methanol is necessary to maintain a volume of 100 μL added to the syringe.
(b) Dilute an aliquot of the methanol extract and then take 100 μL for analysis.

QUALITY CONTROL   Demonstrate, through the analysis of a reagent water blank, that interferences from the analytical system, glassware, and reagents are under control. Blank samples should be carried through all stages of the sample preparation and measurement steps. For each analytical batch (up to 20 samples), a reagent blank, matrix spike, and matrix spike duplicate must be analyzed (the frequency of the spikes may be different for different monitoring programs). The blank and spiked samples must be carried through all stages of the sample preparation and measurement steps. QC samples mentioned in the section on Interferences will also be needed as appropriate to those situations.

Matrix spiking standards should be prepared from volatile organic compounds which will be representative of the compounds being investigated. The recommended internal standards are chlorobenzene-d5, 1,4-difluorobenzene, 1,4-dichlorobenzene-d4, and pentafluorobenzene. Using stock standard solutions, prepare secondary dilution standards containing the compounds of interest, either singly or mixed together in methanol. Store them in a vial with no headspace for no more than one week. Surrogates recommended are toluene-d8, 4-bromofluorobenzene, and dibromofluoromethane. Each sample undergoing GC/MS analysis must be spiked with 10 µL of the surrogate spiking solution prior to analysis.

REFERENCE   Test Methods for Evaluating Solid Waste (SW-846). U.S. EPA 1983, Method 8260A, Rev. 1, Nov. 1990. Office of Solid Waste, Washington, DC.

1,1,2,2-Tetrachloroethane   **EPA Method 601**
CAS #79-34-5

TITLE   Purgeable Halocarbons

MATRIX   Wastewater.

APPLICATION   Method covers 29 purgeable halocarbons. (Method 624 provides GC/MS conditions appropriate for the qualitative and quantitative confirmation of results). Method describes conditions for a 2nd GC column to confirm measurements made with primary column.

INTERFERENCES   Impurities in the purge gas and organic compounds outgassing from the plumbing ahead of the trap. With high- and low-level samples, there can be carryover contamination. Diffusion of volatile organics through the septum seal into the sample.

INSTRUMENTATION   GC-equipped with halide-specific detector. (With purge-and-trap unit).

RANGE   8.0–500 µg/L.

MDL   0.03 µg/L.

PRECISION   0.23X + 2.79 µg/L (overall precision).

ACCURACY   0.95C + 0.19 µg/L (as recovery).

SAMPLING METHOD   25-mL glass vial. Teflon®-lined septum.

STABILITY   Cool, 4°C, 0.008% Sodium thiosulfate.

MHT   14 days.

QUALITY CONTROL   The lab must on an ongoing basis, spike at least 10% of the samples from each sample site being monitored to assess accuracy.

REFERENCE   Method 601, *Federal Register* Part VIII 40 CFR Part 136, Oct 26, 1984.

1,1,2,2-Tetrachloroethane   **EPA Method 624**
CAS #79-34-5

TITLE   Purgeables

MATRIX   Wastewater.

APPLICATION   Method covers 31 purgeable organics. An inert gas is bubbled through a 5-mL water sample in a specially designed purging chamber. Here, purgeables are transferred from aqueous to gaseous phase, passed onto a sorbent column, and trapped. Trap is heated and backflushed with inert gas to desorb purgeables onto a GC column, where purgeables are separated.

INTERFERENCES   Impurities in the purge gas, organic compounds outgassing from the plumbing ahead of the trap, and solvent vapors in the lab. With high- and low-level samples, there can be carryover contamination.

INSTRUMENTATION   GC/MS with purge-and-trap unit.

RANGE   5–600 µg/L

MDL   6.9 µg/L

PRECISION   0.20X + 0.41 µg/L (overall precision).

ACCURACY   0.93C + 1.76 µg/L (as recovery).

SAMPLING METHOD   25-mL glass vial. Teflon®-lined septum.

STABILITY   Cool, 4°C, 0.008% Sodium thiosulfate.

MHT   14 days.

QUALITY CONTROL   The lab must on an ongoing basis, spike at least 5% of the samples from each sample site being monitored to assess accuracy.

REFERENCE   Method 624, *Federal Register* Part VIII 40 CFR Part 136, Oct 26, 1984.

1,1,2,2-Tetrachloroethane   **EPA Method 8010**
CAS #79-34-5

TITLE   Halogenated Volatile Organics

MATRIX   Groundwater, soils, sludges, water miscible liquid wastes, and non-water miscible wastes.

APPLICATION   This method is used for the analysis of 39 halogenated VOCs. Samples are analyzed using direct injection or purge-and-trap methods. Groundwater must be analyzed by the purge-and-trap method. The method provides an optional

GC column which is used for analyte confirmation and that may help resolve analytes from interferences.

INTERFERENCES There can be carryover contamination with high- and low-level samples. Impurities may come from the purge-and-trap apparatus, organic compounds outgassing from the plumbing ahead of trap, diffusion of VOCs through the sample bottle septum during shipping or storage, or from solvent vapors in the lab.

INSTRUMENTATION GC capable of on-column injections or purge-and-trap sample introduction and a halogen specific detector. Column 1: 8 ft by 0.1 in 1%. SP-1000 on Carbopack-B. Column 2: 6 ft by 0.1 in bonded n-octane on Porasil-C.

RANGE 8–500 µg/L (reagent water).

MDL 0.03 µg/L (reagent water).

PQL FACTORS FOR MULTIPLYING × MDL VALUE

Matrix	Multiplication Factor
Groundwater	10
Low-level soil	10
Water miscible liquid waste	500
High-level soil and sludge	1250
Non-water miscible waste	1250

PRECISION 0.23X + 2.79 µg/L (overall precision).

ACCURACY 0.95C + 0.19 µg/L (as recovery).

SAMPLING METHOD For water and liquid samples; use glass 40-mL vials with Teflon®-lined septum caps and collect two vials per sample location with no headspace. For solids and concentrated waste samples; use widemouth glass bottles with Teflon® liners.

STABILITY For concentrated wastes, soils, sediments, or sludges: cool to 4°C. For liquids: add 4 drops of concentrated hydrochloric acid and cool to 4°C.

MHT 14 days.

QUALITY CONTROL Analyze a reagent blank, matrix spike, and matrix spike duplicate/duplicate for each analytical batch (up to 20 samples). Demonstrate the purity of glassware and reagents by analyzing a reagent water method blank. Internal, surrogate, and five concentration level calibration standards are used.

REFERENCE Test Methods for Evaluating Solid Waste (SW-846), U.S. EPA Office of Solid Waste, Washington, DC, Method 8010B, Rev. 2, Nov. 1992.

Tetrachloroethylene   **EPA Method 1624**
CAS #127-18-4

TITLE Volatile Organic Compounds by Isotope Dilution GC/MS

MATRIX Compounds may be determined in waters, soils, and municipal sludges by this method.

METHOD SUMMARY This method is used to determine 58 volatile toxic organic pollutants associated with the CWA (as amended 1987); the RCRA (as amended 1986); the CERCLA (as amended 1986); and other compounds amenable to purge-and-trap gas chromatography-mass spectrometry (GC/MS).

If the solids content is less than 1%, stable isotopically-labeled analogs of the compounds of interest are added to a 5-mL sample and the sample is purged with an inert gas at 20–25°C in a chamber designed for soil or water samples. If the solids content is greater than 1%, 5 mL of reagent water and the labeled compounds are added to a 5-g aliquot of sample and the mixture is purged at 40°C. Compounds that will not purge at 20–25°C or at 40°C are purged at 78–85°C. In the purging process, the volatile compounds are transferred from the aqueous phase into the gaseous phase where they are passed into a sorbent column, and trapped. After purging is completed, the trap is backflushed and heated rapidly to desorb the compounds into a GC. The compounds are separated by the GC and detected by a MS. The labeled compounds serve to correct the variability of the analytical technique.

INTERFERENCES Impurities in the purge gas, organic compounds outgassing from the plumbing upstream of the trap, and solvent vapors in the lab account for most problems. Samples can be contaminated by diffusion of volatile organic compounds (particularly methylene chloride) through the bottle seal during shipment and storage. Contamination by carryover can occur when high-level and low-level samples are analyzed sequentially. When an unusually concentrated sample is encountered, follow it by analysis of a reagent water blank to check for carryover.

INSTRUMENTATION Major equipment includes a GC with linear temperature programming and a glass jet separator as the MS interface, a MS with 70 eV electron impact ionization, and a data system to collect and record response factors.

Column: 2.8 m × 2 mm I.D. glass, packed with 1% SP-1000 on Carbopak B, 60/80 mesh, or equivalent.

PRECISION & ACCURACY The detection limits of the method are usually dependent on the level of interferences rather than instrumental limitations. The method detection limits were determined in digested sludge (low solids) and in filter cake or compost (high solids).

The MDL (in µg/kg) for low solids is 106 and for high solids is 10.

Labeled and native compound precision (in µg/L) as standard deviation was 6.6.

Labeled and native compound accuracy (in µg/L) as average recovery was 15–29.

Acceptance criteria are at 20 µg/L for this compound.

SAMPLE COLLECTION, PRESERVATION & HANDLING Grab samples are collected in glass containers having a total volume greater than 20 mL. Fill and seal each bottle so that no air bubbles are entrapped. Samples are maintained at 0 to 4°C from the time of collection until analysis. If an aqueous sample contains residual chlorine, add sodium thiosulfate preservative (10 mg/40 mL) to the empty sample bottles just prior to shipment

to the sample site. All samples must be analyzed within 14 days of collection.

SAMPLE PREPARATION Samples containing less than 1% solids are analyzed directly as aqueous samples. Samples containing 1% solids or greater are analyzed as solid samples utilizing one of two methods, depending on the levels of pollutants, in the sample. Samples containing 1% solids or greater, and low to moderate levels of pollutants are analyzed by purging a known weight of sample added to 5 mL of reagent water. Samples containing 1% solids or greater, and high levels of pollutants, are extracted with methanol, and an aliquot of the methanol extract is added to reagent water and purged.

QUALITY CONTROL A field blank prepared from reagent water and carried through the sampling and handling protocol may serve as a check on contamination from shipment and storage.

The analyst is permitted to modify this method to improve separations or lower the costs of measurements, provided all performance specifications are met. Analyses of blanks are required. When results of spikes indicate atypical method performance for samples, the samples are diluted to bring method performance within acceptable limits. Analyze two sets of four 5-mL aliquots (8 aliquots total) of the aqueous performance standard. Spike all samples with labeled compounds to assess method performance on the sample matrix. Compute the percent recovery of the labeled compounds using the internal standard method. Compare the percent recovery for each compound with the corresponding labeled compound recovery. Reagent water blanks are analyzed to demonstrate freedom from carryover contamination. Field replicates may be collected to determine the precision of the sampling technique, and spiked samples may be required to determine the accuracy of the analysis when the internal method is used.

REFERENCE Volatile Organic Compounds by Isotope Dilution GC/MS. Office of Water Regulation and Standards, U.S. EPA Industrial Technology Division, Washington, DC, EPA Method 1624, Rev. C, June 1989 (contact W.A. Telliard, U.S. EPA, Office of Water Regulations and Standards, 401 M St., SW, Washington, DC, 20460. Phone: 202-382-7131).

Tetrachloroethylene   **EPA Method 502**
CAS #127-18-4

TITLE Volatile Organic Compounds in Water By Purge and Trap Capillary Column Gas Chromatography with Photoionization and Electrolytic Conductivity Detectors in Series. U.S. EPA Method 502.2, Rev. 2.0, 1989.

MATRIX Drinking water and raw source water. The latter should include most surface water and groundwater sources.

METHOD SUMMARY This method covers 60 volatile organic compounds that contain halogen atoms and/or that are aromatic. An inert gas (zero grade nitrogen or helium) is bubbled through a 25-mL or a 5-mL water sample (depending on the expected concentration of the analytes). Purged sample components are trapped in a tube of sorbent materials. When purging is complete, the sorbent tube is heated and backflushed with helium to desorb the trapped sample onto a capillary GC column. The column is temperature programmed to separate the method analytes which are then detected with a photoionization detector (PID) and a Hall electrolytic conductivity (HECD) placed in series. The PID is selective for aromatic compounds and the HECD is selective for halogenated compounds.

INTERFERENCES Impurities in the purge gas and from organic compounds outgassing from the plumbing ahead of the trap account for many contamination problems. Interferences purged or coextracted from the samples will vary considerably from source to source, depending upon the particular sample or extract being tested. Cross-contamination can occur whenever high-level and low-level samples are analyzed sequentially. Samples also can be contaminated by diffusion of volatile organics (particularly methylene chloride and fluorocarbons) through the septum seal into the sample during shipment and storage. The lab where volatile analysis is performed and also the refrigerated storage area should be completely free of solvents.

INSTRUMENTATION A GC containing a series configuration of a high temperature photoionization detector (PID) equipped with 10.0 eV (nominal) lamp and Hall electrolytic conductivity detector (HECD) is required. Also required is an all-glass 5-mL purging device, a sorbent trap, and a thermal desorption apparatus which is connected to the GC system.

Column 1: VOCOL glass wide-bore capillary column.
Column 2: RTX–502.2 mega-bore capillary column.
Column 3: DB-62 mega-bore capillary column.

PRECISION & ACCURACY Method detection limits are dependent upon the characteristics of the gas chromatographic system used. Analytes that are not separated chromatographically cannot be individually identified and used in the same calibration mixture or water samples unless an alternative technique for identification and quantification, such as mass spectrometry, is used.

Electrolytic conductivity detetor (c) range in µg/L (a) was 0.02–200.
Electrolytic conductivity detetor (c) MDL in µg/L (b) was 0.04.
Electrolytic conductivity detetor (c) accuracy as % recovery was 97.
Electrolytic conductivity detetor (c) precision as % RSD was 2.5.
Photoionization detector (d) range in µg/L (a) was 0.02–200.
Photoionization detector (d) MDL in µg/L (b) was 0.05.
Photoionization detector (d) accuracy as % recovery was 101.
Photoionization detector (d) precision as % RSD was 1.8.

(a) *The applicable concentration range of this method is compound, instrument, and matrix-dependent. It is listed as being approximately 0.02 to 200 µg/L but no specific information is provided so caution should be observed.*
(b) *The method detection limits reports with this method are compound, instrument, and matrix-dependent. The values reported were calculated using reagent water fortified with the corresponding compounds at 10 µg/L and a GC-equipped with a 60 m × 0.75 mm VOLCOL wide bore*

capillary column with 1.5 μm film thickness and using helium carrier gas.

(c) *Recoveries and relative standard deviations were determined from seven samples of reagent water fortified with 10 μg/L of each compound. 2-Bromo-1-chloropropane was used as the internal standard for calculating average recoveries.*

(d) *Recoveries and relative standard deviations were determined from seven samples of reagent water fortified with 10 μg/L of each compound. Fluorobenzene was used as the internal standard for calculating average recoveries.*

SAMPLING METHOD Collect samples using a 40- to 120-mL screw-cap vial (prewashed with detergent, rinsed with distilled water and oven dried at 105°C) with a Teflon®-faced silicone septum. Collect bubble-free samples and place the septum with the Teflon® side down on the water.

SAMPLE PRESERVATION If residual chlorine is present in the water add about 25 mg of ascorbic acid to each vial before samples are collected to remove the chlorine. Add hydrochloric acid to reduce pH to <2, immediately cool samples to 4°C, and store them in a solvent-free refrigerator at 4°C until analysis.

MHT The maximum holding time for samples is 14 days from the time they were collected.

SAMPLE PREPARATION Remove the plungers from two 5-mL syringes and attach a closed syringe valve to each. Warm the sample to room temperature, open the sample bottle, and carefully pour the sample into one of the syringe barrels to just short of overflowing. Replace the syringe plunger, invert the syringe, and compress the sample. Open the syringe valve and vent any residual air while adjusting the sample volume to 5.0 mL. Add 10 μL of the internal calibration standard to the sample through the syringe valve. Close the valve. Fill the second syringe in an identical manner from the same sample bottle. Reserve this second syringe for a reanalysis if necessary.

QUALITY CONTROL As an initial demonstration of lab accuracy and precision, analyze 4 to 7 replicates of a lab fortified blank containing analyte at 0.1–5 μg/L. Collect all samples in duplicate. Surrogate analytes (similar to those of the analytes of interest), whose concentration is known in every sample, are measured using the same internal standard calibration procedure. Duplicate field reagent water blanks (trip blanks) must be analyzed with each set of samples, lab reagent blanks (method blanks) must be analyzed with each batch of samples processed as a group within a work shift. Also, a single lab-fortified blank that contains each of the analytes of interest should be analyzed with each batch of samples processed as a group within a work shift. A 3- to 5-point calibration curve is needed depending on the calibration range factor required.

EPA CONTACT & HOTLINE For technical questions contact Dr. Baldev Bathija, U.S. EPA, Office of Ground Water and Drinking Water (WH-550D), 401 M St. SW, Washington, DC 20460. Tel. (202) 260-3040. For further information the EPA Safe Drinking Water Hotline may be called at: (800) 426-4791.

REFERENCE Methods for the Determination of Organic Compounds in Drinking Water, EPA/600/4-88/039 (revised July 1991; Final Rule for determination of compliance with the MCL for Total Trihalomethanes under 141.30, in 40 CFR Part 141, Vol. 58, No. 147, Fed. Reg., Tuesday Aug. 3, 1993). U.S. EPA Environmental Monitoring Systems Laboratory, Cincinnati, OH, 45268, U.S.A. Available from the National Technical Information Service (NTIS), 5285 Port Royal Road, Springfield, VA 22161; Tel. 800-553-6847. NTIS Order Number is PB91-231480.

Tetrachloroethylene   **EPA Method 524**
CAS #127-18-4

TITLE Measurement of Purgeable Organic Compounds in Water by Capillary Column GC/MS.

MATRIX Drinking water and raw source water; the latter should include most surface water and groundwater sources.

METHOD SUMMARY Method 524.2 covers 60 volatile organic compounds. An inert gas (zero grade nitrogen or helium) is bubbled through a 25-mL or a 5-mL water sample (depending on the expected concentration of the analytes). Purged sample components are trapped in a tube of sorbent materials. When purging is complete, the sorbent tube is heated and backflushed with helium to desorb the trapped sample onto a capillary GC column.

INTERFERENCES Impurities in the purge gas and from organic compounds outgassing from the plumbing ahead of the trap account for many contamination problems. Interferences purged or coextracted from the samples will vary considerably from source to source, depending upon the particular sample or extract being tested. Cross-contamination can occur whenever high-level and low-level samples are analyzed sequentially. Samples also can be contaminated by diffusion of volatile organics (particularly methylene chloride and fluorocarbons) through the septum seal into the sample during shipment and storage.

INSTRUMENTATION A GC/MS with a data system equipped with one of the following capillary GC columns:

Column 1: VOCOL glass wide bore capillary column.
Column 2: DB-624 fused silica capillary column.
Column 3: DB-5 fused silica capillary column.

Also required is an all-glass 25 mL or 5-mL purging device, a sorbent trap, and a thermal desorption apparatus which is connected to the GC/MS system.

PRECISION & ACCURACY Method detection limits are compound- and instrument-dependent, and may vary from approximately 0.02–0.35 μg/L. Note in the table below that the "true" concentration range used for accuracy and precision measurements was quite narrow. However, the applicable concentration range of this method is primarily column dependent and is approximately 0.02 to 200 μg/L for the wide-bore thick-film columns. Narrow-bore thin-film columns may have a capacity which limits the range to about 0.02 to 20 μg/L. Analytes that are inefficiently purged from water will not be detected when present at low concentrations, but they can be measured with acceptable accuracy and precision when present in sufficient amounts.

Analytes that are not separated chromatographically, but which have different mass spectra and non-interfering quantification ions, can be identified and measured in the same calibration mixture or water sample. Analytes which have very similar mass spectra cannot be individually identified and measured in the same calibration mixture or water samples unless they have different retention times. Co-eluting compounds with very similar mass spectra, typically many structural isomers, must be reported as an isomeric group or pair.

The range (in μg/L) was 0.5–10.
The Method Detection Limig (in μg/L) was 0.14.
The accuracy (as % recovery) was 89.
The precision (in %) was 6.8.

Note: Data were obtained from 16–31 determinations using a wide-bore capillary column and a jet separator interfaced to a quadrupole mass spectrometer. All analytes were in a reagent water matrix.

SAMPLING METHOD Collect samples using a 40- to 120-mL screw-cap vial (prewashed with detergent, rinsed with distilled water and oven dried at 105°C) with a Teflon®-faced silicone septum. Collect bubble-free samples and place the septum with the Teflon® side down on the water.

SAMPLE PRESERVATION If residual chlorine is present in the water add about 25 mg of ascorbic acid to each vial before samples are collected to remove the chlorine. Add hydrochloric acid to reduce pH to <2, and immediately cool samples to 4°C, and store them in a solvent-free refrigerator at 4°C until analysis.

MHT The maximum holding time for samples is 14 days from the time they were collected.

SAMPLE PREPARATION Remove the plungers from two 25-mL (or 5-mL depending on sample size) syringes and attach a closed syringe valve to each. Warm the sample to room temperature, open the sample bottle, and carefully pour the sample into one of the syringe barrels to just short of overflowing. Replace the syringe plunger, invert the syringe, and compress the sample. Open the syringe valve and vent any residual air while adjusting the sample volume to 25.0 mL (or 5 mL). For samples and blanks, add 5 μL of the fortification solution containing the internal standard and the surrogates to the sample through the syringe valve. For calibration standards and lab fortified blanks, add 5 μL of the fortification solution containing the internal standard only. Close the valve. Fill the second syringe in an identical manner from the same sample bottle. Reserve this second syringe for a reanalysis if necessary.

QUALITY CONTROL As an initial demonstration of lab accuracy and precision, analyze 4 to 7 replicates of a lab fortified blank containing analyte at 0.2–5 μg/L. Collect all samples in duplicate. Surrogate analytes (similar to those of the analytes of interest), whose concentration is known in every sample, are measured using the same internal standard calibration procedure. Duplicate field reagent water blanks (trip blanks) must be analyzed with each set of samples, lab reagent blanks (method blanks) must be analyzed with each batch of samples processed as a group within a work shift. Also, a single lab-fortified blank that contains each of the analytes of interest should be analyzed with each batch of samples processed as a group within a work shift. A 3- to 5-point calibration curve is needed depending on the calibration range factor required.

EPA CONTACT & HOTLINE For technical questions contact Dr. Baldev Bathija, U.S. EPA, Office of Ground Water and Drinking Water (WH-550D), 401 M St. SW, Washington, DC 20460. Tel. (202) 260-3040. For further information the EPA Safe Drinking Water Hotline may be called at: (800) 426-4791.

REFERENCE Methods for the Determination of Organic Compounds in Drinking Water, EPA/600/4-88/039 (revised July 1991; Final Rule for determination of compliance with the MCL for Total Trihalomethanes under 141.30, in 40 CFR Part 141, Vol. 58, No. 147, Fed. Reg., Tuesday Aug. 3, 1993). U.S. EPA Environmental Monitoring Systems Laboratory, Cincinnati, OH, 45268, U.S.A. Available from the National Technical Information Service (NTIS), 5285 Port Royal Road, Springfield, VA 22161; Tel. 800-553-6847. NTIS Order Number is PB91-231480.

Tetrachloroethylene   **EPA Method 8021**
CAS #127-18-4

TITLE Halogenated Volatile by Gas Chromatography Using Photoionization and Electrolytic Conductivity Detectors in Series: Capillary Column Technique

MATRIX This method is applicable to nearly all types of samples, regardless of water content, including groundwater, aqueous sludges, caustic liquors, acid liquors, waste solvents, oily wastes, mousses, tars, fibrous wastes, polymeric emulsions, filter cakes, spent carbons, spent catalysts, soils, and sediments.

METHOD SUMMARY This method is used to determine 60 volatile organic compounds in a variety of solid waste matrices. It provides GC conditions for the detection of halogenated and aromatic volatile organic compounds. Samples can be analyzed using direct injection or purge-and-trap (EPA Method 5030). Groundwater samples must be analyzed using EPA Method 5030 (where applicable). A temperature program is used with the GC. Detection is achieved by a photoionization detector (PID) and a Hall electrolytic conductivity detector (HECD) in series.

INTERFERENCES Samples can be contaminated by diffusion of volatile organics (particularly chlorofluorocarbons and methylene chloride) through the sample container septum during shipment and storage.

INSTRUMENTATION A GC-equipped with variable-constant differential flow controllers, subambient oven controller, PID and HECD detectors connected with a short piece of uncoated capillary tubing and a data system.

Column: 60 m × 0.75 mm I.D. VOCOL wide-bore capillary column with 1.5 μm film thickness.

PRECISION & ACCURACY MDLs are compound-dependent and vary with purging efficiency and concentration. The applicable concentration range of this method is compound- and instrument-dependent but is approximately 0.1 to 200 μg/L. Analytes that are inefficiently purged from water will not be detected when present at low concentrations, but they can be measured with acceptable accuracy and precision when present in sufficient amounts. The estimated quantitation limit (EQL) for an individual compound is approximately 1 µg/kg (wet weight) for soil/sediment samples, 100 µg/kg (wet weight) for wastes, and 1 µg/L for groundwater. EQLs will be proportionately higher for sample extracts and samples that require dilution or reduced sample size to avoid saturation of the detector.

MULTIPLICATION FACTORS FOR OTHER MATRICES (a)

Matrix	Factor (b)
Groundwater	10
Low-concentration soil	10
Water miscible liquid waste	500
High-concentration soil and sludge	1250
Non-water miscible waste	1250

(a) Sample EQLs are highly matrix-dependent. The EQLs listed herein are provided for guidance and may not always be achievable. (b) EQL = [Method detection limit] × [Factor]. For non-aqueous samples, the factor is on a wet-weight basis.

SINGLE LABORATORY ACCURACY & PRECISION DATA FOR VOCs IN WATER

This method was tested in a single lab using water spiked at 10 µg/L and the following data was reported:

Recoveries and standard deviations were determined from seven samples and spiked at 10 µg/L of each analyte. Recoveries were determined by the internal standard method. Internal standards were: Fluorobenzene for PID and 2-Bromo-1-chloropropane for HECD.

The average recovery (in percent) for the PID was 101.
The standard deviation of the recovery for the PID was 1.8.
The MDL (in µg/mL) for the PID was 0.05.
The average recovery (in percent) for the HECD was 97.
The standard deviation of the recovery for the HECD was 2.4.
The MDL (in µg/mL) for the HECD was 0.04.

SAMPLE COLLECTION, PRESERVATION & HANDLING

Volatile organics — Standard 40-mL glass screw-cap VOA vials with Teflon®-faced silicone septum may be used for both liquid and solid matrices. When collecting samples, liquids and solids should be introduced into the vials gently to reduce agitation which might drive off volatile compounds. If there are any air bubbles present the sample must be retaken. Tap slightly as they are filled to try and eliminate as much free air space as possible. The two vials from each sampling locations should be sealed in separate plastic bags to prevent cross-contamination between samples particularly if the sampled waste is suspected of containing high levels of volatile organics.

Semivolatile organics — Containers used to collect samples for the determination of semivolatile organic compounds should be soap and water washed followed by methanol (or isopropanol) rinsing. The sample containers should be of glass or Teflon® and have screw-top covers with Teflon® liners.

Preservation for volatile organics — No preservation is used with concentrated waste samples. With liquid samples containing no residual chlorine, 4 drops of concentrated hydrochloric acid are added and the samples are immediately cooled to 4°C.

When liquid samples contain residual chlorine, they are treated as above and, in addition, 4 drops of 4% aqueous sodium thiosulfate are added. Soil, sediment, and sludge samples are only cooled to 4°C.

Preservation for semivolatile organics — No preservation is used with concentrated waste samples. With liquid samples containing no residual chlorine and with soil, sediment, and sludge samples, immediately cooling to 4°C is the only preservation used. When residual chlorine is present then 3 mL of 10% aqueous sodium sulfate is added for each gallon of sample collected, followed by cooling to 4°C.

MHT The holding time for all volatile organics samples is 14 days. Liquid samples must be extracted within 7 days and their extracts analyzed within 40 days. Concentrated waste, soil, sediment, and sludge samples must be extracted within 14 days and their extracts analyzed within 40 days.

SAMPLE PREPARATION Volatile compounds are introduced into the gas chromatograph either by direct injector or purge-and-trap (EPA Method 5030). EPA Method 5030 may be used directly on groundwater samples or low-concentration contaminated soils and sediments. For medium-concentration soils or sediments, methanolic extraction, as described in EPA Method 5030, may be necessary prior to purge-and-trap analysis.

QUALITY CONTROL Calculate surrogate standard recovery on all samples, blanks, and spikes. A trip blank is recommended to check on sampling, storage, and handling contamination. Calibration standards, at a minimum of five concentration levels, are prepared in organic-free reagent water. One of the concentration levels should be at a concentration near, but above, the method detection limit.

A combination of bromochloromethane, 2-bromo-1-chloropropane, 1,4-dichlorobutane, and bromochlorobenzene are recommended as surrogate standards to encompass the range of the temperature program used in this method.

REFERENCE Test Methods for Evaluating Solid Waste, Physical/Chemical Methods, SW-846, 3rd Edition, U.S. EPA, Office of Solid Waste, Washington, DC, EPA Method 8021A, Rev. 1, Nov. 1992.

Tetrachloroethylene   **EPA Method 8240**
CAS #127-18-4

TITLE Volatile Organics By GC/MS: Packed Column Technique

MATRIX Nearly all types of sample matarices, regardless of water content, can be analyzed using this method. This includes groundwater, aqueous sludges, caustic liquors, acid liquors, waste solvents, oily wastes, mousses, tars, fibrous wastes, polymetric emulsions, filter cakes, spent carbons, spent catalysts, soils, and sediments.

METHOD SUMMARY Method 8240B covers 80 volatile organic compounds that are introduced into a gas chromatograph by the purge-and-trap method or by direct injection (in limited applications). For the purge-and-trap method an inert gas (zero grade nitrogen or helium) is bubbled through a 5-mL solution at ambient temperature. Purged sample components are trapped in a tube of sorbent materials. When purging is complete, the sorbent tube is heated and backflushed with inert gas to desorb the trapped components onto a GC column.

INTERFERENCES Impurities in the purge gas and from organic compounds outgassing from the plumbing ahead of the trap account for many contamination problems. Interferences purged or coextracted from the samples will vary considerably from source to source. Cross-contamination can occur whenever high-level and low-level samples are analyzed sequentially. Whenever an unusually concentrated sample is analyzed, it should be followed by the analysis of organic-free reagent water to check for cross-contamination. Samples also can be contaminated by diffusion of volatile organics (particularly methylene chloride and fluorocarbons) through the septum seal into the sample during shipment and storage. A trip blank can serve as a check on such contamination. The lab where volatile analysis is performed and also the refrigerated storage area should be completely free of solvents.

INSTRUMENTATION A gas chromatograph/mass spectrometry/data system (GC/MS) equipped with a 6 ft × 0.1 in I.D. glass column packed with 1% SP-1000 on Carbopack-B (60/80 mesh) is required. Also needed is a 5-mL purging device, a sorbent trap, and a thermal desorption apparatus.

PRECISION & ACCURACY This method is reported to have been tested by 15 laboratories using organic-free reagent water, drinking water, surface water, and industrial wastewaters (not specified) fortified at six concentrations over the range 5–600 µg/L.

Sample estimated quantitation limits (EQLs) are highly matrix-dependent. The EQLs listed may not always be achievable. EQLs listed for soils or sediments are based on wet weight. Normally, data is reported on a dry-weight basis; therefore, EQLs will be higher, based on the percent dry weight of each sample. Note that EQLs are even more variable than MDLs and that they are highly variable depending on the matrix being analyzed.

EQL in groundwater in µg/L was 5.
EQL in low soil or sediment in µg/kg was 5.
Accuracy (a) in µg/L was 1.06C + 0.60.
Precision (b) in µg/L was 0.16x-0.45.

(a) *Average recovery found for measurements of samples containing a concentration of C, in µg/L.*
(b) *Overall precision found for measurements of samples with average recovery X for samples containing a concentration of C in µg/L.*
X = *Average recovery found for measurement of samples containing a concentration of C in µg/L.*

MULTIPLICATION FACTORS FOR OTHER MATRICES

Other Matrices	Factor (a)
Waste miscible liquid waste	50
High-concentration soil and sludge	125
Non-water miscible waste	500

(a) *EQL = [EQL for low soil sediment X [Factor]. For non-aqueous samples, the factor is on a wet-weight basis.*

SAMPLING METHOD
Liquid samples — Use a 40-mL glass screw-cap VOA vial with a Teflon®-faced silicone septum that has been prewashed, rinsed with distilled deionized water, and oven dried. However, if residual chlorine is present, collect sample in a 40-oz. soil VOA container which has been pre-preserved with 4 drops of 10% sodium thiosulfate, mix gently, and then transfer the sample to a 40-mL VOA vial. Collect bubble-free samples in duplicate and seal them in separate plastic bags.

Soils or sediments, and sludges — Use an 8-oz. widemouth glass bottle with a Teflon®-faced silicone septum that has been prewashed with detergent, rinsed with distilled deionized water, and oven dried. Tap slightly to eliminate free air space. Collect samples in duplicate and seal them in separate plastic bags.

SAMPLE PRESERVATION
Liquid samples — Add 4 drops of concentrated HCL and immediately cool samples to 4°C and store in a solvent-free refrigerator.

Soils or sediments, and sludges — Cool samples to 4°C and store in a solvent-free refrigerator.

MHT Maximum holding time is 14 days from the date of sample collection.

SAMPLE PREPARATION
Liquid samples — Remove the plunger from a 5-mL syringe and carefully pour the sample into the syringe barrel to just short of overflowing. Replace the syringe plunger and compress the sample. Open the syringe valve and vent any residual air while adjusting the sample volume to 5.0 mL. If there is only one volatile organic analysis (VOA) vial, a second syringe should be filled at this time to protect against possible loss of sample integrity. Add 10 µL of surrogate spiking solution and 10 µL of internal standard spiking solution through the valve bore of the 5-mL syringe, then close the valve. The surrogate and internal standards may be mixed and added as a single spiking solution.

Sediments, soils, and waste samples — All samples of this type should be screened by GC analysis using a headspace method (EPA Method 3810) or the hexadecane extraction and screening method (EPA Method 3820). Use the screening data to determine whether to use the low-concentration method (0.005–1 mg/kg) or the high-concentration method (>1 mg/kg).

Low-concentration method — The low-concentration method is based on purging a heated sediment or soil sample mixed with organic-free reagent water containing the surrogate and internal standards. Analyze all reagent blanks and standards under the same conditions as the samples.

Use a 5-g sample if the expected concentration is <0.1 mg/kg or a 1-g sample for expected concentrations between 0.1 and 1 mg/kg. Mix the contents of the sample container with a narrow metal spatula. Weigh the amount of the sample into a tared purge device. Add the spiked water to the purge device, which contains the weighed amount of sample, and connect the device to the purge-and-trap system.

High-concentration method — This method is based on extracting the sediment or soil with methanol. A waste sample

is either extracted or diluted, depending on its solubility in methanol. Wastes that are insoluble in methanol are diluted with reagent tetraglyme or possibly polyethylene glycol (PEG). An aliquot of the extract is added to organic-free reagent water containing surrogate and internal standards. This is purged at ambient temperature. All samples with an expected concentration of >1.0 mg/kg should be analyzed by this method.

Mix the contents of the sample container with a narrow metal spatula. For sediments or soils and solid wastes that are insoluble in methanol, weigh 4 g (wet weight) of sample into a tared 20-mL vial. For waste that is soluble in methanol, tetraglyme, or PEG, weigh 1 g (wet weight) into a tared scintillation vial or culture tube or a 10-mL volumetric flask. Quickly add 9.0 mL of appropriate solvent then add 1.0 mL of a surrogate spiking solution to the vial, cap it, and shake it for 2 min.

METHANOL EXTRACT REQUIRED FOR ANALYSIS OF HIGH-CONCENTRATION SOILS OR SEDIMENTS

Approximate Concentration Range	Volume of Methanol Extract (a)
500–10,000 µg/kg	100 µL
1,000–20,000 µg/kg	50 µL
5,000–100,000 µg/kg	10 µL
25,000–500,000 µg/kg	100 µL of 1/50 dilution (b)

Calculate appropriate dilution factor for concentrations exceeding this table.

(a) The volume of methanol added to 5 mL of water being purged should be kept constant. Therefore, add to the 5-mL syringe whatever volume of methanol is necessary to maintain a volume of 100 µL added to the syringe.

(b) Dilute an aliquot of the methanol extract and then take 100 µL for analysis.

QUALITY CONTROL Demonstrate, through the analysis of a reagent water blank, that interferences from the analytical system, glassware, and reagents are under control. Blank samples should be carried through all stages of the sample preparation and measurement steps. For each analytical batch (up to 20 samples), a reagent blank, matrix spike, and matrix spike duplicate must be analyzed (the frequency of the spikes may be different for different monitoring programs). The blank and spiked samples must be carried through all stages of the sample preparation and measurement steps. QC samples mentioned in the section on Interferences will also be needed as appropriate to those situations.

REFERENCE Test Methods for Evaluating Solid Waste (SW-846). U.S. EPA. 1983. Method 8240B, Rev. 2, Nov. 1990. Office of Solid Wastes, Washington, DC.

Tetrachloroethylene   **EPA Method 8260**
CAS #127-18-4

TITLE Volatile Organic Compounds by GC/MS: Capillary Column Technique

MATRIX This method is applicable to nearly all types of samples, regardless of water content, including groundwater, soils, and sediments.

METHOD SUMMARY Method 8260A covers 58 volatile organic compounds that are introduced into a gas chromatograph by the purge-and-trap method or by direct injection (in limited applications). Zero-grade helium is bubbled through a 5-mL solution at ambient temperature. Purged sample components are trapped in a tube containing suitable sorbent materials. When purging is complete, the sorbent tube is heated and backflushed with helium to desorb trapped sample components. The analytes are desorbed directly to a large bore capillary or cryofocussed on a capillary precolumn before being flash evaporated to a narrow bore capillary for analysis.

INTERFERENCES Major contaminant sources are volatile materials in the lab and impurities in the inert purging gas and in the sorbent trap. Interfering contamination may occur when a sample containing low concentrations of volatile organic compounds is analyzed immediately after a sample containing high concentrations of volatile organic compounds. After analysis of a sample containing high concentrations of volatile organic compounds, one or more calibration blanks should be analyzed to check for cross-contamination. Screening of the samples prior to purge-and-trap GC/MS analysis is highly recommended to prevent contamination of the system. This is especially true for soil and waste samples.

Special precautions must be taken to analyze for methylene chloride. The analytical and sample storage area should be isolated from all atmospheric sources of methylene chloride. All gas chromatography carrier gas lines and purge gas plumbing should be constructed from stainless steel or copper tubing. Laboratory clothing previously exposed to methylene chloride fumes during liquid-liquid extraction procedures can contribute to sample contamination.

Samples can also be contaminated by diffusion of volatile organics (particularly methylene chloride and fluorocarbons) through the septum seal during shipment and storage. A trip blank can serve as a check on such contamination.

INSTRUMENTATION GC/MS with a temperature-programmable chromatograph suitable for splitless injection equipped with variable constant differential flow controllers, a subambient oven controller, a purging device, sorbent trap, a thermal desorption apparatus and a capillary precolumn interface when using cryogenic cooling will be needed. The following GC columns may be used:

Column 1: 60 m × 0.75mm I.D. capillary column coated with VOCOL, 1.5 µm film thickness.
Column 2: 30 m × 0.53mm capillary column coated with DB-624 or VOCOL, 3 µm film thickness.
Column 3: 30 m × 0.32mm I.D. capillary column coated with DB-5 or SE-54, 1-µm film thickness.

PRECISION & ACCURACY This method has been tested in a single lab using spiked water. Using a wide-bore capillary column, water was spiked at concentrations between 0.5 and 10 µg/L. Single lab accuracy and precision data are presented.

The MDL actually achieved in a given analysis will vary depending on instrument sensitivity and matrix effects.

The MDL (a) in µg/L was 0.14.
The concentration range in µg/L was 0.5–10.
The mean accuracy (% of true value) was 89.
The precision as relative standard deviation was 6.8.

Note: The MDL is based on a 25-mL sample volume instead of a 5-mL sample volume.

SAMPLING METHOD
Liquid samples — Use a 40-mL glass screw-cap VOA vial with a Teflon®-faced silicone septum that has been prewashed, rinsed with distilled deionized water, and oven dried. If residual chlorine is present, collect the sample in a 4-oz soil VOA container which has been pre-preserved with 4 drops of 10% sodium thiosulfate. Mix gently and transfer the sample to a 40-mL VOA vial. Collect bubble-free samples in duplicate and seal each sample in a separate plastic bag.

Soils, sediments and sludges — Use an 8-oz widemouth glass bottle with Teflon®-faced silicone septum that has been prewashed, rinsed with distilled deionized water, and oven dried. **Do not** heat the septum for more than 1 h. Tap slightly to eliminate any free air space. Collect samples in duplicate and seal each one in a separate plastic bag.

SAMPLE PRESERVATION
Liquid samples — Add 4 drops of concentrated HCL, cool to 4°C and store in a solvent-free refrigerator.

Soils, sediments and sludges — Cool samples to 4°C and store in a solvent-free refrigerator.

MHT The maximum holding time of any sample (liquids, soils, sediments, and sludges) is 14 days.

SAMPLE PREPARATION
Liquid samples — Remove the plunger from a 5-mL syringe and carefully pour the sample into the syringe barrel to just short of overflowing. Replace the syringe plunger and compress the sample. Open the syringe valve and vent any residual air while adjusting the sample volume to 5.0 mL. If there is only one volatile organic analysis (VOA) vial, a second syringe should be filled at this time to protect against possible loss of sample integrity. Add 10 µL of surrogate spiking solution and 10 µL of internal standard spiking solution through the valve bore of the 5-mL syringe, then close the valve. The surrogate and internal standards may be mixed and added as a single spiking solution.

Sediments, soils, and waste samples — All samples of this type should be screened by GC analysis using a headspace method (EPA Method 3810) or the hexadecane extraction and screening method (EPA Method 3820). Use the screening data to determine whether to use the low-concentration method (0.005–1 mg/kg) or the high-concentration method (>1 mg/kg).

Low-concentration method — The low-concentration method is based on purging a heated sediment or soil sample mixed with organic-free reagent water containing the surrogate and internal standards. Analyze all reagent blanks and standards under the same conditions as the samples.

Use a 5-g sample if the expected concentration is <0.1 mg/kg or a 1-g sample for expected concentrations between 0.1 and 1 mg/kg. Mix the contents of the sample container with a narrow metal spatula. Weigh the amount of the sample into a tared purge device. Add the spiked water to the purge device, which contains the weighed amount of sample, and connect the device to the purge-and-trap system.

High-concentration method — This method is based on extracting the sediment or soil with methanol. A waste sample is either extracted or diluted, depending on its solubility in methanol. Wastes that are insoluble in methanol are diluted with reagent tetraglyme or possibly polyethylene glycol (PEG). An aliquot of the extract is added to organic-free reagent water containing surrogate and internal standards. This is purged at ambient temperature. All samples with an expected concentration of >1.0 mg/kg should be analyzed by this method.

Mix the contents of the sample container with a narrow metal spatula. For sediments or soils and solid wastes that are insoluble in methanol, weigh 4 g (wet weight) of sample into a tared 20-mL vial. For waste that is soluble in methanol, tetraglyme, or PEG, weigh 1 g (wet weight) into a tared scintillation vial or culture tube or a 10-mL volumetric flask. Quickly add 9.0 mL of appropriate solvent then add 1.0 mL of a surrogate spiking solution to the vial, cap it, and shake it for 2 min.

METHANOL EXTRACT REQUIRED FOR ANALYSIS OF HIGH-CONCENTRATION SOILS OR SEDIMENTS

Approximate Concentration Range	Volume of Methanol Extract (a)
500–10,000 µg/kg	100 µL
1,000–20,000 µg/kg	50 µL
5,000–100,000 µg/kg	10 µL
25,000–500,000 µg/kg	100 µL of 1/50 dilution (b)

Calculate appropriate dilution factor for concentrations exceeding this table.

(a) The volume of methanol added to 5 mL of water being purged should be kept constant. Therefore, add to the 5-mL syringe whatever volume of methanol is necessary to maintain a volume of 100 µL added to the syringe.
(b) Dilute an aliquot of the methanol extract and then take 100 µL for analysis.

QUALITY CONTROL Demonstrate, through the analysis of a reagent water blank, that interferences from the analytical system, glassware, and reagents are under control. Blank samples should be carried through all stages of the sample preparation and measurement steps. For each analytical batch (up to 20 samples), a reagent blank, matrix spike, and matrix spike duplicate must be analyzed (the frequency of the spikes may be different for different monitoring programs). The blank and spiked samples must be carried through all stages of the sample preparation and measurement steps. QC samples mentioned in the section on Interferences will also be needed as appropriate to those situations.

Matrix spiking standards should be prepared from volatile organic compounds which will be representative of the compounds being investigated. The recommended internal standards

are chlorobenzene-d5, 1,4-difluorobenzene, 1,4-dichlorobenzene-d4, and pentafluorobenzene. Using stock standard solutions, prepare secondary dilution standards containing the compounds of interest, either singly or mixed together in methanol. Store them in a vial with no headspace for no more than one week. Surrogates recommended are toluene-d8, 4-bromofluorobenzene, and dibromofluoromethane. Each sample undergoing GC/MS analysis must be spiked with 10 µL of the surrogate spiking solution prior to analysis.

REFERENCE Test Methods for Evaluating Solid Waste (SW-846). U.S. EPA 1983, Method 8260A, Rev. 1, Nov. 1990. Office of Solid Waste, Washington, DC.

Tetrachloroethylene — EPA Method 503.1
CAS #127-18-4

TITLE Aromatic & Unsaturated VOCs Water

MATRIX Drinking water (finished or in any treatment stage) and raw source water.

APPLICATION Method covers 28 aromatic and unsaturated VOCs. An inert gas is bubbled through a 5-mL water sample. Purged sample components are trapped in tube of sorbent materials. When purging is complete, sorbent tube is heated and backflushed with inert gas to desorb trapped sample onto a packed GC column.

INTERFERENCES During analysis, major contaminant sources are volatile materials in the lab and impurities in purging gas and sorbent trap. With high and low level samples, there can be carryover contamination. Excess water causes a negative baseline deflection.

INSTRUMENTATION Purge and Trap GC w/photoionization detector. (Two GC columns are recommended); Column 1: 5% SP-1200 and 1.75% Bentone 34 on Supelcoport; 5% 1,2,3-tris(2-cyanoethoxy)propane on Chromosorb W.

RANGE 2.2–600 µg/L (Drinking water)

MDL 0.01 µg/L in water

PRECISION RSD = 7.8% at 0.50 µg/L; 19 samples

ACCURACY Average recovery = 97% at 0.50 µg/L; 19 samples

SAMPLING METHOD Use a 40–120-mL screw-cap vial (prewashed with detergent, rinsed with distilled water and oven dried at 105°C) with a PTFE-faced silicone septum. If residual chlorine is in the water add about 25 mg of ascorbic acid to each vial before sample collection. Collect bubble-free samples.

STABILITY Cool to 4°C; HCl to pH <2.

MHT 14 days.

QUALITY CONTROL As an initial demonstration of lab accuracy and precision, analyze 4 to 7 replicates of a lab fortified blank containing analyte at 0.1–5 µg/L. Collect all samples in duplicate.

REFERENCE Method 503.1, Volatile Aromatic & Unsaturated Organic Compounds in H2O by Purge and Trap GC, EPA 600/4-88/039.

Tetrachloroethylene — EPA Method 601
CAS #127-18-4

TITLE Purgeable Halocarbons

MATRIX Wastewater.

APPLICATION Method covers 29 purgeable halocarbons. (Method 624 provides GC/MS conditions appropriate for the qualitative and quantitative confirmation of results). Method describes conditions for a 2nd GC column to confirm measurements made with primary column.

INTERFERENCES Impurities in the purge gas and organic compounds outgassing from the plumbing ahead of the trap. With high- and low-level samples, there can be carryover contamination. Diffusion of volatile organics through the septum seal into the sample.

INSTRUMENTATION GC-equipped with halide-specific detector. (With purge-and-trap unit).

RANGE 8.0–500 µg/L.

MDL 0.03 µg/L.

PRECISION 0.18X + 2.21 µg/L (overall precision).

ACCURACY 0.94C + 0.06 µg/L (as recovery).

SAMPLING METHOD 25-mL glass vial. Teflon®-lined septum.

STABILITY Cool, 4°C, 0.008% Sodium thiosulfate.

MHT 14 days.

QUALITY CONTROL The lab must on an ongoing basis, spike at least 10% of the samples from each sample site being monitored to assess accuracy.

REFERENCE Method 601, *Federal Register* Part VIII 40 CFR Part 136, Oct 26, 1984.

Tetrachloroethylene — EPA Method 624
CAS #127-18-4

TITLE Purgeables

MATRIX Wastewater.

APPLICATION Method covers 31 purgeable organics. An inert gas is bubbled through a 5-mL water sample in a specially designed purging chamber. Here, purgeables are transferred from aqueous to gaseous phase, passed onto a sorbent column, and trapped. Trap is heated and backflushed with inert gas to desorb purgeables onto a GC column, where purgeables are separated.

INTERFERENCES Impurities in the purge gas, organic compounds outgassing from the plumbing ahead of the trap, and solvent vapors in the lab. With high- and low-level samples, there can be carryover contamination.

INSTRUMENTATION GC/MS with purge-and-trap unit.

RANGE 5–600 µg/L

MDL 4.1 µg/L

PRECISION 0.16X–0.45 µg/L (overall precision).

ACCURACY 1.06C + 0.60 µg/L (as recovery).

SAMPLING METHOD 25-mL glass vial. Teflon®-lined septum.

STABILITY Cool, 4°C, 0.008% Sodium thiosulfate.

MHT 14 days.

QUALITY CONTROL The lab must on an ongoing basis, spike at least 5% of the samples from each sample site being monitored to assess accuracy.

REFERENCE Method 624, *Federal Register* Part VIII 40 CFR Part 136, Oct 26, 1984.

Tetrachloroethylene   EPA Method 8010
CAS #127-18-4

TITLE Halogenated Volatile Organics

MATRIX Groundwater, soils, sludges, water miscible liquid wastes, and non-water miscible wastes.

APPLICATION This method is used for the analysis of 39 halogenated VOCs. Samples are analyzed using direct injection or purge-and-trap methods. Groundwater must be analyzed by the purge-and-trap method. The method provides an optional GC column which is used for analyte confirmation and that may help resolve analytes from interferences.

INTERFERENCES There can be carryover contamination with high- and low-level samples. Impurities may come from the purge-and-trap apparatus, organic compounds outgassing from the plumbing ahead of trap, diffusion of VOCs through the sample bottle septum during shipping or storage, or from solvent vapors in the lab.

INSTRUMENTATION GC capable of on-column injections or purge-and-trap sample introduction and a halogen specific detector. Column 1: 8 ft by 0.1 in 1%. SP-1000 on Carbopack-B. Column 2: 6 ft by 0.1 in bonded n-octane on Porasil-C.

RANGE 8–500 µg/L (reagent water).

MDL 0.03 µg/L (reagent water).

PQL FACTORS FOR MULTIPLYING × MDL VALUE

Matrix	Multiplication Factor
Groundwater	10
Low-level soil	10
Water miscible liquid waste	500
High-level soil and sludge	1250
Non-water miscible waste	1250

PRECISION 0.18X + 2.21 µg/L (overall precision).

ACCURACY 0.94C + 0.06 µg/L (as recovery).

SAMPLING METHOD For water and liquid samples; use glass 40-mL vials with Teflon®-lined septum caps and collect two vials per sample location with no headspace. For solids and concentrated waste samples; use widemouth glass bottles with Teflon® liners.

STABILITY For concentrated wastes, soils, sediments, or sludges: cool to 4°C. For liquids: add 4 drops of concentrated hydrochloric acid and cool to 4°C.

MHT 14 days.

QUALITY CONTROL Analyze a reagent blank, matrix spike, and matrix spike duplicate/duplicate for each analytical batch (up to 20 samples). Demonstrate the purity of glassware and reagents by analyzing a reagent water method blank. Internal, surrogate, and five concentration level calibration standards are used.

REFERENCE Test Methods for Evaluating Solid Waste (SW-846), U.S. EPA Office of Solid Waste, Washington, DC, Method 8010B, Rev. 2, Nov. 1992.

2,3,4,6-Tetrachlorophenol   EPA Method 1625
CAS #58-90-2

TITLE Semivolatile Organic Compounds by Isotope Dilution GC/MS

MATRIX The compounds may be determined in waters, soils, and municipal sludges by this method.

METHOD SUMMARY This method is used to determine 176 semivolatile toxic organic pollutants associated with the CWA (as amended 1987); the RCRA (as amended 1986); the CERCLA (as amended 1986); and other compounds amenable to extraction and analysis by capillary column gas chromatography-mass spectrometry (GC/MS).

Stable isotopically-labeled analogs of the compounds of interest are added to the sample. If the solids content is less than 1%, a 1-L sample is extracted at pH 12–13, then at pH <2 with methylene chloride using continuous extraction techniques.

If the solids content is 30% or less, the sample is diluted to 1% solids with reagent water, homogenized ultrasonically, and extracted at pH 12–13, then at pH <2 with methylene chloride using continuous extraction techniques. If the solids content is greater than 30%, the sample is extracted using ultrasonic techniques.

Each extract is dried over sodium sulfate, concentrated to a volume of 5 mL, cleaned up using GPC, if necessary, and concentrated. Extracts are concentrated to 1 mL if GPC is not performed, and to 0.5 mL if GPC is performed.

An internal standard is added to the extract, and a 1-mL aliquot of the extract is injected into the GC. The compounds are

separated by GC and detected by a MS. The labeled compounds serve to correct the variability of the analytical technique.

INTERFERENCES Solvents, reagents, glassware, and other sample processing hardware may yield artifacts and/or elevated baselines causing misinterpretation of chromatograms and spectra. Materials used in the analysis must be demonstrated to be free from interferences under the conditions of analysis by running method blanks initially and with each sample lot (sample started through the extraction process on a given 8-h shift, to a maximum of 20). Specific selection of reagents and purification of solvents by distillation in all glass systems may be required. Glassware and, where possible, reagents are cleaned by solvent rinse and baking at 450°C for 1-h minimum. Interferences coextracted from samples will vary considerably from source to source, depending on the diversity of the site being sampled.

INSTRUMENTATION Major instrumentation includes a GC with a splitless or on-column injection port for capillary column, a MS with 70 eV electron impact ionization, and a data system to collect and record MS data, and process it. A K-D apparatus is used to concentrate extracts.

GC Column: 30 m × 0.25 mm I.D. 5% phenyl, 94% methyl, 1% vinyl silicone bonded phased fused silica capillary column.

PRECISION & ACCURACY The detection limits of the method are usually dependent on the level of interferences rather than instrumental limitations. The limits typify the minimum quantities that can be detected with no interferences present.

The minimum level (in µg/mL) was not listed. This is defined as a minimum level at which the analytical system shall give recognizable mass spectra (background corrected) and acceptable calibration points.

The MDL (in µg/kg) in low solids was not listed and in high solids was not listed; these were determined in digested sludge (low solids) and in filter cake or compost (high solids).

The labeled and native compound initial precision as standard deviation (in µg/L) was not listed.
The labeled and native compound initial accuracy as average recovery (in µg/L) was not listed.

SAMPLE COLLECTION, PRESERVATION & HANDLING Collect samples in glass containers. Aqueous samples which flow freely are collected in refrigerated bottles using automatic sampling equipment. Solid samples are collected as grab samples using widemouth jars. Maintain samples at 0 to 4°C from the time of collection until extraction. If residual chlorine is present in aqueous samples, add 80 mg sodium thiosulfate/L of water. Begin sample extraction within 7 days of collection, and analyze all extracts within 40 days of extraction.

SAMPLE PREPARATION Samples containing 1% solids or less are extracted directly using continuous liquid-liquid extraction techniques. Samples containing 1 to 30% solids are diluted to the 1% level with reagent water and extracted using continuous liquid-liquid extraction techniques. Samples containing greater than 30% solids are extracted using ultrasonic techniques.

- Base/neutral extraction — Adjust the pH of the waters in the extractors to 12–13 with 6 *N* NaOH. Extract with methylene chloride for 24–48 h.
- Acid extraction — Adjust the pH of the waters in the extractors to 2 or less using 6 *N* sulfuric acid. Extract with methylene chloride for 24–48 h.
- Ultrasonic extraction of high solids samples — Add anhydrous sodium sulfate to the sample and QC aliquot(s). Add acetone:methylene chloride (1:1) to the sample and mix thoroughly

Concentrate extracts using a K-D apparatus.

QUALITY CONTROL The analyst is permitted to modify this method to improve separations or lower the costs of measurements, provided all performance specifications are met. Analyses of blanks are required to demonstrate freedom from contamination. When results of spikes indicate atypical method performance for samples, the samples are diluted to bring method performance within acceptable limits.

For low solids (aqueous samples), extract, concentrate, and analyze two sets of four 1-L aliquots (8 aliquots total) of the precision and recovery standard. For high solids samples, two sets of four 30-g aliquots of the high solids reference matrix are used.

Spike all samples with labeled compounds to assess method performance. Compute percent recovery of the labeled compounds using the internal standard method. Compare the labeled compound recovery for each compound with the corresponding labeled compound recovery.

Reagent water and high solids reference matrix blanks are analyzed to demonstrate freedom from contamination. Extract and concentrate a 1-L reagent water blank or a high solids reference matrix blank with each sample's lot (samples started through the extraction process on the same 8-h shift, to a maximum of 20 samples).

Field replicates may be collected to determine the precision of the sampling technique, and spiked samples may be required to determine the accuracy of the analysis when the internal standard method is used.

REFERENCE Semivolatile Organic Compounds by Isotope Dilution GC/MS. Office of Water Regulation and Standards, U.S. EPA Industrial Technology Division, Washington, DC, EPA Method 1625, Rev. C, June 1989 (contact W.A. Telliard, U.S. EPA, Office of Water Regulations and Standards, 401 M St., SW, Washington, DC, 20460. Phone: 202-382-7131).

2,3,4,6-Tetrachlorophenol   EPA Method 8270
CAS #58-90-2

TITLE Semivolatile Organic Compounds by GC/MS

MATRIX This method is used to determine the concentration of semivolatile organic compounds in extracts prepared from all types of solid waste matrices, soils, and groundwater. Although surface waters are not specifically mentioned, this

method should be applicable to water samples from rivers, lakes, etc.

METHOD SUMMARY This method covers 259 semivolatile organic compounds. In very limited applications direct injection of the sample into the GC/MS system may be appropriate, but this results in very high detection limits (approximately 10,000 µg/L). Typically, a 1-L liquid sample, containing surrogate, and matrix spiking standards, is extracted in a continuous extractor first under acid conditions and then under basic conditions. Typically 30 g of a solid sample, containing surrogate, and matrix spiking standards, is extracted ultrasonically. After concentrating the extract to 1 mL it is spiked with 10 µL of an internal standard solution just prior to analysis by GC/MS. The volume injected should contain about 100 ng of base/neutral and 200 ng of acid surrogates (for a 1-µL injection). Analysis is performed by GC/MS using a capillary GC column.

INTERFERENCES Raw GC/MS data from all blanks, samples, and spikes must be evaluated for interferences. Contamination by carryover can occur whenever high-concentration and low-concentration samples are sequentially analyzed. To reduce carryover, the sample syringe must be rinsed out between samples with solvent. Whenever an unusually concentrated sample is encountered, it should be followed by the analysis of blank solvent to check for cross-contamination.

INSTRUMENTATION A GC/MS and a data system are required. The GC column used is a 30 m × 0.25 mm I.D. (or 0.32 mm I.D.) 1um film thickness silicone-coated fused silica capillary column. A continuous liquid-liquid extractor equipped with Teflon® or glass connection joints and stopcocks requiring no lubrication, a K-D concentrating apparatus, water bath, and an ultrasonic disrupter with a minimum power of 300 W and with pulsing capability are also required.

PRECISION & ACCURACY The estimated quantitation limit (EQL) of Method 8270B for determining an individual compound is approximately 1 mg/kg (wet weight) for soil or sediment samples, 1–200 mg/kg for wastes (dependent on matrix and method of preparation), and 10 µg/L for groundwater samples. EQLs will be proportionately higher for sample extracts that require dilution to avoid saturation of the detector.

The EQL(b) for groundwater in µg/L is 10.
The EQL (a, b) for low concentrations in soil and sediment in µg/kg is not determined.
Accuracy as µg/L is not listed.
Overall precision in µg/L is not listed.

(a) *EQLs listed for soil/sediment are based on wet weight. Normally data is reported in a dry-weight basis; therefore, EQLs will be higher based on the % dry weight of each sample. This calculation is based on a 30 g sample and gel permeation chromatography cleanup.*
(b) *Sample EQLs are highly matrix-dependent. The EQLs are provided for guidance and may not always be achievable.*
C = *True value for concentration, in µg/L.*
X = *Average recovery found for measurements of samples containing a concentration of C, in µg/L.*

ESTIMATED QUANTITATION LIMIT

Other Matrices	Factor (a)
High-concentration soil and sludges by sonicator	7.5
Non-water miscible waste	75

(a) *EQL for other matrices = [EQL for low soil/sediment × [Factor]. This estimated EQL is similar to an EPA "Practical Quantitation Limit."*

SAMPLING METHOD
Liquid samples — Use a 1 or 2½ gallon amber glass bottle with a screw-top Teflon®-lined cover that has been prewashed with detergent and rinsed with distilled water and methanol (or isopropanol).

Soils, sediments, or sludges — Use an 8-oz. widemouth glass with a screw-top Teflon®-lined cover that has been prewashed with detergent and rinsed with distilled water and methanol (or isopropanol).

SAMPLE PRESERVATION
Liquid samples — If residual chlorine is present, add 3 mL of 10% sodium thiosulfate per gallon, cool to 4°C and store in a solvent-free refrigerator until analysis; if chlorine is not present, then eliminate the sodium thiosulfate addition.

Soils, sediments, or sludges — Cool samples to 4°C and store in a solvent-free refrigerator.

MHT Liquid samples must be extracted within 7 days and the extracts analyzed within 40 days. Soils, sediments, or sludges may be stored for a maximum of 14 days and the extracts analyzed within 40 days.

SAMPLE PREPARATION
Liquid samples — Transfer 1 L quantitatively to a continuous extractor. If high concentrations are anticipated, a smaller volume may be used and then diluted with organic-free reagent water to 1 L. Adjust pH, if necessary, to pH <2 using 1:1 (V/V) sulfuric acid. Pipette 1.0 mL of a surrogate standard spiking solution into each sample. For the sample in each analytical batch selected for spiking, add 1.0 mL of a matrix spiking standard. For base/neutral acid analysis, the amount of the surrogates and matrix spiking compounds added to the sample should result in a final concentration of 100 ng/µL of each analyte in the extract to be analyzed (assuming a 1-µL injection). Extract with methylene chloride for 18–24 h. Next, adjust the pH of the aqueous phase to pH >11 using 10 N sodium hydroxide and extract it with methylene chloride again for 18–24 h. Dry the extract through a column containing anhydrous sodium sulfate and concentrate it to 1 mL using a K-D concentrator.

Soils, sediments, or sludges — Use 30 g of sample. Nonporous or wet samples (gummy or clay type) that do not have a free-flowing sandy texture must be mixed with anhydrous sodium sulfate until the sample is free flowing. Add 1 mL of surrogate standards to all samples, spikes, standards, and blanks. For the sample in each analytical batch selected for spiking, add 1.0 mL of a matrix spiking standard. For base/neutral acid analysis, the amount added of the surrogates and matrix spiking compounds should result in a final concentration of 100 ng/µL of each base/neutral analyte and 200 ng/µL of each acid analyte

in the extract to be analyzed (assuming a 1-µL injection). Immediately add a 100-mL mixture of 1:1 methylene chloride:acetone and extract the sample ultrasonically for 3 min and then decant or filter the extracts. Repeat the extraction two or more times. Dry the extract using a column with anhydrous sodium sulfate and concentrate it to 1 mL in a K-D concentrator.

QUALITY CONTROL A methylene chloride solution containing 50 ng/µL of decafluorotriphenylphosphine (DFTPP) is used for tuning the GC/MS system each 12-h shift. A system performance check also must be made during every 12-h shift. A standard containing 50 ng/µL each of 4,4'-DDT, pentachlorophenol, and benzidine is required to verify injection port inertness and GC column performance. A calibration standard at mid-concentration, containing each compound of interest, including all required surrogates, must be performed every 12 h during analysis. After the system performance check is met, calibration check compounds (CCCs) are used to check the validity of the initial calibration.

The internal standard responses and retention times in the calibration check standard must be evaluated immediately after or during data acquisition. If the retention time for any internal standard changes by more than 30 seconds from the last check calibration (12 h), the chromatographic system must be inspected for malfunctions and corrections must be made, as required. If the electron ionization current plot (EICP) area for any of the internal standards changes by a factor of two from the last daily calibration standard check, the mass spectrometer must be inspected for malfunctions and corrections must be made, as appropriate.

Demonstrate, through the analysis of a reagent water blank, that interferences from the analytical system, glassware, and reagents are under control. The blank samples should be carried through all stages of the sample preparation and measurement steps. For each analytical batch (up to 20 samples), a reagent blank, matrix spike, and matrix spike duplicate/duplicate must be analyzed (the frequency of the spikes may be different for different monitoring programs). The blank and spiked samples must be carried through all stages of the sample preparation and measurement steps. A QC reference sample concentrate containing each analyte at a concentration of 100 mg/L in methanol is required.

REFERENCE Test Methods for Evaluating Solid Waste (SW-846). U.S. EPA 1983, Method 8270B, Rev. 2, Nov. 1990. Office of Solid Waste, Washington, DC.

Tetrachlorvinphos   **EPA Method 8270**
CAS #961-11-5

TITLE Semivolatile Organic Compounds by GC/MS

MATRIX This method is used to determine the concentration of semivolatile organic compounds in extracts prepared from all types of solid waste matrices, soils, and groundwater. Although surface waters are not specifically mentioned, this method should be applicable to water samples from rivers, lakes, etc.

METHOD SUMMARY This method covers 259 semivolatile organic compounds. In very limited applications direct injection of the sample into the GC/MS system may be appropriate, but this results in very high detection limits (approximately 10,000 µg/L). Typically, a 1-L liquid sample, containing surrogate, and matrix spiking standards, is extracted in a continuous extractor first under acid conditions and then under basic conditions. Typically 30 g of a solid sample, containing surrogate, and matrix spiking standards, is extracted ultrasonically. After concentrating the extract to 1 mL it is spiked with 10 µL of an internal standard solution just prior to analysis by GC/MS. The volume injected should contain about 100 ng of base/neutral and 200 ng of acid surrogates (for a 1-µL injection). Analysis is performed by GC/MS using a capillary GC column.

INTERFERENCES Raw GC/MS data from all blanks, samples, and spikes must be evaluated for interferences. Contamination by carryover can occur whenever high-concentration and low-concentration samples are sequentially analyzed. To reduce carryover, the sample syringe must be rinsed out between samples with solvent. Whenever an unusually concentrated sample is encountered, it should be followed by the analysis of blank solvent to check for cross-contamination.

INSTRUMENTATION A GC/MS and a data system are required. The GC column used is a 30 m × 0.25 mm I.D. (or 0.32 mm I.D.) 1um film thickness silicone-coated fused silica capillary column. A continuous liquid-liquid extractor equipped with Teflon® or glass connection joints and stopcocks requiring no lubrication, a K-D concentrating apparatus, water bath, and an ultrasonic disrupter with a minimum power of 300 W and with pulsing capability are also required.

PRECISION & ACCURACY The estimated quantitation limit (EQL) of Method 8270B for determining an individual compound is approximately 1 mg/kg (wet weight) for soil or sediment samples, 1–200 mg/kg for wastes (dependent on matrix and method of preparation), and 10 µg/L for groundwater samples. EQLs will be proportionately higher for sample extracts that require dilution to avoid saturation of the detector.

The EQL(b) for groundwater in µg/L is 20.
The EQL (a, b) for low concentrations in soil and sediment in µg/kg is not determined.
Accuracy as µg/L is not listed.
Overall precision in µg/L is not listed.

(a) *EQLs listed for soil/sediment are based on wet weight. Normally data is reported in a dry-weight basis; therefore, EQLs will be higher based on the % dry weight of each sample. This calculation is based on a 30 g sample and gel permeation chromatography cleanup.*
(b) *Sample EQLs are highly matrix-dependent. The EQLs are provided for guidance and may not always be achievable.*
C = *True value for concentration, in µg/L.*
X = *Average recovery found for measurements of samples containing a concentration of C, in µg/L.*

ESTIMATED QUANTITATION LIMIT

Other Matrices	Factor (a)
High-concentration soil and sludges by sonicator	7.5
Non-water miscible waste	75

(a) EQL for other matrices = [EQL for low soil/sediment × [Factor]. This estimated EQL is similar to an EPA "Practical Quantitation Limit."

SAMPLING METHOD

Liquid samples — Use a 1 or 2½ gallon amber glass bottle with a screw-top Teflon®-lined cover that has been prewashed with detergent and rinsed with distilled water and methanol (or isopropanol).

Soils, sediments, or sludges — Use an 8-oz. widemouth glass with a screw-top Teflon®-lined cover that has been prewashed with detergent and rinsed with distilled water and methanol (or isopropanol).

SAMPLE PRESERVATION

Liquid samples — If residual chlorine is present, add 3 mL of 10% sodium thiosulfate per gallon, cool to 4°C and store in a solvent-free refrigerator until analysis; if chlorine is not present, then eliminate the sodium thiosulfate addition.

Soils, sediments, or sludges — Cool samples to 4°C and store in a solvent-free refrigerator.

MHT Liquid samples must be extracted within 7 days and the extracts analyzed within 40 days. Soils, sediments, or sludges may be stored for a maximum of 14 days and the extracts analyzed within 40 days.

SAMPLE PREPARATION

Liquid samples — Transfer 1 L quantitatively to a continuous extractor. If high concentrations are anticipated, a smaller volume may be used and then diluted with organic-free reagent water to 1 L. Adjust pH, if necessary, to pH <2 using 1:1 (V/V) sulfuric acid. Pipette 1.0 mL of a surrogate standard spiking solution into each sample. For the sample in each analytical batch selected for spiking, add 1.0 mL of a matrix spiking standard. For base/neutral acid analysis, the amount of the surrogates and matrix spiking compounds added to the sample should result in a final concentration of 100 ng/µL of each analyte in the extract to be analyzed (assuming a 1-µL injection). Extract with methylene chloride for 18–24 h. Next, adjust the pH of the aqueous phase to pH >11 using 10 N sodium hydroxide and extract it with methylene chloride again for 18–24 h. Dry the extract through a column containing anhydrous sodium sulfate and concentrate it to 1 mL using a K-D concentrator.

Soils, sediments, or sludges — Use 30 g of sample. Nonporous or wet samples (gummy or clay type) that do not have a free-flowing sandy texture must be mixed with anhydrous sodium sulfate until the sample is free flowing. Add 1 mL of surrogate standards to all samples, spikes, standards, and blanks. For the sample in each analytical batch selected for spiking, add 1.0 mL of a matrix spiking standard. For base/neutral acid analysis, the amount added of the surrogates and matrix spiking compounds should result in a final concentration of 100 ng/µL of each base/neutral analyte and 200 ng/µL of each acid analyte in the extract to be analyzed (assuming a 1-µL injection). Immediately add a 100-mL mixture of 1:1 methylene chloride:acetone and extract the sample ultrasonically for 3 min and then decant or filter the extracts. Repeat the extraction two or more times. Dry the extract using a column with anhydrous sodium sulfate and concentrate it to 1 mL in a K-D concentrator.

QUALITY CONTROL A methylene chloride solution containing 50 ng/µL of decafluorotriphenylphosphine (DFTPP) is used for tuning the GC/MS system each 12-h shift. A system performance check also must be made during every 12-h shift. A standard containing 50 ng/µL each of 4,4'-DDT, pentachlorophenol, and benzidine is required to verify injection port inertness and GC column performance. A calibration standard at mid-concentration, containing each compound of interest, including all required surrogates, must be performed every 12 h during analysis. After the system performance check is met, calibration check compounds (CCCs) are used to check the validity of the initial calibration.

The internal standard responses and retention times in the calibration check standard must be evaluated immediately after or during data acquisition. If the retention time for any internal standard changes by more than 30 seconds from the last check calibration (12 h), the chromatographic system must be inspected for malfunctions and corrections must be made, as required. If the electron ionization current plot (EICP) area for any of the internal standards changes by a factor of two from the last daily calibration standard check, the mass spectrometer must be inspected for malfunctions and corrections must be made, as appropriate.

Demonstrate, through the analysis of a reagent water blank, that interferences from the analytical system, glassware, and reagents are under control. The blank samples should be carried through all stages of the sample preparation and measurement steps. For each analytical batch (up to 20 samples), a reagent blank, matrix spike, and matrix spike duplicate/duplicate must be analyzed (the frequency of the spikes may be different for different monitoring programs). The blank and spiked samples must be carried through all stages of the sample preparation and measurement steps. A QC reference sample concentrate containing each analyte at a concentration of 100 mg/L in methanol is required.

REFERENCE Test Methods for Evaluating Solid Waste (SW-846). U.S. EPA 1983, Method 8270B, Rev. 2, Nov. 1990. Office of Solid Waste, Washington, DC.

---

n-Tetracosane — EPA Method 1625
CAS #646-31-1

TITLE Semivolatile Organic Compounds by Isotope Dilution GC/MS

MATRIX The compounds may be determined in waters, soils, and municipal sludges by this method.

METHOD SUMMARY This method is used to determine 176 semivolatile toxic organic pollutants associated with the CWA (as amended 1987); the RCRA (as amended 1986); the

CERCLA (as amended 1986); and other compounds amenable to extraction and analysis by capillary column gas chromatography-mass spectrometry (GC/MS).

Stable isotopically-labeled analogs of the compounds of interest are added to the sample. If the solids content is less than 1%, a 1-L sample is extracted at pH 12–13, then at pH <2 with methylene chloride using continuous extraction techniques.

If the solids content is 30% or less, the sample is diluted to 1% solids with reagent water, homogenized ultrasonically, and extracted at pH 12–13, then at pH <2 with methylene chloride using continuous extraction techniques. If the solids content is greater than 30%, the sample is extracted using ultrasonic techniques.

Each extract is dried over sodium sulfate, concentrated to a volume of 5 mL, cleaned up using GPC, if necessary, and concentrated. Extracts are concentrated to 1 mL if GPC is not performed, and to 0.5 mL if GPC is performed.

An internal standard is added to the extract, and a 1-mL aliquot of the extract is injected into the GC. The compounds are separated by GC and detected by a MS. The labeled compounds serve to correct the variability of the analytical technique.

INTERFERENCES Solvents, reagents, glassware, and other sample processing hardware may yield artifacts and/or elevated baselines causing misinterpretation of chromatograms and spectra. Materials used in the analysis must be demonstrated to be free from interferences under the conditions of analysis by running method blanks initially and with each sample lot (sample started through the extraction process on a given 8-h shift, to a maximum of 20). Specific selection of reagents and purification of solvents by distillation in all glass systems may be required. Glassware and, where possible, reagents are cleaned by solvent rinse and baking at 450°C for 1-h minimum. Interferences coextracted from samples will vary considerably from source to source, depending on the diversity of the site being sampled.

INSTRUMENTATION Major instrumentation includes a GC with a splitless or on-column injection port for capillary column, a MS with 70 eV electron impact ionization, and a data system to collect and record MS data, and process it. A K-D apparatus is used to concentrate extracts.

GC Column: 30 m × 0.25 mm I.D. 5% phenyl, 94% methyl, 1% vinyl silicone bonded phased fused silica capillary column.

PRECISION & ACCURACY The detection limits of the method are usually dependent on the level of interferences rather than instrumental limitations. The limits typify the minimum quantities that can be detected with no interferences present.

The minimum level (in µg/mL) was 10. This is defined as a minimum level at which the analytical system shall give recognizable mass spectra (background corrected) and acceptable calibration points.

The MDL (in µg/kg) in low solids was not listed and in high solids was not listed; these were determined in digested sludge (low solids) and in filter cake or compost (high solids).

The labeled and native compound initial precision as standard deviation (in µg/L) was 11.

The labeled and native compound initial accuracy as average recovery (in µg/L) was 80–139.

SAMPLE COLLECTION, PRESERVATION & HANDLING
Collect samples in glass containers. Aqueous samples which flow freely are collected in refrigerated bottles using automatic sampling equipment. Solid samples are collected as grab samples using widemouth jars. Maintain samples at 0 to 4°C from the time of collection until extraction. If residual chlorine is present in aqueous samples, add 80 mg sodium thiosulfate/L of water. Begin sample extraction within 7 days of collection, and analyze all extracts within 40 days of extraction.

SAMPLE PREPARATION Samples containing 1% solids or less are extracted directly using continuous liquid-liquid extraction techniques. Samples containing 1 to 30% solids are diluted to the 1% level with reagent water and extracted using continuous liquid-liquid extraction techniques. Samples containing greater than 30% solids are extracted using ultrasonic techniques.

Base/neutral extraction — Adjust the pH of the waters in the extractors to 12–13 with 6 N NaOH. Extract with methylene chloride for 24–48 h.

Acid extraction — Adjust the pH of the waters in the extractors to 2 or less using 6 N sulfuric acid. Extract with methylene chloride for 24–48 h.

Ultrasonic extraction of high solids samples — Add anhydrous sodium sulfate to the sample and QC aliquot(s). Add acetone:methylene chloride (1:1) to the sample and mix thoroughly

Concentrate extracts using a K-D apparatus.

QUALITY CONTROL The analyst is permitted to modify this method to improve separations or lower the costs of measurements, provided all performance specifications are met. Analyses of blanks are required to demonstrate freedom from contamination. When results of spikes indicate atypical method performance for samples, the samples are diluted to bring method performance within acceptable limits.

For low solids (aqueous samples), extract, concentrate, and analyze two sets of four 1-L aliquots (8 aliquots total) of the precision and recovery standard. For high solids samples, two sets of four 30-g aliquots of the high solids reference matrix are used.

Spike all samples with labeled compounds to assess method performance. Compute percent recovery of the labeled compounds using the internal standard method. Compare the labeled compound recovery for each compound with the corresponding labeled compound recovery.

Reagent water and high solids reference matrix blanks are analyzed to demonstrate freedom from contamination. Extract and concentrate a 1-L reagent water blank or a high solids reference matrix blank with each sample's lot (samples started through the extraction process on the same 8-h shift, to a maximum of 20 samples).

Field replicates may be collected to determine the precision of the sampling technique, and spiked samples may be required to determine the accuracy of the analysis when the internal standard method is used.

REFERENCE Semivolatile Organic Compounds by Isotope Dilution GC/MS. Office of Water Regulation and Standards, U.S. EPA Industrial Technology Division, Washington, DC, EPA Method 1625, Rev. C, June 1989 (contact W.A. Telliard, U.S. EPA, Office of Water Regulations and Standards, 401 M St., SW, Washington, DC, 20460. Phone: 202-382-7131).

---

n-Tetradecane   EPA Method 1625
CAS #629-59-4

TITLE Semivolatile Organic Compounds by Isotope Dilution GC/MS

MATRIX The compounds may be determined in waters, soils, and municipal sludges by this method.

METHOD SUMMARY This method is used to determine 176 semivolatile toxic organic pollutants associated with the CWA (as amended 1987); the RCRA (as amended 1986); the CERCLA (as amended 1986); and other compounds amenable to extraction and analysis by capillary column gas chromatography-mass spectrometry (GC/MS).

Stable isotopically-labeled analogs of the compounds of interest are added to the sample. If the solids content is less than 1%, a 1-L sample is extracted at pH 12–13, then at pH <2 with methylene chloride using continuous extraction techniques.

If the solids content is 30% or less, the sample is diluted to 1% solids with reagent water, homogenized ultrasonically, and extracted at pH 12–13, then at pH <2 with methylene chloride using continuous extraction techniques. If the solids content is greater than 30%, the sample is extracted using ultrasonic techniques.

Each extract is dried over sodium sulfate, concentrated to a volume of 5 mL, cleaned up using GPC, if necessary, and concentrated. Extracts are concentrated to 1 mL if GPC is not performed, and to 0.5 mL if GPC is performed.

An internal standard is added to the extract, and a 1-mL aliquot of the extract is injected into the GC. The compounds are separated by GC and detected by a MS. The labeled compounds serve to correct the variability of the analytical technique.

INTERFERENCES Solvents, reagents, glassware, and other sample processing hardware may yield artifacts and/or elevated baselines causing misinterpretation of chromatograms and spectra. Materials used in the analysis must be demonstrated to be free from interferences under the conditions of analysis by running method blanks initially and with each sample lot (sample started through the extraction process on a given 8-h shift, to a maximum of 20). Specific selection of reagents and purification of solvents by distillation in all glass systems may be required. Glassware and, where possible, reagents are cleaned by solvent rinse and baking at 450°C for 1-h minimum. Interferences coextracted from samples will vary considerably from source to source, depending on the diversity of the site being sampled.

INSTRUMENTATION Major instrumentation includes a GC with a splitless or on-column injection port for capillary column, a MS with 70 eV electron impact ionization, and a data system to collect and record MS data, and process it. A K-D apparatus is used to concentrate extracts.

GC Column: 30 m × 0.25 mm I.D. 5% phenyl, 94% methyl, 1% vinyl silicone bonded phased fused silica capillary column.

PRECISION & ACCURACY The detection limits of the method are usually dependent on the level of interferences rather than instrumental limitations. The limits typify the minimum quantities that can be detected with no interferences present.

The minimum level (in µg/mL) was 10. This is defined as a minimum level at which the analytical system shall give recognizable mass spectra (background corrected) and acceptable calibration points.

The MDL (in µg/kg) in low solids was 256 and in high solids was 3533; these were determined in digested sludge (low solids) and in filter cake or compost (high solids).

Note: Background levels of this compound were present in the sludge tested, resulting in higher than expected MDLs. The MDL for this compound is expected to be approximately 50 µg/kg with no interferences present.

The labeled and native compound initial precision as standard deviation (in µg/L) was 109.
The labeled and native compound initial accuracy as average recovery (in µg/L) was ns-ns.

ns = no specification; the limit was outside the range that could be reliably measured.

SAMPLE COLLECTION, PRESERVATION & HANDLING Collect samples in glass containers. Aqueous samples which flow freely are collected in refrigerated bottles using automatic sampling equipment. Solid samples are collected as grab samples using widemouth jars. Maintain samples at 0 to 4°C from the time of collection until extraction. If residual chlorine is present in aqueous samples, add 80 mg sodium thiosulfate/L of water. Begin sample extraction within 7 days of collection, and analyze all extracts within 40 days of extraction.

SAMPLE PREPARATION Samples containing 1% solids or less are extracted directly using continuous liquid-liquid extraction techniques. Samples containing 1 to 30% solids are diluted to the 1% level with reagent water and extracted using continuous liquid-liquid extraction techniques. Samples containing greater than 30% solids are extracted using ultrasonic techniques.

Base/neutral extraction — Adjust the pH of the waters in the extractors to 12–13 with 6 N NaOH. Extract with methylene chloride for 24–48 h.

Acid extraction — Adjust the pH of the waters in the extractors to 2 or less using 6 N sulfuric acid. Extract with methylene chloride for 24–48 h.

Ultrasonic extraction of high solids samples — Add anhydrous sodium sulfate to the sample and QC aliquot(s). Add acetone:methylene chloride (1:1) to the sample and mix thoroughly

Concentrate extracts using a K-D apparatus.

QUALITY CONTROL The analyst is permitted to modify this method to improve separations or lower the costs of measurements, provided all performance specifications are met. Analyses of blanks are required to demonstrate freedom from contamination. When results of spikes indicate atypical method performance for samples, the samples are diluted to bring method performance within acceptable limits.

For low solids (aqueous samples), extract, concentrate, and analyze two sets of four 1-L aliquots (8 aliquots total) of the precision and recovery standard. For high solids samples, two sets of four 30-g aliquots of the high solids reference matrix are used.

Spike all samples with labeled compounds to assess method performance. Compute percent recovery of the labeled compounds using the internal standard method. Compare the labeled compound recovery for each compound with the corresponding labeled compound recovery.

Reagent water and high solids reference matrix blanks are analyzed to demonstrate freedom from contamination. Extract and concentrate a 1-L reagent water blank or a high solids reference matrix blank with each sample's lot (samples started through the extraction process on the same 8-h shift, to a maximum of 20 samples).

Field replicates may be collected to determine the precision of the sampling technique, and spiked samples may be required to determine the accuracy of the analysis when the internal standard method is used.

REFERENCE Semivolatile Organic Compounds by Isotope Dilution GC/MS. Office of Water Regulation and Standards, U.S. EPA Industrial Technology Division, Washington, DC, EPA Method 1625, Rev. C, June 1989 (contact W.A. Telliard, U.S. EPA, Office of Water Regulations and Standards, 401 M St., SW, Washington, DC, 20460. Phone: 202-382-7131).

Tetraethyl dithiopyrophosphate **EPA Method 8270**
CAS #3689-24-5

TITLE Semivolatile Organic Compounds by GC/MS

MATRIX This method is used to determine the concentration of semivolatile organic compounds in extracts prepared from all types of solid waste matrices, soils, and groundwater. Although surface waters are not specifically mentioned, this method should be applicable to water samples from rivers, lakes, etc.

METHOD SUMMARY This method covers 259 semivolatile organic compounds. In very limited applications direct injection of the sample into the GC/MS system may be appropriate, but this results in very high detection limits (approximately 10,000 µg/L). Typically, a 1-L liquid sample, containing surrogate, and matrix spiking standards, is extracted in a continuous extractor first under acid conditions and then under basic conditions. Typically 30 g of a solid sample, containing surrogate, and matrix spiking standards, is extracted ultrasonically. After concentrating the extract to 1 mL it is spiked with 10 µL of an internal standard solution just prior to analysis by GC/MS. The volume injected should contain about 100 ng of base/neutral and 200 ng of acid surrogates (for a 1-µL injection). Analysis is performed by GC/MS using a capillary GC column.

INTERFERENCES Raw GC/MS data from all blanks, samples, and spikes must be evaluated for interferences. Contamination by carryover can occur whenever high-concentration and low-concentration samples are sequentially analyzed. To reduce carryover, the sample syringe must be rinsed out between samples with solvent. Whenever an unusually concentrated sample is encountered, it should be followed by the analysis of blank solvent to check for cross-contamination.

INSTRUMENTATION A GC/MS and a data system are required. The GC column used is a 30 m × 0.25 mm I.D. (or 0.32 mm I.D.) 1um film thickness silicone-coated fused silica capillary column. A continuous liquid-liquid extractor equipped with Teflon® or glass connection joints and stopcocks requiring no lubrication, a K-D concentrating apparatus, water bath, and an ultrasonic disrupter with a minimum power of 300 W and with pulsing capability are also required.

PRECISION & ACCURACY The estimated quantitation limit (EQL) of Method 8270B for determining an individual compound is approximately 1 mg/kg (wet weight) for soil or sediment samples, 1–200 mg/kg for wastes (dependent on matrix and method of preparation), and 10 µg/L for groundwater samples. EQLs will be proportionately higher for sample extracts that require dilution to avoid saturation of the detector.

The EQL(b) for groundwater in µg/L is not listed.
The EQL (a, b) for low concentrations in soil and sediment in µg/kg is not listed.
Accuracy as µg/L is not listed.
Overall precision in µg/L is not listed.

(a) *EQLs listed for soil/sediment are based on wet weight. Normally data is reported in a dry-weight basis; therefore, EQLs will be higher based on the % dry weight of each sample. This calculation is based on a 30 g sample and gel permeation chromatography cleanup.*

(b) *Sample EQLs are highly matrix-dependent. The EQLs are provided for guidance and may not always be achievable.*

C = *True value for concentration, in µg/L.*
X = *Average recovery found for measurements of samples containing a concentration of C, in µg/L.*

ESTIMATED QUANTITATION LIMIT

Other Matrices	Factor (a)
High-concentration soil and sludges by sonicator	7.5
Non-water miscible waste	75

(a) *EQL for other matrices = [EQL for low soil/sediment × [Factor]. This estimated EQL is similar to an EPA "Practical Quantitation Limit."*

SAMPLING METHOD

Liquid samples — Use a 1 or 2½ gallon amber glass bottle with a screw-top Teflon®-lined cover that has been prewashed with detergent and rinsed with distilled water and methanol (or isopropanol).

Soils, sediments, or sludges — Use an 8-oz. widemouth glass with a screw-top Teflon®-lined cover that has been prewashed with detergent and rinsed with distilled water and methanol (or isopropanol).

SAMPLE PRESERVATION

Liquid samples — If residual chlorine is present, add 3 mL of 10% sodium thiosulfate per gallon, cool to 4°C and store in a solvent-free refrigerator until analysis; if chlorine is not present, then eliminate the sodium thiosulfate addition.

Soils, sediments, or sludges — Cool samples to 4°C and store in a solvent-free refrigerator.

MHT Liquid samples must be extracted within 7 days and the extracts analyzed within 40 days. Soils, sediments, or sludges may be stored for a maximum of 14 days and the extracts analyzed within 40 days.

SAMPLE PREPARATION

Liquid samples — Transfer 1 L quantitatively to a continuous extractor. If high concentrations are anticipated, a smaller volume may be used and then diluted with organic-free reagent water to 1 L. Adjust pH, if necessary, to pH <2 using 1:1 (V/V) sulfuric acid. Pipette 1.0 mL of a surrogate standard spiking solution into each sample. For the sample in each analytical batch selected for spiking, add 1.0 mL of a matrix spiking standard. For base/neutral acid analysis, the amount of the surrogates and matrix spiking compounds added to the sample should result in a final concentration of 100 ng/µL of each analyte in the extract to be analyzed (assuming a 1-µL injection). Extract with methylene chloride for 18–24 h. Next, adjust the pH of the aqueous phase to pH >11 using 10 N sodium hydroxide and extract it with methylene chloride again for 18–24 h. Dry the extract through a column containing anhydrous sodium sulfate and concentrate it to 1 mL using a K-D concentrator.

Soils, sediments, or sludges — Use 30 g of sample. Nonporous or wet samples (gummy or clay type) that do not have a free-flowing sandy texture must be mixed with anhydrous sodium sulfate until the sample is free flowing. Add 1 mL of surrogate standards to all samples, spikes, standards, and blanks. For the sample in each analytical batch selected for spiking, add 1.0 mL of a matrix spiking standard. For base/neutral acid analysis, the amount added of the surrogates and matrix spiking compounds should result in a final concentration of 100 ng/µL of each base/neutral analyte and 200 ng/µL of each acid analyte in the extract to be analyzed (assuming a 1-µL injection). Immediately add a 100-mL mixture of 1:1 methylene chloride:acetone and extract the sample ultrasonically for 3 min and then decant or filter the extracts. Repeat the extraction two or more times. Dry the extract using a column with anhydrous sodium sulfate and concentrate it to 1 mL in a K-D concentrator.

QUALITY CONTROL A methylene chloride solution containing 50 ng/µL of decafluorotriphenylphosphine (DFTPP) is used for tuning the GC/MS system each 12-h shift. A system performance check also must be made during every 12-h shift. A standard containing 50 ng/µL each of 4,4'-DDT, pentachlorophenol, and benzidine is required to verify injection port inertness and GC column performance. A calibration standard at mid-concentration, containing each compound of interest, including all required surrogates, must be performed every 12 h during analysis. After the system performance check is met, calibration check compounds (CCCs) are used to check the validity of the initial calibration.

The internal standard responses and retention times in the calibration check standard must be evaluated immediately after or during data acquisition. If the retention time for any internal standard changes by more than 30 seconds from the last check calibration (12 h), the chromatographic system must be inspected for malfunctions and corrections must be made, as required. If the electron ionization current plot (EICP) area for any of the internal standards changes by a factor of two from the last daily calibration standard check, the mass spectrometer must be inspected for malfunctions and corrections must be made, as appropriate.

Demonstrate, through the analysis of a reagent water blank, that interferences from the analytical system, glassware, and reagents are under control. The blank samples should be carried through all stages of the sample preparation and measurement steps. For each analytical batch (up to 20 samples), a reagent blank, matrix spike, and matrix spike duplicate/duplicate must be analyzed (the frequency of the spikes may be different for different monitoring programs). The blank and spiked samples must be carried through all stages of the sample preparation and measurement steps. A QC reference sample concentrate containing each analyte at a concentration of 100 mg/L in methanol is required.

REFERENCE Test Methods for Evaluating Solid Waste (SW-846). U.S. EPA 1983, Method 8270B, Rev. 2, Nov. 1990. Office of Solid Waste, Washington, DC.

Tetraethyl pyrophosphate   **EPA Method 8270**
CAS #107-49-3

TITLE Semivolatile Organic Compounds by GC/MS

MATRIX This method is used to determine the concentration of semivolatile organic compounds in extracts prepared from all types of solid waste matrices, soils, and groundwater. Although surface waters are not specifically mentioned, this method should be applicable to water samples from rivers, lakes, etc.

METHOD SUMMARY This method covers 259 semivolatile organic compounds. In very limited applications direct injection of the sample into the GC/MS system may be appropriate, but this results in very high detection limits (approximately 10,000 µg/L). Typically, a 1-L liquid sample, containing surrogate, and matrix spiking standards, is extracted in a continuous extractor first under acid conditions and then under basic conditions. Typically 30 g of a solid sample, containing surrogate, and matrix spiking standards, is extracted ultrasonically. After concentrating the extract to 1 mL it is spiked with 10 µL of an internal standard solution just prior to analysis by GC/MS. The volume injected should contain about 100 ng of base/neutral and 200 ng of acid surrogates (for a 1-µL injection). Analysis is performed by GC/MS using a capillary GC column.

INTERFERENCES   Raw GC/MS data from all blanks, samples, and spikes must be evaluated for interferences. Contamination by carryover can occur whenever high-concentration and low-concentration samples are sequentially analyzed. To reduce carryover, the sample syringe must be rinsed out between samples with solvent. Whenever an unusually concentrated sample is encountered, it should be followed by the analysis of blank solvent to check for cross-contamination.

INSTRUMENTATION   A GC/MS and a data system are required. The GC column used is a 30 m × 0.25 mm I.D. (or 0.32 mm I.D.) 1um film thickness silicone-coated fused silica capillary column. A continuous liquid-liquid extractor equipped with Teflon® or glass connection joints and stopcocks requiring no lubrication, a K-D concentrating apparatus, water bath, and an ultrasonic disrupter with a minimum power of 300 W and with pulsing capability are also required.

PRECISION & ACCURACY   The estimated quantitation limit (EQL) of Method 8270B for determining an individual compound is approximately 1 mg/kg (wet weight) for soil or sediment samples, 1–200 mg/kg for wastes (dependent on matrix and method of preparation), and 10 µg/L for groundwater samples. EQLs will be proportionately higher for sample extracts that require dilution to avoid saturation of the detector.

The EQL(b) for groundwater in µg/L is 40.
The EQL (a, b) for low concentrations in soil and sediment in µg/kg is not determined.
Accuracy as µg/L is not listed.
Overall precision in µg/L is not listed.

(a) EQLs listed for soil/sediment are based on wet weight. Normally data is reported in a dry-weight basis; therefore, EQLs will be higher based on the % dry weight of each sample. This calculation is based on a 30 g sample and gel permeation chromatography cleanup.
(b) Sample EQLs are highly matrix-dependent. The EQLs are provided for guidance and may not always be achievable.
C =   True value for concentration, in µg/L.
X =   Average recovery found for measurements of samples containing a concentration of C, in µg/L.

ESTIMATED QUANTITATION LIMIT

Other Matrices	Factor (a)
High-concentration soil and sludges by sonicator	7.5
Non-water miscible waste	75

(a) EQL for other matrices = [EQL for low soil/sediment × [Factor]. This estimated EQL is similar to an EPA "Practical Quantitation Limit."

SAMPLING METHOD
Liquid samples — Use a 1 or 2½ gallon amber glass bottle with a screw-top Teflon®-lined cover that has been prewashed with detergent and rinsed with distilled water and methanol (or isopropanol).

Soils, sediments, or sludges — Use an 8-oz. widemouth glass with a screw-top Teflon®-lined cover that has been prewashed with detergent and rinsed with distilled water and methanol (or isopropanol).

SAMPLE PRESERVATION
Liquid samples — If residual chlorine is present, add 3 mL of 10% sodium thiosulfate per gallon, cool to 4°C and store in a solvent-free refrigerator until analysis; if chlorine is not present, then eliminate the sodium thiosulfate addition.

Soils, sediments, or sludges — Cool samples to 4°C and store in a solvent-free refrigerator.

MHT   Liquid samples must be extracted within 7 days and the extracts analyzed within 40 days. Soils, sediments, or sludges may be stored for a maximum of 14 days and the extracts analyzed within 40 days.

SAMPLE PREPARATION
Liquid samples — Transfer 1 L quantitatively to a continuous extractor. If high concentrations are anticipated, a smaller volume may be used and then diluted with organic-free reagent water to 1 L. Adjust pH, if necessary, to pH <2 using 1:1 (V/V) sulfuric acid. Pipette 1.0 mL of a surrogate standard spiking solution into each sample. For the sample in each analytical batch selected for spiking, add 1.0 mL of a matrix spiking standard. For base/neutral acid analysis, the amount of the surrogates and matrix spiking compounds added to the sample should result in a final concentration of 100 ng/µL of each analyte in the extract to be analyzed (assuming a 1-µL injection). Extract with methylene chloride for 18–24 h. Next, adjust the pH of the aqueous phase to pH >11 using 10 N sodium hydroxide and extract it with methylene chloride again for 18–24 h. Dry the extract through a column containing anhydrous sodium sulfate and concentrate it to 1 mL using a K-D concentrator.

Soils, sediments, or sludges — Use 30 g of sample. Nonporous or wet samples (gummy or clay type) that do not have a free-flowing sandy texture must be mixed with anhydrous sodium sulfate until the sample is free flowing. Add 1 mL of surrogate standards to all samples, spikes, standards, and blanks. For the sample in each analytical batch selected for spiking, add 1.0 mL of a matrix spiking standard. For base/neutral acid analysis, the amount added of the surrogates and matrix spiking compounds should result in a final concentration of 100 ng/µL of each base/neutral analyte and 200 ng/µL of each acid analyte in the extract to be analyzed (assuming a 1-µL injection). Immediately add a 100-mL mixture of 1:1 methylene chloride:acetone and extract the sample ultrasonically for 3 min and then decant or filter the extracts. Repeat the extraction two or more times. Dry the extract using a column with anhydrous sodium sulfate and concentrate it to 1 mL in a K-D concentrator.

QUALITY CONTROL   A methylene chloride solution containing 50 ng/µL of decafluorotriphenylphosphine (DFTPP) is used for tuning the GC/MS system each 12-h shift. A system performance check also must be made during every 12-h shift. A standard containing 50 ng/µL each of 4,4'-DDT, pentachlorophenol, and benzidine is required to verify injection port inertness and GC column performance. A calibration standard at mid-concentration, containing each compound of interest, including all required surrogates, must be performed every 12 h during analysis. After the system performance check is met, calibration check compounds (CCCs) are used to check the validity of the initial calibration.

The internal standard responses and retention times in the calibration check standard must be evaluated immediately after or during data acquisition. If the retention time for any internal standard changes by more than 30 seconds from the last check calibration (12 h), the chromatographic system must be inspected for malfunctions and corrections must be made, as required. If the electron ionization current plot (EICP) area for any of the internal standards changes by a factor of two from the last daily calibration standard check, the mass spectrometer must be inspected for malfunctions and corrections must be made, as appropriate.

Demonstrate, through the analysis of a reagent water blank, that interferences from the analytical system, glassware, and reagents are under control. The blank samples should be carried through all stages of the sample preparation and measurement steps. For each analytical batch (up to 20 samples), a reagent blank, matrix spike, and matrix spike duplicate/duplicate must be analyzed (the frequency of the spikes may be different for different monitoring programs). The blank and spiked samples must be carried through all stages of the sample preparation and measurement steps. A QC reference sample concentrate containing each analyte at a concentration of 100 mg/L in methanol is required.

REFERENCE Test Methods for Evaluating Solid Waste (SW-846). U.S. EPA 1983, Method 8270B, Rev. 2, Nov. 1990. Office of Solid Waste, Washington, DC.

Thallium   EPA Method 6010
CAS #7440-28-0

TITLE Inductively Coupled Plasma-Atomic Emission Spectroscopy

MATRIX This method is applicable to the determination of trace elements, including metals, in groundwater, soils, sludges, sediments, and other solid wastes. All matrices require digestion prior to analysis. The method of standard addition must be used for the analysis of all sample digests unless either serial dilution or matrix spike addition demonstrates it is not required.

METHOD SUMMARY Method 6010 covers 25 elements using ICP analysis. It measures element-emitted light by optical spectrometry. Samples, following an appropriate acid digestion, are nebulized and the resulting aerosol is transported to the plasma torch. Element-specific atomic line emission spectra are produced by a radio-frequency inductively coupled plasma.

INTERFERENCES Interferences may be categorized as spectral or non-spectral. Spectral interferences are caused by overlap of a spectral line from another element, unresolved overlap of molecular band spectra, background contribution from continuous or recombination phenomenon, and stray light from the line emission of high concentration elements. Non-spectral interferences include physical and chemical interferences. Physical interferences are effects associated with the sample nebulization and transport processes. Changes in viscosity and surface tension can cause significant inaccuracies. Chemical interferences include molecular compound formation, ionization effects, and solute vaporization effects. Normally these effects are not significant and can be minimized by careful selection of operating conditions. Chemical interferences are highly dependent on matrix type and the specific analyte element.

INSTRUMENTATION An inductively coupled argon plasma emission spectrometer (ICP) capable of background correction is required.

PRECISION & ACCURACY Detection limits, sensitivity, and optimum ranges of the metals will vary with the matrices and model of the spectrometer. In a single lab evaluation, seven wastes were analyzed for 22 elements. The mean percent relative standard deviation from triplicate analyses for all elements and wastes was 9 ± 2%. The mean percent recovery of spiked elements for all wastes was 93 ± 6%. Spike levels ranged from 100 µg/L to 100 mg/L. The wastes included sludges and industrial wastewaters.

Estimated instrument detection limit in µg/L is 40.
Spiked concentration in µg/L is not listed.
Mean reported value in µg/L is not listed.
Precision as RSD % is not listed.

SAMPLING METHOD Samples should be collected in borosilicate glass, linear polyethylene, polypropylene, or Teflon® bottles that have been prewashed with detergent and tap water, and rinsed with 1:1 nitric acid and tap water or 1:1 hydrochloric acid and tap water. Collect at least 2 g of solids and 200 mL of aqueous samples.

SAMPLE PRESERVATION Add nitric acid to make the samples pH <2.

MHT The maximum holding time for properly preserved samples is 6 months.

SAMPLE PREPARATION Preliminary treatment of most matrices is necessary because of the complexity and variability of sample matrices. Water samples that have been prefiltered and acidified will not need acid digestion. Methods for acid digestion of waters for total recoverable or dissolved metals, acid digestions of aqueous samples and extracts for total metals, and acid digestion of sediments, sludges, and soils are summarized below.

Total recoverable or dissolved metals in water — To prepare surface and groundwater samples for determination of total recoverable and dissolved metals, a 100-mL aliquot of well-mixed sample is acidified with concentrated nitric acid and concentrated hydrochloric acid, then heated until the volume is reduced to 15–20 mL. Adjust the final volume to 100 mL with reagent water.

Total metals in aqueous samples, soil and sediment extracts — To prepare aqueous samples, soil and sediment extracts, and wastes that contain suspended solids, a 100-mL aliquot is made acidic with concentrated nitric acid and the solution is evaporated to about 5 mL on a hot plate. Continue heating and adding additional acid until sample digestion is complete, which is usually indicated when the digestate is light in color or does not change in appearance. Evaporate the solution to about 3 mL and cool it and add a small quantity of 1:1 hydrochloric acid (10 mL/100 mL of final solution). Cover the beaker and reflux for 15 min. Wash down the beaker walls and filters or centrifuge the sample to remove silicates and other insoluble material. Filter the sample and adjust the final volume to 100 mL with reagent water and the final acid concentration to 10%.

Sediments, sludges, and soils — To prepare sediments, sludges and soil samples, transfer 1–2 g to a conical beaker and add 10 mL of 1:1 nitric acid, mix the slurry, and cover it with a watch glass. Heat the sample and reflux for 10–15 min without boiling. Allow it to cool, then add 5 mL of concentrated nitric acid and reflux for 30 min. Repeat last step and then allow the solution to evaporate to 5 mL without boiling. Cool and add 2 mL of water and 3 mL of 30% hydrogen peroxide. Cover and place the beaker on the hot plate. Heat and add 30% hydrogen peroxide in 1-mL aliquots with warming until the effervescence is minimal but do not add more than a total of 10 mL of 30% hydrogen peroxide. If the sample is being prepared for the analysis of Ag, Al, As, Ba, Be, Ca, Cd, Co, Cr, Cu, Fe, K, Mg, Mn, Mo, Na, Ni, Os, Pb, Se, Tl, V, and Zn, then add 5 mL of concentrated hydrochloric acid and 10 mL of water and return the covered beaker to a hot plate for 15 min of additional refluxing without boiling. Dilute the sample to a 100 mL volume with water after cooling and filter or centrifuge to remove particulates.

QUALITY CONTROL Laboratory control samples must be analyzed for each analytical method. A method blank should be analyzed with each batch of samples. The effect of the matrix on method performance must be demonstrated: when appropriate, there should be at least one matrix spike and either one matrix duplicate or one matrix spike duplicate per analytical batch. The bias and precision of the method, as well as the method detection limit for each specific matrix type, must be measured.

Dilute and reanalyze samples that are more concentrated than the linear calibration limit. Employ a minimum of one reagent blank per sample batch to determine if contamination or any memory effects are occurring. Whenever a new or unusual sample matrix is encountered, perform either a serial dilution test or a matrix spike addition test to ensure that neither positive or negative interferences are operating on any of the analyte elements. Check the instrument standardization by verifying calibration every 10 samples using a calibration blank and a check standard.

REFERENCE Test Methods for Evaluating Solid Waste (SW-846). U.S. EPA. 1983. Method 6010, Rev. 0, Sept. 1986. Office of Solid Wastes, Washington, DC.

Thallium — EPA Method 200.7
CAS #7440-28-0

TITLE Inductively Coupled Plasma

MATRIX Dissolved, suspended or (ICP) total element in drinking and surface waters and in domestic and industrial wastewaters.

APPLICATION The method covers the determination of 25 metals. Dissolved elements are determined in filtered and acidified samples after appropriate digestion (which increases dissolved solids). Its primary advantage is that ICP instruments allow simultaneous or rapid sequential determination of many elements in a short time. Samples are first nebulized and the aerosol is transported to a plasma torch in which element specific atomic line emission spectra are produced by a radio frequency inductively coupled plasma. Background correction is required for trace element detection except in the case of line broadning.

INTERFERENCES There are spectral, physical, and chemical interferences. The primary disadvantage of ICP instruments is background radiation from other elements and the plasma gases (spectral interferences). Changes in sample viscosity and surface tension with samples containing high dissolved solids (especially those exceeding 1500 mg/L) or high acid concentrations can cause physical interferences. Ionization effects, solute vaporization and molecular compound formation can cause chemical interferences. Aluminum can cause interference at the 100 mg/L level.

INSTRUMENTATION Inductively coupled argon plasma emission spectroscopy. 190.864 nm wavelength

RANGE Not listed.

MDL 40 µg/L.

PRECISION Not listed.

ACCURACY Mean recovery = 93% ± 6% of spiked elements for all wastes.

SAMPLING METHOD Wash sample container with detergent and tap water, rinse with 1 + 1 nitric acid and tap water, then rinse with 1 + 1 hydrochloric acid and tap water, then rinse with deionized, distilled water in that order. Perform any filtration or acid preservation steps when the sample is collected or as soon as possible thereafter.

STABILITY Cool samples to 4°C.

MHT 24 h.

QUALITY CONTROL Mixed calibration standards, an instrument check standard, and an interference check solution are used in addition to a quality control sample. The quality control sample should be prepared in the same acid matrix as the calibration standards at 10 times the instrumental detection limits and in accordance with the instructions provided by the supplier. Furthermore, two types of blanks are required: a calibration blank and a reagent blank.

REFERENCE Method 200.7, U.S. EPA, EMSL-Cincinnati, OH, Nov. 1980

Thallium **EPA Method 7840**
CAS #7440-28-0

TITLE Atomic Absorption (AA)

MATRIX Drinking, surface and direct aspiration saline waters. Wastewater.

APPLICATION Sample is aspirated and atomized in a flame. A light beam from a Tl hollow cathode lamp is directed through the flame into a monochromator and onto a detector. Since wavelength of light beam is specific for Tl, light energy absorbed by detector is measure of thallium.

INTERFERENCES The most troublesome type is chemical, caused by lack of absorption of atoms bound in molecular combination in the flame. High dissolved solids in sample may result in nonatomic absorbance interference. Background correction is required. Hydrochloric acid should not be used.

INSTRUMENTATION Atomic absorption spectrometer. Thallium hollow cathode lamp. (276.8 nm wavelength)

RANGE 1–20 mg/L

MDL 0.1 mg/L

PRECISION Standard deviation = ±0.018, 0.05, 0.20 at 0.60, 3.0, 15 mg Tl/L

ACCURACY Recoveries = 100, 98, 98% at 0.60, 3.0, 15 mg Tl/L

SAMPLING METHOD Use glass or plastic containers. Collect 200 g of solids and 600 mL of liquid samples.

STABILITY Cool solid samples to 4°C and analyze as soon as possible. Add nitric acid to liquid samples to pH <2.

MHT 6 months.

QUALITY CONTROL At least one duplicate and one spike sample should be run every 20 samples or with each matrix type to verify precision of the method. For 20 or more samples per day, verify working standard curve. Run an additional standard at or near mid-range every 10 samples.

REFERENCE Method 7840, SW-846, 3rd ed., Nov.1986.

Thallium **EPA Method 7841**
CAS #7440-28-0

TITLE Atomic Absorption (AA) Furnace Technique

MATRIX Wastes, mobility procedure, extracts, soils and groundwater

APPLICATION Aqueous samples, EP extracts, industrial wastes, soils, sludges, sediments, and solid wastes require digestion before analysis. An aliquot of sample is placed in the graphite tube in the furnace and slowly evaporated, charred and atomized. Absorption of lamp radiation during atomization is proportional to thallium concentration.

INTERFERENCES The furnace technique is subject to chemical interferences. Composition of sample matrix can effect analysis. Hydrochloric acid or excessive chloride will cause thallium volatilization at low temperatures. Palladium is a suitable matrix modifier. Background correction is required.

INSTRUMENTATION Atomic absorption spectrometer. Thallium hollow cathode lamp or electrodeless discharge lamp. Graphite furnace. Strip-chart recorder.

RANGE 5–100 µg/L

MDL 1 µg/L (276.8 nm wavelength)

PRECISION Not listed.

ACCURACY Not listed.

SAMPLING METHOD Use glass or plastic containers. Collect 200 g of solids and 600 mL of liquid samples.

STABILITY Cool solid samples to 4°C and analyze as soon as possible. Add nitric acid to liquid samples to pH <2.

MHT 6 months.

QUALITY CONTROL At least one duplicate and one spike sample should be run every 20 samples, or with each matrix type to verify method precision. If 20 or more samples are run a day, run a standard (at or near mid-range) every 10 samples.

REFERENCE Method 7841, SW-846, 3rd ed., Nov.1986.

Thianaphthene(2,3-benzothiophene) **EPA Method 1625**
CAS #95-15-8

TITLE Semivolatile Organic Compounds by Isotope Dilution GC/MS

MATRIX The compounds may be determined in waters, soils, and municipal sludges by this method.

METHOD SUMMARY This method is used to determine 176 semivolatile toxic organic pollutants associated with the CWA (as amended 1987); the RCRA (as amended 1986); the CERCLA (as amended 1986); and other compounds amenable to extraction and analysis by capillary column gas chromatography-mass spectrometry (GC/MS).

Stable isotopically-labeled analogs of the compounds of interest are added to the sample. If the solids content is less than 1%, a 1-L sample is extracted at pH 12–13, then at pH <2 with methylene chloride using continuous extraction techniques.

If the solids content is 30% or less, the sample is diluted to 1% solids with reagent water, homogenized ultrasonically, and extracted at pH 12–13, then at pH <2 with methylene chloride using continuous extraction techniques. If the solids content is greater than 30%, the sample is extracted using ultrasonic techniques.

Each extract is dried over sodium sulfate, concentrated to a volume of 5 mL, cleaned up using GPC, if necessary, and concentrated. Extracts are concentrated to 1 mL if GPC is not performed, and to 0.5 mL if GPC is performed.

An internal standard is added to the extract, and a 1-mL aliquot of the extract is injected into the GC. The compounds are separated by GC and detected by a MS. The labeled compounds serve to correct the variability of the analytical technique.

INTERFERENCES Solvents, reagents, glassware, and other sample processing hardware may yield artifacts and/or elevated baselines causing misinterpretation of chromatograms and spectra. Materials used in the analysis must be demonstrated to be free from interferences under the conditions of analysis by running method blanks initially and with each sample lot (sample started through the extraction process on a given 8-h shift, to a maximum of 20). Specific selection of reagents and purification of solvents by distillation in all glass systems may be required. Glassware and, where possible, reagents are cleaned by solvent rinse and baking at 450°C for 1-h minimum. Interferences coextracted from samples will vary considerably from source to source, depending on the diversity of the site being sampled.

INSTRUMENTATION Major instrumentation includes a GC with a splitless or on-column injection port for capillary column, a MS with 70 eV electron impact ionization, and a data system to collect and record MS data, and process it. A K-D apparatus is used to concentrate extracts.

GC Column: 30 m × 0.25 mm I.D. 5% phenyl, 94% methyl, 1% vinyl silicone bonded phased fused silica capillary column.

PRECISION & ACCURACY The detection limits of the method are usually dependent on the level of interferences rather than instrumental limitations. The limits typify the minimum quantities that can be detected with no interferences present.

The minimum level (in µg/mL) was not listed. This is defined as a minimum level at which the analytical system shall give recognizable mass spectra (background corrected) and acceptable calibration points.

The MDL (in µg/kg) in low solids was not listed and in high solids was not listed; these were determined in digested sludge (low solids) and in filter cake or compost (high solids).

The labeled and native compound initial precision as standard deviation (in µg/L) was not listed.

The labeled and native compound initial accuracy as average recovery (in µg/L) was not listed.

SAMPLE COLLECTION, PRESERVATION & HANDLING Collect samples in glass containers. Aqueous samples which flow freely are collected in refrigerated bottles using automatic sampling equipment. Solid samples are collected as grab samples using widemouth jars. Maintain samples at 0 to 4°C from the time of collection until extraction. If residual chlorine is present in aqueous samples, add 80 mg sodium thiosulfate/L of water. Begin sample extraction within 7 days of collection, and analyze all extracts within 40 days of extraction.

SAMPLE PREPARATION Samples containing 1% solids or less are extracted directly using continuous liquid-liquid extraction techniques. Samples containing 1 to 30% solids are diluted to the 1% level with reagent water and extracted using continuous liquid-liquid extraction techniques. Samples containing greater than 30% solids are extracted using ultrasonic techniques.

Base/neutral extraction — Adjust the pH of the waters in the extractors to 12–13 with 6 N NaOH. Extract with methylene chloride for 24–48 h.

Acid extraction — Adjust the pH of the waters in the extractors to 2 or less using 6 N sulfuric acid. Extract with methylene chloride for 24–48 h.

Ultrasonic extraction of high solids samples — Add anhydrous sodium sulfate to the sample and QC aliquot(s). Add acetone:methylene chloride (1:1) to the sample and mix thoroughly

Concentrate extracts using a K-D apparatus.

QUALITY CONTROL The analyst is permitted to modify this method to improve separations or lower the costs of measurements, provided all performance specifications are met. Analyses of blanks are required to demonstrate freedom from contamination. When results of spikes indicate atypical method performance for samples, the samples are diluted to bring method performance within acceptable limits.

For low solids (aqueous samples), extract, concentrate, and analyze two sets of four 1-L aliquots (8 aliquots total) of the precision and recovery standard. For high solids samples, two sets of four 30-g aliquots of the high solids reference matrix are used.

Spike all samples with labeled compounds to assess method performance. Compute percent recovery of the labeled compounds using the internal standard method. Compare the labeled compound recovery for each compound with the corresponding labeled compound recovery.

Reagent water and high solids reference matrix blanks are analyzed to demonstrate freedom from contamination. Extract and concentrate a 1-L reagent water blank or a high solids reference matrix blank with each sample's lot (samples started through the extraction process on the same 8-h shift, to a maximum of 20 samples).

Field replicates may be collected to determine the precision of the sampling technique, and spiked samples may be required to determine the accuracy of the analysis when the internal standard method is used.

REFERENCE Semivolatile Organic Compounds by Isotope Dilution GC/MS. Office of Water Regulation and Standards, U.S. EPA Industrial Technology Division, Washington, DC, EPA Method 1625, Rev. C, June 1989 (contact W.A. Telliard, U.S. EPA, Office of Water Regulations and Standards, 401 M St., SW, Washington, DC, 20460. Phone: 202-382-7131).

Thioacetamide
CAS #62-55-5

EPA Method 1625

TITLE Semivolatile Organic Compounds by Isotope Dilution GC/MS

MATRIX The compounds may be determined in waters, soils, and municipal sludges by this method.

METHOD SUMMARY This method is used to determine 176 semivolatile toxic organic pollutants associated with the CWA (as amended 1987); the RCRA (as amended 1986); the CERCLA (as amended 1986); and other compounds amenable to extraction and analysis by capillary column gas chromatography-mass spectrometry (GC/MS).

Stable isotopically-labeled analogs of the compounds of interest are added to the sample. If the solids content is less than 1%, a 1-L sample is extracted at pH 12–13, then at pH <2 with methylene chloride using continuous extraction techniques.

If the solids content is 30% or less, the sample is diluted to 1% solids with reagent water, homogenized ultrasonically, and extracted at pH 12–13, then at pH <2 with methylene chloride using continuous extraction techniques. If the solids content is greater than 30%, the sample is extracted using ultrasonic techniques.

Each extract is dried over sodium sulfate, concentrated to a volume of 5 mL, cleaned up using GPC, if necessary, and concentrated. Extracts are concentrated to 1 mL if GPC is not performed, and to 0.5 mL if GPC is performed.

An internal standard is added to the extract, and a 1-mL aliquot of the extract is injected into the GC. The compounds are separated by GC and detected by a MS. The labeled compounds serve to correct the variability of the analytical technique.

INTERFERENCES Solvents, reagents, glassware, and other sample processing hardware may yield artifacts and/or elevated baselines causing misinterpretation of chromatograms and spectra. Materials used in the analysis must be demonstrated to be free from interferences under the conditions of analysis by running method blanks initially and with each sample lot (sample started through the extraction process on a given 8-h shift, to a maximum of 20). Specific selection of reagents and purification of solvents by distillation in all glass systems may be required. Glassware and, where possible, reagents are cleaned by solvent rinse and baking at 450°C for 1-h minimum. Interferences coextracted from samples will vary considerably from source to source, depending on the diversity of the site being sampled.

INSTRUMENTATION Major instrumentation includes a GC with a splitless or on-column injection port for capillary column, a MS with 70 eV electron impact ionization, and a data system to collect and record MS data, and process it. A K-D apparatus is used to concentrate extracts.

GC Column: 30 m × 0.25 mm I.D. 5% phenyl, 94% methyl, 1% vinyl silicone bonded phased fused silica capillary column.

PRECISION & ACCURACY The detection limits of the method are usually dependent on the level of interferences rather than instrumental limitations. The limits typify the minimum quantities that can be detected with no interferences present.

The minimum level (in µg/mL) was not listed. This is defined as a minimum level at which the analytical system shall give recognizable mass spectra (background corrected) and acceptable calibration points.

The MDL (in µg/kg) in low solids was not listed and in high solids was not listed; these were determined in digested sludge (low solids) and in filter cake or compost (high solids).

The labeled and native compound initial precision as standard deviation (in µg/L) was not listed.
The labeled and native compound initial accuracy as average recovery (in µg/L) was not listed.

SAMPLE COLLECTION, PRESERVATION & HANDLING
Collect samples in glass containers. Aqueous samples which flow freely are collected in refrigerated bottles using automatic sampling equipment. Solid samples are collected as grab samples using widemouth jars. Maintain samples at 0 to 4°C from the time of collection until extraction. If residual chlorine is present in aqueous samples, add 80 mg sodium thiosulfate/L of water. Begin sample extraction within 7 days of collection, and analyze all extracts within 40 days of extraction.

SAMPLE PREPARATION Samples containing 1% solids or less are extracted directly using continuous liquid-liquid extraction techniques. Samples containing 1 to 30% solids are diluted to the 1% level with reagent water and extracted using continuous liquid-liquid extraction techniques. Samples containing greater than 30% solids are extracted using ultrasonic techniques.

Base/neutral extraction — Adjust the pH of the waters in the extractors to 12–13 with 6 N NaOH. Extract with methylene chloride for 24–48 h.
Acid extraction — Adjust the pH of the waters in the extractors to 2 or less using 6 N sulfuric acid. Extract with methylene chloride for 24–48 h.
Ultrasonic extraction of high solids samples — Add anhydrous sodium sulfate to the sample and QC aliquot(s). Add acetone:methylene chloride (1:1) to the sample and mix thoroughly

Concentrate extracts using a K-D apparatus.

QUALITY CONTROL The analyst is permitted to modify this method to improve separations or lower the costs of measurements, provided all performance specifications are met. Analyses of blanks are required to demonstrate freedom from contamination. When results of spikes indicate atypical method performance for samples, the samples are diluted to bring method performance within acceptable limits.

For low solids (aqueous samples), extract, concentrate, and analyze two sets of four 1-L aliquots (8 aliquots total) of the precision and recovery standard. For high solids samples, two sets of four 30-g aliquots of the high solids reference matrix are used.

Spike all samples with labeled compounds to assess method performance. Compute percent recovery of the labeled compounds using the internal standard method. Compare the labeled compound recovery for each compound with the corresponding labeled compound recovery.

Reagent water and high solids reference matrix blanks are analyzed to demonstrate freedom from contamination. Extract and concentrate a 1-L reagent water blank or a high solids reference matrix blank with each sample's lot (samples started through the extraction process on the same 8-h shift, to a maximum of 20 samples).

Field replicates may be collected to determine the precision of the sampling technique, and spiked samples may be required to determine the accuracy of the analysis when the internal standard method is used.

REFERENCE Semivolatile Organic Compounds by Isotope Dilution GC/MS. Office of Water Regulation and Standards, U.S. EPA Industrial Technology Division, Washington, DC, EPA Method 1625, Rev. C, June 1989 (contact W.A. Telliard, U.S. EPA, Office of Water Regulations and Standards, 401 M St., SW, Washington, DC, 20460. Phone: 202-382-7131).

Thionazine   **EPA Method 8270**
CAS #297-97-2

TITLE Semivolatile Organic Compounds by GC/MS

MATRIX This method is used to determine the concentration of semivolatile organic compounds in extracts prepared from all types of solid waste matrices, soils, and groundwater. Although surface waters are not specifically mentioned, this method should be applicable to water samples from rivers, lakes, etc.

METHOD SUMMARY This method covers 259 semivolatile organic compounds. In very limited applications direct injection of the sample into the GC/MS system may be appropriate, but this results in very high detection limits (approximately 10,000 µg/L). Typically, a 1-L liquid sample, containing surrogate, and matrix spiking standards, is extracted in a continuous extractor first under acid conditions and then under basic conditions. Typically 30 g of a solid sample, containing surrogate, and matrix spiking standards, is extracted ultrasonically. After concentrating the extract to 1 mL it is spiked with 10 µL of an internal standard solution just prior to analysis by GC/MS. The volume injected should contain about 100 ng of base/neutral and 200 ng of acid surrogates (for a 1-µL injection). Analysis is performed by GC/MS using a capillary GC column.

INTERFERENCES Raw GC/MS data from all blanks, samples, and spikes must be evaluated for interferences. Contamination by carryover can occur whenever high-concentration and low-concentration samples are sequentially analyzed. To reduce carryover, the sample syringe must be rinsed out between samples with solvent. Whenever an unusually concentrated sample is encountered, it should be followed by the analysis of blank solvent to check for cross-contamination.

INSTRUMENTATION A GC/MS and a data system are required. The GC column used is a 30 m × 0.25 mm I.D. (or 0.32 mm I.D.) 1um film thickness silicone-coated fused silica capillary column. A continuous liquid-liquid extractor equipped with Teflon® or glass connection joints and stopcocks requiring no lubrication, a K-D concentrating apparatus, water bath, and an ultrasonic disrupter with a minimum power of 300 W and with pulsing capability are also required.

PRECISION & ACCURACY The estimated quantitation limit (EQL) of Method 8270B for determining an individual compound is approximately 1 mg/kg (wet weight) for soil or sediment samples, 1–200 mg/kg for wastes (dependent on matrix and method of preparation), and 10 µg/L for groundwater samples. EQLs will be proportionately higher for sample extracts that require dilution to avoid saturation of the detector.

The EQL(b) for groundwater in µg/L is 20.
The EQL (a, b) for low concentrations in soil and sediment in µg/kg is not determined.
Accuracy as µg/L is not listed.
Overall precision in µg/L is not listed.

(a) EQLs listed for soil/sediment are based on wet weight. Normally data is reported in a dry-weight basis; therefore, EQLs will be higher based on the % dry weight of each sample. This calculation is based on a 30 g sample and gel permeation chromatography cleanup.
(b) Sample EQLs are highly matrix-dependent. The EQLs are provided for guidance and may not always be achievable.
C = True value for concentration, in µg/L.
X = Average recovery found for measurements of samples containing a concentration of C, in µg/L.

ESTIMATED QUANTITATION LIMIT

Other Matrices	Factor (a)
High-concentration soil and sludges by sonicator	7.5
Non-water miscible waste	75

(a) EQL for other matrices = [EQL for low soil/sediment × [Factor]. This estimated EQL is similar to an EPA "Practical Quantitation Limit."

SAMPLING METHOD
Liquid samples — Use a 1 or 2½ gallon amber glass bottle with a screw-top Teflon®-lined cover that has been prewashed with detergent and rinsed with distilled water and methanol (or isopropanol).

Soils, sediments, or sludges — Use an 8-oz. widemouth glass with a screw-top Teflon®-lined cover that has been prewashed with detergent and rinsed with distilled water and methanol (or isopropanol).

SAMPLE PRESERVATION
Liquid samples — If residual chlorine is present, add 3 mL of 10% sodium thiosulfate per gallon, cool to 4°C and store in a solvent-free refrigerator until analysis; if chlorine is not present, then eliminate the sodium thiosulfate addition.

Soils, sediments, or sludges — Cool samples to 4°C and store in a solvent-free refrigerator.

MHT Liquid samples must be extracted within 7 days and the extracts analyzed within 40 days. Soils, sediments, or sludges may be stored for a maximum of 14 days and the extracts analyzed within 40 days.

SAMPLE PREPARATION

Liquid samples — Transfer 1 L quantitatively to a continuous extractor. If high concentrations are anticipated, a smaller volume may be used and then diluted with organic-free reagent water to 1 L. Adjust pH, if necessary, to pH <2 using 1:1 (V/V) sulfuric acid. Pipette 1.0 mL of a surrogate standard spiking solution into each sample. For the sample in each analytical batch selected for spiking, add 1.0 mL of a matrix spiking standard. For base/neutral acid analysis, the amount of the surrogates and matrix spiking compounds added to the sample should result in a final concentration of 100 ng/μL of each analyte in the extract to be analyzed (assuming a 1-μL injection). Extract with methylene chloride for 18–24 h. Next, adjust the pH of the aqueous phase to pH >11 using 10 *N* sodium hydroxide and extract it with methylene chloride again for 18–24 h. Dry the extract through a column containing anhydrous sodium sulfate and concentrate it to 1 mL using a K-D concentrator.

Soils, sediments, or sludges — Use 30 g of sample. Nonporous or wet samples (gummy or clay type) that do not have a free-flowing sandy texture must be mixed with anhydrous sodium sulfate until the sample is free flowing. Add 1 mL of surrogate standards to all samples, spikes, standards, and blanks. For the sample in each analytical batch selected for spiking, add 1.0 mL of a matrix spiking standard. For base/neutral acid analysis, the amount added of the surrogates and matrix spiking compounds should result in a final concentration of 100 ng/μL of each base/neutral analyte and 200 ng/μL of each acid analyte in the extract to be analyzed (assuming a 1-μL injection). Immediately add a 100-mL mixture of 1:1 methylene chloride:acetone and extract the sample ultrasonically for 3 min and then decant or filter the extracts. Repeat the extraction two or more times. Dry the extract using a column with anhydrous sodium sulfate and concentrate it to 1 mL in a K-D concentrator.

QUALITY CONTROL A methylene chloride solution containing 50 ng/μL of decafluorotriphenylphosphine (DFTPP) is used for tuning the GC/MS system each 12-h shift. A system performance check also must be made during every 12-h shift. A standard containing 50 ng/μL each of 4,4'-DDT, pentachlorophenol, and benzidine is required to verify injection port inertness and GC column performance. A calibration standard at mid-concentration, containing each compound of interest, including all required surrogates, must be performed every 12 h during analysis. After the system performance check is met, calibration check compounds (CCCs) are used to check the validity of the initial calibration.

The internal standard responses and retention times in the calibration check standard must be evaluated immediately after or during data acquisition. If the retention time for any internal standard changes by more than 30 seconds from the last check calibration (12 h), the chromatographic system must be inspected for malfunctions and corrections must be made, as required. If the electron ionization current plot (EICP) area for any of the internal standards changes by a factor of two from the last daily calibration standard check, the mass spectrometer must be inspected for malfunctions and corrections must be made, as appropriate.

Demonstrate, through the analysis of a reagent water blank, that interferences from the analytical system, glassware, and reagents are under control. The blank samples should be carried through all stages of the sample preparation and measurement steps. For each analytical batch (up to 20 samples), a reagent blank, matrix spike, and matrix spike duplicate/duplicate must be analyzed (the frequency of the spikes may be different for different monitoring programs). The blank and spiked samples must be carried through all stages of the sample preparation and measurement steps. A QC reference sample concentrate containing each analyte at a concentration of 100 mg/L in methanol is required.

REFERENCE Test Methods for Evaluating Solid Waste (SW-846). U.S. EPA 1983, Method 8270B, Rev. 2, Nov. 1990. Office of Solid Waste, Washington, DC.

Thiophenol (Benzenethiol)   **EPA Method 8270**
CAS #108-98-5

TITLE Semivolatile Organic Compounds by GC/MS

MATRIX This method is used to determine the concentration of semivolatile organic compounds in extracts prepared from all types of solid waste matrices, soils, and groundwater. Although surface waters are not specifically mentioned, this method should be applicable to water samples from rivers, lakes, etc.

METHOD SUMMARY This method covers 259 semivolatile organic compounds. In very limited applications direct injection of the sample into the GC/MS system may be appropriate, but this results in very high detection limits (approximately 10,000 μg/L). Typically, a 1-L liquid sample, containing surrogate, and matrix spiking standards, is extracted in a continuous extractor first under acid conditions and then under basic conditions. Typically 30 g of a solid sample, containing surrogate, and matrix spiking standards, is extracted ultrasonically. After concentrating the extract to 1 mL it is spiked with 10 μL of an internal standard solution just prior to analysis by GC/MS. The volume injected should contain about 100 ng of base/neutral and 200 ng of acid surrogates (for a 1-μL injection). Analysis is performed by GC/MS using a capillary GC column.

INTERFERENCES Raw GC/MS data from all blanks, samples, and spikes must be evaluated for interferences. Contamination by carryover can occur whenever high-concentration and low-concentration samples are sequentially analyzed. To reduce carryover, the sample syringe must be rinsed out between samples with solvent. Whenever an unusually concentrated sample is encountered, it should be followed by the analysis of blank solvent to check for cross-contamination.

INSTRUMENTATION A GC/MS and a data system are required. The GC column used is a 30 m × 0.25 mm I.D. (or 0.32 mm I.D.) 1um film thickness silicone-coated fused silica

capillary column. A continuous liquid-liquid extractor equipped with Teflon® or glass connection joints and stopcocks requiring no lubrication, a K-D concentrating apparatus, water bath, and an ultrasonic disrupter with a minimum power of 300 W and with pulsing capability are also required.

PRECISION & ACCURACY The estimated quantitation limit (EQL) of Method 8270B for determining an individual compound is approximately 1 mg/kg (wet weight) for soil or sediment samples, 1–200 mg/kg for wastes (dependent on matrix and method of preparation), and 10 µg/L for groundwater samples. EQLs will be proportionately higher for sample extracts that require dilution to avoid saturation of the detector.

The EQL(b) for groundwater in µg/L is 20.
The EQL (a, b) for low concentrations in soil and sediment in µg/kg is not determined.
Accuracy as µg/L is not listed.
Overall precision in µg/L is not listed.

(a) EQLs listed for soil/sediment are based on wet weight. Normally data is reported in a dry-weight basis; therefore, EQLs will be higher based on the % dry weight of each sample. This calculation is based on a 30 g sample and gel permeation chromatography cleanup.
(b) Sample EQLs are highly matrix-dependent. The EQLs are provided for guidance and may not always be achievable.
$C =$ True value for concentration, in µg/L.
$X =$ Average recovery found for measurements of samples containing a concentration of C, in µg/L.

ESTIMATED QUANTITATION LIMIT

Other Matrices	Factor (a)
High-concentration soil and sludges by sonicator	7.5
Non-water miscible waste	75

(a) EQL for other matrices = [EQL for low soil/sediment × [Factor]. This estimated EQL is similar to an EPA "Practical Quantitation Limit."

SAMPLING METHOD

Liquid samples — Use a 1 or 2½ gallon amber glass bottle with a screw-top Teflon®-lined cover that has been prewashed with detergent and rinsed with distilled water and methanol (or isopropanol).

Soils, sediments, or sludges — Use an 8-oz. widemouth glass with a screw-top Teflon®-lined cover that has been prewashed with detergent and rinsed with distilled water and methanol (or isopropanol).

SAMPLE PRESERVATION

Liquid samples — If residual chlorine is present, add 3 mL of 10% sodium thiosulfate per gallon, cool to 4°C and store in a solvent-free refrigerator until analysis; if chlorine is not present, then eliminate the sodium thiosulfate addition.

Soils, sediments, or sludges — Cool samples to 4°C and store in a solvent-free refrigerator.

MHT Liquid samples must be extracted within 7 days and the extracts analyzed within 40 days. Soils, sediments, or sludges may be stored for a maximum of 14 days and the extracts analyzed within 40 days.

SAMPLE PREPARATION

Liquid samples — Transfer 1 L quantitatively to a continuous extractor. If high concentrations are anticipated, a smaller volume may be used and then diluted with organic-free reagent water to 1 L. Adjust pH, if necessary, to pH <2 using 1:1 (V/V) sulfuric acid. Pipette 1.0 mL of a surrogate standard spiking solution into each sample. For the sample in each analytical batch selected for spiking, add 1.0 mL of a matrix spiking standard. For base/neutral acid analysis, the amount of the surrogates and matrix spiking compounds added to the sample should result in a final concentration of 100 ng/µL of each analyte in the extract to be analyzed (assuming a 1-µL injection). Extract with methylene chloride for 18–24 h. Next, adjust the pH of the aqueous phase to pH >11 using 10 N sodium hydroxide and extract it with methylene chloride again for 18–24 h. Dry the extract through a column containing anhydrous sodium sulfate and concentrate it to 1 mL using a K-D concentrator.

Soils, sediments, or sludges — Use 30 g of sample. Nonporous or wet samples (gummy or clay type) that do not have a free-flowing sandy texture must be mixed with anhydrous sodium sulfate until the sample is free flowing. Add 1 mL of surrogate standards to all samples, spikes, standards, and blanks. For the sample in each analytical batch selected for spiking, add 1.0 mL of a matrix spiking standard. For base/neutral acid analysis, the amount added of the surrogates and matrix spiking compounds should result in a final concentration of 100 ng/µL of each base/neutral analyte and 200 ng/µL of each acid analyte in the extract to be analyzed (assuming a 1-µL injection). Immediately add a 100-mL mixture of 1:1 methylene chloride:acetone and extract the sample ultrasonically for 3 min and then decant or filter the extracts. Repeat the extraction two or more times. Dry the extract using a column with anhydrous sodium sulfate and concentrate it to 1 mL in a K-D concentrator.

QUALITY CONTROL A methylene chloride solution containing 50 ng/µL of decafluorotriphenylphosphine (DFTPP) is used for tuning the GC/MS system each 12-h shift. A system performance check also must be made during every 12-h shift. A standard containing 50 ng/µL each of 4,4′-DDT, pentachlorophenol, and benzidine is required to verify injection port inertness and GC column performance. A calibration standard at mid-concentration, containing each compound of interest, including all required surrogates, must be performed every 12 h during analysis. After the system performance check is met, calibration check compounds (CCCs) are used to check the validity of the initial calibration.

The internal standard responses and retention times in the calibration check standard must be evaluated immediately after or during data acquisition. If the retention time for any internal standard changes by more than 30 seconds from the last check calibration (12 h), the chromatographic system must be inspected for malfunctions and corrections must be made, as required. If the electron ionization current plot (EICP) area for any of the internal standards changes by a factor of two from the last daily calibration standard check, the mass spectrometer

must be inspected for malfunctions and corrections must be made, as appropriate.

Demonstrate, through the analysis of a reagent water blank, that interferences from the analytical system, glassware, and reagents are under control. The blank samples should be carried through all stages of the sample preparation and measurement steps. For each analytical batch (up to 20 samples), a reagent blank, matrix spike, and matrix spike duplicate/duplicate must be analyzed (the frequency of the spikes may be different for different monitoring programs). The blank and spiked samples must be carried through all stages of the sample preparation and measurement steps. A QC reference sample concentrate containing each analyte at a concentration of 100 mg/L in methanol is required.

REFERENCE Test Methods for Evaluating Solid Waste (SW-846). U.S. EPA 1983, Method 8270B, Rev. 2, Nov. 1990. Office of Solid Waste, Washington, DC.

Thioxanthone   **EPA Method 1625**
CAS #492-22-8

TITLE Semivolatile Organic Compounds by Isotope Dilution GC/MS

MATRIX The compounds may be determined in waters, soils, and municipal sludges by this method.

METHOD SUMMARY This method is used to determine 176 semivolatile toxic organic pollutants associated with the CWA (as amended 1987); the RCRA (as amended 1986); the CERCLA (as amended 1986); and other compounds amenable to extraction and analysis by capillary column gas chromatography-mass spectrometry (GC/MS).

Stable isotopically-labeled analogs of the compounds of interest are added to the sample. If the solids content is less than 1%, a 1-L sample is extracted at pH 12–13, then at pH <2 with methylene chloride using continuous extraction techniques.

If the solids content is 30% or less, the sample is diluted to 1% solids with reagent water, homogenized ultrasonically, and extracted at pH 12–13, then at pH <2 with methylene chloride using continuous extraction techniques. If the solids content is greater than 30%, the sample is extracted using ultrasonic techniques.

Each extract is dried over sodium sulfate, concentrated to a volume of 5 mL, cleaned up using GPC, if necessary, and concentrated. Extracts are concentrated to 1 mL if GPC is not performed, and to 0.5 mL if GPC is performed.

An internal standard is added to the extract, and a 1-mL aliquot of the extract is injected into the GC. The compounds are separated by GC and detected by a MS. The labeled compounds serve to correct the variability of the analytical technique.

INTERFERENCES Solvents, reagents, glassware, and other sample processing hardware may yield artifacts and/or elevated baselines causing misinterpretation of chromatograms and spectra. Materials used in the analysis must be demonstrated to be free from interferences under the conditions of analysis by running method blanks initially and with each sample lot (sample started through the extraction process on a given 8-h shift, to a maximum of 20). Specific selection of reagents and purification of solvents by distillation in all glass systems may be required. Glassware and, where possible, reagents are cleaned by solvent rinse and baking at 450°C for 1-h minimum. Interferences coextracted from samples will vary considerably from source to source, depending on the diversity of the site being sampled.

INSTRUMENTATION Major instrumentation includes a GC with a splitless or on-column injection port for capillary column, a MS with 70 eV electron impact ionization, and a data system to collect and record MS data, and process it. A K-D apparatus is used to concentrate extracts.

GC Column: 30 m × 0.25 mm I.D. 5% phenyl, 94% methyl, 1% vinyl silicone bonded phased fused silica capillary column.

PRECISION & ACCURACY The detection limits of the method are usually dependent on the level of interferences rather than instrumental limitations. The limits typify the minimum quantities that can be detected with no interferences present.

The minimum level (in μg/mL) was not listed. This is defined as a minimum level at which the analytical system shall give recognizable mass spectra (background corrected) and acceptable calibration points.

The MDL (in μg/kg) in low solids was not listed and in high solids was not listed; these were determined in digested sludge (low solids) and in filter cake or compost (high solids).

The labeled and native compound initial precision as standard deviation (in μg/L) was not listed.
The labeled and native compound initial accuracy as average recovery (in μg/L) was not listed.

SAMPLE COLLECTION, PRESERVATION & HANDLING Collect samples in glass containers. Aqueous samples which flow freely are collected in refrigerated bottles using automatic sampling equipment. Solid samples are collected as grab samples using widemouth jars. Maintain samples at 0 to 4°C from the time of collection until extraction. If residual chlorine is present in aqueous samples, add 80 mg sodium thiosulfate/L of water. Begin sample extraction within 7 days of collection, and analyze all extracts within 40 days of extraction.

SAMPLE PREPARATION Samples containing 1% solids or less are extracted directly using continuous liquid-liquid extraction techniques. Samples containing 1 to 30% solids are diluted to the 1% level with reagent water and extracted using continuous liquid-liquid extraction techniques. Samples containing greater than 30% solids are extracted using ultrasonic techniques.

Base/neutral extraction — Adjust the pH of the waters in the extractors to 12–13 with 6 N NaOH. Extract with methylene chloride for 24–48 h.
Acid extraction — Adjust the pH of the waters in the extractors to 2 or less using 6 N sulfuric acid. Extract with methylene chloride for 24–48 h.

Ultrasonic extraction of high solids samples — Add anhydrous sodium sulfate to the sample and QC aliquot(s). Add acetone:methylene chloride (1:1) to the sample and mix thoroughly

Concentrate extracts using a K-D apparatus.

QUALITY CONTROL The analyst is permitted to modify this method to improve separations or lower the costs of measurements, provided all performance specifications are met. Analyses of blanks are required to demonstrate freedom from contamination. When results of spikes indicate atypical method performance for samples, the samples are diluted to bring method performance within acceptable limits.

For low solids (aqueous samples), extract, concentrate, and analyze two sets of four 1-L aliquots (8 aliquots total) of the precision and recovery standard. For high solids samples, two sets of four 30-g aliquots of the high solids reference matrix are used.

Spike all samples with labeled compounds to assess method performance. Compute percent recovery of the labeled compounds using the internal standard method. Compare the labeled compound recovery for each compound with the corresponding labeled compound recovery.

Reagent water and high solids reference matrix blanks are analyzed to demonstrate freedom from contamination. Extract and concentrate a 1-L reagent water blank or a high solids reference matrix blank with each sample's lot (samples started through the extraction process on the same 8-h shift, to a maximum of 20 samples).

Field replicates may be collected to determine the precision of the sampling technique, and spiked samples may be required to determine the accuracy of the analysis when the internal standard method is used.

REFERENCE Semivolatile Organic Compounds by Isotope Dilution GC/MS. Office of Water Regulation and Standards, U.S. EPA Industrial Technology Division, Washington, DC, EPA Method 1625, Rev. C, June 1989 (contact W.A. Telliard, U.S. EPA, Office of Water Regulations and Standards, 401 M St., SW, Washington, DC, 20460. Phone: 202-382-7131).

Tin   **EPA Method 7870**
CAS #7440-31-5

TITLE Tin (Atomic Absorption, Direct Aspiration)

MATRIX This method is applicable to a large number of metals in drinking, surface, and saline waters, and domestic and industrial wastes, as well as groundwater, EP extracts, soils, sludges, sediments and other solid wastes which require digestion prior to analysis.

METHOD SUMMARY Solid samples are digested and acidified liquid samples are analyzed directly with an AA. A series of standards of tin must be run and a calibration curve is constructed from this data.

INTERFERENCES The most troublesome type of interference in atomic absorption spectrophotometry is usually termed "chemical" and is caused by lack of absorption of atoms bound in molecular combination in the flame. This phenomenon can occur when the flame is not sufficiently hot to dissociate the molecule or when the dissociated atom is immediately oxidized to a compound that will not dissociate further at the temperature of the flame. The presence of high dissolved solids in the sample may result in an interference from nonatomic absorbance such as light scattering. Preferably, samples containing high solids should be extracted. Ionization interferences occur when the flame temperature is sufficiently high to generate the removal of an electron from a neutral atom, giving a positively charged ion. All metals are not equally stable in the digestate, and tin should be analyzed as soon as possible.

Spectral interference can occur when an absorbing wavelength of an element present in the sample but not being determined falls within the width of the absorption line of the element of interest. The results of the determination will then be erroneously high, due to the contribution of the interfering element to the atomic absorption signal. Interference can also occur when resonant energy from another element in a multi-element lamp, or from a metal impurity in the lamp cathode, falls within the bandpass of the slit setting when that other metal is present in the sample. This type of interference may sometimes be reduced by narrowing the slit width.

INSTRUMENTATION An AA spectrophotometer: single or dual-channel, single or double beam instrument having a grating monochromator, photomultiplier detector, adjustable slits, a wavelength range of 190–800 nm and provisions for interfacing with a strip-chart recorder, and a hollow cathode lamp for tin are required.

PRECISION & ACCURACY Detection limits, sensitivity, and optimum ranges of the metals will vary with the matrices and models of atomic absorption spectrophotometers. A detection limit of 0.8 mg/L and a sensitivity of 4 mg/L have been obtained by direct aspiration. For clean aqueous samples, this detection limit may be extended downward with scale expansion and upward by using a less sensitive wavelength or by rotating the burner head. Detection limits by direct aspiration may also be extended through concentration of the sample and/or through solvent extraction techniques.

SAMPLING METHOD Collect samples in 1-L glass or plastic bottles (previously washed with detergent, rinsed with tap water, 1:1 nitric acid, tap water, 1:1 hydrochloric acid, tap water and Type II water). The same sample containers are used for liquid and for solid samples.

SAMPLE PRESERVATION

Liquid samples — Add nitric acid to adjust the pH of the sample to <2 and immediately cool it to 4°C.

Soils/Sediments: Solid samples usually require no preservation except refrigeration at 4°C during storage; therefore do not adjust the pH of solid samples.

MHT The maximum holding time for samples with this method is 6 months.

SAMPLE PREPARATION

Digestion of soils, sludges and sediments — Mix the sample thoroughly to achieve homogeneity. For each digestion procedure, weigh to the nearest 0.01 g and transfer to a conical beaker a 1.0 to 2.0 g portion of sample. Add 10 mL of a 1:1 water and nitric acid solution, mix the slurry, and cover with a watch glass. Heat the sample to 95°C and reflux for 10–15 min without boiling. Allow the sample to cool, add 5 mL of concentrated nitric acid, replace the watch glass, and reflux for 30 min. Repeat this last step to ensure complete oxidation.

Using a ribbed watch glass, allow the solution to evaporate to 5 mL without boiling, while maintaining a covering of solution over the bottom of the beaker. Allow the sample to cool. Add 2 mL of Type II water and 3 mL of 30% hydrogen peroxide. Cover the beaker with a watch glass and return it to the hot plate for warming and to start the peroxide reaction. Care must be taken to ensure that losses do not occur due to excessively vigorous effervescence. Heat until effervescence subsides and cool the beaker. Continue to add 30% hydrogen peroxide in 1-mL aliquots with warming until the effervescence is minimal or until the general appearance is unchanged. **Do not** add more than a total of 10 mL of 30% hydrogen peroxide.

Add 5 mL of concentrated hydrochloric acid and 10 mL of Type II water, return the covered beaker to the hot plate, and reflux for an additional 15 min without boiling. After cooling, dilute to 100 mL with Type II water. Particulates in the digestate should be removed by filtration, centrifugation or by allowing the sample to settle. Filter the solution through Whatman #41 paper and dilute to 100 mL with Type II water. Centrifuge at 2,000–3,000 rpm for 10 min to clear the supernatant. The diluted sample has an approximate acid concentration of 5% (v/v) hydrochloric acid and 5% (v/v) nitric acid. Drinking water, free of particulates, may be analyzed directly.

QUALITY CONTROL A calibration curve must be prepared each day with a minimum of a reagent blank and three standards, verified by use of at least a reagent blank and one standard at or near the mid-range. Checks throughout the day must be within 20% of original curve. If 20 or more samples per day are analyzed, the working standard curve must be verified by running an additional standard at or near the mid-range every 10 samples. Checks must be within ±20% of the true value. At least one duplicate and one spike sample should be run every 20 samples, or with each matrix type to verify precision of the method.

REFERENCE Test Methods for Evaluating Solid Waste (SW-846). U.S. EPA. 1983. Method 7870, Rev. 0, Sept. 1986. Office of Solid Wastes, Washington, DC.

Tin   **EPA Method 7870**
CAS #7440-31-5

TITLE Atomic Absorption (AA)

MATRIX Drinking, surface and direct aspiration saline waters. Wastewater.

APPLICATION Sample is aspirated and atomized in a flame. A light beam from a Sn hollow cathode lamp is directed through the flame into a monochromator and onto a detector. Since wavelength of light beam is specific for Sn, light energy absorbed by detector is measure of tin.

INTERFERENCES The most troublesome type is chemical, caused by lack of absorption of atoms bound in molecular combination in the flame. High dissolved solids in sample may result in nonatomic absorbance interference. Ionization and spectral interferences can occur.

INSTRUMENTATION Atomic absorption spectrometer. Tin hollow cathode lamp. (286.3 nm wavelength)

RANGE 10–300 mg/L

MDL 0.8 mg/L

PRECISION Standard deviation = ±0.25, 0.50, 0.50 at 4.0, 20, 60 mg Sn/L

ACCURACY Recoveries = 96, 101, 101% at 4.0, 20, 60 mg Sn/L

SAMPLING METHOD Use glass or plastic containers. Collect 200 g of solids and 600 mL of liquid samples.

STABILITY Cool solid samples to 4°C and analyze as soon as possible. Add nitric acid to liquid samples to pH <2.

MHT 6 months.

QUALITY CONTROL At least one duplicate and one spike sample should be run every 20 samples or with each matrix type to verify precision of the method. For 20 or more samples per day, verify working standard curve. Run an additional standard at or near mid-range every 10 samples.

REFERENCE Method 7870, SW-846, 3rd ed., Nov.1986.

Tokuthion (Protothiofos)   **EPA Method 8140**
CAS #34643-46-4

TITLE Organophosphorus Pesticides

MATRIX Groundwater, soils, sludges, water miscible liquid wastes, and non-water miscible wastes.

APPLICATION This method is used for the analysis of 21 organophosphorus pesticides. Samples are extracted, concentrated, and analyzed using direct injection of both neat and diluted organic liquid into a gas chromatograph (GC).

INTERFERENCES Solvents, reagents, and glassware may introduce artifacts. Other interferences may come from coextracted compounds from samples. The use of Florisil cleanup materials may produce low recoveries. Elemental sulfur may interfere with some compounds when using a flame photometric detector. Sulfur cleanup (Method 3660) may alleviate sulfur interference.

INSTRUMENTATION GC capable of on-column injections and a flame photometric detector (FPD) or a thermionic detector. Column 1: 1.8 m by 2 mm with 5% SP-2401 on Supelcoport.

Column 2: 1.8 m by 2 mm with 3% SP-2401 on Supelcoport. Column 3: 50 cm by ⅛ in Teflon® with 15% SE-54 on Gas Chrom Q. The preferred column is Column Number 1.

RANGE 5.3–64 µg/L

MDL 0.5 µg/L (in reagent water).

PQL FACTORS FOR MULTIPLYING × FID MDL VALUE

Matrix	Multiplication Factor
Groundwater	10
Low-level soil by sonication with GPC cleanup	670
High-level soil and sludge by sonication	10,000
Non-water miscible waste	100,000

PRECISION 6.8% (single operator standard deviation)

ACCURACY 64.6% (single operator average recovery)

SAMPLING METHOD Use 8-oz. widemouth glass bottles with Teflon®-lined caps for concentrated waste samples, soils, sediments, and sludges. Use 1 or 2½ gallon amber glass bottles with Teflon®-lined caps for liquid (water) samples.

STABILITY Cool soil, sediment, sludge, and liquid samples to 4°C. If residual chlorine is present in liquid samples add 3 mL of 10% sodium thiosulfate per gallon of sample and cool to 4°C.

MHT 14 days for concentrated waste, soil, sediment, or sludge; 7 days for liquid samples; all extracts must be analyzed within 40 days.

QUALITY CONTROL A quality control check sample concentrate containing this compound in acetone at a concentration 1,000 times more concentrated than the selected spike concentration is required. The QC check sample concentrate may be prepared from pure standard materials or purchased as certified solutions. Use appropriate trip, matrix, control site, method, reagent, and solvent blanks. Internal, surrogate, and five concentration level calibration standards are used.

REFERENCE Method 8140, SW-846, 3rd ed., Sept. 1986.

Tokuthion (Protothiofos)   **EPA Method 8141**
CAS #34643-46-4

TITLE Organophosphorus Compounds by Gas Chromatography: Capillary Column Technique

MATRIX This method covers aqueous and solid matrices. This includes a wide variety such as drinking water, groundwater, industrial wastewaters, surface waters, soils, solids, and sediments.

METHOD SUMMARY This is a GC method used to determine the concentration of 28 organophosphorus pesticides.

The use of Gel Permeation Cleanup (EPA Method 3640) for sample cleanup has been demonstrated to yield recoveries of less than 85% for many method analytes and is therefore not recommended for use with this method.

This method provides GC conditions for the detection of ppb concentrations of organophosphorus compounds. Prior to the use of this method, appropriate sample preparation techniques must be used. Water samples are extracted at a neutral pH with methylene chloride as a solvent by using a separatory funnel (EPA Method 3510) or a continuous liquid-liquid extractor (EPA Method 3520). Soxhlet extraction (EPA Method 3540) or ultrasonic extraction (EPA Method 3550) using methylene chloride/acetone (1:1) are used for solid samples. Both neat and diluted organic liquids (EPA Method 3580) may be analyzed by direct injection. Spiked samples are used to verify the applicability of the chosen extraction technique to each new sample type. A GC with a flame photometric (FPD) or nitrogen-phosphorus detector (NPD) is used for this multiresidue procedure.

INTERFERENCES The use of Florisil cleanup materials (EPA Method 3620) for some of the compounds in this method has been demonstrated to yield recoveries less than 85% and is therefore not recommended for all compounds. Use of phosphorus or halogen specific detectors, however, often obviates the necessity for cleanup for relatively clean sample matrices. If particular circumstances demand the use of an alternative cleanup procedure, the analyst must determine the elution profile and demonstrate that the recovery of each analyte is no less than 85%.

Use of a flame photometric detector (FPD) in the phosphorus mode will minimize interferences from materials that do not contain phosphorus. Elemental sulfur, however, may interfere with the determination of certain organophosphorus compounds by flame photometric gas chromatography. Sulfur cleanup using EPA Method 3660 may alleviate this interference. A nitrogen phosphorus detector (NPD) is also recommended.

A few analytes coelute on certain columns. Therefore, select a second column for confirmation where coelution of the analytes of interest does not occur.

Method interferences may be caused by contaminants in solvents, reagents, glassware, and other sample processing hardware that lead to discrete artifacts or elevated baselines in gas chromatograms. All these materials must be routinely demonstrated to be free from interferences under the conditions of the analysis by analyzing reagent blanks.

INSTRUMENTATION A GC with a NPD or a FPD will be needed. A data system or integrator is recommended for measuring peak areas and/or peak heights. A Kuderna-Danish (K-D) apparatus will be needed for extract concentration.

Column 1: 15 m × 0.53 mm megabore capillary column, 1.0 µm film thickness, DB-210.

Column 2: 15 m × 0.53 mm megabore capillary column, 1.5 µm film thickness, SPB-608.

Column 3: 15 m × 0.53 mm megabore capillary column, 1.0 µm film thickness, DB-5.

Three megabore capillary columns are included for analysis of organophosphates by this method. Column 1 (DB-210 or equivalent) and Column 2 (SPB-608 or equivalent) are recommended if a large number of organophosphorus analytes are to be determined. If the superior resolution offered by

Column 1 and Column 2 is not required, Column 3 (DB-5 or equivalent) may be used. For megabore capillary columns, automatic injections of 1 µL are recommended.

PRECISION & ACCURACY The MDL actually achieved in a given analysis will vary, as it is dependent on instrument sensitivity and matrix effects. Single operator accuracy and precision studies have been conducted with spiked water and soil samples.

MULTIPLICATION FACTORS FOR OTHER MATRICES (a)

Matrix	Factor (b)
Groundwater (EPA Method 3510 or EPA Method 3520)	10
Low-concentration soil by Soxhlet and no cleanup	10 (c)
Low-concentration soil by ultrasonic extraction with GPC cleanup	6.7 (c)
High-concentration soil and sludges by ultrasonic extraction	500 (c)
Non-Water miscible waste (EPA Method 3580)	1000 (c)

(a) Sample EQLs are highly matrix-dependent. The EQLs listed here are provided for guidance and may not always be achievable.
(b) EQL = [Method detection limit] × [Factor]. For non-aqueous samples the factor is on a wet-weight basis.
(c) Multiply this factory times the soil MDL.

The MDL (in µg/L) when reagent water was extracted using a separatory funnel was 0.07.

The MDL (in µg/kg) when soil was extracted using Soxhlet extraction (EPA Method 3540) was 5.5.

Accuracy (as % recovery) with separatory funnel extraction ranged from not recovered (with low spikes) to 90 (with high spikes).

Accuracy (as % recovery) with continuous liquid-liquid extraction ranged from 132 (with low spikes) to 90 (with high spikes).

Accuracy (as % recovery) with Soxhlet extraction of soils ranged from not recovered (with low spikes to 89 (with high spikes).

Accuracy (as % recovery) with ultrasonic extraction of soils ranged from not recovered (with low spikes) to 82 (with high spikes).

SAMPLE COLLECTION, PRESERVATION & HANDLING
Containers used to collect samples for the determination of semivolatile organic compounds should be soap and water washed followed by methanol (or isopropanol) rinsing. The sample containers should be of glass or Teflon® and have screw-top covers with Teflon® liners.

No preservation is used with concentrated waste samples. With liquid samples containing no residual chlorine and with soil, sediment, and sludge samples, immediately cooling to 4°C is the only preservation used. When residual chlorine is present then 3 mL of 10% aqueous sodium sulfate is added for each gallon of sample collected, followed by cooling to 4°C.

Liquid samples must be extracted within 7 days and their extracts analyzed within 40 days. Concentrated waste, soil, sediment, and sludge samples must be extracted within 14 days and their extracts analyzed within 40 days.

SAMPLE PREPARATION In general, water samples are extracted at a neutral pH with methylene chloride, using either EPA Method 3510 or EPA Method 3520. Solid samples are extracted using either EPA Method 3540 or EPA Method 3550 with methylene chloride/acetone (1:1) as the extraction solvent.

Prior to GC analysis, the extraction solvent may be exchanged to hexane. Single lab data indicates that samples should not be transferred with 100% hexane during sample workup as the more water soluble organophosphorus compounds may be lost.

If cleanup is performed on the samples, the analyst should analyze the samples by GC. This will confirm elution patterns and the absence of interferences from the reagents. If peak detection and identification is prevented by the presence of interferences, further cleanup is required.

QUALITY CONTROL The analyst should monitor the performance of the extraction, cleanup (when used), and analytical system and the effectiveness of the method in dealing with each sample matrix by spiking each sample, standard, and blank with one or two surrogates (e.g., organophosphorus compounds not expected to be present in the sample). Deuterated analogs of analytes should not be used as surrogates for gas chromatographic analysis due to coelution problems.

A minimum of five concentrations for each analyte of interest should be prepared through dilution of the stock standards with isooctane. One of the concentrations should be at a concentration near, but above, the MDL.

Include a mid-level check standard after each group of 10 samples in the analysis sequence. GC/MS techniques should be judiciously employed to support qualitative identifications made with this method. Follow the GC/MS operating requirements specified in EPA Method 8270.

When available, chemical ionization mass spectra may be employed to aid in the qualitative identification process. To confirm an identification of a compound, the background-corrected mass spectrum of the compound must be obtained from the sample extract and must be compared with a mass spectrum from a stock or calibration standard analyzed under the same chromatographic conditions. The molecular ion and all other ions present above 20% relative abundance in the mass spectrum of the standard must be present in the mass spectrum of the sample with agreement to ±20%. The retention time of the compound in the sample must be within six seconds of the retention time for the same compound in the standard solution.

Should the MS procedure fail to provide satisfactory results, additional steps may be taken before reanalysis. These steps may include the use of alternate packed or capillary GC columns or additional sample cleanup.

REFERENCE Test Methods for Evaluating Solid Waste, Physical/Chemical Methods, SW-846, 3rd Edition, U.S. EPA, Office of Solid Waste, Washington, DC, EPA Method 8141 July 1992.

Toluene
CAS #108-88-3

EPA Method 1624

TITLE Volatile Organic Compounds by Isotope Dilution GC/MS

MATRIX Compounds may be determined in waters, soils, and municipal sludges by this method.

METHOD SUMMARY This method is used to determine 58 volatile toxic organic pollutants associated with the CWA (as amended 1987); the RCRA (as amended 1986); the CERCLA (as amended 1986); and other compounds amenable to purge-and-trap gas chromatography-mass spectrometry (GC/MS).

If the solids content is less than 1%, stable isotopically-labeled analogs of the compounds of interest are added to a 5-mL sample and the sample is purged with an inert gas at 20–25°C in a chamber designed for soil or water samples. If the solids content is greater than 1%, 5 mL of reagent water and the labeled compounds are added to a 5-g aliquot of sample and the mixture is purged at 40°C. Compounds that will not purge at 20–25°C or at 40°C are purged at 78–85°C. In the purging process, the volatile compounds are transferred from the aqueous phase into the gaseous phase where they are passed into a sorbent column, and trapped. After purging is completed, the trap is backflushed and heated rapidly to desorb the compounds into a GC. The compounds are separated by the GC and detected by a MS. The labeled compounds serve to correct the variability of the analytical technique.

INTERFERENCES Impurities in the purge gas, organic compounds outgassing from the plumbing upstream of the trap, and solvent vapors in the lab account for most problems. Samples can be contaminated by diffusion of volatile organic compounds (particularly methylene chloride) through the bottle seal during shipment and storage. Contamination by carryover can occur when high-level and low-level samples are analyzed sequentially. When an unusually concentrated sample is encountered, follow it by analysis of a reagent water blank to check for carryover.

INSTRUMENTATION Major equipment includes a GC with linear temperature programming and a glass jet separator as the MS interface, a MS with 70 eV electron impact ionization, and a data system to collect and record response factors.

Column: 2.8 m × 2 mm I.D. glass, packed with 1% SP-1000 on Carbopak B, 60/80 mesh, or equivalent.

PRECISION & ACCURACY The detection limits of the method are usually dependent on the level of interferences rather than instrumental limitations. The method detection limits were determined in digested sludge (low solids) and in filter cake or compost (high solids).

The MDL (in µg/kg) for low solids is 27 and for high solids is 4.
Labeled and native compound precision (in µg/L) as standard deviation was 6.3.
Labeled and native compound accuracy (in µg/L) as average recovery was 15–29.

Acceptance criteria are at 20 µg/L for this compound.

SAMPLE COLLECTION, PRESERVATION & HANDLING Grab samples are collected in glass containers having a total volume greater than 20 mL. Fill and seal each bottle so that no air bubbles are entrapped. Samples are maintained at 0 to 4°C from the time of collection until analysis. If an aqueous sample contains residual chlorine, add sodium thiosulfate preservative (10 mg/40 mL) to the empty sample bottles just prior to shipment to the sample site. All samples must be analyzed within 14 days of collection.

SAMPLE PREPARATION Samples containing less than 1% solids are analyzed directly as aqueous samples. Samples containing 1% solids or greater are analyzed as solid samples utilizing one of two methods, depending on the levels of pollutants, in the sample. Samples containing 1% solids or greater, and low to moderate levels of pollutants are analyzed by purging a known weight of sample added to 5 mL of reagent water. Samples containing 1% solids or greater, and high levels of pollutants, are extracted with methanol, and an aliquot of the methanol extract is added to reagent water and purged.

QUALITY CONTROL A field blank prepared from reagent water and carried through the sampling and handling protocol may serve as a check on contamination from shipment and storage.

The analyst is permitted to modify this method to improve separations or lower the costs of measurements, provided all performance specifications are met. Analyses of blanks are required. When results of spikes indicate atypical method performance for samples, the samples are diluted to bring method performance within acceptable limits. Analyze two sets of four 5-mL aliquots (8 aliquots total) of the aqueous performance standard. Spike all samples with labeled compounds to assess method performance on the sample matrix. Compute the percent recovery of the labeled compounds using the internal standard method. Compare the percent recovery for each compound with the corresponding labeled compound recovery. Reagent water blanks are analyzed to demonstrate freedom from carryover contamination. Field replicates may be collected to determine the precision of the sampling technique, and spiked samples may be required to determine the accuracy of the analysis when the internal method is used.

REFERENCE Volatile Organic Compounds by Isotope Dilution GC/MS. Office of Water Regulation and Standards, U.S. EPA Industrial Technology Division, Washington, DC, EPA Method 1624, Rev. C, June 1989 (contact W.A. Telliard, U.S. EPA, Office of Water Regulations and Standards, 401 M St., SW, Washington, DC, 20460. Phone: 202-382-7131).

Toluene
CAS #108-88-3

EPA Method 502

TITLE Volatile Organic Compounds in Water By Purge and Trap Capillary Column Gas Chromatography with Photoionization and Electrolytic Conductivity Detectors in Series. U.S. EPA Method 502.2, Rev. 2.0, 1989.

MATRIX Drinking water and raw source water. The latter should include most surface water and groundwater sources.

METHOD SUMMARY This method covers 60 volatile organic compounds that contain halogen atoms and/or that are aromatic. An inert gas (zero grade nitrogen or helium) is bubbled through a 25-mL or a 5-mL water sample (depending on the expected concentration of the analytes). Purged sample components are trapped in a tube of sorbent materials. When purging is complete, the sorbent tube is heated and backflushed with helium to desorb the trapped sample onto a capillary GC column. The column is temperature programmed to separate the method analytes which are then detected with a photoionization detector (PID) and a Hall electrolytic conductivity (HECD) placed in series. The PID is selective for aromatic compounds and the HECD is selective for halogenated compounds.

INTERFERENCES Impurities in the purge gas and from organic compounds outgassing from the plumbing ahead of the trap account for many contamination problems. Interferences purged or coextracted from the samples will vary considerably from source to source, depending upon the particular sample or extract being tested. Cross-contamination can occur whenever high-level and low-level samples are analyzed sequentially. Samples also can be contaminated by diffusion of volatile organics (particularly methylene chloride and fluorocarbons) through the septum seal into the sample during shipment and storage. The lab where volatile analysis is performed and also the refrigerated storage area should be completely free of solvents.

INSTRUMENTATION A GC containing a series configuration of a high temperature photoionization detector (PID) equipped with 10.0 eV (nominal) lamp and Hall electrolytic conductivity detector (HECD) is required. Also required is an all-glass 5-mL purging device, a sorbent trap, and a thermal desorption apparatus which is connected to the GC system.

Column 1: VOCOL glass wide-bore capillary column.
Column 2: RTX–502.2 mega-bore capillary column.
Column 3: DB-62 mega-bore capillary column.

PRECISION & ACCURACY Method detection limits are dependent upon the characteristics of the gas chromatographic system used. Analytes that are not separated chromatographically cannot be individually identified and used in the same calibration mixture or water samples unless an alternative technique for identification and quantification, such as mass spectrometry, is used.

Electrolytic conductivity detetor (c) range in μg/L (a) was 0.02–200.
Electrolytic conductivity detetor (c) MDL in μg/L (b) was not listed.
Electrolytic conductivity detetor (c) accuracy as % recovery was not listed.
Electrolytic conductivity detetor (c) precision as % RSD was not listed.
Photoionization detector (d) range in μg/L (a) was 0.02–200.
Photoionization detector (d) MDL in μg/L (b) was 0.01.
Photoionization detector (d) accuracy as % recovery was 99.
Photoionization detector (d) precision as % RSD was 0.8.

(a) The applicable concentration range of this method is compound, instrument, and matrix-dependent. It is listed as being approximately 0.02 to 200 μg/L but no specific information is provided so caution should be observed.
(b) The method detection limits reports with this method are compound, instrument, and matrix-dependent. The values reported were calculated using reagent water fortified with the corresponding compounds at 10 μg/L and a GC-equipped with a 60 m × 0.75 mm VOLCOL wide bore capillary column with 1.5 μm film thickness and using helium carrier gas.
(c) Recoveries and relative standard deviations were determined from seven samples of reagent water fortified with 10 μg/L of each compound. 2-Bromo-1-chloropropane was used as the internal standard for calculating average recoveries.
(d) Recoveries and relative standard deviations were determined from seven samples of reagent water fortified with 10 μg/L of each compound. Fluorobenzene was used as the internal standard for calculating average recoveries.

SAMPLING METHOD Collect samples using a 40- to 120-mL screw-cap vial (prewashed with detergent, rinsed with distilled water and oven dried at 105°C) with a Teflon®-faced silicone septum. Collect bubble-free samples and place the septum with the Teflon® side down on the water.

SAMPLE PRESERVATION If residual chlorine is present in the water add about 25 mg of ascorbic acid to each vial before samples are collected to remove the chlorine. Add hydrochloric acid to reduce pH to <2, immediately cool samples to 4°C, and store them in a solvent-free refrigerator at 4°C until analysis.

MHT The maximum holding time for samples is 14 days from the time they were collected.

SAMPLE PREPARATION Remove the plungers from two 5-mL syringes and attach a closed syringe valve to each. Warm the sample to room temperature, open the sample bottle, and carefully pour the sample into one of the syringe barrels to just short of overflowing. Replace the syringe plunger, invert the syringe, and compress the sample. Open the syringe valve and vent any residual air while adjusting the sample volume to 5.0 mL. Add 10 μL of the internal calibration standard to the sample through the syringe valve. Close the valve. Fill the second syringe in an identical manner from the same sample bottle. Reserve this second syringe for a reanalysis if necessary.

QUALITY CONTROL As an initial demonstration of lab accuracy and precision, analyze 4 to 7 replicates of a lab fortified blank containing analyte at 0.1–5 μg/L. Collect all samples in duplicate. Surrogate analytes (similar to those of the analytes of interest), whose concentration is known in every sample, are measured using the same internal standard calibration procedure. Duplicate field reagent water blanks (trip blanks) must be analyzed with each set of samples, lab reagent blanks (method blanks) must be analyzed with each batch of samples processed as a group within a work shift. Also, a single lab-fortified blank that contains each of the analytes of interest should be analyzed with each batch of samples processed as a group within a work shift. A 3- to 5-point calibration curve is needed depending on the calibration range factor required.

EPA CONTACT & HOTLINE For technical questions contact Dr. Baldev Bathija, U.S. EPA, Office of Ground Water and Drinking Water (WH-550D), 401 M St. SW, Washington, DC 20460. Tel. (202) 260-3040. For further information the EPA Safe Drinking Water Hotline may be called at: (800) 426-4791.

REFERENCE Methods for the Determination of Organic Compounds in Drinking Water, EPA/600/4-88/039 (revised July 1991; Final Rule for determination of compliance with the MCL for Total Trihalomethanes under 141.30, in 40 CFR Part 141, Vol. 58, No. 147, Fed. Reg., Tuesday Aug. 3, 1993). U.S. EPA Environmental Monitoring Systems Laboratory, Cincinnati, OH, 45268, U.S.A. Available from the National Technical Information Service (NTIS), 5285 Port Royal Road, Springfield, VA 22161; Tel. 800-553-6847. NTIS Order Number is PB91-231480.

Toluene — EPA Method 524
CAS #108-88-3

TITLE Measurement of Purgeable Organic Compounds in Water by Capillary Column GC/MS.

MATRIX Drinking water and raw source water; the latter should include most surface water and groundwater sources.

METHOD SUMMARY Method 524.2 covers 60 volatile organic compounds. An inert gas (zero grade nitrogen or helium) is bubbled through a 25-mL or a 5-mL water sample (depending on the expected concentration of the analytes). Purged sample components are trapped in a tube of sorbent materials. When purging is complete, the sorbent tube is heated and backflushed with helium to desorb the trapped sample onto a capillary GC column.

INTERFERENCES Impurities in the purge gas and from organic compounds outgassing from the plumbing ahead of the trap account for many contamination problems. Interferences purged or coextracted from the samples will vary considerably from source to source, depending upon the particular sample or extract being tested. Cross-contamination can occur whenever high-level and low-level samples are analyzed sequentially. Samples also can be contaminated by diffusion of volatile organics (particularly methylene chloride and fluorocarbons) through the septum seal into the sample during shipment and storage.

INSTRUMENTATION A GC/MS with a data system equipped with one of the following capillary GC columns:

Column 1: VOCOL glass wide bore capillary column.
Column 2: DB-624 fused silica capillary column.
Column 3: DB-5 fused silica capillary column.

Also required is an all-glass 25 mL or 5-mL purging device, a sorbent trap, and a thermal desorption apparatus which is connected to the GC/MS system.

PRECISION & ACCURACY Method detection limits are compound- and instrument-dependent, and may vary from approximately 0.02–0.35 µg/L. Note in the table below that the "true" concentration range used for accuracy and precision measurements was quite narrow. However, the applicable concentration range of this method is primarily column dependent and is approximately 0.02 to 200 µg/L for the wide-bore thick-film columns. Narrow-bore thin-film columns may have a capacity which limits the range to about 0.02 to 20 µg/L. Analytes that are inefficiently purged from water will not be detected when present at low concentrations, but they can be measured with acceptable accuracy and precision when present in sufficient amounts.

Analytes that are not separated chromatographically, but which have different mass spectra and non-interfering quantification ions, can be identified and measured in the same calibration mixture or water sample. Analytes which have very similar mass spectra cannot be individually identified and measured in the same calibration mixture or water samples unless they have different retention times. Co-eluting compounds with very similar mass spectra, typically many structural isomers, must be reported as an isomeric group or pair.

The range (in µg/L) was 0.5–10.
The Method Detection Limig (in µg/L) was 0.11.
The accuracy (as % recovery) was 102.
The precision (in %) was 8.0.

Note: Data were obtained from 16–31 determinations using a wide-bore capillary column and a jet separator interfaced to a quadrupole mass spectrometer. All analytes were in a reagent water matrix.

SAMPLING METHOD Collect samples using a 40- to 120-mL screw-cap vial (prewashed with detergent, rinsed with distilled water and oven dried at 105°C) with a Teflon®-faced silicone septum. Collect bubble-free samples and place the septum with the Teflon® side down on the water.

SAMPLE PRESERVATION If residual chlorine is present in the water add about 25 mg of ascorbic acid to each vial before samples are collected to remove the chlorine. Add hydrochloric acid to reduce pH to <2, and immediately cool samples to 4°C, and store them in a solvent-free refrigerator at 4°C until analysis.

MHT The maximum holding time for samples is 14 days from the time they were collected.

SAMPLE PREPARATION Remove the plungers from two 25-mL (or 5-mL depending on sample size) syringes and attach a closed syringe valve to each. Warm the sample to room temperature, open the sample bottle, and carefully pour the sample into one of the syringe barrels to just short of overflowing. Replace the syringe plunger, invert the syringe, and compress the sample. Open the syringe valve and vent any residual air while adjusting the sample volume to 25.0 mL (or 5 mL). For samples and blanks, add 5 µL of the fortification solution containing the internal standard and the surrogates to the sample through the syringe valve. For calibration standards and lab fortified blanks, add 5 µL of the fortification solution containing the internal standard only. Close the valve. Fill the second syringe in an identical manner from the same sample bottle. Reserve this second syringe for a reanalysis if necessary.

QUALITY CONTROL As an initial demonstration of lab accuracy and precision, analyze 4 to 7 replicates of a lab fortified

blank containing analyte at 0.2–5 µg/L. Collect all samples in duplicate. Surrogate analytes (similar to those of the analytes of interest), whose concentration is known in every sample, are measured using the same internal standard calibration procedure. Duplicate field reagent water blanks (trip blanks) must be analyzed with each set of samples, lab reagent blanks (method blanks) must be analyzed with each batch of samples processed as a group within a work shift. Also, a single lab-fortified blank that contains each of the analytes of interest should be analyzed with each batch of samples processed as a group within a work shift. A 3- to 5-point calibration curve is needed depending on the calibration range factor required.

EPA CONTACT & HOTLINE For technical questions contact Dr. Baldev Bathija, U.S. EPA, Office of Ground Water and Drinking Water (WH-550D), 401 M St. SW, Washington, DC 20460. Tel. (202) 260-3040. For further information the EPA Safe Drinking Water Hotline may be called at: (800) 426-4791.

REFERENCE Methods for the Determination of Organic Compounds in Drinking Water, EPA/600/4-88/039 (revised July 1991; Final Rule for determination of compliance with the MCL for Total Trihalomethanes under 141.30, in 40 CFR Part 141, Vol. 58, No. 147, Fed. Reg., Tuesday Aug. 3, 1993). U.S. EPA Environmental Monitoring Systems Laboratory, Cincinnati, OH, 45268, U.S.A. Available from the National Technical Information Service (NTIS), 5285 Port Royal Road, Springfield, VA 22161; Tel. 800-553-6847. NTIS Order Number is PB91-231480.

Toluene   **EPA Method 8021**
CAS #108-88-3

TITLE Halogenated Volatile by Gas Chromatography Using Photoionization and Electrolytic Conductivity Detectors in Series: Capillary Column Technique

MATRIX This method is applicable to nearly all types of samples, regardless of water content, including groundwater, aqueous sludges, caustic liquors, acid liquors, waste solvents, oily wastes, mousses, tars, fibrous wastes, polymeric emulsions, filter cakes, spent carbons, spent catalysts, soils, and sediments.

METHOD SUMMARY This method is used to determine 60 volatile organic compounds in a variety of solid waste matrices. It provides GC conditions for the detection of halogenated and aromatic volatile organic compounds. Samples can be analyzed using direct injection or purge-and-trap (EPA Method 5030). Groundwater samples must be analyzed using EPA Method 5030 (where applicable). A temperature program is used with the GC. Detection is achieved by a photoionization detector (PID) and a Hall electrolytic conductivity detector (HECD) in series.

INTERFERENCES Samples can be contaminated by diffusion of volatile organics (particularly chlorofluorocarbons and methylene chloride) through the sample container septum during shipment and storage.

INSTRUMENTATION A GC-equipped with variable-constant differential flow controllers, subambient oven controller, PID and HECD detectors connected with a short piece of uncoated capillary tubing and a data system.

Column: 60 m × 0.75 mm I.D. VOCOL wide-bore capillary column with 1.5 µm film thickness.

PRECISION & ACCURACY MDLs are compound-dependent and vary with purging efficiency and concentration. The applicable concentration range of this method is compound- and instrument-dependent but is approximately 0.1 to 200 µg/L. Analytes that are inefficiently purged from water will not be detected when present at low concentrations, but they can be measured with acceptable accuracy and precision when present in sufficient amounts. The estimated quantitation limit (EQL) for an individual compound is approximately 1 µg/kg (wet weight) for soil/sediment samples, 100 µg/kg (wet weight) for wastes, and 1 µg/L for groundwater. EQLs will be proportionately higher for sample extracts and samples that require dilution or reduced sample size to avoid saturation of the detector.

MULTIPLICATION FACTORS FOR OTHER MATRICES (a)

Matrix	Factor (b)
Groundwater	10
Low-concentration soil	10
Water miscible liquid waste	500
High-concentration soil and sludge	1250
Non-water miscible waste	1250

(a) Sample EQLs are highly matrix-dependent. The EQLs listed herein are provided for guidance and may not always be achievable. (b) EQL = [Method detection limit] × [Factor]. For non-aqueous samples, the factor is on a wet-weight basis.

SINGLE LABORATORY ACCURACY & PRECISION DATA FOR VOCs IN WATER
This method was tested in a single lab using water spiked at 10 µg/L and the following data was reported:

Recoveries and standard deviations were determined from seven samples and spiked at 10 µg/L of each analyte. Recoveries were determined by the internal standard method. Internal standards were: Fluorobenzene for PID and 2-Bromo-1-chloropropane for HECD.

The average recovery (in percent) for the PID was 99.
The standard deviation of the recovery for the PID was 0.8.
The MDL (in µg/mL) for the PID was 0.01.
The average recovery (in percent) for the HECD was none (no response for this detector).
The standard deviation of the recovery for the HECD was none (no response for this detector)-.
The MDL (in µg/mL) for the HECD was none (no response for this detector).

SAMPLE COLLECTION, PRESERVATION & HANDLING
Volatile organics — Standard 40-mL glass screw-cap VOA vials with Teflon®-faced silicone septum may be used for both liquid and solid matrices. When collecting samples, liquids and solids should be introduced into the vials gently to reduce agitation which might drive off volatile compounds. If there are any air bubbles present the sample must be retaken. Tap slightly as they are filled to try and eliminate as much free air space as possible. The two vials from each sampling locations should

be sealed in separate plastic bags to prevent cross-contamination between samples particularly if the sampled waste is suspected of containing high levels of volatile organics.

Semivolatile organics — Containers used to collect samples for the determination of semivolatile organic compounds should be soap and water washed followed by methanol (or isopropanol) rinsing. The sample containers should be of glass or Teflon® and have screw-top covers with Teflon® liners.

Preservation for volatile organics — No preservation is used with concentrated waste samples. With liquid samples containing no residual chlorine, 4 drops of concentrated hydrochloric acid are added and the samples are immediately cooled to 4°C. When liquid samples contain residual chlorine, they are treated as above and, in addition, 4 drops of 4% aqueous sodium thiosulfate are added. Soil, sediment, and sludge samples are only cooled to 4°C.

Preservation for semivolatile organics — No preservation is used with concentrated waste samples. With liquid samples containing no residual chlorine and with soil, sediment, and sludge samples, immediately cooling to 4°C is the only preservation used. When residual chlorine is present then 3 mL of 10% aqueous sodium sulfate is added for each gallon of sample collected, followed by cooling to 4°C.

MHT The holding time for all volatile organics samples is 14 days. Liquid samples must be extracted within 7 days and their extracts analyzed within 40 days. Concentrated waste, soil, sediment, and sludge samples must be extracted within 14 days and their extracts analyzed within 40 days.

SAMPLE PREPARATION Volatile compounds are introduced into the gas chromatograph either by direct injector or purge-and-trap (EPA Method 5030). EPA Method 5030 may be used directly on groundwater samples or low-concentration contaminated soils and sediments. For medium-concentration soils or sediments, methanolic extraction, as described in EPA Method 5030, may be necessary prior to purge-and-trap analysis.

QUALITY CONTROL Calculate surrogate standard recovery on all samples, blanks, and spikes. A trip blank is recommended to check on sampling, storage, and handling contamination. Calibration standards, at a minimum of five concentration levels, are prepared in organic-free reagent water. One of the concentration levels should be at a concentration near, but above, the method detection limit.

A combination of bromochloromethane, 2-bromo-1-chloropropane, 1,4-dichlorobutane, and bromochlorobenzene are recommended as surrogate standards to encompass the range of the temperature program used in this method.

REFERENCE Test Methods for Evaluating Solid Waste, Physical/Chemical Methods, SW-846, 3rd Edition, U.S. EPA, Office of Solid Waste, Washington, DC, EPA Method 8021A, Rev. 1, Nov. 1992.

Toluene EPA Method 8240
CAS #108-88-3

TITLE Volatile Organics By GC/MS: Packed Column Technique

MATRIX Nearly all types of sample matarices, regardless of water content, can be analyzed using this method. This includes groundwater, aqueous sludges, caustic liquors, acid liquors, waste solvents, oily wastes, mousses, tars, fibrous wastes, polymetric emulsions, filter cakes, spent carbons, spent catalysts, soils, and sediments.

METHOD SUMMARY Method 8240B covers 80 volatile organic compounds that are introduced into a gas chromatograph by the purge-and-trap method or by direct injection (in limited applications). For the purge-and-trap method an inert gas (zero grade nitrogen or helium) is bubbled through a 5-mL solution at ambient temperature. Purged sample components are trapped in a tube of sorbent materials. When purging is complete, the sorbent tube is heated and backflushed with inert gas to desorb the trapped components onto a GC column.

INTERFERENCES Impurities in the purge gas and from organic compounds outgassing from the plumbing ahead of the trap account for many contamination problems. Interferences purged or coextracted from the samples will vary considerably from source to source. Cross-contamination can occur whenever high-level and low-level samples are analyzed sequentially. Whenever an unusually concentrated sample is analyzed, it should be followed by the analysis of organic-free reagent water to check for cross-contamination. Samples also can be contaminated by diffusion of volatile organics (particularly methylene chloride and fluorocarbons) through the septum seal into the sample during shipment and storage. A trip blank can serve as a check on such contamination. The lab where volatile analysis is performed and also the refrigerated storage area should be completely free of solvents.

INSTRUMENTATION A gas chromatograph/mass spectrometry/data system (GC/MS) equipped with a 6 ft × 0.1 in I.D. glass column packed with 1% SP-1000 on Carbopack-B (60/80 mesh) is required. Also needed is a 5-mL purging device, a sorbent trap, and a thermal desorption apparatus.

PRECISION & ACCURACY This method is reported to have been tested by 15 laboratories using organic-free reagent water, drinking water, surface water, and industrial wastewaters (not specified) fortified at six concentrations over the range 5–600 µg/L.

Sample estimated quantitation limits (EQLs) are highly matrix-dependent. The EQLs listed may not always be achievable. EQLs listed for soils or sediments are based on wet weight. Normally, data is reported on a dry-weight basis; therefore, EQLs will be higher, based on the percent dry weight of each sample. Note that EQLs are even more variable than MDLs and that they are highly variable depending on the matrix being analyzed.

EQL in groundwater in µg/L was 5.
EQL in low soil or sediment in µg/kg was 5.
Accuracy (a) in µg/L was $0.98C + 2.03$.
Precision (b) in µg/L was $0.22x - 1.71$.

(a) *Average recovery found for measurements of samples containing a concentration of C, in µg/L.*
(b) *Overall precision found for measurements of samples with average recovery X for samples containing a concentration of C in µg/L.*

$X =$ *Average recovery found for measurement of samples containing a concentration of C in µg/L.*

MULTIPLICATION FACTORS FOR OTHER MATRICES

Other Matrices	Factor (a)
Waste miscible liquid waste	50
High-concentration soil and sludge	125
Non-water miscible waste	500

(a) EQL = [EQL for low soil sediment X [Factor]. For non-aqueous samples, the factor is on a wet-weight basis.

SAMPLING METHOD

Liquid samples — Use a 40-mL glass screw-cap VOA vial with a Teflon®-faced silicone septum that has been prewashed, rinsed with distilled deionized water, and oven dried. However, if residual chlorine is present, collect sample in a 40-oz. soil VOA container which has been pre-preserved with 4 drops of 10% sodium thiosulfate, mix gently, and then transfer the sample to a 40-mL VOA vial. Collect bubble-free samples in duplicate and seal them in separate plastic bags.

Soils or sediments, and sludges — Use an 8-oz. widemouth glass bottle with a Teflon®-faced silicone septum that has been prewashed with detergent, rinsed with distilled deionized water, and oven dried. Tap slightly to eliminate free air space. Collect samples in duplicate and seal them in separate plastic bags.

SAMPLE PRESERVATION

Liquid samples — Add 4 drops of concentrated HCL and immediately cool samples to 4°C and store in a solvent-free refrigerator.

Soils or sediments, and sludges — Cool samples to 4°C and store in a solvent-free refrigerator.

MHT Maximum holding time is 14 days from the date of sample collection.

SAMPLE PREPARATION

Liquid samples — Remove the plunger from a 5-mL syringe and carefully pour the sample into the syringe barrel to just short of overflowing. Replace the syringe plunger and compress the sample. Open the syringe valve and vent any residual air while adjusting the sample volume to 5.0 mL. If there is only one volatile organic analysis (VOA) vial, a second syringe should be filled at this time to protect against possible loss of sample integrity. Add 10 µL of surrogate spiking solution and 10 µL of internal standard spiking solution through the valve bore of the 5-mL syringe, then close the valve. The surrogate and internal standards may be mixed and added as a single spiking solution.

Sediments, soils, and waste samples — All samples of this type should be screened by GC analysis using a headspace method (EPA Method 3810) or the hexadecane extraction and screening method (EPA Method 3820). Use the screening data to determine whether to use the low-concentration method (0.005–1 mg/kg) or the high-concentration method (>1 mg/kg).

Low-concentration method — The low-concentration method is based on purging a heated sediment or soil sample mixed with organic-free reagent water containing the surrogate and internal standards. Analyze all reagent blanks and standards under the same conditions as the samples.

Use a 5-g sample if the expected concentration is <0.1 mg/kg or a 1-g sample for expected concentrations between 0.1 and 1 mg/kg. Mix the contents of the sample container with a narrow metal spatula. Weigh the amount of the sample into a tared purge device. Add the spiked water to the purge device, which contains the weighed amount of sample, and connect the device to the purge-and-trap system.

High-concentration method — This method is based on extracting the sediment or soil with methanol. A waste sample is either extracted or diluted, depending on its solubility in methanol. Wastes that are insoluble in methanol are diluted with reagent tetraglyme or possibly polyethylene glycol (PEG). An aliquot of the extract is added to organic-free reagent water containing surrogate and internal standards. This is purged at ambient temperature. All samples with an expected concentration of >1.0 mg/kg should be analyzed by this method.

Mix the contents of the sample container with a narrow metal spatula. For sediments or soils and solid wastes that are insoluble in methanol, weigh 4 g (wet weight) of sample into a tared 20-mL vial. For waste that is soluble in methanol, tetraglyme, or PEG, weigh 1 g (wet weight) into a tared scintillation vial or culture tube or a 10-mL volumetric flask. Quickly add 9.0 mL of appropriate solvent then add 1.0 mL of a surrogate spiking solution to the vial, cap it, and shake it for 2 min.

METHANOL EXTRACT REQUIRED FOR ANALYSIS OF HIGH-CONCENTRATION SOILS OR SEDIMENTS

Approximate Concentration Range	Volume of Methanol Extract (a)
500–10,000 µg/kg	100 µL
1,000–20,000 µg/kg	50 µL
5,000–100,000 µg/kg	10 µL
25,000–500,000 µg/kg	100 µL of 1/50 dilution (b)

Calculate appropriate dilution factor for concentrations exceeding this table.

(a) The volume of methanol added to 5 mL of water being purged should be kept constant. Therefore, add to the 5-mL syringe whatever volume of methanol is necessary to maintain a volume of 100 µL added to the syringe.

(b) Dilute an aliquot of the methanol extract and then take 100 µL for analysis.

QUALITY CONTROL Demonstrate, through the analysis of a reagent water blank, that interferences from the analytical system, glassware, and reagents are under control. Blank samples should be carried through all stages of the sample preparation and measurement steps. For each analytical batch (up to 20 samples), a reagent blank, matrix spike, and matrix spike duplicate must be analyzed (the frequency of the spikes may be different for different monitoring programs). The blank and spiked samples must be carried through all stages of the sample preparation and measurement steps. QC samples mentioned in the section on Interferences will also be needed as appropriate to those situations.

REFERENCE Test Methods for Evaluating Solid Waste (SW-846). U.S. EPA. 1983. Method 8240B, Rev. 2, Nov. 1990. Office of Solid Wastes, Washington, DC.

Toluene
CAS #108-88-3
EPA Method 8260

TITLE Volatile Organic Compounds by GC/MS: Capillary Column Technique

MATRIX This method is applicable to nearly all types of samples, regardless of water content, including groundwater, soils, and sediments.

METHOD SUMMARY Method 8260A covers 58 volatile organic compounds that are introduced into a gas chromatograph by the purge-and-trap method or by direct injection (in limited applications). Zero-grade helium is bubbled through a 5-mL solution at ambient temperature. Purged sample components are trapped in a tube containing suitable sorbent materials. When purging is complete, the sorbent tube is heated and backflushed with helium to desorb trapped sample components. The analytes are desorbed directly to a large bore capillary or cryofocussed on a capillary precolumn before being flash evaporated to a narrow bore capillary for analysis.

INTERFERENCES Major contaminant sources are volatile materials in the lab and impurities in the inert purging gas and in the sorbent trap. Interfering contamination may occur when a sample containing low concentrations of volatile organic compounds is analyzed immediately after a sample containing high concentrations of volatile organic compounds. After analysis of a sample containing high concentrations of volatile organic compounds, one or more calibration blanks should be analyzed to check for cross-contamination. Screening of the samples prior to purge-and-trap GC/MS analysis is highly recommended to prevent contamination of the system. This is especially true for soil and waste samples.

Special precautions must be taken to analyze for methylene chloride. The analytical and sample storage area should be isolated from all atmospheric sources of methylene chloride. All gas chromatography carrier gas lines and purge gas plumbing should be constructed from stainless steel or copper tubing. Laboratory clothing previously exposed to methylene chloride fumes during liquid-liquid extraction procedures can contribute to sample contamination.

Samples can also be contaminated by diffusion of volatile organics (particularly methylene chloride and fluorocarbons) through the septum seal during shipment and storage. A trip blank can serve as a check on such contamination.

INSTRUMENTATION GC/MS with a temperature-programmable chromatograph suitable for splitless injection equipped with variable constant differential flow controllers, a subambient oven controller, a purging device, sorbent trap, a thermal desorption apparatus and a capillary precolumn interface when using cryogenic cooling will be needed. The following GC columns may be used:

Column 1: 60 m × 0.75mm I.D. capillary column coated with VOCOL, 1.5 μm film thickness.
Column 2: 30 m × 0.53mm capillary column coated with DB-624 or VOCOL, 3 μm film thickness.
Column 3: 30 m × 0.32mm I.D. capillary column coated with DB-5 or SE-54, 1-μm film thickness.

PRECISION & ACCURACY This method has been tested in a single lab using spiked water. Using a wide-bore capillary column, water was spiked at concentrations between 0.5 and 10 μg/L. Single lab accuracy and precision data are presented. The MDL actually achieved in a given analysis will vary depending on instrument sensitivity and matrix effects.

The MDL (a) in μg/L was 0.11.
The concentration range in μg/L was 0.5–10.
The mean accuracy (% of true value) was 102.
The precision as relative standard deviation was 8.0.

Note: The MDL is based on a 25-mL sample volume instead of a 5-mL sample volume.

SAMPLING METHOD

Liquid samples — Use a 40-mL glass screw-cap VOA vial with a Teflon®-faced silicone septum that has been prewashed, rinsed with distilled deionized water, and oven dried. If residual chlorine is present, collect the sample in a 4-oz soil VOA container which has been pre-preserved with 4 drops of 10% sodium thiosulfate. Mix gently and transfer the sample to a 40-mL VOA vial. Collect bubble-free samples in duplicate and seal each sample in a separate plastic bag.

Soils, sediments and sludges — Use an 8-oz widemouth glass bottle with Teflon®-faced silicone septum that has been prewashed, rinsed with distilled deionized water, and oven dried. **Do not** heat the septum for more than 1 h. Tap slightly to eliminate any free air space. Collect samples in duplicate and seal each one in a separate plastic bag.

SAMPLE PRESERVATION

Liquid samples — Add 4 drops of concentrated HCL, cool to 4°C and store in a solvent-free refrigerator.

Soils, sediments and sludges — Cool samples to 4°C and store in a solvent-free refrigerator.

MHT The maximum holding time of any sample (liquids, soils, sediments, and sludges) is 14 days.

SAMPLE PREPARATION

Liquid samples — Remove the plunger from a 5-mL syringe and carefully pour the sample into the syringe barrel to just short of overflowing. Replace the syringe plunger and compress the sample. Open the syringe valve and vent any residual air while adjusting the sample volume to 5.0 mL. If there is only one volatile organic analysis (VOA) vial, a second syringe should be filled at this time to protect against possible loss of sample integrity. Add 10 μL of surrogate spiking solution and 10 μL of internal standard spiking solution through the valve bore of the 5-mL syringe, then close the valve. The surrogate and internal standards may be mixed and added as a single spiking solution.

Sediments, soils, and waste samples — All samples of this type should be screened by GC analysis using a headspace method (EPA Method 3810) or the hexadecane extraction and screening method (EPA Method 3820). Use the screening data to determine whether to use the low-concentration method (0.005–1 mg/kg) or the high-concentration method (>1 mg/kg).

Low-concentration method — The low-concentration method is based on purging a heated sediment or soil sample mixed with organic-free reagent water containing the surrogate and internal standards. Analyze all reagent blanks and standards under the same conditions as the samples.

Use a 5-g sample if the expected concentration is <0.1 mg/kg or a 1-g sample for expected concentrations between 0.1 and 1 mg/kg. Mix the contents of the sample container with a narrow metal spatula. Weigh the amount of the sample into a tared purge device. Add the spiked water to the purge device, which contains the weighed amount of sample, and connect the device to the purge-and-trap system.

High-concentration method — This method is based on extracting the sediment or soil with methanol. A waste sample is either extracted or diluted, depending on its solubility in methanol. Wastes that are insoluble in methanol are diluted with reagent tetraglyme or possibly polyethylene glycol (PEG). An aliquot of the extract is added to organic-free reagent water containing surrogate and internal standards. This is purged at ambient temperature. All samples with an expected concentration of >1.0 mg/kg should be analyzed by this method.

Mix the contents of the sample container with a narrow metal spatula. For sediments or soils and solid wastes that are insoluble in methanol, weigh 4 g (wet weight) of sample into a tared 20-mL vial. For waste that is soluble in methanol, tetraglyme, or PEG, weigh 1 g (wet weight) into a tared scintillation vial or culture tube or a 10-mL volumetric flask. Quickly add 9.0 mL of appropriate solvent then add 1.0 mL of a surrogate spiking solution to the vial, cap it, and shake it for 2 min.

METHANOL EXTRACT REQUIRED FOR ANALYSIS OF HIGH-CONCENTRATION SOILS OR SEDIMENTS

Approximate Concentration Range	Volume of Methanol Extract (a)
500–10,000 µg/kg	100 µL
1,000–20,000 µg/kg	50 µL
5,000–100,000 µg/kg	10 µL
25,000–500,000 µg/kg	100 µL of 1/50 dilution (b)

Calculate appropriate dilution factor for concentrations exceeding this table.

(a) The volume of methanol added to 5 mL of water being purged should be kept constant. Therefore, add to the 5-mL syringe whatever volume of methanol is necessary to maintain a volume of 100 µL added to the syringe.
(b) Dilute an aliquot of the methanol extract and then take 100 µL for analysis.

QUALITY CONTROL Demonstrate, through the analysis of a reagent water blank, that interferences from the analytical system, glassware, and reagents are under control. Blank samples should be carried through all stages of the sample preparation and measurement steps. For each analytical batch (up to 20 samples), a reagent blank, matrix spike, and matrix spike duplicate must be analyzed (the frequency of the spikes may be different for different monitoring programs). The blank and spiked samples must be carried through all stages of the sample preparation and measurement steps. QC samples mentioned in the section on Interferences will also be needed as appropriate to those situations.

Matrix spiking standards should be prepared from volatile organic compounds which will be representative of the compounds being investigated. The recommended internal standards are chlorobenzene-d5, 1,4-difluorobenzene, 1,4-dichlorobenzene-d4, and pentafluorobenzene. Using stock standard solutions, prepare secondary dilution standards containing the compounds of interest, either singly or mixed together in methanol. Store them in a vial with no headspace for no more than one week. Surrogates recommended are toluene-d8, 4-bromofluorobenzene, and dibromofluoromethane. Each sample undergoing GC/MS analysis must be spiked with 10 µL of the surrogate spiking solution prior to analysis.

REFERENCE Test Methods for Evaluating Solid Waste (SW-846). U.S. EPA 1983, Method 8260A, Rev. 1, Nov. 1990. Office of Solid Waste, Washington, DC.

Toluene   EPA Method 503.1
CAS #108-88-3

TITLE Aromatic & Unsaturated VOCs Water

MATRIX Drinking water (finished or in any treatment stage) and raw source water.

APPLICATION Method covers 28 aromatic and unsaturated VOCs. An inert gas is bubbled through a 5-mL water sample. Purged sample components are trapped in tube of sorbent materials. When purging is complete, sorbent tube is heated and backflushed with inert gas to desorb trapped sample onto a packed GC column.

INTERFERENCES During analysis, major contaminant sources are volatile materials in the lab and impurities in purging gas and sorbent trap. With high and low level samples, there can be carryover contamination. Excess water causes a negative baseline deflection.

INSTRUMENTATION Purge and Trap GC w/photoionization detector. (Two GC columns are recommended); Column 1: 5% SP-1200 and 1.75% Bentone 34 on Supelcoport; Column 2: 1,2,3-tris(2-cyanoethoxy)propane on Chromosorb W.

RANGE 2.2–600 µg/L. (Drinking water)

MDL 0.02 µg/L in water

PRECISION RSD = 6.6% at 0.40 µg/L conc.; 13 samples

ACCURACY Average recovery = 94% at 0.40 µg/L conc.; 13 samples

SAMPLING METHOD Use a 40–120-mL screw-cap vial (prewashed with detergent, rinsed with distilled water and oven dried at 105°C) with a PTFE-faced silicone septum. If residual chlorine is in the water add about 25 mg of ascorbic acid to each vial before sample collection. Collect bubble-free samples.

STABILITY Cool to 4°C; HCl to pH <2.

MHT 14 days.

QUALITY CONTROL As initial demonstration of lab accuracy and precision, analyze 4 to 7 replicates of a lab fortified blank containing the analyte at 0.1–5 µg/L. Collect all samples in duplicate.

REFERENCE Method 503.1, Volatile Aromatic & Unsaturated Organic Compounds in H2O by Purge and Trap GC, EPA 600/4-88/039.

Toluene   **EPA Method 602**
CAS #108-88-3

TITLE Purgeable Aromatics

MATRIX Wastewater.

APPLICATION Method covers 7 purgeable aromatics. (Method 624 provides GC/MS conditions appropriate for the qualitative and quantitative confirmation of results). Method describes conditions for a 2nd GC column to confirm measurements made with primary column.

INTERFERENCES Impurities in the purge gas and organic compounds outgassing from the plumbing ahead of the trap. With high- and low-level samples, there can be carryover contamination. Diffusion of volatile organics through the septum seal into the sample.

INSTRUMENTATION GC-equipped with photoionization detector. (With purge-and-trap unit)

RANGE 2.1–550 µg/L.

MDL 0.2 µg/L.

PRECISION 0.18X + 0.71 µg/L (overall precision).

ACCURACY 0.94C + 0.65 µg/L (as recovery).

SAMPLING METHOD 25-mL glass vial. Teflon®-lined septum.

STABILITY Cool,4°C, 0.008% Na2S2O3. HCl to pH 2.

MHT 14 days.

QUALITY CONTROL The lab must on an ongoing basis, spike at least 10% of the samples from each sample site being monitored to assess accuracy.

REFERENCE Method 602, *Federal Register* Part VIII 40 CFR Part 136, Oct 26, 1984.

Toluene   **EPA Method 624**
CAS #108-88-3

TITLE Purgeables

MATRIX Wastewater.

APPLICATION Method covers 31 purgeable organics. An inert gas is bubbled through a 5-mL water sample in a specially designed purging chamber. Here, purgeables are transferred from aqueous to gaseous phase, passed onto a sorbent column, and trapped. Trap is heated and backflushed with inert gas to desorb purgeables onto a GC column, where purgeables are separated.

INTERFERENCES Impurities in the purge gas, organic compounds outgassing from the plumbing ahead of the trap, and solvent vapors in the lab. With high- and low-level samples, there can be carryover contamination.

INSTRUMENTATION GC/MS with purge-and-trap unit.

RANGE 5–600 µg/L

MDL 6.0 µg/L

PRECISION 0.22X–1.71 µg/L (overall precision).

ACCURACY 0.98C + 2.03 µg/L (as recovery).

SAMPLING METHOD 25-mL glass vial. Teflon®-lined septum.

STABILITY Cool,4°C, 0.008% Na2S2O3. HCl to pH 2.

MHT 14 days.

QUALITY CONTROL The lab must on an ongoing basis, spike at least 5% of the samples from each sample site being monitored to assess accuracy.

REFERENCE Method 624, *Federal Register* Part VIII 40 CFR Part 136, Oct 26, 1984.

Toluene   **EPA Method 8020**
CAS #108-88-3

TITLE Aromatic Volatile Organics

MATRIX Groundwater, soils, sludges water miscible liquid wastes, and non-water miscible wastes.

APPLICATION This method is used to analyze for 8 aromatic VOCs. Samples are analyzed using direct injection or purge-and-trap methods. Groundwater must be analyzed by the purge-and-trap method. The method provides an optional GC column that is used for analyte confirmation and may also help resolve analytes from interferences.

INTERFERENCES There can be carryover contamination with high- and low-level samples. Impurities may come from the purge-and-trap apparatus, organic compounds outgassing from the plumbing ahead of trap, diffusion of VOCs through the sample bottle septum during shipping or storage, or from solvent vapors in the lab.

INSTRUMENTATION GC capable of on-column injections or purge-and-trap sample introduction and a photoionization detector (PID). Column 1: 6 ft by 0.082 in with 5% SP-1200 and 1.75% Bentone-34 on Supelcoport. Column 2: 8 ft by 0.1 in with 5% 1,2,3-tris(2-cyanoethoxy)propane on Chromosorb W-AW.

RANGE   2.1–500 µg/L

MDL   0.2 µg/L (reagent water).

PQL FACTORS FOR MULTIPLYING × MDL VALUE

Matrix	Multiplication Factor
Groundwater	10
Low-level soil	10
Water miscible liquid waste	500
High-level soil and sludge	1250
Non-water miscible waste	1250

PRECISION   0.18X — 0.71 µg/L (overall precision).

ACCURACY   0.94C + 0.65 µg/L (as recovery).

SAMPLING METHOD   For water and liquid samples use glass 40-mL vials with Teflon®-lined septum caps and collect two vials per sample location with no headspace. For solids and concentrated waste samples use widemouth glass bottles with Teflon® liners. Cool all samples to 4°C

STABILITY   For concentrated wastes, soils, sediments, or sludges cool to 4°C. For liquids, add 4 drops of concentrated hydrochloric acid and cool to 4°C.

MHT   14 days.

QUALITY CONTROL   Analyze a reagent blank, matrix spike, and matrix spike duplicate/duplicate for each analytical batch (up to 20 samples). Demonstrate the purity of glassware and reagents by analyzing a reagent water method blank. Internal, surrogate, and five concentration level calibration standards are used. The QC check sample concentrate should contain this compound at 10 µg/mL in methanol.

REFERENCE   Test Methods for Evaluating Solid Waste (SW-846), U.S. EPA Office of Solid Waste, Washington, DC, Method 8020A, Rev. 1, Nov. 1992.

Toluene diisocyanate   **EPA Method 8270**
CAS #584-84-9

TITLE   Semivolatile Organic Compounds by GC/MS

MATRIX   This method is used to determine the concentration of semivolatile organic compounds in extracts prepared from all types of solid waste matrices, soils, and groundwater. Although surface waters are not specifically mentioned, this method should be applicable to water samples from rivers, lakes, etc.

METHOD SUMMARY   This method covers 259 semivolatile organic compounds. In very limited applications direct injection of the sample into the GC/MS system may be appropriate, but this results in very high detection limits (approximately 10,000 µg/L). Typically, a 1-L liquid sample, containing surrogate, and matrix spiking standards, is extracted in a continuous extractor first under acid conditions and then under basic conditions. Typically 30 g of a solid sample, containing surrogate, and matrix spiking standards, is extracted ultrasonically. After concentrating the extract to 1 mL it is spiked with 10 µL of an internal standard solution just prior to analysis by GC/MS. The volume injected should contain about 100 ng of base/neutral and 200 ng of acid surrogates (for a 1-µL injection). Analysis is performed by GC/MS using a capillary GC column.

INTERFERENCES   Raw GC/MS data from all blanks, samples, and spikes must be evaluated for interferences. Contamination by carryover can occur whenever high-concentration and low-concentration samples are sequentially analyzed. To reduce carryover, the sample syringe must be rinsed out between samples with solvent. Whenever an unusually concentrated sample is encountered, it should be followed by the analysis of blank solvent to check for cross-contamination.

INSTRUMENTATION   A GC/MS and a data system are required. The GC column used is a 30 m × 0.25 mm I.D. (or 0.32 mm I.D.) 1um film thickness silicone-coated fused silica capillary column. A continuous liquid-liquid extractor equipped with Teflon® or glass connection joints and stopcocks requiring no lubrication, a K-D concentrating apparatus, water bath, and an ultrasonic disrupter with a minimum power of 300 W and with pulsing capability are also required.

PRECISION & ACCURACY   The estimated quantitation limit (EQL) of Method 8270B for determining an individual compound is approximately 1 mg/kg (wet weight) for soil or sediment samples, 1–200 mg/kg for wastes (dependent on matrix and method of preparation), and 10 µg/L for groundwater samples. EQLs will be proportionately higher for sample extracts that require dilution to avoid saturation of the detector.

The EQL(b) for groundwater in µg/L is 100.
The EQL (a, b) for low concentrations in soil and sediment in µg/kg is not determined.
Accuracy as µg/L is not listed.
Overall precision in µg/L is not listed.

(a)   *EQLs listed for soil/sediment are based on wet weight. Normally data is reported in a dry-weight basis; therefore, EQLs will be higher based on the % dry weight of each sample. This calculation is based on a 30 g sample and gel permeation chromatography cleanup.*

(b)   *Sample EQLs are highly matrix-dependent. The EQLs are provided for guidance and may not always be achievable.*

C =   *True value for concentration, in µg/L.*

X =   *Average recovery found for measurements of samples containing a concentration of C, in µg/L.*

ESTIMATED QUANTITATION LIMIT

Other Matrices	Factor (a)
High-concentration soil and sludges by sonicator	7.5
Non-water miscible waste	75

(a) *EQL for other matrices = [EQL for low soil/sediment × [Factor]. This estimated EQL is similar to an EPA "Practical Quantitation Limit."*

SAMPLING METHOD

Liquid samples — Use a 1 or 2½ gallon amber glass bottle with a screw-top Teflon®-lined cover that has been prewashed with

detergent and rinsed with distilled water and methanol (or isopropanol).

Soils, sediments, or sludges — Use an 8-oz. widemouth glass with a screw-top Teflon®-lined cover that has been prewashed with detergent and rinsed with distilled water and methanol (or isopropanol).

SAMPLE PRESERVATION
Liquid samples — If residual chlorine is present, add 3 mL of 10% sodium thiosulfate per gallon, cool to 4°C and store in a solvent-free refrigerator until analysis; if chlorine is not present, then eliminate the sodium thiosulfate addition.

Soils, sediments, or sludges — Cool samples to 4°C and store in a solvent-free refrigerator.

MHT Liquid samples must be extracted within 7 days and the extracts analyzed within 40 days. Soils, sediments, or sludges may be stored for a maximum of 14 days and the extracts analyzed within 40 days.

SAMPLE PREPARATION
Liquid samples — Transfer 1 L quantitatively to a continuous extractor. If high concentrations are anticipated, a smaller volume may be used and then diluted with organic-free reagent water to 1 L. Adjust pH, if necessary, to pH <2 using 1:1 (V/V) sulfuric acid. Pipette 1.0 mL of a surrogate standard spiking solution into each sample. For the sample in each analytical batch selected for spiking, add 1.0 mL of a matrix spiking standard. For base/neutral acid analysis, the amount of the surrogates and matrix spiking compounds added to the sample should result in a final concentration of 100 ng/µL of each analyte in the extract to be analyzed (assuming a 1-µL injection). Extract with methylene chloride for 18–24 h. Next, adjust the pH of the aqueous phase to pH >11 using 10 N sodium hydroxide and extract it with methylene chloride again for 18–24 h. Dry the extract through a column containing anhydrous sodium sulfate and concentrate it to 1 mL using a K-D concentrator.

Soils, sediments, or sludges — Use 30 g of sample. Nonporous or wet samples (gummy or clay type) that do not have a free-flowing sandy texture must be mixed with anhydrous sodium sulfate until the sample is free flowing. Add 1 mL of surrogate standards to all samples, spikes, standards, and blanks. For the sample in each analytical batch selected for spiking, add 1.0 mL of a matrix spiking standard. For base/neutral acid analysis, the amount added of the surrogates and matrix spiking compounds should result in a final concentration of 100 ng/µL of each base/neutral analyte and 200 ng/µL of each acid analyte in the extract to be analyzed (assuming a 1-µL injection). Immediately add a 100-mL mixture of 1:1 methylene chloride:acetone and extract the sample ultrasonically for 3 min and then decant or filter the extracts. Repeat the extraction two or more times. Dry the extract using a column with anhydrous sodium sulfate and concentrate it to 1 mL in a K-D concentrator.

QUALITY CONTROL A methylene chloride solution containing 50 ng/µL of decafluorotriphenylphosphine (DFTPP) is used for tuning the GC/MS system each 12-h shift. A system performance check also must be made during every 12-h shift. A standard containing 50 ng/µL each of 4,4'-DDT, pentachlorophenol, and benzidine is required to verify injection port inertness and GC column performance. A calibration standard at mid-concentration, containing each compound of interest, including all required surrogates, must be performed every 12 h during analysis. After the system performance check is met, calibration check compounds (CCCs) are used to check the validity of the initial calibration.

The internal standard responses and retention times in the calibration check standard must be evaluated immediately after or during data acquisition. If the retention time for any internal standard changes by more than 30 seconds from the last check calibration (12 h), the chromatographic system must be inspected for malfunctions and corrections must be made, as required. If the electron ionization current plot (EICP) area for any of the internal standards changes by a factor of two from the last daily calibration standard check, the mass spectrometer must be inspected for malfunctions and corrections must be made, as appropriate.

Demonstrate, through the analysis of a reagent water blank, that interferences from the analytical system, glassware, and reagents are under control. The blank samples should be carried through all stages of the sample preparation and measurement steps. For each analytical batch (up to 20 samples), a reagent blank, matrix spike, and matrix spike duplicate/duplicate must be analyzed (the frequency of the spikes may be different for different monitoring programs). The blank and spiked samples must be carried through all stages of the sample preparation and measurement steps. A QC reference sample concentrate containing each analyte at a concentration of 100 mg/L in methanol is required.

REFERENCE Test Methods for Evaluating Solid Waste (SW-846). U.S. EPA 1983, Method 8270B, Rev. 2, Nov. 1990. Office of Solid Waste, Washington, DC.

---

o-Toluidine EPA Method 1625
CAS #95-53-4

TITLE Semivolatile Organic Compounds by Isotope Dilution GC/MS

MATRIX The compounds may be determined in waters, soils, and municipal sludges by this method.

METHOD SUMMARY This method is used to determine 176 semivolatile toxic organic pollutants associated with the CWA (as amended 1987); the RCRA (as amended 1986); the CERCLA (as amended 1986); and other compounds amenable to extraction and analysis by capillary column gas chromatography-mass spectrometry (GC/MS).

Stable isotopically-labeled analogs of the compounds of interest are added to the sample. If the solids content is less than 1%, a 1-L sample is extracted at pH 12–13, then at pH <2 with methylene chloride using continuous extraction techniques.

If the solids content is 30% or less, the sample is diluted to 1% solids with reagent water, homogenized ultrasonically, and extracted at pH 12–13, then at pH <2 with methylene chloride using continuous extraction techniques. If the solids content is greater than 30%, the sample is extracted using ultrasonic techniques.

Each extract is dried over sodium sulfate, concentrated to a volume of 5 mL, cleaned up using GPC, if necessary, and concentrated. Extracts are concentrated to 1 mL if GPC is not performed, and to 0.5 mL if GPC is performed.

An internal standard is added to the extract, and a 1-mL aliquot of the extract is injected into the GC. The compounds are separated by GC and detected by a MS. The labeled compounds serve to correct the variability of the analytical technique.

INTERFERENCES Solvents, reagents, glassware, and other sample processing hardware may yield artifacts and/or elevated baselines causing misinterpretation of chromatograms and spectra. Materials used in the analysis must be demonstrated to be free from interferences under the conditions of analysis by running method blanks initially and with each sample lot (sample started through the extraction process on a given 8-h shift, to a maximum of 20). Specific selection of reagents and purification of solvents by distillation in all glass systems may be required. Glassware and, where possible, reagents are cleaned by solvent rinse and baking at 450°C for 1-h minimum. Interferences coextracted from samples will vary considerably from source to source, depending on the diversity of the site being sampled.

INSTRUMENTATION Major instrumentation includes a GC with a splitless or on-column injection port for capillary column, a MS with 70 eV electron impact ionization, and a data system to collect and record MS data, and process it. A K-D apparatus is used to concentrate extracts.

GC Column: 30 m × 0.25 mm I.D. 5% phenyl, 94% methyl, 1% vinyl silicone bonded phased fused silica capillary column.

PRECISION & ACCURACY The detection limits of the method are usually dependent on the level of interferences rather than instrumental limitations. The limits typify the minimum quantities that can be detected with no interferences present.

The minimum level (in µg/mL) was not listed. This is defined as a minimum level at which the analytical system shall give recognizable mass spectra (background corrected) and acceptable calibration points.

The MDL (in µg/kg) in low solids was not listed and in high solids was not listed; these were determined in digested sludge (low solids) and in filter cake or compost (high solids).

The labeled and native compound initial precision as standard deviation (in µg/L) was not listed.
The labeled and native compound initial accuracy as average recovery (in µg/L) was not listed.

SAMPLE COLLECTION, PRESERVATION & HANDLING Collect samples in glass containers. Aqueous samples which flow freely are collected in refrigerated bottles using automatic sampling equipment. Solid samples are collected as grab samples using widemouth jars. Maintain samples at 0 to 4°C from the time of collection until extraction. If residual chlorine is present in aqueous samples, add 80 mg sodium thiosulfate/L of water. Begin sample extraction within 7 days of collection, and analyze all extracts within 40 days of extraction.

SAMPLE PREPARATION Samples containing 1% solids or less are extracted directly using continuous liquid-liquid extraction techniques. Samples containing 1 to 30% solids are diluted to the 1% level with reagent water and extracted using continuous liquid-liquid extraction techniques. Samples containing greater than 30% solids are extracted using ultrasonic techniques.

Base/neutral extraction — Adjust the pH of the waters in the extractors to 12–13 with 6 N NaOH. Extract with methylene chloride for 24–48 h.
Acid extraction — Adjust the pH of the waters in the extractors to 2 or less using 6 N sulfuric acid. Extract with methylene chloride for 24–48 h.
Ultrasonic extraction of high solids samples — Add anhydrous sodium sulfate to the sample and QC aliquot(s). Add acetone:methylene chloride (1:1) to the sample and mix thoroughly

Concentrate extracts using a K-D apparatus.

QUALITY CONTROL The analyst is permitted to modify this method to improve separations or lower the costs of measurements, provided all performance specifications are met. Analyses of blanks are required to demonstrate freedom from contamination. When results of spikes indicate atypical method performance for samples, the samples are diluted to bring method performance within acceptable limits.

For low solids (aqueous samples), extract, concentrate, and analyze two sets of four 1-L aliquots (8 aliquots total) of the precision and recovery standard. For high solids samples, two sets of four 30-g aliquots of the high solids reference matrix are used.

Spike all samples with labeled compounds to assess method performance. Compute percent recovery of the labeled compounds using the internal standard method. Compare the labeled compound recovery for each compound with the corresponding labeled compound recovery.

Reagent water and high solids reference matrix blanks are analyzed to demonstrate freedom from contamination. Extract and concentrate a 1-L reagent water blank or a high solids reference matrix blank with each sample's lot (samples started through the extraction process on the same 8-h shift, to a maximum of 20 samples).

Field replicates may be collected to determine the precision of the sampling technique, and spiked samples may be required to determine the accuracy of the analysis when the internal standard method is used.

REFERENCE Semivolatile Organic Compounds by Isotope Dilution GC/MS. Office of Water Regulation and Standards, U.S. EPA Industrial Technology Division, Washington, DC, EPA Method 1625, Rev. C, June 1989 (contact W.A. Telliard, U.S. EPA, Office of Water Regulations and Standards, 401 M St., SW, Washington, DC, 20460. Phone: 202-382-7131).

o-Toluidine
CAS #95-53-4
EPA Method 8270

TITLE Semivolatile Organic Compounds by GC/MS

MATRIX This method is used to determine the concentration of semivolatile organic compounds in extracts prepared from all types of solid waste matrices, soils, and groundwater. Although surface waters are not specifically mentioned, this method should be applicable to water samples from rivers, lakes, etc.

METHOD SUMMARY This method covers 259 semivolatile organic compounds. In very limited applications direct injection of the sample into the GC/MS system may be appropriate, but this results in very high detection limits (approximately 10,000 µg/L). Typically, a 1-L liquid sample, containing surrogate, and matrix spiking standards, is extracted in a continuous extractor first under acid conditions and then under basic conditions. Typically 30 g of a solid sample, containing surrogate, and matrix spiking standards, is extracted ultrasonically. After concentrating the extract to 1 mL it is spiked with 10 µL of an internal standard solution just prior to analysis by GC/MS. The volume injected should contain about 100 ng of base/neutral and 200 ng of acid surrogates (for a 1-µL injection). Analysis is performed by GC/MS using a capillary GC column.

INTERFERENCES Raw GC/MS data from all blanks, samples, and spikes must be evaluated for interferences. Contamination by carryover can occur whenever high-concentration and low-concentration samples are sequentially analyzed. To reduce carryover, the sample syringe must be rinsed out between samples with solvent. Whenever an unusually concentrated sample is encountered, it should be followed by the analysis of blank solvent to check for cross-contamination.

INSTRUMENTATION A GC/MS and a data system are required. The GC column used is a 30 m × 0.25 mm I.D. (or 0.32 mm I.D.) 1um film thickness silicone-coated fused silica capillary column. A continuous liquid-liquid extractor equipped with Teflon® or glass connection joints and stopcocks requiring no lubrication, a K-D concentrating apparatus, water bath, and an ultrasonic disrupter with a minimum power of 300 W and with pulsing capability are also required.

PRECISION & ACCURACY The estimated quantitation limit (EQL) of Method 8270B for determining an individual compound is approximately 1 mg/kg (wet weight) for soil or sediment samples, 1–200 mg/kg for wastes (dependent on matrix and method of preparation), and 10 µg/L for groundwater samples. EQLs will be proportionately higher for sample extracts that require dilution to avoid saturation of the detector.

The EQL(b) for groundwater in µg/L is 10.
The EQL (a, b) for low concentrations in soil and sediment in µg/kg is not determined.
Accuracy as µg/L is not listed.
Overall precision in µg/L is not listed.

(a) *EQLs listed for soil/sediment are based on wet weight. Normally data is reported in a dry-weight basis; therefore, EQLs will be higher based on the % dry weight of each sample.*
This calculation is based on a 30 g sample and gel permeation chromatography cleanup.
(b) *Sample EQLs are highly matrix-dependent. The EQLs are provided for guidance and may not always be achievable.*
$C =$ *True value for concentration, in µg/L.*
$X =$ *Average recovery found for measurements of samples containing a concentration of C, in µg/L.*

ESTIMATED QUANTITATION LIMIT

Other Matrices	Factor (a)
High-concentration soil and sludges by sonicator	7.5
Non-water miscible waste	75

(a) *EQL for other matrices = [EQL for low soil/sediment × [Factor]. This estimated EQL is similar to an EPA "Practical Quantitation Limit."*

SAMPLING METHOD
Liquid samples — Use a 1 or 2½ gallon amber glass bottle with a screw-top Teflon®-lined cover that has been prewashed with detergent and rinsed with distilled water and methanol (or isopropanol).

Soils, sediments, or sludges — Use an 8-oz. widemouth glass with a screw-top Teflon®-lined cover that has been prewashed with detergent and rinsed with distilled water and methanol (or isopropanol).

SAMPLE PRESERVATION
Liquid samples — If residual chlorine is present, add 3 mL of 10% sodium thiosulfate per gallon, cool to 4°C and store in a solvent-free refrigerator until analysis; if chlorine is not present, then eliminate the sodium thiosulfate addition.

Soils, sediments, or sludges — Cool samples to 4°C and store in a solvent-free refrigerator.

MHT Liquid samples must be extracted within 7 days and the extracts analyzed within 40 days. Soils, sediments, or sludges may be stored for a maximum of 14 days and the extracts analyzed within 40 days.

SAMPLE PREPARATION
Liquid samples — Transfer 1 L quantitatively to a continuous extractor. If high concentrations are anticipated, a smaller volume may be used and then diluted with organic-free reagent water to 1 L. Adjust pH, if necessary, to pH <2 using 1:1 (V/V) sulfuric acid. Pipette 1.0 mL of a surrogate standard spiking solution into each sample. For the sample in each analytical batch selected for spiking, add 1.0 mL of a matrix spiking standard. For base/neutral acid analysis, the amount of the surrogates and matrix spiking compounds added to the sample should result in a final concentration of 100 ng/µL of each analyte in the extract to be analyzed (assuming a 1-µL injection). Extract with methylene chloride for 18–24 h. Next, adjust the pH of the aqueous phase to pH >11 using 10 N sodium hydroxide and extract it with methylene chloride again for 18–24 h. Dry the extract through a column containing anhydrous sodium sulfate and concentrate it to 1 mL using a K-D concentrator.

Soils, sediments, or sludges — Use 30 g of sample. Nonporous or wet samples (gummy or clay type) that do not have a free-flowing

sandy texture must be mixed with anhydrous sodium sulfate until the sample is free flowing. Add 1 mL of surrogate standards to all samples, spikes, standards, and blanks. For the sample in each analytical batch selected for spiking, add 1.0 mL of a matrix spiking standard. For base/neutral acid analysis, the amount added of the surrogates and matrix spiking compounds should result in a final concentration of 100 ng/µL of each base/neutral analyte and 200 ng/µL of each acid analyte in the extract to be analyzed (assuming a 1-µL injection). Immediately add a 100-mL mixture of 1:1 methylene chloride:acetone and extract the sample ultrasonically for 3 min and then decant or filter the extracts. Repeat the extraction two or more times. Dry the extract using a column with anhydrous sodium sulfate and concentrate it to 1 mL in a K-D concentrator.

QUALITY CONTROL A methylene chloride solution containing 50 ng/µL of decafluorotriphenylphosphine (DFTPP) is used for tuning the GC/MS system each 12-h shift. A system performance check also must be made during every 12-h shift. A standard containing 50 ng/µL each of 4,4'-DDT, pentachlorophenol, and benzidine is required to verify injection port inertness and GC column performance. A calibration standard at mid-concentration, containing each compound of interest, including all required surrogates, must be performed every 12 h during analysis. After the system performance check is met, calibration check compounds (CCCs) are used to check the validity of the initial calibration.

The internal standard responses and retention times in the calibration check standard must be evaluated immediately after or during data acquisition. If the retention time for any internal standard changes by more than 30 seconds from the last check calibration (12 h), the chromatographic system must be inspected for malfunctions and corrections must be made, as required. If the electron ionization current plot (EICP) area for any of the internal standards changes by a factor of two from the last daily calibration standard check, the mass spectrometer must be inspected for malfunctions and corrections must be made, as appropriate.

Demonstrate, through the analysis of a reagent water blank, that interferences from the analytical system, glassware, and reagents are under control. The blank samples should be carried through all stages of the sample preparation and measurement steps. For each analytical batch (up to 20 samples), a reagent blank, matrix spike, and matrix spike duplicate/duplicate must be analyzed (the frequency of the spikes may be different for different monitoring programs). The blank and spiked samples must be carried through all stages of the sample preparation and measurement steps. A QC reference sample concentrate containing each analyte at a concentration of 100 mg/L in methanol is required.

REFERENCE Test Methods for Evaluating Solid Waste (SW-846). U.S. EPA 1983, Method 8270B, Rev. 2, Nov. 1990. Office of Solid Waste, Washington, DC.

Total Organic Halides (TOX)   EPA Method 450.1

TITLE Inorganics, Non-Metallics

MATRIX Drinking and ground waters.

APPLICATION Method determines TOX as Cl by carbon absorption. Organic halides are defined as all organic species containing chlorine, bromine, and iodine that are adsorbed by granular activated carbon under method conditions. Fluorine containing species are not determined by this method.

INTERFERENCES Method interferences may be caused by contaminants, reagents, glassware, and other sample processing hardware. Improperly handled (or stored) activated carbon (or carbon samples which register >1000 ng/40 mg) can interfere; also halogenated organic vapors.

INSTRUMENTATION Dohrmann microcoulometric-titration and adsorption systems. Strip-chart recorder

RANGE Sensitivity limit = 5 µg/L.

PRECISION SD = 4.3 at 71 Average halide µg Cl/L. (Tap water).

ACCURACY Avg recovery = 89% at dose as 88 µg Cl/L (chloroform)

SAMPLING METHOD 250 mL amber glass fitted with Teflon®-lined cap.

STABILITY Minimize volatile loss. Store at 4°C without headspace.

QUALITY CONTROL The lab must develop and maintain a statement of method accuracy for their lab. Before analysis, analyst must demonstrate ability to generate acceptable.

PRECISION & ACCURACY On appropriate QC check samples. Run all samples in duplicate.

REFERENCE Total Organic Halide Method 405.1-Interim, EPA-600/4-81-056, U.S. EPA, EMSL, 1980.

Total Solids (TS)   EPA Method 160.3

TITLE Physical Properties

MATRIX Drinking, surface and saline waters. Wastewater.

APPLICATION Date issued 1971. Also referred to as residue, total. A well mixed sample is evaporated in a weighed dish and dried to constant weight in an oven at 103–105c. Increase in weight over empty dish represents the total solids. Total solids is sum of homogenous suspended and dissolved materials in a sample.

INTERFERENCES Non-representative particulates such as leaves, sticks, fish and lumps of fecal matter should be excluded from the sample if it is determined that their inclusion is not desired in the final result.

INSTRUMENTATION Drying oven at 103–105°C. Porcelain Evaporating Dish (100 mL).

RANGE 10 mg/L to 20,000 mg/L.

MDL Not listed.

PRECISION Not listed.

ACCURACY Not listed.

SAMPLING METHOD plastic or glass. (100 mL).

STABILITY Cool, 4°C.

MHT 7 Days.

QUALITY CONTROL Floating oil and grease, if present, should be included in the sample and dispersed by a blender device before aliquoting.

REFERENCE Methods for the Chemical Analysis of Water and Wastes, EPA-600/4-79-020, U.S. EPA, EMSL, 1979.

Total Suspended Solids (TSS) — EPA Method 160.2

TITLE Physical Properties

MATRIX Drinking, surface and saline waters. Wastewater.

APPLICATION Date issued 1971. Also referred to as residue, non filterable. A well mixed sample is filtered through a weighed glass fiber filter, and the residue on the filter is dried to constant weight at 103–105 c: Increase in weight of the filter represents TSS. The filtrate may be used for residue, filterable.

INTERFERENCES Samples high in dissolved solids; saline waters, brines, and some wastes may be subject to a positive interference. Select filtering apparatus with care, so that washing of filter and dissolved solids in the filter minimizes this potential interference.

INSTRUMENTATION Glass fiber filter discs (Reeves Angel 934-AH, or equiv). Drying oven@103–105c

RANGE 4–20,000 mg/L.

MDL Not listed.

PRECISION Not listed.

ACCURACY Not listed.

SAMPLING METHOD plastic or glass. (100 mL).

STABILITY Cool, 4°C.

MHT 7 Days.

QUALITY CONTROL Non representative particulates such as leaves, sticks, fish, and lumps of fecal matter should be excluded from the sample if it is determined that their inclusion is not desired in the final result.

REFERENCE EPA Methods for the Chemical Analysis of Water and Wastes, EPA-600/4-79-020, U.S. EPA, EMSL, 1979.

Toxaphene — EPA Method 505
CAS #8001-35-2

TITLE Analysis of Organohalide Pesticides and Commercial Polychlorinated Biphenyl (PCB) Products in Water by Microextraction and Gas Chromatography. U.S. EPA Method 505, Rev. 2.0, 1989.

MATRIX This method is applicable to drinking water and raw source water. The latter should include most surface water and groundwater sources.

METHOD SUMMARY Method 505 covers 25 pesticides and commercial PCB products. This is a very sensitive method that is more useful for monitoring than for exploratory analyses. 5-mL of water are saturated with sodium chloride and then extracted by shaking with 2 mL of hexane. The sample extracts are transferred to an autosampler setup to inject 1–2 µL portions into a gas chromatograph (GC) for analysis. Alternatively, 1–2 µL portions of samples, blanks, and standards may be manually injected. Each extract is analyzed by capillary GC/ECD with confirmation using either a second capillary column or GC/MS. The electron capture detector is easy to use, but it is a nonselective detector. The microextraction technique also eliminates the expensive sample preparation costs of other methods, but it has the disadvantage of being less sensitive than most because the extracts are not concentrated.

INTERFERENCES Method interferences may be caused by contaminants in solvents, reagents, glassware, and other sample processing apparatus that lead to discrete artifacts or elevated baselines. Interfering contamination may occur when a sample containing low concentrations of analytes is analyzed immediately following a sample containing relatively high concentrations of the analytes. Matrix interferences also may be caused by contaminants that are coextracted from the sample; cleanup of sample extracts may be necessary in these cases. Some pesticides and commercial PCB products from aqueous solutions adhere to glass surfaces, so sample transfers and contact with glass surfaces should be minimized. Some pesticides are rapidly oxidized by chlorine so dechlorination with sodium thiosulfate at the time of sample collection is important. Also, splitless injectors may cause degradation of some pesticides.

INSTRUMENTATION A gas chromatograph/electron capture detector/data system, with temperature programming and split/splitless injector suitable for use with capillary columns is needed.

Column 1: 0.32 mm I.D. × 30 m fused silica capillary with chemically bond methyl polysiloxane phase (DB-1, 1.0 µm film, or equivalent).

Column 2: 0.32 mm I.D. × 30 m fused silica capillary with 1:1 mixed phase of dimethyl silicone and polyethylene glycol (Durawax-DX3, 0.25 µm film, or equivalent).

Column 3: 0.32 mm I.D. × 25 m fused silica capillary with chemically bonded 50:50 methyl-phenyl silicone (OV-17, 1.5 µm film, or equivalent).

Column 1 should be used as the primary analytical column. Columns 2 and 3 are recommended for use as confirmatory columns when GC/MS confirmation is not available.

PRECISION & ACCURACY Method detection limits are dependent upon the characteristics of the gas chromatographic system used. Analytes that are not separated chromatographically cannot be individually identified and used in the same calibration mixture or water samples unless an alternative technique

for identification and quantification, such as mass spectrometry, is used.

The concentration(s) (in µg/L) used for these QC measurements was 10 and 80.

The MDL (in µg/L) was 1.0.

The accuracy (% recovery) for reagent water at the above concentration(s) was not available and not available and the precision (%) was 12.6 and 15.3.

The accuracy (% recovery) for groundwater at the above concentration(s) was not listed and not listed and the precision (%) was not listed and not listed.

The accuracy (% recovery) for tap water at the above concentration(s) was 110 and 114 and the precision (5) was 9.5 and 13.5.

Note: No range of concentrations is provided with this method.

SAMPLING METHOD Collect samples using a 40-mL screw-cap vial (prewashed with detergent, rinsed with distilled water and oven dried at 400°C for one h) with a Teflon®-faced silicone septum. Collect bubble-free samples and place the septum with the Teflon® side down on the water.

SAMPLE PRESERVATION If residual chlorine is present in the water add about 3 mg of sodium thiosulfate to each vial before samples are collected to remove the chlorine. Alternatively, add 75 µL of 0.04 g/mL solution of sodium thiosulfate to each vial just prior to sampling. Immediately cool samples to 4°C, and store them in a solvent-free refrigerator at 4°C until analysis.

MHT The maximum holding time is 14 days from the time the sample was collected until it must be analyzed.

SAMPLE PREPARATION Remove the sample from storage and allow it to come to room temperature. Remove a 5-mL volume from each container and weigh the container to the nearest 0.1 g. Add 6 g of sodium chloride and 2.0 mL of hexane to each sample bottle. Recap the sample and shake it vigorously for one min. Allow the water and hexane phases to separate, remove the cap, and transfer 0.5 mL of hexane into an autosampler vial using a disposable glass pipette. Transfer the remaining hexane phase into a second autosampler vial and store at 4°C for reanalysis, if necessary. Discard the remaining sample/hexane mixture and reweigh the empty container to determine net weight of sample.

QUALITY CONTROL Minimum quality control requirements are initial demonstration of lab capability, analysis of lab reagent blanks, fortified blanks, fortified sample matrix, and quality control samples. The lab must analyze at least one fortified blank per sample set, or at least one for every 20 samples. The fortifying concentration of each analyte should be 10 times the method detection limit or the maximum calibration limit (MCL), whichever is less. Calculate accuracy as percent recovery and develop control limits from the mean percent recovery and standard deviation.

The lab must add a known concentration of the analytes to a minimum of 10% of the routine samples, or one lab fortified sample matrix per sample set. Calculate the percent recovery for each analyte and compare to the control limits established from the analyses of the fortified blanks.

EPA CONTACT & HOTLINE For technical questions contact Dr. Baldev Bathija, U.S. EPA, Office of Ground Water and Drinking Water (WH-550D), 401 M St. SW, Washington, DC 20460. Tel. (202) 260-3040. For further information the EPA Safe Drinking Water Hotline may be called at: (800) 426-4791.

REFERENCE Methods for the Determination of Organic Compounds in Drinking Water, EPA/600/4-88/039 (revised July 1991). U.S. EPA Environmental Monitoring Systems Laboratory, Cincinnati, OH, 45268, U.S.A. Available from the National Technical Information Service (NTIS), 5285 Port Royal Road, Springfield, VA 22161; Tel. 800-553-6847. NTIS Order Number is PB91-231480.

Toxaphene   **EPA Method 625**
CAS #8001-35-2

TITLE Base/Neutrals and Acids, U.S. EPA Method 625

MATRIX This methods covers municipal and industrial wastewaters.

METHOD SUMMARY Approximately 1 L of sample is serially extracted with methylene chloride at a pH greater than 11 and again at a pH less than 2 using a separatory funnel or a continuous extractor. The methylene chloride extract is dried, concentrated to a volume of 1 mL, and analyzed by GC/MS. Qualitative identification of the parameters in the extract is performed using the retention time and the relative abundance of three characteristic masses (m/z). Qualitative analysis is performed using either external or internal standard techniques with a single characteristic m/z.

INTERFERENCES Method interferences may be caused by contaminants in solvents, reagents, glassware, and other sample processing hardware. Glassware must be scrupulously cleaned. Glassware should be heated in a muffle furnace at 400°C for 5 to 30 min. Some thermally stable materials, such as PCBs, may not be eliminated by this treatment. Solvent rinses with acetone and pesticide quality hexane may be substituted for the muffle furnace heating. Matrix interferences may be caused by contaminants that are coextracted from the sample. The base-neutral extraction may cause significantly reduced recovery of phenols. The packed gas chromatographic columns recommended for the basic fraction may not exhibit sufficient resolution for some analytes.

INSTRUMENTATION A GC/MS system with an injection port designed for on-column injection when using packed columns and for splitless injection when using capillary columns.

Column for base/neutrals: 1.8 m long × 2 mm I.D. glass, packed with 3% SP-2550 on Supelcoport (100/120 mesh) or equivalent.

Column for acids: 1.8 m long × 2 mm I.D. glass, packed with 1% SP-1240DA on Supelcoport (100/120 mesh) or equivalent.

PRECISION & ACCURACY The MDL concentrations were obtained using reagent water. The MDL actually achieved in a given analysis will vary depending on instrument sensitivity and matrix effects. This method was tested by 15 laboratories using reagent water, drinking water, surface water, and industrial wastewaters spiked at six concentrations over the range 5 to 100 µg/L. Single operator precision, overall precision, and method accuracy were found to be directly related to the concentration of the parameter matrix.

The MDL (in µg/L) in reagent water was not detected.
The standard deviation (in µg/L based on 4 recovery measurements) was not reported.
The range (in µg/L) for average recovery for 4 measurements was not reported.
The range (in %) for percent recovery was not reported.
Accuracy (in µg/L) as expected recovery for one or more measurements of a sample containing a true concentration of C was not reported.
Precision (in µg/L) as expected single analyst standard deviation of measurements at an average concentration found at X was not reported.
Overall precision (in µg/L) as expected interlaboratory standard deviation of measurements in an average concentration found at X was not reported.

C = *True value of the concentration in µg/L.*
X = *Average recovery found for measurements of samples containing a concentration at C in µg/L.*

SAMPLE PREPARATION Adjust the pH to >11 with sodium hydroxide and serially extract in a separatory funnel with methylene chloride or else in a continuous extractor. Next, adjust the pH to <2 with sulfuric acid and serially extract in a separatory funnel with methylene chloride or else in a continuous extractor. Dry the extracts separately through a column of anhydrous sodium sulfate and then concentrate each of the extracts to 1.0 mL using a K-D apparatus.

SAMPLE COLLECTION, PRESERVATION & HANDLING
Grab samples must be collected in glass containers. All samples must be refrigerated at 4°C from the time of collection until extraction. If residual chlorine is present, add 80 mg of sodium thiosulfate/L of sample and mix well. All samples must be extracted within 7 days of collection and completely analyzed within 40 days of extraction.

QUALITY CONTROL Make an initial, one-time, demonstration of the ability to generate acceptable accuracy and precision with this method. Before processing any samples, the analyst must analyze a reagent water blank to demonstrate that interferences from the analytical system and glassware are under control. Each time a set of samples is extracted or reagents are changed, a reagent water blank must be processed. Spike and analyze a minimum of 5% of all samples to monitor and evaluate lab data quality. A QC check sample concentrate that contains each parameter of interest at a concentration of 100 µg/mL in acetone is required. PCBs and multicomponent pesticides may be omitted from this test.

After analysis of five spiked wastewater samples, calculate the average percent recovery and the standard deviation of the percent recovery. Spike all samples with the surrogate standard spiking solution and calculate the percent recovery of each surrogate compound.

REFERENCE *Federal Register*, Vol. 49, No. 209. Friday, Oct. 26, 1984.

Toxaphene **EPA Method 8080**
CAS #8001-35-2

TITLE Organochlorine Pesticides and Polychlorinated Biphenyls By Gas Chromatography

MATRIX This method is used to determine the concentration of various organochlorine pesticides and polychlorinated biphenyls in extracts prepared from water, groundwater, soils, and sediments.

METHOD SUMMARY This method covers 26 pesticides and Aroclor (PCB) mixtures and it is suitable for monitoring-type analyses. After extraction, concentration and solvent exchange to hexane, a 2- to 5-µL sample aliquot is injected into a GC using the solvent flush technique, and the analytes are detected by an electron capture detector (ECD) or an electrolytic conductivity detector in the halogen mode (HECD). Both neat and diluted organic liquids may be analyzed by direct injection.

INTERFERENCES Interferences coextracted from the samples will vary considerably from source to source. Interferences by phthalate esters can pose a major problem in pesticide determinations when using the ECD. Cross-contamination of clean glassware routinely occurs when plastics are handled during extraction steps, especially when solvent-wetted surfaces are handled. The contamination from phthalate esters can be completely eliminated with a microcoulometric or electrolytic conductivity detector. Solvents, reagent, glassware, and other sample processing hardware may yield artifacts and/or interferences to sample analysis.

INSTRUMENTATION A gas chromatograph capable of on-column injections is needed. It must be equipped with an ECD or a HECD and one of the following GC columns:

Column 1: Supelcoport (100/120 mesh) coated with 1.5% SP-2250/1.95% SP-2401 packed in a 1.8 m × 4 mm I.D. glass column.
Column 2: Supelcoport (100/120 mesh) coated with 3% OV-1 in a 1.8 m × 4 mm I.D. glass column.

PRECISION & ACCURACY The method was tested by 20 laboratories using organic-free reagent water, drinking water, surface water, and three industrial wastewaters spiked at six concentrations. Concentrations used in the study ranged from 0.5 to 30 µg/L for single-component pesticides and from 8.5 to 400 µg/L for multicomponent parameters. Overall precision and method accuracy were found to be directly related to the concentration of the analyte and essentially independent of the sample matrix. The sensitivity of this method usually depends on the concentration of interferences rather than on instrumental limitations.

MDL in µg/L was 0.24.
Concentration range in µg/L was 8.5–40.

Accuracy as recovery (x*) in µg/L was 0.80C + 1.74.
Overall precision (S*) in µg/L was 0.20x + 0.22.

x* Expected recovery for one or more measurements of a sample containing concentration C, in µg/L.
S* = Expected interlaboratory standard deviation of measurements at an average concentration found of the analyte in µg/L.
C = True value for the concentration, in µg/L.
X = Average recovery found for measurements of samples containing a concentration of C, in µg/L.

SAMPLING METHOD
Liquid samples — Use a 1 or 2½ gallon amber glass bottle with a screw-top Teflon®-lined cover. Pre-wash the bottle with detergent, rinse with distilled water and methanol (or isopropanol).

Soil, sediments, and sludges — Use an 8-oz. widemouth glass with a screw-top Teflon®-lined cover. Pre-wash the bottle with detergent, rinse with distilled water and methanol (or isopropanol).

SAMPLE PRESERVATION   Cool water, soil, sediment, or sludge samples immediately to 4°C.

Water samples — If residual chlorine is present, add 3 mL of 10% sodium thiosulfate per gallon and cool to 4°C. All extracts and samples should be stored under refrigeration.

MHT   Liquid samples must be extracted within 7 days and the extracts must be analyzed within 40 days. Soils, sediments, and sludges may be stored for a maximum of 14 days prior to extraction.

SAMPLE PREPARATION
Liquid samples — Extract 1 L samples in a continuous extractor at pH 5–9 with methylene chloride after adding 1.0 mL of surrogate spiking solution to each sample. Pass the extract through a column of anhydrous sodium sulfate to dry and concentrate it in a K-D apparatus to 1 mL volume.

Soils, sediments and sludges — Rapidly weigh approximately 30 g of sample into a 400-mL beaker to avoid loss of the more volatile extractables. Nonporous or wet samples (gummy or clay type) that do not have a free-flowing sandy texture must be mixed with anhydrous sodium sulfate until the sample is free flowing. Add 1 mL of surrogate standards to all samples, spikes, standards, and blanks. Add 100 mL of 1:1 methylene chloride:acetone and extract ultrasonically. Decant and filter extracts, dry the extract by passing it through a drying column containing anhydrous sodium sulfate and concentrate to 1 mL in a K-D apparatus.

Hexane solvent exchange — Add 50 mL of hexane, a new boiling chip, and concentrate until the apparent volume of liquid reaches 1 mL. Adjust the extract volume to 10.0 mL. Stopper the concentration tube and store refrigerated at 4°C if further processing will not be performed immediately. If the extract will be stored longer than two days, transfer it to a vial with Teflon®-lined screw-cap or crimp top.

QUALITY CONTROL   Demonstrate through the analysis of a reagent water blank, that all glassware and reagents are interference free. Each time a set of samples is processed, a method blank should be processed as a safeguard against chronic lab contamination. A reagent blank, a matrix spike, and a duplicate or matrix spike duplicate must be performed for each analytical batch (up to a maximum of 20 samples) analyzed.

Analytical system performance must be verified by analyzing QC check samples. The QC check sample concentration should contain each single-component analyte at the following concentrations in acetone: 4,4'-DDD, 10 µg/mL; 4,4'-DDT, 10 µg/mL; endosulfan II, 10 µg/mL; endosulfan sulfate, 10 µg/mL; and any other single-component pesticide at 2 µg/mL. If the method is only to be used to analyze PCBs, Chlordane, or Toxaphene, the QC check sample concentrate should contain the most representative multicomponent parameter at a concentration of 50 µg/mL in acetone.

REFERENCE   Test Methods for Evaluating Solid Waste (SW-846). U.S. EPA. 1983. Method 8080B, Rev. 2, Nov. 1990. Office of Solid Wastes, Washington, DC.

Toxaphene EPA Method 8270
CAS #8001-35-2

TITLE   Semivolatile Organic Compounds by GC/MS

MATRIX   This method is used to determine the concentration of semivolatile organic compounds in extracts prepared from all types of solid waste matrices, soils, and groundwater. Although surface waters are not specifically mentioned, this method should be applicable to water samples from rivers, lakes, etc.

METHOD SUMMARY   This method covers 259 semivolatile organic compounds. In very limited applications direct injection of the sample into the GC/MS system may be appropriate, but this results in very high detection limits (approximately 10,000 µg/L). Typically, a 1-L liquid sample, containing surrogate, and matrix spiking standards, is extracted in a continuous extractor first under acid conditions and then under basic conditions. Typically 30 g of a solid sample, containing surrogate, and matrix spiking standards, is extracted ultrasonically. After concentrating the extract to 1 mL it is spiked with 10 µL of an internal standard solution just prior to analysis by GC/MS. The volume injected should contain about 100 ng of base/neutral and 200 ng of acid surrogates (for a 1-µL injection). Analysis is performed by GC/MS using a capillary GC column.

INTERFERENCES   Raw GC/MS data from all blanks, samples, and spikes must be evaluated for interferences. Contamination by carryover can occur whenever high-concentration and low-concentration samples are sequentially analyzed. To reduce carryover, the sample syringe must be rinsed out between samples with solvent. Whenever an unusually concentrated sample is encountered, it should be followed by the analysis of blank solvent to check for cross-contamination.

INSTRUMENTATION   A GC/MS and a data system are required. The GC column used is a 30 m × 0.25 mm I.D. (or 0.32 mm I.D.) 1um film thickness silicone-coated fused silica

capillary column. A continuous liquid-liquid extractor equipped with Teflon® or glass connection joints and stopcocks requiring no lubrication, a K-D concentrating apparatus, water bath, and an ultrasonic disrupter with a minimum power of 300 W and with pulsing capability are also required.

PRECISION & ACCURACY The estimated quantitation limit (EQL) of Method 8270B for determining an individual compound is approximately 1 mg/kg (wet weight) for soil or sediment samples, 1–200 mg/kg for wastes (dependent on matrix and method of preparation), and 10 µg/L for groundwater samples. EQLs will be proportionately higher for sample extracts that require dilution to avoid saturation of the detector.

The EQL(b) for groundwater in µg/L is not listed.
The EQL (a, b) for low concentrations in soil and sediment in µg/kg is not listed.
Accuracy as µg/L is not listed.
Overall precision in µg/L is not listed.

(a) *EQLs listed for soil/sediment are based on wet weight. Normally data is reported in a dry-weight basis; therefore, EQLs will be higher based on the % dry weight of each sample. This calculation is based on a 30 g sample and gel permeation chromatography cleanup.*
(b) *Sample EQLs are highly matrix-dependent. The EQLs are provided for guidance and may not always be achievable.*
C = *True value for concentration, in µg/L.*
X = *Average recovery found for measurements of samples containing a concentration of C, in µg/L.*

ESTIMATED QUANTITATION LIMIT

Other Matrices	Factor (a)
High-concentration soil and sludges by sonicator	7.5
Non-water miscible waste	75

(a) *EQL for other matrices = [EQL for low soil/sediment × [Factor]. This estimated EQL is similar to an EPA "Practical Quantitation Limit."*

SAMPLING METHOD
Liquid samples — Use a 1 or 2½ gallon amber glass bottle with a screw-top Teflon®-lined cover that has been prewashed with detergent and rinsed with distilled water and methanol (or isopropanol).

Soils, sediments, or sludges — Use an 8-oz. widemouth glass with a screw-top Teflon®-lined cover that has been prewashed with detergent and rinsed with distilled water and methanol (or isopropanol).

SAMPLE PRESERVATION
Liquid samples — If residual chlorine is present, add 3 mL of 10% sodium thiosulfate per gallon, cool to 4°C and store in a solvent-free refrigerator until analysis; if chlorine is not present, then eliminate the sodium thiosulfate addition.

Soils, sediments, or sludges — Cool samples to 4°C and store in a solvent-free refrigerator.

MHT Liquid samples must be extracted within 7 days and the extracts analyzed within 40 days. Soils, sediments, or sludges may be stored for a maximum of 14 days and the extracts analyzed within 40 days.

SAMPLE PREPARATION
Liquid samples — Transfer 1 L quantitatively to a continuous extractor. If high concentrations are anticipated, a smaller volume may be used and then diluted with organic-free reagent water to 1 L. Adjust pH, if necessary, to pH <2 using 1:1 (V/V) sulfuric acid. Pipette 1.0 mL of a surrogate standard spiking solution into each sample. For the sample in each analytical batch selected for spiking, add 1.0 mL of a matrix spiking standard. For base/neutral acid analysis, the amount of the surrogates and matrix spiking compounds added to the sample should result in a final concentration of 100 ng/µL of each analyte in the extract to be analyzed (assuming a 1-µL injection). Extract with methylene chloride for 18–24 h. Next, adjust the pH of the aqueous phase to pH >11 using 10 N sodium hydroxide and extract it with methylene chloride again for 18–24 h. Dry the extract through a column containing anhydrous sodium sulfate and concentrate it to 1 mL using a K-D concentrator.

Soils, sediments, or sludges — Use 30 g of sample. Nonporous or wet samples (gummy or clay type) that do not have a free-flowing sandy texture must be mixed with anhydrous sodium sulfate until the sample is free flowing. Add 1 mL of surrogate standards to all samples, spikes, standards, and blanks. For the sample in each analytical batch selected for spiking, add 1.0 mL of a matrix spiking standard. For base/neutral acid analysis, the amount added of the surrogates and matrix spiking compounds should result in a final concentration of 100 ng/µL of each base/neutral analyte and 200 ng/µL of each acid analyte in the extract to be analyzed (assuming a 1-µL injection). Immediately add a 100-mL mixture of 1:1 methylene chloride:acetone and extract the sample ultrasonically for 3 min and then decant or filter the extracts. Repeat the extraction two or more times. Dry the extract using a column with anhydrous sodium sulfate and concentrate it to 1 mL in a K-D concentrator.

QUALITY CONTROL A methylene chloride solution containing 50 ng/µL of decafluorotriphenylphosphine (DFTPP) is used for tuning the GC/MS system each 12-h shift. A system performance check also must be made during every 12-h shift. A standard containing 50 ng/µL each of 4,4′-DDT, pentachlorophenol, and benzidine is required to verify injection port inertness and GC column performance. A calibration standard at mid-concentration, containing each compound of interest, including all required surrogates, must be performed every 12 h during analysis. After the system performance check is met, calibration check compounds (CCCs) are used to check the validity of the initial calibration.

The internal standard responses and retention times in the calibration check standard must be evaluated immediately after or during data acquisition. If the retention time for any internal standard changes by more than 30 seconds from the last check calibration (12 h), the chromatographic system must be inspected for malfunctions and corrections must be made, as required. If the electron ionization current plot (EICP) area for any of the internal standards changes by a factor of two from the last daily calibration standard check, the mass spectrometer must be inspected for malfunctions and corrections must be made, as appropriate.

Demonstrate, through the analysis of a reagent water blank, that interferences from the analytical system, glassware, and reagents are under control. The blank samples should be carried through all stages of the sample preparation and measurement steps. For each analytical batch (up to 20 samples), a reagent blank, matrix spike, and matrix spike duplicate/duplicate must be analyzed (the frequency of the spikes may be different for different monitoring programs). The blank and spiked samples must be carried through all stages of the sample preparation and measurement steps. A QC reference sample concentrate containing each analyte at a concentration of 100 mg/L in methanol is required.

REFERENCE Test Methods for Evaluating Solid Waste (SW-846). U.S. EPA 1983, Method 8270B, Rev. 2, Nov. 1990. Office of Solid Waste, Washington, DC.

2,4,5-TP (Silvex)   **EPA Method 8150**
CAS #93-72-1

TITLE Chlorinated Herbicides

MATRIX Groundwater, soils, sludges, water miscible liquid wastes, and non-water miscible wastes.

APPLICATION This method is used for the analysis of 10 chlorinated herbicides. Samples are extracted, hydrolyzed with potassium hydroxide, and extraneous organics are removed by a solvent wash. After acidification, the acids are extracted, concentrated and converted to their methyl esters using diazomethane. They are then analyzed using direct injection into a gas chromatograph (GC). Be very careful because diazomethane can explode under certain conditions and it is also a carcinogen.

INTERFERENCES Organic acids and phenols (especially chlorinated acids and phenols) may cause interferences. Phthalate esters are not as significant an interference as with other GC-ECD methods if an electron capture detector is used. The herbicides may react readily with alkaline substances and be lost during analysis so all glassware and glass wool must be acid rinsed and sodium sulfate must be acidified with sulfuric acid prior to use. Sensitivity usually depends on the level of interferences rather than on instrumentation.

INSTRUMENTATION GC capable of on-column injections and an electron capture detector (ECD)or a halogen specific detector. Column 1: 1.8 m by 4 mm with 1.5% SP-2250/1.95% SP-2401 on Supelcoport. Column 2: 1.8 m by 4 mm with 5% OV-210 on Gas Chrom Q. Column 3: 1.98 m by 2 mm with 0.1%. SP-1000 on Carbopack C. The preferred column is Column Number 1 or 2.

RANGE Not listed.

MDL 0.17 µg/L (in reagent water; ECD)

PQL FACTORS FOR MULTIPLYING × FID MDL VALUE

Matrix	Multiplication Factor
Groundwater	10
Low-level soil by sonication with GPC cleanup	670
High-level soil and sludge by sonication	10,000
Non-water miscible waste	100,000

PRECISION (as standard deviation) 5% with 1.0 µg/L spike in drinking water; 4% with 1.3 µg/L in municipal water.

ACCURACY (as mean recovery) 88% with 1.0 µg/L spike in drinking water; 88% with 1.3 µg/L in municipal water.

SAMPLING METHOD Use 8-oz. widemouth glass bottles with Teflon®-lined caps for concentrated waste samples, soils, sediments, and sludges. Use 1 or 2½ gallon amber glass bottles with Teflon®-lined caps for liquid (water) samples.

STABILITY Cool soil, sediment, sludge, and liquid samples to 4°C. If residual chlorine is present in liquid samples add 3 mL of 10% sodium thiosulfate per gallon of sample and cool to 4°C.

MHT 14 days for concentrated waste, soil, sediment, or sludge; 7 days for liquid samples; all extracts must be analyzed within 40 days.

QUALITY CONTROL A quality control check sample concentrate containing this compound in acetone at a concentration 1,000 times more concentrated than the selected spike concentration is required. The QC check sample concentrate may be prepared from pure standard materials or purchased as certified solutions. Use appropriate trip, matrix, control site, method, reagent, and solvent blanks. Internal, surrogate, and five concentration level calibration standards are used.

REFERENCE Method 8150, SW-846, 3rd ed., Sept. 1986.

2,4,5-TP (Silvex)   **EPA Method 8151**
CAS #93-72-1

TITLE Chlorinated Herbicides by GC Using Methylation or Pentafluorobenzylation Derivatization: Capillary Column Technique.

MATRIX This method covers aqueous and solid matrices. This includes a wide variety such as drinking water, groundwater, industrial wastewaters, surface waters, soils, solids, and sediments.

METHOD SUMMARY This is a GC method for determining 19 chlorinated acid herbicides in aqueous, soil, and waste matrices. Because these compounds are produced and used in various forms (i.e., acid, salt, ester, etc.) a hydrolysis step is included to convert the herbicide to the acid form prior to analysis. This method provides hydrolysis, extraction, derivatization and GC conditions for the analysis of chlorinated acid herbicides in water, soil, and waste samples. Water samples are hydrolyzed *in situ*, extracted with diethyl ether, and then esterified with either diazomethane or pentafluorobenzyl bromide. The derivatives are determined by gas chromatography with an electron capture detector (GC/ECD). The results are reported as acid equivalents. The sensitivity of this method depends on the level of interferences in addition to instrumental limitations.

INTERFERENCES Method interferences may be caused by contaminants in solvents, reagents, glassware, and other sample processing hardware. Immediately prior to use, glassware should be rinsed with the next solvent to be used. Matrix interferences may be caused by contaminants that are coextracted from the sample. Organic acids, especially chlorinated acids, cause the most direct interference with the determination by methylation. Phenols, including chlorophenols, may also interfere with this procedure. The determination using pentafluorobenzylation is more sensitive, and more prone to interferences from the presence of organic acids of phenols than by methylation. Alkaline hydrolysis and subsequent extraction of the basic solution removes many chlorinated hydrocarbons and phthalate esters that might otherwise interfere with the ECD analysis. The herbicides, being strong organic acids, react readily with alkaline substances and may be lost during analysis. Therefore, glassware must be acid-rinsed and then rinsed to constant pH with organic-free reagent water.

INSTRUMENTATION A GC suitable for Grob-type injection using capillary columns. A data system for measuring peak heights and/or peak areas is recommended. An electron capture detector (ECD) is used. Also a K-D apparatus, a diazomethane generator, a centrifuge and an ultrasonic disrupter will be required.

Narrow Bore Columns:
Primary Column 1: 30 m × 0.25 mm, 5% phenyl/95% methyl silicone (DB-5), 0.25 μm film thickness.
Primary Column 1a (GC/MS): 30 m × 0.32 mm, 5% phenyl/95% methyl silicone (DB-5), 1-μm film thickness.
Column 2: 30 m × 0.25 mm DB-608 with a 25 μm film thickness.
Confirmation Column: 30 m × 0.25 mm, 14% cyanopropyl phenyl silicone (DB-1701), 0.25 μm film thickness.

Megabore Columns:
Primary Column: 30 m × 0.53 mm DB-608 with 0.83 μm film thickness.
Confirmation Column: 30 m × 0.53 mm, 14% cyanopropyl phenyl silicone (DB-1701), 1.0 μm film thickness.

PRECISION & ACCURACY Method detection limits (MDLs) are compound-dependent and vary with derivitization efficiency, derivative recovery, the matrix sampled, and herbicide concentration.

The estimated MDL (in μg/L) was 0.075 for aqueous samples using GC/ECD.

The estimated MDL (in μg/kg) was 0.28 for soil samples using GC/ECD when corrected back to 50 g samples extracted and concentrated to 10 mL with 5-μL injections.

The estimated GC/MS identification limit (in ng) was 4.5 for soil samples using GC/MS.

Mean percent recovery, calculated from 7–8 determinations of spiked reagent water, after diazomethane derivatization, from a spike concentration (in μg/L) of 0.4 was 117 with a standard deviation of the percent recovery of 16.4.

Mean percent recovery, calculated from 10 determinations of spiked clay and clay/still bottom samples over the linear concentration range (in ng/g) of 0.42–828 was 94.5 with a percent relative standard deviation of 5.7. The RSD % was calculated on 10 samples high in the linear concentration range and 10 low in the range. The linear concentration range was determined using standard solutions and corrected to 50 g soil samples.

SAMPLE COLLECTION, PRESERVATION & HANDLING
Containers used to collect samples for the determination of semivolatile organic compounds should be soap and water washed followed by methanol (or isopropanol) rinsing. The sample containers should be of glass or Teflon® and have screw-top covers with Teflon® liners.

No preservation is used with concentrated waste samples. With liquid samples containing no residual chlorine and with soil, sediment, and sludge samples, immediately cooling to 4°C is the only preservation used. When residual chlorine is present then 3 mL of 10% aqueous sodium sulfate is added for each gallon of sample collected, followed by cooling to 4°C.

The holding time for all volatile organics samples is 14 days. Liquid samples must be extracted within 7 days and their extracts analyzed within 40 days. Concentrated waste, soil, sediment, and sludge samples must be extracted within 14 days and their extracts analyzed within 40 days.

SAMPLE PREPARATION
Preparation of soil, sediment, and other solid samples — Acidify 30 g (dry weight) solids with 0.1 M phosphate buffer (pH = 2.5) and thoroughly mix the contents. Spike the sample with surrogate compound(s). The ultrasonic extraction of solids must be optimized for each type of sample. In order for the ultrasonic extractor to efficiently extract solid samples, the sample must be free flowing when the solvent is added. Acidified anhydrous sodium sulfate should be added to clay-type soils, or any other solid that is not a free-flowing sandy texture, until a free flowing mixture is obtained. Add methylene chloride and perform ultrasonic extraction. Combine organic extracts from the repetitive extractings of the sample and centrifuge. Add aqueous potassium hydroxide, water, and methanol to the extract and reflux the mixture on a water bath. Extract the solution three times with methylene chloride and discard the methylene chloride phase. The basic solution contains the herbicide salts. Adjust the pH of the solution to <2 with cold sulfuric acid and extract three times with methylene chloride. Combine the extracts and pour them through a prerinsed drying column containing acidified anhydrous sodium sulfate. Collect the dried extracts in a K-D flask and concentrate them.

Preparation of aqueous samples — Measure 1 L of sample into a 2 L separatory funnel and spike it with surrogate compound(s). Add NaCl to the sample, then add 6 N NaOH to the sample to a pH of 12 or more and let the sample sit at room temperature for 1 h to hydrolyze esters. Extract the sample three times with methylene chloride and discard the extracts. Then add cold 12 N sulfuric acid to a pH less than or equal to 2, and extract the sample three times with ethyl ether. Collect the ether phase in a flask containing acidified anhydrous sodium sulfate and allow it to remain in contact with the sodium sulfate for a minimum of 2 h. The drying step is very critical to ensuring complete esterification; any moisture remaining in the ether will result in low herbicide recoveries.

Extract concentration and derivatization — The combined ether extract is concentrated to about 1 mL using a K-D apparatus followed by using a micro Snyder column or nitrogen gas blowdown. If methyl esters are to be produced, then dilute the concentrated ether extract with 1 mL of isooctane and 0.5 mL of methanol, dilute to a final volume of 4 mL, and esterify with diazomethane. If pentafluorobenzene esters are to be produced, then dilute concentrated ether extract with acetone to a final volume of 4 mL and esterify with pentafluorobenzyl bromide.

QUALITY CONTROL Select a representative spike concentration for each compound (acid or ester) to be measured. Using stock standard, prepare a quality control check sample concentrate, in acetone, that is 1000 times more concentrated than the selected concentrations. Use this quality control check sample concentrate to prepare quality control check samples. Calculate surrogate standard recovery on all standards, samples, blanks, and spikes. GC/MS techniques should be judiciously employed to support qualitative identifications made with this method. When available, chemical ionization mass spectra may be employed to aid the qualitative identification process.

REFERENCE Test Methods for Evaluating Solid Waste, Physical/Chemical Methods, SW-846, 3rd Edition, U.S. EPA, Office of Solid Waste, Washington, DC, EPA Method 8151, Nov. 1990.

Tri-p-tolyl phosphate   **EPA Method 8270**
CAS #78-32-0

TITLE Semivolatile Organic Compounds by GC/MS

MATRIX This method is used to determine the concentration of semivolatile organic compounds in extracts prepared from all types of solid waste matrices, soils, and groundwater. Although surface waters are not specifically mentioned, this method should be applicable to water samples from rivers, lakes, etc.

METHOD SUMMARY This method covers 259 semivolatile organic compounds. In very limited applications direct injection of the sample into the GC/MS system may be appropriate, but this results in very high detection limits (approximately 10,000 μg/L). Typically, a 1-L liquid sample, containing surrogate, and matrix spiking standards, is extracted in a continuous extractor first under acid conditions and then under basic conditions. Typically 30 g of a solid sample, containing surrogate, and matrix spiking standards, is extracted ultrasonically. After concentrating the extract to 1 mL it is spiked with 10 μL of an internal standard solution just prior to analysis by GC/MS. The volume injected should contain about 100 ng of base/neutral and 200 ng of acid surrogates (for a 1-μL injection). Analysis is performed by GC/MS using a capillary GC column.

INTERFERENCES Raw GC/MS data from all blanks, samples, and spikes must be evaluated for interferences. Contamination by carryover can occur whenever high-concentration and low-concentration samples are sequentially analyzed. To reduce carryover, the sample syringe must be rinsed out between samples with solvent. Whenever an unusually concentrated sample is encountered, it should be followed by the analysis of blank solvent to check for cross-contamination.

INSTRUMENTATION A GC/MS and a data system are required. The GC column used is a 30 m × 0.25 mm I.D. (or 0.32 mm I.D.) 1um film thickness silicone-coated fused silica capillary column. A continuous liquid-liquid extractor equipped with Teflon® or glass connection joints and stopcocks requiring no lubrication, a K-D concentrating apparatus, water bath, and an ultrasonic disrupter with a minimum power of 300 W and with pulsing capability are also required.

PRECISION & ACCURACY The estimated quantitation limit (EQL) of Method 8270B for determining an individual compound is approximately 1 mg/kg (wet weight) for soil or sediment samples, 1–200 mg/kg for wastes (dependent on matrix and method of preparation), and 10 μg/L for groundwater samples. EQLs will be proportionately higher for sample extracts that require dilution to avoid saturation of the detector.

The EQL(b) for groundwater in μg/L is 10.
The EQL (a, b) for low concentrations in soil and sediment in μg/kg is not determined.
Accuracy as μg/L is not listed.
Overall precision in μg/L is not listed.

(a) EQLs listed for soil/sediment are based on wet weight. Normally data is reported in a dry-weight basis; therefore, EQLs will be higher based on the % dry weight of each sample. This calculation is based on a 30 g sample and gel permeation chromatography cleanup.

(b) Sample EQLs are highly matrix-dependent. The EQLs are provided for guidance and may not always be achievable.

$C =$ True value for concentration, in μg/L.
$X =$ Average recovery found for measurements of samples containing a concentration of C, in μg/L.

ESTIMATED QUANTITATION LIMIT

Other Matrices	Factor (a)
High-concentration soil and sludges by sonicator	7.5
Non-water miscible waste	75

(a) EQL for other matrices = [EQL for low soil/sediment × [Factor]. This estimated EQL is similar to an EPA "Practical Quantitation Limit."

SAMPLING METHOD

Liquid samples — Use a 1 or 2½ gallon amber glass bottle with a screw-top Teflon®-lined cover that has been prewashed with detergent and rinsed with distilled water and methanol (or isopropanol).

Soils, sediments, or sludges — Use an 8-oz. widemouth glass with a screw-top Teflon®-lined cover that has been prewashed with detergent and rinsed with distilled water and methanol (or isopropanol).

SAMPLE PRESERVATION

Liquid samples — If residual chlorine is present, add 3 mL of 10% sodium thiosulfate per gallon, cool to 4°C and store in a

solvent-free refrigerator until analysis; if chlorine is not present, then eliminate the sodium thiosulfate addition.

Soils, sediments, or sludges — Cool samples to 4°C and store in a solvent-free refrigerator.

MHT Liquid samples must be extracted within 7 days and the extracts analyzed within 40 days. Soils, sediments, or sludges may be stored for a maximum of 14 days and the extracts analyzed within 40 days.

SAMPLE PREPARATION
Liquid samples — Transfer 1 L quantitatively to a continuous extractor. If high concentrations are anticipated, a smaller volume may be used and then diluted with organic-free reagent water to 1 L. Adjust pH, if necessary, to pH <2 using 1:1 (V/V) sulfuric acid. Pipette 1.0 mL of a surrogate standard spiking solution into each sample. For the sample in each analytical batch selected for spiking, add 1.0 mL of a matrix spiking standard. For base/neutral acid analysis, the amount of the surrogates and matrix spiking compounds added to the sample should result in a final concentration of 100 ng/µL of each analyte in the extract to be analyzed (assuming a 1-µL injection). Extract with methylene chloride for 18–24 h. Next, adjust the pH of the aqueous phase to pH >11 using 10 N sodium hydroxide and extract it with methylene chloride again for 18–24 h. Dry the extract through a column containing anhydrous sodium sulfate and concentrate it to 1 mL using a K-D concentrator.

Soils, sediments, or sludges — Use 30 g of sample. Nonporous or wet samples (gummy or clay type) that do not have a free-flowing sandy texture must be mixed with anhydrous sodium sulfate until the sample is free flowing. Add 1 mL of surrogate standards to all samples, spikes, standards, and blanks. For the sample in each analytical batch selected for spiking, add 1.0 mL of a matrix spiking standard. For base/neutral acid analysis, the amount added of the surrogates and matrix spiking compounds should result in a final concentration of 100 ng/µL of each base/neutral analyte and 200 ng/µL of each acid analyte in the extract to be analyzed (assuming a 1-µL injection). Immediately add a 100-mL mixture of 1:1 methylene chloride:acetone and extract the sample ultrasonically for 3 min and then decant or filter the extracts. Repeat the extraction two or more times. Dry the extract using a column with anhydrous sodium sulfate and concentrate it to 1 mL in a K-D concentrator.

QUALITY CONTROL A methylene chloride solution containing 50 ng/µL of decafluorotriphenylphosphine (DFTPP) is used for tuning the GC/MS system each 12-h shift. A system performance check also must be made during every 12-h shift. A standard containing 50 ng/µL each of 4,4'-DDT, pentachlorophenol, and benzidine is required to verify injection port inertness and GC column performance. A calibration standard at mid-concentration, containing each compound of interest, including all required surrogates, must be performed every 12 h during analysis. After the system performance check is met, calibration check compounds (CCCs) are used to check the validity of the initial calibration.

The internal standard responses and retention times in the calibration check standard must be evaluated immediately after or during data acquisition. If the retention time for any internal standard changes by more than 30 seconds from the last check calibration (12 h), the chromatographic system must be inspected for malfunctions and corrections must be made, as required. If the electron ionization current plot (EICP) area for any of the internal standards changes by a factor of two from the last daily calibration standard check, the mass spectrometer must be inspected for malfunctions and corrections must be made, as appropriate.

Demonstrate, through the analysis of a reagent water blank, that interferences from the analytical system, glassware, and reagents are under control. The blank samples should be carried through all stages of the sample preparation and measurement steps. For each analytical batch (up to 20 samples), a reagent blank, matrix spike, and matrix spike duplicate/duplicate must be analyzed (the frequency of the spikes may be different for different monitoring programs). The blank and spiked samples must be carried through all stages of the sample preparation and measurement steps. A QC reference sample concentrate containing each analyte at a concentration of 100 mg/L in methanol is required.

REFERENCE Test Methods for Evaluating Solid Waste (SW-846). U.S. EPA 1983, Method 8270B, Rev. 2, Nov. 1990. Office of Solid Waste, Washington, DC.

---

n-Triacontane   **EPA Method 1625**
CAS #638-68-6

TITLE Semivolatile Organic Compounds by Isotope Dilution GC/MS

MATRIX The compounds may be determined in waters, soils, and municipal sludges by this method.

METHOD SUMMARY This method is used to determine 176 semivolatile toxic organic pollutants associated with the CWA (as amended 1987); the RCRA (as amended 1986); the CERCLA (as amended 1986); and other compounds amenable to extraction and analysis by capillary column gas chromatography-mass spectrometry (GC/MS).

Stable isotopically-labeled analogs of the compounds of interest are added to the sample. If the solids content is less than 1%, a 1-L sample is extracted at pH 12–13, then at pH <2 with methylene chloride using continuous extraction techniques.

If the solids content is 30% or less, the sample is diluted to 1% solids with reagent water, homogenized ultrasonically, and extracted at pH 12–13, then at pH <2 with methylene chloride using continuous extraction techniques. If the solids content is greater than 30%, the sample is extracted using ultrasonic techniques.

Each extract is dried over sodium sulfate, concentrated to a volume of 5 mL, cleaned up using GPC, if necessary, and concentrated. Extracts are concentrated to 1 mL if GPC is not performed, and to 0.5 mL if GPC is performed.

An internal standard is added to the extract, and a 1-mL aliquot of the extract is injected into the GC. The compounds are

separated by GC and detected by a MS. The labeled compounds serve to correct the variability of the analytical technique.

INTERFERENCES Solvents, reagents, glassware, and other sample processing hardware may yield artifacts and/or elevated baselines causing misinterpretation of chromatograms and spectra. Materials used in the analysis must be demonstrated to be free from interferences under the conditions of analysis by running method blanks initially and with each sample lot (sample started through the extraction process on a given 8-h shift, to a maximum of 20). Specific selection of reagents and purification of solvents by distillation in all glass systems may be required. Glassware and, where possible, reagents are cleaned by solvent rinse and baking at 450°C for 1-h minimum. Interferences coextracted from samples will vary considerably from source to source, depending on the diversity of the site being sampled.

INSTRUMENTATION Major instrumentation includes a GC with a splitless or on-column injection port for capillary column, a MS with 70 eV electron impact ionization, and a data system to collect and record MS data, and process it. A K-D apparatus is used to concentrate extracts.

GC Column: 30 m × 0.25 mm I.D. 5% phenyl, 94% methyl, 1% vinyl silicone bonded phased fused silica capillary column.

PRECISION & ACCURACY The detection limits of the method are usually dependent on the level of interferences rather than instrumental limitations. The limits typify the minimum quantities that can be detected with no interferences present.

The minimum level (in µg/mL) was 10. This is defined as a minimum level at which the analytical system shall give recognizable mass spectra (background corrected) and acceptable calibration points.

The MDL (in µg/kg) in low solids was 252 and in high solids was 658; these were determined in digested sludge (low solids) and in filter cake or compost (high solids).

Note: Background levels of this compound were present in the sludge tested, resulting in higher than expected MDLs. The MDL for this compound is expected to be approximately 50 µg/kg with no interferences present.

The labeled and native compound initial precision as standard deviation (in µg/L) was 32.
The labeled and native compound initial accuracy as average recovery (in µg/L) was 61–200.

SAMPLE COLLECTION, PRESERVATION & HANDLING Collect samples in glass containers. Aqueous samples which flow freely are collected in refrigerated bottles using automatic sampling equipment. Solid samples are collected as grab samples using widemouth jars. Maintain samples at 0 to 4°C from the time of collection until extraction. If residual chlorine is present in aqueous samples, add 80 mg sodium thiosulfate/L of water. Begin sample extraction within 7 days of collection, and analyze all extracts within 40 days of extraction.

SAMPLE PREPARATION Samples containing 1% solids or less are extracted directly using continuous liquid-liquid extraction techniques. Samples containing 1 to 30% solids are diluted to the 1% level with reagent water and extracted using continuous liquid-liquid extraction techniques. Samples containing greater than 30% solids are extracted using ultrasonic techniques.

Base/neutral extraction — Adjust the pH of the waters in the extractors to 12–13 with 6 N NaOH. Extract with methylene chloride for 24–48 h.
Acid extraction — Adjust the pH of the waters in the extractors to 2 or less using 6 N sulfuric acid. Extract with methylene chloride for 24–48 h.
Ultrasonic extraction of high solids samples — Add anhydrous sodium sulfate to the sample and QC aliquot(s). Add acetone:methylene chloride (1:1) to the sample and mix thoroughly

Concentrate extracts using a K-D apparatus.

QUALITY CONTROL The analyst is permitted to modify this method to improve separations or lower the costs of measurements, provided all performance specifications are met. Analyses of blanks are required to demonstrate freedom from contamination. When results of spikes indicate atypical method performance for samples, the samples are diluted to bring method performance within acceptable limits.

For low solids (aqueous samples), extract, concentrate, and analyze two sets of four 1-L aliquots (8 aliquots total) of the precision and recovery standard. For high solids samples, two sets of four 30-g aliquots of the high solids reference matrix are used.

Spike all samples with labeled compounds to assess method performance. Compute percent recovery of the labeled compounds using the internal standard method. Compare the labeled compound recovery for each compound with the corresponding labeled compound recovery.

Reagent water and high solids reference matrix blanks are analyzed to demonstrate freedom from contamination. Extract and concentrate a 1-L reagent water blank or a high solids reference matrix blank with each sample's lot (samples started through the extraction process on the same 8-h shift, to a maximum of 20 samples).

Field replicates may be collected to determine the precision of the sampling technique, and spiked samples may be required to determine the accuracy of the analysis when the internal standard method is used.

REFERENCE Semivolatile Organic Compounds by Isotope Dilution GC/MS. Office of Water Regulation and Standards, U.S. EPA Industrial Technology Division, Washington, DC, EPA Method 1625, Rev. C, June 1989 (contact W.A. Telliard, U.S. EPA, Office of Water Regulations and Standards, 401 M St., SW, Washington, DC, 20460. Phone: 202-382-7131).

Triademefon **EPA Method 507**
CAS #43121-43-3

TITLE Determination of Nitrogen and Phosphorus-Containing Pesticides in Water by GC/NPD

MATRIX This method is applicable to the determination of certain nitrogen and phosphorus-containing pesticides in finished drinking water and groundwater.

METHOD SUMMARY Method 507 covers 46 nitrogen- and phosphorus-containing pesticides. A 1-L sample is fortified with a surrogate standard, salted, buffered, extracted with methylene chloride, and concentrated; then the solvent is exchanged with methyl tert-butyl ether (MTBE) and concentrated again, and a 2-µL aliquot of a sample extract is injected into a GC system equipped with a selective nitrogen-phosphorus detector and a capillary column for analysis.

INTERFERENCES Method interferences may be caused by contaminants in solvents, reagents, glassware, and other sample processing apparatus. Interfering contamination may occur when a sample containing low concentrations of analytes is analyzed immediately following a sample containing relatively high concentrations. One or more injections of MTBE should be made following the analysis of a sample with high concentrations of analytes to check for analyte carryover. Matrix interferences may be caused by contaminants that are coextracted from the sample. The extent of matrix interferences will vary considerably from source to source, depending upon the water sampled.

INSTRUMENTATION A gas chromatograph system (GC) equipped with a nitrogen-phosphorus detector (NPD) is needed.

Column 1: 30 m × 0.25 mm I.D. DB-5 bonded fused silica column, 0.25 µm film thickness, or equivalent.
Column 2: 30 m × 0.25 mm I.D. DB-1701 bonded fused silica column, 0.25 µm film thickness, or equivalent.

PRECISION & ACCURACY This method has been validated in a single lab and estimated detection limits (EDLs) have been determined for each analyte. Observed detection limits may vary among waters, depending upon the nature of the interferences in the sample matrix and the specific instrumentation used. Analytes that are not separated chromatographically cannot be individually identified and measured unless an alternative technique for identification and quantification exist.

The estimated detection limit (in µg/L) was 0.65. The EDL is defined as either method detection limit or a level of compound in a sample yielding a peak in the final extract with signal-to-noise ratio of approximately 5, whichever value is higher.

The concentration used for these measurements (in µg/L) was 6.5.
The accuracy (as % recovery) was 93.
The precision (% RSD) was 8.

SAMPLING METHOD Grab samples are collected in 1-L glass sample bottles (prewashed with detergent and hot tap water, rinsed with reagent water, and dried in an oven at 400°C for 1 h) with screw caps lined with PTFE-fluorocarbon.

SAMPLE PRESERVATION Add mercuric chloride to the sample bottle in amounts to produce a concentration of 10 mg/L. If residual chlorine is present, add 80 mg of sodium thiosulfate/L of sample to the sample bottle prior to collection. After collection, seal bottle and shake vigorously for 1 min, then cool the sample to 4°C immediately and store it at 4°C in the dark until extraction.

MHT Maximum holding time of the samples, and in some cases the extracts, is 14 days.

SAMPLE PREPARATION Fortify the sample with 50 µL of the surrogate standard solution, adjust to pH 7 with phosphate buffer, add 100 g NaCl to the sample, and seal and shake to dissolve the salt; then extract with methylene chloride in a separatory funnel or in a mechanical tumbler bottle. Dry the extract by pouring it through a solvent-rinsed drying column containing about 10 cm of anhydrous sodium sulfate. Collect the extract in a Kuderna-Danish (K-D) concentrator and rinse the column with 20–30 mL methylene chloride. Concentrate the extract to about 2 mL and rinse the flask and its lower joint into the concentrator tube with 1 to 2 mL of methyl t-butyl ether (MTBE). Add 5–10 mL of MTBE and concentrate the extract twice (adding more MTBE) to a final volume of 5.0 mL and store it at 4°C until analysis.

Note: If methylene chloride is not completely removed from the final extract, it may cause detector problems.

QUALITY CONTROL Minimum quality control requirements are initial demonstration of lab capability, determination of surrogate compound recoveries in each sample and blank, monitoring internal standard peak area or height in each sample and blank, analysis of lab reagent blanks, lab fortified samples, lab fortified blanks, and other QC samples. A lab reagent blank is analyzed to demonstrate that all glassware and reagent interferences are under control.

Initial demonstration of capability is fulfilled by analyzing four fortified reagent water samples with the recovery value for each analyte falling within the acceptable range (±30% average recovery). Surrogate recoveries from samples or method blanks must be 70–130%. The internal standard response for any sample chromatogram should not deviate from the daily calibration check standard's internal standard response by more than 30% or lab fortified blanks and sample matrices are used to assess lab performance and analyte recovery, respectively.

If the response for the target analyte peak exceeds the working range of the system, dilute the extract and reanalyze. Alternative techniques such as an alternate detector or second chromatography column should be used to confirm peak identification when sample components are not resolved adequately.

EPA CONTACT & HOTLINE For technical questions contact Dr. Baldev Bathija, U.S. EPA, Office of Ground Water and Drinking Water (WH-550D), 401 M St. SW, Washington, DC 20460. Tel. (202) 260-3040. For further information the EPA Safe Drinking Water Hotline may be called at: (800) 426-4791.

REFERENCE Methods for the Determination of Organic Compounds in Drinking Water, EPA/600/4-88/039 (revised July 1991). U.S. EPA Environmental Monitoring Systems Laboratory, Cincinnati, OH, 45268, U.S.A. Available from the National Technical Information Service (NTIS), 5285 Port Royal Road, Springfield, VA 22161; Tel. 800-553-6847. NTIS Order Number is PB91-231480.

1,2,3-Trichlorobenzene
CAS #87-61-6
EPA Method 1625

TITLE Semivolatile Organic Compounds by Isotope Dilution GC/MS

MATRIX The compounds may be determined in waters, soils, and municipal sludges by this method.

METHOD SUMMARY This method is used to determine 176 semivolatile toxic organic pollutants associated with the CWA (as amended 1987); the RCRA (as amended 1986); the CERCLA (as amended 1986); and other compounds amenable to extraction and analysis by capillary column gas chromatography-mass spectrometry (GC/MS).

Stable isotopically-labeled analogs of the compounds of interest are added to the sample. If the solids content is less than 1%, a 1-L sample is extracted at pH 12–13, then at pH <2 with methylene chloride using continuous extraction techniques.

If the solids content is 30% or less, the sample is diluted to 1% solids with reagent water, homogenized ultrasonically, and extracted at pH 12–13, then at pH <2 with methylene chloride using continuous extraction techniques. If the solids content is greater than 30%, the sample is extracted using ultrasonic techniques.

Each extract is dried over sodium sulfate, concentrated to a volume of 5 mL, cleaned up using GPC, if necessary, and concentrated. Extracts are concentrated to 1 mL if GPC is not performed, and to 0.5 mL if GPC is performed.

An internal standard is added to the extract, and a 1-mL aliquot of the extract is injected into the GC. The compounds are separated by GC and detected by a MS. The labeled compounds serve to correct the variability of the analytical technique.

INTERFERENCES Solvents, reagents, glassware, and other sample processing hardware may yield artifacts and/or elevated baselines causing misinterpretation of chromatograms and spectra. Materials used in the analysis must be demonstrated to be free from interferences under the conditions of analysis by running method blanks initially and with each sample lot (sample started through the extraction process on a given 8-h shift, to a maximum of 20). Specific selection of reagents and purification of solvents by distillation in all glass systems may be required. Glassware and, where possible, reagents are cleaned by solvent rinse and baking at 450°C for 1-h minimum. Interferences coextracted from samples will vary considerably from source to source, depending on the diversity of the site being sampled.

INSTRUMENTATION Major instrumentation includes a GC with a splitless or on-column injection port for capillary column, a MS with 70 eV electron impact ionization, and a data system to collect and record MS data, and process it. A K-D apparatus is used to concentrate extracts.

GC Column: 30 m × 0.25 mm I.D. 5% phenyl, 94% methyl, 1% vinyl silicone bonded phased fused silica capillary column.

PRECISION & ACCURACY The detection limits of the method are usually dependent on the level of interferences rather than instrumental limitations. The limits typify the minimum quantities that can be detected with no interferences present.

The minimum level (in μg/mL) was 10. This is defined as a minimum level at which the analytical system shall give recognizable mass spectra (background corrected) and acceptable calibration points.

The MDL (in μg/kg) in low solids was 260 and in high solids was 164; these were determined in digested sludge (low solids) and in filter cake or compost (high solids).

Note: Background levels of this compound were present in the sludge tested, resulting in higher than expected MDLs. The MDL for this compound is expected to be approximately 50 μg/kg with no interferences present.

The labeled and native compound initial precision as standard deviation (in μg/L) was 69.

The labeled and native compound initial accuracy as average recovery (in μg/L) was 15–229.

SAMPLE COLLECTION, PRESERVATION & HANDLING Collect samples in glass containers. Aqueous samples which flow freely are collected in refrigerated bottles using automatic sampling equipment. Solid samples are collected as grab samples using widemouth jars. Maintain samples at 0 to 4°C from the time of collection until extraction. If residual chlorine is present in aqueous samples, add 80 mg sodium thiosulfate/L of water. Begin sample extraction within 7 days of collection, and analyze all extracts within 40 days of extraction.

SAMPLE PREPARATION Samples containing 1% solids or less are extracted directly using continuous liquid-liquid extraction techniques. Samples containing 1 to 30% solids are diluted to the 1% level with reagent water and extracted using continuous liquid-liquid extraction techniques. Samples containing greater than 30% solids are extracted using ultrasonic techniques.

Base/neutral extraction — Adjust the pH of the waters in the extractors to 12–13 with 6 N NaOH. Extract with methylene chloride for 24–48 h.

Acid extraction — Adjust the pH of the waters in the extractors to 2 or less using 6 N sulfuric acid. Extract with methylene chloride for 24–48 h.

Ultrasonic extraction of high solids samples — Add anhydrous sodium sulfate to the sample and QC aliquot(s). Add acetone:methylene chloride (1:1) to the sample and mix thoroughly

Concentrate extracts using a K-D apparatus.

QUALITY CONTROL The analyst is permitted to modify this method to improve separations or lower the costs of measurements, provided all performance specifications are met. Analyses of blanks are required to demonstrate freedom from contamination. When results of spikes indicate atypical method performance for samples, the samples are diluted to bring method performance within acceptable limits.

For low solids (aqueous samples), extract, concentrate, and analyze two sets of four 1-L aliquots (8 aliquots total) of the

precision and recovery standard. For high solids samples, two sets of four 30-g aliquots of the high solids reference matrix are used.

Spike all samples with labeled compounds to assess method performance. Compute percent recovery of the labeled compounds using the internal standard method. Compare the labeled compound recovery for each compound with the corresponding labeled compound recovery.

Reagent water and high solids reference matrix blanks are analyzed to demonstrate freedom from contamination. Extract and concentrate a 1-L reagent water blank or a high solids reference matrix blank with each sample's lot (samples started through the extraction process on the same 8-h shift, to a maximum of 20 samples).

Field replicates may be collected to determine the precision of the sampling technique, and spiked samples may be required to determine the accuracy of the analysis when the internal standard method is used.

REFERENCE Semivolatile Organic Compounds by Isotope Dilution GC/MS. Office of Water Regulation and Standards, U.S. EPA Industrial Technology Division, Washington, DC, EPA Method 1625, Rev. C, June 1989 (contact W.A. Telliard, U.S. EPA, Office of Water Regulations and Standards, 401 M St., SW, Washington, DC, 20460. Phone: 202-382-7131).

1,2,4-Trichlorobenzene   **EPA Method 1625**
CAS #120-82-1

TITLE Semivolatile Organic Compounds by Isotope Dilution GC/MS

MATRIX The compounds may be determined in waters, soils, and municipal sludges by this method.

METHOD SUMMARY This method is used to determine 176 semivolatile toxic organic pollutants associated with the CWA (as amended 1987); the RCRA (as amended 1986); the CERCLA (as amended 1986); and other compounds amenable to extraction and analysis by capillary column gas chromatography-mass spectrometry (GC/MS).

Stable isotopically-labeled analogs of the compounds of interest are added to the sample. If the solids content is less than 1%, a 1-L sample is extracted at pH 12–13, then at pH <2 with methylene chloride using continuous extraction techniques.

If the solids content is 30% or less, the sample is diluted to 1% solids with reagent water, homogenized ultrasonically, and extracted at pH 12–13, then at pH <2 with methylene chloride using continuous extraction techniques. If the solids content is greater than 30%, the sample is extracted using ultrasonic techniques.

Each extract is dried over sodium sulfate, concentrated to a volume of 5 mL, cleaned up using GPC, if necessary, and concentrated. Extracts are concentrated to 1 mL if GPC is not performed, and to 0.5 mL if GPC is performed.

An internal standard is added to the extract, and a 1-mL aliquot of the extract is injected into the GC. The compounds are separated by GC and detected by a MS. The labeled compounds serve to correct the variability of the analytical technique.

INTERFERENCES Solvents, reagents, glassware, and other sample processing hardware may yield artifacts and/or elevated baselines causing misinterpretation of chromatograms and spectra. Materials used in the analysis must be demonstrated to be free from interferences under the conditions of analysis by running method blanks initially and with each sample lot (sample started through the extraction process on a given 8-h shift, to a maximum of 20). Specific selection of reagents and purification of solvents by distillation in all glass systems may be required. Glassware and, where possible, reagents are cleaned by solvent rinse and baking at 450°C for 1-h minimum. Interferences coextracted from samples will vary considerably from source to source, depending on the diversity of the site being sampled.

INSTRUMENTATION Major instrumentation includes a GC with a splitless or on-column injection port for capillary column, a MS with 70 eV electron impact ionization, and a data system to collect and record MS data, and process it. A K-D apparatus is used to concentrate extracts.

GC Column: 30 m × 0.25 mm I.D. 5% phenyl, 94% methyl, 1% vinyl silicone bonded phased fused silica capillary column.

PRECISION & ACCURACY The detection limits of the method are usually dependent on the level of interferences rather than instrumental limitations. The limits typify the minimum quantities that can be detected with no interferences present.

The minimum level (in μg/mL) was 10. This is defined as a minimum level at which the analytical system shall give recognizable mass spectra (background corrected) and acceptable calibration points.

The MDL (in μg/kg) in low solids was 49 and in high solids was 24; these were determined in digested sludge (low solids) and in filter cake or compost (high solids).

The labeled and native compound initial precision as standard deviation (in μg/L) was 19.
The labeled and native compound initial accuracy as average recovery (in μg/L) was 82–136.

SAMPLE COLLECTION, PRESERVATION & HANDLING Collect samples in glass containers. Aqueous samples which flow freely are collected in refrigerated bottles using automatic sampling equipment. Solid samples are collected as grab samples using widemouth jars. Maintain samples at 0 to 4°C from the time of collection until extraction. If residual chlorine is present in aqueous samples, add 80 mg sodium thiosulfate/L of water. Begin sample extraction within 7 days of collection, and analyze all extracts within 40 days of extraction.

SAMPLE PREPARATION Samples containing 1% solids or less are extracted directly using continuous liquid-liquid extraction techniques. Samples containing 1 to 30% solids are diluted to the 1% level with reagent water and extracted using continuous

liquid-liquid extraction techniques. Samples containing greater than 30% solids are extracted using ultrasonic techniques.

- Base/neutral extraction — Adjust the pH of the waters in the extractors to 12–13 with 6 N NaOH. Extract with methylene chloride for 24–48 h.
- Acid extraction — Adjust the pH of the waters in the extractors to 2 or less using 6 N sulfuric acid. Extract with methylene chloride for 24–48 h.
- Ultrasonic extraction of high solids samples — Add anhydrous sodium sulfate to the sample and QC aliquot(s). Add acetone:methylene chloride (1:1) to the sample and mix thoroughly

Concentrate extracts using a K-D apparatus.

QUALITY CONTROL The analyst is permitted to modify this method to improve separations or lower the costs of measurements, provided all performance specifications are met. Analyses of blanks are required to demonstrate freedom from contamination. When results of spikes indicate atypical method performance for samples, the samples are diluted to bring method performance within acceptable limits.

For low solids (aqueous samples), extract, concentrate, and analyze two sets of four 1-L aliquots (8 aliquots total) of the precision and recovery standard. For high solids samples, two sets of four 30-g aliquots of the high solids reference matrix are used.

Spike all samples with labeled compounds to assess method performance. Compute percent recovery of the labeled compounds using the internal standard method. Compare the labeled compound recovery for each compound with the corresponding labeled compound recovery.

Reagent water and high solids reference matrix blanks are analyzed to demonstrate freedom from contamination. Extract and concentrate a 1-L reagent water blank or a high solids reference matrix blank with each sample's lot (samples started through the extraction process on the same 8-h shift, to a maximum of 20 samples).

Field replicates may be collected to determine the precision of the sampling technique, and spiked samples may be required to determine the accuracy of the analysis when the internal standard method is used.

REFERENCE Semivolatile Organic Compounds by Isotope Dilution GC/MS. Office of Water Regulation and Standards, U.S. EPA Industrial Technology Division, Washington, DC, EPA Method 1625, Rev. C, June 1989 (contact W.A. Telliard, U.S. EPA, Office of Water Regulations and Standards, 401 M St., SW, Washington, DC, 20460. Phone: 202-382-7131).

1,2,3-Trichlorobenzene   **EPA Method 502**
CAS #87-61-6

TITLE Volatile Organic Compounds in Water By Purge and Trap Capillary Column Gas Chromatography with Photoionization and Electrolytic Conductivity Detectors in Series. U.S. EPA Method 502.2, Rev. 2.0, 1989.

MATRIX Drinking water and raw source water. The latter should include most surface water and groundwater sources.

METHOD SUMMARY This method covers 60 volatile organic compounds that contain halogen atoms and/or that are aromatic. An inert gas (zero grade nitrogen or helium) is bubbled through a 25-mL or a 5-mL water sample (depending on the expected concentration of the analytes). Purged sample components are trapped in a tube of sorbent materials. When purging is complete, the sorbent tube is heated and backflushed with helium to desorb the trapped sample onto a capillary GC column. The column is temperature programmed to separate the method analytes which are then detected with a photoionization detector (PID) and a Hall electrolytic conductivity (HECD) placed in series. The PID is selective for aromatic compounds and the HECD is selective for halogenated compounds.

INTERFERENCES Impurities in the purge gas and from organic compounds outgassing from the plumbing ahead of the trap account for many contamination problems. Interferences purged or coextracted from the samples will vary considerably from source to source, depending upon the particular sample or extract being tested. Cross-contamination can occur whenever high-level and low-level samples are analyzed sequentially. Samples also can be contaminated by diffusion of volatile organics (particularly methylene chloride and fluorocarbons) through the septum seal into the sample during shipment and storage. The lab where volatile analysis is performed and also the refrigerated storage area should be completely free of solvents.

INSTRUMENTATION A GC containing a series configuration of a high temperature photoionization detector (PID) equipped with 10.0 eV (nominal) lamp and Hall electrolytic conductivity detector (HECD) is required. Also required is an all-glass 5-mL purging device, a sorbent trap, and a thermal desorption apparatus which is connected to the GC system.

Column 1: VOCOL glass wide-bore capillary column.
Column 2: RTX–502.2 mega-bore capillary column.
Column 3: DB-62 mega-bore capillary column.

PRECISION & ACCURACY Method detection limits are dependent upon the characteristics of the gas chromatographic system used. Analytes that are not separated chromatographically cannot be individually identified and used in the same calibration mixture or water samples unless an alternative technique for identification and quantification, such as mass spectrometry, is used.

Electrolytic conductivity detetor (c) range in µg/L (a) was 0.02–200.
Electrolytic conductivity detetor (c) MDL in µg/L (b) was 0.03.
Electrolytic conductivity detetor (c) accuracy as % recovery was 98.
Electrolytic conductivity detetor (c) precision as % RSD was 3.1.
Photoionization detector (d) range in µg/L (a) was 0.02–200.
Photoionization detector (d) MDL in µg/L (b) was not detected.
Photoionization detector (d) accuracy as % recovery was 106.
Photoionization detector (d) precision as % RSD was 1.8.

(a) The applicable concentration range of this method is compound, instrument, and matrix-dependent. It is listed as being approximately 0.02 to 200 µg/L but no specific information is provided so caution should be observed.
(b) The method detection limits reports with this method are compound, instrument, and matrix-dependent. The values reported were calculated using reagent water fortified with the corresponding compounds at 10 µg/L and a GC-equipped with a 60 m × 0.75 mm VOLCOL wide bore capillary column with 1.5 µm film thickness and using helium carrier gas.
(c) Recoveries and relative standard deviations were determined from seven samples of reagent water fortified with 10 µg/L of each compound. 2-Bromo-1-chloropropane was used as the internal standard for calculating average recoveries.
(d) Recoveries and relative standard deviations were determined from seven samples of reagent water fortified with 10 µg/L of each compound. Fluorobenzene was used as the internal standard for calculating average recoveries.

SAMPLING METHOD Collect samples using a 40- to 120-mL screw-cap vial (prewashed with detergent, rinsed with distilled water and oven dried at 105°C) with a Teflon®-faced silicone septum. Collect bubble-free samples and place the septum with the Teflon® side down on the water.

SAMPLE PRESERVATION If residual chlorine is present in the water add about 25 mg of ascorbic acid to each vial before samples are collected to remove the chlorine. Add hydrochloric acid to reduce pH to <2, immediately cool samples to 4°C, and store them in a solvent-free refrigerator at 4°C until analysis.

MHT The maximum holding time for samples is 14 days from the time they were collected.

SAMPLE PREPARATION Remove the plungers from two 5-mL syringes and attach a closed syringe valve to each. Warm the sample to room temperature, open the sample bottle, and carefully pour the sample into one of the syringe barrels to just short of overflowing. Replace the syringe plunger, invert the syringe, and compress the sample. Open the syringe valve and vent any residual air while adjusting the sample volume to 5.0 mL. Add 10 µL of the internal calibration standard to the sample through the syringe valve. Close the valve. Fill the second syringe in an identical manner from the same sample bottle. Reserve this second syringe for a reanalysis if necessary.

QUALITY CONTROL As an initial demonstration of lab accuracy and precision, analyze 4 to 7 replicates of a lab fortified blank containing analyte at 0.1–5 µg/L. Collect all samples in duplicate. Surrogate analytes (similar to those of the analytes of interest), whose concentration is known in every sample, are measured using the same internal standard calibration procedure. Duplicate field reagent water blanks (trip blanks) must be analyzed with each set of samples, lab reagent blanks (method blanks) must be analyzed with each batch of samples processed as a group within a work shift. Also, a single lab-fortified blank that contains each of the analytes of interest should be analyzed with each batch of samples processed as a group within a work shift. A 3- to 5-point calibration curve is needed depending on the calibration range factor required.

EPA CONTACT & HOTLINE For technical questions contact Dr. Baldev Bathija, U.S. EPA, Office of Ground Water and Drinking Water (WH-550D), 401 M St. SW, Washington, DC 20460. Tel. (202) 260-3040. For further information the EPA Safe Drinking Water Hotline may be called at: (800) 426-4791.

REFERENCE Methods for the Determination of Organic Compounds in Drinking Water, EPA/600/4-88/039 (revised July 1991; Final Rule for determination of compliance with the MCL for Total Trihalomethanes under 141.30, in 40 CFR Part 141, Vol. 58, No. 147, Fed. Reg., Tuesday Aug. 3, 1993). U.S. EPA Environmental Monitoring Systems Laboratory, Cincinnati, OH, 45268, U.S.A. Available from the National Technical Information Service (NTIS), 5285 Port Royal Road, Springfield, VA 22161; Tel. 800-553-6847. NTIS Order Number is PB91-231480.

1,2,4-Trichlorobenzene   EPA Method 502
CAS #120-82-1

TITLE Volatile Organic Compounds in Water By Purge and Trap Capillary Column Gas Chromatography with Photoionization and Electrolytic Conductivity Detectors in Series. U.S. EPA Method 502.2, Rev. 2.0, 1989.

MATRIX Drinking water and raw source water. The latter should include most surface water and groundwater sources.

METHOD SUMMARY This method covers 60 volatile organic compounds that contain halogen atoms and/or that are aromatic. An inert gas (zero grade nitrogen or helium) is bubbled through a 25-mL or a 5-mL water sample (depending on the expected concentration of the analytes). Purged sample components are trapped in a tube of sorbent materials. When purging is complete, the sorbent tube is heated and backflushed with helium to desorb the trapped sample onto a capillary GC column. The column is temperature programmed to separate the method analytes which are then detected with a photoionization detector (PID) and a Hall electrolytic conductivity (HECD) placed in series. The PID is selective for aromatic compounds and the HECD is selective for halogenated compounds.

INTERFERENCES Impurities in the purge gas and from organic compounds outgassing from the plumbing ahead of the trap account for many contamination problems. Interferences purged or coextracted from the samples will vary considerably from source to source, depending upon the particular sample or extract being tested. Cross-contamination can occur whenever high-level and low-level samples are analyzed sequentially. Samples also can be contaminated by diffusion of volatile organics (particularly methylene chloride and fluorocarbons) through the septum seal into the sample during shipment and storage. The lab where volatile analysis is performed and also the refrigerated storage area should be completely free of solvents.

INSTRUMENTATION A GC containing a series configuration of a high temperature photoionization detector (PID) equipped with 10.0 eV (nominal) lamp and Hall electrolytic conductivity detector (HECD) is required. Also required is an

all-glass 5-mL purging device, a sorbent trap, and a thermal desorption apparatus which is connected to the GC system.

Column 1: VOCOL glass wide-bore capillary column.
Column 2: RTX–502.2 mega-bore capillary column.
Column 3: DB-62 mega-bore capillary column.

PRECISION & ACCURACY Method detection limits are dependent upon the characteristics of the gas chromatographic system used. Analytes that are not separated chromatographically cannot be individually identified and used in the same calibration mixture or water samples unless an alternative technique for identification and quantification, such as mass spectrometry, is used.

Electrolytic conductivity detetor (c) range in µg/L (a) was 0.02–200.
Electrolytic conductivity detetor (c) MDL in µg/L (b) was 0.03.
Electrolytic conductivity detetor (c) accuracy as % recovery was 102.
Electrolytic conductivity detetor (c) precision as % RSD was 2.1.
Photoionization detector (d) range in µg/L (a) was 0.02–200.
Photoionization detector (d) MDL in µg/L (b) was 0.02.
Photoionization detector (d) accuracy as % recovery was 104.
Photoionization detector (d) precision as % RSD was 2.2.

(a) *The applicable concentration range of this method is compound, instrument, and matrix-dependent. It is listed as being approximately 0.02 to 200 µg/L but no specific information is provided so caution should be observed.*

(b) *The method detection limits reports with this method are compound, instrument, and matrix-dependent. The values reported were calculated using reagent water fortified with the corresponding compounds at 10 µg/L and a GC-equipped with a 60 m × 0.75 mm VOLCOL wide bore capillary column with 1.5 µm film thickness and using helium carrier gas.*

(c) *Recoveries and relative standard deviations were determined from seven samples of reagent water fortified with 10 µg/L of each compound. 2-Bromo-1-chloropropane was used as the internal standard for calculating average recoveries.*

(d) *Recoveries and relative standard deviations were determined from seven samples of reagent water fortified with 10 µg/L of each compound. Fluorobenzene was used as the internal standard for calculating average recoveries.*

SAMPLING METHOD Collect samples using a 40- to 120-mL screw-cap vial (prewashed with detergent, rinsed with distilled water and oven dried at 105°C) with a Teflon®-faced silicone septum. Collect bubble-free samples and place the septum with the Teflon® side down on the water.

SAMPLE PRESERVATION If residual chlorine is present in the water add about 25 mg of ascorbic acid to each vial before samples are collected to remove the chlorine. Add hydrochloric acid to reduce pH to <2, immediately cool samples to 4°C, and store them in a solvent-free refrigerator at 4°C until analysis.

MHT The maximum holding time for samples is 14 days from the time they were collected.

SAMPLE PREPARATION Remove the plungers from two 5-mL syringes and attach a closed syringe valve to each. Warm the sample to room temperature, open the sample bottle, and carefully pour the sample into one of the syringe barrels to just short of overflowing. Replace the syringe plunger, invert the syringe, and compress the sample. Open the syringe valve and vent any residual air while adjusting the sample volume to 5.0 mL. Add 10 µL of the internal calibration standard to the sample through the syringe valve. Close the valve. Fill the second syringe in an identical manner from the same sample bottle. Reserve this second syringe for a reanalysis if necessary.

QUALITY CONTROL As an initial demonstration of lab accuracy and precision, analyze 4 to 7 replicates of a lab fortified blank containing analyte at 0.1–5 µg/L. Collect all samples in duplicate. Surrogate analytes (similar to those of the analytes of interest), whose concentration is known in every sample, are measured using the same internal standard calibration procedure. Duplicate field reagent water blanks (trip blanks) must be analyzed with each set of samples, lab reagent blanks (method blanks) must be analyzed with each batch of samples processed as a group within a work shift. Also, a single lab-fortified blank that contains each of the analytes of interest should be analyzed with each batch of samples processed as a group within a work shift. A 3- to 5-point calibration curve is needed depending on the calibration range factor required.

EPA CONTACT & HOTLINE For technical questions contact Dr. Baldev Bathija, U.S. EPA, Office of Ground Water and Drinking Water (WH-550D), 401 M St. SW, Washington, DC 20460. Tel. (202) 260-3040. For further information the EPA Safe Drinking Water Hotline may be called at: (800) 426-4791.

REFERENCE Methods for the Determination of Organic Compounds in Drinking Water, EPA/600/4-88/039 (revised July 1991; Final Rule for determination of compliance with the MCL for Total Trihalomethanes under 141.30, in 40 CFR Part 141, Vol. 58, No. 147, Fed. Reg., Tuesday Aug. 3, 1993). U.S. EPA Environmental Monitoring Systems Laboratory, Cincinnati, OH, 45268, U.S.A. Available from the National Technical Information Service (NTIS), 5285 Port Royal Road, Springfield, VA 22161; Tel. 800-553-6847. NTIS Order Number is PB91-231480.

1,2,3-Trichlorobenzene   **EPA Method 524**
CAS #87-61-6

TITLE Measurement of Purgeable Organic Compounds in Water by Capillary Column GC/MS.

MATRIX Drinking water and raw source water; the latter should include most surface water and groundwater sources.

METHOD SUMMARY Method 524.2 covers 60 volatile organic compounds. An inert gas (zero grade nitrogen or helium) is bubbled through a 25-mL or a 5-mL water sample (depending on the expected concentration of the analytes). Purged sample components are trapped in a tube of sorbent materials. When purging is complete, the sorbent tube is heated and backflushed with helium to desorb the trapped sample onto a capillary GC column.

INTERFERENCES Impurities in the purge gas and from organic compounds outgassing from the plumbing ahead of the trap account for many contamination problems. Interferences purged or coextracted from the samples will vary considerably from source to source, depending upon the particular sample or extract being tested. Cross-contamination can occur whenever high-level and low-level samples are analyzed sequentially. Samples also can be contaminated by diffusion of volatile organics (particularly methylene chloride and fluorocarbons) through the septum seal into the sample during shipment and storage.

INSTRUMENTATION A GC/MS with a data system equipped with one of the following capillary GC columns:

Column 1: VOCOL glass wide bore capillary column.
Column 2: DB-624 fused silica capillary column.
Column 3: DB-5 fused silica capillary column.

Also required is an all-glass 25 mL or 5-mL purging device, a sorbent trap, and a thermal desorption apparatus which is connected to the GC/MS system.

PRECISION & ACCURACY Method detection limits are compound- and instrument-dependent, and may vary from approximately 0.02–0.35 µg/L. Note in the table below that the "true" concentration range used for accuracy and precision measurements was quite narrow. However, the applicable concentration range of this method is primarily column dependent and is approximately 0.02 to 200 µg/L for the wide-bore thick-film columns. Narrow-bore thin-film columns may have a capacity which limits the range to about 0.02 to 20 µg/L. Analytes that are inefficiently purged from water will not be detected when present at low concentrations, but they can be measured with acceptable accuracy and precision when present in sufficient amounts.

Analytes that are not separated chromatographically, but which have different mass spectra and non-interfering quantification ions, can be identified and measured in the same calibration mixture or water sample. Analytes which have very similar mass spectra cannot be individually identified and measured in the same calibration mixture or water samples unless they have different retention times. Co-eluting compounds with very similar mass spectra, typically many structural isomers, must be reported as an isomeric group or pair.

The range (in µg/L) was 0.5–10.
The Method Detection Limig (in µg/L) was 0.03.
The accuracy (as % recovery) was 109.
The precision (in %) was 8.6.

Note: Data were obtained from 16–31 determinations using a wide-bore capillary column and a jet separator interfaced to a quadrupole mass spectrometer. All analytes were in a reagent water matrix.

SAMPLING METHOD Collect samples using a 40- to 120-mL screw-cap vial (prewashed with detergent, rinsed with distilled water and oven dried at 105°C) with a Teflon®-faced silicone septum. Collect bubble-free samples and place the septum with the Teflon® side down on the water.

SAMPLE PRESERVATION If residual chlorine is present in the water add about 25 mg of ascorbic acid to each vial before samples are collected to remove the chlorine. Add hydrochloric acid to reduce pH to <2, and immediately cool samples to 4°C, and store them in a solvent-free refrigerator at 4°C until analysis.

MHT The maximum holding time for samples is 14 days from the time they were collected.

SAMPLE PREPARATION Remove the plungers from two 25-mL (or 5-mL depending on sample size) syringes and attach a closed syringe valve to each. Warm the sample to room temperature, open the sample bottle, and carefully pour the sample into one of the syringe barrels to just short of overflowing. Replace the syringe plunger, invert the syringe, and compress the sample. Open the syringe valve and vent any residual air while adjusting the sample volume to 25.0 mL (or 5 mL). For samples and blanks, add 5 µL of the fortification solution containing the internal standard and the surrogates to the sample through the syringe valve. For calibration standards and lab fortified blanks, add 5 µL of the fortification solution containing the internal standard only. Close the valve. Fill the second syringe in an identical manner from the same sample bottle. Reserve this second syringe for a reanalysis if necessary.

QUALITY CONTROL As an initial demonstration of lab accuracy and precision, analyze 4 to 7 replicates of a lab fortified blank containing analyte at 0.2–5 µg/L. Collect all samples in duplicate. Surrogate analytes (similar to those of the analytes of interest), whose concentration is known in every sample, are measured using the same internal standard calibration procedure. Duplicate field reagent water blanks (trip blanks) must be analyzed with each set of samples, lab reagent blanks (method blanks) must be analyzed with each batch of samples processed as a group within a work shift. Also, a single lab-fortified blank that contains each of the analytes of interest should be analyzed with each batch of samples processed as a group within a work shift. A 3- to 5-point calibration curve is needed depending on the calibration range factor required.

EPA CONTACT & HOTLINE For technical questions contact Dr. Baldev Bathija, U.S. EPA, Office of Ground Water and Drinking Water (WH-550D), 401 M St. SW, Washington, DC 20460. Tel. (202) 260-3040. For further information the EPA Safe Drinking Water Hotline may be called at: (800) 426-4791.

REFERENCE Methods for the Determination of Organic Compounds in Drinking Water, EPA/600/4-88/039 (revised July 1991; Final Rule for determination of compliance with the MCL for Total Trihalomethanes under 141.30, in 40 CFR Part 141, Vol. 58, No. 147, Fed. Reg., Tuesday Aug. 3, 1993). U.S. EPA Environmental Monitoring Systems Laboratory, Cincinnati, OH, 45268, U.S.A. Available from the National Technical Information Service (NTIS), 5285 Port Royal Road, Springfield, VA 22161; Tel. 800-553-6847. NTIS Order Number is PB91-231480.

1,2,4-Trichlorobenzene   EPA Method 524
CAS #120-82-1

TITLE Measurement of Purgeable Organic Compounds in Water by Capillary Column GC/MS.

MATRIX Drinking water and raw source water; the latter should include most surface water and groundwater sources.

METHOD SUMMARY Method 524.2 covers 60 volatile organic compounds. An inert gas (zero grade nitrogen or helium) is bubbled through a 25-mL or a 5-mL water sample (depending on the expected concentration of the analytes). Purged sample components are trapped in a tube of sorbent materials. When purging is complete, the sorbent tube is heated and backflushed with helium to desorb the trapped sample onto a capillary GC column.

INTERFERENCES Impurities in the purge gas and from organic compounds outgassing from the plumbing ahead of the trap account for many contamination problems. Interferences purged or coextracted from the samples will vary considerably from source to source, depending upon the particular sample or extract being tested. Cross-contamination can occur whenever high-level and low-level samples are analyzed sequentially. Samples also can be contaminated by diffusion of volatile organics (particularly methylene chloride and fluorocarbons) through the septum seal into the sample during shipment and storage.

INSTRUMENTATION A GC/MS with a data system equipped with one of the following capillary GC columns:

Column 1: VOCOL glass wide bore capillary column.
Column 2: DB-624 fused silica capillary column.
Column 3: DB-5 fused silica capillary column.

Also required is an all-glass 25 mL or 5-mL purging device, a sorbent trap, and a thermal desorption apparatus which is connected to the GC/MS system.

PRECISION & ACCURACY Method detection limits are compound- and instrument-dependent, and may vary from approximately 0.02–0.35 µg/L. Note in the table below that the "true" concentration range used for accuracy and precision measurements was quite narrow. However, the applicable concentration range of this method is primarily column dependent and is approximately 0.02 to 200 µg/L for the wide-bore thick-film columns. Narrow-bore thin-film columns may have a capacity which limits the range to about 0.02 to 20 µg/L. Analytes that are inefficiently purged from water will not be detected when present at low concentrations, but they can be measured with acceptable accuracy and precision when present in sufficient amounts.

Analytes that are not separated chromatographically, but which have different mass spectra and non-interfering quantification ions, can be identified and measured in the same calibration mixture or water sample. Analytes which have very similar mass spectra cannot be individually identified and measured in the same calibration mixture or water samples unless they have different retention times. Co-eluting compounds with very similar mass spectra, typically many structural isomers, must be reported as an isomeric group or pair.

The range (in µg/L) was 0.5–10.
The Method Detection Limig (in µg/L) was 0.04.
The accuracy (as % recovery) was 108.
The precision (in %) was 8.3.

Note: Data were obtained from 16–31 determinations using a wide-bore capillary column and a jet separator interfaced to a quadrupole mass spectrometer. All analytes were in a reagent water matrix.

SAMPLING METHOD Collect samples using a 40- to 120-mL screw-cap vial (prewashed with detergent, rinsed with distilled water and oven dried at 105°C) with a Teflon®-faced silicone septum. Collect bubble-free samples and place the septum with the Teflon® side down on the water.

SAMPLE PRESERVATION If residual chlorine is present in the water add about 25 mg of ascorbic acid to each vial before samples are collected to remove the chlorine. Add hydrochloric acid to reduce pH to <2, and immediately cool samples to 4°C, and store them in a solvent-free refrigerator at 4°C until analysis.

MHT The maximum holding time for samples is 14 days from the time they were collected.

SAMPLE PREPARATION Remove the plungers from two 25-mL (or 5-mL depending on sample size) syringes and attach a closed syringe valve to each. Warm the sample to room temperature, open the sample bottle, and carefully pour the sample into one of the syringe barrels to just short of overflowing. Replace the syringe plunger, invert the syringe, and compress the sample. Open the syringe valve and vent any residual air while adjusting the sample volume to 25.0 mL (or 5 mL). For samples and blanks, add 5 µL of the fortification solution containing the internal standard and the surrogates to the sample through the syringe valve. For calibration standards and lab fortified blanks, add 5 µL of the fortification solution containing the internal standard only. Close the valve. Fill the second syringe in an identical manner from the same sample bottle. Reserve this second syringe for a reanalysis if necessary.

QUALITY CONTROL As an initial demonstration of lab accuracy and precision, analyze 4 to 7 replicates of a lab fortified blank containing analyte at 0.2–5 µg/L. Collect all samples in duplicate. Surrogate analytes (similar to those of the analytes of interest), whose concentration is known in every sample, are measured using the same internal standard calibration procedure. Duplicate field reagent water blanks (trip blanks) must be analyzed with each set of samples, lab reagent blanks (method blanks) must be analyzed with each batch of samples processed as a group within a work shift. Also, a single lab-fortified blank that contains each of the analytes of interest should be analyzed with each batch of samples processed as a group within a work shift. A 3- to 5-point calibration curve is needed depending on the calibration range factor required.

EPA CONTACT & HOTLINE For technical questions contact Dr. Baldev Bathija, U.S. EPA, Office of Ground Water and Drinking Water (WH-550D), 401 M St. SW, Washington, DC 20460. Tel. (202) 260-3040. For further information the EPA Safe Drinking Water Hotline may be called at: (800) 426-4791.

REFERENCE Methods for the Determination of Organic Compounds in Drinking Water, EPA/600/4-88/039 (revised July 1991; Final Rule for determination of compliance with the MCL for Total Trihalomethanes under 141.30, in 40 CFR Part 141, Vol. 58, No. 147, Fed. Reg., Tuesday Aug. 3, 1993). U.S. EPA Environmental Monitoring Systems Laboratory, Cincinnati, OH,

45268, U.S.A. Available from the National Technical Information Service (NTIS), 5285 Port Royal Road, Springfield, VA 22161; Tel. 800-553-6847. NTIS Order Number is PB91-231480.

1,2,4-Trichlorobenzene   EPA Method 625
CAS #120-82-1

TITLE   Base/Neutrals and Acids, U.S. EPA Method 625

MATRIX   This methods covers municipal and industrial wastewaters.

METHOD SUMMARY   Approximately 1 L of sample is serially extracted with methylene chloride at a pH greater than 11 and again at a pH less than 2 using a separatory funnel or a continuous extractor. The methylene chloride extract is dried, concentrated to a volume of 1 mL, and analyzed by GC/MS. Qualitative identification of the parameters in the extract is performed using the retention time and the relative abundance of three characteristic masses (m/z). Qualitative analysis is performed using either external or internal standard techniques with a single characteristic m/z.

INTERFERENCES   Method interferences may be caused by contaminants in solvents, reagents, glassware, and other sample processing hardware. Glassware must be scrupulously cleaned. Glassware should be heated in a muffle furnace at 400°C for 5 to 30 min. Some thermally stable materials, such as PCBs, may not be eliminated by this treatment. Solvent rinses with acetone and pesticide quality hexane may be substituted for the muffle furnace heating. Matrix interferences may be caused by contaminants that are coextracted from the sample. The base-neutral extraction may cause significantly reduced recovery of phenols. The packed gas chromatographic columns recommended for the basic fraction may not exhibit sufficient resolution for some analytes.

INSTRUMENTATION   A GC/MS system with an injection port designed for on-column injection when using packed columns and for splitless injection when using capillary columns.

Column for base/neutrals: 1.8 m long × 2 mm I.D. glass, packed with 3% SP-2550 on Supelcoport (100/120 mesh) or equivalent.

Column for acids: 1.8 m long × 2 mm I.D. glass, packed with 1% SP-1240DA on Supelcoport (100/120 mesh) or equivalent.

PRECISION & ACCURACY   The MDL concentrations were obtained using reagent water. The MDL actually achieved in a given analysis will vary depending on instrument sensitivity and matrix effects. This method was tested by 15 laboratories using reagent water, drinking water, surface water, and industrial wastewaters spiked at six concentrations over the range 5 to 100 µg/L. Single operator precision, overall precision, and method accuracy were found to be directly related to the concentration of the parameter matrix.

The MDL (in µg/L) in reagent water was not reported.
The standard deviation (in µg/L based on 4 recovery measurements) was 26.1.
The range (in µg/L) for average recovery for 4 measurements was 57.3–129.2.
The range (in %) for percent recovery was 44–142.
Accuracy (in µg/L) as expected recovery for one or more measurements of a sample containing a true concentration of C was 0.94C–0.79.
Precision (in µg/L) as expected single analyst standard deviation of measurements at an average concentration found at X was 0.15X + 0.85.
Overall precision (in µg/L) as expected interlaboratory standard deviation of measurements in an average concentration found at X was 0.21X + 0.38.

C = *True value of the concentration in µg/L.*
X = *Average recovery found for measurements of samples containing a concentration at C in µg/L.*

SAMPLE PREPARATION   Adjust the pH to >11 with sodium hydroxide and serially extract in a separatory funnel with methylene chloride or else in a continuous extractor. Next, adjust the pH to <2 with sulfuric acid and serially extract in a separatory funnel with methylene chloride or else in a continuous extractor. Dry the extracts separately through a column of anhydrous sodium sulfate and then concentrate each of the extracts to 1.0 mL using a K-D apparatus.

SAMPLE COLLECTION, PRESERVATION & HANDLING   Grab samples must be collected in glass containers. All samples must be refrigerated at 4°C from the time of collection until extraction. If residual chlorine is present, add 80 mg of sodium thiosulfate/L of sample and mix well. All samples must be extracted within 7 days of collection and completely analyzed within 40 days of extraction.

QUALITY CONTROL   Make an initial, one-time, demonstration of the ability to generate acceptable accuracy and precision with this method. Before processing any samples, the analyst must analyze a reagent water blank to demonstrate that interferences from the analytical system and glassware are under control. Each time a set of samples is extracted or reagents are changed, a reagent water blank must be processed. Spike and analyze a minimum of 5% of all samples to monitor and evaluate lab data quality. A QC check sample concentrate that contains each parameter of interest at a concentration of 100 µg/mL in acetone is required. PCBs and multicomponent pesticides may be omitted from this test.

After analysis of five spiked wastewater samples, calculate the average percent recovery and the standard deviation of the percent recovery. Spike all samples with the surrogate standard spiking solution and calculate the percent recovery of each surrogate compound.

REFERENCE   Federal Register, Vol. 49, No. 209. Friday, Oct. 26, 1984.

1,2,3-Trichlorobenzene   EPA Method 8021
CAS #87-61-6

TITLE   Halogenated Volatile by Gas Chromatography Using Photoionization and Electrolytic Conductivity Detectors in Series: Capillary Column Technique

MATRIX This method is applicable to nearly all types of samples, regardless of water content, including groundwater, aqueous sludges, caustic liquors, acid liquors, waste solvents, oily wastes, mousses, tars, fibrous wastes, polymeric emulsions, filter cakes, spent carbons, spent catalysts, soils, and sediments.

METHOD SUMMARY This method is used to determine 60 volatile organic compounds in a variety of solid waste matrices. It provides GC conditions for the detection of halogenated and aromatic volatile organic compounds. Samples can be analyzed using direct injection or purge-and-trap (EPA Method 5030). Groundwater samples must be analyzed using EPA Method 5030 (where applicable). A temperature program is used with the GC. Detection is achieved by a photoionization detector (PID) and a Hall electrolytic conductivity detector (HECD) in series.

INTERFERENCES Samples can be contaminated by diffusion of volatile organics (particularly chlorofluorocarbons and methylene chloride) through the sample container septum during shipment and storage.

INSTRUMENTATION A GC-equipped with variable-constant differential flow controllers, subambient oven controller, PID and HECD detectors connected with a short piece of uncoated capillary tubing and a data system.

Column: 60 m × 0.75 mm I.D. VOCOL wide-bore capillary column with 1.5 μm film thickness.

PRECISION & ACCURACY MDLs are compound-dependent and vary with purging efficiency and concentration. The applicable concentration range of this method is compound- and instrument-dependent but is approximately 0.1 to 200 μg/L. Analytes that are inefficiently purged from water will not be detected when present at low concentrations, but they can be measured with acceptable accuracy and precision when present in sufficient amounts. The estimated quantitation limit (EQL) for an individual compound is approximately 1 μg/kg (wet weight) for soil/sediment samples, 100 μg/kg (wet weight) for wastes, and 1 μg/L for groundwater. EQLs will be proportionately higher for sample extracts and samples that require dilution or reduced sample size to avoid saturation of the detector.

MULTIPLICATION FACTORS FOR OTHER MATRICES (a)

Matrix	Factor (b)
Groundwater	10
Low-concentration soil	10
Water miscible liquid waste	500
High-concentration soil and sludge	1250
Non-water miscible waste	1250

(a) Sample EQLs are highly matrix-dependent. The EQLs listed herein are provided for guidance and may not always be achievable. (b) EQL = [Method detection limit] × [Factor]. For non-aqueous samples, the factor is on a wet-weight basis.

SINGLE LABORATORY ACCURACY & PRECISION DATA FOR VOCs IN WATER
This method was tested in a single lab using water spiked at 10 μg/L and the following data was reported:

Recoveries and standard deviations were determined from seven samples and spiked at 10 μg/L of each analyte. Recoveries were determined by the internal standard method. Internal standards were: Fluorobenzene for PID and 2-Bromo-1-chloropropane for HECD.

The average recovery (in percent) for the PID was 106.
The standard deviation of the recovery for the PID was 1.9.
The MDL (in μg/mL) for the PID was not determined.
The average recovery (in percent) for the HECD was 98.
The standard deviation of the recovery for the HECD was 3.1.
The MDL (in μg/mL) for the HECD was 0.03.

SAMPLE COLLECTION, PRESERVATION & HANDLING
Volatile organics — Standard 40-mL glass screw-cap VOA vials with Teflon®-faced silicone septum may be used for both liquid and solid matrices. When collecting samples, liquids and solids should be introduced into the vials gently to reduce agitation which might drive off volatile compounds. If there are any air bubbles present the sample must be retaken. Tap slightly as they are filled to try and eliminate as much free air space as possible. The two vials from each sampling locations should be sealed in separate plastic bags to prevent cross-contamination between samples particularly if the sampled waste is suspected of containing high levels of volatile organics.

Semivolatile organics — Containers used to collect samples for the determination of semivolatile organic compounds should be soap and water washed followed by methanol (or isopropanol) rinsing. The sample containers should be of glass or Teflon® and have screw-top covers with Teflon® liners.

Preservation for volatile organics — No preservation is used with concentrated waste samples. With liquid samples containing no residual chlorine, 4 drops of concentrated hydrochloric acid are added and the samples are immediately cooled to 4°C. When liquid samples contain residual chlorine, they are treated as above and, in addition, 4 drops of 4% aqueous sodium thiosulfate are added. Soil, sediment, and sludge samples are only cooled to 4°C.

Preservation for semivolatile organics — No preservation is used with concentrated waste samples. With liquid samples containing no residual chlorine and with soil, sediment, and sludge samples, immediately cooling to 4°C is the only preservation used. When residual chlorine is present then 3 mL of 10% aqueous sodium sulfate is added for each gallon of sample collected, followed by cooling to 4°C.

MHT The holding time for all volatile organics samples is 14 days. Liquid samples must be extracted within 7 days and their extracts analyzed within 40 days. Concentrated waste, soil, sediment, and sludge samples must be extracted within 14 days and their extracts analyzed within 40 days.

SAMPLE PREPARATION Volatile compounds are introduced into the gas chromatograph either by direct injector or purge-and-trap (EPA Method 5030). EPA Method 5030 may be used directly on groundwater samples or low-concentration contaminated soils and sediments. For medium-concentration soils or sediments, methanolic extraction, as described in EPA Method 5030, may be necessary prior to purge-and-trap analysis.

QUALITY CONTROL Calculate surrogate standard recovery on all samples, blanks, and spikes. A trip blank is recommended to check on sampling, storage, and handling contamination. Calibration standards, at a minimum of five concentration levels, are prepared in organic-free reagent water. One of the concentration levels should be at a concentration near, but above, the method detection limit.

A combination of bromochloromethane, 2-bromo-1-chloropropane, 1,4-dichlorobutane, and bromochlorobenzene are recommended as surrogate standards to encompass the range of the temperature program used in this method.

REFERENCE Test Methods for Evaluating Solid Waste, Physical/Chemical Methods, SW-846, 3rd Edition, U.S. EPA, Office of Solid Waste, Washington, DC, EPA Method 8021A, Rev. 1, Nov. 1992.

1,2,4-Trichlorobenzene   **EPA Method 8021**
CAS #120-82-1

TITLE Halogenated Volatile by Gas Chromatography Using Photoionization and Electrolytic Conductivity Detectors in Series: Capillary Column Technique

MATRIX This method is applicable to nearly all types of samples, regardless of water content, including groundwater, aqueous sludges, caustic liquors, acid liquors, waste solvents, oily wastes, mousses, tars, fibrous wastes, polymeric emulsions, filter cakes, spent carbons, spent catalysts, soils, and sediments.

METHOD SUMMARY This method is used to determine 60 volatile organic compounds in a variety of solid waste matrices. It provides GC conditions for the detection of halogenated and aromatic volatile organic compounds. Samples can be analyzed using direct injection or purge-and-trap (EPA Method 5030). Groundwater samples must be analyzed using EPA Method 5030 (where applicable). A temperature program is used with the GC. Detection is achieved by a photoionization detector (PID) and a Hall electrolytic conductivity detector (HECD) in series.

INTERFERENCES Samples can be contaminated by diffusion of volatile organics (particularly chlorofluorocarbons and methylene chloride) through the sample container septum during shipment and storage.

INSTRUMENTATION A GC-equipped with variable-constant differential flow controllers, subambient oven controller, PID and HECD detectors connected with a short piece of uncoated capillary tubing and a data system.

Column: 60 m × 0.75 mm I.D. VOCOL wide-bore capillary column with 1.5 µm film thickness.

PRECISION & ACCURACY MDLs are compound-dependent and vary with purging efficiency and concentration. The applicable concentration range of this method is compound- and instrument-dependent but is approximately 0.1 to 200 µg/L. Analytes that are inefficiently purged from water will not be detected when present at low concentrations, but they can be measured with acceptable accuracy and precision when present in sufficient amounts. The estimated quantitation limit (EQL) for an individual compound is approximately 1 µg/kg (wet weight) for soil/sediment samples, 100 µg/kg (wet weight) for wastes, and 1 µg/L for groundwater. EQLs will be proportionately higher for sample extracts and samples that require dilution or reduced sample size to avoid saturation of the detector.

MULTIPLICATION FACTORS FOR OTHER MATRICES (a)

Matrix	Factor (b)
Groundwater	10
Low-concentration soil	10
Water miscible liquid waste	500
High-concentration soil and sludge	1250
Non-water miscible waste	1250

(a) Sample EQLs are highly matrix-dependent. The EQLs listed herein are provided for guidance and may not always be achievable. (b) EQL = [Method detection limit] × [Factor]. For non-aqueous samples, the factor is on a wet-weight basis.

SINGLE LABORATORY ACCURACY & PRECISION DATA FOR VOCs IN WATER
This method was tested in a single lab using water spiked at 10 µg/L and the following data was reported:

Recoveries and standard deviations were determined from seven samples and spiked at 10 µg/L of each analyte. Recoveries were determined by the internal standard method. Internal standards were: Fluorobenzene for PID and 2-Bromo-1-chloropropane for HECD.

The average recovery (in percent) for the PID was 104.
The standard deviation of the recovery for the PID was 2.2.
The MDL (in µg/mL) for the PID was 0.02.
The average recovery (in percent) for the HECD was 102.
The standard deviation of the recovery for the HECD was 2.1.
The MDL (in µg/mL) for the HECD was 0.03.

SAMPLE COLLECTION, PRESERVATION & HANDLING
Volatile organics — Standard 40-mL glass screw-cap VOA vials with Teflon®-faced silicone septum may be used for both liquid and solid matrices. When collecting samples, liquids and solids should be introduced into the vials gently to reduce agitation which might drive off volatile compounds. If there are any air bubbles present the sample must be retaken. Tap slightly as they are filled to try and eliminate as much free air space as possible. The two vials from each sampling locations should be sealed in separate plastic bags to prevent cross-contamination between samples particularly if the sampled waste is suspected of containing high levels of volatile organics.

Semivolatile organics — Containers used to collect samples for the determination of semivolatile organic compounds should be soap and water washed followed by methanol (or isopropanol) rinsing. The sample containers should be of glass or Teflon® and have screw-top covers with Teflon® liners.

Preservation for volatile organics — No preservation is used with concentrated waste samples. With liquid samples containing no residual chlorine, 4 drops of concentrated hydrochloric acid are added and the samples are immediately cooled to 4°C. When liquid samples contain residual chlorine, they are treated as above and, in addition, 4 drops of 4% aqueous sodium thiosulfate are added. Soil, sediment, and sludge samples are only cooled to 4°C.

Preservation for semivolatile organics — No preservation is used with concentrated waste samples. With liquid samples containing no residual chlorine and with soil, sediment, and sludge samples, immediately cooling to 4°C is the only preservation used. When residual chlorine is present then 3 mL of 10% aqueous sodium sulfate is added for each gallon of sample collected, followed by cooling to 4°C.

MHT The holding time for all volatile organics samples is 14 days. Liquid samples must be extracted within 7 days and their extracts analyzed within 40 days. Concentrated waste, soil, sediment, and sludge samples must be extracted within 14 days and their extracts analyzed within 40 days.

SAMPLE PREPARATION Volatile compounds are introduced into the gas chromatograph either by direct injector or purge-and-trap (EPA Method 5030). EPA Method 5030 may be used directly on groundwater samples or low-concentration contaminated soils and sediments. For medium-concentration soils or sediments, methanolic extraction, as described in EPA Method 5030, may be necessary prior to purge-and-trap analysis.

QUALITY CONTROL Calculate surrogate standard recovery on all samples, blanks, and spikes. A trip blank is recommended to check on sampling, storage, and handling contamination. Calibration standards, at a minimum of five concentration levels, are prepared in organic-free reagent water. One of the concentration levels should be at a concentration near, but above, the method detection limit.

A combination of bromochloromethane, 2-bromo-1-chloropropane, 1,4-dichlorobutane, and bromochlorobenzene are recommended as surrogate standards to encompass the range of the temperature program used in this method.

REFERENCE Test Methods for Evaluating Solid Waste, Physical/Chemical Methods, SW-846, 3rd Edition, U.S. EPA, Office of Solid Waste, Washington, DC, EPA Method 8021A, Rev. 1, Nov. 1992.

1,2,4-Trichlorobenzene   **EPA Method 8120**
CAS #120-82-1

TITLE Chlorinated Hydrocarbons by Gas Chromatography

MATRIX This method covers aqueous and solid matrices. This includes a wide variety such as drinking water, groundwater, industrial wastewaters, surface waters, soils, solids, and sediments.

METHOD SUMMARY This method is used to determine the concentration of 14 chlorinated hydrocarbons. It provides gas chromatographic conditions for the detection of ppb concentrations of certain chlorinated hydrocarbons. Prior to use of this method, appropriate sample extraction techniques must be used. Both neat and diluted organic liquids (EPA Method 3580, Waste Dilution) may be analyzed by direct injection. A 2 to 5 µg/mL aliquot of the extract is injected into a gas chromatograph (GC) using the solvent flush technique, and compounds in the GC effluent are detected by an electron capture detector (ECD).

INTERFERENCES Solvents, reagents, glassware, and other sample processing hardware may yield discrete artifacts and/or elevated baselines causing misinterpretation of gas chromatograms. Interferences coextracted from samples will vary considerably from source to source, depending upon the waste being sampled.

INSTRUMENTATION An analytical system complete with GC suitable for on-column injections and accessories, including detectors, column supplies, recorder, gases and syringes is required. A data system for measuring peak areas and/or peak heights is recommended. The GC is equipped with an electron capture detector (ECD). A K-D apparatus is needed for sample preparation.

Column 1: 1.8 m × 2 mm I.D. glass column packed with 1% SP-1000 on Supelcoport (100/120 mesh) or equivalent.

Column 2: 1.8 m × 2 mm I.D. glass column packed with 1.5% OV-1/2.4% OV-225 on Supelcoport (80/100 mesh) or equivalent.

PRECISION & ACCURACY The method was tested by 20 laboratories using organic-free reagent water, drinking water, surface water, and three industrial wastewaters spiked at six concentrations over the range 1.0 to 356 µg/L. Single operator precision, overall precision, and method accuracy were found to be directly related to the concentration of the parameter and essentially independent of the sample matrix.

MULTIPLICATION FACTORS FOR OTHER MATRICES (a)

Matrix	Factor (b)
Groundwater	10
Low-concentration soil by ultrasonic extraction with GPC cleanup	670
High-concentration soil and sludges by ultrasonic extraction	10,000
Waste not miscible with water	100,000

(a) Sample EQLs are highly matrix-dependent. The EQLs listed herein are provided for guidance and may not always be achievable.
(b) EQL = [Method detection limit] × [Factor]. For nonaqueous samples, the factor is on a wet-weight basis.

PRECISION & ACCURACY The estimates below are based upon the performance in a single lab.

The accuracy (in µg/L) as expected recovery for one or more measurements of a sample containing a concentration of C was 0.76C + 0.98.

The precision (in µg/L) as expected single analyst standard deviation of measurements at an average concentration of x" was 0.23 x"-0.44.

The precision (in µg/L) as expected interlaboratory standard deviation measurements at an average concentration found of x" was 0.40 x"-1.37.

$C =$ *True value for the concentration, in µg/L.*
$x" =$ *Average recovery found for measurements of samples containing a concentration of C, in µg/L.*

SAMPLE COLLECTION, PRESERVATION & HANDLING
Extracts must be stored under refrigeration at 4°C and analyzed within 40 days of extraction.

SAMPLE PREPARATION In general, water samples are extracted at a neutral, or as is, pH with methylene chloride using either EPA Method 3510 or EPA Method 3520. Solid samples are extracted using either EPA Method 3540 or EPA Method 3550. Prior to gas chromatographic analysis, the extraction solvent must be exchanged to hexane.

QUALITY CONTROL The quality control check concentrate (EPA Method 8000) should contain each parameter of interest in acetone at the following concentrations: hexachloro-substituted hydrocarbon, 10 µg/mL; and any other chlorinated hydrocarbon, 100 µg/mL. Calculate surrogate standard recovery on all samples, blanks, and spikes.

Prepare stock standard solutions in isooctane or hexane. Calibration standards at a minimum of five concentrations should be prepared through dilution of the stock standards with isooctane or hexane. Internal standards and surrogate standards are also needed.

REFERENCE Test Methods for Evaluating Solid Waste, Physical/Chemical Methods, SW-846, 3rd Edition, U.S. EPA, Office of Solid Waste, Washington, DC, 1990. EPA Method 8120 A Rev. 1, Nov. 1990.

1,2,3-Trichlorobenzene **EPA Method 8121**
CAS #87-61-6

TITLE Chlorinated Hydrocarbons by GC: Capillary Column Technique

MATRIX This method covers aqueous and solid matrices. This includes a wide variety such as drinking water, groundwater, industrial wastewaters, surface waters, soils, solids, and sediments.

METHOD SUMMARY This method provides procedures for the determination of 22 chlorinated hydrocarbons in water, soil/sediment, and waste matrices. A measured volume or weight of sample is extracted by using one of the appropriate sample extraction techniques specified in EPA Method 3510, EPA Method 3520, EPA Method 3540, or EPA Method 3550, or diluted using EPA Method 3580. Aqueous samples are extracted at neutral pH with methylene chloride by using either a separatory funnel (EPA Method 3510) or a continuous liquid-liquid extractor (EPA Method 3520). Solid samples are extracted with hexane/acetone (1:1) by using a Soxhlet extractor (EPA Method 3540) or with methylene chloride/acetone (1:1) by using an ultrasonic extractor (EPA Method 3550). After cleanup, the extract or diluted sample is analyzed by gas chromatography with electron capture detection (GC/ECD).

The sensitivity level of this method usually depends on the level of interferences rather than on instrumental limitations. This method may be used in conjunction with EPA Method 3620, Florisil Column Cleanup, EPA Method 3660, Sulfur Cleanup, and EPA Method 3640, Gel Permeation Chromatography, to aid in the elimination of interferences.

INTERFERENCES Solvents, reagents, glassware, and other hardware used in sample processing may introduce artifacts which may result in elevated baselines, causing misinterpretation of gas chromatograms. Interferants coextracted from the samples will vary considerably from waste to waste. Glassware must be scrupulously clean. Phthalate esters, if present in a sample, will interfere only with the BHC isomers. The presence of elemental sulfur will result in large peaks, and can often mask the region of compounds eluting after 1,2,4,5-tetrachlorobenzene. The tetrabutylammonium (TBA)-sulfite procedure (EPA Method 3660) works well for the removal of elemental sulfur. Waxes and lipids can be removed by gel permeation chromatography (EPA Method 3640).

INSTRUMENTATION A GC suitable for on-column injections and all required accessories, including and electron capture detector (ECD), analytical columns, recorder, gases, and syringes are needed. A data system for measuring peak heights and/or peak areas is recommended. A Kuderna-Danish (K-D) apparatus will also be needed to concentrate extracts.

Column 1: 30 m × 0.53 mm I.D. fused-silica capillary column chemically bonded with trifluoropropyl methyl silicone (DB-210 or equivalent).
Column 2: 30 m × 0.53 mm I.D. fused-silica capillary column chemically bonded with polyethylene glycol (DB-WAX or equivalent).

PRECISION & ACCURACY This method has been tested in a single lab by using organic-free reagent water, sandy loam samples, and extracts which were spiked with the test compounds at one concentration. Single-operator precision and method accuracy were found to be related to the concentration of compound and the type of matrix. The accuracy and precision technique will be determined by the sample matrix, sample preparation technique, optional cleanup techniques, and calibration procedures used.

MULTIPLICATION FACTORS FOR OTHER MATRICES (a)

Matrix	Factor (b)
Groundwater	10
Low-concentration soil by ultrasonic extraction with GPC cleanup	670
High-concentration soil and sludges by ultrasonic extraction	10,000
Waste not miscible with water	100,000

(a) Sample EQLs are highly matrix-dependent. The EQLs listed herein are provided for guidance and may not always be achievable. (b) EQL = [Method detection limit] × [Factor]. For nonaqueous samples, the factor is on a wet-weight basis.

PRECISION & ACCURACY MDL is the method detection limit for organic-free reagent water. MDL was determined from the analysis of eight replicate aliquots processed through the entire analytical method (extraction, Florisil cartridge cleanup, and GC/ECD analysis).

The MDL (in ng/L) was 39.

The accuracy (as average % recovery using 5 determinations and no Florisil cleanup) from a spike concentration of 10 µg/L and separatory funnel extraction was 95% with a final volume of 10 mL.

The precision (as RSD% using 5 determinations and no Florisil cleanup) from a spike concentration of 10 µg/L and separatory funnel extraction was 4.4% with a final volume of 10 mL.

The accuracy (as average % recovery using 5 determinations and no Florisil cleanup), from a spike concentration of 3300µg/L and ultrasonic extraction of solid samples using 1:1 methylene chloride and acetone, was 79% with a final volume of 10 mL.

The precision (as RSD% using 5 determinations and no Florisil cleanup), from a spike concentration of 3300µg/L and ultrasonic extraction of solid samples using 1:1 methylene chloride and acetone, was 4.3% with a final volume of 10 mL.

SAMPLE COLLECTION, PRESERVATION & HANDLING
Volatile organics — Standard 40-mL glass screw-cap VOA vials with Teflon®-faced silicone septum may be used for both liquid and solid matrices. When collecting samples, liquids and solids should be introduced into the vials gently to reduce agitation which might drive off volatile compounds. If there are any air bubbles present the sample must be retaken. The vials with solids should be tapped slightly as they are filled to try and eliminate as much free air space as possible. Two vials from each sampling location should be sealed in separate plastic bags to prevent cross-contamination between samples.

Semivolatile organics — Containers used to collect samples for the determination of semivolatile organic compounds should be soap and water washed followed by methanol (or isopropanol) rinsing. The sample containers should be of glass or Teflon® and have screw-top covers with Teflon® liners.

Preservation for volatile organics — No preservation is used with concentrated waste samples. With liquid samples containing no residual chlorine, 4 drops of concentrated hydrochloric acid are added and the samples are immediately cooled to 4°C. When liquid samples contain residual chlorine, they are treated as above and, in addition, 4 drops of 4% aqueous sodium thiosulfate are added to remove the residual chlorine. Soil, sediment, and sludge samples are only cooled to 4°C.

Preservation for semivolatile organics — No preservation is used with concentrated waste samples. With liquid samples containing no residual chlorine and with soil, sediment, and sludge samples, immediately cooling to 4°C is the only preservation used. When residual chlorine is present then 3 mL of 10% aqueous sodium sulfate is added for each gallon of sample collected, followed by cooling to 4°C.

Holding times — The holding time for all volatile organics samples is 14 days. Liquid samples must be extracted within 7 days and their extracts analyzed within 40 days. Concentrated waste, soil, sediment, and sludge samples must be extracted within 14 days and their extracts analyzed within 40 days.

SAMPLE PREPARATION Prepare stock standard solutions in hexane. Calibration standards at a minimum of five concentrations should be prepared through dilution of the stock standards with hexane. The suggested internal standards are: 2,5-dibromotoluene, 1,3,5-tribromobenzene, and α, α-dibromo-m-xylene. The analyst can use any of the three compounds provided that they are resolved from matrix interferences. Recommended surrogate compounds are α-2,6-trichlorotoluene, 1,4-dichloronaphthalene, and 2,3,4,5,6-pentachlorotoluene.

In general, water samples are extracted at a neutral pH with methylene chloride using a separatory funnel (EPA Method 3510) or a continuous liquid-liquid extractor (EPA Method 3520). Solid samples are extracted with hexane/acetone (1:1 v:v) using a Soxhlet extractor (EPA Method 3540) or with methylene chloride/acetone (1:1 v:v) using an ultrasonic extractor (EPA Method 3550). Non-aqueous waste samples may be diluted using EPA Method 3580. Prior to Florisil cleanup or gas chromatographic analysis, the extraction solvent must be exchanged to hexane. Sample extracts that will be subjected to gel permeation chromatography do not need solvent exchange.

Cleanup procedures may not be necessary for a relatively clean matrix. If removal of interferences such as chlorinated phenols, phthalate esters, etc., is required, proceed with the procedure outlined in EPA Method 3620.

QUALITY CONTROL Analyze a quality control check standard to demonstrate that the operation of the GC is in control. The frequency of the check standard analysis is equivalent to 10% of the samples analyzed. If the recovery of any compound found in the check standard is less than 80% of the certified value, the problem must be corrected and a new set of calibration standards must be prepared and analyzed. Calculate surrogate standard recoveries for all samples, blanks, and spikes. An internal standard peak area check must be performed on all samples. The internal standard must be evaluated for acceptance by determining whether the measured area for the internal standard deviates by more than 30% from the average area for the internal standard in the calibration standards. When the internal standard peak area is outside that limit, all samples that fall outside the QC criteria must be reanalyzed. Any compound confirmed by two columns may also be confirmed by GC/MS (EPA Method 8270). The GC/MS would normally require a minimum concentration of 1 ng/µL in the final extract for each compound. Include a mid-concentration calibration standard after each group of 20 samples in the analysis sequence. The response factors for the mid-concentration calibration must be within 15% of the average values for the multiconcentration calibration.

REFERENCE Test Methods for Evaluating Solid Waste, Physical/Chemical Methods, SW-846, 3rd Edition, U.S. EPA, Office of Solid Waste, Washington, DC, 1990. EPA Method 8121, Rev. 0, Nov. 1990.

1,2,4-Trichlorobenzene   **EPA Method 8121**
CAS #120-82-1

TITLE Chlorinated Hydrocarbons by GC: Capillary Column Technique

MATRIX This method covers aqueous and solid matrices. This includes a wide variety such as drinking water, groundwater, industrial wastewaters, surface waters, soils, solids, and sediments.

METHOD SUMMARY This method provides procedures for the determination of 22 chlorinated hydrocarbons in water, soil/sediment, and waste matrices. A measured volume or weight of sample is extracted by using one of the appropriate sample extraction techniques specified in EPA Method 3510, EPA Method 3520, EPA Method 3540, or EPA Method 3550, or diluted using EPA Method 3580. Aqueous samples are extracted at neutral pH with methylene chloride by using either a separatory funnel (EPA Method 3510) or a continuous liquid-liquid extractor (EPA Method 3520). Solid samples are extracted with hexane/acetone (1:1) by using a Soxhlet extractor (EPA Method 3540) or with methylene chloride/acetone (1:1) by using an ultrasonic extractor (EPA Method 3550). After cleanup, the extract or diluted sample is analyzed by gas chromatography with electron capture detection (GC/ECD).

The sensitivity level of this method usually depends on the level of interferences rather than on instrumental limitations. This method may be used in conjunction with EPA Method 3620, Florisil Column Cleanup, EPA Method 3660, Sulfur Cleanup, and EPA Method 3640, Gel Permeation Chromatography, to aid in the elimination of interferences.

INTERFERENCES Solvents, reagents, glassware, and other hardware used in sample processing may introduce artifacts which may result in elevated baselines, causing misinterpretation of gas chromatograms. Interferants coextracted from the samples will vary considerably from waste to waste. Glassware must be scrupulously clean. Phthalate esters, if present in a sample, will interfere only with the BHC isomers. The presence of elemental sulfur will result in large peaks, and can often mask the region of compounds eluting after 1,2,4,5-tetrachlorobenzene. The tetrabutylammonium (TBA)-sulfite procedure (EPA Method 3660) works well for the removal of elemental sulfur. Waxes and lipids can be removed by gel permeation chromatography (EPA Method 3640).

INSTRUMENTATION A GC suitable for on-column injections and all required accessories, including and electron capture detector (ECD), analytical columns, recorder, gases, and syringes are needed. A data system for measuring peak heights and/or peak areas is recommended. A Kuderna-Danish (K-D) apparatus will also be needed to concentrate extracts.

Column 1: 30 m × 0.53 mm I.D. fused-silica capillary column chemically bonded with trifluoropropyl methyl silicone (DB-210 or equivalent).
Column 2: 30 m × 0.53 mm I.D. fused-silica capillary column chemically bonded with polyethylene glycol (DB-WAX or equivalent).

PRECISION & ACCURACY This method has been tested in a single lab by using organic-free reagent water, sandy loam samples, and extracts which were spiked with the test compounds at one concentration. Single-operator precision and method accuracy were found to be related to the concentration of compound and the type of matrix. The accuracy and precision technique will be determined by the sample matrix, sample preparation technique, optional cleanup techniques, and calibration procedures used.

MULTIPLICATION FACTORS FOR OTHER MATRICES (a)

Matrix	Factor (b)
Groundwater	10
Low-concentration soil by ultrasonic extraction with GPC cleanup	670
High-concentration soil and sludges by ultrasonic extraction	10,000
Waste not miscible with water	100,000

(a) Sample EQLs are highly matrix-dependent. The EQLs listed herein are provided for guidance and may not always be achievable. (b) EQL = [Method detection limit] × [Factor]. For nonaqueous samples, the factor is on a wet-weight basis.

PRECISION & ACCURACY MDL is the method detection limit for organic-free reagent water. MDL was determined from the analysis of eight replicate aliquots processed through the entire analytical method (extraction, Florisil cartridge cleanup, and GC/ECD analysis).

The MDL (in ng/L) was 130.

The accuracy (as average % recovery using 5 determinations and no Florisil cleanup) from a spike concentration of 10 µg/L and separatory funnel extraction was 95% with a final volume of 10 mL.

The precision (as RSD% using 5 determinations and no Florisil cleanup) from a spike concentration of 10 µg/L and separatory funnel extraction was 3.0% with a final volume of 10 mL.

The accuracy (as average % recovery using 5 determinations and no Florisil cleanup), from a spike concentration of 3300µg/L and ultrasonic extraction of solid samples using 1:1 methylene chloride and acetone, was 89% with a final volume of 10 mL.

The precision (as RSD% using 5 determinations and no Florisil cleanup), from a spike concentration of 3300µg/L and ultrasonic extraction of solid samples using 1:1 methylene chloride and acetone, was 2.7% with a final volume of 10 mL.

SAMPLE COLLECTION, PRESERVATION & HANDLING
Volatile organics — Standard 40-mL glass screw-cap VOA vials with Teflon®-faced silicone septum may be used for both liquid and solid matrices. When collecting samples, liquids and solids should be introduced into the vials gently to reduce agitation which might drive off volatile compounds. If there are any air bubbles present the sample must be retaken. The vials with solids should be tapped slightly as they are filled to try and eliminate as much free air space as possible. Two vials from each sampling location should be sealed in separate plastic bags to prevent cross-contamination between samples.

Semivolatile organics — Containers used to collect samples for the determination of semivolatile organic compounds should be soap and water washed followed by methanol (or isopropanol) rinsing. The sample containers should be of glass or Teflon® and have screw-top covers with Teflon® liners.

Preservation for volatile organics — No preservation is used with concentrated waste samples. With liquid samples containing no residual chlorine, 4 drops of concentrated hydrochloric acid are added and the samples are immediately cooled to 4°C. When liquid samples contain residual chlorine, they are treated as above and, in addition, 4 drops of 4% aqueous sodium thiosulfate are added to remove the residual chlorine. Soil, sediment, and sludge samples are only cooled to 4°C.

Preservation for semivolatile organics — No preservation is used with concentrated waste samples. With liquid samples containing no residual chlorine and with soil, sediment, and sludge samples, immediately cooling to 4°C is the only preservation used. When residual chlorine is present then 3 mL of 10% aqueous sodium sulfate is added for each gallon of sample collected, followed by cooling to 4°C.

Holding times — The holding time for all volatile organics samples is 14 days. Liquid samples must be extracted within 7 days and their extracts analyzed within 40 days. Concentrated waste, soil, sediment, and sludge samples must be extracted within 14 days and their extracts analyzed within 40 days.

SAMPLE PREPARATION Prepare stock standard solutions in hexane. Calibration standards at a minimum of five concentrations should be prepared through dilution of the stock standards with hexane. The suggested internal standards are: 2,5-dibromotoluene, 1,3,5-tribromobenzene, and α, α-dibromo-m-xylene. The analyst can use any of the three compounds provided that they are resolved from matrix interferences. Recommended surrogate compounds are α-2,6-trichlorotoluene, 1,4-dichloronaphthalene, and 2,3,4,5,6-pentachlorotoluene.

In general, water samples are extracted at a neutral pH with methylene chloride using a separatory funnel (EPA Method 3510) or a continuous liquid-liquid extractor (EPA Method 3520). Solid samples are extracted with hexane/acetone (1:1 v:v) using a Soxhlet extractor (EPA Method 3540) or with methylene chloride/acetone (1:1 v:v) using an ultrasonic extractor (EPA Method 3550). Non-aqueous waste samples may be diluted using EPA Method 3580. Prior to Florisil cleanup or gas chromatographic analysis, the extraction solvent must be exchanged to hexane. Sample extracts that will be subjected to gel permeation chromatography do not need solvent exchange.

Cleanup procedures may not be necessary for a relatively clean matrix. If removal of interferences such as chlorinated phenols, phthalate esters, etc., is required, proceed with the procedure outlined in EPA Method 3620.

QUALITY CONTROL Analyze a quality control check standard to demonstrate that the operation of the GC is in control. The frequency of the check standard analysis is equivalent to 10% of the samples analyzed. If the recovery of any compound found in the check standard is less than 80% of the certified value, the problem must be corrected and a new set of calibration standards must be prepared and analyzed. Calculate surrogate standard recoveries for all samples, blanks, and spikes. An internal standard peak area check must be performed on all samples. The internal standard must be evaluated for acceptance by determining whether the measured area for the internal standard deviates by more than 30% from the average area for the internal standard in the calibration standards. When the internal standard peak area is outside that limit, all samples that fall outside the QC criteria must be reanalyzed. Any compound confirmed by two columns may also be confirmed by GC/MS (EPA Method 8270). The GC/MS would normally require a minimum concentration of 1 ng/µL in the final extract for each compound. Include a mid-concentration calibration standard after each group of 20 samples in the analysis sequence. The response factors for the mid-concentration calibration must be within 15% of the average values for the multiconcentration calibration.

REFERENCE Test Methods for Evaluating Solid Waste, Physical/Chemical Methods, SW-846, 3rd Edition, U.S. EPA, Office of Solid Waste, Washington, DC, 1990. EPA Method 8121, Rev. 0, Nov. 1990.

1,3,5-Trichlorobenzene **EPA Method 8121**
CAS #108-70-3

TITLE Chlorinated Hydrocarbons by GC: Capillary Column Technique

MATRIX This method covers aqueous and solid matrices. This includes a wide variety such as drinking water, groundwater, industrial wastewaters, surface waters, soils, solids, and sediments.

METHOD SUMMARY This method provides procedures for the determination of 22 chlorinated hydrocarbons in water, soil/sediment, and waste matrices. A measured volume or weight of sample is extracted by using one of the appropriate sample extraction techniques specified in EPA Method 3510, EPA Method 3520, EPA Method 3540, or EPA Method 3550, or diluted using EPA Method 3580. Aqueous samples are extracted at neutral pH with methylene chloride by using either a separatory funnel (EPA Method 3510) or a continuous liquid-liquid extractor (EPA Method 3520). Solid samples are extracted with hexane/acetone (1:1) by using a Soxhlet extractor (EPA Method 3540) or with methylene chloride/acetone (1:1) by using an ultrasonic extractor (EPA Method 3550). After cleanup, the extract or diluted sample is analyzed by gas chromatography with electron capture detection (GC/ECD).

The sensitivity level of this method usually depends on the level of interferences rather than on instrumental limitations. This method may be used in conjunction with EPA Method 3620, Florisil Column Cleanup, EPA Method 3660, Sulfur Cleanup, and EPA Method 3640, Gel Permeation Chromatography, to aid in the elimination of interferences.

INTERFERENCES Solvents, reagents, glassware, and other hardware used in sample processing may introduce artifacts which may result in elevated baselines, causing misinterpretation of gas chromatograms. Interferants coextracted from the samples will vary considerably from waste to waste. Glassware must be scrupulously clean. Phthalate esters, if present in a sample, will interfere only with the BHC isomers. The presence of elemental sulfur will result in large peaks, and can often mask the region of compounds eluting after 1,2,4,5-tetrachlorobenzene. The tetrabutylammonium (TBA)-sulfite procedure (EPA Method 3660) works well for the removal of elemental sulfur. Waxes and lipids can be removed by gel permeation chromatography (EPA Method 3640).

INSTRUMENTATION A GC suitable for on-column injections and all required accessories, including and electron capture detector (ECD), analytical columns, recorder, gases, and syringes are needed. A data system for measuring peak heights and/or peak areas is recommended. A Kuderna-Danish (K-D) apparatus will also be needed to concentrate extracts.

Column 1: 30 m × 0.53 mm I.D. fused-silica capillary column chemically bonded with trifluoropropyl methyl silicone (DB-210 or equivalent).
Column 2: 30 m × 0.53 mm I.D. fused-silica capillary column chemically bonded with polyethylene glycol (DB-WAX or equivalent).

PRECISION & ACCURACY This method has been tested in a single lab by using organic-free reagent water, sandy loam samples, and extracts which were spiked with the test compounds at one concentration. Single-operator precision and method accuracy were found to be related to the concentration of compound and the type of matrix. The accuracy and precision technique will be determined by the sample matrix, sample preparation technique, optional cleanup techniques, and calibration procedures used.

MULTIPLICATION FACTORS FOR OTHER MATRICES (a)

Matrix	Factor (b)
Groundwater	10
Low-concentration soil by ultrasonic extraction with GPC cleanup	670
High-concentration soil and sludges by ultrasonic extraction	10,000
Waste not miscible with water	100,000

(a) Sample EQLs are highly matrix-dependent. The EQLs listed herein are provided for guidance and may not always be achievable.
(b) EQL = [Method detection limit] × [Factor]. For nonaqueous samples, the factor is on a wet-weight basis.

PRECISION & ACCURACY MDL is the method detection limit for organic-free reagent water. MDL was determined from the analysis of eight replicate aliquots processed through the entire analytical method (extraction, Florisil cartridge cleanup, and GC/ECD analysis).

The MDL (in ng/L) was 12.

The accuracy (as average % recovery using 5 determinations and no Florisil cleanup) from a spike concentration of 10 µg/L and separatory funnel extraction was 93% with a final volume of 10 mL.

The precision (as RSD% using 5 determinations and no Florisil cleanup) from a spike concentration of 10 µg/L and separatory funnel extraction was 6.2% with a final volume of 10 mL.

The accuracy (as average % recovery using 5 determinations and no Florisil cleanup), from a spike concentration of 3300µg/L and ultrasonic extraction of solid samples using 1:1 methylene chloride and acetone, was 75% with a final volume of 10 mL.

The precision (as RSD% using 5 determinations and no Florisil cleanup), from a spike concentration of 3300µg/L and ultrasonic extraction of solid samples using 1:1 methylene chloride and acetone, was 5.3% with a final volume of 10 mL.

SAMPLE COLLECTION, PRESERVATION & HANDLING
Volatile organics — Standard 40-mL glass screw-cap VOA vials with Teflon®-faced silicone septum may be used for both liquid and solid matrices. When collecting samples, liquids and solids should be introduced into the vials gently to reduce agitation which might drive off volatile compounds. If there are any air bubbles present the sample must be retaken. The vials with solids should be tapped slightly as they are filled to try and eliminate as much free air space as possible. Two vials from each sampling location should be sealed in separate plastic bags to prevent cross-contamination between samples.

Semivolatile organics — Containers used to collect samples for the determination of semivolatile organic compounds should be soap and water washed followed by methanol (or isopropanol) rinsing. The sample containers should be of glass or Teflon® and have screw-top covers with Teflon® liners.

Preservation for volatile organics — No preservation is used with concentrated waste samples. With liquid samples containing no residual chlorine, 4 drops of concentrated hydrochloric acid are added and the samples are immediately cooled to 4°C. When liquid samples contain residual chlorine, they are treated as above and, in addition, 4 drops of 4% aqueous sodium thiosulfate are added to remove the residual chlorine. Soil, sediment, and sludge samples are only cooled to 4°C.

Preservation for semivolatile organics — No preservation is used with concentrated waste samples. With liquid samples containing no residual chlorine and with soil, sediment, and sludge samples, immediately cooling to 4°C is the only preservation used. When residual chlorine is present then 3 mL of 10% aqueous sodium sulfate is added for each gallon of sample collected, followed by cooling to 4°C.

Holding times — The holding time for all volatile organics samples is 14 days. Liquid samples must be extracted within 7 days and their extracts analyzed within 40 days. Concentrated waste, soil, sediment, and sludge samples must be extracted within 14 days and their extracts analyzed within 40 days.

SAMPLE PREPARATION Prepare stock standard solutions in hexane. Calibration standards at a minimum of five concentrations should be prepared through dilution of the stock standards with hexane. The suggested internal standards are: 2,5-dibromotoluene, 1,3,5-tribromobenzene, and α, α-dibromo-m-xylene. The analyst can use any of the three compounds provided that they are resolved from matrix interferences. Recommended surrogate compounds are α-2,6-trichlorotoluene, 1,4-dichloronaphthalene, and 2,3,4,5,6-pentachlorotoluene.

In general, water samples are extracted at a neutral pH with methylene chloride using a separatory funnel (EPA Method 3510) or a continuous liquid-liquid extractor (EPA Method

3520). Solid samples are extracted with hexane/acetone (1:1 v:v) using a Soxhlet extractor (EPA Method 3540) or with methylene chloride/acetone (1:1 v:v) using an ultrasonic extractor (EPA Method 3550). Non-aqueous waste samples may be diluted using EPA Method 3580. Prior to Florisil cleanup or gas chromatographic analysis, the extraction solvent must be exchanged to hexane. Sample extracts that will be subjected to gel permeation chromatography do not need solvent exchange.

Cleanup procedures may not be necessary for a relatively clean matrix. If removal of interferences such as chlorinated phenols, phthalate esters, etc., is required, proceed with the procedure outlined in EPA Method 3620.

QUALITY CONTROL Analyze a quality control check standard to demonstrate that the operation of the GC is in control. The frequency of the check standard analysis is equivalent to 10% of the samples analyzed. If the recovery of any compound found in the check standard is less than 80% of the certified value, the problem must be corrected and a new set of calibration standards must be prepared and analyzed. Calculate surrogate standard recoveries for all samples, blanks, and spikes. An internal standard peak area check must be performed on all samples. The internal standard must be evaluated for acceptance by determining whether the measured area for the internal standard deviates by more than 30% from the average area for the internal standard in the calibration standards. When the internal standard peak area is outside that limit, all samples that fall outside the QC criteria must be reanalyzed. Any compound confirmed by two columns may also be confirmed by GC/MS (EPA Method 8270). The GC/MS would normally require a minimum concentration of 1 ng/µL in the final extract for each compound. Include a mid-concentration calibration standard after each group of 20 samples in the analysis sequence. The response factors for the mid-concentration calibration must be within 15% of the average values for the multiconcentration calibration.

REFERENCE Test Methods for Evaluating Solid Waste, Physical/Chemical Methods, SW-846, 3rd Edition, U.S. EPA, Office of Solid Waste, Washington, DC, 1990. EPA Method 8121, Rev. 0, Nov. 1990.

1,2,3-Trichlorobenzene   **EPA Method 8260**
CAS #87-61-6

TITLE Volatile Organic Compounds by GC/MS: Capillary Column Technique

MATRIX This method is applicable to nearly all types of samples, regardless of water content, including groundwater, soils, and sediments.

METHOD SUMMARY Method 8260A covers 58 volatile organic compounds that are introduced into a gas chromatograph by the purge-and-trap method or by direct injection (in limited applications). Zero-grade helium is bubbled through a 5-mL solution at ambient temperature. Purged sample components are trapped in a tube containing suitable sorbent materials. When purging is complete, the sorbent tube is heated and backflushed with helium to desorb trapped sample components. The analytes are desorbed directly to a large bore capillary or cryofocussed on a capillary precolumn before being flash evaporated to a narrow bore capillary for analysis.

INTERFERENCES Major contaminant sources are volatile materials in the lab and impurities in the inert purging gas and in the sorbent trap. Interfering contamination may occur when a sample containing low concentrations of volatile organic compounds is analyzed immediately after a sample containing high concentrations of volatile organic compounds. After analysis of a sample containing high concentrations of volatile organic compounds, one or more calibration blanks should be analyzed to check for cross-contamination. Screening of the samples prior to purge-and-trap GC/MS analysis is highly recommended to prevent contamination of the system. This is especially true for soil and waste samples.

Special precautions must be taken to analyze for methylene chloride. The analytical and sample storage area should be isolated from all atmospheric sources of methylene chloride. All gas chromatography carrier gas lines and purge gas plumbing should be constructed from stainless steel or copper tubing. Laboratory clothing previously exposed to methylene chloride fumes during liquid-liquid extraction procedures can contribute to sample contamination.

Samples can also be contaminated by diffusion of volatile organics (particularly methylene chloride and fluorocarbons) through the septum seal during shipment and storage. A trip blank can serve as a check on such contamination.

INSTRUMENTATION GC/MS with a temperature-programmable chromatograph suitable for splitless injection equipped with variable constant differential flow controllers, a subambient oven controller, a purging device, sorbent trap, a thermal desorption apparatus and a capillary precolumn interface when using cryogenic cooling will be needed. The following GC columns may be used:

Column 1: 60 m × 0.75mm I.D. capillary column coated with VOCOL, 1.5 µm film thickness.
Column 2: 30 m × 0.53mm capillary column coated with DB-624 or VOCOL, 3 µm film thickness.
Column 3: 30 m × 0.32mm I.D. capillary column coated with DB-5 or SE-54, 1-µm film thickness.

PRECISION & ACCURACY This method has been tested in a single lab using spiked water. Using a wide-bore capillary column, water was spiked at concentrations between 0.5 and 10 µg/L. Single lab accuracy and precision data are presented. The MDL actually achieved in a given analysis will vary depending on instrument sensitivity and matrix effects.

The MDL (a) in µg/L was 0.03.
The concentration range in µg/L was 0.5–10.
The mean accuracy (% of true value) was 109.
The precision as relative standard deviation was 8.6.

Note: The MDL is based on a 25-mL sample volume instead of a 5-mL sample volume.

SAMPLING METHOD

Liquid samples — Use a 40-mL glass screw-cap VOA vial with a Teflon®-faced silicone septum that has been prewashed, rinsed with distilled deionized water, and oven dried. If residual chlorine is present, collect the sample in a 4-oz soil VOA container which has been pre-preserved with 4 drops of 10% sodium thiosulfate. Mix gently and transfer the sample to a 40-mL VOA vial. Collect bubble-free samples in duplicate and seal each sample in a separate plastic bag.

Soils, sediments and sludges — Use an 8-oz widemouth glass bottle with Teflon®-faced silicone septum that has been prewashed, rinsed with distilled deionized water, and oven dried. **Do not** heat the septum for more than 1 h. Tap slightly to eliminate any free air space. Collect samples in duplicate and seal each one in a separate plastic bag.

SAMPLE PRESERVATION

Liquid samples — Add 4 drops of concentrated HCL, cool to 4°C and store in a solvent-free refrigerator.

Soils, sediments and sludges — Cool samples to 4°C and store in a solvent-free refrigerator.

MHT The maximum holding time of any sample (liquids, soils, sediments, and sludges) is 14 days.

SAMPLE PREPARATION

Liquid samples — Remove the plunger from a 5-mL syringe and carefully pour the sample into the syringe barrel to just short of overflowing. Replace the syringe plunger and compress the sample. Open the syringe valve and vent any residual air while adjusting the sample volume to 5.0 mL. If there is only one volatile organic analysis (VOA) vial, a second syringe should be filled at this time to protect against possible loss of sample integrity. Add 10 µL of surrogate spiking solution and 10 µL of internal standard spiking solution through the valve bore of the 5-mL syringe, then close the valve. The surrogate and internal standards may be mixed and added as a single spiking solution.

Sediments, soils, and waste samples — All samples of this type should be screened by GC analysis using a headspace method (EPA Method 3810) or the hexadecane extraction and screening method (EPA Method 3820). Use the screening data to determine whether to use the low-concentration method (0.005–1 mg/kg) or the high-concentration method (>1 mg/kg).

Low-concentration method — The low-concentration method is based on purging a heated sediment or soil sample mixed with organic-free reagent water containing the surrogate and internal standards. Analyze all reagent blanks and standards under the same conditions as the samples.

Use a 5-g sample if the expected concentration is <0.1 mg/kg or a 1-g sample for expected concentrations between 0.1 and 1 mg/kg. Mix the contents of the sample container with a narrow metal spatula. Weigh the amount of the sample into a tared purge device. Add the spiked water to the purge device, which contains the weighed amount of sample, and connect the device to the purge-and-trap system.

High-concentration method — This method is based on extracting the sediment or soil with methanol. A waste sample is either extracted or diluted, depending on its solubility in methanol. Wastes that are insoluble in methanol are diluted with reagent tetraglyme or possibly polyethylene glycol (PEG). An aliquot of the extract is added to organic-free reagent water containing surrogate and internal standards. This is purged at ambient temperature. All samples with an expected concentration of >1.0 mg/kg should be analyzed by this method.

Mix the contents of the sample container with a narrow metal spatula. For sediments or soils and solid wastes that are insoluble in methanol, weigh 4 g (wet weight) of sample into a tared 20-mL vial. For waste that is soluble in methanol, tetraglyme, or PEG, weigh 1 g (wet weight) into a tared scintillation vial or culture tube or a 10-mL volumetric flask. Quickly add 9.0 mL of appropriate solvent then add 1.0 mL of a surrogate spiking solution to the vial, cap it, and shake it for 2 min.

METHANOL EXTRACT REQUIRED FOR ANALYSIS OF HIGH-CONCENTRATION SOILS OR SEDIMENTS

Approximate Concentration Range	Volume of Methanol Extract (a)
500–10,000 µg/kg	100 µL
1,000–20,000 µg/kg	50 µL
5,000–100,000 µg/kg	10 µL
25,000–500,000 µg/kg	100 µL of 1/50 dilution (b)

Calculate appropriate dilution factor for concentrations exceeding this table.

(a) The volume of methanol added to 5 mL of water being purged should be kept constant. Therefore, add to the 5-mL syringe whatever volume of methanol is necessary to maintain a volume of 100 µL added to the syringe.

(b) Dilute an aliquot of the methanol extract and then take 100 µL for analysis.

QUALITY CONTROL

Demonstrate, through the analysis of a reagent water blank, that interferences from the analytical system, glassware, and reagents are under control. Blank samples should be carried through all stages of the sample preparation and measurement steps. For each analytical batch (up to 20 samples), a reagent blank, matrix spike, and matrix spike duplicate must be analyzed (the frequency of the spikes may be different for different monitoring programs). The blank and spiked samples must be carried through all stages of the sample preparation and measurement steps. QC samples mentioned in the section on Interferences will also be needed as appropriate to those situations.

Matrix spiking standards should be prepared from volatile organic compounds which will be representative of the compounds being investigated. The recommended internal standards are chlorobenzene-d5, 1,4-difluorobenzene, 1,4-dichlorobenzene-d4, and pentafluorobenzene. Using stock standard solutions, prepare secondary dilution standards containing the compounds of interest, either singly or mixed together in methanol. Store them in a vial with no headspace for no more than one week. Surrogates recommended are toluene-d8, 4-bromofluorobenzene, and dibromofluoromethane. Each sample undergoing GC/MS analysis must be spiked with 10 µL of the surrogate spiking solution prior to analysis.

REFERENCE Test Methods for Evaluating Solid Waste (SW-846). U.S. EPA 1983, Method 8260A, Rev. 1, Nov. 1990. Office of Solid Waste, Washington, DC.

1,2,4-Trichlorobenzene   **EPA Method 8260**
CAS #120-82-1

TITLE Volatile Organic Compounds by GC/MS: Capillary Column Technique

MATRIX This method is applicable to nearly all types of samples, regardless of water content, including groundwater, soils, and sediments.

METHOD SUMMARY Method 8260A covers 58 volatile organic compounds that are introduced into a gas chromatograph by the purge-and-trap method or by direct injection (in limited applications). Zero-grade helium is bubbled through a 5-mL solution at ambient temperature. Purged sample components are trapped in a tube containing suitable sorbent materials. When purging is complete, the sorbent tube is heated and backflushed with helium to desorb trapped sample components. The analytes are desorbed directly to a large bore capillary or cryofocussed on a capillary precolumn before being flash evaporated to a narrow bore capillary for analysis.

INTERFERENCES Major contaminant sources are volatile materials in the lab and impurities in the inert purging gas and in the sorbent trap. Interfering contamination may occur when a sample containing low concentrations of volatile organic compounds is analyzed immediately after a sample containing high concentrations of volatile organic compounds. After analysis of a sample containing high concentrations of volatile organic compounds, one or more calibration blanks should be analyzed to check for cross-contamination. Screening of the samples prior to purge-and-trap GC/MS analysis is highly recommended to prevent contamination of the system. This is especially true for soil and waste samples.

Special precautions must be taken to analyze for methylene chloride. The analytical and sample storage area should be isolated from all atmospheric sources of methylene chloride. All gas chromatography carrier gas lines and purge gas plumbing should be constructed from stainless steel or copper tubing. Laboratory clothing previously exposed to methylene chloride fumes during liquid-liquid extraction procedures can contribute to sample contamination.

Samples can also be contaminated by diffusion of volatile organics (particularly methylene chloride and fluorocarbons) through the septum seal during shipment and storage. A trip blank can serve as a check on such contamination.

INSTRUMENTATION GC/MS with a temperature-programmable chromatograph suitable for splitless injection equipped with variable constant differential flow controllers, a subambient oven controller, a purging device, sorbent trap, a thermal desorption apparatus and a capillary precolumn interface when using cryogenic cooling will be needed. The following GC columns may be used:

Column 1: 60 m × 0.75mm I.D. capillary column coated with VOCOL, 1.5 μm film thickness.
Column 2: 30 m × 0.53mm capillary column coated with DB-624 or VOCOL, 3 μm film thickness.
Column 3: 30 m × 0.32mm I.D. capillary column coated with DB-5 or SE-54, 1-μm film thickness.

PRECISION & ACCURACY This method has been tested in a single lab using spiked water. Using a wide-bore capillary column, water was spiked at concentrations between 0.5 and 10 μg/L. Single lab accuracy and precision data are presented. The MDL actually achieved in a given analysis will vary depending on instrument sensitivity and matrix effects.

The MDL (a) in μg/L was 0.04.
The concentration range in μg/L was 0.5–10.
The mean accuracy (% of true value) was 108.
The precision as relative standard deviation was 8.3.

Note: The MDL is based on a 25-mL sample volume instead of a 5-mL sample volume.

SAMPLING METHOD
Liquid samples — Use a 40-mL glass screw-cap VOA vial with a Teflon®-faced silicone septum that has been prewashed, rinsed with distilled deionized water, and oven dried. If residual chlorine is present, collect the sample in a 4-oz soil VOA container which has been pre-preserved with 4 drops of 10% sodium thiosulfate. Mix gently and transfer the sample to a 40-mL VOA vial. Collect bubble-free samples in duplicate and seal each sample in a separate plastic bag.

Soils, sediments and sludges — Use an 8-oz widemouth glass bottle with Teflon®-faced silicone septum that has been prewashed, rinsed with distilled deionized water, and oven dried. **Do not** heat the septum for more than 1 h. Tap slightly to eliminate any free air space. Collect samples in duplicate and seal each one in a separate plastic bag.

SAMPLE PRESERVATION
Liquid samples — Add 4 drops of concentrated HCL, cool to 4°C and store in a solvent-free refrigerator.

Soils, sediments and sludges — Cool samples to 4°C and store in a solvent-free refrigerator.

MHT The maximum holding time of any sample (liquids, soils, sediments, and sludges) is 14 days.

SAMPLE PREPARATION
Liquid samples — Remove the plunger from a 5-mL syringe and carefully pour the sample into the syringe barrel to just short of overflowing. Replace the syringe plunger and compress the sample. Open the syringe valve and vent any residual air while adjusting the sample volume to 5.0 mL. If there is only one volatile organic analysis (VOA) vial, a second syringe should be filled at this time to protect against possible loss of sample integrity. Add 10 μL of surrogate spiking solution and 10 μL of internal standard spiking solution through the valve bore of the 5-mL syringe, then close the valve. The surrogate and internal standards may be mixed and added as a single spiking solution.

Sediments, soils, and waste samples — All samples of this type should be screened by GC analysis using a headspace method (EPA Method 3810) or the hexadecane extraction and screening method (EPA Method 3820). Use the screening data to determine whether to use the low-concentration method (0.005–1 mg/kg) or the high-concentration method (>1 mg/kg).

Low-concentration method — The low-concentration method is based on purging a heated sediment or soil sample mixed with organic-free reagent water containing the surrogate and internal standards. Analyze all reagent blanks and standards under the same conditions as the samples.

Use a 5-g sample if the expected concentration is <0.1 mg/kg or a 1-g sample for expected concentrations between 0.1 and 1 mg/kg. Mix the contents of the sample container with a narrow metal spatula. Weigh the amount of the sample into a tared purge device. Add the spiked water to the purge device, which contains the weighed amount of sample, and connect the device to the purge-and-trap system.

High-concentration method — This method is based on extracting the sediment or soil with methanol. A waste sample is either extracted or diluted, depending on its solubility in methanol. Wastes that are insoluble in methanol are diluted with reagent tetraglyme or possibly polyethylene glycol (PEG). An aliquot of the extract is added to organic-free reagent water containing surrogate and internal standards. This is purged at ambient temperature. All samples with an expected concentration of >1.0 mg/kg should be analyzed by this method.

Mix the contents of the sample container with a narrow metal spatula. For sediments or soils and solid wastes that are insoluble in methanol, weigh 4 g (wet weight) of sample into a tared 20-mL vial. For waste that is soluble in methanol, tetraglyme, or PEG, weigh 1 g (wet weight) into a tared scintillation vial or culture tube or a 10-mL volumetric flask. Quickly add 9.0 mL of appropriate solvent then add 1.0 mL of a surrogate spiking solution to the vial, cap it, and shake it for 2 min.

METHANOL EXTRACT REQUIRED FOR ANALYSIS OF HIGH-CONCENTRATION SOILS OR SEDIMENTS

Approximate Concentration Range	Volume of Methanol Extract (a)
500–10,000 μg/kg	100 μL
1,000–20,000 μg/kg	50 μL
5,000–100,000 μg/kg	10 μL
25,000–500,000 μg/kg	100 μL of 1/50 dilution (b)

Calculate appropriate dilution factor for concentrations exceeding this table.

(a) The volume of methanol added to 5 mL of water being purged should be kept constant. Therefore, add to the 5-mL syringe whatever volume of methanol is necessary to maintain a volume of 100 μL added to the syringe.

(b) Dilute an aliquot of the methanol extract and then take 100 μL for analysis.

QUALITY CONTROL Demonstrate, through the analysis of a reagent water blank, that interferences from the analytical system, glassware, and reagents are under control. Blank samples should be carried through all stages of the sample preparation and measurement steps. For each analytical batch (up to 20 samples), a reagent blank, matrix spike, and matrix spike duplicate must be analyzed (the frequency of the spikes may be different for different monitoring programs). The blank and spiked samples must be carried through all stages of the sample preparation and measurement steps. QC samples mentioned in the section on Interferences will also be needed as appropriate to those situations.

Matrix spiking standards should be prepared from volatile organic compounds which will be representative of the compounds being investigated. The recommended internal standards are chlorobenzene-d5, 1,4-difluorobenzene, 1,4-dichlorobenzene-d4, and pentafluorobenzene. Using stock standard solutions, prepare secondary dilution standards containing the compounds of interest, either singly or mixed together in methanol. Store them in a vial with no headspace for no more than one week. Surrogates recommended are toluene-d8, 4-bromofluorobenzene, and dibromofluoromethane. Each sample undergoing GC/MS analysis must be spiked with 10 μL of the surrogate spiking solution prior to analysis.

REFERENCE Test Methods for Evaluating Solid Waste (SW-846). U.S. EPA 1983, Method 8260A, Rev. 1, Nov. 1990. Office of Solid Waste, Washington, DC.

1,2,4-Trichlorobenzene   **EPA Method 8270**
CAS #120-82-1

TITLE Semivolatile Organic Compounds by GC/MS

MATRIX This method is used to determine the concentration of semivolatile organic compounds in extracts prepared from all types of solid waste matrices, soils, and groundwater. Although surface waters are not specifically mentioned, this method should be applicable to water samples from rivers, lakes, etc.

METHOD SUMMARY This method covers 259 semivolatile organic compounds. In very limited applications direct injection of the sample into the GC/MS system may be appropriate, but this results in very high detection limits (approximately 10,000 μg/L). Typically, a 1-L liquid sample, containing surrogate, and matrix spiking standards, is extracted in a continuous extractor first under acid conditions and then under basic conditions. Typically 30 g of a solid sample, containing surrogate, and matrix spiking standards, is extracted ultrasonically. After concentrating the extract to 1 mL it is spiked with 10 μL of an internal standard solution just prior to analysis by GC/MS. The volume injected should contain about 100 ng of base/neutral and 200 ng of acid surrogates (for a 1-μL injection). Analysis is performed by GC/MS using a capillary GC column.

INTERFERENCES Raw GC/MS data from all blanks, samples, and spikes must be evaluated for interferences. Contamination by carryover can occur whenever high-concentration and low-concentration samples are sequentially analyzed. To reduce carryover, the sample syringe must be rinsed out between samples with solvent. Whenever an unusually concentrated

sample is encountered, it should be followed by the analysis of blank solvent to check for cross-contamination.

INSTRUMENTATION A GC/MS and a data system are required. The GC column used is a 30 m × 0.25 mm I.D. (or 0.32 mm I.D.) 1um film thickness silicone-coated fused silica capillary column. A continuous liquid-liquid extractor equipped with Teflon® or glass connection joints and stopcocks requiring no lubrication, a K-D concentrating apparatus, water bath, and an ultrasonic disrupter with a minimum power of 300 W and with pulsing capability are also required.

PRECISION & ACCURACY The estimated quantitation limit (EQL) of Method 8270B for determining an individual compound is approximately 1 mg/kg (wet weight) for soil or sediment samples, 1–200 mg/kg for wastes (dependent on matrix and method of preparation), and 10 µg/L for groundwater samples. EQLs will be proportionately higher for sample extracts that require dilution to avoid saturation of the detector.

The EQL(b) for groundwater in µg/L is 10.
The EQL (a, b) for low concentrations in soil and sediment in µg/kg is 660.
Accuracy as µg/L is $0.94C - 0.79$.
Overall precision in µg/L is $0.21X + 0.39$.

(a) *EQLs listed for soil/sediment are based on wet weight. Normally data is reported in a dry-weight basis; therefore, EQLs will be higher based on the % dry weight of each sample. This calculation is based on a 30 g sample and gel permeation chromatography cleanup.*
(b) *Sample EQLs are highly matrix-dependent. The EQLs are provided for guidance and may not always be achievable.*
C = *True value for concentration, in µg/L.*
X = *Average recovery found for measurements of samples containing a concentration of C, in µg/L.*

ESTIMATED QUANTITATION LIMIT

Other Matrices	Factor (a)
High-concentration soil and sludges by sonicator	7.5
Non-water miscible waste	75

(a) *EQL for other matrices = [EQL for low soil/sediment × [Factor]. This estimated EQL is similar to an EPA "Practical Quantitation Limit."*

SAMPLING METHOD
Liquid samples — Use a 1 or 2½ gallon amber glass bottle with a screw-top Teflon®-lined cover that has been prewashed with detergent and rinsed with distilled water and methanol (or isopropanol).

Soils, sediments, or sludges — Use an 8-oz. widemouth glass with a screw-top Teflon®-lined cover that has been prewashed with detergent and rinsed with distilled water and methanol (or isopropanol).

SAMPLE PRESERVATION
Liquid samples — If residual chlorine is present, add 3 mL of 10% sodium thiosulfate per gallon, cool to 4°C and store in a solvent-free refrigerator until analysis; if chlorine is not present, then eliminate the sodium thiosulfate addition.

Soils, sediments, or sludges — Cool samples to 4°C and store in a solvent-free refrigerator.

MHT Liquid samples must be extracted within 7 days and the extracts analyzed within 40 days. Soils, sediments, or sludges may be stored for a maximum of 14 days and the extracts analyzed within 40 days.

SAMPLE PREPARATION
Liquid samples — Transfer 1 L quantitatively to a continuous extractor. If high concentrations are anticipated, a smaller volume may be used and then diluted with organic-free reagent water to 1 L. Adjust pH, if necessary, to pH <2 using 1:1 (V/V) sulfuric acid. Pipette 1.0 mL of a surrogate standard spiking solution into each sample. For the sample in each analytical batch selected for spiking, add 1.0 mL of a matrix spiking standard. For base/neutral acid analysis, the amount of the surrogates and matrix spiking compounds added to the sample should result in a final concentration of 100 ng/µL of each analyte in the extract to be analyzed (assuming a 1-µL injection). Extract with methylene chloride for 18–24 h. Next, adjust the pH of the aqueous phase to pH >11 using 10 N sodium hydroxide and extract it with methylene chloride again for 18–24 h. Dry the extract through a column containing anhydrous sodium sulfate and concentrate it to 1 mL using a K-D concentrator.

Soils, sediments, or sludges — Use 30 g of sample. Nonporous or wet samples (gummy or clay type) that do not have a free-flowing sandy texture must be mixed with anhydrous sodium sulfate until the sample is free flowing. Add 1 mL of surrogate standards to all samples, spikes, standards, and blanks. For the sample in each analytical batch selected for spiking, add 1.0 mL of a matrix spiking standard. For base/neutral acid analysis, the amount added of the surrogates and matrix spiking compounds should result in a final concentration of 100 ng/µL of each base/neutral analyte and 200 ng/µL of each acid analyte in the extract to be analyzed (assuming a 1-µL injection). Immediately add a 100-mL mixture of 1:1 methylene chloride:acetone and extract the sample ultrasonically for 3 min and then decant or filter the extracts. Repeat the extraction two or more times. Dry the extract using a column with anhydrous sodium sulfate and concentrate it to 1 mL in a K-D concentrator.

QUALITY CONTROL A methylene chloride solution containing 50 ng/µL of decafluorotriphenylphosphine (DFTPP) is used for tuning the GC/MS system each 12-h shift. A system performance check also must be made during every 12-h shift. A standard containing 50 ng/µL each of 4,4'-DDT, pentachlorophenol, and benzidine is required to verify injection port inertness and GC column performance. A calibration standard at mid-concentration, containing each compound of interest, including all required surrogates, must be performed every 12 h during analysis. After the system performance check is met, calibration check compounds (CCCs) are used to check the validity of the initial calibration.

The internal standard responses and retention times in the calibration check standard must be evaluated immediately after or during data acquisition. If the retention time for any internal standard changes by more than 30 seconds from the last check calibration (12 h), the chromatographic system must be

inspected for malfunctions and corrections must be made, as required. If the electron ionization current plot (EICP) area for any of the internal standards changes by a factor of two from the last daily calibration standard check, the mass spectrometer must be inspected for malfunctions and corrections must be made, as appropriate.

Demonstrate, through the analysis of a reagent water blank, that interferences from the analytical system, glassware, and reagents are under control. The blank samples should be carried through all stages of the sample preparation and measurement steps. For each analytical batch (up to 20 samples), a reagent blank, matrix spike, and matrix spike duplicate/duplicate must be analyzed (the frequency of the spikes may be different for different monitoring programs). The blank and spiked samples must be carried through all stages of the sample preparation and measurement steps. A QC reference sample concentrate containing each analyte at a concentration of 100 mg/L in methanol is required.

REFERENCE Test Methods for Evaluating Solid Waste (SW-846). U.S. EPA 1983, Method 8270B, Rev. 2, Nov. 1990. Office of Solid Waste, Washington, DC.

1,2,3-Trichlorobenzene — EPA Method 503.1
CAS #87-61-6

TITLE Aromatic & Unsaturated VOCs Water

MATRIX Drinking water (finished or in any treatment stage) and raw source water.

APPLICATION Method covers 28 aromatic and unsaturated VOCs. An inert gas is bubbled through a 5-mL water sample. Purged sample components are trapped in tube of sorbent materials. When purging is complete, sorbent tube is heated and backflushed with inert gas to desorb trapped sample onto a packed GC column.

INTERFERENCES During analysis, major contaminant sources are volatile materials in the lab and impurities in purging gas and sorbent trap. With high and low level samples, there can be carryover contamination. Excess water causes a negative baseline deflection.

INSTRUMENTATION Purge and Trap GC w/photoionization detector. (Two GC columns are recommended); Column 1: 5% SP-1200 and 1.75% Bentone 34 on Supelcoport; 5% 1,2,3-tris(2-cyanoethoxy)propane on Chromosorb W.

RANGE 2.2–600 µg/L (Drinking water)

MDL 0.03 µg/L in water

PRECISION RSD = 10.4% at 0.50 µg/L

ACCURACY Average recovery = 85% at 0.50 µg/L

SAMPLING METHOD Use a 40–120-mL screw-cap vial (prewashed with detergent, rinsed with distilled water and oven dried at 105°C) with a PTFE-faced silicone septum. If residual chlorine is in the water add about 25 mg of ascorbic acid to each vial before sample collection. Collect bubble-free samples.

STABILITY Cool to 4°C; HCl to pH <2.

MHT 14 days.

QUALITY CONTROL As an initial demonstration of lab accuracy and precision, analyze 4 to 7 replicates of a lab fortified blank containing analyte at 0.1–5 µg/L. Collect all samples in duplicate.

REFERENCE Method 503.1, Volatile aromatic & unsaturated organic compounds in H2O by Purge and Trap GC, EPA 600/4-88/039.

1,2,4-Trichlorobenzene — EPA Method 503.1
CAS #120-82-1

TITLE Aromatic & Unsaturated VOCs Water

MATRIX Drinking water (finished or in any treatment stage) and raw source water.

APPLICATION Method covers 28 aromatic and unsaturated VOCs. An inert gas is bubbled through a 5-mL water sample. Purged sample components are trapped in tube of sorbent materials. When purging is complete, sorbent tube is heated and backflushed with inert gas to desorb trapped sample onto a packed GC column.

INTERFERENCES During analysis, major contaminant sources are volatile materials in the lab and impurities in purging gas and sorbent trap. With high and low level samples, there can be carryover contamination. Excess water causes a negative baseline deflection.

INSTRUMENTATION Purge and Trap GC w/photoionization detector. (Two GC columns are recommended); Column 1: 5% SP-1200 and 1.75% Bentone 34 on Supelcoport; 5% 1,2,3-tris(2-cyanoethoxy)propane on Chromosorb W.

RANGE 2.2–600 µg/L (Drinking water)

MDL 0.03 µg/L in water

PRECISION RSD = 14.3% at 80.8 µg/L

ACCURACY Average recovery = 96% at 80.8 µg/L

SAMPLING METHOD Use a 40–120-mL screw-cap vial (prewashed with detergent, rinsed with distilled water and oven dried at 105°C) with a PTFE-faced silicone septum. If residual chlorine is in the water add about 25 mg of ascorbic acid to each vial before sample collection. Collect bubble-free samples.

STABILITY Cool to 4°C; HCl to pH <2.

MHT 14 days.

QUALITY CONTROL As an initial demonstration of lab accuracy and precision, analyze 4 to 7 replicates of a lab fortified blank containing analyte at 0.1–5 µg/L. Collect all samples in duplicate.

REFERENCE Method 503.1, Volatile Aromatic & Unsaturated Organic Compounds in H2O by Purge and Trap GC, EPA 600/4-88/039.

1,1,1-Trichloroethane   EPA Method 1624
CAS #71-55-6

TITLE Volatile Organic Compounds by Isotope Dilution GC/MS

MATRIX Compounds may be determined in waters, soils, and municipal sludges by this method.

METHOD SUMMARY This method is used to determine 58 volatile toxic organic pollutants associated with the CWA (as amended 1987); the RCRA (as amended 1986); the CERCLA (as amended 1986); and other compounds amenable to purge-and-trap gas chromatography-mass spectrometry (GC/MS).

If the solids content is less than 1%, stable isotopically-labeled analogs of the compounds of interest are added to a 5-mL sample and the sample is purged with an inert gas at 20–25°C in a chamber designed for soil or water samples. If the solids content is greater than 1%, 5 mL of reagent water and the labeled compounds are added to a 5-g aliquot of sample and the mixture is purged at 40°C. Compounds that will not purge at 20–25°C or at 40°C are purged at 78–85°C. In the purging process, the volatile compounds are transferred from the aqueous phase into the gaseous phase where they are passed into a sorbent column, and trapped. After purging is completed, the trap is backflushed and heated rapidly to desorb the compounds into a GC. The compounds are separated by the GC and detected by a MS. The labeled compounds serve to correct the variability of the analytical technique.

INTERFERENCES Impurities in the purge gas, organic compounds outgassing from the plumbing upstream of the trap, and solvent vapors in the lab account for most problems. Samples can be contaminated by diffusion of volatile organic compounds (particularly methylene chloride) through the bottle seal during shipment and storage. Contamination by carryover can occur when high-level and low-level samples are analyzed sequentially. When an unusually concentrated sample is encountered, follow it by analysis of a reagent water blank to check for carryover.

INSTRUMENTATION Major equipment includes a GC with linear temperature programming and a glass jet separator as the MS interface, a MS with 70 eV electron impact ionization, and a data system to collect and record response factors.

Column: 2.8 m × 2 mm I.D. glass, packed with 1% SP-1000 on Carbopak B, 60/80 mesh, or equivalent.

PRECISION & ACCURACY The detection limits of the method are usually dependent on the level of interferences rather than instrumental limitations. The method detection limits were determined in digested sludge (low solids) and in filter cake or compost (high solids).

The MDL (in μg/kg) for low solids is 16 and for high solids is 4.
Labeled and native compound precision (in μg/L) as standard deviation was 5.9.
Labeled and native compound accuracy (in μg/L) as average recovery was 11–33.

Acceptance criteria are at 20 μg/L for this compound.

SAMPLE COLLECTION, PRESERVATION & HANDLING Grab samples are collected in glass containers having a total volume greater than 20 mL. Fill and seal each bottle so that no air bubbles are entrapped. Samples are maintained at 0 to 4°C from the time of collection until analysis. If an aqueous sample contains residual chlorine, add sodium thiosulfate preservative (10 mg/40 mL) to the empty sample bottles just prior to shipment to the sample site. All samples must be analyzed within 14 days of collection.

SAMPLE PREPARATION Samples containing less than 1% solids are analyzed directly as aqueous samples. Samples containing 1% solids or greater are analyzed as solid samples utilizing one of two methods, depending on the levels of pollutants, in the sample. Samples containing 1% solids or greater, and low to moderate levels of pollutants are analyzed by purging a known weight of sample added to 5 mL of reagent water. Samples containing 1% solids or greater, and high levels of pollutants, are extracted with methanol, and an aliquot of the methanol extract is added to reagent water and purged.

QUALITY CONTROL A field blank prepared from reagent water and carried through the sampling and handling protocol may serve as a check on contamination from shipment and storage.

The analyst is permitted to modify this method to improve separations or lower the costs of measurements, provided all performance specifications are met. Analyses of blanks are required. When results of spikes indicate atypical method performance for samples, the samples are diluted to bring method performance within acceptable limits. Analyze two sets of four 5-mL aliquots (8 aliquots total) of the aqueous performance standard. Spike all samples with labeled compounds to assess method performance on the sample matrix. Compute the percent recovery of the labeled compounds using the internal standard method. Compare the percent recovery for each compound with the corresponding labeled compound recovery. Reagent water blanks are analyzed to demonstrate freedom from carryover contamination. Field replicates may be collected to determine the precision of the sampling technique, and spiked samples may be required to determine the accuracy of the analysis when the internal method is used.

REFERENCE Volatile Organic Compounds by Isotope Dilution GC/MS. Office of Water Regulation and Standards, U.S. EPA Industrial Technology Division, Washington, DC, EPA Method 1624, Rev. C, June 1989 (contact W.A. Telliard, U.S. EPA, Office of Water Regulations and Standards, 401 M St., SW, Washington, DC, 20460. Phone: 202-382-7131).

1,1,2-Trichloroethane   EPA Method 1624
CAS #79-00-5

TITLE Volatile Organic Compounds by Isotope Dilution GC/MS

MATRIX Compounds may be determined in waters, soils, and municipal sludges by this method.

METHOD SUMMARY This method is used to determine 58 volatile toxic organic pollutants associated with the CWA (as amended 1987); the RCRA (as amended 1986); the CERCLA (as amended 1986); and other compounds amenable to purge-and-trap gas chromatography-mass spectrometry (GC/MS).

If the solids content is less than 1%, stable isotopically-labeled analogs of the compounds of interest are added to a 5-mL sample and the sample is purged with an inert gas at 20–25°C in a chamber designed for soil or water samples. If the solids content is greater than 1%, 5 mL of reagent water and the labeled compounds are added to a 5-g aliquot of sample and the mixture is purged at 40°C. Compounds that will not purge at 20–25°C or at 40°C are purged at 78–85°C. In the purging process, the volatile compounds are transferred from the aqueous phase into the gaseous phase where they are passed into a sorbent column, and trapped. After purging is completed, the trap is backflushed and heated rapidly to desorb the compounds into a GC. The compounds are separated by the GC and detected by a MS. The labeled compounds serve to correct the variability of the analytical technique.

INTERFERENCES Impurities in the purge gas, organic compounds outgassing from the plumbing upstream of the trap, and solvent vapors in the lab account for most problems. Samples can be contaminated by diffusion of volatile organic compounds (particularly methylene chloride) through the bottle seal during shipment and storage. Contamination by carryover can occur when high-level and low-level samples are analyzed sequentially. When an unusually concentrated sample is encountered, follow it by analysis of a reagent water blank to check for carryover.

INSTRUMENTATION Major equipment includes a GC with linear temperature programming and a glass jet separator as the MS interface, a MS with 70 eV electron impact ionization, and a data system to collect and record response factors.

Column: 2.8 m × 2 mm I.D. glass, packed with 1% SP-1000 on Carbopak B, 60/80 mesh, or equivalent.

PRECISION & ACCURACY The detection limits of the method are usually dependent on the level of interferences rather than instrumental limitations. The method detection limits were determined in digested sludge (low solids) and in filter cake or compost (high solids).

The MDL (in µg/kg) for low solids is 26 and for high solids is 1.
Labeled and native compound precision (in µg/L) as standard deviation was 7.1.
Labeled and native compound accuracy (in µg/L) as average recovery was 12–30.

Acceptance criteria are at 20 µg/L for this compound.

SAMPLE COLLECTION, PRESERVATION & HANDLING Grab samples are collected in glass containers having a total volume greater than 20 mL. Fill and seal each bottle so that no air bubbles are entrapped. Samples are maintained at 0 to 4°C from the time of collection until analysis. If an aqueous sample contains residual chlorine, add sodium thiosulfate preservative (10 mg/40 mL) to the empty sample bottles just prior to shipment to the sample site. All samples must be analyzed within 14 days of collection.

SAMPLE PREPARATION Samples containing less than 1% solids are analyzed directly as aqueous samples. Samples containing 1% solids or greater are analyzed as solid samples utilizing one of two methods, depending on the levels of pollutants, in the sample. Samples containing 1% solids or greater, and low to moderate levels of pollutants are analyzed by purging a known weight of sample added to 5 mL of reagent water. Samples containing 1% solids or greater, and high levels of pollutants, are extracted with methanol, and an aliquot of the methanol extract is added to reagent water and purged.

QUALITY CONTROL A field blank prepared from reagent water and carried through the sampling and handling protocol may serve as a check on contamination from shipment and storage.

The analyst is permitted to modify this method to improve separations or lower the costs of measurements, provided all performance specifications are met. Analyses of blanks are required. When results of spikes indicate atypical method performance for samples, the samples are diluted to bring method performance within acceptable limits. Analyze two sets of four 5-mL aliquots (8 aliquots total) of the aqueous performance standard. Spike all samples with labeled compounds to assess method performance on the sample matrix. Compute the percent recovery of the labeled compounds using the internal standard method. Compare the percent recovery for each compound with the corresponding labeled compound recovery. Reagent water blanks are analyzed to demonstrate freedom from carryover contamination. Field replicates may be collected to determine the precision of the sampling technique, and spiked samples may be required to determine the accuracy of the analysis when the internal method is used.

REFERENCE Volatile Organic Compounds by Isotope Dilution GC/MS. Office of Water Regulation and Standards, U.S. EPA Industrial Technology Division, Washington, DC, EPA Method 1624, Rev. C, June 1989 (contact W.A. Telliard, U.S. EPA, Office of Water Regulations and Standards, 401 M St., SW, Washington, DC, 20460. Phone: 202-382-7131).

1,1,1-Trichloroethane   **EPA Method 502**
CAS #71-55-6

TITLE Volatile Organic Compounds in Water By Purge and Trap Capillary Column Gas Chromatography with Photoionization and Electrolytic Conductivity Detectors in Series. U.S. EPA Method 502.2, Rev. 2.0, 1989.

MATRIX Drinking water and raw source water. The latter should include most surface water and groundwater sources.

METHOD SUMMARY This method covers 60 volatile organic compounds that contain halogen atoms and/or that are aromatic. An inert gas (zero grade nitrogen or helium) is bubbled through a 25-mL or a 5-mL water sample (depending on the expected concentration of the analytes). Purged sample components are trapped in a tube of sorbent materials. When purging is complete, the sorbent tube is heated and backflushed with helium to desorb the trapped sample onto a capillary GC column. The column is temperature programmed to separate the method analytes which are then detected with a photoionization detector (PID) and a Hall electrolytic conductivity (HECD) placed in series. The PID is selective for aromatic compounds and the HECD is selective for halogenated compounds.

INTERFERENCES Impurities in the purge gas and from organic compounds outgassing from the plumbing ahead of the trap account for many contamination problems. Interferences purged or coextracted from the samples will vary considerably from source to source, depending upon the particular sample or extract being tested. Cross-contamination can occur whenever high-level and low-level samples are analyzed sequentially. Samples also can be contaminated by diffusion of volatile organics (particularly methylene chloride and fluorocarbons) through the septum seal into the sample during shipment and storage. The lab where volatile analysis is performed and also the refrigerated storage area should be completely free of solvents.

INSTRUMENTATION A GC containing a series configuration of a high temperature photoionization detector (PID) equipped with 10.0 eV (nominal) lamp and Hall electrolytic conductivity detector (HECD) is required. Also required is an all-glass 5-mL purging device, a sorbent trap, and a thermal desorption apparatus which is connected to the GC system.

Column 1: VOCOL glass wide-bore capillary column.
Column 2: RTX–502.2 mega-bore capillary column.
Column 3: DB-62 mega-bore capillary column.

PRECISION & ACCURACY Method detection limits are dependent upon the characteristics of the gas chromatographic system used. Analytes that are not separated chromatographically cannot be individually identified and used in the same calibration mixture or water samples unless an alternative technique for identification and quantification, such as mass spectrometry, is used.

Electrolytic conductivity detetor (c) range in µg/L (a) was 0.02–200.
Electrolytic conductivity detetor (c) MDL in µg/L (b) was 0.03.
Electrolytic conductivity detetor (c) accuracy as % recovery was 104.
Electrolytic conductivity detetor (c) precision as % RSD was 3.3.
Photoionization detector (d) range in µg/L (a) was 0.02–200.
Photoionization detector (d) MDL in µg/L (b) was not listed.
Photoionization detector (d) accuracy as % recovery was not listed.
Photoionization detector (d) precision as % RSD was not listed.

(a) *The applicable concentration range of this method is compound, instrument, and matrix-dependent. It is listed as being approximately 0.02 to 200 µg/L but no specific information is provided so caution should be observed.*
(b) *The method detection limits reports with this method are compound, instrument, and matrix-dependent. The values reported were calculated using reagent water fortified with the corresponding compounds at 10 µg/L and a GC-equipped with a 60 m × 0.75 mm VOLCOL wide bore capillary column with 1.5 µm film thickness and using helium carrier gas.*
(c) *Recoveries and relative standard deviations were determined from seven samples of reagent water fortified with 10 µg/L of each compound. 2-Bromo-1-chloropropane was used as the internal standard for calculating average recoveries.*
(d) *Recoveries and relative standard deviations were determined from seven samples of reagent water fortified with 10 µg/L of each compound. Fluorobenzene was used as the internal standard for calculating average recoveries.*

SAMPLING METHOD Collect samples using a 40- to 120-mL screw-cap vial (prewashed with detergent, rinsed with distilled water and oven dried at 105°C) with a Teflon®-faced silicone septum. Collect bubble-free samples and place the septum with the Teflon® side down on the water.

SAMPLE PRESERVATION If residual chlorine is present in the water add about 25 mg of ascorbic acid to each vial before samples are collected to remove the chlorine. Add hydrochloric acid to reduce pH to <2, immediately cool samples to 4°C, and store them in a solvent-free refrigerator at 4°C until analysis.

MHT The maximum holding time for samples is 14 days from the time they were collected.

SAMPLE PREPARATION Remove the plungers from two 5-mL syringes and attach a closed syringe valve to each. Warm the sample to room temperature, open the sample bottle, and carefully pour the sample into one of the syringe barrels to just short of overflowing. Replace the syringe plunger, invert the syringe, and compress the sample. Open the syringe valve and vent any residual air while adjusting the sample volume to 5.0 mL. Add 10 µL of the internal calibration standard to the sample through the syringe valve. Close the valve. Fill the second syringe in an identical manner from the same sample bottle. Reserve this second syringe for a reanalysis if necessary.

QUALITY CONTROL As an initial demonstration of lab accuracy and precision, analyze 4 to 7 replicates of a lab fortified blank containing analyte at 0.1–5 µg/L. Collect all samples in duplicate. Surrogate analytes (similar to those of the analytes of interest), whose concentration is known in every sample, are measured using the same internal standard calibration procedure. Duplicate field reagent water blanks (trip blanks) must be analyzed with each set of samples, lab reagent blanks (method blanks) must be analyzed with each batch of samples processed as a group within a work shift. Also, a single lab-fortified blank that contains each of the analytes of interest should be analyzed with each batch of samples processed as a group within a work shift. A 3- to 5-point calibration curve is needed depending on the calibration range factor required.

EPA CONTACT & HOTLINE For technical questions contact Dr. Baldev Bathija, U.S. EPA, Office of Ground Water and Drinking Water (WH-550D), 401 M St. SW, Washington, DC 20460. Tel. (202) 260-3040. For further information the EPA Safe Drinking Water Hotline may be called at: (800) 426-4791.

REFERENCE Methods for the Determination of Organic Compounds in Drinking Water, EPA/600/4-88/039 (revised July 1991; Final Rule for determination of compliance with the MCL for Total Trihalomethanes under 141.30, in 40 CFR Part

141, Vol. 58, No. 147, Fed. Reg., Tuesday Aug. 3, 1993). U.S. EPA Environmental Monitoring Systems Laboratory, Cincinnati, OH, 45268, U.S.A. Available from the National Technical Information Service (NTIS), 5285 Port Royal Road, Springfield, VA 22161; Tel. 800-553-6847. NTIS Order Number is PB91-231480.

1,1,2-Trichloroethane   EPA Method 502
CAS #79-00-5

TITLE Volatile Organic Compounds in Water By Purge and Trap Capillary Column Gas Chromatography with Photoionization and Electrolytic Conductivity Detectors in Series. U.S. EPA Method 502.2, Rev. 2.0, 1989.

MATRIX Drinking water and raw source water. The latter should include most surface water and groundwater sources.

METHOD SUMMARY This method covers 60 volatile organic compounds that contain halogen atoms and/or that are aromatic. An inert gas (zero grade nitrogen or helium) is bubbled through a 25-mL or a 5-mL water sample (depending on the expected concentration of the analytes). Purged sample components are trapped in a tube of sorbent materials. When purging is complete, the sorbent tube is heated and backflushed with helium to desorb the trapped sample onto a capillary GC column. The column is temperature programmed to separate the method analytes which are then detected with a photoionization detector (PID) and a Hall electrolytic conductivity (HECD) placed in series. The PID is selective for aromatic compounds and the HECD is selective for halogenated compounds.

INTERFERENCES Impurities in the purge gas and from organic compounds outgassing from the plumbing ahead of the trap account for many contamination problems. Interferences purged or coextracted from the samples will vary considerably from source to source, depending upon the particular sample or extract being tested. Cross-contamination can occur whenever high-level and low-level samples are analyzed sequentially. Samples also can be contaminated by diffusion of volatile organics (particularly methylene chloride and fluorocarbons) through the septum seal into the sample during shipment and storage. The lab where volatile analysis is performed and also the refrigerated storage area should be completely free of solvents.

INSTRUMENTATION A GC containing a series configuration of a high temperature photoionization detector (PID) equipped with 10.0 eV (nominal) lamp and Hall electrolytic conductivity detector (HECD) is required. Also required is an all-glass 5-mL purging device, a sorbent trap, and a thermal desorption apparatus which is connected to the GC system.

Column 1: VOCOL glass wide-bore capillary column.
Column 2: RTX–502.2 mega-bore capillary column.
Column 3: DB-62 mega-bore capillary column.

PRECISION & ACCURACY Method detection limits are dependent upon the characteristics of the gas chromatographic system used. Analytes that are not separated chromatographically cannot be individually identified and used in the same calibration mixture or water samples unless an alternative technique for identification and quantification, such as mass spectrometry, is used.

Electrolytic conductivity detetor (c) range in µg/L (a) was 0.02–200.
Electrolytic conductivity detetor (c) MDL in µg/L (b) was not detected.
Electrolytic conductivity detetor (c) accuracy as % recovery was 109.
Electrolytic conductivity detetor (c) precision as % RSD was 5.6.
Photoionization detector (d) range in µg/L (a) was 0.02–200.
Photoionization detector (d) MDL in µg/L (b) was not listed.
Photoionization detector (d) accuracy as % recovery was not listed.
Photoionization detector (d) precision as % RSD was not listed.

(a) *The applicable concentration range of this method is compound, instrument, and matrix-dependent. It is listed as being approximately 0.02 to 200 µg/L but no specific information is provided so caution should be observed.*
(b) *The method detection limits reports with this method are compound, instrument, and matrix-dependent. The values reported were calculated using reagent water fortified with the corresponding compounds at 10 µg/L and a GC-equipped with a 60 m × 0.75 mm VOLCOL wide bore capillary column with 1.5 µm film thickness and using helium carrier gas.*
(c) *Recoveries and relative standard deviations were determined from seven samples of reagent water fortified with 10 µg/L of each compound. 2-Bromo-1-chloropropane was used as the internal standard for calculating average recoveries.*
(d) *Recoveries and relative standard deviations were determined from seven samples of reagent water fortified with 10 µg/L of each compound. Fluorobenzene was used as the internal standard for calculating average recoveries.*

SAMPLING METHOD Collect samples using a 40- to 120-mL screw-cap vial (prewashed with detergent, rinsed with distilled water and oven dried at 105°C) with a Teflon®-faced silicone septum. Collect bubble-free samples and place the septum with the Teflon® side down on the water.

SAMPLE PRESERVATION If residual chlorine is present in the water add about 25 mg of ascorbic acid to each vial before samples are collected to remove the chlorine. Add hydrochloric acid to reduce pH to <2, immediately cool samples to 4°C, and store them in a solvent-free refrigerator at 4°C until analysis.

MHT The maximum holding time for samples is 14 days from the time they were collected.

SAMPLE PREPARATION Remove the plungers from two 5-mL syringes and attach a closed syringe valve to each. Warm the sample to room temperature, open the sample bottle, and carefully pour the sample into one of the syringe barrels to just short of overflowing. Replace the syringe plunger, invert the syringe, and compress the sample. Open the syringe valve and vent any residual air while adjusting the sample volume to 5.0 mL. Add 10 µL of the internal calibration standard to the sample through the syringe valve. Close the valve. Fill the second syringe in an identical manner from the same sample bottle. Reserve this second syringe for a reanalysis if necessary.

QUALITY CONTROL As an initial demonstration of lab accuracy and precision, analyze 4 to 7 replicates of a lab fortified blank containing analyte at 0.1–5 µg/L. Collect all samples in duplicate. Surrogate analytes (similar to those of the analytes of interest), whose concentration is known in every sample, are measured using the same internal standard calibration procedure. Duplicate field reagent water blanks (trip blanks) must be analyzed with each set of samples, lab reagent blanks (method blanks) must be analyzed with each batch of samples processed as a group within a work shift. Also, a single lab-fortified blank that contains each of the analytes of interest should be analyzed with each batch of samples processed as a group within a work shift. A 3- to 5-point calibration curve is needed depending on the calibration range factor required.

EPA CONTACT & HOTLINE For technical questions contact Dr. Baldev Bathija, U.S. EPA, Office of Ground Water and Drinking Water (WH-550D), 401 M St. SW, Washington, DC 20460. Tel. (202) 260-3040. For further information the EPA Safe Drinking Water Hotline may be called at: (800) 426-4791.

REFERENCE Methods for the Determination of Organic Compounds in Drinking Water, EPA/600/4-88/039 (revised July 1991; Final Rule for determination of compliance with the MCL for Total Trihalomethanes under 141.30, in 40 CFR Part 141, Vol. 58, No. 147, Fed. Reg., Tuesday Aug. 3, 1993). U.S. EPA Environmental Monitoring Systems Laboratory, Cincinnati, OH, 45268, U.S.A. Available from the National Technical Information Service (NTIS), 5285 Port Royal Road, Springfield, VA 22161; Tel. 800-553-6847. NTIS Order Number is PB91-231480.

1,1,1-Trichloroethane   EPA Method 524
CAS #71-55-6

TITLE Measurement of Purgeable Organic Compounds in Water by Capillary Column GC/MS.

MATRIX Drinking water and raw source water; the latter should include most surface water and groundwater sources.

METHOD SUMMARY Method 524.2 covers 60 volatile organic compounds. An inert gas (zero grade nitrogen or helium) is bubbled through a 25-mL or a 5-mL water sample (depending on the expected concentration of the analytes). Purged sample components are trapped in a tube of sorbent materials. When purging is complete, the sorbent tube is heated and backflushed with helium to desorb the trapped sample onto a capillary GC column.

INTERFERENCES Impurities in the purge gas and from organic compounds outgassing from the plumbing ahead of the trap account for many contamination problems. Interferences purged or coextracted from the samples will vary considerably from source to source, depending upon the particular sample or extract being tested. Cross-contamination can occur whenever high-level and low-level samples are analyzed sequentially. Samples also can be contaminated by diffusion of volatile organics (particularly methylene chloride and fluorocarbons) through the septum seal into the sample during shipment and storage.

INSTRUMENTATION A GC/MS with a data system equipped with one of the following capillary GC columns:

Column 1: VOCOL glass wide bore capillary column.
Column 2: DB-624 fused silica capillary column.
Column 3: DB-5 fused silica capillary column.

Also required is an all-glass 25 mL or 5-mL purging device, a sorbent trap, and a thermal desorption apparatus which is connected to the GC/MS system.

PRECISION & ACCURACY Method detection limits are compound- and instrument-dependent, and may vary from approximately 0.02–0.35 µg/L. Note in the table below that the "true" concentration range used for accuracy and precision measurements was quite narrow. However, the applicable concentration range of this method is primarily column dependent and is approximately 0.02 to 200 µg/L for the wide-bore thick-film columns. Narrow-bore thin-film columns may have a capacity which limits the range to about 0.02 to 20 µg/L. Analytes that are inefficiently purged from water will not be detected when present at low concentrations, but they can be measured with acceptable accuracy and precision when present in sufficient amounts.

Analytes that are not separated chromatographically, but which have different mass spectra and non-interfering quantification ions, can be identified and measured in the same calibration mixture or water sample. Analytes which have very similar mass spectra cannot be individually identified and measured in the same calibration mixture or water samples unless they have different retention times. Co-eluting compounds with very similar mass spectra, typically many structural isomers, must be reported as an isomeric group or pair.

The range (in µg/L) was 0.5–10.
The Method Detection Limig (in µg/L) was 0.08.
The accuracy (as % recovery) was 98.
The precision (in %) was 8.1.

Note: Data were obtained from 16–31 determinations using a wide-bore capillary column and a jet separator interfaced to a quadrupole mass spectrometer. All analytes were in a reagent water matrix.

SAMPLING METHOD Collect samples using a 40- to 120-mL screw-cap vial (prewashed with detergent, rinsed with distilled water and oven dried at 105°C) with a Teflon®-faced silicone septum. Collect bubble-free samples and place the septum with the Teflon® side down on the water.

SAMPLE PRESERVATION If residual chlorine is present in the water add about 25 mg of ascorbic acid to each vial before samples are collected to remove the chlorine. Add hydrochloric acid to reduce pH to <2, and immediately cool samples to 4°C, and store them in a solvent-free refrigerator at 4°C until analysis.

MHT The maximum holding time for samples is 14 days from the time they were collected.

SAMPLE PREPARATION Remove the plungers from two 25-mL (or 5-mL depending on sample size) syringes and attach a closed syringe valve to each. Warm the sample to room temperature, open the sample bottle, and carefully pour the sample into one of the syringe barrels to just short of overflowing. Replace the syringe plunger, invert the syringe, and compress the sample. Open the syringe valve and vent any residual air while adjusting the sample volume to 25.0 mL (or 5 mL). For samples and blanks, add 5 µL of the fortification solution containing the internal standard and the surrogates to the sample through the syringe valve. For calibration standards and lab fortified blanks, add 5 µL of the fortification solution containing the internal standard only. Close the valve. Fill the second syringe in an identical manner from the same sample bottle. Reserve this second syringe for a reanalysis if necessary.

QUALITY CONTROL As an initial demonstration of lab accuracy and precision, analyze 4 to 7 replicates of a lab fortified blank containing analyte at 0.2–5 µg/L. Collect all samples in duplicate. Surrogate analytes (similar to those of the analytes of interest), whose concentration is known in every sample, are measured using the same internal standard calibration procedure. Duplicate field reagent water blanks (trip blanks) must be analyzed with each set of samples, lab reagent blanks (method blanks) must be analyzed with each batch of samples processed as a group within a work shift. Also, a single lab-fortified blank that contains each of the analytes of interest should be analyzed with each batch of samples processed as a group within a work shift. A 3- to 5-point calibration curve is needed depending on the calibration range factor required.

EPA CONTACT & HOTLINE For technical questions contact Dr. Baldev Bathija, U.S. EPA, Office of Ground Water and Drinking Water (WH-550D), 401 M St. SW, Washington, DC 20460. Tel. (202) 260-3040. For further information the EPA Safe Drinking Water Hotline may be called at: (800) 426-4791.

REFERENCE Methods for the Determination of Organic Compounds in Drinking Water, EPA/600/4-88/039 (revised July 1991; Final Rule for determination of compliance with the MCL for Total Trihalomethanes under 141.30, in 40 CFR Part 141, Vol. 58, No. 147, Fed. Reg., Tuesday Aug. 3, 1993). U.S. EPA Environmental Monitoring Systems Laboratory, Cincinnati, OH, 45268, U.S.A. Available from the National Technical Information Service (NTIS), 5285 Port Royal Road, Springfield, VA 22161; Tel. 800-553-6847. NTIS Order Number is PB91-231480.

1,1,2-Trichloroethane   EPA Method 524
CAS #79-00-5

TITLE Measurement of Purgeable Organic Compounds in Water by Capillary Column GC/MS.

MATRIX Drinking water and raw source water; the latter should include most surface water and groundwater sources.

METHOD SUMMARY Method 524.2 covers 60 volatile organic compounds. An inert gas (zero grade nitrogen or helium) is bubbled through a 25-mL or a 5-mL water sample (depending on the expected concentration of the analytes). Purged sample components are trapped in a tube of sorbent materials. When purging is complete, the sorbent tube is heated and backflushed with helium to desorb the trapped sample onto a capillary GC column.

INTERFERENCES Impurities in the purge gas and from organic compounds outgassing from the plumbing ahead of the trap account for many contamination problems. Interferences purged or coextracted from the samples will vary considerably from source to source, depending upon the particular sample or extract being tested. Cross-contamination can occur whenever high-level and low-level samples are analyzed sequentially. Samples also can be contaminated by diffusion of volatile organics (particularly methylene chloride and fluorocarbons) through the septum seal into the sample during shipment and storage.

INSTRUMENTATION A GC/MS with a data system equipped with one of the following capillary GC columns:

Column 1: VOCOL glass wide bore capillary column.
Column 2: DB-624 fused silica capillary column.
Column 3: DB-5 fused silica capillary column.

Also required is an all-glass 25 mL or 5-mL purging device, a sorbent trap, and a thermal desorption apparatus which is connected to the GC/MS system.

PRECISION & ACCURACY Method detection limits are compound- and instrument-dependent, and may vary from approximately 0.02–0.35 µg/L. Note in the table below that the "true" concentration range used for accuracy and precision measurements was quite narrow. However, the applicable concentration range of this method is primarily column dependent and is approximately 0.02 to 200 µg/L for the wide-bore thick-film columns. Narrow-bore thin-film columns may have a capacity which limits the range to about 0.02 to 20 µg/L. Analytes that are inefficiently purged from water will not be detected when present at low concentrations, but they can be measured with acceptable accuracy and precision when present in sufficient amounts.

Analytes that are not separated chromatographically, but which have different mass spectra and non-interfering quantification ions, can be identified and measured in the same calibration mixture or water sample. Analytes which have very similar mass spectra cannot be individually identified and measured in the same calibration mixture or water samples unless they have different retention times. Co-eluting compounds with very similar mass spectra, typically many structural isomers, must be reported as an isomeric group or pair.

The range (in µg/L) was 0.5–10.
The Method Detection Limig (in µg/L) was 0.10.
The accuracy (as % recovery) was 104.
The precision (in %) was 7.3.

Note: Data were obtained from 16–31 determinations using a wide-bore capillary column and a jet separator interfaced to a quadrupole mass spectrometer. All analytes were in a reagent water matrix.

SAMPLING METHOD Collect samples using a 40- to 120-mL screw-cap vial (prewashed with detergent, rinsed with distilled water and oven dried at 105°C) with a Teflon®-faced silicone septum. Collect bubble-free samples and place the septum with the Teflon® side down on the water.

SAMPLE PRESERVATION If residual chlorine is present in the water add about 25 mg of ascorbic acid to each vial before samples are collected to remove the chlorine. Add hydrochloric acid to reduce pH to <2, and immediately cool samples to 4°C, and store them in a solvent-free refrigerator at 4°C until analysis.

MHT The maximum holding time for samples is 14 days from the time they were collected.

SAMPLE PREPARATION Remove the plungers from two 25-mL (or 5-mL depending on sample size) syringes and attach a closed syringe valve to each. Warm the sample to room temperature, open the sample bottle, and carefully pour the sample into one of the syringe barrels to just short of overflowing. Replace the syringe plunger, invert the syringe, and compress the sample. Open the syringe valve and vent any residual air while adjusting the sample volume to 25.0 mL (or 5 mL). For samples and blanks, add 5 µL of the fortification solution containing the internal standard and the surrogates to the sample through the syringe valve. For calibration standards and lab fortified blanks, add 5 µL of the fortification solution containing the internal standard only. Close the valve. Fill the second syringe in an identical manner from the same sample bottle. Reserve this second syringe for a reanalysis if necessary.

QUALITY CONTROL As an initial demonstration of lab accuracy and precision, analyze 4 to 7 replicates of a lab fortified blank containing analyte at 0.2–5 µg/L. Collect all samples in duplicate. Surrogate analytes (similar to those of the analytes of interest), whose concentration is known in every sample, are measured using the same internal standard calibration procedure. Duplicate field reagent water blanks (trip blanks) must be analyzed with each set of samples, lab reagent blanks (method blanks) must be analyzed with each batch of samples processed as a group within a work shift. Also, a single lab-fortified blank that contains each of the analytes of interest should be analyzed with each batch of samples processed as a group within a work shift. A 3- to 5-point calibration curve is needed depending on the calibration range factor required.

EPA CONTACT & HOTLINE For technical questions contact Dr. Baldev Bathija, U.S. EPA, Office of Ground Water and Drinking Water (WH-550D), 401 M St. SW, Washington, DC 20460. Tel. (202) 260-3040. For further information the EPA Safe Drinking Water Hotline may be called at: (800) 426-4791.

REFERENCE Methods for the Determination of Organic Compounds in Drinking Water, EPA/600/4-88/039 (revised July 1991; Final Rule for determination of compliance with the MCL for Total Trihalomethanes under 141.30, in 40 CFR Part 141, Vol. 58, No. 147, Fed. Reg., Tuesday Aug. 3, 1993). U.S. EPA Environmental Monitoring Systems Laboratory, Cincinnati, OH, 45268, U.S.A. Available from the National Technical Information Service (NTIS), 5285 Port Royal Road, Springfield, VA 22161; Tel. 800-553-6847. NTIS Order Number is PB91-231480.

1,1,1-Trichloroethane   EPA Method 8021
CAS #71-55-6

TITLE Halogenated Volatile by Gas Chromatography Using Photoionization and Electrolytic Conductivity Detectors in Series: Capillary Column Technique

MATRIX This method is applicable to nearly all types of samples, regardless of water content, including groundwater, aqueous sludges, caustic liquors, acid liquors, waste solvents, oily wastes, mousses, tars, fibrous wastes, polymeric emulsions, filter cakes, spent carbons, spent catalysts, soils, and sediments.

METHOD SUMMARY This method is used to determine 60 volatile organic compounds in a variety of solid waste matrices. It provides GC conditions for the detection of halogenated and aromatic volatile organic compounds. Samples can be analyzed using direct injection or purge-and-trap (EPA Method 5030). Groundwater samples must be analyzed using EPA Method 5030 (where applicable). A temperature program is used with the GC. Detection is achieved by a photoionization detector (PID) and a Hall electrolytic conductivity detector (HECD) in series.

INTERFERENCES Samples can be contaminated by diffusion of volatile organics (particularly chlorofluorocarbons and methylene chloride) through the sample container septum during shipment and storage.

INSTRUMENTATION A GC-equipped with variable-constant differential flow controllers, subambient oven controller, PID and HECD detectors connected with a short piece of uncoated capillary tubing and a data system.

Column: 60 m × 0.75 mm I.D. VOCOL wide-bore capillary column with 1.5 µm film thickness.

PRECISION & ACCURACY MDLs are compound-dependent and vary with purging efficiency and concentration. The applicable concentration range of this method is compound- and instrument-dependent but is approximately 0.1 to 200 µg/L. Analytes that are inefficiently purged from water will not be detected when present at low concentrations, but they can be measured with acceptable accuracy and precision when present in sufficient amounts. The estimated quantitation limit (EQL) for an individual compound is approximately 1 µg/kg (wet weight) for soil/sediment samples, 100 µg/kg (wet weight) for wastes, and 1 µg/L for groundwater. EQLs will be proportionately higher for sample extracts and samples that require dilution or reduced sample size to avoid saturation of the detector.

MULTIPLICATION FACTORS FOR OTHER MATRICES (a)

Matrix	Factor (b)
Groundwater	10
Low-concentration soil	10
Water miscible liquid waste	500
High-concentration soil and sludge	1250
Non-water miscible waste	1250

(a) Sample EQLs are highly matrix-dependent. The EQLs listed herein are provided for guidance and may not always be achievable. (b) EQL = [Method detection limit] × [Factor]. For non-aqueous samples, the factor is on a wet-weight basis.

SINGLE LABORATORY ACCURACY & PRECISION DATA FOR VOCs IN WATER

This method was tested in a single lab using water spiked at 10 µg/L and the following data was reported:

Recoveries and standard deviations were determined from seven samples and spiked at 10 µg/L of each analyte. Recoveries were determined by the internal standard method. Internal standards were: Fluorobenzene for PID and 2-Bromo-1-chloropropane for HECD.

The average recovery (in percent) for the PID was none (no response for this detector).
The standard deviation of the recovery for the PID was none (no response for this detector)-.
The MDL (in µg/mL) for the PID was none (no response for this detector).
The average recovery (in percent) for the HECD was 104.
The standard deviation of the recovery for the HECD was 3.4.
The MDL (in µg/mL) for the HECD was 0.03.

SAMPLE COLLECTION, PRESERVATION & HANDLING

Volatile organics — Standard 40-mL glass screw-cap VOA vials with Teflon®-faced silicone septum may be used for both liquid and solid matrices. When collecting samples, liquids and solids should be introduced into the vials gently to reduce agitation which might drive off volatile compounds. If there are any air bubbles present the sample must be retaken. Tap slightly as they are filled to try and eliminate as much free air space as possible. The two vials from each sampling locations should be sealed in separate plastic bags to prevent cross-contamination between samples particularly if the sampled waste is suspected of containing high levels of volatile organics.

Semivolatile organics — Containers used to collect samples for the determination of semivolatile organic compounds should be soap and water washed followed by methanol (or isopropanol) rinsing. The sample containers should be of glass or Teflon® and have screw-top covers with Teflon® liners.

Preservation for volatile organics — No preservation is used with concentrated waste samples. With liquid samples containing no residual chlorine, 4 drops of concentrated hydrochloric acid are added and the samples are immediately cooled to 4°C. When liquid samples contain residual chlorine, they are treated as above and, in addition, 4 drops of 4% aqueous sodium thiosulfate are added. Soil, sediment, and sludge samples are only cooled to 4°C.

Preservation for semivolatile organics — No preservation is used with concentrated waste samples. With liquid samples containing no residual chlorine and with soil, sediment, and sludge samples, immediately cooling to 4°C is the only preservation used. When residual chlorine is present then 3 mL of 10% aqueous sodium sulfate is added for each gallon of sample collected, followed by cooling to 4°C.

MHT The holding time for all volatile organics samples is 14 days. Liquid samples must be extracted within 7 days and their extracts analyzed within 40 days. Concentrated waste, soil, sediment, and sludge samples must be extracted within 14 days and their extracts analyzed within 40 days.

SAMPLE PREPARATION Volatile compounds are introduced into the gas chromatograph either by direct injector or purge-and-trap (EPA Method 5030). EPA Method 5030 may be used directly on groundwater samples or low-concentration contaminated soils and sediments. For medium-concentration soils or sediments, methanolic extraction, as described in EPA Method 5030, may be necessary prior to purge-and-trap analysis.

QUALITY CONTROL Calculate surrogate standard recovery on all samples, blanks, and spikes. A trip blank is recommended to check on sampling, storage, and handling contamination. Calibration standards, at a minimum of five concentration levels, are prepared in organic-free reagent water. One of the concentration levels should be at a concentration near, but above, the method detection limit.

A combination of bromochloromethane, 2-bromo-1-chloropropane, 1,4-dichlorobutane, and bromochlorobenzene are recommended as surrogate standards to encompass the range of the temperature program used in this method.

REFERENCE Test Methods for Evaluating Solid Waste, Physical/Chemical Methods, SW-846, 3rd Edition, U.S. EPA, Office of Solid Waste, Washington, DC, EPA Method 8021A, Rev. 1, Nov. 1992.

1,1,2-Trichloroethane   EPA Method 8021
CAS #79-00-5

TITLE Halogenated Volatile by Gas Chromatography Using Photoionization and Electrolytic Conductivity Detectors in Series: Capillary Column Technique

MATRIX This method is applicable to nearly all types of samples, regardless of water content, including groundwater, aqueous sludges, caustic liquors, acid liquors, waste solvents, oily wastes, mousses, tars, fibrous wastes, polymeric emulsions, filter cakes, spent carbons, spent catalysts, soils, and sediments.

METHOD SUMMARY This method is used to determine 60 volatile organic compounds in a variety of solid waste matrices. It provides GC conditions for the detection of halogenated and aromatic volatile organic compounds. Samples can be analyzed using direct injection or purge-and-trap (EPA Method 5030). Groundwater samples must be analyzed using EPA Method 5030 (where applicable). A temperature program is used with the GC. Detection is achieved by a photoionization detector (PID) and a Hall electrolytic conductivity detector (HECD) in series.

INTERFERENCES Samples can be contaminated by diffusion of volatile organics (particularly chlorofluorocarbons and methylene chloride) through the sample container septum during shipment and storage.

INSTRUMENTATION A GC-equipped with variable-constant differential flow controllers, subambient oven controller, PID and HECD detectors connected with a short piece of uncoated capillary tubing and a data system.

Column: 60 m × 0.75 mm I.D. VOCOL wide-bore capillary column with 1.5 µm film thickness.

PRECISION & ACCURACY MDLs are compound-dependent and vary with purging efficiency and concentration. The applicable concentration range of this method is compound- and instrument-dependent but is approximately 0.1 to 200 µg/L. Analytes that are inefficiently purged from water will not be detected when present at low concentrations, but they can be measured with acceptable accuracy and precision when present in sufficient amounts. The estimated quantitation limit (EQL) for an individual compound is approximately 1 µg/kg (wet weight) for soil/sediment samples, 100 µg/kg (wet weight) for wastes, and 1 µg/L for groundwater. EQLs will be proportionately higher for sample extracts and samples that require dilution or reduced sample size to avoid saturation of the detector.

MULTIPLICATION FACTORS FOR OTHER MATRICES (a)

Matrix	Factor (b)
Groundwater	10
Low-concentration soil	10
Water miscible liquid waste	500
High-concentration soil and sludge	1250
Non-water miscible waste	1250

(a) Sample EQLs are highly matrix-dependent. The EQLs listed herein are provided for guidance and may not always be achievable.
(b) EQL = [Method detection limit] × [Factor]. For non-aqueous samples, the factor is on a wet-weight basis.

SINGLE LABORATORY ACCURACY & PRECISION DATA FOR VOCs IN WATER
This method was tested in a single lab using water spiked at 10 µg/L and the following data was reported:

Recoveries and standard deviations were determined from seven samples and spiked at 10 µg/L of each analyte. Recoveries were determined by the internal standard method. Internal standards were: Fluorobenzene for PID and 2-Bromo-1-chloropropane for HECD.

The average recovery (in percent) for the PID was none (no response for this detector).
The standard deviation of the recovery for the PID was none (no response for this detector).
The MDL (in µg/mL) for the PID was none (no response for this detector).
The average recovery (in percent) for the HECD was 109.
The standard deviation of the recovery for the HECD was 6.2.
The MDL (in µg/mL) for the HECD was not determined.

SAMPLE COLLECTION, PRESERVATION & HANDLING
Volatile organics — Standard 40-mL glass screw-cap VOA vials with Teflon®-faced silicone septum may be used for both liquid and solid matrices. When collecting samples, liquids and solids should be introduced into the vials gently to reduce agitation which might drive off volatile compounds. If there are any air bubbles present the sample must be retaken. Tap slightly as they are filled to try and eliminate as much free air space as possible. The two vials from each sampling locations should be sealed in separate plastic bags to prevent cross-contamination between samples particularly if the sampled waste is suspected of containing high levels of volatile organics.

Semivolatile organics — Containers used to collect samples for the determination of semivolatile organic compounds should be soap and water washed followed by methanol (or isopropanol) rinsing. The sample containers should be of glass or Teflon® and have screw-top covers with Teflon® liners.

Preservation for volatile organics — No preservation is used with concentrated waste samples. With liquid samples containing no residual chlorine, 4 drops of concentrated hydrochloric acid are added and the samples are immediately cooled to 4°C. When liquid samples contain residual chlorine, they are treated as above and, in addition, 4 drops of 4% aqueous sodium thiosulfate are added. Soil, sediment, and sludge samples are only cooled to 4°C.

Preservation for semivolatile organics — No preservation is used with concentrated waste samples. With liquid samples containing no residual chlorine and with soil, sediment, and sludge samples, immediately cooling to 4°C is the only preservation used. When residual chlorine is present then 3 mL of 10% aqueous sodium sulfate is added for each gallon of sample collected, followed by cooling to 4°C.

MHT The holding time for all volatile organics samples is 14 days. Liquid samples must be extracted within 7 days and their extracts analyzed within 40 days. Concentrated waste, soil, sediment, and sludge samples must be extracted within 14 days and their extracts analyzed within 40 days.

SAMPLE PREPARATION Volatile compounds are introduced into the gas chromatograph either by direct injector or purge-and-trap (EPA Method 5030). EPA Method 5030 may be used directly on groundwater samples or low-concentration contaminated soils and sediments. For medium-concentration soils or sediments, methanolic extraction, as described in EPA Method 5030, may be necessary prior to purge-and-trap analysis.

QUALITY CONTROL Calculate surrogate standard recovery on all samples, blanks, and spikes. A trip blank is recommended to check on sampling, storage, and handling contamination. Calibration standards, at a minimum of five concentration levels, are prepared in organic-free reagent water. One of the concentration levels should be at a concentration near, but above, the method detection limit.

A combination of bromochloromethane, 2-bromo-1-chloropropane, 1,4-dichlorobutane, and bromochlorobenzene are recommended as surrogate standards to encompass the range of the temperature program used in this method.

REFERENCE Test Methods for Evaluating Solid Waste, Physical/Chemical Methods, SW-846, 3rd Edition, U.S. EPA, Office of Solid Waste, Washington, DC, EPA Method 8021A, Rev. 1, Nov. 1992.

1,1,1-Trichloroethane   EPA Method 8240
CAS #71-55-6

TITLE Volatile Organics By GC/MS: Packed Column Technique

MATRIX Nearly all types of sample matarices, regardless of water content, can be analyzed using this method. This includes groundwater, aqueous sludges, caustic liquors, acid liquors, waste solvents, oily wastes, mousses, tars, fibrous wastes, polymetric emulsions, filter cakes, spent carbons, spent catalysts, soils, and sediments.

METHOD SUMMARY Method 8240B covers 80 volatile organic compounds that are introduced into a gas chromatograph by the purge-and-trap method or by direct injection (in limited applications). For the purge-and-trap method an inert gas (zero grade nitrogen or helium) is bubbled through a 5-mL solution at ambient temperature. Purged sample components are trapped in a tube of sorbent materials. When purging is complete, the sorbent tube is heated and backflushed with inert gas to desorb the trapped components onto a GC column.

INTERFERENCES Impurities in the purge gas and from organic compounds outgassing from the plumbing ahead of the trap account for many contamination problems. Interferences purged or coextracted from the samples will vary considerably from source to source. Cross-contamination can occur whenever high-level and low-level samples are analyzed sequentially. Whenever an unusually concentrated sample is analyzed, it should be followed by the analysis of organic-free reagent water to check for cross-contamination. Samples also can be contaminated by diffusion of volatile organics (particularly methylene chloride and fluorocarbons) through the septum seal into the sample during shipment and storage. A trip blank can serve as a check on such contamination. The lab where volatile analysis is performed and also the refrigerated storage area should be completely free of solvents.

INSTRUMENTATION A gas chromatograph/mass spectrometry/data system (GC/MS) equipped with a 6 ft × 0.1 in I.D. glass column packed with 1% SP-1000 on Carbopack-B (60/80 mesh) is required. Also needed is a 5-mL purging device, a sorbent trap, and a thermal desorption apparatus.

PRECISION & ACCURACY This method is reported to have been tested by 15 laboratories using organic-free reagent water, drinking water, surface water, and industrial wastewaters (not specified) fortified at six concentrations over the range 5–600 µg/L.

Sample estimated quantitation limits (EQLs) are highly matrix-dependent. The EQLs listed may not always be achievable. EQLs listed for soils or sediments are based on wet weight. Normally, data is reported on a dry-weight basis; therefore, EQLs will be higher, based on the percent dry weight of each sample. Note that EQLs are even more variable than MDLs and that they are highly variable depending on the matrix being analyzed.

EQL in groundwater in µg/L was 5.
EQL in low soil or sediment in µg/kg was 5.
Accuracy (a) in µg/L was 1.06C + 0.73.
Precision (b) in µg/L was 0.21x-0.39.

(a) *Average recovery found for measurements of samples containing a concentration of C, in µg/L.*
(b) *Overall precision found for measurements of samples with average recovery X for samples containing a concentration of C in µg/L.*

$X =$ *Average recovery found for measurement of samples containing a concentration of C in µg/L.*

MULTIPLICATION FACTORS FOR OTHER MATRICES

Other Matrices	Factor (a)
Waste miscible liquid waste	50
High-concentration soil and sludge	125
Non-water miscible waste	500

(a) EQL = [EQL for low soil sediment X [Factor]. For non-aqueous samples, the factor is on a wet-weight basis.

SAMPLING METHOD
Liquid samples — Use a 40-mL glass screw-cap VOA vial with a Teflon®-faced silicone septum that has been prewashed, rinsed with distilled deionized water, and oven dried. However, if residual chlorine is present, collect sample in a 40-oz. soil VOA container which has been pre-preserved with 4 drops of 10% sodium thiosulfate, mix gently, and then transfer the sample to a 40-mL VOA vial. Collect bubble-free samples in duplicate and seal them in separate plastic bags.

Soils or sediments, and sludges — Use an 8-oz. widemouth glass bottle with a Teflon®-faced silicone septum that has been prewashed with detergent, rinsed with distilled deionized water, and oven dried. Tap slightly to eliminate free air space. Collect samples in duplicate and seal them in separate plastic bags.

SAMPLE PRESERVATION
Liquid samples — Add 4 drops of concentrated HCL and immediately cool samples to 4°C and store in a solvent-free refrigerator.

Soils or sediments, and sludges — Cool samples to 4°C and store in a solvent-free refrigerator.

MHT Maximum holding time is 14 days from the date of sample collection.

SAMPLE PREPARATION
Liquid samples — Remove the plunger from a 5-mL syringe and carefully pour the sample into the syringe barrel to just short of overflowing. Replace the syringe plunger and compress the sample. Open the syringe valve and vent any residual air while adjusting the sample volume to 5.0 mL. If there is only one volatile organic analysis (VOA) vial, a second syringe should be filled at this time to protect against possible loss of sample integrity. Add 10 µL of surrogate spiking solution and 10 µL of internal standard spiking solution through the valve bore of the 5-mL syringe, then close the valve. The surrogate and internal standards may be mixed and added as a single spiking solution.

Sediments, soils, and waste samples — All samples of this type should be screened by GC analysis using a headspace method (EPA Method 3810) or the hexadecane extraction and screening method (EPA Method 3820). Use the screening data to determine whether to use the low-concentration method (0.005–1 mg/kg) or the high-concentration method (>1 mg/kg).

Low-concentration method — The low-concentration method is based on purging a heated sediment or soil sample mixed with organic-free reagent water containing the surrogate and

internal standards. Analyze all reagent blanks and standards under the same conditions as the samples.

Use a 5-g sample if the expected concentration is <0.1 mg/kg or a 1-g sample for expected concentrations between 0.1 and 1 mg/kg. Mix the contents of the sample container with a narrow metal spatula. Weigh the amount of the sample into a tared purge device. Add the spiked water to the purge device, which contains the weighed amount of sample, and connect the device to the purge-and-trap system.

High-concentration method — This method is based on extracting the sediment or soil with methanol. A waste sample is either extracted or diluted, depending on its solubility in methanol. Wastes that are insoluble in methanol are diluted with reagent tetraglyme or possibly polyethylene glycol (PEG). An aliquot of the extract is added to organic-free reagent water containing surrogate and internal standards. This is purged at ambient temperature. All samples with an expected concentration of >1.0 mg/kg should be analyzed by this method.

Mix the contents of the sample container with a narrow metal spatula. For sediments or soils and solid wastes that are insoluble in methanol, weigh 4 g (wet weight) of sample into a tared 20-mL vial. For waste that is soluble in methanol, tetraglyme, or PEG, weigh 1 g (wet weight) into a tared scintillation vial or culture tube or a 10-mL volumetric flask. Quickly add 9.0 mL of appropriate solvent then add 1.0 mL of a surrogate spiking solution to the vial, cap it, and shake it for 2 min.

METHANOL EXTRACT REQUIRED FOR ANALYSIS OF HIGH-CONCENTRATION SOILS OR SEDIMENTS

Approximate Concentration Range	Volume of Methanol Extract (a)
500–10,000 µg/kg	100 µL
1,000–20,000 µg/kg	50 µL
5,000–100,000 µg/kg	10 µL
25,000–500,000 µg/kg	100 µL of 1/50 dilution (b)

Calculate appropriate dilution factor for concentrations exceeding this table.

(a) The volume of methanol added to 5 mL of water being purged should be kept constant. Therefore, add to the 5-mL syringe whatever volume of methanol is necessary to maintain a volume of 100 µL added to the syringe.
(b) Dilute an aliquot of the methanol extract and then take 100 µL for analysis.

QUALITY CONTROL Demonstrate, through the analysis of a reagent water blank, that interferences from the analytical system, glassware, and reagents are under control. Blank samples should be carried through all stages of the sample preparation and measurement steps. For each analytical batch (up to 20 samples), a reagent blank, matrix spike, and matrix spike duplicate must be analyzed (the frequency of the spikes may be different for different monitoring programs). The blank and spiked samples must be carried through all stages of the sample preparation and measurement steps. QC samples mentioned in the section on Interferences will also be needed as appropriate to those situations.

REFERENCE Test Methods for Evaluating Solid Waste (SW-846). U.S. EPA. 1983. Method 8240B, Rev. 2, Nov. 1990. Office of Solid Wastes, Washington, DC.

1,1,2-Trichloroethane   EPA Method 8240
CAS #79-00-5

TITLE Volatile Organics By GC/MS: Packed Column Technique

MATRIX Nearly all types of sample matarices, regardless of water content, can be analyzed using this method. This includes groundwater, aqueous sludges, caustic liquors, acid liquors, waste solvents, oily wastes, mousses, tars, fibrous wastes, polymetric emulsions, filter cakes, spent carbons, spent catalysts, soils, and sediments.

METHOD SUMMARY Method 8240B covers 80 volatile organic compounds that are introduced into a gas chromatograph by the purge-and-trap method or by direct injection (in limited applications). For the purge-and-trap method an inert gas (zero grade nitrogen or helium) is bubbled through a 5-mL solution at ambient temperature. Purged sample components are trapped in a tube of sorbent materials. When purging is complete, the sorbent tube is heated and backflushed with inert gas to desorb the trapped components onto a GC column.

INTERFERENCES Impurities in the purge gas and from organic compounds outgassing from the plumbing ahead of the trap account for many contamination problems. Interferences purged or coextracted from the samples will vary considerably from source to source. Cross-contamination can occur whenever high-level and low-level samples are analyzed sequentially. Whenever an unusually concentrated sample is analyzed, it should be followed by the analysis of organic-free reagent water to check for cross-contamination. Samples also can be contaminated by diffusion of volatile organics (particularly methylene chloride and fluorocarbons) through the septum seal into the sample during shipment and storage. A trip blank can serve as a check on such contamination. The lab where volatile analysis is performed and also the refrigerated storage area should be completely free of solvents.

INSTRUMENTATION A gas chromatograph/mass spectrometry/data system (GC/MS) equipped with a 6 ft × 0.1 in I.D. glass column packed with 1% SP-1000 on Carbopack-B (60/80 mesh) is required. Also needed is a 5-mL purging device, a sorbent trap, and a thermal desorption apparatus.

PRECISION & ACCURACY This method is reported to have been tested by 15 laboratories using organic-free reagent water, drinking water, surface water, and industrial wastewaters (not specified) fortified at six concentrations over the range 5–600 µg/L.

Sample estimated quantitation limits (EQLs) are highly matrix-dependent. The EQLs listed may not always be achievable. EQLs listed for soils or sediments are based on wet weight. Normally, data is reported on a dry-weight basis; therefore, EQLs will be higher, based on the percent dry weight of each sample. Note that EQLs are even more variable than MDLs and

that they are highly variable depending on the matrix being analyzed.

EQL in groundwater in μg/L was 5.
EQL in low soil or sediment in μg/kg was 5.
Accuracy (a) in μg/L was 0.95C + 1.71.
Precision (b) in μg/L was 0.18x + 0.00.

(a) *Average recovery found for measurements of samples containing a concentration of C, in μg/L.*
(b) *Overall precision found for measurements of samples with average recovery X for samples containing a concentration of C in μg/L.*
X = *Average recovery found for measurement of samples containing a concentration of C in μg/L.*

MULTIPLICATION FACTORS FOR OTHER MATRICES

Other Matrices	Factor (a)
Waste miscible liquid waste	50
High-concentration soil and sludge	125
Non-water miscible waste	500

(a) *EQL = [EQL for low soil sediment X [Factor]. For non-aqueous samples, the factor is on a wet-weight basis.*

SAMPLING METHOD
Liquid samples — Use a 40-mL glass screw-cap VOA vial with a Teflon®-faced silicone septum that has been prewashed, rinsed with distilled deionized water, and oven dried. However, if residual chlorine is present, collect sample in a 40-oz. soil VOA container which has been pre-preserved with 4 drops of 10% sodium thiosulfate, mix gently, and then transfer the sample to a 40-mL VOA vial. Collect bubble-free samples in duplicate and seal them in separate plastic bags.

Soils or sediments, and sludges — Use an 8-oz. widemouth glass bottle with a Teflon®-faced silicone septum that has been prewashed with detergent, rinsed with distilled deionized water, and oven dried. Tap slightly to eliminate free air space. Collect samples in duplicate and seal them in separate plastic bags.

SAMPLE PRESERVATION
Liquid samples — Add 4 drops of concentrated HCL and immediately cool samples to 4°C and store in a solvent-free refrigerator.

Soils or sediments, and sludges — Cool samples to 4°C and store in a solvent-free refrigerator.

MHT Maximum holding time is 14 days from the date of sample collection.

SAMPLE PREPARATION
Liquid samples — Remove the plunger from a 5-mL syringe and carefully pour the sample into the syringe barrel to just short of overflowing. Replace the syringe plunger and compress the sample. Open the syringe valve and vent any residual air while adjusting the sample volume to 5.0 mL. If there is only one volatile organic analysis (VOA) vial, a second syringe should be filled at this time to protect against possible loss of sample integrity. Add 10 μL of surrogate spiking solution and 10 μL of internal standard spiking solution through the valve bore of the 5-mL syringe, then close the valve. The surrogate and internal standards may be mixed and added as a single spiking solution.

Sediments, soils, and waste samples — All samples of this type should be screened by GC analysis using a headspace method (EPA Method 3810) or the hexadecane extraction and screening method (EPA Method 3820). Use the screening data to determine whether to use the low-concentration method (0.005–1 mg/kg) or the high-concentration method (>1 mg/kg).

Low-concentration method — The low-concentration method is based on purging a heated sediment or soil sample mixed with organic-free reagent water containing the surrogate and internal standards. Analyze all reagent blanks and standards under the same conditions as the samples.

Use a 5-g sample if the expected concentration is <0.1 mg/kg or a 1-g sample for expected concentrations between 0.1 and 1 mg/kg. Mix the contents of the sample container with a narrow metal spatula. Weigh the amount of the sample into a tared purge device. Add the spiked water to the purge device, which contains the weighed amount of sample, and connect the device to the purge-and-trap system.

High-concentration method — This method is based on extracting the sediment or soil with methanol. A waste sample is either extracted or diluted, depending on its solubility in methanol. Wastes that are insoluble in methanol are diluted with reagent tetraglyme or possibly polyethylene glycol (PEG). An aliquot of the extract is added to organic-free reagent water containing surrogate and internal standards. This is purged at ambient temperature. All samples with an expected concentration of >1.0 mg/kg should be analyzed by this method.

Mix the contents of the sample container with a narrow metal spatula. For sediments or soils and solid wastes that are insoluble in methanol, weigh 4 g (wet weight) of sample into a tared 20-mL vial. For waste that is soluble in methanol, tetraglyme, or PEG, weigh 1 g (wet weight) into a tared scintillation vial or culture tube or a 10-mL volumetric flask. Quickly add 9.0 mL of appropriate solvent then add 1.0 mL of a surrogate spiking solution to the vial, cap it, and shake it for 2 min.

METHANOL EXTRACT REQUIRED FOR ANALYSIS OF HIGH-CONCENTRATION SOILS OR SEDIMENTS

Approximate Concentration Range	Volume of Methanol Extract (a)
500–10,000 μg/kg	100 μL
1,000–20,000 μg/kg	50 μL
5,000–100,000 μg/kg	10 μL
25,000–500,000 μg/kg	100 μL of 1/50 dilution (b)

Calculate appropriate dilution factor for concentrations exceeding this table.

(a) *The volume of methanol added to 5 mL of water being purged should be kept constant. Therefore, add to the 5-mL syringe whatever volume of methanol is necessary to maintain a volume of 100 μL added to the syringe.*
(b) *Dilute an aliquot of the methanol extract and then take 100 μL for analysis.*

QUALITY CONTROL Demonstrate, through the analysis of a reagent water blank, that interferences from the analytical system, glassware, and reagents are under control. Blank samples should be carried through all stages of the sample preparation and measurement steps. For each analytical batch (up to 20 samples), a reagent blank, matrix spike, and matrix spike duplicate must be analyzed (the frequency of the spikes may be different for different monitoring programs). The blank and spiked samples must be carried through all stages of the sample preparation and measurement steps. QC samples mentioned in the section on Interferences will also be needed as appropriate to those situations.

REFERENCE Test Methods for Evaluating Solid Waste (SW-846). U.S. EPA. 1983. Method 8240B, Rev. 2, Nov. 1990. Office of Solid Wastes, Washington, DC.

1,1,1-Trichloroethane   **EPA Method 8260**
CAS #71-55-6

TITLE Volatile Organic Compounds by GC/MS: Capillary Column Technique

MATRIX This method is applicable to nearly all types of samples, regardless of water content, including groundwater, soils, and sediments.

METHOD SUMMARY Method 8260A covers 58 volatile organic compounds that are introduced into a gas chromatograph by the purge-and-trap method or by direct injection (in limited applications). Zero-grade helium is bubbled through a 5-mL solution at ambient temperature. Purged sample components are trapped in a tube containing suitable sorbent materials. When purging is complete, the sorbent tube is heated and backflushed with helium to desorb trapped sample components. The analytes are desorbed directly to a large bore capillary or cryofocussed on a capillary precolumn before being flash evaporated to a narrow bore capillary for analysis.

INTERFERENCES Major contaminant sources are volatile materials in the lab and impurities in the inert purging gas and in the sorbent trap. Interfering contamination may occur when a sample containing low concentrations of volatile organic compounds is analyzed immediately after a sample containing high concentrations of volatile organic compounds. After analysis of a sample containing high concentrations of volatile organic compounds, one or more calibration blanks should be analyzed to check for cross-contamination. Screening of the samples prior to purge-and-trap GC/MS analysis is highly recommended to prevent contamination of the system. This is especially true for soil and waste samples.

Special precautions must be taken to analyze for methylene chloride. The analytical and sample storage area should be isolated from all atmospheric sources of methylene chloride. All gas chromatography carrier gas lines and purge gas plumbing should be constructed from stainless steel or copper tubing. Laboratory clothing previously exposed to methylene chloride fumes during liquid-liquid extraction procedures can contribute to sample contamination.

Samples can also be contaminated by diffusion of volatile organics (particularly methylene chloride and fluorocarbons) through the septum seal during shipment and storage. A trip blank can serve as a check on such contamination.

INSTRUMENTATION GC/MS with a temperature-programmable chromatograph suitable for splitless injection equipped with variable constant differential flow controllers, a subambient oven controller, a purging device, sorbent trap, a thermal desorption apparatus and a capillary precolumn interface when using cryogenic cooling will be needed. The following GC columns may be used:

Column 1: 60 m × 0.75mm I.D. capillary column coated with VOCOL, 1.5 µm film thickness.
Column 2: 30 m × 0.53mm capillary column coated with DB-624 or VOCOL, 3 µm film thickness.
Column 3: 30 m × 0.32mm I.D. capillary column coated with DB-5 or SE-54, 1-µm film thickness.

PRECISION & ACCURACY This method has been tested in a single lab using spiked water. Using a wide-bore capillary column, water was spiked at concentrations between 0.5 and 10 µg/L. Single lab accuracy and precision data are presented. The MDL actually achieved in a given analysis will vary depending on instrument sensitivity and matrix effects.

The MDL (a) in µg/L was 0.08.
The concentration range in µg/L was 0.5–10.
The mean accuracy (% of true value) was 98.
The precision as relative standard deviation was 8.1.

Note: The MDL is based on a 25-mL sample volume instead of a 5-mL sample volume.

SAMPLING METHOD
Liquid samples — Use a 40-mL glass screw-cap VOA vial with a Teflon®-faced silicone septum that has been prewashed, rinsed with distilled deionized water, and oven dried. If residual chlorine is present, collect the sample in a 4-oz soil VOA container which has been pre-preserved with 4 drops of 10% sodium thiosulfate. Mix gently and transfer the sample to a 40-mL VOA vial. Collect bubble-free samples in duplicate and seal each sample in a separate plastic bag.

Soils, sediments and sludges — Use an 8-oz widemouth glass bottle with Teflon®-faced silicone septum that has been prewashed, rinsed with distilled deionized water, and oven dried. **Do not** heat the septum for more than 1 h. Tap slightly to eliminate any free air space. Collect samples in duplicate and seal each one in a separate plastic bag.

SAMPLE PRESERVATION
Liquid samples — Add 4 drops of concentrated HCL, cool to 4°C and store in a solvent-free refrigerator.

Soils, sediments and sludges — Cool samples to 4°C and store in a solvent-free refrigerator.

MHT The maximum holding time of any sample (liquids, soils, sediments, and sludges) is 14 days.

SAMPLE PREPARATION
Liquid samples — Remove the plunger from a 5-mL syringe and carefully pour the sample into the syringe barrel to just

short of overflowing. Replace the syringe plunger and compress the sample. Open the syringe valve and vent any residual air while adjusting the sample volume to 5.0 mL. If there is only one volatile organic analysis (VOA) vial, a second syringe should be filled at this time to protect against possible loss of sample integrity. Add 10 μL of surrogate spiking solution and 10 μL of internal standard spiking solution through the valve bore of the 5-mL syringe, then close the valve. The surrogate and internal standards may be mixed and added as a single spiking solution.

Sediments, soils, and waste samples — All samples of this type should be screened by GC analysis using a headspace method (EPA Method 3810) or the hexadecane extraction and screening method (EPA Method 3820). Use the screening data to determine whether to use the low-concentration method (0.005–1 mg/kg) or the high-concentration method (>1 mg/kg).

Low-concentration method — The low-concentration method is based on purging a heated sediment or soil sample mixed with organic-free reagent water containing the surrogate and internal standards. Analyze all reagent blanks and standards under the same conditions as the samples.

Use a 5-g sample if the expected concentration is <0.1 mg/kg or a 1-g sample for expected concentrations between 0.1 and 1 mg/kg. Mix the contents of the sample container with a narrow metal spatula. Weigh the amount of the sample into a tared purge device. Add the spiked water to the purge device, which contains the weighed amount of sample, and connect the device to the purge-and-trap system.

High-concentration method — This method is based on extracting the sediment or soil with methanol. A waste sample is either extracted or diluted, depending on its solubility in methanol. Wastes that are insoluble in methanol are diluted with reagent tetraglyme or possibly polyethylene glycol (PEG). An aliquot of the extract is added to organic-free reagent water containing surrogate and internal standards. This is purged at ambient temperature. All samples with an expected concentration of >1.0 mg/kg should be analyzed by this method.

Mix the contents of the sample container with a narrow metal spatula. For sediments or soils and solid wastes that are insoluble in methanol, weigh 4 g (wet weight) of sample into a tared 20-mL vial. For waste that is soluble in methanol, tetraglyme, or PEG, weigh 1 g (wet weight) into a tared scintillation vial or culture tube or a 10-mL volumetric flask. Quickly add 9.0 mL of appropriate solvent then add 1.0 mL of a surrogate spiking solution to the vial, cap it, and shake it for 2 min.

METHANOL EXTRACT REQUIRED FOR ANALYSIS OF HIGH-CONCENTRATION SOILS OR SEDIMENTS

Approximate Concentration Range	Volume of Methanol Extract (a)
500–10,000 μg/kg	100 μL
1,000–20,000 μg/kg	50 μL
5,000–100,000 μg/kg	10 μL
25,000–500,000 μg/kg	100 μL of 1/50 dilution (b)

Calculate appropriate dilution factor for concentrations exceeding this table.

(a) The volume of methanol added to 5 mL of water being purged should be kept constant. Therefore, add to the 5-mL syringe whatever volume of methanol is necessary to maintain a volume of 100 μL added to the syringe.
(b) Dilute an aliquot of the methanol extract and then take 100 μL for analysis.

QUALITY CONTROL Demonstrate, through the analysis of a reagent water blank, that interferences from the analytical system, glassware, and reagents are under control. Blank samples should be carried through all stages of the sample preparation and measurement steps. For each analytical batch (up to 20 samples), a reagent blank, matrix spike, and matrix spike duplicate must be analyzed (the frequency of the spikes may be different for different monitoring programs). The blank and spiked samples must be carried through all stages of the sample preparation and measurement steps. QC samples mentioned in the section on Interferences will also be needed as appropriate to those situations.

Matrix spiking standards should be prepared from volatile organic compounds which will be representative of the compounds being investigated. The recommended internal standards are chlorobenzene-d5, 1,4-difluorobenzene, 1,4-dichlorobenzene-d4, and pentafluorobenzene. Using stock standard solutions, prepare secondary dilution standards containing the compounds of interest, either singly or mixed together in methanol. Store them in a vial with no headspace for no more than one week. Surrogates recommended are toluene-d8, 4-bromofluorobenzene, and dibromofluoromethane. Each sample undergoing GC/MS analysis must be spiked with 10 μL of the surrogate spiking solution prior to analysis.

REFERENCE Test Methods for Evaluating Solid Waste (SW-846). U.S. EPA 1983, Method 8260A, Rev. 1, Nov. 1990. Office of Solid Waste, Washington, DC.

1,1,2-Trichloroethane **EPA Method 8260**
CAS #79-00-5

TITLE Volatile Organic Compounds by GC/MS: Capillary Column Technique

MATRIX This method is applicable to nearly all types of samples, regardless of water content, including groundwater, soils, and sediments.

METHOD SUMMARY Method 8260A covers 58 volatile organic compounds that are introduced into a gas chromatograph by the purge-and-trap method or by direct injection (in limited applications). Zero-grade helium is bubbled through a 5-mL solution at ambient temperature. Purged sample components are trapped in a tube containing suitable sorbent materials. When purging is complete, the sorbent tube is heated and backflushed with helium to desorb trapped sample components. The analytes are desorbed directly to a large bore capillary or cryofocussed on a capillary precolumn before being flash evaporated to a narrow bore capillary for analysis.

INTERFERENCES Major contaminant sources are volatile materials in the lab and impurities in the inert purging gas and in the sorbent trap. Interfering contamination may occur when a sample containing low concentrations of volatile organic compounds is analyzed immediately after a sample containing high concentrations of volatile organic compounds. After analysis of a sample containing high concentrations of volatile organic compounds, one or more calibration blanks should be analyzed to check for cross-contamination. Screening of the samples prior to purge-and-trap GC/MS analysis is highly recommended to prevent contamination of the system. This is especially true for soil and waste samples.

Special precautions must be taken to analyze for methylene chloride. The analytical and sample storage area should be isolated from all atmospheric sources of methylene chloride. All gas chromatography carrier gas lines and purge gas plumbing should be constructed from stainless steel or copper tubing. Laboratory clothing previously exposed to methylene chloride fumes during liquid-liquid extraction procedures can contribute to sample contamination.

Samples can also be contaminated by diffusion of volatile organics (particularly methylene chloride and fluorocarbons) through the septum seal during shipment and storage. A trip blank can serve as a check on such contamination.

INSTRUMENTATION GC/MS with a temperature-programmable chromatograph suitable for splitless injection equipped with variable constant differential flow controllers, a subambient oven controller, a purging device, sorbent trap, a thermal desorption apparatus and a capillary precolumn interface when using cryogenic cooling will be needed. The following GC columns may be used:

Column 1: 60 m × 0.75mm I.D. capillary column coated with VOCOL, 1.5 µm film thickness.
Column 2: 30 m × 0.53mm capillary column coated with DB-624 or VOCOL, 3 µm film thickness.
Column 3: 30 m × 0.32mm I.D. capillary column coated with DB-5 or SE-54, 1-µm film thickness.

PRECISION & ACCURACY This method has been tested in a single lab using spiked water. Using a wide-bore capillary column, water was spiked at concentrations between 0.5 and 10 µg/L. Single lab accuracy and precision data are presented. The MDL actually achieved in a given analysis will vary depending on instrument sensitivity and matrix effects.

The MDL (a) in µg/L was 0.10.
The concentration range in µg/L was 0.5–10.
The mean accuracy (% of true value) was 104.
The precision as relative standard deviation was 7.3.

Note: The MDL is based on a 25-mL sample volume instead of a 5-mL sample volume.

SAMPLING METHOD

Liquid samples — Use a 40-mL glass screw-cap VOA vial with a Teflon®-faced silicone septum that has been prewashed, rinsed with distilled deionized water, and oven dried. If residual chlorine is present, collect the sample in a 4-oz soil VOA container which has been pre-preserved with 4 drops of 10% sodium thiosulfate. Mix gently and transfer the sample to a 40-mL VOA vial. Collect bubble-free samples in duplicate and seal each sample in a separate plastic bag.

Soils, sediments and sludges — Use an 8-oz widemouth glass bottle with Teflon®-faced silicone septum that has been prewashed, rinsed with distilled deionized water, and oven dried. **Do not** heat the septum for more than 1 h. Tap slightly to eliminate any free air space. Collect samples in duplicate and seal each one in a separate plastic bag.

SAMPLE PRESERVATION

Liquid samples — Add 4 drops of concentrated HCL, cool to 4°C and store in a solvent-free refrigerator.

Soils, sediments and sludges — Cool samples to 4°C and store in a solvent-free refrigerator.

MHT The maximum holding time of any sample (liquids, soils, sediments, and sludges) is 14 days.

SAMPLE PREPARATION

Liquid samples — Remove the plunger from a 5-mL syringe and carefully pour the sample into the syringe barrel to just short of overflowing. Replace the syringe plunger and compress the sample. Open the syringe valve and vent any residual air while adjusting the sample volume to 5.0 mL. If there is only one volatile organic analysis (VOA) vial, a second syringe should be filled at this time to protect against possible loss of sample integrity. Add 10 µL of surrogate spiking solution and 10 µL of internal standard spiking solution through the valve bore of the 5-mL syringe, then close the valve. The surrogate and internal standards may be mixed and added as a single spiking solution.

Sediments, soils, and waste samples — All samples of this type should be screened by GC analysis using a headspace method (EPA Method 3810) or the hexadecane extraction and screening method (EPA Method 3820). Use the screening data to determine whether to use the low-concentration method (0.005–1 mg/kg) or the high-concentration method (>1 mg/kg).

Low-concentration method — The low-concentration method is based on purging a heated sediment or soil sample mixed with organic-free reagent water containing the surrogate and internal standards. Analyze all reagent blanks and standards under the same conditions as the samples.

Use a 5-g sample if the expected concentration is <0.1 mg/kg or a 1-g sample for expected concentrations between 0.1 and 1 mg/kg. Mix the contents of the sample container with a narrow metal spatula. Weigh the amount of the sample into a tared purge device. Add the spiked water to the purge device, which contains the weighed amount of sample, and connect the device to the purge-and-trap system.

High-concentration method — This method is based on extracting the sediment or soil with methanol. A waste sample is either extracted or diluted, depending on its solubility in methanol. Wastes that are insoluble in methanol are diluted with reagent tetraglyme or possibly polyethylene glycol (PEG). An aliquot of the extract is added to organic-free reagent water containing surrogate and internal standards. This is purged at

ambient temperature. All samples with an expected concentration of >1.0 mg/kg should be analyzed by this method.

Mix the contents of the sample container with a narrow metal spatula. For sediments or soils and solid wastes that are insoluble in methanol, weigh 4 g (wet weight) of sample into a tared 20-mL vial. For waste that is soluble in methanol, tetraglyme, or PEG, weigh 1 g (wet weight) into a tared scintillation vial or culture tube or a 10-mL volumetric flask. Quickly add 9.0 mL of appropriate solvent then add 1.0 mL of a surrogate spiking solution to the vial, cap it, and shake it for 2 min.

METHANOL EXTRACT REQUIRED FOR ANALYSIS OF HIGH-CONCENTRATION SOILS OR SEDIMENTS

Approximate Concentration Range	Volume of Methanol Extract (a)
500–10,000 μg/kg	100 μL
1,000–20,000 μg/kg	50 μL
5,000–100,000 μg/kg	10 μL
25,000–500,000 μg/kg	100 μL of 1/50 dilution (b)

Calculate appropriate dilution factor for concentrations exceeding this table.

(a) The volume of methanol added to 5 mL of water being purged should be kept constant. Therefore, add to the 5-mL syringe whatever volume of methanol is necessary to maintain a volume of 100 μL added to the syringe.

(b) Dilute an aliquot of the methanol extract and then take 100 μL for analysis.

QUALITY CONTROL Demonstrate, through the analysis of a reagent water blank, that interferences from the analytical system, glassware, and reagents are under control. Blank samples should be carried through all stages of the sample preparation and measurement steps. For each analytical batch (up to 20 samples), a reagent blank, matrix spike, and matrix spike duplicate must be analyzed (the frequency of the spikes may be different for different monitoring programs). The blank and spiked samples must be carried through all stages of the sample preparation and measurement steps. QC samples mentioned in the section on Interferences will also be needed as appropriate to those situations.

Matrix spiking standards should be prepared from volatile organic compounds which will be representative of the compounds being investigated. The recommended internal standards are chlorobenzene-d5, 1,4-difluorobenzene, 1,4-dichlorobenzene-d4, and pentafluorobenzene. Using stock standard solutions, prepare secondary dilution standards containing the compounds of interest, either singly or mixed together in methanol. Store them in a vial with no headspace for no more than one week. Surrogates recommended are toluene-d8, 4-bromofluorobenzene, and dibromofluoromethane. Each sample undergoing GC/MS analysis must be spiked with 10 μL of the surrogate spiking solution prior to analysis.

REFERENCE Test Methods for Evaluating Solid Waste (SW-846). U.S. EPA 1983, Method 8260A, Rev. 1, Nov. 1990. Office of Solid Waste, Washington, DC.

1,1,1-Trichloroethane   **EPA Method 601**
CAS #71-55-6

TITLE Purgeable Halocarbons

MATRIX Wastewater.

APPLICATION Method covers 29 purgeable halocarbons. (Method 624 provides GC/MS conditions appropriate for the qualitative and quantitative confirmation of results). Method describes conditions for a 2nd GC column to confirm measurements made with primary column.

INTERFERENCES Impurities in the purge gas and organic compounds outgassing from the plumbing ahead of the trap. With high- and low-level samples, there can be carryover contamination. Diffusion of volatile organics through the septum seal into the sample.

INSTRUMENTATION GC-equipped with halide-specific detector. (With purge-and-trap unit).

RANGE 8.0–500 μg/L.

MDL 0.03 μg/L.

PRECISION 0.20X + 0.37 μg/L (overall precision).

ACCURACY 0.90C–0.16 μg/L (as recovery).

SAMPLING METHOD 25-mL glass vial. Teflon®-lined septum.

STABILITY Cool, 4°C, 0.008% Sodium thiosulfate.

MHT 14 days.

QUALITY CONTROL The lab must on an ongoing basis, spike at least 10% of the samples from each sample site being monitored to assess accuracy.

REFERENCE Method 601, *Federal Register* Part VIII 40 CFR Part 136, Oct 26, 1984.

1,1,2-Trichloroethane   **EPA Method 601**
CAS #79-00-5

TITLE Purgeable Halocarbons

MATRIX Wastewater.

APPLICATION Method covers 29 purgeable halocarbons. (Method 624 provides GC/MS conditions appropriate for the qualitative and quantitative confirmation of results). Method describes conditions for a 2nd GC column to confirm measurements made with primary column.

INTERFERENCES Impurities in the purge gas and organic compounds outgassing from the plumbing ahead of the trap. With high- and low-level samples, there can be carryover contamination. Diffusion of volatile organics through the septum seal into the sample.

INSTRUMENTATION GC-equipped with halide-specific detector. (With purge-and-trap unit).

RANGE 8.0–500 μg/L.

MDL 0.02 µg/L.

PRECISION 0.19X + 0.67 µg/L (overall precision).

ACCURACY 0.86C + 0.30 µg/L (as recovery).

SAMPLING METHOD 25-mL glass vial. Teflon®-lined septum.

STABILITY Cool, 4°C, 0.008% Sodium thiosulfate.

MHT 14 days.

QUALITY CONTROL The lab must on an ongoing basis, spike at least 10% of the samples from each sample site being monitored to assess accuracy.

REFERENCE Method 601, *Federal Register* Part VIII 40 CFR Part 136, Oct 26, 1984.

1,1,1-Trichloroethane　　　　　　　**EPA Method 624**
CAS #71-55-6

TITLE Purgeables

MATRIX Wastewater.

APPLICATION Method covers 31 purgeable organics. An inert gas is bubbled through a 5-mL water sample in a specially designed purging chamber. Here, purgeables are transferred from aqueous to gaseous phase, passed onto a sorbent column, and trapped. Trap is heated and backflushed with inert gas to desorb purgeables onto a GC column, where purgeables are separated.

INTERFERENCES Impurities in the purge gas, organic compounds outgassing from the plumbing ahead of the trap, and solvent vapors in the lab. With high- and low-level samples, there can be carryover contamination.

INSTRUMENTATION GC/MS with purge-and-trap unit.

RANGE 5–600 µg/L

MDL 3.8 µg/L

PRECISION 0.21X–0.39 µg/L (overall precision).

ACCURACY 1.06C + 0.73 µg/L (as recovery).

SAMPLING METHOD 25-mL glass vial. Teflon®-lined septum.

STABILITY Cool, 4°C, 0.008% Sodium thiosulfate.

MHT 14 days.

QUALITY CONTROL The lab must on an ongoing basis, spike at least 5% of the samples from each sample site being monitored to assess accuracy.

REFERENCE Method 624, *Federal Register* Part VIII 40 CFR Part 136, Oct 26, 1984.

1,1,2-Trichloroethane　　　　　　　**EPA Method 624**
CAS #79-00-5

TITLE Purgeables

MATRIX Wastewater.

APPLICATION Method covers 31 purgeable organics. An inert gas is bubbled through a 5-mL water sample in a specially designed purging chamber. Here, purgeables are transferred from aqueous to gaseous phase, passed onto a sorbent column, and trapped. Trap is heated and backflushed with inert gas to desorb purgeables onto a GC column, where purgeables are separated.

INTERFERENCES Impurities in the purge gas, organic compounds outgassing from the plumbing ahead of the trap, and solvent vapors in the lab. With high- and low-level samples, there can be carryover contamination.

INSTRUMENTATION GC/MS with purge-and-trap unit.

RANGE 5–600 µg/L

MDL 5.0 µg/L

PRECISION 0.18X + 0.00 µg/L (overall precision).

ACCURACY 0.95C + 1.71 µg/L (as recovery).

SAMPLING METHOD 25-mL glass vial. Teflon®-lined septum.

STABILITY Cool, 4°C, 0.008% Sodium thiosulfate.

MHT 14 days.

QUALITY CONTROL The lab must on an ongoing basis, spike at least 5% of the samples from each sample site being monitored to assess accuracy.

REFERENCE Method 624, *Federal Register* Part VIII 40 CFR Part 136, Oct 26, 1984.

1,1,1-Trichloroethane　　　　　　　**EPA Method 8010**
CAS #71-55-6

TITLE Halogenated Volatile Organics

MATRIX Groundwater, soils, sludges, water miscible liquid wastes, and non-water miscible wastes.

APPLICATION This method is used for the analysis of 39 halogenated VOCs. Samples are analyzed using direct injection or purge-and-trap methods. Groundwater must be analyzed by the purge-and-trap method. The method provides an optional GC column which is used for analyte confirmation and that may help resolve analytes from interferences.

INTERFERENCES There can be carryover contamination with high- and low-level samples. Impurities may come from the purge-and-trap apparatus, organic compounds outgassing from the plumbing ahead of trap, diffusion of VOCs through the sample bottle septum during shipping or storage, or from solvent vapors in the lab.

INSTRUMENTATION GC capable of on-column injections or purge-and-trap sample introduction and a halogen specific detector. Column 1: 8 ft by 0.1 in 1%. SP-1000 on Carbopack-B. Column 2: 6 ft by 0.1 in bonded n-octane on Porasil-C.

RANGE 8–500 µg/L (reagent water)

MDL 0.03 µg/L (reagent water).

PQL FACTORS FOR MULTIPLYING × MDL VALUE

Matrix	Multiplication Factor
Groundwater	10
Low-level soil	10
Water miscible liquid waste	500
High-level soil and sludge	1250
Non-water miscible waste	1250

PRECISION 0.20X + 0.37 µg/L (overall precision).

ACCURACY 0.90C–0.16 µg/L (as recovery).

SAMPLING METHOD For water and liquid samples; use glass 40-mL vials with Teflon®-lined septum caps and collect two vials per sample location with no headspace. For solids and concentrated waste samples; use widemouth glass bottles with Teflon® liners.

STABILITY For concentrated wastes, soils, sediments, or sludges: cool to 4°C. For liquids: add 4 drops of concentrated hydrochloric acid and cool to 4°C.

MHT 14 days.

QUALITY CONTROL Analyze a reagent blank, matrix spike, and matrix spike duplicate/duplicate for each analytical batch (up to 20 samples). Demonstrate the purity of glassware and reagents by analyzing a reagent water method blank. Internal, surrogate, and five concentration level calibration standards are used.

REFERENCE Test Methods for Evaluating Solid Waste (SW-846), U.S. EPA Office of Solid Waste, Washington, DC, Method 8010B, Rev. 2, Nov. 1992.

1,1,2-Trichloroethane **EPA Method 8010**
CAS #79-00-5

TITLE Halogenated Volatile Organics

MATRIX Groundwater, soils, sludges, water miscible liquid wastes, and non-water miscible wastes.

APPLICATION This method is used for the analysis of 39 halogenated VOCs. Samples are analyzed using direct injection or purge-and-trap methods. Groundwater must be analyzed by the purge-and-trap method. The method provides an optional GC column which is used for analyte confirmation and that may help resolve analytes from interferences.

INTERFERENCES There can be carryover contamination with high- and low-level samples. Impurities may come from the purge-and-trap apparatus, organic compounds outgassing from the plumbing ahead of trap, diffusion of VOCs through the sample bottle septum during shipping or storage, or from solvent vapors in the lab.

INSTRUMENTATION GC capable of on-column injections or purge-and-trap sample introduction and a halogen specific detector. Column 1: 8 ft by 0.1 in 1%. SP-1000 on Carbopack-B. Column 2: 6 ft by 0.1 in bonded n-octane on Porasil-C.

RANGE 8–500 µg/L (reagent water)

MDL 0.02 µg/L (reagent water).

PQL FACTORS FOR MULTIPLYING × MDL VALUE

Matrix	Multiplication Factor
Groundwater	10
Low-level soil	10
Water miscible liquid waste	500
High-level soil and sludge	1250
Non-water miscible waste	1250

PRECISION 0.19X + 0.67 µg/L (overall precision).

ACCURACY 0.86C + 0.30 µg/L (as recovery).

SAMPLING METHOD For water and liquid samples; use glass 40-mL vials with Teflon®-lined septum caps and collect two vials per sample location with no headspace. For solids and concentrated waste samples; use widemouth glass bottles with Teflon® liners.

STABILITY For concentrated wastes, soils, sediments, or sludges: cool to 4°C. For liquids: add 4 drops of concentrated hydrochloric acid and cool to 4°C.

MHT 14 days.

QUALITY CONTROL Analyze a reagent blank, matrix spike, and matrix spike duplicate/duplicate for each analytical batch (up to 20 samples). Demonstrate the purity of glassware and reagents by analyzing a reagent water method blank. Internal, surrogate, and five concentration level calibration standards are used.

REFERENCE Test Methods for Evaluating Solid Waste (SW-846), U.S. EPA Office of Solid Waste, Washington, DC, Method 8010B, Rev. 2, Nov. 1992.

Trichloroethylene **EPA Method 1624**
CAS #79-01-6

TITLE Volatile Organic Compounds by Isotope Dilution GC/MS

MATRIX Compounds may be determined in waters, soils, and municipal sludges by this method.

METHOD SUMMARY This method is used to determine 58 volatile toxic organic pollutants associated with the CWA (as amended 1987); the RCRA (as amended 1986); the CERCLA (as amended 1986); and other compounds amenable to purge-and-trap gas chromatography-mass spectrometry (GC/MS).

If the solids content is less than 1%, stable isotopically-labeled analogs of the compounds of interest are added to a 5-mL sample and the sample is purged with an inert gas at 20–25°C in a chamber designed for soil or water samples. If the solids content is greater than 1%, 5 mL of reagent water and the labeled compounds are added to a 5-g aliquot of sample and the mixture is purged at 40°C. Compounds that will not purge at 20–25°C or at 40°C are purged at 78–85°C. In the purging process, the volatile compounds are transferred from the aqueous phase into the gaseous phase where they are passed into a sorbent column, and trapped. After purging is completed, the trap is backflushed and heated rapidly to desorb the compounds into a GC. The compounds are separated by the GC and detected by a MS. The labeled compounds serve to correct the variability of the analytical technique.

INTERFERENCES Impurities in the purge gas, organic compounds outgassing from the plumbing upstream of the trap, and solvent vapors in the lab account for most problems. Samples can be contaminated by diffusion of volatile organic compounds (particularly methylene chloride) through the bottle seal during shipment and storage. Contamination by carryover can occur when high-level and low-level samples are analyzed sequentially. When an unusually concentrated sample is encountered, follow it by analysis of a reagent water blank to check for carryover.

INSTRUMENTATION Major equipment includes a GC with linear temperature programming and a glass jet separator as the MS interface, a MS with 70 eV electron impact ionization, and a data system to collect and record response factors.

Column: 2.8 m × 2 mm I.D. glass, packed with 1% SP-1000 on Carbopak B, 60/80 mesh, or equivalent.

PRECISION & ACCURACY The detection limits of the method are usually dependent on the level of interferences rather than instrumental limitations. The method detection limits were determined in digested sludge (low solids) and in filter cake or compost (high solids).

The MDL (in μg/kg) for low solids is 41 and for high solids is 2.
Labeled and native compound precision (in μg/L) as standard deviation was 8.9.
Labeled and native compound accuracy (in μg/L) as average recovery was 17–30.

Acceptance criteria are at 20 μg/L for this compound.

SAMPLE COLLECTION, PRESERVATION & HANDLING Grab samples are collected in glass containers having a total volume greater than 20 mL. Fill and seal each bottle so that no air bubbles are entrapped. Samples are maintained at 0 to 4°C from the time of collection until analysis. If an aqueous sample contains residual chlorine, add sodium thiosulfate preservative (10 mg/40 mL) to the empty sample bottles just prior to shipment to the sample site. All samples must be analyzed within 14 days of collection.

SAMPLE PREPARATION Samples containing less than 1% solids are analyzed directly as aqueous samples. Samples containing 1% solids or greater are analyzed as solid samples utilizing one of two methods, depending on the levels of pollutants, in the sample. Samples containing 1% solids or greater, and low to moderate levels of pollutants are analyzed by purging a known weight of sample added to 5 mL of reagent water. Samples containing 1% solids or greater, and high levels of pollutants, are extracted with methanol, and an aliquot of the methanol extract is added to reagent water and purged.

QUALITY CONTROL A field blank prepared from reagent water and carried through the sampling and handling protocol may serve as a check on contamination from shipment and storage.

The analyst is permitted to modify this method to improve separations or lower the costs of measurements, provided all performance specifications are met. Analyses of blanks are required. When results of spikes indicate atypical method performance for samples, the samples are diluted to bring method performance within acceptable limits. Analyze two sets of four 5-mL aliquots (8 aliquots total) of the aqueous performance standard. Spike all samples with labeled compounds to assess method performance on the sample matrix. Compute the percent recovery of the labeled compounds using the internal standard method. Compare the percent recovery for each compound with the corresponding labeled compound recovery. Reagent water blanks are analyzed to demonstrate freedom from carryover contamination. Field replicates may be collected to determine the precision of the sampling technique, and spiked samples may be required to determine the accuracy of the analysis when the internal method is used.

REFERENCE Volatile Organic Compounds by Isotope Dilution GC/MS. Office of Water Regulation and Standards, U.S. EPA Industrial Technology Division, Washington, DC, EPA Method 1624, Rev. C, June 1989 (contact W.A. Telliard, U.S. EPA, Office of Water Regulations and Standards, 401 M St., SW, Washington, DC, 20460. Phone: 202-382-7131).

Trichloroethylene   **EPA Method 502**
CAS #79-01-6

TITLE Volatile Organic Compounds in Water By Purge and Trap Capillary Column Gas Chromatography with Photoionization and Electrolytic Conductivity Detectors in Series. U.S. EPA Method 502.2, Rev. 2.0, 1989.

MATRIX Drinking water and raw source water. The latter should include most surface water and groundwater sources.

METHOD SUMMARY This method covers 60 volatile organic compounds that contain halogen atoms and/or that are aromatic. An inert gas (zero grade nitrogen or helium) is bubbled through a 25-mL or a 5-mL water sample (depending on the expected concentration of the analytes). Purged sample components are trapped in a tube of sorbent materials. When purging is complete, the sorbent tube is heated and backflushed with helium to desorb the trapped sample onto a capillary GC column. The column is temperature programmed to separate the method analytes which are then detected with a photoionization detector (PID) and a Hall electrolytic conductivity (HECD) placed in series. The PID is selective for aromatic compounds and the HECD is selective for halogenated compounds.

INTERFERENCES Impurities in the purge gas and from organic compounds outgassing from the plumbing ahead of the trap account for many contamination problems. Interferences purged or coextracted from the samples will vary considerably from source to source, depending upon the particular sample or extract being tested. Cross-contamination can occur whenever high-level and low-level samples are analyzed sequentially. Samples also can be contaminated by diffusion of volatile organics (particularly methylene chloride and fluorocarbons) through the septum seal into the sample during shipment and storage. The lab where volatile analysis is performed and also the refrigerated storage area should be completely free of solvents.

INSTRUMENTATION A GC containing a series configuration of a high temperature photoionization detector (PID) equipped with 10.0 eV (nominal) lamp and Hall electrolytic conductivity detector (HECD) is required. Also required is an all-glass 5-mL purging device, a sorbent trap, and a thermal desorption apparatus which is connected to the GC system.

Column 1: VOCOL glass wide-bore capillary column.
Column 2: RTX–502.2 mega-bore capillary column.
Column 3: DB-62 mega-bore capillary column.

PRECISION & ACCURACY Method detection limits are dependent upon the characteristics of the gas chromatographic system used. Analytes that are not separated chromatographically cannot be individually identified and used in the same calibration mixture or water samples unless an alternative technique for identification and quantification, such as mass spectrometry, is used.

Electrolytic conductivity detetor (c) range in µg/L (a) was 0.02–200.
Electrolytic conductivity detetor (c) MDL in µg/L (b) was 0.01.
Electrolytic conductivity detetor (c) accuracy as % recovery was 96.
Electrolytic conductivity detetor (c) precision as % RSD was 3.6.
Photoionization detector (d) range in µg/L (a) was 0.02–200.
Photoionization detector (d) MDL in µg/L (b) was 0.02.
Photoionization detector (d) accuracy as % recovery was 100.
Photoionization detector (d) precision as % RSD was 0.78.

(a) The applicable concentration range of this method is compound, instrument, and matrix-dependent. It is listed as being approximately 0.02 to 200 µg/L but no specific information is provided so caution should be observed.
(b) The method detection limits reports with this method are compound, instrument, and matrix-dependent. The values reported were calculated using reagent water fortified with the corresponding compounds at 10 µg/L and a GC-equipped with a 60 m × 0.75 mm VOLCOL wide bore capillary column with 1.5 µm film thickness and using helium carrier gas.
(c) Recoveries and relative standard deviations were determined from seven samples of reagent water fortified with 10 µg/L of each compound. 2-Bromo-1-chloropropane was used as the internal standard for calculating average recoveries.
(d) Recoveries and relative standard deviations were determined from seven samples of reagent water fortified with 10 µg/L of each compound. Fluorobenzene was used as the internal standard for calculating average recoveries.

SAMPLING METHOD Collect samples using a 40- to 120-mL screw-cap vial (prewashed with detergent, rinsed with distilled water and oven dried at 105°C) with a Teflon®-faced silicone septum. Collect bubble-free samples and place the septum with the Teflon® side down on the water.

SAMPLE PRESERVATION If residual chlorine is present in the water add about 25 mg of ascorbic acid to each vial before samples are collected to remove the chlorine. Add hydrochloric acid to reduce pH to <2, immediately cool samples to 4°C, and store them in a solvent-free refrigerator at 4°C until analysis.

MHT The maximum holding time for samples is 14 days from the time they were collected.

SAMPLE PREPARATION Remove the plungers from two 5-mL syringes and attach a closed syringe valve to each. Warm the sample to room temperature, open the sample bottle, and carefully pour the sample into one of the syringe barrels to just short of overflowing. Replace the syringe plunger, invert the syringe, and compress the sample. Open the syringe valve and vent any residual air while adjusting the sample volume to 5.0 mL. Add 10 µL of the internal calibration standard to the sample through the syringe valve. Close the valve. Fill the second syringe in an identical manner from the same sample bottle. Reserve this second syringe for a reanalysis if necessary.

QUALITY CONTROL As an initial demonstration of lab accuracy and precision, analyze 4 to 7 replicates of a lab fortified blank containing analyte at 0.1–5 µg/L. Collect all samples in duplicate. Surrogate analytes (similar to those of the analytes of interest), whose concentration is known in every sample, are measured using the same internal standard calibration procedure. Duplicate field reagent water blanks (trip blanks) must be analyzed with each set of samples, lab reagent blanks (method blanks) must be analyzed with each batch of samples processed as a group within a work shift. Also, a single lab-fortified blank that contains each of the analytes of interest should be analyzed with each batch of samples processed as a group within a work shift. A 3- to 5-point calibration curve is needed depending on the calibration range factor required.

EPA CONTACT & HOTLINE For technical questions contact Dr. Baldev Bathija, U.S. EPA, Office of Ground Water and Drinking Water (WH-550D), 401 M St. SW, Washington, DC 20460. Tel. (202) 260-3040. For further information the EPA Safe Drinking Water Hotline may be called at: (800) 426-4791.

REFERENCE Methods for the Determination of Organic Compounds in Drinking Water, EPA/600/4-88/039 (revised July 1991; Final Rule for determination of compliance with the MCL for Total Trihalomethanes under 141.30, in 40 CFR Part 141, Vol. 58, No. 147, Fed. Reg., Tuesday Aug. 3, 1993). U.S. EPA Environmental Monitoring Systems Laboratory, Cincinnati, OH, 45268, U.S.A. Available from the National Technical Information Service (NTIS), 5285 Port Royal Road, Springfield, VA 22161; Tel. 800-553-6847. NTIS Order Number is PB91-231480.

Trichloroethylene   **EPA Method 524**
CAS #79-01-6

TITLE Measurement of Purgeable Organic Compounds in Water by Capillary Column GC/MS.

MATRIX Drinking water and raw source water; the latter should include most surface water and groundwater sources.

METHOD SUMMARY Method 524.2 covers 60 volatile organic compounds. An inert gas (zero grade nitrogen or helium) is bubbled through a 25-mL or a 5-mL water sample (depending on the expected concentration of the analytes). Purged sample components are trapped in a tube of sorbent materials. When purging is complete, the sorbent tube is heated and backflushed with helium to desorb the trapped sample onto a capillary GC column.

INTERFERENCES Impurities in the purge gas and from organic compounds outgassing from the plumbing ahead of the trap account for many contamination problems. Interferences purged or coextracted from the samples will vary considerably from source to source, depending upon the particular sample or extract being tested. Cross-contamination can occur whenever high-level and low-level samples are analyzed sequentially. Samples also can be contaminated by diffusion of volatile organics (particularly methylene chloride and fluorocarbons) through the septum seal into the sample during shipment and storage.

INSTRUMENTATION A GC/MS with a data system equipped with one of the following capillary GC columns:

Column 1: VOCOL glass wide bore capillary column.
Column 2: DB-624 fused silica capillary column.
Column 3: DB-5 fused silica capillary column.

Also required is an all-glass 25 mL or 5-mL purging device, a sorbent trap, and a thermal desorption apparatus which is connected to the GC/MS system.

PRECISION & ACCURACY Method detection limits are compound- and instrument-dependent, and may vary from approximately 0.02–0.35 µg/L. Note in the table below that the "true" concentration range used for accuracy and precision measurements was quite narrow. However, the applicable concentration range of this method is primarily column dependent and is approximately 0.02 to 200 µg/L for the wide-bore thick-film columns. Narrow-bore thin-film columns may have a capacity which limits the range to about 0.02 to 20 µg/L. Analytes that are inefficiently purged from water will not be detected when present at low concentrations, but they can be measured with acceptable accuracy and precision when present in sufficient amounts.

Analytes that are not separated chromatographically, but which have different mass spectra and non-interfering quantification ions, can be identified and measured in the same calibration mixture or water sample. Analytes which have very similar mass spectra cannot be individually identified and measured in the same calibration mixture or water samples unless they have different retention times. Co-eluting compounds with very similar mass spectra, typically many structural isomers, must be reported as an isomeric group or pair.

The range (in µg/L) was 0.5–10.
The Method Detection Limig (in µg/L) was 0.19.
The accuracy (as % recovery) was 90.

The precision (in %) was 7.3.

Note: Data were obtained from 16–31 determinations using a wide-bore capillary column and a jet separator interfaced to a quadrupole mass spectrometer. All analytes were in a reagent water matrix.

SAMPLING METHOD Collect samples using a 40- to 120-mL screw-cap vial (prewashed with detergent, rinsed with distilled water and oven dried at 105°C) with a Teflon®-faced silicone septum. Collect bubble-free samples and place the septum with the Teflon® side down on the water.

SAMPLE PRESERVATION If residual chlorine is present in the water add about 25 mg of ascorbic acid to each vial before samples are collected to remove the chlorine. Add hydrochloric acid to reduce pH to <2, and immediately cool samples to 4°C, and store them in a solvent-free refrigerator at 4°C until analysis.

MHT The maximum holding time for samples is 14 days from the time they were collected.

SAMPLE PREPARATION Remove the plungers from two 25-mL (or 5-mL depending on sample size) syringes and attach a closed syringe valve to each. Warm the sample to room temperature, open the sample bottle, and carefully pour the sample into one of the syringe barrels to just short of overflowing. Replace the syringe plunger, invert the syringe, and compress the sample. Open the syringe valve and vent any residual air while adjusting the sample volume to 25.0 mL (or 5 mL). For samples and blanks, add 5 µL of the fortification solution containing the internal standard and the surrogates to the sample through the syringe valve. For calibration standards and lab fortified blanks, add 5 µL of the fortification solution containing the internal standard only. Close the valve. Fill the second syringe in an identical manner from the same sample bottle. Reserve this second syringe for a reanalysis if necessary.

QUALITY CONTROL As an initial demonstration of lab accuracy and precision, analyze 4 to 7 replicates of a lab fortified blank containing analyte at 0.2–5 µg/L. Collect all samples in duplicate. Surrogate analytes (similar to those of the analytes of interest), whose concentration is known in every sample, are measured using the same internal standard calibration procedure. Duplicate field reagent water blanks (trip blanks) must be analyzed with each set of samples, lab reagent blanks (method blanks) must be analyzed with each batch of samples processed as a group within a work shift. Also, a single lab-fortified blank that contains each of the analytes of interest should be analyzed with each batch of samples processed as a group within a work shift. A 3- to 5-point calibration curve is needed depending on the calibration range factor required.

EPA CONTACT & HOTLINE For technical questions contact Dr. Baldev Bathija, U.S. EPA, Office of Ground Water and Drinking Water (WH-550D), 401 M St. SW, Washington, DC 20460. Tel. (202) 260-3040. For further information the EPA Safe Drinking Water Hotline may be called at: (800) 426-4791.

REFERENCE Methods for the Determination of Organic Compounds in Drinking Water, EPA/600/4-88/039 (revised July 1991; Final Rule for determination of compliance with the

MCL for Total Trihalomethanes under 141.30, in 40 CFR Part 141, Vol. 58, No. 147, Fed. Reg., Tuesday Aug. 3, 1993). U.S. EPA Environmental Monitoring Systems Laboratory, Cincinnati, OH, 45268, U.S.A. Available from the National Technical Information Service (NTIS), 5285 Port Royal Road, Springfield, VA 22161; Tel. 800-553-6847. NTIS Order Number is PB91-231480.

Trichloroethylene   **EPA Method 8021**
CAS #79-01-6

TITLE Halogenated Volatile by Gas Chromatography Using Photoionization and Electrolytic Conductivity Detectors in Series: Capillary Column Technique

MATRIX This method is applicable to nearly all types of samples, regardless of water content, including groundwater, aqueous sludges, caustic liquors, acid liquors, waste solvents, oily wastes, mousses, tars, fibrous wastes, polymeric emulsions, filter cakes, spent carbons, spent catalysts, soils, and sediments.

METHOD SUMMARY This method is used to determine 60 volatile organic compounds in a variety of solid waste matrices. It provides GC conditions for the detection of halogenated and aromatic volatile organic compounds. Samples can be analyzed using direct injection or purge-and-trap (EPA Method 5030). Groundwater samples must be analyzed using EPA Method 5030 (where applicable). A temperature program is used with the GC. Detection is achieved by a photoionization detector (PID) and a Hall electrolytic conductivity detector (HECD) in series.

INTERFERENCES Samples can be contaminated by diffusion of volatile organics (particularly chlorofluorocarbons and methylene chloride) through the sample container septum during shipment and storage.

INSTRUMENTATION A GC-equipped with variable-constant differential flow controllers, subambient oven controller, PID and HECD detectors connected with a short piece of uncoated capillary tubing and a data system.

Column: 60 m × 0.75 mm I.D. VOCOL wide-bore capillary column with 1.5 µm film thickness.

PRECISION & ACCURACY MDLs are compound-dependent and vary with purging efficiency and concentration. The applicable concentration range of this method is compound- and instrument-dependent but is approximately 0.1 to 200 µg/L. Analytes that are inefficiently purged from water will not be detected when present at low concentrations, but they can be measured with acceptable accuracy and precision when present in sufficient amounts. The estimated quantitation limit (EQL) for an individual compound is approximately 1 µg/kg (wet weight) for soil/sediment samples, 100 µg/kg (wet weight) for wastes, and 1 µg/L for groundwater. EQLs will be proportionately higher for sample extracts and samples that require dilution or reduced sample size to avoid saturation of the detector.

MULTIPLICATION FACTORS FOR OTHER MATRICES (a)

Matrix	Factor (b)
Groundwater	10
Low-concentration soil	10
Water miscible liquid waste	500
High-concentration soil and sludge	1250
Non-water miscible waste	1250

(a) Sample EQLs are highly matrix-dependent. The EQLs listed herein are provided for guidance and may not always be achievable. (b) EQL = [Method detection limit] × [Factor]. For non-aqueous samples, the factor is on a wet-weight basis.

SINGLE LABORATORY ACCURACY & PRECISION DATA FOR VOCs IN WATER
This method was tested in a single lab using water spiked at 10 µg/L and the following data was reported:

Recoveries and standard deviations were determined from seven samples and spiked at 10 µg/L of each analyte. Recoveries were determined by the internal standard method. Internal standards were: Fluorobenzene for PID and 2-Bromo-1-chloropropane for HECD.

The average recovery (in percent) for the PID was 100.
The standard deviation of the recovery for the PID was 0.78.
The MDL (in µg/mL) for the PID was 0.02.
The average recovery (in percent) for the HECD was 96.
The standard deviation of the recovery for the HECD was 3.5.
The MDL (in µg/mL) for the HECD was 0.01.

SAMPLE COLLECTION, PRESERVATION & HANDLING
Volatile organics — Standard 40-mL glass screw-cap VOA vials with Teflon®-faced silicone septum may be used for both liquid and solid matrices. When collecting samples, liquids and solids should be introduced into the vials gently to reduce agitation which might drive off volatile compounds. If there are any air bubbles present the sample must be retaken. Tap slightly as they are filled to try and eliminate as much free air space as possible. The two vials from each sampling locations should be sealed in separate plastic bags to prevent cross-contamination between samples particularly if the sampled waste is suspected of containing high levels of volatile organics.

Semivolatile organics — Containers used to collect samples for the determination of semivolatile organic compounds should be soap and water washed followed by methanol (or isopropanol) rinsing. The sample containers should be of glass or Teflon® and have screw-top covers with Teflon® liners.

Preservation for volatile organics — No preservation is used with concentrated waste samples. With liquid samples containing no residual chlorine, 4 drops of concentrated hydrochloric acid are added and the samples are immediately cooled to 4°C. When liquid samples contain residual chlorine, they are treated as above and, in addition, 4 drops of 4% aqueous sodium thiosulfate are added. Soil, sediment, and sludge samples are only cooled to 4°C.

Preservation for semivolatile organics — No preservation is used with concentrated waste samples. With liquid samples containing no residual chlorine and with soil, sediment, and sludge samples, immediately cooling to 4°C is the only preservation used. When residual chlorine is present then 3 mL of 10% aqueous sodium sulfate is added for each gallon of sample collected, followed by cooling to 4°C.

MHT The holding time for all volatile organics samples is 14 days. Liquid samples must be extracted within 7 days and their extracts analyzed within 40 days. Concentrated waste, soil, sediment, and sludge samples must be extracted within 14 days and their extracts analyzed within 40 days.

SAMPLE PREPARATION Volatile compounds are introduced into the gas chromatograph either by direct injector or purge-and-trap (EPA Method 5030). EPA Method 5030 may be used directly on groundwater samples or low-concentration contaminated soils and sediments. For medium-concentration soils or sediments, methanolic extraction, as described in EPA Method 5030, may be necessary prior to purge-and-trap analysis.

QUALITY CONTROL Calculate surrogate standard recovery on all samples, blanks, and spikes. A trip blank is recommended to check on sampling, storage, and handling contamination. Calibration standards, at a minimum of five concentration levels, are prepared in organic-free reagent water. One of the concentration levels should be at a concentration near, but above, the method detection limit.

A combination of bromochloromethane, 2-bromo-1-chloropropane, 1,4-dichlorobutane, and bromochlorobenzene are recommended as surrogate standards to encompass the range of the temperature program used in this method.

REFERENCE Test Methods for Evaluating Solid Waste, Physical/Chemical Methods, SW-846, 3rd Edition, U.S. EPA, Office of Solid Waste, Washington, DC, EPA Method 8021A, Rev. 1, Nov. 1992.

Trichloroethylene   **EPA Method 8240**
CAS #79-01-6

TITLE Volatile Organics By GC/MS: Packed Column Technique

MATRIX Nearly all types of sample matarices, regardless of water content, can be analyzed using this method. This includes groundwater, aqueous sludges, caustic liquors, acid liquors, waste solvents, oily wastes, mousses, tars, fibrous wastes, polymetric emulsions, filter cakes, spent carbons, spent catalysts, soils, and sediments.

METHOD SUMMARY Method 8240B covers 80 volatile organic compounds that are introduced into a gas chromatograph by the purge-and-trap method or by direct injection (in limited applications). For the purge-and-trap method an inert gas (zero grade nitrogen or helium) is bubbled through a 5-mL solution at ambient temperature. Purged sample components are trapped in a tube of sorbent materials. When purging is complete, the sorbent tube is heated and backflushed with inert gas to desorb the trapped components onto a GC column.

INTERFERENCES Impurities in the purge gas and from organic compounds outgassing from the plumbing ahead of the trap account for many contamination problems. Interferences purged or coextracted from the samples will vary considerably from source to source. Cross-contamination can occur whenever high-level and low-level samples are analyzed sequentially. Whenever an unusually concentrated sample is analyzed, it should be followed by the analysis of organic-free reagent water to check for cross-contamination. Samples also can be contaminated by diffusion of volatile organics (particularly methylene chloride and fluorocarbons) through the septum seal into the sample during shipment and storage. A trip blank can serve as a check on such contamination. The lab where volatile analysis is performed and also the refrigerated storage area should be completely free of solvents.

INSTRUMENTATION A gas chromatograph/mass spectrometry/data system (GC/MS) equipped with a 6 ft × 0.1 in I.D. glass column packed with 1% SP-1000 on Carbopack-B (60/80 mesh) is required. Also needed is a 5-mL purging device, a sorbent trap, and a thermal desorption apparatus.

PRECISION & ACCURACY This method is reported to have been tested by 15 laboratories using organic-free reagent water, drinking water, surface water, and industrial wastewaters (not specified) fortified at six concentrations over the range 5–600 µg/L.

Sample estimated quantitation limits (EQLs) are highly matrix-dependent. The EQLs listed may not always be achievable. EQLs listed for soils or sediments are based on wet weight. Normally, data is reported on a dry-weight basis; therefore, EQLs will be higher, based on the percent dry weight of each sample. Note that EQLs are even more variable than MDLs and that they are highly variable depending on the matrix being analyzed.

EQL in groundwater in µg/L was 5.
EQL in low soil or sediment in µg/kg was 5.
Accuracy (a) in µg/L was $1.04C + 2.27$.
Precision (b) in µg/L was $0.12x + 0.59$.

(a) *Average recovery found for measurements of samples containing a concentration of C, in µg/L.*
(b) *Overall precision found for measurements of samples with average recovery X for samples containing a concentration of C in µg/L.*
$X =$ *Average recovery found for measurement of samples containing a concentration of C in µg/L.*

MULTIPLICATION FACTORS FOR OTHER MATRICES

Other Matrices	Factor (a)
Waste miscible liquid waste	50
High-concentration soil and sludge	125
Non-water miscible waste	500

(a) *EQL = [EQL for low soil sediment X [Factor]. For non-aqueous samples, the factor is on a wet-weight basis.*

SAMPLING METHOD
Liquid samples — Use a 40-mL glass screw-cap VOA vial with a Teflon®-faced silicone septum that has been prewashed, rinsed with distilled deionized water, and oven dried. However, if residual chlorine is present, collect sample in a 40-oz. soil VOA container which has been pre-preserved with 4 drops of

10% sodium thiosulfate, mix gently, and then transfer the sample to a 40-mL VOA vial. Collect bubble-free samples in duplicate and seal them in separate plastic bags.

Soils or sediments, and sludges — Use an 8-oz. widemouth glass bottle with a Teflon®-faced silicone septum that has been prewashed with detergent, rinsed with distilled deionized water, and oven dried. Tap slightly to eliminate free air space. Collect samples in duplicate and seal them in separate plastic bags.

SAMPLE PRESERVATION

Liquid samples — Add 4 drops of concentrated HCL and immediately cool samples to 4°C and store in a solvent-free refrigerator.

Soils or sediments, and sludges — Cool samples to 4°C and store in a solvent-free refrigerator.

MHT Maximum holding time is 14 days from the date of sample collection.

SAMPLE PREPARATION

Liquid samples — Remove the plunger from a 5-mL syringe and carefully pour the sample into the syringe barrel to just short of overflowing. Replace the syringe plunger and compress the sample. Open the syringe valve and vent any residual air while adjusting the sample volume to 5.0 mL. If there is only one volatile organic analysis (VOA) vial, a second syringe should be filled at this time to protect against possible loss of sample integrity. Add 10 µL of surrogate spiking solution and 10 µL of internal standard spiking solution through the valve bore of the 5-mL syringe, then close the valve. The surrogate and internal standards may be mixed and added as a single spiking solution.

Sediments, soils, and waste samples — All samples of this type should be screened by GC analysis using a headspace method (EPA Method 3810) or the hexadecane extraction and screening method (EPA Method 3820). Use the screening data to determine whether to use the low-concentration method (0.005–1 mg/kg) or the high-concentration method (>1 mg/kg).

Low-concentration method — The low-concentration method is based on purging a heated sediment or soil sample mixed with organic-free reagent water containing the surrogate and internal standards. Analyze all reagent blanks and standards under the same conditions as the samples.

Use a 5-g sample if the expected concentration is <0.1 mg/kg or a 1-g sample for expected concentrations between 0.1 and 1 mg/kg. Mix the contents of the sample container with a narrow metal spatula. Weigh the amount of the sample into a tared purge device. Add the spiked water to the purge device, which contains the weighed amount of sample, and connect the device to the purge-and-trap system.

High-concentration method — This method is based on extracting the sediment or soil with methanol. A waste sample is either extracted or diluted, depending on its solubility in methanol. Wastes that are insoluble in methanol are diluted with reagent tetraglyme or possibly polyethylene glycol (PEG). An aliquot of the extract is added to organic-free reagent water containing surrogate and internal standards. This is purged at ambient temperature. All samples with an expected concentration of >1.0 mg/kg should be analyzed by this method.

Mix the contents of the sample container with a narrow metal spatula. For sediments or soils and solid wastes that are insoluble in methanol, weigh 4 g (wet weight) of sample into a tared 20-mL vial. For waste that is soluble in methanol, tetraglyme, or PEG, weigh 1 g (wet weight) into a tared scintillation vial or culture tube or a 10-mL volumetric flask. Quickly add 9.0 mL of appropriate solvent then add 1.0 mL of a surrogate spiking solution to the vial, cap it, and shake it for 2 min.

METHANOL EXTRACT REQUIRED FOR ANALYSIS OF HIGH-CONCENTRATION SOILS OR SEDIMENTS

Approximate Concentration Range	Volume of Methanol Extract (a)
500–10,000 µg/kg	100 µL
1,000–20,000 µg/kg	50 µL
5,000–100,000 µg/kg	10 µL
25,000–500,000 µg/kg	100 µL of 1/50 dilution (b)

Calculate appropriate dilution factor for concentrations exceeding this table.

(a) The volume of methanol added to 5 mL of water being purged should be kept constant. Therefore, add to the 5-mL syringe whatever volume of methanol is necessary to maintain a volume of 100 µL added to the syringe.

(b) Dilute an aliquot of the methanol extract and then take 100 µL for analysis.

QUALITY CONTROL Demonstrate, through the analysis of a reagent water blank, that interferences from the analytical system, glassware, and reagents are under control. Blank samples should be carried through all stages of the sample preparation and measurement steps. For each analytical batch (up to 20 samples), a reagent blank, matrix spike, and matrix spike duplicate must be analyzed (the frequency of the spikes may be different for different monitoring programs). The blank and spiked samples must be carried through all stages of the sample preparation and measurement steps. QC samples mentioned in the section on Interferences will also be needed as appropriate to those situations.

REFERENCE Test Methods for Evaluating Solid Waste (SW-846). U.S. EPA. 1983. Method 8240B, Rev. 2, Nov. 1990. Office of Solid Wastes, Washington, DC.

Trichloroethylene   EPA Method 8260
CAS #79-01-6

TITLE Volatile Organic Compounds by GC/MS: Capillary Column Technique

MATRIX This method is applicable to nearly all types of samples, regardless of water content, including groundwater, soils, and sediments.

METHOD SUMMARY Method 8260A covers 58 volatile organic compounds that are introduced into a gas chromatograph by the purge-and-trap method or by direct injection (in limited applications). Zero-grade helium is bubbled through a 5-mL solution at ambient temperature. Purged sample components are trapped in a tube containing suitable sorbent materials. When purging is complete, the sorbent tube is heated and backflushed with helium to desorb trapped sample components. The analytes are desorbed directly to a large bore capillary or cryofocussed on a capillary precolumn before being flash evaporated to a narrow bore capillary for analysis.

INTERFERENCES Major contaminant sources are volatile materials in the lab and impurities in the inert purging gas and in the sorbent trap. Interfering contamination may occur when a sample containing low concentrations of volatile organic compounds is analyzed immediately after a sample containing high concentrations of volatile organic compounds. After analysis of a sample containing high concentrations of volatile organic compounds, one or more calibration blanks should be analyzed to check for cross-contamination. Screening of the samples prior to purge-and-trap GC/MS analysis is highly recommended to prevent contamination of the system. This is especially true for soil and waste samples.

Special precautions must be taken to analyze for methylene chloride. The analytical and sample storage area should be isolated from all atmospheric sources of methylene chloride. All gas chromatography carrier gas lines and purge gas plumbing should be constructed from stainless steel or copper tubing. Laboratory clothing previously exposed to methylene chloride fumes during liquid-liquid extraction procedures can contribute to sample contamination.

Samples can also be contaminated by diffusion of volatile organics (particularly methylene chloride and fluorocarbons) through the septum seal during shipment and storage. A trip blank can serve as a check on such contamination.

INSTRUMENTATION GC/MS with a temperature-programmable chromatograph suitable for splitless injection equipped with variable constant differential flow controllers, a subambient oven controller, a purging device, sorbent trap, a thermal desorption apparatus and a capillary precolumn interface when using cryogenic cooling will be needed. The following GC columns may be used:

Column 1: 60 m × 0.75mm I.D. capillary column coated with VOCOL, 1.5 μm film thickness.
Column 2: 30 m × 0.53mm capillary column coated with DB-624 or VOCOL, 3 μm film thickness.
Column 3: 30 m × 0.32mm I.D. capillary column coated with DB-5 or SE-54, 1-μm film thickness.

PRECISION & ACCURACY This method has been tested in a single lab using spiked water. Using a wide-bore capillary column, water was spiked at concentrations between 0.5 and 10 μg/L. Single lab accuracy and precision data are presented. The MDL actually achieved in a given analysis will vary depending on instrument sensitivity and matrix effects.

The MDL (a) in μg/L was 0.19.
The concentration range in μg/L was 0.5–10.
The mean accuracy (% of true value) was 90.
The precision as relative standard deviation was 7.3.

Note: The MDL is based on a 25-mL sample volume instead of a 5-mL sample volume.

SAMPLING METHOD
Liquid samples — Use a 40-mL glass screw-cap VOA vial with a Teflon®-faced silicone septum that has been prewashed, rinsed with distilled deionized water, and oven dried. If residual chlorine is present, collect the sample in a 4-oz soil VOA container which has been pre-preserved with 4 drops of 10% sodium thiosulfate. Mix gently and transfer the sample to a 40-mL VOA vial. Collect bubble-free samples in duplicate and seal each sample in a separate plastic bag.

Soils, sediments and sludges — Use an 8-oz widemouth glass bottle with Teflon®-faced silicone septum that has been prewashed, rinsed with distilled deionized water, and oven dried. **Do not** heat the septum for more than 1 h. Tap slightly to eliminate any free air space. Collect samples in duplicate and seal each one in a separate plastic bag.

SAMPLE PRESERVATION
Liquid samples — Add 4 drops of concentrated HCL, cool to 4°C and store in a solvent-free refrigerator.

Soils, sediments and sludges — Cool samples to 4°C and store in a solvent-free refrigerator.

MHT The maximum holding time of any sample (liquids, soils, sediments, and sludges) is 14 days.

SAMPLE PREPARATION
Liquid samples — Remove the plunger from a 5-mL syringe and carefully pour the sample into the syringe barrel to just short of overflowing. Replace the syringe plunger and compress the sample. Open the syringe valve and vent any residual air while adjusting the sample volume to 5.0 mL. If there is only one volatile organic analysis (VOA) vial, a second syringe should be filled at this time to protect against possible loss of sample integrity. Add 10 μL of surrogate spiking solution and 10 μL of internal standard spiking solution through the valve bore of the 5-mL syringe, then close the valve. The surrogate and internal standards may be mixed and added as a single spiking solution.

Sediments, soils, and waste samples — All samples of this type should be screened by GC analysis using a headspace method (EPA Method 3810) or the hexadecane extraction and screening method (EPA Method 3820). Use the screening data to determine whether to use the low-concentration method (0.005–1 mg/kg) or the high-concentration method (>1 mg/kg).

Low-concentration method — The low-concentration method is based on purging a heated sediment or soil sample mixed with organic-free reagent water containing the surrogate and internal standards. Analyze all reagent blanks and standards under the same conditions as the samples.

Use a 5-g sample if the expected concentration is <0.1 mg/kg or a 1-g sample for expected concentrations between 0.1 and 1 mg/kg. Mix the contents of the sample container with a narrow metal spatula. Weigh the amount of the sample into a tared purge device. Add the spiked water to the purge device, which contains the weighed amount of sample, and connect the device to the purge-and-trap system.

High-concentration method — This method is based on extracting the sediment or soil with methanol. A waste sample is either extracted or diluted, depending on its solubility in methanol. Wastes that are insoluble in methanol are diluted with reagent tetraglyme or possibly polyethylene glycol (PEG). An aliquot of the extract is added to organic-free reagent water containing surrogate and internal standards. This is purged at ambient temperature. All samples with an expected concentration of >1.0 mg/kg should be analyzed by this method.

Mix the contents of the sample container with a narrow metal spatula. For sediments or soils and solid wastes that are insoluble in methanol, weigh 4 g (wet weight) of sample into a tared 20-mL vial. For waste that is soluble in methanol, tetraglyme, or PEG, weigh 1 g (wet weight) into a tared scintillation vial or culture tube or a 10-mL volumetric flask. Quickly add 9.0 mL of appropriate solvent then add 1.0 mL of a surrogate spiking solution to the vial, cap it, and shake it for 2 min.

METHANOL EXTRACT REQUIRED FOR ANALYSIS OF HIGH-CONCENTRATION SOILS OR SEDIMENTS

Approximate Concentration Range	Volume of Methanol Extract (a)
500–10,000 µg/kg	100 µL
1,000–20,000 µg/kg	50 µL
5,000–100,000 µg/kg	10 µL
25,000–500,000 µg/kg	100 µL of 1/50 dilution (b)

Calculate appropriate dilution factor for concentrations exceeding this table.

(a) The volume of methanol added to 5 mL of water being purged should be kept constant. Therefore, add to the 5-mL syringe whatever volume of methanol is necessary to maintain a volume of 100 µL added to the syringe.
(b) Dilute an aliquot of the methanol extract and then take 100 µL for analysis.

QUALITY CONTROL Demonstrate, through the analysis of a reagent water blank, that interferences from the analytical system, glassware, and reagents are under control. Blank samples should be carried through all stages of the sample preparation and measurement steps. For each analytical batch (up to 20 samples), a reagent blank, matrix spike, and matrix spike duplicate must be analyzed (the frequency of the spikes may be different for different monitoring programs). The blank and spiked samples must be carried through all stages of the sample preparation and measurement steps. QC samples mentioned in the section on Interferences will also be needed as appropriate to those situations.

Matrix spiking standards should be prepared from volatile organic compounds which will be representative of the compounds being investigated. The recommended internal standards are chlorobenzene-d5, 1,4-difluorobenzene, 1,4-dichlorobenzene-d4, and pentafluorobenzene. Using stock standard solutions, prepare secondary dilution standards containing the compounds of interest, either singly or mixed together in methanol. Store them in a vial with no headspace for no more than one week. Surrogates recommended are toluene-d8, 4-bromofluorobenzene, and dibromofluoromethane. Each sample undergoing GC/MS analysis must be spiked with 10 µL of the surrogate spiking solution prior to analysis.

REFERENCE Test Methods for Evaluating Solid Waste (SW-846). U.S. EPA 1983, Method 8260A, Rev. 1, Nov. 1990. Office of Solid Waste, Washington, DC.

Trichloroethylene   **EPA Method 503.1**
CAS #79-01-6

TITLE Aromatic & Unsaturated VOCs Water

MATRIX Drinking water (finished or in any treatment stage) and raw source water.

APPLICATION Method covers 28 aromatic and unsaturated VOCs. An inert gas is bubbled through a 5-mL water sample. Purged sample components are trapped in tube of sorbent materials. When purging is complete, sorbent tube is heated and backflushed with inert gas to desorb trapped sample onto a packed GC column.

INTERFERENCES During analysis, major contaminant sources are volatile materials in the lab and impurities in purging gas and sorbent trap. With high and low level samples, there can be carryover contamination. Excess water causes a negative baseline deflection.

INSTRUMENTATION Purge and Trap GC w/photoionization detector. (Two GC columns are recommended); Column 1: 5% SP-1200 and 1.75% Bentone 34 on Supelcoport; 5% 1,2,3-tris(2-cyanoethoxy)propane on Chromosorb W.

RANGE 2.2–600 µg/L (Drinking water)

MDL 0.01 µg/L in water

PRECISION RSD = 6.8% at 0.50 µg/L; 19 samples

ACCURACY Average recovery = 97% at 0.50 µg/L; 19 samples

SAMPLING METHOD Use a 40–120-mL screw-cap vial (prewashed with detergent, rinsed with distilled water and oven dried at 105°C) with a PTFE-faced silicone septum. If residual chlorine is in the water add about 25 mg of ascorbic acid to each vial before sample collection. Collect bubble-free samples.

STABILITY Cool to 4°C; HCl to pH <2.

MHT 14 days.

QUALITY CONTROL As an initial demonstration of lab accuracy and precision, analyze 4 to 7 replicates of a lab fortified blank containing analyte at 0.1–5 µg/L. Collect all samples in duplicate.

REFERENCE Method 503.1, Volatile Aromatic & Unsaturated Organic Compounds in H2O by Purge and Trap GC, EPA 600/4-88/039.

Trichloroethylene   **EPA Method 601**
CAS #79-01-6

TITLE Purgeable Halocarbons

MATRIX Wastewater.

APPLICATION Method covers 29 purgeable halocarbons. (Method 624 provides GC/MS conditions appropriate for the qualitative and quantitative confirmation of results). Method describes conditions for a 2nd GC column to confirm measurements made with primary column.

INTERFERENCES Impurities in the purge gas and organic compounds outgassing from the plumbing ahead of the trap. With high- and low-level samples, there can be carryover contamination. Diffusion of volatile organics through the septum seal into the sample.

INSTRUMENTATION GC-equipped with halide-specific detector. (With purge-and-trap unit).

RANGE 8.0–500 µg/L.

MDL 0.12 µg/L.

PRECISION 0.23X + 0.30 µg/L (overall precision).

ACCURACY 0.87C + 0.48 µg/L (as recovery).

SAMPLING METHOD 25-mL glass vial. Teflon®-lined septum.

STABILITY Cool, 4°C, 0.008% Sodium thiosulfate.

MHT 14 days.

QUALITY CONTROL The lab must on an ongoing basis, spike at least 10% of the samples from each sample site being monitored to assess accuracy.

REFERENCE Method 601, *Federal Register* Part VIII 40 CFR Part 136, Oct 26, 1984.

Trichloroethylene   **EPA Method 624**
CAS #79-01-6

TITLE Purgeables

MATRIX Wastewater.

APPLICATION Method covers 31 purgeable organics. An inert gas is bubbled through a 5-mL water sample in a specially designed purging chamber. Here, purgeables are transferred from aqueous to gaseous phase, passed onto a sorbent column, and trapped. Trap is heated and backflushed with inert gas to desorb purgeables onto a GC column, where purgeables are separated.

INTERFERENCES Impurities in the purge gas, organic compounds outgassing from the plumbing ahead of the trap, and solvent vapors in the lab. With high- and low-level samples, there can be carryover contamination.

INSTRUMENTATION GC/MS with purge-and-trap unit.

RANGE 5–600 µg/L

MDL 1.9 µg/L

PRECISION 0.12X + 0.59 µg/L (overall precision).

ACCURACY 1.04C + 2.27 µg/L (as recovery).

SAMPLING METHOD 25-mL glass vial. Teflon®-lined septum.

STABILITY Cool, 4°C, 0.008% Sodium thiosulfate.

MHT 14 days.

QUALITY CONTROL The lab must on an ongoing basis, spike at least 5% of the samples from each sample site being monitored to assess accuracy.

REFERENCE Method 624, *Federal Register* Part VIII 40 CFR Part 136, Oct 26, 1984.

Trichloroethylene   **EPA Method 8010**
CAS #79-01-6

TITLE Halogenated Volatile Organics

MATRIX Groundwater, soils, sludges, water miscible liquid wastes, and non-water miscible wastes.

APPLICATION This method is used for the analysis of 39 halogenated VOCs. Samples are analyzed using direct injection or purge-and-trap methods. Groundwater must be analyzed by the purge-and-trap method. The method provides an optional GC column which is used for analyte confirmation and that may help resolve analytes from interferences.

INTERFERENCES There can be carryover contamination with high- and low-level samples. Impurities may come from the purge-and-trap apparatus, organic compounds outgassing from the plumbing ahead of trap, diffusion of VOCs through the sample bottle septum during shipping or storage, or from solvent vapors in the lab.

INSTRUMENTATION GC capable of on-column injections or purge-and-trap sample introduction and a halogen specific detector. Column 1: 8 ft by 0.1 in 1%. SP-1000 on Carbopack-B. Column 2: 6 ft by 0.1 in bonded n-octane on Porasil-C.

RANGE 8–500 µg/L (reagent water).

MDL 0.12 µg/L (reagent water).

PQL FACTORS FOR MULTIPLYING × MDL VALUE

Matrix	Multiplication Factor
Groundwater	10
Low-level soil	10
Water miscible liquid waste	500
High-level soil and sludge	1250
Non-water miscible waste	1250

PRECISION 0.23X + 0.30 µg/L (overall precision).

ACCURACY 0.87C + 0.48 µg/L (as recovery).

SAMPLING METHOD For water and liquid samples; use glass 40-mL vials with Teflon®-lined septum caps and collect two vials per sample location with no headspace. For solids and concentrated waste samples; use widemouth glass bottles with Teflon® liners.

STABILITY For concentrated wastes, soils, sediments, or sludges: cool to 4°C. For liquids: add 4 drops of concentrated hydrochloric acid and cool to 4°C.

MHT 14 days.

QUALITY CONTROL Analyze a reagent blank, matrix spike, and matrix spike duplicate/duplicate for each analytical batch (up to 20 samples). Demonstrate the purity of glassware and reagents by analyzing a reagent water method blank. Internal, surrogate, and five concentration level calibration standards are used.

REFERENCE Test Methods for Evaluating Solid Waste (SW-846), U.S. EPA Office of Solid Waste, Washington, DC, Method 8010B, Rev. 2, Nov. 1992.

Trichlorofluoromethane   **EPA Method 1624**
CAS #75-69-4

TITLE Volatile Organic Compounds by Isotope Dilution GC/MS

MATRIX Compounds may be determined in waters, soils, and municipal sludges by this method.

METHOD SUMMARY This method is used to determine 58 volatile toxic organic pollutants associated with the CWA (as amended 1987); the RCRA (as amended 1986); the CERCLA (as amended 1986); and other compounds amenable to purge-and-trap gas chromatography-mass spectrometry (GC/MS).

If the solids content is less than 1%, stable isotopically-labeled analogs of the compounds of interest are added to a 5-mL sample and the sample is purged with an inert gas at 20–25°C in a chamber designed for soil or water samples. If the solids content is greater than 1%, 5 mL of reagent water and the labeled compounds are added to a 5-g aliquot of sample and the mixture is purged at 40°C. Compounds that will not purge at 20–25°C or at 40°C are purged at 78–85°C. In the purging process, the volatile compounds are transferred from the aqueous phase into the gaseous phase where they are passed into a sorbent column, and trapped. After purging is completed, the trap is backflushed and heated rapidly to desorb the compounds into a GC. The compounds are separated by the GC and detected by a MS. The labeled compounds serve to correct the variability of the analytical technique.

INTERFERENCES Impurities in the purge gas, organic compounds outgassing from the plumbing upstream of the trap, and solvent vapors in the lab account for most problems. Samples can be contaminated by diffusion of volatile organic compounds (particularly methylene chloride) through the bottle seal during shipment and storage. Contamination by carryover can occur when high-level and low-level samples are analyzed sequentially. When an unusually concentrated sample is encountered, follow it by analysis of a reagent water blank to check for carryover.

INSTRUMENTATION Major equipment includes a GC with linear temperature programming and a glass jet separator as the MS interface, a MS with 70 eV electron impact ionization, and a data system to collect and record response factors.

Column: 2.8 m × 2 mm I.D. glass, packed with 1% SP-1000 on Carbopak B, 60/80 mesh, or equivalent.

PRECISION & ACCURACY The detection limits of the method are usually dependent on the level of interferences rather than instrumental limitations. The method detection limits were determined in digested sludge (low solids) and in filter cake or compost (high solids).

The MDL (in µg/kg) for low solids is not listed and for high solids is not listed.

Labeled and native compound precision (in µg/L) as standard deviation was not listed.

Labeled and native compound accuracy (in µg/L) as average recovery was not listed.

Acceptance criteria are at 20 µg/L for this compound.

SAMPLE COLLECTION, PRESERVATION & HANDLING Grab samples are collected in glass containers having a total volume greater than 20 mL. Fill and seal each bottle so that no air bubbles are entrapped. Samples are maintained at 0 to 4°C from the time of collection until analysis. If an aqueous sample contains residual chlorine, add sodium thiosulfate preservative (10 mg/40 mL) to the empty sample bottles just prior to shipment to the sample site. All samples must be analyzed within 14 days of collection.

SAMPLE PREPARATION Samples containing less than 1% solids are analyzed directly as aqueous samples. Samples containing 1% solids or greater are analyzed as solid samples utilizing one of two methods, depending on the levels of pollutants, in the sample. Samples containing 1% solids or greater, and low to moderate levels of pollutants are analyzed by purging a known weight of sample added to 5 mL of reagent water. Samples containing 1% solids or greater, and high levels of pollutants, are extracted with methanol, and an aliquot of the methanol extract is added to reagent water and purged.

QUALITY CONTROL A field blank prepared from reagent water and carried through the sampling and handling protocol may serve as a check on contamination from shipment and storage.

The analyst is permitted to modify this method to improve separations or lower the costs of measurements, provided all performance specifications are met. Analyses of blanks are required. When results of spikes indicate atypical method performance for samples, the samples are diluted to bring method performance within acceptable limits. Analyze two sets of four 5-mL aliquots (8 aliquots total) of the aqueous performance standard. Spike all samples with labeled compounds to assess method performance on the sample matrix. Compute the percent recovery of the labeled compounds using the internal standard method. Compare the percent recovery for each compound with the corresponding labeled compound recovery. Reagent water blanks are analyzed to demonstrate freedom from carryover contamination. Field replicates may be collected to determine the precision of the sampling technique, and spiked samples may be required to determine the accuracy of the analysis when the internal method is used.

REFERENCE Volatile Organic Compounds by Isotope Dilution GC/MS. Office of Water Regulation and Standards, U.S. EPA Industrial Technology Division, Washington, DC, EPA Method 1624, Rev. C, June 1989 (contact W.A. Telliard, U.S.

EPA, Office of Water Regulations and Standards, 401 M St., SW, Washington, DC, 20460. Phone: 202-382-7131).

Trichlorofluoromethane   **EPA Method 502**
CAS #75-69-4

TITLE Volatile Organic Compounds in Water By Purge and Trap Capillary Column Gas Chromatography with Photoionization and Electrolytic Conductivity Detectors in Series. U.S. EPA Method 502.2, Rev. 2.0, 1989.

MATRIX Drinking water and raw source water. The latter should include most surface water and groundwater sources.

METHOD SUMMARY This method covers 60 volatile organic compounds that contain halogen atoms and/or that are aromatic. An inert gas (zero grade nitrogen or helium) is bubbled through a 25-mL or a 5-mL water sample (depending on the expected concentration of the analytes). Purged sample components are trapped in a tube of sorbent materials. When purging is complete, the sorbent tube is heated and backflushed with helium to desorb the trapped sample onto a capillary GC column. The column is temperature programmed to separate the method analytes which are then detected with a photoionization detector (PID) and a Hall electrolytic conductivity (HECD) placed in series. The PID is selective for aromatic compounds and the HECD is selective for halogenated compounds.

INTERFERENCES Impurities in the purge gas and from organic compounds outgassing from the plumbing ahead of the trap account for many contamination problems. Interferences purged or coextracted from the samples will vary considerably from source to source, depending upon the particular sample or extract being tested. Cross-contamination can occur whenever high-level and low-level samples are analyzed sequentially. Samples also can be contaminated by diffusion of volatile organics (particularly methylene chloride and fluorocarbons) through the septum seal into the sample during shipment and storage. The lab where volatile analysis is performed and also the refrigerated storage area should be completely free of solvents.

INSTRUMENTATION A GC containing a series configuration of a high temperature photoionization detector (PID) equipped with 10.0 eV (nominal) lamp and Hall electrolytic conductivity detector (HECD) is required. Also required is an all-glass 5-mL purging device, a sorbent trap, and a thermal desorption apparatus which is connected to the GC system.

Column 1: VOCOL glass wide-bore capillary column.
Column 2: RTX–502.2 mega-bore capillary column.
Column 3: DB-62 mega-bore capillary column.

PRECISION & ACCURACY Method detection limits are dependent upon the characteristics of the gas chromatographic system used. Analytes that are not separated chromatographically cannot be individually identified and used in the same calibration mixture or water samples unless an alternative technique for identification and quantification, such as mass spectrometry, is used.

Electrolytic conductivity detetor (c) range in µg/L (a) was 0.02–200.
Electrolytic conductivity detetor (c) MDL in µg/L (b) was 0.03.
Electrolytic conductivity detetor (c) accuracy as % recovery was 96.
Electrolytic conductivity detetor (c) precision as % RSD was 3.5.
Photoionization detector (d) range in µg/L (a) was 0.02–200.
Photoionization detector (d) MDL in µg/L (b) was not listed.
Photoionization detector (d) accuracy as % recovery was not listed.
Photoionization detector (d) precision as % RSD was not listed.

(a) *The applicable concentration range of this method is compound, instrument, and matrix-dependent. It is listed as being approximately 0.02 to 200 µg/L but no specific information is provided so caution should be observed.*
(b) *The method detection limits reports with this method are compound, instrument, and matrix-dependent. The values reported were calculated using reagent water fortified with the corresponding compounds at 10 µg/L and a GC-equipped with a 60 m × 0.75 mm VOLCOL wide bore capillary column with 1.5 µm film thickness and using helium carrier gas.*
(c) *Recoveries and relative standard deviations were determined from seven samples of reagent water fortified with 10 µg/L of each compound. 2-Bromo-1-chloropropane was used as the internal standard for calculating average recoveries.*
(d) *Recoveries and relative standard deviations were determined from seven samples of reagent water fortified with 10 µg/L of each compound. Fluorobenzene was used as the internal standard for calculating average recoveries.*

SAMPLING METHOD Collect samples using a 40- to 120-mL screw-cap vial (prewashed with detergent, rinsed with distilled water and oven dried at 105°C) with a Teflon®-faced silicone septum. Collect bubble-free samples and place the septum with the Teflon® side down on the water.

SAMPLE PRESERVATION If residual chlorine is present in the water add about 25 mg of ascorbic acid to each vial before samples are collected to remove the chlorine. Add hydrochloric acid to reduce pH to <2, immediately cool samples to 4°C, and store them in a solvent-free refrigerator at 4°C until analysis.

MHT The maximum holding time for samples is 14 days from the time they were collected.

SAMPLE PREPARATION Remove the plungers from two 5-mL syringes and attach a closed syringe valve to each. Warm the sample to room temperature, open the sample bottle, and carefully pour the sample into one of the syringe barrels to just short of overflowing. Replace the syringe plunger, invert the syringe, and compress the sample. Open the syringe valve and vent any residual air while adjusting the sample volume to 5.0 mL. Add 10 µL of the internal calibration standard to the sample through the syringe valve. Close the valve. Fill the second syringe in an identical manner from the same sample bottle. Reserve this second syringe for a reanalysis if necessary.

QUALITY CONTROL As an initial demonstration of lab accuracy and precision, analyze 4 to 7 replicates of a lab fortified blank containing analyte at 0.1–5 µg/L. Collect all samples in

duplicate. Surrogate analytes (similar to those of the analytes of interest), whose concentration is known in every sample, are measured using the same internal standard calibration procedure. Duplicate field reagent water blanks (trip blanks) must be analyzed with each set of samples, lab reagent blanks (method blanks) must be analyzed with each batch of samples processed as a group within a work shift. Also, a single lab-fortified blank that contains each of the analytes of interest should be analyzed with each batch of samples processed as a group within a work shift. A 3- to 5-point calibration curve is needed depending on the calibration range factor required.

EPA CONTACT & HOTLINE For technical questions contact Dr. Baldev Bathija, U.S. EPA, Office of Ground Water and Drinking Water (WH-550D), 401 M St. SW, Washington, DC 20460. Tel. (202) 260-3040. For further information the EPA Safe Drinking Water Hotline may be called at: (800) 426-4791.

REFERENCE Methods for the Determination of Organic Compounds in Drinking Water, EPA/600/4-88/039 (revised July 1991; Final Rule for determination of compliance with the MCL for Total Trihalomethanes under 141.30, in 40 CFR Part 141, Vol. 58, No. 147, Fed. Reg., Tuesday Aug. 3, 1993). U.S. EPA Environmental Monitoring Systems Laboratory, Cincinnati, OH, 45268, U.S.A. Available from the National Technical Information Service (NTIS), 5285 Port Royal Road, Springfield, VA 22161; Tel. 800-553-6847. NTIS Order Number is PB91-231480.

Trichlorofluoromethane   EPA Method 524
CAS #75-69-4

TITLE Measurement of Purgeable Organic Compounds in Water by Capillary Column GC/MS.

MATRIX Drinking water and raw source water; the latter should include most surface water and groundwater sources.

METHOD SUMMARY Method 524.2 covers 60 volatile organic compounds. An inert gas (zero grade nitrogen or helium) is bubbled through a 25-mL or a 5-mL water sample (depending on the expected concentration of the analytes). Purged sample components are trapped in a tube of sorbent materials. When purging is complete, the sorbent tube is heated and backflushed with helium to desorb the trapped sample onto a capillary GC column.

INTERFERENCES Impurities in the purge gas and from organic compounds outgassing from the plumbing ahead of the trap account for many contamination problems. Interferences purged or coextracted from the samples will vary considerably from source to source, depending upon the particular sample or extract being tested. Cross-contamination can occur whenever high-level and low-level samples are analyzed sequentially. Samples also can be contaminated by diffusion of volatile organics (particularly methylene chloride and fluorocarbons) through the septum seal into the sample during shipment and storage.

INSTRUMENTATION A GC/MS with a data system equipped with one of the following capillary GC columns:

Column 1: VOCOL glass wide bore capillary column.
Column 2: DB-624 fused silica capillary column.
Column 3: DB-5 fused silica capillary column.

Also required is an all-glass 25 mL or 5-mL purging device, a sorbent trap, and a thermal desorption apparatus which is connected to the GC/MS system.

PRECISION & ACCURACY Method detection limits are compound- and instrument-dependent, and may vary from approximately 0.02–0.35 µg/L. Note in the table below that the "true" concentration range used for accuracy and precision measurements was quite narrow. However, the applicable concentration range of this method is primarily column dependent and is approximately 0.02 to 200 µg/L for the wide-bore thick-film columns. Narrow-bore thin-film columns may have a capacity which limits the range to about 0.02 to 20 µg/L. Analytes that are inefficiently purged from water will not be detected when present at low concentrations, but they can be measured with acceptable accuracy and precision when present in sufficient amounts.

Analytes that are not separated chromatographically, but which have different mass spectra and non-interfering quantification ions, can be identified and measured in the same calibration mixture or water sample. Analytes which have very similar mass spectra cannot be individually identified and measured in the same calibration mixture or water samples unless they have different retention times. Co-eluting compounds with very similar mass spectra, typically many structural isomers, must be reported as an isomeric group or pair.

The range (in µg/L) was 0.5–10.
The Method Detection Limig (in µg/L) was 0.08.
The accuracy (as % recovery) was 89.
The precision (in %) was 8.1.

Note: Data were obtained from 16–31 determinations using a wide-bore capillary column and a jet separator interfaced to a quadrupole mass spectrometer. All analytes were in a reagent water matrix.

SAMPLING METHOD Collect samples using a 40- to 120-mL screw-cap vial (prewashed with detergent, rinsed with distilled water and oven dried at 105°C) with a Teflon®-faced silicone septum. Collect bubble-free samples and place the septum with the Teflon® side down on the water.

SAMPLE PRESERVATION If residual chlorine is present in the water add about 25 mg of ascorbic acid to each vial before samples are collected to remove the chlorine. Add hydrochloric acid to reduce pH to <2, and immediately cool samples to 4°C, and store them in a solvent-free refrigerator at 4°C until analysis.

MHT The maximum holding time for samples is 14 days from the time they were collected.

SAMPLE PREPARATION Remove the plungers from two 25-mL (or 5-mL depending on sample size) syringes and attach a closed syringe valve to each. Warm the sample to room temperature, open the sample bottle, and carefully pour the sample into one of the syringe barrels to just short of overflowing. Replace the syringe plunger, invert the syringe, and compress the sample. Open the syringe valve and vent any residual air

while adjusting the sample volume to 25.0 mL (or 5 mL). For samples and blanks, add 5 µL of the fortification solution containing the internal standard and the surrogates to the sample through the syringe valve. For calibration standards and lab fortified blanks, add 5 µL of the fortification solution containing the internal standard only. Close the valve. Fill the second syringe in an identical manner from the same sample bottle. Reserve this second syringe for a reanalysis if necessary.

QUALITY CONTROL As an initial demonstration of lab accuracy and precision, analyze 4 to 7 replicates of a lab fortified blank containing analyte at 0.2–5 µg/L. Collect all samples in duplicate. Surrogate analytes (similar to those of the analytes of interest), whose concentration is known in every sample, are measured using the same internal standard calibration procedure. Duplicate field reagent water blanks (trip blanks) must be analyzed with each set of samples, lab reagent blanks (method blanks) must be analyzed with each batch of samples processed as a group within a work shift. Also, a single lab-fortified blank that contains each of the analytes of interest should be analyzed with each batch of samples processed as a group within a work shift. A 3- to 5-point calibration curve is needed depending on the calibration range factor required.

EPA CONTACT & HOTLINE For technical questions contact Dr. Baldev Bathija, U.S. EPA, Office of Ground Water and Drinking Water (WH-550D), 401 M St. SW, Washington, DC 20460. Tel. (202) 260-3040. For further information the EPA Safe Drinking Water Hotline may be called at: (800) 426-4791.

REFERENCE Methods for the Determination of Organic Compounds in Drinking Water, EPA/600/4-88/039 (revised July 1991; Final Rule for determination of compliance with the MCL for Total Trihalomethanes under 141.30, in 40 CFR Part 141, Vol. 58, No. 147, Fed. Reg., Tuesday Aug. 3, 1993). U.S. EPA Environmental Monitoring Systems Laboratory, Cincinnati, OH, 45268, U.S.A. Available from the National Technical Information Service (NTIS), 5285 Port Royal Road, Springfield, VA 22161; Tel. 800-553-6847. NTIS Order Number is PB91-231480.

Trichlorofluoromethane   **EPA Method 8021**
CAS #75-69-4

TITLE Halogenated Volatile by Gas Chromatography Using Photoionization and Electrolytic Conductivity Detectors in Series: Capillary Column Technique

MATRIX This method is applicable to nearly all types of samples, regardless of water content, including groundwater, aqueous sludges, caustic liquors, acid liquors, waste solvents, oily wastes, mousses, tars, fibrous wastes, polymeric emulsions, filter cakes, spent carbons, spent catalysts, soils, and sediments.

METHOD SUMMARY This method is used to determine 60 volatile organic compounds in a variety of solid waste matrices. It provides GC conditions for the detection of halogenated and aromatic volatile organic compounds. Samples can be analyzed using direct injection or purge-and-trap (EPA Method 5030). Groundwater samples must be analyzed using EPA Method 5030 (where applicable). A temperature program is used with the GC. Detection is achieved by a photoionization detector (PID) and a Hall electrolytic conductivity detector (HECD) in series.

INTERFERENCES Samples can be contaminated by diffusion of volatile organics (particularly chlorofluorocarbons and methylene chloride) through the sample container septum during shipment and storage.

INSTRUMENTATION A GC-equipped with variable-constant differential flow controllers, subambient oven controller, PID and HECD detectors connected with a short piece of uncoated capillary tubing and a data system.

Column: 60 m × 0.75 mm I.D. VOCOL wide-bore capillary column with 1.5 µm film thickness.

PRECISION & ACCURACY MDLs are compound-dependent and vary with purging efficiency and concentration. The applicable concentration range of this method is compound- and instrument-dependent but is approximately 0.1 to 200 µg/L. Analytes that are inefficiently purged from water will not be detected when present at low concentrations, but they can be measured with acceptable accuracy and precision when present in sufficient amounts. The estimated quantitation limit (EQL) for an individual compound is approximately 1 µg/kg (wet weight) for soil/sediment samples, 100 µg/kg (wet weight) for wastes, and 1 µg/L for groundwater. EQLs will be proportionately higher for sample extracts and samples that require dilution or reduced sample size to avoid saturation of the detector.

MULTIPLICATION FACTORS FOR OTHER MATRICES (a)

Matrix	Factor (b)
Groundwater	10
Low-concentration soil	10
Water miscible liquid waste	500
High-concentration soil and sludge	1250
Non-water miscible waste	1250

(a) Sample EQLs are highly matrix-dependent. The EQLs listed herein are provided for guidance and may not always be achievable. (b) EQL = [Method detection limit] × [Factor]. For non-aqueous samples, the factor is on a wet-weight basis.

SINGLE LABORATORY ACCURACY & PRECISION DATA FOR VOCs IN WATER
This method was tested in a single lab using water spiked at 10 µg/L and the following data was reported:

Recoveries and standard deviations were determined from seven samples and spiked at 10 µg/L of each analyte. Recoveries were determined by the internal standard method. Internal standards were: Fluorobenzene for PID and 2-Bromo-1-chloropropane for HECD.

The average recovery (in percent) for the PID was none (no response for this detector).

The standard deviation of the recovery for the PID was none (no response for this detector)-.

The MDL (in µg/mL) for the PID was none (no response for this detector).

The average recovery (in percent) for the HECD was 96.

The standard deviation of the recovery for the HECD was 3.4. The MDL (in µg/mL) for the HECD was 0.03.

SAMPLE COLLECTION, PRESERVATION & HANDLING

Volatile organics — Standard 40-mL glass screw-cap VOA vials with Teflon®-faced silicone septum may be used for both liquid and solid matrices. When collecting samples, liquids and solids should be introduced into the vials gently to reduce agitation which might drive off volatile compounds. If there are any air bubbles present the sample must be retaken. Tap slightly as they are filled to try and eliminate as much free air space as possible. The two vials from each sampling locations should be sealed in separate plastic bags to prevent cross-contamination between samples particularly if the sampled waste is suspected of containing high levels of volatile organics.

Semivolatile organics — Containers used to collect samples for the determination of semivolatile organic compounds should be soap and water washed followed by methanol (or isopropanol) rinsing. The sample containers should be of glass or Teflon® and have screw-top covers with Teflon® liners.

Preservation for volatile organics — No preservation is used with concentrated waste samples. With liquid samples containing no residual chlorine, 4 drops of concentrated hydrochloric acid are added and the samples are immediately cooled to 4°C. When liquid samples contain residual chlorine, they are treated as above and, in addition, 4 drops of 4% aqueous sodium thiosulfate are added. Soil, sediment, and sludge samples are only cooled to 4°C.

Preservation for semivolatile organics — No preservation is used with concentrated waste samples. With liquid samples containing no residual chlorine and with soil, sediment, and sludge samples, immediately cooling to 4°C is the only preservation used. When residual chlorine is present then 3 mL of 10% aqueous sodium sulfate is added for each gallon of sample collected, followed by cooling to 4°C.

MHT The holding time for all volatile organics samples is 14 days. Liquid samples must be extracted within 7 days and their extracts analyzed within 40 days. Concentrated waste, soil, sediment, and sludge samples must be extracted within 14 days and their extracts analyzed within 40 days.

SAMPLE PREPARATION Volatile compounds are introduced into the gas chromatograph either by direct injector or purge-and-trap (EPA Method 5030). EPA Method 5030 may be used directly on groundwater samples or low-concentration contaminated soils and sediments. For medium-concentration soils or sediments, methanolic extraction, as described in EPA Method 5030, may be necessary prior to purge-and-trap analysis.

QUALITY CONTROL Calculate surrogate standard recovery on all samples, blanks, and spikes. A trip blank is recommended to check on sampling, storage, and handling contamination. Calibration standards, at a minimum of five concentration levels, are prepared in organic-free reagent water. One of the concentration levels should be at a concentration near, but above, the method detection limit.

A combination of bromochloromethane, 2-bromo-1-chloropropane, 1,4-dichlorobutane, and bromochlorobenzene are recommended as surrogate standards to encompass the range of the temperature program used in this method.

REFERENCE Test Methods for Evaluating Solid Waste, Physical/Chemical Methods, SW-846, 3rd Edition, U.S. EPA, Office of Solid Waste, Washington, DC, EPA Method 8021A, Rev. 1, Nov. 1992.

Trichlorofluoromethane EPA Method 8240
CAS #75-69-4

TITLE Volatile Organics By GC/MS: Packed Column Technique

MATRIX Nearly all types of sample matarices, regardless of water content, can be analyzed using this method. This includes groundwater, aqueous sludges, caustic liquors, acid liquors, waste solvents, oily wastes, mousses, tars, fibrous wastes, polymetric emulsions, filter cakes, spent carbons, spent catalysts, soils, and sediments.

METHOD SUMMARY Method 8240B covers 80 volatile organic compounds that are introduced into a gas chromatograph by the purge-and-trap method or by direct injection (in limited applications). For the purge-and-trap method an inert gas (zero grade nitrogen or helium) is bubbled through a 5-mL solution at ambient temperature. Purged sample components are trapped in a tube of sorbent materials. When purging is complete, the sorbent tube is heated and backflushed with inert gas to desorb the trapped components onto a GC column.

INTERFERENCES Impurities in the purge gas and from organic compounds outgassing from the plumbing ahead of the trap account for many contamination problems. Interferences purged or coextracted from the samples will vary considerably from source to source. Cross-contamination can occur whenever high-level and low-level samples are analyzed sequentially. Whenever an unusually concentrated sample is analyzed, it should be followed by the analysis of organic-free reagent water to check for cross-contamination. Samples also can be contaminated by diffusion of volatile organics (particularly methylene chloride and fluorocarbons) through the septum seal into the sample during shipment and storage. A trip blank can serve as a check on such contamination. The lab where volatile analysis is performed and also the refrigerated storage area should be completely free of solvents.

INSTRUMENTATION A gas chromatograph/mass spectrometry/data system (GC/MS) equipped with a 6 ft × 0.1 in I.D. glass column packed with 1% SP-1000 on Carbopack-B (60/80 mesh) is required. Also needed is a 5-mL purging device, a sorbent trap, and a thermal desorption apparatus.

PRECISION & ACCURACY This method is reported to have been tested by 15 laboratories using organic-free reagent water, drinking water, surface water, and industrial wastewaters (not specified) fortified at six concentrations over the range 5–600 µg/L.

Sample estimated quantitation limits (EQLs) are highly matrix-dependent. The EQLs listed may not always be achievable. EQLs listed for soils or sediments are based on wet weight.

Normally, data is reported on a dry-weight basis; therefore, EQLs will be higher, based on the percent dry weight of each sample. Note that EQLs are even more variable than MDLs and that they are highly variable depending on the matrix being analyzed.

EQL in groundwater in µg/L was not listed.
EQL in low soil or sediment in µg/kg was not listed.
Accuracy (a) in µg/L was 0.99C + 0.39.
Precision (b) in µg/L was 0.34x-0.39.

(a) Average recovery found for measurements of samples containing a concentration of C, in µg/L.
(b) Overall precision found for measurements of samples with average recovery X for samples containing a concentration of C in µg/L.
X = Average recovery found for measurement of samples containing a concentration of C in µg/L.

MULTIPLICATION FACTORS FOR OTHER MATRICES

Other Matrices	Factor (a)
Waste miscible liquid waste	50
High-concentration soil and sludge	125
Non-water miscible waste	500

(a) EQL = [EQL for low soil sediment X [Factor]. For non-aqueous samples, the factor is on a wet-weight basis.

SAMPLING METHOD

Liquid samples — Use a 40-mL glass screw-cap VOA vial with a Teflon®-faced silicone septum that has been prewashed, rinsed with distilled deionized water, and oven dried. However, if residual chlorine is present, collect sample in a 40-oz. soil VOA container which has been pre-preserved with 4 drops of 10% sodium thiosulfate, mix gently, and then transfer the sample to a 40-mL VOA vial. Collect bubble-free samples in duplicate and seal them in separate plastic bags.

Soils or sediments, and sludges — Use an 8-oz. widemouth glass bottle with a Teflon®-faced silicone septum that has been prewashed with detergent, rinsed with distilled deionized water, and oven dried. Tap slightly to eliminate free air space. Collect samples in duplicate and seal them in separate plastic bags.

SAMPLE PRESERVATION

Liquid samples — Add 4 drops of concentrated HCL and immediately cool samples to 4°C and store in a solvent-free refrigerator.

Soils or sediments, and sludges — Cool samples to 4°C and store in a solvent-free refrigerator.

MHT Maximum holding time is 14 days from the date of sample collection.

SAMPLE PREPARATION

Liquid samples — Remove the plunger from a 5-mL syringe and carefully pour the sample into the syringe barrel to just short of overflowing. Replace the syringe plunger and compress the sample. Open the syringe valve and vent any residual air while adjusting the sample volume to 5.0 mL. If there is only one volatile organic analysis (VOA) vial, a second syringe should be filled at this time to protect against possible loss of sample integrity. Add 10 µL of surrogate spiking solution and 10 µL of internal standard spiking solution through the valve bore of the 5-mL syringe, then close the valve. The surrogate and internal standards may be mixed and added as a single spiking solution.

Sediments, soils, and waste samples — All samples of this type should be screened by GC analysis using a headspace method (EPA Method 3810) or the hexadecane extraction and screening method (EPA Method 3820). Use the screening data to determine whether to use the low-concentration method (0.005–1 mg/kg) or the high-concentration method (>1 mg/kg).

Low-concentration method — The low-concentration method is based on purging a heated sediment or soil sample mixed with organic-free reagent water containing the surrogate and internal standards. Analyze all reagent blanks and standards under the same conditions as the samples.

Use a 5-g sample if the expected concentration is <0.1 mg/kg or a 1-g sample for expected concentrations between 0.1 and 1 mg/kg. Mix the contents of the sample container with a narrow metal spatula. Weigh the amount of the sample into a tared purge device. Add the spiked water to the purge device, which contains the weighed amount of sample, and connect the device to the purge-and-trap system.

High-concentration method — This method is based on extracting the sediment or soil with methanol. A waste sample is either extracted or diluted, depending on its solubility in methanol. Wastes that are insoluble in methanol are diluted with reagent tetraglyme or possibly polyethylene glycol (PEG). An aliquot of the extract is added to organic-free reagent water containing surrogate and internal standards. This is purged at ambient temperature. All samples with an expected concentration of >1.0 mg/kg should be analyzed by this method.

Mix the contents of the sample container with a narrow metal spatula. For sediments or soils and solid wastes that are insoluble in methanol, weigh 4 g (wet weight) of sample into a tared 20-mL vial. For waste that is soluble in methanol, tetraglyme, or PEG, weigh 1 g (wet weight) into a tared scintillation vial or culture tube or a 10-mL volumetric flask. Quickly add 9.0 mL of appropriate solvent then add 1.0 mL of a surrogate spiking solution to the vial, cap it, and shake it for 2 min.

METHANOL EXTRACT REQUIRED FOR ANALYSIS OF HIGH-CONCENTRATION SOILS OR SEDIMENTS

Approximate Concentration Range	Volume of Methanol Extract (a)
500–10,000 µg/kg	100 µL
1,000–20,000 µg/kg	50 µL
5,000–100,000 µg/kg	10 µL
25,000–500,000 µg/kg	100 µL of 1/50 dilution (b)

Calculate appropriate dilution factor for concentrations exceeding this table.

(a) The volume of methanol added to 5 mL of water being purged should be kept constant. Therefore, add to the 5-mL syringe whatever volume of methanol is necessary to maintain a volume of 100 µL added to the syringe.

(b) Dilute an aliquot of the methanol extract and then take 100 µL for analysis.

QUALITY CONTROL Demonstrate, through the analysis of a reagent water blank, that interferences from the analytical system, glassware, and reagents are under control. Blank samples should be carried through all stages of the sample preparation and measurement steps. For each analytical batch (up to 20 samples), a reagent blank, matrix spike, and matrix spike duplicate must be analyzed (the frequency of the spikes may be different for different monitoring programs). The blank and spiked samples must be carried through all stages of the sample preparation and measurement steps. QC samples mentioned in the section on Interferences will also be needed as appropriate to those situations.

REFERENCE Test Methods for Evaluating Solid Waste (SW-846). U.S. EPA. 1983. Method 8240B, Rev. 2, Nov. 1990. Office of Solid Wastes, Washington, DC.

Trichlorofluoromethane **EPA Method 8260**
CAS #75-69-4

TITLE Volatile Organic Compounds by GC/MS: Capillary Column Technique

MATRIX This method is applicable to nearly all types of samples, regardless of water content, including groundwater, soils, and sediments.

METHOD SUMMARY Method 8260A covers 58 volatile organic compounds that are introduced into a gas chromatograph by the purge-and-trap method or by direct injection (in limited applications). Zero-grade helium is bubbled through a 5-mL solution at ambient temperature. Purged sample components are trapped in a tube containing suitable sorbent materials. When purging is complete, the sorbent tube is heated and backflushed with helium to desorb trapped sample components. The analytes are desorbed directly to a large bore capillary or cryofocussed on a capillary precolumn before being flash evaporated to a narrow bore capillary for analysis.

INTERFERENCES Major contaminant sources are volatile materials in the lab and impurities in the inert purging gas and in the sorbent trap. Interfering contamination may occur when a sample containing low concentrations of volatile organic compounds is analyzed immediately after a sample containing high concentrations of volatile organic compounds. After analysis of a sample containing high concentrations of volatile organic compounds, one or more calibration blanks should be analyzed to check for cross-contamination. Screening of the samples prior to purge-and-trap GC/MS analysis is highly recommended to prevent contamination of the system. This is especially true for soil and waste samples.

Special precautions must be taken to analyze for methylene chloride. The analytical and sample storage area should be isolated from all atmospheric sources of methylene chloride. All gas chromatography carrier gas lines and purge gas plumbing should be constructed from stainless steel or copper tubing. Laboratory clothing previously exposed to methylene chloride fumes during liquid-liquid extraction procedures can contribute to sample contamination.

Samples can also be contaminated by diffusion of volatile organics (particularly methylene chloride and fluorocarbons) through the septum seal during shipment and storage. A trip blank can serve as a check on such contamination.

INSTRUMENTATION GC/MS with a temperature-programmable chromatograph suitable for splitless injection equipped with variable constant differential flow controllers, a subambient oven controller, a purging device, sorbent trap, a thermal desorption apparatus and a capillary precolumn interface when using cryogenic cooling will be needed. The following GC columns may be used:

Column 1: 60 m × 0.75mm I.D. capillary column coated with VOCOL, 1.5 µm film thickness.
Column 2: 30 m × 0.53mm capillary column coated with DB-624 or VOCOL, 3 µm film thickness.
Column 3: 30 m × 0.32mm I.D. capillary column coated with DB-5 or SE-54, 1-µm film thickness.

PRECISION & ACCURACY This method has been tested in a single lab using spiked water. Using a wide-bore capillary column, water was spiked at concentrations between 0.5 and 10 µg/L. Single lab accuracy and precision data are presented. The MDL actually achieved in a given analysis will vary depending on instrument sensitivity and matrix effects.

The MDL (a) in µg/L was 0.08.
The concentration range in µg/L was 0.5–10.
The mean accuracy (% of true value) was 89.
The precision as relative standard deviation was 8.1.

Note: The MDL is based on a 25-mL sample volume instead of a 5-mL sample volume.

SAMPLING METHOD

Liquid samples — Use a 40-mL glass screw-cap VOA vial with a Teflon®-faced silicone septum that has been prewashed, rinsed with distilled deionized water, and oven dried. If residual chlorine is present, collect the sample in a 4-oz soil VOA container which has been pre-preserved with 4 drops of 10% sodium thiosulfate. Mix gently and transfer the sample to a 40-mL VOA vial. Collect bubble-free samples in duplicate and seal each sample in a separate plastic bag.

Soils, sediments and sludges — Use an 8-oz widemouth glass bottle with Teflon®-faced silicone septum that has been prewashed, rinsed with distilled deionized water, and oven dried. **Do not** heat the septum for more than 1 h. Tap slightly to eliminate any free air space. Collect samples in duplicate and seal each one in a separate plastic bag.

SAMPLE PRESERVATION

Liquid samples — Add 4 drops of concentrated HCL, cool to 4°C and store in a solvent-free refrigerator.

Soils, sediments and sludges — Cool samples to 4°C and store in a solvent-free refrigerator.

MHT The maximum holding time of any sample (liquids, soils, sediments, and sludges) is 14 days.

SAMPLE PREPARATION
Liquid samples — Remove the plunger from a 5-mL syringe and carefully pour the sample into the syringe barrel to just short of overflowing. Replace the syringe plunger and compress the sample. Open the syringe valve and vent any residual air while adjusting the sample volume to 5.0 mL. If there is only one volatile organic analysis (VOA) vial, a second syringe should be filled at this time to protect against possible loss of sample integrity. Add 10 µL of surrogate spiking solution and 10 µL of internal standard spiking solution through the valve bore of the 5-mL syringe, then close the valve. The surrogate and internal standards may be mixed and added as a single spiking solution.

Sediments, soils, and waste samples — All samples of this type should be screened by GC analysis using a headspace method (EPA Method 3810) or the hexadecane extraction and screening method (EPA Method 3820). Use the screening data to determine whether to use the low-concentration method (0.005–1 mg/kg) or the high-concentration method (>1 mg/kg).

Low-concentration method — The low-concentration method is based on purging a heated sediment or soil sample mixed with organic-free reagent water containing the surrogate and internal standards. Analyze all reagent blanks and standards under the same conditions as the samples.

Use a 5-g sample if the expected concentration is <0.1 mg/kg or a 1-g sample for expected concentrations between 0.1 and 1 mg/kg. Mix the contents of the sample container with a narrow metal spatula. Weigh the amount of the sample into a tared purge device. Add the spiked water to the purge device, which contains the weighed amount of sample, and connect the device to the purge-and-trap system.

High-concentration method — This method is based on extracting the sediment or soil with methanol. A waste sample is either extracted or diluted, depending on its solubility in methanol. Wastes that are insoluble in methanol are diluted with reagent tetraglyme or possibly polyethylene glycol (PEG). An aliquot of the extract is added to organic-free reagent water containing surrogate and internal standards. This is purged at ambient temperature. All samples with an expected concentration of >1.0 mg/kg should be analyzed by this method.

Mix the contents of the sample container with a narrow metal spatula. For sediments or soils and solid wastes that are insoluble in methanol, weigh 4 g (wet weight) of sample into a tared 20-mL vial. For waste that is soluble in methanol, tetraglyme, or PEG, weigh 1 g (wet weight) into a tared scintillation vial or culture tube or a 10-mL volumetric flask. Quickly add 9.0 mL of appropriate solvent then add 1.0 mL of a surrogate spiking solution to the vial, cap it, and shake it for 2 min.

METHANOL EXTRACT REQUIRED FOR ANALYSIS OF HIGH-CONCENTRATION SOILS OR SEDIMENTS

Approximate Concentration Range	Volume of Methanol Extract (a)
500–10,000 µg/kg	100 µL
1,000–20,000 µg/kg	50 µL
5,000–100,000 µg/kg	10 µL
25,000–500,000 µg/kg	100 µL of 1/50 dilution (b)

Calculate appropriate dilution factor for concentrations exceeding this table.

(a) The volume of methanol added to 5 mL of water being purged should be kept constant. Therefore, add to the 5-mL syringe whatever volume of methanol is necessary to maintain a volume of 100 µL added to the syringe.
(b) Dilute an aliquot of the methanol extract and then take 100 µL for analysis.

QUALITY CONTROL Demonstrate, through the analysis of a reagent water blank, that interferences from the analytical system, glassware, and reagents are under control. Blank samples should be carried through all stages of the sample preparation and measurement steps. For each analytical batch (up to 20 samples), a reagent blank, matrix spike, and matrix spike duplicate must be analyzed (the frequency of the spikes may be different for different monitoring programs). The blank and spiked samples must be carried through all stages of the sample preparation and measurement steps. QC samples mentioned in the section on Interferences will also be needed as appropriate to those situations.

Matrix spiking standards should be prepared from volatile organic compounds which will be representative of the compounds being investigated. The recommended internal standards are chlorobenzene-d5, 1,4-difluorobenzene, 1,4-dichlorobenzene-d4, and pentafluorobenzene. Using stock standard solutions, prepare secondary dilution standards containing the compounds of interest, either singly or mixed together in methanol. Store them in a vial with no headspace for no more than one week. Surrogates recommended are toluene-d8, 4-bromofluorobenzene, and dibromofluoromethane. Each sample undergoing GC/MS analysis must be spiked with 10 µL of the surrogate spiking solution prior to analysis.

REFERENCE Test Methods for Evaluating Solid Waste (SW-846). U.S. EPA 1983, Method 8260A, Rev. 1, Nov. 1990. Office of Solid Waste, Washington, DC.

Trichlorofluoromethane   **EPA Method 601**
CAS #75-69-4

TITLE Purgeable Halocarbons

MATRIX Wastewater.

APPLICATION Method covers 29 purgeable halocarbons. (Method 624 provides GC/MS conditions appropriate for the qualitative and quantitative confirmation of results). Method describes conditions for a 2nd GC column to confirm measurements made with primary column.

INTERFERENCES Impurities in the purge gas and organic compounds outgassing from the plumbing ahead of the trap. With high- and low-level samples, there can be carryover contamination. Diffusion of volatile organics through the septum seal into the sample.

INSTRUMENTATION GC-equipped with halide-specific detector. (With purge-and-trap unit).

RANGE 8.0–500 µg/L.

MDL Not determined.

PRECISION 0.26X + 0.91 µg/L (overall precision).

ACCURACY 0.89X–0.07 µg/L (as recovery).

SAMPLING METHOD 25-mL glass vial. Teflon®-lined septum.

STABILITY Cool, 4°C, 0.008% $Na_2S_2O_3$.

MHT 14 days.

QUALITY CONTROL The lab must on an ongoing basis, spike at least 10% of the samples from each sample site being monitored to assess accuracy.

REFERENCE Method 601, *Federal Register* Part VIII 40 CFR Part 136, Oct 26, 1984.

Trichlorofluoromethane   EPA Method 624
CAS #75-69-4

TITLE Purgeables

MATRIX Wastewater.

APPLICATION Method covers 31 purgeable organics. An inert gas is bubbled through a 5-mL water sample in a specially designed purging chamber. Here, purgeables are transferred from aqueous to gaseous phase, passed onto a sorbent column, and trapped. Trap is heated and backflushed with inert gas to desorb purgeables onto a GC column, where purgeables are separated.

INTERFERENCES Impurities in the purge gas, organic compounds outgassing from the plumbing ahead of the trap, and solvent vapors in the lab. With high- and low-level samples, there can be carryover contamination.

INSTRUMENTATION GC/MS with purge-and-trap unit.

RANGE 5–600 µg/L

MDL Not determined.

PRECISION 0.34X–0.39 µg/L (overall precision).

ACCURACY 0.99C + 0.39 µg/L (as recovery).

SAMPLING METHOD 25-mL glass vial. Teflon®-lined septum.

STABILITY Cool, 4°C, 0.008% $Na_2S_2O_3$.

MHT 14 days.

QUALITY CONTROL The lab must on an ongoing basis, spike at least 5% of the samples from each sample site being monitored to assess accuracy.

REFERENCE Method 624, *Federal Register* Part VIII 40 CFR Part 136, Oct 26, 1984.

Trichlorofluoromethane   EPA Method 8010
CAS #75-69-4

TITLE Halogenated Volatile Organics

MATRIX Groundwater, soils, sludges, water miscible liquid wastes, and non-water miscible wastes.

APPLICATION This method is used for the analysis of 39 halogenated VOCs. Samples are analyzed using direct injection or purge-and-trap methods. Groundwater must be analyzed by the purge-and-trap method. The method provides an optional GC column which is used for analyte confirmation and that may help resolve analytes from interferences.

INTERFERENCES There can be carryover contamination with high- and low-level samples. Impurities may come from the purge-and-trap apparatus, organic compounds outgassing from the plumbing ahead of trap, diffusion of VOCs through the sample bottle septum during shipping or storage, or from solvent vapors in the lab.

INSTRUMENTATION GC capable of on-column injections or purge-and-trap sample introduction and a halogen specific detector. Column 1: 8 ft by 0.1 in 1%. SP-1000 on Carbopack-B. Column 2: 6 ft by 0.1 in bonded n-octane on Porasil-C.

RANGE 8–500 µg/L (reagent water).

MDL Not determined.

PQL FACTORS FOR MULTIPLYING × MDL VALUE

Matrix	Multiplication Factor
Groundwater	10
Low-level soil	10
Water miscible liquid waste	500
High-level soil and sludge	1250
Non-water miscible waste	1250

PRECISION 0.26X ׀ 0.91 µg/L (overall precision).

ACCURACY 0.89C–0.07 µg/L (as recovery).

SAMPLING METHOD For water and liquid samples; use glass 40-mL vials with Teflon®-lined septum caps and collect two vials per sample location with no headspace. For solids and concentrated waste samples; use widemouth glass bottles with Teflon® liners.

STABILITY For concentrated wastes, soils, sediments, or sludges: cool to 4°C. For liquids: add 4 drops of concentrated hydrochloric acid and cool to 4°C.

MHT 14 days.

QUALITY CONTROL Analyze a reagent blank, matrix spike, and matrix spike duplicate/duplicate for each analytical batch (up to 20 samples). Demonstrate the purity of glassware and reagents by analyzing a reagent water method blank. Internal, surrogate, and five concentration level calibration standards are used.

REFERENCE Test Methods for Evaluating Solid Waste (SW-846), U.S. EPA Office of Solid Waste, Washington, DC, Method 8010B, Rev. 2, Nov. 1992.

Trichloronate   **EPA Method 8141**
CAS #327-98-0

TITLE Organophosphorus Compounds by Gas Chromatography: Capillary Column Technique

MATRIX This method covers aqueous and solid matrices. This includes a wide variety such as drinking water, groundwater, industrial wastewaters, surface waters, soils, solids, and sediments.

METHOD SUMMARY This is a GC method used to determine the concentration of 28 organophosphorus pesticides.

The use of Gel Permeation Cleanup (EPA Method 3640) for sample cleanup has been demonstrated to yield recoveries of less than 85% for many method analytes and is therefore not recommended for use with this method.

This method provides GC conditions for the detection of ppb concentrations of organophosphorus compounds. Prior to the use of this method, appropriate sample preparation techniques must be used. Water samples are extracted at a neutral pH with methylene chloride as a solvent by using a separatory funnel (EPA Method 3510) or a continuous liquid-liquid extractor (EPA Method 3520). Soxhlet extraction (EPA Method 3540) or ultrasonic extraction (EPA Method 3550) using methylene chloride/acetone (1:1) are used for solid samples. Both neat and diluted organic liquids (EPA Method 3580) may be analyzed by direct injection. Spiked samples are used to verify the applicability of the chosen extraction technique to each new sample type. A GC with a flame photometric (FPD) or nitrogen-phosphorus detector (NPD) is used for this multiresidue procedure.

INTERFERENCES The use of Florisil cleanup materials (EPA Method 3620) for some of the compounds in this method has been demonstrated to yield recoveries less than 85% and is therefore not recommended for all compounds. Use of phosphorus or halogen specific detectors, however, often obviates the necessity for cleanup for relatively clean sample matrices. If particular circumstances demand the use of an alternative cleanup procedure, the analyst must determine the elution profile and demonstrate that the recovery of each analyte is no less than 85%.

Use of a flame photometric detector (FPD) in the phosphorus mode will minimize interferences from materials that do not contain phosphorus. Elemental sulfur, however, may interfere with the determination of certain organophosphorus compounds by flame photometric gas chromatography. Sulfur cleanup using EPA Method 3660 may alleviate this interference. A nitrogen phosphorus detector (NPD) is also recommended.

A few analytes coelute on certain columns. Therefore, select a second column for confirmation where coelution of the analytes of interest does not occur.

Method interferences may be caused by contaminants in solvents, reagents, glassware, and other sample processing hardware that lead to discrete artifacts or elevated baselines in gas chromatograms. All these materials must be routinely demonstrated to be free from interferences under the conditions of the analysis by analyzing reagent blanks.

INSTRUMENTATION A GC with a NPD or a FPD will be needed. A data system or integrator is recommended for measuring peak areas and/or peak heights. A Kuderna-Danish (K-D) apparatus will be needed for extract concentration.

Column 1: 15 m × 0.53 mm megabore capillary column, 1.0 μm film thickness, DB-210.
Column 2: 15 m × 0.53 mm megabore capillary column, 1.5 μm film thickness, SPB-608.
Column 3: 15 m × 0.53 mm megabore capillary column, 1.0 μm film thickness, DB-5.

Three megabore capillary columns are included for analysis of organophosphates by this method. Column 1 (DB-210 or equivalent) and Column 2 (SPB-608 or equivalent) are recommended if a large number of organophosphorus analytes are to be determined. If the superior resolution offered by Column 1 and Column 2 is not required, Column 3 (DB-5 or equivalent) may be used. For megabore capillary columns, automatic injections of 1 μL are recommended.

PRECISION & ACCURACY The MDL actually achieved in a given analysis will vary, as it is dependent on instrument sensitivity and matrix effects. Single operator accuracy and precision studies have been conducted with spiked water and soil samples.

MULTIPLICATION FACTORS FOR OTHER MATRICES (a)

Matrix	Factor (b)
Groundwater (EPA Method 3510 or EPA Method 3520)	10
Low-concentration soil by Soxhlet and no cleanup	10 (c)
Low-concentration soil by ultrasonic extraction with GPC cleanup	6.7 (c)
High-concentration soil and sludges by ultrasonic extraction	500 (c)
Non-Water miscible waste (EPA Method 3580)	1000 (c)

(a) Sample EQLs are highly matrix-dependent. The EQLs listed here are provided for guidance and may not always be achievable.
(b) EQL = [Method detection limit] × [Factor]. For non-aqueous samples the factor is on a wet-weight basis.
(c) Multiply this factory times the soil MDL.

The MDL (in μg/L) when reagent water was extracted using a separatory funnel was 0.80.
The MDL (in μg/kg) when soil was extracted using Soxhlet extraction (EPA Method 3540) was 40.0.

Accuracy (as % recovery) with separatory funnel extraction ranged from not recovered (with low spikes) to 94 (with high spikes).

Accuracy (as % recovery) with continuous liquid-liquid extraction ranged from not recovered (with low spikes) to 21 (with high spikes).

Accuracy (as % recovery) with Soxhlet extraction of soils ranged from 56 (with low spikes to 53 (with high spikes).

Accuracy (as % recovery) with ultrasonic extraction of soils ranged from not recovered (with low spikes) to 31 (with high spikes).

SAMPLE COLLECTION, PRESERVATION & HANDLING
Containers used to collect samples for the determination of semivolatile organic compounds should be soap and water washed followed by methanol (or isopropanol) rinsing. The sample containers should be of glass or Teflon® and have screw-top covers with Teflon® liners.

No preservation is used with concentrated waste samples. With liquid samples containing no residual chlorine and with soil, sediment, and sludge samples, immediately cooling to 4°C is the only preservation used. When residual chlorine is present then 3 mL of 10% aqueous sodium sulfate is added for each gallon of sample collected, followed by cooling to 4°C.

Liquid samples must be extracted within 7 days and their extracts analyzed within 40 days. Concentrated waste, soil, sediment, and sludge samples must be extracted within 14 days and their extracts analyzed within 40 days.

SAMPLE PREPARATION In general, water samples are extracted at a neutral pH with methylene chloride, using either EPA Method 3510 or EPA Method 3520. Solid samples are extracted using either EPA Method 3540 or EPA Method 3550 with methylene chloride/acetone (1:1) as the extraction solvent.

Prior to GC analysis, the extraction solvent may be exchanged to hexane. Single lab data indicates that samples should not be transferred with 100% hexane during sample workup as the more water soluble organophosphorus compounds may be lost.

If cleanup is performed on the samples, the analyst should analyze the samples by GC. This will confirm elution patterns and the absence of interferences from the reagents. If peak detection and identification is prevented by the presence of interferences, further cleanup is required.

QUALITY CONTROL The analyst should monitor the performance of the extraction, cleanup (when used), and analytical system and the effectiveness of the method in dealing with each sample matrix by spiking each sample, standard, and blank with one or two surrogates (e.g., organophosphorus compounds not expected to be present in the sample). Deuterated analogs of analytes should not be used as surrogates for gas chromatographic analysis due to coelution problems.

A minimum of five concentrations for each analyte of interest should be prepared through dilution of the stock standards with isooctane. One of the concentrations should be at a concentration near, but above, the MDL.

Include a mid-level check standard after each group of 10 samples in the analysis sequence. GC/MS techniques should be judiciously employed to support qualitative identifications made with this method. Follow the GC/MS operating requirements specified in EPA Method 8270.

When available, chemical ionization mass spectra may be employed to aid in the qualitative identification process. To confirm an identification of a compound, the background-corrected mass spectrum of the compound must be obtained from the sample extract and must be compared with a mass spectrum from a stock or calibration standard analyzed under the same chromatographic conditions. The molecular ion and all other ions present above 20% relative abundance in the mass spectrum of the standard must be present in the mass spectrum of the sample with agreement to ±20%. The retention time of the compound in the sample must be within six seconds of the retention time for the same compound in the standard solution.

Should the MS procedure fail to provide satisfactory results, additional steps may be taken before reanalysis. These steps may include the use of alternate packed or capillary GC columns or additional sample cleanup.

REFERENCE Test Methods for Evaluating Solid Waste, Physical/Chemical Methods, SW-846, 3rd Edition, U.S. EPA, Office of Solid Waste, Washington, DC, EPA Method 8141 July 1992.

Trichloronate	EPA Method 8140
CAS #327-98-0	

TITLE Organophosphorus Pesticides

MATRIX Groundwater, soils, sludges, water miscible liquid wastes, and non-water miscible wastes.

APPLICATION This method is used for the analysis of 21 organophosphorus pesticides. Samples are extracted, concentrated, and analyzed using direct injection of both neat and diluted organic liquid into a gas chromatograph (GC).

INTERFERENCES Solvents, reagents, and glassware may introduce artifacts. Other interferences may come from coextracted compounds from samples. The use of Florisil cleanup materials may produce low recoveries. Elemental sulfur may interfere with some compounds when using a flame photometric detector. Sulfur cleanup (Method 3660) may alleviate sulfur interference.

INSTRUMENTATION GC capable of on-column injections and a flame photometric detector (FPD) or a thermionic detector. Column 1: 1.8 m by 2 mm with 5% SP-2401 on Supelcoport. Column 2: 1.8 m by 2 mm with 3% SP-2401 on Supelcoport. Column 3: 50 cm by ⅛ in Teflon® with 15% SE-54 on Gas Chrom Q. The preferred column is Column Number 1.

RANGE 20 µg/L only.

MDL 0.15 µg/L (in reagent water).

PQL FACTORS FOR MULTIPLYING × FID MDL VALUE

Matrix	Multiplication Factor
Groundwater	10
Low-level soil by sonication with GPC cleanup	670
High-level soil and sludge by sonication	10,000
Non-water miscible waste	100,000

PRECISION   18.6% (single operator standard deviation)

ACCURACY   105.0% (single operator average recovery)

SAMPLING METHOD   Use 8-oz. widemouth glass bottles with Teflon®-lined caps for concentrated waste samples, soils, sediments, and sludges. Use 1 or 2½ gallon amber glass bottles with Teflon®-lined caps for liquid (water) samples.

STABILITY   Cool soil, sediment, sludge, and liquid samples to 4°C. If residual chlorine is present in liquid samples add 3 mL of 10% sodium thiosulfate per gallon of sample and cool to 4°C.

MHT   14 days for concentrated waste, soil, sediment, or sludge; 7 days for liquid samples; all extracts must be analyzed within 40 days.

QUALITY CONTROL   A quality control check sample concentrate containing this compound in acetone at a concentration 1,000 times more concentrated than the selected spike concentration is required. The QC check sample concentrate may be prepared from pure standard materials or purchased as certified solutions. Use appropriate trip, matrix, control site, method, reagent, and solvent blanks. Internal, surrogate, and five concentration level calibration standards are used.

REFERENCE   Method 8140, SW-846, 3rd ed., Sept. 1986.

2,3,6-Trichlorophenol **EPA Method 1625**
CAS #933-75-5

TITLE   Semivolatile Organic Compounds by Isotope Dilution GC/MS

MATRIX   The compounds may be determined in waters, soils, and municipal sludges by this method.

METHOD SUMMARY   This method is used to determine 176 semivolatile toxic organic pollutants associated with the CWA (as amended 1987); the RCRA (as amended 1986); the CERCLA (as amended 1986); and other compounds amenable to extraction and analysis by capillary column gas chromatography-mass spectrometry (GC/MS).

Stable isotopically-labeled analogs of the compounds of interest are added to the sample. If the solids content is less than 1%, a 1-L sample is extracted at pH 12–13, then at pH <2 with methylene chloride using continuous extraction techniques.

If the solids content is 30% or less, the sample is diluted to 1% solids with reagent water, homogenized ultrasonically, and extracted at pH 12–13, then at pH <2 with methylene chloride using continuous extraction techniques. If the solids content is greater than 30%, the sample is extracted using ultrasonic techniques.

Each extract is dried over sodium sulfate, concentrated to a volume of 5 mL, cleaned up using GPC, if necessary, and concentrated. Extracts are concentrated to 1 mL if GPC is not performed, and to 0.5 mL if GPC is performed.

An internal standard is added to the extract, and a 1-mL aliquot of the extract is injected into the GC. The compounds are separated by GC and detected by a MS. The labeled compounds serve to correct the variability of the analytical technique.

INTERFERENCES   Solvents, reagents, glassware, and other sample processing hardware may yield artifacts and/or elevated baselines causing misinterpretation of chromatograms and spectra. Materials used in the analysis must be demonstrated to be free from interferences under the conditions of analysis by running method blanks initially and with each sample lot (sample started through the extraction process on a given 8-h shift, to a maximum of 20). Specific selection of reagents and purification of solvents by distillation in all glass systems may be required. Glassware and, where possible, reagents are cleaned by solvent rinse and baking at 450°C for 1-h minimum. Interferences coextracted from samples will vary considerably from source to source, depending on the diversity of the site being sampled.

INSTRUMENTATION   Major instrumentation includes a GC with a splitless or on-column injection port for capillary column, a MS with 70 eV electron impact ionization, and a data system to collect and record MS data, and process it. A K-D apparatus is used to concentrate extracts.

GC Column: 30 m × 0.25 mm I.D. 5% phenyl, 94% methyl, 1% vinyl silicone bonded phased fused silica capillary column.

PRECISION & ACCURACY   The detection limits of the method are usually dependent on the level of interferences rather than instrumental limitations. The limits typify the minimum quantities that can be detected with no interferences present.

The minimum level (in µg/mL) was 50. This is defined as a minimum level at which the analytical system shall give recognizable mass spectra (background corrected) and acceptable calibration points.

The MDL (in µg/kg) in low solids was not listed and in high solids was not listed; these were determined in digested sludge (low solids) and in filter cake or compost (high solids).

The labeled and native compound initial precision as standard deviation (in µg/L) was 30.
The labeled and native compound initial accuracy as average recovery (in µg/L) was 58–137.

SAMPLE COLLECTION, PRESERVATION & HANDLING
Collect samples in glass containers. Aqueous samples which flow freely are collected in refrigerated bottles using automatic sampling equipment. Solid samples are collected as grab samples using widemouth jars. Maintain samples at 0 to 4°C from the time of collection until extraction. If residual chlorine is present in aqueous samples, add 80 mg sodium thiosulfate/L

of water. Begin sample extraction within 7 days of collection, and analyze all extracts within 40 days of extraction.

SAMPLE PREPARATION Samples containing 1% solids or less are extracted directly using continuous liquid-liquid extraction techniques. Samples containing 1 to 30% solids are diluted to the 1% level with reagent water and extracted using continuous liquid-liquid extraction techniques. Samples containing greater than 30% solids are extracted using ultrasonic techniques.

Base/neutral extraction — Adjust the pH of the waters in the extractors to 12–13 with 6 N NaOH. Extract with methylene chloride for 24–48 h.

Acid extraction — Adjust the pH of the waters in the extractors to 2 or less using 6 N sulfuric acid. Extract with methylene chloride for 24–48 h.

Ultrasonic extraction of high solids samples — Add anhydrous sodium sulfate to the sample and QC aliquot(s). Add acetone:methylene chloride (1:1) to the sample and mix thoroughly

Concentrate extracts using a K-D apparatus.

QUALITY CONTROL The analyst is permitted to modify this method to improve separations or lower the costs of measurements, provided all performance specifications are met. Analyses of blanks are required to demonstrate freedom from contamination. When results of spikes indicate atypical method performance for samples, the samples are diluted to bring method performance within acceptable limits.

For low solids (aqueous samples), extract, concentrate, and analyze two sets of four 1-L aliquots (8 aliquots total) of the precision and recovery standard. For high solids samples, two sets of four 30-g aliquots of the high solids reference matrix are used.

Spike all samples with labeled compounds to assess method performance. Compute percent recovery of the labeled compounds using the internal standard method. Compare the labeled compound recovery for each compound with the corresponding labeled compound recovery.

Reagent water and high solids reference matrix blanks are analyzed to demonstrate freedom from contamination. Extract and concentrate a 1-L reagent water blank or a high solids reference matrix blank with each sample's lot (samples started through the extraction process on the same 8-h shift, to a maximum of 20 samples).

Field replicates may be collected to determine the precision of the sampling technique, and spiked samples may be required to determine the accuracy of the analysis when the internal standard method is used.

REFERENCE Semivolatile Organic Compounds by Isotope Dilution GC/MS. Office of Water Regulation and Standards, U.S. EPA Industrial Technology Division, Washington, DC, EPA Method 1625, Rev. C, June 1989 (contact W.A. Telliard, U.S. EPA, Office of Water Regulations and Standards, 401 M St., SW, Washington, DC, 20460. Phone: 202-382-7131).

2,4,5-Trichlorophenol   EPA Method 1625
CAS #95-95-4

TITLE Semivolatile Organic Compounds by Isotope Dilution GC/MS

MATRIX The compounds may be determined in waters, soils, and municipal sludges by this method.

METHOD SUMMARY This method is used to determine 176 semivolatile toxic organic pollutants associated with the CWA (as amended 1987); the RCRA (as amended 1986); the CERCLA (as amended 1986); and other compounds amenable to extraction and analysis by capillary column gas chromatography-mass spectrometry (GC/MS).

Stable isotopically-labeled analogs of the compounds of interest are added to the sample. If the solids content is less than 1%, a 1-L sample is extracted at pH 12–13, then at pH <2 with methylene chloride using continuous extraction techniques.

If the solids content is 30% or less, the sample is diluted to 1% solids with reagent water, homogenized ultrasonically, and extracted at pH 12–13, then at pH <2 with methylene chloride using continuous extraction techniques. If the solids content is greater than 30%, the sample is extracted using ultrasonic techniques.

Each extract is dried over sodium sulfate, concentrated to a volume of 5 mL, cleaned up using GPC, if necessary, and concentrated. Extracts are concentrated to 1 mL if GPC is not performed, and to 0.5 mL if GPC is performed.

An internal standard is added to the extract, and a 1-mL aliquot of the extract is injected into the GC. The compounds are separated by GC and detected by a MS. The labeled compounds serve to correct the variability of the analytical technique.

INTERFERENCES Solvents, reagents, glassware, and other sample processing hardware may yield artifacts and/or elevated baselines causing misinterpretation of chromatograms and spectra. Materials used in the analysis must be demonstrated to be free from interferences under the conditions of analysis by running method blanks initially and with each sample lot (sample started through the extraction process on a given 8-h shift, to a maximum of 20). Specific selection of reagents and purification of solvents by distillation in all glass systems may be required. Glassware and, where possible, reagents are cleaned by solvent rinse and baking at 450°C for 1-h minimum. Interferences coextracted from samples will vary considerably from source to source, depending on the diversity of the site being sampled.

INSTRUMENTATION Major instrumentation includes a GC with a splitless or on-column injection port for capillary column, a MS with 70 eV electron impact ionization, and a data system to collect and record MS data, and process it. A K-D apparatus is used to concentrate extracts.

GC Column: 30 m × 0.25 mm I.D. 5% phenyl, 94% methyl, 1% vinyl silicone bonded phased fused silica capillary column.

PRECISION & ACCURACY The detection limits of the method are usually dependent on the level of interferences

rather than instrumental limitations. The limits typify the minimum quantities that can be detected with no interferences present.

The minimum level (in µg/mL) was 10. This is defined as a minimum level at which the analytical system shall give recognizable mass spectra (background corrected) and acceptable calibration points.

The MDL (in µg/kg) in low solids was 32 and in high solids was 55; these were determined in digested sludge (low solids) and in filter cake or compost (high solids).

The labeled and native compound initial precision as standard deviation (in µg/L) was 30.

The labeled and native compound initial accuracy as average recovery (in µg/L) was 58–137.

SAMPLE COLLECTION, PRESERVATION & HANDLING

Collect samples in glass containers. Aqueous samples which flow freely are collected in refrigerated bottles using automatic sampling equipment. Solid samples are collected as grab samples using widemouth jars. Maintain samples at 0 to 4°C from the time of collection until extraction. If residual chlorine is present in aqueous samples, add 80 mg sodium thiosulfate/L of water. Begin sample extraction within 7 days of collection, and analyze all extracts within 40 days of extraction.

SAMPLE PREPARATION

Samples containing 1% solids or less are extracted directly using continuous liquid-liquid extraction techniques. Samples containing 1 to 30% solids are diluted to the 1% level with reagent water and extracted using continuous liquid-liquid extraction techniques. Samples containing greater than 30% solids are extracted using ultrasonic techniques.

- Base/neutral extraction — Adjust the pH of the waters in the extractors to 12–13 with 6 N NaOH. Extract with methylene chloride for 24–48 h.
- Acid extraction — Adjust the pH of the waters in the extractors to 2 or less using 6 N sulfuric acid. Extract with methylene chloride for 24–48 h.
- Ultrasonic extraction of high solids samples — Add anhydrous sodium sulfate to the sample and QC aliquot(s). Add acetone:methylene chloride (1:1) to the sample and mix thoroughly

Concentrate extracts using a K-D apparatus.

QUALITY CONTROL

The analyst is permitted to modify this method to improve separations or lower the costs of measurements, provided all performance specifications are met. Analyses of blanks are required to demonstrate freedom from contamination. When results of spikes indicate atypical method performance for samples, the samples are diluted to bring method performance within acceptable limits.

For low solids (aqueous samples), extract, concentrate, and analyze two sets of four 1-L aliquots (8 aliquots total) of the precision and recovery standard. For high solids samples, two sets of four 30-g aliquots of the high solids reference matrix are used.

Spike all samples with labeled compounds to assess method performance. Compute percent recovery of the labeled compounds using the internal standard method. Compare the labeled compound recovery for each compound with the corresponding labeled compound recovery.

Reagent water and high solids reference matrix blanks are analyzed to demonstrate freedom from contamination. Extract and concentrate a 1-L reagent water blank or a high solids reference matrix blank with each sample's lot (samples started through the extraction process on the same 8-h shift, to a maximum of 20 samples).

Field replicates may be collected to determine the precision of the sampling technique, and spiked samples may be required to determine the accuracy of the analysis when the internal standard method is used.

REFERENCE Semivolatile Organic Compounds by Isotope Dilution GC/MS. Office of Water Regulation and Standards, U.S. EPA Industrial Technology Division, Washington, DC, EPA Method 1625, Rev. C, June 1989 (contact W.A. Telliard, U.S. EPA, Office of Water Regulations and Standards, 401 M St., SW, Washington, DC, 20460. Phone: 202-382-7131).

2,4,6-Trichlorophenol   EPA Method 1625
CAS #88-06-2

TITLE Semivolatile Organic Compounds by Isotope Dilution GC/MS

MATRIX The compounds may be determined in waters, soils, and municipal sludges by this method.

METHOD SUMMARY This method is used to determine 176 semivolatile toxic organic pollutants associated with the CWA (as amended 1987); the RCRA (as amended 1986); the CERCLA (as amended 1986); and other compounds amenable to extraction and analysis by capillary column gas chromatography-mass spectrometry (GC/MS).

Stable isotopically-labeled analogs of the compounds of interest are added to the sample. If the solids content is less than 1%, a 1-L sample is extracted at pH 12–13, then at pH <2 with methylene chloride using continuous extraction techniques.

If the solids content is 30% or less, the sample is diluted to 1% solids with reagent water, homogenized ultrasonically, and extracted at pH 12–13, then at pH <2 with methylene chloride using continuous extraction techniques. If the solids content is greater than 30%, the sample is extracted using ultrasonic techniques.

Each extract is dried over sodium sulfate, concentrated to a volume of 5 mL, cleaned up using GPC, if necessary, and concentrated. Extracts are concentrated to 1 mL if GPC is not performed, and to 0.5 mL if GPC is performed.

An internal standard is added to the extract, and a 1-mL aliquot of the extract is injected into the GC. The compounds are separated by GC and detected by a MS. The labeled compounds serve to correct the variability of the analytical technique.

INTERFERENCES Solvents, reagents, glassware, and other sample processing hardware may yield artifacts and/or elevated baselines causing misinterpretation of chromatograms and spectra. Materials used in the analysis must be demonstrated to be free from interferences under the conditions of analysis by running method blanks initially and with each sample lot (sample started through the extraction process on a given 8-h shift, to a maximum of 20). Specific selection of reagents and purification of solvents by distillation in all glass systems may be required. Glassware and, where possible, reagents are cleaned by solvent rinse and baking at 450°C for 1-h minimum. Interferences coextracted from samples will vary considerably from source to source, depending on the diversity of the site being sampled.

INSTRUMENTATION Major instrumentation includes a GC with a splitless or on-column injection port for capillary column, a MS with 70 eV electron impact ionization, and a data system to collect and record MS data, and process it. A K-D apparatus is used to concentrate extracts.

GC Column: 30 m × 0.25 mm I.D. 5% phenyl, 94% methyl, 1% vinyl silicone bonded phased fused silica capillary column.

PRECISION & ACCURACY The detection limits of the method are usually dependent on the level of interferences rather than instrumental limitations. The limits typify the minimum quantities that can be detected with no interferences present.

The minimum level (in µg/mL) was 10. This is defined as a minimum level at which the analytical system shall give recognizable mass spectra (background corrected) and acceptable calibration points.

The MDL (in µg/kg) in low solids was not listed and in high solids was not listed; these were determined in digested sludge (low solids) and in filter cake or compost (high solids).

The labeled and native compound initial precision as standard deviation (in µg/L) was 57.

The labeled and native compound initial accuracy as average recovery (in µg/L) was 59–205.

SAMPLE COLLECTION, PRESERVATION & HANDLING Collect samples in glass containers. Aqueous samples which flow freely are collected in refrigerated bottles using automatic sampling equipment. Solid samples are collected as grab samples using widemouth jars. Maintain samples at 0 to 4°C from the time of collection until extraction. If residual chlorine is present in aqueous samples, add 80 mg sodium thiosulfate/L of water. Begin sample extraction within 7 days of collection, and analyze all extracts within 40 days of extraction.

SAMPLE PREPARATION Samples containing 1% solids or less are extracted directly using continuous liquid-liquid extraction techniques. Samples containing 1 to 30% solids are diluted to the 1% level with reagent water and extracted using continuous liquid-liquid extraction techniques. Samples containing greater than 30% solids are extracted using ultrasonic techniques.

- Base/neutral extraction — Adjust the pH of the waters in the extractors to 12–13 with 6 N NaOH. Extract with methylene chloride for 24–48 h.
- Acid extraction — Adjust the pH of the waters in the extractors to 2 or less using 6 N sulfuric acid. Extract with methylene chloride for 24–48 h.
- Ultrasonic extraction of high solids samples — Add anhydrous sodium sulfate to the sample and QC aliquot(s). Add acetone:methylene chloride (1:1) to the sample and mix thoroughly

Concentrate extracts using a K-D apparatus.

QUALITY CONTROL The analyst is permitted to modify this method to improve separations or lower the costs of measurements, provided all performance specifications are met. Analyses of blanks are required to demonstrate freedom from contamination. When results of spikes indicate atypical method performance for samples, the samples are diluted to bring method performance within acceptable limits.

For low solids (aqueous samples), extract, concentrate, and analyze two sets of four 1-L aliquots (8 aliquots total) of the precision and recovery standard. For high solids samples, two sets of four 30-g aliquots of the high solids reference matrix are used.

Spike all samples with labeled compounds to assess method performance. Compute percent recovery of the labeled compounds using the internal standard method. Compare the labeled compound recovery for each compound with the corresponding labeled compound recovery.

Reagent water and high solids reference matrix blanks are analyzed to demonstrate freedom from contamination. Extract and concentrate a 1-L reagent water blank or a high solids reference matrix blank with each sample's lot (samples started through the extraction process on the same 8-h shift, to a maximum of 20 samples).

Field replicates may be collected to determine the precision of the sampling technique, and spiked samples may be required to determine the accuracy of the analysis when the internal standard method is used.

REFERENCE Semivolatile Organic Compounds by Isotope Dilution GC/MS. Office of Water Regulation and Standards, U.S. EPA Industrial Technology Division, Washington, DC, EPA Method 1625, Rev. C, June 1989 (contact W.A. Telliard, U.S. EPA, Office of Water Regulations and Standards, 401 M St., SW, Washington, DC, 20460. Phone: 202-382-7131).

2,4,6-Trichlorophenol   **EPA Method 625**
CAS #88-06-2

TITLE Base/Neutrals and Acids, U.S. EPA Method 625

MATRIX This methods covers municipal and industrial wastewaters.

METHOD SUMMARY Approximately 1 L of sample is serially extracted with methylene chloride at a pH greater than 11 and again at a pH less than 2 using a separatory funnel or a continuous extractor. The methylene chloride extract is dried, concentrated to a volume of 1 mL, and analyzed by GC/MS. Qualitative identification of the parameters in the extract is performed using the retention time and the relative abundance of three characteristic masses (m/z). Qualitative analysis is performed using either external or internal standard techniques with a single characteristic m/z.

INTERFERENCES Method interferences may be caused by contaminants in solvents, reagents, glassware, and other sample processing hardware. Glassware must be scrupulously cleaned. Glassware should be heated in a muffle furnace at 400°C for 5 to 30 min. Some thermally stable materials, such as PCBs, may not be eliminated by this treatment. Solvent rinses with acetone and pesticide quality hexane may be substituted for the muffle furnace heating. Matrix interferences may be caused by contaminants that are coextracted from the sample. The base-neutral extraction may cause significantly reduced recovery of phenols. The packed gas chromatographic columns recommended for the basic fraction may not exhibit sufficient resolution for some analytes.

INSTRUMENTATION A GC/MS system with an injection port designed for on-column injection when using packed columns and for splitless injection when using capillary columns.

Column for base/neutrals: 1.8 m long × 2 mm I.D. glass, packed with 3% SP-2550 on Supelcoport (100/120 mesh) or equivalent.

Column for acids: 1.8 m long × 2 mm I.D. glass, packed with 1% SP-1240DA on Supelcoport (100/120 mesh) or equivalent.

PRECISION & ACCURACY The MDL concentrations were obtained using reagent water. The MDL actually achieved in a given analysis will vary depending on instrument sensitivity and matrix effects. This method was tested by 15 laboratories using reagent water, drinking water, surface water, and industrial wastewaters spiked at six concentrations over the range 5 to 100 µg/L. Single operator precision, overall precision, and method accuracy were found to be directly related to the concentration of the parameter matrix.

The MDL (in µg/L) in reagent water was 2.7.

The standard deviation (in µg/L based on 4 recovery measurements) was 31.7.

The range (in µg/L) for average recovery for 4 measurements was 52.4–129.2.

The range (in %) for percent recovery was 37–144.

Accuracy (in µg/L) as expected recovery for one or more measurements of a sample containing a true concentration of C was $0.91C - 0.18$.

Precision (in µg/L) as expected single analyst standard deviation of measurements at an average concentration found at X was $0.16X + 2.22$.

Overall precision (in µg/L) as expected interlaboratory standard deviation of measurements in an average concentration found at X was $0.22X + 1.81$.

C = True value of the concentration in µg/L.
X = Average recovery found for measurements of samples containing a concentration at C in µg/L.

SAMPLE PREPARATION Adjust the pH to >11 with sodium hydroxide and serially extract in a separatory funnel with methylene chloride or else in a continuous extractor. Next, adjust the pH to <2 with sulfuric acid and serially extract in a separatory funnel with methylene chloride or else in a continuous extractor. Dry the extracts separately through a column of anhydrous sodium sulfate and then concentrate each of the extracts to 1.0 mL using a K-D apparatus.

SAMPLE COLLECTION, PRESERVATION & HANDLING Grab samples must be collected in glass containers. All samples must be refrigerated at 4°C from the time of collection until extraction. If residual chlorine is present, add 80 mg of sodium thiosulfate/L of sample and mix well. All samples must be extracted within 7 days of collection and completely analyzed within 40 days of extraction.

QUALITY CONTROL Make an initial, one-time, demonstration of the ability to generate acceptable accuracy and precision with this method. Before processing any samples, the analyst must analyze a reagent water blank to demonstrate that interferences from the analytical system and glassware are under control. Each time a set of samples is extracted or reagents are changed, a reagent water blank must be processed. Spike and analyze a minimum of 5% of all samples to monitor and evaluate lab data quality. A QC check sample concentrate that contains each parameter of interest at a concentration of 100 µg/mL in acetone is required. PCBs and multicomponent pesticides may be omitted from this test.

After analysis of five spiked wastewater samples, calculate the average percent recovery and the standard deviation of the percent recovery. Spike all samples with the surrogate standard spiking solution and calculate the percent recovery of each surrogate compound.

REFERENCE Federal Register, Vol. 49, No. 209. Friday, Oct. 26, 1984.

2,4,5-Trichlorophenol **EPA Method 8270**
CAS #95-95-4

TITLE Semivolatile Organic Compounds by GC/MS

MATRIX This method is used to determine the concentration of semivolatile organic compounds in extracts prepared from all types of solid waste matrices, soils, and groundwater. Although surface waters are not specifically mentioned, this method should be applicable to water samples from rivers, lakes, etc.

METHOD SUMMARY This method covers 259 semivolatile organic compounds. In very limited applications direct injection of the sample into the GC/MS system may be appropriate, but this results in very high detection limits (approximately 10,000 µg/L). Typically, a 1-L liquid sample, containing surrogate, and matrix spiking standards, is extracted in a continuous extractor first under acid conditions and then under basic conditions. Typically 30 g of a solid sample, containing surrogate, and matrix spiking standards, is extracted ultrasonically. After concentrating the extract to 1 mL it is spiked with 10 µL of an

internal standard solution just prior to analysis by GC/MS. The volume injected should contain about 100 ng of base/neutral and 200 ng of acid surrogates (for a 1-μL injection). Analysis is performed by GC/MS using a capillary GC column.

INTERFERENCES Raw GC/MS data from all blanks, samples, and spikes must be evaluated for interferences. Contamination by carryover can occur whenever high-concentration and low-concentration samples are sequentially analyzed. To reduce carryover, the sample syringe must be rinsed out between samples with solvent. Whenever an unusually concentrated sample is encountered, it should be followed by the analysis of blank solvent to check for cross-contamination.

INSTRUMENTATION A GC/MS and a data system are required. The GC column used is a 30 m × 0.25 mm I.D. (or 0.32 mm I.D.) 1um film thickness silicone-coated fused silica capillary column. A continuous liquid-liquid extractor equipped with Teflon® or glass connection joints and stopcocks requiring no lubrication, a K-D concentrating apparatus, water bath, and an ultrasonic disrupter with a minimum power of 300 W and with pulsing capability are also required.

PRECISION & ACCURACY The estimated quantitation limit (EQL) of Method 8270B for determining an individual compound is approximately 1 mg/kg (wet weight) for soil or sediment samples, 1–200 mg/kg for wastes (dependent on matrix and method of preparation), and 10 μg/L for groundwater samples. EQLs will be proportionately higher for sample extracts that require dilution to avoid saturation of the detector.

The EQL(b) for groundwater in μg/L is 10.
The EQL (a, b) for low concentrations in soil and sediment in μg/kg is 660.
Accuracy as μg/L is not listed.
Overall precision in μg/L is not listed.

(a) EQLs listed for soil/sediment are based on wet weight. Normally data is reported in a dry-weight basis; therefore, EQLs will be higher based on the % dry weight of each sample. This calculation is based on a 30 g sample and gel permeation chromatography cleanup.
(b) Sample EQLs are highly matrix-dependent. The EQLs are provided for guidance and may not always be achievable.
C = True value for concentration, in μg/L.
X = Average recovery found for measurements of samples containing a concentration of C, in μg/L.

ESTIMATED QUANTITATION LIMIT

Other Matrices	Factor (a)
High-concentration soil and sludges by sonicator	7.5
Non-water miscible waste	75

(a) EQL for other matrices = [EQL for low soil/sediment × [Factor]. This estimated EQL is similar to an EPA "Practical Quantitation Limit."

SAMPLING METHOD
Liquid samples — Use a 1 or 2½ gallon amber glass bottle with a screw-top Teflon®-lined cover that has been prewashed with detergent and rinsed with distilled water and methanol (or isopropanol).

Soils, sediments, or sludges — Use an 8-oz. widemouth glass with a screw-top Teflon®-lined cover that has been prewashed with detergent and rinsed with distilled water and methanol (or isopropanol).

SAMPLE PRESERVATION
Liquid samples — If residual chlorine is present, add 3 mL of 10% sodium thiosulfate per gallon, cool to 4°C and store in a solvent-free refrigerator until analysis; if chlorine is not present, then eliminate the sodium thiosulfate addition.

Soils, sediments, or sludges — Cool samples to 4°C and store in a solvent-free refrigerator.

MHT Liquid samples must be extracted within 7 days and the extracts analyzed within 40 days. Soils, sediments, or sludges may be stored for a maximum of 14 days and the extracts analyzed within 40 days.

SAMPLE PREPARATION
Liquid samples — Transfer 1 L quantitatively to a continuous extractor. If high concentrations are anticipated, a smaller volume may be used and then diluted with organic-free reagent water to 1 L. Adjust pH, if necessary, to pH <2 using 1:1 (V/V) sulfuric acid. Pipette 1.0 mL of a surrogate standard spiking solution into each sample. For the sample in each analytical batch selected for spiking, add 1.0 mL of a matrix spiking standard. For base/neutral acid analysis, the amount of the surrogates and matrix spiking compounds added to the sample should result in a final concentration of 100 ng/μL of each analyte in the extract to be analyzed (assuming a 1-μL injection). Extract with methylene chloride for 18–24 h. Next, adjust the pH of the aqueous phase to pH >11 using 10 N sodium hydroxide and extract it with methylene chloride again for 18–24 h. Dry the extract through a column containing anhydrous sodium sulfate and concentrate it to 1 mL using a K-D concentrator.

Soils, sediments, or sludges — Use 30 g of sample. Nonporous or wet samples (gummy or clay type) that do not have a free-flowing sandy texture must be mixed with anhydrous sodium sulfate until the sample is free flowing. Add 1 mL of surrogate standards to all samples, spikes, standards, and blanks. For the sample in each analytical batch selected for spiking, add 1.0 mL of a matrix spiking standard. For base/neutral acid analysis, the amount added of the surrogates and matrix spiking compounds should result in a final concentration of 100 ng/μL of each base/neutral analyte and 200 ng/μL of each acid analyte in the extract to be analyzed (assuming a 1-μL injection). Immediately add a 100-mL mixture of 1:1 methylene chloride:acetone and extract the sample ultrasonically for 3 min and then decant or filter the extracts. Repeat the extraction two or more times. Dry the extract using a column with anhydrous sodium sulfate and concentrate it to 1 mL in a K-D concentrator.

QUALITY CONTROL A methylene chloride solution containing 50 ng/μL of decafluorotriphenylphosphine (DFTPP) is used for tuning the GC/MS system each 12-h shift. A system performance check also must be made during every 12-h shift. A standard containing 50 ng/μL each of 4,4'-DDT, pentachlorophenol, and benzidine is required to verify injection port inertness and GC column performance. A calibration standard

at mid-concentration, containing each compound of interest, including all required surrogates, must be performed every 12 h during analysis. After the system performance check is met, calibration check compounds (CCCs) are used to check the validity of the initial calibration.

The internal standard responses and retention times in the calibration check standard must be evaluated immediately after or during data acquisition. If the retention time for any internal standard changes by more than 30 seconds from the last check calibration (12 h), the chromatographic system must be inspected for malfunctions and corrections must be made, as required. If the electron ionization current plot (EICP) area for any of the internal standards changes by a factor of two from the last daily calibration standard check, the mass spectrometer must be inspected for malfunctions and corrections must be made, as appropriate.

Demonstrate, through the analysis of a reagent water blank, that interferences from the analytical system, glassware, and reagents are under control. The blank samples should be carried through all stages of the sample preparation and measurement steps. For each analytical batch (up to 20 samples), a reagent blank, matrix spike, and matrix spike duplicate/duplicate must be analyzed (the frequency of the spikes may be different for different monitoring programs). The blank and spiked samples must be carried through all stages of the sample preparation and measurement steps. A QC reference sample concentrate containing each analyte at a concentration of 100 mg/L in methanol is required.

REFERENCE Test Methods for Evaluating Solid Waste (SW-846). U.S. EPA 1983, Method 8270B, Rev. 2, Nov. 1990. Office of Solid Waste, Washington, DC.

2,4,6-Trichlorophenol   EPA Method 8270
CAS #88-06-2

TITLE Semivolatile Organic Compounds by GC/MS

MATRIX This method is used to determine the concentration of semivolatile organic compounds in extracts prepared from all types of solid waste matrices, soils, and groundwater. Although surface waters are not specifically mentioned, this method should be applicable to water samples from rivers, lakes, etc.

METHOD SUMMARY This method covers 259 semivolatile organic compounds. In very limited applications direct injection of the sample into the GC/MS system may be appropriate, but this results in very high detection limits (approximately 10,000 µg/L). Typically, a 1-L liquid sample, containing surrogate, and matrix spiking standards, is extracted in a continuous extractor first under acid conditions and then under basic conditions. Typically 30 g of a solid sample, containing surrogate, and matrix spiking standards, is extracted ultrasonically. After concentrating the extract to 1 mL it is spiked with 10 µL of an internal standard solution just prior to analysis by GC/MS. The volume injected should contain about 100 ng of base/neutral and 200 ng of acid surrogates (for a 1-µL injection). Analysis is performed by GC/MS using a capillary GC column.

INTERFERENCES Raw GC/MS data from all blanks, samples, and spikes must be evaluated for interferences. Contamination by carryover can occur whenever high-concentration and low-concentration samples are sequentially analyzed. To reduce carryover, the sample syringe must be rinsed out between samples with solvent. Whenever an unusually concentrated sample is encountered, it should be followed by the analysis of blank solvent to check for cross-contamination.

INSTRUMENTATION A GC/MS and a data system are required. The GC column used is a 30 m × 0.25 mm I.D. (or 0.32 mm I.D.) 1um film thickness silicone-coated fused silica capillary column. A continuous liquid-liquid extractor equipped with Teflon® or glass connection joints and stopcocks requiring no lubrication, a K-D concentrating apparatus, water bath, and an ultrasonic disrupter with a minimum power of 300 W and with pulsing capability are also required.

PRECISION & ACCURACY The estimated quantitation limit (EQL) of Method 8270B for determining an individual compound is approximately 1 mg/kg (wet weight) for soil or sediment samples, 1–200 mg/kg for wastes (dependent on matrix and method of preparation), and 10 µg/L for groundwater samples. EQLs will be proportionately higher for sample extracts that require dilution to avoid saturation of the detector.

The EQL(b) for groundwater in µg/L is 10.
The EQL (a, b) for low concentrations in soil and sediment in µg/kg is 660.
Accuracy as µg/L is 0.91C–0.18.
Overall precision in µg/L is 0.22X + 1.81.

(a) *EQLs listed for soil/sediment are based on wet weight. Normally data is reported in a dry-weight basis; therefore, EQLs will be higher based on the % dry weight of each sample. This calculation is based on a 30 g sample and gel permeation chromatography cleanup.*
(b) *Sample EQLs are highly matrix-dependent. The EQLs are provided for guidance and may not always be achievable.*
C = *True value for concentration, in µg/L.*
X = *Average recovery found for measurements of samples containing a concentration of C, in µg/L.*

ESTIMATED QUANTITATION LIMIT

Other Matrices	Factor (a)
High-concentration soil and sludges by sonicator	7.5
Non-water miscible waste	75

(a) *EQL for other matrices = [EQL for low soil/sediment × [Factor]. This estimated EQL is similar to an EPA "Practical Quantitation Limit."*

SAMPLING METHOD

Liquid samples — Use a 1 or 2½ gallon amber glass bottle with a screw-top Teflon®-lined cover that has been prewashed with detergent and rinsed with distilled water and methanol (or isopropanol).

Soils, sediments, or sludges — Use an 8-oz. widemouth glass with a screw-top Teflon®-lined cover that has been prewashed

with detergent and rinsed with distilled water and methanol (or isopropanol).

SAMPLE PRESERVATION

Liquid samples — If residual chlorine is present, add 3 mL of 10% sodium thiosulfate per gallon, cool to 4°C and store in a solvent-free refrigerator until analysis; if chlorine is not present, then eliminate the sodium thiosulfate addition.

Soils, sediments, or sludges — Cool samples to 4°C and store in a solvent-free refrigerator.

MHT Liquid samples must be extracted within 7 days and the extracts analyzed within 40 days. Soils, sediments, or sludges may be stored for a maximum of 14 days and the extracts analyzed within 40 days.

SAMPLE PREPARATION

Liquid samples — Transfer 1 L quantitatively to a continuous extractor. If high concentrations are anticipated, a smaller volume may be used and then diluted with organic-free reagent water to 1 L. Adjust pH, if necessary, to pH <2 using 1:1 (V/V) sulfuric acid. Pipette 1.0 mL of a surrogate standard spiking solution into each sample. For the sample in each analytical batch selected for spiking, add 1.0 mL of a matrix spiking standard. For base/neutral acid analysis, the amount of the surrogates and matrix spiking compounds added to the sample should result in a final concentration of 100 ng/µL of each analyte in the extract to be analyzed (assuming a 1-µL injection). Extract with methylene chloride for 18–24 h. Next, adjust the pH of the aqueous phase to pH >11 using 10 N sodium hydroxide and extract it with methylene chloride again for 18–24 h. Dry the extract through a column containing anhydrous sodium sulfate and concentrate it to 1 mL using a K-D concentrator.

Soils, sediments, or sludges — Use 30 g of sample. Nonporous or wet samples (gummy or clay type) that do not have a free-flowing sandy texture must be mixed with anhydrous sodium sulfate until the sample is free flowing. Add 1 mL of surrogate standards to all samples, spikes, standards, and blanks. For the sample in each analytical batch selected for spiking, add 1.0 mL of a matrix spiking standard. For base/neutral acid analysis, the amount added of the surrogates and matrix spiking compounds should result in a final concentration of 100 ng/µL of each base/neutral analyte and 200 ng/µL of each acid analyte in the extract to be analyzed (assuming a 1-µL injection). Immediately add a 100-mL mixture of 1:1 methylene chloride:acetone and extract the sample ultrasonically for 3 min and then decant or filter the extracts. Repeat the extraction two or more times. Dry the extract using a column with anhydrous sodium sulfate and concentrate it to 1 mL in a K-D concentrator.

QUALITY CONTROL A methylene chloride solution containing 50 ng/µL of decafluorotriphenylphosphine (DFTPP) is used for tuning the GC/MS system each 12-h shift. A system performance check also must be made during every 12-h shift. A standard containing 50 ng/µL each of 4,4′-DDT, pentachlorophenol, and benzidine is required to verify injection port inertness and GC column performance. A calibration standard at mid-concentration, containing each compound of interest, including all required surrogates, must be performed every 12 h during analysis. After the system performance check is met, calibration check compounds (CCCs) are used to check the validity of the initial calibration.

The internal standard responses and retention times in the calibration check standard must be evaluated immediately after or during data acquisition. If the retention time for any internal standard changes by more than 30 seconds from the last check calibration (12 h), the chromatographic system must be inspected for malfunctions and corrections must be made, as required. If the electron ionization current plot (EICP) area for any of the internal standards changes by a factor of two from the last daily calibration standard check, the mass spectrometer must be inspected for malfunctions and corrections must be made, as appropriate.

Demonstrate, through the analysis of a reagent water blank, that interferences from the analytical system, glassware, and reagents are under control. The blank samples should be carried through all stages of the sample preparation and measurement steps. For each analytical batch (up to 20 samples), a reagent blank, matrix spike, and matrix spike duplicate/duplicate must be analyzed (the frequency of the spikes may be different for different monitoring programs). The blank and spiked samples must be carried through all stages of the sample preparation and measurement steps. A QC reference sample concentrate containing each analyte at a concentration of 100 mg/L in methanol is required.

REFERENCE Test Methods for Evaluating Solid Waste (SW-846). U.S. EPA 1983, Method 8270B, Rev. 2, Nov. 1990. Office of Solid Waste, Washington, DC.

2,4,6-Trichlorophenol **EPA Method 8040**
CAS #88-06-2

TITLE Phenols

MATRIX Groundwater, soils, sludges, water miscible liquid wastes, and non-water miscible wastes.

APPLICATION This method is used for the analysis of 17 phenols. Samples are extracted, concentrated, and analyzed using direct injection of both neat and diluted organic liquids. Pentafluorobenzylbromide (PFB) derivatives also may be made to increase sensitivity of the method.

INTERFERENCES There can be carryover contamination with high- and low-level samples. Solvents, reagents, and glassware may introduce artifacts. Other interferences may come from coextracted compounds from samples.

INSTRUMENTATION GC capable of on-column injections and a flame with detector (FID) or electron capture detector (ECD). Column for underivatized phenol: 1.8 m by 2.0 mm with 1% SP-1240DA on Supelcoport. Column for derivatized phenols: 1.8 m by 2.0 mm with 5% OV-17 on Chromosorb W-AW-DMCS.

RANGE 12–450 µg/L

MDL 0.64 µg/L (FID) and 0.58 µg/L (ECD)

PQL FACTORS FOR MULTIPLYING × FID MDL VALUE

Matrix	Multiplication Factor
Groundwater	10
Low-level soil by sonication with GPC cleanup	670
High-level soil and sludge by sonication	10,000
Non-water miscible waste	100,000

PRECISION 0.13X + 2.40 µg/L (overall precision using FID)

ACCURACY 0.86C–0.40 µg/L (as recovery using FID)

SAMPLING METHOD Use 8-oz. widemouth glass bottles with Teflon®-lined caps for concentrated waste samples, soils, sediments, and sludges. Use 1 or 2½ gallon amber glass bottles with Teflon®-lined caps for liquid (water) samples.

STABILITY Cool soil, sediment, sludge, and liquid samples to 4°C. If residual chlorine is present in liquid samples add 3 mL of 10% sodium thiosulfate per gallon of sample and cool to 4°C.

MHT 14 days for concentrated waste, soil, sediment, or sludge; 7 days for liquid samples; all extracts must be analyzed within 40 days.

QUALITY CONTROL A quality control check sample concentrate containing each analyte of interest is required. The QC check sample concentrate may be prepared from pure standard materials or purchased as certified solutions Use appropriate trip, matrix, control site, method, reagent, and solvent blanks. Internal, surrogate, and five concentration level calibration standards are used. The QC check sample concentrate should contain this compound at 100 µg/mL in 2-propanol.

REFERENCE Test Methods for Evaluating Solid Waste (SW-846), U.S. EPA Office of Solid Waste, Washington, DC, Method 8040A, Rev. 1, Nov. 1990.

1,2,3-Trichloropropane   EPA Method 1624
CAS #96-18-4

TITLE Volatile Organic Compounds by Isotope Dilution GC/MS

MATRIX Compounds may be determined in waters, soils, and municipal sludges by this method.

METHOD SUMMARY This method is used to determine 58 volatile toxic organic pollutants associated with the CWA (as amended 1987); the RCRA (as amended 1986); the CERCLA (as amended 1986); and other compounds amenable to purge-and-trap gas chromatography-mass spectrometry (GC/MS).

If the solids content is less than 1%, stable isotopically-labeled analogs of the compounds of interest are added to a 5-mL sample and the sample is purged with an inert gas at 20–25°C in a chamber designed for soil or water samples. If the solids content is greater than 1%, 5 mL of reagent water and the labeled compounds are added to a 5-g aliquot of sample and the mixture is purged at 40°C. Compounds that will not purge at 20–25°C or at 40°C are purged at 78–85°C. In the purging process, the volatile compounds are transferred from the aqueous phase into the gaseous phase where they are passed into a sorbent column, and trapped. After purging is completed, the trap is backflushed and heated rapidly to desorb the compounds into a GC. The compounds are separated by the GC and detected by a MS. The labeled compounds serve to correct the variability of the analytical technique.

INTERFERENCES Impurities in the purge gas, organic compounds outgassing from the plumbing upstream of the trap, and solvent vapors in the lab account for most problems. Samples can be contaminated by diffusion of volatile organic compounds (particularly methylene chloride) through the bottle seal during shipment and storage. Contamination by carryover can occur when high-level and low-level samples are analyzed sequentially. When an unusually concentrated sample is encountered, follow it by analysis of a reagent water blank to check for carryover.

INSTRUMENTATION Major equipment includes a GC with linear temperature programming and a glass jet separator as the MS interface, a MS with 70 eV electron impact ionization, and a data system to collect and record response factors.

Column: 2.8 m × 2 mm I.D. glass, packed with 1% SP-1000 on Carbopak B, 60/80 mesh, or equivalent.

PRECISION & ACCURACY The detection limits of the method are usually dependent on the level of interferences rather than instrumental limitations. The method detection limits were determined in digested sludge (low solids) and in filter cake or compost (high solids).

The MDL (in µg/kg) for low solids is not listed and for high solids is not listed.

Labeled and native compound precision (in µg/L) as standard deviation was not listed.

Labeled and native compound accuracy (in µg/L) as average recovery was not listed.

Acceptance criteria are at 20 µg/L for this compound.

SAMPLE COLLECTION, PRESERVATION & HANDLING Grab samples are collected in glass containers having a total volume greater than 20 mL. Fill and seal each bottle so that no air bubbles are entrapped. Samples are maintained at 0 to 4°C from the time of collection until analysis. If an aqueous sample contains residual chlorine, add sodium thiosulfate preservative (10 mg/40 mL) to the empty sample bottles just prior to shipment to the sample site. All samples must be analyzed within 14 days of collection.

SAMPLE PREPARATION Samples containing less than 1% solids are analyzed directly as aqueous samples. Samples containing 1% solids or greater are analyzed as solid samples utilizing one of two methods, depending on the levels of pollutants, in the sample. Samples containing 1% solids or greater, and low to moderate levels of pollutants are analyzed by purging a known weight of sample added to 5 mL of reagent water. Samples containing 1% solids or greater, and high levels of pollutants, are extracted with methanol, and an aliquot of the methanol extract is added to reagent water and purged.

QUALITY CONTROL A field blank prepared from reagent water and carried through the sampling and handling protocol may serve as a check on contamination from shipment and storage.

The analyst is permitted to modify this method to improve separations or lower the costs of measurements, provided all performance specifications are met. Analyses of blanks are required. When results of spikes indicate atypical method performance for samples, the samples are diluted to bring method performance within acceptable limits. Analyze two sets of four 5-mL aliquots (8 aliquots total) of the aqueous performance standard. Spike all samples with labeled compounds to assess method performance on the sample matrix. Compute the percent recovery of the labeled compounds using the internal standard method. Compare the percent recovery for each compound with the corresponding labeled compound recovery. Reagent water blanks are analyzed to demonstrate freedom from carryover contamination. Field replicates may be collected to determine the precision of the sampling technique, and spiked samples may be required to determine the accuracy of the analysis when the internal method is used.

REFERENCE Volatile Organic Compounds by Isotope Dilution GC/MS. Office of Water Regulation and Standards, U.S. EPA Industrial Technology Division, Washington, DC, EPA Method 1624, Rev. C, June 1989 (contact W.A. Telliard, U.S. EPA, Office of Water Regulations and Standards, 401 M St., SW, Washington, DC, 20460. Phone: 202-382-7131).

1,2,3-Trichloropropane   **EPA Method 502**
CAS #96-18-4

TITLE Volatile Organic Compounds in Water By Purge and Trap Capillary Column Gas Chromatography with Photoionization and Electrolytic Conductivity Detectors in Series. U.S. EPA Method 502.2, Rev. 2.0, 1989.

MATRIX Drinking water and raw source water. The latter should include most surface water and groundwater sources.

METHOD SUMMARY This method covers 60 volatile organic compounds that contain halogen atoms and/or that are aromatic. An inert gas (zero grade nitrogen or helium) is bubbled through a 25-mL or a 5-mL water sample (depending on the expected concentration of the analytes). Purged sample components are trapped in a tube of sorbent materials. When purging is complete, the sorbent tube is heated and backflushed with helium to desorb the trapped sample onto a capillary GC column. The column is temperature programmed to separate the method analytes which are then detected with a photoionization detector (PID) and a Hall electrolytic conductivity (HECD) placed in series. The PID is selective for aromatic compounds and the HECD is selective for halogenated compounds.

INTERFERENCES Impurities in the purge gas and from organic compounds outgassing from the plumbing ahead of the trap account for many contamination problems. Interferences purged or coextracted from the samples will vary considerably from source to source, depending upon the particular sample or extract being tested. Cross-contamination can occur whenever high-level and low-level samples are analyzed sequentially. Samples also can be contaminated by diffusion of volatile organics (particularly methylene chloride and fluorocarbons) through the septum seal into the sample during shipment and storage. The lab where volatile analysis is performed and also the refrigerated storage area should be completely free of solvents.

INSTRUMENTATION A GC containing a series configuration of a high temperature photoionization detector (PID) equipped with 10.0 eV (nominal) lamp and Hall electrolytic conductivity detector (HECD) is required. Also required is an all-glass 5-mL purging device, a sorbent trap, and a thermal desorption apparatus which is connected to the GC system.

Column 1: VOCOL glass wide-bore capillary column.
Column 2: RTX–502.2 mega-bore capillary column.
Column 3: DB-62 mega-bore capillary column.

PRECISION & ACCURACY Method detection limits are dependent upon the characteristics of the gas chromatographic system used. Analytes that are not separated chromatographically cannot be individually identified and used in the same calibration mixture or water samples unless an alternative technique for identification and quantification, such as mass spectrometry, is used.

Electrolytic conductivity detetor (c) range in µg/L (a) was 0.02–200.
Electrolytic conductivity detetor (c) MDL in µg/L (b) was 0.4.
Electrolytic conductivity detetor (c) accuracy as % recovery was 99.
Electrolytic conductivity detetor (c) precision as % RSD was 2.3.
Photoionization detector (d) range in µg/L (a) was 0.02–200.
Photoionization detector (d) MDL in µg/L (b) was not listed.
Photoionization detector (d) accuracy as % recovery was not listed.
Photoionization detector (d) precision as % RSD was not listed.

(a) The applicable concentration range of this method is compound, instrument, and matrix-dependent. It is listed as being approximately 0.02 to 200 µg/L but no specific information is provided so caution should be observed.
(b) The method detection limits reports with this method are compound, instrument, and matrix-dependent. The values reported were calculated using reagent water fortified with the corresponding compounds at 10 µg/L and a GC-equipped with a 60 m × 0.75 mm VOLCOL wide bore capillary column with 1.5 µm film thickness and using helium carrier gas.
(c) Recoveries and relative standard deviations were determined from seven samples of reagent water fortified with 10 µg/L of each compound. 2-Bromo-1-chloropropane was used as the internal standard for calculating average recoveries.
(d) Recoveries and relative standard deviations were determined from seven samples of reagent water fortified with 10 µg/L of each compound. Fluorobenzene was used as the internal standard for calculating average recoveries.

SAMPLING METHOD Collect samples using a 40- to 120-mL screw-cap vial (prewashed with detergent, rinsed with distilled water and oven dried at 105°C) with a Teflon®-faced silicone septum. Collect bubble-free samples and place the septum with the Teflon® side down on the water.

SAMPLE PRESERVATION If residual chlorine is present in the water add about 25 mg of ascorbic acid to each vial before samples are collected to remove the chlorine. Add hydrochloric acid to reduce pH to <2, immediately cool samples to 4°C, and store them in a solvent-free refrigerator at 4°C until analysis.

MHT The maximum holding time for samples is 14 days from the time they were collected.

SAMPLE PREPARATION Remove the plungers from two 5-mL syringes and attach a closed syringe valve to each. Warm the sample to room temperature, open the sample bottle, and carefully pour the sample into one of the syringe barrels to just short of overflowing. Replace the syringe plunger, invert the syringe, and compress the sample. Open the syringe valve and vent any residual air while adjusting the sample volume to 5.0 mL. Add 10 µL of the internal calibration standard to the sample through the syringe valve. Close the valve. Fill the second syringe in an identical manner from the same sample bottle. Reserve this second syringe for a reanalysis if necessary.

QUALITY CONTROL As an initial demonstration of lab accuracy and precision, analyze 4 to 7 replicates of a lab fortified blank containing analyte at 0.1–5 µg/L. Collect all samples in duplicate. Surrogate analytes (similar to those of the analytes of interest), whose concentration is known in every sample, are measured using the same internal standard calibration procedure. Duplicate field reagent water blanks (trip blanks) must be analyzed with each set of samples, lab reagent blanks (method blanks) must be analyzed with each batch of samples processed as a group within a work shift. Also, a single lab-fortified blank that contains each of the analytes of interest should be analyzed with each batch of samples processed as a group within a work shift. A 3- to 5-point calibration curve is needed depending on the calibration range factor required.

EPA CONTACT & HOTLINE For technical questions contact Dr. Baldev Bathija, U.S. EPA, Office of Ground Water and Drinking Water (WH-550D), 401 M St. SW, Washington, DC 20460. Tel. (202) 260-3040. For further information the EPA Safe Drinking Water Hotline may be called at: (800) 426-4791.

REFERENCE Methods for the Determination of Organic Compounds in Drinking Water, EPA/600/4-88/039 (revised July 1991; Final Rule for determination of compliance with the MCL for Total Trihalomethanes under 141.30, in 40 CFR Part 141, Vol. 58, No. 147, Fed. Reg., Tuesday Aug. 3, 1993). U.S. EPA Environmental Monitoring Systems Laboratory, Cincinnati, OH, 45268, U.S.A. Available from the National Technical Information Service (NTIS), 5285 Port Royal Road, Springfield, VA 22161; Tel. 800-553-6847. NTIS Order Number is PB91-231480.

1,2,3-Trichloropropane   EPA Method 524
CAS #96-18-4

TITLE Measurement of Purgeable Organic Compounds in Water by Capillary Column GC/MS.

MATRIX Drinking water and raw source water; the latter should include most surface water and groundwater sources.

METHOD SUMMARY Method 524.2 covers 60 volatile organic compounds. An inert gas (zero grade nitrogen or helium) is bubbled through a 25-mL or a 5-mL water sample (depending on the expected concentration of the analytes). Purged sample components are trapped in a tube of sorbent materials. When purging is complete, the sorbent tube is heated and backflushed with helium to desorb the trapped sample onto a capillary GC column.

INTERFERENCES Impurities in the purge gas and from organic compounds outgassing from the plumbing ahead of the trap account for many contamination problems. Interferences purged or coextracted from the samples will vary considerably from source to source, depending upon the particular sample or extract being tested. Cross-contamination can occur whenever high-level and low-level samples are analyzed sequentially. Samples also can be contaminated by diffusion of volatile organics (particularly methylene chloride and fluorocarbons) through the septum seal into the sample during shipment and storage.

INSTRUMENTATION A GC/MS with a data system equipped with one of the following capillary GC columns:

Column 1: VOCOL glass wide bore capillary column.
Column 2: DB-624 fused silica capillary column.
Column 3: DB-5 fused silica capillary column.

Also required is an all-glass 25 mL or 5-mL purging device, a sorbent trap, and a thermal desorption apparatus which is connected to the GC/MS system.

PRECISION & ACCURACY Method detection limits are compound- and instrument-dependent, and may vary from approximately 0.02–0.35 µg/L. Note in the table below that the "true" concentration range used for accuracy and precision measurements was quite narrow. However, the applicable concentration range of this method is primarily column dependent and is approximately 0.02 to 200 µg/L for the wide-bore thick-film columns. Narrow-bore thin-film columns may have a capacity which limits the range to about 0.02 to 20 µg/L. Analytes that are inefficiently purged from water will not be detected when present at low concentrations, but they can be measured with acceptable accuracy and precision when present in sufficient amounts.

Analytes that are not separated chromatographically, but which have different mass spectra and non-interfering quantification ions, can be identified and measured in the same calibration mixture or water sample. Analytes which have very similar mass spectra cannot be individually identified and measured in the same calibration mixture or water samples unless they have different retention times. Co-eluting compounds with very similar mass spectra, typically many structural isomers, must be reported as an isomeric group or pair.

The range (in µg/L) was 0.5–10.
The Method Detection Limig (in µg/L) was 0.32.
The accuracy (as % recovery) was 108.
The precision (in %) was 14.4.

Note: Data were obtained from 16–31 determinations using a wide-bore capillary column and a jet separator interfaced to a quadrupole mass spectrometer. All analytes were in a reagent water matrix.

SAMPLING METHOD Collect samples using a 40- to 120-mL screw-cap vial (prewashed with detergent, rinsed with distilled water and oven dried at 105°C) with a Teflon®-faced silicone septum. Collect bubble-free samples and place the septum with the Teflon® side down on the water.

SAMPLE PRESERVATION If residual chlorine is present in the water add about 25 mg of ascorbic acid to each vial before samples are collected to remove the chlorine. Add hydrochloric acid to reduce pH to <2, and immediately cool samples to 4°C, and store them in a solvent-free refrigerator at 4°C until analysis.

MHT The maximum holding time for samples is 14 days from the time they were collected.

SAMPLE PREPARATION Remove the plungers from two 25-mL (or 5-mL depending on sample size) syringes and attach a closed syringe valve to each. Warm the sample to room temperature, open the sample bottle, and carefully pour the sample into one of the syringe barrels to just short of overflowing. Replace the syringe plunger, invert the syringe, and compress the sample. Open the syringe valve and vent any residual air while adjusting the sample volume to 25.0 mL (or 5 mL). For samples and blanks, add 5 µL of the fortification solution containing the internal standard and the surrogates to the sample through the syringe valve. For calibration standards and lab fortified blanks, add 5 µL of the fortification solution containing the internal standard only. Close the valve. Fill the second syringe in an identical manner from the same sample bottle. Reserve this second syringe for a reanalysis if necessary.

QUALITY CONTROL As an initial demonstration of lab accuracy and precision, analyze 4 to 7 replicates of a lab fortified blank containing analyte at 0.2–5 µg/L. Collect all samples in duplicate. Surrogate analytes (similar to those of the analytes of interest), whose concentration is known in every sample, are measured using the same internal standard calibration procedure. Duplicate field reagent water blanks (trip blanks) must be analyzed with each set of samples, lab reagent blanks (method blanks) must be analyzed with each batch of samples processed as a group within a work shift. Also, a single lab-fortified blank that contains each of the analytes of interest should be analyzed with each batch of samples processed as a group within a work shift. A 3- to 5-point calibration curve is needed depending on the calibration range factor required.

EPA CONTACT & HOTLINE For technical questions contact Dr. Baldev Bathija, U.S. EPA, Office of Ground Water and Drinking Water (WH-550D), 401 M St. SW, Washington, DC 20460. Tel. (202) 260-3040. For further information the EPA Safe Drinking Water Hotline may be called at: (800) 426-4791.

REFERENCE Methods for the Determination of Organic Compounds in Drinking Water, EPA/600/4-88/039 (revised July 1991; Final Rule for determination of compliance with the MCL for Total Trihalomethanes under 141.30, in 40 CFR Part 141, Vol. 58, No. 147, Fed. Reg., Tuesday Aug. 3, 1993). U.S. EPA Environmental Monitoring Systems Laboratory, Cincinnati, OH, 45268, U.S.A. Available from the National Technical Information Service (NTIS), 5285 Port Royal Road, Springfield, VA 22161; Tel. 800-553-6847. NTIS Order Number is PB91-231480.

1,2,3-Trichloropropane   EPA Method 8021
CAS #96-18-4

TITLE Halogenated Volatile by Gas Chromatography Using Photoionization and Electrolytic Conductivity Detectors in Series: Capillary Column Technique

MATRIX This method is applicable to nearly all types of samples, regardless of water content, including groundwater, aqueous sludges, caustic liquors, acid liquors, waste solvents, oily wastes, mousses, tars, fibrous wastes, polymeric emulsions, filter cakes, spent carbons, spent catalysts, soils, and sediments.

METHOD SUMMARY This method is used to determine 60 volatile organic compounds in a variety of solid waste matrices. It provides GC conditions for the detection of halogenated and aromatic volatile organic compounds. Samples can be analyzed using direct injection or purge-and-trap (EPA Method 5030). Groundwater samples must be analyzed using EPA Method 5030 (where applicable). A temperature program is used with the GC. Detection is achieved by a photoionization detector (PID) and a Hall electrolytic conductivity detector (HECD) in series.

INTERFERENCES Samples can be contaminated by diffusion of volatile organics (particularly chlorofluorocarbons and methylene chloride) through the sample container septum during shipment and storage.

INSTRUMENTATION A GC-equipped with variable-constant differential flow controllers, subambient oven controller, PID and HECD detectors connected with a short piece of uncoated capillary tubing and a data system.

Column: 60 m × 0.75 mm I.D. VOCOL wide-bore capillary column with 1.5 µm film thickness.

PRECISION & ACCURACY MDLs are compound-dependent and vary with purging efficiency and concentration. The applicable concentration range of this method is compound- and instrument-dependent but is approximately 0.1 to 200 µg/L. Analytes that are inefficiently purged from water will not be detected when present at low concentrations, but they can be measured with acceptable accuracy and precision when present in sufficient amounts. The estimated quantitation limit (EQL) for an individual compound is approximately 1 µg/kg (wet weight) for soil/sediment samples, 100 µg/kg (wet weight) for wastes, and 1 µg/L for groundwater. EQLs will be proportionately higher for sample extracts and samples that require dilution or reduced sample size to avoid saturation of the detector.

MULTIPLICATION FACTORS FOR OTHER MATRICES (a)

Matrix	Factor (b)
Groundwater	10
Low-concentration soil	10
Water miscible liquid waste	500
High-concentration soil and sludge	1250
Non-water miscible waste	1250

(a) Sample EQLs are highly matrix-dependent. The EQLs listed herein are provided for guidance and may not always be achievable. (b) EQL = [Method detection limit] × [Factor]. For non-aqueous samples, the factor is on a wet-weight basis.

SINGLE LABORATORY ACCURACY & PRECISION DATA FOR VOCs IN WATER

This method was tested in a single lab using water spiked at 10 µg/L and the following data was reported:

Recoveries and standard deviations were determined from seven samples and spiked at 10 µg/L of each analyte. Recoveries were determined by the internal standard method. Internal standards were: Fluorobenzene for PID and 2-Bromo-1-chloropropane for HECD.

The average recovery (in percent) for the PID was none (no response for this detector).

The standard deviation of the recovery for the PID was none (no response for this detector).

The MDL (in µg/mL) for the PID was none (no response for this detector).

The average recovery (in percent) for the HECD was 99.

The standard deviation of the recovery for the HECD was 2.3.

The MDL (in µg/mL) for the HECD was 0.4.

SAMPLE COLLECTION, PRESERVATION & HANDLING

Volatile organics — Standard 40-mL glass screw-cap VOA vials with Teflon®-faced silicone septum may be used for both liquid and solid matrices. When collecting samples, liquids and solids should be introduced into the vials gently to reduce agitation which might drive off volatile compounds. If there are any air bubbles present the sample must be retaken. Tap slightly as they are filled to try and eliminate as much free air space as possible. The two vials from each sampling locations should be sealed in separate plastic bags to prevent cross-contamination between samples particularly if the sampled waste is suspected of containing high levels of volatile organics.

Semivolatile organics — Containers used to collect samples for the determination of semivolatile organic compounds should be soap and water washed followed by methanol (or isopropanol) rinsing. The sample containers should be of glass or Teflon® and have screw-top covers with Teflon® liners.

Preservation for volatile organics — No preservation is used with concentrated waste samples. With liquid samples containing no residual chlorine, 4 drops of concentrated hydrochloric acid are added and the samples are immediately cooled to 4°C. When liquid samples contain residual chlorine, they are treated as above and, in addition, 4 drops of 4% aqueous sodium thiosulfate are added. Soil, sediment, and sludge samples are only cooled to 4°C.

Preservation for semivolatile organics — No preservation is used with concentrated waste samples. With liquid samples containing no residual chlorine and with soil, sediment, and sludge samples, immediately cooling to 4°C is the only preservation used. When residual chlorine is present then 3 mL of 10% aqueous sodium sulfate is added for each gallon of sample collected, followed by cooling to 4°C.

MHT The holding time for all volatile organics samples is 14 days. Liquid samples must be extracted within 7 days and their extracts analyzed within 40 days. Concentrated waste, soil, sediment, and sludge samples must be extracted within 14 days and their extracts analyzed within 40 days.

SAMPLE PREPARATION Volatile compounds are introduced into the gas chromatograph either by direct injector or purge-and-trap (EPA Method 5030). EPA Method 5030 may be used directly on groundwater samples or low-concentration contaminated soils and sediments. For medium-concentration soils or sediments, methanolic extraction, as described in EPA Method 5030, may be necessary prior to purge-and-trap analysis.

QUALITY CONTROL Calculate surrogate standard recovery on all samples, blanks, and spikes. A trip blank is recommended to check on sampling, storage, and handling contamination. Calibration standards, at a minimum of five concentration levels, are prepared in organic-free reagent water. One of the concentration levels should be at a concentration near, but above, the method detection limit.

A combination of bromochloromethane, 2-bromo-1-chloropropane, 1,4-dichlorobutane, and bromochlorobenzene are recommended as surrogate standards to encompass the range of the temperature program used in this method.

REFERENCE Test Methods for Evaluating Solid Waste, Physical/Chemical Methods, SW-846, 3rd Edition, U.S. EPA, Office of Solid Waste, Washington, DC, EPA Method 8021A, Rev. 1, Nov. 1992.

1,2,3-Trichloropropane   **EPA Method 8240**
CAS #96-18-4

TITLE Volatile Organics By GC/MS: Packed Column Technique

MATRIX Nearly all types of sample matarices, regardless of water content, can be analyzed using this method. This includes groundwater, aqueous sludges, caustic liquors, acid liquors, waste solvents, oily wastes, mousses, tars, fibrous wastes, polymetric emulsions, filter cakes, spent carbons, spent catalysts, soils, and sediments.

METHOD SUMMARY Method 8240B covers 80 volatile organic compounds that are introduced into a gas chromatograph by the purge-and-trap method or by direct injection (in limited applications). For the purge-and-trap method an inert gas (zero grade nitrogen or helium) is bubbled through a 5-mL solution at ambient temperature. Purged sample components are trapped in a tube of sorbent materials. When purging is complete, the sorbent tube is heated and backflushed with inert gas to desorb the trapped components onto a GC column.

INTERFERENCES Impurities in the purge gas and from organic compounds outgassing from the plumbing ahead of the trap account for many contamination problems. Interferences purged or coextracted from the samples will vary considerably from source to source. Cross-contamination can occur whenever high-level and low-level samples are analyzed sequentially. Whenever an unusually concentrated sample is analyzed, it should be followed by the analysis of organic-free reagent water to check for cross-contamination. Samples also can be contaminated by diffusion of volatile organics (particularly methylene chloride and fluorocarbons) through the septum seal into the sample during shipment and storage. A trip blank can serve as a check on such contamination. The lab where volatile analysis is performed and also the refrigerated storage area should be completely free of solvents.

INSTRUMENTATION A gas chromatograph/mass spectrometry/data system (GC/MS) equipped with a 6 ft × 0.1 in I.D. glass column packed with 1% SP-1000 on Carbopack-B (60/80 mesh) is required. Also needed is a 5-mL purging device, a sorbent trap, and a thermal desorption apparatus.

PRECISION & ACCURACY This method is reported to have been tested by 15 laboratories using organic-free reagent water, drinking water, surface water, and industrial wastewaters (not specified) fortified at six concentrations over the range 5–600 µg/L.

Sample estimated quantitation limits (EQLs) are highly matrix-dependent. The EQLs listed may not always be achievable. EQLs listed for soils or sediments are based on wet weight. Normally, data is reported on a dry-weight basis; therefore, EQLs will be higher, based on the percent dry weight of each sample. Note that EQLs are even more variable than MDLs and that they are highly variable depending on the matrix being analyzed.

EQL in groundwater in µg/L was 5.
EQL in low soil or sediment in µg/kg was 5.
Accuracy (a) in µg/L was not listed.
Precision (b) in µg/L was not listed.

(a) *Average recovery found for measurements of samples containing a concentration of C, in µg/L.*
(b) *Overall precision found for measurements of samples with average recovery X for samples containing a concentration of C in µg/L.*
X = *Average recovery found for measurement of samples containing a concentration of C in µg/L.*

MULTIPLICATION FACTORS FOR OTHER MATRICES

Other Matrices	Factor (a)
Waste miscible liquid waste	50
High-concentration soil and sludge	125
Non-water miscible waste	500

(a) *EQL = [EQL for low soil sediment X [Factor]. For non-aqueous samples, the factor is on a wet-weight basis.*

SAMPLING METHOD
Liquid samples — Use a 40-mL glass screw-cap VOA vial with a Teflon®-faced silicone septum that has been prewashed, rinsed with distilled deionized water, and oven dried. However, if residual chlorine is present, collect sample in a 40-oz. soil VOA container which has been pre-preserved with 4 drops of 10% sodium thiosulfate, mix gently, and then transfer the sample to a 40-mL VOA vial. Collect bubble-free samples in duplicate and seal them in separate plastic bags.

Soils or sediments, and sludges — Use an 8-oz. widemouth glass bottle with a Teflon®-faced silicone septum that has been prewashed with detergent, rinsed with distilled deionized water, and oven dried. Tap slightly to eliminate free air space. Collect samples in duplicate and seal them in separate plastic bags.

SAMPLE PRESERVATION
Liquid samples — Add 4 drops of concentrated HCL and immediately cool samples to 4°C and store in a solvent-free refrigerator.

Soils or sediments, and sludges — Cool samples to 4°C and store in a solvent-free refrigerator.

MHT Maximum holding time is 14 days from the date of sample collection.

SAMPLE PREPARATION
Liquid samples — Remove the plunger from a 5-mL syringe and carefully pour the sample into the syringe barrel to just short of overflowing. Replace the syringe plunger and compress the sample. Open the syringe valve and vent any residual air while adjusting the sample volume to 5.0 mL. If there is only one volatile organic analysis (VOA) vial, a second syringe should be filled at this time to protect against possible loss of sample integrity. Add 10 µL of surrogate spiking solution and 10 µL of internal standard spiking solution through the valve bore of the 5-mL syringe, then close the valve. The surrogate and internal standards may be mixed and added as a single spiking solution.

Sediments, soils, and waste samples — All samples of this type should be screened by GC analysis using a headspace method (EPA Method 3810) or the hexadecane extraction and screening method (EPA Method 3820). Use the screening data to determine whether to use the low-concentration method (0.005–1 mg/kg) or the high-concentration method (>1 mg/kg).

Low-concentration method — The low-concentration method is based on purging a heated sediment or soil sample mixed with organic-free reagent water containing the surrogate and internal standards. Analyze all reagent blanks and standards under the same conditions as the samples.

Use a 5-g sample if the expected concentration is <0.1 mg/kg or a 1-g sample for expected concentrations between 0.1 and 1 mg/kg. Mix the contents of the sample container with a narrow metal spatula. Weigh the amount of the sample into a tared purge device. Add the spiked water to the purge device, which contains the weighed amount of sample, and connect the device to the purge-and-trap system.

High-concentration method — This method is based on extracting the sediment or soil with methanol. A waste sample is either extracted or diluted, depending on its solubility in methanol. Wastes that are insoluble in methanol are diluted with reagent tetraglyme or possibly polyethylene glycol (PEG). An aliquot of the extract is added to organic-free reagent water

containing surrogate and internal standards. This is purged at ambient temperature. All samples with an expected concentration of >1.0 mg/kg should be analyzed by this method.

Mix the contents of the sample container with a narrow metal spatula. For sediments or soils and solid wastes that are insoluble in methanol, weigh 4 g (wet weight) of sample into a tared 20-mL vial. For waste that is soluble in methanol, tetraglyme, or PEG, weigh 1 g (wet weight) into a tared scintillation vial or culture tube or a 10-mL volumetric flask. Quickly add 9.0 mL of appropriate solvent then add 1.0 mL of a surrogate spiking solution to the vial, cap it, and shake it for 2 min.

METHANOL EXTRACT REQUIRED FOR ANALYSIS OF HIGH-CONCENTRATION SOILS OR SEDIMENTS

Approximate Concentration Range	Volume of Methanol Extract (a)
500–10,000 µg/kg	100 µL
1,000–20,000 µg/kg	50 µL
5,000–100,000 µg/kg	10 µL
25,000–500,000 µg/kg	100 µL of 1/50 dilution (b)

Calculate appropriate dilution factor for concentrations exceeding this table.

(a) The volume of methanol added to 5 mL of water being purged should be kept constant. Therefore, add to the 5-mL syringe whatever volume of methanol is necessary to maintain a volume of 100 µL added to the syringe.
(b) Dilute an aliquot of the methanol extract and then take 100 µL for analysis.

QUALITY CONTROL Demonstrate, through the analysis of a reagent water blank, that interferences from the analytical system, glassware, and reagents are under control. Blank samples should be carried through all stages of the sample preparation and measurement steps. For each analytical batch (up to 20 samples), a reagent blank, matrix spike, and matrix spike duplicate must be analyzed (the frequency of the spikes may be different for different monitoring programs). The blank and spiked samples must be carried through all stages of the sample preparation and measurement steps. QC samples mentioned in the section on Interferences will also be needed as appropriate to those situations.

REFERENCE Test Methods for Evaluating Solid Waste (SW-846). U.S. EPA. 1983. Method 8240B, Rev. 2, Nov. 1990. Office of Solid Wastes, Washington, DC.

1,2,3-Trichloropropane   **EPA Method 8260**
CAS #96-18-4

TITLE Volatile Organic Compounds by GC/MS: Capillary Column Technique

MATRIX This method is applicable to nearly all types of samples, regardless of water content, including groundwater, soils, and sediments.

METHOD SUMMARY Method 8260A covers 58 volatile organic compounds that are introduced into a gas chromatograph by the purge-and-trap method or by direct injection (in limited applications). Zero-grade helium is bubbled through a 5-mL solution at ambient temperature. Purged sample components are trapped in a tube containing suitable sorbent materials. When purging is complete, the sorbent tube is heated and backflushed with helium to desorb trapped sample components. The analytes are desorbed directly to a large bore capillary or cryofocussed on a capillary precolumn before being flash evaporated to a narrow bore capillary for analysis.

INTERFERENCES Major contaminant sources are volatile materials in the lab and impurities in the inert purging gas and in the sorbent trap. Interfering contamination may occur when a sample containing low concentrations of volatile organic compounds is analyzed immediately after a sample containing high concentrations of volatile organic compounds. After analysis of a sample containing high concentrations of volatile organic compounds, one or more calibration blanks should be analyzed to check for cross-contamination. Screening of the samples prior to purge-and-trap GC/MS analysis is highly recommended to prevent contamination of the system. This is especially true for soil and waste samples.

Special precautions must be taken to analyze for methylene chloride. The analytical and sample storage area should be isolated from all atmospheric sources of methylene chloride. All gas chromatography carrier gas lines and purge gas plumbing should be constructed from stainless steel or copper tubing. Laboratory clothing previously exposed to methylene chloride fumes during liquid-liquid extraction procedures can contribute to sample contamination.

Samples can also be contaminated by diffusion of volatile organics (particularly methylene chloride and fluorocarbons) through the septum seal during shipment and storage. A trip blank can serve as a check on such contamination.

INSTRUMENTATION GC/MS with a temperature-programmable chromatograph suitable for splitless injection equipped with variable constant differential flow controllers, a subambient oven controller, a purging device, sorbent trap, a thermal desorption apparatus and a capillary precolumn interface when using cryogenic cooling will be needed. The following GC columns may be used:

Column 1: 60 m × 0.75mm I.D. capillary column coated with VOCOL, 1.5 µm film thickness.
Column 2: 30 m × 0.53mm capillary column coated with DB-624 or VOCOL, 3 µm film thickness.
Column 3: 30 m × 0.32mm I.D. capillary column coated with DB-5 or SE-54, 1-µm film thickness.

PRECISION & ACCURACY This method has been tested in a single lab using spiked water. Using a wide-bore capillary column, water was spiked at concentrations between 0.5 and 10 µg/L. Single lab accuracy and precision data are presented. The MDL actually achieved in a given analysis will vary depending on instrument sensitivity and matrix effects.

The MDL (a) in µg/L was 0.32.
The concentration range in µg/L was 0.5–10.
The mean accuracy (% of true value) was 108.
The precision as relative standard deviation was 14.4.

Note: The MDL is based on a 25-mL sample volume instead of a 5-mL sample volume.

SAMPLING METHOD

Liquid samples — Use a 40-mL glass screw-cap VOA vial with a Teflon®-faced silicone septum that has been prewashed, rinsed with distilled deionized water, and oven dried. If residual chlorine is present, collect the sample in a 4-oz soil VOA container which has been pre-preserved with 4 drops of 10% sodium thiosulfate. Mix gently and transfer the sample to a 40-mL VOA vial. Collect bubble-free samples in duplicate and seal each sample in a separate plastic bag.

Soils, sediments and sludges — Use an 8-oz widemouth glass bottle with Teflon®-faced silicone septum that has been prewashed, rinsed with distilled deionized water, and oven dried. **Do not** heat the septum for more than 1 h. Tap slightly to eliminate any free air space. Collect samples in duplicate and seal each one in a separate plastic bag.

SAMPLE PRESERVATION

Liquid samples — Add 4 drops of concentrated HCL, cool to 4°C and store in a solvent-free refrigerator.

Soils, sediments and sludges — Cool samples to 4°C and store in a solvent-free refrigerator.

MHT The maximum holding time of any sample (liquids, soils, sediments, and sludges) is 14 days.

SAMPLE PREPARATION

Liquid samples — Remove the plunger from a 5-mL syringe and carefully pour the sample into the syringe barrel to just short of overflowing. Replace the syringe plunger and compress the sample. Open the syringe valve and vent any residual air while adjusting the sample volume to 5.0 mL. If there is only one volatile organic analysis (VOA) vial, a second syringe should be filled at this time to protect against possible loss of sample integrity. Add 10 µL of surrogate spiking solution and 10 µL of internal standard spiking solution through the valve bore of the 5-mL syringe, then close the valve. The surrogate and internal standards may be mixed and added as a single spiking solution.

Sediments, soils, and waste samples — All samples of this type should be screened by GC analysis using a headspace method (EPA Method 3810) or the hexadecane extraction and screening method (EPA Method 3820). Use the screening data to determine whether to use the low-concentration method (0.005–1 mg/kg) or the high-concentration method (>1 mg/kg).

Low-concentration method — The low-concentration method is based on purging a heated sediment or soil sample mixed with organic-free reagent water containing the surrogate and internal standards. Analyze all reagent blanks and standards under the same conditions as the samples.

Use a 5-g sample if the expected concentration is <0.1 mg/kg or a 1-g sample for expected concentrations between 0.1 and 1 mg/kg. Mix the contents of the sample container with a narrow metal spatula. Weigh the amount of the sample into a tared purge device. Add the spiked water to the purge device, which contains the weighed amount of sample, and connect the device to the purge-and-trap system.

High-concentration method — This method is based on extracting the sediment or soil with methanol. A waste sample is either extracted or diluted, depending on its solubility in methanol. Wastes that are insoluble in methanol are diluted with reagent tetraglyme or possibly polyethylene glycol (PEG). An aliquot of the extract is added to organic-free reagent water containing surrogate and internal standards. This is purged at ambient temperature. All samples with an expected concentration of >1.0 mg/kg should be analyzed by this method.

Mix the contents of the sample container with a narrow metal spatula. For sediments or soils and solid wastes that are insoluble in methanol, weigh 4 g (wet weight) of sample into a tared 20-mL vial. For waste that is soluble in methanol, tetraglyme, or PEG, weigh 1 g (wet weight) into a tared scintillation vial or culture tube or a 10-mL volumetric flask. Quickly add 9.0 mL of appropriate solvent then add 1.0 mL of a surrogate spiking solution to the vial, cap it, and shake it for 2 min.

METHANOL EXTRACT REQUIRED FOR ANALYSIS OF HIGH-CONCENTRATION SOILS OR SEDIMENTS

Approximate Concentration Range	Volume of Methanol Extract (a)
500–10,000 µg/kg	100 µL
1,000–20,000 µg/kg	50 µL
5,000–100,000 µg/kg	10 µL
25,000–500,000 µg/kg	100 µL of 1/50 dilution (b)

Calculate appropriate dilution factor for concentrations exceeding this table.

(a) The volume of methanol added to 5 mL of water being purged should be kept constant. Therefore, add to the 5-mL syringe whatever volume of methanol is necessary to maintain a volume of 100 µL added to the syringe.

(b) Dilute an aliquot of the methanol extract and then take 100 µL for analysis.

QUALITY CONTROL Demonstrate, through the analysis of a reagent water blank, that interferences from the analytical system, glassware, and reagents are under control. Blank samples should be carried through all stages of the sample preparation and measurement steps. For each analytical batch (up to 20 samples), a reagent blank, matrix spike, and matrix spike duplicate must be analyzed (the frequency of the spikes may be different for different monitoring programs). The blank and spiked samples must be carried through all stages of the sample preparation and measurement steps. QC samples mentioned in the section on Interferences will also be needed as appropriate to those situations.

Matrix spiking standards should be prepared from volatile organic compounds which will be representative of the compounds being investigated. The recommended internal standards are chlorobenzene-d5, 1,4-difluorobenzene, 1,4-dichlorobenzene-d4, and pentafluorobenzene. Using stock standard solutions, prepare secondary dilution standards containing the compounds of interest, either singly or mixed together in methanol. Store them in a vial with no headspace for no more than one week. Surrogates recommended are toluene-d8, 4-bromofluorobenzene, and dibromofluoromethane. Each sample undergoing GC/MS analysis must be spiked with 10 µL of the surrogate spiking solution prior to analysis.

REFERENCE Test Methods for Evaluating Solid Waste (SW-846). U.S. EPA 1983, Method 8260A, Rev. 1, Nov. 1990. Office of Solid Waste, Washington, DC.

Tricyclozole
CAS #41814-78-2

EPA Method 507

TITLE Determination of Nitrogen and Phosphorus-Containing Pesticides in Water by GC/NPD

MATRIX This method is applicable to the determination of certain nitrogen and phosphorus-containing pesticides in finished drinking water and groundwater.

METHOD SUMMARY Method 507 covers 46 nitrogen- and phosphorus-containing pesticides. A 1-L sample is fortified with a surrogate standard, salted, buffered, extracted with methylene chloride, and concentrated; then the solvent is exchanged with methyl tert-butyl ether (MTBE) and concentrated again, and a 2-µL aliquot of a sample extract is injected into a GC system equipped with a selective nitrogen-phosphorus detector and a capillary column for analysis.

INTERFERENCES Method interferences may be caused by contaminants in solvents, reagents, glassware, and other sample processing apparatus. Interfering contamination may occur when a sample containing low concentrations of analytes is analyzed immediately following a sample containing relatively high concentrations. One or more injections of MTBE should be made following the analysis of a sample with high concentrations of analytes to check for analyte carryover. Matrix interferences may be caused by contaminants that are coextracted from the sample. The extent of matrix interferences will vary considerably from source to source, depending upon the water sampled.

INSTRUMENTATION A gas chromatograph system (GC) equipped with a nitrogen-phosphorus detector (NPD) is needed.

Column 1: 30 m × 0.25 mm I.D. DB-5 bonded fused silica column, 0.25 µm film thickness, or equivalent.

Column 2: 30 m × 0.25 mm I.D. DB-1701 bonded fused silica column, 0.25 µm film thickness, or equivalent.

PRECISION & ACCURACY This method has been validated in a single lab and estimated detection limits (EDLs) have been determined for each analyte. Observed detection limits may vary among waters, depending upon the nature of the interferences in the sample matrix and the specific instrumentation used. Analytes that are not separated chromatographically cannot be individually identified and measured unless an alternative technique for identification and quantification exist.

The estimated detection limit (in µg/L) was 1. The EDL is defined as either method detection limit or a level of compound in a sample yielding a peak in the final extract with signal-to-noise ratio of approximately 5, whichever value is higher.

The concentration used for these measurements (in µg/L) was 10.
The accuracy (as % recovery) was 86.
The precision (% RSD) was 7.

SAMPLING METHOD Grab samples are collected in 1-L glass sample bottles (prewashed with detergent and hot tap water, rinsed with reagent water, and dried in an oven at 400°C for 1 h) with screw caps lined with PTFE-fluorocarbon.

SAMPLE PRESERVATION Add mercuric chloride to the sample bottle in amounts to produce a concentration of 10 mg/L. If residual chlorine is present, add 80 mg of sodium thiosulfate/L of sample to the sample bottle prior to collection. After collection, seal bottle and shake vigorously for 1 min, then cool the sample to 4°C immediately and store it at 4°C in the dark until extraction.

MHT Maximum holding time of the samples, and in some cases the extracts, is 14 days.

SAMPLE PREPARATION Fortify the sample with 50 µL of the surrogate standard solution, adjust to pH 7 with phosphate buffer, add 100 g NaCl to the sample, and seal and shake to dissolve the salt; then extract with methylene chloride in a separatory funnel or in a mechanical tumbler bottle. Dry the extract by pouring it through a solvent-rinsed drying column containing about 10 cm of anhydrous sodium sulfate. Collect the extract in a Kuderna-Danish (K-D) concentrator and rinse the column with 20–30 mL methylene chloride. Concentrate the extract to about 2 mL and rinse the flask and its lower joint into the concentrator tube with 1 to 2 mL of methyl t-butyl ether (MTBE). Add 5–10 mL of MTBE and concentrate the extract twice (adding more MTBE) to a final volume of 5.0 mL and store it at 4°C until analysis.

Note: If methylene chloride is not completely removed from the final extract, it may cause detector problems.

QUALITY CONTROL Minimum quality control requirements are initial demonstration of lab capability, determination of surrogate compound recoveries in each sample and blank, monitoring internal standard peak area or height in each sample and blank, analysis of lab reagent blanks, lab fortified samples, lab fortified blanks, and other QC samples. A lab reagent blank is analyzed to demonstrate that all glassware and reagent interferences are under control.

Initial demonstration of capability is fulfilled by analyzing four fortified reagent water samples with the recovery value for each analyte falling within the acceptable range (±30% average recovery). Surrogate recoveries from samples or method blanks must be 70–130%. The internal standard response for any sample chromatogram should not deviate from the daily calibration check standard's internal standard response by more than 30% or lab fortified blanks and sample matrices are used to assess lab performance and analyte recovery, respectively.

If the response for the target analyte peak exceeds the working range of the system, dilute the extract and reanalyze. Alternative techniques such as an alternate detector or second chromatography column should be used to confirm peak identification when sample components are not resolved adequately.

EPA CONTACT & HOTLINE For technical questions contact Dr. Baldev Bathija, U.S. EPA, Office of Ground Water and Drinking Water (WH-550D), 401 M St. SW, Washington, DC 20460. Tel. (202) 260-3040. For further information the EPA Safe Drinking Water Hotline may be called at: (800) 426-4791.

REFERENCE Methods for the Determination of Organic Compounds in Drinking Water, EPA/600/4-88/039 (revised July 1991). U.S. EPA Environmental Monitoring Systems Laboratory, Cincinnati, OH, 45268, U.S.A. Available from the National Technical Information Service (NTIS), 5285 Port Royal Road, Springfield, VA 22161; Tel. 800-553-6847. NTIS Order Number is PB91-231480.

O,O,O-Triethyl phosphorothioate   EPA Method 8270
CAS #126-68-1

TITLE Semivolatile Organic Compounds by GC/MS

MATRIX This method is used to determine the concentration of semivolatile organic compounds in extracts prepared from all types of solid waste matrices, soils, and groundwater. Although surface waters are not specifically mentioned, this method should be applicable to water samples from rivers, lakes, etc.

METHOD SUMMARY This method covers 259 semivolatile organic compounds. In very limited applications direct injection of the sample into the GC/MS system may be appropriate, but this results in very high detection limits (approximately 10,000 µg/L). Typically, a 1-L liquid sample, containing surrogate, and matrix spiking standards, is extracted in a continuous extractor first under acid conditions and then under basic conditions. Typically 30 g of a solid sample, containing surrogate, and matrix spiking standards, is extracted ultrasonically. After concentrating the extract to 1 mL it is spiked with 10 µL of an internal standard solution just prior to analysis by GC/MS. The volume injected should contain about 100 ng of base/neutral and 200 ng of acid surrogates (for a 1-µL injection). Analysis is performed by GC/MS using a capillary GC column.

INTERFERENCES Raw GC/MS data from all blanks, samples, and spikes must be evaluated for interferences. Contamination by carryover can occur whenever high-concentration and low-concentration samples are sequentially analyzed. To reduce carryover, the sample syringe must be rinsed out between samples with solvent. Whenever an unusually concentrated sample is encountered, it should be followed by the analysis of blank solvent to check for cross-contamination.

INSTRUMENTATION A GC/MS and a data system are required. The GC column used is a 30 m × 0.25 mm I.D. (or 0.32 mm I.D.) 1um film thickness silicone-coated fused silica capillary column. A continuous liquid-liquid extractor equipped with Teflon® or glass connection joints and stopcocks requiring no lubrication, a K-D concentrating apparatus, water bath, and an ultrasonic disrupter with a minimum power of 300 W and with pulsing capability are also required.

PRECISION & ACCURACY The estimated quantitation limit (EQL) of Method 8270B for determining an individual compound is approximately 1 mg/kg (wet weight) for soil or sediment samples, 1–200 mg/kg for wastes (dependent on matrix and method of preparation), and 10 µg/L for groundwater samples. EQLs will be proportionately higher for sample extracts that require dilution to avoid saturation of the detector.

The EQL(b) for groundwater in µg/L is Not Tested.
The EQL (a, b) for low concentrations in soil and sediment in µg/kg is not determined.
Accuracy as µg/L is not listed.
Overall precision in µg/L is not listed.

(a)   *EQLs listed for soil/sediment are based on wet weight. Normally data is reported in a dry-weight basis; therefore, EQLs will be higher based on the % dry weight of each sample. This calculation is based on a 30 g sample and gel permeation chromatography cleanup.*
(b)   *Sample EQLs are highly matrix-dependent. The EQLs are provided for guidance and may not always be achievable.*
$C =$ *True value for concentration, in µg/L.*
$X =$ *Average recovery found for measurements of samples containing a concentration of C, in µg/L.*

ESTIMATED QUANTITATION LIMIT

Other Matrices	Factor (a)
High-concentration soil and sludges by sonicator	7.5
Non-water miscible waste	75

(a) *EQL for other matrices = [EQL for low soil/sediment × [Factor]. This estimated EQL is similar to an EPA "Practical Quantitation Limit."*

SAMPLING METHOD

Liquid samples — Use a 1 or 2½ gallon amber glass bottle with a screw-top Teflon®-lined cover that has been prewashed with detergent and rinsed with distilled water and methanol (or isopropanol).

Soils, sediments, or sludges — Use an 8-oz. widemouth glass with a screw-top Teflon®-lined cover that has been prewashed with detergent and rinsed with distilled water and methanol (or isopropanol).

SAMPLE PRESERVATION

Liquid samples — If residual chlorine is present, add 3 mL of 10% sodium thiosulfate per gallon, cool to 4°C and store in a solvent-free refrigerator until analysis; if chlorine is not present, then eliminate the sodium thiosulfate addition.

Soils, sediments, or sludges — Cool samples to 4°C and store in a solvent-free refrigerator.

MHT Liquid samples must be extracted within 7 days and the extracts analyzed within 40 days. Soils, sediments, or sludges may be stored for a maximum of 14 days and the extracts analyzed within 40 days.

SAMPLE PREPARATION

Liquid samples — Transfer 1 L quantitatively to a continuous extractor. If high concentrations are anticipated, a smaller volume may be used and then diluted with organic-free reagent water to 1 L. Adjust pH, if necessary, to pH <2 using 1:1 (V/V) sulfuric acid. Pipette 1.0 mL of a surrogate standard spiking

solution into each sample. For the sample in each analytical batch selected for spiking, add 1.0 mL of a matrix spiking standard. For base/neutral acid analysis, the amount of the surrogates and matrix spiking compounds added to the sample should result in a final concentration of 100 ng/μL of each analyte in the extract to be analyzed (assuming a 1-μL injection). Extract with methylene chloride for 18–24 h. Next, adjust the pH of the aqueous phase to pH >11 using 10 N sodium hydroxide and extract it with methylene chloride again for 18–24 h. Dry the extract through a column containing anhydrous sodium sulfate and concentrate it to 1 mL using a K-D concentrator.

Soils, sediments, or sludges — Use 30 g of sample. Nonporous or wet samples (gummy or clay type) that do not have a free-flowing sandy texture must be mixed with anhydrous sodium sulfate until the sample is free flowing. Add 1 mL of surrogate standards to all samples, spikes, standards, and blanks. For the sample in each analytical batch selected for spiking, add 1.0 mL of a matrix spiking standard. For base/neutral acid analysis, the amount added of the surrogates and matrix spiking compounds should result in a final concentration of 100 ng/μL of each base/neutral analyte and 200 ng/μL of each acid analyte in the extract to be analyzed (assuming a 1-μL injection). Immediately add a 100-mL mixture of 1:1 methylene chloride:acetone and extract the sample ultrasonically for 3 min and then decant or filter the extracts. Repeat the extraction two or more times. Dry the extract using a column with anhydrous sodium sulfate and concentrate it to 1 mL in a K-D concentrator.

QUALITY CONTROL A methylene chloride solution containing 50 ng/μL of decafluorotriphenylphosphine (DFTPP) is used for tuning the GC/MS system each 12-h shift. A system performance check also must be made during every 12-h shift. A standard containing 50 ng/μL each of 4,4′-DDT, pentachlorophenol, and benzidine is required to verify injection port inertness and GC column performance. A calibration standard at mid-concentration, containing each compound of interest, including all required surrogates, must be performed every 12 h during analysis. After the system performance check is met, calibration check compounds (CCCs) are used to check the validity of the initial calibration.

The internal standard responses and retention times in the calibration check standard must be evaluated immediately after or during data acquisition. If the retention time for any internal standard changes by more than 30 seconds from the last check calibration (12 h), the chromatographic system must be inspected for malfunctions and corrections must be made, as required. If the electron ionization current plot (EICP) area for any of the internal standards changes by a factor of two from the last daily calibration standard check, the mass spectrometer must be inspected for malfunctions and corrections must be made, as appropriate.

Demonstrate, through the analysis of a reagent water blank, that interferences from the analytical system, glassware, and reagents are under control. The blank samples should be carried through all stages of the sample preparation and measurement steps. For each analytical batch (up to 20 samples), a reagent blank, matrix spike, and matrix spike duplicate/duplicate must be analyzed (the frequency of the spikes may be different for different monitoring programs). The blank and spiked samples must be carried through all stages of the sample preparation and measurement steps. A QC reference sample concentrate containing each analyte at a concentration of 100 mg/L in methanol is required.

REFERENCE Test Methods for Evaluating Solid Waste (SW-846). U.S. EPA 1983, Method 8270B, Rev. 2, Nov. 1990. Office of Solid Waste, Washington, DC.

Trifluralin EPA Method 8270
CAS #1582-09-8

TITLE Semivolatile Organic Compounds by GC/MS

MATRIX This method is used to determine the concentration of semivolatile organic compounds in extracts prepared from all types of solid waste matrices, soils, and groundwater. Although surface waters are not specifically mentioned, this method should be applicable to water samples from rivers, lakes, etc.

METHOD SUMMARY This method covers 259 semivolatile organic compounds. In very limited applications direct injection of the sample into the GC/MS system may be appropriate, but this results in very high detection limits (approximately 10,000 μg/L). Typically, a 1-L liquid sample, containing surrogate, and matrix spiking standards, is extracted in a continuous extractor first under acid conditions and then under basic conditions. Typically 30 g of a solid sample, containing surrogate, and matrix spiking standards, is extracted ultrasonically. After concentrating the extract to 1 mL it is spiked with 10 μL of an internal standard solution just prior to analysis by GC/MS. The volume injected should contain about 100 ng of base/neutral and 200 ng of acid surrogates (for a 1-μL injection). Analysis is performed by GC/MS using a capillary GC column.

INTERFERENCES Raw GC/MS data from all blanks, samples, and spikes must be evaluated for interferences. Contamination by carryover can occur whenever high-concentration and low-concentration samples are sequentially analyzed. To reduce carryover, the sample syringe must be rinsed out between samples with solvent. Whenever an unusually concentrated sample is encountered, it should be followed by the analysis of blank solvent to check for cross-contamination.

INSTRUMENTATION A GC/MS and a data system are required. The GC column used is a 30 m × 0.25 mm I.D. (or 0.32 mm I.D.) 1um film thickness silicone-coated fused silica capillary column. A continuous liquid-liquid extractor equipped with Teflon® or glass connection joints and stopcocks requiring no lubrication, a K-D concentrating apparatus, water bath, and an ultrasonic disrupter with a minimum power of 300 W and with pulsing capability are also required.

PRECISION & ACCURACY The estimated quantitation limit (EQL) of Method 8270B for determining an individual compound is approximately 1 mg/kg (wet weight) for soil or sediment samples, 1–200 mg/kg for wastes (dependent on matrix and method of preparation), and 10 μg/L for groundwater

samples. EQLs will be proportionately higher for sample extracts that require dilution to avoid saturation of the detector.

The EQL(b) for groundwater in µg/L is 10.
The EQL (a, b) for low concentrations in soil and sediment in µg/kg is not determined.
Accuracy as µg/L is not listed.
Overall precision in µg/L is not listed.

(a) EQLs listed for soil/sediment are based on wet weight. Normally data is reported in a dry-weight basis; therefore, EQLs will be higher based on the % dry weight of each sample. This calculation is based on a 30 g sample and gel permeation chromatography cleanup.
(b) Sample EQLs are highly matrix-dependent. The EQLs are provided for guidance and may not always be achievable.
$C =$ True value for concentration, in µg/L.
$X =$ Average recovery found for measurements of samples containing a concentration of C, in µg/L.

ESTIMATED QUANTITATION LIMIT

Other Matrices	Factor (a)
High-concentration soil and sludges by sonicator	7.5
Non-water miscible waste	75

(a) EQL for other matrices = [EQL for low soil/sediment × [Factor]. This estimated EQL is similar to an EPA "Practical Quantitation Limit."

SAMPLING METHOD

Liquid samples — Use a 1 or 2½ gallon amber glass bottle with a screw-top Teflon®-lined cover that has been prewashed with detergent and rinsed with distilled water and methanol (or isopropanol).

Soils, sediments, or sludges — Use an 8-oz. widemouth glass with a screw-top Teflon®-lined cover that has been prewashed with detergent and rinsed with distilled water and methanol (or isopropanol).

SAMPLE PRESERVATION

Liquid samples — If residual chlorine is present, add 3 mL of 10% sodium thiosulfate per gallon, cool to 4°C and store in a solvent-free refrigerator until analysis; if chlorine is not present, then eliminate the sodium thiosulfate addition.

Soils, sediments, or sludges — Cool samples to 4°C and store in a solvent-free refrigerator.

MHT Liquid samples must be extracted within 7 days and the extracts analyzed within 40 days. Soils, sediments, or sludges may be stored for a maximum of 14 days and the extracts analyzed within 40 days.

SAMPLE PREPARATION

Liquid samples — Transfer 1 L quantitatively to a continuous extractor. If high concentrations are anticipated, a smaller volume may be used and then diluted with organic-free reagent water to 1 L. Adjust pH, if necessary, to pH <2 using 1:1 (V/V) sulfuric acid. Pipette 1.0 mL of a surrogate standard spiking solution into each sample. For the sample in each analytical batch selected for spiking, add 1.0 mL of a matrix spiking standard. For base/neutral acid analysis, the amount of the surrogates and matrix spiking compounds added to the sample should result in a final concentration of 100 ng/µL of each analyte in the extract to be analyzed (assuming a 1-µL injection). Extract with methylene chloride for 18–24 h. Next, adjust the pH of the aqueous phase to pH >11 using 10 N sodium hydroxide and extract it with methylene chloride again for 18–24 h. Dry the extract through a column containing anhydrous sodium sulfate and concentrate it to 1 mL using a K-D concentrator.

Soils, sediments, or sludges — Use 30 g of sample. Nonporous or wet samples (gummy or clay type) that do not have a free-flowing sandy texture must be mixed with anhydrous sodium sulfate until the sample is free flowing. Add 1 mL of surrogate standards to all samples, spikes, standards, and blanks. For the sample in each analytical batch selected for spiking, add 1.0 mL of a matrix spiking standard. For base/neutral acid analysis, the amount added of the surrogates and matrix spiking compounds should result in a final concentration of 100 ng/µL of each base/neutral analyte and 200 ng/µL of each acid analyte in the extract to be analyzed (assuming a 1-µL injection). Immediately add a 100-mL mixture of 1:1 methylene chloride:acetone and extract the sample ultrasonically for 3 min and then decant or filter the extracts. Repeat the extraction two or more times. Dry the extract using a column with anhydrous sodium sulfate and concentrate it to 1 mL in a K-D concentrator.

QUALITY CONTROL A methylene chloride solution containing 50 ng/µL of decafluorotriphenylphosphine (DFTPP) is used for tuning the GC/MS system each 12-h shift. A system performance check also must be made during every 12-h shift. A standard containing 50 ng/µL each of 4,4'-DDT, pentachlorophenol, and benzidine is required to verify injection port inertness and GC column performance. A calibration standard at mid-concentration, containing each compound of interest, including all required surrogates, must be performed every 12 h during analysis. After the system performance check is met, calibration check compounds (CCCs) are used to check the validity of the initial calibration.

The internal standard responses and retention times in the calibration check standard must be evaluated immediately after or during data acquisition. If the retention time for any internal standard changes by more than 30 seconds from the last check calibration (12 h), the chromatographic system must be inspected for malfunctions and corrections must be made, as required. If the electron ionization current plot (EICP) area for any of the internal standards changes by a factor of two from the last daily calibration standard check, the mass spectrometer must be inspected for malfunctions and corrections must be made, as appropriate.

Demonstrate, through the analysis of a reagent water blank, that interferences from the analytical system, glassware, and reagents are under control. The blank samples should be carried through all stages of the sample preparation and measurement steps. For each analytical batch (up to 20 samples), a reagent blank, matrix spike, and matrix spike duplicate/duplicate must be analyzed (the frequency of the spikes may be different for different monitoring programs). The blank and spiked samples must be carried through all stages of the sample

preparation and measurement steps. A QC reference sample concentrate containing each analyte at a concentration of 100 mg/L in methanol is required.

REFERENCE Test Methods for Evaluating Solid Waste (SW-846). U.S. EPA 1983, Method 8270B, Rev. 2, Nov. 1990. Office of Solid Waste, Washington, DC.

1,2,3-Trimethoxybenzene   EPA Method 1625
CAS #634-36-6

TITLE Semivolatile Organic Compounds by Isotope Dilution GC/MS

MATRIX The compounds may be determined in waters, soils, and municipal sludges by this method.

METHOD SUMMARY This method is used to determine 176 semivolatile toxic organic pollutants associated with the CWA (as amended 1987); the RCRA (as amended 1986); the CERCLA (as amended 1986); and other compounds amenable to extraction and analysis by capillary column gas chromatography-mass spectrometry (GC/MS).

Stable isotopically-labeled analogs of the compounds of interest are added to the sample. If the solids content is less than 1%, a 1-L sample is extracted at pH 12–13, then at pH <2 with methylene chloride using continuous extraction techniques.

If the solids content is 30% or less, the sample is diluted to 1% solids with reagent water, homogenized ultrasonically, and extracted at pH 12–13, then at pH <2 with methylene chloride using continuous extraction techniques. If the solids content is greater than 30%, the sample is extracted using ultrasonic techniques.

Each extract is dried over sodium sulfate, concentrated to a volume of 5 mL, cleaned up using GPC, if necessary, and concentrated. Extracts are concentrated to 1 mL if GPC is not performed, and to 0.5 mL if GPC is performed.

An internal standard is added to the extract, and a 1-mL aliquot of the extract is injected into the GC. The compounds are separated by GC and detected by a MS. The labeled compounds serve to correct the variability of the analytical technique.

INTERFERENCES Solvents, reagents, glassware, and other sample processing hardware may yield artifacts and/or elevated baselines causing misinterpretation of chromatograms and spectra. Materials used in the analysis must be demonstrated to be free from interferences under the conditions of analysis by running method blanks initially and with each sample lot (sample started through the extraction process on a given 8-h shift, to a maximum of 20). Specific selection of reagents and purification of solvents by distillation in all glass systems may be required. Glassware and, where possible, reagents are cleaned by solvent rinse and baking at 450°C for 1-h minimum. Interferences coextracted from samples will vary considerably from source to source, depending on the diversity of the site being sampled.

INSTRUMENTATION Major instrumentation includes a GC with a splitless or on-column injection port for capillary column, a MS with 70 eV electron impact ionization, and a data system to collect and record MS data, and process it. A K-D apparatus is used to concentrate extracts.

GC Column: 30 m × 0.25 mm I.D. 5% phenyl, 94% methyl, 1% vinyl silicone bonded phased fused silica capillary column.

PRECISION & ACCURACY The detection limits of the method are usually dependent on the level of interferences rather than instrumental limitations. The limits typify the minimum quantities that can be detected with no interferences present.

The minimum level (in µg/mL) was not listed. This is defined as a minimum level at which the analytical system shall give recognizable mass spectra (background corrected) and acceptable calibration points.

The MDL (in µg/kg) in low solids was not listed and in high solids was not listed; these were determined in digested sludge (low solids) and in filter cake or compost (high solids).

The labeled and native compound initial precision as standard deviation (in µg/L) was not listed.
The labeled and native compound initial accuracy as average recovery (in µg/L) was not listed.

SAMPLE COLLECTION, PRESERVATION & HANDLING Collect samples in glass containers. Aqueous samples which flow freely are collected in refrigerated bottles using automatic sampling equipment. Solid samples are collected as grab samples using widemouth jars. Maintain samples at 0 to 4°C from the time of collection until extraction. If residual chlorine is present in aqueous samples, add 80 mg sodium thiosulfate/L of water. Begin sample extraction within 7 days of collection, and analyze all extracts within 40 days of extraction.

SAMPLE PREPARATION Samples containing 1% solids or less are extracted directly using continuous liquid-liquid extraction techniques. Samples containing 1 to 30% solids are diluted to the 1% level with reagent water and extracted using continuous liquid-liquid extraction techniques. Samples containing greater than 30% solids are extracted using ultrasonic techniques.

Base/neutral extraction — Adjust the pH of the waters in the extractors to 12–13 with 6 N NaOH. Extract with methylene chloride for 24–48 h.
Acid extraction — Adjust the pH of the waters in the extractors to 2 or less using 6 N sulfuric acid. Extract with methylene chloride for 24–48 h.
Ultrasonic extraction of high solids samples — Add anhydrous sodium sulfate to the sample and QC aliquot(s). Add acetone:methylene chloride (1:1) to the sample and mix thoroughly

Concentrate extracts using a K-D apparatus.

QUALITY CONTROL The analyst is permitted to modify this method to improve separations or lower the costs of measurements, provided all performance specifications are met. Analyses of blanks are required to demonstrate freedom from contamination. When results of spikes indicate atypical method performance for samples, the samples are diluted to bring method performance within acceptable limits.

For low solids (aqueous samples), extract, concentrate, and analyze two sets of four 1-L aliquots (8 aliquots total) of the precision and recovery standard. For high solids samples, two sets of four 30-g aliquots of the high solids reference matrix are used.

Spike all samples with labeled compounds to assess method performance. Compute percent recovery of the labeled compounds using the internal standard method. Compare the labeled compound recovery for each compound with the corresponding labeled compound recovery.

Reagent water and high solids reference matrix blanks are analyzed to demonstrate freedom from contamination. Extract and concentrate a 1-L reagent water blank or a high solids reference matrix blank with each sample's lot (samples started through the extraction process on the same 8-h shift, to a maximum of 20 samples).

Field replicates may be collected to determine the precision of the sampling technique, and spiked samples may be required to determine the accuracy of the analysis when the internal standard method is used.

REFERENCE Semivolatile Organic Compounds by Isotope Dilution GC/MS. Office of Water Regulation and Standards, U.S. EPA Industrial Technology Division, Washington, DC, EPA Method 1625, Rev. C, June 1989 (contact W.A. Telliard, U.S. EPA, Office of Water Regulations and Standards, 401 M St., SW, Washington, DC, 20460. Phone: 202-382-7131).

Trimethyl phosphate	EPA Method 8270
CAS #512-56-1	

TITLE Semivolatile Organic Compounds by GC/MS

MATRIX This method is used to determine the concentration of semivolatile organic compounds in extracts prepared from all types of solid waste matrices, soils, and groundwater. Although surface waters are not specifically mentioned, this method should be applicable to water samples from rivers, lakes, etc.

METHOD SUMMARY This method covers 259 semivolatile organic compounds. In very limited applications direct injection of the sample into the GC/MS system may be appropriate, but this results in very high detection limits (approximately 10,000 µg/L). Typically, a 1-L liquid sample, containing surrogate, and matrix spiking standards, is extracted in a continuous extractor first under acid conditions and then under basic conditions. Typically 30 g of a solid sample, containing surrogate, and matrix spiking standards, is extracted ultrasonically. After concentrating the extract to 1 mL it is spiked with 10 µL of an internal standard solution just prior to analysis by GC/MS. The volume injected should contain about 100 ng of base/neutral and 200 ng of acid surrogates (for a 1-µL injection). Analysis is performed by GC/MS using a capillary GC column.

INTERFERENCES Raw GC/MS data from all blanks, samples, and spikes must be evaluated for interferences. Contamination by carryover can occur whenever high-concentration and low-concentration samples are sequentially analyzed. To reduce carryover, the sample syringe must be rinsed out between samples with solvent. Whenever an unusually concentrated sample is encountered, it should be followed by the analysis of blank solvent to check for cross-contamination.

INSTRUMENTATION A GC/MS and a data system are required. The GC column used is a 30 m × 0.25 mm I.D. (or 0.32 mm I.D.) 1um film thickness silicone-coated fused silica capillary column. A continuous liquid-liquid extractor equipped with Teflon® or glass connection joints and stopcocks requiring no lubrication, a K-D concentrating apparatus, water bath, and an ultrasonic disrupter with a minimum power of 300 W and with pulsing capability are also required.

PRECISION & ACCURACY The estimated quantitation limit (EQL) of Method 8270B for determining an individual compound is approximately 1 mg/kg (wet weight) for soil or sediment samples, 1–200 mg/kg for wastes (dependent on matrix and method of preparation), and 10 µg/L for groundwater samples. EQLs will be proportionately higher for sample extracts that require dilution to avoid saturation of the detector.

The EQL(b) for groundwater in µg/L is 10.
The EQL (a, b) for low concentrations in soil and sediment in µg/kg is not determined.
Accuracy as µg/L is not listed.
Overall precision in µg/L is not listed.

(a) *EQLs listed for soil/sediment are based on wet weight. Normally data is reported in a dry-weight basis; therefore, EQLs will be higher based on the % dry weight of each sample. This calculation is based on a 30 g sample and gel permeation chromatography cleanup.*
(b) *Sample EQLs are highly matrix-dependent. The EQLs are provided for guidance and may not always be achievable.*
C = *True value for concentration, in µg/L.*
X = *Average recovery found for measurements of samples containing a concentration of C, in µg/L.*

ESTIMATED QUANTITATION LIMIT

Other Matrices	Factor (a)
High-concentration soil and sludges by sonicator	7.5
Non-water miscible waste	75

(a) *EQL for other matrices = [EQL for low soil/sediment × [Factor]. This estimated EQL is similar to an EPA "Practical Quantitation Limit."*

SAMPLING METHOD

Liquid samples — Use a 1 or 2½ gallon amber glass bottle with a screw-top Teflon®-lined cover that has been prewashed with detergent and rinsed with distilled water and methanol (or isopropanol).

Soils, sediments, or sludges — Use an 8-oz. widemouth glass with a screw-top Teflon®-lined cover that has been prewashed with detergent and rinsed with distilled water and methanol (or isopropanol).

SAMPLE PRESERVATION

Liquid samples — If residual chlorine is present, add 3 mL of 10% sodium thiosulfate per gallon, cool to 4°C and store in a solvent-free refrigerator until analysis; if chlorine is not present, then eliminate the sodium thiosulfate addition.

Soils, sediments, or sludges — Cool samples to 4°C and store in a solvent-free refrigerator.

MHT Liquid samples must be extracted within 7 days and the extracts analyzed within 40 days. Soils, sediments, or sludges may be stored for a maximum of 14 days and the extracts analyzed within 40 days.

SAMPLE PREPARATION

Liquid samples — Transfer 1 L quantitatively to a continuous extractor. If high concentrations are anticipated, a smaller volume may be used and then diluted with organic-free reagent water to 1 L. Adjust pH, if necessary, to pH <2 using 1:1 (V/V) sulfuric acid. Pipette 1.0 mL of a surrogate standard spiking solution into each sample. For the sample in each analytical batch selected for spiking, add 1.0 mL of a matrix spiking standard. For base/neutral acid analysis, the amount of the surrogates and matrix spiking compounds added to the sample should result in a final concentration of 100 ng/µL of each analyte in the extract to be analyzed (assuming a 1-µL injection). Extract with methylene chloride for 18–24 h. Next, adjust the pH of the aqueous phase to pH >11 using 10 N sodium hydroxide and extract it with methylene chloride again for 18–24 h. Dry the extract through a column containing anhydrous sodium sulfate and concentrate it to 1 mL using a K-D concentrator.

Soils, sediments, or sludges — Use 30 g of sample. Nonporous or wet samples (gummy or clay type) that do not have a free-flowing sandy texture must be mixed with anhydrous sodium sulfate until the sample is free flowing. Add 1 mL of surrogate standards to all samples, spikes, standards, and blanks. For the sample in each analytical batch selected for spiking, add 1.0 mL of a matrix spiking standard. For base/neutral acid analysis, the amount added of the surrogates and matrix spiking compounds should result in a final concentration of 100 ng/µL of each base/neutral analyte and 200 ng/µL of each acid analyte in the extract to be analyzed (assuming a 1-µL injection). Immediately add a 100-mL mixture of 1:1 methylene chloride:acetone and extract the sample ultrasonically for 3 min and then decant or filter the extracts. Repeat the extraction two or more times. Dry the extract using a column with anhydrous sodium sulfate and concentrate it to 1 mL in a K-D concentrator.

QUALITY CONTROL

A methylene chloride solution containing 50 ng/µL of decafluorotriphenylphosphine (DFTPP) is used for tuning the GC/MS system each 12-h shift. A system performance check also must be made during every 12-h shift. A standard containing 50 ng/µL each of 4,4'-DDT, pentachlorophenol, and benzidine is required to verify injection port inertness and GC column performance. A calibration standard at mid-concentration, containing each compound of interest, including all required surrogates, must be performed every 12 h during analysis. After the system performance check is met, calibration check compounds (CCCs) are used to check the validity of the initial calibration.

The internal standard responses and retention times in the calibration check standard must be evaluated immediately after or during data acquisition. If the retention time for any internal standard changes by more than 30 seconds from the last check calibration (12 h), the chromatographic system must be inspected for malfunctions and corrections must be made, as required. If the electron ionization current plot (EICP) area for any of the internal standards changes by a factor of two from the last daily calibration standard check, the mass spectrometer must be inspected for malfunctions and corrections must be made, as appropriate.

Demonstrate, through the analysis of a reagent water blank, that interferences from the analytical system, glassware, and reagents are under control. The blank samples should be carried through all stages of the sample preparation and measurement steps. For each analytical batch (up to 20 samples), a reagent blank, matrix spike, and matrix spike duplicate/duplicate must be analyzed (the frequency of the spikes may be different for different monitoring programs). The blank and spiked samples must be carried through all stages of the sample preparation and measurement steps. A QC reference sample concentrate containing each analyte at a concentration of 100 mg/L in methanol is required.

REFERENCE Test Methods for Evaluating Solid Waste (SW-846). U.S. EPA 1983, Method 8270B, Rev. 2, Nov. 1990. Office of Solid Waste, Washington, DC.

2,4,5-Trimethylaniline   EPA Method 1625
CAS #137-17-7

TITLE Semivolatile Organic Compounds by Isotope Dilution GC/MS

MATRIX The compounds may be determined in waters, soils, and municipal sludges by this method.

METHOD SUMMARY This method is used to determine 176 semivolatile toxic organic pollutants associated with the CWA (as amended 1987); the RCRA (as amended 1986); the CERCLA (as amended 1986); and other compounds amenable to extraction and analysis by capillary column gas chromatography-mass spectrometry (GC/MS).

Stable isotopically-labeled analogs of the compounds of interest are added to the sample. If the solids content is less than 1%, a 1-L sample is extracted at pH 12–13, then at pH <2 with methylene chloride using continuous extraction techniques.

If the solids content is 30% or less, the sample is diluted to 1% solids with reagent water, homogenized ultrasonically, and extracted at pH 12–13, then at pH <2 with methylene chloride using continuous extraction techniques. If the solids content is greater than 30%, the sample is extracted using ultrasonic techniques.

Each extract is dried over sodium sulfate, concentrated to a volume of 5 mL, cleaned up using GPC, if necessary, and concentrated. Extracts are concentrated to 1 mL if GPC is not performed, and to 0.5 mL if GPC is performed.

An internal standard is added to the extract, and a 1-mL aliquot of the extract is injected into the GC. The compounds are separated by GC and detected by a MS. The labeled compounds serve to correct the variability of the analytical technique.

INTERFERENCES Solvents, reagents, glassware, and other sample processing hardware may yield artifacts and/or elevated baselines causing misinterpretation of chromatograms and spectra. Materials used in the analysis must be demonstrated to be free from interferences under the conditions of analysis by running method blanks initially and with each sample lot (sample started through the extraction process on a given 8-h shift, to a maximum of 20). Specific selection of reagents and purification of solvents by distillation in all glass systems may be required. Glassware and, where possible, reagents are cleaned by solvent rinse and baking at 450°C for 1-h minimum. Interferences coextracted from samples will vary considerably from source to source, depending on the diversity of the site being sampled.

INSTRUMENTATION Major instrumentation includes a GC with a splitless or on-column injection port for capillary column, a MS with 70 eV electron impact ionization, and a data system to collect and record MS data, and process it. A K-D apparatus is used to concentrate extracts.

GC Column: 30 m × 0.25 mm I.D. 5% phenyl, 94% methyl, 1% vinyl silicone bonded phased fused silica capillary column.

PRECISION & ACCURACY The detection limits of the method are usually dependent on the level of interferences rather than instrumental limitations. The limits typify the minimum quantities that can be detected with no interferences present.

The minimum level (in µg/mL) was not listed. This is defined as a minimum level at which the analytical system shall give recognizable mass spectra (background corrected) and acceptable calibration points.

The MDL (in µg/kg) in low solids was not listed and in high solids was not listed; these were determined in digested sludge (low solids) and in filter cake or compost (high solids).

The labeled and native compound initial precision as standard deviation (in µg/L) was not listed.
The labeled and native compound initial accuracy as average recovery (in µg/L) was not listed.

SAMPLE COLLECTION, PRESERVATION & HANDLING Collect samples in glass containers. Aqueous samples which flow freely are collected in refrigerated bottles using automatic sampling equipment. Solid samples are collected as grab samples using widemouth jars. Maintain samples at 0 to 4°C from the time of collection until extraction. If residual chlorine is present in aqueous samples, add 80 mg sodium thiosulfate/L of water. Begin sample extraction within 7 days of collection, and analyze all extracts within 40 days of extraction.

SAMPLE PREPARATION Samples containing 1% solids or less are extracted directly using continuous liquid-liquid extraction techniques. Samples containing 1 to 30% solids are diluted to the 1% level with reagent water and extracted using continuous liquid-liquid extraction techniques. Samples containing greater than 30% solids are extracted using ultrasonic techniques.

- Base/neutral extraction — Adjust the pH of the waters in the extractors to 12–13 with 6 *N* NaOH. Extract with methylene chloride for 24–48 h.
- Acid extraction — Adjust the pH of the waters in the extractors to 2 or less using 6 *N* sulfuric acid. Extract with methylene chloride for 24–48 h.
- Ultrasonic extraction of high solids samples — Add anhydrous sodium sulfate to the sample and QC aliquot(s). Add acetone:methylene chloride (1:1) to the sample and mix thoroughly

Concentrate extracts using a K-D apparatus.

QUALITY CONTROL The analyst is permitted to modify this method to improve separations or lower the costs of measurements, provided all performance specifications are met. Analyses of blanks are required to demonstrate freedom from contamination. When results of spikes indicate atypical method performance for samples, the samples are diluted to bring method performance within acceptable limits.

For low solids (aqueous samples), extract, concentrate, and analyze two sets of four 1-L aliquots (8 aliquots total) of the precision and recovery standard. For high solids samples, two sets of four 30-g aliquots of the high solids reference matrix are used.

Spike all samples with labeled compounds to assess method performance. Compute percent recovery of the labeled compounds using the internal standard method. Compare the labeled compound recovery for each compound with the corresponding labeled compound recovery.

Reagent water and high solids reference matrix blanks are analyzed to demonstrate freedom from contamination. Extract and concentrate a 1-L reagent water blank or a high solids reference matrix blank with each sample's lot (samples started through the extraction process on the same 8-h shift, to a maximum of 20 samples).

Field replicates may be collected to determine the precision of the sampling technique, and spiked samples may be required to determine the accuracy of the analysis when the internal standard method is used.

REFERENCE Semivolatile Organic Compounds by Isotope Dilution GC/MS. Office of Water Regulation and Standards, U.S. EPA Industrial Technology Division, Washington, DC, EPA Method 1625, Rev. C, June 1989 (contact W.A. Telliard, U.S. EPA, Office of Water Regulations and Standards, 401 M St., SW, Washington, DC, 20460. Phone: 202-382-7131).

2,4,5-Trimethylaniline   **EPA Method 8270**
CAS #137-17-7

TITLE Semivolatile Organic Compounds by GC/MS

MATRIX This method is used to determine the concentration of semivolatile organic compounds in extracts prepared

from all types of solid waste matrices, soils, and groundwater. Although surface waters are not specifically mentioned, this method should be applicable to water samples from rivers, lakes, etc.

METHOD SUMMARY This method covers 259 semivolatile organic compounds. In very limited applications direct injection of the sample into the GC/MS system may be appropriate, but this results in very high detection limits (approximately 10,000 µg/L). Typically, a 1-L liquid sample, containing surrogate, and matrix spiking standards, is extracted in a continuous extractor first under acid conditions and then under basic conditions. Typically 30 g of a solid sample, containing surrogate, and matrix spiking standards, is extracted ultrasonically. After concentrating the extract to 1 mL it is spiked with 10 µL of an internal standard solution just prior to analysis by GC/MS. The volume injected should contain about 100 ng of base/neutral and 200 ng of acid surrogates (for a 1-µL injection). Analysis is performed by GC/MS using a capillary GC column.

INTERFERENCES Raw GC/MS data from all blanks, samples, and spikes must be evaluated for interferences. Contamination by carryover can occur whenever high-concentration and low-concentration samples are sequentially analyzed. To reduce carryover, the sample syringe must be rinsed out between samples with solvent. Whenever an unusually concentrated sample is encountered, it should be followed by the analysis of blank solvent to check for cross-contamination.

INSTRUMENTATION A GC/MS and a data system are required. The GC column used is a 30 m × 0.25 mm I.D. (or 0.32 mm I.D.) 1um film thickness silicone-coated fused silica capillary column. A continuous liquid-liquid extractor equipped with Teflon® or glass connection joints and stopcocks requiring no lubrication, a K-D concentrating apparatus, water bath, and an ultrasonic disrupter with a minimum power of 300 W and with pulsing capability are also required.

PRECISION & ACCURACY The estimated quantitation limit (EQL) of Method 8270B for determining an individual compound is approximately 1 mg/kg (wet weight) for soil or sediment samples, 1–200 mg/kg for wastes (dependent on matrix and method of preparation), and 10 µg/L for groundwater samples. EQLs will be proportionately higher for sample extracts that require dilution to avoid saturation of the detector.

The EQL(b) for groundwater in µg/L is 10.
The EQL (a, b) for low concentrations in soil and sediment in µg/kg is not determined.
Accuracy as µg/L is not listed.
Overall precision in µg/L is not listed.

(a) *EQLs listed for soil/sediment are based on wet weight. Normally data is reported in a dry-weight basis; therefore, EQLs will be higher based on the % dry weight of each sample. This calculation is based on a 30 g sample and gel permeation chromatography cleanup.*
(b) *Sample EQLs are highly matrix-dependent. The EQLs are provided for guidance and may not always be achievable.*
C = *True value for concentration, in µg/L.*
X = *Average recovery found for measurements of samples containing a concentration of C, in µg/L.*

ESTIMATED QUANTITATION LIMIT

Other Matrices	Factor (a)
High-concentration soil and sludges by sonicator	7.5
Non-water miscible waste	75

(a) *EQL for other matrices = [EQL for low soil/sediment × [Factor]. This estimated EQL is similar to an EPA "Practical Quantitation Limit."*

SAMPLING METHOD
Liquid samples — Use a 1 or 2½ gallon amber glass bottle with a screw-top Teflon®-lined cover that has been prewashed with detergent and rinsed with distilled water and methanol (or isopropanol).

Soils, sediments, or sludges — Use an 8-oz. widemouth glass with a screw-top Teflon®-lined cover that has been prewashed with detergent and rinsed with distilled water and methanol (or isopropanol).

SAMPLE PRESERVATION
Liquid samples — If residual chlorine is present, add 3 mL of 10% sodium thiosulfate per gallon, cool to 4°C and store in a solvent-free refrigerator until analysis; if chlorine is not present, then eliminate the sodium thiosulfate addition.

Soils, sediments, or sludges — Cool samples to 4°C and store in a solvent-free refrigerator.

MHT Liquid samples must be extracted within 7 days and the extracts analyzed within 40 days. Soils, sediments, or sludges may be stored for a maximum of 14 days and the extracts analyzed within 40 days.

SAMPLE PREPARATION
Liquid samples — Transfer 1 L quantitatively to a continuous extractor. If high concentrations are anticipated, a smaller volume may be used and then diluted with organic-free reagent water to 1 L. Adjust pH, if necessary, to pH <2 using 1:1 (V/V) sulfuric acid. Pipette 1.0 mL of a surrogate standard spiking solution into each sample. For the sample in each analytical batch selected for spiking, add 1.0 mL of a matrix spiking standard. For base/neutral acid analysis, the amount of the surrogates and matrix spiking compounds added to the sample should result in a final concentration of 100 ng/µL of each analyte in the extract to be analyzed (assuming a 1-µL injection). Extract with methylene chloride for 18–24 h. Next, adjust the pH of the aqueous phase to pH >11 using 10 N sodium hydroxide and extract it with methylene chloride again for 18–24 h. Dry the extract through a column containing anhydrous sodium sulfate and concentrate it to 1 mL using a K-D concentrator.

Soils, sediments, or sludges — Use 30 g of sample. Nonporous or wet samples (gummy or clay type) that do not have a free-flowing sandy texture must be mixed with anhydrous sodium sulfate until the sample is free flowing. Add 1 mL of surrogate standards to all samples, spikes, standards, and blanks. For the sample in each analytical batch selected for spiking, add 1.0 mL of a matrix spiking standard. For base/neutral acid analysis, the amount added of the surrogates and matrix spiking compounds should result in a final concentration of 100 ng/µL of each base/neutral analyte and 200 ng/µL of each acid analyte

in the extract to be analyzed (assuming a 1-μL injection). Immediately add a 100-mL mixture of 1:1 methylene chloride:acetone and extract the sample ultrasonically for 3 min and then decant or filter the extracts. Repeat the extraction two or more times. Dry the extract using a column with anhydrous sodium sulfate and concentrate it to 1 mL in a K-D concentrator.

QUALITY CONTROL A methylene chloride solution containing 50 ng/μL of decafluorotriphenylphosphine (DFTPP) is used for tuning the GC/MS system each 12-h shift. A system performance check also must be made during every 12-h shift. A standard containing 50 ng/μL each of 4,4'-DDT, pentachlorophenol, and benzidine is required to verify injection port inertness and GC column performance. A calibration standard at mid-concentration, containing each compound of interest, including all required surrogates, must be performed every 12 h during analysis. After the system performance check is met, calibration check compounds (CCCs) are used to check the validity of the initial calibration.

The internal standard responses and retention times in the calibration check standard must be evaluated immediately after or during data acquisition. If the retention time for any internal standard changes by more than 30 seconds from the last check calibration (12 h), the chromatographic system must be inspected for malfunctions and corrections must be made, as required. If the electron ionization current plot (EICP) area for any of the internal standards changes by a factor of two from the last daily calibration standard check, the mass spectrometer must be inspected for malfunctions and corrections must be made, as appropriate.

Demonstrate, through the analysis of a reagent water blank, that interferences from the analytical system, glassware, and reagents are under control. The blank samples should be carried through all stages of the sample preparation and measurement steps. For each analytical batch (up to 20 samples), a reagent blank, matrix spike, and matrix spike duplicate/duplicate must be analyzed (the frequency of the spikes may be different for different monitoring programs). The blank and spiked samples must be carried through all stages of the sample preparation and measurement steps. A QC reference sample concentrate containing each analyte at a concentration of 100 mg/L in methanol is required.

REFERENCE Test Methods for Evaluating Solid Waste (SW-846). U.S. EPA 1983, Method 8270B, Rev. 2, Nov. 1990. Office of Solid Waste, Washington, DC.

1,2,4-Trimethylbenzene   **EPA Method 502**
CAS #95-63-6

TITLE Volatile Organic Compounds in Water By Purge and Trap Capillary Column Gas Chromatography with Photoionization and Electrolytic Conductivity Detectors in Series. U.S. EPA Method 502.2, Rev. 2.0, 1989.

MATRIX Drinking water and raw source water. The latter should include most surface water and groundwater sources.

METHOD SUMMARY This method covers 60 volatile organic compounds that contain halogen atoms and/or that are aromatic. An inert gas (zero grade nitrogen or helium) is bubbled through a 25-mL or a 5-mL water sample (depending on the expected concentration of the analytes). Purged sample components are trapped in a tube of sorbent materials. When purging is complete, the sorbent tube is heated and backflushed with helium to desorb the trapped sample onto a capillary GC column. The column is temperature programmed to separate the method analytes which are then detected with a photoionization detector (PID) and a Hall electrolytic conductivity (HECD) placed in series. The PID is selective for aromatic compounds and the HECD is selective for halogenated compounds.

INTERFERENCES Impurities in the purge gas and from organic compounds outgassing from the plumbing ahead of the trap account for many contamination problems. Interferences purged or coextracted from the samples will vary considerably from source to source, depending upon the particular sample or extract being tested. Cross-contamination can occur whenever high-level and low-level samples are analyzed sequentially. Samples also can be contaminated by diffusion of volatile organics (particularly methylene chloride and fluorocarbons) through the septum seal into the sample during shipment and storage. The lab where volatile analysis is performed and also the refrigerated storage area should be completely free of solvents.

INSTRUMENTATION A GC containing a series configuration of a high temperature photoionization detector (PID) equipped with 10.0 eV (nominal) lamp and Hall electrolytic conductivity detector (HECD) is required. Also required is an all-glass 5-mL purging device, a sorbent trap, and a thermal desorption apparatus which is connected to the GC system.

Column 1: VOCOL glass wide-bore capillary column.
Column 2: RTX–502.2 mega-bore capillary column.
Column 3: DB-62 mega-bore capillary column.

PRECISION & ACCURACY Method detection limits are dependent upon the characteristics of the gas chromatographic system used. Analytes that are not separated chromatographically cannot be individually identified and used in the same calibration mixture or water samples unless an alternative technique for identification and quantification, such as mass spectrometry, is used.

Electrolytic conductivity detetor (c) range in μg/L (a) was 0.02–200.
Electrolytic conductivity detetor (c) MDL in μg/L (b) was not listed.
Electrolytic conductivity detetor (c) accuracy as % recovery was not listed.
Electrolytic conductivity detetor (c) precision as % RSD was not listed.
Photoionization detector (d) range in μg/L (a) was 0.02–200.
Photoionization detector (d) MDL in μg/L (b) was 0.05.
Photoionization detector (d) accuracy as % recovery was 99.
Photoionization detector (d) precision as % RSD was 1.2.

(a) *The applicable concentration range of this method is compound, instrument, and matrix-dependent. It is listed as* being approximately 0.02 to 200 µg/L but no specific information is provided so caution should be observed.

(b) The method detection limits reports with this method are compound, instrument, and matrix-dependent. The values reported were calculated using reagent water fortified with the corresponding compounds at 10 µg/L and a GC-equipped with a 60 m × 0.75 mm VOLCOL wide bore capillary column with 1.5 µm film thickness and using helium carrier gas.

(c) Recoveries and relative standard deviations were determined from seven samples of reagent water fortified with 10 µg/L of each compound. 2-Bromo-1-chloropropane was used as the internal standard for calculating average recoveries.

(d) Recoveries and relative standard deviations were determined from seven samples of reagent water fortified with 10 µg/L of each compound. Fluorobenzene was used as the internal standard for calculating average recoveries.

SAMPLING METHOD Collect samples using a 40- to 120-mL screw-cap vial (prewashed with detergent, rinsed with distilled water and oven dried at 105°C) with a Teflon®-faced silicone septum. Collect bubble-free samples and place the septum with the Teflon® side down on the water.

SAMPLE PRESERVATION If residual chlorine is present in the water add about 25 mg of ascorbic acid to each vial before samples are collected to remove the chlorine. Add hydrochloric acid to reduce pH to <2, immediately cool samples to 4°C, and store them in a solvent-free refrigerator at 4°C until analysis.

MHT The maximum holding time for samples is 14 days from the time they were collected.

SAMPLE PREPARATION Remove the plungers from two 5-mL syringes and attach a closed syringe valve to each. Warm the sample to room temperature, open the sample bottle, and carefully pour the sample into one of the syringe barrels to just short of overflowing. Replace the syringe plunger, invert the syringe, and compress the sample. Open the syringe valve and vent any residual air while adjusting the sample volume to 5.0 mL. Add 10 µL of the internal calibration standard to the sample through the syringe valve. Close the valve. Fill the second syringe in an identical manner from the same sample bottle. Reserve this second syringe for a reanalysis if necessary.

QUALITY CONTROL As an initial demonstration of lab accuracy and precision, analyze 4 to 7 replicates of a lab fortified blank containing analyte at 0.1–5 µg/L. Collect all samples in duplicate. Surrogate analytes (similar to those of the analytes of interest), whose concentration is known in every sample, are measured using the same internal standard calibration procedure. Duplicate field reagent water blanks (trip blanks) must be analyzed with each set of samples, lab reagent blanks (method blanks) must be analyzed with each batch of samples processed as a group within a work shift. Also, a single lab-fortified blank that contains each of the analytes of interest should be analyzed with each batch of samples processed as a group within a work shift. A 3- to 5-point calibration curve is needed depending on the calibration range factor required.

EPA CONTACT & HOTLINE For technical questions contact Dr. Baldev Bathija, U.S. EPA, Office of Ground Water and Drinking Water (WH-550D), 401 M St. SW, Washington, DC 20460. Tel. (202) 260-3040. For further information the EPA Safe Drinking Water Hotline may be called at: (800) 426-4791.

REFERENCE Methods for the Determination of Organic Compounds in Drinking Water, EPA/600/4-88/039 (revised July 1991; Final Rule for determination of compliance with the MCL for Total Trihalomethanes under 141.30, in 40 CFR Part 141, Vol. 58, No. 147, Fed. Reg., Tuesday Aug. 3, 1993). U.S. EPA Environmental Monitoring Systems Laboratory, Cincinnati, OH, 45268, U.S.A. Available from the National Technical Information Service (NTIS), 5285 Port Royal Road, Springfield, VA 22161; Tel. 800-553-6847. NTIS Order Number is PB91-231480.

1,3,5-Trimethylbenzene   EPA Method 502
CAS #108-67-8

TITLE Volatile Organic Compounds in Water By Purge and Trap Capillary Column Gas Chromatography with Photoionization and Electrolytic Conductivity Detectors in Series. U.S. EPA Method 502.2, Rev. 2.0, 1989.

MATRIX Drinking water and raw source water. The latter should include most surface water and groundwater sources.

METHOD SUMMARY This method covers 60 volatile organic compounds that contain halogen atoms and/or that are aromatic. An inert gas (zero grade nitrogen or helium) is bubbled through a 25-mL or a 5-mL water sample (depending on the expected concentration of the analytes). Purged sample components are trapped in a tube of sorbent materials. When purging is complete, the sorbent tube is heated and backflushed with helium to desorb the trapped sample onto a capillary GC column. The column is temperature programmed to separate the method analytes which are then detected with a photoionization detector (PID) and a Hall electrolytic conductivity (HECD) placed in series. The PID is selective for aromatic compounds and the HECD is selective for halogenated compounds.

INTERFERENCES Impurities in the purge gas and from organic compounds outgassing from the plumbing ahead of the trap account for many contamination problems. Interferences purged or coextracted from the samples will vary considerably from source to source, depending upon the particular sample or extract being tested. Cross-contamination can occur whenever high-level and low-level samples are analyzed sequentially. Samples also can be contaminated by diffusion of volatile organics (particularly methylene chloride and fluorocarbons) through the septum seal into the sample during shipment and storage. The lab where volatile analysis is performed and also the refrigerated storage area should be completely free of solvents.

INSTRUMENTATION A GC containing a series configuration of a high temperature photoionization detector (PID) equipped with 10.0 eV (nominal) lamp and Hall electrolytic conductivity detector (HECD) is required. Also required is an all-glass 5-mL purging device, a sorbent trap, and a thermal desorption apparatus which is connected to the GC system.

Column 1: VOCOL glass wide-bore capillary column.
Column 2: RTX–502.2 mega-bore capillary column.
Column 3: DB-62 mega-bore capillary column.

PRECISION & ACCURACY Method detection limits are dependent upon the characteristics of the gas chromatographic system used. Analytes that are not separated chromatographically cannot be individually identified and used in the same calibration mixture or water samples unless an alternative technique for identification and quantification, such as mass spectrometry, is used.

Electrolytic conductivity detetor (c) range in μg/L (a) was 0.02–200.
Electrolytic conductivity detetor (c) MDL in μg/L (b) was not listed.
Electrolytic conductivity detetor (c) accuracy as % recovery was not listed.
Electrolytic conductivity detetor (c) precision as % RSD was not listed.
Photoionization detector (d) range in μg/L (a) was 0.02–200.
Photoionization detector (d) MDL in μg/L (b) was 0.01.
Photoionization detector (d) accuracy as % recovery was 101.
Photoionization detector (d) precision as % RSD was 1.4.

(a) *The applicable concentration range of this method is compound, instrument, and matrix-dependent. It is listed as being approximately 0.02 to 200 μg/L but no specific information is provided so caution should be observed.*
(b) *The method detection limits reports with this method are compound, instrument, and matrix-dependent. The values reported were calculated using reagent water fortified with the corresponding compounds at 10 μg/L and a GC-equipped with a 60 m × 0.75 mm VOLCOL wide bore capillary column with 1.5 μm film thickness and using helium carrier gas.*
(c) *Recoveries and relative standard deviations were determined from seven samples of reagent water fortified with 10 μg/L of each compound. 2-Bromo-1-chloropropane was used as the internal standard for calculating average recoveries.*
(d) *Recoveries and relative standard deviations were determined from seven samples of reagent water fortified with 10 μg/L of each compound. Fluorobenzene was used as the internal standard for calculating average recoveries.*

SAMPLING METHOD Collect samples using a 40- to 120-mL screw-cap vial (prewashed with detergent, rinsed with distilled water and oven dried at 105°C) with a Teflon®-faced silicone septum. Collect bubble-free samples and place the septum with the Teflon® side down on the water.

SAMPLE PRESERVATION If residual chlorine is present in the water add about 25 mg of ascorbic acid to each vial before samples are collected to remove the chlorine. Add hydrochloric acid to reduce pH to <2, immediately cool samples to 4°C, and store them in a solvent-free refrigerator at 4°C until analysis.

MHT The maximum holding time for samples is 14 days from the time they were collected.

SAMPLE PREPARATION Remove the plungers from two 5-mL syringes and attach a closed syringe valve to each. Warm the sample to room temperature, open the sample bottle, and carefully pour the sample into one of the syringe barrels to just short of overflowing. Replace the syringe plunger, invert the syringe, and compress the sample. Open the syringe valve and vent any residual air while adjusting the sample volume to 5.0 mL. Add 10 μL of the internal calibration standard to the sample through the syringe valve. Close the valve. Fill the second syringe in an identical manner from the same sample bottle. Reserve this second syringe for a reanalysis if necessary.

QUALITY CONTROL As an initial demonstration of lab accuracy and precision, analyze 4 to 7 replicates of a lab fortified blank containing analyte at 0.1–5 μg/L. Collect all samples in duplicate. Surrogate analytes (similar to those of the analytes of interest), whose concentration is known in every sample, are measured using the same internal standard calibration procedure. Duplicate field reagent water blanks (trip blanks) must be analyzed with each set of samples, lab reagent blanks (method blanks) must be analyzed with each batch of samples processed as a group within a work shift. Also, a single lab-fortified blank that contains each of the analytes of interest should be analyzed with each batch of samples processed as a group within a work shift. A 3- to 5-point calibration curve is needed depending on the calibration range factor required.

EPA CONTACT & HOTLINE For technical questions contact Dr. Baldev Bathija, U.S. EPA, Office of Ground Water and Drinking Water (WH-550D), 401 M St. SW, Washington, DC 20460. Tel. (202) 260-3040. For further information the EPA Safe Drinking Water Hotline may be called at: (800) 426-4791.

REFERENCE Methods for the Determination of Organic Compounds in Drinking Water, EPA/600/4-88/039 (revised July 1991; Final Rule for determination of compliance with the MCL for Total Trihalomethanes under 141.30, in 40 CFR Part 141, Vol. 58, No. 147, Fed. Reg., Tuesday Aug. 3, 1993). U.S. EPA Environmental Monitoring Systems Laboratory, Cincinnati, OH, 45268, U.S.A. Available from the National Technical Information Service (NTIS), 5285 Port Royal Road, Springfield, VA 22161; Tel. 800-553-6847. NTIS Order Number is PB91-231480.

1,2,4-Trimethylbenzene   **EPA Method 524**
CAS #95-63-6

TITLE Measurement of Purgeable Organic Compounds in Water by Capillary Column GC/MS.

MATRIX Drinking water and raw source water; the latter should include most surface water and groundwater sources.

METHOD SUMMARY Method 524.2 covers 60 volatile organic compounds. An inert gas (zero grade nitrogen or helium) is bubbled through a 25-mL or a 5-mL water sample (depending on the expected concentration of the analytes). Purged sample components are trapped in a tube of sorbent materials. When purging is complete, the sorbent tube is heated and backflushed with helium to desorb the trapped sample onto a capillary GC column.

INTERFERENCES Impurities in the purge gas and from organic compounds outgassing from the plumbing ahead of

the trap account for many contamination problems. Interferences purged or coextracted from the samples will vary considerably from source to source, depending upon the particular sample or extract being tested. Cross-contamination can occur whenever high-level and low-level samples are analyzed sequentially. Samples also can be contaminated by diffusion of volatile organics (particularly methylene chloride and fluorocarbons) through the septum seal into the sample during shipment and storage.

INSTRUMENTATION A GC/MS with a data system equipped with one of the following capillary GC columns:

Column 1: VOCOL glass wide bore capillary column.
Column 2: DB-624 fused silica capillary column.
Column 3: DB-5 fused silica capillary column.

Also required is an all-glass 25 mL or 5-mL purging device, a sorbent trap, and a thermal desorption apparatus which is connected to the GC/MS system.

PRECISION & ACCURACY Method detection limits are compound- and instrument-dependent, and may vary from approximately 0.02–0.35 µg/L. Note in the table below that the "true" concentration range used for accuracy and precision measurements was quite narrow. However, the applicable concentration range of this method is primarily column dependent and is approximately 0.02 to 200 µg/L for the wide-bore thick-film columns. Narrow-bore thin-film columns may have a capacity which limits the range to about 0.02 to 20 µg/L. Analytes that are inefficiently purged from water will not be detected when present at low concentrations, but they can be measured with acceptable accuracy and precision when present in sufficient amounts.

Analytes that are not separated chromatographically, but which have different mass spectra and non-interfering quantification ions, can be identified and measured in the same calibration mixture or water sample. Analytes which have very similar mass spectra cannot be individually identified and measured in the same calibration mixture or water samples unless they have different retention times. Co-eluting compounds with very similar mass spectra, typically many structural isomers, must be reported as an isomeric group or pair.

The range (in µg/L) was 0.5–10.
The Method Detection Limig (in µg/L) was 0.13.
The accuracy (as % recovery) was 99.
The precision (in %) was 8.1.

Note: Data were obtained from 16–31 determinations using a wide-bore capillary column and a jet separator interfaced to a quadrupole mass spectrometer. All analytes were in a reagent water matrix.

SAMPLING METHOD Collect samples using a 40- to 120-mL screw-cap vial (prewashed with detergent, rinsed with distilled water and oven dried at 105°C) with a Teflon®-faced silicone septum. Collect bubble-free samples and place the septum with the Teflon® side down on the water.

SAMPLE PRESERVATION If residual chlorine is present in the water add about 25 mg of ascorbic acid to each vial before samples are collected to remove the chlorine. Add hydrochloric acid to reduce pH to <2, and immediately cool samples to 4°C, and store them in a solvent-free refrigerator at 4°C until analysis.

MHT The maximum holding time for samples is 14 days from the time they were collected.

SAMPLE PREPARATION Remove the plungers from two 25-mL (or 5-mL depending on sample size) syringes and attach a closed syringe valve to each. Warm the sample to room temperature, open the sample bottle, and carefully pour the sample into one of the syringe barrels to just short of overflowing. Replace the syringe plunger, invert the syringe, and compress the sample. Open the syringe valve and vent any residual air while adjusting the sample volume to 25.0 mL (or 5 mL). For samples and blanks, add 5 µL of the fortification solution containing the internal standard and the surrogates to the sample through the syringe valve. For calibration standards and lab fortified blanks, add 5 µL of the fortification solution containing the internal standard only. Close the valve. Fill the second syringe in an identical manner from the same sample bottle. Reserve this second syringe for a reanalysis if necessary.

QUALITY CONTROL As an initial demonstration of lab accuracy and precision, analyze 4 to 7 replicates of a lab fortified blank containing analyte at 0.2–5 µg/L. Collect all samples in duplicate. Surrogate analytes (similar to those of the analytes of interest), whose concentration is known in every sample, are measured using the same internal standard calibration procedure. Duplicate field reagent water blanks (trip blanks) must be analyzed with each set of samples, lab reagent blanks (method blanks) must be analyzed with each batch of samples processed as a group within a work shift. Also, a single lab-fortified blank that contains each of the analytes of interest should be analyzed with each batch of samples processed as a group within a work shift. A 3- to 5-point calibration curve is needed depending on the calibration range factor required.

EPA CONTACT & HOTLINE For technical questions contact Dr. Baldev Bathija, U.S. EPA, Office of Ground Water and Drinking Water (WH-550D), 401 M St. SW, Washington, DC 20460. Tel. (202) 260-3040. For further information the EPA Safe Drinking Water Hotline may be called at: (800) 426-4791.

REFERENCE Methods for the Determination of Organic Compounds in Drinking Water, EPA/600/4-88/039 (revised July 1991; Final Rule for determination of compliance with the MCL for Total Trihalomethanes under 141.30, in 40 CFR Part 141, Vol. 58, No. 147, Fed. Reg., Tuesday Aug. 3, 1993). U.S. EPA Environmental Monitoring Systems Laboratory, Cincinnati, OH, 45268, U.S.A. Available from the National Technical Information Service (NTIS), 5285 Port Royal Road, Springfield, VA 22161; Tel. 800-553-6847. NTIS Order Number is PB91-231480.

1,3,5-Trimethylbenzene   EPA Method 524
CAS #108-67-8

TITLE Measurement of Purgeable Organic Compounds in Water by Capillary Column GC/MS.

MATRIX Drinking water and raw source water; the latter should include most surface water and groundwater sources.

METHOD SUMMARY Method 524.2 covers 60 volatile organic compounds. An inert gas (zero grade nitrogen or helium) is bubbled through a 25-mL or a 5-mL water sample (depending on the expected concentration of the analytes). Purged sample components are trapped in a tube of sorbent materials. When purging is complete, the sorbent tube is heated and backflushed with helium to desorb the trapped sample onto a capillary GC column.

INTERFERENCES Impurities in the purge gas and from organic compounds outgassing from the plumbing ahead of the trap account for many contamination problems. Interferences purged or coextracted from the samples will vary considerably from source to source, depending upon the particular sample or extract being tested. Cross-contamination can occur whenever high-level and low-level samples are analyzed sequentially. Samples also can be contaminated by diffusion of volatile organics (particularly methylene chloride and fluorocarbons) through the septum seal into the sample during shipment and storage.

INSTRUMENTATION A GC/MS with a data system equipped with one of the following capillary GC columns:

Column 1: VOCOL glass wide bore capillary column.
Column 2: DB-624 fused silica capillary column.
Column 3: DB-5 fused silica capillary column.

Also required is an all-glass 25 mL or 5-mL purging device, a sorbent trap, and a thermal desorption apparatus which is connected to the GC/MS system.

PRECISION & ACCURACY Method detection limits are compound- and instrument-dependent, and may vary from approximately 0.02–0.35 µg/L. Note in the table below that the "true" concentration range used for accuracy and precision measurements was quite narrow. However, the applicable concentration range of this method is primarily column dependent and is approximately 0.02 to 200 µg/L for the wide-bore thick-film columns. Narrow-bore thin-film columns may have a capacity which limits the range to about 0.02 to 20 µg/L. Analytes that are inefficiently purged from water will not be detected when present at low concentrations, but they can be measured with acceptable accuracy and precision when present in sufficient amounts.

Analytes that are not separated chromatographically, but which have different mass spectra and non-interfering quantification ions, can be identified and measured in the same calibration mixture or water sample. Analytes which have very similar mass spectra cannot be individually identified and measured in the same calibration mixture or water samples unless they have different retention times. Co-eluting compounds with very similar mass spectra, typically many structural isomers, must be reported as an isomeric group or pair.

The range (in µg/L) was 0.5–10.
The Method Detection Limig (in µg/L) was 0.05.
The accuracy (as % recovery) was 92.

The precision (in %) was 7.4.

Note: Data were obtained from 16–31 determinations using a wide-bore capillary column and a jet separator interfaced to a quadrupole mass spectrometer. All analytes were in a reagent water matrix.

SAMPLING METHOD Collect samples using a 40- to 120-mL screw-cap vial (prewashed with detergent, rinsed with distilled water and oven dried at 105°C) with a Teflon®-faced silicone septum. Collect bubble-free samples and place the septum with the Teflon® side down on the water.

SAMPLE PRESERVATION If residual chlorine is present in the water add about 25 mg of ascorbic acid to each vial before samples are collected to remove the chlorine. Add hydrochloric acid to reduce pH to <2, and immediately cool samples to 4°C, and store them in a solvent-free refrigerator at 4°C until analysis.

MHT The maximum holding time for samples is 14 days from the time they were collected.

SAMPLE PREPARATION Remove the plungers from two 25-mL (or 5-mL depending on sample size) syringes and attach a closed syringe valve to each. Warm the sample to room temperature, open the sample bottle, and carefully pour the sample into one of the syringe barrels to just short of overflowing. Replace the syringe plunger, invert the syringe, and compress the sample. Open the syringe valve and vent any residual air while adjusting the sample volume to 25.0 mL (or 5 mL). For samples and blanks, add 5 µL of the fortification solution containing the internal standard and the surrogates to the sample through the syringe valve. For calibration standards and lab fortified blanks, add 5 µL of the fortification solution containing the internal standard only. Close the valve. Fill the second syringe in an identical manner from the same sample bottle. Reserve this second syringe for a reanalysis if necessary.

QUALITY CONTROL As an initial demonstration of lab accuracy and precision, analyze 4 to 7 replicates of a lab fortified blank containing analyte at 0.2–5 µg/L. Collect all samples in duplicate. Surrogate analytes (similar to those of the analytes of interest), whose concentration is known in every sample, are measured using the same internal standard calibration procedure. Duplicate field reagent water blanks (trip blanks) must be analyzed with each set of samples, lab reagent blanks (method blanks) must be analyzed with each batch of samples processed as a group within a work shift. Also, a single lab-fortified blank that contains each of the analytes of interest should be analyzed with each batch of samples processed as a group within a work shift. A 3- to 5-point calibration curve is needed depending on the calibration range factor required.

EPA CONTACT & HOTLINE For technical questions contact Dr. Baldev Bathija, U.S. EPA, Office of Ground Water and Drinking Water (WH-550D), 401 M St. SW, Washington, DC 20460. Tel. (202) 260-3040. For further information the EPA Safe Drinking Water Hotline may be called at: (800) 426-4791.

REFERENCE Methods for the Determination of Organic Compounds in Drinking Water, EPA/600/4-88/039 (revised July 1991; Final Rule for determination of compliance with the

MCL for Total Trihalomethanes under 141.30, in 40 CFR Part 141, Vol. 58, No. 147, Fed. Reg., Tuesday Aug. 3, 1993). U.S. EPA Environmental Monitoring Systems Laboratory, Cincinnati, OH, 45268, U.S.A. Available from the National Technical Information Service (NTIS), 5285 Port Royal Road, Springfield, VA 22161; Tel. 800-553-6847. NTIS Order Number is PB91-231480.

1,2,4-Trimethylbenzene	EPA Method 8021
CAS #95-63-6	

TITLE Halogenated Volatile by Gas Chromatography Using Photoionization and Electrolytic Conductivity Detectors in Series: Capillary Column Technique

MATRIX This method is applicable to nearly all types of samples, regardless of water content, including groundwater, aqueous sludges, caustic liquors, acid liquors, waste solvents, oily wastes, mousses, tars, fibrous wastes, polymeric emulsions, filter cakes, spent carbons, spent catalysts, soils, and sediments.

METHOD SUMMARY This method is used to determine 60 volatile organic compounds in a variety of solid waste matrices. It provides GC conditions for the detection of halogenated and aromatic volatile organic compounds. Samples can be analyzed using direct injection or purge-and-trap (EPA Method 5030). Groundwater samples must be analyzed using EPA Method 5030 (where applicable). A temperature program is used with the GC. Detection is achieved by a photoionization detector (PID) and a Hall electrolytic conductivity detector (HECD) in series.

INTERFERENCES Samples can be contaminated by diffusion of volatile organics (particularly chlorofluorocarbons and methylene chloride) through the sample container septum during shipment and storage.

INSTRUMENTATION A GC-equipped with variable-constant differential flow controllers, subambient oven controller, PID and HECD detectors connected with a short piece of uncoated capillary tubing and a data system.

Column: 60 m × 0.75 mm I.D. VOCOL wide-bore capillary column with 1.5 µm film thickness.

PRECISION & ACCURACY MDLs are compound-dependent and vary with purging efficiency and concentration. The applicable concentration range of this method is compound- and instrument-dependent but is approximately 0.1 to 200 µg/L. Analytes that are inefficiently purged from water will not be detected when present at low concentrations, but they can be measured with acceptable accuracy and precision when present in sufficient amounts. The estimated quantitation limit (EQL) for an individual compound is approximately 1 µg/kg (wet weight) for soil/sediment samples, 100 µg/kg (wet weight) for wastes, and 1 µg/L for groundwater. EQLs will be proportionately higher for sample extracts and samples that require dilution or reduced sample size to avoid saturation of the detector.

MULTIPLICATION FACTORS FOR OTHER MATRICES (a)

Matrix	Factor (b)
Groundwater	10
Low-concentration soil	10
Water miscible liquid waste	500
High-concentration soil and sludge	1250
Non-water miscible waste	1250

(a) Sample EQLs are highly matrix-dependent. The EQLs listed herein are provided for guidance and may not always be achievable.
(b) EQL = [Method detection limit] × [Factor]. For non-aqueous samples, the factor is on a wet-weight basis.

SINGLE LABORATORY ACCURACY & PRECISION DATA FOR VOCs IN WATER
This method was tested in a single lab using water spiked at 10 µg/L and the following data was reported:

Recoveries and standard deviations were determined from seven samples and spiked at 10 µg/L of each analyte. Recoveries were determined by the internal standard method. Internal standards were: Fluorobenzene for PID and 2-Bromo-1-chloropropane for HECD.

The average recovery (in percent) for the PID was 99.
The standard deviation of the recovery for the PID was 1.2.
The MDL (in µg/mL) for the PID was 0.05.
The average recovery (in percent) for the HECD was none (no response for this detector).
The standard deviation of the recovery for the HECD was none (no response for this detector)-.
The MDL (in µg/mL) for the HECD was none (no response for this detector).

SAMPLE COLLECTION, PRESERVATION & HANDLING
Volatile organics — Standard 40-mL glass screw-cap VOA vials with Teflon®-faced silicone septum may be used for both liquid and solid matrices. When collecting samples, liquids and solids should be introduced into the vials gently to reduce agitation which might drive off volatile compounds. If there are any air bubbles present the sample must be retaken. Tap slightly as they are filled to try and eliminate as much free air space as possible. The two vials from each sampling locations should be sealed in separate plastic bags to prevent cross-contamination between samples particularly if the sampled waste is suspected of containing high levels of volatile organics.

Semivolatile organics — Containers used to collect samples for the determination of semivolatile organic compounds should be soap and water washed followed by methanol (or isopropanol) rinsing. The sample containers should be of glass or Teflon® and have screw-top covers with Teflon® liners.

Preservation for volatile organics — No preservation is used with concentrated waste samples. With liquid samples containing no residual chlorine, 4 drops of concentrated hydrochloric acid are added and the samples are immediately cooled to 4°C. When liquid samples contain residual chlorine, they are treated as above and, in addition, 4 drops of 4% aqueous sodium thiosulfate are added. Soil, sediment, and sludge samples are only cooled to 4°C.

Preservation for semivolatile organics — No preservation is used with concentrated waste samples. With liquid samples containing no residual chlorine and with soil, sediment, and sludge samples, immediately cooling to 4°C is the only preservation used. When residual chlorine is present then 3 mL of 10% aqueous sodium sulfate is added for each gallon of sample collected, followed by cooling to 4°C.

MHT The holding time for all volatile organics samples is 14 days. Liquid samples must be extracted within 7 days and their extracts analyzed within 40 days. Concentrated waste, soil, sediment, and sludge samples must be extracted within 14 days and their extracts analyzed within 40 days.

SAMPLE PREPARATION Volatile compounds are introduced into the gas chromatograph either by direct injector or purge-and-trap (EPA Method 5030). EPA Method 5030 may be used directly on groundwater samples or low-concentration contaminated soils and sediments. For medium-concentration soils or sediments, methanolic extraction, as described in EPA Method 5030, may be necessary prior to purge-and-trap analysis.

QUALITY CONTROL Calculate surrogate standard recovery on all samples, blanks, and spikes. A trip blank is recommended to check on sampling, storage, and handling contamination. Calibration standards, at a minimum of five concentration levels, are prepared in organic-free reagent water. One of the concentration levels should be at a concentration near, but above, the method detection limit.

A combination of bromochloromethane, 2-bromo-1-chloropropane, 1,4-dichlorobutane, and bromochlorobenzene are recommended as surrogate standards to encompass the range of the temperature program used in this method.

REFERENCE Test Methods for Evaluating Solid Waste, Physical/Chemical Methods, SW-846, 3rd Edition, U.S. EPA, Office of Solid Waste, Washington, DC, EPA Method 8021A, Rev. 1, Nov. 1992.

1,3,5-Trimethylbenzene   **EPA Method 8021**
CAS #108-67-8

TITLE Halogenated Volatile by Gas Chromatography Using Photoionization and Electrolytic Conductivity Detectors in Series: Capillary Column Technique

MATRIX This method is applicable to nearly all types of samples, regardless of water content, including groundwater, aqueous sludges, caustic liquors, acid liquors, waste solvents, oily wastes, mousses, tars, fibrous wastes, polymeric emulsions, filter cakes, spent carbons, spent catalysts, soils, and sediments.

METHOD SUMMARY This method is used to determine 60 volatile organic compounds in a variety of solid waste matrices. It provides GC conditions for the detection of halogenated and aromatic volatile organic compounds. Samples can be analyzed using direct injection or purge-and-trap (EPA Method 5030). Groundwater samples must be analyzed using EPA Method 5030 (where applicable). A temperature program is used with the GC. Detection is achieved by a photoionization detector (PID) and a Hall electrolytic conductivity detector (HECD) in series.

INTERFERENCES Samples can be contaminated by diffusion of volatile organics (particularly chlorofluorocarbons and methylene chloride) through the sample container septum during shipment and storage.

INSTRUMENTATION A GC-equipped with variable-constant differential flow controllers, subambient oven controller, PID and HECD detectors connected with a short piece of uncoated capillary tubing and a data system.

Column: 60 m × 0.75 mm I.D. VOCOL wide-bore capillary column with 1.5 μm film thickness.

PRECISION & ACCURACY MDLs are compound-dependent and vary with purging efficiency and concentration. The applicable concentration range of this method is compound- and instrument-dependent but is approximately 0.1 to 200 μg/L. Analytes that are inefficiently purged from water will not be detected when present at low concentrations, but they can be measured with acceptable accuracy and precision when present in sufficient amounts. The estimated quantitation limit (EQL) for an individual compound is approximately 1 μg/kg (wet weight) for soil/sediment samples, 100 μg/kg (wet weight) for wastes, and 1 μg/L for groundwater. EQLs will be proportionately higher for sample extracts and samples that require dilution or reduced sample size to avoid saturation of the detector.

MULTIPLICATION FACTORS FOR OTHER MATRICES (a)

Matrix	Factor (b)
Groundwater	10
Low-concentration soil	10
Water miscible liquid waste	500
High-concentration soil and sludge	1250
Non-water miscible waste	1250

(a) Sample EQLs are highly matrix-dependent. The EQLs listed herein are provided for guidance and may not always be achievable. (b) EQL = [Method detection limit] × [Factor]. For non-aqueous samples, the factor is on a wet-weight basis.

SINGLE LABORATORY ACCURACY & PRECISION DATA FOR VOCs IN WATER

This method was tested in a single lab using water spiked at 10 μg/L and the following data was reported:

Recoveries and standard deviations were determined from seven samples and spiked at 10 μg/L of each analyte. Recoveries were determined by the internal standard method. Internal standards were: Fluorobenzene for PID and 2-Bromo-1-chloropropane for HECD.

The average recovery (in percent) for the PID was 101.
The standard deviation of the recovery for the PID was 1.4.
The MDL (in μg/mL) for the PID was 0.004.
The average recovery (in percent) for the HECD was none (no response for this detector).
The standard deviation of the recovery for the HECD was none (no response for this detector).
The MDL (in μg/mL) for the HECD was none (no response for this detector).

SAMPLE COLLECTION, PRESERVATION & HANDLING
Volatile organics — Standard 40-mL glass screw-cap VOA vials with Teflon®-faced silicone septum may be used for both liquid and solid matrices. When collecting samples, liquids and solids should be introduced into the vials gently to reduce agitation which might drive off volatile compounds. If there are any air bubbles present the sample must be retaken. Tap slightly as they are filled to try and eliminate as much free air space as possible. The two vials from each sampling locations should be sealed in separate plastic bags to prevent cross-contamination between samples particularly if the sampled waste is suspected of containing high levels of volatile organics.

Semivolatile organics — Containers used to collect samples for the determination of semivolatile organic compounds should be soap and water washed followed by methanol (or isopropanol) rinsing. The sample containers should be of glass or Teflon® and have screw-top covers with Teflon® liners.

Preservation for volatile organics — No preservation is used with concentrated waste samples. With liquid samples containing no residual chlorine, 4 drops of concentrated hydrochloric acid are added and the samples are immediately cooled to 4°C. When liquid samples contain residual chlorine, they are treated as above and, in addition, 4 drops of 4% aqueous sodium thiosulfate are added. Soil, sediment, and sludge samples are only cooled to 4°C.

Preservation for semivolatile organics — No preservation is used with concentrated waste samples. With liquid samples containing no residual chlorine and with soil, sediment, and sludge samples, immediately cooling to 4°C is the only preservation used. When residual chlorine is present then 3 mL of 10% aqueous sodium sulfate is added for each gallon of sample collected, followed by cooling to 4°C.

MHT The holding time for all volatile organics samples is 14 days. Liquid samples must be extracted within 7 days and their extracts analyzed within 40 days. Concentrated waste, soil, sediment, and sludge samples must be extracted within 14 days and their extracts analyzed within 40 days.

SAMPLE PREPARATION Volatile compounds are introduced into the gas chromatograph either by direct injector or purge-and-trap (EPA Method 5030). EPA Method 5030 may be used directly on groundwater samples or low-concentration contaminated soils and sediments. For medium-concentration soils or sediments, methanolic extraction, as described in EPA Method 5030, may be necessary prior to purge-and-trap analysis.

QUALITY CONTROL Calculate surrogate standard recovery on all samples, blanks, and spikes. A trip blank is recommended to check on sampling, storage, and handling contamination. Calibration standards, at a minimum of five concentration levels, are prepared in organic-free reagent water. One of the concentration levels should be at a concentration near, but above, the method detection limit.

A combination of bromochloromethane, 2-bromo-1-chloropropane, 1,4-dichlorobutane, and bromochlorobenzene are recommended as surrogate standards to encompass the range of the temperature program used in this method.

REFERENCE Test Methods for Evaluating Solid Waste, Physical/Chemical Methods, SW-846, 3rd Edition, U.S. EPA, Office of Solid Waste, Washington, DC, EPA Method 8021A, Rev. 1, Nov. 1992.

1,2,4-Trimethylbenzene **EPA Method 8260**
CAS #95-63-6

TITLE Volatile Organic Compounds by GC/MS: Capillary Column Technique

MATRIX This method is applicable to nearly all types of samples, regardless of water content, including groundwater, soils, and sediments.

METHOD SUMMARY Method 8260A covers 58 volatile organic compounds that are introduced into a gas chromatograph by the purge-and-trap method or by direct injection (in limited applications). Zero-grade helium is bubbled through a 5-mL solution at ambient temperature. Purged sample components are trapped in a tube containing suitable sorbent materials. When purging is complete, the sorbent tube is heated and backflushed with helium to desorb trapped sample components. The analytes are desorbed directly to a large bore capillary or cryofocussed on a capillary precolumn before being flash evaporated to a narrow bore capillary for analysis.

INTERFERENCES Major contaminant sources are volatile materials in the lab and impurities in the inert purging gas and in the sorbent trap. Interfering contamination may occur when a sample containing low concentrations of volatile organic compounds is analyzed immediately after a sample containing high concentrations of volatile organic compounds. After analysis of a sample containing high concentrations of volatile organic compounds, one or more calibration blanks should be analyzed to check for cross-contamination. Screening of the samples prior to purge-and-trap GC/MS analysis is highly recommended to prevent contamination of the system. This is especially true for soil and waste samples.

Special precautions must be taken to analyze for methylene chloride. The analytical and sample storage area should be isolated from all atmospheric sources of methylene chloride. All gas chromatography carrier gas lines and purge gas plumbing should be constructed from stainless steel or copper tubing. Laboratory clothing previously exposed to methylene chloride fumes during liquid-liquid extraction procedures can contribute to sample contamination.

Samples can also be contaminated by diffusion of volatile organics (particularly methylene chloride and fluorocarbons) through the septum seal during shipment and storage. A trip blank can serve as a check on such contamination.

INSTRUMENTATION GC/MS with a temperature-programmable chromatograph suitable for splitless injection equipped with variable constant differential flow controllers, a subambient oven controller, a purging device, sorbent trap, a thermal desorption apparatus and a capillary precolumn interface when using cryogenic cooling will be needed. The following GC columns may be used:

Column 1: 60 m × 0.75mm I.D. capillary column coated with VOCOL, 1.5 µm film thickness.
Column 2: 30 m × 0.53mm capillary column coated with DB-624 or VOCOL, 3 µm film thickness.
Column 3: 30 m × 0.32mm I.D. capillary column coated with DB-5 or SE-54, 1-µm film thickness.

PRECISION & ACCURACY This method has been tested in a single lab using spiked water. Using a wide-bore capillary column, water was spiked at concentrations between 0.5 and 10 µg/L. Single lab accuracy and precision data are presented. The MDL actually achieved in a given analysis will vary depending on instrument sensitivity and matrix effects.

The MDL (a) in µg/L was 0.13.
The concentration range in µg/L was 0.5–10.
The mean accuracy (% of true value) was 99.
The precision as relative standard deviation was 8.1.

Note: The MDL is based on a 25-mL sample volume instead of a 5-mL sample volume.

SAMPLING METHOD
Liquid samples — Use a 40-mL glass screw-cap VOA vial with a Teflon®-faced silicone septum that has been prewashed, rinsed with distilled deionized water, and oven dried. If residual chlorine is present, collect the sample in a 4-oz soil VOA container which has been pre-preserved with 4 drops of 10% sodium thiosulfate. Mix gently and transfer the sample to a 40-mL VOA vial. Collect bubble-free samples in duplicate and seal each sample in a separate plastic bag.

Soils, sediments and sludges — Use an 8-oz widemouth glass bottle with Teflon®-faced silicone septum that has been prewashed, rinsed with distilled deionized water, and oven dried. **Do not** heat the septum for more than 1 h. Tap slightly to eliminate any free air space. Collect samples in duplicate and seal each one in a separate plastic bag.

SAMPLE PRESERVATION
Liquid samples — Add 4 drops of concentrated HCL, cool to 4°C and store in a solvent-free refrigerator.

Soils, sediments and sludges — Cool samples to 4°C and store in a solvent-free refrigerator.

MHT The maximum holding time of any sample (liquids, soils, sediments, and sludges) is 14 days.

SAMPLE PREPARATION
Liquid samples — Remove the plunger from a 5-mL syringe and carefully pour the sample into the syringe barrel to just short of overflowing. Replace the syringe plunger and compress the sample. Open the syringe valve and vent any residual air while adjusting the sample volume to 5.0 mL. If there is only one volatile organic analysis (VOA) vial, a second syringe should be filled at this time to protect against possible loss of sample integrity. Add 10 µL of surrogate spiking solution and 10 µL of internal standard spiking solution through the valve bore of the 5-mL syringe, then close the valve. The surrogate and internal standards may be mixed and added as a single spiking solution.

Sediments, soils, and waste samples — All samples of this type should be screened by GC analysis using a headspace method (EPA Method 3810) or the hexadecane extraction and screening method (EPA Method 3820). Use the screening data to determine whether to use the low-concentration method (0.005–1 mg/kg) or the high-concentration method (>1 mg/kg).

Low-concentration method — The low-concentration method is based on purging a heated sediment or soil sample mixed with organic-free reagent water containing the surrogate and internal standards. Analyze all reagent blanks and standards under the same conditions as the samples.

Use a 5-g sample if the expected concentration is <0.1 mg/kg or a 1-g sample for expected concentrations between 0.1 and 1 mg/kg. Mix the contents of the sample container with a narrow metal spatula. Weigh the amount of the sample into a tared purge device. Add the spiked water to the purge device, which contains the weighed amount of sample, and connect the device to the purge-and-trap system.

High-concentration method — This method is based on extracting the sediment or soil with methanol. A waste sample is either extracted or diluted, depending on its solubility in methanol. Wastes that are insoluble in methanol are diluted with reagent tetraglyme or possibly polyethylene glycol (PEG). An aliquot of the extract is added to organic-free reagent water containing surrogate and internal standards. This is purged at ambient temperature. All samples with an expected concentration of >1.0 mg/kg should be analyzed by this method.

Mix the contents of the sample container with a narrow metal spatula. For sediments or soils and solid wastes that are insoluble in methanol, weigh 4 g (wet weight) of sample into a tared 20-mL vial. For waste that is soluble in methanol, tetraglyme, or PEG, weigh 1 g (wet weight) into a tared scintillation vial or culture tube or a 10-mL volumetric flask. Quickly add 9.0 mL of appropriate solvent then add 1.0 mL of a surrogate spiking solution to the vial, cap it, and shake it for 2 min.

METHANOL EXTRACT REQUIRED FOR ANALYSIS OF HIGH-CONCENTRATION SOILS OR SEDIMENTS

Approximate Concentration Range	Volume of Methanol Extract (a)
500–10,000 µg/kg	100 µL
1,000–20,000 µg/kg	50 µL
5,000–100,000 µg/kg	10 µL
25,000–500,000 µg/kg	100 µL of 1/50 dilution (b)

Calculate appropriate dilution factor for concentrations exceeding this table.

(a) The volume of methanol added to 5 mL of water being purged should be kept constant. Therefore, add to the 5-mL syringe whatever volume of methanol is necessary to maintain a volume of 100 µL added to the syringe.
(b) Dilute an aliquot of the methanol extract and then take 100 µL for analysis.

QUALITY CONTROL Demonstrate, through the analysis of a reagent water blank, that interferences from the analytical system, glassware, and reagents are under control. Blank samples should be carried through all stages of the sample preparation and measurement steps. For each analytical batch (up to 20 samples), a reagent blank, matrix spike, and matrix spike duplicate must be analyzed (the frequency of the spikes may be different for different monitoring programs). The blank and spiked samples must be carried through all stages of the sample preparation and measurement steps. QC samples mentioned in the section on Interferences will also be needed as appropriate to those situations.

Matrix spiking standards should be prepared from volatile organic compounds which will be representative of the compounds being investigated. The recommended internal standards are chlorobenzene-d5, 1,4-difluorobenzene, 1,4-dichlorobenzene-d4, and pentafluorobenzene. Using stock standard solutions, prepare secondary dilution standards containing the compounds of interest, either singly or mixed together in methanol. Store them in a vial with no headspace for no more than one week. Surrogates recommended are toluene-d8, 4-bromofluorobenzene, and dibromofluoromethane. Each sample undergoing GC/MS analysis must be spiked with 10 µL of the surrogate spiking solution prior to analysis.

REFERENCE Test Methods for Evaluating Solid Waste (SW-846). U.S. EPA 1983, Method 8260A, Rev. 1, Nov. 1990. Office of Solid Waste, Washington, DC.

1,3,5-Trimethylbenzene **EPA Method 8260**
CAS #108-67-8

TITLE Volatile Organic Compounds by GC/MS: Capillary Column Technique

MATRIX This method is applicable to nearly all types of samples, regardless of water content, including groundwater, soils, and sediments.

METHOD SUMMARY Method 8260A covers 58 volatile organic compounds that are introduced into a gas chromatograph by the purge-and-trap method or by direct injection (in limited applications). Zero-grade helium is bubbled through a 5-mL solution at ambient temperature. Purged sample components are trapped in a tube containing suitable sorbent materials. When purging is complete, the sorbent tube is heated and backflushed with helium to desorb trapped sample components. The analytes are desorbed directly to a large bore capillary or cryofocussed on a capillary precolumn before being flash evaporated to a narrow bore capillary for analysis.

INTERFERENCES Major contaminant sources are volatile materials in the lab and impurities in the inert purging gas and in the sorbent trap. Interfering contamination may occur when a sample containing low concentrations of volatile organic compounds is analyzed immediately after a sample containing high concentrations of volatile organic compounds. After analysis of a sample containing high concentrations of volatile organic compounds, one or more calibration blanks should be analyzed to check for cross-contamination. Screening of the samples prior to purge-and-trap GC/MS analysis is highly recommended to prevent contamination of the system. This is especially true for soil and waste samples.

Special precautions must be taken to analyze for methylene chloride. The analytical and sample storage area should be isolated from all atmospheric sources of methylene chloride. All gas chromatography carrier gas lines and purge gas plumbing should be constructed from stainless steel or copper tubing. Laboratory clothing previously exposed to methylene chloride fumes during liquid-liquid extraction procedures can contribute to sample contamination.

Samples can also be contaminated by diffusion of volatile organics (particularly methylene chloride and fluorocarbons) through the septum seal during shipment and storage. A trip blank can serve as a check on such contamination.

INSTRUMENTATION GC/MS with a temperature-programmable chromatograph suitable for splitless injection equipped with variable constant differential flow controllers, a subambient oven controller, a purging device, sorbent trap, a thermal desorption apparatus and a capillary precolumn interface when using cryogenic cooling will be needed. The following GC columns may be used:

Column 1: 60 m × 0.75mm I.D. capillary column coated with VOCOL, 1.5 µm film thickness.
Column 2: 30 m × 0.53mm capillary column coated with DB-624 or VOCOL, 3 µm film thickness.
Column 3: 30 m × 0.32mm I.D. capillary column coated with DB-5 or SE-54, 1-µm film thickness.

PRECISION & ACCURACY This method has been tested in a single lab using spiked water. Using a wide-bore capillary column, water was spiked at concentrations between 0.5 and 10 µg/L. Single lab accuracy and precision data are presented. The MDL actually achieved in a given analysis will vary depending on instrument sensitivity and matrix effects.

The MDL (a) in µg/L was 0.05.
The concentration range in µg/L was 0.5–10.
The mean accuracy (% of true value) was 92.
The precision as relative standard deviation was 7.4.

Note: The MDL is based on a 25-mL sample volume instead of a 5-mL sample volume.

SAMPLING METHOD
Liquid samples — Use a 40-mL glass screw-cap VOA vial with a Teflon®-faced silicone septum that has been prewashed, rinsed with distilled deionized water, and oven dried. If residual chlorine is present, collect the sample in a 4-oz soil VOA container which has been pre-preserved with 4 drops of 10% sodium thiosulfate. Mix gently and transfer the sample to a 40-mL VOA vial. Collect bubble-free samples in duplicate and seal each sample in a separate plastic bag.

Soils, sediments and sludges — Use an 8-oz widemouth glass bottle with Teflon®-faced silicone septum that has been prewashed, rinsed with distilled deionized water, and oven dried. **Do not** heat the septum for more than 1 h. Tap slightly to eliminate any free air space. Collect samples in duplicate and seal each one in a separate plastic bag.

SAMPLE PRESERVATION
Liquid samples — Add 4 drops of concentrated HCL, cool to 4°C and store in a solvent-free refrigerator.

Soils, sediments and sludges — Cool samples to 4°C and store in a solvent-free refrigerator.

MHT The maximum holding time of any sample (liquids, soils, sediments, and sludges) is 14 days.

SAMPLE PREPARATION

Liquid samples — Remove the plunger from a 5-mL syringe and carefully pour the sample into the syringe barrel to just short of overflowing. Replace the syringe plunger and compress the sample. Open the syringe valve and vent any residual air while adjusting the sample volume to 5.0 mL. If there is only one volatile organic analysis (VOA) vial, a second syringe should be filled at this time to protect against possible loss of sample integrity. Add 10 μL of surrogate spiking solution and 10 μL of internal standard spiking solution through the valve bore of the 5-mL syringe, then close the valve. The surrogate and internal standards may be mixed and added as a single spiking solution.

Sediments, soils, and waste samples — All samples of this type should be screened by GC analysis using a headspace method (EPA Method 3810) or the hexadecane extraction and screening method (EPA Method 3820). Use the screening data to determine whether to use the low-concentration method (0.005–1 mg/kg) or the high-concentration method (>1 mg/kg).

Low-concentration method — The low-concentration method is based on purging a heated sediment or soil sample mixed with organic-free reagent water containing the surrogate and internal standards. Analyze all reagent blanks and standards under the same conditions as the samples.

Use a 5-g sample if the expected concentration is <0.1 mg/kg or a 1-g sample for expected concentrations between 0.1 and 1 mg/kg. Mix the contents of the sample container with a narrow metal spatula. Weigh the amount of the sample into a tared purge device. Add the spiked water to the purge device, which contains the weighed amount of sample, and connect the device to the purge-and-trap system.

High-concentration method — This method is based on extracting the sediment or soil with methanol. A waste sample is either extracted or diluted, depending on its solubility in methanol. Wastes that are insoluble in methanol are diluted with reagent tetraglyme or possibly polyethylene glycol (PEG). An aliquot of the extract is added to organic-free reagent water containing surrogate and internal standards. This is purged at ambient temperature. All samples with an expected concentration of >1.0 mg/kg should be analyzed by this method.

Mix the contents of the sample container with a narrow metal spatula. For sediments or soils and solid wastes that are insoluble in methanol, weigh 4 g (wet weight) of sample into a tared 20-mL vial. For waste that is soluble in methanol, tetraglyme, or PEG, weigh 1 g (wet weight) into a tared scintillation vial or culture tube or a 10-mL volumetric flask. Quickly add 9.0 mL of appropriate solvent then add 1.0 mL of a surrogate spiking solution to the vial, cap it, and shake it for 2 min.

METHANOL EXTRACT REQUIRED FOR ANALYSIS OF HIGH-CONCENTRATION SOILS OR SEDIMENTS

Approximate Concentration Range	Volume of Methanol Extract (a)
500–10,000 μg/kg	100 μL
1,000–20,000 μg/kg	50 μL
5,000–100,000 μg/kg	10 μL
25,000–500,000 μg/kg	100 μL of 1/50 dilution (b)

Calculate appropriate dilution factor for concentrations exceeding this table.

(a) The volume of methanol added to 5 mL of water being purged should be kept constant. Therefore, add to the 5-mL syringe whatever volume of methanol is necessary to maintain a volume of 100 μL added to the syringe.
(b) Dilute an aliquot of the methanol extract and then take 100 μL for analysis.

QUALITY CONTROL Demonstrate, through the analysis of a reagent water blank, that interferences from the analytical system, glassware, and reagents are under control. Blank samples should be carried through all stages of the sample preparation and measurement steps. For each analytical batch (up to 20 samples), a reagent blank, matrix spike, and matrix spike duplicate must be analyzed (the frequency of the spikes may be different for different monitoring programs). The blank and spiked samples must be carried through all stages of the sample preparation and measurement steps. QC samples mentioned in the section on Interferences will also be needed as appropriate to those situations.

Matrix spiking standards should be prepared from volatile organic compounds which will be representative of the compounds being investigated. The recommended internal standards are chlorobenzene-d5, 1,4-difluorobenzene, 1,4-dichlorobenzene-d4, and pentafluorobenzene. Using stock standard solutions, prepare secondary dilution standards containing the compounds of interest, either singly or mixed together in methanol. Store them in a vial with no headspace for no more than one week. Surrogates recommended are toluene-d8, 4-bromofluorobenzene, and dibromofluoromethane. Each sample undergoing GC/MS analysis must be spiked with 10 μL of the surrogate spiking solution prior to analysis.

REFERENCE Test Methods for Evaluating Solid Waste (SW-846). U.S. EPA 1983, Method 8260A, Rev. 1, Nov. 1990. Office of Solid Waste, Washington, DC.

1,2,4-Trimethylbenzene   **EPA Method 503.1**
CAS #95-63-6

TITLE Aromatic & Unsaturated VOCs Water

MATRIX Drinking water (finished or in any treatment stage) and raw source water.

APPLICATION Method covers 28 aromatic and unsaturated VOCs. An inert gas is bubbled through a 5-mL water sample. Purged sample components are trapped in tube of sorbent materials. When purging is complete, sorbent tube is heated and backflushed with inert gas to desorb trapped sample onto a packed GC column.

INTERFERENCES During analysis, major contaminant sources are volatile materials in the lab and impurities in purging gas and sorbent trap. With high and low level samples, there can be carryover contamination. Excess water causes a negative baseline deflection.

INSTRUMENTATION Purge and Trap GC w/photoionization detector. (Two GC columns are recommended); Column 1: 5% SP-1200 and 1.75% Bentone 34 on Supelcoport; Column 2: 1,2,3-tris(2-cyanoethoxy)propane on Chromosorb W.

RANGE 2.2–600 µg/L. (Drinking water)

MDL 0.006 µg/L. in water

PRECISION RSD = 8.7% at 0.40 µg/L conc. 7 samples

ACCURACY Average recovery = 75% at 0.40 µg/L conc.; 7 saples

SAMPLING METHOD Use a 40–120-mL screw-cap vial (prewashed with detergent, rinsed with distilled water and oven dried at 105°C) with a PTFE-faced silicone septum. If residual chlorine is in the water add about 25 mg of ascorbic acid to each vial before sample collection. Collect bubble-free samples.

STABILITY Cool to 4°C; HCl to pH <2.

MHT 14 days.

QUALITY CONTROL As initial demonstration of lab accuracy and precision, analyze 4 to 7 replicates of a lab fortified blank containing the analyte at 0.1–5 µg/L. QC criteria. Collect all samps in duplicate.

REFERENCE Method 503.1, Volatile Aromatic & Unsaturated Organic Compounds in H2O by Purge and Trap GC, EPA 600/4-88/039.

1,3,5-Trimethylbenzene   **EPA Method 503.1**
CAS #108-67-8

TITLE Aromatic & Unsaturated VOCs Water

MATRIX Drinking water (finished or in any treatment stage) and raw source water.

APPLICATION Method covers 28 aromatic and unsaturated VOCs. An inert gas is bubbled through a 5-mL water sample. Purged sample components are trapped in tube of sorbent materials. When purging is complete, sorbent tube is heated and backflushed with inert gas to desorb trapped sample onto a packed GC column.

INTERFERENCES During analysis, major contaminant sources are volatile materials in the lab and impurities in purging gas and sorbent trap. With high and low level samples, there can be carryover contamination. Excess water causes a negative baseline deflection.

INSTRUMENTATION Purge and Trap GC w/photoionization detector. (Two GC columns are recommended); Column 1: 5% SP-1200 and 1.75% Bentone 34 on Supelcoport; Column 2: 1,2,3-tris(2-cyanoethoxy)propane on Chromosorb W.

RANGE 2.2–600 µg/L. (Drinking water)

MDL 0.003 µg/L in water

PRECISION RSD = 8.7% at 0.50 µg/L conc.; 10 samples

ACCURACY Average recovery = 92% at 0.50 µg/L conc.; 10 samples

SAMPLING METHOD Use a 40–120-mL screw-cap vial (prewashed with detergent, rinsed with distilled water and oven dried at 105°C) with a PTFE-faced silicone septum. If residual chlorine is in the water add about 25 mg of ascorbic acid to each vial before sample collection. Collect bubble-free samples.

STABILITY Cool to 4°C; HCl to pH <2.

MHT 14 days.

QUALITY CONTROL As initial demonstration of lab accuracy and precision, analyze 4 to 7 replicates of a lab fortified blank containing the analyte at 0.1–5 µg/L. QC criteria. Collect all samples in duplicate.

REFERENCE Method 503.1, Volatile Aromatic & Unsaturated Organic Compounds in H2O by Purge and Trap GC, EPA 600/4-88/039.

1,3,5-Trinitrobenzene   **EPA Method 8270**
CAS #99-35-4

TITLE Semivolatile Organic Compounds by GC/MS

MATRIX This method is used to determine the concentration of semivolatile organic compounds in extracts prepared from all types of solid waste matrices, soils, and groundwater. Although surface waters are not specifically mentioned, this method should be applicable to water samples from rivers, lakes, etc.

METHOD SUMMARY This method covers 259 semivolatile organic compounds. In very limited applications direct injection of the sample into the GC/MS system may be appropriate, but this results in very high detection limits (approximately 10,000 µg/L). Typically, a 1-L liquid sample, containing surrogate, and matrix spiking standards, is extracted in a continuous extractor first under acid conditions and then under basic conditions. Typically 30 g of a solid sample, containing surrogate, and matrix spiking standards, is extracted ultrasonically. After concentrating the extract to 1 mL it is spiked with 10 µL of an internal standard solution just prior to analysis by GC/MS. The volume injected should contain about 100 ng of base/neutral and 200 ng of acid surrogates (for a 1-µL injection). Analysis is performed by GC/MS using a capillary GC column.

INTERFERENCES Raw GC/MS data from all blanks, samples, and spikes must be evaluated for interferences. Contamination by carryover can occur whenever high-concentration

and low-concentration samples are sequentially analyzed. To reduce carryover, the sample syringe must be rinsed out between samples with solvent. Whenever an unusually concentrated sample is encountered, it should be followed by the analysis of blank solvent to check for cross-contamination.

INSTRUMENTATION A GC/MS and a data system are required. The GC column used is a 30 m × 0.25 mm I.D. (or 0.32 mm I.D.) 1um film thickness silicone-coated fused silica capillary column. A continuous liquid-liquid extractor equipped with Teflon® or glass connection joints and stopcocks requiring no lubrication, a K-D concentrating apparatus, water bath, and an ultrasonic disrupter with a minimum power of 300 W and with pulsing capability are also required.

PRECISION & ACCURACY The estimated quantitation limit (EQL) of Method 8270B for determining an individual compound is approximately 1 mg/kg (wet weight) for soil or sediment samples, 1–200 mg/kg for wastes (dependent on matrix and method of preparation), and 10 μg/L for groundwater samples. EQLs will be proportionately higher for sample extracts that require dilution to avoid saturation of the detector.

The EQL(b) for groundwater in μg/L is 10.
The EQL (a, b) for low concentrations in soil and sediment in μg/kg is not determined.
Accuracy as μg/L is not listed.
Overall precision in μg/L is not listed.

(a) EQLs listed for soil/sediment are based on wet weight. Normally data is reported in a dry-weight basis; therefore, EQLs will be higher based on the % dry weight of each sample. This calculation is based on a 30 g sample and gel permeation chromatography cleanup.
(b) Sample EQLs are highly matrix-dependent. The EQLs are provided for guidance and may not always be achievable.
C = True value for concentration, in μg/L.
X = Average recovery found for measurements of samples containing a concentration of C, in μg/L.

ESTIMATED QUANTITATION LIMIT

Other Matrices	Factor (a)
High-concentration soil and sludges by sonicator	7.5
Non-water miscible waste	75

(a) EQL for other matrices = [EQL for low soil/sediment × [Factor]. This estimated EQL is similar to an EPA "Practical Quantitation Limit."

SAMPLING METHOD
Liquid samples — Use a 1 or 2½ gallon amber glass bottle with a screw-top Teflon®-lined cover that has been prewashed with detergent and rinsed with distilled water and methanol (or isopropanol).

Soils, sediments, or sludges — Use an 8-oz. widemouth glass with a screw-top Teflon®-lined cover that has been prewashed with detergent and rinsed with distilled water and methanol (or isopropanol).

SAMPLE PRESERVATION
Liquid samples — If residual chlorine is present, add 3 mL of 10% sodium thiosulfate per gallon, cool to 4°C and store in a solvent-free refrigerator until analysis; if chlorine is not present, then eliminate the sodium thiosulfate addition.

Soils, sediments, or sludges — Cool samples to 4°C and store in a solvent-free refrigerator.

MHT Liquid samples must be extracted within 7 days and the extracts analyzed within 40 days. Soils, sediments, or sludges may be stored for a maximum of 14 days and the extracts analyzed within 40 days.

SAMPLE PREPARATION
Liquid samples — Transfer 1 L quantitatively to a continuous extractor. If high concentrations are anticipated, a smaller volume may be used and then diluted with organic-free reagent water to 1 L. Adjust pH, if necessary, to pH <2 using 1:1 (V/V) sulfuric acid. Pipette 1.0 mL of a surrogate standard spiking solution into each sample. For the sample in each analytical batch selected for spiking, add 1.0 mL of a matrix spiking standard. For base/neutral acid analysis, the amount of the surrogates and matrix spiking compounds added to the sample should result in a final concentration of 100 ng/μL of each analyte in the extract to be analyzed (assuming a 1-μL injection). Extract with methylene chloride for 18–24 h. Next, adjust the pH of the aqueous phase to pH >11 using 10 N sodium hydroxide and extract it with methylene chloride again for 18–24 h. Dry the extract through a column containing anhydrous sodium sulfate and concentrate it to 1 mL using a K-D concentrator.

Soils, sediments, or sludges — Use 30 g of sample. Nonporous or wet samples (gummy or clay type) that do not have a free-flowing sandy texture must be mixed with anhydrous sodium sulfate until the sample is free flowing. Add 1 mL of surrogate standards to all samples, spikes, standards, and blanks. For the sample in each analytical batch selected for spiking, add 1.0 mL of a matrix spiking standard. For base/neutral acid analysis, the amount added of the surrogates and matrix spiking compounds should result in a final concentration of 100 ng/μL of each base/neutral analyte and 200 ng/μL of each acid analyte in the extract to be analyzed (assuming a 1-μL injection). Immediately add a 100-mL mixture of 1:1 methylene chloride:acetone and extract the sample ultrasonically for 3 min and then decant or filter the extracts. Repeat the extraction two or more times. Dry the extract using a column with anhydrous sodium sulfate and concentrate it to 1 mL in a K-D concentrator.

QUALITY CONTROL A methylene chloride solution containing 50 ng/μL of decafluorotriphenylphosphine (DFTPP) is used for tuning the GC/MS system each 12-h shift. A system performance check also must be made during every 12-h shift. A standard containing 50 ng/μL each of 4,4'-DDT, pentachlorophenol, and benzidine is required to verify injection port inertness and GC column performance. A calibration standard at mid-concentration, containing each compound of interest, including all required surrogates, must be performed every 12 h during analysis. After the system performance check is met, calibration check compounds (CCCs) are used to check the validity of the initial calibration.

The internal standard responses and retention times in the calibration check standard must be evaluated immediately after

or during data acquisition. If the retention time for any internal standard changes by more than 30 seconds from the last check calibration (12 h), the chromatographic system must be inspected for malfunctions and corrections must be made, as required. If the electron ionization current plot (EICP) area for any of the internal standards changes by a factor of two from the last daily calibration standard check, the mass spectrometer must be inspected for malfunctions and corrections must be made, as appropriate.

Demonstrate, through the analysis of a reagent water blank, that interferences from the analytical system, glassware, and reagents are under control. The blank samples should be carried through all stages of the sample preparation and measurement steps. For each analytical batch (up to 20 samples), a reagent blank, matrix spike, and matrix spike duplicate/duplicate must be analyzed (the frequency of the spikes may be different for different monitoring programs). The blank and spiked samples must be carried through all stages of the sample preparation and measurement steps. A QC reference sample concentrate containing each analyte at a concentration of 100 mg/L in methanol is required.

REFERENCE Test Methods for Evaluating Solid Waste (SW-846). U.S. EPA 1983, Method 8270B, Rev. 2, Nov. 1990. Office of Solid Waste, Washington, DC.

Triphenylene **EPA Method 1625**
CAS #217-59-4

TITLE Semivolatile Organic Compounds by Isotope Dilution GC/MS

MATRIX The compounds may be determined in waters, soils, and municipal sludges by this method.

METHOD SUMMARY This method is used to determine 176 semivolatile toxic organic pollutants associated with the CWA (as amended 1987); the RCRA (as amended 1986); the CERCLA (as amended 1986); and other compounds amenable to extraction and analysis by capillary column gas chromatography-mass spectrometry (GC/MS).

Stable isotopically-labeled analogs of the compounds of interest are added to the sample. If the solids content is less than 1%, a 1-L sample is extracted at pH 12–13, then at pH <2 with methylene chloride using continuous extraction techniques.

If the solids content is 30% or less, the sample is diluted to 1% solids with reagent water, homogenized ultrasonically, and extracted at pH 12–13, then at pH <2 with methylene chloride using continuous extraction techniques. If the solids content is greater than 30%, the sample is extracted using ultrasonic techniques.

Each extract is dried over sodium sulfate, concentrated to a volume of 5 mL, cleaned up using GPC, if necessary, and concentrated. Extracts are concentrated to 1 mL if GPC is not performed, and to 0.5 mL if GPC is performed.

An internal standard is added to the extract, and a 1-mL aliquot of the extract is injected into the GC. The compounds are separated by GC and detected by a MS. The labeled compounds serve to correct the variability of the analytical technique.

INTERFERENCES Solvents, reagents, glassware, and other sample processing hardware may yield artifacts and/or elevated baselines causing misinterpretation of chromatograms and spectra. Materials used in the analysis must be demonstrated to be free from interferences under the conditions of analysis by running method blanks initially and with each sample lot (sample started through the extraction process on a given 8-h shift, to a maximum of 20). Specific selection of reagents and purification of solvents by distillation in all glass systems may be required. Glassware and, where possible, reagents are cleaned by solvent rinse and baking at 450°C for 1-h minimum. Interferences coextracted from samples will vary considerably from source to source, depending on the diversity of the site being sampled.

INSTRUMENTATION Major instrumentation includes a GC with a splitless or on-column injection port for capillary column, a MS with 70 eV electron impact ionization, and a data system to collect and record MS data, and process it. A K-D apparatus is used to concentrate extracts.

GC Column: 30 m × 0.25 mm I.D. 5% phenyl, 94% methyl, 1% vinyl silicone bonded phased fused silica capillary column.

PRECISION & ACCURACY The detection limits of the method are usually dependent on the level of interferences rather than instrumental limitations. The limits typify the minimum quantities that can be detected with no interferences present.

The minimum level (in µg/mL) was not listed. This is defined as a minimum level at which the analytical system shall give recognizable mass spectra (background corrected) and acceptable calibration points.

The MDL (in µg/kg) in low solids was not listed and in high solids was not listed; these were determined in digested sludge (low solids) and in filter cake or compost (high solids).

The labeled and native compound initial precision as standard deviation (in µg/L) was not listed.
The labeled and native compound initial accuracy as average recovery (in µg/L) was not listed.

SAMPLE COLLECTION, PRESERVATION & HANDLING Collect samples in glass containers. Aqueous samples which flow freely are collected in refrigerated bottles using automatic sampling equipment. Solid samples are collected as grab samples using widemouth jars. Maintain samples at 0 to 4°C from the time of collection until extraction. If residual chlorine is present in aqueous samples, add 80 mg sodium thiosulfate/L of water. Begin sample extraction within 7 days of collection, and analyze all extracts within 40 days of extraction.

SAMPLE PREPARATION Samples containing 1% solids or less are extracted directly using continuous liquid-liquid extraction techniques. Samples containing 1 to 30% solids are diluted to the 1% level with reagent water and extracted using continuous liquid-liquid extraction techniques. Samples containing greater than 30% solids are extracted using ultrasonic techniques.

Base/neutral extraction — Adjust the pH of the waters in the extractors to 12–13 with 6 N NaOH. Extract with methylene chloride for 24–48 h.

Acid extraction — Adjust the pH of the waters in the extractors to 2 or less using 6 N sulfuric acid. Extract with methylene chloride for 24–48 h.

Ultrasonic extraction of high solids samples — Add anhydrous sodium sulfate to the sample and QC aliquot(s). Add acetone:methylene chloride (1:1) to the sample and mix thoroughly

Concentrate extracts using a K-D apparatus.

QUALITY CONTROL The analyst is permitted to modify this method to improve separations or lower the costs of measurements, provided all performance specifications are met. Analyses of blanks are required to demonstrate freedom from contamination. When results of spikes indicate atypical method performance for samples, the samples are diluted to bring method performance within acceptable limits.

For low solids (aqueous samples), extract, concentrate, and analyze two sets of four 1-L aliquots (8 aliquots total) of the precision and recovery standard. For high solids samples, two sets of four 30-g aliquots of the high solids reference matrix are used.

Spike all samples with labeled compounds to assess method performance. Compute percent recovery of the labeled compounds using the internal standard method. Compare the labeled compound recovery for each compound with the corresponding labeled compound recovery.

Reagent water and high solids reference matrix blanks are analyzed to demonstrate freedom from contamination. Extract and concentrate a 1-L reagent water blank or a high solids reference matrix blank with each sample's lot (samples started through the extraction process on the same 8-h shift, to a maximum of 20 samples).

Field replicates may be collected to determine the precision of the sampling technique, and spiked samples may be required to determine the accuracy of the analysis when the internal standard method is used.

REFERENCE Semivolatile Organic Compounds by Isotope Dilution GC/MS. Office of Water Regulation and Standards, U.S. EPA Industrial Technology Division, Washington, DC, EPA Method 1625, Rev. C, June 1989 (contact W.A. Telliard, U.S. EPA, Office of Water Regulations and Standards, 401 M St., SW, Washington, DC, 20460. Phone: 202-382-7131).

Tripropyleneglycolmethyl ether   **EPA Method 1625**
CAS #20324-33-8

TITLE Semivolatile Organic Compounds by Isotope Dilution GC/MS

MATRIX The compounds may be determined in waters, soils, and municipal sludges by this method.

METHOD SUMMARY This method is used to determine 176 semivolatile toxic organic pollutants associated with the CWA (as amended 1987); the RCRA (as amended 1986); the CERCLA (as amended 1986); and other compounds amenable to extraction and analysis by capillary column gas chromatography-mass spectrometry (GC/MS).

Stable isotopically-labeled analogs of the compounds of interest are added to the sample. If the solids content is less than 1%, a 1-L sample is extracted at pH 12–13, then at pH <2 with methylene chloride using continuous extraction techniques.

If the solids content is 30% or less, the sample is diluted to 1% solids with reagent water, homogenized ultrasonically, and extracted at pH 12–13, then at pH <2 with methylene chloride using continuous extraction techniques. If the solids content is greater than 30%, the sample is extracted using ultrasonic techniques.

Each extract is dried over sodium sulfate, concentrated to a volume of 5 mL, cleaned up using GPC, if necessary, and concentrated. Extracts are concentrated to 1 mL if GPC is not performed, and to 0.5 mL if GPC is performed.

An internal standard is added to the extract, and a 1-mL aliquot of the extract is injected into the GC. The compounds are separated by GC and detected by a MS. The labeled compounds serve to correct the variability of the analytical technique.

INTERFERENCES Solvents, reagents, glassware, and other sample processing hardware may yield artifacts and/or elevated baselines causing misinterpretation of chromatograms and spectra. Materials used in the analysis must be demonstrated to be free from interferences under the conditions of analysis by running method blanks initially and with each sample lot (sample started through the extraction process on a given 8-h shift, to a maximum of 20). Specific selection of reagents and purification of solvents by distillation in all glass systems may be required. Glassware and, where possible, reagents are cleaned by solvent rinse and baking at 450°C for 1-h minimum. Interferences coextracted from samples will vary considerably from source to source, depending on the diversity of the site being sampled.

INSTRUMENTATION Major instrumentation includes a GC with a splitless or on-column injection port for capillary column, a MS with 70 eV electron impact ionization, and a data system to collect and record MS data, and process it. A K-D apparatus is used to concentrate extracts.

GC Column: 30 m × 0.25 mm I.D. 5% phenyl, 94% methyl, 1% vinyl silicone bonded phased fused silica capillary column.

PRECISION & ACCURACY The detection limits of the method are usually dependent on the level of interferences rather than instrumental limitations. The limits typify the minimum quantities that can be detected with no interferences present.

The minimum level (in μg/mL) was not listed. This is defined as a minimum level at which the analytical system shall give recognizable mass spectra (background corrected) and acceptable calibration points.

The MDL (in µg/kg) in low solids was not listed and in high solids was not listed; these were determined in digested sludge (low solids) and in filter cake or compost (high solids).

The labeled and native compound initial precision as standard deviation (in µg/L) was not listed.

The labeled and native compound initial accuracy as average recovery (in µg/L) was not listed.

SAMPLE COLLECTION, PRESERVATION & HANDLING
Collect samples in glass containers. Aqueous samples which flow freely are collected in refrigerated bottles using automatic sampling equipment. Solid samples are collected as grab samples using widemouth jars. Maintain samples at 0 to 4°C from the time of collection until extraction. If residual chlorine is present in aqueous samples, add 80 mg sodium thiosulfate/L of water. Begin sample extraction within 7 days of collection, and analyze all extracts within 40 days of extraction.

SAMPLE PREPARATION Samples containing 1% solids or less are extracted directly using continuous liquid-liquid extraction techniques. Samples containing 1 to 30% solids are diluted to the 1% level with reagent water and extracted using continuous liquid-liquid extraction techniques. Samples containing greater than 30% solids are extracted using ultrasonic techniques.

- Base/neutral extraction — Adjust the pH of the waters in the extractors to 12–13 with 6 N NaOH. Extract with methylene chloride for 24–48 h.
- Acid extraction — Adjust the pH of the waters in the extractors to 2 or less using 6 N sulfuric acid. Extract with methylene chloride for 24–48 h.
- Ultrasonic extraction of high solids samples — Add anhydrous sodium sulfate to the sample and QC aliquot(s). Add acetone:methylene chloride (1:1) to the sample and mix thoroughly

Concentrate extracts using a K-D apparatus.

QUALITY CONTROL The analyst is permitted to modify this method to improve separations or lower the costs of measurements, provided all performance specifications are met. Analyses of blanks are required to demonstrate freedom from contamination. When results of spikes indicate atypical method performance for samples, the samples are diluted to bring method performance within acceptable limits.

For low solids (aqueous samples), extract, concentrate, and analyze two sets of four 1-L aliquots (8 aliquots total) of the precision and recovery standard. For high solids samples, two sets of four 30-g aliquots of the high solids reference matrix are used.

Spike all samples with labeled compounds to assess method performance. Compute percent recovery of the labeled compounds using the internal standard method. Compare the labeled compound recovery for each compound with the corresponding labeled compound recovery.

Reagent water and high solids reference matrix blanks are analyzed to demonstrate freedom from contamination. Extract and concentrate a 1-L reagent water blank or a high solids reference matrix blank with each sample's lot (samples started through the extraction process on the same 8-h shift, to a maximum of 20 samples).

Field replicates may be collected to determine the precision of the sampling technique, and spiked samples may be required to determine the accuracy of the analysis when the internal standard method is used.

REFERENCE Semivolatile Organic Compounds by Isotope Dilution GC/MS. Office of Water Regulation and Standards, U.S. EPA Industrial Technology Division, Washington, DC, EPA Method 1625, Rev. C, June 1989 (contact W.A. Telliard, U.S. EPA, Office of Water Regulations and Standards, 401 M St., SW, Washington, DC, 20460. Phone: 202-382-7131).

Tris(2,3-dibromopropyl) phosphate   **EPA Method 8270**
CAS #126-72-7

TITLE Semivolatile Organic Compounds by GC/MS

MATRIX This method is used to determine the concentration of semivolatile organic compounds in extracts prepared from all types of solid waste matrices, soils, and groundwater. Although surface waters are not specifically mentioned, this method should be applicable to water samples from rivers, lakes, etc.

METHOD SUMMARY This method covers 259 semivolatile organic compounds. In very limited applications direct injection of the sample into the GC/MS system may be appropriate, but this results in very high detection limits (approximately 10,000 µg/L). Typically, a 1-L liquid sample, containing surrogate, and matrix spiking standards, is extracted in a continuous extractor first under acid conditions and then under basic conditions. Typically 30 g of a solid sample, containing surrogate, and matrix spiking standards, is extracted ultrasonically. After concentrating the extract to 1 mL it is spiked with 10 µL of an internal standard solution just prior to analysis by GC/MS. The volume injected should contain about 100 ng of base/neutral and 200 ng of acid surrogates (for a 1-µL injection). Analysis is performed by GC/MS using a capillary GC column.

INTERFERENCES Raw GC/MS data from all blanks, samples, and spikes must be evaluated for interferences. Contamination by carryover can occur whenever high-concentration and low-concentration samples are sequentially analyzed. To reduce carryover, the sample syringe must be rinsed out between samples with solvent. Whenever an unusually concentrated sample is encountered, it should be followed by the analysis of blank solvent to check for cross-contamination.

INSTRUMENTATION A GC/MS and a data system are required. The GC column used is a 30 m × 0.25 mm I.D. (or 0.32 mm I.D.) 1um film thickness silicone-coated fused silica capillary column. A continuous liquid-liquid extractor equipped with Teflon® or glass connection joints and stopcocks requiring no lubrication, a K-D concentrating apparatus, water bath, and an ultrasonic disrupter with a minimum power of 300 W and with pulsing capability are also required.

PRECISION & ACCURACY The estimated quantitation limit (EQL) of Method 8270B for determining an individual compound is approximately 1 mg/kg (wet weight) for soil or sediment samples, 1–200 mg/kg for wastes (dependent on matrix and method of preparation), and 10 μg/L for groundwater samples. EQLs will be proportionately higher for sample extracts that require dilution to avoid saturation of the detector.

The EQL(b) for groundwater in μg/L is 200.
The EQL (a, b) for low concentrations in soil and sediment in μg/kg is not determined.
Accuracy as μg/L is not listed.
Overall precision in μg/L is not listed.

(a) *EQLs listed for soil/sediment are based on wet weight. Normally data is reported in a dry-weight basis; therefore, EQLs will be higher based on the % dry weight of each sample. This calculation is based on a 30 g sample and gel permeation chromatography cleanup.*
(b) *Sample EQLs are highly matrix-dependent. The EQLs are provided for guidance and may not always be achievable.*
C = *True value for concentration, in μg/L.*
X = *Average recovery found for measurements of samples containing a concentration of C, in μg/L.*

ESTIMATED QUANTITATION LIMIT

Other Matrices	Factor (a)
High-concentration soil and sludges by sonicator	7.5
Non-water miscible waste	75

(a) *EQL for other matrices = [EQL for low soil/sediment × [Factor]. This estimated EQL is similar to an EPA "Practical Quantitation Limit."*

SAMPLING METHOD
Liquid samples — Use a 1 or 2½ gallon amber glass bottle with a screw-top Teflon®-lined cover that has been prewashed with detergent and rinsed with distilled water and methanol (or isopropanol).

Soils, sediments, or sludges — Use an 8-oz. widemouth glass with a screw-top Teflon®-lined cover that has been prewashed with detergent and rinsed with distilled water and methanol (or isopropanol).

SAMPLE PRESERVATION
Liquid samples — If residual chlorine is present, add 3 mL of 10% sodium thiosulfate per gallon, cool to 4°C and store in a solvent-free refrigerator until analysis; if chlorine is not present, then eliminate the sodium thiosulfate addition.

Soils, sediments, or sludges — Cool samples to 4°C and store in a solvent-free refrigerator.

MHT Liquid samples must be extracted within 7 days and the extracts analyzed within 40 days. Soils, sediments, or sludges may be stored for a maximum of 14 days and the extracts analyzed within 40 days.

SAMPLE PREPARATION
Liquid samples — Transfer 1 L quantitatively to a continuous extractor. If high concentrations are anticipated, a smaller volume may be used and then diluted with organic-free reagent water to 1 L. Adjust pH, if necessary, to pH <2 using 1:1 (V/V) sulfuric acid. Pipette 1.0 mL of a surrogate standard spiking solution into each sample. For the sample in each analytical batch selected for spiking, add 1.0 mL of a matrix spiking standard. For base/neutral acid analysis, the amount of the surrogates and matrix spiking compounds added to the sample should result in a final concentration of 100 ng/μL of each analyte in the extract to be analyzed (assuming a 1-μL injection). Extract with methylene chloride for 18–24 h. Next, adjust the pH of the aqueous phase to pH >11 using 10 N sodium hydroxide and extract it with methylene chloride again for 18–24 h. Dry the extract through a column containing anhydrous sodium sulfate and concentrate it to 1 mL using a K-D concentrator.

Soils, sediments, or sludges — Use 30 g of sample. Nonporous or wet samples (gummy or clay type) that do not have a free-flowing sandy texture must be mixed with anhydrous sodium sulfate until the sample is free flowing. Add 1 mL of surrogate standards to all samples, spikes, standards, and blanks. For the sample in each analytical batch selected for spiking, add 1.0 mL of a matrix spiking standard. For base/neutral acid analysis, the amount added of the surrogates and matrix spiking compounds should result in a final concentration of 100 ng/μL of each base/neutral analyte and 200 ng/μL of each acid analyte in the extract to be analyzed (assuming a 1-μL injection). Immediately add a 100-mL mixture of 1:1 methylene chloride:acetone and extract the sample ultrasonically for 3 min and then decant or filter the extracts. Repeat the extraction two or more times. Dry the extract using a column with anhydrous sodium sulfate and concentrate it to 1 mL in a K-D concentrator.

QUALITY CONTROL A methylene chloride solution containing 50 ng/μL of decafluorotriphenylphosphine (DFTPP) is used for tuning the GC/MS system each 12-h shift. A system performance check also must be made during every 12-h shift. A standard containing 50 ng/μL each of 4,4′-DDT, pentachlorophenol, and benzidine is required to verify injection port inertness and GC column performance. A calibration standard at mid-concentration, containing each compound of interest, including all required surrogates, must be performed every 12 h during analysis. After the system performance check is met, calibration check compounds (CCCs) are used to check the validity of the initial calibration.

The internal standard responses and retention times in the calibration check standard must be evaluated immediately after or during data acquisition. If the retention time for any internal standard changes by more than 30 seconds from the last check calibration (12 h), the chromatographic system must be inspected for malfunctions and corrections must be made, as required. If the electron ionization current plot (EICP) area for any of the internal standards changes by a factor of two from the last daily calibration standard check, the mass spectrometer must be inspected for malfunctions and corrections must be made, as appropriate.

Demonstrate, through the analysis of a reagent water blank, that interferences from the analytical system, glassware, and reagents are under control. The blank samples should be carried through all stages of the sample preparation and measurement

steps. For each analytical batch (up to 20 samples), a reagent blank, matrix spike, and matrix spike duplicate/duplicate must be analyzed (the frequency of the spikes may be different for different monitoring programs). The blank and spiked samples must be carried through all stages of the sample preparation and measurement steps. A QC reference sample concentrate containing each analyte at a concentration of 100 mg/L in methanol is required.

REFERENCE Test Methods for Evaluating Solid Waste (SW-846). U.S. EPA 1983, Method 8270B, Rev. 2, Nov. 1990. Office of Solid Waste, Washington, DC.

1,3,5-Trithiane	EPA Method 1625
CAS #291-21-4	

TITLE Semivolatile Organic Compounds by Isotope Dilution GC/MS

MATRIX The compounds may be determined in waters, soils, and municipal sludges by this method.

METHOD SUMMARY This method is used to determine 176 semivolatile toxic organic pollutants associated with the CWA (as amended 1987); the RCRA (as amended 1986); the CERCLA (as amended 1986); and other compounds amenable to extraction and analysis by capillary column gas chromatography-mass spectrometry (GC/MS).

Stable isotopically-labeled analogs of the compounds of interest are added to the sample. If the solids content is less than 1%, a 1-L sample is extracted at pH 12–13, then at pH <2 with methylene chloride using continuous extraction techniques.

If the solids content is 30% or less, the sample is diluted to 1% solids with reagent water, homogenized ultrasonically, and extracted at pH 12–13, then at pH <2 with methylene chloride using continuous extraction techniques. If the solids content is greater than 30%, the sample is extracted using ultrasonic techniques.

Each extract is dried over sodium sulfate, concentrated to a volume of 5 mL, cleaned up using GPC, if necessary, and concentrated. Extracts are concentrated to 1 mL if GPC is not performed, and to 0.5 mL if GPC is performed.

An internal standard is added to the extract, and a 1-mL aliquot of the extract is injected into the GC. The compounds are separated by GC and detected by a MS. The labeled compounds serve to correct the variability of the analytical technique.

INTERFERENCES Solvents, reagents, glassware, and other sample processing hardware may yield artifacts and/or elevated baselines causing misinterpretation of chromatograms and spectra. Materials used in the analysis must be demonstrated to be free from interferences under the conditions of analysis by running method blanks initially and with each sample lot (sample started through the extraction process on a given 8-h shift, to a maximum of 20). Specific selection of reagents and purification of solvents by distillation in all glass systems may be required. Glassware and, where possible, reagents are cleaned by solvent rinse and baking at 450°C for 1-h minimum.

Interferences coextracted from samples will vary considerably from source to source, depending on the diversity of the site being sampled.

INSTRUMENTATION Major instrumentation includes a GC with a splitless or on-column injection port for capillary column, a MS with 70 eV electron impact ionization, and a data system to collect and record MS data, and process it. A K-D apparatus is used to concentrate extracts.

GC Column: 30 m × 0.25 mm I.D. 5% phenyl, 94% methyl, 1% vinyl silicone bonded phased fused silica capillary column.

PRECISION & ACCURACY The detection limits of the method are usually dependent on the level of interferences rather than instrumental limitations. The limits typify the minimum quantities that can be detected with no interferences present.

The minimum level (in µg/mL) was not listed. This is defined as a minimum level at which the analytical system shall give recognizable mass spectra (background corrected) and acceptable calibration points.

The MDL (in µg/kg) in low solids was not listed and in high solids was not listed; these were determined in digested sludge (low solids) and in filter cake or compost (high solids).

The labeled and native compound initial precision as standard deviation (in µg/L) was not listed.
The labeled and native compound initial accuracy as average recovery (in µg/L) was not listed.

SAMPLE COLLECTION, PRESERVATION & HANDLING Collect samples in glass containers. Aqueous samples which flow freely are collected in refrigerated bottles using automatic sampling equipment. Solid samples are collected as grab samples using widemouth jars. Maintain samples at 0 to 4°C from the time of collection until extraction. If residual chlorine is present in aqueous samples, add 80 mg sodium thiosulfate/L of water. Begin sample extraction within 7 days of collection, and analyze all extracts within 40 days of extraction.

SAMPLE PREPARATION Samples containing 1% solids or less are extracted directly using continuous liquid-liquid extraction techniques. Samples containing 1 to 30% solids are diluted to the 1% level with reagent water and extracted using continuous liquid-liquid extraction techniques. Samples containing greater than 30% solids are extracted using ultrasonic techniques.

Base/neutral extraction — Adjust the pH of the waters in the extractors to 12–13 with 6 N NaOH. Extract with methylene chloride for 24–48 h.
Acid extraction — Adjust the pH of the waters in the extractors to 2 or less using 6 N sulfuric acid. Extract with methylene chloride for 24–48 h.
Ultrasonic extraction of high solids samples — Add anhydrous sodium sulfate to the sample and QC aliquot(s). Add acetone:methylene chloride (1:1) to the sample and mix thoroughly

Concentrate extracts using a K-D apparatus.

QUALITY CONTROL The analyst is permitted to modify this method to improve separations or lower the costs of measurements, provided all performance specifications are met. Analyses of blanks are required to demonstrate freedom from contamination. When results of spikes indicate atypical method performance for samples, the samples are diluted to bring method performance within acceptable limits.

For low solids (aqueous samples), extract, concentrate, and analyze two sets of four 1-L aliquots (8 aliquots total) of the precision and recovery standard. For high solids samples, two sets of four 30-g aliquots of the high solids reference matrix are used.

Spike all samples with labeled compounds to assess method performance. Compute percent recovery of the labeled compounds using the internal standard method. Compare the labeled compound recovery for each compound with the corresponding labeled compound recovery.

Reagent water and high solids reference matrix blanks are analyzed to demonstrate freedom from contamination. Extract and concentrate a 1-L reagent water blank or a high solids reference matrix blank with each sample's lot (samples started through the extraction process on the same 8-h shift, to a maximum of 20 samples).

Field replicates may be collected to determine the precision of the sampling technique, and spiked samples may be required to determine the accuracy of the analysis when the internal standard method is used.

REFERENCE Semivolatile Organic Compounds by Isotope Dilution GC/MS. Office of Water Regulation and Standards, U.S. EPA Industrial Technology Division, Washington, DC, EPA Method 1625, Rev. C, June 1989 (contact W.A. Telliard, U.S. EPA, Office of Water Regulations and Standards, 401 M St., SW, Washington, DC, 20460. Phone: 202-382-7131).

Turbidity (Nephelometric) EPA Method 180.1

TITLE Physical Properties

MATRIX Drinking, surface, and saline waters.

APPLICATION Date issued 1971. Editorial Rev.s 1974 and 1978. EPA Method based on comparison of light scattered by the sample under defined conditions with a standard reference suspension. The higher the intensity, the higher the turbidity.

INTERFERENCES Presence of floating debris and coarse sediments which settle out rapidly will give low readings. Finely divided air bubbles will affect results in a positive manner. Presence of true color, dissolved substances which absorb light, cause low turbidities.

INSTRUMENTATION Nephelometer (with light source) and one or more photo-electric detectors

RANGE 0–40 NTU.

MDL Not listed.

PRECISION SD = ±0.60 and 1.2 Units at NTU levels of 26 and 75.

ACCURACY Not listed.

SAMPLING METHOD plastic or glass (100 mL).

STABILITY Cool, 4°C.

MHT 48 h.

QUALITY CONTROL Use turbidity free water Sample tubes must be clear, colorless glass. Ntu = nephelometric turbidity units.

REFERENCE Methods for the Chemical Analysis of Water and Wastes, EPA-600/4-79-020, U.S. EPA, EMSL, 1979.

V

Vanadium
CAS #7440-62-2

EPA Method 6010

TITLE Inductively Coupled Plasma-Atomic Emission Spectroscopy

MATRIX This method is applicable to the determination of trace elements, including metals, in groundwater, soils, sludges, sediments, and other solid wastes. All matrices require digestion prior to analysis. The method of standard addition must be used for the analysis of all sample digests unless either serial dilution or matrix spike addition demonstrates it is not required.

METHOD SUMMARY Method 6010 covers 25 elements using ICP analysis. It measures element-emitted light by optical spectrometry. Samples, following an appropriate acid digestion, are nebulized and the resulting aerosol is transported to the plasma torch. Element-specific atomic line emission spectra are produced by a radio-frequency inductively coupled plasma.

INTERFERENCES Interferences may be categorized as spectral or non-spectral. Spectral interferences are caused by overlap of a spectral line from another element, unresolved overlap of molecular band spectra, background contribution from continuous or recombination phenomenon, and stray light from the line emission of high concentration elements. Non-spectral interferences include physical and chemical interferences. Physical interferences are effects associated with the sample nebulization and transport processes. Changes in viscosity and surface tension can cause significant inaccuracies. Chemical interferences include molecular compound formation, ionization effects, and solute vaporization effects. Normally these effects are not significant and can be minimized by careful selection of operating conditions. Chemical interferences are highly dependent on matrix type and the specific analyte element.

INSTRUMENTATION An inductively coupled argon plasma emission spectrometer (ICP) capable of background correction is required.

PRECISION & ACCURACY Detection limits, sensitivity, and optimum ranges of the metals will vary with the matrices and model of the spectrometer. In a single lab evaluation, seven wastes were analyzed for 22 elements. The mean percent relative standard deviation from triplicate analyses for all elements and wastes was 9 ± 2%. The mean percent recovery of spiked elements for all wastes was 93 ± 6%. Spike levels ranged from 100 µg/L to 100 mg/L. The wastes included sludges and industrial wastewaters.

Estimated instrument detection limit in µg/L is 8.
Spiked concentration in µg/L is 70.
Mean reported value in µg/L is 69.
Precision as RSD % is 2.9.

SAMPLING METHOD Samples should be collected in borosilicate glass, linear polyethylene, polypropylene, or Teflon® bottles that have been prewashed with detergent and tap water, and rinsed with 1:1 nitric acid and tap water or 1:1 hydrochloric acid and tap water. Collect at least 2 g of solids and 200 mL of aqueous samples.

SAMPLE PRESERVATION Add nitric acid to make the samples pH <2.

MHT The maximum holding time for properly preserved samples is 6 months.

SAMPLE PREPARATION Preliminary treatment of most matrices is necessary because of the complexity and variability of sample matrices. Water samples that have been prefiltered and acidified will not need acid digestion. Methods for acid digestion of waters for total recoverable or dissolved metals, acid digestions of aqueous samples and extracts for total metals, and acid digestion of sediments, sludges, and soils are summarized below.

Total recoverable or dissolved metals in water — To prepare surface and groundwater samples for determination of total recoverable and dissolved metals, a 100-mL aliquot of well-mixed sample is acidified with concentrated nitric acid and concentrated hydrochloric acid, then heated until the volume is reduced to 15–20 mL. Adjust the final volume to 100 mL with reagent water.

Total metals in aqueous samples, soil and sediment extracts — To prepare aqueous samples, soil and sediment extracts, and wastes that contain suspended solids, a 100-mL aliquot is made acidic with concentrated nitric acid and the solution is evaporated to about 5 mL on a hot plate. Continue heating and adding additional acid until sample digestion is complete, which is usually indicated when the digestate is light in color or does not change in appearance. Evaporate the solution to about 3 mL and cool it and add a small quantity of 1:1 hydrochloric acid (10 mL/100 mL of final solution). Cover the beaker and reflux for 15 min. Wash down the beaker walls and filters or centrifuge the sample to remove silicates and other insoluble material. Filter the sample and adjust the final volume to 100 mL with reagent water and the final acid concentration to 10%.

Sediments, sludges, and soils — To prepare sediments, sludges and soil samples, transfer 1–2 g to a conical beaker and add 10 mL of 1:1 nitric acid, mix the slurry, and cover it with a watch glass. Heat the sample and reflux for 10–15 min without boiling. Allow it to cool, then add 5 mL of concentrated nitric acid and reflux for 30 min. Repeat last step and then allow the solution to evaporate to 5 mL without boiling. Cool and add 2 mL of water and 3 mL of 30% hydrogen peroxide. Cover and place the beaker on the hot plate. Heat and add 30% hydrogen peroxide in 1-mL aliquots with warming until the effervescence is minimal but do not add more than a total of 10 mL of 30% hydrogen peroxide. If the sample is being prepared for the analysis of Ag, Al, As, Ba, Be, Ca, Cd, Co, Cr, Cu, Fe, K, Mg, Mn, Mo, Na, Ni, Os, Pb, Se, Tl, V, and Zn, then add 5 mL of concentrated hydrochloric acid and 10 mL of water and return the covered beaker to a hot plate for 15 min of additional refluxing without boiling. Dilute the sample to a 100 mL volume with

water after cooling and filter or centrifuge to remove particulates.

QUALITY CONTROL Laboratory control samples must be analyzed for each analytical method. A method blank should be analyzed with each batch of samples. The effect of the matrix on method performance must be demonstrated: when appropriate, there should be at least one matrix spike and either one matrix duplicate or one matrix spike duplicate per analytical batch. The bias and precision of the method, as well as the method detection limit for each specific matrix type, must be measured.

Dilute and reanalyze samples that are more concentrated than the linear calibration limit. Employ a minimum of one reagent blank per sample batch to determine if contamination or any memory effects are occurring. Whenever a new or unusual sample matrix is encountered, perform either a serial dilution test or a matrix spike addition test to ensure that neither positive or negative interferences are operating on any of the analyte elements. Check the instrument standardization by verifying calibration every 10 samples using a calibration blank and a check standard.

REFERENCE Test Methods for Evaluating Solid Waste (SW-846). U.S. EPA. 1983. Method 6010, Rev. 0, Sept. 1986. Office of Solid Wastes, Washington, DC.

Vanadium   **EPA Method 200.7**
CAS #7440-62-2

TITLE Inductively Coupled Plasma

MATRIX Dissolved, suspended or (ICP) total element in drinking and surface waters and in domestic and industrial wastewaters.

APPLICATION The method covers the determination of 25 metals. Dissolved elements are determined in filtered and acidified samples after appropriate digestion (which increases dissolved solids). Its primary advantage is that ICP instruments allow simultaneous or rapid sequential determination of many elements in a short time. Samples are first nebulized and the aerosol is transported to a plasma torch in which element specific atomic line emission spectra are produced by a radio frequency inductively coupled plasma. Background correction is required for trace element detection except in the case of line broadning.

INTERFERENCES There are spectral, physical, and chemical interferences. The primary disadvantage of ICP instruments is background radiation from other elements and the plasma gases (spectral interferences). Changes in sample viscosity and surface tension with samples containing high dissolved solids (especially those exceeding 1500 mg/L) or high acid concentrations can cause physical interferences. Ionization effects, solute vaporization and molecular compound formation can cause chemical interferences. Chromium, iron and thallium can cause interference at the 100 mg/L level.

INSTRUMENTATION Inductively coupled argon plasma emission spectroscopy. 292.402 nm wavelength

RANGE Not listed.

MDL 8 µg/L.

PRECISION SD = 1.8% Mean at true value 750 µg/L.

ACCURACY Mean recovery = 93% ± 6% of spiked elements for all wastes.

SAMPLING METHOD Wash sample container with detergent and tap water, rinse with 1 + 1 nitric acid and tap water, then rinse with 1 + 1 hydrochloric acid and tap water, then rinse with deionized, distilled water in that order. Perform any filtration or acid preservation steps when the sample is collected or as soon as possible thereafter.

STABILITY Cool samples to 4°C.

MHT 24 h.

QUALITY CONTROL Mixed calibration standards, an instrument check standard, and an interference check solution are used in addition to a quality control sample. The quality control sample should be prepared in the same acid matrix as the calibration standards at 10 times the instrumental detection limits and in accordance with the instructions provided by the supplier. Furthermore, two types of blanks are required: a calibration blank and a reagent blank.

REFERENCE Method 200.7, U.S. EPA, EMSL-Cincinnati, OH, Nov. 1980

Vanadium   **EPA Method 7911**
CAS #7440-62-2

TITLE Atomic Absorption (AA) Furnace Technique

MATRIX Wastes, mobility procedure, extracts, soils and groundwater.

APPLICATION Aqueous samples, EP extracts, industrial wastes, soils, sludges, sediments, and solid wastes require digestion before analysis. An aliquot of sample is placed in the graphite tube in the furnace and slowly evaporated, charred and atomized. Absorption of lamp radiation during atomization is proportional to vanadium concentration.

INTERFERENCES The furnace technique is subject to chemical interferences. Composition of sample matrix can effect analysis. Vanadium is refractory and prone to form carbides, so memory effects are common. Clean the furnace before and after analysis. Background correction is required.

INSTRUMENTATION Atomic absorption spectrometer. Vanadium hollow cathode lamp or electrodeless discharge lamp. Graphite furnace. Strip-chart recorder.

RANGE 10–200 µg/L

MDL 4 µg/L (318.4 nm wavelength)

PRECISION Not listed.

ACCURACY Not listed.

SAMPLING METHOD Use glass or plastic containers. Collect 200 g of solids and 600 mL of liquid samples.

STABILITY Cool solid samples to 4°C and analyze as soon as possible. Add nitric acid to liquid samples to pH <2.

MHT 6 months.

QUALITY CONTROL At least one duplicate and one spike sample should be run every 20 samples, or with each matrix type to verify method precision. If 20 or more samples are run a day, run a standard (at or near mid-range) every 10 samples.

REFERENCE Method 7911, SW-846, 3rd ed., Nov.1986.

Vanadium   EPA Method 7910
CAS #7440-62-2

TITLE Atomic Absorption (AA)

MATRIX Drinking, surface, and direct aspiration saline waters. Wastewater.

APPLICATION Sample is aspirated and atomized in a flame. A light beam from a vanadium hollow cathode lamp is directed through the flame into monochromator and onto detector. Since wavelength of light beam is specific for vanadium, light energy absorbed by detector is measure of vanadium.

INTERFERENCES The most troublesomee type is chemical,caused by lack of absorption of atoms bound in molecular combination in the flame. High dissolved solids in sample may result in nonatomic absorbance interference. Adding aluminum to samples and standards controls interferences.

INSTRUMENTATION Atomic absorption spectrometer. Vanadium hollow cathode lamp. (318.4 nm wavelength)

RANGE 2–100 mg/L

MDL 0.2 mg/L

PRECISION Standard deviation = ±0.10, 0.10, 0.20 at 2.0, 10, 50 mg V/L

ACCURACY Recoveries = 100, 95, 97% at 2.0, 10, 50 mg V/L

SAMPLING METHOD Use glass or plastic containers. Collect 200 g of solids and 600 mL of liquid samples.

STABILITY Cool solid samples to 4°C and analyze as soon as possible. Add nitric acid to liquid samples to pH <2.

MHT 6 months.

QUALITY CONTROL At least one duplicate and one spike sample should be run every 20 samples or with each matrix type to verify precision of the method. For 20 or more samples per day, verify working standard curve. Run an additional standard at or near mid-range every 10 samples.

REFERENCE Method 7910, SW-846, 3rd ed., Nov.1986.

Vernolate   EPA Method 507
CAS #1929-77-7

TITLE Determination of Nitrogen and Phosphorus-Containing Pesticides in Water by GC/NPD

MATRIX This method is applicable to the determination of certain nitrogen and phosphorus-containing pesticides in finished drinking water and groundwater.

METHOD SUMMARY Method 507 covers 46 nitrogen- and phosphorus-containing pesticides. A 1-L sample is fortified with a surrogate standard, salted, buffered, extracted with methylene chloride, and concentrated; then the solvent is exchanged with methyl tert-butyl ether (MTBE) and concentrated again, and a 2-µL aliquot of a sample extract is injected into a GC system equipped with a selective nitrogen-phosphorus detector and a capillary column for analysis.

INTERFERENCES Method interferences may be caused by contaminants in solvents, reagents, glassware, and other sample processing apparatus. Interfering contamination may occur when a sample containing low concentrations of analytes is analyzed immediately following a sample containing relatively high concentrations. One or more injections of MTBE should be made following the analysis of a sample with high concentrations of analytes to check for analyte carryover. Matrix interferences may be caused by contaminants that are coextracted from the sample. The extent of matrix interferences will vary considerably from source to source, depending upon the water sampled.

INSTRUMENTATION A gas chromatograph system (GC) equipped with a nitrogen-phosphorus detector (NPD) is needed.

Column 1: 30 m × 0.25 mm I.D. DB-5 bonded fused silica column, 0.25 µm film thickness, or equivalent.

Column 2: 30 m × 0.25 mm I.D. DB-1701 bonded fused silica column, 0.25 µm film thickness, or equivalent.

PRECISION & ACCURACY This method has been validated in a single lab and estimated detection limits (EDLs) have been determined for each analyte. Observed detection limits may vary among waters, depending upon the nature of the interferences in the sample matrix and the specific instrumentation used. Analytes that are not separated chromatographically cannot be individually identified and measured unless an alternative technique for identification and quantification exist.

The estimated detection limit (in µg/L) was 0.13. The EDL is defined as either method detection limit or a level of compound in a sample yielding a peak in the final extract with signal-to-noise ratio of approximately 5, whichever value is higher.

The concentration used for these measurements (in µg/L) was 1.3.
The accuracy (as % recovery) was 93.
The precision (% RSD) was 6.

SAMPLING METHOD Grab samples are collected in 1-L glass sample bottles (prewashed with detergent and hot tap

water, rinsed with reagent water, and dried in an oven at 400°C for 1 h) with screw caps lined with PTFE-fluorocarbon.

SAMPLE PRESERVATION Add mercuric chloride to the sample bottle in amounts to produce a concentration of 10 mg/L. If residual chlorine is present, add 80 mg of sodium thiosulfate/L of sample to the sample bottle prior to collection. After collection, seal bottle and shake vigorously for 1 min, then cool the sample to 4°C immediately and store it at 4°C in the dark until extraction.

MHT Maximum holding time of the samples, and in some cases the extracts, is 14 days.

SAMPLE PREPARATION Fortify the sample with 50 µL of the surrogate standard solution, adjust to pH 7 with phosphate buffer, add 100 g NaCl to the sample, and seal and shake to dissolve the salt; then extract with methylene chloride in a separatory funnel or in a mechanical tumbler bottle. Dry the extract by pouring it through a solvent-rinsed drying column containing about 10 cm of anhydrous sodium sulfate. Collect the extract in a Kuderna-Danish (K-D) concentrator and rinse the column with 20–30 mL methylene chloride. Concentrate the extract to about 2 mL and rinse the flask and its lower joint into the concentrator tube with 1 to 2 mL of methyl t-butyl ether (MTBE). Add 5–10 mL of MTBE and concentrate the extract twice (adding more MTBE) to a final volume of 5.0 mL and store it at 4°C until analysis.

NOTE If methylene chloride is not completely removed from the final extract, it may cause detector problems.

QUALITY CONTROL Minimum quality control requirements are initial demonstration of lab capability, determination of surrogate compound recoveries in each sample and blank, monitoring internal standard peak area or height in each sample and blank, analysis of lab reagent blanks, lab fortified samples, lab fortified blanks, and other QC samples. A lab reagent blank is analyzed to demonstrate that all glassware and reagent interferences are under control.

Initial demonstration of capability is fulfilled by analyzing four fortified reagent water samples with the recovery value for each analyte falling within the acceptable range (±30% average recovery). Surrogate recoveries from samples or method blanks must be 70–130%. The internal standard response for any sample chromatogram should not deviate from the daily calibration check standard's internal standard response by more than 30% or lab fortified blanks and sample matrices are used to assess lab performance and analyte recovery, respectively.

If the response for the target analyte peak exceeds the working range of the system, dilute the extract and reanalyze. Alternative techniques such as an alternate detector or second chromatography column should be used to confirm peak identification when sample components are not resolved adequately.

EPA Contact and Hotline For technical questions contact Dr. Baldev Bathija, U.S. EPA, Office of Ground Water and Drinking Water (WH-550D), 401 M St. SW, Washington, DC 20460. Tel. (202) 260-3040. For further information the EPA Safe Drinking Water Hotline may be called at: (800) 426-4791.

REFERENCE Methods for the Determination of Organic Compounds in Drinking Water, EPA/600/4-88/039 (revised July 1991). U.S. EPA Environmental Monitoring Systems Laboratory, Cincinnati, OH, 45268, U.S.A. Available from the National Technical Information Service (NTIS), 5285 Port Royal Road, Springfield, VA 22161; Tel. 800-553-6847. NTIS Order Number is PB91-231480.

Vinyl acetate — EPA Method 1624
CAS #108-05-4

TITLE Volatile Organic Compounds by Isotope Dilution GC/MS

MATRIX Compounds may be determined in waters, soils, and municipal sludges by this method.

METHOD SUMMARY This method is used to determine 58 volatile toxic organic pollutants associated with the CWA (as amended 1987); the RCRA (as amended 1986); the CERCLA (as amended 1986); and other compounds amenable to purge-and-trap gas chromatography-mass spectrometry (GC/MS).

If the solids content is less than 1%, stable isotopically-labeled analogs of the compounds of interest are added to a 5-mL sample and the sample is purged with an inert gas at 20–25°C in a chamber designed for soil or water samples. If the solids content is greater than 1%, 5 mL of reagent water and the labeled compounds are added to a 5-g aliquot of sample and the mixture is purged at 40°C. Compounds that will not purge at 20–25°C or at 40°C are purged at 78–85°C. In the purging process, the volatile compounds are transferred from the aqueous phase into the gaseous phase where they are passed into a sorbent column, and trapped. After purging is completed, the trap is backflushed and heated rapidly to desorb the compounds into a GC. The compounds are separated by the GC and detected by a MS. The labeled compounds serve to correct the variability of the analytical technique.

INTERFERENCES Impurities in the purge gas, organic compounds outgassing from the plumbing upstream of the trap, and solvent vapors in the lab account for most problems. Samples can be contaminated by diffusion of volatile organic compounds (particularly methylene chloride) through the bottle seal during shipment and storage. Contamination by carryover can occur when high-level and low-level samples are analyzed sequentially. When an unusually concentrated sample is encountered, follow it by analysis of a reagent water blank to check for carryover.

INSTRUMENTATION Major equipment includes a GC with linear temperature programming and a glass jet separator as the MS interface, a MS with 70 eV electron impact ionization, and a data system to collect and record response factors.

Column: 2.8 m × 2 mm I.D. glass, packed with 1% SP-1000 on Carbopak B, 60/80 mesh, or equivalent.

PRECISION & ACCURACY The detection limits of the method are usually dependent on the level of interferences rather than instrumental limitations. The method detection

limits were determined in digested sludge (low solids) and in filter cake or compost (high solids).

The MDL (in µg/kg) for low solids is not listed and for high solids is not listed.
Labeled and native compound precision (in µg/L) as standard deviation was not listed.
Labeled and native compound accuracy (in µg/L) as average recovery was not listed.

Acceptance criteria are at 20 µg/L for this compound.

SAMPLE COLLECTION, PRESERVATION & HANDLING Grab samples are collected in glass containers having a total volume greater than 20 mL. Fill and seal each bottle so that no air bubbles are entrapped. Samples are maintained at 0 to 4°C from the time of collection until analysis. If an aqueous sample contains residual chlorine, add sodium thiosulfate preservative (10 mg/40 mL) to the empty sample bottles just prior to shipment to the sample site. All samples must be analyzed within 14 days of collection.

SAMPLE PREPARATION Samples containing less than 1% solids are analyzed directly as aqueous samples. Samples containing 1% solids or greater are analyzed as solid samples utilizing one of two methods, depending on the levels of pollutants, in the sample. Samples containing 1% solids or greater, and low to moderate levels of pollutants are analyzed by purging a known weight of sample added to 5 mL of reagent water. Samples containing 1% solids or greater, and high levels of pollutants, are extracted with methanol, and an aliquot of the methanol extract is added to reagent water and purged.

QUALITY CONTROL A field blank prepared from reagent water and carried through the sampling and handling protocol may serve as a check on contamination from shipment and storage.

The analyst is permitted to modify this method to improve separations or lower the costs of measurements, provided all performance specifications are met. Analyses of blanks are required. When results of spikes indicate atypical method performance for samples, the samples are diluted to bring method performance within acceptable limits. Analyze two sets of four 5-mL aliquots (8 aliquots total) of the aqueous performance standard. Spike all samples with labeled compounds to assess method performance on the sample matrix. Compute the percent recovery of the labeled compounds using the internal standard method. Compare the percent recovery for each compound with the corresponding labeled compound recovery. Reagent water blanks are analyzed to demonstrate freedom from carryover contamination. Field replicates may be collected to determine the precision of the sampling technique, and spiked samples may be required to determine the accuracy of the analysis when the internal method is used.

REFERENCE Volatile Organic Compounds by Isotope Dilution GC/MS. Office of Water Regulation and Standards, U.S. EPA Industrial Technology Division, Washington, DC, EPA Method 1624, Rev. C, June 1989 (contact W.A. Telliard, U.S. EPA, Office of Water Regulations and Standards, 401 M St., SW, Washington, DC, 20460. Phone: 202-382-7131).

Vinyl acetate   **EPA Method 8240**
CAS #108-05-4

TITLE Volatile Organics By GC/MS: Packed Column Technique

MATRIX Nearly all types of sample matarices, regardless of water content, can be analyzed using this method. This includes groundwater, aqueous sludges, caustic liquors, acid liquors, waste solvents, oily wastes, mousses, tars, fibrous wastes, polymetric emulsions, filter cakes, spent carbons, spent catalysts, soils, and sediments.

METHOD SUMMARY Method 8240B covers 80 volatile organic compounds that are introduced into a gas chromatograph by the purge-and-trap method or by direct injection (in limited applications). For the purge-and-trap method an inert gas (zero grade nitrogen or helium) is bubbled through a 5-mL solution at ambient temperature. Purged sample components are trapped in a tube of sorbent materials. When purging is complete, the sorbent tube is heated and backflushed with inert gas to desorb the trapped components onto a GC column.

INTERFERENCES Impurities in the purge gas and from organic compounds outgassing from the plumbing ahead of the trap account for many contamination problems. Interferences purged or coextracted from the samples will vary considerably from source to source. Cross-contamination can occur whenever high-level and low-level samples are analyzed sequentially. Whenever an unusually concentrated sample is analyzed, it should be followed by the analysis of organic-free reagent water to check for cross-contamination. Samples also can be contaminated by diffusion of volatile organics (particularly methylene chloride and fluorocarbons) through the septum seal into the sample during shipment and storage. A trip blank can serve as a check on such contamination. The lab where volatile analysis is performed and also the refrigerated storage area should be completely free of solvents.

INSTRUMENTATION A gas chromatograph/mass spectrometry/data system (GC/MS) equipped with a 6 ft × 0.1 in I.D. glass column packed with 1% SP-1000 on Carbopack-B (60/80 mesh) is required. Also needed is a 5-mL purging device, a sorbent trap, and a thermal desorption apparatus.

PRECISION & ACCURACY This method is reported to have been tested by 15 laboratories using organic-free reagent water, drinking water, surface water, and industrial wastewaters (not specified) fortified at six concentrations over the range 5–600 µg/L.

Sample estimated quantitation limits (EQLs) are highly matrix-dependent. The EQLs listed may not always be achievable. EQLs listed for soils or sediments are based on wet weight. Normally, data is reported on a dry-weight basis; therefore, EQLs will be higher, based on the percent dry weight of each sample. Note that EQLs are even more variable than MDLs and that they are highly variable depending on the matrix being analyzed.

EQL in groundwater in µg/L was 50.
EQL in low soil or sediment in µg/kg was 50.
Accuracy (a) in µg/L was not listed.

Precision (b) in µg/L was not listed.

(a) *Average recovery found for measurements of samples containing a concentration of C, in µg/L.*

(b) *Overall precision found for measurements of samples with average recovery X for samples containing a concentration of C in µg/L.*

X = *Average recovery found for measurement of samples containing a concentration of C in µg/L.*

MULTIPLICATION FACTORS FOR OTHER MATRICES

Other Matrices	Factor (a)
Waste miscible liquid waste	50
High-concentration soil and sludge	125
Non-water miscible waste	500

(a) *EQL = [EQL for low soil sediment X [Factor]. For non-aqueous samples, the factor is on a wet-weight basis.*

SAMPLING METHOD

Liquid samples — Use a 40-mL glass screw-cap VOA vial with a Teflon®-faced silicone septum that has been prewashed, rinsed with distilled deionized water, and oven dried. However, if residual chlorine is present, collect sample in a 40-oz. soil VOA container which has been pre-preserved with 4 drops of 10% sodium thiosulfate, mix gently, and then transfer the sample to a 40-mL VOA vial. Collect bubble-free samples in duplicate and seal them in separate plastic bags.

Soils or sediments, and sludges — Use an 8-oz. widemouth glass bottle with a Teflon®-faced silicone septum that has been prewashed with detergent, rinsed with distilled deionized water, and oven dried. Tap slightly to eliminate free air space. Collect samples in duplicate and seal them in separate plastic bags.

SAMPLE PRESERVATION

Liquid samples — Add 4 drops of concentrated HCL and immediately cool samples to 4°C and store in a solvent-free refrigerator.

Soils or sediments, and sludges — Cool samples to 4°C and store in a solvent-free refrigerator.

MHT Maximum holding time is 14 days from the date of sample collection.

SAMPLE PREPARATION

Liquid samples — Remove the plunger from a 5-mL syringe and carefully pour the sample into the syringe barrel to just short of overflowing. Replace the syringe plunger and compress the sample. Open the syringe valve and vent any residual air while adjusting the sample volume to 5.0 mL. If there is only one volatile organic analysis (VOA) vial, a second syringe should be filled at this time to protect against possible loss of sample integrity. Add 10 µL of surrogate spiking solution and 10 µL of internal standard spiking solution through the valve bore of the 5-mL syringe, then close the valve. The surrogate and internal standards may be mixed and added as a single spiking solution.

Sediments, soils, and waste samples — All samples of this type should be screened by GC analysis using a headspace method (EPA Method 3810) or the hexadecane extraction and screening method (EPA Method 3820). Use the screening data to determine whether to use the low-concentration method (0.005–1 mg/kg) or the high-concentration method (>1 mg/kg).

Low-concentration method — The low-concentration method is based on purging a heated sediment or soil sample mixed with organic-free reagent water containing the surrogate and internal standards. Analyze all reagent blanks and standards under the same conditions as the samples.

Use a 5-g sample if the expected concentration is <0.1 mg/kg or a 1-g sample for expected concentrations between 0.1 and 1 mg/kg. Mix the contents of the sample container with a narrow metal spatula. Weigh the amount of the sample into a tared purge device. Add the spiked water to the purge device, which contains the weighed amount of sample, and connect the device to the purge-and-trap system.

High-concentration method — This method is based on extracting the sediment or soil with methanol. A waste sample is either extracted or diluted, depending on its solubility in methanol. Wastes that are insoluble in methanol are diluted with reagent tetraglyme or possibly polyethylene glycol (PEG). An aliquot of the extract is added to organic-free reagent water containing surrogate and internal standards. This is purged at ambient temperature. All samples with an expected concentration of >1.0 mg/kg should be analyzed by this method.

Mix the contents of the sample container with a narrow metal spatula. For sediments or soils and solid wastes that are insoluble in methanol, weigh 4 g (wet weight) of sample into a tared 20-mL vial. For waste that is soluble in methanol, tetraglyme, or PEG, weigh 1 g (wet weight) into a tared scintillation vial or culture tube or a 10-mL volumetric flask. Quickly add 9.0 mL of appropriate solvent then add 1.0 mL of a surrogate spiking solution to the vial, cap it, and shake it for 2 min.

METHANOL EXTRACT REQUIRED FOR ANALYSIS OF HIGH-CONCENTRATION SOILS OR SEDIMENTS

Approximate Concentration Range	Volume of Methanol Extract (a)
500–10,000 µg/kg	100 µL
1,000–20,000 µg/kg	50 µL
5,000–100,000 µg/kg	10 µL
25,000–500,000 µg/kg	100 µL of 1/50 dilution (b)

Calculate appropriate dilution factor for concentrations exceeding this table.

(a) *The volume of methanol added to 5 mL of water being purged should be kept constant. Therefore, add to the 5-mL syringe whatever volume of methanol is necessary to maintain a volume of 100 µL added to the syringe.*

(b) *Dilute an aliquot of the methanol extract and then take 100 µL for analysis.*

QUALITY CONTROL Demonstrate, through the analysis of a reagent water blank, that interferences from the analytical system, glassware, and reagents are under control. Blank samples should be carried through all stages of the sample preparation and measurement steps. For each analytical batch (up to 20 samples), a reagent blank, matrix spike, and matrix spike duplicate must be analyzed (the frequency of the spikes may be different for different monitoring programs). The blank and spiked samples must be carried through all stages of the sample preparation and measurement steps. QC samples mentioned in the section on Interferences will also be needed as appropriate to those situations.

REFERENCE Test Methods for Evaluating Solid Waste (SW-846). U.S. EPA. 1983. Method 8240B, Rev. 2, Nov. 1990. Office of Solid Wastes, Washington, DC.

Vinyl chloride   **EPA Method 1624**
CAS #75-01-4

TITLE Volatile Organic Compounds by Isotope Dilution GC/MS

MATRIX Compounds may be determined in waters, soils, and municipal sludges by this method.

METHOD SUMMARY This method is used to determine 58 volatile toxic organic pollutants associated with the CWA (as amended 1987); the RCRA (as amended 1986); the CERCLA (as amended 1986); and other compounds amenable to purge-and-trap gas chromatography-mass spectrometry (GC/MS).

If the solids content is less than 1%, stable isotopically-labeled analogs of the compounds of interest are added to a 5-mL sample and the sample is purged with an inert gas at 20–25°C in a chamber designed for soil or water samples. If the solids content is greater than 1%, 5 mL of reagent water and the labeled compounds are added to a 5-g aliquot of sample and the mixture is purged at 40°C. Compounds that will not purge at 20–25°C or at 40°C are purged at 78–85°C. In the purging process, the volatile compounds are transferred from the aqueous phase into the gaseous phase where they are passed into a sorbent column, and trapped. After purging is completed, the trap is backflushed and heated rapidly to desorb the compounds into a GC. The compounds are separated by the GC and detected by a MS. The labeled compounds serve to correct the variability of the analytical technique.

INTERFERENCES Impurities in the purge gas, organic compounds outgassing from the plumbing upstream of the trap, and solvent vapors in the lab account for most problems. Samples can be contaminated by diffusion of volatile organic compounds (particularly methylene chloride) through the bottle seal during shipment and storage. Contamination by carryover can occur when high-level and low-level samples are analyzed sequentially. When an unusually concentrated sample is encountered, follow it by analysis of a reagent water blank to check for carryover.

INSTRUMENTATION Major equipment includes a GC with linear temperature programming and a glass jet separator as the MS interface, a MS with 70 eV electron impact ionization, and a data system to collect and record response factors.

Column: 2.8 m × 2 mm I.D. glass, packed with 1% SP-1000 on Carbopak B, 60/80 mesh, or equivalent.

PRECISION & ACCURACY The detection limits of the method are usually dependent on the level of interferences rather than instrumental limitations. The method detection limits were determined in digested sludge (low solids) and in filter cake or compost (high solids).

The MDL (in µg/kg) for low solids is 190 and for high solids is 11.
Background levels of this compound were present in the sludge with low solids, resulting in a higher than expected MDL.
Labeled and native compound precision (in µg/L) as standard deviation was 28.0.
Labeled and native compound accuracy (in µg/L) as average recovery was detected to 59.

Acceptance criteria are at 20 µg/L for this compound.

SAMPLE COLLECTION, PRESERVATION & HANDLING Grab samples are collected in glass containers having a total volume greater than 20 mL. Fill and seal each bottle so that no air bubbles are entrapped. Samples are maintained at 0 to 4°C from the time of collection until analysis. If an aqueous sample contains residual chlorine, add sodium thiosulfate preservative (10 mg/40 mL) to the empty sample bottles just prior to shipment to the sample site. All samples must be analyzed within 14 days of collection.

SAMPLE PREPARATION Samples containing less than 1% solids are analyzed directly as aqueous samples. Samples containing 1% solids or greater are analyzed as solid samples utilizing one of two methods, depending on the levels of pollutants, in the sample. Samples containing 1% solids or greater, and low to moderate levels of pollutants are analyzed by purging a known weight of sample added to 5 mL of reagent water. Samples containing 1% solids or greater, and high levels of pollutants, are extracted with methanol, and an aliquot of the methanol extract is added to reagent water and purged.

QUALITY CONTROL A field blank prepared from reagent water and carried through the sampling and handling protocol may serve as a check on contamination from shipment and storage.

The analyst is permitted to modify this method to improve separations or lower the costs of measurements, provided all performance specifications are met. Analyses of blanks are required. When results of spikes indicate atypical method performance for samples, the samples are diluted to bring method performance within acceptable limits. Analyze two sets of four 5-mL aliquots (8 aliquots total) of the aqueous performance standard. Spike all samples with labeled compounds to assess method performance on the sample matrix. Compute the percent recovery of the labeled compounds using the internal standard method. Compare the percent recovery for each compound with the corresponding labeled compound recovery. Reagent water blanks are analyzed to demonstrate freedom from carryover contamination. Field replicates may be collected to determine the precision of the sampling technique, and spiked samples may be required to determine the accuracy of the analysis when the internal method is used.

REFERENCE Volatile Organic Compounds by Isotope Dilution GC/MS. Office of Water Regulation and Standards, U.S. EPA Industrial Technology Division, Washington, DC, EPA Method 1624, Rev. C, June 1989 (contact W.A. Telliard, U.S.

EPA, Office of Water Regulations and Standards, 401 M St., SW, Washington, DC, 20460. Phone: 202-382-7131).

Vinyl Chloride — EPA Method 502
CAS #75-01-4

TITLE Volatile Organic Compounds in Water By Purge and Trap Capillary Column Gas Chromatography with Photoionization and Electrolytic Conductivity Detectors in Series. U.S. EPA Method 502.2, Rev. 2.0, 1989.

MATRIX Drinking water and raw source water. The latter should include most surface water and groundwater sources.

METHOD SUMMARY This method covers 60 volatile organic compounds that contain halogen atoms and/or that are aromatic. An inert gas (zero grade nitrogen or helium) is bubbled through a 25-mL or a 5-mL water sample (depending on the expected concentration of the analytes). Purged sample components are trapped in a tube of sorbent materials. When purging is complete, the sorbent tube is heated and backflushed with helium to desorb the trapped sample onto a capillary GC column. The column is temperature programmed to separate the method analytes which are then detected with a photoionization detector (PID) and a Hall electrolytic conductivity (HECD) placed in series. The PID is selective for aromatic compounds and the HECD is selective for halogenated compounds.

INTERFERENCES Impurities in the purge gas and from organic compounds outgassing from the plumbing ahead of the trap account for many contamination problems. Interferences purged or coextracted from the samples will vary considerably from source to source, depending upon the particular sample or extract being tested. Cross-contamination can occur whenever high-level and low-level samples are analyzed sequentially. Samples also can be contaminated by diffusion of volatile organics (particularly methylene chloride and fluorocarbons) through the septum seal into the sample during shipment and storage. The lab where volatile analysis is performed and also the refrigerated storage area should be completely free of solvents.

INSTRUMENTATION A GC containing a series configuration of a high temperature photoionization detector (PID) equipped with 10.0 eV (nominal) lamp and Hall electrolytic conductivity detector (HECD) is required. Also required is an all-glass 5-mL purging device, a sorbent trap, and a thermal desorption apparatus which is connected to the GC system.

Column 1: VOCOL glass wide-bore capillary column.
Column 2: RTX–502.2 mega-bore capillary column.
Column 3: DB-62 mega-bore capillary column.

PRECISION & ACCURACY Method detection limits are dependent upon the characteristics of the gas chromatographic system used. Analytes that are not separated chromatographically cannot be individually identified and used in the same calibration mixture or water samples unless an alternative technique for identification and quantification, such as mass spectrometry, is used.

Electrolytic conductivity detetor (c) range in µg/L (a) was 0.02–200.
Electrolytic conductivity detetor (c) MDL in µg/L (b) was 0.04.
Electrolytic conductivity detetor (c) accuracy as % recovery was 95.
Electrolytic conductivity detetor (c) precision as % RSD was 5.9.
Photoionization detector (d) range in µg/L (a) was 0.02–200.
Photoionization detector (d) MDL in µg/L (b) was 0.02.
Photoionization detector (d) accuracy as % recovery was 109.
Photoionization detector (d) precision as % RSD was 5.0.

(a) The applicable concentration range of this method is compound, instrument, and matrix-dependent. It is listed as being approximately 0.02 to 200 µg/L but no specific information is provided so caution should be observed.
(b) The method detection limits reports with this method are compound, instrument, and matrix-dependent. The values reported were calculated using reagent water fortified with the corresponding compounds at 10 µg/L and a GC-equipped with a 60 m × 0.75 mm VOLCOL wide bore capillary column with 1.5 µm film thickness and using helium carrier gas.
(c) Recoveries and relative standard deviations were determined from seven samples of reagent water fortified with 10 µg/L of each compound. 2-Bromo-1-chloropropane was used as the internal standard for calculating average recoveries.
(d) Recoveries and relative standard deviations were determined from seven samples of reagent water fortified with 10 µg/L of each compound. Fluorobenzene was used as the internal standard for calculating average recoveries.

SAMPLING METHOD Collect samples using a 40- to 120-mL screw-cap vial (prewashed with detergent, rinsed with distilled water and oven dried at 105°C) with a Teflon®-faced silicone septum. Collect bubble-free samples and place the septum with the Teflon® side down on the water.

SAMPLE PRESERVATION If residual chlorine is present in the water add about 25 mg of ascorbic acid to each vial before samples are collected to remove the chlorine. Add hydrochloric acid to reduce pH to <2, immediately cool samples to 4°C, and store them in a solvent-free refrigerator at 4°C until analysis.

MHT The maximum holding time for samples is 14 days from the time they were collected.

SAMPLE PREPARATION Remove the plungers from two 5-mL syringes and attach a closed syringe valve to each. Warm the sample to room temperature, open the sample bottle, and carefully pour the sample into one of the syringe barrels to just short of overflowing. Replace the syringe plunger, invert the syringe, and compress the sample. Open the syringe valve and vent any residual air while adjusting the sample volume to 5.0 mL. Add 10 µL of the internal calibration standard to the sample through the syringe valve. Close the valve. Fill the second syringe in an identical manner from the same sample bottle. Reserve this second syringe for a reanalysis if necessary.

QUALITY CONTROL As an initial demonstration of lab accuracy and precision, analyze 4 to 7 replicates of a lab fortified blank containing analyte at 0.1–5 µg/L. Collect all samples in duplicate. Surrogate analytes (similar to those of the analytes

of interest), whose concentration is known in every sample, are measured using the same internal standard calibration procedure. Duplicate field reagent water blanks (trip blanks) must be analyzed with each set of samples, lab reagent blanks (method blanks) must be analyzed with each batch of samples processed as a group within a work shift. Also, a single lab-fortified blank that contains each of the analytes of interest should be analyzed with each batch of samples processed as a group within a work shift. A 3- to 5-point calibration curve is needed depending on the calibration range factor required.

EPA CONTACT & HOTLINE For technical questions contact Dr. Baldev Bathija, U.S. EPA, Office of Ground Water and Drinking Water (WH-550D), 401 M St. SW, Washington, DC 20460. Tel. (202) 260-3040. For further information the EPA Safe Drinking Water Hotline may be called at: (800) 426-4791.

REFERENCE Methods for the Determination of Organic Compounds in Drinking Water, EPA/600/4-88/039 (revised July 1991; Final Rule for determination of compliance with the MCL for Total Trihalomethanes under 141.30, in 40 CFR Part 141, Vol. 58, No. 147, Fed. Reg., Tuesday Aug. 3, 1993). U.S. EPA Environmental Monitoring Systems Laboratory, Cincinnati, OH, 45268, U.S.A. Available from the National Technical Information Service (NTIS), 5285 Port Royal Road, Springfield, VA 22161; Tel. 800-553-6847. NTIS Order Number is PB91-231480.

Vinyl chloride EPA Method 524
CAS #75-01-4

TITLE Measurement of Purgeable Organic Compounds in Water by Capillary Column GC/MS.

MATRIX Drinking water and raw source water; the latter should include most surface water and groundwater sources.

METHOD SUMMARY Method 524.2 covers 60 volatile organic compounds. An inert gas (zero grade nitrogen or helium) is bubbled through a 25-mL or a 5-mL water sample (depending on the expected concentration of the analytes). Purged sample components are trapped in a tube of sorbent materials. When purging is complete, the sorbent tube is heated and backflushed with helium to desorb the trapped sample onto a capillary GC column.

INTERFERENCES Impurities in the purge gas and from organic compounds outgassing from the plumbing ahead of the trap account for many contamination problems. Interferences purged or coextracted from the samples will vary considerably from source to source, depending upon the particular sample or extract being tested. Cross-contamination can occur whenever high-level and low-level samples are analyzed sequentially. Samples also can be contaminated by diffusion of volatile organics (particularly methylene chloride and fluorocarbons) through the septum seal into the sample during shipment and storage.

INSTRUMENTATION A GC/MS with a data system equipped with one of the following capillary GC columns:

Column 1: VOCOL glass wide bore capillary column.
Column 2: DB-624 fused silica capillary column.
Column 3: DB-5 fused silica capillary column.

Also required is an all-glass 25 mL or 5-mL purging device, a sorbent trap, and a thermal desorption apparatus which is connected to the GC/MS system.

PRECISION & ACCURACY Method detection limits are compound- and instrument-dependent, and may vary from approximately 0.02–0.35 µg/L. Note in the table below that the "true" concentration range used for accuracy and precision measurements was quite narrow. However, the applicable concentration range of this method is primarily column dependent and is approximately 0.02 to 200 µg/L for the wide-bore thick-film columns. Narrow-bore thin-film columns may have a capacity which limits the range to about 0.02 to 20 µg/L. Analytes that are inefficiently purged from water will not be detected when present at low concentrations, but they can be measured with acceptable accuracy and precision when present in sufficient amounts.

Analytes that are not separated chromatographically, but which have different mass spectra and non-interfering quantification ions, can be identified and measured in the same calibration mixture or water sample. Analytes which have very similar mass spectra cannot be individually identified and measured in the same calibration mixture or water samples unless they have different retention times. Co-eluting compounds with very similar mass spectra, typically many structural isomers, must be reported as an isomeric group or pair.

The range (in µg/L) was 0.5–10.
The Method Detection Limig (in µg/L) was 0.17.
The accuracy (as % recovery) was 98.
The precision (in %) was 6.7.

Note: Data were obtained from 16–31 determinations using a wide-bore capillary column and a jet separator interfaced to a quadrupole mass spectrometer. All analytes were in a reagent water matrix.

SAMPLING METHOD Collect samples using a 40- to 120-mL screw-cap vial (prewashed with detergent, rinsed with distilled water and oven dried at 105°C) with a Teflon®-faced silicone septum. Collect bubble-free samples and place the septum with the Teflon® side down on the water.

SAMPLE PRESERVATION If residual chlorine is present in the water add about 25 mg of ascorbic acid to each vial before samples are collected to remove the chlorine. Add hydrochloric acid to reduce pH to <2, and immediately cool samples to 4°C, and store them in a solvent-free refrigerator at 4°C until analysis.

MHT The maximum holding time for samples is 14 days from the time they were collected.

SAMPLE PREPARATION Remove the plungers from two 25-mL (or 5-mL depending on sample size) syringes and attach a closed syringe valve to each. Warm the sample to room temperature, open the sample bottle, and carefully pour the sample into one of the syringe barrels to just short of overflowing. Replace the syringe plunger, invert the syringe, and compress the sample. Open the syringe valve and vent any residual air

while adjusting the sample volume to 25.0 mL (or 5 mL). For samples and blanks, add 5 μL of the fortification solution containing the internal standard and the surrogates to the sample through the syringe valve. For calibration standards and lab fortified blanks, add 5 μL of the fortification solution containing the internal standard only. Close the valve. Fill the second syringe in an identical manner from the same sample bottle. Reserve this second syringe for a reanalysis if necessary.

QUALITY CONTROL As an initial demonstration of lab accuracy and precision, analyze 4 to 7 replicates of a lab fortified blank containing analyte at 0.2–5 μg/L. Collect all samples in duplicate. Surrogate analytes (similar to those of the analytes of interest), whose concentration is known in every sample, are measured using the same internal standard calibration procedure. Duplicate field reagent water blanks (trip blanks) must be analyzed with each set of samples, lab reagent blanks (method blanks) must be analyzed with each batch of samples processed as a group within a work shift. Also, a single lab-fortified blank that contains each of the analytes of interest should be analyzed with each batch of samples processed as a group within a work shift. A 3- to 5-point calibration curve is needed depending on the calibration range factor required.

EPA CONTACT & HOTLINE For technical questions contact Dr. Baldev Bathija, U.S. EPA, Office of Ground Water and Drinking Water (WH-550D), 401 M St. SW, Washington, DC 20460. Tel. (202) 260-3040. For further information the EPA Safe Drinking Water Hotline may be called at: (800) 426-4791.

REFERENCE Methods for the Determination of Organic Compounds in Drinking Water, EPA/600/4-88/039 (revised July 1991; Final Rule for determination of compliance with the MCL for Total Trihalomethanes under 141.30, in 40 CFR Part 141, Vol. 58, No. 147, Fed. Reg., Tuesday Aug. 3, 1993). U.S. EPA Environmental Monitoring Systems Laboratory, Cincinnati, OH, 45268, U.S.A. Available from the National Technical Information Service (NTIS), 5285 Port Royal Road, Springfield, VA 22161; Tel. 800-553-6847. NTIS Order Number is PB91-231480.

Vinyl chloride   **EPA Method 8021**
CAS #75-01-4

TITLE Halogenated Volatile by Gas Chromatography Using Photoionization and Electrolytic Conductivity Detectors in Series: Capillary Column Technique

MATRIX This method is applicable to nearly all types of samples, regardless of water content, including groundwater, aqueous sludges, caustic liquors, acid liquors, waste solvents, oily wastes, mousses, tars, fibrous wastes, polymeric emulsions, filter cakes, spent carbons, spent catalysts, soils, and sediments.

METHOD SUMMARY This method is used to determine 60 volatile organic compounds in a variety of solid waste matrices. It provides GC conditions for the detection of halogenated and aromatic volatile organic compounds. Samples can be analyzed using direct injection or purge-and-trap (EPA Method 5030). Groundwater samples must be analyzed using EPA Method 5030 (where applicable). A temperature program is used with the GC. Detection is achieved by a photoionization detector (PID) and a Hall electrolytic conductivity detector (HECD) in series.

INTERFERENCES Samples can be contaminated by diffusion of volatile organics (particularly chlorofluorocarbons and methylene chloride) through the sample container septum during shipment and storage.

INSTRUMENTATION A GC-equipped with variable-constant differential flow controllers, subambient oven controller, PID and HECD detectors connected with a short piece of uncoated capillary tubing and a data system.

Column: 60 m × 0.75 mm I.D. VOCOL wide-bore capillary column with 1.5 μm film thickness.

PRECISION & ACCURACY MDLs are compound-dependent and vary with purging efficiency and concentration. The applicable concentration range of this method is compound- and instrument-dependent but is approximately 0.1 to 200 μg/L. Analytes that are inefficiently purged from water will not be detected when present at low concentrations, but they can be measured with acceptable accuracy and precision when present in sufficient amounts. The estimated quantitation limit (EQL) for an individual compound is approximately 1 μg/kg (wet weight) for soil/sediment samples, 100 μg/kg (wet weight) for wastes, and 1 μg/L for groundwater. EQLs will be proportionately higher for sample extracts and samples that require dilution or reduced sample size to avoid saturation of the detector.

MULTIPLICATION FACTORS FOR OTHER MATRICES (a)

Matrix	Factor (b)
Groundwater	10
Low-concentration soil	10
Water miscible liquid waste	500
High-concentration soil and sludge	1250
Non-water miscible waste	1250

(a) Sample EQLs are highly matrix-dependent. The EQLs listed herein are provided for guidance and may not always be achievable. (b) EQL = [Method detection limit] × [Factor]. For non-aqueous samples, the factor is on a wet-weight basis.

SINGLE LABORATORY ACCURACY & PRECISION DATA FOR VOCs IN WATER
This method was tested in a single lab using water spiked at 10 μg/L and the following data was reported:

Recoveries and standard deviations were determined from seven samples and spiked at 10 μg/L of each analyte. Recoveries were determined by the internal standard method. Internal standards were: Fluorobenzene for PID and 2-Bromo-1-chloropropane for HECD.

The average recovery (in percent) for the PID was 109.
The standard deviation of the recovery for the PID was 5.4.
The MDL (in μg/mL) for the PID was 0.02.
The average recovery (in percent) for the HECD was 95.
The standard deviation of the recovery for the HECD was 5.6.
The MDL (in μg/mL) for the HECD was 0.04.

SAMPLE COLLECTION, PRESERVATION & HANDLING

Volatile organics — Standard 40-mL glass screw-cap VOA vials with Teflon®-faced silicone septum may be used for both liquid and solid matrices. When collecting samples, liquids and solids should be introduced into the vials gently to reduce agitation which might drive off volatile compounds. If there are any air bubbles present the sample must be retaken. Tap slightly as they are filled to try and eliminate as much free air space as possible. The two vials from each sampling locations should be sealed in separate plastic bags to prevent cross-contamination between samples particularly if the sampled waste is suspected of containing high levels of volatile organics.

Semivolatile organics — Containers used to collect samples for the determination of semivolatile organic compounds should be soap and water washed followed by methanol (or isopropanol) rinsing. The sample containers should be of glass or Teflon® and have screw-top covers with Teflon® liners.

Preservation for volatile organics — No preservation is used with concentrated waste samples. With liquid samples containing no residual chlorine, 4 drops of concentrated hydrochloric acid are added and the samples are immediately cooled to 4°C. When liquid samples contain residual chlorine, they are treated as above and, in addition, 4 drops of 4% aqueous sodium thiosulfate are added. Soil, sediment, and sludge samples are only cooled to 4°C.

Preservation for semivolatile organics — No preservation is used with concentrated waste samples. With liquid samples containing no residual chlorine and with soil, sediment, and sludge samples, immediately cooling to 4°C is the only preservation used. When residual chlorine is present then 3 mL of 10% aqueous sodium sulfate is added for each gallon of sample collected, followed by cooling to 4°C.

MHT The holding time for all volatile organics samples is 14 days. Liquid samples must be extracted within 7 days and their extracts analyzed within 40 days. Concentrated waste, soil, sediment, and sludge samples must be extracted within 14 days and their extracts analyzed within 40 days.

SAMPLE PREPARATION
Volatile compounds are introduced into the gas chromatograph either by direct injector or purge-and-trap (EPA Method 5030). EPA Method 5030 may be used directly on groundwater samples or low-concentration contaminated soils and sediments. For medium-concentration soils or sediments, methanolic extraction, as described in EPA Method 5030, may be necessary prior to purge-and-trap analysis.

QUALITY CONTROL
Calculate surrogate standard recovery on all samples, blanks, and spikes. A trip blank is recommended to check on sampling, storage, and handling contamination. Calibration standards, at a minimum of five concentration levels, are prepared in organic-free reagent water. One of the concentration levels should be at a concentration near, but above, the method detection limit.

A combination of bromochloromethane, 2-bromo-1-chloropropane, 1,4-dichlorobutane, and bromochlorobenzene are recommended as surrogate standards to encompass the range of the temperature program used in this method.

REFERENCE Test Methods for Evaluating Solid Waste, Physical/Chemical Methods, SW-846, 3rd Edition, U.S. EPA, Office of Solid Waste, Washington, DC, EPA Method 8021A, Rev. 1, Nov. 1992.

Vinyl chloride — EPA Method 8240
CAS #75-01-4

TITLE Volatile Organics By GC/MS: Packed Column Technique

MATRIX Nearly all types of sample matarices, regardless of water content, can be analyzed using this method. This includes groundwater, aqueous sludges, caustic liquors, acid liquors, waste solvents, oily wastes, mousses, tars, fibrous wastes, polymetric emulsions, filter cakes, spent carbons, spent catalysts, soils, and sediments.

METHOD SUMMARY Method 8240B covers 80 volatile organic compounds that are introduced into a gas chromatograph by the purge-and-trap method or by direct injection (in limited applications). For the purge-and-trap method an inert gas (zero grade nitrogen or helium) is bubbled through a 5-mL solution at ambient temperature. Purged sample components are trapped in a tube of sorbent materials. When purging is complete, the sorbent tube is heated and backflushed with inert gas to desorb the trapped components onto a GC column.

INTERFERENCES Impurities in the purge gas and from organic compounds outgassing from the plumbing ahead of the trap account for many contamination problems. Interferences purged or coextracted from the samples will vary considerably from source to source. Cross-contamination can occur whenever high-level and low-level samples are analyzed sequentially. Whenever an unusually concentrated sample is analyzed, it should be followed by the analysis of organic-free reagent water to check for cross-contamination. Samples also can be contaminated by diffusion of volatile organics (particularly methylene chloride and fluorocarbons) through the septum seal into the sample during shipment and storage. A trip blank can serve as a check on such contamination. The lab where volatile analysis is performed and also the refrigerated storage area should be completely free of solvents.

INSTRUMENTATION A gas chromatograph/mass spectrometry/data system (GC/MS) equipped with a 6 ft × 0.1 in I.D. glass column packed with 1% SP-1000 on Carbopack-B (60/80 mesh) is required. Also needed is a 5-mL purging device, a sorbent trap, and a thermal desorption apparatus.

PRECISION & ACCURACY This method is reported to have been tested by 15 laboratories using organic-free reagent water, drinking water, surface water, and industrial wastewaters (not specified) fortified at six concentrations over the range 5–600 µg/L.

Sample estimated quantitation limits (EQLs) are highly matrix-dependent. The EQLs listed may not always be achievable. EQLs listed for soils or sediments are based on wet weight. Normally, data is reported on a dry-weight basis; therefore, EQLs will be higher, based on the percent dry weight of each sample. Note that EQLs are even more variable than MDLs and

that they are highly variable depending on the matrix being analyzed.

EQL in groundwater in µg/L was 10.
EQL in low soil or sediment in µg/kg was 10.
Accuracy (a) in µg/L was 1.00C.
Precision (b) in µg/L was 0.65x.

(a) Average recovery found for measurements of samples containing a concentration of C, in µg/L.
(b) Overall precision found for measurements of samples with average recovery X for samples containing a concentration of C in µg/L.
X = Average recovery found for measurement of samples containing a concentration of C in µg/L.

MULTIPLICATION FACTORS FOR OTHER MATRICES

Other Matrices	Factor (a)
Waste miscible liquid waste	50
High-concentration soil and sludge	125
Non-water miscible waste	500

(a) EQL = [EQL for low soil sediment X [Factor]. For non-aqueous samples, the factor is on a wet-weight basis.

SAMPLING METHOD

Liquid samples — Use a 40-mL glass screw-cap VOA vial with a Teflon®-faced silicone septum that has been prewashed, rinsed with distilled deionized water, and oven dried. However, if residual chlorine is present, collect sample in a 40-oz. soil VOA container which has been pre-preserved with 4 drops of 10% sodium thiosulfate, mix gently, and then transfer the sample to a 40-mL VOA vial. Collect bubble-free samples in duplicate and seal them in separate plastic bags.

Soils or sediments, and sludges — Use an 8-oz. widemouth glass bottle with a Teflon®-faced silicone septum that has been prewashed with detergent, rinsed with distilled deionized water, and oven dried. Tap slightly to eliminate free air space. Collect samples in duplicate and seal them in separate plastic bags.

SAMPLE PRESERVATION

Liquid samples — Add 4 drops of concentrated HCL and immediately cool samples to 4°C and store in a solvent-free refrigerator.

Soils or sediments, and sludges — Cool samples to 4°C and store in a solvent-free refrigerator.

MHT Maximum holding time is 14 days from the date of sample collection.

SAMPLE PREPARATION

Liquid samples — Remove the plunger from a 5-mL syringe and carefully pour the sample into the syringe barrel to just short of overflowing. Replace the syringe plunger and compress the sample. Open the syringe valve and vent any residual air while adjusting the sample volume to 5.0 mL. If there is only one volatile organic analysis (VOA) vial, a second syringe should be filled at this time to protect against possible loss of sample integrity. Add 10 µL of surrogate spiking solution and 10 µL of internal standard spiking solution through the valve bore of the 5-mL syringe, then close the valve. The surrogate and internal standards may be mixed and added as a single spiking solution.

Sediments, soils, and waste samples — All samples of this type should be screened by GC analysis using a headspace method (EPA Method 3810) or the hexadecane extraction and screening method (EPA Method 3820). Use the screening data to determine whether to use the low-concentration method (0.005–1 mg/kg) or the high-concentration method (>1 mg/kg).

Low-concentration method — The low-concentration method is based on purging a heated sediment or soil sample mixed with organic-free reagent water containing the surrogate and internal standards. Analyze all reagent blanks and standards under the same conditions as the samples.

Use a 5-g sample if the expected concentration is <0.1 mg/kg or a 1-g sample for expected concentrations between 0.1 and 1 mg/kg. Mix the contents of the sample container with a narrow metal spatula. Weigh the amount of the sample into a tared purge device. Add the spiked water to the purge device, which contains the weighed amount of sample, and connect the device to the purge-and-trap system.

High-concentration method — This method is based on extracting the sediment or soil with methanol. A waste sample is either extracted or diluted, depending on its solubility in methanol. Wastes that are insoluble in methanol are diluted with reagent tetraglyme or possibly polyethylene glycol (PEG). An aliquot of the extract is added to organic-free reagent water containing surrogate and internal standards. This is purged at ambient temperature. All samples with an expected concentration of >1.0 mg/kg should be analyzed by this method.

Mix the contents of the sample container with a narrow metal spatula. For sediments or soils and solid wastes that are insoluble in methanol, weigh 4 g (wet weight) of sample into a tared 20-mL vial. For waste that is soluble in methanol, tetraglyme, or PEG, weigh 1 g (wet weight) into a tared scintillation vial or culture tube or a 10-mL volumetric flask. Quickly add 9.0 mL of appropriate solvent then add 1.0 mL of a surrogate spiking solution to the vial, cap it, and shake it for 2 min.

METHANOL EXTRACT REQUIRED FOR ANALYSIS OF HIGH-CONCENTRATION SOILS OR SEDIMENTS

Approximate Concentration Range	Volume of Methanol Extract (a)
500–10,000 µg/kg	100 µL
1,000–20,000 µg/kg	50 µL
5,000–100,000 µg/kg	10 µL
25,000–500,000 µg/kg	100 µL of 1/50 dilution (b)

Calculate appropriate dilution factor for concentrations exceeding this table.

(a) The volume of methanol added to 5 mL of water being purged should be kept constant. Therefore, add to the 5-mL syringe whatever volume of methanol is necessary to maintain a volume of 100 µL added to the syringe.
(b) Dilute an aliquot of the methanol extract and then take 100 µL for analysis.

QUALITY CONTROL Demonstrate, through the analysis of a reagent water blank, that interferences from the analytical system, glassware, and reagents are under control. Blank samples should be carried through all stages of the sample preparation and measurement steps. For each analytical batch (up to 20 samples), a reagent blank, matrix spike, and matrix spike duplicate must be analyzed (the frequency of the spikes may be different for different monitoring programs). The blank and spiked samples must be carried through all stages of the sample preparation and measurement steps. QC samples mentioned in the section on Interferences will also be needed as appropriate to those situations.

REFERENCE Test Methods for Evaluating Solid Waste (SW-846). U.S. EPA. 1983. Method 8240B, Rev. 2, Nov. 1990. Office of Solid Wastes, Washington, DC.

Vinyl chloride **EPA Method 8260**
CAS #75-01-4

TITLE Volatile Organic Compounds by GC/MS: Capillary Column Technique

MATRIX This method is applicable to nearly all types of samples, regardless of water content, including groundwater, soils, and sediments.

METHOD SUMMARY Method 8260A covers 58 volatile organic compounds that are introduced into a gas chromatograph by the purge-and-trap method or by direct injection (in limited applications). Zero-grade helium is bubbled through a 5-mL solution at ambient temperature. Purged sample components are trapped in a tube containing suitable sorbent materials. When purging is complete, the sorbent tube is heated and backflushed with helium to desorb trapped sample components. The analytes are desorbed directly to a large bore capillary or cryofocussed on a capillary precolumn before being flash evaporated to a narrow bore capillary for analysis.

INTERFERENCES Major contaminant sources are volatile materials in the lab and impurities in the inert purging gas and in the sorbent trap. Interfering contamination may occur when a sample containing low concentrations of volatile organic compounds is analyzed immediately after a sample containing high concentrations of volatile organic compounds. After analysis of a sample containing high concentrations of volatile organic compounds, one or more calibration blanks should be analyzed to check for cross-contamination. Screening of the samples prior to purge-and-trap GC/MS analysis is highly recommended to prevent contamination of the system. This is especially true for soil and waste samples.

Special precautions must be taken to analyze for methylene chloride. The analytical and sample storage area should be isolated from all atmospheric sources of methylene chloride. All gas chromatography carrier gas lines and purge gas plumbing should be constructed from stainless steel or copper tubing. Laboratory clothing previously exposed to methylene chloride fumes during liquid-liquid extraction procedures can contribute to sample contamination.

Samples can also be contaminated by diffusion of volatile organics (particularly methylene chloride and fluorocarbons) through the septum seal during shipment and storage. A trip blank can serve as a check on such contamination.

INSTRUMENTATION GC/MS with a temperature-programmable chromatograph suitable for splitless injection equipped with variable constant differential flow controllers, a subambient oven controller, a purging device, sorbent trap, a thermal desorption apparatus and a capillary precolumn interface when using cryogenic cooling will be needed. The following GC columns may be used:

Column 1: 60 m × 0.75mm I.D. capillary column coated with VOCOL, 1.5 μm film thickness.
Column 2: 30 m × 0.53mm capillary column coated with DB-624 or VOCOL, 3 μm film thickness.
Column 3: 30 m × 0.32mm I.D. capillary column coated with DB-5 or SE-54, 1-μm film thickness.

PRECISION & ACCURACY This method has been tested in a single lab using spiked water. Using a wide-bore capillary column, water was spiked at concentrations between 0.5 and 10 μg/L. Single lab accuracy and precision data are presented. The MDL actually achieved in a given analysis will vary depending on instrument sensitivity and matrix effects.

The MDL (a) in μg/L was 0.17.
The concentration range in μg/L was 0.5–10.
The mean accuracy (% of true value) was 98.
The precision as relative standard deviation was 6.7.

Note: The MDL is based on a 25-mL sample volume instead of a 5-mL sample volume.

SAMPLING METHOD
Liquid samples — Use a 40-mL glass screw-cap VOA vial with a Teflon®-faced silicone septum that has been prewashed, rinsed with distilled deionized water, and oven dried. If residual chlorine is present, collect the sample in a 4-oz soil VOA container which has been pre-preserved with 4 drops of 10% sodium thiosulfate. Mix gently and transfer the sample to a 40-mL VOA vial. Collect bubble-free samples in duplicate and seal each sample in a separate plastic bag.

Soils, sediments and sludges — Use an 8-oz widemouth glass bottle with Teflon®-faced silicone septum that has been prewashed, rinsed with distilled deionized water, and oven dried. **Do not** heat the septum for more than 1 h. Tap slightly to eliminate any free air space. Collect samples in duplicate and seal each one in a separate plastic bag.

SAMPLE PRESERVATION
Liquid samples — Add 4 drops of concentrated HCL, cool to 4°C and store in a solvent-free refrigerator.

Soils, sediments and sludges — Cool samples to 4°C and store in a solvent-free refrigerator.

MHT The maximum holding time of any sample (liquids, soils, sediments, and sludges) is 14 days.

SAMPLE PREPARATION
Liquid samples — Remove the plunger from a 5-mL syringe and carefully pour the sample into the syringe barrel to just

short of overflowing. Replace the syringe plunger and compress the sample. Open the syringe valve and vent any residual air while adjusting the sample volume to 5.0 mL. If there is only one volatile organic analysis (VOA) vial, a second syringe should be filled at this time to protect against possible loss of sample integrity. Add 10 µL of surrogate spiking solution and 10 µL of internal standard spiking solution through the valve bore of the 5-mL syringe, then close the valve. The surrogate and internal standards may be mixed and added as a single spiking solution.

Sediments, soils, and waste samples — All samples of this type should be screened by GC analysis using a headspace method (EPA Method 3810) or the hexadecane extraction and screening method (EPA Method 3820). Use the screening data to determine whether to use the low-concentration method (0.005–1 mg/kg) or the high-concentration method (>1 mg/kg).

Low-concentration method — The low-concentration method is based on purging a heated sediment or soil sample mixed with organic-free reagent water containing the surrogate and internal standards. Analyze all reagent blanks and standards under the same conditions as the samples.

Use a 5-g sample if the expected concentration is <0.1 mg/kg or a 1-g sample for expected concentrations between 0.1 and 1 mg/kg. Mix the contents of the sample container with a narrow metal spatula. Weigh the amount of the sample into a tared purge device. Add the spiked water to the purge device, which contains the weighed amount of sample, and connect the device to the purge-and-trap system.

High-concentration method — This method is based on extracting the sediment or soil with methanol. A waste sample is either extracted or diluted, depending on its solubility in methanol. Wastes that are insoluble in methanol are diluted with reagent tetraglyme or possibly polyethylene glycol (PEG). An aliquot of the extract is added to organic-free reagent water containing surrogate and internal standards. This is purged at ambient temperature. All samples with an expected concentration of >1.0 mg/kg should be analyzed by this method.

Mix the contents of the sample container with a narrow metal spatula. For sediments or soils and solid wastes that are insoluble in methanol, weigh 4 g (wet weight) of sample into a tared 20-mL vial. For waste that is soluble in methanol, tetraglyme, or PEG, weigh 1 g (wet weight) into a tared scintillation vial or culture tube or a 10-mL volumetric flask. Quickly add 9.0 mL of appropriate solvent then add 1.0 mL of a surrogate spiking solution to the vial, cap it, and shake it for 2 min.

METHANOL EXTRACT REQUIRED FOR ANALYSIS OF HIGH-CONCENTRATION SOILS OR SEDIMENTS

Approximate Concentration Range	Volume of Methanol Extract (a)
500–10,000 µg/kg	100 µL
1,000–20,000 µg/kg	50 µL
5,000–100,000 µg/kg	10 µL
25,000–500,000 µg/kg	100 µL of 1/50 dilution (b)

Calculate appropriate dilution factor for concentrations exceeding this table.

(a) The volume of methanol added to 5 mL of water being purged should be kept constant. Therefore, add to the 5-mL syringe whatever volume of methanol is necessary to maintain a volume of 100 µL added to the syringe.
(b) Dilute an aliquot of the methanol extract and then take 100 µL for analysis.

QUALITY CONTROL Demonstrate, through the analysis of a reagent water blank, that interferences from the analytical system, glassware, and reagents are under control. Blank samples should be carried through all stages of the sample preparation and measurement steps. For each analytical batch (up to 20 samples), a reagent blank, matrix spike, and matrix spike duplicate must be analyzed (the frequency of the spikes may be different for different monitoring programs). The blank and spiked samples must be carried through all stages of the sample preparation and measurement steps. QC samples mentioned in the section on Interferences will also be needed as appropriate to those situations.

Matrix spiking standards should be prepared from volatile organic compounds which will be representative of the compounds being investigated. The recommended internal standards are chlorobenzene-d5, 1,4-difluorobenzene, 1,4-dichlorobenzene-d4, and pentafluorobenzene. Using stock standard solutions, prepare secondary dilution standards containing the compounds of interest, either singly or mixed together in methanol. Store them in a vial with no headspace for no more than one week. Surrogates recommended are toluene-d8, 4-bromofluorobenzene, and dibromofluoromethane. Each sample undergoing GC/MS analysis must be spiked with 10 µL of the surrogate spiking solution prior to analysis.

REFERENCE Test Methods for Evaluating Solid Waste (SW-846). U.S. EPA 1983, Method 8260A, Rev. 1, Nov. 1990. Office of Solid Waste, Washington, DC.

Vinyl chloride   **EPA Method 8260**
CAS #75-01-4

TITLE Volatile Organic Compounds by GC/MS: Capillary Column Technique

MATRIX This method is applicable to nearly all types of samples, regardless of water content, including groundwater, soils, and sediments.

METHOD SUMMARY Method 8260A covers 58 volatile organic compounds that are introduced into a gas chromatograph by the purge-and-trap method or by direct injection (in limited applications). Zero-grade helium is bubbled through a 5-mL solution at ambient temperature. Purged sample components are trapped in a tube containing suitable sorbent materials. When purging is complete, the sorbent tube is heated and backflushed with helium to desorb trapped sample components. The analytes are desorbed directly to a large bore capillary or cryofocussed on a capillary precolumn before being flash evaporated to a narrow bore capillary for analysis.

INTERFERENCES Major contaminant sources are volatile materials in the lab and impurities in the inert purging gas and in the sorbent trap. Interfering contamination may occur when a sample containing low concentrations of volatile organic compounds is analyzed immediately after a sample containing high concentrations of volatile organic compounds. After analysis of a sample containing high concentrations of volatile organic compounds, one or more calibration blanks should be analyzed to check for cross-contamination. Screening of the samples prior to purge-and-trap GC/MS analysis is highly recommended to prevent contamination of the system. This is especially true for soil and waste samples.

Special precautions must be taken to analyze for methylene chloride. The analytical and sample storage area should be isolated from all atmospheric sources of methylene chloride. All gas chromatography carrier gas lines and purge gas plumbing should be constructed from stainless steel or copper tubing. Laboratory clothing previously exposed to methylene chloride fumes during liquid-liquid extraction procedures can contribute to sample contamination.

Samples can also be contaminated by diffusion of volatile organics (particularly methylene chloride and fluorocarbons) through the septum seal during shipment and storage. A trip blank can serve as a check on such contamination.

INSTRUMENTATION GC/MS with a temperature-programmable chromatograph suitable for splitless injection equipped with variable constant differential flow controllers, a subambient oven controller, a purging device, sorbent trap, a thermal desorption apparatus and a capillary precolumn interface when using cryogenic cooling will be needed. The following GC columns may be used:

Column 1: 60 m × 0.75mm I.D. capillary column coated with VOCOL, 1.5 μm film thickness.
Column 2: 30 m × 0.53mm capillary column coated with DB-624 or VOCOL, 3 μm film thickness.
Column 3: 30 m × 0.32mm I.D. capillary column coated with DB-5 or SE-54, 1-μm film thickness.

PRECISION & ACCURACY This method has been tested in a single lab using spiked water. Using a wide-bore capillary column, water was spiked at concentrations between 0.5 and 10 μg/L. Single lab accuracy and precision data are presented. The MDL actually achieved in a given analysis will vary depending on instrument sensitivity and matrix effects.

The MDL (a) in μg/L was 0.17.
The concentration range in μg/L was 0.5–10.
The mean accuracy (% of true value) was 98.
The precision as relative standard deviation was 6.7.

Note: The MDL is based on a 25-mL sample volume instead of a 5-mL sample volume.

SAMPLING METHOD

Liquid samples — Use a 40-mL glass screw-cap VOA vial with a Teflon®-faced silicone septum that has been prewashed, rinsed with distilled deionized water, and oven dried. If residual chlorine is present, collect the sample in a 4-oz soil VOA container which has been pre-preserved with 4 drops of 10% sodium thiosulfate. Mix gently and transfer the sample to a 40-mL VOA vial. Collect bubble-free samples in duplicate and seal each sample in a separate plastic bag.

Soils, sediments and sludges — Use an 8-oz widemouth glass bottle with Teflon®-faced silicone septum that has been prewashed, rinsed with distilled deionized water, and oven dried. **Do not** heat the septum for more than 1 h. Tap slightly to eliminate any free air space. Collect samples in duplicate and seal each one in a separate plastic bag.

SAMPLE PRESERVATION

Liquid samples — Add 4 drops of concentrated HCL, cool to 4°C and store in a solvent-free refrigerator.

Soils, sediments and sludges — Cool samples to 4°C and store in a solvent-free refrigerator.

MHT The maximum holding time of any sample (liquids, soils, sediments, and sludges) is 14 days.

SAMPLE PREPARATION

Liquid samples — Remove the plunger from a 5-mL syringe and carefully pour the sample into the syringe barrel to just short of overflowing. Replace the syringe plunger and compress the sample. Open the syringe valve and vent any residual air while adjusting the sample volume to 5.0 mL. If there is only one volatile organic analysis (VOA) vial, a second syringe should be filled at this time to protect against possible loss of sample integrity. Add 10 μL of surrogate spiking solution and 10 μL of internal standard spiking solution through the valve bore of the 5-mL syringe, then close the valve. The surrogate and internal standards may be mixed and added as a single spiking solution.

Sediments, soils, and waste samples — All samples of this type should be screened by GC analysis using a headspace method (EPA Method 3810) or the hexadecane extraction and screening method (EPA Method 3820). Use the screening data to determine whether to use the low-concentration method (0.005–1 mg/kg) or the high-concentration method (>1 mg/kg).

Low-concentration method — The low-concentration method is based on purging a heated sediment or soil sample mixed with organic-free reagent water containing the surrogate and internal standards. Analyze all reagent blanks and standards under the same conditions as the samples.

Use a 5-g sample if the expected concentration is <0.1 mg/kg or a 1-g sample for expected concentrations between 0.1 and 1 mg/kg. Mix the contents of the sample container with a narrow metal spatula. Weigh the amount of the sample into a tared purge device. Add the spiked water to the purge device, which contains the weighed amount of sample, and connect the device to the purge-and-trap system.

High-concentration method — This method is based on extracting the sediment or soil with methanol. A waste sample is either extracted or diluted, depending on its solubility in methanol. Wastes that are insoluble in methanol are diluted with reagent tetraglyme or possibly polyethylene glycol (PEG). An aliquot of the extract is added to organic-free reagent water containing surrogate and internal standards. This is purged at

ambient temperature. All samples with an expected concentration of >1.0 mg/kg should be analyzed by this method.

Mix the contents of the sample container with a narrow metal spatula. For sediments or soils and solid wastes that are insoluble in methanol, weigh 4 g (wet weight) of sample into a tared 20-mL vial. For waste that is soluble in methanol, tetraglyme, or PEG, weigh 1 g (wet weight) into a tared scintillation vial or culture tube or a 10-mL volumetric flask. Quickly add 9.0 mL of appropriate solvent then add 1.0 mL of a surrogate spiking solution to the vial, cap it, and shake it for 2 min.

METHANOL EXTRACT REQUIRED FOR ANALYSIS OF HIGH-CONCENTRATION SOILS OR SEDIMENTS

Approximate Concentration Range	Volume of Methanol Extract (a)
500–10,000 µg/kg	100 µL
1,000–20,000 µg/kg	50 µL
5,000–100,000 µg/kg	10 µL
25,000–500,000 µg/kg	100 µL of 1/50 dilution (b)

Calculate appropriate dilution factor for concentrations exceeding this table.

(a) The volume of methanol added to 5 mL of water being purged should be kept constant. Therefore, add to the 5-mL syringe whatever volume of methanol is necessary to maintain a volume of 100 µL added to the syringe.
(b) Dilute an aliquot of the methanol extract and then take 100 µL for analysis.

QUALITY CONTROL Demonstrate, through the analysis of a reagent water blank, that interferences from the analytical system, glassware, and reagents are under control. Blank samples should be carried through all stages of the sample preparation and measurement steps. For each analytical batch (up to 20 samples), a reagent blank, matrix spike, and matrix spike duplicate must be analyzed (the frequency of the spikes may be different for different monitoring programs). The blank and spiked samples must be carried through all stages of the sample preparation and measurement steps. QC samples mentioned in the section on Interferences will also be needed as appropriate to those situations.

Matrix spiking standards should be prepared from volatile organic compounds which will be representative of the compounds being investigated. The recommended internal standards are chlorobenzene-d5, 1,4-difluorobenzene, 1,4-dichlorobenzene-d4, and pentafluorobenzene. Using stock standard solutions, prepare secondary dilution standards containing the compounds of interest, either singly or mixed together in methanol. Store them in a vial with no headspace for no more than one week. Surrogates recommended are toluene-d8, 4-bromofluorobenzene, and dibromofluoromethane. Each sample undergoing GC/MS analysis must be spiked with 10 µL of the surrogate spiking solution prior to analysis.

REFERENCE Test Methods for Evaluating Solid Waste (SW-846). U.S. EPA 1983, Method 8260A, Rev. 1, Nov. 1990. Office of Solid Waste, Washington, DC.

Vinyl chloride   **EPA Method 601**
CAS #75-01-4

TITLE Purgeable Halocarbons

MATRIX Wastewater.

APPLICATION Method covers 29 purgeable halocarbons. (Method 624 provides GC/MS conditions appropriate for the qualitative and quantitative confirmation of results). Method describes conditions for a 2nd GC column to confirm measurements made with primary column.

INTERFERENCES Impurities in the purge gas and organic compounds outgassing from the plumbing ahead of the trap. With high- and low-level samples, there can be carryover contamination. Diffusion of volatile organics through the septum seal into the sample.

INSTRUMENTATION GC-equipped with halide-specific detector. (With purge-and-trap unit).

RANGE 8.0–500 µg/L

MDL 0.18 µg/L

PRECISION 0.27X + 0.40 µg/L (overall precision).

ACCURACY 0.97C–0.36 µg/L (as recovery).

SAMPLING METHOD 25-mL glass vial. Teflon®-lined septum.

STABILITY Cool, 4°C, 0.008% Sodium thiosulfate.

MHT 14 days.

QUALITY CONTROL The lab must on an ongoing basis, spike at least 10% of the samples from each sample site being monitored to assess accuracy.

REFERENCE Method 601, *Federal Register* Part VIII 40 CFR Part 136, Oct 26, 1984.

Vinyl chloride   **EPA Method 624**
CAS #75-01-4

TITLE Purgeables

MATRIX Wastewater.

APPLICATION Method covers 31 purgeable organics. An inert gas is bubbled through a 5-mL water sample in a specially designed purging chamber. Here, purgeables are transferred from aqueous to gaseous phase, passed onto a sorbent column, and trapped. Trap is heated and backflushed with inert gas to desorb purgeables onto a GC column, where purgeables are separated.

INTERFERENCES Impurities in the purge gas, organic compounds outgassing from the plumbing ahead of the trap, and solvent vapors in the lab. With high- and low-level samples, there can be carryover contamination.

INSTRUMENTATION GC/MS with purge-and-trap unit.

RANGE 5–600 µg/L

MDL Not determined.

PRECISION 0.65X µg/L (overall precision).

ACCURACY 1.00C µg/L (as recovery).

SAMPLING METHOD 25-mL glass vial. Teflon®-lined septum.

STABILITY Cool, 4°C, 0.008% Sodium thiosulfate.

MHT 14 days.

QUALITY CONTROL The lab must on an ongoing basis, spike at least 5% of the samples from each sample site being monitored to assess accuracy.

REFERENCE Method 624, *Federal Register* Part VIII 40 CFR Part 136, Oct 26, 1984.

Vinyl chloride EPA Method 8010
CAS #75-01-4

TITLE Halogenated Volatile Organics

MATRIX Groundwater, soils, sludges, water miscible liquid wastes, and non-water miscible wastes.

APPLICATION This method is used for the analysis of 39 halogenated VOCs. Samples are analyzed using direct injection or purge-and-trap methods. Groundwater must be analyzed by the purge-and-trap method. The method provides an optional GC column which is used for analyte confirmation and that may help resolve analytes from interferences.

INTERFERENCES There can be carryover contamination with high- and low-level samples. Impurities may come from the purge-and-trap apparatus, organic compounds outgassing from the plumbing ahead of trap, diffusion of VOCs through the sample bottle septum during shipping or storage, or from solvent vapors in the lab.

INSTRUMENTATION GC capable of on-column injections or purge-and-trap sample introduction and a halogen specific detector. Column 1: 8 ft by 0.1 in 1%. SP-1000 on Carbopack-B. Column 2: 6 ft by 0.1 in bonded n-octane on Porasil-C.

RANGE 8–500 µg/L (reagent water).

MDL 0.18 µg/L (reagent water).

PQL FACTORS FOR MULTIPLYING × MDL VALUE

Matrix	Multiplication Factor
Groundwater	10
Low-level soil	10
Water miscible liquid waste	500
High-level soil and sludge	1250
Non-water miscible waste	1250

PRECISION 0.27X + 0.40 µg/L (overall precision).

ACCURACY 0.97C–0.36 µg/L (as recovery).

SAMPLING METHOD For water and liquid samples; use glass 40-mL vials with Teflon®-lined septum caps and collect two vials per sample location with no headspace. For solids and concentrated waste samples; use widemouth glass bottles with Teflon® liners.

STABILITY For concentrated wastes, soils, sediments, or sludges: cool to 4°C. For liquids: add 4 drops of concentrated hydrochloric acid and cool to 4°C.

MHT 14 days.

QUALITY CONTROL Analyze a reagent blank, matrix spike, and matrix spike duplicate/duplicate for each analytical batch (up to 20 samples). Demonstrate the purity of glassware and reagents by analyzing a reagent water method blank. Internal, surrogate, and five concentration level calibration standards are used.

REFERENCE Test Methods for Evaluating Solid Waste (SW-846), U.S. EPA Office of Solid Waste, Washington, DC, Method 8010B, Rev. 2, Nov. 1992.

m-Xylene   **EPA Method 1624**
CAS #108-38-3

TITLE Volatile Organic Compounds by Isotope Dilution GC/MS

MATRIX Compounds may be determined in waters, soils, and municipal sludges by this method.

METHOD SUMMARY This method is used to determine 58 volatile toxic organic pollutants associated with the CWA (as amended 1987); the RCRA (as amended 1986); the CERCLA (as amended 1986); and other compounds amenable to purge-and-trap gas chromatography-mass spectrometry (GC/MS).

If the solids content is less than 1%, stable isotopically-labeled analogs of the compounds of interest are added to a 5-mL sample and the sample is purged with an inert gas at 20–25°C in a chamber designed for soil or water samples. If the solids content is greater than 1%, 5 mL of reagent water and the labeled compounds are added to a 5-g aliquot of sample and the mixture is purged at 40°C. Compounds that will not purge at 20–25°C or at 40°C are purged at 78–85°C. In the purging process, the volatile compounds are transferred from the aqueous phase into the gaseous phase where they are passed into a sorbent column, and trapped. After purging is completed, the trap is backflushed and heated rapidly to desorb the compounds into a GC. The compounds are separated by the GC and detected by a MS. The labeled compounds serve to correct the variability of the analytical technique.

INTERFERENCES Impurities in the purge gas, organic compounds outgassing from the plumbing upstream of the trap, and solvent vapors in the lab account for most problems. Samples can be contaminated by diffusion of volatile organic compounds (particularly methylene chloride) through the bottle seal during shipment and storage. Contamination by carryover can occur when high-level and low-level samples are analyzed sequentially. When an unusually concentrated sample is encountered, follow it by analysis of a reagent water blank to check for carryover.

INSTRUMENTATION Major equipment includes a GC with linear temperature programming and a glass jet separator as the MS interface, a MS with 70 eV electron impact ionization, and a data system to collect and record response factors.

Column: 2.8 m × 2 mm I.D. glass, packed with 1% SP-1000 on Carbopak B, 60/80 mesh, or equivalent.

PRECISION & ACCURACY The detection limits of the method are usually dependent on the level of interferences rather than instrumental limitations. The method detection limits were determined in digested sludge (low solids) and in filter cake or compost (high solids).

The MDL (in µg/kg) for low solids is not listed and for high solids is not listed.

Labeled and native compound precision (in µg/L) as standard deviation was not listed.

Labeled and native compound accuracy (in µg/L) as average recovery was not listed.

Acceptance criteria are at 20 µg/L for this compound.

SAMPLE COLLECTION, PRESERVATION & HANDLING Grab samples are collected in glass containers having a total volume greater than 20 mL. Fill and seal each bottle so that no air bubbles are entrapped. Samples are maintained at 0 to 4°C from the time of collection until analysis. If an aqueous sample contains residual chlorine, add sodium thiosulfate preservative (10 mg/40 mL) to the empty sample bottles just prior to shipment to the sample site. All samples must be analyzed within 14 days of collection.

SAMPLE PREPARATION Samples containing less than 1% solids are analyzed directly as aqueous samples. Samples containing 1% solids or greater are analyzed as solid samples utilizing one of two methods, depending on the levels of pollutants, in the sample. Samples containing 1% solids or greater, and low to moderate levels of pollutants are analyzed by purging a known weight of sample added to 5 mL of reagent water. Samples containing 1% solids or greater, and high levels of pollutants, are extracted with methanol, and an aliquot of the methanol extract is added to reagent water and purged.

QUALITY CONTROL A field blank prepared from reagent water and carried through the sampling and handling protocol may serve as a check on contamination from shipment and storage.

The analyst is permitted to modify this method to improve separations or lower the costs of measurements, provided all performance specifications are met. Analyses of blanks are required. When results of spikes indicate atypical method performance for samples, the samples are diluted to bring method performance within acceptable limits. Analyze two sets of four 5-mL aliquots (8 aliquots total) of the aqueous performance standard. Spike all samples with labeled compounds to assess method performance on the sample matrix. Compute the percent recovery of the labeled compounds using the internal standard method. Compare the percent recovery for each compound with the corresponding labeled compound recovery. Reagent water blanks are analyzed to demonstrate freedom from carryover contamination. Field replicates may be collected to determine the precision of the sampling technique, and spiked samples may be required to determine the accuracy of the analysis when the internal method is used.

REFERENCE Volatile Organic Compounds by Isotope Dilution GC/MS. Office of Water Regulation and Standards, U.S. EPA Industrial Technology Division, Washington, DC, EPA Method 1624, Rev. C, June 1989 (contact W.A. Telliard, U.S. EPA, Office of Water Regulations and Standards, 401 M St., SW, Washington, DC, 20460. Phone: 202-382-7131).

o-Xylene
CAS #95-47-6
EPA Method 1624

TITLE Volatile Organic Compounds by Isotope Dilution GC/MS

MATRIX Compounds may be determined in waters, soils, and municipal sludges by this method.

METHOD SUMMARY This method is used to determine 58 volatile toxic organic pollutants associated with the CWA (as amended 1987); the RCRA (as amended 1986); the CERCLA (as amended 1986); and other compounds amenable to purge-and-trap gas chromatography-mass spectrometry (GC/MS).

If the solids content is less than 1%, stable isotopically-labeled analogs of the compounds of interest are added to a 5-mL sample and the sample is purged with an inert gas at 20–25°C in a chamber designed for soil or water samples. If the solids content is greater than 1%, 5 mL of reagent water and the labeled compounds are added to a 5-g aliquot of sample and the mixture is purged at 40°C. Compounds that will not purge at 20–25°C or at 40°C are purged at 78–85°C. In the purging process, the volatile compounds are transferred from the aqueous phase into the gaseous phase where they are passed into a sorbent column, and trapped. After purging is completed, the trap is backflushed and heated rapidly to desorb the compounds into a GC. The compounds are separated by the GC and detected by a MS. The labeled compounds serve to correct the variability of the analytical technique.

INTERFERENCES Impurities in the purge gas, organic compounds outgassing from the plumbing upstream of the trap, and solvent vapors in the lab account for most problems. Samples can be contaminated by diffusion of volatile organic compounds (particularly methylene chloride) through the bottle seal during shipment and storage. Contamination by carryover can occur when high-level and low-level samples are analyzed sequentially. When an unusually concentrated sample is encountered, follow it by analysis of a reagent water blank to check for carryover.

INSTRUMENTATION Major equipment includes a GC with linear temperature programming and a glass jet separator as the MS interface, a MS with 70 eV electron impact ionization, and a data system to collect and record response factors.

Column: 2.8 m × 2 mm I.D. glass, packed with 1% SP-1000 on Carbopak B, 60/80 mesh, or equivalent.

PRECISION & ACCURACY The detection limits of the method are usually dependent on the level of interferences rather than instrumental limitations. The method detection limits were determined in digested sludge (low solids) and in filter cake or compost (high solids).

- The MDL (in µg/kg) for low solids is not listed and for high solids is not listed.
- Labeled and native compound precision (in µg/L) as standard deviation was not listed.
- Labeled and native compound accuracy (in µg/L) as average recovery was not listed.

Acceptance criteria are at 20 µg/L for this compound.

SAMPLE COLLECTION, PRESERVATION & HANDLING Grab samples are collected in glass containers having a total volume greater than 20 mL. Fill and seal each bottle so that no air bubbles are entrapped. Samples are maintained at 0 to 4°C from the time of collection until analysis. If an aqueous sample contains residual chlorine, add sodium thiosulfate preservative (10 mg/40 mL) to the empty sample bottles just prior to shipment to the sample site. All samples must be analyzed within 14 days of collection.

SAMPLE PREPARATION Samples containing less than 1% solids are analyzed directly as aqueous samples. Samples containing 1% solids or greater are analyzed as solid samples utilizing one of two methods, depending on the levels of pollutants, in the sample. Samples containing 1% solids or greater, and low to moderate levels of pollutants are analyzed by purging a known weight of sample added to 5 mL of reagent water. Samples containing 1% solids or greater, and high levels of pollutants, are extracted with methanol, and an aliquot of the methanol extract is added to reagent water and purged.

QUALITY CONTROL A field blank prepared from reagent water and carried through the sampling and handling protocol may serve as a check on contamination from shipment and storage.

The analyst is permitted to modify this method to improve separations or lower the costs of measurements, provided all performance specifications are met. Analyses of blanks are required. When results of spikes indicate atypical method performance for samples, the samples are diluted to bring method performance within acceptable limits. Analyze two sets of four 5-mL aliquots (8 aliquots total) of the aqueous performance standard. Spike all samples with labeled compounds to assess method performance on the sample matrix. Compute the percent recovery of the labeled compounds using the internal standard method. Compare the percent recovery for each compound with the corresponding labeled compound recovery. Reagent water blanks are analyzed to demonstrate freedom from carryover contamination. Field replicates may be collected to determine the precision of the sampling technique, and spiked samples may be required to determine the accuracy of the analysis when the internal method is used.

REFERENCE Volatile Organic Compounds by Isotope Dilution GC/MS. Office of Water Regulation and Standards, U.S. EPA Industrial Technology Division, Washington, DC, EPA Method 1624, Rev. C, June 1989 (contact W.A. Telliard, U.S. EPA, Office of Water Regulations and Standards, 401 M St., SW, Washington, DC, 20460. Phone: 202-382-7131).

p-Xylene
CAS #106-42-3
EPA Method 1624

TITLE Volatile Organic Compounds by Isotope Dilution GC/MS

MATRIX Compounds may be determined in waters, soils, and municipal sludges by this method.

METHOD SUMMARY This method is used to determine 58 volatile toxic organic pollutants associated with the CWA (as amended 1987); the RCRA (as amended 1986); the CERCLA (as amended 1986); and other compounds amenable to purge-and-trap gas chromatography-mass spectrometry (GC/MS).

If the solids content is less than 1%, stable isotopically-labeled analogs of the compounds of interest are added to a 5-mL sample and the sample is purged with an inert gas at 20–25°C in a chamber designed for soil or water samples. If the solids content is greater than 1%, 5 mL of reagent water and the labeled compounds are added to a 5-g aliquot of sample and the mixture is purged at 40°C. Compounds that will not purge at 20–25°C or at 40°C are purged at 78–85°C. In the purging process, the volatile compounds are transferred from the aqueous phase into the gaseous phase where they are passed into a sorbent column, and trapped. After purging is completed, the trap is backflushed and heated rapidly to desorb the compounds into a GC. The compounds are separated by the GC and detected by a MS. The labeled compounds serve to correct the variability of the analytical technique.

INTERFERENCES Impurities in the purge gas, organic compounds outgassing from the plumbing upstream of the trap, and solvent vapors in the lab account for most problems. Samples can be contaminated by diffusion of volatile organic compounds (particularly methylene chloride) through the bottle seal during shipment and storage. Contamination by carryover can occur when high-level and low-level samples are analyzed sequentially. When an unusually concentrated sample is encountered, follow it by analysis of a reagent water blank to check for carryover.

INSTRUMENTATION Major equipment includes a GC with linear temperature programming and a glass jet separator as the MS interface, a MS with 70 eV electron impact ionization, and a data system to collect and record response factors.

Column: 2.8 m × 2 mm I.D. glass, packed with 1% SP-1000 on Carbopak B, 60/80 mesh, or equivalent.

PRECISION & ACCURACY The detection limits of the method are usually dependent on the level of interferences rather than instrumental limitations. The method detection limits were determined in digested sludge (low solids) and in filter cake or compost (high solids).

The MDL (in µg/kg) for low solids is not listed and for high solids is not listed.
Labeled and native compound precision (in µg/L) as standard deviation was not listed.
Labeled and native compound accuracy (in µg/L) as average recovery was not listed.

Acceptance criteria are at 20 µg/L for this compound.

SAMPLE COLLECTION, PRESERVATION & HANDLING Grab samples are collected in glass containers having a total volume greater than 20 mL. Fill and seal each bottle so that no air bubbles are entrapped. Samples are maintained at 0 to 4°C from the time of collection until analysis. If an aqueous sample contains residual chlorine, add sodium thiosulfate preservative (10 mg/40 mL) to the empty sample bottles just prior to shipment to the sample site. All samples must be analyzed within 14 days of collection.

SAMPLE PREPARATION Samples containing less than 1% solids are analyzed directly as aqueous samples. Samples containing 1% solids or greater are analyzed as solid samples utilizing one of two methods, depending on the levels of pollutants, in the sample. Samples containing 1% solids or greater, and low to moderate levels of pollutants are analyzed by purging a known weight of sample added to 5 mL of reagent water. Samples containing 1% solids or greater, and high levels of pollutants, are extracted with methanol, and an aliquot of the methanol extract is added to reagent water and purged.

QUALITY CONTROL A field blank prepared from reagent water and carried through the sampling and handling protocol may serve as a check on contamination from shipment and storage.

The analyst is permitted to modify this method to improve separations or lower the costs of measurements, provided all performance specifications are met. Analyses of blanks are required. When results of spikes indicate atypical method performance for samples, the samples are diluted to bring method performance within acceptable limits. Analyze two sets of four 5-mL aliquots (8 aliquots total) of the aqueous performance standard. Spike all samples with labeled compounds to assess method performance on the sample matrix. Compute the percent recovery of the labeled compounds using the internal standard method. Compare the percent recovery for each compound with the corresponding labeled compound recovery. Reagent water blanks are analyzed to demonstrate freedom from carryover contamination. Field replicates may be collected to determine the precision of the sampling technique, and spiked samples may be required to determine the accuracy of the analysis when the internal method is used.

REFERENCE Volatile Organic Compounds by Isotope Dilution GC/MS. Office of Water Regulation and Standards, U.S. EPA Industrial Technology Division, Washington, DC, EPA Method 1624, Rev. C, June 1989 (contact W.A. Telliard, U.S. EPA, Office of Water Regulations and Standards, 401 M St., SW, Washington, DC, 20460. Phone: 202-382-7131).

---

m-Xylene   **EPA Method 502**
CAS #108-38-3

TITLE Volatile Organic Compounds in Water By Purge and Trap Capillary Column Gas Chromatography with Photoionization and Electrolytic Conductivity Detectors in Series. U.S. EPA Method 502.2, Rev. 2.0, 1989.

MATRIX Drinking water and raw source water. The latter should include most surface water and groundwater sources.

METHOD SUMMARY This method covers 60 volatile organic compounds that contain halogen atoms and/or that are aromatic. An inert gas (zero grade nitrogen or helium) is bubbled through a 25-mL or a 5-mL water sample (depending on the expected concentration of the analytes). Purged sample components are trapped in a tube of sorbent materials. When purging is complete, the sorbent tube is heated and backflushed with helium to desorb the trapped sample onto a capillary GC column. The column is temperature programmed to separate the method analytes which are then detected with a photoionization detector (PID) and a Hall electrolytic conductivity (HECD) placed in series. The PID is selective for aromatic compounds and the HECD is selective for halogenated compounds.

INTERFERENCES Impurities in the purge gas and from organic compounds outgassing from the plumbing ahead of the trap account for many contamination problems. Interferences purged or coextracted from the samples will vary considerably from source to source, depending upon the particular sample or extract being tested. Cross-contamination can occur whenever high-level and low-level samples are analyzed sequentially. Samples also can be contaminated by diffusion of volatile organics (particularly methylene chloride and fluorocarbons) through the septum seal into the sample during shipment and storage. The lab where volatile analysis is performed and also the refrigerated storage area should be completely free of solvents.

INSTRUMENTATION A GC containing a series configuration of a high temperature photoionization detector (PID) equipped with 10.0 eV (nominal) lamp and Hall electrolytic conductivity detector (HECD) is required. Also required is an all-glass 5-mL purging device, a sorbent trap, and a thermal desorption apparatus which is connected to the GC system.

Column 1: VOCOL glass wide-bore capillary column.
Column 2: RTX–502.2 mega-bore capillary column.
Column 3: DB-62 mega-bore capillary column.

PRECISION & ACCURACY Method detection limits are dependent upon the characteristics of the gas chromatographic system used. Analytes that are not separated chromatographically cannot be individually identified and used in the same calibration mixture or water samples unless an alternative technique for identification and quantification, such as mass spectrometry, is used.

Electrolytic conductivity detetor (c) range in µg/L (a) was 0.02–200.
Electrolytic conductivity detetor (c) MDL in µg/L (b) was not listed.
Electrolytic conductivity detetor (c) accuracy as % recovery was not listed.
Electrolytic conductivity detetor (c) precision as % RSD was not listed.
Photoionization detector (d) range in µg/L (a) was 0.02–200.
Photoionization detector (d) MDL in µg/L (b) was 0.01.
Photoionization detector (d) accuracy as % recovery was 100.
Photoionization detector (d) precision as % RSD was 1.4.

(a) *The applicable concentration range of this method is compound, instrument, and matrix-dependent. It is listed as being approximately 0.02 to 200 µg/L but no specific information is provided so caution should be observed.*
(b) *The method detection limits reports with this method are compound, instrument, and matrix-dependent. The values reported were calculated using reagent water fortified with the corresponding compounds at 10 µg/L and a GC-equipped with a 60 m × 0.75 mm VOLCOL wide bore capillary column with 1.5 µm film thickness and using helium carrier gas.*
(c) *Recoveries and relative standard deviations were determined from seven samples of reagent water fortified with 10 µg/L of each compound. 2-Bromo-1-chloropropane was used as the internal standard for calculating average recoveries.*
(d) *Recoveries and relative standard deviations were determined from seven samples of reagent water fortified with 10 µg/L of each compound. Fluorobenzene was used as the internal standard for calculating average recoveries.*

SAMPLING METHOD Collect samples using a 40- to 120-mL screw-cap vial (prewashed with detergent, rinsed with distilled water and oven dried at 105°C) with a Teflon®-faced silicone septum. Collect bubble-free samples and place the septum with the Teflon® side down on the water.

SAMPLE PRESERVATION If residual chlorine is present in the water add about 25 mg of ascorbic acid to each vial before samples are collected to remove the chlorine. Add hydrochloric acid to reduce pH to <2, immediately cool samples to 4°C, and store them in a solvent-free refrigerator at 4°C until analysis.

MHT The maximum holding time for samples is 14 days from the time they were collected.

SAMPLE PREPARATION Remove the plungers from two 5-mL syringes and attach a closed syringe valve to each. Warm the sample to room temperature, open the sample bottle, and carefully pour the sample into one of the syringe barrels to just short of overflowing. Replace the syringe plunger, invert the syringe, and compress the sample. Open the syringe valve and vent any residual air while adjusting the sample volume to 5.0 mL. Add 10 µL of the internal calibration standard to the sample through the syringe valve. Close the valve. Fill the second syringe in an identical manner from the same sample bottle. Reserve this second syringe for a reanalysis if necessary.

QUALITY CONTROL As an initial demonstration of lab accuracy and precision, analyze 4 to 7 replicates of a lab fortified blank containing analyte at 0.1–5 µg/L. Collect all samples in duplicate. Surrogate analytes (similar to those of the analytes of interest), whose concentration is known in every sample, are measured using the same internal standard calibration procedure. Duplicate field reagent water blanks (trip blanks) must be analyzed with each set of samples, lab reagent blanks (method blanks) must be analyzed with each batch of samples processed as a group within a work shift. Also, a single lab-fortified blank that contains each of the analytes of interest should be analyzed with each batch of samples processed as a group within a work shift. A 3- to 5-point calibration curve is needed depending on the calibration range factor required.

EPA CONTACT & HOTLINE For technical questions contact Dr. Baldev Bathija, U.S. EPA, Office of Ground Water and Drinking Water (WH-550D), 401 M St. SW, Washington, DC 20460. Tel. (202) 260-3040. For further information the EPA Safe Drinking Water Hotline may be called at: (800) 426-4791.

REFERENCE Methods for the Determination of Organic Compounds in Drinking Water, EPA/600/4-88/039 (revised July 1991; Final Rule for determination of compliance with the

MCL for Total Trihalomethanes under 141.30, in 40 CFR Part 141, Vol. 58, No. 147, Fed. Reg., Tuesday Aug. 3, 1993). U.S. EPA Environmental Monitoring Systems Laboratory, Cincinnati, OH, 45268, U.S.A. Available from the National Technical Information Service (NTIS), 5285 Port Royal Road, Springfield, VA 22161; Tel. 800-553-6847. NTIS Order Number is PB91-231480.

---

o-Xylene **EPA Method 502**
CAS #95-47-6

TITLE Volatile Organic Compounds in Water By Purge and Trap Capillary Column Gas Chromatography with Photoionization and Electrolytic Conductivity Detectors in Series. U.S. EPA Method 502.2, Rev. 2.0, 1989.

MATRIX Drinking water and raw source water. The latter should include most surface water and groundwater sources.

METHOD SUMMARY This method covers 60 volatile organic compounds that contain halogen atoms and/or that are aromatic. An inert gas (zero grade nitrogen or helium) is bubbled through a 25-mL or a 5-mL water sample (depending on the expected concentration of the analytes). Purged sample components are trapped in a tube of sorbent materials. When purging is complete, the sorbent tube is heated and backflushed with helium to desorb the trapped sample onto a capillary GC column. The column is temperature programmed to separate the method analytes which are then detected with a photoionization detector (PID) and a Hall electrolytic conductivity (HECD) placed in series. The PID is selective for aromatic compounds and the HECD is selective for halogenated compounds.

INTERFERENCES Impurities in the purge gas and from organic compounds outgassing from the plumbing ahead of the trap account for many contamination problems. Interferences purged or coextracted from the samples will vary considerably from source to source, depending upon the particular sample or extract being tested. Cross-contamination can occur whenever high-level and low-level samples are analyzed sequentially. Samples also can be contaminated by diffusion of volatile organics (particularly methylene chloride and fluorocarbons) through the septum seal into the sample during shipment and storage. The lab where volatile analysis is performed and also the refrigerated storage area should be completely free of solvents.

INSTRUMENTATION A GC containing a series configuration of a high temperature photoionization detector (PID) equipped with 10.0 eV (nominal) lamp and Hall electrolytic conductivity detector (HECD) is required. Also required is an all-glass 5-mL purging device, a sorbent trap, and a thermal desorption apparatus which is connected to the GC system.

Column 1: VOCOL glass wide-bore capillary column.
Column 2: RTX–502.2 mega-bore capillary column.
Column 3: DB-62 mega-bore capillary column.

PRECISION & ACCURACY Method detection limits are dependent upon the characteristics of the gas chromatographic system used. Analytes that are not separated chromatographically cannot be individually identified and used in the same calibration mixture or water samples unless an alternative technique for identification and quantification, such as mass spectrometry, is used.

Electrolytic conductivity detetor (c) range in µg/L (a) was 0.02–200.
Electrolytic conductivity detetor (c) MDL in µg/L (b) was not listed.
Electrolytic conductivity detetor (c) accuracy as % recovery was not listed.
Electrolytic conductivity detetor (c) precision as % RSD was not listed.
Photoionization detector (d) range in µg/L (a) was 0.02–200.
Photoionization detector (d) MDL in µg/L (b) was 0.02.
Photoionization detector (d) accuracy as % recovery was 99.
Photoionization detector (d) precision as % RSD was 0.8.

(a) *The applicable concentration range of this method is compound, instrument, and matrix-dependent. It is listed as being approximately 0.02 to 200 µg/L but no specific information is provided so caution should be observed.*
(b) *The method detection limits reports with this method are compound, instrument, and matrix-dependent. The values reported were calculated using reagent water fortified with the corresponding compounds at 10 µg/L and a GC-equipped with a 60 m × 0.75 mm VOLCOL wide bore capillary column with 1.5 µm film thickness and using helium carrier gas.*
(c) *Recoveries and relative standard deviations were determined from seven samples of reagent water fortified with 10 µg/L of each compound. 2-Bromo-1-chloropropane was used as the internal standard for calculating average recoveries.*
(d) *Recoveries and relative standard deviations were determined from seven samples of reagent water fortified with 10 µg/L of each compound. Fluorobenzene was used as the internal standard for calculating average recoveries.*

SAMPLING METHOD Collect samples using a 40- to 120-mL screw-cap vial (prewashed with detergent, rinsed with distilled water and oven dried at 105°C) with a Teflon®-faced silicone septum. Collect bubble-free samples and place the septum with the Teflon® side down on the water.

SAMPLE PRESERVATION If residual chlorine is present in the water add about 25 mg of ascorbic acid to each vial before samples are collected to remove the chlorine. Add hydrochloric acid to reduce pH to <2, immediately cool samples to 4°C, and store them in a solvent-free refrigerator at 4°C until analysis.

MHT The maximum holding time for samples is 14 days from the time they were collected.

SAMPLE PREPARATION Remove the plungers from two 5-mL syringes and attach a closed syringe valve to each. Warm the sample to room temperature, open the sample bottle, and carefully pour the sample into one of the syringe barrels to just short of overflowing. Replace the syringe plunger, invert the syringe, and compress the sample. Open the syringe valve and vent any residual air while adjusting the sample volume to 5.0 mL. Add 10 µL of the internal calibration standard to the sample through the syringe valve. Close the valve. Fill the second syringe in an identical manner from the same sample bottle. Reserve this second syringe for a reanalysis if necessary.

QUALITY CONTROL As an initial demonstration of lab accuracy and precision, analyze 4 to 7 replicates of a lab fortified blank containing analyte at 0.1–5 µg/L. Collect all samples in duplicate. Surrogate analytes (similar to those of the analytes of interest), whose concentration is known in every sample, are measured using the same internal standard calibration procedure. Duplicate field reagent water blanks (trip blanks) must be analyzed with each set of samples, lab reagent blanks (method blanks) must be analyzed with each batch of samples processed as a group within a work shift. Also, a single lab-fortified blank that contains each of the analytes of interest should be analyzed with each batch of samples processed as a group within a work shift. A 3- to 5-point calibration curve is needed depending on the calibration range factor required.

EPA CONTACT & HOTLINE For technical questions contact Dr. Baldev Bathija, U.S. EPA, Office of Ground Water and Drinking Water (WH-550D), 401 M St. SW, Washington, DC 20460. Tel. (202) 260-3040. For further information the EPA Safe Drinking Water Hotline may be called at: (800) 426-4791.

REFERENCE Methods for the Determination of Organic Compounds in Drinking Water, EPA/600/4-88/039 (revised July 1991; Final Rule for determination of compliance with the MCL for Total Trihalomethanes under 141.30, in 40 CFR Part 141, Vol. 58, No. 147, Fed. Reg., Tuesday Aug. 3, 1993). U.S. EPA Environmental Monitoring Systems Laboratory, Cincinnati, OH, 45268, U.S.A. Available from the National Technical Information Service (NTIS), 5285 Port Royal Road, Springfield, VA 22161; Tel. 800-553-6847. NTIS Order Number is PB91-231480.

---

p-Xylene   **EPA Method 502**
CAS #106-42-3

TITLE Volatile Organic Compounds in Water By Purge and Trap Capillary Column Gas Chromatography with Photoionization and Electrolytic Conductivity Detectors in Series. U.S. EPA Method 502.2, Rev. 2.0, 1989.

MATRIX Drinking water and raw source water. The latter should include most surface water and groundwater sources.

METHOD SUMMARY This method covers 60 volatile organic compounds that contain halogen atoms and/or that are aromatic. An inert gas (zero grade nitrogen or helium) is bubbled through a 25-mL or a 5-mL water sample (depending on the expected concentration of the analytes). Purged sample components are trapped in a tube of sorbent materials. When purging is complete, the sorbent tube is heated and backflushed with helium to desorb the trapped sample onto a capillary GC column. The column is temperature programmed to separate the method analytes which are then detected with a photoionization detector (PID) and a Hall electrolytic conductivity (HECD) placed in series. The PID is selective for aromatic compounds and the HECD is selective for halogenated compounds.

INTERFERENCES Impurities in the purge gas and from organic compounds outgassing from the plumbing ahead of the trap account for many contamination problems. Interferences purged or coextracted from the samples will vary considerably from source to source, depending upon the particular sample or extract being tested. Cross-contamination can occur whenever high-level and low-level samples are analyzed sequentially. Samples also can be contaminated by diffusion of volatile organics (particularly methylene chloride and fluorocarbons) through the septum seal into the sample during shipment and storage. The lab where volatile analysis is performed and also the refrigerated storage area should be completely free of solvents.

INSTRUMENTATION A GC containing a series configuration of a high temperature photoionization detector (PID) equipped with 10.0 eV (nominal) lamp and Hall electrolytic conductivity detector (HECD) is required. Also required is an all-glass 5-mL purging device, a sorbent trap, and a thermal desorption apparatus which is connected to the GC system.

Column 1: VOCOL glass wide-bore capillary column.
Column 2: RTX–502.2 mega-bore capillary column.
Column 3: DB-62 mega-bore capillary column.

PRECISION & ACCURACY Method detection limits are dependent upon the characteristics of the gas chromatographic system used. Analytes that are not separated chromatographically cannot be individually identified and used in the same calibration mixture or water samples unless an alternative technique for identification and quantification, such as mass spectrometry, is used.

Electrolytic conductivity detetor (c) range in µg/L (a) was 0.02–200.
Electrolytic conductivity detetor (c) MDL in µg/L (b) was not listed.
Electrolytic conductivity detetor (c) accuracy as % recovery was not listed.
Electrolytic conductivity detetor (c) precision as % RSD was not listed.
Photoionization detector (d) range in µg/L (a) was 0.02–200.
Photoionization detector (d) MDL in µg/L (b) was 0.01.
Photoionization detector (d) accuracy as % recovery was 99.
Photoionization detector (d) precision as % RSD was 0.9.

(a) The applicable concentration range of this method is compound, instrument, and matrix-dependent. It is listed as being approximately 0.02 to 200 µg/L but no specific information is provided so caution should be observed.
(b) The method detection limits reports with this method are compound, instrument, and matrix-dependent. The values reported were calculated using reagent water fortified with the corresponding compounds at 10 µg/L and a GC-equipped with a 60 m × 0.75 mm VOLCOL wide bore capillary column with 1.5 µm film thickness and using helium carrier gas.
(c) Recoveries and relative standard deviations were determined from seven samples of reagent water fortified with 10 µg/L of each compound. 2-Bromo-1-chloropropane was used as the internal standard for calculating average recoveries.

(d) Recoveries and relative standard deviations were determined from seven samples of reagent water fortified with 10 µg/L of each compound. Fluorobenzene was used as the internal standard for calculating average recoveries.

SAMPLING METHOD Collect samples using a 40- to 120-mL screw-cap vial (prewashed with detergent, rinsed with distilled water and oven dried at 105°C) with a Teflon®-faced silicone septum. Collect bubble-free samples and place the septum with the Teflon® side down on the water.

SAMPLE PRESERVATION If residual chlorine is present in the water add about 25 mg of ascorbic acid to each vial before samples are collected to remove the chlorine. Add hydrochloric acid to reduce pH to <2, immediately cool samples to 4°C, and store them in a solvent-free refrigerator at 4°C until analysis.

MHT The maximum holding time for samples is 14 days from the time they were collected.

SAMPLE PREPARATION Remove the plungers from two 5-mL syringes and attach a closed syringe valve to each. Warm the sample to room temperature, open the sample bottle, and carefully pour the sample into one of the syringe barrels to just short of overflowing. Replace the syringe plunger, invert the syringe, and compress the sample. Open the syringe valve and vent any residual air while adjusting the sample volume to 5.0 mL. Add 10 µL of the internal calibration standard to the sample through the syringe valve. Close the valve. Fill the second syringe in an identical manner from the same sample bottle. Reserve this second syringe for a reanalysis if necessary.

QUALITY CONTROL As an initial demonstration of lab accuracy and precision, analyze 4 to 7 replicates of a lab fortified blank containing analyte at 0.1–5 µg/L. Collect all samples in duplicate. Surrogate analytes (similar to those of the analytes of interest), whose concentration is known in every sample, are measured using the same internal standard calibration procedure. Duplicate field reagent water blanks (trip blanks) must be analyzed with each set of samples, lab reagent blanks (method blanks) must be analyzed with each batch of samples processed as a group within a work shift. Also, a single lab-fortified blank that contains each of the analytes of interest should be analyzed with each batch of samples processed as a group within a work shift. A 3- to 5-point calibration curve is needed depending on the calibration range factor required.

EPA CONTACT & HOTLINE For technical questions contact Dr. Baldev Bathija, U.S. EPA, Office of Ground Water and Drinking Water (WH-550D), 401 M St. SW, Washington, DC 20460. Tel. (202) 260-3040. For further information the EPA Safe Drinking Water Hotline may be called at: (800) 426-4791.

REFERENCE Methods for the Determination of Organic Compounds in Drinking Water, EPA/600/4-88/039 (revised July 1991; Final Rule for determination of compliance with the MCL for Total Trihalomethanes under 141.30, in 40 CFR Part 141, Vol. 58, No. 147, Fed. Reg., Tuesday Aug. 3, 1993). U.S. EPA Environmental Monitoring Systems Laboratory, Cincinnati, OH, 45268, U.S.A. Available from the National Technical Information Service (NTIS), 5285 Port Royal Road, Springfield, VA 22161; Tel. 800-553-6847. NTIS Order Number is PB91-231480.

---

m-Xylene EPA Method 524
CAS #108-38-3

TITLE Measurement of Purgeable Organic Compounds in Water by Capillary Column GC/MS.

MATRIX Drinking water and raw source water; the latter should include most surface water and groundwater sources.

METHOD SUMMARY Method 524.2 covers 60 volatile organic compounds. An inert gas (zero grade nitrogen or helium) is bubbled through a 25-mL or a 5-mL water sample (depending on the expected concentration of the analytes). Purged sample components are trapped in a tube of sorbent materials. When purging is complete, the sorbent tube is heated and backflushed with helium to desorb the trapped sample onto a capillary GC column.

INTERFERENCES Impurities in the purge gas and from organic compounds outgassing from the plumbing ahead of the trap account for many contamination problems. Interferences purged or coextracted from the samples will vary considerably from source to source, depending upon the particular sample or extract being tested. Cross-contamination can occur whenever high-level and low-level samples are analyzed sequentially. Samples also can be contaminated by diffusion of volatile organics (particularly methylene chloride and fluorocarbons) through the septum seal into the sample during shipment and storage.

INSTRUMENTATION A GC/MS with a data system equipped with one of the following capillary GC columns:

Column 1: VOCOL glass wide bore capillary column.
Column 2: DB-624 fused silica capillary column.
Column 3: DB-5 fused silica capillary column.

Also required is an all-glass 25 mL or 5-mL purging device, a sorbent trap, and a thermal desorption apparatus which is connected to the GC/MS system.

PRECISION & ACCURACY Method detection limits are compound- and instrument-dependent, and may vary from approximately 0.02–0.35 µg/L. Note in the table below that the "true" concentration range used for accuracy and precision measurements was quite narrow. However, the applicable concentration range of this method is primarily column dependent and is approximately 0.02 to 200 µg/L for the wide-bore thick-film columns. Narrow-bore thin-film columns may have a capacity which limits the range to about 0.02 to 20 µg/L. Analytes that are inefficiently purged from water will not be detected when present at low concentrations, but they can be measured with acceptable accuracy and precision when present in sufficient amounts.

Analytes that are not separated chromatographically, but which have different mass spectra and non-interfering quantification ions, can be identified and measured in the same calibration mixture or water sample. Analytes which have very similar mass spectra cannot be individually identified and measured in the same calibration mixture or water samples unless they have different retention times. Co-eluting compounds with very

similar mass spectra, typically many structural isomers, must be reported as an isomeric group or pair.

The range (in µg/L) was 0.1–10.
The Method Detection Limig (in µg/L) was 0.05.
The accuracy (as % recovery) was 97.
The precision (in %) was 6.5.

Note: Data were obtained from 16–31 determinations using a wide-bore capillary column and a jet separator interfaced to a quadrupole mass spectrometer. All analytes were in a reagent water matrix.

SAMPLING METHOD Collect samples using a 40- to 120-mL screw-cap vial (prewashed with detergent, rinsed with distilled water and oven dried at 105°C) with a Teflon®-faced silicone septum. Collect bubble-free samples and place the septum with the Teflon® side down on the water.

SAMPLE PRESERVATION If residual chlorine is present in the water add about 25 mg of ascorbic acid to each vial before samples are collected to remove the chlorine. Add hydrochloric acid to reduce pH to <2, and immediately cool samples to 4°C, and store them in a solvent-free refrigerator at 4°C until analysis.

MHT The maximum holding time for samples is 14 days from the time they were collected.

SAMPLE PREPARATION Remove the plungers from two 25-mL (or 5-mL depending on sample size) syringes and attach a closed syringe valve to each. Warm the sample to room temperature, open the sample bottle, and carefully pour the sample into one of the syringe barrels to just short of overflowing. Replace the syringe plunger, invert the syringe, and compress the sample. Open the syringe valve and vent any residual air while adjusting the sample volume to 25.0 mL (or 5 mL). For samples and blanks, add 5 µL of the fortification solution containing the internal standard and the surrogates to the sample through the syringe valve. For calibration standards and lab fortified blanks, add 5 µL of the fortification solution containing the internal standard only. Close the valve. Fill the second syringe in an identical manner from the same sample bottle. Reserve this second syringe for a reanalysis if necessary.

QUALITY CONTROL As an initial demonstration of lab accuracy and precision, analyze 4 to 7 replicates of a lab fortified blank containing analyte at 0.2–5 µg/L. Collect all samples in duplicate. Surrogate analytes (similar to those of the analytes of interest), whose concentration is known in every sample, are measured using the same internal standard calibration procedure. Duplicate field reagent water blanks (trip blanks) must be analyzed with each set of samples, lab reagent blanks (method blanks) must be analyzed with each batch of samples processed as a group within a work shift. Also, a single lab-fortified blank that contains each of the analytes of interest should be analyzed with each batch of samples processed as a group within a work shift. A 3- to 5-point calibration curve is needed depending on the calibration range factor required.

EPA CONTACT & HOTLINE For technical questions contact Dr. Baldev Bathija, U.S. EPA, Office of Ground Water and Drinking Water (WH-550D), 401 M St. SW, Washington, DC 20460. Tel. (202) 260-3040. For further information the EPA Safe Drinking Water Hotline may be called at: (800) 426-4791.

REFERENCE Methods for the Determination of Organic Compounds in Drinking Water, EPA/600/4-88/039 (revised July 1991; Final Rule for determination of compliance with the MCL for Total Trihalomethanes under 141.30, in 40 CFR Part 141, Vol. 58, No. 147, Fed. Reg., Tuesday Aug. 3, 1993). U.S. EPA Environmental Monitoring Systems Laboratory, Cincinnati, OH, 45268, U.S.A. Available from the National Technical Information Service (NTIS), 5285 Port Royal Road, Springfield, VA 22161; Tel. 800-553-6847. NTIS Order Number is PB91-231480.

o-Xylene — EPA Method 524
CAS #95-47-6

TITLE Measurement of Purgeable Organic Compounds in Water by Capillary Column GC/MS.

MATRIX Drinking water and raw source water; the latter should include most surface water and groundwater sources.

METHOD SUMMARY Method 524.2 covers 60 volatile organic compounds. An inert gas (zero grade nitrogen or helium) is bubbled through a 25-mL or a 5-mL water sample (depending on the expected concentration of the analytes). Purged sample components are trapped in a tube of sorbent materials. When purging is complete, the sorbent tube is heated and backflushed with helium to desorb the trapped sample onto a capillary GC column.

INTERFERENCES Impurities in the purge gas and from organic compounds outgassing from the plumbing ahead of the trap account for many contamination problems. Interferences purged or coextracted from the samples will vary considerably from source to source, depending upon the particular sample or extract being tested. Cross-contamination can occur whenever high-level and low-level samples are analyzed sequentially. Samples also can be contaminated by diffusion of volatile organics (particularly methylene chloride and fluorocarbons) through the septum seal into the sample during shipment and storage.

INSTRUMENTATION A GC/MS with a data system equipped with one of the following capillary GC columns:

Column 1: VOCOL glass wide bore capillary column.
Column 2: DB-624 fused silica capillary column.
Column 3: DB-5 fused silica capillary column.

Also required is an all-glass 25 mL or 5-mL purging device, a sorbent trap, and a thermal desorption apparatus which is connected to the GC/MS system.

PRECISION & ACCURACY Method detection limits are compound- and instrument-dependent, and may vary from approximately 0.02–0.35 µg/L. Note in the table below that the "true" concentration range used for accuracy and precision measurements was quite narrow. However, the applicable concentration range of this method is primarily column dependent and is approximately 0.02 to 200 µg/L for the wide-bore thick-film

columns. Narrow-bore thin-film columns may have a capacity which limits the range to about 0.02 to 20 µg/L. Analytes that are inefficiently purged from water will not be detected when present at low concentrations, but they can be measured with acceptable accuracy and precision when present in sufficient amounts.

Analytes that are not separated chromatographically, but which have different mass spectra and non-interfering quantification ions, can be identified and measured in the same calibration mixture or water sample. Analytes which have very similar mass spectra cannot be individually identified and measured in the same calibration mixture or water samples unless they have different retention times. Co-eluting compounds with very similar mass spectra, typically many structural isomers, must be reported as an isomeric group or pair.

The range (in µg/L) was 0.1–31.
The Method Detection Limig (in µg/L) was 0.11.
The accuracy (as % recovery) was 103.
The precision (in %) was 7.2.

Note: Data were obtained from 16–31 determinations using a wide-bore capillary column and a jet separator interfaced to a quadrupole mass spectrometer. All analytes were in a reagent water matrix.

SAMPLING METHOD   Collect samples using a 40- to 120-mL screw-cap vial (prewashed with detergent, rinsed with distilled water and oven dried at 105°C) with a Teflon®-faced silicone septum. Collect bubble-free samples and place the septum with the Teflon® side down on the water.

SAMPLE PRESERVATION   If residual chlorine is present in the water add about 25 mg of ascorbic acid to each vial before samples are collected to remove the chlorine. Add hydrochloric acid to reduce pH to <2, and immediately cool samples to 4°C, and store them in a solvent-free refrigerator at 4°C until analysis.

MHT   The maximum holding time for samples is 14 days from the time they were collected.

SAMPLE PREPARATION   Remove the plungers from two 25-mL (or 5-mL depending on sample size) syringes and attach a closed syringe valve to each. Warm the sample to room temperature, open the sample bottle, and carefully pour the sample into one of the syringe barrels to just short of overflowing. Replace the syringe plunger, invert the syringe, and compress the sample. Open the syringe valve and vent any residual air while adjusting the sample volume to 25.0 mL (or 5 mL). For samples and blanks, add 5 µL of the fortification solution containing the internal standard and the surrogates to the sample through the syringe valve. For calibration standards and lab fortified blanks, add 5 µL of the fortification solution containing the internal standard only. Close the valve. Fill the second syringe in an identical manner from the same sample bottle. Reserve this second syringe for a reanalysis if necessary.

QUALITY CONTROL   As an initial demonstration of lab accuracy and precision, analyze 4 to 7 replicates of a lab fortified blank containing analyte at 0.2–5 µg/L. Collect all samples in duplicate. Surrogate analytes (similar to those of the analytes of interest), whose concentration is known in every sample, are measured using the same internal standard calibration procedure. Duplicate field reagent water blanks (trip blanks) must be analyzed with each set of samples, lab reagent blanks (method blanks) must be analyzed with each batch of samples processed as a group within a work shift. Also, a single lab-fortified blank that contains each of the analytes of interest should be analyzed with each batch of samples processed as a group within a work shift. A 3- to 5-point calibration curve is needed depending on the calibration range factor required.

EPA CONTACT & HOTLINE   For technical questions contact Dr. Baldev Bathija, U.S. EPA, Office of Ground Water and Drinking Water (WH-550D), 401 M St. SW, Washington, DC 20460. Tel. (202) 260-3040. For further information the EPA Safe Drinking Water Hotline may be called at: (800) 426-4791.

REFERENCE   Methods for the Determination of Organic Compounds in Drinking Water, EPA/600/4-88/039 (revised July 1991; Final Rule for determination of compliance with the MCL for Total Trihalomethanes under 141.30, in 40 CFR Part 141, Vol. 58, No. 147, Fed. Reg., Tuesday Aug. 3, 1993). U.S. EPA Environmental Monitoring Systems Laboratory, Cincinnati, OH, 45268, U.S.A. Available from the National Technical Information Service (NTIS), 5285 Port Royal Road, Springfield, VA 22161; Tel. 800-553-6847. NTIS Order Number is PB91-231480.

---

o-Xylene EPA Method 524
CAS #95-47-6

TITLE   Measurement of Purgeable Organic Compounds in Water by Capillary Column GC/MS.

MATRIX   Drinking water and raw source water; the latter should include most surface water and groundwater sources.

METHOD SUMMARY   Method 524.2 covers 60 volatile organic compounds. An inert gas (zero grade nitrogen or helium) is bubbled through a 25-mL or a 5-mL water sample (depending on the expected concentration of the analytes). Purged sample components are trapped in a tube of sorbent materials. When purging is complete, the sorbent tube is heated and backflushed with helium to desorb the trapped sample onto a capillary GC column.

INTERFERENCES   Impurities in the purge gas and from organic compounds outgassing from the plumbing ahead of the trap account for many contamination problems. Interferences purged or coextracted from the samples will vary considerably from source to source, depending upon the particular sample or extract being tested. Cross-contamination can occur whenever high-level and low-level samples are analyzed sequentially. Samples also can be contaminated by diffusion of volatile organics (particularly methylene chloride and fluorocarbons) through the septum seal into the sample during shipment and storage.

INSTRUMENTATION   A GC/MS with a data system equipped with one of the following capillary GC columns:

Column 1: VOCOL glass wide bore capillary column.

Column 2: DB-624 fused silica capillary column.
Column 3: DB-5 fused silica capillary column.

Also required is an all-glass 25 mL or 5-mL purging device, a sorbent trap, and a thermal desorption apparatus which is connected to the GC/MS system.

PRECISION & ACCURACY Method detection limits are compound- and instrument-dependent, and may vary from approximately 0.02–0.35 µg/L. Note in the table below that the "true" concentration range used for accuracy and precision measurements was quite narrow. However, the applicable concentration range of this method is primarily column dependent and is approximately 0.02 to 200 µg/L for the wide-bore thick-film columns. Narrow-bore thin-film columns may have a capacity which limits the range to about 0.02 to 20 µg/L. Analytes that are inefficiently purged from water will not be detected when present at low concentrations, but they can be measured with acceptable accuracy and precision when present in sufficient amounts.

Analytes that are not separated chromatographically, but which have different mass spectra and non-interfering quantification ions, can be identified and measured in the same calibration mixture or water sample. Analytes which have very similar mass spectra cannot be individually identified and measured in the same calibration mixture or water samples unless they have different retention times. Co-eluting compounds with very similar mass spectra, typically many structural isomers, must be reported as an isomeric group or pair.

The range (in µg/L) was 0.1–31.
The Method Detection Limig (in µg/L) was 0.11.
The accuracy (as % recovery) was 103.
The precision (in %) was 7.2.

Note: Data were obtained from 16–31 determinations using a wide-bore capillary column and a jet separator interfaced to a quadrupole mass spectrometer. All analytes were in a reagent water matrix.

SAMPLING METHOD Collect samples using a 40- to 120-mL screw-cap vial (prewashed with detergent, rinsed with distilled water and oven dried at 105°C) with a Teflon®-faced silicone septum. Collect bubble-free samples and place the septum with the Teflon® side down on the water.

SAMPLE PRESERVATION If residual chlorine is present in the water add about 25 mg of ascorbic acid to each vial before samples are collected to remove the chlorine. Add hydrochloric acid to reduce pH to <2, and immediately cool samples to 4°C, and store them in a solvent-free refrigerator at 4°C until analysis.

MHT The maximum holding time for samples is 14 days from the time they were collected.

SAMPLE PREPARATION Remove the plungers from two 25-mL (or 5-mL depending on sample size) syringes and attach a closed syringe valve to each. Warm the sample to room temperature, open the sample bottle, and carefully pour the sample into one of the syringe barrels to just short of overflowing. Replace the syringe plunger, invert the syringe, and compress the sample. Open the syringe valve and vent any residual air while adjusting the sample volume to 25.0 mL (or 5 mL). For samples and blanks, add 5 µL of the fortification solution containing the internal standard and the surrogates to the sample through the syringe valve. For calibration standards and lab fortified blanks, add 5 µL of the fortification solution containing the internal standard only. Close the valve. Fill the second syringe in an identical manner from the same sample bottle. Reserve this second syringe for a reanalysis if necessary.

QUALITY CONTROL As an initial demonstration of lab accuracy and precision, analyze 4 to 7 replicates of a lab fortified blank containing analyte at 0.2–5 µg/L. Collect all samples in duplicate. Surrogate analytes (similar to those of the analytes of interest), whose concentration is known in every sample, are measured using the same internal standard calibration procedure. Duplicate field reagent water blanks (trip blanks) must be analyzed with each set of samples, lab reagent blanks (method blanks) must be analyzed with each batch of samples processed as a group within a work shift. Also, a single lab-fortified blank that contains each of the analytes of interest should be analyzed with each batch of samples processed as a group within a work shift. A 3- to 5-point calibration curve is needed depending on the calibration range factor required.

EPA CONTACT & HOTLINE For technical questions contact Dr. Baldev Bathija, U.S. EPA, Office of Ground Water and Drinking Water (WH-550D), 401 M St. SW, Washington, DC 20460. Tel. (202) 260-3040. For further information the EPA Safe Drinking Water Hotline may be called at: (800) 426-4791.

REFERENCE Methods for the Determination of Organic Compounds in Drinking Water, EPA/600/4-88/039 (revised July 1991; Final Rule for determination of compliance with the MCL for Total Trihalomethanes under 141.30, in 40 CFR Part 141, Vol. 58, No. 147, Fed. Reg., Tuesday Aug. 3, 1993). U.S. EPA Environmental Monitoring Systems Laboratory, Cincinnati, OH, 45268, U.S.A. Available from the National Technical Information Service (NTIS), 5285 Port Royal Road, Springfield, VA 22161; Tel. 800-553-6847. NTIS Order Number is PB91-231480.

---

p-Xylene   **EPA Method 524**
CAS #106-42-3

TITLE Measurement of Purgeable Organic Compounds in Water by Capillary Column GC/MS.

MATRIX Drinking water and raw source water; the latter should include most surface water and groundwater sources.

METHOD SUMMARY Method 524.2 covers 60 volatile organic compounds. An inert gas (zero grade nitrogen or helium) is bubbled through a 25-mL or a 5-mL water sample (depending on the expected concentration of the analytes). Purged sample components are trapped in a tube of sorbent materials. When purging is complete, the sorbent tube is heated and backflushed with helium to desorb the trapped sample onto a capillary GC column.

INTERFERENCES Impurities in the purge gas and from organic compounds outgassing from the plumbing ahead of the trap account for many contamination problems. Interferences purged or coextracted from the samples will vary considerably from source to source, depending upon the particular sample or extract being tested. Cross-contamination can occur whenever high-level and low-level samples are analyzed sequentially. Samples also can be contaminated by diffusion of volatile organics (particularly methylene chloride and fluorocarbons) through the septum seal into the sample during shipment and storage.

INSTRUMENTATION A GC/MS with a data system equipped with one of the following capillary GC columns:

Column 1: VOCOL glass wide bore capillary column.
Column 2: DB-624 fused silica capillary column.
Column 3: DB-5 fused silica capillary column.

Also required is an all-glass 25 mL or 5-mL purging device, a sorbent trap, and a thermal desorption apparatus which is connected to the GC/MS system.

PRECISION & ACCURACY Method detection limits are compound- and instrument-dependent, and may vary from approximately 0.02–0.35 µg/L. Note in the table below that the "true" concentration range used for accuracy and precision measurements was quite narrow. However, the applicable concentration range of this method is primarily column dependent and is approximately 0.02 to 200 µg/L for the wide-bore thick-film columns. Narrow-bore thin-film columns may have a capacity which limits the range to about 0.02 to 20 µg/L. Analytes that are inefficiently purged from water will not be detected when present at low concentrations, but they can be measured with acceptable accuracy and precision when present in sufficient amounts.

Analytes that are not separated chromatographically, but which have different mass spectra and non-interfering quantification ions, can be identified and measured in the same calibration mixture or water sample. Analytes which have very similar mass spectra cannot be individually identified and measured in the same calibration mixture or water samples unless they have different retention times. Co-eluting compounds with very similar mass spectra, typically many structural isomers, must be reported as an isomeric group or pair.

The range (in µg/L) was 0.5–10.
The Method Detection Limig (in µg/L) was 0.13.
The accuracy (as % recovery) was 104.
The precision (in %) was 7.7.

Note: Data were obtained from 16–31 determinations using a wide-bore capillary column and a jet separator interfaced to a quadrupole mass spectrometer. All analytes were in a reagent water matrix.

SAMPLING METHOD Collect samples using a 40- to 120-mL screw-cap vial (prewashed with detergent, rinsed with distilled water and oven dried at 105°C) with a Teflon®-faced silicone septum. Collect bubble-free samples and place the septum with the Teflon® side down on the water.

SAMPLE PRESERVATION If residual chlorine is present in the water add about 25 mg of ascorbic acid to each vial before samples are collected to remove the chlorine. Add hydrochloric acid to reduce pH to <2, and immediately cool samples to 4°C, and store them in a solvent-free refrigerator at 4°C until analysis.

MHT The maximum holding time for samples is 14 days from the time they were collected.

SAMPLE PREPARATION Remove the plungers from two 25-mL (or 5-mL depending on sample size) syringes and attach a closed syringe valve to each. Warm the sample to room temperature, open the sample bottle, and carefully pour the sample into one of the syringe barrels to just short of overflowing. Replace the syringe plunger, invert the syringe, and compress the sample. Open the syringe valve and vent any residual air while adjusting the sample volume to 25.0 mL (or 5 mL). For samples and blanks, add 5 µL of the fortification solution containing the internal standard and the surrogates to the sample through the syringe valve. For calibration standards and lab fortified blanks, add 5 µL of the fortification solution containing the internal standard only. Close the valve. Fill the second syringe in an identical manner from the same sample bottle. Reserve this second syringe for a reanalysis if necessary.

QUALITY CONTROL As an initial demonstration of lab accuracy and precision, analyze 4 to 7 replicates of a lab fortified blank containing analyte at 0.2–5 µg/L. Collect all samples in duplicate. Surrogate analytes (similar to those of the analytes of interest), whose concentration is known in every sample, are measured using the same internal standard calibration procedure. Duplicate field reagent water blanks (trip blanks) must be analyzed with each set of samples, lab reagent blanks (method blanks) must be analyzed with each batch of samples processed as a group within a work shift. Also, a single lab-fortified blank that contains each of the analytes of interest should be analyzed with each batch of samples processed as a group within a work shift. A 3- to 5-point calibration curve is needed depending on the calibration range factor required.

EPA CONTACT & HOTLINE For technical questions contact Dr. Baldev Bathija, U.S. EPA, Office of Ground Water and Drinking Water (WH-550D), 401 M St. SW, Washington, DC 20460. Tel. (202) 260-3040. For further information the EPA Safe Drinking Water Hotline may be called at: (800) 426-4791.

REFERENCE Methods for the Determination of Organic Compounds in Drinking Water, EPA/600/4-88/039 (revised July 1991; Final Rule for determination of compliance with the MCL for Total Trihalomethanes under 141.30, in 40 CFR Part 141, Vol. 58, No. 147, Fed. Reg., Tuesday Aug. 3, 1993). U.S. EPA Environmental Monitoring Systems Laboratory, Cincinnati, OH, 45268, U.S.A. Available from the National Technical Information Service (NTIS), 5285 Port Royal Road, Springfield, VA 22161; Tel. 800-553-6847. NTIS Order Number is PB91-231480.

---

p-Xylene   EPA Method 524
CAS #106-42-3

TITLE Measurement of Purgeable Organic Compounds in Water by Capillary Column GC/MS.

MATRIX Drinking water and raw source water; the latter should include most surface water and groundwater sources.

METHOD SUMMARY Method 524.2 covers 60 volatile organic compounds. An inert gas (zero grade nitrogen or helium) is bubbled through a 25-mL or a 5-mL water sample (depending on the expected concentration of the analytes). Purged sample components are trapped in a tube of sorbent materials. When purging is complete, the sorbent tube is heated and backflushed with helium to desorb the trapped sample onto a capillary GC column.

INTERFERENCES Impurities in the purge gas and from organic compounds outgassing from the plumbing ahead of the trap account for many contamination problems. Interferences purged or coextracted from the samples will vary considerably from source to source, depending upon the particular sample or extract being tested. Cross-contamination can occur whenever high-level and low-level samples are analyzed sequentially. Samples also can be contaminated by diffusion of volatile organics (particularly methylene chloride and fluorocarbons) through the septum seal into the sample during shipment and storage.

INSTRUMENTATION A GC/MS with a data system equipped with one of the following capillary GC columns:

Column 1: VOCOL glass wide bore capillary column.
Column 2: DB-624 fused silica capillary column.
Column 3: DB-5 fused silica capillary column.

Also required is an all-glass 25 mL or 5-mL purging device, a sorbent trap, and a thermal desorption apparatus which is connected to the GC/MS system.

PRECISION & ACCURACY Method detection limits are compound- and instrument-dependent, and may vary from approximately 0.02–0.35 µg/L. Note in the table below that the "true" concentration range used for accuracy and precision measurements was quite narrow. However, the applicable concentration range of this method is primarily column dependent and is approximately 0.02 to 200 µg/L for the wide-bore thick-film columns. Narrow-bore thin-film columns may have a capacity which limits the range to about 0.02 to 20 µg/L. Analytes that are inefficiently purged from water will not be detected when present at low concentrations, but they can be measured with acceptable accuracy and precision when present in sufficient amounts.

Analytes that are not separated chromatographically, but which have different mass spectra and non-interfering quantification ions, can be identified and measured in the same calibration mixture or water sample. Analytes which have very similar mass spectra cannot be individually identified and measured in the same calibration mixture or water samples unless they have different retention times. Co-eluting compounds with very similar mass spectra, typically many structural isomers, must be reported as an isomeric group or pair.

The range (in µg/L) was 0.5–10.
The Method Detection Limig (in µg/L) was 0.13.
The accuracy (as % recovery) was 104.
The precision (in %) was 7.7.

Note: Data were obtained from 16–31 determinations using a wide-bore capillary column and a jet separator interfaced to a quadrupole mass spectrometer. All analytes were in a reagent water matrix.

SAMPLING METHOD Collect samples using a 40- to 120-mL screw-cap vial (prewashed with detergent, rinsed with distilled water and oven dried at 105°C) with a Teflon®-faced silicone septum. Collect bubble-free samples and place the septum with the Teflon® side down on the water.

SAMPLE PRESERVATION If residual chlorine is present in the water add about 25 mg of ascorbic acid to each vial before samples are collected to remove the chlorine. Add hydrochloric acid to reduce pH to <2, and immediately cool samples to 4°C, and store them in a solvent-free refrigerator at 4°C until analysis.

MHT The maximum holding time for samples is 14 days from the time they were collected.

SAMPLE PREPARATION Remove the plungers from two 25-mL (or 5-mL depending on sample size) syringes and attach a closed syringe valve to each. Warm the sample to room temperature, open the sample bottle, and carefully pour the sample into one of the syringe barrels to just short of overflowing. Replace the syringe plunger, invert the syringe, and compress the sample. Open the syringe valve and vent any residual air while adjusting the sample volume to 25.0 mL (or 5 mL). For samples and blanks, add 5 µL of the fortification solution containing the internal standard and the surrogates to the sample through the syringe valve. For calibration standards and lab fortified blanks, add 5 µL of the fortification solution containing the internal standard only. Close the valve. Fill the second syringe in an identical manner from the same sample bottle. Reserve this second syringe for a reanalysis if necessary.

QUALITY CONTROL As an initial demonstration of lab accuracy and precision, analyze 4 to 7 replicates of a lab fortified blank containing analyte at 0.2–5 µg/L. Collect all samples in duplicate. Surrogate analytes (similar to those of the analytes of interest), whose concentration is known in every sample, are measured using the same internal standard calibration procedure. Duplicate field reagent water blanks (trip blanks) must be analyzed with each set of samples, lab reagent blanks (method blanks) must be analyzed with each batch of samples processed as a group within a work shift. Also, a single lab-fortified blank that contains each of the analytes of interest should be analyzed with each batch of samples processed as a group within a work shift. A 3- to 5-point calibration curve is needed depending on the calibration range factor required.

EPA CONTACT & HOTLINE For technical questions contact Dr. Baldev Bathija, U.S. EPA, Office of Ground Water and Drinking Water (WH-550D), 401 M St. SW, Washington, DC 20460. Tel. (202) 260-3040. For further information the EPA Safe Drinking Water Hotline may be called at: (800) 426-4791.

REFERENCE Methods for the Determination of Organic Compounds in Drinking Water, EPA/600/4-88/039 (revised July 1991; Final Rule for determination of compliance with the

MCL for Total Trihalomethanes under 141.30, in 40 CFR Part 141, Vol. 58, No. 147, Fed. Reg., Tuesday Aug. 3, 1993). U.S. EPA Environmental Monitoring Systems Laboratory, Cincinnati, OH, 45268, U.S.A. Available from the National Technical Information Service (NTIS), 5285 Port Royal Road, Springfield, VA 22161; Tel. 800-553-6847. NTIS Order Number is PB91-231480.

---

m-Xylene   EPA Method 8021
CAS #108-38-3

TITLE Halogenated Volatile by Gas Chromatography Using Photoionization and Electrolytic Conductivity Detectors in Series: Capillary Column Technique

MATRIX This method is applicable to nearly all types of samples, regardless of water content, including groundwater, aqueous sludges, caustic liquors, acid liquors, waste solvents, oily wastes, mousses, tars, fibrous wastes, polymeric emulsions, filter cakes, spent carbons, spent catalysts, soils, and sediments.

METHOD SUMMARY This method is used to determine 60 volatile organic compounds in a variety of solid waste matrices. It provides GC conditions for the detection of halogenated and aromatic volatile organic compounds. Samples can be analyzed using direct injection or purge-and-trap (EPA Method 5030). Groundwater samples must be analyzed using EPA Method 5030 (where applicable). A temperature program is used with the GC. Detection is achieved by a photoionization detector (PID) and a Hall electrolytic conductivity detector (HECD) in series.

INTERFERENCES Samples can be contaminated by diffusion of volatile organics (particularly chlorofluorocarbons and methylene chloride) through the sample container septum during shipment and storage.

INSTRUMENTATION A GC-equipped with variable-constant differential flow controllers, subambient oven controller, PID and HECD detectors connected with a short piece of uncoated capillary tubing and a data system.

Column: 60 m × 0.75 mm I.D. VOCOL wide-bore capillary column with 1.5 μm film thickness.

PRECISION & ACCURACY MDLs are compound-dependent and vary with purging efficiency and concentration. The applicable concentration range of this method is compound- and instrument-dependent but is approximately 0.1 to 200 μg/L. Analytes that are inefficiently purged from water will not be detected when present at low concentrations, but they can be measured with acceptable accuracy and precision when present in sufficient amounts. The estimated quantitation limit (EQL) for an individual compound is approximately 1 μg/kg (wet weight) for soil/sediment samples, 100 μg/kg (wet weight) for wastes, and 1 μg/L for groundwater. EQLs will be proportionately higher for sample extracts and samples that require dilution or reduced sample size to avoid saturation of the detector.

MULTIPLICATION FACTORS FOR OTHER MATRICES (a)

Matrix	Factor (b)
Groundwater	10
Low-concentration soil	10
Water miscible liquid waste	500
High-concentration soil and sludge	1250
Non-water miscible waste	1250

(a) Sample EQLs are highly matrix-dependent. The EQLs listed herein are provided for guidance and may not always be achievable.
(b) EQL = [Method detection limit] × [Factor]. For non-aqueous samples, the factor is on a wet-weight basis.

SINGLE LABORATORY ACCURACY & PRECISION DATA FOR VOCs IN WATER
This method was tested in a single lab using water spiked at 10 μg/L and the following data was reported:

Recoveries and standard deviations were determined from seven samples and spiked at 10 μg/L of each analyte. Recoveries were determined by the internal standard method. Internal standards were: Fluorobenzene for PID and 2-Bromo-1-chloropropane for HECD.

The average recovery (in percent) for the PID was 100.
The standard deviation of the recovery for the PID was 1.4.
The MDL (in μg/mL) for the PID was 0.01.
The average recovery (in percent) for the HECD was none (no response for this detector).
The standard deviation of the recovery for the HECD was none (no response for this detector)-.
The MDL (in μg/mL) for the HECD was none (no response for this detector).

SAMPLE COLLECTION, PRESERVATION & HANDLING
Volatile organics — Standard 40-mL glass screw-cap VOA vials with Teflon®-faced silicone septum may be used for both liquid and solid matrices. When collecting samples, liquids and solids should be introduced into the vials gently to reduce agitation which might drive off volatile compounds. If there are any air bubbles present the sample must be retaken. Tap slightly as they are filled to try and eliminate as much free air space as possible. The two vials from each sampling locations should be sealed in separate plastic bags to prevent cross-contamination between samples particularly if the sampled waste is suspected of containing high levels of volatile organics.

Semivolatile organics — Containers used to collect samples for the determination of semivolatile organic compounds should be soap and water washed followed by methanol (or isopropanol) rinsing. The sample containers should be of glass or Teflon® and have screw-top covers with Teflon® liners.

Preservation for volatile organics — No preservation is used with concentrated waste samples. With liquid samples containing no residual chlorine, 4 drops of concentrated hydrochloric acid are added and the samples are immediately cooled to 4°C. When liquid samples contain residual chlorine, they are treated as above and, in addition, 4 drops of 4% aqueous sodium thiosulfate are added. Soil, sediment, and sludge samples are only cooled to 4°C.

Preservation for semivolatile organics — No preservation is used with concentrated waste samples. With liquid samples containing no residual chlorine and with soil, sediment, and sludge samples, immediately cooling to 4°C is the only preservation used. When residual chlorine is present then 3 mL of 10% aqueous sodium sulfate is added for each gallon of sample collected, followed by cooling to 4°C.

MHT The holding time for all volatile organics samples is 14 days. Liquid samples must be extracted within 7 days and their extracts analyzed within 40 days. Concentrated waste, soil, sediment, and sludge samples must be extracted within 14 days and their extracts analyzed within 40 days.

SAMPLE PREPARATION Volatile compounds are introduced into the gas chromatograph either by direct injector or purge-and-trap (EPA Method 5030). EPA Method 5030 may be used directly on groundwater samples or low-concentration contaminated soils and sediments. For medium-concentration soils or sediments, methanolic extraction, as described in EPA Method 5030, may be necessary prior to purge-and-trap analysis.

QUALITY CONTROL Calculate surrogate standard recovery on all samples, blanks, and spikes. A trip blank is recommended to check on sampling, storage, and handling contamination. Calibration standards, at a minimum of five concentration levels, are prepared in organic-free reagent water. One of the concentration levels should be at a concentration near, but above, the method detection limit.

A combination of bromochloromethane, 2-bromo-1-chloropropane, 1,4-dichlorobutane, and bromochlorobenzene are recommended as surrogate standards to encompass the range of the temperature program used in this method.

REFERENCE Test Methods for Evaluating Solid Waste, Physical/Chemical Methods, SW-846, 3rd Edition, U.S. EPA, Office of Solid Waste, Washington, DC, EPA Method 8021A, Rev. 1, Nov. 1992.

---

o-Xylene   **EPA Method 8021**
CAS #95-47-6

TITLE Halogenated Volatile by Gas Chromatography Using Photoionization and Electrolytic Conductivity Detectors in Series: Capillary Column Technique

MATRIX This method is applicable to nearly all types of samples, regardless of water content, including groundwater, aqueous sludges, caustic liquors, acid liquors, waste solvents, oily wastes, mousses, tars, fibrous wastes, polymeric emulsions, filter cakes, spent carbons, spent catalysts, soils, and sediments.

METHOD SUMMARY This method is used to determine 60 volatile organic compounds in a variety of solid waste matrices. It provides GC conditions for the detection of halogenated and aromatic volatile organic compounds. Samples can be analyzed using direct injection or purge-and-trap (EPA Method 5030). Groundwater samples must be analyzed using EPA Method 5030 (where applicable). A temperature program is used with the GC. Detection is achieved by a photoionization detector (PID) and a Hall electrolytic conductivity detector (HECD) in series.

INTERFERENCES Samples can be contaminated by diffusion of volatile organics (particularly chlorofluorocarbons and methylene chloride) through the sample container septum during shipment and storage.

INSTRUMENTATION A GC-equipped with variable-constant differential flow controllers, subambient oven controller, PID and HECD detectors connected with a short piece of uncoated capillary tubing and a data system.

Column: 60 m × 0.75 mm I.D. VOCOL wide-bore capillary column with 1.5 μm film thickness.

PRECISION & ACCURACY MDLs are compound-dependent and vary with purging efficiency and concentration. The applicable concentration range of this method is compound- and instrument-dependent but is approximately 0.1 to 200 μg/L. Analytes that are inefficiently purged from water will not be detected when present at low concentrations, but they can be measured with acceptable accuracy and precision when present in sufficient amounts. The estimated quantitation limit (EQL) for an individual compound is approximately 1 μg/kg (wet weight) for soil/sediment samples, 100 μg/kg (wet weight) for wastes, and 1 μg/L for groundwater. EQLs will be proportionately higher for sample extracts and samples that require dilution or reduced sample size to avoid saturation of the detector.

MULTIPLICATION FACTORS FOR OTHER MATRICES (a)

Matrix	Factor (b)
Groundwater	10
Low-concentration soil	10
Water miscible liquid waste	500
High-concentration soil and sludge	1250
Non-water miscible waste	1250

(a) Sample EQLs are highly matrix-dependent. The EQLs listed herein are provided for guidance and may not always be achievable.
(b) EQL = [Method detection limit] × [Factor]. For non-aqueous samples, the factor is on a wet-weight basis.

SINGLE LABORATORY ACCURACY & PRECISION DATA FOR VOCs IN WATER
This method was tested in a single lab using water spiked at 10 μg/L and the following data was reported:

Recoveries and standard deviations were determined from seven samples and spiked at 10 μg/L of each analyte. Recoveries were determined by the internal standard method. Internal standards were: Fluorobenzene for PID and 2-Bromo-1-chloropropane for HECD.

The average recovery (in percent) for the PID was 99.
The standard deviation of the recovery for the PID was 0.8.
The MDL (in μg/mL) for the PID was 0.02.
The average recovery (in percent) for the HECD was none (no response for this detector).
The standard deviation of the recovery for the HECD was none (no response for this detector)-.

The MDL (in μg/mL) for the HECD was none (no response for this detector).

SAMPLE COLLECTION, PRESERVATION & HANDLING

Volatile organics — Standard 40-mL glass screw-cap VOA vials with Teflon®-faced silicone septum may be used for both liquid and solid matrices. When collecting samples, liquids and solids should be introduced into the vials gently to reduce agitation which might drive off volatile compounds. If there are any air bubbles present the sample must be retaken. Tap slightly as they are filled to try and eliminate as much free air space as possible. The two vials from each sampling locations should be sealed in separate plastic bags to prevent cross-contamination between samples particularly if the sampled waste is suspected of containing high levels of volatile organics.

Semivolatile organics — Containers used to collect samples for the determination of semivolatile organic compounds should be soap and water washed followed by methanol (or isopropanol) rinsing. The sample containers should be of glass or Teflon® and have screw-top covers with Teflon® liners.

Preservation for volatile organics — No preservation is used with concentrated waste samples. With liquid samples containing no residual chlorine, 4 drops of concentrated hydrochloric acid are added and the samples are immediately cooled to 4°C. When liquid samples contain residual chlorine, they are treated as above and, in addition, 4 drops of 4% aqueous sodium thiosulfate are added. Soil, sediment, and sludge samples are only cooled to 4°C.

Preservation for semivolatile organics — No preservation is used with concentrated waste samples. With liquid samples containing no residual chlorine and with soil, sediment, and sludge samples, immediately cooling to 4°C is the only preservation used. When residual chlorine is present then 3 mL of 10% aqueous sodium sulfate is added for each gallon of sample collected, followed by cooling to 4°C.

MHT The holding time for all volatile organics samples is 14 days. Liquid samples must be extracted within 7 days and their extracts analyzed within 40 days. Concentrated waste, soil, sediment, and sludge samples must be extracted within 14 days and their extracts analyzed within 40 days.

SAMPLE PREPARATION Volatile compounds are introduced into the gas chromatograph either by direct injector or purge-and-trap (EPA Method 5030). EPA Method 5030 may be used directly on groundwater samples or low-concentration contaminated soils and sediments. For medium-concentration soils or sediments, methanolic extraction, as described in EPA Method 5030, may be necessary prior to purge-and-trap analysis.

QUALITY CONTROL Calculate surrogate standard recovery on all samples, blanks, and spikes. A trip blank is recommended to check on sampling, storage, and handling contamination. Calibration standards, at a minimum of five concentration levels, are prepared in organic-free reagent water. One of the concentration levels should be at a concentration near, but above, the method detection limit.

A combination of bromochloromethane, 2-bromo-1-chloropropane, 1,4-dichlorobutane, and bromochlorobenzene are recommended as surrogate standards to encompass the range of the temperature program used in this method.

REFERENCE Test Methods for Evaluating Solid Waste, Physical/Chemical Methods, SW-846, 3rd Edition, U.S. EPA, Office of Solid Waste, Washington, DC, EPA Method 8021A, Rev. 1, Nov. 1992.

p-Xylene   EPA Method 8021
CAS #106-42-3

TITLE Halogenated Volatile by Gas Chromatography Using Photoionization and Electrolytic Conductivity Detectors in Series: Capillary Column Technique

MATRIX This method is applicable to nearly all types of samples, regardless of water content, including groundwater, aqueous sludges, caustic liquors, acid liquors, waste solvents, oily wastes, mousses, tars, fibrous wastes, polymeric emulsions, filter cakes, spent carbons, spent catalysts, soils, and sediments.

METHOD SUMMARY This method is used to determine 60 volatile organic compounds in a variety of solid waste matrices. It provides GC conditions for the detection of halogenated and aromatic volatile organic compounds. Samples can be analyzed using direct injection or purge-and-trap (EPA Method 5030). Groundwater samples must be analyzed using EPA Method 5030 (where applicable). A temperature program is used with the GC. Detection is achieved by a photoionization detector (PID) and a Hall electrolytic conductivity detector (HECD) in series.

INTERFERENCES Samples can be contaminated by diffusion of volatile organics (particularly chlorofluorocarbons and methylene chloride) through the sample container septum during shipment and storage.

INSTRUMENTATION A GC-equipped with variable-constant differential flow controllers, subambient oven controller, PID and HECD detectors connected with a short piece of uncoated capillary tubing and a data system.

Column: 60 m × 0.75 mm I.D. VOCOL wide-bore capillary column with 1.5 μm film thickness.

PRECISION & ACCURACY MDLs are compound-dependent and vary with purging efficiency and concentration. The applicable concentration range of this method is compound- and instrument-dependent but is approximately 0.1 to 200 μg/L. Analytes that are inefficiently purged from water will not be detected when present at low concentrations, but they can be measured with acceptable accuracy and precision when present in sufficient amounts. The estimated quantitation limit (EQL) for an individual compound is approximately 1 μg/kg (wet weight) for soil/sediment samples, 100 μg/kg (wet weight) for wastes, and 1 μg/L for groundwater. EQLs will be proportionately higher for sample extracts and samples that require dilution or reduced sample size to avoid saturation of the detector.

MULTIPLICATION FACTORS FOR OTHER MATRICES (a)

Matrix	Factor (b)
Groundwater	10
Low-concentration soil	10
Water miscible liquid waste	500
High-concentration soil and sludge	1250
Non-water miscible waste	1250

(a) Sample EQLs are highly matrix-dependent. The EQLs listed herein are provided for guidance and may not always be achievable. (b) EQL = [Method detection limit] × [Factor]. For non-aqueous samples, the factor is on a wet-weight basis.

SINGLE LABORATORY ACCURACY & PRECISION DATA FOR VOCs IN WATER

This method was tested in a single lab using water spiked at 10 µg/L and the following data was reported:

Recoveries and standard deviations were determined from seven samples and spiked at 10 µg/L of each analyte. Recoveries were determined by the internal standard method. Internal standards were: Fluorobenzene for PID and 2-Bromo-1-chloropropane for HECD.

The average recovery (in percent) for the PID was 99.
The standard deviation of the recovery for the PID was 0.9.
The MDL (in µg/mL) for the PID was 0.01.
The average recovery (in percent) for the HECD was none (no response for this detector).
The standard deviation of the recovery for the HECD was none (no response for this detector).
The MDL (in µg/mL) for the HECD was none (no response for this detector).

SAMPLE COLLECTION, PRESERVATION & HANDLING

Volatile organics — Standard 40-mL glass screw-cap VOA vials with Teflon®-faced silicone septum may be used for both liquid and solid matrices. When collecting samples, liquids and solids should be introduced into the vials gently to reduce agitation which might drive off volatile compounds. If there are any air bubbles present the sample must be retaken. Tap slightly as they are filled to try and eliminate as much free air space as possible. The two vials from each sampling locations should be sealed in separate plastic bags to prevent cross-contamination between samples particularly if the sampled waste is suspected of containing high levels of volatile organics.

Semivolatile organics — Containers used to collect samples for the determination of semivolatile organic compounds should be soap and water washed followed by methanol (or isopropanol) rinsing. The sample containers should be of glass or Teflon® and have screw-top covers with Teflon® liners.

Preservation for volatile organics — No preservation is used with concentrated waste samples. With liquid samples containing no residual chlorine, 4 drops of concentrated hydrochloric acid are added and the samples are immediately cooled to 4°C. When liquid samples contain residual chlorine, they are treated as above and, in addition, 4 drops of 4% aqueous sodium thiosulfate are added. Soil, sediment, and sludge samples are only cooled to 4°C.

Preservation for semivolatile organics — No preservation is used with concentrated waste samples. With liquid samples containing no residual chlorine and with soil, sediment, and sludge samples, immediately cooling to 4°C is the only preservation used. When residual chlorine is present then 3 mL of 10% aqueous sodium sulfate is added for each gallon of sample collected, followed by cooling to 4°C.

MHT The holding time for all volatile organics samples is 14 days. Liquid samples must be extracted within 7 days and their extracts analyzed within 40 days. Concentrated waste, soil, sediment, and sludge samples must be extracted within 14 days and their extracts analyzed within 40 days.

SAMPLE PREPARATION Volatile compounds are introduced into the gas chromatograph either by direct injector or purge-and-trap (EPA Method 5030). EPA Method 5030 may be used directly on groundwater samples or low-concentration contaminated soils and sediments. For medium-concentration soils or sediments, methanolic extraction, as described in EPA Method 5030, may be necessary prior to purge-and-trap analysis.

QUALITY CONTROL Calculate surrogate standard recovery on all samples, blanks, and spikes. A trip blank is recommended to check on sampling, storage, and handling contamination. Calibration standards, at a minimum of five concentration levels, are prepared in organic-free reagent water. One of the concentration levels should be at a concentration near, but above, the method detection limit.

A combination of bromochloromethane, 2-bromo-1-chloropropane, 1,4-dichlorobutane, and bromochlorobenzene are recommended as surrogate standards to encompass the range of the temperature program used in this method.

REFERENCE Test Methods for Evaluating Solid Waste, Physical/Chemical Methods, SW-846, 3rd Edition, U.S. EPA, Office of Solid Waste, Washington, DC, EPA Method 8021A, Rev. 1, Nov. 1992.

---

m-Xylene — EPA Method 8260
CAS #108-38-3

TITLE Volatile Organic Compounds by GC/MS: Capillary Column Technique

MATRIX This method is applicable to nearly all types of samples, regardless of water content, including groundwater, soils, and sediments.

METHOD SUMMARY Method 8260A covers 58 volatile organic compounds that are introduced into a gas chromatograph by the purge-and-trap method or by direct injection (in limited applications). Zero-grade helium is bubbled through a 5-mL solution at ambient temperature. Purged sample components are trapped in a tube containing suitable sorbent materials. When purging is complete, the sorbent tube is heated and backflushed with helium to desorb trapped sample components. The analytes are desorbed directly to a large bore capillary or cryofocussed on a capillary precolumn before being flash evaporated to a narrow bore capillary for analysis.

INTERFERENCES Major contaminant sources are volatile materials in the lab and impurities in the inert purging gas and in the sorbent trap. Interfering contamination may occur when a sample containing low concentrations of volatile organic compounds is analyzed immediately after a sample containing high concentrations of volatile organic compounds. After analysis of a sample containing high concentrations of volatile organic compounds, one or more calibration blanks should be analyzed to check for cross-contamination. Screening of the samples prior to purge-and-trap GC/MS analysis is highly recommended to prevent contamination of the system. This is especially true for soil and waste samples.

Special precautions must be taken to analyze for methylene chloride. The analytical and sample storage area should be isolated from all atmospheric sources of methylene chloride. All gas chromatography carrier gas lines and purge gas plumbing should be constructed from stainless steel or copper tubing. Laboratory clothing previously exposed to methylene chloride fumes during liquid-liquid extraction procedures can contribute to sample contamination.

Samples can also be contaminated by diffusion of volatile organics (particularly methylene chloride and fluorocarbons) through the septum seal during shipment and storage. A trip blank can serve as a check on such contamination.

INSTRUMENTATION GC/MS with a temperature-programmable chromatograph suitable for splitless injection equipped with variable constant differential flow controllers, a subambient oven controller, a purging device, sorbent trap, a thermal desorption apparatus and a capillary precolumn interface when using cryogenic cooling will be needed. The following GC columns may be used:

Column 1: 60 m × 0.75mm I.D. capillary column coated with VOCOL, 1.5 µm film thickness.
Column 2: 30 m × 0.53mm capillary column coated with DB-624 or VOCOL, 3 µm film thickness.
Column 3: 30 m × 0.32mm I.D. capillary column coated with DB-5 or SE-54, 1-µm film thickness.

PRECISION & ACCURACY This method has been tested in a single lab using spiked water. Using a wide-bore capillary column, water was spiked at concentrations between 0.5 and 10 µg/L. Single lab accuracy and precision data are presented. The MDL actually achieved in a given analysis will vary depending on instrument sensitivity and matrix effects.

The MDL (a) in µg/L was 0.05.
The concentration range in µg/L was 0.1–10.
The mean accuracy (% of true value) was 97.
The precision as relative standard deviation was 6.5.

Note: The MDL is based on a 25-mL sample volume instead of a 5-mL sample volume.

SAMPLING METHOD
Liquid samples — Use a 40-mL glass screw-cap VOA vial with a Teflon®-faced silicone septum that has been prewashed, rinsed with distilled deionized water, and oven dried. If residual chlorine is present, collect the sample in a 4-oz soil VOA container which has been pre-preserved with 4 drops of 10% sodium thiosulfate. Mix gently and transfer the sample to a 40-mL VOA vial. Collect bubble-free samples in duplicate and seal each sample in a separate plastic bag.

Soils, sediments and sludges — Use an 8-oz widemouth glass bottle with Teflon®-faced silicone septum that has been prewashed, rinsed with distilled deionized water, and oven dried. **Do not** heat the septum for more than 1 h. Tap slightly to eliminate any free air space. Collect samples in duplicate and seal each one in a separate plastic bag.

SAMPLE PRESERVATION
Liquid samples — Add 4 drops of concentrated HCL, cool to 4°C and store in a solvent-free refrigerator.

Soils, sediments and sludges — Cool samples to 4°C and store in a solvent-free refrigerator.

MHT The maximum holding time of any sample (liquids, soils, sediments, and sludges) is 14 days.

SAMPLE PREPARATION
Liquid samples — Remove the plunger from a 5-mL syringe and carefully pour the sample into the syringe barrel to just short of overflowing. Replace the syringe plunger and compress the sample. Open the syringe valve and vent any residual air while adjusting the sample volume to 5.0 mL. If there is only one volatile organic analysis (VOA) vial, a second syringe should be filled at this time to protect against possible loss of sample integrity. Add 10 µL of surrogate spiking solution and 10 µL of internal standard spiking solution through the valve bore of the 5-mL syringe, then close the valve. The surrogate and internal standards may be mixed and added as a single spiking solution.

Sediments, soils, and waste samples — All samples of this type should be screened by GC analysis using a headspace method (EPA Method 3810) or the hexadecane extraction and screening method (EPA Method 3820). Use the screening data to determine whether to use the low-concentration method (0.005–1 mg/kg) or the high-concentration method (>1 mg/kg).

Low-concentration method — The low-concentration method is based on purging a heated sediment or soil sample mixed with organic-free reagent water containing the surrogate and internal standards. Analyze all reagent blanks and standards under the same conditions as the samples.

Use a 5-g sample if the expected concentration is <0.1 mg/kg or a 1-g sample for expected concentrations between 0.1 and 1 mg/kg. Mix the contents of the sample container with a narrow metal spatula. Weigh the amount of the sample into a tared purge device. Add the spiked water to the purge device, which contains the weighed amount of sample, and connect the device to the purge-and-trap system.

High-concentration method — This method is based on extracting the sediment or soil with methanol. A waste sample is either extracted or diluted, depending on its solubility in methanol. Wastes that are insoluble in methanol are diluted with reagent tetraglyme or possibly polyethylene glycol (PEG). An aliquot of the extract is added to organic-free reagent water containing surrogate and internal standards. This is purged at

ambient temperature. All samples with an expected concentration of >1.0 mg/kg should be analyzed by this method.

Mix the contents of the sample container with a narrow metal spatula. For sediments or soils and solid wastes that are insoluble in methanol, weigh 4 g (wet weight) of sample into a tared 20-mL vial. For waste that is soluble in methanol, tetraglyme, or PEG, weigh 1 g (wet weight) into a tared scintillation vial or culture tube or a 10-mL volumetric flask. Quickly add 9.0 mL of appropriate solvent then add 1.0 mL of a surrogate spiking solution to the vial, cap it, and shake it for 2 min.

METHANOL EXTRACT REQUIRED FOR ANALYSIS OF HIGH-CONCENTRATION SOILS OR SEDIMENTS

Approximate Concentration Range	Volume of Methanol Extract (a)
500–10,000 µg/kg	100 µL
1,000–20,000 µg/kg	50 µL
5,000–100,000 µg/kg	10 µL
25,000–500,000 µg/kg	100 µL of 1/50 dilution (b)

Calculate appropriate dilution factor for concentrations exceeding this table.

(a) The volume of methanol added to 5 mL of water being purged should be kept constant. Therefore, add to the 5-mL syringe whatever volume of methanol is necessary to maintain a volume of 100 µL added to the syringe.
(b) Dilute an aliquot of the methanol extract and then take 100 µL for analysis.

QUALITY CONTROL Demonstrate, through the analysis of a reagent water blank, that interferences from the analytical system, glassware, and reagents are under control. Blank samples should be carried through all stages of the sample preparation and measurement steps. For each analytical batch (up to 20 samples), a reagent blank, matrix spike, and matrix spike duplicate must be analyzed (the frequency of the spikes may be different for different monitoring programs). The blank and spiked samples must be carried through all stages of the sample preparation and measurement steps. QC samples mentioned in the section on Interferences will also be needed as appropriate to those situations.

Matrix spiking standards should be prepared from volatile organic compounds which will be representative of the compounds being investigated. The recommended internal standards are chlorobenzene-d5, 1,4-difluorobenzene, 1,4-dichlorobenzene-d4, and pentafluorobenzene. Using stock standard solutions, prepare secondary dilution standards containing the compounds of interest, either singly or mixed together in methanol. Store them in a vial with no headspace for no more than one week. Surrogates recommended are toluene-d8, 4-bromofluorobenzene, and dibromofluoromethane. Each sample undergoing GC/MS analysis must be spiked with 10 µL of the surrogate spiking solution prior to analysis.

REFERENCE Test Methods for Evaluating Solid Waste (SW-846). U.S. EPA 1983, Method 8260A, Rev. 1, Nov. 1990. Office of Solid Waste, Washington, DC.

---

m-Xylene **EPA Method 8260**
CAS #108-38-3

TITLE Volatile Organic Compounds by GC/MS: Capillary Column Technique

MATRIX This method is applicable to nearly all types of samples, regardless of water content, including groundwater, soils, and sediments.

METHOD SUMMARY Method 8260A covers 58 volatile organic compounds that are introduced into a gas chromatograph by the purge-and-trap method or by direct injection (in limited applications). Zero-grade helium is bubbled through a 5-mL solution at ambient temperature. Purged sample components are trapped in a tube containing suitable sorbent materials. When purging is complete, the sorbent tube is heated and backflushed with helium to desorb trapped sample components. The analytes are desorbed directly to a large bore capillary or cryofocussed on a capillary precolumn before being flash evaporated to a narrow bore capillary for analysis.

INTERFERENCES Major contaminant sources are volatile materials in the lab and impurities in the inert purging gas and in the sorbent trap. Interfering contamination may occur when a sample containing low concentrations of volatile organic compounds is analyzed immediately after a sample containing high concentrations of volatile organic compounds. After analysis of a sample containing high concentrations of volatile organic compounds, one or more calibration blanks should be analyzed to check for cross-contamination. Screening of the samples prior to purge-and-trap GC/MS analysis is highly recommended to prevent contamination of the system. This is especially true for soil and waste samples.

Special precautions must be taken to analyze for methylene chloride. The analytical and sample storage area should be isolated from all atmospheric sources of methylene chloride. All gas chromatography carrier gas lines and purge gas plumbing should be constructed from stainless steel or copper tubing. Laboratory clothing previously exposed to methylene chloride fumes during liquid-liquid extraction procedures can contribute to sample contamination.

Samples can also be contaminated by diffusion of volatile organics (particularly methylene chloride and fluorocarbons) through the septum seal during shipment and storage. A trip blank can serve as a check on such contamination.

INSTRUMENTATION GC/MS with a temperature-programmable chromatograph suitable for splitless injection equipped with variable constant differential flow controllers, a subambient oven controller, a purging device, sorbent trap, a thermal desorption apparatus and a capillary precolumn interface when using cryogenic cooling will be needed. The following GC columns may be used:

Column 1: 60 m × 0.75mm I.D. capillary column coated with VOCOL, 1.5 µm film thickness.
Column 2: 30 m × 0.53mm capillary column coated with DB-624 or VOCOL, 3 µm film thickness.

Column 3: 30 m × 0.32mm I.D. capillary column coated with DB-5 or SE-54, 1-μm film thickness.

PRECISION & ACCURACY This method has been tested in a single lab using spiked water. Using a wide-bore capillary column, water was spiked at concentrations between 0.5 and 10 μg/L. Single lab accuracy and precision data are presented. The MDL actually achieved in a given analysis will vary depending on instrument sensitivity and matrix effects.

The MDL (a) in μg/L was 0.05.
The concentration range in μg/L was 0.1–10.
The mean accuracy (% of true value) was 97.
The precision as relative standard deviation was 6.5.

Note: The MDL is based on a 25-mL sample volume instead of a 5-mL sample volume.

SAMPLING METHOD

Liquid samples — Use a 40-mL glass screw-cap VOA vial with a Teflon®-faced silicone septum that has been prewashed, rinsed with distilled deionized water, and oven dried. If residual chlorine is present, collect the sample in a 4-oz soil VOA container which has been pre-preserved with 4 drops of 10% sodium thiosulfate. Mix gently and transfer the sample to a 40-mL VOA vial. Collect bubble-free samples in duplicate and seal each sample in a separate plastic bag.

Soils, sediments and sludges — Use an 8-oz widemouth glass bottle with Teflon®-faced silicone septum that has been prewashed, rinsed with distilled deionized water, and oven dried. **Do not** heat the septum for more than 1 h. Tap slightly to eliminate any free air space. Collect samples in duplicate and seal each one in a separate plastic bag.

SAMPLE PRESERVATION

Liquid samples — Add 4 drops of concentrated HCL, cool to 4°C and store in a solvent-free refrigerator.

Soils, sediments and sludges — Cool samples to 4°C and store in a solvent-free refrigerator.

MHT The maximum holding time of any sample (liquids, soils, sediments, and sludges) is 14 days.

SAMPLE PREPARATION

Liquid samples — Remove the plunger from a 5-mL syringe and carefully pour the sample into the syringe barrel to just short of overflowing. Replace the syringe plunger and compress the sample. Open the syringe valve and vent any residual air while adjusting the sample volume to 5.0 mL. If there is only one volatile organic analysis (VOA) vial, a second syringe should be filled at this time to protect against possible loss of sample integrity. Add 10 μL of surrogate spiking solution and 10 μL of internal standard spiking solution through the valve bore of the 5-mL syringe, then close the valve. The surrogate and internal standards may be mixed and added as a single spiking solution.

Sediments, soils, and waste samples — All samples of this type should be screened by GC analysis using a headspace method (EPA Method 3810) or the hexadecane extraction and screening method (EPA Method 3820). Use the screening data to determine whether to use the low-concentration method (0.005–1 mg/kg) or the high-concentration method (>1 mg/kg).

Low-concentration method — The low-concentration method is based on purging a heated sediment or soil sample mixed with organic-free reagent water containing the surrogate and internal standards. Analyze all reagent blanks and standards under the same conditions as the samples.

Use a 5-g sample if the expected concentration is <0.1 mg/kg or a 1-g sample for expected concentrations between 0.1 and 1 mg/kg. Mix the contents of the sample container with a narrow metal spatula. Weigh the amount of the sample into a tared purge device. Add the spiked water to the purge device, which contains the weighed amount of sample, and connect the device to the purge-and-trap system.

High-concentration method — This method is based on extracting the sediment or soil with methanol. A waste sample is either extracted or diluted, depending on its solubility in methanol. Wastes that are insoluble in methanol are diluted with reagent tetraglyme or possibly polyethylene glycol (PEG). An aliquot of the extract is added to organic-free reagent water containing surrogate and internal standards. This is purged at ambient temperature. All samples with an expected concentration of >1.0 mg/kg should be analyzed by this method.

Mix the contents of the sample container with a narrow metal spatula. For sediments or soils and solid wastes that are insoluble in methanol, weigh 4 g (wet weight) of sample into a tared 20-mL vial. For waste that is soluble in methanol, tetraglyme, or PEG, weigh 1 g (wet weight) into a tared scintillation vial or culture tube or a 10-mL volumetric flask. Quickly add 9.0 mL of appropriate solvent then add 1.0 mL of a surrogate spiking solution to the vial, cap it, and shake it for 2 min.

METHANOL EXTRACT REQUIRED FOR ANALYSIS OF HIGH-CONCENTRATION SOILS OR SEDIMENTS

Approximate Concentration Range	Volume of Methanol Extract (a)
500–10,000 μg/kg	100 μL
1,000–20,000 μg/kg	50 μL
5,000–100,000 μg/kg	10 μL
25,000–500,000 μg/kg	100 μL of 1/50 dilution (b)

Calculate appropriate dilution factor for concentrations exceeding this table.

(a) The volume of methanol added to 5 mL of water being purged should be kept constant. Therefore, add to the 5-mL syringe whatever volume of methanol is necessary to maintain a volume of 100 μL added to the syringe.
(b) Dilute an aliquot of the methanol extract and then take 100 μL for analysis.

QUALITY CONTROL Demonstrate, through the analysis of a reagent water blank, that interferences from the analytical system, glassware, and reagents are under control. Blank samples should be carried through all stages of the sample preparation and measurement steps. For each analytical batch (up to 20 samples), a reagent blank, matrix spike, and matrix spike duplicate must be analyzed (the frequency of the spikes may be different for different monitoring programs). The blank and

spiked samples must be carried through all stages of the sample preparation and measurement steps. QC samples mentioned in the section on Interferences will also be needed as appropriate to those situations.

Matrix spiking standards should be prepared from volatile organic compounds which will be representative of the compounds being investigated. The recommended internal standards are chlorobenzene-d5, 1,4-difluorobenzene, 1,4-dichlorobenzene-d4, and pentafluorobenzene. Using stock standard solutions, prepare secondary dilution standards containing the compounds of interest, either singly or mixed together in methanol. Store them in a vial with no headspace for no more than one week. Surrogates recommended are toluene-d8, 4-bromofluorobenzene, and dibromofluoromethane. Each sample undergoing GC/MS analysis must be spiked with 10 µL of the surrogate spiking solution prior to analysis.

REFERENCE Test Methods for Evaluating Solid Waste (SW-846). U.S. EPA 1983, Method 8260A, Rev. 1, Nov. 1990. Office of Solid Waste, Washington, DC.

---

o-Xylene   EPA Method 8260
CAS #95-47-6

TITLE Volatile Organic Compounds by GC/MS: Capillary Column Technique

MATRIX This method is applicable to nearly all types of samples, regardless of water content, including groundwater, soils, and sediments.

METHOD SUMMARY Method 8260A covers 58 volatile organic compounds that are introduced into a gas chromatograph by the purge-and-trap method or by direct injection (in limited applications). Zero-grade helium is bubbled through a 5-mL solution at ambient temperature. Purged sample components are trapped in a tube containing suitable sorbent materials. When purging is complete, the sorbent tube is heated and backflushed with helium to desorb trapped sample components. The analytes are desorbed directly to a large bore capillary or cryofocussed on a capillary precolumn before being flash evaporated to a narrow bore capillary for analysis.

INTERFERENCES Major contaminant sources are volatile materials in the lab and impurities in the inert purging gas and in the sorbent trap. Interfering contamination may occur when a sample containing low concentrations of volatile organic compounds is analyzed immediately after a sample containing high concentrations of volatile organic compounds. After analysis of a sample containing high concentrations of volatile organic compounds, one or more calibration blanks should be analyzed to check for cross-contamination. Screening of the samples prior to purge-and-trap id/liquid extraction procedures can contribute to sample contamination.

Samples can also be contaminated by diffusion of volatile organics (particularly methylene chloride and fluorocarbons) through the septum seal during shipment and storage. A trip blank can serve as a check on such contamination.

INSTRUMENTATION GC/MS with a temperature-programmable chromatograph suitable for splitless injection equipped with variable constant differential flow controllers, a subambient oven controller, a purging device, sorbent trap, a thermal desorption apparatus and a capillary precolumn interface when using cryogenic cooling will be needed. The following GC columns may be used:

Column 1: 60 m × 0.75mm I.D. capillary column coated with VOCOL, 1.5 µm film thickness.
Column 2: 30 m × 0.53mm capillary column coated with DB-624 or VOCOL, 3 µm film thickness.
Column 3: 30 m × 0.32mm I.D. capillary column coated with DB-5 or SE-54, 1-µm film thickness.

PRECISION & ACCURACY This method has been tested in a single lab using spiked water. Using a wide-bore capillary column, water was spiked at concentrations between 0.5 and 10 µg/L. Single lab accuracy and precision data are presented. The MDL actually achieved in a given analysis will vary depending on instrument sensitivity and matrix effects.

The MDL (a) in µg/L was 0.11.
The concentration range in µg/L was 0.1–31.
The mean accuracy (% of true value) was 103.
The precision as relative standard deviation was 7.2.

Note: The MDL is based on a 25-mL sample volume instead of a 5-mL sample volume.

SAMPLING METHOD
Liquid samples — Use a 40-mL glass screw-cap VOA vial with a Teflon®-faced silicone septum that has been prewashed, rinsed with distilled deionized water, and oven dried. If residual chlorine is present, collect the sample in a 4-oz soil VOA container which has been pre-preserved with 4 drops of 10% sodium thiosulfate. Mix gently and transfer the sample to a 40-mL VOA vial. Collect bubble-free samples in duplicate and seal each sample in a separate plastic bag.

Soils, sediments and sludges — Use an 8-oz widemouth glass bottle with Teflon®-faced silicone septum that has been prewashed, rinsed with distilled deionized water, and oven dried. **Do not** heat the septum for more than 1 h. Tap slightly to eliminate any free air space. Collect samples in duplicate and seal each one in a separate plastic bag.

SAMPLE PRESERVATION
Liquid samples — Add 4 drops of concentrated HCL, cool to 4°C and store in a solvent-free refrigerator.

Soils, sediments and sludges — Cool samples to 4°C and store in a solvent-free refrigerator.

MHT The maximum holding time of any sample (liquids, soils, sediments, and sludges) is 14 days.

SAMPLE PREPARATION
Liquid samples — Remove the plunger from a 5-mL syringe and carefully pour the sample into the syringe barrel to just short of overflowing. Replace the syringe plunger and compress the sample. Open the syringe valve and vent any residual air while adjusting the sample volume to 5.0 mL. If there is only one volatile organic analysis (VOA) vial, a second syringe

should be filled at this time to protect against possible loss of sample integrity. Add 10 µL of surrogate spiking solution and 10 µL of internal standard spiking solution through the valve bore of the 5-mL syringe, then close the valve. The surrogate and internal standards may be mixed and added as a single spiking solution.

Sediments, soils, and waste samples — All samples of this type should be screened by GC analysis using a headspace method (EPA Method 3810) or the hexadecane extraction and screening method (EPA Method 3820). Use the screening data to determine whether to use the low-concentration method (0.005–1 mg/kg) or the high-concentration method (>1 mg/kg).

Low-concentration method — The low-concentration method is based on purging a heated sediment or soil sample mixed with organic-free reagent water containing the surrogate and internal standards. Analyze all reagent blanks and standards under the same conditions as the samples.

Use a 5-g sample if the expected concentration is <0.1 mg/kg or a 1-g sample for expected concentrations between 0.1 and 1 mg/kg. Mix the contents of the sample container with a narrow metal spatula. Weigh the amount of the sample into a tared purge device. Add the spiked water to the purge device, which contains the weighed amount of sample, and connect the device to the purge-and-trap system.

High-concentration method — This method is based on extracting the sediment or soil with methanol. A waste sample is either extracted or diluted, depending on its solubility in methanol. Wastes that are insoluble in methanol are diluted with reagent tetraglyme or possibly polyethylene glycol (PEG). An aliquot of the extract is added to organic-free reagent water containing surrogates and internal standards. This is purged at ambient temperature. All samples with an expected concentration of >1.0 mg/kg should be analyzed by this method.

Mix the contents of the sample container with a narrow metal spatula. For sediments or soils and solid wastes that are insoluble in methanol, weigh 4 g (wet weight) of sample into a tared 20-mL vial. For waste that is soluble in methanol, tetraglyme, or PEG, weigh 1 g (wet weight) into a tared scintillation vial or culture tube or a 10-mL volumetric flask. Quickly add 9.0 mL of appropriate solvent then add 1.0 mL of a surrogate spiking solution to the vial, cap it, and shake it for 2 min.

METHANOL EXTRACT REQUIRED FOR ANALYSIS OF HIGH-CONCENTRATION SOILS OR SEDIMENTS

Approximate Concentration Range	Volume of Methanol Extract (a)
500–10,000 µg/kg	100 µL
1,000–20,000 µg/kg	50 µL
5,000–100,000 µg/kg	10 µL
25,000–500,000 µg/kg	100 µL of 1/50 dilution (b)

Calculate appropriate dilution factor for concentrations exceeding this table.

(a) The volume of methanol added to 5 mL of water being purged should be kept constant. Therefore, add to the 5-mL syringe whatever volume of methanol is necessary to maintain a volume of 100 µL added to the syringe.
(b) Dilute an aliquot of the methanol extract and then take 100 µL for analysis.

QUALITY CONTROL Demonstrate, through the analysis of a reagent water blank, that interferences from the analytical system, glassware, and reagents are under control. Blank samples should be carried through all stages of the sample preparation and measurement steps. For each analytical batch (up to 20 samples), a reagent blank, matrix spike, and matrix spike duplicate must be analyzed (the frequency of the spikes may be different for different monitoring programs). The blank and spiked samples must be carried through all stages of the sample preparation and measurement steps. QC samples mentioned in the section on Interferences will also be needed as appropriate to those situations.

Matrix spiking standards should be prepared from volatile organic compounds which will be representative of the compounds being investigated. The recommended internal standards are chlorobenzene-d5, 1,4-difluorobenzene, 1,4-dichlorobenzene-d4, and pentafluorobenzene. Using stock standard solutions, prepare secondary dilution standards containing the compounds of interest, either singly or mixed together in methanol. Store them in a vial with no headspace for no more than one week. Surrogates recommended are toluene-d8, 4-bromofluorobenzene, and dibromofluoromethane. Each sample undergoing GC/MS analysis must be spiked with 10 µL of the surrogate spiking solution prior to analysis.

REFERENCE Test Methods for Evaluating Solid Waste (SW-846). U.S. EPA 1983, Method 8260A, Rev. 1, Nov. 1990. Office of Solid Waste, Washington, DC.

---

o-Xylene EPA Method 8260
CAS #95-47-6

TITLE Volatile Organic Compounds by GC/MS: Capillary Column Technique

MATRIX This method is applicable to nearly all types of samples, regardless of water content, including groundwater, soils, and sediments.

METHOD SUMMARY Method 8260A covers 58 volatile organic compounds that are introduced into a gas chromatograph by the purge-and-trap method or by direct injection (in limited applications). Zero-grade helium is bubbled through a 5-mL solution at ambient temperature. Purged sample components are trapped in a tube containing suitable sorbent materials. When purging is complete, the sorbent tube is heated and backflushed with helium to desorb trapped sample components. The analytes are desorbed directly to a large bore capillary or cryofocussed on a capillary precolumn before being flash evaporated to a narrow bore capillary for analysis.

INTERFERENCES Major contaminant sources are volatile materials in the lab and impurities in the inert purging gas and in the sorbent trap. Interfering contamination may occur when a sample containing low concentrations of volatile organic compounds is analyzed immediately after a sample containing high concentrations of volatile organic compounds. After analysis of a sample containing high concentrations of volatile organic compounds, one or more calibration blanks should be analyzed to check for cross-contamination. Screening of the samples prior to purge-and-trap GC/MS analysis is highly recommended to prevent contamination of the system. This is especially true for soil and waste samples.

Special precautions must be taken to analyze for methylene chloride. The analytical and sample storage area should be isolated from all atmospheric sources of methylene chloride. All gas chromatography carrier gas lines and purge gas plumbing should be constructed from stainless steel or copper tubing. Laboratory clothing previously exposed to methylene chloride fumes during liquid-liquid extraction procedures can contribute to sample contamination.

Samples can also be contaminated by diffusion of volatile organics (particularly methylene chloride and fluorocarbons) through the septum seal during shipment and storage. A trip blank can serve as a check on such contamination.

INSTRUMENTATION   GC/MS with a temperature-programmable chromatograph suitable for splitless injection equipped with variable constant differential flow controllers, a subambient oven controller, a purging device, sorbent trap, a thermal desorption apparatus and a capillary precolumn interface when using cryogenic cooling will be needed. The following GC columns may be used:

Column 1: 60 m × 0.75mm I.D. capillary column coated with VOCOL, 1.5 µm film thickness.
Column 2: 30 m × 0.53mm capillary column coated with DB-624 or VOCOL, 3 µm film thickness.
Column 3: 30 m × 0.32mm I.D. capillary column coated with DB-5 or SE-54, 1-µm film thickness.

PRECISION & ACCURACY   This method has been tested in a single lab using spiked water. Using a wide-bore capillary column, water was spiked at concentrations between 0.5 and 10 µg/L. Single lab accuracy and precision data are presented. The MDL actually achieved in a given analysis will vary depending on instrument sensitivity and matrix effects.

The MDL (a) in µg/L was 0.11.
The concentration range in µg/L was 0.1–31.
The mean accuracy (% of true value) was 103.
The precision as relative standard deviation was 7.2.

Note: The MDL is based on a 25-mL sample volume instead of a 5-mL sample volume.

SAMPLING METHOD

Liquid samples — Use a 40-mL glass screw-cap VOA vial with a Teflon®-faced silicone septum that has been prewashed, rinsed with distilled deionized water, and oven dried. If residual chlorine is present, collect the sample in a 4-oz soil VOA container which has been pre-preserved with 4 drops of 10% sodium thiosulfate. Mix gently and transfer the sample to a 40-mL VOA vial. Collect bubble-free samples in duplicate and seal each sample in a separate plastic bag.

Soils, sediments and sludges — Use an 8-oz widemouth glass bottle with Teflon®-faced silicone septum that has been prewashed, rinsed with distilled deionized water, and oven dried. **Do not** heat the septum for more than 1 h. Tap slightly to eliminate any free air space. Collect samples in duplicate and seal each one in a separate plastic bag.

SAMPLE PRESERVATION

Liquid samples — Add 4 drops of concentrated HCL, cool to 4°C and store in a solvent-free refrigerator.

Soils, sediments and sludges — Cool samples to 4°C and store in a solvent-free refrigerator.

MHT   The maximum holding time of any sample (liquids, soils, sediments, and sludges) is 14 days.

SAMPLE PREPARATION

Liquid samples — Remove the plunger from a 5-mL syringe and carefully pour the sample into the syringe barrel to just short of overflowing. Replace the syringe plunger and compress the sample. Open the syringe valve and vent any residual air while adjusting the sample volume to 5.0 mL. If there is only one volatile organic analysis (VOA) vial, a second syringe should be filled at this time to protect against possible loss of sample integrity. Add 10 µL of surrogate spiking solution and 10 µL of internal standard spiking solution through the valve bore of the 5-mL syringe, then close the valve. The surrogate and internal standards may be mixed and added as a single spiking solution.

Sediments, soils, and waste samples — All samples of this type should be screened by GC analysis using a headspace method (EPA Method 3810) or the hexadecane extraction and screening method (EPA Method 3820). Use the screening data to determine whether to use the low-concentration method (0.005–1 mg/kg) or the high-concentration method (>1 mg/kg).

Low-concentration method — The low-concentration method is based on purging a heated sediment or soil sample mixed with organic-free reagent water containing the surrogate and internal standards. Analyze all reagent blanks and standards under the same conditions as the samples.

Use a 5-g sample if the expected concentration is <0.1 mg/kg or a 1-g sample for expected concentrations between 0.1 and 1 mg/kg. Mix the contents of the sample container with a narrow metal spatula. Weigh the amount of the sample into a tared purge device. Add the spiked water to the purge device, which contains the weighed amount of sample, and connect the device to the purge-and-trap system.

High-concentration method — This method is based on extracting the sediment or soil with methanol. A waste sample is either extracted or diluted, depending on its solubility in methanol. Wastes that are insoluble in methanol are diluted with reagent tetraglyme or possibly polyethylene glycol (PEG). An aliquot of the extract is added to organic-free reagent water containing surrogate and internal standards. This is purged at ambient temperature. All samples with an expected concentration of >1.0 mg/kg should be analyzed by this method.

Mix the contents of the sample container with a narrow metal spatula. For sediments or soils and solid wastes that are insoluble in methanol, weigh 4 g (wet weight) of sample into a tared 20-mL vial. For waste that is soluble in methanol, tetraglyme, or PEG, weigh 1 g (wet weight) into a tared scintillation vial or culture tube or a 10-mL volumetric flask. Quickly add 9.0 mL of appropriate solvent then add 1.0 mL of a surrogate spiking solution to the vial, cap it, and shake it for 2 min.

METHANOL EXTRACT REQUIRED FOR ANALYSIS OF HIGH-CONCENTRATION SOILS OR SEDIMENTS

Approximate Concentration Range	Volume of Methanol Extract (a)
500–10,000 µg/kg	100 µL
1,000–20,000 µg/kg	50 µL
5,000–100,000 µg/kg	10 µL
25,000–500,000 µg/kg	100 µL of 1/50 dilution (b)

Calculate appropriate dilution factor for concentrations exceeding this table.

(a) The volume of methanol added to 5 mL of water being purged should be kept constant. Therefore, add to the 5-mL syringe whatever volume of methanol is necessary to maintain a volume of 100 µL added to the syringe.
(b) Dilute an aliquot of the methanol extract and then take 100 µL for analysis.

QUALITY CONTROL Demonstrate, through the analysis of a reagent water blank, that interferences from the analytical system, glassware, and reagents are under control. Blank samples should be carried through all stages of the sample preparation and measurement steps. For each analytical batch (up to 20 samples), a reagent blank, matrix spike, and matrix spike duplicate must be analyzed (the frequency of the spikes may be different for different monitoring programs). The blank and spiked samples must be carried through all stages of the sample preparation and measurement steps. QC samples mentioned in the section on Interferences will also be needed as appropriate to those situations.

Matrix spiking standards should be prepared from volatile organic compounds which will be representative of the compounds being investigated. The recommended internal standards are chlorobenzene-d5, 1,4-difluorobenzene, 1,4-dichlorobenzene-d4, and pentafluorobenzene. Using stock standard solutions, prepare secondary dilution standards containing the compounds of interest, either singly or mixed together in methanol. Store them in a vial with no headspace for no more than one week. Surrogates recommended are toluene-d8, 4-bromofluorobenzene, and dibromofluoromethane. Each sample undergoing GC/MS analysis must be spiked with 10 µL of the surrogate spiking solution prior to analysis.

REFERENCE Test Methods for Evaluating Solid Waste (SW-846). U.S. EPA 1983, Method 8260A, Rev. 1, Nov. 1990. Office of Solid Waste, Washington, DC.

---

p-Xylene **EPA Method 8260**
CAS #106-42-3

TITLE Volatile Organic Compounds by GC/MS: Capillary Column Technique

MATRIX This method is applicable to nearly all types of samples, regardless of water content, including groundwater, soils, and sediments.

METHOD SUMMARY Method 8260A covers 58 volatile organic compounds that are introduced into a gas chromatograph by the purge-and-trap method or by direct injection (in limited applications). Zero-grade helium is bubbled through a 5-mL solution at ambient temperature. Purged sample components are trapped in a tube containing suitable sorbent materials. When purging is complete, the sorbent tube is heated and backflushed with helium to desorb trapped sample components. The analytes are desorbed directly to a large bore capillary or cryofocussed on a capillary precolumn before being flash evaporated to a narrow bore capillary for analysis.

INTERFERENCES Major contaminant sources are volatile materials in the lab and impurities in the inert purging gas and in the sorbent trap. Interfering contamination may occur when a sample containing low concentrations of volatile organic compounds is analyzed immediately after a sample containing high concentrations of volatile organic compounds. After analysis of a sample containing high concentrations of volatile organic compounds, one or more calibration blanks should be analyzed to check for cross-contamination. Screening of the samples prior to purge-and-trap GC/MS analysis is highly recommended to prevent contamination of the system. This is especially true for soil and waste samples.

Special precautions must be taken to analyze for methylene chloride. The analytical and sample storage area should be isolated from all atmospheric sources of methylene chloride. All gas chromatography carrier gas lines and purge gas plumbing should be constructed from stainless steel or copper tubing. Laboratory clothing previously exposed to methylene chloride fumes during liquid-liquid extraction procedures can contribute to sample contamination.

Samples can also be contaminated by diffusion of volatile organics (particularly methylene chloride and fluorocarbons) through the septum seal during shipment and storage. A trip blank can serve as a check on such contamination.

INSTRUMENTATION GC/MS with a temperature-programmable chromatograph suitable for splitless injection equipped with variable constant differential flow controllers, a subambient oven controller, a purging device, sorbent trap, a thermal desorption apparatus and a capillary precolumn interface when using cryogenic cooling will be needed. The following GC columns may be used:

Column 1: 60 m × 0.75mm I.D. capillary column coated with VOCOL, 1.5 µm film thickness.
Column 2: 30 m × 0.53mm capillary column coated with DB-624 or VOCOL, 3 µm film thickness.
Column 3: 30 m × 0.32mm I.D. capillary column coated with DB-5 or SE-54, 1-µm film thickness.

PRECISION & ACCURACY This method has been tested in a single lab using spiked water. Using a wide-bore capillary column, water was spiked at concentrations between 0.5 and 10 µg/L. Single lab accuracy and precision data are presented. The MDL actually achieved in a given analysis will vary depending on instrument sensitivity and matrix effects.

The MDL (a) in µg/L was 0.13.
The concentration range in µg/L was 0.5–10.
The mean accuracy (% of true value) was 104.
The precision as relative standard deviation was 7.7.

Note: The MDL is based on a 25-mL sample volume instead of a 5-mL sample volume.

SAMPLING METHOD

Liquid samples — Use a 40-mL glass screw-cap VOA vial with a Teflon®-faced silicone septum that has been prewashed, rinsed with distilled deionized water, and oven dried. If residual chlorine is present, collect the sample in a 4-oz soil VOA container which has been pre-preserved with 4 drops of 10% sodium thiosulfate. Mix gently and transfer the sample to a 40-mL VOA vial. Collect bubble-free samples in duplicate and seal each sample in a separate plastic bag.

Soils, sediments and sludges — Use an 8-oz widemouth glass bottle with Teflon®-faced silicone septum that has been prewashed, rinsed with distilled deionized water, and oven dried. **Do not** heat the septum for more than 1 h. Tap slightly to eliminate any free air space. Collect samples in duplicate and seal each one in a separate plastic bag.

SAMPLE PRESERVATION

Liquid samples — Add 4 drops of concentrated HCL, cool to 4°C and store in a solvent-free refrigerator.

Soils, sediments and sludges — Cool samples to 4°C and store in a solvent-free refrigerator.

MHT The maximum holding time of any sample (liquids, soils, sediments, and sludges) is 14 days.

SAMPLE PREPARATION

Liquid samples — Remove the plunger from a 5-mL syringe and carefully pour the sample into the syringe barrel to just short of overflowing. Replace the syringe plunger and compress the sample. Open the syringe valve and vent any residual air while adjusting the sample volume to 5.0 mL. If there is only one volatile organic analysis (VOA) vial, a second syringe should be filled at this time to protect against possible loss of sample integrity. Add 10 µL of surrogate spiking solution and 10 µL of internal standard spiking solution through the valve bore of the 5-mL syringe, then close the valve. The surrogate and internal standards may be mixed and added as a single spiking solution.

Sediments, soils, and waste samples — All samples of this type should be screened by GC analysis using a headspace method (EPA Method 3810) or the hexadecane extraction and screening method (EPA Method 3820). Use the screening data to determine whether to use the low-concentration method (0.005–1 mg/kg) or the high-concentration method (>1 mg/kg).

Low-concentration method — The low-concentration method is based on purging a heated sediment or soil sample mixed with organic-free reagent water containing the surrogate and internal standards. Analyze all reagent blanks and standards under the same conditions as the samples.

Use a 5-g sample if the expected concentration is <0.1 mg/kg or a 1-g sample for expected concentrations between 0.1 and 1 mg/kg. Mix the contents of the sample container with a narrow metal spatula. Weigh the amount of the sample into a tared purge device. Add the spiked water to the purge device, which contains the weighed amount of sample, and connect the device to the purge-and-trap system.

High-concentration method — This method is based on extracting the sediment or soil with methanol. A waste sample is either extracted or diluted, depending on its solubility in methanol. Wastes that are insoluble in methanol are diluted with reagent tetraglyme or possibly polyethylene glycol (PEG). An aliquot of the extract is added to organic-free reagent water containing surrogate and internal standards. This is purged at ambient temperature. All samples with an expected concentration of >1.0 mg/kg should be analyzed by this method.

Mix the contents of the sample container with a narrow metal spatula. For sediments or soils and solid wastes that are insoluble in methanol, weigh 4 g (wet weight) of sample into a tared 20-mL vial. For waste that is soluble in methanol, tetraglyme, or PEG, weigh 1 g (wet weight) into a tared scintillation vial or culture tube or a 10-mL volumetric flask. Quickly add 9.0 mL of appropriate solvent then add 1.0 mL of a surrogate spiking solution to the vial, cap it, and shake it for 2 min.

METHANOL EXTRACT REQUIRED FOR ANALYSIS OF HIGH-CONCENTRATION SOILS OR SEDIMENTS

Approximate Concentration Range	Volume of Methanol Extract (a)
500–10,000 µg/kg	100 µL
1,000–20,000 µg/kg	50 µL
5,000–100,000 µg/kg	10 µL
25,000–500,000 µg/kg	100 µL of 1/50 dilution (b)

Calculate appropriate dilution factor for concentrations exceeding this table.

(a) The volume of methanol added to 5 mL of water being purged should be kept constant. Therefore, add to the 5-mL syringe whatever volume of methanol is necessary to maintain a volume of 100 µL added to the syringe.
(b) Dilute an aliquot of the methanol extract and then take 100 µL for analysis.

QUALITY CONTROL Demonstrate, through the analysis of a reagent water blank, that interferences from the analytical system, glassware, and reagents are under control. Blank samples should be carried through all stages of the sample preparation and measurement steps. For each analytical batch (up to 20 samples), a reagent blank, matrix spike, and matrix spike duplicate must be analyzed (the frequency of the spikes may be different for different monitoring programs). The blank and spiked samples must be carried through all stages of the sample preparation and measurement steps. QC samples mentioned in the section on Interferences will also be needed as appropriate to those situations.

Matrix spiking standards should be prepared from volatile organic compounds which will be representative of the compounds being investigated. The recommended internal standards are chlorobenzene-d5, 1,4-difluorobenzene, 1,4-dichlorobenzene-d4, and pentafluorobenzene. Using stock standard solutions, prepare secondary dilution standards containing the compounds of interest, either singly or mixed together in methanol. Store them in a vial with no headspace for no more than one week. Surrogates recommended are toluene-d8, 4-bromofluorobenzene, and dibromofluoromethane. Each sample undergoing GC/MS analysis must be spiked with 10 μL of the surrogate spiking solution prior to analysis.

REFERENCE Test Methods for Evaluating Solid Waste (SW-846). U.S. EPA 1983, Method 8260A, Rev. 1, Nov. 1990. Office of Solid Waste, Washington, DC.

p-Xylene — EPA Method 8260
CAS #106-42-3

TITLE Volatile Organic Compounds by GC/MS: Capillary Column Technique

MATRIX This method is applicable to nearly all types of samples, regardless of water content, including groundwater, soils, and sediments.

METHOD SUMMARY Method 8260A covers 58 volatile organic compounds that are introduced into a gas chromatograph by the purge-and-trap method or by direct injection (in limited applications). Zero-grade helium is bubbled through a 5-mL solution at ambient temperature. Purged sample components are trapped in a tube containing suitable sorbent materials. When purging is complete, the sorbent tube is heated and backflushed with helium to desorb trapped sample components. The analytes are desorbed directly to a large bore capillary or cryofocussed on a capillary precolumn before being flash evaporated to a narrow bore capillary for analysis.

INTERFERENCES Major contaminant sources are volatile materials in the lab and impurities in the inert purging gas and in the sorbent trap. Interfering contamination may occur when a sample containing low concentrations of volatile organic compounds is analyzed immediately after a sample containing high concentrations of volatile organic compounds. After analysis of a sample containing high concentrations of volatile organic compounds, one or more calibration blanks should be analyzed to check for cross-contamination. Screening of the samples prior to purge-and-trap GC/MS analysis is highly recommended to prevent contamination of the system. This is especially true for soil and waste samples.

Special precautions must be taken to analyze for methylene chloride. The analytical and sample storage area should be isolated from all atmospheric sources of methylene chloride. All gas chromatography carrier gas lines and purge gas plumbing should be constructed from stainless steel or copper tubing. Laboratory clothing previously exposed to methylene chloride fumes during liquid-liquid extraction procedures can contribute to sample contamination.

Samples can also be contaminated by diffusion of volatile organics (particularly methylene chloride and fluorocarbons) through the septum seal during shipment and storage. A trip blank can serve as a check on such contamination.

INSTRUMENTATION GC/MS with a temperature-programmable chromatograph suitable for splitless injection equipped with variable constant differential flow controllers, a subambient oven controller, a purging device, sorbent trap, a thermal desorption apparatus and a capillary precolumn interface when using cryogenic cooling will be needed. The following GC columns may be used:

Column 1: 60 m × 0.75mm I.D. capillary column coated with VOCOL, 1.5 μm film thickness.
Column 2: 30 m × 0.53mm capillary column coated with DB-624 or VOCOL, 3 μm film thickness.
Column 3: 30 m × 0.32mm I.D. capillary column coated with DB-5 or SE-54, 1-μm film thickness.

PRECISION & ACCURACY This method has been tested in a single lab using spiked water. Using a wide-bore capillary column, water was spiked at concentrations between 0.5 and 10 μg/L. Single lab accuracy and precision data are presented. The MDL actually achieved in a given analysis will vary depending on instrument sensitivity and matrix effects.

The MDL (a) in μg/L was 0.13.
The concentration range in μg/L was 0.5–10.
The mean accuracy (% of true value) was 104.
The precision as relative standard deviation was 7.7.

Note: The MDL is based on a 25-mL sample volume instead of a 5-mL sample volume.

SAMPLING METHOD
Liquid samples — Use a 40-mL glass screw-cap VOA vial with a Teflon®-faced silicone septum that has been prewashed, rinsed with distilled deionized water, and oven dried. If residual chlorine is present, collect the sample in a 4-oz soil VOA container which has been pre-preserved with 4 drops of 10% sodium thiosulfate. Mix gently and transfer the sample to a 40-mL VOA vial. Collect bubble-free samples in duplicate and seal each sample in a separate plastic bag.

Soils, sediments and sludges — Use an 8-oz widemouth glass bottle with Teflon®-faced silicone septum that has been prewashed, rinsed with distilled deionized water, and oven dried. **Do not** heat the septum for more than 1 h. Tap slightly to eliminate any free air space. Collect samples in duplicate and seal each one in a separate plastic bag.

SAMPLE PRESERVATION
Liquid samples — Add 4 drops of concentrated HCL, cool to 4°C and store in a solvent-free refrigerator.

Soils, sediments and sludges — Cool samples to 4°C and store in a solvent-free refrigerator.

MHT The maximum holding time of any sample (liquids, soils, sediments, and sludges) is 14 days.

SAMPLE PREPARATION
Liquid samples — Remove the plunger from a 5-mL syringe and carefully pour the sample into the syringe barrel to just short of overflowing. Replace the syringe plunger and compress the sample. Open the syringe valve and vent any residual air while adjusting the sample volume to 5.0 mL. If there is only one volatile organic analysis (VOA) vial, a second syringe

should be filled at this time to protect against possible loss of sample integrity. Add 10 μL of surrogate spiking solution and 10 μL of internal standard spiking solution through the valve bore of the 5-mL syringe, then close the valve. The surrogate and internal standards may be mixed and added as a single spiking solution.

Sediments, soils, and waste samples — All samples of this type should be screened by GC analysis using a headspace method (EPA Method 3810) or the hexadecane extraction and screening method (EPA Method 3820). Use the screening data to determine whether to use the low-concentration method (0.005–1 mg/kg) or the high-concentration method (>1 mg/kg).

Low-concentration method — The low-concentration method is based on purging a heated sediment or soil sample mixed with organic-free reagent water containing the surrogate and internal standards. Analyze all reagent blanks and standards under the same conditions as the samples.

Use a 5-g sample if the expected concentration is <0.1 mg/kg or a 1-g sample for expected concentrations between 0.1 and 1 mg/kg. Mix the contents of the sample container with a narrow metal spatula. Weigh the amount of the sample into a tared purge device. Add the spiked water to the purge device, which contains the weighed amount of sample, and connect the device to the purge-and-trap system.

High-concentration method — This method is based on extracting the sediment or soil with methanol. A waste sample is either extracted or diluted, depending on its solubility in methanol. Wastes that are insoluble in methanol are diluted with reagent tetraglyme or possibly polyethylene glycol (PEG). An aliquot of the extract is added to organic-free reagent water containing surrogate and internal standards. This is purged at ambient temperature. All samples with an expected concentration of >1.0 mg/kg should be analyzed by this method.

Mix the contents of the sample container with a narrow metal spatula. For sediments or soils and solid wastes that are insoluble in methanol, weigh 4 g (wet weight) of sample into a tared 20-mL vial. For waste that is soluble in methanol, tetraglyme, or PEG, weigh 1 g (wet weight) into a tared scintillation vial or culture tube or a 10-mL volumetric flask. Quickly add 9.0 mL of appropriate solvent then add 1.0 mL of a surrogate spiking solution to the vial, cap it, and shake it for 2 min.

METHANOL EXTRACT REQUIRED FOR ANALYSIS OF HIGH-CONCENTRATION SOILS OR SEDIMENTS

Approximate Concentration Range	Volume of Methanol Extract (a)
500–10,000 μg/kg	100 μL
1,000–20,000 μg/kg	50 μL
5,000–100,000 μg/kg	10 μL
25,000–500,000 μg/kg	100 μL of 1/50 dilution (b)

Calculate appropriate dilution factor for concentrations exceeding this table.

(a) The volume of methanol added to 5 mL of water being purged should be kept constant. Therefore, add to the 5-mL syringe whatever volume of methanol is necessary to maintain a volume of 100 μL added to the syringe.
(b) Dilute an aliquot of the methanol extract and then take 100 μL for analysis.

QUALITY CONTROL Demonstrate, through the analysis of a reagent water blank, that interferences from the analytical system, glassware, and reagents are under control. Blank samples should be carried through all stages of the sample preparation and measurement steps. For each analytical batch (up to 20 samples), a reagent blank, matrix spike, and matrix spike duplicate must be analyzed (the frequency of the spikes may be different for different monitoring programs). The blank and spiked samples must be carried through all stages of the sample preparation and measurement steps. QC samples mentioned in the section on Interferences will also be needed as appropriate to those situations.

Matrix spiking standards should be prepared from volatile organic compounds which will be representative of the compounds being investigated. The recommended internal standards are chlorobenzene-d5, 1,4-difluorobenzene, 1,4-dichlorobenzene-d4, and pentafluorobenzene. Using stock standard solutions, prepare secondary dilution standards containing the compounds of interest, either singly or mixed together in methanol. Store them in a vial with no headspace for no more than one week. Surrogates recommended are toluene-d8, 4-bromofluorobenzene, and dibromofluoromethane. Each sample undergoing GC/MS analysis must be spiked with 10 μL of the surrogate spiking solution prior to analysis.

REFERENCE Test Methods for Evaluating Solid Waste (SW-846). U.S. EPA 1983, Method 8260A, Rev. 1, Nov. 1990. Office of Solid Waste, Washington, DC.

---

m-Xylene   EPA Method 503.1
CAS #108-38-3

TITLE Aromatic & Unsaturated VOCs Water

MATRIX Drinking water (finished or in any treatment stage) and raw source water.

APPLICATION Method covers 28 aromatic and unsaturated VOCs. An inert gas is bubbled through a 5-mL water sample. Purged sample components are trapped in tube of sorbent materials. When purging is complete, sorbent tube is heated and backflushed with inert gas to desorb trapped sample onto a packed GC column.

INTERFERENCES During analysis, major contaminant sources are volatile materials in the lab and impurities in purging gas and sorbent trap. With high and low level samples, there can be carryover contamination. Excess water causes a negative baseline deflection.

INSTRUMENTATION Purge and Trap GC w/photoionization detector. (Two GC columns are recommended); Column 1: 5% SP-1200 and 1.75% Bentone 34 on Supelcoport; Column 2: 1,2,3-tris(2-cyanoethoxy)propane on Chromosorb W.

RANGE 2.2–600 μg/L. (Drinking water)

MDL 0.004 µg/L in water

PRECISION RSD = 7.7% at 0.40 µg/L conc.; 7 samples

ACCURACY Average recovery = 90% at 0.40 µg/L conc.; 7 samples

SAMPLING METHOD Use a 40–120-mL screw-cap vial (prewashed with detergent, rinsed with distilled water and oven dried at 105°C) with a PTFE-faced silicone septum. If residual chlorine is in the water add about 25 mg of ascorbic acid to each vial before sample collection. Collect bubble-free samples.

STABILITY Cool to 4°C; HCl to pH <2.

MHT 14 days.

QUALITY CONTROL As initial demonstration of lab accuracy and precision, analyze 4 to 7 replicates of a lab fortified blank containing the analyte at 0.1–5 µg/L. Collect all samples in duplicate.

REFERENCE Method 503.1, Volatile Aromatic & Unsaturated Organic Compounds in H2O by Purge and Trap GC, EPA 600/4-88/039.

---

o-Xylene EPA Method 503.1
CAS #95-47-6

TITLE Aromatic & Unsaturated VOCs Water

MATRIX Drinking water (finished or in any treatment stage) and raw source water.

APPLICATION Method covers 28 aromatic and unsaturated VOCs. An inert gas is bubbled through a 5-mL water sample. Purged sample components are trapped in tube of sorbent materials. When purging is complete, sorbent tube is heated and backflushed with inert gas to desorb trapped sample onto a packed GC column.

INTERFERENCES During analysis, major contaminant sources are volatile materials in the lab and impurities in purging gas and sorbent trap. With high and low level samples, there can be carryover contamination. Excess water causes a negative baseline deflection.

INSTRUMENTATION Purge and Trap GC w/photoionization detector. (Two GC columns are recommended); Column 1: 5% SP-1200 and 1.75% Bentone 34 on Supelcoport; Column 2: 1,2,3-tris(2-cyanoethoxy)propane on Chromosorb W.

RANGE 2.2–600 µg/L. (Drinking water)

MDL 0.004 µg/L in water

PRECISION RSD = 7.2% at 0.40 µg/L conc.; 7 samples

ACCURACY Average recovery = 90% at 0.40 µg/L conc.; 7 samples

SAMPLING METHOD Use a 40–120-mL screw-cap vial (prewashed with detergent, rinsed with distilled water and oven dried at 105°C) with a PTFE-faced silicone septum. If residual chlorine is in the water add about 25 mg of ascorbic acid to each vial before sample collection. Collect bubble-free samples.

STABILITY Cool to 4°C; HCl to pH <2.

MHT 14 days.

QUALITY CONTROL As initial demonstration of lab accuracy and precision, analyze 4 to 7 replicates of a lab fortified blank containing the analyte at 0.1–5 µg/L. Collect all samples in duplicate.

REFERENCE Method 503.1, Volatile Aromatic & Unsaturated Organic Compounds in H2O by Purge and Trap GC, EPA 600/4-88/039.

---

p-Xylene EPA Method 503.1
CAS #106-42-3

TITLE Aromatic & Unsaturated VOCs Water

MATRIX Drinking water (finished or in any treatment stage) and raw source water.

APPLICATION Method covers 28 aromatic and unsaturated VOCs. An inert gas is bubbled through a 5-mL water sample. Purged sample components are trapped in tube of sorbent materials. When purging is complete, sorbent tube is heated and backflushed with inert gas to desorb trapped sample onto a packed GC column.

INTERFERENCES During analysis, major contaminant sources are volatile materials in the lab and impurities in purging gas and sorbent trap. With high and low level samples, there can be carryover contamination. Excess water causes a negative baseline deflection.

INSTRUMENTATION Purge and Trap GC w/photoionization detector. (Two GC columns are recommended); Column 1: 5% SP-1200 and 1.75% Bentone 34 on Supelcoport; Column 2: 1,2,3-tris(2-cyanoethoxy)propane on Chromosorb W.

RANGE 2.2–600 µg/L. (Drinking water)

MDL 0.002 µg/L in water

PRECISION RSD = 8.7% at 0.40 µg/L conc.; 7 samples

ACCURACY Average recovery = 85% at 0.40 µg/L conc.; 7 samples

SAMPLING METHOD Use a 40–120-mL screw-cap vial (prewashed with detergent, rinsed with distilled water and oven dried at 105°C) with a PTFE-faced silicone septum. If residual chlorine is in the water add about 25 mg of ascorbic acid to each vial before sample collection. Collect bubble-free samples.

STABILITY Cool to 4°C; HCl to pH <2.

MHT 14 days.

QUALITY CONTROL As initial demonstration of lab accuracy and precision, analyze 4 to 7 replicates of a lab fortified blank containing the analyte at 0.1–5 µg/L. Collect all samples in duplicate.

REFERENCE Method 503.1, Volatile Aromatic & Unsaturated Organic Compounds in H2O by Purge and Trap GC, EPA 600/4-88/039.

Xylenes   EPA Method 8240
CAS #1330-20-7

TITLE Volatile Organics By GC/MS: Packed Column Technique

MATRIX Nearly all types of sample matarices, regardless of water content, can be analyzed using this method. This includes groundwater, aqueous sludges, caustic liquors, acid liquors, waste solvents, oily wastes, mousses, tars, fibrous wastes, polymetric emulsions, filter cakes, spent carbons, spent catalysts, soils, and sediments.

METHOD SUMMARY Method 8240B covers 80 volatile organic compounds that are introduced into a gas chromatograph by the purge-and-trap method or by direct injection (in limited applications). For the purge-and-trap method an inert gas (zero grade nitrogen or helium) is bubbled through a 5-mL solution at ambient temperature. Purged sample components are trapped in a tube of sorbent materials. When purging is complete, the sorbent tube is heated and backflushed with inert gas to desorb the trapped components onto a GC column.

INTERFERENCES Impurities in the purge gas and from organic compounds outgassing from the plumbing ahead of the trap account for many contamination problems. Interferences purged or coextracted from the samples will vary considerably from source to source. Cross-contamination can occur whenever high-level and low-level samples are analyzed sequentially. Whenever an unusually concentrated sample is analyzed, it should be followed by the analysis of organic-free reagent water to check for cross-contamination. Samples also can be contaminated by diffusion of volatile organics (particularly methylene chloride and fluorocarbons) through the septum seal into the sample during shipment and storage. A trip blank can serve as a check on such contamination. The lab where volatile analysis is performed and also the refrigerated storage area should be completely free of solvents.

INSTRUMENTATION A gas chromatograph/mass spectrometry/data system (GC/MS) equipped with a 6 ft × 0.1 in I.D. glass column packed with 1% SP-1000 on Carbopack-B (60/80 mesh) is required. Also needed is a 5-mL purging device, a sorbent trap, and a thermal desorption apparatus.

PRECISION & ACCURACY This method is reported to have been tested by 15 laboratories using organic-free reagent water, drinking water, surface water, and industrial wastewaters (not specified) fortified at six concentrations over the range 5–600 µg/L.

Sample estimated quantitation limits (EQLs) are highly matrix-dependent. The EQLs listed may not always be achievable. EQLs listed for soils or sediments are based on wet weight. Normally, data is reported on a dry-weight basis; therefore, EQLs will be higher, based on the percent dry weight of each sample. Note that EQLs are even more variable than MDLs and that they are highly variable depending on the matrix being analyzed.

EQL in groundwater in µg/L was 5.
EQL in low soil or sediment in µg/kg was 5.
Accuracy (a) in µg/L was not listed.
Precision (b) in µg/L was not listed.

(a) *Average recovery found for measurements of samples containing a concentration of C, in µg/L.*
(b) *Overall precision found for measurements of samples with average recovery X for samples containing a concentration of C in µg/L.*
X = *Average recovery found for measurement of samples containing a concentration of C in µg/L.*

MULTIPLICATION FACTORS FOR OTHER MATRICES

Other Matrices	Factor (a)
Waste miscible liquid waste	50
High-concentration soil and sludge	125
Non-water miscible waste	500

(a) *EQL = [EQL for low soil sediment X [Factor]. For non-aqueous samples, the factor is on a wet-weight basis.*

SAMPLING METHOD
Liquid samples — Use a 40-mL glass screw-cap VOA vial with a Teflon®-faced silicone septum that has been prewashed, rinsed with distilled deionized water, and oven dried. However, if residual chlorine is present, collect sample in a 40-oz. soil VOA container which has been pre-preserved with 4 drops of 10% sodium thiosulfate, mix gently, and then transfer the sample to a 40-mL VOA vial. Collect bubble-free samples in duplicate and seal them in separate plastic bags.

Soils or sediments, and sludges — Use an 8-oz. widemouth glass bottle with a Teflon®-faced silicone septum that has been prewashed with detergent, rinsed with distilled deionized water, and oven dried. Tap slightly to eliminate free air space. Collect samples in duplicate and seal them in separate plastic bags.

SAMPLE PRESERVATION
Liquid samples — Add 4 drops of concentrated HCL and immediately cool samples to 4°C and store in a solvent-free refrigerator.

Soils or sediments, and sludges — Cool samples to 4°C and store in a solvent-free refrigerator.

MHT Maximum holding time is 14 days from the date of sample collection.

SAMPLE PREPARATION
Liquid samples — Remove the plunger from a 5-mL syringe and carefully pour the sample into the syringe barrel to just short of overflowing. Replace the syringe plunger and compress the sample. Open the syringe valve and vent any residual air while adjusting the sample volume to 5.0 mL. If there is only one volatile organic analysis (VOA) vial, a second syringe should be filled at this time to protect against possible loss of sample integrity. Add 10 µL of surrogate spiking solution and 10 µL of internal standard spiking solution through the valve bore of the 5-mL syringe, then close the valve. The surrogate

and internal standards may be mixed and added as a single spiking solution.

Sediments, soils, and waste samples — All samples of this type should be screened by GC analysis using a headspace method (EPA Method 3810) or the hexadecane extraction and screening method (EPA Method 3820). Use the screening data to determine whether to use the low-concentration method (0.005–1 mg/kg) or the high-concentration method (>1 mg/kg).

Low-concentration method — The low-concentration method is based on purging a heated sediment or soil sample mixed with organic-free reagent water containing the surrogate and internal standards. Analyze all reagent blanks and standards under the same conditions as the samples.

Use a 5-g sample if the expected concentration is <0.1 mg/kg or a 1-g sample for expected concentrations between 0.1 and 1 mg/kg. Mix the contents of the sample container with a narrow metal spatula. Weigh the amount of the sample into a tared purge device. Add the spiked water to the purge device, which contains the weighed amount of sample, and connect the device to the purge-and-trap system.

High-concentration method — This method is based on extracting the sediment or soil with methanol. A waste sample is either extracted or diluted, depending on its solubility in methanol. Wastes that are insoluble in methanol are diluted with reagent tetraglyme or possibly polyethylene glycol (PEG). An aliquot of the extract is added to organic-free reagent water containing surrogate and internal standards. This is purged at ambient temperature. All samples with an expected concentration of >1.0 mg/kg should be analyzed by this method.

Mix the contents of the sample container with a narrow metal spatula. For sediments or soils and solid wastes that are insoluble in methanol, weigh 4 g (wet weight) of sample into a tared 20-mL vial. For waste that is soluble in methanol, tetraglyme, or PEG, weigh 1 g (wet weight) into a tared scintillation vial or culture tube or a 10-mL volumetric flask. Quickly add 9.0 mL of appropriate solvent then add 1.0 mL of a surrogate spiking solution to the vial, cap it, and shake it for 2 min.

METHANOL EXTRACT REQUIRED FOR ANALYSIS OF HIGH-CONCENTRATION SOILS OR SEDIMENTS

Approximate Concentration Range	Volume of Methanol Extract (a)
500–10,000 µg/kg	100 µL
1,000–20,000 µg/kg	50 µL
5,000–100,000 µg/kg	10 µL
25,000–500,000 µg/kg	100 µL of 1/50 dilution (b)

Calculate appropriate dilution factor for concentrations exceeding this table.

(a) The volume of methanol added to 5 mL of water being purged should be kept constant. Therefore, add to the 5-mL syringe whatever volume of methanol is necessary to maintain a volume of 100 µL added to the syringe.
(b) Dilute an aliquot of the methanol extract and then take 100 µL for analysis.

QUALITY CONTROL Demonstrate, through the analysis of a reagent water blank, that interferences from the analytical system, glassware, and reagents are under control. Blank samples should be carried through all stages of the sample preparation and measurement steps. For each analytical batch (up to 20 samples), a reagent blank, matrix spike, and matrix spike duplicate must be analyzed (the frequency of the spikes may be different for different monitoring programs). The blank and spiked samples must be carried through all stages of the sample preparation and measurement steps. QC samples mentioned in the section on Interferences will also be needed as appropriate to those situations.

REFERENCE Test Methods for Evaluating Solid Waste (SW-846). U.S. EPA. 1983. Method 8240B, Rev. 2, Nov. 1990. Office of Solid Wastes, Washington, DC.

Xylenes — EPA Method 8020
CAS #1330-20-7

TITLE Aromatic Volatile Organics

MATRIX Groundwater, soils, sludges, water miscible liquid wastes, and non-water miscible wastes.

APPLICATION This method is used to analyze for 8 aromatic VOCs. Samples are analyzed using direct injection or purge-and-trap methods. Groundwater must be analyzed by the purge-and-trap method. The method provides an optional GC column that is used for analyte confirmation and may also help resolve analytes from interferences.

INTERFERENCES There can be carryover contamination with high- and low-level samples. Impurities may come from the purge-and-trap apparatus, organic compounds outgassing from the plumbing ahead of trap, diffusion of VOCs through the sample bottle septum during shipping or storage, or from solvent vapors in the lab.

INSTRUMENTATION GC capable of on-column injections or purge-and-trap sample introduction and a photoionization detector (PID). Column 1: 6 ft by 0.082 in with 5% SP-1200 and 1.75% Bentone-34 on Supelcoport. Column 2: 8 ft by 0.1 in with 5% 1,2,3-tris(2-cyanoethoxy)propane on Chromosorb W-AW.

RANGE 2.1–500 µg/L.

MDL Not available.

PQL FACTORS FOR MULTIPLYING × MDL VALUE

Matrix	Multiplication Factor
Groundwater	10
Low-level soil	10
Water miscible liquid waste	500
High-level soil and sludge	1250
Non-water miscible waste	1250

PRECISION Not available.

ACCURACY Not available.

SAMPLING METHOD For water and liquid samples use glass 40-mL vials with Teflon®-lined septum caps and collect two vials per sample location with no headspace. For solids and concentrated waste samples use widemouth glass bottles with Teflon® liners. Cool all samples to 4°C

STABILITY For concentrated wastes, soils, sediments, or sludges cool to 4°C. For liquids, add 4 drops of concentrated hydrochloric acid and cool to 4°C.

MHT 14 days.

QUALITY CONTROL Analyze a reagent blank, matrix spike, and matrix spike duplicate/duplicate for each analytical batch (up to 20 samples). Demonstrate the purity of glassware and reagents by analyzing a reagent water method blank. Internal, surrogate, and five concentration level calibration standards are used. The QC check sample concentrate should contain this compound at 10 µg/mL in methanol.

REFERENCE Test Methods for Evaluating Solid Waste (SW-846), U.S. EPA Office of Solid Waste, Washington, DC, Method 8020A, Rev. 1, Nov. 1992.

Z

Zinc
CAS #7440-66-6

EPA Method 6010

TITLE Inductively Coupled Plasma-Atomic Emission Spectroscopy

MATRIX This method is applicable to the determination of trace elements, including metals, in groundwater, soils, sludges, sediments, and other solid wastes. All matrices require digestion prior to analysis. The method of standard addition must be used for the analysis of all sample digests unless either serial dilution or matrix spike addition demonstrates it is not required.

METHOD SUMMARY Method 6010 covers 25 elements using ICP analysis. It measures element-emitted light by optical spectrometry. Samples, following an appropriate acid digestion, are nebulized and the resulting aerosol is transported to the plasma torch. Element-specific atomic line emission spectra are produced by a radio-frequency inductively coupled plasma.

INTERFERENCES Interferences may be categorized as spectral or non-spectral. Spectral interferences are caused by overlap of a spectral line from another element, unresolved overlap of molecular band spectra, background contribution from continuous or recombination phenomenon, and stray light from the line emission of high concentration elements. Non-spectral interferences include physical and chemical interferences. Physical interferences are effects associated with the sample nebulization and transport processes. Changes in viscosity and surface tension can cause significant inaccuracies. Chemical interferences include molecular compound formation, ionization effects, and solute vaporization effects. Normally these effects are not significant and can be minimized by careful selection of operating conditions. Chemical interferences are highly dependent on matrix type and the specific analyte element.

INSTRUMENTATION An inductively coupled argon plasma emission spectrometer (ICP) capable of background correction is required.

PRECISION & ACCURACY Detection limits, sensitivity, and optimum ranges of the metals will vary with the matrices and model of the spectrometer. In a single lab evaluation, seven wastes were analyzed for 22 elements. The mean percent relative standard deviation from triplicate analyses for all elements and wastes was 9 ± 2%. The mean percent recovery of spiked elements for all wastes was 93 ± 6%. Spike levels ranged from 100 µg/L to 100 mg/L. The wastes included sludges and industrial wastewaters.

Estimated instrument detection limit in µg/L is 2.
Spiked concentration in µg/L is 16.
Mean reported value in µg/L is 19.
Precision as RSD % is 45.

SAMPLING METHOD Samples should be collected in borosilicate glass, linear polyethylene, polypropylene, or Teflon® bottles that have been prewashed with detergent and tap water, and rinsed with 1:1 nitric acid and tap water or 1:1 hydrochloric acid and tap water. Collect at least 2 g of solids and 200 mL of aqueous samples.

SAMPLE PRESERVATION Add nitric acid to make the samples pH <2.

MHT The maximum holding time for properly preserved samples is 6 months.

SAMPLE PREPARATION Preliminary treatment of most matrices is necessary because of the complexity and variability of sample matrices. Water samples that have been prefiltered and acidified will not need acid digestion. Methods for acid digestion of waters for total recoverable or dissolved metals, acid digestions of aqueous samples and extracts for total metals, and acid digestion of sediments, sludges, and soils are summarized below.

Total recoverable or dissolved metals in water — To prepare surface and groundwater samples for determination of total recoverable and dissolved metals, a 100-mL aliquot of well-mixed sample is acidified with concentrated nitric acid and concentrated hydrochloric acid, then heated until the volume is reduced to 15–20 mL. Adjust the final volume to 100 mL with reagent water.

Total metals in aqueous samples, soil and sediment extracts — To prepare aqueous samples, soil and sediment extracts, and wastes that contain suspended solids, a 100-mL aliquot is made acidic with concentrated nitric acid and the solution is evaporated to about 5 mL on a hot plate. Continue heating and adding additional acid until sample digestion is complete, which is usually indicated when the digestate is light in color or does not change in appearance. Evaporate the solution to about 3 mL and cool it and add a small quantity of 1:1 hydrochloric acid (10 mL/100 mL of final solution). Cover the beaker and reflux for 15 min. Wash down the beaker walls and filters or centrifuge the sample to remove silicates and other insoluble material. Filter the sample and adjust the final volume to 100 mL with reagent water and the final acid concentration to 10%.

Sediments, sludges, and soils — To prepare sediments, sludges and soil samples, transfer 1–2 g to a conical beaker and add 10 mL of 1:1 nitric acid, mix the slurry, and cover it with a watch glass. Heat the sample and reflux for 10–15 min without boiling. Allow it to cool, then add 5 mL of concentrated nitric acid and reflux for 30 min. Repeat last step and then allow the solution to evaporate to 5 mL without boiling. Cool and add 2 mL of water and 3 mL of 30% hydrogen peroxide. Cover and place the beaker on the hot plate. Heat and add 30% hydrogen peroxide in 1-mL aliquots with warming until the effervescence is minimal but do not add more than a total of 10 mL of 30% hydrogen peroxide. If the sample is being prepared for the analysis of Ag, Al, As, Ba, Be, Ca, Cd, Co, Cr, Cu, Fe, K, Mg, Mn, Mo, Na, Ni, Os, Pb, Se, Tl, V, and Zn, then add 5 mL of concentrated hydrochloric acid and 10 mL of water and return the covered beaker to a hot plate for 15 min of additional refluxing without boiling. Dilute the sample to a 100 mL volume with

water after cooling and filter or centrifuge to remove particulates.

QUALITY CONTROL Laboratory control samples must be analyzed for each analytical method. A method blank should be analyzed with each batch of samples. The effect of the matrix on method performance must be demonstrated: when appropriate, there should be at least one matrix spike and either one matrix duplicate or one matrix spike duplicate per analytical batch. The bias and precision of the method, as well as the method detection limit for each specific matrix type, must be measured.

Dilute and reanalyze samples that are more concentrated than the linear calibration limit. Employ a minimum of one reagent blank per sample batch to determine if contamination or any memory effects are occurring. Whenever a new or unusual sample matrix is encountered, perform either a serial dilution test or a matrix spike addition test to ensure that neither positive or negative interferences are operating on any of the analyte elements. Check the instrument standardization by verifying calibration every 10 samples using a calibration blank and a check standard.

REFERENCE Test Methods for Evaluating Solid Waste (SW-846). U.S. EPA. 1983. Method 6010, Rev. 0, Sept. 1986. Office of Solid Wastes, Washington, DC.

Zinc   **EPA Method 200.7**
CAS #7440-66-6

TITLE Inductively Coupled Plasma

MATRIX Dissolved, suspended or (ICP) total element in drinking and surface waters and in domestic and industrial wastewaters.

APPLICATION The method covers the determination of 25 metals. Dissolved elements are determined in filtered and acidified samples after appropriate digestion (which increases dissolved solids). Its primary advantage is that ICP instruments allow simultaneous or rapid sequential determination of many elements in a short time. Samples are first nebulized and the aerosol is transported to a plasma torch in which element specific atomic line emission spectra are produced by a radio frequency inductively coupled plasma. Background correction is required for trace element detection except in the case of line broadning.

INTERFERENCES There are spectral, physical, and chemical interferences. The primary disadvantage of ICP instruments is background radiation from other elements and the plasma gases (spectral interferences). Changes in sample viscosity and surface tension with samples containing high dissolved solids (especially those exceeding 1500 mg/L) or high acid concentrations can cause physical interferences. Ionization effects, solute vaporization and molecular compound formation can cause chemical interferences. Copper and nickel can cause interference at the 100 mg/L level.

INSTRUMENTATION Inductively coupled argon plasma emission spectroscopy.

213.856 nm wavelength

RANGE Not listed.

MDL 2 µg/L.

PRECISION SD = 5.6% Mean at true value 200 µg/L.

ACCURACY Mean recovery = 93% ± 6% of spiked elements for all wastes.

SAMPLING METHOD Wash sample container with detergent and tap water, rinse with 1 + 1 nitric acid and tap water, then rinse with 1 + 1 hydrochloric acid and tap water, then rinse with deionized, distilled water in that order. Perform any filtration or acid preservation steps when the sample is collected or as soon as possible thereafter.

STABILITY Cool samples to 4°C.

MHT 24 h.

QUALITY CONTROL Mixed calibration standards, an instrument check standard, and an interference check solution are used in addition to a quality control sample. The quality control sample should be prepared in the same acid matrix as the calibration standards at 10 times the instrumental detection limits and in accordance with the instructions provided by the supplier. Furthermore, two types of blanks are required: a calibration blank and a reagent blank.

REFERENCE Method 200.7, U.S. EPA, EMSL-Cincinnati, OH, Nov. 1980

Zinc   **EPA Method 7950**
CAS #7440-66-6

TITLE Atomic Absorption (AA)

MATRIX Drinking, surface and direct aspiration saline waters. Wastewater.

APPLICATION Sample is aspirated and atomized in a flame. A light beam from a Zn hollow cathode lamp is directed through the flame into a monochromator and onto a detector. Since wavelength of light beam is specific for Zn, light energy absorbed by detector is measure of zinc.

INTERFERENCES The most troublesomee type is chemical, caused by lack of absorption of atoms bound in molecular combination in the flame. High dissolved solids in sample may result in nonatomic absorbance interference. Zinc is a universal contaminent so use the method with great care.

INSTRUMENTATION Atomic absorption spectrometer. Zinc hollow cathode lamp. (213.9 nm wavelength)

RANGE 0.05–1 mg/L.

MDL 0.005 mg/L

PRECISION Standard deviation = 114 µg/L at 310 µg/L (true value) 89 labs

ACCURACY As bias = –0.7% at 310 µg/L (true value) 89 labs

SAMPLING METHOD Use glass or plastic containers. Collect 200 g of solids and 600 mL of liquid samples.

STABILITY Cool solid samples to 4°C and analyze as soon as possible. Add nitric acid to liquid samples to pH <2.

MHT 6 months.

QUALITY CONTROL At least one duplicate and one spike sample should be run every 20 samples or with each matrix type to verify precision of the method. For 20 or more samples per day, verify working standard curve. Run an additional standard at or near mid-range every 10 samples.

REFERENCE Method 7950, SW-846, 3rd ed., Nov.1986.

Zinc **EPA Method 7951**
CAS #7440-66-6

TITLE Atomic Absorption (AA) Furnace Technique

MATRIX Wastes, mobility procedure, extracts, soils and groundwater.

APPLICATION Aqueous samples, EP extracts, industrial wastes, soils, sludges, sediments, and solid wastes require digestion before analysis. An aliquot of sample is placed in the graphite tube in the furnace and slowly evaporated, charred and atomized. Absorption of lamp radiation during atomization is proportional to zinc concentration.

INTERFERENCES The furnace technique is subject to chemical interferences. Composition of sample matrix can effect analysis. Modify matrix to remove interferences. Background correction correction must be used. Zinc is universal contaminant so use the method with great care.

INSTRUMENTATION Atomic absorption spectrometer. Zinc hollow cathode lamp or electrodeless discharge lamp. Graphite furnace. Strip-chart recorder

RANGE 0.2–4 µg/L.

MDL 0.05 µg/L (213.9 nm wavelength)

PRECISION Not listed.

ACCURACY Not listed.

SAMPLING METHOD Use glass or plastic containers. Collect 200 g of solids and 600 mL of liquid samples.

STABILITY Cool solid samples to 4°C and analyze as soon as possible. Add nitric acid to liquid samples to pH <2.

MHT 6 months.

QUALITY CONTROL At least one duplicate and one spike sample should be run every 20 samples, or with each matrix type to verify method precision. If 20 or more samples are run a day, run a standard (at or near mid-range) every 10 samples.

REFERENCE Method 7951, SW-846, 3rd ed., (Included as Rev. 0, Dec. 1987)

Abbreviations

<	less than	ICP	inductively coupled plasma emission spectrometry
>	greater than	kg	kilogram(s)
@	at	KI	potassium iodide
AA	atomic absorption	L	liter(s)
AAS	atomic absorption spectrometer	MDL	method detection limit
BOD	biological oxygen demand	mg	milligram(s)
cm	centimeter(s)	MHT	maximum holding time for a sample
C	true value (when used with accuracy)	mL	milliliter(s)
°C	degrees Centigrade	mm	millimeter(s)
$CaCO_3$	calcium carbonate	$Na_2S_2O_3$	sodium thiosulfate
COD	carbon oxygen demand	NH3	ammonia
DO	dissolved oxygen	NH4OH	ammonium hydroxide
EC	electron capture	nm	nanometer
ECD	electron capture detector	PID	photoionization detector
EP	extraction procedure	PQL	practical quantitation limit
FLAA	flame atomic absorption	RSD	relative standard deviation
GC	gas chromatograph or gas chromatography	SD	standard deviation
GC/MS	gas chromatograph/mass spectrometer	TDS	total dissolved solids
GFAA	graphite furnace atomic absorption	TKN	total Kjedahl nitrogen
H_2O_2	hydrogen peroxide	TS	total suspended solids
HCl	hydrochloric acid	mg	microgram(s)
HNO_3	nitric acid	VOC	volatile organic compounds
H_2SO_4	sulfuric acid	X	mean recovery (used with accuracy)

Definitions

accuracy
The nearness of a result or the mean (X) of a set of results to the true value. Accuracy is assessed by means of reference samples and percent recoveries.

ACS reagent grade
See reagent grade.

analytical batch
The basic unit for analytical quality control. The analytical batch is defined as samples that are analyzed together with the same method sequence and the same lots of reagents and with the manipulations common to each sample within the same time period or in continuous sequential time periods. Samples in each batch should be of similar composition.

analytical reagent grade (AR)
See reagent grade.

background samples
Matrices minus the analytes of interest (matrix blanks) that are carried through all steps of the analytical procedure. All reagents, glassware, preparations, and instrumental analyses are included. These are necessary to account for the presence of spurious analytes, interferences, and background concentrations of the analyte of interest.

blank
An artificial sample designed to monitor the introduction of artifacts into the process. For aqueous samples, reagent water is used as a blank matrix. However, a universal blank matrix does not exist for solid samples. The blank is taken through all the appropriate steps of the sample preparation and analysis process.

calibration check
Verification of the ratio of instrument response to analyte amount. A calibration check is performed by analyzing for analyte standards in an appropriate solvent. Calibration check solutions are made from a stock solution that is different from the stock used to prepare standards.

environmental sample (or field sample)
A representative sample of any material — aqueous, nonaqueous, or multimedia — collected from any source for which determination of composition or contamination is requested or required. In this book, environmental samples are classified as:

>**drinking water**
>Delivered (treated and untreated) water designed as potable water.

>**groundwater**
>Water from wells.

>**sludge**
>Municipal and industrial sludges.

>**surface water**
>Water from lakes, rivers, and streams.

>**waste**
>Aqueous and nonaqueous liquid wastes, chemical solids, contaminated soils, and industrial liquid and solid wastes.

>**water/wastewater**
>Raw source water for public drinking water supplies, groundwater, municipal influents, and industrial influents and effluents.

equipment blanks
Samples of analyte-free media that have been used to rinse the sample equipment. They are used to document adequate decontamination of sampling equipment after its use. These blanks are collected after equipment decontamination and before using the equipment for sampling again.

field blanks
Samples of analyte-free media similar to the sample matrix that are transferred from one vessel to another or exposed to the sampling environment at the sampling site. They are used to measure incidental or accidental contamination of a sample during the whole process (sampling, transport, sample preparation, and analysis). Capped and cleaned containers are taken to the sample collection site. Usually at least one field blank with a matrix compatible to the sample of interest is collected with each batch of samples.

field sample
See environmental sample.

instrument blanks
Solvent or reagent blanks used to measure interference or contamination from an analytical instrument by cycling matrices containing materials that are normal to the analysis, but minus the analytes of interest, through the instrument.

laboratory blanks
See background samples.

limit of detection
See method detection limit.

material blanks
Samples of construction materials such as those used in groundwater wells, pump and flow testing, etc. They are used to document decontamination (or measure artifacts) from use of these materials.

matrix samples
See background samples.

matrix spike
Samples to which predetermined quantities of stock solutions of certain analytes are added before sample

preparation (extraction, digestion) and analysis. Samples are split into duplicates, spiked, and analyzed. Percent recoveries are calculated for each of the analytes detected. The relative percent difference between the samples is calculated and used to assess analytical accuracy in terms of recovery.

MDL
See method detection limit.

method blanks
See background samples.

method detection limit (or limit of detection)
The minimum concentration of an analyte above true zero that can be detected and reported. The MDL does not take into account concentrations of the analyte in a background sample and may be equal to or even less than the *method quantitation limit*.

method quantitation limit
The minimum concentration of a substance that can be measured and reported.

MQL
See method quantitation limit.

PQL
See practical quantitation limit.

practical quantitation limit
The lowest level that reliably can be achieved within specified limits of precision and accuracy during routine laboratory operating conditions.

precision
The measurement of agreement of a set of replicate results among themselves without assumption of any prior information as to the true result. Precision is assessed by means of duplicate or replicate sample analysis.

QC check sample
A blank that has been spiked with the analyte(s) from an independent source in order to monitor the execution of the analytical method. The level of the spike is at the regulatory action level when applicable. Otherwise, the spike usually is 5 times the estimate of the quantitation limit. The matrix is phase matched with the samples and well characterized; for example, reagent grade water is appropriate for an aqueous sample.

RCRA
The Resource Conservation and Recovery Act.

reagent blank
An aliquot of analyte-free water or solvent analyzed with the analytical batch.

reagent grade
Reagents that conform to the current specifications of the Committee for Analytical Reagents of the American Chemical Society.

replicate sample
A sample prepared by dividing a sample into two or more separate aliquots. Duplicate samples are considered to be two replicates. Replicate samples are used to measure the precision of the analytical methods used.

rinsate blanks
See equipment blanks.

solvent blanks
Blanks consisting only of the solvent used to dilute or extract the sample. They are used to identify and/or correct for signals produced by the solvent or by impurities in the solvent.

spiked field blanks
Field blanks that have known amounts of the analytes of interest added to them. They are used to measure effects that the sample matrix may have on the analytical methods (usually analyte recovery).

spiked laboratory blanks
Laboratory blanks that have known amounts of the analytes of interest added to them. They are used to measure systematic bias from all laboratory sources.

spiked test samples
Samples used to measure effects that the sample matrix may have on the analytical methods (usually analyte recovery).

standard curve
A curve that plots concentrations of known analyte standard vs. the instrument response to the analyte.

surrogate
Organic compounds that are similar to analytes of interest in chemical composition, extraction, and chromatography, but which normally are not found in environmental samples. These compounds are spiked into all blanks, standards, samples, and spiked samples before analysis. Percent recoveries are calculated for each surrogate.

trip blanks
Samples of analyte-free media taken from the laboratory to the sampling site and returned to the laboratory unopened. They are used to measure cross contamination from the container and preservative during transport, field handling, and storage.